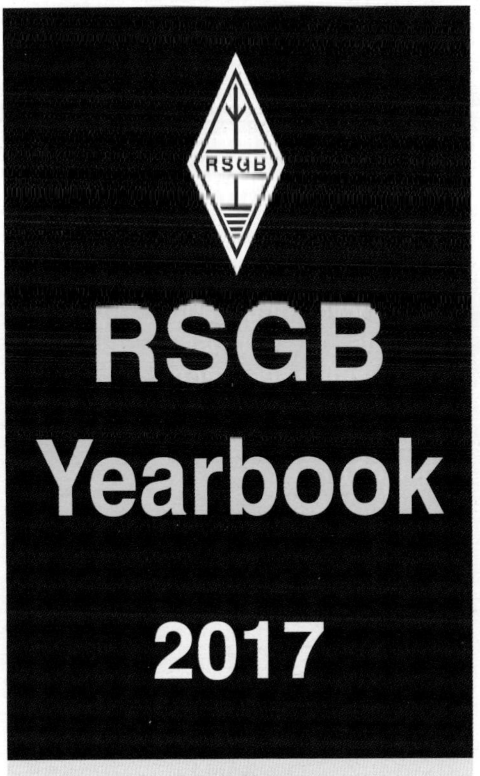


**Editor**
Mike Browne, G3DIH

**Advertising**
Chris Danby, G0DWV, Danby Advertising

**Front Cover**
Kevin Williams, M6CYB

**Production**
Mark Allgar, M1MPA

*Published by the*
Radio Society of Great Britain,
3 Abbey Court, Fraser Road,
Priory Business Park, Bedford MK44 3WH
Website: www.rsgb.org

Tel: 01234 832700
Fax: 01234 831496



ISBN: 9781 9101 9323 5

ISSN: 1460-454X

Printed in Great Britain
by Hobbs the Printers Ltd
of Totton, Hampshire

# Contents

## Information Section



## Operating Section



## UK & Ireland Callbook 2017

# Foreword

As I start my two year period as RSGB President I am delighted to write a foreword for the RSGB Yearbook for 2017. For many, part of the fascination of amateur radio is the breadth of activities available. This excellent and comprehensive publication is rich in information about them and is an invaluable reference document for our 21,000 members. It is also a good read.

Quite rightly, members want to know what the RSGB does and the Yearbook answers that question in some detail. In particular I would encourage you to look at the "Services" map on page 8. Part of my personal preparation for becoming President was to find out more about those things which I did not know much about, even after so many years as an amateur. The Yearbook has been an invaluable reference document.

At the 2016 AGM in Glasgow my predecessor, John Gould, gave some early feedback from the 2015 Amateur Radio survey. The full report to members is at a late stage of preparation and will be a lively topic for discussion and engagement amongst members. The Society is keen to respond to the findings of the survey.

Any RSGB publication owes a huge debt to the volunteers who work to make the RSGB such a successful organisation. Many members will be largely unaware of the extent of volunteer commitment – through the regional organisation, committees, as individuals and in many other ways. All give freely their time to keep amateur radio alive and well. We owe them our heartfelt thanks. It is good to know that the membership has continued to show a gentle increase in recent years.

Callsign listings remain an important feature of the Yearbook and take up about two thirds of the publication. It remains a valuable database but revalidation of licences is essential to maintain its accuracy. The amateur licensing section of Ofcom is there to help you with questions about your own licence or the licensing of your club.

2016 has been a great year for encouraging young people to engage with and enjoy amateur radio. The highlight was the opportunity for pupils in ten schools to talk directly with Tim Peake on the International Space Station as it orbited above them, using amateur radio links provided by ARISS (Amateur Radio on the International Space Station). This and other initiatives have been fully reported in RadCom. The focus on young people will continue into 2017 when the RSGB will follow Austria in hosting the international 'Youth on the Air' event (YOTA).

While there is an important focus on youth, the Society is keenly aware of the needs and interests of the whole age range of members.

I hope you will find this edition of the RSGB Yearbook both helpful and enjoyable. We all owe sincere thanks for this latest edition to Mike Browne, G3DIH, and his team, and also to the staff at RSGB HQ.

*Nick Henwood, G3RWF*
*President*

# Introduction

As far as possible, the information is accurate as of July 2016. As with a publication lasting until December 2017, changes will occur during the lifetime of this edition – so please check the various web sites or the pages of RadCom for information which became available after we closed for press.

This particularly applies to items such as QSL sub-managers, GB2RS newsreaders, local and national clubs and societies, repeaters, beacons and the composition of the regions and volunteer posts.

The 'Operating' section of the *Yearbook* starts on page 88 making it easier to find repeaters, HF beacons, prefix listings and band plans etc all in one place close to the Callsign Listings.

As in previous years, we have had an overwhelming response from the clubs with information for entry within the 'Featured Clubs' section.

I am sorry that we could not include all of the club entries that were received, and also had to edit some copy received so as to allow space for as many club entries as possible. I have however tried to include them all (unedited) on the disk that is included with this *Yearbook*.

## UK Callsign entries

The Callsigns listed in this *Yearbook* reflect the official records held by Spectrum Licensing on behalf of Ofcom. They show correspondence addresses, **not** licence addresses, so do not be misled into thinking that if a foreign address is quoted the station is operating from there. The amateur concerned will have a different station address when in the UK.

The callsign data is obtained from Spectrum Licensing every year and is, therefore, as up to date as it can possibly be at the time of going to press. Note that some callbooks, CDs and websites rely entirely on updates being notified to them by individual licensees, so the information can be many years out of date. To make this book readable and compact, we make a few standard abbreviations, but no changes are made to the substance of any entries.

## Validating your Licence

### Revalidate your licence to avoid revocation

Ofcom has advised the Society that plans will be drawn up to revoke licences that have not been revalidated as required by the licence conditions at intervals of not more than 5 years.

Therefore it is important, so as to retain your licence, and to remain legal to operate on the amateur bands, to re-validate your licence every 5 years.
To validate, log in to *https://services.ofcom.org.uk/* and amend any details as soon as possible, this will automatically validate your licence or by email: *amateur.validations@ofcom.org.uk* If you need assistance in the process, Ofcom staff are available to help, but please be patient during times of heavy workload.

If you have not yet registered to use the Ofcom Online Licensing Service you will need to do so in order to access your licence online. When registering for the first time you will need to have details of your lifetime licence number, which can be found on page 1 of your 23 page Licence document.

## Details withheld

The words "Details withheld at licensee's request" mean exactly what they say, because entries are not within the control of the Society. In some cases there is a perfectly good reason for a licensee not to want his or her address publishing, but in others it is possible that the 'no publicity' box was ticked in error.

### 'Withheld' entries

Should you wish your details to be withheld from publication, or released if they are not, or there is an error in the substance of your entry, you can update the entry online or write to Spectrum Licensing (*and not to the RSGB*) and request them to take the necessary action. Their address is:

Spectrum Licensing,
Riverside House,
2a Southwark Bridge Road,
London SE1 9HA
Tel: 020 7981 3131
Email: spectrum.licensingenquiries@ofcom.org.uk
Web: www.ofcom.org.uk/licensing/olc/

There are several other publishers of the data, so it is vital that you contact the source of data so that it is either blocked or released in *all* callbooks. The Society cannot accept direct input regarding an individual's entry.

### International listings

Amateurs sometimes ask about their entry in the *International Listings* of the Radio Amateur Call Book, sometimes known as the *DX Listings*. This is an entirely separate work published annually by an independent company. If you want to be listed in their publication, please write direct to them, not to the RSGB. Corrections are free, but they do make a small charge for special entries. Their address is:

Radio Amateur Callbook,
ITfM GmbH
PO Box 1170
34216 Baunatal,
Germany.
Web: www.callbook.biz

## Acknowledgments

My thanks go to all the contributors to this book, this includes a number of RSGB Staff, Committee Members, and Club Secretaries.

Also my thanks go to Ann Stevens, G8NVI, for her help, and for proof reading the *Yearbook*, finally thanks also go to Joe Ryan of the IRTS for supplying the Irish callsign listings.

*Mike Browne, G3DIH, Editor*

# At Your Service!

The Society provides a broad range of services for its members through its professional staff and management at Headquarters and a country-wide force of skilled, dedicated and knowledgeable volunteers. To get the best out of the RSGB it is important that you approach the correct part of the Society. On these pages you will find a practical guide to finding the right person for your enquiry.

## YOUR RSGB

The Society's affairs are directed by the Board, supported by the Leadership Team comprising the Regional Managers, Committee Chairs and Honorary Officers. The Board Members and the Regional Managers are elected by the membership in a postal and electronic ballot.

The day to day running of the Society is the responsibility of the General Manager, supported by the Board Members who liaise with each of the committees.

There are 8 portfolios:

1. Amateur Radio Development, Education and Training
2. Technical (Environment)
3. Technical (Technical / Propagation)
4. Business
5. Spectrum (Including sport radio, IOTA and QSL)
6. International and Regulatory
7. Membership Services
8. Public Services / RAYNET

Regional Managers are responsible for Society matters within their respective Regions. Details of the Regional Structure, the Regional Managers and their deputies can be found in this Yearbook.

Full details of Board and Committees' terms of reference can be found on our website.

For HQ staff, both email addresses and telephone details are provided, including the option to select when dialling through the RSGB switchboard (01234 832 700).

### Chairmen and Honorary Officers

These are all volunteers and give their time freely to support the Society. Members should respect the fact that many also have full time day jobs, and so email is the appropriate method of communication.

### General Manager

Steve Thomas, M1ACB,
email: steve.thomas@rsgb.org.uk

### Honorary Treasurer (Acting)

Richard Horton, G4AOJ,
email: g4aoj@rsgb.org.uk

### Company Secretary (Acting)

Steve Hartley, G0FUW,
email: g0fuw@rsgb.org.uk

## WEBSITE

Main website: www.rsgb.org

Members Pages: Log in using your callsign as the user name and your Membership number, without the leading zeros (see your *RadCom* address label) as the password.

If you need to update your Membership details, please visit www.rsgb.org/myaccount/

## THE RSGB BOARD

### President

Nick Henwood, G3RWF (RSGB President)
email: president@rsgb.org.uk

Steve Hartley, G0FUW (Board Chairman)
email: g0fuw@rsgb.org.uk

Stewart Bryant, G3YSX,
email: g3ysx@rsgb.org.uk

Alan Messenger, G0TLK,
email: g0tlk@rsgb.org.uk

Graham Murchie, G4FSG,
email: g4fsg@rsgb.org.uk

Len Paget, GM0ONX,
email: gm0onx@rsgb.org.uk

Ian Shepherd, G4EVK,
email: g4evk@rsgb.org.uk

**Note:** The General Manager, Company Secretary and Acting Honorary Treasurer are not Directors, but are in attendance at Board Meetings.

## REGIONAL MANAGERS

Information on the Regional and Sub Regional Managers, may be found elsewhere in the RSGB Yearbook and on the RSGB website, www.rsgb.org

| | |
|---|---|
| **Region 1** | M Hazel-McGown, MM0ZIF, rm1@rsgb.org.uk |
| **Region 2** | Denny Morrison, GM1BAN, rm2@rsgb.org.uk |
| **Region 3** | K A Wilson, M1CNY, rm3@rsgb.org.uk |
| **Region 4** | Ian Douglas, G7MFN, rm4@rsgb.org.uk |
| **Region 5** | M Vincent, G3UKV, rm5@rsgb.org.uk |
| **Region 6** | C Jones, 2W0LJC, rm6@rsgb.org.uk |
| **Region 7** | Glyn Jones, GW0ANA, rm7@rsgb.org.uk |
| **Region 8** | P Hosey, MI0MSO, rm8@rsgb.org.uk |
| **Region 9** | T O'Reilly, G0NSY rm9@rsgb.org.uk |
| **Region 10** | M Senior, G4EFO, rm10@rsgb.org.uk |
| **Region 11** | P Helliwell, G7SME, rm11@rsgb.org.uk |
| **Region 12** | Keith Haynes, G3WRO, rm12@rsgb.org.uk |
| **Region 13** | J Stevenson, G0EJQ, rm13@rsgb.org.uk |

## SPECIALIST AREAS – CHAIRMEN & HONORARY OFFICERS

The many different activities of the Society are run by its committees, honorary officers and full-time staff. If you wish to take advantage of one of these services or have an administrative enquiry about any one of them, contact details are listed below.

### Abuse and Poor Operating

Amateur Radio Observation Service (AROS), Mark Jones, G0MGX, AROS coordinator, email: aros@rsgb.org.uk, www.rsgb.org/aros/

### Amateur Radio Direction Finding

Bob Titterington, G3ORY,
Chairman, ARDF Committee,
email: ardf.chairman@rsgb.org.uk,
www.rsgb.org/ardf/

### Audio-Visual Library

Videos etc, for short term loan to RSGB affiliated clubs. List published in RSGB Yearbook.
email: ar.dept@rsgb.org.uk

### Contests

Ian Pawson, G0FCT, Chairman, Contest Support, email: csc.chair@rsgb.org.uk, www.rsgb.org/radiosport/
Nick Totterdell, G4FAL, HF Contest Committee email: hfcc.chair@rsgb.org.uk
Andy Cook, G4PIQ, VHF Contest Committee email: vhfcc.chair@rsgb.org.uk

### EMC

Advice on dealing with all types of radio interference. EMC information sheets are available free of charge from the RSGB. They are also on the EMC Committee's website. For specific advice from your local EMC Co-ordinator, see this Yearbook.

To contact the EMC Committee directly, John Rogers, M0JAV, Chairman, EMC Committee, email: emc.chairman@rsgb.org.uk, www.rsgb.org/emc/

### General Technical Matters

Andy Talbot, G4JNT, Chairman, Technical Forum, email: tech.chair@rsgb.org.uk, www.rsgb.org/technicalmatters/

### General Spectrum & Regulatory Matters

Murray Niman, G6JYB, Chairman, Spectrum Forum, email: spectrum.chairman@rsgb.org.uk www.rsgb.org/committees/spectrumforum/

### GB2RS News Service Management

Ken Hatton, G3VBA, GB2RS Manager, email: gb2rs.manager@rsgb.org.uk (GB2RS news items should be sent to radcom@rsgb.org.uk)

### HF Matters

Ian Greenshields, G4FSU, HF Manager, email: hf.manager@rsgb.org.uk

### Intruders to the Amateur Bands

email: iw@rsgb.org.uk
www.rsgb.org/intruders/

### IOTA Activity Programme

Roger Balister, G3KMA, IOTA Manager, email: iota.manager@rsgb.org.uk, www.rsgbiota.org/

**Microwave Matters**
Barry Lewis, G4SJH, Microwave Manager, email: microwave.manager@rsgb.org.uk

**Planning Advice**
John Mattocks, G4TFQ Chairman, Planning Advisory Committee
email: pac.chairman@rsgb.org.uk,
www.rsgb.org/planning/

**Propagation Studies**
Weekly forecasts for GB2RS (for explanation of terms used, see RSGB Yearbook). Monthly predictions in *RadCom*.
Steve Nichols, G0KYA, Chairman, Propagation Studies Committee, email: psc.chairman@rsgb.org.uk,
www.rsgb.org/psc/

**Repeater and Data Communications**
A list of current Repeaters can be found in this Yearbook. Full Repeater information is also available on the RSGB Website.
For all Policy and other matters, contact the Emerging Technology Chairman:
John McCullagh, GI4BWM, Chairman, ETCC, email: etcc.chairman@rsgb.org.uk,
www.ukrepeater.net

**RSGB Awards**
Marcus Hazel-McGown, MM0ZIF, Awards Manager, email: awards@rsgb.org.uk,
www.rsgb.org/awards/

**Training & Education**
Philip Willis, M0PHI, Chairman, Training & Education Committee,
email: tec.chair@rsgb.org.uk,
www.rsgb.org/clubsandtraining/

**VHF Matters**
John Regnault, G4SWX, VHF Manager
email: vhf.manager@rsgb.org.uk

**Youth Committee**
Mike Jones, 2E0MLJ, Chairman, Youth Committee
email:youth.chairman@rsgb.org.uk
www.rsgb.org/youth-committee

Details of the Society's volunteer officers can be found on the RSGB website, www.rsgb.org

## HEADQUARTERS STAFF

*For HQ staff below, both email addresses and telephone details are provided, including the option to select when dialling through the RSGB switchboard (01234 832 700).*

**Technical Amateur Radio Enquiries**
email: AR.dept@rsgb.org.uk
Telephone: 01234 832 700, Option 4

**Amateur Radio Examinations**
email: exams@rsgb.org.uk
Telephone: 01234 832 700, Option 3

*RadCom* **(news items, feature submissions, etc)**
Elaine Richards, G4LFM / Giles Read, G1MFG
email: radcom@rsgb.org.uk
Telephone: 01234 832 700, Option 8

**GB2RS and Club News**
email: GB2RS@rsgb.org.uk
Telephone: 01234 832 700, Option 8

*RadCom* Members Ads
email: memads@rsgb.org.uk
web: tinyurl.com/memadsinfo

**Amateur Radio Licensing Enquiries**
email: AR.dept@rsgb.org.uk
Telephone: 01234 832 700, Option 5

**Sales Department**
(Membership, books and other products)
email: sales@rsgb.org.uk
Telephone: 01234 832 700, Option 1

**Subscription renewals**
Telephone: 01234 832 700, Option 2

**IOTA**
email: IOTA_HQ@rsgb.org.uk

**General Manager**
email: GM.dept@rsgb.org.uk
Telephone: 01234 832 700, Option 9

## HEADQUARTERS AND REGISTERED OFFICE

3 Abbey Court, Fraser Road,
Priory Business Park, Bedford MK44 3WH
Telephone: 01234 832 700
Fax: 01234 831 496

Main website: www.rsgb.org
Log in using your callsign as the user name and your membership number, without the leading zeros (see your *RadCom* address label) as the password.

## QSL BUREAU ADDRESS

PO Box 5, Halifax HX1 9JR, England
Telephone: 01422 359 362
email: qsl@rsgb.org.uk, www.rsgb.org/qsl

## PLAY YOUR PART IN YOUR RSGB

**Have Your Say**
Let us know how we're doing! Through 'Have Your Say' you can let us know your views and you will receive a reply from the General Manager or a Board Member.
email: haveyoursay@rsgb.org.uk
www.rsgb.org/haveyoursay

**Consultations**
From time to time you will find we are consulting the Membership on aspects of Society policy. You can find current consultations at www.rsgb.org/consultations/

**National Radio Centre**
Don't forget to tell your friends about the National Radio Centre at Bletchley Park. Full details at www.nationalradiocentre.com

**Licensing & Special Event Stations**
Licensing and Notices of Variation (NoVs) for special event stations are handled by Ofcom, 0207 981 3131, www.ofcom.org.uk, email: Spectrum.Licensing@ofcom.org.uk

**FAQs**
The RSGB has compiled the questions most frequently asked by Members at: www.rsgb.org/faq/

**Band Plan**
Band Plans are available in this Yearbook, and the very latest version of the band plans are always available on the website at:
www.rsgb.org/committees/spectrumforum/band-plans.php

**Good Operating Practice**
The RSGB fully supports the code of conduct and encourages all amateurs to read the advice at www.rsgb.org/main/clubs-training/training-resources/advanced-resources/ & www.rsgb.org/main/operating/operating-guidelines/

**RSGB Tech**
The purpose of this service is to be the first port of call for technical queries on amateur radio matters. It is open to all radio amateurs. See http://groups.yahoo.com/group/rsgbtech/

**RSGB Shop**
All RSGB goods - books, filters, clothing etc - can be purchased online at:
www.rsgbshop.org/

**Club Finder**
Use the website to find your nearest radio club and check out the facilities they have to offer.
www.rsgb.org/clubsandtraining/

**Yearbook**
If you have moved home, if you would like your name and address to be withheld from future editions of the Yearbook (or released, for use in it), or if your callsign is not listed, you can **only** make the necessary changes to the database via Ofcom, direct by phone or via the internet at Ofcom's website below:

Tel: 020 7981 3131
www.ofcom.org.uk

## DISCOVER RADIO COMMUNICATIONS

# Radio Society of Great Britain

*Benefits that keep you on the air*

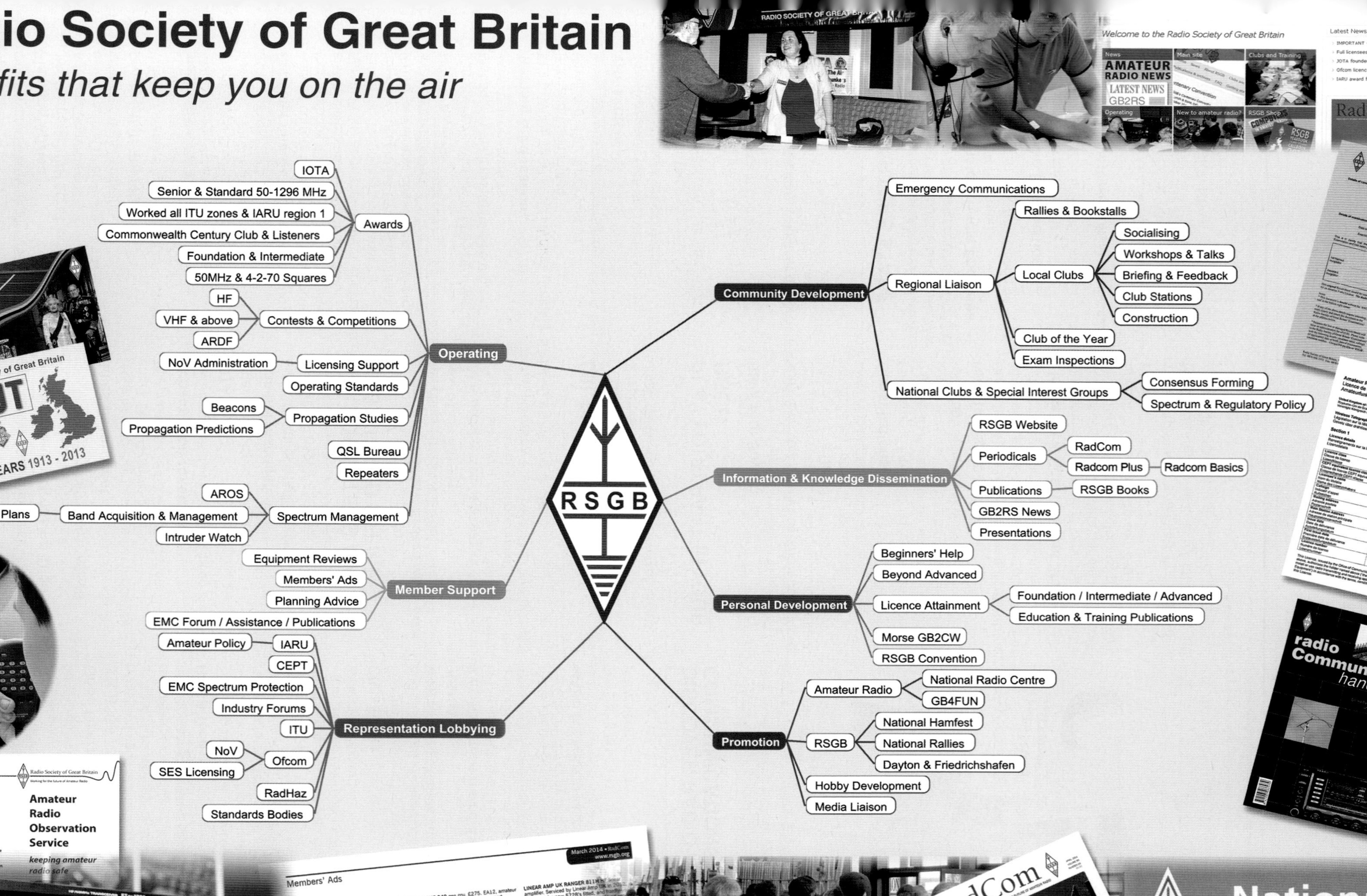

# National Radio Centre

The National Radio Centre, created by the RSGB, is a public showcase for radio communications technology - a technology powering the 21st century economy. The Centre provides the opportunity for members of the public to get up close and personal with the history and technology of radio communications.

This world-class radio communications education centre is situated at Bletchley Park in Buckinghamshire. From the first inventors in the late 19th century through to future radio developments, visitors will find films, interactive displays, hands-on experiments and even the opportunity to go 'on air' in our state-of-the-art amateur radio station.

Visitors learn about the basic principles of radio and discover the history of radio communication. They see how different parts of the radio spectrum have different uses and can explore the different uses of radio and experiment with the building blocks of a radio system.

The NRC also allows visitors to find out about the role of radio amateurs - who push technology to the limits and have fun at the same time.

## The NRC Experience

Starting with a short film *Wireless Communication Powers our Lives*, visitors gain an overview of the role radio communication plays in our lives today - the vital driving force of the 21st century.

### Interactive Displays

The National Radio Centre boasts a collection of interactive displays - both hardware and software - and experiments that get visitors 'up-close and personal' with the workings of a radio communications system.

Interactive touch-screen presentations take visitors through key areas of radio technology, while interactive hardware displays allow visitors to explore and discover the technologies that come together to make radio work.

### Wall of Radio

Then, there comes the 'Wall of Radio'. This fascinating display shows the history of radio from its early beginnings in the 1890s through such milestones as the first regular radio broadcasts in the 1920s and the invention of radar in 1938.

There is much more, including the role of radio amateurs in developing technology over the years and in serving their country during WWII, the development of the mobile phone, amateur radio space satellites and the amateur radio station on the International Space Station.

### GB3RS Live Demonstration

Then visitors can go on air themselves, seeing who they can speak to around the world using our state-of-the-art amateur radio station. Qualified operators are on hand to help you experience something you will never forget.

*The Lord Lieutenant, Sir Henry Aubrey-Fletcher Bt JP being shown some of the Interactive Displays by former RSGB President Bob Whelan G3PJT on July 5th 2013, RSGB Centenary Day.*

### Future Zone

Finally, the Future Zone at the NRC (sponsored by the University of Bedfordshire) looks at the latest developments in radio technology. Radio is the technology that will continue to have a huge impact on our lives in the future.

The NRC is located at Bletchley Park Heritage Site. Entry to the whole site is free for RSGB members on production of a downloadable voucher from the RSGB website: *www.rsgb.org*

*The Wall of Radio shows the history of radio communication.*

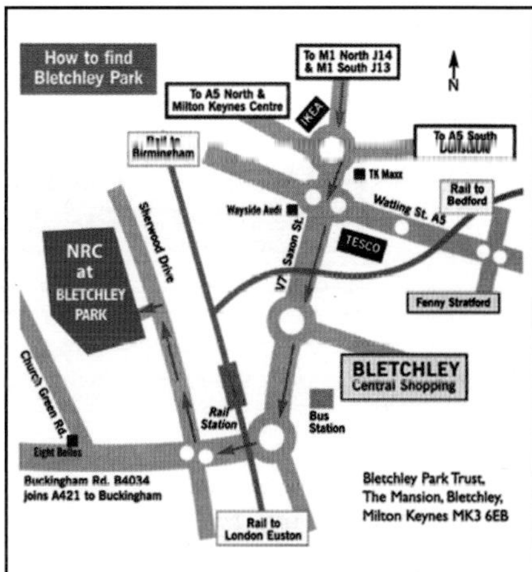


The National Radio Centre located at Bletchley Park, Bletchley MK3 6EB.

For opening times and other details *see*: *www.nationalradiocentre.com*

# RSGBTech

RSGB Tech is an independently-run Yahoo Group. It aims to provide technical help and discussion of amateur radio related technical matters, and currently has around 1,500 members from a range of different interests within amateur radio and around the world. It is a useful source of independent discussion of technical matters raised in RSGB publications, periodicals, website and other communication channels.

## If you have just started in amateur radio, can you use it?

Yes, of course! RSGBTech is meant for everyone, the newcomer and the old hand alike. The biggest asset to the hobby is radio amateurs themselves. There is a wide range of backgrounds, knowledge and experience on the site. Even if you require advice on setting up a station for the first time, need to get to the next stage of the licence or want to source that special component which you have not been able to find for ages then RSGBTech could be the place for you. There is always someone willing to advise.

## Is the site moderated?

Yes, it is moderated in order to ensure that the questions are on topic, ie they have to be technical and related to amateur radio.

## What else you can do on the site?

You can post photographs, files, drawings etc, but please ensure that they are not subject to copyright. If they are, then please ask permission from the person who owns the copyright in the first instance.

## Can you use it to sell equipment?

Sorry, the site is not meant for that, so private sales and ebay are not permitted. Its main purpose is technical queries and discussions.

## How to join the site

Simply go to a search engine, typically Google, and type in "rsgbtech" or the following full URL: *http://groups.yahoo.com/group/rsgbtech/* Once you see the RSGB logo you know that you are on the right track. If you are new to Yahoo! then you need to click on 'new user' to set up your account (at no cost). Once that is done, click on the box that says 'join this group'. Confirmation will then be received very quickly. The whole process only takes a few minutes.

## Sending a technical query

Once you have set up your Yahoo! account and logged in, simply click on the box that says 'start topic', type in the subject details and your technical query. Finally, click on the box that says 'send'. Your message may take a little while before it appears on the site, as it is now in the process of being moderated. You may get a few replies or lots! It may indeed start a flurry of activity and interest from around the globe on a particular topic that keeps the moderators very busy!

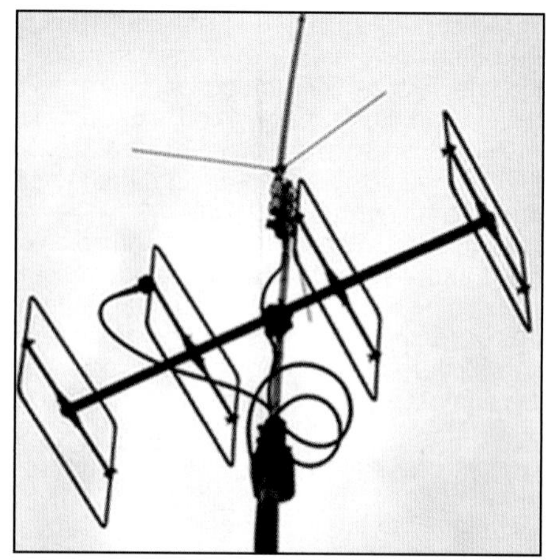

*G7DSU's 'problem quad'.*

## Recently discussed topics

*The topics discussed recently include*:

- Airspy SDR
- 50 Ohm balanced feeder
- Weather Stations
- New PSK31 Satellites
- Wiring Regulations
- EMC and USB Chargers
- Gamma Matches
- KW202 Circuit Diagram
- PIC Processor Enhanced Instruction Set
- The Innards of power transistors

## Answering a technical query or otherwise responding

If you want to respond to a technical query, click on the message and then on the box marked 'reply'. Ensure that it replies to rsgbtech@yahoogroups.com then everyone will be able to see the message.

Keep comments as brief as possible, and if reponding to a long message, delete as much of the previous text as possible to avoid readers having to scroll though masses of repeated informtion to get to the relevent bits.

RSGBTech is one of the most dynamic ways for the modern amateur radio enthusiast to have a technical query dealt with.

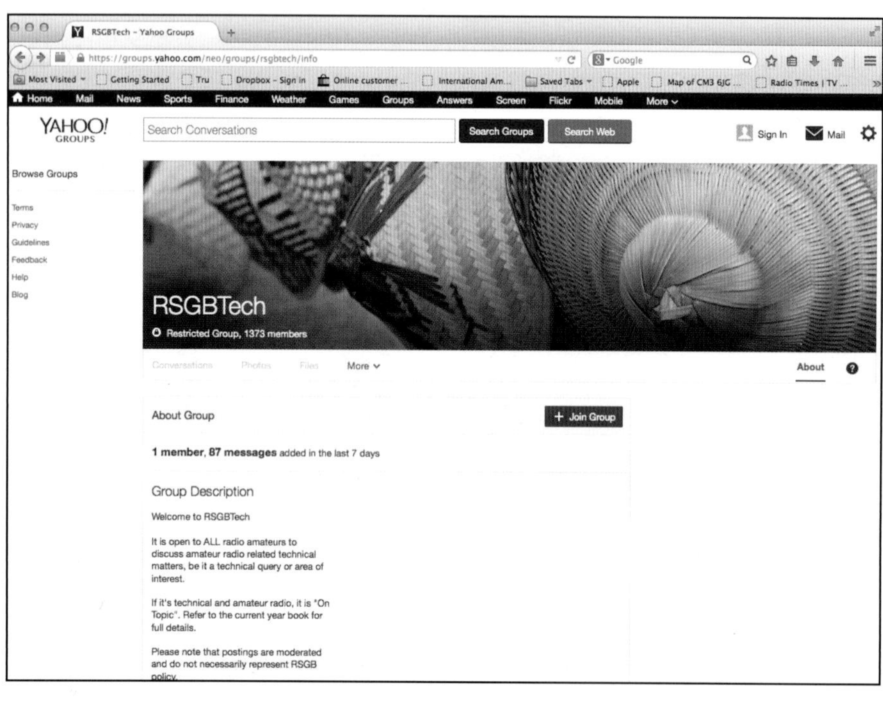


*Typical front page of RSGBTech, the Yahoo! site where you can ask technical questions and give advice.*

Join now and good luck!

**RSGBTech**: *www.rsgb.org/rsgbtech/*

# RSGB online

Home News Events About Us Clubs Training Operating Technical Radio Sport Join Renew Shop
Forums Publications Archives Consultations FAQs Get Started Contact Login Search

Over 80,000 unique visitors come to rsgb.org every month. There are hundreds of pages of information and links to resources from around the world, plus the very latest news from the world of amateur radio.

## PORTAL

***rsgb.org***

Go straight to key areas of the website on our tablet and mobile-friendly front page. Access the latest news headlines, the main site index, guidance for newcomers and information on training and operating, plus the latest edition of RadCom Plus.

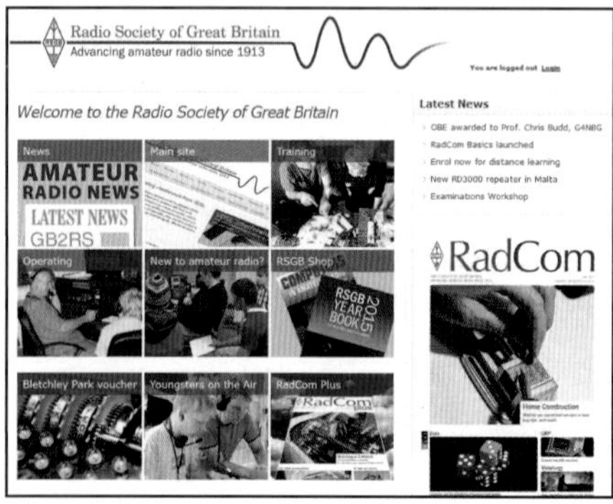


## MAIN SITE

***rsgb.org/main***

The main site provides access to all the content in rsgb.org. The latest updates to the website are listed on this page along with our most important current events and activities.

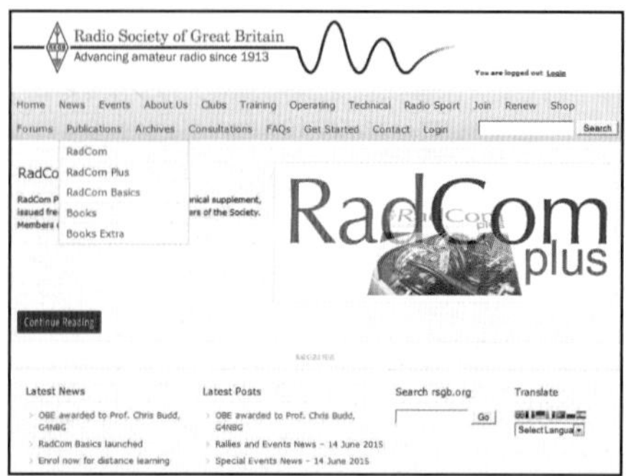


## NEWS

***rsgb.org/news***

The Society's flagship news bulletin GB2RS is published here every Friday, and we add further news items throughout the week. Visit this section for regional, national and world news, plus upcoming special events and all the latest from the world of contesting.

## EVENTS

***rsgb.org/eventsplanner***

UK Events Planner displays a map of forthcoming amateur radio events. Click on the markers to display essential information about each event.

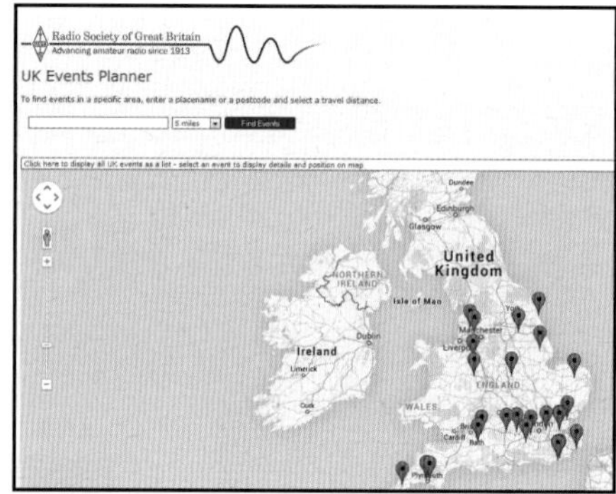


## ABOUT US

***rsgb.org/main/about-us***

A summary of who we are and what we do. You will also find important documents here such as our Code of Conduct, plus a link to the National Radio Centre, the RSGB's world-class educational facility at Bletchley Park.

## CLUBS

***rsgb.org/main/clubs***

Club Finder is the UK's most comprehensive and up-to-date listing of amateur radio clubs. Enter your location or postcode and select a travel distance to display clubs near you. Click the markers to display meeting, contact and training information and links to club websites. This section also includes information on how to affiliate your club to the RSGB and a club media guide, as well as information on the Society's insurance scheme for affiliated clubs.

## TRAINING

***rsgb.org/main/clubs-training***

UK Course and Exam Finder allows you to search for courses and exams in your area. You will also find educational resources for your studies, support material for trainers and useful information about exam fees, together with the forms to download.

## OPERATING

***rsgb.org/main/operating***

Information on band plans, awards, beacons and repeaters, emergency comms, the QSL bureau, planning, CW and NoVs, including online applications for selected NoVs.

## TECHNICAL

***rsgb.org/main/technical***

Home to the EMC and propagation pages and a wide range of other specialist information, including space and satellites and microwave operation. There are also links to the RSGB Technical Forum and useful apps.

## SHOP

***rsgbshop.org***

Here you can join the Society, renew your Membership and buy RSGB and other publications with a Members' discount. The shop also supplies EMC-related components, RSGB-themed polo shirts and baseball caps, and IOTA merchandise. All major cards accepted.

## JOIN

***rsgb.org/join***

Sign up for RSGB Membership and become part of our 20,000-plus community working for the future of amateur radio.

## RSGB FORUMS

***forums.thersgb.org***

Express your views on a wide range of amateur radio topics. There are two permanent forums on EMC and Radio Propagation, as well as occasional consultations on matters of importance to the amateur radio community.

## PUBLICATIONS

***rsgb.org/main/publications-archives***

Members can read the digital edition of RadCom and search the RadCom archive, as well as read our new supplements RadCom Plus and RadCom Basics. You can also browse and purchase the full range of RSGB books.

## ARCHIVES

***rsgb.org/main/archive***

The Photo Archive contains a century of fascinating amateur radio photography, whilst the Events Archive includes preserved material from major RSGB events. Go to the Publications Archive for back issues of current and discontinued publications, including RadCom

Basics and RadCom Plus. You'll find material relating to closed consultations in the Consultations Archive. Our new Video Archive contains both promotional videos about the hobby and also some presentations from last year's RSGB convention.

### CONSULTATIONS

***rsgb.org/main/rsgb-consultations***

A list of current active consultations on matters of importance to the amateur radio community. You are invited to participate. Contains links to relevant forums.

### FAQs

***rsgb.org/main/faq-2***

If you have an amateur radio-related question, chances are it has been asked and answered before. In this section we answer your most common questions on amateur radio, DBS checking, exams, IOTA, RSGB Membership and how to become a radio amateur.

### GET STARTED

***rsgb.org/main/get-started-in-amateur-radio***

Everything you need to know in one place if you are new to amateur radio, from getting licensed to setting up your first radio shack.

### CONTACT

***rsgb.org/main/contact***

The RSGB's address, phone and fax numbers, and departmental email addresses.

### LOGIN

***thersgb.org/members/login***

The Membership Services portal is the place to update your RSGB account details, renew Membership, reset your login password, update your roles and preferences, read

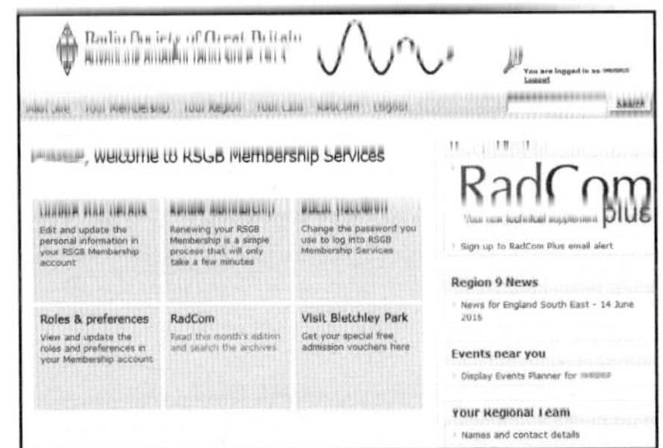

our digital publications and download a free admission voucher to Bletchley Park. You can also browse your region's news and view events taking place in your area. Affiliated clubs and Regional Managers can post events on the UK Events Planner.

### SOCIAL MEDIA

***facebook.com/theRSGB***

***twitter.com/theRSGB***

***youtube.com/theRSGB***

Visit the RSGB's Facebook and Twitter pages for breaking news and extra material not available on rsgb.org. Like us, follow us, ask a question and share news of your amateur radio special events.

### VIDEOS

We've created a variety of videos ranging from a celebration of the Tim Peake school amateur radio contacts, to promotional films that explain more about the hobby and some inspiring stories from some of our current volunteers. Find them – and a selection of videos from last year's RSGB convention – in our video archives or go to our YouTube channel:
*youtube.com/theRSGB*

Join the RSGB at: *www.rsgb.org/joinus*

# Abbey Court

The Radio Society of Great Britain continues to be one of a few radio societies in the world to maintain a full time staff. The Society is now administered from a modern, two-storey, open-plan office situated on the prestigious Priory Business Park in Bedford. No. 3 Abbey Court houses the General Manager's Department, Sales and Accounts, Examinations Department, Website Management, IT, *RadCom* and *RadCom Plus*.

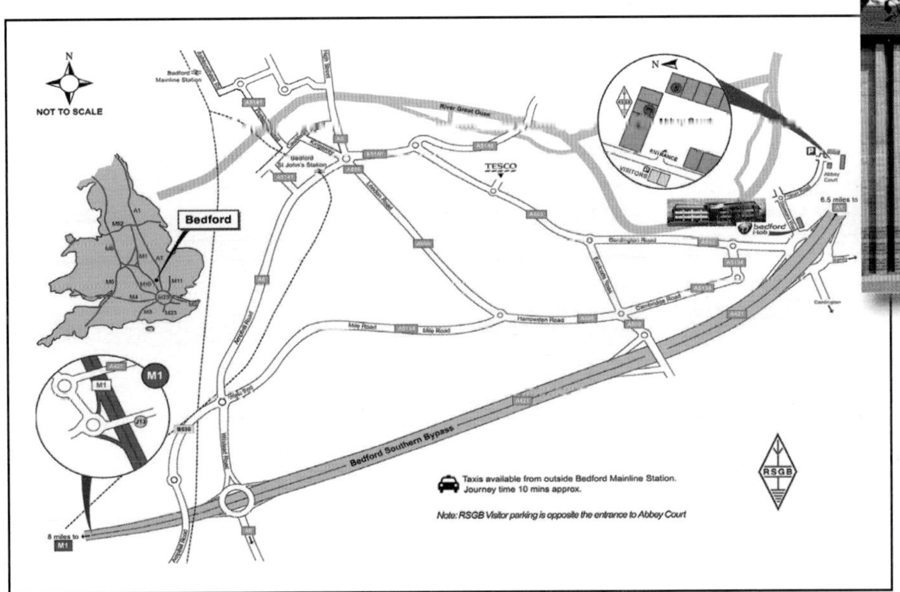

**RSGB**

3 Abbey Court, Fraser Road, Priory Business Park, Bedford MK44 3WH
**Tel**: 01234 832 700
**Fax**: 01234 831 496
**Web**: *www.rsgb.org*

Office hours are Monday to Friday, 8.30am to 4.30pm

# GB1SS: Schools speaking to Tim Peake

AMATEUR RADIO
CONNECTING SCHOOLS
WITH THE
INTERNATIONAL
SPACE STATION
https://principia.ariss.org

*by Heather Parsons**

"Golf Bravo 1 Sierra Sierra, GB1SS, this is GB1SAN calling and standing by, over". And so began the scheduled amateur radio contact between Sandringham School and British ESA astronaut Tim Peake on the International Space Station (ISS) on Friday 8 January 2016. Led by Year 10 pupil Jessica Leigh M6LPJ, a newly-licensed radio amateur, this was the first of ten school contacts as part of the Principia mission, that have caught the imagination of a nation and inspired a new generation of radio amateurs.

### How it all began

In 2013, Ciaran Morgan, M0XTD was appointed RSGB lead for this ARISS (Amateur Radio on the International Space Station) operation. He began discussions with the European Space Station (ESA) and the UK Space Agency, demonstrating what ARISS could do, and in early 2015 ARISS UK Operations was officially brought on board as a partner with the agencies and began the detailed preparations for the school contacts.

Every school in the country was invited to apply to take part and the sixty that answered the call had to go through a rigorous process. Not only did they have to show they had the facilities and capability to host an event that would attract global attention, but also how they would embed the contact into a wide range of STEM-related activities for the pupils at their school and at schools in the surrounding area. This was an important part of the process for the contacts – ARISS in the UK didn't want the contacts to be a one-off event, but the springboard for inspiring young people into careers in the STEM (science, technology, engineering and maths) sector.

This embedding of the contact into such a focused STEM programme was significant and unusual, as Ciaran explained: "It is the first time that a national space agency programme has been so heavily intertwined with the ESA outreach programmes, and also where a core team of ARISS volunteers have been so incorporated into the delivery of that outreach. The key aspect was that the contacts were heavily integrated into the astronaut's own educational programme; consistency was achieved through having the same team of people dealing with both the school amateur radio contacts and the overall astronaut educational programme."

The final ten schools were shortlisted for an amateur radio contact with Tim Peake, but at this stage they knew that their contact still wasn't guaranteed.

*Some of the ARISS equipment used for the contacts*

### Lift off!

On 15 December 2015 Tim Peake launched successfully to the ISS. The launch was celebrated across the UK as millions of people watched the live feed and over 15,000 people gathered at launch events across the country to wave Tim on his way. The RSGB was delighted to be part of the launch events in London, Cardiff and Belfast where we were able to explain more about amateur radio to school pupils and the public, and to share in the excitement of this special day.

### Building up to the contacts

The first four contacts were confirmed soon after launch but, as the project developed, Ciaran was able to deliver the unexpected and exciting news that all ten schools would have a contact with Tim Peake. Working with other schools and educational organisations in their area, each of the ten schools created a wonderful programme of space-related events that involved every aspect of the curriculum. Expert speakers explained the mysteries of space, pupils were involved in science experiments and built models of the ISS, food suitable for astronauts was produced, musicians joined together to

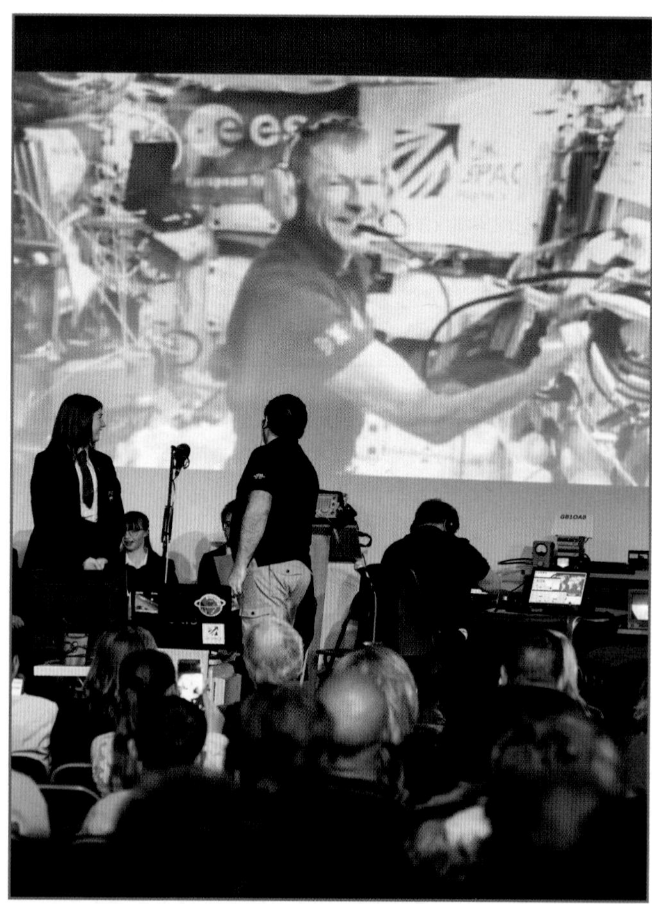

*Tim Peake contact with Oasis Academy Brightstowe, Bristol*

*Pupils building a Rodway receiver*

play space-themed music and to create musical backdrops for artistic displays. Pupils of every age were involved in sporting challenges and dance extravaganzas that reflected the rigors and joys of space travel and exploration.

### Amateur radio buildathons

As part of the build-up, the RSGB worked with local clubs to organise buildathons at the schools. Pupils worked individually, in pairs or small groups to build RSGB/Walford Electronics Rodway receiver kits.

RSGB Board Chair Steve Hartley, G0FUW was the driving force behind the buildathons and was keen to show pupils some amateur radio construction that would help them to understand the technology behind the contact with Tim Peake: "It has been a real team effort putting these events together. It is a very simple receiver that can be built in a few hours and receives medium wave broadcasts to give an instant 'wow' factor, but at the flick of a switch it can receive amateur radio top-band contacts. The Radio Communications Foundation funded 200 of the kits to enable all of the schools to experience this exciting event."

Linking the receiver to a small antenna the pupils were very excited to hear radio transmissions on something that they had just built; it also inspired some girls to think about their career choices. As one pupil said, "It has been a very new experience and quite challenging. Not enough girls want a job like this so it has been good to build this today so more women might start to be interested in this type of career."

The teachers were also so impressed by the kits and the way in which pupils were engaged in the task, that many are planning further radio-building activities in the future.

### Contacts

The contacts began on Friday 8 January 2016 at Sandringham School, St Albans and ended on 9 May at King's School, Ottery St Mary. In the intervening four months, the ARISS team travelled all round the country to set up and run sometimes two contacts in a week, to fit in with Tim's availability: from Royal Masonic School for Girls, Rickmansworth across to Oasis Academy Brightstowe, Bristol, from the City of Norwich Schools over to Powys Combined Schools in Wales; to the south east for St Richard's Catholic College, Bexhill on Sea and Wellesley House in Broadstairs; and then back up north to the Derby High School, Bury and Ashfield Primary School, Otley.

Each contact had its own unique atmosphere, from musicians dressed in space suits to a four-year-old asking Tim a question whilst clutching her teddy bear for moral support. Regardless of age, the excitement that everyone felt when they heard or saw Tim was palpable and spontaneous applause rang out. It was the first time that Ham TV had been used for school amateur radio contacts and seeing Tim wave and smile added a very special element to the events.

Special event call signs had been obtained for each of the ARISS school contacts and the call sign was related to the school so that it had a special meaning for them e.g. GB1SAN for Sandringham School. Nearly all of the schools had a newly-licensed pupil leading the contact, which was nerve-wracking for them but also showed how accessible and exciting amateur radio is for people of all ages. The UK Space Agency headed up expert panels at each contact, answering all sorts of questions from pupils and parents about space, living and working as an astronaut, the possibilities for future exploration and habitation on other planets and even whether they preferred earth or space! The local clubs continued their support for the schools by putting on a special event station (SES) at each event to let people see amateur radio in a more relaxed way after the tension and excitement of the contact with Tim Peake. The clubs, RSGB Regional Team and Youth Committee also played an important role explaining to staff, pupils and sometimes the media, all about amateur radio and the huge variety of activities it encompasses.

From the start of the ISS programme in 2001 until 1 January 2016, in the UK there had been 12 direct contacts with schools via amateur radio and six telebridge contacts. Thanks to the ARISS UK Operations contacts this year under the Principia mission, an additional ten direct contacts have been added to that total. Ciaran sums it up: "This has been a fabulous project and a great team effort that has reached and inspired thousands of young people across the UK. Not only that, but the schools have been left with skills, contacts and the experience of interacting with the communications and space sectors, which was an important goal for creating a legacy from these events."

### Looking ahead

Whilst the contacts are over, in many ways this is just the beginning of making the most of their legacy. The Society has set up the Schools Link Project (***www.rsgb.org/schools-link-project***) to develop wireless-related science and its technologies into activities for use in everyday teaching. Initially we are working with the ten ARISS contact schools and Ian, G3YNU from the Training and Education Committee is leading the project.

Some schools have already set up amateur radio clubs as a result of pupils getting their Foundation Licence to lead the ARISS contacts. Others are being very active with their local clubs and are looking for ways in which to include amateur radio activities into their busy schedules, or are running Foundation courses for other pupils. We will keep sharing news, but if you're interested in getting involved, contact Ian via: ***schools.link@rsgb.org.uk***

At the final contact Libby Jackson, Astronaut Flight Education Programme Manager, UK Space Agency, summed up the amateur radio contacts:

"The UK Space Agency has been thrilled to work with ARISS UK and the RSGB in these calls. You've been wonderful people to work with. It has been really important that we've built this partnership together to deliver such a successful programme to ten schools across the country. Many tens of thousands children have participated in this and I am very confident that they will remember this for the rest of their lives. That will help some of them consider careers in the space sector, in the electronics sector, in all these wonderful careers that can come out of space and amateur radio."

*ARISS and UK Space Agency Teams*

*Heather Parsons, is the RSGB Communications Manager. email: heather.parsons@rsgb.org.uk*

# RSGB QSL Bureau

Whilst sending cards for a much-prized contact will always be quicker by direct mail, QSLing via the RSGB Bureau remains an extremely cost effective option, indeed the RSGB QSL Bureau enables members to exchange cards worldwide in the cheapest practical way.

## How it works

QSL cards arriving at the central bureau are initially separated into UK and Foreign destinations. Overseas cards are sent in bulk to other member societies of the International Amateur Radio Union (IARU). Cards for stations within the UK are sorted into separate callsign groups and sent to the appropriate volunteer collection managers, on a quarterly schedule. They place cards in stamped addressed envelopes (SAEs) provided to them by the call holders.

## Who can use The Bureau?

Unlike the RSGB, many other national societies make extra charges for using their QSL service. The RSGB QSL Bureau is an inclusive membership service and operates as follows:

- **UK RSGB members** can send and receive their personal cards without additional charges, subject to the conditions shown here.
- **UK non-RSGB members** can collect their personal cards only by using the '*Pay-to-Receive*' service but cannot send cards via the bureau. See RSGB website for details
- **Overseas RSGB members** can send their outgoing cards to the RSGB QSL bureau for distribution. UK call holders should collect in the normal way, via their UK call-sign. Non-UK call holders should arrange collection via a UK-based QSL manager, who should also be a member.
- **Overseas non-RSGB members** may send cards addressed to UK-based stations only.
- **Affiliated Societies and independent QSL Managers** can send their own cards and those for club members or stations for whom they act, but should include current RSGB membership details for every station whose cards they wish to send. Cards included from overseas stations and intended for delivery outside the UK will not processed without proof of membership and will not be returned.

## Available Destinations

A full list of IARU partner QSL bureaus can be found at: *www.iaru.org/iaruqsl.html* Keeping an up-to-date copy to hand is vital when deciding which route to send your card. For example, there are currently no bureaus in, Egypt, Kazakhstan, Morocco, and Mauritius, Sudan and several other African and Caribbean countries, plus many more smaller destinations.

Activity also relates to the frequency with which cards can be dispatched to a particular destination. This may range from monthly to annually, according to demand and is something to consider before sending your card via the bureau.

## Responsible QSLing

The Bureau handles approximately 1.5 million cards per year and is one of the busiest in the world. The Society has a policy of discouraging the sending of cards when they are not wanted or cannot be received.

Active Amateurs, GB and Special events, all Clubs and DXpeditions are strongly advised that 100% QSL outgoing is no longer desirable or cost effective.

Transporting large volumes of cards between bureaus, only to have them ultimately destroyed, returned or uncollected, is disappointing and not eco-friendly.

**Tip:** Ask yourself… Do I need to send a card for every contact before QSLing? Always ask the other station if they can receive a bureau card, before sending.

### Log Book of the World (LOTW)

Receiving a nice card for a memorable contact is always a thrill, never matched by an electronic confirmation via the Internet. However, do consider the alternatives, uploading your logs to Logbook of The World can automatically confirm some contacts, such as for contests and award purposes etc.

Confirmations via LOTW are easy and work well for everyone, if a few simple steps are followed. See: www.arrl.org/logbook-of-the-world

### Responsible QSLer...

*Help us to speed up processing and cut waste for everyone,*

- 10 Simple things you can do.....
- Ask new contacts, "If I send you a QSL card, do you collect and how? – every time
- If you don't QSL, be polite but honest "Thanks but no thanks" is all it takes.
- Please don't say you do when you don't, or ignore the other guy's kind offer,
- QSLing 100% outgoing is costly for everyone. Half your cards could be wasted, please check before you send.
- Create your own OQRS system or use ClubLog www.clublog.org to reply to incoming requests with a real card, save time and money for you or your club.
- Make your QRZ.com QSL details clear, honest and visible
- Amend your on-line details if you change your QSL status - don't leave it.
- QSL info– Direct or via...' is confusing. Please clarify what you really mean .
- Clubs Calls – Collect only from the calls sub group. For, 'QSL-Direct,' show an address not a call-sign as this can change, it's often confusing .
- Always collect your cards. - Even if you never send one, they will arrive.

***Be Responsible, it only costs a stamp!***

## OQRS systems - the future of QSLing?

Many stations and most DXpeditions and rare calls are now using the worldwide OQRS network and only responding to requests for QSL cards. This online system means there is now no need to automatically send a card, to receive one via the bureau, or direct.

Using OQRS also speeds up the system so that it can now be only half the time it presently takes to send and receive a card, with the added benefit of not needing to send yours. Simply put cards are only sent in response to OQRS requests for a card. So if you are sending QSL cards you or your QSL manager will receive an email to generate a genuinely wanted card. This saves time and waste for both the user and the QSL system in general and is therefore recommended as good practice.

In the UK we are fortunate to have the free to use ClubLog, courtesy of Michael Wells, G7VJR and his team. Simply go to, www.clublog.org for more information or to register your call, club, GB station or event and start uploading your logs

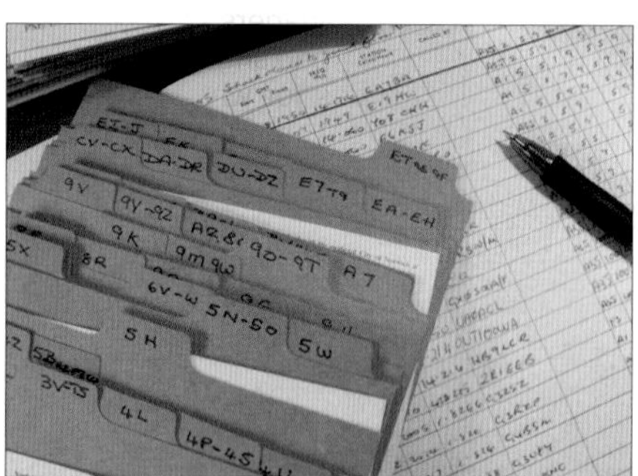

*Hand written cards can be easily pre-sorted, using a simple card index box.*

## Sending cards via the Bureau

Cards from RSGB members for both UK and worldwide should be sent, suitably packed, to the main UK bureau address: **RSGB QSL Bureau PO Box 5, Halifax HX1 9JR, England.**

Members, clubs or DX groups wishing to send large or heavy packages to the Bureau via carriers other than Royal Mail should contact the Bureau for an alternative delivery address.

The bureau operates a 24-hour message line for members' QSL enquiries. Tel: 01422 359362.
When calling, please leave contact details, a brief message and if possible an email address.
email: qsl@rsgb.org.uk (please put your callsign series in the subject box)

### Fair Usage Policy

As part of their subscription, each Member can send up to 15 Kgs of cards through the Bureau each year (about 5000 standard cards).

Each Affiliated Club can send up to 20 Kgs through the Bureau each year. Additional cards will be charged at £6 per kilo or part thereof.

In the interests of fairness to others, members should only send to the Bureau a maximum of 1kgs of cards (approx 300) to any single DXCC entity per month (larger quantities should be sent directly to the bureau in the relevant country - *see* IARU list online).

Heavy users such as DXpeditions, some clubs, etc, will be required to send the bulk of their outgoing cards direct to the destination countries.

### Each batch of cards should contain...

- Proof of current membership; that is an original *RadCom* address label showing address, callsign and membership number, not more than three months old. Leaving out this information will likely result in a delay, while membership status is checked.
  As Clubs receive the RSGB Yearbook each year in lieu of RadCom, they should include sufficient information for a check to be made against the Affiliated society's register. Ideally, in the form of a club letterhead, showing the membership number and renewal date. To speed status checking, clubs and groups are asked to ensure that they register club call and contact details at *My Account* directly in the group's name and not as secondary to a personal call sign, or qsl manager.
- Special event stations (GB) and single letter Abbreviated/Contest callsigns should include the membership number and call of the NoV license holder or affiliated club, for contact purposes.

### Other important points

- Clubs and QSL Managers sending a bulk dispatch to the bureau should ensure that all callsign holders for whom they send cards are current members of RSGB and should enclose current membership details for every callsign with every batch of cards.
- Members who operate from another station, typically a foreign club call or that of an individual overseas amateur, may send cards for contacts made from that station, provided they clearly identify themselves as the operator and state their UK callsign and membership number on each card.
- Listener report QSLs need sufficient information to be of genuine value to the transmitting amateurs. Reception reports relating to broadcasting stations cannot be accepted.
  The bureau system accepts standard cards only no letters, SAE's or money orders.
- All cards, whatever the quantity, should be pre-sorted into alphabetical and numerical country DXCC order (see the Prefix List pages and/or the *RSGB Prefix Guide* which also contains a complete cross reference and awards section).

## Checklist for sending cards

***We need your help to sort more than a million cards each year and reduce delays.***

**A** *First, place your cards into three piles...*

**1. UK destinations**

Pre sort G, M and 2 as per the Sub Managers list.

**2. USA destinations**

Sort by number only, 0-9, regardless of prefix. Separate cards for Alaska, Hawaii and Puerto Rica.

**3. Rest of the World**

In DXCC callsign prefix order.

**4. Calls with numbers first**

Sort in digit order, i.e. 3A, 4X, 8P, 9H etc.

**5. Calls with one letter then one number**

These come before two letter prefixes, i.e. S5 before SM, etc.

**B** *Check ALL cards for possible 'Via' destinations*

Re-sort if necessary, and never rely on your computer print log, for example: F5/G3UGF isn't a French destination.

Africa, Caribbean and DX destinations are mostly QSL Direct only, or via a QSL Manager. Check *www.qrz.com*

**C** *Pack your cards securely and don't forget*

A recent RadCom address wrapper as proof of membership.
Your callsign and return address on the package.
If you put more than ten cards in a C5 envelope, check the dimensions and weight at the Post Office, before sending don't just post.

***Whatever the quantity, never send unsorted cards!***

- Countries with more than one prefix should be placed together. For example, JA and 7J cards (destined for Japan), F and TK-TM cards (destined for France) and SP, HF and 3Z (destined for Poland) may be grouped together.
- Cards for the USA need be sorted separately into call areas (numbers 0-9), regardless of the prefix letters.

**NB**: Exceptionally, cards in the number 4 series with either one or two letters before the number are handled by different bureaus and need to be separated, as are cards for Alaska (KL), Hawaii (KH6-7) and Puerto Rico (KP3-4).

- Cards for Russian Federation and former Soviet countries were traditionally grouped together. They now need to be separated into five individual groups; RA-RZ, UA-UI, UJ-UM, UN-UQ and UR-UZ, as they are no longer sent to a single destination in Moscow. See: *www.iaru.org/iaruqsl.html*
- Cards for UK delivery need to be provided separately to foreign destinations. Our UK sorters currently have to split cards into some 112 alphanumeric categories. For this reason cards should be supplied to us pre-sorted order as per the Sub Manager list on the following pages.
- Envelopes, paper or card dividers to separate countries or call groups are not required, as removing these can sometimes slow down the distribution process.
- Cards sent in date/time/logbook or random order are not acceptable, as they typically take up to five times longer to process. Similarly, those with small print or hand written callsigns can be very difficult to process, resulting in delays for other users. The bureau reserves the right, at its discretion, to reject unsorted cards or those with callsign or routing information in small and difficult to read print. The minimum print size requirement is 12 point.

**Tip:** If you are unsure about your handwriting, why not ask someone else to check the cards to see if they can easily read the callsigns?

## Card issues and some good advice

To avoid possible transit damage and in fairness to others, all cards should be single page, standard postcard sized (140 x 90mm). Card weight/thickness is important and needs to be in the range 130-330gm paper board for easy processing.

## Before you send, please check that we can deliver!

**Check all Vias before you send, as your card may come back to you, or it may never arrive.**

**There are many world destinations, but only 190 IARU member and associate Bureaus worldwide.**

**The following IARU Bureaus are currently closed.**

| | |
|---|---|
| 3B Mauritius | D4 Cape Verde |
| 3DA Swaziland | HH 4V Haiti |
| 4J-K Azerbaijan | HV Vatican City |
| 7P Lesthoto | PZ Suriname |
| 9L Sierra Leone | ST Sudan |
| A3 Tonga | SU, 6A-B Egypt |
| C2 Nauru | V3 Belize |
| C5 Nauru | V4 St Kitts and Nevis |
| C6 Bahamas | XY-XZ Myanmar |
| CN, 5C-G Morocco | Z2 Zimbabwe |

Download your own Bureau list from: *www.iaru.org/iaruqsl.html*

Large or unusual shaped cards are extremely difficult to process and most easily damaged when packed, or folded with others.

Thin, small, and paper cards are slow and extremely difficult to handle. They often stick to cards for other destinations, as do homemade cards using photo print or heat laminated paper. This type of card does not travel well, is difficult to write on and is very easily damaged when subjected to humidity or damp – not recommended!

Multiple page and non-standard cards should be avoided, as they increase the Society's workload and overheads, at the expense of other members. They can also significantly reduce the numbers of individual cards per consignment to overseas destinations. In quantity they should be sent direct to overseas bureaus. Therefore, in the interest of fairness to other members, they are sometimes spread over several shipments.

If multiple page cards are used they should weigh no more than a single page standard card typically 3gm maximum and should be pre folded to clearly show the destination callsign.

## QSL Routing & QSL Direct

It sounds obvious, but the Bureau can only process outgoing cards if there is a destination to which they can be sent. Before sending cards, therefore, (particularly to rare stations or DXpeditions) please check the recipient's QSL policy. This is usually available on QRZ.com or via a websearch.

Many DXpeditions and rare call signs only QSL direct, or respond to an OQRS request often via a QSL Manager who may be in another country. These stations are most often located where there is no bureau service and are operated by visiting non-resident Amateurs from another country. Some stations do not QSL at all, so it is vital to check before sending, whatever route you choose.

Please note that outgoing bureau cards where no destination bureau is available, or no clear 'Via' route is indicated, will be recycled.

**Tip**:

- Ask for the other station's QSL details at the time of the QSO, or by an Internet search before posting.
- Consider posting your most wanted cards direct or to an overseas bureau, if it's active. This helps to speed replies as most bureaux world-wide have backlogs. The IARU world-bureau list can be downloaded at: *www.iaru.org/qsl-bureas.html*. It's good practice to check the listings for changes at least twice a year.
- Always search the web and check *www.qrz.com* first before posting.
- Make sure that any "via" information on your cards appears directly below, or next to, the station call sign, to avoid being missed. Using a different coloured ink for this purpose is a great help.

For guidance on what information to include on your card (and where), *see the example card on page 21*.

### Using printed labels

Avoid cramming too much information on small printed labels. For health and safety reasons all callsigns should be a minimum of 12 point print size and in common, easy to read fonts such as Arial, Times New Roman, or clear, block capital hand written letters. The bureau system is for the exchange of QSL cards only. Envelopes containing letters, photographs, IRCs, stamped addressed envelopes, awards, certificates and other items will not be processed and should be sent by other means.

## Heavy users

Those sending more than a few thousand cards per year should send their largest volumes of cards directly to their top ten destination countries. The remaining balance can be sent via the normal bureau system.

The aim of this is to share some of the burden of cost, without penalising others who may only occasionally send a few more cards than normal. The bureau weighs and notes regular large consignments. Members or clubs may be contacted if their usage becomes excessive, with a request to follow the guidelines above. The IARU bureau list can be found at: *www.iaru.org/iaruqsl.html*

## Packing and posting your cards

The bureau receives many damaged envelopes and packages from both UK and foreign amateurs. It also receives a significant number of requests each month from Royal Mail for payment of additional postage, which are always rejected.

Having first separated pre-sorted, UK destination cards from the rest of the world, please read on...

- Never post loose cards in lightweight or thin envelopes, as they will often cut through the edge of the envelope in transit.
- Always print return address details and callsign on your package, in case it arrives damaged.
- Secure batches of cards with a rubber band or - better still - a banknote style band of thin paper strip, folded around the cards.
- Never place two or more packs of cards side by side in a C5, A4 or larger envelope, as it will fold in transit and split down the middle, allowing the cards to spill out.
- Using lightweight 'Mail-Tuff' style plastic or 'Mail-Lite' style padded bags or Post-Pack envelopes usually avoids this problem.
- Always check the size and weight of your envelopes and packages, before posting as the Post Office now charge by volume as well as by weight.
- The current weight limit for a First Class stamp is 100g, but the package size is limited to 240mm x 165mm and the package should fit through a postal slot only 5mm in height.
- It is possible to send a large envelope A4, or a smaller envelope over 5mm in thickness). This type of envelope is considered to be a 'Large Letter.' Large Letter stamps costs more, but allows the letter thickness to be up to 25mm. 'Second Class Large' offers better value
- The Post Office can supply a paper/card copy of their pricing slot guide for a small charge. Frequent users are advised to obtain a plastic Helix HP5 'Pricing in Proportion' Ruler. It has postal slots built in, to check your packets.

Sending small numbers of cards in separate envelopes is not cost effective for the sender and means much more time spent opening, sorting and checking in the bureau. Sending not more than one pack per month, with your *RadCom* label, resolves many issues and can save you money .

*Part of the overseas side of the bureau*

Recorded delivery is not cost effective. We receive many packages and we are not always asked to sign for individual items, secure packing and a return address offers better value.

## Receiving cards from the Bureau

RSGB is extremely fortunate to have around 80 dedicated volunteer Sub Managers who give freely of their time to support the work of the Bureau and in the service of their fellow radio amateurs. Members' cards are sent to the Sub Managers for onward distribution.

Sub Managers details are subject to change, so it a good idea to check the QSL section of the RSGB website from time to time for the latest information. From the RSGB Home Page click, 'Operating' and follow the links.

Our system relies upon those wishing to receive cards depositing Stamped Addressed Envelopes with Sub Managers, ready for each quarterly despatch.

Members should use SAEs, as Sub Managers are not authorised or insured to accept money in lieu of postage stamps. RSGB is not liable in case of any loss or dispute.

The scheme is open to all RSGB members plus UK-based, pay-to-receive subscribers.

### Collection Envelopes

- Envelopes need to be C5 size (160mm x 230mm) and of strong material (*see earlier*).
- Callsign or Listener number should be printed in the top left hand corner, followed by the a current membership number, immediately below.
- Print the name, delivery address and postcode clearly, as normal.
- Number each envelope sent to the Manager (eg '1 of 6', '2 of 6', '3 of 6', etc) always mark one of them 'Last', so that you will know when a fresh batch should be sent.

*RSGB's new feedback card, is designed to help speed QSL throughput at one of the world's busiest Bureaux. If you receive one... "Please help us to help you.*

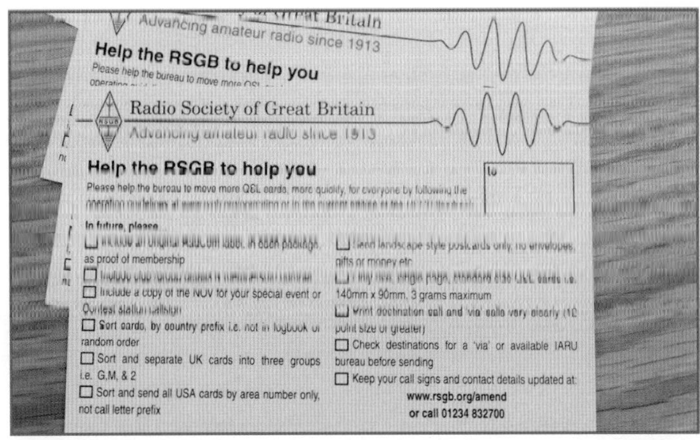


- Envelopes are normally despatched every quarter, subject to card availability.
- Always use stamps worded Second or First Class, rather than a numerical amount, as these will be honoured if the postal rate changes.

No delivery in that quarter means 'no cards waiting'.

Cards for amateurs who have not lodged envelopes are not returned to sender and will, at the Sub Manager's discretion, be recycled after a period of three months. Always keep your envelopes up to date. Many volunteer Sub Managers now operate their own websites, with links from the RSGB website, giving cards waiting, envelope status and next anticipated delivery details. RSGB requires these lists to be confidential. Members permission to display their callsign and details to others, is a condition of inclusion on any such listing, operated by a volunteer.

It is a good idea to note in your diary to check your Manager's list every quarter.

UK amateurs who do not wish to collect cards or those who use a separate QSL Manager are asked to notify the appropriate Sub Manager as a matter of courtesy and also make this clear at www.qrz.com

## More than one callsign?

Stations changing their callsign as a result of a licence upgrade etc should inform RSGB of their change of status, using the 'Amend my Details' form accessed from the home page of the RSGB website, (see 'Operating' listings) or by telephoning headquarters. They also need to maintain envelopes with both the new and old QSL Sub Managers. Typically, envelopes for the old callsigns and membership number need to be available for up to five years after the old call is no longer the primary call.

Club stations should enter their callsign details in the club's name and not as a secondary call of a member or QSL manager, as this gives rise to confusion. Please avoid registering calls using optional club identifiers, such as X,S,C,N

Stations operating from a different prefix, for example G9ABC as GW9ABC/P or GU9ABC/P, need to lodge envelopes with the appropriate Sub Managers for every area of operation, as cards may not be forwarded to the home call.

UK mainland stamps are not valid when sent from the Isle of Man or Channel Islands. Local stamps should be obtained during the period of operation, for use later. When operating outside the UK under CEPT rules, e.g. F/G9ABC, or more importantly with another callsign, it is vital to tell the QSO partner to 'QSL via G9ABC' and not simply state 'via home call'.

Registering any foreign calls separately, together with the QSL route and contact email address at: *www.qrz.com* is extremely helpful to others in these cases.

### Requesting a 'Via' call route

In recent years there has been an explosion in the use of 'Via' requests, where amateurs use a QSL Manager, or wish to have cards sent c/o another callsign. Advising your contacts to send cards via the personal callsign of the RSGB volunteer Sub Manager is not appropriate, as he/she may change.

In many instances the incoming card does not contain the 'Via' information given during the QSO, the expectation being that the bureau sorters will instinctively know the routing.

With so many cards passing through the bureau each week - and the passage of time - this is simply not practical or possible. Finding the routing is a time-consuming process and no longer a realistic or reliable option for our staff.

This problem can be easily avoided by lodging SAEs with the correct Sub Manager, for the actual call used but bearing any alternative delivery address, i.e. that of the station's QSL Manager. This should be considered as a more effective solution and is much to be preferred over giving out a QSL Manager or 'Via' details to every contact.

For example, 'G9ABC, QSL via M8ZZZ' can simply be replaced by sending all cards to the G9ABC RSGB Sub Manager, who holds envelopes marked with the street address for M8ZZZ.

In the case of special event (GB calls), together with all Abbreviated/Contest (single letter suffix) callsigns and personal Special Prefix calls (GR. MQ, 2O. MV etc), no Vias are accepted.

All bureau cards for these groups are sent directly to speciallist Sub Managers - see list. These Managers will only send cards to the NoV holder, unless an authorised alternative destination is confirmed in writing by the callsign holder and registered via the website.

Remember: Even if you never send a QSL card, someone somewhere, sometime, will send you a card. It would be a shame not to receive it, so please send an SAE to your RSGB volunteer sub manager!

## Card design

Whether you are designing and making your own QSL or having it made professionally, ***size, quality and design are the most important factors*** if you are hoping for a reply.

Gone are the days when cards were printed in a single colour (black), with only a callsign and basic information and which took several weeks to produce. The advent of high resolution digital photography and computers has changed everything. High quality commercial QSL cards are now more interesting, colourful, easier, quicker and much cheaper to produce or change than ever before. What's more, professionally printed cards can and often do work out cheaper than making your own.

All the more reason to consider having a distinctive card that gives not just your station details, but perhaps reflects your radio and other interests, family, pets, location or some other part of your life. Cards can be

### Checklist for receiving cards

**1. Register all your callsigns, past/present**

Do this via the RSGB website at 'amend my details', or telephone 01234 832700.

**2. Send C5 stamped envelopes to each Sub Manager**

In addition to your name and address...
Write your callsign and RSGB membership number at the top left.
Number each envelope at the bottom left.
Mark 'Last envelope' at the bottom left of the last envelope.
N.B. Sub Managers are not authorised to accept cash in lieu of SAEs.

**3. Holidays and portable activations**

We don't automatically divert cards to your home call.
For temporary prefix operation (e.g. GM, MW, 2I etc), lodge separate envelopes with the relevant Sub Manager to collect your cards.
N.B. The Channel Islands and Isle of Man use different stamps.

**4. Special event (GB) and abbreviated contest calls (G1A, etc)**

No diversions apply, so please see the Sub Manager list.

**5. Special Prefix NoV callsigns**

For GR, MQ, 2O etc, no diversions apply. See the Sub Manager details.
Multiple callsigns can be listed on the same envelope.

simple, beautiful, artistic, funny, technical or even something completely unexpected. They make a statement about you - so what does your card say?

The range of choice has never been greater, so just use your imagination. Above all, make your QSL card something of quality that stands out; something that the other station will want to keep and display. If you are sending or receiving a 'gift', make it memorable. It's now possible to collect special interest cards showing planes, trains, ships, cars, families, pets, castles, churches, windmills, lighthouses, motorcycles and many other things, in addition to antenna farms, radios, vintage gear and shack interiors.

**Tip:** Remember to tidy up before you take a photo of your station!

The business side of the card is also very important. Here, simple clarity is the key to a good card and to receiving a reply. Use a clear type face that is easy to read. Don't put too much information or too many logos on the card, unless it's a special event when background information is always nice to see. Remember that English is not always a first language, so keep it simple, keep it relevant. Allow enough space to write or print the contact information clearly on the card, ensuring that the destination call is at the ***top right, with any via routing details immediately below.***

Many cards now have space to log more than one contact. This is a great eco-friendly idea. *See example card below from G4EZT.*

## Where to buy cards

The RSGB doesn't endorse any particular producer. Take a look at the cards you receive, as they will often include maker's details.

Apart from your local printer and checking with friends, there are now a whole range of specialist on-line makers offering superb, correctly sized card. We regularly see cards from UK stations being sent to us that have been designed on-line, some produced in other countries, and many are simply stunning. It is possible to download card making software from the Internet, but so much depends on the actual equipment used to make the card that the results are often disappointing or uneconomic, unless you have access to specialist print and cutting machinery. However, where practical, they do make possible one-off special, individual and personalised cards, for QSLing direct.

**Remember:** If you have invested time and energy on your station, isn't it right to do the same with your QSL card? Send something you would be pleased to receive.

## Sub Managers

### Abbreviated & Contest NoV calls

Mr S Imms M0IMM.
26, Greenwood Avenue, Rowley Regis, West Midlands B65 9NJ
contestmgr@gmail.com

### G0 Series

Mr N P Roberts, G4KZZ
13 Rosemoor Close, Hunmanby, Filey, North Yorks YO14 0NB
nipro@btinternet.com

### G1 & G2 Series

Mr. C. Tuckley G8TMV.
98 Woodland Road, Cambridge. 22 3DU
g1g2mgr@tuckley.org

### GW/MW/2W Series

Mr L Thomas, 2W0LLT
15 Blaenwern, Newcastle Emlyn, Carmarthen SA38 9BE
2w0llt@talktalk.net

### G3AAA-DZZ

Mr P J Pasquet, G4RRA
Honey Blossom Cottage Spreyton
Devon EX17 5AL
g4rra@hotmail.com

### G3EAA-HZZ

Mr J G Holland, G3GHS
Tanglewood, Portheast Way, Gorran Haven, St Austell Cornwall PL26 6JA
jgholland@totease.eclipse.co.uk

### G3IAA-LZZ

Mr J R A Fogden, G4WSX
38 Green Lane, Chichester. West Sussex PO19 7NP bambad@g4wsx.talktalk.co.uk

### G3MAA-PZZ

Mr G Brown, G3MZV
'The Rowans.' 52a, Rivelands Road, Swindon Village. Cheltenham GL51 9RF
g3mg3pqsl@virginmedia.com

### G3RAA-RZZ

Mr D S Moffatt, G3RAU
Mill House Middle Street, Glentworth, Gainsborough, Lincs DN21 5BZ
g3rau@aol.com

### G3SAA-SZZ

Mr R K Western, G3SXW
7 Field Close
Chessington Surrey KT9 2QD
g3sxw@btinternet.com

### G3TAA-VZZ

Mr N S Cawthorne, G3TXF
Falcons St George's Avenue Weybridge
Surrey KT13 0BS nigel@G3TXF.com

### G3WAA-XZZ

Mr J. Peden G3ZQQ
51A, Bewdley Road, Kidderminster.
Worcestershire DY11 6RL
jim@peden.orangehome.co.uk

### G3YAA-ZZZ

Mr J Peden, G3ZQQ
51A Bewdley Road, Kidderminster
Worcestershire DY11 6RL
jim@jimpeden.orangehome.co.uk

### G4AAA-FZZ

Mr. J. J. Pascoe, G4ELZ
3, Aller Brake Road, Newton Abbot.
Devon. TQ12 4NJ
g4elz@blueyonder.co.uk

### G4GAA-LZZ

Mr I N Fugler, G4IIY
Lees Hill Farm, Lees Hill Brampton.
Cumbria CA8 2BB ian.g4iiy@zen.co.uk

### G4MAA-NZZ

Mr C G Rowe, G4MAR
29, Lucknow Road, Willenhall West Midlands WV12 4QF cliff.g4mar@gmail.com

### G4OAA-OZZ

Mr S R Harwood, G4OWT
24 Firle Crescent, Lewes, East Sussex BN7 1QG g4owt@btinternet.com
*www.g4owt.btinternet.co.uk*

### G4PAA-PZZ

Mr K T Hutt, G0TSH
The Crossing House, Fenwick. Doncaster DN6 0EZ keith.hutt@virgin.net

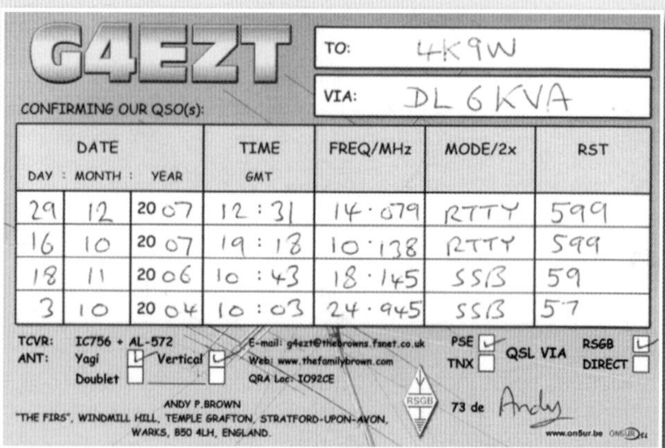
G4EZT

TO: 4K9W

VIA: DL 6 KVA

CONFIRMING OUR QSO(s):

| DATE DAY | MONTH | YEAR | TIME GMT | FREQ/MHz | MODE/2x | RST |
|---|---|---|---|---|---|---|
| 29 | 12 | 2007 | 12:31 | 14·079 | RTTY | 599 |
| 16 | 10 | 2007 | 19:18 | 10·138 | RTTY | 599 |
| 18 | 11 | 2006 | 10:43 | 18·145 | SSB | 59 |
| 3 | 10 | 2004 | 10:03 | 24·945 | SSB | 57 |

TCVR: IC756 + AL-572
ANT: Yagi ☑ Vertical ☑ Doublet ☐
E-mail: g4ezt@thebrowns.fsnet.co.uk
Web: www.thefamilybrown.com
QRA Loc: IO92CE
PSE ☑ TNX ☐ QSL VIA RSGB ☑ DIRECT ☐
ANDY P.BROWN
"THE FIRS", WINDMILL HILL, TEMPLE GRAFTON, STRATFORD-UPON-AVON, WARKS, B50 4LH, ENGLAND.
RSGB
73 de Andy
www.on5ur.be

*Example of a QSL card that's well laid out, easy to read and easy to sort.*

**G4RAA-RZZ**

Mr W Thomas M0WAY
5 Thornfield Road, Ashmore Park, Wednesfield, Wolverhampton WV11 2HR
m0way@hotmail.co.uk

**G4SAA-SZZ**

Mr S L Shenstone, M5BFL
35, Nuffield Road, Hextable, Swanley. Kent BR8 7SL m5bflsteve@aol.com

**G4TAA-TZZ**

Vacant at time of press , see web site for details

**G4UAA-UZZ**

Mr S L Shenstone, M5BFL - 35, Nuffield Road, Hextable, Swanley. Kent BR8 7SL.
m5bflsteve@aol.com

**G4VAA-VZZ**

Mr G R Marley, G0VFV
41 Scalby Road, Burniston Scarborough YO13 0HN g0vfv@amsat.org

**G4WAA-XZZ**

Mr J R A Fogden, G4WSX
38 Green Lane, Chichester, West Sussex PO19 7NP bambad@g4wsx.freeserve.co.uk

**G4YAA-ZZZ**

Mr A F Roberts, G4ZIB
25 Sebright Road, Wolverley, Kidderminster Worcs DY11 5TZ
tonyroberts1948@live.co.uk

**G5 Series**

Mr P J Pasquet, G4RRA
Honey Blossom Cottage, Spreyton, Devon EX17 5AL g4rra@hotmail.com

**G6 Series**

Mr S Wellon, G6DMG
71 Toftdale Green, Lyppard Bourne, Worcester WR4 0PE g6dmg@hotmail.co.uk

**G7 Series**

Mr C. Flanagan G7NRO 19a, High Street, Wolviston. Stockton-on-Tees Tyne & Wear TS22 5JY g77nro@gmail.com

**G8 Series**

Mr D Helliwell, G6FSP
1 Beechfield Avenue Torquay TQ2 8HU
dave@g6fsp.com

**GBxAAA-ZZZ**

Mrs D Williams, M0LXT
20 Neale Close Wollaston Northamptonshire NN29 7UT
qsltrek@hotmail.co.uk
*www.gb-special-event-qsl-status.webs.com*

**GD, MD & 2D Series**

Mr M J H Parnell, GD3YUM
1 Derwent Drive Onchan Isle of Man IM3 2DF martyn@wm.im

**GI, MI & 2I Series**

Dr E H Squance, GI4JTF
11 Ballymenoch Road, Holywood Co Down Northern Ireland BT18 0HH
gi4jtf@gmx.com

**GJ, MJ & 2J Series**

Mr M Roche, MJ0ASP
Flat 1 Stratscombe House. Le Quai Bisson, St Brelade, Jersey JE3 8JT
mathieu.roche@hotmail.com

**GM0 & GM1 Series**

Mr F A Roe, GM0ALS
74 Willow Grove, Livingston, West Lothian EH54 5NP
fred.roe190@googlemail.com

**GM2 & GM3 Series**

Mr C O'Hennessy, GM4VVX
Savalbeg Lairg Sutherland Scotland IV27 4ED cohennessy@madasafish.com

**GM4-8 Series**

Mr. A. Hood GM7GDE
26 Annan Avenue East Kilbride Scotland G75 8XT
gm-qsl@mail.com

**GU, MU & 2U Series**

Mr P F H Cooper, GU0SUP
1 Clos au Pre, Hougue du Pommier, Castel, Guernsey GY5 7FQ
pcooper@guernsey.net

**GW- Series**

Mr. J. L. Lewis. GW0RAD. 189,Heol y Gors, Cwmgors, Ammanford. Carmarthenshire. Wales SA18 1RF gwmanager@sky.com

**2E Series**

Mr R Maltby, 2E1DFI
1 Briar Close Southfields, London Road, Sleaford, Lincolnshire NG34 7NT
ray2e1dfi@aol.com

**2M Series**

Mr S Gill, MM0SGQ
5 Ramornie Place Kingskettle. Fife. Scotland KY15 7PT 2m0sgq@gmail.com

**2W Series**

Mr S J Smith, 2W0VAG – 36, Jones Street, Blaenclydech, Tonypandy , Mid Glamorgan, CF40 2BY
simon@dxcc.co.uk
*www.dxcc.co.uk*

**M0AAA-AZZ**

Mr D W Heard MBE, G6YEK
103 Moorland Road, Weston Super Mare, Somerset BS23 4HU
desheardmbe@hotmail.co.uk

**M0BAA-BZZ**

Mr S J Smith, 2W0VAG
36 Jones Street, Blaenclydach, Tonypandy, Mid Glamorgan, CF40 2BY
simon@dxcc.co.uk www.dxcc.co.uk

**M0CAA-CZZ**

Mr R P Ashman, G4JVJ
44 Conan Doyle Walk, Liden, Swindon SN3 6JB
the.brummie@fsmail.net

**M0DAA-M0FZZ**

Mr J Steel, M0ZAK
6 Central Avenue, Shepshed. Loughborough Leicestershire LE12 9HP
m0zak@ntlworld.com

**M0GAA-M0LZZ**

Mr D E Mappin, G4EDR
13 Willow Close Filey North Yorks YO14 9NY radioham73-qsl@yahoo.co.uk

**M0MAA-M0ZZZ**

Mr S Adaway M0TNG
20 Foundary Street, Barnsley S70 1PL
stu@m0tng.info *www.m0tng.info*

**M1 Series**

Mr R Taylor M0RRV.
2 Chadwick Road, Moorends, Thorne, DN8 4NG South Yorkshire
roy024@tiscali.co.uk
*groups.yahoo.com/group/M6_QSL/*

**M3 Series**

Mr R Taylor, M0RRV
2 Chadwick Road, Moorends, Thorne, DN8 4NG South Yorkshire
roy024@tiscali.co.uk
*groups.yahoo.com/group/M6_QSL/*

**M5 Series**

Mr S L Shenstone, M5BFL
35, Nuffield Road, Hextable, Swanley. Kent BR8 7SL. m5bflsteve@aol.com

**M6 Series**
Mr R Taylor, M0RRV
2 Chadwick Road, Moorends, Thorne, DN8 4NG South Yorkshire
roy024@tiscali.co.uk
*groups.yahoo.com/group/M6_QSL/*

**MM Series**
Mr S Gill MM0SGQ.
5, Ramornie Place, Kingskettle. Fife
2m0sgq@gmail.com

**MW Series**
Mr S J Smith, 2W0VAG – 36, Jones Street, Blaenclydech, Tonypandy , Mid Glamorgan, CF40 2BY
simon@dxcc.co.uk
*www.dxcc.co.uk*

**RS Receiving stations**
Mr R Small, RS8841 13 Rydall Close, Stowmarket, Suffolk IP14 1QX
rob@g3ali.co.uk

**Special UK Prefixes - NoV Call Holders**
**R**. Royal Wedding. **Q**. Queen's Jubilee.
**O**. Olympic Games. **V**. RSGB Centenary.

Mr J. Peden, G3ZQQ 51A, Bewdley Road, Kidderminster. Worcestershire DY11 6RL
jim@peden.orangehome.co.uk

**Note 1**. The sub group for 2 letter suffix G Call signs, has closed. All 2 letter calls are now sorted and distributed with 3 letter calls.

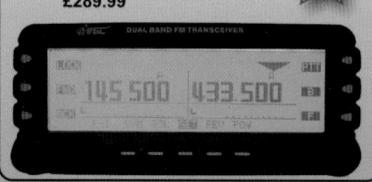

# Intruder Watch

The RSGB Monitoring System, more popularly known as the Intruder Watch is a small team of volunteer observers and forms part of the IARU Monitoring System. As such it submits reports of non amateur transmissions heard on the exclusive HF amateur bands to both the Ofcom Monitoring Station at Baldock and the IARU Region 1. While most of Intruder Watch activities is centred around the HF bands, Intruder Watch also assists leasing with Ofcom and AROS for reports of non-amateur transmissions in the VHF/UHF bands.

Intruders removed from our exclusive amateur bands include broadcast stations, military data transmissions, faulty positioning installations, coast stations, embassies, fax stations, faulty set-top boxes and numerous others. For data transmissions a 'zero beat' frequency will be accurate enough for our observers without decoders. Many software data decoders are available as well as equipment manufactured by companies such as 'Hoka', 'Wavecom' and 'Universal' these are used for analysis of data signals.

Most information received by the co-ordinator arrives from regular observers, but occasional reports are also welcome from anyone who finds what may be an intruding signal on one of our exclusive amateur bands. This information can then be passed on to a suitably equipped observer for further investigation. All reports are welcome and will be acknowledged.

Data communications is by far the most common intruder into the HF amateur bands and it is an area where we could use more support. Other non-data categories of intruding signals include CW, broadcast stations, speech and over-the-horizon radar (OTHR). Any report should include as much information as possible, but preferably frequency, date, time (UTC), mode of transmission, any identification signal or callsign, language used, text (where appropriate) and beam heading where possible.

Intruder Watch is always looking for more volunteer observers so if you think you might like to join our team do please send an email to the Intruder Watch Co-ordinator
Email: *iw@rsgb.org.uk*

Email Intruder Watch: *iw@rsgb.org.uk*

# Youngsters on the Air 2017

The RSGB Youth Committee is delighted to announce that the bid to host the IARU YOTA 2017 summer camp has been successful and we are proud to be hosting this prestigious international event.

Around 100 young people under the age of 26 from all over IARU Region 1 will be coming to the UK to take part. There will be a week-long set of wireless technology activities including a Special Event Station, a buildathon, antenna building, ARDF, a SOTA activation and the opportunity to visit some landmark 'tourist' sites to get the cultural experience.

A project team has been set up and after investigating a range of sites across the UK it has been decided that the event will be hosted at Gilwell Park, the UK Scouting HQ, during the week 5-12 August 2017.

Sara, 2I0SSW, Nick, M6NJR, Arthur, 2E0RTY and Kieran, M6RZR are representing the UK at this year's event in Austria and we will be advertising for the UK team for 2017 in January. Watch this space!

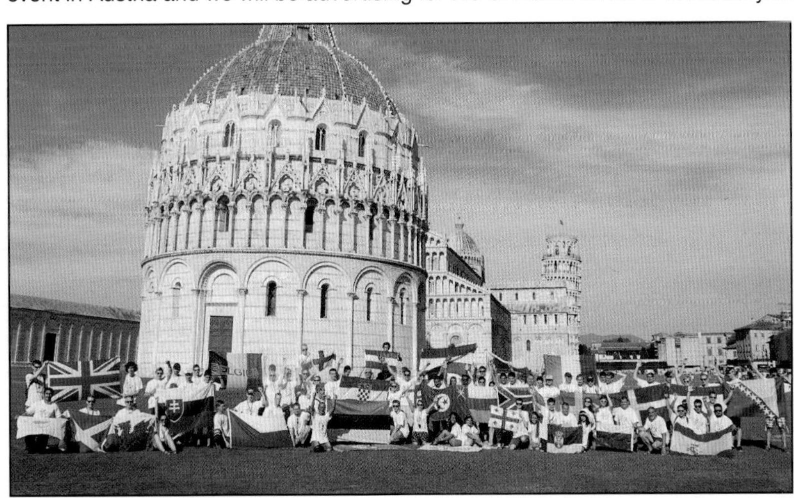

We have already had offers of help for YOTA 2017 from a number of clubs, including Loughton & Epping Forest ARS, Radio Society of Harrow, Camb-Hams and the Radio Scouting leaders from Gilwell Park. If you believe you have any skills to offer during the event please contact *yota@rsgb.org.uk* We would be particularly interested to hear from young Members or Affiliated Societies who could lead an activity or talk that will educate and/or inspire the YOTA teams.

Running an event such as YOTA 2017 is not without some costs. The IARU makes a significant contribution but we need to find the rest. The RSGB is therefore seeking supporters from the amateur radio community who wish to be associated with this event and help financially. Supporters of all sizes are welcome and if you would like to know more please visit ***www.rsgb.org/yotasupporter***

# Morse Tuition, Practice and Assessment ...GB2CW

Whilst a Morse qualification is not needed by the present day licence, amateurs are realising that they are missing out on a lot of fun and DX by not using Morse and limiting themselves to part of the total amateur bandwidth available.

## Introduction

Since the RSGB introduced the Certificate of Competency, there have been a number awarded. Some have been publicised in the *RadCom*, so the scheme is enjoying a considerable degree of success. Making the Certificate an 'award' rather than something mandatory in order to gain a licence has encouraged a lot of newly licensed amateurs to take up the challenge and achieve a skill that will enable them to work more DX, plus pass on their achieved skills to others. It can be compared to chasing DX certificates where effort is rewarded for the time spent in achieving the necessary qualifications.

The Morse certificate is no different - some will wish to obtain it and some won't. In the same vein, some clubs will embrace the idea and others will not. However, the skill will stay with you for the rest of your life.

International recognition is not an aspiration of the initial scheme, but if that develops later in a non-contentious manner it could be a clear additional benefit for some (provided that it is not abused to gain a higher class of licence overseas than is held here in the UK).

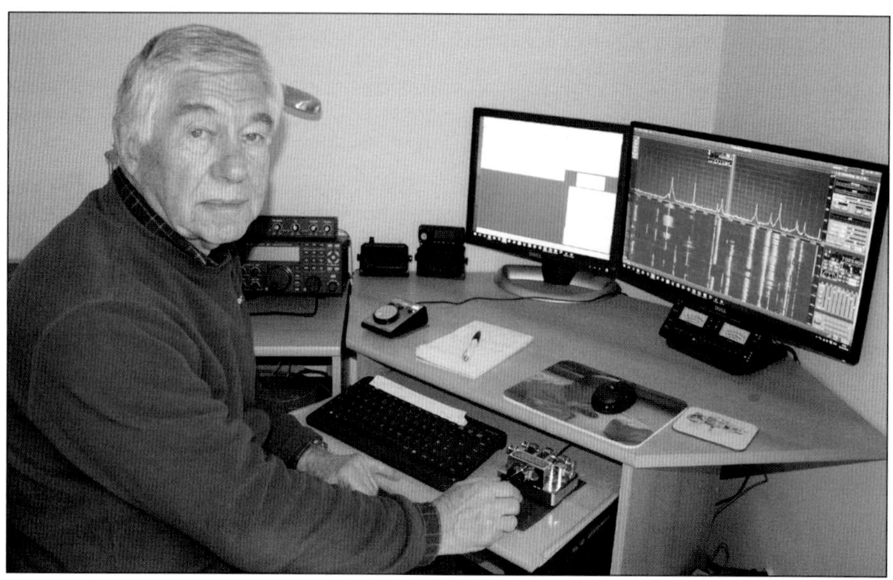

*GB2CW volunteer, Malcolm Prestwood, G3PDH*

## Certificate of Competency

The lowest speed for which the Certificate will be issued is 5WPM. However, experience has shown that most people prefer to obtain a Certificate of at least 15WPM. The low speed of 5WPM was chosen as the threshold to encourage learning the code, and as a step towards inspiring and achieving confidence and moving forward to higher speeds. It is specifically designed for those who are most comfortable with this rate of progress. There is, however, no barrier to those who wish to enter the scheme at a higher level of (say) 12WPM.

The initial Certificate of Competency Morse assessment will require the candidate to receive and send text, including some punctuation, for 3 minutes with no more than three

### GB2CW Schedule

**HF Transmissions**

| Day | Time | Frequency | Call | Location |
|---|---|---|---|---|
| **Monday** | 20.00 | 3.550 | G0WKL | Hampshire |
| | 20.15* | 3.555 | G0IBN | Essex |
| | 20.15** | 7.035 | G0IBN | Essex |
| **Tuesday** | 20.00 | 3.555 | GW0KZW | Prestatyn |
| **Thursday** | 09.00 | 3.605 | G3UKV | Telford |
| **Friday** | 10.00 | 5.301.0 | G0ORH | West Berkshire |
| | 20.00 | 3.563 | GW0KZW | Prestatyn |

**VHF Transmissions**

| Day | Time | Frequency | Call | Location |
|---|---|---|---|---|
| **Monday** | 18.30 | 145.250 | M0APY | Leeds |
| | 20.00 | 145.550 | G3KAF | Stockport |
| **Tuesday** | 09.00 | 145.250 | G4OOC | Pontefract |
| | 18.00 | 145.250 | M0HAZ | Skegness*** |
| | 18.30 | 145.250 | M0APY | Leeds |
| | 19.15 | 145.250 | G0BYA | Stafford |
| | 19.00 | 145.250 | G3LDI | Norwich (Advanced speeds) |
| | 20.00 | 145.250 | G3YLA/G3XLG | Norwich (Intermediate speeds) |
| **Wednesday** | 09.00 | 145.250 | G4OOC | Pontefract |
| | 19.00 | 145.450 | G4ERV | Dorchester, Salisbury and Southampton/Portsmouth |
| | 20.00 | 144.250 | G3XVL | Ipswich |
| | 20.00 | 145.250 | M0PCB | Gloucester |
| **Thursday** | 09.00 | 145.250 | G4OOC | Pontefract |
| | 18.00 | 145.250 | M0HAZ | Skegness |
| | 18.30 | 145.250 | M0APY | Leeds |
| | 19.30 | 145.250 | G0TDJ | Crayford (Kent) |
| | 20.00 | 145.250 | G3PDH | Norwich (Beginners only) |
| **Friday** | 18.30 | 145.250 | M0APY | Leeds |
| **Saturday** | 08.00 | 145.250 | G4OOC | Pontefract |
| **Sunday** | 20.00 | 145.250 Voice / 145.250 CW | G4PVB | St Albans |

**Modes**: A1A/J3E 144.2508 and all HF transmissions
F2A/F3E All VHF transmissions

*GMT ** BST *** Excluding first Tuesday in each month

*Co-ordinator*: Roger Cooke, G3LDI, The Old Nursery, The Drift, Swardeston, Norwich, Norfolk NR14 8LQ. email: g3ldi@yahoo.co.uk

*GB2CW volunteer, Martyn Vincent, G3UKV*

uncorrected errors and will also include some figure groups (receiving and sending). Success in this will merit issuing the Certificate of Competency, after which endorsements (or a new certificate) may be obtained at 12, 15, 20, 25, 30WPM, etc.

To obtain an endorsement or a new certificate at higher speed, further assessments will also include receiving and sending proficiency in a basic rubber-stamp type QSO. This applies also to those wishing to take their initial assessment at a speed higher than 5WPM.

Regardless of speed, a requirement of every assessment taken will be that all sending will be pre-recorded, to guarantee the speed and ensure integrity of the assessment process.

All assessments may be taken using equipment chosen by the candidate and appropriate for the speed being examined, including straight keys, paddles, bugs and semi automatic keys.

## Training

The Society is not prescriptive about the method of training used to achieve the Certificate. There are numerous methods of learning Morse code and it is a personal choice as to the method used. Instructors and students will have preferences and individual teaching and learning styles. No written rules are made, but once the code has been learned it is absolutely necessary for candidates to practice regularly. This may be done in a group, such as at a club, by listening to Morse on the bands or by using one of the numerous computer programs available.

The student can supplement individual or group training at clubs; and (ideally) be further supported by the use of an active and well promoted GB2CW broadcast schedule. Regular attendance to a weekly tutorial on the air using GB2CW in an interactive way is extremely beneficial. 2m FM is preferred to achieve this activity and it is normally a lot of fun, especially with mutual competition with other students in the same class.

There is more comprehensive information which adds to and builds on this in the RSGB book *Morse Code for Radio Amateurs*, by Roger Cooke, G3LDI.

## Assessments

Assessments are conducted under the auspices of any RSGB affiliated club or society.

When a candidate is ready to be assessed, application should be made by the candidate to the local RSGB Regional Manager, stating the speed at which they wish to be assessed.

The assessment will be conducted by an Approved Assessor for the speed of assessment to be conducted (see Approved Assessors). The assessment will be adjudicated by the Regional Manager, Deputy Regional Manager, any other elected RSGB volunteer or an elected member of the club committee.

A successful assessment will be confirmed using a form that can be downloaded from the RSGB website and will be signed by the Approved Assessor and Adjudicator. This form will be retained by the Regional Manager for checking and audit purposes. On completion of a successful assessment, the Regional Manager will issue a formal Certificate of Competency.

When an assessment is requested, it is the Regional Manager's responsibility to contact the Approved Assessor and make arrangements for the assessment to take place at a convenient time and place for all concerned. As your skill improves, so you can apply for an upgrade to a Certificate with a higher speed.

## Consistency and integrity

Consistency, integrity and development of the scheme is monitored by a joint committee of the Regional Team and Amateur Radio Development Committee. The terms of reference for the ARDC encompass training and testing, and ensuring that the scheme is conducted in a thoroughly professional and competent manner. This is intended to guarantee that the Certificate is both desirable and that its reputation is respected both in the UK and internationally.

*List of RSGB-appointed, Approved Morse Assessors on next page*

It may also be desirable that the ARDC will in due time extend its focus to further encourage the use of Morse.

## Learning Morse

Unlike the Foundation licence, where a course of a few hours learning will probably produce a pass for the candidate, learning Morse Code and becoming a proficient operator is akin to learning a musical instrument. Attending a class once a week will not produce results. Occasional listening on the air is also a waste of time. The student has to be motivated and disciplined. Learning a musical instrument requires constant practice, and that does not mean just ten minutes a day. If you aspire to become a top-notch CW operator, consider at least one or two hours per day practice, EVERY day, not just once a week. This must carry on for a few months to reach acceptable speeds. If you cannot meet those requirements, then Morse is not for you. This cannot be stressed enough. The results you obtain will be well worth that effort.

## Volunteers for the GB2CW scheme

Ken Chandler G0ORH volunteered his services and has pioneered a new frequency, 5MHz.

Ken was given permission to operate for four weeks to see if there were any problems. None materialised so he now operates with the following schedule:

| | | | |
|---|---|---|---|
| Friday | 10:00 | local time | 5301.0 |
| | 19:00 | local time | 5301.0 |

Ken reports that 5MHz is very successful and he already has established a class there.

*"Here in Norwich we have three classes running now, a raw beginner's class taken by me. The Intermediate class taking the students up to around 18 wpm is handled by two new volunteers here. One is the well known local BBC TV weatherman Jim Bacon, G3YLA and the other is Ray Spreadbury, G3XLG.*

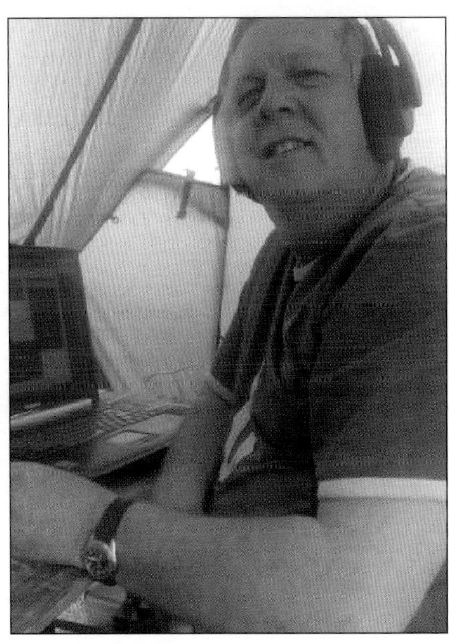

*GB2CW volunteer, Ken Chandler, G0ORH*

*Morse tests on demand*:
G3NCN provides Morse transmissions on demand in the Bracknell area.
Call: John Ellerton on 01344 425666 to arrange a transmission.

*The Intermediate classes are new and once the students attain 18 wpm they progress to the Advanced class run by Malcolm Prestwood, G3PDH who takes them up to 30 wpm".*

The expansion of the GB2CW network is slow but gradually taking more and more volunteers. This is good progress and very satisfying to see. However, additional volunteers are always needed to run GB2CW broadcasts, especially in some of the more remote parts of the UK. Broadcasts can take place on several bands, ranging from 3.5MHz to 50MHz. It may only take an hour of your time per week to ensure that amateur radio continues to have a flourishing pool of CW operators to ensure the future of the mode, so please consider helping.

As interest grows, more tutors are needed. It would be very nice to see volunteer instructors in every Club in the UK, and that is what I would like to see as Coordinator. There is a long way to go to achieve anything like that but in order to maintain this quintessential mode used in amateur radio we need a lot more tutors. Remember, some Elmer taught you, so now it's your turn to be an Elmer!

Also, there are gaps in the coverage of Assessors. Assessors are needed in Regions 8 (Northern Ireland), and 11 (South West England and the Channel Islands). If you live in one of these areas, please consider joining this most worthwhile scheme. We have been fortunate in filling a few areas in the last year or so. However, more are always needed, not only for the vacant areas, but all areas, to act as backup. Full details, including an application form, can be found on the RSGB website at:

*www.rsgb.org/main/operating/morse/certificate-of-competency/approved-assessors/*

Volunteers are scarce and are perceived by some to be those with super human skills and speed in excess of 30 wpm. This is far from the truth and if you have a good average skill level of around 15 to 20 wpm, you could take on the role of instructor to that level anyway.

Computer programs are used for instruction so therefore the Morse sent is perfectly formed so it is completely straightforward to implement on the air. To offer your assistance, please contact the scheme's co-ordinator, Roger Cooke, G3LDI (*details below*).

## Approved Morse Assessors by RSGB Region

| RSGB Region | Location | Speed (WPM) |
|---|---|---|
| **Region 1** | | |
| Ray Evens, GM0CDV | Kelso | up to 12 |
| Derrick Dance, GM4CXP | Kelso | up to 15 |
| Dick Hodge, GM4PPT | Coylton | up to 15 |
| Douglas Panton, MM0MPA | Polmont | up to 25 |
| **Region 2** | | |
| William Cecil, GM3KHH | Moray | up to 25 |
| Bernie Macintosh, GM4WZG | Dalgetty Bay, Fife | up to 25 |
| James Mackinnon, GM4EKC | Aberdeen | up to 30 |
| Norman Mackenzie, GM3WIJ | Aberdeen | up to 25 |
| Thomas Brown, MM0TGB | Kirkcaldy, Fife | up to 30 |
| **Region 3** | | |
| Albert Heyes, G3ZHE | Warrington | up to 20 |
| Colwyn Baillie-Searle, GD4EIP | Isle of Man | up to 25 |
| James France, G3KAF | Stockport | up to 30 |
| Brian Gale, G3UJE | Cheshire | up to 30 |
| Ken Randall, G3RFH | Thornton Clevelys | up to 30 |
| **Region 4** | | |
| Malcolm Brass, G4YMB | Guisborough | up to 12 |
| Michael Taylor, G3WTA | Shieldfield, | |
| Newcastle upon Tyne | up to 30 | |
| David Jackson, G4HYY | Withernsea | up to 15 |
| Terry Bucknell, G4AFS | Bingley, West Yorkshire | up to 40 |
| Tom Sandilands, G0HBV | Berwick upon Tweed) | up to 15 |
| **Region 5** | | |
| Martyn Vincent, G3UKV | Telford, Shropshire | up to 25 |
| Eric Arkinstall, M0KZB | Shrewsbury | up to 20 |
| Iain Kelly, M0PCB | Cheltenham | up to 20 |
| Martin Hallard, M0AJN | Stourbridge | up to 12 |
| **Region 6** | | |
| Anthony Allen Chalk, MW0BXJ | Colwyn Bay | up to 20 |

| RSGB Region | Location | Speed (WPM) |
|---|---|---|
| **Region 7** | | |
| Robert Evans, MW0CVT | | 20+ |
| George Bodley, MW0RZC | | up to 20 |
| David Mead, MW0MWL | | up to 20 |
| Curtis Burke, MW0USK | | up to 12 |
| **Region 9** | | |
| Bob Leask, G3XNG | | up to 30 |
| **Region 10** | | |
| Mick Puttick, G3LIK | | 25 |
| Ray Ezra, G3KOJ | | 25 |
| **Region 11** | | |
| Ken Selleck, G3SNU, | Dartington, Totnes | up to 12 |
| G.W. Davis, G3ICO | Yeovil | up to 12 |
| Alan Hydes, G3XSV | Bristol | up to 20 |
| Robin Thompson, G3TKF | Bath | up to 25 |
| David Barlow, G3PLE | Helston, Cornwall | up to 25 |
| **Region 12** | | |
| Malcolm Prestwood, G3PDH | Norwich | up to 30 |
| John Francis Bonner, G0GKP | Cambridge | up to 15 |
| Bob Leask, G3XNG | | up to 30 |
| Andrew Kersey, G0IBN | Maldon, Essex | up to 35 |
| Dr Malc Williams, G0EGA | Halesworth, Suffolk | up to 20 |
| Bob Whelan, G3PJT | Cambridge | up to 25 |
| **Region 13** | | |
| Martin Farmer, M0MDF | Lincoln | up to 20 |
| Peter Kendall, M0EJL | Lincoln | up to 12 |
| James Nichol, G0EUN | Lincoln | up to 12 |
| Ian Fulton, G4XFC | Lincoln | up to 20 |
| Robert Topliss, G0OTH | Skegness | up to 12 |
| Anthony Freeman, M0HAZ | Boston, Lincs | up to 20 |
| Roy Fretsome, G4WPW | N.Notts | up to 20 |
| Harry Hall, G4ZRL | N.Notts | up to 20 |
| Ken Francom, G3OCA | S Derbyshire | up to 30 |

**If your region is not listed, it does not currently have any assessors.**

***For more information regarding the Morse Code and Abbreviations used by radio amateurs please go to page 153 in this Yearbook***

*Co-ordinator*: Roger Cooke, G3LDI, The Old Nursery, The Drift, Swardeston, Norwich, Norfolk NR14 8LQ. email: g3ldi@yahoo.co.uk

Latest GB2CW broadcast schedule:
*www.rsgb.org/main/operating/morse/certificate-of-competency/gb2cw-broadcast-schedule/*

# Featured Clubs

The Stow Maries Aerodrome, Flying Day, Dengie Hundred ARS

Throughout the UK there are over 500 local clubs and societies affiliated to the RSGB. They vary from small local clubs, through repeater groups to the large contest groups. Below are the details of some of these clubs, telling you a little about what they do for amateur radio and their members, and what they did in 2015/16 or plan to do in 2016/17.

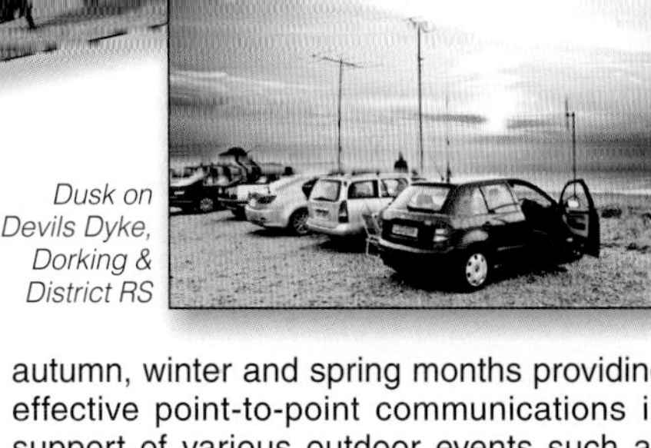

Dusk on Devils Dyke, Dorking & District RS

We are sorry that we cannot include every club that submitted information or every photo from those that did, but we have tried to include everything that every club sent to us unedited, on the disk that is included with this Yearbook.

## Appledore and District Amateur Radio Club

The members of Appledore and District Amateur Radio Club have had another active year with monthly Club meetings at which there is generally a presentation by a club member on a technical subject or an invited guest giving a non-radio talk. The meetings take place every third Monday of the month at the clubhouse in Appledore, just outside Bideford, and a Tuesday coffee morning in Barnstaple.

All club meetings start with the opportunity for members and visitors to mingle and chat over tea, before the presentation. The club's talks programme consist of a varied mix of technical subjects directly related to amateur radio, from 'back to basics' to the more complex areas such as APRS, antennas, radio astronomy and more.

Club members are very active on the following frequencies and operate daily, with the exception of Saturday. Our Zepp Net starting at 1600 local time: Monday, Tuesday and Thursday on 145.450MHz

On Wednesday via the local North Devon repeater GB3DN at 1600 local time.

HF Net: Friday at 1500 on 7.145MHz ± qrm.

On Sunday the 70cm Net via GB3ND, 1100 - local time. Available on Echolink node 221334.

All visitors are very welcome to pop into the club or give us a call on air if visiting North Devon; otherwise please try our Sunday net via Echolink, details above.

The club's Callsigns are G2FKO and GX2FKO and our Web Site can be found at:

http://www.adarc.co.uk/ .

Alternatively contact Alan Fisher (M6CCH) Club Secretary on 01237 422833 or email: fisheralan.af@gmail.com

## Bangor and District Amateur Radio Society GI3XRQ

Bangor & DARS was originally formed over 40 years ago in the seaside town of Bangor, County Down, Northern Ireland. We are one of the largest radio clubs in Northern Ireland with over 60 members.

We meet on the first Thursday of every month September through June mainly at "The Boathouse" in Groomsport. We have a talk at most meetings by either a guest speaker or from one of our own members. The subjects are many and varied but are mainly of a specific radio or a more general technology type area.

Keep an ear out for us operating Field Day contests, Mills on the Air, Railways on the Air, 80m Club Championship and others.

Training courses for the Foundation and Intermediate Licences are run when we have enough numbers. We hope to run Full licence courses in the future if requested.

Check out our web site at www.bdars.com for contact details.

## Braintree and District ARS GX3XG

2014/15 saw the club successfully celebrating its 40th anniversary year starting with a most enjoyable coach outing for members, partners and friends to Bletchley Park. This was followed by a busy bank holiday special event station callsign GB4OAY hosted by a local Scout group and located adjacent to a popular public leisure location. T.

Our aim is to encourage new members into the hobby and to this end we are welcoming both at our outside functions and to all visitors at our meetings. We have forged strong links with the volunteers from the water mill and steam heritage railway that we regularly operate from. At JOTA we are fortunate to attract much interest from lively groups of Scouts, Guides, Brownies and Cubs, and support from their leaders, some of whom are also club members. We produce a full and varied monthly magazine which is circulated both among members and to other local clubs and the press. We also have a Twitter presence @GB6RH.

Our schedule for the coming year is fast filling up with functions ranging from fun (DX hunt and special event stations) to technical (construction and talks) and practical (operating evenings) and we would look forward to seeing new visitors at, and attracting new members from these and other events.

Anyone seeking further information about the club please visit our website www.badars.co.uk and to make contact, follow the link to our Chairman John M5AJB on the Committee page.

## Calderdale Raynet ARG

Formed in the 1970s, is active throughout the autumn, winter and spring months providing effective point-to-point communications in support of various outdoor events such as long-distance walks, fell races and Scout hikes. With our assistance, organisers are able to keep a close eye over the progress of their events and more effectively manage the safety of the participants and any adverse developments. Our ability at providing effective communications consequently draws gratitude from event organisers.

The local terrain in the Pennine hills brings its own challenges to radio signal propagation including lack of mobile phone coverage and many radio shadows on the hillsides and in the valleys. These test the expertise of the Group members and give us opportunities to develop innovative methods for providing ad hoc wireless networks. Applications including APRS are under investigation for their effectiveness aiding well-attended moorland walking events.

In a developing cooperation with Halifax Amateur Radio Society, local people are being encouraged to participate in amateur radio and radio technology and to pursue opportunities for practical experience at the microphone.

We welcome anyone who would like to join with us and we can provide support and training for all the levels of the amateur radio licence examinations. The reward is the satisfaction that comes from tasks performed well – sometimes under difficult conditions – and the knowledge that our participation enhances the safety of members of the local community. For further information please contact the Calderdale Raynet Group via http://www.yorkshireraynet.org/

## CHELMSFORD ARS (CARS)

The Society holds two meetings a month and actively promotes radio at various special events.

CARS sets up special stations at such as Langford's Museum of Power as well as Chelmsford's historic Sandford Mill. This former waterworks opened in 1930, and is now a heritage site which houses the original 1920's 2MT Writtle hut used by Marconi engineers in the early days of radio broadcasting. Transmitting from the hut for both International Marconi and Science Days are always popular events with Club members and also the many ex-Marconi employees who make contact with the station. The Society also supports SOS Radio Week from the Marconi Sailing

Club at Stansgate and field events aimed at introducing newcomers to the hobby.

CARS Skills Night, a monthly free-of-charge initiative by the training team, is now in its third year and continues to draw people from clubs across Essex, Kent, London, Suffolk, and beyond. Its reputation has spread and now many other clubs have copied the successful format of a mix of show-and-tell activities, demonstrations and hands-on practicals. Attractions include satellite reception, test equipment and data demonstrations, antenna construction, call sign badge making, etc.

The Society's long experienced training team continue to have successful years; Foundation, Intermediate and Full examination sessions usually return a near 100% pass rate and their CW classes are as popular as ever. CARS have been actively involved with training since the RAE course was started at Marconi's New Street works in 1960 and continued through the inception of Foundation Licence. The society's reputation for quality training is well-deserved.

Chelmsford Amateur Radio Society is one of the oldest clubs in Essex, if not the Country and with nearly 120 members it is probably one of the largest. Membership is drawn from a wide range of occupations and includes professionals and others from the electronics, electrical and communication fields. Dating from 1936 CARS' roots can be traced back to its formation by engineers from the Marconi Wireless Telegraph Co., Louis Varney, G5RV being one of them. Our well known callsign, G0MWT is part of the tradition of keeping the Marconi legacy alive in the birthplace of radio.

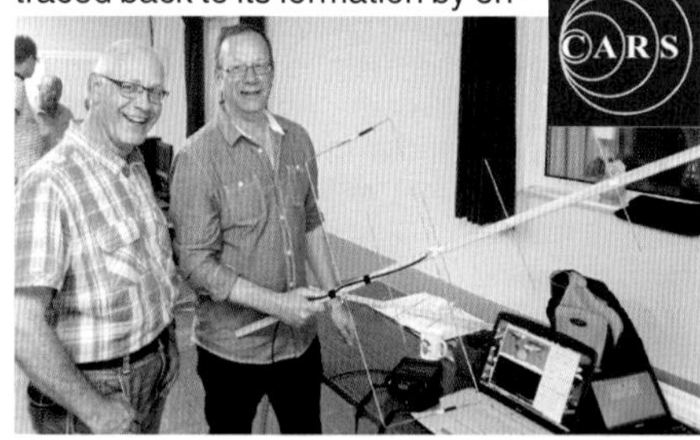


Chelmsford is steeped in radio history, and CARS is proud to be keeping the spirit of the early pioneering days of radio alive. To this end, Society members have very recently been involved in a three month long Marconi historical exhibition at the World's first radio factory at Hall Street in Chelmsford. Their role at the organisers' request was to show visitors young and old, the joys and the art of using Morse code.

Our monthly newsletter gives excellent coverage of all our events and much more.

For information regarding activities, venues, dates and times go to www.g0mwt.org.uk

Training and Skills Nights take place at Danbury Village Hall, Danbury, Essex CM3 4NQ

Facebook: *www.facebook.com/ChelmsfordARS*

Follow us:

@ChelmsfordARS, @TrainWithCARS,

## Cornish Radio Amateur Club - GX4CRC/GB4IMD

Our club has been running successfully for over 70 years. The Cornish Radio Amateur Club meets on the 1st and 3rd Thursday of the month and membership is open to anyone with an interest in radio communications, computing or indeed electronics. We have a wide age range in the club from 14 to over 80years. We all enjoy various aspects of the hobby from traditional methods such as CW and voice to the more modern day digital modes etc.

We try to put on as many special event stations to promote our hobby and make the club meetings a truly sociable evening for everyone.

Our biggest and proudest event we run is International Marconi Day. In 1988 a group of amateurs in our club decided to start up an event to Celebrate Guglielmo Marconi's birthday. This was to be held on the weekend closest as possible to 25th April. This year sadly saw the passing of one of International Marconi Days original founders, Norman Pascoe G4USB. Norman organised the event every year and during 2016 we had 64 stations registered with us from all over the world, which truly made the event a success. We as a club will continue his legacy each year by running what is now a major event in the radio amateur's calendar.

Throughout recent years we have been involved with the Scouts and put on a station for JOTA at Nine Ashes in Bodmin. Each year we have had over 300 Scouts coming into our tent and making a contact on HF or VHF. We also attended local events when possible. We have put stations on at the West of England Steam Engine Rally and also the Cornwall Volkswagen Owners Club Jamboree. Another enjoyable event has been for the Cornwall Air Ambulance Trust and using this opportunity to promote the fact that it was Cornwall who had the very 1st Air Ambulance.

We are also very proud to run Cornish Radio Amateur Club Rally once a year at Penair School in Truro. Whilst rallies nowadays seem to be getting smaller we always have a good turnout each year.

To find out more about our club please visit our new website www.gx4crc.com

## The Crawley ARC

The Crawley ARC has about 50 members and benefits from having sole use of its own premises offering a meeting room, tea bar, modern lecture facilities and well equipped contest standard radio shack with tower, beams and wire antennas.

Club members, some of whom are professional engineers, bring their combined technical expertise in RF engineering, electronics design & construction, antenna design, microwave engineering and electronic servicing to the membership. This experience is invaluable to members starting in the hobby.

Within the Crawley area there are numerous companies engaged in engineering and technology. We appreciate that industry is desperately short of engineers and that young people can gain a professional interest through amateur radio. Similarly, it appears that those retiring also wish to take up the hobby. Our current training programme has been running for six years and covers all three licence levels. Please contact us if we can help with your training.

The Ashdown Forest Repeater Group is an associated organisation well supported by our members who provide a wealth of experience with design updates and ongoing maintenance for both the D-Star and FM repeaters. Call signs GB7MH and GB3MH respectively are co-located in the nearby village of Turners Hill and each have internet linking capability. We are looking at the possibility of replacing the D-Star repeater with one on the DMR network.

CARC hosts one of the four annual Microwave Round Tables bringing together microwave enthusiasts from a wide area with lectures and sale of components and general items of surplus equipment. For the first time in 2016 we also hosted the Broadband HamNet inaugural meeting, which we hope will become an annual event.

The Club's biggest outside event of every year is participation in the VHF National Field Day Contest in association with another local club, which involves the establishment of multiple stations to cover all bands and modes.

On a smaller scale, we are entering the RSGB UKAC regular Tuesday evening contests with some "first time" contesters getting involved as a learning experience. We are also active during the 80m Club Call contests.

We have a lot to offer, and we are a friendly helpful bunch! Please contact us on secretary@carc.org.uk or come and visit on a Wednesday (2000-2200) or Sunday (1100-1330).

## Cray Valley Radio Society

Cray Valley Radio Society celebrates its 70th anniversary in 2016. Throughout its history, the club has been at the forefront of amateur radio in south-east London, and nationally. Being located in the Royal Borough of Greenwich, CVRS was ideally placed to mount a series of high-profile special event stations starting with M2000A celebrating the millennium, and ending with 2O12L for the London Olympics. The Society now focuses on more locally-based events such as Meopham

Windmill GB6MW and Crossness Pumping Station GB2CM.

The Society runs a full programme of training courses from Foundation right through to the Full licence, and is careful to offer ongoing help and support to amateurs after they receive their callsigns, including making equipment available on short term loan to new licensees.

Planning permission was recently obtained from Greenwich Council for a Tennamast at the Club's meeting place, and the mast now supports an A3S HF tribander and 2m yagi together with wire antennas. The shack is opened regularly so that any member who wishes can operate, and members also use the club shack for a number of contest entries. All three RSGB field day contests will also be entered in 2016 from locations within driving distance of Eltham.

Meetings are held on the first and third Thursdays of each month at the 1st Royal Eltham Scout Hall, Rear of 61-71 Southend Crescent, Eltham, London SE9 2SD. A varied range of lectures and activities are held throughout the year, and the Club's magazine 'QUA' is published monthly. There are club nets weekly on 10m and 4m with additional nets and activity planned for the anniversary month of October 2016.

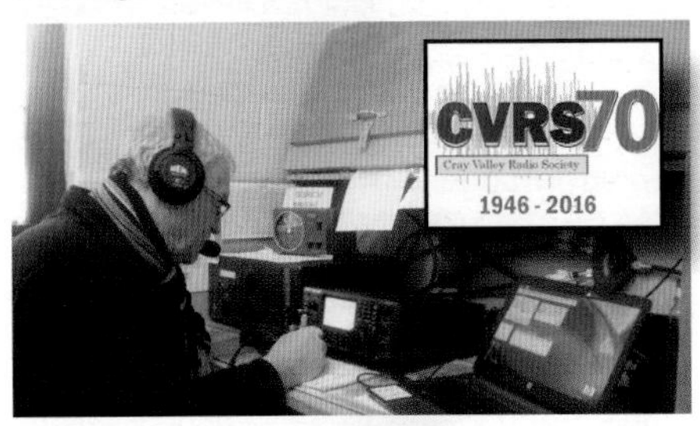


Please visit the Club's web site www.cvrs.org for further information, or contact Richard Cains G7GLW secretary@cvrs.org or 07831 715797.

## Dartmoor Radio Club

The Club was formed in November 1983 at Princetown, Dartmoor by 11 members studying to take the Radio Amateurs examination in May 1984. We meet on the first Thursday of each month at Yelverton War Memorial Hall, Meavy Lane and extend a warm and friendly welcome to a variety of visiting speakers, with talks given by club members.

We run regular nets on 2m with the call sign G1RCD from 20:00 to 21:00HRS on Thursday evenings, apart from Club nights. On Wednesday evenings we run an HF net from 20:00 hrs depending upon conditions with the callsign G0DRC.

The Club has always had the tradition of running Field Days; these provide newcomers with the opportunity of working HF. In summer we locate to Dartmoor and successfully make contacts from North America in the west to Australia and Japan in the east. Our usual location is on the western slopes of Cox Tor. In winter we move indoors to the Sports & Community Centre at Lamerton.

The main event of the year is the Rally held in Tavistock on the May Day Bank Holiday. This Rally has a good reputation with traders and visitors and we are pleased that everyone finds the situation very friendly. This has been our 32nd year!

We are very pleased to support two local community activities. In June we help with communications for the 32 Mile Ultra Marathon Road Race run over a gruelling course in the middle of Dartmoor. Then in October we help with the Abbotsway Walk which sees hundreds of young people crossing the Abbotsway route across Dartmoor ending in Tavistock. This is the 54th year this event has been running; the 300+ runners finish at Kelly College Outdoor Centre, Tavistock.

For full details contact our secretary Viv Watson: Vivwatsondrc@aol.com.

Phone: 01752 823427

## Dengie Hundred Amateur Radio Society

The Dengie Hundred Amateur Radio Society (http://www.dhars.org.uk/) is located about 9 miles from the town of Maldon in Essex, and approximately 3/4 a mile from Althorne railway station. The club is one of the few in the country to have a permanent shack, with permanently installed antennas. Being near the top of a ridge, with views to the rivers Crouch, Blackwater and Thames, the site has an excellent takeoff. Since the shack is permanently set up, members of the club have the option of meeting far more often than with other clubs.

Meetings of the club include:

- The 2nd and 4th Mondays evenings of each month, are the "official" club meeting nights. A speaker from either inside or outside the club is normally arranged. The evenings are quite informal, and very sociable.
- The first Monday of each month is where members meet to just operate amateur radio equipment.
- Members who are retired, or otherwise have some spare time, regularly meet on a Thursday during the day, to chat, perform maintenance, use amateur radio equipment or computers, and discuss future projects.

The club welcomes new members, and particularly welcomes those with other interests. Potential members should contact either the Secretary or Chairman by telephone or email.

Secretary – Stephen Hedgecock (M0SHQ) steve.m0shq@gmail.com /Telephone 07725597622

Chairman – Dr. David Kirkby (G8WRB) drkirkby@kirkbymicrowave.co.uk

Telephone 07910 441670

## Durham And District Amateur Radio Society

The Durham and District Amateur Radio Society meets every Wednesday evening at the Bowburn Community Centre, starting at 7.00 PM. The Society was formed in April 2010. The Bowburn Community Centre is located in Bowburn next to the A1 motorway. The Community Centre has full disabled access. We have a fully equipped shack with rigs and aerials for HF, VHF and UHF. and the clubs call sign is G4EUZ. The Society runs the D-Star repeater GB7PB. We are an RSGB. Registered Examination Centre and have run successful Foundation and Intermediate courses. We have a web site at: *www.radioclubs.net/durhamanddistrictars*

## Edgware & District Radio Society

Edgware & District Radio Society is one of the oldest radio clubs in London. It was founded in 1937 by a group of local enthusiasts as the Edgware Short Wave Society which later changed to the present day name of the Edgware & District Radio Society. We celebrate our 80th anniversary during 2017.

We meet at the Watling Community Centre (WCC), 145 Orange Hill Road, Burnt Oak, Edgware, Middlesex HA8 0TR on the second and fourth Thursdays of each month between 20:00 and 22:00 local time. There is a large free car park and London Transport buses numbered 302, 303, 305, 628 and 642 pass very near WCC and the Burnt Oak Underground station on the Northern Line is about 10 minute walk away.

Some of our people are involved in electronics in their working lives and some for whom amateur radio is just a hobby. But whatever your interest and knowledge base is, we try and help you to enjoy Amateur Radio much more. Please contact Mike GRNW on 020 8950 0658 or *Michael.stewart5@ntlworld.com* for further details or visit our website at: *www.g3asr.co.uk* for the latest news and details of the club programme.

## Felixstowe & DARS

Formed in 1983, the Felixstowe & District ARS provides a varied programme of talks, visits and events roughly fortnightly throughout the year. Annual special events include GB2WTM from Woodbridge Tide Mill for National Mills Weekend and GB2FX from Landguard Museum, next to Landguard Fort at Felixstowe, for Darrell Day; the commemoration of Captain Nathaniel Darrell, who repelled the Dutch invasion of July 1663. The Society also co-hosts the annual East Suffolk Wireless Revival - the Ipswich Rally. Several members are very keen on contesting and are always looking for new operators. Other activities include various operating evenings, talks by some excellent speakers, junk sales and the ever popular annual dinner.

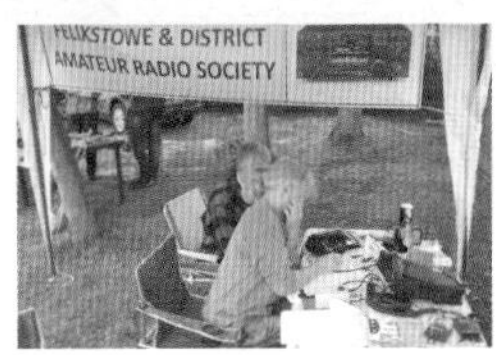


The main thrust of the Society, however, has always been radio training. Since the amateur radio training programme began within the Felixstowe & District ARS in 1989, over 500 people have been successfully trained for the various Morse, Foundation, Intermediate and Advanced licence exams. Over 250 people were successful in the former City & Guilds RAE and Novice RAE exams, and latterly

another 280 people have successfully tackled the RSGB/RCF examinations.

For further details contact the Secretary, Paul G4YQC, QTHR, or checkout the website at: *www.fdars.org.uk*

## Furness Amateur Radio Society

Furness Amateur Radio Society has been representing amateur radio in Furness and South Cumbria for over 100 years and 2016 was another busy year for our small club. The annual mills on the air event is always a popular event for FARS, where we operate GB2GW from Gleaston Water Mill – owned by club member G8ALE and his XLY. The year continued with stations for Lighthouse on the Air, Railways on the Air and YOTA. Our weekly meetings continued to be well supported with social / natter evenings as well as talks and demos. Some of the most popular talks included an introduction to the Raspberry Pi, talks on Contesting, Club members took part on a year long HF ladder contest to encourage HF operating, and various outdoor events were held including ARDF, antenna construction and testing, and portable operating.

A number of Foundation, Intermediate and Advanced exams were held, with successful candidates at all three levels.

For further details please visit: *www.fars.org.uk* or email us at: *info@fars.org.uk*

## Cambridge and District ARC

Cambridge has been associated with wireless development and manufacturing for the best part of a century so it's no surprise that it is a centre of ham radio activity. There are three good amateur radio clubs in the area which means that our club has to work hard to provide something a little different to the others.

We hold regular club evenings, twice a month. On these evenings there is always a special event, it may be a presentation on a technical topic or a talk about a bit of radio history. Constructors' Evenings are a popular item where members each spend a few minutes talking about what they've been making and Surplus Sales never fail to attract the bargain hunters. But, whatever the core of the evening's activities, there's always some time left for members to have a chat, swap ideas or just catch up on each other's news.

For those who enjoy operating we have a shack which is well equipped for both hf and vhf/uhf. We hold regular shack sessions outside of club evenings when members can take part in contests or, for those who prefer, just have a rag-chew on the bands.

New members are always welcome and you can try the club out without any obligation. You'll find lots more information on the website at *www.cdarc.co.uk* or look us up on Facebook or Twitter (Cambridge DARC).

## Guildford & DRS

Founded in 1918 as the Guildford Wireless Alliance, club meetings are on the second, fourth and fifth Fridays of the month and are held at the Guildford Model Engineer's building, Burchett's Gate entrance, Stoke Park, Guildford GU1 1TU.

GDRS can also be heard operating during various contests as "The Guildford Contest Group" using the call-sign G5RS. In 2011, 2012 & 2014 we had success in winning the Practical Wireless 4m contest and came 2nd in 2015.

Members gather from about 7:30 pm (but not much before) and meetings start at 8:00 pm. We are a very friendly club and visitors are very welcome. We have plenty of free parking at the club house.

Coffee, tea and a selection of soft drinks and confectionary are available at most meetings.

Following a rule change GDRS can now offer Associate Membership for those who live a significant distance from the club house and attend meetings infrequently. For full details see our updated Constitution and Rules document, which can be found on our web site *www.gdrs.net*

For enquiries please contact the secretary, Timothy Dabbs, G7JYQ Tel: 020 8241 9396 or via email: *sec.gdrs@virginmedia.com*

## Hastings Electronics and RC

There's been a radio club in Hastings since the 1920s and the present club was founded in 1976 to promote interest in all aspects of electronics, including amateur radio. We've over 70 members, whose interests range from VLF to 1.3GHz ATV, QRP to DX, home-build to vintage restoration, digital modes, RAYNET and Optical Comms.

The Club generally meets on the 4th. Wednesday of each month at the Taplin Centre, named after the late John Taplin, G3HRI. Our June meeting is a Field Day & BBQ, usually held in the Hastings Country Park at Fairlight

In April 2016 we supported St Richards College with their ARISS contact, providing a display at their STEM Space Conference and supervising their Build-a-thon of the RSGB Rodway receiver.

Full details of our programme & licence training are on our exciting new website, www.herc-hastings.org.uk/. If you'd like to get more from Amateur Radio, why not join our Club? There really is something for everyone. Current membership is only £12 a year, with concessions for students & families and a £2 discount for receiving our monthly magazine, Vital Spark, by email as a PDF - fantastic value. Just come along to one of our meetings where you'll be made most welcome, or email us at G6HHherc@yahoo.co.uk.

## Staffordshire Portable ARC

We are a group of men and women that like to operate amateur radio around the country as well as our own county of Staffordshire. At the moment we consist of 6 main operators, and 5 other members that come along as support. Our aim is to promote amateur radio to as many people as possible. This year 2015/2016 we were awarded small club of the year for our region (region 5). Since our club was formed in august 2014 we have participated in Biwota at Fradley junction, and,we also work with the Lichfield and Hatherton canal trust to promote the regeneration of the canal in Lichfield, we did castles on the air from Gwrych Castle in Abergele North Wales, we have done Beaches on the Air this year from Mablethorpe Beach, in Lincolnshire, we work one weekend a month from Chasewater Country Park, this year we worked from the largest carnival in Staffordshire the Lichfield Bower, we work once a month from the Wall Roman Site in Lichfield which is also where we do Museums on the Air, we also do Museums on The Air from Amerton Railway in Stafford, we have also done Scout Fetes and all through the winter we work from our club qth but this is to be changed in the next few weeks as we are moving, To round this up we hope to be out every weekend of the year and we try to encourage other people to come along, we try to be as family friendly as possible. Hopefully next year we will be having a week away in Tenby, South Wales as we want to work St Catherines Island, we hope to be doing Castles on the Air from Stafford Castle, and as with this year we will be doing all the usual events and also we will be trying to get into working with the Scouts as this is something we haven't ventured into yet ie. Jota.

Contact: Neville, 01922 449668

Website : *www.M0SPA,co,uk*

email : *sparc,2014@hotmail.com*

## VERULAM ARC (ST ALBANS)

The club's name is derived from that for an ancient settlement which is now the city of St Albans, and which the Romans called Verulamium. The river Ver flows through the town. It is no surprise therefore that the club's call sign is G3VER and its short call sign G3V. The club is now 55 years old and has over 50 members. It holds monthly meetings in a school in St Albans for discussions, demonstrations and talks from members and invited speakers. The club was delighted to be involved with Sandringham School in its preparation for its contact, in early January 2016, with Tim Peake aboard the International Space Station. As part of the school's preparations, students from the school attended a Foundation level course and examination held by the club. The school

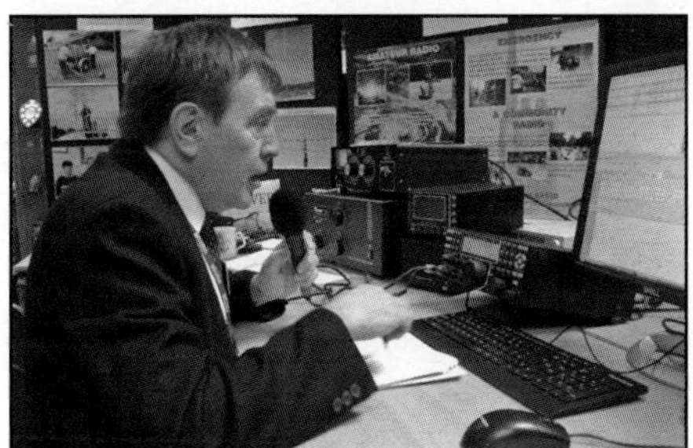

was the first of ten schools selected to make contact with Tim and it attracted much interest from the media, both national and local. The contact with Tim was the high point of the schools four day 'Festival of Space'. As part of the festival, club members helped students build radio receivers from kits supplied by the RSGB.

It has a well-equipped operating cabin high on Dunstable Downs and its facilities and location are ideal for contesting. The club runs courses and holds exams for all three licence levels. Information and news about the club's activities can be found at its website at: *http://www.verulam-arc.org.uk/.*

## West of Scotland (Glasgow) ARS

West of Scotland Club Meets 7:30 for 8.00pm on Friday at the Multi Cultural Centre 21 Rose Street Glasgow G3 6RE

Contact: Sam Liddel, GM4BGS,

Tel: 07831486620

*gm4bgs@yahoo.co.uk*

website: *www.wosars.org.uk*

## Worksop ARS

Worksop Amateur Radio Society is a very active club with over 100 members. We are very fortunate to have our own building in which we have a club room/café bar, a dedicated training/construction/committee room equipped with HF and VHF/UHF stations, two HF shacks for CW, Voice & Data and a VHF/UHF shack fully equipped for 6m, 4m, 2m, 70cm & 23cm for voice, data and D-Star with vertical antennas for 4m, 2m & 70 cm as well as Yagi antennas for all bands.

We are open two nights a week (sometimes three when training courses are running), Tuesday evening is mainly a social night plus CW operation/training and VHF/UHF contesting in RSGB UKAC. Thursday evening is a technical radio night with tuition, talks, demonstrations and construction projects plus more CW operation/training.

Members of the club have a broad range of interests which includes (but is not limited to) CW Operation and Training, VHF/UHF Contesting, HF contesting , IOTA, SOTA, Repeaters, APRS, Digital Modes, D Star, DMR, Antenna Construction and the use of technology linked with Amateur Radio. eg Raspberry Pi, Arduino, etc.

You can find an up to date programme of events on our web site at *www.g3rcw.org.uk*. You can also follow us on Twitter at *www.twitter.com/g3rcw* and or like us on Facebook at *www.facebook.com/g3rcw*

## POLDHU ARC

Located above Poldhu Cove in South West Cornwall, Poldhu ARC is located in the Marconi Centre on the famous site of Marconi's transmitter from where the first transatlantic signal sent on 12th December 1901. Using the three well equipped radio rooms the Club celebrates this anniversary annually, clocking up over 500 QSO's this year.

We have two 60ft. masts with dedicated dipole aerials and a Hex Beam, all making for excellent operating conditions.

Club members run the Marconi Centre, a museum and display of artefacts from the site, demonstrations of spark transmission and magnetic detector receivers; the Centre not only demonstrates items of historic interest but also brings communications into the 21st century by demonstrating data modes including PSK reception and transmission. Youngsters are encouraged to send their name on a Morse key and receive a certificate if it is received correctly. The Centre is visited by over 3000 members of the public and radio enthusiasts from around the world every year.

Visiting radio amateurs are welcomed and

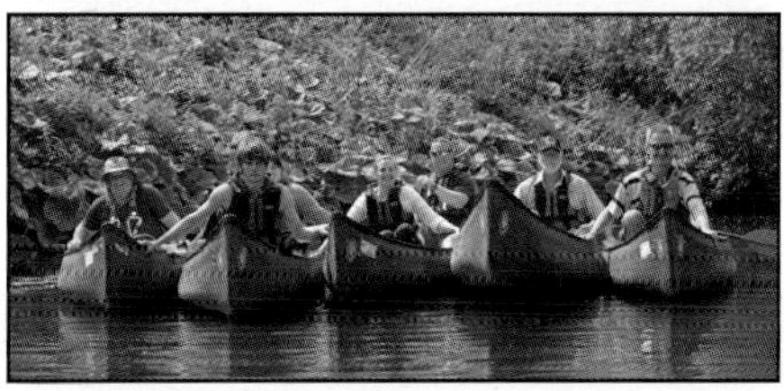

often are delighted to use the permanent special call sign GB2GM which signifies the important role that Guglielmo Marconi played here.

This year, to celebrate the granting of National Minority Status to the Cornish, Cornish amateurs have been able to use 'K' as a regional secondary identifier (Kernow is the Cornish word for Cornwall). Another of our calls is G3MPD (MPD was one of the calls used pre WW1 at Poldhu) and GK3MPD has made over 2000 QSO's to date.

The club has organised a 'Kernow Award' for stations working a number of 'K' prefixes. The age of club members ranges from 11 to 80+ years and members receive a quarterly newsletter and full access to the club facilities. The Club welcomes visitors from other Amateur Radio Clubs, schools and other organisations. Princess Elettra Marconi has visited us on a number of occasions and sent us greetings by radio from the Villa Griffone on International Marconi Day.

The Club net is on Wednesdays at 19.30pm local time 3725kHz + or – QRM and a full members meeting takes place on the second Tuesday of the month, often with first class speakers on radio related subjects.

Details of Club events, the Kernow Award and opening times of the Marconi Centre can be found at *www.gb2gm.org.*

# Online Clubs

## Essex Ham

Essex Ham has two simple aims – to promote the hobby, and to support the work being done by local groups and amateurs. At the centre of what we do, is the Essex Ham website, which forms an online community of over 500 members. The site also contains an ever-growing collection of resources.

We make use of social media, live video and audio streams from events to raise the awareness of amateur radio. Our weekly net continues to be the most popular local on-air meeting place, and in the last 12 months, our Monday Night Net has been joined by the monthly Essex YL Net.

As is highlighted by the RSGB's Train The Trainers, people learn in different ways, and we have recently launched "Foundation Online", a free online course for those looking for to study for their first amateur radio licence outside of the traditional club classroom. The course uses an industry-standard Virtual Learning Environment, and complements the work being done by RSGB assessors and clubs around the UK. In the first 12 months, over 270 potential amateurs enrolled, with special courses being run for local clubs unable to provide online training, as well as a local air cadet group. We also offer a regularly-updated set of Foundation training slides, video tutorials and handouts, which are in use by multiple clubs around the UK.

We've completed several successful field events, including a charity event with the Essex Air Ambulance, a special event station set up to promote a YMCA fundraiser, World Thinking Day, and an activation of the RAF Earls Colne airfield. Essex Ham has also been active in raising the profile of the hobby with youngsters, including demonstrations of Tim Peake ISS contacts.

Members of our team attend as many local events as possible, and we're very active on Twitter, Facebook and YouTube. Regardless of where you live, take a look at our busy website, and sign up as a supporter. Membership is free, and open to all. You can find out more at: ***www.essexham.co.uk***

# GB2RS News

After many years of negotiations with the General Post Office, the RSGB was authorised to broadcast the first news bulletin at 10.00 hours UTC on Sunday 25 September 1955. This broadcast was on 3600kHz from the home of Frank Hicks-Arnold, G6MB, in Walton-on-Thames, using the special callsign GB2RS. Broadcasts have continued on Sundays ever since, and the Society now has nearly 100 volunteer news readers who take it in turns to operate over 60 separate schedules every Sunday. These go out in nine different amateur frequency bands and can be heard throughout the UK and in parts of Western Europe. Listeners in Western Europe should try listening on 7150kHz or after dark on 1990kHz. At 09.00 hours (UK local time) the broadcasts on 3650kHz may also be heard in countries bordering the North Sea and the English Channel. For those in Southern England there are two ATV news transmissions in the 1.3GHz band.

## Broadcast philosophy

Ever since the service began, the news readings have taken place on Sundays, with the morning being the more popular time for listeners. GB2RS News may be likened to a weekly newspaper. The bulk of the editorial work is carried out on Thursdays and Fridays at RSGB HQ, with the completed script being released to the RSGB website later on Friday afternoons.

All readings include a weekly propagation report and forecast, prepared by the RSGB's Propagation Studies Committee. Recipients of these broadcasts can also see the script, containing some graphics, on the RSGB website where it is published on Friday afternoons. Surveys are carried out at selected radio exhibitions and rallies from time-to-time, which show that the majority of UK news recipients still prefer to hear the news on-the-air, rather than to obtain it from the RSGB website. This indicates that amateur radio is still a flourishing hobby, in spite of present day Internet addiction! However, the broadcast is also produced in MP3 format for the RSGB by Jeremy Boot, G4NJH. This material is also made available to other English speaking countries' amateur radio news services. To listen to the news online, go to G4NJH's website (see next page), or download it direct in MP3 format from the RSGB's GB2RS News web pages.

## News readers

The organisation of the GB2RS News Service and the network of newsreaders is the responsibility of the GB2RS News Manager, Contact: Ken Hatton, G3VBA, gb2rs.manager@rsgb.org.uk

The GB2RS News Manager is always in need of volunteers who are willing to assist with the broadcasting of GB2RS News. The form of presentation has evolved gradually over the years, but it still embodies a need for a presenter at the microphone who can give a reasonably fluent delivery.

The newsreaders enjoy making their contribution to the News Service. They often conduct after-news nets, when listeners are invited to call in, exchange information and give reports on the reception.

## Submitting news

News items for GB2RS should be sent to RSGB HQ as far in advance as possible. The deadline is 10am on Thursdays. The RSGB News Desk may be contacted by email. Please send all news stories for *RadCom*, as well as GB2RS, to: *radcom@rsgb.org.uk*. Entries for Around Your Region or GB2RS local news should also be emailed to *radcom@rsgb.org.uk* HQ may be telephoned via 01234 832700 and then selecting 'Editorial' from the automated options.

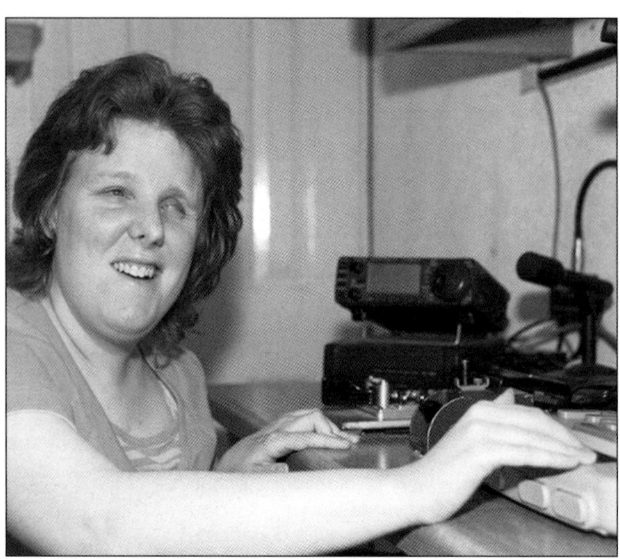

*Blind GB2RS newsreader Annick Morris, M0HDE.*

## GB2RS broadcast schedule (Sundays only, all times local)

| Time | Freq | Mode | Reader(s) | Location |
|---|---|---|---|---|
| **National Transmissions** | | | | |
| 10.00 | 7.127 | LSB | G3ISB/DJ0OK | Roetgen |
| 10.00 | 7.127 | LSB | M0DXM/DJ2XB | Simmerath |
| 10.30 | 7.150 | LSB | GM3JIJ | Stornaway |
| 16.00 | 5.3985 | USB | G4HPE M0RIF | Royston Newton Abbey |
| 16.00 | 5.3985 | USB | GM4NTL G0APM | Sanquhar S. Ockendon |
| 16.00 | 5.3985 | USB | G4MWO G8DQZ | St.Helens Bury St.Edmunds |
| 21.30 | 1.990 | LSB | G4HPE | Royston |
| 21.30 | 1.990 | LSB | GM4NTL | Sanquhar |
| 21.30 | 1.990 | LSB | GM4JET | Johnstone |
| **England South East and East Anglia** | | | | |
| 09.00 | 3.650 | LSB | G8ROG,G4TRN,G4IWS | Reading, Bristol, Bures |
| 09.00 | 3.650 | LSB | G4RDC,G6WPJ | Reading, Bures also SW & Wales |
| 09.00 | 70.425 | FM | G8ROG,G4IWS, | Reading, Bristol also SW news |
| 09.00 | 70.425 | FM | G4RDC | Reading also SW news |
| 09.30 | 145.525 | FM | G8CKN,G8LES | Alton |
| 09.30 | 145.525 | FM | G8MSQ,M0KEL | Lancing, Worthing |
| 09.30 | 145.525 | FM | G0TLU | Lancing |
| 09.30 | 145.525 | FM | G0OZS,M0AKK, | Manningtree, Ipswich |
| 09.30 | 145.525 | FM | G4YQC,G0DVJ, | Felixstowe, Ipswich |
| 09.30 | 433.000 | FM | G8CKN via GB3BN | Bracknell |
| 09.30 | 433.225 | FM | G8CKN via GB3IW | Isle of Wight |
| 09.30 | 1308 | ATV | G8CKN, G8LES via GB3HV | High Wycombe |
| 09.30 | 1316 | ATV | G8CKN, G8LES via GB3IV | Isle of Wight |
| 09.30 | 1316 | ATV | G3NDJ via GB3VR | Brighton |

**For GB2RS News Online, there are links from the RSGB main page.*

| Time | Freq | Mode | Reader(s) | Location |
|---|---|---|---|---|
| 09.00 | 145.525 | FM | M0MBD | Hainault |
| 09.00 | 433.525 | FM | M0MBD | Hainault |
| 10.00 | 70.425 | FM | G4OXY ,G1GSN | Biggleswade |
| 10.00 | 145.525 | FM | G4OXY, GIGSN,G4OXD | Biggleswade, Hitchin |
| 10.30 | 51.530 | FM | G3EKJ, G6DGK | Uckfield |
| 10.30 | 145.525 | FM | G3EKJ, G6DGK | Uckfield |
| 10.00 | 433.525 | FM | G4IMP | Folkestone & Dover |
| 19.00 | 145.625 | FM | G4NZQ,G7URP,G3LDI via GB3NB | Norwich |
| **England South West and Channel Islands** | | | | |
| 09.00 | 145.525 | FM | GJ0PDJ, 2J0SZI | Jersey |
| 09.30 | 51.530 | FM | GU1HTY, GU6EFB | Guernsey |
| 09.30 | 145.525 | FM | GU1HTY, GU0SUP, GU4RUK | Guernsey |
| 09.30 | 145.725 | FM | G3NPB, G4BHD, G2KF via GB3NC | St. Austell |
| 09.30 | 433.525 | FM | G2HDR,G4TRN | Bristol |
| 10.30 | 145.525 | FM | G2HDR,G4TRN | Bristol |
| **England Midlands** | | | | |
| 09.00 | 145.600 | FM | G0ATR,G4AFJ via GB3CF | Leicester |
| 09.30 | 3.650 | LSB | G3STG,G8BGT | Melton Mowbray, Reading |
| 10.30 | 145.725 | FM | M0DWE,M5ZZZ,G0FVI via GB3LM | Lincoln |
| 18.00 | 145.525 | FM | G3USF,G0VVT | Keele, Stoke on Trent |
| 18.30 | 50.790 | FM | G0VVT | Stoke on Trent Via GB3SX |
| 18.30 | 433.525 | FM | G0VVT | Stoke on Trent |
| 19.00 | 145.6375 | FM | G4TSN,G0LCG via GB3IN | Huthwaite Notts |
| **England North** | | | | |
| 9.00 | 50.800 | FM | G4CLI,G0TKF via GB3WY | Wakefield |
| 09.00 | 145.525 | FM | G4OLK, G7GJU | Tyne Tees, Durham |
| 09.00 | 145.525 | FM | G7MFN | Sunderland |
| 09.30 | 145.525 | FM | G4IOD, G4KFP | Gawthorpe,Brighouse |
| 09.30 | 145.525 | FM | M0WIT, G6NTI G6YGV | Cleckheaton, Halifax |
| 09.30 | 145.775 | FM | G0VOF, G4PF via GB3RF | Accrington |
| 10.00 | 145.525 | FM | G3GJA | Hull |
| 10.30 | 3.640 | LSB | G0VOF, G0LYZ | Blackburn, Driffield |
| 10.30 | 3.640 | LSB | G0MRL | Bolton |
| 10.30 | 70.425 | FM | G3VBA,G4GSY | Frodsham, Bury |
| 10.30 | 70.425 | FM | G0NAJ,G1JPV | Dukinfield, Frodsham |
| 10.30 | 145.525 | FM | G3VBA,M0HDE | Frodsham, Wigan |
| 10.30 | 145.525 | FM | G4GSY,G0NAJ | Bury, Dukinfield |
| 10.30 | 145.525 | FM | G1JPV | Frodsham |
| 21.00 | 70.425 | FM | G3VBA,G4GSY | Frodsham, Bury |
| 2100 | 70.425 | FM | G0NAJ,G1JPV | Dukinfield, Frodsham |
| **Scotland** | | | | |
| 09.00 | 145.525 | FM | GM6MEN | Perth |
| 09.30 | 70.425 | FM | GM4DTH | Edinburgh |
| 09.30 | 145.525 | FM | GM4DTH,MM0TSS | Edinburgh |
| 09.30 | 145.650 | FM | GM7GMC,GM1BAN via GB3OC | Kirkwall |
| 09.30 | 145.650 | FM | MM3YMU,MM3YHA via GB3OC | Kirkwall |
| 09.30 | 433.525 | FM | GM4DTH | Edinburgh |
| 10.00 | 51.530 | FM | GM4ILS | Elgin |
| 10.00 | 145.525 | FM | GM4ILS | Elgin |
| 10.00 | 145.525 | FM | GM3VTB | Glasgow |
| 10.00 | 145.525 | FM | GM4COX,GM7GDE | Carluke,E.Kilbride |
| 10.30 | 145.525 | FM | GM3JIJ,MM0DCQ | Stornaway, Skye |
| 11.30 | 3.640 | LSB | GM8MHU | Aberdeen |
| 12.30 | 3.650 | LSB | | |
| 12.30 | 3.640 | LSB | GM3JIJ | Stornaway |
| 19.00 | 145.700 | FM | 2M0CDO,G0AXJ,GM0CDV via GB3BT | Berwick |
| **Northern Ireland and Isle of Man** | | | | |
| 09.30 | 145.525 | FM | GI3WEM,MI0AWL | Banbridge, Carrickfergus |
| 10.00 | 3.640 | LSB | GI4FUM,MI0RYL | Muckamore, Craigavon |
| 10.00 | 433.050 | FM | GI0VTS,MI0AWLGB3UL, | Belfast |
| 11.30 | 70.425 | FM | GI0VTS | Bangor Co. Down |
| **Wales** | | | | |
| 18.30 145.525 | | FM | GW0AQR,GW4KAZ | Caernarfon |

*Above schedule updated by Gordon L Adams G3LEQ - GB2RS News Manager effective June 2015*

*Got a news item?* **Tel:** 01234 832700 **email:** *radcom@rsgb.org.uk*
*Got a network enquiry?* **email:** *gb2rs.manager@ntlworld.com*

## When submitting news items...

***Do:***

- Submit items by email, to: radcom@rsgb.org.uk
- Always give a contact name, callsign (if any) and phone number.
- Say if a phone number is daytime only or evening only.
- Send GB2RS any last-minute details of your rally.
- Give the proper name of your club – there is a Wirral and District Amateur Radio Club and a Wirral Amateur Radio Society, so saying "the Wirral club" could lead to confusion.
- Always give the callsign of a speaker if he/she has one, not just 'Talk by John on aerials'.
- Provide full dates, not 'Last Friday in month'.
- Listen to GB2RS to hear for yourself the format used.

***Don't:***

- Forget to include the venue and opening time of a rally, details of talk-in (if any) and a contact name, callsign, and phone number.
- Send in 'to be confirmed' items. If they are not confirmed we assume that they are not taking place and they will therefore not be broadcast. Only send your item in when it has been confirmed.
- Give more than one contact person or telephone number. There is only time to broadcast one, so you decide which one to use.
- Mix regular club meetings with main news items, such as rallies and special event stations.
- Use GB2RS and *RadCom* as the only means of publicising club events to your own members – the main purpose of GB2RS should be to inform casual listeners and members of other clubs of the exciting things your club is doing.
- Use cryptic titles for talks, or 'in-jokes'. If we don't know what you mean, it is unlikely that anyone else will.

*Read the news online at: www.rsgb.org/news/*

*Listen to the news on-line at G4NJH's website: http://homepage.ntlworld.com/g4njh2/rsgb.html*

# Planning Advice

Many, if not most, radio amateurs never see the need to apply for planning permission for their aerials. After all the aerials work just as well without it and there is a school of thought that if you don't ask for planning permission the Planning Department can't be tempted to say no. This might seem an attractive argument if you use small visually unobtrusive wire aerials, but if you have aspirations of anything more substantial you are likely to fall foul of the local Planning Department.

## Urban Myths

Unfortunately, holding an amateur radio licence in the United Kingdom does not convey any special 'rights' under planning legislation to have an aerial and there are a number of urban myths circulating regarding the need for planning permission.

Amateur radio aerials and masts are generally treated as residential development, exactly the same as a garage or conservatory, and will require planning permission unless they come under one of the following categories:

### Temporary

Unlike non-residential land which has a limit of 28 days, there are no specific time limits on how long a mast or aerial can be present and still be classed as temporary. It is the degree of permanence that is the deciding factor. The fact that the mast or aerial is installed in a ground socket and can be easily removed is not enough for it to be classed as temporary if it is in regular use.

### De minimalist

Visual impact is too small to be a concern to the planning process. There is no legal definition of what is 'de minimalist' and it is left in the first instance to the interpretation of the Planning Department, but it has been successfully argued that a single wire dipole can be classed as de minimalist when it uses existing structures such as a tree to support it. Should you receive an enforcement notice claiming that your installation is not de minimalist and you disagree, you can appeal the enforcement notice.

### Permitted Development

The Town and Country Planning (General Permitted Development) (Amendments) (No.2) (England) Order 2008 permits certain alterations and/or improvements to existing dwelling houses without the need for planning permission. Although no references are made in this Order to amateur radio aerials and masts, some radio amateurs have successfully argued under Part 1, Class A of the Order that an aerial, mast or pole to the rear of and attached to a dwelling house is an 'enlargement, improvement or alteration of a dwellinghouse' *provided the aerial or mast does not protrude above the ridge of the roof.* Similarly it has been successfully argued that a freestanding mast up to 3m in height to the rear of the property is also permitted development.

There is currently no legal ruling on whether this type of installation is actually covered by the provision and it is left in the first instance to the interpretation of individual Planning Departments, but it is known that some Planning Departments do accept this argument whilst others don't.

Some planners will seek to limit the size of aerials attached to masts under this Order, but it should be noted that the size restrictions for aerials in this Order refers only to *satellite and microwave aerials.* The Order gives no guidance on HF or other aerials. Similar legislation (often verbatim) exists in other parts of the United Kingdom.

### Mobile installation

The legal position regarding mobile masts is uncertain if the mast is used for more than 28 days per year in the one location. Some radio amateurs have successfully argued that mobile masts are plant and do not need planning permission, whilst others have failed and had enforcement notices served on them.

### 4 year rule

If your house is not a listed building and you have had your aerials and masts present and unchanged for 4 years or more, no enforcement action can be taken against you. You may be required to prove that the installation has been there for 4 years or more, but this need only be a letter of confirmation from your immediate neighbours or a receipt if it was commercially installed. It also makes sense to take some dated digital photographs of the new system and to note the log. Remember, if you change any part of the installation, e.g. the aerial, the clock starts again for the part you have changed.

A Certificate of Lawfulness for your aerials and mast can be obtained from the Planning Department after four years if you want one, but there is no legal requirement to do this.

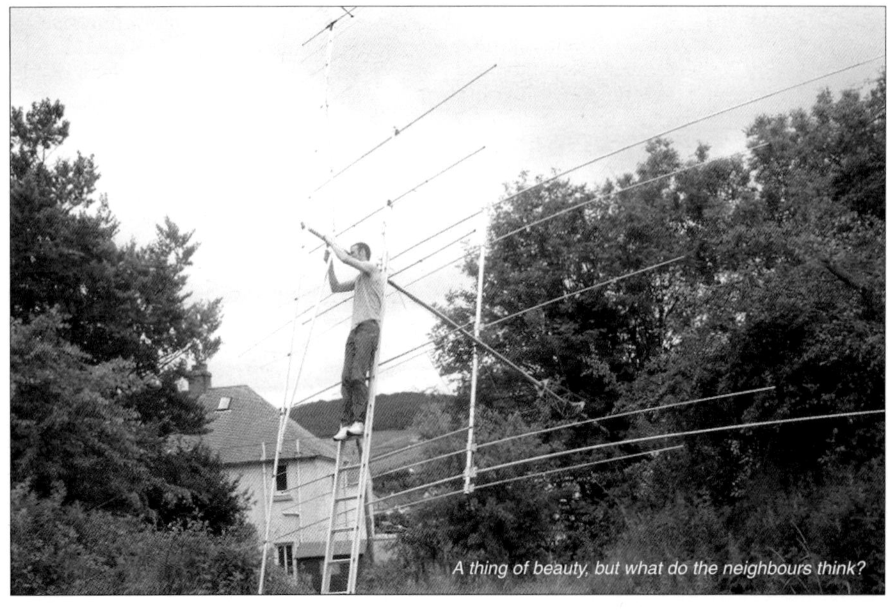

*A thing of beauty, but what do the neighbours think?*

## Applying for planning permission

Each local authority will have their own planning permission application forms, but they generally follow a similar style. They will typically require you to complete a Householders Planning Application form, a site location plan(s) and a development plan(s) showing the dimensions of the proposed aerial and/or mast and the distances to your property and the boundary with neighbouring properties. The number of copies and scale for these plans will be specified by the Planning Department in their planning pack. The drawings need not be professionally prepared, as long as it is clear what your proposals are and they are to the scale specified by the Planning Department. If you forget to show the aerial on your planning drawings you may receive planning permission for the mast only, without permission to attach any aerials.

You will also need to complete a neighbourhood notification form, detailing your 'notifiable neighbours'. A notifiable neighbour is someone who shares a boundary with your property or directly face any part your property from across the road. It is worth discussing your proposals with them before making your submission, so that when the official notice comes through their door it will not be a surprise. If you have TVI issues get these resolved first, as although TVI is not part of the planning process experience has shown neighbours will just object on other grounds, usually visual amenity.

Before formally submitting your planning application, ask if you can discuss the submission with your Case Officer. Minor changes at this stage may alleviate any concerns he/she may have, giving your application a better chance of success. You can also contact RSGB HQ to ask to be put in contact with a member of the Planning Advisory Committee, to discuss your proposals prior to submission. A letter of support from the RSGB for your proposed aerial or mast is also available on request.

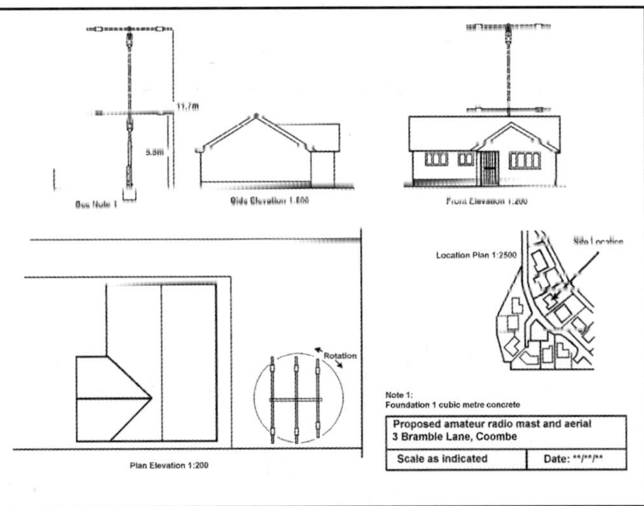


*The kind of drawing that a council will want you to submit with your application.*

## Refusal to grant planning permission

Sadly, not all planning applications are successful and there is sometimes no apparent reason why one Planning Department will grant planning permission for an aerial and mast in one area and another in a neighbouring area will refuse planning permission for a near identical installation.

You will be told why your application was refused. Usually it's on the grounds of visual amenity. Consider if the Planning Department has a valid point. To a radio amateur a large beam is a thing of beauty and a joy to own, but what do your neighbours think? Does it overly dominate the area? The Planning Department has to weigh-up the rights of all involved, not simply take sides. You will usually be able to resubmit a revised application free of charge if it is less than 12 months from the original application. If appropriate, reconsider a less ambitious proposal.

If however you believe the Planning Department has treated your application unfairly you have the right of appeal to the Planning Inspectorate (England and Wales), the Planning Appeals Commission (Northern Ireland) or The Directorate for Planning and Environmental Appeals (DPEA), (Scotland).

The appeal must be made within six months from the planning decision and is usually made in the form of 'Written Submission'. No charge is made for the appeal it is simply a matter of filling in the appropriate form and submitting your evidence in writing. It is also possible to submit your planning appeal electronically, but all documentation must be supplied in an electronic form..

To be successful you must state why you believe the original decision was unsound. Simply saying you disagree or that it will curtail your operations as a licensed radio amateur is not enough. You must establish that the Planning Department has failed to comply with planning law, policy or guidelines, or has sought to impose a different standard on your application than it has done for others.

*An 18m mast granted planning permission on appeal.*

The RSGB's Planning Advisory Committee can assist members in the preparation of a planning appeal if required. If you require assistance, contact RSGB HQ who will put you in contact with your nearest Committee member

If your appeal is not upheld and you have not used up your free resubmission, you can submit a revised proposal free of change if it is still less than 12 months from the original application.

## Enforcement notices

The Planning Department are likely to take enforcement action against you in two circumstances:

1. Where you have erected an aerial or mast which, in the Planning Department's opinion, requires permission and you have not obtained it.
2. Where the Planning Department alleges that you have breached a condition attached to the planning permission they have issued (for example, to keep a mast wound down when not in use).

The first is the most common. If you have not already submitted an application and had it refused the Planning Department will normally write to invite you to submit an application. It is usually worth doing so unless you want to argue that you have permitted development rights for the aerial or they are de minimalist.

The Planning Department may serve on you a Planning Contravention Notice. This requires you to give certain information as to ownership or to attend the Planning Department's Offices at a specific date and time to give details of your installation and why you believe it does not need planning permission (for example, because it's permitted development or de minimalist). You must comply with the Notice, because if you fail to do so you may be prosecuted.

If the Planning Department is not satisfied with your explanation they may elect to issue you an Enforcement Notice. Planning Departments can only do this if they can give reasons why they would not consider granting planning permission and may have to justify their decision to the Planning Inspectorate.

If an Enforcement Notice is issued it will set out what the Planning Department want you to do. Usually this will require you to remove the aerial and/or mast.

Should you be served an Enforcement Notice you have two choices:

1. Comply by removing the offending aerial, mast, etc.
2. Appeal.

You must appeal within 28 days of receiving the Notice. Details on how to appeal are available from the Planning Inspectorate, Scottish Government and the Northern Ireland Planning Appeals Commission websites listed below.

If the notice relates to a breach of conditions, the Planning Department may serve on you an ordinary enforcement notice, (against which you can appeal as above), or alternatively a Breach of Condition Notice, against which there is no appeal.

Failure to comply with an Enforcement Notice quickly can lead to legal action being taken against you, so don't ignore them. If the Planning Department considers that the aerial/mast has a severe environmental concern which requires immediate action they can apply to the Court for an injunction. If such an injunction is granted, you must comply or you will be prosecuted.

## Planning Advisory Committee

The Planning Advisory Committee exists to assist RSGB members with planning applications, enforcement notices and planning appeals. Committee members will not actually prepare your planning application or submit an appeal on your behalf, but can check your application or provide you with a suggested appeal strategy.

The Committee also provides a guide to the planning process. This is available free of charge from RSGB HQ, or as a download from the member's only website.

## Tenancy matters

For tenants, both planning permission and landlord consent are likely to be needed. Obtaining planning permission does not mean the landlord has to agree, so you could find yourself having incurred the expense of obtaining planning permission only to find that the landlord does not agree and so you cannot implement the permission.

Especially for private sector tenants, failure to obtain consent may give the landlord grounds to terminate the tenancy.

The Society cannot become involved in legal disputes between landlords and tenants, but will try to provide advice or to signpost members to other bodies who can help. You may therefore want to contact the PAC Chair before starting - contact details below.

| | Online planning information | Appeals and enforcement notices |
|---|---|---|
| England/Wales | www.planningportal.gov.uk/ | www.planning-inspectorate.gov.uk |
| Scotland | www.eplanning.scotland.gov.uk | www.dpea.scotland.gov.uk |
| Northern Ireland | www.planningni.gov.uk | www.pacni.gov.uk |

# RadCom Reviews

Items of amateur radio equipment are frequently reviewed in RadCom. The types of equipment range from antennas, through budget 'handies', software and ancilliary equipment, to top-of-the-line transceivers. Listed below are the reviews that have taken place in RadCom since January 1994. Please refer to the relevant edition, if you need more information on any of the items.

## A

| | |
|---|---|
| AA&A AMA-3 and AMA-5 (HF loop antennas) | Jul 1994 |
| AADE DFD (digital frequency display kit) | Apr 1998 |
| AADE IIB (digital L/C meter kit) | Apr 2005 |
| AAlog (logging software) | May 2006 |
| ACE-HF (propagation prediction software) | Oct 2006 |
| ACE-HF PRO propagation prediction software | May 2016 |
| Acom 1000 (160m-6m linear) | Mar 2001 |
| Acom 1010 (HF linear amplifier) | Aug 2005 |
| Acom 1500 (160m-6m linear) | Dec 2012 |
| Acom 2000A (HF auto-tuning linear) | Mar 2001 |
| ADI AR-446 (70cm FM mobile) | Mar 1997 |
| ADI AT-600D (2m+70cm handheld) | Apr 1997 |
| ADI Sender-450 (70cm FM handheld) | Apr 1993 |
| Adonis AM-308 (desk microphone) | Jan 2006 |
| Adonis AM-508 (desk microphone) | Jan 2006 |
| Adonis AM-708 (desk microphone) | Jan 2006 |
| AEA Isoloop (HF loop antenna) | Jul 1994 |
| AirNav RadarBox (virtual radar) | Mar 2009 |
| Airspy SDR Dongle | Sep 2015 |
| AKD-6001 (6m FM mobile) | Feb 1994 |
| AKD-7003 (70cm FM mobile) | Sep 1994 |
| AKD HF3E (communications receiver) | Jun 1998 |
| AKD Target HF3 (VLF-HF receiver) | Nov 1996 |
| Albrecht AE485S (10m multimode mobile) | Jan 2001 |
| Alinco DJ-596 (2m+70cm FM handheld) | Dec 2004 |
| Alinco DJ-C1 (2m FM micro handheld) | Feb 1998 |
| Alinco DJ-C4 (70cm FM micro handheld) | Feb 1998 |
| Alinco DJ-F1E (2m FM handheld) | Jun 1992 |
| Alinco DJ-C5 (2m+70cm FM mini handheld) | Jul 1998 |
| Alinco DJ-C7 (2m+70cm FM handheld) | Jan 2005 |
| Alinco DJ-G5 (2m+70cm FM handheld) | Nov 1995 |
| Alinco DJ-G5 (2m+70cm FM handheld) | Apr 1997 |
| Alinco DJ-G7 (2m+70cm+23cm FM handheld) | Aug 2009 |
| Alinco DJ-V5E (2m+70cm FM handheld) | Sep 2000 |
| Alinco DJ-V17 (2m FM handheld) | Jul 2008 |
| Alinco DJ-580SP (2m+70cm FM handheld) | Aug 1993 |
| Alinco DR-150E (2m FM mobile) | May 1995 |
| Alinco DR-610E (2m+70cm FM mobile) | Feb 1999 |
| Alinco DR-M06 (6m FM mobile) | Apr 1995 |
| Alinco DX-70 (HF+6m multimode mobile) | Aug 1995 |
| Alinco DX-70TH (HF+6m multimode mobile) | Aug 1999 |
| Alinco DX-SR8E(100W HF transceiver) | Jan 2014 |
| Alinco switch mode power supplies | Jan 2016 |
| Allsop Helikite (hybrid helium balloon / kite) | Jul 1995 |
| Alpha 4510 (HF power/SWR meter) | Nov 2006 |
| Alpha DX-Jr (lighweight portable 40m - 6m ant) | Feb 2013 |
| Alpha SPID Rak (azimuth rotator) | Mar 2007 |
| Alpin 100 (HF+6m linear amplifier) | Apr 2011 |
| Ameritron ALS-500M (HF linear) | Dec 2005 |
| Amlog 3 (logging software) | Sep 1996 |
| Analyser III Linear Cct Simulator (software) | Jul 1994 |
| AOR AR-DV1 Digital Voice Receiver | May 2016 |
| AOR AR3030 (HF communication receiver) | Feb 1995 |
| AOR AR7030 (VLF-HF receiver) | Jul 1996 |
| AOR ARD9000 Digital Voice (digital voice adapter) | Oct 2005 |
| AOR ARD9800 Fast Data Modem (digital voice adapter) | Jul 2004 |
| Arno Elettronica E-H Antennas (small HF antennas) | Sep 2003 |
| Arno Elettronica Venus-80 (small HF antenna) | Aug 2005 |
| Arno Elettronica Venus-160 (small HF antenna) | Aug 2005 |
| Array Solutions Bandmaster (universal band decoder) | Jan 2011 |
| Array Solutions PowerMaster (wattmeter) | Feb 2006 |
| ARRL Radio Designer (software) | Sep 1995 |
| ATM Motion Picture (video grabber) | Nov 1996 |
| Autek RF-1 RF Analyst (antenna analyser) | Oct 1994 |
| Avair AV-20 (HF power/SWR meter) | Oct 2006 |
| Avair AV-40 (2m-70cm SWR/power meter) | Jun 2006 |
| Aztex TVTX (24cm FM transmitter) | Sep 1991 |

## B

| | |
|---|---|
| Badger Boards Receiver (kit for Novice course) | Apr 1998 |
| Basicomm CW Touch Paddle | Jan 2016 |
| Begali HST (single lever paddle key) | Jul 2009 |
| Begali Sculpture (iambic paddle key) | Feb 2008 |
| Begali Simplex Mono (single lever paddle key) | Feb 2008 |
| bhi Compact In-Line Noise Eliminating Module | Dec 2015 |
| bhi DSPKR (noise reduction speaker) | Feb 2010 |
| bhi DSPKR 10W (noise reduction speaker) | Jan 2011 |
| bhi NEDSP1061 (add-on DSP module for FT-817) | Dec 2003 |
| bhi NEDSP1062 (add-on DSP module) | Jul 2005 |
| bhi NES 10-2 (noise eliminating speaker) | Dec 2002 |
| bhi NES 10-2 Mk2 (noise eliminating speaker) | Nov 2005 |
| bhi NEIM1031 (noise eliminating inline module) | Mar 2004 |
| bhi 1042 (switch box) | Mar 2004 |
| Bilal Isotron (small antennas for 80m and 40m) | Aug 2010 |
| Buddipole (portable HF-VHF antenna) | Mar 2005 |
| Butternut HF2V (HF multiband vertical antenna) | Mar 2005 |

## C

| | |
|---|---|
| Cloud IQ SDR | June 2016 |
| Comet CHA-250B (wideband vertical antenna) | Dec 2006 |
| Comet CSW-201G (antenna switch) | Jul 2007 |
| CommSlab µ-Modem (multimode radio modem) | Sep 1995 |
| Crazy Daisy (Mag loop Antenna) | May 2014 |
| Cross Country Wireless SDR (single band receivers) | Aug 2010 |
| CRT SS6900 (10m multimode mobile) | Aug 2011 |
| CT (contest logging software) | Nov 1999 |
| Cushcraft 13B2 (2m beam) | Mar 2005 |
| Cushcraft MA5B (compact HF beam) | Nov 1999 |
| Cushcraft MA8040V (80m/40m vertical antenna) | Sep 2005 |
| Cushcraft R7 (HF vertical antenna) | Jul 1992 |
| Cushcraft R8 (40m-6m vertical antenna) | Jun 2000 |
| Cushcraft R7000 (HF vertical antenna) | Jan 1997 |
| Cushcraft X7 ('Big Thunder' HF beam) | Nov 1998 |

## D

| | |
|---|---|
| Daiwa CS-201A (antenna switch) | Jul 2007 |
| DB6NT 13cm transverter (kit) | Jan 1998 |
| Derek Stillwell Morse Key (handmade straight key) | Jun 1995 |
| Diamond CA-35RS (lightning arrestor) | May 2006 |
| Diamond CX-210A (antenna switch) | Jul 2007 |
| Diamond SD300 (3-30MHz screwdriver antenna) | Nov 2011 |
| Diamond SX20C (HF power/SWR meter) | Oct 2006 |
| Diamond SX40C (2m-70cm SWR/power meter) | Jun 2006 |
| DG8SAQ VNWA3 (vector network analyser) | Dec 2011 |
| DK2DB 13cm PA (kit) | Jan 1998 |
| DPRE4-6VL (cavity filters for 2m repeater) | Jun 2009 |
| DSP-10 (DSP 2m multimode transceiver kit) | Feb 2000 |
| DX4Win (station logkeeping software) | Aug 2000 |
| DV Dongle (D-Star adapter) | Dec 2008 |
| DV Access Point Dongle (2m 10mW D-Star node) | Mar 2011 |
| DXAID 5.0 (propagation prediction software) | May 2004 |
| DX Engineering HEXX-5TAP-2 (5-band 2-ele hex beam) | Mar 2011 |

## E

| | |
|---|---|
| Elad FDM77 (software defined radio) | Nov 2005 |
| Elecraft K1 (HF QRP CW transceiver kit) | Sep 2001 |
| Elecraft K2 (HF QRP CW transceiver kit) | Mar 2003 |
| Elecraft K3S | April 2016 |
| Elecraft KX3 (HF/6m allmode transceiver) | Apr 2013 |
| Elecraft KPA500 (Solid State Amplifier) | Jan 2013 |
| Elecraft KRC2 (band decoder kit) | Jan 2005 |
| Elecraft N-gen (wideband noise generator) | Nov 2006 |

For a 10-year index of reviews, 1985-1994, see RadCom December 1995.

| | |
|---|---|
| Elecraft P3 Panadapter & K3 Transceiver revisited | Oct 2012 |
| Elecraft T1 (HF-6m QRP auto ATU) | Mar 2007 |
| Elecraft XG2 (receiver test osc. / S-meter calibrator) | Nov 2006 |
| ETO/Alpha 91B (HF linear) | Feb 1997 |
| ETO/Alpha 87A (HF linear) | Feb 1997 |
| EZMaster (SO2R interface) | Mar 2006 |
| EZNEC 2 (antenna modelling software) | Sep 1998 |

## F

| | |
|---|---|
| Flexradio SDR-1000 (software defined transceiver) | Jun 2006 |
| Flexradio Flex 1500 (HF-6m software defined xcvr) | Apr 2011 |
| Flexradio Flex-3000 (HF-6m software defined xcvr) | Aug 2009 |
| Flexradio Flex-5000A (HF-6m software defined xcvr) | Jan 2008 |
| Flexradio Flex-5000A upgrades (auto ATU and 2nd rx) | Mar 2009 |
| Force 12 XR6/XR6C Yagi (11 ele six band ant) | May 2015 |
| FunCube Dongle Pro + | Feb 2013 |

## G

| | |
|---|---|
| G1MFG ATV modules | May 2003 |
| G3LIV Isoterm (data interfaces) | Sep 2009 |
| G3PPD CW501 Insect Filter (audio filter) | Feb 1994 |
| G3WDG (23cm transverter kit) | Jun 2000 |
| G4HUP L-C Meter Kit (test instrument) | Jul 2009 |
| G4TPH Magnetic Loop Antennas (QRP loops) | Oct 2008 |
| G4ZPY 3-in-1 Combo Keyer (electronic keyer) | Jun 1998 |
| Garth 70cm and 23cm bandpass filters | Sep 2007 |
| Gemini 23 (1296MHz linear amplifier) | Feb2016 |
| GH Engineering PA1.6-16 (1.3GHz PA kit) | Jul 2006 |
| Global CX201 (antenna switch) | Jul 2007 |
| Goodwinch TDS (electric tower winches) | Jun 2011 |
| Green Heron RT-21 (digital rotator controller) | Oct 2009 |

## H

| | |
|---|---|
| HamGadgets MasterKeyer MK-1 (Morse keyer) | Oct 2011 |
| HamGadgets PicoKeyer-Plus (Morse keyer kit) | Oct 2011 |
| Ham Radio Deluxe (station control & logging software) | Apr 2005 |
| Hamware AT-502 (remote auto ATU) | Mar 2009 |
| Hamware AT-515 (remote auto ATU) | Mar 2009 |
| Hands RTX/AMP (broadband HF linear amp kit) | Jan 1995 |
| Hatley Crossed-Field Loop (antenna) | May 2002 |
| Heil Proset Headphones (headphones with boom mics) | Aug 1996 |
| Heil Proset 5 (headphones with boom mic) | Mar 2006 |
| Heil Pro 7 communications headset | June 2015 |
| Hexbeam Folding Antennas (20-10m multiband ant) | April 2015 |
| HFx (propagation prediction software) | Oct 1999 |
| HFx (propagation prediction software) | Jun 2000 |
| High Sierra 1800/Pro (HF mobile antenna) | Aug 2004 |
| Hilberling PT-8000A (HF-2m transceiver) | Nov 2013 |
| Howes ASL5 (audio filter kit) | Jun 1995 |
| Howes AT160, VF160 & MA4 (kits) | May 1997 |
| Howes DC2000 (receiver kit) | Jul 1997 |
| Howes DXR20 (receiver kit) | May 1995 |
| Howes Tx2000 (transmitter kit) | Mar 1998 |

## I

| | |
|---|---|
| ICEPAK (propagation prediction software) | Jun 2000 |
| Icom IC-2SET (2m FM handheld) | Jan 1990 |
| Icom IC-207H (2m+70cm FM mobile) | Feb 1999 |
| Icom IC-703 (HF-6m multimode portable) | Oct 2003 |
| Icom IC-706 (HF-2m multimode mobile) | Nov 1995 |
| Icom IC-706 Mk2 (differences from Mk1 version) | Jun 1997 |
| Icom IC-707 (HF multimode base station) | Dec 1993 |
| Icom IC-707 (HF multimode base station) | Apr 1994 |
| Icom IC-7100 (HF-70cm multimode) | Feb 2014 |
| Icom IC-726 (HF+6m multimode base station) | Feb 1990 |
| Icom IC-729 (HF+6m multimode base station) | Apr 1993 |
| Icom IC-736 (HF+6m multimode base station) | May 1995 |
| Icom IC-737 (HF multimode base station) | Sep 1993 |
| Icom IC-738 (HF multimode base station) | May 1995 |
| Icom IC-746 (HF-2m multimode base station) | Mar 1998 |
| Icom IC-756 (HF+6m multimode base station) | May 1997 |
| Icom IC-756 Pro (HF-6m multimode base station) | Mar 2000 |
| Icom IC-756 Pro II (HF-6m multimode base station) | Jun 2002 |
| Icom IC-756 Pro III (HF-6m multimode base station) | Feb 2005 |
| Icom IC-775DSP (HF multimode base station) | Jan 1996 |
| Icom IC-821H (2m+70cm multimode base station) | Jan 1998 |
| Icom IC-910 (2m-23cm multimode base station) | Jul 2001 |
| Icom IC-3230H (2m+70cm mobile) | Feb 1993 |
| Icom IC-7000 (HF-70cm multimode mobile) | Apr 2006 |
| Icom IC-7200 (HF-6m multimode rugged /P/base) | Jan 2009 |
| Icom IC-7400 (HF-2m multimode base station) | Oct 2002 |
| Icom IC-7410 (HF-6m multimode base station) | Jan 2012 |
| Icom IC-7600 (HF-6m multimode base station) | Jun 2009 |
| Icom IC-7700 (HF-6m multimode base station) | Jun 2008 |
| Icom IC-7800 (HF-6m multimode base station) | May & Aug 2004 |
| Icom IC-7851 HF-6m transceiver | Nov 2015 |
| Icom IC-9100 (HF-23cm base station) | Apr 2011 |
| Icom IC-9100 (HF-23cm base station) | May 2012 |
| Icom IC-Delta 1E (2m+70cm+23cm handheld) | Nov 1993 |
| Icom IC-E7 (2m+70cm handheld) | Jul 2006 |
| Icom IC-E80 (2m+70cm FM + D-Star handheld) | Feb 2011 |
| Icom IC-E90 (6m+2m+70cm FM handheld) | Dec 2004 |
| Icom IC-E91 (2m+70cm FM handheld) | Jan 2007 |
| Icom IC-E92 (2m+70cm FM + D-Star handheld) | Nov 2008 |
| Icom IC-E2820 (2m+70cm FM + D-Star mobile) | Mar 2008 |
| Icom IC-R3 (HF-microwave handheld receiver/TV) | Nov 2001 |
| Icom IC-R20 (HF-microwave handheld receiver) | Sep 2004 |
| Icom IC-T3H (2m handheld) | Jul 2002 |
| Icom IC-T7E (2m+70cm handheld) | Apr 1997 |
| Icom IC-T70E (2m+70cm handheld) | Nov 2010 |
| Icom IC-T81E (6m-23cm FM handheld) | Sep 2000 |
| Icom IC-V80E (2m handheld) | Nov 2010 |
| Icom IC-V82 (2m FM/digital handheld) | Jan 2006 |
| Icom ID-E880 (2m+70cm D-Star mobile) | Feb 2011 |
| Icom PCR-1000 (remotable receiver) | Dec 1997 |
| ICS AMT-3 (AMTOR terminal unit) | Jan 1993 |
| Idiom Press Rotor-EZ (rotator controller) | May 2001 |
| IK Telecom DPRE4-6VL (cavity filters) | Jun 2009 |
| Index Labs QRP+ (HF SSB/CW QRP transceiver) | Nov 1994 |
| Index Labs QRP+ (HF SSB/CW QRP transceiver) | Nov 1995 |
| InnovAntennas 9-ele 2m LFA Yagi (antenna) | Mar 2012 |
| InnovAnetnnas 5-element 15m OP-DES Yagi (antenna) | Aug 2012 |
| Innovantennas 20/15/10 DESpole (antenna) | Dec 2013 |
| International Radio Roofing Filter (for FT-1000MP) | Jan 2005 |
| Ionsound (propagation software) | Aug 1994 |
| I-PRO Home (multiband vertical dipole) | Jul 2011 |
| I-PRO Traveller (portable vertical HF dipole) | Jun 2010 |

## J

| | |
|---|---|
| JPS ANC-4 (antenna noise canceller) | Aug 1996 |
| JPS NTR-1 (DSP audio filter) | Sep 1994 |
| JPS NIR-10 (DSP audio filter) | Sep 1994 |
| JRC JST-245 (HF+6m multimode base station) | Oct 1997 |
| JRC NRD-630 (professional MF/HF receiver) | Feb 2008 |
| J-Com W9GR DSP-II (DSP audio filter) | Sep 1994 |

## K

| | |
|---|---|
| Kanga Finningley (80m SDR kit) | Aug 2011 |
| Kanga Foxx3 (QRP transceiver kit) | Sep 2010 |
| Kenwood LF-30A (HF low pass filter) | May 2007 |
| Kenwood TH-79E (2m+70cm handheld) | Apr 1997 |
| Kenwood TH-D7E (2m+70cm FM handheld) | Sep 2000 |
| Kenwood TH-F7E (2m+70cm FM handheld) | Dec 2004 |
| Kenwood TM-D710E (2m+70cm FM mobile) | Nov 2007 |
| Kenwood TM-D710E + AvMap Geosat 5 Blu | Feb 2009 |
| Kenwood TM-G707E (2m+70cm FM mobile) | Feb 1999 |
| Kenwood TS-50 (HF mobile) | May 1993 |
| Kenwood TS-60S (HF multimode mobile) | Aug 1994 |
| Kenwood TS-480HX (HF+6m multimode mobile/base stn) | Mar 2004 |
| Kenwood TS-570D (HF multimode base station) | Dec 1996 |
| Kenwood TS-590S (HF+6m multimode base station) | Jan 2011 |
| Kenwood TS-590SG (HF and 50MHz) | Mar 2015 |
| Kenwood TS-690S (HF+6m multimode base station) | Nov 1992 |
| Kenwood TS-870S (HF multimode base station) | Apr 1996 |
| Kenwood TS-990S (HF+6m Flagship base station) | Jun 2013 |
| Kenwood TS-2000 (HF-23cm multimode base station) | Apr 2001 |
| Kinetics SBS-1eR (virtual radar) | Nov 2009 |
| Kinetic SBS-3 (virtual radar) | Aug 2012 |
| KK7P DSPx & KDSP-10 (DSP kit) | Jul 2005 |
| Kuhne 23cm transverter | Dec 2007 |
| Kuhne MKU LNA 131A HEMT (23cm preamp) | Jan 2008 |
| Kuhne MKU10 G3 (10GHz transverter) | May 2008 |
| Kuhne MKU 432 G2 (70cm transverter) | Apr 2011 |

## L

Lake DTR7-5 (HF QRP transceiver) Oct 1995
Lake Novice Receiver (kit) Feb 2001
LAMCO DU1500L (HF ATU) Feb 2012
LAMCO DU1500T (HF ATU) Feb 2012
LDG AT-100 (automatic ATU) Feb 2006
LDG Z11 (automatic ATU kit) Aug 2002
LDG Z11 Pro (automatic ATU) Jun 2009
Linear Amp UK Challenger (HF linear) Feb 1997
Linear Amp UK Discovery 64 (6m/4m linear) Apr 2010
Linear Amp UK Explorer 1200 (HF linear) Feb 1997
Linear Amp UK Ranger 811 kit (HF linear) Sep 2004
Little Tarheel II (80m-6m mobile antenna) Aug 2008
Logger (station logkeeping software) Aug 2000
Log-EQF (station logkeeping software) Aug 2000
Lowe HF-150 (HF receiver) Dec 1992

## M

Maha MH-C777 Plus-II (intelligent battery charger) Jun 2007
Maldol HF, VHF & UHF mobiles (antennas) Jan 2003
Maldol HVU-8 (80m-70cm vertical antenna) Jan 2004
Maldol MFB-300 (broadband vertical antenna) Apr 2007
MFJ-226 (graphical impedance analyser) Nov 2015
MFJ-249 & MFJ-259 (SWR analysers) May 1994
MFJ-260C (1-650MHz 300W dummy load) Feb 2008
MFJ-269 (HF-UHF SWR analyser) May 2000
MFJ-270 (lightning arrestor) May 2006
MFJ-393 (headphones with boom mic) Mar 2006
MFJ-461 (pocket Morse code reader) May 2002
MFJ-492 (memory keyer) Aug 1994
MFJ-704 (HF low pass filter) May 2007
MFJ-781 (multimode DSP data filter) Sep 1997
MFJ-784B (tunable DSP filter) Feb 1996
MFJ-805 (RF current meter) Jul 2006
MFJ-817C (2m-70cm SWR/power meter) Jun 2006
MFJ-854 (RF current meter) Jul 2006
MFJ-860 (HF power/SWR meter) Oct 2006
MFJ-890 (DX beacon monitor) Jul 2003
MFJ-935 (small loop tuner) Apr 2005
MFJ-939 Mar 2016
MFJ-935B (small loop tuner) Apr 2006
MFJ-936 (small loop tuner) Apr 2005
MFJ-989D (HF QRO manual ATU) Jan 2007
MFJ-991B (automatic ATU) Feb 2006
MFJ-993B (automatic ATU) Jun 2009
MFJ-1702C (antenna switch) Jul 2007
MFJ-1786 Super Hi-Q Loop (HF loop antenna) Jul 1994
MFJ-1786X Super Hi-Q Loop (HF loop antenna) May 2007
MFJ-1897 (HF vertical antenna) Sep 1995
MFJ-8100 (HF regenerative receiver kit) Oct 1993
MFJ-9020 (20m CW transceiver) Mar 1993
MFJ 9020, 9420, 9140 & 9040 (HF QRP transceivers) Oct 1995
MFJ 'Cub' (HF QRP transceiver kit) Feb 2001
Microham Micro Keyer II (multimode digital interface) Jan 2008
Microham Station Master & Six Switch (antenna switch) Apr 2009
MicroKeyer (PIC-based electronic keyer) Mar 1996
Microset CF-300 (DC-1GHz 300W dummy load) Feb 2008
Microset PTS-124 (13.8V DC power supply) Dec 2005
Microtelecom Perseus (software defined receiver) Mar 2008
Microtelecom Perseus (software defined receiver) May 2010
Miniprop Plus (propagation prediction software) Jun 2000
miniVNA Tiny (network analyser) Feb 2015
Miracle Ducker (antenna for FT-817 etc) Nov 2004
Miracle Whip (antenna for FT-817 etc) Feb 2002
Mizuho MX-14S (HF QRP transceiver) Oct 1995
Moonraker HT-90E (2m FM handheld) Sep 2011
Moonraker MT-270M (2m/70cm FM Mobile) Drc 2015
Moonraker SPX-200 (HF-6m mobile antenna) Jan 2007
Morphy Richards 27024 (domestic DAB+DRM receiver) May 2008
MW0JZE seven-band wide-spaced Hexbeam July 2015
MyDEL AnyTone AT-5189 (4m FM mobile) Apr 2011
MyDEL CG3000 (HF auto ATU) Nov 2006
MyDEL HB-1A (compact 3-band QRP CW transceiver) Oct 2010
MyDEL Multi-Trap Dipole (HF antenna) Dec 1995
MyDEL SB-2000 (radio interface) Sep 2010
M0CVO HW-20HP (off-centre fed dipole) Jan 2012

## N

NA (contest logging software) Nov 1999
Netset Pro-44 (VHF scanning receiver) Jun 1994

## O

Optibeam OB9-5 (9-ele 5-band HF beam) Aug 2003
Optibeam OB10-3W (10-element 20/17/15m beam) Mar 2005
Optibeam OB10-5 (10-element 5-band HF beam) Mar 2007
Outbacker Joey (QRP HF portable antenna) May 2008

## P

Palstar AT1KM (HF ATU) May 2005
Palstar AT1500CV (HF QRO manual ATU) May 2005
Palstar AT1500CV (HF QRO manual ATU) Jan 2007
Palstar AT-AUTO (HF auto ATU) Sep 2006
Palstar KH-6 (6m FM handheld) Oct 1997
Palstar ZM30 (digital antenna impedance bridge) Sep 2005
Peak Electronics 'Atlas' Analysers (L, C & R meters) Sep 2005
Peak Atlas DCA PRO (semiconductor analyser) Mar 2013
Peak Atlas ZEN50 intelligent Zener diode tester July 2015
Peet Bros Ultimeter 2100 (weather station) Jun 2008
Piccolo 6m Transceiver Kit (synthesised FM) Jun 1996
Pico Balun and Pico Tuner kits July 2016
Pixie QRP transceiver kit Jan 2016
PJ-80 (DF receiver) Feb 2006
PJ-80 (DF receiver) Apr 2007
PK-4 (Auto CW Pocket Keyer) Nov 2015
PocketDigi (datamode software for PDA/smartphone) Dec 2006
PolyPhaser IS-50UX-C0 (lightning arrestor) May 2006
PolyPhaser IS-B50HN-C2 (lightning arrestor) May 2006
PolyPhaser VHF50HN (lightning arrestor) May 2006
Powerex MH-C9000 (charger for AA and AAA cells) May 2010
Procom DPF 2/33 & DPF 70/6 (2m & 70cm duplexers) Sep 2009
Primetec Primesat Controller (satellite stn controller) Dec 2005
Pro-Am HF Mobile Antennas (mobile whips) Aug 1995
Pro-Am MM-3401 (mobile antenna mag-mount) Aug 1996
Pro Antennas DMV Pro (portable antenna system) May 2009
Pro Antennas Dual Beam Pro (5-band non-resonant beam) May 2011
Pro Antennas I-pro Traveller (portable antenna system) Jun 2010

## Q

Qpak Precision Tuner (mini ATU) Oct 2004
Quickroute PCB Designer (software) May 1993

## R

Ranger RCI-2950DX (10m/12m multimode mobile) Feb 2012
Rexon RL-102 (2m handheld) Apr 1994
RFSpace SDR-IQ (software defined receiver) Mar 2008
RigExpert AA-1000 antenna analyser (Antennas) Aug 2012
RIGblaster Advantage (PC-radio interface) Mar 2012
RIGrunner 4005 (12V distribution panel) Mar 2012
Rigol DS2000 series (oscilloscope) Aug 2013
Rigol DSA815-TG (spectrum analyser) May 2013
Rig Expert AA200 (antenna analyser) May 2008
Roberts RC818 (portable receiver) Jul 1993
Rock-Mite QRP (40m Xtal controlled Kit) May 2014
Rohde & Schwarz FSH3 (spectrum analyser) Sep 2004
R&D WM-BDSTR (weather monitor) Jul 1994

## S

Samlex SEC-1223 (13.8V DC power supply) Dec 2005
Sangean ATS-803A (portable receiver) Jun 1992
SD (contest logging software) Nov 1999
SDR play March 2016
SGC ADSP2 (noise eliminating speaker) Nov 2003
SGC ADSP2 Mk2 (noise eliminating speaker) Nov 2005
SGC SG-211 (internally powered HF-6m auto ATU) Jun 2005
SGC SG-231 (HF-6m auto ATU) Feb 2000
SGC SG-239 (budget HF auto ATU) Jun 2005
SGC SG-500 (HF linear) Dec 2005
SGC SG-2020 (HF 20W SSB/CW transceiver) Mar 1999
SGC Stealth Kit (antenna) May 2003
Shacklog (logging software) Jan 1995
Shacklog (station logkeeping software) Aug 2000
Shure 522 (desk microphone) Jun 2007
Shure 550L (desk microphone) Jun 2007
Shure 572B (fist microphone) Jun 2007
Signal Hound SA44B (spectrum analyser) Sep 2011

| | |
|---|---|
| Signal Hound TG44 (tracking generator) | Sep 2011 |
| SkySweeper (datamodes software) | Apr 2005 |
| SkySweeper professional (datamodes software) | Feb 2008 |
| Softrock V6.2 (HF SDR receiver) | Mar 2007 |
| Sony ICF-SW100E (mini communications receiver) | Jul 1996 |
| SOTA Beams 2m Portable Yagi (antenna) | Jul 2004 |
| SOTAbeams aerials, log book & battery monitor | Oct 2013 |
| SPE Expert 1K-FA (linear amplifier) | Jun 2007 |
| SPE Expert 1.3K-FA linear amplifier | July 2015 |
| Spectran HF-6085 (hand-held spectrum analyser) | Jan 2010 |
| Spiderbeam (5-band antenna kit) | Mar 2010 |
| Spiderbeam 160-18-4WTH (160m vertical antenna) | Sep 2013 |
| Spiderbeam Aerial-51 Model 404-UL (port ant 40m-6m) | May 2015 |
| SRW CobbWebb (HF antenna) | Jun 1993 |
| Standard C108 (2m FM handheld) | Jan 1997 |
| Standard C408 (70cm FM handheld) | Jan 1997 |
| Standard C568 (2m+70cm handheld) | Apr 1997 |
| Standard C5900D (6m+2m+70cm FM mobile) | Jul 1997 |
| Startek ATH-30 (frequency counter) | Sep 1994 |
| StationMaster (station logkeeping software) | Aug 2000 |
| SteppIR 3-element Yagi (14-52MHz beam) | Feb 2004 |
| SteppIR 4-element Yagi (14-52MHz beam) | Oct 2007 |
| sunSDR2 PRO (HF to VHF transceiver) | Dec 2015 |
| Super Antenna MP-1 (portable HF-VHF dipole) | Jun 2008 |
| Super Antenna MP-1 (compact portable antenna) | Jan 2013 |
| Super-Duper Contest Log (software) | Sep 1993 |
| Super Keyer 3 (electronic keyer) | Jan 1997 |
| Super Antenna MP1B Super-Stick | Jan 2013 |

## T

| | |
|---|---|
| Talksafe (Bluetooth adapter) | Sep 2007 |
| Telecom 23CM150 (23cm linear amplifier) | Oct 2009 |
| Tennadyne T10 (13-30MHz log periodic HF beam) | Jan 2002 |
| TenTec Argo (HF QRP transceiver) | Oct 1995 |
| TenTec Eagle 599 (HF-6m compact multimode base stn) | Jul 2011 |
| TenTec Jupiter 538 (HF multimode base station) | Jan 2004 |
| TenTec Omvi VII (HF-6m multimode base station) | Sep 2007 |
| TenTec Orion 565 (HF multimode base station) | Jun 2004 |
| TenTec Orion 2 566 (HF multimode base station) | Aug 2006 |
| TenTec 506 Rebel (CW QRP transceiver) | Dec 2015 |
| TenTec RX340 (professional HF DSP receiver) | Mar 2002 |
| TenTec Scout 555 (HF SSB/CW transceiver) | Nov 1993 |
| TenTec 1320 (20m CW transceiver kit) | Aug 2000 |
| Thamway TX-2200A (136kHz transmitter) | Feb 2010 |
| Timewave DPS-9 & DSP-9+ (DSP audio filter) | Sep 1994 |
| Timewave DSP-59+ (tunable DSP filter) | Feb 1996 |
| Timewave TZ-900 (antenna analyser) | Nov 2009 |
| Tokyo Hy-Power HL-1KFX (HF linear amplifier) | Oct 2005 |
| Tokyo Hy-Power HL-2KFX (HF linear amplifier) | Oct 2005 |
| Tokyo Hy-Power HL-50B (linear amplifier) | Sep 2002 |
| Tokyo Hy-Power HL-100BDX (HF linear amplifier) | Oct 2005 |
| Toyocom MS-5 (mobile hands-free kit) | Mar 2009 |
| Trident 6M5L (6m long yagi antenna) | Jun 2003 |
| TROPIC T-R (sequencer) | Aug 2012 |
| TRlog (contest logging software) | Nov 1999 |
| Turbolog (logging software) | Apr 1992 |
| Turbolog III (software) | Nov 1997 |
| Turbolog (station logkeeping software) | Aug 2000 |
| TYT TH-UVF1 (2m+70cm handheld) | Dec 2010 |
| Tytera MD380 DMR handheld | July 2016 |
| TYT TH-UFV9 (dual band handheld) | Dec 2012 |

## U

| | |
|---|---|
| Ulna 23-24 (GaAsFET Pre-amp) | Sep 1991 |

## V

| | |
|---|---|
| Vargarda 11EL2 (2m beam) | Mar 2005 |
| Vectronics 1010K (10m FM receiver kit) | Feb 2001 |
| Vectronics DL-300M (DC-150MHz 300W dummy load) | Feb 2008 |
| Vectronics LP-30 (HF low pass filter) | May 2007 |
| Vibroplex Original Deluxe (mechanical bug key) | May 1994 |
| Videologic DRX-601E/ES (digital radio tuner) | Oct 2001 |
| Vine Antennas LFA Yagis (loop fed VHF antennas) | Nov 2009 |
| Vortex Whirlwind 6M4(6m delto loop) | Dec 2011 |
| Voyager DX-IV (HF vertical antenna) | Jul 1992 |

## W

| | |
|---|---|
| Walford Berrow (QRP transceiver kit) | Apr 2014 |
| Walford Brent (single band CW transceiver kit) | Feb 2005 |
| Walford Compton (direct conversion receiver kit) | Jun 2002 |
| Walford Langport (80m+20m CW+SSB transceiver kit) | Apr 2000 |
| Walford Radio Today Chedzoy (receiver kit) | Feb 2001 |
| Watson AT-715 (five-in-one power station) | Nov 2009 |
| Watson CS-600 (antenna switch) | Jul 2007 |
| Watson PDX 100 (portable HF antenna) | Nov 2002 |
| Watson Power-Mite-NF (power supply) | Jun 2008 |
| Watson SP-350V (lightning arrestor) | May 2006 |
| Watson VAA-1 (antenna analyser) | Oct 2015 |
| Watson W-25SM (13.8V DC power supply) | Dec 2005 |
| Watson W-184 (headphones with boom mic) | Mar 2006 |
| Watson W-8682 (radio controlled weather centre) | Sep 2008 |
| Watson WM-S (mobile microphone system) | Nov 2005 |
| Wavecom W-Code (datamodes decoding software) | Jul 2012 |
| Wedmore 80m QRP Transceiver (kit) | Aug 1997 |
| Wellbrook ALA1530 (active receiving loop) | Jan 2012 |
| WinCAP Wizard II (propagation prediction software) | Jun 2000 |
| WinCAP Wizard III (propagation prediction software) | Dec 2002 |
| WinRadio WR-G1DDC Excalibur (SDR receiver) | Oct 2010 |
| WinRadio WR-G313i (PC-controlled receiver) | Mar 2005 |
| Wimo Big-Wheel antenna | Jan 2016 |
| Wonder Wand (antenna for FT-817 etc) | Jun 2004 |
| Wouxun KG-699E (4m handheld) | Sep 2010 |
| Wouxun KG-UV6D Pro Pack (2m/70cm handheld) | Jun 2012 |
| Wouxun KG-UVD1P (2m/70cm handheld) | Sep 2010 |
| WriteLog (contest logging software) | Nov 1999 |
| W2IYH (range of audio products) | Jan 2009 |
| W9GR DSP-II (audio filter) | Feb 1994 |

## Y

| | |
|---|---|
| Yaesu ATAS-100 (mobile antenna) | May 1999 |
| Yaesu FRG-100 (base station receiver) | Jul 1993 |
| Yaesu FT-11R (2m handheld) | May 1994 |
| Yaesu FT-41R (70cm handheld) | May 1994 |
| Yaesu FT-50R (2m+70cm handheld) | Apr 1997 |
| Yaesu FT-60R (2m+70cm handheld) | Dec 2004 |
| Yaesu FT-100 (HF-70cm multimode mobile) | Jun 1999 |
| Yaesu FT-450 (HF-6m compact multimode base stn) | Oct 2007 |
| Yaesu FT-450D (HF-6m compact multimode base stn) | Nov 2011 |
| Yaesu FT-817 (160m-70cm multimode portable) | Jun 2001 |
| Yaesu FT-840 (HF multimode base station) | Feb 1994 |
| Yaesu FT-847 (HF-70cm multimode base station) | Aug 1998 |
| Yaesu FT-857 (HF-70cm mobile) | Jun 2003 |
| Yaesu FT-890 (HF multimode base station) | Sep 1992 |
| Yaesu FT-897 (HF-70cm multimode /P/base station) | Apr 2003 |
| Yaesu FT-900 (HF multimode base station) | Nov 1994 |
| Yaesu FT-920 (HF+6m multimode base station) | Aug 1997 |
| Yaesu FT-950 (HF+6m multimode base station) | Dec 2007 |
| Yaesu FT-991 (HF/VHF/UHF transceiver) | Feb 2016 |
| Yaesu FT-1000 (HF multimode base station) | Jun 1991 |
| Yaesu FT-1000MP (HF multimode base station) | Jan 1996 |
| Yaesu FT-1000MP Mark-V (HF multimode base stn) | Oct 2000 |
| Yaesu FT-1000MP Mark-V Field (HF multimode base) | Oct 2002 |
| Yaesu FT-2000 (firmware upgrade) | May 2009 |
| Yaesu FT-2000D (HF-6m multimode base station) | Mar 2008 |
| Yaesu FT-2200 (2m FM mobile) | Dec 1993 |
| Yaesu FT-2500M (2m FM mobile) | Nov 1994 |
| Yaesu FT-8100R (2m+70cm FM mobile) | Feb 1999 |
| Yaesu FT-DX1200 (HF-6m multimode base station) | Mar 2014 |
| Yaesu FT-DX3000 (HF-6m multimode base station) | Jan 2014 |
| Yaesu FT-DX5000 (HF-6m multimode base station) | Jun 2010 |
| Yaesu FT-DX9000D (HF-6m multimode base station) | Oct 2005 |
| Yaesu FT-DX9000D (HF-6m multimode base station) | Dec 2006 |
| Yaesu FTM-10R (2m+70cm FM mobile) | Feb 2008 |
| Yaesu FTM-400DE dual band, dual mode mobile | Jan 2015 |
| Yaesu VR-5000 (HF-microwave multimode receiver) | Aug 2001 |
| Yaesu VX-5R (6m/2m/70cm FM handheld) | Sep 2000 |
| Yaesu VX-7R (6m/2m/70cm FM handheld) | Oct 2003 |
| YouKits FG-01 Antenna Analyser | Sep 2012 |
| YP-3 (6-band 3-ele portable yagi) | May 2009 |

## Z

| | |
|---|---|
| Zeus ZS-1 (SDR HF transceiver) | Dec 2013 |

# Audio Visual Library

The Audio Visual Library has been established as a special facility for societies affiliated to the RSGB in preparing programmes for their club meetings (please understand that it is not available to individual members). The items available to borrow include DVDs and CD-ROM presentations.

## Using the Service

1. Request your choices(s), giving the number(s) shown in brackets (eg S1). Many items are duplicated, but please give alternatives because your first choice may already be out on loan.
2. State the date for when it is required. Please give at least 14 days notice. If something is required at short notice, please telephone HQ to discuss.
3. The charge is £3.00 for each item, which includes packing and postage to you. The return postal cost is your responsibility. Cheques or Postal Orders should be made payable to RSGB. Please send payment with order.
4. Please return items immediately after use (within three days at most).
5. If damage is incurred or quality is poor, please attach a note to this effect to the relevant item on return.

## General Interest

(G2) *ARMADA GB400A*
Account of Plymouth ARC establishing defeat of the Spanish Armada. (40 min)

(G3) *BATTLE OF BRITAIN*
Explains the truth on how the battle was really waged and won. (1994, 60 min)

(G4) *CLASSIC MANOEUVRES*
The Red Arrows on their North American tour 1983 – the world's greatest aerial display team. (1992, 40 min)

(G11) *MELBOURNE RADIO CLUB*
Video made by this famous VK club, showing the city and amateurs' stations. (1982, 65 min)

(G12) *MY AMATEUR RADIO*
Biography G2DPQ – lifetime of amateur radio – recorded just before he died. (1996, 60 min)

(G14) *PASSPORT TO FRIENDSHIP*
World Goodwill Games 1990. Top Ham Operators 'Go for Gold' contest. (1991, 25 min)

(G15) *PJ9W – 1990 CQ WW SSB CONTEST*
Fascinating video on preparing and running a winning contest station, by a keen team of Finnish operators. (1992, 45 min)

(G22) *VHF – ALL YOU NEED TO KNOW*
VHF marine operator's examination manual. (1992, 45 min)

(G23) *VHF THEN AND NOW*
By Jack Hum, G5UM (1987, 85 min)

(G25) *WINNING ON THE HILL*
Superb video of VHF NFD winning station in Rochester, NY, USA. (1995, 58 min)

(G26) *WORLD AT THEIR FINGERTIPS –* RSGB
Growth of the amateur movement in the UK, by John Clarricoats. (45 min)

(G28) *M2000A – AMATEUR RADIO INTO THE NEW MILLENNIUM*
The planning and set-up of the Millennium special special event station M2000A. Includes many highlights of the two month operation (31 December 1999 - 29 February 2000).

(G30) *HUNTERS IN THE SKY*
Fighter aces of WWII.

(G31) *GB100MAR MARCONI CENTENERY CELEBRATIONS*
(Location: Dover, Kent)
i. The exhibition.
ii. The early days of wireless.
iii. The celebrations with special guests and re-enactments.
(1999 138 mins)

(G34) *MOBILE TELEPHONY & HEALTH*
Video made by the National Radiological Protection Board (NRPB). Provides a factual explanation of mobile phone technology and any possible health effects. (27 mins).

(G35) *OT9A "THE STORY" CQ WW DX CONTEST 1999 SSB MULTI-MULTI*
Discover the unique atmosphere and results by one of the biggest multi-multi contest stations in Europe.

(G36) *WRTC FINLAND 2002*
Coverage of the 4th World Radiosport Team Championship. (60 min).

(G37) *BUYING AMATEUR RADIO EQUIPMENT IN HONG KONG*
By Mike, M0AWD. (1997).

(G38) *UCANDO – "DIGITAL 3"*

(G40) *RSGB TODAY*
Meet the people behind *RadCom*; the General Manager; the Amateur Radio Dept; an M3 licence course in Cheshire; HQ facilities; the Commercial Dept; the Accounts Dept; the Annual General Meeting. (2005, 22 min)

(G41) *INSIDE RADCOM*
*RadCom* Editor Steve Telenius-Lowe and Technical Editor George Brown explain how the RSGB's house journal is created.

(G42) *THE M3 COURSE AT FRODSHAM*
Dave and Kath Wilson describe how the Foundation course is run at Frodsham.

(G43) *GB4FUN AT GRESWOLD SCHOOL*
The RSGB's mobile shack visits a school in Solihull. Co-ordinator Carlos Eavis explains amateur radio and telecommunications to the pupils, making a fun lesson out of the classroom.

(G44) *VALVEMAN*
The stroy of one man's lifetime obsession collecting valves. (30 min)

(G45) *HISTORY OF AMATEUR RADIO G2BTO*
First-class introduction to - and history of - amateur radio. 75 mins. CD audio.

## Specially for the Beginner

(B6) *NEW WORLD OF AMATEUR RADIO – ARRL*
(30 min).

(B7) *WHAT IS AMATEUR RADIO?*
A new look at amateur radio today and some of the important uses of this worldwide hobby. (10 min).

## Technical

(T2) *AMATEUR TELEVISION*
A series of short programmes on amateur TV here and in Australia. (60 min)

(T3) *BASIC RADIO MEASUREMENTS*
Practical demonstration of key station measurements, by G3NYK. (Amateur made video quality, 1993, 80 min)

(T9) *SECRET LIFE OF RADIO*
An easy-to-watch video on radio techniques from crystal set to complex rig. Very 'watchable' for all ranges of skill and knowledge. (1992, 35 min)

(T10) *SILICON GLEN*
Electronics industry video. (30 min)

(T11) *SKYWATCHING*
Good guide to daytime sky, explaining sun, clouds, wind, precipitation, weather hazards, etc. Made in USA. (1993, 40 min)

## DXpeditions etc

(D1) *AIRING AILSA*
Donated by the Ayr Amateur Radio Group. (1999, 27 min, 2 DVDs)

(D3) *DXPEDITION CAMPBELL ISLAND ZL9CI* By 9V1YC. (1999, 58 min).
(D5) *DXPEDITION FASTNET*
First-class account of preparing for and executing a successful Dxpedition to the Fastnet Rock Lighthouse. Donated by the IRTS (1991, 55 min).

(D6) *DXPEDITION 4J1FS MALIYSOTSKII IS* (25 min)

(D7) *DXPEDITION 7J1RL OKINO TORISHIMA* (35 min)

(D8) *DXPEDITION TO HEARD ISLAND* (1997). VK0IR by ON877 (53 min)

(D9) *DXPEDITION TO HOWLAND ISLAND – NO1Z/KH1* (22 min). *DXPEDITION TO BOUVET ISLAND – 3Y5X* (30 min). Donated by the NCDXF 1990/91.

(D10) *DXPEDITION TO HOWLAND ISLAND*
The latest expedition to KH1. (1993, 45 min)

(D13) *DXPEDITION VIETNAM, 3W*
New IOTA expedition to AS-130, Con Son Island. (1997, 30 mins)

(D15) *EXPEDITION BORNEO*
Presence Radioamateur (1995, 35 min). French spoken.

(D16) *JARL VISIT TO CHINA*

(D17) *LADY ISLE 2 - THE RETURN*
Donated by the Ayr Amateur Radio Group. (1997, 20 min)

(D19) *DXPEDITION S0ARSD*
(40 min)

(D20) *DXPEDITION VP8ANT*
(47 min)

(D21) *CQ DX PACIFIC - VK2EKY*
(1990)

(D22) *LEBIAZHI ISLAND EM5UIA, EU-180*

(D23) *A52A BHUTAN 2000*
The official DXpedition video. (60 mins)

(D24) *K5K DXPEDITION KINGMAN REEF*
(2000)

(D25) *W6IXP/K6ST ALASKA EXPEDITION*
(2000)

(D26) *P40V* (12 mins) *IS0XV* (120 mins)

(D27) *VK9RS ROWLEY SHOALS*

(D28) *VP8THU - SOUTH SANDWICH ISLANDS* (2002). The Micro-lite DXpedition, Part 1.

(D29) *VP8GEO - SOUTH GEORGIA*
(2002). The Micro-lite DXpedition, Part 2 (60 min).

(D30) *PERU 1999*
IOTA visit to some of the Islands off the coast of Peru.
(D31) *FO0AAA CLIPPERTON ISLAND*
(2000). The official DXpedition video.

(D32) *DXPEDITION: PATAGONIA / ARGENTINA / CHILE*
Including visits to IOTA islands near Ushuaia.
(D33) *YAMBE ISLAND*
(2000). IOTA Expedition. (CD ROM, 27 min).

(D34) *EAST TIMOR*
Powerpoint presentation on activity from East Timor ). (CD ROM)

(D35) *INSELN DOWN UNDER*
DXpedition to Norfolk Island and Lord Howe Island. (2004). Commentary in German.

(D36) *[illegible], MACEDONIA Z38Z, TEMUTO H40*
Four expedition films, 12-20 min each. German commentary.

(D37) *AGALEGA EXPEDITION 3B6RF*
CD Rom contains a number of still photos, a 15 min video clip and text describing the expedition. The video has no soundtrack but the text will help prepare a club member to explain what is happening. S(CD ROM )

(D38) *Z38Z MACEDONIA*
CD requires MPEG2 software to play. (12min). (CD ROM )

(D39) *VP6DI*
DXpedition to Ducie Island. No commentary.

(D40) *KIRIBATI 1999 T30/T33Y/CW – DL7DF*

(D41) *TWO EXPEDITIONS 2004 – GREENLAND NA-220.*(CD ROM)

(D42) *V15BR BAUDIN ROCKS*
Easter 2004 IOTA expedition to Baudin Rocks – good quality video.

(D43) V15BR OC-228, V13JPI OC251, V15WCP OC-261
DXpedition videos to Baudin Rocks, Ladia Julia Percy Island and Waldegrave Island.

(D44) *3B9C - RODRIGUS, D68C - COMOROS*
FSDXA Expeditions.

(D45) *THE LOST ISLANDS*
DXpedition to the 'Lost Islands' in the Central Arctic. (2001)

(D46) *T33C BANABA ISLAND*
The largest ever DXpedition to Banaba, April 2004.

(D47) *HEREHERETUE ATOLL DXPEDITION*
DXpedition to IOTA island OC-052, September 21-23 2004.

(D48) *MEXIQUE 2004*
DXpedition to Mexico. English commentary (some broken English).

(D50) *FT5XO KERGUELEN 2005*
FT5XO Expedition. (54 min)

(D51) *3B7C 2007 St Brandon 2007 Inc Rodrigues – 3B9C*

## Space, Satellites etc.

(S1) *AMATEUR RADIO'S NEWEST FRONTIER*
W5LFL Space Shuttle.

(S4) SATELLITE COMMUNICATION TELECOM
(1979, 60 mins)

(S7) *SATELLITE 'FUJI'*
Professional video (by JARL), showing the construction, launch and use of FUJI. (1986, 30 min)

(S8) *SPACE SHUTTLE WOORE TONY ENGLAND*
Account of the first amateur in space by the man himself. (53 min)

# Radio Communications Foundation

The RCF was established in 2002 and formally incorporated as a Registered Charity in 2003. Although the Charity was established via the RSGB it is independent from the Society.

## Mission

The RCF mission is to increase the engagement of people, especially young people, in radio communications technology. The RCF therefore encourages and assists students to pursue relevant higher education courses, leading to them being employed in the radio communications sector, and also works to raise public interest and involvement in radio communications and its associated science and technologies, including amateur radio.

The interested young person of today is the radio amateur of tomorrow and the engineer of the future.

### Near and medium term objectives:

The objectives are to advance the RCF mission by engaging with four key groups:

- Those at school (and in uniformed groups such as the Scouts, Guides, Army Cadet Force, Royal Air Force Air Cadets, Air Training Corps and Ambulance Brigades) in order to develop an interest in radio communications
- Those planning or undertaking a university or higher education course to encourage the study of radio communications options
- Those planning or considering employment in radio communications
- The public more generally, including those with an interest in amateur radio

Radio communication is so widely practised that it is almost taken for granted, but a greater public understanding of it is vital for the UK economy. There is a serious shortage of radio communications scientists, engineers and technicians, all needed to exploit a myriad of commercial opportunities. The RCF is a charity set up by - but independent from - the RSGB, to create a fund which can support efforts to heighten awareness of the importance of radio communications in classrooms and universities.

Radio communication is one of the vital technologies for the 21st century. It provides the backbone technology for the information economy. Every member of the public uses it. Many innovations were developed by scientists and engineers who had their interest aroused by a hands-on demonstration, perhaps at school, perhaps at an exhibition, perhaps through a demonstration by a radio amateur.

## Fund Raising

To the end of 2015, the Foundation has raised over £341k. The money comes from:

- Members of the RSGB. Many members already make donations, some with their membership renewals, and donations can be increased when Gift Aid is applied to them.
- Through bequests. People can make bequests to the Foundation in their Wills and the Foundation rigorously respects any instructions made in a legacy.
- From fund raising or club events.
- By approaches to industry and public sector sources of grant aided money. The RCF plays its part with industry by raising public awareness of the opportunities for jobs and careers in radio.

## Projects

The following are the sort of projects and activities the Foundation has supported in the past:

### Grants to individual clubs or educational institutions

Examples of such funding include the refurbishment of a mobile training vehicle for a local club, a projector to help with amateur radio training classes in a local area, a contribution towards a portable mast and related equipment for a school of science and technology and help with a project to stream closed radio amateur television broadcasts over the internet.

During the past year the Foundation donated £2,000 to fund the receivers for the Principia project and £500 to the RSGB to sponsor an Arkwright Day at the National Radio Centre. This was an open invitation to all current Arkwright scholars giving them an introduction to amateur radio with excellent feedback from the students and the Arkwright Trust. This was such a successful day that there are plans to host more such days in the future.

### Bursaries and scholarships

In conjunction with the Arkwright Trust, the Foundation has given support for annual scholarships for students who are actively considering higher education in engineering, product or industrial design. The Foundation would consider supporting young licensed amateurs with a bursary to help them through university or college if the courses involve radio communications.

During 2016 the Foundation established with UKESF (UK Electronic Skills Foundation), a competition to promote RF engineering in final year university projects. The first competition will commence in 2017.

*www.ukesf.org/schools/rf-engineering-and-communications/*

### Disbursement of a legacy

The Foundation was charged with managing a bequest where the legacy stipulates that the funding must be used for the development of a suitable amateur satellite project. The outcome was the highly successful FUNCube and the Foundation is keen to involve itself with other high profile projects that helps it to achieve its objectives.

## Radio Communications Examinations

On 1st October 2015, in agreement with Ofcom, responsibility for the UK Amateur Radio exam system transferred from the Radio Communications Foundation to the RSGB.

## How to donate

**There are several ways:**

- Through RSGB membership renewals
- Through the RSGB shop
- By one-off donations
- Via club or other fund raising events
- By the payroll giving scheme for Charities, which some employers offer as a route for regular donations
- Bequest in a Will, so that an interest in amateur radio lives on for the benefit of others

*http://www.rsgbshop.org/acatalog/RCF_Donations.html*

**Gift Aid**

For every £1 you give to us, we get an extra 25p from the Inland Revenue. You must pay an amount of income tax and/or capital gains tax at least equal to the tax that the charity reclaims on your donations in the tax year for the RCF to receive this.



**Registered charity number 1100694**
*http://commsfoundation.org/*

Radio Communications Foundation, 3 Abbey Court, Fraser Road, Bedford MK44 3WH
Web site: *www.commsfoundation.org*
For more information, email:
*secretary@commsfoundation.org*

# Amateur Radio Observation Service

The Amateur Radio Observation Service (AROS) is an advisory and reporting service of the RSGB which is intended to assist radio amateurs and others who may be affected by problems which occur within the amateur bands or which develop on other frequencies as a result of amateur transmissions.

The service investigates reports of licence infringements and instances of poor operating practice which might bring amateur radio as a hobby into disrepute. Reports, complaints and associated supplementary information are accepted from any source and the contents of each communication is regarded as confidential material. The source of any report or supplementary information is not disclosed without the permission of the originator. The originator of any report or complaint should be prepared to respond to further enquiries. Requests for further details may be made to the originator and, in addition, independent verification on an individual case basis may be supplied by AROS Observers.

## Information

A report to AROS should contain details of the alleged infringement and should include: dates, times, modes and details of what was heard, corroborated by a third party if possible. The originator should also state where he/she considers the 'offender' to have infringed the terms of the licence or where the 'offender' has acted in a manner contrary to codes of operational practice which have been agreed nationally or internationally. The identity of the 'offender' or his/her location, where this is known positively, should be included. However, reports where the identity or the location of the offender is not known, or not known with certainty, are still of value and are required.

## Investigation

In terms of investigation and at OFCOM's insistence, AROS serves both RSGB members and non-members alike and is not solely a benefit of RSGB membership. The service however does not include investigations regarding Citizen Band (CB) radio.

When a complaint of poor or abusive behaviour is received by AROS the complainant will normally receive an acknowledgment within 72 hours and will be issued with an AROS reference number. If warranted, an observer in the complainant's area will be alerted to investigate. The observers' identity will not be known to the complainant.

AROS observers are recruited with their identities only known to the AROS Coordinator & his deputy. The observers' identities are also unknown to each other unless prior arrangement is made for them to work as a team. After investigation, and where there is positive information of deliberate malpractice or malicious abuse of amateur radio facilities, a formal report may be made to Ofcom's Enforcement Policy Unit. This report will contain sufficient detail and information to enable further investigations to be made and the authorities may take such action as is appropriate, probably involving Ofcom. However, AROS prefers to settle problems – great or small – within the amateur service. Problems arising are referred to the authorities as a last resort.

Volunteers to become AROS Observers are welcome and should apply directly to AROS.

## Prefixes and suffixes

### Prefixes

1st Character: Foundation: M3, M6
Intermediate: 2
Full: G or M (except M3 and M6)

2nd Character:

| | Intermediate | Foundation/ Full | Club (Full) |
|---|---|---|---|
| England | E | (none) | X |
| Isle of Man | D | D | T |
| Northern Ireland | I | I | N |
| Jersey | J | J | H |
| Scotland | M | M | S |
| Guernsey | U | U | P |
| Wales | W | W | C |

| | |
|---|---|
| GB3 + 2 letters: | Repeaters |
| GB3 + 3 letters: | Beacons |
| GB7 + 2 letters: | Data repeaters |
| GB7 + 3 letters: | Data mailboxes |
| GB + other digits: | Special event stations (class of call sign normally corresponds to the appropriate M format) |
| M/foreign call: | Reciprocal CEPT |

### Suffixes

- -/A Alternative address
- -/M Mobile (includes inland waterways and pedestrian)
- -/P Portable (temporary location)
- -/MM Maritime Mobile

### Approximate Issue Dates for UK Callsigns

| | | | | | | | | | | | |
|---|---|---|---|---|---|---|---|---|---|---|---|
| **Two letters** | | G3MAA | 1957-58 | G4IAA | 1979 | M0CAA | 1998-00 | G6CAA | 1981 | M1CAA | 1997 |
| G2AA | 1920-39 | G3NAA | 1958-60 | G4MAA | 1981 | M5AAA | 1999-00 | G6QAA | not issued | M1DAA | 1998-99 |
| G3AA | 1937-38 | G3OAA | 1960-61 | G4QAA | not issued | G8AAA | 1964-67 | G6RAA | 1982 | M1EAA | 1999-00 |
| G4AA | 1938-39 | G3PAA | 1961-62 | G4RAA | 1982 | G8BAA | 1967-68 | G1AAA | 1983 | | |
| G5AA | 1921-39 | G3QAA | not issued | G4SAA | 1983 | G8CAA | 1968-69 | G1DAA | 1984 | **Foundation** | |
| G6AA | 1921-39 | G3RAA | 1962-63 | G4WAA | 1984 | G8DAA | 1969-70 | G1LAA | 1985 | M3AAA | 2002 |
| G8AA | 1936-37 | G3SAA | 1963-64 | G0AAA | 1985 | G8EAA | 1970-71 | G1QAA | not issued | M6AAA | 2008 |
| | | G3TAA | 1964-65 | G0EAA | 1986 | G8FAA | 1971-72 | G1SAA | 1986 | | |
| **Three letters** | | G3UAA | 1965-66 | G0HAA | 1987 | G8HAA | 1973 | G1XAA | 1987 | **Intermediate** | |
| G2AAA | Pre-war | G3VAA | 1966-67 | G0JAA | 1988 | G8IAA | 1973-74 | G7AAA | 1988 | 2E0AAA | 1991 |
| G3AAA | 1946 | G3WAA | 1967 | G0LAA | 1989 | G8JAA | 1974-75 | G7EAA | 1989 | 2E1AAA | 1991 |
| G3CAA | 1947 | G3XAA | 1967-68 | G0MAA | 1990 | G8KAA | 1975 | G7FAA | 1990 | 2E1BAA | 1992 |
| G3EAA | 1948 | G3YAA | 1968-69 | G0NAA | 1991 | G8MAA | 1976-77 | G7HAA | 1991 | 2E1CAA | 1993 |
| G3GAA | 1949-50 | G3ZAA | 1969-71 | G0SAA | 1992 | G8NAA | 1977 | G7MAA | 1992 | 2E1DAA | 1994 |
| G3HAA | 1950-51 | G4AAA | 1971-72 | G0TAA | 1993 | G8OAA | 1977-78 | G7OAA | 1993 | 2E1EAA | 1995-97 |
| G3IAA | 1951-52 | G4BAA | 1972-73 | G0VAA | 1994 | G8PAA | 1978 | G7SAA | 1994 | 2E1GAA | 1997-99 |
| G3JAA | 1952-54 | G4DAA | 1974-75 | G0WAA | 1995 | G8QAA | not issued | G7TAA | 1995 | 2E1HAA | 1999-00 |
| G3KAA | 1954-56 | G4EAA | 1975-76 | M0AAA | 1996 | G8TAA | 1979 | G7WAA | 1996 | | |
| G3LAA | 1956-57 | G4GAA | 1977 | M0BAA | 1997-98 | G8ZAA | 1981 | M1AAA | 1996 | | |

**Note:** *From April 2000, out-of-sequence callsigns could be requested, so calls in later series may be heard.*

AROS: 3 Abbey Court, Fraser Road, Bedford MK44 3WH. Email: aros@rsgb.org.uk

# Public Service and Emergency Comms

The RSGB Emergency Communications Committee, ECC, was formed in September 2013 with the broad remit to improve the support given to RSGB affiliated RAYNET groups and to improve relations with RAEN (the Network). It has four members spread across the UK and the appropriate RSGB Board member also attends the ECC meetings.

Separate from the RSGB but affiliated to it is the Radio Amateurs' Emergency Network which represents the interests of its RAYNET membership throughout the United Kingdom. It is administered by a Committee of Management, which is formed by Zonal Co-ordinators who are also responsible for their own geographical area. Each Zonal Co-ordinator looks after an area based on the former national civil defence Zones. The Network is the recognised voice of voluntary emergency communications in the UK, and liaises with national voluntary agencies and government bodies. To cover administration costs a small annual charge is levied on each member, and in return offers combined liability and personal accident insurance, discounted supplies, as well as technical and training backup. Further details can be found on the Network's website: *www.raynet-uk.net* The Radio Amateurs' Emergency Network provides a wide range of services to support Members, their Groups and liaisons with our User Services. This support is delivered at Member, Group and National level and is organised by the Committee of Management. Details can be found at: www. raynet-uk.net/main/services.asp

Many amateurs like to contribute to society by using their skills in these ways, thus demonstrating the usefulness of amateur radio, and promoting the Amateur Service.

The amateur radio licence normally restricts the use of a station to self-training in radio communications and as a leisure activity. The latter is interpreted to allow the use of amateur radio in support of community events. The licence also allows any amateur to use his/her station to send traffic on behalf of specified bodies, referred to as 'User Services' in paragraph 17(1)(qq) of the licence.

However, there is a limit to what an individual amateur can achieve, so most will join local clubs or groups and pool their equipment and skills. All amateur radio operators are responsible for the same operation of their station when in a public area.

RSGB and Network groups enjoy the protection of public liability insurance, but such cover is not to be taken for granted and may be invalidated if carelessness is proven.

Please note that RSGB affiliated groups must give the Society a list of all members for the Society's insurance to be valid for that particular member. Non-affiliated groups do not enjoy the benefits of the Society's insurance cover. The Network has its own insurance facilities in place and it is possible that members with dual affiliation will be covered by one or the other.

## Emergencies

Amateur Radio emergency communications in the UK are usually provided by organised groups under the general term 'RAYNET'. Some groups affiliate as societies to the RSGB; others join RAEN; some join both. The 'Network' is a Registered Charity, affiliated to RSGB and there is regular liaison between the two, promoting a coordinated approach to emergency communications. There are around 150 groups around the country.

Natural disasters on various scales occur all the time. Unfortunately, man-made events are increasing in frequency, but the government has recognised this and established 'Resilience Forums' through the country. They have the responsibility of planning 'resilience' or 'the ability to bounce back quickly' in the event of any type of disaster.

One of the first things to come under pressure in an emergency will be communications. All the usual links are liable to damage, either by deliberate targeting or by secondary damage from failing power supplies, gale damage, etc. RAYNET groups are an attractive option to those responsible for emergency planning - because they are quick to respond, flexible, technically skilled, and, perhaps above all at a time of diminishing resource, FREE!

## Non-emergency use

These may be events in their own right, often handled by groups with no desire to join RAYNET groups, or by RAYNET groups as a means of practising procedures, testing equipment and honing skills, etc. For example, the organisers of a charity cycle run or similar event may approach a group and ask for help, or a pro-active group may offer its services.

A RAYNET group may do the same, but will often be better placed to provide the service due to having a close relationship with an Emergency Planning Officer or a Resilience Officer who recognise how useful these events are as part of ongoing training and also include RAYNET in their exercises and plans. RAYNET groups thus become aware of how the User Services work and are used to their procedures.

## Technically minded?

RAYNET is not limited to voice modes. Our user services increasingly look for more advanced communications modes. If you have an interest in data, video or microwaves, please get in touch with a local group and bring your specialist skills to emergency communications.

RAYNET Groups will not be used in positions of danger - Emergency Planning Officers also have to consider H&S - but usually as a second-line resource. For example, when the Doncaster groups were called out in flooding in summer 2007, they were not placed on or near the walls of the dams which were under pressure, but in rest centres and town halls, handling the equally important (but perhaps less dramatic) traffic. An increasing collaboration is being seen where RAYNET Groups are working with 4x4 Responders', and some 4x4 group members are taking out amateur licenses to increase their skill level and the service they can offer. UK band plans allocate the following frequencies to emergency communications, and all Amateurs are requested to respect these:

| | |
|---|---|
| HF: | 3663, 5,278.5 |
| 6m: | 51.65-51.75, 51.77 & 51.79MHz |
| 4m: | 70.350, 70.375, 70.400MHz |
| 2m: | 144.625-144.675MHz, 144.775, 145.200, 145.225, 144.800 (APRS), 144.260 (SSB) |
| 70cm: | 433.700-433.775MHz |
| In-Band | 7.6MHz split temp talk-through 430.800 (mobile) 438.400 (base) |
| In-Band | 1.6MHz split temp talk-through 434.375 (mobile) 432.775 (base) |

Most RAYNET activity takes place on those bands for which equipment is readily available, namely 2m and 70cm, although use is also made of 4m, 6m and 23cm. More use of HF is also being made, to supplement VHF & UHF, particularly for distance and hostile terrain. Increasingly, both 10m and 6m are being used for cross-band talk-through, in addition to the more usual 2m and 70cm. bands. These temporary repeaters are authorised under special Notices of Variation - TT Permits - issued on behalf of Ofcom by the RSGB and the Network.

RAYNET is not restricted to UK operation. There is a growing international organisation, holding regular conferences (GAREC - Global Amateur Radio Emergency Communications), and their proceedings may be found on the Internet.

**International frequencies**
21360kHz - Global
18160kHz - Global
14300kHz - Global
7110kHz & 3760kHz - Region 1

## Becoming Involved

If you feel you'd like to become involved on a regular basis, then you should consider joining a RAYNET Group. First of all, find a group near to you, contact them and arrange to meet. Groups usually suggest that you come along to a few events and observe, and then perhaps apply to join. In fact you do not necessarily need to have an amateur radio licence, as there are often plenty of jobs that

non-licensed persons can do and you may get radio experience by using a PMR446 licence-free radio or even CB. ID cards are issued to members to identify them to user services as a member of RAYNET. Some groups adopt a smart dress code, which is often appreciated by user services, giving them a quick means of identifying members on site at an incident. The Network operates a supply service for all RAYNET members.

## RAYNET Groups in the UK

The RSGB website has a list of RSGB Affiliated Groups (http://rsgb.org/main/operating/emergency-communications/rsgb-affiliated-emergency-groups/), where details are updated whenever they are received.

The Network main page is to be found at *www.raynet-uk.net*

## Affiliated RSGB Emergency Groups

As part of the continued RSGB commitment to Emergency Communications it gives affiliated clubs and groups benefits over and above that available to other affiliated clubs and groups. These extra benefits are at no extra cost and come as part of the usual RSGB club affiliation fee. These benefits include:

### Personal Accident Insurance

The RSGB provides free of charge a specialist personal accident insurance for all the ' members of the Raynet affiliated group regardless of if they are RSGB members or not. This insurance is aligned to that offered by the Network and covers the wide range of risks that members may encounter in conducting Raynet activities. Full details of this policy are available online in the RSGB members' area of the website: *www.rsgb.org/membersonly*

### Raynet Identity Cards

Provided free of charge with a simple online application, these cards provide affiliated Raynet Group members with a proof of membership and insurance that can be used in many circumstances. Again covering all members regardless of whether they are individual RSGB members or not. These cards can be applied for via your Group secretary and you will need to provide an electronic passport style picture that will be used on the card. These services are provided in addition to the usual benefits accorded to RSGB clubs such as Third Party Public and Products Liability insurance, an annual copy of this Yearbook, use of the QSL Bureau, etc. If you would like to take advantage of these benefits you can simply contact RSGB sales on **01234 832700** or by email: sales@rsgb.org.uk for full details.

## On the web

Lots of additional information from the RSGB and the Network may be found on the Internet. There's also the RAYNET Forum, where RAYNET matters such as technical topics, news and views, can be discussed, plus the RAYNET Diary, where all Groups can register their future events

# National Affiliated Clubs & Societies

### Amateur Radio Caravan & Camping Club

If you have an interest in caravanning or camping the ARCC could be for you. We hold rallies from spring to autumn on sites exclusive to us. This is often a farmer's field or similar with basic facilities such as drinking water and chemical toilet emptying point. There is often plenty of room to erect aerials offering the opportunity to experiment, working without electrical interference can often be a revelation.

We cater for the whole family, not just the radio amateur, children are especially welcome (tomorrow's radio amateurs). Some of our current members came to rallies with their parents when they were children.

The club was founded over 30 years ago by three radio amateurs who were also caravanners and one founder member still attends many rallies throughout the year.

It's not all amateur radio, often some kind of get together in the evening is arranged which may be a pub meal or a "pot luck" supper when people bring something to share. All such events are entirely optional of course.

If you think the ARCC may be of interest to you and your family, we would love to let you have more information.

E-mail: *membership@arcc.org.uk* or visit: *www.arcc.org.uk*

### AMSAT-UK (G0AUK)

This is the UK national society specialising in amateur radio satellite matters. It has approximately 400 paid-up members, produces a regular publication, *Oscar News*, for its members four times per year, and holds weekly nets on 3.780kHz ±QRM on Sundays at 10.00 local time. You do not have to be a member of AMSAT-UK to join the net - anyone with an interest in satellites is welcome.

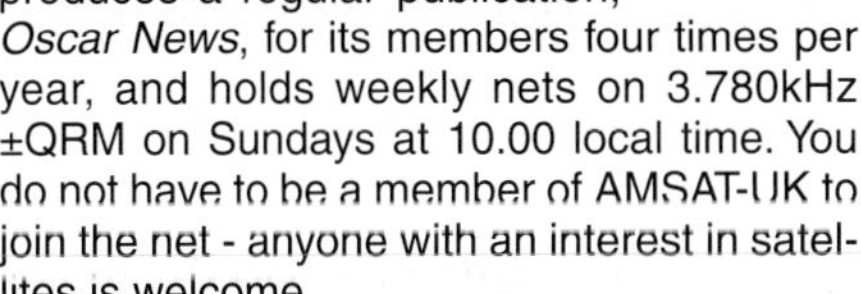

In addition, AMSAT-UK organises an annual Colloquium, normally during the last weekend of July. This has traditionally been held in Guildford and you do not have to be a member to attend. It is an excellent chance to rub shoulders with experienced satellite operators and satellite designers. A series of talks and demonstrations is arranged over the weekend, as well as social events.

Membership of AMSAT-UK is by donation, for which there is a suggested minimum. Extra donations are always welcome and can be sent anonymously. Funds raised are used to build satellites for all to use.

Enquiries and application forms for membership should be sent with an SASE to: Jim Heck, G3WGM, Badgers, Letton Close, Blandford, Dorset DT11 7SS. email: *g3wgm@amsat.org* website: *www.amsat-uk.org*

If you use amateur radio satellites, be prepared to pay something to the organisation which designs, builds and launches them AMSAT!

### Blind Veterans UK ARS *(M0SBV)*

Formerly St Dunstan's, the charity and Society changed its name on 21 February 2012 to be more descriptive. The benefits of the charity are available to any ex-Service person who has a substantial loss of sight in both eyes.

The Society was formed 35 years ago as St Dunstan's to foster the hobby amongst its visually handicapped ex-Service members and encourage new members to join. The Society meets twice a year alternately at their Llandudno or Brighton Centre to operate their station (M0SBV or GB0BVU) as a special event station. Members keep in touch via a daily net on 7.125MHz.

Society Secretary: Ray Hazan, OBE (G0PQQ), 5 Micklefield Way, Seaford, East Sussex, BN25 4EU. email: *ray.hazan@gmail.co*

### British Amateur Radio Teledata Group *(BARTG – G4ATG, GB2ATG)*

Our aim is to encourage and promote the use all types of datacoms within amateur radio including RTTY at various speeds, PSK at all speeds, WSPR, Olivia etc. We are web-based with a world-wide membership and you can join us for FREE.

We run an awards scheme open to both licensed amateurs and listeners. The entry level will appeal to the occasional operator and the upgrades on offer will appeal to the serious DX chaser.

Our Quarter Century award is for making contact with 25 different ARRL DXCC countries. Our PSK31 award is for contacting 40 different ARRL DXCC countries using PSK. We have a series of Continent awards (African, Asian, European, North American, South American and Oceania) for contacting countries in those continents.

All of these awards can be upgraded by contacting more countries and many can be endorsed for bands and modes. We also have a Golden Jubilee award based on our golden jubilee in 2009 and we aim to run a Diamond Jubilee award in 2019. Logs submitted to our contests can be used to apply for any of our awards.

We run a variety of contests each year currently ranging from a 4 hour sprint contest to a 48 hour weekend contest. They are serious contests but are also regarded as being great fun. We also encourage BARTG Members and Friends to enter international data mode contests so that BARTG can form one or more team entries and thus get greater worldwide exposure.

Although membership is free, we welcome donations to our DXpedition fund. This fund enables us to make modest donations to selected DXpeditions.

The 'For Sale/Wanted' page on our website isn't only for buying and selling amateur radio datacoms equipment. 'For Sale' includes equipment being given away and 'Wanted' includes not only equipment but also technical help.

Our website is central to everything we do so please visit it for more information about us, including our contests and awards. If you would like to join us then please go to the 'Friends of BARTG' page on our website. There is no charge and we welcome both licensed amateurs and listeners.

Secretary: Ian Brothwell G4EAN, 56 Arnot Hill Road, Arnold, Nottingham, NG5 6LQ.

email: *bartg@bartg.org.uk*

### British Amateur Television Club *(BATC – RS38114)*

The BATC is the world's largest television technology club with almost 1000 members.

It is based in the UK and has a worldwide membership. The BATC is open to all who are interested in television from programme making and production to making equipment and transmission.

Internet Video Streaming

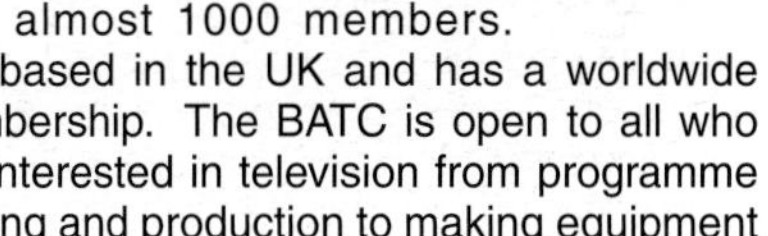

On our streaming site *www.batc.tv*, ATV repeater outputs can be viewed via the internet. This enables repeaters that are out of transmission range to be viewed anywhere in the world, individual members, live conferences and events such as the AMSAT annual conference may be seen, vintage videos are available from the archive and the weekly RSGB news read by Roy, G8CKN is shown live on Sunday mornings.

For technical discussions about television our forum is available at:

*http://www.batc.org.uk/forum/*

We have 'hard to get' components for television in our online shop.

### CQTV

Our 40-page colour magazine is published four times a year and is available in hard copy or as an internet download.

Our BATC 2-day convention is held annually at various venues around the UK and provides technical lectures, products for sale, bring and buy and a chance to meet other club members.

For more information or to join our club see our website: *http://www.batc.org.uk* or email the secretary: *secretary@batc.org.uk*

British Railways ARS *(G4LMR)*

If you are interested in railways (as an enthusiast or volunteer or professionally) and amateur radio (licensed or listener) then we may be precisely the society you are looking for. We are a small and friendly society with enthusiastic members and we would be delighted to welcome you as a member.

Our "Rails & Radio" newsletter is published and posted to our members several times a year. Articles and photos from our members are always welcomed by our editor. We run HF nets depending on band conditions. Our AGM is usually not far from a railway station and our members are often involved in running amateur radio stations at or near railways (often heritage railways). We organise an annual weekend meeting with a railway theme.

We are affiliated to FIRAC (Federation Internationale des Radio Amateurs Cheminots - International Federation of Railway Radio Amateurs) which, amongst other things, runs a congress annually in Europe, organised by its member societies.

BRARS was founded in 1966 and is celebrating its Golden Anniversary in 2016.

To find out more about us, and to get an application form, please go to our website or contact our Treasurer: Mr John Wellard, M0ZAA 19 South Motto, Park Farm, Kingsnorth, Ashford, TN23 3NJ email: *M0ZAA@BRARS.info*

Secretary: Ian Brothwell, G4EAN 56 Arnot Hill Road, Arnold, Nottingham, NG5 6LQ

*www.brars.info*

### British Railways ARS (G4LMR)

If you are interested in railways [as an enthusiast or volunteer or professionally] and amateur radio [licensed or listener] then we may be precisely the society you are looking for. We are a small and friendly society with enthusiastic members and we would be delighted to welcome you as a member.

Our "Rails & Radio" newsletter is published and posted to our members several times a year. Articles and photos from our members are always welcomed by our editor. We run HF nets depending on band conditions. Our AGM is usually not far from a railway station and our members are often involved in running amateur radio stations at or near railways (often heritage railways). We organise an annual weekend meeting with a railway theme.

We are affiliated to FIRAC (Federation Internationale des Radio Amateurs Cheminots - International Federation of Railway Radio Amateurs) which, amongst other things, runs a congress annually in Europe, organised by its member societies.

BRARS was founded in 1966 and is celebrating its Golden Anniversary in 2016.

To find out more about us, and to get an application form, please go to our web site or contact our Treasurer: Mr John Wellard, M0ZAA 19 South Motto, Park Farm, Kingsnorth, Ashford, TN23 3NJ email: *M0ZAA@BRARS.info*

Secretary: Ian Brothwell, G4EAN 56 Arnot Hill Road, Arnold, Nottingham, NG5 6LQ

*www.brars.info*

### British Top Band DF Association *(BTBDFA)*

The BTBDFA was formed in 2000 to centralise the organisation of Direction Finding (DF) on the 160m band in the UK. The association organises the qualifying rounds and the final of the National DF championship. The winner receives the handsome RSGB Trophy, which dates back to 1951. The main aim of the association is to increase the popularity of Top Band DF, which it does by organising events, providing lectures to radio clubs, putting interested people in touch with each other, and by writing articles for magazines, etc.

website: *www.TopBandDF.org.uk*

Details from the secretary, Bill Pechey, G4CUE. Tel: **01491 680552**.

email: *secretary@TopBandDF.org.uk*

### British Young Ladies Amateur Radio Association *(BYLARA – M0BYL)*

BYLARA was formed on 29th April 1979 to further YL operation in Great Britain and to promote friendships with YLs and OMs worldwide.

The association aims to encourage good operating, as well as persuade all YLs across the Great Britain and the World on to the air and to explore more of the hobby renewing old friendships and making new ones, across the airwaves.

We would also like to encourage good operating techniques and courtesy to other operators at all times and also aim to encourage and introduce up and coming Guides, Brownies, Scouts as well as other young people who would like to go into the Hobby. We now have Members and Friendships worldwide and are growing all the time.

We have had our Callsign (M0BYL) since 1998 and more information can be found on our Web Page.

Our Members take part in various Special Events, Dxpeditions, Contests and have our own Net Night.

We are looking forward to 2016 as we have decided to hold an 'International Young Ladies' (IYL) Convention. It will be held in Milton Keynes in conjunction with the RSGB Conference that is also held in October. The Convention will see the arrival of the Ladies from many Countries on Monday, 3rd October, 2016 and the Convention will end on Monday, 10th October 2016. For full details please see BYLARA Website and YL Face Book (where you will find information and will also be able to chat to others on a closed site dedicated to IYL). For any queries please contact me on the details below.

We are always open to new members, OMs and family groups.

For more information, contact The Chairmam – Carol, 2E1RBH on **01305 820400** or email: *carolhodges1@btinternet.com*

*www.bylara.org* or

### CDXC – The UK DX Foundation *(M0CDX)*

CDXC is the UK's premier DX Foundation. We are dedicated to encouraging excellence in DXing and contest operating and have almost 1,000 members – a quarter of whom are overseas.

CDXC began in the 1980s, formed by a small group of keen UK DXers and has grown substantially with our members sharing a common interest in HF DX and contesting. CDXC has become one of the largest and most respected DX groups in the world.

Characterised by a supportive and encouraging approach to all members, CDXC is an influential organisation which develops and enriches the world of HF amateur radio for its members and DXers worldwide.

If your main interest in radio is HF, and you are interested in working DX - rare or distant countries - and competitive radio sport, then you will find yourself amongst like-minded radio amateurs in CDXC.

CDXC members have many interests, but all share an interest in DXing, one of the largest world-wide movements in amateur radio. CDXC members are often extremely active in contests and can be heard working or running pile-ups on the bands. Indeed, many members have been on a significant number of DXpeditions to far-flung DX entities, and very many have "worked" 300 or more DXCC entities.

Every year, many DXpeditions add significant enjoyment to the HF hobby. CDXC members support DXpeditions financially through their membership fees. On average CDXC supports between fifteen and twenty DXpeditions each year. By being a member of CDXC, you can directly encourage DXing in this way. So, when you receive a QSL card bearing the CDXC logo, as most of the high-profile DXpeditions do, remember that your QSO was partly thanks to being a member of one of the biggest HF DX groups in the world!

The CDXC Digest covers everything that's going on in the world of DXing and Contesting and is sent to our members every second month. Traditionally, a fifty to sixty page printed magazine, from 2016 it additionally became available on-line for individual member download.

Timely reports on every significant DXpedition, together with pages of full colour pictures, are a central part of Digest content. Regular features cover DXing, ClubLog news, happenings in the Digital World, Contesting, QRP and IOTA news. You will also find original technical articles, book reviews, features on operating techniques and much more.

You can download a sample copy here: *www.cdxc.org.uk/Archive_Pages*

To join the CDXC World go to: *www.cdxc.org.uk* @cdxcuk

### Duxford Radio Society

*(RS92589)*

The Society now forms the Radio Section of the Imperial War Museums, Duxford. The Society was originally founded in 1986. It consists of a dedicated group of volunteers who research, conserve, restore and display historic military radio and electronic equipment at the Imperial War Museum, Duxford, Cambridgeshire, England. The Society also operates the radio station GB2IWM from the Duxford airfield site, using both modern and vintage equipment. DRS is based in Buildings 177 and 178, adjacent to the American Air Museum. These are open every Sunday and also on many other days of the week when DRS volunteers are present on the site. Other opening times are by prior arrangement by writing to: Duxford Radio Society, c/o The Imperial War Museum, Duxford, Cambs CB22 4QR, England; or contact the Secretary, Beryl Pope, on 01279 656149 email: *secretary@duxfordradiosociety.org* website: *www.duxfordradiosociety.org*

The Society welcomes new members who share an interest in conserving, restoring or operating military radio equipment.

### Essex CW Amateur Radio Club

*(ECWARC – G1FCW, G4C)*

Based and formed in Essex, but open to all licensed radio amateurs worldwide who are interested in Morse code (CW), Essex CW ARC has as its mission to encourage the use and preservation of the CW mode on the amateur bands. It further aims to support those who strive to learn and improve their CW skills (associate members) as well as operators who are already competent at any level (full members). Membership is free and most on-air activities are publicised via the club's website and via a regular email bulletin to members. Activities include club CW nets on the HF bands, ad-hoc communication using recommended ECWARC frequencies on each HF band and team efforts in various CW contests throughout the year. Learning opportunities for those starting out includes a weekly GB2CW QRS transmission on 80m as well as advice for those who are running classes or practice sessions for others in their own areas. For further information please visit the club's website at: *http://www.essexcw.org.uk*

### First Class CW Operators' Club

*(FOC – G4FOC)*

The First Class CW Operators' Club (FOC) was founded in 1938. The aim of the Club is to promote good CW (Morse code) operating, activity, friendship and socialising. The Club is UK-based, with many international members.

FOC has approximately 500 members, who will typically meet both on air and in person. The Club holds a wide range of social gatherings throughout the year, in many different countries. The character of the club is best expressed in its motto, "A man should keep his friendship in constant repair" (Samuel Johnson, 1755).

Membership of FOC is achieved through a process of nomination by existing members, based on on-air activity and other criteria which are explained on the Club's website at *www.g4foc.org*. The Secretary is Michael Wells, G7VJR (email: *michael@g7vjr.org*

### FISTS CW Club

*(GX0IPX, GX3ZQS & MX5IPX)*

FISTS was founded in 1987 by the late Geo Longdon, G3ZQS. It is now an International club with chapters for North America, New Zealand / Australia and East Asia. It actively promotes the use of Morse code on the amateur bands and runs many friendly contests and challenges each year. We have a club website: *www.fists.co.uk* and attend many rallies each year.

We are always looking for new members, the only criteria for membership being a love of Morse. No code proficiency levels are required and we encourage novice and foundation licence holders and SWLs to join. We produce a quarterly newsletter in an A5 booklet form. We have many thousands of members worldwide and would welcome you!

Details from Paul Webb, M0BMN, 40 Links Road, Penn, Wolverhampton, WV4 5RF. email: *m0bmn@fists.co.uk* or *members@fists.co.uk*

### The Five Star DXers Ltd FSDXL – Previously known as The Five Star DXers Association (FSDXA)

The Five Star DXers Limited is a leading global group that organises major DXpeditions. UK based, they hold a number of world records and are known for both the scale and success of their DXpeditions. They take a professional and measured approach to everything they do. DXpeditions to date have been Spratly, 9M0C 1998; Comoros D68C 2001; Rodrigues 3B9C 2004; St Brandon Reef 3B7C 2007 and Christmas Island T32C 2011. For further information contact Neville Cheadle, G3NUG, Chairman – *g3nug@btinternet.com*

### GMDX Group - Scotland's DX Association *(GM5A, GS8VL)*

The GMDX Group, based in Scotland, caters for all radio amateurs who have an interest in DXing, contesting and award chasing. The group has its own award scheme and produces a quarterly magazine, The GMDX Digest, which is sent out to members electronically. The Group has supported most of the major international DXpeditions and criteria for sponsorship is outlined on the group website.

There are usually two meetings per year, including a DX Convention held in April in the Stirling area. Membership is open to all radio amateurs or SWLs with an interest in Contesting and DXing.

Full details from Robert W Ferguson, GM3YTS, 19 Leighton Avenue, Dunblane, Perth FK15 0EB. Tel: 01786 824199.

email: *gm3yts@btinternet.com*

website: *www.gmdx.org.uk*

### G-QRP Club

The club has recently passed its 40th year in existence. The club specialises in low power operation (hence the QRP), primarily on the HF bands but can be found on LF/VHF/UHF and SHF as well. The club produces a quarterly magazine for its members called Sprat that focuses on construction and operation but that also includes articles and comments from members about their activities. Membership is growing and is in excess of 4,000 and contains many active members from DX locations not only the UK. Many members take part in all sorts of radio related activities, especially DXpeditions, buildathons, contests and field day activities to name but a few. There are many, many members who are active SOTA, IOTA and DXCC activators and chasers. Activities cover the whole radio spectrum and many different modes. It is not uncommon to find members active on bands from LF to SHF. There are many members involved in helping others enjoy the hobby through training and construction, several members hold and have held NoV for novel operations.

Further details can be obtained from the membership secretary by emailing: *membership@gqrp.co.uk* or from their website: *www.gqrp.com*

"It is vain to do with more what can be done with less." *William of Occum*, 1290-1350

### International Short Wave League *(ISWL – G4BJC, M1SWL)*

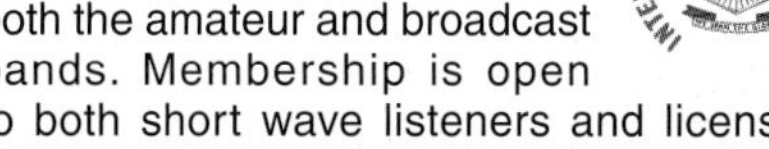

Known as the ISWL, the League was founded in October 1946 and caters for all interested in both the amateur and broadcast bands. Membership is open to both short wave listeners and licensed amateurs.

All members receive monthly the league journal Monitor which includes columns of interest to both Short Wave Listeners and Licensed Amateurs. The pages include Members Mailbag; reception reports for HF/VHF/UHF; SW BC Bands; transmitting topics, "In Days Gone By" a regular feature reviewing receivers of the past, articles of technical interest submitted by members, additional member's contributions include Meet the Member and articles of interest, stations to be looked for during the month, QSL Managers and Addresses and information of general interest world-wide.

A full contests programme and a comprehensive awards programme is open to all members, the latter free of charge to members and also available to non-members at a charge.

Weekly Nets are held on 80m and 40m SSB Sundays; Tuesdays and Saturdays; 160m SSB Sundays; 80m or 40m CW Mondays.

Full details and a membership application can be found on the website: *www.iswl.org.uk* or, are available from the Hon. Secretary; Peter Lewis, G-20322/G4VFG, 18 Bittaford Wood, Ivybridge, Devon, PL21 0ET.

email: *vfgnsu@yahoo.co.uk*

### Midland Amateur Radio & Caravan Club

The group came together back in 2005 as a recreational group who were all interested in ham radio. We are based in the Midlands, Dudley and Stourbridge in fact.

Initially, we met on several weekends during the year and limited ourselves to erecting a small tent and limited antennas. We used our own callsigns and equipment and had such a good time that we invited others to join us.

Over time our numbers have increased to the point that we now operate as a club with the benefits of club insurance etc. We are also members of the ACCEO which allows allows us to have rallies in private fields (as opposed to commercial sites) as well as on special rally sites in the area.

Chairman G4CGB. contact: *info@marcc.co.uk*

### Open University ARC

*(OUARC – G0OUR)*

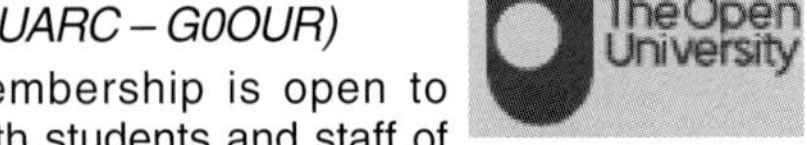


Membership is open to both students and staff of the university. Meetings are held in the shack every Thursday lunchtime on the Walton Hall campus, Milton Keynes, MK7 6AA. The club station is currently active on HF, VHF and UHF.

Further details from the secretary: Adrian Rawlings, M0ANS, 57 High Street, Nash MK17 0EP. Tel: **01908 503355**. email: *adrianrawlings@gmail.com*

website: *g0our.open.ac.uk*

### Prudential Amateur Radio Society

*(G0PRU, G0PPS, G8PRU)*

This is open to all current, retired and pensioned employees of the Prudential Group of companies, together with any SWLs worldwide. The society sends out a special QSL card for those contacts that it makes and for those who QSL it. If you don't want this special card, then please let the operator know at the time of the contact.

Chairman: Gerald Haines, G4SXY. Tel: 020 8657 8494. Secretary: Dennis Egan, GW4XKE. Tel: 029 2051 2959. Publicity Officer: John Wimble, G4TGK. Tel: 01797 362295. Callsign Manager: Mike Butler, G0NRK. Tel: 01833 690515. Overseas Liaison: Alan McCulloch, ZS6KU.

Details about the society can be obtained from the membership secretary: David Dyer, G4DNX, 'The Burlington', 85 Queens Road, Felixstowe, Suffolk IP11 7PE. Tel: **01394 276034**. email: *davidbobtail@me.com*

website: *www.radioclubs.net/prudentialars*

### Radio Amateur Old Timers' Association *(RAOTA – G2OT)*

The association seeks to keep alive the pioneer spirit and traditions of the past in today's amateur radio by means of personal and radio contact, whilst being mindful of any special needs. Our motto is 'Honour the Past, Enjoy the Present, Ensure the Future'.

There are two classes of membership. Full membership is open to anyone who has been actively involved in amateur radio for 25 years or more. There is no requirement to have held an amateur radio licence for the whole of that period, or even to have held one at all. Associate membership is for those who have been actively involved in amateur radio for a shorter period. It carries all the benefits of full membership but without the voting rights. All members receive our quarterly magazine, *OT News*, which is professionally printed using digital techniques. *OT News* is also available on CD or cassette.

Regular nets are held on HF under the callsign G2OT. Awards are available for contacts with other RAOTA members. RAOTA representatives man a display stand at rallies throughout the UK. Details of these events can be found on the website. Applications for membership and requests for a sample copy of *OT News* should be addressed to: RAOTA, 65 Montgomery Avenue, Hove, East Sussex BN3 5BE. website: *www.raota.org*

email: *memsec@raota.org*

### RAIBC *(G4IBC, GB0IBC, GB1IBC)*

RAIBC - the charity working for radio amateurs with disabilities - exists to help and support potential and existing radio amateurs and short wave listeners with disabilities. The club's journal, Radial, contains articles with a view to inform, share common experiences and challenges in the vast range of interests that is amateur radio today. It is available in print, MP3 audio and PDF formats. The RAIBC produces and circulates The Reading Rattle, a compilation of recordings including periodicals like RadCom, PW and Radio User, plus audio versions of key publications such as the licence manuals for all three stages of examination plus CW courses. These recordings are available for download and on CD in MP3 format for the use of qualifying members, together with plenty of help to assist blind and disabled students who may be starting out in the hobby and progressing through the licence stages. We can also offer help and advice in the training of potential radio amateurs with disabilities.

RAIBC also loans amateur radio equipment to members if they meet certain criteria. These can include adapted equipment, i.e. receivers and transceivers with synthesised voice output, or modifications to facilitate ease of use.

There are regular RAIBC nets on HF and VHF using the club's callsigns. Anyone is welcome to call in or visit the RAIBC website for a full list.

We value the interest of amateur radio clubs throughout the UK by way of training and mentoring the SWL's, the prospective and progressing radio amateurs with disabilities in their local areas and you could help the group by visiting a disabled amateur, and maybe setting up their loaned radio. Out of pocket expenses are paid and we're only asking for your time. If you have equipment you no longer require, perhaps we can loan it to one of our members, or use the proceeds of its sale to purchase adapted equipment.

RAIBC relies on the goodwill of fellow amateurs.

Almost all of the charity's income is from donations of money and equipment, and from the legacies of amateurs. As always, a very big thank you goes to everyone who has given their support in the past year!

We attend many rallies throughout the year. Feel free to come up to our stall and have a chat.

It may be possible to provide speakers for radio clubs to talk about RAIBC, its history and activities, plus specific topics in accessibility and use of equipment with modifications.

For more information call the RAIBC Secretary on **08000 141743**,

email: *secretary@raibc.org.uk* visit our

website: *www.raibc.org.uk* or write to the Secretary, Russell Bradley G0OKD, 4 Paddocks Close, Pinxton, Nottingham NG16 6JR

### Radio Officers Association

*(ROARS - M0ROA)*

ROARS is the only affiliated society that comprises members with a professional radio qualification who have been employed as Radio Officers in the Merchant Navy, Civil Aircraft, Coast Stations and Covert Services.

ROARS is part of the Radio Officers Association (ROA) which publishes a quarterly journal called QSO. This contains a number of articles from members illustrating their experiences at sea and ashore in war and peacetime as well as archive and technical material plus an amateur radio section.

ROA members get together twice a year at suitable venues with a maritime connection for a reunion, usually in April and November. Members also keep in touch with a regular CW net each Thursday evening at 19.30 local time on 3538kHz or 7017kHz +/- depending on conditions and time of the year.

New members are always made very welcome. What could be better than to share experiences of the old days with people who have a similar background in professional radio communications?

Website : *www.radioofficers.com*

Chairman : Peter Gavin M0URL

*peter.gavin@btinternet.com*

Secretary : Geoff Valentine G0UVX, 15 Kingsway, Stotford, Hitchin, Herts. SG5 4EL. *sparksatsea@googlemail.com*

### Royal Air Force Amateur Radio Society

*(RAFARS – G8FC, G8RAF, G3RAF, G6RAF)*

Formed in 1938, RAFARS, with over 900 members worldwide, is an international society that aims to promote amateur radio activities within the Royal Air Force. Through amateur radio it maintains and fosters the existing close bonds between radio amateurs still serving and those who have retired from, or have close associations with, the Royal Air Force, Commonwealth, NATO or Allied Air Forces.

Additionally, all radio amateurs and short wave listeners are welcome to join as Associate members if they have an interest in military aviation and the RAF, and are willing to assist the Society in achieving its published aims.

The society runs its own QSL bureau and publishes its own call book, QRZ. An inhouse Journal, QRV, is published twice a

year. A monthly Newsletter will be found on our website and can also be accessed via social networking sites. The Newsletter is also emailed or posted to subscribers.

The RAFARS website (*www.rafars.org*) contains information about the Society, including an abridged membership list, an application form to join, net schedules, monthly newsletters, details of awards and competitions and much, much more.

Prospective members are welcome to join one of our daily or weekly nets.

Further information is available from:

The General Secretary, G7JPN, HQ RAFARS, DCAE Cosford, Wolverhampton WV7 3EX.

Tel: 01902 372722 (answerphone).

email: *rafars.gen.sec@googlemail.com*

### Royal Naval Amateur Radio Society *(RNARS – GB3RN, GB0SUB, G3CRS, G1BZU, G3BZU, G7DOL)*

The RNARS was formed in 1960 to promote amateur radio as an aid to technical education within the Royal Naval Service. In 1964 the society was considered to be sufficiently established to invite the Captain of HMS Mercury to become its first president. When HMS Mercury closed down the Signal School was transferred to HMS Collingwood. The last Commanding Officer of HMS Mercury, Commodore Paul Sutermeister DL RN is our current President. The society headquarters is HMS Collingwood in Fareham and the HQ Shack callsign is GB3RN.

Membership is open to all Radio Amateurs with an interest in maritime affairs but particularly those who have served or are serving in: Royal Navy, Royal Marines, Women's Royal Naval Service, Royal Naval Reserve, Royal Naval Auxiliary Service, Merchant Service, Sea Cadet Corps, Nautical Training Corps, Ministry of Defence in a civilian capacity, Commonwealth & other Navies

We publish a newsletter three times a year, have our own QSL bureau and a number of radio nets are run on 2m and HF. Prospective members are welcome to join in. A number of awards are also available. All the details are shown on the Society's website: *www.rnars.org.uk*

More information on joining is available from the Membership Secretary: Marc Litchman, G0TOC, *g0toc@gb2rn.org.uk*, address QTHR

General enquiries to the Secretary, Joe Kirk, G3ZDF. email: *g3zdf@btinternet.com*

*Patron*: Admiral Sir Philip Jones, KCB RN

*President*: Commodore Paul Sutermeister DL Royal Navy. *Chairman*: Lt Cdr G D Hotchkiss MBE QCB RN, G4BEQ.

### Royal Signals Amateur Radio Society *(RSARS – G4RS)*

Formed in 1961 under the chairmanship of the late Major General Eric Cole CB CBE, G2EC, membership is open to: serving and past members of the British Regular and Territorial Army, civilian staff who have supported army telecommunications, Commonwealth Army signallers and licensed amateurs from other countries who have proven military connections, subject to status.

Members receive a high quality magazine, *Mercury*, three times a year, and the society runs its own contests and awards scheme. More information from: The Membership Secretary, (G0VDE), RSARS, 65 Montgomery Street, Hove, Sussex BN3 5BE.

Tel: **01273 703680**.

email: *memsec@rsars.org.uk*

website: *www.rsars.org.uk*

### The Radio Amateurs' Emergency Network *(G4NRC)*

The 'Network' is a national organisation whose members provide emergency communication for specified 'User Services' in times of emergency and disaster. It was formed in 1992 and became a registered charity in England and Wales (No.1047725) in 1995, and in Scotland (SC046184) in 2015.

Further details can be found on the Internet at: *www.raynet*-uk.net Alternatively, contact the Chairman at: *chair@raynet*-uk.net

Postal correspondence to: 9 Conigre, Chinnor, Oxon OX39 4JY. All enquiries about the Network can be made by telephoning 0303 040 1080 and selecting whom to speak to from the menu. For 24-hour emergency contact, select the* option.

### The UK Microwave Group *(UKuG ) GX3EEZ*

The UK Microwave Group acts as the focal point for radio amateurs and SWLs interested in frequencies above 1GHz. This includes optical communication - "nanowaves". The primary aim of the Group is to develop and promote all forms of amateu interest in the microwave spectrum. This is not limited to traditional amateur radio activities, members are involved in a wide range of interests ranging from casual hilltopping, amateur television, data networking,propagation research, tropo and EME dxing, to monitoring deep space probes and radio astronomy.

All levels of UK licence allow operation on some, or all, of our microwave allocations.

Microwaving, in all its forms, can be a fulfilling amateur radio activity for newcomers and experienced radio amateurs alike.

As a member of the RSGB Spectrum Forum, the Group works closely with the Society on bandplanning and licensing issues.

To focus activity, the Group organises Microwave Contests, and runs an Award Scheme which recognises all levels of achievement. It encourages microwave meetings (Roundtables) a mixture of fleamarket, social event, equipment test facility, and lecture stream. A regionally based Technical Support Team is available, in which experienced members spread across the UK, with excellent test gear, provide facilities to those either beginning their exploration of the microwave spectrum, or wishing to improve existing equipment.

The Group's e-newsletter, 'Scatterpoint' has a worldwide reputation for providing excellent practical articles on microwave techniques, construction and operation, as well as news of recent activities.

'Scatterpoint' is accessible to all: contributors represent a cross section of the amateur microwave community, and the magazine reflects that.

The *www.beaconspot.eu* website is a unique facility supported and funded by UKuG. This aggregates 'spots' on the DXCluster, and inputs from other sources to provide a remarkably complete and up-to-date resource of information on beacons on the bands above 30MHz.

Beacons are an important feature of microwave operation, and the Group offers financial support on a 'payment by results' basis to individuals and groups building beacons. Technical assistance is also available.

Building equipment is aided by the Group's 'Chipbank' service which provides a means of obtaining some specialised components at zero cost. This service is definitely for members only!

Membership of the Group is open to all those interested in amateur microwaves for a remarkably small subscription – currently £6.00 annually. For members under 21years, there is no subscription.

For further information, please visit the Group's website: *www.microwavers.org* or contact our – John Quarmby G3XDY 12 Chestnut Close Rushmere St Andrew Ipswich IP5 1ED Tel **01473 717830** or email: *secretary@microwavers.org*

### UK Six Metre Group *(G5KW)*

The UKSMG was formed in 1982 with the primary aim of promoting 50MHz activity by amateurs worldwide. Today the group is the largest organisation in the world dedicated to 50MHz.

Through its quarterly journal Six News, the group provides comprehensive information on all aspects of the band, including DX and beacon news, propagation, awards, contests, technical articles, equipment reviews, QSL information and DXpedition reports. For further information on "Six News" and to submit items for publication, contact Peter Bacon, G3ZSS (email: *editor@uksmg.org* ).

A major objective of the group is to actively support 50MHz operation from new countries and from DX locations, as well as promoting the establishment of beacons in various parts of the world. Recent DXpedition sponsorship from UKSMG includes: 5Z0L Kenya, 9Q0HQ Zaire, VU4KV Andaman and Nicobar Is., VK9XSP Christmas Is., OZ7IGY Next Gen Beacon and S01A Western Sahara.

Further information on sponsorship can be obtained from Trevor Day, G3ZYY (email: *sponsorship@uksmg.org* ).

The UKSMG website at *www.uksmg.org* carries a wealth of information about the 50MHz band and is guaranteed to provide interesting browsing for newcomers and old hands alike.

Further information about the group and membership details can be obtained from the website or from Tim Hugill, G4FJK, Swandhams House, Sampford Peverell, Tiverton, Devon EX16 7ED or email: *secretary@uksmg.org*

**Vintage & Military Amateur Radio Society**
*(VMARS - M0VMW)*

The VMARS is an international society based in the UK with nearly 400 members. Its main aims are to restore and preserve historic communication and electronic equipment, from military, commercial and amateur sources, encourage and enable the use of historic equipment on the amateur bands, encourage the use of historic modes such as AM on the amateur bands and encourage research into radio history.

These aims are realised through the very wide range of VMARS members' interests. Whereas many are engaged in the operation of "vintage" military and amateur equipment on the air (19-sets, T1154/R1155, T1509, 52-sets, TCS TX/RX, LG300, Heathkit, KW equipment, RA17, AR88 etc) with many a rare item in evidence, and much of which is lovingly restored, there is a very strong interest in military man-packs, military wireless vehicles, spy-sets, radar, history and museums and not least in amateur radio construction of valved equipment.

The Society runs a weekly AM net on 80m at 08.30 local time on Saturdays on 3615kHz ± QRM. All suitably licensed amateurs (whether VMARS members or not) are welcome. During the week, members may be found around 3615kHz at almost any time. There is also SSB activity, on Wednesdays, around 3615kHz and starting at 20.00 local time, primarily for those with ex-military USB-only equipment and on Fridays, a general SSB net, again around 3615kHz and starting at 19.30 local time.

The Society publishes a monthly newsletter and a quarterly 'technical' journal Signal, both of which are free to members. The journal aims to cover the scope of members' interests with constructional, restoration, general interest articles and comment. The Society is represented at a number of amateur radio and vintage rallies and events which include military vehicle shows at which vintage equipment is often demonstrated.

The Society's website includes a virtual library of manuals for vintage equipment, many of which can be downloaded at no charge. This library is supplemented by many paper documents, for which copies are available to members. Further details may be obtained from the website or the Hon. Secretary, John Keeley G6RAV, 93 Park Crescent, Abergavenny NP7 5TL, +44(0)1873 850164. Membership information may be obtained from the Membership Secretary, Ron Swinburne M0WSN, 32 Hollywell Road, Sheldon, Birmingham B26 3BX +44(0)1217 421808

website: *www.vmars.org.uk.*

email: *honsec@vmars.org.uk*

Vintage Operating Group *(M0VOG)*

The Vintage Operating Group was formed with the broad aims of preserving our wireless heritage and assisting like minded individuals with restoration and operation of army radio equipment, through the medium of the Royal Artillery Firepower Museum in the former Woolwich Arsenal, SE London.

Members are interested in early amateur, and of course military communications sets, many using amplitude modulation, making the promotion and continued use of AM, as well as other early modes such as CW, one of the primary aims. The Group, also operate former Military equipment in support of the Museum.

Initially we operated inside the museum. We now operate an annual Armed Forces Day radio station GB2AFD and participate in Military Vehicles Shows in the South East of England, demonstrating and displaying former military radio sets.

Under the terms of our licence and under the careful supervision of a Group member, visitors are most welcome to take a hand at operating, at any of these public shows, we shall be at the Overlord Military Show at Denmead, Waterlooville over the May Bank Holiday weekend, The War and Peace Revival Show located on the former RAF Westenhanger (Folkestone Racecourse) in late July and on the Kent County Show Ground, in Detling, Maidstone over the late August Bank Holiday. You may look out for our special calls of GB2SOE in May, GB4WP in late July and GB4MO in August. Please contact us through Mike Buckley, M1CCF - **0208 654 2582**

**Worked All Britain Awards Group**
*(WAB - G4WAB, G7WAB)*

The group was founded in 1969 by the late John Morris, G3ABG, to encourage greater amateur radio interest in Britain. The group promotes an award programme, contests and activity weekends and makes regular donations to organisations such as the RAIBC who help the less fortunate members of the amateur radio fraternity.

The award scheme, which is open to licensed amateurs and SWLs, is based on the geographical and administrative division of the UK. QSL cards are not required, only log entries, and special record books are available to assist in the claiming of awards. Full details and checklists of all the areas and counties for all the WAB awards are contained in the *WAB Book* or CD ROM. For further details please write to the membership secretary: Steve Mayer, G6TEL 453 Wimborne Road, Poole, Dorset BH15 3EE Phone: 07770415646

Email: *g6tel@worked*-all-britain.org.uk

website: *www.worked-all-britain.co.uk*

**World Association of Christian Radio Amateurs & Listeners**
*(WACRAL – M1CRA)*

WACRAL celebrated their 50th Anniversary in 2008. Dedicated to 'Friendship and Christian Fellowship among Radio Amateurs World-wide', membership is open to committed Christians of all denominations and all nationalities.

Founded by the late Rev. Arthur Shepherd, G3NGF, members are encouraged to follow his example in spreading the 'Good News', fostering international relations, supporting disadvantaged amateurs and SWLs and providing a good role model for the new generations.

WACRAL publish a newsletter four times a year, featuring the news and activities of members around the globe. A members' QSL Bureau is provided and a variety of awards are available.

An annual activity period is open to all on HF and VHF, including CHOTA (Churches and Chapels On The Air) and conferences take place each year.

The popular WACRAL 'Good News' Net meets at 07:30 UK local time each Sunday morning on 3.747kHz SSB and a schedule of national and international Christian nets is promoted, subject to propagation.

You are invited to go to their website at *www.wacral.org* for the latest news and full details of membership, or contact the Membership Secretary: Ian Horsefield, G4OPP, 61 Lewis Court Drive, Boughton Monchelsea, Maidstone, Kent ME17 4LG.

Tel: **01622 746650** Email: *g4opp@wacral.org*

# RSGB Club Affiliation

Many amateur radio societies and clubs choose to affiliate with the Radio Society of Great Britain because they see it as an effective way of demonstrating their support for the aims and aspirations of the National Society. The Society welcomes this support because it can only strengthen its claim to speak on behalf of amateurs and amateur radio.

The Society recognises that much of the vitality of amateur radio lies within clubs and wishes to encourage all clubs, societies and groups to join us to advance amateur radio.

There are tangible benefits for the affiliated society. These include:

- Publicity for club activities through 'club news' in RadCom, via broadcasts on the Society's news service GB2RS, and on the RSGB website..
- Full facilities of the RSGB QSL Bureau for cards bearing the club station call.
- Purchase of publications at a discount with the RSGB.
- Receipt of the *RSGB Yearbook*.
- Freedom to participate in RSGB Affiliated Societies' contests.
- Third party liability insurance up to £10m whilst taking part in club events such as field days.
- Freedom to borrow RSGB DVDs, tapes and display materials. (This facility is also available to certain non-affiliated groups such as schools).

## How to affiliate your club, group or society to the RSGB

Any UK club, group, society or emergency communication group may affiliate to the RSGB, provided it fulfills just a few requirements. The affiliation fee is currently £51.00 (with a £4 discount for those who pay by Direct Debit) and includes receipt of monthly email newsletters.

### Procedure

(i) Please contact RSGB HQ or visit the website for an Affiliated Club membership application form. If your organisation has a callsign, please let us know on the application form. If it does not, we will issue a receiving station number for reference purposes.

Note that for UK clubs, groups and societies the RSGB region and RSGB district will be determined by the address given on the application form. Clubs, groups and societies near to, or spanning county boundaries should decide carefully with which county they wish to be associated and insert the appropriate choice in the address of the club they are entering on the application form.

Please also note that once your club address is on our files, we will regard it as information that can be freely given out to those seeking to contact clubs.

(ii) Please send to the appropriate Regional Manager the following:

- Your completed application form signed by the chairman or secretary
- A copy of the club's constitution or rules. There is no prescribed form of constitution but an example is provided for guidance if required.
- A list of current officers of the club
- A statement of the number of members, and the proportion who are RSGB members.

A list of Regional Managers is published on the RSGB website and in the RSGB Yearbook. The Regional Manager will vet your constitution/ rules and if suitable will countersign your application form. He or she will then return the form and your constitution or rules to you.

Note that only the Regional Manager may countersign an application for a club, group or society. Overseas organisations should send their application form and constitution/ rules direct to RSGB HQ addressed to 'The RSGB Secretary'.

(iii) Finally, please send your countersigned application form, constitution and remittance to:

Membership Secretary, 3 Abbey Court, Fraser Road, Priory Business Park, Bedford, MK44 3WH.

## Notes on Example Constitution (overleaf)

(1) It is recommended that the words 'amateur radio' appear in its title.

(2) It is useful to specify which groups have voting rights and whether reduced subscriptions apply, particularly for students in full time education.

(3) Alternatively, the subscription may be recommended by the Committee for ratification at the AGM.

(4) This period perhaps should not exceed one to three years to avoid placing an undue burden on future Committees

(5) There are great advantages in running the Club's finances on a strict basis, although a less formal arrangement may still be effective.

(6) There are two methods for electing the Committee: the more common is for the meeting to elect the Committee members and for the latter in turn to elect the officers from within the Committee; alternatively, the members may elect individuals to specific offices. The method adopted will need to be specified.

(7) The number of Ordinary Committee members should be related to the size of the Club. Remember that being a committee member is an essential part of the training of the future officers of the Club.

(8) These can replace elected Committee members who have left the Committee.

(9) These can be people who need to be familiar within the work of the Committee such as the editor of the Club magazine or the press officer.

(10) This can be expressed either as a fixed number or, for example, as at least half or two-thirds of the full membership of the committee.

(11) This can be set either as a fixed number or a fixed percentage of the membership (state which members are to be included), or both "whichever is the smaller/greater". It is probably safer to make the numbers on the small side so as to ensure that the meeting can take place.

(12) Such as, among its members, to a charity, or to a club of similar interest.

Downloadable application form: www.rsgb.org/main/files/2012/05/RSGB_ClubAffiliationForm.pdf

# Example Constitution for Affiliated Clubs*

Guidance intended for those writing a constitution for their local club or society which will be acceptable for RSGB affiliation.

## 1. Name

The Club (1) shall be known as the .......

## 2. Aims

The aims of the Club shall be to further the interests of its members in aspects of amateur radio and directly associated activities.

## 3. Membership

Membership shall be open, subject to the discretion of the Committee, to all persons interested in the aims of the Club

(a) **Full members** Full members must be 16 years of age or over.

(b) **Honorary members** Honorary Life Membership may be granted to any person, who, in the opinion of the Committee, has rendered outstanding service to the Club, either directly or indirectly. Such membership shall carry the rights of full membership but shall be free from subscriptions.

(c) **Guests** Members may invite guests to meetings. No visitor may attend more than three meetings in each year.

All members shall abide by the constitution of the Club. The Committee shall have power to expel any member whose conduct, in the opinion of at least three-quarters of the full Committee, renders that person unfit to be a member of the Club. No Member shall be expelled without first having been given an opportunity to appear before the Committee.

## 4. Subscriptions

(a) The annual subscriptions for membership shall be set by the Committee (3).
(b) All subscriptions shall be due and payable at the beginning of the financial year. Members in arrears have no voting rights.
(c) The financial year shall be determined by the Committee
(d) A member shall be deemed to have resigned from the Club, if, by the end of the financial year, the subscription has not been paid.
(e) The Committee shall have the power to waive or reduce subscriptions in special circumstances for a period not exceeding...years at a time (4).

## 5. Finance

All money received by the club shall be promptly deposited in the Club's bank account. Withdrawals require the signature of the Club's Treasurer and one other nominated officer of the Club (5).

## 6. Membership of the Club's Committee

The Club's affairs shall be administered by a Committee elected at the Annual General Meeting (6). The Committee, in whom the Club's property shall be vested, shall consist of:

(a) A Chairman who will preside at all meetings at which he is present.

(b) A Vice-Chairman who will act as chairman in the absence of the Chairman.

(c) A Secretary who will be responsible for:
- (i) keeping the minutes of all meetings of the Club.
- (ii) ensuring that all correspondence is correctly handled.
- (iii) maintaining a master roll of members and honorary members.
- (iv) maintaining a register of Club equipment.

(d) A Treasurer, who will be responsible for:
- (i) keeping the Club's accounts.
- (ii) advising the Committee on all financial matters.
- (iii) preparing the accounts for audit and presenting them at the AGM.

(e) ......Ordinary Committee Members (8).

(f) Not more than......co-opted members who have full voting powers (8), and not more than......who are not permitted to vote (9).

## 7. Committee Standing Orders

(a) The quorum for the Committee shall be..... (10).

In the absence of a quorum, business may be dealt with but any decisions taken only become valid after ratification at the next meeting at which a quorum exists.
(b) Committee meetings may be called by the Chairman, the Secretary or any vote.

## 8. Annual General Meeting

(a) The Annual General Meeting shall normally be held at the beginning of each financial year. At least 21 days notice shall be given to each member in writing.

(b) The quorum for the meeting shall be...... (11).

(c) The agenda for the meeting shall be:
- (i) Apologies for absence
- (ii) Minutes of the previous AGM
- (iii) Chairman's report
- (iv) Secretary's report
- (v) Treasurer's report
- (vi) Election of the new Committee
- (vii) Election of auditors
- (viii) Other business

(d) Items (i) to (v) shall be chaired by the outgoing Chairman, item (vi) by an acting Chairman who is not standing for election to office, and the remaining business by the newly elected Chairman.

(e) Nominations for Committee members will only be valid if confirmed by the nominee at the meeting or previously in writing.

(f) Items to be raised by members under other business must be notified to the Secretary not less than 21 days before the AGM.

## 9. Extraordinary General Meeting

(a) Extraordinary General Meetings may be called by the Committee or not less than......members of the Club, the date of the meeting being the earliest convenient as decided by the Committee. At least 28 days notice in writing must be given to the Secretary, who in turn shall give members at least 14 days notice in writing of the agenda. No other business may be transacted at the EGM.

(b) The quorum for the EGM shall be......(11).

## 10. Amendments to the Constitution

The constitution may be amended only at an EGM called for that purpose.

## 11. Winding-up of the Society

(a) The decision to wind up the Club may be taken only at an EGM.

(b) The funds of the Club shall, after the sale of all assets and the payment of all outstanding debts, be disposed of as directed by members at the final EGM (12).

* Club is used to denote any club or similar organisation wishing to apply for affiliation

Downloadable Model Constitution:
www.rsgb.org/main/files/2012/05/RSGB_ModelClubConstitution.pdf

# Structure and Representation

For administrative purposes the Society divides itself into regions, each comprising four to six districts. Each region has an elected Regional Manager (RM) and each district has an appointed Deputy Regional Manager (DRM). The map below identifies the RSGB Regions. On the following pages, listed by Region, you will find details of the relevant Regional Managers (RM) and the Deputies (DRMs) along with contact information on local clubs and societies, examination centres and Emergency Comms Groups.

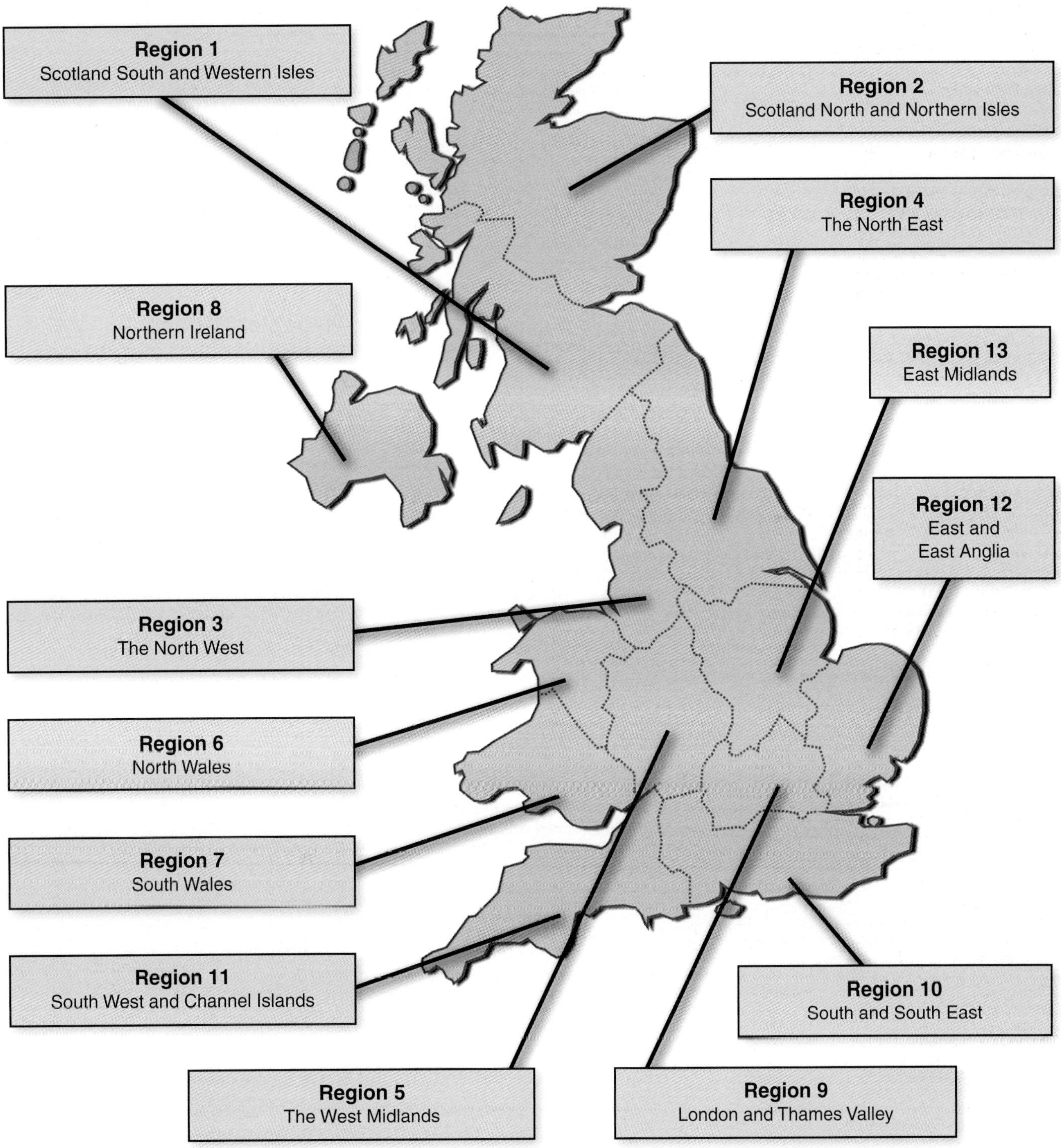

**Region 1**
Scotland South and Western Isles

**Region 2**
Scotland North and Northern Isles

**Region 4**
The North East

**Region 8**
Northern Ireland

**Region 13**
East Midlands

**Region 12**
East and East Anglia

**Region 3**
The North West

**Region 6**
North Wales

**Region 7**
South Wales

**Region 11**
South West and Channel Islands

**Region 10**
South and South East

**Region 5**
The West Midlands

**Region 9**
London and Thames Valley

# Region 1 - Scotland South & Western Isles

**(http://rsgb.org/region1)**

## RM

Marcus Hazel-McGown, MM0ZIF
Tel: 07593 441518
Email: *rm1@rsgb.org.uk*

## DRMs

**District 11** (Central, City of Glasgow)
Vacant Email: *drm11@rsgb.org.uk*
**District 12** (Lanarkshire, Renfrewshire)
Andrew Hood, GM7GDE Tel: 07825 932488
Email: *drm12@rsgb.org.uk*
**District 13** (Ayrshire, Dumfries & Galloway)
Vacant Email: *drm13@rsgb.org.uk*
**District 14** (Dunbartonshire, Argyll & Bute, Western Isles)
Barrie Spink, GM0KZX Tel: 01389 764401
Email: *drm14@rsgb.org.uk*
**District 15** (Lothian)
David Smith, MM0HVU Tel: 07912 871685
Email: *drm15@rsgb.org.uk*
**District 16** (Borders)
Vacant Email: *drm16@rsgb.org.uk*

## Clubs & Societies

**Ayr ARG**
Charles Stewart MM0GNS, email: *mm6ave@gmail.com* Meets 7.15pm-9:30pm on alternate Wednesdays Asda Ayr Superstore Liberator Drive Heathfield Retail Park, Ayr KA8 9BF *www.gm0ayr.org*

**Borders ARS**
Sandy Weddell GM1JFF *a.weddell764@btinternet.com* Meets 2nd Friday in month at 19.30 at the St. John Ambulance Hall Berwick-upon-Tweed TD15 1NG

**Cockenzie & Port Seton ARC**
Mr Bob Glasgow, GM4UYZ. *bob.gm4uyz@talktalk.net* Meets on the 1st Friday of the month at the Thorntree Inn, Lounge Bar Old Cockenzie High Street Cockenzie East Lothian EH32 0DQ *www.cpsarc.com*

**Elderslie ARS**
Alex Jenkins, MM0OIL. *alex@maeatae.com* Meets every thursday night at 19:00pm till 22:00pm at Elderslie Village Hall Stoddart Sq Elserslie Renfrewshire PA5 9AS

**Falkirk & DARS**
P Howson, GM8GAX, *gm8gax@tiscali.co.uk* Meets every Monday 19.30 at 62nd Forth Valley Scouts Hall Denny Road, Larbert, Nr Falkirk, FK5 3AD

**Galashiels & DARS**
Mr Jim Keddie, GM7LUN, *mail@gm7lun.co.uk* Meets at 8.00pm on Wednesdays at the Focus Centre Livingston Place Galashiels Scottish Borders TD1 1DQ *www.galaradioclub.co.uk*

**Greenock and District Scouts and Guides ARC**
Bob Lynch, MM1AWV *mm0tsg@yahoo.com* Meets 7.30 pm every Friday (term time only) at Greenock and District Scout Headquarters 159A Finnart Street Gourock Inverclyde PA16 8HZ http://mm0tsg.wordpress.com

**Kilmarnock & Loudoun ARC**
Len Paget, *klarc@gm0onx.co.uk* Meets 7.30pm on every second and fourth Tuesday of the month at KLARC Clubhouse,EAC Internal Transport 34 Main Street Crookedholm Kilmarnock KA3 6JS *www.klarc.org*

**Livingston & DARS**
Catherine Morris - catherine_*197@msn.com* Meets 7:00 to 9:00 pm every Tuesday evening at Crofthead Centre West Dedridge Livingston EH54 6DG *http://uk.groups.yahoo.com/group/ms0liv*

**Lomond Radio Club**
Mr BP Spink,GM0KZX,*gm0kzx@googlemail.com* Meets 7pm on Thursdays at John Connelly Community Centre Main Street Renton, Dumbarton G82 4PF *www.lomondradioclub.co.uk*

**Lorn ARS**
Stewart McIver MM1AVR Meets on the 1st and 3rd Wednesdays of each month Lancaster Hotel, Oban Tel: 01631 566793 Email: *stewart.mciver@btinternet.com* PA34 5AD *www.gm0lra.freeuk.com*

**Lothians Radio Society**
Mike Burgess MM0MLB *secretary@lothiansradiosociety.com* Meets 7.30pm on second and fourth Wednesday of the month at the Braid Hills Hotel, 134 Braid Road, Edinburgh, EH10 6JD *www.lothiansradiosociety.com*

**Mid Lanark ARS**
Web Admin *info@mlars.org* Meets every Friday 18.30 - 21.30 Newarthill Community Ed. Cent. High Street Newarthill, Motherwell Lanarkshire ML1 5JU *www.mlars.org.uk*

**Na Fir Chlis ARC**
Angus MacLeod, *mm0nfc@hotmail.com* Meets on an irregular basis at Heathbank Hotel Northbay Isle of Barra Outer Hebrides HS9 5YQ *www.ms0nfc.yolasite.com*

**Paisley ARC**
Stuart McKinnon, MM0PAZ *gm0pym@gmail.com* Meets on the 1st and 3rd Thursday at 19:30 at St Ninians Church Hall 81-83 Blackstoun Road Paisley PA3 1NR *http://gm0pym.wix.com/paisleyarc*

**Stirling & DARS**
Myles Williamson, MM0MYL, *chufferfalcon@gmail.com* Meets 7.30pm every Thursday and 11am to 3pm on Sundays at Unit 68, Bandeath Industrial Estate, Throsk, Stirling FK7 7NP

**The Jaggy Thistles ARC**
Charles Stewart MM0GNS, *cgnstewart@hotmail.com* Meets on the first Tuesday of each month between September and May at 46 Rowallan Drive KA3 1TU

**West of Scotland [Glasgow] ARS**
Sam Liddell, GM4BGS,07831 486620 *gm4bgs@yahoo.co.uk* Meets 7.30pm for 8.00pm on Fridays at the Multi Cultural Centre 21 Rose Street Glasgow G3 6RE *www.wosars.org.uk*

**Wigtownshire ARC**
Mr Ellis Gaston, GM0HPK. *ellis@agaston.freeserve.co.uk* Meets 7.30pm on Thursdays at the Aird Unit, Stranraer Academy Stranraer (entrance from Cairnport Road) DG9 8BY *www.gm4riv.org*

## Repeater Groups

**Scottish Bdrs Repeater Grp, RS43855**
Contact: Jim Keddie, GM7LUN Tel: 01896 850245 Email: *mail@gm7lun.com*

## Emergency Comms Groups

**Strathclyde Raynet Group**
Mr Crawford Ross, GM8HBY, *gm8hby@strathclyde*-raynet.com;01236 755177 *www.strathclyde*-raynet.com/

**Dumfries & Galloway Raynet Gp**
Ellis Gaston GM0HPK Group Controller, Tel: 01556 610549, *raynetdg@gmail.com* Meets 8.00pm 2nd Monday in the month at the Emergency Planning Centre Council Offices English Street Dumfries DG12DD *www.dgraynet.co.uk*

**Greater Glasgow Raynet**
Paul Lucas, MM3DDQ, Tel: 01389 499972, *mm3ddq@yahoo.co.uk* Meet 1st Tuesday of each month at 7.30pm at Braehead Shopping Centre Glasgow

## Contest Groups

**Central Scotland FM, RS38728**
9 St. Andrews Crescent DUMBARTON, G82 3ER *info@csfmg.com* 01389 764 401

## Region 1 Examinations

| Club | Place | Exam Secretary | | Telephone | Email |
|---|---|---|---|---|---|
| Greenock and District Scouts and Guides Amateur Radio Club | Gourock | Catherine McClintock | MM1AUF | 01475 635155 | *kai@mcclintock.plus.com* |
| Ayr Amateur Radio Group | Ayr | John Shankland | GM3CSO | 01292 445599 | *johnshankland143@btinternet.com* |
| Ayr Amateur Radio Group | Ayr | John Rankin | 2M0JCG | 01294 469821 | *rankin17@freeuk.com* |
| Cockenzie & Port Seton ARC | Prestonpans | Bob Glasgow | GM4UYZ | 01875 811723 | *bob.gm4uyz@talktalk.net* |
| Elderslie ARS | Johnstone | Alex Jenkins | 2M0LEX | 07742 606101 | *mr.alex.jenkins@btinternet.com* |
| Elderslie ARS | Elderslie | Robert Todd | MM3LVQ | 0141 8893466 | *robert3todd@aol.com* |
| Falkirk & DARS | Falkirk | Kenneth Elliot | GM4NTX | 01324 825914 | |

| | | | | | |
|---|---|---|---|---|---|
| Kilmarnock & Loudoun ARC | Hurlford | Leonard Pagot | GM0ONX | 01563 534383 | gm0onx@gmail.com |
| Zetland Amateur Radio Club | Lerwick | Peter Bruce | GM0CXQ | 01595 880241 | gm0cxq@gmail.com |
| Livingston DARS | Edinburgh | Iain Street | MM0DEC | 01506 417005 | iain.street@btinternet.com |
| Lomond ARC | Dumbarton | Barrie Spink | GM0KZX | 01389 764401 | gm0kzx@googlemail.com |
| Lomond ARC | Dumbarton | William McCue | MM0ELF | 01389 755758 | mm0elf@blueyonder.co.uk |
| Mid Lanark ARS | Motherwell | Denis Barrett | MM0DNX | 0141 573 4847 | midlanark_motherwell@yahoo.co.uk |
| Paisley ARC | Paisley | Stuart McKinnon | MM0PAZ | 01505 341047 | mm0paz@gmail.com |
| Renfrew ARS | Paisley | B McCue | MM0ELF | | mm0elf@blueyonder.co.uk |
| Renfrew ARS | Renfrew | Stewart Magee | 2M0HYZ | 0141 636 9440 | stewartmagee246@btinternet.com |
| Stirling DARS | Stirling | William McTaggart | GM0MZP | | [illegible]@gm0nx.com |
| Stornaway Repeater Group | Stornoway | Carl Taylor | 2M0TYR | 01851 870772 | |
| Wigtownshire ARC | Stranraer | J Hopkins | GM4LPT | 01776 706601 | hopk@btinternet.com |
| WOSARS | Glasgow | John Connelly | MM0SIL | 01417 701532 | joncon1@btinternet.com |
| WOSARS | Glasgow | Kenny Duffus | 2M0ZUN | | kenny@duffus.org |
| HUB Radio Group | Paisley | Robert Markle | 2M0IZC | 07826 411585 | |
| Strathclyde 4x4 Response | Clydebank | Barclay Bannister | MM6MCX | 01389 879423 | bannister12@btinternet.com |
| DX-Interceptors | Airdrie | William McFarland | GM0OBX | 07525 399635 | gm0obx@yahoo.co.uk |
| 122 Squadron ATC | | James McMorland | MM0GUE | 01419 466641 | j.mcmorland225@btinternet.com |
| 1777 Squadron Air Cadets | Dumbarton | William McCue | MM0ELF | 01389 755758 | mm0elf@blueyonder.co.uk |
| Ardrossan Scout Group | Ardrossan | Austen Brown | MM6OOP | 07831 324981 | phthirusp1@aol.com |

# Region 2 - Scotland North & Northern Isles

**(http://rsgb.org/region2)**

## RM

Denny Morrison, GM1BAN 07711 682756 *rm2@rsgb.org.uk*

## DRMs

**District 21** (Highlands)
Doug Fraser, GM6JRX Tel: 01847 821397 Email: *drm21@rsgb.org.uk*

**District 22** (Aberdeenshire, Moray)
Peter Thomson, GM1XEA Tel: 01224 740091 Email: *drm22@rsgb.org.uk*

**District 23** (Angus, Fife. Perth & Kinross)
Andrew Burns, MM0CXA Tel: 07720 321824 Email: *drm23@rsgb.org.uk*

**District 24** (Orkney)
David Wishart, MM5DWW Tel: 01856 721422 Email: *drm24@rsgb.org.uk*

**District 25** (Shetland)
Peter Bruce, GM0CXQ Tel: 01595 880241 Email: *drm25@rsgb.org.uk*

## Clubs & Societies

**Aberdeen ARS**
Fred Gordon GM3ALZ *fred_gordon@btinternet.com* Meets 19.30 - 21.30 on Thursdays at 25th Scout Group, Oakhill Crescent Lane, Aberdeen AB15 5HY *www.aars.org.uk*

**Caithness ARS**
Leslie Thomas (GM0TKB) *gm0tkb@farnorth.org.uk* Meets at 10:30 to 21:30 Craigwell Farm Skirza Caithness KW1 4XX *www.radioclubs.net/c.a.r.s.*

**Dundee ARC**
Jim Wilson Meets every Tuesday during Term Time 19.00-21.00 Dundee and Angus College Old Glamis Road Dundee DD3 8LE *www.dundee*-amateur-radio.co.uk

**Glenrothes & DARC**
Laurie Auchterlonie MM0LJA *mm0ljasecgc@btinternet* Meets in New Football Pavilion Station Road Thornton Fife KY1 4AX *www.gdarc.org.uk*

**Montrose Air Station Heritage Centre**
Ewan Cameron, MM0BIX, *rafmontrose@aol.com* Meets Sundays at 12.00noon at Montrose Air Museum, Waldron Road, Broomfield, Montrose, Angus DD10 9BB *www.rafmontrose.org.uk*

**Moray Firth ARS**
Steve Jackson,2M0SKJ,*MM6SKJ@Yahoo.couk* Meets first Thursday in every month at 7.30pm Ashgrove Community Hall Land Street New Elgin IV30 6BL *www.mfars.co.uk*

**Museum of Communication ARS**
Ken Horne GM3YBQ Meets Wednesday and Saturday at 11am - 4am at 131 High Street Burntisland Fife KY3 9AA

**Orkney ARC**
Mrs Terry Penna. *mm3poi@btinternet.com* Meets on the first Tuesday of each month at the Orkney Club Harbour Street Kirkwall Orkney Islands KW15 1LE *http://eu009.webplus.net*

**Pentland Firth Radio Hams**
Denny Morrison, GM1BAN *gm1ban@fsmail.net* Meets last Sunday of the month 15:00 Wirane West Murkle Thurso KW14 8YT

**Perth & DARG**
Dr Ron Harkess, GM3THI Meets 8.00pm Wednesdays at the Perth Sports & Social Club 18 Leonard Street Perth PH2 8ES

**Sutherland & District A.R.C**
Frank Dinger GM0CSZ. *sadarc@sutherland-arc.org.uk* Meets evey Friday at 7:00pm at Dunrobin Farm Golspie Highland KW10 6RH

## Repeater Groups

**Grampian Repeater Group, GB3GN**
Contact: George Anderson.
Tel: 01467 625967
Email: *repeatercommittee@wkfr.org.uk*

## Emergency Comms Groups

**Grampian Raynet Group**
Mr P. Thomson, GM1XEA. , 01224 740091, *sec@grampian-raynet.org.uk* No regular meetings Grampiam Raynet Secretary 13 Westwood Drive Westhill Aberdeenshire AB32 6WW *www.radioclubs.net/grampianraynet/*

**Orkney Raynet Group**
Dave Wakefield GM0KTH, 01856 731811, *dave.gm0kth.com*

## Region 2 Examinations

| Club | Place | Exam Secretary | | Telephone | Email |
|---|---|---|---|---|---|
| Aberdeen ARC | Aberdeen | Robert Duncan | 2M1HRS | 01224 896142 | robertduncan90@talktalk.net |
| Banff & Buchan ARC | New Pitsligo | Martin Andrew | GM6VXB | 01346 582061 | martin.andrew@btinternet.com |
| Caithness ARS | Caithness | Colin Mair | GM7NUQ | 01847 896151 | colin@mairnet.co.uk |
| Caithness ARS | Caithness | Donald Morrison | GM1BAN | 01847 893590 | denny@west-murkle.freeserve.co.uk |
| Caithness ARS | Caithness | Leslie Thomas | GM0TKB | 01847 851729 | gm0tkb@farnorth.ortg.uk |
| Dundee ARC | Dundee | Peter Carnegie | GM1CMF | 01382 580578 | member@muirheed.freeserve.co.uk |
| Dundee ARC | Dundee | Peter Carnegie | GM1CMF | 01382 582875 | pete@muirheed.freeserve.co.uk |
| Galashiels DARS | Galashiels | Jim Keddie | GM7LUN | 01896 850245 | mail@gm7lun.co.uk |
| Glenrothes DARC | Kirkcaldy | Ken Horne | GM3YBQ | 01592 265789 | kenmarg.horne@btinternet.com |
| Inverness Radio Club | Inverness | Neil Moir | GM7RVR | 01463 225508 | neil.moir@dsl.pipex.com |
| Moray Firth ARS | Moray | John Brown | MM0JMB | 01343 541653 | john@jbrown82.orangehome.co.uk |
| Orkney ARC | | Robert Duncan | MM0RDD | 01856 872102 | |
| Strathmore ARC | Forfar | Graham Scattergood | MM0BSX | 01307 468824 | |
| Ellenroad RC | Lancs | Doreen Hensby | 2E1TKD | 01706 344247 | danddhensby@btinternet.com |
| Montrose Air Station Heritage Centre | Montrose | Ewan Cameron | MM0BIX | 01674 676740 | |
| Na Fir Chlis ARC | | James Cameron | MM0CWJ | | mm0nfc@hotmail.com |

# Region 3 - North West

**(http://rsgb.org/region3)**

## RM

Kath Wilson, M1CNY. Tel: 01270 761608. Email: *rm3@rsgb.org.uk*

## DRMs

**District 31** (Cumbria)
Barry Easdon, G0RZI Tel: 01946 812092
Email: *drm31@rsgb.org.uk*

**District 32** (Lancashire)
Steve Roberts M0SJR Tel: 01514 801764
Email: *drm32@rsgb.org.uk*

**District 33** (Greater Manchester)
Dave Wilson, M0OBW Tel: 07720 656542
Email: *drm33@rsgb.org.uk*

**District 34** (Cheshire)
Julian Woolvin M0JPW Tel: 0151 523 4464
Email: *drm34@rsgb.org.uk*

**District 35** (Isle of Man)
Stuart Hill, GD0OUD Tel: 01624 613226
Email: *drm35@rsgb.org.uk*

**District 36** (Merseyside)
John Clark, 2E0JPC Tel: 07856 260816
Email: *drm36@rsgb.org.uk*

## Clubs & Societies

**Ashton In Makerfield ARC**
Peter Williams M0RGN, *mx0htr@gmail.com* Meets every Monday of each week from 19.30 at Ashton Town Football Club Edge Green Street Ashton-in-Makerfield Wigan WN4 8SL *www.aimarc.co.uk*

**Bolton Wireless Club**
Ian G0CTO Meets at 7:30pm every 2 weeks on a Monday from the The Britannia Hotel Beaumont Road Bolton BL3 4TA *www.boltonwireless.org.uk*

**Burnley & DARS**
Vince Greatwood M0VHG, *burnleyradioclub@gmail.com* Meets on Thursday 7:30PM @ Higham Village Hall Higham Hall Road Higham Burnley BB12 9EZ *www.burnleyradio.club*

**Bury RS**
Mr Mike Bainbridge,G4GSY *mail@buryradiosociety.org.uk* Meets 7:30 pm on Tuesdays at the Mosses Centre Cecil Street Bury Lancs BL9 0SB *www.buryradiosociety.org.uk*

**Central Lancs ARC**
Peter Sinclair G3UCA, *g3uca@blueyonder.co.uk* Meets each Sat or Sun from Easter to end of Oct when the railway is open to visitors at Ribble Steam Railway Museum Chain Caul Road Aston On Ribble Preston PR2 2PD *www.clarc.webs.com*

**Chester & DRS**
Bruce Sutherland, M0CVP, *bfcsuth@gmail.com* Meets 1st, 3rd, 4th (&5th) Tues of each month except August at The Burley Memorial Hall Common Lane, Waverton, Chester or The Waverton Institute, Waverton, Chester CH3 7QN *www.chesterdars.org.uk*

**Chorley & DARS**
Nessie (M0NES) *nessie68@hotmail.com* Meets every Wednesday, 6.30pm at Tatton Community Centre Silverdale Road Chorley PR6 0PR *www.cadars.org*

**Fleetwood Radio Enthusiasts Group**
Mr Robert Baines, *gb0frg@hotmail.com* 0794 0815 659 Meets Tuesday's at 19.30 at Various locations across Fleetwood Please contact us for details of venue *www.fleetwoodradiogroup.moonfruit.com*

**Furness ARS**
Chris Leviston, M0KPW, *info@fars.org.uk* Meets first, third and fifth (if applicable) Monday of each month at 20.00, Farmers Arms Hotel, Newton-in-Furness LA13 0NB Meets second and fourth Wednesdays at 19.30 for talks at Hawcoat Park Sports Club Barrow-in-Furness LA14 4HF *www.fars.org.uk*

**Fylde ARS**
Mr Ken Randall, G3RFH *ken.g3rfh@fsmail.net* Meets every 1st and 3rd Thursdays in the month at South Shore Tennis Club Midgeland Road Blackpool FY4 5HZ

**Isle of Man ARS**
Andy Morgan (Secretary) GD1MIP , *iomars@namx.net* Meets every 2nd Tuesday from 19.30 onwards in the Sea Cadets Hall Tromode Road Tromode Douglas IM9 2AJ *https://iomars.wordpress.com*

**Macclesfield and District ARS**
Adrian Dodd, M0PAI, *m0pai@hotmail.co.uk* Meets 19.30p on Mondays at The Pack Horse Bowling Club Abbey Road Macclesfield SK10 3AU *www.gx4mws.com*

**Manchester Wireless Society**
*secretary@g5ms.com* Meets on the first Monday at 8pm Cleveland Public House Wilton Road Crumpsall M8 4WQ *www.g5ms.com*

**Mid Cheshire ARS**
Peter Paul Fox, G8HAV,*midcars@woollysheep.org* Meets every Wednesday evening 19:30-22:30 Cotebrook Village Hall Stable Lane, Cotebrook nr Tarporley Cheshire (NGR: SJ 571 655) CW6 0JJ *www.midcars.org*

**Morecambe Bay ARS**
Andrew Scarr (Sec) G0LWU, *andrewscarr@uk2.net* Meets 8:00pm on every Tuesday at the Trimpell Sports & Social Club Outmoss Lane Morecambe Lancs LA4 4UP *www.mbars*-g1mbr.co.uk

**Newton-le-Willows ARC**
Lee Leland *m0lgl@nlwarc.co.uk* Meets Tuesdays 19.00 - 21.00 Newton Boys and Girls Club 19 Haydock Street Newton-le-Willows Merseyside WA12 8JF *www.nlwarc.co.uk*

**North Cheshire RC**
Terry Roeves, G3RKF *roeves@talktalk.net* Meets 8.00pm on Sundays at the Morley Green Club Mobberley Road Wilmslow Cheshire SK9 5NT g0vie.co.uk/ncradioclub

**North West Amateur Radio Club**
Simon Skirving, 2E0SIA, *simon2e0sia@gmail.com* Meets every Wednesday at 19.00 pm The Brooklyn Hotel Green Lane Bolton Lancs BL3 2EF *www.nwarc.org.uk*

**Oldham ARC**
Christopher Cunliffe G7OOD, *president@oarc.org.uk* Meets every Thursday at 7:30pm the No 1855 (Royton) Squadron Air Training Corps Park Lane, Royton Oldham OL2 6RE *www.oarc.org.uk*

**Preston ARS**
John Knight M0NDU, *john@engineer.com* Meets every Thursday 7pm at The Lonsdale Club Fulwood Hall Lane Fullwood Preston PR2 8DB *www.facebook.com/groups/prestonamateurradiosociety*

**Radio Millennium Lodge**
Mr N Stackhouse, email: *g1scl@ntlworld.com* Meets the first Friday of Feb, Apr, Jun, Oct & Dec at 6:00pm 15 Westbourne Road Urmston Manchester M41 0XQ *www.rml9709.org.uk*

**Rochdale & DARS (R.A.D.A.R.S)**
Dave Carden, G3RIK,*dave@radars.me.uk* Meets every Wednesday 19.30-21.30 Norden Old Library 617 Edenfield Road Rochdale Lancashire OL11 5XE *www.radars.me.uk*

**Salford University Radio Society**
Vincent Lynch, *enquiries@suradiosociety.org* Meets Technology House, 2 Lissadel Street Salford M6 6AP *www.suradiosociety.org*

**Sands Amateur Radio Contest Group**
Brian Watson, G0RDH, *info@m0scg.org.uk* Meets every other Monday at 8pm at The Owls Nest Bare Lane Morcambe LA4 6DD *www.m0scg.org.uk*

**South Cheshire ARS**
Alan Lewis, *alan@a*-lewis.co.uk 01782 510856 Meets every 2nd & 4th Thursday at 7.30pm Wilson House Ford Lane Crewe CW1 3EH *www.g6tw.co.uk*

**South Lancashire ARC**
Jason Bridson, M0HOY, *jay.brid44@gmail.com* Meets every Wednesday evening 20:00 to 22:00 Bickershaw Village Community Club Bickershaw Lane, Bickershaw Wigan WN2 5TE *www.slarc.co.uk*

**South Manchester Radio & Comp Club**
Ron, G3SVW. *chairman@smrcc.org.uk* Meets every Thursday. 20.15 to 22.00. The Woodheys Club. 299 Washway Road Sale Cheshire M33 4EE *www.smrcc.org.uk*

**Southport & DARC**
Alison Hughes, *secretary@sadarc.org.uk* Meets 8.00pm on 3rd Monday in the month at St Marks Church Hall Scarisbrick Lancs L40 9RE *www.sadarc.org.uk*

**Stockport Radio Society**
Heather Stanley, M6HNS *info@g8srs.co.uk* Meets 1st, 3rd and 4th Tuesday at 19.00-22.00 1st & 3rd Tues - Morse Class 4th Tues - Skills night Walthew House, Shaw Heath Stockport SK2 6QS *www.g8srs.co.uk*

**The Simpson Amateur Radio Society**
Ian Ridings, M0IPR Email: *simpson.radio@ntlworld.com* Meets from 7.00pm

an onwards on Mondays at Heathfield Resource Centre Heathfield Street, Newton Heath Manchester M40 1LF *www.m0ora.com*

**Thornton Cleveleys Amateur Radio Society**

Paul Hughes, G3OSR *paulboserup@tiscali.co.uk* Meets 8pm every Monday evening Except Bank Holidays Cleveleys Community Centre and Church Kensington Road Cleveleys Lancashire FY5 1ER *www.tcars.moonfruit.com*

**Warrington ARC**

Bill Rabbitt, G0PZP *william.rabbitt@btinternet.com* Meets Tuesday 8 pm Thursday 10am Sunday 1pm Grappenhall Community Centre Bellhouse Lane Grappenhall Warrington, Cheshire WA4 2SG *www.warc.org.uk*

**West Manchester RC**

Trevor L Speight, G0TEE. *secretary@wmrc.org.uk* Meets 8.00pm on Thursdays at the Astley & Tyldesley Miners Welfare Club, Meanly Road Astley, Tyldesley Manchester M29 7DW *www.wmrc.org.uk*

**Widnes & Runcorn ARC**

Mr Julian Woolvin, M0JPW *jwoolvin@aol.com* Meets alternate Wednesdays evenings at 7:30pm in The Lostock Sports and Social Club Works Lane Northwich Cheshire CW9 7NW *www.wararc.co.uk*

**Wigan-Douglas Valley ARS**

Mr D Snape, G4GWG, *daveg4gwg@gmail.com* Meets 8.00pm on 1st and 3rd Wednesday in the month at the 30 Culcross Avenue Highfield WN3 6AA *www.radioclubs.net/wigandvars*

**Wirral & DARC**

Simon Richards G0XHF Email: *secretary@wadarc.com* Meets 8.00pm 2nd & 4th Wednesdays of month at the Irby Cricket Club Mill Hill Road Wirral CH61 4XQ *www.wadarc.com*

**Wirral ARS**

Gordon Hunter, G8WWD, *g8wwd@yahoo.co.uk* Meets Tuesday, Wednesday and Thursday every week Club Room Ivy Farm Arrowe Park Road Wirral CH49 5LW *www.g3nwr.org.uk*

**Workington and District ARC**

Alex Hill, G7KSE *mx0wrc@gmail.com* Meets fortnightly at 19.00 at Helena Thompson Museum Park End Road Workington Cumbria CA14 4DE *www.mx0wrc.org*

## Repeater Groups

**UKFM Group Western, GB3MP**

Contact: Julian Wolvern M0JPW Tel: 01515 234464 Email: *dwilson@btinternet.com*

## Emergency Comms Groups

**Southport & District Raynet Group**

Harold Taylor G3LWK, *g3lwk@southport*-raynet.org.uk, 07708 252866 Meets 1st Wednesday in month at the St Teresa's Club Upper Aughton Road Birkdale *www.southport*-raynet.org.uk

**Central Lancs Raynet**

E Wane, G0OFY, *raynet@rfradiofitter.co.uk*, 01257 274368

**Fylde Coast Raynet Group, G7FCR**

Mr R Knighton, 01253 824319, *rayknighton@btinternet.com* 2nd Wednesday of each month 7:30pm Frank Townend Centre, Beach Road, Thornton Cleveleyes Tel: 01253 824319

**Ribble Valley Raynet Group, MQ0RVR**

Lee Roe, M0LMP, 07772 444422, *lee@ribblevalleyraynet.co.uk* Meets every quarter at the SJA Building, King Lane Clitheroe, 01200 453823 or 07772444422

**North Lancashire Raynet Group**

Chaz Warr G0AWM *chaswarr@btinternet.com*, 01524 63060

**Cumbria Raynet Group**

Paul Woodburn M0PWD, 07721 457257, *controller@cumbria*-raynet.org *www.cumbriaraynet.org.uk*

## Contest Groups

**S Wirral Contest Grp, G3CSA**

18 Ploughmans Way Great Sutton South Wirral, CH66 2YJ 0151 339 0842

**Tall Trees Contest Group, M0TTG**

Tall Trees Farm' Noahs Ark Lane Mobberley Great Warford KNUTSFORD, Cheshire WA16 7AX *Bgale111s@googlemail.com* 01565 873205

## Region 3 Examinations

| Club | Place | Exam Secretary | | Telephone | Email |
|---|---|---|---|---|---|
| Alderley Explorer Scout Radio Unit | Chelford | T I Webster | G4ZVA | 01477 537190 | *ian@g4zva.co.uk* |
| Bolton Wireless Club | Bolton | Paul Sephton | M0KDM | 01942 730790 | *psephton@gmail.com* |
| Burnley & DARC | Burnley | Clive Cosgrif | G0DZC | 01282 454371 | *clive.cosgrif@hotmail.com* |
| Bury Radio Society | Bury | Peter E Smith | G2DPL | 0161 797 6736 | *peter.g2dpl@btinternet.com* |
| Castle Rushen RC | Castletown | Daniel Wood | GD0VIK | 01624 826500 | *d.wood@crhs.sch.im* |
| Chester & DARS | Chester | B Sutherland | M0CVP | 01244 343825 | *bfcsuth@gotadsl.co.uk* |
| Chorley & DARS | Bury | Denise Croasdale | M0NES | 01257 413113 | *nessie68@hotmail.com* |
| East Cheshire RG | | Stephen Sparkes | M0DFD | 01625 528462 | *m0dfd@f2s.com* |
| Furness ARS | Ulverston | M Brereton | G8ALE | 01229 869244 | *mxb@watermill.co.uk* |
| Halton Radio Club | Runcorn | Derek Robbinson | G0MQH | 01928 711194 | |
| Isle of Man ARS | Peel | Stuart Hill | GD0OUD | 01624 613226 | *gd0oud@manx.net* |
| Isle of Man College | Douglas | Colin Baillie-Searle | GD4EIP | 01624 801592 | *gd4eip@yahoo.com* |
| Mid-Cheshire ARS | Tarporley | Peter Fox | G0IRA | 01606 553401 | *g8hav@msn.com* |
| Morecambe Bay ARS | Morecambe | Tim Upstone | G4DPT | 01524 425770 | *trupsto@gmail.com* |
| North Cheshire RC | Wilmslow | Jill Gourley | G0OZJ | 0161 4855036 | |
| Northwest ARC | Wigan | David Prior | M0HPT | 01257 252094 | *davidprior@blueyonder.co.uk* |
| Northwest ARC | Bolton | Simon Skirving | 2E0SIA | 01204 432964 | *simon2e0sia@gmail.com* |
| Oldham ARC | Oldham | Geoffrey Oliver | G0BJR | 0161 6524164 | *president@oarc.org.uk* |
| Preston ARS | Preston | John Knight | M0NDU | 01772 722195 | *john@engineer.com* |
| Rochdale & DARS | Rochdale | Bryan Turner | G3RLE | 01706 345732 | *bhturner@btinternet.com* |
| The Simpson ARS | Manchester | Ian Ridings | M0IPR | 0161 2887301 | *simpson.radio@ntlworld.com* |
| Solway DX Group | Cumbria | Pauline English | M0SOL | 01228 527181 | *t.english2@homecall.co.uk* |
| South Lancashire ARC | Wigan | Rina Horner | 2E0RIN | 01204 571627 | *rina.horner@ntlworld.com* |
| Southport & District ARC | Ormskirk | Paul Harvey | M0SET | 01512 222918 | *courses@sadarc.org.uk* |
| Southport & District ARC | Southport | Robert Harwood | G0HRT | 01704 220775 | *rob@g0hrt.co.uk* |
| Stockport RS | Stockport | John Marsh | M0JFM | 01625 859690 | *courses@g8srs.co.uk* |
| Thornton Cleveleys ARS | Thornton Cleveleys | John Webb | G8RDP | 01253 876313 | *john.webb2@o2email.co.uk* |
| Thornton Cleveleys ARS | Fleetwood | John Webb | G8RDP | 01253 876313 | *mail@tcars.org.uk* |
| Warrington ARC | Warrington | Paul Middlehurst | G1DVA | 01925 756595 | *paul@g1dva.net* |
| West Manchester Radio Club | Manchester | Peter Nutt | G4WLI | 01616 431671 | *peter@g4wli.com* |
| Widnes & Runcorn ARC | Northwich | Kath Wilson | M1CNY | 01270 761608 | *dwilson@btinternet.com* |
| Workington & District AR & IT Group | Cumbria | Peter Webster | G8RZ | 07789 622165 | *prw123@live.co.uk* |
| Macclesfield and District ARS | Macclesfield | Arthur C Randles | M0GWF | 01625 876489 | *arthur.randles@btinternet.com* |
| Greater Manchester North Scout County | Manchester | Paul Raine | | 01204 494512 | *paulraine@uku.co.uk* |
| Wirral & District ARC | Wirral | G Brown | | 0151 678 7807 | geoffrey_*brown@o2.co.uk* |
| The Sands AR Contest Group | Morecambe | Damien Davies | G0LLG | 01524 61305 | *damien.davies@btopenworld.com* |
| HMPSARC | | Mrs J Lewer | | 01254 234089 | |
| Ashton in Makerfield ARC | Wigan | Peter Williams | M0RGN | 01942 740486 | *peter.m0rgn@gmail.com* |
| East Lancs Radio Club | | Peter Martin | M0NWI | 01254 368598 | peter-*martin@outlook.com* |
| South Manchester Radio & CC | Sale | Peter Fambely | G0BHP | 0161 9730617 | *pfambely@gmail.com* |
| Bent & Bongs Explorer Scout Unit | | Adam Jordan | M6CRM | 07837 183908 | adam_*jordan15@hotmail.co.uk* |
| Newton-Le-Willows ARC | | Allan Green | 2E0RAG | 01942 389562 | *2E0RAG@nlwarc.co.uk* |
| Salford University Radio Society | Salford | Stephen | Pettitt | 07903 720250 | *steve@scorfluflus.co.uk* |
| Standish Amateur Radio Society | Wigan | David Prior | M0HPT | 01257 252094 | *davidprior@blueyonder.co.uk* |
| Wigan UTC Academy | Wigan | M O'Boyle | | 01942 614440 | *moboyle@wiganutc.org* |

# Region 4 - North East

**(http://rsgb.org/region4)**

## RM

Ian Douglas, G7MFN Tel: 07581 068065 Email: *rm4@rsgb.org.uk*

## DRMs

**District 41** (Northumberland, Tyne & Wear, Cleveland, Co Durham, North Yorkshire, Northallerton) Email: *drm41@rsgb.org.uk*
**District 42** (East Yorkshire)
John Baines, M0JBA Tel: 01482 842430 Email: *drm42@rsgb.org.uk*
**District 43** (West Yorkshire)
Gerald Edinburgh, G3SDY Tel: 01484 602905 Email: *drm43@rsgb.org.uk*
**District 44** (South Yorkshire, NE Lincolnshire)
Adrian Patton, G1BRB Tel: 01472 501138 Email: *drm44@rsgb.org.uk*
**District 45**
Anthony Bonney, M0RHJ Tel: 07764254789 Email: *drm45@rsgb.org.uk*
**District 46** (Five Bridges)
Nancy Bone, G7UUR Tel: 07990 760920 Email: *drm46@rsgb.org.uk*

## Clubs & Societies

**Angel of the North ARC**
Nancy Bone, G7UUR, *nancybone2001@yahoo.co.uk* Meets on Mondays 7pm - 9pm at the Whitehall Road Methodist Church Hall Corner of Coatsworth Road and Whitehall Road Bensham Gateshead NE8 4LH *www.anarc.net*

**Axholme Radio Club**
Mr J R Fennell, G4HOY. *john.fennell471@btinternet.* Meets on Wednesdays 1000 - 1600hrs, Thursdays 1900 - 2100hrs and Saturdays 1000 - 1600hrs, Other times by arrangement. Hollytree Farm, Westend Road Sandtoft Epworth, S.Yorks DN9 1LB

**Bishop Auckland RC**
John West G4LRG *g4ttf@yahoo.co.uk* Meets 8.00pm Thursday evenings at the Stanley Village Hall Rear High Road Stanley, Crook Co Durham DL15 9SN *www.qsl.net/g4ttf*

**Blyth ARC**
Mr K Stewart. *kenwendy1@talktalk.net* Meets each week 7pm Wednesday at the New Delaval Welfare Centre Beatrice Avenue Blyth Northunberland NE24 4BP

**Brigg and District Amateur Radio Club**
David Ogg (M0OGY), mailto:*m0ogy1@tiscali.co.uk* Meets fortnightly at 8pm Thursday Brigg and District Servicemens Club Coney Court Brigg North Lincolnshire DN20 8EX *www.bdarc.co.uk*

**Brimham Contest Group**
Neil Clark G8MC Meets 1st Tuesday of every Month at 7.30pm Brimham Lodge, Brimham Rocks Road Ripley, Harrogate HG3 3HE *www.stevebb.co/brimham.htm*

**Colburn & Richmondshire DARS**
Colin Lyne, 2E0TCN - Tel 01748876391 Meets every other Thursday 7.30pm Colburn Village Hall Colburn Lane Colburn DL9 4LZ

**Denby Dale & DARS**
Darran, G0BWB. *g0bwb@g0bwb.com* Meets 1st & 3rd Wednesdays at Meetings start at 20.00 Pie Hall Denby Dale West Yorkshire HD8 8RX *www.g4cdd.net*

**Durham& District AR Society**
David Barclay, M0BPM *dadars@gmx.com* or Macolm Brooks, M1CKU, 0191 440 1792 Meets 7.00pm - 9.00pm on Wednesdays at the Finchale Training College Pity Me, Durham, DH1 5RX *www.radioclubs.net/dadars*

**East Cleveland ARC**
Mr Alistair G Mackay, G4OLK *alistair.mackay@talk21* Meets Friday evenings 7.00pm at the The New Marske Institute Club Gurney Street, New Marske Near Redcar TS11 8EG

**Finningley ARS**
Stuart Boast G3WDL email *stuartboast@yahoo.co.uk* Meets Tuesdays 6-9pm, Sat 12-6pm at The Hurst Communications Centre Belton Road Sandtoft North Lincolnshire DN8 5SX *www.g0ghk.co.uk*

**Goole R & ES**
Mr K McCann, G6YYN or Mr R Sugden, G0GLZ Tel: 01405 769894 Email: *rpsmail1973@gmail.com* Meets Wednesdays at a variety of locations, includ The Courtyard Centre, Boothferry Rd, Goole and the Barnes Wallis Inn, Howden, Google DN14 6AE *www.gooleradioclub.btck.co.uk*

**Grimsby ARS**
Brian Siddle M6LZX, email *secretary@gars.org.uk* Meets 8.00pm on Thursdays at Cromwell Social Club, Cromwell Road, Grimsby, DN31 2BA Also on the second Thursday of the month at The New Sunnyside Club, Grant Street, Cleethorpes, DN35 8AT *www.gars.org.uk*

**Halifax District Amateur Radio Society**
Martin, M0GQB (Chairman) and Darren, M0WIT (Sec.) Meets every Tuesday between 19.00-21.00 at the Church of the Good Shepherd New Road Mytholmroyd Halifax, West Yorkshire HX7 5EA *www.hadars.org.uk*

**Hambleton ARS**
Tony Wilson G3MAE, *tony55@clannet.co.uk*, Meets at 7.30pm on alternative Wednesdays at The Mencap Centre Northallerton N Yorks DL6 1EG *www.radioclubs.net/hambletonars*

**Hornsea ARS**
Gordon McNaught G3WOV, *gmacnaughtwov@yahoo.co.uk*, Meets 7.30pm on Wednesdays at The Old Bakery, behind the New Inn, Hornsea, East Yorkshire HU18 1BG *www.hornseaarc.co.uk*

**Houghton-le-Spring ARC**
Mr George Thompson, M5GHT *m5ght@hotmail.co.uk* Meets Weekly on Tuesdays from 18.30 at the Dubmire Royal British Legion Dubmire Fencehouses Tyne & Wear DH4 6LJ

**Hull & DARS**
Philip Booth, G4PAA Tel: 01964 502775 Meets 7:30pm every Monday and Thursday at the The Old School Dairycoates Avenue Hull HU3 5DB *www.had*-ars.org.uk

**Humber Fortress DX**
Robert Lane, M0RWL. Email: *m0rwl@rwlane.karoo.co.* Meets Tuesdays and Fridays of each week at 7pm Fort Paul, Battery Road, Paull East Yorkshire HU12 8FP http://hfdxarc.co.uk

**Maltby & DARS**
Bryan Ferris, G7HAR Meets Wednesdays 1930-2130 at the The Centenary Hall Bateman Rd
Clifford Rd Hellaby ROTHERHAM S66 8HA *www.maltbyradio.org.uk*

**Mexborough & DARS**
Mrs Sharon Saiger M0BOH Meets Fridays 19.00-22.00 at The Place, Castle Street, Conisborough Doncaster DN12 3HH *www.madars.net*

**North Wakefield RC**
Ruth Moseley 2E0MHU Meets 8.00pm Thursday at the East Ardsley Cricket Club Nr Wakefield WF2 6DT *www.g4nok.org*

**Northeast Amateur Radio Society**
Gary Cockburn, M0TYN *cgt179@msn.com* Meets every 1st Monday of the month 7pm - 9.30pm Wardles Bar, Albert Street, Hebburn Tyne and Wear NE31 1DW

**Northumbria ARC**
Mike Smith M0ZVX *mandmsmith5@talktalk.net*, 01670 861751 Meets Thursday evenings at the Old Telephone Exchange, Cresswell Road, Ellington, Morpeth, Northumberland NE61 5HR *www.g4aax.org.uk*

**Otley ARS**
Mr Paul Watson, M0PKW. *paul@otleyradio.org* Tel: 07768 996370 Meets 8.00pm Tuesdays at The Clifton with Newall Village Hall, Newall Carr Road, Newall with Clifton, Otley West Yorkshire LS21 2ES *www.otleyradio.org*

**Peterlee Radio Club**
Mr Andrew Pennell, G0NSK. *g0nsk@telinco.co.uk* Meets Every Wednesday Evening at 7.30pm St John Headquarters Armstrong Road North East Industrial Estate Peterlee SR8 5AJ *www.peterlee-radio-club.co.uk*

**PDARS**
*colin.g0nqe@btinternet.com* or via website Meets Thursdays at the Carleton Community Centre Pontefract West Yorkshire WF8 3RJ *www.pdars.com*

**Ripon & DARS**
Mr David Cutter G3UNA. *d.cutter@ntlworld.com* Meets 7.30pm Thursdays at The Bunker, rear of Ripon Town Hall 21 Water Skellgate Ripon North Yorkshire HG4 1BH *www.ripon.org.uk*

**Scarborough Amateur Radio Society**
Janet Porter 2E0SCN Tel 01723 054502 Mobile 07717 063751 *seasidejan@googlemail.com* Meets 7.30pm on Mondays at The Pavillion Scarborough Cricket Club North Marine Road, Scarborough, North Yorks, YO12 7TJ *www.g4bp.org*

**Scarborough Special Events Grp**
Roy Clayton, G4SSH, *g4ssh@tiscali.co.uk* Meets as and when required for AGM and organisation of Special Events *www.sseg.co.uk*

**Scunthorpe Steel ARC**
Mr Alistair Butler, M1ECF. *alistair.butler@btinternet.com* Meets 7.30-8pm Meets every Tuesday in the bar at Brumby Hall, Ashby Road, Scunthorpe, North East Lincolnshire DN16 2AB *www.g4fuh.co.uk*

**Sheffield and District Wireless Society**
Krystyna Haywood *info@sheffieldwireless.org* Meets every 1st and 3rd Wednesday of each month at 7:30pm at Rutland Hotel Dovedale Suite 452 Glossop Road Sheffield S10 2PY

**Sheffield ARC**
Dave Littlewood, *G6DCT.littlewood20@btinternet.com* Meets 7.00pm on Mondays at The Sheffield Transport Sports Club Greenhill Main Road Sheffield S8 7RH *www.sheffieldarc.org.uk*

**South East Northumberland ARC**
John Malia M0JAQ, 01914 526 124 Meets every Wednesday 7pm to 9pm at Seaton Sluice Community Centre Albert Road Seaton Sluice Northumberland NE26 4QX

**South Tyneside ARS**
Darren Raine, M0NCB Email: *d_j_raine@hotmail.com* Meets 7.30pm - 9.00pm Mondays (except bank holidays) at St Peters Church York Avenue Jarrow Tyne & Wear NE32 5LP *www.starsradioclub.co.uk*

**Spen Valley ARS**
Mr J R Wilde, G0FOI, *russell@wildegardens.co.uk* Meets first and third Thursday in the month at Old Bank Club Old Bank Road MIRFIELD West Yorkshire WF14 0HY *www.svars.org.uk*

**Stockton & DARG**
Tony Bonney, M0RHJ, *m0rhj@radioclub.co.uk* Meets Wednesday evenings at 19.00 to 21.00 in the Billingham Community Centre The Causeway Billingham Cleveland TS23 2DA

**The Radio Club**
David Jones, Email: *info@theradioclub.co.uk* Meets 19:30 Thursdays St Mary's Centre Badsworth Pontefract WF9 1AJ *www.theradioclub.co.uk*

**Tynemouth ARC**
Club Secretary. Email: *mail@g0nwm.com* Meets 7pm to 9pm Fridays at St Hildas Church Stanton Road North Shields Tyne & Wear NE29 9QB *www.g0nwm.com*

**Tyneside ARS**
T Gardner G3YVZ Meets Wednesday evenings at the St Teresa's Club 200b Heaton Road Newcastle-upon-Tyne NE6 5HP *www.qsl.net/g3zqm*

**Wakefield & DRS**
Mr D Lockwood, G4CLI. *g4cli@wdrs.org.uk* Meets 8.00pm Thursday at the The Scout HQ, 253 Barnsley Rd Wakefield WF1 5NU *www.wdrs.org.uk*

**Wearside Electronics and Amateur Radio Society**
Bill Smith *radio@wears.co.uk* Meets every Wednesday 19.00 to 22.00 Bowburn Community Centre Durham Road Co.Durham DH6 5AT

**Withernsea Lighthouse Amateur Radio Club (WLARC)**
David Jackson, G4HYY, *g4hyy@tiscali.co.uk* Please contact G4HYY to check meeting venue & date Meets 2nd Monday of alternate months at 7.30pm at The Tea Room, Withernsea Lighthouse Hull Rd Withernsea East Yorks HU19 2DY

**York ARS**
Mr C R Rouse *crouse258@gmail.com* Meets 8pm Fridays at the Guppy's Enterprise Club, 17 Nunnery Lane, York YO23 1AB

**York Radio Club (Amateur)**
Mr Gareth Foster, *G1DRG.g1drg@btopenworld.com* Meets 7.30pm Thursdays at the Bishopthorpe Social Club, Bishopthorpe Main Street, York YO23 2RB

**Yorkshire Radio Friends**
Steve Hall, M0HTH, *s.hall21@btinternet.com* Meets Tuesday 6.30 pm Hatfield Woodhouse Village Hall Hatfield Woodhouse Doncaster DN7 6BP

## Repeater Groups

**South Yorkshire Repeater Group, GB7YD**
Contact: Mr Ernie Bailey, G4LUE Tel: Email: *ernest.bailey@hotmail.co.uk*

## Emergency Comms Groups

**Calderdale RAYNET ARG, G7RRC**
Mr Malcolm Spencer, 01422 883867, *mal.spencer@googlemail.com* *www.calderdaleraynet.net*

**South West Durham Raynet ARG**
Mr Ian Bowman, G7ESY, 01388 812104, *ian.bowman70@yahoo.co.uk* Meets 8.00pm 2nd Mon in the month at the Stanley Crook Village Hall Stanley Crook Co. Durham *http://g4ttf.uhp.me.uk/raynet.php*

**Sheffield & Rotherham Raynet**
Mark Harrison, G6NVT, 0700 349 6787, *g6nvt@btinternet.com* Meets 1st Tuesday of every month at 8 pm atNiagara Sports & Social Club Wadsley Bank Sheffield S6 5BU *www.g6aen.net*

**South Humber Radio ECG**
Carl Flynn G7EOG *groupcontroller@southhumberraynet*, 07654 642558 *www.southhumberraynet.org.uk*

## Contest Groups

**Craven Radio Amateur Group, MX0BCQ**
Nutter Cote Cottage Thornton-in-Craven Skipton, North Yorkshire BD23 3TT *www.mxobcq.weebly.com mx0bcq@gmail.com* 01282 842244

**Ossett Amateur Radio Operators, M0ORO**
Homestead Chancery Road OSSETT, West Yorkshire WF5 9RZ *m0oto@aim.com* 07765 192106

**Northern Fells Contest Group, M0NFD**
28 Neville Road DARLINGTON, County Durham DL3 8HY *clive_davies@ntlworld.com* 01325 467834

**93 Contest Group**
John Spurgeon, G4LKD Tel: 01405 704136 Meets last Saturday of each 14.00 Whitgift House Whitgift Goole DN14 8HL

## Region 4 Examinations

| Club | Place | Exam Secretary | | Telephone | Email |
|---|---|---|---|---|---|
| Sheffield & Rotherham Raynet | Sheffield | Karen Marks | 2E0KMW | 08456 445285 | *km.2e0kmw@btinternet.com* |
| Angel of the North ARC | Gateshead | Nancy Bone | G7UUR | 0191 477 0038 | *nancybone2001@yahoo.co.uk* |
| Axholme RC | Nr Doncaster | Brian Spittlehouse | G7IMD | 01427 872354 | *brianspittlehouse@btinternet.com* |
| Bishop Auckland RAC | Crook | Ian Bowman | G7ESY | 01388 812104 | *ian.bowman70@yahoo.co.uk* |
| Bishop Auckland RAC | Bishop Auckland | Timothy Bevan | M0ACV | 01388 832948 | *m0acv@timothybevan.wanadoo.co.uk* |
| Brigg & District ARC | South Humberside | Graham Dawes | M0AEP | 01652 632806 | *training@mx0gbb.org.uk* |
| Danum School Technology College | Doncaster | Daniel Wood | G0VIK | 01302 370767 | *g0vik@qsl.net* |
| Denby Dale (Pie Hall) ARS | Huddersfield | Geoffrey Brierley | M0OYZ | 01484 313909 | *smogg@ntlworld.com* |
| Durham & District ARS | Bowburn | Malcolm Brooks | M1CKU | 01914 401792 | *dadars@gmx.com* |
| Durham & District ARS | | Malcolm Brooks | M1CKU | 0191 440 1792 | *mac.pat@blueyonder.co.uk* |
| East Cleveland ARC | Cleveland | Stephen Pybus | M1SPY | | *fletch2008@btinternet.com* |
| East of Greenwich RAC | Holmpton | Richard Callan | G8NDP | 01964 650400 | |
| Gateshead AR | Tyne & Wear | Keith Morrison | M1VHT | 01670 858817 | *m3vht@virgin.net* |
| Great Lumley AR & ES | Great Lumley | Malcolm Brooks | M1CKU | 0191 440 1792 | *mac.pat@blueyonder.co.uk* |
| Halifax & District ARS | Hebden Bridge | Anthony Vinters | G0WFG | 01422 822636 | *tony@g0wfg.demon.co.uk* |
| Hambleton ARS | Northallerton | John Richardson | M0JWR | 01748 833309 | *john@richardsons-3.freeserve.co.uk* |
| Hornsea ARC | Hornsea | Richard Guttridge | G4YTV | 01964 562498 | *richard@guttridge.karoo.co.uk* |
| Houghton Le Spring ARC | Tyne & Wear | George Thompson | M5GHT | 0191 5488995 | *m5ght@hotmail.co.uk* |
| Hull & DARS | Hull | Philip Booth | G4PAA | 01964 502775 | *philipbooth2002@hotmail.com* |
| Hull & DARS | Hull | Brian Penn | M3EOV | 01482 643475 | *tykepenn@yahoo.co.uk* |
| Humber Fortress DX ARC | Hull | Bob Lane | M0RWL | 01482 444567 | *m0rwl@rwlane.karoo.co.uk* |
| Keighley ARS | Keighley | Shirley Kendrick | M6SAK | 01535652781 | *shirljohnkendrick@aol.com* |
| Keighley College Radio Club | Keighley | Louise Nilon | M3TLL | 01535 618294 | *louise.nilon@keighley.ac.uk* |
| Kirklees College - Hudd Centre ARC | Huddersfield | Selwyn Horner | G4LNF | 01484 363498 | *s.horner4@ntlworld.com* |
| Mexborough DARS | Doncaster | Alan Farrar | M0GVX | 01709 898475 | *m0gvx@hotmail.com* |
| Mexborough DARS | Doncaster | Darrell Harrop | G0FUO | 01709 898577 | *radiohamis@btinternet.com* |

| | | | | | |
|---|---|---|---|---|---|
| Maltby & District ARS | Maltby | Bryan Ferris | G7HAR | | *bryan.ferris@btinternet.com* |
| Mirfield ATC | Mirfield | James Thornton | G3YDL | 01924 517538 | *g3ydl@ntlworld.com* |
| Nestle UK Employee Radio Club | York | Michael Brown | M0APC | 07802 915523 | *m0apl@hotmail.com* |
| North Wakefield RC | Wakefield | Karen Darwin | M6KSD | 07835 350420 | *karen.darwin@hotmail.co.uk* |
| Northumbria ARC | Northumberland | Gordon Emmerson | G8PNN | 01670 790458 | *g8pnn@aol.com* |
| Otley ARS | Otley | David Foley | G3XNO | 01423 522618 | *davidfol@gmail.com* |
| Peterlee Radio Club | Peterlee | Andrew Pennell | G0NSK | 0191 5675760 | |
| Phoenix Comms ARC | Co. Durham | Janice Bolton | M0ETP | 0191 4151693 | |
| Pontefract and District ARS | Pontefract | Chris Pearson | M0JRQ | 01977 620872 | *chris@m0jrq.com* |
| Radio Jcom | Leeds | Tony Kessler | G4DXA | 0113 2323299 | *info@tonykessler.co.uk* |
| Richmond School ARS | Richmond | C Dennis | | 01748 850111 | *cdennis@richmondschool.net* |
| Ripon & District ARS | Ripon | Rob Hall | M0RBY | 01677 460449 | *m0rby@waylock.co.uk* |
| Rose & Crown ARC | | A Garthwaite | M0RTL | 01226 287872 | *mortl@hotmail.com* |
| Scarborough ARS | Seamer | Peter Freeman | M0HQO | 01751 477418 | *pickeringpete@btinternet.com* |
| SEAREG | Doncaster | John Williams | G8LGC | 01709 860769 | John-*Williams@Tinyonline.co.uk* |
| Sheffield ARC | Sheffield | Selwyn James | M0ZEL | 01246 418704 | *m0zel@prospect1952.plus.com* |
| Silcoates Radio Club | Wakefield | Lucy Townsend | | 01924 499182 | *radioclub@silcoates.org.uk* |
| South Tyneside ARS | Tyne & Wear | David B Harbron | G7PHG | 0191 5191464 | |
| South Yorks Repeater Grp. | Rotherham | Jill Bailey | G8PLJ | 01226 716339 | *ernest.bailey@hotmail.co.uk* |
| South Yorkshire ARS | Barnsley | Andrew Lomas | M0ALA | 01226 237916 | *syars@blueyonder.co.uk* |
| St. Davids RC | Middlesborough | F Clarkson | G7TWU | 01642 292364 | *g7twu@fandbc.co.uk* |
| South Humb Raynet | Cleethorpes | Andy Carlile | G0MNI | 07752 530954 | *andy@andyglassman.me.uk* |
| St Cyprian's | Sheffield | Sonia Joan Howard | | 01142 557875 | *sonia@furd.org* |
| Stockton & District ARG | Cleveland | David London | G0VGB | 01642 290350 | *davidlondon55@hotmail.com* |
| Tynemouth ARC | Tyne & Wear | Ian Bennett | M0IGB | 0191 2531566 | *ianmarybennett@talktalk.net* |
| University of Third Age | Bridlington | Peter Masters | M1DBB | 01262 674524 | *m1dbb@aol.com* |
| Wakefield & DARS | Wakefield | Colin Greenwood | M0HRM | 01924 500394 | *colinandsandy@aol.com* |
| Yarm School ARS | Cleveland | J Doherty | N/A | 01642 784685 | |
| York ARS | York | K R Cass | G3WVO | 01904 422084 | Mr Cass has sight difficulties - No email |
| York Radio Club | York | A Palfrey | G8IMZ | 01904 413342 | *apalg8@aol.com* |
| Barnsley & District ARC | Barnsley | Andrew Garthwaite | M0RTL | 01226 231517 | *m0rtl@hotmail.com* |
| Barnsley & District ARC | Barnsley | Stephen Cook | 2E0ETP | 07835 350420 | *teresastevecook@hotmail.co.uk* |
| Harrogate Ladies College | Harrogate | S Rouse | | 01423 504543 | *mr.rouse@hlc.org.uk* |
| Spen Valley ARS | | Timothy Clough | G4PHR | 01924 499397 | *tclough49@btinternet.com* |
| Thorne Sea Cadets | | Phil Ormsby | G7SUD | 07776 001700 | *randomoldgeezer@live.co.uk* |
| Rotherham District Scouts | Rotherham | Paul Archer | M0PJA | 01909 774106 | *paul@resu.org.uk* |
| Southeast Northumberland ARC | Northumberland | Anthony Brown | M0HLR | 0191 2375060 | *brownt2603@sky.com* |
| Southeast Northumberland ARC | Northumberland | John Malia | M0JAQ | 0191 4526124 | *johnmalia74@hotmail.co.uk* |
| The Radio Club | Badsworth | Chris Pearson | G5VZ | 01977 620812 | *chris@g5VZ.co.uk* |
| Wearside Electronics & ARS | Sunderland | Tara Douglas | M3TNI | 07977 134869 | |
| Sheffield & District Wireless S | | Andrew Bennett | G0HSA | 01142 740165 | *andrew.bennett21@btinternet.com* |
| Colburn & Richmondshire DARS | Colburn | Colin Lyne | 2E0TCN | 01788 76391 | *colinlyne@btinternet.com* |
| Yorkshire Amateur Radio Friends | Hatfield Woodhouse | Stephen Hall | M0HTH | 07900 911782 | *s.hall21@btinternet.com* |
| Newsham Amateur Training Centre | Blyth | John Hurlbutt | 2E0DCV | 01912 371729 | *john.hurlbutt@sky.com* |

# Region 5 - West Midlands

**(http://rsgb.org/region5)**

## RM

Martyn Vincent, G3UKV Tel: 01952 255416
Email: *rm5@rsgb.org.uk*

## DRMs

**District 51** (Staffordshire, Warwickshire)
Robert Williams G1BCZ Tel: 07777 694415
Email: *drm51@rsgb.org.uk*

**District 52** (Central & East Birmingham)
Steven Edwards, G3AGW Tel: 07934 682592 Email: *drm52@rsgb.org.uk*

**District 53** (Shropshire, North Worcestershire & West Birmingham)
Jim Wakenell, G8UGL Tel: 07722 380953
Email: *drm53@rsgb.org.uk*

**District 54** (Gloucestershire, Hereford & South Worcestershire)
Giles Herbert, G0NXA Tel: 07769 658041
Email: *drm54@rsgb.org.uk*

## Clubs & Societies

**Bromsgrove & DARC**
Mr Chris Margetts, M0BQE. *m0bqe@hotmail.com* Meets 8pm on Fridays at the Avoncroft Arts Centre Redditch Road, Bromsgrove, Worcestershire. B60 4JS *www.radioclubs.net/bdarc*

**Burton ARC**
Maria Hancock M6KSW *maria.m6ksw@gmail.com* Meets on each Wednesday of the month at 7.30pm at Stapenhill Institute Club 23 Main Street Stapenhill Burton Upon Trent DE15 9AP *www.burtonarc.co.uk*

**Central Radio Amateur Circle**
Martin Hallard G1TYV. *radio-circle@live.co.uk* Meets one Thursday of each month at 7.30pm at The Sir Robert Peel Inn, 104 Bell Lane Bloxwich West Midlands WS3 2JS *www.radioclubs.net/crac*

**Cheltenham AR Assn**
Derek Thom, G3NKS *secretary@caranet.org* Meets 7.30 for 8pm on the third Thursday of the month at Brizen Young People's Centre Up Hatherley Way Cheltenham GL51 4BB *www.caranet.org*

**Coventry ARS G7ASF**
Mr John Beech, G8SEQ. *john@g8seq.com* 3rd Fri are outdoor events or 2m net see website Meet 1st, 2nd & 4th Friday in month at 20.00 St Bartholomews Church Hall Brinklow Road Coventry CV3 2DT *www.coventryradio.org.uk*

**Dudley and District ARS**
Carl Roberts M0ZCR *m0zcr@live.co.uk* Meets 7.30pm every Tuesday at Ruiton Windmill Vale Street Dudley West Midlands DY3 3XF *www.dadars.co.uk*

**Gloucester A R & E S**
Les Harris *g4aym@aol.com* Meets every Monday except bank holidays & school closures 19.30-22.00hrs Churchdown School Academy Winston Road Churchdown Gloucester GL3 2RB *www.g4aym.org.uk*

**Hereford Amateur Radio Society**
Tim Bridgland Taylor, G0JWJ *timbt@btconnect.com* Meets first Friday of month at 19:30 Hill House Newton HR6 0PF *http://hars.wagnet.co.uk*

**Malvern Hills RAC**
Mike Allenson G3TGD *mike.g3tgd@gmail.com* Meets 8.00pm on 2nd Tuesday in the month at the The Town Club 30 Worcester Road Great Malvern Worcestershire WR14 4QW *www.mhrac.org*

**Mid Warwickshire ARS**
Don Darkes, G4CYG. *midwarwicks@gmail.com* Meets on 2nd and 4th Tuesday of the month 19.30 in Spring Summer and 14.00 in Autumn Winter Warwick Ambulance Assn HQ 61 Emscote Road Warwick CV34 5QR

**Midlands ARS**
Ron Swinburne *M0WSN@aol.com* Meets every Wednesday evening from 7 - 9pm Selly Park Baptist Church 1041 Pershore Rd Stirchley Birmingham B29 7PS *www.radioclubs.net/mars*

**Moorlands & DARS**
Ian King 2E0IDK *m0idk@yahoo.com* Meets 8.30pm on Thursdays at the Foxfield Railway, Caverswall Road Station Caverswall Road, Blythe Bridge Stoke-on-Trent Staffs ST11 9BG

**Nuneaton and DARC**
Neil Yorke M0NKE *[illegible]@ndarc.co.uk* Meets [illegible] Friday of each month at [illegible]pm till 21.00 at Stockingford Community Centre, Haunchwood Road Nuneaton Warwickshire CV10 8DY *www.ndarc.co.uk*

**Riverway Amateur Radio Society**
Robert Fullagar M0RPF, *rfullagar@worldonline.co.uk* Meets every Wednesday at 7:30pm at Stafford & Rugeley Sea Coaotes, Riverway Stafford ST16 3TH

**Rugby ATS**
Mr Stephen Tompsett G8LYB *stephen@tompsett.net* Meets every Tuesday from 20.00 to 22.30 and every Saturday from 14.00 to 18.00 12th Rugby Scout Headquarters Broughton-Leigh Community Junior School Wetherell Way Brownsover, Rugby CV21 1LT *www.rugbyats.co.uk*

**Salop ARS**
Mrs Glenda Evans G1YJB, *salopamaterradio@gmail.com* Meets 8.00pm on Thursdays at The Telepost Club Railway Lane Abbey Foregate Shrewsbury SY2 6BT *www.salopradiosociety.org*

**Sandwell Amateur Radio Club**
Martin Prestidge G2BXP Meets Monday evening at 7:00-9.30pm at Sandwell ARC, R O 55 The Broadway, Oldbury, West Midlands B68 9DP *www.sandwellarc.co.uk*

**Solihull ARS**
Mr Paul Gaskin, G8AYY, *pg012h0844@blueyonder.co.uk* Meets at 8pm on 3rd Thursday in the month at The Shirley Centre 274 Stratford Road Shirley Solihull, West Midlands B90 3AD *http:homepages.which.net/~r_a.hancock/sars.htm*

**South Birmingham RS**
*gemmagordon.m6gkg@gmail.com* Meets 8.00pm every Mon, Wed and Fri nights at West Heath Community Centre, Corner of Fairfax Road and Condover Road,

**Stafford DARS**
Nick Barnes G4KQK, *cnb@doctors.org.uk* Meets 7.30pm Wednesday at the Wildwood Community Centre Wildwood Gate Stafford ST17 4RA *www.g3sbl.org.uk*

**Staffordshire Portable ARC**
Neville Briggs, M0VSP, 01922 449668 *[illegible]@hotmail.co.uk, [illegible]@hotmail.com* Meet on 2nd Mon of each month at 7.30pm at the Horse and Jockey Freeford Bridge Tamworth Road Lichfield, Staffs WS14 9JE

**Stourbridge & DRS**
Mr John Clarke Meets 8.00pm 1st & 3rd Mondays in month at the Old Swinford Hospital School Stourbridge West Midlands DY8 1QX *www.g6oi.org.uk*

**Stratford-upon-Avon and DRS**
Clive Ousbey *sdrsinfo@cousbey.freeserve.co.uk* Meets second and fourth Mondays of most months 7.30pm for 8pm Home Guard Club Tiddington Stratford-upon-Avon Warwickshire CV37 7AY *www.stratfordradiosociety.freeserve.co.uk*

**Sutton Coldfield ARS**
Mr Robert Bird 2E0ZAP, *spirit.guide@hotmail.co.uk* Meets from 7 30pm till 10 30pm 2nd & 4th Mondays in the month (except bank hoildays) Sutton Coldfield Rugby Club, 160 Walmley Road Nr Sutton Coldfield Birmingham B76 2QA *www.g3rsc.co.uk*

**Tamworth ARS**
Richard Redmond 2E0LLE *richard.redmond@outlook.com* Meets Thursdays at 7.30pm at Drayton Village Club Drayton Lane Drayton Bassett Tamworth Staffordshire B78 3TX *www.tamworth*-ars.org.uk

**Telford & DARS G6ZME**
Mr John Humphreys, M0JZH *M0JZH@yahoo.co.uk*, 01952 457234 Meets every Wednesday at 7.00pm events are at 8:00 Village Hall Malthouse Bank Little Wenlock Telford TF6 5BG *www.tdars.org.uk*

**The Vulture Squadron C.G**
Iain Kelly M0PCB *iain@m0pcb.co.uk* Meets 2nd Monday of each month at 7:30pm Meeting venue varies

**Wolverhampton ARS**
Mr V Ravenscroft *secretary@wolverhmaptonars.co.uk* Meets 8.00pm every Thursdays Electricity Board Sports Club St Marks Road Chapel Ash Wolverhampton WV3 0QH *www.wolverhamptonars.co.uk*

**Worcester Radio Amateurs Association**
Rich Moles M0UVA, *secretary@m0zoo.co.uk* Meets 1st & 4th Tuesdays of the month at 19.30 at 3rd Worcester Scout HQ Vicar street, Rainbow Hill Worcester WR3 8EU *www.wraa.co.uk*

**Wythall Radio Club**
Chris Pettitt, *g0eyo@blueyonder.co.uk* 07710 412819 Meets weekly Tuesday and Friday 19.30 to 22.30 Wythall House Silver Street Wythall Birmingham B47 6LZ *www.wythallradioclub.co.uk*

## Repeater Groups

**Kidderminster Repeater Group, M0KRG**
Contact: Ms P Dowie, G8PZT, *g8pzt@gb3kd.org.uk* Tel: 01562 68734 Email: *g8pzt@gb3kd.org.uk*

## Emergency Comms Groups

**Gloucestershire County Raynet**
Max White, M0VNG, 01905 724509 Email: *bmews@hotmail.com* Meets 2nd Tuesday in the month at Ullenwood *www.radioclubs.net/glosraynet*

## Contest Groups

**Cheltenham Cluster, GB7DXC**
c/o 11 Gravel Pits Close Bredon Tewkesbury, Glos GL20 7QL *g0hdb@amdavies.demon.co.uk* 01684 772178

**Travelling Wave Contest Group, MW0TWC**
68 Coalway Road Wolverhampton, WV3 7LZ *peter.burden@gmail.com* 01902 655220

**Bad Weather DX Group, M0WXB**
4 Dunster Grove Cheltenham, Gloucestershire GL51 0PE *www.badwxdx.club M0WXB.radio@virginmedia.co.uk* 07515 340505

## Region 5 Examinations

| Club | Place | Exam Secretary | | Telephone | Email |
|---|---|---|---|---|---|
| Charlie Delta ARC | Bilston | Andrew Bedford | 2E0YDA | 07951 711955 | |
| Bromsgrove & DARC | Bromsgrove | Chris Margetts | M0BQE | | *m0bqe@hotmail.com* |
| Burton upon Trent ARC | Burton upon Trent | Roger Smith | 2E0CLP | 01827 383553 | *remoteswitching@btopenworld.com* |
| Burton upon Trent ARC | Burton upon Trent | Sharon Warren | 2E0NBR | 01283 534580 | *steve@ukspeedtraps.co.uk* |
| Coventry ARS | Coventry | John Beech | G8SEQ | 02476 273190 | *john@g8seq.com* |
| Midland ARS | Birmingham | Ron Swinburne | M0WSN | 0121 7421808 | *M0WSN@aol.com* |
| Cheltenham ARS | Cheltenham | Barry Eames | M0HFY | 01452 720080 | *barry.eames@blueyonder.co.uk* |
| Cheltenham ARS | Rhondda | Nick Booth | M0RAR | 01242 673314 | *Nick@caranet.co.uk* |
| Cotswold Amateur Radio Group | Stroud | D Chatterton | G4PLE/G4VPT | 01453 758311 | *don.chat@virgin.net* |
| Gloucester AR & ES | Churchdown | Anne Reed | 2E1GKY | 01242 699595 | *hamreed@blueyonder.co.uk* |
| South Birmingham RS | Birmingham | Joseph Murphy | G8OWL | 0121 4750908 | *g8owl@hotmail.com* |
| Dudley and District ARS | Kingswinford | Drew Belcher | G7DMO | 01384 378040 | *drew@g7dmo.co.uk* |
| Dudley and District ARS | Dudley | Richard Whyton | M0YYC | 07835 024455 | *exams@dadars.co.uk* |
| Rugby Amateur RTS | Rugby | Stephen Tompsett | G8LYB | 01788 578940 | *stephen@tompsett.net* |
| Malvern Hills RAC | Malvern | M Revell | G7KPR | 01886 830277 | *mrevell@freeuk.com* |
| Moorlands & DARS | Stoke-On-Trent | David Brunt | G6KTE | 01538 750242 | *g6kte@btinternet.com* |
| Phoenix School | Telford | Peter Wallace | M0OAR | 01952 613080 | *p.wallace@blueyonder.co.uk* |
| Salop ARS | Shrewsbury | Glenda Evans | G1YJB | 01939 235412 | *glendaevans@live.co.uk* |
| Solihull ARS | Solihull | R A Hancock | G4BBT | 0121 743 7277 | *r_a.hancock@which.net* |
| South Staffordshire AR Tutor Group | Cannock | E A Matthews | G3FZW | 01543 262495 | *arnold.g3fzw@gmail.com* |

| | | | | | |
|---|---|---|---|---|---|
| St John's Welcome Centre | | John Adlington | M0DVT | 01782 533370 | *m0dvt@qsl.net* |
| Stafford & District ARS | Stafford | Anthony Bairstow | G4RSW | 01785 614771 | *anthony.bairstow@ntlworld.com* |
| Stafford & District ARS | Stafford | Derek Southey | G0EYX | 01785 604904 | *g0eyx.derek@ntlworld.com* |
| Stafford & Rugeley Sea Cadets | Riverway | W Reynolds | M6VIX | 01785 222658 | *m6vix.wend@gmail.com* |
| Staffordshire DARS | Keele | Theocharis Kyriacou | M0TKS | 01782 534313 | *theo@colourexposure.com* |
| Stelar | Keele | Anthony Vinters | G0WFG | 01422 822636 | *tony@g0wfg.demon.co.uk* |
| Stratford Radio Society | Stratford-Upon Avon | John Harris | G8HJS | 01789 293508 | *john.harris1@mypostoffice.co.uk* |
| Tamworth ARS | Tamworth | Steve Smith | M0TSR | 01827 317577 | *steve@spectrumservices.uk.net* |
| Telford & DARS | Telford | Mike Street | G3JKX | 01952 299677 | *g3jkx603@gmail.com* |
| Warwick University R.S | Coventry | Michael Dixon | G4GHJ | 02476 490771 | *g4ghj@lineone.net* |
| Worcester RAA | Worcester | Peter Badham | G0WXJ | 01905 330113 | *chairman@m0zoo.co.uk* |
| Worcester RAA | Worcester | Peter Badham | G0WXJ | 01905 330113 | *chairman@m0zoo.co.uk* |
| Wythall RC | Wythall | Chris Pettitt | G0EYO | 0121 246 7267 | *g0eyo@blueyonder.co.uk* |
| Sutton Coldfield ARS | Birmingham | B Adkins | M1CQN | 0121 328 0703 | *baza052@hotmail.co.uk* |
| Dudley West Explored | Kingswinford | Paul Barton | G6YKT | 01384 292787 | *paul@jamesdidit.com* |
| Forest of Dean ARG | Dry Brook | Clive Wynn | 2E1IFL | 01594 839427 | *gwynn@btinternet.com* |
| Tamworth Radio Scouting Group | Tamworth | Dale Williams | M0WHR | 01827 57088 | *dale.williams@btopenworld.com* |
| Wolverhampton ARS | Wolverhampton | Ron Wellsted | M0RNW | 01902 655207 | *ron@wellsted.org.uk* |
| Sandwell Raynet | Birmingham | Clive Martin | M0LIT | 0121 5322916 | *clive@20hgr.co.uk* |
| Central Radio Amateur Circle | Walsall | Kevin Merchant | G6KOY | | *keving6koy@tiscali.co.uk* |
| Central Radio Amateur Circle | Walsall | Martin Hallard | G1TYV | 01384 358941 | radio-*circle@live.co.uk* |
| Wychavon Area ARG (WAARG) | Pershore | Derek Floyd | 2E0DRF | 01386 860771 | *derek@waarg.net* |
| Hereford Amateur Radio Society | Leominster | Rodney Archard | G0JWJ/M0JLA | 01432 356079 | *radio@crescent.me.uk* |
| Nuneaton & District ARC | Nuneaton | Ray Dunham | G3ZSQ | 07775 760684 | *ray@gg3zsq.plus.com* |
| Gloucester County Raynet | Gloucester | Mary Large | | 01452 302310 | *mary.large@btinternet.com* |
| Forest Federation Radio club | Rangemore | Roger Ridley | M0AZO | 07936 025084 | *roger.ridley@yahoo.co.uk* |

# Region 6 - North Wales

**(http://rsgb.org/region6)**

## RM

Ceri Jones, 2W0LJC Tel: 07917 197988
Email: *rm6@rsgb.org.uk*

## DRMs

**District 61** (Flintshire, Wrexham & Powys)
Mark Harper, MW1MDH Tel: 07967 517892
Email: *drm61@rsgb.org.uk*
**District 62** (Conwy, Denbigh)
Liz Cabban, GW0ETU Tel: 01690 710257
Email: *drm62@rsgb.org.uk*
**District 63** (Gwynedd, Anglesey (Ynys Mon))
John Martin, MW0VTK Tel: 07772 720099
Email: *drm63@rsgb.org.uk*

## Clubs & Societies

**Brecon and Radnor ARS**
Adam Tofarides Meets first Thursday of the month 7pm Llanddew village Hall Llanddew Brecon Powys LD3 9ST

**Conwy Valley ARC**
Mr R W Evans, GW6PMC. *wynneevans@sky.com* Meets 7.30pm on 1st Wednesday of month April - October (excluding August) only The Studio Penrhos Road Colwyn Bay Conway LL28 4DB

**The Dragon ARC**
Stewart Rolfe, GW0ETF, *gw0etf@btinternet.com* Meets 7.30pm for 8.00pm on 1st and 3rd Monday in the month at the Ebenezer Church Hall Lon Foel Graig Llanfairpwll Isle Of Anglesey LL61 5RX
*www.radioclubs.net/dragonarc*

**Halkyn Radio Group**
Bob Stanton GW4KDI, *bobstanton@mail.com* Meets every Wednesday at 8pm Halkyn Radio Group Halkyn Sports Association Pentre Road Halkyn Flintshire CH8 8BS *www.madarc.org.uk*

**Marches Amateur Radio Society**
Marc Griffiths, MW0MDT *marchesars@hotmail.co.uk* Meets second and fourth Thursdays of each month (Except August) at 19:30hrs at Black Park Community Centre Black Park Halton Chirk LL14 5BB

**Meirion ARS**
R A Smith, GW0AYQ. *monbob.bayq34@btinternet.com* Meets 7.30pm on 1st Wednesday in the month (except August) Dyffryn & Talybont Village Hall Dyffryn Ardudwy LL44 2EF *www.meirion-ars.info*

**North Wales RS**
Liz Cabban GW0ETU *lizcabban@vodafoneemail.co.uk* Meets 7.00pm every Thursday at the Colwyn Bay Town Hall Rhiw Road Colwyn Bay LL29 7TE
*www.nwrs.org.uk*

**Porthmadog & DARS**
Robert Hughes-Burton, Sec, MW0RHD Meets 8.00pm on 3rd Thursday in the month The Yacht Club The Harbour Porthmadog Gwynedd LL49 9AT
*www.radioclubs.net/portdist*

**Powys ARC**
Alan Williams GW6EUT, *alanwilliams2@btinternet.com* Meets 8.00pm 1st Thursday in the month Berriew Community Centre Welshpool SY21 8AZ
*www.parc.care4free.net*

**Wrexham ARS**
Eifion Parry (Secretary) *MW6EYU@gmail.com* Meets on 1st and 3rd Tuesday in the month 7.30 for 8.00 pm Brymbo Sports and Social Complex College Hill Tan-Y-Fron Wrexham LL11 5TF
*www.wrexham*-ars.co.uk

## Repeater Groups

**Stornoway Repeater Group, RS194702**
Contact: Tel: 07980 011964
Email: *dave@rtty.co.uk*

## Emergency Comms Groups

**Wrexham & District Raynet Group**
Mr Peter Higgs, GW4IGF, 01244 570212

## Region 6 Examinations

| Club | Place | Exam Secretary | | Telephone | Email |
|---|---|---|---|---|---|
| Dragon ARC | Ynys Mon | Stewart Rolfe | GW0ETF | 01248 362229 | *gw0etta@btinternet.com* |
| MARCHES ARS | Wrexham | Marc Griffiths | 2W0MDT | 01978 840488 | *marcgriffiths@hotmail.co.uk* |
| Merion ARS | Gwynedd | M J Smith | | 01341 242767 | *monbob.bayq34@btinternet.com* |
| Halkyn Radio Group (formally Mold & DARC) | Halkyn | Derek W Jones | GW0UDJ | 01352 714197 | *derwjones@tiscali.co.uk* |
| Mold & DARC | Mold | Derek W Jones | GW0UDJ | 01352 714197 | *derwjones@tiscali.co.uk* |
| North Wales RS | Conwy | Anthony Chalk | MW0BXJ | 01492 530954 | *tonybxj@btopenworld.com* |
| Wrexham ARS | Wrexham | V Priamo | GW6 ZCS | 01878 362446 | *vincepriamo47@gmail.com* |
| Porthmadog & DARS | Gwynedd | A Meadows | | 01766 523499 | |

# Region 7 - South Wales

**(http://rsgb.org/region7)**

## RM

Glyn Jones, GW0ANA Tel: 01446 774522
Email: rm7@rsgb.org.uk

## DRMs

**District 71** (Ceredigion, Pembrokeshire)
Ray Ricketts, GW7AGG Tel: 01970 611853
Email: *drm71@rsgb.org.uk*

**District 72** (Camarthenshire, West Glamorgan, Swansea)
Trevor Nicholas, GW4RVA Tel: 01267 222916 Email: *drm72@rsgb.org.uk*

**District 73** (Mid Glamorgan, East Glamorgan, Cardiff)
Vacant Email: *drm73@rsgb.org.uk*

**District 74** (Monmouthshire, Newport)
Ken Smith, MW0YAC Tel: 01633 876734
Email: *drm74@rsgb.org.uk*

## Clubs & Societies

**Aberdare and District ARS**
hilip Jones MW0PJJ E-mail *mw0pjj@gmail.com* Meets every other Friday Hirwaun YMCA, Manchester Place, Hirwaun, CF44 9RB *www.radioclubs.net/aadars*

**Aberkenfig ARC**
Gareth Evans 2W0GME *info@amberkenfigradioclub.co.uk* Meets every Thursday 19.30 at Aberkenfig Social and Athletic Club Bridgeend Road Aberkenfig CF32 9AP

**Aberystwyth & DARS**
Dave Mansell, GW8SFT. *dman102092@aol.com* Meets 8.00pm on the 2nd Thursday in the month (except August) at the Waunfawr Community Hall Waunfawr Aberystwyth SY23 3PN

**Barry ARS**
Glyn Jones. *glyndxis@talktalk.net* Meets 7.30pm on Tuesdays at Sully Sports & Leisure Club South Road Sully S Glamorgan CF64 5SP *www.bars.btik.com*

**Blackwood & DARS**
L.W.Wright GW8UAM *wynnwright7@aol.com* Meets 7.00pm on Fridays at the Oakdale Comprehensive School Oakdale Blackwood Gwent NP12 0DT *www.gw6gw.co.uk*

**Carmarthen ARS**
M Twyman, GW6KOA, *matthew.twyman63@btinternet.com* Meets 7.00pm on 1st and 3rd Tuesdays in the month at the Cwmduad Community Centre Carmarthen SA33 6XN *www.carmarthenradioclub.org.uk*

**Chepstow & DARS (CDARS)**
Rod Hawkins, *mail@75ohm.co.uk* Meets first and third Tuesday of the month at 19.30 - 22.00 pm Chepstow Athletic Club Chepstow NP16 5JT *www.gw4lwz.org.uk*

**Cleddau ARS**
Mr Ian Baker MW0IBZ, *mw0ibz@pembs.com* Meets Mondays during School term time (not Bank Holidays) at Neyland Community Learning Centre Neyland SA73 1SH *www.cleddau-ars.org.uk*

**Cwmbran & DARS**
Ken Smith Meets on Thursday at 19.00 at Community Centre Council House Ventnor Road Cwmbran NP44 3JY *www.mc0yad.co.uk*

**Hoover (Merthyr) ARC**
Mrs Aeronwen Sneddon. *mw6mws@googlemail.com* Meets Tuesday evening 7.00pm-9.00pm at Galon Uchaf Residents Centre, 9th Avenue, Galon Uchaf CF47 9TL *www.hoover-arc.co.uk*

**Highfields ARC**
Ian Robinson, GW1AWH *IanRobinson4242@aol.com* Meets 7.00pm Tuesdays at Rhiwbina Sports and Social Club Lon-Y-Dail Rhiwbina CARDIFF CF14 6EA *www.highfields-arc.co.uk*

**Llanelli ARS**
Craig Fisher, MW0MXT. *craig@mw0mxt.co.uk* Meets every Monday at 7pm in the Swiss Valley Community Centre Heol Nant, Swiss Valley Llanelli Carms SA14 8EH *www.llanelli-radio-club.tk*

**Mid Glamorgan ARG**
Lynne Beedle,*GW0VMS.gw0vms@yahoo.co.uk* Meets every Thursday Aberkenfig Sports and Social Club CF32 9AP *www.cobwebs.uk.net/mgarg*

**Newport ARS**
Mr Paul Nicholls. Email: *nars@gw4ezw.fsnet.co.uk* Meets Fridays at 19.00 - 21.00 St Julian's Community Learning and Library Centre Beaufort Road Newport Gwent NP19 7UB *www.gw4ezw.org.uk*

**No1 Welsh Wing ATC ARS**
Mr Chris Stubbs, MW0LZZ, *onewingshotoff@hotmail.co* Meets 1st Wednesday of each month at 7.00pm HQ 1344 Sqn ATC Ty Walter Cleall GC Maindy Barracks Cardiff CF14 3YE

**Pembrokeshire RS**
E. Hollowell, GW0GUY Meets Sundays from 18.00 - 22.00 The Oak, St Thomas Green Haverfordwest Pembrokeshire SA61 1QT

**Pencoed Amateur Radio Club**
Madeline Roberts Meets Tuesdays at 7pm at Pencoed Rugby Club Felindre Road Pencoed Nr. BRIDGEND CF35 6PB

**Rhondda ARS**
Mr John Howells, GW4BUZ Meets Tuesdays at 7.30 pm at St Barnabas Church Hall Penygraig Rhondda Valley Mid-Glamorgan CF40 1TA

**Risca & District ARS**
Clive Jenkins, GW6JPC, *riscaham@gmail.com* Meets every Tuesday except Bank Holidays from 19.00 to 21.00 Tuesdays - weekly St John Ambulance Hall Tredegar Terrace Risca Gwent NP11 6BY *www.qsl.net/mc0rrd*

**St. Tybie ARS**
Gareth Woods, GW4JPC, *gw0jpc@yahoo.co.nz* Meets fortnightly at the Old School community Centre in LLandybie, Carmarthenshire SA18 3HX *http://gc0vpr.clubbz.com*

**Swansea & District Amateur Radio Club**
Jeff Downer, GW6TYJ, *jeffrey.downer@tiscali.co.uk* Meets Fortnightly every Tuesday Forge Fach Community Centre Hebron Road Clydach Swansea SA6 5EJ

**Swansea ARS**
*nick.lewis@btinternet.com* Tel: 01792 402035 Meets 7.30pm 1st & 3rd Thursdays in the month Committee Room Clyne Golf Club 118-120 Owls Lodge Lane Swansea SA3 5DP *www.radioclubs.net/swanseaars*

## Repeater Groups

**Tenby Radio Repeater Group,**
Contact: John Rees, GW0JRF Tel: 01834 842238 Email: *gw0jrf@yahoo.co.uk*

## Emergency Comms Groups

**West Glamorgan Raynet**
Mr Mike Rowles (Secretary) GW4WWN, 01639 639745, *gw4wwn@yahoo.com*

## Contest Groups

**Pembrokeshire Contest Group, GW2OP**
Sunray Pendine CARMARTHEN, Dyfed SA33 4PD *g3xjqc@btinternet.com* 01994 453495

## Region 7 Examinations

| Club | Place | Exam Secretary | | Telephone | Email |
|---|---|---|---|---|---|
| Aberdare DARS | Aberdare | Dean Willis | 2W0XTP | 01685 883706 | *dean@deanw106.plus.com* |
| Aberystwyth & DARS | Aberystwyth | Chris Davies | GW7HAE | 01970 611403 | *gw7hae@googlemail.com* |
| Aberystwyth & DARS | Aberystwyth | Daniel Pugh | MW0ZXY | 07967 482173 | *rsgb@daniel*-pugh.co.uk |
| Barry ARS | Vale Of Glamorgan | Steven Trahearn | MW0VRQ | 01446 701061 | *steven@trahearn.com* |
| Blackwood & DARS | Oakdale | Lewis Wynn Wright | GW8UAM | 02920 889156 | *wynnwright@aol.com* |
| Carmarthen ARS | Cwmduad | Allan Jones | GW4VPX | 01559 395485 | *gw4vpx@gmail.com* |
| Carmarthen ARS | Carmarthen | T A Nicholas | GW4RVA | 01267 222916 | *t.nicholas@btinternet.com* |
| Chepstow & DARS | Chepstow | Hazel Trott | MW3SZW | 1291 623735 | *examsec@gw4lwz.org.uk* |
| Chepstow & DARS | Chepstow | Hazel Trott | MW3SZW | 01291 623735 | *htrott@rocketmail.com* |

| | | | | | |
|---|---|---|---|---|---|
| Cleddau ARS | Pembroke Dock | Heinz Holland | MW0ECY | 01834 843189 | *osnok@talk21.com* |
| Cwmbran ARS | Cwmbran | Ken Smith | MW0YAC | 01633 876734 | *ken@mc0yad.co.uk* |
| Ebbw Vale Scouts | Ebbw Vale | Hydren Harrison | GW1IOT | 01495 352100 | *hydren@harr1961.fsnet.co.uk* |
| Highfields ARC | Llanedeyrn | David Mead | MW0MWL | 01443 491864 | *mw0mwl@yahoo.co.uk* |
| Highfields ARC | Cardiff | Rob Lanen | | 029 20625314 | rob-*jo@talktalk.net* |
| Hoover(Merthyr) ARC | Merthyr Tydfil | Aeronwen Sneddon | 2W0MJA | 01685 350594 | *mw6mws@googlemail.com* |
| Llanelli ARS | Merthyr Tydfil | Brian Davies | MW0DVB | 01554 758275 | *briandavies2008@btinternet.com* |
| Risca and DARS | Risca | C Jenkins | GW6JPC | 01495 309954 | *clivejenkins@talktalk.net* |
| Newport ARS | Newport | Paul Nicholls | GW7RIB | 01633 896326 | *paul.nicholls@ipo.gov.uk* |
| No1 Welsh Wing ARS | Newport | Alison Stubbs | | | stubbs_*chris@hotmail.com* |
| Pembrokeshire RS | Haverfordwest | E Hollowell | GW0GUY | 01437 760026 | *phoenix.johnwatts@btinternet.com* |
| Pencoed ARC | Pencoed | Madeline Roberts | MW0RMC | 01639 885126 | *mcr101@yahoo.com* |
| Rhondda ARS | Clydach Vale | Paul Terrell | 2W0KSA | 01443 777231 | *p.terrell@btconnect.com* |
| Risca and DARS | Newport | C Jenkins | GW6JPC | 01495 309954 | *mc0rrd@qsl.net* |
| St Tybie ARS | Llandybie | John Jensen | MW0ANX | 01269 850511 | |
| Powys ARC | Llandrindod Wells | Evan Jones | GW7UNV | 01547 550633 | *gw7unv@aol.com* |
| Neath & District Sea Cadets | | Philip Denyer | | 01656 713284 | *philip.denyer@sky.com* |
| South Glam Raynet Group | | Stephen Williams | GW6CUR | 02920 634613 | *stephen.williams6@ntlworld.com* |
| Aberkenfig Amateur Radio Club | Aberkenfig | Gareth Evans | 2W0GME | 07771 123148 | *info@aberkenfigradioclub.co.uk* |
| Brecon Amateur Radio Group | Brecon | Owen Williams | MW0GMH | 01874 624432 | *owenwilliamsft950@gmail.com* |
| Brecon & Radnor ARS | Brecon | David Bowen | MW0UAA | 01874 610031 | *mw0uaa@yahoo.com* |
| Swansea & District ARC | Llanamlet | Mark Pilot | GW1 DTA | 01792 537838 | *mark.pilot@ntlworld.com* |

# Region 8 - Northern Ireland

**(http://rsgb.org/region8)**

## RM

Philip Hosey, MI0MSO. Tel: 078 4902 5760 Email: *rm8@rsgb.org.uk*

## DRMs

**District 81** (Co Antrim)
William Campbell, MI0WJC Tel: 07712115791 Email: *drm81@rsgb.org.uk*

**District 82** (Co Down)
Bobby Wadey, MI0RYL Tel: 07841 621816 Email: *drm82@rsgb.org.uk*

**District 83** (Co Londonderry, Co Tyrone)
Trevor Campbell, MI5TCC Tel: 07710 468835 Email: *drm83@rsgb.org.uk*

**District 84** (Belfast)
Sara McGarvey, 2I0SSW Tel: 07595 353166 Email: *drm84@rsgb.org.uk*

**District 85** (Co Fermanagh, Co Armagh)
David Parkinson, 2I0SJV Tel: 07967 043982 Email: *drm85@rsgb.org.uk*

## Clubs & Societies

**Antrim & DARC**
Steven Nash, MI0WWF *stn40@btinternet.com* Meets on the 2nd Friday of each month at 7.30pm at Greystone Community Centre, 30 Ballycraigy Road, Antrim BT41 1PW *www.radioclubs.net/adars*

**Ballymena ARC**
Tom Herbison MI0IOU (chairman) 078162 02474 *hkernohan@aol.com* Meets Thursdays night at 70 Nursery Road Gracehill Ballymena Co. Antrim BT42 2QA *http://gi3fff.synthasite.com*

**Bangor & DARS**
Richard White GI4DOH, *gi4doh@ramfihaz.co.uk* Meets 7.30pm to 9.30pm 1st Thursday of the month at the Groomsport Boat House Harbour Car Park Groomsport BT19 6JP *www.bdars.com*

**Belfast RSGB Group**
David Gillespie MI0FBI, *gwocni@hotmail.com* Meets 8.00pm on 3rd Wednesdays of each month September to June at the Maple Leaf Club Park Avenue Standtown Belfast BT4 1PU

**Bushvalley Amateur Radio Club**
Sam Quigg (Secretary) *bovally@btopenworld.com* Meets every 8pm 3rd Thursday of each month at the United Services Club 8, Roe Mill Road Limavady Co. Londonderry BT49 9DF *www.bushvalley-arc.co.uk*

**Carrickfergus ARG**
Mr John Branagh, *GI3YRL. carrickfergusarg@outlook.c* Meets every Tuesday at the Downshire Community School Downshire Road Carrickfergus BT38 7DA *www.radioclubs.net/carg*

**Causeway Coast & Glens ARC**
John Hutton, MI0LJM, *mi0ljm@hotmail.com* Meets every second Wednesday of each month at 7pm at Portballintrae Villiage Hall Beach Road Portballintrae BT57 8RT

**Causeway R C**
Neil Raymond Bolt, MI0RUC. *mi0ruc@btinternet.com* Meets on the 1st Monday of the month at 7pm at Ballysally Youth and Community Centre Ballysally Road Coleraine Co Londonderry BT52 2QA *www.causewayradioclub1.piczo.com*

**City of Belfast Radio Amateur Society**
Frank Hunter GI4NKB *fthunter@virginmedia.com* Meets first Monday of each month at 8pm at Shorts Social Club Holywood Road Belfast

**Greenisland Electronics ARS**
Kenneth McInnes, GI6KDN *bigmac.747@gmail.com* Meets Mondays at 8pm in Mossley Hockey Club 1 The Glade Newtownabbey BT36 5NN *www.gearsradioclub.org.uk*

**Grey Point Fort ARS**
Stephen McFarland GI4RNP *greypointfort@hotmail.co.uk* Meets every Sunday at 10am at Grey Point Fort, Fort Road Helens Bay Crawfordsburn N Ireland BT19 1LD http: *www.greypointfort.magix.net/public*

**Lagan Valley ARS**
Mr Andrew Mulholland, *MI0BPB.gi4gty@hotmail.com* Meets every Wednesday 8pm at The Society Shack Ballynahinch Road Lisburn Co Antrim BT27 5LX *www.qsl.net/gi4gty*

**Lough Erne ARC**
Michael Clarke, MI5MTC email: *michaelclarke719@btinternet.com* Meets at the SHARE Centre. Smith's Strand Lisnaskea Co.Fermanagh BT92 0EQ *www.learc.eu*

**Marconi Radio Group**
Kevin McAuley. *mail@kevinmcauleyphotography.com* Meets on the first and third Thursday of the month at 8pm at Marconi Radio Clubroom 71a Whitepark Road Ballycastle, Co. Antrim BT54 6LP

**Mid Ulster ARC**
Mr Brian Burns MI0TGO *muarc.secretary@yahoo.co.uk* Meets Second Sunday of each month JH Turkington Ltd, James Park, Mahon Road, Portadown, BT62 3EH *www.muarc.com*

**Strabane ARS**
Mr T White, GI7THH. *terrygi7thh@qthr.fsnet.co.uk* Meets the second Thursday of each month at 20.00hrs at 3a Park Road Strabane Co Tyrone Northern Ireland BT82 8EL

**The Foyle & District ARC**
Gerard McLaughlin EI9JU Meets at 8.00pm on 1st Monday of the month exception of December & January at 159 Victoria Road, Bready, Co Tyrone BT82 0DZ *www.fadarc.co.uk*

**West Tyrone ARC**
Philip Hosey MI0MSO *info@watrc.org.uk* Meets every 2nd Wednesday of each month 20.00 - 22.00 The Basement Omagh Community House 2 Drumragh Ave Omagh BT78 1DP *www.wtarc.org.uk*

## Repeater Groups

**Mr Gordon J Bannister, GI8SKR,**
02890 660158, *raynetni@ntlworld.com*
*www.raynetni.com*

## Emergency Comms Groups

**Raynet Northern Ireland**
Mr Gordon J Bannister, GI8SKR,
02890 660158, *raynetni@ntlworld.com*
*www.raynetni.com*

## Contest Groups

**Orchard County DX Club, MN0OCG**
10 Woodview Park Tandragee
CRAIGAVON, County Armagh BT62 2DD
*www.mn0ocg.co.uk orchardcountydx@googlemail.co.uk* 02838 849433

## Region 8 Examinations

| Club | Place | Exam Secretary | | Telephone | Email |
|---|---|---|---|---|---|
| Antrim & District ARS | Antrim | Robert Robinson | MI0GDO | 028 9446 2395 | *robertrobinson2@btinternet.com* |
| Bangor & District | Donaghadee | David Best | MI0OBC | 02891 468034 | *mi0obc@btopenworld.com* |
| Bushvalley ARC | Limavady | John McNerlin | GI4EBC | 02877 707029 | *john.mcnerlin@btinternet.com* |
| Carrickfergus ARG | Carickfergus | John Branagh | GI3YRL | 02893 367208 | |
| Castlerock ARS | Castlerock | Ernest Kyle | 2I0BKI | 02870 343001 | *ernestkyle@btinternet.com* |
| Causeway RC | Coleraine | Neil R Bolt | MI0RUC | 028 207 30668 | *mi0ruc@btinternet.com* |
| Church Island ARG | Bellaghy | John McVeigh | MI0MIO | 07719 100 595 | *McVeighJohn3@aol.com* |
| Foyle and District ARC | Strabane | Trevor Campbell | MI5TCC | 02871 345 405 | *trevsoup@aol.com* |
| Greenisland Electronics ARS | Carrickfergus | Charlie Morrison | GI4FUE | 02893 351903 | *charlie@gi4fue* |
| Lough Erne ARC | Lisnaskea | Herbie Graham | GI6JPO | 02866 387761 | *herbie.graham@virgin.net* |
| Marconi Radio Group | Bushmills | Robert McCann | MI1WWG | 02820 741169 | *robertmccann4@btinternet.com* |
| Mid Ulster ARC | Banbridge | Noel Jenkinson | 2I0SFA | 02838 832602 | *noodle2020@gmail.com* |
| Newry High School ARC | Dungannon | D Gibson | GI3OQR | 02887 723508 | |
| Strabane ARS | Strabane | Terry White | GI7THH | 02871 883461 | *terrygi7thh@qthr.fsnet.co.uk* |
| The Foyle & DARS | Strabane | Trevor Campbell | GI1XGA | 02871 345405 | |
| West Tyrone ARC | Omagh | John Martin | 2I0OMA | 02882 247330 | *john.omagh@btinternet.com* |
| Ballymena ARC | Cullybackey | Jim Kelso | MI6WJK | 02825 643965 | *jim@jkelso.co.uk* |
| Enniskillen ARS | Co Fermanagh | G McLernan | MI3GHW | 02866 323817 | *earsni@btopenworld.com* |
| Queen's University Belfast | Belfast | David Linton | GI0ITJ | 02890 971761 | *amatuerradio@qub.ac.uk* |
| Causeway & Glens ARC | | Stephen Morrow | MI3ULK | 02870 326774 | *stephen769@talktalk.net* |
| Orchard County DX Club | | Noeleen McCann | MI6NVM | 07547 749945 | *noe1een@yahoo.com* |

# Region 9 - London & Thames Valley

**(http://rsgb.org/region9)**

## RM

Tom O'Reilly, G0NSY Tel: 07951 000630
Email: *rm9@rsgb.org.uk*

## DRMs

**District 91** (Hertfordshire, North & East London)
Vacant Email: *drm91@rsgb.org.uk*
**District 92** (Berkshire)
Alison Johnson, G8ROG Tel: 0118 9545368
Email: *drm92@rsgb.org.uk*
**District 93** (Oxfordshire)
Malcolm Andrew, G8NRP Tel: 01235 524844 Email: *drm93@rsgb.org.uk*
**District 94** (Bedfordshire, Buckinghamshire & Stevenage)
Stephen Richardson, M0SLP Tel: 07710 904653 Email: *drm94@rsgb.org.uk*
**District 95** (West London)
Garo Molozian, G0PZA Tel: 07765 657 542
Email: *drm95@rsgb.org.uk*

## Clubs & Societies

**Aylesbury Vale Repeater Group**
Mr Mike Marsden, G8BQH Meets in March, June, July and December at the Robin Hood on the A422 Buckingham to Brackley Rd. (AGM at Stone Village Hall, nr Aylesbury) MK18 5DN

**Aylesbury Vale RS**
Mr V Gerhardi, G6GDI, *avrs@rakewell.com* Meets every 2nd Wednesday of the month at 20.00 - 22.00 The Dog House Inn Broughton Crossing Broughton Aylesbury HP22 5AR *www.avrs.org.uk*

**Banbury Amateur Radio Society**
John Burrell, G8OZH email: *BARS@g8ozh.com* Meets 1st & 3rd Wednesday of each month at 7pm - 9pm at 169 Bloxham Rd, Banbury, Oxfordshire, OX16 9JU and 2nd and 4th Wednesdays on-the-air on 2 metres FM - 144.515MHz
*www.banburyarc.org*

**Bedford & DARC**
Bob Leask, G3XNG . *g4ceo@talk21.com* Meets every Tuesday at 7.30 pm at The Shack Plantation Cottage Ravensden Bedfordshire MK44 2RL *www.badarc.net*

**Bracknell ARC**
Andy Cunningham, M0HAK *andyc@cunningham.me.uk* Meets 8.00pm on 2nd Wednesday in the month at the Bracknell Methodist Church Shepards Lane Bracknell Berkshire RG42 2BT *www.g4bra.org.uk*

**Burnham Beeches RC**
Dave Chislett G4XDU, *bbrclub@btinternet.com*, Meets 8.00pm on 1st and 3rd Mondays in the month at the Farnham Common Village Hall Victoria Road Farnham Common Bucks SL2 3QG
*http://come.to/bbrc*

**Chesham & DARS**
Mr T J Thirlwell G0VFW *cdars@g3mdg.org.uk* Meets 1st and 3rd Wednesdays of the month 20.15 - 22.00 Mezzanine Room M3 White Hill Community Centre White Hill, Chesham Bucks HP5 1AG
*www.radioclubs.net/c&dars*

**Dacorum ARTS**
Tony Mitchell, G0TPK *g0tpk@hotmail.com* Meets third Tuesday every month except August 20.00 to 22.00 Gadebridge Community Centre Galley Hill Queensway Hemel Hempstead HP1 3LG
*www.darts73.co.uk*

**Drowned Rats Radio Group**
David Dix G8LZE *info@drownedrats.uk* Meets second Wednesday of the month 20.00pm - 22.30pm The Coy Carp Copperhill Lane Harefield Middlesex UB9 6HZ

**Dunstable Downs Radio Club**
Mike Scarlett G4CAK *mikescarl@btinternet.com* Meets every friday at 8pm at Chews House, 77 Hight Street South Dunstable Bedfordshire LU6 3SF

**Edgware & DRS**
Mike, G4RNW, *michael.stewart5@ntlworld.com* Meets 8.00pm on 2nd and 4th Thursdays in the month at the Watling Community Centre 145 Orange Hill Road Burnt Oak Edgware, Middlesex HA8 0TR
*www.g3asr.co.uk*

**Harwell ARS**
Malcolm Andrew, G8NRP *info@g3pia.org.uk* Meets at 19.45 for 20.00hrs on the 2nd Thurday of each month Chilton Village Hall Church Hill Chilton OX11 0SH *www.g3pia.org.uk*

**Home Counties Atv Grp**
Mr Thomas Grady, *info@gb3hv.com* Meets on 4th Tuesday in the month at the The Binfield Club Forest Road, Binfield RG42 4HP
*www.gb3hv.com*

**London Hackspace**
Paul Dart M0OKE *pauldart@gmail.com* Meets first Saturday of the month Open Evenings: Every Tuesday from 19:00pm at

447 Hackney Road London E2 9DY *http://london.hackspace.org.uk*

**Maidenhead & DARC**
Peter Hicks, G4KCX *G4KCX@talktalk.net* Meets 7.45 on 1st Thursday and 3rd Tuesday in the month at the The Friends Meeting House 14. West Street Maidenhead Berkshire SL6 1RL

**Mid Thames DFC**
Mrs Doreen Pechey, G8NMO Meets monthly in the winter at various pubs, currently Dashwood Arms Piddington Bucks HP14 3JG *www.topbanddf.org.uk*

**Milton Keynes ARS**
Dave Barlow Meets first and third Monday of each month 19.3 TS Invincible Bletchley Park Milton Keynes MK3 6EB *www.mkars.org.uk*

**Newbury & DARS**
*phill.morris@live.co.uk* Meets 19.30hrs on 4th Wednesday of the month in the Meeting room, The Travellers Friend pub, Crookham Common, Thatcham RG19 8EA *www.nadars.org.uk*

**Oxford & DARS**
Graham Diacon G8EWT, *radiog8ewt@gmail.com* Meets 7.30 for 8.00pm on the 1st and 3rd Tuesday of every month at The Gladiator Club 263 Iffley Road Oxford OX4 1SJ *www.odars.org*

**Reading & DARC**
Min Standen,G0JM, *G0JMS@radarc.org* Meets at 8pm on 2nd and 4th Thursdays Woodford Park Leisure Centre Haddon Drive Woodley, Reading RG5 4LW *www.radarc.org*

**RNARS London**
J Faxholm, 0208 5811 484 email *info@gb2rn.org.uk* Meets last Thursday of every alternate month at 15.00hrs on HMS Belfast Battleship Lane Tooley Street London SE1 2JH *www.gb2rn.org.uk*

**Radio Society of Harrow**
Linda, G7RJL Meets 1st and 3rd Fridays of the month at 20.00 at the Harrow District Masonic Centre Northwick Circle Kenton HA2 8AL *www.g3efx.org.uk*

**Shefford & DARS**
John Burnett 2E0OAK, *www.sadars.co.uk* Meets at 8pm Thursdays (with summer break) at the Community Hall Ampthill Road Shefford Beds SG17 5AX *www.sadars.co.uk*

**Southgate ARC**
Mr D F Berry, G4DFB Tel: 02083603614 Meets at 7.30pm for 8.00pm on the 2nd Wednesday of each month at the Hazelwood Lawn Tennis and Squash Club Ridge Avenue Winchmore Hill, London N21 2AJ *www.southgatearc.org*

**Stevenage & DARS**
Mr Rob McTait G2BKZ, *rob_g2bkz@talktalk.net* Meets 7.30pm on Tuesdays at the Stevenage Resource Centre Chells Way Stevenage Herts SG2 0LT *www.sadars.org*

**Triple B Amateur Radio Contest Group**
M Goodey, G0GJV - Tel: 07770938478 Meets Fridays 10pm Jack O'Mewbury Terrace Road South Binfield RG42 5PH

**Verulam ARC (St Albans) G3VER, G8VER**
Ralph Nash 01923 265572 Meets 3rd Tuesday of each month at 7.45pm at: Aboyne Lodge School St Albans, Herts AL3 5NL *www.radioclubs.net/verulam*

**Whitton ARG**
Mr Ian Clabon G0OFN 020 8894 9131 Meets at 7.30pm every Friday at the Whitton Community Centre Percy Road Whitton TW2 6JL *www.warg.info*

## Repeater Groups

**Guildford UHF Repeater Group**, GB3GF. Contact: Alex Morris, G6ZPR. Tel: 01483 892348. Website: gb3gf.co.uk

## Emergency Comms Groups

**Mid-Thames Raynet**
Matt Wimpenny-Smith, M1BTF, 07921 260182, *midthamesraynet@yahoo.co.uk* Meets on the 1st Monday in the month (except bank holidays). *www.webspawner.com/users/gerrypgz/*

**North Hertfordshire Raynet Association**
Mr S Donald, G6EDD, 01763 242876, *stuartdonald2010@yahoo.co.uk*

## Contest Groups

**Beds, Mid Contest A, G4MBC**
c/o Sandholm Bridge End Road Red Lodge Bury St Edmunds, Suffolk IP28 8LQ *g4mbc@homeshack.freeserve.co.uk* 07899 064100

## Region 9 Examinations

| Club | Place | Exam Secretary | | Telephone | Email |
|---|---|---|---|---|---|
| Bracknell ARC | Bracknell | David Ferrington | M0XDF | 01344 483922 | m0xdf@alphadene.co.uk |
| Burnham Beeches Radio Club | Farnham Common | Gregory Head | G4EBY | 01494 630525 | gregory.head@ntlworld.com |
| Chesham & DARS | Tring | Ian slaney | G0RTF | 01494 725842 | g0rtf@katzenrolli.eu |
| Chesham & DARS | Chesham | Jeremy Browne | G3XZG | 01494 782244 | jeremy@bpssolicitors.co.uk |
| Cray Valley RS | Eltham | F Parradine | G0FDP | 020 8850 5224 | francisparradine@btinternet.com |
| Eton College ARS | Windsor | Michael Wilcockson | G7KYI | 01753 671304 | m.wilcockson@etoncollege.org.uk |
| Hackspace Foundation | London | Samuel Keating-Fry | 2E0SKF | 07756 677992 | sam@keat.in |
| Milton Keynes ARS | Milton Keynes | Graham Parry | G7OSR | 01908 679692 | graham7osr@btinternet.com |
| Milton Keynes ARS | Milton Keynes | Andrew Thomas | G8GNI | 01908 263 758 | shoptilyoudrop@homecall.co.uk |
| Newbury & DARS | Newtown | Steve Elliott | M0SEL | 07707 105558 | se293el@gmail.com |
| Oldfield School ARC | Hampton | A G Fisher | G4VBH | 0208 5720465 | g4vbh@blueyonder.co.uk |
| Harwell ARS | Didcot | Malcolm Andrew | G8NRP | 01235 524844 | info@g3pia.org.uk |
| Radio Scouting Team | Cuffley | Melanie Cross | M1EJQ | 01992 635393 | |
| Radio Society of Harrow | Hatch End | Michael Bruce | M0ITI | 01923 720253 | m0iti@orion.org.uk |
| Reading & DARC | Reading | Harry Hogg | G3NGX | 01491 872919 | g3ngx@radarc.org |
| RSGB | Milton Keynes | Carol Meredith | M6MUP | | |
| Shefford DARS | Sandy | Stuart Baker | G3RXQ | | baker@nildram.co.uk |
| Silverthorn Radio Club | London | Tom Dawson | M5AJK | 0208 967 7621 | m5ajk@blueyonder.co.uk |
| Southgate ARC | Enfield | David Sharp | M0XDS | 01992 422622 | david.sharp1@tesco.net |
| Stevenage & DARS | Stevenage | R McTait | G2BKZ | 01438 489045 | rob_g2bkz@talktalk.net |
| Verulam ARC | St. Albans | Ralph Nash | G1BSZ | 01923 265572 | g1bsz@aol.com |
| Verulam ARC | St. Albans | Peter Baker | G4HSO | | g4hso@btinternet.com |
| Whitton ARG | Twickenham | Ian Clabon | G0OFN | 0208 5728615 | |
| Harpenden & Wheathampstead District Scouts | Hitchin | Mark Hubbard | M6MDH | 01438 833386 | sl@skimptonscouts.org.uk |
| Wokingham Air Cadets | Wokingham | Kathleen Lambert | | 07968 447553 | kfalambert@btinternet.com |
| UK High Altitude Society | | Anthony Stirk | M0UPU | 01274 550910 | anthony@nevis.co.uk |
| Sea Cadets - London North District | Chingford | Marc Litchman | G0TOC | 0208 5021645 | g0toc@hotmail.com |
| Milton Keynes Raynet | | Christopher Sayles | 2E0CVV | 01908 691689 | christopher.sayles@btinternet.com |
| Royal Hospital Chelsea ARS | | Ray Petrie | G0SLL | 07762 195679 | rayg0sll@chelsea-pensioners.co.uk |
| Elstree ARS | Elstree | Ray Snow | G0BSP | 0208 9547044 | raysnow@aol.com |
| Ystrad Mynach College | | Rhys Boulton | MW6YHR | 01443 816888 | rhys.boulton@cymoedd.ac.uk |
| Halton Amateur Radio Society | Aylesbury | William David Green | M5AGW | 02083 668680 | dave.g8bcq@virgin.net |
| Gilwell Park Scouts | Chingford | Steve Bunting | M0BPQ | 02082 812322 | steve@m0bpq.com |

# Region 10 - South & South East

**(http://rsgb.org/region10)**

## RM

Mick Senior, G4EFO. Tel: 07956 107490 Email: *rm10@rsgb.org.uk*

## DRMs

**District 101** (Surrey, London South of the Thames)
Graham Smith, G4NMD Tel: 07971 326589
Email: *drm101@rsgb.org.uk*

**District 102** (Wiltshire)
Vacant Email: *drm102@rsgb.org.uk*

**District 103** (West Sussex)
Adrian Boyd, G4LRP Tel: 07714 664957
Email: *drm103@rsgb.org.uk*

**District 104** (Hampshire)
Paul Steed, G0VEP Tel: 02392 371677
Email: *drm104@rsgb.org.uk*

**District 105** (Isle of Wight)
Chris Brown G0DWE Tel: 01983 529060
Email: *drm105@rsgb.org.uk*

**District 106** (Kent)
Keith Bird, G4JED Tel: 01732 446331 Email: *drm106@rsgb.org.uk*

**District 107** (East Sussex)
Daniel Adkin, M0HOW Tel: 01424 882008
Email: *drm107@rsgb.org.uk*

## Clubs & Societies

**Basingstoke ARC**
Peter, G0KQA *barcsecretary@yahoo.co.uk* Meets 3rd Monday of each month at 7.30 to 8pm start at Mays Bounty Cricket and Sports Club Fairfield Road Basingstoke RG21 3DR *www.basingstokearc.co.uk*

**Brede Steam ARS**
Dan Adkin M0HOW, *Daniel@adkin.net* Meets on the 1st Saturday of the month and every Tuesday 4 Westminster Crescent Hastings East Sussex TN34 2AW *www.bsars.co.uk*

**Bredhurst Rx & Tx Soc**
Kevin Richardson G0PEK *secretary@brats-qth.org* Meets 8.30pm Thursdays at The Parkwood Community Centre Long Catlis Road Rainham Gillingham, Kent ME8 9PN *www.brats*-qth.org

**Brickfields ARS**
Doug Fenna, M0DSF *GX0BAR@yahoo.com* Meets Mondays & Thursdays at 2pm Newchurch Isle of Wight Note: Access to the shack is not easy Contact the secretary to arrange attendance PO36 0NL *www.b-a-r-s.org.uk*

**Bromley & DARS**
Andy Brooker *enquiries@bdars.co.uk* Meets 3rd Tuesday in the month 19:45 for 20.00 Victory Social Club Kechill Gardens Hayes Bromley BR2 7NG *www.bdars.org*

**Caterham RG**
Mr P N Lewis, G4APL. *catrad@skywaves.demon.co.uk* Meets on alternate Fridays evenings at fellow member's houses *www.theskywaves.net*

**Chippenham & DARC**
Dimitris Vainas: email: *hobgoblin24@hotmail.com* Meets every Tuesday from 7pm at the Sea Cadet HQ Long Close Chippenham SN15 3JY *www.g3vro.org.uk*

**Coulsdon Amateur Trans. Soc**
Mike Buckley M1CCF - *MikeB@vmars.co.uk* Meets 8.15pm on 2nd Mon in the month & 4th Mon in month at 8pm Banstead Scout Hut St Swithuns Church Hall Grovelands Road Purley Surrey CR8 4YN *www.catsradio.org*

**Crawley Amateur Radio Club**
Contact: Phil Moore M0TZZ *secretary@carc.org.uk* Meets 8.00pm Wednesdays and 11.00 am Sundays at the Tilgate Forest Rec. Centre Hut 18, Tilgate Forest Crawley, West Sussex RH11 9BQ *www.carc.org.uk*

**Cray Valley RS**
Richard Cains G7GLW *secretary@cvrs.org* Meets 1st & 3rd thursday of the month at 20.00 - 22.00 at the 1st Royal Eltham Scouts HQ Rear of 61 - 71 Southend Crescent Eltham London SE9 2SD *www.cvrs.org*

**Crystal Palace Radio Elec Club**
Mr Bob Burns, G3OOU. *g3oou@aol.com* Meets 7.30pm on 1st Friday each month at the All Saints Church Parish Rooms Beulah Hill London SE19 2QQ *www.g3oou.co.uk*

**Darenth Valley RS**
Mike Wallace G8AXA *g8axa@yahoo.co.uk* Meets 2nd & 4th Wenesday of the month 20.00 - 22.30 Crockenhill Village Hall Stones Cross Road Crockenhill Swanley BR8 8LS *www.g0kdv.com*

**Dorking & DRS**
George Brind G4CMU, *ddrs.secretary@yahoo.co.uk* Meets 4th Tuesday of each Month at 19.45 pm at The Friends Meeting House, Butter Hill, South St, Dorking, Surrey RH4 2LE *www.ddrs.org.uk*

**Dover Radio Club**
Matt Curtis M1CMN *m1cmn@btinternet.com* Meets 7.00pm Wednesdays at The Old Park Community Centre Gordon Road Whitfield Kent CT15 6BL *www.darc.org.uk*

**East Kent Radio Society**
Karl Davies M1DFM *admin@ekrs.co.uk* Meets 8.00pm on the 1st and 3rd Mondays in the month at The Herne Mill Mill Lane Herne Bay Kent CT6 7DR *www.ekrs.co.uk*

**Eastbourne Radio & Electronics Club**
Dave Williams, G8PUO, *dave.williams@ditronix.com* Meets 2nd & 4th Monday of each month 19.30 p.m. Eastbourne Sports Park Cross Levels Way Eastbourne West Sussex BN21 2UF *www.ereclub.org.uk*

**Echelford ARS**
Philip Fauchon, G0JSP, email: *g0jsp@hotmail.com* Meets 7pm with active CW practice on 2nd & 4th Thursdays in the month at Weybridge Vandals Rugby Club Desborough Island Walton on Thames KT12 1QP *http://beam.to/ears*

**Fareham & DARC**
Chris Jenkins-Powell G7MFR *chris@jenkins-powell.com* Meets 8pm on Wednesdays at the Fareham Motorboat & Sailing Club Lower Quay Fareham Hants PO16 0RA *www.fareham-darc.co.uk*

**Farnborough & DRS**
Derek G3OFA, *mail@farnboroughradio.org.uk* Meets every 2nd and 4th Wednesday of the month at 19.30 pm The Community Centre Meudon Avenue Farnborough Hants GU14 7LE *www.farnboroughradio.org.uk*

**Fort Purbrook ARC**
Graham Yoxall, M0CYX, *m0cyx@fparc.org.uk* Meets last Friday of each month (except Dec) at 19.00 - 21.00 pm Fort Purbrook, Meeting Room FP-1 OFF Portsdown Hill Road Cosham Portsmouth PO6 1BJ *www.fparc.org.uk*

**Guildford & DRS**
Timothy Dabbs, G7JYQ *sec.gdrs@virginmedia.com* Meets 7.30pm for 8.00pm, 2nd, 4th & 5th Fridays in the month at the Guildford Model Engineers HQ Stoke Park Guildford Surrey GU1 1TU *www.gdrs.net*

**Hastings Elec & RC**
Gordon Sweet, *gordonsweet2000@yahoo.co.uk* Meets 19.00pm 4th Wednesday each month The John Taplin Centre St Leonards-on-Sea East Sussex TN38 0LQ *www.herc-hastings.org.uk*

**Hilderstone Radio and Electronics Club**
Ian Warnecke 2E0DUE *hilderstoneclub@gmail.com* Meets 2nd and 4th Thursday of the month (not August) at 7:00pm Crampton Tower Museum The Broadway Broadstairs Kent CT10 2AB *http://g0hrs.org*

**Hog's Back ARC**
Frank Heritage M0AEU *enquiries@hogsback*-arc.org.uk Meets 2nd and 4th Mondays of the month (excluding Bank holidays), Scout Centre, Pankridge Street, Crondall, Farnham, Surrey GU10 5RQ *www.hogsback-arc.org.uk*

**Horndean & DARC**
Mr Stuart Swain, *G0FYX@msn.com* Meets 1st & 3rd Thursday in the month at 19.30 Anders Hall off Milton Road Waterlooville Hants PO7 6AW *www.hdarc.co.uk*

**Horsham ARC**
Mr Alister Watt, G3ZBU. *info@harc.org.uk* Meets 8pm 1st Thur in month at the Guide Hall Denne Road Horsham West Sussex RH12 1JF *www.harc.org.uk*

**Isle of Wight RS**
Alan Ash, G3PZB. *alan.ash@talktalk.net* Meets 7.00pm Fridays at Haylands Farm Salters Road Haylands, Ryde Isle of Wight PO33 3HU *www.g3sky.org.uk*

**Itchen Valley ARC**
Mr Quentin Gee, M1ENU, *gg2@ecs.soton.ac.uk* Meets 2nd and 4th Fridays of the month at the Bianchi Suite Otterbourne Village Hall Cranbourne Drive Otterbourne SO21 2ET *www.ivarc.org.uk*

**Kent Weald Radio Club**
Mr P J Blunt, G0UXG *palybl@btinternet.com* Meets on the last Saturday of each month at 19.00 hrs at Headcorn Airfield Briefing Room Headcorn Ashford, Kent TN27 9HX *www.kentwealdradioclub.orangehome.co.uk*

**Maidstone YMCA Amateur Radio Society**
Trevor Collins G6ALJ *trev@dr.com*, *g3trf@mail.com* Meets 7.30pm every Monday evening (except Bank Holidays) at Maidstone YMCA ME15 9AE *www.g3trf.weebly.com*

**Medway ARTS**
Keith Anderson, M0KJA, email *keith@m0kja.co.uk* Tel: 07799 791014 Meets 7.30pm Fridays at Tunbury Hall Catkin Close Tunbury Avenue Walderslade, Chatham ME5 9HP *www.g5mw.org.uk*

**New Forest Amateur Radio Society**
Richard Ferguson, M0RBF *newforestradio@yahoo.co.uk* Meets monthly (venue varies) 19.30 - 19.45 Lymington Community Centre The Filly Inn, Brockenhurst, SO42 7UF Various other New Forest Locations

**Newhaven Fort ARG**
Mike Jones G6GOS, *mike1943@btconnect.com* Meets daily at 10 am at Newhaven Fort Fort Road Newhaven East Sussex BN9 9DS

**North Kent RS**
Mr Dave Collings, G4YIB, *nkrecc@crystaldave.com* Meets 8.00pm on 1st and 3rd Tuesday in the month at The Pop-in-Parlour Graham Road Bexleyheath Kent DA6 7EG

**Reigate ATS**
Mr T Trew, G8JXV Email: *rats@qsl.net* Meets 8.00pm 3rd Tuesday in month in the Conference Room of the RNIB College, Philanthropic Road Redhill Surrey RH1 4DG *www.qsl.net/rats*

**Southampton ARC**
Malcolm Troy, G0WFQ, *malcolm.troy@virgin.net* Meets at 7.30pm, Third Wednesday of each month in Room 1, 1st Floor, Cantell Maths and Computing College, Violet Road, Bassett, Southampton, SO16 3GT *www.southampton-arc.org.uk*

**Southampton University Wireless Society**
Mr P Cromp M0DNY *contact@suws.org.uk* Meets every Thurs at 6pm at Building 59 University Road Southampton SO17 1BJ *www.suws.org.uk*

**Southdown ARS**
Andy Holden, M6GND *committee@sars.club* Meets first Monday of each month WRVS Russell Centre 24 Hyde Road East Sussex BN21 4SX *www.southdown-amateur-radio-society.org.uk*

**Surbiton Heritage AR & Elec Society**
Tony M0SHA email: *info@m0sha.com* Meets 7.45 pm on 1st Wednesday of the month at the The Coffee Bar 1st Floor, Surbiton Hill Methodist Church 39 Ewell Road, Surbiton KT6 6AF *www.m0sha.com*

**University of Surrey E & ARS**
Mr Bob Doran, G4VRC, *ears@surrey.ac.uk* Meets 1.00pm Wednesdays. Open only to University students and staff. For location see notice board outside shack in AC05 GU2 7XH *www.ussu.co.uk/ears*

**Surrey Radio Contact Club**
John Kennedy, G3MCX Meets 7.45pm - 9.45pm on 1st and 3rd Mondays in the month at the Trinity School Shirley Park Croydon CR9 7AT *www.g3src.org.uk*

**Mid Sussex ARS**
Russell Nelson, G7TMR *g7tmr@msars.org.uk* Meets every Friday 19.30 - 22.00 Cyprus Hall Cyprus Road Burgess Hill West Sussex RH15 8DX *www.msars.org.uk*

**Sutton & Cheam RS**
John Puttock, G0BWV, *info@scrs.org.uk* 020 8644 9945 Meets 8pm on 3rd Thur in the month at the Sutton United Football Club Borough Sports Ground Gander Green Lane Sutton, Surrey SM1 2PA *www.scrs.org.uk*

**Swindon & DARC**
Den, M0ACM *secretary@sdarc.net* Meets 7.00 pm every Thursday except August at the Pinetrees Centre Pinehurst Circle Swindon SN2 1RF *www.sdarc.net*

**The QRZ ARG of Sussex**
Stuart Constable, M0CHW Tel: 07719 481072 Meets The Radio Shack (Herstmonceux Megacycles) Herstmonceux Castle Wartling Road, Herstmonceux Hailsham East Sussex (Do not use the postcode to navigate) BN27 1RN

**The Wings Museum**
Barrie Bloomfield G4OKB *bloomfieldbarrie@hotmail.com* Meets first Meets every Wed of each month at 11am Wings Museum Brantridge Lane Balcombe West Sussex RH17 6JT

**Three Counties ARC**
Mr John Rivett M3RRX *jrivett@mail2web.com* Meets 19.30pm for 20.00pm on the 2nd & 4th Mondays in the month at the Crondall Scout Centre Pankridge Street Crondall nr Farnham Surrey GU10 5RQ *www.radioclubs.net/tcarc*

**Trowbridge & DARC**
Mr Ian Carter, G0GRI. *tdarc@btinternet.com* Meets 8.00pm 1st & 3rd Wednesdays in the month at the Southwick Village Hall Southwick Trowbridge Wilts BA14 9QN *www.tdarc.uk*

**Waterside (New Forest) ARS**
Mr Tim Williams, G4YVY, 02380894278 Meets 8pm on 1st & 3rd Tues of month (not Aug) at Applemore Scout HQ near Hythe Southampton SO45 4RQ *www.watersidears.org.uk*

**West Kent ARS G1WKS**
Steven Hardy, 2E0VKH *secretary@wkars.org.uk* Meets 8pm on the 2nd Monday of the month at Bidborough Village Hall Bidborough Kent TN3 0XD *www.wkars.org.uk*

**Wey Valley ARG**
Andrew Vine, M0GJH, *wvarg@btinternet.com* Meets 1st and 3rd Fridays of each month at 19:30 for 20:00, Guildford Rowing Club The Boathouse Shalford Road Guildford, Surrey GU1 3XL *www.weyvalleyarg.org.uk*

**Wimbledon & DARS**
Andrew Maish G4ADM Meets 8pm on the 2nd & last friday of the month at Martin Way Methodist Church (corner) Buckleigh Avenue Merton Park London SW20 9JZ *www.gx3wim.org.uk*

**Worthing & DARC**
Gordon Hutchinson 2E0GTG or 01903 761487 Meets 8.00pm on Wednesdays at the Lancing Parish Hall, South Street, Lancing, Sussex BN15 8AJ *www.wadarc.org.uk*

**Worthing Radio Events Group**
Mark Hillman, M0TVV *markyhillman@yahoo.co.uk* Meets Every 1st Monday of each month at 8pm at The John Selden Public House Salvington Road Worthing BN13 2HN *www.m0reg.co.uk*

## Repeater Groups

**Guildford UHF Repeater Group, GB3GF**
Contact: Mr Alex Morris, G6ZPR Tel: 01483 892348 Email: *morris.alex@btconnect.com*

**Ridgeway Repeater Group, GB3WH**
Contact: Mr R Loss, G4XUT (Secretary) Email: *G4XUT@rrg.org.uk*

## Emergency Comms Groups

**Mid-Thames Raynet**
Matt Wimpenny-Smith, M1BTF, 07921 260182, *midthamesraynet@yahoo.co.uk* Meets on the 1st Monday in the month (except bank holidays). *www.webspawner.com/users/gerrypgz/*

**North Hertfordshire Raynet Association**
Mr S Donald, G6EDD, 01763 242876, *stuartdonald2010@yahoo.co.uk*

## Contest Groups

**Three A's Contest, G0AAA**
9 Golf Close WOKING, Surrey GU22 8PE *g3wvg@btinternet.com* 0181 3973319

**Vecta Contest Group ARS, M0VCT**
33 Oakwood Road Ryde, Isle of Wight PO33 3JU *fred.wp.dawson@googlemail.com* 07885 634518

## Region 10 Examinations

| Club | Place | Exam Secretary | | Telephone | Email |
|---|---|---|---|---|---|
| Alleyn's School Radio Club | Dulwich | David Davies | G4WVK | 02087 711639 | |
| Bromley & District ARS | Bromley | A Brooker | G4WGZ | 01689 878089 | andy *g4wgz@o2.co.uk* |
| Croyden Sea Cadets | Croydon | E Tozeland | M3VRW | 01883 724877 | *oldsod@btinternet.com* |
| Dorking & DARS | Dorking | George Brind | G4CMU | 01306 631115 | *george@brind.org.uk* |
| Epsom Scouts Radio Group | Epsom | John E Kelly | G3YGG | 01372 812776 | *windflowersuk@hotmail.com* |
| Bredhurst Receiving & TS | Gillingham | Charles Darley | G4VSZ | 07982 244788 | *charles.darley@blueyonder.co.uk* |
| Bredhurst Receiving & TS | Gravesend | Michael Jury | 2E0HMJ | 01634 324112 | *mikeanddiane@blueyonder.co.uk* |
| Andover ARC | Wildhern | Andrew Milner | M0ZFX | 07525 176749 | *andymilneruk@gmail.com* |
| Banbury ARS | Banbury | Michael Gritton | M0UJU | [illegible] | [illegible] |
| Basingstoke ARC | Basingstoke | R M Kerswill | M0KEN | 01256 881214 | *baroxam@btinternet.com* |
| Basingstoke ARC | Basingstoke | Alex Salmon | M0ZER | 07905 524175 | *mrasalmon@gmail.com* |
| Brede Steam ARS | Hastings | Steve Stewart | M0SSR | 01424 720815 | *m0nuc@aol.com* |
| Bromley Adult Education Centre | Bromley | Jane Monghan | N/A | 0208 4600020 | *monaghands@bromleyadulteducation.ac.uk* |
| Hog's Back ARC | Farnham | Kevin Wood | G7BCS | 01420 563908 | *courses@hogsback*-arc.org.uk |
| Chippenham & DARC | Chippenham | Ian Carter | G0GRI | 01225 864698 | *ian.g0gri@btinternet.com* |
| Crawley ARC | Crawley | Phil Moore | M0TZZ | 01342 832154 | *secretary@carc.org.uk* |
| Fareham & District ARC | Fareham | Derek Clarkson | G4JLP | 01329 823405 | *g4jlp@lineone.net* |
| Farnborough & District RS | Farnborough | Kevin Wood | G7BCS | 01420 563908 | *kjwood@iee.org* |
| Hastings Electronics & Radio Club | St Leonards-on-Sea | Phil Parkman | G3MGQ | 01580 881028 | *phil.parkman@virgin.net* |
| Horndean and District ARC | Lovedean | Julia Tribe | G0IUY | 02392 785568 | *juliatribe@ntlworld.com* |
| Horsham ARC | Horsham | Adrian Boyd | G4LRP | 01403 733087 | *g4lrp@btinternet.com* |
| Horsham ARC | Horsham | Andrew Vine | M0GJH | 01403 325343 | *m0gjh@btinternet.com* |
| Isle of Wight Radio Society | Ryde | Alan Robinson | G0VPO | 01983 401797 | *alinco9@hotmail.com* |
| Itchen Valley Radio Club | Chandlers Ford | Sheila Williams | G0VNI | 02380 813827 | |
| Lymington Community Association RC | Lymington | Gillian Ferguson | 2E0SEW | 01425 612612 | *gill@hamandchips.com* |
| Richmond upon Thames Scouts ARC | Petersham | Marian Lonsdale | | 0208 9416325 | |
| Mid Sussex ARS | Burgess Hill | Sue Davis | G6YPY | 01273 845103 | *sue@figgerit.co.uk* |
| Bishop Waltham | Bishop's Waltham | Andy Digby | G0JLX | 07768 282880 | *instructor@asel.demon.co.uk* |
| Newhaven Fort ARG | Newhaven | Joan Parish | M0JOA | 01273 612038 | *jeparish13@gmail.com* |
| QRZ Radio Group of Sussex | Hailsham | Stuart Constable | M0CHW | 07719 481072 | *qrzargos@gmail.com* |
| Dover Radio Club | Dover | Anthony Willsher | M0VYW | 01303 220441 | *antony@keybored.plus.com* |
| Hilderstone Radio & Electronics C | Kent | Clive Widdus | G1WCR | 01843 297143 | *clivewi@btinternet.com* |
| Royal Naval ARS | Guildford | G F de Voil | G8SKK | 01483 566962 | *gerald@devoil.wanadoo.co.uk* |
| RNARS | Fareham | J Kirk | G3ZDF | 01243 536586 | *g3zdf@btinternet.com* |
| Rosemary Amateur Radio Group | Ryde | Allan Thornton | M0TIW | 01983 566027 | *m0tiw@aol.com* |
| Sandown Primary School | Hastings | S Stewart | M0SSR | 01424 720815 | *m0ssr@aol.com* |
| SHARES | Surbiton | Tony Fell | G7DGW | 0845 2698971 | *info@m0sha.com* |
| Sutton & Cheam ARS | Banstead | Neil Horton | M0ZEY | 07939 111043 | neil_horton*@ntlworld.com* |
| University of Surrey EARS | Guildford | Philip Handley | M0PVI | 07841 704111 | *wikrok@gmail.com* |
| Southdown ARS | Eastborne | Peter Martin | G6GVM | 01323 731514 | *peterg6gvm@btinternet.com* |
| Swindon & DARC | Swindon | Mike Beale | M5CBS | 01793 480358 | *beale133@btinternet.com* |
| The Portsmouth College | Portsmouth | Paul Steed | G0VEP | 02392 371677 | *paul.steed@ntlworld.com* |
| Trowbridge DARC | Trowbridge | Ian Carter | G0GRI | 01225 864698 | *ian.g0gri@btinternet.com* |
| TS Royal George ARG | Ryde | A Thornton | M0TIW | 01983 566027 | *m0tiw@aol.com* |
| Waterlooville ARC | Waterlooville | A Shillabeer | M3SHI | | *capper1234@ntlworld.com* |
| Worthing and District ARC | Lancing | Alastair Weller | M0OAL | 07966 037314 | *him@alastairweller.co.uk* |
| Worthing and District ARC | Shoreham-by-Sea | Alastair Weller | M0OAL | 07966 037314 | *training@wadarc.org.uk* |
| Worthing Radio Events Group | Worthing | Chris Thrower | | 01903 690069 | *chrissiethrower@gmail.com* |
| Southampton ARC | Southampton | Malcolm Troy | G0WFQ | | *malcolm.troy@virgin.net* |
| Crystal Palace REC | London | Alan O'Donovan | G8NKM | 02087 789660 | *alan.odonovan@btinternet.com* |
| Fort Purbrook ARC | Portsmouth | Graham Yoxall | M0CYX | 02392 297409 | *grahamyoxall@aol.com* |
| Southampton University WS | Southampton | Henry Kennedy | M0HWY | 07522 127595 | *hk1g10@ecs.soton.ac.uk* |
| West Kent ARS | Aylesford | Kevin Jeffery | M0KAO | 01892 532456 | *kjeffery150@gmail.com* |
| Sussex 4x4 Response | Angmering | David Green | G4OTV | 01892 654763 | *dave.w.green@btinternet.com* |
| Sussex 4x4 Response | Worthing | Ian MacDonald | M0IAD | 01903 261972 | *ian@ianm.net* |
| Polish Scouts Radio Club | | Wojtek Bernasinski | G0IDA | 07968 115436 | *bernig0ida@gmail.com* |
| Fareham West Scouts ARC | | Mark Elbourn | 2E0EFA | 01329 317485 | *mark.elbourn@mail.com* |
| Thanet Radio & Electronics Club | Ramsgate | Denis Kirkden | M0ZDE | 01843 836188 | *m6des@live.co.uk* |
| Eastbourne Radio & EC | Pevensey | Daniel Adkin | M0HOW | 01424 882008 | *daniel@adkin.net* |
| University Of Kent | | Fred Barnes | M1FRB | 01227 816168 | *F.R.M.Barnes@kent.ac.uk* |
| G-QRP Club | Bexley Heath | Beverley Maltby | | 02030 455176 | *beverley.maltby@tlcbexley.ac.uk* |
| Magdalen College School ARC | Oxfordshire | Andrew Baker-Munton | G0KPY | 01865 744170 | *abmunton@hotmail.com* |
| Darenth Valley Radio Society | Swanley | Michael Wallace | G0AXA | 01689 [illegible] | *g8axa@yahoo.co.uk* |

# Region 11 - South West & Channel Islands

**(http://rsgb.org/region11)**

## RM

Pam Helliwell, G7SME Tel: 01803 326033
Email: *rm11@rsgb.org.uk*

## DRMs

**District 111** (Cornwall & Plymouth)
Mike Jones, 2E0MLJ Tel: 07577 447754
Email: *drm111@rsgb.org.uk*
**District 112** (Devon)
Martin Sables, G7NTY Tel: 01769 581349
Email: *drm112@rsgb.org.uk*
**District 113** (Somerset & Bristol)
Dick Elford, G0XAY Tel: 01454 218362
Email: *drm113@rsgb.org.uk*
**District 114** (Dorset)
Bill Coombes, G4ERV Tel: 07575 300199
Email: *drm114@rsgb.org.uk*
**District 115** (Jersey)
Peter Bertram, GJ8PVL Tel: 07829 722722
Email: *drm115@rsgb.org.uk*
**District 116** (Guernsey)
Jerry Bligh, MU0ZVV Tel: 01481 243322
Email: *drm116@rsgb.org.uk*

## Clubs & Societies

**Appledore & DARC**
Alan Fisher, *fisheralan.af@gmail.com* Meets 7.30pm on 3rd Monday in the month at the Appledore Football Club EX39 1PA *http://mysite.wanadoo-members.co.uk/gx2fko*

**Blackmore Vale ARS**
Mr Keith Chadwick M0TMO; *keith.m0tmo@btinternet.co* Meets 7.30pm every Tuesday at The New Remembrance Hall Remembrance field Charlton Shaftsbury SP7 0PL *www.radioclubs.net/bvars*

**North Bristol ARC**
Dick Elford G0XAY *g0xay@aol.com* Meets every Friday at 19.00 SHE7 Building Braemar Crescent Northville Filton, Bristol BS7 0TD *www.nbarc.org.uk*

**Burnham Amateur Radio Club**
Brian Mudge, G3MDD, Meets 1st & 3rd Wed of the month 7pm The Bay Centre Cassis Close Burnham on Sea Somerset TA8 1NN *www.barc.co.com*

**Callington ARS**
J E Vivian, G4PBN, lumley85-*cars@yahoo.co.uk* Meets monthly except for August 1st Wed of every month 19.00-21.00 Council Chamber of Callington Town Hall New Road Callington PL17 7BD *www.g1xic.co.uk*

**Christchurch ARS**
Martin Clack M0KZC, *martin@clacksp6.fsnet.co.uk* Meets 8.00pm on Thursdays at The Clubhouse adjacent to East Christchurch Sports and Social Club Grange Road, Christchurch, Dorset BH23 4JE *www.radioclubs.net/christchurchars*

**Cornish RAC**
Steven Holland, G7VOH *g7voh@btinternet.com* Meets on the 1st & 3rd Thur of the month from 7.30pm at Gweal an Top School School Lane Redruth TR15 2ER *www.cornishradioamateurclub.org.uk*

**Dartmoor Radio Club**
Mrs Viv *Watson.vivwatsondrc@aol.com*,01752 823427 Meets 7.30pm on 1st Thursday in the month at the Yelverton War Memorial Village Hall, Meavy Lane, Yelverton Devon PL20 6AL *www.radioclubs.net/g1rcd*

**Exeter ARS**
Mr Nick Johnson, M0NRJ Meets every other Wednesday at 18.30 pm America Hall De La Rue Way Exeter, EX4 8PX *www.exeterars.co.uk*

**Exmouth ARC**
Michael Newport, G1GZG. *michael.newport1@btinternet.com* Meets 1st and 3rd Wednesdays of the month at The Scout Hut Marpool Hill Exmouth EX8 1TD *www.G0XRC.org*

**Flight Refuelling ARS**
Mr John Hart, G4POF. *info@frars.org.uk* Meets twice a week Wednesdays and Sundays 19:00 to 22:00 Cobham Sports and Social Club Merley Park Road Merley, Wimborne Dorset BH21 3DA *www.frars.org.uk*

**Gordano ARG**
Malcolm Pitt *mal@g4kpm.co.uk* Meets fourth Wednesday of the month 20:00 The Ship Down Road Redcliff Bay Portishead BS20 8LD

**Guernsey ARS GU3HFN**
Mr. Richard Robilliard , GU2RS Tel: 01481 23754 Meets 8.00pm the 1st Friday of each month in St Peters Port at various venues *www.gars.org.gg*

**Holsworthy ARC**
Ken Sharman G7VJA, *ken@g7vja.co.uk* Meets the 1st Wednesday of each month at 7.30pm at the Holsworthy Community college EX22 6JD

**Jersey ARS**
Mike Turner.GJ0PDJ email: *tuckshop2@live.co.uk* Meets 8.00pm Mondays and Fridays La Moye Signal Station La Rue Du Sud (just passed the prison at La Moye on the right) *www.radioclubs.net/gj3dvc*

**Mid Somerset ARC**
David Edwards *davidedwa6@talktalk.net* Meets 2nd Tuesday of each month at 7.30pm at Peter Street Rooms Peter Street Shepton Mallet BA4 5BL *www.midsarc.org.uk*

**Plymouth Radio Club, G8PRC**
David Beck, 2E0DTC, *d.beck123@btinternet.com* Meets on the 2nd Tuesday of each month at 7:00pm at Weston Mill Oak Villa Social Club Ferndale Road Weston Mill Plymouth PL2 2EL *www.radioclubs.net/g3prc*

**Poldhu ARC**
Keith Matthew, G0WYS, *gowys@yahoo.co.uk* Meets every Tuesday and Friday from 7pm at the Marconi Centre Poldhu Cove Mullion Cornwall TR12 7JB *www.gb2gm.org.uk*

**Poole Radio Society**
Bill Coombes G4ERV *secretary@g4prs.org.uk* Meets 8pm Friday evenings at St Osmund's Hall Florence Road Parkstone BH14 9JF *www.g4prs.org.uk*

**Riviera ARC**
Ian Nelson M0IDP *rivieraARC@gmail.com* Meets first and third Thursdays of every month Acorn Community Centre 19.30 - 21.00 Lummaton Cross Torquay Devon TQ2 8ET *www.rivieraarc.org.uk*

**Saltash & DARC**
Mark Chanter 2E0MGC, *sadrc2eomgc@gmail.com* Meets first Thursday of each month at 7.45pm at the, plus additional meetings - contact Burraton Community Centre Grenfell Avenue Saltash Cornwall PL12 4JB *www.sadarc.co.uk*

**Shirehampton ARC**
Mr R G Ford, G4GTD. *G4GTD@shirehampton*-arc.org.uk Meets 7.30pm on Fridays at Avonmouth Sea Cadets T.S Enterprise, Station Road Shirehampton Bristol BS11 9XA *www.shirehampton-arc.org.uk*

**Sidmouth Amateur Radio Society**
Dave Lee G6XUV *g6xuv@hotmail.com* Meets 7.00pm every Tuesday at the Thorn Golf Centre Salcombe Regis Sidmouth Devon EX10 0JH *www.sidmouthars.org.uk*

**South Bristol ARC**
Andy Jenner G7KNA *enquiries@sbarc.co.uk* Meets Thursdays at 7.30pm ar Novers Park Community Centre Rear of 122 – 124 Novers Park Road Bristol BS4 1RN *www.sbarc.co.uk*

**South Dorset RS**
Chris Bagley, G4RSL *g4rsl@sdrg.co.uk* Meets 7.30pm on 2nd Tuesday in the month at The Southill Youth and Community Centre Radipole Lane Southill Weymouth, Dorset DT4 9SF *www.sdrg.co.uk.*

**T.S.W.A.R.C**
Nick Burnet, club secretary, 2E0IFC, Meets 8pm every Tuesday at ROC Site 7 Crosses Tiverton EX16 8JR

**Taunton & DARS**
Mr David Rosewarn, M0CIF *daverosewarn@ime.com* Meets on the first Wednesday of each month at 19.30 Tangier Scout and Guide Centre Tangier Castle Street Taunton TA1 4AS *www.tauntonradioclub.org.uk*

**Bristol RSGB Group**
Robin Thompson G3TKF. Email: *robin@g3tkf.co.uk* Meets 7.30pm on the last Monday of the month in The Back Bar Room Bristol Lawn Tennis and Squash, Redland Green, Redland, Bristol BS6 7HF *www.g6yb.org/cms*

**Thornbury & Sth Glos ARC**
Stan Goodwin G0RYM, Paul Smart M0ZMD *tsgarc@gmail.* Meets 7.30pm every Wednesday *www.tsgarc.uk* Chantry TDCA 52 Castle Street Thornbury Bristol BS35 1HB *www.tsgarc.ham-radio-op.net*

**Torbay ARS**
Geoff Coker *sec@tars.org.uk* Meets on Fridays from 19.00 at Teignbridge District Scout Headquarters Wolborough Street Newton Abbot Devon TQ12 1LJ *www.tars.org.uk*

**Watcombe Radio Club**
Mr Pete Worlledge, M0BHJ, *peteworlledge@hotmail.co* Meets 7.30pm every Monday at Torbay Scout Camp Site Easterfield Lane Torquay TQ1 4SW

**Weston-Super-Mare RS**
Martin Jones *martin@g7uwi.wanadoo.co.uk* Meets Mondays at 7:30pm at the Devonshire Road Social Club. BS23 4LG *www.radioclubs.net/wsmrs*

**Yeovil ARC**
W J Harris G8UED, *g8ued@gmail.com* Meets 7.30pm Thursdays at the Abbey Community Centre, The Forum Abbey Manor Park Yeovil Somerset BA20 2BE *http://yeovil-arc.com*

## Repeater Groups

**Devon & Cornwall Repeater Group, GB3PL**
Contact: Dereck White *discosliue@btinternet.com* Tel: 01822 610864 Email: *discosliue@btinternet.com*

**The Wessex Repeater Group, GB3JB**
Contact: Mr Dave Boniface, G3ZXX
Tel: 01963 370279
Email: *peterhillman440@btinternet.com*

**Jersey Amateur Radio Repeater Group, GB3GJ**
Contact: Peter Bertram, GJ8PVL Tel: 07829 722722 Email: *gj8pvl@hotmail.com*

**Mid Cornwall Beacon and Repeater Group, GB3NC**
Contact: Tel: 01726 850363 Email: *g0vdu@yahoo.co.uk* Tel: 01726 850363 Email: *g0vdu@yahoo.co.uk*

## Emergency Comms Groups

**South Devon Raynet**
Mrs Pam Helliwell, G7SME, 01803 326033, *pam@g6fsp.com*

**North Devon Raynet Group**
Steve Smith Tel:01769 574357, *boghay@hotmail.co.uk*

**Taunton Raynet Group**
Mr David L Smith - G4XUR, 01823 324631, *committee@tauntonraynet.co.uk* *www.tauntonraynet.co.uk*

## Contest Groups

**Wessex Contest Group ARS, MX0WCB**
5 Sydenham Buildings BATH, BA2 3BS *www.mx0wcb.com g0fuw@tiscali.co.uk* 01225 464394

## Region 11 Examinations

| Club | Place | Exam Secretary | | Telephone | Email |
|---|---|---|---|---|---|
| Appledore & District ARC | Appledore | John Lovell | G3JKL | 01237 478410 | *jklovell@gmail.com* |
| Bournemouth Radio Society | Kinson | N Higgins | G7VIK | 01202 779905 | *g7vik@ntlworld.com* |
| Callington ARS | Callington | John Vivian | G4PBN | 01822 835834 | lumley85-*cars@yahoo.co.uk* |
| Christchurch ARS | Christchurch | Christine Clack | 2E0DUK | 01425 650396 | *chrisclack@fsmail.net* |
| Cornish ARC | Truro | Ken Tarry | G0FIC | 01209 821073 | *pendennis38@btinternet.com* |
| Exeter ARS | Exeter | Peter Longhurst | G3ZVI | 01392 469405 | *g3zvi@yahoo.co.uk* |
| Exeter ARS | Exeter | Peter Longhurst | G3ZVI | 01392 469045 | *g3zvigh3dv@gmail.com* |
| Flight Refuelling ARS | Wimborne | John Hart | G4POF | 01425 653404 | *g4pof@hotmail.com* |
| Guernsey Amateur Radio Society | St Saviours | Roderick Loveridge | GU1IIW | 01481 257728 | *rml@cwgsy.net* |
| Guernsey Amateur Radio Society | Vale | Bob Beebe | GU4YOX | 01481 256755 | *gu4yox@cwgsy.net* |
| Holsworthy ARC | Holsworthy | Geoff Forster | G7VTE | 01409 253430 | *g7vte@talktalk.net* |
| Holsworthy ARC | Holsworthy | David Horton | M3EOQ | 01288 353561 | *m3eoq@hotmail.com* |
| Jersey ARS | St Brelade | Paul Ahier | MJ0PMA | 01534 855586 | *paul@jerseyars.com* |
| Newquay & District ARS | Nr Newquay | R C R Young | M0BGA | 01637 875848 | *rcry100@yahoo.com* |
| No. 7 Overseas (Jersey) Sqn ATC | St Brelade | Victoria Atherton | | 01534 736664 | *vicki.atherton@7os.org* |
| Sidmouth Amateur Radio Society | Sidmouth | Dave Lee | G6XUV | 07721 436810 | *g6xuv@hotmail.com* |
| North Bristol ARC | Bristol | Anthony Zerafa | M0BUV | 01179 677922 | *a-zerafa@hotmail.com* |
| Plymouth RC | Plymouth | S Hart | 2E0YSH | 01752 346678 | *sheo@fsmail.net* |
| Poldhu ARC | Mullion | Robin Ridge | 2E0RER | 01326 569919 | *robin.ridge@googlemail.com* |
| Poole Radio Society | Poole | Alan Walker | G4UWS | 01202 732912 | *alanandsue@greenbee.net* |
| Poole Radio Society | Poole | Stephen Morris | M0SXM | 01202 576566 | *steve.2384@btinternet.com* |
| Portland ARC | Weymouth | Kerry Morris | G1WIK | 01305 788591 | |
| Shirehampton ARC | | Clive Maby | G4NAQ | 01179 687323 | *clive@themabys.co.uk* |
| Shirehampton ARC | | Ron Ford | G4GTD | 0117 9856353 | *g4gtd@shirehampton-arc.org.uk* |
| South Bristol ARC | Bristol | Peter Hill | G0DRX | 07700 017661 | *pete.drx@raynet.co.uk* |
| Taunton and District ARC | Taunton | Michael Coles | M0CIE | 01823 259125 | *mccoles@sky.com* |
| Taunton Raynet Group | Taunton | R J Bonar | G1ONV | 01643 863462 | *bob.g1onv@btinternet.com* |
| Thornbury & South Glos ARC | Bristol | J Jones | G8AZT | 01454 883912 | *ajones@thornburyarc.org* |
| Thornbury & South Glos ARC | Thornbury | J Jones | G8AZT | 01454 883912 | *ajones4621@hotmail.com* |
| Torbay ARS | Newton Abbot | Derrick Webber | G3LHJ | 01626 354437 | *g3lhj@freeuk.com* |
| Torbay ARS | Torquay | Larry Hill | M1ARW | 01803 616565 | *larryhill@onetel.net* |
| Torbay ARS | Torquay | Pam Helliwell | G7SME | 01803 326033 | *pam@g6fsp.com* |
| Watcombe Radio Club | Torquay | Pete Worlledge | M0BHJ | 01803 552872 | *ron.edinborough@blueyonder.co.uk* |
| West Devon Raynet | Plymouth | I Harley | G6BJJ | 01752 500153 | *west.devon@raynet-*uk.net |
| Weston Super Mare RG | Weston Super Mare | James Denmead | M5AFH | | *courses@gb3we.com* |
| Weston Super-Mare RS | Weston Super Mare | David L Dyer | G4CXQ | 01934 416231 | *g4cxq@btinternet.com* |
| Yeovil ARC | Yeovil | G W Davis | G3ICO | 01935 425669 | *g3ico8@gmail.com* |
| City of Bristol RSGB Group | Bath | Steve Hartley | G0FUW | 01225 464394 | *g0fuw@tiscali.co.uk* |
| Blackmore Vale ARS | Shaftesbury | Keith Chadwick | M0TMO | 01767 851260 | *keith.m0tmo@btinternet.com* |
| Riviera Amateur RC | Torquay | Ann Nelson | M6RWJ | 07970 960497 | *devonannuk@yahoo.co.uk* |
| Riviera Amateur RC | Torquay | Steph Foster | G4XKH | 07758 857529 | *steph.p.foster@gmail.com* |
| Tiverton South West ARC | | Alan Burnett | G0IFC | 01884 251461 | *alan.w.burnett@gmail.com* |
| HCC Radio Club | Helston | S Powell | | 01326 575026 | |
| Lockyer Technology Centre | Sidmouth | Iain Grant | M1OOO | 01395 579941 | *meteorscan@gmail.com* |
| Tintagel & District RAC | Tintagel | John Brooks | M0HJO | 01840 770480 | *m0hjo@btinternet.com* |
| University of Bristol Aerospace RS | | Steve Burrow | | 01173 315542 | *steve.burrow@bristol.ac.uk* |

# Region 12 - East & East Anglia

## (http://rsgb.org/region12)

## RM

Keith Haynes, G3WRO Tel: 07711 102531
Email: *rm12@rsgb.org.uk*

## DRMs

**District 121** (Cambridgeshire )
Colin Tuckley, G8TMV Tel: 07799 143369
Email: *drm121@rsgb.org.uk*
**District 122** (Norfolk)
Mark Taylor, G0LGJ Tel: 01362 691099
Email: *drm122@rsgb.org.uk*
**District 123** (Essex - North & West)
Peter Onion, G0DZB Tel: 01206 792950
Email: *drm123@rsgb.org.uk*
**District 124** (Essex - South & East)
Vic Rogers, G6BHE Tel: 07957 461694
Email: *drm124@rsgb.org.uk*
**District 125** (Suffolk)
Keith Gaunt, G7CIY Tel: 07955 686923
Email: *drm125@rsgb.org.uk*

## Clubs & Societies

**Bishops Stortford ARS**
Tony Judge, G0PQF, *g0pqf@hotmail.com* Meets 8.00pm on third Monday of the month at Farnham Village Hall Rectory Lane Farnham Essex CM23 1HU *www.bsars.org*

**Bittern DX Group G6IPU**
Linda Leavold G0AJJ *secretary@bittern-dxers.org.uk* Meets on 2nd and last Thursday of every month at Pinewood Park I.B.C NR26 8TU *www.bittern*-dxers.org.uk

**Braintree & DARS**
Edwin Hume G0LPO Tel. 01376 324031 Meets 8.00pm on 1st and 3rd Mondays in the month at the Braintree Hockey Club Church Street, Bocking, Braintree CM7 5LJ *www.badars.co.uk*

**Bury St Edmunds ARS**
Mr Jack Myers Meets on the third Wednesday each month at the Rougham Tower Museum, Rougham IP28 6TX (Apr - Oct) The Manger Pub, Bradfield Combust , IP30 0LW *www.bsears.co.uk*

**Cambridge & DARC**
Ian Cooke M0HTA *publicity@cdarc.co.uk* Meets 7.30pm on 2nd & 4th Fridays of the month Coleridge College Radegund Road Cambridge CB1 3RJ *www.cdarc.co.uk*

**Cambridge Univ. Wireless Soc**
Dan McGraw M0WUT, *chairman@g6uw.org* Meets Thursdays 8pm at the Maypole Public House Park Street Cambridge CB5 8AS *www.g6uw.org*

**Chelmsford ARS**
Colin Page, G0TRM *secretary@g0mwt.org.uk* Meets first Tuesday of each month at 19.30 1 The Leeway Danbury Chelmsford Essex CM3 4PS *www.g0mwt.org.uk*

**Coal House Fort Radio Society**
John Parker, M1DUC Tel: 01375 383419 Meets twice a month (contact for details) at Coalhouse Fort East Tilbury Village Essex RM18 8PB

**Colchester Radio Amateurs**
Stefan Borrell M0XLB *secretary@g3co.org.uk* Meets 7.30pm on the third Thursday of each month Wilson Marriage Centre at 19.30 - 21.30 Barrack Street Colchester Essex CO1 2LR *www.g3co.org.uk*

**Dengie Hundred ARS**
Stephen Hedgecock, M0SHQ, *shedgecock516@o2.co.uk* Meets 2nd and 4th Mondays of each month 19:30 to 22:00 Latchingdon Lower Burnham Road Latchingdon Essex CM3 6HF *www.dhars.org.uk*

**Felixstowe & DARS**
Mr Paul Whiting, G4YQC *pjw@btinternet.com* Meets 8pm on roughly alternate Mondays in month Orwell Park School Nacton near Ipswich IP10 0EP *www.fdars.org.uk*

**Great Yarmouth RC**
John Attwood 2E0TWQ *g3yrc.radioclub@gmail.com* Meets second and fourth Friday of the month from 19.00 to 21.30 Bradwell Community Centre Bradwell Great Yarmouth Norfolk NR31 8QH *www.qsl.net/g3yrc*

**Harlow & DARS**
Mike Simkins, G7OBS, *enquires@g6ut.com* Meets 8.00pm every Friday the Mark Hall Barn First (Mandela) Avenue Harlow Essex CM20 2LE *www.g6ut.com*

**Harwich ARIG**
Kevin Francis M0JVC *g0rgh@amsat.org* Meets at 8pm on the 2nd Wednesday of each month at Park Pavillion Barrack Lane Dovercourt, Harwich Essex CO12 3NS *www.harig.org.uk*

**Havering & DARC**
Spencer Tomlinson *g4hrc@haveringradioclub.co.uk* Meets every Wednesday 19.30 to 21.45 Fairkytes Arts Centre 51 Billet Lane Hornchurch Essex RM11 1AX *www.haveringradioclub.co.uk*

**Huntingdonshire ARS**
Mrt Philip Haylock, G7KJW *phinorm@talktalk.net* Meets on 2nd and 4th Thursdays in the month, 7.30pm - 10.00pm at Buckden Village Hall Burberry Road Buckden St Neots PE19 5UY *www.radioclubs.net/huntsars*

**Ipswich Radio Club**
John Gee, G4BAV, *johng4bav@talktalk.net* Meets every Wednesday at 19.30 except 1st Wed of month at our contest site Shrubbery Farm Otley IP6 9PD

**Kings Lynn ARC**
Mr Eric Allison G4JNQ 01485600587 *eric@eric-a.demon.co.uk* Meets on Thursdays at 7.30pm to late at the Scout HQ, Chequers Lane North Runcton Kings Lynn, Norfolk PE33 0QN *www.klarc.org.uk*

**Leiston ARC**
Ian Pryke, M0IAH, *secretary@larc.org.uk* Meets 2nd Tuesday of the month at Quaker Meeting House, Waterloo Avenue Leiston Suffolk IP16 4HE *www.larc.org.uk*

**Loughton & Epping Forest ARS**
Marc Litchman, *info@lefars.org.uk*, 0208502 1645 Meets alternate Fridays at 19:45 at All Saints House, Romford Road, Chigwell Row, Essex IG7 4QD *www.lefars.org.uk*

**March & DRAS**
Mr E S CAMPBELL, G8PHS Meets 7.30pm on Tuesdays at British Legion Club Rookswood Road March Cambs PE15 8DP

**Martlesham RS**
Mr Keith Gaunt G7CIY *g7ciy@raynet-uk.net* Meets last Saturday of every month The Orwell Crossing A14 Ipswich IP10 0DD *www.m6t.net*

**Norfolk ARC**
David Palmer G7URP *radio@dcpmicro.com* Meets 7.00pm and 10:00pm every Wednesday at Sixth Form Common Room (front of school) City of Norwich School Eaton Road Norwich Norfolk NR4 6PP *www.norfolkamateurradio.org*

**Norfolk Coast ARS**
Steve Appleyard, G3PND *info@norfolkcoastamateurs.co.uk* Meets 3rd Thursday in the month at 7 pm East Runton Village Hall Felbrigg Road East Runton Norfolk NR27 9PH *www.norfolkcoastamateurs.co.uk*

**North Norfolk ARG**
01263 512377 *g4nre.uk@gmail.com* Meets Thursdays at 10am 88 Central Road Cromer NR27 9BW *www.gb2mc.co*

**Peterborough and District Amateur Radio Club**
Alan Ralph *g8xlh@ntlworld.com* Meets fourth Wednesday of the month Doors open at 19.00 to 21.30 Southfields Community Centre Southfields Avenue Stanground Peterborough, Cambs PE2 8RY *www.padarc.co.uk*

**Selex Galileo Basildon RC**
Mike Purser G0NEM, *michael.purser@selexgalileo.com* Meets on various dates BAE Systems Social Club, Gardiners Lane, Basildon, Essex. SS14 3AP

**South Essex ARS**
Dave G4UVJ *g4uvj@btinternet.com* Meets second Tuesday of each month 1900 to 21:00hrs Swans Green Hall Hart Road Thundersley Essex SS7 3PE *www.southessex-ars.co.uk*

**Sudbury & DRA**
Kim Edwards, G8GRL, *g8grl@k1m.co.uk* Meets 2nd Wednesday in the month at 8pm at the Wells Hall, Old School Wells Hall Road Great Cornard Sudbury, Suffolk CO10 0NH *www.k1m.co.uk/Radioclub.html*

19:30 hrs Jubilee Hall Waterside Farm Sports Centre Somnes Avenue Canvey Island SS8 9RA *www.thamesarg.org.uk*

**The Martello Tower Group**
Keith Maton G6NHU *g6nhu@me.com* Meets 1st Tuesday of month at 19:30 The Martello Tower, Tower Estate The Orchards Holiday Park, Point Clear Essex CO16 8LJ *www.martellotowergroup.com*

**Thurrock Acorns Amateur Radio Club**
Gordon Hayers 2E0ELI *acorns@taarc.co.uk* Meets every 3rd Tuesday of the month 20:00 to 22.00 at 1st Grays Scout Group HQ Cromwell Road Grays Essex RM17 5HG

**Vange ARS**
Steve Adams G0KVZ, *vars@live.co.uk* Meets 8.00pm on Thursdays at St Gabriels Youth And Community Centre Rectory Road Basildon Essex SS13 2AA *www.vangeradio.org.uk*

**Wisbech AR & Elec. Club**
Alan Bridgeland, M0DUQ *m0duq@talktalk.net* Meets Mondays 7.30 pm Elme Hall Hotel 69 Elm High Rd Wisbech, PE14 0DQ *www.warec.org.uk*

## Repeater Groups

**Cambridgeshire Rep Grp, G3PYE**
Contact: Phil Nice, G8IER Tel: 01353 775298 Email: *secretary@cambridgerepeaters.net*

**Essex Repeater Group, GB3DA**
Contact: Mr Murray Niman, G6JYB Tel: 01245 242617 Email: *secretary@essexrepeatergroup.org.uk*

**Leicestershire Repeater Group, GB3CF**
Contact: Geoff Dover, G4AFJ Tel: 01455 823344 Email: *geoffrey@geoffg4afj.plus.com*

## Emergency Comms Groups

**Suffolk RAYNET**
Russell Keen, *suffolk.county@raynet*-uk.net

**Essex Raynet**
Roland Taylor, *rhwt@btinternet.com*, *www.essexraynet.co.uk*

**Norfolk County RAYNET**
Mike Mobbs G8EEY, 01603 403305, *secretary@norfolkraynet.org.uk*

## Contest Groups

**Blackwater Amateur Radio Contest Group, M0HCY**
The Old Cottage Waterside Road Bradwell-on-Sea Southminster, Essex CM0 7QT *www.barcg.org.uk cbenn10453@aol.com* 01621 776298

**Essex Amateur Radio DX Group, MX0XYD**
*www.eardx.co.uk G0RNU@talktalk.net* 07585 906696

**Magnetic Fields Contest Group, RS305859**
16 Melrose Drive Peterborough, Cambridgeshire PE2 9DN *admin@magneticfieldscg.org* 07958 666580

**Secret Nuclear Bunker Contest Group, M0SNB**
23 Diban Avenue Hornchurch, Essex RM12 4YF *george@george*-smart.co.uk 0771 9631311

## Region 12 Examinations

| Club | Place | Exam Secretary | | Telephone | Email |
|---|---|---|---|---|---|
| All Ages Radio Club of Sth W Norfolk | Stoke Ferry | JC Nicholas-Letch | G3PRU | 01366 500704 | *johnnl@hotmail.com* |
| Barking Radio & Electronics Society | Ilford | W Chewter | G0IQK | 0208 4784758 | *g0iqk@barkingradio.org.uk* |
| Bittern DX Group | Banningham | Kenneth Holloway | M6KAH | 01263 823326 | *examsec@bittern*-dxers.org.uk |
| Cambridge and DARC | Cambridge | Peter Howell | M0DCV | 01223 870665 | *ptr.howell@ntlworld.com* |
| Cambridge University WS | Cambridge | Martin Atherton | G3ZAY | 01223 424714 | *g3zay@btinternet.com* |
| Chelmsford ARS | Chelmsford | David Bradley | M0BQC | 01245 602838 | *davidbradley@blueyonder.co.uk* |
| Chelmsford ARS | Nr Chelmsford | David Ingrey | M0HBV | 01206 243739 | *davidingrey@gmail.com* |
| Colchester ARS | Colchester | Brian Fitzsimmons | G0GGM | 01206 822547 | *brianfitz@aspects.net* |
| Colchester ARS | Colchester | Edward Erbes | M0HDK | 01473 682280 | *ederbes@globalnet.co.uk* |
| Colchester ARS | Colchester | Victor Leppard | M0VLL | 01206 367582 | *m0vll@live.co.uk* |
| Felixstowe & DARS | Ipswich | Paul Whiting | G4YQC | 01394 273507 | *pjw@btinternet.com* |
| Forest Heath | Brandon | Phyllis Seaman | M3GFW | 01379 783842 | *paul@seaman3276.fsnet.co.uk* |
| Peterborough & District ARC | Peterborough | Tracey Ralph | M5ATR | 01733 753477 | *m5atr@ntlworld.com* |
| Happing ARC | Norwich | Jill Bailey | G8PLJ | 01226 716339 | *ernest.bailey@hotmail.co.uk* |
| Harlow & District ARS | Harlow | Mike Simkins | G7OBS | 070 9287 2237 | *training@g6ut.com* |
| Hartismere High School | Eye | G Sessions | 2E1FWX | 07021 129566 | |
| Havering & DARS | Hornchurch | Stephen Lambert | G8PMU | 01708 443516 | *haveringradioclub@gmail.com* |
| Leiston ARC | Leiston | John Francis | G4XVE | 01728 648586 | *pintail@globalnet.co.uk* |
| Loughton & Epping Forest ARS | Chigwell Row | Marc Litchman | G0TOC | 0208 5021645 | *g0toc@lefars.org.uk* |
| Lowestoft & DARC | Lowestoft | Carl Leake | M0XBE | | *m0xbe2@gmail.com* |
| Maidstone YMCA ARS | Maidstone | Ray Davidson | M0RAY | 01795 420129 | *ka1828gg@yahoo.co.uk* |
| Medway ARTS | Chatham | Keith Anderson | M0KJA | 01622 725477 | *keith@m0kja.co.uk* |
| Medway ARTS | Rochester | Linda Reay | G6MXR | 01634 376516 | *linda.reay@virgin.net* |
| Norfolk ARC | Great Ellingham | David Palmer | G7URP | 01953 457322 | *radio@dcpmicro.com* |
| Phoenix Radio Club | Gillingham | Linda Reay | G6MXR | 01634 376516 | *linda.reay@[illegible]* |
| West Suffolk College | Bury St Edmunds | Beverley Burroughs | | 01284 716350 | *beverley.burroughs@wsc.ac.uk* |
| North Norfolk Amateur Radio Group (Norfolk Coast ARS) | Norfolk | Stephen Appleyard | G3PND | 01263 519485 | *sfappleyard@btinternet.com* |
| Bury St Edmunds ARS | Bury St Edmunds | Darren Coe | G7SDC | 01284 701732 | *darren@boydaz.force9.co.uk* |
| Bedford & District ARC | Ravensden | H Ehm | M0ZAE | 01234 306104 | *m0zae@yahoo.co.uk* |
| Bottisham Village College RC | Bottisham | Peter Howell | M0DCV | 01223 870665 | *ptr.howell@ntlworld.com* |
| Essex Amateur RS | Colchester | Chris Fox | M0PXZ | 01206 824561 | *m0pxz@foxearch.org* |
| Thames ARG | Canvey Island | Mark Sanderson | M0IEO | 01268 698304 | *marksanderson5@sky.com* |
| UK High Altitude Society | Cambridge | Philip Crump | M0DNY | 07743 336864 | *phil@philcrump.co.uk* |
| Thurrock Acorns ARC | Grays | Nicholas Wilkinson | G4HCK | 01375 373718 | *nicholas@graduatestudy.eu* |
| Thurrock Acorns ARC | Grays | Nicholas Wilkinson | G4HCK | 01375 373718 | *ttt@taarc.co.uk* |
| Kings Lynn ARC | Kings Lynn | Peter Elms | G0IJU | 01553 671660 | *hwes@talk21.com* |
| South Essex ARS | Thundersley | Stephen Smith | 2E0UEH | 01702 302234 | *steve2e0ueh@outlook.com* |

# Region 13 - East Midlands

## (http://rsgb.org/region13)

## RM

Jim Stevenson, G0EJQ Tel: 07500 061306
Email: *rm13@rsgb.org.uk*

## DRMs

**District 131** (Leicestershire & Rutland)
Mark Burrows, 2E0SBM Tel: 07789929730
Email: *drm131@rsgb.org.uk*
**District 132** (South Derbyshire / South Nottinghamshire)
Amanda Higton, M0HLF Tel: 07722 242292
Email: *drm132@rsgb.org.uk*
**District 133** (North Nottinghamshire)
Dr John Rogers, M0JAV Tel: 07836 731544
Email: *drm133@rsgb.org.uk*
**District 134** (Northamptonshire)
Richard Steele, G6TVB Tel: 07889 692865
Email: *drm134@rsgb.org.uk*
**District 135** (Lincolnshire South )
Graham Boor, G8NWC Tel: 01775 760832
Email: *drm135@rsgb.org.uk*
**District 136** (North Derbyshire )
Paul Archer, M0PJA Tel: 07890 626684
Email: *drm136@rsgb.org.uk*
District 137 (North of South Lincolnshire)
Andrew Gilfillan, G0FVI Tel: 07909 680047
Email: *drm137@rsgb.org.uk*

## Clubs & Societies

**ARC of Nottingham**
Club Secretary *arconradio@virginmedia.com* Meets every 2nd Thursday of the month at various venues NG9 1FY *http://arconradio.webspace.virginmedia.com*

**Bolsover ARS**
Mr Alvey Street, G4KSY. Email: *alvey@talktalk.net* Meets 8pm on Wednesdays at Bolsover Sports and Social Club (former coalite club) Moor Lane, Bolsover Derbyshire S44 6EB *www.g4rsb.org.uk*

**Buxton RA**
Mr Derek Carson, G4IHO. *g4iho@g4iho.co.uk* Meets 8.00pm on 2nd and 4th Tuesday in the month at the Leewood Hotel Buxton SK17 6TQ
*www.radioclubs.net/buxtonra*

**Chesterfield & DARS**
John Kilroy, G4PBC, *g4pbcjohn@ntlworld.com* Meets Wednesdays at 8pm to 9.30pm at Mill Public House 236 Station Road Brimington Chesterfield S43 1LT

**Chesterfield & North Derbyshire ARS**
Suzi Lilly. M6SIY email: *info@cndars.com* Meets alternative Wednesdays of Month from 18.30 to 21.00 Nags Head 47 Market Street Clay Cross Chesterfield, S45 9JE
*www.m0oct.com*

**Sth Derbys & Ashby W ARG**
John Ashmore G7KVZ *ashmore808@btinternet.com* Meets 7.00pm on Wednesdays at the Moira Replan Centre 17 Ashby Road Moira, Swadlincote Derbyshire DE12 6DJ *www.sdawarg.org.uk*

**Derby & DARS**
Chris Gent G4AKE Meets 7.30pm on Tuesdays at United Reformed Church Carlton Road Derby DE23 6HE
*www.dadars.org.uk*

**Derbyshire Spire Radio Club**
Andy Hubbard, 2E0LRM *hubbarda@sky.com* Meets every Friday at 6:45pm at Chesterfield Community Fire Station Braidwood Way Chesterfield Derbyshire S18 1PH

**Friskney and East Lincolnshire Communication Club**
Brendan Sykes 2E0BDS *FELCC@btinternet.com* Meets 1st Tuesday of each month at 19:30 Friskney Village Hall Church Road Friskney Boston PE22 8RD
*http://felcc.co.uk*

**Grantham Amateur Radio Club**
Kevin Burton G6SSN, *g6ssn@btinternet.com* Meets first Tuesday of the month Meet 19.30 for 20.00 start Grantham West Community Center Trent Rd Grantham NG31 7QX *www.garc.org.uk*

**Hinckley ARES**
Colin Brink 07986 045152 Meets 7.45pm on Tuesdays at the Hinckley Sea Cadets Rear of the Wharf Inn Coventry Road Hinckley LE10 0NQ *www.hares.org.uk*

**Hucknall Rolls Royce ARC**
Mark Attenborough, *Secretary@hrrartc.com* Meets at 8.30pm but doors usually open for 8pm Hucknall Rolls Royce Leisure Association Gate 1 Watnall Road Hucknall, Nottingham NG15 6BU
*www.hrrarc.com*

**Kettering & DARS**
Michael Lawrence 2E0XVX *2e0xvx@gmail.com* Meets at 7:00pm on Tuesday evenings and from 10:00am Sundays at Harrington Aviation Museum Sunnyvale Farm Nursery Off Lamport Road Harrington, Northants NN6 9PF *www.g5kn.org*

**Leicester RS**
*john.g0ijm@btinternet.com* Meets every Monday at 19.00 - 22.00 pm Gilroes Cottage Groby Road Leicester LE3 9QJ
*http://g3lrs.org.uk*

**Lincoln Short Wave Club**
Mrs Pam Rose, G4STO Meets 8.00pm Wednesdays (and from 9.00m to 12 noon on Saturdays in shack) LSWC C/o BSA Social Club Village Hall Lane Aisthorpe Lincolnshire LN1 2SG
*www.g5fz.co.uk*

**Loughborough & DARC**
Mr Chris Walker, G1ETZ, *g1etz@aol.com* Meets Tuesday evening from 19.30 Glenmore Community Centre Thorpe Road Shepshed LE12 9LU
*www.radioclubs.net/ladarc*

**Mansfield ARS**
David Peat, G0RDP Meets 7.30pm on Mondays at the King William IV Public House Sutton Road Mansfield NG18 5QE
*www.g3gqc.co.uk*

**Melton Mowbray ARS**
Graham Mason G4PTK *g4ptk99@ntlworld.com* Meets 7.30pm on 3rd Friday of the month at the South Melton Community Centre Dalby Rd Melton Mowbray Leics LE13 0BQ
*www.melton-mowbray-ars.org.uk*

**Northampton Radio Club**
John Cockrill G4CZB, email: *g4czb@hotmail.co.uk* Meets every Thursday 20.00 to 23.00 The Northampton Branch of the British Sub-Aqua Club Crestwood Road Northampton NN3 8JJ
*http://nrc1913.clubbz.com*

**Northampton Scout ARG**
NSARG Team *nsarg@nsarg.co.uk* Meets Saturdays Overstone Scout Activity Cntr. Northampton NN6 0AF

**Nunsfield House ARG**
Mr Adrian Price, G1OXH. *sec@nharg.org.uk* Meets 7.45pm on Fridays at the Nunsfield House 33 Boulton Lane Alvaston Derby DE24 0FD
*www.nharg.org.uk*

**Phoenix Amateur Radio Club**
Alan Clayton G7HZZ *m0phx@outlook.com* Meets once a month (variable) at 7:30pm at Mellish Rugby Fottball Club Mapperly Plains Arnold Nottingham NG11 6QE
*www.m0phx.org.uk*

**RAF Waddington ARC**
Mr Bob Pickles, G3VCA Meets 7.30pm Thursdays at The Pyewipe Fossebank, Saxilby Road Lincoln LN1 2BG
*www.g0raf.co.uk*

**South Kesteven ARS**
Andrew Garratt, M0NRD *enquiry@skars.co.uk* Meets every 1st & 3rd Friday of the month 7.30pm 47F Grantham ATC Triggs Lane Watergate Lincolnshire NG31 6NT
*www.skars.co.uk*

**South Notts ARC**
Robin Carter, G4NDM, *robin.rcarter@ntlworld.com* 07879 860207 Meets 7.00pm Wednesdays at the Greens Mill Belvoir Hill Sneinton Nottingham NG2 4QB
*www.radioclubs.net/snarc*

**Spalding & DARS**
Graham Boor, G8NWC. *secretary@sdars.org.uk* Meets Thursday & Friday weekly at 19.00 Please contact for latest meeting venue *www.sdars.org.uk*

**Sth Normanton, Alfreton & DARC**
L Lawrence, 2E0BQS *adylawri@btinternet.com* Meets 7.00pm on Mondays, except bank holidays at Post Mill Centre (formerly Community Centre) Market Street, South Normanton, Alfreton, Derbyshire DE55 2JE
*www.snadarc.com*

**Welland Valley ARS**
Peter Rivers G4XEX, *g4xex@fsmail.net*
Meets 7.30pm on third monday in the month at the Great Bowden village hall Great Bowden Market Harborough Leics LE16 7EU *www.wvars.com*

**Worksop ARS**
Fred Pickersgill, G3XXN *info@g3rcw.org.uk*
Meets Tuesdays 10.00 22.00 & Thursdays 19.00 to 22.00 59 61 West Street Worksop S80 1JP *www.qsl.net/g3rcw*

## Repeater Groups

**Leicestershire Repeater Group**, GB3CF
Contact: Geoff Dover, G4AFJ Tel: 01455 823344 geoffrey@geoffg4afj.plus.com
*www.leicestershirerepeatergroup.org.uk*

## Emergency Comms Groups

**Daventry Raynet Group**
Malcom Ogle, G1DLH, 01327 301989
*www.daventryraynet.org.uk*

**NE Derbyshire RAYNET Group**
Roger Neale, G4OIE, 01623 811141, *g4oie@perkeo.demon.co.uk*

**North Lincolnshire Raynet Group**
Dr Alan S Clark, G8EVI,
*alanclarklincsraynet@tiscali.co.uk*
*www.radioclubs.net/northlincsraynet*

**Derbyshire Dales Raynet Group**
Robin Carter G4NDM, *robin.rcarter@ntlworld.com* 07879860207
*www.derbyshire-dales-raynet.org*

**Derby Raynet**
Martin Winfield G7MKS, *contactus@derbyraynet.co.uk*, 01332 704144Ascot Drive Fire Station Community Room Ascot Drive Derby DE24 8GZ
*www.derbyraynet.co.uk*

## Contest Groups

**Parallel Lines, G4LIP**
The Barn Main Street Ashby Parva

## Region 13 Examinations

| Club | Place | Exam Secretary | | Telephone | Email |
|---|---|---|---|---|---|
| Air Cadet Radio Society | Sleaford | William Green | M5AGW | 0208 366 8680 | *dave.g8bcq@virgin.net* |
| Bramcote Lorne School | Retford | P Hitchcock | M3BAG | 0177 7838933 | *p.hitchcock@bramcote*-lorne.notts.sch.uk |
| British Red Cross Centre | Cleethorpes | Carl Flynn | G7EOG | 0845 2800083 | *mail@g7eog.co.uk* |
| Chesterfield & District Scouts ARC | Chesterfield | Keith Greatorex | G0THF | 01623 811839 | *g0thf@hotmail.com* |
| Daventry Raynet | Daventry | David Pink | G6EGO | 01327 311113 | |
| Eagle Radio Group | Mablethorpe | Charles Wilkie | G0CBM | 01507 441856 | *charles.wilkie@googlemail.com* |
| Eagle Radio Group | Mablethorpe | Charles Wilkie | G0CBM | 01507 441856 | *s.wilkie@virgin.net* |
| Finningley ARS | Sandtoft | Stuart Boast | G3WDL | 01405 815324 | *stuartboast@yahoo.co.uk* |
| Franklin ARC | Skegness | Robert Topliss | G0OTH | 01754 765408 | *robert.skegness@homecall.co.uk* |
| Friskney & East Lincolnshire Communications Club | Boston | Anthony Freeman | M0HAZ | 07956 654481 | *m0haz@hotmail.co.uk* |
| Grantham ARC | Grantham | Alan Gibson | G0RCI | 01476 402559 | *gm0rci@ntlworld.com* |
| Hinckley ARES | Hinkley | Clive Woodley | G3XPU | 01455 610055 | *clive.woodley@talktalk.net* |
| Hucknall Rolls Royce ARC | Nottingham | Steve Sorockyj | G0LCG | 0115 8400429 | *steveg0lcg@sky.com* |
| Kettering & DARS | Northants | George Christofi | G0JKZ | 07966 422602 | *g0jkz@terminalcomputers.com* |
| Kettering & DARS | Harrington | Lorna Froggatt | 2E1EFT | 01536 762523 | *lornafroggatt@aol.com* |
| Leicester Radio Society | Leicester | Duncan Gunn | M0OFL | 07889 901632 | duncan-a-*gunn@hotmail.com* |
| Lincoln Shortwave Club | Aisthorpe | Pam Rose | G4STO | 01427 788356 | *pamelagrose@tiscali.co.uk* |
| Chesterfield and North Derby ARS | Chesterfield | Suzi Lilley | M6SIY | 01246 862321 | *secretary@m0oct.com* |
| Chesterfield and North Derby ARS | Chesterfield | Suzi Lilley | M6SIY | 01246 862321 | *suziewong7@hotmail.com* |
| M0OCT ARS - North East Derby ARS | Chesterfield | Steve Brown | G6IBQ | 01246 275889 | *exams@m0oct.com* |
| Stockport RS | Melton Mowbray | Sheila Griffiths | | 01664 480733 | *sheila.g3stg@btinternet.com* |
| Northampton Radio Club | Northampton | J Cockrill | G4CZB | | *g4czb@hotmail.co.uk* |
| Nunsfield House ARS | Alvaston | Ken Frankcom | G3OCA | 01332 720976 | *g3oca1@ntlworld.com* |
| Pioneer Explorer Scout Unit | Wigston | Jim Andrews | G1HUL | 01530 249218 | *rce@stuckinthemud.org* |
| RAF Waddington ARC | Lincoln | Robert Pickles | G3VCA | 01522 528708 | *robert@pyewipe.co.uk* |
| Sandwell ARC | Oldbury | M J Prestidge | G2BXP | 0121 552 4902 | |
| Scunthorpe Steel ARC | Scunthorpe | W Jackson | G0DLL | 01724 846441 | |
| St John Ambulance | Chesterfield | Paul Steed | G0VEP/N3SSH | 02392 649471 | *paul.steed@.sja.org.uk* |
| South Normanton Alfreton & DARC | Alfreton | Melville Nutt | 2E0RNU | 01773 834533 | *melville.nutt@btinternet.com* |
| South Notts ARC | Sneinton | Robin Carter | G4NDM | 0115 974 3749 | *robin.pcarter@ntlworld.com* |
| Spalding & DARS | Spalding | John Hill | G4NBR | 01775 680596 | *jrhelectronics@btinternet.com* |
| Spalding & DARS | Spalding | John Hall | M1CDL | 01733 566517 | *m1cdl@btinternet.com* |
| Stenigot Chain Home DX Group | Alford | A Matthews | M0AQC | 01507 451547 | *seaspirit@talk21.com* |
| South Derbyshire & Ashby Woulds ARG | Swadlincote | John Whiten | M0GNO | 01530 416156 | |
| The Priory LSST School ARC | Lincoln | D Mackinder | G4DWP | 01522 595700 | *djm@priorylsst.co.uk* |
| Thorpe Camp Museum ARG | Tattershall Thorpe | Anthony Freeman | M0HAZ | 07956 654481 | *m0haz@hotmail.co.uk* |
| Trent Vale ARC | Nottingham | Paul Ryder | G0TSG | 0115 9135440 | *g0tsg@tvarc.co.uk* |
| WARS | Worksop | Carol Archer | M6ZCA | 01909 774106 | *carol@pjarcher.plus.com* |
| Sherwood Amateur Radio Club | Beeston | Aikaterini Miariti-Rippon | M6EFL | 07429 937983 | *katmiariti@yahoo.com* |
| Sherwood Amateur Radio Club | Linby | Edward Rippon | M0EPR | 07429 937983 | *m0epr.sarc@yahoo.com* |
| Buxton Radio Amateurs | Buxton | June Carson | G8RHT | 01298 25506 | *g8rht@g4iho.co.uk* |
| Stourbridge & DARS | | John Clarke | M1EJG | 01562 700513 | *johnclarke.m1ejg@btinternet.com* |
| Leicester DX | Leicester | Clive Horne | M3CGH | 07712 474864 | *leicesterdx@mail2world.com* |
| North Lindsey College | | Jannine Anderson | | 01724 294647 | *jannine.anderson@northlindsey.ac.uk* |
| South Kesteven ARS | Grantham | Andrew Garratt | M0NRD | 07969 062859 | *chairman@skars.co.uk* |
| Lincolnshire Scout Radio Club | Sutterton | Alan Blackhorse Hull | G0KYD | 01205 481452 | *american.west@yahoo.co.uk* |

# Ofcom

## Regulatory Principles

Ofcom was established as a public corporation by the Office of Communications Act 2002. Ofcom is the regulator for the UK communications industries, with responsibilities across television, radio, telecommunications and wireless communications services.

## Ofcom's Statutory Duties Under the Communications Act 2003

"3(1) It shall be the principal duty of Ofcom, in carrying out their functions;

(a) to further the interests of citizens in relation to communications matters; and

(b) to further the interests of consumers in relevant markets, where appropriate by promoting competition"

## Ofcom's specific duties fall into six areas:

1. Ensuring the optimal use of the electro magnetic spectrum
2. Ensuring that a wide range of electronic communications services - including high speed data services - is available through- out the UK
3. Ensuring a wide range of TV and radio services of high quality and wide appeal
4. Maintaining plurality in the provision of broadcasting
5. Applying adequate protection for audiences against offensive or harmful material
6. Applying adequate protection for audiences against unfairness or the infringement of privacy

## Ofcom's Regulatory Principles

- Ofcom will regulate with a clearly articulated and publicly reviewed annual plan, with stated policy objectives.
- Ofcom will intervene where there is a specific statutory duty to work towards a public policy goal which markets alone cannot achieve.
- Ofcom will operate with a bias against intervention, but with a willingness to intervene firmly, promptly and effectively where required.
- Ofcom will strive to ensure its interventions will be evidence-based, proportionate, consistent, accountable and transparent in both deliberation and outcome.
- Ofcom will always seek the least intrusive regulatory mechanisms to achieve its policy objectives.
- Ofcom will research markets constantly and will aim to remain at the forefront of technological understanding.
- Ofcom will consult widely with all relevant stakeholders and assess the impact of regulatory action before imposing regulation upon a market.

Licences are issued on a lifetime basis. This is subject to the licence being validated with Ofcom at least once every five years. Amateur licensees are therefore reminded that their licence (now 'lifetime') must be validated at least once every 5 years in accordance with the terms, conditions and limitations of their licence. If your licence has been amended (e.g. by notifying us of a change of address) this will count as a validation. It only takes a few minutes to register your details and validate or make any necessary changes to your licence on-line (*https://services.ofcom.org.uk/*) If you are experiencing difficulties or need assistance in processing your licence on-line, you can call: 0300 123 1000 or 020 7981 3131 or Textphone: 020 7981 3043 or 0300 123 2024 (please note that Textphone numbers only work with special equipment used by people who are deaf or hard of hearing).

## Applying for an Amateur radio licence

You can apply for, vary, re-validate or surrender an amateur radio licence by either using the online system (subject to conditions), or by completing a paper based application form (new applications are subject to a small administrative fee, unless you are 75 years of age or over).

The free online system has eased administration and reduced costs for amateur users. This approach makes it easier for users to comply with legal obligations. More information on how to apply for an amateur radio licence is available on the Ofcom website (see below).

If you have any questions about your licence please contact:

Spectrum Licensing
Riverside House,
2a Southwark Bridge Rd,
London SE1 9HA.
Tel: 0300 123 1000 or 020 7981 3131
Fax: 020 7981 3333.
Email: *spectrum.licensing@ofcom.org.uk*
Website: *www.ofcom.org.uk*

## Varying the terms and conditions of a licence

Any licence changes affecting all amateurs are subject to Ofcom publishing a Notice of Variation (NoV) or a consultation (depending on the changes proposed). These are announced by Ofcom in the manner specified in the licence and will also appear on the Ofcom website.

## Ofcom and the RSGB

Ofcom works closely with the RSGB and meets on a regular basis to ensure that where feasible, the amateur licence meets the needs of today's amateur. Ofcom continues to work closely with the Society to facilitate regulatory requirements on a number of ongoing developments in amateur radio.Frequencies in the 5 MHz band have been incorporated into the licence.

Spectrum allocations are listed in the Licence, which is published on the Ofcom website: *http://licensing.ofcom.org.uk/binaries/spectrum/amateur-radio/guidance-for-licensees/nov/licence.pdf*

The temporary extension to the Amateur Radio Special Research Permit to operate in the band 501kHz to 504kHz has now expired. These NoVs will not be renewed and will no longer be available. However, Amateur radio has been given an alternative allocation, on a secondary basis, of 472 479kHz and this band has been included in the frequencies listed in the Licence

Special Contests Call signs are available for individuals and clubs who regularly participate in Special Contests. These are administered for Ofcom by the RSGB and further information is available at: *http://rsgb.org/main/operating/licensing-novs-visitors/online-nov-application/application-for-a-special-contest-call-sign/* Ofcom also supports the Society in the international scene (see below).

## International

There are three main areas of interest; Spectrum, Commonality of licensing conditions and Reciprocal agreements as detailed below.

## Spectrum

Ofcom is responsible for administering the International Telecommunication Union (ITU) Radio Regulations in the UK, and ensures the effective management of spectrum taking into account international obligations.

## Commonality of licensing conditions

Ofcom represents the UK within the European Conference of Postal and Telecommunications Administrations (CEPT) and continually works towards harmonisation of licence conditions and mutual recognition including countries outside Europe. The UK has not adopted any recognition agreements for the Foundation or Intermediate level licence.

## Reciprocal agreements

A number of reciprocal agreements have already been negotiated which allow UK amateurs who have passed the 'Full' examination (or past equivalents) to operate abroad, although there are still a number of countries with which no agreement currently exists. In general, Ofcom prefers reciprocal operation under the CEPT arrangements. Non-participating countries (including those not in CEPT) may apply to CEPT to be included in the CEPT Reciprocal arrangement, governed by Recommendations T/R 61-01 and 61-02. This route is considered as being more effective and efficient than the process of numerous individual bi-lateral arrangements between two administrations. However, Ofcom

will continue to review needs for new agreements as appropriate.

Due to the many differing arrangements in countries at levels below the full licence level, Ofcom does not currently recognise reciprocal agreements at Foundation and Intermediate level.

## Repeater network

The processing of applications and issuing of the Notices of Variation (NoV) is carried out by Ofcom, with initial assessments undertaken by the Society. Clearance involves consultation with Government departments and with Ofcom's local offices where appropriate. The NoV holder remains responsible for compliance with the amateur licence.

## Packet Data and Internet (voice) Linking

Ofcom is responsible for the issuing of Notice of Variations (NoVs) for Packet Data Nodes and Internet Gateways, including those linked to Voice Repeaters. Initial assessments of applications are undertaken by the Society. The NoV may or may not permit unattended operation, but the NoV holder remains responsible for compliance with the amateur licence.

## Special Event Stations

Ofcom is responsible for the issuing of NoVs for Special Event Stations. Please see elsewhere in the Yearbook for full details. For further details regarding NoVs, please contact Spectrum Licensing at Ofcom.

## Enforcement

The number of people who misuse amateur radio is fortunately small. However, there are some individuals who cause considerable problems to other amateurs or to other licensed radio users by transmitting in unauthorised frequency bands, using obscene language or generally using radio in an antisocial way. The majority of abuse is directed at the repeater network.

Repeater keeper responsibilities include taking steps to prevent and stop abuse. Details of repeater abuse should therefore be sent to the ETCC Chairman, c/o RSGB HQ. Other cases of abuse should also be taken up with Ofcom through the Amateur Radio Observation Service (AROS). Subject to priorities, Ofcom endeavours to take action in cases of abuse or deliberate interference involving the amateur service. Having obtained evidence against an offender, Ofcom may issue formal warnings or Conformity Notices, initiate prosecution proceedings and/or revoke a licence, depending on the seriousness of the offence.

## UK amateur radio technical information

The UK Interface Requirement (IR) for Amateur Radio is IR 2028 and includes the technical details of radio equipment for Foundation, Intermediate and Full licensees. The IR is available at: *www.ofcom.org.uk/radiocomms/ifi/tech/interface_req/ir2028.pdf*

### Provision of Information

Ofcom's policy is to provide its information in electronic form via its website, *www.ofcom.org.uk* Some of these information sheets are available free in paper form. The following relate to the amateur service.

**Amateur application forms**

Of346 Amateur Licence Application, amendment, validation, surrender form
Of346a Amateur Licence amendment, validation, surrender form
OfW 287 Application for a Notice of Variation for a Special Event Callsign
OfW 306 Application for an Amateur Radio Special Research Permit

# Licences

There are three levels of qualification for gaining an amateur radio licence; the three-tier structure consisting of Foundation, Intermediate and Advanced Amateur radio licence.

The licensing structure in the United Kingdom is now progressive, which means all new entrants have to enter the hobby at the Foundation level, progressing through the Intermediate licence and finally to the Advanced Amateur radio licence.

## Foundation Amateur Radio Licence

Introduced in January 2002, the Foundation licence is designed as an entry point into amateur radio and forms the first part of the three-tier amateur licensing structure.

A Foundation licence can be obtained after a short course of study lasting some 12-15 hours. During the course, assessments take place in a number of practical areas. At the end of the course the candidate sits a short, multiple-choice exam, which lasts 45 minutes. There are 26 straightforward questions to answer. The pass mark is 19 correctly answered questions. A 'pass' will allow operation of an existing Full licensee's station as a fully qualified Foundation Amateur immediately. You will be able to operate your own station when you receive your licence and M6 callsign.

### Foundation Syllabus

1) Amateur Radio
2) Licensing Conditions
3) Technical Basics
4) Transmitters and Receivers
5) Feeder and Antenna
6) Propagation
7) EMC
8) Operating Practices and Procedures
9) Safety
10) Morse Code

## Intermediate Amateur Radio Licence

The second level in the three-tier structure. Candidates must have completed the requirements of the Foundation licence syllabus and passed the associated examination as a prerequisite to sitting the Intermediate Licence Examination.

To obtain the Intermediate licence it is advisable to take a training course. This is a little longer than the Foundation course, lasting some 20-25 hours. The aim is to teach many of the fundamentals of radio in a stimulating way, by undertaking practical tasks such as soldering, constructing a small project and a variety of other exercises, building on the experience gained as a Foundation licence holder.

After completing the practical assessments, a candidate will be ready to sit the Intermediate amateur radio licence examination. Once again this examination is multiple-choice, this time with 45 questions. Lasting 75 minutes, the pass mark is 27 correctly answered questions.

### Intermediate Syllabus

1) Amateur Radio
2) Licensing Conditions
3) Technical Basics
4) Transmitters and Receivers
5) Feeder and Antenna
6) Propagation
7) EMC
8) Operating Practices & Procedures
9) Safety
10) Components, soldering & colour code

## Licence types and basic privileges

| Type | Foundation | Intermediate | Advanced |
|---|---|---|---|
| **Prefix(es)** | **M3, M6** | **2*0, 2*1** | **G, M0, M1, M5** |
| **Amateur bands** | 136kHz-430MHz, 10GHz | All | All |
| **Maximum output power** | 10W [1] | 50W [1] | 400W [1] |
| **Permitted TX equipment** | Commercial, kit | Commercial, kit, home-made | Commercial, kit, home-made |
| **TX modes** [2] | All permitted | All permitted | All permitted |
| **Act as mailbox/BBS** | No | No | Yes |
| **Act as repeater?** | No | No | Yes |
| **Low power device/beacon?** | Yes | Yes | Yes |
| **CEPT reciprocal operation?** | No | No | Yes (but check) |
| **Other licensees allowed to operate?** | Yes [3] | Yes [3] | Yes |
| **Disaster comms permitted?** | Yes | Yes | Yes |
| **Control station remotely?** | Yes [4] | Yes [4] | Yes |

* Regional identifier letter. [1] Some bands have lower power limits. [2] Not all modes are permitted on all bands. [3] Other licensee must hold UK licence. [4] Restrictions apply.

## Advanced Amateur Radio Licence (formerly known as the RAE – C&G exam No.7650)

This is the highest level of licence that you can obtain. To gain an Advanced licence it is necessary to pass the Advanced radio communications examination, which contains 62 multiple-choice questions and lasts two hours. Once again the examination covers radio theory and licence conditions, but because holding an Advanced licence enables you to use 400 watts power output from your transmitter such subjects as Electro Magnetic Compatability (EMC), antenna design and safety issues are covered in some depth. The licence allows access to all the amateur allocations with full power.

When studying for the Advanced radio communications examination there is no requirement to take a formal training course because there is no practical element. It is possible to study at home on your own if you so wish, but you should recognise that a good understanding of the material is required.

Many local amateur radio clubs and societies and technical colleges run courses specifically for the Advanced radio communications examination. Alternatively, there are some correspondence and Internet courses available

### Advanced Syllabus

1) Amateur Radio
2) Licensing Conditions
3) Technical Aspects
4) Transmitters and Receivers
5) Feeder and Antenna
6) Propagation
7) EMC
8) Operating Practices and Procedures
9) Safety
10) Measurements

## How and where to find a training course

The quickest ways to find a training course is to check on the RSGB website Clubs and Training web page or look at the Local Information section in this Yearbook.

Tutors may run courses using their own personal station. Local amateur radio clubs, Scouts, Guides or other organisations such as the Air Training Corps may also run courses. Youth organisations are likely to run courses solely for their own members, but it is always worth asking. If you are at school and your school does not run a course, suggest to someone that they do so. The RSGB will be happy to advise on how this might be done and point to sources of assistance and training.

The syllabuses for the three examinations are published on the RCF website.

## Costs

There is likely to be a small charge for training course administration. It is best to ask before committing to attend, but the costs should not be high and certainly should not be a reason to miss out. Examinations and assessments cost £27.50 for the Foundation licence, £32.50 for the Intermediate licence and £37.50 for the Advanced licence.

## Study material

The following titles are available from the RSGB Shop.

*Foundation Licence Now!*
*Intermediate Licence*
*– Building on the Foundation*
*Advance - The Full Licence Manual*
*Exam Secrets*

## Audio versions of books

### *Foundation Licence Now!*

Available on four audio cassettes or one data CD. The data CD contains MP3 files and may be listened to on a 'Daisy'* or a 'Symphony'** player, as well as on a PC. The CD can be used to produce five audio CDs by anyone with a CD writer installed in their PC. For copies contact the Royal National Institute for the Blind (RNIB). Tel: 0845 702 3153. Alternatively, Kelvin Marsh, M0AID, of the Radio Amateur Invalid and Blind Club (RAIBC). Tel: 01823 412087.

### *Intermediate Licence – Building on the Foundation*

Available on one data CD. This CD can also be used to produce eight audio CDs. For copies contact Kelvin Marsh, M0AID, of the RAIBC. Tel: 01823 412087.

### *Advance – The Full Licence Manual*

Available on one data CD. For copies contact the RNIB. Tel: 0845 702 3153. Alternatively, Kelvin Marsh, M0AID, of the RAIBC. Tel: 01823 412087.

Please note that the RNIB are re-indexing their files and may not be able to assist at present. If you find that this is the case please contact Gill Cowsill at Ivy Bridge Recording Centre, who will be able to supply copies of the tapes/CDs. Tel: 01752 690092.

## Special needs candidates

Arrangements can be made for candidates with special needs who would otherwise have difficulties taking any of the amateur radio licence examinations. The key requirement is proper advice on what provisions should be made for the candidates concerned. This must be from an appropriate health, educational or other professional. The need is not for a statement of the candidate's circumstances, but what should be done in order to provide equality of access to the examinations.

Where a disability prevents a candidate from carrying out a practical task then advanced permission should be sought, again supported by proper advice, to waive that specific requirement. Not all requirements can be waived and this will be dealt with on a case by case basis.

It is important to start this process well in advance of requesting and exam, as it may take some weeks to obtain appropriate professional advice. Advice should be sought from the RSGB Examination Department

**RSGB Clubs and Training web page:**
***www.rsgb.org/clubsandtraining***

* Daisy players are available from the RNIB's Talking Book service and allow a visually impaired or blind person to listen to many hours of pre-recorded material. The player allows skipping between tracks, or in this case, pages, and bookmarks can be set to allow an easy return to a study point.

** Symphony players are available from the British Wireless for the Blind Fund (BWBF) and allow the listener to pause and return to the same point, provided the CD has not been changed.

# Special Contest Callsigns

The holder of any UK amateur radio Full club licence may apply for a Special Contest Callsign, for use in a number of contests.

## Ofcom Policy and Procedure

The holder of a UK Amateur Radio Full (Club) Licence or an Amateur Full/Reciprocal Licence may apply for a special contests such as those shown in **Table 1**. The callsign will consist of G or M, a regional locator (if appropriate), a chosen digit and a chosen suffix letter, eg G8Z or GW8Z etc. RSLs may be inserted into an SCC. 520 callsigns are available:

G*(number) A-Z, and M*(number) A-Z.

The RSGB administers SCCs for Ofcom. Applicants should apply on the form at *http://rsgb.org/main/operating/licensing-novs-visitors/online-nov-application/application-for-a-special-contest-call-sign*, giving three choices of callsign in order of preference. Current availability may be checked by contacting the RSGB

Allow up to four weeks for processing.

Individual licensees or club licensees (holder of the club licence) will be expected to provide evidence of having entered at least five of the contests listed in Table 1 within the last three years and having achieved at least one third of the number of contacts of the leader in the appropriate table. However, in a contest where the licensee achieves more than one half of the number of contacts of the leader, (rather than one third), this contest will count as two contests towards the requirement to have entered five contests within the last three years. Individual licensees wishing to apply for a special contest callsign should do so with achievements gained under their own individual callsign.

Details of the application will be passed to the RSGB's Contests Committee for a recommendation on issuing a NoV. Ofcom will treat the RSGB Contests Committee's response only as a recommendation. The NoV to the licence is issued by Ofcom to the holder of the relevant licence.

Licences are issued until 31 December, and subsequently every three years. If the licensee wishes to renew the callsign that was issued, a fresh application must be made within the period of two months immediately prior to the expiry date of the NoV. Unless a fresh application has been submitted in this period, the callsign will be withdrawn for a period of two years prior to it being made available for general re-issue. Please note that 'Licensees' who hold a special contests callsigns NoV can use the following (RSGB's Contests Committee) Email address to ask for a reminder about their NoV renewal: *chairman@rsgbcc.org.* Whilst the RSGB has offered this facility, please note that the onus is on the licensee in case of a reminder not being received.

The RSGB will send NoVs directly to successful applicant.

## Extract of Terms relating to the NoV for use of Special Contests Callsigns:

1. Terms and expressions defined in the Licence shall have the same meaning herein except where the context requires otherwise.
2. The special contests callsign shall only be used for the contests specified in Table 1 and at no other times.
3. The special contests callsign shall only be used until the date of expiry. Unless a fresh application is made for renewal within the period of 2 months immediately prior to this date, the callsign will be withdrawn for a period of 2 years and then become available for general re-issue.
4. Where the licence relates to an Amateur Full (Club) Licence, only members of the club may use the special contests callsign, subject to the conditions in this Licence.
5. A separate log must be kept in respect of the special contests callsign.
6. The appropriate regional secondary locator (if any) should be used. For guidance see Section 2 of the Licence - Notes to the licence. - Note (c).
7. The special contests callsign shall not be used outside the United Kingdom.
8. The Notice of Variation together with Annex A (in the application document) shall be read as integral to the Licence and both shall be kept with the Licence.

| Contest | Mode | Month | Contest | Mode | Month |
|---|---|---|---|---|---|
| ARRL DX | CW | February | IOTA | Multi | July |
| ARRL DX | SSB | March | WAE DX | CW | August |
| ARRL 1.8MHz | CW | | WAE DX | RTTY | November |
| ARRL 28MHz | Multi | December | WAE DX | SSB | September |
| CQ WPX | CW | May | ARRL RTTY Roundup | RTTY | January |
| CQ WPX | RTTY | February | BARTG | RTTY | March |
| CQ WPX | SSB | March | Russian DX | Multi | March |
| CQ Worldwide | CW | November | IARU 50MHz Trophy | Multi | June |
| CQ Worldwide | RTTY | September | IARU 144MHz Trophy | Multi | September |
| CQ Worldwide | SSB | October | IARU 432MHz - 248GHz | Multi | October |
| CQ Worldwide 160m | CW | January | RSGB 144MHz & 432MHz | Multi | March |
| CQ Worldwide 160m | SSB | February | RSGB 432MHz - 248GHz | Multi | May |
| IARU HF Championship | Multi | July | 2m Marconi Memorial | CW | November |

**Table 1:** *Events in which special contest callsigns may be used.*

***For a list of current Special Contest Callsigns, see the Callsign Listings section on of this Yearbook***

The application form may be found and printed from the Internet at: *http://licensing.ofcom.org.uk/binaries/spectrum/amateur-radio/apply-for-a-licence/ofw286.pdf*

# Special Event Callsigns

## Do you really need a GB callsign?

All club stations are able to pass greetings messages sent by a non-licensed third party. This means that applying for a GB callsign no longer holds any advantages for a club!

Provided the club uses the prefix letters (see below), the club station is allowed to pass greetings messages and to operate simultaneously on more than one band. The club prefixes are very distinctive and create interest through their rarity. If a club regularly operates a special event station using the club callsign, this will help increase the club's identity. It will also benefit by being able to print its QSL cards in larger, more economic quantities.

Another advantage is that any suitably licensed and authorised club member may operate the club station. This gives greater flexibility over a GB callsign which is just a variation to an individual's licence.

Best of all, you don't have to fill in any forms or give 28 days notice of operation!

| Old club prefix | New club prefix |
|---|---|
| G/M | GX/MX |
| GD/MD | GT/MT |
| GI/MI | GN/MN |
| GJ/MJ | GH/MH |
| GM/MM | GS/MS |
| GU/MU | GP/MP |
| GW/MW | GC/MC |

### GB callsigns

Ofcom issues and despatches Notices of Variation authorising special event (GB) callsigns. Consequently, all enquiries and correspondence should be addressed to Ofcom and not to the Society. Ofcom has stated that GB callsigns are issued for special event stations and that as such, they should normally be open to viewing by members of the public.

### Applying for a GB callsign

No charge is currently made by Ofcom for a special event callsign, but application forms must be sent in at least 28 days prior to the start of the event.

Applications are normally processed shortly after receipt. If nothing has been received 14 days prior to the event, please contact Ofcom immediately. Please note that no authority exists until a Notice of Variation has been received.

A Notice of Variation will only be issued to licensees who hold a current Full, including a Club licence (ie not Foundation or Intermediate). It will be valid for a maximum of 28 consecutive days. The station may only be established and operated at one specified location. This must be the address stated on the application form which must be detailed enough for anyone to find easily. Operation of a special event station from a licensee's home address is not normally permitted.

Only the person responsible for the station need sign the form, as the authorisation is by Notice of Variation to that individual's licence. This person is required to be present to supervise the correct operation of the station. Additional operators need only sign and write their callsigns in the logbook.

If you have not used the callsign before, you can avoid last-minute disappointment by first contacting Ofcom, who can check that it is available and reserve it for you. A GB callsign may be reserved for up to six months in advance. When a GB callsign has been used it will not normally be re-issued to another amateur for use at a different event for a period of two years.

The holder of a Full Licence must apply for a GB callsign.

Subject to availability, special event callsigns are available in the following formats:

| | |
|---|---|
| GB0 + 2 or 3 letters | GB1 + 2 or letters |
| GB2 + 2 or 3 letters | GB5 + 3 letters |
| GB4 + 2 or 3 letters | GB6 + 3 letters |
| GB5 + 2 letters | GB8 + 3 letters |
| GB6 + 2 letters | GB8 + 2 letters |

### Greetings messages

The guidelines agreed with Ofcom are:

1. Each greetings message should not exceed five minutes.
2. Each person may pass only one message to each station with which the originating station is in contact.
3. A non-licensed person may speak into the microphone but the licensed radio amateur must identify the station and operate the transmitter controls at all times.
4. Greetings messages by third parties may only be sent from and received by stations within the UK or the USA, Canada, Falkland Is, Gibraltar, Malta, the Maldives and Pitcairn Is. The licensee may exchange greetings as in any QSO, with any station.

### Charitable events

It is recognised that some special event stations will be established at certain charitable events where a major concern will be the raising of funds.

Ofcom has agreed that the charity (if one is involved) or the reason for establishing the special event station may be mentioned 'on-air' provided that under no circumstances may a donation be requested during the contact, and sending of QSL cards must not be conditional upon the pledge of a donation. It is in the interests of everyone who holds a special event station licence that operators keep within the spirit of this by not asking for any money over the air.

The station may be sponsored per con- tact, ie the licensee may in advance of the event seek from his/her friends and relatives sponsorship assurances under the usual arrangements for sponsorship. You must not seek sponsors 'on-air' at any time.

### QSL information

Special event stations generate many QSL cards, so it is important that you use the QSL Bureau correctly.

For instructions, see the 'QSL Bureau' pages in this edition of the *RSGB Yearbook*.

## JOTA and Radio Scouting

Jamboree on the Air (JOTA) is an annual event designed to allow Scouts to send greetings messages to each other. Started in 1957, it now involves approximately 600,000 Scouts and Guides, with the help of over 23,000 radio amateurs in over 100 countries.

JOTA takes place on the third full weekend of October each year, officially between 00.00 Saturday and 24.00 Sunday, although most stations run for a period within these hours to suit their own requirements. The event is organised by the Scout movement, supported by radio amateurs or clubs. Their aim is to bring Scouts around the world closer together and to introduce them to the capabilities of amateur radio.

All amateur bands are used. Most stations use a special event or a club call, allowing the Scouts to pass greetings messages over the air. JOTA information packs are sent to all participating GB stations at the beginning of October. Any clubs taking part in JOTA wishing to receive this information pack should contact the SES Administrator at RSGB HQ.

The interest fostered by JOTA and World Jamboree has spread and many Scout camps and campsites boast amateur radio facilities. A number of proficiency badges in the radio, electronics and computer fields are available for Scouts. Several countries have permanent Scout Headquarters stations – for example the World Scout Bureau in Geneva has the callsign HB9S and Gilwell Park in the UK operates under the callsign GB2GP.

Many countries run periodic Scout nets. There are regular weekly UK and European nets aimed at Scouters who are also radio amateurs.

## Thinking Day On The Air

## Usual Scout Net Frequencies

| Band | SSB (Phone) | CW |
|---|---|---|
| 80m | 3.740 and 3.940* | 3.590 |
| 40m | 7.090 | 7.030 |
| 20m | 14.290 | 14.070 |
| 17m | 18.140 | 18.080 |
| 15m | 21.360 | 21.140 |
| 12m | 24.960 | 24.910 |
| 10m | 28.990 | 28.190 |

The UK Scout net is on Saturdays, 3.740MHz at 0900 local time. The European Scout net is on Saturdays, 14.290MHz at 09:30GMT.

Thinking Day On The Air (TDOTA) is organised by The Guide Association on the third full weekend in February, to celebrate the birthdays of the founder of the movement, Lord Baden-Powell, and of his wife, Lady Baden-Powell, the World Chief Guide, on 22 February.

The aim of TDOTA is to encourage the girls to make Guiding friendships with members of other units and to introduce them to amateur radio. Station organisers are asked to keep these objectives in mind.

Guide amateur radio stations rely on the goodwill of radio amateurs in setting up stations, though the association has an increasing number of members of all ages holding callsigns.

Guiders interested in organising a TDOTA station can apply for a comprehensive information pack with suggestions for activities, logos for certificates and posters to report forms. Further information is published from time to time in the association's magazines.

Stations are requested to complete a brief report which is sent to Girlguiding UK HQ. All the information is collated into a National Report, which is sent to those who took part and contributed to the report. Copies of the current report are available on receipt of an A4 SASE or from the website (see below).

Amateur radio has a place in the programme for all age groups, encouraging girls to embrace the technical aspects and international perspectives of a world-wide movement. Girlguiding UK supports the revised amateur radio licence structure and particularly welcomes the Foundation Licence.

While the main focus remains TDOTA, Guides can be heard on the air at other times of the year from camps, activity days and leader training courses.

**Info about JOTA and other Scout activities**:
The Scout Association, Gilwell Park, Chingford, London, E4 7QW.
Tel: 020 8524 5246

Website: *www.scouts.org.uk*

**Info about Girlguiding & TDOTA***:*
*Send A4 SASE to*: The Programme Team, Girlguiding UK, 17-19 Buckingham Palace Rd, London SW1V 0PT

Website: *www.girlguiding.org.uk*

***For a list of permanent Special Event Callsigns, see the Callsign Listings section of this Yearbook***

# Your Time

We rely on the active support of a myriad of volunteers to enable the RSGB to provide the range of member services that we do. Indeed, we can only operate with the active support of dedicated and committed people who believe in the future of amateur radio and are prepared to give their time into making that future actually happen.

## People like you

Putting something back into the hobby or passing on your experience, knowledge and specialised skill is a rewarding and fulfilling experience.

The RSGB has many areas in which your skills might be used to help others. Some of them require a level of specialised knowledge: EMC, Planning, Repeater Management and Data Communications. In other areas, enthusiasm and commitment is all that is required: GB2RS newsreading, Emergency Services or Deputy RSGB Regional Manager, to name but a few.

When it comes to training, why not consider becoming an RSGB Registered Instructor? There is a constant need for Instructors for the Foundation, Intermediate and Advanced amateur radio licence courses.

You decide when and where you can help (please note that all applications for Registered Instructors are, for security reasons, subject to a vetting procedure).

If you can't find time to volunteer to work for the Society, remember the other area where you can make a real difference - mentoring.

Too often we hear of newly licenced amateurs who feel 'on their own' after completing their studies. You can pass on your experience and skill by 'taking under your wing' a new or prospective amateur and ensuring that he or she learns appropriate operating practice and behaviour. In that way we can ensure that our bands are properly used, to the greater enjoyment of everyone, and newcomers can make the most of amateur radio.

*For further information on becoming a volunteer for the Society, contact the*

RSGB General Manager,
Tel: **01234 832700** or
email: GM.dept@rsgb.org.uk
website *www.rsgb.org/volunteer*

# EMC (ElectroMagnetic Compatibility)

The RSGB can offer help to members on EMC matters through its EMC Committee, which consists of volunteers who have professional as well as amateur radio experience in the field of EMC.

## Introduction

Operating an amateur radio station in the 21st century in an urban or suburban environment presents particular challenges. Not only may there be limited space for antennas but the presence nearby of other electronic devices can result in emissions raising the noise level on the amateur bands, as well as breakthrough from amateur transmissions into other devices. EMC, or 'ElectroMagnetic Compatibility' is the term used to describe the ability of devices to co-exist without excessive interaction.

Fortunately, cases of breakthrough from amateur transmissions are becoming less frequent, and by following the "Good Radio Housekeeping" guidance can generally be managed. See Avoiding Interference below.

Sadly, however, the combined effect of numerous other electronic devices nearby can add together to form what has been termed 'radio fog', raising local noise floors and impeding communication.

## Particular threats

Almost any electronic device has the potential to cause emissions of some sort. Most are benign and conform to relevant Standards, but some have significant potential to cause problems:

- PLT/Home Plug devices
- xDSL wired internet
- Plasma TVs
- Switch-mode power supplies (SMPSU)
- PV Solar Panels
- Wind Farms

– Plus a plethora of other electronic devices, such as:

- Remote controlled lamps
- LED low voltage lighting modules
- RF-excited lighting modules

Many of the sources of interference are familiar to members. The RSGB, through members of its EMC Committee, is represented on international standards bodies working to achieve standards which should allow coexistence of electronic devices with radio communications systems.

## Summary

Many complaints of EMC problems can only be solved with the active cooperation of both parties. This requires diplomacy and tact. Whilst Ofcom can sometimes help in difficult cases, the responsibility rests with the individual amateur to try to assess causes of interference or breakthrough, and within limits, to effect a cure. Complaints from neighbours of interference may be related to environmental impact of antennas, so see the Planning Advice pages in this Yearbook and discuss planning issues with the Planning Advisory Committee.

Interference problems are not often understood by complainants or the owners of the offending apparatus, so help them to understand - be a good radio neighbour and be sensitive to their point of view.

# Data Transmission Systems Using Telephone Lines & Electricity Cables

Technologies which use the telephone lines and the electricity cables to carry high speed data signals have been a source of concern to radio amateurs for more than a decade. These notes give a brief outline of the radio interference (RFI) threat that may be expected from the various technologies.

### Dial-up modems

These use audio signals on the phone line and have now been almost completely superseded by DSLs and fibre optic links. There are a whole family of DSLs, but the only ones of interest to us are ADSL and VDSL.

## ADSL (Asymmetric Digital Subscriber Line)

Technique and frequencies used generally up to 1.1MHz, but could be up to 2.2MHz. ADSL is usually fed into the phone line at the local exchange, which could be up to 5km from the customer's premises.

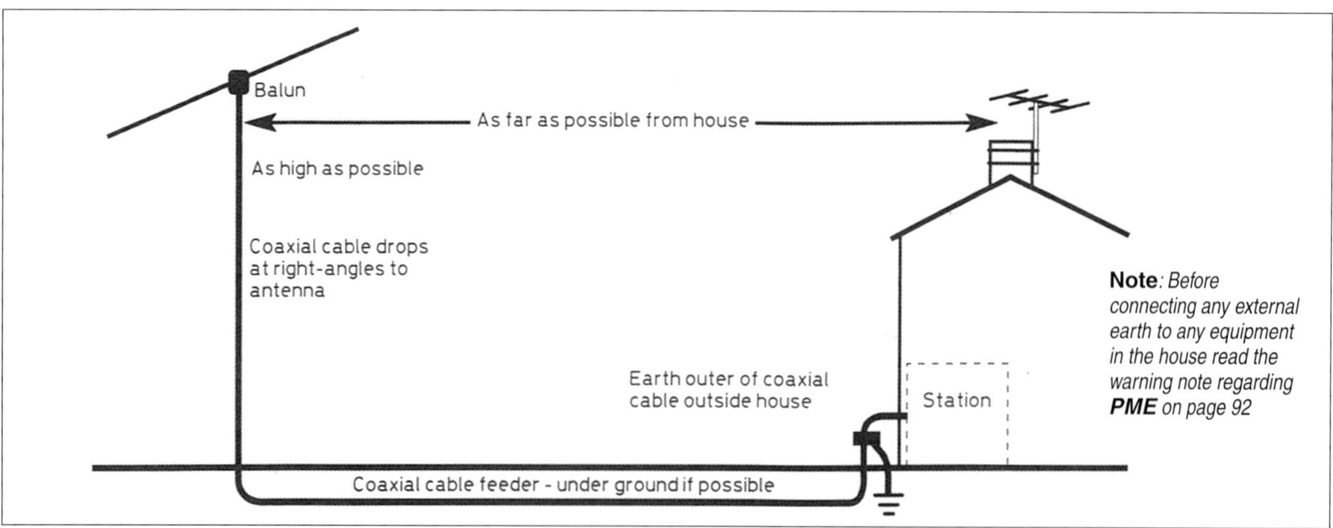


*Good radio housekeeping – site your antenna and feeder system well away from the house.*

RSGB EMC Committee website: www.rsgb.org/main/technical/emc/

### Deployment

It is widely deployed in the UK with many millions of customers. The numbers will decline as customers change to VDSL

### Interference Potential

In general, interference from ADSL is not a problem to amateur radio but there have been a number of reports of breakthrough to ADSL by amateur transmissions. More information can be found in the EMC Columns of RadCom. An index and link to past issues can be found on the EMCC website.

## VDSL (Very High Speed Digital Subscriber Line)

### Techniques and Frequencies used

VDSL operates at up to about 17MHz and is launched into the telephone lines at the street cabinet (It is sometimes called FTTC Fibre-To-The-Cabinet). Since only a relatively short length of telephone cable from the street cabinet to the customer is involved (1km maximum), data speeds of up to 40Mb/s are possible.

### Deployment

Deployment is well advanced in the UK and the service is now available in most urban and many rural locations. The availability will increase rapidly and it will probably be a major factor in the Government's Broadband Britain policy.

### Interference potential

Until recently it had been assumed that VDSL2 would not be a significant RFI problem. This has been brought into question by a number of reports of interference to the amateur bands which have occurred since the upgrading of VDSL2 about the end of 2012. Investigation is in hand to determine the extent of the problem and the best way to tackle individual cases. Installations with underground connections seldom exhibit problems.

Until recently installation practice in the UK has been for a technician to install the splitter and modem using appropriate high quality twisted pair cable, but "self install" options are now becoming available. The Society is actively involved in ensuring that this does not become a major RFI threat.

## Systems using electricity cables

This is known as Power Line Telecommunications (PLT or PLC). In the USA it is usually known as BPL, Broadband over Power Lines. Low frequency signalling on the electricity mains has a long history, but so far as radio amateurs are concerned PLT refers to Internet access and computer networking and also, recently, to Smart Metering. *See below..*

There are two types of PLT; Access PLT and In-house PLT

## Access PLT

### Techniques and frequencies used

This is internet access using the mains supply cables mainly using frequencies at the low end of HF band. There are two obvious problems. Firstly, the signal and consequent RFI enter every house on the circuit, whether they want the PLT service or not. Secondly, the data stream is shared by a large number of customers which is a severe constraint on data speed.

### Deployment

Access PLT has not been deployed beyond the trial stage in the UK. The availability of other, potentially much faster technologies such as the DSLs and fibre would appear to rule out Access PLT as a practical proposition in Europe. This may not apply in other parts of the world where geography and infrastructure are different.

### Interference potential

There is no imperative system requirement to limit launch power, as there is with the DSLs, so ultimately the only limit on radio interference will be the emission regulations.

## In-House PLT

This makes use of modems which plug into a mains socket and communicate with one another via the electricity wiring in the house. The modems are called Power Line Adapters (PLAs).

### Frequencies used

Systems vary, but typically 4 to 28MHz. Some new devices go up to about 70MHz

### Deployment

Apart from computer networking, Power Line Adapters are widely used for video distribution in Internet TV systems (IPTV). Some years ago the devices used by the major provider of this service in the UK emitted serious interference all the time even when not transmitting data.

New standards now require that devices limit their emissions in amateur bands, and are quiescent when no data is being passed.

### Interference potential

All PLAs reduce their launch power in the international amateur bands. This is known as 'notching'. This seems to be reasonably effective, though some filling due to intermodulation has been observed. Without the notching the interference on the amateur bands would be intolerable. Discussions on an EMC Standard specific to PLT have resulted in two new Standards EN50561-1 below 30MHz and EN50561-3 above 30MHz. Both seem to give a reasonable degree of protection to the amateur bands. This is too big a subject for these short notes and further information can be found on the RSGB website.

## Smart Metering PLT vs. PLS

In 2009 the Government announced that it wished to press ahead with plans for Smart Metering, with a full implementation by 2020. This announcement had been long expected; the RSGB was prepared for it and had been active in the relevant BSI committee.

The European Commission has been working on environmental issues for many years and the 'Energy Efficiency Directive' sets out some clear objectives. We are all aware of the removal of tungsten filament lamps from the market and the efforts to reduce carbon emissions.

Smart Metering in the home is another part of this plan. The emotive words which have caught the eye of radio amateurs are 'Power Line Technology'. Whilst the method of remote reading of the millions of meters has still not been finally decided, it is most likely that it will use established technology, so the immediate alarm is unwarranted since in this case it really means Power Line Signalling (PLS). PLS has existed for over 100 years and is widely used for the control of street lighting, the switching of tariff rates and control of the power grid. It uses very low data rates and a frequency below 150kHz, so this is not the same as broadband PLT. That is not to say that the PLC industry will not press for its technology to be employed. There are many hurdles for either scheme not least of which is the noisy nature of the domestic mains supply.

### Smart Network

Smart Networks and Smart Metering complement one another in their aims to improve the overall efficiency of the power distribution and usage. There are considerable challenges that electricity suppliers face with optimisation of the electricity grid, and the Smart Networks initiative seeks to improve things. It will use signalling below 150kHz in what is referred to in Europe as the Cenelec Bands. Whilst there are many political ramifications raised by the proposals, they should have little or no impact on the spectrum assigned to amateur radio.

RSGB has been fully engaged with the development of existing Standards for PLS and will continue to work with the BSI and IEC committees that influence future development as it happens. Members will be informed as developments occur in the EMC column in RadCom or the EMC pages of the RSGB website.

# Other Potential Sources of Interference

The emphasis on preservation of the Environment has resulted in many schemes aimed at reducing the use of energy and harvesting of renewable sources. These have inevitably resulted in consequential environmental impact.

## Energy Harvesting Systems

### Solar panels

The Government incentives offered to house-holders and industrial users has encouraged many electrical power users to install Photo Voltaic (PV) panels on their roofs. These installations are a potential source of RFI. From the outset, it must be said that there are good RFI-free installations, and of course the converse is true.

An installation consists of the solar panels on the roof, and much more importantly an inverter, usually placed somewhere below the roof, which are connected by cabling. The inverter is the source of RFI and the cables are potentially the antenna that radiates the energy. The current UK Government and Ofcom view is that solar PV installations are comprised of separate items of apparatus rather than being an integrated fixed installation. The RSGB's view is that even so, an installer is responsible for ensuring the apparatus meets the EMC compliance requirements when the apparatus is first taken into service (see the section on the EMC Directive). In any case any member contemplating a Solar Energy Harvesting system should check that the installer understands the requirements of the EMC regulations. The industry has given some recognition to the potential RFI problems and lightweight invertors which can be installed within the roof space have been introduced. The interconnection between these is usually quite short and results in very low antenna efficiency, and low radiation. At the same time, much greater care has been taken to ensure that the leakage of RFI from the units is minimised.

However, it is also true to say that the move towards so called 'transformerless' invertors has presented new challenges. These invertors, using solid state commutation to created the 50Hz AC signal, produce high frequency spikes which leak more readily from the unit housing. The EMC Committee is continuing to gather information on PV systems, and this information will be published in David Lauder's regular RadCom EMC column, and subsequently in the Yearbook.

### Wind Farms

It is not necessary to travel very far in the UK to see a hilltop wind farm installation, and members have expressed concerns regarding how these will affect the amateur bands. An installation consists of the wind turbine itself and a complex control system at the base of the mast. There are a number of arrangements available for feeding 50Hz energy into the National Grid. Almost all of these involve complex electrical conversion of the voltages and current, with the inevitable switched mode power convertor playing an important part.

The most probable cause of RFI from a wind farm is from the electrical control systems at the base of the tower, with once again the cables connected to the top acting as an antenna. Although these are usually screened within the metal structure, at ground level there may be feeds to the control systems that radiate.

The EMC Committee is gathering information from members and David Lauder will be making measurements on actual installations. These will be reported in his regular column.

### Utility Services

A regular source of complaints to the EMC Committee comes from members who live in a rural, normally quiet location. An unexpected high noise level appears on the lower HF bands. In almost all cases the mains power feed is overhead.

As well as the possibility of arcing on the power line itself, frequently the cause has been found to be thyristor controlled motors installed in pumping stations operated by the water or sewage utility. The problem seems to be that the installing electricians have little or no knowledge of RF grounding. Overhead power lines accentuate the radiation, acting as long wire antennas. Fortunately, the RFI is evident on MW broadcast stations, and is easily demonstrated with a portable radio as coming from an enclosure housing a pump. The advice from the EMCC has been to contact the Utility, who will usually be sympathetic to the problem.

### Solid State Lighting

LED lamp modules are tumbling in price and becoming more attractive as an energy saving method. Reservations regarding the suitability in the domestic environment are being overcome, with units available with more acceptable colour temperature characteristics. This has led more homeowners to consider using LEDs in down-lighters in kitchens and bathrooms.

There are two types of LED modules available for the consumer.

1) Replacement for the GU16 and MU11 series, which operate from 230V AC. These units incorporate the electronics to convert the mains to a constant current DC supply, suitable for the LED array.
2) GU16 and MU11 Series, which operate on 12V DC. These contain the necessary electronics to provide the constant current supply for the LEDs.

The LED modules which operate on 230V are tested to existing standards, which ensure that the emissions are generally well-controlled and cause few known problems. The LED modules that operate on 12V do not fall under any specific standards regime and are a very variable bunch. Tests have identified some modules that create very high levels of RFI.

One of the issues with the 12V units is that they are often considered as replacements in installations which are already equipped with so called 'electronic transformers'. These are in fact SMPSUs which convert 230V mains to 12V DC.

When a 12V LED replacement is connected to these with one SMPSU operating on one frequency, and the power convertor in the lamp module operating at a different frequency, and with different peak load demands, there are problems that compound to produce RFI. Work by DARC in Germany has resulted in the submission of a paper to CISPR, requesting a programme of work aimed at introducing a standard for 12V LED modules. In the meantime work is continuing to identify rogue modules and to make the membership aware.

### RF-excited Lamp Modules

Fortunately the threat posed by the introduction of RF Excited Plasma lighting modules has diminished, as the work on producing even more efficient LED modules has taken over.

These modules are produced mainly for street lighting and are still being installed and pushed by some suppliers. The EMC Committee has received very few complaints regarding these lamps in recent times.

## Putting the RFI in context

### Background noise on the HF bands

How much noise would one expect in a typical residential location?

Situations where there are continuous, high level broadband sources of interference are unusual in residential areas, though they are common in industrial/commercial premises. In residential locations broadband noise is usually relatively low, with occasional periods of high level noise. In addition there may be high levels of narrowband interference on specific frequencies. Where there is continuous broadband noise in a residential location it is likely to be something specific like an alarm system or some device such as a switch mode power supply.

The RSGB EMC Committee rule of thumb, for the HF amateur bands, is that in a residential area the ambient noise should not exceed 0dBuV/m measured on a horizontal dipole 10m away from the house (measured in a 9kHz bandwidth, quasi peak). This figure is a bit optimistic below 7MHz but is definitely pessimistic above 14MHz. Measurements at 28MHz should show a noise level well below 0dBuV/m. See 'Notes on the RSGB Observations of the HF Ambient Noise' on the EMC Committee website.

Links to RSGB and Ofcom EMC problem reporting web pages:
*www.rsgb.org/main/technical/emc/*

The natural noise on HF can be as high as 30 or 40dB above thermal, and this defines the ambient noise floor This means that fairly high S-meter readings can be expected on say 80m, even when no man-made noise is present. It is possible that readings as high as S5 to S6 may be seen at certain times particularly at night.

Vertical and poorly balanced horizontal antennas tend to pick up more noise than a well balanced horizontal dipole. This really comes under the heading of Good Radio Housekeeping (EMC leaflet **EMC 10**).

# Are You Getting Interference?

The RSGB wants to establish the extent of interference to the radio spectrum from switched mode PSUs, data-over-mains, xDSL, wind farms, Solar PV and other interfering devices. We are therefore asking everyone who is experiencing local interference to help us.

Most data-over-mains devices are 'notched', so that they do not cause high levels of emissions on amateur bands. However, elsewhere in the shortwave spectrum there remains the possibility of interference.

We should all remember that there are other forms of interference to the enjoyment of our frequencies - from local switch-mode power supplies, some plasma televisions, electrical machinery, etc. Please also remember that there is a high level of natural noise on the lower frequency bands,. The EMC Committee needs feedback from members on the radio frequency interference (RFI) they are experiencing. Please report problems through the RSGB EMC forum *http://forums.thersgb.org/index.php?forums/emcmatters/* or to one of the advisors

*http://rsgb.org/main/about-us/committees/electromagnetic-compatibility-committee/emc-technical-advisors/.* For the latest information see the EMCC Section of the RSGB website.

*The following will help you identify different types of RFI.*

**There are three simple steps to take:**

### 1. Identifying the form of interference

You should first check that the equipment in your own house is not the source of the interference. There are 'hidden' sources in many everyday pieces of electronic equipment so we recommend that, if you can, you turn off all your power circuits at the circuit-breaker except the one powering your receiver (or use a battery powered receiver) and double check that there is nothing in your own home contributing to the interference. It is easy to overlook a small device that could be the source of the problem. If you switch off all the breakers except one, make sure there is nothing connected on the remaining live circuit other than the receiver you are using. Be sure that you warn all other people in the household that the power, including lighting, may go off.

You should now try to determine what the interference is caused by, because there are many other possible interfering signal sources. SMPSUs are high on the list, because of the millions of them that are now in use, so tune across the affected bands using the AM detector position on your receiver and turn off the AGC. Compare the sound of your interference with the sound clips available on the website.

If you are reasonably sure that the interference comes from a PLA device (the sound clips may help), then tune outside the amateur band, and check if the level of the interference increases. What you should hear is that the interference increases rapidly as you tune outside of the amateur bands. If it does, it is more likely than not that you have identified a PLA device, whilst if it is more broadly observed across the spectrum a SMPSU may be the culprit.

Make a note of the amateur bands affected, and the strength of the interference as indicated on your S-meter including any frequencies where a step change is evident and the size of the step. You might also like to record the interference level on some of the shortwave broadcast bands. If you have access to an SDR or band scope a capture using this is useful evidence and helps in diagnosis.

### 2. Tell RSGB

Post a description of the interference and any additional information on the RSGB EMC forum. The RSGB will try to offer advice and in some cases will assist in preparing the information for a formal complaint to Ofcom.

Just as important is the database which we build up from complaints will help us in our discussions with Ofcom. Remember we are interested in knowing of all forms of interference.

### 3. Making a Complaint to OFCOM

A complaint of interference can be made directly to OFCOM but this should not be done without careful thought. Recently OFCOM have advised the RSGB that they have received a number of complaints where the Interference was within the limits one would expect at a particular location. This was put down partly to the wide publicity surrounding RFI issues and also to the popularity of SDRs (Software Defined Radios) which, while giving a fascinating picture of the radio spectrum, are open to misinterpretation where interference is concerned. This, it is claimed, is stretching their limited resources which should be going into "serious" cases. OFCOMs position is that interference should be judged by its effect in actual operating conditions. Without taking any position on the validity or prevalence of this situation, the RSGB has agreed to advise members not to complain to OFCOM before discussing the situation with the EMCC.

Contact the EMCC by any of the means listed under "Getting Help and Advice". Your query will be dealt with directly or if necessary passed on to a committee member who deals specifically with your type of problem.

Of course this does not prevent any radio amateur (or any other person) from contacting OFCOM directly if they so wish, but it is hoped that it will at least help to facilitate official action where it is really needed..

# Complaints Procedure

### TV and Radio interference

The BBC has responsibility for investigating complaints of interference to domestic radio and television. All complaints should be made to the BBC. You can find the BBC's diagnostic guidance at the following address: *www.radioandtvhelp.co.uk/interference/rtis_tv/radcom_tools*

This page also carries useful commentary for any of your neighbours who may be affected by your transmissions. There is also a facility to contact the RTIS where the basic diagnostic guidelines have not helped. If, following the investigation by the BBC, there is evidence of interference caused by something which is unlawful, the BBC may refer your case back to Ofcom for possible enforcement action.

### Interference to amateur radio

Amateurs often mistakenly believe that the 'non-protected' status of the Amateur Radio Service means they are not entitled to any action in the case of interference caused to them. In fact, 'non-protected' is only in respect of interference from other authorised services operating in the same bands. Amateurs are as entitled to protection from external interference as any other radio user, although we must accept that Ofcom will have to give priority to safety of life and business radio users.

# I'm Causing Breakthrough

Nowadays it is unusual for interference to be caused by a faulty transmitter and though interference from harmonics and other spurious emissions is occasionally encountered, by far the most common cause of interference from amateur stations is breakthrough.

Information on this type of problem can be found in a number of EMC Leaflets available on-line. This section deals with breakthrough caused by the fundamental of the transmitter getting into nearby electronic equipment and causing it to malfunction. Most modern electronic equipment is designed to have a reasonable immunity to radio frequency (RF) fields, but this may not be enough to cope with the large fields which can arise from a nearby amateur transmitter.

The most important factor in reducing breakthrough into local radio

and electronic equipment is good radio practice and particularly the siting of antennas. EMC Leaflet 10 sets out the factors you should consider if you live in close proximity to neighbours and wish to minimize breakthrough problems. In the paragraphs below we provide a suggested approach in dealing with complaints of interference. More details can be found on the RSGB website.

## The first step

You must confirm that no equipment in your own house is affected by your transmissions. Put simply, if you cannot resolve breakthrough into your own domestic electronic equipment, you are starting from a poor position in discussions with your neighbours.

Where your transmissions are affecting your other equipment, try to resolve the problem through application of the relevant filters or ferrite rings available from the RSGB Shop.

Look for the obvious sources of problems. Ask yourself 'What is acting as an antenna to pick up my signals?' Very often this will be long loudspeaker leads to audio systems (fit ferrite toroids near the audio amplifier output), the coaxial down-lead to the television (try a "braid-breaker" ), or the mains cable to the affected equipment (again, try a toroid at the point that the mains cable enters the equipment).. Breakthrough onto telephones can be tricky - see EMC leaflet 5. Again, ferrite toroids fitted to the telephone cable near the handset can help, but some telephones are much more susceptible to breakthrough than others. It may be helpful to consider having a very basic handset (with minimum amount of electronics) to confirm the source of the problem. Disconnect all telephones, modems etc from your internal telephone network and connect the simple phone. Confirm that the breakthrough is cleared. If not, fit toroids to that handset until it is. Then progressively add back telephones on the network, one at a time, checking each time that the breakthrough is cleared. Remember that one telephone that is susceptible to RF can add back interference on to the whole network. When one handset added causes the breakthrough to return, try toroids, but be prepared to discard the offending handset if all else fails. Ultimately you should have a complete telephone network free from breakthrough.

Breakthrough onto security alarms is another cause of problem, and EMC leaflet 3 provides some useful guidance.

### Now to the real work

Once you have cleared all cases of break- through in your own home, the difficult part begins! Before you speak with your neighbour, you may find it useful to consult an EMC Advisor. See: *www.rsgb.org/main/about-us/committees/electromagnetic-compatibility-committee/emc-technical-advisors/* The resolving of an EMC problem calls for great diplomacy with your neighbour. Often the initial reaction is that the problem must be with your transmitter, as without it reception is faultless. Read EMC leaflet **1** and let your neighbour have a copy to read. Explain that you are willing to help and that a few simple steps involving no modification to his equipment, may resolve the issue. Add that your own TV/radio/alarm/telephone (as appropriate) is unaffected by your transmissions. You should not consider any internal modifications to your neighbours' equipment. But you could offer to try toroids on the downlead, speaker leads, telephone leads and mains leads of affected equipment, or a braid-breaker on the TV antenna feeder. If the problem is a security alarm, read again EMC leaflet **3**.

If these basic steps fail it may be a case of asking the retailer to provide some help. Equally, there is the option of the neighbour complaining to Ofcom. EMC leaflets that will help you in your work are listed in 'Tackling EMC Problems' pages and are available on the EMC Committee web site.

Please note: These leaflets are intended for members of the Radio Society of Great Britain, but are available to non-members on the understanding that any information is given in good faith and the Society cannot be held responsible for any misuse or misunderstanding. An up-to-date list of all of the available leaflets may be found on the RSGB EMC web site.

# Avoiding Interference

## Avoiding interference from the transmitter

### Spurious Emissions

At one time complaints of interference to TV from harmonics of amateur transmitters were a major concern in amateur radio. Nowadays complaints of this type are rare mainly because TV has moved up to UHF but also because transceivers, whether home brew or commercial, are designed with reduction of harmonics and other spurious emissions in mind.

Spurious emissions do still occasionally cause problems, for instance when harmonics of a 2m transmitter fall onto a UHF TV frequency or a harmonic of an HF or 50MHz transmitter might fall on a VHF radio frequency. Such cases are easy to identify by considering the frequency of the station being interfered with and the operating frequency of the amateur station. The solution is to check that the transmitter is working correctly and if necessary fit a low pass or band pass filter. Further information on spurious emissions can be found in the Radio Communication Handbook.

It is worth noting that interference to digital TV will not cause the typical picture and audio degradation which was associated with analogue TV, but will cause the picture to 'freeze', appear as blocks, or possibly disappear altogether until the receiver re-synchronises. These effects can also be caused by a number of signal degradation situations not related to amateur radio.

### Breakthrough

When the fundamental signal from the transmitter gets into radio and electronic devices and causes interference it is usually called "breakthrough" to emphasise the fact that it not caused by a fault at the transmitter but a lack of immunity of the victim equipment. Breakthrough can be to either radio or non-radio equipment such as telephones or audio units. Most cases of HF interference to radio and TV are actually breakthrough, with the fundamental of the amateur signal getting in via the braid of the antenna coax or the mains lead, and causing overloading and inter-modulation effects.

**There are two ways of tackling breakthrough problems.**

1. By taking care to operate the amateur station so as to minimise RF energy getting into nearby radio and electronic equipment. This has been called good radio housekeeping and is covered in more detail in EMC leaflet **10**.
2. By increasing the immunity of the affected equipment. This aspect is covered separately in 'Tackling EMC Problems'.

## Avoiding interference to amateur radio reception

There are three ways of dealing with interference to reception

### 1. Tackling the interference at source

This is the best option and should always be considered first. The object is to track down the source of interference and then persuade the owner to take action to suppress it or modify the use of the offending device so as to minimise the effect on your amateur operation.

This will probably not be too much of a problem if the device is in your own home, but may be much more difficult if it is in a neighbouring property. Possible actions depend on whether the device is compliant with EMC regulations or not, but the golden rule is that any approach to neighbours should be diplomatic. It is not possible in these notes to do justice to this difficult subject. Further information can be found in EMC Leaflets **04** and EMC **09**. Post on to the RSGB EMC forum or contact an EMC advisors if you need specific help.

### 2. Reducing the coupling

The term good radio housekeeping was coined to cover breakthrough situations and especially to publicise the need to operate an amateur station with 'due care and attention' and to bear in mind the reasonable expectations of neighbours *see* EMC Leaflet **10**. For the purposes

of these notes, good radio housekeeping has been expanded to include a discussion of the application of these principles to minimising received interference.

When the station is located in a residential area, siting the antenna in relation to surrounding properties is of major importance. Antennas should be as far from your own and neighbouring houses as possible, and as high as practical. This applies to both transmitting and receiving, since situations which cause breakthrough will also couple noise from the same wiring back to the antenna.

Some HF antennas can function near ground level, but this is not a good policy from the EMC point of view.

On HF there is, however, one big difference between transmission and reception. This is that, regardless of any local noise, there is an ambient noise level on the HF band, which greatly outweighs the thermal noise generated in the receiver front end. So, unless the antenna is very inefficient, the ambient noise dictates the received noise level. This means that it might be better for reception to mount a small, relatively inefficient, antenna in a place where local interference is least high up and far from buildings. In special circumstances it might be worth considering an active antenna. Apart from this, good housekeeping rules for HF receiving and transmitting antennas are the same. They should be:

**a – Horizontally Polarised**. House wiring tends to look like an earthed vertical antenna and is more susceptible to vertical radiation. Likewise - but for rather more complex reasons - vertical receiving antennas tend to be noisier than horizontal ones.

**b – Balanced**. Out-of-balance currents on feeders generate vertically polarised radiation and likewise tend to pick-up vertically polarised noise.

**c – Compact**. So that one end is not much closer to the house than the other.

For most of us it is not possible to fulfil both conditions (b) and (c) at the lower HF frequencies, unless we have a very large or oddly shaped garden. However they illustrate what to consider when making a compromise.

With VHF antennas there is a trade-off between antenna siting and feeder loss.

Where a high gain antenna is used, careful consideration must be given to the effective radiated power (ERP) and the proximity of nearby houses.

### 3. Actions to reduce the effects of interference at the receiver

It is usually better to tackle the interference at source, but if this is not possible the only option is to attempt to minimise the effect of the interference at the receiving end. First look at your radio housekeeping and at the same time check the whole antenna/earth and feeder installation for corroded joints. These can cause passive intermodulation products (PIPs), which, though not really interference, have the effect of increasing background noise. Interference can enter a receiver from the mains by unexpected common impedances and tests with a battery-operated receiver may give clues to what is happening. Don't forget that, in the absence of a signal, the receiver AGC will pull up the interference to a more or less constant level. This often leads to false conclusions.

If all else fails there are anti-interference measures which can be used at the receiver itself. Most amateurs are familiar with the function of the noise blanker and its much less effective grandfather the noise limiter. Modern transceivers include digital signal processing (DSP) which can be very effective with some types of interference.

In difficult cases it might be worth considering interference cancelling.This can be tricky to set up and operate but when functioning correctly is remarkably effective.

# EMC Help and Advice

This is a support service for RSGB members who may require technical assistance with an EMC or RF interference issue. Amateur radio encompasses a wide range of technical interests and within the ranks of the membership there is a strong pool of knowledge that can be made available to others.
The EMCC is reorganising its Help and Advice service to make it more flexible and easy to use. Look on the EMCC website for the latest information

## How to get help from the RSGB EMC Committee

1. The EMC Matters forum. This can be used to seek advice on a specific EMC problem but it may also be used to report issues of interest which may be of help to other amateurs.
2 The EMC Problem Reporting Form. This can be found on the EMCC website. It provides information needed to guide you through diagnosis of the problem
3 Email or telephone the advisor/coordinator using contact details given on the EMCC website.

Whichever method you use it is important to give as much information as possible. The EMC Problem Reporting form on the website can be used as a check list.

### Other Sources of technical help

It is important that members be aware of all possible sources of technical help.

### Amateur radio clubs

All radio amateurs are encouraged to join and regularly support their local clubs. You will find a list of RSGB affiliated clubs elsewhere in the RSGB Yearbook. If you join one you will find that your fellow members will be a valuable source of help and assistance with almost any technical problem.

### Technical publications

The RSGB produces a number of excellent publications, available to members and non-members alike. These cover a range of subjects and are readily available by post. New entrants to the hobby are particularly encouraged to improve their level of technical knowledge by reading them. Members of the RSGB receive RadCom and can access RadCom Plus, which also covers a wide range of technical subjects. Many RadCom columnists offer technical assistance but, in general, they can only give a reply via their future articles, and this may not be appropriate if a speedy response is needed. RadCom columnists often list a number of helpful websites in their articles. It may be worth checking these out before going any further.

Remember also the EMC Leaflets listed in "Tackling EMC Problems" later in this section.

### Course instructors

If you have recently successfully graduated from a course, your former instructors may be able to provide assistance.

### Amateur radio suppliers

Many of these regularly advertise in RadCom and are usually very helpful in giving general technical advice in relation to any product that you have purchased from them. Members should bear in mind that suppliers may not always be free to spend time on the telephone and their priority will be to attend to customers in the shop. Please be discerning if you call on one of them for assistance.

### Regional Managers

Regional Managers and their deputies will be pleased to give assistance. A list is published regularly in RadCom and also in the Yearbook. They can suggest the best local contact to help you with your problems.

### *WARNING: Protective Multiple Earthing (PME)*

Many houses in the UK are wired on what is known as the PME system. In this system the earth conductor of the consumer's installation is bonded to the neutral close to where the supply enters the premises, and there is no separate earth conductor going back to the sub-station. Under certain rare supply fault conditions a shock or fire risk could occur where external conductors such as antennas or earths are connected. For this reason the supply regulations require additional bonding and similar precautions in PME systems.

Many houses in UK were wired on the old TN-S system, where a separate earth goes back to the sub-station. In such systems there were no problems with connecting an external radio antenna or earth. It has recently become evident that changes to maintenance and installation practice mean that the inherent safety of the old TNS systems cannot always be guaranteed. The situation is being reviewed. Until further information is available all installations should be treated as if they were PME.

**Read EMC lealet 07 Earthing and the Radio Amateur before connecting any earth or antenna system to equipment inside the house.**

**If in doubt consult a qualified electrician**

EMC Leaflet 07 is available on request from RSGB, or from the RSGB EMC Committee website.

# Tackling EMC Problems

A look at the information and filters available.

## RSGB Information

The RSGB EMC Committee has produced a series of information sheets, listed below. Copies are available on the EMC Committee website.

**EMC 01** Radio Transmitters and Domestic Electronic Equipment
**EMC 02** Radio Transmitters and Home Security Systems
**EMC 03** Dealing with Alarm EMC Problems Advice to RSGB Members
**EMC 04** Locating Sources of Interference to Amateur Radio Reception
**EMC 05** Radio Transmitters and Telephones
**EMC 07** Earthing and the Radio Amateur
**EMC 08** TV Distribution Amplifiers
**EMC 09** Handling Inbound Interference
**EMC 10** Avoiding Interference to Nearby Electronic Equipment
**EMC 12** Part P and the Radio Amateur
**EMC 14** Interference from in-house PLT
**EMC 15** VDSL Interference to HF Radio

## Use of Filters and Ferrite Rings

### TV and video recorder

In recent years interference to TV (TVI) has become much less of a problem. There are several reasons for this including improved TV set design and the phasing out of analogue TV in the UK. As a result, the once popular specific TVI filters are now rarely encountered. On the other hand, the ferrite ring choke remains an essential weapon in the EMC armoury.

With a TV set alone, a suitable ferrite ring 'braid breaker' in the coaxial aerial cable at 'X' in **Fig 1** may be all that is required to cure breakthrough. If this does not cure the problem, then a ferrite ring may also be required on the mains cable at 'Y'.

**Fig 2** shows a video recorder, hard disc recorder or DVD recorder with its RF output connected to the aerial socket of a TV. In practice this arrangement is unlikely to be found in a modern digital TV installation, where connection by SCART or HDMI would be expected. A filter and/or 'braid breaker' should be fitted in the coaxial cable at 'X1' but if this is not sufficient, another should be fitted at either 'X2' or at 'X3'. In some cases, ferrite rings are required on the mains cables at 'Y1' and/or at 'Y2'.

### Audio systems

RF breakthrough in an audio system can often be cured by fitting ferrite rings to the loudspeaker cables as shown by 'Y1' and 'Y2' in **Fig 3**. In some cases, additional ferrite rings may be required such as at 'Y3' on the mains cable. They may also be required on audio cables or mains cables of other units such as a separate CD player or a 'surround sound' audio connection to a DVD player or TV. The use of an optical digital audio cable if possible should improve RF immunity.

### Telephones

In cases where a wired telephone suffers from audible breakthrough of signals from an amateur radio transmitter, an ADSL 'microfilter' can make an effective RF filter, whether or not the telephone line has ADSL. There are large variations between the performance (and price) of different ADSL microfilters however. It is advisable to obtain a good quality model with a relatively large case containing four or more ferrite-cored inductors plus a ferrite cored bifilar wound choke.

### Characteristics of filters available from RSGB

All the filters listed below allow UHF TV signals to pass through but, as with any filter, there is a small loss in the pass band.

In many cases, particularly on the HF bands, amateur signals are picked up by the TV aerial downlead rather than by the TV aerial itself. If this causes breakthrough, a so-called 'braid breaker' is required. This could be part of a filter or it could be a separate unit. There are several types of 'braid breaker' although some of these do not actually break the braid at DC.

### Ferrite ring choke

Winding a length of coaxial cable onto one or more ferrite rings (see **Fig 4**) has the advantage that it introduces very little loss or impedance mismatch for the wanted UHF signal and also maintains the integrity of the coaxial cable screening.

### 1:1 transformer braid breaker

Included in the HPFS filter, this gives excellent rejection of common mode signals on the HF and lower VHF bands, plus it can be more effective than ferrite rings on the HF bands below 10MHz. However, it generally introduces greater loss of the wanted UHF TV signal.

### HPFS High pass filter (special)

RSGB order code: Filter 3
Pass band: UHF TV
Stop band: All bands up to and including 144MHz
Braid breaking: Includes 1:1 transformer type braid breaker

The HPFS is a combined high pass filter and braid breaker. It is a good all-round filter for all HF and VHF amateur bands but is not effective on 430-440MHz. Due to the relatively high pass-band loss (up to 4dB), it is not suitable for areas where the TV signal strength is low and may not be suitable for digital terrestrial TV.

### Ferrite rings (RSGB Order code: FRIN)

A ferrite ring made in the USA by Fair-Rite Corporation in type 43 material, part number 2643802702. The inside diameter is 22.85mm (0.9in) and the width is 12.7mm (0.5in). It can accommodate twelve turns of 5mm diameter cable and is equivalent to FT140-43. Details of the characteristics are given below.

## Use of Ferrite Chokes

RF breakthrough into electronic equipment can be caused by signals being picked up on external cables such as loudspeaker cables, mains cables. This effect can often be reduced or eliminated by winding the affected cable onto a suitable ferrite core, which presents a high impedance to unwanted RF signals without affecting the wanted signals (see **Fig 4**). For the best chance of success, it is important

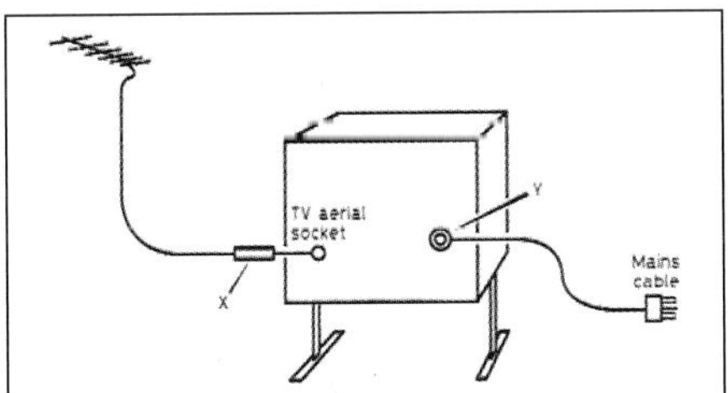

*Fig 1: Fitting filters to TV set alone*

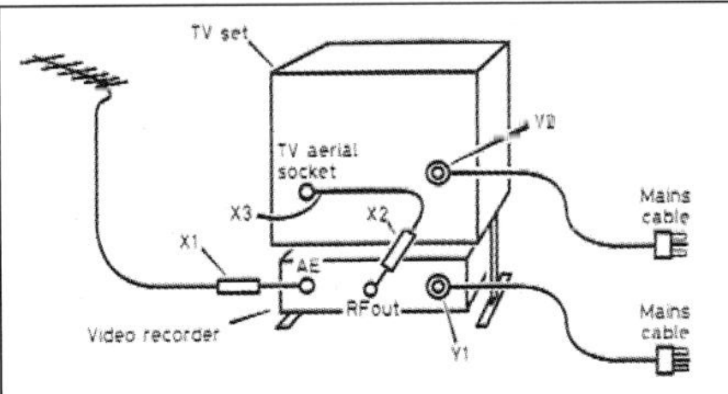

*Fig 2: Fitting filters to a TV set with a video recorder*

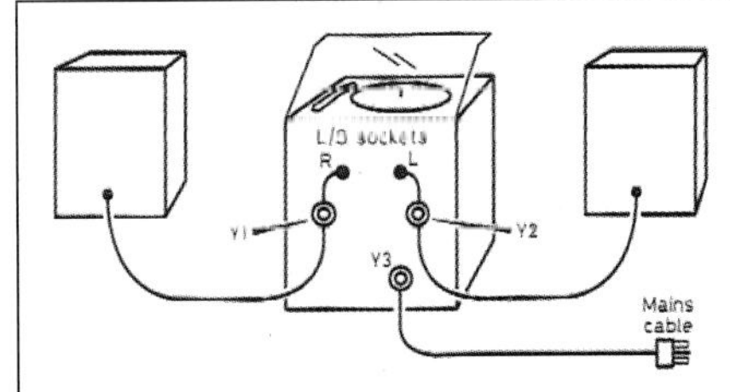

*Fig 3: Fitting common mode chokes to an audio system*

to use a suitable grade of ferrite with enough turns in order to obtain the highest possible impedance at the frequencies of interest (3 - 5kΩ or more is possible).

### Winding a choke

Above about 10MHz, it becomes increasingly important to minimise the stray capacitance between the ends of the winding. Stray capacitance can be reduced by using the winding method shown in **Fig 4**. It is vital that the end of the cable is always threaded through the core in the same direction as shown by the arrows. Finally, the ends should be secured with self-locking cable ties as shown.

If the cable is very thick, if it has connectors which cannot easily be removed or if it is not long enough, the best solution is usually to make up a short plug-in extension lead using the thinnest suitable cable. This is wound through a ferrite core and then fitted with suitable connectors. Normal TV coax should not be wound tightly through a ferrite ring or other core, otherwise it may collapse internally and short-circuit. Instead, a one metre length of miniature 75  coaxial cable, such as Maplin Electronics' XR88V, should be used and fitted with coaxial connectors.

In some cases, fitting a clip-on ferrite core onto the cable may be the only feasible solution, more than one clip-on may be needed.

### Ferrite ring core characteristics

There is no single grade of ferrite which is ideal for EMC use on all HF and VHF amateur bands. Fair-Rite type 43 material gives good results from 7MHz upwards, but for the 1.8MHz and 3.5MHz bands better results are obtained with a higher permeability ferrite such as Fair-Rite type 31 or type 73 material if available.

Minimising stray capacitance becomes particularly important at VHF, so for 144MHz the recommended number of turns is 6 or 7 on a ring core or 3 for a clip-on core. Further details of various cores and their characteristics are given below.

**Fig 5** shows the approximate impedance of various chokes wound on ferrite ring cores, measured using a network analyser.

Curve 'A' shows the characteristics of a 12-turn winding on a single Fair-Rite 2643802702 ring core which gives good results from 7 to 28MHz. For cables thicker than 5mm, similar results can be achieved using 8-9 turns on two rings stacked together or 6-turns on four rings stacked together. (**Note**: Half the number of turns gives one quarter of the inductance.) Curve 'B' shows the characteristics of a 12-turn winding on two Fair-Rite 2643802702 ring cores wound together. This gives excellent performance from 3.5 to 14MHz but it can be seen that the impedance is likely to peak at around 8.5MHz.

Curve 'C' shows the characteristics of a 12-turn winding on two Fair-Rite 2643802702 ring cores wound separately, which reduces the overall stray capacitance but requires more cable. This gives excellent performance from 3.5 to 50MHz and it can be seen that the impedance is likely to peak at around 15MHz.

Around 80-144MHz, curves B and C both show unwanted resonances that depend on the exact winding configuration.

Curve 'D' shows the characteristics of a 6-turn winding on a single Fair-Rite 2643802702 ring core. This gives good results from 50-144MHz.

### Other types of ferrite core

Clip-on ferrite cores can be fitted to a cable without access to the ends. **Fig 6** shows the characteristics of two types of clip-on ferrite core:

TDK type ZCAT3035-1330 with a 13mm diameter hole. The ferrite is TDK Type 30, which is primarily intended for EMC use at VHF.

Fair-Rite type 0431176451 with an 18mm diameter hole. The ferrite is Fair-Rite Type 31, which has higher permeability at HF than the TDK Type 30.

In **Fig 6**, Curve 'A' shows the measured characteristics of a TDK clip-on core type ZCAT3035-1330 with a single turn winding (passing the cable through the core aperture once). This gives an impedance of typically 110-220  from 10-144MHz.

Curve 'B' shows the characteristics of a Fair- Rite clip-on core type 0431176451 with a single turn winding (derived from manufacturer's published data). This gives an impedance of typically 135-300  from 10-144MHz.

Curve 'C' shows the measured characteris- tics of a three turn winding on a TDK clip-on core type ZCAT3035-1330. This gives an impedance of 870-1750  from 10-144MHz.

Curve 'D' shows the typical characteristics of a three turn winding on a Fair-Rite clip-on core type 0431176451 (derived from manufacturer's published data). This gives an impedance of 400-2250  from 10-144MHz.

In practice, the impedance of a single turn winding is relatively small and may be insufficient in many cases. Below resonance, two turns should give four times as much impedance and three turns should give nine times as much. For frequencies up to 70MHz, aim for three turns and for frequencies up to 30MHz, four, five or more turns would be better if possible, but the two halves of the core must close together without the slightest air gap.

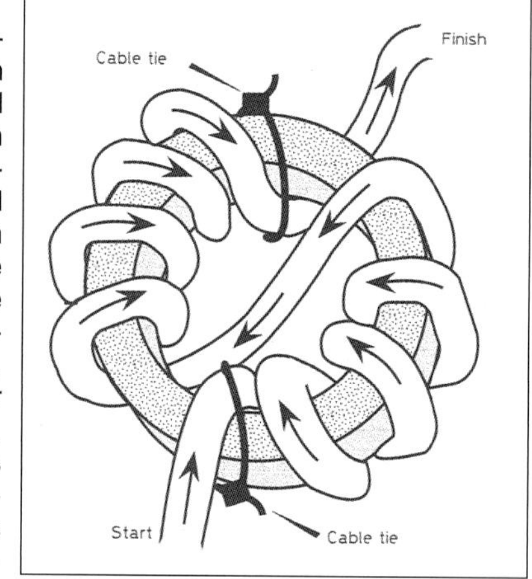


*Fig 4: Recommended winding method for ring cores*

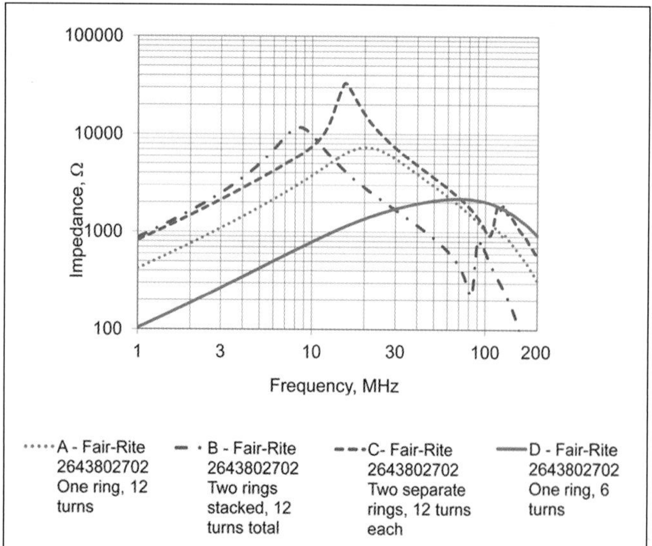


*Fig 5: Impedance of various ferrite rings at different frequencies*

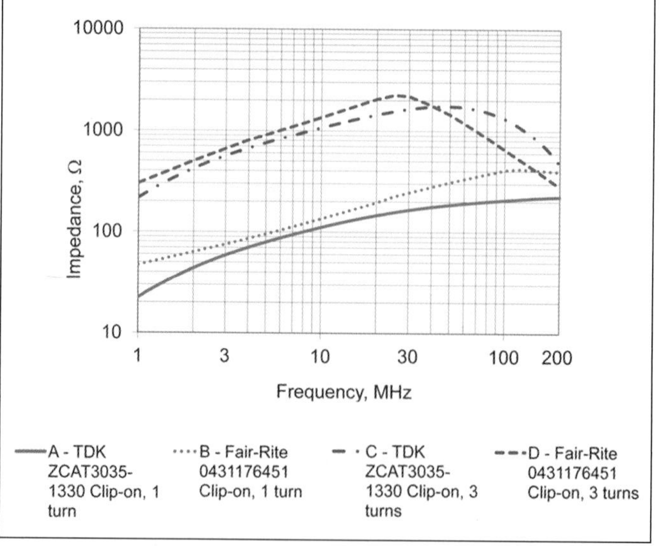


*Fig 6: Impedance of some other types of ferrite cores at various frequencies*

## EMC Directive

The existing EU Directive on EMC, 2004/108/EC, was superseded by Directive 2014/30. This should have been implemented in UK law by 19 April 2016 but the Society understands this will now take place later in the year. The main differences in the new Directive are enhanced market surveillance and safeguard procedures. In the meantime Statutory Instrument 2006/3418 will continue to apply. The EMC Directive specifically cites the protection of the amateur service as one of its aims.

The Directive includes Fixed Installations as well as single items of electrical or electronic apparatus. All things that are within the scope however are still required to meet Essential Requirements in respect of their emissions and immunity to electromagnetic interference in order that they can be declared compliant with the Directive and placed on the EU market or taken into service. As well as technical requirements there are administrative requirements to be met. Only compliant apparatus may carry the CE mark. To meet the Essential Requirements a technical assessment has to be carried out on the apparatus. This can be done against an EU Harmonised Standard or by carrying out measurements which may rely on existing relevant technical specifications.

Fixed Installations are not required to carry the CE mark and there are no Harmonised Standards for them, although individual pieces of equipment within the installation may need to meet the requirements for apparatus. Those responsible for installations must however carry out tests using good engineering practice to prove compliance with the Essential Requirements when the installation is first put into service and must keep documentation to show this.

There is some doubt about the precise scope of Fixed Installations and The Society is currently engaged in discussions with BIS and Ofcom about the compliance of domestic solar panel installations that have caused interference to amateurs at first switch-on. The current BIS but not, apparently, the EU Commission's, view is that these are not Fixed Installations but apparatus rather like items in a domestic video or HiFi set up. Although RSGB has taken up with Ofcom cases where such apparatus does not comply when first taken into service, which the Directive and UK regulations require, Ofcom has maintained it does not have the power to act.

The main enforcement authorities in the UK are Trading Standards and Ofcom. Both have a statutory duty of enforcement, but Ofcom's is specifically in respect of protection of the radio spectrum so it is clearly to them that we should take complaints of non-compliance where interference is caused. The Secretary of State (through the Department of Business, Innovation and Skills – BIS) has a residual, discretionary, power to enforce. The EMC committee continues to watch for cases of non-compliance and will continue to take up worthy cases.

### RTTE Directive

As with the EMCD the current Radio Equipment and Telecommunications Terminal Equipment (RTTE) Directive 1999/5 has also been superseded. The new Radio Equipment Directive (2014/53) will also need to be transposed into UK law but in the meantime the existing UK Regulations apply. The new Directive will only apply to radio equipment, fixed line telecommunications equipment only needing now to be compliant individually with the Low Voltage and EMC Directives. So far as amateurs are concerned, the Directive covers commercially available equipment but not kits intended for amateurs or home built equipment not intended to be placed on the market.

Amateurs are expected to use their expertise when self-building equipment to ensure it does not cause EMC problems.

A similar enforcement regime is in place to that of the EMC Regulations. RTTE compliance does not infer a right to use radio equipment; national rules still apply, so that in the UK exemption or licensing is still required under the Wireless Telegraphy Act 2006.

### Interference Legislation

Both the EMC and RTTE Directives are designed to facilitate the free movement of goods in the EU and compliance is required when apparatus is either first placed on the market and/or first taken into service. They do not apply to in-use situations. For radio interference, in-use situations should be covered by interference Regulations under the Wireless Telegraphy Act 2006.

However, such Regulations as existed were based on long outdated standards. The Society's repeated requests to Ofcom to bring in new Regulations, resulted in Ofcom consulting on new legislation. The new Regulations, SI2016/426, came into force in April 2016. While the Society broadly welcomed their introduction it is taking up with Ofcom what it sees as substantial unjustified omissions.

So far as interference from TVs is concerned (plasma screen TV, masthead amps and the like), the TV receiving licence contains a condition that the user's TV receiving apparatus must not cause undue interference to other radio use. Ofcom have powers to enforce this and they have said that they will remind users of their obligations in this respect in appropriate circumstances. The Society hopes however, that the new interference Regulations described above will also cover such cases, but the boundary between the two regimes is not yet clear.

# Operating Abroad

Assuming that the country allows Amateur operation, there are generally two routes that may apply to operating in it. For temporary operating a number of countries allow this under a European Conference of Postal and Telecommunication Administrations (CEPT) agreement. The second, that allows permanent operating, is to obtain a reciprocal licence.

## CEPT agreement

CEPT is a group of countries from Europe, Scandinavia and countries near to Europe.

Amongst their many actions, many but not all have agreed a common standard of amateur radio licence (*T/R 61-01*) so as to facilitate temporary operation when visiting a fellow CEPT country, as well as a number of countries outside CEPT who have agreed and met the conditions to use *T/R 61-01*. Once each CEPT member's country has confirmed that their amateur radio licence conforms to the *T/R 61-01* minimum standard, then its amateurs may operate in other countries which have also confirmed the recommendation.

CEPT *T/R 61-01* operation does not replace reciprocal licensing - rather it supplements it. Only temporary operation is permitted under CEPT rules, eg from holiday accommodation or mobile. Therefore, if you seek either long-term (over three months) residence or additional facilities, you will still need to apply for a reciprocal licence. Operating under CEPT regulations means that you are restricted by the regulations of the foreign country; thus the RSGB recommends that you get a copy of the licensing regulations for the countries in which you plan to operate.

Providing you hold a **Full licence** you can operate in countries which have implemented *CEPT Recommendation T/R 61-01* in accordance with the terms of that recommendation.

To underline the point licences ***excluded*** from this arrangement are **Full (Reciprocal)**, **Full (Temporary Reciprocal)** and **Full (Club)** along with the **Intermediate** and the **Foundation** licence holders.

If you satisfy the above then you must also ensure that at all times you comply with the requirements applicable to the use of amateur radio equipment at the location in the country concerned. Some vary from ours, so it is vital to check before your journey. Also, remember that if requested you must present your licence documention plus any other documentation required to the relevant authorities in that country. Unless told otherwise you must use the callsign specified in section 1 of your licence *after* the appropriate host country callsign prefix.

To get the up-to-date list of countries (both CEPT and non-CEPT countries) that have implemented *CEPT T/R 61-01*, and to be sure you have the latest information check out the ERO website which has at the top of the page a downloadable version of *CEPT Recommendation T/R 61-01*. Another useful source of up-to-date information on licensing is OH2MCN's website.

## Reciprocal licensing

A reciprocal licence is a licence issued by a foreign country to you because that country recognises the standards of the UK licence. Some countries do this unilaterally, others require a two-way recognition - a reciprocal licence. The callsign issued is sometimes your own callsign with the foreign country's suffix or prefix. In other countries it is allocated in their normal series of callsigns.

At the moment not all countries have followed the UK by abandoning the Morse test requirement for an HF licence. If you held a **Class-B Full licence**, check that you will be able to operate on HF in any country you are travelling to.

Due to overseas post and administration delays, it is generally best to allow at least two to three months for your application to be processed - longer if it is a Third World country where amateur radio is not so sympathetically regarded. The use of air mail certainly helps in this regard.

Like operating under the *CEPT T/R 61-01* agreement, reciprocal licences are **not** available for **Foundation** or **Intermediate** licence holders.

## Other information

Most countries have a national society which looks after the well being of that country's amateurs. A list may be found on the IARU website. Few are of the size of the RSGB - indeed many are staffed entirely by volunteers. Nevertheless, they will all give you as much assistance as they can.

There is usually little problem with customs. It certainly helps to be able to show that the equipment was purchased abroad and is not being exported. Unfortunately, neither a reciprocal licence nor operation under CEPT regulations is deemed an exemption from customs formalities. If in doubt, you should seek additional advice about importing/exporting equipment. Information is available on 020 7202 4227.

Useful resources on operating in foreign countries can be found on the the ARRL website at:
*www.arrl.org/reciprocal-permit* and OH2MCN's website at:
*www.qsl.net/oh2mcn/license.htm* and also at: European Radiocommunications Office website: *www.ero.dk/*

Details of IARU Societies (on the IARU website): www.iaru.org/iaru-soc.html
European Radiocommunications Office website: www.ero.dk/

# Amateur Radio Direction Finding

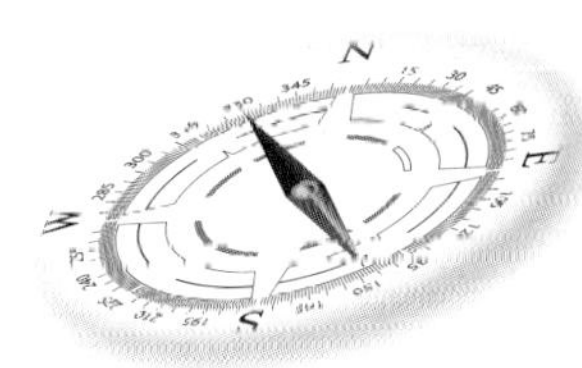

Amateur direction finding in the UK goes back to the time between the wars when competitions were run using the 1.8MHz band. This tradition continues to the present day with competitions organised by the British Top Band DF Association, a society affiliated to the RSGB, which was formed in 2000. These competitions generally involve two hidden transmitters and take place across most of an Ordnance Survey 1:50,000 map sheet. There are eight qualifying rounds each year prior to a three transmitter National Final in September. The 1950 RSGB Council Cup is awarded to the winner although the competitions are no longer promoted or run directly by the RSGB. The British Top Band DF Association organises other smaller events, some of which take place after dark.

The arrival of the commercial VHF hand-held radio in the 1970s spawned a different kind of direction finding competition, generally using the 144MHz band. These events are organised by local clubs, many of which are affiliated to the RSGB. Usually a club member parks up in the countryside and makes a series of transmissions using a mobile 144MHz radio. Other club members attempt to locate him and the evening often concludes comparing notes in a local hostelry. There are potential issues concerning the Road Traffic Act and the wording of motor insurance policies for these car-based competitions. The Road Traffic Act makes it an offence in the primary legislation to engage in racing or trials of speed on the public highway and Clubs will wish to frame their rules to avoid any possibility of infringing this. The Act makes both the participant and the organiser responsible in the event of an infringement. An increasing number of insurers now include a clause which invalidates cover if the insured enters any kind of competition irrespective of whether racing or trials of speed are involved or not. It is worthwhile checking the small print on the policy to avoid driving while uninsured.

In continental Europe, things developed along a different track after the end of WW2 in 1945. Back then, only in Great Britain, Eire and Czechoslovakia were amateurs allowed to operate on 1.8MHz, so countries wishing to introduce surface wave direction finding simply used the lowest frequency band available to them, which was 3.5MHz. Today, this choice is embodied in the set of rules supported by the IARU and also in thousands of DF receivers for this band across the world.

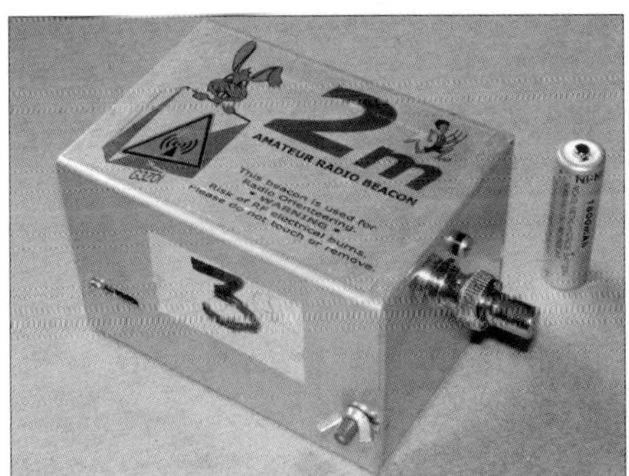


*Fig 1: A typical 2m ARDF transmitter with an AA cell for size comparison.*

| TX | Minute 1 | Minute 2 | Minute 3 | Minute 4 | Minute 5 | Minute 6 |
|---|---|---|---|---|---|---|
| No.1 | MOE (one dot) | Silent | Silent | Silent | Silent | MOE (one dot) |
| No.2 | Silent | MOI (two dots) | Silent | Silent | Silent | Silent |
| No.3 | Silent | Silent | MOS (three dots) | Silent | Silent | Silent |
| No.4 | Silent | Silent | Silent | MOH (four dots) | Silent | Silent |
| No.5 | Silent | Silent | Silent | Silent | MO5 (five dots) | Silent |
| | 1 x 5-minute cycle (all five transmitters operate on the same freq) | | | | | |

*Table 1: The timing sequence of the five hidden transmitters.*

Region 1 of the IARU has an ARDF Working Group responsible for the formulation of rules, since by far the greatest interest is in Europe. It is the custom for Regions 2 and 3 to adopt the rules originating in Region 1.

The situation today is that ARDF is extremely vibrant in Europe. There have been seventeen bi-annual World Championships and twenty Region 1 Championships, also a bi-annual event but in odd numbered years.

Competitions are organised using the 3.5MHz band, where propagation is predictable and good bearings are generally obtained, and the 144MHz band, which exhibits significant multi-path propagation, sometimes leading to misleading bearings being obtained. The competitions take place entirely on foot and no motor vehicles are involved.

In Great Britain the RSGB made rather a late start to this international style of ARDF, with the first UK event being held in 2002. Many European countries have over 50 years of experience, especially in the old Soviet Bloc countries where there used to be significant state support for activities like ARDF that were also useful militarily. State support has ebbed away since 1989 but the Eastern European countries have inherited a long tradition of direction finding that makes them dominant at World level.

## Introduction to the IARU Rules

### Transmitters and timing

Five low power transmitters (3W output if 3.5MHz is being used and 800mW if 144MHz is chosen) are deployed in the area to be used. A typical transmitter is shown in Fig 1. All the transmitters operate on the same frequency, but not all at the same time. They transmit in sequence and send an identifier in Morse code for one minute each. Before the reader freaks out at the mention of Morse code, it should be pointed out that the identifier is simply a matter of dot counting.

The first transmitter sends the letters MOE in Morse. The first two letters are long ones in Morse and serve to keep the transmitter on the air for a while, to allow the competitor to swing the aerial carried and assess the direction of the transmitter. The last letter is a single dot, so one dot denotes transmitter 1. This transmitter sends for one minute before shutting down.

The second transmitter then radiates the

morse sequence MOI. The last letter (I) is two dots, to denote transmitter number 2. Transmitters 3, 4 and 5 transmit MOS, MOH and MO5 respectively. The sequence is shown pictorially in Table 1.

In the UK it is now a licence condition that the callsign of the supervising licensed amateur is radiated at the end of each transmission, so the one minute transmission terminates with a burst of higher speed Morse, which is this callsign.

In addition to the five hidden transmitters there is a beacon transmitter operating on a different frequency, which radiates the letters MO repeatedly in Morse and is interrupted at intervals with the callsign of the supervising licensed amateur. This transmission is continuous and enables competitors who get hopelessly lost to simply DF the beacon to find their way to the finish.

All the transmitters use some form of omnidirectional antenna. For 3.5MHz an 8m vertical wire with an 8m counterpoise is frequently deployed, while on 144MHz a pair of crossed horizontal dipoles (aka a turnstile antenna) at a height of about 3m is commonplace.

## Proof of finding the transmitters

Clearly it is necessary for the competitor to demonstrate that each assigned transmitter has been visited. This can be done in one of two ways:

1. The competitor carries a control card (see Fig 2) with a space for each of the five hidden transmitters plus one for the beacon if the latter is to be registered. At each transmitter there is a needle punch (see Fig 3) which is used to mark a unique pattern of needle holes in the card. In international competition the beacon will also be 'punched' but practice varies in domestic races.
2. Electronic timing equipment may be used. Each competitor carries a microchip (see Fig 4), which is inserted into a unit at each transmitter. The transmitter writes its identity plus the time of the visit to the microchip. On completion of the course, the competitor punches at the finish and then downloads all the data to a computer, which is able to print the time taken, the transmitters visited and all the split times.

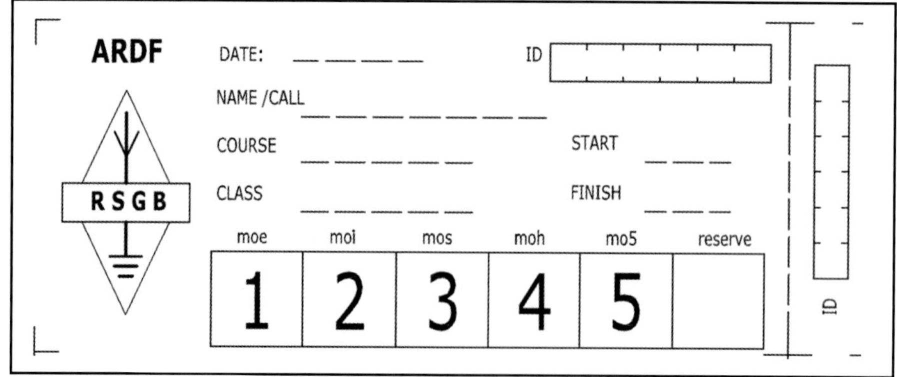
ARDF DATE: ID
RSGB
NAME /CALL
COURSE START
CLASS FINISH

| moe | moi | mos | moh | mo5 | reserve |
|---|---|---|---|---|---|
| 1 | 2 | 3 | 4 | 5 | |

ID

*Fig 2: Control Card for a five transmitter DF hunt.*

## Age categories

ARDF is organised into a series of age categories and the adult age categories in force are shown in Table 2.

| Men | Women |
|---|---|
| M19 | W19 |
| M21 | W21 |
| M40 | W35 |
| M50 | W50 |
| M60 | W60 |
| M70 | |

*Table 2: Age categories for competitors.*

To explain how the system works, consider the age group M21. The M denotes a male age group. A man enters the M21 class on 1 January of the year in which he becomes 21 and leaves it on 1 January of the year in which he becomes 40 (M40 being the next age group).

There are a total of eleven adult age groups, with the older age groups hunting fewer transmitters over shorter distances than the younger age groups.

The result of this is that competition is against one's peers and this considerably broadens the appeal of this radio sport.

## Transmitter placement and the time limit

There are three further rules, which can be of great significance, depending on the shape and size of the area used for the competition. No transmitters can be placed within 750m of the start. In domestic competition this distance is frequently reduced to 400m, to avoid 'sterilising' a large part of a small wood as far as transmitter placement is concerned.

The second restriction is that there can be no transmitter within 400m of the finish.

Finally, transmitters must be placed at least 400m apart.

A time limit is rigorously enforced for competitions, with the rule that any competitor over time is placed below a competitor who has found at least one transmitter and is within the time limit. It can be rather galling to find all five transmitters and finish one minute outside the time, to be beaten by someone who took nearly two hours to find just one transmitter. The time limit is decided by the course planner, but two hours is a frequent choice, with 90 minutes for easier areas. The object of this rule is to constrain those competitors who are determined to find all the transmitters at any cost, even if this sees them still hunting as darkness falls.

## Equipment

It is obvious that competitors will require a receiver for the frequency band being used and a directional antenna for that band. These two items are normally combined into one unit and at UK events there is usually equipment available on loan, although it is sensible to confirm this with the organiser beforehand. The receiver should be an AM receiver, since the direction to the hidden transmitters will be determined by swinging the antenna from side to side and noting changes in signal amplitude. The amplitude limiter in an FM receiver makes it less suitable for this task, although it can still be made to work in this application.

There is a need to plot bearings on the map provided at the start. Beginners usually plot more bearings than experienced competitors. To do this, the map should be taped to a lightweight board of some kind and either a spirit pen or a chinagraph (wax) pencil can then be used to mark the map. Lines can be drawn by both of these markers on any clear plastic covering used to protect the map. Neither of them will run if they later become wet, but only the chinagraph will make a satisfactory mark if the plastic is already wet.

A compass will be required to measure the bearings. The type with a rectangular base plate also doubles as a protractor. The compass is best looped round the wrist with the cord normally provided.

Proving the visit made to each transmitter involves carrying either a Control Card or an SI 'dibber'. This 'dibber' is a plastic encased

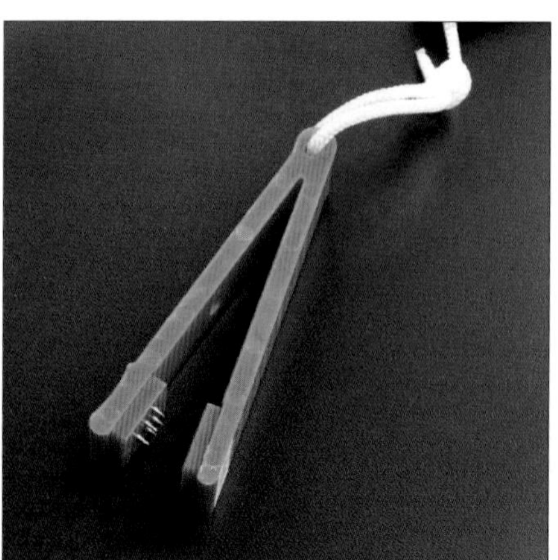

*Fig 3: Pin punch found at the transmitter and used to mark the appropriate box on the control card. The punches at each transmitter carry a unique pattern of pins.*

*Fig 4: The electronic punching 'dibber' is carried on a finger of the competitor by an elastic strap. This is inserted into the unit at each transmitter to register that the competitor has been there.*

microchip used for electronic 'punching'. It is on a small elastic strap which allows it to be attached to the index finger. The control card is best pinned with safety pins to the front of the clothing. Many control cards are printed on tough, waterproof Tyvek paper and require no protection or strengthening. Control cards that are printed on thin card should be covered with Sellotape to both waterproof and strengthen them.

Moving to the desirable rather than the essential, a whistle should be carried. In some competitions is mandatory with a 'no whistle – no run' policy. The emergency signal is six blasts of the whistle at one minute intervals. Also in the desirable category is a circle stencil to mark the circles around the start and the finish, in which no controls can be placed. Obviously a different stencil is required for a map at 1:10,000 scale to one at 1:15,000 scale.

A check list is shown in Table 3.

| Item carried | Note(s) |
|---|---|
| Receiver | Usually fixed to the antenna. |
| Antenna | Usually fixed to the receiver. |
| Map | Normally issued at the start line, 5 or 10 minutes before starting. |
| Lightweight rigid board for the map | To deal with wet weather conditions, the competitor will need waterproofing (a plastic folder or sticky backed plastic film) to cover the map and possibly tape to fix the map to the board. |
| Compass | The type with a rectangular backplate doubles as a protractor. |
| Spirit pens and/or wax pencils | Will not run if it rains, but note that only wax pencils will write satisfactorily on plastic film that is already wet. |
| Circle stencil | 750 and 400m circles at the map scale in use. |
| Control Card or SI 'dibber' | To register that the competitor has visited each assigned transmitter. |
| Whistle | Emergency signal is 6 blasts at 1 minute intervals. |

*Table 3: A checklist of the items you require for a competition.*

## Competition hints

### Pre-start

There is normally five minutes after being given the map and before getting the signal to start, in which the competitor is able to:

a. waterproof and protect the map as deemed appropriate for the weather conditions,
b. on the map, draw a 750m circle around the start and a 400m circle around the finish (in domestic competition the 750m start circle is often reduced to 400m),
c. study the map to identify height features in particular.

### Start + 5 minutes

After being given the start signal, the rules oblige the competitor to keep moving to the end of the start funnel. Once at the end the aim should be to listen to each transmitter in turn, assess the strength of the signal and plot the bearing – all within the 60 seconds that it is on the air. Prior practice at this procedure will pay big dividends for the beginner. If the 144MHz band is being used, bearings taken from high spots are more accurate than those taken from valleys. It may pay to sacrifice a complete transmitter cycle and climb to the top of a nearby hill or spur from which more accurate bearings can be obtained.

### Decision time

Based on the information gained by listening just once to each of the transmitters, the most important decision of the day must be made. This is the choice of the first transmitter to be visited. In the case of a co-located start/finish, the penalties for getting it wrong are not as severe compared to a split start/finish. With a co-located start and finish, a poor choice can often be rectified on the route back from the furthest transmitters to the finish. When the start and finish are at separate locations, a bad decision may mean a lot of 'back tracking' and hence wasted time.

### Bearing quality

The surface wave propagation on 3.5MHz during daytime leads to bearings which are generally pretty accurate. While it is wise to avoid wire fences and overhead power lines when taking bearings, the accuracy of the plotted bearing is determined by the equipment and skill of the competitor. This results in fast runners being able to get to the transmitters first, assuming that they also have reasonable direction finding skills.

On 144MHz there is a lot of multi-path propagation, with the signal being reflected or scattered from steep hillsides, rock outcrops and even the edges of wooded areas. The bearings obtained vary greatly in 'quality'. A sharp, clear peak in the signal as the antenna is swung from side to side is indicative of a single path signal and this is often the direct path from the transmitter. Multi-path propagation most often reveals itself as a rather diffuse bearing as the antenna is swung. Sometimes there may be more than one distinct peak to the signal and this is where an antenna with very low side and back responses comes into its own, to differentiate between the direct and the multi-path signals.

*Andrew, G4KWQ 'flies the flag' at the World ARDF Championships.*

All this interpretation of bearings coupled with the need to view the bearings against the background of any high ground in the vicinity; leads to the winner needing to process all this information quickly and accurately. Hence, being able to run fast is no longer such a key quality to gain victory.

## Your first event

### Event information

Finding out about competitions is clearly the first step and there are currently around 15-20 events held in the UK each year. There is usually a break around Christmas and January, with the 'season' generally commencing in February and the last event in November or December.

The RSGB website is the gateway to information about ARDF events. The URL is given below.

### Clothing and equipment

The issue of equipment has already been covered above. As far as clothing is concerned, for a first outing, stout shoes and outdoor attire is sufficient. If you become more committed, then studded orienteering shoes, gaiters to protect the lower leg against brambles and nettles and an orienteering suit are more appropriate. On occasions when the weather is particularly inclement (thankfully few) then a cagoule or other waterproof garment will be needed.

### Registration

Events that are run in conjunction with an orienteering event will benefit from the direction signs to that event. ARDF events that are freestanding are not likely to be extensively signed, so the competitor should ensure that a copy of the Grid Reference and of the map extract that is frequently given with the event details, are carried on the journey to the venue.

RSGB ARDF page: www.rsgb.org/main/radio-sport/

Once there, it is necessary to register and pay the event fee (usually of the order of £5-£6). This provides for entry for all the competitions taking place on the day. There is usually a full scale event in the morning, followed by a more relaxed competition after lunch. All of this provides an excellent day of radio sport and makes it well worthwhile to travel a fair distance for a full day of radio in the open air.

At registration you may have to make a choice regarding the number of transmitters you wish to hunt. Details of the frequencies of the transmitters (hidden transmitters on one frequency and the homing beacon on a second frequency), the radius of the zone around the start in which transmitters may not be placed and your individual start time will be given to you. Finally, the time limit for the event should be noted.

If electronic timing is being used, it will probably be necessary for you to hire an electronic 'chip' (colloquially known as a dibber).

If pin punching is in use, then you will be given a control card. Fill this out with your details. If it is made of plain paper or thin card, protect and strengthen it with Sellotape and finally pin it to the front of your clothing with safety pins.

### The map

An orienteering style map will be in use and the most common scale in domestic competition is 1:10,000. These maps show much more ground detail than an Ordnance Survey map. The first thing to note is that the white bits are trees – quite the opposite to an Ordnance Survey map. The white parts of the map denote runnable forest and various shades of green show less runnable areas, with dark green being really impenetrable and well worth avoiding. Fortunately you are very unlikely to have transmitters located in these latter areas.

Open and semi-open areas are shown in a yellow ochre colour.

Only the start (a triangle) and the finish (a double circle with a smaller circle inside a larger one) will be marked on the map.

### After the start

In simple terms, listen to all of the transmitters to get a bearing and an idea of the signal strength of each one, decide which transmitter you wish to visit first and then head for the one selected.

Finding your very first hidden transmitter is a great moment and one to be remembered for a very long time. Keep an eye on the clock, so that you get back to the finish inside the time, and see if more transmitters can be located.

Newcomers are usually a bit erratic in their first events, as is to be expected. For some a brilliant performance at the first outing can be followed by poor and disappointing results at subsequent events. Experience tells us that it takes about six outings for the majority of competitors to settle in and be able to locate all the assigned transmitters inside the time on a reliable basis. In other words don't get discouraged by a few poor results; it will all come together for you with a bit of experience.

After finishing there is the opportunity to compare notes with other competitors and to get some tips on how to avoid any mistakes at future events.

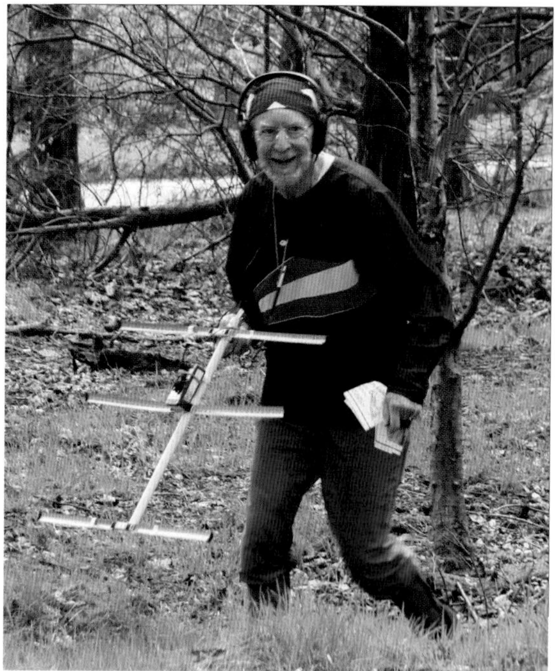

*Now competing in the 70+ age category and not a scrap of lycra in sight; Robert Vickers G3ORI shows that ARDF can be enjoyed by amateurs of any age.*

## Other Formats

There are two variants of the basic format. The first is Foxoring, which is a hybrid of Orienteering and Radio Direction Finding. A large number of very low power 3.5MHz transmitters are deployed and a circle is marked on the map within which each one will be audible. The competitor uses orienteering techniques to navigate to the area of the circle. Once the signals from the transmitter are picked up, direction finding enables the transmitter to be located.

The second variant is the sprint format. This provides two clusters of five transmitters operating on two different frequencies in the 3.5MHz band and keyed at different speeds. Each transmission lasts just 12 seconds and competitors have to return to a spectator beacon after finding all the transmitters in the first group and before they set out to find transmitters in the second group. The transmitters are keyed with the usual MOE, MOI etc. The format is very fast and furious and winning times of less than 15 minutes are not unknown.

## Further Information

A description of competitive direction finding cannot be exhaustive in the space available here. The RSGB book 'Radio Orienteering – The ARDF Handbook', goes into much fuller detail and is essential reading for the beginner.

Event information is available on the RSGB website: www.rsgb.org/main/radio-sport/ARDF/events

The ARDF pages also include the results of competitions, details of the big international events and information about sources of suitable equipment.

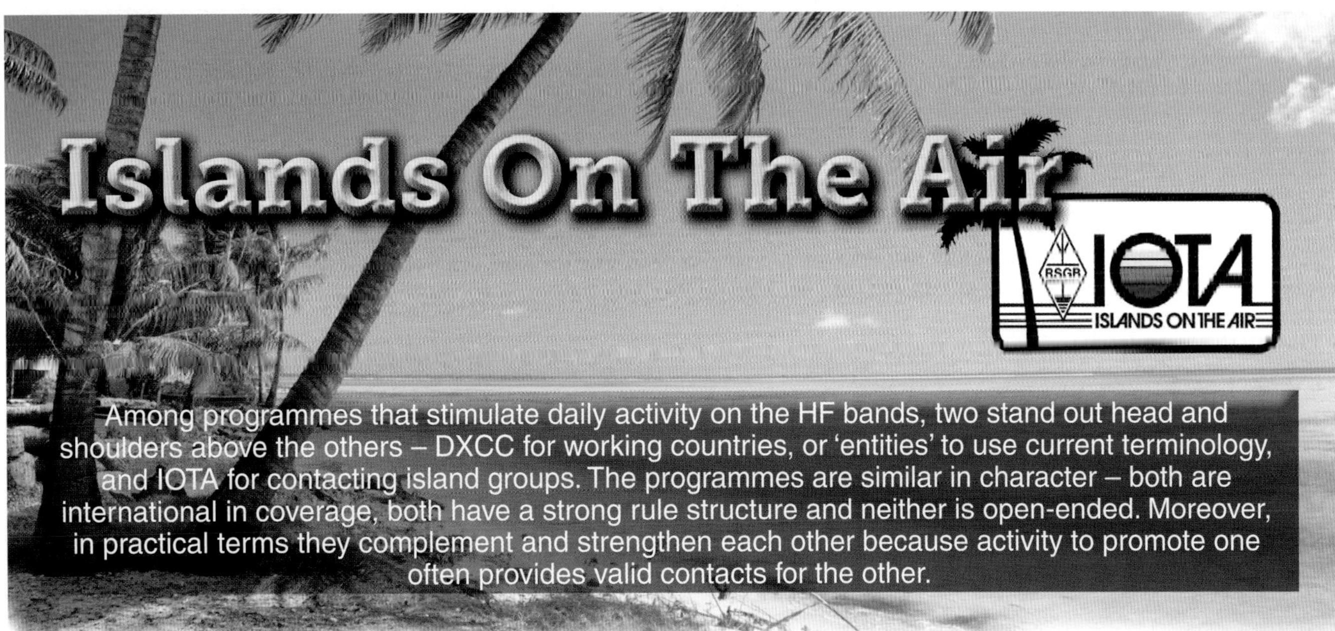

Islands On The Air
RSGB
IOTA
ISLANDS ON THE AIR
Among programmes that stimulate daily activity on the HF bands, two stand out head and shoulders above the others – DXCC for working countries, or 'entities' to use current terminology, and IOTA for contacting island groups. The programmes are similar in character – both are international in coverage, both have a strong rule structure and neither is open-ended. Moreover, in practical terms they complement and strengthen each other because activity to promote one often provides valid contacts for the other.


IOTA, or the Islands On The Air Programme to give it its full title, was created in 1964 by the late Geoff Watts, a leading British short wave listener and the only SWL in the DX Hall of Fame. When the programme was taken over, at Geoff's request, by the RSGB in 1985 it was already a favourite for many DXers. Its popularity has since grown each year, not only among ever-increasing numbers of island chasers but also among a rapidly expanding band of amateurs attracted by the possibilities for operating portable from islands. For both it is a fun pastime adding much enjoyment to on-the-air activity.

The basic building block for IOTA is the IOTA Group. The oceans' islands have been corralled into some 1200 IOTA Groups with, for reasons of geography, varying numbers of 'counters', i.e. qualifying islands, in each. Only in very few cases do the rules of IOTA allow single islands to count separately, DXCC island entities such as Barbados being one. The number of groups is now capped and further changes are expected to be minimal.

Each group activated has been issued with an IOTA reference number, for example EU-005 for Great Britain. Part of the fun of IOTA is that it is an evolving programme with new groups being activated for the first time. Currently some 1122 of the 1200 groups have confirmed numbers.

The objective, for the island chaser, is to make radio contact with at least one counter in as many of these groups as possible and, for the DX-peditioner, to provide island contacts. A wide range of separate certificates, graded in difficulty, is currently available for island chasers as well as two prestigious awards for high achievement (see the table overleaf). Applicants may be any licensed radio amateur (or SWL on a 'heard' basis) who has had confirmed contacts with the required number of IOTA Groups listed.

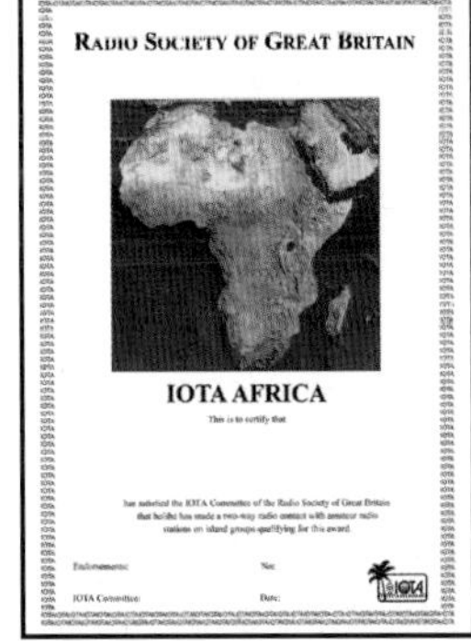


## IOTA Directory

The latest *RSGB IOTA Directory, 17th Edition, published in May 2016,* gives a full listing of IOTA Groups together with the names of 15,000 qualifying islands. You can order a copy on-line on the Society website at: *www.rsgb.org/shop* or direct from RSGB.

## Applying for an Award

IOTA is in the process of introducing a method of electronic confirmation of contact by QSO matching with logs on Club Log. The system will however continue to accept confirmation by QSL cards. Award applicants should prepare and submit their applications electronically on the Internet. Full details of the application procedure and a list of checkpoints can be found on the IOTA website at www.rsgbiota.org After signing off your application on-line for processing by your checkpoint, you should immediately send him by post your application print-out, the appropriate checking fee and any cards required for confirmation.

## Island-chasing

1000 or more IOTA Groups may seem an enormous target. If you are a long time DXer who has worked it all and are looking for something new, you will already have amassed a very respectable IOTA score from among your DXCC contacts. If, however, you are new to the bands or one of the many amateurs who adopt a more relaxed approach to their operating, you can take full advantage of a very high level of IOTA activity, comprising easy and semi-rare groups, to launch you on your way. Well over 600 IOTA groups are usually activated over a three year period with, during a typical summer weekend, some 20/25 IOTA Groups being heard around the IOTA meeting frequencies. An enthusiast should be able to gain the IOTA Plaque of Excellence for working 750 groups in about six years, operating mainly at weekends. This must be a reasonable target to go for – after all, how long does it take to get to the top of the DXCC Honour Roll?

IOTA is one of the few award programmes that has an annual Honour Roll and other performance listings. These create a great deal of interest when they are published each spring. Many IOTA enthusiasts are more interested in participating in these listings than in collecting the certificates. All you need to enable you to participate is a registered score of at least 100 Islands.

## Operating from an island

Many amateurs are fortunate enough to live on an island and to be able to give out an IOTA every time they make a contact. Others are not so lucky. For both there is the lure of operating portable from a rare or rarer group – the fun of being at the other end of a pile-up for a few days. Many islands lie within a few hours' reach and, subject to the availability of suitable equipment, could be put on the air relatively easily.

Those amateurs lucky enough to be able to activate a rare or semi-rare IOTA Group can expect to generate huge pile-ups with thousands of contacts during even a short 2/3 day period. Rare groups are not all remote and difficult to access. Even in Europe and North America there are many that are needed by the chasers. For those interested, a list of most wanted IOTA Groups in each continent, ranked by rarity, can be viewed on the IOTA website.

## Categories of application

IOTA began as an award for single operators working on the HF bands (1.8 to 30MHz). However, in response to demand, the IOTA Committee subsequently introduced categories specifically for club stations and for working on VHF/UHF (50MHz and above).

Address: Islands on the Air (IOTA) Ltd, 5 Morton, Tadworth Park, Tadworth, KT20 5UA.
Email: islandsontheair@outlook.com

## IOTA meeting frequencies

Nobody and no group in amateur radio is entitled to reserved frequencies, but the IOTA community has adopted a number of 'meeting frequencies' which island stations are encouraged to use when they are free – and to operate close to, without causing interference, if they are occupied. The frequencies are 3755, 7055, 14260, 18128, 21260, 24950, 28460 and 28560kHz on SSB and 3530, 10115, 14040, 18098, 21040, 24920 and 28040kHz on CW. No specific frequency has been nominated for 7MHz CW, but it is recommended that operations should include a frequency above 7025kHz when the band is open to North America.

## IOTA contest

The IOTA Contest, first held in July 1993, has become enormously popular and now regularly attracts more than 2000 entries. It provides an opportunity annually, at the end of July, to work large numbers of rare and semi-rare IOTA Groups. Contest rules and results are available from the RSGB HF Contest Committee website.

## IOTA Annual Listings

The following pages show the IOTA Annual Listings as of February 2016. The lists are divided as follows:

**The Honour Roll** is a list of the callsigns of stations with a checked score equalling or exceeding 50% of the total of numbered IOTA groups, excluding those with provisional numbers, at the time of preparation.

**The Annual Listing** is a list of the callsigns of stations with a checked score of 100 or more IOTA groups but less than the qualifying threshold for entry into the Honour Roll.

**The Club Listing** is a list of the callsigns of club or multi-operator stations with a checked score of 100 or more IOTA groups.

**The VHF/UHF Listing** is a list of the callsigns of stations with a checked score of 100 or more IOTA groups on the VHF/UHF bands.

**The SWL Listing** is a list of SWLs with a checked score of 100 or more IOTA groups.

Listing in the 2016 tables was restricted to those participants who had updated their scores since February 2011. IOTA rules limit inclusion in the listings to those participants who have updated their scores at least once in the preceding five years and have opted to have their scores published.

All participants should be reminded that the final decision on acceptance of credits is made at IOTA HQ and that this can mean downward adjustments to scores at any time to reflect corrections of one sort or another. Data-cleansing work is on-going and covers every participant's complete record, not just the latest credits added. Although efforts are made to alert participants to score changes, this cannot be guaranteed to happen in each case. Remember, the line of communication is via your checkpoint, so please do not route queries direct to IOTA HQ or the IOTA Manager. Always check first to see if the answer is in the IOTA Directory (any edition since 2000).

## IOTA Awards

| Award | All Band Categories | VHF /UHF Categories |
|---|---|---|
| IOTA 100 Islands of the World Certificate | 100 Confirmed IOTA Groups including 1 from all 7 continents | 100 Confirmed IOTA Groups including 5 continents |
| IOTA 200 Islands of the World Certificate | 200 Confirmed IOTA Groups including 1 from all 7 continents | 200 Confirmed IOTA Groups including 5 continents |
| IOTA 300 Islands of the World Certificate | 300 Confirmed IOTA Groups including 1 from all 7 continents | |
| IOTA 400 Islands of the World Certificate | 400 Confirmed IOTA Groups including 1 from all 7 continents | |
| IOTA 500 Islands of the World Certificate | 500 Confirmed IOTA Groups including 1 from all 7 continents | |
| IOTA 600 Islands of the World Certificate | 600 Confirmed IOTA Groups including 1 from all 7 continents | |
| IOTA 700 Islands of the World Certificate | 700 Confirmed IOTA Groups including 1 from all 7 continents | |
| IOTA 800 Islands of the World Certificate | 800 Confirmed IOTA Groups including 1 from all 7 continents | |
| IOTA 900 Islands of the World Certificate | 900 Confirmed IOTA Groups including 1 from all 7 continents | |
| IOTA 1000 Islands of the World Certificate | 1000 Confirmed IOTA Groups including 1 from all 7 continents | |
| IOTA 1100 Islands of the World Certificate | 1100 Confirmed IOTA Groups including 1 from all 7 continents | |
| IOTA 750 Islands Plaque (with shields for each additional 25 IOTA Groups) | 750 Confirmed IOTA Groups including 1 from all 7 continents | 300 Confirmed IOTA Groups including 5 continents |
| IOTA 1000 Islands Trophy (with shields for each additional 25 IOTA Groups) | 1000 Confirmed IOTA Groups including 1 from all 7 continents | |
| IOTA Africa Certificate | 75 African IOTA Groups | 50 African IOTA Groups |
| IOTA Antarctica Certificate | 75% of Antarctic IOTA Groups | 50% of Antarctic IOTA Groups |
| IOTA Asia Certificate | 75 Asian IOTA Groups | 50 Asian IOTA Groups |
| IOTA Europe Certificate | 75 European IOTA Groups | 50 European IOTA Groups |
| IOTA North America Cert. | 75 North American IOTA Groups | 50 North American IOTA Groups |
| IOTA Oceania Certificate | 75 Oceanian IOTA Groups | 50 Oceanian IOTA Groups |
| IOTA South America Certificate | 75% of South American IOTA Groups or 75 South American Groups, whichever is the lesser number at the time of application | 50% of South American IOTA Groups or 50 South American Groups, whichever is the lesser number at the time of application |
| IOTA World Diploma | 50% of the IOTA Groups in all 7 continents or 50 IOTA Groups for the continents where there are more than 100 IOTA Groups | |
| IOTA Arctic Islands Certificate | 75 Arctic Island Groups | 50 Arctic Island Groups |
| IOTA British Isles Certificate | 75% of the British Isles Groups | 50% of the British Isles Groups |
| IOTA West Indies Certificate | 75% of the West Indies Groups | 50% of the West Indies Groups |

## 2016 Honour Roll

| Pos. | Callsign | Total |
|---|---|---|
| 1 | 9A2AA | 1115 |
| 1 | I2YDX | 1115 |
| 3 | I8ACB | 1114 |
| 4 | I1JQJ | 1113 |
| 5 | K9PPY | 1112 |
| 5 | VE6VK | 1112 |
| 7 | G3KMA | 1111 |
| 7 | I1SNW | 1111 |
| 9 | F2DS | 1110 |
| 10 | I8XTX | 1109 |
| 11 | ON6HE | 1108 |
| 12 | HB9AFI | 1107 |
| 12 | I8KNT | 1107 |
| 12 | VE3XN | 1107 |
| 15 | I4LCK | 1106 |
| 15 | N8JV | 1106 |
| 15 | W9DC | 1106 |
| 18 | G3NDC | 1104 |
| 18 | W5BOS | 1104 |
| 20 | OM3JW | 1103 |
| 21 | CT1ZW | 1102 |
| 21 | F6DLM | 1102 |
| 21 | W1NG | 1102 |
| 24 | DL8NU | 1100 |
| 24 | F6BFH | 1100 |
| 24 | HA0DU | 1100 |
| 27 | ON4AAC | 1099 |
| 27 | YT7DX | 1099 |
| 29 | EA8AKN | 1098 |
| 29 | IK1JJB | 1098 |
| 29 | OE3WWB | 1098 |
| 29 | ON4XL | 1098 |
| 33 | IK8FIQ | 1094 |
| 33 | W1DIG | 1094 |
| 35 | DF2NS | 1093 |
| 35 | K6DT | 1093 |
| Pos'n | Callsign | Total |
| 35 | N5JR | 1093 |
| 38 | SM0AJU | 1092 |
| 39 | I2YBC | 1091 |
| 39 | WD8MGQ | 1091 |
| 41 | IK8DDN | 1090 |
| 41 | K8NA | 1090 |
| 43 | K9AJ | 1089 |
| 43 | VE3LDT | 1089 |
| 45 | 4Z4DX | 1088 |
| 46 | AA5AT | 1087 |
| 46 | G3ZAY | 1087 |
| 46 | K7SO | 1087 |
| 46 | SM5DJZ | 1087 |
| 50 | DL8USA | 1086 |
| 51 | GM3ITN | 1085 |
| 51 | SM3EVR | 1085 |
| 53 | OE3SGA | 1084 |
| 53 | W4DKS | 1084 |
| 55 | N5UR | 1083 |
| 55 | SM6CVX | 1083 |
| 57 | F9GL | 1082 |
| 57 | IK1ADH | 1082 |
| 57 | ON4IZ | 1082 |
| 60 | F6AJA | 1081 |
| 61 | I4EAT | 1080 |
| 62 | IK2MLY | 1079 |
| 63 | F6FHO | 1078 |
| 64 | K8SIX | 1077 |
| 65 | CT1EEB | 1076 |
| 65 | IK5IWU | 1076 |
| 67 | N6VR | 1075 |
| 67 | UA9YE | 1075 |
| 69 | VE7IG | 1073 |
| 69 | WB9EEE | 1073 |
| 71 | F6CKH | 1072 |
| 72 | 4X4JU | 1071 |
| 73 | F5IL | 1070 |
| 73 | IK4WMA | 1070 |
| 73 | PY7ZZ | 1070 |
| 76 | K5MT | 1069 |
| 77 | I4GAS | 1067 |
| 78 | JE1DXC | 1065 |
| 78 | SM3CXS | 1065 |
| 80 | F6AXP | 1064 |
| 80 | OK1ADM | 1064 |
| 82 | OZ4RT | 1062 |
| 83 | DK1RV | 1061 |
| 83 | I2FUG | 1061 |
| 85 | N5ET | 1060 |
| 85 | VE3LYC | 1060 |

| Pos. | Callsign | Total |
|---|---|---|
| 87 | AD5A | 1059 |
| 87 | N7RO | 1059 |
| 87 | PA3EXX | 1059 |
| 90 | EA4MY | 1058 |
| 90 | G3OCA | 1058 |
| 90 | G3RUV | 1058 |
| 93 | OH2BLD | 1057 |
| 93 | SP8BOW | 1057 |
| 95 | G0ANH | 1056 |
| 95 | HB9BZA | 1056 |
| 96 | N6AWD | 1055 |
| 98 | VE7QCR | 1054 |
| 99 | F5NPS | 1053 |
| 100 | IK8RGC | 1052 |
| 101 | K1OA | 1051 |
| 102 | VE7YL | 1050 |
| 103 | DL1BKK | 1049 |
| 103 | EA3JL | 1049 |
| 103 | G3OAG | 1049 |
| 103 | JF1SEK | 1049 |
| 107 | EA3KB | 1048 |
| 108 | GJ3LFJ | 1047 |
| 108 | VK4MA | 1047 |
| 110 | I4MKN | 1046 |
| 111 | DK6IP | 1045 |
| 111 | W1JR | 1045 |
| 113 | F5XL | 1044 |
| 113 | S52KM | 1044 |
| 115 | DK2PR | 1043 |
| 115 | DL1BDD | 1043 |
| 115 | G3HTA | 1043 |
| 115 | ON4ON | 1043 |
| 115 | R6AF | 1043 |
| 115 | WB2YQH | 1043 |
| 121 | VE7DP | 1042 |
| 122 | JA8RJE | 1041 |
| 123 | DL8FL | 1040 |
| 123 | G4SOZ | 1040 |
| 125 | JA1QXY | 1037 |
| 126 | G0APV | 1036 |
| 127 | IT9DAA | 1035 |
| 127 | S51RU | 1035 |
| 129 | G4WFZ | 1034 |
| 129 | HA5KG | 1034 |
| 131 | IK1AIG | 1032 |
| 132 | G0DQS | 1030 |
| 133 | K3FN | 1029 |
| 134 | JA1EY | 1028 |
| 134 | RU6K | 1028 |
| 136 | DL4MCF | 1027 |
| 136 | F6DZU | 1027 |
| 136 | ON4BAV | 1027 |
| 136 | RZ1OA | 1027 |
| 140 | OK1JKM | 1026 |
| 141 | UA4HBW | 1025 |
| 142 | SP9FKQ | 1024 |
| 143 | HA5AGS | 1023 |
| 144 | DK6NJ | 1022 |
| 144 | WC6DX | 1022 |
| 146 | OZ1BUR | 1021 |
| 147 | DL5ME | 1020 |
| 147 | DL6MST | 1020 |
| 147 | G3SJX | 1020 |
| 150 | 9A5CY | 1019 |
| 150 | N4AH | 1019 |
| 152 | DF9ZN | 1018 |
| 153 | HB9RG | 1017 |
| 153 | K1HTV | 1017 |
| 153 | VE3JV | 1017 |
| 153 | ZL1ARY | 1017 |
| 157 | DL8MLD | 1016 |
| 157 | F5PAC | 1016 |
| 157 | G3NUG | 1016 |
| 157 | JA5IU | 1016 |
| 157 | KD1CT | 1016 |
| 162 | RA9YN | 1014 |
| 163 | K5MK | 1013 |
| 164 | SM3NXS | 1012 |
| 165 | IK4HLU | 1011 |
| 165 | KD6WW | 1011 |
| 165 | SM5FWW | 1011 |
| 168 | JF4VZT | 1010 |
| 168 | PT7WA | 1010 |
| 170 | DJ3XG | 1008 |
| 170 | JA4UQY | 1008 |
| 172 | UA4SKW | 1007 |
| 173 | JA2KVB | 1006 |

| Pos. | Callsign | Total |
|---|---|---|
| 173 | W1PKU | 1006 |
| 175 | DF6EX | 1005 |
| 175 | DL5CT | 1005 |
| 175 | JA8MS | 1005 |
| 178 | DL7CM | 1003 |
| 178 | R3OK | 1003 |
| 180 | DL1BKI | 1001 |
| 180 | EA7DUD | 1001 |
| 180 | JO1WKO | 1001 |
| 180 | UR3IFD | 1001 |
| 184 | [illegible] | 1000 |
| 184 | JR7TEQ | 1000 |
| 184 | UA0ZC | 1000 |
| 187 | [illegible] | [illegible] |
| 188 | IT9EJW | 994 |
| 188 | JA9IFF | 994 |
| 188 | W1CU | 994 |
| 191 | JA1SKE | 993 |
| 191 | WI8A | 993 |
| 193 | IK5ACO | 992 |
| 194 | DL8DSL | 991 |
| 195 | G4BWP | 990 |
| 195 | N6JV | 990 |
| 197 | VE7KDU | 985 |
| 198 | DK8UH | 984 |
| 199 | I1FY | 983 |
| 199 | SM4CTT | 983 |
| 199 | VE7SMP | 983 |
| 202 | IK8TWV | 982 |
| 203 | IZ4BEZ | 981 |
| 204 | UT7QF | 980 |
| 204 | W1OX | 980 |
| 206 | AB5EU | 979 |
| 207 | PT7BZ | 978 |
| 208 | DL6ATM | 977 |
| 208 | IT9FXY | 977 |
| 208 | UA3AKO | 977 |
| 211 | DL5DSM | 976 |
| 212 | DL6ZXG | 975 |
| 212 | OE6IMD | 975 |
| 212 | RZ3EC | 975 |
| 212 | W4ABW | 975 |
| 216 | UA9YJO | 974 |
| 217 | UR5LCV | 973 |
| 218 | F5TJC | 972 |
| 218 | K2VV | 972 |
| 218 | RA6AR | 972 |
| 221 | UT5JAJ | 971 |
| 222 | EU7A | 969 |
| 223 | HA5WA | 968 |
| 223 | K9MUF | 968 |
| 225 | DJ5AV | 967 |
| 225 | DJ9HX | 967 |
| 225 | N4WW | 967 |
| 228 | HA1RW | 966 |
| 228 | HA5DA | 966 |
| 230 | 9A2EU | 965 |
| 231 | IK2QPR | 962 |
| 231 | JA3FGJ | 962 |
| 233 | AG9S | 961 |
| 234 | JR0DLU | 960 |
| 235 | EI7CC | 959 |
| 235 | UA4CC | 959 |
| 237 | VE3EXY | 957 |
| 238 | HA9PP | 956 |
| 239 | DL2RNS | 955 |
| 239 | RA3DX | 955 |
| 239 | UA9LP | 955 |
| 242 | HK3JJH | 954 |
| 242 | LZ1HA | 954 |
| 244 | DH5VK | 953 |
| 244 | SP7GAQ | 953 |
| 244 | SP8HXN | 953 |
| 244 | UR3HC | 953 |
| 248 | 9A7W | 952 |
| 249 | RY7G | 951 |
| 250 | DL2VPF | 950 |
| 250 | DL5MX | 950 |
| 250 | N7GR | 950 |
| 250 | R7DX | 950 |
| 250 | W5ZPA | 950 |
| 250 | W7MO | 950 |
| 256 | I5HOR | 948 |
| 256 | N9BX | 948 |
| 258 | DJ5AI | 946 |
| 258 | HA8IB | 946 |
| 258 | I1BUP | 946 |

| Pos. | Callsign | Total |
|---|---|---|
| 261 | HA1AG | 945 |
| 261 | JH4IFF | 945 |
| 263 | G3RTE | 944 |
| 263 | G3UAS | 944 |
| 265 | DL6XK | 943 |
| 266 | IK2WAL | 942 |
| 266 | OZ1ACB | 942 |
| 268 | HB9CEX | 939 |
| 268 | VK3UY | 939 |
| 270 | JA7DOT | 935 |
| 270 | K8CW | 935 |
| 270 | UT5URW | 935 |
| 273 | PA0ZH | 934 |
| 273 | [illegible] | 934 |
| 273 | W8WFN | 934 |
| 276 | SP5TZC | 933 |
| 276 | UY5XE | 933 |
| 278 | 9A3NM | 932 |
| 278 | DL2DXA | 932 |
| 280 | K9RR | 931 |
| 281 | W5RQ | 930 |
| 282 | JA1BPA | 929 |
| 282 | OH2BF | 929 |
| 282 | OM3XX | 929 |
| 285 | JA3UCO | 928 |
| 285 | UY9IF | 928 |
| 287 | JH1QVW | 927 |
| 287 | YO7LCB | 927 |
| 289 | DL3APO | 926 |
| 289 | UA3AGW | 926 |
| 291 | F5HNQ | 925 |
| 291 | F6EOO | 925 |
| 291 | IT9YRE | 925 |
| 291 | OZ1HPS | 925 |
| 295 | CT4NH | 923 |
| 295 | JA7BWT | 923 |
| 297 | HA7UW | 922 |
| 298 | DL2CHN | 921 |
| 298 | JA1GHH | 921 |
| 300 | IK4HPU | 920 |
| 300 | IN3ASW | 920 |
| 302 | G3XPO | 919 |
| 303 | VE3VHB | 918 |
| 304 | HA0IH | 917 |
| 304 | LY5A | 917 |
| 306 | 9A2NO | 916 |
| 306 | JA6LCJ | 916 |
| 308 | SM3NRY | 915 |
| 309 | HB9BHY | 914 |
| 309 | R0FA | 914 |
| 311 | 7N1GMK | 913 |
| 311 | DK1FW | 913 |
| 311 | DL3EA | 913 |
| 314 | G3XTT | 911 |
| 315 | DL6KVA | 910 |
| 316 | I5CRL | 909 |
| 317 | SM6CMU | 908 |
| 317 | UR5ZEL | 908 |
| 319 | I0SYQ | 907 |
| 320 | LA2PA | 906 |
| 321 | DL1JIU | 905 |
| 321 | ON4CD | 905 |
| 323 | AH6HY | 904 |
| 323 | K0DEQ | 904 |
| 325 | SM5JE | 903 |
| 326 | N4MM | 902 |
| 326 | UY5ZZ | 902 |
| 326 | W2YC | 902 |
| 329 | JH4GJR | 901 |
| 330 | IK2WXZ | 900 |
| 331 | EA3BT | 898 |
| 331 | OK1DH | 898 |
| 333 | DK2BR | 897 |
| 333 | JQ1ALQ | 897 |
| 333 | VK7BC | 897 |
| 336 | KD3CQ | 896 |
| 337 | CT1BXX | 895 |
| 337 | RZ6LY | 895 |
| 339 | JN3SAC | 894 |
| 339 | SM6DHU | 894 |
| 341 | DJ4GJ | 893 |
| 341 | IZ4CZE | 893 |
| 343 | IK8JVG | 888 |
| 344 | G0WRE | 886 |
| 344 | RA3RGQ | 886 |
| 346 | HA6NF | 884 |
| 347 | W5PF | 883 |

| Pos. | Callsign | Total |
|---|---|---|
| 348 | JA0DNP | 882 |
| 348 | N6FX | 882 |
| 350 | IK8CVZ | 880 |
| 350 | IZ8DBJ | 880 |
| 352 | R7KM | 879 |
| 353 | JH8JYV | 878 |
| 354 | DL2HU | 877 |
| 354 | RM0F | 877 |
| 354 | SM6CAS | 877 |
| 357 | CT1CJJ | 876 |
| 357 | SP6CIK | 876 |
| 359 | 9A3JB | 875 |
| 359 | RZ3FW | 875 |
| 361 | DL6AWI | 874 |
| 361 | EA1EAU | 874 |
| 363 | UA0SFN | 873 |
| 364 | I8DVJ | 872 |
| 364 | JM1PXG | 872 |
| 366 | DJ8QP | 871 |
| 367 | JE8TGI | 870 |
| 368 | DL4FDM | 869 |
| 369 | I2PQW | 867 |
| 370 | JR2UJT | 866 |
| 371 | IK5PWQ | 863 |
| 372 | ON4CAS | 862 |
| 372 | S55SL | 862 |
| 374 | RJ3AA | 861 |
| 375 | JL7BRH | 860 |
| 375 | ON7DR | 860 |
| 377 | SM5BFJ | 859 |
| 378 | IK6DLK | 857 |
| 379 | LZ1BJ | 856 |
| 380 | R9OK | 855 |
| 381 | AB5EB | 854 |
| 381 | SM5ARL | 854 |
| 383 | W2KKZ | 852 |
| 384 | HB9BIN | 851 |
| 384 | K6VVA | 851 |
| 384 | UY0ZG | 851 |
| 387 | 5B4MF | 850 |
| 387 | I1CAW | 850 |
| 389 | JR6SVM | 846 |
| 390 | JH2IEE | 845 |
| 391 | I5ZGQ | 844 |
| 391 | R7KC | 844 |
| 393 | I2MQP | 843 |
| 394 | G4DUW | 842 |
| 394 | OZ8BZ | 842 |
| 394 | SV1FJA | 842 |
| 397 | OE3JHC | 838 |
| 397 | W3TN | 838 |
| 399 | JA9BEK | 837 |
| 400 | DL5BUT | 836 |
| 400 | I2VGW | 836 |
| 400 | JA1WPX | 836 |
| 403 | JK1OPL | 834 |
| 403 | K2SHZ | 834 |
| 405 | IK4DRR | 833 |
| 406 | SM5BMB | 832 |
| 407 | DL1CL | 831 |
| 408 | US4EX | 829 |
| 409 | JL1BYZ | 827 |
| 410 | EA3WL | 826 |
| 410 | EA7TV | 826 |
| 410 | JE7JIS | 826 |
| 410 | RK6AM | 826 |
| 414 | PY4OY | 824 |
| 415 | CT1EKY | 823 |
| 415 | IZ5BAM | 823 |
| 417 | DK6AO | 822 |
| 417 | N4QQ | 822 |
| 417 | RG4F | 822 |
| 420 | JE3GUG | 821 |
| 420 | OE6GRG | 821 |
| 420 | UT5UGR | 821 |
| 423 | RU3FM | 819 |
| 423 | SM6BZV | 819 |
| 423 | W5FKX | 819 |
| 426 | HB9ICC | 818 |
| 426 | R7NB | 818 |
| 428 | OM7CA | 817 |
| 429 | JA9GPG | 816 |
| 430 | OE3EVA | 815 |
| 430 | UA6MF | 815 |
| 432 | DK5WL | 813 |
| 432 | K0AP | 813 |
| 432 | W6YOO | 813 |

| Pos. | Callsign | Total |
|---|---|---|
| 435 | DF7GK | 812 |
| 435 | OZ7DN | 812 |
| 437 | HA5UK | 810 |
| 437 | WB5JID | 810 |
| 439 | DF0QP | 808 |
| 440 | DL7VSN | 807 |
| 440 | G4NXG/M | 807 |
| 440 | IV3ZOF | 807 |
| 440 | SM3TLG | 807 |
| 444 | IZ8EFB | 806 |
| 444 | JA2CEJ | 806 |
| 446 | AI9Y | 805 |
| 446 | IK2ZJN | 805 |
| 446 | JG3LGD | 805 |
| 446 | UA3TCJ | 805 |
| 450 | I4KMN | 804 |
| 451 | AB5C | 803 |
| 451 | JJ0NCC | 803 |
| 451 | K8AJK | 803 |
| 451 | UT7WZ | 803 |
| 451 | W6RLL | 803 |
| 456 | HA0HW | 802 |
| 456 | JA1AML | 802 |
| 456 | RN3QN | 802 |
| 459 | CU3EJ | 801 |
| 459 | SV1GYG | 801 |
| 459 | YT2AA | 801 |
| 462 | BA4DW | 800 |
| 462 | G0RCI | 800 |
| 462 | JA1NLX | 800 |
| 462 | JE1LFX | 800 |
| 466 | SM3DMP | 799 |
| 467 | IK2ILH | 794 |
| 468 | JH1IAQ | 793 |
| 469 | G3SWH | 792 |
| 470 | F6HQP | 790 |
| 471 | I5OYY | 789 |
| 471 | ON5NT | 789 |
| 471 | SA3ANZ | 789 |
| 474 | IK7MXB | 788 |
| 475 | DL1EJA | 787 |
| 475 | VA3DXA | 787 |
| 477 | JH1IED | 786 |
| 477 | RW3XZ | 786 |
| 479 | AC0A | 785 |
| 480 | JM1XCW | 781 |
| 480 | SV1DPI | 781 |
| 482 | SM7NGH | 780 |
| 483 | I0MOM | 779 |
| 484 | UA0CW | 778 |
| 484 | UR7GW | 778 |
| 486 | F5CQ | 777 |
| 486 | JF6WTY | 777 |
| 488 | SM7DXQ | 776 |
| 489 | ON7TK | 775 |
| 490 | IK2OVC | 771 |
| 490 | UA3DPM | 771 |
| 492 | W3AWU | 770 |
| 493 | G3KHZ | 769 |
| 493 | JA1BNW | 769 |
| 495 | KA2ZJE | 768 |
| 495 | R7KW | 768 |
| 497 | 5B4AHJ | 766 |
| 497 | OE2VEL | 766 |
| 499 | SM7CQY | 764 |
| 500 | JG1OWV | 759 |
| 500 | VE3ZZ | 759 |
| 502 | CT1AHU | 755 |
| 502 | JA7MGP | 755 |
| 502 | WA5VGI | 755 |
| 505 | DK3GG | 754 |
| 505 | UA9YF | 754 |
| 507 | HB9BGV | 753 |
| 507 | KC6AWX | 753 |
| 509 | HB9AMO | 752 |
| 509 | JA2AH | 752 |
| 509 | ON5SY | 752 |
| 509 | R7AA | 752 |
| 509 | SM1TDE | 752 |
| 509 | UA9FAR | 752 |
| 509 | UY5AA | 752 |
| 509 | VA7ZT | 752 |
| 517 | N9GKE | 751 |
| 517 | SM5AQD | 751 |
| 519 | UA4PT | 750 |
| 520 | JH1XUP | 745 |
| 520 | NI6T | 745 |

**2016 Honour Roll**

| Pos. | Callsign | Total |
|---|---|---|
| 520 | W5ZE | 745 |
| 523 | VK8NSB | 744 |
| 524 | N6KZ | 743 |
| 525 | IK8BQE | 742 |
| 526 | 9A1DX | 741 |
| 527 | K3PT | 739 |
| 528 | DL3JON | 738 |
| 529 | DK1BX | 737 |
| 530 | KJ3L | 735 |
| 531 | UA4PF | 733 |
| 532 | SM6CUK | 732 |
| 533 | PR7FB | 731 |
| 534 | DL6ZFG | 730 |
| 534 | DL9RCF | 730 |
| 534 | JA1HP | 730 |
| 537 | ON7WW | 729 |
| 538 | IF9ZWA | 727 |
| 538 | K2XF | 727 |
| 540 | NA5AR | 724 |
| 541 | EA1AUS | 722 |
| 541 | ON5JV | 722 |
| 543 | F5PAL | 721 |
| 543 | IK2VUC | 721 |
| 543 | IZ1ANU | 721 |
| 543 | JF1RDH | 721 |
| 547 | UY5BC | 720 |
| 548 | HA0IS | 717 |
| 548 | JA1GRM | 717 |
| 548 | JA6TMU | 717 |
| 551 | CT1EGW | 716 |
| 552 | JH2KXN | 715 |
| 552 | JH4BTI | 715 |
| 552 | K8GI | 715 |
| 555 | RA3CQ | 713 |
| 555 | W9IXX | 713 |
| 557 | K6PJ | 712 |
| 558 | F8GB | 711 |
| 559 | UA0FO | 710 |
| 559 | W7BEM | 710 |
| 561 | F8AMV | 709 |

| Pos. | Callsign | Total |
|---|---|---|
| 561 | WD8ANZ | 709 |
| 563 | GM7TUD | 708 |
| 564 | GW0IWD | 704 |
| 565 | EA6LP | 703 |
| 565 | IN3QCI | 703 |
| 565 | KD7H | 703 |
| 568 | DL1FU | 702 |
| 568 | G3TXF | 702 |
| 568 | I4GAD | 702 |
| 568 | VK4BUI | 702 |
| 572 | DL3KZA | 701 |
| 572 | I8IHG | 701 |
| 572 | IN3NJB | 701 |
| 572 | RA1OD | 701 |
| 576 | F6GCP | 700 |
| 576 | M0OXO | 700 |
| 576 | R0AZ | 700 |
| 579 | DL8AAV | 698 |
| 580 | JH4DYP | 696 |
| 580 | JN6RZM | 696 |
| 580 | R2DO | 696 |
| 583 | N4NX | 695 |
| 583 | UA9LBQ | 695 |
| 585 | IK2RPE | 693 |
| 586 | G4XRX | 692 |
| 587 | OE3RPB | 688 |
| 588 | JH1MXV | 687 |
| 588 | WW8W | 687 |
| 590 | K4JP | 686 |
| 591 | KF8UN | 684 |
| 592 | JA1PUK | 683 |
| 593 | CT1DKS | 682 |
| 594 | G3LUW | 681 |
| 594 | IK4MSV | 681 |
| 596 | 9A2Y | 678 |
| 596 | EA3GHZ | 678 |
| 596 | ON8BN | 678 |
| 599 | I7PXV | 677 |
| 599 | JJ1CZR | 677 |
| 601 | DL5KUD | 674 |

| Pos. | Callsign | Total |
|---|---|---|
| 601 | IK8YTA | 674 |
| 603 | AA8LL | 673 |
| 603 | DL2BQV | 673 |
| 603 | EA1YY | 673 |
| 603 | F5BOY | 673 |
| 603 | F6ACV | 673 |
| 603 | RA9CMO | 673 |
| 609 | MD0CCE | 672 |
| 609 | RW0LT | 672 |
| 611 | JA3KZV | 671 |
| 611 | K0KG | 671 |
| 613 | N1KC | 670 |
| 614 | DS5ACV | 669 |
| 615 | G4IUF | 668 |
| 616 | K6FW | 665 |
| 617 | DL3JPN | 664 |
| 618 | W1KSZ | 662 |
| 618 | W4UM | 662 |
| 620 | RN8W | 660 |
| 620 | W1OW | 660 |
| 622 | DL2VPO | 655 |
| 622 | SV1OZ | 655 |
| 624 | ND7K | 654 |
| 624 | OE2LCM | 654 |
| 624 | WD8PKF | 654 |
| 627 | JO3AXC | 652 |
| 627 | OZ3SK | 652 |
| 629 | DL3BRE | 651 |
| 630 | EA7TG | 649 |
| 630 | JA2FGL | 649 |
| 630 | JA6FIO | 649 |
| 630 | ON7LX | 649 |
| 634 | EA1ABS | 642 |
| 635 | IZ5JMZ | 641 |
| 636 | HB9AGH | 639 |
| 637 | I2BUH | 638 |
| 638 | AA4V | 637 |
| 639 | DL7UKA | 635 |
| 639 | HA0IT | 635 |
| 641 | DH2PC | 634 |

| Pos. | Callsign | Total |
|---|---|---|
| 641 | DL1XE | 634 |
| 643 | US5QR | 632 |
| 644 | W5VFO | 631 |
| 645 | RA0FF | 630 |
| 645 | UT7UW | 630 |
| 647 | JA7DHJ | 627 |
| 647 | SM5ELV | 627 |
| 649 | HA5FA | 626 |
| 649 | LA5HE | 626 |
| 651 | DJ9IN | 622 |
| 651 | F8NAN | 622 |
| 651 | UA9JLL | 622 |
| 651 | UX2KA | 622 |
| 655 | DL6MHG | 621 |
| 655 | R3AW | 621 |
| 657 | W2FB | 620 |
| 658 | DL4AO | 619 |
| 658 | JK1TCV | 619 |
| 660 | DL3NM | 618 |
| 661 | HL4CBX | 617 |
| 662 | DL2OE | 615 |
| 662 | I2ZBX | 615 |
| 662 | IK0CNA | 615 |
| 662 | JF2OZH | 615 |
| 666 | HB9DKZ | 613 |
| 667 | DJ1OJ | 612 |
| 667 | DL8UAT | 612 |
| 667 | EC1AIJ | 612 |
| 667 | WA1ZIC | 612 |
| 671 | DL9UBF | 611 |
| 671 | G3KWK | 611 |
| 671 | HA1DAE | 611 |
| 671 | RA6YJ | 611 |
| 671 | UA3ECJ | 611 |
| 676 | UA4SJS | 610 |
| 677 | JA1FGB | 609 |
| 678 | SV2DGH | 607 |
| 678 | W0GLG | 607 |
| 680 | CT1BOY | 606 |
| 680 | I5KG | 606 |

| Pos. | Callsign | Total |
|---|---|---|
| 680 | K8NKQ | 606 |
| 680 | N6PF | 606 |
| 684 | DF6TC | 605 |
| 684 | K9RHY | 605 |
| 684 | VE1JS | 605 |
| 687 | RL6M | 604 |
| 687 | RW9LL | 604 |
| 689 | F4BKV | 603 |
| 689 | G4KFT | 603 |
| 691 | IZ8EJB | 602 |
| 691 | N6VS | 602 |
| 691 | PA7RA | 602 |
| 694 | EA1EDF | 601 |
| 694 | HB9DDZ | 601 |
| 694 | IK8CNT | 601 |
| 694 | W5WP | 601 |
| 698 | G4VMX | 600 |
| 698 | RA6ATZ | 600 |
| 700 | JI3MJK | 599 |
| 701 | ON5JE | 597 |
| 702 | SM7DAY | 595 |
| 703 | DL2YY | 594 |
| 703 | SM7BHH | 594 |
| 705 | K2AJY | 590 |
| 706 | F6FYD | 589 |
| 707 | DJ8VC | 587 |
| 708 | JG1UKW | 586 |
| 708 | SP5APW | 586 |
| 710 | HB9BXE | 584 |
| 710 | HB9TKS | 584 |
| 710 | KE4DH | 584 |
| 713 | DL2VFR | 582 |
| 713 | DL3EEE | 582 |
| 713 | JA4GXS | 582 |
| 713 | UR5WBQ | 582 |
| 717 | HA5VZ | 580 |
| 718 | JA8COE | 578 |
| 719 | DL7VOX | 576 |
| 720 | DK7MD | 575 |
| 720 | KE5K | 575 |

| Pos. | Callsign | Total |
|---|---|---|
| 722 | DL7VKD | 574 |
| 722 | S58MU | 574 |
| 724 | SM4AZQ | 573 |
| 724 | WX4G | 573 |
| 726 | 7K3QPL | 572 |
| 727 | DL4MN | 571 |
| 727 | N6AR | 571 |
| 729 | JH1OCC | 570 |
| 729 | JO1CRA | 570 |
| 729 | RA1QX | 570 |
| 732 | N6UK | 569 |
| 733 | AA0MZ | 567 |
| 733 | DS4DRE | 567 |
| 733 | N3RW | 567 |
| 733 | UN7JX | 567 |
| 737 | JL3CRS | 566 |
| 738 | PT7ZT | 565 |
| 739 | I3ZSX | 563 |
| 740 | AA0FT | 561 |

## Annual Update Deadline

IOTA enthusiasts are reminded that the last date for submitting applications or updates to checkpoints (and mailing cards and fees) for inclusion in the 2017 Honour Roll and other performance tables is 31 January 2017. If submitted/postmarked after that date, they will be processed in the normal way but the scores will be held over to the following year's listing. It is important that members who have not updated since the 2012 annual listings and wish to remain listed should make a submission on or before 31 January 2017.

## Official sources of information

RSGB IOTA's website (www.rsgbiota.org), provides the following:

- A List of Frequently Asked Questions (FAQ) on IOTA
- IOTA Programme Rules
- A list of authorised IOTA checkpoints
- A detailed listing of IOTA groups by continent, region and country
- A listing of IOTA groups by short title only
- A listing of the most wanted IOTA groups by continent
- Information on how to obtain the IOTA directory (link to the RSGB website: www.rsgb.org)

The RSGB IOTA website allows you to search the island listings by IOTA group number or island name – and returns the relevant listing together with details of rarity and past and future operations.

*Radio Communication (RadCom),* the monthly journal of the RSGB (see its website *www.rsgb.org*), covers IOTA in the monthly HF column.

## Sources of information on IOTA activity

### RSGB IOTA website (www.rsgbiota.org)

Containing, among so much else, details of forthcoming DXpeditions taken from reliable sources. Check the bottom of the home page for expeditions currently active and follow the link for future activity. Intending DXpeditioners are encouraged to enter their own information via a web-form. This service is moderated so there may be a delay before the information appears. DXpeditioners should also send their information direct to the news editors below.

### DX-World.net

A very attractive website run by a dedicated team of DXers and IOTA enthusiasts, which is updated daily by Col McGowan, MM0NDX. It features a web-form for readers to submit their own DX or IOTA information.

### 425 DX News

A weekly email DX bulletin issued by IOTA Committee Member (Europe) Mauro Pregliasco, I1JQJ (i1jqj@425dxn.org). This carries IOTA news – official news releases and listings, DXpedition activity reports, a calendar of forthcoming events – as well as a host of other useful material. For more information, check the 425 DX News website (www.425dxn.org).

### The Daily DX

The first subscription-based email DX bulletin, published by Bernie McClenny, W3UR (email: bernie@dailydx.com). This provides a comprehensive daily commentary on the DX scene with a prominent IOTA section listing island activity. You may prefer to subscribe to The Weekly DX, which is also available. For full details, check Bernie's home page (www.dailydx.com).

### QRZ DX

Another long established weekly bulletin, published by Carl Smith, N4AA (email: QRZDX@dxpub.com), in paper and Email formats. For more information check Carl's website: *www.dxpub.com*

## Internet reflectors and forums

### IOTA-chasers Yahoo Forum

This provides a meeting-place for IOTA enthusiasts to exchange information and share views about different aspects of the IOTA Programme including island operations, QSL queries, etc.

### Other information sources

The DX Summit web cluster *www.dxsummit.fi* is an invaluable way of keeping an eye on the bands when you are away from your QTH. It offers a listing of the last 25, 250, etc spots logged from clusters worldwide, as well as WWV reports and announcements. The search facility is a particularly valuable aid to IOTA DXing as you can check for spots for callsigns and IOTA numbers.

An increasing number of IOTA operators and expeditioners maintain home pages with IOTA information and features. Try the pages of W9DC (www.w9dc.com).

## The Launch of a New Company, Islands On The Air (IOTA) Ltd

Late April 2016 saw the start of an exciting new development in IOTA's narrative with the launch of Islands on the Air (IOTA) Ltd, a not-for-profit company, limited by guarantee. With the agreement of the RSGB this body now has full responsibility for all aspects of the programme, its day to day management, strategy, policy, finance, development, promotion and marketing. Currently its directors are Roger Balister G3KMA, Stan Lee G4XXI and Cezar Trifu VE3LYC, but a full Board of Directors will be formed shortly. The new IOTA Directory, published by the RSGB in April, refers to the newly set-up company under a different name but this has now been changed in the way always intended. It will take a little time to carry through all aspects of the changed governance but IOTA enthusiasts should be assured that the new company is fully committed to carrying on the management of the programme in accordance with the traditions established over the last five decades.

For more information: *https://www.rsgbiota.org*

### 2016 Annual Listing

| Pos. | Callsign | Total | Pos. | Callsign | Total | Pos. | Callsign | Total | Pos. | Callsign | Total | Pos. | Callsign | Total | Pos. | Callsign | Total |
|---|---|---|---|---|---|---|---|---|---|---|---|---|---|---|---|---|---|
| 741 | SM0BSB | 556 | 804 | EW1LM | 485 | 867 | UA9UDX | 419 | 930 | M0KCM | 372 | 990 | NG7Z | 320 | 1054 | HB9FAZ | 269 |
| 742 | KN6KI | 554 | 805 | DL8ZBA | 484 | 868 | DL4BBH | 418 | 931 | NL7V | 370 | 993 | DJ2DA | 319 | 1056 | F4GDI | 268 |
| 743 | DK3DUA | 551 | 806 | G3VKW | 482 | 869 | W9ILY | 417 | 932 | VK2DX | 368 | 994 | 7K1CPT | 318 | 1057 | RU3EQ | 267 |
| 744 | R9TO | 550 | 807 | DM1TT | 481 | 870 | RA3TAR | 416 | 933 | G1VDP | 367 | 994 | UA1OIW | 318 | 1058 | GM0DEQ | 266 |
| 745 | DL6JZ | 549 | 808 | DL3TC | 480 | 870 | US0YA | 416 | 933 | GW4TSG | 367 | 996 | K4WES | 317 | 1058 | SM5AFU | 266 |
| 745 | K9UQN | 549 | 808 | KN7D | 480 | 872 | DL5MHQ | 415 | 933 | K1NU | 367 | 997 | R3BM | 316 | 1060 | G3LAA | 264 |
| 747 | K1ZN | 547 | 810 | IK2DOT | 478 | 872 | F5LIW | 415 | 933 | VE1WT | 367 | 998 | G4FVK | 314 | 1061 | DK5DC | 263 |
| 748 | E73Y | 545 | 810 | RX3MX | 478 | 872 | K4MIJ | 415 | 937 | DL4NN | 366 | 998 | I1YDT | 314 | 1062 | RM2A | 261 |
| 749 | IK1NLZ | 543 | 812 | EA3LS | 477 | 875 | JH3GCN | 414 | 938 | DG5LAC | 365 | 998 | PT7YV | 314 | 1063 | 4X6KJ | 260 |
| 749 | MM0ABJ | 543 | 813 | DL6MKA | 474 | 875 | OK2QA | 414 | 938 | VK5CE | 365 | 1001 | DL9MRF | 312 | 1063 | G4FFN | 260 |
| 751 | IK3OYU | 542 | 813 | I8INW | 474 | 875 | RN0SRR | 414 | 940 | SV9AHZ | 362 | 1001 | N4AL | 312 | 1065 | G3JTO | 259 |
| 752 | OE3KKA | 540 | 813 | JI3DST | 474 | 878 | SV1BTK | 413 | 941 | F9CI | 360 | 1001 | RN3FT | 312 | 1065 | JA3AVO | 259 |
| 753 | G3LHJ | 538 | 816 | JA1FVS | 472 | 879 | F8DZY | 412 | 941 | SM7BHM | 360 | 1004 | CU3AC | 311 | 1065 | JA4CZM | 259 |
| 754 | G0GKY | 532 | 817 | W6OUL | 470 | 879 | JP3AYQ | 412 | 943 | DJ6OI | 359 | 1004 | DL3LBM | 311 | 1065 | VE3RNH | 259 |
| 754 | G4JFS | 532 | 818 | DL7UXG | 466 | 879 | XE1RBV | 412 | 943 | EA3CCN | 359 | 1004 | NN7A | 311 | 1069 | DL8DXF | 258 |
| 756 | F1TXI | 531 | 819 | 4Z4BS | 464 | 882 | EI3IO | 411 | 943 | H44MS | 359 | 1004 | VA7CRZ | 311 | 1069 | IT9ABN | 258 |
| 757 | A65BR | 528 | 819 | DF1BN | 464 | 882 | IZ8FFA | 411 | 946 | IZ4MJP | 358 | 1008 | UT5UA | 310 | 1071 | DM3ZF | 256 |
| 757 | DJ4EY | 528 | 819 | HB9DOT | 464 | 882 | OZ1ADL | 411 | 947 | W3LL | 357 | 1009 | N3RC | 309 | 1071 | JH1FVE | 256 |
| 759 | VE2BR | 527 | 822 | DK1YP | 463 | 885 | G3NPA | 410 | 948 | EA5HEU | 355 | 1009 | PT7CB | 309 | 1071 | K3VAT | 256 |
| 760 | SM6BZE | 525 | 822 | DL6MIG | 463 | 885 | JA8ZO | 410 | 948 | G3USR | 355 | 1011 | I5GKS | 308 | 1074 | EA5IY | 254 |
| 761 | IT9RTA | 524 | 824 | JF6XQJ | 461 | 885 | JF1MTV | 410 | 950 | K4MM | 354 | 1011 | NQ7R | 308 | 1075 | DJ6XG | 253 |
| 761 | SM4DDS | 524 | 825 | JE3AGN | 460 | 885 | K7LAY | 410 | 951 | G0TRB | 352 | 1011 | RA9HM | 308 | 1075 | IK4YCQ | 253 |
| 763 | EA1AST | 523 | 826 | DL5DF | 458 | 885 | N0ODK | 410 | 951 | IK2PZC | 352 | 1014 | AA1QD | 307 | 1075 | JA2MNB | 253 |
| 763 | I4YCE | 523 | 827 | DF2FZ | 454 | 885 | UR8IDX | 410 | 953 | G0THF | 350 | 1014 | YO9FLD | 307 | 1075 | XE1J | 253 |
| 765 | I5JFG | 522 | 827 | JH3GFA | 454 | 891 | DH6DAO | 409 | 954 | RA3BL | 349 | 1016 | DM3PKK | 306 | 1079 | K3CH | 252 |
| 766 | DL1ASA | 520 | 827 | UA4PCM | 454 | 891 | DK6CQ | 409 | 955 | N7QU | 348 | 1016 | PY2VA | 306 | 1080 | G0AHC | 251 |
| 767 | HB9AAA | 518 | 830 | G4MFX | 452 | 893 | I8IGS | 408 | 955 | ON4RO | 348 | 1018 | HB9BQB | 305 | 1080 | SA7AUH | 251 |
| 768 | JE3SSL | 515 | 831 | DF8HS | 451 | 893 | VA2WT | 408 | 957 | F5EOT | 346 | 1018 | SQ1EIX | 305 | 1082 | JI1LAT | 250 |
| 769 | W1OK | 514 | 832 | I5PLS | 449 | 895 | UA3FX | 407 | 958 | VA3NQ | 345 | 1020 | DL3OV | 304 | 1083 | DG1ASA | 248 |
| 769 | W4HG | 514 | 833 | JR3QHQ | 444 | 896 | WB3LHD | 406 | 959 | SV1JG | 344 | 1020 | W0KEU | 304 | 1084 | G4DJC | 247 |
| 771 | IW9HII | 513 | 833 | OH2BN | 444 | 897 | K6OHM | 405 | 960 | HA1ZH | 343 | 1022 | DJ7YM | 303 | 1084 | HA8TI | 247 |
| 771 | OH8US | 513 | 833 | RA6FG | 444 | 898 | IK2CMN | 404 | 961 | DL8AAB | 342 | 1022 | EA5ARC | 303 | 1084 | N1HOQ | 247 |
| 773 | GW4BKG | 512 | 836 | CT1CQK | 443 | 898 | JS3OSI | 404 | 962 | DL6DQW | 340 | 1022 | JG3WCZ | 303 | 1087 | DL1BSH | 245 |
| 774 | UT4EK | 511 | 836 | DJ6UP | 443 | 898 | OE8TLK | 404 | 963 | HB9CWA | 339 | 1022 | KA3CRC | 303 | 1087 | DL4FAP | 245 |
| 775 | DL3SUG | 510 | 836 | JH3CUL | 443 | 901 | IK2RLS | 403 | 964 | JR3ADB | 338 | 1022 | RA4FEA | 303 | 1089 | IK4NZD | 244 |
| 776 | JA9TWN | 509 | 836 | W4OX | 443 | 901 | IW2FND | 403 | 965 | DJ5FZ | 337 | 1027 | DL7HKL | 301 | 1090 | IW5AOT | 243 |
| 776 | RK9UN | 509 | 840 | HB9KT | 442 | 901 | IZ3ETU | 403 | 965 | G0PCF | 337 | 1027 | F8VOE | 301 | 1091 | XE1MW | 242 |
| 776 | RT0F | 509 | 841 | UA1OMS | 441 | 904 | DL6CNG | 402 | 967 | VK4CAG | 335 | 1027 | JA3APU | 301 | 1092 | JE6HCL | 241 |
| 776 | W2SM | 509 | 842 | JH6JMM | 440 | 904 | N2NVH | 402 | 968 | JA2ACI | 333 | 1027 | K2MHE | 301 | 1093 | DL5XL | 240 |
| 780 | IK2ANI | 508 | 842 | SP6MLX | 440 | 906 | JA4LKB | 401 | 968 | K3VAR | 333 | 1027 | KM6HB | 301 | 1094 | JA1GQV | 239 |
| 780 | S52OT | 508 | 844 | EA3IM | 439 | 906 | VE1AI | 401 | 968 | VR2XLN | 333 | 1027 | OH6JKW | 301 | 1094 | W0MF | 239 |
| 782 | DL5ZL | 507 | 844 | G3NKC | 439 | 908 | DL1TRK | 400 | 971 | JJ3FRB | 331 | 1027 | WB1ATZ | 301 | 1094 | W9HR | 239 |
| 783 | DL1ROJ | 505 | 846 | DB3LO | 438 | 908 | WA2VQV | 400 | 971 | SV1IW | 331 | 1034 | 7N1NXF | 300 | 1097 | UR7UT | 238 |
| 784 | DK7YY | 504 | 847 | DL5JK | 437 | 908 | WG5M | 400 | 973 | DL5XAT | 330 | 1034 | 7N2JZT | 300 | 1098 | DL4ALI | 236 |
| 784 | EV1R | 504 | 847 | DS5FNE | 437 | 908 | YO9FNP | 400 | 973 | G4ZCS | 330 | 1034 | DL7UGO | 300 | 1098 | OE7LVI | 236 |
| 784 | JP1EWY | 504 | 849 | DL2MDZ | 433 | 912 | CT1AVR | 397 | 973 | IV3BSF | 330 | 1034 | KK9M | 300 | 1100 | G4FCI | 234 |
| 784 | KJ6P | 504 | 849 | DL3AWB | 433 | 912 | G3LPS | 397 | 973 | MM0EAX | 330 | 1034 | M5KJM | 300 | 1101 | DL8UVG | 233 |
| 788 | JH0JQS | 502 | 849 | RU9LA | 433 | 914 | K7ACZ | 394 | 977 | DJ9ER | 329 | 1039 | YO3APJ | 295 | 1101 | IZ2GNQ | 233 |
| 788 | N3NT | 502 | 849 | UT3IW | 433 | 915 | IW0GBU | 393 | 977 | DL9LF | 329 | 1040 | DL2VM | 293 | 1103 | KD8MQY | 232 |
| 790 | IK2GPQ | 501 | 853 | EI9FBB | 432 | 916 | N1RR | 392 | 977 | ER3DX | 329 | 1041 | DK1AX | 292 | 1103 | MM0BQI | 232 |
| 790 | NN1N | 501 | 853 | IZ4BBF | 432 | 916 | ON4CCN | 392 | 980 | 4Z1UF | 328 | 1042 | DK4MX | 288 | 1103 | R0TR | 232 |
| 790 | VE3ESE | 501 | 853 | UX3IA | 432 | 918 | DL7GN | 389 | 980 | RA6MQ | 328 | 1042 | G4POF | 288 | 1106 | HA5ARX | 230 |
| 793 | CT1BLE | 500 | 856 | DL2DWC | 430 | 919 | SM0KRN | 383 | 980 | UY1HY | 328 | 1044 | 7M3IYU | 286 | 1106 | I6TIH | 230 |
| 793 | G0FYX | 500 | 856 | G4UZN | 430 | 920 | JE3GRQ | 382 | 983 | DL2ASB | 326 | 1045 | G3TTC | 283 | 1106 | JI1IXW | 230 |
| 793 | R8MC | 500 | 856 | N8AGU | 430 | 921 | DL1AY | 380 | 983 | JG3SKK | 326 | 1046 | RN0C | 282 | 1109 | CT1JOH | 229 |
| 796 | IN3XUG | 498 | 859 | NQ3A | 427 | 921 | EA2WD | 380 | 983 | W8OP | 326 | 1047 | HB9ARF | 279 | 1109 | OZ1HX | 229 |
| 797 | VE2ACP | 494 | 859 | UW5IM | 427 | 923 | DK3DG | 379 | 986 | IZ3KNK | 325 | 1048 | PP1CZ | 278 | 1111 | KS6A | 228 |
| 798 | DL3ZAI | 493 | 861 | R9SG | 426 | 923 | JG4OOU | 379 | | | | 1049 | AA6RE | 277 | 1112 | UT4NY | 227 |
| 798 | R3DG | 493 | 861 | SM5CBM | 426 | 925 | DL8DZV | 376 | 987 | RU6AI | 324 | 1049 | G0DRM | 277 | 1113 | PS8ACL | 226 |
| 800 | IZ2AMW | 491 | 863 | RD0L | 425 | 926 | G3YJQ | 375 | 988 | DL2DQL | 323 | 1051 | N2SQW | 274 | 1113 | W1GWN | 226 |
| 800 | OE1PMU | 491 | 863 | W8JRK | 425 | 926 | UA0LCZ | 375 | 989 | AA3TH | 322 | 1052 | G0PHY | 273 | 1115 | G3PEM | 225 |
| 802 | DL5KUR | 486 | 865 | M0URX | 421 | 928 | KW0U | 374 | 990 | IZ5FSA | 320 | 1053 | IW8EPH | 271 | 1115 | IZ8XQC | 225 |
| 802 | RU9HM | 486 | 866 | JA6CBG | 420 | 929 | DL2GAC | 373 | 990 | JF2AXT | 320 | 1054 | DH5JG | 269 | 1117 | CT1IZW | 224 |

## 2016 Annual Listing

| Pos. | Callsign | Total |
|---|---|---|
| 1117 | JA2AYP | 224 |
| 1117 | N1WQ | 224 |
| 1120 | IK2YGZ | 223 |
| 1120 | JE2CPI | 223 |
| 1120 | K5KUA | 223 |
| 1120 | WB0YEA | 223 |
| 1124 | HB9AJK | 222 |
| 1124 | RV9MA | 222 |
| 1126 | HB9RUZ | 221 |
| 1126 | JA1ODB | 221 |
| 1128 | DJ8WO | 220 |
| 1128 | G3LDI | 220 |
| 1128 | N2JTO | 220 |
| 1128 | UT0NN | 220 |
| 1128 | W1UL | 220 |
| 1128 | YO5CRQ | 220 |
| 1134 | DJ8OB | 218 |
| 1134 | IZ7ECL | 218 |
| 1134 | IZ8FQI | 218 |
| 1134 | JR3CNQ | 218 |
| 1134 | K9AAA | 218 |
| 1139 | N5WR | 216 |
| 1139 | RW9QA | 216 |
| 1141 | DH0JAE | 215 |
| 1141 | DL9WO | 215 |
| 1143 | IK5TSZ | 214 |
| 1143 | UA9YPS | 214 |
| 1143 | UR5ZTH | 214 |
| 1143 | WF2S | 214 |
| 1147 | DF5BX | 213 |
| 1147 | DL2RZG | 213 |
| 1147 | WA4FLZ | 213 |
| 1150 | DD6UDD | 212 |
| 1150 | DL1JPF | 212 |
| 1150 | G4AXX | 212 |
| 1150 | K5CX | 212 |
| 1154 | OM5NL | 211 |
| 1154 | RL3AA | 211 |
| 1154 | SV1MO | 211 |
| 1154 | WD8NVN | 211 |
| 1158 | DL6DH | 210 |
| 1158 | JG1GCO | 210 |
| 1158 | JH7CFX | 210 |
| 1158 | KD4POJ | 210 |
| 1158 | N4CBS | 210 |
| 1158 | W0NB | 210 |
| 1164 | JN1RFY | 209 |
| 1165 | DL2QT | 208 |
| 1165 | JE1UMG | 208 |
| 1165 | JE2RBK | 208 |
| 1168 | GM4CHX | 207 |
| 1168 | KG4JSZ | 207 |
| 1168 | N7TY | 207 |
| 1168 | UT5IP | 207 |
| 1172 | XE1R | 206 |
| 1173 | DF1PY | 205 |
| 1173 | E73XL | 205 |
| 1173 | IZ2KXC | 205 |
| 1173 | M0AID | 205 |
| 1173 | RA3AOS | 205 |
| 1178 | DM1LM | 204 |
| 1178 | F5AAR | 204 |
| 1178 | R9CAC | 204 |
| 1178 | US3LR | 204 |
| 1178 | WA3FRP | 204 |
| 1183 | DJ3CQ | 203 |
| 1183 | DL9ZWG | 203 |
| 1183 | I2YPY | 203 |
| 1183 | IK8IPL | 203 |
| 1187 | N1LID | 202 |
| 1187 | OZ1DYI | 202 |
| 1187 | W6IYS | 202 |
| 1190 | JN1FRL | 201 |
| 1190 | W8WV | 201 |
| 1192 | AB1OC | 200 |
| 1192 | DJ1VT | 200 |
| 1192 | JA6CMQ | 200 |
| 1192 | M0DDT | 200 |
| 1192 | RA4DAR | 200 |
| 1192 | UA6CEY | 200 |
| 1198 | OE3CHC | 199 |
| 1198 | SM1LF | 199 |
| 1200 | DL1HTW | 197 |
| 1201 | K7WK | 195 |
| 1202 | IW1AZJ | 194 |
| 1203 | DL6MHW | 193 |
| 1204 | M0BUI | 189 |
| 1205 | HG6IA | 185 |
| 1205 | IS0UWX | 185 |
| 1207 | EA5UJ | 182 |
| 1207 | IK0TRV | 182 |
| 1209 | DK8PX | 181 |
| 1209 | UA1AFZ | 181 |
| 1211 | HB9MXY | 179 |
| 1211 | I1ANP | 179 |
| 1213 | CU7AA | 178 |
| 1213 | SM6IWT | 178 |
| 1215 | HA5OV | 177 |
| 1216 | R6FAA | 176 |
| 1217 | IZ2KPE | 175 |
| 1217 | JA1TBX | 175 |
| 1219 | EA2RY | 174 |
| 1220 | DK6HD | 173 |
| 1221 | W9HBH | 171 |
| 1222 | HB9TRR | 169 |
| 1223 | DF9VJ | 167 |
| 1223 | RK9DM | 167 |
| 1225 | DL1ZBO | 164 |
| 1226 | WU1U | 163 |
| 1227 | IT9XUA | 162 |
| 1228 | N9EAJ | 161 |
| 1229 | UT2HC | 160 |
| 1230 | CE3FZ | 156 |
| 1230 | DL3MR | 156 |
| 1230 | US0MS | 156 |
| 1233 | JE1XXT | 155 |
| 1234 | DF4ZY | 154 |
| 1234 | R7HL | 154 |
| 1236 | F1VEV | 153 |
| 1236 | GM4SSA | 153 |
| 1236 | N7HD | 153 |
| 1239 | 7L2PDJ | 152 |
| 1239 | DL4NAZ | 152 |
| 1239 | EI8JX | 152 |
| 1239 | UR6LEY | 152 |
| 1243 | EA5EOR | 150 |
| 1243 | K0YY | 150 |
| 1245 | AA4FL | 149 |
| 1245 | N1IA | 149 |
| 1245 | UA0QPA | 149 |
| 1245 | UA6XT | 149 |
| 1245 | UT1AO | 149 |
| 1250 | JA0CJK | 148 |
| 1251 | XE2AA | 147 |
| 1252 | DH4BAZ | 146 |
| 1252 | G4WGE | 146 |
| 1254 | G6OKU | 145 |
| 1255 | DB8PZ | 144 |
| 1256 | IW5EIJ | 143 |
| 1257 | DL2DQN | 142 |
| 1257 | IT9YVO | 142 |
| 1257 | JF0EBM | 142 |
| 1257 | PY2CX | 142 |
| 1261 | ON6LY | 141 |
| 1261 | PP5BZ | 141 |
| 1263 | 9A3PM | 139 |
| 1263 | HA9MDN | 139 |
| 1265 | CT1AVC | 138 |
| 1265 | UA6LP | 138 |
| 1267 | DL6UAA | 137 |
| 1267 | EB3JT | 137 |
| 1267 | VE3NEA | 137 |
| 1267 | W4JDS | 137 |
| 1271 | AG0A | 136 |
| 1271 | SV9COL | 136 |
| 1271 | UN7FW | 136 |
| 1274 | DL2YBG | 135 |
| 1274 | G4MPK | 135 |
| 1276 | AE4WG | 134 |
| 1276 | JQ3MWA | 134 |
| 1278 | G4OTV | 133 |
| 1278 | IK3JLV | 133 |
| 1278 | JA7ARM | 133 |
| 1278 | NH6T/W4 | 133 |
| 1282 | EI7JZ | 132 |
| 1282 | I4VJC | 132 |
| 1282 | JA1KEB | 132 |
| 1282 | OE1WEU | 132 |
| 1286 | DG8HJ | 131 |
| 1286 | JA7FVA | 131 |
| 1288 | RA9UGU | 129 |
| 1289 | AC8HN | 128 |
| 1289 | DL9MWG | 128 |
| 1289 | G5CL | 128 |
| 1292 | AA2TH | 127 |
| 1292 | DL1DWL | 127 |
| 1292 | K4PWS | 127 |
| 1292 | NH6T | 127 |
| 1296 | IK2RGT | 126 |
| 1296 | PP5VB | 126 |
| 1296 | UK7AL | 126 |
| 1299 | UW1HM | 125 |
| 1300 | KM2O | 124 |
| 1300 | UA4HDB | 124 |
| 1300 | W7GSV | 124 |
| 1300 | WX2CX | 124 |
| 1304 | G4HUN | 123 |
| 1304 | K4KGG | 123 |
| 1304 | RZ1O | 123 |
| 1304 | UA9CNX | 123 |
| 1308 | C31US | 122 |
| 1308 | I8IEQ | 122 |
| 1308 | IZ5VYP | 122 |
| 1311 | DL1JGA | 121 |
| 1311 | IZ2USP | 121 |
| 1311 | OM3YCY | 121 |
| 1314 | EA5DPL | 120 |
| 1314 | IZ5KDD | 120 |
| 1314 | K6UM | 120 |
| 1314 | K9BQL | 120 |
| 1314 | R7AX | 120 |
| 1314 | UX3MZ | 120 |
| 1320 | IW0HQE | 119 |
| 1320 | RW9TP | 119 |
| 1320 | W1EEB | 119 |
| 1320 | W5MJ | 119 |
| 1324 | 2E0AOZ | 118 |
| 1324 | DM3PYA | 118 |
| 1324 | IK5QPS | 118 |
| 1324 | JH1IHO | 118 |
| 1324 | PY8AZT | 118 |
| 1324 | R7NA | 118 |
| 1324 | UR5UEY | 118 |
| 1324 | XE3D | 118 |
| 1332 | M0RYB | 117 |
| 1332 | PY2SRL | 117 |
| 1332 | UA1TGQ | 117 |
| 1335 | 9A2GA | 116 |
| 1335 | HK3W | 116 |
| 1335 | KC0BMF | 116 |
| 1335 | UA3TCQ | 116 |
| 1339 | N3KEM | 115 |
| 1339 | W5DQ | 115 |
| 1341 | DB1WT | 114 |
| 1341 | IZ8GUQ | 114 |
| 1341 | PY2DA | 114 |
| 1341 | RA1TL | 114 |
| 1341 | UA9CIM | 114 |
| 1341 | VE7TJF | 114 |
| 1341 | W0OVM | 114 |
| 1348 | MU0GSY | 113 |
| 1348 | UX2IB | 113 |
| 1348 | WH7DX | 113 |
| 1351 | LA6VQ | 112 |
| 1351 | N2SO | 112 |
| 1351 | W0WLL | 112 |
| 1351 | W2GBY | 112 |
| 1351 | W7TLV | 112 |
| 1351 | XE1FAS | 112 |
| 1357 | JA1UAV | 111 |
| 1357 | JG3KMT | 111 |
| 1357 | KF6ILA | 111 |
| 1357 | PA0KVA | 111 |
| 1357 | ZL3PAH | 111 |
| 1362 | G0BPK | 110 |
| 1362 | G4RHR | 110 |
| 1362 | IK1UWL | 110 |
| 1362 | IZ1QLT | 110 |
| 1362 | JM1LRA | 110 |
| 1362 | JS3CTQ | 110 |
| 1362 | K6UIP | 110 |
| 1362 | KY6J | 110 |
| 1362 | LZ1MDU | 110 |
| 1362 | WA6AEE | 110 |
| 1372 | G0WSC | 109 |
| 1373 | AE6RR | 108 |
| 1373 | EU4CQ | 108 |
| 1373 | JA5NSR | 108 |
| 1373 | UR4UT | 108 |
| 1377 | DL6MLA | 107 |
| 1377 | K1LOG | 107 |
| 1377 | KG2U | 107 |
| 1377 | RW0LD | 107 |
| 1377 | UR7LY | 107 |
| 1377 | W7OXB | 107 |
| 1383 | DM2GON | 106 |
| 1383 | K5HM | 106 |
| 1383 | PA9JO | 106 |
| 1383 | RA7KW | 106 |
| 1383 | UX0HO | 106 |
| 1383 | W4HVW | 106 |
| 1389 | AB1QB | 105 |
| 1389 | DL2AJB | 105 |
| 1389 | F5VHQ | 105 |
| 1389 | HB9EXU | 105 |
| 1389 | K0LDS | 105 |
| 1389 | K6VXI | 105 |
| 1395 | IK2MLS | 104 |
| 1395 | IK2SAV | 104 |
| 1395 | N4YHC | 104 |
| 1395 | NS6E | 104 |
| 1395 | PY1ON | 104 |
| 1400 | PA0QRB | 103 |
| 1400 | RX3DFW | 103 |
| 1400 | RX3X | 103 |
| 1400 | UF2F/1 | 103 |
| 1400 | VE7JH | 103 |
| 1400 | W1WBB | 103 |
| 1400 | XE1EE | 103 |
| 1407 | AG6OX | 102 |
| 1407 | G4AYU | 102 |
| 1407 | HA1RJ | 102 |
| 1407 | UT8IU | 102 |
| 1411 | EA6SB | 101 |
| 1411 | IV3AOL | 101 |
| 1411 | IZ1KGY | 101 |
| 1411 | JI3GME | 101 |
| 1411 | K5KWR | 101 |
| 1411 | PT7AK | 101 |
| 1411 | RU3U | 101 |
| 1418 | EA2BCJ | 100 |
| 1418 | IK4VFB | 100 |
| 1418 | JH1GBO | 100 |
| 1418 | N4NDX | 100 |
| 1418 | RW5CW | 100 |
| 1418 | W3GLL | 100 |
| 1418 | YO4AUL | 100 |

## 2016 Club Listing

| Pos. | Callsign | Total |
|---|---|---|
| 1 | UT7WZA | 1077 |
| 2 | DK0EE | 925 |
| 3 | 9A1CCY | 916 |
| 4 | SK6PJ | 803 |
| 5 | SL0ZG | 799 |
| 6 | DK0PM | 739 |
| 7 | DL0IOA | 705 |
| 8 | UR4LWC | 589 |
| 9 | SK7DX | 584 |
| 10 | SK3BG | 532 |
| 11 | 9A1HBC | 509 |
| 12 | IQ2VA | 404 |
| 13 | VE3IC | 402 |
| 14 | RO2E | 305 |
| 15 | HA5KFV | 235 |
| 16 | DL0KB | 228 |
| 17 | UR4CWQ | 182 |
| 18 | 9A1CFR | 108 |
| 18 | DL1RMJ | 108 |
| 20 | PW7T | 106 |

## 2016 VHF-UHF Listing

| Pos | Name/callsign | Total |
|---|---|---|
| 1 | IW1AZJ | 194 |
| 2 | I4EAT | 192 |
| 3 | I1ANP | 179 |
| 4 | SM6CMU | 155 |
| 5 | IW9HII | 138 |
| 6 | G3KMA | 130 |
| 7 | IK4WMA | 128 |
| 8 | IZ4BEZ | 120 |
| 9 | HB9RUZ | 119 |
| 10 | OZ1BUR | 118 |
| 11 | EI3IO | 116 |
| 12 | DJ2DA | 111 |
| 12 | JA1UAV | 111 |
| 12 | SM0AJU | 111 |
| 15 | IK1UWL | 110 |
| 16 | JI1IXW | 109 |
| 17 | W1JR | 106 |
| 18 | JE3GRQ | 103 |
| 18 | N4MM | 103 |

## 2016 SWL Listing

| Pos. | Callsign | Total |
|---|---|---|
| 1 | I1-21171 | 1109 |
| 2 | UA3-147-412 | 1089 |
| 3 | BRS8841 | 1074 |
| 4 | I1-12387 | 986 |
| 5 | W1-7897 | 884 |
| 6 | F-59706 | 750 |
| 7 | SM4-3434 | 631 |
| 8 | DE0RFR | 617 |
| 9 | DE3EAR | 613 |
| 10 | JH8GAU | 536 |
| 11 | DE0DKR | 520 |
| 12 | UA1-136-644/MM | 503 |
| 13 | JA4-4665 | 451 |
| 14 | DE0ASS | 346 |
| 15 | F10437 | 315 |
| 16 | PS7AB-SWL | 309 |
| 17 | DE2EBF | 305 |
| 18 | SM7-8190 | 106 |

# Trophy Winners

**The Society is fortunate to have a large number of trophies.**

Many of these are awarded to winners of various contests, whilst others give public recognition to some particular aspect of Society work. They are presented at a number of events - typically the RSGB Convention and AGM.

*Below you will find details of some of the trophy winners, presented during 2015/2016*

## The Board

**Calcutta Key**
*(For work associated with international friendship through amateur radio)*
**Colin Thomas, G3PSM**

**Founder's Trophy**
*(For outstanding service to the Society)*
**Colin Dalziel, GM8LBC**

**Special RSGB Award for Long Service**
*(for 35 years service as GB2RS Manager)*
**Dave Sumner, K1ZZ**

## Training & Education Committee

**Kenwood Trophy**
*(For making a significant contribution to training and development in amateur radio within the UK)*
**Train the Trainers Team - Paul Whatton, G4DCV, David Evans, G0EVA, Derek Hughes, G7LFC**

## EMC Committee

**G5RV Trophy**
*(For outstanding contributions in the EMC field)*
**Three As CG**

## ARDF Committee

**3.5MHz Trophy**
**Steve Chalk SWL**

**144MHz Plate**
**Andrew Soltysik, G4KWQ**

**Sprint**
**Andrew Soltysik, G4KWQ**

## Spectrum Forum (HF Manager)

**ROTAB Trophy**
*(For outstanding and consistent DX work)*
**Michael Wells, G7VJR**

**G5RP Shield**
*(For greatest progress in the DX field made by an RSGB member resident in the UK during the year)*
**Jamie Williams, 2E0SDV**

## HF Contests

**BERU Senior Rose Bowl**
*(Winner of Commonwealth Contest)*
**John Sluymer, VE3EJ**

**BERU Junior Rose Bowl**
*(Runner-up in Commonwealth Contest)*
**Bob Whelan, J34G (G3PJT Op)**

**BERU Receiving Rose Bowl**
*(Winner of Commonwealth Contest receiving section)*
**Not awarded**

**The Lilliput Cup**
*(New - Commonwealth - Single Op Low Power)*
**Brian Campbell, VE3MGY**

**The Rosebery Shield**
*(New - Commonwealth - Single Op Assisted)*
**Gabor Horvath, VE7UF**

**Commonwealth Medal**
*(Commonwealth - Contribution To Contest/ Most Improved)*
**Brian Coyne, C4Z**

**Braaten Trophy**
*(Leading G station in the ARRL DX CW Contest)*
**Justin Snow, G4A (G4TSH OP)**

**Bristol Trophy**
*(Highest score in the 'other' section of NFD)*
**Stockport RS G3LX/P**

**CDXC Geoff Watts Trophy**
*(IOTA winners multi-op high power section)*
**DL1KZA Contest Team**

**Colonel Thomas Rose Bowl**
*(Highest placed G in Commonwealth Contest)*
**Don Beattie, G3BJ**

**Cyril Leyden G4RYY - Memorial Trophy**
*(Leading single operator Island High Power, 12hr CW, UK, in the IOTA contest)*
**Ed Taylor, GM7O (GW3SQX OP)**

**David Hill G4IQM Memorial Trophy**
*(The club having the highest aggregate score in Club Calls Contest)*
**Bristol CG**

**David King G3PFS Trophy**
*(Leading single operator, Island, High Power, 12hr SSB, UK, in the IOTA Contest)*
**Steve Cole, GW4BLE**

**Edgware Trophy**
*(Winning team in AFS 80m CW Contest)*
**Three As CG**

**Flight Refuelling ARS Trophy**
*(Winning team in AFS 80m SSB Contest)*
**Bristol CG**

**Frank Hoosen G3YF Trophy**
*(Leading 14MHz score in NFD)*
**Three As CG, GW0AAA/P**

**G2QT Cup**
*(Winner of the RSGB HF Contest Championship)*
**Nick Totterdell, G4FAL**

**G3DYY Memorial Trophy**
*(Leading single operator Island High Power, 24hr CW, UK, in the IOTA Contest)*
**Mike Chamberlain, G3P**

**G3PSH Memorial Trophy**
*(Winner of the Restricted section of SSB Field Day)*
**Adrian Rees, MW1LCR**

**G3XTJ Memorial Trophy**
*(Most accurate log in RoPoCo 2 Contest)*
**Graham Bubloz, G4FNL**

**G5MY Trophy**
*(Highest aggregate score in the RoPoCo Contests)*
**Graham Bubloz, G4FNL**

**G5RV Memorial Shield**
*(Winning team in 80m Club Championships)*
**Three As CG**

**G6ZR Memorial Cup**
*(Runner-up in the Open section of NFD)*
**Newbury & DARS, G3KLH/P**

**G8KW Trophy**
*(Leading UK entrant in the CQWW CW Contest)*
**Mark Haynes, G9W (M0DXR OP)**

**G4RYY - Memorial Trophy**
*(IOTA leading UK SO 12hr CW Section)*
**Ed Taylor, GW7O (GW3SQX OP)**

**GM5VG Trophy**
*(IOTA leading UK multi-op station)*
**Bristol CG**

**Gravesend Trophy**
*(Runner-up in the Restricted section of NFD)*
**Sussex Downs CG**

**GMDX Group Trophy**
*(Leading single operator Island High Power, 24hr SSB station in the IOTA Contest)*
**Steve Wilson, M2X (G3VMW OP)**

**GMDX Group Trophy**
*(Leading single operator Island Low Power, 24hr SSB station in the IOTA Contest)*
**Jim Martin, GA1J (MM0BQI OP)**

**Henry Lewis G3GIQ Memorial Cup** *(IOTA leading UK entrant 24h SSB)*
**Mike Clark, M0ZDZ**

**Horace Freeman Trophy**
*(80m Club Championship - winner "other" section)*
**Tall Trees CG**

**Houston Fergus Trophy**
*(Winner of the 10W section of Low Pwr Field Day)*
**Melton Mowbray ARS G4FOX/P**

**John Dunnington Trophy**
*(Commonwealth - highest UK restricted)*
**Bob Hammond, G4DBW**

**IOTA Contest Manager Trophy**
*(Leading Multi-operator, Island, Low Power, Expedition in the IOTA contest)*
**Radio Klub Dubrovnik 9A5D/P**

**IOTA Trophy**
*(Leading multi-operator, Island, High Power, Expedition in the IOTA contest)*

**Lichfield Trophy**
*(Highest individual score in AFS SSB Contest)*
**Mark Haynes, M0DXR**

**Maitland Trophy**
*(Scottish station with highest aggregate score in both Top Band Contests)*
**Allan Duncan, GM4ZUK**

**Marconi Trophy**
*(Highest individual score in AFS CW Contest)*
**Three As CG**

**Milne Cup**
*(Leading GD, GI, GJ, GM, GU or GW station in the ARRL DX CW Contest)*
**Jim Fisher, GM7R**

**MM0BQI Summer Isles Trophy**
*(Leading Single op Island, Expedition station in IOTA Contest)*
**Norbert Moravansky SV8/OM6NM**

**Newbury Trophy**
*(80m club Sprints -- [Local]Club winner)*
**Torbay Amateur Radio Society**

**NFD Shield**
*(Winner / overall highest score in NFD)*
**Brimham CG G6MC/P**

**Northumbria Cup**
*(Winner of the Open Section of SSB Field Day)*
**Northern Lights CG GD0EMG**

**Powditch Transmitting Trophy**
*(Leading single operator station in the 28MHz section of the 21/28MHz Phone Contest)*
**Mark Haynes, M0DXR**

**Reading QRP Trophy**
*(Leading station in the Low Power Section of NFD)*
**Horsham ARC G4HRS/P**

**Ross Cary Rose Bowl**
*(Leading G entry in the Restricted Section of Commonwealth Contest)*
**Andy Cook, G4PIQ**

**RSGB CDXC Cup**
*(Leading British Isles entrant in the unassisted category of CQWW SSB)*
**Keith Kerr, GM5X**

**RSGB HF Contest Committee Trophy**
*(Low Power Contest - Winner 3W Fixed)*
**De Montfort University ARS G4ARI**

**Scottish NFD Trophy**
*(Leading GM station in NFD)*
**North of Scotland CG GM2MP/P**

**Senior Rose Bowl**
*(Commonwealth - Overall winner)*
**John Sluymer, VE3EJ**

**Somerset Trophy**
*(Winner of first 1.8MHz CW Contest)*
**Bryn Tinton, G3SWC**

**Southgate Trophy**
*(Winner of the 3W section of Low Power Field Day)*
**Northampton RC G3GWB/P**

**T E Wilson G6VQ Cup**
*(Winner of the 21/28MHz CW Contest)*
**David Cree, G3TBK**

**Verulam Silver Jubilee Trophy**
*(Most accurate log in RoPoCo 1 Contest)*
**Graham Bubloz, G4FNL**

**Victor Desmond Cup**
*(Winner of the 2nd 1.8MHz Contest)*
**Graham Bubloz, G4FNL**

**VP8GQ Trophy**
*(Commonwealth - top non-UK 12hr)*
**Jeffrey Briggs, VY2ZM**

**Whitworth Cup**
*(Winner of the 21/28MHz Contest)*
**Nigel Brown, G0GDA**

**1930 Committee Cup**
*(Winner of the Low Power 80m Contest, single operator section)*
**Newbury & DARS G0ORH**

## VHF Contests

**144MHz Backpackers Trophy**
*(Leading station in the 144MHz Backpackers series of contests)*
**Andrew Vare, GW4XZL/P**

**Arthur Watts Trophy**
*(Winner of the Low Power section of VHF NFD)*
**Warrington CG G3CKR/P**

**Bolton Wireless Club Trophy**
*(Winning club in the UK Activity Contest)*
**Travelling Wave CG**

**Cockenzie Quaich**
*(Leading resident Scottish station in Restricted section of VHF NFD. All ops to have been resident in Scotland for at least 6 months prior to contest.)*
**Lothians RS, GM3HAM/P**

**Four Metre Cup**
*(Winner of the Single Operator section of the 70MHz Trophy contest)*
**Erik Gedvilas, G8XVJ/P**

**G0ODQ Trophy**
*(Winning club in the 432MHz AFS Contest)*
**Blacksheep CG**

**G3JYP, Memorial Award**
*(70MHz CW)*
**Tiverton SW RC G4TSW**

**G3MEH Trophy**
*(Winning club in the 144MHz AFS Contest)*
**Bristol Contest Group**

**John Pilags Memorial Trophy**
*(Awarded to the leading Single Operator Fixed station in the VHF Contests Championship)*
**Roger Piper, G3MEH**

**Martlesham Trophy**
*(Winner of the Restricted section of VHF NFD)*
**South Birmingham RS**

**Mitchell-Milling Trophy**
*(Winner of the Multi Operator section of the 144MHz Trophy contest)*
**Parallel Lines CG, G8P**

**Racal Radio Cup**
*(Winner of the Open section of the VHF Contests Championship)*
**Five Bells Contest Group**

**Scottish Trophy**
*(Leading Scottish station in the Low Power section of VHF NFD)*
**Loch Fyne Kippers**

**SMC Six Metre Cup**
*(Highest scoring Single Operator UK entry in the 50MHz Trophy contest)*
**2015 Erik Gedvilas G8XVJ/P**

**Surrey Trophy**
*(Winner of the Open Section of VHF NFD)*
**Blacksheep CG, M0BAA/P**

**Turton Trophy**
*(Leading Scottish station in VHF NFD - all operators to have been resident in GM for at least six months prior to the contest)*
**Cockenzie & Port Seton ARC MM0CPS/P**

**Telford Trophy**
*(Leading Fixed station in the 50MHz Trophy contest)*
**Blacksheep CG, G8T**

**Thorogood Trophy**
*(144MHz Trophy, SO Section)*
**Roger Piper, G3MEH**

**VHF Contests Committee Cup**
*(Overall winner of the 1.3GHz Trophy contest)*
**Colchester RA M1CRO/P**

**VHF Manager's Trophy**
*(Winner of the 70MHz Trophy contest)*
**Blacksheep CG M0BAA/P**

**G6ZR Memorial Trophy**
*(Winner of the 2.3GHz Trophy contest)*
**South Birmingham RS, G3OHM/P**

**1951 Council Cup**
*(Winner of the 432MHz Trophy contest)*
**ParallelLines CG, G8P**

**Low Power VHF Championship Trophy**
*(Winner of the Low Power Section of the VHF Contests Championship)*
**Richard Staples, G4HGI**

**G5BY Trophy**
*(Winner of the Mix and Match section of VHF NFD)*
**Trowbridge & DARC G2BQY/P**

**G6NB Trophy**
*(Winner of the 144MHz Club Championship)*
**Travelling Wave CG**

**10GHz Trophy**
*(Winner of the May 10GHz Trophy contest)*
**Colchester RA M1CRO/P**

**The Foundation Shield**
*(Leading Foundation Licensee in the 144MHz UK Activity Contest)*
**Andrew Power, M6RET/P**

**The Intermediate Shield**
*(Leading Intermediate Licensee in the 144MHz UK Activity Contest)*
**Pete Millard, 2E0NEY**

## Spectrum Forum (VHF Manager)

**Harold Rose Trophy**
*(To the person making an outstanding contribution to 50MHz)*
**Alain Stievenart, ON4KST**

**Don Cameron G4SST Memorial**
*(For outstanding contribution to low power amateur radio communicaton)*
**Paul Darlington, M0XPD**

**Fraser Shepherd Award**
*(For advances in space communication)*
**Ian Lamb, G8KQW / John Hazell, G8ACE**

**Louis Varney Cup**
*(For advances in space communication)*
**Wouter Weggerlaar, PA3WEG**

**1962 Committee Cup**
*(Awarded for the best VHF equipment development)*
**Justin Johnson, G0KSC**

## Technical Forum

**Courteney-Price Trophy**
*(For the most outstanding published technical contribution to amateur radio)*
**Peter Dodd, G3LDO** (posthumous award)

**Bennett Prize**
*(For any innovation which furthers the art of radio communications )*
**Steve Ireland, VK6VZ**

**Ostermeyer Trophy**
*(For most meritorious description of a piece of home-constructed or electronic equipment published in* RadCom*)*
**Barry Chambers, G8AGN**

**Norman Keith Adams Prize**
*(For the most original article published in* RadCom*)*
**Stuart Wisher, G8CYW**

**Wortley-Talbot Trophy**
*(For outstanding experimental work in amateur radio)*
**Paul Darlington, M0XPD**

## UK Microwave Group

**Dain Evans, G3RPE Memorial Cup**
*(Winner of the 10GHz Cumulative Contest)*
**Ian Lamb G8KQW/P**

**Dave Cox, G0RRJ Memorial Trophy**
*(Winner of the 24GHz Cumulative Contest)*
**Ian Lamb G8KQW/P**

**Jack Brooker, G3JMB Trophy**
*(Winner of the 10GHz Cumulative Contest, restricted section)*
**Stewart Wilkinson G0LGS/P**

**Tim Leighfield, G3KEU Trophy**
*(Winner of the 5.7GHz Cumulative Contest)*
**Ian Lamb, G8KQW**

**24GHz Cumulative Trophy**
*(Winner of the 24GHz Cumulative Contest)*
**Ian Lamb G8KQW/P**

**47GHz Cumulative Trophy**
*(Winner of the 47GHz Cumulative Contest)*
**Ian Lamb G8KQW/P**

**G3VVB Trophy**
*(For Home Construction)*
**Roger Ray, G8CUB**

**Fraser Shepherd Award**
*(For research into microwave applications to radio communication)*
**Ian Lamb, G8KQW / John Hazell, G8ACE**

**Les Sharrock, G3BNL Trophy**
*(For innovation or technical development of microwave equipment or techniques)*
**Chris Bartram GW4DGU**

**G3EEZ Memorial Trophy**
*(For outstanding contributions to Amateur Microwave Communication)*
**Charlie Suckling G3WDG**

## Other trophies awarded at the AGM

**Jack Wylie Trophy**
**Tom Wylie, GM4FDM**

**Jock Kyle Memorial**
**Colin Dalziel, GM8LBC**

**Don Cameron G4SST Memorial Trophy**
**Paul Darlington, M0XPD**

**Life Vice Presidency**
**Not awarded**

**Club of The Year - Large Club**
*(over 25 members)*
**Not awarded**

**Club of The Year - Small Club**
*(under 25 members)*
**Not awarded**

**ICOM UK SPONSOR**
*(IOTA leading UK MO Fixed Station)*
**Brimham CG G6MC**

**ICOM UK SPONSOR**
*(IOTA UK DXpedition MO Runner-up)*
**Stockport RS GM5O**

**ICOM UK SPONSOR**
*(IOTA UK MO LP Expedition)*
**Cray Valley RC MJ8C**

**ICOM UK SPONSOR**
*(IOTA SO HP Fixed statiion)*
**Victor Zagaynov, RU4SO**

**ICOM UK SPONSOR**
*(IOTA SO QRP station)*
**Antonin Beckyna, OK7CM**

**Pat Hawker, G3VA Award**
*(Overall winner in Construction Competition)*
**Anthony Watts, M6KWH**

# HF Awards

## General Rules

The following general rules and conditions apply to HF certificates and awards issued by the Radio Society of Great Britain and should be read in conjunction with the conditions which govern the particular award programme:

### Applicant eligibility

1. Claimants from the UK, Channel Isles and Isle of Man must be members of the RSGB and, as proof of membership, should provide a recent address label from RadCom. Applicants from elsewhere need not be members of the RSGB.
2. Claimants may be either licensed radio amateurs or short wave listeners. All certificates, but not special plaques, are available on a 'heard' basis to listeners.

### Claim eligibility

3. Each claim must be submitted in a form acceptable to the HF Awards Manager. Where application forms are provided for particular award programmes, these should be used although a computer generated form including the same headings will generally be accepted. Each claim must include the following signed declaration: "I declare that all the contacts were made by me personally from the same DXCC country and in accordance with the terms of my radio transmitting license, and that none of the QSLs have been amended in any way since receipt. I accept that a breach of these rules may result in disqualification from the awards programme. I further accept that the decision of the HF Committee shall be final in all cases of dispute."
4. All claims from within the UK, Channel Isles and Isle of Man must be accompanied by QSL cards. Claims from elsewhere must also be accompanied by QSL cards but only in the case of those categories of award attracting a plaque. In all other cases a statement from the applicant's national society that the necessary cards have been checked will be accepted although the HF Awards Manager reserves the right to ask to see some or all of the cards. For IOTA claims special rules apply (see the IOTA Directory).
5. Each claim must be accompanied by a fee per certificate or class of certificate. Applicants submitting cards for checking must include sufficient payment to cover their return. Cards will only be returned by air, recorded delivery (UK only), or registered mail (overseas) if adequate postage is enclosed (for current fees please check on the RSGB website or with the Awards Manager).

### Contact eligibility

6. All contacts must be made by the holder of the callsign.
7. Contacts may be made from any location in the same DXCC country.
8. Except where otherwise indicated, credit will be given for contacts made on or after 15 November 1945 on any of the amateur bands below 30MHz.
9. Contacts with land mobile stations will be accepted, provided the location at the time of contact is clearly stated on the QSL card.
10. Credit will be given for two-way contacts on the same mode and band, ie not cross-mode or cross-band. Certificate endorsements for single mode transmission and/or single band may be made on the submission of cards clearly confirming the mode or frequency of transmission, but the request must be made at the time of application. Special rules apply for IOTA.
11. Please note that e-QSLs are not acceptable as proof of QSOs.

### Disqualification

12. The submission for credit of any altered or forged confirmations or, equally, bad behaviour on or off the air, which is judged by the HF Committee to bring a particular programme into disrepute, may result in disqualification of the applicant from all RSGB's award programmes. The decision of the HF Committee on this and other matters of dispute will be final.

## Applications For Awards

All claims, except IOTA, should be sent to the HF Awards Manager (details below).

Prepare your application in accordance with the requirements of the award being claimed. Send QSL cards when required, and do not forget to enclose details of your name, callsign and full address as well as the certificate fee, adequate postage for the return of QSL cards, and for UK applicants, proof of RSGB membership. Payment may be made by cheque drawn on a UK bank in Pounds Sterling and payable to 'Radio Society of Great Britain'.

## Further Information

The HF Awards Manager may be contacted by email, and information about all awards may be found on the Internet (details below).

## Commonwealth Century Club (CCC)

This award may be claimed by any licensed radio amateur eligible under the general rules who can produce evidence of having contacted, since 15 November 1945, amateur radio stations in at least 100 Commonwealth call areas on the list current at the time of application.

The certificate holder may claim, on payment of a contributory charge, a plaque with a plate detailing name, callsign, date and number of the award. Additionally, an amateur providing evidence of having contacted all the Commonwealth call areas on the list current at the time of application may claim the Supreme Plaque in recognition of the magnitude of the achievement, again on payment of a contributory charge.

*New style Award for all Class of Certificate*

### Notes

(a) Credit for South Georgia and the South Sandwich Is will only be given for contacts with stations using a VP8 callsign. Credit for Antarctica and the South Orkney and South Shetland Is will only be given for contacts with stations using a callsign issued by a Commonwealth government.

(b) Where, very occasionally, a contact is made with a station using a callsign legitimately outside the geographical area to which the prefix normally applies, it will count for the actual area from which the operation took place. The evidence submitted will need to be clear.

(c) Contacts verified by entry to the Annual Commonwealth Contest are acceptable as valid confirmation that a QSO took place. The Awards Manager maintains files on all call areas worked by all entrants on each band therefore if you work a new one during the contest please email awards@rsgb.org.uk should you wish to have it included on your Award Check Sheet Record. Records are not updated automatically, as not all entrants are working towards the CCC Award Program. QSOs from 2005-2007 can be used.

## DX Listeners' Century Award (DXLCA)

This award may be claimed by any short wave listener eligible under the general rules who can produce evidence of having heard amateur radio stations located in at least 100 DXCC countries. Stickers are available for every 25 additional countries confirmed. Submit a list in radio prefix order with the callsign and country name.

Endorsements are available for hearing 100 countries on 5, 6, 7, 8 and 9 bands (they need not be the same countries on each band).

## 5 Band Commonwealth Century Club (5BCCC)

This award, available in five classes, may be claimed by any licensed radio amateur under the general rules who can produce evidence of having made two-way communication, since 15 November 1945, with the requisite number of amateur radio stations located in

RSGB Awards Manager: Marcus Hazel-McGown, MM0ZIF,
27 Ashdale Avenue, Saltcoats, North Ayrshire, KA21 6AA.
Email: awards@rsgb.org.uk Web site: *http://www.rsgb.org/awards*

the call areas listed, using all five bands, 3.5, 7, 14, 21 and 28 MHz. Each station should be located in a different call area per band.

The five classes are for contacts as follows:

| | |
|---|---|
| 5BCCC Supreme | 500 stations |
| 5BCCC Class 1 | 450 stations |
| 5BCCC Class 2 | 400 stations, with a minimum of 50 on each band |
| 5BCCC Class 3 | 300 stations, with a minimum of 40 on each band |
| 5BCCC Class 4 | 200 stations, with a minimum of 30 on each band |

Certificates will be issued to winners of all classes. Additionally, as in the case of the CCC, winners of the Class 1 award will be eligible to claim a plaque suitably inscribed on payment of a contributory charge, while winners of the Supreme Award will be able to claim an engraved plaque on payment of a contributory charge.

### WARC Bands Endorsement

A holder of the CCC award who can provide evidence of contact with the required number of call areas on the 10, 18 and 24MHz bands may claim a WARC band 'sticker' endorsement. This is available in five classes as follows:

| | |
|---|---|
| 5BCCC (WARC) Supreme | 300 call areas |
| 5BCCC (WARC) Class 1 | 275 call areas |
| 5BCCC (WARC) Class 2 | 250 call areas, with a minimum of 50 on each band |
| 5BCCC (WARC) Class 3 | 200 call areas, with a minimum of 40 on each band |
| 5BCCC (WARC) Class 4 | 150 call areas, with a minimum of 30 on each band |

Note: On 10MHz credit will only be given for contacts on CW and datamodes.

### Top band endorsement

A holder of the CCC award who can produce evidence of contact with the required number of call areas on the 1.8MHz band may claim a Top Band 'sticker' endorsement. This is available in five classes for 30, 40, 50, 60 and 70 call areas.

### Credit for deleted call areas

Credit will be given for contacts with stations in Commonwealth countries at the time of contact, and additionally, with stations using a Commonwealth callsign in the Antarctic and the S Orkney and S Shetland Is. Since 1945 some countries have left the Commonwealth, while others have joined. For 5BCCC (including endorsements) contacts with 'deleted' call areas made at a time when the countries concerned were in the Commonwealth may count in place of 'missing' credits on any band up to the maximum possible for that band (at July 1996 this was 132). Deleted call areas do not count for CCC.

### Applications for CCC and 5BCCC

Applications for 5BCCC should use the special application form available from the HF

**Commonwealth Century Club Supreme Award**

| Award Number | All Areas Supreme | Date Issued |
|---|---|---|
| 1 | G3AAE | ? |
| 2 | G3NOF | ? |
| 3 | G3GIQ | 12/07/1995 |
| 4 | G3ZDA | 21/12/1996 |
| 5 | G4LVQ | 20/11/1997 |
| 6 | G3KYF | 01/03/1999 |
| 7 | G0OJX | 13/11/2001 |
| 8 | G4BWP | 16/01/2007 |
| 9 | G3LZQ | 25/02/2007 |
| 10 | G3VKW | [illegible]/2010 |

**Notes:**

1. Computer records began when G4BWP took over the awards programme in 1994. Early records are limited, hence the lack of date information and the odd question mark.
2. Sequential numbers for all certificates were used, irrespective of CCC Award classification.
3. Corrections or confirmations would be most helpful to keep the database accurate.

**Commonwealth Century Call Area Standard Award**

| Award Number | 100 Areas Standard | Date Issued | Award Number | 100 Areas Standard | Date Issued | Award Number | 100 Areas Standard | Date Issued |
|---|---|---|---|---|---|---|---|---|
| 1 | G4STH | | 32 | A92BE | | 63 | G0EHO | 05/10/1999 |
| 2 | G3AAE | Supreme-1 ? | 33 | G4KBX | | 64 | G3LAS | 28/12/1999 |
| 3 | G4ADD | | 34 | G0LRX | | 65 | KT4BW | 28/06/2000 |
| 4 | G3NOF | Supreme-2 ? | 35 | GM4XLU | Jul-93 | 66 | JA1CKE | 03/01/2001 |
| 5 | 9H4G | | 36 | G3KYF | Jul-93 | 67 | DK7YY | 10/01/2001 |
| 6 | G3IFB | Jul-88 ? | 37 | G4BWP | | 68 | G4ZOY | 20/03/2001 |
| 7 | GM3CIX | | 38 | UA9-154-1289 | | 69 | VE7SMP | 09/05/2001 |
| 8 | GW4RHW | | 39 | UA3-142-1925 | | 70 | G3HSL | 09/05/2001 |
| 9 | G0ANH | | 40 | KI6PG | 12/06/1994 | 71 | GI4SNC | 09/05/2001 |
| 10 | ? | | 41 | SM7HV/HK7 | 20/06/1994 | 72 | G0TYV | 13/11/2001 |
| 11 | Y78SL | | 42 | VE3MS | 09/07/1994 | 73 | GI3VAW | 13/11/2001 |
| 12 | CX4HS | | 43 | AB4DU | 16/07/1994 | 74 | WC6DX | 25/02/2002 |
| 13 | DL3RK | | 44 | JE1VTZ | 14/08/1994 | 75 | VK2DPD | 24/06/2002 |
| 14 | GM3WIL | | 45 | G4MVA | 29/08/1994 | 76 | JH4BTI | 24/07/2003 |
| 15 | UB4WZA | Jul-89 | 46 | ON5KL | 08/10/1994 | 77 | 7M3ESJ | 24/07/2003 |
| 16 | G3VOF | | 47 | G3EZZ | 09/11/1994 | 78 | G3LPS | 02/05/2004 |
| 17 | K3FN | | 48 | G0ARF | 30/12/1994 | 79 | G4KHM | 18/05/2004 |
| 18 | G3SJX | | 49 | G3TBK | 22/04/1995 | 80 | G3EKJ | 06/03/2005 |
| 19 | G4OBK | | 50 | G4IUF | 09/09/1995 | 81 | JA2EPW* | 21/03/2005 |
| 20 | ORS43992 | | 51 | JA8XDM | 17/11/1995 | 82 | JA8AWR | 28/07/2005 |
| 21 | G4NXG/M | | 52 | G3KWK | 14/10/1995 | 83 | G3IAF | 04/11/2005 |
| 22 | AA6VC | | 53 | W4ZYT | 06/01/1996 | 84 | G3NXT | 03/10/2007 |
| 23 | UA3-142-7453 | | 54 | G4XRX | 13/01/1996 | 85 | G3AKU | 30/10/2007 |
| 24 | G4IJW | | 55 | AA6WJ | 28/02/1996 | 86 | G3VKW | 30/11/2008 |
| 25 | UA3DRB | | 56 | VK4SS | 19/06/1996 | 87 | G3CWW* | 31/05/2009 |
| 26 | I8SAT | | 57 | BRS10167 | 18/01/1997 | 88 | VK2RO | 31/08/2009 |
| 27 | G0AHC | | 58 | VU2SMN | 19/02/1997 | 89 | EA3IM | 30/11/2010 |
| 28 | KA4GYU | | 59 | KQ4WD | 01/04/1997 | 90 | JA7OXR | 01/05/2012 |
| 29 | G4ZYQ | | 60 | BRS47426 | 28/01/1998 | 91 | G0HIO* | 01/05/2012 |
| 30 | KA1LMR | | 61 | G2FSR | 23/01/1999 | | | |
| 31 | SM0DJZ | | 62 | ONL7681 | 23/01/1999 | | (* = all CW) | |

**5-Band CCC Awards (not WARC bands)**

| Award Number | 500 Areas Supreme | 450 Areas Class 1 | 400 (min 50/band) Class 2 | 300 (min 40) Class 3 | 200 (min 30) Class 4 |
|---|---|---|---|---|---|
| 1 | G3TJW | G3UML | K2SHZ | G3UML | G8JR |
| 2 | G3GIQ | UP1BZZ | G4GIR | K2SHZ | K2SHZ |
| 3 | OK3EY | G4GIR | - | G3SWH | G3SJX |
| 4 | VK9NS | I8SAT | G3TBK | G3KWK | YU2OB |
| 5 | A92BE | G3TBK | CT3FT | G0EHO | G3IFB |
| 6 | G3UML | G3SWH | G3SWH | VK2NU | G3EZZ |
| 7 | G4BWP | G3IAF | G3KWK | G0EHO | G3SWH |
| 8 | LY2ZZ | G3VKW | G3AKU | G3LPS | G3KWK |
| 9 | G3IFB | - | G3VKW | G3VKW | G0EHO |
| 10 | G3TBK | - | - | G3AKU | G3LPS |
| 11 | G3SJX | - | - | - | G3WW |
| 12 | G3KWK | - | - | - | - |
| 13 | G3LZQ | - | - | - | - |
| 14 | G3TXF | - | - | - | - |
| 15 | G3SWH | - | - | - | - |
| 16 | G3IAF | - | - | - | - |
| 17 | G3VKW | - | - | - | - |

**WARC CCC Endorsements**

| Award Number | 300 Call Areas Supreme | 275 Class 1 | 250 (min 50) Class 2 | 200 (min 40) Class 3 | 150 (min 30) Class 4 |
|---|---|---|---|---|---|
| 1 | G3TBK | G3TBK | G3SWH | G3SWH | G3EZZ |
| 2 | G4BWP | G3SWH | G3TBK | G3TBK | G3SWH |
| 3 | G3SWH | G4BWP | G4BWP | G3SJX | G3TBK |
| 4 | G3TXF | G3IAF | G3AKU | G4BWP | G4BWP |
| 5 | G3SJX | - | G3LZQ | G3LZQ | G0EHO |
| 6 | G3IAF | - | - | G3VKW | G3KWK |
| 7 | - | - | - | - | VK2NU |
| 8 | - | - | - | - | G3VKW |

**160m CCC Endorsements**

| Award Number | 70 Areas Supreme | 60 Areas Class 1 | 50 Areas Class 2 | 40 Areas Class 3 | 30 Areas Class 4 |
|---|---|---|---|---|---|
| 1 | G4BWP | G4BWP | G4BWP | G4BWP | SM5HV/HK7 |
| 2 | G3LZQ | G3LZQ | G3LZQ | G3KWK | G3SJX |
| 3 | - | - | - | G3LZQ | G3TBK |
| 4 | - | - | - | G3SJX | G4BWP |
| 5 | - | - | - | - | G3KWK |
| 6 | - | - | - | - | G3TXF |

Some IARU, WAS and RSGB HF Award application forms are available in PDF format by sending an Email request specifying the award to: *awards@rsgb.org.uk*

Awards Manager. The form allows space for callsigns to be recorded for contacts on each band. A checklist of current Commonwealth call areas is also available and can be used for applications for the CCC Award. Please send an A5 size SASE for the checklist and application forms.

## IARU Region 1 Award

This award, available in three classes, may be claimed by any licensed radio amateur eligible under the general rules who can produce evidence of having contacted amateur radio stations located in the required number of countries whose national societies are members of the Region 1 Division of the International Amateur Radio Union (IARU)

The three classes are for contacts as follows:

| | |
|---|---|
| Class 1 | All member countries on the current list |
| Class 2 | 60 member countries |
| Class 3 | 40 member countries |

## Worked ITU Zones (WITUZ)

This award may be claimed by any licensed radio amateur eligible under the general rules who can produce evidence of having contacted, since 15 November 1945, land-based amateur radio stations in at least 70 of the 75 broadcasting zones as defined by the International Telecommunications Union (ITU).

The certificate holder may claim, on payment of a contributory charge, a plaque with a plate detailing name, callsign, date and number of the award. Additionally, an amateur providing evidence of having contacted all 75 ITU zones may claim the Supreme Plaque in recognition of the magnitude of the achievement, again on payment of a contributory charge.

## IARU Region 1 Award Recent Issues

| Call | Class | Mode | Countries | Cert No. |
|---|---|---|---|---|
| G3UFO | 2 | MIXED | 64 | 6918 |
| YO3VU | 2 | MIXED | 78 | 6919 |
| G4FFN | 2 | SSB | 63 | 6920 |
| OH8US | 1 | CW | 95 | 6921 |
| SM7ZDC | 2 | SSB | 64 | 6922 |
| SP3FGQ | 2 | CW | 64 | 6923 |
| M5LRO | 2 | MIXED | 85 | 6924 |
| EA3IM | 2 | MIXED | 85 | 6925 |
| 9W2ESM | 2 | SSB | 89 | 6926 |
| GI4CWZ | 3 | CW | 40 | 6927 |
| IW0HDU | 2 | MIXED 10m | 68 | 6928 |
| AJ4FM | 3 | MIXED | 51 | 6929 |
| EA7NA | 2 | SSB | 81 | 6930 |
| JA1RUR | 2 | SSB | 60 | 6931 |
| WA0JZK | 2 | MIXED | 62 | 6932 |
| AJ4FM | 2 | MIXED | 60 | 6933 |
| PA0QRB | 2 | CW | 60 | 6934 |
| K4JC | 2 | MIXED | 77 | 6935 |
| W4JU | 3 | MIXED | 40 | 6936 |
| MI0AHH | 3 | MIXED | 40 | 6937 |
| IV3GOW | 1 | MIXED | 95 | 6938 |
| TA1DX | 2 | MIXED | 94 | 6939 |
| TA1DX | 2 | RTTY | 78 | 6940 |
| G0THF | 1 | SSB | 95 | 6941 |
| G5CL | 2 Update | QRP CW | 81 | 6893 |

## 5 Band Worked ITU Zones (5BWITUZ)

This award, available in five classes, may be claimed by any licensed radio amateur eligible under the general rules who can produce evidence of having contacted, since 15 November 1945, the required number of land-based amateur radio stations located in the 75 ITU broadcasting zones, using all 5 bands: 3.5, 7, 14, 21 and 28MHz. Each station should be located in a different ITU zone per band. The five classes are for contacts as follows:

| | |
|---|---|
| 5BWITUZ Supreme | 350 zones |
| 5BWITUZ Class 1 | 325 zones |
| 5BWITUZ Class 2 | 300 zones, with a minimum of 50 on each band |
| 5BWITUZ Class 3 | 250 zones, with a minimum of 40 on each band |
| 5BWITUZ Class 4 | 200 zones, with a minimum of 30 on each band |

Certificates will be issued to winners of all classes. Also, as in the case of the WITUZ, winners of the Class 1 award may claim a plaque suitably inscribed, while winners of the Supreme Award will be eligible for the Supreme Plaque, both on payment of a contributory charge.

### WARC band endorsement

A holder of the WITUZ award who can produce evidence of contact with the required number of ITU zones on the 10, 18 and 24MHz bands may claim a WARC band 'sticker' endorsement. This is available in five classes as follows:

| | |
|---|---|
| 5BWITUZ (WARC) Supreme | 210 zones |
| 5BWITUZ (WARC) Class 1 | 195 zones |
| 5BWITUZ (WARC) Class 2 | 180 zones, with a minimum of 50 on each band |
| 5BWITUZ (WARC) Class 3 | 150 zones, with a minimum of 40 on each band |
| 5BWITUZ (WARC) Class 4 | 120 zones, with a minimum of 30 on each band |

Note: On 10MHz credit will only be given for contacts on CW and datamodes.

### Top Band endorsement

A holder of the WITUZ award who can produce evidence of contact with the required number of call areas on the 1.8MHz band may claim a Top Band 'sticker' endorsement. This is available in five classes: 20, 30, 40, 50 and 60 zones.

### Notes

(a) The number of ITU broadcasting (ie land) zones recognized by the ITU is 75 zones (Zones 1 to 75) and this therefore is the maximum score which can be claimed per band. However, the islands of Minami Torishima (JD1) and Salas-y-Gomez (CE0) lie outside the 75 zones in sea zones 90 and 85 respectively. For 5BWITUZ (including endorsements) contacts with these islands may count in place of one 'missing' credit on any band up to the maximum 75 for that band. Contacts with Minami Torishima and Salas-y-Gomez do not count for WITUZ.

## Worked ITU Zones Award Holders

### 5-Band WITUZ

| Award Number | 350 Zones Supreme | 325 Zones Class 1 | 300 (min 50/band) Class 2 | 250 (min 40/band) Class 3 | 200 (min 30/band) Class 4 |
|---|---|---|---|---|---|
| 1 | OK3EY | UP1BZZ | G4GIR | YU1OB | Y37XJ |
| 2 | UP1BZZ | RT5UN | A92BE | UY5XE | UY5XE |
| 3 | I8SAT | UA6AD | G3SWH | A92BE | OE1-1040 |
| 4 | UA0FZ | UA6JW | G3LZQ | RB0HZ | G3SJX |
| 5 | UY0MM | UB4WZA | IV3GOW | - | G4MVA |
| 6 | G3SJX | OK1ADM | - | - | A92BE |
| 7 | - | G3VOF | - | - | G3RHM |
| 8 | - | A92BE | - | - | (missed) |
| 9 | - | UA4PO | - | - | G3EZZ |
| 10 | - | DE0DXM | - | - | F-10095 |
| 11 | - | G3SJX | - | - | SM5HV/HK7 |
| 12 | - | K2SHZ | - | - | G3WW |
| 13 | - | UY5AA | - | - | - |
| 14 | - | G3LZQ | - | - | - |

### WARC 5-Band WITUZ Endorsements

| Award Number | 210-Zones Supreme | 195-Zones Class 1 | 180 (50/band) Class 2 | 150 (40/band) Class 3 | 120 (30/band) Class 4 |
|---|---|---|---|---|---|
| 1 | UY0MM | - | G3SWH | SM5HV/HK7 | G3EZZ |
| 2 | - | - | G3SJX | G3SJX | SM5HV/HK7 |
| 3 | - | - | G3LZQ | - | G3SJX |
| 4 | - | - | - | - | G3LZQ |

### 160m WITUZ Endorsements

| Award Number | 70 Zones Supreme | 60 Zones Class 1 | 50 Zones Class 2 | 40 Zones Class 3 | 30 Zones Class 4 |
|---|---|---|---|---|---|
| 1 | - | - | G3SJX | G3SJX | G3SWH |
| 2 | - | - | G3LZQ | - | - |

## Worked All ITU Zones Series Supreme Award 75-Zones

| Call | Date |
|---|---|
| EA5AT | August 1994 |
| ON5KL | October 1994 |
| LY2ZZ | May 1995 |
| DE0DXM (SWL) | October 1995 |
| VK4SS | August 1996 |
| HB9CMZ | November 1996 |
| JA7JI | February 1998 |
| G3SJX | November 2001 |
| WC6DX | February 2002 |
| K2SHZ | April 2006 |

## Members of IARU Region 1 (from February 2012)

| Prefix | Country | Nat. Soc. | Prefix | Country | Nat. Soc. |
|---|---|---|---|---|---|
| ZA | Albania | AARA | EL | Liberia | LRAA |
| 7X | Algeria | ARA | HB0 | Liechtenstein | AFVL |
| C3 | Andorra | URA | LY | Lithuania | LRMD |
| EK | Armenia | FRRA | LX | Luxembourg | RL |
| OE | Austria | OEVSV | TZ | Mali | CRAM |
| A9 | Bahrain | ARAB | 9H | Malta | MARL |
| EU | Belarus | BFRR | 3B | Mauritius | MARS |
| ON | Belgium | UBA | ER | Moldova | ARDM |
| T9/E7 | Bosnia & Herzegovina | ARABIH | 3A | Monaco | ARM |
| A2 | Botswana | BAR | JT | Mongolia | MRSF |
| LZ | Bulgaria | BFRA | 4O | Montenegro | MARP |
| XT | Burkina Faso | ARBF | CN | Morocco | ARRAM |
| TJ | Cameroon | ARTJ | C9 | Mozambique | LREM |
| TN | Congo | URAC | V5 | Namibia | NARL |
| 9A | Croatia | HRS | PA | Netherlands | VERON |
| 5B | Cyprus inc UK Sov Bases | CARS | 5N | Nigeria | NARS |
| OK | Czech Republic | CRC | LA | Norway | NRRL |
| 9Q | Democratic Republic of Congo | ARAC | A4 | Oman | ROARS |
| OZ | Denmark | EDR | SP | Poland | PZK |
| J2 | Djibouti | ARAD | CT | Portugal (inc. CU, CT3) | REP |
| SU | Egypt | EARA | A7 | Qatar | QARS |
| ES | Estonia | EARU | 3X | Republic of Guinee | ARGUI |
| ET | Ethiopia | EARS | YO | Romania | FRR |
| OY | Faeroe Islands | FRA | R | Russian Federation | SRR |
| OH | Finland | SARL | T7 | San Marino | ARRSM |
| Z3 | F.Y.R. Macedonia | RSM | 6W | Senegal | ARAS |
| F | France (inc. TK) | REF | YU | Serbia | SRS |
| TR | Gabon | AGRA | 9L | Sierra Leone | SLARS |
| C5 | Gambia | RSTG | OM | Slovakia | SARA |
| 4L | Georgia | NARG | S5 | Slovenia | ZRS |
| DL | Germany | DARC | ZS | South Africa | SARL |
| 9G | Ghana | GARS | EA | Spain | URE |
| ZB2 | Gibraltar | GARS | 3DA | Swaziland | RSS |
| SV | Greece | RAAG | SM | Sweden | SSA |
| HA | Hungary | MRASZ | HB9 | Switzerland | USKA |
| TF | Iceland | IRA | YK | Syria | SSTARS |
| YI | Iraq | IARS | EY | Tadjikstan (ex UJ) | TARL |
| EI | Ireland | IRTS | 5H | Tanzania | TARC |
| 4X | Israel | IARC | 3V | Tunisia | CAST |
| I | Italy | ARI | TA | Turkey | TRAC |
| TU | Ivory Coast | ARAI | EZ | Turkmenistan (ex UH) | LRT |
| JY | Jordan | RJARS | 5X | Uganda | UARS |
| UN | Kazakstan | KFRR | UR | Ukraine | UARL |
| 5Z | Kenya | ARSK | A6 | United Arab Emirates | EARS |
| 9K | Kuwait | KARS | G | UK (G GD GI GJ GM GU GW) | RSGB |
| YL | Latvia | LRAL | 9J | Zambia | RSZ |
| OD | Lebanon | RAL | Z2 | Zimbabwe | ZARS |
| 7P | Lesotho | LARS | | | |

(b) In the case of WITUZ and 5BWITUZ, confirmations need not bear the appropriate ITU zone number but in order to count for credit they should give the location of the station in sufficient detail to place it clearly within one particular zone. Doubtful cases indicating possible overlap across two zones will not be given credit.

(c) The HF Awards Manager will use as his reference a list which is based on the Radio Amateurs' Prefix Map of the World published by Radio Amateur Call Book Inc, PO Box 2013, Lakewood, New Jersey 08701, USA. In the case of countries which encompass two or more ITU zones, eg USA, Russia and Brazil, zonal boundaries will generally follow the longitude/latitude grid lines as shown in the map. In the few instances of discrepancy between the map and the accompanying prefix/country list the decision of the HF Awards Manager will be final.

### Applications

Applications for WITUZ and 5BWITUZ should use the special application form available from the Awards Manager. A check list of ITU Zones and a map are included with the application form.

Please send an A5 size SASE for the check list and application form.

## Worked All Continents (WAC)

This award, issued by IARU headquarters, may be obtained by any licensed radio amateur in the UK, Channel Islands or Isle of Man who is a member of the RSGB and can produce evidence of having made two-way contact with amateur radio stations located in each of the six continents: North America, South America, Europe, Africa, Asia and Oceania.

### 136kHz Certificates

| Cert. No. | 5 Countries | 10 Countries | 15 Countries | 20 Countries |
|---|---|---|---|---|
| 1 | G4GVC | G4GVC | G4GVC | PA0BWL |
| 2 | G0MIN | DJ7RD | PA0BWL | F6BWO |
| 3 | G3XTZ | F6BWO | DK1IS | DK1IS |
| 4 | OK1FIG | PA0BWL | S52AB | OK2BVG |
| 5 | G0AKY | S52AB | | |
| 6 | F6BWO | F6CWA | | |
| 7 | G3YMC | | | |
| 8 | PA0BWL | | | |
| 9 | EA1PX | | | |
| 10 | YO2IS | | | |
| 11 | SP2KDS | | | |
| 12 | F4DTL | | | |
| 13 | SO5AS | | | |

Applicants should send QSL cards to the RSGB HF Awards Manager who will certify the claim to the IARU headquarters society (ARRL) for issuance of the award. They should also enclose an SASE for return of the cards, and proof of RSGB membership.

All contacts must be made from the same country or separate territory within the same continent. Various endorsements including 'all 1.8MHz' are available. In addition both a 5 and 6 Band WAC may be claimed.

## 136kHz Award

This award is to recognize achievements in both transmission and reception on the 136kHz band, and to stimulate experimentation and station improvement.

The award is available in three categories, with endorsements for additional countries heard/worked.

The basic award is for confirmed two-way QSOs on 136kHz with five countries from the ARRL DXCC/WAE country list.

The SWL award is for confirmation of SWL reports from five countries. The SWL award may also be claimed by amateurs working crossband to stations transmitting in the 136kHz band.

The third category is for cross-band contacts where the station claiming the award has worked five countries by transmitting on the 136kHz band and receiving stations on other amateur bands.

Cross-mode contacts will be allowed for this award. The categories of the award may not be mixed but awards from some or all of the categories may be claimed and endorsed concurrently.

Once the basic award has been claimed, it may be endorsed in steps of each additional five countries worked or heard.

## American award checkpoints

The UK checkpoints for CQ awards (WAZ, WPX and CQ DX) are G3LZQ and GM3YTS (email: gm3yts@btinternet.com). Both are QTHR. The UK checkpoint for the ARRL WAS Award is G3LZQ. For claims for CQ Worked All Zones, a new Microsoft Word document and Excel Spreadsheet are available from G3LZQ.

## DXCC field checking

In the UK, the approved DXCC checkers are GM3YTS, G0KRL, G3RTE and G4BWP. Requests for information, application forms etc should be sent with a large SASE to Fred Handscombe, G4BWP, QTHR, or emailed to fredch@homeshack.freeserve.co.uk

Details on QSL submission, costs etc can be obtained from g0krl@arrl.net (if you live in England and have a G, M, GX, MX or 2E prefix), gm3yts@btinternet.com (if you live in the rest of the UK ie GM, GD, GI, GJ, GU, GW plus MD, MI, MJ etc etc + 2 and associated Club Prefixes).

For more information on the IOTA Programme, see the separate feature in this edition of the RSGB Yearbook.

# VHF Awards

## General Rules

1. Awards are available to licensed amateurs and listeners (on a heard basis). All claims must be fully supported by QSL cards. For the various squares awards, these cards must also bear the IARU (Maidenhead) locator details. A card without an IARU locator originally printed on it is acceptable provided that it bears some adequate form of positional information (for example old QTH locator or latitude and longitude), in which case the IARU locator square designation should be clearly added to the card by the award claimant.
2. For all awards with a fixed station category, the applicant must state that all the contacts were made from the same location. In the case of an amateur moving home location he/she may apply for the award using QSL cards gained from more than one location, but the award will be endorsed as "gained from more than one location".
3. Endorsements such as all CW, all SSB, all auroral contacts, or all contacts made during the first year of being licensed may be made on application. The appropriate information must be contained on the QSL cards and a declaration signed when applying for endorsements.
4. The charges for awards are: RSGB members £4, US$7, 6 Euros. UK residents who are not RSGB members £7, US$13, 8 Euros. Overseas applicants who are members of their national society £7, US$13 or 8 Euros. Other overseas applicants £9, US$18. Where applicable, proof of membership of their national (IARU approved) society is required, eg recent RadCom label or photocopy of membership card. There is no charge for stickers to update levels of achievement, but, if a new certificate is requested, the charges above will apply.
5. All claims must be submitted to the Awards Manager, RSGB Awards Manager: Marcus Hazel-McGown, MM0ZIF, 27 Ashdale Avenue, Saltcoats, North Ayrshire, KA21 6AA. email: awards@rsgb.org.uk

Website: *http://www.rsgb.org/awards*

Application forms may be obtained from this address by sending an SASE (A4 or A5 size preferred).

6. For the safe return of the QSL cards, adequate postage and an SASE must be sent with the application. If Recorded Delivery is required within the UK, then adequate postage must be included and a completed Recorded Delivery sticker enclosed for attachment.

## RSGB 50MHz Countries Award

The initial qualification for this certificate is proof of completed two-way QSOs on 50MHz with 10 countries. Stickers will be provided for increments of every ten countries worked. Only contacts with countries permitting 50MHz operation can be considered.

### Rules

1. All contacts must have been made on or after 1 June 1987.
2. QSL cards submitted must be arranged in alphabetical order of the countries claimed, and a checklist enclosed.
3. Stations are eligible for awards in the following categories:
(a) Fixed stations
(b) Temporary location or portable (/P).
Categories cannot be mixed.

## RSGB 50MHz DX Certificate

This certificate takes into account the considerable potential for cross-band working when transmitting in the 50MHz band. There is therefore no stipulation on the band used for the incoming signal. The initial qualification is confirmation from 25 different countries of a successful QSO with transmission from the applicant's country taking place within the 50MHz band. Stickers will be provided for increments of 25 countries confirmed.

### Rules

1. All contacts must have been made on or after 1 June 1987.
2. QSL cards submitted must be arranged in alphabetical order of the countries claimed, and a checklist enclosed.
3. Stations are eligible for awards in the following categories:
   (a) Fixed stations
   (b) Temporary location or portable (/P).

Categories cannot be mixed.

## RSGB 50MHz Squares Award

The 50MHz Squares Award is intended to mark successful VHF achievement. The initial qualification needed for this certificate is proof that 25 different locator squares have been worked with complete two-way QSOs within the 50MHz band. Squares in any country will qualify provided that operation from that country is formally authorised. Additional stickers will be provided when proof is submitted for increments of 25 squares.

### Rules

1. All contacts must have been made on or after 1 June 1987.
2. QSL cards submitted must be arranged in alphabetical order of the QTH squares claimed, and a checklist enclosed.

*New style Award for all Class of Certificate*

3. Stations are eligible for awards in the following categories:
(a) Fixed stations
(b) Temporary location or portable (/P).

Categories cannot be mixed.

## 4-2-70 Squares Award

These are intended to mark successful VHF/UHF achievement. Initially, a certificate and one sticker will be issued. Further stickers will be issued as additional locator squares are claimed. The title of each award gives the number of locator squares and countries needed to qualify for the award. For example, to obtain the 144MHz 40/10 award you must have QSL cards confirming contact with 40 locator squares including 10 countries on 144MHz. The following awards are available:

| **4m band** | |
|---|---|
| 70MHz 20/4 | 70MHz 25/6 |
| 70MHz 30/8 | 70MHz 35/8 |
| 70MHz 40/8 | 70MHz 45/8 |
| 70MHz 50/8 | |

| **2m band** | **70cm band** |
|---|---|
| 144MHz 40/10 | 432MHz 30/6 |
| 144MHz 60/15 | 432MHz 40/10 |
| 144MHz 80/18 | 432MHz 50/13 |
| 144MHz 100/20 | 432MHz 60/15 |
| 144MHz 125/20 | 432MHz 70/15 |
| 144MHz 150/20 | 432MHz 80/15 |
| 144MHz 175/20 | 432MHz 90/15 |
| 144MHz 200/30 | 432MHz 100/15 |
| 144MHz 225/30 | 432MHz 110/15 |
| 144MHz 250/35 | 432MHz 120/18 |
| 144MHz 275/35 | 432MHz 130/18 |
| 144MHz 300/40 | 432MHz 140/20 |
| 144MHz 325/40 | 432MHz 150/20 |
| 144MHz 350/45 | 432MHz 160/20 |
| 144MHz 375/45 | 432MHz 170/23 |
| 144MHz 400/50 | 432MHz 180/25 |
| 144MHz 425/50 | |
| 144MHz 450/50 | |

### Rules

1. All contacts must have been made after 31 December 1978.
2. Eligible countries are those shown in the countries list printed elsewhere in this Yearbook.
3. Stations are eligible for awards in the following categories:
   (a) Fixed stations
   (b) Portable stations, any location
   (c) Mobile stations, any location.

Categories cannot be mixed

4. QSL cards submitted must be arranged in alphabetical order of the QTH squares claimed, and a checklist enclosed.

## VHF Countries and Postal Districts Awards

Awards intended to mark successful VHF/UHF achievements.

### Requirement

| Title of award | Countries/ Districts |
|---|---|
| 50MHz Standard Transmitting | 12/60 |
| 50MHz Senior Transmitting | 25/90 |
| 70MHz Standard Transmitting | 3/45 |
| 70MHz Senior Transmitting | 6/80 |
| 144MHz Standard Transmitting | 9/65 |
| 144MHz Senior Transmitting | 15/100 |
| 432MHz Standard Transmitting | 3/40 |
| 432MHz Senior Transmitting | 9/70 |
| 1296MHz Standard Transmitting | 3/30 |
| 1296MHz Senior Transmitting | 6/60 |

**Listener Awards** to be on an 'as heard' basis.

**Supreme Award** (for fixed stations only). For holding: 3 Senior awards or 2 Senior + one 1296MHz Awards.

### Rules

1. Starting date for this award is 1 January 2000.
2. All contacts made after 1 January 1990 are eligible.
3. Eligible districts are listed in the RSGB Yearbook.
4. Eligible countries are listed in the Prefix section of the RSGB Yearbook.
5. For QSL cards not showing a 'district code', the claimant may add the appropriate code providing suitable other location information is on the card.
6. Stations are eligible for awards in the following categories:
   (a) Fixed stations
   (b) Portable stations (/P any location)
   (c) Mobile stations (/M any location).

Categories cannot be mixed.

7. Each different confirmed contact with a station in a Scottish district may count up to a maximum of three per district. Belfast (BT) may count up to six contacts. (See page 103 of this Yearbook for a checklist.)
8. The 'county and country' based awards (available until 31 December 2002, the rules for which were published in the 1999 RSGB Yearbook) continue to be valid towards the Supreme Award.
9. VHFC reserves the right to modify these criteria as necessary

## RSGB Foundation Award

Award for holders of 'Foundation' callsign only. Contacts to claim this award must be made during the first year of being licensed. Band and/or mode endorsements will be available.

### Rules

1. This award is available only to holders of UK Foundation class callsign. The contacts claimed must be made during the first year of being licensed.
2. Starting date is 1 January 2002.
3. Modes: CW, SSB, AM and FM. No repeater contacts permitted.
4. Bands: 50, 70, 144, 432 and 1296MHz.
5. Qualification for this award will be:
   Bronze Award: 50 points.
   Silver Award: 75 points
   Gold Award: 100 points.
6. Points scoring.
   Bands: 50 & 144MHz - 1pt per QSO
   70 & 432MHz - 2pts per QSO
7. All claims must include a log extract certified by a Novice Instructor, an RSGB affiliated club official or two licensed amateurs. Multiple contacts with the same amateur on a single band when fixed, /P and /M only count for a single point although the same amateur may be worked on each of the bands for qualifying points. For endorsements the log must show the appropriate details.
8. The charge for this award is: RSGB members, £4. UK residents who are not RSGB members, £7. Where applicable, proof of RSGB membership is required, eg a recent RadCom label.
9. All claims for the award must be submitted to the RSGB VHF/UHF Awards Manager.
10. State clearly on the application your name, (as you wish it to appear on the certificate) and sign a declaration that you have acted in abidance of the award rules and by the conditions of your amateur radio license.

## RSGB Intermediate Award

This award is for holders of Intermediate callsign only and is designed to encourage activity in the CW, SSB & FM simplex sections of the bands.

There are three levels to the award and claimants may take advantage of a 'multiplier system' if they wish to acquire the points required, especially for the higher levels.

### Rules

1. This award is available only to holders of UK Intermediate callsigns.
2. Starting date is 1 January 2000.
3. Modes: CW, SSB, AM and FM. No repeater contacts permitted.
4. Bands: 50, 144, 432 and 1296MHz.
5. Award Levels:
   Basic [Bronze]: 1000 points.
   Intermediate [Silver]: 10,000 points.
   Advanced [Gold]: 25,000 points.
6. Scoring.
   50MHz - 1pt / QSO
   144MHz - 1pt / QSO
   432MHz - 2pts / QSO
   1296MHz - 4pts / QSO

   *Multipliers*:

   May be claimed as follows: Locator Squares (eg JO01), Postal Districts and Countries - 1pt.

   Distance Bonuses:

   50MHz for contacts in excess of 2000km - 1pt

   144MHz for contacts in excess of 1000km - 1pt.

   432MHz for contacts in excess of 500km - 3pt.

   1296MHz for contacts in excess of 500km - 5pt.
7. All claims must include a log extract certified either by a Intermediate Instructor, an RSGB affiliated club official or two licensed amateurs. A summary sheet of the points claimed per contact listed by callsign and if multipliers are claimed then a summary sheet of these as well. Multiple contacts with the same amateur on a single band when fixed,

   /P and /M only count for a single point although the same amateur may be worked on each of the bands for qualifying points.
8. The charges for each level of this award are: RSGB members £3.00. UK residents who are not RSGB members £6.00. Where applicable, proof of membership of the RSGB is required, eg a recent RadCom label.
9. All claims for the award must be submitted to the RSGB Awards Manager.
10. State clearly on the application your name, (as you wish it to appear on the certificate) and sign a declaration that you have acted in abidance of the award rules and by the conditions of your amateur radio license.

Website:
*http://www.rsgb.org/awards*

# Microwave Awards

As of 2010, RSGB microwave awards were transferred to the UK Microwave Group (UKuG).

The following awards are intended to mark achievement on the microwave bands. Successful applicants will initially receive a certificate and one sticker. Further stickers will be issued as later claims are received.

## Microwave Squares Awards

Existing RSGB records have been transferred so that claims for extra stickers to add to existing RSGB Squares Awards can be accepted. The existing sticker design has been retained.

Awards are available in 5 square increments on the following bands:

1.3GHz: 5 to 150 Locator squares
2.3GHz: 5 to 75 Locator squares
3.4GHz: 5 to 75 Locator squares
5.7GHz: 5 to 75 Locator squares
10GHz: 5 to 75 Locator squares
24GHz: 5 to 25 Locator squares

- Initial claims will require submission of QSL cards for all contacts claimed.
- Subsequent increments will need the additional cards and the countersigned check list from the previous claim.
- QSL cards must include the IARU QTH locator of the station worked.
- eQSL confirmations can be accepted from stations that have completed eQSL verification
- Contacts must have been made after 31 December 1978.
- All contacts must be two-way on the band in question, cross-band contacts are not eligible.
- Awards are available for fixed or portable/mobile stations but these categories cannot be mixed.
- Claims for portable operation must be for contacts made from one site, defined as anywhere within a 5km radius of the point operated from.
- Cards should be listed and sorted in IARU QTH Locator alphanumeric order.
- A self-addressed envelope must be included for return of the QSL cards, with sufficient postage value in UK stamps or IRCs.
- Certificates are free for members of the UK Microwave Group.
- Claims from non-members must include payment of £3 or $5 when an initial certificate is requested.
- Subsequent stickers will be enclosed with returned QSL cards and checklists.

Claims should be made using the application form and squares checklist available on the UK Microwave Group website: *www.microwavers.org* and sent to the address given below.

## UKuG/SOTA Microwave Distance Awards

These new awards are jointly issued by UKuG and SOTA, and recognise achievement in working distance. They replace the previous RSGB/UKuG distance awards, and are available in 50km endorsements for all bands 1.3GHz and above.

For further information, please see: *http://www.sota-shop.co.uk/microwave.html*

## Recording and recognising 'Firsts'

The UK Microwave Group has an award that recognises the achievements of British stations in making first contacts with other countries, and maintains a list of firsts as a historic record of the development of microwave operating techniques in Great Britain.

Should you not wish to claim a certificate we would still be very grateful for details of your First to ensure the records are accurate.

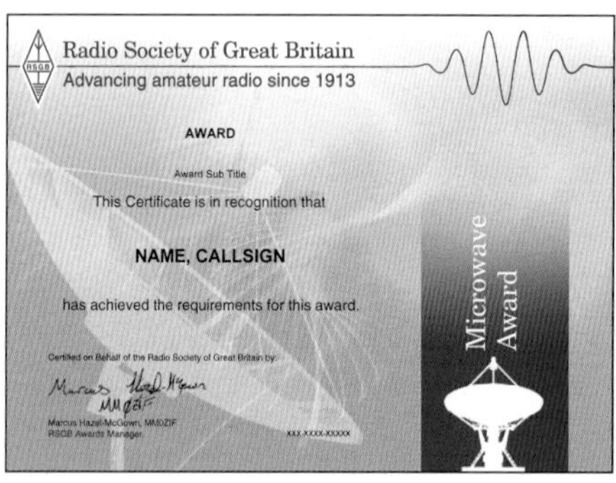


Certificates will be awarded to stations that can demonstrate that they completed the first contact between one of the countries in Great Britain and Northern Ireland (Prefixes M, MM, MI, MW, MU, MJ, MD - and the G and 2E equivalents) and any other country on a particular band above 1GHz. Contacts within Great Britain are valid for this award (eg Scotland to Guernsey).

Only one 'First' will be recognised per band, irrespective of the propagation mode (eg tropo, rainscatter or EME).

Claims should be made using the application form available on the UK Microwave Group website.

A QSL will be required for award of a certificate (but are not required to provisionally enter a First in the database). QSL cards should not be sent until requested. eQSL confirmation will be accepted from stations that have completed eQSL verification.

All claims should be sent to John Quarmby, G3XDY. Details below.

Please note that data provided for the above awards will be held by the UK Microwave Group for the purposes of providing a published list on the UKuG website and in the group's newsletter *Scatterpoint*, of achievements by stations in Great Britain and Northern Ireland. Data will not be used for any other purposes.

You do not have to be a member of the UK Microwave Group to lodge a claim.

The decision of the UK Microwave Group committee is final on the validity of claims.

Prior to making a claim, please check the Firsts database on the UK Microwave Group website at: *www.microwavers.org.uk* for existing contacts.

## Trophies & Awards

In addition to the certificates and distance/squares awards, and the trophies awarded for microwave contests by the RSGB Contest Committee, the UK Microwave Group also awards a number of cups and trophies, both for contest operating successes and for contribution to the facet of the hobby in both technical and supportive aspects. Most of these awards are presented annually at the UK Microwave Group Round Table event held at Martlesham, Suffolk, each April.

Photos of all the trophies can be found online at: *www.microwavers.org/trophies.htm*

## Operating Trophies

### 5.7GHz - G3KEU

In memory of Tim Leighfield, (SK 2002), this cup is awarded annually to the winner of the UKuG 5.7GHz cumulative contest. See: *www.g3pho.free-online.co.uk/microwaves/keu.html*

### 10GHz - G3JMB

Awarded in memoriam to Jack Brooker MBE, (SK 2004) is awarded annually to the winner of the Restricted section of the 10GHz Cumulative contest. *www.r-type.org/g3jmb/*

This trophy is an attractive glass plaque, but due to the inadvisability of engraving it, each winner is awarded a small plaque which they retain in perpetuity.

### 10GHz - G3RPE

Awarded to the winner overall of the 10GHz Cumulative contest, this cup commemorates Dain S Evans, BSc, PhD, FIM and RSGB President in 1978. Dain was one of the prime movers of the expansion in 10GHz construction and activity in the UK and was the first, with G3ZGO, to break the 150km distance barrier, setting the bar for those who followed.

### 24GHz - G0RRJ

This cup, in memory of Dave Cox (SK 2009), is awarded to the winner of the 24GHz Cumulative contest.

Send microwave award applications to: John Quarmby, G3XDY, 12 Chestnut Close, Rushmere St Andrew, Ipswich IP5 1ED email: g3xdy@btinternet.com

**24GHz and 47GHz - Trophy**

These two cups are awarded to the winners of the 24 and 47GHz Trophy events (as distinct from the 24GHz Cumulative sessions)

## Contribution Awards

There are four awards in this area:

**G3BNL**

In memory of Les Sharrock, this ornate decanter trophy is awarded by the RSGB on the nomination of the UK Microwave Group committee for innovation or technical development of microwave equipment or techniques.

**Fraser-Shepherd Award**

This award, for research into microwave applications for radio communication, is presented by the RSGB on the nomination of the UK Microwave Group committee. Fraser Shepherd, GM3FGW was an early pioneer of UHF and microwave operation and techniques, and was a participant in the first GM 432MHz moonbounce activity.

**G3EEZ**

Awarded for contributions to Microwave Communications, this cup is in memory of Alan Wakeman.

**G3VVB**

Presented for the best piece of home constructed microwave equipment exhibited at a Round Table, this magnificent trophy is in memory of Cyril James, a keen home constructor who produced microwave artefacts of very high quality workmanship. The entries and judging of this event are a regular feature of Microwave Round Tables, with the trophy awarded at the RSGB Convention in October.

Send microwave award applications to: John Quarmby, G3XDY, 12 Chestnut Close, Rushmere St Andrew, Ipswich IP5 1ED email: g3xdy@btinternet.com

# Microwaves

This an aspect of amateur radio where experimentation and construction are still alive and well. And it's never been easier to get started.

## What are microwaves?

In amateur radio terms, frequencies above 1000MHz (1GHz).

## Are microwaves only for line-of-sight communication?

Absolutely not! Try telling those operators who routinely work hundreds of kilometres with relatively modest antennas and power that they can't do it. They can, and so could you!

Microwaves is just another aspect of amateur radio which can be learned. You can take it in so many directions ...

All amateur radio equipment costs money, but a simple system of antenna and transverter could cost less than a new headset! Less if you are prepared to build your own from a kit.

## So what can we do with microwaves?

Lots! That question is probably better answered by asking what we can't do! The microwave bands support a wide variety of propagation modes, some of which will be completely unfamiliar to people who haven't ventured above VHF.

On most bands troposcatter – scattering from the upper part of the lowest layer of the atmosphere, the troposphere, is important. This will reliably support contacts over several hundred kilometres with good modern equipment. With modest kit, perhaps a couple of watts to a discreet yagi antenna a couple of metres long on 1.3 or 2.3GHz, from a site with open horizons – troposcatter path losses are very dependent on the vertical angle of your horizon - will produce regular contacts up to around 200km without much effort. A similar power level to a repurposed satellite TV dish will provide contacts on 10GHz over the same range, or further.

The troposphere is involved in another way. At times the masses of air from different sources, some from the cold dry Arctic, some, perhaps, from the warm, wet Atlantic, move over each other. In regions sometimes thousands of kilometres across, these boundaries are able to 'duct' microwave signals over long distances. Ducting produces openings on the microwave bands just like those on VHF/UHF, and signals are often propagated over paths in excess of 1000km.

Other ways of propagating microwave signals include scatter from rainfall (and other forms of precipitation) and scattering from aircraft. Although rainscatter can occasionally be heard on 1.3 and 2.3GHz, it really starts to be useful at 3.4GHz and above. 'Aircraft scatter' can allow all year propagation up to about 700km.

On the higher frequency bands, at 24GHz and above absorption by water in the atmosphere becomes a challenge but it has not deterred UK microwavers from having a go. The UK DX record on the 24GHz band is almost 400km. Contacts have been made by UK radioamateurs on all of our allocated bands up to, and including 241GHz.

Microwave moonbounce (EME) currently produces worldwide DX from the UK on all of our bands up to 24GHz. Some operators apply for HF-style awards, such as WAC. Some UK stations are advancing towards DXCC on 1.3GHz. On the microwave bands, EME is practical with quite modest antennas. Significant numbers of contacts can be made on 23cms with single yagis or 2m diameter dishes. On 13cms and up even smaller antennas can be used. 10GHz has become a favourite for EME. There 50W and a 2.4m dish will allow a station to hear its own Moon echoes on SSB. 24GHz is starting to make headway as a moonbounce band. Bands up to 76GHz have been used experimentally.

*A modern 10 and 24GHz home system*

Amateur microwavers can be found throughout the world, especially in the Americas, Western Europe, Australasia and Japan. Regular EME contacts are made between all of these areas, and there even DXpeditions bringing activity to other areas.

Amateur television has found its way into earth orbit with DATV transmissions being made from the ISS at 2.4GHz, while the geostationary Es'hail2 satellite is due to launch in 2017 with digital and linear transponders using 2.4GHz uplinks, and 10GHz downlinks.

A few UK radioamateurs have heard deep space satellites such as Voyager at the very edge of our Solar System. Others have discovered the joy of amateur Radio Astronomy.

Microwaving can be challenging, it can be fun, it can be educational, it can be social. Finally it's what you make it. It's a rapidly developing area of our hobby.

## How did amateur microwaves come about?

As early as 1894 to 1896, Jagadish Chandra Bose, an Indian physicist, experimented on 60GHz over a one mile distance using

primitive semiconductors! He was a truly remarkable man who is now seen as the 'Father of Microwaves'. It may seem strange, but microwaves pre-date HF radio!

In 1946 the first amateur microwave contacts were recorded in the USA, when W1LZV/2 worked W2JN over two miles on 10GHz and W1NVL/2 worked W9SAD/2 on 21.9GHz (800 feet!). The World 10GHz record was then set at 7.6 miles by W4HPJ/4 and W6IFE/3.

The first UK amateur microwave contact was made in 1949 by G3BAK and G3LZ over 27 miles, at the time a new world 10GHz record.

For the next twenty years operation on the bands above 3.4GHz centred mainly on the use of tubes, such as klystrons as local oscillators and transmitters with waveguide mounted diode mixers in wideband FM systems. Gunn diodes replaced the klystrons during the '70s, and the '80s saw a sharp spike in activity on 10GHz. Activity on the lower frequency microwave bands also grew, with many people using NBFM, or even AM, generated by solid-state varactor multipliers from 70cm.

At the beginning of the '80s, more advanced microwavers started to move to an approach more like that on the lower frequency bands, with CW and SSB becoming more common. The problem with low-power, wideband systems is that it is very difficult, particularly above 3.4GHz, to cover many non-line-of-sight paths.

More recently, the field has continued to develop, with narrowband standards in use on all bands. All DX operation is now on narrowband, often using weak-signal modulation schemes, such as K1JT's WSJT. Transmitters capable of several watts output are commonplace on 10GHz.

In addition commercial kits and ready-made modules have become available.

In the last decade, much focus has been on increasing frequency stability by locking oscillators to GPS stabilised standards. SDR technology operating directly at microwave frequencies is becoming more common.

The current State of Play: UK Microwave Terrestrial Records

| Band (GHz) | Distance (km) |
|---|---|
| 1.3 | 2617 |
| 2.3 | 1389 |
| 3.4 | 1137 |
| 5.7 | 1244 |
| 10 | 1429 |
| 24 | 408 |
| 47 | 203 |
| 76 | 129 |
| 134 | 35.6 |
| 145 | 1.29 |
| 241GHz | 0.03 |
| Light (red) | 129.1 |

Although much has been made above about the DX potential of the microwave bands, and many enthusiasts concentrate on this aspect of the hobby, there are other important user groups.

Conventional voice repeaters/beacons operate in the 23cm band. These have recently been joined by a number of digital voice repeaters using the 'Dstar' modulation scheme.

Amateur Television is very well established above 1GHz, with a network of repeaters. Many now employ Digital Video technology, although many analogue FM TV repeaters remain in service.

Digital Networking is an interest of an increasing number amateur radio operators, and several long haul networks exist using IEEE802 standards at both 2.4 and 5.6GHz. In many countries on the European mainland extensive amateur owned and operated high-speed data networks, linked to the Internet, operate on 1.3 and 2.3GHz.

M0EYT portable.

## More about the Microwave Bands

Many bands are shared with other users and are increasingly under threat from commercial interests. For example parts of the 2.3 and 3.4GHz bands have been subject to Public Sector Spectrum Release (PSSR) which has seen loss of access to some frequencies and other changes to the main UK amateur licence schedule. Users of parts of 2.3GHz (shown by ***) are now also specifically required to register with Ofcom. There are both limits to times of activity and to location, which make continued sharing with the main User possible. NoVs for new additional bands such as 2300-2302MHz and >275GHz are also available on request. (N)

Many of the bands shown in the following list contain amateur beacons and repeaters which are useful for frequency calibration and as indicators of propagation conditions. A more detailed spectrum allocation can be found in the Band plans Section of this Yearbook and on the RSGB Spectrum Forum website (URL below).

| | |
|---|---|
| 23cm | 1240.0-1325.0MHz |
| 13cm | 2300.0-2302.0MHz(N) |
| 13cm | 2310.0-2350.0MHz*** |
| 13cm | 2390.0-2450.0MHz |
| 9cm | 3400.0-3410.0MHz |
| 6cm | 5650.0-5680.0MHz |
| 6cm | 5755.0-5765.0MHz |
| 6cm | 5820.0-5850.0MHz |
| 3cm | 10.000-10.125GHz |
| 3cm | 10.225-10.500GHz |
| 12mm | 24.000-24.250GHz |
| 6mm | 47.00-47.2GHz |
| 4mm | 75.50-81.0GHz |
| 2.5mm | 122.25-123, |
| 2.2mm | 134-141 |
| 1.25mm | 241-250GHz |
| also | >250GHz (N) |

## Ways into microwaves

The 1.3GHz (23cm) and the 10GHz (3cm) bands are the easiest ones on which to make a start. There is a lot of ready-made equipment, including antennas for these bands. The on-line auction and flea-market sites can be a great source of microwave surplus.

## Simple 10GHz wideband gear

The days of simple Gunn oscillator transceivers are now past, if only because the intruder alarm modules which many people used in the past are no longer easily and cheaply available. The Gunn oscillator diodes used are obsolescent as low cost items. Modern intruder alarm modules are much less easy to modify. However, if old wideband gear is available, perhaps lurking in someone's attic, it can still be used, if you can find someone to cooperate with. In this case, it would be sensible to check that the gear is operating in a part of the band to which we still have access.

Satellite TV LNBs make very acceptable receive converters when combined with a cheap DATV "Dongle" or the more expensive Funcube Dongle. Many are good enough for narrowband use, but there is no obvious easily available companion transmitter. If you have a beacon or ATV repeater locally, a LNB, particularly if mounted in a dish, would make a great tool for initially exploring microwave propagation at 10GHz. You could also take part in UkuG contests, as 'one-way' contacts are allowed for 50% of the points of a two-way!

Another way of getting going, assuming that your licence allows it, is to consider using 5.6GHz and to exploit (cheap!) video-sender equipment. These often consist of a wideband, synthesised FM transmitter and a companion

A more modern satellite TV LNB.

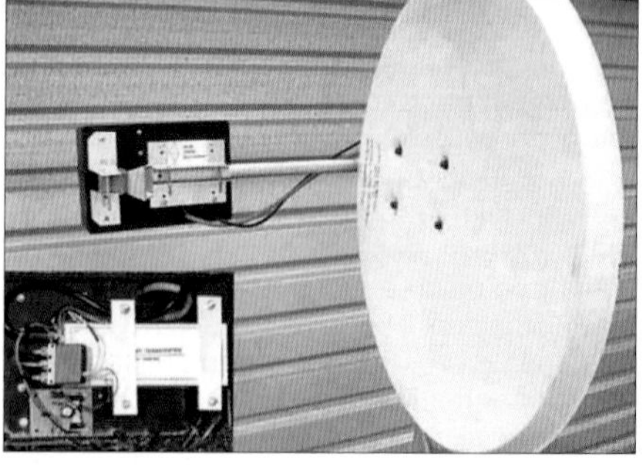
The 'Quickstart' microwave system

A small dish microwave EME system for 13cms

receiver, which can be tuned to a channel within the amateur band. Combine these with a suitable antenna, also available cheaply, and you have a simple transceiver which is capable of surprisingly good performance over line-of-sight paths.

There are groups around the UK using video sender technology, but if you can't locate one (try asking on 'ukmicrowaves') or they are too far away and you know someone close-by who is interested in experimenting, go ahead and try!

As with all worthwhile activities, there is a learning curve. You might not work very far initially, so it's worth finding someone with similar interests to experiment with. It's still an excellent, fun and cheap way to 'wet your feet' in microwaves. Don't forget you will also need something for talkback liaison, be it a mobile phone, computer with internet access, or even 2m/70cm!

## Other bands for a beginner

1296MHz is the other band which can be recommended for people taking their first steps into the microwaving world. The technology is not too dissimilar to that used on VHF/UHF, and equipment needed to make a start is easily available. Recently, the Company owned by Bulgarian microwaver, Hristyiyan, LZ5HP, has been producing a simple 2W transverter – not a kit - of good performance, at a very attractive price *www.sg-lab.com/TR1300/tr1300.html.* This is small enough, and light enough to be mounted in a waterproof box close to the feed point of a 1296MHz antenna. From many locations this will allow regular QSOs up to around 250km – much further in a tropo opening.

## Microwave Antennas

On 1296MHz and 2.3GHz, yagi or sometimes dish antennas are the norm. High gain dishes and horns are the most commonly used antennas at 3.4GHz and above.

Microwave antennas give much higher gains than HF ones. 100 microwatts of 10GHz CW into a 60cm dish will be heard over any line-of-sight path in the UK by a receiver using a similar antenna.

On 10GHz, a 60cm ex-Satellite TV dish with a suitable feed horn will have a gain of around 35dB. This will give a potential range greater than 144MHz, for similar transmit power.

The extra gain comes from concentrating the transmitted energy into a narrow beam; an antenna with 35dB gain will have a -3dB beamwidth of about ±1.5°, so you have to point the dish more accurately than you would, say, a 2m yagi. The ability to go to a site, and to work out where to accurately point a dish is another skill which has to be learned as a microwaver!

## Equipment

Ready built transverters are available from a number of sources, such as Kuhne Electronik (DB6NT) in Germany, DEMI in the USA, and LZ5BP in Bulgaria. Kuhne and DEMI, along with Mini-Kits in Australia also market kits.

There is a market for second-user microwave equipment. A request on one of the on-line groups, such as 'ukmicrowaves' <groups.yahoo.com/ukmicrowaves> could well produce results.

## Making QSOs, Is it like the lower frequency bands?

No! Perhaps on 1.3GHz, it can be, but at higher frequencies the beamwidth of antennas is so small that the chances of hearing another station randomly are not large. So, the use of talkback has developed. Historically, in the UK, stations liased on 144.175MHz SSB. In most of the rest of Europe, 432MHz SSB was used. More recently, the ON4KST chat server <*www.on4kst..com*> has become very popular. While there are those who regret the way in which the Internet has replaced a 'pure' amateur radio talkback channel, it has its advantages.

The first is that as 2m propagation is often not the same as that at higher frequencies: the microwave bands are often 'open' when 2m is not, for well understood reasons. More prosaically, the use of a public talkback channel, rather than a semi-private VHF link, discourages people from trying to complete QSOs or even make QSOs over talkback.

On the microwave (and VHF/UHF) bands the completion of a QSO is defined rather differently to the usual practice on the lower frequencies. On the band in which you are operating you must exchange callsigns, some piece of unknown information, like a report or a QTH locator. Both stations must acknowledge the receipt of of the callsigns and unknown information from the other station. If you don't keep to that procedure, you haven't made a contact which would be seen as valid.

There are three sets of parameters you need to get right, if you are to work another station; the beam heading, the frequency, and who is to transmit and when.

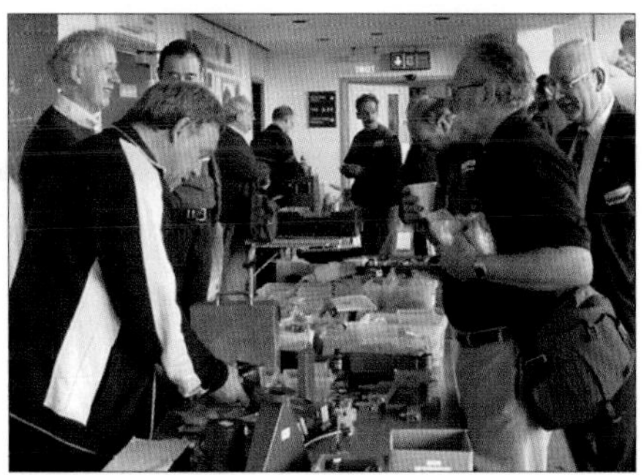

The Martlesham Round Table Fleamarket.

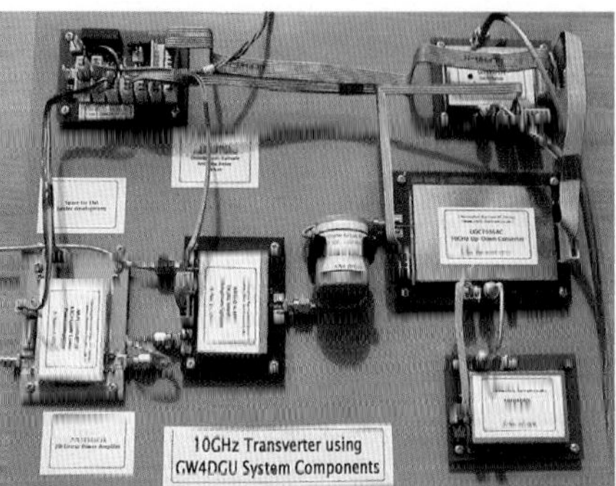


New-technology 10GHz PCB modules from GW4DGU

The beam heading is easily obtained from a number of sources. Most conveniently this can be found by a couple of clicks in the ON4KST interface. But there are a number of apps available for all of the major operating systems which will convert an input in the shape of two QTH locators into a bearing and distance. The majority of antenna rotators – particularly those aimed at HF operators - do not have particularly accurate azimuth indications and some modification will be needed. Fortunately, some European manufacturers of rotators understand the need for more accurate calibration. For portable operation, a simple compass rose which can be calibrated by reference to a local landmark can produce an elegant solution.

Frequency is very important. Older transverters used either a free-running crystal oscillator or a crystal oscillator partially stabilised with a simple temperature control oven. That resulted in the frequent need to tune over perhaps 20kHz to find another station, even if both stations were fairly confident of their frequency. Many weak and transient signals were missed because of this. Modern practice is to phase-lock the transverter local oscillator to a GPS stabilised frequency reference, and even at 10GHz frequency accuracies of a very few Hz can be obtained.

## If this all sounds very different to what you're used to, you can get help?'

The UK Microwave Group (UKuG) has a policy of supporting newcomers. It runs a Technical Support service, covering a large part of the UK. This is a voluntary service to members – not a right of membership – and depends on the goodwill of the volunteers freely to provide (within sensible limits) their expertise, and often very extensive test equipment to support other members. Many of those providing this service are involved professionally with microwaves and have excellent laboratory facilities. In the commercial world the time they give would cost lots of money!

To find out more about UK Microwave Group you can find out more information by looking on the National Affiliated Clubs & Societies pages in this Yearbook or by searching their website:

*www.microwavers.org*

***Have fun on microwaves!***

# Propagation

An explanation of the solar and geophysical information published by the RSGB.

Each week the GB2RS news bulletin includes a brief solar and propagation report and forecast prepared by members of the Society's Propagation Studies Committee. As carried on most GB2RS broadcasts and as posted on the Internet, these reports look back on the week up to the Thursday before transmission, and ahead to the week from the Sunday of transmission.

The usual format for the bulletins is a review of solar activity, geomagnetic activity and ionospheric data during the week. This is followed by the forecast.

Spacecraft data is used to compile the weekly propagation report, including imagery from the STEREO (Solar TErrestrial RElations Observatory) twin orbiters, which were launched in 2006. Unfortunately one of them has since developed a fault and imagery is currently only available from the STEREO Ahead spacecraft.

The Solar Dynamics Observatory (SDO), which was launched in February 2010, also helps us understand the Sun's influence on the Earth and near-Earth space.

Although a lot can happen in a few days (new regions appear, old ones decay), from now on more accurate forecasting will be possible.

## Solar Activity

The general level of solar activity for a 24-hour period from midnight to midnight is described as:

**Very low**
Either no solar flares or only A- or B-class flares.

**Low**
C-class flares, which usually have little or no impact on propagation.

**Moderate**
Between one and four isolated M-class flares.

**High**
Five or more M-class flares or isolated (1-4) M5 or greater flares, including X-class solar flares.

**Very high**
Several flares of M5 or greater magnitude, including X-class flares.

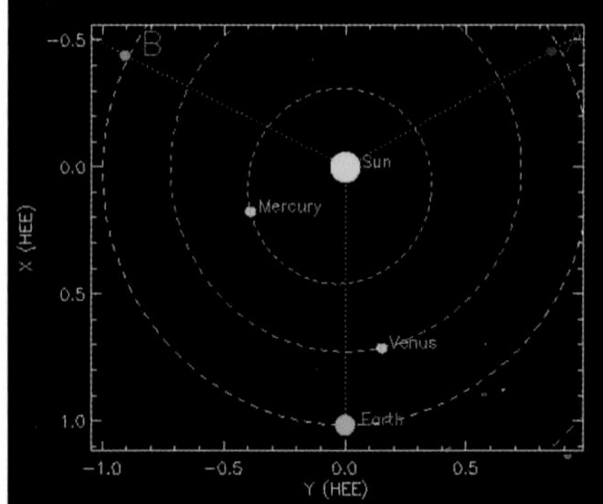


**Fig 1:** *Present locations of the STEREO spacecraft.*

A flare is a sudden eruption of energy on the solar disc emitting radiation and particles, which can last anything from a few minutes to some hours. Flares are classified as 'A', 'B', 'C', 'M' or 'X' according to their X-ray energy level. This is measured by satellites in terms of megaelectron volts (MeV).

There are four energy thresholds: 2, 10, 50 and 100MeV, and flares are classified by numbers, ie M3, X4 etc. Generally the broadcasts mention only M- or X-class events, as they are the most likely to have an impact on propagation.

However, flares are also classified by their optical importance, which gives a measure of their size and brilliance. Size is indicated by a number between 1 and 4, while brilliance is either 'F' = faint, 'N' = normal, 'B' = bright or 'S' meaning sub (so 'SF' indicates sub-faint). The energy and optical indicators combine to give the complete flare data, eg 'M3/2B'.

## Major Flares

Flares above M9/3B and all X-class flares can be very disruptive to the ionosphere. They can lead to severely-degraded HF propagation and any associated Earth-facing coronal mass ejections (CMEs) can cause auroral events, usually after a lapse of between 30 and 50 hours, particularly if they are on the Sun's limbs or have just passed the central meridian. Flare effects may be reported as:

**Sudden Ionospheric Disturbance (SID) or Short-Wave Fadeout (SWF)**
HF propagation blacked out or degraded for between a few minutes and many hours, with the lower bands affected first and most severely, the higher bands affected less and recovering first, but LF (<500kHz) and VLF DX signals may be enhanced.

**(Sudden) Storm Commencement**
Increase or decrease in the northward component of the geomagnetic field, marking the beginning of a geomagnetic storm. The onset may be very sudden (SSC) or more gradual (SC).

**Noise Bursts or Noise Storms**
Enhanced emissions from the Sun at radio wavelengths, associated with major flare events or complex solar active regions. They may last only a few minutes or for many hours.

**Proton Events**
These may be mentioned if they have an energy level exceeding 10MeV. Proton events cause high absorption in the D region of the ionosphere, particularly affecting transpolar propagation due to polar cap absorption, which can be degraded for days or even weeks following such an event.

## Coronal Holes

Coronal holes are holes in the Sun's outer corona through which material is ejected by various means. There are always holes at the Sun's polar regions but tongues sometimes extend to the equatorial regions, or small holes can form. The passage of these can cause a magnetic disturbance. This is particularly so if the interplanetary magnetic field is southerly, as this couples to the Earth's northward field. What have become known as 'Scottish' type auroras can generally be attributed to the passage of a coronal hole. If known about, coronal holes are always referred to in the text due to their importance. Coronal holes become more geoeffective when at low solar latitudes and are more numerous around the time of solar maximum and during the first few declining years after solar maximum.

## Other Solar Events

Now and again reference is made to solar filaments. They appear as prominences on the solar limbs and as dark snaking strings of material against the limb as viewed in the light of hydrogen alpha. Occasionally the magnetic fields that hold filaments together break apart and fling the filamentary material into space. Filaments can last for several solar rotations before fading or erupting. These events can be sudden and are mostly unpredictable and can cause widespread auroras, ionospheric blackouts, and worldwide disruption to radio communication. Sometimes eruptive prominences are reported, but because they are located on the solar limbs they are not so geoeffective and therefore not so disruptive.

## Satellite Data

Increasingly, new forms of data from satellites are supplementing or replacing traditional forms of ground-gathered data in explaining and predicting the 'propagation weather'. Some or all of the following may feature in a bulletin:

**X-ray background flux**
A more sensitive indicator than solar flux. It is reported on a rising scale of A1-9, B1-9 and C1-9. GB2RS reports the weekly average and any unusual levels.

**>2MeV electron fluence**
Referred to as 'high', 'normal' or 'low.' High levels adversely affect the HF bands.

**Solar wind speed**
The ACE satellite measures the speed of the flux of solar particles and magnetic fields moving outwards from the sun. Normal velocities are around 350-400km/s, though speeds exceeding 1000km/s have been recorded.

**Particle densities**
Under 10 per $cm^3$ are 'low'; 10-25 'moderate' and above 25 'high'.

**Bz**
The orientation of the interplanetary magnetic field measured in nanoteslas. A southerly (-ve) orientation, coupling with Earth's northern orientation, results in HF disturbances and auroras. A northerly (+ve) orientation has little or no effect.

| K | 0 | 1 | 2 | 3 | 4 | 5 | 6 | 7 | 8 | 9 |
|---|---|---|---|---|---|---|---|---|---|---|
| A | 0 | 3 | 7 | 15 | 27 | 48 | 80 | 140 | 240 | 400 |

**Table 1:** *Geomagnetic activity look-up table for a typical middle-latitude magnetic observatory.*

## Solar Flux

This is the 2800MHz radio noise output from the Sun at midday. This frequency is chosen because the radio Sun looks the same size as the visible Sun. The figure given is that obtained at Penticton (BC, Canada), which is the world standard. The level varies from (at the cycle's minimum) about 64 units to a maximum of around 300 units. The higher the level, the more intense is the Sun's ionising radiation and the higher the frequency that can be reflected from the ionosphere. Good HF band conditions require a high solar flux but the level of magnetic activity will also be a crucial factor. A 90-day average of flux levels is given, as this has been found to be best for home computer prediction programs.

## Geomagnetic Activity

References to geomagnetic activity are made in terms of the worldwide or 'planetary' A index, expressed as 'Ap' units. During magnetic storms, the A-index may reach levels as high as 100. During severe storms, the A-index may exceed 200. Great 'rogue' storms may produce index values in excess of 300, although storms associated with indices this high are very rare indeed. Generally, an Ap index of 0-10 is considered 'quiet', 11-20 is 'unsettled', 21-50 as 'sub-storm', 51-80 'storm', 81 and above 'severe storm' or 'major storm'. High levels of geomagnetic activity - say roughly over 30 - are associated with poor HF conditions, especially on the higher bands. The greater the index over about 50, the greater the likelihood of aurora. And the higher the index, the further south auroral working may be possible. While auroral events, including auroral-E, are most likely to be found at 50, 70 and 144MHz, they can sometimes be found at 28MHz and, during major disturbances, contacts may be possible at 432MHz.

The Ap index is linear, unlike the alternative K index, used by WWV among others, which is quasi-logarithmic and open ended. Each observatory uses a look-up table created for that specific location, to convert an amplitude range into an associated K-index value. The table above shows the look-up table for a typical middle-latitude magnetic observatory.

## Ionospheric Data

The critical frequency is the highest frequency reflected back from the ionosphere from a signal sent vertically upwards by an ionosonde. The maximum frequency that can be used for normal communication at equal latitudes is very roughly about three times the critical frequency (so a critical frequency of 7.0MHz indicates that it should be possible to work due east at single hop distance - 3000km - at 21.0MHz). For southerly working the multiplication factor would be higher but over northerly paths it would be lower. However, the actual level attainable also depends on the level of geomagnetic activity, the season and the time of day.

During the hours of darkness, the normally relatively smooth ionospheric layers can break up during magnetically disturbed periods. This is referred to as **Spread F**. The break-up can be vertical or horizontal, or both at once. The resulting holes give rise to deep fading. Northern circuits are more prone to this effect, which is more likely to occur during the early morning. References in the bulletins usually refer to the number of hours when Spread F has been present, or if it was very bad, on any particular day.

**Blanketing E** means that the E layer is so intensely ionised that the ionosonde cannot see through it. This effect is often associated with summertime sporadic-E or, for northern stations, with auroral conditions.

## Absorption

Sometimes, for northern stations, complete absorption of the ionosonde signal occurs. This suggests that the D region is so heavily ionised that it is absorbing all but the strongest radio signals. These events can be associated with proton events, or high energy 'M' or 'X' flare events, or by electron precipitation from the Earth's radiation belts.

## Seasonal Changes

The daily highs tend to be higher in winter and lower in summer. The darkness hour lows vary in the opposite way – high in summer and low in winter. The weekly average variations are balanced against these seasonal changes, and reference is made to any discrepancy when this applies. For the HF bands, the higher the daily 'highs', the better is the chance of DX on the higher HF bands.

The average times of the highest and lowest frequency recorded are given. The high times vary with the season, being around midday in the winter and early evening (about 2000UTC) in the summer. The low times do not vary much, being usually about 0400UTC.

## Forecasts

Each week the bulletins include a forecast of expected events for the seven days following the Sunday of broadcast. This includes expected levels of solar flux, geomagnetic activity and the passage of any expected coronal holes. Maximum Usable Frequencies (MUFs) during daylight hours at equal latitudes are estimated for southern England. Scotland and Northern England will generally be down on these levels by around 3MHz at equal latitudes due to factors such as geomagnetic activity, which may affect northern areas more. In general, north-south paths will tend to be more readily workable than east-west, especially around the equinoxes. Bulletins usually include a forecast for one or more path.

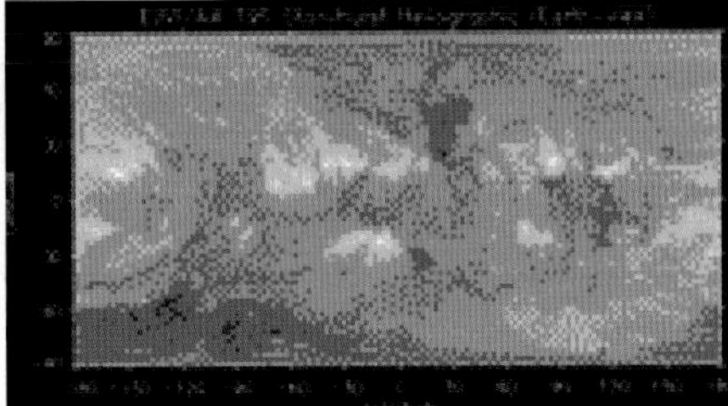

**Fig 2:** *A whole Sun image, created by combining images from Earth, STEREO-A and STEREO-B.*

The MUF given in these forecasts indicates the frequency up to which the path should be workable on 50% of days. On the better days, therefore, this value should be exceeded - at times by a considerable margin.

It also indicates a lower and more consistently reliable Optimum Working Frequency. This is the frequency on which it should be possible to operate on 90% of days. During months when sporadic-E propagation is prevalent - roughly May to August and around Christmas and the New Year - reminders are given of the likelihood of openings on 2, 4, 6 and 10m, and brief reports of major openings may be included.

Solar activity may be mentioned in the form "the quiet side of the Sun" or "the active side of the Sun"; which refers to the best chance of flare activity. For instance, the forecasts will attempt to predict the best chance of solar activity reaching moderate or high levels, therefore that being the active side of the Sun. This works in reverse for the quiet side.

The table published each month in *RadCom* shows path predictions from the UK to 31 locations around the world, 27 being short path and four being long path. All are F-layer predictions. The numbers indicate the expected reliability of the circuit; with a '1' representing between 1 and 19% of days, '2' between 20 and 29% of days, etc. The colours represent relative signal strength; a dot being no signal, black being a weak signal, blue being a fair signal, and red being a strong signal.

While every effort is made to ensure the table is as accurate as possible, it should always be remembered that propagation prediction is not an exact science. There will be times, for example during magnetically disturbed periods, when the table may be unduly optimistic. Alternatively, and especially during the peak of the solar cycle, conditions may for a time considerably exceed predicted values.

It should also be noted that these HF predictions are for F-layer propagation. They do not take into account either sporadic-E which, particularly during the summer, enlivens 28MHz, or auroral-E which occurs occasionally at HF (and VHF) during geomagnetic disturbances. They also do not include 'greyline' long-distance openings, which are a feature of the lower bands during the periods around dusk and dawn, as these are of too short a duration to appear in the two-hourly steps used in the tables. Neither can they take into the reckoning other short-lived phenomena like back-scatter and side-scatter. All these make the task of the forecaster more complex - but also make day-to-day working on the bands more varied and challenging. Fuller explanations of these and other solar and propagation phenomena, as well as near-real-time data, can be found on the Internet, notably at the Propagation Studies Committee (PSC) page.

RSGB PSC: www.rsgb.org/psc/
GB2RS propagation enquiries: Steve Nichols, G0KYA. Email: psc.chairman@rsgb.org.uk (or QTHR)

# HF Propagation in 2017

HF propagation prediction is an imprecise art, offering only a measure of the probability of a particular path being open at a particular time and on a particular band. However, the broad trends of what we can expect are reasonably clear, and understanding them raises the chances of our using the amateur bands productively.

First, let's take a look back at 2016.

Last year was characterised by declining sunspot activity. The continued slide down from the peak of Solar Cycle 24 also brought a host of solar flares and associated coronal mass ejections (CMEs), which were good for aurora enthusiasts on 10m, 6m and 2m, but not so good for lovers of the lower HF bands.

The declining solar flux index (SFI) was also not so good for the upper HF bands At the time this was being prepared in July 2016 we had already seen a few days with absolutely no sunspots and the SFI had hit a low of 73.

Considering that at sunspot minimum it is never worse than 64-65 it shows how the cycle had declined quite rapidly. Current thinking puts the next sunspot minimum some time around 2018-2019.

We also had problems with plasma flowing towards Earth from solar coronal holes - vast areas of the sun with open magnetic fields that allow large quantities of hot plasma to flow out. If the coronal hole is earth-facing we say it is geo-effective and as the plasma hits we often see the K index soar, with visible and radio aurora.

These coronal holes can be seen as black patches on the sun's surface when it is viewed in extreme UV light via the SDO spacecraft.

Coronal holes are a feature of the declining phase that occurs after solar maximum and it will interesting to see if they continue in 2017.

So the best of solar cycle 24 has long gone, but what can we expect in 2017?

We can probably expect to see average SFI figures around 70-80 at the beginning of the year and more towards 70 towards the end, although it is very hard to be precise. If you are using a VOACAP-based propagation prediction program, such as HAMCAP or ACE HF, you will need to know the Smoothed Sunspot Number (SSN) rather than the SFI. This can be found from a link at: *www.voacap.com*

Expect solar flares, coronal mass ejections (CMEs) and coronal holes to continue at times, which as I said earlier are more common on the downward leg of a solar cycle. These may therefore bring continued unsettled band conditions in 2017.

In terms of frequencies, 20m (14MHz) will remain the HF DX band of choice, with 17m and 15m perhaps struggling to open to DX at times.

The 12m and 10m (24/28MHz) bands may fail to open much at all, apart from during the main Sporadic E season from May to late August in the northern hemisphere.

Now let's look at the whole year, season-by-season, band-by-band. From a propagation perspective, conditions are dependent upon the angle the sun makes with the ionosphere, so the periods around both equinoxes are likely to be similar. There is a gradual change from one season-type to another, so the periods listed below should not be taken too literally.

## Winter period

### (Jan-Feb / Nov-Dec)

These periods are when the low bands (160m, 80m and 40m) come into their own. Generally, winter is a good time for East-West paths on HF too.

### 160m (1.8MHz or Top Band)

Solar absorption will prevent skip during daylight hours. You should be able to work other UK stations out to about 50-80 miles via ground wave. The band will start to come alive around sunset and openings up to around 1,300 miles should be possible, with frequent openings up 2,300 miles. DX openings to the east from the UK should be possible around midnight and to the west before sunrise for well-equipped stations.

### 80m (3.5MHz)

Expect a similar pattern to Top Band, with DX openings at night with peaks at midnight and around sunrise (greyline openings). Openings around the UK and out to around 500 miles should be possible during the day and between 750-2,300 miles at night. A low, horizontal antenna will be useful for relatively local, NVIS (Near Vertical Incidence Skywave) signals, but lower angle radiation, such as obtained with a vertical, will be required for DX.

### 40m (7MHz)

Another great DX band at this time of year. 40m should open for DX in an easterly direction during the late afternoon and towards the south at

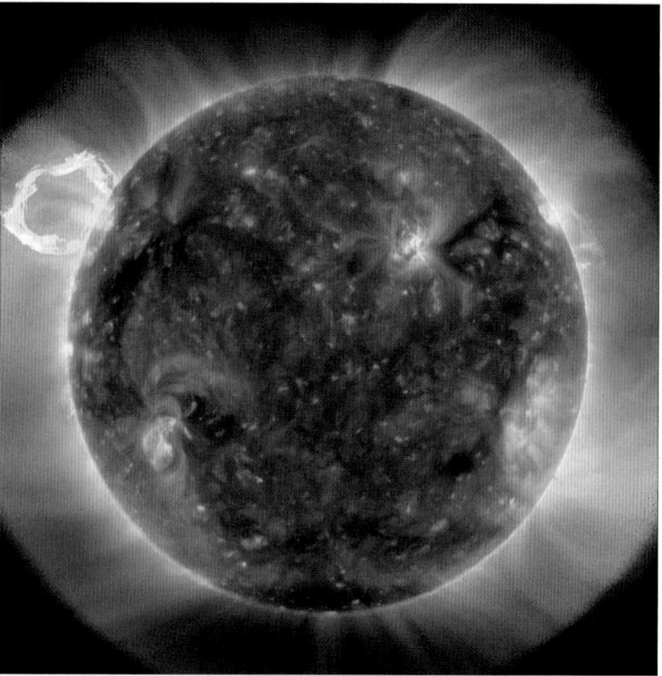

**Above:** *The Sun as seen through an Extreme Ultraviolet Imaging Telescope*

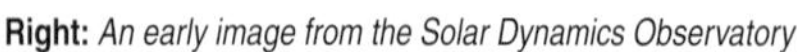

**Right:** *An early image from the Solar Dynamics Observatory*

sunset. Paths during the afternoon may also include W6 (west coast USA) in mid winter. Openings to the west, including long path to VK/ZL, should be possible after midnight and should peak just before sunrise. Relatively local contacts may still be possible during the day if the daytime critical frequency remains above 7MHz.

### 20m (14MHz)

This is likely to provide great DX openings during the hours of daylight. Peak conditions will be a couple of hours after sunrise for paths to the east and a couple of hours before sunset for paths to the west. Contacts up to 2,300 miles should be possible during daylight hours, and the band may remain open after sunset if the solar flux remains high. Occasional DX openings towards South America may be possible after nightfall. Watch out for higher levels of D layer absorption around local noon, which may mean that you are better off heading for the higher bands at this time, if they are open.

### 17m/15m (18MHz/21MHz)

These could provide some good DX openings during daylight hours at times. The period from noon to late afternoon may be best, but both bands are likely to close soon after sunset and remain closed until some time after sunrise the following day. If the SFI remains low (around 70) both bands might struggle to open during daylight.

### 12m/10m (24MHz/28MHz)

If the solar flux index remains below about 90 both bands might struggle to open at all. A brief spell of Sporadic-E can sometimes occur in the New Year, resulting in very strong, but short-lived propagation on 10m out to around 1,300 miles.

## Equinox periods

### (Feb-May / Aug-Nov)

The equinox periods provide longer daytime periods than winter, but logically, shorter night-time periods too. These tend to be the best months for working North-South paths, such as UK to South Africa.

### 160m (1.8MHz or Top Band)

Look for short-skip and DX openings at night. Again, no daylight skip is possible due to absorption, but openings out to 1,300 miles and occasionally further afield can be expected at night with conditions peaking around midnight and again at sunrise (greyline).

### 80m (3.5MHz)

This band will generally follow the characteristics of Top Band at night, but will also provide good openings out to around 250 miles during the day. These will lengthen to around 500-2,300 miles at night with fairly good DX opportunities at times.

### 40m (7MHz)

Forty metres should open to DX in an easterly direction at sunset. Openings to the west should be possible after midnight and should peak just before sunrise. Contacts should be possible during the day, although lower critical frequencies may mean that it is difficult to work other UK stations while perfectly possible to talk to European stations.

### 20m (14MHz)

This is likely to be the best DX band between sunrise and sunset. The bands may occasionally open after dark, mainly to the southern hemisphere. Good openings will be possible during daylight hours out to around 2,300 miles.

### 17m/15m (18MHz/21MHz)

These bands could provide fairly good DX openings during daylight hours, especially to Africa and South America, with 17m being open more often than 15m. Once again, both bands may struggle to open during times of low solar flux. Both bands are likely to close after sunset.

### 12m/10m (24MHz/28MHz)

These bands will continue to be disappointing at this point in the solar cycle. There may be many days where there are no signals at all, although occasional brief openings to DX may be possible, especially if the solar flux heads towards the high 80s/90s.

## Summer Solstice period

### (May-Aug)

Daytime MUFs are likely to be lower than those of winter. The so-called 'Seasonal Anomaly' is thought to be due to a large summer electron loss rate caused by an increase in the molecular/atomic ratio of the ionosphere and the reaction rates being temperature sensitive.

It is not all bad news though. Night-time MUFs may be higher in summer than those in winter. Note that DX on the low bands, if possible, is unlikely to occur until around midnight or the early hours, due to the late sunset.

### 160m (1.8MHz or Top Band)

High levels of static and solar absorption mean that the band will not really support sky-wave contacts during the day. During darkness, short-skip openings may occur, but DX may be a rarity. Occasional openings can occur during the hours of darkness, especially around local midnight/early hours. Not the best season for Top Band.

### 80m (3.5MHz)

Will generally follow the characteristics of Top Band with high levels of static. Absorption will grow to a maximum at midday for inter-G contacts, so you may be better going to 40m. DX capabilities will be poor to fair during the hours of darkness, compared with the winter.

### 40m (7MHz)

Will suffer from high static, caused by high numbers of thunderstorms. Nevertheless, night-time openings should be reliable from sunset to sunrise. Local daytime openings will be possible on the whole with 40m being the band of choice for contacts around the UK. Night-time skip distances are likely to be between 300 and 2,300 miles.

### 20m (14MHz)

Still likely to be a good DX band around the clock, although the band will be noisier than the winter period and perhaps not as reliable for long-haul contacts in the summer. The higher MUFs at night mean that 20m may remain open during the evening and night to DX. Short skip may also be possible due to summer Sporadic-E.

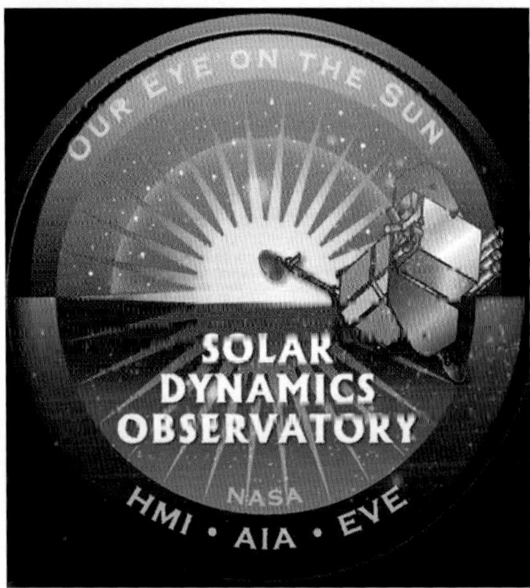

*The Solar Dynamics Observatory mission logo.*

### 17m/15m (18MHz/21MHz)

May provide some DX openings during daylight hours, especially to the southern hemisphere, but only if the SFI is high enough. Both bands are likely to close after sunset. Sporadic-E will provide good short-skip openings, predominantly in the May-June period.

### 12m/10m (24MHz/28MHz)

Sporadic-E openings will provide regular openings out to around 1,300 miles. Multi-hop Sporadic-E openings are possible, providing relatively good but short-lived paths to DX beyond this range. Propagation via the F2 layer is likely to be less reliable with the decreasing solar flux index and the seasonal summer doldrums.

**Steve Nichols, G0KYA**
*Chairman, RSGB Propagation Studies Committee*

# Amateur Satellites

Since soon after the launch of the first artificial satellite, "Sputnik-1" radio amateurs have been constructing and operating amateur satellites. There have been more than one hundred amateur satellites launched since then, of which more than twenty are currently operating and available to amateurs. In the early days, only a few well-equipped stations with operators well versed in orbital mechanics were able to make QSOs consistently. Nowadays, modern technology takes the strain of the mathematics and the physics, making listening to and communicating via amateur satellites far easier than it used to be. Indeed it's often the case that amateurs find that they already have all the equipment and knowledge required to start operating amateur satellites.

***All radio amateurs licensed in the UK can communicate via satellites, Foundation licensees are actively encouraged to join in!***

## How far can I communicate via an amateur satellite?

How far a station can expect to communicate using an amateur satellite depends on a number of factors, but it is possible to operate almost globally using amateur satellites. One benefit of using amateur satellites as a communication medium is that it is possible to predict very accurately and consistently when a QSO can be made. Because satellite communication almost always demands line of sight between the groundstation and the satellite, the vagaries of propagation present in many other DX modes can generally be ignored

## What's onboard an amateur satellite?

Different amateur satellites have differing capabilities in their payloads. Analogue satellites provide voice and often CW transponders. Some analogue transponders simply receive a single channel FM signal on one band and retransmit the channel on another band. These are known as 'bent-pipe' transponders. Others carry linear transponders designed for CW and SSB QSOs. Some satellites support an international version of the APRS network. Additionally there are a rapidly increasing number of very small 'CubeSats' which generally provide simple CW telemetry. These are often only 10cm cubes and have a mass of only 1kg. As technology progresses these small spacecraft are becoming more capable.

As mentioned, because of the limitation of a single FM channel, a linear transponder receives a range of frequencies (or 'passband') on one amateur band and re-transmits this passband on another band.

There are several amateur satellites providing digital capabilities. Some may simply provide telemetry information, while others provide two-way communications from simple single-channel digipeating to store-and-forward BBSs. Over the years, there have been a large number of different hardware modems required for digital satellite operation. Thanks to software DSP techniques, almost all of these modems are now emulated using a PC and a soundcard.

Unlike a terrestrial repeater, almost all amateur satellites receive on one band and retransmit on another. This saves on payload weight on the satellite where potentially expensive, large and heavy filters would be needed. At the groundstation end, it allows operators to listen easily to their own signals without de-sensing their receivers. Operating full duplex like this is recommended practice on amateur satellites. If you can't hear your own signal on the satellite's downlink, it's difficult to know whether you are on frequency or even making it to the satellite at all.

## ARISS

As well as unmanned spacecraft, amateur radio also has a permanent presence in space on board the manned International Space Station. ARISS (Amateur Radio on the International Space Station) is an international organisation comprised of national amateur societies, national AMSAT societies and various space agencies. It is designed to be an educational tool to encourage young people to become interested and involved in science, space and amateur radio. Primarily, this takes the part of scheduled live QSOs between amateur stations and astronauts onboard the International Space Stations. It is also possible for individual amateurs to hold random QSOs with astronauts - surely a highlight for any amateur's career! When the astronauts are busy, there is also an unattended packet station on board.

During 2016, Tim Peake, the first British ESA Astronaut, made contact with ten schools throughout the UK. Most of these also include live video from the HamTV installation that is now operational in the Columbus module. These were the first ever ARISS contacts to have video as well as audio downlinks and this facility contributed greatly to impact of the contacts. Great support was given by both the UK Space Agency and the STEM outreach teams of the RSGB itself.

## Amateur satellite orbits

Although, these days, amateur satellite operators certainly don't need an understanding of orbital mechanics, knowing some basics of a satellite's orbit can help when operating.

Some satellites orbit higher in space than others. The low earth orbit (LEO) satellites have coverage areas (or 'footprints') of some 3,000 miles diameter, just allowing transatlantic QSOs. Because of their low orbit, they complete an entire orbit in roughly 93 minutes, so they are only visible for a maximum of about fifteen minutes each orbit. Because of the short 'pass', QSOs and overs tend to be of short duration.

The 'Phase 3' high earth orbit (HEO) satel-

*Tim Peake live video from the ISS using the HamTV system*

*Before the contact at Kings School Ottery St Mary*

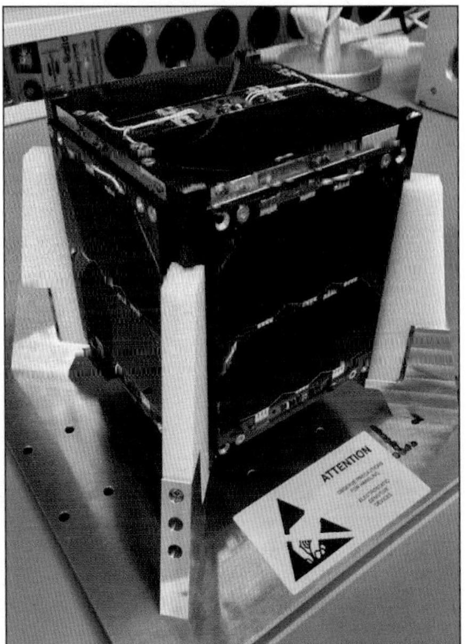


*The Nayif-1 CubeSat Flight Model in the cleanroom*

lites had the benefit of a much larger footprint because they were designed to operate from a much higher altitude. They also appeared to the observer on the ground to be hovering around for several hours at a time, perfect for a ragchew. These spacecraft are in a highly elliptical orbit but, at the time of writing, there are no operational Phase 3 orbiting satellites, although at least one is currently in development.

The great news is however, that there are currently two "Phase 4" satellites under construction. These are intended to operate from a geosynchronous or geostationary (GSO) orbit. One of these, a hosted payload on the Es'Hail-2 satellite, is expected to provide coverage of Europe, Africa and much of Asia and will use the 2.4GHz band for uplinks and the 10GHz band for downlinks.

To figure out when a satellite is passing over, and where it will be in the sky at a given moment in time, prediction software is used. If the station has directional antennas, most prediction software can also steer the antennas in real time. It's important to ensure that prediction software has up-to-date orbital parameters, called Keplerian Elements, or often simply 'Keps'. These parameters are downloadable from the Internet, and most software can be configured to obtain these updates automatically.

Because satellites are moving relatively rapidly compared to the groundstation, there will be some degree of Doppler shift in the receiving and transmitting frequencies. Doppler shift is experienced in daily life when, for example, the tone of an emergency vehicle's siren drops as it passes. When operating amateur satellites, Doppler shift may be adjusted by manual tuning, or alternatively most prediction software can update the radio's frequency in real time.

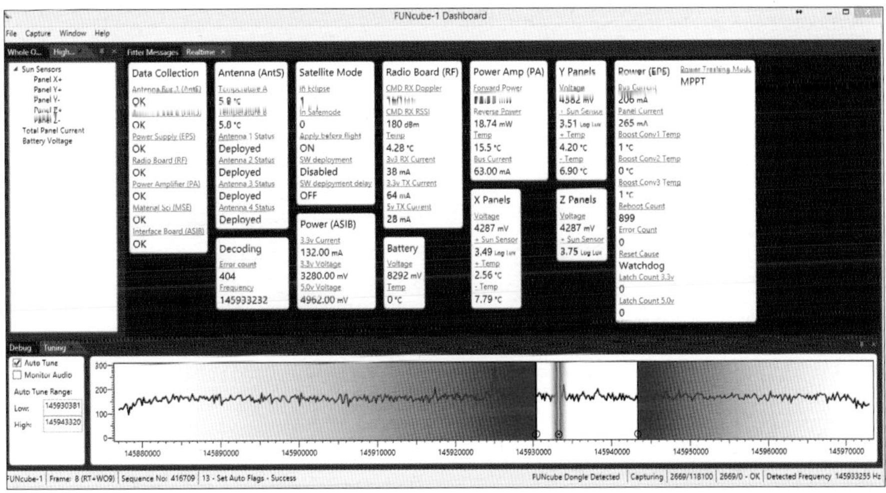


*FUNcube-1_telemetry display*

## Groundstations for operating amateur satellites

It is not necessary to have a large or expensive station to operate satellites. In its simplest form, a dual band handheld FM transceiver can be used to make satellite contacts. When used with a handheld Yagi antenna, this simple configuration can make an effective station for portable contacts.

At the other end of the scale, a station may have a steerable antenna array under automatic computer control and a radio with satellite-specific functionality such as full duplex operation, SSB capability and tracking uplink and downlink VFOs. The benefit of the larger station is that it allows the operator to use more satellites more consistently.

A well-equipped satellite station has antennas of similar size to a standard domestic TV and FM Band II antenna configuration, with perhaps eight elements on 70cm and four on 2m. This makes it surprisingly easy to make a neighbour-friendly and capable amateur satellite station. These antennas are usually crossed dipoles and can either be steerable in azimuth only, or, for more consistent QSOs, in elevation too.

As technology progresses there are more satellites operating on the higher bands. For microwave use a small dish is employed and because of their narrower beamwidths it's essential to be able to point the antennas accurately both in azimuth and in elevation. Rather than purchasing expensive Az/El rotators, antenna pointing is very often done manually with the antennas at ground level, especially in temporary or portable configurations. Because the satellite is above the horizon in the sky, locations often considered ineffective for terrestrial radio communication can be effective for amateur satellites.

Perhaps the most important rule of thumb in satellite operation is to concentrate on the

station's receiving equipment before investing time, money and effort in the transmitting side. The nature of satellite communications means that the old adage 'if you can't hear them, you can't work them' is especially true. It is often tempting to improve your signal by increasing your station's ERP. It is likely that the transponder may already be limiting (eg, in linear transponders, the transponder's AGC has started to attenuate the passband so that its output can be maintained in the linear region), so more ERP is not going to be beneficial. Masthead preamps are always beneficial.

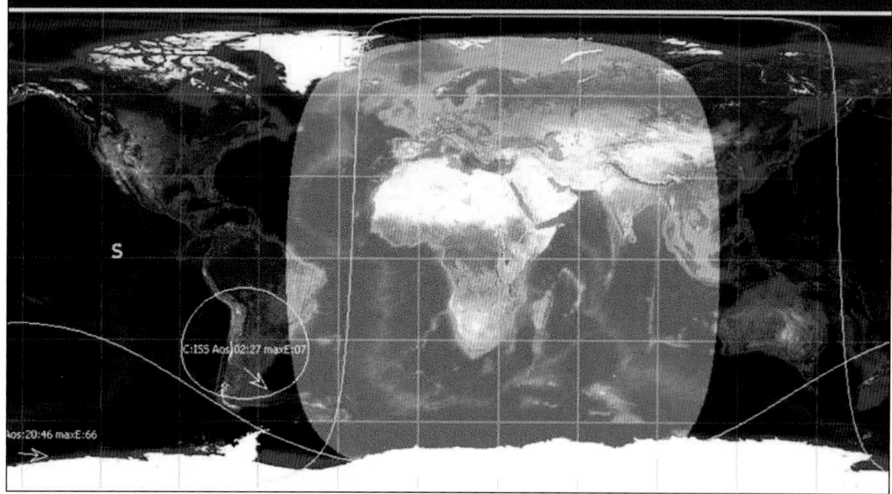


## How are amateur satellites launched?

Historically, amateur satellites have been launched by generous space agencies with space available on their rockets. Often this space would otherwise have been dead weight ballast. Lately most of these free rides have dried up, but the launch costs for CubeSats can often be managed. The collaboration between radio amateurs and universities is also leading to having shared missions as described below.

## Who makes amateur satellites?

Amateur satellites continue to be made by many organisations throughout the world, by individual national AMSAT societies, or a collection of national AMSAT societies. Many educational establishments have also launched amateur satellites.

AMSAT-UK is at the forefront of satellite building. They are presently working with AMSAT-NL on another CubeSat project called Nayif-1. This is based on the Funcube-1 system and is planned for launch in late 2016. Members are also continuing the organisation's link with the European Space Agency (ESA) and are developing a new transponder for the new low earth orbit European Student Earth Orbiter microsat. (ESEO). This will carry a single channel FM 23cms to 2 metre transponder and educational outreach telemetry for schools using the highly successful FUNcube format. This is now expected to launch in 2017.

AMSAT-UK has already created the FUNcube-1 CubeSat in collaboration with AMSAT-NL and this was successfully launched in late 2013. This also carries the name Oscar AO73 and acts as a linear 70cms to 2 metre transponder at night and during weekends and holidays. At other times it provides telemetry data for educational outreach for schools and colleges. The sharing of this resource is intended to encourage the uptake of STEM (Science Technology Engineering & Mathematics) subjects as well as increasing knowledge about and interest in amateur radio. This spacecraft continues to operate nominally and has provided more than 900MB of data which is stored on a central Data Warehouse and which is available for research purposes.

AMSAT-UK has also provided a similar FUNcube-2 payload for the UKube-1 spacecraft which was developed for the UK Space Agency. This spacecraft is also in orbit and completing its science mission. There is also FUNcube transponder on another CubeSat known as Oscar EO79. Again this is already in orbit awaiting the completion of its science mission before being placed into regular service.

All artificial satellites have a limited lifetime. Solar panels gradually become less efficient as they are bombarded by the radiation in space, from which we are protected here on Earth. On other occasions it might be that the rechargeable batteries fail first. Because of their inherent limited lifetime, there is a continual need for replacement satellites.

The longest surviving amateur satellite is AO-7, launched in 1974. AO-7 went silent for over two decades until Pat Gowen, G3IOR, heard it again in 2002. It is believed that AO-7 originally stopped functioning due to the battery going short circuit. After many years of the cumulative electrical and chemical stresses on the battery from the solar cells attempting to charge them, one of the cells in the battery went open circuit and now AO-7 is available again - although only when it is in sunlight.

## Where to find more information on amateur satellites

The AMSAT-UK and AMSAT-NA websites are very good sources of the most up-to-date information about satellites, and there is also a very active AMSAT reflector available on the internet. To join, send an Email with the following text on the first line of your message: subscribe AMSAT-BB Send your request to: majordomo@amsat.org

## Who pays for amateur satellites?

Funding for amateur satellites comes from a number of sources, including the national radio societies and national AMSAT societies.

In the United Kingdom, AMSAT-UK represents the interests of amateur satellite operators, and, as discussed above, AMSAT-UK members have been instrumental in the development, manufacture and funding of several amateur satellites. With the generous donations and subscriptions of its members, AMSAT-UK helps to keep amateur satellites in space. So if you use amateur satellites, remember to contribute by joining AMSAT-UK!

## Which amateur satellites are currently operational?

The list can change very quickly with new launches taking place and some failing or burning up and de-orbiting... It is therefore not possible to publish a sensibly up to date list here. The *http://www.amsat.org/status/* page shows a large amount of detail for each spacecraft and this is updated by the hour. More details of each spacecraft can be found here *http://www.dk3wn.info/p/?page_id=29535* As will be seen, most satellites presently use frequencies in 2m or 70cms bands for which equipment is very readily available. A simple on-line satellite prediction service is provided here *http://www.heavens-above.com/* Be sure to enter your location first and then select the "amateur satellites" tab

**AMSAT-UK represents the amateur satellite community in the UK whose members not only operate amateur satellites, but also help to design, build and fund them**

# Try Amateur Satellites for yourself!

Before trying an amateur satellite for the first time, it's highly recommended that reference is made to the current status of the satellite on the AMSAT web pages and the Oscar status page at: http://www.amsat.org/status/ Both operating schedules and modes of satellite operation regularly change. Over time, batteries, solar panels and other hardware can and do fail, and there's little opportunity to repair satellites once they're in space.

For a first timer, SO-50 (also known as Saudisat 1C) is a satellite that can be operated with fairly minimal equipment. SO-50 is a satellite in the Amateur Satellite Service providing functions very similar to a traditional terrestrial FM repeater. It was developed by King Abdulaziz City for Science & Technology (KACST) in Saudi Arabia on 20 December 2002. It is a single channel FM bent-pipe transponder with a 2m uplink and 70cm downlink, and requires a CTCSS tone on the 2m uplink. So if you have a way of transmitting on 2m FM and receiving on 70cm FM, you already have the equipment to try operating satellites. Particularly useful are the full duplex handheld transceivers which have been available since the 1980s.

For an antenna, it's very beneficial to have a dual band antenna with some directional gain such as the Arrow or Elk antennae (these are available from AMSAT-UK). Alternatively, a simple dual band whip can be used, preferably one which is an end-fed $\lambda/2$ whip at 70cm: there is some benefit to be gained on the receive side from the $\lambda/2$ antenna because a ground plane is not required.

If you have Internet access, print out some pass predictions from the Heavens Above website, where it's known as Saudisat 1C. Alternatively, there are many software prediction packages available on a variety of platforms. Take the opportunity to print out these predictions.

The passes of SO-50 are about 15 minutes in duration, so it's worthwhile planning the pass. Take the radio(s) and antenna outside, and ensure that you will have a line of sight path to the satellite as it goes past. A compass and an accurate watch are also essential aids when planning the pass.

Practice receiving the satellite for a few passes first. Remember to open the squelch of your receiver. Although it's not very weak, SO-50 will appear to suffer fading, and you will find that leaving the squelch open will aid reception. For SO-50, the downlink frequency is 436.800MHz. It's worth having a receiver which can be tuned in 5kHz steps to correct for Doppler. At the beginning of the pass, you'll find the frequency on the ground is about 10kHz high, starting at 436.810, and throughout the pass it will slowly decrease in frequency. When you hear the satellite move out of the receiver's FM passband (it sounds distorted, just like any other off-frequency FM signal), decrease the frequency in 5kHz steps. At the end of the pass, the receiver will be down to 436.790MHz.

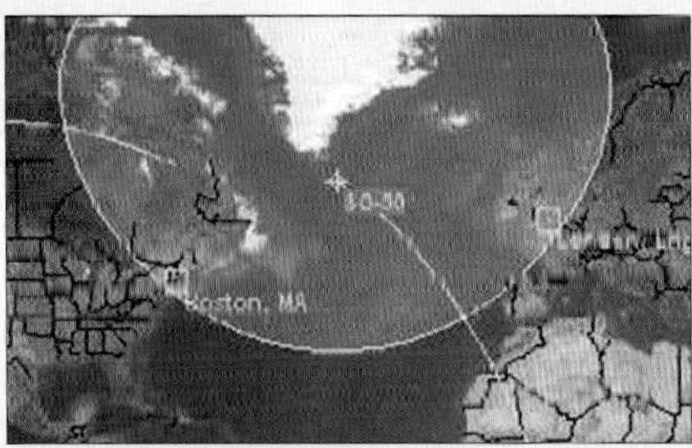


*The footprint of Saudisat 1C (SO-50) showing co-visibility in both north east USA and the UK*

If your radio allows it, consider programming five consecutive memories: 436.810, 436.805, 436.800, 436.795, 436.790. This way you can leave the radio's VFO on the standard 25kHz spacing and jump straight to the memories for operating SO-50 without having to keep adjusting the radio's channel step.

The biggest knack to learn is how to manoeuvre the antenna for the best signal. There is no hard or fast rule here, but keep in mind that you should know approximately where the satellite is in the sky, and that you are aiming to orient your antenna at the satellite. Also consider that because satellites spin like a gyroscope, their antenna polarisation changes. You'll find turning the antenna to try to match polarisations will prevent deep fades. It takes a couple of passes to get the hang of it. Keeping the pass listing, compass and your watch readily available is very useful here.

Note that the transponder on SO-50 is on a ten minute timer, and it must be switched on first. To switch on, or 'arm' the transponder, an operator sends a very short transmission together with a CTCSS tone of 74.4Hz on the uplink frequency, 145.850MHz. After ten minutes, the transponder switches off. Listen carefully, and if you can already hear the transponder, another operator has already switched the transponder on and of course there's no need to arm it yourself.

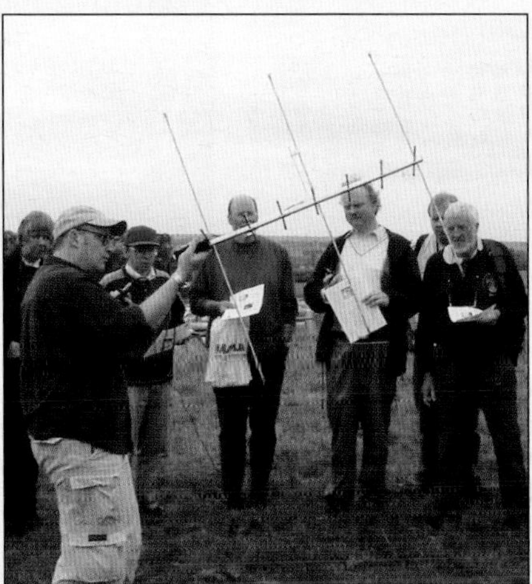

*Howard Long, G6LVB, demonstrates operation of amateur satellite operation with nothing more than a dual-band handheld transceiver and an 'Arrow' antenna, so-called because its elements are made from arrow shafts.*

When it comes to conducting a QSO, you'll also need to be able to transmit a 67.0Hz CTCSS tone along with your FM transmission on 145.850MHz. 5W will easily be sufficient as long as the 'alligators' aren't about. This term is often used in the amateur satellite community to describe an operator who's all mouth and no ears. It is obvious from monitoring the downlink that these operators usually cannot hear the satellite at all! The author has worked SO-50 on 25mW during a quiet time, so 5W is certainly enough if the satellite is not too busy, or a gain antenna is used.

Because you'll be operating full duplex, take along some headphones. This will prevent feedback via the satellite when you're transmitting. If you have one, take a tape recorder or digital recording device with you too for logging purposes. There's a lot to do, and recording the QSOs will allow you to keep your hands available for things other than logging. After the pass, you can write the QSOs into your log book.

Doppler shift is proportional to frequency as well as relative velocity, so although it is almost essential to be able to adjust for Doppler at 70cm, it is usually unnecessary to move from the SO-50 uplink frequency of 145.850 because the Doppler shift is only about ±3kHz on 2m.

From your practice receiving sessions, you will have already discovered that the nature of QSOs on SO-50 can be quick, almost contest-style exchanges. SO-50's footprint comfortably encompasses the whole of Europe, so there are often a large number of amateurs in the footprint. The footprint is also just large enough to cover both the UK and the North East coast of North America when the pass allows it. Because there are relatively few operators in the footprint on these transatlantic passes, it can sometimes be easier to have QSO with Canada or the United States than it is with Europe - all on your handheld transceiver!

## SO-50 quick facts

- Downlink 436.800MHz +/- 10kHz Doppler.
- Uplink 145.850MHz.
- Short 74.4Hz CTCSS transmission to arm transponder for ten minutes.
- 67.0Hz CTCSS while conducting QSOs.

AMSAT-UK address: 'Badgers', Letton Close, Blandford, Dorset DT11 7SS, UK.
email: g3wgm@amsat-uk.org (Jim Heck) website: www.amsat-uk.org
AMSAT-NA website: www.amsat.org

# Datacommunications

The Emerging Technology Co-ordination Committee (ETCC) exists to deal with all matters concerning amateur radio repeaters and data communications on behalf of the RSGB. It assists Ofcom in the processing of applications for Notices of Variation (NoV's) for repeaters, Internet Gateways and mailboxes. It is also responsible for the coordination of all requests for site and frequency clearances prior to submission to Ofcom.

From 1990 to 2007 the Data Communications Committee acted as the body responsible for facilitating (by means of frequency co-ordination) simplex internet voice gateways. At the time of writing there are operational gateways on 29MHz, 51MHz, 70MHz, 145MHz, 430MHz, 431MHz, 434MHz and 1297MHz. Full details of the currently licensed gateways plus the innovative online NoV application system can be found at the ETCC website (see below) under 'Internet Linking'.

ETCC recommends general operating practices for data communications. A soft copy of this is available in *The Guide to Repeater and Internet Gateway Licensing in UK*. Visit the ETCC main site Documents section to download it. The website not only contains the latest information about most digital modes, but also a comprehensive list of links to datacomms related sites.

Please see the separate repeaters section covering all aspects of our work in respect of repeaters both analogue and digital.

## What is Packet Radio?

Packet radio is digital communications via radio. It began in 1978 in Canada and was introduced into the UK in the 1980s. Packet radio mailboxes (BBS) were first licensed in the UK in about 1988. The numbers grew until the late 1990's, when numbers then started to decline due to the widespread availability of broadband Internet to most of the UK population. However, some mailboxes do still operate in UK, so the following information should help any newcomers to the mode.

## What can I do on Packet?

### Live Contacts

Like RTTY, packet radio can be used to talk to other amateurs, chatting keyboard to keyboard. Some mailboxes also have a conference mode, so people can log on and chat with many people at once just like an HF net.

### Mailboxes

Mailboxes allow amateurs to connect with their local mailbox and send and receive text messages. These messages can be sent as personal messages to another amateur anywhere in the world (or in space!). Alternatively, messages can be sent as a bulletin for any amateur to read.

Within the UK your messages will normally be relayed via other mailboxes using Internet gateways. These are used to forward mail to more distant continental mailboxes, so it is possible to exchange mail with amateurs on the other side of the world by simply logging into your local mailbox using a low power VHF transceiver and a simple aerial system, as well as a Terminal Node Controller (TNC) and computer.

### File Transfer

Packet radio also allows you to be able to transfer files between amateur packet stations in both text and binary format.

### DXCluster

There are a few DXCluster stations around the UK and, alongside, the 'automatic position reporting' component of the network. The provision of near real-time on a truly global network exchange of DX, band condition and stations heard/worked information is highly valued by the world's DX community.

### APRS

Automatic Packet Reporting System (APRS) was developed by Bob Bruninga, WB4APR, to track mobile GPS stations with two-way radio. APRS can be used in a number of applications for data, communications and telemetry.

In the UK, the late Roger Barker, G4IDE, authored a protocol-compatible variant called 'UI-View'. This is extremely popular and is well supported with regular updates and third-party extensions that add increased functionality to the base product. Examples of add-ons include 'DXCluster spy', satellite telemetry decoding, rig control, rotator control the list is long! For full details see the website at www.ui-view.org

APRS remains one of the most active packet radio modes. ETCC issue NoV's for the operation of APRS digipeaters and internet gateways.

NoV's are now required for any unattended packet radio operation. Please see the ETCC website for details, as well as the on-line forms.

### What will I need?

As well as a VHF or UHF FM radio transceiver, you will also need a TNC and a terminal or computer with some form of terminal software program or a specialist packet radio software program. And finally... a great deal of patience and willingness to learn.

## Other data modes

### RTTY or Radio Teletype

A frequency-shift-keying mode that has been in use longer than any other digital mode (except Morse). It uses a simple five-bit code to represent the alphabet, numbers, a few punctuation marks and a few control codes. As there is no error-correction, QRM and QRN can have seriously detrimental effects on copy. Despite all that, it is still the most popular digital mode. The bandwidth of a RTTY signal is 230Hz, and at 45.45 Baud it gives about 60 WPM throughput.

### PSK31

This mode has the advantage of having a very narrow bandwidth and was designed for 'real-time' keyboard-to-keyboard QSOs. It can be used to send almost all of the characters shown on your keyboards and has even been used to send small pictures. If only lower case letters are used, you get about 53 WPM, but with all capitals, this is reduced to about 39 WPM. The bandwidth is as the name suggests 31Hz.

### MFSK16

Uses 16 tones and has forward error correction, where it sends all data twice with an interleaving technique to reduce errors from such things as static crashes. It has a comparatively wide bandwidth of 316Hz, which allows faster baud rates, and has greater immunity to multipath phase shifts. This wider bandwidth gives you around 42 WPM.

### Hellschrieber

Uses facsimile techniques to transmit and receive characters. It has been in use since the 1920s but has come on in leaps and bounds thanks to modern day DSP soundcard processing. Characters are painted on the screen in ticker-tape like fashion and are read directly from the screen, as opposed to being decoded and printed. Although this mode has a relatively small bandwidth of about 75Hz, it can handle about 35 WPM.

### MT63

An excellent mode for sending text over propagation paths that suffer from fading and interference from other signals. It works by encoding text with a matrix of 64 tones over time and frequency. Although this is rather complicated, it does provide error correction at the receiving end, and gives about 100WPM. MT63 has a wide bandwidth of 1kHz.

### Throb

Uses either 5 or 9 tones, depending on the version in use. The latest is the 9-tone version, and has speeds of 1, 2 and 4 Baud, enabling data rates of 10, 20 and 40WPM respectively. It appears to be quite good under poor propagation conditions and although it isn't commonly heard, it does seem to be gaining popularity.

### 23Hz RTTY

A relatively new thing, and hasn't proved as popular as hoped. It sounds similar to PSK31 but is a bit narrower in bandwidth. Although several programs now include this mode, it hasn't taken off well.

## Where to find the other data communication signals

In recent years there has been an explosion in the number of digital modes, with many different people writing software to decode them. Some of the modes may be familiar to readers, others perhaps not. Many of you will know the sound of RTTY, and maybe that of PSK31, but do you know what Throb or MFSK sound like or where to find their signals in the amateur bands?

Within the amateur bands the data communications signals are to be found around the frequencies shown in the bandplans.

ETCC website: www.rsgb.org/main/about-us/committees/emerging-technology-co-ordination-committee/

Throb, Hellschrieber and MFSK16 tend to congregate just below the RTTY segment.

These frequencies are approximate, but will give you an idea of where to start looking. A full set of datacommunications bandplans is available on the DCC website.

## Getting started

The advent of soundcard software for the PC heralded a new era in the digital modes, especially since most of the software is free!

The advice given here is intended for people who are newcomers to the data modes and is therefore presented at a beginner's level. It does not explore the theory behind the data modes in any depth. Neither does it cover all aspects. Once you have understood the basic principles, you can start to learn the more advanced features at your own pace.

It is simply not possible to include instructions for connecting a particular make or model of radio here; for that you should consult the manual that came with it. The same goes for installing a soundcard in your computer. If you are not confident in wiring-up cables then you should either buy one of the commercial interfaces or get someone to do it for you.

These explanations will refer mainly to RTTY and PSK31, as they are the most popular of the digital modes in use at present. However, details are given of other modes present on the bands.

### Background

Computer soundcards use Digital Signal Processing (DSP) to handle sound, and these techniques lend themselves very well to the processing and decoding of the audio data signals from the output of your radio.

Although RTTY, AmTOR and PacTOR have been around for quite some time, it is the general feeling that a renewed interest in the digital modes really came about when Peter Martinez, G3PLX, created the PSK31 program for the Windows operating system.

At first, it proved to be quite difficult to tune in a PSK signal, so much so that many gave up before they had even watched a QSO in progress.

For those of us who persevered, however, it proved to be something of an enlightenment. At that point, tuning was aided only by turning the tuning knob, with the aid of the phase scope, and it took a great deal of patience and careful finger work to tune in one of these new sounds.

You only needed to be off by as little as 5Hz to get garbage on the screen. After several months, Peter came up with a version that included a waterfall display, and that really made a big difference. It was probably a key turning point. Now you could see which way to turn the dial, and you had a fair chance of getting it right quite quickly. Later versions allowed you to point and click on a signal on the waterfall and it was tuned in instantly.

On the bands one was quite likely to meet up with G3PLX, and although he wouldn't hesitate to tell you that you were over-driving the soundcard, if indeed you were, he would offer suggestions, and between you, it was possible to adjust the levels during the QSO to an optimum state. Since then the number of operators has increased considerably. Unfortunately, many have not taken the advice offered by those with greater experience, especially with respect to the adjustments of sound levels and transmitter power. One common mistake is to leave the speech processor switched on, which will definitely cause you problems with the transmitted tones.

In the early days it was the norm to use between 5 and 10 watts of transmitted power and many used much less than that. This level of signal is perfectly adequate for world-wide communications, although these days many stations seem to use several hundred watts into large beams. Apart from the fact that it is not necessary, it also tends to reduce the bandwidth available to others.

Some new modes, such as MFSK16, have been developed in the past few years with the idea of replacing RTTY. Although they do have a following, RTTY has remained at the forefront of the digital modes. PSK31 has gained respect mainly because it is so good at low power levels, making it ideal for QRP work.

## Operating the Data Modes

When you are ready to begin using the digital modes, if you are new to data communications it is suggested that initially you spend some time just listening and watching QSOs in progress. This will give you an idea of how they are conducted, what sort of phrases are commonly used and general etiquette.

When you are set up and as you begin operating you may find that RTTY and PSK give you greater scope for learning, mainly because they are more common, but also because they are easier to operate.

The world is waiting for you. It is quite feasible to achieve DXCC on digital modes and the good news is that there are many rare DX countries out there that regularly appear on one of the digital modes. A lot of the DXpeditions these days include RTTY and/or PSK31.

Remember that RTTY is 100% duty-cycle, so please watch your output power! PSK31 is about 80 per cent duty cycle, but as it is used at much lower power this is of less importance. Curiously, Hellschrieber has a duty cycle of only about 21 per cent, so is a lot more 'equipment friendly'.

If you change the output power when you change mode, you may need to adjust the mic gain, as the ALC may well have altered. Just hit the transmit key and adjust the gain so that the ALC is showing slightly, then back it off a touch. It shouldn't make any difference to the indicated output power, but your signal will be cleaner.

If you use the DX Cluster while you are using RTTY, you may well find that the spots are not where they are listed. This is because operators in Europe tend to use different standards from the rest of the world. In Europe, the norm is to use 'low tones' (1275 and 1445Hz) on USB whilst everyone else seems to use 'high tones' (2125 and 2295Hz) on LSB. Although the tone values won't make any difference to your receiving, the sideband will.

Many of the RTTY programs cater for the American market, which means that 'Normal' equates to LSB. If you intend using USB, you may need to find the 'Reverse' button to invert the tones. If you have a RTTY signal with the audio tuned in nicely but only get garbage on the screen, try hitting the 'Reverse' button and you will probably get clear text after that. It can be quite common to see someone operating 'Inverted', and although it will work equally well providing the other station is also inverted, it can reduce the chance of getting a reply to a CQ call.

If you want to work DX and increase your country count, a good way to do this is to enter one of the many RTTY contests. There are about 14 or 15 such contests per year, and many have sections for single operators with low power. No matter how many contacts you make, always try and submit your log, as this helps the contest organiser get an idea of popularity (it also allows them to verify other logs). If you are unsure about your log, then simply send it as a checklog.

One thing you will regularly see in PSK31 is the use of all capitals in the transmitted text. PSK uses an alphabet called 'Varicode', which has shorter codes for the more common letters (similar to Morse) and which also includes both upper and lower case letters. The lower case letters, being more common, have the shorter codes, so any given sentence takes longer to send in capitals. There really isn't any need to use capital letters at all in PSK, even for callsigns. RTTY uses ITC2 (5-element) code that doesn't include lower case, so all capitals is the only option.

Many operators use abbreviations much like in a CW QSO, and it really depends on the type of QSO you are having as to how you will operate. Don't worry about typing mistakes, as these are normal, and anyway who can tell if it wasn't a burst of static that caused his screen to mis-represent what you just typed?

If you are trying to contact a rare DX station, listen for a while and see if the operator is working in any kind of pattern. Listen to the callers and see how they are working. It may be that timing is the key to making the contact. Unless you have big antennas way up high, plus a big linear, and can drown everyone else out, don't try to be first in the pile; wait and let your call be the last one to be heard. That can often get you a response!

Don't forget that with most of the digital modes, if the transmissions of two or more stations overlap, all you get on screen is garbage. It is no good transmitting over the top of a QSO that is in progress, even though it may be nothing to do with the DX station, you will not be read and you will disrupt the QSO in progress.

When you do get through, the other station really doesn't need to know your working conditions or your life history, so keep your QSO short and leave space for others to have a go. It's a good idea to create a macro just for responding to a DX station, perhaps something like this: HISCALL DE MYCALL - TNX - UR ALSO 599 599, VY 73 ES GD DX DE MYCALL. That is all that is needed and anything more is only likely to get lost in the pile-up that will follow.

PSK31 is an excellent mode if you have a limited set-up and can operate only low power. 5 to 10 watts is quite practical with this mode and you can easily obtain DXCC at QRP levels.

## To Sum Up:

1. Check and re-check your volume settings.
2. Listen BEFORE you transmit.
3. Watch your power levels.
4. Don't worry too much about spelling mistakes.

# Contest Calendars

Contests are sporting events between amateur stations on specific bands and modes, conducted according to published rules. The activity appeals mainly to those with a competitive instinct, but construction and station optimisation are also important.

## 2017 RSGB HF Contest Calendar

| Date | Time (UTC) | Contest | Sections |
|---|---|---|---|
| 8 Jan | 1400-1800 | AFS Contest CW | ALL |
| | | AFS Super League | |
| 14 Jan | 1400-1800 | AFS Contest PHONE | ALL |
| | | AFS Super League | |
| 6 Feb | 2000-2130 | 80m CC SSB | LOW QRP |
| 11 Feb | 2100-0100 | **1st 1.8MHz Contest** | UK-Assisted UK-Unassisted Non-UK-Assisted Non-UK-Unassisted |
| 15 Mar | 2000-2130 | 80m CC DATA | LOW QRP |
| 23 Feb | 2000-2130 | 80m CC CW | LOW QRP |
| 6 Mar | 2000-2130 | 80m CC DATA | LOW QRP |
| 11-12 Mar | 1000-1000 | **Commonwealth Contest** | Open-SOA Open-SOU Restricted-SOA restricted-HF SOU Multi-Op Hq |
| 15 Mar | 2000-2130 | 80m CC CW | LOW QRP |
| 23 Mar | 2000-2130 | 80m CC SSB | LOW QRP |
| 2 Apr | 1900-2030 | **RoPoCo SSB** | ALL |
| 3 Apr | 1900-2030 | 80m CC CW | LOW QRP |
| 8 Apr | 1700-2100 | International Sprint CW | ALL |
| 12 Apr | 1900-2030 | 80m CC SSB | LOW QRP |
| 15 Apr | 1700-2100 | International Sprint CW | ALL |
| 20 Apr | 1900-2030 | 80m CC DATA | LOW QRP |
| 1 May | 1900-2030 | 80m CC SSB | LOW QRP |
| 10 May | 1900-2030 | 80m CC DATA | LOW QRP |
| 18 May | 1900-2030 | 80m CC CW | LOW QRP |
| 3-4 Jun | 1500-1500 | NFD | Open Restricted LowPower |
| 5 Jun | 1900-2030 | 80m CC DATA | LOW QRP |
| 14 Jun | 1900-2030 | 80m CC CW | LOW QRP |
| 22 Jun | 1900-2030 | 80m CC SSB | LOW QRP |
| 3 Jul | 1900-2030 | 80m CC CW | LOW QRP |
| 12 Jul | 1900-2030 | 80m CC SSB | LOW QRP |
| 16 Jul | 0900-1600 | Low Power Contest | A B C D |
| 20 Jul | 1900-2030 | 80m CC DATA | LOW QRP |
| 29-30 Jul | 1200-1200 | IOTA Contest | Single Operator Unassisted Single Operator Assisted Multi Operator |
| 09 Aug | 1900-2000 | 80m Club Sprint CW | LOW QRP |
| 24 Aug | 1900-2000 | 80m Club Sprint SSB | LOW QRP |
| 2-3 Sep | 1300-1300 | SSB Field Day | Open (SSBFD) Restricted (SSBFD) |
| 13 Sep | 1900-2000 | 80m Club Sprint SSB | LOW QRP |
| 28 Sep | 1900-2000 | 80m Club Sprint CW | LOW QRP |
| 8 Oct | 0700-1900 | 21/28MHz Contest | UK Open UK Restricted UK QRP Non-UK Open Non-UK Restricted Non-UK QRP |
| | **HF Championship** | | |
| 11 Oct | 1900-2000 | 80m Club Sprint CW | LOW QRP |
| 26 Oct | 1900-2000 | 80m Club Sprint SSB | LOW QRP |
| 8 Nov | 2000-2100 | 80m Club Sprint SSB | LOW QRP |
| 11 Nov | 2000-2300 | Club Calls (1.8MHz AFS) | ALL |
| | | AFS Super League | |
| 18 Nov Non- | 2100-0100 | **2nd 1.8MHz Contest** | UK-Assisted UK-Unassisted Non-UK-Assisted UK-Unassisted |
| 23 Nov | 2000-2100 | 80m Club Sprint CW | LOW QRP |
| | | AFS Super League 2014-15 | |
| | | **AFS Super League 2015-16** | |

Events in **bold** qualify towards the HF Championship

The Contest calendars in these pages are provisional, and before entering you should check the website before getting on the air to participate at: www.rsgbcc.org

### HF Key to Sections

| | |
|---|---|
| A | 10W Fixed |
| ALL | ALL |
| B | 10W Portable |
| C | 3W Fixed |
| D | 3W Portable |
| HQ | HQ |
| LOW | 100W maximum output power |
| LowPower | Low Power |
| Multi Operator | |
| Multi-Op | Multi Operator |
| Non-UK Open | Non-UK Open |
| Non-UK QRP | Non-UK QRP |
| Non-UK Restricted | Non-UK Restricted |
| Non-UK-Assisted | (b) Non-UK |
| Non-UK-Unassisted | (b) Non-UK |
| OPEN-SOA | Open - Single Operator Assisted |
| OPEN-SOU | Open - Single Operator Unassisted |
| Open | Open |
| Open (SSBFD) | Open |
| QRP | 10W maximum output power |
| RESTRICTED-SOA | Restricted - Single Operator Assisted |
| RESTRICTED-SOU | Restricted - Single Operator Unassisted |
| Restricted | Restricted |
| Restricted (SSBFD) | Restricted |
| Single Operator Assisted | |
| Single Operator Unassisted | |
| UK Open | UK Open |
| UK QRP | UK QRP |
| UK Restricted | UK Restricted |
| UK-Assisted | (a) UK stations |
| UK-Unassisted | (a) UK stations |

### VHF/UHF Key to Sections

| | |
|---|---|
| 10H | 10W Hill Toppers |
| 3B | 2m 3W Backpacker |
| 6O | 6 hours others |
| 6S | 6 hours Single Op Fixed |
| A | All |
| AL | UKAC Low Power |
| ALL | All |
| AO | UKAC Open |
| AR | UKAC Restricted |
| AX | UKAC DXers |
| FSO | Fixed Station Sweepers Open |
| FSR | Fixed Station Sweepers Restricted |
| L | Low Power Section of VHF NFD |
| LP | Single Operator, Fixed, 25W - Single Antenna |
| M | Mix and Match Section of VHF NFD |
| MS | Single Transmitter Section of VHF NFD |
| O | Open |
| Open | Open section of VHF NFD |
| Overseas | Overseas |
| R | Restricted Section of VHF NFD |
| SAO | SHF UKAC Open |
| SAR | SHF UKAC Restricted |
| SF | Single Op Fixed |
| SO | Single Op Others |
| Sweeper | VHF NFD Overall Sweeper Results |

### VHF/UHF Key to Multipliers

| | |
|---|---|
| M1 | Post Codes and Countries |
| M2 | QTH Locators |
| M3 | Post Codes, Countries and Locators |
| M4 | Countries and Locators |
| M5 | UK QTH Locators |
| M6 | UK Locators plus non-UK Countries |
| M7 | All large square locators count as multipliers; A locator in which a UK station has been worked counts as a double multiplier |

### VHF/UHF Key to Special Rules

| | |
|---|---|
| Backpacker | Special Backpackers Rules |
| VHF Championship | VHF Championship |
| VHF CW Championship | VHF CW Championship |
| VHFNFD | Special Rules for VHF NFD |
| Low Power Contest | 25W max. transmit o/p power, |
| S1 | 1 Point per QSO |
| S2 | Deleted |
| S3 | Affiliated Societies contest |
| S4 | Runs concurrently with a backpackers contest |
| S5 | Cumulative contest |
| S6 | Runs concurrently with the first few hours of an RSGB 24 hour event |
| S7 | Runs concurrently with all or part of an IARU co-ordinated contest |
| S8 | Activity contest |
| S9 | Club Championship |
| S10 | Overall UKAC Club Championship |
| S11 | Activity contest |
| AFS Super League 2013-14 | AFS Super League |
| AFS Super League 2014-15 | AFS Super League |

## 2017 RSGB VHF Contest Calendar

| Date | Time (UTC) | Contest Name | Sections |
|---|---|---|---|
| Every Tuesday of the Month | | | |
| 1st | 2000-2230L | 144MHz UKAC | AO AR AL AX |
| 2nd | 2000-2230L | 432MHz UKAC | AO AR AL AX |
| 3rd | 2000-2230L | 1.3GHz UKAC | AO AR AL |
| 4th (Jan-Nov) | 2000-2230L* | SHF UKAC (Jan-Nov Only) | SAO SAR |
| 4th (Jan-Nov) | 2000-2230L | 50MHz UKAC (Jan-Nov Only) | AO AR AL AX |
| 5th | 2000-2230L | 70MHz UKAC | AO AR AL AX |
| 5 Feb | 0900-1300 | 432MHz AFS | O SF |
| | | AFS Super League | |
| 26 Feb | 1000-1200 | 70MHz Cumulatives #1 | O SF |
| 4-5 Mar | 1400-1400 | March 144 432MHz | O 6O SF SO 6S |
| | | VHF Championship | |
| 12 Mar | 1000-1200 | 70MHz Cumulatives #2 | O SF |
| 2 Apr | 0900-1200 | First 70MHz Contest | O SF |
| 9 Apr | 0900-1200 | First 50MHz Contest | O SF |
| 6-7 May | 1400-1400 | May 432MHz-248GHz Contest | O SF |
| 6 May | 1400-2200 | 432MHz Trophy Contest | O SF |
| | | VHF Championship | |
| 6 May | 1400-2200 | 10GHz Trophy Contest | A |
| 14 May | 0900-1200 | 70MHz Contest CW | O SF |
| | | VHF CW Championship | |
| 20-21 May | 1400-1400 | 144MHz May Contest | O SF SO 6S 6O |
| | | VHF Championship | |
| 21 May | 1100-1500 | 1st 144MHz Backpackers | 3B 10H |
| 28 May | 1400-1600 | 70MHz Cumulatives #3 | O SF |
| 11 Jun | 0900-1300 | 2nd 144MHz Backpackers | 3B 10H |
| 17-18 Jun | 1400-1400 | 50MHz Trophy Contest | O SF SO 6O 6S Overseas |
| | | VHF Championship | |
| 25 Jun | 0900-1200 | 50MHz Contest CW | O SF |
| | | VHF CW Championship | |
| 25 Jun | 1400-1600 | 70MHz Cumulatives #4 | O SF |
| 1-2 Jul | 1400-1400* | VHF NFD | Open R L M MS FSO FSR |
| 2 Jul | 1100-1500 | 3rd 144MHz Backpackers | 3B 10H |
| 16 Jul | 1000-1600 | 70MHz Trophy Contest | O SO SF |
| | | VHF Championship | |
| 5 Aug | 1400-2000 | 144MHz Low Power Contest | O SF SO |
| | | VHF Championship | |
| 5 Aug | 1300-1700 | 4th 144MHz Backpackers | 3B 10H |
| 6 Aug | 0800-1200 | 432MHz Low Power Contest | O SF SO |
| | | VHF Championship | |
| 13 Aug | 1400-1600 | 70MHz Cumulatives #5 | O SF |
| 2-3Sep | 1400-1400 | 144MHz Trophy Contest | O SF SO 6O 6S |
| | | VHF Championship | |
| 3 Sep | 1100-1500 | 5th 144MHz Backpackers | 3B 10H |
| 17 Sep | 0900-1200 | Second 70MHz Contest | O SF |
| 7 Oct | 1400-2200 | 1.2GHz Trophy / 2.3GHz Trophy | O SF |
| | | VHF Championship | |
| 7-8 Oct | 1400-1400 | Oct 432MHz-248GHz Contest | O SF |
| 15 Oct | 0900-1300 | 50MHz AFS Contest | O SF |
| | | AFS Super League | |
| 4-5 Nov | 1400-1400 | 144MHz CW Marconi | SF O 6S 6O |
| | | VHF CW Championship | |
| 3 Dec | 1000-1600 | 144MHz AFS | SF O |
| | | AFS Super League | |
| 26-29 Dec | 1400-1600 | 50/70/144/432MHz Christmas | |
| | | Cumulatives Contest | SF O |
| | | 144MHz Backpackers Championship | |
| | | AFS Super League 2014-15 | |
| | | AFS Super League 2015-16 | |
| | | VHF CW Championship | |
| | | UKAC Overall Results | AO AR AL |
| | | VHF Championship | SF O LPt |

*Times marked with a * denote times vary per band L = Local*

### HF Key to Special Rules

| | |
|---|---|
| HF Championship | Special Rules for HF Championship |
| S1 | Affiliated Societies contest |
| S2 | Commonwealth Contest |
| AFS Super League 2014-15 | Special Rules for AFS Super League 2014-15 |
| AFS Super League 2015-16 | Special Rules for AFS Super League 2015-16 |

### HF Key to Multipliers

| | |
|---|---|
| M1 | DXCCs worked on each band |
| M2 | UK Districts per band and mode |
| M3 | DXCC & UK District Bonus |
| M4 | IOTA Points |

# Beacons

Beacons are intended mainly as propagation indicators although, especially on the microwave bands, they may also serve as signal sources for alignment purposes. The table lists a selection, some of which may be heard regularly in the UK, so that variation in strength gives an indication of conditions. Others may be heard occasionally, and the appearance of one can indicate exceptional propagation.

For example, the 144MHz beacon GB3VHF can always be heard over much of the UK so, if its strength is above average, then there is a 'lift' on. If the 50MHz beacon in Newfoundland, not usually audible in the UK, appears then there is a path to North America. Conversely, the 28MHz beacon GB3RAL is of little interest to UK stations but can indicate to overseas operators the presence of propagation to the UK.

## Co-ordination

On HF, beacons are co-ordinated by the IARU. At 28MHz, 28.190-28.199 is reserved for regional networks, 28.200 is shared by the International Beacon Project beacons and 28.201-28.225 is allocated to approved continuous cycle beacons.

## Setting up a beacon

The UK licence permits a private station to operate as a low-power unattended beacon, but only in some bands above 2.3GHz, together with 70MHz and part of 432MHz.

Establishing a permanent and reliable beacon at a remote site can be a complex undertaking, maybe more suited to a group than an individual. Site clearance by Ofcom is required before a licence (which will be GB3 + three letters) can be issued. Full details are given in *Guide to Beacon Licensing* available from RSGB on receipt of a large SASE.

## HF Beacons

| Freq | Call | Nearest Town | Locator | ERPw | Antenna | Direction | Mode | Status |
|---|---|---|---|---|---|---|---|---|
| IARU Region 1 discourages beacon operation on 1.8 MHz | | | | | | | | |
| 1.815.00 | EW1OZ | Minsk | | 0.025 | | | | ? |
| 1.836.20 | SK2AU | Skelleftea | KP04LQ | 0.4 | LW | | A1 | ? |
| 1.837.00 | IW3FZQ | Monselice PD | JN55VF | 8 | Dipole | | | TEST |
| 1.840.00 | OK0EK | Kromeriz | JN89QG | 4 | Vertical | Omni | A1 | T NonOp |
| 1.853.00 | OK0EV | Near Prague | JN79EV | 0.1 | 25m Vert | Omni | A1 | PT |
| 1.875.00 | DL3KR | | JO63LV | 5 | Dipole | | A1 | |
| IARU Region 1 discourages beacon operation on 3.5MHz | | | | | | | | |
| 3.525.00 | PY2DSB | GH28XL | | 5 | Vertical | | A1 | 24 |
| 3.548.67 | ER1AAZ | Chisinau | | 4 | Dipole | | | ? |
| 3.576.80 | IZ3DVW | Monselice | JN55VF | 0.5 | Inv V | | A1 | IRREG |
| 3.578.70 | OK2PYA | | IN89TI | 1mw | | | | ? |
| 3.579.00 | DK0WCY | Scheggerott | JO44VQ | 30 | Dipole | | A1 | PT/zz |
| 3.579.80 | SM2IUF | Kalix | KP15NU | QRPP | | | | 15-07UT |
| 3.580.30 | OK1IF | | JO70MS | | | | | ? |
| 3.594.50 | OK0EU | Panska Ves | JO70GM | 100mw | Mag Loop | N-S | A1 | 24qq |
| 3.600.00 | OK0EN | Nr Kladno | JO70AC | 150mw | LW 41m | SE-NW | A1 | 24 |
| 3.694.00 | VK6SH | E.Carrington WA | OF77XX | 3 | Horiz Loop | | A1 | 10-22UTC |
| 5.195.00 | DRA5 | Scheggerott | JO44VQ | 30 | dipole | | A1 | 06-24 LT |
| | | | | | | | psk.rtty | 24 |
| 5.250.00 | ZS6KTS | Johannesburg | KG43CW | 15 | Inv. Vee | | psk.WSPR | 24 |
| 5.289.50 | OV1BCN | 10k S Soroe | JO55SI | 30 | 32m F.Dip@1m | USB/MT63 | H+4.19.34.49 | |
| 5.290.50 | OV1BCN | " | " | " | | | A1 | 24 |
| 5.290.00 | GB3RAL | Nr Didcot | IO91IN | 10<158uw | Inv. Vee | | A1+psk | zzz |
| 5.290.00 | GB3WES | Cumbria | IO84QN | 10<158uw | Inv. Vee | | A1+psk | zzz |
| 5.290.00 | GB3ORK | Orkney | IO89JA | 10<158uw | Inv. Vee | | A1+psk | zzz |
| 5.291.00 | HB9AW | Sursee | JN43BA | 10/5/1/.1/.001 | 1/2Dip | | A1+psk | 5m cycle |
| 5.398.50 | SZ1SV | | KM17UX | 30/15/3 | | | A1.psk31 | H+0+15+30+45 |
| IARU Region 1 discourages beacon operation on 7MHz | | | | | | | | |
| 7.020.00 | LW6HBU | Cordoba | | | Dipole | N-S | A1 | ? |
| 7023} | ZS6SRL | Johannesburg | KG33WV | 0.4 | | | | IRREG |
| 7023} | ZS6YI | Johannesburg | KG33XH | 0.4 | | | | IRREG |
| 7023} | ZS4BFN | Bloemfontein | KG30DV | 0.4 | | | | IRREG |
| 7023} | ZS1AFU | Simonstownj | KF07HN | 0.4 | | | | IRREG |
| 7023} | ZS2LAW | Grahamstown | KF36GQ | 0.4 | | | | IRREG |
| 7023} | ZS1HMO | Somerset West | JF95PN | 0.4 | | | | IRREG |
| 7.025.00 | ZS1AGI | George Airport | KF16EA | 0.2 | 1/2 Dipole | E-W | A1 | OP? |
| 7.026.00 | PP8AA | | | | | | ? | ? |
| 7.030.00 | PY5ZW | Medianeira PR | GG24WQ | | | | A1 | ? |
| 7.032.00 | PU2NJL | Guarulhos SP | GG66RM | | | | | ? |
| 7.033.00 | PY2LL | Guarulhos SP | GH40JU | 4 | | | A1 | ? |
| [illegible] | PY2RFF | San Pedro SP | | 3 | | | A1 | IRREG |
| 7034V | PT9BCN | | GG29RN | 12 | | | A1 | 24? |
| 7.035.00 | VA3NDO | Toronto ON | FN03HR | qrp | | | | ? |
| 7.035.00 | PS8RF | Teresina PI | GI84OV | | | | A1 | ? |
| 7.036.40 | IK0UWF | Sardinia | | | | | | ? |
| 7037.5V | SK7OB | Saxtorp | JO65LU | | | | A1 | TEST? |
| 7.038.50 | OK0EU | Panska Ves | JO70GM | 1 | Mag Loop | N-S | A1 | OP? |
| 7.039.10 | SK7CQ | | | | | | | TEST |
| 7.039.20 | IK1HGI | Cerano | | 0.1 | | | | QRSS |
| 7.039.40 | OK0EPB | Prague | JO70EC | 10 | Dipole@22m | | A1 | 24 |
| 7.039.60 | IZ3DVW | Monselice Padua | JN55VF | 0.5 | Inv V | | A1 | 24 |
| 7.040.00 | HC1AKP | | | | | | | ? |
| 7.040.00 | CE3DNP | | | | | | | ? |
| 7.047.50 | YD0MWK | | OI33MQ | 5 | Inv Vee@15m | A1 | Planned | |
| 7.048.00 | ZU6DD | | | | | | | |
| 7.050.00 | HG4FC | | JN97EC | 0.4 | | | | |
| 7.082.65 | VHK3QQ | Bogota | FJ24XR | 10 | Dipole | | A1 | PT |
| IARU Region 1 discourages beacon operation on 10MHz (DK0WCY excepted) | | | | | | | | |
| 10.123.00 | HP1AVS | Cerro Jefe | FJ09HD | 2.5 | Slope Dip | Omni | A1 | 24 |

| Freq | Call | Nearest Town | Locator | ERPw | Antenna | Direction | Mode | Status |
|---|---|---|---|---|---|---|---|---|
| 10.129.50 | W0ERE | HighlandsvilleMO | EM36HX | 3 | G5RV | E-W | A1 | INT |
| 10.130.00 | OK1IF | Liberec | JO70HG | 0.5 | | | A1 | ? |
| 10.133.00 | SK6RUD | Oxaback | JO67KI | 0.5 | 1/4 GP | Omni | A1 | 24 |
| 10.134.00 | OK0EF | Nr Kladno | JO70BC | .1/.2/.5 1/2 Vert | Omni | A1 | 24 | |
| 10.136.70 | IQ2UL | Sondrio | JN46WE | 150mw | | | | ? |
| 10.137.20 | IK3NWX | Nr MonselicePD | JN55VB | 4.2 | Rot. Dip. | E-W | A1 | 24 |
| 10.138.70 | WSPR beacons here | | | | | | | |
| 10139V | IZ0NHW | Nea Smirni | JN61VL | 0.2 | | | | ? |
| 10.139.40 | YO8BRX | | KN37FW | 10mw | | | | |
| 10.139.60 | PY3PSI | Porto Alegre | GF49KX | 1.6 | Hor. Dip | N-S | A1 | IRREG |
| 10.139.70 | SV8GXC | | KM17UW | 0.08 | | | | ? |
| 10.140.00 | WSPR beacons around here | | | | | | | |
| 10.140.10 | IK0IXI | Civitavecchia | | 45mw | Dipole | | | ? |
| 10.140.07 | IQ2DP | San Donate MI | JN45PJ | 0.4 | Vertical | Omni | | 24 |
| 10.140.60 | DL5KZ | Numbrecht | JO30SU | 0.1 | Dipole | | A1 | ? |
| 10.142.51 | IK1HGI | Tricati | JN45IK | 0.1 | Dipole | | QRSS3 | 24 |
| 10.144.00 | DK0WCY | Scheggerott | JO44VQ | 30 | Dipole | | A1.psk. rtty | 24zz |
| 10.144.60 | YO8RIX | | | 20mw | | | | |
| 10.149.70 | IZ8BZX | Torre del Greco | JN70ES | .1/.5/1 whip | | Omni | QRSS | EXP |
| 10.150.00 | IZ5ILH | Firenze | JN53PS | 2 | | | | ? |
| 14.020.00 | I2QIL | | JN55CN | 70mw | | | | ? |
| 14.062.00 | UA1AVA | St Petersburg | KO59EW | 0.1 | 1/2 Dip | | A1 | 06-15UTC |
| 14.095.00 | RN3RHO | | | 0.5 | | | | ? |
| 14.097.00 | WSPR beacons around here | | | | | | | |
| 14.098.00 | IW3ICH | S.Martino di Venezze | JN55WD | | | | | ? |
| 14.099.00 | IZ0HCC | Rome | JN61FT | | | | | ? |
| 14.100.00 | 4U1UN | UN NY | FN30AS | 100-0.1 | Vertical | Omni | A1 | TNonOp |
| 14.100.00 | VE8AT | Eureka.Nunavut | EQ79AX | 100-0.1 | Vertical | Omni | A1 | IBP cycle |
| 14.100.00 | W6WX | Mt Umunhum CA | CM97BD | 100-0.1 | Vertical | Omni | A1 | IBP cycle |
| 14.100.00 | KH6WO | Laie. Oahu | BL11AP | 100-0.1 | Vertical | Omni | A1 | IBP cycle |
| 14.100.00 | ZL6B | Nr Masterton | RE78TW | 100-0.1 | Vertical | Omni | A1 | IBP cycle |
| 14.100.00 | VK6RBP | Rolystone | OF87AV | 100-0.1 | Vertical | Omni | A1 | IBP cycle |
| 14.100.00 | JA2IGY | Mt Asama | PM84JK | 100-0.1 | Vertical | Omni | A1 | IBP cycle |
| 14.100.00 | RR9O | Novosibirsk | NO14KX | 100-0.1 | Vertical | Omni | A1 | IBP cycle |
| 14.100.00 | VR2B | Hong Kong | OL72CQ | 100-0.1 | Vertical | Omni | A1 | IBP cycle |
| 14.100.00 | 4S7B | Colombo | NJ06CR | 100-0.1 | Vertical | Omni | A1 | NonOp |
| 14.100.00 | ZS6DN | Pretoria | KG44DC | 100-0.1 | Vertical | Omni | A1 | IBP cycle |
| 14.100.00 | 5Z4B | Kiambu | KI88MX | 100-0.1 | Vertical | Omni | A1 | IBP cycle |
| 14.100.00 | 4X6TU | Tel Aviv | KM72JB | 100-0.1 | Vertical | Omni | A1 | IBP cycle |
| 14.100.00 | OH2B | Lohja | KP20BM | 100-0.1 | Vertical | Omni | A1 | IBP cycle |
| 14.100.00 | CS3B | Santo da Serra | IM12OR | 100-0.1 | Vertical | Omni | A1 | IBP cycle |
| 14.100.00 | LU4AA | Buenos Aires | GF05TJ | 100-0.1 | Vertical | Omni | A1 | IBP cycle |
| 14.100.00 | OA4B | Lima | FH17MW | 100-0.1 | Vertical | Omni | A1 | IBP cycle |
| 14.100.00 | YV5B | Caracas | FK06NK | 100-0.1 | Vertical | Omni | A1 | IBP cycle |
| 14.101.00 | R1ANF | S.Shetland ANT | GC07 | 100-0.1 | | | | ? |
| 14.101.10 | RI1AND | Mirny Base ANT | | | | | | ? |
| 14.101.00 | IN3UFW | | JN56NF | | | | | ? |
| 14.103.00 | IV3VOU | Verzegnis UD | JN66LJ | | | | | ? |
| 14.161.00 | I1YRB | Torre Bert | JN35UB | 0.2 | Vertical | Omni | QRSS3 | 24 |
| 18.095.50 | HP1AVS | Cerro Jefe | FJ09HD | 2.5 | Inv Vee | | A1 | 24 |
| 18.098.00 | SM7ZFB | Loddekpinge | JO65LS | 0.1 | | | | ? |
| 18.100.00 | IK6BAK | Montefelcino | JN63KR | 1 | Inv Vee | Omni | A1 | 24 |
| 18.100.00 | 9H1LO | Hosta. Malta | JM75FV | 0.5 | G5RV | | psk | 24 |
| 18.101.00 | VE3RAT | Thornhill ONT | FN03GL | 1 | Vertical | Omni | A1 | 24 |
| 18.102.00 | I1M | Bordighera | JN33UT | 10 | 5/8 Vert | Omni | A1 | 24 |
| 18.104.60 | Many WSPR beacons around here | | | | | | | |
| 18.109.25 | IQ3VO | Verona | JN55LL | 5 | GP | Omni | A1 | 24 |
| 18.109.80 | LU6YCB | Neuquen | FF51WB | 2 | Dipole | | A1 | 24 |
| 18.110.00 | 4U1UN | U. Nations NY | FN30AS | 100-0.1 | Vertical | Omni | A1 | TNonOp |
| 18.110.00 | VE8AT | Eureka. Nunavut | EQ79AX | 100-0.1 | Vertical | Omni | A1 | IBP cycle |
| 18.110.00 | W6WX | Mt Umunhum CA | CM97BD | 100-0.1 | Vertical | Omni | A1 | IBP cycle |
| 18.110.00 | KH6WO | Laie. Oahu | BL11AP | 100-0.1 | Vertical | Omni | A1 | IBP cycle |
| 18.110.00 | ZL6B | Nr Masterton | RE78TW | 100-0.1 | Vertical | Omni | A1 | IBP cycle |
| 18.110.00 | VK6RBP | 28k SE Perth | OF87AV | 100-0.1 | Vertical | Omni | A1 | IBP cycle |
| 18.110.00 | JA2IGY | Mt Asama | PM84JK | 100-0.1 | Vertical | Omni | A1 | IBP cycle |
| 18.110.00 | RR9O | Novosibirsk | NO14KX | 100-0.1 | Vertical | Omni | A1 | IBP cycle |
| 18.110.00 | VR2B | Hong Kong | OL72CQ | 100-0.1 | Vertical | Omni | A1 | IBP cycle |
| 18.110.00 | 4S7B | Colombo | NJ06CC | 100-0.1 | Vertical | Omni | A1 | NonOp |
| 18.110.00 | ZS6DN | Pretoria | KG44DC | 100-0.1 | Vertical | Omni | A1 | IBP cycle |
| 18.110.00 | 5Z4B | Kiambu | KI88MX | 100-0.1 | Vertical | Omni | A1 | IBP cycle |
| 18.110.00 | 4X6TU | Tel Aviv | KM72JB | 100-0.1 | Vertical | Omni | A1 | IBP cycle |
| 18.110.00 | OH2B | Lohja | KP20 | 100-0.1 | Vertical | Omni | A1 | IBP cycle |
| 18.110.00 | CS3B | Santo da Serra | IM12OR | 100-0.1 | Vertical | Omni | A1 | IBP cycle |
| 18.110.00 | LU4AA | Buenos Aires | GF05TJ | 100-0.1 | Vertical | Omni | A1 | IBP cycle |
| 18.110.00 | OA4B | Lima | FH17MW | 100-0.1 | Vertical | Omni | A1 | IBP cycle |
| 18.110.00 | YV5B | Caracas | FK60NK | 100-0.1 | Vertical | Omni | A1 | IBP cycle |
| 18.112.40 | IQ0LT | Latina LT | JN61KL | | | | | ? |
| 18.139.90 | EA1GIB | | | | | | | ? |
| 18.140.10 | 9H1LO | Malta | JM75GE | 2 | Horiz Dip | | psk31 | OP? |
| 21.052.30 | DL5KZ | Numbrecht | JO30SU | 0.25 | 3-el Yagi | 180 | A1 | EXP 24 |
| 21.068.00 | SK7CQ | | | | H.Dipole | | | EXP |
| 21.068.00 | SK7OB | | | | V.Dipole | | | EXP |
| 21.094.60 | WSPR beacons around here | | | | | | | |
| 21.133.50 | LU1DWE | La Plata BA | | 4 | 1/2 Dip | | A1 | 24 |
| 21.145.70 | IZ3DVW | Nr Monselice PD | JN55VF | 2.6 | Inv. V Dip | | A1 | 24 |
| 21.149.00 | F5ZHL | | JO10SI | | | | | ? |
| 21.149.50 | IW9HMQ | Nr Catania | JM77MN | | | | | wkend |
| 21.150.00 | 4U1UN | UN New York | FN30AS | 100-0.1 | Vertical | Omni | A1 | TNonOp |
| 21.150.00 | VE8AT | Eureka. Nunavut | EQ79AX | 100-0.1 | Vertical | Omni | A1 | IBP cycle |
| 21.150.00 | W6WX | Mt Umunhum | CM97BD | 100-0.1 | Vertical | Omni | A1 | IBP cycle |
| 21.150.00 | KH6WO | Laie. Oahu HI | BL11AP | 100-0.1 | Vertical | Omni | A1 | IBP cycle |
| 21.150.00 | ZL6B | Masterton | RE78TW | 100-0.1 | Vertical | Omni | A1 | IBP cycle |
| 21.150.00 | VK6RBP | 28k SE Perth | OF87AV | 100-0.1 | Vertical | Omni | A1 | IBP cycle |
| 21.150.00 | JA2IGY | Mt Asama | PM84JK | 100-0.1 | Vertical | Omni | A1 | IBP cycle |

Please notify errors/changes to: **Martin Harrison, G3USF**, HF Beacon Coordinator, Region 1 of the IARU, 1 Church Fields, Keele, Staffs ST5 5HP, England. Tel (home): **+44 (0)1782 627396** Fax (work): **+44 (0)1782 58359**2 Email:hf.beacons@rsgb.org.uk

| Freq | Call | Nearest Town | Locator | ERPw | Antenna | Direction | Mode | Status |
|---|---|---|---|---|---|---|---|---|
| 21.150.00 | RR9O | Novosibirsk | NO14KX | 100-0.1 | Vertical | Omni | A1 | IBP cycle |
| 21.150.00 | VR2B | Hong Kong | OL72CQ | 100-0.1 | Vertical | Omni | A1 | IBP cycle |
| 21.150.00 | 4S7B | Colombo | NJ06CC | 100-0.1 | Vertical | Omni | A1 | NonOp |
| 21.150.00 | ZS6DN | Pretoria | KG44DC | 100-0.1 | Vertical | Omni | A1 | IBP cycle |
| 21.150.00 | 5Z4B | Kiambu | KI88MX | 100-0.1 | Vertical | Omni | A1 | IBP cycle |
| 21.150.00 | 4X6TU | Tel Aviv | KM72JB | 100-0.1 | Vertical | Omni | A1 | IBP cycle |
| 21.150.00 | OH2B | Lohja | KP20 | 100-0.1 | Vertical | Omni | A1 | IBP cycle |
| 21.150.00 | CS3B | Santo da Serra | IM12OR | 100 0.1 | Vertical | Omni | A1 | IBP cycle |
| 21.150.00 | LU4AA | Buenos Aires | GF05TJ | 100 0.1 | Vertical | Omni | A1 | IBP cycle |
| 21.150.00 | OA4B | Lima | FH17MW | 100 0.1 | Vertical | Omni | A1 | IBP cycle |
| 21.150.00 | YV5B | Caracas | FK60NK | 100 0.1 | Vertical | Omni | A1 | IBP cycle |
| 21.151.00 | I1M | Bordighera | JN33UT | 10 | 2 5/8 Vert | Omni | A1 | 24 |
| 21.155.50 | IQ0LT | Latina LT | JN61KL | | | | | ? |
| 21.171.00 | KA5FYI | Austin TX | EM10 | 1.25 | 1/4 Vert | Omni | A1 | INT |
| 21.241.50 | I1YRB | Torre Bert (TO) | JN35UB | 0.2 | Vertical | Omni | QRSS3 | 24 |
| 21.265.00 | XE1FAS | Puebla PU | EK09UB | | | | A1 | ? |
| 21.393.80 | PY3PSI | Porto Alegre | GF49KX | 4 | Dipole | N-S | A1 | INT |
| 24.912.00 | SK6RUD | Oxaback | JO67KT | 0.5 | GP | Omni | A1 | 24 |
| 24.915.00 | IQ6FU | Fano PU | JN63MU | 5 | Inv Vee | Omni | A1 | 24 |
| 24.920.00 | IY4M | Bologna | JN54QK | | | | A1 | H+30>H+60m |
| 24.929.50 | XE1FAS | Puebla PU | EK09UB | | | | A1 | ? |
| 24.930.00 | 4U1UN | UN NY | FN30AS | 100-0.1 | Vertical | Omni | A1 | TNonOp |
| 24.930.00 | VE8AT | Eureka. Nunavut | EQ79AX | 100-0.1 | Vertical | Omni | A1 | IBP cycle |
| 24.930.00 | W6WX | Mt Umunhum CA | CM97BD | 100-0.1 | Vertical | Omni | A1 | IBP cycle |
| 24.930.00 | KH6WO | Laie. Oahu HI | BL11AP | 100-0.1 | Vertical | Omni | A1 | IBP cycle |
| 24.930.00 | ZL6B | Nr Masterton | RE78TW | 100-0.1 | Vertical | Omni | A1 | IBP cycle |
| 24.930.00 | VK6RBP | 28k SE Perth | OF87AV | 100-0.1 | Vertical | Omni | A1 | IBP cycle |
| 24.930.00 | JA2IGY | Mt Asama | PM84JK | 100-0.1 | Vertical | Omni | A1 | IBP cycle |
| 24.930.00 | RR9O | Novosibirsk | NO14KX | 100-0.1 | Vertical | Omni | A1 | IBP cycle |
| 24.930.00 | VR2B | Hong Kong | OL72CQ | 100-0.1 | Vertical | Omni | A1 | IBP cycle |
| 24.930.00 | 4S7B | Colombo | NJ06CC | 100-0.1 | Vertical | Omni | A1 | NonOp |
| 24.930.00 | ZS6DN | Pretoria | KG44DC | 100-0.1 | Vertical | Omni | A1 | IBP cycle |
| 24.930.00 | 5Z4B | Kiambu | KI88MX | 100-0.1 | Vertical | Omni | A1 | IBP cycle |
| 24.930.00 | 4X6TU | Tel Aviv | KM72JB | 100-0.1 | Vertical | Omni | A1 | IBP cycle |
| 24.930.00 | OH2B | Lohja | KP20 | 100-0.1 | Vertical | Omni | A1 | IBP cycle |
| 24.930.00 | CS3B | Santo da Serra | IM12OR | 100-0.1 | Vertical | Omni | A1 | IBP cycle |
| 24.930.00 | LU4AA | Buenos Aires | GF05TJ | 100-0.1 | Vertical | Omni | A1 | IBP cycle |
| 24.930.00 | OA4B | Lima | FH17MW | 100-0.1 | Vertical | Omni | A1 | IBP cycle |
| 24.930.00 | YV5B | Caracas | FK06NK | 100-0.1 | Vertical | Omni | A1 | IBP cycle |
| 24.931.00 | 7Z1CQ | Jeddah | KL91ON | 5 | Vert.Dip | Omni | A1 | 24 |
| 24.986.00 | JE7YNQ 0 | Fukushima | QM07 | 5 | Dipole | | A1 | QRT? |
| 24.990.00 | I1YRB | Torre Bert(TO) | JN35UB | 0.2 | Vertical | Omni | QRSS3 | 24 |
| 28.164.00 | VE3DJI | Burlington ON | FN03CI | | | | A1 | 24? |
| 28.166.00 | XE2O | Monterey NL | DL95UR | 5 | 1/4 Vert | Omni | A1 | 24 |
| 28.169.00 | ZB2TEN | Gibraltar | | 4 | 1/4 Vert | Omni | A1 | 24 |
| 28.171.00 | XE1FAS | Publa PU | EK09UB | 12 | Dipole | | A1 | 24 |
| 28.173.00 | IZ1EPM | 27k NE Turin | JN35WD | 20 | 1/2 Vert | Omni | A1 | 24 |
| 28.175.00 | VE3TEN | Ottawa ON | FN25 | 10 | GP | Omni | A1 | 24 |
| 28.176.90 | HP1RCP | Cerro Jefe | FJ09HD | 5 | Slope Dip | | A1 | 24 |
| 28.180.30 | I1M | Bordighera | JN33UT | May-20 | 2x5/8 Vert | Omni | A1 | 30/60min |
| 28.182.40 | SV3AQR | Amalias | KM07QS | 4 | GP | Omni | A1 | 24 |
| 28.183.20 | XE1RCS | Cerro Gordo | EK09OS | 8 | AR10 | Omni | A1 | 24 |
| 28.184.00 | VE2REA | Quebec QC | FN46IT | | | | A1 | 24 |
| 28.185.00 | VA3SRC | Burlington ON | FN03BH | 5 | Dipole | | A1 | PT |
| 28.187.60 | VE7KC | Penticton BC | DN09EL | | | | A1 | 24 |
| 28.188.00 | OE3XAC | Kaiserkogel | JN78SB | 20 | 7/8 GP@750m | Omni | A1 | 24 |
| 28.188.10 | JE7YNQ | Fukushima | QM07 | | | | | 24 |
| 28.188.90 | SV5TEN | Raad | KM46CK | 5 | Vertical | Omni | | 24 |
| 28.189.50 | LU2DT | Mar del Plata | GF12 | 5 | Vert. Dip. | Omni | A1/psk | |
| 28.189.80 | LU8XW | Ushuaia | FD55UE | | | | | ? |
| 28.190.00 | LU3HFA | Cordoba CD | FF78UP | 5 | Vertical | Omni | A1 | 24? |
| 28.190 50 | VA3ROR | Orillia ON | FN04EQ | 5 | Whip | Omni | A1 | 24 |
| 28.191.50 | A62ER | Sharjah UAE | LL75QI | | | | A1 | OP? |
| 28.192.00 | EP4HR | | LL69GP | 20/2/0.2 | Dipole | | A1 | 24 |
| 28.192.90 | VE4ARM | Austin MB | EM09HW | 5 | GP | Omni | A1 | 24 |
| 28.192.00 | LU8EML | Ensenada BA | GF18AD | | | | A1 | 24 |
| 28.193.50 | A47RB | Oman | LL93FO | 10 | Vertical | Omni | A1 | OP? |
| 28.193.50 | LU2XPK | | FF66DE | | | | A1 | ? |
| 28.195.10 | IY4M | Bologna | JN54QK | 20 | 5/8 GP | Omni | A1 | H>H+30m |
| 28.196.00 | VA3ITA | Bramton ON | FN03CW | | | | A1 | IRREG |
| 28.196.10 | LU4JJ | Concordia ER | GF08XO | | | | A1 | 24 |
| 28.196.70 | LU5FB | Rosario SF | FF97PB | | | | A1 | 24 |
| 28.197.00 | VE7MTY | Vancouver BC | CN09 | 5 | Vertical | Omni | A1 | 24 |
| 28.198.00 | LU2ERC | Ensenada | GF15AD | 10 | Vertical | Omni | A1 | 24 |
| 28.198.00 | LU9FE | | FF98GR | | | | | ? |
| 28.199.30 | LU1FHH | El Trebol SF | | | | | A1 | 24 |
| 28.200.00 | 4U1UN | UN New York | FN30AS | 100-0.1 | Vertical | Omni | A1 | TNonOp |
| 28.200.00 | VE8AT | Eureka.Nunavut | EQ79AX | 100-0.1 | Vertical | Omni | A1 | IBP cycle |
| 28.200.00 | W6WX | Mt Umunhum CA | CM97BD | 100-0.1 | Vertical | Omni | A1 | IBP cycle |
| 28.200.00 | KH6WO | Laie. Oahu HI | BL11AP | 100-0.1 | Vertical | Omni | A1 | IBP cycle |
| 28.200.00 | ZL6B | Nr Masterton | RE78TW | 100-0.1 | Vertical | Omni | A1 | IBP cycle |
| 28.200.00 | VK6RBP | 28k SE Perth | OF87AV | 100-0.1 | Vertical | Omni | A1 | IBP cycle |
| 28.200.00 | JA2IGY | Mt Asama | PM84JK | 100-0.1 | Vertical | Omni | A1 | IBP cycle |
| 28.200.00 | RR9O | Novosibirsk | NO14KX | 100-0.1 | Vertical | Omni | A1 | IBP cycle |
| 28.200.00 | VR2B | Hong Kong | OL72CQ | 100-0.1 | Vertical | Omni | A1 | IBP cycle |
| 28.200.00 | 4S7B | Colombo | NJ06CC | 100-0.1 | Vertical | Omni | A1 | NonOP |
| 28.200.00 | ZS6DN | Pretoria | KG44DC | 100-0.1 | Vertical | Omni | A1 | IBP cycle |
| 28.200.00 | 5Z4B | Kiambu | KI88MX | 100-0.1 | Vertical | Omni | A1 | IBP cycle |
| 28.200.00 | 4X6TU | Tel Aviv | KM72JB | 100-0.1 | Vertical | Omni | A1 | IBP cycle |
| 28.200.00 | OH2B | Lohja | KP20 | 100-0.1 | Vertical | Omni | A1 | IBP cycle |
| 28.200.00 | CS3B | Santo da Serra | IM12OR | 100-0.1 | Vertical | Omni | A1 | IBP cycle |
| 28.200.00 | LU4AA | Buenos Aires | GF05TJ | 100-0.1 | Vertical | Omni | A1 | IBP cycle |
| 28.200.00 | OA4B | Lima | FH17MW | 100-0.1 | Vertical | Omni | A1 | IBP cycle |
| 28.200.00 | YV5B | Caracas | FK60NK | 100-0.1 | Vertical | Omni | A1 | IBP cycle |

For the latest HF beacon listing, see: *www.keele.ac.uk/depts/por/28.htm*

| Freq | Call | Nearest Town | Locator | ERPw | Antenna | Direction | Mode | Status |
|---|---|---|---|---|---|---|---|---|
| 28.200.50 | VA3GMT | Toronto | | | | | A1 | ? |
| 28.200.80 | AC7AV | Oak Harbor WA | | | | | A1 | ? |
| 28.201.30 | PU2SUT | Sao Paulo | GG66TB | 20 | Dipole | | A1 | ? |
| 28.202.00 | WN2WNC | New Berlin NY | FN22IO | | | | A1 | |
| 28.202.50 | KA3BWP | Stafford VA | FM18GK | | | | A1 | 24 |
| 28.203.00 | PY2WFG | Ipisanga SP | GG77FF | | | | A1 | 24 |
| 28.203.00 | KB1QZY | Springfield MA | FN32QC | 2 | Imax2000 V | Omni | A1 | 24 |
| 28.203.00 | KG8CO | Clinton MI | EN82AB | 5 | Vertical | Omni | A1 | 24 |
| 28.203.00 | N6DXX | Sacramento CA | CM98FM | | | | A1 | ? |
| 28.203.50 | K6LLL | MissionViejo CA | DM13EO | | | | A1 | 24 |
| 28.204.00 | WA2NTK | Big Flats NY | FN12NE | | | E-W | A1 | 24 |
| 28.204V | WL7N | Ward Cove AK | CO45KK | | | | A1 | ? |
| 28.204.00 | W6CF | San Francisco | CM87UU | | | | A1 | 24 |
| 28.204.00 | KA1KNW | Windsor CT | FN31RU | 10 | | | A1 | ? |
| 28.205.00 | DL0IGI | Hohenpeissenb'g | JN57MT | Varies | 1/4 Vert | Omni | A1 | 24 |
| 28.205.20 | VN3NIA | Nr Ridgway PA | FN01PK | 4 | Dipole | | A1 | 24 |
| 28.205.90 | HS0BBD | Bangkok | OK13 | | | | | ? |
| 28.206.10 | VA3GRR | Brampton ON | FN03 | 1.75 | 1/2 Vert | Omni | A1 | 24 |
| 28.206.30 | K9EJ | Toledo | EM59UG | 2 | Vert @4m | Omni | A1 | 24 |
| 28.206.50 | HP1RIS | Panama City | FJ09GA | 3 | Vert Dip | Omni | A1 | 24 |
| 28.207.00 | ON0RY | Binche | JO20CK | 5 | Vertical | Omni | A1 | 24 |
| 28.207.30 | KW7HR | Pasco WA | DN06KG | | | | A1 | 24? |
| 28.207.80 | W4CND | Jemison AL | EM63QA | 2 | Vertical | Omni | A1 | 24 |
| 28.208.00 | KE6TE | Elk Grove CA | CM98HK | | | | A1 | OP? |
| 28.208.00 | AK2F | Randolph NJ | FN20QT | | | | A1 | share |
| 28.208.00 | WN2A | Budd Lake NJ | FN20OU | | | | A1 | AK2F |
| 28.208.10 | JR0YAN | Nr Toyama | PM86JW | 25 | Hor.Loop | | A1 | 24+++ |
| 28.208.20 | IZ3LCJ | S.Lucia di Piave | JN65DU | 5 | 1/4GP@15m | Omni | A1 | QRT? |
| 28.208.50 | NB7A | Reno NV | | | | | A1 | 24 |
| 28.208.70 | N8PVL | Livonia MI | EN82GJ | | | | A1 | 24? |
| 28.209.00 | KH6AP | Kikei Maui HI | BL10SS | 20 | 3/8 Vert | Omni | A1 | 24 |
| 28.209.50 | K9CW | Thomasboro IL | EN50WF | 2 | AR99@15' | Omni | A1 | 24 |
| 28.209.80 | KV6Q | Sab Diego | DN12JS | 3 | | | A1 | ? |
| 28.210.00 | PT2SSB | Brasilia | GH64CI | | | | A1 | ? |
| 28.210.00 | KB9UGA | Egg Harbor WI | | | | | A1 | 24 |
| 28.210.40 | NT4F | Wilmington NC | FM14AE | 5 | | | A1 | 24 |
| 28.210.50 | VE4TEN | Kelowna BC | | 2 | 1/2 Vert | Omni | A1 | 24 |
| 28.211.00 | DB0FKS | Nr Frankfurt/M J | N40IT | 0.2 | DV27 Vert | Omni | A1 | OP? |
| 28.211.00 | K5ARC | Galvez LA | EM40 | 20 | Vertical | Omni | A1 | 24? |
| 28.211.00 | CE1TUW | Antofagasta | FF46RQ | | | | A1 | ? |
| 28.211.10 | LA4TEN | Nr Hellvik | JO28WL | 250 | Vertical | Omni | A1 | 24 |
| 28.211.80 | AC7GZ | Chandler AZ | DM43 | | | | A1 | ? |
| 28.212.00 | 7Z1AL | Dammam | LL56BK | 10 | Vertical | | A1 | OP? |
| 28.212.50 | K0KP | Fredenberg MN | EN36VW | 0.5 | GP | Omni | A1 | OP? |
| 28.212.50 | KJ4QYB | RainbowCity AL | EM63WO | | | | A1 | ? |
| 28.212.60 | LU7DQP | Lanos Oeste BA | F05TH | | | | A1 | 24 |
| 28.213.30 | KD8RKJ | Cleveland OH | EN91CK | 2 | Vertical | Omni | A1 | 24 |
| 28.213.50 | KE4KAA | Big Stone Gap VA | EM86OV | 5 | | | A1 | 24 |
| 28.213.50 | W3IK | Gray TN | EM86 | | | | A1 | 24 |
| 28.213.80 | KF5KBZ | Austin TX | EM10FB | | | | A1 | ? |
| 28.214.00 | N4PAL | Longwood FL | EL98HQ | 5 | Vert@4.5m | Omni | A1 | 24? |
| 28.214.00 | LA9TEN | Snertingdal | JP50EV | 10 | 5/8 GP | Omni | A1 | 24 |
| 28.214.80 | FR1GZ | Reunion I. | LG79RC | | | | | irreg? |
| 28.215.00 | LU5EGY | Buenos Aires | GF05QI | | | | A1 | 24 |
| 28.215.00 | YV5LIX | | | | | | | |
| 28.215.00 | KA9SZX | Paxton IL | EN50VD | 1 | Antron99 | Omni | A1 | 24 |
| 28.215.00 | W4JPL | Liberty NC | FM05FV | | | | A1 | 24 |
| 28.215.00 | GB3RAL | Nr Didcot | IO91IN | 25 | HorizDip | Omni | F1 | 24 |
| 28.215.30 | XE3D | Merida YUC | EL50EG | | | | A1 | 24 |
| 28.215.50 | KD5CKP | Olive Branch MS | EM54BW | 3 | Vertical | Omni | A1 | 24 |
| 28.215.80 | K6WKX | Santa Cruz CA | CM86XX | 10 | Horiz. Dip | Omni | A1 | 24 |
| 28.216.00 | K3FX | Neptune City NJ | FN20XE | 7 | 1/2 Vert | Omni | A1 | 24 |
| 28.216.00 | N7MA | Cataldo ID | DN17SN | 5 | Vertical | Omni | A1 | 24 |
| 28.217.00 | WB0FTL | Alden MN | EN33FP | 5 | AR10@25' | Omni | A1 | 24 |
| 28.217.00 | W6GY | Star ID | DN13RP | | | | A1 | 24 |
| 28.217.50 | WA1LAD | West Warwick RI | FN41FQ | 4.5 | J-Pole | Omni | A1 | 24 |
| 28.218.00 | IQ5MS | Marina di Massa | JN54AA | | Vertical | Omni | A1 | 24 |
| 28.218.30 | AC0KC | Fort Lupton CO | DN70OA | qrpp | | | A1 | ? |
| 28.218.30 | KD8RKJ | Cleveland OH | | 2 | Vertical | Omni | A1 | ? |
| 28.218.30 | KJ4LAA | Decatur AL | EM64LN | 3 | Vertical | Omni | A1 | ? |
| 28.218.50 | W5RDW | Murphy TX | EM12QX | 5 | AR10@25' | Omni | A1 | 24 |
| 28.218.60 | KA4RRU | Catlett VA | FM18EN | | | | A1 | 24 |
| 28.218.80 | KN8DMK | Amanda OH | EM89OO | 3 | Slope Dip. | NW/SE | A1 | ? |
| 28.219.60 | PY2UEP | | GG58GA | | | | A1 | ? |
| 28.219.90 | KB9DJA | Mooresville IN | EM69RO | 35 | GP | Omni | A1 | 24 |
| 28.220.00 | 5B4CY | Mandria | KM64KU | 15 | GP | Omni | F1 | 24 |
| 28.220.00 | W8VO | Sterling Hts MI | EN82NU | 5 | Vert@50' | Omni | A1 | OP? |
| 28.220.30 | KF5VXU | Batesville AR | | | | | | ? |
| 28.220.40 | WK4DS | Trenton GA | EM74FU | 2 | Dipole | NE-SW | A1 | 24 |
| 28.220.50 | YM7KK | Giresun | KN90IV | 4 | | | A1 | 24 |
| 28.220.50 | N5FUN | Caroltton TX | EM12 | | | | A1 | ? |
| 28.220.80 | SV2MCG | Thessaloniki | KN10FC | | | | A1 | ? |
| 28.221.00 | OZ7IGY | Jystrup | JO55WM | 16 | Halo | Omni | F1a | 24 |
| 28.221.00 | WE4S | Rock Springs CA | | | | | A1 | 24 |
| 28.221.50 | KC0TKS | Sedalia MO | EM38IQ | 5 | J-poleVert Omni | A1 | 24 | |
| 28.221.90 | W1DLO | Calais ME | | | | | A1 | 24 |
| 28.222.00 | TP2CE | Strasbourg | JN38VO | 450mw | GP R-7000 | Omni | A1 | 24 |
| 28.222.00 | K6JCA | Carmel Valley CA | CM96DN | | | | A1 | QRT? |
| 28.222.00 | IZ0KBA | Castel Madama | JN61KX | 4 | GP 350m aslOmni | A1 | 24 | |
| 28.222.80 | N4QDK | Sauratown Mt NC | EM95 | 3 | dipole | | A1 | 24 |
| 28.222.80 | N3PV | Chula Vista CA | | | | | A1 | OP? |
| 28.223.00 | WY5B | Biloxi MS | EM50NK | | | | A1 | 24 |
| 28.223.00 | UR3OMP | | KN77KU | QRPP | | | | OP? |
| 28.223.00 | 9H1LO | Malta | JM75 | 8 | GP | | A1 | OP? |
| 28.223.00 | KP3FT | Ponce PR | FK68 | 4 | Vertical | Omni | A1 | 24 |

| Freq | Call | Nearest Town | Locator | ERPw | Antenna | Direction | Mode | Status |
|---|---|---|---|---|---|---|---|---|
| 28.223.30 | XE3ACB | Hecelchakan | EL40 | | | | A1 | |
| 28.223.60 | PB5A | | JO21DE | | | | A1 | ? |
| 28.224.00 | WD0AKX | Albert Lea MN | EN33 | 5 | Vertical | OMNI | A1 | 24 |
| 28.224.20 | YB9BWN | Denpasar. Bali | OJ13FK | 2 | Dipole | EU/VK | A1 | 24 |
| 28.224.70 | HA5BHA | Nr Budapest | JN97KO | 5 | Omni | | | 24 |
| 28.224.80 | KW7Y | Marysville WA | CN88SD | 4 | Vertical | Omni | A1 | 24 |
| 28.225.00 | YM7TEN | | KN91RB | 1 | Vertical | Omni | A1 | 24 |
| 28.225.00 | K6FRC | Angels Common CA | CM97GP | | Vertical | Omni | A1 | 24 |
| 28.225.00 | K5GJR | CorpusChristi TX | EL17HR | | | | A1 | 24 |
| 28.225.50 | W2DLL | Nr Buffalo NY | FN02PP | 8 | 1/2 Vert. | Omni | A1 | 24 |
| 28.225.60 | WB0LYV | Beatrice NE | EN10 | | V DeltaLoop | | A1 | ? |
| 28.226.00 | ED1YCA | | IN73AL | 10 | 5/b V 3dBd | Omni | A1 | 24 |
| 28.226.00 | PU4CDX | Barao de Cocais | GH80UB | | | | A1 | Irreg |
| 28.226.20 | WA6HXW | WestnCovina CA | DM14AB | | | | A1 | irreg |
| 28.226.50 | N7MSH | North Powder OR | DN15 | | | | A1 | 24 |
| 28.226.60 | KC6WGN | Las Vegas NV | DM26LD | 10 | | Omni | A1 | 24 |
| 28.226.80 | PY2RFF | Sao Pedro SP | GG67AL | 1.5 | 1/4 Vert | Omni | A1 | 24 |
| 28.226.70 | KU4A | Lexngton KY | EM78SB | | | | A1 | ? |
| 28.227.00 | VE9AT | White Head I. | | 0.1 | Dipole | | A1 | OP? |
| 28.227.00 | KJ4HYV | Zellwood FL | | | | | A1 | 24 |
| 28.227.50 | KC5MO | Austin TX | EM10BF | 2 | Dipole | | A1 | 24 |
| 28.227.70 | IW3FZQ | Monselice PD | JN55VF | 5 | J-Pole@20m Omni | A1 | OP? | |
| 28.228.00 | ZL3TEN | Rolliston | RE66 | 10 | 1/2 Vert | Omni | A1 | 24 |
| 28.228.30 | TG9TEN | | EK44 | | | | A1 | OP? |
| 28.228.80 | N3PV | Chula Vista CA | | | | | A1 | |
| 28.229.00 | ZL2MHF | Nr Wellington | RE78NU | 10 | 1/2 Vert | Omni | F1 | 24 |
| 28.229.30 | KA2LIM | Pine Valley NY | FN12NP | | | | A1 | ? |
| 28.229.70 | IQ8CZ | Catanzo | JM88HV | 10 | GP | Omni | A1 | 24 |
| 28.230.00 | OH5SHF | Kouvola | KP30HV | 6 | H Dipole | 20/200 | F1 | 24 |
| 28.230.00 | WA4ZKO | Dry Ridge KY | EM78PP | 4 | A99 | Omni | A1 | 24 |
| 28.230.50 | AA0RQ | Pine CO | | | | | A1 | ? |
| 28230? | KI4AED | Ocoee FL | EL98FN | 5 | Antron 99 | Omni | A1 | 24 |
| 28.230.00 | KQ4TG | Leland NC | FM40XF | | | | A1 | 24 |
| 28.230.50 | HP6RCP | Santiago | EJ98MB | 3 | AR99 | Omni | A1 | IRREG |
| 28.231.00 | F5ZEH | | IN88VA | .5/5/50 | 1/2V. 3-elY | | A1 | 24 |
| 28.231.80 | WA4FC | PrinceGeorge VA | FM17HD | 5 | Ringo@200' | Omni | A1 | 24 |
| 28.232.00 | N1FSX | Simla CO | DM89AC | 5 | Vertical | Omni | A1 | 24 |
| 28.232.60 | N9BPE | Tuscaloosa IL | EM59UT | 2 | 1/2Dip@20' Omni | A1 | 24 | |
| 28.232.70 | N2MH | West Orange NJ | FN20UT | | | | A1 | 24 |
| 28.233.00 | N2UHC | St Paul KS | EM27JM | 4 | Vert Dip | Omni | A1 | 24 |
| 28.231.60 | SV2AHT | Hortiatis | KN10NO | | | | A1 | 24 |
| 28.232.30 | W7SWL | Tucson AZ | | 5 | Vertical | Omni | A1 | ? |
| 28.233.00 | I0KNQ | Genzano di Roma | JN61FU | 2 | Turnstile | Omni | A1 | 24 |
| 28.233.00 | KB9GSY | Hammond IN | EN61FP | | | | A1 | 24 |
| 28.234.00 | K4DP | Covington VA | FM07AR | | | | A1 | IRREG |
| 28.234.30 | K4DXY | Birmingham AL | EM63PP | 2 | | | A1 | ? |
| 28.234.80 | VE1CBZ | KeswickRidge NB | FN65 | 2 | A99 Vert | Omni | A1 | 24 |
| 28.235.00 | KI4AED | Ocoee FL | EL98FN | 5 | Antron 99 | Omni | A1 | 24 |
| 28.235.00 | NP4LW | San Sebastian | FK68MI | 15 | Vertical | | A1 | ? |
| 28.235.00 | KQ4FM | Southlake TX | EM12JW | 5 | GP | | A1 | 24 |
| 28.235.10 | KI4HOZ | Pickens SC | EM83XM | | | | A1 | ? |
| 28.235.60 | VE3GOP | Mississauga ON | FN03GD | 0.2 | A99 Vert | Omni | A1 | 24 |
| 28.236.00 | W8YT | Martinsburg WV | FM19AJ | 5 | Vertical | Omni | A1 | 24 |
| 28.236.50 | W0KIZ | Nr Denver CO | DM79 | 5 | Vertical | Omni | A1 | 24 |
| 28.237.00 | K7TIA | Houston TX | | | | | A1 | ? |
| 28.237.50 | WA2NEW | Beach Haven NJ | | | | | A1 | ? |
| 28.237.60 | LA5TEN | Nr Oslo | JO59JP | 15 | 1/2 Vert | Omni | A1 | 24 |
| 28.237.80 | K7ZSA | Alger WA | CN88UO | 5 | Vertical | Omni | A1 | 24? |
| 28.238.00 | KB2SEO | Eton GA | EM74OT | 5 | 1/4 GP | Omni | A1 | 24 |
| 28.239.00 | VA7PL | Crystal Mountn | DM09 | 5 | GP | Omni | A1 | 24 |
| 28.239.00 | PP6AJM | Nosso Senhora | HH19LD | | | | | ? da Socorro |
| 28.239.20 | AL7FS | Anchorage AK | BP51BD | 3 | 1/2 Vert | Omni | A1 | PT |
| 28.239.50 | WA3HGT | Montoursville PA | FN11MG | | | | | |
| 28.239.80 | N4LEM | Cocoa FL | EL98 | 50 | Vertical | Omni | A1 | ? |
| 28.239.80 | IZ8RVA | Agropoli SA | JN70LI | | | | A1 | 24? |
| 28.240.00 | I0KNQ | Rome | JN61FU | | | | A1 | 24 |
| 28.240.00 | XE3OAX | Ocotlan. OAX | EK17PA | 0.5? | #NAME? | | A1 | 24 |
| 28.240.11 | WE6Z | Granite Bay CA | CM98JS | 4 | Vertical | Omni | A1 | 24 |
| 28.240.50 | N2DWS | PortRepublic NJ | FM29SM | | | | A1 | 24 |
| 28.240.50 | W4RKC | Winchester VA | FM09VD | 5 | 1/2 Vert | Omni | A1 | 24 |
| 28.240.60 | YO2X | Timisoara | KN05PS | 2 | GP | Omni | A1 | 09-15UT |
| 28.240.70 | AJ8T | Sturgis MI | EN71HS | | | | A1 | 24 |
| 28.241.50 | F5ZUU | Malataverne | JN24IL | 5 | 1/2 Vert | Omni | A1 | 24 |
| 28.241.50 | K5DZE | Erlanger KY | EM78QS | 5 | | | A1 | 24 |
| 28.242.00 | IZ8DXB | Naples | JN70LN | 6 | | | A1 | 24? |
| 28.242.50 | WD9CVP | Elgin IL | EN52UA | | | | A1 | 24 |
| 28.242.70 | F5ZWE | Foix | JN02TW | 15 | Vert Dip | Omni | A1 | 24 |
| 28.243.50 | KG4TIM | Guantanamo ? | | | | | A1 | IRREG |
| 28.244.00 | WA6APQ | Long Beach CA | DM13 | 30 | Vertical | Omni | A1 | 24 |
| 28.244.00 | GB3TEN | Fleetwood | IO83LV | 3 | Vertical | Omni | F1A | 24 |
| 28.245.00 | N8RT | Blairsville GA | EM84 | 40 | Ringo | Omni | A1 | 24 |
| 28.245.00 | PU2KXC | | | | | | | ? |
| 28.245.00 | DB0TEN | Bomlitz | JO42TW | 2 | 1/2 GP | Omni | A1 | 24? |
| 28.245.00 | KG4TIM | Guantanamo | | | | | | |
| 28.245.60 | SV2AHT | Hortiatis | KN1ONO | | | | | 24 |
| 28.245.80 | PP6AJM | NS do Socorro SE | HH19LD | 10 | Dipole | | A1 | 24 |
| 28.246.00 | VE9BEA | Crabbe Mntn NB | FN66 | 6 | AR10@43' | Omni | A1 | 24 |
| 28.246.00 | KG2GL | Nutley NJ | FN20UT | 5 | R5Vert @40' | Omni | A1 | INT |
| 28.246.20 | KI4LEV | Clarksville TN | EM66 | 5 | | | A1 | ? |
| 28.247.90 | N1ME | Bangor ME | FN54PS | 5 | | | A1 | 24 |
| 28.248.50 | K5DDJ | San Antonio TX | EL09 | 0.5 | GP | Omni | A1 | 24 |
| 28.249.00 | N7LT | Bozeman MT | DN45LQ | 4/.4/.04 1/2 | GP | Omni | A1 | 24 |
| 28.249.00 | KA3JOE | Bensalem PA | FN29MB | | | | A1 | ? |
| 28.249.10 | ER1TEN | Chisinau | KN47IB | 4 | Vertical | Omni | A1 | 24 |
| 28.249.50 | PY3PSI | Porto Alegre | GF49KX | 2.8 | GP | Omni | A1 | IRREG |

| Freq | Call | Nearest Town | Locator | ERPw | Antenna | Direction | Mode | Status |
|---|---|---|---|---|---|---|---|---|
| 26.249.80 | W4CJB | Santa Rosa Beach FL | EM60WR | | | | A1 | ? |
| 28.249.90 | W3ATV | Trevose PA | FN20 | 1 | Dipole | | A1 | 24 |
| 28.250.00 | K1GND | Johnston RI | FN41FT | | | | A1 | |
| 28.250.00 | K8NDB | Somerton AZ | DM22QQ | 4 | 1/4 Vert | Omni | A1 | 24 |
| 28.250.00 | N4ES | Tampa FL | EL88TA | | 20/2/.2/.02 | Horiz | A1 | SYNCH)n |
| 28.250.00 | N4ESS | Zephyrhills FL | EL88VG | | | | A1 | SYNCH)n |
| 28.250.00 | WB4WOR | Greensboro NC | FM06BT | | 20/2/.2/.02 | Horiz | A1 | SYNCH)n |
| 28.250.00 | K7EK | Graham WA | CN87TB | 25 | 1/2 Vert | Omni | A1 | SYNCH)n |
| 28250.00 | N4ES | Clearwater FL | EL88PA | 20/2/.2/.02 | Horiz | | A1 | SYNCH)n |
| 28.250.00 | K6FRC | SutterButtes CA | CM97GP | 10 | Vertical | Omni | A1 | 24 |
| 28.250.00 | K0HTF | Des Moines IA | EN31DO | 3 | Inv.V@10' | | A1 | 24? |
| 28.250.10 | Z21ANB | Bulawayo | KG47 | 8 | GP | Omni | F1 | TNonOp |
| 28.251.00 | N5ZTW | DrippingSprings TX | EM00VF | | | | A1 | 24? |
| 28.251.15 | WA4GEH | Clayton NC | FM05SN | | | | A1 | 24 |
| 28.251.30 | ED4YAK | Shell Knob MO | IN80FK | 5 | Vertical | Omni | A1 | ? |
| 28.252.00 | KX5TX | Whitney TX | EM11HX | | | | A1 | ? |
| 28.252.00 | WA2DVU | Cape May NJ | FM29NC | | | | A1 | ? |
| 28.252.00 | K7OC | Oregon City OR | | | | | A1 | |
| 28.252.00 | WW9EE | Tremont IL | EN50GK | | | | A1 | ? |
| 28.252.50 | W6PC/4 | Ocala FL | EL89VD | 10 | Dipole | Omni | A1 | 24 |
| 28.253.00 | N3BSQ | Bethel Park PA | | | | | A1 | 24? |
| 28.253.20 | SR7IHM | | | 5 | | | | 24? |
| 28.253.00 | KG4YUV | Crandall GA | EM74OW | 7 | A99 Vert | Omni | A1 | 24 |
| 28.253.00 | K8HWW | Warren MI | EN82MN | 3 | | | A1 | 24? |
| 28.253.80 | XE1USG | Puebla | EK09VB | | | | A1 | ? |
| 28.254.00 | W4CJB | Pt WashingtonFL | | | | | A1 | 24? |
| 28.254.30 | N1FCU | Windham ME | FN43ST | | | | A1 | 24? |
| 28.254.50 | K4JEE | Louisville KY | EM78RD | | | | A1 | 24? |
| 28.254.50 | K5AHH | Broken Bow OK | | | | | A1 | ? |
| 28.255.00 | N0AR | St Paul MN | EM73SW | 0.5 | 1/2 Vert | Omni | A1 | 24? |
| 28.255.50 | K8HWW | StirlingHeights MI | EN82MN 3 | | Vertical | Omni | A1 | 24 |
| 28.256.00 | C30P | Andorra | JN02SM | 10 | R5 Vert@2m | Omni | A1 | 24 |
| 28.256.00 | WI5V | Oklahoma City | | | | | | |
| 28.256.50 | VK3RMH | 25kNEMelbourne | QF22OH | 20-Feb | 1/2 Vert | Omni | A1 | 24 |
| 28.256.50 | K9JHQ | O'Fallon IL | EM58AM | 10 | Vertical | Omni | A1 | 24? |
| 28257V | KB4UPI | Bessemer AL | EM63MG | | | | A1 | 24 |
| 28.257.30 | WA2DVU | Cape May NJ | | 10 | Mosley 57 | 45 | A1 | 24 |
| 28.257.50 | N5WYN | Seven Points TX | EM12VI | | | | A1 | |
| 28.257.80 | WY5I | Port St KucieFL | EL97TF | 5 | 7db Coll. | Omni | A1 | 07-2200LT |
| 28.258.00 | EA7JNC | La Linea de la | IM76IE | 8 | | | | 24 Conception |
| 28.258.00 | NM5TW | Albuquerque NM | DM65RD | 5 | Vert Dip | Omni | A1 | 24 |
| 28.258.30 | N1YPM | Corea ME | EM19FA | | | | A1 | ? |
| 28.259.00 | K5TLL | Hattiesburg MS | EM51GG | | | | A1 | ? |
| 28.259.00 | AB8CL | Arcanum OH | EN79RA | | | | A1 | ? |
| 28.259.00 | AA4AN | Brentwood TN | EM65NW | 4 | Vertical | Omni | A1 | 24 |
| 28.259.00 | F5ZVM | Valenciennes 59 | JO10PA | 5 | 3dbi Vert | Omni | A1 | 24 |
| 28.259.30 | VK5W1 | Adelaide | PF95HG | 10 | GP | Omni | A1 | 24 |
| 28.259.30 | AF6PI | Indio CA | | | | | A1 | |
| 28.260.00 | AD5KO | Mena AR | EM24VS | 20 | Vert@20' | Omni | A1 | 24 |
| 28.260.10 | W7LFD/0 | Shell Knob MO | | 5 | Vertical | Omni | A1 | ? |
| 28.260.80 | NJ3T | Somerset PA | FM09LX | | | | A1 | 24? |
| 28.260.80 | W5TXR | Schertz TX | EL09VP | | | | A1 | ? |
| 28.261.00 | N7LF | Corbett OR | CN85VI | | | | A1 | 24 |
| 28.261.50 | N4VBV | Sumter SC | EM93TW | 5 | Attic Dip | | A1 | 24 |
| 28.261.60 | RK3XWA | Kaluga | KO84DM | | | | | 24 |
| 28.261.80 | VK2RSY | Sydney | QF56MH | 25 | 1/2 Vert | Omni | A1 | 24 |
| 28.262.30 | K8TK | Clarklake MI | EN72TC | 2 | GP@15' | Omni | A1 | 24 |
| 28.262.50 | WF4HAM | Altamonte Springs FL | EL95HP | 6 | A99@40' | Omni | A1 | INT |
| 28.263.00 | VK3RRU | Mildura | QF15AT | 20 | | | A1 | ? |
| 28.263.00 | ED4YBA | Cuneca | IN80WC | 5 | GP | Omni | A1 | 24 |
| 28.263.50 | N5YEY | Kilgore TX | EM22OJ | | | | A1 | 24 |
| 28.263.50 | W4JPL | Liberty NC | FM05EW | 4 | | | A1 | 24 |
| 28.264.00 | AB8Z | Parma OH | EN91DJ | 5 | 5/8 Vert | Omni | A1 | 24 |
| 28.264.00 | VK6RWA | Carine WA | OF78WB | 20 | 5/8 Vert | Omni | A1 | 24 |
| 28.264.50 | K7NWS | Kent WA | CN87TK | 1 | GP | Omni | A1 | 24 |
| 28.264.50 | W5ZA | Shreveport LA | EM32DJ | 3 | Vert Dip | Omni | A1 | 24 |
| 28.265.00 | DF0ANN | Moritzberg Hill | JN59PL | 5 | Dipole | E-W | A1 | 24 |
| 28.265.00 | VK4RC | Woody Point QL | DQG62NS | 10 | Vertical | Omni | A1 | ? |
| 28.265.00 | KJ3P | Schwenksville PA | | 5 | | | A1 | ? |
| 28.265.00 | PT9BCN | CampoGrande MS | GG29RN | 12 | 1/2 Dipole | | A1 | 24 |
| 28.265.40 | KR4HO | Lake City FL | EM80QG | 1 | Vertical | Omni | A1 | ? |
| 28.265.00 | NC4SW | Zebulon NC | FM05 | | | | A1 | 24 |
| 28.265.00 | N7SCQ | Dixon CA | CM98CK | | | | A1 | 24 |
| 28.266.20 | KB3ZI | Bloomsberg PA | | | | | A1 | 24 |
| 28.266.50 | KA1EKS | Millinocket ME | FN55OO | 4 | A99 GP | Omni | A1 | 24 |
| 28.266.50 | W5DJT | Pocola OK | EM25SH | | | | A1 | 24 |
| 28.266.60 | WN5KNY | RadiumSprings NM | DM62LP | | | | A1 | 24 |
| 28.267.00 | VK7RAE | TAS | QE38DT | 10 | Vartical | Omni | A1 | 24 |
| 28.267.50 | W5EFR | Houston Tx | EL29EW | 2.75 | | | A1 | TQRT |
| 28.267.60 | OH9TEN | Pirttikoski | KP36OI | 20 | 1/2 GP | Omni | A1 | 24 |
| 28.268.00 | KB0QZ | Centralia MO | EM39WE | 5 | Vertical | Omni | A1 | 24 |
| 28.268.00 | NM0R | St Genevieve MO | EM47UV | | | | A1 | |
| 28.268.30 | VK8VF | Darwin NT | PH57KP | 25 | 1/4 Vert | Omni | A1 | 24 |
| 28.268.50 | KG4GXS | CoralSpringsFL | EL96UG | 3 | Dip@23' | E-W | A1 | 24 |
| 28.268.60 | K7ZS | Hillsboro OR | CN85MM | | | | A1 | 24 |
| 28.268.80 | KD5ITM | Spring TX | | 4 | G5RV@50' | | A1 | |
| 28.268.90 | AA1TT | Claremont NH | FN33 | 5 | | | A1 | 24 |
| 28.268.90 | SV6DBG | Ioannina | KM09KQ | 2 | 5/8Vertical | Omni | A1/rtty | 24 |
| 28.269.50 | W3HH | Nr Ocala FL | EL89VB | 6 | Hamstick | Omni | A1 | 24 |
| 28.270.00 | VK4RTL | Townsville | QH30JS | 5 | Vertical | Omni | F1 | 24 |
| 28.271.70 | W4TIY | Dallas GA | EM73OW | 4 | 1/4over5/8 | Omni | A1 | 24 |
| 28.271.80 | SV2HQL | Katakali-Grevens | KM09UV | 5 | 5/8 GP | Omni | A1 | 24 |
| 28.272.30 | N1KON | Centerville IN | EM79LT | 5 | Vertical | Omni | A1 | 24 |
| 28.271.90 | AC0RR | Springfield MO | EM37IE | | | | A1 | ? |

| Freq | Call | Nearest Town | Locator | ERPw | Antenna | Direction | Mode | Status |
|---|---|---|---|---|---|---|---|---|
| 28.272.00 | PY1RJ | San Goncalo RJ | GG87ME | 4 | | | A1 | 24 |
| 28.272.50 | K5BTV | Cumming GA | EM74 | 0.25 HF6V | Vertical | Omni | A1 | 24 |
| 28.273.00 | PY4MAB | Pocos de Caldas | GG68RE | 2 | Dipole | | A1 | 24 |
| 28.273.00 | AC4DJ | Eustis FL | EL98EU | 20 | Ringo | Omni | A1 | 24 |
| 28.273.00 | WF4HAM | Altamont Spr.FL | | 10 | Ringo | | A1 | ? |
| 28.273.00 | DL3RTL | Berlin | | 1 | GP | Omni | A1 | OP? |
| 28.274.00 | LW1DZ | Escobar BA | GF05OQ | 10 | Loop | | A1 | 24 |
| 28.274.00 | WI4L | Dalton GA | EM74NQ | | | | A1 | 24 |
| 28.274.70 | N0UD | Halliday ND | DN87SH | | | | A1 | 24 |
| 28.275.00 | PY2EMG | Jacarel SP | GG76AO | | | | A1 | ? |
| 28.276.00 | NP2EH | St John VI | FK78OI | | | | A1 | ? |
| 28.275.00 | KG4GVV | Summerville SC | EM93 | | Vertical | Omni | A1 | 24 |
| 28.275.70 | XE1AKM | Alvarez(Colima) | | | | | A1 | 24 |
| 28276} | K4UKB | Danville KY | EM77NP | 10 | 5/8 Vert | Omni | A1 | SYNCHx |
| 28276} | K4FUM | Stone Mntn GA | EM73WU | | | | A1 | SYNCHx |
| 28.276.50 | XE2YBG | Victoria Tamaulipas | EL03 | | | | A1 | ? |
| 28.277.00 | WB7RBN | Pasco WA | DN06IG | | | | A1 | 24 |
| 28.277.00 | WI4L | Dalton GA | EM74MS | | | | A1 | 24 |
| 28.277.00 | WD8AQS | Fremont MI | EN73AL | | | | A1 | 24 |
| 28.277.30 | KD4MZM | Sarasota FL | EL87RG | 3 | Ringo@15' | Omni | A1 | 24 |
| 28.277.60 | DM0AAB | Kiel | JO54GH | 10 | GP | Omni | F1 | 24 |
| 28.278.00 | WA4OTD | Carmel IN | EM69 | 5 | Indoor Dip | | A1 | 24 |
| 28.278.20 | KE4IFI | Lexington SC | | | | | A1 | 24 |
| 28.278.50 | WA6MHZ | Crest CA | | | Ringo@20' | Omni | A1 | 24 |
| 28.279.00 | DB0UM | Schwedt | JO73CE | 2 | SlopeDipV | Omni | A1 | 24 |
| 28.280.00 | KA3NXN | Charlottesville | FM08SA | | | | A1 | 24 |
| 28.280.00 | K5AB | Goldthwaite TX | EM10DH | 20 | 5/8 GP@45' | Omni | A1 | 24 |
| 28.280.00 | PU5AAD | Nova Brasilia | GG51PS | 10 | | | A1 | 24 |
| 28.280.00 | N6SPP | Concord CA | CM97 | 10 | Vert Dip | Omni | A1 | 24 |
| 28.280.50 | WB6FYR | Quartz Hill CA | DM04VP | 10 | 5/8 Vert | Omni | A1 | 24 |
| 28.281.00 | W8EH | Middletown OH | EM79TL | 7.5 | Vert@40' | Omni | A1 | 24 |
| 28.281.20 | IK6ZEW | Pescara | JN72OK | | | | A1 | ? |
| 28.281.50 | W4HEW | MillegevilleGA | EM94LX | | | | A1 | 24 |
| 28.282.00 | LA6TEN | Kirkenes | KP49XQ | 10/1/.1 | | Omni | A1 | OP? |
| 28.282.00 | HP1ATM | Santiago | | | | | A1 | ? |
| 28.282.00 | XE2ES | Mexicali | DM22RP | | | | A1 | 24 |
| 28.282.00 | N2IFC | Allamuchy NM | | | | | A1 | ? |
| 28.282.40 | KD0GZJ | Loveland CO | DN70KJ | 1 | Vertical | | A1 | 24 |
| 28.282.60 | OK0EG | Hradec Kralove | JO70WE | 10 | GP | Omni | F1 | 24 |
| 28.282.80 | W0ERE | Fordlan MO | EM36 | 5 | Vertical | Omni | A1 | 24 |
| 28.283.00 | K7YSP | Gainsville GA | EM84AH | | | | A1 | 24 |
| 28.283.60 | KC9GNK | Madison WI | EN53 | 4 | Inv Vee | | A1 | 24 |
| 28.283.80 | W5OM | | DN28 | 98 | 3-el | | A1 | IRREG |
| 28.284.00 | K2XG | Monticello KY | EM76NU | 5 | V.Dip@25' | Omni | A1 | 24 |
| 28.284.80 | WD8AQS | Fremont MI | EN73AL | 5 | | | A1 | 24 |
| 28.284.80 | WA3IIA | Bloomsberg PA | FN11TA | | | | A1 | 24 |
| 28.285.00 | VP8ADE | Adelaide I. | FC52WK | 10 | 1/4 Vert | Omni | A1 | 24 |
| 28.284.50 | KB9NK | Hudsonville MI | EN72BU | | | | A1 | 24? |
| 28.285.00 | PT9BCN | CampoGrande MS | GG29RN | 1 | Vertical | Omni | A1 | 24 |
| 28.285.80 | WA4ROX | Largo FL | EL87 | 0.75 | | | A1 | 24 |
| 28.285.80 | W0ILO | Fargo ND | EN16 | | | | A1 | |
| 28.286.00 | WI6J | Bakersfield CA | DM05JJ | 5 | Vertical | Omni | A1 | 24 |
| 28.286.00 | N5AQM | Chandler AZ | DM43AH | 2 | Vertical | Omni | A1 | 24 |
| 28.286.00 | N2PD | Middletown NY | FN21 | 5 | | | A1 | ? |
| 28.286.70 | K3XR | SinkingSpringPA | FN10 | | | | A1 | ? |
| 28.270.00 | GB3XMB | Nr Waddington | IO83SV | 10 | V.Dip@800' | Omni | F1A | 24 |
| 28.287.00 | W6WTG | Bakersfield CA | DM05MJ | | | | A1 | 24 |
| 28.287.30 | W2SDX | Buffalo NY | FM18LV | | | | A1 | |
| 28.287.50 | N8PUM | Ishpeming MI | | 0.5 | Loop | | A1 | 24 |
| 28.288.00 | WA7LNW | Harmony Mesa UT | DM37 | 5 | FWave Loop | EU/PAC | A1 | 24 |
| 28.288.00 | K4LJP | W Palm Beach FL | EL96 | 5 | AR99@30' | Omni | A1 | 24 |
| 28.288.00 | RA3ATX | | KO85NX | | | | | OP? |
| 28.288.50 | ND3E | Newcastle DE | | | | | A1 | 24 |
| 28.289.00 | KB9WA | Egg Harbor WI | | | | | A1 | |
| 28.289.00 | WJ5O | Nr Columbus AL | EM72NE | 2 | Yagi | NE | A1 | 24 |
| 28.289.50 | N1KXR | Medway MA | FN32 | 5 | | | A1 | 24 |
| 28.289.80 | VR2TEN | Hong Kong | OL72 | 5 | 1/4 GP | Omni | A1 | 24 |
| 28.289.80 | W0ERE | Highlandsville MO | EM36HX | | Varies | Varies | A1 | 24 |
| 28289V | PS8RF | Teresina PI | GI84OW | | | | A1 | 24 |
| 28.290.00 | N6UN | San Diego CA | DM12 | 5 | | | A1 | 24 |
| 28.290.30 | WB4WOR | Randleman NC | FM06BT | 3 | Vertical | Omni | A1 | 24 |
| 28.291.00 | K5TLJ | Trumann AR | EM45RQ | 20 | | | A1 | ? |
| 28.291.80 | N5MAV | Midland TX | | | | | A1 | |
| 28.292.00 | VA3VA | Windsor ON | EN82 | 5 | Horiz Dip | | A1 | 24 |
| 28.292.50 | KM4GS | KentuckyLake KY | EM56 | 0.5 | Vertical | Omni | A1 | 24 |
| 28.292.50 | SK0CT | Sollentuna | JO89XK | 10 | GP | Omni | A1 | 24 |
| 28.292.80 | K7GFH | Damascus OR | CN85SJ | 3 | Attic Dip | | A1 | 24 |
| 28.293.40 | ND4Z | Gilbert SC | EM94JA | 5 | 5/8@40' | Omni | A1 | 24 |
| 28.293.70 | W4DJD | Woodbridge VA | FM18HP | | | | A1 | 24 |
| 28.294.00 | K7RON | Peoria AZ | DM33 | | | | A1 | 24 |
| 28.294.00 | KE4IAP | Woodbridge VA | FM18HP | | | | A1 | 24 |
| 28.295.00 | KD1ZX | CentralFalls RI | FN41HV | 4 | Vert@10' | Omni | A1 | 24 |
| 28.294.80 | NR8O | Fort Thomas KY | EM79KY | | | | A1 | |
| 28.295.00 | PU5ATX | Santa Catarina G | G51PR | | Dipole | | A1 | Jun-50 |
| 28.295.10 | SK2TEN | Kristineberg | KP08FC | 5 | Vertical | Omni | A1 | OP? |
| 28.295.40 | K1SPD | La Vergne TN | EM65RX | | | | A1 | ? |
| 28.295.50 | K4IT | Flatwoods KY | EM88PM | 3.5 | Dipole | | A1 | 24 |
| 28.295.50 | IZ0CWW | Cervaro FR | JN61VL | 3 | | | | 24 |
| 28.295.80 | W3APL | Laurel MD | FM19NE | 10 | Hor.Dipole | NE/SW | A1 | 24 |
| 28.296.00 | KA7BGR | CentralPoint OR | CN82 | | | | A1 | 24 |
| 28.296.20 | W5JDG | Washington TX | | | | | A1 | 24 |
| 28.297.10 | NS9RC | Evanston IL | EN62DC | 3 | 1/2 Vert | Omni | A1 | 24 |
| 28.297.30 | K9EEI | Tuscola FL | | | | | A1 | |
| 28.297.80 | WA3BM | Valencia PA | FN00SQ | | | | A1 | 24 |
| 28.298.00 | V73TEN | Roi Namur I | RJ39RJ | | Horiz OA50 | Omni | | 24? |

| Freq | Call | Nearest Town | Locator | ERPw | Antenna | Direction | Mode | Status |
|---|---|---|---|---|---|---|---|---|
| 28.298.00 | K5TLL | Hattiesburg MS | EM51GG | 5 | Vertical | Omni | A1 | 24 |
| 28.298.10 | 5WZ8D | Blachester OH | EM89OO | | | | A1 | 24 |
| 28.298.10 | SK7GH | Bor | JO77BF | 5 | | | A1 | 24 |
| 28.298.10 | K7PO | Tenopah AZ | DM32 | 0.5 | Vertical | Omni | A1 | 24 |
| 28.298.50 | K7FL | Battle Ground WA | CN85SS | 4 | Horiz Loop | | A1 | 24 |
| 28.298.60 | K4JDR | Raleigh NC | FM05 | 10 | Vertical | Omni | A1 | 24 |
| 28.299.00 | N1SCA | Palm Bay FL | EL97QX | | | | A1 | 24 |
| 28.300.00 | K6FRC | Sutters Mntn CA | CM99 | | | | A1 | 24 |
| 28.300.00 | IK6ZEW | Pescara | JN72CK | 0.4 | Vertical | | A1 | ? |
| 28.300.10 | IW0HBY | Rome | JN61DM | 10 | | | | ? |
| 28.300.00 | KF4MS | Tallahassee FL | EM70VM | 10 | A99@25' | Omni | A1 | 24 |
| 28.300.00 | VE3CKN | Gloucester ONT | | | | | A1 | |
| 28.300.30 | PU5ZAA | Novegentes | GG53QE | 3 | | | A1 | 24 |
| 28.301.00 | PI7ETE | Amersfoort | JO22QD | 0.5 | Vertical | Omni | F1A | 24 |
| 28.320.97 | IZ8HUJ | Pignola PZ | JN70VN | 0.4 | Windom | | | ? |
| 28.321.00 | I1YRB | Torre Bert (TO) | JN35UB | 0.2 | Vertical | Omni | QRSS3 | 24 |
| 28.321.20 | IS0GOV | Cagliari | JM49NF | 0.3 | Vertical | Omni | QRSS3 | 24 |
| 28.321.20 | IK3ERY | Vittorio Veneto | JN65DX | 0.1 | Vertical | OMNI | QRSS3 | 24 |
| 28.321.47 | IZ1GJH | Casarza Lig GE | JN44RG | 0.1 | Vertical | Omni | QRSS3 | 24 |
| 28.321.65 | IN3KLQ | Nr Trento | JN56RG | 0.3 | Vertical | Omni | QRSS3 | 24 |
| 28.321.70 | IW4EMG | Ferrara | JN54RU | 0.1 | | | QRSS3 | 24 |
| 28.321.80 | I3GNQ | Tencarola PD | JN55VJ | 0.4 | GP | Omni | QRSS3 | 24 |
| 28.321.85 | IK0IXI | Aprilia LT | JN62VB | 0.1 | Dipole@30m | NE/SW | A1/F1 | 24 |
| 28.321.80 | I1SXT | | JN44PH | | | | | ? |
| 28.321.86 | IZ8JFA | Cosenza | JM89CH | 0.3 | | N/NE | QRSS3 | 24 |
| 28.321.77 | I1SXT | | | | | | | ? |
| 28.321.94 | IW9BAJ | Sicily | JM77NO | 1 | | | QRSS3 | 24 |
| 28.321.95 | IW0HK | Civitavecchia | JN52WD | 1/.1 Dipole | Omni | | QRSS3 | 24 |
| 28.321.70 | IZ1TAA | Nr La Spezia | JN54AC | 0.1 | Dipole | NW | QRSS3 | 24 |
| 28.322.00 | IQ3QR | | JN55XR | 0.5 | | | QRSS3 | 24 |
| 28.322.00 | IZ5ILK | | | 0.1 | GP | | | ? |
| 28.322.00 | IZ0ANE | Cassino | | 0.1 | | | | ? |
| 28.322.01 | DJ6GT | | | 0.1 | | | QRSS3 | ? |
| 28.322.01 | IW1QIF | Davagna-Genova | JN44LL | 0.1 | Long Wire | | QRSS3 | 24 |
| 28.322.04 | IT9YAF | Canicatti | JM67WI | 0.1 | Vertical | Omni | QRSS3 | 24 |
| 28.322.05 | IK1HGI | Trecate | JN45IK | 0.1 | Dipole | NNW/SSE | QRSS3 | 24 |
| 28.322.08 | IS0GSR | Nr Cagliari | JM49IN | 0.1 | Dipole | N-S | QRSS3 | 24 |
| 28.322.11 | IK8YTN | Salerno | JN70LG | 0.1 | Inv V | E-W | QRSS3 | 24 |
| 28.322.18 | IZ0HCC | Roma | JN61FT | 0.1 | Vertical | Omni | QRSS3 | ? |
| 28.322.20 | IK8SUT | Salerno | JN70JQ | 0.1 | Inv V | E-W | QRSS3 | 24 |
| 28.322.20 | IW7DEC | Nr Bari | JN81GF | 0.1 | 1/2 Vert | Omni | QRSS3 | 24 |
| 28.322.32 | IW9FRA | Trapani.Sicily | JM68HA | 0.1 | Dipole | N-S | QRSS3 | 24 |
| 38.322.36 | IW3SGT | Trieste | JN65VP | | | | QRSS3 | 24 |
| 28.322.45 | IW4EMG | Nr Ferrara | JN54RW | | Vertical | Omni | QRSS3 | ? |
| 28.322.60 | F1VJT | StGermain/Puoh | JN16XX | 0.1 | | | | ? |
| 28.322.62 | IK1BPL | Novara | JN45HK | 0.1 | GP | NNW/SSE | QRSS3 | 24 |
| 28.322.63 | IK0VVE | Nr Latina | JN61KN | 0.1 | Dipole | NNW/SSE | QRSS3 | 24 |
| 28.322.70 | G3ZJG | Nr Leicester | | | | | QRSS3 | PT |

**Notes:**

? = Activity uncertain. INT = Intermittent. PT = Part-time. IRREG = Irregular. UC = Under Construction. (T)NonOp = (Temporarily)NotOperational. qq = see OK0EU paragraph below. OP? = Operational? V = Varies. EXP = Experimental. Wkend = Weekends during daylight. LT=Local Time. +++ = 10 minutes constant carrier followed by ID.

SYNCH: frequency sharing synchronised beacons

SYNCHn current sequence: N4ESS(0.00seconds), N4ES(0.20) WB4WOR (0.30) K7EK(0.40) N4ES(0.50).

SYNCHx: K4UKB ID at 00 seconds and K4FUM at 30 seconds. ££ K6FRC transmits on 28245, 28250, 28300

zzz These beacons transmit in sequence at 15-minute intervals: GB3RAL on the hour and H+15m, H+30, H+45. GB3WES at H+1, H+16 etc and GB3ORK at H+2, H+17 etc. Transmissions have a stepped power sequence and a 30-second sounder sequence of 0.5ms pulses at 40Hz prf

### OK0EU

OK0EU is a cluster of five high-stability transmitters with 5Hz separation, running for multi-point measurement. The callsign is sent 40Hz up to identify ionospheric echoes. Output power 1 watt to a magnetic loop radiating N-S.

Transmitters are on:

- 3.594.492 Vackov (JO60EF),
- 3.594.496 Diouha Louka (JO60TP),
- 3.594.500 Panksa Ves (JO70GM),
- 3.594.504 Pruhonice (JN79GX),
- 3.594.508 Kasperske Hory (JN69SD).

### DK0WCY/DRA5

zz DK0WCY 3579 operates 0720-0900UTC and 1600-1900UTC LOCAL time (ie UTC+1 winter and UTC+2 summer) DRA5 and DK-0WCY on 10144 are 24/7. Transmissions on 3579 are cw only, with beacon id interrupted at ten minute intervals by a datagram giving the Kiel K, current fof2 and MUF at Juliusruh and indications of current solar-geophysical events. DRA5 and DK0WCY on 10MHz carry the CW datagrams but, at 10 minutes after the hour the datagram transmission is on RTTY and contains additional information on X-ray background and hf2, using RTTY. The H+50 transmission is PSK31(BPSK).

During auroral events the carrier at the end of the ID changes to a series of dots. See also: www.dk0wcy.de

### Schedule of IBP/NCDXF Beacon Transmissions

International Beacon Project beacons transmit for ten seconds on each frequency in turn in the sequence shown below. The whole cycle takes three minutes and then repeats. They send callsign at 22WPM and 100 watts, then four 1-second dashes at 100W, 10W, 1W and 0.1W. Equipment is a Kenwood TS-50, Cushcraft R-5 multiband vertical and a Trimble Navigation GPS receiver to ensure sychronization, with a control unit built by NCDXF.

## Schedule of IBP/NCDXF Beacon Transmissions

| | | Frequency (kHz) | | | | |
|---|---|---|---|---|---|---|
| Location | Call | 14100 | 18110 | 21150 | 24930 | 28200 |
| United Nations NY | 4U1UN | 00.00 | 00.10 | 00.20 | 00.30 | 00.40 |
| Northern Canada | VE8AT | 00.10 | 00.20 | 00.30 | 00.40 | 00.50 |
| USA (CA) | W6WX | 00:20 | 00.30 | 00:40 | 00.50 | 01:00 |
| Hawaii | KH6WO | 00.30 | 00.40 | 00.50 | 01.00 | 01.10 |
| New Zealand | ZL6B | 00.40 | 00.50 | 01.00 | 01.10 | 01.20 |
| West Australia | VK6RBP | 00.50 | 01.00 | 01.10 | 01.20 | 01.30 |
| Japan | JA2IGY | 01.00 | 01.10 | 01.20 | 01.30 | 01.40 |
| Siberia | RR9O | 01.10 | 01.20 | 01.30 | 01.40 | 01.50 |
| China | VR2HK | 01.20 | 01.30 | 01.40 | 01.50 | 02.00 |
| Sri Lanka | 4S7B | 01.30 | 01.40 | 01.50 | 02.00 | 02.10 |
| South Africa | ZS6DN | 01:40 | 01.50 | 02:00 | 02:10 | 02:20 |
| Kenya | 5Z4B | 01.50 | 02.00 | 02.10 | 02.20 | 02.30 |
| Israel | 4X6TU | 02:00 | 02:10 | 02:20 | 02.30 | 02:40 |
| Finland | OH2B | 02:10 | 02:20 | 02:30 | 02:40 | 02:50 |
| Madeira | CS3B | 02.20 | 02.30 | 02.40 | 02.50 | 00.00 |
| Argentina | LU4AA | 02:30 | 02:40 | 02:50 | 00.00 | 00:10 |
| Peru | OA4B | 02.40 | 02.50 | 00.00 | 00.10 | 00.20 |
| Venezuela | YV5B | 02:50 | 00.00 | 00:10 | 00:20 | 00:30 |

DK0WCY: *www.dk0wcy.de* NCDXF: *www.ncdxf.org/beacons.html*

# IARU Region 1 VHF/UHF & Microwave Beacons

This list of VHF/UHF/Microwave Beacons is compiled for IARU Region 1 by G0RDI and builds upon the valuable work contributed by G3UUT. Thanks are also due to the VHF/UHF/Microwave Managers of radio societies across Region 1, beacon keepers, beacon co-ordinators and VHF/UHF DXers too numerous to mention.

If the Status column for a beacon is blank, the beacon can be assumed to be on-air. However, the reliability of this information can be judged by the source and date in the Last Update column.

Entries which are shown in italic text are for beacons which have been reported as operational, but for which there is no IARU R1 coordination information on file.

Note that this list includes information regarding beacons notified to the R1 coordinator which operate (or are Planned to operate) outside of the IARU R1 band Planned beacon segments due to local licensing requirements.

Also, at an Interim C5 Meeting (Vienna, 02/2010) it was proposed that the 6m beacon band will be relocated to 50.400-50.500MHz, with all beacons moved or closed down no later that 01/01/2013. Again, some member Societies may be unable to comply for local regulatory reasons. As a general principle, all beacons should move to band Planned frequencies as soon as possible.

All inputs are welcome, and should be sent to g0rdi@77hz.com

You are free to use information from this beacon list, but please acknowledge IARU Region 1 and G0RDI if you do so.

| Freq | Call | Nearest Town | Locator | m ASL | Antenna | Direction | Power | Info | Status | Last update |
|---|---|---|---|---|---|---|---|---|---|---|
| 24.9305 | GB3RAL | | IO91EN | 140 | | | 20 | G0MJW | Planning | 02/07 G6JYB |
| 28.191 | GB3RAL | | IO91EN | 140 | | | 20 | G0MJW | Planning | 02/07 G6JYB |
| 28.271 | OZ7IGY | Jystrup | JO55WM | 98 | Turnstile | Omni | 10 | OZ7IS | | 01/14 OZ7IS |
| 40.021 | OZ7IGY | Jystrup | JO55WM | 94 | Halo | Omni | 10 | OZ7IS | | 01/14 OZ7IS |
| 40.050 | GB3RAL | | IO91EN | 140 | | | | G0MJW | QRV | 06/08 G6JYB |
| 50.000 | GB3BUX | Buxton, Derbys | IO93BF | 457 | Turnstile | Omni | 20 | G7EKY | | 06/14 G6JYB |
| 50.000 | 9A1CAL | | JN86EL | | Turnstile | Omni | 1 | | QRT since 2006? | 04/13 G0RDI |
| 50.001 | IW3FZQ/B | Monselice PD | JN55VF | 25 | 5/8 Vertiical | Omni | 8 | | | 04/13 IK1YWB |
| 50.001 | VE1SMU | Halifax NS | FN84 | | 3 el Yagi | 90° | 40 | | | |
| 50.0025 | 7Q7SIX | Malawi | KF75 | | | | | | | 05/02 G3USF |
| 50.004 | I0JX/B | Roma | JN61HV | 100 | 5/8 Vertical | Omni | 5 | | | 04/13 IK1YWB |
| 50.0047 | 4N0SIX | Belgrade | KN04FU | | Dipole | Omni | 1 | | | |
| 50.007 | HG1BVB | Hörman-forrás | JN87FI | 725 | Cross Dipoles | Omni | 20 | HA1YA | | 03/13 HA5NF |
| 50.008 | I5MXX/B | Pieve a Nievole PT | JN53JU | | 5/8 Vertical | Omni | | | | 04/13 IK1YWB |
| 50.008 | EA3RCC | Barcelona | JN11BP | 580 | GP | Omni | 5 | EA3ABN | QRV | 09/07 EA3ABN |
| 50.0108 | SV9SIX | Iraklio | KM25NH | | Vertical dipole | Omni | 30 | | | 05/02 G3USF |
| 50.011 | OK0EK | Kromeriz | JN89QG | 300 | 2 x Dipole | Omni | 10 | OK2PWM | QRT | 11/12 OK1HH |
| 50.012 | OH1SIX | Ikaalinen | KP11QU | | | | | | PLANNED | 07/11 OH6DD |
| 50.012 | OX3SIX | Kulesuk | GP15EO | 1200 | Vert Dipole | Omni | 100 | | | 11/05 OZ2TG |
| 50.013 | LZ1JH | | KN22TK | | GP | Omni | 1 | | QRT | 10/13 LZ1NY |
| 50.013 | CU3URA | Terceira, Azores | HM68 | | 5/8 Vertical | | 5 | | | 06/98 EA7KW |
| 50.014 | OK0SIX | Nr. Milevsko | JN79FM | 700 | Vertical | Omni | 2.5 | OK5VTR | QSY -> 50.414 | 11/12 OK1HH |
| 50.016 | GB3BAA | Nr Tring | IO91PS | 272 | Vertical Dipole | Omni | 10 | G0RDI | Will QSY -> .416 | 05/14 G0RDI |
| 50.017 | OH0SIX | Stalsby | JP90XI | 65 | Dipole | 0°/180° | 3 | | | 07/11 OH6DD |
| 50.019 | IZ1EPM/B | Saronsella TO | JN35WD | 385 | GH50 5/8 | Omni | 10 | | | 04/13 IK1YWB |
| 50.020 | IW8RSB/B | Simeri CZ | JN88HW | 8 | 5/8 Vertical | Omni | 5 | | | 04/13 IK1YWB |
| 50.020 | CS5BSIX | | IM58IS | | Yagi | Transatlantic | 5 | | PLANNED | 07/08 CT1END |
| 50.0207 | IK5ZUL/B | Follonica GR | JN52JW | | | Omni | | | | 04/13 IK1YWB |
| 50.022 | HG8BVB | Gerla | KN06OQ | 91 | Ground Plane | Omni | 5 | HA8BS | | 03/13 HA5NF |
| 50.0225 | FR5SIX | Reunion Is | LG78RQ | 1700 | Vertical | Omni | 0.7 | FR4LI | WILL QSY 50.430 | 03/14 F6HTJ |
| 50.023 | SR5FHX | Slubianka | KO02LL | 115 | 3el Yagi | 240° | 3 | SP5XMU | | 04/05 SP6LB |
| 50.023 | LX0SIX | Bourscheid | JN39BF | 500 | Horizontal Dipole | 0°/180° | 10 | LX1JX | | 05/02 G3USF |
| 50.0247 | UN1SIX | Kazakhstan | MN83KE | 3400 | GP | Omni | 12 | | | 05/02 G3USF |
| 50.025 | OH2SIX | Lohja | KP20DH | 63 | Dipole | Omni | 50 | | | 07/11 OH6DD |
| 50.025 | 9H1SIX | Attard, Malta | JM75FV | 75 | Ground Plane | Omni | 7 | 9H1ES | | |
| 50.026 | IQ4FA/B | Ferrara | JN54TU | | Ground Plane | Omni | 5 | | Will QSY -> .426 | 04/13 IK1YWB |
| 50.026 | SR9FHA | Chorgawica | KN09AS | 700 | 5/8 Vertical | Omni | 5 | SP9SVH | | 04/07 SP6LB |
| 50.027 | SK7SIX | Hultsfred | JO77UM | | | Omni | 15 | | | 12/01 SM0KAK |
| 50.027 | CN8LI | Rabat | IM64 | | J Pole | Omni | 8 | CN8LI | | 03/99 CN8LI |
| 50.0276 | SR6SIX | Sztobno / Wolow | JO81HH | | Ground Plane | Omni | 10 | SP6GZZ | | 07/94 SP6LB |
| 50.028 | SR8SIX | Sanok | KN19CN | | | | | | Non op | 05/02 G3USF |
| 50.028 | SR3FHB | Chelmce | JO91CQ | 180 | Dipole | Omni | 5 | SP3DRT | | 04/07 SP6LB |
| 50.031 | HG7BVA | Gyömro | JN97QJ | 180 | Ground Plane | Omni | 5 | HA7CR | | 03/13 HA5NF |
| 50.0315 | CT0SIX | Tavarde | IM59QM | | H. Dipole | 90°/270° | | | | 09/06 CT2IRJ |
| 50.033 | OH5RAC | Kuusankoski | KP30HV | 157 | 2dBd | 200° | 20 | | | 08/06 OH6DD |
| 50.033 | OH5SHF | Kouvola | KP30HV | 145 | 2dBd | 180° | 20 | | | 07/11 OH6DD |
| 50.035 | OY6SMC | Faroe Is | IP62MB | | | | | | | 11/05 OZ2TG |
| 50.035 | CQ3SIX | Madeira Is | IM21IP | | Halo | Omni | 10 | | | 05/07 G0LGS |
| 50.0366 | SR2SIX | Bydgoszcz | JO93BC | | | | | | | 05/02 G3USF |
| 50.037 | CT1ART | Serra do Caldeirao | IM67AH | 510 | 6el Yagi | | 40 | CT1EPS | | 01/08 CT1HZE |
| 50.0375 | ES0SIX | Hiiuma Island | KO18CW | 105 | Hor. dipole | 90°/270° | 15 | ES1CW | | 12/01 SM0KAK |
| 50.040 | SV1SIX | Athens | KM17UX | 130 | Vertical Dipole | Omni | 30 | | | 08/00 G3UUT |
| 50.041 | ON0SIX | Vieux Genappe | JO20EP | 178 | | | 50 | ON4KVJ | | 05/07 ON4PC |
| 50.0424 | GB3MCB | St Austell | IO70OJ | 320 | Dipole | 90°/270° | 40 | M0PSB | | 06/14 G6JYB |
| 50.044 | ZS6TWB | Haenertsburg | KG46XA | | 5/8 vertical | Omni | 15 | | | 11/00 ZS6PJS |
| 50.045 | SR2FHM | Gdansk | JO94HI | 5 | Dipole | | 7 | SQ2BXI | | 04/07 SP6LB |
| 50.045 | OX3VHF | Qaqortoq | GP60XR | 15 | Ground Plane | Omni | 20 | OX3JUL | | 11/05 OZ2TG |
| 50.047 | YO2S | Timisoara | KN05PS | | Dipole | | 1 | YO2IS | | 06/99 YO2IS |
| 50.0472 | 4N1SIX | Belgrade | KN04OO | | Vee | Omni | 10 | | Non op | 05/02 G3USF |
| 50.0485 | TR0A | Libreville | JJ40 | | 5 el Yagi | 0° | 15 | | | 05/02 G3USF |
| 50.050 | LZ0SJB | Silven | KN32DR | 1096 | Ground Plane | Omni | 2 | LZ1SJ | QRV / Not Coordi | 10/13 LZ1NY |
| 50.050 | IARU IBP | | | | | | | | | 06/02 G3UUT |
| 50.050 | GB3RAL | | IO91IN | 140 | | | 17 | G0MJW | QRV | 06/08 G6JYB |
| 50.052 | LZ0SIX | Silven | KN22XS | 1500 | Dipole | | 1 | LZ1GHT | QRV / Not Coordi | 10/13 LZ1NY |
| 50.052 | SK2CP | Kiruna | KP07MV | 630 | | Omni | 30 | SM2UHF | | 05/10 SM2UHF |
| 50.053 | PI7SIX | Utrecht | JO22NC | 40 | Dipole | 0°/180° | 12 | PA3FYM | | 07/03 PE2KP |
| 50.054 | LZ0MON | Montana | KN13OJ | 160 | Ground Plane | Omni | 0.5 | LZ2CM | QRV / Not Coordi | 10/13 LZ1NY |
| 50.054 | OZ6VHF | Oestervraa | JO57EI | 84 | Turnstile | Omni | 25 | OZ1IPU | | 11/05 OZ2TG |

| Freq | Call | Nearest Town | Locator | m ASL | Antenna | Direction | Power | Info | Status | Last update |
|---|---|---|---|---|---|---|---|---|---|---|
| 50.057 | TF3SIX | Reykjavik | HP94BC | 120 | | Omni | 20 | TF3GW | | 12/07 OZ7IS |
| 50.057 | IT9X/B | Messina | JM78SG | 586 | Loop | Omni | 10 | | | 04/13 IK1YWB |
| 50.058 | HB9SIX | Nr. St. Gall | JN47QF | 2502 | J-pole | Omni | 6 | HB9RUZ | | 07/08 HB9RUZ |
| 50.058 | IQ4AD/B | Parma | JN54DT | | 5/8 Vertical | Omni | | | | 04/13 IK1YWB |
| 50.058 | IW0DTK/B | Latina | JN61TG | 10 | 5/8 Vertical | Omni | 10 | | | 04/13 IK1YWB |
| 50.060 | GB3RMK | Inverness | IO77UO | 242 | Dipole | 0°/180° | 32 | GM3WOJ | | 08/09 G6JYB |
| 50.061 | EA3VHF | Gerona | JN11MV | | Vertical | Omni | 1 | | Intermittant | 05/02 G3USF |
| 50.062 | OZ2VHF | Esbjerg | JO45FL | | Horiz Dipole | 0°/180° | 1 | | | 06/99 G3USF |
| 50.0625 | GB3NGI | Ballymena | IO65VB | 518 | Halo | Omni | 30 | GI6ATZ | QRV | 06/10 G6JYB |
| 50.063 | LY0SIX | | KO24PS | | 6 el Yagi | 250° | 7 | LY2MW | | 03/04 LY2MW |
| 50.064 | GB3LER | Lerwick | IP90JD | 104 | Dipole | 0°/180° | 30 | GM4WMM | QRV | 04/09 G6JYB |
| 50.065 | GB3IOJ | Jersey Island | IN89VE | 74 | Halo | Omni | 25 | GJ8PVL | QRV | 01/13 G6JYB |
| 50.066 | OE3XAC | Kaiserkogel | JN78SB | 792 | 5/8 vertical GP | Omni | 10 | OE3KLU | QRV | 01/10 OE1MCU |
| 50.067 | OH9SIX | Pirttikoski | KP36OI | 192 | 2 x X dipole | Omni | 35 | | | 07/11 OH6DD |
| 50.068 | HG8BVD | Kisráta | KN06HT | 85 | Ground Plane | Omni | 5 | HA8MV | | 03/13 HA5NF |
| 50.070 | SK3SIX | Edsbyn | JP71XF | 505 | Hor X dipole | Omni | 10 | SM3EQY | | 07/98 SM6CEN |
| 50.0735 | EA8SIX | | IL28GD | | | Omni | 14 | | | 05/02 G3USF |
| 50.0745 | ED7YAD | Malaga | IM76qo | 925 | Loop | Omni | 15 | EA7UU | QSY --> 50.475 | 11/14 EA7KW |
| 50.075 | YO3KWJ | | KN34BJ | | Ground Plane | Omni | 5 | Y03JW | | 05/99 YO3JW |
| 50.076 | CS1RLA/B | Aldeia de Chaos | IM57PX | 295 | Dipole | Omni | 2.5 | CT4RK | | 07/07 CT1FBF |
| 50.076 | CS5BLA | Aldeia de Chaos | IM57PX | 295 | Horizontal | | 2.5 | CT4RK | QRV | 04/09 CT1FBF |
| 50.077 | DL???? | | | | | | | | Planned | 01/01 DJ3TF |
| 50.078 | OD5SIX | Lebanon | KM74WK | | 1/4 Vertical | Omni | 7 | OD5SB | | 01/96 OD5SB |
| 50.080 | 4X4SIX | Tel Aviv | KN72JB | | Dipole | | 3 | | | 05/02 G3USF |
| 50.080 | UU5SIX | Nr Yalta | KN74AL | | Dipole | | 10 | | | 05/02 G3USF |
| 50.084 | UT5G | Petri UKR | KN66LS | 105 | Ground plane | Omni | 10 | UT7GA | | 04/00 UR4LL |
| 50.0847 | UR4LL | | KO70XG | 106 | H dipole | Omni | 8 | UR4LL | | 04/00 UR4LL |
| 50.406 | F5ZHQ | Hyeres | JN23BD | 240 | Halo | Omni | 5 | | TESTING | 03/14 F6HTJ |
| 50.413 | C30SIX | | | | | | | C31US | PLANNING | 10/15 C31US |
| 50.415 | IW3ICH/B | Rovigo | JN55WD | 20 | I-Pole | Omni | 1 | | | 04/13 IK1YWB |
| 50.418 | F1ZFE | Sarreguemines | JN39OC | 392 | Vertical | Omni | 2 | | PLANNING | 06/13 F6HTJ |
| 50.422 | S55ZRS | Mt. Kum | JN76MC | 1219 | Ground Plane | Omni | 1 | | | 08/15 S51FB |
| 50.445 | ED1YCA | Pico Gorfoli | IN73AL | 650 | | Omni | 10 | EB1TR | LICENSED | 05/14 EB1TR |
| 50.445 | ED1YCE | Laredo | IN83GK | 18 | Loop | Omni | 5 | EA1DPP | IN PROGRESS | 03/14 EA1AL |
| 50.446 | JW5SIX | Hopen | KQ26MM | 10 | Vertical Dipole | Omni | 10 | LA5RIA | | 12/14 LA0BY |
| 50.447 | JW7SIX | Isfjord Radio | JQ68TB | 10 | 3 el Yagi | 180° | 30 | LA0BY | | 12/14 LA0BY |
| 50.448 | FX4SIX | Neuville | JN06CQ | 153 | 2 x dipole | Omni | 25 | F5GTW | QRV | 03/14 F6HTJ |
| 50.449 | JW9SIX | Bjornoya | JQ94LM | 25 | Dipole | Omni | 10 | LA5RIA | Temp. QRT | 12/14 LA0BY |
| 50.450 | IARU IBP | | | | | | | | | 06/02 G3UUT |
| 50.451 | LA7SIX | Malselv | JP99EC | 320 | 4 el Yagi | 190° | 100 | LA5TFA | | 12/14 LA0BY |
| 50.457 | IW0DAQ/B | Albano Laziale RM | JN61HQ | 258 | QuadLoop | Omni | 2 | | | 04/13 IK1YWB |
| 50.457 | IT9X/B | Messina | JM78SG | 586 | Loop | Omni | 10 | | | 04/13 IK1YWB |
| 50.458 | IW0DTK/B | Latina | JN61TG | 10 | 5/8 Vertical | Omni | 10 | | | 04/13 IK1YWB |
| 50.459 | LA9SIX | | JP50EV | 430 | 5/8 Vertical | Omni | 25 | | PLANNING | 11/12 LA0BY |
| 50.459 | F5ZHI | Valenciennes | JO10PH | 90 | Halo | Omni | 3 | | QRV | 04/14 F6HTJ |
| 50.463 | LA2SIX | | JP53EG | | | | | | PLANNED | 02/10 LA0BY |
| 50.471 | OZ7IGY | Jystrup | JO55WM | 100 | Big Wheel | Omni | 40 | OZ7IS | | 01/14 OZ7IS |
| 50.473 | OK0EQ | Namest n./O | JN89BE | 430 | Dipole | Omni | 4 | OK2ZI | PLANNED | 11/12 OK1HH |
| 50.475 | OK0NCC | Praha | JN79EW | 531 | Dipoles | Omni | 5 | OK1IMJ | PLANNED | 11/12 OK1HH |
| 50.477 | OK0EXD | Ricany | JN79IW | 503 | Vertical | Omni | 10 | OK1DXD | PLANNED | 11/12 OK1HH |
| 50.479 | JX7SIX | Jan Mayen | IQ50RX | 269 | 3 el yagi | 160° | 40 | LA7DFA | Temp. QRT | 12/14 LA0BY |
| 50.481 | F1ZFB | Archiac | IN95TM | 103 | Dipole | 0°/180° | 10 | | QRV | 03/14 F6HTJ |
| 50.4995 | 5B4CY | Zyghi, Cyprus | KM64PR | 30 | Ground Plane | Omni | 20 | 5B4BBC | | 07/92 5B4JE |
| 50.5203 | SZ2DF | | KM25 | | 4 x 16 el | 30°/330° | 1000 | | | 05/02 G3USF |
| 52.450 | VK5VF | Mt Lofty Adelaide | PF95 | | Turnstile | Omni | 10 | | | 05/02 G3USF |
| 60.050 | GB3RAL | | IO91EN | 140 | | | | G0MJW | QRV | 06/08 G6JYB |
| 70.000 | GB3BUX | Buxton, Derbys | IO93BF | 456 | 2 x Turnstile | Omni | 20 | G7EKY | | 06/14 G6JYB |
| 70.002 | ZS1FOR/B | Western Cape | JF96FB | | Horiz. Dipole | | 15 | | | 09/99 ZS5JF |
| 70.005 | ZS5MTL | | KG50IG | | | Omni | 50 | | | 09/99 ZS5JF |
| 70.007 | GB3WSX | Yeovil | IO80QW | 105 | Yagi | 70° | 150 | G3ZXX | | 06/05 G0RDI |
| 70.010 | J5FOUR | | IK21EV | | 4el Yagi | 020° | 20 | J5JUA | QRV | 03/08 CT1FFU |
| 70.012 | OX4M | | GP15EO | 1200 | Dipole | Omni | 25 | | | 11/05 OZ2TG |
| 70.014 | S55ZRS | Mt. Kum | JN76MC | 1219 | | | | | Planned | 06/98 S57C |
| 70.015 | ZR6FOR | | | | | | | | Planned | 09/99 ZS5JF |
| 70.016 | GB3BAA | Nr Tring | IO91PS | 272 | 2 x Dipole | Omni | 25 | G0RDI | Temp. off air | 05/14 G0RDI |
| 70.01628 | CS3BFM | Santo de Serra / Madeira | | IM12OR | | 736 | Dipole | | 4 | ARRM 03/14 CT1END |
| 70.018 | OH2FOUR | Lohja | KP20DH | 83 | Dipole | Omni | 15 | | | 07/11 OH6DD |
| 70.020 | GB3ANG | Dundee | IO86MN | 370 | 3 el Yagi | 160° | 100 | GM4ZUK | | 06/02 G3UUT |
| 70.021 | OZ7IGY | Jystrup | JO55WM | 102 | Big Wheel | Omni | 40 | OZ7IS | | 01/14 OZ7IS |
| 70.023 | PI7EPO | Rijswijk | JO22EA | | Halo | Omni | 20 | PA5DD | QRT | 10/09 PA0EZ |
| 70.025 | GB3MCB | St Austell | IO70OJ | 320 | 2 el Yagi | 45° | 40 | M0PSB | QRV | 06/14 G6JYB |
| 70.027 | GB3CFG | Carrickfergus | IO74CR | 271 | 2 x 3 el Yagi | 45°/135° | 65 | GI0GDP | QRV | 05/09 G0HIK |
| 70.029 | S55ZMB | | JN76VK | 250 | 4 el Yagi | 310° | 5 | S51DI | | 09/99 70MHz.org |
| 70.030 | S56A | | JN76GB | | | | 1 | | Personal | 06/98 G3NKS |
| 70.030 | G Personal Beacons | | | | | | | | | |
| 70.033 | OH5RBG | Kouvla | KP30HV | 140 | 7 dBi | 232° | 15 | | | 07/11 OH6DD |
| 70.035 | OY6BEC | | IP62OA | 300 | | 225° | 25 | | | 11/05 OZ2TG |
| 70.040 | SV1FOUR | Athens | KM17UX | | Halo | | 10 | SV1DH | QRV | 10/06 OD5TE |
| 70.045 | OE5QL | | JN78CJ | 840 | Vertical | Omni | See info | Click Here | 01-Jun-09 | 04/09 OE5MPL |
| 70.050 | GB3RAL | | IO91EN | 140 | | | 15 | G0MJW | QRV | 06/08 G6JYB |
| 70.050 | Proposed for IARU IBP | | | | | | | | | 06/02 G3UUT |
| 70.060 | HG1BVC | Hörman-forrás | JN87FI | 725 | 2el Yagi | 315° | 10 | HA1YA | | 03/13 HA5NF |

| Freq | Call | Nearest Town | Locator | m ASL | Antenna | Direction | Power | Info | Status | Last update |
|---|---|---|---|---|---|---|---|---|---|---|
| 70.063 | LA2VHF | | JP53EG | | 4 el | 015° | 25 | LA0BY | PLANNED | 01/10 LA0BY |
| 70.065 | LA5VHF | | JP48AD | 85 | GP | Omni | 20 | LA4YGA | PLANNED | 11/10 LA0BY |
| 70.001 | HG8BVC | Kisrata | KN06HT | 05 | Crossed Dipoles | Omni | 5 | HA8MV | | 03/13 HA5NF |
| 70.061 | LA7VHF | Maalselv | JP99EC | | 4 el | 190° | 40 | LA0BY | QRV | 12/14 LA0BY |
| 70.085 | A92C/B | Mina Salman | LL56HE | | Dipole | Omni | 10 | | | 06/12 A92IO |
| 70.091 | IW9GDC/B | Messina | JM78SD | | Big Wheel | Omni | 10 | IW9GDC | | 08/13 IW9GDC |
| 70.109 | IZ1DYE/B | Giaveno TO | JN35PA | 1350 | Dipole | Omni | 5 | | | 04/13 IK1YWB |
| 70.114 | 5B4CY | Zyyi, Cyprus | KM64PR | 30 | 4 el Yagi | 315° | 15 | 5B4BBC | | 06/98 PA3BFM |
| 70.117 | OK0EE | Zdar nad Sazavou | JN89CK | 520 | Crossed Dipoles | Omni | 10 | OK1CDJ | | 11/13 OK1HH |
| 70.130 | EI4RF | Dublin | IO63WD | 120 | 2 x 5 el Yagi | 45°/135° seq | 25 | EI9GK | | 00/95 G3NKS |
| 70.159 | CS5ALG | Serra do Caldeirao | IM67AH | | Dipole | | 5 | DCI | | [illegible] CT1END |
| 70.160 | CS4BEM | Ilha Terceira / Acores | [illegible] | | Dipole | | 5 | URA | | 03/14 CT1END |
| 70.164 | CS5BFM | Fazendas de Almeirim | IM59RD | 12 | Dipole | 45°/225° | 8 | ARR | | 03/14 CT1END |
| 144.282 | W1RJA/B | Rhode Is | FN41CJ | 140 | 5 el Yagi | | 500 | W1RJA | QRT | 06/98 K1ZZ |
| 144.300 | VE1SMU/B | Nova Scotia | FN84CM | | 4 x 9 el Yagi | 61° | 4.8kW | VE1KG | | 06/98 VE1KG |
| 144.400 | VO1?? | Transatlantic beacon | | | Ground plane | Omni | 1 | VO1NA | | 03/00 VO1NA |
| 144.401 | CU????? | | | | | | | CU2IF | Planning | 06/05 G0RDI |
| 144.402 | OY6BEC | | IP62OA | 300 | 2 x 4 el Yagi | 45°/135° | 50 | | QRT | 03/04 OZ2TG |
| 144.402 | EA8VHF | Grand Canary Is | IL28GC | | | Omni | 10 | | | 06/98 EA2SG |
| 144.403 | EI2WRB | Portlaw | IO62IG | 248 | 5 el Yagi | 95° | 200 | EI6GY | Not operational | 04/14 EI8JA |
| 144.404 | IW1AVR/B | Giaveno TO | JN35PA | 1350 | Big Wheel | Omni | 5 | | | 04/13 IK1YWB |
| 144.404 | ED1ZAG | Cerceda | IN53RE | 615 | Big Wheel | Omni | 25 | EB1IVY | | 10/13 EB1IVY |
| 144.405 | F5ZRB | Quistinic | IN87KW | 165 | 7 ele Yagi | 215° | 40 TX | F6ETI | Trans Atl. | 03/13 F6HTJ |
| 144.406 | CT1ART | Serra do Caldeirao | IM67AH | 510 | Loop | Omni | 40 | CT1EPS | | 01/08 CT1HZE |
| 144.407 | GB3SSS | Poldhu | IO70IA | 50 | 8/8 slot Yagi | 300° | 200 | G7THT | Trans Atl | 06/14 G6JYB |
| 144.408 | CU7VHF | Faial | HM58QM | 1000 | 6 el Yagi | 290° | 25 | | PLANNED | 07/08 CT1HZE |
| 144.408 | OZ0FOX | | JO55IL | 25 | Clover Leaf | Omni | 0.25 | | | 11/05 OZ2TG |
| 144.409 | F5ZSF | Lannion | IN88GS | 145 | 9 el Yagi | 90° | 5 TX | F6DBI | | 06/12 F6HTJ |
| 144.410 | DB0MFI | DOK T 21 | JN58KR | | | Omni | 10 | DG1MFI | PLANNING | 07/04 DJ3HW |
| 144.410 | DB0SI | Schwerin DOK V 14 | JO53QP | 90 | Big wheel | Omni | 10 TX | DL1SUZ | | 07/04 DJ3HW |
| 144.410 | ZS2VHF | Port Elizabeth | KF25UX | | 5 el Yagi | 45° | 160 | ZS2FM | | 09/99 ZS5JF |
| 144.410 | YO2X | Timisoara | KN05PS | | Turnstile | | 3/0.5 | YO2IS | | 06/99 YO2IS |
| 144.410 | CS5BLA | Aldeia de Chaos | IM57PX | 295 | Vertical | | 2.5 | CT4RK | QRV | 04/09 CT1FBF |
| 144.412 | SK4MPI | Borlaenge | JP70NJ | 520 | 4 x 6 el Yagi | 45°/315° | 1500 | SM4HFI | | 04/06 SM6CEN |
| 144.4125 | SV1SV-VHF | | KM18UE | | 1/4 wave | | 0.5 | | Not coordinated | 06/10 OD5TE |
| 144.413 | F5ZXT | Var | JN33 | | | | | F5PVX | PLANNED | 06/12 F6HTJ |
| 144.414 | DB0JW | Wurselen DOK G 05 | JO30DU | 238 | 7 el Yagi | 22° | 50 | DL9KAS | | 07/04 DJ3HW |
| 144.415 | IQ2MI/B | M.Te Rosa-Vercelli | JN35WW | 4559 | 5 el Yagi | 110° | 0.5 | | | 04/13 IK1YWB |
| 144.415 | ED7YAD | | | | | | | EA7UU | PLANNED | 01/09 EA7UU |
| 144.416 | PI7CIS | Scheveningen | JO22DC | 40 | Dipole | 90/270° | 50 | PA0CIS | | 07/03 PE2KP |
| 144.417 | OH9VHF | Pirttikoski | KP36OI | 310 | 7 dBd gain | 200° | 160 | | | 07/11 OH6DD |
| 144.417 | CU3??? | Azores | HM68LL | | | | | CU3EQ | Planning | 02/07 CT1HZE |
| 144.418 | ON0VHF | Louvain-La-Neuve | JO20HP | 141 | | | 25 | ON7ZV | | 05/07 ON4PC |
| 144.419 | IQ2CY/B | Cremona | JN55AD | 46 | Big wheel | Omni | 10 | | | 04/13 IK1YWB |
| 144.4195 | LZ0DVK | Sofia | KN12QO | 711 | HB9CV | 320° | 10 | LZ1DVK | QRV / Not Coordi | 10/13 LZ1NY |
| 144.420 | HG5BVB | János-hegy | JN97LM | 485 | Big Wheel | Omni | 3.5 | HA5BDJ | Temp. QRT | 03/13 HA5NF |
| 144.420 | OD5KU/B | | KM73UW | 750 | 3 el Yagi | 300° | 80 | OD5KU | QRV | 09/11 OD5TE |
| 144.420 | DB0RTL | DOK P 60 | JN48PL | 480 | Big wheel | Omni | 15 | DL8SDL | PLANNING | 07/04 DJ3HW |
| 144.421 | OZ???? | | | | | | QRO | Planned | Transatlantic | 11/03 OZ7IS |
| 144.421 | CS3DUB | Madeira | IM12OP | | | | | CT3HF | Testing | 02/07 CT3HF |
| 144.422 | DB0TAU | DOK F 11 | JO40HG | 326 | 4 x 4 el Yagi | Omni | 15 | DL4FCS | | 07/04 DJ3HW |
| 144.423 | PI7HVN | Heerenveen | JO22WW | 52 | | Omni | 10 | PA3FHY | | 07/03 PE2KP |
| 144.425 | F5ZAM | Blaringhem | JO10EQ | 99 | Big wheel | Omni | 10 | F6BPB | | 06/12 F6HTJ |
| 144.425 | SR9VHK | Siemianowice | JO90MH | 350 | Dipole | 150°/300° | 3 | SP9BGS | | 04/07 SP6LB |
| 144.426 | EA6VHF | San Jose, Ibiza | JM08PV | 150 | | Omni | 20 | EA6FB | | 06/98 EA2SG |
| 144.427 | OK0EJ | Frydek-Mistek | JN99FN | 1323 | 4 el Yagi | 270° | 0.3 | OK2UWF | | 08/01 OK1HH |
| 144.427 | PI7PRO | Nieuwegein | JO22NA | 10 | Halo | Omni | 10 | PI4VRZ | | 07/03 PE2KP |
| 144.428 | DB0JT | Oberndorf DOK C 16 | JN67JT | 785 | 4 x Dipole | 0° | 30 | DJ8QP | | 07/04 DJ3HW |
| 144.428 | SR3VHC | Chelmce | JO91CQ | 180 | 2 x Dipole | Omni | 1 | SP3DRT | | 04/05 SP6LB |
| 144.429 | IQ3MF/B | Cormons GO | JN65RW | 130 | 2 x Turnstile | Omni | 10 | IV3HWT | | 04/13 IK1YWB |
| 144.430 | GB3VHF | Fairseat, Kent | JO01EH | 205 | 2 x 3 el Yagi | 288°/348° | 30 | G0FDZ | | 05/10 G6JYB |
| 144.431 | 9A0BVH | | JN85JO | 489 | V Dipole | Omni | 1 | | QRT ? | 03/95 9A2MP |
| 144.432 | 9H1A | Malta | JM75FV | 160 | Turnstile | Omni | 1.5 | 9H1BT | QRT ? | 05/97 9H1PA |
| 144.433 | OH7VHF | Tuupovaara | KP52IJ | 165 | 2 x Big Wheel | Omni | 50 | | | 07/11 OH6DD |
| 144.433 | TF? | | | | | | | | Planned | 03/98 TF3AOT |
| 144.433 | IZ3DVW/B | Monselice PD | JN55VF | 25 | 3 el Yagi | 30° | 10 | | | 04/13 IK1YWB |
| 144.434 | DB0LBV | DOK S 30 | JO61EH | 232 | 2 x Dipole | Omni | 0.4 TX | DL1LWM | | 07/04 DJ3HW |
| 144.435 | HB9OK | Monte Generoso | JN45MW | | | | | | | 10/09 IK7UXW |
| 144.435 | SK2VHG | Kiruna | KP07MV | 535 | 16 el Yagi | 180° | 800 | SK2CP | | 05/10 SM2UHF |
| 144.435 | IQ5MS/B | Massa MS | JN54AB | | Big Wheel | Omni | 3 | | | 04/13 IK1YWB |
| 144.436 | DM0DUB | Salzburger Kopf | JO40AQ | 700 | | | | DM8MM | Was DA5DUB ? | 03/14 PA0O |
| 144.437 | LA1VHF | Oslo | JO59MS | 358 | X300 (Vert) | Omni | 20 | LA4PE | Temp QRT | 04/04 LA0BY |
| 144.437 | F1ZXK | Preaux | JN18KF | 166 | | Omni | 30/10/3/1.3 | F4BUC | | 03/13 F6HTJ |
| 144.438 | 3A2B | Monaco | JN33RR | 50 | Ground Plane | Omni | 50 | 3A2LF | QRT ??? | 08/03 F6HTJ |
| 144.438 | OK0EO | Olomouc | JN89QQ | 602 | Ring dipole | Omni | 0.05 | OK2VI X | Planned | 00/01 OK1HH |
| 144.438 | F1ZXK | Preaux | JN18KF | 100 | Big Wheel | Omni | 10 | F4BUC | QRV | 05/09 F6HTJ |
| 144.439 | OZ3VHF | | JO55HM | 78 | Clover Leaf | Omni | 0.5 | | Temp QTH | 11/05 OZ2TG |
| 144.440 | DB0RG | DOK Z 47 | JO51GO | | | Omni | 1 | DL4OAN | | 07/04 DJ3HW |
| 144.440 | DL0UH | Melsungen DOK Z 25 | JO41RD | 385 | V Dipole | Omni | 1 | DJ3KO | | 03/99 DJ3TF |
| 144.440 | ZS0VST | Orange Free State | | | 8 el Yagi | 0° | 30 | ZS4NS | Planned | 03/00 ZS5JF |
| 144.441 | LA4VHF | Egersund | JO28WL | 50 | 2 x 8 el Yagi | 200°/0° | 200 | LA1YCA | Planned | 05/06 LA0BY |
| 144.442 | IK4PNJ/B | Pianoro BO | JN54QK | 375 | 2 x Cross Dipole | Omni | 10 | | | 04/13 IK1YWB |
| 144.443 | OH2VHF | Inkoo | KP20BB | 65 | 9/9 dBi | 20°/230° | 130/130 | OH2LNM | | 07/11 OH6DD |

| Freq | Call | Nearest Town | Locator | m ASL | Antenna | Direction | Power | Info | Status | Last update |
|---|---|---|---|---|---|---|---|---|---|---|
| 144.444 | DB0KI | Bayreuth DOK Z42 | JO50WC | 1025 | Dipole | Omni | 2.5 | DC9NL | | 07/04 DJ3HW |
| 144.444 | IQ5LU/B | Lucca | JN53GW | 40 | Big wheel | Omni | 6 | | | 04/13 IK1YWB |
| 144.445 | GB3LER | Lerwick | IP90JD | 108 | 2 x 6 el Yagi | 45°/135° | 500/500 | G6JYB | QRT | 04/09 G6JYB |
| 144.445 | ED1YCA | | IN73AL | | Loop | Omni | 10 | EB1TR | QRV | 06/13 EB1TR |
| 144.446 | OK0EB | Ceske Budejovice | JN78DU | 1084 | Miniwheel | Omni | 0.07/0.007 | OK1APG | | 08/01 OK1HH |
| 144.447 | SK1VHF | Klintehamn | JO97CJ | 65 | 2 x Cloverleaf | Omni | 10 | | | 04/06 SM6CEN |
| 144.448 | SK6VHF | Tjorn Island | JO57TX | 120 | Loop | Omni | 10 | | | 04/06 SM6CEN |
| 144.448 | HB9HB | Biel | JN37OE | 1300 | 3 el Yagi | 345° | 120 | HB9AMH | | 09/99 HB9/G4KLX |
| 144.450 | DL0UB | Trebbin DOK Z 20 | JO62KK | 120 | 4 x Dipole | Omni | 10 TX | DL7ACG | | 07/04 DJ3HW |
| 144.450 | F5ZVJ | Remoulins | JN24GB | 300 | Big wheel | Omni | 5 | F5IHN | | 06/12 F6HTJ |
| 144.451 | LA7VHF | Maalselv | JP99EC | 320 | 10 el Yagi | 190° | 100 | LA5TFA | QRV | 12/14 LA0BY |
| 144.452 | OK0EC | As | JO60CF | 778 | 3 el Yagi | 90° | 0.7 | OK1MO | | 08/01 OK1HH |
| 144.453 | GB3ANG | Dundee | IO86MN | 370 | 4 el Yagi | 160° | 20 | GM4ZUK | | 06/02 G3UUT |
| 144.454 | IQ0AH/B | Olbia SS | JN40QW | 350 | Turnstile | Omni | 1 | | | 04/13 IK1YWB |
| 144.455 | F5ZXV | Nancy | JN38CO | 238 | Halo | Omni | 2.5 | F5OOM | | 06/12 F6HTJ |
| 144.455 | OH5ADB | Hamina | KP30NN | 65 | Dipole | 135°/315° | 0.1 | | | 07/11 OH6DD |
| 144.456 | DB0RHN | Rhoen DOK Z 62 | JO50AL | 930 | Dipole | 0°/180° | 1 TX | DG3NAE | | 07/04 DJ3HW |
| 144.456 | OM0MVC | Zvolen | JN98NO | 400 | 5el Yagi | 270° | 5 TX | OM7AC | | 10/10 OM7AC |
| 144.457 | SK2VHF | Vindeln | JP94TF | 300 | 2 x 10 el Yagi | 0°/225° | 100 | | | 04/06 SM6CEN |
| 144.458 | F1ZAT | Brive | JN05VE | 578 | Big wheel | Omni | 3 | F1HSU | | 06/12 F6HTJ |
| 144.459 | LA5VHF | Bodo | JP77KI | 260 | 2 x 6 el Quad | 15°/180° | 100 | LA1UG | QRT | 04/01 LA0BY |
| 144.460 | HG1BVA | Kétvölgy | JN86CW | 370 | 2 x Big Wheel | Omni | 5 | HA1YA | | 03/13 HA5NF |
| 144.460 | TF3VVV | Reykjavik | HP94GF | 140 | | Omni | 25 | TF3GW | | 12/07 OZ7IS |
| 144.461 | IQ0OS/B | Ostia RM | JN61DR | | 2 x Hentenna | Omni | | | | 04/13 IK1YWB |
| 144.461 | SK7VHF | Falsterbo | JO65KJ | 25 | 2 x Cloverleaf | Omni | 10 | | | 04/06 SM6CEN |
| 144.462 | SR5VHW | Warsaw | KO02GH | 130 | | Omni | 10 | SQ5MX | QRV | 01/10 SQ5MX |
| 144.4625 | LZ0TWO | Silven | KN22XS | 1500 | Dipole | | 1 | LZ1GHT | QRV / Not Coordi | 10/13 LZ1NY |
| 144.463 | LA2VHF | Melhus | JP53EG | 710 | 12 el Yagi | 15° | 500 | LA1K | | 05/06 LA0BY |
| 144.464 | F1ZDU | Pierre St. Martin | IN92OX | 1700 | 1 / 0.25 | 0° / 180° | | F5FGP | TESTING | 06/12 F6HTJ |
| 144.464 | IK7XLW/B | Gravina BA | JN80FT | | Big Wheel | Omni | 8 | | | 04/13 IK1YWB |
| 144.465 | DF0ANN | DOK B 25 | JN59PL | 630 | V Dipole | Omni | 0.3 TX | DL8ZX | | 07/04 DJ3HW |
| 144.465 | CN8LI | Rabat | IM64 | | 5 el Yagi | 25° | 30 TX | CN8LI | | 05/01 CN8LI |
| 144.466 | OZ4UHF | Osterlars Bornholm Is | JO75LD | 130 | Clover Leaf | Omni | 10 | OZ1HTB | | 11/05 OZ2TG |
| 144.467 | IK7UXW/B | Brindisi | JN80XP | | 5el Yagi | 320° | 10 | | | 04/13 IK1YWB |
| 144.467 | OK0ED | Frydek-Mistek | JN99DQ | 290 | 2 x Dipole | Omni | 0.1 | OK2UWF | | 08/01 OK1HH |
| 144.467 | HB9RR | Zurich | JN47FI | 871 | 4 x Dipole | Omni | | | | 09/99 HB9/G4KLX |
| 144.468 | F1ZAW | Beaune | JN26IX | 561 | Big wheel | Omni | 10 | F1RXC | QRV | 06/12 F6HTJ |
| 144.468 | LA6VHF | Kirkenes | KP59AL | 70 | 9 el Yagi | 210° | 300 | LA4OO | Non op | 05/06 LA0BY |
| 144.469 | GB3MCB | St Austell | IO70OJ | 320 | 3 el Yagi | 45° | 40 | M0PSB | | 06/14 G6JYB |
| 144.470 | OH2VHH | Vantaa | KP20MH | 75 | Halo | Omni | 2 | | Temp. QRT | 07/11 OH6DD |
| 144.470 | OK0EZ | Chrudim | JN79VV | 350 | X dipoles | Omni | 2/0.5 | OK1DXF | TEMP QRT | 07/10 OK1HH |
| 144.471 | OZ7IGY | Jystrup | JO55WM | 104 | Big Wheel | Omni | 40 | OZ7IS | | 01/14 OZ7IS |
| 144.472 | TF? | | | | | | | | Plan ? | 03/98 TF3AOT |
| 144.473 | ED2YAC | Durango | IN83RD | 1016 | 2 el Yagi | 010° | 10 | EB1RL | Planning | 02/09 EB1RL |
| 144.473 | OK0EQ | Namest n./O | JN89BE | 430 | Dipole | Omni | 4 | OK2ZI | PLANNED | 06/10 OK1HH |
| 144.474 | OK0EL | Benecko | JO70SQ | 1030 | Dipole | 90°/270° | 0.005 | OK1AIY | | 08/01 OK1HH |
| 144.475 | DL0SG | DOK U 14 | JN69KA | 1024 | 4 x 4 el Yagi | Omni | 5 TX | DJ4YJ | | 07/04 DJ3HW |
| 144.475 | LY2WN | Jonava | KO25GC | | 2 x Dipole | Omni | 15 | LY2WN | | 01/98 LY2IC |
| 144.475 | YU1VHF | Pozarevac | KN04OO | 200 | 2 x QQ | 135°/337° | 10 | YU1AU | | 03/98 YT1MM |
| 144.476 | F5ZAL | Pic Neulos | JN12LL | 1100 | Big wheel | Omni | 8 | F6HTJ | | 06/12 F6HTJ |
| 144.477 | DB0ABG | DOK U 01 | JN59WI | 522 | Big Wheel | Omni | 4 TX | DJ3TF | Temp.QRT | 07/04 DJ3HW |
| 144.478 | S55ZRS | Mt. Kum | JN76MC | 1219 | Crossed dipoles | Omni | 3 | | | 06/12 S51FB |
| 144.478 | LA3VHF | Mandal | JO38RA | 30 | 9 el Yagi | 180° | 120 | LA3BAA | | 05/06 LA0BY |
| 144.478 | OM0MVA | Bratislava | JN88NE | 570 | Dipole | 90°/270° | 0.11 | OM3ID | | 08/01 OK1HH |
| 144.479 | SR5VHF | Wesola | KO02OF | 130 | Turnstile | Omni | 0.75 | SP5TAT | | 02/95 SP6LB |
| 144.480 | EA3VHF | | JN11MV | | | | | | QRT ?? | 08/03 F6HTJ |
| 144.480 | F0QAD | Paris | JN18BW | 104 | | | 10 | F1FPP | PLANNING | 05/13 F6HTJ |
| 144.480 | LA8VHF | Stavern | JO48XX | 30 | 3x2 el Yagi | 150° | 100 | LA6LCA | | 05/06 LA0BY |
| 144.481 | HG8BVA | Vészto | KN06PW | 85 | 5el Yagi | 295° | 3 | HA8MV | | 03/13 HA5NF |
| 144.481 | SR3VHX | | JO82KL | | | | | | | 08/09 SQ3FYK |
| 144.482 | GB3NGI | Ballymena | IO65VB | 528 | 2 x 4 el Yagi | 45°/135° | 120/120 | GI6ATZ | QRV | 04/09 G6JYB |
| 144.484 | EA8??? | La Palma | IL18 | | | | | DL6FAW | Planned | 03/04 DL6FAW |
| 144.484 | LA9VHF | Egersund | JO28WL | | | | 200 | LA1YCA | Planned | 05/05 LA0BY |
| 144.485 | TK5ZMK | Coti Chiavari | JN41JS | 635 | Big Wheel | Omni | 5 | | QRV | 03/13 F6HTJ |
| 144.485 | IW0DTK/B | Minturno LT | JN61TG | 10 | Halo | Omni | 3 | | | 04/13 IK1YWB |
| 144.486 | DL0PR | Garding DOK Z 69 | JO44JH | 75 | 6 el Yagi | 0°/180° | 200 TX | DL8LD | | 07/04 DJ3HW |
| 144.487 | SR2VHM | Gdansk | JO94HI | 5 | 9el Yagi | 240° | 5 | SQ2BXI | | 04/07 SP6LB |
| 144.487 | 5T5VHF | | IK28AC | | Ground Plane | Omni | 30 | 5T5SN | 144.306 ? | 02/05 5T5SN |
| 144.488 | SV3AQR/B | Amalias | KM07QS | 50 | 3 el Yagi | 320° | 5 | SV3AQR | | 12/04 SV3AQR |
| 144.488 | EI2DKH | Bantry | IO51DN | 56 | 2 x 5el LFA-Q | 270° | 75 | EI8JK | QRV | 10/14 EI8JK |
| 144.490 | DB0FAI | Langerringn DOK T01 | JN58IC | 590 | 16 el Yagi | 305° | 1000 | DL5MCG | | 03/99 DJ3TF |
| 144.855 | LZ1VHF | | KN12PO | | GP | | 1 | | QRT | 10/13 LZ1NY |
| 144.902 | OX3VHF | | GP60QQ | 70 | 2 x 4 el | 45°/90° | 80 | | | 03/04 OZ2TG |
| 144.922 | ZS6TLB | Peitersburg | KG46RC | | 2 x 5 el Yagis | 215° | 10 | | QRT | 09/99 ZS5JF |
| 145.250 | RB9FA | Perm | LO88DA | | Vertical | Omni | 2 | RV9FF | F1A | 09/03 RV9FF |
| | | | | | | | | | | |
| 431.999 | LZ1UHF | | KN12PO | | GP | Omni | 1 | | QRT | 10/13 LZ1NY |
| 432.000 | YO2U | Timisoara | KN05PS | | 4 el Yagi | 315° | 0.2 | YO2IS | | 06/99 YO2IS |
| 432.128 | S55ZNG | Trstelj | JN65UU | 643 | Horizontal Loop | Omni | 0.1 | S50M | | 06/98 S57C |
| 432.400 | OE3XMB | Muckenkogel | JN77TX | 1154 | 9 el Yagi | 337° | 2 | OE3FFC | | 02/05 OE3FFC |
| 432.401 | SK2UHF | Vindeln | JP94WG | 448 | 2 x 2 x 6 el Yagi | 0°/200° | 30 | SK2AT | | 04/06 SM6CEN |
| 432.401 | F5ZBU | Preaux | JN18KF | 166 | 5 | 4 x 6 el | 5 | Omni | TEMP. QRT | 03/13 F6HTJ |
| 432.402 | OY6BEC | | IP62OA | 300 | 7 dB group | 135° | | | QRT | 03/04 OZ2TG |
| 432.404 | F5ZZI | Hyeres | JN33BD | 240 | Big wheel | Omni | 5 | F5PVX | | 11/10 F6HTJ |

| Freq | Call | Nearest Town | Locator | m ASL | Antenna | Direction | Power | Info | Status | Last update |
|---|---|---|---|---|---|---|---|---|---|---|
| 432.405 | SK1UHF | Klintehamn | JO07CJ | [illegible] | Alford Slot | Omni | 50 | | | 04/06 SM6CEN |
| 432.405 | PI7QHN | Zandvoort | JO22KH | 20 | 3 dB Gain | Omni | 2 | PA2EAR | | [illegible] |
| 432.406 | OK0EO | Olomouc | [illegible] | [illegible] | Ring Dipole | Omni | 0.05 | OK2VLX | Planned | 08/01 OK1HH |
| 432.407 | CT1ART | Serra do Caldeirao | IM67AH | 510 | Vertical | Omni | 40 | CT1EPS | | 01/08 CT1HZE |
| 432.407 | PI7YSS | Zutphen | JO32CD | 60 | Halo | Omni | 2 | PA0JAZ | | 03/05 PA0EZ |
| 432.407 | HG8BUA | Vészto | KN06PW | 85 | Big Wheel | Omni | 3 | HA8MV | | 03/13 HA5NF |
| 432.408 | F5ZPH | Quistinic | IN87KW | 105 | [illegible] | 100° | 20 | [illegible] | | 02/09 F6HTJ |
| 432.410 | IW1AVH/B | Giaveno TO | JN35PA | 1350 | Big Wheel | Omni | 3 | | | 04/13 IK1YWB |
| 432.411 | HB9OK | Monte Generoso | JN46MW | | | | | | QRV | 10/00 IK7UXW |
| 432.412 | SK0UHF | Vurberg | JO67EH | 175 | Clover Leaf | Omni | 10 | SM6ESG | | 04/06 SM6CEN |
| 432.413 | F5ZTX | Lacapelle | JN14EB | 625 | 2 x 3 el | 0° / 90° | 40 | [illegible] | [illegible] | [illegible] |
| [illegible] | [illegible] | [illegible] | [illegible] | 307 | 6 dBd gain | 200° | 70 | | | 07/11 OH6DD |
| 432.418 | F1ZQT | Moragne | IN95OX | 80 | Big Wheel | Omni | 1 | F1MMR | | 02/09 F6HTJ |
| 432.420 | F5ZAS | Eyne | JN12BL | 2400 | Big wheel | Omni | 15 | F6HTJ | TEMP QRT | 04/12 F6HTJ |
| 432.425 | CS5BLA | Aldeia do Chaos | IM57PX | 295 | Vertical | | 2.5 | CT4RK | PLANNED | 04/09 CT1FBF |
| 432.425 | LY2WN | Jonava | KO25GC | 96 | 2 x dipole | Omni | 32 | LY2SA | | 03/05 LY2SA |
| 432.428 | SK5BN/B | | JO88LL | 49 | 2 x H. Dipole | Omni | 10-Jan | SM5YLG | | 01/09 SM5YLG |
| 432.428 | HG7BUA | Dobogóko | JN97KR | 700 | Slot | Omni | 2 | HA5NF | | 03/13 HA5NF |
| 432.430 | GB3UHF | Fairseat, Kent | JO01EH | 205 | 2 x 3el Yagi | 288°/348° | | G0FDZ | Planning | 02/15 G6JYB |
| 432.432 | HB9F | Interlaken | JN36XN | 3573 | Corner reflector | 0° | 15 | HB9MHS | | 01/04 PA0EZ |
| 432.432 | OH6UHF | Uusikaarlepyy | KP13GM | 55 | 3 x Big Wheel | Omni | 7 | OH6UH | | 07/11 OH6DD |
| 432.435 | OH5SHF | Kuovola | KP30HV | 147 | 5 dBd | 217° | 20 | | | 07/11 OH6DD |
| 432.435 | IQ5MS/B | Massa MS | JN54AB | | Big Wheel | Omni | 2.5 | | | 04/13 IK1YWB |
| 432.436 | F5ZAA | Nerignac | JN06IH | 205 | Big wheel | Omni | 20 | F5EAN | QRV | 04/12 F6HTJ |
| 432.437 | LA1UHF | Oslo | JO59MS | 380 | X300 (Vert) | Omni | 30 | LA4PE | Temp QRT | 04/04 LA0BY |
| 432.440 | F1ZTV | Cloutons | JN24WX | 2120 | Loop | Omni | 2 | F1LCE | QRV | 03/12 F6HTJ |
| 432.440 | SK7MHH | Faerjestaden | JO86GP | 45 | Alford slot | Omni | 100 | | | 04/06 SM6CEN |
| 432.441 | LA5UHF | Jaeren | JO28UO | 150 | 6 el Yagi | 220° | 300 | LA3EQ | | 05/06 LA0BY |
| 432.442 | CT???? | | | | | | | | Proposed | 06/03 CT1DDN |
| 432.443 | OH2UHF | Inkoo | KP20BB | 65 | 2 x 10 dBi | 20°/230° | 60/60 | | | 07/11 OH6DD |
| 432.444 | F5ZBU | Preaux | JN18KF | 166 | 4 x HB9CV | Omni | 5 | F2AI | | 05/10 F6HTJ |
| 432.445 | IQ5BA/B | Monte Quegna SI | JN53OI | 515 | Cloverleaf | Omni | 3 | | | 04/13 IK1YWB |
| 432.447 | S55ZRS | Mt Kum | JN76MC | 1219 | Slot | Omni | 1 | | | 04/14 S51FB |
| 432.448 | HG3BUA | Misinateto | JN96CC | 585 | Slot | Omni | 2 | HG5AZB | | 03/13 HA5NF |
| 432.449 | OZ1UHF | Frederikshavn | JO57FJ | 150 | Big wheel | Omni | 10 | OZ9NT | | 11/05 OZ2TG |
| 432.450 | ON???? | Leuven | JO20HU | | | | 19 | | Planning | 07/07 ON4PC |
| 432.450 | I5WBE/B | Cornocchio FI | JN53LK | 610 | 2 x Loop | Omni | 6 | | | 04/13 IK1YWB |
| 432.452 | OK0EC | As | JO60CF | 778 | 10 el Yagi | 90° | 1 | OK1MO | PLANNED | 07/10 OK1HH |
| 432.453 | GB3ANG | Dundee | IO86MN | 370 | 9 el Yagi | 170° | 100 | GM4ZUK | | 02/15 G6JYB |
| 432.454 | F5ZZY | Nancy | JN38CO | 238 | Halo | Omni | 3.5 | F5OOM | | 04/12 F6HTJ |
| 432.455 | SK3UHF | Ravson | JP92FW | 200 | 2 x dipole | Omni | 50 | SM3AFT | | 04/06 SM6CEN |
| 432.455 | IZ3KLB/B | Conco VI | JN55TT | 850 | 8 x 5/8 Vertical | Omni | 5 | | | 04/13 IK1YWB |
| 432.460 | SK4BX/B | Garphyttan | JO79LH | 270 | 4 x log periodic | N/E/S/W | 50 | | | 04/06 SM6CEN |
| 432.460 | HG1BUA | Hörman-forrás | JN87FI | 725 | Slot | Omni | 3 | HA1YA | | 03/13 HA5NF |
| 432.462 | SR5UHW | Warsaw | KO02GH | 130 | | Omni | 10 | SQ5MX | PLANNED | 01/10 SQ5MX |
| 432.463 | LA2UHF | Melhus | JP53EG | 710 | 10 el Yagi | 15° | 300 | LA1K | Non op | 05/06 LA0BY |
| 432.465 | DF0ANN | Altdorf | JN59PL | 630 | Big wheel | Omni | 1 TX | DL8ZX | | 04/13 OK1HH |
| 432.466 | OK0EA | Trutnov | JO70UP | 1355 | 2 x 15 el Yagi | 180°/270° | 10 erp | OK1IA | QSY -> .468 | 04/13 OK1HH |
| 432.467 | ON0UHF | Bruxelles | JO20ET | | | | | ON4LC | | 05/07 ON4PC |
| 432.468 | LA6UHF | Kirkenes | KP59AL | 70 | 15 el Yagi | 210° | 40 | LA4OO | Non op | 05/06 LA0BY |
| 432.470 | HG5BUA | János-hegy | JN97LM | 485 | Big Wheel | Omni | 1 | HA5BDJ | Temp. QRT | 03/13 HA5NF |
| 432.470 | GB3MCB | St Austell | IO70OJ | 320 | 4 el Yagi | 45° | 12 | M0PSB | | 02/15 G6JYB |
| 432.471 | OZ7IGY | Jystrup | JO55WM | 105 | Big Wheel | Omni | 40 | OZ7IS | | 01/14 OZ7IS |
| 432.473 | ED2YAE | Durango | IN83RD | 1016 | 2 el Yagi | 010° | 1 | EB1RL | Planning | 02/09 EB1RL |
| 432.473 | OK0EQ | Namest n./O | JN89BE | 430 | Dipole | Omni | 4 | OK2ZI | PLANNED | 06/10 OK1HH |
| 432.477 | DB0ABG | DOK U 01 | JN59WI | 522 | Big Wheel | Omni | 1 TX | DJ3TF | QRT | 03/99 DJ3TF |
| 432.478 | LA3UHF | Mandal | JO38RA | 12 | 13 el Yagi | 180° | 50 | LA3BAA | | 05/06 LA0BY |
| 432.480 | LA8UHF | Tonsberg | JO59FB | 30 | 8 el Yagi | 90°/180° | 50 | LA6LCA | Keys "LA1UHG" | 05/06 LA0BY |
| 432.482 | GB3NGI | Ballymena | IO65VB | | 12 el Yagi | 125° | 250 | GI6ATZ | QRV | 11/14 G6JYB |
| 432.483 | OZ2ALS | Sonderborg | JO45WA | 65 | 4 x dipole | Omni | 40 | OZ9DT | Non op? | 11/05 OZ2TG |
| 432.485 | LA4UHF | Haugesund | JO29PJ | 75 | 3 el Yagi | 250° | 15 | LA9NQ | Non op | 05/06 LA0BY |
| 432.485 | IW0DTK/B | Minturno LT | JN61TG | 10 | Turnstile | Omni | 5 | | | 04/13 IK1YWB |
| 432.486 | OE????? | Salzburg | JN67NT | 1280 | 10 dB | 320° | 50 | OE2WPO | PLANNING | 11/12 OE2WPO |
| 432.487 | F1ZBY | Roc Blanc | JN13TV | 942 | Big wheel | Omni | 5 | F4DVR | | 04/12 F6HTJ |
| 432.488 | DB0AD | | JO40AQ | | | | | | | 02/05 DL7AJA |
| 432.489 | SK7MHL | Lund | JO65OR | 100 | Alford slot | Omni | 40 | SM7ECM | | 04/06 SM6CEN |
| 432.800 | DB0GD | Rhoen | JO50AL | 930 | Dipole | 0°/180° | 1 TX | DG6ZX | --> 432.456 | 03/99 DJ3TF |
| 432.810 | DB0ZW | DOK U 17 | JN69EQ | 825 | Schlitz | Omni | 1 TX | DC9RK | --> 432.410 | 03/99 DJ3TF |
| 432.812 | SV1SV-UHF | | KM18UE | | 1/4 wave | | 0.5 | | Not coordinated | 06/10 OD5TE |
| 432.416 | PI7CIS | Scheveningen | JO22DC | 40 | | Omni | 100 | PA0CIS | | 05/07 GW4DGU |
| 432.830 | LA7UHF | Bergen | JP20LG | 30 | 4 el Yagi | 0° | 200 | LA6LU | --> 432.441 | 04/04 LA0BY |
| 432.835 | ES0UHF | Hiiumaa Island | KO18CW | 105 | Alford slot | Omni | 50 | ES0NW | --> 432.475 | 12/01 SM0KAK |
| 432.840 | DB0KI | Bayreuth | JO50WC | 925 | Dipole | Omni | 10 | DC9NL | --> 432.444 | 03/99 DJ3TF |
| 432.845 | DB0LBV | DOK S 30 | JO61EH | 234 | Schlitz | Omni | 2 TX | DL1LWM | --> 432.434 | 03/99 DJ3TF |
| 432.847 | 9A0BUH | | JN85JO | 489 | V dipole | Omni | 1 | | --> 432.431 | 03/95 9A2MP |
| 432.850 | DL0UB | DOK Z 20 | JO62KK | 120 | Malteser | Omni | 10 TX | DL7ACG | --> 432.450 | 03/99 DJ3TF |
| 432.870 | EI2WRB | Portlaw | IO62IJ | 248 | 5 el Yagi | 95° | 250 | EI9GO | --> 432.403 | 07/96 G8GXP |
| 432.870 | OK0EZ | Chrudim | JN79VV | 350 | X dipoles | Omni | 2.5 | OK1DXF | TEMP QRT | 07/10 OK1HH |
| 432.423 | PI7HVN | Heerenveen | JO22WW | 50 | Horizontal | Omni | 0.5 | PE1HUE | | 12/04 PA0EZ |
| 432.875 | OH7UHF | Kuopio | KP32TW | 215 | 6 dBd | 225° | 15/1.5/.15 | | --> 432.490 | 08/06 OH6DD |
| 432.886 | OK0EP | Sumperk | JO80OB | 1505 | 2 x 4 el Yagi | 150°/280° | 2 x 5 | OK1VPZ | Unable to QSY | 10/13 OK1VPZ |
| 432.888 | OM0MUA | Bratislava | JN88NE | 570 | Dipole | 90°/270° | 0.08 | OM3ID | --> 432.478 | 08/01 OK1HH |
| 432.890 | GB3SUT | Sutton Coldfield | IO92CO | 270 | 2 x 8 el Yagi | 0°/135° | 10 | G6JYB | QRT | 01/09 G6JYB |
| 432.900 | DB0YI | Hildesheim Z 35 | JO42XC | 480 | Big wheel | Omni | 3 TX | DL4AS | --> 432.425 | 03/99 DJ3TF |

| Freq | Call | Nearest Town | Locator | m ASL | Antenna | Direction | Power | Info | Status | Last update |
|---|---|---|---|---|---|---|---|---|---|---|
| 432.900 | ZS6UHF | Pietersburg | KG46RC | | 13 el Yagi | 215° | 10 | | QRT | 09/99 ZS5JF |
| 432.908 | EA8UHF | Grand Canary Is | IL28GC | | | Omni | 10 | | --> 432.488 | 06/98 EA2SG |
| 432.910 | GB3MLY | Emley Moor | IO93EO | 600 | 6 el Yagi | 150° | 40 | G3PYB | QRT | 03/09 G6JYB |
| 432.918 | EA6UHF | Ibiza Is | JM08PV | | | Omni | 10 | QRT | --> 432.425 | 08/03 F6HTJ |
| 432.920 | DB0UBI | DOK N 59 | JO42GE | 125 | 8el Coll | 45° | 12 | DD8QA | --> 432.415 | 03/99 DJ3TF |
| 432.925 | DB0JG | Bocholt DOK N17 | JO31GT | 45 | Clover Leaf | Omni | 1 TX | DL3QP | --> 432.412 | 03/99 DJ3TF |
| 432.934 | GB3BSL | Bristol | IO81QJ | 252 | 4 x 3 el Yagi | 90° | 250 | GW8AWM | NOW QRT | 06/13 G6JYB |
| 432.940 | DB0RG | Melsungen DOK Z25 | JO41RD | 385 | V?Dipole | Omni | 1 | DJ3KO | --> 432.440 | 03/99 DJ3TF |
| 432.945 | DB0OS | Erndtebruck DOK N32 | JO40CW | 730 | 2 el Yagi | 270° | 0.3 | DG6YW | --> 432.417 | 03/99 DJ3TF |
| 432.945 | DB0LB | DOK P06 | JN48NV | 367 | Corner dipole | 0°/180° | 0.2 TX | DK3PS | --> 432.417 | 03/99 DJ3TF |
| 432.950 | DB0IH | Oberthal DOK Q 18 | JN39ML | 630 | Big wheel | Omni | 1 | DC8DV | --> 432.447 | 03/99 DJ3TF |
| 432.965 | GB3LER | Lerwick | IP90JD | 104 | 12 el Yagi | 165° | 675 | G6JYB | QRT | 04/09 G6JYB |
| 432.970 | OK0EB | Ceske Budejovice | JN78DU | 1084 | Mini Wheel | Omni | 0.03/0.16 | OK1APG | --> 432.446 | 08/01 OK1HH |
| 432.975 | DB0JW | Aachen DOK G 05 | JO30DU | 238 | 2 x 11 el Yagi | 45° | 50 | DL9KAS | --> 432.414 | 03/99 DJ3TF |
| 432.975 | DB0SGA | DOK U 14 | JN69KA | 1024 | 4 x 11 Yagi | Omni | 5 TX | DJ4YJ | --> 432.475 | 03/99 DJ3TF |
| 432.980 | S55ZCE | Sv. Jungert | JN76OH | 574 | Ground plane (V) | Omni | 0.07 | S51KQ | --> 432.448 | 06/98 S57C |
| 432.982 | SR5UHF | Wesola | KO02OF | 130 | Turnstile | Omni | 0.25 | SP5TAT | --> 432.479 | 02/95 SP6LB |
| 432.990 | DB0VC | DOK Z 10 | JO54IF | 300 | 4 x DQ | Omni | 10 | DL8LAO | --> 432.420 | 03/99 DJ3TF |
| 1296.000 | W2ETI | EME Beacon | FN21TA | | 4x RH helix | Tracks moon | 64 dBmi | www.setileague.org | | 06/02 G3UUT |
| 1296.000 | YO2U | Timisoara | KN05PS | | 10 el Yagi | 0° | 0.05 | YO2IS | | 06/99 YO2IS |
| 1296.050 | HB9BBD/P | Goldau | JN47GA | 1674 | 3 x Stacked Dipole | 0° | 9 | HB9BBD | | 03/07 HB9BBD |
| 1296.050 | S55ZSE | Kokos | JN65WP | 620 | Slot | Omni | 0.3 | | S53MV | 06/12 S51FB |
| 1296.063 | S55ZNG | Trstelj | JN65UU | 643 | V-J Slot | Omni | 0.1 | S50M | | 06/98 S57C |
| 1296.090 | S55ZMS | Dolina | JN86CR | 350 | Slot | Omni | 0.3 | | S53M | 06/12 S51FB |
| 1296.380 | S55ZRS | Mt. Kum | JN76MC | 1219 | Slot | Omni | 0.2 | | | 06/12 S51FB |
| 1296.457 | IZ1ERR/B | Bagnolo CN | JN34OS | 1550 | Quad | 45° | 0.2 | | | 04/13 IK1YWB |
| 1296.739 | F5ZBS | Strasbourg | JN38PJ | 1070 | Big wheel | Omni | 4 | F6BUF | | 06/12 F6HTJ |
| 1296.749 | IK5CON/B | Camaiore LU | JN53CW | 30 | Big Wheel | Omni | 10 | | | 04/13 IK1YWB |
| 1296.790 | IQ7FG/B | Rigano Garganicio FG | JN71TQ | 590 | 16 Slot | | 1 | | | 04/13 IK1YWB |
| 1296.800 | CS5BCT | Cadaval - Montejunto | IM59KF | 660 | Alford slot | Omni | 20 | CT1ARR | PLANNED | 03/11 CT1END |
| 1296.800 | DB0GD | DOK Z 62 | JO50AL | 930 | Dipol | Omni | 1 TX | DG6ZX | | 03/99 DJ3TF |
| 1296.800 | DB0HEG | Wassertrudingen | JN59GB | 700 | 4 x Slot | Omni | 0.5 TX | DL2QQ | | 01/06 DF9IC |
| 1296.800 | SK6MHI | Hoenoe | JO57TQ | 40 | Alford slot | Omni | 30 | SM6CEN | | 04/06 SM6CEN |
| 1296.800 | OE3XMB | Muckenhogel | JN77TX | 1154 | 2 x Quad | 337° | 0.1 | OE3FFC | | 02/05 OE3FFC |
| 1296.805 | SK6UHI | Tjorn | JO57TX | 120 | Alford slot | Omni | 30 | SM6CEN | | 04/06 SM6CEN |
| 1296.805 | DB0RIG | DOK P 17 | JN48WQ | 780 | 4 x Yagi Box | Omni | 50 | DG9SQ | | 03/99 DJ3TF |
| 1296.810 | SK7MHF | Nassjo | JO77IP | 505 | Alford slot | Omni | 30 | SM5GEP | | 04/06 SM6CEN |
| 1296.810 | PI7DIJ | Dokkum | JO33AI | 20 | 9 el Yagi | 200° | 1 | PA3DIJ | QRT | 10/09 PA0EZ |
| 1296.810 | DB0ZW | Weiden | JN69EQ | 825 | Slot | Omni | 1 | DC9RK | | 03/99 DJ3TF |
| 1296.810 | LZ0VVV | Sofia | KN12PQ | 597 | Big Wheel | Omni | 1 | LZ1NY | QRV / Not Coordi | 10/13 LZ1NY |
| 1296.812 | F1ZBI | Le Petit Ballon | JN37NX | 1278 | Quad | 180° | 0.8 | F5AHO | | 06/12 F6HTJ |
| 1296.815 | DB0VI | Saarbrucken | JN39NF | 400 | 13 el Yagi | | 0.2 | DL4VCG | | 07/04 PE1IWT |
| 1296.816 | F1ZTF | Segonzac | IN95VO | 125 | Big wheel | Omni | 10 | F1MMR | | 06/12 F6HTJ |
| 1296.818 | IQ0RM/B | Roma | JN61FW | 110 | Alford Slot | Omni | 1 | | | 04/13 IK1YWB |
| 1296.820 | LA5SHF | Stavanger | JO28UX | 50 | 8 el Yagi | 240° | 50 | LA3EQ | | 05/06 LA0BY |
| 1296.820 | DB0OT | Lathen | JO32QR | 80 | Big wheel | Omni | 1 | DL1BFZ | | 02/06 SM7LCB |
| 1296.825 | F5ZRS | Chamrousse | JN25UD | 1700 | Horn | 337° | 0.1 | F5LGJ | | 06/12 F6HTJ |
| 1296.825 | IQ4AD/B | Monte Cassio PR | JN54AO | 1010 | 2 el Yagi | 35° | 2 | | | 04/13 IK1YWB |
| 1296.825 | CS5BLA | Aldeia de Chaos | IM57px | 295 | Horizontal | | 2.5 | CT4RK | PLANNED | 04/09 CT1FBF |
| 1296.825 | DB0HF | Wandsbek | JO53BO | 65 | Big wheel | Omni | 0.3 TX | DK2NH | | 06/05 GM4OGI |
| 1296.825 | OE1XTB | Vienna | JN88EE | 170 | 4 x dipole | Omni | 10 | OE1MOS | QRT | 09/99 OE1MCU |
| 1296.825 | DB0ABG | DOK U 01 | JN59WI | 522 | Slot | Omni | 0.5 TX | DJ3TF | | 03/99 DJ3TF |
| 1296.830 | GB3MHZ | Martlesham | JO02PB | 80 | 2 x 16 Slot wg | 90°/270° | 700 | G7OCD | | 06/14 G6JYB |
| 1296.830 | SR6LHZ | Sniezne Kotly | JO70SS | 1490 | Dipole | 40° | 1 | SP6LB | | 04/07 SP6LB |
| 1296.830 | SV3GKE/B | Lefkada | KM08HQ | 1150 | 17 el | 320° | 5 | | | 10/09 IK7UXW |
| 1296.835 | DB0AJ | DOK C 09 | JN57 | 620 | 12 el Yagi | 0° | 50 | DK2RV | QRT | 03/99 DJ3TF |
| 1296.835 | SK0UHG | Vaellingby | JO89WI | 60 | Horizontal | Omni | 10 | | | 04/06 SM6CEN |
| 1296.840 | OH6SHF | Uusikaarlepyy | KP13GM | 56 | 4 x EIA | 180° | 30 | OH6DD | | 07/11 OH6DD |
| 1296.840 | DB0KI | Bayreuth | JO50WC | 925 | Slot | Omni | 80 | DC9NL | | 02/06 G4EAT |
| 1296.845 | DB0LBV | Leipzig | JO61EH | 234 | 4 x Slot | Omni | 2 | DL1LWM | | 02/06 OZ1FF |
| 1296.845 | SR3LHE | Chelmce | JO91CQ | 180 | 2 x Helix | Omni | 2 | SP3DRT | Ex SR3SHF | 04/07 SP6LB |
| 1296.847 | F5ZBM | Faviers | JN18JS | 160 | Alford Slot | Omni | 10 | F6ACA | | 06/12 F6HTJ |
| 1296.850 | DL0UB | Berlin | JO62KK | 120 | 4 x Box | Omni | 10 | DL7ACG | | 12/05 DL3YEE |
| 1296.850 | GB3FRS | Farnborough | IO91OG | 36 | Wimo PA-23R | 340° | 25 | G8ATK | | 06/08 G6JYB |
| 1296.854 | F1ZBK | Nancy | JN38BP | 420 | | Omni | 5 | F1DND | TESTING | 06/12 F6HTJ |
| 1296.854 | DB0JO | Witten | JO31SL | 312 | 4 x 15 el Yagi | 270° | 350 | DG8DCI | | 02/06 OZ1CTZ |
| 1296.855 | OZ3UHF | | JO56CE | 150 | 5 el Yagi | 180° | 6 | OZ1GMP | QRT Q2 99 | 09/99 OZ7IS |
| 1296.855 | SK3UHG | Nordingra | JP92FW | 200 | Horizontal | Omni | 10 | SM6CEN | Temp QRT | 04/06 SM6CEN |
| 1296.857 | ON0NR | Namur | JO20KR | 272 | 16 x slot | Omni | 100 | | PLANNING | 02/10 ON5QI |
| 1296.860 | LA8SHF | Tonsberg | JO59FB | 30 | 13 dB Horn | 180° | 60 | LA6LCA | Keys "LA1UHG" | 05/06 LA0BY |
| 1296.860 | IK0XUH/B | Ardea RM | JN61GO | 27 | Colinear | Omni | 0.5 | | | 04/13 IK1YWB |
| 1296.860 | DB0LB | | JN48NV | 367 | Big Wheel | Omni | 0.3 | DK3PS | | 01/06 OE5VRL |
| 1296.860 | GB3MCB | St Austell | IO70OJ | 300 | 23 el Yagi | 45° | 48 | M0PSB | QRV | 06/14 G6JYB |
| 1296.862 | SR5LHW | Warsaw | KO02GH | 130 | | Omni | 5 | SQ5MX | PLANNED | 01/10 SQ5MX |
| 1296.862 | F1ZAK | Istres | JN23MM | 114 | Slotted WG | Omni | 20 | F1AAM | | 06/12 F6HTJ |
| 1296.863 | HG5BUB | János-hegy | JN97LM | 485 | Slot | Omni | 0.8 | HA5BDJ | Temp. QRT | 03/13 HA5NF |
| 1296.865 | DB0JK | Koln | JO30LX | 260 | 4 x 8 el Yagi | Omni | 40 | DK2KA | | 02/06 OZ1FF |
| 1296.865 | HB9EME | Neuchatel | JN37KB | 1145 | 10 dBi | Omni | 12 | HB9HLM | | 02/06 F5PEJ |
| 1296.870 | DB0IBB | Westernkappeln | JO32VG | 200 | 4 x Slot | Omni | 170 | DB7QW | | 10/05 G4EAT |
| 1296.870 | LA2SHF | Trondheim | | | | | | LA1BFA | Planned | 05/06 LA0BY |
| 1296.872 | F1ZMT | Le Mans | JN07CX | 85 | Panel | 180° | 10 | F1BJD | | 06/12 F6HTJ |
| 1296.875 | GB3USK | Bristol | IO81QJ | 235 | 2 x 15el Yagi | 90° | 250 | GW8AWM | NOW QRT | 06/13 G6JYB |
| 1296.875 | OH6??? | Viitasaari | KP23WC | | | | | | PLANNED | 07/11 OH6DD |

| Freq | Call | Nearest Town | Locator | m ASL | Antenna | Direction | Power | Info | Status | Last update |
|---|---|---|---|---|---|---|---|---|---|---|
| 1296.875 | HB9[illegible] | Monte Generoso | JN45MW | | | | | | QRV | 10/09 IK7UXW |
| 1296.875 | F6CGJ | Landornoqu | IN70UK | 121 | Quad | 00° | 8 | F6CGJ | | 00/12 F6HTJ |
| 1296.878 | DB0FAI | DOK F 01 | JN58IC | 610 | | Omni | 10 | DL5MCG | | 03/99 DJ3TF |
| 1296.880 | ON0SHF | Ellignies Saint Anne | JO10UN | 63 | | | 5 | ON5PX | | 05/07 ON4PC |
| 1296.880 | LA3SHF | Flekkeroy | JO38XB | 5 | 2 x 15 el Yagi | 180° | 10 | LA8AK | QRT | 04/04 LA0BY |
| 1296.883 | DB0INN | DOK C 15 | JN68GI | 504 | Schlitz | Omni | 1 TX | DL3MBG | | 03/99 DJ3TF |
| 1296.885 | OY6BEC | | IP62OA | 300 | | | 20 | | Non op | 11/05 OZ2TG |
| 1296.885 | DB0TUD | Dresden | JO61UA | 260 | Quad | Omni | | DL4DTU | | 02/05 DG1BHA |
| 1296.885 | OE3XEA | Kaiserkogel | JN78SB | 725 | | | 1 | OE3EFS | QRT | 09/99 OE1MCU |
| 1296.885 | SR1LHX | Czluchow | JO83QP | 255 | DL7KN | N - S | 1 | SQ1BVJ | | 04/07 SP6LB |
| 1296.886 | F1ZBC | Adriers | JN06IO | [illegible] | [illegible] | Omni | 10 | F1AFJ | | 06/12 F6HTJ |
| 1296.888 | OM0MSA | Bratislava | JN88NE | 570 | Dipole | 90°/270° | 0.045 | OM3ID | | 06/05 HA1YA |
| 1296.890 | GB3DUN | Dunstable, Beds | IO91SV | 263 | Alford slot | Omni | 40 | G3ZFP | QRV / QRP | 06/08 G6JYB |
| 1296.890 | LA4SHF | Jaeren | JO20UO | 175 | Colinear | 180° | 40 | LA9VFA | | 05/08 LA0BY |
| 1296.895 | ON0RUG | Gent | JO11UB | 90 | | | 10 | ON6UG | | 05/07 ON4PC |
| 1296.900 | HG9BUA | Kis-kohát | KN08FD | 900 | Slot | Omni | 1 | HA9MDP | | 03/13 HA5NF |
| 1296.900 | OH1SHF | Salo | KP11NJ | 25 | 6 x 5/8 | Omni | 2 | | Temp. QRT | 07/11 OH6DD |
| 1296.900 | DB0AN | Muenster-Nienberger | JO31SX | 100 | Big wheel | Omni | 1 TX | DF1QE | | 06/98 DL9QJ |
| 1296.900 | IQ3ZB/B | Monte Cesen TV | JN65AW | 1100 | Alford Slot | Omni | 10 | | | 04/13 IK1YWB |
| 1296.900 | GB3IOW | Newport, IOW | IO90IO | 250 | Alford Slot | Omni | 100 | G8MBU | | 06/06 G6JYB |
| 1296.900 | OZ5SHF | Horsens | JO45WX | 205 | Dipole | N/S | 2 | | | 02/06 DL4DTU |
| 1296.902 | LX0SHF | Walferdange | JN39BP | 420 | 2 x Big wheel | Omni | 3 | LX1JX | | 07/92 LX1JX |
| 1296.902 | OK0EA | Trutnov | JO70UP | 1355 | 4 x 15 el Yagi | S/SW/W/NW | 1.6 | OK1IA | | 04/13 OK1HH |
| 1296.902 | F5ZAN | Pic Neulos | JN12LL | 1100 | Slotted WG | Omni | 10 | F6HTJ | | 06/12 F6HTJ |
| 1296.905 | SK4BX/B | Garphyttan | JO79LI | 270 | Horizontal | Omni | 10 | SM4RWI | | 04/06 SM6CEN |
| 1296.905 | DB0AD | DOK R14 | JO40AQ | 693 | V dipole | Omni | 1 | DL7AJA | | 03/99 DJ3TF |
| 1296.905 | GB3CFG | Carrickfergus | IO74CR | 271 | Slot | Omni | 30 | GI0GDP | QRV | 04/08 G6JYB |
| 1296.910 | DB0UX | Karlsruhe DOK A 35 | JN48FX | 275 | Big wheel | Omni | 1 | DK2DB | | 03/99 DJ3TF |
| 1296.910 | GB3CLE | Clee Hill, Salop | IO82RL | 540 | 2 x 15/15 Yagi | 0° | 40 | G8DIR | QRV again | 09/06 G7IEI |
| 1296.915 | DB0UBI | DOK N 59 | JO42GE | 165 | Horn | 45° | 2.5 | DD8QA | | 03/99 DJ3TF |
| 1296.915 | ES0SHF | Saarema | KO18DN | | 2 x Double diamond | 90°/270° | 60/60 | ES5PC | | 08/00 ES5PC |
| 1296.916 | PI7QHN | Zandvoort | JO22FH | 20 | 6 dB Gain | Omni | 4 erp | PA0QNH | | 05/07 PA0EZ |
| 1296.917 | TK5ZMV | Coti Chiavari | JN41JS | 635 | Yagi | 0° | 10 | TK5EP | | 06/12 F6HTJ |
| 1296.918 | DB0VC | Lutjenberg | JO54IF | 300 | 2 x Big wheel | Omni | 12 | DL8LAO | | 05/07 PA0EZ |
| 1296.920 | TK5ZMV | Ajaccio | JN41JS | 635 | Yagi | 315° | 50 | TK5EP | | 13/08 F6HTJ |
| 1296.920 | 9A0BLB | | JN83HG | 778 | Dipole | | 1 | | | 03/95 9A2MP |
| 1296.924 | HG3BUB | Misinateto | JN96CC | 585 | Slot | Omni | 0.8 | HG5AZB | | 03/13 HA5NF |
| 1296.925 | ES0SHF | Saarema | KO18DN | | 2 x Double Diamond | 90°/270° | 60 | ES5PC | | 01/06 SM0DFP |
| 1296.925 | DB0KME | DOK C 35 | JN67HT | 800 | Vertical | Omni | 1 TX | DL8MCG | | 03/99 DJ3TF |
| 1296.928 | OH2SHF | Inkoo | KP20BB | 63 | 10 / 10 dBd | 230°/20° | 150/150 | | | 07/11 OH6DD |
| 1296.930 | OZ7IGY | Jystrup | JO55WM | 96 | 4 x Big wheel | Omni | 40 | OZ7IS | | 01/14 OZ7IS |
| 1296.930 | OK0EL | Benecko | JO70SQ | 1030 | Horn | 270° | 0.8 | OK1AIY | | 08/01 OK1HH |
| 1296.930 | GB3MLE | Emley Moor | IO93EO | 600 | Corner Reflector | 160° | 50 | G3TSA | QRT | 03/09 G6JYB |
| 1296.933 | F5ZBT | Pessac | IN94QT | 90 | 2 x Big wheel | Omni | 10 | F6DBP | | 03/13 F6HTJ |
| 1296.935 | DB0YI | Hucksheim | JO42XC | 480 | Big wheel | Omni | 3 | DL4AS | | 11/05 DK1KR |
| 1296.935 | OH5SHF | Kuusankoski | KP30HV | 142 | Alford Slot | Omni | 25 | | | 07/11 OH6DD |
| 1296.940 | DL0UH | Melsungen DOK Z 25 | JO41RD | 385 | V?Dipole | Omni | 1 | DJ3KO | | 03/99 DJ3TF |
| 1296.940 | SK7MHH | Farjestaden | JO86GP | 45 | Alford slot | Omni | 10 | | | 02/01 SM6CEN |
| 1296.945 | HB9F | Bern | JN46SW | 1015 | Corner reflector | 0° | 15 | HB9MHS | | 08/94 HB9DX |
| 1296.945 | DB0AJA | | JN59AS | 364 | FX 2304 V | 285° - 315° | 20 | DF6NA | QRV | 06/08 DF6NA |
| 1296.945 | LA7SHF | Bergen | JP20QJ | 565 | | 180° | 15 | LA3QMA | Planning | 12/06 LA0BY |
| 1296.945 | OH9SHF | Pirttikoski | KP36OI | 236 | 10 dBd | 200° | 30 | OH6DD | | 07/11 OH6DD |
| 1296.945 | DB0OS | Hitchembach | JO40CW | 730 | 6 el array | 270° | 1 | DG6YW | | 02/06 G3XDY |
| 1296.948 | IZ1DYE/B | Giaveno TO | JN35PA | 1350 | 5 el Yagi | | 3 | | | 04/13 IK1YWB |
| 1296.950 | DB0HG | DOK F11 | JO40HG | 300 | Big wheel | Omni | 3 | DL3DC | | 03/99 DJ3TF |
| 1296.950 | OZ5UHF | Kobenhavn | JO65GQ | 35 | Colinear | Omni | 1 | OZ3TZ | | 11/05 OZ2TG |
| 1296.955 | OZ1UHF | Fredrikshavn | JO57FJ | 150 | Big wheel | Omni | 10 | OZ9NT | | 02/06 GM4OGI |
| 1296.955 | OK0EB | Ceske Budejovice | JN78DU | 1084 | | | | OK1VHB | PLANNED | 10/07 OK1HH |
| 1296.957 | SK3GW/B | Osterfamebo | JP80JH | 90 | 32 el | 210° | 10 | | | 04/06 SM6CEN |
| 1296.960 | OK0EJ | Frydek Msitek | JN99FN | 1323 | Horn | Omni | 0.1 | | | 09/05 OK2BFH |
| 1296.960 | HG7BUB | Dobogóko | JN97KR | 700 | Slot | Omni | 1 | HA5NF | | 03/13 HA5NF |
| 1296.965 | DF0ANN | Lauf DOK B 25 | JN59PL | 630 | 4 x DQ | Omni | 0.5 | DL8ZX | | 03/99 DJ3TF |
| 1296.965 | OK0EO | Olomouc | JN89QQ | 602 | 2 el Yagi | SW | 0.1 | OK2VLK | Planned | 06/06 G6JYB |
| 1296.965 | GB3ANG | Dundee | IO86MN | 319 | Slot Yagi | 170° | 40 | GM4ZUK | | 06/06 G6JYB |
| 1296.970 | GB3RAI | | IO91EN | 140 | | | 10 | G0MJW | Planning | 02/07 G6JYB |
| 1296.970 | SK7MHL | Lund | JO65OR | 100 | Alford slot | Omni | 15 | SM7ECM | | 04/06 SM6CEN |
| 1296.973 | OK0EQ | Namest n./O | JN89BE | 430 | Big Wheel | Omni | 4 | OK2ZI | QRV | 06/10 OK1HH |
| 1296.975 | DL0SG | DOK U 14 | JN69KA | 1024 | 4 x DQ | Omni | 5 TX | DJ4YJ | | 03/99 DJ3TF |
| 1296.975 | OH3RNE | Tampere | KP11UM | 247 | Big Wheel | Omni | 25 | | FSK 1800 Hz | 07/11 OH6DD |
| 1296.975 | ON0AZ | Antwerpen | JO21FE | 75 | Clover leaf | Omni | 10 | ON7BPS | | 05/07 ON4PC |
| 1296.978 | HG1BUB | Hörman-forrás | JN87FI | 725 | Slot | Omni | 1.5 | HA1YA | | 03/13 HA5NF |
| 1296.980 | DB0JU | Kleve | JO31CV | 150 | Helical | Omni | 2.4 | DF5EO | | 03/99 DJ3TF |
| 1296.983 | F5ZWX | Grand Cap | JN23XE | 780 | | Slot | 0.5 | F5PVX | TESTING | 03/13 F6HTJ |
| 1296.983 | OZ2ALS | | JO45WA | 65 | 2 x slot | Omni | 10 | OZ9DT | QRT | 11/05 OZ2TG |
| 1296.985 | SR1LHX | | | | | | | | | 06/06 SP3BEK |
| 1296.985 | GB3CSB | Kilsyth | IO76XA | 259 | PA-23R | 150° | 100 | GM6BIG | QRV | 05/10 G6JYB |
| 1296.990 | GB3EDN | Edinburgh | IO85HW | 117 | Slotted WG | Omni | 25 | GM8BJF | | 06/06 G6JYB |
| 1296.990 | DB0FB | DOK Z 06 | JN47AU | 1495 | 8el Group | 45° | 5 TX | DJ3EN | | 03/99 DJ3TF |
| 1296.990 | DB0AS | Peiss | JN67CR | 1565 | Dipolfeld | 10° | 0.5 | DL2AS | | 03/06 DL3LFA |
| 1296.995 | DB0WOS | DOK U 16 | JN68ST | 850 | 4 x DQ | Omni | 5 | DF8RU | | 03/99 DJ3TF |
| 1296.090 | S55ZMS | Dolina | JN86CR | 350 | Slot | Omni | 0.5 | | S53M | 06/12 S51FB |
| 1297.010 | DB0JW | Aachen | JO30DU | 225 | 4 x 12 el Yagi | 45° | 70 | DL9KAS | | 11/05 DL3YEE |
| 1297.040 | DB0LB | DOK P 06 | JN48NV | 367 | Big wheel | Omni | 0.3 TX | DK3PS | | 03/99 DJ3TF |

| Freq | Call | Nearest Town | Locator | m ASL | Antenna | Direction | Power | Info | Status | Last update |
|---|---|---|---|---|---|---|---|---|---|---|
| 2304.040 | S55ZNG | Trstelj | JN65UU | 643 | V-J Slot | Omni | 0.1 | S50M | | 10/96 S51KQ |
| 2304.050 | S55ZSE | Kokos | JN65WP | 620 | Slot | Omni | 0.5 | | S53MV | 06/12 S51FB |
| 2304.795 | HB9OK | Monte Generoso | JN45MW | | | | | | QRV | 10/09 IK7UXW |
| 2320.050 | S55ZSE | Kokos | JN65WP | 620 | Slot | Omni | 0.5 | | S53MV | 06/12 S51FB |
| 2320.800 | SK6MHI | Goteborg | JO57XQ | 135 | Slotted WG | Omni | 10 | SM6EAN | | 04/06 SM6CEN |
| 2320.810 | SK7MHF | Nassjo | JO77IP | 505 | 2 x Big wheel | Omni | 0.1 | SM7MXO | | 04/06 SM6CEN |
| 2320.810 | DB0ZW | DOK U 17 | JN69EQ | 825 | 6 x Slot | Omni | 1 | DC9RK | | 03/99 DJ3TF |
| 2320.814 | SK0UHH | Taeby | JO99BM | 90 | Horizontal | Omni | 25 | | | 04/06 SM6CEN |
| 2320.815 | DB0IH | Nohfelden DOK Q 18 | JN39ML | 630 | Big wheel | Omni | 5 | DC8DV | | 03/99 DJ3TF |
| 2320.816 | F1ZQU | Segonzac | IN95VO | 125 | Slot | Omni | 18 | F1MMR | | 06/12 F6HTJ |
| 2320.819 | IQ0RM/B | Roma | JN61FW | 110 | Alford Slot | Omni | 1 | | | 04/13 IK1YWB |
| 2320.820 | DB0OT | Esterwegen DOK I 26 | JO32QR | 80 | Big wheel | Omni | 1 TX | DL1BFZ | | 03/99 DJ3TF |
| 2320.825 | OE1XTB | Vienna | JN88EE | 170 | 4 x dipole | Omni | 1 | OE1MOS | QRT | 09/99 OE1MCU |
| 2320.825 | DB0HF | Harksheide DOK E27 | JO53BO | 65 | Big wheel | Omni | 0.3 TX | DK2NH | | 03/99 DJ3TF |
| 2320.830 | DB0JX | Willich DOK R21 | JO31FF | 115 | Double helical | Omni | 0.1 TX | DK4TJ | | 03/99 DJ3TF |
| 2320.830 | GB3MHZ | Martlesham | JO02PB | 85 | Slotted WG | Omni | 25 | G7OCD | | 06/14 G6JYB |
| 2320.832 | OH6SHF | Uusikaarleypyy | KP13GM | 56 | 7 dB | 180° | 0.2 | | Temp. QRT | 07/11 OH6DD |
| 2320.833 | DB0FGB | DOK B 09 | JO50WB | 1150 | Slot | Omni | 12 | DB8UY | | 03/99 DJ3TF |
| 2320.835 | F5ZAC | Cerdagne | JN12BL | 2400 | Panel | 45° | 5 | F6HTJ | | 06/12 F6HTJ |
| 2320.840 | DB0KI | Bayreuth DOK Z42 | JO50WC | 925 | Slot | Omni | 40 | DC9NL | | 03/99 DJ3TF |
| 2320.840 | F1ZYY | Mugron | IN93PS | 100 | Panel | 023° | 4 | F1MOZ | QRV | 06/12 F6HTJ |
| 2320.842 | OH3SHF | Tampere | KP11VK | 222 | 6 dBi | Omni | 125 | | Accuracy 10^-8 | 07/11 OH6DD |
| 2320.845 | DB0LBV | DOK S 30 | JO61EH | 234 | DQ | 135°/225° | 1.5 TX | DL1LWM | | 03/99 DJ3TF |
| 2320.845 | SR3SHF | Chelmce | JO91CQ | 180 | 2 x Helix | Omni | 1 | SP3DRT | | 04/07 SP6LB |
| 2320.850 | DB0GW | DOK L 01 | JO31JK | 80 | 2 x Helix | Omni | 8 | DL4JK | | 03/99 DJ3TF |
| 2320.850 | DL0UB | DOK Z 20 | JO62KK | 120 | 5 x Dipole | Omni | 10 TX | DL7ACG | | 03/99 DJ3TF |
| 2320.850 | LA4SHF | Jaeren | JO28UO | 170 | Log periodic | 225° | 10 | LA3EQ | QRV | 12/06 LA0BY |
| 2320.855 | DB0SHF | DOK Z 46 | JN48XS | 800 | 6 x Dipole | 260° | 0.2 | DL1SBE | | 03/99 DJ3TF |
| 2320.855 | F1ZUM | Orleans | JN07WV | 170 | Slot | Omni | 2 | F1JGP | | 06/12 F6HTJ |
| 2320.857 | PI7GHG | Europoort | JO21CV | 30 | 10 el Yagi | 270° | 30 | PE1GHG | QRT | 10/09 PA0EZ |
| 2320.860 | LA8SHF | Tonsberg | JO59FB | 30 | 13 dB Horn | 180° | 50 | LA6LCA | Keys "LA1UHG" | 05/06 LA0BY |
| 2320.860 | HG9BUB | Kis-kohát | KN08FB | 930 | Slot | Omni | 0.7 | HA9MDP | | 03/13 HA5NF |
| 2320.864 | F5ZVY | | IN93EK | 65 | 23el Yagi | 023° | 0.7 | F2CT | Testing | 06/12 F6HTJ |
| 2320.865 | PI7TGA | Nijmegen | JO21VT | 75 | | 135°/270° | 50 | PA0TGA | | 07/03 PE2KP |
| 2320.868 | HG5BUC | János-hegy | JN97LM | 485 | Slot | Omni | 1 | HA5BDJ | | 03/13 HA5NF |
| 2320.870 | DB0IBB | DOK N 49 | JO32VG | 200 | 10 x Slot | Omni | 4 | DB7QW | | 03/99 DJ3TF |
| 2320.872 | F1ZRI | Le Mans | IN98WE | 260 | 14 el Yagi | 190° | 80 | F1BJD | | 06/12 F6HTJ |
| 2320.880 | LA3SHF | Flekkeroy | JO38XB | 5 | 2 x 6 dB Horn | 90°/180° | 1 | LA8AK | | 04/04 LA0BY |
| 2320.880 | DB0YI | Hildesheim DOK Z 35 | JO42XC | 480 | Big Wheel | Omni | 3 TX | DL4AS | | 03/99 DJ3TF |
| 2320.880 | DB0GO | DOK N 32 | JO41ED | 738 | 10 x Slot | Omni | 50 | DB1DI | | 03/99 DJ3TF |
| 2320.883 | DB0INN | DOK C 15 | JN68GI | 504 | Slot | Omni | 1 TX | DL3MBG | | 03/99 DJ3TF |
| 2320.885 | DB0TUD | DOK S07 | JO61UA | 260 | Slot | Omni | | DL4DTU | | 03/99 DJ3TF |
| 2320.885 | PI7RMD | Roermond | JO31AE | 100 | 2 x Quad | 180° | 10 | PE1KXH | QRT | 10/09 PA0EZ |
| 2320.886 | F5ZMF | Adriers | JN06JG | 230 | Slot | Omni | 5 | F5BJL | | 06/12 F6HTJ |
| 2320.888 | OM0MTA | Bratislava | JN88EE | 570 | | 90°/270° | 0.012 | OM3ID | | 08/01 OK1HH |
| 2320.889 | IZ1DYE/B | Giaveno TO | JN35PA | 1350 | | | | | | 04/13 IK1YWB |
| 2320.890 | GB3ANT | Norwich | JO02PP | 75 | Alford slot | Omni | 5 | G8VLL | | 06/06 G6JYB |
| 2320.900 | DB0JW | DOK G 05 | JO30DU | 238 | 6 el Array | 45° | 25 | DL9KAS | | 03/99 DJ3TF |
| 2320.900 | HG3BUC | Misinateto | JN96CC | 585 | Slot | Omni | 1 | HG5AZB | | 03/13 HA5BF |
| 2320.900 | DB0UX | Grotzingen DOK A 35 | JN48FX | 275 | Big wheel | Omni | 1 | DK2DB | | 03/99 DJ3TF |
| 2320.900 | F6DWG | Beauvais | JN19FK | 140 | Slot | Omni | 2 | F6DWG | | 06/12 F6HTJ |
| 2320.902 | LX0THF | Walferdange | JN39BP | 420 | Double quad | Omni | 0.5 | LX1JX | | 11/93 LX1KQ |
| 2320.902 | F6DPH | Chartrette | JN18IM | | Panel | 180° | 5 | F6DPH | | 06/12 F6HTJ |
| 2320.905 | GB3SCS | Bell Hill, Dorset | IO80UU69 | 274 | Alford slot | Omni | 70 | G4JNT | | 06/14 G6JYB |
| 2320.910 | GB3ZME | Telford | IO82RP | 218 | Slot | Omni | 80 | G3UKV | QRV | 06/14 G6JYB |
| 2320.912 | DL0UH | DOK Z 25 | JO41RD | 385 | 6 x Dipole | 0° | 2 | DJ3K0 | | 03/99 DJ3TF |
| 2320.915 | DB0UBI | DOK N 59 | JO42GE | 165 | Collinear | 45° | 0.5 | DD8QA | | 03/99 DJ3TF |
| 2320.920 | DB0VC | Albersdorf DOK Z 10 | JO54IF | 300 | Big wheel | Omni | 3 | DL8LAO | | 03/99 DJ3TF |
| 2320.920 | PI7QHN | Zandvoort | JO22KH | 20 | | Omni | 0.2 TX | PA0QHN | QRT | 10/09 PA0EZ |
| 2320.925 | GB3BSS | Stroud | IO81SR | 70 | Slotted WG | Omni | 25 | G4CJZ | LIC/AWAIT QRV | 06/13 G6JYB |
| 2320.928 | OH2SHF | Inkoo | KP20BB | 63 | 10/10dB | 230°/20° | 230/20 | | Now QRV | 06/07 OH6DD |
| 2320.928 | OH2SHF | Inkoo | KP20BB | 63 | 10 / 10 dB | 230°/20° | 250 | | | 07/11 OH6DD |
| 2320.930 | F5EJZ | | IN99IO | 120 | 2 x quad | 090° / 135° | 2 | F5EJZ | | 06/12 F6HTJ |
| 2320.930 | OZ7IGY | Jystrup | JO55WM | 100 | Alford Slot | Omni | 15 | OZ7IS | | 01/14 OZ7IS |
| 2320.930 | OK0EL | Benecko | JO70SQ | 1030 | Horn | 270° | 0.8 | OK1AIY | | 08/01 OK1HH |
| 2320.930 | PI7PLA | Zuidlaren | JO33IC | 50 | | Omni | 0.15 TX | PA0PLA | QRT | 10/09 PA0EZ |
| 2320.933 | F5ZEN | Pessac | IN94QT | 83 | | 130° | | F6CBC | | 03/13 F6HTJ |
| 2320.935 | OH5SHF | Kouvola | KP30HV | 142 | 15.5 dBd | 240° | 25 | | | 07/11 OH6DD |
| 2320.935 | OH5SHF | Kuusankoski | KP30HV | 146 | 10 dBd | 240° | 11 | | Now QRV | 10/06 OH6DD |
| 2320.937 | DB0JO | Kamp-Lintfort DOK Z03 | JO31SL | 312 | Horn | 270° | 0.2 TX | DG8DCI | | 03/99 DJ3TF |
| 2320.940 | DB0DON | DOK T21 | JN58KR | 532 | Slot | Omni | 1 | DL5MEL | | 03/99 DJ3TF |
| 2320.940 | SK7MHH | Farjestaden | JO86GP | 45 | 30 cm dish | 0° | 50 | | | 04/06 SM6CEN |
| 2320.945 | DB0OS | Hitchinbach DOK N 32 | JO40CW | 730 | 8 el array | 270° | 2 | DG6YW | | 03/99 DJ3TF |
| 2320.950 | DB0KP | DOK P 09 | JN47TS | 435 | Slot | Omni | 0.1 TX | DL1GBQ | | 03/99 DJ3TF |
| 2320.950 | OZ9UHF | | JO65HP | 30 | Slot | Omni | 5 | OZ2TG | QRT | 11/05 OZ2TG |
| 2320.955 | GB3LES | Leicester | IO92IQ | 220 | Slot | 160° | 30 | G3TQF | | 06/06 G6JYB |
| 2320.955 | OZ1UHF | | JO57FJ | 150 | Slot | Omni | 8 | OZ9NT | | 11/05 OZ2TG |
| 2320.960 | SK4BX/B | Garphyttan | JO79LI | 265 | Slotted WG | Omni | 250 | | Temp QRT | 04/06 SM6CEN |
| 2320.960 | DB0AJA | | JN59AS | 364 | 10 dB | 315° | 10 | DF6NA | QRV | 06/08 DF6NA |
| 2320.965 | DF0ANN | Lauf DOK B 25 | JN59PL | 630 | 4 x D Q | Omni | 5 TX | DL8ZX | | 03/99 DJ3TF |
| 2320.966 | OZ4UHF | | JO75LD | 130 | 4 x patch | 45°/225° | 200 | | Planned | 11/05 OZ2TG |
| 2320.967 | DB0AS | Rosenheim DOK C 14 | JN67CR | 1560 | 28 el Yagi | 337° | 0.5 TX | DL2AS | | 03/99 DJ3TF |
| 2320.970 | HG1BUC | Hörman-forrás | JN87FI | 725 | Slot | Omni | 0.8 | HA1YA | Temp QRT | 03/13 HA5NF |
| 2320.970 | SK7MHL | Lund | JO65OR | 100 | WG slot | Omni | 25 | SM7ECM | | 04/06 SM6CEN |

| Freq | Call | Nearest Town | Locator | m ASL | Antenna | Direction | Power | Info | Status | Last update |
|---|---|---|---|---|---|---|---|---|---|---|
| 2320.973 | OK0EQ | Namest n./O | JN89BE | 430 | Big Wheel | Omni | 5 | OK2ZI | QRV | 06/10 OK1HH |
| 2320.975 | ON0KUL | Tielt Winge | JO20KV | 100 | | | 1 | ON4IY | | 05/07 ON4BG |
| 2320.975 | DB0JL | DOK R 25 | JO31MC | 195 | Slot | Omni | 2 | DF1EQ | | 03/99 DJ3TF |
| 2320.980 | DB0JU | Duesburg DOK L 04 | JO31CV | 150 | Helical | Omni | 1 TX | DF5EO | | 03/99 DJ3TF |
| 2320.985 | GB3CSB | Kilsyth | IO76XA | 259 | PA-13R | 150° | 25 | GM6BIG | QRV | 01/13 G6JYB |
| 2320.987 | F1ZSO | | JN05MP | 517 | | | | F1DXP | Planned | 04/06 F6HTJ |
| 2320.990 | PI7EHG | Schipol | JO22JH | 80 | | Omni | 10 | PA0EHG | QRT | 10/09 PA0EZ |
| | | | | | | | | | | |
| 3400.015 | PI7SHF | Schipol | JO22JH | 80 | 10 dB Slot | Omni | 2 TX | PA0EHG | QRT | 10/09 PA0EZ |
| 3400.020 | DB0AS | DOK C 29 | JN67CR | 1565 | Double 0 | 10 | 0.5 TX | DL2AS | | 03/99 DJ3TF |
| 3400.025 | DB0HI | DOK E 27 | JO53BO | 65 | | 202° | | [illegible] | | 00/00 DJ3TF |
| 3400.040 | DB0KI | Bayreuth DOK Z 42 | [illegible] | [illegible] | Slot | Omni | 50 | DC9NL | | 03/99 DJ3TF |
| 3400.050 | S55ZSE | Kokos | JN65WP | 620 | Slot | Omni | 0.5 | | S53MV | 06/12 S51FB |
| 3400.050 | DB0JL | DOK R 25 | JO31MC | 195 | Helical | Omni | 1 | DF1EQ | | 03/00 DJ3TF |
| 3400.090 | S55ZMS | Dolina | JN86CR | 350 | Slot | Omni | 0.5 | | S53M | 06/12 S51FB |
| 3400.160 | PA0JCA | | JO22NB | 40 | | Omni | 25 erp | | QRV | 10/09 PA0EZ |
| 3400.165 | PI7CKK | Groningen | JO33GE | 55 | 10 dB Slot | Omni | 5/30 erp | PA0CKK | QRT | 10/09 PA0EZ |
| 3400.400 | OK0EL | Benecko | JO70SQ | 1030 | Horn | 270° | 0.2 | OK1AIY | Planned | 08/01 OK1HH |
| 3400.800 | OH3SHF | Tampere | KP11VK | 222 | 6 dBi | Omni | 60 | | | 07/11 OH6DD |
| 3400.830 | GB3MHZ | Martlesham | JO02PB | 85 | 2 x slotted lines | 90°/270° | 7 | G7OCD | | 06/14 G6JYB |
| 3400.850 | LA4SHF | Jaeren | JO28UO | 170 | Log periodic | 225° | 2 | LA3EQ | QRV | 12/06 LA0BY |
| 3400.850 | DB0GW | Duisburg DOK L 01 | JO31JK | 80 | Double Helical | Omni | 8 | DL4JK | | 03/99 DJ3TF |
| 3400.900 | GB3OHM | Birmingham | IO92AJ | 171 | 16 Slot waveguide | Omni | 100 | G8SH | | 06/06 G6JYB |
| 3400.905 | GB3SCF | Bell Hill, Dorset | IO80UU69 | 274 | Alford slot | Omni | 70 | G4JNT | | 06/14 G6JYB |
| 3400.910 | GB3ZME | Telford | IO82RP | 218 | Slotted WG | Omni | 80 | G3UKV | | 06/08 G6JYB |
| 3400.920 | PA3CRX | Rotterdam | JO21FV | | Horizontal | Omni | 14.5 | PA3CRX | PROPOSED | 09/15 PA3CRX |
| 3400.930 | OZ7IGY | Jystrup | JO55WM | 97 | WG Slot | Omni | 25 | OZ7IS | | 01/14 OZ7IS |
| 3400.945 | DB0AJA | | JN59AS | 364 | 3 x 120° sector | Omni | 20 | DF6NA | QRV | 06/08 DF6NA |
| 3400.955 | GB3LEF | Leicester | IO92IQ | | | | 8 | G3TQF | QRV / QRP | 06/08 G6JYB |
| 3400.955 | OZ1UHF | | JO57FJ | 150 | | | | | | 11/05 OZ2TG |
| 3400.970 | GB3RAL | | IO91EN | 140 | | | 10 | G0MJW | Planning | 02/07 G6JYB |
| 3400.973 | OK0EQ | Namest n./O | JN89BE | 430 | Slot | Omni | 1 | OK2ZI | PLANNED | 06/10 OK1HH |
| 3400.985 | GB3CSB | Kilsyth | IO76XA | 259 | Slotted WG | 150° | 25 | GM6BIG | QRV | 01/13 G6JYB |
| 3456.800 | DB0KHT | DOK F 13 | JO40FE | 247 | Horn | Omni | 10 | DJ1RV | | 03/99 DJ3TF |
| 3456.830 | DB0JX | DOK R 21 | JO31FF | 115 | Helical | Omni | 0.1 TX | DK4TJ | | 03/99 DJ3TF |
| 3456.850 | DL0UB | DOK Z 20 | JO62KK | 120 | 12 x Slot | Omni | 10 TX | DL7ACG | | 03/99 DJ3TF |
| 3456.855 | DB0SHF | DOK Z 46 | JN48XS | 800 | Horn | 260° | 0.5 TX | DL1SBE | | 03/99 DJ3TF |
| 3456.883 | DB0INN | DOK C 15 | JN68GI | 504 | Slot | Omni | 1 TX | DL3MBG | | 03/99 DJ3TF |
| 3456.885 | DB0TUD | DOK S07 | JO61UA | 260 | Slot | Omni | | DL4DTU | | 03/99 DJ3TF |
| | | | | | | | | | | |
| 5668.880 | GB3MAN | Rochdale | IO83WO | 153 | Slotted WG | Omni | 25 | G6GXK | Planned | 09/07 G6GXK |
| 5760.010 | S55ZSE | Kokos | JN65WP | 620 | Slot | Omni | 0.5 | | S53MV | 06/12 S51FB |
| 5760.030 | OK0EL | Benecko | JO70SQ | 1030 | Horn | 270° | 0.08 | OK1AIY | | 08/01 OK1HH |
| 5760.030 | PI7EHG | Schipol | JO22JH | 80 | Horizontal | Omni | 2 TX | PA0EHG | QRT | 10/09 PA0EZ |
| 5760.045 | S55ZRS | Mt. Kum | JN76MC | 1219 | Slot | Omni | | | | 06/12 S51FB |
| 5760.050 | OK0EA | Trutnov | JO70UP | 1355 | 12 el Slot | 180°/270° | 0.5 | OK1IA | | 04/13 OK1HH |
| 5760.060 | F1ZAO | Plougonver | IN88HL | 326 | Slotted WG | Omni | 1 | F1LHC | | 04/12 F6HTJ |
| 5760.070 | DB0JL | DOK R 25 | JO31MC | 195 | Slot | Omni | 0.8 | DF1EQ | | 03/99 DJ3TF |
| 5760.090 | S55ZMS | Dolina | JN86CR | 350 | Slot | Omni | 0.5 | | S53M | 06/12 S51FB |
| 5760.100 | DB0AS | DOK C 29 | JN67CR | 1565 | Double 8 | 10° | 0.5 TX | DL2AS | | 03/99 DJ3TF |
| 5760.177 | IK1YWB/B | Bagnolo CN | JN34OS | 1550 | Slot 12 | 65° | 1 | | | 04/13 IK1YWB |
| 5760.455 | HB9OK | Monte Generoso | JN45MW | | | | | | QRV | 10/09 IK7UXW |
| 5760.800 | IQ4AD/B | Monte Cassio PR | JN54AO | 1010 | Slot 6 | 35° | 0.8 | | | 04/13 IK1YWB |
| 5760.800 | DB0KHT | DOK F 13 | JO40FE | 247 | Horn | Omni | 0.5 TX | DJ1RV | | 03/99 DJ3TF |
| 5760.800 | OH3SHF | Tampere | KP11VK | 222 | 15 dBi | Omni | 150 | | Accuracy 10^-8 | 07/11 OH6DD |
| 5760.800 | SK6MHI | Goteborg | JO57XQ | 135 | Sectoral Horn | 270° | 5 | SM6EAN | | 04/06 SM6CEN |
| 5760.805 | DB0RIG | DOK P 17 | JN48WQ | 780 | | Omni | 15 | DG9SQ | | 03/99 DJ3TF |
| 5760.820 | F5ZBE | Favières | JN18JS | 160 | Slot | Omni | 12 | F5HRY | | 04/12 F6HTJ |
| 5760.830 | GB3MHZ | Martlesham | JO02PB | 85 | Slotted WG | Omni | 2 | G7OCD | | 06/14 G6JYB |
| 5760.830 | DB0JX | DOK R 21 | JO31FF | 115 | Slot | Omni | 0.08 TX | DK4TJ | | 03/99 DJ3TF |
| 5760.833 | DB0FGB | DOK B 09 | JO50WB | 1150 | Slot | Omni | 12 | DB8UY | | 03/99 DJ3TF |
| 5760.840 | DB0KI | Bayreuth DOK Z 42 | JO50WC | 925 | Slot | Omni | 20 | DC9NL | | 03/99 DJ3TF |
| 5760.845 | F1ZBD | Orleans | JN07WV | 170 | Slot | Omni | 10 | F1JGP | | 04/12 F6HTJ |
| 5760.850 | LA4SHF | Jaeren | JO28UO | 170 | Log periodic | 225° | 2 | LA3EQ | QRV | 12/06 LA0BY |
| 5760.850 | DL0UB | DOK Z 20 | JO62KK | 120 | 12 x Slot | Omni | 0.2 TX | DL7ACG | | 03/99 DJ3TF |
| 5760.850 | I3EME/B | Monte Tomba TV | JN55WV | 890 | Slot 8 | 170° | 0.22 | | | 04/13 IK1YWB |
| 5760.855 | F5ZPR | Talence | IN94QT | 60 | Slot | Omni | 10 | F6CBC | | 04/06 F6HTJ |
| 5760.855 | DB0SHF | DOK Z 46 | JN48XS | 800 | Array | 260° | 0.4 TX | DL1SBE | | 03/99 DJ3TF |
| 5760.860 | DB0ARB | DOK U 02 | JN69NC | 1456 | Slot | Omni | 3 | DJ4YJ | | 03/99 DJ3TF |
| 5760.860 | LA8SHF | Tonsberg | JO59FB | 30 | 13 dB Horn | 180° | 25 | LA6LCA | Keys "LA1UHG" | 05/06 LA0BY |
| 5760.862 | F5ZUO | Pic Neulos | JN12LL | 1100 | Slot | Omni | 1 | F6HTJ | | 04/12 F6HTJ |
| 5760.865 | OE1XVB | Vienna Simmering | JN88EF | 191 | Slotted WG | Omni | 4 | OE1WRS | QRT | 09/99 OE1MCU |
| 5760.866 | F5ZUO | Pic Naulos | JN12LL | 1100 | Slot | Omni | 10 | F6HTJ | QRV | 06/07 F6HTJ |
| 5760.870 | GB3MAN | Rochdale | IO83WO | 153 | Slotted WG | Omni | 25 | G6GXK | QRV | 01/13 G6JYB |
| 5760.875 | HG5BSA | János-hegy | JN97LM | 485 | Slot | Omni | 0.2 | HA5BDJ | Temp. QRT | 03/13 HA5NF |
| 5760.883 | F5ZWY | Grand Cap | JN23XE | 780 | Slot | Omni | 1 | F5PVX | | 04/12 F6HTJ |
| 5760.883 | DB0INN | DOK C 15 | JN68GI | 504 | Slot | Omni | 1 TX | DL3MBG | | 03/99 DJ3TF |
| 5760.885 | DB0TUD | DOK S07 | JO61UA | 260 | Slot | Omni | | DL4DTU | | 03/99 DJ3TF |
| 5760.890 | HG3BSA | Misinateto | JN96CC | 585 | Slot | Omni | 0.5 | HG5AZB | | 03/13 HA5NF |
| 5760.900 | OZ5SHF | | JO45WX | 205 | WG | Omni | 2 | | | 11/05 OZ2TG |
| 5760.900 | GB3OHM | Birmingham | IO92AJ | 185 | Alford Slot | Omni | 50 | G8SH | QRV | 08/09 G6JYB |
| 5760.900 | DB0CU | DOK A 28 | JN48BI | 970 | Slot | Omni | 5 | DJ7FJ | | 03/99 DJ3TF |
| 5760.904 | F6DWG | Beauvais | JN19FK | 140 | Slot | Omni | 8 | F6DWG | | 04/12 F6HTJ |
| 5760.912 | SK0UX/B | Taby | JO99BM | 80 | Slotted WG | Omni | 100 | | | 04/06 SM6CEN |

| Freq | Call | Nearest Town | Locator | m ASL | Antenna | Direction | Power | Info | Status | Last update |
|---|---|---|---|---|---|---|---|---|---|---|
| 5760.905 | GB3SCC | Bell Hill, Dorset | IO80UU69 | 274 | Slotted WG | Omni | 7 | G4JNT | | 06/14 G6JYB |
| 5760.910 | GB3ZME | Telford | IO82RP | 218 | Slotted WG | Omni | 80 | G3UKV | | 06/08 G6JYB |
| 5760.920 | GB3FNM | Farnham | IO91OF | 207 | Slotted WG | Omni | 25 | G4EPX | | 04/09 G6JYB |
| 5760.925 | GB3KEU | Sheffield | IO93GH | 198 | Slotted WG | Omni | 25 | G3PHO | QRV | 08/09 G6JYB |
| 5760.928 | OH2SHF | Helsinki | KP20LE | 107 | 10 dBd | Omni | 60 | | | 07/11 OH6DD |
| 5760.930 | OZ7IGY | Jystrup | JO55WM | 97 | IKEA dish | 35° | 400 | OZ7IS | | 01/14 OZ7IS |
| 5760.930 | OZ7IGY | Jystrup | JO55WM | 99 | WG Slot | Omni | 50 | OZ7IS | | 01/14 OZ7IS |
| 5760.933 | F5ZPR | Pessac | IN94QT | 83 | Horn | 130° | 8 | F6CBC | | 03/13 F6HTJ |
| 5760.945 | OE2XRO | Sonnblick | JN67LA | 3105 | Slotted WG | Omni | 35 | OE1MCU | | 09/99 OE1MCU |
| 5760.945 | DB0AJA | | JN59AS | 364 | 2 x 10 slots | Omni | 10 | DF6NA | QRV | 06/08 DF6NA |
| 5760.949 | F5ZYK | Angers | IN97RL | 60 | Slot | Omni | 3 | F6APE | | 03/13 F6HTJ |
| 5760.950 | OZ9UHF | | JO65HP | 30 | Slotted Waveguide | Omni | 50 | OZ2TG | | 11/05 OZ2TG |
| 5760.951 | F1ZWJ | Lacapelle | JN14EB | 625 | Slot | Omni | 0.2 | F1BOH | QRT | 04/12 F6HTJ |
| 5760.955 | OZ1UHF | | JO57FJ | 150 | Slotted Waveguide | Omni | 8 | OZ9NT | | 11/05 OZ2TG |
| 5760.970 | SK7MHL | Lund | JO65OR | 100 | WG slot | Omni | 10 | SM7ECM | | 04/06 SM6CEN |
| 5760.970 | GB3RAL | | IO91EN | 140 | | | 10 | G0MJW | Planning | 02/07 G6JYB |
| 5760.973 | OK0EQ | Namest n./O | JN89BE | 430 | Slot | Omni | 1 | OK2ZI | PLANNED | 06/10 OK1HH |
| 5760.975 | HG1BSA | Hörman-forrás | JN87FI | 725 | Slot | Omni | 0.4 TX | HA1YA | | 03/13 HA5NF |
| 5760.975 | ON0KUL | Tielte- Winge | JO20KV | 100 | | | 5 | ON4IY | | 05/07 ON4PC |
| 10100.000 | GB3IOW | Newport, IOW | IO90IP | 250 | Slotted waveguide | Omni | 1 | G8MBU | | 06/06 G6JYB |
| 10368.033 | I3CLZ/B | Cima Carega VI | JN55NR | 2217 | Slot | Omni | 0.6 | | | 04/13 IK1YWB |
| 10368.037 | PI7SHY | Utrecht | JO22NB | 40 | Slotted WG (10 dB) | Omni | 10 erp | PA0EZ | QRV | 10/09 PA0EZ |
| 10368.050 | LX0DU | Soleuvre | JN29XM | 280 | 1.3m Dish | 63° | 20 kW | | | 02/98 LX1SC |
| 10368.050 | OK0EL | Benecko | JO70SQ | 1030 | 12 el slot WG | 90°/270° | 0.15 | OK1AIY | TEMP QRT | 07/10 OK1HH |
| 10368.050 | OZ9UHF | | JO65HP | 30 | Slotted WG | Omni | 3 | OZ2TG | QRT | 11/05 OZ2TG |
| 10368.053 | F5XBD | Favieres | JN18JS | 160 | Slot | Omni | 60 | F5HRY | | 06/07 F6HTJ |
| 10368.070 | I4BER/B | Piane di Mocogno MO | JN54IG | 1350 | Slot 16 | | 0.07 | | | 04/13 IK1YWB |
| 10368.072 | F5ZBB | Favieres | JN18JS | 160 | Slot | Omni | 3 | F5HRY | | 05/10 F6HTJ |
| 10368.073 | S55ZRS | Mt. Kum | JN76MC | 1219 | Slot | Omni | 0.27 | | | 06/12 S51FB |
| 10368.075 | PI7GOE | Kapelle | JO11XL | | Slotted WG | Omni | 0.1 | PE1CIG | QRT | 10/09 PA0EZ |
| 10368.080 | OK0EA | Trutnov | JO70UP | 1355 | 12 el Slot | 180°/270° | 0.5 | OK1IA | | 04/13 OK1HH |
| 10368.090 | PA0TGA/A | Nijmegen | JO21VT | 75 | 6 dB | Omni | 4 erp | PA0TGA | | 05/07 PA0EZ |
| 10368.108 | F1ZAP | Plougonver | IN88HL | 326 | Slotted WG | Omni | 0.5 | F1LHC | | 06/12 F6HTJ |
| 10368.120 | DB0JL | DOK R 25 | JO31MC | 195 | Slot | Omni | 0.15 | DF1EQ | | 03/99 DJ3TF |
| 10368.142 | IT9CIT/B | Alcamo TP | JM67LX | 400 | Slot 12 | 0° | 1.16 | | | 04/13 IK1YWB |
| 10368.150 | OE8XXQ | Dobratsch | JN76UO | 2166 | Horn | 0° | 1 | OE8MI | | 09/99 OE1MCU |
| 10368.170 | HB9OK | Monte Generoso | JN45MW | | | | | | QRV | 10/09 IK7UXW |
| 10368.175 | DB0AS | DOK C 29 | JN67CR | 1565 | Horn | 10° | 0.5 TX | DL2AS | | 03/99 DJ3TF |
| 10368.165 | PI7EHG | | JO22HC | 10 | 13 dBi Slot | Omni | 50 erp | PA0EHG | | 10/09 PA0EZ |
| 10368.187 | IZ1ERR/B | Bagnolo CN | JN34OS | 1550 | Horn | 55° | 0.5 | | | 04/13 IK1YWB |
| 10368.270 | DL0WY | Rosenheim DOK C29 | JN67AQ | 1838 | 10 dB Slot horn | 45°/270° | 0.1 TX | DJ8VY | | 03/99 DJ3TF |
| 10368.270 | I1TEX/B | Rossana CN | JN34QM | 1450 | Horn | 65° | 1 | | | 04/13 IK1YWB |
| 10368.333 | F5ZEP | Talence | IN94QT | 83 | Horn | 130° | 5 | F6CBC | | 06/12 F6HTJ |
| 10368.300 | F5ZPS | Talence | IN94QT | 83 | Horn | 25° | 8 | F6CBC | | 06/12 F6HTJ |
| 10368.320 | F5ELY | | IN99IO | 120 | Horn | 147° | 1.2 | F5ELY | | 06/12 F6HTJ |
| 10368.755 | F1XAE | Mt Ventoux | JN24PE | 1910 | Horn | 270° | 5 | F1UNA | Temp. QRT | 04/06 F6HTJ |
| 10368.755 | GB3CAM | Wyton, Cambs. | IO92WI | 65 | 10dB slot | Omni | 5 | G4HJW | | 06/14 G6JYB |
| 10368.796 | IQ0RM/B | Roma | JN61FW | 110 | Double Slot | N/S | 0.2 | | | 04/13 IK1YWB |
| 10368.800 | OH3SHF | Tampere | KP11VK | 222 | 15 dBi | Omni | 80 | | Accuracy 10^-8 | 07/11 OH6DD |
| 10368.800 | SK6MHI | Goteborg | JO57XQ | 135 | Slotted WG | Omni | 5 | SM6EAN | | 04/06 SM6CEN |
| 10368.805 | DB0XL | DOK E-IG | JO53HU | 45 | Slot | Omni | 1 | DK1KR | | 03/99 DJ3TF |
| 10368.810 | SK6YH/B | Goteborg | JO57XQ | 135 | Dish 33 dB | 60° | 10000 | | | 04/06 SM6CEN |
| 10368.810 | GB3MAN | Rochdale | IO83WO | 153 | Slotted WG | Omni | 20 | G6GXK | QRV | 06/14 G6JYB |
| 10368.815 | DB0MAX | DOK B 41 | JN58SP | 420 | | | | DL4MDQ | | 03/99 DJ3TF |
| 10368.820 | LA5SHF | Stavanger | JO28UX | 50 | 15 dB Horn | 240° | 25 | LA3EQ | | 05/06 LA0BY |
| 10368.820 | DB0KHT | DOK F 13 | JO40FE | 247 | Horn | Omni | 3 | DJ1RV | | 03/99 DJ3TF |
| 10368.822 | SK3SHH | Rafson | JP92FW | 200 | Dish 33 dB | 180° | 10000 | | | 04/06 SM6CEN |
| 10368.824 | IQ5FI/B | M.te Secchieta AR | JN53SR | 1400 | Slot 16 | Omni | 1.5 | | | 04/13 IK1YWB |
| 10368.825 | DB0HRO | DOK V 09 | JO64AD | 185 | Slot | Omni | 0.2 TX | DL5CC | | 03/99 DJ3TF |
| 10368.825 | F1XAU | Sombernon | JN27IH | 516 | Slot WG | Omni | 13 | F1MPE | | 04/06 F6HTJ |
| 10368.830 | SR6XHZ | Snienze Kloty | JO70SS | 1490 | Horn | 40° | 1 | SP6LB | | 04/07 SP6LB |
| 10368.830 | DB0JX | Wickrath DOK R 21 | JO31FF | 115 | 10 dB Slot | Omni | 0.09 TX | DK4TJ | | 03/99 DJ3TF |
| 10368.830 | GB3MHZ | Martlesham | JO02PB | 80 | 12 Slot waveguide | Omni | 5 | G7OCD | | 06/14 G6JYB |
| 10368.833 | DB0FGB | DOK B 09 | JO50WB | 1150 | Slot | Omni | 7 | DB8UY | | 03/99 DJ3TF |
| 10368.840 | SK0SHI | Edsberg | JO89XK | | Horizontal | Omni | 1 | | | 02/01 SM6CEN |
| 10368.840 | DB0KI | Bayreuth DOK Z 42 | JO50WC | 925 | Slot | Omni | 13 | DC9NL | | 03/99 DJ3TF |
| 10368.840 | DB0JO | Kamp-Lintfort DOK Z03 | JO31SL | 312 | 6 x Slot | Omni | 1 | DG8DCI | | 03/99 DJ3TF |
| 10368.842 | F5ZTR | Beauvais | JN19FK | 140 | Slot | Omni | 10 | F6DWG | | 06/12 F6HTJ |
| 10368.842 | SK0SHH | Kista | JO89XJ | 60 | Dish 33 dB | 237° | 10000 | | | 04/06 SM6CEN |
| 10368.845 | SR3XHR | Jarocin | JO81SX | | Slot | Omni | 0.2 | SP3WYP | Temp QRT | 04/07 SP6LB |
| 10368.845 | DB0SZB | DOK S 45 | JO60JM | 767 | Slot | Omni | 15 | DG0YC | | 03/99 DJ3TF |
| 10368.846 | SK0SHI | Edsberg | JO89XK | 70 | | Omni | 1 | | | 04/06 SM6CEN |
| 10368.850 | I3EME/B | Monte Tomba TV | JN55WV | 890 | Slot 12 | 170° | 2.2 | | | 04/13 IK1YWB |
| 10368.850 | DL0UB | DOK Z 20 | JO62KK | 120 | 12 x Slot | Omni | 0.1 TX | DL7ACG | | 03/99 DJ3TF |
| 10368.850 | DB0GG | DOK P 24 | JN48NS | 400 | Slot | Omni | 0.05 TX | DL5AAP | | 03/99 DJ3TF |
| 10368.850 | GB3SEE | Reigate | IO91VG | 250 | Slotted waveguide | Omni | 3 | G0OLX | | 06/06 G6JYB |
| 10368.850 | SR0CWK | Czestochowa | JO90NS | 282 | Slot | Omni | 4.5 | SP9NLY | | 04/07 SP6LB |
| 10368.850 | LA4SHF | Jaeren | JO28UO | 150 | 2 x 16 dB Horn | 180°/0° | 8 | LA3EQ | | 05/06 LA0BY |
| 10368.853 | SK1SHH | Klintehamn | JO97CJ | 52 | Dish 33 dB | 0° | 10000 | | | 04/06 SM6CEN |
| 10368.855 | IQ6AN/B | Ancona | JN63QN | 250 | Horn | 350° | 10 | | | 04/13 IK1YWB |
| 10368.855 | DB0SHF | DOK Z 46 | JN48XS | 800 | Horn | 260° | 0.1 TX | DL1SBE | | 03/99 DJ3TF |
| 10368.855 | F1ZCL | Mt. Doublier | JN33KQ | 1200 | Slot | Omni | 0.1 | F1BDB | | 06/12 F6HTJ |
| 10368.859 | F1DLT | La Roche | JN27UR | | Corner | 0°/270° | 30 | F1DLT | | 04/06 F6HTJ |
| 10368.860 | DB0ARB | DOK U 02 | JN69NC | 1456 | Slot | Omni | 3 | DJ4YJ | | 03/99 DJ3TF |

| Freq | Call | Nearest Town | Locator | m ASL | Antenna | Direction | Power | Info | Status | Last update |
|---|---|---|---|---|---|---|---|---|---|---|
| 10368.860 | LA8SHF | Tonsberg | JO59FB | 30 | 13 dB Horn | 180° | 10 | LA6LCA | Keys "LA1UHG" | 05/06 LA0BY |
| 10368.860 | F5ZAE | Pic Neulos | JN12LL | 1100 | Slotted WG | Omni | 1 | F2CF | | 06/12 F6HTJ |
| 10368.865 | DB0JK | Köln DOK Z 12 | JO30LX | 200 | Slot | Omni | 200 | DK2KA | | 03/99 DJ3TF |
| 10368.865 | F1ZAI | Orleans | JN07WV | 170 | Slot | Omni | 1 | F1JGP | | 06/12 F6HTJ |
| 10368.870 | DB0IBB | DOK N 49 | JO32VG | 245 | Slot | Omni | 2 | DB7QW | | 03/99 DJ3TF |
| 10368.870 | GB3KBQ | Taunton | IO80LW | 167 | Slotted waveguide | Omni | 0.2 | G4UVZ | | 06/06 G6JYB |
| 10368.870 | OE8XGQ | Gerlitze | JN66WQ | 1909 | Slotted WG | Omni | 1,5 | OE8MI | | 09/99 OE1MCU |
| 10368.875 | ON0AZ | Antwerpen | JO21FE | 78 | | | 10 | ON7BPS | | 05/07 ON4PC |
| 10368.875 | OE5XBM | Breitenstein | JN78DJ | 985 | Slotted WG | Omni | 10 | OE5VRL | | 08/09 OE1MCU |
| 10368.877 | HG5BSB | János-hegy | JN97LM | 405 | Slot | Omni | 0.5 | HA5BDJ | | 03/13 HA5NF |
| 10368.880 | GB3CEM | Wolverhampton | IO82WO | 165 | Slotted waveguide | Omni | 25 | QRT | | [illegible] |
| 10368.880 | OE1XVB | [illegible] | [illegible] | 100 | Slotted WG | Omni | 1.5 | OE1WRS | | 09/99 OE1MCU |
| 10368.883 | DB0INN | DOK C 15 | JN68GI | 504 | Slot | Omni | 1 TX | DL3MBG | | 03/99 DJ3TF |
| 10368.884 | HB9G | Geneva | JN36BK | 1600 | Slotted waveguide | Omni | 2 | HB9PBD | | 12/05 HB9PBD |
| 10368.885 | DB0TUD | DOK C 07 | JO61UA | 286 | Slot | Omni | 5 | DL4DTU | | 03/99 DJ3TF |
| 10368.890 | DB0KLX | DOK K 16 | JN39VK | 350 | Slot | Omni | 1 TX | DC2UG | | 03/99 DJ3TF |
| 10368.890 | LA9SHF | Hoggodal | JO59FS | 160 | 10 dB Horn | 180° | 100 | LA8GKA | | 05/06 LA0BY |
| 10368.890 | ON0RUG | Gent | JO11UB | 90 | | | 1 | ON6UG | | 05/07 ON4PC |
| 10368.892 | S55ZKP | Slavnik | JN65XM | 1028 | Slot | Omni | 0.4 | | | 06/12 S51FB |
| 10368.892 | F5EJZ/B | Percy | IN98JW | 300 | Horn | 090° / 135° | 0.21 | F5EJZ | | 06/12 F6HTJ |
| 10368.895 | GB3NGI | Ballymena | IO65VB | 518 | Slotted WG | Omni | 20 | GI6ATZ | QRV | 11/14 G6JYB |
| 10368.895 | DB0ECA | DOK C 08 | JN57UV | 705 | Slot | Omni | 10 | DC8EC | | 03/99 DJ3TF |
| 10368.900 | DB0CU | DOK A 28 | JN48BI | 970 | Slot | Omni | 5 | DJ7FJ | | 03/99 DJ3TF |
| 10368.900 | DB0UX | DOK A 35 | JN48FX | 275 | Slot | Omni | 1 | DK2DB | | 03/99 DJ3TF |
| 10368.900 | IQ2CF/B | Bovezzo BS | JN55DO | 962 | Horn | 350° | 10 | | | 04/13 IK1YWB |
| 10368.900 | OH1SHF | Salo | KP10NJ | 25 | Corner Reflector | 100° | 1 | | Temp. QRT | 07/11 OH6DD |
| 10368.900 | OZ5SHF | | JO45WX | 210 | Slotted WG | Omni | 4 | OZ2OE | | 11/05 OZ2TG |
| 10368.900 | GB3AZA | Scarborough | IO94TF | 75 | Single 18" dish | | 50 | G8AZA | QRV | 04/08 G6JYB |
| 10368.900 | F5ZBA | Gueret | JN06WD | 700 | Slot | Omni | 2 | F1NYN | | 06/12 F6HTJ |
| 10368.905 | GB3SCX | Bell Hill, Dorset | IO80UU69 | 274 | Slotted waveguide | Omni | 0.9 | G4JNT | | 06/14 G6JYB |
| 10368.910 | HG3BSB | Misinateto | JN96CC | 585 | Slot | Omni | 1 | HG5AZB | | 03/13 HA5NF |
| 10368.910 | GB3RPE | Swansea | IO81AO | 60 | Slotted Waveguide | Omni | 4 | GW4ADL | | 06/06 G6JYB |
| 10368.910 | DB0HEX | DOK Z 85 | JO51HT | 1341 | Slot | Omni | 8 | DG0CBP | | 03/99 DJ3TF |
| 10368.915 | OZ4SHF | | JO65BV | 22 | Slotted WG | Omni | 10 | OZ1UM | | 11/05 OZ2TG |
| 10368.916 | SR1XHX | Czluchow | JO83QP | 255 | Slot | Omni | 1 | SQ1BVJ | | 04/07 SP6LB |
| 10368.919 | F5ZWM | Ste-Fortunade | JN05VE | 578 | Slot | Omni | 2 | F6ETI | | 06/12 F6HTJ |
| 10368.920 | OE2XBO | Haunsberg | JN67MW | 740 | Slotted WG | Omni | 1.5 | OE2HFO | | 09/99 OE1MCU |
| 10368.920 | DB0VC | DOK Z 10 | JO54IF | 291 | Slot | Omni | 1 | DL8LAO | | 03/99 DJ3TF |
| 10368.920 | SK7MHF | Naessjoe | JO77IP | 505 | Slotted WG | Omni | 4 | SM7MXO | | 09/00 SM7MXO |
| 10368.924 | F1DLT | La Roche | JN27UR | | Horn | 315° | 30 | F1DLT | | 05/05 F6HTJ |
| 10368.928 | OH2SHF | Kauniainen | KP20IF | 85 | 10 dB | Omni | 15 | | | 08/06 OH6DD |
| 10368.928 | OH2SHF | Helsinki | KP20LE | | 10 dB | Omni | 15 | | | 07/11 OH6DD |
| 10368.928 | F1URI | via Mt Blanc | JN35FU | 1660 | Dish | JN35KT | 0.7 | F1URI | | 06/12 F6HTJ |
| 10368.930 | DB0HO | DOK Z 49 | JN47QT | 487 | Slot | Omni | 10 | DF6TK | | 03/99 DJ3TF |
| 10368.930 | OZ7IGY | Jystrup | JO55WM | 98 | WG Slot | Omni | 100 | OZ7IS | PLANNED | 01/14 OZ7IS |
| 10368.930 | OE3XMB | Muckenkogel | JN77TX | 1154 | Slot | Omni | 0.2 | OE3FFC | | 02/05 OE3FFC |
| 10368.930 | OZ7IGY | Jystrup | JO55WM | 96 | 27 dB Horn | 35° | 1000 | OZ7IS | | 01/14 OZ7IS |
| 10368.935 | OH5SHF | Kouvola | KP30HV | 144 | 12 dBd | 240° | 15 | | | 07/11 OH6DD |
| 10368.935 | OH5TEN | Kuusankoshi | KP30HV | 94 | 43 dB | 220° | 15000 | | New | 06/07 OH6DD |
| 10368.940 | GB3CCX | Cheltenham | IO81XW | 342 | SlottedWG | Omni | 3 | G4SGI | | 06/14 G6JYB |
| 10368.940 | SK7MHH | Faerjestaden | JO86GP | 45 | | | | | | 02/01 SM6CEN |
| 10368.940 | DB0DON | DOK T21 | JN58KR | 532 | Slot | Omni | 1 | DL5MEL | | 03/99 DJ3TF |
| 10368.945 | DB0AJA | | JN59AS | 364 | 2 x 10 slots | Omni | 1 | DF6NA | QRV | 06/08 DF6NA |
| 10368.945 | OE2XRO | Sonnblick | JN67LA | 3105 | Slotted WG | Omni | 12.5 | OE1MCU | | 09/99 OE1MCU |
| 10368.950 | DB0FHR | DOK C 31 | JN67BU | 474 | Slot | Omni | 1 | DL5MEA | | 03/99 DJ3TF |
| 10368.950 | F5ZTT | Lacapelle | JN14EB | 625 | Slot | Omni | 1 | F6CXO | | 06/12 F6HTJ |
| 10368.953 | SK1SHF | Klintehamn | JO97CJ | 52 | Horizontal | | | | | 02/01 SM6CEN |
| 10368.955 | GB3LEX | Leicester | IO92IQ | 213 | Slotted WG | Omni | 1 | G3TQF | QRV | 08/09 G6JYB |
| 10368.955 | OZ1UHF | | JO57FJ | 150 | Slotted WG | Omni | 0.8 | OZ9NT | | 11/05 OZ2TG |
| 10368.957 | F1ZXJ | Forbach | JN39KD | 300 | Slot | Omni | 0.2 | F1ULQ | | 06/12 F6HTJ |
| 10368.960 | GB3CMS | Chelmsford | JO01GR | 107 | Slotted WG | Omni | 0.3 | G1EUC | QRV / QRP | 06/13 G6JYB |
| 10368.960 | SK4BX/B | Garphyttan | JO79LI | 265 | Slotted WG | Omni | 8 | | | 04/06 SM6CEN |
| 10368.965 | DF0ANN | DOK B 25 | JN59PL | 630 | 12 x Slot | Omni | 0.2 TX | DL8ZX | | 03/99 DJ3TF |
| 10368.970 | SK7MHL | Lund | JO65OR | 100 | WG slot | Omni | 10 | SM7ECM | | 04/06 SM6CEN |
| 10368.970 | GB3RAL | | IO91EN | 140 | | | 10 | G0MJW | Planning | 02/07 G6JYB |
| 10368.973 | OK0EO | Namest n./O | JN89BE | 430 | Slot | Omni | 1 | OK2ZI | PLANNED | 06/10 OK1HH |
| 10368.957 | F1ZXJ | | JN39KD | | Slot | Omni | 10 | F1ULQ | | 06/08 F6HTJ |
| 10368.975 | ON0KUL | Tielte- Winge | JO20KV | 100 | | | 1 | ON4IY | | 05/07 ON4PC |
| 10368.975 | OZ3SHF | | JO45NL | 58 | Slotted WG | Omni | 2 | OZ1IN | | 09/99 OZ7IS |
| 10368.980 | GB3MCB | St Austell | IO70OJ | 300 | Sectoral Horn | 67° | 10 | M0P3B | QRV | 06/14 G6JYB |
| 10368.981 | HG1BSD | Hörman-forrás | JN87FI | 725 | Slot | Omni | 0.25 | HA1YA | | 03/13 HA5NF |
| 10368.983 | F5ZWZ | Grand Cap | JN23XE | 780 | Slot | Omni | 1 | F5PVX | | 06/12 F6HTJ |
| 10368.987 | SR6NCI | Czarna Gora | JO80JG | 1150 | Slot | Omni | 1 | SP6GWB | | 04/07 SP6LB |
| 10368.994 | F5ZAD | Chalon sur Saône | JN26KT | | Slot | Omni | 0.2 | F6FAT | | 06/12 F6HTJ |
| 10369.000 | F1XAN | Bus St Remy | JN09TD | 300 | Slotted WG | Omni | 1.5 | F1PBZ | Temp. QRT | 04/06 F6HTJ |
| 10386.090 | S55ZMS | Dolina | JN86CR | 350 | Slot | Omni | 0.2 | | S53M | 06/12 S51FB |
| 10435.000 | GB3JET | Dukinfield | IO83XL | | | | 0.2 | G3WFK | Non op | 06/06 G6JYB |
| | | | | | | | | | | |
| 24000.860 | LA1SHG | Toensberg | JO59FB | 30 | 13dB Horn | 180° | ? | LA6LCA | | 04/01 PA0EZ |
| 24025.000 | GB3IOW | Newport, IOW | IO90IO | 250 | Sectoral horn | | 8 | G8MBU | | 06/05 G8MBU |
| 24048.100 | HB9OK | Monte Generoso | JN45MW | | | | | | QRV | 10/09 IK7UXW |
| 24048.050 | ON0KUL | Tielte- Winge | JO20KV | 100 | | | 1 | ON4IY | | 05/07 ON4PC |
| 24048.073 | OK0EQ | Namest n./O | JN89BE | 430 | Slot | Omni | 2 | OK2ZI | PLANNED | 07/10 OK1HH |

| Freq | Call | Nearest Town | Locator | m ASL | Antenna | Direction | Power | Info | Status | Last update |
|---|---|---|---|---|---|---|---|---|---|---|
| 24048.135 | I3EME/B | Monte Tomba TV | JN55WV | 890 | Slot 12 | 170° | 0.18 | | | 04/13 IK1YWB |
| 24048.152 | HG8BSC | Szentes | KN06DP | 90 | Slot | Omni | 0.17 | HA8MV | | 03/13 HA5NF |
| 24048.170 | F5ZTS | Beauvais | JN19FK | 140 | Dish | 50° | 0.5 | F6DWG | | 09/11 F6HTJ |
| 24048.180 | F6DKW | Velizy | JN18CS | 230 | Slot | Omni | 15 | F6DKW | | 04/06 F6HTJ |
| 24048.200 | PI7EHG | Schiphol | JO22JH | 90 | 30cm dish | 266° | 2 | PA0EHG | QRT | 10/09 PA0EZ |
| 24048.233 | F5ZEG | Talence | IN94QT | 83 | Horn | 130° | 0.5 | F6CBC | | 09/11 F6HTJ |
| 24048.250 | PA0JCA | | JO22NB | 40 | | 030° | 100 erp | | QRV | 10/09 PA0EZ |
| 24048.252 | F1ZAQ | Plougonver | IN88HL | 326 | Slotted WG | Omni | 0.08 | F1LHC | | 09/11 F6HTJ |
| 24048.300 | F5ZYA | Lacapelle | JN14EB | 625 | Slot | Omni | 0.5 | F6CXO | | 03/13 F6HTJ |
| 24048.373 | IK1YWB/B | Bagnolo CN | JN34OS | 1550 | Horn | 65° | 1 | | | 04/13 IK1YWB |
| 24048.392 | F6DKW | Velizy | JN18CS | 230 | Slot | Omni | 0.5 | F6DKW | | 09/11 F6HTJ |
| 24048.550 | F1ZPE | Orleans | JN07WV | 170 | Horn / Slot | 0° / Omni | 0.35 | F1JGP | | 09/11 F6HTJ |
| 24048.738 | F1ZSE | Foix | JN02TW | 1200 | Slot | Omni | 0.1 | F1AAM | | 03/13 F6HTJ |
| 24048.800 | OH3SHF | Tampere | KP11VK | 222 | 18 dBi | Omni | 30 | | | 07/11 OH6DD |
| 24048.829 | HG2BSC | Kab-hegy | JN87TB | 635 | Slot | Omni | 0.15 | HG5AZB | | 03/13 HA5NF |
| 24048.830 | GB3MHZ | Martlesham | JO02PB | 85 | SlottedWG | Omni | 25 | G7OCD | | 06/14 G6JYB |
| 24048.850 | GB3MAN | Rochdale | IO83WO | 153 | Slotted WG | Omni | 3 | G6GXK | QRV | 09/07 G6GXK |
| 24048.860 | F1ZSE | Mt Ventoux | JN24PE | 1910 | Slot | Omni | 1 | F1AAM | Planned | 04/06 F6HTJ |
| 24048.870 | GB3CAM | Wyton, Cambs. | IO92WI | 65 | 10dB slot | Omni | 2.5 | G4HJW | | 06/14 G6JYB |
| 24048.884 | HG5BSE | János-hegy | JN97LM | 485 | Slot | Omni | 0.18 | HA5BDJ | Temp. QRT | 03/13 HA5NF |
| 24048.885 | DB0TUD | DOK S 07 | JO61UA | 260 | Slot | Omni | | DL4DTU | | 02/07 DL4DTU |
| 24048.890 | GB3DUN | Dunstable | IO91SV | 260 | Slotted WG | Omni | 1 | G3ZFP | | 06/06 G6JYB |
| 24048.890 | SK6SHG | Tjorn | JO57TX | 110 | 2 x sector horn | N/S | 2 x 1 | | | 04/06 SM6CEN |
| 24048.900 | F1DFY | Grand Cap | JN23XE | 780 | Slot | Omni | 0.9 | F1DFY | | 03/13 F6HTJ |
| 24048.900 | OZ5SHF | | JO45WX | 205 | WG | Omni | 0.5 | | | 11/05 OZ2TG |
| 24048.905 | GB3SCK | Bell Hill, Dorset | IO80UU69 | 274 | Slotted WG | Omni | 3 | G4JNT | | 06/14 G6JYB |
| 24048.910 | GB3ZME | Telford | IO82RP | 218 | Slotted WG | Omni | 8 | G3UKV | | 06/08 G6JYB |
| 24048.920 | GB3FNM | Farnham | IO91OF | 207 | Slotted WG | Omni | 25 | G4EPX | | 04/09 G6JYB |
| 24048.921 | HG4BSE | Meleg-hegy | JN97HG | 352 | Slot | Omni | 0.1 | HA5KHC | | 03/13 HA5NF |
| 24048.930 | OZ7IGY | Jystrup | JO55WM | 98 | WG Slot | Omni | 10 | OZ7IS | | 01/14 OZ7IS |
| 24048.930 | HG3BSC | Misinateto | JN96CC | 585 | Slot | Omni | 0.18 | HG5AZB | Temp. QRT | 03/13 HA5NF |
| 24048.945 | DB0AJA | | JN59AS | 364 | 2 x 8 slots | Omni | 0.5 | DF6NA | QRV | 06/08 DF6NA |
| 24048.950 | OZ9UHF | | JO65HP | 30 | Slot | Omni | 0.5 | | | 11/05 OZ2TG |
| 24048.950 | HB9G | | JN46KE | | | | | | | 04/03 HB9AHL |
| 24048.970 | SK7MHL | Lund | JO65OR | 100 | Slotted WG | Omni | 1 | | | 04/06 SM6CEN |
| 24050.000 | OK0EL | Benecko | JO70SQ | 1030 | 12 el slot WG | 90°/270° | 0.015 | OK1AIY | | 08/01 OK1HH |
| 24192.000 | ON0RUG | Gent | JO11UB | 90 | | | 1 | ON6UG | | 05/07 ON4PC |
| 24192.050 | DB0KHT | DOK F 13 | JO40FE | | Horn | Omni | 0.02 TX | DJ1RV | | 03/99 DJ3TF |
| 24192.055 | DB0JO | DOK Z 03 | JO31SL | 312 | 6 x Slot | Omni | 0.6 | DG8DCI | | 03/99 DJ3TF |
| 24192.120 | DB0JL | DOK R 25 | JO31MC | 195 | Slot | Omni | 0.01 | DF1EQ | | 03/99 DJ3TF |
| 24192.200 | LX0DUF | Soleuvre | JN29XM | 280 | 0.4m Dish | 63° | 1.2 kW | | | 02/98 LX1SC |
| 24192.405 | DB0AS | DOK C 29 | JN67CR | 1565 | Horn | 10° | 0.5 TX | DL2AS | | 03/99 DJ3TF |
| 24192.800 | SK6MHI | Goteborg | JO57XQ | 135 | 2 x Sectoral Horn | 225°/315° | 1 | SM6EAN | | 02/01 SM6CEN |
| 24192.830 | F5XAF | Paris | JN18DU | | Dish | 90° | 0.1 | F5ORF | QRT | 04/06 F6HTJ |
| 24192.833 | DB0FGB | DOK B 09 | JO50WB | 1150 | Slot | Omni | 0.6 | DB8UY | | 03/99 DJ3TF |
| 24192.840 | DB0KI | Bayreuth DOK Z 42 | JO50WC | 925 | Slot | 0° | 0.5 | DC9NL | | 03/99 DJ3TF |
| 24192.853 | DL0WY | DOK C 29 | JN67AQ | 1838 | Sectored horn | 45°/270° | 0.01 | DJ8VY | | 03/99 DJ3TF |
| 24192.860 | LA8SHF | Tonsberg | JO59FB | 30 | 13 dB Horn | 180° | | LA6LCA | Keys "LA1UHG" | 05/06 LA0BY |
| 24192.860 | DB0ARB | DOK U 02 | JN69NC | 1456 | Parabola | 225° | 0.03 | DJ4YJ | | 03/99 DJ3TF |
| 24192.865 | DB0JK | DOK Z 12 | JO30LX | 260 | 2 x H-Horn | Omni | 1 | DK2KA | | 03/99 DJ3TF |
| 24192.875 | ON0AZ | Antwerpen | JO21FE | 75 | | | 1 | ON7BPS | | 05/07 ON4PC |
| 24192.875 | OE5XBM | Breitenstein | JN78DJ | 985 | Slotted WG | Omni | 0.5 | OE5VRL | | 09/99 OE1MCU |
| 24192.875 | DB0HW | DOK H46 | JO51GT | 1016 | Slot | Omni | 1 | DL3AAS | | 03/99 DJ3TF |
| 24192.895 | DB0ECA | DOK C 08 | JN57UV | 705 | Slot | 0° | | DC8EC | QRT | 03/99 DJ3TF |
| 24192.900 | DB0CU | DOK A 28 | JN48BI | 970 | Horn | 180° | 5 | DJ7FJ | | 03/99 DJ3TF |
| 24192.900 | SK0UX/B | Taeby | JO99BM | | Horizontal | | | | | 02/01 SM6CEN |
| 24192.910 | DB0HEX | DOK Z 85 | JO51HT | 1341 | Slot | Omni | 8 | DG0CBP | | 03/99 DJ3TF |
| 24192.915 | OZ4SHF | | JO65BV | 22 | Slotted WG | Omni | 10 | OZ1UM | | 11/05 OZ2TG |
| 24048.940 | GB3AMU | Cardiff | IO81JN | 266 | Sectorial Horn | 135° | 0.5 | GW3PPF | | 06/06 G6JYB |
| 24192.945 | OE2XRO | Sonnblick | JN67LA | 3105 | Slotted WG | Omni | 5 | OE1MCU | | 09/99 OE1MCU |
| 24192.955 | | | | | | | | | | |
| 24192.955 | OZ1UHF | | JO57FJ | 150 | Slot | Omni | 0.5 | | QRT | 11/05 OZ2TG |
| 24192.970 | SK7MHL | Lund | JO65OR | 100 | WG slot | Omni | 1 | SM7ECM | | 04/02 SM7ECM |
| | | | | | | | | | | |
| 47048.200 | F5ZEF | Pessac | IN94QT | 83 | Dish | 50° | 0.03 | F6CBC | | 03/13 F6HTJ |
| 47088.100 | DB0AS | DOK C 29 | JN67CR | 1565 | Horn | 10° | 0.5 TX | DL2AS | | 03/99 DJ3TF |
| 47088.833 | DB0FGB | DOK B09 | JO50WB | 1150 | Slot | Omni | 0.2 | DB8UY | | 03/99 DJ3TF |
| 47088.853 | DL0WY | DOK C29 | JN67AQ | 1830 | Horn | 0°/90°/270° | 0.1 | DJ8VY | | 03/99 DJ3TF |
| 47088.865 | DB0JK | DOK Z 12 | JO30LX | 260 | 2 x H-Horn | Omni | 0.1 TX | DK2KA | | 03/99 DJ3TF |
| 47088.875 | OE5XBM | Hellmonsoedt | JN78DK | 855 | Slotted WG | Omni | 0.25 | OE5VRL | | 09/99 OE1MCU |
| 47088.895 | DB0ECA | DOK C 08 | JN57UV | 705 | Horn | 0° | | DC8EC | | 03/99 DJ3TF |
| 47088.920 | GB3FNM | Farnham | IO91OF | 207 | Slotted WG | Omni | 10 | G4EPX | | 04/09 G6JYB |
| 76032.833 | DB0FGB | DOK B 09 | JO50WB | 1150 | Slot | Omni | 0.01 | DB8UY | QRT | 03/99 DJ3TF |
| 76032.895 | DB0ECA | DOK C 08 | JN57UV | 705 | Horn | 0° | | DC8EC | QRT | Mar-99 |

**NOTES** * Licensed/Awaiting ** Not coord

**Direction Column**

45°/315° Transmits at 45° and 315° continuously

0°/225° 1.5 Transmits at 0° and 225° alternatively at 1.5 minute intervals

**Power Column**

Unless otherwise stated, power is given as the Estimate Radiated Power in Watts relative to a dipole. The letters TX after a power level indicate the power level at the transmitter output.

**Status Column**

No Information means 'On Air'. The reliability of this information can be judged by the source and date in the 'Last update' column.

# Abbreviations & Codes

Numerous abbreviations and codes are used by radio amateurs, especially when using Morse or datamodes. Many cross language barriers, enabling people without a common language to communicate, but some also find their way into spoken conversations when it would be just as simple to use plain language. Listed here are many of the common ones, but this list is by no means exhaustive and new abbreviations are being introduced all the time.

## Abbreviations

|  |  |
|---|---|
| ABT | About |
| AGN | Again |
| AM | Amplitude Modulation |
| ANI | Any |
| ANT | Antenna |
| BCI | Broadcast Interference |
| BCNU | Be Seeing You |
| BK | Break |
| BTW | By The Way |
| BURO | (QSL) Bureau |
| B4 | Before |
| CCT | Circuit |
| CFM | Confirm |
| CHK | Check |
| CLD | Called |
| CLG | Calling |
| CONDX | Conditions |
| CPI | Copy |
| CPY | Copy |
| CQ | General call ('Seek You') |
| CS | Call Sign |
| CU | See You |
| CU AGN | See You Again |
| CUD | Could |
| CUL | See You Later |
| CUZ | Because |
| CW | Continuous Wave (Morse) |
| DE | From |
| DN | Down |
| DR | Dear |
| DX | Long Distance |
| EL | Element |
| ENUF | Enough |
| ES | And |
| EU | Europe |
| EVE | Evening |
| FB | Fine Business (excellent) |
| FER | For |
| FM | Frequency Modulation |
| FM | From |
| FONE | Phone (telephony) |
| FREQ | Frequency |
| GA | Go Ahead |
| GA | Good Afternoon |
| GB | Good Bye |
| GD | Good |
| GD | Good Day |
| GE | Good Evening |
| GG | Going |
| GLD | Glad |
| GM | Good Morning |
| GN | Good Night |
| GND | Ground |
| GP | Ground Plane |
| GUD | Good |
| HI | Laughter |
| HI | High |
| HPE | Hope |
| HR | Here |
| HR | Hear |
| HRD | Heard |
| HV | Have |
| HVE | Have |
| HVG | Having |
| HVY | Heavy |
| HW | How |
| HW | How Copy? |
| IMI | Repeat (question mark) |
| INFO | Information |
| K | Invitation to Transmit |
| LID | A Bad Operator |
| LNG | Long |
| LP | Long Path |
| LSN | Listen |
| LTR | Later |
| LW | Long Wire |
| LW | Long Wave |
| MA | Millamperes |
| MGR | Manager |
| MI | My |
| MILS | Millamperes |
| MNI | Many |
| MOM | Moment |
| MSG | Message |
| MULT | Multiplier |
| N | No |
| NIL | Nothing |
| NR | Number |
| NR | Near |
| NW | Now |
| OB | Old Boy |
| OC | Old Chap |
| OK | Correct |
| OM | Old Man |
| OP | Operator |
| OT | Old Timer |
| OW | Old Woman |
| PLS | Please |
| PSE | Please |
| PWR | Power |
| R | Received |
| R | Are |
| RC | Ragchew |
| RCD | Received |
| RCVR | Receiver |
| RE | Regarding |
| REF | Referring to |
| RFI | Radio Frequency Interference |
| RIG | Station Equipment |
| RPRT | Report |
| RPT | Repeat |
| RPT | Report |
| RTTY | Radio Teletype |
| RST | Readability, Strength, Tone (report) |
| RX | Receive |
| RX | Receiver |
| SA | Say |
| SED | Said |
| SEZ | Says |
| SHUD | Should |
| SIG | Signal |
| SK | Silent Key (deceased) |
| SKED | Schedule |
| SN | Soon |
| SP | Short Path |
| SRI | Sorry |
| SSB | Single Side Band |
| STN | Station |
| SUM | Some |
| SWL | Short Wave Listener |
| TEMP | Temperature |
| TEST | Testing |
| TEST | Contest |
| THRU | Through |
| TKS | Thanks |
| TMW | Tomorrow |
| TNX | Thanks |
| TR | Transmit |
| T/R | Transmit/Receive |
| TRBL | Trouble |
| TRX | Transceiver |
| TT | That |
| TU | Thank You |
| TVI | Television Interference |
| TX | Transmitter; Transmit |
| U | You |
| UFB | Ultra Fine Business |
| UR | Your |
| UR | You're |
| URS | Yours |
| VERT | Vertical |
| VFB | Very Fine Business |
| VFO | Variable Frequency Oscillator |
| VY | Very |
| W | Watts |
| WID | With |
| WKD | Worked |
| WKG | Working |
| WL | Well |
| WL | Will |
| WPM | Words Per Minute |
| WRD | Word |
| WRK | Work |
| WUD | Would |
| WX | Weather |

| | |
|---|---|
| XCVR | Transceiver |
| XMAS | Christmas |
| XMTR | Transmitter |
| XTAL | Crystal |
| XYL | Wife |
| YF | Wife |
| YL | Young Lady (girlfriend) |
| YR | Year |
| Z | Zulu (Time) |
| 55 | Best Success |
| 73 | Best Regards |
| 88 | Love and Kisses |

## Five-unit Code

The so-called Murray Code is used by RTTY operators and telex machines. It consists of one 'start' bit, five 'data' bits, then 1.5 'stop' bits. As there are only 32 possible combinations of code, a limited character set is available (e.g. no lower case). Traditionally, amateur RTTY is sent at 45.45 Bauds, which results in an element length of 22ms. It is known officially as the International Telegraphic Alphabet No.2.

| Binary | Dec | Hex | Octal | Letter | Figure |
|---|---|---|---|---|---|
| 00000 | 0 | 00 | 00 | [Blank] | |
| 00001 | 1 | 01 | 01 | T | 5 |
| 00010 | 2 | 02 | 02 | [Carr. Return] | |
| 00011 | 3 | 03 | 03 | O | 9 |
| 00100 | 4 | 04 | 04 | [Space] | |
| 00101 | 5 | 05 | 05 | H | # (note) |
| 00110 | 6 | 06 | 06 | N | , |
| 00111 | 7 | 07 | 07 | M | . |
| 01000 | 8 | 08 | 10 | [Line Feed] | |
| 01001 | 9 | 09 | 11 | L | ) |
| 01010 | 10 | 0A | 12 | R | 4 |
| 01011 | 11 | 0B | 13 | G | & (note) |
| 01100 | 12 | 0C | 14 | I | 8 |
| 01101 | 13 | 0D | 15 | P | 0 |
| 01110 | 14 | 0E | 16 | C | : |
| 01111 | 15 | 0F | 17 | V | ; |
| 10000 | 16 | 10 | 20 | E | 3 |
| 10001 | 17 | 11 | 21 | Z | " |
| 10010 | 18 | 12 | 22 | D | $ |
| 10011 | 19 | 13 | 23 | B | ? |
| 10100 | 20 | 14 | 24 | S | Bell |
| 10101 | 21 | 15 | 25 | Y | 6 |
| 10110 | 22 | 16 | 26 | F | ! (note) |
| 10111 | 23 | 17 | 27 | X | / |
| 11000 | 24 | 18 | 30 | A | - |
| 11001 | 25 | 19 | 31 | W | 2 |
| 11010 | 26 | 1A | 32 | J | ' |
| 11011 | 27 | 1B | 33 | [Figure Shift] | |
| 11100 | 28 | 1C | 34 | U | 7 |
| 11101 | 29 | 1D | 35 | Q | 1 |
| 11110 | 30 | 1E | 36 | K | ( |
| 11111 | 31 | 1F | 37 | [Letter Shift] | |

Note:
The letters F, G and H in the Figures mode are not allocated internationally. Each country is free to use them as they see fit. The American usage is shown in the table above.

## Morse Code

The standard alphabet and numbers, as required for the Foundation Licence Morse Assessment.

| | | | |
|---|---|---|---|
| A | • ▬ | N | ▬ • |
| B | ▬ • • • | O | ▬ ▬ ▬ |
| C | ▬ • ▬ • | P | • ▬ ▬ • |
| D | ▬ • • | Q | ▬ ▬ • ▬ |
| E | • | R | • ▬ • |
| F | • • ▬ • | S | • • • |
| G | ▬ ▬ • | T | ▬ |
| H | • • • • | U | • • ▬ |
| I | • • | V | • • • ▬ |
| J | • ▬ ▬ ▬ | W | • ▬ ▬ |
| K | ▬ • ▬ | X | ▬ • • ▬ |
| L | • ▬ • • | Y | ▬ • ▬ ▬ |
| M | ▬ ▬ | Z | ▬ ▬ • • |

| | | | |
|---|---|---|---|
| 1 | • ▬ ▬ ▬ ▬ | 6 | ▬ • • • • |
| 2 | • • ▬ ▬ ▬ | 7 | ▬ ▬ • • • |
| 3 | • • • ▬ ▬ | 8 | ▬ ▬ ▬ • • |
| 4 | • • • • ▬ | 9 | ▬ ▬ ▬ ▬ • |
| 5 | • • • • • | 0 | ▬ ▬ ▬ ▬ ▬ |

## Special Morse Characters

There are numerous procedural and punctuation characters. The following are those most commonly used by Morse operators.

| | | |
|---|---|---|
| $\overline{\text{AR}}$ (+) | [di-dah-di-dah-dit] | End of message (used before the final calls and is written as 'AR' or '+') |
| $\overline{\text{CT}}$ | [dah-di-dah-di-dah] | Preliminary call |
| $\overline{\text{BT}}$ (=) | [dah-di-di-di-dah] | Separation signal (used in text and is written as 'BT' or '=') |
| $\overline{\text{KN}}$ | [dah-di-dah-dah-dit] | Transmit only the station called (used after the final calls and is written as 'KN') |
| $\overline{\text{VA}}$ | [di-di-di-dah-di-dah] | Transmission ends (written as 'VA' or 'SK') |
| ? | [di-di-dah-dah-di-dit] | Question mark (written as 'IMI' or '?') |
| / | [dah-di-di-dah-dit] | Oblique stroke (can be used as part of a callsign and is written as '/') |
| . | [di-dah-di-dah-di-dah] | Full stop |
| Error | [di-di-di-di-di-di-di-dit] | Erases the word in which a mistake has been made |
| @ | [di-dah-dah-di-dah-dit] | As used in Email addresses |

## Phonetic Alphabet

The Phonetic Alphabet used by radio amateurs today was developed by NATO in the 1950s to be intelligible (and pronounceable) to all NATO allies. It replaced several other phonetic alphabets and is now widely used in business and telecommunications across Europe and North America.

| | | | |
|---|---|---|---|
| A | Alpha | N | November |
| B | Bravo | O | Oscar |
| C | Charlie | P | Papa |
| D | Delta | Q | Quebec |
| E | Echo | R | Romeo |
| F | Foxtrot | S | Sierra |
| G | Golf | T | Tango |
| H | Hotel | U | Uniform |
| I | India | V | Victor |
| J | Juliet | W | Whiskey |
| K | Kilo | X | X-ray |
| L | Lima | Y | Yankee |
| M | Mike | Z | Zulu |

## Q Codes

There are a huge number of three-letter Q codes. Radio amateurs use only a small percentage of them, as many are of relevance only to shipping, aircraft, the police etc. They fall into the following pattern.

| | |
|---|---|
| QAA-QNZ | Aeronautical |
| QOA-QQZ | Maritime |
| QRA-QUZ | All services |
| QZA-QZZ | Other |

All Q codes follow the form of a question and answer. The list below gives details of those in common use by radio amateurs.

## ASCII Code

The American Standard Code for Information Interchange is the code used in computing and for packet radio. Standard ASCII consists of seven data bits, which gives 128 possible combinations. The first 32 characters are used for control and the remaining 96 are representable characters.

| Dec | Hex | Chr | Ctrl | Dec | Hex | Chr | Dec | Hex | Chr | Dec | Hex | Chr |
|---|---|---|---|---|---|---|---|---|---|---|---|---|
| 0 | 0 | NUL | ^@ | 32 | 20 | SP | 64 | 40 | @ | 96 | 60 | ‘ |
| 1 | 1 | SOH | ^A | 33 | 21 | ! | 65 | 41 | A | 97 | 61 | a |
| 2 | 2 | STX | ^B | 34 | 22 | " | 66 | 42 | B | 98 | 62 | b |
| 3 | 3 | ETX | ^C | 35 | 23 | # | 67 | 43 | C | 99 | 63 | c |
| 4 | 4 | EOT | ^D | 36 | 24 | $ | 68 | 44 | D | 100 | 64 | d |
| 5 | 5 | ENQ | ^E | 37 | 25 | % | 69 | 45 | E | 101 | 65 | e |
| 6 | 6 | ACK | ^F | 38 | 26 | & | 70 | 46 | F | 102 | 66 | f |
| 7 | 7 | BEL | ^G | 39 | 27 | ‘ | 71 | 47 | G | 103 | 67 | g |
| 8 | 8 | BS | ^H | 40 | 28 | ( | 72 | 48 | H | 104 | 68 | h |
| 9 | 9 | HT | ^I | 41 | 29 | ) | 73 | 49 | I | 105 | 69 | i |
| 10 | 0A | LF | ^J | 42 | 2A | * | 74 | 4A | J | 106 | 6A | j |
| 11 | 0B | VT | ^K | 43 | 2B | + | 75 | 4B | K | 107 | 6B | k |
| 12 | 0C | FF | ^L | 44 | 2C | , | 76 | 4C | L | 108 | 6C | l |
| 13 | 0D | CR | ^M | 45 | 2D | - | 77 | 4D | M | 109 | 6D | m |
| 14 | 0E | SO | ^N | 46 | 2E | . | 78 | 4E | N | 100 | 6E | n |
| 15 | 0F | SI | ^O | 47 | 2F | / | 79 | 4F | O | 111 | 6F | o |
| 16 | 10 | DLE | ^P | 48 | 30 | 0 | 80 | 50 | P | 112 | 70 | p |
| 17 | 11 | DC1 | ^Q | 49 | 31 | 1 | 81 | 51 | Q | 113 | 71 | q |
| 18 | 12 | DC2 | ^R | 50 | 32 | 2 | 82 | 52 | R | 114 | 72 | r |
| 19 | 13 | DC3 | ^S | 51 | 33 | 3 | 83 | 53 | S | 115 | 73 | s |
| 20 | 14 | DC4 | ^T | 52 | 34 | 4 | 84 | 54 | T | 116 | 74 | t |
| 21 | 15 | NAK | ^U | 53 | 35 | 5 | 85 | 55 | U | 117 | 75 | u |
| 22 | 16 | SYN | ^V | 54 | 36 | 6 | 86 | 56 | V | 118 | 76 | v |
| 23 | 17 | ETB | ^W | 55 | 37 | 7 | 87 | 57 | W | 119 | 77 | w |
| 24 | 18 | CAN | ^X | 56 | 38 | 8 | 88 | 58 | X | 120 | 78 | x |
| 25 | 19 | EM | ^Y | 57 | 39 | 9 | 89 | 59 | Y | 121 | 79 | y |
| 26 | 1A | SUB | ^Z | 58 | 3A | : | 90 | 5A | Z | 122 | 7A | z |
| 27 | 1B | ESC | | 59 | 3B | ; | 91 | 5B | [ | 123 | 7B | { |
| 28 | 1C | FS | | 60 | 3C | < | 92 | 5C | \ | 124 | 7C | | |
| 29 | 1D | GS | | 61 | 3D | = | 93 | 5D | ] | 125 | 7D | } |
| 30 | 1E | RS | | 62 | 3E | > | 94 | 5E | ^ | 126 | 7E | ~ |
| 31 | 1F | US | | 63 | 3F | ? | 95 | 5F | _ | 127 | 7F | DEL |

| Q-code | Question | Answer | Colloquial use (if different)/explanation |
|---|---|---|---|
| QRB | How far are you from my station? | The distance between our stations is ... | |
| QRG | What is my exact frequency? | Your frequency is ... | Frequency of operation |
| QRH | Does my frequency vary? | Your frequency varies | |
| QRI | How is the tone of my transmission? | The tone of your transmission is ... | |
| QRL | Are you (or is the frequency) busy? | I am (or the frequency is) busy | |
| QRM | Are you suffering interference? | I am suffereing interference | Man-made interference |
| QRN | Are you troubled by static? | I am troubled by static | Natural interference (atmospherics) |
| QRO | Shall I increase power? | Increase power | High power |
| QRP | Shall I decrease power? | Decrease power | Low power |
| QRQ | Shall I send faster? | Send faster | High speed |
| QRS | Shall I send more slowly? | Send slower | Low speed |
| QRT | Shall I stop transmitting? | Stop transmitting | To close down |
| QRU | Have you anything for me? | I have nothing for you | |
| QRV | Are you ready to transmit? | I am ready to transmit | |
| QRX | When will you call again? | I will call you again at ... | To stand-by |
| QRZ | Who is calling me? | You are being called by ... | |
| QSA | What is the strength of my signal? | The strength of your signal is ... | |
| QSB | Does the strength of my signals vary? | The strength of your signals varies | Fading |
| QSD | Is my keying defective? | Your keying is defective | |
| QSK | Can you hear me between your signals (and if so can I break-in)? | I can hear you between my signals (and it is OK to break-in on my transmission) | Break-in Morse operation |
| QSL | Can you acknowledge receipt? | I am acknowledging receipt | A card to confirm contact |
| QSO | Can you communicate with ... ? | I can communicate with ... | A contact |
| QSP | Will you relay to ...? | I will relay to ... | |
| QSX | Will you listen for ... (callsign) on ...? | I am listening for ... on ... | Split frequency operation |
| QSY | Shall I change frequency? | Change frequency to ... | |
| QTF | Will you give me the position of my station according to bearings taken? | The position of your station according to bearings taken is... | Beam heading (in degrees) |
| QTH | What is your position (location)? | My position (location) is ... | |
| QTR | What is the exact time? | The exact time is ... | |

# Repeater Listings

There are over 700 repeaters licensed in the United Kingdom. They range in frequency from 28MHz to 10GHz. Many are traditional FM units, but there are also several for amateur television. Repeaters are increasingly being connected to the Internet and D-STAR digital repeaters are increasing in number.

## Repeaters (alphabetical)

| Callsign | Channel | Locator | Location |
|---|---|---|---|
| GB3AA | RV53 | IO81RO | Bristol |
| GB3AC | RU78 | IO81SR | Lydney Glos |
| GB3AE | R50-01 | IO71PR | Tenby |
| GB3AG | RV58 | IO86ON | Forfar |
| GB3AH | RB11 | JO02KP | East Dereham |
| GB3AK | RM14A | IO81RO | Bristol |
| GB3AL | RV59 | IO91QP | Amersham |
| GB3AM | R50-13 | IO91QP | Amersham |
| GB3AN | RB08 | IO73UJ | Amlwch |
| GB3AP | RV49 | IO71OV | Haverfordwest |
| GB3AR | RV56 | IO73VC | Caernarfon |
| GB3AS | RV48 | IO84LS | Carlisle |
| GB3AU | RB07 | IO91QP | Amersham |
| GB3AV | RB02 | IO91OT | Aylesbury |
| GB3AW | RB10 | IO91HH | Newbury |
| GB3AY | RV52 | IO75OR | Dalry |
| GB3BB | RV56 | IO81HW | Brecon |
| GB3BC | RV60 | IO81IO | Pontypridd |
| GB3BE | RU69 | IO85SS | Duns |
| GB3BF | RV63 | IO92SD | Bedford |
| GB3BG | RB09 | IO81KS | Blaenavon |
| GB3BI | RV58 | IO77WO | Inverness |
| GB3BK | RM0A | JO01AK | Bromley |
| GB3BL | RB07 | IO92SD | Bedford |
| GB3BM | RV57 | IO92BL | Birmingham |
| GB3BN | RB0 | IO91OJ | Bracknell |
| GB3BP | RU71 | IO93GA | Belper |
| GB3BR | RB06 | IO90WT | Brighton |
| GB3BS | RU68 | IO81TK | Bristol City |
| GB3BT | RV56 | IO85WT | Berwick On Tweed |
| GB3BV | RB01 | IO91SR | Hemel Hempstead |
| GB3BX | RV54 | IO82QN | Much Wenlock |
| GB3BZ | RU68 | JO01GW | Braintree |
| GB3CA | RB13 | IO84OT | Carlisle |
| GB3CB | RB14 | IO92BL | Birmingham |
| GB3CC | RU77 | IO90RM | Chichester |
| GB3CD | RV55 | IO94DR | Crook |
| GB3CE | RB14 | JO01KV | Colchester |
| GB3CF | RV48 | IO92IQ | Markfield |
| GB3CG | RV58 | IO81VU | Gloucester |
| GB3CH | RB02 | IO70SM | Liskeard |
| GB3CI | RB02 | IO92PM | Corby |
| GB3CJ | 10M | IO92NF | Northampton |
| GB3CK | RB0 | JO01JF | Charing Kent |
| GB3CL | RB09 | JO01OT | Clacton |
| GB3CM | RB08 | IO71VW | Carmarthen |
| GB3CO | RV53 | IO92PM | Corby |
| GB3CP | RV59 | IO64IG | Fermanagh |
| GB3CR | RB06 | IO83LC | Caergwrle |
| GB3CS | RV60 | IO85AU | Motherwell |
| GB3CT | R50-14 | IO92MG | Northampton |
| GB3CV | RB09 | IO92GK | Coventry |
| GB3CW | RB04 | IO82HL | Newtown Powys |
| GB3DA | RV58 | JO01GR | Danbury |
| GB3DB | R50-06 | JO01HR | Danbury |
| GB3DC | RV55 | IO92GW | Derby |
| GB3DD | RB10 | IO86MM | Dundee |
| GB3DE | RB07 | JO02NF | Ipswich |
| GB3DG | RV62 | IO74UV | Newton Stewart |
| GB3DI | RB06 | IO91IN | Didcot |
| GB3DM | RU74 | IO75RW | Dumbarton |
| GB3DN | RV51 | IO70UW | Stibb Cross |
| GB3DQ | RU78 | IO70RI | Polperro |
| GB3DR | RV59 | IO80QR | Dorchester |
| GB3DS | RB13 | IO93KH | Worksop |
| GB3DT | RB0 | IO80WU | Blandford Camp |
| GB3DU | RV61 | IO85SS | Duns |
| GB3DV | RB01 | IO93JK | Maltby |
| GB3DW | RV53 | IO72VW | Criccieth |
| GB3DX | RB12 | IO65HA | Londonderry |
| GB3DY | RB10 | IO93FB | Wirksworth |
| GB3EA | RV55 | JO02GE | Wickhambrook |
| GB3EB | RV63 | JO00BW | Uckfield |
| GB3ED | RB14 | IO85JW | Edinburgh |
| GB3EE | RB12 | IO93FE | Chesterfield |
| GB3EF | R50-01 | JO02NF | Stowmarket |
| GB3EH | RB08 | IO92GC | Edge Hill |

| Callsign | Channel | Locator | Location |
|---|---|---|---|
| GB3EK | RU71 | JO01QJ | Margate |
| GB3EL | RV48 | JO01AM | London |
| GB3ER | RB03 | JO01GR | Danbury |
| GB3ES | RV54 | JO00HV | Hastings |
| GB3EV | RV56 | IO84SQ | Dufton |
| GB3EW | RV49 | IO80FR | Exeter |
| GB3EX | RU76 | IN89GW | Silverton |
| GB3EZ | RU78 | JO02GE | Wickhambrook |
| GB3FC | RU68 | IO83LU | Blackpool |
| GB3FE | RV53 | IO86BC | Stirling |
| GB3FF | RV48 | IO86GC | Kelty |
| GB3FG | RV57 | IO71WW | Carmarthen |
| GB3FH | R50-06 | IO81OH | Somerset |
| GB3FI | RU74 | IO81OH | Cheddar |
| GB3FJ | RU76 | IO93XE | Asgarby |
| GB3FK | RV60 | JO01OC | Folkestone |
| GB3FM | RM2 | IO91OF | Farnham |
| GB3FN | RB15 | IO91OF | Farnham |
| GB3FR | RV62 | JO03AE | Spilsby Lincs. |
| GB3FX | R50-10 | IO91OF | Farnham |
| GB3GB | RB12 | IO92BN | Birmingham |
| GB3GC | R50-02 | IO70WN | Gunnislake |
| GB3GD | RV50 | IO74SG | Snaefell Iom |
| GB3GF | RB12 | IO91RF | Guildford |
| GB3GH | RB05 | IO81VU | Gloucester |
| GB3GJ | RV51 | IN89WE | St Helier |
| GB3GL | RU72 | IO75WU | Glasgow |
| GB3GN | RV62 | IO87TA | Banchory |
| GB3GO | RV63 | IO83BH | Llandudno |
| GB3GR | RU75 | IO92RV | Grantham |
| GB3GT | R50-12 | IO82QJ | Ludlow |
| GB3GU | RB13 | IN89RK | Guernsey |
| GB3GX | 10M | IO82QJ | Ludlow |
| GB3GY | RB11 | IO93XN | Grimsby |
| GB3HA | RV60 | IO85XA | Corbridge |
| GB3HB | RB15 | IO70OJ | St Austell |
| GB3HC | RB06 | IO82PB | Hereford |
| GB3HD | RB09 | IO93BP | Huddersfield |
| GB3HE | RB14 | JO00HV | Hastings |
| GB3HE | RU74 | JO00HV | Hastings |
| GB3HF | R50-05 | IO93HO | Barnsley |
| GB3HG | RV50 | IO94JF | Thirsk |
| GB3HH | RV56 | IO93BF | Buxton |
| GB3HI | RV56 | IO76DK | Isle Of Mull |
| GB3HJ | RB01 | IO94EB | Harrogate |
| GB3HK | RB02 | IO85ON | Selkirk |
| GB3HM | R50-11 | IO93GA | Belper |
| GB3HN | RB11 | IO91VW | Hitchin |
| GB3HR | RB14 | IO91TO | Harrow |
| GB3HS | RV52 | IO93RT | Walkington |
| GB3HT | RB11 | IO92HN | Hinckley |
| GB3HW | RB13 | JO01CO | Romford |
| GB3HY | RU72 | IO90WX | Haywards Heath |
| GB3IC | RB05 | IO82WO | Wolverhampton |
| GB3IE | RU68 | IO70XJ | Plymouth |
| GB3IG | RV62 | IO68QE | Stornoway |
| GB3IH | RB04 | JO02OB | Ipswich |
| GB3IK | RV61 | JO01FJ | Rochester |
| GB3IM-C | RU66 | IO74SD | Douglas Iom |
| GB3IM-P | RU70 | IO74PF | Peel |
| GB3IM-R | RU66 | IO74TI | Ramsey Iom |
| GB3IM-S | RB05 | IO74SG | Douglas Iom |
| GB3IN | RV51 | IO93GD | Matlock |
| GB3IP | RV61 | IO82WS | Stafford |
| GB3IR | RV61 | IO94DJ | Richmond Yorks |
| GB3IS | RV60 | IO77BF | Broadford |
| GB3IW | RB09 | IO90JQ | Ryde |
| GB3JB | RV63 | IO81VC | Mere Wiltshire |
| GB3JC | RU72 | JO02QP | Norwich |
| GB3JL | RV63 | IO70SM | Liskeard |
| GB3JS | RB01 | JO02UO | Great Yarmouth |
| GB3JU | RU72 | IN89WE | St Helier |
| GB3JX | R50-12 | JO02QP | Norwich |
| GB3KC | RU72 | IO82WK | Stourbridge |
| GB3KD | RV63 | IO82UJ | Kidderminster |
| GB3KE | RV55 | IO75UV | Glasgow |
| GB3KI | RV53 | JO01NI | Herne Bay |

| Callsign | Channel | Locator | Location |
|---|---|---|---|
| GB3KK | RU78 | IO65VE | Ballycastle |
| GB3KL | RB04 | JO02FR | Kings Lynn |
| GB3KN | RV56 | JO01GH | Maidstone |
| GB3KR | RB03 | IO82UJ | Kidderminster |
| GB3KS | RV50 | JO01PA | Dover |
| GB3KU | RB03 | IO83XM | Ashton-Under-Lyne |
| GB3KV | RU78 | IO75XX | Kilsyth |
| GB3KY | RV57 | JO02FS | Kings Lynn |
| GB3LA | RV57 | IO85CJ | Sanquhar |
| GB3LB | RV63 | IO85NS | Lauder |
| GB3LC | RB09 | IO93WH | Louth |
| GB3LD | RV54 | IO84OA | Lancaster |
| GB3LE | RB04 | IO92IQ | Markfield |
| GB3LF | RB14 | IO84PH | Kendal |
| GB3LH | RB15 | IO82OP | Shrewsbury |
| GB3LI | RB10 | IO83LL | Liverpool |
| GB3LL | RB0 | IO83BH | Llandudno |
| GB3LM | RV58 | IO93RF | Lincoln |
| GB3LP | R50-04 | IO83MK | Liverpool |
| GB3LR | RU69 | JO00AS | Newhaven |
| GB3LS | RB02 | IO93RF | Lincoln |
| GB3LT | RB10 | IO91SV | Luton |
| GB3LU | RV54 | IP90JD | Shetland |
| GB3LV | RB02 | IO91XP | North London |
| GB3LW | RU72 | IO91WM | London |
| GB3LY | RV48 | IO65NC | Limavady N.I |
| GB3MA | RB01 | IO83UO | Bury |
| GB3MD | RU68 | IO93KD | Mansfield |
| GB3ME | RB06 | IO92JJ | Rugby |
| GB3MF | RB02 | IO83WG | Macclesfield |
| GB3MG | RU72 | IO81IO | Pontypridd |
| GB3MH | RV50 | IO91WC | East Grinstead |
| GB3MI | RV57 | IO83VM | Manchester |
| GB3MK | RB0 | IO92OB | Milton Keynes |
| GB3MM | RM6 | IO82XP | Wolverhampton |
| GB3MN | RV52 | IO83XH | Disley |
| GB3MP | RV60 | IO83IF | Denbigh |
| GB3MR | RB14 | IO83XH | Stockport |
| GB3MS | RB07 | IO82VE | Worcester |
| GB3MT | RU72 | IO64QS | Magherafelt |
| GB3MW | RB10 | IO92FH | Leamington Spa |
| GB3NA | RV49 | IO93GK | Sheffield |
| GB3NB | RV50 | JO02PN | Norwich |
| GB3NC | RV58 | IO70OJ | St Austell |
| GB3ND | RB14 | IO71VA | Bideford |
| GB3NE | RV61 | IO91HJ | Newbury |
| GB3NF | RV50 | IO92KX | Nottingham |
| GB3NG | RV50 | IO87XO | Fraserburgh |
| GB3NH | RU70 | IO92NF | Northampton |
| GB3NI | RV58 | IO74CO | Belfast N.I. |
| GB3NK | RB04 | JO01BL | Erith |
| GB3NL | RV62 | IO91XP | North London |
| GB3NM | RB07 | IO92KX | Nottingham |
| GB3NN | RB02 | JO02JV | Wells Norfolk |
| GB3NO | RM0 | JO02PP | Norwich |
| GB3NP | RU71 | IO92LD | Towcester |
| GB3NR | RB0 | JO02PP | Norwich |
| GB3NS | DVU54 | IO91WG | Caterham |
| GB3NT | RB0 | IO94FW | Newcastle/Tyne |
| GB3NU | RU71 | JO02OW | Sheringham |
| GB3NW | RV50 | IO82VE | Worcester |
| GB3NX | RU68 | IO91WB | Crawley Sussex |
| GB3NY | RU66 | IO94VC | Bridlington |
| GB3OA | RV49 | IO83LP | Southport |
| GB3OC | RV52 | IO88LX | Kirkwall |
| GB3OH | RU76 | IO85EX | Linlithgow |
| GB3OM | RU76 | IO64JQ | Omagh |
| GB3OV | RB05 | IO92VG | St Neots |
| GB3OY | RU76 | JO01AO | Buckhurst Hill |
| GB3PA | RV50 | IO75QV | Paisley |
| GB3PC | RV62 | IO90KT | Portsmouth |
| GB3PE | RV54 | IO92TN | Peterborough |
| GB3PI | RV60 | IO92XA | Royston |
| GB3PK | RV53 | IO65WE | Ballycastle Ni |
| GB3PL | RV61 | IO70VM | E.Cornwall |
| GB3PO | RV52 | JO02OB | Ipswich |
| GB3PP | RB15 | IO83PS | Preston |

# Repeaters (alphabetical)

| Callsign | Channel | Locator | Location |
|---|---|---|---|
| GB3PH | RV54 | IO86GI | Perth |
| GB3PI | RM3 | IO92XA | Royston |
| GB3PU | RB0 | IO86GI | Perth |
| GB3PW | RV62 | IO82HL | Newtown Powys |
| GB3PX | R50-07 | IO92XA | Royston |
| GB3PY | RB08 | JO02AF | Cambridge |
| GB3PZ | RU72 | IO83XI | Dukinfield |
| GB3RA | RV51 | IO82JH | Llandrindod Wells |
| GB3RD | RB08 | IO93IF | Bolsover |
| GB3RD | RV54 | IO91JM | Reading |
| GB3RE | RB11 | JO01HH | Maidstone |
| GB3RF | RV62 | IO83TR | Accrington |
| GB3RH | RB11 | IO80MS | Axminster |
| GB3RT | RB0 | IO81LP | Cwmbran |
| GB3RU | RB11 | IO91LK | Reading |
| GB3SB | RV52 | IO85ON | Selkirk |
| GB3SC | RV50 | IO90BR | Bournemouth |
| GB3SD | RB14 | IO80SQ | Weymouth |
| GB3SE | RM3 | IO83WA | Stoke On Trent |
| GB3SF | RU75 | IO81DA | South Molton |
| GB3SI | RV50 | IO70JB | Helston |
| GB3SK | RB06 | JO01MH | Canterbury |
| GB3SL | R50-02 | IO75XX | Kilsyth |
| GB3SM | RB13 | IO93BA | Stoke On Trent |
| GB3SN | RV58 | IO91LC | Alton Hants |
| GB3SO | RB0 | JO03DH | Sutton On Sea |
| GB3SP | RB04 | IO71OQ | Pembroke |
| GB3SS | RV48 | IO87MO | Elgin Moray |
| GB3ST | RB09 | IO83WA | Stoke On Trent |
| GB3SW | RV57 | IO80JR | Sidmouth |
| GB3SX | R50-08 | IO93BA | Stoke On Trent |
| GB3SY | RB06 | IO93HN | Barnsley |
| GB3SZ | RB15 | IO90AS | Bournemouth |
| GB3TA | RV59 | IO92DP | Tamworth |
| GB3TD | RB03 | IO91DL | Swindon |
| GB3TE | RV49 | JO01MT | Clacton On Sea |
| GB3TF | RB08 | IO82RP | Telford |
| GB3TH | RB15 | IO92DP | Tamworth |
| GB3TJ | RB12 | IO85XA | Corbridge |
| GB3TJ | RB12 | IO85XA | Corbridge |
| GB3TO | RV49 | IO92NE | Northampton |
| GB3TP | RV58 | IO93BV | Keighley |
| GB3TR | RV52 | IO80FM | Torquay |
| GB3TS | RB07 | IO94GX | Sunderland |
| GB3TU | RB09 | IO91PS | Tring Herts |
| GB3TW | RV58 | IO94EW | Gateshead |
| GB3TW | RV58 | IO94EW | Gateshead |
| GB3TX | 10M | JO02PP | Norwich |
| GB3TY | R50-07 | IO74BS | Carrickfergus |
| GB3UB | RB04 | IO81UJ | Bath Avon |
| GB3UI | RB01 | IO92CK | Birmimgham |
| GB3UK | RU69 | IO81XW | Cheltenham |
| GB3UL | RB02 | IO74CO | Belfast N.I. |
| GB3UM | R50-03 | IO92IQ | Markfield Leic |
| GB3UO | RU66 | IO82LW | Chirk |
| GB3US | RB0 | IO93GJ | Sheffield |
| GB3VA | RV56 | IO91LT | Brill |
| GB3VE | RB04 | IO84SQ | Dufton |
| GB3VH | RB13 | IO91VT | Welwyn G City |
| GB3VI | R50-15 | IO92BJ | Birmingham |
| GB3VM | RV49 | IO82PH | Ludlow |
| GB3VN | RU74 | IO82QI | Ludlow |
| GB3VS | RB03 | IO80LX | Taunton |
| GB3VT | RV58 | IO83WA | Stoke On Trent |
| GB3VW | RV48 | JO02QP | Norwich |
| GB3WA | RU70 | IO75OR | Dalry |
| GB3WB | RB12 | IO81MH | Weston-Sup-Mare |
| GB3WC | RM15 | IO93EP | Wakefield |
| GB3WD | RV56 | IO70WJ | Plymouth |
| GB3WE | RV55 | IO81MH | Weston-Sup-Mare |
| GB3WF | RB14 | IO93DV | Otley |
| GB3WG | RB06 | IO81CP | Port Talbot |
| GB3WH | RV52 | IO91EM | Swindon |
| GB3WI | RB15 | JO02CP | Wisbech |
| GB3WJ | RB05 | IO93QN | Scunthorpe |
| GB3WK | RV62 | IO92FH | Leamington Spa |
| GB3WL | RU76 | IO92BJ | Birmingham |
| GB3WN | RB0 | IO82XP | Wolverhampton |
| GB3WP | RU75 | IO83XL | Hyde |
| GB3WR | RV48 | IO81PH | Wells |
| GB3WS | RV60 | IO91WB | Crawley Sussex |
| GB3WT | RV62 | IO64JQ | Omagh N.I |
| GB3WU | RU66 | IO82VE | Worcester |
| GB3WW | RV62 | IO81CP | Port Talbot |
| GB3WX-10 | | 10M | Mere Wiltshire |
| GB3WX-6 | 6M | IO81VC | Mere Wiltshire |
| GB3WY | R50-09 | IO93EP | Wakefield |
| GB3XD | R50-02 | IO93WH | Louth |
| GB3XL | RU71 | IO93CT | Shipley |
| GB3XN | [illegible] | [illegible] | Worksop |
| GB3XP | RV55 | IO91VJ | Morden |
| GB3XX | RB13 | IO92KG | Daventry |
| GB3YC | RV60 | IO94SC | Driffield |
| GB3YL | RB14 | JO02VI | Lowestoft |
| GB3YS | RB03 | IO80QW | Yeovil |
| GB3YW | RV63 | IO93EP | Wakefield |
| GB3ZA | RV59 | IO82PB | Hereford |
| GB3ZB | RU66 | IO81QJ | Bristol |
| GB3ZI | RU78 | IO90WT | Stafford |
| GB3ZW | R50-10 | IO82HL | Newtown Powys |
| GB3ZX | RU66 | JO00CT | Eastbourne |
| GB3ZY | R50-09 | IO81QJ | Bristol |
| GB4DPY | RD00 | JO02AF | Cambridge |
| GB7AA | DVU54 | IO81RO | Bristol |
| GB7AD | DVU73 | IO81RO | Bristol |
| GB7AK | DVU42 | JO00AX | Barking |
| GB7AU | DVU56 | IO91QP | Amersham |
| GB7AV | DVU35 | IO91OT | Aylesbury |
| GB7BD | DVU36 | IO81PJ | Bristol |
| GB7BE | DVU53 | JO02TK | Beccles |
| GB7BH | DVU46 | IO90WU | Brighton |
| GB7BK | DVU59 | IO91JM | Reading |
| GB7BM | DVU32 | IO92BL | Birmingham |
| GB7BP | DVU36 | IO91PX | Milton Keynes |
| GB7BR | RU74 | IO74TI | Ramsey |
| GB7BS | DVU13 | IO81TK | Bristol City |
| GB7BX | DVU35 | IO82QN | Much Wenlock |
| GB7CA | RU74 | IO74SD | Douglas |
| GB7CD | DVU56 | IO81JL | Cardiff |
| GB7CF | DVU41 | JO01OC | Folkestone |
| GB7CH | DVU38 | IO93WH | Louth |
| GB7CK | DVU59 | JO01OC | Folkestone |
| GB7CL | DVU51 | JO01MT | Clacton On Sea |
| GB7CM | RV56 | IO80XS | Wimbourne |
| GB7CT | RV51 | IO91PS | Tring |
| GB7CW | DVU32 | IO81EN | Bridgend |
| GB7DA-B | DVU50 | IO85AV | Airdrie |
| GB7DA-C | RV62 | IO85AV | Airdrie |
| GB7DB | DVU53 | IO92RA | Ampthill |
| GB7DC | DVU39 | IO92GW | Derby |
| GB7DD | DVU53 | IO86NL | Dundee |
| GB7DE-B | DVU48 | IO86NF | Upper Largo |
| GB7DE-C | RV51 | IO86NF | Upper Largo |
| GB7DG | DVU43 | IO82WI | Bromsgrove |
| GB7DK-B | DVU73 | IO74LV | Stranraer |
| GB7DK-C | RV55 | IO74LV | Stranraer |
| GB7DL | DVU54 | JO33BB | Friskney |
| GB7DN | DVU33 | IO64MW | Dungiven |
| GB7DP-B | DVU40 | IO86LL | Dundee |
| GB7DR | DVU34 | IO90AR | Poole |
| GB7DS | DVU34 | JO02PP | Norwich |
| GB7DV | DVU49 | IO83QJ | St. Helens |
| GB7DX | DVU56 | JO01IA | Wittersham |
| GB7DZ | DVU42 | IO94FX | Newcastle-Upon-Tyne |
| GB7EB | DVU46 | JO02TK | Beccles |
| GB7ED | DVU54 | IO80GR | Exeter |
| GB7EE | DVU57 | IO85JW | Edinburgh |
| GB7EK | DVU50 | JO01NI | Whitstable |
| GB7EL | DVU57 | IO83VT | Nelson |
| GB7EN | DVU43 | IO92PH | Wellingborough |
| GB7EP | DVU32 | IO91UI | Epsom |
| GB7ER | RV62 | IO80GR | Exeter |
| GB7ES | DVU35 | JO00DT | Eastbourne |
| GB7EX | DVU57 | JO01GN | Southend-On-Sea |
| GB7FC | RU78 | IO83LT | Blackpool |
| GB7FH | DVU49 | IO90JU | Fareham |
| GB7FI | RU71 | IO81OH | Axbridge |
| GB7FK-C | RV63 | JO01OC | Folkestone |
| GB7FU | DVU13 | IO92TU | Pointon |
| GB7FW | DVU53 | IO92BL | Birmingham |
| GB7GB | DVU34 | IO91WM | Birmingham |
| GB7GC | DVU35 | IO92WN | Grimsby |
| GB7GD | RV55 | IO87VD | Aberdeen |
| GB7GF | DVU55 | IO91RF | Guildford |
| GB7GL | DVU57 | IO81SU | Mitcheldean |
| GB7GT | DVU33 | IO82QJ | Ludlow |
| GB7HD-B | DVU36 | IO93DP | Huddersfield |
| GB7HE | DVU43 | JO00HV | Hastings |
| GB7HM | DVU51 | IO83LC | Caergwrie |
| GB7HR | DVU36 | IO91SL | Heathrow |
| GB7HS | DVU34 | IO93BR | Batley |
| GB7HU | DVU39 | IO93RS | South Cave |
| GB7HX | DVU46 | IO93BP | Huddersfield |
| GB7IC-A | 23CM | JO01NI | Herne Bay |
| GB7IC-B | DVU36 | JO01NI | Herne Bay |
| GB7IC-C | RV53 | JO01NI | Herne Bay |
| GB7ID | RU78 | IO91WF | Redhill |
| GB7IK | DVU43 | JO01FJ | Rochester |
| GB7IQ | DVU49 | IO82WS | Stafford |
| GB7IT | DVU41 | IO81MH | Weston-Sup-Mare |
| GB7JB | DVU37 | IO81VC | Mere Wiltshire |
| GB7JF | DVU51 | IO92NE | Northampton |
| GB7JH-B | DVU57 | IO90TT | Shoreham-By-Sea |
| GB7JH-C | RV52 | IO90TT | Shoreham-By-Sea |
| GB7JL | RU77 | IO83QL | Wigan |
| GB7KH | DVU49 | JO01DQ | Kelvedon Hatch |
| GB7KM | DVU53 | IO81XQ | Cirencester |
| GB7KT | DVU40 | IO91GF | Andover |
| GB7LE | DVU53 | IO93FU | Leeds |
| GB7LF | DVU43 | IO84OA | Lancaster |
| GB7LN | DVU02 | IO93RF | Lincoln |
| GB7LO | DVU41 | JO01BJ | Bromley |
| GB7LP | DVU32 | IO83MK | Liverpool |
| GB7LY | DVU53 | IO64IX | Londonderry |
| GB7MA | RV48 | IO83UO | Bury |
| GB7MB | DVU56 | IO84NA | Heysham |
| GB7MC | DVU43 | IO70OJ | St Austell |
| GB7MH | DVU51 | IO91WC | East Grinstead |
| GB7MK | DVU48 | JO02OB | Ipswich |
| GB7MR | DVU59 | IO83XN | Manchester |
| GB7NB | DVU55 | JO02PN | Norwich |
| GB7NE | DVU36 | IO95FE | Ashington |
| GB7NI | RV60 | IO74BS | Carrickfergus |
| GB7NL | DVU37 | IO80FR | Exeter |
| GB7NM | DVU37 | IO83UN | Manchester |
| GB7NS | DVU13 | IO91WG | Caterham |
| GB7NU | DVU43 | JO02OW | Sheringham |
| GB7OK | RV57 | JO01BJ | Bromley |
| GB7PB | RV53 | IO94FT | Durham |
| GB7PD-B | RB05 | IO71MQ | Pembroke Dock |
| GB7PD-C | RV59 | IO71MQ | Pembroke Dock |
| GB7PE | DVU40 | IO92XO | Peterborough |
| GB7PI | DVU73 | IO92XA | Royston |
| GB7PN | DVU34 | IO83HH | Prestatyn |
| GB7PP | DVU35 | JO02OB | Ipswich |
| GB7PT | DVU57 | IO92XA | Royston |
| GB7RB-C | RV54 | IO81GK | Cowbridge |
| GB7RE | DVU57 | IO93MH | Retford |
| GB7RN | RV51 | IO90JT | Fareham |
| GB7RR | DVU48 | IO93KA | Nottingham |
| GB7RW | RV48 | IO94SO | Whitby |
| GB7RY | RU76 | JO00HW | Rye |
| GB7SB | RV48 | IO90WT | Brighton |
| GB7SC | DVU38 | IO90QT | Bognor Regis |
| GB7SD | DVU33 | IO80SQ | Weymouth |
| GB7SE | DVU38 | JO01DM | Thurrock |
| GB7SF | RV59 | IO93GK | Sheffield |
| GB7SH | RU77 | IO93GK | Sheffield |
| GB7SI | DVU42 | IO83WA | Stoke-On-Trent |
| GB7SJ | RU70 | IO83RF | Northwich |
| GB7SN | DVU54 | IO93GH | Sheffield |
| GB7SO | DVU46 | IO92BJ | Solihull |
| GB7SR | DVU51 | IO93GK | Sheffield |
| GB7SS | DVU59 | JO01HO | Hockley |
| GB7SU | DVU36 | IO90HV | Southampton |
| GB7SX | DVU62 | IO90QT | Bognor Regis |
| GB7TC | DVU42 | IO91DL | Swindon |
| GB7TD | DVU13 | IO93EP | Wakefield |
| GB7TE | RV62 | JO01OT | Clacton On Sea |
| GB7TP | DVU55 | IO93BV | Keighley |
| GB7TQ | RU71 | IO80FM | Torquay |
| GB7TV | DVU38 | IO94LN | New Marske |
| GB7UL | DVU42 | IO74BS | Carrickfergus |
| GB7VO | DVU56 | IO82PE | Leominster |
| GB7WB | DVU39 | IO81MH | Weston-Sup-Mare |
| GB7WC | DVU39 | IO83QI | Warrington |
| GB7WF | DVU39 | IO82UJ | Bewdley |
| GB7WI | DVU33 | IO93QP | Winterton |
| GB7WL | DVU37 | IO91QP | Amersham |
| GB7WP | DVU73 | IO83KJ | Birkenhead |
| GB7WT | DVU48 | IO64HQ | Omagh N.I |
| GB7XX | DVU37 | IO94FW | Felling |
| GB7YD-A | RM12 | IO93HO | Barnsley |
| GB7YD-B | DVU41 | IO93HO | Barnsley |
| GB7YD-C | RV54 | IO93HO | Barnsley |
| GB7ZI | DVU55 | IO82XR | Stafford |
| GB7ZP | DVU39 | JO01HR | Danbury |

# Repeaters (by output frequency)

| Channel | F (out) | F (in) | Callsign | CTCSS | Location | Keeper |
|---|---|---|---|---|---|---|
| 10M | 29.2100 | 50.5200 | GB3WX-10 | 77Hz | Mere Wiltshire | G3ZXX |
| 10M | 29.6400 | 29.5400 | GB3CJ | 77Hz | Northampton | G1IRG |
| 10M | 29.6500 | 29.5500 | GB3TX | 94.8Hz | Norwich | M0ZAH |
| 10M | 29.6900 | 29.5900 | GB3GX | 103.5Hz | Ludlow | G1MAW |
| 6M | 50.5200 | 29.2100 | GB3WX-6 | 77Hz | Mere Wiltshire | G3ZXX |
| R50-01 | 50.7200 | 51.2200 | GB3AE | 94.8Hz | Tenby | GW4AKZ |
| R50-01 | 50.7200 | 51.2200 | GB3EF | 110.9Hz | Stowmarket | G1YFF |
| R50-02 | 50.7300 | 51.2300 | GB3GC | 77Hz | Gunnislake | M0YDW |
| R50-02 | 50.7300 | 51.2300 | GB3SL | 103.5Hz | Kilsyth | GM4COX |
| R50-02 | 50.7300 | 51.2300 | GB3XD | 71.9Hz | Louth | G7AJP |
| R50-03 | 50.7400 | 51.2400 | GB3UM | 77Hz | Markfield Leicester | M1NAS |
| R50-04 | 50.7500 | 51.2500 | GB3LP | 77Hz | Liverpool | M1SWB |
| R50-05 | 50.7600 | 51.2600 | GB3HF | 71.9Hz | Barnsley | G4LUE |
| R50-06 | 50.7700 | 51.2700 | GB3DB | 110.9Hz | Danbury | G6JYB |
| R50-06 | 50.7700 | 51.2700 | GB3FH | 77Hz | Somerset | G4RKY |
| R50-07 | 50.7800 | 51.2800 | GB3PX | 77Hz | Royston | G4NBS |
| R50-07 | 50.7800 | 51.2800 | GB3TY | 110.9Hz | Carrickfergus | GI6DKQ |
| R50-08 | 50.7900 | 51.2900 | GB3SX | 103.5Hz | Stoke On Trent | G4SCY |
| R50-09 | 50.8000 | 51.3000 | GB3WY | 82.5Hz | Wakefield | G1XCC |
| R50-09 | 50.8000 | 51.3000 | GB3ZY | 77Hz | Bristol | G4RKY |
| R50-10 | 50.8100 | 51.3100 | GB3FX | 82.5Hz | Farnham | G4EPX |
| R50-10 | 50.8100 | 51.3100 | GB3ZW | 103.5Hz | Newtown Powys | GW4NQJ |
| R50-11 | 50.8200 | 51.3200 | GB3HM | 71.9Hz | Belper | G8IQP |
| R50-12 | 50.8300 | 51.3300 | GB3GT | 103.5Hz | Ludlow | G1MAW |
| R50-12 | 50.8300 | 51.3300 | GB3JX | 94.8Hz | Norwich | M0ZAH |
| R50-13 | 50.8400 | 51.3400 | GB3AM | 77Hz | Amersham | G0RDI |
| R50-14 | 50.8500 | 51.3500 | GB3CT | 77Hz | Northampton | G7SYT |
| R50-15 | 50.8600 | 51.3600 | GB3VI | 67Hz | Birmingham | G8NDT |
| RV48 | 145.6000 | 145.0000 | GB3AS | 77Hz | Carlisle | G1XSZ |
| RV48 | 145.6000 | 145.0000 | GB3CF | 77Hz | Markfield | G4AFJ |
| RV48 | 145.6000 | 145.0000 | GB3EL | 82.5Hz | London | G4RZZ |
| RV48 | 145.6000 | 145.0000 | GB3FF | 103.5Hz | Kelty | GM7LUN |
| RV48 | 145.6000 | 145.0000 | GB3LY | 110.9Hz | Limavady N.I | GI3USS |
| RV48 | 145.6000 | 145.0000 | GB3SS | 67Hz | Elgin Moray | GM4ILS |
| RV48 | 145.6000 | 145.0000 | GB3VW | 94.8Hz | Norwich | M0ZAH |
| RV48 | 145.6000 | 145.0000 | GB3WR | 94.8Hz | Wells | G4RKY |
| RV48 | 145.6000 | 145.0000 | GB7MA | | Bury | G7LWT |
| RV48 | 145.6000 | 145.0000 | GB7RW | | Whitby | G4EQS |
| RV48 | 145.6000 | 145.0000 | GB7SB | | Brighton | G4PAP |
| RV49 | 145.6125 | 145.0125 | GB3AP | 94.8Hz | Haverfordwest | GW4AKZ |
| RV49 | 145.6125 | 145.0125 | GB3EW | 77Hz | Exeter | G8XQQ |
| RV49 | 145.6125 | 145.0125 | GB3NA | 71.9Hz | Sheffield | G4LUE |
| RV49 | 145.6125 | 145.0125 | GB3OA | 82.5Hz | Southport | G4EID |
| RV49 | 145.6125 | 145.0125 | GB3TE | 103.5Hz | Clacton On Sea | G0MBA |
| RV49 | 145.6125 | 145.0125 | GB3TO | 77Hz | Northampton | G6NYH |
| RV49 | 145.6125 | 145.0125 | GB3VM | 103.5Hz | Ludlow | G4AIJ |
| RV50 | 145.6250 | 145.0250 | GB3GD | 110.9Hz | Snaefell Iom | GD4HOZ |
| RV50 | 145.6250 | 145.0250 | GB3HG | 88.5Hz | Thirsk | G8PXB |
| RV50 | 145.6250 | 145.0250 | GB3KS | 103.5Hz | Dover | M1CMN |
| RV50 | 145.6250 | 145.0250 | GB3MH | 88.5Hz | East Grinstead | G3NZP |
| RV50 | 145.6250 | 145.0250 | GB3NB | 94.8Hz | Norwich | G8VLL |
| RV50 | 145.6250 | 145.0250 | GB3NF | 77Hz | Nottingham | G4NRZ |
| RV50 | 145.6250 | 145.0250 | GB3NG | 67Hz | Fraserburgh | GM4ZUK |
| RV50 | 145.6250 | 145.0250 | GB3NW | 67Hz | Worcester | G4IDF |
| RV50 | 145.6250 | 145.0250 | GB3PA | 103.5Hz | Paisley | GM7OAW |
| RV50 | 145.6250 | 145.0250 | GB3SC | 71.9Hz | Bournemouth | M0AUY |
| RV50 | 145.6250 | 145.0250 | GB3SI | 77Hz | Helston | M1ERD |
| RV51 | 145.6375 | 145.0375 | GB3DN | 77Hz | Stibb Cross | G1BHM |
| RV51 | 145.6375 | 145.0375 | GB3GJ | | St Helier | GJ8PVL |
| RV51 | 145.6375 | 145.0375 | GB3IN | 71.9Hz | Matlock | G4TSN |
| RV51 | 145.6375 | 145.0375 | GB3RA | 103.5Hz | Llandrindod Wells | GW7UNV |
| RV51 | 145.6375 | 145.0375 | GB7CT | 3Hz | Tring | G0RDI |
| RV51 | 145.6375 | 145.0375 | GB7DE-C | | Upper Largo | MM0DXE |
| RV51 | 145.6375 | 145.0375 | GB7RN | | Fareham | G3ZDF |
| RV52 | 145.6500 | 145.0500 | GB3AY | 103.5Hz | Dalry | MM0YET |
| RV52 | 145.6500 | 145.0500 | GB3HS | 88.5Hz | Walkington | G0VRM |
| RV52 | 145.6500 | 145.0500 | GB3MN | 82.5Hz | Disley | G8LZO |
| RV52 | 145.6500 | 145.0500 | GB3OC | 77Hz | Kirkwall | GM0HQG |
| RV52 | 145.6500 | 145.0500 | GB3PO | 110.9Hz | Ipswich | G7CIY |
| RV52 | 145.6500 | 145.0500 | GB3SB | 118.8Hz | Selkirk | GM0FTJ |
| RV52 | 145.6500 | 145.0500 | GB3TR | 94.8Hz | Torquay | G8XST |
| RV52 | 145.6500 | 145.0500 | GB3WH | 118.8Hz | Swindon | G4LDL |
| RV52 | 145.6500 | 145.0500 | GB7JH-C | | Shoreham-By-Sea | G8MUQ |
| RV53 | 145.6625 | 145.0625 | GB3AA | 94.8Hz | Bristol | G4CJZ |
| RV53 | 145.6625 | 145.0625 | GB3CO | 77Hz | Corby | G1DIW |
| RV53 | 145.6625 | 145.0625 | GB3DW | Hz | Criccieth | GW4KAZ |
| RV53 | 145.6625 | 145.0625 | GB3FE | 103.5Hz | Stirling | GM0MZB |
| RV53 | 145.6625 | 145.0625 | GB3KI | 103.5Hz | Herne Bay | G4TKR |
| RV53 | 145.6625 | 145.0625 | GB3PK | 110.9Hz | Ballycastle Ni | MI0CRR |
| RV53 | 145.6625 | 145.0625 | GB7IC-C | | Herne Bay | G4TKR |
| RV53 | 145.6625 | 145.0625 | GB7PB | | Durham | G4EBN |
| RV54 | 145.6750 | 145.0750 | GB3BX | 146.2Hz | Much Wenlock | M1GIZ |
| RV54 | 145.6750 | 145.0750 | GB3ES | 103.5Hz | Hastings | G6ZZX |
| RV54 | 145.6750 | 145.0750 | GB3LD | 110.9Hz | Lancaster | G3VVT |
| RV54 | 145.6750 | 145.0750 | GB3LU | 77Hz | Shetland | GM4SLV |
| RV54 | 145.6750 | 145.0750 | GB3PE | 94.8Hz | Peterborough | M0ZPU |
| RV54 | 145.6750 | 145.0750 | GB3PR | 94.8Hz | Perth | GM8KPH |
| RV54 | 145.6750 | 145.0750 | GB3RD | 118.8Hz | Reading | G8DOR |
| RV54 | 145.6750 | 145.0750 | GB7RB-C | | Cowbridge | GW6CUR |
| RV54 | 145.6750 | 145.0750 | GB7YD-C | | Barnsley | G4LUE |
| RV55 | 145.6875 | 145.0875 | GB3CD | 118.8Hz | Crook | G0OCB |
| RV55 | 145.6875 | 145.0875 | GB3DC | 71.9Hz | Derby | G7NPW |
| RV55 | 145.6875 | 145.0875 | GB3EA | 110.9Hz | Wickhambrook | G1YFF |
| RV55 | 145.6875 | 145.0875 | GB3KE | 103.5Hz | Glasgow | GM7SVK |
| RV55 | 145.6875 | 145.0875 | GB3WE | 94.8Hz | Weston-Super-Mare | G4SZM |
| RV55 | 145.6875 | 145.0875 | GB3XP | 82.5Hz | Morden | M0ZEY |
| RV55 | 145.6875 | 145.0875 | GB7DK-C | | Stranraer | GM0HPK |
| RV55 | 145.6875 | 145.0875 | GB7GD | | Aberdeen | GM0MYQ |
| RV56 | 145.7000 | 145.1000 | GB3AR | 110.9Hz | Caernarfon | GW4KAZ |
| RV56 | 145.7000 | 145.1000 | GB3BB | 94.8Hz | Brecon | MW0UAA |
| RV56 | 145.7000 | 145.1000 | GB3BT | 118.8Hz | Berwick On Tweed | GM1JFF |
| RV56 | 145.7000 | 145.1000 | GB3EV | 77Hz | Dufton | G7ITT |
| RV56 | 145.7000 | 145.1000 | GB3HH | 71.9Hz | Buxton | G7EKY |
| RV56 | 145.7000 | 145.1000 | GB3HI | 103.5Hz | Isle Of Mull | MM0JRM |
| RV56 | 145.7000 | 145.1000 | GB3KN | 103.5Hz | Maidstone | G3VFC |
| RV56 | 145.7000 | 145.1000 | GB3VA | 118.8Hz | Brill | G8BQH |
| RV56 | 145.7000 | 145.1000 | GB3WD | 77Hz | Plymouth | G7LUL |
| RV56 | 145.7000 | 145.1000 | GB7CM | 71.9Hz | Wimbourne | M0MRP |
| RV57 | 145.7125 | 145.1125 | GB3BM | 67Hz | Birmingham | G8AMD |
| RV57 | 145.7125 | 145.1125 | GB3FG | 94.8Hz | Carmarthen | GW8KCY |
| RV57 | 145.7125 | 145.1125 | GB3KY | 94.8Hz | Kings Lynn | G1SCQ |
| RV57 | 145.7125 | 145.1125 | GB3LA | 103.5Hz | Sanquhar | GM3SAN |
| RV57 | 145.7125 | 145.1125 | GB3MI | 82.5Hz | Manchester | G0TOG |
| RV57 | 145.7125 | 145.1125 | GB3SW | 77Hz | Sidmouth | G6XUV |
| RV57 | 145.7125 | 145.1125 | GB7OK | | Bromley | G1HIG |
| RV58 | 145.7250 | 145.1250 | GB3AG | 94.8Hz | Forfar | GM1CMF |
| RV58 | 145.7250 | 145.1250 | GB3BI | 67Hz | Inverness | GM1VAD |
| RV58 | 145.7250 | 145.1250 | GB3CG | 118.8Hz | Gloucester | G3LVP |
| RV58 | 145.7250 | 145.1250 | GB3DA | 110.9Hz | Danbury | G6JYB |
| RV58 | 145.7250 | 145.1250 | GB3LM | 71.9Hz | Lincoln | G7AVU |
| RV58 | 145.7250 | 145.1250 | GB3NC | 77Hz | St Austell | G3IGV |
| RV58 | 145.7250 | 145.1250 | GB3NI | 110.9Hz | Belfast N.I. | GI3USS |
| RV58 | 145.7250 | 145.1250 | GB3SN | 71.9Hz | Alton Hants | G4EPX |
| RV58 | 145.7250 | 145.1250 | GB3TP | 82.5Hz | Keighley | G8ZMG |
| RV58 | 145.7250 | 145.1250 | GB3TW | 118.8Hz | Gateshead | G7UUR |
| RV58 | 145.7250 | 145.1250 | GB3TW | 118.8Hz | Gateshead | G7UUR |
| RV58 | 145.7250 | 145.1250 | GB3VT | 103.5Hz | Stoke On Trent | G8NSS |
| RV59 | 145.7375 | 145.1375 | GB3AL | 77Hz | Amersham | G0RDI |
| RV59 | 145.7375 | 145.1375 | GB3CP | 110.9Hz | Fermanagh | GI8RLE |
| RV59 | 145.7375 | 145.1375 | GB3DR | 71.9Hz | Dorchester | G8NIC |
| RV59 | 145.7375 | 145.1375 | GB3TA | 67Hz | Tamworth | M0TSD |
| RV59 | 145.7375 | 145.1375 | GB3ZA | 118.8Hz | Hereford | G0JWJ |
| RV59 | 145.7375 | 145.1375 | GB7PD-C | | Pembroke Dock | MW0HVB |
| RV59 | 145.7375 | 145.1375 | GB7SF | | Sheffield | M1ERS |
| RV60 | 145.7500 | 145.1500 | GB3BC | 94.8Hz | Pontypridd | GW6CUR |
| RV60 | 145.7500 | 145.1500 | GB3CS | 103.5Hz | Motherwell | GM8HBY |
| RV60 | 145.7500 | 145.1500 | GB3FK | 103.5Hz | Folkestone | M1CMN |
| RV60 | 145.7500 | 145.1500 | GB3HA | | Corbridge | G7UUR |
| RV60 | 145.7500 | 145.1500 | GB3IS | 88.5Hz | Broadford | GM8RBR |
| RV60 | 145.7500 | 145.1500 | GB3MP | 110.9Hz | Denbigh | M0OBW |
| RV60 | 145.7500 | 145.1500 | GB3PI | 77Hz | Royston | G4NBS |
| RV60 | 145.7500 | 145.1500 | GB3WS | 88.5Hz | Crawley Sussex | G4EFO |
| RV60 | 145.7500 | 145.1500 | GB3YC | 88.5Hz | Driffield | M0KXQ |
| RV60 | 145.7500 | 145.1500 | GB7NI | | Carrickfergus | GI6DKQ |
| RV61 | 145.7625 | 145.1625 | GB3DU | 118.8Hz | Duns | GM7LUN |
| RV61 | 145.7625 | 145.1625 | GB3IK | 103.5Hz | Rochester | G6CKK |
| RV61 | 145.7625 | 145.1625 | GB3IP | 103.5Hz | Stafford | G7PFT |
| RV61 | 145.7625 | 145.1625 | GB3IR | 88.5Hz | Richmond Yorks | G4FZN |
| RV61 | 145.7625 | 145.1625 | GB3NE | 118.8Hz | Newbury | G8JIP |
| RV61 | 145.7625 | 145.1625 | GB3PL | 77Hz | E.Cornwall | M0YDW |
| RV62 | 145.7750 | 145.1750 | GB3DG | 103.5Hz | Newton Stewart | MM1BHO |
| RV62 | 145.7750 | 145.1750 | GB3FR | 71.9Hz | Spilsby Lincs. | G8SFU |
| RV62 | 145.7750 | 145.1750 | GB3GN | 67Hz | Banchory | GM4ZUK |
| RV62 | 145.7750 | 145.1750 | GB3IG | 88.5Hz | Stornoway | GM0LZE |
| RV62 | 145.7750 | 145.1750 | GB3NL | 82.5Hz | North London | G4DFB |
| RV62 | 145.7750 | 145.1750 | GB3PC | 71.9Hz | Portsmouth | G8UCY |
| RV62 | 145.7750 | 145.1750 | GB3PW | 103.5Hz | Newtown Powys | GW4NQJ |
| RV62 | 145.7750 | 145.1750 | GB3RF | 82.5Hz | Accrington | G0BMH |
| RV62 | 145.7750 | 145.1750 | GB3WK | 67Hz | Leamington Spa | G6FEO |
| RV62 | 145.7750 | 145.1750 | GB3WT | 110.9Hz | Omagh N.I | GI3NVW |
| RV62 | 145.7750 | 145.1750 | GB3WW | 94.8Hz | Port Talbot | GW4FOI |
| RV62 | 145.7750 | 145.1750 | GB7DA-C | Hz | Airdrie | GM4AUP |
| RV62 | 145.7750 | 145.1750 | GB7ER | 77Hz | Exeter | M0ZZT |
| RV62 | 145.7750 | 145.1750 | GB7TE | | Clacton On Sea | G7HJK |
| RV63 | 145.7875 | 145.1875 | GB3BF | 77Hz | Bedford | G8MGP |
| RV63 | 145.7875 | 145.1875 | GB3EB | 88.5Hz | Uckfield | G8PUO |
| RV63 | 145.7875 | 145.1875 | GB3GO | 110.9Hz | Llandudno | GW6STK |
| RV63 | 145.7875 | 145.1875 | GB3JB | 103.5Hz | Mere Wiltshire | G3ZXX |
| RV63 | 145.7875 | 145.1875 | GB3JL | 77Hz | Liskeard | G4RKY |
| RV63 | 145.7875 | 145.1875 | GB3KD | 118.8Hz | Kidderminster | G8PZT |
| RV63 | 145.7875 | 145.1875 | GB3LB | 118.8Hz | Lauder | GM7LUN |

# Repeaters (by output frequency)

| Channel | F (out) | F (in) | Callsign | CTCSS | Location | Keeper |
|---|---|---|---|---|---|---|
| RV63 | 145.7875 | 145.1875 | GB3YW | 82.5Hz | Wakefield | G1XCC |
| RV63 | 145.7875 | 145.1875 | GB7FK-C | | Folkestone | M1CMN |
| RU66 | 430.8250 | 438.4250 | GB3IM-C | 110.9Hz | Douglas Iom | GD4HOZ |
| RU66 | 430.8250 | 438.4250 | GB3IM-R | 71.9Hz | Ramsey Iom | GD4HOZ |
| RU66 | 430.8250 | 438.4250 | GB3NY | 88.5Hz | Bridlington | M0DPH |
| RU66 | 430.8250 | 438.4250 | GB3UO | 110.9Hz | Chirk | GW0WZZ |
| RU66 | 430.8250 | 438.4250 | GB3WU | 110.9Hz | Worcester | G8TIC |
| RU66 | 430.8250 | 438.4250 | GB3ZB | 77Hz | Bristol | G4RKY |
| RU66 | 430.8250 | 438.4250 | [illegible] | 88.5Hz | Eastbourne | [illegible] |
| RU68 | 430.8500 | 438.4500 | GB3BS | 118.8Hz | Bristol City | G4SDR |
| RU68 | 430.8500 | 438.4500 | GB3BZ | 110.9Hz | Braintree | G0DEC |
| RU68 | 430.8500 | 438.4500 | GB3FC | 82.5Hz | Blackpool | G6AOS |
| RU68 | 430.8500 | 438.4500 | GB3IE | 77Hz | Plymouth | G7DQC |
| RU68 | 430.8500 | 438.4500 | GB3MD | 71.9Hz | Mansfield | G0UYQ |
| RU68 | 430.8500 | 438.4500 | GB3NX | 88.5Hz | Crawley Sussex | G4EFO |
| RU69 | 430.8625 | 438.4625 | GB3BE | 118.8Hz | Duns | GM7LUN |
| RU69 | 430.8625 | 438.4625 | GB3LR | 88.5Hz | Newhaven | G0TJH |
| RU69 | 430.8625 | 438.4625 | GB3UK | 103.5Hz | Cheltenham | G4FPV |
| RU70 | 430.8750 | 438.4750 | GB3IM-P | 110.9Hz | Peel | GD4HOZ |
| RU70 | 430.8750 | 438.4750 | GB3NH | 77Hz | Northampton | G4YKE |
| RU70 | 430.8750 | 438.4750 | GB3WA | 103.5Hz | Dalry | GM7GDE |
| RU70 | 430.8750 | 438.4750 | GB7SJ | 103.5Hz | Northwich | M0WTX |
| RU71 | 430.8875 | 438.4875 | GB3BP | 71.9Hz | Belper | G0MGX |
| RU71 | 430.8875 | 438.4875 | GB3EK | 103.5Hz | Margate | G4TKR |
| RU71 | 430.8875 | 438.4875 | GB3NP | 77Hz | Towcester | G4YKE |
| RU71 | 430.8875 | 438.4875 | GB3NU | 94.8Hz | Sheringham | G8SAU |
| RU71 | 430.8875 | 438.4875 | GB3XL | 82.5Hz | Shipley | M0IRK |
| RU71 | 430.8875 | 438.4875 | GB7FI | 77Hz | Axbridge | G4RKY |
| RU71 | 430.8875 | 438.4875 | GB7TQ | 94.8Hz | Torquay | G8XST |
| RU72 | 430.9000 | 438.5000 | GB3GL | 103.5Hz | Glasgow | GM3SAN |
| RU72 | 430.9000 | 438.5000 | GB3HY | Hz | Haywards Heath | G6DGK |
| RU72 | 430.9000 | 438.5000 | GB3JC | 94.8Hz | Norwich | M0ZAH |
| RU72 | 430.9000 | 438.5000 | GB3JU | 88.5Hz | St Helier | GJ8PVL |
| RU72 | 430.9000 | 438.5000 | GB3KC | 67Hz | Stourbridge | G0EWH |
| RU72 | 430.9000 | 438.5000 | GB3LW | 82.5Hz | London | G4RFC |
| RU72 | 430.9000 | 438.5000 | GB3MG | 94.8Hz | Pontypridd | GW6CUR |
| RU72 | 430.9000 | 438.5000 | GB3MT | 110.9Hz | Magherafelt | MI0GRN |
| RU72 | 430.9000 | 438.5000 | GB3PZ | 82.5Hz | Dukinfield | G4ZPZ |
| RU74 | 430.9250 | 438.5250 | GB3DM | 103.5Hz | Dumbarton | GM0RTY |
| RU74 | 430.9250 | 438.5250 | GB3FI | 77Hz | Cheddar | G4RKY |
| RU74 | 430.9250 | 438.5250 | GB3HE | 103.5Hz | Hastings | G8PUO |
| RU74 | 430.9250 | 438.5250 | GB3VN | 103.5Hz | Ludlow | G4OYX |
| RU74 | 430.9250 | 438.5250 | GB3XN | 71.9Hz | Worksop | G3XXN |
| RU74 | 430.9250 | 438.5250 | GB7BR | 3Hz | Ramsey | GD4HOZ |
| RU74 | 430.9250 | 438.5250 | GB7CA | 2Hz | Douglas | GD4HOZ |
| RU75 | 430.9375 | 438.5375 | GB3GR | 71.9Hz | Grantham | G8SAU |
| RU75 | 430.9375 | 438.5375 | GB3SF | 77Hz | South Molton | G6SQX |
| RU75 | 430.9375 | 438.5375 | GB3WP | 82.5Hz | Hyde | G6YRK |
| RU76 | 430.9500 | 438.5500 | GB3EX | 77Hz | Silverton | G7NBU |
| RU76 | 430.9500 | 438.5500 | GB3FJ | 71.9Hz | Asgarby | G3ZPU |
| RU76 | 430.9500 | 438.5500 | GB3OH | 94.8Hz | Linlithgow | GM0MZB |
| RU76 | 430.9500 | 438.5500 | GB3OM | 1Hz | Omagh | GI4SXV |
| RU76 | 430.9500 | 438.5500 | GB3OY | 82.5Hz | Buckhurst Hill | G7UZN |
| RU76 | 430.9500 | 438.5500 | GB3WL | 67Hz | Birmingham | G3YXM |
| RU76 | 430.9500 | 438.5500 | GB7RY | | Rye | M0HOW |
| RU77 | 430.9625 | 438.5625 | GB3CC | 88.5Hz | Chichester | G3UEQ |
| RU77 | 430.9625 | 438.5625 | GB7JL | 82.5Hz | Wigan | G6TPG |
| RU77 | 430.9625 | 438.5625 | GB7SH | 71.9Hz | Sheffield | M1ERS |
| RU78 | 430.9750 | 438.5750 | GB3AC | 94.8Hz | Lydney Glos | G4CJZ |
| RU78 | 430.9750 | 438.5750 | GB3DQ | 77Hz | Polperro | G1YDQ |
| RU78 | 430.9750 | 438.5750 | GB3EZ | 110.9Hz | Wickhambrook | G1YFF |
| RU78 | 430.9750 | 438.5750 | GB3KK | 110.9Hz | Ballycastle | MI0MRV |
| RU78 | 430.9750 | 438.5750 | GB3KV | 103.5Hz | Kilsyth | GM3SAN |
| RU78 | 430.9750 | 438.5750 | GB3ZI | | Stafford | G8VPR |
| RU78 | 430.9750 | 438.5750 | GB7FC | 82.5Hz | Blackpool | M0AUT |
| RU78 | 430.9750 | 438.5750 | GB7ID | 3Hz | Redhill | M0IDD |
| RB0 | 433.0000 | 434.6000 | GB3BN | 118.8Hz | Bracknell | G8DOR |
| RB0 | 433.0000 | 434.6000 | GB3CK | 103.5Hz | Charing Kent | M0ZAA |
| RB0 | 433.0000 | 434.6000 | GB3DT | 71.9Hz | Blandford Camp | G8BXQ |
| RB0 | 433.0000 | 434.6000 | GB3LL | 110.9Hz | Llandudno | GW6SIX |
| RB0 | 433.0000 | 434.6000 | GB3MK | 77Hz | Milton Keynes | G4CAK |
| RB0 | 433.0000 | 434.6000 | GB3NR | 94.8Hz | Norwich | G8VLL |
| RB0 | 433.0000 | 434.6000 | GB3NT | 118.8Hz | Newcastle/Tyne | G7UUR |
| RB0 | 433.0000 | 434.6000 | GB3PU | 94.8Hz | Perth | GM8KPH |
| RB0 | 433.0000 | 434.6000 | GB3RT | 94.8Hz | Cwmbran | MW0YAC |
| RB0 | 433.0000 | 434.6000 | GB3SO | 71.9Hz | Sutton On Sea | M0MFP |
| RB0 | 433.0000 | 434.6000 | GB3US | 103.5Hz | Sheffield | G4CUI |
| RB0 | 433.0000 | 434.6000 | GB3WN | 67Hz | Wolverhampton | G4OKE |
| RB01 | 433.0250 | 434.6250 | GB3BV | 118.8Hz | Hemel Hempstead | G8BQH |
| RB01 | 433.0250 | 434.6250 | GB3DV | 71.9Hz | Maltby | G0EPX |
| RB01 | 433.0250 | 434.6250 | GB3HJ | 118.8Hz | Harrogate | G4MEM |
| RB01 | 433.0250 | 434.6250 | GB3JS | 94.8Hz | Great Yarmouth | M0JGX |
| RB01 | 433.0250 | 434.6250 | GB3MA | 82.5Hz | Bury | G7LWT |
| RB01 | 433.0250 | 434.6250 | GB3UI | 67Hz | Birmingham | G4LCH |
| RB02 | 433.0500 | 434.6500 | GB3AV | 118.8Hz | Aylesbury | G8BQH |
| RB02 | 433.0500 | 434.6500 | [illegible] | 77Hz | Liskeard | G4RKY |
| RB02 | 433.0500 | 434.6500 | GB3CI | 77Hz | Corby | G7HPE |
| RB02 | 433.0500 | 434.6500 | GB3HK | 118.8Hz | Selkirk | GM0FTJ |
| RB02 | 433.0500 | 434.6500 | GB3LS | 71.9Hz | Lincoln | G7AVU |
| RB02 | 433.0500 | 434.6500 | GB3LV | 82.5Hz | North London | G4DFB |
| RB02 | 433.0500 | 434.6500 | GB3MF | 103.5Hz | Macclesfield | G1JVI |
| RB02 | 433.0500 | 434.6500 | GB3NN | 94.8Hz | Wells Norfolk | G0FVF |
| RB02 | 433.0500 | 434.6500 | GB3UI | 110.9Hz | Belfast N.I. | GI3USS |
| RB02 | 433.0500 | 434.6500 | GB3YS | 77Hz | Yeovil | G8NIC |
| RB03 | 433.0750 | 434.6750 | GB3ER | 110.9Hz | Danbury | G6JYB |
| RB03 | 433.0750 | 434.6750 | GB3KR | 67Hz | Kidderminster | G8NTU |
| RB03 | 433.0750 | 434.6750 | GB3KU | 82.5Hz | Ashton-Under-Lyne | M0NCZ |
| RB03 | 433.0750 | 434.6750 | GB3TD | 118.8Hz | Swindon | G4XUT |
| RB03 | 433.0750 | 434.6750 | GB3VS | 94.8Hz | Taunton | G4UVZ |
| RB04 | 433.1000 | 434.7000 | GB3CW | 103.5Hz | Newtown Powys | GW4NQJ |
| RB04 | 433.1000 | 434.7000 | GB3IH | 110.9Hz | Ipswich | G7CIY |
| RB04 | 433.1000 | 434.7000 | GB3KL | 94.8Hz | Kings Lynn | G0IJU |
| RB04 | 433.1000 | 434.7000 | GB3LE | 77Hz | Markfield | M1NAS |
| RB04 | 433.1000 | 434.7000 | GB3NK | 103.5Hz | Erith | G4EGU |
| RB04 | 433.1000 | 434.7000 | GB3SP | 94.8Hz | Pembroke | GW4VRO |
| RB04 | 433.1000 | 434.7000 | GB3UB | 118.8Hz | Bath Avon | G4KVI |
| RB04 | 433.1000 | 434.7000 | GB3VE | 77Hz | Dufton | G7ITT |
| RB05 | 433.1250 | 434.7250 | GB3GH | 118.8Hz | Gloucester | G3LVP |
| RB05 | 433.1250 | 434.7250 | GB3IC | 67Hz | Wolverhampton | M0VRR |
| RB05 | 433.1250 | 434.7250 | GB3IM-S | 110.9Hz | Douglas Iom | GD4HOZ |
| RB05 | 433.1250 | 434.7250 | GB3OV | 94.8Hz | St Neots | M1JUL |
| RB05 | 433.1250 | 434.7250 | GB3WJ | 88.5Hz | Scunthorpe | G3TMD |
| RB05 | 433.1250 | 434.7250 | GB7PD-B | | Pembroke Dock | MW0HVB |
| RB06 | 433.1500 | 434.7500 | GB3BR | 88.5Hz | Brighton | G4PAP |
| RB06 | 433.1500 | 434.7500 | GB3CR | 110.9Hz | Caergwrle | M0OBW |
| RB06 | 433.1500 | 434.7500 | GB3DI | 118.8Hz | Didcot | G8CUL |
| RB06 | 433.1500 | 434.7500 | GB3HC | 118.8Hz | Hereford | G0JWJ |
| RB06 | 433.1500 | 434.7500 | GB3ME | 67Hz | Rugby | G7BQM |
| RB06 | 433.1500 | 434.7500 | GB3SK | 103.5Hz | Canterbury | G6DIK |
| RB06 | 433.1500 | 434.7500 | GB3SY | 71.9Hz | Barnsley | G4LUE |
| RB06 | 433.1500 | 434.7500 | GB3WG | 94.8Hz | Port Talbot | GW4FOI |
| RB07 | 433.1750 | 434.7750 | GB3AU | 82.5Hz | Amersham | G0RDI |
| RB07 | 433.1750 | 434.7750 | GB3BL | 77Hz | Bedford | G8MGP |
| RB07 | 433.1750 | 434.7750 | GB3DE | 110.9Hz | Ipswich | G1NRL |
| RB07 | 433.1750 | 434.7750 | GB3MS | 67Hz | Worcester | G7WIG |
| RB07 | 433.1750 | 434.7750 | GB3NM | 71.9Hz | Nottingham | G4IRX |
| RB07 | 433.1750 | 434.7750 | GB3TS | 118.8Hz | Sunderland | G7MFN |
| RB08 | 433.2000 | 434.8000 | GB3AN | 110.9Hz | Amlwch | GW6DOK |
| RB08 | 433.2000 | 434.8000 | GB3CM | 94.8Hz | Carmarthen | GW8KCY |
| RB08 | 433.2000 | 434.8000 | GB3EH | 67Hz | Edge Hill | G4OHB |
| RB08 | 433.2000 | 434.8000 | GB3PY | 77Hz | Cambridge | G4NBS |
| RB08 | 433.2000 | 434.8000 | GB3RB | 71.9Hz | Bolsover | G1SLE |
| RB08 | 433.2000 | 434.8000 | GB3TF | 103.5Hz | Telford | G3UKV |
| RB08 | 433.2000 | 434.8000 | GB40PY | 77Hz | Cambridge | G4NBS |
| RB09 | 433.2250 | 434.8250 | GB3BG | 94.8Hz | Blaenavon | GW7LOP |
| RB09 | 433.2250 | 434.8250 | GB3CL | 103.5Hz | Clacton | G7HJK |
| RB09 | 433.2250 | 434.8250 | GB3CV | 67Hz | Coventry | G7TRJ |
| RB09 | 433.2250 | 434.8250 | GB3HD | 82.5Hz | Huddersfield | G0ISX |
| RB09 | 433.2250 | 434.8250 | GB3IW | 71.9Hz | Ryde | G4IKI |
| RB09 | 433.2250 | 434.8250 | GB3LC | 71.9Hz | Louth | G7AJP |
| RB09 | 433.2250 | 434.8250 | GB3ST | 103.5Hz | Stoke On Trent | G8NSS |
| RB09 | 433.2250 | 434.8250 | GB3TU | 77Hz | Tring Herts | G0RDI |
| RB10 | 433.2500 | 434.8500 | GB3AW | 71.9Hz | Newbury | G8DOR |
| RB10 | 433.2500 | 434.8500 | GB3DD | 94.8Hz | Dundee | GM4UGF |
| RB10 | 433.2500 | 434.8500 | GB3DY | 71.9Hz | Wirksworth | G3ZYC |
| RB10 | 433.2500 | 434.8500 | GB3LI | 82.5Hz | Liverpool | G3WIC |
| RB10 | 433.2500 | 434.8500 | GB3LT | 77Hz | Luton | G8XTW |
| RB10 | 433.2500 | 434.8500 | GB3MW | 67Hz | Leamington Spa | G6FEO |
| RB11 | 433.2750 | 434.8750 | GB3AH | 94.8Hz | East Dereham | G0LGJ |
| RB11 | 433.2750 | 434.8750 | GB3GY | 88.5Hz | Grimsby | G0ATW |
| RB11 | 433.2750 | 434.8750 | GB3HN | 82.5Hz | Hitchin | G4LOO |
| RB11 | 433.2750 | 434.8750 | GB3HT | 77Hz | Hinckley | G4ALB |
| RB11 | 433.2750 | 434.8750 | GB3RE | 103.5Hz | Maidstone | G6RVS |
| RB11 | 433.2750 | 434.8750 | GB3RH | 94.8Hz | Axminster | G6WWY |
| RB11 | 433.2750 | 434.8750 | GB3RU | 118.8Hz | Reading | G8DOR |
| RB12 | 433.3000 | 434.9000 | GB3DX | 110.9Hz | Londonderry | GI4YWT |
| RB12 | 433.3000 | 434.9000 | GB3EE | 71.9Hz | Chesterfield | G1SLE |
| RB12 | 433.3000 | 434.9000 | GB3GB | 67Hz | Birmingham | G8NDT |
| RB12 | 433.3000 | 434.9000 | GB3GF | 88.5Hz | Guildford | G4EML |
| RB12 | 433.3000 | 434.9000 | GB3TJ | 118.8Hz | Corbridge | G7UUR |
| RB12 | 433.3000 | 434.9000 | GB3TJ | 118.8Hz | Corbridge | G7UUR |
| RB12 | 433.3000 | 434.9000 | GB3WB | 94.8Hz | Weston-Super-Mare | G4TBD |
| RB13 | 433.3250 | 434.9250 | GB3CA | 77Hz | Carlisle | G1XSZ |
| RB13 | 433.3250 | 434.9250 | GB3DS | 71.9Hz | Worksop | G3XXN |

# Repeaters (by output frequency)

| Channel | F (out) | F (in) | Callsign | CTCSS | Location | Keeper |
|---|---|---|---|---|---|---|
| RB13 | 433.3250 | 434.9250 | GB3GU | 71.9Hz | Guernsey | GU6EFB |
| RB13 | 433.3250 | 434.9250 | GB3HW | 110.9Hz | Romford | G4GBW |
| RB13 | 433.3250 | 434.9250 | GB3SM | 103.5Hz | Stoke On Trent | G4SCY |
| RB13 | 433.3250 | 434.9250 | GB3VH | 82.5Hz | Welwyn G City | G4THF |
| RB13 | 433.3250 | 434.9250 | GB3XX | 77Hz | Daventry | G1ZJK |
| RB14 | 433.3500 | 434.9500 | GB3CB | 67Hz | Birmingham | G8VIQ |
| RB14 | 433.3500 | 434.9500 | GB3CE | 110.9Hz | Colchester | G7HJK |
| RB14 | 433.3500 | 434.9500 | GB3ED | 94.8Hz | Edinburgh | GM4GZW |
| RB14 | 433.3500 | 434.9500 | GB3HE | 103.5Hz | Hastings | G8PUO |
| RB14 | 433.3500 | 434.9500 | GB3HR | 82.5Hz | Harrow | G3YXZ |
| RB14 | 433.3500 | 434.9500 | GB3LF | 110.9Hz | Kendal | G3VVT |
| RB14 | 433.3500 | 434.9500 | GB3MR | 82.5Hz | Stockport | G8LZO |
| RB14 | 433.3500 | 434.9500 | GB3ND | 77Hz | Bideford | G4SOF |
| RB14 | 433.3500 | 434.9500 | GB3SD | 71.9Hz | Weymouth | G0EVW |
| RB14 | 433.3500 | 434.9500 | GB3WF | 82.5Hz | Otley | M0JSE |
| RB14 | 433.3500 | 434.9500 | GB3YL | 94.8Hz | Lowestoft | G4RKP |
| RB15 | 433.3750 | 434.9750 | GB3FN | 82.5Hz | Farnham | G4EPX |
| RB15 | 433.3750 | 434.9750 | GB3HB | 77Hz | St Austell | G3IGV |
| RB15 | 433.3750 | 434.9750 | GB3LH | 103.5Hz | Shrewsbury | G8DIR |
| RB15 | 433.3750 | 434.9750 | GB3PP | 82.5Hz | Preston | M0NED |
| RB15 | 433.3750 | 434.9750 | GB3SZ | 71.9Hz | Bournemouth | M0AUY |
| RB15 | 433.3750 | 434.9750 | GB3TH | 67Hz | Tamworth | G8MKT |
| RB15 | 433.3750 | 434.9750 | GB3WI | 94.8Hz | Wisbech | M0DUQ |
| DVU13 | 439.1625 | 430.1625 | GB7BS | 3Hz | Bristol City | G4SDR |
| DVU13 | 439.1625 | 430.1625 | GB7FU | 3Hz | Pointon | G8SJP |
| DVU13 | 439.1625 | 430.1625 | GB7NS | 3Hz | Caterham | G0OLX |
| DVU13 | 439.1625 | 430.1625 | GB7TD | 1Hz | Wakefield | G1XCC |
| DVU32 | 439.4000 | 430.4000 | GB7BM | | Birmingham | G8VIQ |
| DVU32 | 439.4000 | 430.4000 | GB7CW | 3Hz | Bridgend | GW0UZK |
| DVU32 | 439.4000 | 430.4000 | GB7EP | 3Hz | Epsom | G0OXZ |
| DVU32 | 439.4000 | 430.4000 | GB7LN | 1Hz | Lincoln | G0RZR |
| DVU32 | 439.4000 | 430.4000 | GB7LP | 1Hz | Liverpool | M1SWB |
| DVU33 | 439.4125 | 430.4125 | GB7DN | | Dungiven | GI0AZB |
| DVU33 | 439.4125 | 430.4125 | GB7GT | 13Hz | Ludlow | G1MAW |
| DVU33 | 439.4125 | 430.4125 | GB7SD | 1Hz | Weymouth | G0EVW |
| DVU33 | 439.4125 | 430.4125 | GB7WI | 1Hz | Winterton | G0UZJ |
| DVU34 | 439.4250 | 430.4250 | GB7DR | 5Hz | Poole | G0ECX |
| DVU34 | 439.4250 | 430.4250 | GB7DS | 1Hz | Norwich | M0ZAH |
| DVU34 | 439.4250 | 430.4250 | GB7GB | 8Hz | Birmingham | G8NDT |
| DVU34 | 439.4250 | 430.4250 | GB7HS | 2Hz | Batley | G1XCC |
| DVU34 | 439.4250 | 430.4250 | GB7PN | 1Hz | Prestatyn | G4NOY |
| DVU35 | 439.4375 | 430.4375 | GB7AV | 3Hz | Aylesbury | G0RAS |
| DVU35 | 439.4375 | 430.4375 | GB7BX | 5Hz | Much Wenlock | M1GIZ |
| DVU35 | 439.4375 | 430.4375 | GB7ES | | Eastbourne | M0LRE |
| DVU35 | 439.4375 | 430.4375 | GB7GC | | Grimsby | G7EOG |
| DVU35 | 439.4375 | 430.4375 | GB7PP | | Ipswich | G8HUE |
| DVU36 | 439.4500 | 430.4500 | GB7BD | | Bristol | G4RKY |
| DVU36 | 439.4500 | 430.4500 | GB7BP | | Milton Keynes | G0NBI |
| DVU36 | 439.4500 | 430.4500 | GB7HD-B | | Huddersfield | G6OCD |
| DVU36 | 439.4500 | 430.4500 | GB7HR | 3Hz | Heathrow | G0OXZ |
| DVU36 | 439.4500 | 430.4500 | GB7IC-B | | Herne Bay | G4TKR |
| DVU36 | 439.4500 | 430.4500 | GB7NE | | Ashington | G7RWC |
| DVU36 | 439.4500 | 430.4500 | GB7SU | 8Hz | Southampton | G6IGA |
| DVU37 | 439.4625 | 430.4625 | GB7JB | 1Hz | Mere Wiltshire | G3ZXX |
| DVU37 | 439.4625 | 430.4625 | GB7NL | | Exeter | M0NLO |
| DVU37 | 439.4625 | 430.4625 | GB7NM | | Manchester | G7LWT |
| DVU37 | 439.4625 | 430.4625 | GB7WL | 3Hz | Amersham | G0RDI |
| DVU37 | 439.4625 | 430.4625 | GB7XX | 10Hz | Felling | G4MSF |
| DVU38 | 439.4750 | 430.4750 | GB7CH | | Louth | G7AJP |
| DVU38 | 439.4750 | 430.4750 | GB7SC | 3Hz | Bognor Regis | G0AFN |
| DVU38 | 439.4750 | 430.4750 | GB7SE | 3Hz | Thurrock | M0PFX |
| DVU38 | 439.4750 | 430.4750 | GB7TV | | New Marske | M0RIG |
| DVU39 | 439.4875 | 430.4875 | GB7DC | | Derby | G7NPW |
| DVU39 | 439.4875 | 430.4875 | GB7HU | | South Cave | G0VRM |
| DVU39 | 439.4875 | 430.4875 | GB7WB | | Weston-Super-Mare | G4SZM |
| DVU39 | 439.4875 | 430.4875 | GB7WC | | Warrington | G4VSS |
| DVU39 | 439.4875 | 430.4875 | GB7WF | | Bewdley | G8OXG |
| DVU39 | 439.4875 | 430.4875 | GB7ZP | | Danbury | G6JYB |
| DVU40 | 439.5000 | 430.5000 | GB7DP-B | | Dundee | GM0ROU |
| DVU40 | 439.5000 | 430.5000 | GB7KT | 1Hz | Andover | G3ZXX |
| DVU40 | 439.5000 | 430.5000 | GB7PE | 3Hz | Peterborough | M0ZPU |
| DVU41 | 439.5125 | 430.5125 | GB7CF | | Folkestone | M1CMN |
| DVU41 | 439.5125 | 430.5125 | GB7IT | 1Hz | Weston-Super-Mare | G4SZM |
| DVU41 | 439.5125 | 430.5125 | GB7LO | 3Hz | Bromley | G1HIG |
| DVU41 | 439.5125 | 430.5125 | GB7YD-B | | Barnsley | G4LUE |
| DVU42 | 439.5250 | 430.5250 | GB7AK | 3Hz | Barking | G8YPK |
| DVU42 | 439.5250 | 430.5250 | GB7DZ | 7Hz | Newcastle-Upon-Tyne | M0MBA |
| DVU42 | 439.5250 | 430.5250 | GB7SI | 1Hz | Stoke-On-Trent | G8NSS |
| DVU42 | 439.5250 | 430.5250 | GB7TC | 2Hz | Swindon | G8VRI |
| DVU42 | 439.5250 | 430.5250 | GB7UL | 1Hz | Carrickfergus | GI6DKQ |
| DVU43 | 439.5375 | 430.5375 | GB7DG | | Bromsgrove | M1JSS |
| DVU43 | 439.5375 | 430.5375 | GB7EN | | Wellingborough | G7HIF |
| DVU43 | 439.5375 | 430.5375 | GB7HE | | Hastings | M0JBR |
| DVU43 | 439.5375 | 430.5375 | GB7IK | 3Hz | Rochester | G6CKK |
| DVU43 | 439.5375 | 430.5375 | GB7LF | | Lancaster | G3VVT |
| DVU43 | 439.5375 | 430.5375 | GB7MC | | St Austell | M1DNS |
| DVU43 | 439.5375 | 430.5375 | GB7NU | | Sheringham | G8SAU |
| DVU46 | 439.5750 | 430.5750 | GB7BH | 4Hz | Brighton | G7TXU |
| DVU46 | 439.5750 | 430.5750 | GB7EB | 2Hz | Beccles | M0JGX |
| DVU46 | 439.5750 | 430.5750 | GB7HX | 1Hz | Huddersfield | G0ISX |
| DVU46 | 439.5750 | 430.5750 | GB7SO | | Solihull | G7DDN |
| DVU48 | 439.6000 | 430.6000 | GB7DE-B | | Upper Largo | MM0DXE |
| DVU48 | 439.6000 | 430.6000 | GB7MK | 13Hz | Ipswich | M1NIZ |
| DVU48 | 439.6000 | 430.6000 | GB7RR | 1Hz | Nottingham | G0LCG |
| DVU48 | 439.6000 | 430.6000 | GB7WT | | Omagh N.I | GI3NVW |
| DVU49 | 439.6125 | 430.6125 | GB7DV | | St. Helens | G1DVA |
| DVU49 | 439.6125 | 430.6125 | GB7FH | | Fareham | G6ORL |
| DVU49 | 439.6125 | 430.6125 | GB7IQ | | Stafford | G7PFT |
| DVU49 | 439.6125 | 430.6125 | GB7KH | | Kelvedon Hatch | M1GEO |
| DVU50 | 439.6250 | 430.6250 | GB7DA-B | | Airdrie | GM4AUP |
| DVU50 | 439.6250 | 430.6250 | GB7EK | 3Hz | Whitstable | G6MRI |
| DVU51 | 439.6375 | 430.6375 | GB7CL | 3Hz | Clacton On Sea | G0MBA |
| DVU51 | 439.6375 | 430.6375 | GB7HM | 1Hz | Caergwrie | G1SYG |
| DVU51 | 439.6375 | 430.6375 | GB7JF | | Northampton | G7HIF |
| DVU51 | 439.6375 | 430.6375 | GB7MH | | East Grinstead | G3NZP |
| DVU51 | 439.6375 | 430.6375 | GB7SR | 2Hz | Sheffield | M0GAV |
| DVU53 | 439.6625 | 430.6625 | GB7BE | | Beccles | M0JGX |
| DVU53 | 439.6625 | 430.6625 | GB7DB | | Ampthill | G3YQO |
| DVU53 | 439.6625 | 430.6625 | GB7DD | 1Hz | Dundee | MM0DUN |
| DVU53 | 439.6625 | 430.6625 | GB7FW | 1Hz | Birmingham | G8VIQ |
| DVU53 | 439.6625 | 430.6625 | GB7KM | 3Hz | Cirencester | G0RMA |
| DVU53 | 439.6625 | 430.6625 | GB7LE | 2Hz | Leeds | G1XCC |
| DVU53 | 439.6625 | 430.6625 | GB7LY | 1Hz | Londonderry | GI4YWT |
| DVU54 | 439.6750 | 430.6750 | GB3NS | 82.5Hz | Caterham | G0OLX |
| DVU54 | 439.6750 | 430.6750 | GB7AA | 1Hz | Bristol | G4CJZ |
| DVU54 | 439.6750 | 430.6750 | GB7DL | | Friskney | M0VBR |
| DVU54 | 439.6750 | 430.6750 | GB7ED | 5Hz | Exeter | M0ZZT |
| DVU54 | 439.6750 | 430.6750 | GB7SN | 1Hz | Sheffield | M1ERS |
| DVU55 | 439.6875 | 430.6875 | GB7GF | 3Hz | Guildford | G4EML |
| DVU55 | 439.6875 | 430.6875 | GB7NB | | Norwich | G0LGJ |
| DVU55 | 439.6875 | 430.6875 | GB7TP | 1Hz | Keighley | M0IRK |
| DVU55 | 439.6875 | 430.6875 | GB7ZI | | Stafford | G8VPR |
| DVU56 | 439.7000 | 430.7000 | GB7AU | | Amersham | G0RDI |
| DVU56 | 439.7000 | 430.7000 | GB7CD | | Cardiff | GW6CUR |
| DVU56 | 439.7000 | 430.7000 | GB7DX | | Wittersham | G0GCQ |
| DVU56 | 439.7000 | 430.7000 | GB7MB | 1Hz | Heysham | G4TUZ |
| DVU56 | 439.7000 | 430.7000 | GB7VO | | Leominster | G8XYJ |
| DVU57 | 439.7125 | 430.7125 | GB7EE | 1Hz | Edinburgh | GM7RYR |
| DVU57 | 439.7125 | 430.7125 | GB7EL | 2Hz | Nelson | G4BLH |
| DVU57 | 439.7125 | 430.7125 | GB7EX | 3Hz | Southend-On-Sea | G8YPK |
| DVU57 | 439.7125 | 430.7125 | GB7GL | | Mitcheldean | G4SGI |
| DVU57 | 439.7125 | 430.7125 | GB7JH-B | | Shoreham-By-Sea | G8MUQ |
| DVU57 | 439.7125 | 430.7125 | GB7PT | | Royston | G4NBS |
| DVU57 | 439.7125 | 430.7125 | GB7RE | 1Hz | Retford | M0CMN |
| DVU59 | 439.7375 | 430.7375 | GB7BK | 3Hz | Reading | G8DOR |
| DVU59 | 439.7375 | 430.7375 | GB7CK | 3Hz | Folkestone | M1CMN |
| DVU59 | 439.7375 | 430.7375 | GB7MR | 2Hz | Manchester | G8UVC |
| DVU59 | 439.7375 | 430.7375 | GB7SS | | Hockley | G6XYU |
| DVU62 | 439.7750 | 430.7750 | GB7SX | 88.5Hz | Bognor Regis | G0AFN |
| DVU73 | 439.9125 | 430.9125 | GB7AD | | Bristol | G4CJZ |
| DVU73 | 439.9125 | 430.9125 | GB7DK-B | | Stranraer | GM0HPK |
| DVU73 | 439.9125 | 430.9125 | GB7PI | | Royston | M0ZPU |
| DVU73 | 439.9125 | 430.9125 | GB7WP | | Birkenhead | G4BKF |
| 23CM | 1290.650 | 1270.650 | GB7IC-A | | Herne Bay | G4TKR |
| RM0 | 1297.000 | 1291.000 | GB3NO | 94.8Hz | Norwich | G8VLL |
| RM2 | 1297.050 | 1291.050 | GB3FM | 100Hz | Farnham | G4EPX |
| RM3 | 1297.075 | 1291.075 | GB3PS | 77Hz | Royston | G4NBS |
| RM3 | 1297.075 | 1291.075 | GB3SE | 103.5Hz | Stoke On Trent | G8NSS |
| RM6 | 1297.150 | 1291.150 | GB3MM | 67Hz | Wolverhampton | G4OKE |
| RM12 | 1297.300 | 1291.300 | GB7YD-A | | Barnsley | G4LUE |
| RM14A | 1297.350 | 1277.350 | GB3AK | 94.8Hz | Bristol | G4CJZ |
| RM15 | 1297.375 | 1291.375 | GB3WC | 82.5Hz | Wakefield | G1XCC |
| RM0A | 1299.850 | 1293.850 | GB3BK | 103.5Hz | Bromley | G0WYG |

## Internet-linked (alphabetical)

| Callsign | Ch | Location | Echolink | IRLP |
|---|---|---|---|---|
| GB3AE | R50-01 | Tenby | 117406 | |
| GB3AG | RV58 | Forfar | 117931 | |
| GB3AL | RV59 | Amersham | 19063 | |
| GB3AM | R50-13 | Amersham | 4125 | |
| GB3AR | RV58 | Caernarfon | 208003 | |
| GB3BC | RV60 | Pontypridd | 39300 | |
| GB3BM | RV57 | Birmingham | | 5702 |
| GB3BN | RB0 | Bracknell | 1938 | |
| GB3CA | RB13 | Carlisle | 412685 | 5280 |
| GB3CG | RV58 | Gloucester | 190502 | |
| GB3CH | RB02 | Liskeard | | 5992 |
| GB3DC | RV55 | Derby | 92369 | |
| GB3DQ | RU78 | Polperro | 418341 | 5612 |
| GB3DU | RV61 | Duns | 276441 | |
| GB3DV | RB01 | Malling | 120618 | 6130 |
| GB3DX | RB12 | Londonderry | 7125 | |
| GB3EE | RB12 | Chesterfield | | 5046 |
| GB3EK | RU71 | Margate | 48360 | |
| GB3FH | R50-06 | Somerset | 228585 | 5361 |
| GB3FK | RV60 | Folkestone | 235976 | |
| GB3GN | RV62 | Banchory | 19583 | |
| GB3HE | RB14 | Hastings | 71066 | |
| GB3HH | RV56 | Buxton | 97616 | |
| GB3IE | RU68 | Plymouth | 27871 | |
| GB3IK | RV61 | Rochester | 263025 | |
| GB3IM-C | RU66 | Douglas Iom | 464453 | |
| GB3IM-P | RU70 | Peel | 464453 | |
| GB3IM-R | RU68 | Ramsey Iom | 464453 | |
| GB3IM-S | RB05 | Douglas Iom | 464453 | |
| GB3IN | RV51 | Matlock | 98258 | |
| GB3IR | RV61 | Richmond Yorks | 1353 | 5562 |
| GB3JS | RB01 | Great Yarmouth | 246617 | |
| GB3JX | R50-12 | Norwich | 040059 | |
| GB3KC | RU72 | Stourbridge | 430900 | |
| GB3KD | RV63 | Kidderminster | 78750 | |
| GB3KE | RV55 | Glasgow | 5411 | 5110 |
| GB3KL | RB04 | Kings Lynn | 77288 | |
| GB3KR | RB03 | Kidderminster | 4304 | |
| GB3KS | RV50 | Dover | 346463 | |
| GB3KU | RB03 | Ashton-Under-Lyne | | 1234567 |
| GB3LF | RB14 | Kendal | 184457 | 5140 |
| GB3LR | RU69 | Newhaven | 494669 | |
| GB3LS | RB02 | Lincoln | 268511 | |
| GB3LV | RB02 | North London | 155403 | 5600 |
| GB3MH | RV50 | East Grinstead | | 5877 |
| GB3MI | RV57 | Manchester | 197681 | |
| GB3NC | RV58 | St Austell | 282184 | |
| GB3ND | RB14 | Bideford | 221334 | |
| GB3NK | RB04 | Frith | 54760 | |
| GB3NU | RU71 | Sheringham | 388653 | |
| GB3OA | RV49 | Southport | 5302 | 5302 |
| GB3PA | RV50 | Paisley | 116678 | |
| GB3PY | RB08 | Cambridge | 222303 | |
| GB3PZ | RU72 | Dukinfield | 2591 | 5400 |
| GB3SB | RV52 | Selkirk | 116678 | |
| GB3SD | RB14 | Weymouth | 112689 | |
| GB3TD | RB03 | Swindon | 43307 | |
| GB3TF | RV58 | Keighley | 257033 | |
| GB3TR | RV52 | Torquay | | 5582 |
| GB3UB | RB04 | Bath Avon | 201135 | |
| GB3US | RB0 | Sheffield | 223073 | 5150 |
| GB3XN | RU74 | Worksop | 153126 | 5708 |
| GB3YL | | Lowestoft | 227697 | |
| GB3ZB | RU66 | Bristol | | 5429 |
| GB7SJ | RU70 | Northwich | 41360 | |

## Digital Nodes including D-Star (by Channel)

| Ch No | Callsign | Location | Type |
|---|---|---|---|
| RV48 | GB3SS | Elgin Moray | FUSION |
| RV48 | GB3VW | Norwich | dPMR |
| RV48 | GB3WR | Wells | FUSION |
| RV48 | GB7MA | Bury | D-STAR |
| RV48 | GB7RW | Whitby | D-STAR |
| RV48 | GB7SB | Brighton | D-STAR |
| RV49 | GB3NA | Sheffield | FUSION |
| RV49 | GB3TO | Northampton | D-STAR |
| RV49 | GB3VM | Ludlow | FUSION |
| RV51 | GB3IN | Matlock | DMR/? |
| RV51 | GB7CT | Tring | DMR/3 |
| RV51 | GB7DE-C | Upper Largo | D-STAR |
| RV51 | GB7RN | Fareham | D-STAR |
| RV52 | GB3HS | Walkington | FUSION |
| RV52 | GB3MN | Disley | FUSION |
| RV52 | GB7JH-C | Shoreham-By-Sea | D-STAR |
| RV53 | GB3AA | Bristol | FUSION |
| RV53 | GB3KI | Herne Bay | D-STAR |
| RV53 | GB7IC-C | Herne Bay | D-STAR |
| RV53 | GB7PB | Durham | D-STAR |
| RV54 | GB3PR | Perth | FUSION |
| RV54 | GB7IE | Plymouth | FUSION |
| RV54 | GB7RB-C | Cowbridge | D-STAR |
| RV54 | GB7YD-C | Barnsley | D-STAR |
| RV55 | GB3CD | Crook | FUSION |
| RV55 | GB3KE | Glasgow | FUSION |
| RV55 | GB3WE | Weston-Super-Mare | DMR/1 |
| RV55 | GB7DK-C | Stranraer | D-STAR |
| RV55 | GB7GD | Aberdeen | D-STAR |
| RV56 | GB7CM | Wimbourne | FUSION |
| RV57 | GB3MI | Manchester | D-STAR |
| RV57 | GB7OK | Bromley | D-STAR |
| RV58 | GB3LM | Lincoln | FUSION |
| RV58 | GB3TW | Gateshead | FUSION |
| RV58 | GB3VT | Stoke On Trent | D-STAR |
| RV59 | GB3ZA | Hereford | FUSION |
| RV59 | GB7PD-C | Pembroke Dock | D-STAR |
| RV59 | GB7SF | Sheffield | D-STAR |
| RV60 | GB3HA | Corbridge | FUSION |
| RV60 | GB7NI | Carrickfergus | D-STAR |
| RV61 | GB3IP | Stafford | DMR/1 |
| RV62 | GB3RF | Accrington | FUSION |
| RV62 | GB7DA-C | Airdrie | D-STAR |
| RV62 | GB7ER | Exeter | FUSION |
| RV62 | GB7TE | Clacton On Sea | D-STAR |
| RV63 | GB7FK-C | Folkestone | D-STAR |
| RU66 | GB3KW | Glasgow | FUSION |
| RU70 | GB7SJ | Northwich | FUSION |
| RU71 | GB3DP | Belper | FUSION |
| RU71 | GB3NP | Towcester | FUSION |
| RU71 | GB3XL | Shipley | DMR/1 |
| RU71 | GB7FI | Axbridge | DMR/3 |
| RU71 | GB7TQ | Torquay | FUSION |
| RU72 | GB3MG | Pontypridd | FUSION |
| RU74 | GB3DM | Dumbarton | FUSION |
| RU74 | GB3FI | Cheddar | DMR/3 |
| RU74 | GB3VN | Ludlow | FUSION |
| RU74 | GB7BR | Ramsey | DMR/3 |
| RU74 | GB7CA | Douglas | DMR/2 |
| RU76 | GB3OM | Omagh | DMR/1 |
| RU76 | GB7RY | Rye | FUSION |
| RU77 | GB7JL | Wigan | DMR/1 |
| RU77 | GB7SH | Sheffield | FUSION |
| RU78 | GB3ZI | Stafford | FUSION |
| RU78 | GB7FC | Blackpool | DMR/1 |
| RU78 | GB7ID | Redhill | DMR/3 |
| RB0 | GB3DT | Blandford | FUSION |
| RB0 | GB3MK | Milton Keynes | FUSION |
| RB0 | GB3PU | Perth | FUSION |
| RB04 | GB7HF | Welham Green | FUSION |
| RB05 | GB7PD-B | Pembroke Dock | D-STAR |
| RB06 | GB3HC | Hereford | FUSION |
| RB07 | GB3TS | Sunderland | FUSION |
| RB08 | GB3AN | Amlwch | FUSION |
| RB08 | GB3TF | Telford | FUSION |
| RB09 | GB3HD | Huddersfield | FUSION |
| RB10 | GB3DD | Dundee | FUSION |
| RB12 | GB3TJ | Corbridge | FUSION |
| RB14 | GB3WF | Otley | FUSION |
| RB15 | GB3LH | Shrewsbury | FUSION |
| RB15 | GB3WI | Wisbech | FUSION |
| DVU13 | GB7BS | Bristol City | DMR/3 |
| DVU13 | GB7FU | Pointon | DMR/3 |
| DVU13 | GB7NS | Caterham | DMR/3 |
| DVU13 | GB7TD | Wakefield | DMR/1 |
| DVU32 | GB7BM | Birmingham | D-STAR |
| DVU32 | GB7CW | Bridgend | DMR/3 |
| DVU32 | GB7EP | Epsom | DMR/3 |
| DVU32 | GB7LN | Lincoln | DMR/1 |
| DVU32 | GB7LP | Liverpool | DMR/1 |
| DVU33 | GB7DN | Dungiven | D-STAR |
| DVU33 | GB7GT | Ludlow | DMR/13 |
| DVU33 | GB7SD | Weymouth | DMR/1 |
| DVU33 | GB7VS | Shropham | DMR/1 |
| DVU33 | GB7WI | Winterton | DMR/1 |
| DVU34 | GB7DR | Poole | DMR/5 |
| DVU34 | GB7DS | Norwich | DMR/1 |
| DVU34 | GB7GB | Birmingham | DMR/8 |
| DVU34 | GB7HS | Batley | DMR/2 |
| DVU34 | GB7KB | Welham Green | D-STAR |
| DVU34 | GB7PN | Prestatyn | DMR/1 |
| DVU35 | GB7AV | Aylesbury | DMR/3 |
| DVU35 | GB7BX | Much Wenlock | DMR/5 |
| DVU35 | GB7ES | Eastbourne | D-STAR |
| DVU35 | GB7FO | Blackpool | DMR/1 |
| DVU35 | GB7GC | Grimsby | D-STAR |
| DVU35 | GB7PP | Ipswich | D-STAR |
| DVU36 | GB7BD | Bristol | D-STAR |
| DVU36 | GB7BP | Milton Keynes | D-STAR |
| DVU36 | GB7HD-B | Huddersfield | D-STAR |
| DVU36 | GB7HR | Heathrow | DMR/3 |
| DVU36 | GB7IC-B | Herne Bay | D-STAR |
| DVU36 | GB7NE | Ashington | D-STAR |
| DVU36 | GB7SU | Southampton | DMR/8 |
| DVU37 | GB7JB | Mere Wiltshire | DMR/1 |
| DVU37 | GB7NL | Exeter | D-STAR |
| DVU37 | GB7NM | Manchester | |
| DVU37 | GB7SK | Leicester | DMR/1 |
| DVU37 | GB7WL | Amersham | DMR/3 |
| DVU37 | GB7XX | Felling | DMR/10 |
| DVU38 | GB7CH | Louth | D-STAR |
| DVU38 | GB7LR | Leicester | DMR/1 |
| DVU38 | GB7SC | Bognor Regis | DMR/3 |
| DVU38 | GB7SE | Thurrock | DMR/3 |
| DVU38 | GB7TV | New Marske | D-STAR |
| DVU39 | GB7DC | Derby | D-STAR |
| DVU39 | GB7HU | South Cave | D-STAR |
| DVU39 | GB7NF | Newhaven | DMR/1 |
| DVU39 | GB7WB | Weston-Super-Mare | D-STAR |
| DVU39 | GB7WC | Warrington | D-STAR |
| DVU39 | GB7WF | Bewdley | D-STAR |
| DVU39 | GB7ZP | Chelmsford | D-STAR |
| DVU40 | GB7DP-B | Dundee | D-STAR |
| DVU40 | GB7FR | Worthing | DMR/4 |
| DVU40 | GB7KT | Andover | DMR/1 |
| DVU40 | GB7PE | Peterborough | DMR/3 |
| DVU40 | GB7UZ | Lancaster | DMR/1 |
| DVU41 | GB7CF | Folkestone | FUSION |
| DVU41 | GB7IT | Weston-Super-Mare | DMR/1 |
| DVU41 | GB7LO | Bromley | DMR/3 |
| DVU41 | GB7YD-B | Barnsley | D-STAR |
| DVU42 | GB7AK | Barking | DMR/3 |
| DVU42 | GB7DZ | Newcastle-Upon-Tyne | DMR/7 |
| DVU42 | GB7SI | Stoke-On-Trent | DMR/1 |
| DVU42 | GB7TC | Swindon | DMR/2 |
| DVU42 | GB7UL | Carrickfergus | DMR/1 |
| DVU43 | GB7DG | Bromsgrove | D-STAR |
| DVU43 | GB7EN | Wellingborough | D-STAR |
| DVU43 | GB7HE | Hastings | D-STAR |
| DVU43 | GB7IK | Rochester | DMR/3 |
| DVU43 | GB7IS | Weston-Super-Mare | FUSION |
| DVU43 | GB7LF | Lancaster | D-STAR |
| DVU43 | GB7MC | St Austell | D-STAR |
| DVU43 | GB7NU | Sheringham | D-STAR |
| DVU46 | GB7BH | Brighton | DMR/4 |
| DVU46 | GB7EB | Beccles | DMR/2 |
| DVU46 | GB7HX | Huddersfield | DMR/1 |
| DVU46 | GB7SO | Solihull | D-STAR |
| DVU48 | GB7DE-B | Upper Largo | D-STAR |
| DVU48 | GB7MK | Ipswich | DMR/13 |
| DVU48 | GB7RR | Nottingham | DMR/1 |
| DVU48 | GB7WT | Omagh N.I | D-STAR |
| DVU49 | GB7DV | St. Helens | D-STAR |
| DVU49 | GB7FH | Fareham | D-STAR |
| DVU49 | GB7IQ | Stafford | D-STAR |
| DVU49 | GB7IQ | Stafford | DMR/1 |
| DVU49 | GB7KH | Kelvedon Hatch | D-STAR |
| DVU50 | GB7DA-B | Airdrie | D-STAR |
| DVU50 | GB7EK | Whitstable | DMR/3 |
| DVU50 | GB7HB | Tandragee | DMR/1 |
| DVU50 | GB7OZ | Oswestry | DMR/8 |
| DVU51 | GB7CL | Clacton On Sea | DMR/3 |
| DVU51 | GB7HM | Caergwrie | DMR/1 |
| DVU51 | GB7JF | Northampton | D-STAR |
| DVU51 | GB7MH | East Grinstead | D-STAR |
| DVU51 | GB7MJ | Romsey | DMR/5 |
| DVU51 | GB7SR | Sheffield | DMR/2 |
| DVU53 | GB7BE | Beccles | D-STAR |
| DVU53 | GB7DB | Ampthill | D-STAR |
| DVU53 | GB7DD | Dundee | DMR/1 |
| DVU53 | GB7FW | Birmingham | DMR/1 |
| DVU53 | GB7KM | Cirencester | DMR/3 |
| DVU53 | GB7LE | Leeds | DMR/2 |
| DVU53 | GB7LY | Londonderry | DMR/1 |
| DVU54 | GB7AA | Bristol | DMR/1 |
| DVU54 | GB7DL | Friskney | D-STAR |
| DVU54 | GB7ED | Exeter | DMR/5 |
| DVU54 | GB7SN | Sheffield | DMR/1 |
| DVU55 | GB7GF | Guildford | DMR/3 |
| DVU55 | GB7NB | Norwich | D-STAR |
| DVU55 | GB7NY | Newry | DMR/1 |
| DVU55 | GB7TP | Keighley | DMR/1 |
| DVU55 | GB7ZI | Stafford | D-STAR |
| DVU56 | GB7AU | Amersham | D-STAR |
| DVU56 | GB7CD | Cardiff | D-STAR |
| DVU56 | GB7MB | Heysham | DMR/1 |
| DVU56 | GB7VO | Leominster | D-STAR |
| DVU57 | GB7EE | Edinburgh | DMR/1 |
| DVU57 | GB7EL | Nelson | DMR/2 |
| DVU57 | GB7EX | Southend-On-Sea | DMR/3 |
| DVU57 | GB7GL | Mitcheldean | D-STAR |
| DVU57 | GB7JH-B | Shoreham-By-Sea | D-STAR |
| DVU57 | GB7KA | Kilrea | DMR/3 |
| DVU57 | GB7PT | Royston | FUSION |
| DVU57 | GB7RE | Retford | DMR/1 |
| DVU59 | GB7BJ | Bromyard | DMR/13 |
| DVU59 | GB7BK | Reading | DMR/3 |
| DVU59 | GB7CK | Folkestone | DMR/3 |
| DVU59 | GB7MR | Manchester | DMR/2 |
| DVU59 | GB7SS | Hockley | D-STAR |
| DVU62 | GB7CS | East Kilbride | FUSION |
| DVU62 | GB7SX | Bognor Regis | FUSION |
| DVU73 | GB7DK-B | Stranraer | D-STAR |
| DVU73 | GB7WP | Birkenhead | D-STAR |
| | GB7YD-A | Barnsley | D-STAR |
| | GB7IC-A | Herne Bay | D-STAR |
| RM12 | GB7YD-A | Barnsley | D-STAR |

As new applications are being made on a regular basis please visit the RSGB website and you will find more information on Repeaters at: www.ukrepeater.net

# RSGB Band Plan

**EFFECTIVE FROM 1ST JUNE 2016 UNLESS OTHERWISE SHOWN**

The combination of the IARU Region 1 Conference at Varna, spectrum release and Ofcom licence changes are incorporated into the 2016 band plan on these pages. The result is quite a contrast between relatively few changes at HF, as opposed to Microwave bands.

### HF

The addition of a usage note in the 472kHz band and a wider bandwidth all-modes segment in 29.0-29.1MHz that were agreed at Varna. In addition the licence notes that referred to NoVs for 472kHz their incorporation into the new licence and

### VHF/UHF

The most noticeable feature is that 146-147MHz has been included. However IARU changes and the Ofcom-ETCC packet review also result in changes to the main 145MHz band. Several packet channels have been cleared whilst the bottom of the band is now shared with new narrowband amateur satellite downlinks.

Both 145MHz and other VHF bands see the deletion of old RTTY and FAX channels room for all-modes usage. 70cm also sees some change including a more consistent designation for the 12.5kHz operation of Internet gateways. A landmark change is the removal of the UK beacon segment, IARU beacon frequencies.

### MICROWAVE

The Ofcom spectrum release changes see the 2350-2390 and 3410-3475MHz ranges removed from the appropriate band plans and some of the remaining frequencies being reset to all modes. In future this may change further as new data and DATV developments become clear. The new 2300-2302MHz segment (if you have the NoV) is incorporated as a separate table. The 10GHz band also sees a clearout of old designations and updates for repeater and wideband usage. A new shaded warning zone in the bottom 1010.125GHz section indicates where the Primary User now has increased use, having been pressured out of other spectrum.

### GENERAL NOTES

These have also been updated, including the need to refer to for certain bands. New notes provide an usage. Another new note, agreed by IARU Region 1 at Varna emphasises that all VHF WSPR frequencies in the band plans are transmitted centre frequencies and not ambiguous dial settings.

### FINALLY...

As we have said before, band plans are living entities and do evolve over time. Please ensure you only refer or link to the current ones on the RSGB website and remove any older ones you have locally.

The band plan including the master section of the RSGB website – and if you are unsure, by all means contact the HF, VHF or Microwave Spectrum Manager via:

*hf.manager@rsgb.org.uk or*

*vhf.manager@rsgb.org.uk or*

*mw.manager@rsgb.org.uk*

| 136kHz | NECESSARY BANDWIDTH | UK USAGE |
|---|---|---|
| 135.7-137.8kHz | 200Hz | CW, QRSS and Narrowband Digital Modes |

**Licence Notes:** Amateur Service – Secondary User. 1 watt (0dBW) ERP.
**R.R. 5.67B.** The use of the band 135.7-137.8kHz in Algeria, Egypt, Iran (Islamic Republic of), Iraq, Lebanon, Syrian Arab Republic Sudan, South Sudan and Tunisia is limited to fixed and maritime mobile services. The amateur service shall not be used in the above-mentioned countries in the band 135.7-137.8kHz, and this should be taken into account by the countries authorising such use. (WRC-12).

| 472kHz (600m) | NECESSARY BANDWIDTH | UK USAGE |
|---|---|---|

IARU Region 1 does not have a formal band plan for this allocation but has a usage recommendation (Note 1).

| 472-479kHz | 500Hz | CW, QRSS and Narrowband Digital Modes |
|---|---|---|

**Note 1:** Usage recommendation – 472-475kHz CW only 200Hz maximum bandwidth, 475-479kHz CW and Digimodes.
**Note 2:** It should be emphasised that this band is available on a non-interference basis to existing services. UK amateurs should be aware that some overseas stations may be restricted in terms of transmit frequency in order to avoid interference to nearby radio navigation service Non-Directional Beacons.
**Licence Notes:** Amateur Service – Secondary User. Full Licensees only, **5 watts EIRP maximum**. Note that conditions regarding this band are specified by the Licence Schedule notes.
**R.R. 5.80B.** The use of the frequency band 472-479kHz in Algeria, Saudi Arabia, Azerbaijan, Bahrain, Belarus, China, Comoros, Djibouti, Egypt, United Arab Emirates, the Russian Federation, Iraq, Jordan, Kazakhstan, Kuwait, Lebanon, Libya, Mauritania, Oman, Uzbekistan, Qatar, Syrian Arab Republic, Kyrgyzstan, Somalia, Sudan, Tunisia and Yemen is limited to the maritime mobile and aeronautical radionavigation services. The amateur service shall not be used in the above-mentioned countries in this frequency band, and this should be taken into account by the countries authorising such use. (WRC 12).

| 1.8MHz (160m) | NECESSARY BANDWIDTH | UK USAGE |
|---|---|---|
| 1,810-1,838kHz | 200Hz | Telegraphy |
| 1,838-1,840 | 500Hz | Narrowband Modes |
| 1,840-1,843 | 2.7kHz | All Modes |
| 1,843-2,000 | 2.7kHz | Telephony (Note 1), Telegraphy<br>1,836kHz – QRP (low power) Centre of Activity<br>1,960kHz – DF Contest Beacons (14dBW) |

**Note 1:** Lowest LSB carrier frequency (dial setting) should be 1,843kHz. AX25 packet should not be used on the 1.8MHz band.
**Licence Notes:** 1,810-1,850kHz – Primary User: 1,810-1,830kHz on a non-interference basis to stations outside of the UK. 1,850-2,000kHz – Secondary User. 32W (15dBW) maximum.
**Notes to the Band Plan:** As on page 167.

| 3.5MHz (80m) | NECESSARY BANDWIDTH | UK USAGE |
|---|---|---|
| 3,500-3,510kHz | 200Hz | Telegraphy – Priority for Inter-Continental Operation |
| 3,510-3,560 | 200Hz | Telegraphy – Contest Preferred. 3,555kHz – QRS (slow telegraphy) Centre of Activity |
| 3,560-3,580 | 200Hz | Telegraphy 3,560kHz – QRP (low power) Centre of Activity |
| 3,570-3,580 | 200 Hz | Narrow band modes |
| 3,580-3,590 | 500Hz | Narrowband Modes |
| 3,590-3,600 | 500Hz | Narrowband Modes – Automatically Controlled Data Stations (unattended) |
| 3,600-3,620 | 2.7kHz | All Modes – Automatically Controlled Data Stations (unattended), (Note 1) |
| 3,600-3,650 | 2.7kHz | All Modes – Phone Contest Preferred, (Note 1).<br>3,630kHz – Digital Voice Centre of Activity |
| 3,650-3,700 | 2.7kHz | All Modes – Telephony, Telegraphy<br>3,663kHz May Be Used For UK Emergency Comms Traffic<br>3,690kHz SSB QRP (low power) Centre of Activity |
| 3,700-3,775 | 2.7kHz | All Modes – Phone Contest Preferred<br>3,735kHz – Image Mode Centre of Activity<br>3,760kHz – IARU Region 1 Emergency Centre of Activity |
| 3,775-3,800 | 2.7kHz | Priority for Inter-Continental Telephony (SSB) Operation |

**Note 1.** Lowest LSB carrier frequency (dial setting) should be 3,603kHz.
**Licence Notes:** Primary User: Shared with other user services.
**Notes to the Band Plan:** As on page 167.

| 5MHz (60m) | AVAILABLE WIDTH | UK USAGE |
|---|---|---|
| 5,258.5-5,264kHz | 5.5kHz | 5,262kHz CW QRP Centre of Activity |
| 5,276-5,284 | 8kHz | 5,278.5kHz – May be used for UK Emergency Comms Traffic |
| 5,288.5-5,292 | 3.5kHz | Beacons on 5290kHz (Note 2), WSPR |
| 5,298-5,307 | 9kHz | |
| 5,313-5,323 | 10kHz | 5,317kHz – AM 6kHz maximum bandwidth |
| 5,333-5,338 | 5kHz | |
| 5,354-5,358 | 4kHz | |
| 5,362-5,374.5 | 12.5kHz | 5,362-5,370kHz – Digital Mode Activity in the UK |
| 5,378-5,382 | 4kHz | |
| 5,395-5,401.5 | 6.5kHz | |
| 5,403.5-5,406.5 | 3kHz | 5,403.5kHz USB Common International Frequency |

Unless indicated, usage is All Modes (necessary bandwidth to be within channel limits).
**Note 1:** Upper Sideband is recommended for SSB activity.
**Note 2:** Activity should avoid interference to the experimental beacons on 5290kHz.
**Note 3:** Amplitude Modulation is permitted with a maximum bandwidth of 6kHz, on frequencies with at least 6kHz available width.
**Licence Notes:** Full Licensees only, **Secondary User, 100 watts maximum**. Note that conditions on transmission bandwidth, power and antennas are specified in the Licence.
**Notes to the Band Plan.** As on page 167.

| 7MHz (40m) | NECESSARY BANDWIDTH | UK USAGE |
|---|---|---|
| 7,000-7,040kHz | 200Hz | Telegraphy – 7,030kHz QRP (low power) Centre of Activity |
| 7,040-7,047 | 500Hz | Narrowband Modes (Note 2) |
| 7,047-7,050 | 500Hz | Narrowband Modes, Automatically Controlled Data Stations (unattended) |
| 7,050-7,053 | 2.7kHz | All Modes, Automatically Controlled Data Stations (unattended), (Note 1) |
| 7,053-7,060 | 2.7kHz | All Modes, Digimodes |
| 7,060-7,100 | 2.7kHz | All Modes, SSB Contest Preferred Segment Digital Voice 7,070kHz; SSB QRP Centre of Activity 7,090kHz |
| 7,100-7,130 | 2.7kHz | All Modes, 7,110kHz – Region 1 Emergency Centre of Activity |
| 7,130-7,200 | 2.7kHz | All Modes, SSB Contest Preferred Segment; 7,165kHz – Image Centre of Activity |
| 7,175-7,200 | 2.7kHz | All Modes, Priority For Inter-Continental Operation |

Lowest LSB carrier frequency (dial setting) should be 7,053kHz.
**Note 2:** PSK31 activity starts from 7,040kHz. Since 2009, the narrowband modes segment starts at 7,040kHz.
**Licence Notes:** 7,000-7,100kHz Amateur and Amateur Satellite Service – Primary User. 7,100-7,200kHz Amateur Service – Primary User.
**Notes to the Band Plan:** As on page 167.

| 10MHz (30m) | NECESSARY BANDWIDTH | UK USAGE |
|---|---|---|
| 10,100-10,140kHz | 200Hz | Telegraphy (CW)<br>10,116kHz – QRP (low power) Centre of Activity |
| 10.130-10.150 | 500Hz | Narrowband Modes<br>Automatically Controlled Data Stations (unattended) should avoid the use of the 10MHz band |

**Licence Notes:** Amateur Service – Secondary User.
**Notes to the Band Plan:** As on page 167.
The 10MHz band is allocated to the amateur service only on a secondary basis. The IARU has agreed that only CW and other narrow bandwidth modes are to be used on this band. Likewise the band is not to be used for contests and bulletins. SSB may be used on the 10MHz band during emergencies involving the immediate safety of life and property, and only by stations actually involved with the handling of emergency traffic. The band segment 10,120-10,140kHz may only be used for SSB transmissions in the area of Africa south of the equator during local daylight hours.

| 14MHz (20m) | NECESSARY BANDWIDTH | UK USAGE |
|---|---|---|
| 14,000-14,060kHz | 200Hz | Telegraphy – Contest Preferred<br>14,055kHz – QRS (slow telegraphy) Centre of Activity |
| 14,060-14,070 | 200Hz | Telegraphy<br>14,060kHz – QRP (low power) Centre of Activity |
| 14,070-14,089 | 500Hz | Narrowband Modes |
| 14,089-14,099 | 500Hz | Narrowband Modes – Automatically Controlled Data Stations (unattended) |
| 14,099-14,101 | | IBP – Reserved Exclusively for Beacons |
| 14,101-14,112 | 2.7kHz | All Modes – Automatically Controlled Data Stations (unattended) |
| 14,112-14,125 | 2.7kHz | All Modes (excluding digimodes) |
| 14,125-14,300 | 2.7kHz | All Modes – SSB Contest Preferred Segment<br>14,130kHz – Digital Voice Centre of Activity<br>14,195 ±5kHz – Priority for DXpeditions<br>14,230kHz – Image Centre of Activity<br>14,285kHz – QRP Centre of Activity |
| 14,300-14,350 | 2.7kHz | All Modes<br>14,300kHz – Global Emergency Centre of Activity |

**Licence Notes:** Amateur Service – Primary User. 14,000-14,250kHz Amateur Satellite Service – Primary User.
**Notes to the Band Plan:** As on page 167.

| 18MHz (17m) | NECESSARY BANDWIDTH | UK USAGE |
|---|---|---|
| 18,068-18,095kHz | 200Hz | Telegraphy – 18,086kHz QRP (low power) Centre of Activity |
| 18,095-18,105 | 500Hz | Narrowband Modes |
| 18,105-18,109 | 500Hz | Narrowband Modes – Automatically Controlled Data Stations (unattended) |
| 18,109-18,111 | | IBP – Reserved Exclusively for Beacons |
| 18,111-18,120 | 2.7kHz | All Modes – Automatically Controlled Data Stations (unattended) |
| 18,120-18,168 | 2.7kHz | All Modes, 18,130kHz – SSB QRP Centre of Activity<br>18,150kHz – Digital Voice Centre of Activity<br>18,160kHz – Global Emergency Centre of Activity |

**Licence Notes:** Amateur and Amateur Satellite Service – Primary User. The band is not to be used for contests or bulletins.
**Notes to the Band Plan:** As on page 167.

| 21MHz (15m) | NECESSARY BANDWIDTH | UK USAGE |
|---|---|---|
| 21,000-21,070kHz | 200Hz | Telegraphy<br>21,055kHz – QRS (slow telegraphy) Centre of Activity<br>21,060kHz – QRP (low power) Centre of Activity |
| 21,070-21,090 | 500Hz | Narrowband Modes |
| 21,090-21,110 | 500Hz | Narrowband Modes – Automatically Controlled Data Stations (unattended) |
| 21,110-21,120 | 2.7kHz | All Modes (excluding SSB) – Automatically Controlled Data Stations (unattended) |
| 21,120-21,149 | 500Hz | Narrowband Modes |
| 21,149-21,151 | | IBP – Reserved Exclusively For Beacons |
| 21,151-21,450 | 2.7kHz | All Modes<br>21,180kHz – Digital Voice Centre of Activity<br>21,285kHz – QRP Centre of Activity<br>21,340kHz – Image Centre of Activity<br>21,360kHz – Global Emergency Centre of Activity |

**Licence Notes:** Amateur and Amateur Satellite Service – Primary User.
**Notes to the Band Plan:** As on page 167.

| 24MHz (12m) | NECESSARY BANDWIDTH | UK USAGE |
|---|---|---|
| 24,890-24,915kHz | 200Hz | Telegraphy<br>24,906kHz – QRP (low power) Centre of Activity |
| 24,915-24,925 | 500Hz | Narrowband Modes |
| 24,925-24,929 | 500Hz | Narrowband Modes – Automatically Controlled Data Stations (unattended) |
| 24.929-24.931 | | IBP – Reserved Exclusively For Beacons |
| 24,931-24,940 | 2.7kHz | All Modes – Automatically Controlled Data Stations (unattended) |
| 24,940-24,990 | 2.7kHz | All Modes, 24,950kHz – SSB QRP Centre of Activity<br>24,960kHz – Digital Voice Centre of Activity |

**Licence Notes:** Amateur and Amateur Satellite Service – Primary User. The band is not to be used for contests or bulletins.
**Notes to the Band Plan:** As on page 167.

| 28MHz (10m) | NECESSARY BANDWIDTH | UK USAGE |
|---|---|---|
| 28,000-28,070kHz | 200Hz | Telegraphy<br>28,055kHz – QRS (slow telegraphy) Centre of Activity<br>28,060kHz – QRP (low power) Centre of Activity |
| 28,070-28,120 | 500Hz | Narrowband Modes |
| 28,120-28,150 | 500Hz | Narrowband Modes – Automatically Controlled Data Stations (unattended) |
| 28,150-28,190 | 500Hz | Narrowband Modes |
| 28,190-28,199 | | IBP – Regional Time Shared Beacons |
| 28,199-28,201 | | IBP – World Wide Time Shared Beacons |
| 28,201-28,225 | | IBP – Continuous-Duty Beacons |
| 28,225-28,300 | 2.7kHz | All Modes – Beacons |
| 28,300-28,320 | 2.7kHz | All Modes – Automatically Controlled Data Stations (unattended) |
| 28,320-29,000 | 2.7kHz | All Modes –<br>28,330kHz – Digital Voice Centre of Activity<br>28,360kHz – QRP Centre of Activity<br>28,680kHz – Image Centre of Activity |
| 29,000-29,100 | 6kHz | All Modes |
| 29,100-29,200 | 6kHz | All Modes – FM Simplex – 10kHz Channels |
| 29,200-29,300 | 6kHz | All Modes – Automatically Controlled Data Stations (unattended)<br>29,270kHz Internet Gateways Channel<br>29,280kHz UK Internet Voice Gateway (unattended)<br>29,290kHz – UK Internet Voice Gateway (unattended) |
| 29,300-29,510 | 6kHz | Satellite Links |
| 29,510-29,520 | Guard Channel | |
| 29,520-29,590 | 6kHz | All Modes – FM Repeater Inputs (RH1-RH8) |
| 29,600 | 6kHz | All Modes – FM Calling Channel |
| 29,610 | 6kHz | All Modes – FM Simplex Repeater (parrot) – input and output |
| 29,620-29,700 | 6kHz | All Modes – FM Repeater Outputs (RH1-RH8) |

**Licence Notes:** Amateur and Amateur Satellite Service – Primary User: 26dBW permitted. Beacons may be established for DF competitions except within 50km of NGR SK985640 (Waddington).
**Notes to the Band Plan:** As on page 167.

| 50MHz (6m) | NECESSARY BANDWIDTH | UK USAGE |
|---|---|---|
| 50.000-50.100MHz | 500Hz | Telegraphy Only (except for Beacon Project) (Note 2) |
| | | 50.000-50.030MHz reserved for future Synchronised Beacon Project (Note 2) |
| | | Region 1: 50.000-50.010; Region 2: 50.010-50.020; Region 3: 50.020-50.030 |
| | | 50.050MHz – Future International Centre of Activity |
| | | 50.090MHz – Inter-Continental DX Centre of Activity (Note 1) |
| 50.100-50.200 | 2.7kHz | SSB/Telegraphy – International Preferred |
| | | 50.100-50.130MHz – Inter-Continental DX Telegraphy & SSB (Note 1) |
| | | 50.110MHz – Inter-Continental DX Centre of Activity |
| | | 50.130-50.200MHz – General International Telegraphy & SSB |
| | | 50.150MHz – International Centre of Activity |
| 50.200-50.300 | 2.7kHz | SSB/Telegraphy – General Usage |
| | | 50.285MHz – Crossband Centre of Activity |
| 50.300-50.400 | 2.7kHz | MGM/Narrowband/Telegraphy |
| | | 50.305MHz – PSK Centre of Activity |
| | | 50.310-50.320MHz – EME |
| | | 50.320-50.380MHz – MS |
| 50.400-50.500 | | Propagation Beacons only |
| 50.500-52.000 | 12.5kHz | All Modes |
| | | 50.510MHz – SSTV (AFSK) |
| | | 50.520MHz – Internet Voice Gateway (10kHz channels), (IARU common channel) |
| | | 50.530MHz – Internet Voice Gateway (10kHz channels), (IARU common channel) |
| | | 50.540MHz – Internet Voice Gateway (10kHz channels), (IARU common channel) |
| | | 50.550MHz – Image/Fax working frequency |
| | | 50.600MHz – RTTY (FSK) |
| | | 50.620-50.750MHz – Digital communications |
| | | 50.630MHz – Digital Voice (DV) calling |
| | | 50.710-50.890MHz – FM/DV Repeater Outputs (10kHz channel spacing) |
| | | 51.210-51.390MHz – FM/DV Repeater Inputs (10kHz channel spacing) (Note 4) |
| | | 51.410-51.590MHz – FM/DV Simplex (Note 3) (Note 4) |
| | | 51.510MHz – FM Calling Frequency |
| | | 51.530MHz – GB2RS News Broadcast and Slow Morse |
| | | 51.650 & 51.750MHz – See Note 5 (25kHz aligned) |
| | | 51.770 & 51.790MHz – See Note 5 |
| | | 51.810-51.990MHz – FM/DV Repeater Outputs (IARU aligned channels) |

**Note 1:** Only to be used between stations in different continents (not for intra-European QSOs).
**Note 2:** 50.0-50.1MHz is currently shared with Propagation Beacons. These were due to be migrated by Aug 2014 to 50.4-50.5MHz, to create more space for Telegraphy and a new Synchronised Beacon Project.
**Note 3:** 20kHz channel spacing. Channel centre frequencies start at 51.430MHz.
**Note 4:** Embedded data traffic is allowed with digital voice (DV).
**Note 5:** May be used for Emergency Communications and Community Events.
**Licence Notes:** Amateur Service 50.0-51.0MHz – Primary User. Amateur Service 51.0-52.0MHz – Secondary User. 100W (20dBW) maximum. Available on the basis on non-interference to other services (inside or outside the UK).
**Notes to the Band Plan:** As on page 167.

| 70MHz (4m) | NECESSARY BANDWIDTH | UK USAGE (NOTE 1) |
|---|---|---|
| 70.000-70.090MHz | 1kHz | Propagation Beacons Only |
| 70.090-70.100 | 1kHz | Personal Beacons |
| 70.100-70.250 | 2.7kHz | Narrowband Modes |
| | | 70.185MHz – Cross-band Activity Centre |
| | | 70.200MHz – CW/SSB Calling |
| | | 70.250MHz – MS Calling |
| 70.250-70.294 | 12kHz | All Modes |
| | | 70.260MHz – AM/FM Calling |
| | | 70.270MHz MGM Centre of Activity |
| 70.294-70.500 | 12kHz | All Modes Channelised Operations Using 12.5kHz Spacing |
| | | 70.3000MHz |
| | | 70.3125MHz – Digital Modes |
| | | 70.3250MHz – DX Cluster |
| | | 70.3375MHz – Digital Modes |
| | | 70.3500MHz – Internet Voice Gateway (Note 2) |
| | | 70.3625MHz – Internet Voice Gateway |
| | | 70.3750MHz – See Note 2 |
| | | 70.3875MHz – Internet Voice Gateway |
| | | 70.4000MHz – See Note 2 |
| | | 70.4125MHz – Internet Voice Gateway |
| | | 70.4250MHz – FM Simplex – used by GB2RS news broadcast |
| | | 70.4375MHz – Digital Modes (special projects) |
| | | 70.4500MHz – FM Calling |
| | | 70.4625MHz – Digital Modes |
| | | 70.4750MHz |
| | | 70.4875MHz – Digital Modes |

**Note 1:** Usage by operators in other countries may be influenced by restrictions in their national allocations.
**Note 2:** May be used for Emergency Communications and Community Events.
**Licence Notes:** Amateur Service 70.0-70.5MHz – Secondary User: 160W (22dBW) maximum. Available on the basis of non-interference to other services (inside or outside the UK).
**Notes to the Band Plan:** As on page 167.

| 144MHz (2m) | NECESSARY BANDWIDTH | UK USAGE |
|---|---|---|
| 144.000-144.025MHz | 2700Hz | All Modes – including Satellite Downlinks |
| 144.025-144.110 | 500Hz | Telegraphy (including EME CW) |
| | | 144.050MHz – Telegraphy Centre of Activity |
| | | 144.100MHz – Random MS Telegraphy Calling, (Note 1) |
| 144.110-144.150 | 500Hz | Telegraphy and MGM |
| | | 144.138MHz – PSK31 Centre of Activity |
| | | EME MGM Activity (Note 7) |
| 144.150-144.180 | 2700Hz | Telegraphy, MGM and SSB |
| 144.180-144.360 | 2700Hz | Telegraphy and SSB |
| | | 144.175MHz – Microwave Talk-back |
| | | 144.195-144.205MHz – Random MS SSB |
| | | 144.200MHz – Random MS SSB Calling Frequency |
| | | 144.250MHz – GB2RS News Broadcast and Slow Morse |
| | | 144.260MHz – USB. (Note 10) |
| | | 144.300MHz – SSB Centre of Activity |
| 144.360-144.399 | 2700Hz | Telegraphy, MGM, SSB |
| | | 144.370MHz – MGM Calling Frequency |
| 144.400-144.490 | | Propagation Beacons only |
| 144.490-144.500 | | Beacon Guard Band |
| 144.500-144.794 | 20kHz | All Modes (Note 8) |
| | | 144.500MHz – Image Modes Centre (SSTV, FAX, etc) |
| | | 144.600MHz – Data Centre of Activity (MGM, RTTY, etc) |
| | | 144.6125MHz – UK Digital Voice (DV) Calling (Note 9) |
| | | 144.625-144.675MHz – See Note 10 |
| | | 144.750MHz – ATV Talk-back |
| | | 144.775-144.794MHz – See Note 10 |
| 144.794-144.990 | 12kHz | MGM Digital Communications (Note 15) |
| | | 144.800-144.9875MHz – MGM/Digital Communications |
| | | 144.8000MHz – Unconnected Nets – APRS, UiView etc (Note 14) |
| | | 144.8125MHz – DV Internet Voice Gateway |
| | | 144.8250MHz – DV Internet Voice Gateway |
| | | 144.8375MHz – DV Internet Voice Gateway |
| | | 144.8500MHz – DV Internet Voice Gateway |
| | | 144.8625MHz – DV Internet Voice Gateway |
| | | 144.9250MHz – TCP/IP Usage |
| | | 144.9375MHz – AX25 Usage |
| | | 144.9500MHz – AX25 Usage |
| | | 144.9625MHz – FM Internet Voice Gateway |
| | | 144.9750MHz, 144.9875MHz To Be Decided (Note 11) |
| 144.990-145.1935 | 12kHz | FM/DV RV48-RV63 Repeater Input Exclusive (Note 2 & 5) |
| 145.200 | 12kHz | FM/DV Space Communications (eg ISS) – Earth-to-Space |
| | | 145.2000MHz – (Note 4 & 10) |
| 145.200-145.5935 | 12kHz | FM/DV V16-V48 – FM/DV Simplex (Note 3, 5 & 6) |
| | | 145.2250MHz – See Note 10 |
| | | 145.2375MHz – FM Internet Voice Gateway (IARU common channel) |
| | | 145.2500MHz – Used for Slow Morse Transmissions |
| | | 145.2875MHz – FM Internet Voice Gateway (IARU common channel) |
| | | 145.3375MHz – FM Internet Voice Gateway (IARU common channel) |
| | | 145.5000MHz – FM Calling (Note 12) |
| | | 145.5250MHz – Used for GB2RS News Broadcast. |
| | | 145.5500MHz – Used for Rally/exhibition Talk-in |
| | | 145.5750MHz, 145.5875MHz (Note 11) |
| 145.5935-145.7935 | 12kHz | FM/DV RV48-RV63 – Repeater Output (Note 2) |
| 145.800 | 12kHz | FM/DV Space Communications (eg ISS) – Space-Earth |
| 145.806-146.000 | 12kHz | All Modes – Satellite Exclusive |

**Note 1:** Meteor scatter operation can take place up to 26kHz higher than the reference frequency.
**Note 2:** 12.5kHz channels numbered RV48-RV63. RV48 input = 145.000MHz, output = 145.600MHz.
**Note 3:** 12.5kHz simplex channels numbered V16-V46. V16 = 145.200MHz.
**Note 4:** Emergency Communications Groups utilising this frequency should take steps to avoid interference to ISS operations in non-emergency situations.
**Note 5:** Embedded data traffic is allowed with digital voice (DV).
**Note 6:** Simplex use only – no DV gateways.
**Note 7:** EME activity using MGM is commonly practiced between 144.110-144.160MHz.
**Note 8:** Amplitude Modulation (AM) is acceptable within the All Modes segment. AM usage is typically found on 144.550MHz. Users should consider adjacent channel activity when selecting operating frequencies.
**Note 9:** In other countries IARU Region 1 recommends 145.375MHz.
**Note 10:** May be used for Emergency Communications and Community Events.
**Note 11:** May be used for repeaters in other IARU Region 1 countries.
**Note 12:** DV users are asked not to use this channel, and use 144.6125MHz for calling.
**Note 13:** Not used.
**Note 14:** 144.800 use should be NBFM to avoid interference to 144.8125 DV Gateways.
**Licence Notes:** Amateur Service and Amateur Satellite Service – Primary User. Beacons may be established for DF competitions except within 50km of TA 012869 (Scarborough).
**Notes to the Band Plan**: As on page 167.

| 146MHz | NECESSARY BANDWIDTH | UK USAGE |
|---|---|---|
| 146.000-146.900MHz | 500kHz | Wideband Digital Modes (High speed data, DATV etc) |
| | | 146.500MHz Centre frequency for wideband modes (Note 1) |
| 146.900-147.000MHz | 12kHz | Narrowband Digital Modes including Digital Voice |
| | | 146.900 |
| | | 146.9125 |
| | | 146.925 |
| | | 146.9375 Not available in/near Scotland (see Licence Notes & NoV terms) |
| | | 146.9500 |
| | | 146.9625 |
| | | 146.9750 |
| | | 146.9875 |

**Note 1:** Users of wideband modes must ensure their spectral emissions are contained with the band limits.
**Licence Notes:** Full Licensees only, with NoV, **25W ERP max** – not available in the Isle of Man or Channel Isles. Note that additional restrictions on geographic location, antenna height and upper frequency limit are specified by the NoV terms.
It should be emphasised that this band is UK-specific and is available on a non-interference basis to existing services. Upper Band limit 147.000MHz (or 146.93750 where applicable) are absolute limits and not centre frequencies. The absolute band frequency limit in or within 40km of Scotland is 146.93750MHz – see NoV schedule
**Notes to the Band Plan:** As on page 167.

| 430MHz (70cm) IARU Recommendation | NECESSARY BANDWIDTH | UK USAGE |
|---|---|---|
| 430.0000-431.9810MHz | 20kHz | 430.0125-430.0750MHz – FM Internet Voice Gateways (Notes 7, 8) |
| All Modes | | |
| | | 430.4000-430.7750 – UK DV 9MHz Split Repeaters – inputs |
| Digital Links | | |
| 430.6000-430.9250 | | 430.8000MHz – 7.6MHz Talk-through (Note 10) |
| Digital Repeaters | | 430.8250-430.9750MHz – RU66-RU78 7.6MHz Split Repeaters – outputs |
| | | See Licence Exclusion Note; 431-432MHz |
| | | 430.9900-431.9000MHz – Digital Communications |
| | | 431.0750-431.1750MHz – DV Internet Voice Gateways (Note 8) |
| 432.0000-432.1000 | 500Hz | 432.0000-432.0250MHz – Moonbounce (EME) |
| Telegraphy | | 432.0500MHz – Telegraphy Centre of Activity |
| MGM | | 432.0880MHz – PSK31 Centre of Activity |
| 432.1000-432.4000 | 2700Hz | 432.2000MHz – SSB Centre of Activity |
| SSB, Telegraphy | | 432.3500MHz – Microwave Talk-back (Europe) |
| MGM | | 432.3700MHz – FSK441 Calling Frequency |
| 432.4000-432.5000 | 500Hz | Propagation Beacons only |
| Beacons Exclusive | | |
| 432.5000-432.9940 | 25kHz | 432.5000MHz – Narrowband SSTV Activity |
| Centre | | |
| All Modes | (Note 11) | 432.6250-432.6750MHz Digital Communications (25kHz channels) |
| Non-channelised | | 432.7750MHz 1.6MHz Talk-through – Base TX (Note 10) |
| 432.9940-433.3810 | 25kHz | 433.0000-433.3750MHz (RB0-RB15) – RU240-RU270 |
| FM repeater outputs in UK only (Note 1) | (Note 11) | FM/DV Repeater Outputs (25kHz channels) in UK Only |
| 433.3940-433.5810 | 25kHz | 433.4000MHz U272 – IARU Region 1 SSTV (FM/AFSK) |
| | (Note 11) | 433.4250MHz U274 |
| FM/DV (Notes 12, 13) | | 433.450MHz U276 (Note 5) |
| Simplex Channels | | 433.4750MHz U278 |
| | | 433.5000MHz U280 – FM Calling Channel |
| | | 433.5250MHz U282 |
| | | 433.5500MHz U284 – Used for Rally/Exhibition Talk-in |
| | | 433.5750MHz U286 |
| 433.6000-434.0000 | 25kHz | |
| All Modes | (Note 11) | 433.6250-6750MHz – Digital Communications (25kHz channels) |
| 433.800MHz for APRS where 144.800MHz cannot be used | | 433.700MHz (Note 10) |
| | | 433.7250-433.7750MHz (Note 10) |
| | | 433.8000-434.2500MHz – Digital Communications |
| 434.000-434.5940 | 25kHz | 433.9500-434.0500MHz – Internet Voice Gateways (Note 8) |
| | (Note 11) | 434.3750MHz 1.6MHz Talk-through – Mobile TX (Note 10) |
| | | 434.4750-434.5250MHz – Internet Voice Gateways (Note 8) |
| 434.5940-434.9810 | 25kHz | 434.6000-434.9750MHz (RB0-RB15) RU240-RU270 |
| FM repeater inputs in UK only & ATV (Note 4) | (Note 11) | FM/DV Repeater Inputs (25kHz channels) in UK Only (Note 12) |
| 435.0000-438.0000 | 20kHz | Satellites and Fast Scan TV (Note 4) |
| | | 437.0000 – Experimental DATV Centre of Activity (Note 14) |
| 438.0000-440.0000 | 25kHz | 438.0250-438.1750MHz – IARU Region 1 Digital Communications |
| All Modes | (Note 11) | 438.2000-439.4250MHz (Note 1) |
| | | 438.4000MHz – 7.6MHz Talk-through (Note 10) |
| | | 438.4250-438.5750MHz RU66-RU78 – 7.6MHz Split Repeaters – inputs |
| | | Contd |
| | | 438.6125MHz – UK DV calling (Note 12) (Note 13) |
| | | 439.6000-440.0000MHz – Digital Communications |
| | | 439.400-439.775MHz – UK DV 9MHz split repeaters - Outputs |

**Note 1.** In Switzerland, Germany and Austria, repeater Inputs are 431.050-431.825MHz with 25kHz spacing and outputs 438.650-439.425MHz. In Belgium, France and the Netherlands repeater outputs are 430.025-430.375MHz with 12.5kHz spacing and inputs at 431.625-431.975MHz. In other European countries repeater inputs are 433.000-433.375MHz with 25kHz spacing and outputs at 434.600-434.975MHz, in the middle of the UK allocation.
**Note 4:** ATV carrier frequencies shall be chosen to avoid interference to other users, in particular the satellite service and repeater inputs.
**Note 5:** In other countries IARU Region 1 recommends 433.450MHz for DV calling.
**Note 7:** Users must accept interference from repeater output channels in France and the Netherlands at 430.025-430.575MHz. Users with sites that allow propagation to other countries (notably France and the Netherlands) must survey the proposed frequency before use to ensure that they will not cause interference to users in those countries.
**Note 8:** All internet voice gateways: 12.5kHz channels, maximum deviation ±2.4kHz, maximum effective radiated power 5W (7dBW), attended only operation in the presence of the NoV holder.
**Note 10:** May be used for Emergency Communications and Community Events.
**Note 11:** IARU Region 1 recommended maximum bandwidths are 12.5 or 20kHz.
**Note 12:** Embedded data traffic is allowed with digital voice (DV).
**Note 13:** Simplex use only - no DV gateways.
**Note 14:** QPSK 2 Mega-symbols/second maximum recommended.
**Licence Notes:** Amateur Service – Secondary User. Amateur Satellite Service: 435-438MHz – Secondary User. Exclusion: 431-432MHz not available within 100km radius of Charing Cross, London. Power Restriction 430-432MHz is 40 watts effective radiated power maximum.
**Notes to the Band Plan:** As on page 167.

| 1.3GHz (23cm) | NECESSARY BANDWIDTH | UK USAGE |
|---|---|---|
| 1240.000-1240.500MHz | 2700Hz | Alternative Narrowband Segment – see Note 7 – 1240.000-1240.750MHz |
| 1240.500-1240.750 | | Alternative Propagation Beacon Segment |
| 1240.750-1241.000 | 20kHz | FM/DV Repeater Inputs |
| 1241.000-1241.750 | 150kHz | DD High Speed Digital Data – 5 x 150kHz channels |
| All Modes | | 1241.075, 1241.225, 1241.375, 1241.525, 1241.675MHz (±75kHz) |
| 1241.750-1242.000 | 20kHz | 25kHz Channels available for FM/DV use |
| All Modes | | 1241.775-1241.975MHz |
| 1242.000-1249.000 | | TV Repeaters (Note 9) |
| ATV | | New DATV Repeater Inputs |
| | | Original ATV Repeater Inputs: 1248, 1249 |
| 1249.000-1249.250 | 20kHz | FM/DV Repeater Outputs, 25kHz Channels (Note 9) |
| | | 1249.025-1249.225MHz |
| 1250.00 | | In order to prevent interference to Primary Users, caution must be exercised prior to using 1250-1290MHz in the UK |
| 1260.000-1270.000 | | Amateur Satellite Service – Earth to Space Uplinks Only |
| Satellites | | |
| 1290.000 | | |
| 1290.994-1291.481 | 20kHz | FM/DV Repeater Inputs (Note 5) |
| | | 1291.000-1291.375MHz (RM0-RM15) 25kHz spacing |
| 1291.494-1296.000 | All Modes | |
| All Modes | | Preferred Narrowband segment |
| 1296.000-1296.150 | 500Hz | 1296.000-1296.025MHz – Moonbounce |
| Telegraphy, MGM | | 1296.138MHz – PSK31 Centre of Activity |
| 1296.150-1296.800 | 2700Hz | 1296.200MHz – Narrowband Centre of Activity |
| Telegraphy, SSB & MGM | | 1296.400-1296.600MHz – Linear Transponder Input |
| (Note 1) | | 1296.500MHz – Image Mode Centre of Activity (SSTV, FAX etc) |
| | | 1296.600MHz – Narrowband Data Centre of Activity (MGM, RTTY etc) |
| | | 1296.600-1296.700MHz – Linear Transponder Output |
| | | 1296.750-1296.800MHz – Local Beacons, 10W ERP max |
| 1296.800-1296.994 | | 1296.800-1296.990MHz – Propagation Beacons only |
| | | Beacons exclusive |
| 1296.994-1297.481 | 20kHz | FM/DV Repeater Outputs (Note 5) |
| | | 1297.000-1297.375MHz (RM0-RM15) |
| 1297.494-1297.981 | 20kHz | FM/DV Simplex ((Notes 2, 5 & 6)) 25kHz spacing |
| | | 1297.500-1297.750MHz (SM20-SM30) |
| FM/DV simplex 2, 5, 6) | | 1297.725MHz – Digital Voice (DV) Calling (Notes (IARU recommended) |
| | | 1297.900-1297.975MHz – FM Internet Voice Gateways (IARU common channels, 25kHz) |
| 1298.000-1299.000 | 20kHz | All Modes |
| All Modes | | General mixed analogue or digital use in channels |
| | | 1298.025-1298.975MHz (RS1-RS39) |
| 1299.000-1299.750 | 150kHz | DD High Speed Digital Data – 5 x 150kHz channels |
| All Modes | | 1299.075, 1299.225, 1299.375, 1299.525, 1299.675MHz (±75kHz) |

| 1.3GHz (23cm) IARU Recommendation | NECESSARY BANDWIDTH | UK USAGE Contd |
|---|---|---|
| 1299.750-1300.000 All Modes | 20kHz | 25kHz Channels Available for FM/DV use 1299.775-1299.975MHz |
| 1300.000-1325.000 ATV | | TV Repeaters (UK only) (Note 9) New DATV Repeater Outputs Original ATV Repeater Outputs: 1308.0, 1310.0, 1311.5, 1312.0, 1316.0, 1318.5MHz |

**Note 1:** Local traffic using narrowband modes should operate between 1296.500-1296.800MHz during contests and band openings.
**Note 2:** Stations in countries that do not have access to 1298-1300MHz may also use the FM simplex segment for digital communications.
**Note 3:** IARU Region 1 recommended maximum bandwidth is 20kHz. See also Note 7.
**Note 4:** deleted.
**Note 5:** Embedded data traffic is allowed with digital voice (DV).
**Note 6:** Simplex use only – no DV gateways.
**Note 7:** 1240.000-1240.750 has been designated by IARU as an alternative centre for narrowband activity and beacons. Operations in this range should be on a flexible basis to enable coordinated activation of this alternate usage.
**Note 8:** The band 1240-1300MHz is subject to major replanning. Contact the Microwave Manager for further information.
**Note 9:** Repeaters and Migration to DATV, inc option for new DATV simplex are subject to further development and coordination.
**Note 10:** QPSK 4 Mega-symbols/second maximum recommended.
**Licence Notes:** Amateur Service – Secondary User. Amateur Satellite Service: 1,260-1,270MHz – Secondary User Earth to Space only. In the sub-band 1,298-1,300MHz unattended operation is not allowed within 50km of SS206127 (Bude), SE202577 (Harrogate), or in Northern Ireland.
**Notes to the Band Plan:** As on page 167.

| 2.3-2.302GHz IARU Recommendation | NECESSARY BANDWIDTH | UK USAGE |
|---|---|---|

Access to this band requires an appropriate NoV, which is available to Full licensees only. Please note that the current NoVs last for up to three years prior to expiry.

| | | |
|---|---|---|
| 2300.000-2300.400MHz | 2.7kHz | Narrowband Modes (including CW, SSB, MGM) 2300.350-2300.400MHz Attended Beacons |
| 2300.400-2301.800MHz | 500kHz | Wideband Modes (NBFM, DV, Data, DATV, etc) Note 1 |
| 2301.800-2302.000MHz | 2.7kHz | Narrowband modes (including CW, SSB, MGM) EME Usage |

**Note 1:** Users of wideband modes must ensure their spectral emissions are contained within the band limits.
**Note 2:** Full licensees only with NoV, 400 watts maximum, not available in the Isle of Man or Channel isles. Note additional restrictions on usage are specified by the NoV terms. It should be emphasised that this is UK-specific and is available on a non interference basis to exisiting services.
**Notes to the Band Plan:** As on page 167.

| 2.3GHz (13cm) IARU Recommendation | NECESSARY BANDWIDTH | UK USAGE |
|---|---|---|
| 2,310.000-2,320.000MHz (National band plans) | 200kHz | 2,310.000-2,310.500MHz – Repeater links 2,311.000-2,315.000MHz – High speed data |
| | | Preferred Narrowband Segment |
| 2,320.000-2,320.150 | 500Hz | 2,320.000-2,320.025MHz – Moonbounce |
| 2,320.150-2,320.800 | 2.7kHz | 2,320.200MHz – SSB Centre of Activity |
| | | 2,320.750-2,320.800MHz – Local Beacons, 10W ERP max |
| 2,320.800-2,321.000 Beacons exclusive | | 2,320.800-2,320.990MHz – Propagation Beacons Only |
| 2321.000-2322.000 | 20kHz | FM/DV. See also Note 1 |
| 2,322.000-2,350.000 | | Wideband Modes including Data, ATV |
| 2,390.000-2,400.000 | | All Modes |
| 2,400.000-2,450.000MHz Satellites | | 2,435.000MHz ATV Repeater Outputs 2,440.000MHz ATV Repeater Outputs |

**Note 1:** Stations in countries which do not have access to the All Modes section 2,322-2,390MHz, use the simplex and repeater segment 2,320-2,322MHz for data transmission.
**Note 2:** Stations in countries that do not have access to the narrowband segment 2,320-2,322MHz, use the alternative narrowband segment 2,304-2,306MHz and 2,308-2,310MHz.
**Note 3:** The segment 2,433-2,443MHz may be used for ATV if no satellite is using the segment.
**Licence Notes:** Amateur Service – Secondary User. Users must accept interference from ISM users. Amateur Satellite Service: 2,400-2,450MHz – Secondary User. Users must accept interference from ISM users. Operation in 2310-2350 and 2390-2400 MHz are subject to specific conditions and guidance. In the sub-bands 2,310.000-2,310.4125 and 2,392-2,450MHz unattended operation is not allowed within 50km of SS206127 (Bude) or SE202577 (Harrogate). ISM = Industrial, scientific and medical.
**Notes to the Band Plan:** As on page 167.

| 3.4GHz (9cm) IARU Recommendation | NECESSARY BANDWIDTH | UK USAGE |
|---|---|---|
| 3,400.000-3,401.000MHz | 2.7kHz | Narrowband Modes (including CW, SSB, MGM, EME) 3,400.100MHz – Centre of Activity (Note 1) |
| | | 3,400.750-3,400.800MHz – Local Beacons, 10W ERP max |
| 3,400.800-3,400.995 Propagation Beacons | | 3,400.800-3,400.995MHz – Propagation Beacons Only |
| 3,400.000-3,401.000MHz | 200kHz | 3,401.000-3,402.000MHz Data, Remote Control |
| 3,402.000-3,410.000 Outputs All Modes (Notes 2, 3) | | Wideband Modes including DATV Repeater |

**Note 1:** EME has migrated from 3456MHz to 3400MHz to promote harmonised usage and activity.
**Note 2:** Stations in many European countries have access to 3400-3410MHz as permitted by ECA Table Footnote EU17.
**Note 3:** Amateur Satellite downlinks planned.
**Licence Notes:** Amateur Service – Secondary User. Subject to specific conditions and guidance.
**Notes to the Band Plan:** As on page 167.

| 5.7GHz (6cm) IARU Recommendation | UK USAGE |
|---|---|
| 5,650.000-5,668.000MHz Satellite Uplinks | Amateur Satellite Service – Earth to Space Only |
| 5,650.000-5,670.000 Narrowband CW/EME/SSB | 5,668.200MHz – Alternative Centre of Activity 5,668.8MHz – Beacons |
| 5,670.000-5,680.000 All Modes | |
| 5.755.000-5,760.000 All Modes | |
| 5,760.000-5,762.000 Narrowband CW/EME/SSB | 5,760.100MHz – Current Centre of Activity 5,760.750-5,760.800MHz – Local Beacons, 10W ERP max |
| 5760.800-5760.995 Propagation Beacons | 5,760.800-5,760.995MHz – Propagation Beacons only |
| 5,762.000-5,765.000 All Modes | |
| 5,820.000-5,830.000 All Modes | |
| 5,830.000-5,850.000 Satellite Downlinks | Amateur Satellite Service – Space to Earth Only |

**Licence Notes:** Amateur Service: 5,650-5,680MHz – Secondary User. 5,755-5,765 and 5,820-5,850MHz – Secondary User. Users must accept interference from ISM users. Amateur Satellite Service: 5,650-5,670MHz and 5,830-5,850MHz – Secondary User. Users must accept interference from ISM users. Unattended operation is permitted for remote control, digital modes and beacons, except in the sub-bands 5,670-5,680MHz within 50km of SS206127 (Bude) and SE202577 (Harrogate). ISM = Industrial, scientific and medical.
**Notes to the Band Plan:** As on page 167.

| 10GHz (3cm) IARU Recommendation | NECESSARY BANDWIDTH | UK USAGE |
|---|---|---|
| 10,000.000-10,125.000MHz All Modes | | Note 4 10,065MHz ATV Repeater Outputs |
| 10,225.000-10,250.000 All Modes | | 10,240MHz ATV Repeaters |
| 10,250.000-10,350.000 Digital Modes | | |
| 10,350.000-10,368.000 All Modes | | 10,352.5-10,368MHz Wideband Modes (Note 2) |
| 10,368-10,370MHz Narrowband Telegraphy EME/SSB | 2.7kHz | 10,368-10,370 Narrowband Modes (Note 3) 10,368.1MHz Centre of Activity |
| | | 10,368.750-10,368.800MHz – Local Beacons, 10W ERP max |
| 10,368.800-10,368.995 Propagation Beacons | | 10,368.800-10,368.995MHz – Propagation Beacons Only |
| 10,370.000-10,450.000 All Modes | | 10,371MHz Voice Repeaters Rx 10,425 ATV Repeaters |
| 10,450.000-10,475.000 All Modes & Satellites | | 10,400-10,475MHz Unattended Operation 10,450-10,452MHz Alternative Narrowband Segment (Note 3) 10,471MHz Voice Repeaters Tx |
| 10,475.000-10,500.000 All Modes and satellites | | Amateur Satellite Service ONLY |

**Note 1:** Deleted.
**Note 2:** Wideband FM is preferred between 10,350-10,400MHz to encourage compatibility between narrowband systems.
**Note 3:** 10,450MHz is used as an alternative narrowband segment in countires where 10,368MHz is not available.
**Note 4:** 10,000-10,125MHz is subject to increased Primary user utilisation and NoV restrictions.
**Note 5:** 10,475-10,500MHz is allocated ONLY to the Amateur Satellite Service and NOT to the Amateur Service.
**Licence Notes:** Amateur Service – Secondary User. Foundation licensees 1 watt maximum. Amateur Satellite Service: 10,450-10,500MHz – Secondary User. Unattended operation is permitted for remote control, digital modes and beacons except in the sub-bands 10,000-10,125MHz within 50km of SO916223 (Cheltenham), SS206127 (Bude), SK985640 (Waddington) and SE202577 (Harrogate).
**Notes to the Band Plan:** As on page 167.

| 24GHz (12mm) IARU Recommendation | UK USAGE |
|---|---|
| 24,000.000-24,050.000MHz Satellites | 24,025MHz Preferred Operating Frequency for Wideband Equipment |
| | 24,048.2MHz – Narrowband Centre of Activity |
| | 24,048.750-24,048.800MHz Local Beacons, 10W ERP max |
| Propagation Beacons 24,050.000-24,250.000 All Modes | |

**Licence Notes:** Amateur Service: 24,000-24,050MHz – Primary User: Users must accept interference from ISM users. 24,050-24,150MHz – Secondary User. May only be used with written permission of Ofcom. Users must accept interference from ISM users. 24,150-24,250MHz – Secondary User. Users must accept interference from ISM users. Amateur Satellite Service: 24,000-24,050MHz Primary User: Users must accept Interference from ISM users. Unattended operation is permitted for remote control, digital modes and beacons, except in the sub bands 24,000-24,050MHz within 50km of SK985640 (Waddington) and SE202577 (Harrogate).
ISM = Industrial, scientific and medical.
**Notes to the Band Plan:** As on page 167.

| 47GHz (6mm) IARU Recommendation | UK USAGE |
|---|---|
| 47,000.000-47,200.000MHz | 47,088.2MHz – Centre of Narrowband Activity |
| 47,088.000-47,090.000 Narrowband Segment | 47,088.8-47,089.0MHz – Propagation Beacons Only |

**Licence Notes:** Amateur Service and Amateur Satellite Service – Primary User. Unattended operation is permitted for remote control, digital modes and beacons, except within 50km of SK985640 (Waddington) and SE202577 (Harrogate).
**Notes to the Band Plan:** As on page 167.

| 76GHz (4mm) IARU Recommendation | UK USAGE |
|---|---|
| 75,500-76,000MHz All Modes (preferred) | 75,976.200MHz – IARU Region 1 Preferred Centre of Activity |
| 76,000.000-77,500.000 All Modes | |
| 77,500-78,000 | 77,500.200MHz – Alternative IARU Recommended Narrowband Segment |
| All Modes (preferred) | |
| 78,000-81,000 All Modes | |

**Licence Notes:**
75,500-75,875MHz Amateur Service and Amateur Satellite Service – Secondary User.
75,875-76,000MHz Amateur Service and Amateur Satellite Service – Primary User.
76,000-77,500MHz Amateur Service and Amateur Satellite Service – Secondary User.
77,500-78,000MHz Amateur Service and Amateur Satellite Service – Primary User.
78,000-81,000MHz Amateur service and Amateur Satellite Service – Secondary User.
Unattended operation is permitted for remote control, digital modes and beacons, except within 50km of SK985640 (Waddington) and SE202577 (Harrogate).
**Notes to the Band Plan:** As on page 167.

| 134GHz (2mm) IARU Recommendation | UK USAGE |
|---|---|
| 134,000-134,928MHz All Modes | |
| 134,928 -134,930 Narrowband Modes | IARU Region 1 Preferred Centre of Activity |
| | 134,928.800-134,928.990 – Propagation Beacons Only |
| 134,930 -136,000 All Modes | |

**Licence Notes:** 134,000-136,000MHz Amateur Service and Amateur Satellite Service – Primary User. Unattended operation is permitted for remote control, digital modes and beacons, except within 50km of SK985640 (Waddington) and SE202577 (Harrogate).

**THE FOLLOWING BANDS ARE ALSO ALLOCATED TO THE AMATEUR SERVICE AND THE AMATEUR SATELLITE SERVICE**

122,250-123,000MHz – Amateur Service only, Secondary User
136,000-141,000MHz – Secondary User
241,000-248,000MHz – Secondary User
248,000-250,000MHz – Primary User
**Notes to the Band Plan:** As on page 167.

## NOTES TO THE BAND PLAN

**ITU-R Recommendation SM.328 (extract)**

**Necessary bandwidth:** For a given class of emission, the width of the frequency band which is just sufficient to ensure the transmission of information at the rate and with the quality required under specified conditions.

**Foundation and Intermediate** Licence holders are advised to check their Licences for the permitted power limits and conditions applicable to their class of Licence.

**All Modes:** CW, SSB and those modes listed as Centres of Activity, plus AM. Consideration should be given to adjacent channel users.

**Image Modes:** Any analogue or digital image modes within the appropriate bandwidth, for example SSTV and FAX.

**Narrowband Modes:** All modes using up to 500Hz bandwidth, including CW, RTTY, PSK, etc.

**Digimodes:** Any digital mode used within the appropriate bandwidth, for example RTTY, PSK, MT63, etc.

**Sideband usage:** Below 10MHz use lower sideband (LSB), above 10MHz use upper sideband (USB). Note the lowest dial settings for LSB Voice modes are 1843, 3603 and 7043kHz on 160, 80 and 40m. Note that on (5MHz) USB is used.

**Amplitude Modulation (AM):** AM with a bandwidth greater than 2.7kHz is acceptable in the All Modes segments provided users consider adjacent channel activity when selecting operating frequencies (Davos 2005).

**Extended SSB (eSSB):** Extended SSB (eSSB) is only acceptable in the All Modes segments provided users consider adjacent channel activity when selecting operating frequencies.

**Digital Voice (DV):** Users of Digital Voice (DV) should check that the channel is not in use by other modes (CT08_C5_Rec20).

**FM Repeater & Gateway Access:** CTCSS Access is recommended. Toneburst access is being withdrawn in line with IARU-R1 recommendations.

**Beacons** Propagation Beacon Sub-bands are highlighted – please avoid transmitting in them!

**MGM:** Machine Generated Modes indicates those transmission modes relying fully on computer processing such as RTTY, AMTOR, PSK31, JTxx, FSK441 and the like. This does not include Digital Voice (DV) or Digital Data (DD).

**WSPR:** Above 30MHz, WSPR frequencies in the band plan are the centre of the transmitted frequency (not the suppressed carrier frequency or the VFO dial setting).

CW QSOs are accepted across all bands, except within beacon segments (Recommendation DV05_C4_Rec_13).

Contest activity shall not take place on the 10, 18 and 24MHz (30, 17 and 12m) bands.

-----------

Non-contesting radio amateurs are recommended to use the contest-free HF bands (30, 17 and 12m) during the largest international contests (DV05_C4_Rev_07).

The term 'automatically controlled data stations' include Store and Forward stations.

**Transmitting Frequencies:** The announced frequencies in the band plan are understood as 'transmitted frequencies' (not those of the suppressed carrier!).

**Unmanned transmitting stations:** IARU member societies are requested to limit this activity on the HF bands. It is recommended that any unmanned transmitting stations on HF shall only be activated under operator control except for beacons agreed with the IARU Region 1 Beacon Coordinator, or specially licensed experimental stations.

**472-479kHz:** Access is available to Full licensees only - see licence schedule for additional conditions.

**1.8MHz:** Radio amateurs in countries that have a SSB allocation ONLY below 1840kHz, may continue to use it, but the National Societies in those countries are requested to take all necessary steps with their licence administrations to adjust phone allocations in accordance with the Region 1 Band Plan (UBA – Davos 2005).

**3.5MHz:** Inter-Continental operations should be given priority in the segments 3500-3510kHz and 3775- 3800kHz. Where no DX traffic is involved, the contest segments should not include 3500-3510kHz or 3775-3800kHz. Member societies will be permitted to set other (lower) limits for national contests (within these limits). 3510-3600kHz may be used for unmanned ARDF beacons (CW, A1A) (Recommendation DV05_C4_Rec_12). Member societies should approach their national telecommunication authorities and ask them not to allocate frequencies other than amateur stations in the band segment that IARU has assigned to Inter-Continental long distance traffic.

**5MHz:** Access is available to Full licensees only- see licence schedule for additional conditions.

**7MHz:** The band segment 7040-7060kHz may be used for automatic controlled data stations (unattended) traffic in the areas of Africa south from the equator during local daylight hours. Where no DX traffic is involved, the contest segment should not include 7,175-7,200kHz.

**10MHz:** SSB may be used during emergencies involving the immediate safety of life and property and only by stations actually involved in the handling of emergency traffic. The band segment 10120kHz to 10140kHz may be used for SSB transmissions in the area of Africa south of the equator during local daylight hours.
News bulletins on any mode should not be transmitted on the 10MHz band.

**28MHz:** Member societies should advise operators not to transmit on frequencies between 29.3 and 29.51MHz to avoid interference to amateur satellite downlinks.

**Experimentation with NBFM Packet Radio on 29MHz band:** Preferred operating frequencies on each 10kHz from 29.210 to 29.290MHz inclusive should be used. A deviation of ±2.5kHz being used with 2.5kHz as maximum modulation frequency.

**146-147MHz & 2300-2302MHz**
Access to these bands requires an appropriate NoV, which is available to Full licensees only.

**430MHz**
The use of Amplitude Modulation (AM) is acceptable in the all modes segments but users are asked to consider

**1.3GHz**
The band is subject to re-planning. It is also shared with air traffic radar.

**2.3GHz (2310-2350 & 2390-2400MHz)**
Operation is subject to specific licence conditions and guidance - see also the Ofcom PSSR statement.

**3.4GHz (3400-3410MHz)**
Operation is subject to specific licence conditions and guidance - see also the Ofcom PSSR statement.

# Locators

The IARU Locator System, usually just called 'Locator', provides a means of pinpointing stations throughout the world. It is most often used by operators above 30MHz, as a means of calculating the distance between two stations. It is also used on the 136kHz band for the same reason. For use by operators on the upper microwave bands, it can have eight digits, though only the first six are dealt with here. The system is based upon latitude and longitude.

As the map and diagrams show, there are three sizes of 'rectangle'. The largest, known as a 'field', is 20° of longitude (east-west) by 10° latitude (north-south), and is designated by two letters. Most of Britain is in IO field. The next rectangle, known as a 'square' (though it is actually neither truly square nor rectangular!) is 2° of longitude by 1° of latitude. One hundred squares make up one field and, as the map shows, these are given numbers 00 in the south-west corner to 99 in the north-east. Dublin is in IO63. Finally, each square is divided into 576 'sub-squares', 5 minutes of longitude by 2.5 minutes of latitude, and given letters from AA to XX.

To find out your locator, first use a map of your area to determine your exact latitude and longitude, then use the map on this page and the squares diagram opposite to pinpoint your locator. Computer programs and online calculators are available to do this more easily, especially for those who operate from various locations.

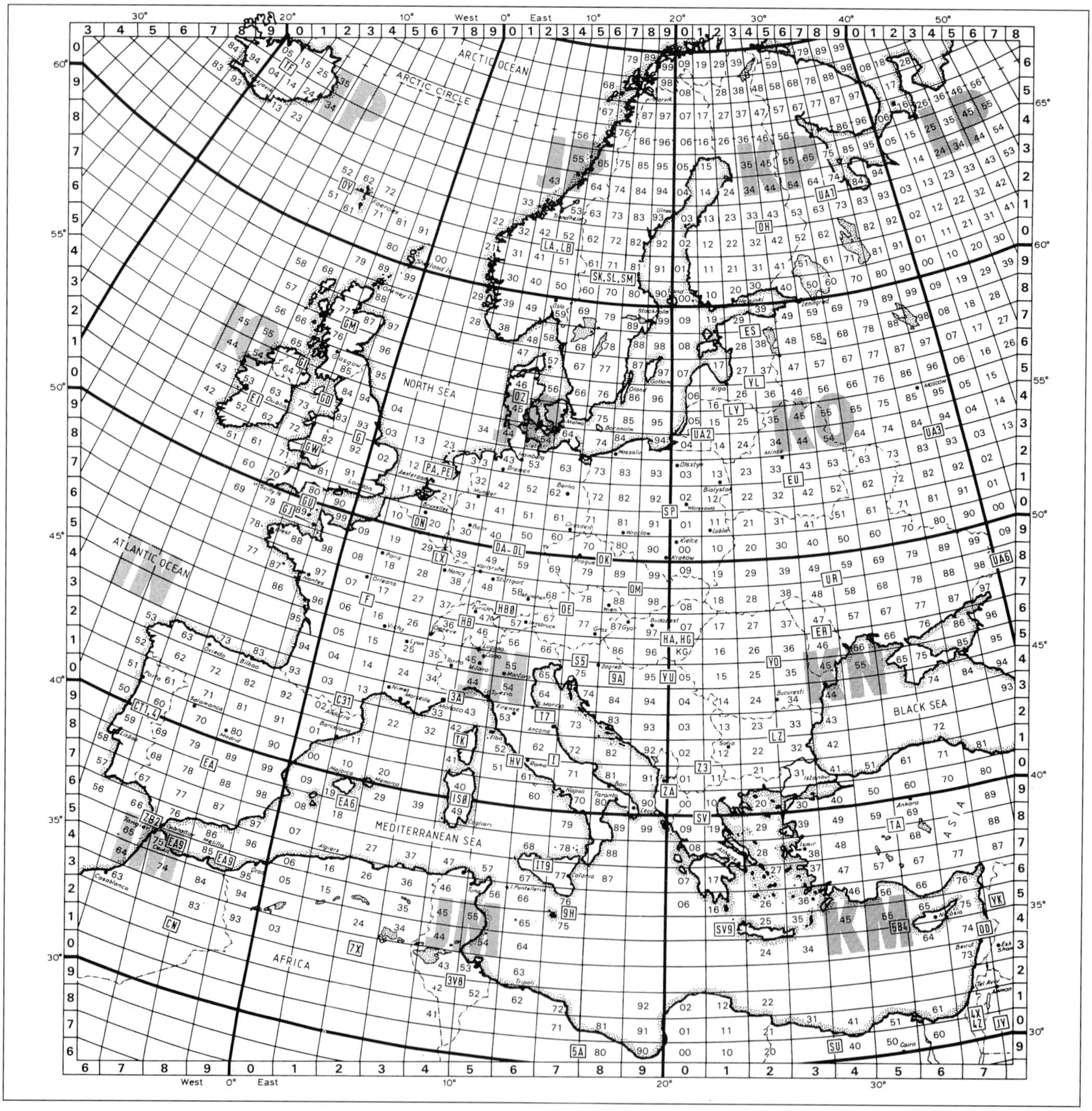

On-line Lat+Long to/from Locator calculators: www.arrl.org/locate/grid.html and www.amsat.org/cgi-bin/gridconv

On-line NGR to Locator calculator: www.ntay.com/contest/NGR2Loc.html

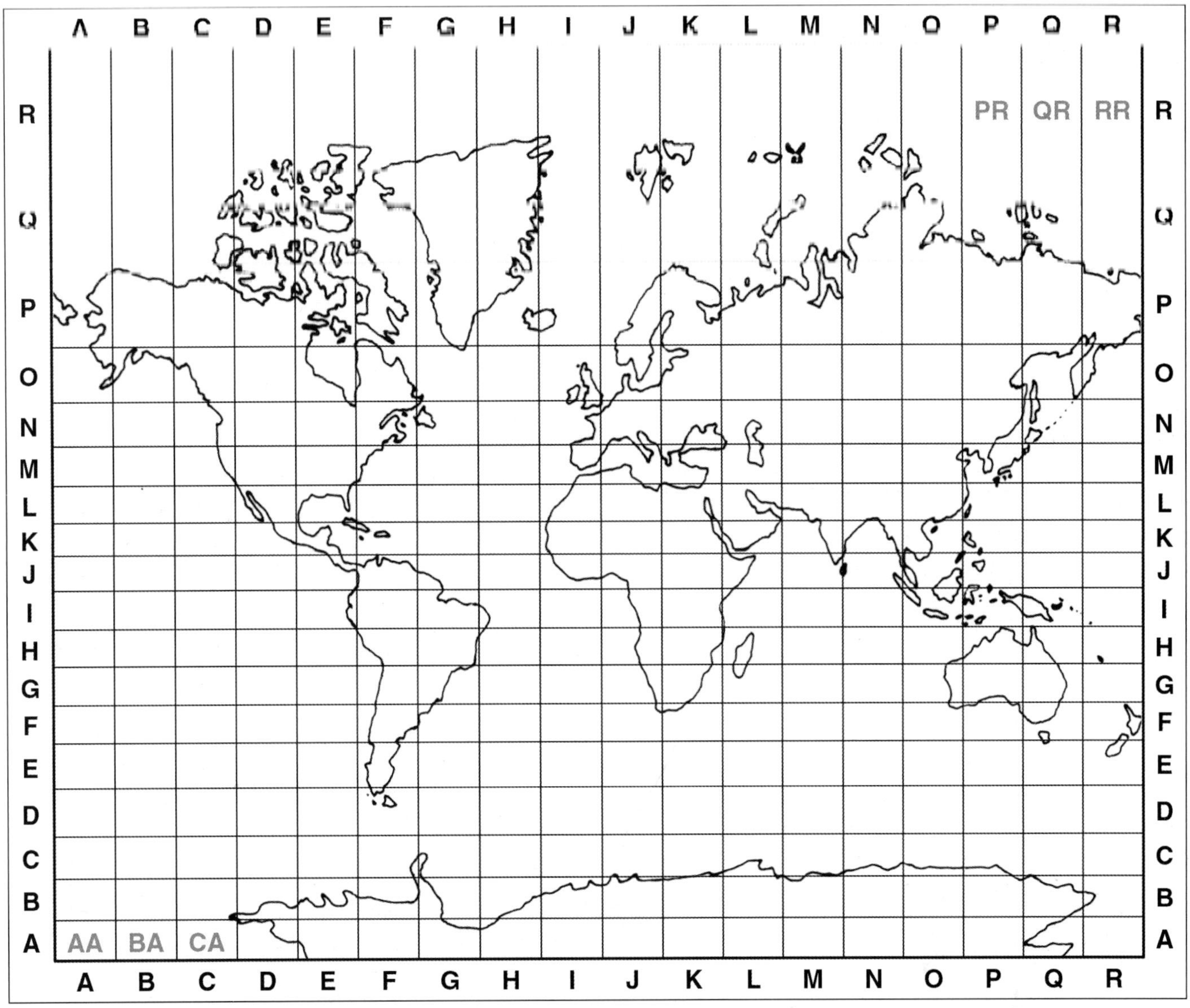

The IARU Locator system may be used throughout the world without repeats. The map above shows the fields that make up the first two letters of the Locator. Examples are shown at two of the corners. The map left shows numbering of squares within the fields.

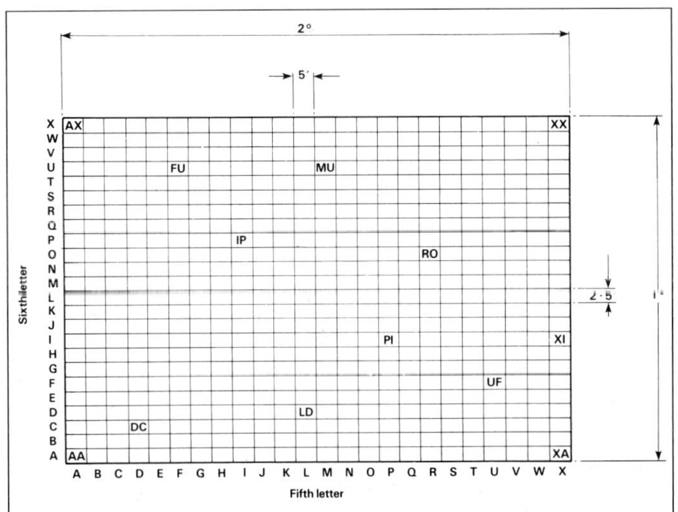


A square (the numbered part of the Locator) is divided into 576 sub-squares, designated AA to XX. Each sub-square is 5' W-E and 2.5' N-S.

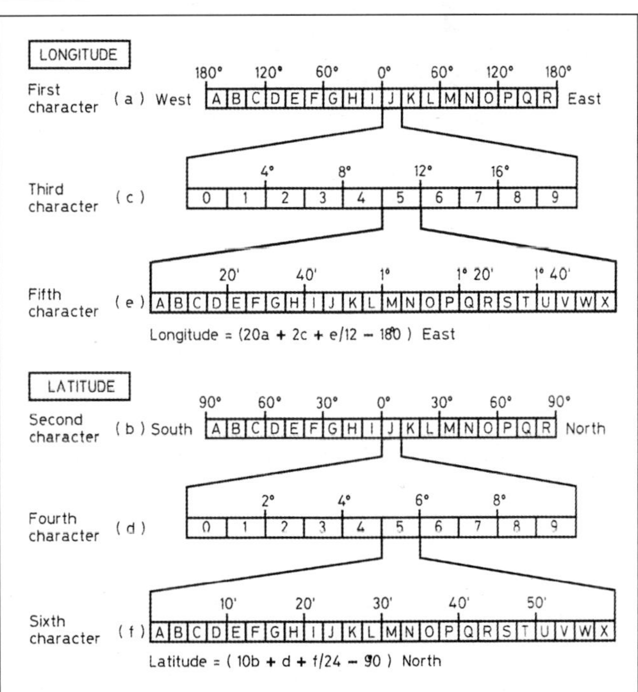


The final two letters may be calculated thus.

A large Locator map of Europe is available from RSGB.
See www.rsgbshop.org or phone 01234 832700.

# Prefix List

Callsigns for the world's nations are determined by the International Telecommunications Union (ITU). This is the United Nations agency that co-ordinates radio activity for all spectrum users. The prefixes used by a country for both commercial and amateur radio purposes are determined from one or more ITU allocation blocks issued to that country. The amateur radio callsigns in use for a particular country might use one or a number of combinations derived from the authorised ITU allocation(s) for that country. The following list shows callsign prefixes currently in use. Most are derived from the callsign blocks allocated to administrations by the ITU for use within the countries, territories and dependencies for which a country is responsible. Also shown are some unauthorised prefixes which may be heard and which may or may not be recognised as a DXCC entity, eg 1A0 (SMOM). 1B (the Turkish area of North Cyprus) and 1Z (Karea State - Myanmar) are unofficial and are not recognised for DXCC purposes, so these are not shown.

Full information on prefixes is contained in the ***RSGB Prefix Guide***.

| Prefix | Entity | Cont. | ITU | CQ | 1.8 | 3.5 | 7.0 | 10.1 | 14 | 18 | 21 | 24 | 28 | 50 | 144 | oth. |
|---|---|---|---|---|---|---|---|---|---|---|---|---|---|---|---|---|
| 1A | Sov. Mil. Order of Malta | EU | 28 | 15 | | | | | | | | | | | | |
| 3A | Monaco | EU | 27 | 14 | | | | | | | | | | | | |
| 3B6, 7 | Agalega & St. Brandon Is. | AF | 53 | 39 | | | | | | | | | | | | |
| 3B8 | Mauritius | AF | 53 | 39 | | | | | | | | | | | | |
| 3B9 | Rodriguez I. | AF | 53 | 39 | | | | | | | | | | | | |
| 3C | Equatorial Guinea | AF | 47 | 36 | | | | | | | | | | | | |
| 3C0 | Annobon I. | AF | 52 | 36 | | | | | | | | | | | | |
| 3D2 | Fiji | OC | 56 | 32 | | | | | | | | | | | | |
| 3D2 | Conway Reef | OC | 56 | 32 | | | | | | | | | | | | |
| 3D2 | Rotuma I. | OC | 56 | 32 | | | | | | | | | | | | |
| 3DA | Swaziland | AF | 57 | 38 | | | | | | | | | | | | |
| 3V | Tunisia | AF | 37 | 33 | | | | | | | | | | | | |
| 3W, XV | Vietnam | AS | 49 | 26 | | | | | | | | | | | | |
| 3X | Guinea | AF | 46 | 35 | | | | | | | | | | | | |
| 3Y | Bouvet | AF | 67 | 38 | | | | | | | | | | | | |
| 3Y | Peter 1 I. | AN | 72 | 12 | | | | | | | | | | | | |
| 4J, 4K | Azerbaijan | AS | 29 | 21 | | | | | | | | | | | | |
| 4L | Georgia | AS | 29 | 21 | | | | | | | | | | | | |
| 4O | Montenegro | EU | 28 | 15 | | | | | | | | | | | | |
| 4S | Sri Lanka | AS | 41 | 22 | | | | | | | | | | | | |
| 4U_ITU | ITU HQ | EU | 28 | 14 | | | | | | | | | | | | |
| 4U_UN | United Nations HQ | NA | 08 | 05 | | | | | | | | | | | | |
| 4W | Timor - Leste | OC | 54 | 28 | | | | | | | | | | | | |
| 4X, 4Z | Israel | AS | 39 | 20 | | | | | | | | | | | | |
| 5A | Libya | AF | 38 | 34 | | | | | | | | | | | | |
| 5B, C4, P3 | Cyprus | AS | 39 | 20 | | | | | | | | | | | | |
| 5H-5I | Tanzania | AF | 53 | 37 | | | | | | | | | | | | |
| 5N | Nigeria | AF | 46 | 35 | | | | | | | | | | | | |
| 5R | Madagascar | AF | 53 | 39 | | | | | | | | | | | | |
| 5T | Mauritania | AF | 46 | 35 | | | | | | | | | | | | |
| 5U | Niger | AF | 46 | 35 | | | | | | | | | | | | |
| 5V | Togo | AF | 46 | 35 | | | | | | | | | | | | |
| 5W | Samoa | OC | 62 | 32 | | | | | | | | | | | | |
| 5X | Uganda | AF | 48 | 37 | | | | | | | | | | | | |
| 5Y-5Z | Kenya | AF | 48 | 37 | | | | | | | | | | | | |
| 6V-6W | Senegal | AF | 46 | 35 | | | | | | | | | | | | |
| 6Y | Jamaica | NA | 11 | 08 | | | | | | | | | | | | |
| 7O5 | Yemen | AS | 39 | 21 | | | | | | | | | | | | |
| 7P | Lesotho | AF | 57 | 38 | | | | | | | | | | | | |
| 7Q | Malawi | AF | 53 | 37 | | | | | | | | | | | | |
| 7T-7Y | Algeria | AF | 37 | 33 | | | | | | | | | | | | |
| 8P | Barbados | NA | 11 | 08 | | | | | | | | | | | | |
| 8Q | Maldives | AS/AF | 41 | 22 | | | | | | | | | | | | |
| 8R | Guyana | SA | 12 | 09 | | | | | | | | | | | | |
| 9A | Croatia | EU | 28 | 15 | | | | | | | | | | | | |
| 9G | Ghana | AF | 46 | 35 | | | | | | | | | | | | |
| 9H | Malta | EU | 28 | 15 | | | | | | | | | | | | |
| 9I-9J | Zambia | AF | 53 | 36 | | | | | | | | | | | | |
| 9K | Kuwait | AS | 39 | 21 | | | | | | | | | | | | |
| 9L | Sierra Leone | AF | 46 | 35 | | | | | | | | | | | | |
| 9M2, 4 | West Malaysia | AS | 54 | 28 | | | | | | | | | | | | |
| 9M6, 8 | East Malaysia | OC | 54 | 28 | | | | | | | | | | | | |
| 9N | Nepal | AS | 42 | 22 | | | | | | | | | | | | |
| 9Q-9T | Dem. Rep. of Congo | AF | 52 | 36 | | | | | | | | | | | | |

Key to abbreviations of continents:
**AF** = Africa, **AN** = Antarctica, **AS** = Asia, **EU** = Europe,
**NA** = North America, **OC** = Oceania, **SA** = South America

| Prefix | Entity | Cont. | ITU | CQ | 1.8 | 3.5 | 7.0 | 10.1 | 14 | 18 | 21 | 24 | 28 | 50 | 144 | oth. |
|---|---|---|---|---|---|---|---|---|---|---|---|---|---|---|---|---|
| 9U | Burundi | AF | 52 | 36 | | | | | | | | | | | | |
| 9V | Singapore | AS | 54 | 28 | | | | | | | | | | | | |
| 9X | Rwanda | AF | 52 | 36 | | | | | | | | | | | | |
| 9Y-9Z | Trinidad & Tobago | SA | 11 | 09 | | | | | | | | | | | | |
| A2 | Botswana | AF | 57 | 38 | | | | | | | | | | | | |
| A3 | Tonga | OC | 62 | 32 | | | | | | | | | | | | |
| A4 | Oman | AS | 39 | 21 | | | | | | | | | | | | |
| A5 | Bhutan | AS | 41 | 22 | | | | | | | | | | | | |
| A6 | United Arab Emirates | AS | 39 | 21 | | | | | | | | | | | | |
| A7 | Qatar | AS | 39 | 21 | | | | | | | | | | | | |
| A9 | Bahrain | AS | 39 | 21 | | | | | | | | | | | | |
| AP | Pakistan | AS | 41 | 21 | | | | | | | | | | | | |
| B | China | AS | 33, 42-44 | 23, 24 | | | | | | | | | | | | |
| BS7 | Scarborough Reef | AS | 50 | 27 | | | | | | | | | | | | |
| BU-BX | Taiwan | AS | 44 | 24 | | | | | | | | | | | | |
| BV9P | Pratas I. | AS | 44 | 24 | | | | | | | | | | | | |
| C2 | Nauru | OC | 65 | 31 | | | | | | | | | | | | |
| C3 | Andorra | EU | 27 | 14 | | | | | | | | | | | | |
| C5 | The Gambia | AF | 46 | 35 | | | | | | | | | | | | |
| C6 | Bahamas | NA | 11 | 08 | | | | | | | | | | | | |
| C8-9 | Mozambique | AF | 53 | 37 | | | | | | | | | | | | |
| CA-CE | Chile | SA | 14,16 | 12 | | | | | | | | | | | | |
| CE0 | Easter I. | SA | 63 | 12 | | | | | | | | | | | | |
| CE0 | Juan Fernandez Is. | SA | 14 | 12 | | | | | | | | | | | | |
| CE0 | San Felix & San Ambrosio | SA | 14 | 12 | | | | | | | | | | | | |
| CE9/KC4 | Antarctica | AN | 67,69-74 | S | | | | | | | | | | | | |
| CM, CO | Cuba | NA | 11 | 08 | | | | | | | | | | | | |
| CN | Morocco | AF | 37 | 33 | | | | | | | | | | | | |
| CP | Bolivia | SA | 12,14 | 10 | | | | | | | | | | | | |
| CT | Portugal | EU | 37 | 14 | | | | | | | | | | | | |
| CT3 | Madeira Is. | AF | 36 | 33 | | | | | | | | | | | | |
| CU | Azores | EU | 36 | 14 | | | | | | | | | | | | |
| CV-CX | Uruguay | SA | 14 | 13 | | | | | | | | | | | | |
| CY0 | Sable I. | NA | 09 | 05 | | | | | | | | | | | | |
| CY9 | St. Paul I. | NA | 09 | 05 | | | | | | | | | | | | |
| D2-3 | Angola | AF | 52 | 36 | | | | | | | | | | | | |
| D4 | Cape Verde | AF | 46 | 35 | | | | | | | | | | | | |
| D6 | Comoros | AF | 53 | 39 | | | | | | | | | | | | |
| DA-DR | Fed. Rep. of Germany | EU | 28 | 14 | | | | | | | | | | | | |
| DU-DZ | Philippines | OC | 50 | 27 | | | | | | | | | | | | |
| E3 | Eritrea | AF | 48 | 37 | | | | | | | | | | | | |
| E4 | Palestine | AS | 39 | 20 | | | | | | | | | | | | |
| E5 | N. Cook Is. | OC | 62 | 32 | | | | | | | | | | | | |
| E5 | S. Cook Is. | OC | 62 | 32 | | | | | | | | | | | | |
| E6 | Niue | OC | 62 | 32 | | | | | | | | | | | | |
| E7 | Bosnia-Herzegovina | EU | 28 | 15 | | | | | | | | | | | | |
| EA-EH | Spain | EU | 37 | 14 | | | | | | | | | | | | |
| EA6-EH6 | Balearic Is. | EU | 37 | 14 | | | | | | | | | | | | |
| EA8-EH8 | Canary Is. | AF | 36 | 33 | | | | | | | | | | | | |
| EA9-EH9 | Ceuta & Melilla | AF | 37 | 33 | | | | | | | | | | | | |
| EI-EJ | Ireland | EU | 27 | 14 | | | | | | | | | | | | |
| EK | Armenia | AS | 29 | 21 | | | | | | | | | | | | |
| EL | Liberia | AF | 46 | 35 | | | | | | | | | | | | |
| EP-EQ | Iran | AS | 40 | 21 | | | | | | | | | | | | |
| ER | Moldova | EU | 29 | 16 | | | | | | | | | | | | |
| ES | Estonia | EU | 29 | 15 | | | | | | | | | | | | |
| ET | Ethiopia | AF | 48 | 37 | | | | | | | | | | | | |
| EU-EW | Belarus | EU | 29 | 16 | | | | | | | | | | | | |
| EX | Kyrgyzstan | AS | 30, 31 | 17 | | | | | | | | | | | | |
| EY | Tajikistan | AS | 30 | 17 | | | | | | | | | | | | |
| EZ | Turkmenistan | AS | 30 | 17 | | | | | | | | | | | | |
| F | France | EU | 27 | 14 | | | | | | | | | | | | |
| FG, TO | Guadeloupe | NA | 11 | 08 | | | | | | | | | | | | |
| FH, TO | Mayotte | AF | 53 | 39 | | | | | | | | | | | | |
| FJ, TO | Saint Barthelemy | NA | 11 | 08 | | | | | | | | | | | | |
| FK, TX | New Caledonia | OC | 56 | 32 | | | | | | | | | | | | |
| FK, TX | Chesterfield Is. | OC | 56 | 30 | | | | | | | | | | | | |
| FM, TO | Martinique | NA | 11 | 08 | | | | | | | | | | | | |
| FO, TX | Austral I. | OC | 63 | 32 | | | | | | | | | | | | |
| FO, TX | Clipperton I. | NA | 10 | 07 | | | | | | | | | | | | |
| FO, TX | French Polynesia | OC | 63 | 32 | | | | | | | | | | | | |
| FO, TX | Marquesas Is. | OC | 63 | 31 | | | | | | | | | | | | |

The letter ‘S’ against an ITU or CQ Zone indicates that the entity is split across several.

| Prefix | Entity | Cont. | ITU | CQ | 1.8 | 3.5 | 7.0 | 10.1 | 14 | 18 | 21 | 24 | 28 | 50 | 144 | oth. |
|---|---|---|---|---|---|---|---|---|---|---|---|---|---|---|---|---|
| FP | St. Pierre & Miquelon | NA | 09 | 05 | | | | | | | | | | | | |
| FR, TO | Reunion I. | AF | 53 | 39 | | | | | | | | | | | | |
| FT/G, TO | Glorioso Is. | AF | 53 | 39 | | | | | | | | | | | | |
| FT/J,E, TO | Juan de Nova, Europa | AF | 53 | 39 | | | | | | | | | | | | |
| FT/T, TO | Tromelin I. | AF | 53 | 39 | | | | | | | | | | | | |
| FS, TO | Saint Martin | NA | 11 | 08 | | | | | | | | | | | | |
| FT/W | Crozet I. | AF | 68 | 39 | | | | | | | | | | | | |
| FT/X | Kerguelen Is. | AF | 68 | 39 | | | | | | | | | | | | |
| FT/Z | Amsterdam & St. Paul Is. | AF | 68 | 39 | | | | | | | | | | | | |
| FW | Wallis & Futuna Is. | OC | 62 | 32 | | | | | | | | | | | | |
| FY | French Guiana | SA | 12 | 09 | | | | | | | | | | | | |
| G, GX, M, MX, 2E | England | EU | 27 | 14 | | | | | | | | | | | | |
| GD, GT, MD, MT, 2D | Isle of Man | EU | 27 | 14 | | | | | | | | | | | | |
| GI, GN, MI, MN, 2I | Northern Ireland | EU | 27 | 14 | | | | | | | | | | | | |
| GJ, GH, MJ, MH, 2J | Jersey | EU | 27 | 14 | | | | | | | | | | | | |
| GM, GS, MM, MS, 2M | Scotland | EU | 27 | 14 | | | | | | | | | | | | |
| GU, GP, MU, MP, 2U | Guernsey | EU | 27 | 14 | | | | | | | | | | | | |
| GW, GC, MW, MC, 2W | Wales | EU | 27 | 14 | | | | | | | | | | | | |
| H4 | Solomon Is. | OC | 51 | 28 | | | | | | | | | | | | |
| H40 | Temotu Province | OC | 51 | 32 | | | | | | | | | | | | |
| HA, HG | Hungary | EU | 28 | 15 | | | | | | | | | | | | |
| HB | Switzerland | EU | 28 | 14 | | | | | | | | | | | | |
| HB0 | Liechtenstein | EU | 28 | 14 | | | | | | | | | | | | |
| HC-HD | Ecuador | SA | 12 | 10 | | | | | | | | | | | | |
| HC8-HD8 | Galapagos Is. | SA | 12 | 10 | | | | | | | | | | | | |
| HH | Haiti | NA | 11 | 08 | | | | | | | | | | | | |
| HI | Dominican Republic | NA | 11 | 08 | | | | | | | | | | | | |
| HJ-HK, 5J-5K | Colombia | SA | 12 | 09 | | | | | | | | | | | | |
| HK0 | Malpelo I. | SA | 12 | 09 | | | | | | | | | | | | |
| HK0 | San Andres & Providencia | NA | 11 | 07 | | | | | | | | | | | | |
| HL, 6K-6N | Republic of Korea | AS | 44 | 25 | | | | | | | | | | | | |
| HO-HP | Panama | NA | 11 | 07 | | | | | | | | | | | | |
| HQ-HR | Honduras | NA | 11 | 07 | | | | | | | | | | | | |
| HS, E2 | Thailand | AS | 49 | 26 | | | | | | | | | | | | |
| HV | Vatican | EU | 28 | 15 | | | | | | | | | | | | |
| HZ | Saudi Arabia | AS | 39 | 21 | | | | | | | | | | | | |
| I | Italy | EU | 28 | 15,33 | | | ✓ | | | | | | | | | |
| IS0, IM0 | Sardinia | EU | 28 | 15 | | | | | | | | | | | | |
| J2 | Djibouti | AF | 48 | 37 | | | | | | | | | | | | |
| J3 | Grenada | NA | 11 | 08 | | | | | | | | | | | | |
| J5 | Guinea-Bissau | AF | 46 | 35 | | | | | | | | | | | | |
| J6 | St. Lucia | NA | 11 | 08 | | | | | | | | | | | | |
| J7 | Dominica | NA | 11 | 08 | | | | | | | | | | | | |
| J8 | St. Vincent | NA | 11 | 08 | | | | | | | | | | | | |
| JA-JS, 7J-7N | Japan | AS | 45 | 25 | | | | | | | | | | | | |
| JD | Minami Torishima | OC | 90 | 27 | | | | | | | | | | | | |
| JD | Ogasawara | AS | 45 | 27 | | | | | | | | | | | | |
| JT-JV | Mongolia | AS | 32,33 | 23 | | | | | | | | | | | | |
| JW | Svalbard | EU | 18 | 40 | | | | | | | | | | | | |
| JX | Jan Mayen | EU | 18 | 40 | | | | | | | | | | | | |
| JY | Jordan | AS | 39 | 20 | | | | | | | | | | | | |
| K, W, N, AA-AK | United States of America | NA | 6,7,8 | 3,4,5 | | | | | | | | | | | | |
| KG4 | Guantanamo Bay | NA | 11 | 08 | | | | | | | | | | | | |
| KH0 | Mariana Is. | OC | 64 | 27 | | | | | | | | | | | | |
| KH1 | Baker & Howland Is. | OC | 61 | 31 | | | | | | | | | | | | |
| KH2 | Guam | OC | 64 | 27 | | | | | | | | | | | | |
| KH3 | Johnston I. | OC | 61 | 31 | | | | | | | | | | | | |
| KH4 | Midway I. | OC | 61 | 31 | | | | | | | | | | | | |
| KH5 | Palmyra & Jarvis Is. | OC | 61, 62 | 31 | | | | | | | | | | | | |
| KH5K | Kingman Reef | OC | 61 | 31 | | | | | | | | | | | | |
| KH6,7 | Hawaii | OC | 61 | 31 | | | | | | | | | | | | |
| KH7K | Kure I. | OC | 61 | 31 | | | | | | | | | | | | |
| KH8 | American Samoa | OC | 62 | 32 | | | | | | | | | | | | |
| KH8 | Swains I. | OC | 62 | 32 | | | | | | | | | | | | |
| KH9 | Wake I. | OC | 65 | 31 | | | | | | | | | | | | |
| KL, AL, NL, WL | Alaska | NA | 1, 2 | 1 | | | | | | | | | | | | |
| KP1 | Navassa I. | NA | 11 | 08 | | | | | | | | | | | | |
| KP2 | Virgin Is. | NA | 11 | 08 | | | | | | | | | | | | |
| KP3, 4 | Puerto Rico | NA | 11 | 08 | | | | | | | | | | | | |
| KP5 | Desecheo I. | NA | 11 | 08 | | | | | | | | | | | | |
| LA-LN | Norway | EU | 18 | 14 | | | | | | | | | | | | |
| LO-LW | Argentina | SA | 14,16 | 13 | | | | | | | | | | | | |

There are numerous instances of entities that do not count for DXCC before (or after) certain dates. Check the ARRL website for details: *www.arrl.org/country-lists-prefixes*

| Prefix | Entity | Cont. | ITU | CQ | 1.8 | 3.5 | 7.0 | 10.1 | 14 | 18 | 21 | 24 | 28 | 50 | 144 | oth. |
|---|---|---|---|---|---|---|---|---|---|---|---|---|---|---|---|---|
| LX | Luxembourg | EU | 27 | 14 | | | | | | | | | | | | |
| LY | Lithuania | EU | 29 | 15 | | | | | | | | | | | | |
| LZ | Bulgaria | EU | 28 | 20 | | | | | | | | | | | | |
| OA-OC | Peru | SA | 12 | 10 | | | | | | | | | | | | |
| OD | Lebanon | AS | 39 | 20 | | | | | | | | | | | | |
| OE | Austria | EU | 28 | 15 | | | | | | | | | | | | |
| OF-OI | Finland | EU | 18 | 15 | | | | | | | | | | | | |
| OH0 | Aland Is. | EU | 18 | 15 | | | | | | | | | | | | |
| OJ0 | Market Reef | EU | 18 | 15 | | | | | | | | | | | | |
| OK-OL | Czech Republic | EU | 28 | 15 | | | | | | | | | | | | |
| OM | Slovak Republic | EU | 28 | 15 | | | | | | | | | | | | |
| ON-OT | Belgium | EU | 27 | 14 | | | | | | | | | | | | |
| OU-OW, OZ | Denmark | EU | 18 | 14 | | | | | | | | | | | | |
| OX | Greenland | NA | 5, 75 | 40 | | | | | | | | | | | | |
| OY | Faroe Is. | EU | 18 | 14 | | | | | | | | | | | | |
| P2 | Papua New Guinea | OC | 51 | 28 | | | | | | | | | | | | |
| P4 | Aruba | SA | 11 | 09 | | | | | | | | | | | | |
| P5 | DPR of Korea | AS | 44 | 25 | | | | | | | | | | | | |
| PA-PI | Netherlands | EU | 27 | 14 | | | | | | | | | | | | |
| PJ2 | Curacao | SA | 11 | 09 | | | | | | | | | | | | |
| PJ4 | Bonaire | SA | 11 | 09 | | | | | | | | | | | | |
| PJ5, 6 | Saba & St. Eustatius | NA | 11 | 08 | | | | | | | | | | | | |
| PJ7 | St Maarten | NA | 11 | 08 | | | | | | | | | | | | |
| PP-PY, ZV-ZZ | Brazil | SA | 12, 13, 15 | 11 | | | | | | | | | | | | |
| PP0-PY0F | Fernando de Noronha | SA | 13 | 11 | | | | | | | | | | | | |
| PP0-PY0S | St. Peter & St. Paul Rocks | SA | 13 | 11 | | | | | | | | | | | | |
| PP0-PY0T | Trindade & Martim Vaz Is. | SA | 15 | 11 | | | | | | | | | | | | |
| PZ | Suriname | SA | 12 | 09 | | | | | | | | | | | | |
| R1/F | Franz Josef Land | EU | 75 | 40 | | | | | | | | | | | | |
| S0 | Western Sahara | AF | 46 | 33 | | | | | | | | | | | | |
| S2 | Bangladesh | AS | 41 | 22 | | | | | | | | | | | | |
| S5 | Slovenia | EU | 28 | 15 | | | | | | | | | | | | |
| S7 | Seychelles | AF | 53 | 39 | | | | | | | | | | | | |
| S9 | Sao Tome & Principe | AF | 47 | 36 | | | | | | | | | | | | |
| SA-SM, 7S-8S | Sweden | EU | 18 | 14 | | | | | | | | | | | | |
| SN-SR | Poland | EU | 28 | 15 | | | | | | | | | | | | |
| ST | Sudan | AF | 47, 48 | 34 | | | | | | | | | | | | |
| SU | Egypt | AF | 38 | 34 | | | | | | | | | | | | |
| SV-SZ, J4 | Greece | EU | 28 | 20 | | | | | | | | | | | | |
| SV/A | Mount Athos | EU | 28 | 20 | | | | | | | | | | | | |
| SV5, J45 | Dodecanese | EU | 28 | 20 | | | | | | | | | | | | |
| SV9, J49 | Crete | EU | 28 | 20 | | | | | | | | | | | | |
| T2 | Tuvalu | OC | 65 | 31 | | | | | | | | | | | | |
| T30 | W. Kiribati (Gilbert Is. ) | OC | 65 | 31 | | | | | | | | | | | | |
| T31 | C. Kiribati (British Phoenix Is) | OC | 62 | 31 | | | | | | | | | | | | |
| T32 | E. Kiribati (Line Is.) | OC | 61, 63 | 31 | | | | | | | | | | | | |
| T33 | Banaba I. (Ocean I.) | OC | 65 | 31 | | | | | | | | | | | | |
| T5, 6O | Somalia | AF | 48 | 37 | | | | | | | | | | | | |
| T7 | San Marino | EU | 28 | 15 | | | | | | | | | | | | |
| T8 | Palau | OC | 64 | 27 | | | | | | | | | | | | |
| TA-TC | Turkey | EU/AS | 39 | 20 | | | | | | | | | | | | |
| TF | Iceland | EU | 17 | 40 | | | | | | | | | | | | |
| TG, TD | Guatemala | NA | 12 | 07 | | | | | | | | | | | | |
| TI, TE | Costa Rica | NA | 11 | 07 | | | | | | | | | | | | |
| TI9 | Cocos I. | NA | 12 | 07 | | | | | | | | | | | | |
| TJ | Cameroon | AF | 47 | 36 | | | | | | | | | | | | |
| TK | Corsica | EU | 28 | 15 | | | | | | | | | | | | |
| TL | Central Africa | AF | 47 | 36 | | | | | | | | | | | | |
| TN | Congo (Republic of the) | AF | 52 | 36 | | | | | | | | | | | | |
| TR | Gabon | AF | 52 | 36 | | | | | | | | | | | | |
| TT | Chad | AF | 47 | 36 | | | | | | | | | | | | |
| TU | Cote d'Ivoire | AF | 46 | 35 | | | | | | | | | | | | |
| TY | Benin | AF | 46 | 35 | | | | | | | | | | | | |
| TZ | Mali | AF | 46 | 35 | | | | | | | | | | | | |
| UA-UI1-7, RA-RZ | European Russia | EU | S | 16 | | | | | | | | | | | | |
| UA2, RA2 | Kaliningrad | EU | 29 | 15 | | | | | | | | | | | | |
| UA-UI8, 9, 0, RA-RZ | Asiatic Russia | AS | S | S | | | | | | | | | | | | |
| UJ-UM | Uzbekistan | AS | 30 | 17 | | | | | | | | | | | | |
| UN-UQ | Kazakhstan | AS | 29-31 | 17 | | | | | | | | | | | | |
| UR-UZ, EM-EO | Ukraine | EU | 29 | 16 | | | | | | | | | | | | |
| V2 | Antigua & Barbuda | NA | 11 | 08 | | | | | | | | | | | | |
| V3 | Belize | NA | 11 | 07 | | | | | | | | | | | | |

| Prefix | Entity | Cont. | ITU | CQ | 1.8 | 3.5 | 7.0 | 10.1 | 14 | 18 | 21 | 24 | 28 | 50 | 144 | oth. |
|---|---|---|---|---|---|---|---|---|---|---|---|---|---|---|---|---|
| V4 | St. Kitts & Nevis | NA | 11 | 08 | | | | | | | | | | | | |
| V5 | Namibia | AF | 57 | 38 | | | | | | | | | | | | |
| V6 | Micronesia | OC | 65 | 27 | | | | | | | | | | | | |
| V7 | Marshall Is. | OC | 65 | 31 | | | | | | | | | | | | |
| V8 | Brunei Darussalam | OC | 54 | 28 | | | | | | | | | | | | |
| VA-VG, VO,VY | Canada | NA | 2-4, 9, 75 | 1-5 | | | | | | | | | | | | |
| VK, AX | Australia | OC | 55, 58, 59 | 29, 30 | | | | | | | | | | | | |
| VK0 | Heard I. | AF | 68 | 39 | | | | | | | | | | | | |
| VK0 | Macquarie I. | OC | 60 | 30 | | | | | | | | | | | | |
| VK9C | Cocos (Keeling) Is. | OC | 54 | 29 | | | | | | | | | | | | |
| VK9L | Lord Howe I. | OC | 60 | 30 | | | | | | | | | | | | |
| VK9M | Mellish Reef | OC | 56 | 30 | | | | | | | | | | | | |
| VK9N | Norfolk I. | OC | 60 | 32 | | | | | | | | | | | | |
| VK9W | Willis I. | OC | 55 | 30 | | | | | | | | | | | | |
| VK9X | Christmas I. | OC | 54 | 29 | | | | | | | | | | | | |
| VP2E | Anguilla | NA | 11 | 08 | | | | | | | | | | | | |
| VP2M | Montserrat | NA | 11 | 08 | | | | | | | | | | | | |
| VP2V | British Virgin Is. | NA | 11 | 08 | | | | | | | | | | | | |
| VP5 | Turks & Caicos Is. | NA | 11 | 08 | | | | | | | | | | | | |
| VP6 | Pitcairn I. | OC | 63 | 32 | | | | | | | | | | | | |
| VP646 | Ducie I. | OC | 63 | 32 | | | | | | | | | | | | |
| VP8 | Falkland Is. | SA | 16 | 13 | | | | | | | | | | | | |
| VP8, LU | South Georgia I. | SA | 73 | 13 | | | | | | | | | | | | |
| VP8, LU | South Orkney Is. | SA | 73 | 13 | | | | | | | | | | | | |
| VP8, LU | South Sandwich Is. | SA | 73 | 13 | | | | | | | | | | | | |
| VP8, LU, CE9, HF0, 4K1 | South Shetland Is. | SA | 73 | 13 | | | | | | | | | | | | |
| VP9 | Bermuda | NA | 11 | 05 | | | | | | | | | | | | |
| VQ9 | Chagos Is. | AF | 41 | 39 | | | | | | | | | | | | |
| VR | Hong Kong | AS | 44 | 24 | | | | | | | | | | | | |
| VU | India | AS | 41 | 22 | | | | | | | | | | | | |
| VU4 | Andaman & Nicobar Is. | AS | 49 | 26 | | | | | | | | | | | | |
| VU7 | Lakshadweep Is. | AS | 41 | 22 | | | | | | | | | | | | |
| XA-XI | Mexico | NA | 10 | 06 | | | | | | | | | | | | |
| XA4-XI4 | Revillagigedo | NA | 10 | 06 | | | | | | | | | | | | |
| XT | Burkina Faso | AF | 46 | 35 | | | | | | | | | | | | |
| XU | Cambodia | AS | 49 | 26 | | | | | | | | | | | | |
| XW | Laos | AS | 49 | 26 | | | | | | | | | | | | |
| XX9 | Macao | AS | 44 | 24 | | | | | | | | | | | | |
| XY-XZ | Myanmar | AS | 49 | 26 | | | | | | | | | | | | |
| YA, T6 | Afghanistan | AS | 40 | 21 | | | | | | | | | | | | |
| YB-YH | Indonesia | OC | 51,54 | 28 | | | | | | | | | | | | |
| YI | Iraq | AS | 39 | 21 | | | | | | | | | | | | |
| YJ | Vanuatu | OC | 56 | 32 | | | | | | | | | | | | |
| YK | Syria | AS | 39 | 20 | | | | | | | | | | | | |
| YL | Latvia | EU | 29 | 15 | | | | | | | | | | | | |
| YN,H6-7,HT | Nicaragua | NA | 11 | 07 | | | | | | | | | | | | |
| YO-YR | Romania | EU | 28 | 20 | | | | | | | | | | | | |
| YS, HU | El Salvador | NA | 11 | 07 | | | | | | | | | | | | |
| YT-YU | Serbia | EU | 28 | 15 | | | | | | | | | | | | |
| YV-YY, 4M | Venezuela | SA | 12 | 09 | | | | | | | | | | | | |
| YV0 | Aves I. | NA | 11 | 08 | | | | | | | | | | | | |
| Z2 | Zimbabwe | AF | 53 | 38 | | | | | | | | | | | | |
| Z3 | Macedonia | EU | 28 | 15 | | | | | | | | | | | | |
| Z8 | South Sudan (Rep of) | AF | 48 | 34 | | | | | | | | | | | | |
| ZA | Albania | EU | 28 | 15 | | | | | | | | | | | | |
| ZB2 | Gibraltar | EU | 37 | 14 | | | | | | | | | | | | |
| ZC4 | UK Sov. Base Areas on Cyprus | | AS | 3 9 | | | | | | | | | | | | |
| 20 | | | | | | | | | | | | | | | | |
| ZD7 | St. Helena | AF | 66 | 36 | | | | | | | | | | | | |
| ZD8 | Ascension I. | AF | 66 | 36 | | | | | | | | | | | | |
| ZD9 | Tristan da Cunha & Gough I. | AF | 66 | 38 | | | | | | | | | | | | |
| ZF | Cayman Is. | NA | 11 | 08 | | | | | | | | | | | | |
| ZK3 | Tokelau Is. | OC | 62 | 31 | | | | | | | | | | | | |
| ZL-ZM | New Zealand | OC | 60 | 32 | | | | | | | | | | | | |
| ZL7 | Chatham Is. | OC | 60 | 32 | | | | | | | | | | | | |
| ZL8 | Kermadec Is. | OC | 60 | 32 | | | | | | | | | | | | |
| ZL9 | Auckland & Campbell Is. | OC | 60 | 32 | | | | | | | | | | | | |
| ZP | Paraguay | SA | 14 | 11 | | | | | | | | | | | | |
| ZR-ZU | South Africa | AF | 57 | 38 | | | | | | | | | | | | |
| ZS8 | Prince Edward & Marion Is. | AF | 57 | 38 | | | | | | | | | | | | |

For the latest DXCC list: www.arrl.org/country-lists-prefixes

# Postcodes

When taking part in a VHF contest which includes Postcodes as part of the exchange, it is useful to know the approximate direction in which to beam to make contact with areas that have not already been worked.

For contesting purposes, single letter Postcodes are padded-out to two letters.

The map here shows the relative positions only (ie no boundaries).

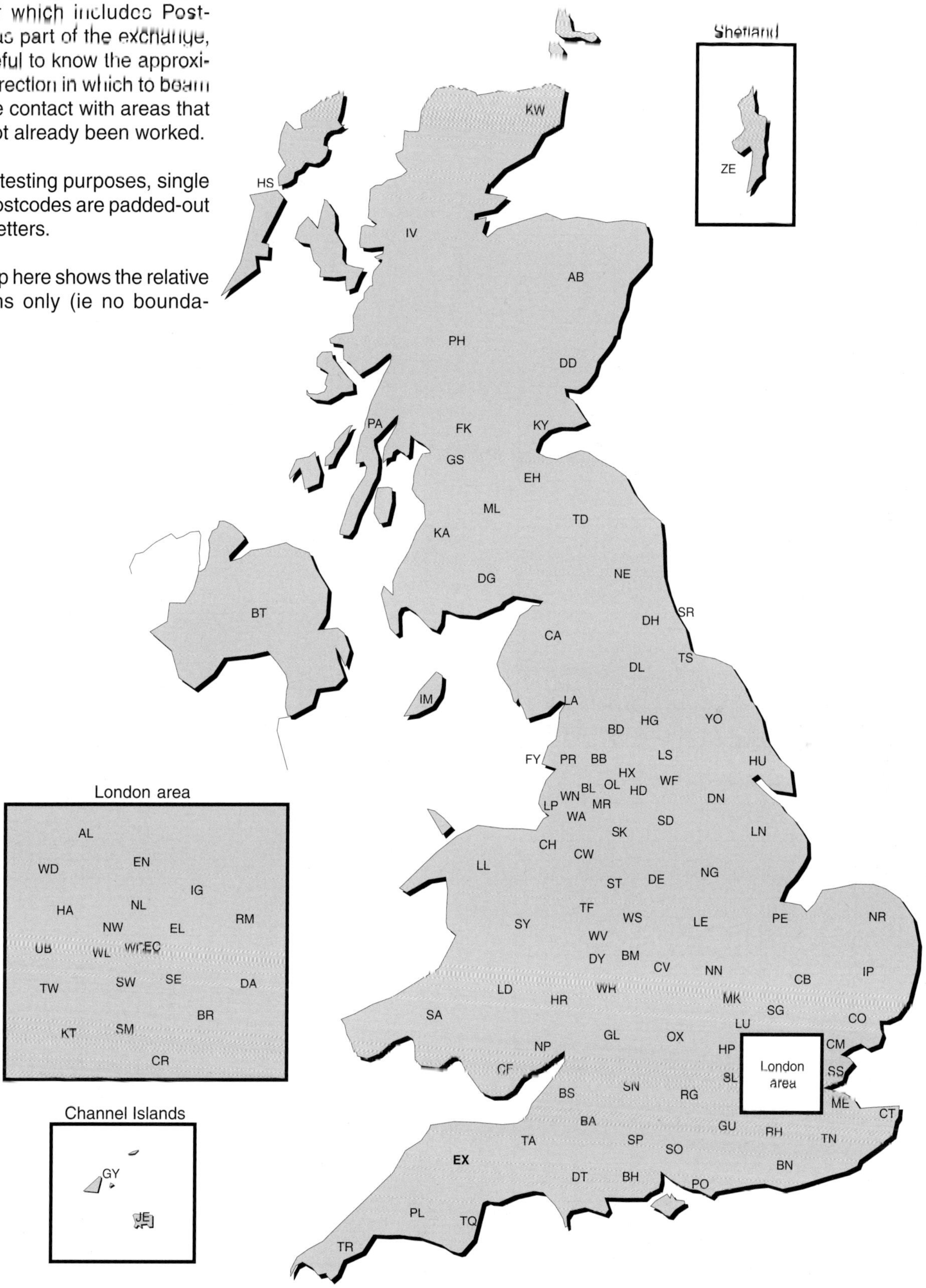

# Postcode Areas Check List

| Area | Location | | 1st QSO | 2nd QSO |
|---|---|---|---|---|
| AB | Aberdeen | (1) | | |
| | | (2) | | |
| | | (3) | | |
| AL | St Albans | | | |
| BA | Bath | | | |
| BB | Blackburn | | | |
| BD | Bradford | | | |
| BH | Bournemouth | | | |
| BL | Bolton | | | |
| BM | Birmingham | | | |
| BN | Brighton | | | |
| BR | Bromley | | | |
| BS | Bristol | | | |
| BT | Belfast | (1) | | |
| | | (2) | | |
| | | (3) | | |
| | | (4) | | |
| | | (5) | | |
| | | (6) | | |
| CA | Carlisle | | | |
| CB | Cambridge | | | |
| CF | Cardiff | | | |
| CH | Chester | | | |
| CM | Chelmsford | | | |
| CO | Colchester | | | |
| CR | Croydon | | | |
| CT | Canterbury | | | |
| CV | Coventry | | | |
| CW | Crewe | | | |
| DA | Dartford | | | |
| DD | Dundee | (1) | | |
| | | (2) | | |
| | | (3) | | |
| DE | Derby | | | |
| DG | Dumfries | (1) | | |
| | | (2) | | |
| | | (3) | | |
| DH | Durham | | | |
| DL | Darlington | | | |
| DN | Doncaster | | | |
| DT | Dorchester | | | |
| DY | Dudley | | | |
| EC | London EC1 - 4 | | | |
| EH | Edinburgh | (1) | | |
| | | (2) | | |
| | | (3) | | |
| EL | London E1 - 18 | | | |
| EN | Enfield | | | |
| EX | Exeter | | | |
| FK | Falkirk | (1) | | |
| | | (2) | | |
| | | (3) | | |
| FY | Blackpool | | | |
| GL | Gloucester | | | |
| GS | Glasgow | (1) | | |
| | | (2) | | |
| | | (3) | | |
| GU | Guildford | | | |
| GY | Guernsey | | | |
| HA | Harrow | | | |
| HD | Huddersfield | | | |
| HG | Harrogate | | | |
| HP | Hemel Hempstead | | | |
| HR | Hereford | | | |
| HS | Scottish Islands | (1) | | |
| | | (2) | | |
| | | (3) | | |
| HU | Hull | | | |
| HX | Halifax | | | |
| IG | Ilford | | | |
| IM | Isle of Man | | | |
| IP | Ipswich | | | |
| IV | Inverness | (1) | | |
| | | (2) | | |
| | | (3) | | |
| JE | Jersey | | | |
| KA | Kilmarnock | (1) | | |
| | | (2) | | |
| | | (3) | | |
| KT | Kingston upon Thames | | | |

| Area | Location | | 1st QSO | 2nd QSO |
|---|---|---|---|---|
| KW | Orkney | (1) | | |
| | | (2) | | |
| | | (3) | | |
| KY | Kirkcaldy | (1) | | |
| | | (2) | | |
| | | (3) | | |
| LA | Lancaster | | | |
| LD | Llandrindod Wells | | | |
| LE | Leicester | | | |
| LL | Llandudno | | | |
| LN | Lincoln | | | |
| LP | Liverpool | | | |
| LS | Leeds | | | |
| LU | Luton | | | |
| ME | Medway | | | |
| MK | Milton Keynes | | | |
| ML | Motherwell | (1) | | |
| | | (2) | | |
| | | (3) | | |
| MR | Manchester | | | |
| NE | Newcastle upon Tyne | | | |
| NG | Nottingham | | | |
| NL | London N1 – 22 | | | |
| NN | Northampton | | | |
| NP | Newport | | | |
| NR | Norwich | | | |
| NW | London NW1 – 11 | | | |
| OL | Oldham | | | |
| OX | Oxford | | | |
| PA | Paisley | (1) | | |
| | | (2) | | |
| | | (3) | | |
| PE | Peterborough | | | |
| PH | Perth | (1) | | |
| | | (2) | | |
| | | (3) | | |
| PL | Plymouth | | | |
| PO | Portsmouth | | | |
| PR | Preston | | | |
| RG | Reading | | | |
| RH | Redhill | | | |
| RM | Romford | | | |
| SA | Swansea | | | |
| SD | Sheffield | | | |
| SE | London SE1 – 28 | | | |
| SG | Stevenage | | | |
| SK | Stockport | | | |
| SL | Slough | | | |
| SM | Sutton | | | |
| SN | Swindon | | | |
| SO | Southampton | | | |
| SP | Salisbury | | | |
| SR | Sunderland | | | |
| SS | Southend-on-Sea | | | |
| ST | Stoke-on-Trent | | | |
| SW | London SW1 – 20 | | | |
| SY | Shrewsbury | | | |
| TA | Taunton | | | |
| TD | Tweed | (1) | | |
| | | (2) | | |
| | | (3) | | |
| TF | Telford | | | |
| TN | Tonbridge | | | |
| TQ | Torquay | | | |
| TR | Truro | | | |
| TS | Teesside | | | |
| TW | Twickenham | | | |
| UB | Uxbridge | | | |
| WA | Warrington | | | |
| WC | London WC1 – 2 | | | |
| WD | Watford | | | |
| WF | Wakefield | | | |
| WL | London W1 – 14 | | | |
| WN | Wigan | | | |
| WR | Worcester | | | |
| WS | Walsall | | | |
| WV | Wolverhampton | | | |
| YO | York | | | |
| ZE | Shetland Isles | (1) | | |
| | | (2) | | |
| | | (3) | | |

# Callsign Listings





## Frequently Asked Questions

Throughout the year the RSGB receives a number of standard queries about an individual's entry in the *Yearbook*. Here are the answers given:

**Q: I have just gained my new licence, please include my new details in the next Yearbook.**
A: *Your details will be passed on to us by Spectrum Licensing from the data you supply on your licence application form. There is no need to contact the Society directly.*

**Q: I am an RSGB member. Thank you for sending *RadCom* to my new address, but why didn't you change my *Yearbook* entry?**
A: *The Yearbook lists all UK amateurs, not just RSGB members, and the records are entirely separate. Your callbook address details come from Spectrum Licensing. Did you re-validate your licence with your up-to-date details or let them know you moved? (***read Important Note below***)*

**Q: My entry shows me as being 'details withheld', but I want my full address to be listed.**
A: *Please contact Spectrum Licensing and ask them to release your details.* **The Society cannot accept direct input from licensees**.

**Q: You have published my address, but I would like it withheld.**
A: *As above, but ask Spectrum Licensing to withhold it from all callbook publishers.*

**Q: My callsign is not shown at all, please include it.**
A: *If a callsign is not shown in the Yearbook it was not licensed at the time the data was supplied to the RSGB. Sometimes, through an administrative error or misunderstanding, an amateur believes that he or she is licensed but the licensing records show that the licence has lapsed. If this is the case, contact Spectrum Licensing and discuss the matter with them as it is important that you re-validate your licence if you intend using it.*

## Important Note

We do not make amendments to the basic callsign data. Should the details that appear in the callsign section of this *Yearbook* need amending, you should contact the licensing authority direct. Your submission should be prior to the **1st June** to allow sufficient time for the amendment to be processed.

If you have not validated your licence, which you must do every 5 years, or if any of your details have changed, ie change of address, then your licence may be void and will not appear in the Callsign listings section of this *Yearbook*, and you may not be licensed to operate.

To validate your licence, log in to *https://services.ofcom.org.uk/* and amend any details as soon as possible, this will automatically validate your licence.

# United Kingdom

## 2*0

2E0 AAC Gary Ockett, 15 Deep Dale Street, Houghton le Spring, Dh5 0dq
2E0 AAF Steve Rhenius, Baythorne Cottage Baythorne End, Halstead, CO9 4AB
2E0 AAI David Simmons, 23 Fairey Street, Cofton Hackett, Birmingham, B45 8GU
2E0 AAJ Richard Yarrow, 27 Staplers Road, Newport, PO30 2DB
2E0 AAK Richard Berridge, Bracklyn, St. Clare Road, Deal, CT14 7QB
2E0 AAN Mark Keilty, 17 Cliff Road, Wallasey, Wirral, CH44 3DJ
2E0 AAO STEPHEN THIRLWALL, 2 Crossfield Avenue Blythe Bridge, Stoke-on-Trent, ST11 9PL
2M0 AAQ William Ferguson, 58o Fraser Walk, Kilmarnock, KA3 7NT
2E0 AAX E Wills, 1 Scott Close, High Street, Marlborough, SN8 3AF
2E0 AAZ W Elston, 49 Langdale Road, Kingsley, Northampton, NN2 7QQ
2E0 ABD R Mold, 134 Kipling Avenue, Brighton, BN2 6UE
2E0 ABL F Hayes, 88 Johns Road, Fareham, PO16 0RX
2E0 ABT L Simons, Westwood, Faris Lane, Addlestone, KT15 3DJ
2E0 ACA Ian Craig, Ferndown, Tilley Lane, Hailsham, BN27 4UT
2W0 ACD P Bennett, 13 Thornbury Close, Baglan, Port Talbot, SA12 8EU
2E0 ACE A Backhouse, 2 Brook Cottages, Garrigill, Alston, CA9 3EA
2E0 ACQ H Barnes, 24 Burleigh Place, Oakley, Bedford, MK43 7SG
2E0 ACR H Smallwood, 14 Shaftesbury Avenue, Bedford, MK40 3SA
2E0 ACV E Barnes, 3 Layton Road, Ashton-on-Ribble, Preston, PR2 1PB
2E0 ADA R Gaskell, 18 Woodcroft, Kennington, Oxford, OX1 5NH
2E0 ADL John Wresdell, Bracey Bridge Farm, Harpham, Driffield, YO25 4DE
2E0 ADN M Holmes, 55 Fitzpain Road, West Parley, Ferndown, BH22 8RZ
2E0 AED A Sumner, 78 Woodlands Way, Southwater, Horsham, RH13 9HZ
2E0 AES B Hyde, 54 The Byway, Darlington, DL1 1EQ
2E0 AFL M Coxhead, Baronsway, Parker Lane, Preston, PR4 4JX
2E0 AFP Robert Forrest-Webb, 1 Trelasdee Farm Cottages, St. Weonards, Hereford, HR2 8PU
2E0 AFT Andrew Rowe, 27 Silver Street, Cheddar, BS27 3LE
2E0 AGI E McLusky, 11 Ripon Road, Killinghall, Harrogate, HG3 2DG
2E0 AGQ B Read, 10 Shamblers Road, Cowes, PO31 7HF
2M0 AGS G J Robbins, 39 Locheil Gardens, Glenrothes, KY7 6YL
2M0 AIE Matthew Hooks, Braehead Mains, Main Street, Braehead Forth, Lanark, ML11 8HA
2E0 AIK M Footring, 26 Ernest Road, Wivenhoe, Colchester, CO7 9LG
2E0 AIL G Harper, 48 Norlands Lane, Widnes, WA8 5AS
2E0 AIM J Cowley, 39 Alpine Way, Tow Law, Bishop Auckland, DL13 4DS
2E0 AIT K Lloyd, 2 Bishopstone Drive, Beltinge, Herne Bay, CT6 6RE
2E0 AJK David Forsyth, 20 Chapel View, Rowlands Gill, NE39 2PN
2E0 AJP L Humphrey, 133 Hocombe Road, Chandler's Ford, Eastleigh, SO53 5QD
2E0 AJQ Andrew Nelson, 29 Coxford Road, Southampton, SO16 5FG
2E0 AJT William Atkins, 55 Park Grange Croft, Norfolk Park, Sheffield, S2 3QJ
2E0 AJX Christopher Overson, The Studio 6 Grenville Street, Bideford, EX39 2EA
2M0 AKI Darren Hague, 13 North Dell, Ness, Isle of Lewis, HS2 0SW
2E0 AKJ S Davin, 39 Nags Head Hill, Bristol, BS5 8LN
2E0 AKK E Jones, 43 Wesley Road, Wimborne, BH21 2QB
2E0 AKL R Mcdermott, 2 Monument Close, Wellington, TA21 9AL
2E0 AKN P Newton, 61 Ashbourne Crescent, Taunton, TA1 2RA
2E0 AKO Malcolm Newton, 43 hayfield road, Minehead, TA246AD
2E0 AKQ Jeremy Paul, Mimosa Lodge, 59 Baring Road, Cowes, PO31 8DW
2E0 AKR Andrew Nesbitt, 9 Manor Road, Richmond, TW9 1YD
2M0 AKS A Curlis, 94 Kirkhill Road, Aberdeen, AB11 8FX
2M0 AKT Keith Ross, 9 Bennan Place, East Kilbride, Glasgow, G75 9NR
2E0 AKU R BRISLEY, 15 Elm Fields, Old Romney, Romney Marsh, TN29 9SN
2E0 AKY Eric Curling, 919 Oxford Road, Tilehurst, Reading, RG30 6TP
2E0 ALA Nigel Kind, 18 Cunningham Road, Bentley, Walsall, WS2 0AY
2E0 ALB Maaruf Ali, 12 Hazeleigh Gardens, Woodford Green, IG8 8DX
2E0 ALD P Girling, The Gorse, Leiston Road, Aldeburgh, IP15 5QE
2E0 ALF A Taylor, 113 Queensway, Grantham, NG31 9RG
2E0 ALH J Side, Railway Crossing Cottage, Ash Road, Sandwich, CT13 9JB
2E0 ALJ Geoffrey Brierley, 35 Ochrewell Avenue, Deighton, Huddersfield, HD2 1LL
2E0 ALL A Sawyer, 26 Dallas Brett Crescent, Folkestone, CT19 6NE
2M0 ALS M Gill, Easter Templand, Fortrose, IV10 8RA
2E0 ALW R Hatcher, 61 Holland Road, Oxted, RH8 9AU
2E0 ALZ D Gunning, 19 Gordon Avenue, Prestatyn, LL19 8RU
2E0 AMB J Lloyd-Owen, 34 Maes Y Castell, Llandudno, LL30 1NG
2E0 AMW Geoffrey Fowle, 12 Lytham Road, Broadstone, BH18 8JS
2M0 ANE G Steele, 1 James Street, Bannockburn, Whins of Milton, Stirling, FK7 0NQ
2E0 AOK Paul Godolphin, Shepherds Cottage, Flakebridge, Appleby-in-Westmorland, CA16 6JZ
2E0 AOL T Money, 119 Twyford Way, Canford Heath, Poole, BH17 8SR
2E0 AOS E Hedley, 17 Chowdene Bank, Gateshead, NE9 6JJ
2E0 AOU Stuart Tilly, 24 Whinham Way, Morpeth, NE61 2TF
2E0 AOZ I Marshall, 44 Cromwell Crescent, Pontefract, WF8 2EJ
2M0 APB John Macdonald, 63 Perceval Road South, Stornoway, HS1 2TL
2E0 APG G Wilson, 19 Ercall Close, Trench, Telford, TF2 6RR
2E0 APJ M France, 10 Grampian Avenue, Wakefield, WF2 8JZ
2E0 APN T Humble, 43 Whiteley Crescent, Bletchley, Milton Keynes, MK3 5DQ
2W0 APT G Davies, 27 Twyniago, Pontarddulais, Swansea, SA4 8HX
2M0 APX Shelley Barbour, 40 Mannerston Holdings, Linlithgow, EH49 7ND
2E0 APY J Rayson, 18 Station Street, South Wigston, Wigston, LE18 4TH
2E0 APZ Martin Meyer, 72 Trevelyan, Bracknell, RG12 8YD
2E0 AQE J Brown, 55 Barrington Road, Rubery, Birmingham, B45 9EU
2E0 AQI D Black, 8 Cornwood Close, Finchley, London, N2 0HP
2E0 AQU Paul Webb, 40 Links Road, Penn, Wolverhampton, WV4 5RF
2E0 ARA John Griffin, 35 Cottage Street, Kingswinford, DY6 7QE
2E0 ARJ Robin Birch, 15 Chester Street, Cirencester, GL7 1HF
2E0 ARV I Strachan, 1 Gantley Avenue, Billinge, Wigan, WN5 7AF
2E0 ASU Juna Bell, 122 Howard Drive, Letchworth Garden City, SG6 2DE
2E0 ASW S Pearson, 8 The Pastures, Edlesborough, Dunstable, LU6 2HL
2E0 ASX R Bannister, 60 St. Johns Avenue, Bridlington, YO16 4NL
2E0 ATB L Coneley, 157 Gordon Road, Fareham, PO16 7TB
2E0 ATF J Constable, 9 Ridgeway Close, Heathfield, TN21 8NS
2E0 ATY Ralph Browne, 2 Martham Close, London, SE28 8NF
2E0 ATZ Robin MacDonald, 4 Throwley Drive, Herne Bay, CT6 8LP
2W0 AUC R Henderson, 150 Heol Maes Eglwys, Pantlasau, Swansea, SA6 6NW
2E0 AUI David Tyson, 21 Providence Place, Filey, YO14 9DU
2E0 AUM A Peach, 14 Northfield Avenue, Rocester, Uttoxeter, ST14 5LE
2E0 AUU P Christopoulos, Cosenda, Tye Green Village, Harlow, CM18 6QY
2E0 AUV C Brice, 10 Swan Close, Weston-Super-Mare, BS22 8XR
2E0 AVA T Milne, 3 Birch Road, New Ollerton, Newark, NG22 9PU
2E0 AVB M Orr, 42 Mayfield Road, London, W12 9LU
2E0 AVD Wendy Jukes, 22 Hazelmere Road, Creswell, Worksop, S80 4HS
2E0 AVK R Harrall, 47 Hawksmoor Road, Stafford, ST17 9DS
2M0 AVL A Fraser, 29 Seafield Gardens, Aberdeen, AB15 7YB
2E0 AVP A Dixon, 15 Maple Crescent, Basingstoke, RG21 5SX
2E0 AVQ J Paul, 38 Avon Road, Upminster, RM14 1QU
2E0 AVS S Panczel, 11 Chauncy Road, Manchester, M40 3GG
2E0 AVW Richard Smith, London Road, Shardlow, Derby, DE72 2GL
2E0 AVZ N Heyne, Croxdale, Chiddingly Road, Heathfield, TN21 0JH
2E0 AWC K Necchi, 59 Winters Way, Waltham Abbey, EN9 3HP
2E0 AWE J Sanders, 219 Station Road, Drayton, Portsmouth, PO6 1PY
2E0 AWG S Rope, 51 Medeswell Close, Brundall, Norwich, NR13 5QG
2E0 AWI R Desbois, 12 Dove Walk, Hornchurch, RM12 5HH
2E0 AWK J Jackson, 48 Whitefield Avenue, Rochdale, OL11 5YG
2E0 AWM I Reynolds, 98 Priestley Road, Stevenage, SG2 0BP
2E0 AWR J Gibson, 14 Douglas Bank Drive, Wigan, WN6 7NH
2E0 AWS A Storer, 40 Danetre Drive, Daventry, NN11 4GY
2E0 AWT G Griffiths, 16 Back Lane, Winteringham, Scunthorpe, DN15 9NW
2E0 AWU G Dunne, Three Ways, Northbourne Road, Deal, CT14 0HJ
2E0 AWW R Evans, 4 Green Terrace, Deiniolen, Caernarfon, LL55 3LE
2M0 AWY J Fulton, 35 Heatherbank, Livingston, EH54 6EE
2E0 AWZ R Turton, 31 Second Avenue, Woodlands, Doncaster, DN6 7QQ
2E0 AXA K Kerridge, 31 Second Avenue, Woodlands, Doncaster, DN6 7QQ
2E0 AXB M Costello, 2 Flintham Close, Birmingham, B27 6RL
2E0 AXN S Jenner, 4 Christie Close, Chatham, ME5 7NG
2E0 AXT A Smith, 186 Longmead Drive, Nottingham, NG5 6DJ
2E0 AXZ C Cook, 112 Waterford Road, Ipswich, IP1 5NJ
2E0 AYI K Williams, 36 Castle Street, Tiverton, EX16 6RG
2E0 AYK M Longbottom, 32 Anns Hill Road, Gosport, PO12 3JY
2E0 AYQ M Walsh, 25 Felstead Road, Orpington, BR6 9AA
2E0 AYT R Cole, 41 Ninehams Road, Caterham, CR3 5LN
2M0 AYU J Bruce, 33 Kintail Place, Dingwall, IV15 9RL
2E0 AYW M Pedley, 70 Dane Avenue, Barrow-in-Furness, LA14 4JY
2E0 AYX Cecil Penfold, 14 Romney Road, Tetbury, GL8 8JU
2E0 AYY Paul Rainer, 3 St. Martins Road, Folkestone, CT20 3LA
2M0 AYZ Colin Gajewski, 7/2 Brunswick Road, Edinburgh, EH7 5NG
2E0 AZA P Livsey, 33 Aldingham Walk, Morecambe, LA4 4EW
2E0 AZB M Livsey, 26 Aldingham Walk, Morecambe, LA4 4EW
2E0 AZD F Russell, 157 Durnford Street, Plymouth, PL1 3QR
2E0 AZF D Whitworth, 570 Dereham Road, Norwich, NR5 8TE
2E0 AZJ V Hocking, 80 Barton Tors, Bideford, EX39 4HA
2E0 AZK A Cartwright, 17 Greenheath Way, Wirral, CH46 3RX
2E0 AZL L Murphy, 22 Shenley Fields Drive, Birmingham, B31 1XH
2E0 AZM R Harper, 17 Conway Drive, Wrexham, LL13 9HR
2W0 AZP D Edwards, 9 Ffos Y Cerridden, Nelson, Treharris, CF46 6HQ
2E0 AZS Anthony Elliott, 32 Tredegar Walk, Hartlepool, TS26 0TP
2E0 AZU A Groat, Rose Cottage, Ivy Dene Lane, East Grinstead, RH19 3TN
2M0 AZW T Galt, 14 Colwood Avenue, Glasgow, G53 7XT
2E0 AZZ A Dalzell, 9 Pyms Lane, Crewe, CW1 3PJ
2E0 BAB G Grayson, 2 Greenacres Grove, Shelf, Halifax, HX3 7RN
2I0 BAC C McConnell, 41 Moyra Road, Doagh, Ballyclare, BT39 0SQ
2I0 BAD A Hayes, 9 Ashlea Avenue, Ballymoney, BT53 7BZ
2E0 BAF Debbie Joanne Davies, 31 Beverley Close, Rainworth, Mansfield, NG21 0LW
2E0 BAH P Mannering, 101 Burford, Brookside, Telford, TF3 1LJ
2E0 BAJ F Cappleman, 8 The Woodlands, Lilleshall, Newport, TF10 9EN
2E0 BAK M Verrall, Flat 4, 84 Briar Way, Skegness, PE25 3PU
2W0 BAO T Mulcuck, 61 Hillside View, Graigwen, Pontypridd, CF37 2LG
2E0 BAP S Savage, 450 Locking Road, Milton, Weston-Super-Mare, BS22 8PS
2E0 BAU C Young, 15 Shelton Avenue, East Ayton, Scarborough, YO13 9HB
2E0 BAV A Yorke, 33 Avon Crescent, Stratford-upon-Avon, CV37 7EX
2E0 BAX Sean Amesbury, 13 Haddon Close, Macclesfield, SK11 7YG
2E0 BAY A Harris, 32 King Edward Road, Gillingham, ME7 2RE
2E0 BBA I Wengraf, 45 School Lane, Higham, Rochester, ME3 7JR
2E0 BBG D Skinner, 77 Rolleston Avenue, Petts Wood, Orpington, BR5 1AL
2E0 BBI A Hayward, 23 Greenbank Close, Grampound Road, Truro, TR2 4TD
2E0 BBL Carl Lishman, 6 Clarence Road, Accrington, BB5 0NA
2E0 BBM Simon Beales, 49 Greenacres, Old Newton, Stowmarket, IP14 4EJ
2E0 BBN S Woodcock, 3 Lower House Road, Leyland, PR25 1HT
2W0 BBO L Powell, 2 Gelliderw, Pontardawe SA8 4NB, Swansea, SA8 4NB
2E0 BBQ Rupert Gash, 2 The Marsh, Benington, Boston, PE22 0DH
2E0 BBS M Janes, 1 Cheswell Gardens, Church Crookham, Fleet, GU51 5NJ
2W0 BBT S christie-powell, 2 Gelliderw Pontardawe, Swansea, SA8 4NB
2E0 BBX John Akinin, 70 Valley Road, West Bridgford, Nottingham, NG2 6HQ
2E0 BBY F Coles, 8 Moore Close, Church Crookham, Fleet, GU52 6JD
2E0 BBZ M Thyer, 8 Roman Road, Taunton, TA1 2BD
2E0 BCB Dean HOLLAND, 13 Linley Drive, Boston, PE21 7EJ
2E0 BCD Barry Grice, 17 Albion Field Drive, West Bromwich, B71 4HN
2E0 BCE J Ferris, 39 Gladstone St., Beeston, Nottingham, NG9 1EU
2E0 BCF H Donnelly, 6 Famet Walk, Purley, CR8 2DY
2E0 BCG W Walker, 9 Malthouse Lane, Dorchester-on-Thames, Wallingford, OX10 7LF
2E0 BCJ D Neville, 21n Alderwood Avenue, Liverpool, L24 2UB
2E0 BCK T Gale, Flat 15 Browning Apartments, 140 Hamlets Way, London, E3 4GS
2M0 BCL Thomas McCall, 119 Claremont, Alloa, FK10 2ER
2E0 BCM B Marston, 38 Bulstrode Road, Ipswich, IP2 8HA
2E0 BCO S norman, 27 Ashburton Road, Ickburgh, Thetford, IP26 5JA
2E0 BCQ R Britt, Thoroughfare House, South Burlingham Road, Norwich, NR13 4FA
2D0 BCR Robert Cunningham, 3 Kellets Cottage, Lhergy Cripperty, Union Mills, Isle of Man, IM4 4NF
2E0 BCS J Schofield, 6 Robin Royd Avenue, Mirfield, WF14 0LF
2E0 BCW S Bywater, Birch Wood, Norwich Road, Cromer, NR27 0HG
2E0 BCX J Allen, 20 Spa Hill, Kirton Lindsey, Gainsborough, DN21 4BA
2E0 BDB Anthony Grosvenor, 10 Neves Close, Lingwood, Norwich, NR13 4AW
2E0 BDD A Morrison, 43 Alexandra Road, Northampton, NN1 5QP
2E0 BDI S Moakes, 46 Parsonage Street, Stockport, SK4 1HZ
2E0 BDJ D Anthony, 20 Parkfield Close, Leyland, PR26 7XJ
2M0 BDN J Walker, 49 Great King Street, Edinburgh, EH3 6RP
2E0 BDO S Grainger, 15 Carr House Lane, Wirral, CH46 6EN
2E0 BDP M Saunders, 70 Underwood Lane, Crewe, CW1 3LE
2E0 BDQ William Taylor, 99 St. Marys Close, Littlehampton, BN17 5QQ
2M0 BDR B Keiller, Da Cro, Branchiclate, Burra Isle, ZE2 9LA
2E0 BDS Brendan Derbin-Sykes, 1 Lentons Lane, Friskney, Boston, PE22 8RR
2M0 BDT Andrew Halcrow, Da Cro, Branchiclate, Burra Isle, ŽE2 9LA
2E0 BDV J Goulding, 79 Dalston Drive, Manchester, M20 5LQ
2W0 BDW B Perrett, 43 Glyn Collen, Cardiff, CF23 7ER
2E0 BEA J Rushton, 25 Garth Meadow, Catterick, Richmond, DL10 7RT
2M0 BEB G Bourhill, 30c Salters Road, Wallyford, Musselburgh, EH21 8AA
2M0 BEC M Mitchell, Raithburn Farm, Glasgow Road, Kilmarnock, KA3 6ES
2W0 BED C Rosser, 16 Thomas St., Penygraig, Tonypandy, CF40 1EU
2E0 BEE Alistair Bell, 2 Croft Foot, Sandwith, Whitehaven, CA28 9UG
2E0 BEF Mary Staton, 52 School Road, Newborough, Peterborough, PE6 7RG
2E0 BEG Michael Davidson, 19 Mason Street, Workington, CA14 3EH
2E0 BEH C loughran, 8 Douglas Road, Dover, CT17 0BD
2E0 BEI Clive brett, 60 Pine Tree Avenue, Canterbury, CT2 7TJ
2M0 BEL P McCluskey, 119 Tower Drive, Gourock, PA19 1SG
2E0 BEP Alison Holmes, 127 Lower Oxford Street, Castleford, WF10 4AG
2M0 BET D Boden, 42 Kirkwynd, Maybole, KA19 7AE
2E0 BEV I Lowe, 1 Hazelby Road, Creswell, Worksop, S80 4BB
2E0 BEW D Powis, Fircroft, Pound Lane, Woodbridge, IP13 0LN
2E0 BFA J Mullarkey, 41 Foyle Avenue, Chaddesden, Derby, DE21 6TZ
2I0 BFB H McErlean, 24 Mullaghboy Heights, Magherafelt, BT45 5NU
2E0 BFC Michael Williams, 30 Elm Drive, Risca, Newport, NP11 6HJ
2E0 BFD W Corbett, 32 Fairview Avenue, Risca, Newport, NP11 6HU
2E0 BFF Anthony Bateman, 154 Arnold Road, Mangotsfield, Bristol, BS16 9LB
2E0 BFG Peter Davies, 53 Lammas Road, Cheddington, Leighton Buzzard, LU7 0RY
2E0 BFJ Gary Swain, 3 Flaxfield Drive, Crewkerne, TA18 8DF
2E0 BFM Andrew Knights, 81 Green Lane, Barnard Castle, DL12 8LF
2E0 BFN Wayne Carty, 49 Princess Gardens, Blackburn, BB2 5EJ
2E0 BFS B Townshend, 9 Norfolk Place, Boston, PE21 9JJ
2E0 BFT David Doroba, Flat 3, 305a London Road South, Lowestoft, NR33 0DX
2W0 BFV W Harries, 18 Bro Teify, Alltyblacca, Llanybydder, SA40 9SR
2E0 BFZ Peter Hinde, 19 Tadcaster Road, Sheffield, S8 0RA
2E0 BGC Norman Speight, Flat 14, Cranbrook, London, NW1 0LJ
2U0 BGE J Bligh, The Bounty, Salines Lane, St. Sampson, Guernsey, GY2 4FL
2W0 BGI M Lewis, 4 Coldwell Terrace, Pembroke, SA71 4QL
2E0 BGJ M Whitaker, 5 Horns Drove, Rownhams, Southampton, SO16 8AH
2E0 BGL B Lloyd, 104 Wootton Street, Bedworth, CV12 9DZ
2E0 BGM S Russon, 165 Billington Avenue, Newton-le-Willows, WA12 0AU
2E0 BGO J Woodruff, 62 Burton St., Rishton, Blackburn, BB1 4PD
2E0 BGP E Neil, 5 Winsford Hill, Furzton, Milton Keynes, MK4 1BJ
2W0 BGQ I Canterbury, Brynllethryd Bungalow, Senghenydd, Caerphilly, CF83 4HJ
2W0 BGS John Brydges, 9 Twynygarreg, Treharris, CF46 5RL
2E0 BGV R Johnson, 30 Thorpe Downs Road, Church Gresley, Swadlincote, DE11 9FB
2E0 BGX P Sarll, 81 Austendyke Road, Weston Hills, Spalding, PE12 6BX
2E0 BGZ Ronald Dale, 17 Spencer Gardens, Brackley, NN13 6AQ
2E0 BHA P Bennett, 1 Queens Road, Carterton, OX18 3YB
2E0 BHB R Fallows, Loughrigg, Cranmore Avenue, Yarmouth, PO41 0XS
2E0 BHC L Heenan, 1 Howard Close, Daventry, NN11 4TD
2E0 BHD David Thorn, 10 Orchard Road, Basingstoke, RG22 6NU
2E0 BHH Rod Bullen, 2 Redlands Cottages, East Coker, Yeovil, BA22 9HF
2E0 BHJ J List, 41 Westbury Crescent, Dover, CT17 9QQ
2E0 BHN James McColl, 6 Grenville Close, Bodmin, PL31 2FB
2E0 BHP Gary Suter, 11 Summerdown Close, Durrington, Worthing, BN13 3QG
2E0 BHQ David George, 9 winscombe court, Frome, BA11 2DZ
2E0 BHS N Tideswell, 19 Wish Court, Ingram Crescent West, Hove, BN3 5NY
2I0 BHT Samuel Quigg, 100 Whispering Pines, Limavady, BT49 0UF
2E0 BHU John Hickman, Ardoch, Harlestone Road, Northampton, NN6 8AW
2E0 BHY Steven Preston, 22 Baddesley Close, North Baddesley, Southampton, SO52 9DR
2E0 BHZ T Mildenhall, 58 Montrose Avenue, Datchet, Slough, SL3 9NJ
2E0 BIB Russell Orton, 18 Clarel Street, Penistone, Sheffield, S36 6AU
2E0 BIC Tony Humphries, 10 Cropthorne Avenue, Leicester, LE5 4QL
2I0 BID A Jamison, 11 Richmond Gardens, Newtownabbey, BT36 5LA
2E0 BII Robin Hathaway, 38 Windsor Walk, Lindford, Bordon, GU35 0SG
2M0 BIK Leslie Tombe, Benview, The Quarters, Fochabers, IV327HJ
2M0 BIL William McCue, 188 Redburn, Alexandria, G83 9BU
2E0 BIM Kevin Hawke, 111 Dorchester Avenue, Plymouth, PL5 4AZ
2M0 BIN Frederick Coombes, 44 Lochfield ROAD, Paisley, PA2 7RL
2E0 BIO R Aston, 8 Parliament Road, Lea Park Estate, Thame, OX9 3TE
2I0 BIQ K Blake, 3 Red Row, Tandragee Road, Craigavon, BT63 6BB

| | | |
|---|---|---|
| 2I0 | BIN | James Smyth, 37 Ardfreelin, Newry, BT34 1JG |
| 2E0 | BIS | J Cosson, 25 Fox Brook, St. Neots, PE19 6AL |
| 2E0 | BIU | P Rushby, 16 Foxhill Lane, Selby, YO8 9AD |
| 2E0 | BIY | P Lewis, 16 Valley Road, St. Albans, AL3 6LR |
| 2E0 | BJA | B Johnson, 15 Oak Avenue, Willington, Crook, DL15 0BJ |
| 2E0 | BJB | Anthony Chaplin, 33 The Crofts, Little Wakering, Southend-on-Sea, SS3 0JS |
| 2E0 | BJD | R Rowe, 16 Orchard Road, Plymouth, PL2 2QY |
| 2W0 | BJE | Sidney Merrifield, 37 South View Drive, Rumney, Cardiff, CF3 3LX |
| 2E0 | BJK | Royston Robertson, 2 Dunsley Close, Middlesbrough, TS3 7DR |
| 2E0 | BJL | [illegible] Flat [illegible] Avenue, Epsom, KT19 9HD |
| 2E0 | BJM | B Maggs, 44 Coldharbour Road, Hungerford, RG17 0AZ |
| 2E0 | BJ[illegible] | Philip Harman, [illegible] Close, West Row, Bury St Edmunds, IP28 8QD |
| 2E0 | BJ[illegible] | M [illegible], [illegible] Jubilee Road, Halifax, HX[illegible] 0LD |
| 2W0 | BJR | Robert Johnson, 25 Lon Tyrhaul, Llansamlet, Swansea, SA7 9SF |
| 2E0 | BJS | Nigel Roberts, 40 Armour Road, Tilehurst, Reading, RG31 6HN |
| 2E0 | BJT | R Carter, 43 Sheldon Avenue, Standish, Wigan, WN6 0LW |
| 2M0 | BJU | G Craig, 1 Butt Avenue, Helensburgh, G84 0DA |
| 2E0 | BJV | T Gabriel, 57 West Down Road, Delabole, PL33 9DT |
| 2E0 | BJW | B White, 9 Springfield Close, Wirral, CH49 7NJ |
| 2E0 | BJX | Graham Dyson, 6 Twynersh Avenue, Chertsey, KT16 9DE |
| 2E0 | BJY | Simon Billingham, 6 Swinford Leys Wombourne, Wolverhampton, WV5 8HT |
| 2W0 | BKA | Mark Mainwaring, 36 Oak Street, Gilfach Goch, Porth, CF39 8UG |
| 2E0 | BKB | Brian Hall, 6 Marshall Close, Parkgate, Rotherham, S62 6DB |
| 2E0 | BKD | Graham Flack, 20 The Pastures, Hardwick, Cambridge, CB23 7XA |
| 2E0 | BKE | Adrian Anderson, 89a Malmesbury Park Road, Bournemouth, BH8 8PS |
| 2I0 | BKI | Ernest Kyle, 2, Wattstown, Coleraine, BT521SP |
| 2E0 | BKJ | Gillian Douch, High House, Ulverston Road, Ulverston, LA12 0PZ |
| 2M0 | BKL | Steve Fradley, 30 Polmont Park, Polmont, Falkirk, FK2 0XT |
| 2W0 | BKM | Emlyn Thomas, 29 Maes y Wern, Carway, Kidwelly, SA17 4HF |
| 2E0 | BKN | Max Stokes, 15 Parc Terrace, Newlyn, Penzance, TR18 5AS |
| 2E0 | BKO | Aaron Oxlade, 27 Spenfield Court, Northampton, NN3 8LZ |
| 2E0 | BKP | S Martin, 70 Moorlands Drive, Stainburn, Workington, CA14 4UJ |
| 2E0 | BKQ | David Griffin, 4 Willowside Way, Royston, SG8 5ET |
| 2E0 | BKT | Gordon Milsom, Flat 8, Sovereign Court, High Wycombe, HP13 6XL |
| 2E0 | BKV | S Bird, 9 Almery Drive, Carlisle, CA2 4EX |
| 2E0 | BKY | Francis Goodall, 1 Parkfield Grove, Leeds, LS11 7LS |
| 2E0 | BKZ | William Owen, 8 Sandhurst Avenue, Lytham St. Annes, FY8 2DA |
| 2W0 | BLA | J Jones, Bronydd, Blaenffos, Boncath, SA37 0HZ |
| 2E0 | BLC | John Farina, 9 Mallards Close, Alveley, Bridgnorth, WV15 6JL |
| 2E0 | BLD | V Parton, 51 Marston Grove, Stoke-on-Trent, ST1 6EF |
| 2E0 | BLF | A Spaxman, 12 Stanhope Gardens, Barnsley, S75 2QB |
| 2E0 | BLG | R Williams, Plaen Cottage, Bodfari, Denbigh, LL16 4BS |
| 2E0 | BLI | Jordan Skittrall, 14 Tamarin Gardens, Cambridge, CB1 9GH |
| 2E0 | BLJ | Alan Marks, grosvenor hotel, 51 grosvenor road, Scarborough, YO112LZ |
| 2E0 | BLK | Stuart Heaton, 2 Hunmanby Road, Reighton, Filey, YO14 9RT |
| 2E0 | BLL | Mike Green, 11 Moorhead Gardens, Warton, Preston, PR4 1WA |
| 2E0 | BLN | D Blake, 3 Carrside, Eastfield, Scarborough, YO11 3DE |
| 2E0 | BLQ | Jason Brewster, 44 Beaulieu Close, Hounslow, TW4 5EW |
| 2E0 | BLR | Robert Payne, 74 Churchill Avenue, Newmarket, CB8 0BY |
| 2E0 | BLS | Leslie Turner, 16 Woodland Place, Scarborough, YO12 6EP |
| 2E0 | BLT | John Owen, 90 Granville Street, Kingswinford, DY6 8LW |
| 2E0 | BLW | Robert Silcox, 103 Oakdale Road, Downend, Bristol, BS16 6EG |
| 2E0 | BLX | Gordon Thorpe, 81 knoll drive, Coventry, CV3 5PJ |
| 2E0 | BLY | A Bailey, 58 Billy Buns Lane, Wombourne, Wolverhampton, WV5 9BP |
| 2E0 | BLZ | David Forster, 23 Field Street, Padiham, Burnley, BB12 7AU |
| 2E0 | BMB | A Bolla, 11 Shelley Crescent, Blyth, NE24 5RH |
| 2E0 | BMD | John Turner, 17 Beechwood Road, Dronfield, S18 1PW |
| 2E0 | BME | Colin King, 5 Manor Road, Tamworth, B77 3PE |
| 2M0 | BMF | Fergus Thomson, 11 Carmichael Place, Irvine, KA12 0XH |
| 2E0 | BMG | W Mcgill, 49 Anthony Close, Colchester, CO4 0LD |
| 2E0 | BMH | Brian Hawes, 3 Orchard Close, Cassington, Witney, OX29 4BU |
| 2E0 | BMI | James Colderwood, 34 Desborough Way, Norwich, NR7 0RR |
| 2E0 | BMJ | R Burrows, 62 Fletcher Road, Burbage, Hinckley, LE10 2PS |
| 2M0 | BMK | J cosgrove, The Cottages, Kirkinch, Blairgowrie, PH12 8SL |
| 2W0 | BMM | N Hughes, Mountain Farm Cottage, Clynderwen, Llandissilio, SA66 7PX |
| 2M0 | BMN | C Lewis, 9 Cessnock Road, Troon, KA10 6NJ |
| 2E0 | BMO | Roger Rimmer, 8 Greensward Close, Standish, Wigan, WN6 0RY |
| 2E0 | BMP | John Redfern, 19 Stanton Green, Shrewsbury, SY1 4PL |
| 2W0 | BMR | Spencer Williams, 11 Beyron Clos, Pontypridd, CF37 5HW |
| 2E0 | BMU | Philip Wright, 34 Coles Lane, West Bromwich, B71 2QJ |
| 2E0 | BMW | A Rowe, Southern Point, Grange View, Houghton le Spring, DH4 4HU |
| 2E0 | BMY | Allan Farrar, 8 Wensley Street, Thurnscoe, Rotherham, S63 0PX |
| 2E0 | BNA | Neal Anderson, 3 Phelipps Road, Corfe Mullen, Wimborne, BH21 3NN |
| 2E0 | BNB | Brian Hayes, 22 Urban Way, Biggleswade, SG18 0HT |
| 2E0 | BNC | Cameron Murray, 7 Pilmoor Drive, Richmond, DL10 5BJ |
| 2E0 | BND | L Woollard, 46 Woodhill Lane, Morecambe, LA4 4NN |
| 2E0 | BNE | Stuart Sanderson, 65 Holm Flatt Street, Parkgate, Rotherham, S62 6HJ |
| 2E0 | BNF | Andrew Webb, 47 Granville Street, Gloucester, GL1 5HL |
| 2E0 | BNH | J Hawkes, 183 Borden Lane, Sittingbourne, ME10 1DA |
| 2E0 | BNI | George Marshall, 12 Arthur Avenue, Caister-on-Sea, Great Yarmouth, NR30 5PQ |
| 2E0 | BNJ | S Scott, Watercrook Bungalow, Natland, Kendal, LA9 7QB |
| 2E0 | BNK | John Stoppard, 14 Brookside Bar, Chesterfield, S40 3PJ |
| 2E0 | BNO | S Pearson, 29 Darwin Road, Chesterfield, S40 4RX |
| 2E0 | BNT | Bruce Trayhurn, 15 Wight Drive, Caister-on-Sea, Great Yarmouth, NR30 5UN |
| 2E0 | BNV | K Peel, 123 Cunningham Road, Tamerton Foliot, Plymouth, PL5 4PU |
| 2E0 | BNW | Peter Ashton, 14 Poppy Close, Boston, PE21 7TJ |
| 2E0 | BNZ | William Whitcher, 17 Watermead, Stratton St. Margaret, Swindon, SN3 4WE |
| 2E0 | BOB | R Hastings, 2 Boleyn Way, Boreham, Chelmsford, CM3 3JJ |
| 2E0 | BOC | Owen Williams, 39 Camden Road, Maes-Y-Coed, Brecon, LD3 7RT |
| 2E0 | BOD | Jon Tusler, Kilima, Batts Corner, Farnham, GU10 4EX |
| 2E0 | BOF | P Callaghan, 41 Higher Ash Road, Talke, Stoke-on-Trent, ST7 1JN |
| 2E0 | BOI | Rosario Massimino, 115 Trelowarren Street, Camborne, TR14 8AW |
| 2E0 | BOJ | Mark Anthony, The Bungalow, Magpies Cottage, Redruth, TR16 5JL |
| 2W0 | BOK | A Budding, 54 Wern Isaf, Dowlais, Merthyr Tydfil, CF48 3NY |
| 2E0 | BON | Jason Ball, Manor House, Tolgus Hill, Redruth, TR15 1AX |
| 2D0 | BOR | Robert Gee, Flat 1D, Quarmby Road, Huddersfield, HD3 4HQ |
| 2M0 | BOS | S McCurdy, 5 Kestrel Place, Greenock, PA16 7BL |
| 2E0 | BOT | A Canning, The Mount, Birmingham Road, [illegible] |
| 2E0 | BOV | Ting-Yueh Liu, 52 Tenison Road, Cambridge, CB1 2DW |
| 2W0 | BOX | Colin Davis, Denant Mill, Dreenhill, Haverfordwest, SA62 3TS |
| 2E0 | BOY | Shaun Kirkpatrick, Homeland, Rowanshill Crescent, Stranraer, DG9 0HL |
| 2E0 | BOZ | David Tidswell, 1 Cherrytree Grove, Spalding, PE11 2NA |
| 2E0 | BPE | James Freeman, 12 Norfolk Terrace, Cambridge, CB1 2NG |
| 2E0 | BPF | Frederick Banner, 14 Dale Road, Barnard Castle, DL12 8LQ |
| 2E0 | BPG | B Dixon, 21 Pankhurst Road, Hoo, Rochester, ME3 9DF |
| 2E0 | BPI | Adrian Schuler, 6 Tatham Court, [illegible] |
| 2W0 | BPJ | Richard Jones, 36 Heol-y-Cae, Cefn Coed, Merthyr Tydfil, CF48 2NT |
| 2E0 | BPL | Dennis Golding, Windrush Cottage, [illegible], Chippenham, SN15 1EL |
| 2M0 | BPM | Martin Dickson, 41 Hossack Drive, Elgin, IV30 6JY |
| 2E0 | BPN | Jonathan Barker, 20 Ardley Road, Fewcott, Bicester, OX27 7PA |
| 2I0 | BPO | J Rice, 12 The Crescent, Ballymoney, BT53 6ES |
| 2E0 | BPP | Colin Bell, 20 Kingfisher Close, Blackburn, BB1 8NS |
| 2E0 | BPS | John Murray, 2 The Cuttings, Hampstead Norreys, Thatcham, RG18 0RR |
| 2E0 | BPT | Lorne Clark, 16 Kibblewhite Crescent, Twyford, Reading, RG10 9AX |
| 2E0 | BPU | Nicholas Phillips, First Floor Flat, 116 Lodge Road, Croydon, CR0 2PF |
| 2M0 | BPV | Neil Davidson, 25 Hopetoun Court, Bucksburn, Aberdeen, AB21 9QS |
| 2E0 | BPW | J Hall, 1 Nash Close, Earley, Reading, RG6 5SL |
| 2E0 | BPX | Dave Adshead, 16 Moat Way, Swavesey, Cambridge, CB24 4TR |
| 2E0 | BPY | Mark Rogers, 22 Robson Drive, Hoo, Rochester, ME3 9EA |
| 2E0 | BQB | Theophilus Horsoo, 5 Kelmarsh Court, Great Holm, Milton Keynes, MK8 9EN |
| 2E0 | BQC | Kevin Bindley, 56 Iona Close, Beaumont Leys, Leicester, LE4 0QY |
| 2E0 | BQE | John Callis, 51 Pipistrelle Way, Oadby, Leicester, LE2 4QA |
| 2E0 | BQF | Duncan Gunn, 40 The Pastures, Oadby, Leicester, LE2 4QD |
| 2E0 | BQG | Sam Turner, 12 Park Street, Morecambe, LA4 6BN |
| 2E0 | BQH | Michael Bailey, 17 Sparrowhawk Way, Hartford, Huntingdon, PE29 1XE |
| 2E0 | BQJ | Daniel Trudgian, 18 Hart Close, Wootton Bassett, Swindon, SN4 7FN |
| 2E0 | BQK | Michael Carr, 25 Malvern Avenue, Fareham, PO14 1QF |
| 2E0 | BQL | Ben Hall, 64 Synehurst Crescent, Badsey, Evesham, WR11 7XX |
| 2E0 | BQM | Nick Jewitt, 45 North Road, Kirkburton, Huddersfield, HD8 0QH |
| 2E0 | BQN | Roger Suffling, 44 Belgrave Manor, Woking, GU22 7TW |
| 2E0 | BQO | D Green, 12 Nostell Road, Ashton-in-Makerfield, Wigan, WN4 9XD |
| 2E0 | BQP | Kevin Mills, 106 Goodwin Crescent, Swinton, Mexborough, S64 8QR |
| 2E0 | BQQ | John Addy, 12 Wortley Avenue, Swinton, Mexborough, S64 8PT |
| 2E0 | BQR | Stephen James, Wardens House, Kirkstone Close, Doncaster, DN5 9QZ |
| 2E0 | BQU | Christopher Langmaid, Flat 4, Woodlawn High Street, Partridge Green, Horsham, RH13 8HR |
| 2E0 | BQV | Jonny Parrett, 6 Shelley Road, East Grinstead, RH19 1TA |
| 2E0 | BQW | Anton Chapman, 24 Eaton Grange Drive, Long Eaton, Nottingham, NG10 3QE |
| 2E0 | BQX | N Groat, Rose Cottage, Ivy Dene Lane, East Grinstead, RH19 3TN |
| 2E0 | BQY | Andre Clinchant, 14 Taylor Close, Norton Fitzwarren, Taunton, TA2 6TA |
| 2E0 | BQZ | Keith Puttock, 12 Beechfields, School Lane, Petworth, GU28 9DH |
| 2E0 | BRC | Elaine Smith, 13 Eagle Avenue, Waterlooville, PO8 9UB |
| 2M0 | BRD | B Donnelly, 19 Douglas Drive, Dunfermline, KY12 9YG |
| 2E0 | BRF | Colin Dawson, 9 Mulberry Close, Poringland, Norwich, NR14 7WF |
| 2M0 | BRH | Christopher Jones, Croy Lodge, Shandon, Helensburgh, G84 8NN |
| 2E0 | BRI | B Bott, 15 Lansdowne Crescent, Darton, Barnsley, S75 5PW |
| 2E0 | BRJ | Paul Shirlaw, 32 West Street, Faversham, ME13 7JG |
| 2E0 | BRK | B Kemp, 4 Creek View, Basildon, SS16 4RU |
| 2E0 | BRL | Roy Dewis, 6 St Nicolas Close, Pevensey, BN245LB |
| 2E0 | BRP | Graham Armitage, Windmill Cottage, Greens Gardens, Nottingham, NG2 4QD |
| 2E0 | BRQ | K Shaw, 2 Montrose Avenue, Montrose Street, Hull, HU8 7RY |
| 2E0 | BRS | B Simmonds, 55 Pepys Road, St. Neots, PE19 2EN |
| 2E0 | BRT | Peter Allen, 21 Chase Vale, Burntwood, WS7 3GD |
| 2E0 | BRU | Bruce Pickering, 7 Front Street Grindale, Bridlington, YO16 4XU |
| 2E0 | BRY | M EDMOND, 12 Yeoman Close, Worksop, S80 2RR |
| 2I0 | BSA | D Cooke, 7 Killyclooney Road, Dunamanagh, Strabane, BT82 0LZ |
| 2E0 | BSB | Tim Wooldridge, 12 Redwood Avenue, Leyland, PR25 1RN |
| 2M0 | BSE | R Ewing, 7 Middlemas Drive, Kilmarnock, KA1 3DZ |
| 2E0 | BSF | Ben Streeter, Fairway, West Chiltington Road, Pulborough, RH20 2EE |
| 2E0 | BSG | Brian Holland, 11 Silverlands Park, Buxton, SK17 6QX |
| 2I0 | BSH | Robin Vage, 80 Chinauley Park, Banbridge, BT32 4JL |
| 2E0 | BSI | Catreena Ferguson, Royd Moor, Royd Moor Lane, Pontefract, WF9 1AZ |
| 2E0 | BSJ | M Croxford Simmons, 37 Queens Road, Askern, Doncaster, DN6 0LU |
| 2E0 | BSK | Stephen Everson, 41 Westminster Lane, Newport, PO30 5ZF |
| 2E0 | BSM | Terence Hall, 18 Common Lane, New Haw, Addlestone, KT15 3LH |
| 2E0 | BSN | Michael Blagg, 17 Flint Avenue, Forest Town, Mansfield, NG19 0DS |
| 2E0 | BSQ | Duncan Gray, 68 Endeavour Way, Hythe Marina Village, Southampton, SO45 6LA |
| 2E0 | BSR | Robert Gay, 47 Egerton Street, Flat C, Chester, CH1 3ND |
| 2E0 | BSS | C Windsor, 44 Paragon Place, Norwich, NR2 4BL |
| 2E0 | BST | S Clay, Akers Lodge, 6 Penn Way, Rickmansworth, WD3 5HQ |
| 2E0 | BSU | J Grosvenor, 10 Neves Close, Lingwood, Norwich, NR13 4AW |
| 2E0 | BSW | Colin Godwin, 0, RED EARL LANE, Malvern, WR142ST |
| 2E0 | BSX | Ian Colvin, 80 Silvester Road, Cowplain, Waterlooville, PO8 8TS |
| 2E0 | BTA | C Wynne, 43 Lansdown Road, Broughton, Chester, CH4 0NZ |
| 2E0 | BTB | B Bowen, 2 Veronica Road, Manchester, M20 6SU |
| 2E0 | BTD | J Wynne, 43 Lansdown Road, Broughton, Chester, CH4 0NZ |
| 2W0 | BTE | J Davies, 122 Heol Frank, Penlan, Swansea, SA5 7EG |
| 2E0 | BTF | Timothy Banks, 18 Leicester Road, Newport, NP19 7ER |
| 2E0 | BTO | Alan King, 6 Dunsfold Close, Crawley, RH11 8EY |
| 2W0 | BTP | B Page, 9 De Braose Close, Cardiff, CF5 2DH |
| 2E0 | BTQ | Tomos Rogers, 9 Maryland Road, Risca, Newport, NP11 6BB |
| 2E0 | BTR | Albert Passey, 3 The Yard, Bayton, Kidderminster, DY14 9LH |
| 2E0 | BTS | G Tyler, Crofton, Stoney Ley, Worcester, WR6 5NG |
| 2I0 | BTT | Roger Laverty, 23 Hyacinth Avenue, Ballykelly, Limavady, BT49 9HT |
| 2E0 | BTU | Julius Katz, 8 Astor Drive, Birmingham, B13 9QR |
| 2E0 | BTV | Jose Barbieri, 20 Gilbard Court, Chineham, Basingstoke, RG24 8RG |
| 2E0 | BTW | David Richards, 73 Greenfields Avenue, Alton, GU34 2EW |
| 2E0 | BTX | Peter Chamberlain, 22 Stanedge Grove, Wigan, WN3 5PL |
| 2E0 | BTZ | David Shipton, Hillside, Tintern, Chepstow, NP16 6TF |
| 2E0 | BUA | J Watson, 20 St. Marys Gardens, Hilperton Marsh, Trowbridge, BA14 7PG |
| 2E0 | BUD | Andrew Hanson, Pilgrim Cottage, South Road, Truro, TR3 7AD |
| 2E0 | BUE | Stephen Mather, 104 Appley Lane North, Appley Bridge, Wigan, WN6 9DG |
| 2E0 | BUF | Nigel Ham, 59 Thorpe Gardens, Alton, GU34 2BQ |
| 2E0 | BUI | S Chuter, C/O 94 Westfield, Bognor Regis, PO22 9HF |
| 2E0 | BUJ | Christopher Cherry, 12 Scarisbrick New Road, Southport, PR8 6PY |
| 2E0 | BUK | M Buckland, 7 Heath Close, Newport, PO30 1HN |
| 2E0 | BUN | Lee Knight, Flat 4, Barrington House, Portsmouth, PO2 7DD |
| 2E0 | BUO | B Jenson, 10 Tintern Close, Portsmouth, PO6 4LS |
| 2E0 | BUP | Gary Knight, 131 Washington Road, Portsmouth, PO2 7DF |
| 2W0 | BUQ | Stephen Gibbon, 1 Graig Terrace, Senghenydd, Caerphilly, [illegible] |
| 2M0 | BUT | Frederick Pudsey, 21/2 Bathfield, Edinburgh, EH6 4DU |
| 2E0 | BUV | Athanasios Markettos, 98 Foster Road, Trumpington, Cambridge, CB2 9JR |
| 2M0 | BUX | Arthur Young, 4/4 Prestonfield Terrace, Edinburgh, EH16 5EE |
| 2M0 | BUY | Jonathan Henderson, 7 Rowanhill Close, Port Seton, Prestonpans, EH32 0SY |
| 2E0 | BUZ | David Burrows, 10 Fleming Avenue, Bottesford, Nottingham, NG13 0ED |
| 2E0 | BVB | John Roberts, Worlds Wonder, Warehorne, Ashford, TN26 2LU |
| 2E0 | BVD | Anthony Burns, 76, 76, Morprth, NE65 0TF |
| 2E0 | BVE | Joseph McBride, 65 Dunston Road, Hull, HU5 5ES |
| 2E0 | BVG | Jack Thomas, 15 Brant Road, Lincoln, LN5 8RL |
| 2E0 | BVH | Mike Norris, 35 Sudbrooke Road, London, SW12 8TQ |
| 2E0 | BVJ | Wayne Tunstall, 89 Lever Street, Little Lever, Bolton, BL3 1BA |
| 2E0 | BVK | Amanda Bennett, Erwenni, Llanbedrog, Pwllheli, LL53 7PA |
| 2M0 | BVN | Ian Hepworth, Bronte Cottage, Inverugie, Peterhead, AB42 3DN |
| 2E0 | BVO | Andrew Little, 60a Murray Road, Horndean, Waterlooville, PO8 9JL |
| 2E0 | BVQ | G Turner, 28 Chapel Close, Needingworth, St. Ives, PE27 4SH |
| 2E0 | BVR | N Breckons, 5 Berrybut Way, Stamford, PE9 1DS |
| 2W0 | BVS | John Smith, 15 Harvey Crescent, Aberavon, Port Talbot, SA12 6DF |
| 2E0 | BVT | Mark Elkington, Warner Leys, Iron Hill Farm, Daventry, NN11 6YJ |
| 2E0 | BVU | Timothy Loker, 24 St. Albans Hill, Hemel Hempstead, HP3 9NG |
| 2E0 | BVV | T Leaworthy, 7 Maesderwen Rise, Stafford Road, Pontypool, NP4 5SS |
| 2E0 | BVZ | James Harris, Flat 3, Herstmonceux Place, Church Road, Herstmonceux, Hailsham, BN27 1RL |
| 2E0 | BWA | Laura Garnett, 32 George Fox Way, Norwich, NR5 8BJ |
| 2E0 | BWD | Errol Mehmet, 8 Hailsham Road, London, SW17 9EN |
| 2E0 | BWH | B Harrison, 24 Alderton Road, Nottingham, NG5 6DX |
| 2E0 | BWI | Adrian Tring, 12 Ainsdale Close, Orpington, BR6 8DJ |
| 2D0 | BWJ | John Swift, 56 Leymoor Road, Huddersfield, HD3 4SW |
| 2E0 | BWK | Kevin Bushell, 4 Birch Grove, Harrogate, HG1 4HR |
| 2E0 | BWL | Robert Lane, 9 Hartoft Road, Hull, HU5 4JZ |
| 2E0 | BWM | Annas Alamudi, 18 Silverthorne Loft Apartments, 400 Albany Road, London, SE5 0DJ |
| 2E0 | BWN | C Mason, Riding Lea, Middleton Road, Barnard Castle, DL12 0AQ |
| 2E0 | BWP | Derek Le Mare, The Sycamore, Church Bank, Barnard Castle, DL12 0AH |
| 2E0 | BWQ | Brian North, 54 Parklands, Mablethorpe, LN12 1BY |
| 2E0 | BWT | Alasdair Cockett, 10 San Marcos Drive, Chafford Hundred, Grays, RM16 6LT |
| 2E0 | BWU | Peter Moule, 30 Hillview Road, Chelmsford, CM1 7RX |
| 2E0 | BWV | David Jones, 77 Brinkburn Grove, Banbury, OX16 3WX |
| 2E0 | BWX | Thomas Ward, 20 Ollerton Road, Edwinstowe, Mansfield, NG21 9QG |
| 2E0 | BWY | Artur Ferenc, 2a Rosedene Avenue, London, SW16 2LT |
| 2E0 | BXA | John Oakley, 59 Bewsey Street, Warrington, WA2 7JQ |
| 2E0 | BXC | Michael Wilkins, 11 Brockwell Lane, Kelvedon, Colchester, CO5 9BB |
| 2E0 | BXD | Joe Bell, 8 Firsleigh Park, Roche, St. Austell, PL26 8JN |
| 2E0 | BXE | Roy Steele, 175 Vale Road, Seaford, BN25 3HH |
| 2E0 | BXF | P Freeman, 24 Roe Green Close, Hatfield, AL10 9PE |
| 2E0 | BXG | Harry Hughes, 27 The Holt, Hailsham, BN27 3ND |
| 2I0 | BXJ | J Steele, 46 Circular Road, Newtownards, BT23 4BN |
| 2E0 | BXK | Richard Hope, 32 Winstanley Place, Rugeley, WS15 2QB |
| 2E0 | BXM | Kenneth Foster, 10 Bleaswood Road Oxenholme, Kendal, LA9 7EY |
| 2M0 | BXN | Glen Moir, 38 Forest Park, Stonehaven, AB39 2GF |
| 2E0 | BXQ | Lee Denham, 8 Kings Field, Bursledon, Southampton, SO31 8EN |
| 2E0 | BXS | Benjamin Sims, 4 New Cottages, Cranwich Road, Thetford, IP26 5EQ |
| 2E0 | BXV | Mervin Browne, 60 Lindsay Avenue, Abington, Northampton, NN3 2LP |
| 2E0 | BXW | Steven Johnston, 67 Eversfield Road, Horsham, RH13 5JS |
| 2E0 | BXX | Jason Searle, 2 Tukes Avenue, Gosport, PO13 0SE |
| 2E0 | BXY | William Taylor, Garth Wood, Fishers Brae, Eyemouth, TD14 5NJ |
| 2E0 | BXZ | Jude Reynolds, 64 Albury Road, Merstham, Redhill, RH1 3LL |
| 2E0 | BYA | David Passey, Blue House Cottage, Blue House Lane Albrighton, Wolverhampton, WV7 3AA |
| 2E0 | BYC | M Matthews, 28 Kempson Drive, Great Cornard, Sudbury, CO10 0YE |
| 2E0 | BYF | J Milne, 9 Roman Road, Colchester, CO1 1UR |
| 2E0 | BYG | Peter Bailey, 103 Jarden, Letchworth Garden City, SG6 2NZ |
| 2E0 | BYH | Sidney Leadbetter, 11 Cogos Park, Mylor Bridge, Falmouth, TR11 5SF |
| 2M0 | BYI | Charles Williamson, 31 Medrox Gardens, Cumbernauld, Glasgow, G67 4AJ |
| 2E0 | BYJ | Emilio Isaac, 162b Hitchin Road, Stotfold, Hitchin, SG5 4JE |
| 2M0 | BYK | Sascha Troscheit, 20 James Street, St. Andrews, KY16 8YA |
| 2I0 | BYL | B Crozier, 33 Cullentragh Road, Poyntzpass, Newry, BT35 6SD |
| 2E0 | BYM | M Brough, 12 Longfield Court, Barnoldswick, BB18 6LP |
| 2E0 | BYN | Ian Patterson, 63 Orchard Road, South Ockendon, RM15 6HP |
| 2E0 | BYO | John Morris, 15 New Wanstead, London, E11 2SH |
| 2E0 | BYQ | James Summerhill, 43 Rangers Walk, Bristol, BS15 3PW |
| 2M0 | BYT | John Cairney, 5 James Street, Bannockburn, Stirling, FK7 0NQ |
| 2E0 | BYW | Mark Bradley, 55 Queensway, Penwortham, Preston, PR1 0DT |
| 2E0 | BZA | K Moulder, 51a Aston Cantlow Road, Wilmcote, Stratford-upon-Avon, CV37 9XN |
| 2M0 | BZB | Stewart Campbell, 22 Golf View Crescent, New Elgin, Elgin, IV30 6JP |
| 2E0 | BZC | Paul Davies, 67a James Street, Stoke-on-Trent, ST4 5HR |
| 2E0 | BZE | Thomas Munro, 71 Zig Zag Road, Liverpool, L12 9EQ |
| 2E0 | BZG | Clifford Warwick, 104 Church Road, Formby, Liverpool, L37 3NH |
| 2E0 | BZH | Richard Hopkins, 17 Springside Court, Josephs Road, Guildford, GU1 1BT |
| 2E0 | BZI | John Donnelly, 23 Pitts Croft, Neston, Corsham, SN13 9ST |
| 2E0 | BZJ | B Cooke, 2 Harvey Place, Andover, SP10 2BU |

UK Callsigns

**IMPORTANT NOTE**

**Revalidate licence to avoid revocation** – Ofcom has advised the Society that plans will be drawn up to revoke licences that have not been revalidated as required by the licence conditions. The quickest way to revalidate is to do so online via the Ofcom website: *https://services.ofcom.org.uk/* or by email: *amateur.validations@ofcom.org.uk* Ofcom staff are available to help, but please be patient during times of heavy workload.

2M0 BZL Charles Watkinson, The Hillock Farmhouse, Lumphanan, Banchory, AB31 4QL

2E0 BZM Michael Tew, Willowell, Spring Valley Lane, Colchester, CO7 7SD

2E0 BZO Dimitrios Chatzikos, 53 Benbow Court, Shenley Church End, Milton Keynes, MK5 6JE

2E0 BZS Philip Burt, 56 Winslade Road, Sidmouth, EX10 9EX

2E0 BZT P Burgess, Tally Ho Cottage, High Street, Swindon, SN4 0AE

2E0 BZU Peter Fenney, 44 Birch Avenue, Oldham, OL8 3TX

2E0 BZV Irene Govan, 9 Willowbank, Sandwich, CT13 9QA

2E0 BZX R Crockford, 17 Tadcroft Walk, Calcot, Reading, RG31 7JR

2M0 BZZ W Anderson, Hillview, North Street, Johnstone, PA6 7HJ

2E0 CAA Clare Orange, 79 Heath Avenue, Werrington, Stoke-on-Trent, ST9 0HU

2E0 CAD C Dodgson, 16 Jefferson Street, Goole, DN14 6SH

2E0 CAE Carl Frizzell, 85 Gibbon Road, Newhaven, BN9 9ER

2E0 CAG Christopher Ives, 54 Gray Street, Clowne, Chesterfield, S43 4RU

2E0 CAJ Catherine Travis, 4 Kingsdale, Worksop, S81 0XJ

2E0 CAK Andrew Martyn, 6 North Side, New Tupton, Chesterfield, S42 6BW

2E0 CAL Clinton Bowley, 2 Cottage, Middle Battenhall Farm, Worcester, WR5 2JL

2E0 CAO Matilde Castanheira, Lillian Penson Hall, University of London, London, W2 1TT

2E0 CAP Jonathan Williams, 41 Overton Lane, Hammerwich, Burntwood, WS7 0LQ

2E0 CAQ David Arnold, The Chase, Rectory Road, Penzance, TR19 6BB

2E0 CAR C Goulding, 9 Lune Drive, Leyland, PR25 5SX

2E0 CAS C Spence, 30 Chestnut Drive, Shirebrook, Mansfield, NG20 8NH

2E0 CAT E Taylor, 39 Gill Street, Newcastle upon Tyne, NE4 8BH

2E0 CAU Dorne Holmes, 5 The Cottages, Low Road, Dereham, NR20 3DG

2E0 CAV C Vernon, 29 Alice St., Deane, Bolton, BL3 5PJ

2E0 CAW Simon Court, Eastgate Cottage, Perrys Lane, Norwich, NR10 4HJ

2E0 CBA Peter Humphreys, 30 The Chestnuts, Hinstock, Market Drayton, TF9 2SX

2E0 CBB Brian Clements, 23 Croft Terrace, Egremont, CA22 2AT

2M0 CBC Colin Bryson, 29 Roull Road, Edinburgh, EH12 7JW

2M0 CBE Colin Ellison, 1 Newton Road, St. Fergus, Peterhead, AB42 3DD

2E0 CBF Adrian Cunningham, 18 Bradway, Whitwell, Hitchin, SG4 8BE

2E0 CBG Cleeton Gough, 104 Canley Road, Coventry, CV5 6AR

2E0 CBH Julian Wooldridge, 7 Heather Gardens, Belton, Great Yarmouth, NR31 9PP

2W0 CBJ V Holden, 2 Anglesey Close, Tonteg, Pontypridd, CF38 1LY

2E0 CBL Ian Botham, 12 Lairgill, Bentham, Lancaster, LA2 7JZ

2E0 CBN Barry Adby, 26 Love Lane, Watlington, OX49 5RA

2E0 CBO Peter Rogers, 16 Begonia Close, Basingstoke, RG22 5RA

2E0 CBQ Roderick Smith, 5 Elizabeth Place, 13 Heath Road, Haywards Heath, RH16 3AX

2D0 CBT Mohini Hersom, 26 Bourne Court, Station Approach, Ruislip, HA4 6SW

2E0 CBU Joe Horry, 5 Donington Road, Bicker, Boston, PE20 3EF

2I0 CBV Alan Shilliday, 26 Iskymeadow Road, Armagh, BT60 3JS

2E0 CBX Iain Connors, 3 Wheatfield Way, Chelmsford, CM1 2QZ

2E0 CBZ Clifford Nicholls, 26 Maes Geraint, Pentraeth, LL75 8UR

2E0 CCA Christopher Chambers, 21 Sullington Way, Shoreham-by-Sea, BN43 6PJ

2E0 CCB Stuart Connolly, 82 Cheswood Drive, Minworth, Sutton Coldfield, B76 1YE

2E0 CCC J Restall, 1 Johndory, Dosthill, Tamworth, B77 1NY

2E0 CCF Gerald Wright, 2 Hillcrest Drive, Castleford, WF103QN

2E0 CCG G Brierley, 15 First Avenue, Flint, CH6 5LP

2E0 CCJ Christopher Jones, 6 Cleeve Park Cottages, Icknield Road, Reading, RG8 0DJ

2E0 CCK Mark Buxton, 610 SQN Air Traning Corps, Cadet Training Centre, Chester, CH1 4AN

2E0 CCL Susan Gillard, 1 Chevening Close, Stoke Gifford, Bristol, BS34 8NJ

2E0 CCQ Tanvir Panesar, 135a Hargate Way, Hampton Hargate, Peterborough, PE7 8FL

2E0 CCR Paul Alborough, Flat 8, Morey Court, Ryde, PO33 1HA

2E0 CCW Mark Shoyer, 22 St. Andrews Road, Whitehill, Bordon, GU35 9QN

2E0 CCY F Gear, 251 Abington Avenue, Northampton, NN3 2BU

2E0 CDE A Bateson, North Sands, 7 Meadoway, Preston, PR4 5BB

2E0 CDF Bradley Myler-Cook, 11a York Street, Boston, PE21 6JN

2E0 CDI Scott Lee-Ray, The Paddock, Sutton Road, Alford, LN13 9RL

2W0 CDJ C Josey, 157 Waterloo Road, Penygroes, Llanelli, SA14 7PU

2E0 CDK D Barnett, 20 Middlemead Road, Bookham, Leatherhead, KT23 3DA

2E0 CDL C Lote, 8 Warren Place, Walsall, WS8 6BY

2E0 CDM Christopher McNulty, 91 Barn Hey Crescent, Wirral, CH47 9RW

2E0 CDO Daniel Fleming, 4 Oxenrig Farm Cottage, Coldstream, TD12 4EY

2E0 CDQ William Haddock, 5 Bradley Close, Middlewich, CW10 0PF

2E0 CDR Craig Russell, 255 Leeds Road, Shipley, BD18 1EH

2E0 CDS Christopher Small, Riddings Barn, Hope Bagot, Ludlow, SY8 3AE

2E0 CDT Colin Taylor, 1 Jasmine Gardens, Warrington, WA5 1GU

2E0 CDU Michael Hall, 29 The Spinney, Finchampstead, Wokingham, RG40 4UN

2E0 CDV A Drew, 51 Hobart Road, Cambridge, CB1 3PT

2E0 CDW Chris Wade, 31 Melton Green, Wath-upon-Dearne, Rotherham, S63 6AA

2W0 CDZ Paul Gough, 21 Clos Tyclyd, Cardiff, CF14 2HP

2E0 CEA Martin Anderson, 27 Laing Road, Colchester, CO4 3UT

2E0 CEB Brian Smithers, 16 Potters Grove, New Malden, KT3 5DE

2W0 CED Colin Davies, 70 Farm Drive, Port Talbot, SA12 6TF

2M0 CEE Robert Renshaw, Smithy House, Scotscalder, Halkirk, KW12 6XJ

2E0 CEF Frederick Price, 538 Abergele Road, Old Colwyn, Colwyn Bay, LL29 9LD

2E0 CEG I GREATHEAD, 3 Helmington Terrace, Hunwick, Crook, DL15 0LQ

2E0 CEH Philip Blyth, 20 Common Lane, Beccles, NR34 9RH

2I0 CEI Bryan Craney, 8a Drumhoy Drive, Carrickfergus, BT38 8NN

2E0 CEJ R Parrish, 5 Kestrel Lane, Cheadle, Stoke-on-Trent, ST10 1RU

2E0 CEK Douglas McAuslan, Casa Arco Iris, Via Variante Nascente, Santa Barbara de Nex, Portugal, 8005-491

2E0 CEM Michael Miller, Barn Cottage, Wingfield Hall, Manor Road, Alfreton, DE55 7NH

2E0 CEN Duncan McTaggart, 59 Gainsborough Road, Richmond, TW9 2DZ

2E0 CEO Chris Gozzard, Craig Dulas, Rhydyfoel Road, Abergele, LL22 8EG

2E0 CEP Jonathan Pelham, 20 Merchants Court, Bedford, MK42 0AT

2E0 CER Michael Everitt, 10 Morris Close, Hatherleigh, Okehampton, EX20 3NX

2E0 CES John Brawn, 9 Westbury Road, Westbury-on-Trym, Bristol, BS9 3AY

2E0 CEU Neil Hoare, 5 Kelsey Head, Port Solent, Portsmouth, PO6 4TA

2M0 CEX Paul Rice, 36 Namur Road, Penicuik, EH26 0LL

2E0 CEY Tracey Edwards, 5 Sienna Mews, Plumstead Road, Norwich, NR1 4LR

2M0 CFA Royston Mannifield, 2 Plewlands Avenue, Edinburgh, EH10 5JY

2M0 CFB Ian Watson, 10 Christie Place, Elgin, IV30 4HX

2E0 CFD Andrew McNeil, 2 Palmerston Crescent, Liverpool, L19 1RB

2E0 CFE Tim Carpenter, 11 Castle Road, Southwick, Fareham, PO17 6EY

2E0 CFG Fufu Fang, 82 Mill Hill Road, Norwich, NR2 3DS

2E0 CFH Anne Andrew, 80 Hamble Drive, Abingdon, OX14 3TE

2E0 CFI Martin Summers, 21 Quantock Avenue, Caversham, Reading, RG4 6PY

2E0 CFK Paul Rimmington, 28 Skipton Road Swallownest, Sheffield, S26 4NQ

2E0 CFL ADL Norville, 137 Foster Road, Trumpington, Cambridge, CB2 9JW

2E0 CFM Edward Vaughan, 10 Evans Close, Manchester, M20 2SQ

2E0 CFP Johnhenry Hitchens, 6 Oak Road, Alderholt, Fordingbridge, SP63bl

2E0 CFQ John Hunt, 14 Nevill Close, Hanslope, Milton Keynes, MK19 7NY

2E0 CFT A Wells, 186 Manford Way, Chigwell, IG7 4DG

2E0 CFV Edward Beever, 160 Granby Road, Buxton, SK17 7TA

2I0 CFW Desmond McGlone, 10 O'Neill Terrace, Dromore, Omagh, BT78 3AW

2E0 CFX Darryl Powis, 18 Merlin Court, Huddersfield, HD4 7SP

2E0 CFY Andrew Davis, 72 Westbourne Avenue, Clevedon, BS21 7XY

2D0 CFZ Colin Ingles, 1 Hillberry View, Onchan, Isle of Man, IM3 3GB

2D0 CGB Graham Grimshaw, 1 Hardy Close, Pinner, HA5 1NL

2M0 CGE Alistair MacDonald, 1 Edinmore Cottage, Rothesay, Isle of Bute, PA20 0QT

2E0 CGG Steven Brown, 4 Dorado Gardens, Orpington, BR6 7TD

2E0 CGH Wendy Durrant, 19 Rydal Rd, Gosport, PO12 4ES

2E0 CGI James Garrard, 3 West End Gardens, Nafferton, Driffield, YO25 4QE

2E0 CGJ David Ramsell, 36 West Street, Burton-on-Trent, DE15 0BW

2E0 CGK Nigel Allen, 59 Sherborne Road, Chichester, PO19 3AN

2E0 CGL C Lewis, 3 Sovereign Way, Calcot, Reading, RG31 4US

2W0 CGM Nigel Wells, 52 Lowerdale Drive, Llantrisant, Pontyclun, CF72 8DY

2E0 CGP Ian Duffie, Trebeighan Farm, Saltash, PL12 5AE

2E0 CGR Mark Cashman, Flat 3, Linden Court, Romsey, SO51 8BR

2E0 CGS J Goodyear, 30 Ashburton Road, Alresford, SO24 9HH

2E0 CGT Jonathan Russell, 75 Swallow Dale, Thringstone, Coalville, LE67 8LY

2E0 CGU Ian Albrighton, 12 Clewley Road, Branston, Burton-on-Trent, DE14 3JE

2E0 CGV David Ingrey, 1 Ponders Road, Fordham, Colchester, CO6 3LX

2E0 CGW Nicholas Bown, 18a Warley Hill, Warley, Brentwood, CM14 5HA

2E0 CGX J Shufflebotham, 316 Stockport Road, Hyde, SK14 5RU

2I0 CGZ Darryl Mudd, 14 Bloomfield Road, Belfast, BT5 5LT

2E0 CHA Charmain Nakajima, 22 Royds Crescent, Rhodesia, Worksop, S80 3HF

2E0 CHH Christopher Hill, 9 Oliver Road, Newport, NP19 0HU

2E0 CHK Alex Wild, 181 The Strand, Goring-by-Sea, Worthing, BN12 6DY

2E0 CHL Timothy Chapman, 1 East Dean Road, Lockerley, Romsey, SO51 0JL

2E0 CHN Paul Gale, 37 Hazlebury Road, Poole, BH17 7AX

2E0 CHQ Sean Drake-Brockman, 13 St. Johns Place, Bury St. Edmunds, IP33 1SW

2E0 CHT M Cook, 14 Speyside Close, Carterton, OX18 1TT

2E0 CHU George Hatt, 4H Colman House, Earlham Road, Norwich, NR4 7TJ

2E0 CHV S Taylor, 43 Toronnen, Bangor, LL57 4TG

2E0 CHW C West, 1 Willetts Mews, Hoddesdon, EN11 9DX

2E0 CHY Geoffrey Lyon, 1 Eckersley Street, Wigan, WN1 3PP

2E0 CHZ Cheryl Jewell, 3 Marsh Gate, Clee St. Margaret, Craven Arms, SY7 9DU

2E0 CIA Domingo Campanario, 3 Foxearth Hall, Leek Road, Stoke-on-Trent, ST9 0DG

2E0 CID Kevin Percy, 55 Buxton Avenue, Heanor, DE75 7UN

2E0 CIG Peter Holland, 30 Knighton Park Road, London, SE26 5RJ

2E0 CIH Michael Verrechia, 7 Willow Lane, Great Cambourne, Cambridge, CB23 6AB

2E0 CII Thomas Pearsall, 16 Langdale Road, Leyland, PR25 3AR

2E0 CIJ Derek Simpson, 50 Castle Hill, Berkhamsted, HP4 1HF

2E0 CIK Charles Storr, 40 Weelsby Way, Hessle, HU13 0JW

2E0 CIM Steven Preece, 14 Bettespol Meadows, Redbourn, St. Albans, AL3 7EW

2E0 CIN Allan Mawson, 14 Pontop View, Consett, DH8 7JB

2E0 CIO Alan Bulman, 21 Stannington Road, North Shields, NE29 7JY

2E0 CIQ J Chapman, South View, Mill End, Buntingford, SG9 0SU

2E0 CIR I Sapstead, 18 Rib Close, Standon, Ware, SG11 1QS

2E0 CIS David Woodbine, 29 Compass Tower, Munnings Road, Norwich, NR7 9TW

2E0 CIT Jeffrey O'Brian, 83 Bramdean Crescent, London, SE12 0UJ

2E0 CIV Matthew Ireland, Pen y Gadlas, Ffordd Bryniau, Prestatyn, LL19 8RD

2E0 CIX Jennifer Rowsell, 55 Scarlet Oaks, Camberley, GU15 1RD

2E0 CJA Lee McGaughey, 5 Railway Cottages, Station Lane, Brough, HU15 1RQ

2E0 CJB Christopher Beresford, 13 Chaseside Avenue, Twyford, Reading, RG10 9BT

2E0 CJD Christopher Eyre, 23 Nelson Street, Congleton, CW12 4BS

2E0 CJF Kevin Lambert, 38 Whittleford Road, Nuneaton, CV10 9HU

2W0 CJG John Loughlin, 453 Heol-y-Waun, Penrhys, Ferndale, CF43 3NW

2W0 CJI Max Day, 11 Troedrhiw-Trwyn, Pontypridd, CF37 2SE

2E0 CJJ Richard Squires, Hillcrest, Pontfadog, Llangollen, LL20 7AS

2E0 CJK Peter Mullen, 14 Anderson Road, Hemswell Cliff, Gainsborough, DN21 5XP

2E0 CJM Keith Nicholson, 11 Lancaster Way, Skellingthorpe, Lincoln, LN6 5UF

2E0 CJO Martin Reynolds, 24 Burton Close, Corringham, Stanford-le-Hope, SS17 7SB

2E0 CJP C Price, 10 St. James Park, Lower Milkwall, Coleford, GL16 7LG

2E0 CJQ Thomas Willis, 143 Clarence House, Leeds, LS10 1LH

2E0 CJS Philip Hanman, 7 Tremenheere Road, Penzance, TR18 2AH

2E0 CJV Adam Sharam, 30 Heywood Avenue, Maidenhead, SL6 3JA

2E0 CJW Chris Wright, 16 Baseley Way Longford, Coventry, CV6 6QA

2E0 CJZ Joshua Smith, 38 Marshfield Street, Newport, NP19 0GX

2I0 CKB Cormac Kelly, 7a Bancran Road, Draperstown, Magherafelt, BT45 7DT

2E0 CKC Alan Bradshaw, 130 Low Lane, Morecambe, LA4 6PS

2E0 CKE Adam Cooke, Iolanthe, Chidham Lane, Chichester, PO18 8TH

2E0 CKI Andrew Pickles, 87a Laburnum Road, Waterlooville, PO7 7EW

2E0 CKJ Paul Wright, 16 Hainault Avenue, Giffard Park, Milton Keynes, MK14 5PA

2E0 CKL Mark Atherton, 39 Fairmead Road, Wirral, CH46 8TU

2E0 CKM Antony Pawlak, 8 Healey Close, Crewe, CW1 4RS

2I0 CKN Malcolm Allen, 48 Kevlin Gardens, Omagh, BT78 1QS

2E0 CKO Brian Le Page, 88 Reading Road, Finchampstead, Wokingham, RG40 4RA

2E0 CKP Clarence Prior, 38 Windmill Road, Wombwell, Barnsley, S73 8PP

2E0 CKQ John Legrain, 22 Cromwell Drive, Didcot, OX11 9RB

2E0 CKR Bryan Cox, 7 Wolsey Avenue, London, E6 6HG

2E0 CKS Anthony Hickey, 144 Gisburn Road, Barnoldswick, BB18 5LQ

2E0 CKT Barry Cunningham, 33 Barry Street, Burnley, BB12 6DT

2W0 CKV Elizabeth Price, 1 Brynderi, Pontyates, Llanelli, SA15 5SU

2E0 CKW Richard Pringle, 14 Marjorie Street, Cramlington, NE23 6XQ

2E0 CKX Lars Schuy, 32 Sudeley Street, Brighton, BN2 1HE

2E0 CKY Anthony Barrett, 4 Wood Cottages, Cummings Cross, Newton Abbot, TQ12 6HJ

2E0 CLD Mark Shopland, 128 Whitewood Park, Liverpool, L9 7LG

2E0 CLE C Edgson, 59 Gilmour Crescent, Worcester, WR3 7PJ

2M0 CLF C Forsyth, 14b Osborne Terrace, Edinburgh, EH12 5HG

2E0 CLH Simon Saunders, 5 Park Court, Woking, GU22 7NW

2E0 CLI Michael Randle, 21 Grazebrook Croft, Birmingham, B32 3NL

2E0 CLJ Clive Jones, 33 Graig Ebbw, Rassau, Ebbw Vale, NP23 5SF

2E0 CLL C Lee, 46 Little Lane, Huthwaite, Sutton-in-Ashfield, NG17 2RA

2M0 CLN Colin Cosgrove, The Cottages, Kirkinch, Blairgowrie, PH12 8SL

2E0 CLP Roger Smith, Five Elms, Lullington Road, Tamworth, B79 9JA

2I0 CLS Gaibriel O'Neill, 46 Ashgrove Road, Newtownabbey, BT36 6LJ

2W0 CLT Gerald Williams, 36 Park Street, Taibach, Port Talbot, SA13 1TD

2M0 CLU Robert Adamson, 6 Camdean Crescent, Rosyth, Dunfermline, KY11 2TJ

2E0 CLW David Clewer, 45 Ashfield Road, Andover, SP10 3PE

2E0 CLX Nigel Royan, 23 Durham Terrace, London, W2 5PB

2E0 CLY Andrew Page, 207 Brooklyn Road, Cheltenham, GL51 8DZ

2E0 CLZ Iain Jones, 8a Orchard Close, Longford, Gloucester, GL2 9BB

2M0 CMA Anne Marie Campbell, 1b Craig Road, Troon, KA10 6DA

2E0 CMC Alan Applegate, 13 Deacons Close, KINGS STANLEY, Stonehouse, GL10 3JA

2E0 CMD Barry Eames, 22 Ashgrove Close, Hardwicke, Gloucester, GL2 4RT

2E0 CME Reginald Boardman, 12 St. Margarets Road, Alderton, Tewkesbury, GL20 8NN

2E0 CMF Stewart Mason, 8 Barrowby Gate, Grantham, NG31 7LT

2E0 CMH Anthony Brown, 3 Alston Road, New Hartley, Whitley Bay, NE25 0ST

2E0 CMK Chris Norris, 53 Station Road, Castlethorpe, Milton Keynes, MK19 7HF

2E0 CMN Neil Curtis, 15a High Street, Headcorn, Ashford, TN27 9NH

2E0 CMO C Overton, 99 Hope Avenue, Goldthorpe, Rotherham, S63 9DZ

2E0 CMP Christopher Pegrum, 3 Bretland Road, Tunbridge Wells, TN4 8PS

2E0 CMQ Paul Chester Chester, 33 Salehurst Road, London, SE4 1AS

2E0 CMR Carl Millar, 60 Rodney Way, Ilkeston, DE7 8PW

2E0 CMT Sean Metcalfe, 55 Coventry Close, Corfe Mullen, Wimborne, BH21 3UW

2E0 CMY Frederick Southgate, 52 Jeffrey Lane, Belton, Doncaster, DN9 1LT

2E0 CMZ Clement Rawlin, 5 Japonica Hill, Immingham, DN40 1LT

2E0 CNA Peter Davies, 2 Lynfords Drive, Runwell, Wickford, SS11 7PP

2E0 CNB David James, Bramble Cottage, Tray Lane, Atherington, Umberleigh, EX37 9HY

2E0 CNC Tony Ward, 1 Darrismere Villas, Edinburgh Street, Hull, HU3 5AS

2E0 CND Steven Bradley, 6 Downing Street, South Normanton, Alfreton, DE55 2HE

2E0 CNE Michael Shepherd, Oak View, Church Road, Halstead, CO9 4PR

2E0 CNG Christopher Smith, 44 Brooksfield, Bildeston, Ipswich, IP7 7EJ

2E0 CNH Neil Sinclair, 8 Milton Avenue, Scarborough, YO12 4ES

2E0 CNL C Lockyear, 26 Wentworth Gardens, Exeter, EX4 1NH

2E0 CNN Darren Turnbull, 63 Brecklands, Mundford, Thetford, IP26 5EG

2E0 CNP F Hatfull, 16b Church Street, Easton on the Hill, Stamford, PE9 3LL

2E0 CNQ Colin Greenwood, 44 Fountain Street, Heckmondwike, WF16 9HS

2M0 CNR Samuel Burnside, Woodend Farm, Buchlyvie, Stirling, FK8 3PD

2E0 CNS Maxim Hatfull, 16b Church Street, Easton on the Hill, Stamford, PE9 3LL

2E0 CNU Timothy Cocks, 9 Mountfield Way, Westgate-on-Sea, CT8 8HR

2E0 CNV Timothy Pelham, 172a Gloucester Road, Patchway, Bristol, BS34 5BG

2E0 CNW Andrew Palmer, 127 Gloucester Road, Brighton, BN1 4AF

2E0 CNX I Buckton, 67 Tennyson Avenue, Middlesbrough, TS6 7ND

2E0 CNY Gordon Fryer, 9 Church Road, Chepstow, NP16 5HP

2M0 CNZ John Rayne, 8 Bankton Grove, Livingston, EH54 9DW

2E0 COA Lee Layland, 3 Thirlmere Road, Golborne, Warrington, WA3 3HH

2E0 COB J Cobbold, 2 The Green, Blencogo, Wigton, CA7 0DF

2E0 COD G Glasgow, 4 Beech Avenue, Culcheth, Warrington, WA3 4JF

2E0 COF John Murray, 65 Sincil Bank, Lincoln, LN5 7TH

2E0 COH Andrew Nisbet, 65a Hamilton Road, Felixstowe, IP11 7BE

2E0 COI Peter Bailey, 16 Manchester Road, Holland-on-Sea, Clacton-on-Sea, CO15 5PL

2E0 COJ Kevin Cartwright, 53 Sedgley Road, Dudley, DY1 4NE

2E0 COL Stephen Britt-Hazard, 9 Serbin Close, London, E10 6JL

2E0 COM Roger Hammond, 3 Hunt Road, Earls Colne, Colchester, CO6 2NX

2E0 CON John Middleton, 16 Kyme Road, Boston, PE21 8NQ

2J0 COQ Leslie Langlois, Farleyer, La Rue Des Platons, Jersey, JE3 5AA

2M0 COT Fred Gordon, Croft of Torrancroy, Strathdon, AB36 8US

2E0 COV Vincent Hopkins, 109 Smith Street, Coventry, CV6 5EH

2E0 COX Alan Cox, 1 Low House Cottages, Coniston, LA21 8ER

2E0 COY Arthur Pointon, 9 Parkwood Avenue, Stoke-on-Trent, ST4 8PD

2E0 CPB Christopher Bond, Tryfan, Vicarage Lane, Neston, CH64 5TJ

2E0 CPD Richard Briant, Talarvor, Llanon, SY23 5HG

2E0 CPE Michael Phillips, 59 Bradeley Road, Haslington, Crewe, CW1 5PX

2E0 CPF George Evans, 13 Lydgate Road, Sale, M33 3LW

2E0 CPG Christopher Leviston, 13 Pryors Walk, Askam-in-Furness, LA16 7JG

2E0 CPK Kenneth Jackson, 4 Milfoil Close, Marton-in-Cleveland, Middlesbrough, TS7 8SE

2E0 CPL Clive Luckett, 257 Folkestone Road, Dover, CT17 9LL

2M0 CPN Alan Woodford, Nordkette, Levenwick, Shetland, ZE2 9GY

2E0 CPO David Jones, 2 Trewen, Llandinam, SY17 5BU

2E0 CPQ Bruce Saunders, 88 Bramwoods Road, Chelmsford, CM2 7LT

2E0 CPR Ronald Packman, 42 Shad Thames, London, SE1 2YD

2E0 CPS William Coburn, 42 Hinton Wood Avenue, Christchurch, BH23 5AH

2M0 CPV Jonathan Hutchinson, Hawthorn Cottage, Addiewell, West Calder, EH55 8NL

2E0 CPX Paul Strickland, 7 School Lane, Offley, Hitchin, SG5 3AZ

2E0 CPZ Patrick Kirkden, 22 Leas Green, Broadstairs, CT10 2PL

2E0 CQB Ian McKean, 142b High Street, Cranfield, Bedford, MK43 0EL

2E0 CQC David Clavey, 32 Apollo Close, Dunstable, LU5 4AQ

2E0 CQD Alistair Kennington, 63 Emmanuel Court, Scunthorpe, DN16 2LR

2E0 CQG Philip Jones, 10 Moulton Road, Tivetshall St. Margaret, Norwich, NR15 2AJ

2E0 CQH Simon Baynton, 50 Briton Way, Wymondham, NR18 0TT
2M0 CQI Jeffery Browne, 33 Pilgrims Hill, Linlithgow, EH49 7LN
2E0 CQJ Nicholas Parry, Worlingham Court, Mill Lane, [illegible], NR34 7DE
2E0 CQL Kevin Jeffery, 9 Gordon Road, Tunbridge Wells, TN4 9DL
2E0 CQM C Harding, 27 Eston Avenue, Malvern, WR14 2SR
2E0 CQN Barry McGlynn, 22 Bracken Bank Way, Keighley, BD22 7AB
2E0 CQO Malcolm Mutkin, 13 The Grove, Radlett, WD7 7NF
2E0 CQQ Ian Pryke, 9 Charles Avenue, Grundisburgh, Woodbridge, IP13 6TH
2E0 CQR Philip Moye, 13 Post Mill Gardens, Grundisburgh, Woodbridge, IP13 6UP
2E0 CQS P Elsey, 62b Coleraine Road, London, SE3 7PE
2D0 CQT Florence Agbolma, Flat 6, [illegible]
2E0 CQV Patrick Chong, 320 Glenalmond Avenue, Cambridge, [illegible]
2E0 CQX Alan Davis, Old Malt Kiln House, [illegible], DL8 5JS
2E0 CQZ John Street, 22 Roman [illegible], [illegible]
2E0 CRB Kym Dutfield Cooke, Tan yr Ffail, Segurinside, Llandudno Junction, LL31 9QE
2E0 CRC Ashish Dhakoo, 4 Bryden Cottages, High Street, Uxbridge, UB8 2NY
2E0 CRD C Densham, 27 Lloyds Crescent, Exeter, EX1 3JQ
2E0 CRH Paul Bull, 87 Braemor Road, Calne, SN11 9DU
2E0 CRI John Parris, Starsmead Farmhouse, Haresfield, Stonehouse, GL10 3EG
2E0 CRM C Murly, C/O 1 Mount Pleasant, Middleton, Leeds, LS10 3TB
2E0 CRN John Cranston, 7 Cowen Gardens, Gateshead, NE9 7TY
2M0 CRQ Carrie Welsh, 28 Peacock Wynd, Motherwell, ML1 4ZL
2M0 CRR Colin Hodger, 23 Harrysmuir Road, Pumpherston, Livingston, EH53 0NT
2E0 CRS Christopher Marsh, 16 Drake Head Lane, Conisbrough, Doncaster, DN12 2AA
2E0 CRU C Redmond, 6 Apsley Road, Southsea, PO4 8RH
2E0 CRV S Hodder, 32 Stubbs Close, Wellingborough, NN8 4UQ
2E0 CRX S Cross, 31 Parkfields, Abram, Wigan, WN2 5XR
2E0 CSD Craig Davies, 3 King Alfreds Green, Leeds, LS6 4PZ
2E0 CSE Terence Chapman, 12 Greenways, Chilcompton, Radstock, BA3 4HT
2E0 CSF C Finnis, 44 Disraeli Road, Christchurch, BH23 3NB
2E0 CSG D Pollard, 8 Drammen Avenue, Burnley, BB11 5EA
2E0 CSH Graham Webster, 15 Bridge Road, Chichester, PO19 7NW
2E0 CSJ Mark Catchpole, Woodcote, Five Oaks Road, Horsham, RH13 0RQ
2E0 CSK David Rudling, Rose Cottage, Ludwells Lane, Southampton, SO32 2NP
2E0 CSL Christopher Lester, 21 Mortimer Way, Witham, CM8 1SZ
2E0 CSN Gordon Flinn, 38 Fir Grove, Whitehill, Bordon, GU35 9ED
2E0 CSO Colin Opie, 354 Beaumont Road, Plymouth, PL4 9EN
2E0 CSQ Neil Saunders, 41 Drivers Mead, Lingfield, RH7 6EX
2E0 CSU Graham Wilman, 8 Oldfield Drive, Mobberley, Knutsford, WA16 7HB
2E0 CSV Alan Parker, 9 Milecastle Court, Newcastle upon Tyne, NE5 2PA
2M0 CSX A Thomson, 5 Gib Grove, Dunfermline, KY11 8DH
2E0 CSY Simon Bradley, 9 Crofton Road, Southsea, PO4 8NX
2M0 CSZ Stephen Leighton, 75 Dovecot Road, Tullibody, Alloa, FK10 2QU
2E0 CTA Andrew Cross, 12 Appleby Drive, Langdon Hills, Basildon, SS16 6NU
2M0 CTB William Scott, 6f Drumcross Road, Bathgate, EH48 4HG
2E0 CTC C Castle, 32 Inglefield Road, Ilkeston, DE7 5AP
2E0 CTD Francis Davison, 137 Hellis Wartha, Helston, TR13 8WF
2E0 CTE Christopher Etchells, 7 Woodlands Drive, Sandford, Wareham, BH20 7QA
2E0 CTF Phillip Evans, 144 Grenville Street, Stockport, SK3 9ET
2M0 CTI Michael Jamieson, 5 Straid Bheag, Barremman, Helensburgh, G84 0QX
2E0 CTJ Craig Johnson, 2 Dawber Street, Worksop, S81 7DS
2E0 CTK John Schleswick, Flat 46, Fitzroy House, 55-59 Great Pulteney Street, Bath, BA2 4DW
2E0 CTL Kevin Baker, 27 St. Matthews Close, Cherry Willingham, Lincoln, LN3 4LS
2E0 CTM Charles Meakin, 102 Ryknield Road, Kilburn, Belper, DE56 0PF
2M0 CTN Richard Tait, 9 Fifth Avenue, Glasgow, G12 0AS
2E0 CTO James Killman, 19 Moorland Avenue, Walkeringham, Doncaster, DN10 4LG
2E0 CTQ Tim Crew, 1 Wood End, Farnborough, GU14 7BA
2E0 CTR Trevor Wright, 11 Ash Close, Daventry, NN11 0XH
2E0 CTT M Ward, 39 Neil Avenue, Holt, NR25 6TG
2E0 CTU David Foley, 1 Hill Rise Close, Harrogate, HG2 0DQ
2E0 CTW Jonathan Hobbs, 82 Perry's Lane, Wroughton, Swindon, SN4 9AP
2D0 CTX Piotr Sniezek, 204 Quadrant Court, Empire Way, Wembley, HA9 0EY
2E0 CTZ Michael Carr, 51 Langton Road, Holton-le-Clay, Grimsby, DN36 5BH
2E0 CUA Jose Valle Espin, 203 Broadway, Horsforth, Leeds, LS18 4HL
2E0 CUB Adam Clements, 49 Canberra Court, West Avenue, Huntingdon, PE26 1EY
2E0 CUC Arthur Hunter, 22 Lindsay Court, Whitburn, Sunderland, SR6 7LN
2E0 CUE Richard Weaver, 15 Sharps Field, Headcorn, Ashford, TN27 9UF
2E0 CUH James Farr, 64 Parc an Tansys, Pengegon, Camborne, TR14 7PH
2E0 CUJ J Barry, 7 Rockfield Rise, Undy, Caldicot, NP26 3FG
2E0 CUK Ian Troughton, Rhiwbina, Pentre Lane, Cwmbran, NP44 3AP
2E0 CUL Karl Richards, 44 Curly Bridge Close, Farnborough, GU14 9AU
2E0 CUO Harry Hope, 51 Margravine Gardens, London, W6 8RN
2E0 CUS Anthony Cussen, The Poplars, Mill End, Southminster, CM0 7LJ
2E0 CUU Malcolm Boon, 45 Carlton Road, Wickford, SS11 7ND
2E0 CUV James Stewart, 19 Salisbury Road, Dover, CT16 1EX
2E0 CUW Richard Tofts, Elmcroft, Redhill Road, Ross-on-Wye, HR9 5AU
2E0 CUY Samuel Bache, 82 Whittingham Road, Halesowen, B63 3TP
2D0 CVB David Boardman, 57 Sunningdale Road, Huddersfield, HD4 5DX
2E0 CVC Andrew Chaplin, 10 St. Leonards Road, Malinslee, Telford, TF4 2EB
2E0 CVD Cameron Herd, 182 Hungerhill Road, Nottingham, NG3 3LL
2E0 CVE Clive Wilkinson, High Breck, Dolwyd, Colwyn Bay, LL28 5HS
2E0 CVF Andrew Davies, 4 Capella Path, Hailsham, BN27 2JY
2E0 CVG C Green, 4 Lyme Grove, Knott End-on-Sea, Poulton-le-Fylde, FY6 0AJ
2E0 CVJ Robert Fidler, 44 Windermere Avenue, Ramsgate, CT11 0PF
2M0 CVK Scott Muir, Cairnside, Burnhead, Dundee, DD3 0QN
2E0 CVL Nathan Edwards, 37 Rifle Green, Blaenavon, Pontypool, NP4 9QN
2E0 CVM Steven Broderick, 179 Malpas Road, Newport, NP20 5PP
2E0 CVN S Treacher, 8 Berman Court, Back Road, Newton Abbot, TQ12 1FY
2E0 CVO James Pauline, 54 Laurel Road, Bassaleg, Newport, NP10 8NY
2E0 CVP James Mason, 77 Albutts Road, Walsall, WS8 7ND
2E0 CVQ Steven Knott, 15 Meadowlands Drive, Haslemere, GU27 2FD
2I0 CVR James Allen, 192 Joanmount Gardens, Belfast, BT14 6PA
2E0 CVU Paul Swingewood, 9 Goodall Grove, Great Barr, Birmingham, B43 7PQ
2E0 CVV Christopher Sayles, 11 Malton Close, Monkston, Milton Keynes, MK10 9HR
2E0 CVW Daniel Levy, Flat 36, Claydon House, London, NW4 1LS
2E0 CVX Graham Spicer, 44 Cowley Lane, Chapeltown, Sheffield, S35 1SY
2E0 CVY Robert Smith, 15 Hollybush Road, North Walsham, NR28 9XT
2E0 CVZ Paul Herron, 102 Garden City Villas, Ashington, [illegible]
2E0 CWB Maurice Fletcher, 7 Richard Street, Bacup, [illegible]
2E0 CWC Carl Lewis, 9 Chatsworth Gardens, Sydenham, Leamington Spa, CV31 1WA
2E0 CWE Phillip Stuart, 5 Welbeck Gardens, Woodthorpe, Nottingham, NG5 4NX
2E0 CWF Alan Angus, 51 Osprey Drive, Blyth, NE24 3QS
2E0 CWI Robert Last, 30 Abbot Road, Bury St. Edmunds, IP33 3UB
2E0 CWJ John Taylor, 90 Village Road, Gosport, PO12 2LG
2E0 CWK James Deery, Flat 6, [illegible], London, SW5 9ER
2W0 CWL K Saltmarsh, 15 Colbourne Road, Beddau, Pontypridd, CF38 2LN
2E0 CWM George Moore, 40 Main Street, [illegible], Sleaford, NG34 8QG
2E0 CWO Brian Ledson, [illegible]
2E0 CWQ John Clarke, 160 Hall Lane Estate, Willington, Crook, DL15 0PP
2E0 CWR Colin Ralphson, 20 Moncal Grove, Buxton, SK17 7TF
2E0 CWS K Roberts, Burnwithian Cottage, Burnwithian, Redruth, TR16 5LG
2M0 CWV Neil Ford, 11 Kenmore Way, Coatbridge, ML5 4FN
2E0 CWW Andrew Rowland-Stuart, 86 Wiltshire House, Lavender Street, Brighton, BN2 1LE
2E0 CWZ Kevin Legg, Bennetts, High Street, Clacton-on-Sea, CO16 0EG
2E0 CXA Douglas Storer, 13 The Square, Lower Burraton, Saltash, PL12 4SH
2E0 CXD James Johnson, 226 Preston New Road, Southport, PR9 8NY
2E0 CXE Praful Naik, 82 Misbourne Road, Uxbridge, UB10 0HW
2M0 CXG Alistair Donald, 10 Fraser Road, Burghead, Elgin, IV30 5YN
2E0 CXH Phillip Arnold, 20 Upper Seagry, Chippenham, SN15 5EX
2M0 CXI David Plummer, 39 St. Nicholas Drive, Banchory, AB31 5YG
2E0 CXJ Colin Norman, 15 Maple Close, Sedbergh, LA10 5JE
2E0 CXK Ian Nicholls, 34 West Close, Bath, BA2 1PY
2E0 CXL Robert Chandler, 12 Lyndhurst Drive, Hale, Altrincham, WA15 8EA
2E0 CXM Brian Cullen, 4 Bree Close, Weston-Super-Mare, BS22 7JX
2E0 CXN James Bligh-Wall, 3 George Street, Elworth, Sandbach, CW11 3BL
2E0 CXO K Ralph, 11 Burrard Road, London, E16 3QL
2E0 CXP R Gill, 45 Biggin Lane, Ramsey, Huntingdon, PE26 1NB
2E0 CXQ Tim Pettis, 11 Curlew Drive, Hythe, Southampton, SO45 3GB
2E0 CXU Craig Ashworth, 12 North Terrace, Tebay, Penrith, CA10 3XH
2E0 CXV David Morgan, Ty Bettws, Kilgwrrwg, Chepstow, NP16 6PN
2E0 CXW Colin Wilson, 87 Levensgarth Avenue, Fulwood, Preston, PR2 9FP
2E0 CYC Neil Marley, Penstemons, Chapel Lane Pen Selwood, Wincanton, BA9 8LY
2E0 CYE John Jones, Isfryn Bungalow, Glan-y-Nant, Llanidloes, SY18 6PQ
2W0 CYF Simon Phillips, 58 Taff Embankment, Cardiff, CF11 7BG
2E0 CYL Robert Horvath, 3 Back Knowl Road, Mirfield, WF14 9SA
2E0 CYM A Rowlands, Lluest Wen, Penygarth, Caernarfon, LL55 1EY
2E0 CYO Peter Gregory, 2 Lennox Close, Hunmanby, Filey, YO14 0PY
2E0 CYP Stewart Challis, 73 Rivenhall Way, Hoo, Rochester, ME3 9GF
2E0 CYR Michael Woodruff, 21 Cross Green Close, Formby, Liverpool, L37 4BP
2E0 CYS Peter Martin, 108 Headlands Grove, Swindon, SN2 7HP
2E0 CYT Liam Starrett, 50 Danes Road, Bicester, OX26 2LP
2E0 CYU Jeremy Powell, 23 Park Road, Norton, Malton, YO17 9DZ
2E0 CYV Philip Richards, 2 The Mayflowers, Norwich, NR11 6FZ
2I0 CYW Chaoyang Wang, 32 Broadlands, Carrickfergus, BT38 7BL
2E0 CYX Simon Keith Rogers Rogers, 30 Coed Celynen Drive, Abercarn, Newport, NP11 5AU
2W0 CYY Charlotte Smith, 29 Heol Cwarrel Clark, Caerphilly, CF83 2NE
2E0 CYZ Mark Tarrant, Wayside Cottage, Gabber Lane, Plymouth, PL9 0AW
2E0 CZA Gary Youll, 4 Shaftsbury Court, Barnstaple Road, Scunthorpe, DN17 1YB
2E0 CZC Stephen Ingledew, 34 Sunningbrook Road, Tiverton, EX16 6EB
2E0 CZD John Crill, 92 Waterside, Exeter, EX2 8GZ
2E0 CZE Michael Reed, 2 Fir Close, Willand, Cullompton, EX15 2PZ
2E0 CZG Stephen Sissens, 20 Fallow Drive, Eaton Socon, St. Neots, PE19 8QL
2E0 CZI Paul Patterson, 3 Barnes Close, Southampton, SO18 5FE
2E0 CZJ Jonathan Chalmers, 19 Brettenham Crescent, Ipswich, IP4 2UB
2E0 CZK Robert Silcock, 18 Saxon Road, Southampton, SO15 1JJ
2E0 CZM Colin Rose, 132 Golf Green Road, Jaywick, Clacton-on-Sea, CO15 2RW
2E0 CZN Daniel Endean, 11 Forrester Drive, Brackley, NN13 6NE
2W0 CZP Albert Jones, 31 Russell Terrace, Carmarthen, SA31 1SZ
2E0 CZR Dean Rogers, 5 Semple Gardens, Chatham, ME4 6QD
2E0 CZS P Handley, 97 Applegarth Avenue, Guildford, GU2 8LX
2E0 CZT Rhodri Morgan, 14 Ash Road, Ashurst, Southampton, SO40 7AT
2E0 CZU Neil Adam, Tan Ffordd, Mynydd Llandygai, Bangor, LL57 4LX
2E0 CZW Graham Street, 105 Jeals Lane, Sandown, PO36 9NS
2E0 CZZ Paul Hampton, Caretakers Flat, T.A. Centre, Newport, NP20 5XE
2E0 DAA David Wilson, 94 Lon Hedydd, Llanfairpwllgwyngyll, LL61 5JY
2E0 DAB D Bambrook, 18 Vorvain Close, Bicester, OX26 3SR
2M0 DAC Donald Campbell, 10 Dalgate Mill, Kiltarlity, Beauly, IV4 7GL
2E0 DAH David Horner, 21 Ainsworth Road, Little Lever, Bolton, BL3 1RG
2E0 DAI David Holman, 20 Green Drive, Wolverhampton, WV10 6DW
2E0 DAJ James Bowley, 2 Cottage, Middle Battenhall Farm, Worcester, WR5 2JL
2E0 DAL David Stringer, 18 Townfield Close, Ravenglass, CA18 1SL
2E0 DAO David Oddie, 5 The Bridleway, Forest Town, Mansfield, NG19 0QJ
2W0 DAP Terry Woodley, [illegible], Neath, SA10 6AA
2E0 DAQ Gary Clarke, 11 Blackfordby Lane, Moira, Swadlincote, DE12 6EX
2E0 DAR D Robertson, 53 Moor Lane, Weston-Super-Mare, BS22 6RA
2E0 DAT D Toyne, 19 Poachers Rest, Wolton, Lincoln, LN2 3TH
2E0 DAW Don Williams, 18 Lower Greave Road, Meltham, Holmfirth, HD9 4DY
2E0 DAX Donald Sobey, Flat 2 73 Park Road, Blackpool, FY1 4JQ
2E0 DAY Susan Darby, 4 Whately Mews, Whately Road, Lymington, SO41 0XS
2E0 DBA Derek Barnes, 11 Yewside, Gosport, PO13 0ZD
2E0 DBB David Walker, Flat 5, Seward Court, 380-396 Lymington Road, Christchurch, BH23 5HD
2E0 DBH Darren Hoare, 47 High Street, Chalgrove, Oxford, OX44 7SJ
2E0 DBI Paul Marchant, 16 Melrose Drive, Peterborough, PE2 9DN
2I0 DBK David O'Hale, 6 Cochron Road, Newry, BT35 6DD
2E0 DBL P Kimberlee, 24 Jacey Road, Shirley, Solihull, B90 3LJ
2E0 DBM Darren Mellor, 28 Winster Road, Staveley, Chesterfield, S43 3NJ
2E0 DBN Darren Booth, 75 Meynell Road, Sheffield, S5 8GL
2E0 DBO Jonathan Ledger, 18 Claremont Street, Rotherham, S61 2LT
2E0 DBP D Payne, 31 Cockering Road, Canterbury, CT1 3UP
2E0 DBQ Denis Moger, 23 Elmsleigh Road, Paignton, TQ4 5AX
2E0 DBS D Baines, 21 Vera Road, Norwich, NR6 5HU
2E0 DBW Daniel Whyatt, 11 The Perrings, Nailsea, Bristol, BS48 4YD
2E0 DBY Christopher Riley, [illegible], Colchester, [illegible]
2E0 DBZ D Teasdale, 43 Easington Road, Stockton-on-Tees, [illegible]
2E0 DCA C Price, 9 Arlington Avenue, Aston, Sheffield, S26 2AA
2E0 DCD Bruce Savage, 33 Sky End Lane, Hordle, Lymington, SO41 0HG
2E0 DCF Clayton Lonie Jr, 41 De la Hay Avenue, Plymouth, PL3 4HS
2E0 DCH D Hannington, 3 Canadian Avenue, Gillingham, ME7 2DN
2E0 DCJ David Bishop, 62 Brindley Crescent, Hednesford, Cannock, WS12 4DS
2E0 DCK Richard Powell, Gerdd y Don, Llansantffraed, Llanon, SY23 5HS
2E0 DCL David Clark, [illegible], Sandhurst, GU47 0UT
2E0 DCM David Martin, 14 Freeston Terrace, St. Georges, Telford, TF2 9HD
2E0 DCN David Filby, [illegible]
2E0 DCP Derek [illegible], Crewe, CW1 5PB
2E0 DCS D Sharpen, 52 Woodsend Road, Urmston, Manchester, M41 8QT
2E0 DCV John Hurlbutt, 55 Prospect Avenue, Seaton Delaval, Whitley Bay, NE25 0EL
2E0 DCX Denis Cook, 44 Statfold Lane, Fradley, Lichfield, WS13 8NY
2E0 DCY Alastair Kerr, 23 Manor Park, Duloe, Liskeard, PL14 4PT
2E0 DCZ David Mooney, 107 Todder Road, South Croydon, CR2 8AR
2E0 DDB C Massey, 23 Ladymead Lane, Langford, Bristol, BS40 5EG
2E0 DDC David Coe, 199 Newark Road, North Hykeham, Lincoln, LN6 8QS
2E0 DDD Robert Hirst, 106-108 Washerwall Lane, Werrington, Stoke-on-Trent, ST9 0LR
2E0 DDE Michael Smith, 5 Cresswell Road, Ellington, Morpeth, NE61 5HR
2E0 DDF David Pennison, 69 Caneland Court, Waltham Abbey, EN9 3DS
2E0 DDG David Davies, 32 New North Road, Reigate, RH2 8NA
2E0 DDH David De La Haye, 4 Nicola Mews, Ilford, IG6 2QE
2E0 DDJ D Jones, 11 Alma Place, Sebastopol, Pontypool, NP4 5EA
2E0 DDK Brian Lewis, 10 Healey Avenue, Knypersley, Stoke-on-Trent, ST8 6SQ
2E0 DDL David Curtis, 7 Neale Close, Aylsham, Norwich, NR11 6DJ
2E0 DDN Brian Southern, 25 Chilgrove Avenue, Blackrod, Bolton, BL6 5TR
2E0 DDO David Lyons, 2 Goswick Farm Cottages, Berwick-upon-Tweed, TD15 2RW
2E0 DDP Roger Carpenter, 2 Greenfield Terrace, Trinant, Newport, NP11 3LJ
2E0 DDR David Randles, 111 East Pines Drive, Thornton-Cleveleys, FY5 3RY
2E0 DDU Stefan Borrell, Rose Cottage, Colchester Main Road, Colchester, CO7 8DD
2E0 DDW Dominic Webb, 9 Dunsfold Close, Crawley, RH11 8EY
2E0 DDZ Darren May, 12 Marl Crescent, Llandudno Junction, LL31 9HS
2E0 DEC Derek Chebsey, 21 Shortlands Lane, Walsall, WS3 4AG
2E0 DEE D Nicholson, 24 Barnmead, Haywards Heath, RH16 1UZ
2E0 DEG Andrew Titmus, 5 Kithurst Crescent, Goring-by-Sea, Worthing, BN12 6AJ
2E0 DEH Luc Mathlin, 29 Wagtail Drive, Stowmarket, IP14 5GH
2E0 DEI Geoffrey Watson, 88 Avenue Road, Sandown, PO36 8BE
2E0 DEK Derek Gibson, 25 Middleham Close, Ouston, Chester le Street, DH2 1TA
2E0 DEO C Phillips, 14 Laburnum Way, Hatfield Peverel, Chelmsford, CM3 2LP
2E0 DEP Christopher Garner, 30 Pendula Road, Wisbech, PE13 3RR
2E0 DEQ D pritchard, 104 Edgehill Road, Wirral, CH46 6AS
2E0 DER Derlwyn Williams, 10 Bronllys, Gaerwen, LL60 6JN
2M0 DES G Taylor, 15 Ronaldsvoe, Kirkwall, KW15 1XE
2E0 DEU Marc Hanbuerger, Arboris, New Road Hill, Reading, RG7 5RY
2E0 DEV Robert Barter, 17 West Gate, Plumpton Green, Lewes, BN7 3BQ
2E0 DEX D Rigby, 32 Springs Road, Chorley, PR6 7AN
2E0 DEZ Des Turner, 29 Balmoral Road, Castle Bromwich, Birmingham, B36 0JT
2E0 DFA Brian Geall, 129 Jewell Road, Bournemouth, BH8 0JP
2E0 DFB David Grundy, 44 Heathend Road, Alsager, Stoke-on-Trent, ST7 2SH
2E0 DFF Michael Bookham, 116 Clare Gardens, Petersfield, GU31 4EU
2E0 DFG Jeremy Tarrant, 70 Sunnymead, Midsomer Norton, Radstock, BA3 2SD
2E0 DFI Malcolm Philpott, Garmisch, Hazel Road, Aldershot, GU12 6HP
2E0 DFJ David Jacobs, 7 Coppice Close, Ravenstone, Coalville, LE67 2NS
2E0 DFL David Humm, 15 Sherborne Road, Farnborough, GU14 6JS
2E0 DFM Richard Russell, 1 Horeb Cottages, Rhiw Road, Colwyn Bay, LL29 7TL
2E0 DFN Goronwy Edwards, 17 Glan y Mor Road, Penrhyn Bay, Llandudno, LL30 3NL
2E0 DFO Paul Heiney, 6 Arthur Street, Oxford, OX2 0AS
2E0 DFP Darren Parker, 53 Brisbane Way, Cannock, WS12 2GR
2E0 DFQ Patrick Love, 30 Salisbury Road, Canterbury, CT2 7HH
2E0 DFS David Smith, 186 Weekes Drive, Slough, SL1 2YR
2E0 DFT D Fisher, 34 Orange Croft, Tickhill, Doncaster, DN11 9EW
2E0 DFV Roger Rigby, 11 Mallow Close, Thornbury, Bristol, BS35 1UE
2E0 DGA Paul Wilson, 45 Newquay Close, Hartlepool, TS26 0XG
2M0 DGB Duncan Baillie, 126 Main St., Fauldhouse, Bathgate, EH47 9BW
2E0 DGC David Cowling, 11 Shakespeare Avenue, Scunthorpe, DN17 1SA
2E0 DGD Douglas Bailey, 2b Queens Road, Enfield, EN1 1NE
2E0 DGG Mark Tinsell Stanton, 38 Comberton Road, Kidderminster, DY10 3DT
2E0 DGH D Humphrey, 42 Ratcliffe Road, Sileby, Loughborough, LE12 7PZ
2M0 DGI Peter Riddle, Carngeal, Pitlochry, PH16 5JL
2M0 DGJ Zoe Bak, 62/6 North Gyle Loan, Edinburgh, EH12 8LD
2E0 DGL Donald Lock, 22 The Towers, Southgate, Stevenage, SG1 1HE
2E0 DGM Sarah Hopkins, 3 Butts Road, Wolverhampton, WV4 5QD
2E0 DGN Andrew Rackham, 31 Severn Road, Pontllanfraith, Blackwood, NP12 2GA
2E0 DGQ Paul Flanagan, 71 [illegible], Pelton Fell, Chester le Street, DH2 2BY
2E0 DGR Arthur Norton, Flat 9, Brownhill Court, Southampton, SO16 9LB
2E0 DGS Daniel Smith, 48 Shirley Gardens, Tunbridge Wells, TN4 8TH
2E0 DGT Roy Fripp, 41 Sweyns Lease, East Boldre, Brockenhurst, SO42 7WQ
2E0 DGU William Alexander, 81 Cherry Lane, Lymm, WA13 0SY
2E0 DGV Pete Hodson, 21 Green Hill, London Road, Worcester, WR5 2AA
2E0 DHA David Atkins, 20 Nappsbury Road, Luton, LU4 9AL
2E0 DHB D Bisson, 48 Elmsfield Avenue, Rochdale, OL11 5XN
2I0 DHC Paul White, 46 Pine Cross, Dunmurry, Belfast, BT17 9QY
2E0 DHD David Lockyer, 19b Drury Lane, Buckley, CH7 3DU
2E0 DHE Robin Barnard, 3 Heaths Close, Enfield, EN1 3UP
2E0 DHF Jack Jackson, 49 Leafield Rise, Two Mile Ash, Milton Keynes, MK8 8BX
2E0 DHG Raymond Holmes, 29 Whalesmead Road, Eastleigh, SO50 8HJ
2M0 DHI Robert Buchan, 5 Fairview Terrace, Danestone, Aberdeen, AB22 8ZH
2E0 DHJ Terence Thompson, 14 Queen Street, Northwich, CW9 5JL
2E0 DHK Barnaby Davies, 12 Scalebor Gardens, Burley-in-Wharfedale, LS29 7BX



**IMPORTANT NOTE**

**Revalidate licence to avoid revocation** – Ofcom has advised the Society that plans will be drawn up to revoke licences that have not been revalidated as required by the licence conditions. The quickest way to revalidate is to do so online via the Ofcom website: *https://services.ofcom.org.uk/* or by email: *amateur.validations@ofcom.org.uk* Ofcom staff are available to help, but please be patient during times of heavy workload.

2E0 DHO Paul Chadwick, 112 Sandy Lane, Warrington, WA2 9JA
2E0 DHQ David Rimmer, 41 Ashburton Road, Wallasey, CH44 5XB
2I0 DHR David Richards, 70 Cherryhill Avenue, Dundonald, Belfast, BT16 1JD
2E0 DHS D Sherwin, 5 North Road, Buxton, SK17 7EA
2E0 DHT Jason Berry, 4 Millfield Road, Chorley, PR7 1RE
2E0 DHV Peter Walton, 11 Parkfield Road, Northwich, CW9 7AR
2E0 DHW Adam Stabler, 11 Lincolns Avenue, Gedney Hill, Spalding, PE12 0PQ
2E0 DHX Matthew Watkins, 108 Honiton Road, Exeter, EX1 3EQ
2E0 DHY David Ball, 27 Bramble Close, Aston, Birmingham, B6 5HW
2E0 DHZ Andrew Hitchcott, 121 Oakhurst Road, Acocks Green, Birmingham, B27 7PB
2M0 DIB Catherine Morris, 23 Sedgebank, Livingston, EH54 6HE
2E0 DID David Foyston, 10 Ash Grove, Wickersley, Rotherham, S66 2LJ
2M0 DIF Iain Smith, 32 Kaimes Avenue, Kirknewton, EH27 8AU
2E0 DIG Axel Taylor, 130a Hazelwood Avenue, Eastbourne, BN22 0UX
2E0 DIH Edward Coles, 46 Hampshire Court, Upper St. James's Street, Brighton, BN2 1JF
2E0 DII Ian Phillips, 324 The Meadway Tilehurst, Reading, RG30 4PD
2E0 DIJ Duane Yates, 16 Sunnyfield Road, Prestwich, Manchester, M25 2RD
2E0 DIL Dilawar Yakub, 42 Swift Close, Blackburn, BB1 6LF
2E0 DIM Dimitris Vainas, 51 Magister Road, Bowerhill, Melksham, SN12 6FD
2E0 DIN R Noon, Forest Hill Cottage, Rushall Lane, Wimborne, BH21 3RT
2E0 DIP David Webb, 52 Simpkin Close, Eaton Socon, St. Neots, PE19 8PD
2E0 DIQ Andrew Collins, 14 Double Corner, Mendlesham Road, Cotton, Stowmarket, IP14 4RF
2E0 DIT Nicholas Baulf, 1 Lower Chart Cottages, Brasted Chart, Westerham, TN16 1LS
2E0 DIU Ioan Jones, 90 Preston, Cirencester, GL7 5PR
2E0 DIV P Williams, 63 Trem Eryri, Llanfairpwllgwyngyll, LL61 5JF
2E0 DIX Bernadette Smith, 7 Kestrel Avenue, Bransholme, Hull, HU7 4ST
2E0 DJB Darryl Burden, 16 Milnthorpe Lane, Wakefield, WF2 7DE
2E0 DJC Dean Cole, 14 Inner Loop Road, Beachley, Chepstow, NP16 7HF
2E0 DJF Donald Fagg, 62 Hawkins Road, Folkestone, CT19 4JA
2E0 DJH David Harris, 6 Baker Lea, Monkland, Leominster, HR6 9DB
2E0 DJI Darren Oliver, 20 Five Oaks Close, Malvern, WR14 2SW
2E0 DJJ George McCaffery, 7 Cliffe Court, Sunderland, SR6 9NT
2I0 DJM Joe McBride, 22 Birchwood, Omagh, BT79 7RA
2E0 DJP Darren Parvin, 11 Stanhope Way, Sevenoaks, TN13 2DZ
2E0 DJQ M Barnaby, 8 Callowood Croft, Purleigh, Chelmsford, CM3 6NZ
2E0 DJR Julian Rudd, 5 St Andrews Close, Blofield, Norwich, NR13 4JX
2W0 DJS D James, 10 Hafan Deg, Pencoed, Bridgend, CF35 6YG
2M0 DJT David Rodger, 25 Wilson Road, Banchory, AB31 5UY
2E0 DJU Timothy Bannister, 6 Tanners Road, North Baddesley, Southampton, SO52 9FD
2E0 DJV Anthony Wheeler, 8 Elsworth Grove, Birmingham, B25 8EJ
2D0 DJX Stephen Scott, 13 Silver Close, Harrow, HA3 6JT
2E0 DJY David Bennett, 31 Park Road North, Urmston, Manchester, M41 5AT
2E0 DJZ Binoy Issac, 9b Poplar Grove, Stockport, SK2 7JD
2E0 DKA David Carmichael, 22 California Close Great Sankey, Warrington, WA5 8WU
2E0 DKB Darren Banks, 41 East Road, Rotherham, S65 2UX
2E0 DKD Daryn Saxon, 53 Westmorland Road, South Shields, NE34 7JJ
2E0 DKE Christopher Boyle, 8 Westlees Close, North Holmwood, Dorking, RH5 4TN
2M0 DKF Alasdair Gow, 0/1 319 Glasgow Harbour Terraces, Glasgow, G11 6BL
2E0 DKG Ernesto Gomez Lozano, Flat D, 212 Kennington Road, Oxford, OX1 5PG
2E0 DKI David Iveson, 11 Newport Road, North Cave, Brough, HU15 2NU
2E0 DKJ Ellen Musselle, 2 Rectory Crescent, Middle Barton, Chipping Norton, OX7 7BP
2E0 DKM W Molloy, 32 Millers Barn Road, Jaywick, Clacton-on-Sea, CO15 2QB
2E0 DKO Peter Taylor, 32 Heliers Road, Liverpool, L13 4DH
2E0 DKP Andrew Lutley, Springfield, Rookery Hill, Ashtead, KT21 1HY
2I0 DKQ Gregory Gardiner, 60 Limestone Meadows, Moira, Craigavon, BT67 0UT
2E0 DKR Darren Reeves, Flat 2, Challonsleigh, Blandford Forum, DT11 7HB
2E0 DKS C Wilkes, 2 Kings Crescent, Edlington, Doncaster, DN12 1BD
2E0 DKT Mark Bumstead, Windmill Girl III, Riverside Boatyard, Southampton, SO31 1AA
2M0 DKU Stephen Boyd, Gowanbank Chalet, Garelochhead, Helensburgh, G84 0AE
2M0 DKV James Flannigan, 21 Kirkbean Avenue, Rutherglen, Glasgow, G73 4EA
2E0 DKW Craig Nicholl, 36 Eylewood Road, London, SE27 9NA
2M0 DKX Paul Bacon, 12 The Greens, Maddiston, Falkirk, FK2 0FN
2E0 DKY Barry Cross, 22 Park Avenue, Washingborough, Lincoln, LN4 1DB
2E0 DKZ Darren Hyde, 136 Station Road, Woodmancote, Cheltenham, GL52 9HN
2E0 DLA Phillip Booth, 7 Handley Crescent, East Rainton, Houghton le Spring, DH5 9QX
2E0 DLC Barry Tufnell, 1 Moorlands Court Wath-upon-Dearne, Rotherham, S63 6DD
2E0 DLD D Dewsbury, 62 Yew Tree Drive, Leicester, LE3 6PL
2W0 DLE T Edwards, Broadfield House, Vicarage Road, Tonypandy, CF40 1HP
2E0 DLF Andrew Cook, 84 Clent View Road, Birmingham, B32 4LW
2E0 DLG Christopher Gilbert, 63 Lovell Road, Bedford, MK42 0LS
2E0 DLH D Hill, 11 Paddock Lane, Metheringham, Lincoln, LN4 3YG
2E0 DLJ D Johnstone, 14 Carr Hey, Wirral, CH46 6EL
2E0 DLK Lawrence Sargent, 13 Park View, Thornton, Liverpool, L23 4TD
2E0 DLL M Richardson, 34 Clarence Green, Newton Aycliffe, DL5 5HZ
2E0 DLO Simon Strange, 94 Digby Avenue, Nottingham, NG3 6DY
2E0 DLP David Ion, 78 Blackmore Street, Derby, DE23 8AX
2E0 DLR Derek Taylor, Flat 207, The Metropole, Folkestone, CT20 2LU
2E0 DLV Cesar Lombao, 6 Privet Close, Lower Earley, Reading, RG6 4NY
2E0 DLZ D Dunne, 1 Burton Gardens, Brierfield, Nelson, BB9 5DR
2E0 DMA D Aldridge, 5 Alpine Close, Paulton, Bristol, BS39 7SE
2E0 DMB D Browne, 27 Lewis Road, Emsworth, PO10 7RP
2I0 DMC Declan McCloskey, 1 Dernaflaw Cottages, Dernaflaw Road, Londonderry, BT47 4PP
2E0 DMD David McArthur, 7 Gore Avenue, Salford, M5 5LF
2E0 DME D Priestley, 8 Cokefield Avenue, Nuthall, Nottingham, NG16 1AU
2W0 DMG D Griffiths, 8 Heol Cynwyd, Llangynwyd, Maesteg, CF34 9TB
2E0 DMI Alan Yates, 19 Lauriston Park, Cheltenham, GL50 2QL
2E0 DMJ Diane Gegg, 84 Aberconway Crescent, New Rossington, Doncaster, DN11 0JP
2E0 DMM David Moran, 94 Moran Road, Knutton, Newcastle, ST5 6EY
2E0 DMN Daniel Zubrzycki, 16 Oldfield Avenue, Hull, HU6 7UN
2E0 DMP D Powell, 58 Lessingham Avenue, Swinley, Wigan, WN1 2HX
2E0 DMQ K HENNEY, Flat 18, Ash House, 22 Brook Avenue, Ascot, SL5 7SG
2E0 DMS Darren Stewart, 79 Eastfield Road, Driffield, YO25 5EZ
2E0 DMU David Pearson, 37 Elmridge, Leigh, WN7 1HN
2E0 DMV Grenville Weston, 131 Ringwood Road, Eastbourne, BN22 8TQ
2E0 DMW Roger Weir, 130 Alexander Square, Eastleigh, SO50 4BX
2E0 DMX Dominic Moffat, 27 Cinque Ports Way, Seaford, BN25 3UE
2E0 DMY Daniel Humphrey, 11 Colborne Close, Poole, BH15 1UR
2E0 DNB N Brown, 241 Bury Road, Tottington, Bury, BL8 3DY
2E0 DNC Dave Swift, 15 Gloucester Walk, Westbury, BA13 3XF
2E0 DNE Bala Rajagopal, 4 Balliol Road, Caversham, Reading, RG4 7DT
2E0 DNF David Featherby, 14 Station Road, Sutton, Ely, CB6 2RL
2E0 DNH Roy Finch, Garth Cottage, North Cowton, Northallerton, DL7 0HL
2E0 DNI William Oliver, Pwllmeyric, Chepstow, NP16 6LE
2E0 DNJ Darren Jarvice, 15 Meden Avenue, Warsop, Mansfield, NG20 0PS
2E0 DNL Derek Neill, 7 Ashbrow Road, Northampton, NN4 8ST
2M0 DNM D MacKenzie, 4 Nabhar, Laxay, Isle of Lewis, HS2 9PJ
2E0 DNO Dean Close, 22 Station Road, Dodworth, Barnsley, S75 3JE
2E0 DNR Jennifer Hughes, Midfield Farm, Midfield Caravan Site, Aberystwyth, SY23 4DX
2E0 DNS John Cummins, Flat 30, St. Giles, Moor Hall Lane, Chelmsford, CM3 8AR
2E0 DNU Giles Cater, 7 Seymour Street, Chelmsford, CM2 0RX
2W0 DNV Derek Hooper, Nant y Dryslwyn Cottage, Ty Mawr, Llanybydder, SA40 9RD
2E0 DNW Lee Stamper, 22 Douglas Road, Workington, CA14 2QY
2E0 DNX Daniel Smith, 49a Seacroft Drive, St. Bees, CA27 0AF
2M0 DOC John Dock, 75 Ferguslie Park Avenue, Paisley, PA3 1JW
2E0 DOD Adrian Dodd, 68 Windlehurst Road, High Lane, Stockport, SK6 8AE
2E0 DOE Christopher Taylor, 23 Heol Derw, Brynmawr, Ebbw Vale, NP23 4TT
2E0 DOF Scott Black, 7 Harwood Close, Gosport, PO13 0TY
2E0 DOG Ronald Scholefield, 4 Minnie Street, Haworth, Keighley, BD22 8PR
2M0 DOI Denis Speirs, 45 Elmbank Crescent, Arbroath, DD11 4EZ
2E0 DOJ D Jenkins, 1 Green End Road, Sawtry, Huntingdon, PE28 5UX
2M0 DOL Jonathan Marsh, 8 Hazelton Way, Broughty Ferry, Dundee, DD5 3BT
2E0 DOM Dean Maddison, 9 Rowley Way, Sunnyside, Rotherham, S66 3ZY
2E0 DOP Alexander Seelig, Flat 67, Regents Riverside, Reading, RG1 8QS
2E0 DOQ Mohammad Dinally, 208 Ley Hill Farm Road, Birmingham, B31 1UQ
2E0 DOW Brian Lewin, 68 Brackley Square, Woodford Green, IG8 7LS
2E0 DOX David Richards, Flat 40, Leander Court, Teignmouth, TQ14 8AQ
2E0 DOZ Dorian Logan, Cedar House, Reading Road North, Fleet, GU51 4AQ
2E0 DPD Bernard Scannell, 60 Burnside Road, Dagenham, RM8 1XD
2E0 DPF Michael Reaney, Odessa Marine, Little London, Newport, PO30 5BS
2E0 DPH Philip Hughes, 111 Wisbech Road, Littleport, Ely, CB6 1JJ
2E0 DPI Adam Studdart, 11 Degas Close, Connah's Quay, Deeside, CH5 4WQ
2E0 DPL Colin Glass, The Old Homestead, Havikil Lane, Knaresborough, HG5 9HN
2E0 DPO D Greenland, 1 Hilltop, Tuesley Lane, Godalming, GU7 1SB
2E0 DPR Darren Richardson, 25 Comptons Lane, Horsham, RH13 5NL
2E0 DPS Daniel Seedhouse, 5 June Crescent, Amington, Tamworth, B77 3BH
2E0 DPU Daniel Pugh, 8 Clos Deiniol, Llanbadarn Fawr, Aberystwyth, SY23 3TX
2E0 DPX Wil Currie, 11 Boston Road, Ipswich, IP4 4EQ
2E0 DPY David Pye, 12 Buchanan Drive, Hindley Green, Wigan, WN2 4HJ
2E0 DPZ Ronald Lyddall, 102 Chapel Road, Brightlingsea, Colchester, CO7 0HE
2E0 DQB Anthony Smith, 116 Pilling Lane, Preesall, Poulton-le-Fylde, FY6 0HG
2E0 DQD Paul Driver, 68 Ripon Road, Dewsbury, WF12 7LG
2E0 DQG Quentin Wright, 9 Browning Avenue, Warwick, CV34 6JQ
2E0 DQJ Harold Woodfin, 8 Bank Hall Close, Bury, BL8 2UL
2E0 DQL David Coles, 36 York Hill, Loughton, IG10 1HT
2M0 DQN Gordon McLeod, 75 Grange Avenue, Wishaw, ML2 0AH
2E0 DQO James Marks, Chantry End, Oak Hill, Epsom, KT18 7BU
2E0 DQP Anne Lewis, Four Winds Cottage, Main Street, Brough, HU15 1RJ
2E0 DQQ Stuart McLoughlin, 40 Rowlandson Gardens, Bristol, BS7 9UH
2W0 DQT Allan Williams, 62 Wern Road, Llanelli, SA15 1SR
2E0 DQU Ian Lawton, 38 Battershall Close, Plymouth, PL9 9UU
2E0 DQX Jacob Saunders, 123 Medway Road, Ferndown, BH22 8UR
2M0 DQY John Dow, 52 Muirfield Way, Deans, Livingston, EH54 8EN
2E0 DRA Michael Draper, 160 Chanctonbury Road, Burgess Hill, RH15 9HA
2W0 DRB David Barraclough, 6 Bryn Terrace, Llangynwyd, Maesteg, CF34 0EA
2E0 DRE D Dean, 12 Abbeydale Road South, Sheffield, S7 2QN
2E0 DRG Carl Schofield, 1187 Manchester Road, Castleton, Rochdale, OL11 2XZ
2E0 DRI Karl Abel, 7 Foldgate View, Ludlow, SY8 1NB
2E0 DRJ David Farrant, 13 Bramham Down, Guisborough, TS14 7BY
2W0 DRK David Machon, 22 Albert Street, Caerau, Maesteg, CF34 0UF
2D0 DRM P Black, Rock plain, main road, Crosby, Isle of Man, IM4 2DR
2E0 DRN D Rayne, 73 Southfield Road, Hinckley, LE10 1UA
2M0 DRO Robert Drummond, 11 Firwood Drive, Bo'ness, EH51 0NX
2E0 DRQ David Roberts, 3 Heather Avenue, Melksham, SN12 6FX
2E0 DRT Corrina Brock, 3 Morley Street, Norwich, NR3 1ND
2E0 DRW Winton Wightman, 36 Holyoake Avenue, Woking, GU21 4PW
2M0 DRY David Drysder, 37 Farburn Drive, Stonehaven, AB39 2BZ
2E0 DRZ Alison Potts, 103 Etherstone Street, Leigh, WN7 4HY
2E0 DSB Dennis Slade, 22 Oaklands Road, Mangotsfield, Bristol, BS16 9EY
2M0 DSG Derek Gartshore, 85 Springhill Street, Douglas, Lanark, ML11 0NZ
2E0 DSH Darren Hind, 19 Ellington Road, Arnold, Nottingham, NG5 8SJ
2E0 DSI David Stocker, 2 Paull Road, Bodmin, PL31 1QJ
2E0 DSJ Darren Jenkins, 15 Homefield Close, Winscombe, BS25 1JE
2E0 DSK Dawn Manning, 153 Pavilion Road, Worthing, BN14 7EG
2M0 DSL D Latto, 8 Aspen Avenue, Glenrothes, KY7 5TA
2W0 DSO Chris Summerfield, 11 Woodland Park, Penderyn, Aberdare, CF44 9TX
2W0 DSP A Elias, 31 Banc Y Gors, Upper Tumble, Llanelli, SA14 6BR
2E0 DSQ Philip Taylor, 104 Winstanley Drive, Leicester, LE3 1PA
2E0 DSS Walter Stewart, 43 Newlands Drive, Halesowen, B62 9DX
2M0 DSU Ronald Murray, 18 Braids Road, Kirkcaldy, KY2 6JE
2E0 DSV Robert Nicholson, 4 Morris Court, Aylesbury, HP21 9QT
2E0 DSW David Weight, 12 Durrants Path, Chesham, HP5 2LH
2E0 DSX Dean Sullivan, 15 Market Lane, Witham, CM8 1GF
2M0 DSY Sajimon Chacko, 210 Hillington Road South, Glasgow, G52 2BB
2E0 DSZ Vasilije Perovic, Trinity College, Cambridge, CB2 1TQ
2E0 DTB David Bradley, 45 Fourth Avenue, Ketley Bank, Telford, TF2 0AS
2E0 DTC David Beck, 94 Shaldon Crescent, Plymouth, PL5 3RB
2E0 DTD Trevor Davies, 7 Crescent Road, Warley, Brentwood, CM14 5JR
2I0 DTE David Best, 13 Cranley Green, Bangor, BT19 7FE
2E0 DTF Matthew Bostock, 86 Beauvale Drive, Ilkeston, DE7 8SJ
2E0 DTG David Griffiths, 43 Laneside Road, Grange-over-Sands, LA11 7BX
2E0 DTH David Thompson, 17 Sandpiper Close, Blyth, NE24 3QN
2M0 DTJ Geoffrey Lewin, Larch Cottage, Lein Road, Fochabers, IV32 7NW
2E0 DTL C Brink, 138 Brookside, Burbage, Hinckley, LE10 2TN
2W0 DTM Nicholas Sugg, 18 Dolgynog, Penderyn, Aberdare, CF44 9JT
2E0 DTN Dirk Niggemann, 35 Holm Court, Twycross Road, Godalming, GU7 2QT
2E0 DTO E Bray, 28 Henshall Avenue, Latchford, Warrington, WA4 1PY
2M0 DTP Tom Burnett, 45 The Murrays Brae, Edinburgh, EH17 8UF
2E0 DTQ Tiffany Kilfeather, Flat 1, 57 Chalk Hill, Watford, WD19 4DA
2W0 DTR Dean Rosser, 42 Cobden Street, Aberaman, Aberdare, CF44 6EN
2E0 DTS B Oldfield, 4 Creteway Close, Folkestone, CT19 6LH
2E0 DTV Matthew King, 126 Blythsford Road, Hall Green, Birmingham, B28 0UT
2E0 DTW Daniel Williams, 11 Berkeley Gardens, London, N21 2BE
2E0 DTX Laurence Cook, 43 Midge Hall Drive, Rochdale, OL11 4AX
2E0 DTY Martin Pesendorfer, 13 Blake Road, London, N11 2AD
2M0 DTZ John Hogg, 31 Woodlea Court, Crosshouse, Kilmarnock, KA2 0ES
2E0 DUA Keith Lynch, Medindie, Woodside, Ryton, NE40 4SY
2E0 DUB Tim Price, 40 East Street, Kidderminster, DY10 1SE
2E0 DUD Jonathan Storey, 3 Woodside Road, Poole, BH14 9JH
2E0 DUE Ian Warnecke, 12 Caxton Road, Margate, CT9 5NP
2E0 DUF Sam Lo, Upper Maisonette, 41 Park Street, Bath, BA1 2TD
2E0 DUH Justin Shears, 161 Park Road, Keynsham, Bristol, BS31 1AS
2E0 DUI R Sutton, 80 Fishbourne Lane, Ryde, PO33 4EU
2E0 DUJ Steve Bluff, 2 Astor Mews, High Street, Tidworth, SP9 7TR
2E0 DUL Dulyn Davies, 2 Hendre Ddu, Manod, Blaenau Ffestiniog, LL41 4BH
2E0 DUM Ivan Clarke, 86 Charlton Road, Andover, SP10 3JY
2W0 DUN Thomas Dungey, 8 Downs View Close, Aberthin, Cowbridge, CF71 7HG
2E0 DUO Anita Richards, 114 Northleach Close, Redditch, B98 8RD
2E0 DUP Vance Downes, 55 Ashfield Road, Bromborough, Wirral, CH62 7EE
2E0 DUQ Aiffah Ali, 42 Blease Close, Staverton, Trowbridge, BA14 8WD
2I0 DUR Andrew Savage, 469 Old Belfast Road, Bangor, BT19 1RQ
2E0 DUS James Fenton, 4 Forest Hills, Newport, PO30 5NG
2E0 DUU Ian Holdford, 46 Hildreth Road, Prestwood, Great Missenden, HP16 0LY
2E0 DUW Leslie Jones, 44 Allthorpe Drive, Loughborough, LE11 4QU
2E0 DUZ Trevor Hope, 59 Chatsworth Crescent, Walsall, WS4 1QU
2E0 DVD A Swan, 47 Warren Close, Whitehill, Bordon, GU35 9EX
2E0 DVF Leslie Lemmon, 4 Honington Close, Wickford, SS11 8XB
2E0 DVI Graham Ridley, 12 Garforth Avenue, Steeton, Keighley, BD20 6SP
2E0 DVK Richard Steele, 244a Uttoxeter Road, Blythe Bridge, Stoke-on-Trent, ST11 9LY
2E0 DVM Darren Vincelli, 20 Worthington Close, Hyde, SK14 3QX
2E0 DVN James Sanders, 76 Fullerton Road, Plymouth, PL2 3AX
2E0 DVO Kevin Wells, 45 Laburnum Avenue, Yaxley, Peterborough, PE7 3YQ
2E0 DVP David Price, 11 Cefn Melindwr, Capel Bangor, Aberystwyth, SY23 3LS
2E0 DVU Michael Hughes, 183 Station Road, Hednesford, Cannock, WS12 4DP
2E0 DVV Jonathan Allen, Croston, Old Hall Road, Ulverston, LA12 7DL
2E0 DVW Malcolm Roberts, 5 Cliff Cottages, Cliff Road, Hessle, HU13 0HB
2E0 DVX Nick Malyon, 19 Swallow Cliffe, Shoeburyness, Southend-on-Sea, SS3 8BL
2E0 DVY Joby Poriyath, 18 Howard Close, Cambridge, CB5 8QU
2E0 DWA Andrew Dickson, The Rowans, Pwllmeyric, Chepstow, NP16 6LA
2E0 DWB David Belton, 4 Sandown Road, Toton, Nottingham, NG9 6GN
2M0 DWC Dennis Cowie, 69 Broomfield Park, Portlethen, Aberdeen, AB12 4XT
2U0 DWD Adam Prosser, Woodlands, La Vassalerie St. Andrew, St. Andrew, Guernsey, GY6 8XL
2E0 DWE Derek Toller, Field Cottage, Ash Lane, Derby, DE65 6HT
2E0 DWF Darren Leyland, 20 Newton Heath, Middlewich, CW10 9HL
2E0 DWG David Gough, 29 Belvedere Road, Biggin Hill, Westerham, TN16 3HX
2E0 DWJ David Johnston, 5 Moorfield Crescent, Hemsworth, Pontefract, WF9 4EQ
2E0 DWK Craig Duckworth, 261 Barden Lane, Burnley, BB10 1JA
2E0 DWM David Mardlin, 13 Churchill Crescent, Sonning Common, Reading, RG4 9RU
2E0 DWP Daniel Prout, 2 Pine Crest Way, Bream, Lydney, GL15 6HG
2E0 DWR David Rich, 41 Ronald Road, Newport, NP19 7GF
2E0 DWS Derek Sewell, 19 St. Leonards Way, Ashley Heath, Ringwood, BH24 2HS
2E0 DWT David Wells, 34 Bramble Way, Wymondham, NR18 0UN
2E0 DWU Andre Edmonds, 20 Tomline Road, Ipswich, IP3 8BZ
2E0 DWV Emil Preda, 59 Cambridge Street, Stockport, SK2 6PY
2E0 DWW Kevin Gorringe, 36 West Close, Polegate, BN26 6EL
2E0 DWZ Mark Davey, Romea, Long Street, Attleborough, NR17 1LW
2J0 DXA Steve Huelin, Stebezel, La Petite Rue De La Pointe, St Peter, Jersey, JE3 7YZ
2M0 DXC A Brown, 14 Preston Place, Pathhead, EH37 5QS
2E0 DXJ Dominic Barrett, 208 Doncaster Road, Rotherham, S65 2UE
2E0 DXK John Whitworth, 31 Shirley Close, Chesterfield, S40 4RJ
2E0 DXL Ivor Newton, 16 Cross Close, Newquay, TR7 3LB
2E0 DXO Paul Higginson, 9 St. Mildreds Way, Heysham, Morecambe, LA3 2QJ
2E0 DXV Tony Agar, 2 Belsay Close, Ferryhill, DL17 8SX
2E0 DXW Mark Roper, 11 East Close, Beverley, HU17 7JN
2E0 DXZ Lakota Brearley, Ash Tree Lodge, Snaith Road, Goole, DN14 0AT
2I0 DYA Nat Cully, 42 Omerbane Road, Cloughmills, Ballymena, BT44 9PE
2E0 DYB David Brough, 38 Tynedale Avenue, Crewe, CW2 7NY
2E0 DYD Graham Lloyd, 1 Holmside Terrace, Stanley, DH9 6ET
2E0 DYG David Young, 75 Broadlands Road, Southampton, SO17 3AP
2E0 DYJ Mark Bentley, 5 Stokewell Road, Wath-upon-Dearne, Rotherham, S63 6EL
2E0 DYK David McCarthy, 102 Grove Street, New Balderton, Newark, NG24 3AS
2E0 DYM P Cottam, Eastleigh, Kings Nympton, Umberleigh, EX37 9ST
2E0 DYN David Jones, Drove Farm, Sheepdrove, Hungerford, RG17 7UN
2E0 DYP Wayne Hartley, 30 Coltman Avenue, Beverley, HU17 0EY

2E0 DYU Paul Sykes, 18 Hill Fold, South Elmsall, Pontefract, WF9 2BZ
2E0 DYV Peter Collier, Flat 9, Henry House, London, SW11 [illegible]
2E0 DYX Timothy Carpenter, 21a Rails Lane, Hayling Island, PO11 9LG
2E0 DYY leslie hopgood, 5c Cherry Tree Road North, Blackpool, FY4 4NY
2E0 DZC Lee Davison, 58 Priestley Court, South Shields, NE34 9NQ
2E0 DZE Marie White, 9 Hawksworth Close, Rotherham, S65 3JX
2E0 DZF Peter Ridgers, 231 The Greenway, Epsom, KT18 7JE
2E0 DZJ Graham Allen, 14 The Parsonage, Sixpenny Handley, Salisbury, SP5 5QJ
2E0 DZN [illegible], 18 [illegible] Drive, Oxford, OX3 0NW
2E0 DZO paul sargeant, 6 Meldon Way, Blaydon-on-Tyne, NE21 [illegible]
2E0 DZQ Jan Ostapiuk, [illegible] Place, Gateshead, NE8 1XN
2E0 DZR Paul [illegible] Drive, Bamber Bridge, Preston, PR5 [illegible]
2E0 DZS Zoltan Dorzsi, 217 Bensham Road, Bensham, Gateshead, [illegible]
2E0 DZT John Emery, Mulberry Cottage, Quarry Lane, Chard, TA20 [illegible]
2E0 DZV Nick Harris, 10 Pentland Drive, North Hykeham, Lincoln, LN6 9TG
2E0 DZY David Atkins, 32 Draybrook, Orton Goldhay, Peterborough, PE2 5SH
2E0 DZZ Duncan Taylor, 1 Mayfield Farm Cottages, Reston, Eyemouth, TD14 5LQ
2E0 EAA Barry Matthews, 30 Oaklands Drive, Brandon, IP27 0NR
2M0 EAC Jim Woods, 12 Westbank Terrace, MacMerry, Tranent, EH33 1QE
2E0 EAD David Cook, 19 Almond Avenue, Risca, Newport, NP11 6PF
2E0 EAF Russell Bond, 21 Coleridge Close, Bletchley, Milton Keynes, MK3 5AF
2E0 EAI Stewart Bide, 9 Greenway, Watchet, TA23 0BP
2E0 EAL John Oldman, 94 Mornington Road, London, E4 7DT
2E0 EAN Paul Fulbrook, 167 Droitwich Road, Fernhill Heath, Worcester, WR3 7TZ
2E0 EAO Martin Harrison, 91 Rye Road, Hastings, TN35 5DH
2I0 EAR Eneas Rainey, 22 Cherry Gardens, Ballymoney, BT53 7AS
2I0 EAS David Elliott, 15 Derrychara Park, Enniskillen, BT74 6JP
2E0 EAT E Taylor, 14 Sycamore Grove, Doncaster, DN4 6NX
2E0 EAU Edward Aksamit, 14 Popplewell Gardens, Gateshead, NE9 6TU
2E0 EAV David Blake, Pound Farm, Swan Lane, Leigh, Swindon, SN6 6RD
2E0 EAW Edward Cross, 12b Oakridge, Three Rivers Country Park, Clitheroe, BB7 3JW
2E0 EAY Alonza Driver, 12 Almond Tree Avenue, Malton, YO17 7DF
2E0 EAZ Edward Heath, 63 Meadway, Dunstable, LU6 3JT
2E0 EBA Robert Aldridge, 6 Sheppards Court, Horsenden Lane North, Greenford, UB6 7QJ
2I0 EBB Anthony Connolly, 68 Willowbank Gardens, Belfast, BT15 5AJ
2E0 EBD Brian Clayton, 26 Wood Walk, Mexborough, S64 9SG
2I0 EBF Albert Toner, 10 Heatherlea Avenue, Portstewart, BT55 7HF
2E0 EBG Eamonn Bias, 10 Riverdale Road, Shrewsbury, SY2 5TA
2E0 EBJ F Taylor, 55 Leicester Avenue, Horwich, Bolton, BL6 5QX
2E0 EBK David Levett, 11 Love Lane, London, SE25 4NG
2E0 EBL Charles Haynes, 25 Barnards Hill Lane, Seaton, EX12 2EQ
2E0 EBN Peter Singleton, 7 Queen Street, Glossop, SK13 8EL
2E0 EBO J Hann, 2 Leighton Green, Westbury, BA13 3PN
2E0 EBP Trevor Stokes, 1 Hunters Reach, Bradwell, Milton Keynes, MK13 9BT
2E0 EBQ Michael Priest, 35 Albert Road, Chaddesden, Derby, DE21 6SJ
2E0 EBR Andrew Buckland, 21 Malton Close, Monkston, Milton Keynes, MK10 9HR
2I0 EBS Edwin McKnight, 14 Marlacoo Beg Road, Portadown, Craigavon, BT62 3TF
2E0 EBU Robert Clow, 25 Scott Street, Newcastleton, TD9 0QQ
2E0 EBV David Cutts, 38 Berkeley Drive, Hornchurch, RM11 3PY
2E0 EBW D Gemmell, 36 Church Street, Dumfries, DG2 7AS
2E0 EBX Kevin Carter, 50 Elliman Avenue Bottom flat, Slough, SL2 5BG
2E0 EBZ A CHANCE, 24 Doddsfield Road, Slough, SL2 2AD
2E0 ECD Craig Dennis, 1 West Villa, Crathorne, Yarm, TS15 0BA
2E0 ECG Terrance Anthony, 27 Evelyn Avenue, Doncaster, DN2 6LN
2E0 ECI William Canavan, 9 The Ridings, Deanshanger, Milton Keynes, MK19 6JD
2M0 ECK Alex Falconer, 61 Mountcastle Drive North, Edinburgh, EH8 7SP
2E0 ECM Mark Cadman, 7 Horsham Avenue, Stourbridge, DY8 5LU
2E0 ECO Andrew Hawksworth, 17 St. Clements Court, Weston, Crewe, CW2 5NS
2E0 ECW E Williams, 19 Raglans, Exeter, EX2 8XN
2E0 ECZ M Douglas, 486 Malpas Road, Newport, NP20 6NB
2E0 EDE Edward Harman, 53 Anthony Road, Borehamwood, WD6 4NB
2E0 EDG Neil Chamberlain, 8 Southfields, Binbrook, Market Rasen, LN8 6DX
2E0 EDI E Stormes, 1 Meadowbank, Belton, Doncaster, DN9 1NW
2D0 EDL David Simmons, 8 Lower Grange, Huddersfield, HD2 1RU
2E0 EDM Edward Moore, 44 Bridge Street, Oxford, OX2 0BB
2E0 EDP Edward Palo, 13 Welwyn Close, St. Helens, WA9 5HL
2D0 EDQ E Quinney, 69 Clagh Vane, Ballasalla, Isle of Man, IM9 2HF
2E0 EDS Edwin Kaye, 119 St. Bernards Avenue, Louth, LN11 8AS
2E0 EDX Ian Taylor, 37 Wood Green Drive, Thornton-Cleveleys, FY5 3DH
2M0 EDY Edward Higgins, 44a Mossvale Street, Paisley, PA3 2LR
2E0 EEB Joseph Cameron, 29 Webster Road, Stanford-le Hope, SS17 0BE
2E0 EEK John Charlton, Hillside House, Ham Lane, Bristol, BS41 8JA
2E0 EEM Nicholas Bristow, 4 Manor Flats, Rothamsted Estate, Harpenden, AL5 2BE
2E0 EEO Stuart Leask, 1 Collington Street, Beeston, Nottingham, NG9 1FJ
2E0 EER Jason Salter, 20 Burrow Road, Chigwell, IG7 4HQ
2E0 EET Paul Woodburn, 21 The Row, Silverdale, Carnforth, LA5 0UG
2E0 EEU Alexander Bullard, 15 Rowan Drive, Lutterworth, LE17 4SP
2E0 EFA Mark Elbourn, 1 Downside, Gosport, PO13 0JS
2E0 EFC David Owings, 11 Thingwall Road East, Thingwall, Wirral, CH61 3UY
2E0 EFH Edward Hull, Flat 17, Parade Court, Portsmouth, PO2 0RB
2M0 EFI F wenseth, 2 Sunnybank Cottage, Logie Coldstone, Aboyne, AB34 5PQ
2E0 EFN Even Almas, 10, Rindal, Norway, 6657
2E0 EFO Michael Luper, Apartment 14, Thames Point, The Boulevard, London, SW6 2SX
2U0 EFR Denzil Robert, Nos Treis Liberation Drive, 7 Route Des Clos Landais, Guernsey, Guernsey, GY7 9PH
2M0 EGI Edward Ireland, The Steading, Blairmains, Shotts, ML7 5TJ
2W0 EGK K Martin, 12 Heol Fargoed, Bargoed, CF81 8PP
2I0 EGN Robert Nelson, 32 Rallagh Road, Dungiven, Londonderry, BT47 4TT
2E0 EGO Michael Read, 53 West Way, Lancing, BN15 8LX
2E0 EGP John Churchill, West Winds, Brandheath Lane, New End, Redditch, B96 6NG
2E0 EGR William Smith, 29 Peasemore Road, Sunderland, SR4 0HN
2E0 EHB Neil Livingstone, 2 Mickleton, Wilnecote, Tamworth, B77 4QY
2E0 EHD Edward Delasalle, 31 West Hill Road, Hoddesdon, EN11 9DL
2E0 EHT Graeme Clark, 65 Chyvelah Vale, Gloweth, Truro, TR1 3YJ
2E0 EHV Alan MacDonald, Woodside Cottage, Horton Way, Verwood, BH31 6JJ
2I0 EID Michael Hinph, 36 O'Hanagnue Park, Doogbrook, Newry, BT35 7AA
2I0 EIG Jamye McGoldrick, 45 Stewarts Road, Dromara, Dromore, BT25 2AN
2I0 EIR Conor Robinson, 71 Eglantine Road, Lisburn, BT27 5RQ
2E0 EJA E Cole, 24 Patricks Orchard, Uffington, Faringdon, SN7 7RL
2E0 EJK R Keast, 7 The Finches, Newport, PO30 5GU
2U0 EJL James Littlewood, Wayland, LES MARTIN, L'islet, Guernsey, GY2 4XW
2E0 EJM E Marsh, 16 Laurel Close, North Warnborough, Hook, RG29 1BH
2D0 EJR Ray Edwards, 23 Queens Walk, Ruislip, HA4 0LX
2I0 EJT John Billingham, 27 [illegible] Gardens, Portadown, Craigavon, BT62 3BN
2E0 EJW Eric Woodward, [illegible] Manor, Hartfields, Hartlepool, TS26 0NW
2E0 EKB G Fatluk, [illegible] Road, Worthing, BN13 1DX
2I0 EKN William Martin, 01 Camgart Road, Tempo, Tempo, BT94 3EQ
2E0 EKR N harris, 45 Sleigh Road, Sturry, Canterbury, CT2 0HT
2E0 EKT Matthew Evans, 9 Hollis Close, Long Ashton, Bristol, BS41 9AZ
2E0 ELD Emma Dalton, 120 Goodway Road, Birmingham, B44 0NC
2E0 ELI Gordon Hayers, 87 Bradleigh Avenue, Grays, RM17 5RH
2E0 ELK A Hawkins, 5 Ranworth Road, Great Sankey, Warrington, WA5 3EH
2E0 ELO Daniel Smith, 7 Kestrel Avenue, Bransholme, Hull, HU7 4ST
2M0 ELP C Maxwell, 29 Ambleside Rise, Hamilton, ML3 7HJ
2E0 ELT William Foster, 55 Drake Avenue Minster on Sea, Sheerness, ME12 3SA
2E0 EMB Leon Lee, 8 William Avenue, Margate, CT9 3XT
2E0 EME Paul Smith, 5 Olivers Hill, Cherhill, Calne, SN11 8UR
2E0 EMF Wayne Johnson, 10 Archdale Road, Nottingham, NG5 6EB
2E0 EMG Eric MacGurk, 10 Elmore Road, Lee on Solent, PO139DU
2E0 EML David Polley, 6 Coneygear Road, Hartford, Huntingdon, PE29 1QL
2M0 EMM Edward Munro, 55 Abergeldie Road, Aberdeen, AB10 6ED
2E0 EMP Christian Keszei, 2 Blackmore Hill Farm Cottages, Calvert Road, Buckingham, MK18 2HA
2E0 EMX Alan Holt, 36 The Maltings, Malmesbury, SN16 0RN
2E0 END M Townsend, 71 Elm Court, Newbridge, Newport, NP11 5LU
2E0 ENG Matthew Winch, 2 Cranleigh Gardens, Cowes, PO31 8AS
2E0 ENN Sean Burton, 20 Flowerdown Avenue, Cranwell, Sleaford, NG34 8HZ
2E0 ENP Kevin Matthews, St. Helens Cottage, Flimby, Maryport, CA15 8RX
2E0 ENQ Michael McKenna, 25 Heaton Place, Norton Road, Colwyn Bay, LL28 4TL
2E0 ENW Edwin Wilson, 60 Seathorne, Withernsea, HU19 2BB
2E0 ENZ P Joynson, 10 Rothesay Gardens, Prenton Hall Road, Prenton, CH43 3DW
2E0 EOD Aaron Billingham, 6 Kemble Close, Lincoln, LN6 0NR
2E0 EOF Charlie Westcott, 50 Bentinck Street, Sutton-in-Ashfield, NG17 4AZ
2E0 EOK Elwyn Powell, 28 Frederick Avenue, Hereford, HR1 1HL
2E0 EOL David Palmer, 19 Herbert Road, Bath, BA2 3PP
2E0 EOP Nathan Haigh, 10 Moor Park Gardens, Dewsbury, WF12 7AS
2I0 EOS Jim Kelso, 32 Old Park Manor, Ballymena, BT42 1RW
2E0 EOU John Hinds, 69 Carshalton Grove, Wolverhampton, WV2 2QZ
2I0 EPC Trevor Crawford, 21 Ardranny Drive, Newtownabbey, BT36 6BD
2W0 EPE Paul Plummer, 26 Hill Road, Neath Abbey, Neath, SA10 7NR
2E0 EPM Jacqueline Caswell, 10 Beech Close, Scole, Diss, IP21 4EH
2E0 EPR Eric Phiri, 26 The Drove, Andover, SP10 3DL
2E0 EPT R Paris, 14 Jarden, Letchworth Garden City, SG6 2NP
2I0 EQC M McCourt, 52a Moira Road, Crumlin, BT29 4JL
2I0 EQR D McDonnell, 52 Moira Road, Glenavy, Crumlin, BT29 4JL
2I0 EQS T McDonnell, 52 Moira Road, Glenavy, Crumlin, BT29 4JL
2E0 ERD Raymond Overy, 62 Dykelands Road, Sunderland, SR6 8ER
2E0 ERE Michael Bruce, 28 Pheasants Way, Rickmansworth, WD3 7ES
2E0 ERF Garry Sheppard, 32 Bramble Drive, Hailsham, BN27 3EG
2E0 ERG Raymond Harriman, 9 Millers Close, Rushden, NN10 9RP
2E0 ERK Esta Kissin, 115 Aarons Hill, Godalming, GU7 2LJ
2E0 ERM Eric Milner, 16 Spring Valley Court, Bramley, Leeds, LS13 4TT
2E0 ERP Ashley Burton, 12 Munden Grove, Watford, WD24 7EE
2E0 ERS Raymond Springall, 27 Westbourne Park, Scarborough, YO12 4AS
2E0 ERT Rob Smith, 21 Canal Road Crossflatts, Bingley, BD16 2SR
2E0 ERV Steve Slack, 10 Micheldever Road, Whitchurch, RG28 7JG
2I0 ESA James McGoldrick, 45 Stewarts Road, Dromara, Dromore, BT25 2AN
2E0 ESC Matthew Bennett-Blacklock, Theatre View Apartments, 19 Short Street, London, SE1 8LJ
2E0 ESJ N Evans, 38 Cockster Road, Longton, Stoke-on-Trent, ST3 2EG
2M0 ESL Gordon Robinson, 3 Ivy Lane, Dysart, Kirkcaldy, KY1 2XD
2E0 ESO M Bull, 14 Ermin Walk, Thatcham, RG19 3SD
2E0 ESS Edward Slevin, Woodcock Hall, Cobbs Brow Lane, Wigan, WN8 7NB
2E0 ESU M Eade, Roemah-Kita, Strathcona Avenue, Leatherhead, KT23 4HP
2E0 ESX John Hawthorn, 1 Tudor Close, Leigh-on-Sea, SS9 5AR
2E0 ETA Mark Shasby, 19 Crawshaw Grange, Crawshawbooth, Rossendale, BB4 8LY
2I0 ETB William Curry, 7 Ballyversal Road, Coleraine, BT52 2ND
2E0 ETC Laurence Kay, 27 Millbrook Drive, Shawbury, Shrewsbury, SY4 4PQ
2E0 ETD Richard Sindall, 16 Chantrell Road, Wirral, CH48 9XP
2E0 ETE Andrew Beardsley, 10 Moreton Close, Church Crookham, Fleet, GU52 8NS
2E0 ETH Christopher Marshall, 51 Hedgerow Close, Redditch, B98 7QF
2E0 ETN Robert Bradshaw, 272 Councillor Lane, Cheadle Hulme, Cheadle, SK8 5PN
2E0 ETP Steven Cook, 62 Ollerton Road, Barnsley, S71 3DS
2E0 ETT Thomas Horsten, Kastelsvej 4, 2.Tv, Copenhagen E, Denmark, 2100
2E0 ETV Karl Bianchini, 10 St. Leonards Road, Headington, Oxford, OX3 8AA
2I0 ETW Peter Moore, 32 Kinnegar Rocks, Donaghadee, BT21 0EZ
2E0 EUI Damien Nolan, Flat 7, Fonthill Court, London, SE23 3SJ
2E0 EUN Edward Underhill, 61 Goldthorne Avenue, Sheldon, Birmingham, B26 3LA
2W0 EUO Alan Jones, 75 Hollybush Road, Cardiff, CF23 6SZ
2E0 EUR Stephen Milner, Pavilion House, School Lane, Ormskirk, L40 3TG
2E0 EUW Martin Augustus, 84 Wright Way, Stapleton, Bristol, BS16 1WH
2E0 EVA Michael Knowles, 12 Dalestorth Avenue, Mansfield, NG19 6NT
2E0 EVB Joseph Barnes, Glebe Farm, Billington, Stafford, ST18 9DQ
2E0 EVE D Cattermole, Blaxhall Hall Crossing, Little Glemham, Woodbridge, IP13 0BP
2E0 EVP Martin Brasher, 48 Eldertree Road, Thorpe Hesley, Rotherham, S61 2TQ
2E0 EVX S Hall, 12 Lady Jane Grey Road, King's Lynn, PE30 2NW
2E0 EVZ David Lynch, 7 Dollant Avenue, Canvey Island, SS8 9EJ
2E0 EWL John Keefe, 64 Heath Lane, Blackfordby, Swadlincote, DE11 8AA
2E0 EWM Ellie Melman, 177 Grantham Road, London, E12 6NB
2E0 EWS Ernest Unerwin, 1 Warren Place, Barnsley, S70 4LD
2M0 EWY Alan Kinnersley, 5a Regent Terrace, Dunshalt, Cupar, KY14 7HD
2E0 EXC Christopher Wilson, 21 New Road, Hythe, Southampton, SO45 6BN
2E0 EXJ Stephen Baldwin, 143 Oxford Road, Swindon, SN3 4JA
2E0 EXL P Morris, 10 Haslam Avenue, Sutton, SM3 9ND
2E0 EXO Thomas Crocker, 32 Godmanston Close, Poole, BH17 8BU
2I0 EXP Adrian Boyd, 27 The Meadows, Dungannon, BT71 6PW
2E0 EYY Barry Smith, 28 Newhill Road, Wath-upon-Dearne, Rotherham, S63 6JY
2E0 EXY J Sergeant, 16 Green Park, [illegible], BB7 0TJ
2E0 EYO C Rowell, Westview, 1 Chapel Lane, Hull, HU12 0UC
2E0 EYE Fionn Ross, Foundry Cottage, Crowdore Lane, [illegible], TN33 9LP
2E0 EYP James Head, 31 Merebrook [illegible], [illegible], SK11 8RH
2E0 EZK John Power, 12 Campbell Gordon Way, London, NW2 6RS
2E0 EZL T Newton, 1 Brimley Park, Bovey Tracey, Newton Abbot, TQ13 9DE
2E0 EZX William Taylor, 34 Pedley Avenue, Westfield, Sheffield, S20 8EZ
2E0 EZY Harold Foster, 6 Lakeway, Blackpool, FY3 8PF
2E0 FAA Declan McGlone, 32 Shipley Mill Close, Kingsnorth, Ashford, TN23 3NR
2E0 FAB J Whalley, Flat 9, 37 Church Street, Southport, PR9 0QT
2E0 FAC Philip Burke, 38 Bosworth Square, Rochdale, OL11 3QG
2E0 FAE Alan Foote, Flat One, Kimber's Close Kennet Road, Newbury, RG14 5JF
2E0 FAH P Harris, Flat 33, Buckingham Court Shrubbs Drive, Bognor Regis, PO22 7SE
2E0 FAJ Samuel Kingstone, 3 Roman Way, Tamworth, B79 8NF
2E0 FAM Colin Horridge, 6 Back Sreet, East Stockwith, Gainsborough, DN21 3DL
2E0 FAN Paul Pearce, 41 Tennyson Avenue, Boldon Colliery, NE35 9EP
2E0 FAP Alan Phillips, 3 Pen y Llys, Rhyl, LL18 4EH
2E0 FAQ G Austin, 2 Foxfields Way Huntington, Cannock, WS12 4TA
2W0 FAR Adam Burgess, 18 Fairmeadows, Maesteg, CF34 9JL
2E0 FAS Norman Duke, 15 Plover Gardens, Barrow-in-Furness, LA14 3AY
2E0 FAU Barry Cairns, 4 Spence Court, Great Ayton, Middlesbrough, TS9 6DW
2E0 FAV Adrian Barter, 17 West Gate, Plumpton Green, Lewes, BN7 3BQ
2E0 FAY Nigel Fahey, 5 Hillside, Felmingham, North Walsham, NR28 0LE
2E0 FBA Michael McPhee, 63 Mumford Close, West Bergholt, Colchester, CO6 3HY
2E0 FBC Frank Foy, 4 The Square, East Rounton, Northallerton, DL6 2LB
2E0 FBD David Smith, Heath Farm, Heath Road, Bury St. Edmunds, IP30 9RL
2E0 FBE Michael Goodman, 20b Orchard Estate, Little Downham, Ely, CB6 2TU
2E0 FBH Zheng Yao, 56 The Spinney North Cray, Sidcup, DA14 5NF
2E0 FBL Frank Baker, 275 Bye Pass Road, Beeston, Nottingham, NG9 5HS
2E0 FBM Alan Rand, 17 Fairways Drive, Harrogate, HG2 7ES
2E0 FBN Tim Weston, 4 The Pightle, Peasemore, Newbury, RG20 7JS
2I0 FBY Ivan Gillespie, 32 Maghaberry Manor, Moira, Craigavon, BT67 0JZ
2W0 FCF Dean Clark, 137 Llanedeyrn Road, Penylan, Cardiff, CF23 9DW
2E0 FCH Adam Finch, 6 Clover Way, Thetford, IP24 1LQ
2W0 FCM L Paschalis, 45 Pencisely Road, Cardiff, CF5 1DH
2E0 FCS Carl Spencer, 18 Coatsby Road, Kimberley, Nottingham, NG16 2TH
2E0 FCZ Chris Cooper, 25 Waterside Close, Loughborough, LE11 1LP
2E0 FDD Brian Evans, 2 Hastings Road, Eccles, Manchester, M30 8JR
2E0 FDG Brendon Pettit, 35 Lakeside Rise, Blundeston, Lowestoft, NR32 5BE
2E0 FDS Keith Bowyer, 50 Birkdale Gardens, Winsford, CW7 2LE
2M0 FDZ S McLaughlin, 21 Shirrel Road, Motherwell, ML1 4RD
2E0 FEB Stuart Scotching, 26 Newton Way, Leighton Buzzard, LU7 4YU
2E0 FEC George Eycott, 1 Ham Road, Wanborough, Swindon, SN4 0DF
2E0 FEM Samantha Blackham, 5 Rogate Road, Worthing, BN13 2DT
2E0 FEO Raymond Hunter, 3 Sandyway, Croyde, Braunton, EX33 1PP
2I0 FEX D Rantin, 8 Buchanans Road, Newry, BT35 6NS
2W0 FEY Anthony Fey, 28 Bryn Rhedyn, Caerphilly, CF83 3BT
2E0 FFL Frederic Labrosse, 72 Ger y Llan Penrhyncoch, Aberystwyth, SY23 3HQ
2E0 FFM John O'Reilly, 22 Abbey Place, Crewe, CW1 4JR
2E0 FFS Robert Baines, 2 Lower Lune Street, Fleetwood, FY7 6DA
2E0 FFW Paul Drake, Flat 1, Richmond Court, Eagle Close, Yeovil, BA22 8JY
2M0 FFY Christopher Northcott, 14/3 Marytree House, 12 Craigour Green, Edinburgh, EH17 7RP
2E0 FGA Peter Meanwell, 20 Crow Park Avenue, Sutton-on-Trent, Newark, NG23 6QG
2E0 FGH Nicholas Speller, 3 Homestall Close, Oxford, OX2 9SW
2E0 FGM Stephen Haigh, 17 Glebe Street, Swadlincote, DE11 9BW
2E0 FGQ Nicholas Bennett, 35 West Shepton, Shepton Mallet, BA4 5UD
2E0 FGT M Jones, 93 Barrs Road, Cradley Heath, B64 7HH
2D0 FGW David Crouch, 87 Nibthwaite Road, Harrow, HA1 1TD
2E0 FGY Barrie Bestwick, 185 Ashbourne Road, Turnditch, Belper, DE56 2LH
2D0 FHG Frank Goldsmith, 6 Fistard Road, Port St. Mary, Isle of Man, IM9 5HF
2E0 FHR Kelvin Scott, 5 Peel Street, Padiham, Burnley, BB12 8RP
2E0 FIA A Smeed, 5 The Causeway, Sibton, Saxmundham, IP17 2JA
2E0 FIB Simon Watling, 1 Chediston Green, Chediston, Halesworth, IP19 0BB
2E0 FIE James Hobson, 5 Maes Briallen, Llandudno, LL30 1JJ
2E0 FIF Michael Hadfield, 22 Mansfield Road, Clowne, Chesterfield, S43 4DH
2E0 FIJ Donard De-Cogan, 52 Gurney Road, New Costessey, Norwich, NR5 0HL
2E0 FIR Michael Firth, 209 High Street, Wickham Market, Woodbridge, IP13 0RQ
2E0 FIT Robert Lynch, 2 Launceston Close, Oldham, OL8 2XE
2E0 FIZ Alan Brims, 47 Chipchase Avenue, Cramlington, NE23 6TS
2E0 FJA Alexander Ferriroli, 142 Hillbury Road, Warlingham, CR6 9TD
2E0 FJD Karl Dobson, 1 Howarth Road, Ashton-on-Ribble, Preston, PR2 2HH
2E0 FJP John Park, 18 Ladgate Grange, Middlesbrough, TS3 7SL
2E0 FJZ Brian Jones, 1 Edgeworth Road, Hindley Green, Wigan, WN2 4PT
2E0 FKA Jonathan Davies, 5 Beauchamp Road, Kenilworth, CV8 1GH
2E0 FKH S Caddy, 51 Worthington Road, Balderton, Newark, NG24 3RE
2E0 FKS Brian Hoare, 2 St. Peters Close, South Newington, Banbury, OX15 4JL
2E0 FKU Christopher Gibson, 11 Parkside Avenue, Queensbury, Bradford, BD13 2HQ
2E0 FKV Mark Sherrey, 14 The Grove, Hallatrow, Bristol, BS39 6ES
2E0 FLA David Walker, 290 Shannon Road, Hull, HU8 9RY
2E0 FLF Craig Wilson, 12 Desmond Avenue, Hornsea, HU18 1AF
2M0 FLG Mark Bradshaw, 32 Greycraigs, Cairneyhill, Dunfermline, KY12 8XL

UK Callsigns

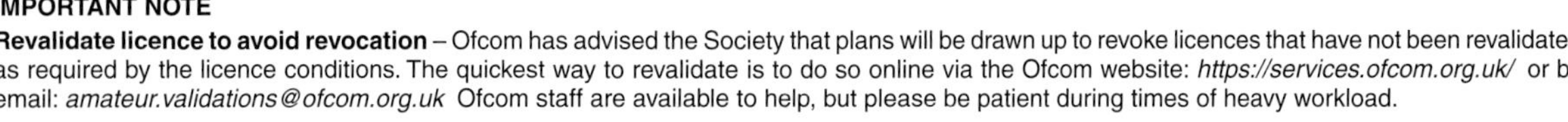
**IMPORTANT NOTE**

**Revalidate licence to avoid revocation** – Ofcom has advised the Society that plans will be drawn up to revoke licences that have not been revalidated as required by the licence conditions. The quickest way to revalidate is to do so online via the Ofcom website: *https://services.ofcom.org.uk/* or by email: *amateur.validations@ofcom.org.uk* Ofcom staff are available to help, but please be patient during times of heavy workload.

2E0 FLI Edwin Flikkema, 7 St. James Mews, Great Darkgate Street, Aberystwyth, SY23 1DW
2M0 FLJ James Morris, 10 Middlemas Road, Dunbar, EH42 1GJ
2E0 FLN Julian Horn, 8 Princess Close, Watton, Thetford, IP25 6XA
2I0 FLO I Nicholl, 7 Killyclooney Road, Dunamanagh, Strabane, BT82 0LZ
2E0 FLR Steve Rhenius, Baythorne Cottage, Baythorne End, Halstead, CO9 4AB
2E0 FMA Andrew West, 33 Mundays Row, Waterlooville, PO8 0HF
2E0 FMB T Huntriss, 1 Threefields, Ingol, Preston, PR2 7BE
2E0 FME D Bennett, 29 Margraten Avenue, Canvey Island, SS8 7JD
2E0 FMG Philip Booker, 17 Colton Copse, Chandler's Ford, Eastleigh, SO53 4HQ
2E0 FML Adam Smith, 6 Rawlinson Avenue, Caistor, Market Rasen, LN7 6NQ
2E0 FMS G Bailey, 61 Great Ranton, Pitsea, Basildon, SS13 1JS
2E0 FMX Nithin Shajan, 19 Sturgess Avenue, London, NW4 3TR
2E0 FMY Florence Masters, 91 Mayfair Avenue, Worcester Park, KT4 7SJ
2E0 FNB F Nuttall, 4 Kingholm Gardens, Bolton, BL1 3DJ
2E0 FNG Matthew Grice, 48 St. Ives Road, Coventry, CV2 5FZ
2I0 FNN Damien Wilson, 81 Parknasilla Way, Aghagallon, Craigavon, BT67 0AU
2E0 FNQ J Mason, 11 Scriven Grove, Haxby, York, YO32 3NW
2E0 FNY David Chatterton, 3 Hunt Close, South Wonston, Winchester, SO21 3HY
2E0 FOD Paul Dekkers, 21 Nodens Way, Lydney, GL15 5NP
2W0 FOG Martin Haywood-Samuel, 38 Tanygraig Road, Llanelli, SA14 9LH
2E0 FOK Dave Sutherland, 78 Holmden Avenue, Wigston, LE18 2EF
2E0 FOL William Bartle, 6 The Cottages, Eccles Road, High Peak, SK23 0EZ
2E0 FOR Ian Forester, 35 Thackeray Street, Sinfin, Derby, DE24 9GY
2E0 FOX B Hopkins, 60 Hales Gardens, Birmingham, B23 5DF
2E0 FPA R Etchells, 6 Woodbank Court, Canterbury Road, Manchester, M41 7DY
2I0 FPB David Neill, 8 Castle Meadows, Carrowdore, Newtownards, BT22 2TZ
2I0 FPK Frank Kearney, 45 Sperrin Park, Omagh, BT78 5BA
2E0 FPO Beverley Bruce, 9 New Road, Ironbridge, Telford, TF8 7AU
2I0 FPT Daniel Hawthorne, 58 Seagahan Road, Collone, Armagh, BT60 2BH
2E0 FQC J Roberts, 41 Mcneill Avenue, Crewe, CW1 3NW
2E0 FQR Aaron Coote, 148 Clarendon Street, Dover, CT17 9RB
2E0 FQT John Mumby, 68 East Common Lane, Scunthorpe, DN16 1QH
2M0 FRA Charles Fraser-Hopewell, 2/1 70 Albert Road, Glasgow, G42 8DW
2E0 FRB D Munday, 29 Coombe Park, Wroxall, Ventnor, PO38 3PH
2E0 FRC Frank Clements, 40 Ellison Fold Terrace, Darwen, BB3 3EB
2E0 FRF A Church, The Willows, Warboys Road, Huntingdon, PE28 3AH
2E0 FRG Fergus Noble, 1045, 45th Street Apartment A, California, United States, 94608
2E0 FRK D Church, The Willows, Warboys Road, Huntingdon, PE28 3AH
2E0 FRO P Froggatt, 11 Goldsmith Road, Walsall, WS3 1DL
2E0 FRY Christopher Fryer, 14 Perks Road, Wolverhampton, WV11 2ND
2M0 FSB David Kelly, 21 Dhailling Road, Dunoon, PA23 8EA
2E0 FSE Lisa Beaney, Penns Cottage, Horsham Road, Steyning, BN44 3LJ
2M0 FSF G Stoddart, 1 Barrs Brae, Kilmacolm, PA13 4DE
2E0 FSG Bernard Jones, 9 St. James Close, Hanslope, Milton Keynes, MK19 7LF
2E0 FSH Callum Winfield, 30 Rodney Way, Ilkeston, DE7 8PW
2E0 FSI Dorian Woolger, 8 Old Cottages, Horsham Road, Worthing, BN14 0TQ
2E0 FSK George Hardill, 107 Leicester Road, Whitwick, Coalville, LE67 5GN
2I0 FSL Eoghan Murray, 33 Orpen Avenue, Belfast, BT10 0BS
2E0 FSM Matthew Bruce, 27 Blaenant, Emmer Green, Reading, RG4 8PH
2E0 FSN Hector Hamilton, Flat B, 9 Cambridge Drive, London, SE12 8AG
2M0 FSP Stephen Paterson, 14-16 New Street, Findochty, Buckie, AB56 4PS
2E0 FSX Frank Riches, 4 Priory Close, Chelmsford, CM1 2SY
2M0 FTA Lukasz Pinkowski, 73 Willow Grove, Livingston, EH54 5NA
2E0 FTC Justin Forbes, Weald Barkfold Farm, Plaistow, Billingshurst, RH14 0PJ
2E0 FTH Sean Macdonald, 157 Delapre Drive, Banbury, OX16 3WS
2E0 FTL Michael Dorrington, 19 Shaftesbury Drive, Wardle, Rochdale, OL12 9LT
2E0 FTM Darren Fletcher, 97 Wallace Crescent, Carshalton, SM5 3SU
2E0 FTQ Anthony Street, 110 Magdalen Street, Colchester, CO1 2LF
2E0 FTV Steven Barber, 82 Prunus Road, Crewe, CW1 4HB
2E0 FTW J Northall, 5 West Winds Road, Winterton, Scunthorpe, DN15 9RU
2E0 FTX Jakub Krol, 40 Hampton Gardens, Southend-on-Sea, SS2 6RW
2E0 FUD Robert Blackwell, Vikings Hall, Baylham, Ipswich, IP6 8JS
2E0 FUH Chris Pomfrett, 17 Manifold Close, Sandbach, CW11 1XP
2E0 FUN Stuart Southern, 37 Conway Road, Calcot, Reading, RG31 4XP
2E0 FUR James Pattinson, 28 Dunley Close, Swindon, SN25 2BL
2I0 FUT Melvyn Crozier, 33 Cullentragh Road, Poyntzpass, Newry, BT35 6SD
2E0 FUZ D King, 215 Hartland Road, Reading, RG2 8DN
2E0 FVL Peter Penycate, 8 Campbell Road, Tangmere, Chichester, PO20 2HX
2E0 FVV Mark Roberts, 463, Brighton Road, Lancing, BN15 8LF
2E0 FWC F Clark, 14 Warwick Road, Bude, EX23 8EU
2E0 FWD Christopher Board, Pinmoor, Moretonhampstead, Newton Abbot, TQ13 8QA
2E0 FWN E Hunter, 38 Blewitt Street, Hednesford, Cannock, WS12 4BD
2E0 FWR Frank Waller, 249 Summer Lane, Wombwell, Barnsley, S73 8QB
2E0 FWY David Smith, 7 Oakley Grove, Wolverhampton, WV4 4LN
2E0 FXD Daniel Sawyer, 2 St. Johns Court, Palmerston Mews, Bournemouth, BH1 4JH
2E0 FXP Stephen Westley, 76 Rockingham Close, Birchwood, Warrington, WA3 6UY
2M0 FXX Michael McGrorty, 17 Fernbank, Stirling, FK9 5AD
2E0 FYE R Fye, 201 North Wing The Residence, Kershaw Drive, Lancaster, LA1 3SY
2M0 FYF John fyfe, 53a Ware Road, Glasgow, G34 9AR
2M0 FYG J Rodger, 7 Heathryfold Circle, Aberdeen, AB16 7DQ
2E0 FYL Phyllis Catton, 97 High Street, South Hiendley, Barnsley, S72 9AN
2E0 FYQ Peter Moore, 22 Audit Hall Road, Empingham, Oakham, LE15 8PH
2E0 FZJ James Foster, 23 High Street, Cumnor, Oxford, OX2 9PE
2E0 FZK A McBirnie, 25 Ulverston Road, Swarthmoor, Ulverston, LA12 0JB
2E0 FZM Clive Ramsdale, 87 Mill Lane, Kirk Ella, Hull, HU10 7JN
2E0 GAC N Bexon, 60 Whitwell Road, Nottingham, NG8 6JT
2E0 GAF George Sole, 16 Beech Crescent, Hythe, Southampton, SO45 3QG
2E0 GAG Graeme Davies, 10 Leaway, Prudhoe, NE42 6QE
2E0 GAH G Hudson, 26 Griffins Brook Lane, Birmingham, B30 1PU
2E0 GAK G Kell, 14 Carisbrooke Lane, Garforth, Leeds, LS25 2LE
2E0 GAL Ruth Hughes, 19 Pendine Crescent, North Hykeham, Lincoln, LN6 8UW
2M0 GAN Gerald Prior, 41 Beechwood, Linlithgow, EH49 6SD
2E0 GAO Geoffrey Bridge, 49 Wilton Gardens, Radcliffe, Manchester, M26 2UP
2E0 GAP Andrew Pilkington, 26 Ryelands Close, Market Harborough, LE16 7XE
2E0 GAQ Ifor Williams, 5 Fron Goch, Llanberis, Caernarfon, LL55 4LE
2E0 GAR G Farrar, 174 Houghton Road, Thurnscoe, Rotherham, S63 0SA
2E0 GAU Gary Cooper, Holmfield, Chelmorton, Buxton, SK17 9SG
2E0 GAV Charles Holmes, 8 Byron Way, Caister-on-Sea, Great Yarmouth, NR30 5RW
2W0 GAY Anthony Ferguson, Mount, Salem, Llandeilo, SA19 7HD
2E0 GAZ G Botterill, 36 Metchley Drive, Harborne, Birmingham, B17 0JX
2E0 GBA Ian Stevenson, 79 Lunedale Road, Darlington, DL3 9AT
2E0 GBB Gordon Foster, 22 Bradley Cottages, Consett, DH8 6JZ
2E0 GBD Owen Cook, 38 Redbourn Way, Scunthorpe, DN16 1NE
2E0 GBE Graham Bell, 83 Coopers Green, Bicester, OX26 4XJ
2E0 GBF Phil Dawes, 49 Altofts Lodge Drive, Altofts, Normanton, WF6 2LB
2E0 GBG David Gillingham, 3 Rosier Close, Thatcham, RG19 4FN
2E0 GBH M Brinnen, 82 Victoria Road, Mablethorpe, LN12 2AJ
2E0 GBI Lance Catterall, 14 Dunham Drive, Whittle-le-Woods, Chorley, PR6 7DN
2E0 GBJ Brian Cave, 2 Beaufort Close, Newcastle upon Tyne, NE5 3XL
2E0 GBK John Bell, 6 Highfields, Fetcham, Leatherhead, KT22 9XA
2D0 GBM Peter Birchall, 7 Richmond Close, Douglas, Isle of Man, IM2 6HR
2E0 GBN Garry Barton, 9 Tees Crescent, Stanley, DH9 6HX
2E0 GBO Martin King, Gate House, Lower Bentham, Lancaster, LA2 7DD
2E0 GBP Philip Lindley, 17 Swallow Lane, Aston, Sheffield, S26 2GR
2E0 GBT Frederick Woods, 30 Hurst Close, Chandler's Ford, Eastleigh, SO53 3PA
2E0 GBU David Torrance, 30 St. Norbert Drive, Ilkeston, DE7 4EH
2E0 GBV Garell Brotherhood, 17 Baldwin Close, Forest Town, Mansfield, NG19 0LR
2E0 GCB G Buxton, 11 The Green, Northfield, Birmingham, B31 5HT
2I0 GCC Gerard O'Reilly, 20 Lower Clonard Street, Belfast, BT12 4NH
2E0 GCD George Bosnyak, 10 Station Road, Stannington, Morpeth, NE61 6DS
2E0 GCE Geoffrey Elsworthy, 40 Moorfield Way, Wilberfoss, York, YO41 5PL
2M0 GCF John Brown, 78 Eglisay Street, Glasgow, G22 7RG
2E0 GCG Henriks Vecenans, 155 Upper Dale Road, Derby, DE23 8BP
2E0 GCH Shaun Shreeves, 6 Bowshaw Avenue, Batemoor, Sheffield, S8 8EZ
2E0 GCI Alexander Clarke, 57 Welland Avenue, Grimsby, DN34 5JP
2E0 GCL Andrew Finn, 202 Northgate Road, Stockport, SK3 9NJ
2E0 GCM Nick Baker, 56 Chalklands, Bourne End, SL8 5TJ
2I0 GCN Stephen Morrow, 769 Farranseer Park, Macosquin, Coleraine, BT51 4NB
2E0 GCO Sharon Lake, 85 Clarkson Road, Norwich, NR5 8ED
2E0 GCP G Piddington, 45 Pleasant View Road, Crowborough, TN6 2UU
2M0 GCS Graham Cochrane, 33 Portland Road, Galston, KA4 8EA
2E0 GCW B Barker, 44 Falcon Crescent, Bilston, WV14 9BE
2E0 GCY Gary Cornish, 78 Kerry Avenue, Ipswich, IP1 5LD
2E0 GDA Richard Davison, 2 Marlow Terrace, Mold, CH7 1HH
2E0 GDB Gary Brookes, Llecyn y Llan, Llanerchymedd, LL71 8EH
2E0 GDF Martin Joynson-Ellis, Roadside Cottage, Craiglemine, Newton Stewart, DG8 8NE
2E0 GDG Gianfranco Di Genova, Flat 2, Elfin Court, Southampton, SO17 1DY
2E0 GDH Steven Hunter, 9 Gelt Burn, Didcot, OX11 7TZ
2E0 GDL G Ludlow, 12 The Paddocks, Middleton on the Wolds, Driffield, YO25 9UN
2E0 GDM Gerry Martin, Flat 16, Sorrel House, Birmingham, B24 0TQ
2E0 GDN Kenneth Young, 51 Barrow Road, Barton-upon-Humber, DN18 5BS
2E0 GDO Geoffrey Cochrane, 133 Cotman Fields, Norwich, NR1 4EP
2E0 GDT Andrew Shaw, 20 Hillcrest Close, Thrapston, Kettering, NN14 4TB
2E0 GDY Garry Dealey, 69 Upper Belmont Road, Chesham, HP5 2DD
2E0 GDZ Gary Dean, 62 Baptist Close, Abbeymead, Gloucester, GL4 5GD
2E0 GEB Stuart Marr, 49 Gallows Hill, Ripon, HG4 1RG
2D0 GEE Nandesh Patel, 78 Wesley Close, South Harrow, Harrow, HA2 0QE
2E0 GEF Geoffrey Winterbottom, 35 Abingdon View, Worksop, S81 7RT
2E0 GEG George Bramham, 1 Watson Avenue, Dewsbury, WF12 8PZ
2M0 GEJ George Jamieson, 6 Maryville Park, Aberdeen, AB15 6DU
2M0 GEK John Wright, 43 Spey Court, Stirling, FK7 7QZ
2E0 GEL James Willetts, 102 Welch Road, Cheltenham, GL51 0EG
2E0 GEM Paul Murray, 45 Commercial Street, Risca, Newport, NP11 6AW
2E0 GEN Andrew Askam, 8 The Pastures, Weston-on-Trent, Derby, DE72 2DQ
2W0 GER T Doak, 27 Hill St., Gilfach Goch, Porth, CF39 8TW
2E0 GET David Baker, 25 Hitherspring, Corsham, SN13 9UT
2E0 GEV Andrew Sherman, 31 Peartree Avenue, Kingsbury, Tamworth, B78 2LG
2E0 GFB Adam Durrant, 22 Supple Close, Norwich, NR1 4PP
2M0 GFC Peter Davis, 2 Virkie Cottages, Virkie, Shetland, ZE3 9JS
2E0 GFE Gary Teale, 97a Wokingham Road, Reading, RG6 1LH
2E0 GFF Andrew Banks, 2 Holt Close, Farnborough, GU14 8DG
2I0 GFO Geoffrey Craig, 103 Moyle Parade, Larne, BT40 1ET
2E0 GFW Graham Watson, The Dell, Nova Scotia Road, Great Yarmouth, NR29 3QD
2E0 GFX Joshua Duffain, 53 Maes y Ffynnon, Brecon, LD3 9PL
2E0 GGG Mark Brady, 24 Gregory Avenue, Colwyn Bay, LL29 7ND
2E0 GGI R Gleave, 52 Cranborne Avenue, Warrington, WA4 6DE
2E0 GGM Andrew Mansfield, 3 The Coppice, Thrapston, Kettering, NN14 4QA
2E0 GGO Gareth Jones, 109 Montgomery Avenue, Lowestoft, NR32 4DU
2E0 GGT Graham Townsend, 19 Landor Crescent, Rugeley, WS15 1LP
2E0 GGW Graham Willard, 4 Varrier Jones Place, Papworth Everard, Cambridge, CB23 3XP
2M0 GGY A Espie, 70 Everard Rise, Livingston, EH54 6JD
2E0 GHA Graham Hill-Adams, 6 Broadleaze Way, Winscombe, BS25 1JX
2E0 GHB G Bourne, 72 Cornish Way, Royton, Oldham, OL2 6JY
2E0 GHD Alan Burleton, 27 Doncaster Road, Bristol, BS10 5PN
2E0 GHK David Hindle, 18 Haig Street, Selby, YO8 4BY
2E0 GHP John Phillips, The Manor, Blackwoods, York, YO61 3ER
2E0 GHR R Gill, 84 Leypark Road, Exeter, EX1 3NT
2E0 GHX Steven Aucoin, 296 Turkey Road, Bexhill-on-Sea, TN39 5HY
2I0 GHY Ian Gibb, 1 Shankill Road, Garvary, Enniskillen, BT94 3DB
2E0 GHZ John Wakefield, Oakhurst, Lower Common Road, Romsey, SO51 6BT
2E0 GIA David Owen, Tanrallt, Blaenpennal, Aberystwyth, SY23 4TP
2I0 GIF Graham Clarke, 12 Church Green, Dromore, BT25 1LL
2E0 GIG Richard Eyre, 123 Baden Powell Road, Chesterfield, S40 2RL
2E0 GIH E Peck, 11 Blake Road, Stapleford, Nottingham, NG9 7HN
2M0 GIL Graeme Gilmour, 100 Main St., Milngavie, Glasgow, G62 6JN
2E0 GIW Graham Williams, 99 Maes Llwyn, Amlwch, LL68 9BG
2E0 GIX Alex Hodgson, Bryngwyn Bach Rhuallt, St. Asaph, LL17 0TH
2E0 GJE Gary Groves, 5 Beech Road, Ashurst, Southampton, SO40 7AY
2E0 GJJ Gareth Johnson, 199 Lynwood, Folkestone, CT19 5TA
2W0 GJR Gareth Reason, 454 Cowbridge Road West, Cardiff, CF5 5BZ
2E0 GKA Peter Houghton, 151c London Road, Calne, SN11 0AQ
2I0 GKB Gareth Black, 41a Meeting House Lane, Lisburn, BT27 5BY
2E0 GKD A McCreadie, 37 Beddie Crescent, Wigtown, Newton Stewart, DG8 9HX
2E0 GKL Kirk Lord, 21 Norfolk Road, Littlehampton, BN17 5PW
2E0 GKM Robin Ley, 23 Heronbridge Close, Westlea, Swindon, SN5 7DR
2E0 GKR Lee Bullen, 2 Rowley Cottages, Hermitage Road, Upton, Langport, TA10 9NP
2E0 GKS George Skea, 15 The Shires, Gilwern, Abergavenny, NP7 0EX
2E0 GLA M Gladders, 2 Albion Mansions, Saltburn-by-The-Sea, TS12 1JP
2I0 GLC Geoff Crabbe, 39 Arran Avenue, Ballymena, BT42 4AP
2E0 GLD A Goold, 6 The Elms, Kempston, Bedford, MK42 7JN
2M0 GLI Graham Irvine, 11 Hazel Road, Cumbernauld, Glasgow, G67 3BN
2E0 GLL David Firth, 7 Martinet Drive, Lee-on-The-Solent, PO13 8GP
2E0 GLR Edwin Rimmer, 26 Kenmore Road, Prenton, CH43 3AS
2E0 GLS Gary Stevens, 17 Manston Close, Ernesettle, Plymouth, PL5 2SN
2E0 GLT Glitsun Cheeran, 201 Eastcombe Avenue, London, SE7 7LH
2E0 GLV Christopher Tanner, Pen y Gogarth, Llaneilian, Amlwch, LL68 9NH
2E0 GLW G Whittle, 22 Warwick Street, Leigh, WN7 2NH
2I0 GLY Nigel Sands, 6 The Granary, Waringstown, Craigavon, BT66 7TG
2E0 GMA Glenn Marsden, 38 Sandhill Road, Rawmarsh, Rotherham, S62 5NT
2M0 GMB I Birse, North Milton Of Corsindae, Midmar, Inverurie, AB51 7QP
2E0 GMD Michael Drury, 19 Cuffley Avenue, Watford, WD25 9RB
2E0 GMG Alexander Andover, Bishoper Farmhouse, Brokenborough, Malmesbury, SN16 9SR
2E0 GMM John Cook, 42 Pampas Close, Colchester, CO4 9ST
2E0 GMN Juan Rufes, Flat 1 & 3-8, 12 Smyrna Road, London, NW6 4LY
2E0 GMO Maurice Boland, 24 Hallam Close, Moulton, Northampton, NN3 7LB
2E0 GMS Graham Brooks, 14 Chalton Crescent, Havant, PO9 4PT
2E0 GMU G Murch, 79 Alderson Crescent, Liverpool, L37 3LY
2E0 GMW Simon Coombs, 34 Mast Drive, Hull, HU9 1ST
2E0 GMZ Huw Hughes, Llecyn y Llan, Llanerchymedd, Llannerch-Y-Medd, LL71 8EH
2W0 GNG Paul Smith, 3 Islington Road, Bridgend, CF31 4QY
2E0 GNI Llyr Mercer, 38 Manadon Drive, Plymouth, PL5 3DJ
2E0 GNN A Glover, 103a Latimer Street, Liverpool, L5 2RF
2E0 GNS Glen Sandell, 20 Kirkby View, Sheffield, S12 2NB
2E0 GNU Edward Brook, 30 Pitchstone Court, Farnley, Leeds, LS12 5SZ
2E0 GNW Adrian Land, 27 Peaks Lane, New Waltham, Grimsby, DN36 4LG
2M0 GOE Tearlach MacDonald, Main Road Farm, Balephuil, Isle of Tiree, PA77 6UE
2E0 GOL Andrew Goldsmith, 7 Fengate Drove, Weeting, Brandon, IP27 0PW
2E0 GOM C Blackburn, 158 Dyas Road, Great Barr, Birmingham, B44 8SW
2E0 GON Peter Gonczarow, 25 Ribchester Avenue, Burnley, BB10 4PD
2E0 GOO Jason Barker, Pearl Bungalow, Killerby Cliff, Scarborough, YO11 3NR
2E0 GOQ Ronald Gibbs, 7 Thornhill, Eastfield, Scarborough, YO11 3LY
2E0 GOS N Gostling, 49 Roundhouse Road, Dudley, DY3 2AX
2E0 GOW C Gowing, Barbosa, Remembrance Road, Newbury, RG14 6BA
2E0 GPA Mark Phillips, 1 The Vale, Oakham, LE15 6JQ
2E0 GPB Gregory Beacher, 22 Trowbridge Gardens, Luton, LU2 7JY
2E0 GPC Gary Coleman, 19 Grunmore Drive, Stretton, Burton-on-Trent, DE13 0GZ
2E0 GPD Philip Dimes, 5 Meadowbrook, Oxted, RH8 9LT
2E0 GPE Glenn Pearson, 41 Myrica Grove, Hoole, Chester, CH2 3EW
2E0 GPF Gerard Fleming, 1 Balmoral Drive, Methley, Leeds, LS26 9LE
2E0 GPG G Bates, 230 Brook Street, Erith, DA8 1DZ
2E0 GPH Gary Hart, 11 Sadlers Ride, West Molesey, KT8 1SU
2E0 GPK Glenn Kendall, 6 Badger Wood, Todmorden, OL14 6BB
2E0 GPL Thomas Arrow, Crystalwood, Stonemans Hill, Newton Abbot, TQ12 5PZ
2E0 GPS G Perkins, Gamekeepers Cottage, Snarehill, Thetford, IP24 2QA
2E0 GPT Richard Hampson, 12 Oakhays, South Molton, EX36 4DB
2E0 GPU Andrew Taylor, 16 Bellmans Road Whittlesey, Peterborough, PE7 1TY
2E0 GPX Gary Matthews, 81 Kipling Avenue, Goring-by-Sea, Worthing, BN12 6LH
2E0 GPY N Shepherd, 12 Barnfield Close, Radcliffe, Manchester, M26 3UA
2E0 GQD J Reynolds, 3 Ardleigh, Basildon, SS16 5RA
2E0 GQT Ian Alderman, 107 Manton Drive, Luton, LU2 7DL
2E0 GQW Edward Tart, Sunnybank Farm, Wattlesborough Heath, Shrewsbury, SY5 9EG
2E0 GRA Graeme Hayward, 129 Nipsells Chase, Mayland, Chelmsford, CM3 6EJ
2M0 GRE Gregory Lailvaux, 4 Oxenfoord Avenue, Pathhead, EH37 5QD
2E0 GRF Griffith Hewis, 10 Albert Road, New Malden, KT3 6BS
2E0 GRH Stephen Walters, 87 Fairbourne Close, Bransholme, Hull, HU7 5DH
2E0 GRI George Reywer, 1 Tiverton Close, Houghton le Spring, DH4 4XR
2M0 GRK James Black, 13 Dunlop Street, Greenock, PA16 9BG
2E0 GRL Luke Spear, 57 Station Road, Melbourne, Derby, DE73 8EB
2E0 GRM N Hallwood, 32 Hawthorne Avenue, Ripley, DE5 3PJ
2E0 GRN Adam Young, 48 Sussex Street, Cleethorpes, DN35 7NP
2E0 GRP Graham Priestley, 53 Millfield Gardens, Crowland, Peterborough, PE6 0HA
2E0 GRR Carol Hebden, Reedecraft, Mill Green Road, Spalding, PE11 3PU
2E0 GRS G Street, Flat 9, Weavers Cottages, Congleton, CW12 1AG
2E0 GRW Matthew Harvey, 129 Goldthorn Hill, Wolverhampton, WV2 4PS
2E0 GRX Graham Kennedy, 4 Calder Crescent, Whitefield, Manchester, M45 8LH
2E0 GRY G Collis, 16 Hill Grove, Barrow Hill, Chesterfield, S43 2NW
2E0 GSA Gary Smith, 2 Hawthorn Rise, Groby, Leicester, LE6 0EX
2E0 GSB Gary Bertola, 17 Caraway Drive, Branston, Burton-on-Trent, DE14 3FQ
2E0 GSC G Chaffey, 63 Underwood Road, Eastleigh, SO50 6FX
2E0 GSF T Gadd, 20 Gladstone Street, Swindon, SN1 2AX
2I0 GSG Gary Gregg, 30 Claremont Avenue, Moira, Craigavon, BT67 0SS
2E0 GSJ K Andrews, 80 Hollywell Road, Lincoln, LN5 9DA
2E0 GSK Michael Silver, 52 Park Crescent, Elstree, Borehamwood, WD6 3PU
2E0 GSL Lee Clark, 30 Warwick Square, London, SW1V 2AD
2E0 GSR Graham Iredale, Ship Cottage, Main Street, Maryport, CA15 7DX
2E0 GST Graham Starling, 4 Three Corner Drive, Norwich, NR6 7HA
2E0 GSW Thomas Rowlands, 7 Northfield Crescent, Beeston, Nottingham, NG9 5GR
2E0 GTA A Holbrook, 6 Birch Tree Way, Maidstone, ME15 7RR
2E0 GTB P Rigden, 11 Railway Cottages, Station Road, Whitstable, CT5 1JZ
2E0 GTE Gary Cockburn, 20 Hexham Avenue, Hebburn, NE31 2HN

2E0 GTL G Taylor, 31 Ashfurlong Crescent, Sutton Coldfield, B75 6EN
2E0 GTM Gordon Moon, 9 Blackstock Court, Bootle, L30 0PN
2I0 GTO George Shaw, 49 Cloughey Road, Portaferry, Newtownards, BT22 1NQ
2M0 [illegible] [illegible], 34 Centre Street, Kelty, KY4 0EQ
2E0 GTT Michael Smith, 24 Fifth Avenue, Portsmouth, PO6 3PE
2E0 GTZ Andrew Blamire, 21 The Laurels, Banstead, SM7 2HG
2E0 GUA John Hammond, 8 Rowntree Way, Saffron Walden, CB11 4DG
2E0 GUH Darren Hughes, 32 Achille Road, Grimsby, DN34 5RR
2E0 GUI Michael Topple, 41 Whitehall Close, Colchester, CO2 8AJ
2M0 GUL James Hume, 8/11 Leslie Place, Edinburgh, EH4 1NH
2E0 GUN A Price, 67 Mansfield Road, Glapwell, Chesterfield, S44 5QA
2E0 GUT Gerhard [illegible], Flat 140, 100 London Street, [illegible]
2E0 GUV P Brown, 1 Octavian Close, Hatch Warren, Basingstoke, RG22 4[illegible]
2E0 GUY Stephen Mellor, 11 Bolton Meadow, Leyland, [illegible]
2E0 GVC Graham Clayton, The Forge, High Street, [illegible]
2E0 GVO G Owen, Flat 3, Osborne House, Lille Lane, [illegible], LL58 8DB
2I0 GWA Andrew Cummings, 19 Bachelors Walk, Keady, Armagh, BT60 2NA
2E0 GWB George Bunting, 31 Hardwick Avenue, Allestree, Derby, DE22 2LN
2E0 GWC Andrew Colman, 5 Burn Heads Road, Hebburn, NE31 2TD
2E0 GWD Nicholas Reeves, Flat 2, Delamore, Ivybridge, PL21 9QT
2E0 GWE Gerald Watson, 20 Windermere Drive, West Auckland, Bishop Auckland, DL14 9LF
2E0 GWF B Whitemore, 24 Rectory Close, Wraxall, Bristol, BS48 1LT
2E0 GWI N Davies, 27 Grafton Road, Ellesmere Port, CH65 2BD
2E0 GWK Keith Jones, Gorswen, Brynrefail, Caernarfon, LL55 3NT
2E0 GWM Michael Martin, 1 Y Gorlan, Bryn Street, Newtown, SY16 2HN
2E0 GWP G Prescott, 3 View Fields, Station Road, Doncaster, DN9 3AE
2E0 GWR Brian Bosson, 1 Broomsgrove, Pewsey, SN9 5LE
2E0 GWS George Salter, 9 Spring Gardens, Malvern Link, Malvern, WR14 1AP
2I0 GWX Gordon Todd, 27 Ardreagh Road, Aghadowey, Coleraine, BT51 4DN
2E0 GXB G Beaver, 23 West Drive Gardens, Soham, Ely, CB7 5EF
2E0 GXE Ben Thomson, 50 Thomson Street, Stockport, SK3 9DR
2E0 GXF Guy Fernando, 1 Rosemary Avenue, West Molesey, KT8 1QF
2E0 GXI David Burt, 2 Cae Masarn, Pentre Halkyn, Holywell, CH8 8JY
2E0 GXK Peter Blagden, 24 Ashgrove Avenue, Gloucester, GL4 4NE
2E0 GXX Ian Bardell, 32 Bridle Road, Watton, Thetford, IP25 6NA
2M0 GXZ George Sinclair, 33 Keptie Road, Arbroath, DD11 3EF
2W0 GYB Laurence Brown, 13 Station Road Loughor, Swansea, SA4 6TR
2E0 GYC Robert McKnight, Gortadrohid, Reengaroga, Co. Cork, Ireland
2I0 GYL Grace McCormick, 46 Lany Road, Moira, Craigavon, BT67 0NZ
2M0 GYM James Branson, East Bank, South Road, Fochabers, IV32 7LU
2M0 GYN Alasdair Connell, 39 Glebe Crescent, Maybole, KA19 7HZ
2E0 GYP Jason Wells, 18 Roewood Road, Holbury, Southampton, SO45 2JH
2E0 GYW Thomas Clark, 15 Braybrook Court, Bradford, BD8 7BH
2E0 GYZ Will Rittman, 70 Market Street, Chapel-en-le-Frith, High Peak, SK23 0HY
2M0 GZA Stephen Hargreaves, 4 Oxenfoord Avenue, Pathhead, EH37 5QD
2E0 GZT Tony Marshall, 63a Newport Road, Ventnor, PO38 1BD
2E0 HAB Hooman Atifeh, 9 Stagshaw Close, East Hunsbury, Northampton, NN4 0WE
2W0 HAC C Thomas, 2 Ffordd Donaldson, Copper Quarter, Swansea, SA1 7FJ
2E0 HAF Jeremy Raehse Felstead, 10 Rubens Close, Aylesbury, HP19 8SW
2E0 HAG Clive Hall, 28 Tidebrook Place, Stoke-on-Trent, ST6 6XF
2E0 HAH Mark McKenna, 1 Kennet Avenue, Jarrow, NE32 4DB
2E0 HAJ Stephen Cordner, 29 Buxton Road, Aylsham, Norwich, NR11 6JD
2W0 HAK K Vaughan, 26 Mount Pleasant, Bargoed, CF81 8UU
2E0 HAL G Smith, 7 Kestrel Avenue, Bransholme, Hull, HU7 4ST
2E0 HAN Hannah Hopkins, 3 Colegrave Road, Bloxham, Banbury, OX15 4NT
2E0 HAP Anthony Craven, 45 Benhams Drive, Horley, RH6 8QT
2E0 HAQ Ayub Malik, 3 Berkeley Street, Nelson, BB9 0SJ
2E0 HAS Royston Williams, 34 Maendu Terrace, Brecon, LD3 9HH
2E0 HAT Richard Hatton, 1 Bowman Mews, Southfields, London, SW18 5TN
2E0 HAV Selwyn James, 35 Prospect Road, Dronfield, S18 2EA
2E0 HAW Christopher Cross, 7 Anne Close, Norwich, NR7 0PH
2M0 HAY Scott Hay, 20 Woodside Way, Glenrothes, KY7 5DF
2E0 HAZ G Hazlewood, 102 Throne Road, Rowley Regis, B65 9JX
2E0 HBB Howard Russell, 4 Dearnsdale Close, Stafford, ST16 1SD
2E0 HBD David Hardy, 61 Westbourne Road, Handsworth, Birmingham, B21 8AU
2E0 HBE David Newman, 78 Clapham Court, Gloucester, GL1 3DE
2I0 HBO Karol Mikicki, 429 Beersbridge Road, Belfast, BT5 5DU
2D0 HBT Thomas Richley, 30 Chicheley Road, Harrow, HA3 6QL
2E0 HBY A Foster, 62 Spa Road, Atherton, Manchester, M46 9NQ
2E0 HCC Gary Belgium, 590 Wells Road, Bristol, BS14 9BD
2M0 HCF Craig Moir, 1/2 26 Kilnside Road, Paisley, PA1 1RH
2E0 HCL Glynn Holland, 6 Moorfield Road, Widnes, WA8 3JE
2E0 HCT Peter Bell, 12a Mill Lane, Carlton, Goole, DN14 9NG
2E0 HCW Cyril Haynes, 4 Thorn Close, Rugby, CV21 1JN
2M0 HDA Francis Parkinson, Garrell Park, Burnbank Terrace, Glasgow, G65 0AE
2W0 HDB Howard Bancroft, Stop and Call, Goodwick, SA64 0EX
2W0 HDC Neils Orchard, The Burrows, Spring Gardens, Whitland, SA34 0HL
2E0 HDE David Edmondson, 21 Hawthorne Close, Heathfield, TN21 8HP
2E0 HDF Andrew Brain, 20 South Street, Spennymoor, DL16 7TU
2E0 HDG Alan Copse, 3 The Limes, Market Overton, Oakham, LE15 7PX
2E0 HDM Paul Allin, 25 Castleton Road, Hope, Hope Valley, S33 6SB
2E0 HDU Peter Loomes, 107 Main Street, Sedgeberrow, Evesham, WR11 7UE
2E0 HDW Matthew Langham, 12 Cornflower Drive, Chelmsford, CM1 6XY
2E0 HDX Simon Hewick, 56 Hemswell Avenue, Hull, HU9 5JZ
2E0 HDY Anthony Hardy, 3 Cornwall Road, Leicester, LE4 0BB
2E0 HEA John Heagren, 84 Avon Drive, Alderbury, Salisbury, SP5 3TH
2D0 HEB H Blackburn, 32 Close Rushen, Castletown, Isle of Man, IM9 1NN
2E0 HEF David Hobinson, Height End Farm, Kirk Hill Road, Rossendale, BB4 8TZ
2E0 HEO Lance Davis-Edmonds, Bladnoch Cottage, Bladnoch, Newton Stewart, DG8 9AB
2E0 HEP J Hobbs, 2 Eccles Road, Wittering, Peterborough, PE8 6AU
2E0 HES Ralph Heslop, 7 Fieldfare Close, Clanfield, Waterlooville, PO8 0NQ
2E0 HEX John Ash, 47 Stein Road, Emsworth, PO10 8LB
2E0 HFA Jennifer Wilson, Flat 5, Blake House, London, SE1 7DX
2D0 HFT David Merridale, THE GRANARY, FALLEDGE LANE, Upper Denby, HD8 8YH
2W0 HFU Gareth Ralls, 2 Yew Close, Merthyr Tydfil, CF47 9SD
2E0 HGE A Hitchens, 16 Harrisons Place, Northwich, CW8 1HX
2E0 HGG D Rennie, 27 Orrell Road, Liverpool, L21 8NQ
2E0 HGO Piotr Niewiadomski, 79a Dartmouth Road, London, SE23 3HT
2E0 [illegible] Stuart Raynor, 108 Rockover Drive, Pudsey, LS28 8EQ
2E0 [illegible] D Clark, 13 Wilson Crescent, Lostock Gralam, Northwich, CW9 7QH
2E0 HHE John Cryan, 12 Stamford Avenue, Sunderland, SR3 4AT
2E0 HHK Steve Collins, 5 Fernleigh Gardens, Stafford, ST16 1HA
2E0 HHU J Newham, 6 Belsfield Gardens, Jarrow, NE32 5QB
2E0 HIG Brian Higgins, 2 Bishops Yard, High Street, Huntingdon, PE28 3JB
2E0 HIP David Slater, 13 Longford Close, Rainham, Gillingham, ME8 8EW
2E0 HIT I Petrie, 88 Vicarage Road, Henley-on-Thames, RG9 1JT
2E0 HIZ [illegible] Easdown, 38 North Street, Barming, Maidstone, ME16 9HF
2E0 [illegible] Barry Halliwell, 18a Rushton Road, Desborough, Kettering, NN14 2RW
2E0 HJD Michael Dixon, 55 Henthorn Road, Clitheroe, BB7 2LD
2E0 HJE [illegible], 31 Fitznain Road, West Parley, Ferndown, BH22 8RZ
2E0 HJJ Henry Jackson, Wingfield Farm, Cublington Road, Leighton Buzzard, LU7 0LB
2M0 HJP Hedley Phillips, Maplebank, Leithen Road, Innerleithen, EH44 6NJ
2M0 HJS John Smith, 10 High Street, Portknockie, Buckie, AB56 4LD
2E0 HJZ M Pridmore, 61 Gillards, Bishops Hull, Taunton, TA1 5HH
2E0 HKC Andrew Chambers, 19 Marina Road, Durrington, Salisbury, SP4 8DB
2E0 HKK P Buick, Poundgate Farm, Uckfield Road, Crowborough, TN6 3TA
2E0 HKR A Hill, 11 Pelham Drive, Hull, HU9 2AS
2E0 HLC Thomas Winter, 8 Thorpe Street, Hartlepool, TS24 0DX
2E0 HLF John Taylor, The Old Mission Hall Orton Avenue, Peterborough, PE2 9HL
2E0 HLH H Hall, 7 Front Street Grindale, Bridlington, YO16 4XU
2E0 HLM Klaus Schmidt, Church, Corner, Mareham-le-Fen, PE22 7RA
2E0 HLO Anthony Rowan, 14 Craven Lea, Liverpool, L12 0NF
2E0 HLP P Hallas, 37 Oakfield Road, Bromborough, Wirral, CH62 7BA
2E0 HMC Heidi Coghlan, Charterhouse, Orchard Road, Salisbury, SP5 2JA
2E0 HMD John Hammond, 57 Phoenix Drive, Leadenhall, Milton Keynes, MK6 5NB
2E0 HMM Daniel Lawson, 2 The Blossoms, Fulwood, Preston, PR2 9RF
2E0 HMS E Barnes, 2 Trem Y Garnedd, Bangor, LL57 1NA
2E0 HNC Howard Clark, 36 Market Oak Lane, Hemel Hempstead, HP3 8JL
2E0 HNF David Glover, 24 Cadeby Road, Sprotbrough, Doncaster, DN5 7SD
2E0 HNJ Peter Harris, 33 Kirklea Road, Houghton le Spring, DH5 8DP
2E0 HNK Harley Baird, 35 St. Peters Road, Wolvercote, Oxford, OX2 8AX
2E0 HNX C Shearan, 59 Arnold Crescent, Mexborough, S64 9JX
2E0 HOG Michael Stroud, 1 Sefton Court, Welwyn Garden City, AL8 6WW
2W0 HOH Peter Gostelow, Coedmor Llangrannog, Llandysul, SA44 6AG
2E0 HOK Ken Hough, 15 Moorside Road, Endmoor, Kendal, LA8 0EN
2E0 HOL K Craner, 46 Meadowhill Crescent, Redditch, B98 8HT
2E0 HOO Hugh Coram, Flat 3 19 Courtland Road, Paignton, TQ3 2AB
2E0 HOQ Neil Lambert, 17 Starcross Road, Weston-Super-Mare, BS22 6NY
2E0 HOS John Wallace, 323 High Street, Dalbeattie, DG5 4DX
2E0 HOT K Jones, 1 Alway Crescent, Newport, NP19 9SX
2E0 HOU Steve Houssart, Flat 3, Virginia Court, London, SE16 6PU
2E0 HPB Paul Brundrett, 114 Ack Lane East, Bramhall, Stockport, SK7 2AB
2E0 HPD Robert David Hodson, 99 Alcester Road, Hollywood, Birmingham, B47 5NR
2E0 HPF J Scannell, 53 Morley Croft, Farington Moss, Leyland, PR26 6QS
2E0 HPI Carl Gorse, 34 Ellison Street, Hartlepool, TS26 9AN
2E0 HPJ James Whiteside, The Old Antique Shop, Bank Street Pulham Market, Diss, IP21 4TG
2E0 HPL Paul Evans, 1 Solent Apartments, 16-17 South Parade, Southsea, PO5 2AZ
2W0 HPM Michael Melody, 7 Tegfan, Pontyclun, CF72 9BP
2E0 HPR Ian Hooper, 25 Honey Lane, Buntingford, SG9 9BQ
2E0 HQD Rob Wilkes, 33 Pembroke Road, Bulwark, Chepstow, NP16 5AF
2E0 HQJ C Sherwood, 1 Savile Road, Elland, HX5 0LA
2E0 HQO Robert Olive, Lorien, The Ridge, Thatcham, RG18 9HZ
2E0 HRD Vincent Greatwood, 11 The Green, Long Preston, Skipton, BD23 4PQ
2E0 HRH Marc Bloore, 6a Lovatt Close, Stretton, Burton-on-Trent, DE13 0HZ
2E0 HRJ Robin Huelin, 15 Hill Chase, Walderslade, Chatham, ME5 9HE
2W0 HRL H Leonard, 11 Newton Road, Grangetown, Cardiff, CF11 8AJ
2I0 HRM Jake Mercer, 32 Templemore Avenue, Belfast, BT5 4FT
2E0 HRS Toby Dunne, 5 Lambsickle Lane, Weston, Runcorn, WA7 4QZ
2I0 HRV Tommy Darrah, 42 Pinewood Avenue, Carrickfergus, BT38 8EW
2E0 HRY Harry Roxbrough, 17 Stanwell Close, Sheffield, S9 1PZ
2E0 HSB Deiniol Murphy, 23 Lowndes Close, Stockport, SK2 6DW
2I0 HSL Neil Davis, 19 Toberhewny Hall, Lurgan, Craigavon, BT66 8JZ
2E0 HST Christian Bolton, 201 Lime Tree Avenue, Crewe, CW1 4HZ
2E0 HSW H Scott Whittle, 92 The Grove, London, W5 5LG
2E0 HTB Henryk Banasiak, 11 Westfield Road, Backwell, Bristol, BS48 3NE
2E0 HTC Gary Carter, 19 Brathay Crescent, Barrow-in-Furness, LA14 2BG
2E0 HTM Alex Taylor, 65 Teign Bank Road, Hinckley, LE10 0ED
2E0 HTR Heather Moore, 62 Limefield Street, Accrington, BB5 2AF
2E0 HTS Simon Davison, 5 Denby Drive, Baildon, Shipley, BD17 7PQ
2E0 HTV Gheorghe Radulescu, 41 Sherard Road, London, SE9 6EX
2E0 HUB Mark Hubbard, 14 Parkfield Crescent, Kimpton, Hitchin, SG4 8EQ
2M0 HUD Aeneas Allan, 117 Bruce Gardens, Inverness, IV3 5BD
2W0 HUL Robert Hulme, 33 Clos Rhandir, Loughor, Swansea, SA4 6UF
2E0 HUM Harry Crabb, 12a Connaught Avenue, Frinton-on-Sea, CO13 9PW
2E0 HUN Alison Instone, 63 Larch Road, New Ollerton, Newark, NG22 9SX
2E0 HUR Jonathan Jefferies, Millfield Cottage, 1 Bolnhurst Road, Bedford, MK44 2LF
2E0 HUT Barry Hudson, 12 Elmfield Road, Hebburn, NE31 2DY
2E0 HUU Patrick Harper, 5 Hermitage, Llangollen, LL20 8BE
2E0 HVE Martin Simonsohn, 5 Pitt Close, Blandford St. Mary, Blandford Forum, DT11 9PS
2E0 HVL Hong Ly, 18 Mullway, Letchworth Garden City, SG6 4DH
2E0 HVN Gareth Edwards, 7 Maple Crescent, Leigh, WN7 5QX
2E0 HVW Stephen Downe, 9 Danesway, Exeter, EX4 9ES
2E0 HVZ James Godfrey, 6 Moor Lane, Croyde, Braunton, EX33 1NN
2E0 HWC Harry Cheesman, 49 Front Street, Chirton, North Shields, NE29 7QN
2E0 HWD Andrew Fugard, Flat 4, 1 Hazelmere Road, London, NW6 6PY
2E0 HWG Cori Haws, 5 Mallow Close, Locks Heath, Southampton, SO31 6XF
2E0 HWK Adrian Brand, 6 Walnut Close, Milton, Cambridge, CB24 6ET
2E0 HWN D Wardman, 2 Silver St., Scruton, Northallerton, DL7 0QR
2I0 HWW P Ford, 25 Carnhill, Londonderry, BT48 8BA
2E0 HXT M Bower, 21 Raglans, Exeter, EX2 8XN
2E0 HYC T Rutt, Granthorpe, Hull Road, Hull, HU11 5RN
2E0 HYE Thomas Byers, 1 Hazelwood Avenue, Sunderland, SR5 5AH
2E0 [illegible] [illegible], 9 Forth Close, [illegible]
2E0 HYK Timothy Ward, 37 High Mead, Royal Wootton Bassett, [illegible]
2E0 HYM A Mears, Flat 248, 5 Charter House, Portsmouth, PO1 2SN
2M0 HYZ Stewart Magee, 30 Burnfield Drive, Mansewood, Glasgow, G43 1BW
2M0 HZL Hazel McKay, 44a Torbane Drive, East Whitburn, Bathgate, EH47 0JQ
2E0 HZS Harry Smith, 11 Kettlebrook Road, Tamworth, B77 1AB
2E0 IAD Jack Fuller, 3 The Russets, Meopham, Gravesend, DA13 0HH
2E0 IAF Simon Pryke, Pately, School Lane, Woodbridge, IP13 6DX
2E0 [illegible] Ian [illegible], Farningham, Dartford, DA4 0HF
2M0 IAH Ian MacDonald, 20 Newbigging Terrace, Auchtertool, Kirkcaldy, KY2 5XL
2E0 [illegible] [illegible], Ramsgate, CT12 6DD
2E0 [illegible] [illegible] TN17 9JD
2E0 IAI Ian MacDonald, Broomhill, Mill Lane, Worthing, BN13 3DH
2E0 IAN I Campbell, 19b Elphinstone Road, Southsea, PO5 3HP
2E0 IAO Anthony OConnell, 31 North Avenue, Tredegar, NP22 3HF
2E0 IAS Elizabeth Alexander, 25 Dunelm Road, Hetton-le-Hole, Houghton le Spring, DH5 9LB
2E0 IAZ Darrin Goldthorpe, 30 Morrissey Close, St. Helens, WA10 4JW
2E0 IBA S Jones, 11 Croft Close, Rowton, Chester, CH3 7QQ
2E0 IBB Bruce Beard, 1 Friars Walk, Newcastle, ST5 2HA
2E0 IBI Alan Douglas, 3 Beech Avenue, Bilsborrow, Preston, PR3 0RH
2E0 IBM A McElroy, 36 Bod Offa Drive, Buckley, CH7 2PB
2E0 IBN L Thacker, Hunters Moon, Reigate Road, Horley, RH6 0HU
2E0 IBT J Ferrol, 29 Westlands, Haltwhistle, NE49 9BS
2E0 IBU Adam Buckley, 25 Queensway, Pilsley, Chesterfield, S45 8EJ
2M0 IBW I Macdonald, The Cottage, High Craigton, Glasgow, G62 7HA
2E0 ICB I Buchner, 52 Hillview, Coldstream, TD12 4ED
2W0 ICD Ivor Davies, 3 Keteringham Close, Sully, Penarth, CF64 5JW
2E0 ICE Paul Hallas, 11 Wensleydale Avenue, Eastham, Wirral, CH62 8ET
2E0 ICI I Lippett, 41 Springvale, Gillingham, ME8 0JG
2E0 ICK A Morris, 71 Lurdin Lane, Standish, Wigan, WN6 0AQ
2E0 ICP Ian Pass, 69 Cotswold Road, Bath, BA2 2DL
2E0 ICT M Gascoyne, 31 Dale View, Hemsworth, Pontefract, WF9 4TA
2E0 ICU D Marsh, 16 Laurel Close, North Warnborough, Hook, RG29 1BH
2E0 ICY Ewen Moore, 23 Woodland Road, Rode Heath, Stoke-on-Trent, ST7 3TJ
2E0 IDA Ian Sharples, 20 Poplar Avenue, Euxton, Chorley, PR7 6BE
2E0 IDC Ian Cosham, 54 Hawkins Crescent, Shoreham-by-Sea, BN43 6TP
2I0 IDJ Kieran McLaverty, 123a Castle Road, Antrim, BT41 4ND
2E0 IDK Ian King, 7 Greenacres Avenue, Blythe Bridge, Stoke-on-Trent, ST11 9HU
2E0 IDN Ian Norfolk, Arwelfa, High Street, Uckfield, TN22 3LP
2E0 IDR Ian Reeve, 36 Stone Pippin Orchard, Badsey, Evesham, WR11 7AA
2W0 IDT Ian Booth, 4 Church Meadow, Boverton, Llantwit Major, CF61 2AT
2E0 IEA Edward Aspden, 9 Cledford Crescent, Middlewich, CW10 0EZ
2M0 IEC E Cohen, 234 Allison Street, Glasgow, G42 8RT
2E0 IED Andy Wedge, 30 Primrose Way, Locks Heath, Southampton, SO31 6WX
2E0 IEE Steven Spice, Ebor Lodge, Udimore Road, Rye, TN31 6BX
2E0 IEI Rees Adams, Aston View, Brownshill, Stroud, GL6 8AG
2E0 IEO M Sanderson, 2 East Crescent, Canvey Island, SS8 9HL
2E0 IET C Blount, 55 Silverthorne Drive, Caversham, Reading, RG4 7NR
2E0 IFC Nick Burnet, 27 Mackenzie Way, Tiverton, EX16 4AW
2E0 IFF Peter Webster, 15 Napier Street, Workington, CA14 2PT
2E0 IFW E Evans, 313 Parkgate Road, Chester, CH1 4BE
2D0 IGL Paul Penn-Bixby, 4 Gold Hill, Edgware, HA8 9BZ
2E0 IGM Gary Benford, 34 Victoria Gardens, Colchester, CO4 9YD
2E0 IGN Robert Buchan-Terrey, Godre'r Coed, Aberhosan, Machynlleth, SY20 8RA
2E0 IGQ I Jameson, 9 Pine Hey, Neston, CH64 3TJ
2E0 IGW G Taylor, 39 Gill Street, Newcastle upon Tyne, NE4 8BH
2E0 IHB Roger Bleaney, 40 Broadstone Road, Harpenden, AL5 1RF
2E0 IHH Ian Hutchinson, Bridgend, Mill Lane, North Hykeham, Lincoln, LN6 9PA
2E0 IHI Catherine Colless, 128 Ditton Lane, Fen Ditton, Cambridge, CB5 8SS
2E0 IHO Keith Brown, 143 Princes Road, Ellesmere Port, CH65 8EP
2E0 IIT Graham Coleman, 2 Chapel House, 2 Alpine Road, Ventnor, PO38 1BT
2E0 IJH Ian Hope, 5 The Crescent, Northfleet, Gravesend, DA11 7EB
2E0 IJK D Barstow, 11 Carlton Street, Featherstone, Pontefract, WF7 6AA
2W0 IJL K Smith, 62 Waterloo Road, Talywain, Pontypool, NP4 7HJ
2E0 IJQ Robert Bedford, 29 Kent Road, Brookenby, Market Rasen, LN8 6EW
2E0 IJW Gary Williamson, 4b Havelock Place, Bridlington, YO16 4JN
2E0 IKB B Wilkes, 2 Kings Crescent, Edlington, Doncaster, DN12 1BD
2E0 IKH Tony Anderson, 45 Dudley Road, Clacton-on-Sea, CO15 3DW
2E0 IKM Michael Moffat, 19 Croftfield Road, Seaton, Workington, CA14 1QW
2E0 IKV Steven Rush, 88 Mountview, Borden, Sittingbourne, ME9 8JZ
2E0 IKW Gary Cannon, 30 Main Street, Flixton, Scarborough, YO11 3UB
2E0 ILN David Dee, 25 Blatcher Close, Minster on Sea, Sheerness, ME12 3PG
2E0 ILO Milo Noblet, 1 Lingdale Road, Wirral, CH48 5DG
2I0 IMB I Barr, 64 Owenreagh Drive, Strabane, BT82 9DT
2E0 IMC I Coggon, 45, Ansten Crescent, Doncaster, DN4 6EZ
2W0 IMD Horia Ilie, Flat 1 30 Alexandra Road, Swansea, SA1 5DQ
2E0 IMG Ian Gough, 14 Currane Road, Nuneaton, CV10 0HY
2E0 IMJ D Coppin, 7 Firtree Close, Rough Common, Canterbury, CT2 9DB
2E0 IMM Charles Milne, 24 Berton Close, Blunsdon, Swindon, SN26 7BE
2D0 IMN T Hardwick, 3 Poplar Terrace, Douglas, Isle of Man, IM2 4AR
2I0 IMO Thomas McNaughter, 36 Elms Park, Coleraine, BT52 2QE
2M0 IMP Seonag Robertson, 20 Knockard Place, Pitlochry, PH16 5JF
2E0 IMS S Edwards, 68 Hampshire Road, Droylsden, Manchester, M43 7PL
2E0 IMT Ian Turner, 1 Elmwood Rise, Dudley, DY3 3QJ
2E0 IMW I Walker, 24 Hawthorn Road, Norwich, NR5 0LP
2E0 IMZ Ian Ross, 48 Henry Drive, Leigh-on-Sea, SS9 3QF
2E0 INC Carl Jenkins, 25 Longmeadow Grove, Birmingham, B31 4SU
2E0 IND Peter Ind, 30 Thompson Road, Stroud, GL5 1SY
2M0 INE Malcolm Scott, 28b Highfield Place, Birkhill, Dundee, DD2 5PZ
2M0 INS Cephas Ralph, 37 Seaview Terrace, Edinburgh, EH15 2HE
2E0 INT Paul Kerton, 21 Appledore, Bracknell, RG12 8QY
2E0 INV Nathan Bookham, 116 Clare Gardens, Petersfield, GU31 4EU
2M0 IOB Roderick Kennedy, 45 Rodney Road, Gourock, PA19 1XG

UK Callsigns

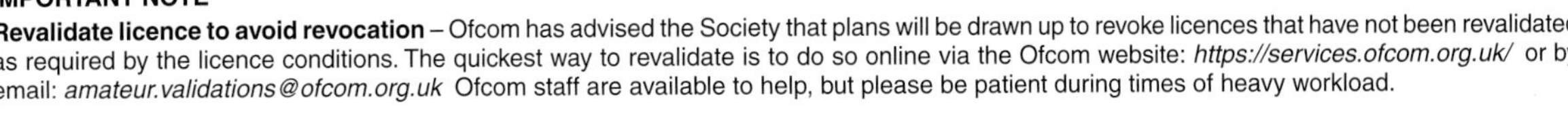

**IMPORTANT NOTE**

**Revalidate licence to avoid revocation** – Ofcom has advised the Society that plans will be drawn up to revoke licences that have not been revalidated as required by the licence conditions. The quickest way to revalidate is to do so online via the Ofcom website: *https://services.ofcom.org.uk/* or by email: *amateur.validations@ofcom.org.uk* Ofcom staff are available to help, but please be patient during times of heavy workload.

2E0 IOG D Mathewson, 33 Thornton Road, Bootle, L20 5AN
2M0 IOK S webb, Fawn View, Kennethmont, Huntly, AB54 4PF
2D0 IOM Darrell Allcote, 18 Maynrys, Castletown, Isle of Man, IM9 1HR
2E0 IOS Roger Smith, 62 Norwich Avenue, Plymouth, PL5 4JQ
2E0 IOU David Savage, 72 Pitmore Road, Eastleigh, SO50 4LW
2E0 IOZ Alexander Nikitits, 270a The Ridgeway, St. Albans, AL4 9XQ
2E0 IPA D Bradley, 37 Leamoor Avenue, Somercotes, Alfreton, DE55 1RL
2I0 IPB Aubrey Kincaid, 428 Cushendall Road, Ballymena, BT43 6QE
2E0 IPC William Catney, 92 Second Avenue, Liversedge, WF15 8JW
2E0 IPW Paul White, 13 Field Court, Oxted, RH8 0PD
2E0 IPX Philip Ashby, 2 Hamilton Road, Felixstowe, IP11 7AU
2E0 IQO M Vardy, 1 Sutton Middle Lane, Kirkby-in-Ashfield, Nottingham, NG17 8FX
2E0 IQT S Magrys, 13 Norfolk Road, Kidsgrove, Stoke-on-Trent, ST7 1EZ
2M0 IQU W Curry, 35 Tarvit Terrace, Springfield, Cupar, KY15 5SE
2E0 IQX Alan Emmerson, 8 Weston Close, Cannock, WS11 7YX
2M0 IRC James Livingstone, 17 Livingstone Drive, Bo'ness, EH51 0BQ
2E0 IRE Morgan O'Donovan, 58 Ocean View, Cirencester, GL76PA
2E0 IRN Joshua Glicklich, 86 Ainsdale Road, Bolton, BL3 3ER
2E0 IRX Gordon Harman, 58 Laurence Avenue, Witham, CM8 1JB
2M0 ISA I Watson, 64 Anstruther Street, Law, Carluke, ML8 5JG
2E0 ISK Keith Hyde, 8 Pennant Close, Birchwood, Warrington, WA3 6RR
2E0 ISO I Butler, 23 Owen Way, Basingstoke, RG24 9GH
2E0 ISQ R Lilley, 3 Coultshead Avenue, Billinge, Wigan, WN5 7HS
2E0 ISS Paul Dann, 4 Middlefields Court, Middlefields, Letchworth Garden City, SG6 4NQ
2E0 ISZ R Priamo, 58 Ffordd Glyn, Coed-y-Glyn, Wrexham, LL13 7QW
2E0 ITC A Checketts, 77 Shenstone Avenue, Stourbridge, DY8 3EH
2E0 ITJ Tony Johnson, 81 Welbeck Street, Whitwell, Worksop, S80 4TN
2E0 ITM Iwan Mitchell, 9 Rhiw Grange, Colwyn Bay, LL29 7TT
2E0 ITN Robert Goodall, 76 Beaconfield Road, Plymouth, PL2 3LF
2E0 ITU Paul James, 25 Brookfield Road, Churchdown, Gloucester, GL3 2PQ
2E0 ITV G Moorhouse, 1 Woodlands Avenue, Spilsby, PE23 5EP
2I0 ITY David Given, 15 Middle Road, Lisburn, BT27 6UU
2E0 IUH A Morris, 22 Dixon Avenue, Newton-le-Willows, WA12 0NE
2E0 IUI David Marshall, 7 Lancaster Close, Newton-le-Willows, WA12 9EY
2E0 IUK David Grayson, 79 Errington Avenue, Sheffield, S2 2EA
2W0 IUN Ieuan Jones, 21 Albert Street, Maesteg, CF34 0UF
2E0 IVO W Gissing, 2 Yeo Moor, Clevedon, BS21 6UQ
2E0 IVR I Ross, 15 Earlswood Drive, Madeley, Telford, TF7 5SF
2E0 IVY David Slee, Turnpike, Brearton, Harrogate, HG3 3BX
2W0 IVZ R Vials, 46-48 Park Place, Gilfach, Bargoed, CF8 8NA
2E0 IWB Ian Bunting, 6 Forster Close, Aylsham, Norwich, NR11 6BD
2M0 IWD I Davidson, 3 Hillcrest Avenue, Kirkcaldy, KY2 5TU
2E0 IWF Ian Francis, 34 Furlong Road, Bourne End, SL8 5AA
2W0 IWM Ian Miles, 40 Seymour Street, Mountain Ash, CF45 4BL
2E0 IWT M Champness, 10 Isaac Square, Great Baddow, Chelmsford, CM2 7PP
2E0 IWZ A Hanna, 35 Orchard Drive, Mayland, Chelmsford, CM3 6EP
2E0 IXC D Watson, 10 Gimson Close, Tuffley, Gloucester, GL4 0YQ
2E0 IXH R Maas, 15 Pine Court, Attleborough, NR17 2HU
2E0 IXI Richard Forss, Lower Conghurst Oast, Conghurst Lane, Cranbrook, TN18 4RW
2E0 IXX D Cattermole, 39 Moor Lea, Braunton, EX33 2PF
2I0 IYH Albert Wilson, 108a Salia Avenue, Carrickfergus, BT38 8NE
2E0 IYY H Burnett, 10 Galton Avenue, Christchurch, BH23 1JU
2I0 IZI Adrian Ismay, 21 Hillsborough Drive, Belfast, BT6 9DS
2E0 IZP Ian Pipe, 8 Glebe Drive, Stottesdon, Kidderminster, DY14 8UF
2E0 IZR Jason Bridson, 10 Clegg Street, Astley, Manchester, M29 7DB
2E0 IZW Isabel Whiteley, 2 The Meade, Manchester, M21 8FA
2E0 JAA J Ahmed, 59 Ramsgate, Lofthouse, Wakefield, WF3 3PX
2E0 JAB J Bradbury, 11 Ravensbourne House, Arlington Road, Twickenham, TW1 2AX
2E0 JAC J Crank, 38 Harley Avenue, Harwood, Bolton, BL2 4NU
2E0 JAF John King, 22 Latchmere Gardens, Leeds, LS16 5DN
2W0 JAI Julie Griffiths, 85 Tudor Estate, Maesteg, CF34 0SW
2E0 JAJ John Johnson, 62a Julien Road, London, W5 4XA
2M0 JAL C MacLean, 16 Glamis Avenue, Elderslie, Johnstone, PA5 9NR
2E0 JAM J Dodds, 8 York Road, Rowley Regis, B65 0RR
2E0 JAN George Martin, 12 Poolside, Phase 1, St. Joseph, Trinidad and Tobago
2E0 JAO Julian Lockyear, Berrow Bank, Bromsberrow, Ledbury, HR8 1SG
2I0 JAP Elizabeth Rantin, 8a Buchanans Road, Newry, BT35 6NS
2E0 JAQ J bosworth, 10 Aston Street, Leeds, LS13 2BJ
2E0 JAR Janet Rance, 11 Orchard Lane, Longton, Preston, PR4 5AX
2M0 JAT Alexander Murray, 119 Carnarc Crescent, Inverness, IV3 8SJ
2E0 JAW Jason Phipps, 42 Cavell Avenue, Peacehaven, BN10 7NS
2E0 JAX John Burdett, 5 Winston Drive, Wainscott, Rochester, ME2 4LJ
2E0 JAZ Jim Sadler, 10 Spindle Warren, Havant, PO9 2PU
2E0 JBA John Woods, 3 Ingle Avenue, Morley, Leeds, LS27 9NP
2E0 JBC David Croft, 33 Roughaw Road, Skipton, BD23 2PY
2E0 JBD J Dukes, 79 Jubilee Avenue, Boston, PE21 9LE
2D0 JBE Joseph McCartney, 5 Riverside, Ramsey, Ramsey, Isle of Man, IM8 3DA
2E0 JBF Gerald Brown, 51 Arncliffe Drive, Knottingley, WF11 8RH
2E0 JBG John Brownbill, 18 Milestone Road, Stratford-upon-Avon, CV37 7HH
2E0 JBI Jonathan Bethell, 47 Montgomery Road, Ipswich, IP2 8QB
2E0 JBJ J Jenkins, 13 Birch Hill, Newport, NP20 6JD
2E0 JBK S Glass, 16 Norman Way, Colchester, CO3 4PS
2E0 JBL John Chatterton, 6 Bayliss Road, Wargrave, Reading, RG10 8DR
2E0 JBM Jacqueline Moppett, 59 piccadilly, Tamworth, B78 2ER
2E0 JBQ Duncan Robinson, 27 Abbeylea Drive, Westhoughton, Bolton, BL5 3ZD
2E0 JBS J Byrne, 45 Jenkins Drive, Sheffield, S9 1AR
2E0 JBW Stanley Pascoe, Trewyn, Carnmenellis, Redruth, TR16 6PG
2E0 JBX John Barratt, 17 Main Road, Collyweston, Stamford, PE9 3PF
2E0 JBY Jason Bibby, 12 James Street, Burton-on-Trent, DE14 3SB
2M0 JBZ James Coubrough, 41 Bridge Court, Alexandria, G83 0BZ
2E0 JCA James Aubury, 27 Gravel Walk, Tewkesbury, GL20 5NH
2E0 JCB James Waring, 12 Mary Street, Farnhill, Keighley, BD20 9AU
2E0 JCC Jake Whitton, 11 Dursley Road, Bristol, BS11 9XB
2E0 JCD J Dalgliesh, 61 Clonners Field, Stapeley, Nantwich, CW5 7GU
2E0 JCE Jason Tompkins, 3 Hartwell Road, Portsmouth, PO3 5TN
2M0 JCF J Forsyth, 14b Osborne Terrace, Edinburgh, EH12 5HG
2M0 JCG John Rankin, 17 Dippin Place, Saltcoats, KA21 6AB
2I0 JCH John Henderson, 1 Brook Lodge Ballinderry Lower, Lisburn, BT28 2GZ
2E0 JCK Eric Beechill, Belleroyd Farm Blackshaw Head, Hebden Bridge, HX7 7JP
2E0 JCM John Clark-McIntyre, 184 Quebec Road, Blackburn, BB2 7DP
2E0 JCN John Pritchard, 1 Tan y Coed, Maesgeirchen, Bangor, LL57 1LU
2E0 JCO Julian Crewe, 22 Myrtle Tree Crescent, Weston-Super-Mare, BS22 9UL
2E0 JCP J Percival, Blue Cedars, Gresford, Wrexham, LL12 8RN
2E0 JCQ James Stevens, 23 Edlyn Close, Berkhamsted, HP4 3PQ
2E0 JCR Jeffrey Seear, 40 Rochester Way, Crowborough, TN6 2DT
2E0 JCS John Sanderson, 54 Kelvedon Close, Chelmsford, CM1 4DG
2E0 JCT John Griffiths, 257 Bolton Road, Ashton-in-Makerfield, Wigan, WN4 8TG
2E0 JCU John Underwood, 12 Forsythia Close, Bicester, OX26 3GA
2W0 JCV John Baldwin, 94 King Street, Abertridwr, Caerphilly, CF83 4BG
2E0 JCW Jack Wright, 19 Halstead Close, Woodley, Reading, RG5 4LD
2E0 JCX Steven Jackson, 5 Duchess Park Close, Shaw, Oldham, OL2 7YN
2E0 JCY John Connolly, 2 Waring Avenue, St. Helens, WA9 2QG
2E0 JCZ John Robb, 37 Wroxham Road, Woodley, Reading, RG5 3AX
2E0 JDA J Dale, Corydon, Church Street, Sevenoaks, TN14 7SW
2E0 JDC L Short, 10 Carville Crescent, Brentford, TW8 9RD
2E0 JDD J Delves, 34 Tatton Road, Crewe, CW2 8QA
2E0 JDE Johnathan Hardingham, 11 All Saints Close, Weybourne, Holt, NR25 7HH
2E0 JDF Andrew Powell, 2 Ormsby Close, Hopton, Great Yarmouth, NR31 9TY
2E0 JDH D HARBRON, 6 West View, Penshaw, Houghton le Spring, DH4 7HP
2E0 JDI Joseph Redhead, 28 Sandfields, Frodsham, WA6 6PT
2E0 JDK John Hazeltine, 21 Hassock Way, Wimblington, March, PE15 0PJ
2E0 JDL Glen Wilkins, 7 Byron Road, Newport, NP20 3HJ
2E0 JDM James Matheson, 24 Grange Road, Dacre Banks, Harrogate, HG3 4HA
2E0 JDO Josh Harriott, 1 Laggan Close, Tamworth, B77 2TZ
2E0 JDP Neil Pipkin, 46 Charles Avenue, Albrighton, Wolverhampton, WV7 3LF
2D0 JEA Jeanie Hill, 54 Wybourn Drive, Onchan, Isle of Man, IM3 4AT
2E0 JED Gerald Maguire, 39 Cooper St., Stretford, Manchester, M32 8NA
2E0 JEE J Egleton, 81 Pinchbeck Road, Spalding, PE11 1QF
2E0 JEH John Hewart, 14 Kestrel Close, Marple, Stockport, SK6 7JS
2E0 JEK John Kay, 36 Winnington Road, Marple, Stockport, SK6 6PT
2E0 JEM J Carvill, THE LODGE, OLDBURY ROAD, Worcester, WR2 6AA
2E0 JEN P Halpin, 50 Celtic Road, Deal, CT14 9EF
2E0 JES Colin Ember, 35 Mattock Lane, Ealing, London, W5 5BH
2E0 JET Ronald Cunliffe, 89 Colyers Lane, Erith, DA8 3NG
2E0 JEZ Jeremy Powell, 46 Woodmancote, Yate, Bristol, BS37 4LL
2E0 JFC James Carroll, 103 Brays Lane, Coventry, CV2 4DS
2E0 JFK J Kirk, 3 Knold Park, Margate, CT9 3BH
2I0 JFO William Forde, 35 Torr Gardens, Larne, BT40 2JH
2E0 JFW J Witchell, 43 Elm Way, Shepton Mallet, BA4 5JX
2E0 JGD James Dinsbier, Little Downs, Sandgate Lane, Pulborough, RH20 3HJ
2E0 JGE John Glover, 12 Willow Street, London, E4 7EG
2E0 JGG J Gibbons, 28 Deveron Gardens, South Ockendon, RM15 5ET
2E0 JGH Jonathan Hunt, 15 Greenway, London, SW20 9BQ
2E0 JGP Jonathan Paradi, 168 Castle Road, Northolt, UB5 4SG
2E0 JGR John Ritson, Granville House, Newton Arlosh, Wigton, CA7 5ET
2E0 JGS Joseph Seaton, 52 Shrubbery Street, Kidderminster, DY10 2QY
2E0 JGW James Whitehead, 12 Polkerris Road, Carharrack, Redruth, TR16 5RJ
2E0 JHC J Clark, 27 The Gabriels, Newbury, RG14 6PZ
2E0 JHD David Capstick, 3 Andrew Close, Dibden Purlieu, Southampton, SO45 4LS
2E0 JHF Philip Attwater, 42 Danescourt Crescent, Sutton, SM1 3EA
2E0 JHG John Ginever, 66 London Road, Maidstone, ME16 8QU
2M0 JHN J Brown, 60 Laburnum Lea, Hamilton, ML3 7LZ
2E0 JHO Brian Hodgson, 28 Grove Drive, Woodhall Spa, LN10 6RT
2E0 JHP D Porter, 105 Shore Road, Littleborough, OL15 9LJ
2E0 JHT James Hill, 45 Venus Street, Congresbury, Bristol, BS49 5HA
2E0 JIA Ian Boddy, 5 Boverton Avenue, Brockworth, Gloucester, GL3 4ER
2I0 JIE Joseph Evans, 404 Foreglen Road, Dungiven, Londonderry, BT47 4PN
2E0 JIF James Isherwood, 11 Manor Crescent, Chesterfield, S40 1HU
2E0 JIG David Harris, 35 Itchenor Road, Hayling Island, PO11 9SN
2E0 JIL Gill Whitehead, 29 Coulsons Road, Bristol, BS14 0NN
2E0 JIR S Buckley, 31 Rose Avenue, Irlam, Manchester, M44 6AQ
2E0 JIW J Biggin, Galadean, Farriers Way, Newport, PO30 3JP
2E0 JIX Jago Packer, 20 Shipman Road, Market Weighton, York, YO43 3RB
2E0 JJA Jonathon Adler, 1 Searles Meadow, Dry Drayton, Cambridge, CB23 8BW
2E0 JJB John Barton, 93 Cardigan Road, Bridlington, YO15 3JU
2E0 JJK James King, 18 Ross Road, Wallington, SM6 8QB
2E0 JJN James Nicholls, 4 Sadler Close, Colchester, CO2 7LU
2E0 JJR R Ellery, No 14 Roughtor View, Planet Park, Delabole, PL33 9BX
2W0 JJW Roger Woodland, 2 Waun Las, Neath, SA10 7RW
2E0 JKB Stuart Lucas, 31 Lilian Close, Norwich, NR6 6RZ
2E0 JKD Keith Winward, 123 Fulbeck Road, Middlesbrough, TS3 0RL
2E0 JKG Justin Gaskin, Badgers Barn, Canterbury Road, Folkestone, CT18 8DF
2E0 JKR Ronald Whatmough, 150 Whelley, Wigan, WN1 3UE
2E0 JKT John Turner, 34 Vaughan Road, Stotfold, Hitchin, SG5 4EH
2D0 JKW John Wardle, Cooyrt Vane, Ballamodha Straight, Ballamodha, Isle of Man, IM9 3AY
2E0 JKY Kevin Young, 48 Sussex Street, Cleethorpes, DN35 7NP
2E0 JLB James Bookham, 116 Clare Gardens, Petersfield, GU31 4EU
2E0 JLD Jamie Drinkell, 64 Pioneer Avenue, Burton Latimer, Kettering, NN15 5LH
2E0 JLK Karl Robinson, 103 Recreation Street, Mansfield, NG18 2HP
2E0 JLM James Meek, 30 Cleveland Square, London, W2 6DD
2E0 JLN Jillian Ullersperger, 60 Reeds Avenue, Earley, Reading, RG6 5SR
2E0 JLR Benjamin Taylor, 12 Clovelly Road, Stockport, SK2 5AZ
2M0 JLS John Leitch, 25 Lime Street, Grangemouth, FK3 8LZ
2E0 JLX John Landless, 2 Aspen Way, Banstead, SM7 1LE
2E0 JMB J Banham, Timandra, Mill Road, Norwich, NR15 2ST
2E0 JMC James McInnes-Boylan, 54 Fernbeck Close, Farnworth, Bolton, BL4 8BR
2E0 JMD Jonathan Davies, 10 Leaway, Prudhoe, NE42 6QE
2E0 JME James Smith, 10 Wayside Road, Bridlington, YO16 4BA
2E0 JMF Mark Hardwick, 34 The Pastures, Long Bennington, Newark, NG23 5EG
2E0 JMH James Hewitt, 1 Highfield, Gloucester Road, Chepstow, NP16 7DF
2W0 JMK Mark Dean, 12 Maes Y Bedol, Garnant, Ammanford, SA18 2EP
2E0 JMR J Rodley, 268 Grovehill Road, Beverley, HU17 0HP
2E0 JMW Julia Walker, 194a Mount Vale, Mount Vale, York, YO24 1DL
2W0 JMX Brendan Preece, 4 Waltham Close, Morriston, Swansea, SA6 7PH
2E0 JMY Jamie Bickers, 3 The Old Brickyard, West Haddon, Northampton, NN6 7GP
2E0 JND James Philps, 130 Turkey Road, Bexhill-on-Sea, TN39 5HH
2E0 JNH James Betteridge, 57 Wood Road, Chaddesden, Derby, DE21 4LY
2E0 JNP John Perry, Flat 7 Lancaster House, Belle Vue Rd, Paignton, TQ4 6HD
2E0 JOC R Denim, 15 Saxon Rise, Collingbourne Ducis, Marlborough, SN8 3HQ
2E0 JOE J Fletcher, 32 Chapel Lane, Barwick in Elmet, Leeds, LS15 4EJ
2E0 JOF J Miles, 11 Enborne Gate, Newbury, RG14 6AZ
2E0 JOG Graham Smith, 559 New Ashby Road, Loughborough, LE11 4EX
2E0 JOH John Hatton, 49 Buxton Street, Morecambe, LA4 5SR
2M0 JOK John Stewart, 1 Barns Park, Dalgety Bay, Dunfermline, KY11 9XX
2E0 JOP Joan Priestman, 198 Felmongers, Harlow, CM20 3DW
2I0 JOS Joshua Millar, 3 Lismurn Park, Ahoghill, Ballymena, BT42 1JN
2E0 JOX J Jones, 15 Kinnaird Road, Sheffield, S5 0NN
2E0 JPA John Wake, 60 Cloverville Approach, Odsal, Bradford, BD6 1ET
2E0 JPC J Clark, 1 Brooklime Road, Liverpool, L11 2YH
2E0 JPD Ronald Young, 26 Silent Woman Park, Coldharbour, Wareham, BH20 7PE
2E0 JPE Joshua Elsmore, 8 Clos Aberconway, Prestatyn, LL19 9HU
2W0 JPF J Freelove, 12 Honeyborough Road, Neyland, Milford Haven, SA73 1RE
2E0 JPH J Hazell, 90 Lichfield Road, St. Annes, Bristol, BS4 4BN
2E0 JPM James Monahan, 48 Church Road, Earley, Reading, RG6 1HS
2I0 JPP Jordan Page, 259 Bridge Street, Portadown, Craigavon, BT63 5AR
2E0 JPR Peeyush Gaur, 34 Queensberry Avenue, Copford, Colchester, CO6 1YN
2E0 JPS J Smith, 26 Hazel Close, Laindon, Basildon, SS15 5GT
2E0 JPT J Taylor, 6 Hawks Close, Walsall, WS6 7LE
2E0 JPU J Patient, 4 Bucklebury Heath, South Woodham Ferrers, Chelmsford, CM3 5ZU
2E0 JPW Jason Websdale, Blacksmith Cottage, Black Street, Lowestoft, NR33 8EG
2E0 JPX Jean-Paul Parkes, 24 Kenilworth Road, Lichfield, WS14 9DP
2E0 JQF S Walrond, 26 Birchwood Avenue, Weston-Super-Mare, BS23 3JE
2E0 JQI E Lightbown, 18 Formby Close, Blackburn, BB2 3JZ
2E0 JQK P Blackie, 30 Queens Avenue, Ilfracombe, EX34 9LS
2E0 JQW Jieqiong Wang, Flat 31, 74 Arlington Avenue, London, N1 7AY
2E0 JRA A Jessop, 4 Katherine St., Thurcroft, Rotherham, S66 9LG
2E0 JRB Jack Barrow, 45 Windermere Road, Seaham, SR7 8HW
2E0 JRD John Dowdeswell, 18 Lechlade Gardens, Fareham, PO15 6HF
2E0 JRL J Lambert, 82 22nd Avenue, Hull, HU6 9LS
2E0 JRN James Lynn, 2 The Fairways, Redhill, RH1 6LP
2E0 JRP J Preston, 5 Bodiam Crescent, Eastbourne, BN22 9HQ
2E0 JRR Joshua Riley, The Thatched House, 18 Bond Street, Norwich, NR9 4HA
2E0 JRS John Stallard, 6 Richmond Crescent, Leominster, HR6 8RX
2E0 JRT Neville Wing, 39 Whittington Road, Hutton, Brentwood, CM13 1JX
2E0 JRW M Williams, 76 Quince, Amington, Tamworth, B77 4EU
2E0 JRZ Barry Handley, 68 Northfield Avenue, Rothwell, Leeds, LS26 0SW
2M0 JSB J Bence, 5 Braeside Gardens, Hamilton, ML3 7PN
2M0 JSF John French, 19 Woodside Avenue, Bridge of Weir, PA11 3PQ
2E0 JSG Justin Godfrey, 29 Ridgewood Drive, Harpenden, AL5 3LJ
2E0 JSH J Harris, 23 Winchester Avenue, Chatham, ME5 9AR
2E0 JSJ Trevor Peck, 7 Byron Road, Mablethorpe, LN12 1JD
2E0 JSK S Killian, 31 Cherry Hills, Watford, WD19 6DH
2E0 JSM Jacob Preston, 35 First Avenue, Kidsgrove, Stoke-on-Trent, ST7 1DN
2E0 JSN John Slattery, 64 Rydal Avenue, Chadderton, Oldham, OL9 0QX
2E0 JSO J O'Shea, 56 Crummock Gardens, London, NW9 0DJ
2I0 JSQ John Quinn, 5 Woodhill Heights, Lurgan, Craigavon, BT66 7DJ
2E0 JSR Zalam Rathore, 7 Ashdale Avenue, Bolton, BL3 4PH
2E0 JSS John Symonds, 45a Waterford Road, Ipswich, IP1 5NL
2E0 JTB D Baker, 65 Madison Street, Tunstall, ST6 5HS
2E0 JTH James Thornhill, 47 Hopton Lane, Mirfield, WF14 8JP
2E0 JTI Stephen Smith, 72 Throstle Lane, Leeds, LS10 4EY
2E0 JTM Denise Bache, 62 Whittingham Road, Halesowen, B63 3TP
2E0 JTN Martin Chivers, 4 Hunters Lodge, Fareham, PO15 5NF
2E0 JTQ Joshua Cook, 37 Leigham Court Drive, Leigh-on-Sea, SS9 1PT
2E0 JTW James Johnson, 4 Wallace Close, Hullbridge, Hockley, SS5 6NE
2D0 JTY Jonathan Hewlett, 28b The Broadway, Joel Street, Northwood, HA6 1PF
2E0 JUL Julia Hardy, LAMBDA HOUSE, SEANOR LANE, Chesterfield, S45 8DH
2E0 JUW Adrian Davis, 34 Novers Park Drive, Bristol, BS4 1RG
2E0 JVA Andrew Vasarhelyi, 1 Eldon Close, Langley Park, Durham, DH7 9FR
2E0 JVM Adrian Mori, 33 Valerian Road, Hedge End, Southampton, SO30 0GR
2E0 JVP Michael Mason, 7 Langland Close, Malvern, WR14 2UY
2M0 JVR James Robertson, 3 Richmond Place, Fochabers, IV32 7HF
2E0 JVV John Waddy, 70 Linden Avenue, Prestbury, Cheltenham, GL52 3DS
2E0 JWC John Cater, 5 Shady Grove, Hilton, Derby, DE65 5FX
2E0 JWE John Elsmore, 8 Clos Aberconway, Prestatyn, LL19 9HU
2E0 JWG James Guess, Flat 257, Helen Gladstone House, London, SE1 0QB
2E0 JWH John Hope, 40 Birch Lane, Rugeley, WS15 1EJ
2E0 JWJ D Shields, 42 Studland Park, Westbury, BA13 3HL
2E0 JWL J Laverie, 3 Kirkland Terrace, Wigtown, Newton Stewart, DG8 9TL
2E0 JWP John Wishart, 19 Chepstow Close, Chippenham, SN14 0XP
2E0 JWS Christopher Shaw, 365 Sopwith Crescent, Wimborne, BH21 1XH
2E0 JWW John Webster, 72 Grosvenor Street, Derby, DE24 8AT
2E0 JWY John Willby, 10 Sunbury Road, Birmingham, B31 4LJ
2E0 JXB John Bischoff, 6 Harton Close, Bromley, BR1 2UD
2E0 JXN Paul Jackson, Langsmead Barn, Eastbourne Road, Lingfield, RH7 6JX
2I0 JXO David Burke, 7 Edinburgh Villas, Omagh, BT79 0DW
2E0 JXX Jonathan Creaser, 8 Millwood Road, Hounslow, TW3 2HH
2E0 JYA James Young, 1 Pen Tye, Gwinear, Hayle, TR27 5HL
2W0 JYC M Coleman, 96 Shelone Road, Briton Ferry, Neath, SA11 2PU
2W0 JYI James Blaxland, 28 Limewood Close, St. Mellons, Cardiff, CF3 0BU
2E0 JYM J Downes, 30 Forestgate, Haxby, York, YO32 2WT
2E0 JYN S Bobby, 56 Ffordd Offa, Rhosllanerchrugog, Wrexham, LL14 2EY
2E0 JYX G Burlington, Podgwell Cottage, Seven Leaze Lane, Stroud, GL6 6NJ
2E0 JZC Jonathan Clark, 26 Heron Way, Sandbach, CW11 3AU
2E0 JZK John Howlett, 29 Little London, Heytesbury, Warminster, BA12 0ES
2E0 JZU Paul Holmes, 82 Moore Avenue, Norwich, NR6 7LG
2E0 KAB Kenneth Bull, 111 Hinksford Mobile Home Park, Kingswinford, DY6 0BB
2E0 KAC David Carr, 19 Kingsmead Walk, Speedwell, Bristol, BS5 7RL

2E0 KAE Roger Dyson, 4 Royston Lane, Royston, Barnsley, S71 4NL
2E0 KAG Michael Hughes, 10 Penrith Crescent, North Hykeham, Lincoln, LN6 8UW
2E0 KAK Robert Hambly, 144 Station Road, Irchester, Wellingborough, NN29 7EW
2E0 KAL K Lott, 6 Centurion Close, College Town, Sandhurst, GU47 0HH
2E0 KAR Abdelghani Mesbah, 121 Hood Street, Nottingham, NG5 4AQ
2E0 KAS Kevin Sharpe, 18 Dudhill Road, Rowley Regis, B65 8HT
2M0 KAU Martin Krawczyk, 19 Wishart Archway, Dundee, DD1 2JA
2E0 KAX Neil Griffiths, 67 Warstones Drive, Wolverhampton, WV4 4PF
2W0 KAY J Harvey, 17 Heol y Plwyf, Ynysybwl, Pontypridd, CF37 3HU
2E0 KBA James Fletcher, 16 Chalcot Drive, Hednesford, Cannock, WS12 4SF
2I0 KBB Jonathan Elliott, 20 North Road, Newtownards, BT23 7AN
2E0 KBD Paul Smith, 10 [illegible] Road, [illegible] on Tees, TS19 0EH
2E0 KBE J Hatton, 37 St. Cuthberts Way, Holystone, Newcastle upon Tyne, NE27 0UZ
2E0 KBF Barry Wiggins, 103 Lowlands Avenue, Sutton Coldfield, B74 3RF
2E0 KBG Keith Glaysher, 66 Talbot Road, Farnham, GU9 8RR
2E0 KBJ S Inman, 9 Colbert Avenue, Ilkley, LS29 8LU
2E0 KBK Roger Skinner, 29 Kenyon Road, Portsmouth, PO2 0JZ
2E0 KBL Sean Kneeshaw, 40 The Broadwalk, Otley, LS21 2RL
2E0 KBN Keith Burness, 4 Fenwick Street, Boldon Colliery, NE35 9HU
2E0 KBO David Jones, 19 Ffordd Hebog, Y Felinheli, LL56 4QZ
2E0 KBP Bisher Al-rawi, Flat 17, Harrow Lodge, London, NW8 8HR
2I0 KBS Kevin Boyle, 764 Springfield Road, Belfast, BT12 7JD
2E0 KBX Kevin Buxey, 17 Gort Crescent, Southampton, SO19 8LH
2E0 KCB Keith Smart, 33 East Street, Littlehampton, BN17 6AU
2E0 KCF Andrew Hankins, 16 Eastwick Barton, Nomansland, Tiverton, EX16 8PP
2E0 KCL John Jopson, 12 Charing Close, Ringwood, BH24 1FA
2E0 KCN Ben Somerville Roberts, 21 Regency Way, Ponteland, Newcastle upon Tyne, NE20 9AU
2E0 KCO Kevin Cornmell, 19 Forest Road, Chandler's Ford, Eastleigh, SO53 1NA
2E0 KCP Robert Hutton, 22a Victoria Road, Maldon, CM9 5HF
2E0 KCW Patrick Walsh, 181 Hermes Close, Hull, HU9 4DR
2E0 KCX David Askew, 19 Oliver Road, Staplehurst, Tonbridge, TN12 0TE
2E0 KCZ Raymond Robinson, 22 Riddings Court, Timperley, Altrincham, WA15 6BG
2E0 KDA Alan Nunn, 6 Carleton Glen, Pontefract, WF8 2RT
2E0 KDB David Barnett, 72a Clough Hall Road, Kidsgrove, Stoke-on-Trent, ST7 1AW
2E0 KDE A Lee, 14 Bernice Avenue, Chadderton, Oldham, OL9 8QJ
2E0 KDF Albert Hamilton, 56 Wyvern, Telford, TF7 5QH
2E0 KDG Chris Harman, 58 Laurence Avenue, Witham, CM8 1JB
2E0 KDH K Hall, 18 Brooklands Drive, Littleover, Derby, DE23 1DN
2E0 KDI Lawrence Azzopardi, 11 Ilynton Avenue, Firsdown, Salisbury, SP5 1SH
2E0 KDR Matthew Kidner, 4 Tonypistyll Road, Newbridge, Newport, NP11 4HJ
2E0 KDT Robert Simpson, 5 Rogate Road, Worthing, BN13 2DT
2E0 KDV D Craven, 69 Markham Avenue, Leeds, LS19 6NE
2E0 KEA T Lupton, 81 Home Farm Lane, Bury St. Edmunds, IP33 2QL
2W0 KED Steven Kedward, 9 Lawrence Avenue, Aberdare, CF44 9EW
2E0 KEG K Stone, 34 Daventry Grove, Birmingham, B32 1JA
2E0 KEI K Hastings, 161 Cottingley Road, Allerton, Bradford, BD15 9LD
2E0 KEP Tim Keep, 119 Radley Road, Abingdon, OX14 3RX
2E0 KEQ Kevin Mogford, 49 Cefn Road, Rogerstone, Newport, NP10 9AQ
2E0 KER R Kerswill, 38 Longbridge Road, Bramley, Tadley, RG26 5AN
2E0 KES Robert Noakes, 31 Pilot Road, Hastings, TN34 2AP
2I0 KEW K McDonald, 37 Ardgarvan Cottages, Limavady, BT49 0NF
2E0 KEZ D Brough, 57 Francis Road, Ashford, TN23 7UP
2E0 KFB Francis Bejoy Kuttikkate, 34 Shetland Crescent, Rochford, SS4 3FJ
2E0 KFH John Rowe, 22 Treaty Road, Glenfield, Leicester, LE3 8LU
2E0 KFK Krzysztof Kozlowski, WOKING HOMES / FLAT 2, ORIENTAL ROAD, Woking, GU22 7BE
2E0 KFM Andrew Burfield, 4 Eastern Crescent, Chelmsford, CM1 4JQ
2E0 KFO Sean Howard, 17 Webdell Court, Norwich, NR1 2NB
2E0 KFR Christopher Pearcey, 17 Peppercorn Close, Christchurch, BH23 3BL
2E0 KFT John Ferrol, 29 Westlands, Haltwhistle, NE49 9BS
2E0 KFX M Hill, 4 Farmland Way, Hailsham, BN27 1SP
2E0 KGB Sergei Moissejev, 50 Filey Road, Reading, RG1 3QQ
2E0 KGC Keith Cossey, 34 Pinewood Road, Hordle, Lymington, SO41 0GP
2E0 KGD Mark Tabberer, 29 Chase Vale, Chasetown, Burntwood, WS7 3GD
2E0 KGJ Kenneth Green, 12 Merton Place, Grays, RM16 4HL
2E0 KGP P Kelly, Arosfa, Westminster Road, Wrexham, LL11 6DN
2E0 KGQ J Kelly, Arosfa, Westminster Road, Wrexham, LL11 6DN
2E0 KGT Gary Tagg, Tinkers Cottage, Nevendon Road, Wickford, SS12 0QB
2E0 KGV Clive Moulding, 28 Queens Avenue, Highworth, Swindon, SN6 7BA
2E0 KHA Anthony Hughes, 8 Bullens Green Lane, Colney Heath, St. Albans, AL4 0QS
2E0 KHG Keith Gibbs, 40a Oakwood Road, Hollywood, Birmingham, B47 5DX
2E0 KHI Keith Hale, 42 Heyes Street, Liverpool, L5 6SG
2E0 KHK Steven Rosser, 25 Clos tir y pwll, Blaenavon, NP11 5GE
2E0 KHO David Plunght, 21 Buscot Drive, Abingdon, OX14 2BJ
2E0 KHW Keith White, 4 Top Birches, St. Neots, PE19 6DD
2E0 KIA Matthew West, 6 Elm Close, Norton, Stourbridge, DY8 0JH
2E0 KIL Terry Lee, 68 Wharton Drive, Springfield, Chelmsford, CM1 6BF
2E0 KIM Kim Matthews, 4 George Croft, Gayton le Marsh, Alford, LN13 0NP
2E0 KIT Alan Gladman, 19 Colchester Road, Wymering, Portsmouth, PO6 3RH
2E0 KJI Matthew Bidwell, 3 Walsingham Place, London, SW4 9RR
2E0 KJJ Jeremy Kent, Meadow Bank, Rye Lane, Sevenoaks, TN14 5JF
2E0 KJK Karol Jan Kolesnik, 15 Steer Road, Swanage, BH19 2RU
2W0 KJO Julia Orchard, The Burrows, Spring Gardens, Whitland, SA34 0HL
2E0 KKC Edward Brown, Flat 8, Jacobs Court, High Street, Colchester, CO7 0AQ
2E0 KKJ Krzysztof Juszczak, 68 College Road, Sandy, SG19 1RH
2E0 KKO Terence Stack, 31a Chester Road South, Kidderminster, DY10 1XJ
2E0 KLD Richard Delrosa, 18 Rangers Avenue, Dursley, GL11 4AS
2E0 KLF Kevin Francis, 203 Colchester Road, Lawford, Manningtree, CO11 2BU
2M0 KLL Robert Bertram, Noroc, 46 Main Street, Pathhead, EH37 5QB
2E0 KLN Kevin Nevins, 4 Ubbanford, Norham, Berwick-upon-Tweed, TD15 2LA
2E0 KLR Antony Henshall, 37 Marlow Drive, Branston, Burton-on-Trent, DE14 3TX
2E0 KLS Peter Elstub, 4 Granville Lodge, Church Street, Telford, TF2 9LX
2E0 KLV Keith Brown, 41 Church Street, Swinton, Mexborough, S64 8EF
2E0 KLW Lee Woods, 193 Wimberley Street, Blackburn, BB1 8HU
2E0 KLY Kenneth Young, 14 Beechwood Avenue, Chatham, ME5 7HH
2E0 KMA Karl Meehan, 55 Aketon Road, Castleford, WF10 5DN
2E0 KMB K Bailey, 50 Billy Buns Lane, Wombourne, Wolverhampton, WV5 0BP
2E0 KMD Keith Deans, 31 Northcroft, Sandy, SG19 1JJ
2E0 KMF Kevin Moysey, 109 Langbrook Cottages, Langbrook, Ivybridge, PL21 9JX
2E0 KMG Kevin Cook, 7 Grenville Terrace, Bideford, EX39 4BE
2E0 KMI K Mills, 6 West Coombe, Bristol, BS9 2BA
2E0 KMP Andrew Adams, 45 Four Oaks Road, Tedburn St. Mary, Exeter, EX6 6AP
2E0 KMS K Shenton, 2 The Croft, Stramshall, Uttoxeter, ST14 5AG
2W0 KMU G Cattle, 39 Park View, Abercynon, Mountain Ash, CF45 4TP
2E0 KMZ Keith Missenden, 47 Roseacre Drive, Elswick, Preston, PR4 3UQ
2E0 KNA T Clarke, 111 Telford Way, High Wycombe, HP13 5DZ
2E0 KNB M Dennon, 49 Dorset Street, Nottingham, [illegible]
2E0 KNC Stephen Henson, 297 Underwood Lane, Crewe, CW1 [illegible]
2E0 KNE J Dale, 37 Bussey Road, Norwich, NR6 6JF
2E0 KNF Kevin Sumner, 18 Grange Road, Fleetwood, FY7 8BH
2E0 KNH Keith Holman, 39 Trellech Court, Yeovil, BA21 3TE
2E0 KNL Glenn Knowles, 29 Delamore Crescent, Cramlington, NE23 3FY
2E0 KNM Mark Hill, 109 Kitchener Street, St. Helens, WA10 4LU
2E0 KNT Kent Royce, 11 Church Lane, Stibbington, Peterborough, PE8 6LP
2E0 KNU Richard Weightman, 17 Shaw Drive, Knutsford, WA16 8JG
2E0 KOI H Walker, 9 Humphries Close, Leicester, LE5 4LU
2E0 KOM Leonard Wilson, 6 Marrick Road, Middlesbrough, TS3 7RX
2W0 KOP Avril Martin, 117 Fforchaman Road, Cwmaman, Aberdare, CF44 6NL
2E0 KOR L Ross, 133 Petersmith Drive, New Ollerton, Newark, NG22 9SG
2E0 KOS Andrew Hollings, 39 Rendham Road, Saxmundham, IP17 1EA
2I0 KPA S Frazer, 2 Cavanballaghy Road, Killylea, Armagh, BT60 4NZ
2E0 KPC Paul Gagliardi, 7 Saxon Way, Jarrow, NE32 3QA
2E0 KPD Paul Davies, 18 Barton Lane, Kingswinford, DY6 9EY
2M0 KPE Kevin Page, 43 Lothian Court, Glenrothes, KY6 1LZ
2W0 KPH Kenneth House, Liddington, Dehewydd Lane, Pontypridd, CF38 2EN
2E0 KPI N Gregg, 84 Aberconway Crescent, New Rossington, Doncaster, DN11 0JP
2E0 KPL Leon Kiddell, 1 Sparham Hill, Sparham, Norwich, NR9 5QT
2W0 KPN Adrian Thomas, 10 Chapel Street, Gorseinon, Swansea, SA4 4DT
2E0 KPO Steven Warren, 1 Hoveley Close, Stapenhill, Burton-on-Trent, DE15 9EW
2E0 KPP D fellows, 147 Olive Lane, Halesowen, B62 8LR
2E0 KPR G Andrews, 151 Great Gregorie, Basildon, SS16 5QQ
2E0 KPT B Harris, 1 Newbridge Cottages, Midgehole, Hebden Bridge, HX7 7AL
2E0 KPX Danny Stephenson, Flat 8, Thornwood Court, 84-88 Hudson Road, Leigh-on-Sea, SS9 5NF
2M0 KPZ Nial Stewart, 35 Newbattle Gardens, Dalkeith, EH22 3DR
2E0 KQV David Harrison, 64 Crosby Road, Grimsby, DN33 1LU
2E0 KRB Andrew Cowan, 217 South Park Road, Wimbledon, London, SW19 8RY
2E0 KRD K Dukes, 127 Carlton Road, Boston, PE21 8LL
2E0 KRK James Birch, 3 Partridge Way, High Wycombe, HP13 5JX
2E0 KRN Karen Pidwell, 2 Welford Road, Chapel Brampton, Northampton, NN6 8AF
2E0 KRR Ram Rao, 3 Weller Mews, Enfield, EN2 8FG
2E0 KRS Kris Tharanee, 525 Burton Road, Littleover, Derby, DE23 6FT
2E0 KRT K Taylor, 3 The Drive, Lichfield, WS14 9QT
2E0 KRX Kevin Rosema, Apartment 801, 25 Goswell Road, London, EC1M 7AJ
2E0 KSG Katrina Stevens, 61a Main Road, Hoo, Rochester, ME3 9AA
2E0 KSH Krystyna Haywood, 126 Derby Street, Sheffield, S2 3NF
2E0 KSO A Jackson, 5 Woodside Lane, London, N12 8RB
2E0 KSW Kevin Sewell, 12 Haylands Square, South Shields, NE34 0JB
2E0 KTD Kate Davidson, 5 Hanover Parc, Indian Queens, St. Columb, TR9 6ER
2E0 KTG K Gribben, 44 Fern Close, Birchwood, Warrington, WA3 7NU
2M0 KTL Kit Lane, 23 Mayfield Avenue, Tillicoultry, FK13 6HB
2E0 KTV Bipin Chauhan, 45 Burnham Drive, Whetstone, Leicester, LE8 6HY
2E0 KTW K Wheeler, 3 Praze Road, Porthleven, Helston, TR13 9LR
2E0 KTX Kevin Browne, 56 Moorhouse Avenue, Wakefield, WF2 9QG
2E0 KUC David Holman, 38 Polyear Close, Polgooth, St. Austell, PL26 7BH
2E0 KUF John Inwood, Flat 359, Hagley Road Retirement Village, 330 Hagley Road, Birmingham, B17 8BP
2E0 KUH Sarah Terry, 201a Urmston Lane, Stretford, Manchester, M32 9EF
2E0 KVA A Kossick, 33 Verney Road, Winslow, Buckingham, MK18 3BN
2E0 KVB Margaret Wall, 227 Wayfield Road, Chatham, ME5 0HJ
2E0 KVE K Waterhouse, 74 Clifford Road, West Bromwich, B70 8JY
2E0 KVF Konrad Emery-Ford, 15 Anson Close, Grantham, NG31 7EN
2E0 KVJ David Boyes, 1 Fluxton Cottages, Fluxton, Ottery St. Mary, EX11 1RL
2E0 KVK Kevin Sim, 49 St. Julians Wells, Kirk Ella, Hull, HU10 7AF
2M0 KVM Kevin Mair, 26 Hawthorn Drive, Wishaw, ML2 8JS
2E0 KVR Alan King, 23 Tower Crescent, Lincoln, LN2 5QF
2E0 KWO K Comben, 9 West Lane, North Baddesley, Southampton, SO52 9GB
2E0 KWG Roy Knight, Thrift, Ashwells Road, Brentwood, CM15 0SG
2E0 KWM David Wright, 203 Winn Street, Lincoln, LN2 5EY
2E0 KWT John Towner, 4 The Copse, Scarborough, YO12 5HG
2E0 KWW Keith Willson, Ludpit Cottage, Ludpit Lane, Etchingham, TN19 7DB
2E0 KXD C Collins, 2 Kew Crescent, Sheffield, S12 3LP
2I0 KXM Keith Mitchell, 80 Markethill Road, Collone, Armagh, BT60 1LF
2E0 KXX Bruce Cook, 11 Arbrook Lane, Esher, KT10 9EG
2E0 KYI Keith Armstrong, 29 Thorntree Avenue, Crofton, Wakefield, WF4 1NU
2E0 KZC Martin Clack, 42a Provost Street, Fordingbridge, SP6 1AY
2E0 KZH David Taylor, 49 Boggart Hill Gardens, Seacroft, Leeds, LS14 1LJ
2M0 KZI A Broom, 140 Green Road, Paisley, PA2 9AJ
2E0 KZJ P Lamb, 13 Pool End, St. Helens, WA9 3RE
2E0 KZL I Hallatt, 11 Cheshire St., Audlem, Crewe, CW3 0AH
2E0 KZM M Osborne, Flat 5, 7 St Catherines Road, Littlehampton, BN17 5HS
2E0 LAA Anthony Lewis, 38 Olive Houses, Leicester Road, Leicester, LE8 8BF
2E0 LAB Luke Bain, 45 Larpool Crescent, Whitby, YO22 4JD
2E0 LAC Sian Llewellyn, 53 Cripps Avenue, Cefn Golau, Tredegar, NP22 3PF
2E0 LAD Michael Leech, 11 Westlake Close, Torpoint, PL11 2BZ
2E0 LAH Brian Beckett, 21 Horseshoes Lane, Langley, Maidstone, ME17 1SR
2E0 LAI Leslie Potter, 18 Elizabeth Gardens, Wakefield, WF1 3SZ
2E0 LAL Lewis Larkins, 34 Guycroft, Otley, LS21 3DS
2M0 LAO Scott Ramsay, 6 Cross Road, Peebles, EH45 8DH
2E0 LAQ P Hickling, 49 Roundhill Close, Syston, Leicester, LE7 1PP
2E0 LAR Andrew Davies, 96 Broad Lane, Kirkby, Liverpool, L32 6QQ
2M0 LAS Frank Davidson, 27 Gordon Way, Livingston, EH54 8JG
2E0 LAV Anthony Loyd, Maple House, Pangbourne Road, Reading, RG8 8LN
2M0 LAW Arthur McCaig, 46 Patterson Drive, Law, Carluke, ML8 5LT
2E0 LAX Andrew [illegible], 0 Marshall Close, Purley on Thames, Reading, RG8 8DQ
2E0 LAY Sidney Lay, 7 Hunt Street, Swindon, SN1 3HW
2W0 LAZ Robin Lasbury, 57 Westbourne Road, Whitchurch, Cardiff, CF14 2BR
2E0 LBA Aleksejs Polakovs, 76 Sandringham Crescent, Leeds, LS17 8DF
2E0 LBB Leigh Brown, 20 Pinfold Lane, Mirfield, WF14 9HZ
2E0 LBE Matthew Ryall, 2 Greenacres, Trinant, Newport, NP11 3AE
2D0 LBG John Shatford, 31 Pinner Park Avenue, Harrow, HA2 6LG
2M0 LBH Alex Hunsley, 42 Waverley Place, Edinburgh, EH7 5SA
2E0 LBI Leslie Bell, 42 Ocean Road, Walney, Barrow-in-Furness, LA14 3DY
2E0 LBJ [illegible], 18 Bridlington Road, Driffield, YO25 5HZ
2E0 LBK [illegible], 17 Manor Close, Abbotts Ann, Andover, [illegible]
2E0 LBL Laurence Lay, 17 Herbert Road, Hornchurch, RM11 [illegible]
2W0 LBN Lee Au-Yeung, 104 Fleet Street, Swansea, SA1 3UX
2I0 LBS L Stewart, 22 Edenmore Court, Newtownabbey, BT37 0ZA
2E0 LBZ David Taylor, 143 Sandhurst Road, London, SE6 1UR
2E0 LCA David Aldred, 14 The Meadows, Radcliffe, Manchester, M26 4NS
2E0 LCE Ian Pilton, Rainbow Barn, Pool Foot Farm, Ulverston, LA12 8AA
2E0 LCJ Lyndon Jones, Ty'r Ysgol, Holland Street, Ebbw Vale, NP23 6HT
2E0 LCM Lawrence Micallef, 132 Kings Hedges Road, Cambridge, CB4 2PB
2E0 LCN T Cass, April Cottage, South Eau Bank, Spalding, PE12 0QR
2E0 LCR Laurence Rimington, 3 Amicombe, Wilnecote, Tamworth, B77 4JJ
2E0 LCW T Wright, 2 Lilyville Road, London, SW6 5DW
2I0 LDC Paul Floyd, 9 Killybrack Mews, Omagh, BT79 7FB
2E0 LDD D Darling, 10a South St., Portslade, Brighton, BN41 2LE
2E0 LDE G Matts, 34 Barry Road, Leicester, LE5 1FA
2E0 LDF Reginald Irving, 2 Wasdale Close, Cockermouth, CA13 9JD
2E0 LDJ Leigh Jepson, 16 Golborne Street, Newton-le-Willows, WA12 9TH
2E0 LDM Leslie Mason, 9 Trenethick Avenue, Helston, TR13 8LU
2E0 LDN Kevin Poulton, 21 East View, London, E4 9JA
2E0 LDQ Liam Dobinson, 20 Newholme Crescent, Evenwood, Bishop Auckland, DL14 9RY
2E0 LDR Lyndon Reynolds, 49 Westborough Way, Anlaby Common, Hull, HU4 7SW
2E0 LDS Simon Wheeldon, 32 Beech Grove Terrace, Garforth, Leeds, LS25 1EG
2E0 LDV D Butterfield, 57 Holmes Road, Retford, DN22 6QU
2W0 LDX S Jones, 14 Lower Cross Road, Llanelli, SA15 1NQ
2E0 LDY D Hayes, 60 Shelby Road, Worthing, BN13 2TT
2E0 LDZ Trevor Clapp, Windrush, One Pin Lane, Slough, SL2 3QY
2E0 LEA D Anderson, 35 Sycamore Road, East Leake, Loughborough, LE12 6PP
2E0 LEE L Tunstall, 8 York Road, Rowley Regis, B65 0RR
2E0 LEF Neville Briggs, 20 Broad Lane, Pelsall, Walsall, WS4 1AP
2E0 LEG James Landless, 2 Aspen Way, Banstead, SM7 1LE
2E0 LEM Joseph Francis, 22 Acre Lane, Carshalton, SM5 3AB
2W0 LEN L Hayes, 56 Snowden Road, Cardiff, CF5 4PR
2E0 LEO L Steer, 51 Kings Chase, East Molesey, KT8 9DG
2E0 LEV Michael Leveridge, 17 Gladstone Court, Dewsbury, WF13 4DQ
2M0 LEW Mark Strachan, 62 Charleston Drive, Dundee, DD2 2EZ
2M0 LEX Alex Jenkins, Buchanan House, 58 Port Dundas Road, 5th Floor, Glasgow, G4 0HF
2E0 LEY Lee Medley, 9 Polyplatt Lane, Scampton, Lincoln, LN1 2TL
2E0 LEZ Leslie Trend, 140 Ardleigh, Basildon, SS16 5RW
2E0 LFE Lee Ansell, 114 Bowleaze, Greenmeadow, Cwmbran, NP44 4LG
2E0 LFI Andrew Birch, 22 Ullswater Road, Burnley, BB10 4HX
2E0 LFK Leonard Brown, 71 Chiltern Way, Nottingham, NG5 5NP
2E0 LFM Paul Dimmick, 17 Keir Hardie Court, 17 Kier Hardie Wood Lee Court, Newbiggin by The Sea, ne646lh
2E0 LFR L Rofix, Birds Hill Cottage, Clopton, Woodbridge, IP13 6SE
2M0 LFS Anthony Miles, 9 Buchanan Drive, Lenzie, Glasgow, G66 5HS
2E0 LFT Anthony Norden, 10 School Lane, Watton at Stone, Hertford, SG14 3SF
2E0 LFX John Lamb, 9a Matlock Road, Canvey Island, SS8 0EW
2W0 LFY Alan Jones, 10 Dan y Bryn, Caerau, Maesteg, CF34 0UW
2E0 LGB G Benson, 2 Guisborough Road, Nunthorpe, Middlesbrough, TS7 0LB
2E0 LGE Richard Samphire, Courtlands, Newport Road, Caldicot, NP26 3BZ
2E0 LGG Louis Martin, 78 Llwyn Ynn, Talybont, LL43 2AG
2E0 LGH Stuart Hallam, 18a Market Street, Hoylake, Wirral, CH47 2AE
2E0 LGL Marius Rusu, 9 Nash Close, Corby, NN18 0QU
2E0 LGO Patrick Frost, 26 Hollies Court, Britannia Road, Banbury, OX16 5DR
2E0 LGR L Rickman, 42 Sycamore Close, Poole, BH17 7YJ
2E0 LGS Peter Schoenmaker, 24 Greenheys Drive, London, E18 2HB
2E0 LGT D Page, 17 Hedge End Walk, Havant, PO9 5LS
2E0 LGV A Meek, 19 Stevenson St. East, Accrington, BB5 0SB
2E0 LGW Algernon O'Connell, 5 Westend Terrace, Gloucester, GL1 2RX
2E0 LGZ G Cash, 00 Vittoria Court, Birkenhead, CH41 3LF
2E0 LHC A Strudwick, 31 Skipper Way, Lee-on-The-Solent, PO13 9EU
2D0 LHR Simon Kapadia, 7 Elms Lane, Wembley, HA0 2NX
2E0 LHS Gareth Carver, 4 Andrews Road, Farnborough, GU14 9RY
2E0 LIT Clive Martin, 20 Hall Green Road, West Bromwich, B71 3LA
2E0 LIV A Nicholson, 24 Barnmead, Haywards Heath, RH16 1UZ
2E0 LIW William Sawyer, 20 Park Terrace, Willington, Crook, DL15 0QL
2E0 LIZ E Greatorex, 22 Marlborough Way, Uttoxeter, ST14 7HL
2E0 LJB L Bedford, 29 Kent Road, Brookenby, Market Rasen, LN8 6EW
2E0 LJC Ceri Jones, 19 Crud y Castell, Denbigh, LL16 4PQ
2E0 LJD John Dyer, 32 Brynystwyth, Penparcau, Aberystwyth, SY23 1SS
2E0 LJG Laura Goldsmith, 7 Fengate Drove, Weeting, Brandon, IP27 0PW
2E0 LJH Laura Halloway, 82 Northwall Road, Deal, CT14 6PP
2E0 LJK Laura Marriott, 94 Lyndhurst Road, Worthing, BN11 2DW
2I0 LJQ William Phair, 98 Cedar Grove, Holywood, BT18 9QB
2E0 LJR Andrew Siddall, 12 Russell Gardens, Sipson, West Drayton, UB7 0LS
2E0 LJS Michael Fearon, 70 George Street, Heywood, OL10 4PW
2E0 LJT Albert Tranter, 122 Summerhill Road, Bristol, BS5 8JU
2I0 LJW G Gorman, 4 Springwell Park, Groomsport, Bangor, BT19 6LF
2E0 LKC Peter Cairns, 16 East Avenue, Heald Green, Cheadle, SK8 3DL
2E0 LKE Leslie Kett, 52 Northgate, Hornsea, HU18 1EU
2E0 LKH K Hull, 3 Enas Crescent, Ena Street, Hull, HU3 2TL
2E0 LKM Luke McDonnell, 108 Long Lane, Garston, Liverpool, L19 6PQ

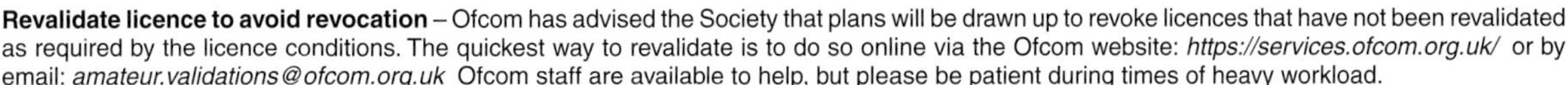

**IMPORTANT NOTE**

**Revalidate licence to avoid revocation** – Ofcom has advised the Society that plans will be drawn up to revoke licences that have not been revalidated as required by the licence conditions. The quickest way to revalidate is to do so online via the Ofcom website: *https://services.ofcom.org.uk/* or by email: *amateur.validations@ofcom.org.uk* Ofcom staff are available to help, but please be patient during times of heavy workload.

2E0 **LKS** Lawrence Spriggs, 19 Mackenzie Square, Stevenage, SG2 9TT
2E0 **LKT** Lee Taylor, Apartment 10, The Church Apartments 47a Seamer Road, Scarborough, YO12 4EF
2E0 **LLA** Andrew Jones, 2 Erw Terrace, Bethel, Caernarfon, LL55 1YT
2E0 **LLC** Lynda Addison, 45 Fir Terraces, Esh Winning, DH7 9JQ
2E0 **LLD** Edwin Daniels, 2 Garstons Close, Fareham, PO14 4EN
2E0 **LLE** Richard Redmond, 28 Common Lane, Polesworth, Tamworth, B78 1LS
2I0 **LLG** Stephen Horner, 10 Meadow Court, Bushmills, BT57 8SD
2E0 **LLI** Alan Foley, 23 Church Lane, Wymington, Rushden, NN10 9LW
2E0 **LLL** Christian Wadsworth, tyn llwyn, Talybont, LL43 2AN
2E0 **LLM** Janet Proudman, 61 Iffley Road, Oxford, OX4 1EB
2E0 **LLO** John Lovelock, Sea Spray, The Lizard, Helston, TR12 7NU
2W0 **LLT** L Thomas, 15 Blaenwern, Newcastle Emlyn, SA38 9BE
2E0 **LLW** Peter Sampson, 16 Rutland Place, Cirencester, GL7 1PR
2E0 **LLX** J Wright, 2 Regent Road, Church, Accrington, BB5 4AR
2W0 **LLY** Paul Kyte, 1 Dan y Bryn, Caerau, Maesteg, CF34 0UW
2E0 **LMD** Anne Bate, 16 East Avenue, Heald Green, Cheadle, SK8 3DL
2E0 **LME** G Beardmore, 9 Ashmore Drive, Gnosall, Stafford, ST20 0RP
2E0 **LMG** Christine Murray, 3 Rookery Dell, Deepcar, Sheffield, S36 2ND
2E0 **LMH** Lee Hudson, 68 Eleanor Road, Harrogate, HG2 7AJ
2E0 **LMK** Shaun Wills, 8 Amherst Road, Newcastle upon Tyne, NE3 2QQ
2W0 **LMM** Peter Jones, 115 Wordsworth Gardens, Pontypridd, CF37 5HH
2E0 **LMR** Gareth Southall, 28 Manor Road, Woodford Halse, Daventry, NN11 3QP
2E0 **LMS** E Spurr, 20 Mannington Way, West Moors, Ferndown, BH22 0JE
2M0 **LNF** I Mcgurk, 40 Beechwood Drive, Alexandria, G83 9NP
2E0 **LNG** Ronald Rider, 25 Kimber Close, Lancing, BN15 8QD
2M0 **LNR** Thomas Couper, 10 Sclandersburn Road, Denny, FK6 5LP
2E0 **LNU** Marian Durban, 62 Westfield Way, Charlton, Wantage, OX12 7EP
2I0 **LNZ** Linzi Craig, 170 Donaghadee Road, Bangor, BT20 4PP
2E0 **LOE** Nicola Chaplin, 5 Maxwell Street, Bury, BL9 7QA
2E0 **LOG** D Whelan, 431 Leeds Road, Huddersfield, HD2 1XT
2E0 **LOL** L Woolley, 4 Robert Street, Warrington, WA5 1TQ
2E0 **LON** Ian Lonsdale, 23 Hunts Field, Clayton-le-Woods, WLC Scout Council, Chorley, PR6 7TT
2I0 **LOR** Andrew Bell, 4 Mount Pleasant View, Newtownabbey, BT37 0ZY
2E0 **LOW** Mark Duchar, 4 Miller Gardens, Pelton Fell, Chester le Street, DH2 2NX
2E0 **LPD** Darren Lester, 171 Glenavon Road, Birmingham, B14 5BT
2I0 **LPG** Billy Bruce, 19 Ashvale Heights, Hillsborough, BT26 6DJ
2E0 **LPJ** Andrew Pomfrey-Jones, 46 Hampton Road, Erdington, Birmingham, B23 7JJ
2I0 **LPO** John McErlean, 24 McCorley Road, Toomebridge, Antrim, BT41 3NH
2E0 **LPR** M Price, 9 Herbarth Close, Liverpool, L9 1JZ
2E0 **LPW** Louis Walker, 55 Silverlands Road, St. Leonards-on-Sea, TN37 7DF
2E0 **LQR** Thomas Longmore, 3 Dairy Farm Cottages, Northlands Road, Gainsborough, DN21 5DN
2E0 **LQW** Rodney Edwards, 46 Lavers Oak, Martock, TA12 6HG
2E0 **LRA** L Ayre, 30 Tithe Lane, Calverton, Nottingham, NG14 6HY
2U0 **LRB** Leslie Bichard, Brise De Mer, Les Rouvets De Bas, Guernsey, Guernsey, GY7 9QF
2E0 **LRD** Karl Bainbridge, 29 Bluebell Grove, Calne, SN11 9QH
2E0 **LRG** Michael Parker, Ridgeways, Mill Common, Halesworth, IP19 8RQ
2E0 **LRJ** A Smith, 101 Chaucer Drive, Lincoln, LN2 4LT
2E0 **LRK** Lee Kelsey, 111-113 George Street, Mablethorpe, LN12 2BS
2I0 **LRN** Kevin Bell, 3 Alexandra Crescent, Larne, BT40 1NE
2M0 **LRO** J Barton, 14 Backmarch Crescent, Rosyth, Dunfermline, KY11 2RW
2E0 **LRP** R Hughes, 7 Willow Place, Darlington, DL1 5LX
2E0 **LSB** A Shepherd, 39 Minehead Road, Dudley, DY1 2NZ
2E0 **LSE** George Bystryakov, 20 Elmhurst Gardens, Leeds, LS17 8BG
2E0 **LSI** Nigel Highfield, 298 Mersea Road, Colchester, CO2 8QY
2E0 **LSL** Andrew Wood, 1 Abbey Close, Cranleigh, GU6 8TP
2E0 **LSR** Kevin Titmarsh, 3 Meadow Cottages, Banningham, Norwich, NR11 7ED
2E0 **LSS** Jack Lusted, 49 Asher Reeds, Langton Green, Tunbridge Wells, TN3 0AN
2E0 **LST** Lee Timmins, 83 Loxdale Sidings, Bilston, WV14 0TN
2E0 **LSV** R Derham, Netherwood, Copse Lane, Hook, RG29 1SX
2E0 **LSW** L Wellington, 5 Pasture Road, London, SE6 1JF
2E0 **LSX** Alan Dale, 37 Bussey Road, Norwich, NR6 6JF
2E0 **LTC** Christopher Anderson, 191 Waveney Road, Hull, HU8 9NA
2E0 **LTD** Steven Boyles, 4 Ervins Lock Road, Wigston, LE18 4NQ
2E0 **LTF** Lee Farrell, 1 Meadway, Ince, Wigan, WN2 2BZ
2E0 **LTH** Lee Thornton, 11 Polruan Road, Truro, TR1 1QR
2E0 **LTJ** Michael Brett, Lindon, Bunkers Hill, Wisbech, PE13 4SQ
2E0 **LTR** Nigel Rice, 30 Oveton Way, Bookham, Leatherhead, KT23 4ND
2E0 **LTS** Laurence Stant, EARS, University of Surrey Student's Union, Guildford, GU2 7XH
2E0 **LTT** Mark Lovatt, 3 Withington Close, Atherton, Manchester, M46 0EZ
2E0 **LTU** Arnoldas Jakstas, Flat 9, Kendal Court, 112 Godstone Road, Kenley, CR8 5GE
2E0 **LTV** Aidan Ascroft, Caretakers Cottage, School Road, Worksop, S81 9PX
2E0 **LTX** David Akerman, The Brick Barn, Coppice Farm, Ross-on-Wye, HR9 7QW
2E0 **LTZ** Reinhard Lenicker, 18 Wellington Grove, Bradford, BD2 3AL
2E0 **LUD** Peter Lloyd, 8 Maydor Avenue, Saltney Ferry, Chester, CH4 0AH
2M0 **LUG** Lewis Affleck, 1 Fank Brae, Mallaig, PH41 4RQ
2E0 **LUK** Celso Cavalcante Pinheiro Filho, Flat 6, 2a Trumans Road, London, N16 8BD
2E0 **LUL** Laura Pearson, 4 Brentwood Close, Thorpe Audlin, Pontefract, WF8 3ES
2E0 **LUY** Lucy Isaac Sneath, 21 Garrick Close, Lincoln, LN5 8TG
2E0 **LVR** Oliver De Peyer, Flat 5, Molasses House, London, SW11 3TN
2M0 **LWB** Leslie Bradley, Amon Sul, Kiltarlity, Beauly, IV4 7HT
2E0 **LWR** A Mundy, 27 Yorke Road, Croxley Green, Rickmansworth, WD3 3DW
2E0 **LWT** Andrew Teed, 21 Sheen Close, Salisbury, SP2 9PJ
2E0 **LXA** Adam Lowery, 21 Westlea Avenue, Riddlesden, Keighley, BD20 5EJ
2E0 **LXD** D Collins, 30 Upham Road, Swindon, SN3 1DN
2E0 **LXF** D Shuttleworth, 27 Union St., Egerton, Bolton, BL7 9SP
2E0 **LXR** Leslie Rowlands, 6 St. Michaels Avenue, Clevedon, BS21 6LL
2I0 **LXS** Andrew Logan, 15 Park Lane, Saintfield, Ballynahinch, BT24 7PR
2I0 **LXW** M Lewis, 7 Liester Park, Ballyrobert, Ballyclare, BT39 9RZ
2M0 **LXX** C Sturgeon, 7 Chalmers Avenue, Ayr, KA7 2NF
2E0 **LYD** Barry Vile, 24 Hudson Close, Dover, CT16 2SG
2E0 **LYN** A Lynn, 32 Ennerdale Road, Newcastle upon Tyne, NE6 4DH
2E0 **LYR** T Martin, 46 Hayes Crescent, Frodsham, WA6 7PG
2E0 **LZB** Adrian Highfield, 38 Brunswick Gardens, Garforth, Leeds, LS25 1HF
2E0 **LZE** Elizabeth Stone, Oakley, Main Road, Salisbury, SP4 6EE
2E0 **LZM** John Roberts, 51 Bradfield Road, Broxtowe, Nottingham, NG8 6GP
2E0 **LZT** David Young, 20 Summerhouse, Tickenham, Clevedon, BS21 6SN
2E0 **MAA** Michael Milne, Flambards, Manor Road, Dunmow, CM6 2JR
2E0 **MAB** M Bridgeland, 17 Oldfield Lane, Wisbech, PE13 2RJ
2E0 **MAC** C McCarthy, 7 Aneurin Avenue, Crumlin, Newport, NP11 5HN
2E0 **MAD** M Davidson, 26 Hurford Drive, Thatcham, RG19 4WA
2E0 **MAF** Matthew Beckett, 59 Broadacre, Caton, Lancaster, LA2 9NH
2E0 **MAH** M Holbrook, 9 Beechwood Mount, Hemsworth, Pontefract, WF9 4ES
2E0 **MAJ** M Jones, 20 Chelsea Drive, Sutton Coldfield, B74 4UG
2E0 **MAL** Malcolm Frame, 125 Park Road, Stanley, DH9 7QE
2E0 **MAN** R Lomax, 11 Sherbourne Drive, Heywood, OL10 4ST
2E0 **MAO** Michael Cowhey, 11 Aspen Way, Newport, NP20 6LB
2E0 **MAP** M Woolley, 84 Bowthorpe Road, Norwich, NR2 3TP
2E0 **MAQ** Michael Finn, 23 Spa Lane, Hinckley, LE10 1JA
2M0 **MAV** John Cattigan, Lunan Home Farm Cottage, Lunan Bay, Arbroath, DD11 5ST
2E0 **MAX** M McSherry, 5 Briery Croft, Stainburn, Workington, CA14 1XJ
2E0 **MAY** D Maydew, 128 Thorne Road, Willenhall, WV13 1AW
2E0 **MAZ** Mario Stevenson, 127 Walton Road, Chesterfield, S40 3BX
2E0 **MBA** H Anderton, 69 Sycamore Grove, Lancaster, LA1 5RS
2E0 **MBB** M Bartley, 19 South Avenue, Shadforth, Durham, DH6 1LB
2E0 **MBD** N Draper, 107 Arkwrights, Harlow, CM20 3LY
2M0 **MBE** C Hebenton, 43 East Avenue, Uddingston, Glasgow, G71 6LG
2E0 **MBG** John Girard, 49 Beech Crescent, Hythe, Southampton, SO45 3QF
2I0 **MBI** Brendan McDonald, 20 Aughan Park, Poyntzpass, Newry, BT35 6TW
2E0 **MBK** Mark Lewis, 73 Addenbrooke Street, Wednesbury, WS108HJ
2E0 **MBO** Michael Hughes, 58 Grange Lane North, Scunthorpe, DN16 1RW
2E0 **MBQ** Andrew Blamires, 2 Foldings Grove, Scholes, Cleckheaton, BD19 6DQ
2E0 **MBR** Paul Threakall, 83 Gregory Avenue, Birmingham, B29 5DG
2E0 **MBS** Martin Strange, 101 Southbroom Road, Devizes, SN10 1LY
2E0 **MBT** Michael Buchanan, 36 Church Lane, Manby, Louth, LN11 8HL
2E0 **MBV** Michael McHugh, 51 Rutland Street, Hyde, SK14 4SY
2E0 **MBW** Martin Rickaby, 57 Hylton Road, Jarrow, NE32 5DN
2E0 **MCA** M Addison, 319 Long Lane, London, N2 8JW
2E0 **MCB** Michael Luxton, 3 The Paddocks, Newgate Street, Brecon, LD3 8DJ
2E0 **MCC** H Shekhdar, Manora Lodge, Sea Bank Road, Skegness, PE24 5QU
2M0 **MCD** Michael McDonald, 106 Stamperland Gardens, Clarkston, Glasgow, G76 8NR
2E0 **MCG** J McGill, 186 Timbrell Avenue, Crewe, CW1 3LZ
2E0 **MCH** Mark Bailey, 34 Jephson Drive, Birmingham, B26 2HW
2E0 **MCJ** Marc Jeffrey, 17 Beweys Park, Lower Burraton, Saltash, PL12 4SW
2E0 **MCK** Michael Bridgehouse, 43 Age Croft, Oldham, OL8 2HG
2E0 **MCL** M Carney, 2 Lilac Meadows, Lawley Village, Telford, TF4 2NX
2E0 **MCM** Billy Cameron, 9 Finchale Road, Hebburn, NE31 2HR
2E0 **MCN** Michael Bridger, 11, Beecham Close, Newcastle upon Tyne, NE15 6LG
2E0 **MCQ** Matthew Ellis, Timbers, Fernhill Park, Woking, GU22 0DL
2E0 **MCS** Lewis Allcock, 26 Castleton Grove, Inkersall, Chesterfield, S43 3HU
2W0 **MCT** A McTaggart, Brick Hall, Hundleton, Pembroke, SA71 5QX
2E0 **MCW** M Carlin, 44 Sileby Road, Barrow upon Soar, Loughborough, LE12 8LR
2E0 **MDA** Michael Dailey, 58 Waincliffe Mount, Leeds, LS11 8AH
2E0 **MDC** Michael Crawley, 16 The Meadows, Herne Bay, CT6 7XF
2E0 **MDE** Christian Cundall, 43 High Street, Great Gonerby, Grantham, NG31 8JR
2E0 **MDG** Marc Griffiths, Mandalay, Bromfield Street, Wrexham, LL14 1NF
2E0 **MDH** Martin Walters, 65 Bannawell Street, Tavistock, PL19 0DP
2E0 **MDJ** Matthew Wilkinson, 1 Oxford Way, Cheltenham, GL51 3HH
2E0 **MDK** Andrew Currie, 20 Portal Road, Eastleigh, SO50 6AY
2E0 **MDN** Michael Bray, 26 South Park Close, Redruth, TR15 3AR
2E0 **MDR** Mark Bradley, 13 Elizabeth Avenue, Bilston, WV14 8EA
2E0 **MDT** Brian Hiley, 9 Pinfold Lane, Harby, Melton Mowbray, LE14 4BU
2E0 **MDU** Nicholas Alders, 14 Forest Rise, Crowborough, TN6 2ES
2E0 **MDV** James Bremner, 21 Embleton Drive, Blyth, NE24 4QJ
2E0 **MDZ** Matthew Smith, 31 Atlantic Crescent, Sheffield, S8 7FW
2E0 **MED** A Medhurst, 44 Battle Road, Hailsham, BN27 1DS
2E0 **MEG** Stewart Ridley, 123 Lanercost Drive, Newcastle upon Tyne, NE5 2DL
2E0 **MEH** Daniel Howarth, 32 Cotswold Drive, Rothwell, Leeds, LS26 0QZ
2E0 **MEI** Michael Bennetts, 2 Chywoone Terrace, Newlyn, Penzance, TR18 5NR
2E0 **MEK** Matthew Prentice, 2 Wickenden Road, Sevenoaks, TN13 3PJ
2E0 **MEL** M McGoldrick, 7 Walnut Drive, Tiverton, EX16 6HE
2E0 **MEO** Alex Bond, 85 Eaves Lane, Chorley, PR6 0PU
2E0 **MEQ** Michael Rea, 15 Wensleydale Close, Royton, Oldham, OL2 5TQ
2E0 **MES** M Skinner, 5 Sycamore Avenue, Upminster, RM14 2HR
2E0 **MET** Gregg Lewis, 33a Richmond Park Road, Kingston upon Thames, KT2 6AQ
2E0 **MEU** Agnes Sharif, 10 The Boundary, Seaford, BN25 1DG
2E0 **MEV** Michael Clarke, 54 Stafford Grove, Shenley Church End, Milton Keynes, MK5 6AZ
2E0 **MEX** Robert Hicks, 1 Shenstone Road, Maypole, Birmingham, B14 4TH
2E0 **MEY** Michael Sadler, 14 Woodlands Avenue, Water Orton, Birmingham, B46 1SA
2E0 **MEZ** Mike Marsh, 25 Southdown Road, Seaford, BN25 4PD
2E0 **MFA** Frank Alfrey, 16 Walls Road, Bembridge, PO35 5RA
2E0 **MFC** Lee Ross, 2 Bedford Street, Blackburn, BB2 4EU
2E0 **MFD** Terrence Heath, 16 Beacons Park, Brecon, LD3 9BR
2E0 **MFF** Mark Clarke, 4 Mill Lane, Brant Broughton, Lincoln, LN5 0RP
2E0 **MFG** Mark McGowan, 48 Alderley Road, Thelwall, Warrington, WA4 2JA
2E0 **MFI** Maxwell Berrisford, 5 Branwell Drive, Haworth, Keighley, BD22 8HG
2I0 **MFJ** B Traynor, 94 Markville, Portadown, Craigavon, BT63 5SZ
2M0 **MFK** Mark Cook, 7 Donald Gardens, Dundee, DD2 2RZ
2E0 **MFM** Ian Ridsdale, 15 Carlton Road, Hough-on-the-Hill, Grantham, NG32 2BG
2E0 **MFN** Peter Roberts, 17 Cannon Hill, Prenton, CH43 4XR
2E0 **MFS** Mark Feast, 10 Brackendale Road, Swanwick, Alfreton, DE55 1DJ
2E0 **MFT** Alan Hill, 1 Rochester Close, Mountsorrel, Loughborough, LE12 7UH
2E0 **MFV** Darren Baker, 39 Taylor Road, Wallington, SM6 0AZ
2E0 **MGA** Mark Greensmith, 14 Fountain Road, Draycott-in-the-Clay, Ashbourne, DE6 5HP
2E0 **MGC** Mark Chanter, 7 Woodford Crescent, Plymouth, PL7 4QY
2E0 **MGI** Matthew Isbell, 20 Woodland Crescent, Wolverhampton, WV3 8AS
2E0 **MGL** M Talbot, 26 Chevalier Grove, Crownhill, Milton Keynes, MK8 0EJ
2M0 **MGM** Matthew Geldart, 13b Greystone Place, Newtonhill, Stonehaven, AB39 3UL
2E0 **MGR** Malcolm Reeks, 33 Madresfield Village, Madresfield, Malvern, WR13 5AA
2E0 **MGT** James Gardiner, 31 Rowdy Road Tilehurst, Reading, RG306EH
2E0 **MGW** Matt Whitticombe, 18 Fairclough Street, Burtonwood, Warrington, WA5 4HJ
2E0 **MGX** Mark Deeley, Unit 8, West Cannock Way, Cannock Chase Enterprise Centre, Tachosoft UK Limited, Cannock, WS12 0QW
2M0 **MGY** J Boag, 182 St. Fillans Road, Dundee, DD3 9LH
2E0 **MHE** Stephen Snelson, 212 Dickson Road, Blackpool, FY1 2JS
2E0 **MHM** Michael Byard, 1 Fieldside, Long Wittenham, Abingdon, OX14 4QB
2M0 **MHN** Neil Morris, 23 sedgebank, Sedgebank, Livingston, EH54 6HE
2E0 **MHR** Michael Haynes, 25 Barnards Hill Lane, Seaton, EX12 2EQ
2E0 **MIB** Vaughan Ball, 24 Carr Lane, Warsop, Mansfield, NG20 0BN
2E0 **MID** Peter Staite, Chestnut Farm, Eastville, Boston, PE22 8LX
2M0 **MIF** Derek Mifsud, 25 Priory Road, Linlithgow, EH49 6BP
2E0 **MIG** Paul Cattermole, Blaxhall Hall Crossing, Little Glemham, Woodbridge, IP13 0BP
2E0 **MIH** Michael Humphries, 5 Coppice Mead, Stotfold, Hitchin, SG5 4JX
2E0 **MIJ** Michael Jones, 29 Highbridge Road, Burnham-on-Sea, TA8 1LL
2E0 **MIL** Ian Millman, 70 Springdale Avenue, Broadstone, BH18 9EX
2E0 **MIS** Dawn Smout, Sunrays, Warbage Lane, Bromsgrove, B61 9BH
2E0 **MIT** Russell Hayward, Brook House, Brook Street, Mitcheldean, GL17 0AU
2E0 **MIU** John Marsh, 14 Eyam Road, Hazel Grove, Stockport, SK7 6HP
2E0 **MIV** Brian Davies, 60 Queensway, Blackburn, BB2 4QT
2E0 **MIX** Derek Edge, 18 Sandringham Avenue, Whitehaven, CA28 6XL
2E0 **MIY** Paul Billingham, 393 Landseer Road, Ipswich, IP3 9LT
2E0 **MIZ** A Bartlett, 62 Kewstoke Road, Bath, BA2 5PU
2W0 **MJA** Aeronwen Sneddon, 3 Marigold Close, Gurnos, Merthyr Tydfil, CF47 9DA
2E0 **MJC** M Churcher, 71 Twyn Road, Abercarn, Newport, NP11 5JY
2E0 **MJD** Martin Juhe, 75 Pondcroft Road, Knebworth, SG3 6DE
2E0 **MJE** Melanie Parker, 1 Ham Road, Wanborough, Swindon, SN4 0DF
2E0 **MJG** Michael Gillingham, Frecheville Rectory, Brackenfield Grove, Sheffield, S12 4XS
2E0 **MJH** Michael Holroyd, 9 Coniston Green, Aylesbury, HP20 2AJ
2E0 **MJJ** John Jones, 19 Southbank Street, Leek, ST13 5LS
2E0 **MJL** M Lee, Up To Date House, Shore Road, Boston, PE22 0NA
2E0 **MJM** Matt Middleditch, 8 Royal Close, Yeovil, BA21 4NX
2E0 **MJO** Simone Marcomini, 18 Chadwick Place, Long Ditton, Surbiton, KT6 5RE
2E0 **MJP** M Marsh, 19 First Avenue, South Kirkby, Pontefract, WF9 3EP
2E0 **MJS** Stuart McMurtrie, 5 Hill Road, Carshalton, SM5 3RA
2E0 **MJT** Matthew Troth, 21 Willow Road, Bromsgrove, B61 8PN
2E0 **MJX** Micheal Cresswell, 44 The Lea, Birmingham, B33 8JP
2M0 **MJY** Martin Yarrow, Lomond Villa, Downies Village, Aberdeen, AB12 4QX
2E0 **MJZ** Dave Cook, 15 Kendricks Fold, Rainhill, Prescot, L35 9LX
2E0 **MKB** M Ballard, 41 Middlefield Avenue, Halesowen, B62 9QJ
2E0 **MKC** A Goodenough, 336 Herne Road, Ramsey St. Marys, Huntingdon, PE26 2TD
2E0 **MKE** Michael Gregory, 65 Nursery Crescent, North Anston, Sheffield, S25 4BR
2E0 **MKF** Malcolm Amphlett, Highbanks, Charnes Road, Market Drayton, TF9 4LQ
2E0 **MKG** Mark Gray, 15 The Circle, Cwmbran, NP44 7JP
2E0 **MKH** Michael Heaton-Bentley, 65 Brookfield Road, Thornton-Cleveleys, FY5 4DR
2E0 **MKI** T Palmer, 29 Field End, Maresfield, Uckfield, TN22 2DJ
2E0 **MKJ** Michael Johnson, 7 Norfolk Wing, Tortington Manor, Arundel, BN18 0FD
2E0 **MKK** M Kilkenny, 23 Hazelhurst Road, Stalybridge, SK15 1HD
2E0 **MKT** Timothy Walker, 11 Banburies Close, Bletchley, Milton Keynes, MK3 6JP
2E0 **MKV** Michael Vardy, 60 Hucklow Avenue, North Wingfield, Chesterfield, S42 5PU
2E0 **MKW** M Wiggins, 2 Cherry Tree Close, Halstead, CO9 2UA
2E0 **MKX** Martin Keyte, 3 Lower High St., Mow Cop, Stoke-on-Trent, ST7 3PB
2E0 **MLA** A Highfield, 29 Blewitt Street, Brierley Hill, DY5 4AW
2E0 **MLD** Matthew Dixon, 37 Carlton Close, Parkgate, Neston, CH64 6RB
2E0 **MLE** Derek Pilkington, 197 Saltings Road, Snodland, ME6 5HP
2E0 **MLF** Michael Raynor, 21 Teversal Avenue, Pleasley, Mansfield, NG19 7QQ
2E0 **MLG** simon Gordon, 8 Maesteg Cymau, Wrexham, LL11 5EP
2E0 **MLH** Merlin Howse, Woodland, Moretonhampstead, Newton Abbot, TQ13 8SD
2E0 **MLJ** Michael Jones, 11 Lower Glen Park, Pensilva, Liskeard, PL14 5PP
2E0 **MLK** Marie Kipling, 12 Jolly Brows, Bolton, BL2 4LZ
2E0 **MLL** Andrew McCall, 40 Nethway Avenue, Blackpool, FY3 8JU
2E0 **MLS** Michael Heywood, 16 Edinburgh Drive, Hindley Green, Wigan, WN2 4HL
2E0 **MLV** M Grantham, 7 Goodwin Close, Sandiacre, Nottingham, NG10 5FF
2E0 **MLX** A Boyes, 12 Leyburn Grove, Stockton-on-Tees, TS18 5NH
2I0 **MMA** John Morrison, 70 Ravenswood, Banbridge, BT32 3RD
2M0 **MMB** M Cleland, 85 Carfin Street, Motherwell, ML1 4JL
2W0 **MMD** D Holloway, 14 Woodbrook Terrace, Burry Port, SA16 0NF
2M0 **MMF** Robert Tripney, 7 Sunnyside St., Camelon, Falkirk, FK1 4BJ
2M0 **MMG** M Gourlay, 14 Holmes Holdings, Broxburn, EH52 5NS
2E0 **MMH** Michael Hall, 67 Darlinghurst Grove, Leigh-on-Sea, SS9 3LF
2E0 **MMJ** Manmeet Majhail, 3 Poynders Hill, Hemel Hempstead, HP2 4PQ
2M0 **MMM** M Greig, 7 St. Ronans Road, Forres, IV36 1BQ
2M0 **MMO** M Overthrow, 63 Primrose Avenue, Larkhall, ML9 1JX
2E0 **MMP** M Porter, 102 Vulcan Close, Padgate, Warrington, WA2 0HN
2I0 **MMT** Michael Torley, 4 Yew Tree Park, Newry, BT34 2QP
2E0 **MMX** Ryan Hewson, Tad-Cu, Almond Avenue, Llandrindod Wells, LD1 6DH
2E0 **MMZ** Scott Phillips, 25 Sunnydene Avenue, London, E4 9RE
2E0 **MNC** M Craner, 46 Meadowhill Crescent, Redditch, B98 8HT
2E0 **MND** Amanda Harrop, 35 Langdale Crescent, Dalton-in-Furness, LA15 8NR
2E0 **MNG** Neal Giuliano, 13 Walton Drive, Derby, DE23 1GN
2E0 **MNJ** Martyn Kenny, 27 Brangwyn Crescent, Newport, NP19 7QY
2E0 **MNP** Matt Pomfret, 5 Malvern Crescent, Ince, Wigan, WN3 4QA
2E0 **MNU** Peter Richardson, 14 Portland Street, Worksop, S80 1RZ
2E0 **MNY** Raymond Parker, 53 Tunstall Road, Canterbury, CT2 7BX
2E0 **MOB** Clive Larner, 98 Allandale, Hemel Hempstead, HP2 5AT
2M0 **MOF** Thomas Moffat, 11 Mansfield Road, Prestwick, KA9 2DL
2M0 **MOK** William Fulton, 15 Staffa Avenue, Port Glasgow, PA14 6DT
2E0 **MOL** Richard Moles, 14 Dorsett Road, Stourport-on-Severn, DY13 8EL

2E0 MOR [illegible], 61 Oaklands, Bunbrooke, Northampton, NN7 3QU
2E0 MOY Richard Moys, 1[illegible] [illegible] Avenue, Fareham, PO16 7DF
2E0 MOZ Maurice Meadowcroft, 8 Lamlash Road, Blackburn, BB1 2AU
2E0 MPA Matthew Ashworth, 123 Forest Road, Liss, GU33 7BP
2E0 MPB Richard Johnson, 24 Fairfields, Upper Denby, Huddersfield, HD8 8UB
2E0 MPC Michael Carter, 113 Old Road, Tintwistle, Glossop, SK13 1JZ
2E0 MPE Ronald Eaton, 31 Pinfold Lane Ruskington, Sleaford, NG34 9EU
2E0 MPG Paul McGrath, 24 Broadoak Drive, Lanchester, Durham, DH7 0QA
2E0 MPJ James Neal, 75 Park Lane, Castle Donington, Derby, DE74 2JG
2E0 MPN M Nolan, 5 Ryeleaze, Potterne, Devizes, SN10 5NJ
2E0 MPU [illegible], 41 Exeter Street, Blackburn, BB2 4AU
2E0 MPR [illegible], 19 St. Bedes, 14 Conduit Road, Bedford, MK40 1ED
2E0 MPX Paul Matthew, 24 Jubilee Close, Pamber Heath, [illegible] 0HP
2E0 MQA David Rogers, 9 Prospect Place, Stafford, ST17 4HZ
2E0 MQC Martin Le Moine, 115 Rothesay Road, Blackburn, BB1 2ER
2E0 MQT F Walker, 54 Burnage Lane, Burnage, Manchester, M19 2NL
2E0 MRA Andrew Armson, 2 Windmill Gardens, St. Helens, WA9 1EN
2E0 MRD Derek Millard, 112 Avenue Road, Sandown, PO36 8DZ
2E0 MRG G Whittle, 22 St. Oswalds Close, Finningley, Doncaster, DN9 3ED
2E0 MRJ M Jarrett, 17 Greenhill Gardens, Minster, Ramsgate, CT12 4EP
2E0 MRM M Tetley, 27 Cunningham Hill Road, St. Albans, AL1 5BX
2M0 MRO Michael Reid, 2 Pinkie Gardens, Newmachar, Aberdeen, AB21 0QF
2E0 MRP Mark Peters, 25 Windsor Court, Falmouth, TR11 3DZ
2E0 MRQ Peter Gibbs, 9 Walton Heath, Darlington, DL1 3HZ
2I0 MRY M Ruddy, C/O 6 Iveagh Park, Greysteel, Londonderry, BT47 3DD
2E0 MSA Michael Statham, Broad Oak Bungalow, Manston, Sturminster Newton, DT10 1EZ
2M0 MSB S Brown, 21 Whiteford Avenue, Dumbarton, G82 3JU
2E0 MSE Mark Edmonds, 60 Shenstone Road, Maypole, Birmingham, B14 4TJ
2E0 MSI Mark Sims, 5 Sandy Leaze, Bradford-on-Avon, BA15 1LX
2W0 MSL Nicholas Berrall, 41 Nantgarw Road, Caerphilly, CF83 3FB
2E0 MSM Malamkunnu Mohammed Shafi, 142 Whitethorn Street, London, E3 4DB
2I0 MSO Philip Hosey, 13 Glenelly Gardens, Omagh, BT79 7XG
2E0 MSW Mark Wyatt, 18 Eastcote Lane, Hampton-in-Arden, Solihull, B92 0AS
2E0 MTB T Beckett, 95 Warrens Hall Road, Dudley, DY2 8DH
2E0 MTC Catherine Mathewson, 33 Thornton Road, Bootle, L20 5AN
2E0 MTD Mark Davies, 11 High Street, Malltraeth, Bodorgan, LL62 5AS
2E0 MTE Eifion Thomas, 13 Cwrt Dolafon, Dolafon Road, Newtown, SY16 2HU
2E0 MTH Matthew Knowles, 11 Thorneycroft Avenue, Birkenhead, CH41 8HJ
2E0 MTL Morris Leach, 64 Grove Street, Wantage, OX12 7BG
2E0 MTM Tomasz Mloduchowski, Flat 4, Gwynne House, London, E1 2AG
2E0 MTN Martyn Newell, 55 Station Road, Brimington, Chesterfield, S43 1JU
2M0 MTO Raymond Foulds, 83 Croftfoot Road, Glasgow, G44 5JU
2E0 MTR Michael Reilly, 26 Roman Way, Folkestone, CT19 4JT
2E0 MTT Matthew Nassau, 1A Burford Road, Bromley, BR1 2EY
2E0 MTX Neil Challis, 48 Brunsfield Close, Wirral, CH46 6HE
2E0 MTY Darren Raine, 91 Lulworth Avenue, Jarrow, NE32 3SB
2E0 MUA James Anderson, 121 Barton Road, Stretford, Manchester, M32 9AF
2E0 MUD S Sparks, 25 Wilwick Lane, Macclesfield, SK11 8RS
2E0 MUN A Munford, 16 Broadhurst Way, Brierfield, Nelson, BB9 5HG
2M0 MUR Gordon Murray, The Barn House, Springfield Farm, Carluke, ML8 4QZ
2E0 MUS Adrian Sutton, 3 Grotes Buildings, London, SE3 0QG
2E0 MUT John Merritt, 41 Great Grove, Bushey, WD23 3BQ
2E0 MUW J Blaylock, 23 Sunnyway, Blakelaw, Newcastle upon Tyne, NE5 3QB
2E0 MUZ Murray Colpman, Kirk House, Goodworth Clatford, Andover, SP11 7RN
2E0 MVD Mark Denham, 2 Shorts Corner, Frithville, Boston, PE22 7EA
2E0 MVH S Smith, 121 Springfield Road, St. Helens, WA10 3RR
2I0 MVP Alexander Simpson, 10 Woodview Park, Tandragee, Craigavon, BT62 2DD
2E0 MVT Darren Harris, 27 Ashley Road, Poole, BH14 9BS
2E0 MWA M Austin, 14 The Green, Brown Edge, Stoke-on-Trent, ST6 8RN
2E0 MWB Mark Bryant, 284 Brantingham Road, Chorlton cum Hardy, Manchester, M21 0QU
2E0 MWC M Caffrey, 40 Kingston Way, Seaford, BN25 4NG
2E0 MWH Michael Hetherington, 18 Wesley Street, Low Fell, Gateshead, NE9 5YN
2E0 MWJ Michael Willis, 51 Barnsdale Close, Loughborough, LE11 5AN
2E0 MWN Michael Singer, 1 Bentley Road, Slough, SL1 5BB
2E0 MWT Richard Woolley, 10 Hazelmoor Fold, Blackley, Elland, HX5 0DR
2E0 MWW M Wheal, 7 Ryecroft Drive, Withernsea, HU19 2LP
2E0 MXC Mark Craven, 78 Connaught Road, Brookwood, Woking, GU24 0HF
2E0 MXM Michael Meehan, 14 Grosvenor Road, Walton, Liverpool, L4 5RB
2E0 MXP Michael Palmer, New Haven, Stoneraise, Carlisle, CA5 7AX
2E0 MXR Andrew Wilson, 28 Langham Road, Bristol, BS4 2LJ
2W0 MXI C Fisher, 22 Troed Y Bryn, Upper Tumble, Llanelli, SA14 6BP
2E0 MXW D Platt, 50 Poplars Road, Stalybridge, SK15 3EN
2E0 MYD H Ibbitson, Tor View, Whitstone, Holsworthy, EX22 6TB
2E0 MYE D Sykes, 2 The Street, Claxton, Norwich, NR14 7AS
2E0 MYH D Morgan, 87 Pool Hayes Lane, Willenhall, WV12 4PX
2E0 MYK Michael Knowles, 86 West Shore Road, Walney, Barrow-in-Furness, LA14 3UD
2E0 MYT Michael Corrigan, 33 Westbourne Road, Knott End-on-Sea, Poulton-le-Fylde, FY6 0BS
2E0 MYX Stephen Elliott, 79 Somerton Road, Bolton, BL2 6LN
2E0 MZB B Gutteridge, 54 Malthouse Road, Southgate, Crawley, RH10 6BG
2E0 MZL Graham Johnson, 22 Beechwood Close, Blythe Bridge, Stoke-on-Trent, ST11 9RH
2E0 MZM Mark O'Loughlin, 2 Hen Ysgol, Forge Road, Crickhowell, NP8 1LU
2E0 MZU Christopher Smith, 105 Netherton Road, Worksop, S80 2SA
2E0 MZZ Emma Reeve, 12 Sime Street, Worksop, S80 1TD
2E0 NAD Oliver Bross, 8 Queens Drive, Buckley, CH7 2LJ
2E0 NAF Nigel Foster, 18 Austen Ave, Sawley, Nottingham, NG103GG
2E0 NAG T Bown, 16 Sandringham Court, Queen Elizabeth Road, Nuneaton, CV10 9AR
2E0 NAI Paul Turner, 43 Nelson Way, Mundesley, Norwich, NR11 8JD
2E0 NAJ Nigel Auckland, Glenfield, The Avenue, Southampton, SO32 1BP
2E0 NAM Neil Carey, 16 Cannamanning Road, Penwithick, St. Austell, PL26 8UX
2M0 NAN Sohan Ram, 28 Craigievar Gardens, Kirkcaldy, KY2 5SD
2E0 NAP Nicholas Deery, 25 Ribblesdale Place, Preston, PR1 3NA
2E0 NAQ Nick Barnard, 10 Whites Lane Kessingland, Lowestoft, NR33 7TF
2E0 NAR Richard Nagy, 40 Oakhampton Road, London, NW7 1NH
2E0 NAS Neil Inglis, 74 Runswick Avenue, Whitby, YO21 3UE
2I0 NAT C Mooney, 12 Curragh Walk, Londonderry, BT48 8HX
2M0 NAX Alfred Anderson, 18 [illegible] Street, Wishaw, ML2 8RA
2E0 NAY M Williams, Jurys, Fore Street, South Molton, EX36 3HL
2E0 NAZ Nathan Azizoff, 7 Spencer Close, London, N3 3TX
2E0 NBC David Waters, Flat 1, Pastors Hill House, Pastors Hill, Lydney, GL15 6NA
2E0 NBE Neil Irvine, 100 Cavendish Road, Sunbury-on-Thames, TW16 7PL
2E0 NBG Nigel Newman, 1 Ockendon Road, North Ockendon, Upminster, RM14 3PT
2E0 NBM N Modi, 90 Hereford Road, Basingstoke, RG23 8QL
2E0 NBR S Warren, 1 Morley Close, Stapenhill, Burton-on-Trent, DE15 9EW
2E0 NBZ N [illegible], 81 [illegible] Croft, Dunstable, LU6 3JZ
2W0 NOA N [illegible], 22 [illegible] Court, Caldy Close, Barry, CF62 [illegible]
2E0 NCB David Lawson, 30 Meadowcroft, St. Helens, WA9 0XJ
2E0 NCC N croft, 22 King Edward Crescent, Leeds, LS18 4BE
2E0 NCE Dorothy Stanley, 58 Wells Gardens, Basildon, SS14 3QS
2E0 NCF Alan Blake, Northfield Cottage, Droxford Road, Fareham, PO17 5AZ
2E0 NCG Mark Dumpleton, 23 Watermans Yard, Norwich, NR2 4SD
2E0 NCI Bernard Dowley, 120 Capel Street, Capel-le-Ferne, Folkestone, CT18 7HB
2E0 NCK Nick Taylor, 212 Plantation Hill, Worksop, S81 0HD
2M0 NCM N Cunningham, 11 Glendoune Street, Girvan, KA26 0AA
2E0 NCN Kieran Clarke, 15 Grig Place, Alsager, Stoke-on-Trent, ST7 2SU
2E0 NCO Kevin Tonge, 98 Trescott Road, Northfield, Birmingham, B31 5QB
2E0 NCS N Sunley, 1 East Lea View, Cayton, Scarborough, YO11 3TN
2E0 NCY Jeanpierre Mooneapillay, 354 Upper Elmers End Road, Beckenham, BR3 3HG
2E0 NDG Nigel Graven, 33 Sheldrake Road, Broadheath, Altrincham, WA14 5LJ
2E0 NDH Neil Hewitt, 36 Kenilworth Road, Doncaster, DN4 0UD
2I0 NDJ Nigel Jameson, 15a Ednagee Road, Castlederg, BT81 7QF
2E0 NDP Neil Plunkett, 11 Stoneleigh Gardens, Grappenhall, Warrington, WA4 3LE
2E0 NDR Nigel Nash, Roann, Bedmond Road, Hemel Hempstead, HP3 8SH
2E0 NDT Rees Thatcher, 83 Westfield Drive, North Greetwell, Lincoln, LN2 4RE
2E0 NDW Peter Mackay, 30 Main Road, Austrey, Atherstone, CV9 3EH
2E0 NDY Anthony Williams, 12 St. Wilfrids Crescent, Brayton, Selby, YO8 9EU
2E0 NDZ Andrew Humphriss, 44 Bishops Close, Stratford-upon-Avon, CV37 9ED
2E0 NEC Carey Humphries, 44 Linksway, Folkestone, CT19 5LS
2E0 NEI Neil Yorke, 21 Braemar Way, Nuneaton, CV10 7LF
2E0 NEL Christopher Nelson, 14 Windy Harbour Road, Southport, PR8 3DU
2E0 NEO Neil Thomson, 2 Doon Terrace, Dumfries, DG2 9EE
2E0 NER Michael Straughan, 71 Silcoates Lane, Wrenthorpe, Wakefield, WF2 0PA
2E0 NEV Neil Griffin, 54 Edinburgh Road, Newmarket, CB8 0QD
2E0 NEY P Millard, Weavern House, Hartham Lane, Chippenham, SN14 7EA
2E0 NFB Neill Bisiker, 31 Lansdowne Avenue, Waterlooville, PO7 5BL
2E0 NFC Alan Cockburn, 52 Devon Road, Hebburn, NE31 2DW
2E0 NFK Richard Gowler, Merlins Lodge, Church Road, Norwich, NR12 0JP
2E0 NFS Nicholas Stephens, 3 Spinney House, College Road, Windermere, LA23 1PX
2E0 NGB Norman Bland, 3 Kennet Road, Newbury, RG14 5JA
2E0 NGC Lee Akred, 25 Kitchener Street, Walney, Barrow-in-Furness, LA14 3QW
2E0 NGF Stephen Dale, 76 Houldsworth Drive, Stoke-on-Trent, ST6 6TJ
2E0 NGG Neil G Clare, 123 Cunningham Road, Tamerton Foliot, Plymouth, PL5 4PU
2I0 NGK James Allen, 3 Malwood Close, Belfast, BT9 6QX
2I0 NGM Andrew McKay, 17 Thorn Hill Road, Banbridge, BT32 3TL
2M0 NGO John Nattress, 44 Broadlands, Carnoustie, DD7 6JY
2E0 NGR Nik Grey, 1 Norwich Road, Little Plumstead, Norwich, NR13 5JQ
2E0 NHB Nigel Barker, 17 Pippin Walk, Hardwick, Cambridge, CB23 7QD
2E0 NHM Nigel Meakin, 60 Canberra Way, Warton, Preston, PR4 1XY
2E0 NHR Steven Sawyer, 19 malvern close, Ashington, NE630TD
2E0 NHS J Kelly, 12 Park Road, Milford on Sea, Lymington, SO41 0QU
2M0 NIA Niamh Hague, 11 Auchriny Circle, Bucksburn, Aberdeen, AB21 9JJ
2E0 NIB Nigel Bennett, 44 Glenmoor Road, Buxton, SK17 7DD
2I0 NIE Chriss Morton, 29 Lackaboy View, Enniskillen, BT74 4DY
2E0 NIF G Calder, 41 Wood End Way, Chandler's Ford, Eastleigh, SO53 4LN
2M0 NIT Diamantino De Freitas, 14 York Street, Clydebank, G81 2PH
2E0 NIX Nicholas Robertson, Craigenveoch Farm, Glenluce, Newton Stewart, DG8 0LD
2E0 NJC Nicholas Long, 25 Blendworth Lane, Southampton, SO18 5GY
2E0 NJE Neil Gonzales, 46 Whitton View, Rothbury, Morpeth, NE65 7QN
2E0 NJJ D Wharlley, 15 Crampton Court, Grosvenor Road, Broadstairs, CT10 2XU
2E0 NJK Nicky Kendall, 19 Clowance Lane, Mount Wise, Plymouth, PL1 4HU
2E0 NJO Nathan Jones, 5 Montgomery Crescent, Quarry Bank, Brierley Hill, DY5 2HB
2E0 NJS Nigel Sheridan, Cemetery Lodge, Lochmaben, Lockerbie, DG11 1RL
2E0 NKC D Ansell, 30 Curzon Avenue, Horsham, RH12 2LB
2E0 NKF S Harrison, 13 Gillsike House, Thornbury Road, Wakefield, WF2 8BN
2E0 NKI Nicola Crabb, 1 Council Houses, Hall Lane, Norwich, NR12 7BB
2E0 NKM Nigel Morse, 33 Tower Close, Bassingbourn, Royston, SG8 5JX
2E0 NKP Nicholas Palin, 21 Ford Lane, Crewe, CW1 3EQ
2E0 NKR Mark Moss, 6 Orchard Close, Watford, WD17 3DU
2E0 NKT Nicholas Kent, Flat 1, Manor House, Redruth, TR15 1AX
2M0 NLA Alan Cunningham, 36 Station Brae Gardens, Dreghorn, Irvine, KA11 4FB
2E0 NLD Neil Brown, 9 Devonshire Avenue, Wigston, LE18 4LP
2E0 NLK N Lake, 64 Womersley Road, Norwich, NR1 4QD
2E0 NLM Paul Maybin, 16 Appleby Road, London, E16 1LQ
2E0 NLP John Watts, 70 Castleway North, Leasowe, Wirral, CH46 1RW
2E0 NLW Stephen Jones, 30 Crown Fields Close, Newton-le-Willows, WA12 0JW
2E0 NLY Brett Plackett, 36 Dartmouth Crescent, Bredbury, Stockport, SK5 8BG
2E0 NMC N Mcintyre, 27 Chapel Close St Ann's Chapel, Gunnislake, PL18 9JB
2M0 NMD Thomas Ormiston, 22 St. Ronans Road, Innerleithen, EH44 6LZ
2E0 NMK Simon Bateson, 2 Green Crescent, Coxhoe, Durham, DH6 4BE
2E0 NNB Alan Hopper, 7 Holmesdale Villas, Swallow Lane, Dorking, RH5 4EY
2E0 NNE David Hanwell, 28 Chipperfield Road, Norwich, NR7 9RR
2E0 NNH Gary Bansil, 32 Nethermead Court, Northampton, NN3 8NE
2E0 NNQ J Blamey, 46 First Avenue, Canvey Island, SS8 9LP
2E0 NNX Daniel Austin, 1002 Marsden House, Marsden Road, Bolton, BL1 2JX
2E0 NOC Colin Arbon, 8 Orchard Avenue, Ashford, TW15 1JB
2E0 NOD N Lightfoot, 4 Prospect Close, Hatfield Peverel, Chelmsford, CM3 2JE
2E0 NOK Colin Smith, 8 Pitts Street, Bradford, BD4 9JJ
2E0 NON Geoffrey Fielding, Chapel Court, Chapel Lane, Malvern, WR13 5HX
2M0 NOP Norman Price, 5 Hallree Cottage, Heriot, EH38 5YD
2D0 NOT D Ali, 25 Sunnydale Avenue, Port Erin, Isle of Man, IM9 6EU
2E0 NOW M Clarke, 14 Tower Court, Haverhill, CB9 [illegible]
2E0 NOZ J Norrington, 32 Fulfen Way, Saffron Walden, CB11 4DW
2E0 NPN Nigel Swift, 59 Milton Avenue, Malton, YO17 7LB
2E0 NPP Peter Hayward, 14 Micklewright Avenue, Crewe, CW1 4DF
2E0 NPS Christopher Kenyon, The Farmhouse, 10, Watermill Lane, Spilsby, PE23 5AG
2U0 NPT N Thomas, 6 Tunstall Terrace, Gibauderie, St. Peter Port, Guernsey, GY1 1XJ
2E0 NPX Neil Paxman, 128 Coggeshall Road, Braintree, CM7 9ES
2W0 NOE S Walmsley, 20 [illegible] Court, Machen, Caerphilly, CF83 8TT
2E0 NQU E Wagner, 7 [illegible] Road, London, NW2 3SN
2W0 NRA M Chell, Mesen Fach, Llanybydder, SA40 9TY
2E0 NRB Matthew Beckett, 4 Sandcross Close, Orrell, Wigan, WN5 7AH
2E0 NRH Nicholas Hickson, 27 Cressing Road, Witham, CM8 2NP
2E0 NRJ Nick Johnson, Delair, Western Road, Crediton, EX17 3NB
2E0 NRW Nicholas Waters, 9 Shirley Road, Droitwich, WR9 8NR
2M0 NSA Ana Custura, 85 Lord Hay's Grove Old Aberdeen, Aberdeen, AB24 1WT
2E0 NSC Neil Smith, 40 Fairdale Drive, Newthorpe, Nottingham, NG16 2FG
2E0 NSG Neville Gregson, 4 Pollard House, Maldwyn Avenue, Bolton, BL3 3RB
2E0 NSQ Stephen Beedham, 27 Malpas Close Bransholme, Hull, HU7 4HH
2E0 NSR Edward Parrish, 89 Delamere Drive, Macclesfield, SK10 2PS
2E0 NSS M Price, 25 School Crescent, Lydney, GL15 5TA
2E0 NSW N White, 2 Appleby Cottages, Whithorn, Newton Stewart, DG8 8DQ
2D0 NSY Stephen O'Riordan, 46 Grange Road, London, HA20LW
2E0 NTA Alan Briscoe, 69 Sharpe Street, Tamworth, B77 3HZ
2E0 NTC Gerard Bull, 9 Kilburn Place, Dudley, DY2 8HP
2I0 NTH Charles McCormick, Flat 4, Legacorry House, Main Street, Armagh, BT61 9RW
2E0 NTJ Nicholas Jones, 1 Olaf Close, Andover, SP10 5NJ
2E0 NTW C Northwood, Apartment 50, 2 Munday Street, Manchester, M4 7BB
2M0 NTY Colin McClymont, 115 Glenavon Road, Flat13/1, Glasgow, G20 0HT
2E0 NUC W Groves, 3 Tetbury Close, Newport, NP20 5HX
2E0 NUG Matthew Wells, 23 Eastmead, Bognor Regis, PO21 4QT
2E0 NUL Jon Unwin, 59 Hempstalls Lane, Newcastle, ST5 0SN
2E0 NUQ Michael Dickenson, 6 The Pavilions, Blandford Forum, DT11 7GF
2E0 NVB N Betts, 12 Sandy Lane, Worksop, S80 1SW
2E0 NVK Laurence Bolton, 59 Picquets Way, Banstead, SM7 1AB
2E0 NVP M Weir, 153 Tyndale Crescent, Birmingham, B43 7HX
2E0 NVS Phillip Rees, 3 Nash Green, Hemel Hempstead, HP3 8AA
2E0 NWA Nicholas Wong, Montefiore House, Wessex Lane, Southampton, SO18 2NU
2E0 NWB John Benbow, 44 Copthorne Park, Shrewsbury, SY3 8TJ
2E0 NWE Owen Price, 1 King Street, Odiham, Hook, RG29 1NN
2I0 NWO Dorothy Adams, 65 Rose Park, Limavady, BT49 0BF
2E0 NWR William Westlake, 2 Chegwin Court, Newquay, TR7 2DE
2E0 NWT N Topping, 7 Beckstone Close, Harrington, Workington, CA14 5QR
2E0 NWY Simon Newhouse, 28 Hillmorton Lane, Lilbourne, Rugby, CV23 0SS
2E0 NYC Stuart Vzor, 40 Henlow Road, Birmingham, B14 5DS
2E0 NYE N Whittaker, 1 Edendale, Hull, HU7 4BX
2E0 NYF Dave Lamble, 4 Laburnum Road, Chorley, PR6 7BG
2E0 NYM Clive Matthews, The Lawns, Ridsale Street, Darlington, DL1 4EG
2E0 NYX L Naylor, 23 Lilla Close, Whitby, YO21 3LY
2E0 NZD Matthew Phillips, 44 Hilderic Crescent, Dudley, DY1 2ET
2E0 OAA Christopher King, 8a Barton Road, Bedford, MK42 0NA
2M0 OAB Sam Milne, 5 Moriston Court, Grangemouth, FK3 0JJ
2W0 OAG A Graham, 2 Heol Undeb, Beddau, Pontypridd, CF38 2LB
2E0 OAH Kevin Johnson, 32 Redmire Close, Bransholme, Hull, HU7 5AQ
2E0 OAI David Saunders, 17 Sandy Lane Prestwich, Manchester, M25 9RU
2E0 OAK John Burnett, 218 High Street, Clapham, Bedford, MK41 6BS
2E0 OAO Abdullah Al-Shakarchi, 17 Fairfax Place, London, NW6 4EJ
2E0 OAP Michael Deary, 7 Newbold Avenue, Sunderland, SR5 1LG
2I0 OAZ Norman Armstrong, 1 Diamond Cottages, Ardmore Road, Crumlin, BT29 4QU
2E0 OBB Owen Boar, 19 Blyford Road, Lowestoft, NR32 4PZ
2E0 OBC Christopher Bridges, 53 St. Margarets London Road, Guildford, GU1 1TL
2E0 OBI Paul Sherratt, 39 vimy road, Leighton Buzzard, LU7 1FQ
2E0 OBL Mark Orbell, 21 Reedings Road, Barrowby, Grantham, NG32 1AU
2E0 OBS Brian Heath, 108 Cow Lane, Bramcote, Nottingham, NG9 3BB
2E0 OBZ D Thomas, 51 Sandringham Avenue, Vicars Cross, Chester, CH3 5JF
2E0 OCB Oliver Carpenter-Beale, 6 Betherinden Cottages, Bodiam Road, Cranbrook, TN18 5LW
2W0 OCF Nigel Smith, 23 Pennyroyal Close, St. Mellons, Cardiff, CF3 0NB
2E0 OCG O Giles, Holly Cottage, Main Road, Crewe, CW4 0LL
2E0 OCH Colin Howard, 1 Beale Road, Cheltenham, GL51 0JN
2E0 OCL Leanne Hendry, 109 Grove Avenue, New Costessey, Norwich, NR5 0HZ
2E0 OCM Ian Johnson, 10 Westbury Road, Shrewsbury, SY1 3HF
2E0 OCV Nigel Powis, 24 Rosemullion Close, Exhall, Coventry, CV7 9NQ
2E0 OCW William Joyce, 2 Palmers Cottage, Main Street, Oakham, LE15 8DH
2E0 ODD David Bambrough, 7 Barnwell View, Herrington Burn, Houghton le Spring, DH4 7FB
2E0 ODF J Lashley, 33 Goodes Avenue, Syston, Leicester, LE7 2JH
2E0 ODL K Bedford, 29 Kent Road Brookenby, Binbrook, Market Rasen, LN8 6EW
2E0 ODO Elwyn White, 3 Davy Drive, Maltby, Rotherham, S66 7EN
2E0 ODS Dale Robins, 12 Kestrel Way, Duffryn, Newport, NP10 8WF
2E0 ODT Belinda Hendry, 109 Grove Avenue, New Costessey, Norwich, NR5 0HZ
2J0 ODX Paul Ahier, Les Trois Carres, La Rue D'Aval, Jersey, JE3 6ER
2E0 OEE Jack Hendry, 109 Grove Avenue, New Costessey, Norwich, NR5 0HZ
2E0 OEM Joe Summers, Little Trembroath, Stithians, Truro, TR3 7DT
2E0 OES Robert Barnes, 275 Oregon Way, Chaddesden, Derby, DE21 6UR
2E0 OEV Martin Cuff, 14 The Mount, Ringwood, BH24 1XX
2E0 OEZ Ian Beresford, 16a Holbeck Hill, Scarborough, YO11 2XD
2E0 OFF George Kenyon, 2 Langdale Terrace, Stalybridge, SK15 1EX
2E0 OFK Gareth Tasker, 49 Grasmere Street, Liverpool, L5 6RH
2E0 OFM Peter Joyce, 92 Essex Road Halling, Rochester, ME2 1AX

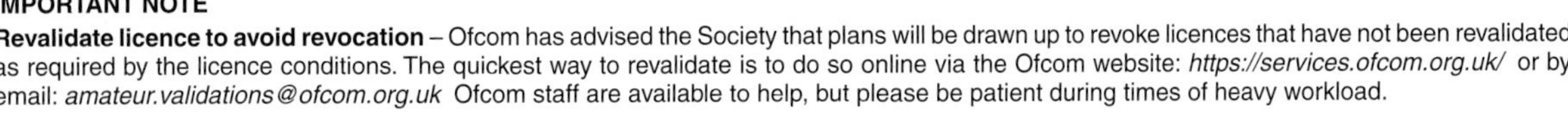

**IMPORTANT NOTE**

**Revalidate licence to avoid revocation** – Ofcom has advised the Society that plans will be drawn up to revoke licences that have not been revalidated as required by the licence conditions. The quickest way to revalidate is to do so online via the Ofcom website: *https://services.ofcom.org.uk/* or by email: *amateur.validations@ofcom.org.uk* Ofcom staff are available to help, but please be patient during times of heavy workload.

2E0 OGY Christopher Hodgetts, 16 Myrtle Drive, Rogerstone, Newport, NP10 9EA
2E0 OGZ G Wilkinson, 50 Sherburn Road, Durham, DH1 2JR
2I0 OHE E Paterson, 1 Sycamore Grove, Belfast, BT4 2RB
2M0 OIC James McArdle, 1 Queen Street, Hamilton, ML3 9JR
2E0 OIL Martin O'Connor, 28 Cardigan Road, Southport, PR8 4SF
2E0 OIN H List, 41 Westbury Crescent, Dover, CT17 9QQ
2E0 OJB J Bayliss, 39 Elms Avenue, Littleover, Derby, DE23 6FB
2E0 OJD Kevin Winton, 130 George V Avenue, Worthing, BN11 5RX
2E0 OJE Andrew Forbes, Flat 10, Denby House, Paignton, TQ4 6ES
2I0 OJK Jonathan Kavanagh, Flat 1, 161 Andersonstown Road, Belfast, BT11 9EA
2E0 OJS Oliver Squire, 91 Victoria Road, London, N22 7XG
2E0 OKC Andrew Londors, 112 Kingston Hill Avenue, Romford, RM6 5QL
2E0 OKH Owain Hopkins, Apartment 17, White Croft Works, Sheffield, S3 7AH
2E0 OKK Ian Day, 137 Tuffley Lane, Tuffley, Gloucester, GL4 0NZ
2E0 OKP Stephen Walsh, 41 The Rydales, Hull, HU5 1QD
2E0 OKS Marek Biadon, 57 Fern Hill Road, Oxford, OX4 2JW
2E0 OKY R Eglington, 33 Bradley Lane, Bilston, WV14 8EW
2E0 OKZ A Cammish, 6 West Vale, Filey, YO14 9AY
2E0 OLE Oliver Rofix, Birds Hill Clopton, Woodbridge, IP13 6SE
2E0 OLF Martin Mills, 17 Hornby Street, Plymouth, PL2 1JD
2E0 OLG Dean Rugen, 19 Jacksons Close, Haskayne, Ormskirk, L39 7LD
2M0 OLK O Keast, 6 Prospecthill Place, Greenock, PA15 4DW
2E0 OLO Christopher Rennie, 28 Foxwell Drive, Hucclecote, Gloucester, GL3 3LF
2E0 OLT Owain Thomas, Garth Celyn, St. Davids Road, Aberystwyth, SY23 1EU
2I0 OMA John Martin, 23 Winters Gardens, Omagh, BT79 0DZ
2E0 OMG Martin Robinson, 10 Bramley Gardens, Poulton-le-Fylde, FY6 7RD
2E0 OMI Hans Kassier, 26 Higher Port View, Saltash, PL12 4BX
2M0 OML T Cockayne, 2b Bogleshole Road, Cambuslang, Glasgow, G72 7PR
2M0 OMS M Scullion, 24 Langmuir Road, Kirkintilloch, Glasgow, G66 2QE
2E0 OMT T Baggley, 16 Seaton Road, Seaton, Workington, CA14 1DT
2E0 OMV J Barton, 37 Lytton Road, Sheffield, S5 8AX
2E0 ONE Paul Greenwood, 1 The Garth, Whitby, YO21 3PD
2E0 ONO Robert James, 55 Kirkwall Road, Plymouth, PL5 3TL
2M0 ONS Donald Anderson, Dail Darach, Monydrain Road, Lochgilphead, PA31 8LG
2E0 ONV John Bonar, 40 Quarry Close, Minehead, TA24 6EE
2M0 ONW Kevin Harper, 93 Craufurdland Road, Kilmarnock, KA3 2HU
2E0 OOC B Cooper, 71 High Street, Birstall, Batley, WF17 9RG
2E0 OOM Michael Buist, 23 St. Chads Drive, Gravesend, DA12 4EL
2E0 OON Cass May, 16 Trelawn Road, London, E10 5QD
2E0 OOO Roy Clayton, 9 Green Island, Irton, Scarborough, YO12 4RN
2M0 OOT J ferrans, 77 Knockinlaw Road, Kilmarnock, KA3 2AS
2E0 OPB O Blackburn, 128 High St., Crigglestone, Wakefield, WF4 3EF
2E0 OPC Owen Campbell, 3 Hillside Close, Helsby, Frodsham, WA6 9LB
2E0 OPM David Whitehouse, 6 Larch Close, Heathfield, TN21 8YW
2E0 OPO Oscar Silva, 1 Grangewood Terrace, London, SE25 6TA
2E0 OPS D Lapham-Crozier, 109 Aylesbury Crescent, Plymouth, PL5 4HX
2E0 OPU K Morris, 80 Bridge Street, Chatteris, PE16 6RN
2E0 OQH Dennis Cooper, 52 Meadow Lane, Birkenhead, CH42 3YE
2E0 OQZ P Hall, 13 Sheard Avenue, Ashton-under-Lyne, OL6 8DS
2E0 ORI Daniel Dart, Ticklebelly Cottage Lower Charlton Trading Estate, Shepton Mallet, BA4 5QE
2M0 ORK Marc Herridge, The Hollies, Petticoat Lane, Orkney, KW17 2RP
2E0 ORP M Orpen, Daymer, Tey Road, Colchester, CO6 3RY
2E0 ORT Barbara Roberts, 6 Trem Y Moelwyn, Tanygrisiau, Blaenau Ffestiniog, LL41 3SS
2M0 OSC Jennifer Hanley, 44 Waverley Crescent, Livingston, EH54 8JN
2E0 OSE Paul Pope, 2 Hale Villas, Honiton, EX14 9TQ
2E0 OSG J Bellis, 32 Broughton Road, Lodge, Wrexham, LL11 5NG
2E0 OSH Sharon Owen, 8 Old Tanymanod Terrace, Blaenau Ffestiniog, LL41 4BU
2M0 OSK Andrew Twort, 17 Balallan, Isle of Lewis, HS2 9PN
2E0 OSO Anna James, 55 Kirkwall Road, Plymouth, PL5 3TL
2E0 OSX A Logan, 23 Cherry Tree Rise, Walkern, Stevenage, SG2 7JL
2E0 OSY David Coppenhall, 56 Vicarage Lane, Elworth, Sandbach, CW11 3BU
2E0 OTB Paul Hateley, 44 Painters Croft, Coseley, Bilston, WV14 8AP
2I0 OTC Bob Emerson, 67 Castlemore Avenue, Belfast, BT6 9RH
2E0 OTE D Barlow, 7 Peter Street, Eccles, Manchester, M30 0JF
2E0 OTI Daniel Scotcher, 17 St. Dominics Square, Luton, LU4 0UN
2E0 OTP Michael Anostalgia, 136 Avenue Road Extension, Leicester, LE2 3EH
2E0 OTT J Slobin, 45 Dale Edge, Eastfield, Scarborough, YO11 3EP
2I0 OTW Paul Fallon, 18 Church View, Killough, Downpatrick, BT30 7RJ
2E0 OTY Peter Lane, 21 Rycroft Avenue, St. Neots, PE19 1DT
2E0 OTZ M Wright, 8 St. Wilfrids Road, Oundle, Peterborough, PE8 4NX
2E0 OUK Ivan Hrynkiewicz, 21 York Road, Cannock, WS11 8ES
2E0 OVB Robert Gowers, 43 Tungstone Way, Market Harborough, LE16 9GA
2M0 OVD Derek Adamson, 5 Central Quadrant, Ardrossan, KA22 7DY
2E0 OVF Stephen Hedgecock, 37 Tennyson Road, Maldon, CM9 6BE
2E0 OVI Ovidiu Popa, 14 Mulberry Close, Horsham, RH12 2NH
2E0 OVL John Hirst, 57 Newgate Street, Doddington, March, PE15 0SR
2E0 OVR Fiona Farrer, 16 High Street, Eagle, Lincoln, LN6 9DH
2E0 OVT J Jones, 40 Ffordd Coed Marion, Caernarfon, LL55 2EF
2M0 OVV S Monaghan, 13 Ballyhennan Crescent, Tarbet, Arrochar, G83 7DB
2E0 OWC Stuart Iles, 12 St. Peters Road, Burntwood, WS7 0DJ
2E0 OWH Trevor Harrison, 58 Ascot Drive, Cannock, WS11 1PE
2E0 OWL S Hurley, 11 Beresford Avenue, Wirral, CH63 7LR
2E0 OXF A Comerford, 21 New Cross Road, Headington, Oxford, OX3 8LP
2E0 OXO D Harden, 59 Violet Avenue, Edlington, Doncaster, DN12 1NW
2M0 OXQ S McKinnon, 8 Rowanlea Avenue, Paisley, PA2 0RP
2M0 OXX Alex Berry, 41 Bruce Drive, Stenhousemuir, Larbert, FK5 4DD
2E0 OYG Brian Nuttall, 102 Highcroft Avenue, Blackpool, FY2 0BW
2E0 OYN R Watson, 60 Beresford Avenue, Surbiton, KT5 9LJ
2E0 OZE Michael Crockford, Centre Cottage Kelk, Driffield, YO25 8HL
2E0 OZH M Flynn, 20 Manwood Avenue, Canterbury, CT2 7AH
2E0 OZI Scott Carpenter, 52 Mewstone Avenue, Wembury, Plymouth, PL9 0JZ
2W0 OZO Sheridan Hayward, 22 Dewsland Street, Milford Haven, SA73 2AU
2E0 OZQ A Sherer, 28 Baroness Road, Grimsby, DN34 4DP
2E0 OZW K Ozwell, 109 Abbey Road, Grimsby, DN32 0HN
2E0 OZY Colin Osborne, Gwinwydden, Tremont Road, Llandrindod Wells, LD1 5BH
2E0 PAA Clive Jago, 20 Glanville Road, Tavistock, PL19 0EA
2E0 PAB Phillip Stone, 40 Highfields, Halstead, CO9 1NH
2I0 PAC Paddy Dallas, 12 Glendun Crescent, Coleraine, BT52 1UJ
2E0 PAD Joe Stainton, 24 Clifton Road, Huddersfield, HD1 4LL
2E0 PAE Paul Illidge, 55 East Park Road, Spofforth, Harrogate, HG3 1BH
2E0 PAF Philip Taylor, 47 Pickhurst Park, Bromley, BR2 0TN
2E0 PAH Paul Hunt, 33 Drakes Close, Bridgwater, TA6 3TD
2E0 PAJ Patrick Kiernan, 63 King Street, Southport, PR8 1LG
2E0 PAK Paul Watson, 10 Whitelands Crescent, Baildon, Shipley, BD17 6NN
2E0 PAN Adrian Paffey, 1 St. Vincent Road, Newport, NP19 0AN
2E0 PAO David Pike, 46 Haymans Close, Cullompton, EX15 1EH
2E0 PAP Peter Woodyard, 65 Raglan Street, Lowestoft, NR32 2JS
2E0 PAT Graham Dobson, 4 Durley Gardens, Orpington, BR6 9LL
2E0 PAU Peter Dossett, 20 Vineyard Close, Southampton, SO19 7DD
2E0 PAV S Richards, 18 Lowfields, Staxton, Scarborough, YO12 4SR
2E0 PAX Edward Goodwin, 55 Twickenham Road, Sunderland, SR3 4JN
2E0 PBB William Ramsell, 8 Didcot Drive, Marchington, Uttoxeter, ST14 8LT
2M0 PBC C Montague, 74 Holmbyre Road, Glasgow, G45 9QD
2E0 PBE Peter Edwards, 791 Windmill Lane, Denton, Manchester, M34 2ER
2E0 PBF Patrick Bigsby, 14 Rutland Avenue, Sidcup, DA15 9DZ
2E0 PBH Alan Billings, 46 Thorley Drive, Cheadle, Stoke-on-Trent, ST10 1SA
2E0 PBJ Paul Mansfield, 56 Sunningdale, Waltham, Grimsby, DN37 0UG
2E0 PBK Paul Parkin, Hawksworth House, Main Street, Frolesworth, Lutterworth, LE17 5EG
2E0 PBL P Ball, 101 Chelwood Drive, Bath, BA2 2PS
2I0 PBM P Bingham, 28 Carnew Road, Katesbridge, Banbridge, BT32 5PS
2E0 PBN Paul Dent, 33 Cavalier Close, Dibden, Southampton, SO45 5TU
2E0 PBO Paul Kay, 30 Broadway, Grange Park, St. Helens, WA10 3RX
2E0 PBP Paul Jones, 50 Clay Lane, Doncaster, DN2 4RJ
2E0 PBR Peter Alley, 58 Osprey Close, Watford, WD25 9AR
2E0 PBS P Sycamore, 17 Markham Avenue, Weymouth, DT4 0QL
2E0 PBT Paul Burgess, 61 Grosvenor Avenue, Torquay, TQ2 7JX
2E0 PBU W Dickson, The Rowans, Pwllmeyric, Chepstow, NP16 6LA
2E0 PBV Michael James, 42 Doone Way, Ilfracombe, EX34 8HS
2D0 PBW Ivelin Yovchev, 11 Beverley Drive, Edgware, HA8 5NQ
2E0 PBY Paul Barker, 65 Cornwall Road, Walmer, Deal, CT14 7SA
2I0 PBZ Philip Bell, 3 Alexandra Crescent, Larne, BT40 1NE
2E0 PCA Peter Kelly, 20 Fareham Close, Walton-le-Dale, Preston, PR5 4JX
2E0 PCC P Garraway, The Poplars, Crowell Road, Chinnor, OX39 4HP
2W0 PCD Paul Day, 15-16 Troedrhiw-Trwyn, Pontypridd, CF37 2SE
2W0 PCE Philip Iles, 150 Pen-y-Bryn, Caerphilly, CF83 2LA
2E0 PCF Paul Faulkner, 32 Manvers Road, Beighton, Sheffield, S20 1AY
2E0 PCG P Green, 8 Grassthorpe Road, Sheffield, S12 2JH
2E0 PCH Phil Haywood, 5 Mayfield Drive, Kenilworth, CV8 2SW
2E0 PCI Paul Collins, 16 Fern Grove, Haverhill, CB9 9ND
2E0 PCL Gareth Owen, 38 Trentham Drive, Bridlington, YO16 6ES
2W0 PCN P Nash, 110 Aberporth Road, Cardiff, CF14 2RY
2E0 PCO Philip Coombes, 2 Bissoe Cottages, Bissoe, Truro, TR4 8SU
2E0 PCQ Peter Morris, 14 Marina Road, Darlington, DL3 0AL
2E0 PCR Carol Vincent, 81 Trethannas Gardens, Praze, Camborne, TR14 0LL
2W0 PCT Stephen Gau, Disgwylfa, The Downs, Cardiff, CF5 6SB
2E0 PCU Peter Roberts, 26 Park Road, Wallasey, CH44 9EB
2E0 PCV D Whiting, 133 Belfield Road, Accrington, BB5 2JD
2M0 PCW Sarah Skerratt, 3/2 18 Mardale Crescent, Edinburgh, EH10 5AG
2E0 PCZ Paul Colyer, 23 Florida Road, Torquay, TQ1 1JY
2E0 PDB Paul Phipps, Meakers Cottage, Long Load, Langport, TA10 9JX
2E0 PDG E Aitken, 20 Plover Drive, Bury, BL9 6JH
2E0 PDH Callum Macleod, 2 Welford Road, Chapel Brampton, Northampton, NN6 8AF
2E0 PDL Michael Garry, 34 Conway Road, Paignton, TQ4 5LH
2E0 PDM Paul March, 46 Christchurch Road, Tilbury, RM18 8XP
2E0 PDO Andrew Dingwall, 48 Village Farm Caravan Site, Bilton Lane, Harrogate, HG1 4DL
2E0 PDP J Clarkson, 56 Edward Bailey Close, Binley, Coventry, CV3 2LZ
2E0 PDQ Denise Carpenter, 34b Carey Park, Killigarth, Looe, PL13 2JP
2E0 PDU Leslie Fuller, Rosemar Lodge Westford, Wellington, TA21 0DX
2E0 PDX G Swindells, 15 Benedict Close, Salford, M7 2GB
2E0 PDZ Peter Harper, 7 Duncan Gardens, Bath, BA1 4NQ
2E0 PEC Barry Clayton, 5 Greycourt Close, Halifax, HX1 3LR
2W0 PEE Neville Tanner, 3 Maes y Tyra, Resolven, Neath, SA11 4NN
2E0 PEF Peter Freeman, 57 Ruffa Lane, Pickering, YO18 7HN
2W0 PEG Jon Reason, 159 Caerau Lane, Cardiff, CF5 5JS
2W0 PEH Brian Sellers, 86 St. John Street, Ogmore Vale, Bridgend, CF32 7BB
2E0 PEL S Peel, 21 Fairfield Avenue, Ormesby, Middlesbrough, TS7 9BB
2E0 PEM Paul Metters, 11 Horton Avenue, South Shields, NE34 8NL
2E0 PEP Stephen Hill, 35 Longs Way, Wokingham, RG40 1QW
2E0 PES Paul Sewell, 17 Chatham Close, Coventry, CV3 1LY
2E0 PEW Paul Woolley, 84 Bowthorpe Road, Norwich, NR2 3TP
2E0 PFA Peter Fernie, 39 North Parade, Falmouth, TR11 2TE
2E0 PFB Peter Browne, 151 North Road, St. Andrews, Bristol, BS6 5AH
2E0 PFD Paul Devlin, Brynteg, Fron Bache, Llangollen, LL20 7BP
2E0 PFF Simon Debenham, 10 Elizabeth Close, Wellingborough, NN8 2JA
2E0 PFG Patrick Goddard, 62, Woodlands Drive, Thetford, IP24 1JJ
2E0 PFH Paul Holmes, 18 Raleigh Avenue, Whiston, Prescot, L35 3PL
2E0 PFL Peter Leng, The Barn, Gildersleets, Settle, BD24 0AH
2E0 PFO Paul Noble, 14 Park Street, Swallownest, Sheffield, S26 4UP
2E0 PFR P Ratcliffe, 2 Newlands Avenue, Whitby, YO21 3DX
2E0 PFT Angus Perfect, 3 Chelmarsh Close, Chellaston, Derby, DE73 6PB
2E0 PFX Paul Fuller, 6 Annalee Road, South Ockendon, RM15 5DJ
2E0 PFY Richard Ball, 77 Old Brumby Street, Scunthorpe, DN16 2AJ
2E0 PGB P Barnard, Hawthorns, Old Church Road, Chelmsford, CM3 8BG
2E0 PGC Philip Challans, Flat 4, Sandringham Court, 2 Chandos Square, Broadstairs, CT10 1QN
2E0 PGH Paul Hill, 14 Drovers Way, Woodlands, Ivybridge, PL21 9XA
2E0 PGI Geoffrey Hartless, 32 Long Acre, Mablethorpe, LN12 1JF
2E0 PGL Philip Lewis, 154 Meadow Head, Sheffield, S8 7UF
2E0 PGM Peter McFadden, Maple Cottage, Great Gap, Leighton Buzzard, LU7 9DZ
2E0 PGP William Dover, Silverdale, Fox Lane, Basingstoke, RG23 7BB
2E0 PGR P Grainger, 36 Orchard Road, Wigton, CA7 9JL
2E0 PGS Peter Stevenson, 6 Dighton Gate, Stoke Gifford, Bristol, BS34 8XA
2E0 PGT Paul Thompson, 25 Pitclose Road, Birmingham, B31 3HU
2E0 PHB P Haslam-Brunt, 488 Klightwood Road, Lightwood, Stoke-on-Trent, ST3 7EW
2E0 PHL Philip Probst, 37 Devonshire Street, Skipton, BD23 2ET
2E0 PHM Philip Meerman, 24 Horseshoe Crescent, Burghfield Common, Reading, RG7 3XW
2W0 PHP Chris Maggs, 15 Stuart Street, Treorchy, CF42 6SN
2E0 PHS Paul Sladen, 25 Linden Grove, Beeston, Nottingham, NG9 2AD
2E0 PHU P Uttley, 55 Dunce Park Close, Elland, HX5 0PF
2E0 PHX Paul Hunter, 160 Pembroke Road, Northampton, NN5 7ER
2E0 PIA Mark Hill, 56 Moorhouse Avenue, Wakefield, WF2 9QG
2M0 PID Victoria Hamilton, 10/3 Fox Street, Edinburgh, EH6 7HN
2E0 PIK Bernard Pike, 19 Cardigan Gardens, Reading, RG1 5QP
2E0 PIO T Nakagawa, 7 Milton Street, Barrowford, Nelson, BB9 6HE
2E0 PIP P Marsh, 16 Laurel Close, North Warnborough, Hook, RG29 1BH
2E0 PIT Conrad Fox, Millstone Cottage, Prior Wath Road, Scarborough, YO13 0AZ
2E0 PIW R Metcalfe, 33 Midland Terrace, Hellifield, Skipton, BD23 4HJ
2E0 PIX Aidan Clayton, 7 Church Terrace, Reading, RG1 6AS
2E0 PJD P Dawson, 88 Urmson Road, Wallasey, CH45 7LQ
2E0 PJE Peter Elmore, 8 Gray Street, Elsecar, Barnsley, S74 8JR
2E0 PJH Phil Holmes, 1 Leonards Place, Bingley, BD16 1AD
2W0 PJJ Philip Jones, 23 Pinecroft Avenue, Aberdare, CF44 0HY
2E0 PJM Philip McLaren, 10 Haulfryn, Ruthin, LL15 1HB
2E0 PJN P Northover, 66 Howard Drive, Letchworth Garden City, SG6 2DQ
2E0 PJR Peter Radford, 43 Bells Lane, Nottingham, NG8 6EX
2E0 PJS P Spilman, 28 Staines Way, Louth, LN11 0DF
2E0 PJT P Tomlinson, 11 Haynes Close, Clifton, Nottingham, NG11 8JN
2E0 PJY Paul Jay, 23 Manor Road, Wendover, Aylesbury, HP22 6HN
2M0 PKA Pavan Akula, 270 Springhill Road, Aberdeen, AB16 7SL
2E0 PKB P Beier, 20 Markham Avenue, Armthorpe, Doncaster, DN3 2AZ
2E0 PKK P Knight, 73 Bramley Crescent, Southampton, SO19 9LJ
2E0 PKL Darren Cadet, 2 Paddockside, Middleton, Ludlow, SY8 3EB
2E0 PKR Steven Parker, 57 Queen Street, Horncastle, LN9 6BH
2E0 PKS Richard Harlow, 28 Dovecliff Crescent, Stretton, Burton-on-Trent, DE13 0JH
2E0 PKU David Winstanley, 43 Florence Street, St. Helens, WA9 5NA
2E0 PKY Brian Collins, 15 Bonds Meadow, Lowestoft, NR32 3QL
2E0 PLA P brown, 11 Booth Crescent, Rossendale, BB4 9BT
2E0 PLC T Kyriacou, 54 Sutton Avenue, Silverdale, Newcastle, ST5 6TB
2E0 PLE G Dennis, 21 Rydal Crescent, Scarborough, YO12 4JJ
2E0 PLH P Holmes, 28 Ackworth Drive, Manchester, M23 1LD
2E0 PLK Artur Jedryka, 71 West Royd Drive, Shipley, BD18 1HL
2E0 PLL Paul Lettington, 21 Bideford Close, Woodley, Reading, RG5 3SE
2E0 PLR Peter Tolcher, 15 Langstone Close, Torquay, TQ1 3TX
2E0 PLS Jacek Walczak, 18 Heathfield, Chippenham, SN15 1BQ
2E0 PLV Paul Le Vallois, 14 London Row, Arlesey, SG15 6RX
2E0 PLX Paul Levy, 43 Conroy Drive, Dawley, Telford, TF4 2RW
2E0 PLY Darren Hensman, 130 Clittaford Road, Plymouth, PL6 6DW
2E0 PMA Nicholas Perkins, 39 Ladychapel Road, Abbeymead, Gloucester, GL4 5FQ
2E0 PMB Peter Browne, Ham Cottage, Hammingden Lane, Haywards Heath, RH17 6SR
2E0 PMC Colin Bowman, 26 Albany Hill, Tunbridge Wells, TN2 3RX
2E0 PMD P Dowling, 22 Chelkar Way, York, YO30 5ZH
2E0 PME Peter Martin, 5 Shropshire Drive, Wilpshire, Blackburn, BB1 9NF
2E0 PML Christopher Suddell, Lynhurst, Littleworth Lane, Horsham, RH13 8JX
2E0 PMM P Mather, 11 Odette Court, Gilstead, Bingley, BD16 3QN
2E0 PMP Prakash Punjabi, 62 Cleveland Road, London, W13 8AJ
2M0 PMR Alastair Graham, 27 Crichton Road, Pathhead, EH37 5RA
2E0 PMV P Mansfield, 27 Popplechurch Drive, Swindon, SN3 5DE
2E0 PMX A Marlow, 66 Woodborough Road, Winscombe, BS25 1BA
2E0 PMZ Paul Eckersley, 18 Mulberry Court, Guildford, GU4 7EQ
2E0 PNA Nigel Hine, 13 Wilton Crescent, Alderley Edge, SK9 7RE
2E0 PNB Panagiotis Bozikis, 336 Higham Hill Road, London, E17 5RG
2E0 PNC G Billington, 47 Smithy Leisure Park, Cabus Nook Lane, Preston, PR3 1AA
2E0 PNG Philip Green, 34 Drydens Close, Titchmarsh, Kettering, NN14 3DD
2E0 PNK Charlotte Taylor, 212 Plantation Hill, Worksop, S81 0HD
2E0 PNN Paul Bowen, 12 Powell Place, Newport, TF10 7BS
2E0 PNP Robert Bartha, 6 Chappell Close, Aylesbury, HP19 9QA
2E0 PNR Anthony Ault, 124 High Street, Aylesbury, HP20 1RB
2E0 PNX Keven Kenton, 24 Penygraig Road, Brymbo, Wrexham, LL11 5AD
2E0 POB Andrew Carden, Hazelgrove, South Allington, Kingsbridge, TQ7 2NB
2I0 POD Alistair Hunter, 38 Robinson Road, Bangor, BT19 6NJ
2E0 POI Lorna Smart, 139 Northumberland Street, Norwich, NR2 4EH
2E0 POQ Richard Finch, 19b Kiln Road, Newbury, RG14 2LS
2E0 POU P Daubaris, 32 Chalcombe Road, Abbey Wood, London, SE2 9QS
2E0 POZ P Stilwell, 23 Edale Moor, Liden, Swindon, SN3 6LT
2E0 PPA Paul Ridley, 218 Lichfield Road, Rushall, Walsall, WS4 1SA
2E0 PPD P Bengey, 3 Millmead Road, Bath, BA2 3JW
2E0 PPH Paul Huband, 11 Barnwood Road, Birmingham, B32 2LZ
2E0 PPK N Green, 11 Wythburn Way, Rugby, CV21 1PX
2W0 PPL Alexander Dighton, 84 Trefelin, Aberdare, CF44 8LF
2E0 PPO John Grint, 10 Paddock Gardens, Attleborough, NR17 2EW
2E0 PPR Paul Rickwood, 8 Bealeys Avenue, Wolverhampton, WV11 1EG
2I0 PPW Jonathan MacFarlane, 1 Main Street, Uttony, Magheraveely, Enniskillen, BT92 6NB
2E0 PPY S Evans, 51 St. Georges Road, Dudley, DY2 8EY
2E0 PPZ L Westwood, 28 Ash Crescent, Kingswinford, DY6 8DJ
2E0 PQR Peter Nathan, 7 Pitt Drive, Seaford, BN25 3JB
2E0 PRC Paul Craig, 4 Poolside, Burston, Stafford, ST18 0DR
2E0 PRD P Denham, 48 Marl Court, Thornhill, Cwmbran, NP44 5TY
2I0 PRL Peter Reid, 1 Nettlehill Mews, Lisburn, BT28 3HN
2I0 PRM Edith Simpson, 10 Woodview Park, Tandragee, Craigavon, BT62 2DD
2E0 PRO Philip Robinson, 13 Carrfield Avenue, Liverpool, L23 9SS
2E0 PRS Peter Shaw, 32 Hardwick Road East, Worksop, S80 2NT
2M0 PSA P Smith, 13 Newmills Grove, Balerno, EH14 5SY
2E0 PSC Philip Croxford, 1 Meteor Close, Bicester, OX26 4YA

2E0 PSD Julian Eames, 6 The Oaklands, Cold Meece, Stone, ST15 0QH

2E0 PSH J Storey, 11 Enderby Road, Sunderland, SR4 6BA

2E0 PSK Stephen [illegible], 117 Victoria Avenue, Hastings, TN35 [illegible]

2E0 PSM Paul Smart, 142 Finch Road, Chipping Sodbury, Bristol, BS37 6JB

2E0 PSN Gareth James, 20 Redcar Road, Romford, RM3 9PT

2E0 PSO Peter Sheffield, 13 St. Winifred Road, Wallasey, CH45 5EJ

2E0 PSP J Anderson, 57 Chapel Lane, Hadfield, Glossop, SK13 1NX

2E0 PSR Philip Shaw, 25 Headcorn Road, Platts Heath, Maidstone, ME17 2NH

2E0 PSW John Godfrey, 4 Cherry Close, Houghton Conquest, Bedford, MK45 3LQ

2E0 PSX Philip Simpson, 7 Hawthorn Close, Wootton, Ulceby, DN39 6RB

2E0 PSZ Susan MacDonald, Woodside Cottage, Horton Way, Verwood, BH31 6JJ

2E0 P[illegible] Paul Chambers, 257 Kings Acre Road, Hereford, HR4 0SR

2M0 PTF [illegible], Willowbank, Kirkton of Tough, Alford, AB33 [illegible]

2E0 PTG Peter Gavin, 11 Campbell Close, Yateley, GU46 [illegible]

2E0 PTI B Parton, 51 Marston Grove, Stoke-on-Trent, ST1 6EF

2E0 PTS Phillip Boultwood, 32 Makepiece Road, Bracknell, RG42 2HJ

2E0 PUB A Fulton, 8 Priest Hill Gardens, Wetherby, LS22 7UD

2E0 PUE A Hannon, 8 Circular Road West, Liverpool, L11 1AZ

2E0 PUG David Cobbold, 65 St. Olaves Road, Bury St. Edmunds, IP32 6HR

2E0 PUL David Pullen, 5 Weldon Close, Shotton Colliery, Durham, DH6 2YJ

2E0 PUN John Rideout, 4 Treetops, Northampton, NN3 8XA

2E0 PUS P Ellis, 40 Grasmere Road Royton, Oldham, OL2 6SR

2E0 PVN Paul Nicholls, 23 Bishops Gate, Birmingham, B31 4AJ

2E0 PVQ Edwin Rhodes, The Old Forge, Stoke Gabriel, Totnes, TQ9 6RL

2E0 PVW Paul Armstrong, 10 Shirdley Avenue, Liverpool, L32 7QG

2E0 PWC Paul Castle, 3 Wye Road, Brockworth, Gloucester, GL3 4PP

2E0 PWD John Shaw, 6 Garth End, Huntington, York, YO32 9QU

2E0 PWF Chris Cousins, 43 Avon Close, Little Dawley, Telford, TF4 3HP

2E0 PWG Paul Green, 15 Dickenson Road, Chesterfield, S41 0RX

2E0 PWI Philip Warwick, 24 Chiltern Close, Berinsfield, Wallingford, OX10 7PZ

2E0 PWK S Platts, 59 Sea View Road, Drayton, Portsmouth, PO6 1EW

2E0 PWL M Mynn, 15 Shearling Drive, Lower Cambourne, Cambridge, CB23 6BZ

2E0 PWM P Mitchell, 13 Ashorne Close, Matchborough, Redditch, B98 0EY

2W0 PWO Peter Oseland, 6 Oaklands Close, Bridgend, CF31 4SJ

2E0 PWP Paul Thompson, 3 Floyers Field, West Stafford, Dorchester, DT2 8FJ

2E0 PWR David Riley-Kydd, 26 Talwrn Road, Wrexham, LL11 3PG

2E0 PXD Paul Donaghy, 67 Brockenhurst Way, Bicknacre, Chelmsford, CM3 4XN

2M0 PXH Paul Holmes, Maraval, Doune Road, Dunblane, FK15 9AT

2E0 PXP John Maudsley, Knight Stainforth Hall, Little Stainforth, Settle, BD24 0DP

2E0 PXW Barry Smith, 25 Lancing Road, Ellesmere Port, CH65 5BB

2E0 PXY Camilla Fox, 45 Park Road, Wivenhoe, Colchester, CO7 9LS

2E0 PXZ Chris Fox, 45 Park Road, Wivenhoe, Colchester, CO7 9LS

2E0 PYA W Allen, 109 Barston Road, Oldbury, B68 0PU

2E0 PYC Richard Watson, 8 Bourne Close, Warminster, BA12 9PT

2E0 PYM Angie Nutt, 77 Exeter Close, Stevenage, SG1 4PW

2E0 PYN Lindsay Scott, 28 Cavendish Place, New Silksworth, Sunderland, SR3 1JW

2E0 PYR Pauline Robinson, 15 Cornelius Drive, Wirral, CH61 9PY

2E0 PZK Peter Kirby, 102 Waterloo Road, Crowthorne, RG45 7NW

2E0 RAA Robert Keeley, 17 Pembroke Avenue, Wirral, CH46 0TP

2W0 RAD Robert Miles, 63 Phillip Street, Caegarw, Mountain Ash, CF45 4BG

2E0 RAF A Woodrup, 8 Hawarden Road, Preston, PR1 4TS

2E0 RAG A Green, 18 Harold Avenue, Ashton-in-Makerfield, Wigan, WN4 9UZ

2E0 RAH R Haynes, 47 Alder Drive, Alderholt, Fordingbridge, SP6 3EP

2E0 RAI Richard Trim, 23 Coleman Road, Bournemouth, BH11 8EQ

2E0 RAJ Philip Penfold, 2 The Leas, Essenden Road, St. Leonards-on-Sea, TN38 0PU

2E0 RAK Paul Graham, 19 Pontop View, Rowlands Gill, NE39 2JP

2E0 RAL A Clewes, 20 Linden Drive, Crewe, CW1 6HN

2E0 RAM R Mason, 73 Edinburgh Road, Chatham, ME4 5BZ

2E0 RAS Ray Shippey, 43 Westbury Street, Bradford, BD4 8PB

2E0 RBA Richard Bowman, 3 Hodders Way, Cargreen, Saltash, PL12 6NY

2E0 RBC Derek Judge, 18 Sheridan Street, Bacup, OL13 8BH

2E0 RBG Robert Rigden, 36a Atherston, Bristol, BS30 8YB

2E0 RBH Michael Clifford, 100 Cromwell Road, Hounslow, TW3 3QJ

2E0 RBI R Gilbert, 61 Coltstead, New Ash Green, Longfield, DA3 8LN

2E0 RBK B Kelly, 21 Hogarth Walk, Bristol, BS7 9XS

2E0 RBN Vincent Steele, 175 Vale Road, Seaford, BN25 3HH

2E0 RBO Ronald Coleman, 5 Meeting Lane, Burton Latimer, Kettering, NN15 5LS

2E0 RBP Robert Cobb, 57 ADAMS DRIVE, Willesborough, Ashford, TN24 0FX

2E0 RBQ Roberta Titmarsh, 38 Cromer Road, Sheringham, NR26 8RR

2E0 RBR Clifford Dunstan, 67 Knights Way, Mount Ambrose, Redruth, TR15 1PA

2E0 RBU Rodney Booth, 142 Heath Lane, Earl Shilton, LE9 7PD

2I0 RBV Ralph Montgomery, 24 Cameron Court, Ballyclare, BT39 9UZ

2E0 RBW Robert Williams, Bardsville, Porthdafarch Road, Holyhead, LL65 2LL

2E0 RBX Robert Jones, 25 Whiteway Drive, Gresford, Wrexham, LL12 8HW

2E0 RBY Robert Hall, Concorde Cottage, Ellingstring, Ripon, HG4 4PW

2E0 RBZ Ronald Booker, 6 Kipling Road, Dursley, GL11 4QD

2E0 RCA B Whiteley, 2a Beechfield Close, Thorpe Willoughby, Selby, YO8 9QJ

2E0 RCB R Brown, 9 Dayleaf Crescent, Oakwood, Derby, DE21 2UG

2E0 RCC R Chadwick, 4 Gleneagles Drive, Haydock, St. Helens, WA11 0YS

2M0 RCD Stuart McKenzie, 0/2 69 Glenkirk Drive, Glasgow, G15 6AU

2E0 RCF Robert Goody, 113 Kenneth Road, Basildon, SS13 2BH

2E0 RCI Shane Lofthouse, 30 Broughton Grove, Skipton, BD23 1TL

2E0 RCL Rodney Buckland, 34 Beechwood Drive, Meopham, Gravesend, DA13 0TX

2E0 RCM R Medland, 5 Bay Tree Cottages, Hospital Road, Bude, EX23 9BP

2E0 RCN Raymond Northway, 8 Dean Close, Wick, Littlehampton, BN17 7ND

2E0 RCO Barry Walker, 53 Barley Cross, Wick St. Lawrence, Weston-Super-Mare, BS22 9TB

2E0 RCR Robert Rawson, 68 Broom Nook, Leeds, LS10 3LR

2E0 RCT Mark Russell, 107 Cambridge Road, Hitchin, SG4 0JH

2E0 RCU Robin Gripp, 23 Edmond Locard Court, Chepstow, NP16 6FA

2E0 RCV R Treacher, 93 Elibank Road, London, SE9 1QJ

2E0 RCW Cliff Wilson, 31 Violet Road, South Woodford, London, E18 1DG

2M0 RCZ A Conlon, Kilrae, Barrpath, Glasgow, G65 0EX

2D0 RDA M Salt, 1 Chantry Close, Harrow, HA3 9QZ

2W0 RDD Robert Cotterell, 49 Graham Court, Caerphilly, CF83 1RF

2E0 RDE Robert Seeley, 2 Church Road, Folkestone, CT20 3LH

2E0 RDF John Bailey, 22 Wilford Drive, Ely, CB6 1TL

2E0 RDG Robert Rogerson, 93 Auchencrieff Road, Locharbriggs, Dumfries, DG1 1UZ

2M0 RDH Robert [illegible], 2 Watson Place, Dunfermline, KY12 0DR

2E0 RDI Reginald Topley, 85 Stuart Road, Aylsham, Norwich, NR11 6UW

2E0 RDJ Robbie Cole, 14 Inner Loop Road, Beachley, Chepstow, NP16 7HF

2M0 RDK Robert Walmsley, 95 Race Road, Bathgate, EH48 2AU

2E0 RDN Robert Newton, 38 Bedford Road, Denton, Northampton, NN7 1DR

2E0 RDP Matthew Payne, The Devonhurst, 13 Eastern Esplanade, Broadstairs, CT10 1DR

2E0 RDQ Ronald Owen, 4 Aldersleigh Drive, Stafford, ST17 4RY

2M0 RDR Richard Rowe, 5 Lever Road, Helensburgh, G84 9DP

2M0 RDT Robin Tourish, 8 Linnpark Gardens, Johnstone, PA5 8LH

2E0 RDU J Crosby, 65 Bradwell Avenue, Stretford, Manchester, M32 9RT

2E0 RDW Richard Wyatt, 207 Weston Road, Stoke-on-Trent, ST3 6HA

2E0 RDX B Wilkes, 9 Barnsley Avenue, Conisbrough, Doncaster, DN12 3LB

2E0 RDZ Richard Shipman, 1 Lledfair Place, Heol Pentrerhedyn, Machynlleth, SY20 8DL

2E0 REB Robert Beardsley, 10 Moreton Close, Church Crookham, Fleet, GU52 8NS

2E0 REC Michael Barker, 18 Nickleby Road, Waterlooville, PO8 0RH

2E0 REG R Davis, 9 Mossdale Road, Liverpool, L33 1UQ

2M0 REH Richard Hay, Roddach Cottage East, Cummingston, Elgin, IV30 5XY

2W0 REJ John Richardson, 15 Calland Street, Plasmarl, Swansea, SA6 8LE

2E0 REK Ruth Kelly, 42 Hinton Wood Avenue, Christchurch, BH23 5AH

2E0 REL Rupert Allen, 44 Overing Avenue, Great Waldingfield, Sudbury, CO10 0RJ

2E0 REM Rory Whalley, 188 Astley Street, Astley, Manchester, M29 7AX

2E0 REN Alan Pendle, 17 Norrington Grove, Birmingham, B31 5NY

2E0 RER Robin Ridge, Roskellan House, Maenlay, Helston, TR12 7QR

2E0 RES Richard Strong, 2 Dean Avenue, Thornbury, Bristol, BS35 1JJ

2E0 REU Brian Robinson, 11 Wimbledon Drive, Stockport, SK3 9RZ

2E0 REX Christopher Moreton, 20 Millbrook Court, Little Mill, Pontypool, NP4 0HT

2E0 REY Ramon Milton, 66 Hoo Marina Park, Vicarage Lane, Rochester, ME3 9TG

2E0 RFD Richard Dell, 18 Greenacres, Fulwood, Preston, PR2 7DA

2E0 RFE R Parkinson, 27 Hopton Avenue, Mirfield, WF14 8JW

2E0 RFF Mark Petchey, 74 Avondale, Ellesmere Port, CH65 6RW

2E0 RFG R Gray, Upper Bisterne Farmhouse, Bisterne, Ringwood, BH24 3BP

2E0 RFH Roger Henderson, 9 Green Mead, South Woodham Ferrers, Chelmsford, CM3 5NL

2E0 RFI Antony Driscoll, 82 Station Road, Langford, Biggleswade, SG18 9PQ

2E0 RFK Robert Eardley, Bridge Cottage, Martin, Fordingbridge, SP6 3LD

2E0 RFL Robert Tongs, 9 Woodland Drive, Winterslow, Salisbury, SP5 1SZ

2E0 RFM Rowan Corney, Lavender Cottage, Worlds End, Waterlooville, PO7 4QU

2E0 RFQ D Brown, 38 Tamworth Road, Sutton Coldfield, B75 6DG

2W0 RFT S Beer, 49 Central Street, Pwllypant, Caerphilly, CF83 2NJ

2E0 RFU Jonathan Byrne, 316 Turncroft Lane, Stockport, SK1 4BP

2W0 RFX R Ashworth, The Vicarage, Crymych, SA41 3RN

2W0 RGA Roy Anderson, 156 Cockett Road, Cockett, Swansea, SA2 0FQ

2E0 RGB F Birtwistle, 27 Southwell Road, Wisbech, PE13 3LF

2E0 RGC Ron Charbonneau, 11 Farriers Close, Billingshurst, RH14 9LT

2I0 RGD Bertie Gilliland, 28 Baird Avenue, Donaghcloney, Craigavon, BT66 7LP

2E0 RGK K Harley, 5 Saltrens Cottages, Monkleigh, Bideford, EX39 5JP

2E0 RGL Gary Lewis, 93 Eastcliff, Portishead, Bristol, BS20 7AD

2I0 RGM Ryan Murphy, 40 Stoneypath, Londonderry, BT47 2AF

2E0 RGO Max White, 7 Overthwart Crescent, Worcester, WR4 0JW

2E0 RGR R Hopwood, 136 Bradbury Lane, Hednesford, Cannock, WS12 4EN

2E0 RGS Richard Simpson, 22 Kenworthy Road, Stocksbridge, Sheffield, S36 1BZ

2I0 RGT R Todd, 3 Granville Manor, Kells, Ballymena, BT42 3JE

2D0 RGW D Corkish, 10 Vicarage Close, Ballabeg, Castletown, Isle of Man, IM9 4LQ

2E0 RGY P Haydon, 101 Blandford Street, Ashton-under-Lyne, OL6 7HG

2E0 RHC Richard Cook, 62 High Hazel Road, Moorends, Doncaster, DN8 4QN

2E0 RHE Richard East, 6 Ashley Road, Worcester, WR5 3AY

2E0 RHI Anthony Johnston, 44 Cradoc Road, Brecon, LD3 9LH

2E0 RHL Rachel Landragin, 101 Linden Gardens, Enfield, EN1 4DY

2E0 RHM Robert Murphy, 23 Lowndes Close, Stockport, SK2 6DW

2I0 RHN Robert Dunwoody, 16 Dernalea Road, Milford, Armagh, BT60 4DZ

2E0 RHO Rhosyn Celyn, 20 Aylesbury Street, Wolverton, Milton Keynes, MK12 5HZ

2I0 RHQ Hilary Selfridge, 147 Culcrum Road, Dunloy, Ballymena, BT44 9DT

2E0 RHS S Hawkins, Forest Edge, Deer Park, Blandford Forum, DT11 0AY

2E0 RHT Ralph Harkness, 4 Cassalands, Dumfries, DG2 7NS

2E0 RIA Maria Landragin, 101 Linden Gardens, Enfield, EN1 4DY

2E0 RIH Raymond Hart, Jays, South Street, Gillingham, SP8 5ET

2E0 RIN Rina Horner, 21 Ainsworth Road, Little Lever, Bolton, BL3 1RG

2E0 RIO Sarah Rutt, Granthorpe, Hull Road, Hull, HU11 5RN

2I0 RIR Edward Stevenson, 25 Woodlands Manor, Portadown, Craigavon, BT62 4JP

2E0 RIS John Edwards, 45 Bramshaw Gardens, Bournemouth, BH8 0BT

2E0 RIW Rowan Cruse Howse, Woodland, Moretonhampstead, Newton Abbot, TQ13 8SD

2E0 RIZ C Chilton, 217 North Dean Road, Keighley, BD22 6QT

2E0 RJA R Ashman, 61 Fairfield Road, Burgess Hill, RH15 8NP

2E0 RJD J Dixon, 23 Dee Way, Winsford, CW7 3JB

2E0 RJE R Earnshaw, 53 Blue Waters Drive, Paignton, TQ4 6JF

2E0 RJH Robert Harrison, 10 20 Hall Lane, Kirkburton, Huddersfield, HD8 0QW

2M0 RJJ Rolfe James, The Garret, Alyth, Blairgowrie, PH11 8HQ

2E0 RJL Robert Lovesey, 33 Ty Isaf Park Avenue, Risca, Newport, NP11 6NB

2E0 RJM Roger Millington, Quaintways, The Avenue, Tarporley, CW6 0BA

2E0 RJO Richard Hart, 4 Glade Mews, Guildford, GU1 2FB

2E0 RJP Robert Pounder, 65 Stubsmead, Swindon, SN3 3TB

2E0 RJR Robert Radley, 131 North Marine Road, Scarborough, YO12 7HU

2E0 RJU Andrew Foster, 7 Moira Dale, Castle Donington, Derby, DE74 2PG

2E0 RJX Richard Miller, 23 Clarendon Road, Sevenoaks, TN13 1EU

2E0 RJZ Roger Duthie, 14 Kettles Close, Oakington, Cambridge, CB24 3XA

2E0 RKB Albert Turner, 140 Thorndon Avenue, West Horndon, Brentwood, CM13 3TR

2W0 RKF William Lewis, 53 Leyshon Road, Gwaun Cae Gurwen, Ammanford, SA18 1EN

2E0 RKK Edward Whiten, 17 Scott Close, Ashby-de-la-Zouch, LE65 1HT

2E0 RKM Mark Beasley, 292 North Road, Yate, Bristol, BS37 7LL

2E0 RKO Norman Jacobs, 43 The Pines, Yapton, Arundel, BN18 0EG

2E0 RKR Graham Hirst, 15 Hengist Close, Horsham, RH12 1SB

2E0 RKS Richard Styles, 4 Coningsby Close, Gainsborough, DN21 1SS

2E0 RKX Mark Hemming, 11 Blackberry Way, Evesham, WR11 3AH

2D0 RLA R Allcote, 77 Ballanorris Crescent, Ballabeg, Castletown, Isle of Man, IM9 4ER

2E0 RLG Jack Berrisford, 8 Streatham Road, Derby, DE22 4AY

2E0 RLJ R Johnson, 23 Friars Dene Road, Gateshead, NE10 0DR

2E0 RLR Roger Hanson, 3 Lower Collier Fold, Cawthorne, Barnsley, S75 4HT

2E0 RLW Richard Wood, 7 Wishart Green, Old Farm Park, Milton Keynes, MK7 8QB

2E0 RLY A Hock, 29 Marley Road, Kingswinford, DY6 8BQ

2E0 RMC [illegible] Campbell, 8 [illegible] Bank, York, YO10 5HH

2I0 RMD Raymond Nelson, [illegible] Road, Rosslea, BT92 7FN

2M0 RMH R Hay, 12 Mitchell Brae, Balmedie, Aberdeen, AB23 8FW

2E0 RMI J Salmon, 25 Helston Road, Chelmsford, CM1 6JF

2E0 RMJ Jez Mitchell, 11 Brookside Drive, Oadby, Leicester, LE2 4PB

2I0 RMK Ann Mackenzie, 30 Dalriada Gardens, Ballycastle, BT54 6DZ

2E0 RML Richard Lunson, 130 Pilgrims Way, Bedford, MK42 9TZ

2E0 RMM John Hodson, 77 Waltham Road, Woodford Green, IG8 8DW

2M0 RMN Ross Nicoll, 15 Redford Walk, Edinburgh, EH13 0AF

2W0 RMO Ryan Orchard, The Burrows, Spring Gardens, Whitland, SA34 0HL

2M0 RMP James Mclelland, 9 Roberts Quadrant, Bellshill, ML4 2BG

2E0 RMR Mark Manley, 20 sorrel drive, Newport, NP20 5DP

2E0 RMS Robert Stevenson, 97 Queen Street, Crewe, CW1 4AL

2E0 RMT Mark Callow, 4 The Firs, Canvey Island, SS8 9TW

2E0 RMV Julian Vine, 2 Chapel Meadow Cottages, Chapel Road, Deal, CT14 0JF

2W0 RMY Adrian Lewis, 3 Aster View, Port Talbot, SA12 7ED

2E0 RMZ Richard Mansfield, 8 Haysoms Drive, Greenham, Thatcham, RG19 8EY

2E0 RNA Peter Bannon, 73 London Road, Worcester, WR5 2DU

2M0 RND M Gerrard, 10 Whinhill Gardens, Aberdeen, AB11 7WD

2E0 RNE R Emery, 67 Victoria Street, Gillingham, ME7 1EW

2E0 RNF Martin Faulkner, 6 Stanley Avenue, Queenborough, ME11 5DT

2E0 RNI D Taylor, 8 Curlew Close, Winsford, CW7 1SW

2E0 RNJ John Reynolds, 38 Spring Lane, Hockley Heath, Solihull, B94 6QY

2E0 RNM M James, 7 Pixey Place, Oxford, OX2 8BB

2E0 RNO Jeremy Franks, 14 The Hamlet, Slades Hill, Templecombe, BA8 0HJ

2E0 RNP Rodney Penaluna, 113 Brocklesby Road, Scunthorpe, DN17 2LW

2D0 RNR Jason Shettler, 504 Leeds Road, Huddersfield, HD2 1YW

2E0 RNS Paul Rainey, 27 School Road, Silver End, Witham, CM8 3RZ

2E0 RNT Richard Taylor, 22 Shakespeare Court, Chaucer Way, Hoddesdon, EN11 9QS

2E0 RNU Melville Nutt, 110 Birkinstyle Lane, Shirland, Alfreton, DE55 6BT

2E0 RNW Nick Rapson, 15 School Close, Bampton, Tiverton, EX16 9NN

2E0 RNX Jeremy Faulks, 11 Fishguard Spur, Slough, SL1 1TS

2E0 ROB R Morgan, 153 Beanfield Avenue, Coventry, CV3 6NY

2I0 ROC Alistair McCann, 6 Bowens Meadow, Lurgan, BT66 7UT

2E0 ROD Roderick Burton, 23 Freston, Paston, Peterborough, PE4 7EN

2E0 ROI W Chorlton, 25 Ash Grove, Orrell, Wigan, WN5 8NG

2E0 ROM Romano Pasika, 192 Longfield Lane, Cheshunt, Waltham Cross, EN7 6AQ

2E0 ROP Paul Bolton, 1 Acorn Rise, Hollesley, Woodbridge, IP12 3JT

2E0 ROS C Ross, 27 The Meadows, Skegness, PE25 2JA

2M0 ROT Stuart McIntosh, 1 Ivybank Road, Port Glasgow, PA14 5LH

2E0 ROV Robert Van-der-Wijst, 6 Willow Street, Romford, RM7 7LJ

2E0 ROY Roy Trzeciak-Hicks, 24 Wolston Meadow, Middleton, Milton Keynes, MK10 9AY

2E0 RPA Roger Ashley, 15 Wimbourne Drive, Gillingham, ME8 9EN

2E0 RPC Roger Cockayne, 20 The Shrubbery, Rugeley, WS15 1JJ

2E0 RPD I Handley, Rosedale, Chapman Street, Market Rasen, LN8 3DS

2E0 RPE Andrew Wallman, 30 Elmsway, Bramhall, Stockport, SK7 2AE

2E0 RPF Robert Fullagar, 6 Locke Way, Stafford, ST16 3RE

2E0 RPH Roger Hann, 29 Sherrell Park, Bere Alston, Yelverton, PL20 7AZ

2I0 RPM R Gault, 7 Gardenmore Place, Larne, BT40 1SE

2E0 RPO W Donnelly, 4 Mayfield Road, Bentham, Lancaster, LA2 7LP

2E0 RPR Martyn Roper, 13 St. Cuthbert Street, Worksop, S80 2HN

2E0 RPT R Thompson, 109 Battle Road, Hailsham, BN27 1UD

2E0 RPY Henry Kennedy, 11 Green Road, High Wycombe, HP13 5BD

2E0 RQK B Dare, 1 St. Johns Villas, Sivell Place, Exeter, EX2 5ES

2E0 RQN Joseph Foster, 23 High Street, Cumnor, Oxford, OX2 9PE

2E0 RRC Richard Wilson, 84 Sir Thomas Whites Road, Coventry, CV5 8DR

2I0 RRE Robert Rantin, 8a Buchanans Road, Newry, BT35 6NS

2E0 RRF Derry White, 2 Birchwood Avenue, Breaston, Derby, DE72 3AQ

2E0 RRL Kathleen Woodhams, 83 Langdale Place, Newton Aycliffe, DL5 7DY

2M0 RRO C Duncan, 131 Croftend Avenue, Glasgow, G44 5PF

2M0 RRS Darren Lee, 11 Linton Road, Dundee, DD2 2SZ

2M0 RRT Mark Batchelor, Flat 11, 21 Albany Terrace, Dundee, DD3 6HR

2W0 RRY Peter Smith, 19 Grandison Street, Neath, SA11 2PG

2D0 RSB Richard Blandford, 60 Benomley Road, Almondbury, Huddersfield, HD5 8LS

2E0 RSD Stephane Ray, 28 Stenbury View, Wroxall, Ventnor, PO38 3DB

2E0 RSG Simon Cowgill, 17 Tooley Street, Boston, PE21 6DP

2E0 RSH R Hodgkinson, 39 Oxford Road, Carlton-in-Lindrick, Worksop, S81 0[illegible]

2E0 RSI R simms, 3 The Byeway, London, SW14 7NL

2E0 RSM M Shortreed, 13 Marshfield Road, Settle, BD24 9DA

2E0 RSV Graham Jones, 31 Liverpool Road, Buckley, CH7 3LH

2M0 RTA Robert Turpie, 11 Ashkirk Place, Dundee, DD4 0TN

2E0 RTC Christopher Ring, 29 Shelley Close, Newport Pagnell, MK16 8JB

2M0 RTD D Robertson, 17 Keswick Drive, Hamilton, ML3 7HN

2E0 RTE Paul Ballington, 7 Links Close, Sinfin, Derby, DE24 9PF

2W0 RTJ Richard Johns, 39 Tyla Coch, Llanharry, Pontyclun, CF72 9LT

2E0 RTM Stephen Key, 159 Launcelot Road, Bromley, BR1 5EA

2E0 RTN Martine Symmonds, 24 Woodville Grove, Stockport, SK5 7HU

2E0 RTP Rael Paster, 8 Rachaels Lake View, Warfield, Bracknell, RG42 3XU

2E0 RTQ Kieron Jones, 1 Mowbray Street, Epworth, Doncaster, DN9 1HR

2E0 RTW R Whitfield, 47 Denchworth Road, Wantage, OX12 9AY

2E0 RTY Arthur Loukes, 14 Batchwood View, St. Albans, AL3 5TD



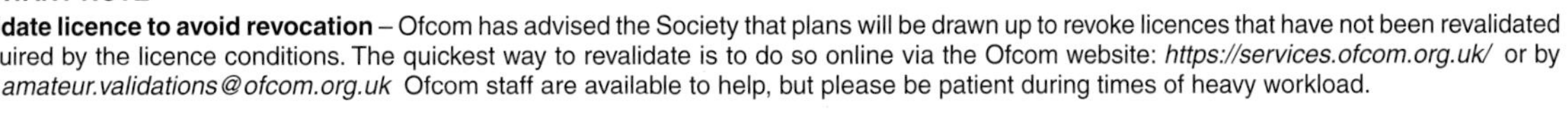

**IMPORTANT NOTE**

**Revalidate licence to avoid revocation** – Ofcom has advised the Society that plans will be drawn up to revoke licences that have not been revalidated as required by the licence conditions. The quickest way to revalidate is to do so online via the Ofcom website: *https://services.ofcom.org.uk/* or by email: *amateur.validations@ofcom.org.uk* Ofcom staff are available to help, but please be patient during times of heavy workload.

2E0 RUD R Rudd, 11 Woodlands Way, Lepton, Huddersfield, HD8 0JA
2E0 RUG Jon Fry, 104 Freemantle Road, Rugby, CV22 7HY
2E0 RUI Rui Lima Matos, Flat 9, 55 Shepherds Hill, London, N6 5QP
2M0 RUP R Hart, Rosalis, Piperhill, Nairn, IV12 5SD
2E0 RUS Russell Garland, 113 The Drive, Feltham, TW14 0AH
2E0 RUZ Russell Brierley, 39 Hatfield Road, Alvaston, Derby, DE24 0BU
2E0 RVE Rob Harvey, 15 Abbey Park Way Weston, Crewe, CW2 5NR
2M0 RVF Steven Wilson, 46 Babylon Road, Bellshill, ML4 2HQ
2I0 RVH Thomas Nelson, 1, 1 Annashanco, Rosslea, BT92 7PT
2E0 RVI Ravi Miranda, Flat 56, Amelia House 11 Boulevard Drive, London, NW9 5JP
2I0 RVT Robert Todd, 58 Kilrea Road, Portglenone, Ballymena, BT44 8JB
2E0 RVV David Kemp, 2 Darwin Close, Elston, Newark, NG23 5PQ
2E0 RWB Ronald Whalley, 65 Stanley Street, Nelson, BB9 7ET
2E0 RWF Tom Mitchell, 9 Rhiw Grange, Colwyn Bay, LL29 7TT
2I0 RWG Richard Gilmore, 86 Lylehill Road, Templepatrick, Ballyclare, BT39 0HL
2M0 RWH Robert Humphrey, 170 Clement Rise, Livingston, EH54 6LP
2E0 RWN R Nock, 83 Coles Lane, West Bromwich, B71 2QW
2E0 RWP R Penny, 93 Mirfield Grove, Hull, HU9 4QR
2E0 RWT A White, 3 Ipswich Place, Thornton-Cleveleys, FY5 1SP
2E0 RWX R Williamson, 23 Harrowdyke, Barton-upon-Humber, DN18 5LN
2M0 RWZ Robert Walker, 10 Westpark Gate, Saline, Dunfermline, KY12 9US
2E0 RXC Ross Codrai, 27 Howard Avenue, West Wittering, Chichester, PO20 8EX
2E0 RXN Julian Cartwright, 37a Station Road, Whitwell, Worksop, S80 4UF
2E0 RXR Roy Rich, 9 Spring Close, Verwood, BH31 6LB
2E0 RXT Matthew Harrison, 2 Rosemount Court, Holly Bank Road, York, YO24 4EG
2E0 RXW Reuben Wells, The Turkey House, Park Farm, Tudeley, Tonbridge, TN11 0NL
2E0 RXX Greg Acton, 39 Craig Road, Macclesfield, SK11 7YH
2M0 RYL R Haynes, 29 Invercauld Road, Aberdeen, AB16 5RP
2E0 RYN Derrick Renshaw, 25 Ashley Road, Worksop, S81 7JS
2E0 RYP C Barrett, 1 Mead Road, Padgate, Warrington, WA1 3TN
2E0 RYS Ryan Sayre, 8 Lorne Road, Richmond, TW10 6DS
2E0 RYX Ian Van Der Linde, 77 Port Vale, Hertford, SG14 3AF
2E0 RZB Ken Slade, 7 Cottage Farm Mews, The Street, King's Lynn, PE33 9JQ
2J0 RZD Robert Luscombe, Flat, 1 Rouge Bouillon, St. Helier, Jersey, JE2 3ZA
2M0 RZE G Smith, 40 Pirleyhill Drive, Shieldhill, Falkirk, FK1 2EA
2E0 RZH M Flynn, 20 Manwood Avenue, Canterbury, CT2 7AH
2E0 RZL Robert Higgins, 44 Maeshyfryd Road, Holyhead, LL65 2AL
2E0 RZM Graham Kingstone, 17 Ullswater Drive, Leighton Buzzard, LU7 2QR
2I0 RZT James Thompson, 119 Rathkyle, Antrim, BT41 1LN
2E0 RZX Ben Forrest, 32 Idonia Road, Perton, Wolverhampton, WV6 7NQ
2E0 SAA Steven Garrett, 44 Wardle Crescent, Leek, ST13 5PW
2E0 SAF Sean Finch, 25 Bluebell Avenue, Wigan, WN6 8NS
2I0 SAI Simon Barnes, 191 Marlacoo Road, Portadown, Craigavon, BT62 3TD
2E0 SAK S Poyser, 80 Derby Road, Long Eaton, NG10 4LB
2E0 SAL M BANCROFT, 1 John Street, Knutton, Newcastle, ST5 6DT
2E0 SAN S Neachell, 59 Gilmour Crescent, Worcester, WR3 7PJ
2E0 SAT Michael Ward, Fair View, Queen Street, Winkleigh, EX19 8JB
2E0 SAU Cyril Mokes, 22 Oxclose Lane, Arnold, Nottingham, NG5 6GA
2E0 SAW Stacy Williams, Flat 35, Winterton House, London, E1 2QR
2M0 SAX G Sproul, 132 Muirdrum Avenue, Glasgow, G52 3AP
2E0 SAY Mark Baker, 20 Centurion Rise, Hastings, TN34 2UL
2E0 SAZ S Waller, 17 Vere Road, Peterborough, PE1 3DZ
2E0 SBB Anthony Southwell, 56 Lambrook Road, Taunton, TA1 2AF
2E0 SBC Kenneth Smith, 7 Rosebery Avenue, Morecambe, LA4 5RU
2E0 SBD Samantha Shailes, 9 Ingham Street, Padiham, Burnley, BB12 8DR
2E0 SBH Stephen Ray, 18 Crescent Way, Cholsey, Wallingford, OX10 9NE
2E0 SBJ Carlo Didcott, 9 Maid Marion Avenue, Selston, Nottingham, NG16 6QH
2E0 SBK Steve Harvey, 40 Thales Drive, Arnold, Nottingham, NG5 7NF
2E0 SBL Steven York, 1 The Cottage, Dogdyke Bank, Lincoln, LN4 4JQ
2E0 SBM M Burrows, 4 Melton Street, Earl Shilton, Leicester, LE9 7FP
2E0 SBN Harry Buckley, 10 Lower Hey Lane, Mossley, Ashton-under-Lyne, OL5 9DE
2E0 SBO Robert Raine, 91 Lulworth Avenue, Jarrow, NE32 3SB
2M0 SBP Keith Verrall, 7 Roshven View, Arisaig, PH39 4NX
2E0 SBS Gerrard Hamilton, 5 Eascott Common, Eastcott, Devizes, SN10 4PL
2E0 SBX Stephen Elliott, 74 Preston Avenue, Alfreton, DE55 7JX
2E0 SBZ David Smith, 105 Princes Street, Dunstable, LU6 3AS
2E0 SCA Sean Waudby, 36 21st Avenue, Hull, HU6 8DJ
2E0 SCB Stephen Elias, 20 Attlee Way, Cefn Golau, Tredegar, NP22 3TA
2E0 SCC David Riman, 22 Princess Road, Hinckley, LE10 1EB
2E0 SCD John Lord, 5 Langworthy Avenue Little Hulton, Manchester, M38 9GQ
2E0 SCE Derek Hagan, 8 Charles Close, Westcliff-on-Sea, SS0 0EU
2E0 SCF Simon Faulkner, Mount Pleasant, Elkstones, Buxton, SK17 0LU
2M0 SCG Stephen Greenland, Flat 12, Weymouth Court, 201 Weymouth Drive, Glasgow, G12 0ER
2E0 SCH Phil Hayes, 4 London Road, Roade, Northampton, NN7 2NL
2D0 SCJ C Short, 8 Whitley Willows, Lepton, Huddersfield, HD8 0GD
2E0 SCK Mark Stillman, 58 Highfield Road, Bognor Regis, PO22 8PH
2W0 SCL Anthony Davies, 25 Llanfair Road, Tonypandy, CF40 1TA
2E0 SCM J Stephenson, 4 Carlow Drive, West Sleekburn, Choppington, NE62 5UT
2E0 SCN Janet Porter, 14 Longwestgate, Scarborough, YO11 1QB
2E0 SCO John Hepburn, 32 Green Croft, Ashington, NE63 8EF
2W0 SCP S Peel, 28 Dan yr Allt, Llanelli, SA14 8AT
2E0 SCQ Steve Chandler, 7 Clinton Road, Redruth, TR15 2LL
2E0 SCR Christopher Petrie, 14 Rotherfield Avenue, Eastbourne, BN23 8JQ
2E0 SCS Arthur Mcallister-Bowditch, 24 Morse Close, Chippenham, SN15 3FY
2E0 SCV Steven Shaw, 2 Daleside, Todmorden, OL14 7NE
2E0 SCW Stuart Whittaker, 25 Cleveleys Road, Blackburn, BB2 3JS
2E0 SCX Stephen Hassall, 21 Bridgnorth Grove, Newcastle, ST5 7QP
2E0 SDA Adam Sanderson, 65 Holm Flatt Street, Parkgate, Rotherham, S62 6HJ
2E0 SDC S Cornwell, 46 School Lane, Lower Cambourne, Cambridge, CB23 5DG
2E0 SDD Steven David, 34 Hardwicke Walk, Kings Heath, Birmingham, B14 5XX
2E0 SDJ Danny Wild, 10 The Green, Lydd, Romney Marsh, TN29 9ES
2E0 SDK S Allington, 137 Marshall Lane, Northwich, CW8 1LA
2E0 SDM Carl Reid, 28 Albion Road, London, N16 9PH
2E0 SDO Malcolm Johns, 151 Somerset Street, Abertillery, NP13 1DR
2E0 SDP Raymond Marco De Vries, Corner Cottage, Hillcrest Close, Sturminster Newton, DT10 2DL
2E0 SDQ Philip Weaver, 1 Madeley Street, Newcastle, ST5 6LS
2E0 SDT S Theaker, 10 Grange Fields Mount, Leeds, LS10 4QN
2E0 SDV Jamie Williams, 41 Overton Lane, Hammerwich, Burntwood, WS7 0LQ
2E0 SDW S Whitehead, 55 Crombie Road, Sidcup, DA15 8AT
2E0 SDY James Popple, 10 Kingsmead Park, Waterbeach, Cambridge, CB25 9PF
2E0 SDZ Eleri Ayre, 1 Spring Gardens, Broadmayne, Dorchester, DT2 8PP
2E0 SEA M Bown, 47 Ullswater Crescent, Weymouth, DT3 5HF
2E0 SEB Scott Gordon, 15 Turnstone Road, Chatham, ME5 8NE
2I0 SEC Seamus Carlin, 9 Mullandra Park, Kilcoo, Newry, BT34 5LS
2E0 SEE Axel Seedig, 8 Barton Court, Cambridge Road West, Farnborough, GU14 6QA
2M0 SEF Stuart Fleming, 20 Park Road, Invergowrie, Dundee, DD2 5AH
2M0 SEG Magnus Henry, Selkie Stanes, Scatness, Shetland, ZE3 9JW
2I0 SEH Stefan Pynappels, 38 Dora Avenue, Newry, BT34 1JW
2E0 SEJ Shane Johnson, 2 North Square, Edlington, Doncaster, DN12 1ED
2I0 SEK Raymond Thomson, 1 Litchfield Park, Coleraine, BT51 3TN
2D0 SEO Simon Bates, 6 Foxdell, Northwood, HA6 2BU
2E0 SEP Andrew Howard, 24 Ladybower Lane, Poulton-le-Fylde, FY6 7FY
2E0 SER Michelle Edmonds, 20 Tomline Road, Ipswich, IP3 8BZ
2E0 SES Marco Cianni, 121 Springfield Park Avenue, Chelmsford, CM2 6EW
2E0 SET Paul Setter, 199 Southbourne Grove, Westcliff-on-Sea, SS0 0AN
2E0 SEW Gillian Ferguson, 31 Barton Court Road, New Milton, BH25 6NW
2E0 SEY Jake Mottram, 196 Queens Drive, Nantwich, CW5 5LA
2E0 SEZ S Ezard, 59 Station Farm, Croesyceiliog, Cwmbran, NP44 2JW
2I0 SFA Noel Jenkinson, 35 Scarvagh Heights, Scarva, Craigavon, BT63 6LY
2E0 SFB Paul Latham, 20 Kenyon Avenue, Wrexham, LL11 2ST
2E0 SFC Russell Bates, 61 Park View, Crowmarsh Gifford, Wallingford, OX10 8BN
2E0 SFS Mark Heenan, 15 Woodacre Green, Bardsey, Leeds, LS17 9AB
2E0 SFT M Shelton, 92 Timberley Drive, Grimsby, DN37 9QZ
2E0 SGG Steve Gibbs, 61a Main Road, Hoo, Rochester, ME3 9AA
2E0 SGH Steven Holt, 14 Fir Street, Cadishead, Manchester, M44 5AU
2E0 SGI A Parker, 4 Park Avenue, Shirebrook, Mansfield, NG20 8JW
2I0 SGM Stephen Murray, 117, KNOCKVIEW DRIVE, Tandragee, BT622BL
2M0 SGQ Stephen Gill, 5 Ramornie Place, Kingskettle, Cupar, KY15 7PT
2E0 SGY Simon Gregory, 9 Croftlands Road, Wythenshawe, M22 9YE
2E0 SHA Jim Ridley, 73 The Markhams, New Ollerton, Newark, NG22 9QY
2E0 SHB S Burrows, 78a Coronation Road, Earl Shilton, Leicester, LE9 7HJ
2E0 SHG David Grindrod, 6 Croft Way, Market Drayton, TF9 3UB
2I0 SHM S Montgomery, 2 Woodland Gardens, Lisburn, BT27 4PL
2E0 SHP Andrew Sharp, 30 Burbidge Close, Calcot, Reading, RG31 7ZU
2E0 SHV Amrit Sidhu-Brar, White Gates, Main Road, Northampton, NN7 3NA
2E0 SHW Adrian Dransfield, 135 Rosedale Grove, Hull, HU5 5DA
2E0 SHY William Shelley, 91 Canterbury House, Stratfield Road, Borehamwood, WD6 1NT
2I0 SHZ Sharon Lewis, 15 Foyle Drive, Ballykelly, Limavady, BT49 9PG
2E0 SIA Simon Skirving, 1 Hallington Close, Bolton, BL3 6YH
2E0 SIB Darren Sibley, 25 Avery Close, Leighton Buzzard, LU7 4UP
2E0 SID Sidney Frampton, 20 Winslow Close, Boldon Colliery, NE35 9LR
2E0 SIJ Jacqueline Simkins, 37 St. Andrews Meadow, Harlow, CM18 6BL
2E0 SIM S Matley, 67 Alexandra Road, Chandlers Ford, SO53 2BP
2E0 SIS Aydin Tanseli, 157 Warwick Road, Rayleigh, SS6 8SG
2E0 SIX S Hannah, 4 Station Road, Minsterley, Shrewsbury, SY5 0BG
2E0 SIY Simon Shaul, Shepherds Cottage, Middle Street, Gainsborough, DN21 5BU
2E0 SJA A Spears, 106 Cleves Way, Ashford, TN23 5DF
2E0 SJD Stephen Down, 1 Dove Close, Honiton, EX14 2GP
2E0 SJF S Farnell, 16 Lily Way, Lowestoft, NR33 8NN
2E0 SJG Stephen Goodridge, Trelane, Pelynt, Looe, PL13 2LF
2E0 SJH S Humphreys, 52 Llys Owain, Bangor, LL57 1SH
2E0 SJI Stuart Ibbotson, 17 Marley Combe Road, Haslemere, GU27 3SN
2E0 SJK Shirley Kendrick, 29 Waterside, Silsden, Keighley, BD20 0LQ
2E0 SJM S Moore, 33 Church Street, Heavitree, Exeter, EX2 5EP
2E0 SJN Steven Nash, 3 Brightside, Waterlooville, PO7 7BA
2E0 SJP Stan Parker, 36 Eton Close, Lincoln, LN6 0YF
2E0 SJQ Steven Carpenter, Field View, Old Lyndhurst Road, Southampton, SO40 2NL
2E0 SJR S Roberts, 7 Alberta Grove, Prescot, L34 1PX
2E0 SJS S Spencer, 55 Witton Lane, West Bromwich, B71 2AA
2E0 SJT Stuart tweddle, 3 Bron Ffinan, Pentraeth, LL75 8UT
2I0 SJV David Parkinson, 16 Beechwood Gardens, Moira, Craigavon, BT67 0LB
2E0 SKA G Brown, 3 Willow Lane, Goostrey, Crewe, CW4 8PP
2E0 SKD Mark Powell, 98 Provan Court, Ipswich, IP3 8GG
2E0 SKE Martin Swan, 35 Colston Close, Plymouth, PL6 6AY
2W0 SKG Wayne Lloyd, 29 Duffryn Street, Mountain Ash, CF45 3NU
2E0 SKI Andre Skarzynski, 93 Richmond Walk, St. Albans, AL4 9BB
2E0 SKK Adam Greig, 3 Fir Grange Avenue, Weybridge, KT13 9AR
2E0 SKL Carl Hare, Flat 6, Elswyn House, 64 Hatherley Road, Sidcup, DA14 4AW
2E0 SKZ Jack Parfitt, 5 Sheridan Road, Frimley, Camberley, GU16 7DU
2E0 SLD D Dash, 36 Rockvilla Close, Varteg, Pontypool, NP4 7QF
2E0 SLE Stephanie Eyre, 123 Baden Powell Road, Chesterfield, S40 2RL
2E0 SLH S Hill, 108 Hitchin Close, Romford, RM3 7EQ
2E0 SLJ Stephen Legg, 98 Shenstone Valley Road, Halesowen, B62 9TF
2E0 SLK James Blezard, 10 North Row, Barrow-in-Furness, LA13 0HE
2E0 SLM John Stringer, 31 Pipit Lane, Birchwood, Warrington, WA3 6NY
2E0 SLN Robert Nixon, Dove Crag, Silverton Lane, Morpeth, NE65 7RJ
2E0 SLO Denise Willingham, 42 Childers Court, Ipswich, IP3 0DU
2E0 SLP S Williams, 63 Trem Eryri, Llanfairpwllgwyngyll, LL61 5JF
2E0 SLR B Robins, 99 Uplands Road, Dudley, DY2 8BB
2E0 SLS Kenneth Holloway, 6 Britons Lane Close Beeston Regis, Sheringham, NR26 8SH
2E0 SLT S Thompson, 80 Mountain Road, Dewsbury, WF12 0BP
2E0 SMA Stephen Harcourt, 71 Ingleby Road, Long Eaton, Nottingham, NG10 3DG
2E0 SMB S Bowkett, 4 May Street, Newport, NP19 0EG
2I0 SMD S Davey, 16 Maritime Drive, Carrickfergus, BT38 8GQ
2E0 SMG S Gearey, 32 Bridgeside, Deal, CT14 9SS
2E0 SMK Susan Kendrick, 103a Latimer Street, Liverpool, L5 2RF
2E0 SML Samuel Weightman, Rose Cottage, Ellingstring, Ripon, HG4 4PW
2M0 SMM Stewart Harvey, 63 Darley Road, Cumbernauld, Glasgow, G68 0JR
2M0 SMN Stewart Nicoll, 15 Redford Walk, Edinburgh, EH13 0AF
2E0 SMO S O'Neill, 1 Avenue Cottages, Winchester Road, Fareham, PO17 5EX
2E0 SMS S Stratford, 23 The Fairway, Banbury, OX16 0RR
2E0 SMX Steven Hogg, 57 The Grange, Burton-on-Trent, DE14 2EX
2I0 SMY Sam Dallas, 101 Coagh Road, Stewartstown, Dungannon, BT71 5JL
2E0 SMZ Simon Edwards, 30 Morrison Road, Tipton, DY4 7PU
2E0 SNA Paul Harvey, 22 Meredale Road, Liverpool, L18 5EX
2E0 SND Andrew Williamson, 25 Manor Road, Rugby, CV21 2SZ
2E0 SNE Michael Bryan, 13 Elmwood Avenue, Sunderland, SR5 5AW
2I0 SNG Stephen Gilmour, 14g Malcolm Road, Lurgan, Craigavon, BT66 8DF
2E0 SNJ Simon Worger, 6 Glendale Terrace, Mornington Road, Whitehill Bordon, GU35 9AJ
2E0 SNM Steven Hindmarsh, 55 Warkworth Crescent, Seaham, SR7 8JT
2E0 SNP S Powell, 3 Tadgedale Avenue, Loggerheads, Market Drayton, TF9 4DD
2E0 SNS Shaun Button, 6 Farfield, Retford, DN22 7TL
2E0 SNU E Collins, 38 Meynell Road, Sheffield, S5 8GN
2W0 SNW S Williams, 56 Heol Llansantffraid, Sarn, Bridgend, CF32 9NH
2E0 SNX Thanawit Lertruengpanya, Flat 1, Mallow Court, London, SE13 7PR
2M0 SOE Raymond Baxter, 24g Bradan Road, Troon, KA10 6DS
2E0 SOF Wayne Soffe, 96 Urban Road, Doncaster, DN4 0EP
2M0 SOP A Gordon, 26 East Millicent Avenue, Golspie, KW10 6TL
2E0 SOT S Gregory, 11 Ribblesdale Avenue, Congleton, CW12 2BS
2E0 SOX R Cook, 46 Wheatsheaf Road, Tividale, Oldbury, B69 1SW
2E0 SOZ Robert Mather, 4 Vimy Road, Wednesbury, WS10 9BQ
2E0 SPA Stuart Etheridge, 50 Pond Road, Horsford, Norwich, NR10 3SW
2E0 SPB D Evans, 21 Quilter Close, Bilston, WV14 9AX
2E0 SPD Stephen Denman, 12 Dyke Vale Road, Sheffield, S12 4ER
2E0 SPG Martin Hennessey, 57 Northern Road, Aylesbury, HP199QT
2E0 SPH Dariusz Janowicz, 20 Salisbury Road, St. Leonards-on-Sea, TN37 6RX
2M0 SPL Scott Ling, Leadburnlea, Leadburn, West Linton, EH46 7BE
2E0 SPM Stephen Morris, 23 De Courtenai Close, Bournemouth, BH11 9PG
2E0 SPN John Daniels, 27 Hammerwater Drive, Warsop, Mansfield, NG20 0DJ
2E0 SPR S Randall, 23 Onslow Road, Plymouth, PL2 3QG
2E0 SPT K Tremain, 26 Longbeech Park, Canterbury Road, Ashford, TN27 0HA
2E0 SPU Paul Saunders, 62 Parkfield Avenue, Eastbourne, BN22 9SF
2E0 SPZ R Hinde, 3 Baunhill Close, Northampton, NN3 3EQ
2E0 SQK Paul Henry, 22 Huddleston Close, Wirral, CH49 8JP
2E0 SQL P Goodhall, 2 Manor Place, Oxford, OX1 3UN
2E0 SQN Ian Franklin, 23 Ingle Drive, Ashby-de-la-Zouch, LE65 2LW
2E0 SRB Stephen Blaikie, 22 Juno Close, Goring-by-Sea, Worthing, BN12 4UB
2E0 SRC Derek Barker, 12 The Weavers, Denstone, Uttoxeter, ST14 5DP
2W0 SRD David Bray, 24 Lon Gwesyn, Birchgrove, Swansea, SA7 9LD
2M0 SRF Robin Farrer, 23 Upper Craigour, Edinburgh, EH17 7SE
2E0 SRJ John Statham, Oakwoods, School Lane, Reading, RG8 8LT
2D0 SRP Paul Skidmore, 36 Princes Drive, Harrow, HA1 1XH
2E0 SRV Stephen Vickers, 35 Lanchester Road, Birmingham, B38 9AG
2M0 SRX Steven Russell, 3 Rankin Road, Wishaw, ML2 8PG
2M0 SRY Sean Young, 6 Ramsey Cottages, Bonnyrigg, EH19 3JG
2E0 SSA Altaf Dossa, 24 Warwick Drive, Cheshunt, Waltham Cross, EN8 0BW
2E0 SSC Steve Charles, 29 Woolford Close, Winchester, SO22 4DN
2E0 SSD Andrew Gillard, 4 Horton Avenue, Thame, OX9 3NJ
2E0 SSG A Butler, 12 South Bank Cottages, South Stoke, Reading, RG8 0HX
2E0 SSJ Scott Shortland, 10 Highclere Court, Knaphill, Woking, GU21 2QP
2E0 SSK Sarder Kamal, 40 Catterick Way, Borehamwood, WD6 4QT
2E0 SSL Graham Sawyer, 432 Rowood Drive, Solihull, B92 9LQ
2E0 SSN Brian Woods, 28 Delph Drive, Burscough, Ormskirk, L40 5BE
2M0 SSO Craig Haldane, 72A Coatbridge Road, Glenmavis, Airdrie, ml6 0nj
2E0 SST Stephen Slapper, 1 Standards Keep, Standards Road, Bridgwater, TA7 0EZ
2E0 SSX K Baker, 64 Pendle Drive, Basildon, SS14 3LZ
2E0 SSY Michael Hossell, 80 Murray Road, Sheffield, S11 7GG
2E0 STA S Collis, 21 Holme Hall Crescent, Chesterfield, S40 4PQ
2M0 STB Iain Learmonth, 85 Lord Hay's Grove Old Aberdeen, Aberdeen, AB24 1WT
2E0 STI Steve Pearce, 20 Barcote Walk, Plymouth, PL6 5QE
2D0 STL S Leslie, 16 Little Meddow, Andreas, Isle of Man, IM7 4HY
2I0 STN Steven Nash, 45 Parkfield Road, Ahoghill, Ballymena, BT42 1LY
2E0 STP Benjamin Freeman, 14 Rhee Spring, Baldock, SG7 6TD
2E0 STQ Edward St Quinton, Mill Cottage, The Thorofare, Woodbridge, IP13 8BB
2E0 STT A Burgess, 1 Laurel Crescent, Long Eaton, Nottingham, NG10 3NL
2M0 STV S McCormick, 4 Birch Way, Renfrew, PA4 8FB
2E0 STX Stuart Jackson, 64 Main Road, Moulton, Northwich, CW9 8PB
2I0 SUB Colin Williams, 15 Gleneden Park, Newtownabbey, BT37 0QL
2E0 SUD A Greenhalgh, 61 Long Meadows, Chorley, PR7 2YB
2E0 SUE Susan Turford, 1 Portland Crescent, Bolsover, Chesterfield, S44 6EG
2E0 SUF Simon Batley, 2 Boulge Road, Hasketon, Woodbridge, IP13 6LA
2E0 SUS Susan Millard, 112 Avenue Road, Sandown, PO36 8DZ
2E0 SUT Rupert Sutton, Yew Tree Farm, Paddol Green, Shrewsbury, SY4 5QZ
2E0 SUU Wendy Malcolm-Brown, Flat 11, Chiltern Court, Harpenden, AL5 5LY
2E0 SUX A Suttle, Delvillewood House, 81 Albert Street, Shildon, DL4 2DN
2E0 SUZ Susan Coombes, 33 Clarence Park Road, Bournemouth, BH7 6LF
2E0 SVG Joshua Savage, 44 Hastings Road, Maidstone, ME15 7SP
2E0 SVK Lubomir Mikolka, Flat, 1 Scotney Court, Romney Marsh, TN29 9JP
2E0 SVT Stephen Tayler, 22 Wheatley Road, Leicester, LE4 2HN
2E0 SVV A Kemp, 2 Darwin Close, Elston, Newark, NG23 5PQ
2E0 SVW D Wilkinson, Bryn Y Mor, North Road, Caernarfon, LL55 1BE
2E0 SVZ Andrew Abson, 117 Lysander Road, Rubery, Birmingham, B45 0EN
2M0 SWA S Anderson, 33 Dryden Avenue, Loanhead, EH20 9JT
2E0 SWB S Bennett, 38 Jowetts Lane, West Bromwich, B71 2QU
2E0 SWE Oscar Hall, 2 Beverley Lodge, Paradise Road, Richmond, TW9 1LL
2E0 SWF R Jenkinson, Esperance, West End Road, Doncaster, DN9 1LB
2M0 SWM Peter Holmes Holmes, 15a Ochlochy Park, Dunblane, FK15 0DU
2W0 SWO Stephen Owen, 500 Cowbridge Road West, Cardiff, CF5 5DA
2E0 SWS S Scotson, 86 Derrydown Road, Birmingham, B42 1RT
2E0 SWT Tamer Akay, 87 Swithland Avenue, Leicester, LE4 5BQ
2M0 SWY Mark Shewry, 110 Forbeshill, Forres, IV36 1JL
2E0 SWZ Ian Swindells, 69 Danby Close, Newton Moor, Hyde, SK14 4AF
2E0 SXC Sean Ward, 22 St. Margarets Close, Horstead, Norwich, NR12 7ER

2E0 SXF Stephen Levsen, 8 Craig Close Broughton, Brigg, DN20 0SE
2E0 SXJ Patrick Duckles, 8 Railway Cottages, Skillings Lane, Brough, HU15 1EN
2I0 SXM Geoffrey Hutton, 13 Meadowbank, Seapatrick, Banbridge, BT32 4[illegible]
2E0 SXP Stephen Perring, 26 Celandine Grove, Thatcham, RG18 4[illegible]
2E0 SXY S Lord, 34 Alsop Street, Leek, ST13 5NZ
2E0 SYD Kenneth Hunt, 13 Beaumaris Court, Spondon, Derby, DE21 7RG
2E0 SYE S Greenheart, 168 Beech Hill Lane, Wigan, WN6 8PL
2M0 SYL A Haynes, 29 Invercauld Road, Aberdeen, AB16 5RP
2E0 SYM C Symons, 7 Cavendish Street, Staveley, Chesterfield, S43 3QZ
2E0 SYN Matthew Hickey, Squirrels Hollow, 3 Castleford Drive, Prestbury, SK10 4BG
2E0 SYS Michael Styles, 6 Kilpale Close, Caerwent, Caldicot, NP26 5AF
2E0 SYW Sonny Ward, 24 [illegible] Road, [illegible], NR7 0DQ
2E0 SYY Colin Gibson, [illegible], Wigan, WN5 7LH
2E0 SZH Colin Hogarty, 4 Whitland Avenue, Bristol, BS13 9QQ
2J0 SZI Michael Drown, 200 Le Marais, St. Clement, Jersey, JE2 6GF
2E0 SZZ Malcolm Smith, 21 Buckden Close, Woodley, Reading, RG5 4HD
2I0 TAA Tom Boyd, 40 Walnut Park, Larne, BT40 2WF
2E0 TAC Anthony Paxton, 20f Green End, Granborough, Buckingham, MK18 3NT
2E0 TAG Joseph Bingham, 31 Wyro Close, Paignton, TQ4 7RU
2W0 TAI Tom Evans, 27 Addison Road, Neath, SA11 2AY
2E0 TAJ Alan Turner, 174 Preston Road, Standish, Wigan, WN6 0NP
2E0 TAK D Heathcote, 154 High St., Harriseahead, Stoke-on-Trent, ST7 4JX
2E0 TAL Anthony Walton, 65 Broadway East, Rotherham, S65 2XA
2E0 TAM A Dodds, 19 Westgate, Oldbury, B69 1BA
2I0 TAN Anthony Kelly, 16 Union Street Mews, Coleraine, BT52 1EN
2E0 TAO Geoffrey Jones, 57 Oxford Road, Banbury, OX16 9AJ
2E0 TAQ Robert Dicker, 38 Inkerman Road, Southampton, SO19 9DA
2E0 TAR Kevin Parry, 80 Cripps Avenue, Cefn Golau, Tredegar, NP22 3PB
2M0 TAS Dennis Branson, Derelochy, Kingsteps, Nairn, IV12 5LF
2E0 TAT A Horton, 11 Hilton Road, Tividale, Oldbury, B69 1JU
2E0 TAU A Smith, 46 Mulberry Close, Goldthorpe, Rotherham, S63 9LB
2E0 TAV Davy Rajanayagam, 87 Riffel Road, London, NW2 4PG
2E0 TAW T Williamson, 286 Glynswood, Chard, TA20 1BX
2E0 TAX Thomas McIntyre, Coach House, Commercial Street, Griffithstown, Pontypool, NP4 5JF
2E0 TAY A Taylor, 15 Woodford Glebe, Welford, Northampton, NN6 6AF
2E0 TAZ A Barron, 80 Primrose Crescent, Norwich, NR7 0SF
2E0 TBD Andrew Marshall, 13 The Markhams, New Ollerton, Newark, NG22 9QX
2E0 TBE Edward Whitehouse, 16 rue Gaston de Caillavet, Paris, France, 75015
2E0 TBF David Bateson, 19 Rothesay Road, Heysham, Morecambe, LA3 2UR
2E0 TBH James Smith, Flat 15, Hastings House, London, W12 7PY
2E0 TBI John Lynch, Beechway, Raddel Lane, Warrington, WA4 4EE
2E0 TBL James Wilson, 35 Lawson Avenue, Jarrow, NE32 5UF
2E0 TBQ Alexander Wong, 43 Northern Road, Aylesbury, HP19 9QT
2E0 TBR Howard Hylton, 214 School Road, Hall Green, Birmingham, B28 8PF
2E0 TBS Christopher Burnham, 2 Barry House, Elmleigh Road, Bristol, BS16 9AG
2E0 TBT Nicholas Hall, 37 Hayfell Avenue Heron Hill, Kendal, LA9 7JH
2E0 TBV A Hodgkinson, 41 Allott Crescent, Jump, Barnsley, S74 0LB
2E0 TBW Ben Fitzgerald-O'Connor, 24 Routh Street, London, E6 5XX
2E0 TBX Charles Pulford, 41 Morris Street, Sheringham, NR26 8JY
2E0 TBZ Christopher Cowen, Rosita, White Street Green, Sudbury, CO10 5JN
2E0 TCB Andrew Bruton, 29 Helyers Green, Wick, Littlehampton, BN17 7HB
2E0 TCC Allan Fleming, 39 Urswick Green, Barrow-in-Furness, LA13 0BH
2E0 TCF Timothy Guy, 16 Cogdeane Road, Poole, BH17 9AS
2E0 TCG Anthony Gilberts, 22 Granby Road, Buxton, SK17 7TW
2E0 TCI Andy Ashton, 46 Kingsland, Harlow, CM18 6XL
2I0 TCJ Trevor McKee, 4 Earlford Heights, Newtownabbey, BT36 5WZ
2E0 TCK Paul Pritchard, 11 Beacon Avenue, Dunstable, LU6 2AD
2W0 TCM M Digby, 40 Waterloo Road, Ammanford, SA18 3SF
2E0 TCN Colin Lyne, 4 Bridge Close, Catterick Garrison, DL9 4PG
2E0 TCO Thomas Corcoran, 191 Queensway, Rochdale, OL11 2NA
2E0 TCQ Michael Nicholas, 34 Doe Quarry Terrace, Dinnington, Sheffield, S25 2NP
2E0 TCU Brian Neal, 6 Canterbury Street, Chaddesden, Derby, DE21 4LG
2E0 TCV Treve Curnow, 5 Bosleake Row, Bosleake, Redruth, TR15 3YG
2E0 TCX Andrew Wilson, 38 Cotleigh Drive, Sheffield, S12 4HU
2E0 TCY Charlie Maddex, 25 Firfield Road, Benfleet, SS7 3UU
2E0 TDD Thomas Dixon, Lawn Cottage, Wyver Lane, Belper, DE56 2EF
2E0 TDE A Dilworth, 808 Liverpool Road, Southport, PR8 3QF
2E0 TDF William Welch, Kenilworth, School Lane, Oswestry, SY11 3LD
2E0 TDH Peter Thorley, 57 Riverside Drive, Hambleton, Poulton-le-Fylde, FY6 9EH
2E0 TDI D Davies, 32 Newlyn Crescent, Puriton, Bridgwater, TA7 8BS
2E0 TDJ Thomas Kelly, 50 Ivanhoe Road, Herne Bay, CT6 6EQ
2I0 TDL Thomas Browne, 7 Hawthorn Park, Greysteel, Londonderry, BT47 3YE
2E0 TDO Tim Digman, 74 Baddlesmere Road, Whitstable, CT5 2LA
2E0 TDP Anthony Pickett, 4 Trembel Road, Mullion, Helston, TR12 7DY
2E0 TDR James McMullen, 25 Dene Road, North Shields, NE30 2JW
2E0 TDS Thomas Stewart, 91 Chequers Field, Welwyn Garden City, AL7 4TX
2E0 TDT Anthony Maclean, 10 Elizabeth Close, West Hallam, Ilkeston, DE7 6LW
2E0 TDV Thomas Millichamp, 9 Wells Close, Bridgnorth, WV16 5JQ
2E0 TED G Cahill, 81 Albemarle Road, Willesborough, Ashford, TN24 0HJ
2E0 TEE Alan Chapman, 1 Fortunes Way, Bedhampton, Havant, PO9 3LX
2E0 TEF Raymond Ashman, Cartref, Salisbury Road, Salisbury, SP4 9QZ
2E0 TEH Terry Harris, 12 Maple Close, Stourport-on-Severn, DY13 8TA
2E0 TEI Martyn Bell, 36 Schneider Road, Barrow in Furness, LA14 5DW
2E0 TEK G Tecklenberg, 32 Emperor Way, Peterborough, PE2 9FE
2E0 TEN Adrian Riddick, 30 Britannia Road, Banbury, OX16 5DW
2E0 TEW Jeremy Woodland, 14 Kelham Green, Nottingham, NG3 2LP
2E0 TEZ Terry Kemp, 30 Tawny Sedge, King's Lynn, PE30 3PW
2E0 TFE Jonathan King, 60 Harwood Avenue, Bromley, BR1 3DU
2E0 TFH Thomas Harrison, 1 Tall Trees, Colchester, CO4 5DU
2E0 TFI Brian May, 10 Farrier Place, Downs Barn, Milton Keynes, MK14 7PJ
2E0 TFM Ian Whiteley, 29 Harvey Avenue, Wirral, CH49 1RT
2E0 TFO Robert Styles, 52 Vernham Grove, Bath, BA2 2TB
2E0 TFT Stuart Mayor, 12 Yealand Avenue, Heysham, Morecambe, LA3 2LT
2E0 TFX Tom Fisk, 2 Hall Farm Cottage, Caston Road, Attleborough, NR17 1BW
2E0 TGB Wayne Millington, 93 Feiashill Road, Trysull, Wolverhampton, WV5 7HT
2E0 TGC G Cooper, 10 Granary Court Magdalene Lawn, Barnstaple, EX32 7FA
2M0 TGD Ian Currie, 4 Greendyke Cottage, Falkirk, FK2 8PP
2E0 TGF A McGoff, 55 Knights End Road, March, PE15 9QA
2E0 TGG Thomas Sutton, Yew Tree Farm, Paddol Green, Shrewsbury, SY4 5QZ
2E0 TGL Andrew Wallace, 17 Dennis Road, Liskeard, PL14 3NS
2M0 TGM Thomas McConnell, 163 Strathaven Road, Stonehouse, Larkhall, ML9 3JN
2M0 TGN Anthony Barclay, 21 Netherlea, Scone, Perth, PH2 6QA
2E0 TGQ T Garcia-Quismondo, 11 Half Moon Lane, Worthing, [illegible]
2E0 TGS A Sayers, 4 Roughley Avenue, Warrington, WA5 1BL
2E0 TGV David Cook, 35 Elmwood Drive, Breadsall, Derby, DE21 4GA
2E0 THS E Graham, 25 South End Road, Cottingham, Hull, HU12 0DP
2E0 THZ Mark Hopewell, 4 Cotes Crescent, Bicton Heath, Shrewsbury, SY3 5AS
2M0 TIA Samuel Martin, 104 The Braes, Tullibody, Alloa, FK10 2TT
2E0 TIF Christopher [illegible] Close, Aston, Sheffield, S26 2GA
2E0 TIG Gary Spiers, 68 Thamesmead, Walton on Thames, KT12 2S[illegible]
2E0 TIL Paul Athersmith, [illegible], Stirchley, Telford, TF3 1RW
2E0 TIN Douglas [illegible], Duns, TD11 3LG
2M0 TIP Timothy Hawes, 100/8 Great Junction Street, Edinburgh, EH6 5JB
2E0 TIV Stephen Bryan, Flat 1, 81 Swallow Way, Cullompton, EX15 1GH
2E0 TJC T Catton, 97 High St., South Hiendley, Barnsley, S72 9AN
2E0 TJE T Ellis, 127 Hillmorton Road, Coventry, CV2 1FY
2I0 TJF Trevor Ferguson, 3 Wheatfield Park, Ballybogy, Ballymoney, BT53 6NT
2E0 TJG Jeremy Wong, St Edmund's College, Mount Pleasant, Cambridge, CB3 0BN
2E0 TJI Stephen Guest, 19 Ellesmere Avenue, Ashton-under-Lyne, OL6 8UT
2I0 TJK John Mackenzie, 30 Dalriada Gardens, Ballycastle, BT54 6DZ
2I0 TJM T Mulholland, 215 Finaghy Road North, Belfast, BT11 9ED
2M0 TJO Tim Johnston, The Old Schoolhouse, Luggate Burn, Haddington, EH41 4QA
2E0 TJP Patrick Neal, 14 Hilltop Close, Desborough, Kettering, NN14 2LQ
2I0 TJR T Ruddell, 30 Ballynacor Meadows, PORTADOWN, Craigavon, BT63 5UU
2E0 TJS A Steer, 51 Kings Chase, East Molesey, KT8 9DG
2E0 TJT Jeremy Thompson, 32 Church Street, Warnham, Horsham, RH12 3QR
2E0 TJU Evan Duffield, 92 Crosby Street, Stockport, SK2 6SP
2E0 TJX T Jones, 15 Kinnaird Road, Sheffield, S5 0NN
2U0 TKB Thomas Burnett, Flat 1, Clairval House, St Peter Port, Guernsey, GY1 1GD
2E0 TKE Russell McKie, 16 silver street, creetown, Newton Stewart, DG8 7HU
2E0 TKF David Long, 25 St. Matthias Road, Deepcar, Sheffield, S36 2SG
2W0 TKS Steven Davies, 5 Maldwyn Street, Cardiff, CF11 9JR
2E0 TKV T King, 24 Royston Avenue, Basildon, SS15 4EW
2E0 TKX Amar Sood, Parima, Sewardstone Road, London, E4 7RA
2E0 TKY Stephen Potter, 93 Church Lane, Bocking, Braintree, CM7 5SD
2E0 TLB F Smith, 13 Heather Walk, Crowborough, TN6 2HA
2E0 TLC Charles Parry, 27 Tynedale Close, Stockport, SK5 7NA
2E0 TLD Nonie Sinclair, 11 Primrose Close, Warton, Preston, PR4 1EN
2M0 TLE Kevin Quillien, 1/2 81 Laurel Street, Glasgow, G11 7QX
2E0 TLG Duncan James, Coombe House, Coombe, Presteigne, LD8 2HL
2E0 TLM Ian Williams, 36 Telford Road, Tamworth, B79 8EY
2I0 TLT W Thompson, 25 Darby Road, Carrickfergus, BT38 7XU
2E0 TLX David Burdsall, 37 Fulmar Walk, Whitburn, Sunderland, SR6 7BW
2E0 TLY Ian Bain, 45 Larpool Crescent, Whitby, YO22 4JD
2E0 TMA Timothy Ahern, 39 Essex Road, Romford, RM7 8BE
2W0 TMB Mark Woodington, 44 Glas y Gors, Aberdare, CF44 0BQ
2E0 TMC Stuart Oram, 9 Springbrook, Eynesbury, St. Neots, PE19 2DT
2E0 TMD Paul Kirby, 36 Durham Road, London, E12 5AX
2I0 TME Timothy Evans, 30 Seagoe Park, Portadown, Craigavon, BT63 5HR
2E0 TMF Andrew Taylor, 11 Fillingfir Drive, Leeds, LS16 5EG
2E0 TMH Terry McElwee, Little Borough, Borough Farm Road, Godalming, GU8 5JZ
2E0 TMN Anthony Macnauton, 27a Lincoln Road, Poole, BH12 2HT
2E0 TMO Thomas Williams, Moor Farm, Moor Lane, Lincoln, LN3 4EG
2E0 TMS Trevor Mackenzie, 2 Newcastle Street, Carlisle, CA2 5UH
2E0 TMX Adam Lee, 181 Tytherington, Warminster, BA12 7AE
2E0 TMY Thomas Haley, 3 Orchard View, Cropredy, Banbury, OX17 1NR
2E0 TNB Neil Williams, 60 Denbigh Close, Wrexham, LL12 7TW
2E0 TNC Thomas Humphries, The Nook, Yeldham Road, Halstead, CO9 3QJ
2E0 TNE Daniel Camp, Kidbrooke Lodge, Lewes Road, Forest Row, RH18 5AF
2E0 TNV Mark Clough, 8 Skeldyke Road, Kirton, Boston, PE20 1LR
2E0 TOF John Morgan, Glas y Dorlan, Pontrhydfendigaid, Ystrad Meurig, SY25 6EJ
2E0 TOG Brian Whelan, 147 Lawsons Road, Thornton-Cleveleys, FY5 4PL
2M0 TOK John McGurk, Pf2, 18 Warriston Road, Edinburgh, EH7 4HN
2E0 TOL Spencer Tomlinson, 8 Levett Road, Stanford-le-Hope, SS17 0BB
2E0 TOP R Styles, 16 Hatton Park, Bromyard, HR7 4EY
2M0 TOR Scott Gibson, 93 Allan Street, Coatbridge, ML5 5RD
2E0 TOT Steven Dix, 8 Beaumont Road, Longlevens, Gloucester, GL2 0EJ
2E0 TOX Paul Tivey, 49 Saxon Street, Burton-on-Trent, DE15 9RL
2E0 TOY Leo Bodnar, 47 Alchester Court, Towcester, NN12 6RL
2E0 TPA Anthony Patrick, The Woodlands, Nantwich Road, Chester, CH3 9JH
2I0 TPC Thomas Crozier, 201d Linn Road, Larne, BT40 2AJ
2E0 TPD Frederick Harwood, 1, South Highall Cottage, Woodhall Spa, LN10 6UH
2E0 TPE Benjamin Angus, Network Rail Advanced Apprenticeship, Faraday Building, Hms Sultan Gosport, PO12 3BY
2E0 TPO Anthony Gravell, 21 Wickridge Close, Stroud, GL5 1ST
2E0 TPH Tim Hazel, 84 Rodwell Avenue, Weymouth, DT4 8SQ
2E0 TPN Gary Grace, 194 Lillechurch Road, Dagenham, RM8 2EW
2E0 TPU Alastair Smith, 50 Wellesbourne Road, Barford, Warwick, CV35 8DS
2E0 TPP G Finney, 4 Cressida Court, Braunstone Lane, Leicester, LE3 3AP
2E0 TPR Thomas Prince, 6 Hyperion Ave, Newcastle, NE34 9AE
2E0 TPY Timothy Pollett, 10 Bridport Road, Poole, BH12 4BS
2E0 TQB Derek Holmes, 4 Council House, Nidds Lane, Boston, PE20 1LZ
2E0 TQF C Bailey, 37 Cherry Tree Drive, Filey, YO14 9UZ
2E0 TQS A Raeper, 40 Springfield Road, Stoke-on-Trent, ST4 6RU
2E0 TRA Tracey Mussell, Whitehouse, Priory Road, Boston, PE21 0RD
2W0 TRD R George, 18 Bryndedwyddfa, Penygroes, Llanelli, SA14 7PR
2E0 TRE Thomas Ellis, 84 Revelstoke Road, London, SW18 5PB
2E0 TRF D Pollington, 30 Rectory Avenue, Rochford, SS4 3AQ
2E0 TRH Antony Hargreaves, 27 Meadow Head Close, Blackburn, BB2 4TY
2E0 TRI Nicholas Rostant, 19 Pentre Jane Morgan, Penglais, Aberystwyth, SY23 3TE
2E0 TRK A Cooper, 23 Ash Street, Manchester, M9 5XY
2D0 TRL T Leece, Colby Croft, Glen Road, Colby, Isle of Man, IM9 4NX
2I0 TRM Stephen Davison, 60 cornation place, Craigavon, BT66 7AN
2E0 TRO Martin Tromans, 10 Crofters View, Little Wenlock, Telford, TF6 5AU
2E0 TRP Trevor Parsons, 5 Blackmoor Road, Auborn, Lincoln, LN5 [illegible]
2W0 TRR Thomas Rees, 7 Heathy Close, Pen-y-Fai, Bridgend, CF31 4[illegible]
2W0 TRS David Humphreys, 99 Park Road, Treorchy, CF42 6LB
2E0 TRU Graham Truman, 9 Riddell Avenue, Langold, Worksop, S81 9SS
2E0 TRW Christopher Jacobs, Flat 33, The Lodge, Waterlooville, PO7 8BX
2E0 TRX Razvan Moldoveanu, 17 Lynchford Road, Farnborough, GU14 6AR
2E0 TRY Herschel Chawdhry, Trinity College, Cambridge, CB2 1TQ
2E0 TSA Tony Stamp, 41 Glamorgan Close, Mitcham, CR4 1XG
2E0 TSC M Hemmings, [illegible], Brierley Hill, DY5 4JF
2E0 TSD Steve Smith, 103 Comberford Road, Tamworth, Tamworth, B79 8PF
2E0 TSJ S Holt, 8 Moulton Drive, [illegible], NR16 [illegible]
2E0 TSM Colin [illegible], [illegible], Reading, RG10 0BL
2E0 TSO Ian Macfarlane, 70 Ashby Drive, Rushden, NN10 9HH
2E0 TSP Rob Connolly, 29 Gayer Street, Coventry, CV6 7EU
2M0 TSR David Murphy, 38 Lothian Road, Stewarton, Kilmarnock, KA3 3BT
2E0 TSU S Watson, 5 Birchwood Avenue, Whickham, Newcastle upon Tyne, NE16 5QS
2E0 TTA T Arscott, 29 Parsonsfield Road, Banstead, SM7 1JW
2E0 TTB Trevor Bate, 87 Dunsheath, Telford, TF3 2BY
2M0 TTF James McMorland, 382 Maryhill Road, Glasgow, G20 7YQ
2E0 TTG Tony Hoyle, 60 Greenbank Crescent, Marple, Stockport, SK6 7PB
2E0 TTH Peter Troth, Beuna Vista, Hawford Wood, Droitwich, WR9 0EZ
2E0 TTI Matti Juvonen, 7 Alphin Brook, Didcot, OX11 7FG
2E0 TTL James Sales, 10 Wolsey Drive, Walton-on-Thames, KT12 3AY
2E0 TTM Toby Moncaster, 4 Crossways Gardens, Cambridge, CB2 9JT
2E0 TTS D Brook, 140 Dearne Hall Road, Barugh Green, Barnsley, S75 1LX
2E0 TTT Alastair Rosenschein, 101 Christchurch Road, London, SW14 7AT
2E0 TTW Robert Glynn, 106 Fairway Avenue, West Drayton, UB7 7AP
2E0 TTY Adrian Smith, 93 Sheriffs Highway, Gateshead, NE9 6QN
2E0 TUB Christine Austen, Fenbank House, Roman Bank, Holbeach Clough, Spalding, PE12 8DH
2E0 TUC Brian Tucker, 12 Alpha Place, Appledore, Bideford, EX39 1QY
2E0 TUF Julian Caswell, 3 Pavilion Court, Roydon, Diss, IP22 5SP
2E0 TUG Andrew Barrett-Sprot, 1 Malting End, Wickhambrook, Newmarket, CB8 8YH
2E0 TUH Richard Taylor, Penhawger Park, Liskeard, PL14 3LW
2I0 TUI Charles Stockdale, 3 Hightown Drive, Newtownabbey, BT36 7TG
2E0 TUK Tania Goddard, 217 Speedwell Road, Bristol, BS5 7SP
2E0 TUN Andrew Tunney, 79 Scott Street, Burnley, BB12 6NJ
2E0 TUR Terence Baines, 10 Croydon Avenue, Leigh, WN7 1TP
2E0 TUS Richard Austen, Fenbank House, Roman Bank, Holbeach Clough, Spalding, PE12 8DH
2E0 TUT Carol Howard, 6 Robinson Way, Burbage, Hinckley, LE10 2EU
2I0 TUV Charles Bailie, 26 Moatview Park, Dundonald, Belfast, BT16 2BE
2E0 TUW Richard Harris, 7 Fosse Lane, Shepton Mallet, BA4 4PS
2E0 TUX N Trangmar, 8 Maxstoke Close, Meriden, Coventry, CV7 7NB
2E0 TVD Stephen Pettit, 88 Abbotsbury, Great Hollands, Bracknell, RG12 8QX
2E0 TVM Thomas Nott, 87 Powney Road, Maidenhead, SL6 6EG
2E0 TVR S Martin, 14 Mount Road, Thatcham, RG18 4LA
2E0 TVS Neil Simmonds, 3 Noneley Hall Barns, Noneley, Shrewsbury, SY4 5SL
2E0 TVV Scott Baggott, 2 Ellison Street, West Bromwich, B70 7ES
2E0 TVW Anthony Wilson, 11 Headland Way, Alton, Stoke-on-Trent, ST10 4AN
2E0 TVX Raymond Taylor, 10 Barnwell Lane, Cromford, Matlock, DE4 3QY
2E0 TVZ David Darby, 8 Mulberry Close, Blackfield, Southampton, SO45 1FH
2E0 TWA Anthony Weatherall, The Old Telephone Exchange, The Street, Canterbury, CT3 1ED
2E0 TWI Christopher Wild, 203 High Street, Saltney, Chester, CH4 8SJ
2E0 TWK Alice Champion, 2 St. Andrews Hill, Waterbeach, Cambridge, CB25 9NA
2E0 TWL Terence Larman, 861 London Road, Westcliff-on-Sea, SS0 9SZ
2E0 TWO Kevin Monaghan, The Bulstone Hotel, Branscombe, Seaton, EX12 3BL
2E0 TWP Paul Miller, 46 Great Brooms Road, Tunbridge Wells, TN4 9DH
2E0 TWQ John Attwood, 13 John Winter Court, Euston Road, Great Yarmouth, NR30 1DU
2E0 TWS Trevor Wood, 44 Wincobank Lane, Sheffield, S4 8AA
2E0 TWT James Colbourne, 43, Westfield Road, Bilston, WV14 6EW
2E0 TWW Timothy Larman, 861b London Road, Westcliff on Sea, SS0 9SZ
2E0 TWZ Noel Houghton, 100 Ellerburn Avenue, Hull, HU6 9RW
2M0 TXA Thomas Dorricott, 11 Dakota Way, Renfrew, PA4 0NP
2I0 TXB Josh Morrison, 70 Ravenswood, Banbridge, BT32 3RD
2E0 TXG Trevor Ruddick, Hazel Gill, Croglin, Carlisle, CA4 9RR
2E0 TXJ John Hopkins, 53 Sprules Road, London, SE4 2NL
2M0 TXK Tim Kerby, 1 St. Marks Lane, Edinburgh, EH15 2PX
2E0 TXL Philip Dunnicliffe, 19 Woodland Road, Chelmsford, CM1 2AT
2I0 TXM Andrew McCarvoy, 66a Scaddy road, Downpatrick, BT30 9BS
2E0 TXP Patrick Cassells, 5 Saxon Way, Liverpool, L33 4DW
2M0 TXR Alexander McNeill, 13 Spinkhill, Laurieston, Falkirk, FK2 9JR
2M0 TXY A Caldwell, 180 Pappert, Alexandria, G83 9LG
2E0 TYC Tyrone Corcoran, 50 Grange Road, Bracebridge Heath, Lincoln, LN4 2PW
2E0 TYE Michael Armstrong, Tan yr Ffail, Segurincide, Llandudno Junction, LL31 9QE
2E0 TYG Carl Morris, 17 Percy Road, Wrexham, LL13 7EA
2E0 TYL M Tyler, 40 Bullards Lane, Woodbridge, IP12 4HE
2E0 TYT Carl Gibson, 17 Clyde Court, Grantham, NG31 7RB
2M0 TZB Martin Lawson, 23 Kirkfield View, Livingston Village, Livingston, EH54 7BP
2E0 TZD Allan Maiden, 70 Green End Road, Manchester, M19 1LE
2E0 TZM Andrew Laister, 2 Warlow Crest, Greenfield, Oldham, OL3 7HD
2E0 TZO Paul Gibson, 7 Greenfields Road, Horley, RH6 8HW
2E0 TZR Tristan Tozer, 9 Bainbridge Court, Plymouth, PL7 4HH
2E0 TZY Steven Crabb, 1 Council Houses, Hall Lane, Norwich, NR12 7BB
2E0 TZZ Philip Moore, 24 Plough Road, Dormansland, Lingfield, RH7 6PS
2W0 UAA Mark Uphill, 1 Brynview Avenue, Ystrad Mynach, Hengoed, CF82 7DB
2E0 UAB Carl Kent, 19 Coppice Rise, Harrogate, HG1 2DP
2E0 UAC M Timms, 29 Sutherland Avenue, Coventry, CV5 7ND



**IMPORTANT NOTE**

**Revalidate licence to avoid revocation** – Ofcom has advised the Society that plans will be drawn up to revoke licences that have not been revalidated as required by the licence conditions. The quickest way to revalidate is to do so online via the Ofcom website: *https://services.ofcom.org.uk/* or by email: *amateur.validations@ofcom.org.uk* Ofcom staff are available to help, but please be patient during times of heavy workload.

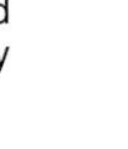

2I0 UAD A Pritchard, 16 Ballymaconnell Road South, Bangor, BT19 6DQ
2E0 UAE John Firth, 36 Howley Grange Road, Halesowen, B62 0HW
2E0 UAK Andrew Vaile, 76 Stanbury Road, London, SE15 2DB
2M0 UAL Kenneth Mackenzie, Alderwood, Braes, Ullapool, IV26 2TB
2E0 UAM C Byrne, 31 Graham Drive, Castleford, WF10 3EY
2E0 UAO Michael Lewis, 6 Remembrance Road, Newbury, RG14 6BA
2E0 UAR Peter Barnes, 13 Lavender Close, Thornbury, Bristol, BS35 1UL
2E0 UAV Tyler Ward, James Barn, Horham, Eye, IP21 5ER
2D0 UAY D Levey, Heriots Wood, The Common, Stanmore, HA7 3HT
2E0 UBH Russell Tolman, 10 Woodcote Way, Abingdon, OX14 5NE
2E0 UBN Benjamin Shephard, 74 Harcourt Street, Kirkby-in-Ashfield, Nottingham, NG17 8DD
2E0 UBT Colin Bowes, 11 burghwallis Lane, Sutton, Doncaster, DN69JU
2E0 UBU Derek Hampton, 17 Hamilton Road, Worcester, WR5 1AG
2E0 UBW John Cardwell, 35 Bush Lane, Freckleton, Preston, PR4 1SB
2E0 UCH William Toher, The Chapel, Station Road, Darlington, DL2 1JG
2I0 UCS M Edwards, 21 Old Grange Avenue, Carrickfergus, BT38 7UE
2E0 UCV Thomas Kilroy, 55 Summerfield Crescent, Brimington, Chesterfield, S43 1HB
2E0 UDA Andrew Cattell, 2 St. James Close, Ruscombe, Reading, RG10 9LJ
2E0 UDE Richard Brown, 62 Charlecote Park, Telford, TF3 5HD
2E0 UDG David Gaut, 14 Bedale Close, Belmont, Durham, DH1 2BB
2E0 UDM Darren Meehan, 47 Clinton Road, Shirley, Solihull, B90 4RN
2M0 UEA E Ewing, Arisaig, Priestland, Darvel, KA17 0LP
2E0 UEH Stephen Smith, 39 Wimborne Road, Southend-on-Sea, SS2 5JG
2E0 UEL Mark Tointon, 13 Ridgeway, Broadstone, BH18 8DY
2E0 UFM Adrian Parkhouse, 3 St. Margarets Avenue, Ashford, TW15 1DR
2E0 UGF Michael Sims, 133 Canterbury Road, Hawkinge, Folkestone, CT18 7BS
2M0 UGL Nigel Rogers, 108 Beechwood Road, Cumbernauld, Glasgow, G67 2NP
2E0 UGM Graham Mountain, 34 Albert Road, Warlingham, CR6 9EP
2E0 UGO Darren Hill, 19 Farren Road, Birmingham, B31 5HH
2E0 UHF Nicholas Booth, Greenfield, Westmancote, Tewkesbury, GL20 7EP
2E0 UHJ Philip Hopkinson, 28 Stockdove Way, Thornton-Cleveleys, FY5 2AR
2E0 UHL Mark Simpson, 32 Underhill Lane, Wolverhampton, WV10 8NS
2E0 UHS Ian Phillpott, 14 Buttercup Close, Paddock Wood, Tonbridge, TN12 6BG
2E0 UIP D Woodhouse, Lund Cotage, Long Acre, Selby, YO8 6PE
2E0 UIQ A Greenhough, 14 The Dale, Wirksworth, Matlock, DE4 4EJ
2E0 UJD Kevin Colman, 10 South Rise, North Walsham, NR28 0EE
2E0 UKA Andrew Reay, 12 Victoria Avenue, South Hylton, Sunderland, SR4 0QZ
2E0 UKD John Parfrey, 47 Ford Lane, Rainham, RM13 7AS
2E0 UKG Geoffery Stratford, 23 The Fairway, Banbury, OX16 0RR
2E0 UKK George Radulescu, 41 Sherard Road, London, SE9 6EX
2E0 UKM Marwan Qassim, Winchester Road, Kings Somborne, Stockbridge, SO20 6NY
2E0 UKX Anthony Fletcher, 2 Brow Crescent, Halfway, Sheffield, S20 4GB
2E0 ULC Paul Power, 16 Mainstone Close, Redditch, B98 0PP
2E0 ULH David Butchart, Flat 26, Seldon House, London, SW8 4DP
2E0 ULY Andrew Ulyatt, 15 Garden Close, Kington, HR5 3FJ
2M0 UMH Leslie Mitchell Hynd, Smithy House Bruichladdich, Isle of Islay, PA49 7UN
2E0 UMR Umar Munir, 100 Ranelagh Road, Southall, UB1 1DG
2E0 UNI Geoff Rigby, Gas House Farm, Shavington Park, Market Drayton, TF9 3SY
2E0 UNY Lee Betts, 12a Maesgwyn, Pontnewydd, Cwmbran, NP44 1BQ
2E0 UOG Tony Cater, 14, britannia road marshgreen, Wigan, WN5OEN
2E0 UOJ Kevin Barry, 25 Delabole Road, Merstham, Redhill, RH1 3PB
2E0 UOK A Abraham, Flat 2, 41 Francis Road, Birmingham, B33 8SL
2E0 UPA Jan Van Der Elsen, SULA Lightship, Llanthony Road, Gloucester, GL2 5HH
2E0 UPH Aled Williams, 8 Old Tanymanod Terrace, Blaenau Ffestiniog, LL41 4BU
2E0 UPT S Thomas, 103 Liverpool Road, Upton, Chester, CH2 1BB
2E0 UPU Anthony Stirk, 5 Hall Stone Court, Shelf, Halifax, HX3 7NY
2E0 URF Richard Freeman, 9 Bramley Road, Wisbech, PE13 3PA
2E0 URJ Robert Jordan, Pheasants Walk, Copyhold Lane, Haslemere, GU27 3DZ
2M0 URP Samantha Gray-Jones, Flat C, 7 Nelson Street, Aberdeen, AB24 5EP
2E0 USA Adrian Jepson, 24 Shortbrook Close, Westfield, Sheffield, S20 8LE
2E0 USC Stephen Coleman, 32 Southwell Road, Wisbech, PE13 3LQ
2W0 USN J Barry, Flat 3, 31 Ely Road, Cardiff, CF5 2JF
2E0 USQ Lawrence King, 1 Kingsbury Drive, Old Windsor, Windsor, SL4 2NQ
2E0 USV Denis Soames, 40 Woodland Drive, North Anston, Sheffield, S25 4EP
2E0 UTC Jonathan Bennette Alincastre, 90 York Crescent, Durham, DH1 5PT
2E0 UTD Anthony McLean, 47 Tarn Drive, Bury, BL9 9QB
2M0 UTH P Dower, 1670 Maryhill Road, Glasgow, G20 0HJ
2D0 UTL Adrian Salt, 1 Chantry Close, Harrow, HA3 9QZ
2E0 UTT Bernard Bull, Swan Cottage, Swan Road, Welshpool, SY21 0RH
2E0 UTX Alex Emmerson, 31 Culver Road, Stockport, SK3 8PG
2E0 UUA Ian Ward, 37 Maes Gwydryn, Abersoch, Pwllheli, LL53 7ED
2E0 UUW John Douch, High House, Ulverston Road, Ulverston, LA12 0PZ
2E0 UVO Andrew Ruocco, 83 Epsom Road, Morden, SM4 5PR
2E0 UVP Darren Bisbey, 17 Benson Close, Lichfield, WS13 6DA
2E0 UVZ K Such, 38 Hornby Grove, Hull, HU9 4PG
2E0 UWI R Wilmot, 41 Milton Brow, Weston-Super-Mare, BS22 8DD
2E0 UWK John Hadley, 75 Glendower Avenue, Coventry, CV5 8BD
2E0 UWT Andrew Chapman, 10 Derwent Road, Seaton Sluice, Whitley Bay, NE26 4JH
2E0 UXS Carol Dutton, 7 Ellery Grove, Lymington, SO41 9DX
2E0 UYB J Smith, 9 Trafalgar Road, Newport, PO30 1QD
2E0 UZK Alan Sweet, 3 Beechwood Grove, Blackpool, FY2 0DZ
2W0 UZO Daniel White, 222 St. Fagans Road, Cardiff, CF5 3EW
2E0 VAA Kent Britain, Blenheim Cottage, Falkenham, Ipswich, IP10 0QU
2E0 VAC Charles Rayment, Brambles, Alltami Road, Mold, CF7 6RT
2E0 VAF John Edmunds, 22 Horsewhim Drive, Kelly Bray, Callington, PL17 8GL
2W0 VAG S Smith, 36 Jones Street, Tonypandy, CF40 2BY
2E0 VAI Thomas Oliver, 17 East Lea, Newbiggin-by-The-Sea, NE64 6BQ
2E0 VAJ Albert Taylor, 90 Coppice Avenue, Eastbourne, BN20 9QJ
2E0 VAM Mark Higham, 30 Broome Road, Southport, PR8 4EQ
2E0 VAN Dominic Stinson, 1 The Croft, Earls Colne, Colchester, CO6 2NH
2E0 VAO Simon Pankhurst, 57 Barley Croft, Harlow, CM18 7QZ
2E0 VAS V Papanikolaou, 104 West Drive Gardens, Soham, Ely, CB7 5EX
2E0 VAT Mark Raynor, 68 Cambridge Street, South Elmsall, Pontefract, WF9 2AR
2E0 VAU Simon Fairbourn, 17 Perry's Lane, Wroughton, Swindon, SN4 9AX
2E0 VAV Allan Burnett-Provan, 17 Courtenay Avenue, Sutton, SM2 5ND
2E0 VAW Ian Kane, 44 Hafod Arthen Estate, Brynithel, Abertillery, NP13 2HY
2W0 VAY Alan Davies, 19 Williams Place, Merthyr Tydfil, CF47 9YH
2E0 VBB S Coulson, 17 Charlton Street, York, YO23 1JN
2E0 VBK Bill Kalogerakis, Inglewood, Madingley Road, Cambridge, CB23 7PH
2E0 VBM Alan Bairstow, 12 Danesfield Avenue, Waltham, Grimsby, DN37 0QE
2E0 VBN Keith Sloan, Woodland Halt, Old Station Road, Winchester, SO21 1BA
2E0 VBQ Paul McDonough, 91 Lever Street, Little Lever, Bolton, BL3 1BA
2E0 VBR Jess Baughan, Chestnut Farm, Eastville, Boston, PE22 8LX
2E0 VBT Martynas Kveksas, 29 Saxon Way, Reigate, RH2 9DH
2E0 VBW Brian Whall, 3 Farrow Close, Great Moulton, Norwich, NR15 2HR
2E0 VBX Peter Otterwell, 50 Hythe Road, Staines-upon-Thames, TW18 3EE
2E0 VBY D Potter, 30 Mersham Gardens, Goring-by-Sea, Worthing, BN12 4TQ
2E0 VCB Barry Ashall, 22 Bloomer Wood View, Sutton-in-Ashfield, NG17 1HA
2E0 VCC Darrell Jacobs, 5 Tor View, Tregadillett, Launceston, PL15 7HB
2E0 VCD David Jones, 102 Bryce Road, Brierley Hill, DY5 4ND
2E0 VCE Brian McGuirk, 9 Almond Crescent, Standish, Wigan, WN6 0AZ
2E0 VCM Charles Mallory, 11 Baymead Meadow, North Petherton, Bridgwater, TA6 6QW
2E0 VCP Colin Pinder, 70 Highfield Road, Beverley, HU17 9QR
2E0 VCU Paul Pritchard, 12 Easton Crescent, Billingshurst, RH14 9TU
2E0 VCY Graham Shakespeare, 6 Waterworks Cottages, Clough Road, Hull, HU6 7QB
2E0 VDC Neil Fairbairn, 15 Hewitt Road, Dover, CT16 1TH
2E0 VDE Richard Ferguson, 31 Barton Court Road, New Milton, BH25 6NW
2E0 VDM Jeffrey Savage, Rufford, Barnes Lane, Lymington, SO41 0RR
2E0 VDP D Powell, 36 Beechwood Drive, Stone, ST15 0EH
2W0 VDW Nigel Thomas, 57 Brynhyfryd Street, Treorchy, CF42 6DT
2E0 VEF P Johnson, 90 North Road, Southport, PR9 8QR
2W0 VEH David Thomas, 57 Brynhyfryd Street, Treorchy, CF42 6DT
2E0 VEK Kevin Lowcock, 43 Larch Street, Nelson, BB9 9RH
2E0 VET Steve Froggatt, 140 Greenlea Court, Huddersfield, HD5 8QB
2E0 VEX Andrew Crawford, 4 Trimpley Drive, Kidderminster, DY11 5LB
2E0 VFA Andrew Colcombe, 217 Church Drive, Quedgeley, Gloucester, GL2 4US
2I0 VFO Joseph Sills, 145 Ballycolman Estate, Strabane, BT82 9AJ
2M0 VFV Simon Young, 103 Feorlin Way, Garelochhead, Helensburgh, G84 0EB
2E0 VGB V Greenway-Brown, 207 Lowe Avenue, Wednesbury, WS10 8NS
2E0 VGC David Carter, 85 Dingle Street, Oldbury, B69 2DZ
2E0 VGK B Lunn, 204a Main Street, Horsley Woodhouse, Ilkeston, DE7 6AX
2E0 VGV Gautham Venugopalan, 3 Southwater Close, London, E14 7TE
2E0 VHA Richard Parker, 29 Hill Lea Gardens, Cheddar, BS27 3JH
2E0 VHC M Sands, Room 3, 8 Upperton Gardens, Eastbourne, BN21 2AH
2E0 VHF Chris Craswell, 49 Alexandria Walk, Cheltenham, GL525LG
2E0 VHV Luke Kelly, 9 Ham Lane, Farrington Gurney, Bristol, BS39 6TW
2E0 VHZ James Telfer, 50 Agraria Road, Guildford, GU2 4LF
2E0 VIA Laurie Kirkcaldy, 62 West Garth Road, Exeter, EX4 5AN
2E0 VIS Anibal Oliveira, 13 A Lakefield Road, London, N22 6RR
2E0 VJB V Bowkett, 9 Gwealmayowe Park, Helston, TR13 0PE
2E0 VJH Ray Heming, Milepost Cottage, Benenden Road, Ashford, TN27 8BY
2D0 VJK Robert Smith, 3 Rheast Barrule, Castletown, Castletown, Isle of Man, IM9 1HW
2E0 VJL Vanessa Lea, 30 Cardiff Road, Pwllheli, LL53 5NU
2E0 VJO Jonathan Sawyer, 9 Waller Court, Caversham, Reading, RG4 6DB
2E0 VJX Bradley Walker, 255 Packington Avenue, Birmingham, B34 7RU
2W0 VKA Anthony Vincent, 88 Lake Street, Ferndale, CF43 4HE
2E0 VKB Kevin Davies, 23 Egmanton Road, Meden Vale, Mansfield, NG20 9QN
2E0 VKG Andrew Smith, 32 Cotswold Drive, Rothwell, Leeds, LS26 0QZ
2E0 VKK Richard Cresswell, Meadow View, Hulver Road, Beccles, NR34 7UW
2E0 VKN Ian Astley, 1 Howard Crescent, Durkar, Wakefield, WF4 3AJ
2M0 VKO S MacDonald, 366 Millcroft Road, Cumbernauld, Glasgow, G67 2QW
2E0 VKQ R Vickerstaff, 16 Sewell Wontner Close, Kesgrave, Ipswich, IP5 2GB
2E0 VKS D Vickers, 178 Bakewell Road, Matlock, DE4 3BA
2E0 VKW Victoria Williams, Moor Farm, Moor Lane, Lincoln, LN3 4EG
2E0 VKY Victoria Fleming, 1 Balmoral Drive, Methley, Leeds, LS26 9LE
2E0 VKZ Mark Hillman, Flat 5, 32 South Terrace, Littlehampton, BN17 5NU
2E0 VLB Vladislav Boev, Flat 4, Camborne House, Sutton, SM2 6RL
2M0 VLF Iain Mcglynn, 25 Fairhill Avenue, Hamilton, ML3 8JS
2E0 VLL Lisa Burbidge, 33 Burcote Fields, Towcester, NN12 6TH
2E0 VLT Paul Honey, 3 Peterswood, Harlow, CM18 7RJ
2E0 VMA Chantel Skupski, 57 Three Nooks, Bamber Bridge, Preston, PR5 8EN
2E0 VMC J Argent, 7 Lloyds Hill, Buckley, CH7 3ER
2D0 VMN Voirrey Matthewman, Monte Rosa, 7 Ballaughton Close, Douglas, Isle of Man, IM2 1JE
2E0 VMV Rod Vale, 611 College Road, Birmingham, B44 0AY
2E0 VNK David Bishop, 1 Charnwood Drive, Barton Seagrave, Kettering, NN15 6TU
2E0 VNL Matthew Harris, 5 Lynmore Close, Northampton, NN4 9QU
2E0 VNN Victor Nikolaidis, 35-46 Ernst Chain Road, Manor Park, Guildford, GU2 7YW
2E0 VNO Darryl Harwood, 36 Seaview Drive, Great Wakering, Southend-on-Sea, SS3 0BE
2M0 VNW Allan Sim, 44 Hillmoss, Kilmaurs, Kilmarnock, KA3 2RS
2W0 VOC Aubrey Parsons, 21 Rectory Drive, St. Athan, Barry, CF62 4PD
2E0 VOD Nicholas McLean, 21 Matlock Avenue, Wigston, LE18 4NA
2I0 VOF Patrick McFadden, 35 West Wind Terrace, Hillsborough, BT26 6BS
2E0 VOK R remnant, 172 Burnham Road, Highbridge, TA9 3EH
2I0 VOQ David Sinton, 34 West Link, Holywood, BT18 9NX
2E0 VPO Robert Dean, 15 Gorge Road, Dudley, DY3 1LF
2E0 VPT Graham Taylor, 57 Edinburgh Avenue, Walsall, WS2 0JD
2M0 VPU Jason Smith, 82 Overton Road, Netherburn, Larkhall, ML9 3BT
2E0 VPW V Williams, 11 Priory Green, Highworth, Swindon, SN6 7NU
2E0 VRC Andrew Newbould, 20 Gorsemoor Road, Heath Hayes, Cannock, WS12 3TG
2E0 VRD Vicky Bowen, 4 Crossley Gardens, Halifax, HX1 5PU
2E0 VRE Douglas Easden, 20 Brunel Way, Calne, SN11 9FN
2E0 VRR Ronald Oxley, 17 Hardhurst Road, Alvaston, Derby, DE24 0LF
2E0 VRT Julian Garwood, 4 Ryedale, Carlton Colville, Lowestoft, NR33 8TB
2E0 VRX Craig Bradley, 22 Park Street, Skipton, BD23 1NS
2E0 VSW Victor Wallace, 10 Maes Llydan, Benllech, Tyn-Y-Gongl, LL74 8RD
2E0 VTA Duncan McNicholl, 186 Coldhams Lane, Cambridge, CB1 3HH
2M0 VTB Alexander Hamilton, 10/3 Fox Street, Edinburgh, EH6 7HN
2E0 VTK John Martin, 62 Llwyn Ynn, Talybont, LL43 2AL
2E0 VTR N Burton, 11 Weldon Avenue, Stoke-on-Trent, ST3 6PN
2E0 VTS Peter Wilkes, 8 Cloverdale, Stafford, ST17 4QJ
2E0 VTT John Owen, 8 Highridge Crescent, Bristol, BS13 8HN
2E0 VTV Ivor Roberts, 15 Broadcroft, Hemel Hempstead, HP2 5YX
2I0 VTZ Stephen Gore, 5 Rosebrook, Dungiven, Londonderry, BT47 4GA
2E0 VUK Brian Lewis, 68 Irwin Avenue, Rednal, Birmingham, B45 8QU
2M0 VUV Robert Fraser, 72 Ferguson Drive, Denny, FK6 5AG
2E0 VVA Andrew Amos, 19 Poets Gate, Cheshunt, Waltham Cross, EN7 6SB
2E0 VVC Matthew Walker, 3 Finch Close, Tadley, RG26 3YJ
2E0 VVE Kenneth Macmanus, 23 Mount Pleasant Residential Park, Bloomhill Road, Doncaster, DN8 4ST
2E0 VVJ Thomas Sandham, 96 South Road, Morecambe, LA4 6JS
2E0 VVL Gareth Davies, 66 Allt-yr-yn View, Newport, NP20 5GG
2W0 VVO Stuart Barry, 7 Redhill Park, Haverfordwest, SA61 2HA
2M0 VVS Iain Lindsay, 265 Stirling Street, Denny, FK6 6QJ
2E0 VWG Michael Cross, 11 Polyplatt Lane, Scampton, Lincoln, LN1 2TL
2E0 VWK M Poole, 15 Roberts Place, Dorchester, DT1 2JJ
2E0 VWL James Campion, 25 Suffield Road, Liverpool, L4 1UL
2E0 VWW Stuart Hegarty, 10 New Street, Ash, Canterbury, CT3 2BH
2E0 VWX Terence McBride, 53 Blackdown Grove, St. Helens, WA9 2BD
2M0 VXB Majed Al Saeed, 9 Appin Place, Edinburgh, EH14 1NJ
2E0 VXI Darren Holden, 24 Penny Gate Close, Hindley, Wigan, WN2 3DP
2M0 VXL Julian May, 12 Clochbar Gardens, Milngavie, Glasgow, G62 7JP
2E0 VXR Anthony Murdoch, 2 Birtwistle Terrace, Langho, Blackburn, BB6 8BT
2E0 VXT Andrew Calvert, 11 Pine Tree Walk, Poole, BH17 7EH
2E0 VXX Tristan Quiney, 20 Britannia Gardens, Stourport-on-Severn, DY13 9NZ
2D0 VYN Martin Roberts, 11 Oakleigh Road, Pinner, HA5 4HB
2E0 VYW Antony Willsher, 1 Tolputt Court, Gladstone Road, Folkestone, CT19 5NE
2E0 VZL Stuart Haycock, 51 South Crescent, Southend-on-Sea, SS2 6TB
2E0 WAA Oliver Prin, 19 The Colliers, Heybridge Basin, Maldon, CM9 4SE
2E0 WAE Andrew Ward, 29 Mainwaring Road, Wallasey, CH44 9DN
2E0 WAF Harold Burch, 46 School Lane, Horton Kirby, Dartford, DA4 9DQ
2E0 WAG D Wagstaff, 68 Braziers Quay, South Street, Bishop's Stortford, CM23 3YW
2I0 WAH Thomas Quin, 165 Marlacoo Road, Portadown, Craigavon, BT62 3TD
2I0 WAI Martin McErlean, 38a Culbane Road, Portglenone, Ballymena, BT44 8NZ
2E0 WAJ Wayne Johnson, 145 Netherton Road, Worksop, S80 2SA
2E0 WAK Peter Holton, 66 Mill Road, Gillingham, ME7 1JB
2E0 WAP Anthony Woodhouse, 4 Grafton Close, St. Albans, AL4 0EX
2I0 WAS William Campbell, 9 Rochester Court, Coleraine, BT52 2JL
2E0 WAT A Watmough, Apartment 31, Wyatville House, Buxton, SK17 6WJ
2E0 WAU Albert Lander, 37 Berry Drive, Paignton, TQ3 3QW
2E0 WAV Alan Snelson, 6 Rayleigh Close, Braintree, CM7 9TX
2E0 WAY W Davies, 17 Oakdale Avenue, Harrogate, HG1 2JN
2I0 WBD William McDonald, 14 Edenmore Park, Limavady, BT49 0RG
2E0 WBE David Buckley, 66 Tharp Road, Wallington, SM6 8LE
2I0 WBF William Turkington, 8a Drummullan Road, Moneymore, Magherafelt, BT45 7XS
2E0 WBG William Bennison, 21 Ashdene Close, Chadderton, Oldham, OL1 2QG
2E0 WBH W howie, 152 Norwood Road, Birkby, Huddersfield, HD2 2YD
2M0 WBJ William Jackson, 3 Annick Road, Dreghorn, Irvine, KA11 4EY
2E0 WBL Andrew Riley, 35 Ross Avenue, Wirral, CH46 2SA
2E0 WBO William Jones, 50 Bridge Place, Croydon, CR0 2BB
2E0 WBQ David Hodgson, 11 Harmony Place, Mountain, Bradford, BD13 1LD
2E0 WBR Wayne Reeves, 33 Pond Bank, Blisworth, Northampton, NN7 3EL
2E0 WBS Adrian Whadcoat, 38 Edwin Panks Road, Hadleigh, Ipswich, IP7 5JL
2E0 WBT Brian Weston, 10 Clement Drive, Peterborough, PE2 9RQ
2I0 WBU William Bradley, 16 Mullaghanagh Lane, Dungannon, BT71 7NY
2E0 WBW Reginald Howard, 13 Top Common, East Runton, Cromer, NR27 9PW
2E0 WBX Alan Coats, 57 Mill Hill, Boulton Moor, Derby, DE24 5AF
2E0 WBZ Keith Hunt, Flat 32, Greenford House, West Bromwich, B70 6DX
2E0 WCB Alan Williams, 74 Broadfield Road, Bristol, BS4 2UW
2E0 WCC Colin Calvert, 1 Moorsholme Avenue, Manchester, M40 9BW
2E0 WCE Mark Hall, 20 Diamond Drive, Oakwood, Derby, DE21 2JP
2E0 WCL Oliver Fallon, 26 Central Avenue, Corfe Mullen, Wimborne, BH21 3JD
2E0 WCM Danny Neumunn, 92 Miner Street, Walsall, WS2 8QL
2E0 WCO Paul Austen, Flat 4, Orchard Court, 13 Barnhorn Road, Bexhill-on-Sea, TN39 4QB
2W0 WCQ Eon Edwards, 1 Brynhyfryd, Sarn, Bridgend, CF32 9UR
2E0 WCX Wayne Dix, 21 Pine Vale Crescent, Bournemouth, BH10 6BG
2E0 WDB William Bull, 117 Walton Road, Wednesbury, WS10 0EU
2I0 WDD David Milligan, 30 Belgrano Ahoghill, Ballymena, BT42 2QQ
2M0 WDG William Goodfellow, 1 Yester Place, Haddington, EH41 3BE
2E0 WDH W Henderson, 14 Highfield Road, Newcastle upon Tyne, NE5 5HS
2E0 WDI I Woollen, 33 The Oaks, Taunton, TA1 2QX
2E0 WDM Stephen Thompson, 64 Church Road, Fordham, Colchester, CO6 3NJ
2E0 WDS Wayne Sargeant, 44a Nelson Street, Buckingham, MK18 1DA
2E0 WDY Colin Johnson, 45 Gordon Road, Chelmsford, CM2 9LN
2E0 WEC C Newton, 7 Moss Close, Bridgwater, TA6 4NA
2E0 WEE Neil Froggatt, 13 Stroudes Close, Worcester Park, KT4 7RB
2E0 WEF I Chambers, 2 Belford Road, Borehamwood, WD6 4HY
2E0 WEG William Gray, 39 Guest Avenue, Poole, BH12 1JA
2E0 WEJ W Jefferies, 26 Norcutt Road, Twickenham, TW2 6SR
2E0 WEK Keith Alabaster, 16 Butlers Road, Horsham, RH13 6AJ
2E0 WEL William Easdown, 38 North Street, Barming, Maidstone, ME16 9HF
2E0 WEO Steven Kiel, 32 Weavers Avenue, Frizington, CA26 3AT
2E0 WES S Weston, 73 Priory Road, Ashton-in-Makerfield, Wigan, WN4 9UP
2E0 WET Alan Forrest, 1 Errington Bungalows, Sacriston, Durham, DH7 6NE
2M0 WEV George Weir, 95 White Street, Whitburn, Bathgate, EH47 0BH
2E0 WEZ J Weston, 29 Langdale Road, Orrell, Wigan, WN5 0EB
2W0 WFB Laurie Bowman, Chanrick, Penderyn Road, Aberdare, CF44 9RU
2E0 WFC Joe Paradas, 13 Meadow Road, Hemel Hempstead, HP3 8AH
2E0 WFD Paul Blundell, 22 Regina Crescent, Havercroft, Wakefield, WF4 2ER
2E0 WFG W Griffiths, 68 Altcar Lane, Formby, Liverpool, L37 6AY
2D0 WFH William Hogg, Medhamstead, Lhergydhoo, Isle of Man, IM5 2AE
2E0 WFM David Silkstone, 169 Otley Road, Harrogate, HG2 0DA

| | | |
|---|---|---|
| 2M0 | WFN | William Noon, 0/1 445 Royston Road, Glasgow, G21 2DE |
| 2E0 | WGB | Gerald Beale, 34 Teville Road, Worthing, BN11 1UG |
| [illegible] | WGC | Christopher Watkins, 25 Citadilla Close, Gatherley Road, Richmond, DL10 7JE |
| 2U0 | WGE | Robert Batiste, Asile de Paix, Clos Des Sablons, Sandy Lane, Guernsey, Guernsey, GY2 4RN |
| 2E0 | WGF | Mark Horn, 105 Wards Hill Road, Minster on Sea, Sheerness, ML12 2LH |
| 2E0 | WGI | S sugihara, Southfield, Park Lane, Wokingham, RG40 4PY |
| 2I0 | WGL | William Leonard, 57 Mullanavehy, Enniskillen, BT92 2EW |
| 2I0 | WGM | Graeme McCusker, 41 The Granary, Waringstown, Craigavon, BT66 7TG |
| 2E0 | WGO | Ian Paterson, 11 Ocho Rios Mews, Eastbourne, BN23 5UB |
| 2E0 | WGP | Wayne Power, 23 Drawbridge Close, Maidstone, ME15 7PD |
| 2E0 | WHA | William Armes, 11 Rutland Road, [illegible] 4HW |
| 2E0 | WHB | Brian Marks, 107 Linnet Drive, [illegible] |
| 2E0 | WHD | James Gardner, Silverdale, Vicarage Lane, Ormskirk, L40 6HQ |
| 2E0 | WHF | David Cracknell, 120 Woodhill, London, SE18 5JL |
| 2E0 | WHH | John Cook, 20 Huntingdon Close, Totton, Southampton, SO40 3NX |
| 2E0 | WHN | William Northcote, 58 Warren Avenue, Wakefield, WF2 7JN |
| 2E0 | WHO | M Wells, 42 Eggesford Road, Stenson Fields, Derby, DE24 3BH |
| 2E0 | WHU | Antony Hodgson, 515 Ashingdon Road, Rochford, SS4 3HE |
| 2M0 | WIC | Roderick Mackay, 12 Robertson Square, Wick, KW1 5NF |
| 2E0 | WIE | Lisa Shasby, 19 Crawshaw Grange, Crawshawbooth, Rossendale, BB4 8LY |
| 2E0 | WIG | J Shaw, 54 Dicconson Street, Wigan, WN1 2AT |
| 2E0 | WIL | William Whyatt, 11 The Perrings, Nailsea, Bristol, BS48 4YD |
| 2E0 | WIS | Dorian Wiskow, 15 Ferndale Close, Sandbach, CW11 4HZ |
| 2E0 | WIZ | Steven Keen, 13 Ivy Road, Kettering, NN16 9TG |
| 2E0 | WJA | William White, 15 St. Walstans Road, Taverham, Norwich, NR8 6NF |
| 2E0 | WJB | William Bradley, 4 forest view avenue, London, E10 6DX |
| 2E0 | WJC | W Cromack, 45 Southroyd Park, Pudsey, LS28 8AX |
| 2E0 | WJE | W Ellis, 16 Furlong Drive, Tean, Stoke-on-Trent, ST10 4LD |
| 2E0 | WJI | John Dale, 47 Mungo Park Road, Rainham, RM13 7PD |
| 2E0 | WJL | W Lloyd, 14 Lewis Grove, Wolverhampton, WV11 3HR |
| 2E0 | WJP | Wilfred Paterson, 1 Burnside Terrace, Stranraer, DG9 8HH |
| 2E0 | WJR | Daryl Howard, 1 Watermill Drive, St. Leonards-on-Sea, TN38 8WD |
| 2M0 | WJS | Stuart Wilson, 2 Kinnear Court, Guardbridge, St. Andrews, KY16 0UE |
| 2E0 | WJT | William Twemlow, Flat 6, 27 Marmion Road, Liverpool, L17 8TT |
| 2I0 | WKE | John Wilkinson, 11 Fairview Park, Dromore, BT25 1PN |
| 2E0 | WKG | Andrew Cole, 104 Newport Road, Cowes, PO31 7PS |
| 2E0 | WKH | Kris Harbour, 43 Falcon Drive, Stanwell, Staines-upon-Thames, TW19 7EU |
| 2E0 | WKT | Geoffrey Williams, 18 Elmsleigh Road, Farnborough, GU14 0ET |
| 2E0 | WKV | Kevin Dale, 26 Warwick Place, Langdon Hills, Basildon, SS16 6DU |
| 2E0 | WKZ | Ben Drury, 6 Ellen Grove, Harrogate, HG1 4RH |
| 2M0 | WLA | William Lawson, 60 Inglis Avenue, Port Seton, Prestonpans, EH32 0AQ |
| 2E0 | WLD | William Daley, 27 Rosebery Street, Manchester, M14 4UR |
| 2E0 | WLK | R Readman, 1 Millside Close, Kilham, Driffield, YO25 4SF |
| 2E0 | WLN | James Wilson, 125 Langroyd Road, Colne, BB8 9ED |
| 2M0 | WLX | Brian Ewart, 94 Kirkness Street, Airdrie, ML6 6ET |
| 2E0 | WLY | Matthew Walters, 39 Portland Place, Coseley, Bilston, WV14 9TB |
| 2W0 | WMB | A Gibbs, 105 Oak Place, Bargoed, CF81 8NT |
| 2I0 | WMC | William McCormick, 6 Church Street, Rosslea, Enniskillen, BT92 7DD |
| 2E0 | WMD | M Peters, 9 Evelyn Close, Twickenham, TW2 7BL |
| 2E0 | WMG | Kevan Pugh, Col Bern, Church Rd, Colchester, CO7 8HS |
| 2I0 | WMH | William Hawkes, 12 Meadow Court, Newtownards, BT23 8YE |
| 2M0 | WMJ | William MacKenzie, 7 Urquhart Grove, Elgin, IV30 8TB |
| 2E0 | WML | William Little, Burnside, Main Street, Lochans, Stranraer, DG9 9AW |
| 2E0 | WMP | Michael Weaver, 16 Avocet Drive, Kidderminster, DY10 4JT |
| 2M0 | WMU | Michael Marino, 35 Niddrie House Park, Edinburgh, EH16 4UH |
| 2E0 | WMY | Christopher Walmsley, 6 Holly Close, Brighton, BN1 6RZ |
| 2E0 | WNI | Robert Karpinski, 55 Cambridge Avenue, New Malden, KT3 4LD |
| 2E0 | WNM | Wayne McCoo, 8 Newlands Road, Parson Drove, Wisbech, PE13 4LB |
| 2E0 | WNT | Tom Corker, North Side, Wingerworth Hall Estate, Chesterfield, S42 6PL |
| 2E0 | WNW | Andrew Walker, 4 Pretymen Crescent, New Waltham, Grimsby, DN36 4NS |
| 2E0 | WOB | Robert Landragin, 101 Linden Gardens, Enfield, EN1 4DY |
| 2E0 | WOL | Wolfgang Walther, 139 East Street, Epsom, KT17 1EJ |
| 2E0 | WOS | Warwick Barnes, Cushendall, Lyngate Road, North Walsham, NR28 0DH |
| 2E0 | WOW | Diane Martin, kiln close, main road, Lincoln, LN4 4QH |
| 2E0 | WOZ | H Greenhalgh, 61 Long Meadows, Chorley, PR7 2YB |
| 2E0 | WPD | Dudley Woodhams, 83 Langdale Place, Newton Aycliffe, DL5 7DY |
| 2E0 | WPE | M Shaw, 10 Beechwood Avenue, Shevington, Wigan, WN6 8EH |
| 2E0 | WPH | Paul Holmquest, 6 Rhyme Hall Mews, Fawley, Southampton, SO45 1FX |
| 2E0 | WPJ | Peter Joyner, 3 Barton Road, Canterbury, CT1 1YG |
| 2E0 | WPS | Wayne Phillips, 38 Beeches Road Great Barr, Birmingham, B42 2HF |
| 2E0 | WPT | Michael Thompson, 35 Princes Avenue, Desborough, Kettering, NN14 2RQ |
| 2E0 | WPZ | D Chilvers, Flint Cottage, Cherrytree Road, Norwich, NR11 7LQ |
| 2E0 | WQK | Drew Blackie, 30 Queens Avenue, Ilfracombe, EX34 9LS |
| 2E0 | WRF | W Fuller, 10 Grasmere Road, Knottingley, WF11 0NQ |
| 2E0 | WRI | Peter Wright, 4a Alma Street, Melbourne, Derby, DE73 8GA |
| 2E0 | WRK | Joseph Erinjeri, 63 Bulls Green, Westbrook, Warrington, WA5 7XT |
| 2I0 | WRR | W Rea, 3 Carwood Way, Newtownabbey, BT36 5JT |
| 2E0 | WRS | S Webber, 59 Mincinglake Road, Exeter, EX4 7DY |
| 2M0 | WRX | K Glacken, 14 Hailes Avenue, Edinburgh, EH13 0NA |
| 2E0 | WRY | Lawrence Curtis, 39 Mount Stewart Street, Seaham, SR7 7NG |
| 2I0 | WSH | William Hamilton, 9 Susan Street, Belfast, BT5 4FE |
| 2E0 | WSJ | J Woodroof, 37 Danefield Road, Northampton, NN3 2LT |
| 2M0 | WSK | John Muchowski, 71 The Braes, Tullibody, Alloa, FK10 2TT |
| 2E0 | WSM | Jullian Claydon, 17 Canterbury Close, Weston-Super-Mare, BS22 7TS |
| 2E0 | WSR | Alan Rigler, 10 The Ball, Dunster, Minehead, TA24 6SD |
| 2E0 | WSS | L Shand, 52 Ten Acre Way, Rainham, Gillingham, ME8 8TL |
| 2E0 | WST | Richard West, 557 East Bank Road, Sheffield, S2 2AG |
| 2E0 | WSW | Susan Thorne, 2 Ellfield Close, Bristol, BS13 8EF |
| 2E0 | WSX | Alan Holmes, 2 Park Farm Cottages, Park Lane, Chichester, PO20 3TL |
| 2E0 | WSZ | Jeffery Hocking, 26 Musket Road, Heathfield, Newton Abbot, TQ12 6SB |
| 2E0 | WTA | Trevor Wood, 61 Berry Avenue, Watford, WD24 6RU |
| 2E0 | WTD | Kevin Metcalfe, 33 Corsican Drive, Hednesford, Cannock, WS12 4SS |
| 2M0 | WTE | P Jackson, 4 Wester Tarbat House, Kildary, Invergordon, IV18 0GF |
| 2E0 | WTG | Duncan Cooper, Little Heath, Bradfield Common, North Walsham, NR28 0QR |
| 2E0 | WTH | Philip Newth, 20 Barrow Close, Redditch, B98 0NL |
| 2M0 | WTN | Alistair Ross, 29 East Banks, Wick, KW1 5NL |
| 2E0 | WTQ | Thomas Walsh, 2 Ashfield Mews, Ashington, NE63 [illegible] |
| 2M0 | WTT | Shaun Paterson, Free Church Manse, Church Street, Golspie, KW10 6TT |
| 2E0 | WTY | Robert Clare, Kimberley, Boston Road, Boston, PE20 3AP |
| 2E0 | WTZ | D Scott, 198 Slade Green Road, Erith, DA8 2JG |
| 2E0 | WUF | Arentas Butkus, 73a Hudson Road, Bexleyheath, DA7 4PQ |
| 2E0 | WUL | William Murdoch, 64 Cotton Street, Castle Douglas, DG7 1AH |
| 2E0 | WUN | Robert Lester, 17 Clarence Road, Capel-le-Ferne, Folkestone, CT18 7LW |
| 2E0 | WVD | M Bradshaw, 118 Queens Road, Vicars Cross, Chester, CH3 5HE |
| 2E0 | WVE | Mervyn Huggett, 12 West View Cottages, Lewes Road, Haywards Heath, RH16 2LJ |
| 2E0 | WVM | M Edwards, [illegible] Farm, [illegible], WA1 4DY |
| 2E0 | WVS | W Symons, Pommel-House, 4 Trevassack Court, Hayle, TR27 4NA |
| 2E0 | WWA | Alexander Kerr, 9 Martindale Way, Sawston, Cambridge, CB22 3BT |
| 2I0 | WWB | William Bradley, 14 Ardmore Grange, Ballygowan, Newtownards, BT23 5TZ |
| 2E0 | WWF | Edward Field, 4 Redhouse Drive, Sonning Common, Reading, RG4 9NT |
| 2M0 | WWM | Roy Jowett, Fearnoch, Ardentallen, Oban, PA34 4SF |
| 2E0 | WWN | W Northover, 13 Dagenham Avenue, Dagenham, RM9 6LD |
| 2E0 | WWR | Paul Abram, 2 Blackthorn Close, Marford, Wrexham, LL12 8LB |
| 2E0 | WWS | Douglas Cox, 9 Northbrook Copse, Bracknell, RG12 0UA |
| 2E0 | WWT | Allan Walls, 7 Waveney Grove, York, YO30 6EQ |
| 2M0 | WWX | Andrew Prentice, 24 Victoria Road, Grangemouth, FK3 9JN |
| 2E0 | WWZ | R Alexander, 14 Ashfield Terrace, Appley Bridge, Wigan, WN6 9AG |
| 2E0 | WXD | Benjamin Wild, 1 Sunnymount, Midsomer Norton, Radstock, BA3 2AS |
| 2E0 | WXM | Eric Lake, 28 Hampsons Grove, Ruabon, Wrexham, LL14 6AN |
| 2M0 | WXS | Allan McCall, 1 Finlayson Drive, Airdrie, ML6 8LU |
| 2E0 | WXT | T Nimash, 185 Hazelbury Road, Bristol, BS14 9EU |
| 2E0 | WYE | A H Ayres, Brynhyfryd, Phocle Green, Ross on Wye, HR9 7TW |
| 2E0 | WYT | Trevor Webster, 1 Fen Close, Newton, Alfreton, DE55 5TD |
| 2E0 | WYZ | Simon Melton, 2 The Orchard, Bishopthorpe, York, YO23 2RX |
| 2E0 | WZT | Mark Fulbrook, 2 Cob Place, Westbury, BA13 3GS |
| 2E0 | XAA | D Pollard, 110 Rowan Way, Malpas, Newport, NP20 6JN |
| 2E0 | XAE | Andrew Oxborrow, 24 Ickworth Crescent, Rushmere St. Andrew, Ipswich, IP4 5PQ |
| 2E0 | XAG | David Robinson, 49 Meldon Drive, Bradley, Bilston, WV14 8BQ |
| 2E0 | XAH | Alfred Harden, 16 Shining Cliff Court, Bawtry, Doncaster, DN10 6SW |
| 2E0 | XAI | Harry Parfitt, 5 Sheridan Road, Frimley, Camberley, GU16 7DU |
| 2E0 | XAM | A Morgan, 18 Keysworth Drive, Wareham, BH20 7BD |
| 2I0 | XAN | William Nelson, 17 Killygullan Drive, Killygullan, Lisnaskea, Enniskillen, BT92 0HJ |
| 2E0 | XAO | James Cooke, Iolanthe, Chidham Lane, Chichester, PO18 8TH |
| 2E0 | XAR | Steven Halliday, 8 Newby Farm Road, Scarborough, YO12 6UN |
| 2E0 | XAV | James Martin, 20 Hall Green Road, West Bromwich, B71 3LA |
| 2E0 | XAW | A Winkley, 77 Lechlade Road, Birmingham, B43 5ND |
| 2E0 | XAY | Steven Greaves, 409 Beaumont Leys Lane, Leicester, LE4 2BH |
| 2E0 | XAZ | James Deacon, 42a Fairfield, Christchurch, BH23 1QX |
| 2W0 | XBC | Christopher Powell, 1 Llwyn-Onn, Penderyn, Aberdare, CF44 9YJ |
| 2M0 | XBD | Stephanie Boyd, 1 St. Marks Lane, Edinburgh, EH15 2PX |
| 2E0 | XBE | Shane Best, 38 Greensway, Abertysswg, Tredegar, NP22 5AR |
| 2E0 | XBG | Joshua Walker, 6 Wellington Terrace, Islip, Kettering, NN14 3LJ |
| 2E0 | XBK | K Hardy, 17 Purcell Cole, Writtle, Chelmsford, CM1 3NB |
| 2E0 | XBM | C Atkinson, 7 Hamilton Road, Grantham, NG31 9QG |
| 2E0 | XBN | Brian Johnson, 6 Trevor Road, Swinton, Manchester, M27 0YH |
| 2E0 | XBT | Sean Pryer, 16 Wayside Avenue, Worthing, BN13 3JU |
| 2E0 | XBW | Bradley Woollett, 24 Earlsworth Road, Willesborough, Ashford, TN24 0DN |
| 2E0 | XBX | M Lee, 46 Little Lane, Huthwaite, Sutton-in-Ashfield, NG17 2RA |
| 2E0 | XCB | D Beech, The Presbytery, Cotswold Avenue, Newcastle, ST5 6HP |
| 2E0 | XCH | Christopher Hayes, 7 Hadstock Close, Sandiacre, Nottingham, NG10 5LQ |
| 2E0 | XCM | Roydan Styles, Padcroft, Weir, OL13 8QL |
| 2E0 | XCO | Matthew MacDonell, 54 Cinque Foil, Peacehaven, BN10 8DZ |
| 2E0 | XCP | Chris Parker, 40 Holman Way, Ivybridge, PL21 9TE |
| 2M0 | XCT | E Whitaker, Breal House, Drumindorsair, Beauly, IV4 7AH |
| 2E0 | XCV | Mark Abberley, 10 Cranesbill Close, Featherstone, Wolverhampton, WV10 7TY |
| 2E0 | XDC | Derek Copsey, Fairview, Mill Lane, Brentwood, CM15 0PP |
| 2E0 | XDD | David Baseden Butt, 24 Lowry Way, Stowmarket, IP14 1UF |
| 2E0 | XDF | David Ferrington, The Redwoods, 20 Innings Lane, Bracknell, RG42 3TR |
| 2E0 | XDG | Geoffrey Welch, Amazonas, Sandy Lane, Liverpool, L38 3RP |
| 2E0 | XDI | Jonathan Peain, 29 Wild Flower Way, Ditchingham, Bungay, NR35 2SF |
| 2E0 | XDM | Daniel Cole, 39 Hillside Road, Southminster, CM0 7AL |
| 2I0 | XDR | Robert Cross, 15 Ballyfore Road, Larne, BT40 3NF |
| 2M0 | XDS | Donald Suttie, 37 Ullapool Crescent, Dundee, DD2 4TT |
| 2W0 | XDT | R Snape, Ashdale, Broadmoor, Kilgetty, SA68 0RN |
| 2E0 | XDX | Alisdair Lark, 20 Lawfield, Coldingham, Eyemouth, TD14 5PB |
| 2E0 | XDY | Craig Bass, 51 Dane Park Road, Ramsgate, CT11 7LP |
| 2E0 | XDZ | Graham Parsons, 2 The Close, East Grinstead, RH19 1DQ |
| 2E0 | XEA | Andrew Welch, 18 Monk Close, Tipton, DY4 7TP |
| 2E0 | XEE | Andy King, 125 Shirley Road, Southampton, SO15 3FF |
| 2E0 | XEN | Jon Hartley, 71 Pullman Lane, Godalming, GU7 1YB |
| 2E0 | XFF | Mark Chamberlain, 79 Riddy Lane, Luton, LU3 2AJ |
| 2E0 | XFH | Mike Ashton, Lodge Farm Bungalow, Wattisham Road, Ipswich, IP7 7LU |
| 2M0 | XFM | Barry Burrows, 27 Bughtknowes Drive, Bathgate, EH48 4DP |
| 2E0 | XFR | Jacob Stamford, 12 Springhead, Tunbridge Wells, TN2 3NY |
| 2E0 | XGA | G White, 89 Kings Drive, Thingwall, Wirral, CH61 9QA |
| 2E0 | XGS | Gary Stanley, 95 Old Vicarage, Westhoughton, Bolton, BL5 2EG |
| 2E0 | XGW | Gareth Whall, 10 Hillcrest Court, Ipswich Road, Diss, IP21 4YJ |
| 2E0 | XHG | Michael Bolton, 11 Silvia Way, Fleetwood, FY7 7JF |
| 2E0 | XHL | Paul Snook, 7 Sandhurst Avenue, Kwazulu Natal, South Africa, 3610 |
| 2E0 | XIK | Joshua Lambert Hurley, 19 Hill Close, West Bridgford, Nottingham, NG2 6GQ |
| 2E0 | XIS | M Morton, 26 Elderberry Gardens, Witham, CM8 2PT |
| 2E0 | XJJ | Jonathan Jones, 2 Lavender Gardens, Warrington, WA5 1BQ |
| 2E0 | XJL | Jonathan Welch, 49 Walshs Manor, Stantonbury, Milton Keynes, MK14 6BU |
| 2E0 | XJP | Christopher Dennis, Hillsdene, Plex Lane, Ormskirk, L39 7JY |
| 2E0 | XJR | John Raffill, 34 Almond Road, Kettering, NN16 9PF |
| 2E0 | XJT | S Ramsden, 76 Brigg Lane, Camblesforth, Selby, YO8 8HD |
| 2E0 | XKC | J Cunningham, Boleyn House, Erwarton, Ipswich, IP9 1LL |
| 2E0 | XKD | Lindsay Booth, 8 Rowthorne Close, Northampton, NN5 4WB |
| [illegible] | XKL | Roger Jones, Flat 2, Ton y Geraint, 33 Princess Street, Llangollen, LL20 8RD |
| 2E0 | XKO | Paul Goodridge, 22 Horefield, Porton, Salisbury, SP4 0LE |
| 2E0 | XKT | Keith Todman, 12 Winscombe, Bracknell, RG12 8UD |
| 2E0 | XKX | Sean Lyon, 10 Sycamore Close, Preston, Hull, HU12 8TZ |
| 2E0 | XLG | Daniel Moore, 2 Queens Garth, Thornton in Craven, Skipton, BD23 3TH |
| 2D0 | XLJ | L Justin, Garth, Park View Road, Pinner, HA5 3YF |
| 2E0 | XLM | Stevan Wing, Flat 32, Wilkinson Drop, Benfleet, SS7 2BG |
| 2E0 | XLX | John Gascoigne, 64 Prestwold Way, Aylesbury, HP19 8GZ |
| 2E0 | XLY | Mark Oxley, 49 Dalton Crescent, Shildon, DL4 2LE |
| 2E0 | XMC | Martin Callis, 1 Webb Close, Letchworth Garden City, SG6 2TY |
| 2W0 | XMG | P Provis, Dingle Gardens, [illegible], CF11 0EJ |
| 2E0 | XMK | Michael Ross, 115 New Street, Brightlingsea, Colchester, CO7 0DJ |
| 2E0 | XMP | Michael Pearce, 1 Hillside Close, Helsby, Frodsham, WA6 9LB |
| 2E0 | XMS | Mark Street, Flat 6, Derwent Court, Solihull, B92 7BU |
| 2E0 | XMT | Tommy Kwiatkowski, 19 Caspian Walk, London, E16 3JD |
| 2E0 | XNF | Nathaniel Ferrington, 20 Innings Lane, Warfield, Bracknell, RG42 3TR |
| 2E0 | XNL | Gijsbert Molendijk, 47 Lodge Road, Scunthorpe, DN15 7EN |
| 2E0 | XOD | Ian Donnelly, 17 Jessop Close, Horncastle, LN9 6RR |
| 2E0 | XOJ | Malcolm Benson, 11 Hield Grove, Aston by Budworth, Northwich, CW9 6LN |
| 2E0 | XOL | Trevor Brownen, 43 Great Rea Road, Brixham, TQ5 9SW |
| 2E0 | XOR | Michael Hauser, 27 Abbey Street Ickleton, Saffron Walden, CB10 1SS |
| 2E0 | XOT | John Messenger, 34 Goylands Close, Llandrindod Wells, LD1 5RB |
| 2E0 | XPJ | Julian Parfitt, 5 Sheridan Road, Frimley, Camberley, GU16 7DU |
| 2E0 | XPK | Colin Park, 197 Occupation Road, Albert Village, Swadlincote, DE11 8HD |
| 2E0 | XPM | Patrick Mullen, 12 Poplar Grove, Conisbrough, Doncaster, DN12 2JG |
| 2E0 | XPP | Paul Jarvis, 24 St. Peters Gardens, Leeds, LS13 3EH |
| 2E0 | XPT | Tom Parfitt, 5 Sheridan Road, Frimley, Camberley, GU16 7DU |
| 2E0 | XQK | Dennis Goodfellow, 60 Pickering Green, Gateshead, NE9 7DX |
| 2E0 | XRG | Gavin Duffy, 34 Twentyfifth Avenue, Blyth, NE24 2QW |
| 2E0 | XRM | Dennis Bingham, 33 Sheffield Road, Creswell, Worksop, S80 4HN |
| 2E0 | XRS | Richard Stevens, Durham House, Cavendish Road, Sudbury, CO10 8PJ |
| 2E0 | XRX | Alexander Wright, Hills Road, Cambridge, CB2 8PH |
| 2E0 | XRZ | Michael Nicholls, Grahams Onsett Farm, Newcastleton, TD9 0TT |
| 2E0 | XSD | Colin Catlin, 27 Main Street, Frizington, CA26 3SA |
| 2E0 | XSG | Craig Braisby, 4 Langmans Way, Woking, GU21 3QY |
| 2E0 | XSL | S Looker, 165 Mollison Drive, Wallington, SM6 9GX |
| 2E0 | XSW | Stuart Whall, 17 Vicarage Road, Deopham, Wymondham, NR18 9DR |
| 2E0 | XTC | Kevin Haworth, 11 Petersfield Close, Bootle, L30 1SG |
| 2E0 | XTL | Matthew Porter, 8 Stanton Drive, Ludlow, SY8 2PH |
| 2E0 | XTM | J Maguire, 14 Botha Road, St. Eval, Wadebridge, PL27 7TS |
| 2W0 | XTP | Dean Willis, 51 Fforchaman Road, Cwmaman, Aberdare, CF44 6NG |
| 2E0 | XTV | Terence Benson, 83 Glovers Road, Birmingham, B10 0LE |
| 2E0 | XUH | Glyn Thomas, Lowside Barn, Rothersyke, Egremont, CA22 2UD |
| 2E0 | XUL | Michael Brooks, Garth, Cemmaes, Machynlleth, SY20 9PR |
| 2E0 | XUU | Ravi Gopan, 84 Hilmanton, Lower Earley, Reading, RG6 4HN |
| 2E0 | XUZ | Nigel Hanson-Collins, 92 Howbury Lane, Slade Green, Erith, DA8 2DR |
| 2E0 | XVF | Jeremy Smith, 8 Mayfields, Spennymoor, DL16 6RN |
| 2E0 | XVT | Christopher Williams, 1 South View, Freeholdland Road, Pontypool, NP4 8LL |
| 2E0 | XVX | Michael Lawrence, 16 Timson Close, Market Harborough, LE16 7UU |
| 2E0 | XVZ | Andrew Sibley, 27 Sherwood Road, Tetbury, GL8 8BU |
| 2E0 | XWD | R Hawkins, Nook Cottage, Common-y-Coed, Caldicot, NP26 3AX |
| 2E0 | XXB | S Bunce, 15 Downs View Road, Bembridge, PO35 5QS |
| 2E0 | XXK | Kam Mitchell, 1 Denstroude Cottages, Denstroude Lane, Canterbury, CT2 9JX |
| 2E0 | XXM | Michael Jennings, Springfield Farm, The Causeway, King's Lynn, PE34 3PP |
| 2E0 | XXO | Stuart Nesling, 64 Ruskin Avenue, Lincoln, LN2 4BT |
| 2M0 | XXP | Alan Pitkethley, 99 Margaretvale Drive, Larkhall, ML9 1EH |
| 2E0 | XXT | Terence Archer, 241 Beaver Lane, Ashford, TN23 5PA |
| 2E0 | XXX | M Davey, 27 Earls Drive, Newcastle, ST5 3QR |
| 2E0 | XYA | Phillip Hodges, 191 Broadstone Road, Stockport, SK4 5HP |
| 2E0 | XYM | Mark Leonard, 5 Nettleton Garth, Burstwick, Hull, HU12 9DY |
| 2E0 | XYT | Edward Wood, 13 Rosedale, Welwyn Garden City, AL7 1DW |
| 2E0 | XYX | Anthony Austin, 4 Cornwall Avenue, Oldbury, B68 0SW |
| 2E0 | XYZ | C Edgar, 61 Winchester Avenue, Lancaster, LA1 4HX |
| 2E0 | XZI | Jason Ball, 83 pendra loweth, Falmouth, TR11 5BJ |
| 2M0 | XZX | Gabriel Queen, G/R 31 Provost Road, Dundee, DD3 8AF |
| 2E0 | XZZ | Simon Rouse, 7 Cranbrook Road, Thurnby, Leicester, LE7 9UA |
| 2E0 | YAB | Shane Morgan, 20 Cwrt y Babell, Cwmfelinfach, Newport, NP11 7NR |
| 2E0 | YAD | R Williams, 92 [illegible], Greenmeadow, Cwmbran, NP44 4LF |
| 2W0 | YAE | Gary Thatcher-Sharp, 20 Dilys Street, Blaencwm, Treorchy, CF42 5DT |
| 2M0 | YAF | Christopher Tait, 21 Mount Avenue, Kilmarnock, KA1 1UF |
| 2M0 | YAG | Neil Hirst, 25 Conifer Road, Mayfield, Dalkeith, EH22 5BY |
| 2E0 | YAL | David Parker, 5 Beam Avenue, Dagenham, RM10 9BS |
| 2E0 | YAO | Lorna Jex, 26 Springdale Crescent, Brundall, Norwich, NR13 5RA |
| 2E0 | YAP | Alexander Kissin, 115 Aarons Hill, Godalming, GU7 2LJ |
| 2E0 | YAR | Frank Lees, 5 St. Winifred Road, Rainhill, Prescot, L35 0PY |
| 2E0 | YAS | John Stacey, 130 Stocks Lane, East Wittering, Chichester, PO20 8NT |
| 2E0 | YAV | William Jones, 8 Oakbrook Close, Ewyas Harold, Hereford, HR2 0NX |
| 2E0 | YAW | Michael Driscoll, 59 Havendale, Hedge End, Southampton, SO30 0FD |
| 2E0 | YAX | Alan Garn, 5 Bassett Street, Walsall, WS2 9PZ |
| 2J0 | YAY | James Bryant, 5 Louiseberg Court Queen's Road, St. Helier, Jersey, JE2 3GQ |
| 2E0 | YBL | Linda Jones, 8 Oakbrook Close, Ewyas Harold, Hereford, HR2 0NX |
| 2E0 | YBR | Gregory Fordyce, 2 Church Street, East End, Earlston, TD4 6HS |
| 2E0 | YBS | B Scroggs, Thatchways, High Street, Banbury, OX15 5HW |
| 2W0 | YBZ | Paul Smith, 29 Heol Cwarrel Clark, Caerphilly, CF83 2NE |
| 2E0 | YCD | David Ashton-Hilton, 14 Weetwood Road, Congresbury, Bristol, BS49 5BN |
| 2M0 | YCG | Callum Graham, 5 Ashkirk Road, Strathaven, ML10 6JT |
| 2M0 | YCJ | Colwyn Jones, 11b Ettrick Road, Edinburgh, EH10 5BJ |
| 2E0 | YDA | Andrew Bedford, 1 Carder Crescent, Bilston, WV14 0JT |

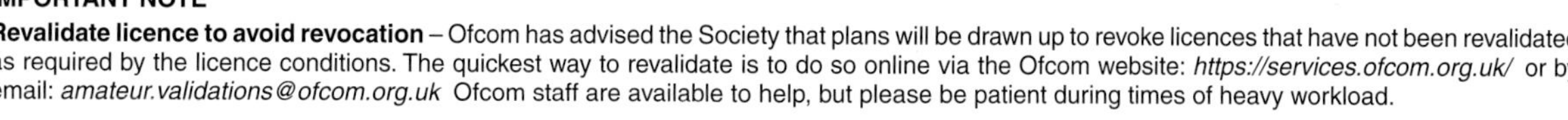

**IMPORTANT NOTE**

**Revalidate licence to avoid revocation** – Ofcom has advised the Society that plans will be drawn up to revoke licences that have not been revalidated as required by the licence conditions. The quickest way to revalidate is to do so online via the Ofcom website: *https://services.ofcom.org.uk/* or by email: *amateur.validations@ofcom.org.uk* Ofcom staff are available to help, but please be patient during times of heavy workload.

2E0 YDB Daniel Bower, 89 Halifax Road, Sheffield, S6 1LA
2I0 YDF D Foley, 14 Chestnut Hall Court, Maghaberry, BT67 0GJ
2E0 YDJ Darran Jackson, 3 Laburnum Road, Cadishead, Manchester, M44 5AS
2E0 YDK D Edwards, 25 Bryn Coed, Gwersyllt, Wrexham, LL11 4UE
2E0 YDM Carl Preece, 14 Dock Street, Widnes, WA8 0QX
2E0 YDT Alan Carney, 9 Hart Square, Sunderland, SR4 8BS
2E0 YDX J leighton, 27 The Pastures, Cayton, Scarborough, YO11 3UU
2E0 YEP Sean Quinn, 17 Cleveland Road, Southampton, SO18 2AP
2M0 YEQ Gordon Pearce, 1 Inchbelle Farm Cottage, Kirkintilloch, Glasgow, G66 1RS
2E0 YES M Casey, 7 Cobham Avenue, Manchester, M40 5QW
2E0 YEW Andrew McEwen, 4 The Pantyles, Nightingale Lane, Sevenoaks, TN14 6BX
2W0 YFC J Evans, 11 Dew Crescent, Cardiff, CF5 5PB
2M0 YFR David Cockburn, 88 Knockmarloch Drive, Kilmarnock, KA1 4QN
2E0 YFZ Joseph Blower, 4 Lamorna Close, Luton, LU3 2TH
2E0 YGB Andrew Birch, 3 Partridge Way, High Wycombe, HP13 5JX
2E0 YGH Francis Armstrong, 38 Dovecote Drive, Haydock, St. Helens, WA11 0SD
2E0 YGS Grahame Moss, 125 Lavender Avenue, Mitcham, CR4 3RS
2M0 YIC Sarah Bellis, 3 Ivy Lane, Dysart, Kirkcaldy, KY1 2XD
2M0 YIO Brian Fullerton, 55 Alexander Avenue, Stevenston, KA20 4BG
2E0 YIP Shaun Chng, Wolfson College, Cambridge, CB3 9BB
2E0 YJF David Moran, 23 Abbotsfield Crescent, Tavistock, PL19 8EY
2E0 YJL John Cairns, 17 Alfred Avenue, Worsley, Manchester, M28 2TX
2E0 YJW Jason Williams, 10 Masefield Avenue Eaton Ford, St. Neots, PE19 7LS
2E0 YJY Michael Kealey, 24 Ben Nevis Road, Birkenhead, CH42 6QY
2E0 YKS Brenda Shackleton, 54a Blueleighs Park Homes, Ipswich, IP6 0ND
2E0 YLE Allan Doyle, 54 Bro Syr Ifor, Tregarth, Bangor, LL57 4AS
2E0 YLH J Haystead, 11 Lumley Close, Maltby, Rotherham, S66 7SG
2E0 YLL Rebecca Bowen, 25 Maendu Terrace, Brecon, LD3 9HH
2E0 YLP Rupert Campbell-Black, 10 Wren Close, Towcester, NN12 6RD
2I0 YLT Summer McCormick, 46 Lany Road, Moira, Craigavon, BT67 0NZ
2D0 YLX David Cain, 7 Cronk y Berry Mews, Douglas, Isle of Man, IM2 6HQ
2E0 YLY Stuart Myland, 2 Willhays Close, Kingsteignton, Newton Abbot, TQ12 3YT
2E0 YME Mark Smith, 2 Tullig, Cahirciveen, County Kerry, Ireland, 111111
2I0 YMF Martin Foley, 44 Gallows Street, Dromore, BT25 1BD
2E0 YMR Kevin Humphreys, 16 Thames Close, Ferndown, BH22 8XA
2E0 YMT Paul Horrox, 39 Wilton Grove, Heywood, OL10 1AS
2E0 YND Bryan Anderson, 41 Lower Meadow, Harlow, CM18 7RE
2E0 YNI Stuart Widdowson, 11 Belmont Drive, Staveley, Chesterfield, S43 3PQ
2E0 YNT Andre Ashby, 5 New Street, Osbournby, Sleaford, NG34 0DL
2E0 YOK Christopher Bietz, 5 Consort House, Brewery Lane, Wymondham, NR18 0BD
2E0 YOM James Thresher, 328 Gospel Lane, Birmingham, B27 7AJ
2E0 YOP Timothy Court, Eastgate Cottage, Perrys Lane, Norwich, NR10 4HJ
2M0 YOY James Moir, 41 Brisbane Terrace, East Kilbride, Glasgow, G75 8DL
2E0 YOZ Jamie Wilson, 5 Queens Road, Hoylake, Wirral, CH47 2AG
2E0 YPG R Irwin, 21 Penn Street, Belper, DE56 1GH
2E0 YPJ Paul Kirby, 30 New Street, Eccleston, Chorley, PR7 5TW
2E0 YPK Eleanor Maddex, 16 The Larches, Benfleet, SS7 4NR
2E0 YPW Paul Woodfin, Laurel Cottage, Barrow Street, Much Wenlock, TF13 6EN
2E0 YQC G Hope, 3 Bean Close, Great Chart, Ashford, TN23 3BG
2E0 YQT John Best, 24 Suggitts Lane, Cleethorpes, DN35 7JJ
2E0 YRM Martin Radulov, 60 St. Marks Avenue, Northfleet, Gravesend, DA11 9LW
2E0 YRW John Woods, 21 Appleyard Crescent, Norwich, NR3 2QN
2E0 YSF Leo Metcalfe, 40 St. Anns Court, Hartlepool, TS24 7HY
2E0 YSO Michael Mayson, Bell Cottage, School Road, Huntingdon, PE28 3AT
2E0 YSP Alan Gobey, Nut Tree Cottage, Valley rd, Ipswich, IP8 4LR
2M0 YSR Christopher Phillips, 8 The Square, Newtongrange, Dalkeith, EH22 4QD
2E0 YSU G Cummings, 18 Castleton Boulevard, Skegness, PE25 2TX
2E0 YTF Gary Garman, 11 Rye Close, Norwich, NR3 2LF
2E0 YTT R Spooner, 45 Shaftesbury Avenue, Southport, PR8 4NH
2E0 YUD Stephen Nutt, 77 Exeter Close, Stevenage, SG1 4PW
2E0 YUN Yunfei Li Song, The Colony, Chesterton Lane, Cambridge, CB4 3AA
2E0 YVR Linda Palir, 116 Carville Crescent, Brentford, TW8 9RD
2E0 YWN R Cook, 17 Baroness Grove, Salford, M7 1LP
2E0 YWP Douglas Spooner, 30 Clover Road, Norwich, NR7 8TF
2E0 YXB Richard Barrett, 18 Bullstake Close, Oxford, OX2 0HN
2E0 YXO Michael Bilverstone, 12 Westlea Road, Sywell, Northampton, NN6 0BY
2E0 YXZ Storm Christofi, 19 Kingsland Avenue, Northampton, NN2 7PP
2E0 YYD P Dumpleton, 20 Cambridge Road North, Mablethorpe, LN12 1QR
2E0 YYF Gerald Finney, 121 School Lane, Caverswall, Stoke-on-Trent, ST11 9EN
2E0 YYG Daren Harwood, 5 Willows Close, Washington, NE38 7BB
2E0 YYK Steven Hoyle, 20 Brandsby Grove, Huntington, York, YO31 9HL
2W0 YYP Paul Griffey, 61 Cottesmore Way, Cross Inn, Pontyclun, CF72 8BG
2E0 YYT Gary Finney, 121 School Lane, Caverswall, Stoke-on-Trent, ST11 9EN
2M0 YYU Amy Anderson, 16 Walker Court, Glasgow, G11 6QP
2E0 YYY Michael Hunter, 126 Turner Street, Stoke-on-Trent, ST1 2NE
2E0 YYZ S Wall, 26 Wallace Lane, Whelley, Wigan, WN1 3XT
2E0 YZA Bethany Wilson, 28 Warren Crescent, Marsh Lane, Sheffield, S21 5RW
2E0 YZC Daniel Matheson, 21 Warren Hill Road, Woodbridge, IP12 4DU
2E0 YZM Graham Brown, 134 Skipper Way, Lee-on-The-Solent, PO13 8HD
2E0 YZQ Alan Kent, 4 Sellerdale Drive, Wyke, Bradford, BD12 9DA
2E0 YZX Mark Galloway, 2 Edendale Terrace, Horden, Peterlee, SR8 4RD
2E0 YZZ Paul Collingham, 1 Wychwood Drive, Trowell, Nottingham, NG9 3RB
2W0 ZAA Stephen Tozer, 110 Glanffornwg, Wild Mill, Bridgend, CF31 1RL
2E0 ZAC M Cotton, 18 St. Oswalds Crescent, Brereton, Sandbach, CW11 1RW
2E0 ZAE Peter Mason, 20 Coronation Road, Six Bells, Abertillery, NP13 2PJ
2E0 ZAF Ricky Amos, 6 Eccles Road, Wittering, Peterborough, PE8 6AU
2E0 ZAH L Almond, 26 Ashbourne Drive, Desborough, Kettering, NN14 2XG
2E0 ZAI Barry Hardy, 10 Spring Farm Road, Burton-on-Trent, DE15 9BN
2E0 ZAJ S Rafter, 30 Monmouth Grove, St. Helens, WA9 1QB
2E0 ZAL John Moore, Moorelake Lodge, Barholm Road, Stamford, PE9 4RJ
2E0 ZAP R Bird, 78 Arden Road, Hockley, Tamworth, B77 5JE
2E0 ZAU Matthew Southgate, 107 Englands Lane, Loughton, IG10 2QL
2M0 ZAX Adam Hutchison, 24 Tanna Drive, Glenrothes, KY7 6FX
2E0 ZBB Mark Palmer, 116 Claverham Road, Yatton, Bristol, BS49 4LE
2W0 ZBC Glyn Jones, 12 Wilson Place, Cardiff, CF5 4LN
2E0 ZBD J Wells, 15 Phillips Crescent, Needham Market, Ipswich, IP6 8TF
2E0 ZBE John Ellery, 7 Midanbury Crescent, Southampton, SO18 4FN
2M0 ZBF Graham Irving, 55 Gillbank Avenue, Carluke, ML8 5UW
2M0 ZBH Paul McLaren, 1 Morayvale, Aberdour, Burntisland, KY3 0XE
2E0 ZBW Robert Weaver, 116 Carville Crescent, Brentford, TW8 9RD
2E0 ZBZ Mike Carvell, 10 Burns Close, Stevenage, SG2 0JN
2E0 ZCB Christopher Button, 37 Smith Square, Harworth, Doncaster, DN11 8HW
2E0 ZCG C Gregory, 81 Fiskerton Way, Oakwood, Derby, DE21 2HY
2E0 ZCJ Charles Jonas, 1 St. Johns Road, Stansted, CM24 8JP
2E0 ZCM Andrew Birkhead, 9 Parkfield Terrace, Branscombe, Seaton, EX12 3DD
2E0 ZCP Chandith Palawinna, 41 Lealands Drive, Uckfield, TN22 1DW
2E0 ZDA D Phillips, 359 Finchampstead Road, Finchampstead, Wokingham, RG40 3JU
2E0 ZDB David Brownsea, 47 Southill Road, Bournemouth, BH9 1SH
2E0 ZDC Dunstan Cooke, Apartment 9, 27 Sheldon Square, London, W2 6DW
2E0 ZDE Denis Kirkden, 57 Crow Hill Road, Margate, CT9 5PF
2E0 ZDH D Hardwick, 30 Halfcot Avenue, Stourbridge, DY9 0YB
2E0 ZDJ Louis Macrides, 5 Apple Farm Lane, Weston-Super-Mare, BS24 7TJ
2E0 ZDM David Mason, 25 Primary Gardens, Hendon, Sunderland, SR2 8QT
2E0 ZDW Darren Whitley, 10 Kenmore Drive, Cleckheaton, BD19 3EJ
2E0 ZDX Phil Jones, 102 Manor House Lane, Preston, PR1 6HP
2M0 ZEB Graeme Barrie, 24 Mauldeth Road, Broxburn, EH52 6FB
2E0 ZED Alan Henderson, 5 Snipe House Cottages, Alnwick, NE66 2JD
2M0 ZEE Paul Russell, 21 St. Andrews Drive, Law, Carluke, ML8 5GB
2E0 ZEH Stephen Hendy, Flat 2, 33 Kingston Road, Leatherhead, KT22 7SL
2E0 ZEN H Haydon, 101 Blandford Street, Ashton-under-Lyne, OL6 7HG
2M0 ZET H Dally, 3 Gremmasgaet, Lerwick, Shetland, ZE1 0NE
2E0 ZEV Barrie Dexter, 13 Guildford Avenue, Sheffield, S2 2PJ
2M0 ZFG Steven Street, 13 Cobbler View, Arrochar, G83 7AD
2E0 ZFV Peter Whiteley, 1 Newton Close, Fareham, PO14 3LF
2E0 ZFX Stephen Pantony, 40 Park Avenue, Redhill, RH1 5DP
2I0 ZFZ Robert McKay, 31 Squires Hill Crescent, Belfast, BT14 8RE
2E0 ZGA Gary Campbell, 10 Welbeck Road, Rochdale, OL16 4XP
2E0 ZGL Adam Lunn, 57 Greets Green Road, West Bromwich, B70 9ES
2E0 ZGS Zbigniew Sznober, 9 MOOR ROAD Dawley, Telford, TF4 2AR
2E0 ZGX P Beltrami, 15 Woodroffe Square, Calne, SN11 8PW
2E0 ZHG Chris Barnes, 23 South Street, Crewe, CW2 6HN
2E0 ZHN Elia Mady, 130 Staveley Gardens, London, W4 2SF
2E0 ZIP S Kiley, 178 Kingfisher Drive, Woodley, Reading, RG5 3LQ
2E0 ZIV Iain Vickers, 3 Nesbit Road, St. Marys Bay, Romney Marsh, TN29 0SF
2E0 ZJA David Bowen, 25 Maendu Terrace, Brecon, LD3 9HH
2E0 ZJB Mark Collier, 8 Masefield Mews, Dereham, NR19 2SY
2E0 ZJO Jonathan Rawlinson, Westfield Farm, Risden Lane, Cranbrook, TN18 5DU
2E0 ZJQ Adam Rawlinson, Westfield Farm, Risden Lane, Hawkhurst, Sandhurst, Cranbrook, TN18 5DU
2E0 ZKD Timothy Newton, Wraysbury, Forestside, Rowland's Castle, PO9 6ED
2E0 ZKT Peter Stone, 32 Worcestershire Lea, Warfield, Bracknell, RG42 3TQ
2E0 ZLA Adam Zeller, Flat 1, 57 Chalk Hill, Watford, WD19 4DA
2E0 ZLD Zofia Dunne, 1 Burton Gardens, Brierfield, Nelson, BB9 5DR
2E0 ZLM Luke Milburn, 55 Hyde Heath Court, Crawley, RH10 3UQ
2E0 ZLO Mark Holmes, 6 Wells Court, Saxilby, Lincoln, LN1 2GY
2E0 ZMB Mark Breslin, 15 Acorn Gardens, East Cowes, PO32 6TD
2E0 ZMI Thomas Kelly, 2 Weaver House, Chester Road, Runcorn, WA7 3EG
2E0 ZML J Marr, Touchstone, Heathfield Road, Bembridge, PO35 5UW
2E0 ZMM David Mainwaring, 1 Buckingham Close, Didcot, OX11 8TX
2E0 ZMO Andrew Maguire, 132 Wigan Road, Ormskirk, L39 2BA
2E0 ZMR Iain Nicholson, 2 Broom Close, Leyland, PR25 5RQ
2E0 ZMS Matthew Strickland, Ancoats, Piercy End, York, YO62 6DQ
2E0 ZMT M Thompson, 133 Redford Avenue, Horsham, RH12 2HH
2E0 ZMZ Michael Coleman, 3 Tummon Road, Sheffield, S2 5FD
2M0 ZNQ William Beaton, 4 Moorfield Gardens, Springfield, Cupar, KY15 5SH
2E0 ZNZ Guy Richardson, Berwick Cottage, Bailes Lane, Guildford, GU3 2AX
2E0 ZOM Jeff Skinner, 36 Milton Road, Waterloo, Liverpool, L22 4RF
2E0 ZOR Alexander McCrystal, 2 Surtees Close, Maltby, Rotherham, S66 8EE
2E0 ZOT Martin Toher, The Chapel, Station Road, Darlington, DL2 1JG
2E0 ZOZ Adrian Hunter, 9 Gelt Burn, Didcot, OX11 7TZ
2E0 ZPA Paul Archer, 31 Stoney Bank Drive, Kiveton Park, Sheffield, S26 6SJ
2E0 ZPM Christopher Williams, 39 Meldrum Court, Temple Herdewyke, Southam, CV47 2UF
2E0 ZPN Lance Gibbs, 37 Oxford Road, Fulwood, Preston, PR2 3JL
2E0 ZPT Mark Atfield, 42 Pauls Croft, Cricklade, Swindon, SN6 6AJ
2E0 ZPY Richard Pyner, 1 Avon Court, 63 Shakespeare Road, Bedford, MK40 2DS
2E0 ZRB Robin Brown, 194 Wymersley Road, Hull, HU5 5LN
2E0 ZRG Rob Greaves, 7 Eller Brook Close, Heath Charnock, Chorley, PR6 9NQ
2E0 ZRL Anthony Briggs, 3 Swallow Avenue, Leeds, LS12 4RD
2E0 ZRM D Morgan, 2 Raymond Way, Plymouth, PL7 4EG
2E0 ZRQ Glenn Crane, 22 Brewery Street, Burgh le Marsh, Skegness, PE24 5LG
2E0 ZRT Timothy Cooper, 9 Websters Close, Shepshed, Loughborough, LE12 9AT
2E0 ZRX C Waterworth, 4 Mossdale Road, Ashton-in-Makerfield, Wigan, WN4 0EQ
2E0 ZSA Simon Airs, 6 The Willows, Culham, Abingdon, OX14 4NN
2E0 ZSB Steven Bannister, 162 Dobcroft Road, Sheffield, S11 9LH
2E0 ZSE Paul Holmes, 53 Bishops Hull Road, Bishops Hull, Taunton, TA1 5EP
2E0 ZSH Shaun Hampson, 12 Flying Fields Drive, Macclesfield, SK11 7GE
2E0 ZSJ John Gibson, 22 Woodburn Drive, Chapeltown, Sheffield, S35 1YS
2E0 ZSK Seth Kneller, 366a Kingsland Road, London, E8 4DA
2E0 ZSR Stuart Robottom-Scott, 73 St. Bernards Road, Solihull, B92 7DF
2E0 ZST Steve Harris, 61 Monks Park Avenue, Bristol, BS7 0UA
2E0 ZSU Peter Westwell, Roden House, Dobsons Bridge, Whitchurch, SY13 2QL
2E0 ZSY Thomas Symons, Southgate, The Commons, Mullion, TR12 7HZ
2E0 ZTD George Berry, 5 Oakholme Rise, Worksop, S81 7LJ
2E0 ZTG A Hill, 5 Park Road, Thurnscoe, Rotherham, S63 0TG
2E0 ZTL Craig Ingamells, 2 St. Mary's Drive, Sutterton, Boston, PE20 2LU
2E0 ZTM T Moore, 16 Warwick Drive, Earby, Barnoldswick, BB18 6LX
2E0 ZUT Mehmet Beyoglu, 177 Nags Head Road, Enfield, EN3 7AD
2E0 ZUX A Remnant, 172 Burnham Road, Highbridge, TA9 3EH
2E0 ZVG Ian Browne, 85 White Eagle Road, Haydon Leigh, Swindon, SN25 1PY
2E0 ZVL Vincent Lynch, 16 Okehampton Crescent, Sale, M33 5HR
2W0 ZVR Barry Bateman, Galltygog Farm, Llwydcoed, Aberdare, CF44 0DJ
2E0 ZWA Douglas Tordoff, 49 Dale Edge, Eastfield, Scarborough, YO11 3EP
2E0 ZWE L Werndle, 46 Buckingham Road, Richmond, TW10 7EQ
2W0 ZWR G Spicer, 6 Cromwell Road, Neath, SA10 8DR
2E0 ZWW William Warwicker, 13 Elm Tree Avenue, Tile Hill, Coventry, CV4 9EU
2I0 ZXD Joseph Baker, 324 Clonmeen, Drumgor, Craigavon, BT65 4AT
2E0 ZXG Gareth Carless, Silver Cottage, Silver Street, South Petherton, TA13 5BY
2E0 ZXJ Jon Wildsmith, 7 Doctors Hill, Stourbridge, DY9 0YE
2I0 ZXM Michael Meagher, 42 Mourne View Park, Newry, BT35 6BZ
2E0 ZXQ Ian Talbot, 41 Elmwood Close, Cannock, WS11 6LX
2E0 ZXR Matthew Holbrook-Bull, 66 Wayman Road, Corfe Mullen, Wimborne, BH21 3PN
2E0 ZXV Ashley Booth, 11 Kinnaird Close, Elland, HX5 9JF
2E0 ZXX Oliver Spurway, 89 Woolpitch Wood, Chepstow, NP16 6DR
2E0 ZYG Colin Richardson, 12 Ingsgarth, Pickering, YO18 8DA
2E0 ZYK Mark Edge, 19 Burton Avenue, Rushall, WS4 1NH
2E0 ZYL Sue Allen, Milverton, Mill Road, Pulborough, RH20 2PZ
2D0 ZYX Stephen Entwisle, 30 Arden Mhor, Pinner, HA5 2HR
2E0 ZZC Simon Alexander, 13 Padgate, Thorpe End, Norwich, NR13 5DG
2E0 ZZF David Jones, 1 Brig y Nant, Llangefni, LL77 7QD
2E0 ZZT Nigel Payne, 19 Sid Park Road, Sidmouth, EX10 9BW
2W0 ZZU Elgan Jones, 39 Ger-y-Llan, Velindre, Llandysul, SA44 5YB

## 2*1

2E1 ABE Hannah Forder, 4 Jackson Drive, Kennington, Oxford, OX1 5LL
2E1 ABN Jon Morrison, 52 Kimberley Close, Dover, CT16 2JW
2E1 ABQ Nicola Harman, 7 Maple Avenue, Torpoint, PL11 2NE
2E1 ABW S Minnock, 32 Sandwood Road, Sandwich, CT13 0AQ
2E1 ABY Michael O'Brien, 14 Westdean Close, Dover, CT17 0NP
2E1 ACG V Hammonds, 22 The Croft Meriden, Coventry, CV7 7NQ
2E1 ACK D Nye, 5 Charles Road, Deal, CT14 9AT
2W1 ACM Darryl Young, 1 Hawthorn Road, Llanharry, Pontyclun, CF72 9JD
2E1 ACS Neville Roberts, 37a Rockley Avenue, Birdwell, Barnsley, S70 5QY
2E1 ACW C Pooler, 18 Johnstone Close, Wrockwardine Wood, Telford, TF2 7DA
2E1 ACZ Roger Stanley, 219 Fartown, Pudsey, LS28 8NH
2E1 ADJ J Bridgman, 5 Drayton Avenue, Mackworth, Derby, DE22 4JU
2E1 ADO Kenneth Jenkins, 79 Beaufort Road, Newport, NP19 7PB
2E1 ADP T Thompson, 19 Park End, Summer Lane Caravan Park, Banwell, BS29 6JD
2E1 ADQ Cheryll Hammett, 63 Treffry Road, Truro, TR1 1WL
2E1 ADR D Palmer, 133 Victoria Road East, Thornton-Cleveleys, FY5 5HH
2E1 ADT Kelvin Barbery, 17 Polbreen Avenue, St. Agnes, TR5 0TR
2E1 AEC Cerys Vincent, 134 Wolds Drive, Keyworth, Nottingham, NG12 5DA
2E1 AED P Weston, 3 Parc Y Llan, Llanfair Dyffryn Clwyd, Ruthin, LL15 2YL
2E1 AEJ E Jones, 19 Foxhollow, Bar Hill, Cambridge, CB23 8EP
2E1 AEQ Viviene Parrish, 89 Delamere Drive, Macclesfield, SK10 2PS
2E1 AFA J Davis, 5 James Way, Camberley, GU15 2RQ
2E1 AFC Jeffrey Rossiter, 7 Valley View, Bodmin, PL31 1BE
2E1 AFH G Tibbett, 16 The Dingle, Fulwood, Preston, PR2 3EX
2E1 AFI Judith Charnley, 30 Dunkirk Avenue, Fulwood, Preston, PR2 3RY
2E1 AFN Lauraine Swindale, 17 Crofton Close, Bracknell, RG12 0UR
2E1 AFR F Batty, 26 Kingsmead Park, Elstead, Godalming, GU8 6DZ
2E1 AFS L Jenkins, 49 Harts Grove, Chiddingfold, Godalming, GU8 4RG
2E1 AGE J Prince, Field House, 25 Chiltern Road, Slough, SL1 7NF
2E1 AGQ J Collins, 61 Albemarle Road, Gorleston, Great Yarmouth, NR31 7AS
2E1 AGV Edward Muircroft, 84 Longley Avenue west, Sheffield, S5 8WF
2E1 AHK J Robinson, 35 Vegal Crescent, Halifax, HX3 5PA
2E1 AHU T Hassall, 5 Ashworth Street, Bacup, OL13 9LS
2W1 AID Simon Williams, 17 Pond Mawr, Maesteg, CF34 0NG
2E1 AII D Swann, 89 Leazes View, Rowlands Gill, NE39 2JT
2E1 AIT J Tonks, 295 Quinton Road West, Quinton, Birmingham, B32 1PG
2E1 AIY Barbara Whalley, 46 Wayside, Woodside, Telford, TF7 5NG
2E1 AKW Michelle Bradley, Flat 20, Crown Court, Portsmouth, PO1 1QN
2I1 ALE Desmond Auld, 37 Castlewellan Road, Rathfriland, Newry, BT34 5EL
2E1 AMB A Collins, 141 Downside Avenue, Findon Valley, Worthing, BN14 0EY
2E1 AMW D Evans, 9 Robin Close, Farnworth, Bolton, BL4 0RG
2E1 ANG P Jackson, 55 Bomers Field, Rednal, Birmingham, B45 8TQ
2E1 ANH Claire Jackson, 55 Bomers Field, Rednal, Birmingham, B45 8TQ
2E1 ANN M Kearney, 18 Wayside Mews, Maidenhead, SL6 7EJ
2E1 ANQ A Bell, 8 Silk Mill Green, Leeds, LS16 6DU
2M1 ANY Lee Waterall, 3 Wavell Street, Grangemouth, FK3 8TG
2W1 AOD David Wanklyn, 21 Upper Street, Maesteg, CF34 9DU
2E1 AOF G Tutt, 46 Heathcroft Avenue, Sunbury-on-Thames, TW16 7TL
2E1 AOG J Menday, 3 Ash Grove, Guildford, GU2 8UT
2E1 AOK R Gill, 45 Biggin Lane, Ramsey, Huntingdon, PE26 1NB
2E1 APW David Jenkinson, 9 Dalton Street, Cockermouth, CA13 0AR
2E1 APX D Johnson, 3 Plantation Avenue, Swalwell, Newcastle upon Tyne, NE16 3JN
2E1 AQH S Allgood, 53 The Avenue, Leighton Bromswold, Huntingdon, PE28 5AW
2E1 ARG M Trigg, 41 Veasey Road, Hartford, Huntingdon, PE29 1TA
2E1 ARS W Hornby, Lindenstrasse 9, Allschwil, Switzerland, 4123
2E1 ASF T Stevens, 20 The Butts, Crudwell, Malmesbury, SN16 9HF
2E1 ATH J Minnock, 32 Sandwood Road, Sandwich, CT13 0AQ
2E1 ATV R Corden, Konrad Cottage, Welburn, York, YO60 7DX
2E1 AUN D Austin, 66 Homewood Avenue, Sittingbourne, ME10 1XJ
2E1 AUQ E Harding, 17 Summerfield Close, Wokingham, RG41 1PH
2E1 AVM S Peacock, 16 Banalog Terrace, Hollybush, Blackwood, NP12 0SF
2E1 AVT J Hoggan, 34 Dickens Road, Crawley, RH10 5AR
2E1 AVX James Preece, 17 Cherry Tree Avenue, Staines, TW18 1JB
2M1 AVZ N Sinclair, 16 Sycamore Glade, Livingston, EH54 9JG
2E1 AWS Nigel Cook, 24 Thornhurst, Churchill Avenue, Herne Bay, CT6 6SQ
2E1 AWZ V Hilton, 232 Hurst Rise, Matlock, DE4 3EW
2E1 AXD A Richmond, 57 The Fairway, Daventry, NN11 4NW
2E1 AXE L Richmond, 57 The Fairway, Daventry, NN11 4NW
2I1 AXH Kenneth Bird, 115 Halftown Road, Lisburn, BT27 5RF
2E1 AXI F Preece, 8 Gregory Road, Hedgerley, Slough, SL2 3XL
2E1 AXL W Rose, 7 Harby Close, Grantham, NG31 7XA
2E1 AYO E Phillips, 2 Oak Street, Newport, NP9 7HW
2E1 AYS Paul Cartwright, 41 Sandgate Drive, Kippax, Leeds, LS25 7EX
2E1 AZA E Williams, 37 Danesby Crescent, Denby, Ripley, DE5 8RF
2E1 AZK C Morris, 10 Hempits Grove, Acton Trussell, Stafford, ST17 0SL

| | | |
|---|---|---|
| 2E1 | AZQ | J Perkins, Highfield House, Newtown, Buxton, SK17 0NF |
| 2E1 | AZU | D Williams, 75 Queens Avenue, [illegible], Bangor, LL57 1N[illegible] |
| 2E1 | AZW | R Roberts, Rose Cottage, Castle Hill, Leyburn, DL8 4QN |
| 2E1 | BAD | B Rowland, Deacons Cottage, Bridleway, Croft, LE9 6EE |
| 2E1 | BAE | T Ladley, Marisdene, London Road, Faversham, ME13 9LF |
| 2E1 | BBO | P Kent, Old Cottage, Hermitage Lane, Maidstone, ME14 0HP |
| 2E1 | BBY | J Duncan, 4 Lady Moss, Tweedbank, Galashiels, TD1 3SB |
| 2E1 | BCC | Nicholas Kluger-Langer, 23 Vernon Walk, Northampton, NN1 5ST |
| 2E1 | BCF | J Brown, 6 South Street High Spen, Rowlands Gill, NE39 2HF |
| 2E1 | BDB | P Hudson, 1 Dean Moore Close, St. Albans, AL1 1DW |
| 2E1 | BDC | P Kennedy, 71 Wilbert Lane, Beverley, HU17 0AJ |
| 2E1 | BDV | P McLusky, 11 Ripon Road, Killinghall, Harrogate, HG3 2DG |
| 2E1 | BEV | M Norman, 11 [illegible], Milton Keynes, MK3 7BP |
| 2E1 | BFF | R Greer, 159 Lucas Avenue, Chelmsford, CM2 9JR |
| 2E1 | BFH | Michael Knight, 350 Shelley Road, Wellingborough, NN8 3EW |
| 2E1 | BFP | J Philpot, 7 Providence Place, Ilkeston, DE7 8AL |
| 2E1 | BFW | P Hyde, 10 Highfield Crescent, Taunton, TA1 5JH |
| 2E1 | BFX | C Mann, 11 North River Road, Runham Vauxhall, Great Yarmouth, NR30 1JY |
| 2E1 | BGQ | M Wynn, 22 Matthews Drive, Wickersley, Rotherham, S66 1NN |
| 2E1 | BHB | Alexander Comis, 178 Lordswood Road, Birmingham, B17 8QH |
| 2E1 | BHC | Paul Comis, 65 Montague Way, Chellaston, Derby, DE73 5AS |
| 2E1 | BHF | J Clifford, 16 Park View Road, Birmingham, B31 5AU |
| 2E1 | BHU | P Gosling, 10 Prospect Road, Carlton, Nottingham, NG4 1LY |
| 2E1 | BIM | C Faulkner, 1 Westland, Martlesham Heath, Ipswich, IP5 3SU |
| 2E1 | BIT | Christopher Swain, Box 404551, Gaborone, Botswana |
| 2E1 | BJD | R Saunders, 31 Greenwood Road, High Green, Sheffield, S35 3GU |
| 2E1 | BJG | Steven Benson, 45 Maple Way, Selston, Nottingham, NG16 6FA |
| 2E1 | BKF | G Muircroft, 62 New Road, Rotherham, S61 2DU |
| 2E1 | BKK | C Berry, Roseneath, Walcote Road, Lutterworth, LE17 6EQ |
| 2E1 | BKP | S Martin, 70 Moorlands Drive, Stainburn, Workington, CA14 4UJ |
| 2E1 | BKT | D Jump, 49 Merryfield Grange, Bolton, BL1 5GS |
| 2E1 | BLA | D Burgess, 67 Fair Close, Beccles, NR34 9QT |
| 2E1 | BLG | M Levinson-Withall, The Bungalows, 20 Moor End Avenue, Salford, M7 3NX |
| 2E1 | BLP | Ben Mulley, 8 Drinkstone Road, Gedding, Bury St. Edmunds, IP30 0QB |
| 2E1 | BLT | A Rayner, 25 Spencer Close, Marske-by-the-Sea, Redcar, TS11 6BD |
| 2E1 | BME | Michael Riley, 2 Keeble Drive, Washingborough, Lincoln, LN4 1DZ |
| 2E1 | BMF | Dawn Carskake, 38 Loppets Road, Crawley, RH10 5DW |
| 2E1 | BMJ | Daniel Peters, 7 Gravel Lane, Drayton, Abingdon, OX14 4HY |
| 2E1 | BMV | B Mycock, 69 Bentley Road, Uttoxeter, ST14 7EN |
| 2E1 | BOG | I Skinner, 4 St. Marys Crescent, Rogiet, Caldicot, NP26 3TB |
| 2E1 | BOM | M Love, 72a Hart Plain Avenue, Waterlooville, PO8 8RX |
| 2E1 | BOO | M Constantine, 18 Hillbeck, Halifax, HX3 5LU |
| 2E1 | BPN | Alexander Collins, Flat 2, 49 Dukes Head Street, Lowestoft, NR32 1JY |
| 2E1 | BPV | D Roberts, 98 Pinkneys Road, Maidenhead, SL6 5DN |
| 2E1 | BRA | M Scotton, 15 Grove Road, Aston, Stone, ST15 0DW |
| 2E1 | BRC | G Thornsby, 25 Kipling Way, Stowmarket, IP14 1TS |
| 2E1 | BRD | M Larcombe, 52 Orchard Road, Burgess Hill, RH15 9PL |
| 2E1 | BRG | C Sanderson, 14 Hazelwood Avenue, York, YO10 3PD |
| 2E1 | BRT | A Cooper, 5 Roman Avenue South Stamford Bridge, York, YO41 1EZ |
| 2E1 | BSC | C Castle, 26 Chestnut Walk, Pulborough, RH20 1AW |
| 2E1 | BST | Sheryl Long, 10 St. George Road, Bulwark, Chepstow, NP16 5LA |
| 2E1 | BTG | M Joyner, Brimar, Nelson Park Road, Dover, CT15 6HL |
| 2E1 | BTK | K Carson, Mandalay, Berrynarbor Park, Sterridge Valley, Ilfracombe, EX34 9TA |
| 2E1 | BUJ | Patricia Stott, 12 Castle View, Ovingham, Prudhoe, NE42 6AT |
| 2E1 | BUM | F Stone, 7 Cherry Tree Close, Hilton, Derby, DE65 5FD |
| 2E1 | BUR | J Izzard, Sunnyside, West End, Hailsham, BN27 4NH |
| 2E1 | BVJ | E Constantine, 18 Hillbeck, Halifax, HX3 5LU |
| 2E1 | BVQ | N Brown, 55 Shakespeare Terrace, Chorley, PR6 7AQ |
| 2E1 | BVS | E Elliott, 8 Edge Lane, Mottram, Hyde, SK14 6SE |
| 2E1 | BVY | Dean Fox, 165b Rossmore Road, Poole, BH12 2HG |
| 2E1 | BWT | Ashley Ross, 42 New Heritage Way, North Chailey, Lewes, BN8 4GD |
| 2E1 | BYI | Sheila Piccavey, 611 Manchester Road, Linthwaite, Huddersfield, HD7 5QX |
| 2E1 | BYK | A Sellors, 12 Morfa View, Bodelwyddan, Rhyl, LL18 5TT |
| 2M1 | BYW | T Conlan, 12 Rowantree Road, Mayfield, Dalkeith, EH22 5ER |
| 2E1 | BYY | Wendy Kennedy, 1 Lynton Road, Hindley, Wigan, WN2 4EH |
| 2E1 | BZB | D Peter, 41 Coleswood Road, Harpenden, AL5 1EF |
| 2E1 | BZH | Geoffrey Low, 24 Meadway, Halton Brook, Runcorn, WA7 2DX |
| 2E1 | BZI | David Low, 29 Mountbatten Road, Malvern, WR14 2YD |
| 2E1 | CAF | Sven Kirkpatrick, 29 Barnfields, Gloucester, GL4 6WE |
| 2E1 | CAH | R Richards, [illegible] North Holme Court, Northampton, NN3 8UX |
| 2E1 | CAJ | J Godding, 70 Rodway Road, Tilehurst, Reading, RG30 6DT |
| 2E1 | CAQ | M Porter, 12 Woodland Crescent, London, SE16 6YP |
| 2E1 | CAT | L O'Ryan, 12 Minton Close, Congleton, CW12 3TD |
| 2W1 | CAU | Andrew Burt, 14 Kenry Street, Tonypandy, CF40 1DE |
| 2E1 | CAW | R Marshall, 15 Whisby Court, Holton-le-Clay, Grimsby, DN36 5RG |
| 2E1 | CBH | S Stretch, 5 Ledwych Road, Droitwich, WR9 9LA |
| 2E1 | CBU | C Richards, 39 North Holme Court, Northampton, NN3 8UX |
| 2E1 | CCF | P Izzard, 7 Yardley Drive, Northampton, NN2 8PE |
| 2E1 | CCG | R Winship, 32 Lytes Cary Road, Keynsham, Bristol, BS31 1XD |
| 2E1 | CCI | A Murphy, 34 Hawkenbury Way, Lewes, BN7 1LT |
| 2E1 | CCN | David Hill, 28 Pendarves Flats, St. Clare Street, Penzance, TR18 2PL |
| 2E1 | CDK | G Langdon, 43 Daniel Street, Ryde, PO33 2BH |
| 2E1 | CDS | B Allen, 78 Bargates, Christchurch, BH23 1QL |
| 2E1 | CDZ | A Woods, 40 Windsor Way, Sandy, SG19 1JL |
| 2E1 | CEE | D Whish, 62 Marion Road, Prestatyn, LL19 7DF |
| 2E1 | CEQ | M Sayers, Flat 6, 24 Knole Road, Bournemouth, BH1 4DH |
| 2E1 | CEU | J Columbine, West Lodge, 166 Tollerton Lane, Nottingham, NG12 4FW |
| 2E1 | CEZ | J Brown, Kingsdown Cottage, Fron, Montgomery, SY15 6SB |
| 2E1 | CFB | Derek Williams, 100 Hills Lane Drive, Madeley, Telford, TF7 4BX |
| 2E1 | CHX | Mark Reavill, 11 Clarence Road, Beeston, Nottingham, NG9 5HY |
| 2E1 | CIK | D Gillatt, 1 City Mills, Skeldergate, York, YO1 6DB |
| 2E1 | CIO | C Brooks, 34 Wentworth Crescent, New Marske, Redcar, TS11 8DB |
| 2E1 | CIP | R Bufton, 7 Laburnum Close, Rassau, Ebbw Vale, NP23 5TS |
| 2E1 | CIR | Richard Bush, Church View, Overcross Banham, Norwich, NR16 2BY |
| 2E1 | CIT | J Watkins-Field, Sharlions, 27 Bosvigo Road, Truro, TR1 3DG |
| 2E1 | CIX | W Scott, 18 Manor Gardens, Killinghall, Harrogate, HG3 2DS |
| 2E1 | CIY | B Harratt, 8 Aaron Wilkinson Court, South Kirkby, Pontefract, WF9 3JT |
| 2E1 | CJB | Kenneth Jordan, 11 Sandringham Place, Hucknall, Nottingham, NG15 8EU |
| 2E1 | CJC | R Richardson, 3 Cautley Drive, Killinghall, Harrogate, HG3 2DJ |
| 2E1 | CJD | Peter Taylor, 13 Mackenzie Crescent, Burncross, Sheffield, S35 1UR |
| 2E1 | CJF | Stephen Curtis, 354 St. Helens Road, Leigh, WN7 3PQ |
| 2E1 | CJJ | Samantha Hull, 1 Occupation Lane, New Bolingbroke, Boston, PE22 7LW |
| 2E1 | CJN | H Hughes, 46 The Boundary, Oldbrook, Milton Keynes, MK6 2HT |
| 2E1 | CJZ | Z Hodges, 12 Linwal Avenue, Houghton-on-the-Hill, Leicester, LE7 9HD |
| 2E1 | CKH | K Riley, 16 King St., Westhoughton, Bolton, BL5 3AX |
| 2E1 | CKQ | E Swain, 11 Blackdown, Fullers Slade, Milton Keynes, MK11 2AA |
| 2W1 | [illegible] | M Holmes, 77 Harlech Drive, Merthyr Tydfil, CF48 1JU |
| [illegible] | [illegible] | Gerald Harvey, 2 Mulberry Way, Sittingbourne, [illegible] |
| 2E1 | CLM | E Woolley, 82 Pennycroft Road, Uttoxeter, ST14 7L |
| 2E1 | CML | R Norris, 20 Laburnum Close, Guildford, GU1 1NA |
| 2E1 | CMZ | D Bracher, 29 Bungalow Park, Holders Road, Salisbury, SP4 7PJ |
| 2E1 | CNM | A Raxworthy, 32 St Marys Avenue, Alverstoke, Gosport, PO12 2HX |
| 2W1 | CNN | A Gray, 36 Heol Pentre Felen, Morriston, Swansea, SA6 6BY |
| 2E1 | CNO | L Call, 3 Southfield, Bramhope, Leeds, LS16 9DR |
| 2E1 | COC | J Taylor-Cram, 7 Hart Plain Avenue, Cowplain, Waterlooville, PO8 8RP |
| 2E1 | COG | D Oakes, 2 Hillcrest, Scotton, Catterick Garrison, DL9 3NJ |
| 2E1 | COM | M Holt, 20 Lingfield Mount, Leeds, LS17 7FP |
| 2E1 | COV | I Cockshoot, 72 Princess Margaret Avenue, Cliftonville, Margate, CT9 3EF |
| 2E1 | CPB | Andrew Cadey, 45 St. Mildreds Road, Westgate-on-Sea, CT8 8RJ |
| 2E1 | CPC | Matthew Sheppard, 11 Parvian Road, Leicester, LE2 6TS |
| 2E1 | CPF | R Kensall, 40 Eskdale Avenue, Ramsgate, CT11 0PB |
| 2E1 | CPI | Beatrice Rowley, 7 Hall Farm Close, Castle Donington, Derby, DE74 2NG |
| 2E1 | CPJ | P Fisk, 38 Bedingfield Crescent, Halesworth, IP19 8EE |
| 2E1 | CPP | Nigel Newman, 89 Sea Place, Goring-by-Sea, Worthing, BN12 4BH |
| 2E1 | CPQ | P Goodwin, 60 Dale Crescent, Congleton, CW12 3EP |
| 2W1 | CPS | David Probert, 32 Heol Penlan, Longford, Neath, SA10 7LB |
| 2E1 | CPV | P Porter, 7 Long Road, Framingham Earl, Norwich, NR14 7RY |
| 2E1 | CQM | Iain Hurst, 234 Nuncargate Road, Kirkby-in-Ashfield, Nottingham, NG17 9AE |
| 2E1 | CQP | V Brightwell, 40 Streete Court Road, Westgate-on-Sea, CT8 8BX |
| 2E1 | CQQ | E Whelan, 54 Boroughbridge Road, Northallerton, DL7 8BN |
| 2E1 | CRA | M Lewis, 15 Highcliffe Avenue, Chester, CH1 5DP |
| 2E1 | CRI | Joann Mosby, 1 School Road, Golcar, Huddersfield, HD7 4NU |
| 2E1 | CSD | Andrew Smith, 30 Lime Grove, Grantham, NG31 9JD |
| 2E1 | CTU | Steven Crawshaw, Quarry Top, 19 Heather Close, Nelson, BB9 5HB |
| 2E1 | CVE | R Trow, 5 Cranberry Drive, Stourport-on-Severn, DY13 8TH |
| 2E1 | CWE | M Skewes, 47 Pentrevah Road, Penwithick, St. Austell, PL26 8UA |
| 2J1 | CWG | J Totty, 34 Clos Paumelle, St. Saviour, Jersey, JE2 7TW |
| 2J1 | CWH | Christopher Totty, 34 Le Clos Paumelle, Bagatelle Road, St. Saviour, Jersey, JE2 7TW |
| 2E1 | CWJ | David Thatcher, 6 Ivel View, Sandy, SG19 1AU |
| 2E1 | CWN | P Kemble, 88 Mayfield Road, Ipswich, IP4 3NG |
| 2E1 | CWP | J Parker, 19 Mayfair Close, Dukinfield, SK16 5HR |
| 2E1 | CWQ | P Millward, 28 Olive Grove, Burton Joyce, Nottingham, NG14 5FG |
| 2E1 | CWX | E Parker, 19 Mayfair Close, Dukinfield, SK16 5HR |
| 2E1 | CXE | J Mortimer, 2 Springdale, Earley, Reading, RG6 5PR |
| 2E1 | CXF | A Whittaker, 62 Ingham Street, Padiham, Burnley, BB12 8DR |
| 2E1 | CXI | R Maunder, 12 Hamble Springs, Bishops Waltham, Southampton, SO32 1SG |
| 2E1 | CXP | E Bradshaw, 41 Sherwood Road, Woodley, Stockport, SK6 1LH |
| 2W1 | CYC | D Tittman, 16 St. Georges Road, Heath, Cardiff, CF14 4AQ |
| 2E1 | CYD | G Reilly, 10 Gloucester Road, Huyton, Liverpool, L36 1XX |
| 2E1 | CYE | R Stenhouse, High Park, Common Road, Norwich, NR16 1HH |
| 2E1 | CYI | D Pomery, 3 Mayfair Park, Minorca Lane, St. Austell, PL26 8QN |
| 2E1 | CYP | J Bick, 45 Gloucester Road, Almondsbury, Bristol, BS32 4HH |
| 2E1 | CYS | I Limbert, 138 Dinas Lane, Liverpool, L36 2NT |
| 2E1 | CYU | Stephen Barker, 26 Rye Court Helmsley, York, YO62 5DY |
| 2E1 | CYZ | M Baxter, 5 Farnborough Street, Farnborough, GU14 8AG |
| 2E1 | CZB | B Fido, 7 Claires Walk, Parklands Mobile Hom, Scunthorpe, DN17 1SW |
| 2E1 | CZF | Anita Pink, 31 The Fairway, Daventry, NN11 4NW |
| 2E1 | CZJ | J Bosworth, 57 Livingstone Road, Derby, DE23 6PS |
| 2E1 | CZO | Ben Coombs, 10 Horseshoe Walk, Widcombe, Bath, BA2 6DE |
| 2E1 | DAK | C Wilderspin, 3 Ferndale, Eaglestone, Milton Keynes, MK6 5AE |
| 2E1 | DAO | J Patrick, 52 Huntsmans Corner, Wrexham, LL12 7UH |
| 2E1 | DAR | Adrian Cottle, 1 Bathite Cottages, Shaft Road, Bath, BA2 7HN |
| 2E1 | DBP | A Gener, 3e Dartmouth Terrace, Greenwich, London, SE10 8AX |
| 2E1 | DBQ | Stephen Walker, 33 Parkside, Somercotes, Alfreton, DE55 4LA |
| 2E1 | DBS | Keith Budd, 20 Marshal Road, Poole, BH17 7HA |
| 2E1 | DBT | R Verma, 43 Farley Road, Derby, DE23 0DW |
| 2E1 | DBZ | S Issatt, 7 Birch Road, Doncaster, DN4 6PD |
| 2E1 | DCV | Brian Shields, 20 Grosley Court, Grantham, NG31 7RH |
| 2E1 | DDJ | John Gallagher, 71 Castle Hill, Beccles, NR34 7BJ |
| 2E1 | DDZ | S Arter, 18 Essex Road, Westgate-on-Sea, CT8 8AP |
| 2E1 | DEA | P Smith, 23 Gainsborough Close, Llantarnam, Cwmbran, NP44 3BX |
| 2E1 | DEM | Neil Humphreys, 4 Rose Cottages, Annscroft, Shrewsbury, SY5 8AU |
| 2E1 | DEN | P Hart, 18 Harewood Road, Rochdale, OL11 5TG |
| 2E1 | DEP | L Darton, 8 Foster Grove, Sandy, SG19 1HP |
| 2E1 | DET | A Whyman, 8 Staplers Close, Great Totham, Maldon, CM9 8UN |
| 2E1 | DFE | R Scott, 315 Ormskirk Road, Wigan, WN5 9DL |
| 2E1 | DFZ | M Axon, 48 Cowslip Road, Broadstone, BH18 9QZ |
| 2E1 | DGL | Paul Lewis, 166 Euston Road, Morecambe, LA4 5LE |
| 2W1 | DGM | John Foster, 56 Hillrise Park, Clydach, Swansea, SA6 5DX |
| 2E1 | DHC | Mark Axworthy, 3Shepherds Court, 1 Shepherds Street, St Leonard's on Sea, TN38 0ET |
| 2M1 | DHG | Garry Russell, 70 Easter Road, Kinloss, Forres, IV36 3FG |
| 2E1 | DHX | D Walker, 17 Pinehurst Avenue, Christchurch, BH23 3NS |
| 2E1 | DIA | J Laffin, 154 Blenheim Drive, Allestree, Derby, DE22 2GN |
| 2E1 | DIG | G Williams, 6 St. Georges Drive, Prestatyn, LL19 8EH |
| 2E1 | DIH | J Bentley, 4 Highway, Crowthorne, RG45 6HE |
| 2E1 | DKU | A Whittle, 9 Dale View, Littleborough, OL15 0BP |
| 2E1 | DLA | Peter Craig, 25 Harts Green Road, Birmingham, B17 9TZ |
| 2E1 | DLD | E Harrison, 55 Hudson Close, Worcester, WR2 4DP |
| 2E1 | DLM | Dean Wilson, 76 Cheadle Road, Uttoxeter, ST14 7BY |
| 2E1 | DLO | B Bush, Church View, Overcross, Norwich, NR16 2BY |
| 2E1 | DLR | Richard Diaper, 30 Holmcroft Road, Kidderminster, DY10 3AG |
| 2E1 | DLS | J Field, 27 Lovelace Road, Barnet, EN4 8EA |
| 2E1 | DLT | P Lilley, 12 Trueman Gardens, Arnold, Nottingham, NG5 [illegible] |
| 2E1 | DLX | P Thackray, 20 Darfield Street, Leeds, LS8 5DB |
| 2E1 | DMH | M Rippin, Gaverne, Welford Road, Stratford-upon-Avon, CV37 8RA |
| 2E1 | DMI | R Dixon, 14 Blackdown Close, Peterlee, SR8 2JW |
| 2E1 | DMU | S Wilson, 218 Bredhurst Road, Gillingham, ME8 0HD |
| 2E1 | DMZ | R Laverick, 12 Greenlands Road, Redcar, TS10 2DG |
| 2E1 | DNB | Peter Lomas, 42 Lane End, Pudsey, LS28 9AD |
| 2E1 | DNC | Micheal Harris, Flat 22, Alexandra Court, The Royal Seabathing, Margate, CT9 5NT |
| 2E1 | DNF | John Foy, 2 Lark Rise, Northampton, NN3 0QT |
| 2E1 | DNK | A Waller, 4 Rose Court, [illegible] |
| 2E1 | DNX | Marilyn MacKenzie, 70 [illegible], Weymouth, DT4 0AS |
| 2E1 | DOA | P Dalby, Windfall, 11 Greensward Lane, Hockley, SS5 5HD |
| 2E1 | DOZ | S Brenchley, 89 Thicket Mead, Midsomer Norton, Radstock, BA3 2SL |
| 2E1 | DPG | D Stone, Bridor, 12 Robertson Avenue, Sleaford, NG34 8NJ |
| 2E1 | DPK | Sylvia Hill, 22b Strait Lane, Hurworth On Tees, Darlington, DL2 2AL |
| 2E1 | DPQ | Raymond Tattersall, 56 Larch Road, New Ollerton, Newark, NG22 9SX |
| 2E1 | DQM | N Edwards, 609 Upper Richmond Road West, Richmond, TW10 5DU |
| 2E1 | DQQ | David Horsley, 1 Mead Close, Swanley, BR8 8DQ |
| 2E1 | DQT | A Charles, 6 Bridewell Street, Wymondham, NR18 0AR |
| 2E1 | DQZ | C Houlden, 29 Court Barton, Weston, Portland, DT5 2HJ |
| 2E1 | DRB | P Waller, 4 Rose Court, Ty Canol, Cwmbran, NP44 6JH |
| 2E1 | DRC | P Hatcher, 32 Slough Road, Iver, SL0 0DT |
| 2E1 | DRU | K McCann, Treverven, Back Lane, Selby, YO8 6QP |
| 2E1 | DRV | J Wohlgemuth, 37 Broadcoombe, South Croydon, CR2 8HR |
| 2E1 | DRX | R Hauxwell, 65 Harleston Way, Heworth, Gateshead, NE10 9BQ |
| 2E1 | DRY | Gary Symonds, Flat 13, Bradbury House, Norwich, NR2 3PT |
| 2E1 | DSU | James Hewitt, 49 Calder Drive, Sutton Coldfield, B76 1YR |
| 2E1 | DSX | C Haddon, 1 Victoria Place, Weston, Portland, DT5 2AA |
| 2E1 | DTE | C Folkerd, The Old Manse, London, W5 5QT |
| 2E1 | DTF | D Folkerd, The Old Manse, London, W5 5QT |
| 2E1 | DTK | R Cleary, 5 Gregson Road, Widnes, WA8 0BX |
| 2E1 | DTN | J Bennett, Rectory Cottage, Broad Street, Bristol, BS40 5LD |
| 2E1 | DTR | George Ellis, 223 The Uplands, Palacefields, Runcorn, WA7 2UE |
| 2E1 | DUW | M Goldby, Waylands Gate, St. Johns Road, New Milton, BH25 5SD |
| 2E1 | DWM | J Bush, Church View, Overcross, Norwich, NR16 2BY |
| 2E1 | DXB | I Humberstone, 20 Kingswood Road, Colchester, CO4 5JX |
| 2E1 | DXV | Andrew Powers, 9 Courtybella Gardens, Newport, NP20 2GN |
| 2E1 | DYL | Malcolm Reid, 15 Joseph Rich Avenue, Madeley, Telford, TF7 5EY |
| 2E1 | DYT | Catherine Block, 10 Beatrice Road, Capel-le-Ferne, Folkestone, CT18 7LL |
| 2E1 | DZH | Jane Richards, 1 Stocks Mead, Washington, Pulborough, RH20 4AU |
| 2E1 | DZJ | R Morris, 7 Chapmans Close, Stirchley, Telford, TF3 1ED |
| 2E1 | DZL | R Neville, 4 Danson Gardens, Blackpool, FY2 0XH |
| 2E1 | DZP | A Cannon, 20 Gladwyn Street, Stoke-on-Trent, ST2 8JZ |
| 2M1 | DZS | C Clark, 3 Old Cottages, Seton Mains, Longniddry, EH32 0PG |
| 2E1 | DZV | J Keegan, The Cottage, 11 Condor Grove, Lytham St. Annes, FY8 2HE |
| 2M1 | DZW | B Waugh, 93 Denholm Road, Musselburgh, EH21 6TU |
| 2M1 | DZX | R Stuart, Tigh Na Coille, Bishop Kinkell, Dingwall, IV7 8AW |
| 2E1 | EAK | Yvette Wood, 67 Bay View Road, Duporth, St. Austell, PL26 6BN |
| 2W1 | EAN | Gary Taylor, 55 Haulfryn, Tregynwr, Carmarthen, SA31 2DT |
| 2E1 | EAS | H Southgate, 70 Bouverie Road, Hardingstone, Northampton, NN4 7EQ |
| 2E1 | EAV | I Duggan, 21 Chetwynd Road, Toton, Nottingham, NG9 6FW |
| 2E1 | EAW | M Barnett, 23 Francis Close, Penkridge, Stafford, ST19 5HP |
| 2E1 | EAX | Russell Edwards, Flat 10, Longford Court, 60 London Road, Sevenoaks, TN13 2UG |
| 2E1 | EAZ | David Finch, 174 Park Road, Chesterfield, S40 2LL |
| 2M1 | EBJ | Derek Martin, Inverleod, Avoch, IV9 8PR |
| 2E1 | EBL | Keith Dignall, 11 Mottershead Road, Widnes, WA8 7LD |
| 2E1 | EBN | S Valvona, 75 Bettesworth Road, Ryde, PO33 3EN |
| 2E1 | EBR | C Smith, 16 Meadowbrook, Ancaster, Grantham, NG32 3RR |
| 2E1 | EBX | L Collinson, 12 Victoria Avenue, Hunstanton, PE36 6BX |
| 2M1 | ECF | J MacKenzie, Rainbows End, Lochside, Lairg, IV27 4EG |
| 2E1 | ECG | A Topping, 30 St. Pauls Avenue, Nottingham, NG7 5EB |
| 2E1 | ECL | C Gray, 19 Marsh View, Newton, Preston, PR4 3SX |
| 2E1 | ECM | Peter Fletcher, 11 The Banks, Long Buckby, Northampton, NN6 7QQ |
| 2E1 | ECN | M Chapman, 15 Norwood Road, Somersham, Huntingdon, PE28 3EY |
| 2E1 | ECV | P Elsey, 129 Kingsway, Chandler's Ford, Eastleigh, SO53 5BX |
| 2E1 | EDA | E Kurtz, 1 Leonard Medler Way, Hevingham, Norwich, NR10 5LE |
| 2E1 | EDB | M Bradwell, 6 Moorfoot Gardens, Gateshead, NE11 9LA |
| 2E1 | EDD | S Lansdell, 42 Marylebone Crescent, Derby, DE22 4JX |
| 2M1 | EDM | D MARTIN, 45 Tiree Place, Newton Mearns, Glasgow, G77 6UJ |
| 2J1 | EDR | Annette Price, Wheatlands, La Rue Des Longchamps, St. Brelade, Jersey, JE3 8BN |
| 2E1 | EDT | Edward Durkin, 30 Douglas Road West, Stafford, ST16 3NX |
| 2M1 | EDT | J Wilson, 4f Langside Street, Clydebank, G81 5HJ |
| 2E1 | EDV | Julie Blanche, 11 Woodside, Southminster, CM0 7RD |
| 2E1 | EDW | James Blanche, 11 Woodside, Southminster, CM0 7RD |
| 2E1 | EEK | Stephen Paffott, 15 Centaury Gardens, Horton Heath, Eastleigh, SO50 7NY |
| 2E1 | EEP | J Jones, Glanrafon Garage, Pontfadog, Llangollen, LL20 7AR |
| 2E1 | EET | L Edwards, 39 Foskitt Court, Northampton, NN3 9AX |
| 2E1 | EFG | S Philpot, 93 Princess Drive, Grantham, NG31 9QA |
| 2E1 | EFQ | D Bushby, 66 Sandy Road, Everton, Sandy, SG19 2JU |
| 2E1 | EFT | L Froggatt, 255 Rushton Road, Desborough, Kettering, NN14 2QB |
| 2E1 | EGI | G Durrant, 51 Raglan Avenue, Waltham Cross, EN8 8DA |
| 2E1 | EGU | W Lupton, Eaton Cottage, Shorts Green Lane, Shaftesbury, SP7 9PA |
| 2E1 | EGV | M Townson, 53 Brompton Road, Bradford, BD4 7JD |
| 2E1 | EHB | I Greenall, 356 Warrington Road, Abram, Wigan, WN2 5XA |
| 2E1 | EHF | John Goodman, St. Francis House Highlands, Banbury, OX16 1FA |
| 2E1 | EHM | D whittaker, 68 Querns Road, Canterbury, CT1 1PZ |
| 2E1 | EHP | D Moses, 121 Badger Avenue, Crewe, CW1 3JN |
| 2E1 | EHY | N Waters, 23 Harold Road, Birchington, CT7 9NA |
| 2E1 | EID | Russell Owens, 96 Gaer Park Road, Newport, NP20 3NT |
| 2W1 | EIN | Christopher Bodley, 38 Bryn Hawddgar, Clydach, Swansea, SA6 5LA |

UK Callsigns

**IMPORTANT NOTE**

**Revalidate licence to avoid revocation** – Ofcom has advised the Society that plans will be drawn up to revoke licences that have not been revalidated as required by the licence conditions. The quickest way to revalidate is to do so online via the Ofcom website: *https://services.ofcom.org.uk/* or by email: *amateur.validations@ofcom.org.uk* Ofcom staff are available to help, but please be patient during times of heavy workload.

2E1 EIO Annie Godolphin, Shepherds Cottage, Flakebridge, Appleby-in-Westmorland, CA16 6JZ
2E1 EIU David Meddings, 42a Argyle Road, Poulton-le-Fylde, FY6 7EW
2E1 EIV A Eyre, St. Michael Mead, The Common, Norwich, NR12 8BA
2E1 EIX A Chruscinski, 39 Sherwood Rise, Mansfield Woodhouse, Mansfield, NG19 7NP
2E1 EJC Terry Bain, 23 Salisbury Crescent, Blandford Forum, DT11 7LX
2E1 EJD W Ling, Valley Farm Equestrian Centre, Wickham Market, Woodbridge, IP13 0ND
2U1 EJF Andrew Scheffer, Foveat, Rue de Jardins, Les Prins, Guernsey, Guernsey, GY6 8EZ
2M1 EJI R Lynch, 21 Carnoustie Avenue, Gourock, PA19 1HF
2E1 EJU David Lumley, 19 Bramley Avenue, Needingworth, St. Ives, PE27 4UD
2E1 EJX Matthew Tullett, The Lodge, York Road, Knaresborough, HG5 0SW
2U1 EKE C Ayres, Rousay, Bailiffs Cross Road, St. Andrew, Guernsey, GY6 8RY
2U1 EKH Ken Johnson, 9 Clos Spurway, Victoria Avenue, Guernsey, Guernsey, GY2 4AH
2M1 EKI Kirstine Hughes, 49 Marmion Drive, Kirkintilloch, Glasgow, G66 2BH
2E1 EKM D Baldwin, 51 Queens Road, Broadstairs, CT10 1PG
2E1 EKQ M Eades, 8 Wellington Street, Gainsborough, DN21 1BX
2W1 EKR D Pollard, 19 The Drive, Bargoed, CF81 8JX
2E1 ELE Peter Williams, 20 Elm Close, Great Haywood, Stafford, ST18 0SP
2M1 ELU Iain Sinclair, Airdaniar, Kilchrenan, Taynuilt, PA35 1HG
2E1 EMH V Newton, 60 The Lynch, Winscombe, BS25 1AR
2E1 EMI George Paterson, 4 Rowallan Drive, Bedford, MK41 8AW
2E1 EMK James Roff, 6 Canal Close, Wilcot, Pewsey, SN9 5NW
2E1 EMN J Thompson, 8 Roman Drive, Leeds, LS8 2DR
2M1 ENI Donald Paterson, Leuchlands Croft, Whitecairns, Aberdeen, AB23 8UT
2M1 ENK Douglas Paterson, 29 Arnothill Gardens, Falkirk, FK1 5BQ
2E1 ENN G Cattle, 50 Oakland Avenue, York, YO31 1DF
2E1 ENZ R Baxter, 20 Thorpe St., Thorpe Hesley, Rotherham, S61 2RP
2E1 EOD N bridges, Acomb, Station Road, Pershore, WR10 3BB
2E1 EOI J Allum, 122 Long Chaulden, Hemel Hempstead, HP1 2HY
2E1 EOK Christine Blackman, 32 Frenchs Gate, Dunstable, LU6 1BQ
2E1 EOO D Webb, 11 Alfriston Road, Worthing, BN14 7QU
2E1 EOQ D Horton, Glen View, New Road, Bude, EX23 9LE
2E1 EOR Katie Horton, The Old School House, Kelly, Lifton, PL16 0HJ
2M1 EOV M MacLeod, 4 Portnaguran, Isle of Lewis, HS2 0HD
2E1 EOW Darryl Forward, 19 Kelly Close, Poole, BH17 8QP
2E1 EOZ Robert Cannon, 31 Moretons Mews, Basildon, SS13 3NB
2E1 EPA P Allaker, 3 Eden Cottages, Watling Street, Consett, DH8 6HZ
2E1 EPD M Freedman, Rivermeade, Irwell Vale, Bury, BL0 0QA
2E1 EPE P Freedman, Rivermeade, Irwell Vale, Bury, BL0 0QA
2E1 EPL Gareth Morris, 8 Llys Heulog, Sisson Street, Rhyl, LL18 2AQ
2E1 EPO Francis Hodge, Flat 7, Windsor Court, Rhyl, LL18 1TF
2E1 EPQ T Fanning, 26 Mandeville Close, Tilehurst, Reading, RG30 4JT
2M1 EPV D Stewart, 19 Arthur Street, Blairgowrie, PH10 6PF
2E1 EQE D Evans, 5 Compton Drive, Streetly, Sutton Coldfield, B74 2DA
2E1 EQI V Collins, 30 Upham Road, Swindon, SN3 1DN
2E1 EQQ K Wood, 52 Ashfield Avenue, Beeston, Nottingham, NG9 1PY
2E1 EQR John Jones, 17 Dunkeld Drive, Shrewsbury, SY2 5UZ
2E1 EQY Stephen Heard, 42 Hallowell Down, South Woodham Ferrers, Chelmsford, CM3 5FS
2E1 ERJ David Lusty, 104 Polstain Road, Threemilestone, Truro, TR3 6DB
2E1 ESK L Hodge, 11 Glebelands, Bampton, OX18 2LH
2E1 ESM K James, 36 Lemon Hill, Mylor Bridge, Falmouth, TR11 5NA
2E1 ESN Catherine McLean, 18 Chatfield Road, Gosport, PO13 0TN
2E1 ESQ A Mould, 95 Stanton Road, Southampton, SO15 4HU
2E1 ESW D Wilkinson, 139 Church Road, Jackfield, Telford, TF8 7ND
2E1 ETB R Moore, Thorpefield Farm, 91 Thorpe Street, Rotherham, S61 2RP
2E1 ETJ A Willis, Kilncroft, Broadlayings, Newbury, RG20 9TS
2M1 ETM S Mackie, 25 Carlaverock Drive, Tranent, EH33 2EE
2W1 ETN Damien Jorgensen, 5 Woodham Road, Barry, CF63 4JE
2W1 ETW Richard Thomas, 24 Heol Innes, Llanelli, SA15 4LA
2E1 EUE Emaleen Hunt, Flat 47, Redbridge Tower, Southampton, SO16 9AU
2W1 EUR A Joynes, 25 St. Annes Gardens, Maesycwmmer, Hengoed, CF82 7QQ
2M1 EUV E Clark, 3 Old Cottages, Seton Mains, Longniddry, EH32 0PG
2E1 EUY W Partridge, 15 Cranbourne Avenue, Wolverhampton, WV4 6RJ
2E1 EVH F Stevenson, 56 Barrows Hill Lane, Westwood, Nottingham, NG16 5HJ
2E1 EVJ R Chipperfield, 5 Lullingstone Close, Hempstead, Gillingham, ME7 3TS
2E1 EVK Richard Chipperfield, 5 Clayton Avenue, Upminster, RM14 2EZ
2E1 EVM John Merrick, 52 Lombardy Close, Hempstead, Gillingham, ME7 3SQ
2E1 EWK B Ashman, 108 Eastwood Drive, Highwoods, Colchester, CO4 9SL
2E1 EWN Ben Storkey, 9 Snatchup, Redbourn, St. Albans, AL3 7HD
2E1 EXA P Johnson, 110 Rachel Clarke Close, Stanford-le-Hope, SS17 7SX
2E1 EXI D Taylor, 188 Walstead Road, Walsall, WS5 4DN
2E1 EXK J Wilkes, 47 Greenwood Park, Hednesford, Cannock, WS12 4DQ
2I1 EXU P Robinson, 20 Harwood Park, Carrickfergus, BT38 7LZ
2E1 EYC M Davies, 57 Ladybrook Lane, Mansfield, NG18 5JF
2E1 EYF E Hedley, 17 Chowdene Bank, Gateshead, NE9 6JJ
2E1 EYI Steven Riddle, 10 Clarendon Avenue, Trowbridge, BA14 7BN
2E1 EYL Mark Harris, 1 Brampton Court, Bowerhill, Melksham, SN12 6TH
2E1 EYS B Egglestone, 4 Lancaster Close, Etherley Dene, Bishop Auckland, DL14 0RP
2W1 EYZ S Gray, 36 Heol Pentre Felen, Morriston, Swansea, SA6 6BY
2M1 EZA J Boyle, Flat 3/2, 33 St. Mungo Avenue, Glasgow, G4 0PH
2E1 FAN K Blanchard, 17 Stephens Way, Sleaford, NG34 7JN
2E1 FAT S Hallsworth, 27 Westfield Avenue, Heanor, DE75 7BN
2E1 FBA R Locke, 12a Kennedy Court, Walesby, Newark, NG22 9PQ
2E1 FBK Christopher Strange, 12 Cricketts Lane, Chippenham, SN15 3EF
2E1 FBS D Seabridge, 31 Charlestown Drive, Allestree, Derby, DE22 2HA
2E1 FBY Ian Roper, 69 duckworth lane, Bradford, BD9 5EX
2E1 FCC Maurice Williams, Cornacres, Dodwell, Stratford upon Avon, CV37 9ST
2E1 FCD Mary Williams, Cornacres, Dodwell, Stratford upon Avon, CV37 9ST
2E1 FCE J Lindsay, 10 New Station Road, Swinton, Mexborough, S64 8AH
2E1 FCO I Brady, 6 Bristow Close, Bletchley, Milton Keynes, MK2 2XP
2E1 FDD M Warren, 145 Shirehall Road, Sheffield, S5 0JL
2E1 FDF G Bilson, 34 Bramlyn Close, Clowne, Chesterfield, S43 4QP
2E1 FDJ K Newnam, 1 Wheatlands Close, Maulden, Bedford, MK45 2AQ
2E1 FDK L Rule, 46 Meadowsweet Road, Poole, BH17 7XT
2E1 FDM Jenny Mew, 1 Council Houses, Norwich Road, Norwich, NR9 4NY
2E1 FDP G Westwood, 6 Monkton Road, Borough Green, Sevenoaks, TN15 8SD
2E1 FDT Phillip Brown, 19 Huxley Close, Nottingham, NG8 4PU
2E1 FDU M Darton, 8 Foster Grove, Sandy, SG19 1HP
2E1 FDY Kenneth Rankin, 31 Gorsey Bank Road, Hockley, Tamworth, B77 5JD
2E1 FEC F Dunmore, 36 Dove Rise, Oadby, Leicester, LE2 4NY
2E1 FEF K Rhodes, 30 Priory Road, Louth, LN11 9AL
2E1 FEG B Rhodes, 2 Kent Avenue, Theddlethorpe, Mablethorpe, LN12 1QE
2E1 FET N Moore, 84 Franklynn Road, Haywards Heath, RH16 4DH
2E1 FFJ Ros Dennis, Chapel House, Farlesthorpe, Alford, LN13 9PH
2E1 FFL P Harbinson, 19 Pennine Road, Dewsbury, WF12 7AW
2E1 FFZ D Fawcett, 6 Wand Hill, Boosbeck, Saltburn-by-The-Sea, TS12 3AW
2E1 FGB N Ash, 35 Fairford Road, Tilehurst, Reading, RG31 6PY
2W1 FGR J Clark, 22 Heol Yr Wylan, Cwmrhydyceirw, Swansea, SA6 6TB
2E1 FHO P Sawyers, 15 Park Terrace, Whitby, YO21 1PN
2E1 FHQ C Ward, 18 Aspen Grove, Aldershot, GU12 4EU
2E1 FHZ D Hebb, 45 Marklew Avenue, Grimsby, DN34 4AD
2E1 FIE M Robertson, 98 Hawthorn Road, Bognor Regis, PO21 2DG
2E1 FIQ H Pook, Beverley Friory, Friars Lane, Beverley, HU17 0DF
2E1 FIV D Dalzell, 9 Pyms Lane, Crewe, CW1 3PJ
2E1 FIX J Shaw, 50 Elmpark Way, Rooley Moor, Rochdale, OL12 7JQ
2E1 FJL M Gray, 19 Marsh View, Newton, Preston, PR4 3SX
2E1 FJN I Pearson, Warren Cottage, Pontfadog, Llangollen, LL20 7AT
2E1 FJP B Melling, 68 Westfield Drive, Ribbleton, Preston, PR2 6TH
2E1 FJV I Page, 84 Beaulieu Close, Toothill, Swindon, SN5 8AH
2E1 FJZ Peter Edwards, Cam O'R Afon, Dolywern, Llangollen, LL20 7AD
2E1 FKD E Merrington, Cartref, Ball Lane, Frodsham, WA6 8HP
2E1 FKJ J Ewing, 130 Uttoxeter Road, Hill Ridware, Rugeley, WS15 3QX
2E1 FKM C Andrews, Flat 2, 1st Floor, 153 Thanet Street, Chesterfield, S45 9JT
2E1 FKT M Boyes, 19 Rowe Ashe Way, Locks Heath, Southampton, SO31 7EY
2E1 FKZ Arthur Woodward, 19 Hazel Grove, Winchester, SO22 4PQ
2E1 FLD P Whittaker, 34 New Road, Newhall, Swadlincote, DE11 0SP
2E1 FLN G McMillan, 10 Thornton Court, Girton, Cambridge, CB3 0NS
2E1 FLW S Burling, Ongar Cottage, 28 Main Road, Macclesfield, SK11 0BU
2E1 FMC R Sanderson, 92 Edge Lane, Dewsbury, WF12 0HB
2E1 FMW A Oughton, 176 South Lodge Drive, Southgate, London, N14 4XN
2E1 FNB F Morgan, 171 Town Road, London, N9 0HJ
2E1 FNJ L Carr, 10 Bonds Road, Hemblington, Norwich, NR13 4QF
2U1 FNQ J Le Page, Heathwick, Les Martins, St. Martin, Guernsey, GY4 6QJ
2E1 FNX J Meredith, 2 Hamilton Road, Dawley, Telford, TF4 3NG
2E1 FNY C Warren, Clifden Farm, Quenchwell ROAD, Truro, TR3 6LN
2E1 FON S Simpson, 20 Staveley Grove, Keighley, BD22 7DH
2E1 FOU Kevin Perry, 238 Sherwood Street, Warsop, Mansfield, NG20 0HJ
2E1 FOW Julia Thacker, 15 Riverside Mews, Hall Yard, Stoke-on-Trent, ST10 4FE
2E1 FOX Jean Camp, 1 Higher Tresillian Cottages, Tresillian, Newquay, TR8 4PL
2E1 FPI G Goodearl, 4 South Terrace, High Street, Dartford, DA4 0DF
2E1 FPK A Cartwright, 7 Pen Parc, Malltraeth, Bodorgan, LL62 5BG
2E1 FPM Layla Noel, 58 Easenhall Lane, Redditch, B98 0BJ
2E1 FPP D Paddon, 21 Oak Park Drive, Havant, PO9 2XE
2E1 FPU J Overland, 73 Butchers Lane, Walton on The Naze, CO14 8UE
2E1 FPV Jack Stringer, The Cottage, High Street, Radstock, BA3 5AL
2E1 FPW J Green, 2 Broadmeadow, Kingswinford, DY6 7HG
2E1 FQB P Elcombe, 16 Blenheim Avenue, Martham, Great Yarmouth, NR29 4TW
2E1 FQE C Carter, 331a Ordnance Road, Enfield, EN3 6HE
2M1 FQI G Robinson, 12 Hannahston Avenue, Drongan, Ayr, KA6 7AU
2E1 FQO Michael Dunthorne, 29 Clarion Way, Cannock, WS11 4NN
2E1 FQY O James, 24 Fryer Avenue, Leamington Spa, CV32 6HY
2E1 FRC H Doyle, Hurst House, Stratford Road, Henley-in-Arden, B95 6AB
2E1 FRE D Franklin, The Old Vicarage, Westville Road, Boston, PE22 7HJ
2E1 FRI Jonathan Wood, 3 Harold Collins Place, Colchester, CO1 2GQ
2E1 FRQ Anne Storry, 99 Swineshead Road, Wyberton Fen, Boston, PE21 7JG
2E1 FRW T Depledge, 16 Pennington Walk, Retford, DN22 6LR
2E1 FRY E Young, 7 The Quadrant, Fordingbridge, SP6 1BW
2E1 FRZ Terry Jennings, 23 De Lacy Court, New Ollerton, Newark, NG22 9RN
2E1 FSF J Kinch, 7 Fox Lane, Oakley, Basingstoke, RG23 7BB
2E1 FSG R Kinch, 7 Fox Lane, Oakley, Basingstoke, RG23 7BB
2E1 FSH C Forber, 32 Larch Avenue, Newton-le-Willows, WA12 8JF
2E1 FSR G Hammond, 50 Fernhill Close, Woodbridge, IP12 1LB
2E1 FSV D Feetenby, 32 Hawkins Close, Daventry, NN11 4JQ
2E1 FSX D Charlton, 20 Bailey Crescent, South Elmsall, Pontefract, WF9 2TL
2E1 FSY M Fey, 37 Winnards Close, West Parley, Ferndown, BH22 8PA
2E1 FSZ T Bashford, 198 Uplands Road, West Moors, Ferndown, BH22 0EY
2E1 FTA S Goodall, 37 Woodfield Road, Bear Cross, Bournemouth, BH11 9EU
2E1 FTE C Bingham, 56 Newgate Lane, Mansfield, NG18 2LQ
2E1 FTF E Sheppard, 4 Lindrick Avenue, Swinton, Mexborough, S64 8TE
2E1 FTH R Megone, 16 Mercer Close, Basingstoke, RG22 6NZ
2E1 FTI M Burrows, 40 Fairmile Road, Christchurch, BH23 2LL
2E1 FTV J Bailey, 13 Newark Road, Mexborough, S64 9EZ
2W1 FUD David Green, 17 Glyn-y-Mel Pencoed, Bridgend, CF35 6YA
2E1 FUH David Hartley, 2 Thirlmere Avenue, Burnley, BB10 1HU
2E1 FUJ J Hartley, 23 Broomfield Road, Fleetwood, FY7 7HA
2E1 FUQ J QUARTERMAINE, 3 Markham Close, Duston, Northampton, NN5 6TW
2W1 FVH R Hallett, Llamedos, 151 Trealaw Road, Tonypandy, CF40 2NX
2E1 FVJ R Halford, 104 Gladstone Avenue, London, N22 6LH
2E1 FVK T Kiely, 192 Morley Avenue, London, N22 6NT
2E1 FVS M Hotchin, 122 Buckingham Avenue, Scunthorpe, DN15 8NS
2E1 FVY D Ripley, 5 Rope Walk, Cranbrook, TN17 3DZ
2E1 FWA C Walker, 12 Bradshaw Crescent, Honley, C/O Mr C Walker, Holmfirth, HD9 6EG
2E1 FWD C Booth, 8 Heathfield Mews, Martlesham Heath, Ipswich, IP5 3UF
2E1 FWM M Foister, 38 Nine Acres, Kennington, Ashford, TN24 9JW
2E1 FWX Graham Sessions, 5 Luxton Court, Cullompton, EX15 1FJ
2E1 FXN D Boland, 9 Clapham Court, Gloucester, GL1 3DE
2E1 FYC D Martyr, 52 Parklawn Avenue, Epsom, KT18 7SL
2E1 FYI Paul Wootton, 56 Smithfield Road, Market Drayton, TF9 1EN
2E1 FYZ T Wilkie, Bramcote, Grange Road, Sutton on Sea, LN12 2RE
2E1 FZC J Charter, 36 Northumberland Avenue, London, E12 5HD
2E1 FZH M Saunders, 105 Raynham Road, Bury St. Edmunds, IP32 6ED
2E1 FZK B Barnett, 7 Holts Lane, Tutbury, Burton-on-Trent, DE13 9LE
2E1 FZO P Geier, 24 Deirdre Avenue, Wickford, SS12 0AX
2E1 FZU D Peters, 25 Corndon Crescent, Shrewsbury, SY1 4LD
2E1 FZY J Doyle, 33 Bodenham Road, Northfield, Birmingham, B31 5DP
2W1 GAC W Hicks, 2 Second Avenue, Clase, Swansea, SA6 7LN
2E1 GAD C Lee, 2 Edward Road, Romford, RM6 6UH
2M1 GBG Murrie Thomson, 194 East Main Street, Broxburn, EH52 5HQ
2E1 GBM J Lake, 25 Erlensee Way, Biggleswade, SG18 8GG
2E1 GBN A Vincent, 9 Broad Park Avenue, Ilfracombe, EX34 8DZ
2E1 GCB Edward Muxlow, 17 Station Road, Grasby, Barnetby, DN38 6AP
2E1 GCC Byron Smith, 18a King Edwards Road, Ware, SG12 7EJ
2E1 GCF C Horn, 12 Melbourne Road, Chichester, PO19 7NE
2E1 GDA C Norris, 8 Sutton Lane, Middlewich, CW10 9AU
2E1 GDB Margaret Jeffery, 14 Holly Mount, Shavington, Crewe, CW2 5AZ
2E1 GDD D Townson, 4 Crawford Street, Bradford, BD4 7JJ
2E1 GDF Shaun Okeefe, 63 Panama Circle, Derby, DE24 1AE
2E1 GDG D Baker, 78 Station Road, Whittlesey, Peterborough, PE7 1UE
2E1 GDK V Potts, 14 Topcliffe Mead, Morley, Leeds, LS27 8UH
2E1 GDM Mark Banner, 23 Astral Grove, Hucknall, Nottingham, NG15 6FY
2E1 GDO Robert Hacker, Flat 7, Dove Court, Ramsgate, CT11 8QA
2E1 GDY Andrew Palmer, 4a Clomendy Road, Cwmbran, NP44 3LS
2E1 GES C Knowlson, 23 Hawthorne Avenue, Shipley, BD18 2JB
2E1 GEZ R Scott, Kirklands, Craigend Road, Galashiels, TD1 2RJ
2M1 GFG R Dempster, 42 Kirkhill Road, Edinburgh, EH16 5DD
2E1 GFW M Jones, 68 Hampton Park Road, Hereford, HR1 1TJ
2E1 GGL E Mills, 70 Crescent Road, Rochdale, OL11 3LG
2E1 GGT A Walker, 12 Bladen Close, Cheadle Hulme, Cheadle, SK8 5RU
2E1 GHE A Carter, 50 Harberd Tye, Chelmsford, CM2 9GJ
2E1 GHF A Richardson, 42 Gypsey Road, Bridlington, YO16 4AZ
2E1 GHI E Mills, 3 Whitfield Close, Wilford, Nottingham, NG11 7AU
2E1 GHX C Lewin, The Hawthorns, Hawthorne Drive, Stafford, ST19 9NQ
2E1 GHZ Philip Mather, 18 Watcombe Cottages, Richmond, TW9 3BD
2I1 GIH L Costford, Aughakeerin, Derrygonnelly, Co. Fermanagh, BT93 6FR
2E1 GIK A Craven, 4 Amanda Drive, Louth, LN11 0AZ
2E1 GIZ Elizabeth Christieson, September Cottage, Rushlake Green, Heathfield, TN21 9PP
2E1 GJC A Brown, Hurst Farm, Ashworth Road, Rochdale, OL11 5UP
2E1 GJD S Burgess, 14 Shrubcote, Tenterden, TN30 7BA
2E1 GJE P Lee, 2 Dennis Street, Worksop, S80 2LL
2E1 GJG Harriet Kennedy, 19 High Street, East Hoathly, Lewes, BN8 6DR
2E1 GJJ Andrew Noon, 12 Stoney Fold, Telford, TF3 5GQ
2E1 GJO S Higgs, 239 Whalley Drive, Bletchley, Milton Keynes, MK3 6PL
2E1 GJP K Turner, 2 Bungalow, Dunston Fen, Lincoln, LN4 3AP
2E1 GJT C Wojcik, 153 Netherton Road, Worksop, S80 2SD
2E1 GKB M Brearley, 136 Elmsfield Avenue, Rochdale, OL11 5XA
2E1 GKE David SHEPHARD, 107 Withywood Drive, Telford, TF3 2HX
2E1 GKF Nigel Salter, 10 Jarvis Place, St. Michaels, Tenterden, TN30 6DQ
2W1 GKJ Matthew Moore, 56 Morfa Street, Bridgend, CF31 1HD
2E1 GKP C Dix, 141a Jerningham Road, London, SE14 5NJ
2E1 GKY Anne Reed, 32 Hollis Garden, Cheltenham, GL51 6JQ
2M1 GLD E Clark, 3 Old Cottages, Seton Mains, Longniddry, EH32 0PG
2E1 GLR John Halsall, 83 Poole Road, Leeds, LS15 7HD
2E1 GLS D Brown, 14 Beech Tree Road, Featherstone, Pontefract, WF7 5EB
2E1 GLT M Simpson, 20 Mount Pleasant Residential Park, Bloomhill Road, Doncaster, DN8 4ST
2E1 GLY C Mills, 26 Goossens Close, Ringland, Newport, NP19 9JN
2E1 GMA S Jordan, 31 Rocky Bank Road, Birkenhead, CH42 7LB
2E1 GMD Matt Buckland-Hoby, 9 Charles Darwin Road, Plymouth, PL1 4GU
2E1 GMM B Richards, 8 East Avenue, Griffithstown, Pontypool, NP4 5AB
2E1 GMO M Hastry, 56 Kilsyth Close, Fearnhead, Warrington, WA2 0SQ
2E1 GMQ Judith Barrett, 114 William Street, Long Eaton, Nottingham, NG10 4GD
2E1 GMT Jennifer Williams, 25 Peghouse Rise, Stroud, GL5 1RU
2E1 GMV S Fletcher, 5 Hayeswood Road, Stanley Common, Ilkeston, DE7 6GB
2E1 GNE R Wright, 3 Ednall Lane, Bromsgrove, B60 2DB
2E1 GNK G Smith, 106 Broadoak Road, Manchester, M22 9PL
2E1 GNN P Saben, Tredinneck Moor, Newmill, Penzance, TR20 8XT
2E1 GNR A Taylor, 16 Penny Lane, Collins Green, Warrington, WA5 4DS
2E1 GNU A Watson, 7 Branksome Drive, Morecambe, LA4 5UJ
2E1 GOC S Bedell, 1 Pheasant Field Drive, Spondon, Derby, DE21 7LR
2E1 GOE C Hopkins, 28 Pitcairn Road, Mitcham, CR4 3LL
2E1 GOK K Phillips, 30 Allendale, Ilkeston, DE7 4LE
2E1 GOM J Constable, 9 Ridgeway Close, Heathfield, TN21 8NS
2E1 GOP B Cartwright, 14 Zealand Close, Hinckley, LE10 1TJ
2E1 GOZ I Press, 19 Banwell Close, Keynsham, Bristol, BS31 1JX
2E1 GPE M Ruff, Brambledown, Camberlot Road, Hailsham, BN27 3QG
2E1 GPG Dominic Howells, Flat 1-2, 130 Essex Road, London, N1 8LX
2E1 GPV M Allott, 1 Charles Street, Lancaster, LA1 4UU
2E1 GQB G Meek, 443 Springfield Road, Chelmsford, CM2 6AP
2E1 GQD Alan Smith, 31 Haywards Place, Easterton, Devizes, SN10 4PP
2E1 GQN Tasmin Stevens, 25 Avenue Road, New Milton, BH25 5JP
2E1 GQU M Forster, 33 Deer Valley Road, Holsworthy, EX22 6DA
2E1 GRA A Merrill, 66 Royal Oak Drive, Selston, Nottingham, NG16 6RJ
2E1 GRG D Brady, 6 Bristow Close, Bletchley, Milton Keynes, MK2 2XP
2E1 GRT Matthew Keeling, 40 Lambseth Street, Eye, IP23 7AQ
2E1 GSC Paul Sherratt, 23 Paulina Avenue, Nottingham, NG15 8JA
2E1 GSM D Ridgway, 145 Bodmin Road, Astley, Manchester, M29 7PE
2E1 GSW J Harratt, 8 The Runcie Building, Ripon College, Oxford, OX44 9EX
2E1 GSX Howard Mustoe, 5 Dartmouth Park Avenue, London, NW5 1JL
2E1 GTB C Barnes, 533 Maidstone Road, Blue Bell Hill, Chatham, ME5 9QP
2E1 GTD S Keene, 4 Blackberry Lane, Four Marks, Alton, GU34 5BN
2E1 GTE S Hall, 122 Norwich Road, New Costessey, Norwich, NR5 0EH
2E1 GTF Stuart Silk, 89 St. Barts Road, Sandwich, CT13 0AS
2E1 GTI A Cadier, 28 Romney Avenue, Folkestone, CT20 3QJ
2E1 GTT A Davies, Little Platt, Bodle Street Green, Hailsham, BN27 4RA
2E1 GTW F James, 1 The Gorseway, Bexhill-on-Sea, TN39 4PP
2E1 GUC A Ibrahim, 14 Dunvegan Road, London, SE9 1SA
2E1 GUN S Kirby, 2 Kneeton Park, Middleton Tyas, Richmond, DL10 6SB
2E1 GVB E Artis, 71 South Quay, Great Yarmouth, NR30 2RW

2E1 GVC G Carlin, 38 Balfour St., East Bowling, Bradford, BD4 7JT
2E1 GVD G Carlin, 38 Balfour St., East Bowling, Bradford, BD4 7JT
2E1 GVJ J Millington, 6 Fentonhouse Lane, Wheaton Aston, Stafford, ST19 9NU
2E1 GV8 G Gallacher, 112 Central Drive, Stoke-on-Trent, [illegible]
2E1 GWX Robert Shepherd, 91 Saxon Road, Hastings, TN35 5HH
2E1 GXE M Rogers, 45 Church Road, Westoning, Bedford, MK45 5LP
2E1 GXH D Webb, 51 Garden Road, Walton on The Naze, CO14 8RR
2E1 GXI J Adams, 28 Greenside, Stoke Prior, Bromsgrove, B60 4EB
2E1 GXL P Travis, 42 Trafalgar Road, Wallasey, CH44 0EB
2E1 GXQ P Walker, 78 Kirkby Road, Desford, Leicester, LE9 9JG
2E1 [illegible] [illegible], Gerrards Cross, SL9 7EB
2E1 GXU Rogers Steven, 20 Collingwood Road, Woodbridge, IP12 1JL
2E1 [illegible] M Rogers, [illegible] Tudor Close, Framlingham, Woodbridge, IP13 9SL
2E1 [illegible] [illegible] Road, Galashiels, TD1 2BJ
2E1 GXY P Batty, 134 Plymouth Road, Scunthorpe, DN17 1TS
2E1 GYB M Peterson, 4 Allen Drive, Mansfield, NG18 3HZ
2E1 GYC S Jenkins, 45 Dorchester Road, Solihull, B91 1LN
2E1 GYD M Hall, 48 Dorchester Road, Solihull, B91 1LN
2E1 GYG D Abbott, 5 Heathcote Gardens, Rudheath, Northwich, CW9 7JB
2E1 GYH S Pearson, 8 The Pastures, Edlesborough, Dunstable, LU6 2HL
2E1 GYN W Ashley, 23 Lavenham Close, Clacton-on-Sea, CO16 8BZ
2E1 GYO C Stanley, 9 Pyramid Caravan Site, Beeston, Nottingham, NG9 1NS
2E1 GYR H Webster, 67 Greenways, Over Kellet, Carnforth, LA6 1DE
2M1 GYX Alistair MacLeod, 13 Wyvern Park, Edinburgh, EH9 2JY
2E1 GYZ L Rowley, 20 Long Leasow, Selly Oak, Birmingham, B29 4LT
2E1 GZF John Walker, 1 Riverside Court, Louth, LN11 7AG
2E1 GZV A Ager, 5 Matthews Close, Bedhampton, Havant, PO9 3NJ
2E1 GZY Alex Wilson, 22 Ormesby Road, RAF Coltishall, Norwich, NR10 5JY
2E1 GZZ M COLES, 29 Sydney Road, Exeter, EX2 9AH
2E1 HAC M Lucas, 48 Sycamore Drive, Ash Vale, Aldershot, GU12 5PR
2E1 HAF R Johnson, 14 Denison Court, Clifton Street, Barnsley, S70 1UA
2E1 HAL Martin Farraway, 199 Ilkeston Road, Nottingham, NG7 3HF
2E1 HAM D Langmead, 38 Milton Grove, London, N11 1AX
2E1 HAQ Kevin Graffham, 15 Hayes Road, Clacton-on-Sea, CO15 1TX
2E1 HAS B Heirene, 9 Ryecroft Crescent, Barnet, EN5 3BP
2E1 HAU M Durrant, 8 Drake Avenue, Great Yarmouth, NR30 4BS
2E1 HAW James Beith, 18 Avenue Road, New Milton, BH25 5JP
2E1 HBF L Buggs, 2 Archway Cottages, Valley Road, Leiston, IP16 4AR
2E1 HBG M Hannaford, 12 Victoria Street, Norwich, NR1 3QX
2E1 HBJ C Masters, 85 Petersham Road, Creekmoor, Poole, BH17 7DW
2E1 HBS D Treacher, Barrons Cottage, Welburn, York, YO60 7DZ
2E1 HBT E Marlow, 17 Fellows Court, Weymouth Terrace, London, E2 8LP
2E1 HCB D Broom, 114 Gammons Lane, Watford, WD2 5HY
2E1 HCG A Docherty, The Flat, Winton House, Stoke-on-Trent, ST4 2RQ
2M1 HCP A Smith, 4 The Terrace, Lhanbryde, Elgin, IV30 8NY
2M1 HCQ A Smith, 4 The Terrace, Lhanbryde, Elgin, IV30 8NY
2E1 HCT R Cowlishaw, 23 Aldrich Drive, Willen, Milton Keynes, MK15 9HP
2E1 HDB M Dimambro, 26 Fetcham Court, Bank Top, Newcastle upon Tyne, NE3 2UL
2E1 HDC P Dimambro, 26 Fetcham Court, Bank Top, Newcastle upon Tyne, NE3 2UL
2E1 HDE A Morris, 71 Lurdin Lane, Standish, Wigan, WN6 0AQ
2E1 HDF Robert Harrison, 36 Windermere Road, Farnworth, Bolton, BL4 0QH
2E1 HDZ T Lockett, 14 Tildsley Crescent, Weston, Runcorn, WA7 4RN
2E1 HEB Ray Currant, 8 Milford Hill, Harpenden, AL5 5BH
2E1 HEE H Merrington, Cartref, Ball Lane, Frodsham, WA6 8HP
2E1 HEF J Hastry, 56 Kilsyth Close, Fearnhead, Warrington, WA2 0SQ
2E1 HEG Edna Mather, 72 Cranleigh Road, Worthing, BN14 7QW
2E1 HEK L Taylor, 127 Dundee Close, Fearnhead, Warrington, WA2 0UJ
2E1 HEM C Kulikovsky, 42 Highlands Grove, Bradford, BD7 4BG
2E1 HEO Andrew Austin, Flat 4, De Cham Court, 33 De Cham Road, St. Leonards-on-Sea, TN37 6JA
2E1 HER D Webster, 3 Eden Avenue, Bare, Morecambe, LA4 6QL
2E1 HES M Davenport, 24 Willow Grove, Belper, DE56 1LX
2E1 HEV S Brandon, 2 Moss Bank, Winsford, CW7 2ED
2E1 HFA Sarah Emerton, 18 Avenue Road, New Milton, BH25 5JP
2M1 HFE C Budas, 20 Oak Avenue, Bearsden, Glasgow, G61 3HD
2E1 HFH Steven OVERALL, Flat 74, Douglas Buildings, London, SE1 1EL
2E1 HFN J Newell, 4 Honeyfield Drive, Ripley, DE5 3JL
2E1 HFS P Dunne, 2 All Saints Road, Wyke Regis, Weymouth, DT4 9EZ
2E1 HFV H Shackleton, Woodroyd, 66 Spring Ave, Keighley, BD21 4TA
2E1 HFW M Hemstock, 4 Tavistock Avenue, Perivale, Greenford, UB6 8AJ
2E1 HFX K Taylor, 46 Hunters Field, Stanford in the Vale, Faringdon, SN7 8LX
2W1 HFZ Leighton Jones, 52 George Street, Aberdare, CF44 6SH
2E1 HGA T Winwood, 2 The Warren, Abingdon, OX14 3XB
2E1 HGE A Higgs, 26 Avon Avenue, Ringwood, BH24 2BH
2E1 HGF Paul Crookes, 11 Degens Way, Hugglescote, Coalville, LE67 2XD
2E1 HGG P Mead, 9 Abraham Drive, Silver End, Witham, CM8 3SP
2E1 HGJ J Jordan, 4 Tuckers Close, Loughborough, LE11 2PG
2E1 HGM K Cronin, 9 Marriott Close, Beeston, Nottingham, NG9 4JB
2E1 HGR D Edwards, 18 Garden House Lane, East Grinstead, RH19 4JT
2E1 HGT Adrian [illegible], 85 St. Catherines Road, Crawley, RH10 3TB
2E1 HGU T Lawrence, 85 West Avenue, Clacton-on-Sea, CO15 1HB
2E1 HGY Marilyn Henson, 1 Eaton Close, Rainworth, Mansfield, NG21 0AR
2E1 HHA A Cittone, Elyria, 22 Charles Lovell Way, Scunthorpe, DN17 1YL
2E1 HHB Daniel Botterell, 12 Selsey Avenue, Clacton-on-Sea, CO15 1NQ
2E1 HHE Tom Carlson, 15 White Hedge Drive, St. Albans, AL3 5TU
2E1 HHG J Bradbury, 192 Greenwood Road, Bakersfield, Nottingham, NG3 7FY
2E1 HHL S Williams, Alwent Farm, Staindrop, Darlington, DL2 3NS
2E1 HHY S Day, 10 Second Avenue, Wolverhampton, WV10 9PP
2E1 HID D Smith, 23 Marlborough Road, Long Eaton, Nottingham, NG10 2BS
2E1 HIL I Finch, 29 Sherwood Road, Grimsby, DN34 5TG
2M1 HIN J McPhee, 101 Birkenside, Gorebridge, EH23 4JF
2E1 HIO WR Roberts, 30 Park Boulevard, Clacton-on-Sea, CO15 5RH
2E1 HIQ Anthony Mayes, 126 Walesby Lane, New Ollerton, Newark, NG22 9UU
2E1 HJA K Mieske, 10 Cowdrey Road, London, SW19 8TU
2E1 HJE Christopher Clifton, 35 Farrowdene Road, Reading, RG2 8SD
2E1 HJO M Hyde, 10 Devonshire Drive, Barnsley, S75 1EE
2E1 HJS R Burns, 130 Kingsway, Mapplewell, Barnsley, S75 6EU
2M1 HKA S Mcintyre-Stewart, 4 Howie Crescent, Rosneath, Helensburgh, G84 0RL
2E1 HKB Kieron Brunning, 70 Cobbold Road, Felixstowe, IP11 7QR
2E1 HKC K Cullum, 22 Orwell View Road, Shotley, Ipswich, IP9 1NP
2E1 HKE R Wilson, Newstead Farm, Clay Lane, Norwich, NR10 4PP
2E1 HKM E Miles, 31 Winnipeg Road, Bentley, Doncaster, DN5 0ED
2E1 HKS S Hall, 1 George Place, Wellington, Telford, TF1 2AJ
2E1 HKY D Green, 89 Upper Ratton Drive, Eastbourne, BN20 9DJ
2E1 HLA J Moran, 27 Mellor Grove, Bolton, BL1 6DA
2M1 HLE Gilbert Stephen, 121 Dunbar Street, Lossiemouth, IV31 6RE
2E1 HLF E Carder, 45 Chalklands, Linton, Cambridge, CB21 4JQ
2E1 HLH E Gardner, New House, Birdbush Avenue, Saffron Walden, CB11 4DJ
2E1 HLL [illegible], 8 [illegible] Avenue, Atherton, Manchester, M46 9RR
2E1 HLO C Mitchell-Watson, 144 Shakespeare Crescent, Dronfield, S18 1ND
2E1 HLT [illegible], 25 West Avenue, Clacton-on-Sea, CO15 1HB
2E1 [illegible] [illegible], Worcester, [illegible]
2E1 [illegible] T Kimm, 99 Midland Road, Bramhall, Stockport, SK7 3DT
2E1 HLW Myles McNeany, 29 Grenfell Road, Manchester, M20 6TG
2E1 HMR Tristan Emmett, 33 St. Brelades Avenue, Poole, BH12 4JR
2E1 HMQ K Vance, 5 Riversdale Close, Birstall, Leicester, LE4 4EH
2E1 HNB K Hughes, High Lane Cottage, Congleton Road, Macclesfield, SK11 9RH
2E1 HNF Joseph Li, 13 Tothill Street, Minster, Ramsgate, CT12 4AG
2E1 HNH Allen Chalk, 42 Erskine Road, Colwyn Bay, LL29 8EU
2E1 HNN T Bennellick, 18 Bailie Close, Abingdon, OX14 5RF
2E1 HNS Terence Middleton, 31 Coltman Avenue, Long Crendon, Aylesbury, HP18 9DP
2I1 HNZ M McCrum, 2 New Road, Donaghadee, BT21 0DR
2E1 HOF J Johnson, 2 Meadow Drive, Canon Pyon, Hereford, HR4 8NT
2E1 HOK S Langley, 100 Pogmoor Road, Barnsley, S75 2EF
2E1 HOO Diane Anstie, 20 Keyes Road, Norwich, NR1 2JX
2E1 HOS T Keating, 65 Shorncliffe Avenue, Norwich, NR3 2HT
2E1 HOT J Tosh, 65 Shorncliffe Avenue, Norwich, NR3 2HT
2E1 HPC S Carr, 10 Bonds Road, Hemblington, Norwich, NR13 4QF
2E1 HPD D Carr, 10 Bonds Road, Hemblington, Norwich, NR13 4QF
2E1 HPM L Hall, 9 Stone Court, South Hiendley, Barnsley, S72 9DL
2E1 HPS J Vandervord, 13 Granville Avenue, Ramsgate, CT12 6DX
2E1 HPT A Spain, 60 The Maples, Broadstairs, CT10 2PE
2E1 HPZ W Bellis, Cliffe Bungalow, Barnsley Road, Barnsley, S72 9JX
2E1 HQA Frederica Kennedy, 19 High Street, East Hoathly, Lewes, BN8 6DR
2E1 HQH J Garner, 30 Birds Avenue, Garlinge, Margate, CT9 5NE
2E1 HQP M Newth, 3 Juniper Road, Southampton, SO18 4EH
2E1 HQV B Cummings, 8 Spey House, Criterion Street, Stockport, SK5 6TD
2E1 HQW S Hackney, 8 Spey House, Criterion Street, Stockport, SK5 6TD
2E1 HQY Pamela Procter, 1 Brow Hey, Bamber Bridge, Preston, PR5 8DS
2E1 HQZ A Webb, 48 Warren Road, Stoke-on-Trent, ST6 6HW
2E1 HRB S Dykes, 15 Sunningdale Road, Chelmsford, CM1 2NH
2E1 HRC P Edwards, 5 Brindley Road, Silsden, Keighley, BD20 0LD
2E1 HRJ A Hollis, 89 Longfield Lane, Cheshunt, Waltham Cross, EN7 6AN
2E1 HRM D Morgan, 171 Town Road, London, N9 0HJ
2E1 HRN Adam Fower, 31 Hillswood Avenue, Leek, ST13 8EQ
2M1 HRS Robert Duncan, 90 Faulds Gate, Aberdeen, AB12 5QT
2E1 HRY P Odle, 24 Longfellow Road, Gillingham, ME7 5QG
2E1 HSA T Newman, Sometimes (The Workshop), South Pew, Dorchester, DT2 9HZ
2E1 HSB Chris Price, 162 Stamshaw Road, Portsmouth, PO2 8LX
2E1 HSD B Manning, Barton Farm, Barton Road, Wisbech, PE13 4TL
2M1 HSG Mark Douglas, 195 Dumbuck Road, Dumbarton, G82 3NU
2E1 HSJ M Tulk, 8 Cleves Close, Weymouth, DT4 9JU
2E1 HSL Terry Brodie, 8 Meadow Way, Plymouth, PL7 4JB
2E1 HSP A Saville, 4 Shannon Court, Downs Barn, Milton Keynes, MK14 7PP
2E1 HSR M Rogers, 47 Tregarrian Road, Tolvaddon, Camborne, TR14 0HD
2E1 HTE Chris Hall, 2 Brambling Lane, Wath-upon-Dearne, Rotherham, S63 7GT
2E1 HTK E Bateman, 32 Park Avenue, Bodelwyddan, Rhyl, LL18 5TB
2E1 HTM J Brice, 10 Swan Close, Weston-Super-Mare, BS22 8XR
2M1 HTR A Pollard, 12 Royfold Crescent, Aberdeen, AB15 6BH
2E1 HTU R Corbett, 11 Old Office Close, Dawley Bank, Telford, TF4 2QA
2E1 HTV C Sargent, 8 Gilwell Grove, Priorslee, Telford, TF2 9SR
2E1 HTY A Semple, Linden Lea, Lydham, Bishops Castle, SY9 5HB
2E1 HUB R Semple, The Inn On The Green, Wentnor, Bishops Castle, SY9 5EF
2E1 HUC K Jones, 24 Brooke Avenue, Margate, CT9 5NG
2E1 HUE D Taylor, 2 Delph Cottages, Barkisland, Halifax, HX4 0BW
2E1 HUJ P Adams, 20 Grosvenor Road, London, W4 4EH
2E1 HUQ K Gerrard, 8 Windsor Crescent, Little Houghton, Barnsley, S72 0HG
2E1 HUX H Huxley, Hen Berllan, Nant Mawr Road, Buckley, CH7 2BS
2E1 HVB A Scothern, 42 Griceson Close, Ollerton, Newark, NG22 9BD
2E1 HVI G Reilly, B I Z Ltd, Millmarsh Lane, Enfield, EN3 7QA
2E1 HVL G Bateman, 141 Bury Street, Ruislip, HA4 7TQ
2E1 HVM C Jackson, 76 Margards Lane, Verwood, BH31 6JP
2M1 HVR Caroline Hextall, 4 Hawthornbank, Cockenzie, Prestonpans, EH32 0HZ
2E1 HVT R Hicks, 31 Arundel Road, Great Yarmouth, NR30 4LD
2E1 HVU S Hicks, 50 Garfield Road, Great Yarmouth, NR30 4JU
2E1 HVZ C Stevens, 82 Bembridge Drive, Alvaston, Derby, DE24 0UQ
2E1 HWI C Zdziech, 200 Kensington Street, Rochdale, OL11 1QS
2E1 HWJ T Winch, Whitehall Barn, Stowmarket Road, Stowmarket, IP14 6BU
2E1 HWQ Sean Cannon, 40 Hawthorne Avenue, Louth, LN11 0LD
2E1 HWU R Noakes, 14 Sackville Road, Immingham, DN40 1EE
2E1 HWV M Brett, 30 Belmont Avenue, London, N13 4HD
2E1 HXB B Wheat, 23 Stead Street, Eckington, Sheffield, S21 4FY
2E1 HXC A Rodger, 4 Burns Nurseries, Wootton Road, King's Lynn, PE30 3BG
2E1 HXD A Brett, 1c Old Park Road, Palmers Green, London, N13 4RG
2E1 HXM L Tunstall, 0 York Road, Rowley Regis, B65 0RR
2E1 HXN J Dodds, 8 York Road, Rowley Regis, B65 0RR
2E1 HXP M Walsh, 23 Moss Fold Road, Darwen, BB3 0AQ
2E1 HXR E Lacey, 6 Weetshaw Close, Shafton, Barnsley, S72 8PZ
2E1 HXT K Jones, 11 St. Davids Close, Gobowen, Oswestry, SY11 3JF
2E1 HXY R Ling, 8 Spa Hill, Kirton Lindsey, Gainsborough, DN21 4NE
2E1 HXZ P Ling, 8 Spa Hill, Kirton Lindsey, Gainsborough, DN21 4NE
2E1 HYD antony wilson, 135 Britannia Road Morley, Leeds, LS27 0DS
2E1 HYE D Bruce, 6 Princes Way, King's Lynn, PE30 2QL
2E1 HYI Paul Redfern, 42 Newton Street, Retford, DN22 7AD
2E1 HYT K Webb, 43 New Road, Hornsea, HU18 1PH
2E1 HYX M Hammond, 10 Collingwood Close, Braintree, CM7 9UG
2E1 HZI E Sinclair, 21 Longport Avenue, Manchester, M20 1EN
2E1 HZM Michael Fleming, 54 Madison Avenue, Bradford, BD4 0JJ
2E1 HZV E Colley, 14 Hawthorne Close, Tyldesley, Manchester, M29 8PH
2E1 HZY P Hartley, 14 Medway Walk, Wigan, WN5 9NQ
2E1 IAB J Jackson, 40 Lightbounds Road, Bolton, BL1 5UN
2E1 IAC Martyn Dix, 10 College Hill, Godalming, GU7 1YA
2E1 IAE Mark Hedges, Oakfield Cooper Lane, Holmfirth, HD9 3DP
2E1 IAI N Revell, 34 Chestnut Walk, Chelmsford, CM1 4JT
2E1 IAS S Loane, 30 St Wilfrids Road, Burgess Hill, RH15 8BD
2E1 IAT R Randell, 32 Windsor Close, Rubery, Birmingham, B45 0DA
2E1 IAX [illegible] Road, Coventry, CV5 6GH
2E1 IAY J Healey, 28 Thatchers Place, Westlands, Droitwich, WR9 9FD
2E1 IAZ [illegible] Place, Westlands, Droitwich, WR9 9ED
2M1 IBL [illegible], Prestonpans, EH32 9BD
2M1 [illegible] G Noonan, 8 Johnston Terrace, Port Seton, Prestonpans, EH32 0BB
2E1 IDJ D Marshall, 143 Middleton Road, Banbury, OX16 3QS
2E1 IBN M Tucker, 21 Pen Y Fan Close, Pentwyn Crumlin, Newport, NP11 3JQ
2E1 IBP J Siddall, 17 Dalebrook Road, Sale, M33 3LD
2E1 IBS I Grahame, 24 Bushby Avenue, Broxbourne, EN10 6QE
2E1 IBT J Jordan, 28 Old Ashby Road, Loughborough, LE11 4PG
2M1 IBX J Millar, 18a Dougall Street, Tayport, DD6 9JD
2E1 ICB Ian Bellhouse, 37 Runnymede Avenue Kingswood, Hull, HU7 3FZ
2E1 ICI C Rengifo, 61 Clarendon Road, Sale, M33 2DY
2E1 ICM L Kerswill, 18c Jewell Road, Bournemouth, BH8 0JQ
2E1 ICO A Ryan, 18 Belmount Avenue, Newcastle upon Tyne, NE3 5QD
2E1 ICT G Mears, 4 Uplands Road, Woodford Green, IG8 8JN
2E1 ICU R Ropinski, 38 The Leys, Little Eaton, Derby, DE21 5AR
2E1 ICW A Spiers, 2 Adeline Cottages, Jacobs Well Road, Guildford, GU4 7PD
2E1 IDC Carl Hillier, 15 Dabbs Hill Lane, Northolt Park, Northolt, UB5 4AQ
2E1 IDE G Swain, 33 Saville St., Blidworth, Mansfield, NG21 0RW
2E1 IDG D Greaves, The White House, 25 London Road, Leicester, LE8 9GF
2E1 IDH K Reynolds, 3 Lilac Close, Chelmsford, CM2 9NY
2E1 IDI C Welsh, 56 Longacres, St. Albans, AL4 0DR
2E1 IDM D Cheetham, 405 Jenkin Road, Sheffield, S9 1AY
2E1 IEK E Howie, 102 Rushdean Road, Rochester, ME2 2QB
2E1 IFA F Maddison, 87 Godolphin Road, London, W12 8JN
2E1 IFL C Wynn, 45 Hillcrest Road, Berry Hill, Coleford, GL16 7RG
2E1 IFM A Gow, 11 Rodley Square, Lydney, GL15 5AZ
2E1 IFN K Gow, 11 Rodley Square, Lydney, GL15 5AZ
2E1 IFW D Hyde, The Grove, 7 Mill Lane, Kidderminster, DY10 3ND
2E1 IGA E Crane, 161 Heath Way, Horsham, RH12 5XX
2E1 IGG K Horsley, 76 Bernwood Road, Headington, Oxford, OX3 9LQ
2E1 IGI D Willimot, 5 Green Lane, Upton, Huntingdon, PE28 5YE
2E1 IGJ H Schnaar, 21 Anglesey Close, Crawley, RH11 9HG
2E1 IGK C Smith, 10 The Row, Weeting, Brandon, IP27 0QG
2E1 IGN G Howe, 22 Freston, Paston, Peterborough, PE4 7EN
2E1 IGP A Wolstencroft, 201 Walton Road, Sale, M33 4ER
2M1 IGQ C Bond, Strathllan House, Doune Road, Dunblane, FK15 9AR
2E1 IGU Richard Ransom, 97 Park Square West, Jaywick, Clacton-on-Sea, CO15 2NU
2E1 IGY James Kerr, 14 Seafield Close, Barton on Sea, New Milton, BH25 7HR
2E1 IHB W Clarke, 41 Upton Road, Atherton, Manchester, M46 9RQ
2E1 IHE D Siviter, 72 Sandfield Road, West Bromwich, B71 3NE
2E1 IHF L Parrish, 5 Kestrel Lane, Cheadle, Stoke-on-Trent, ST10 1RU
2E1 IHJ D Seeby, 59 Dallamoor, Telford, TF3 2EE
2E1 IHK J Smith, 54 Hillside Avenue, Bridgnorth, WV15 6BU
2W1 IHN D Carpenter, Kilkiffeth, Pontfaen, Fishguard, SA65 9TP
2E1 IHO J Willingham, 49 Creek Road, Hayling Island, PO11 9RA
2E1 IHS Alma Hardman, 47 Oatlands Road, Manchester, M22 1AH
2E1 IHT G Large, 11 Ranworth, King's Lynn, PE30 4XD
2E1 IHW J Eaton, 184 Gore Road, New Milton, BH25 5NQ
2E1 IHY C Sneap, 14 Calver Close, Nottingham, NG8 1AT
2E1 IIA C Lee, 154 Grangeway, Runcorn, WA7 5JA
2E1 IIC Gordon Foreman, 41 Winnards Park, Sarisbury Green, Southampton, SO31 7BX
2E1 IID G Gudgeon, 28 Park Close, Stevenage, SG2 8PX
2E1 IIE David Jones, 31 Summerhill Drive, Liverpool, L31 3DN
2E1 IIG H Holmes De Wyvill Sinclair, 27 Mount Crescent, Bridlington, YO16 7HR
2E1 IIJ G Colclough, Little Hallands, Norton, Seaford, BN25 2UN
2E1 IIL T Rodgers, 57 Knowles Hill, Rolleston-on-Dove, Burton-on-Trent, DE13 9DY
2E1 IIM A Bagg, 1 Stone Road, Burnham-on-Sea, TA8 1JU
2E1 IIP Robert Mullen, 58 Jasmin Avenue, Newcastle upon Tyne, NE5 1TL
2E1 IIV G Guinan, 5a Temple Lane, Silver End, Witham, CM8 3QY
2M1 IIW M Smith, 25 Charleston Crescent, Cove, Aberdeen, AB12 3DZ
2E1 IJD C Wilcockson, Conybeare House, Willowbrook, Windsor, SL4 6HL
2E1 IJE A Wilcockson, Conybeare House, Willowbrook, Windsor, SL4 6HL
2E1 IJF J Grubb, Waterloo Farm, Foston-on-the-Wolds, Driffield, YO25 8BH
2E1 IJK David Butler, Church Cottage, Church Road, Badminton, GL9 1HT
2E1 IJL B Gow, 11 Rodley Square, Lydney, GL15 5AZ
2E1 IJM F Lawless, 99 Ribbleton Avenue, Ribbleton, Preston, PR2 6DA
2E1 IJN P Nevard, Millinder House, Westerdale, Whitby, YO21 2DE
2E1 IJO B Blackham, 5 Reedham Drive, Bramley, Rotherham, S66 2SW
2E1 IJQ J Mallichan, 17 Napier Road, Gillingham, ME7 4HB
2E1 IJR R Rendell, 43 Springmead, Chard, TA20 2EW
2E1 IJS A Reed, 139 Wigmore Road, Gillingham, ME8 0TH
2E1 IJU E Heagren, 156 Eastbrooks, Pitsea, Basildon, SS13 3QH
2E1 IJX J Underwood, 12 Forge Lane, Gillingham, ME7 1UG
2E1 IJY A Brown, 24 Greenfields, Langley Mill, Nottingham, NG16 4GJ
2E1 IJZ Russell Meech, 26 Priory Street, Tonbridge, TN9 2AN
2E1 IKA M Flanagan, 33 Ullswater Road, Chorley, PR7 2JB
2E1 IKB Barry Randall, 5 Percival Road, Eastbourne, BN22 9JL
2E1 IKM K Ralph, 15 Hansell Road, Norwich, NR7 0LY
2E1 ILH Dean Ledger, 12 Reedling Drive, Southsea, PO4 8UF
2E1 ILM A Chandoo, 59 Stanton Road, Birmingham, B43 5HH
2E1 INC R Davenport, 10 Woodend Lane, Hyde, SK14 1DT
2E1 INT Daniel Kenyon, 88 Knutsford Road, Wilmslow, SK9 6JD

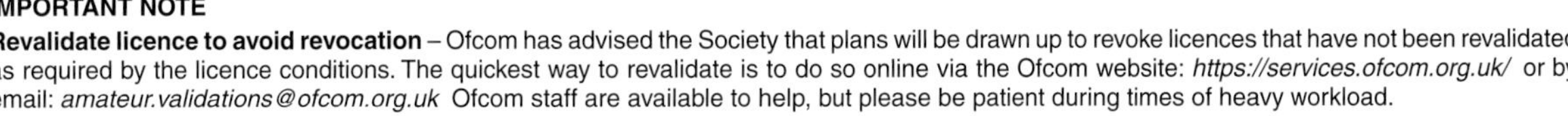

**IMPORTANT NOTE**

**Revalidate licence to avoid revocation** – Ofcom has advised the Society that plans will be drawn up to revoke licences that have not been revalidated as required by the licence conditions. The quickest way to revalidate is to do so online via the Ofcom website: *https://services.ofcom.org.uk/* or by email: *amateur.validations@ofcom.org.uk* Ofcom staff are available to help, but please be patient during times of heavy workload.

2E1 INW I Williamson, 196 Bruntwood Lane, Heald Green, Cheadle, SK8 3AS
2E1 IOS H Handrick, 13 Ormonde Way, New Rossington, Doncaster, DN11 0SB
2E1 ITE G Cahill, 81 Albemarle Road, Willesborough, Ashford, TN24 0HJ
2E1 ITI D Quick, Rosegarth, Woodbine Road, Blackwood, NP12 1QH
2E1 IVT F Handley, 155 Heathcote Street, Longton, Stoke-on-Trent, ST3 1AD
2E1 IWD D Brooks, 61 Carisbrooke High Street, Newport, PO30 1NR
2E1 IWG W Goodwin, 4 Southfields Close, Bishops Waltham, Southampton, SO32 1EY
2E1 JAA J Ahmed, 59 Ramsgate, Lofthouse, Wakefield, WF3 3PX
2E1 JAC J Marter, 4 Meadow Way, Seaford, BN25 4QT
2E1 JBJ A Berry, 4 Newlands Park Way, Newick, Lewes, BN8 4PG
2E1 JCM J Matthews, 9 Clive Green, Shrewsbury, SY2 5QL
2E1 JEF G Hensby, Flat 12, Edel Quinn House, Wirral, CH49 6PN
2E1 JEH J Haimes, 15 The Avenue, Hersden, Canterbury, CT3 4HL
2E1 JGM James Moody, 44 Frensham, Cheshunt, Waltham Cross, EN7 6HB
2E1 JGW J Wright, 2 St. Leonards Park, East Grinstead, RH19 1EE
2E1 JIM J Deacon, Spring Valley, Churt Road, Farnham, GU10 2QU
2E1 JJM J Martin, 1 Collins Lane, West Harting, Petersfield, GU31 5NZ
2E1 JJN J Nicholson, 6 Mill Gardens, West End, Southampton, SO18 3AG
2M1 JKG J Grieve, 39 Kenmount Drive, Kennoway, Leven, KY8 5HA
2E1 JKL J Loader, Furlong Farm, Henley, Langport, TA10 9AX
2E1 JKP J Peggram, Cherry Trees, Broad Lane, Bracknell, RG12 9BY
2E1 JLC John Chaundy, Ambleside, Barnes Lane, Lymington, SO41 0RL
2I1 JMC James McCaw, 62 High Street, Ballymena, BT43 6DT
2E1 JMG J Greaves, The White House, 25 London Road, Leicester, LE8 9GF
2E1 JMW J Williams, 73 Telford Road, Tamworth, B79 8EY
2E1 JOD J Antimano, 23 Nupton Drive, Barnet, EN5 2QU
2W1 JOL Terence Philpott, 6 Hillside, Fochriw, Bargoed, CF81 9LQ
2E1 JON Jonathan Sturgeon, Windyridge, Linkside West, Hindhead, GU26 6PA
2E1 JOY E Nye, 11 Barnhill Close, Marlow, SL7 3HA
2E1 JRB James Bancroft, 20 Hawthorn Road, Sedgefield, Stockton-on-Tees, TS21 3BZ
2E1 JRC B Goodall, 21 Carr Hill Avenue, Calverley, Pudsey, LS28 5QG
2E1 JRS J Stew, 19 Salisbury Close, Sittingbourne, ME10 3BL
2E1 KAJ K Johnson, 22 Prior Road, Greatstone, New Romney, TN28 8SB
2E1 KCC C Coates, 104 Orion Way, Willesborough, Ashford, TN24 0DZ
2E1 KID R Gow, 11 Rodley Square, Lydney, GL15 5AZ
2E1 KIP J Barker, 5 Severn Avenue, Weston-Super-Mare, BS23 4DH
2E1 KJB K Blanch, Sticelett Farm, Rolls Hill, Cowes, PO31 8NE
2E1 KLT G Clark, 28 Manor Road, Woolton, Liverpool, L25 8QG
2E1 KYQ L Fisher, Annstuyvonne, Baghill Green, Wakefield, WF3 1DL
2E1 LAM A Lam, Partridge Cottage, Redpale, Heathfield, TN21 9NR
2E1 LCO Michael Kinsey, Hyfrydle, Cyffylliog, Ruthin, LL15 2DW
2E1 LEC C Cattel, 21 School Hill, Chickerell, Weymouth, DT3 4BA
2E1 LED F Chorlton, 25 Ash Grove, Orrell, Wigan, WN5 8NG
2E1 LEN L Kinley, 100 Withington Lane, Aspull, Wigan, WN2 1JE
2E1 LES L Tomkins, 46 The Boulevard, Great Sutton, Ellesmere Port, CH65 7DZ
2E1 LGA A Edwards, 23 Brittany Avenue, Ashby-de-la-Zouch, LE65 2QY
2E1 LGE L Edwards, 8 St. Andrews Close, High Ham, Langport, TA10 9DD
2E1 LGJ L Gaston-Johnston, 23a Nashleigh Hill, Chesham, HP5 3JQ
2E1 LGV L Verghese, 19 Old Mansion Close, Eastbourne, BN20 9DP
2E1 LIS Elizabeth Crossley, 3 Delibes Road, Basingstoke, RG22 4LZ
2E1 LIZ E Greatorex, 22 Marlborough Way, Uttoxeter, ST14 7HL
2E1 LJL Lee Lewis, Flat 4, 41 Marine Parade, Lowestoft, NR33 0QN
2E1 LJW L Walker, 125 Devereux Road, West Bromwich, B70 6RQ
2E1 LME M Hulme, 13 Cherry Tree Court, Diss, IP22 4QW
2E1 LOZ L Dodman, 30 Cambridge Street, Rugby, CV21 3NQ
2M1 LPT C Macnab, 18 Lochport, North Uist, Western Isles, HS6 5EU
2E1 MAR M Davies, Newton Lodge, Newton, Ellesmere, SY12 0PF
2E1 MAZ Maria Adlington, 21 Newstead Road, Stoke-on-Trent, ST2 8HU
2E1 MDC C Mackay, 665a Edenfield Road, Rochdale, OL11 5XE
2E1 MEL P Cosens, 34 Waterloo Road, Salisbury, SP1 2JX
2E1 MEP M Pearce, 137 Westwood Road, Salisbury, SP2 9HN
2E1 MFC M Clews, 16 Chestnut Street, Worcester, WR1 1PA
2E1 MGB H Knight, 10 Welford Road, Barton, Alcester, B50 4NP
2E1 MGH M Hickey, 1 Preston Way, Christchurch, BH23 4QT
2E1 MIB Damian Clegg, 1 Green Croft, Hereford, HR2 7NT
2M1 MIC M Budas, 20 Oak Avenue, Bearsden, Glasgow, G61 3HD
2E1 MIN M Dawson, 140a Healey Road, Scunthorpe, DN16 1HT
2E1 MJF M Faulkner, 49 Oakfield Road, Shrewsbury, SY3 8AD
2E1 MJH Mark Hickford, 3 Ashen Road, Clare, Sudbury, CO10 8LQ
2E1 MJM M Marter, 4 Meadow Way, Seaford, BN25 4QT
2E1 MPB M Bates, Apartment 801, Imperial Point, Salford, M50 3RB
2E1 MPN M Nurse, 81 Lexden Drive, Seaford, BN25 3JF
2E1 MSC M Cook, 9 Drenewydd, Park Hall, Oswestry, SY11 4AH
2E1 MWS M Southall, 23 Ffordd Elias, Old Colwyn, Colwyn Bay, LL29 9LA
2E1 NAC Nicola Cosens, 1 Bake Farm Cottages, Salisbury Road, Salisbury, SP5 4JT
2E1 NFD N De-Thabrew, 12 Balfour Road, Dover, CT16 2NQ
2E1 NII C Sims, 7 Ainthorpe Lane, Ainthorpe, Whitby, YO21 2JN
2E1 NJC N Cook, Chinthurst, Springfield Road, Woolacombe, EX34 7BX
2E1 NPH Aidan Arnold, 2 Duck Lane, Haddenham, Ely, CB6 3UE
2E1 NRL N Loveridge, 26 Haylands, Portland, DT5 2JZ
2E1 NRQ J Dean, 9 School Hill, Chickerell, Weymouth, DT3 4BA
2E1 OBI B Atkins, 30 Rishworth Rise, Shaw, Oldham, OL2 7QA
2E1 ODG D Goodall, 19 Rossefield Avenue, Leeds, LS13 3SG
2E1 OLI O Bradley, 19 Lincoln Close, Eastbourne, BN20 7TZ
2E1 ORT M Hickford, Conifers, 3 Ashen Road, Sudbury, CO10 8LQ
2E1 OUJ J Walsh, 155 Hunter Drive, Bletchley, Milton Keynes, MK2 3NG
2E1 OZO D Day, Blakeney, Arbor Lane, Lowestoft, NR33 7BQ
2E1 OZY O Morris, 44 Leamington Road, Weymouth, DT4 0EZ
2E1 PAL Annabel Lewis, Sky Waves, 20 Annes Walk, Caterham, CR3 5EL
2E1 PAW P Woodward, 5 Lupin Way, Clacton, CO167DX
2E1 PDM Paul Marney, 137 Thistle Grove, Welwyn Garden City, AL7 4AG
2E1 PDQ A Pennington, 16 Invicta Road, Margate, CT9 3SL
2E1 PEC Peter Cosens, 1 Bake Farm Cottage, Salisbury Road, Salisbury, SP5 4JT
2E1 PGA P Graham, 556 Mather Avenue, Liverpool, L19 4UG
2E1 PGB DT Brierley, 639 Borough Road, Birkenhead, CH42 9QA
2E1 PGL P Lloyd, Pentrip, Wynnstay Yard, Wrexham, LL14 6DP
2E1 PHW P Wade, Fovant Cottage, Sherbourne Street, Bembridge, PO35 5RY
2E1 PJJ Jane Miller, Flat 1 Block 2, St. Phillips Place, Eastbourne, BN22 8LW
2E1 PMT J Green, 10 Holme Dene, Haxey, Doncaster, DN9 2JX
2E1 PPK A Boag, 53 Castlewood Road, London, N16 6DJ
2E1 PPM Christopher Martin, 4 Chaloner Place, Aylesbury, HP21 8NW
2E1 RAD J Elliott, 13 Ormonde Way, New Rossington, Doncaster, DN11 0SB
2E1 RAF Roy Walker, 35 Romany Close, Letchworth Garden City, SG6 4LA
2E1 RAO N BRENT, 53 Middlewich Street, Crewe, CW1 4DA
2E1 RBA Adrian Russell - Bishop, 227 Ardleigh Green Road, Hornchurch, RM11 2ST
2E1 RBH Carol Hodges, Marine Cottage, 1a Clements Lane, Portland, DT5 1AS
2E1 RFS R Starling, 10 School Lane, Lawford, Manningtree, CO11 2HZ
2E1 RIO R Odle, 24 Longfellow Road, Gillingham, ME7 5QG
2E1 RJS R Spevack, 87 Albany Road, Hersham, Walton-on-Thames, KT12 5QG
2E1 RMD Brian Debenham, 80 Stewart Road, Chelmsford, CM2 9BD
2E1 RMS Robert Stevenson, 97 Queen Street, Crewe, CW1 4AL
2E1 RON C Robinson, 11 Poplar Way, Leeds, LS13 4SU
2E1 RSB Raymond Mather, 76, Moreton, CH46 9PS
2E1 RSS R Hark, 5 Victoria Park, Bagillt, CH6 6JS
2E1 RWC R Cornwall, 9 Bishop Close, Dunholme, Lincoln, LN2 3US
2E1 RWN N Rowan, 27 Crieff Road, London, SW18 2EB
2E1 SAM S Martyr, Old Coach House, Westdean, Seaford, BN25 4AL
2E1 SAZ Sarah Greatorex, 54 Lilac Grove, Glapwell, Chesterfield, S44 5NG
2E1 SBF C Noon, 24 Sunset Walk, Bush Estate, Norwich, NR12 0SX
2E1 SCO Robin Jeffrey, Burnbrae, Crocketford, Dumfries, DG2 8QR
2E1 SDI G Baines, 60 Parkdale Road, Thurmaston, Leicester, LE4 8JP
2W1 SDR A Rose, 5 Hunter Street, Cadoxton, Barry, CF63 2HY
2E1 SGK S Knott, 24 John Street, Leek, ST13 8BL
2E1 SHE S Brown, 11 Wordsworth Avenue, Tamworth, B79 8BZ
2E1 SIS L Downes, 6 Greenland Crescent, Beeston, Nottingham, NG9 5LB
2M1 SJB S Budas, 20 Oak Avenue, Bearsden, Glasgow, G61 3HD
2E1 SJG S Grant, 47 Coneyford Road, Shard End, Birmingham, B34 7AY
2E1 SKA William Pitt, 237 Broadway, Dunscroft, Doncaster, DN7 4HS
2E1 SKR A Skrzypecki, The Chestnuts, Birdcage Lane, Halifax, HX3 0JQ
2E1 SKY P Staerck, 42 Plantation Hill, Worksop, S81 0RJ
2E1 SOB A Weatherall, 1 Dean Place, Stoke-on-Trent, ST1 3HS
2E1 SOX A Hughes, 4 Cobden Court, Birkenhead, CH42 3YH
2E1 SPY S Palmer, 21 Ibbett Close, Kempston, Bedford, MK43 9BT
2W1 SRB S Bowen, 41 Bro Dawel, Merthyr Tydfil, CF47 0YU
2E1 STK L Meek, 3 St. Johns Grove, Kirk Hammerton, York, YO26 8DE
2E1 STO Graham Stone, 40 Friars Road, Stoke-on-Trent, ST2 8DS
2E1 SUE S Macpherson, 18 Mountbatten Avenue, Dukinfield, SK16 5BU
2E1 SWB E Jenkins, 8 Ffordd Elias, Old Colwyn, Colwyn Bay, LL29 9LA
2I1 SWD S McAuley, Layde View, 19 Rathlin Avenue, Ballycastle, BT54 6DQ
2E1 TAB T Brierley, 6 Bridle Avenue, Wallasey, CH44 7BJ
2E1 TAG D Taggart, 2 Treewall Gardens, Bromley, BR1 5BT
2W1 TBD T Davies, 72 Eversley Road, Sketty, Swansea, SA2 9DF
2E1 TCP T Clayton, 14 Medway Walk, Wigan, WN5 9NQ
2E1 TDM T Meredith, 18 Hyde Place, Llanhilleth, Abertillery, NP13 2RT
2E1 TIM Timothy Wightman, Laithbutts Farm, Cowan Bridge, Carnforth, LA6 2JL
2E1 TKD Doreen Hensby, 28 Moorland Crescent, Whitworth, Rochdale, OL12 8SU
2E1 TMB Thomas Bolderstone, 7 Teal Close, Snettisham, King's Lynn, PE31 7RE
2E1 TNE A Millard, 6 Connaught Road, Weymouth, DT4 0SA
2E1 TOM T Lake, 77 Grafton Road, King's Lynn, PE30 3EX
2E1 TON A Steele, 15 Roman Meadow, Downton, Salisbury, SP5 3LB
2E1 TWB A brown, 7 Brookfield Road, Wooburn Green, High Wycombe, HP10 0PZ
2E1 UJE A Dalzell, 9 Pyms Lane, Crewe, CW1 3PJ
2E1 UTD Roger Smith, 14 Oakfield Cottages, Brockton, Shrewsbury, SY5 9JA
2E1 VAR Richard Preece, 26 Chineway Gardens, Ottery St. Mary, EX11 1JJ
2M1 VFO Mark Bartlett, 14d St. Andrew Street, Perth, PH2 8SA
2E1 VMR Matthew Sides, 17 Mayville Avenue, Llay, Wrexham, LL12 0PW
2M1 VXB C Andrew, 23 Shore Street, Inverallochy, Fraserburgh, AB43 8WA
2E1 WCD W Denny, 86 Lloyds Avenue, Kessingland, Lowestoft, NR33 7TR
2E1 WEB C Webb, 1 The Square, Eltisley Road, Sandy, SG19 3BT
2E1 WGB G Barber, 35 Lower Park Crescent, Bishop's Stortford, CM23 3PU
2E1 WIN C Wingfield, 35 Causey Farm Road, Hayley Green, Halesowen, B63 1EQ
2E1 WJB B Walsh, 20 Edge Fold Crescent, Worsley, Manchester, M28 7EX
2E1 WNA Ian Jones, 151 Atherton Road, Hindley, Wigan, WN2 3EE
2E1 WPW A Newton, 12 Hewett Street, Warsop Vale, Mansfield, NG20 8XN
2E1 WRC W Chorlton, 25 Ash Grove, Orrell, Wigan, WN5 8NG
2E1 WVF D Fowler, Millhouse, 8 Church Road, Stockport, SK6 5PR
2E1 XDJ Keith Senior, 20a Union Street, Hemsworth, Pontefract, WF9 4AP
2E1 XXX J Day, 98 Steynburg Street, Hull, HU9 2PF
2E1 YAP A Sheppard, 1 Waveney Walk, Crawley, RH10 6RL
2W1 YEG J Thorne, 11 Dowland Road, Penarth, CF64 3QX
2E1 YES J Glover, 31 West Drive, Lancaster, LA1 5BY
2E1 YRK Christopher Wright, 55 Booth Street, Denton, Manchester, M34 3HU
2E1 ZPR W Toomer, 10 Northfield, Tarrant Hinton, Blandford Forum, DT11 8JD
2E1 1BGN Andrew Rollitt-Smith, 9 St Helens Road, Doncaster, DN4 5EQ
2E1 1DLR David Carr, 33 Livingstone Street, Leek, ST13 5JU
2E1 1EXP Catherine Staerck, 8 Cresswell Road, Worksop, S80 1SU
2W1 FJG Adrian Heigh, Janneen, Henry Street, Wrexham, LL14 4DA

## G*0

GM0 0SXQ Marshall Hepburn, Toll House, Cose, Banchory, AB31 4RY
G0 AAA The Three A's Contest Group c/o Kenneth Pritchard, 9 Golf Close, Pyrford, Woking, GU22 8PE
G0 AAM G Willetts, 2 Underlane, Boyton, Launceston, PL15 9RR
G0 AAN L Schnurr, 42 Basin Road, Heybridge Basin, Maldon, CM9 4RQ
G0 AAT P Wheatley, 44 Primrose Crescent, Worcester, WR5 3HT
G0 AAU J Blades, 42 Ellesmere, Burnmoor, Houghton le Spring, DH4 6EA
GM0 AAX Graham Anthony, 10 Cedar Road, Kilmarnock, KA1 2HP
G0 AAY I Brooks, The Breck, 210 Breck Road, Poulton-le-Fylde, FY6 7JZ
G0 ABB Michael Honeywell, 23 Deverell Place, Waterlooville, PO7 5ED
G0 ABE P Hughes, 59 Jeffreys Road, Wrexham, LL12 7PD
G0 ABI P Green, Camellia Cottage, The Challices, Chulmleigh, EX18 7QX
G0 ABM T Johnson, 143 Queens Road, Tunbridge Wells, TN4 9JY
G0 ABN A Samuels, 45 Mermaid Close, Chatham, ME5 7PT
G0 ABP D Orgill, 32 Upland Avenue, Chesham, HP5 2EB
G0 ABT T Thomas, Tymawr Farm, Llanwern, Brecon, LD3 7UW
G0 ABV D Wood, 18 Bankhouse Road, Nelson, BB9 7RA
G0 ABW J Harding, Fen End Farm, High Street, Huntingdon, PE19 4UE
G0 ACA J Kliffen, 8 West Park, Minehead, TA24 8AW
G0 ACD Michael Amos, 41 Jocelyn Road, Richmond, TW9 2TJ
GW0 ACH John Cooper, 157 Bryn Road, Brynmenyn, Bridgend, CF32 9LU
GD0 ACK Donald Lamb, 339 Victoria Road, Ruislip, HA4 0DS
G0 ACQ Mark Godden, 20 Channel View Road, Weston, Portland, DT5 2AY
G0 ADA Nigel Law, Post Office, 2 Priory Close, Norwich, NR13 3AA
G0 ADB C Willis, 5 Gower Drive, Biddenham, Bedford, MK40 4PZ
G0 ADC Kevin Shepherd, 15 Gronant Road, Prestatyn, LL19 9DT
GI0 ADD OBO ARMAGH AND DUNGANNON DISTRICT A.R.C. c/o John Ashe, 49 Deans Walk, Richhill, Armagh, BT61 9LD
GM0 ADF D Mackinnon, 60 Mount Stuart Drive, Wemyss Bay, PA18 6DX
G0 ADG R Bristeir, 94 Burnthwaite Road, Fulham, London, SW6 5BG
G0 ADH R Razey, 2 Park Farm Cottage, 26 St. Georges Road, Wallingford, OX10 8HP
G0 ADJ D Elsworth, 34 Seal Road, Bramhall, Stockport, SK7 2JR
G0 ADK J Saueressig, 8 The Ridgeway, River, Dover, CT17 0NX
G0 ADL B Barlow, 134 Bury Road, Radcliffe, Manchester, M26 2UX
G0 ADO A Hodgson, Arla Burn Farm, Middleton in Teesdale, DL12 0QU
G0 ADP A Wither, 30 Mersey Road, Aigburth, Liverpool, L17 6AD
GW0 ADS J Jenkins, Derwen Las, Llanwnnen, Lampeter, SA48 7LG
G0 ADT Ian Houghton, 28 Beechnut Drive Blackwater, Camberley, GU17 0DJ
G0 ADU B Gibson, 55 Ledward Street, Winsford, CW7 3EN
G0 ADW Peter Radford, 42 Ashbury Drive, Weston-Super-Mare, BS22 9QS
GM0 ADX Kilmarnock and Loudoun ARC c/o Allan McKay, 2 Osprey Drive, Kilmarnock, KA1 3LQ
GW0 ADY M Jones, Stretford End, Dryslwyn, Carmarthen, SA32 8SA
G0 ADZ Dennis Price, Pippins, High Street, Newmarket, CB8 9DQ
G0 AED H Gerard, 18 Hunstanton Road, Dersingham, King's Lynn, PE31 6HQ
GM0 AEG D Greatorex, 40 Robertson Road, Lhanbryde, Elgin, IV30 8PE
G0 AEL K Horton, 16 Linden Close, West Parley, Ferndown, BH22 8RS
G0 AEN S Webb, Fieldways, The Drift, Chard, TA20 4DN
G0 AEP G Roper, 17 Slepe Crescent, Poole, BH12 4DH
G0 AEU P Tietz, 5 Chevin Road, Belper, DE56 2UW
G0 AEV S Reed, Bridlands, Middle Common, Chippenham, SN15 5NN
G0 AEW D Arlette, 12 Polmear, Par, PL24 2AT
G0 AEX E Hannaby, 34 Woodlea Lane, Meanwood, Leeds, LS6 4SX
GM0 AEY D Palmer, 36 Kilsyth Road, Haggs, Bonnybridge, FK4 1HE
G0 AEZ J Howarth, 7a Liddell Drive, Llandudno, LL30 1UH
G0 AFH I Burns, Little Delmar Farm, Leywood Road, Meopham, DA13 0UD
G0 AFJ A Brown, 33 Marion Road, Haydock, St. Helens, WA11 0PY
G0 AFN Peter Howard, 1 Avon Close, Bognor Regis, PO22 6BX
G0 AFP D Rayner, 69 Lovelace Drive, Woking, GU22 8QZ
G0 AFQ J Cook, 71 Richmond Avenue, Burscough, Ormskirk, L40 7RB
G0 AFR Clive Mears, 11 Aberford Close, Reading North, Reading, RG30 2NX
G0 AFT C Bovey, 12 The Mead, Beaconsfield, HP9 1AW
G0 AFU Michael Bousfield, Station House, Kielder, Hexham, NE48 1EG
G0 AFY R Norman, 98 Foxwell Drive, Headington, Oxford, OX3 9QF
G0 AFZ C Wilson, 9-10 Daventry Street, Southam, CV47 1PH
G0 AGB R Ellis, 22 Fern Road, Storrington, Pulborough, RH20 4LW
G0 AGC Allen Core, 1 Partridge Ride, Loggerheads, Market Drayton, TF9 2QX
G0 AGD P Shortland, 69 South Parade, Worksop, S81 0BS
G0 AGJ R Hughes, 90 Oxford Road, Old Marston, Oxford, OX3 0RD
G0 AGL Emyr Williams, Caerbergam Cottage, Llanbedr, LL45 2HT
G0 AGN G Speirs, 43 Sheuchan View, Stranraer, DG9 7TA
G0 AGO Robert White, 137 Fennells, Harlow, CM19 4RR
G0 AGR AYLESBURY RAYNET GRP c/o R Needs, 13 Greenway, Great Horwood, Milton Keynes, MK17 0QR
G0 AGU B Humphries, 22 Leander Close, Burntwood, WS7 1PW
G0 AGV A Rollings, 24 Millburn Court, Sheuchan Street, Stranraer, DG9 0DX
G0 AGZ John Squire, Dyffryn, Llanymynech, SY22 6EW
G0 AHA A Pannell, 3 Nethercourt Gardens, Ramsgate, CT11 0RY
G0 AHB A Bennett, 2 Portland Place, Hertford Heath, Hertford, SG13 7RR
G0 AHC B Hewitt, 32 Pinehurst Drive, Kings Norton, Birmingham, B38 8TH
G0 AHD C Richardson, 122 Elmton Road, Creswell, Worksop, S80 4DE
G0 AHE P Moss, Vivenda, Wick Road, Langham, CO4 5PE
G0 AHI T Deacon, 9 Mulberry Close, Woodley, Reading, RG5 3LR
G0 AHJ J Johnson, 14 Park House Lane, Prestbury, Macclesfield, SK10 4HZ
G0 AHK D Layne, 5 Howe Close, Christchurch, BH23 3JA
G0 AHL R Sears, 19 Shepherds Grove Park, Stanton, Bury St. Edmunds, IP31 2AY
G0 AHM P Gregg, 5 Rosevear Road, Bugle, St. Austell, PL26 8PH
G0 AHO C Grant, 20 Muriel Kenny Court, Hethersett, Norwich, NR9 3EZ
G0 AHR P Cuthbert, 115 Tintern Avenue, Whitefield, Manchester, M45 8WY
G0 AHU John Tolson, 9 Buckingham Drive, Churchdown, Gloucester, GL3 1NG
G0 AHV R Gilling, 24 Bellerby Road, Skellow, Doncaster, DN6 8PD
G0 AIG M Sutherland, 28 Sycamore Way, Littlethorpe, Leicester, LE19 2HT
G0 AIH Richard Baker, 38 The Front, Middleton One Row, Darlington, DL2 1AU
G0 AII D Fairbank, 2 Sandpit Road, Welwyn Garden City, AL7 3TN
GI0 AIJ I Greenwood, Deers Leap, 24 Tullyrusk Road, Crumlin, BT29 4JQ
G0 AIL D Penny, 30 Belvedere Road, Yeovil, BA21 5JB
G0 AIM S Robinson, 164 Leigh Road, Westhoughton, Bolton, BL5 2LE
G0 AIN Steve Withnell, 85 Headroomgate Road, Lytham St. Annes, FY8 3BG
G0 AIO A Owens, 69 Locomotion Way, North Shields, NE29 6XE
GI0 AIQ F Holland, 413 Ballyoran Park, Portadown, Craigavon, BT62 1JX
GM0 AIR D Parker, Devon Cottage, Main Street, Cupar, KY15 7QX
G0 AIS J Borg, 94 Coldershaw Road, London, W13 9DT
G0 AIX David Westlake, Chyvellin, Newmill, Penzance, TR20 8XW
GW0 AIY Robert Gibbons, Cefnysgwyn, Capel Isaac, Llandeilo, SA19 7UA
G0 AIZ W Chesterton, Homelea Farm, Fosse Way, Coventry, CV7 9LR
G0 AJA A Astley, 16 Cedar Avenue, Ellesmere, SY12 9PA
G0 AJB A Botherway, 4 Brodrick Drive, Ilkley, LS29 9SN
G0 AJF S Harrison, 25 High Spring Road, Keighley, BD21 4TF
G0 AJH J Hornsby, 15 Coronation Drive, Hornchurch, RM12 5BL
GW0 AJI S Davies, 20 Ashcroft Crescent, Cardiff, CF5 3RN
G0 AJJ Linda Leavold, 8 Wilkinson Way, North Walsham, NR28 9BB

| | | |
|---|---|---|
| G0 | AJI | R Bromsgrove, 34 Boundary Drive, Hunts Cross, Liverpool, L25 0QD |
| G0 | AJT | D Rollings, 2 Challoch Crescent, Leswalt, Stranraer, DG9 0LN |
| GW0 | AJU | A Underwood, Rock Hill, Llanarthney, Carmarthen, SA32 8LJ |
| G0 | AJW | J Wolcby, 47 Links Road, Knott End-on-Sea, Poulton-le-Fylde, FY6 0DF |
| G0 | AJX | M Coles, 13 Dawnay Road, Dilton, Hull, HU11 4HR |
| G0 | AJZ | Bernard Park, 147 Castle Road, Ings Farm, Redcar, TS10 2LT |
| G0 | AKC | A Chamberlain, 455 Norwich Road, Ipswich, IP1 5DR |
| G0 | AKF | K Farrance, Clarewood, Tabley Road, Knutsford, WA16 0NE |
| G0 | AKH | Kevin Hale, 58 St. Stephens Road, Saltash, PL12 4BJ |
| G0 | AKI | Miquel Eavis, 61 Hitchmead Road, Biggleswade, SG18 0NL |
| [illegible] | [illegible] | D [illegible], 51 Loachkin Avenue, Inverness, IV3 8L[illegible] |
| G0 | AKK | J Oliveiro, 1 Laurel Drive, Bognor Regis, PO21 3ND |
| [illegible] | [illegible] | D King, 4 Ringwood Drive, Horning, Norwich, NR12 8LZ |
| [illegible] | [illegible] | R [illegible], 9 Belfont Walk, London, N7 0SN |
| G0 | AKQ | R Cleveland, 2 Morse Close, Brundall, Norwich, NR13 5LG |
| G0 | AKR | A Poynter, Hill Top Farm, Warren Road, Chatham, ML5 9[illegible] |
| G0 | AKS | David Newell, 7 Edward Road West, Clevedon, BS21 7DY |
| G0 | AKU | R Reanney, 8 Stapleford Court, Ellesmere Port, CH66 1RW |
| GW0 | AKV | James Anderson, 173 Gate Road, Penygroes, Llanelli, SA14 7RW |
| G0 | ALA | Denis Whitehouse, 40 Fernleigh Crescent, Up Hatherley, Cheltenham, GL51 3QL |
| G0 | ALB | C Matcham, 28 Buckingham Drive, Luton, LU2 9RA |
| G0 | ALC | J Richards, 77 Poxon Road, Walsall Wood, Walsall, WS9 9JR |
| G0 | ALE | TATSFIELD ARTS c/o Peter Madagan, 40 Lagham Park, South Godstone, Godstone, RH9 8ER |
| G0 | ALI | Alison Soars, 8 Nomis Park, Congresbury, Bristol, BS49 5HB |
| G0 | ALJ | R Prior, 274 West View Lodge , Canterbury road, Herne, ct6 7hb |
| G0 | ALQ | J Amer, 1 Kingfisher Way, Watton, Thetford, IP25 6SR |
| G0 | ALR | Adrian Wagenaar, La Villette, Folles, France, 87250 |
| GM0 | ALS | F Roe, 74 Willow Grove, Livingston, EH54 5NA |
| G0 | ALW | G Chalmers, 38 Grove Hill, Kelso, TD5 7AS |
| G0 | ALX | Margaret Chalmers, 38 Grovehill, Kelso, TD5 7AS |
| GD0 | AMD | Andrew Dorman, 1 Sprucewood Rise, Foxdale, Douglas, Isle of Man, IM4 3JP |
| G0 | AMO | M Adams, 9 Brancaster Avenue, Charlton, Andover, SP10 4EN |
| G0 | AMP | Robert Senft, 11 Maltravers Drive, Littlehampton, BN17 5EY |
| G0 | AMS | A Sinclair, 4 Blackbrook Park Avenue, Fareham, PO15 5JJ |
| G0 | AMU | Lee Parrott, 3 Fox Gardens, Lymm, WA13 9EY |
| G0 | AMW | Carl Dunn, 90 Mushroom Green, Dudley, DY2 0EE |
| G0 | AMX | P Appleyard, 28 Romany Close, Letchworth Garden City, SG6 4JZ |
| G0 | AMY | J Sheppeck, 25 Kingsleigh Road, Heaton Mersey, Stockport, SK4 3QF |
| G0 | AMZ | Kelvin Fay, 2 Chapel Row, Hartley Wintney, Hook, RG27 8NJ |
| GW0 | ANA | Glynwyn Jones, 2 Castle Precinct, Llandough, Cowbridge, CF71 7LX |
| G0 | ANE | W Burrows, 35 Bamford Avenue, Barnsley, S71 3SJ |
| GM0 | ANG | J Fish, Senekal, Alma Road, Fort William, PH33 6HB |
| G0 | ANH | J Wright, 5 Abbeville Avenue, Whitby, YO21 1JD |
| G0 | ANK | Ian Smallwood, 27 Cormorant Way, Herne Bay, CT6 6HG |
| G0 | ANL | P Ellis, 8 Shropshire Close, Woolston, Warrington, WA1 4DY |
| G0 | ANM | James Mooney, 2 Madford Lane, Launceston, PL15 9EB |
| G0 | ANN | M Viney, 12 Palgrave House, Sherwell Road, Norwich, NR6 6PU |
| G0 | ANO | D Lawton, Grenehurst, Pinewood Road, High Wycombe, HP12 4DD |
| G0 | ANP | D Guy, 7 Park Avenue, Castle Cary, BA7 7HE |
| G0 | ANS | John Vye, 16 Breach Close, Steyning, BN44 3RZ |
| G0 | ANT | Eden Valley RS c/o David Shaw, Cotehouse, Bleatarn, Appleby-in-Westmorland, CA16 6PX |
| G0 | ANV | Daryl Burchell, 31 Thornton Road, Girton, Cambridge, CB3 0NP |
| G0 | ANW | J Hockley, 6 King Edward Road, Birchington, CT7 0EL |
| G0 | ANX | W Tully, 8 The Croft, West Hanney, Wantage, OX12 0LD |
| G0 | AOA | T Marten, 10 Chieveley Drive, Tunbridge Wells, TN2 5HG |
| G0 | AOB | M Beal, Heath Farm Bradworthy, Holsworthy, EX22 7SL |
| G0 | AOC | S Colledge, 32 Wakeman Street, Worcester, WR3 8BQ |
| G0 | AOD | D Heathcote, 8 Ferrers Avenue, Tutbury, Burton-on-Trent, DE13 9JR |
| G0 | AOE | N Evans, 16 Humbledon Park, Sunderland, SR3 4AA |
| GM0 | AOF | R Wallace, 1 Holding West Kincardine, Crieff, PH7 3RP |
| G0 | AOH | John Abbruscato, Whittingham Leeds Hill House, Leeds Hill, Woodbridge, IP12 1LW |
| G0 | AOJ | F Fenwick, 6 School Lane, Bonby, Brigg, DN20 0PP |
| G0 | AOK | D Gall, 2 Norham Close, Wideopen, Newcastle upon Tyne, NE13 7HS |
| G0 | AOL | G Parsons, 248 Filey Road, Scarborough, YO11 3AQ |
| G0 | AOM | R Day, 3 Railway Cottages, Newby Bridge, Ulverston, LA12 8AW |
| G0 | AOO | J Butterwick, 45 Fox Howe, Coulby Newham, Middlesbrough, TS8 0RU |
| G0 | AOP | P Warriner, 36 Eskdaleside, Sleights, Whitby, YO22 5EP |
| G0 | AOQ | Samuel Boyd, 7 Kings Close, Pocklington, York, YO42 2GX |
| G0 | AOS | M Reynolds, Willhay Cottage, Willhay Lane, Axminster, EX13 5RW |
| G0 | AOU | David Armitage, 12 Loughborough Close, Sale, M33 5UF |
| G0 | AOW | Susan Sands, 33 High Street, Kinver, Stourbridge, DY7 6HF |
| G0 | AOX | A Sands, Gabledown, Bridgnorth Road, Stourbridge, DY7 6RW |
| G0 | AOY | T Rudd, Grasmere, Burgh, Woodbridge, IP13 6SU |
| G0 | AOZ | Roger Powell, Town Pond Cottage, Town Pond Lane, Abingdon, OX13 5HS |
| G0 | APB | P Duckley, 7 Callams Close, Rainham, Gillingham, ME8 9ES |
| G0 | API | J Fell, 14 Rectory Avenue, Corfe Mullen, Wimborne, BH21 3EZ |
| GM0 | APN | John Loveday, Crombiebrae, Inverurie, AB51 0JT |
| G0 | APP | Adrian Parmment, 5 New Captains Road, West Mersea, Colchester, CO5 8QP |
| G0 | APV | V Covell-London, 15 St. Nicholas Way, Potter Heigham, Great Yarmouth, NR29 5LG |
| G0 | APY | G Flood, 4 Campbell Crescent, Great Sankey, Warrington, WA5 3DA |
| G0 | AQA | Stuart Baynes, 1 Reeves Paddock, Townsend, Wells, BA5 3FG |
| G0 | AQB | J Ireland, The Grange, Grange Lane, Worcester, WR2 6RW |
| GI0 | AQD | David Burn, 17 Old Shore Road, Newtownards, BT23 8NE |
| G0 | AQF | David Dolphin, 16 Golden Cross Lane, Catshill, Bromsgrove, B61 0LQ |
| G0 | AQH | G Griffin, 23 St. Giles Close, Shoreham-by-Sea, BN43 6GR |
| G0 | AQI | E Greenhalgh, 19 Rooks Nest Lane, Therfield, Royston, SG8 9QX |
| G0 | AQR | John Richards, 21 Cae Gwyn, Caernarfon, LL55 1LL |
| G0 | AQS | P Goldthorpe, 29 Broadoak Road, Ashton-under-Lyne, OL6 8QN |
| G0 | AQT | V Taylor, 5 St. Matthews Drive, St. Leonards-on-Sea, TN38 0TR |
| G0 | AQU | R Mason, 28 Vandyke, Great Hollands, Bracknell, RG12 8UP |
| G0 | AQZ | David McDonald, 24 Wenning Street, Nelson, BB9 0LE |
| GW0 | ARA | Aberystwyth & District RS c/o Robin Clews, Maesygaer, Ciliau Aeron, Lampeter, SA48 7SG |
| GM0 | ARD | J Hoey, 152 Muirhouse Avenue, Motherwell, ML1 2LB |
| G0 | ARF | R Canning, Green Lane Cottage, Eardisland, Leominster, HR6 9BN |
| G0 | ARG | ARIEL RADIO GRP c/o Tom Elliner, 53 Hillside, Banstead, SM7 1HQ |
| GM0 | ARH | A Haxton, Lanimar, Dickson Avenue, Montrose, DD10 9[illegible] |
| G0 | ARK | K Hudspeth, 67 Bloomfield Road, Blackwood, NP12 1LX |
| G0 | ARP | A Price, Brook House, Drury Lane, Shrewsbury, SY4 1DT |
| G0 | ARQ | M Burchmore, 49 School Lane, Horton Kirby, Dartford, DA4 9DQ |
| GM0 | ART | A Gray, 191 Greengairs Road, Greengairs, Airdrie, ML6 7SZ |
| G0 | ARU | John Lumb, 2 Briarwood Avenue, Bury St. Edmunds, IP33 3QF |
| G0 | ARW | J R [illegible], 9 [illegible] Close, Liverpool, L16 [illegible] |
| GM0 | ARY | N Hamilton, Glenwood Cottage, Enzie, Buckie, AB56 5BW |
| G0 | [illegible] | M E[illegible], 9 St. Hild Close, Darlington, DL3 8LD |
| [illegible] | [illegible] | [illegible], The Old Reading Room, Ashwater, Beaworthy, EX21 5EF |
| G0 | ASH | J Coupe, 65 Irongate, Bamber Bridge, Preston, PR5 6[illegible] |
| G0 | ASI | K Rodden, 12 Bonis Crescent, Stockport, SK2 7HH |
| G0 | ASK | S Gray, 29 Verity Walk, Stourbridge, DY8 4XS |
| G0 | ASL | J Gray, 29 Verity Walk, Stourbridge, DY8 4XS |
| G0 | ASM | N Marston, 14 Greystoke Avenue, Sunderland, SR2 0DX |
| G0 | ASN | S Hendry, 5 Harvey Road, Great Totham, Maldon, CM9 8QA |
| G0 | ASP | E Mason, 28 Pendil Close, Wellington, Telford, TF1 2PQ |
| G0 | ASQ | L Williams, 24 Alston Drive, Bare, Morecambe, LA4 6QR |
| G0 | ASX | N Ward, 104 Kingsley Road, Bishops Tachbrook, Leamington Spa, CV33 9RZ |
| G0 | ASZ | E Gamble, 87 Silverdale Drive, Waterlooville, PO7 6DP |
| GM0 | ATA | R Mulholland, 1 Larch Grove, Milton of Campsie, Glasgow, G66 8HG |
| G0 | ATB | Veronica Herbert, 98 Blithdale Road, Abbey Wood, London, SE2 9HL |
| G0 | ATC | 59 (Huddersfield) Squadron Air Cadets c/o Ian Halliwell, 61 Cliffe Road, Shepley, Huddersfield, HD8 8AG |
| G0 | ATD | D Welch, 51 Verbena Way, Worle, Weston-Super-Mare, BS22 6RL |
| G0 | ATG | T Goodyer, 11 Upper Bere Wood, Waterlooville, PO7 7HX |
| G0 | ATK | George Atkins, 97 South Street, Tillingham, Southminster, CM0 7TH |
| GM0 | ATL | P Smith, 29 Rowan Drive, Bearsden, Glasgow, G61 3HQ |
| G0 | ATO | N Gregorian, 449 E Providencia Ave, Burbank, California, United States, 91501 2916 |
| G0 | ATP | C Abela, 14 Warren Road, Barkingside, Ilford, IG6 1BJ |
| GM0 | ATQ | L Morgan, 12 Bayview Road, Gourock, PA19 1XE |
| G0 | ATS | Elaine Green, Chylean, Tintagel, PL34 0HH |
| G0 | ATW | J Ferrier, 30 Grimsby Road, Laceby, Grimsby, DN37 7DB |
| G0 | ATZ | Clive Best, 34 Julius Hill, Warfield, Bracknell, RG42 3UN |
| G0 | AUB | Fred Groves, 14 Edenfield Road, Mobberley, Knutsford, WA16 7HE |
| G0 | AUE | Anthony Knowles, 17 The Deneway, Sompting, Lancing, BN15 9SU |
| G0 | AUF | R Griffiths, 26 Hamilton Road, Morecambe, LA4 6QG |
| G0 | AUG | Thomas Hopkinson, Whitbarrow Hall, Caravan Park, Penrith, CA11 0XB |
| G0 | AUH | M Hopkinson, Whitbarrow Hall, Caravan Park, Penrith, CA11 0XB |
| G0 | AUI | Carroll Stiller, 6 Barn Cottage Lane, Haywards Heath, RH16 3QW |
| G0 | AUJ | Malcolm O'Dell, 19 Redwing Rise, Royston, SG8 7XU |
| G0 | AUK | AMSAT-UK c/o J Heck, Badgers, Letton Close, Blandford Forum, DT11 7SS |
| GM0 | AUL | Robert McKenzie, 26 Gladstone Place, Woodside, Aberdeen, AB24 2RP |
| G0 | AUN | D Taylor, 24 Kingshill Drive, Hoo, Rochester, ME3 9JP |
| G0 | AUR | A Hogg, 1 Champions Way, South Woodham Ferrers, Chelmsford, CM3 5NJ |
| G0 | AUT | A Utting, 9 Sydney Road, Spixworth, Norwich, NR10 3PG |
| G0 | AUV | Anthony Johnson, 125 Charles Street, Sileby, Loughborough, LE12 7SH |
| G0 | AUW | Richard Philpott, 4 Dukeswood, Chestfield, Whitstable, CT5 3PJ |
| G0 | AUX | Kieran Daly, Granarogue, Co. Monaghan, Ireland |
| G0 | AVB | Kenneth Graham, 98 Dalswinton Avenue, Dumfries, DG2 9NR |
| G0 | AVD | Paul Richards, 11 Seabourne Road, Holyhead, LL65 1AL |
| G0 | AVE | David Brannon, 10 Rochester Crescent, Crewe, CW1 5YF |
| G0 | AVH | Thomas Mcguire, 33 Sandy Lane, Hindley, Wigan, WN2 4EJ |
| G0 | AVJ | Malcolm Searley, 49 Hollymount Close, Exmouth, EX8 5PQ |
| G0 | AVP | Paul Raxworthy, 32 St. Marys Avenue, Alverstoke, Gosport, PO12 2HX |
| G0 | AVU | G Logan, Fenton Hill Farm, Wooler, NE71 6JL |
| GW0 | AVW | T Steele, 53 Second Avenue, Trecenydd, Caerphilly, CF83 2SP |
| G0 | AWA | Maurice Hall, 5 Sandringham Way, Ponteland, Newcastle upon Tyne, NE20 9AE |
| G0 | AWG | W McNamara, 22 Cissbury Avenue, Peacehaven, BN10 8TJ |
| G0 | AWH | Nicholas Bergstrom-Allen, Garden Cottage, Thicket Priory, York, YO19 6DE |
| GI0 | AWK | D Payne, 10 Grovemount Court, Altnagelvin, Londonderry, BT47 5JP |
| G0 | AWM | C Warr, 29 Barton Road, Lancaster, LA1 4ER |
| G0 | AWR | D Proud, Willow Cottage, Tresevern, Truro, TR3 7AT |
| GW0 | AWT | Susan Richardson, Glasfryn, Porthyrhyd, Llanwrda, SA19 8DF |
| G0 | AWW | D Elliott, 188 Seaview Road, Wallasey, CH45 5HB |
| G0 | AWY | R Titmuss, 70 Mallards Rise, Church Langley, Harlow, CM17 9PL |
| G0 | AWZ | D Richardson, Becksyde, Cow Moor Bridge, York, YO3 9UA |
| G0 | AXA | A Hargrave, 36 Fromondes Road, Cheam, Sutton, SM3 8QR |
| G0 | AXB | J Moore, 53 Thatchers Court, Westlands, Droitwich, WR9 0EG |
| G0 | AXC | Dave Lee, 188 Manstone Avenue, Sidmouth, EX10 9TJ |
| G0 | AXD | S Waters, 27 Mill Lane, Shepherdswell, Dover, CT15 7LJ |
| G0 | AXE | M Davenport, 119 Gravel Lane, Wilmslow, SK9 6EG |
| G0 | AXI | N Davies, 15 Fyfield Road, Oxford, OX2 6QE |
| G0 | AXJ | Roy McCluskey, 29 Hotspur Avenue, Bedlington, NE22 5TD |
| GM0 | AXM | John Thomson, 16 Ravelstone Terrace, Edinburgh, EH4 3TP |
| G0 | AXO | Rayburn Bainbridge, 9 Hamilton Crescent, North Shields, NE29 8DW |
| G0 | AXQ | Paul Austin, 66 Homewood Avenue, Sittingbourne, ME10 1XJ |
| G0 | AXR | R Ashton, 9 St. Chads Close, Little Haywood, Stafford, ST18 0QW |
| G0 | AXS | S Birch, 29 Manners Road, Southsea, PO4 0BA |
| G0 | AXU | G Woods, Leyland House, Farleton, Lancaster, LA2 9LF |
| GM0 | AXX | F Gilhooly, 26 Arnott Gardens, Edinburgh, EH14 2LB |
| GM0 | AXY | Einar Dons, 37 Ashley Drive, Edinburgh, EH11 1RP |
| G0 | AXZ | W Johnson, 29 Wentworth Park, Allendale, Hexham, NE47 9DR |
| G0 | AYA | M Chown, 15 Hambleden Walk, Maidenhead, SL6 7UH |
| GI0 | AYB | J Throne, Fascadail, 12 Mason Road, Londonderry, BT47 2RY |
| G0 | AYC | C Porter, 10 Cotefield Drive, Leighton Buzzard, LU7 3DS |
| G0 | AYD | D Dixon, 3 Towns End, Wylye, Warminster, BA12 0RN |
| G0 | AYF | W Collier, 8 Douglas Street, Hindley, Wigan, WN2 3HP |
| GI0 | AYG | M Evans, 12 Tullymore Park, Ballymena, BT42 2AU |
| G0 | AYI | B Spencer, 161a West Lane, Hayling Island, PO11 0JW |
| G0 | AYM | Keith Luxton, 2 Trinity Court, Westward Ho, Bideford, EX39 1LT |
| G0 | AYP | D Groves, 185 Rhyl Coast Road, Rhyl, LL18 3US |
| G0 | AYQ | Robert Smith, 4 Glan Ysgethin, Talybont, LL43 2BB |
| GM0 | AYR | AYR AMATEUR RADIO GROUP c/o Geoff Cheesworth, Auchinway, Skares, Cumnock, KA18 2RE |
| GM0 | AYT | A McDougall, Ceol Na Mara, 3 Bowfield Road, West Kilbride, KA[illegible] |
| G0 | AYX | P Towell, 25 Cedar Close, Grafham, Huntingdon, PE28 0DZ |
| G0 | AYY | A Perry, Chala, Woodhouse Lane Hill, Lyme Regis, DT7 3SX |
| GI0 | AZA | E Harper, 404 Foreglen Road, Dungiven, Londonderry, BT47 4PN |
| GI0 | AZB | Henry Evans, 404 Foreglen Road, Dungiven, BT47 4PN |
| GM0 | AZC | J Sherry, 26 Grahamshill Terrace, Fankerton, Denny, FK6 5HX |
| G0 | AZD | [illegible] Road, Beeches Estate, Cirencester, GL7 1BX |
| G0 | AZE | L Owen, 68 Clevedon Road, Tickenham, Clevedon, BS21 6RD |
| G0 | AZG | Timothy Wh[illegible], [illegible], Clitheroe Road, Clitheroe, BB7 3DA |
| G0 | AZH | [illegible] Street, Clitheroe, BB7 1ND |
| G0 | AZM | M Walton, 6 Home Farm Court, Home Farm Close, Leicester, LE4 0SU |
| G0 | AZP | S Tricker, 1 Drewitt Court, 75 Godstow Road, Oxford, OX2 [illegible] |
| G0 | AZQ | Anthony Pearce, 21 Cherry Way, Nafferton, Driffield, YO25 4PA |
| G0 | AZR | Jeremy Norman, The End Peg, The Street, Norwich, NR11 7A[illegible] |
| GM0 | AZU | J Aiken, 48 Kirkwall Avenue, Blantyre, Glasgow, G72 9NX |
| GM0 | AZV | Christopher Norton, 3 Nether Balfour Cottages, Durris, Banchory, AB31 6BL |
| GW0 | AZW | R Dooley, 98 Gelli Aur, Treboeth, Swansea, SA5 9DG |
| G0 | AZX | R Pritchard, 10 Dolphin Crescent, Paignton, TQ3 1AE |
| G0 | BAA | NORTH CHESHIRE RC c/o G Gourley, 6a Longsight Lane, Cheadle Hulme, Cheadle, SK8 6PW |
| G0 | BAF | N Quinn, 15 Newham Lane, Steyning, BN44 3LR |
| G0 | BAG | R Cox, 2 Yardlea Close, Rowland's Castle, PO9 6DQ |
| GW0 | BAH | P Bateman, 48 Ogmore Drive, Nottage, Porthcawl, CF36 3HR |
| G0 | BAI | P Goodger, 125 Mill Hill Wood Way, Ibstock, LE67 6QD |
| G0 | BAJ | R Edinborough, Glendevon, 3 Manor Road, Paignton, TQ3 2HT |
| G0 | BAK | W Schofield, 36 Long Row, Horsforth, Leeds, LS18 5AA |
| G0 | BAM | E Smith, 166 Tudor Way, Dines Green, Worcester, WR2 5QY |
| G0 | BAN | James Emerson, 26 Cardwell Street, Roker, Sunderland, SR6 0JP |
| G0 | BAO | M Heath, Brambles, 1 Bestwall Road, Wareham, BH20 4HY |
| G0 | BAP | R Harman, 42 Newlyn Close, Bransholme, Hull, HU7 4PQ |
| G0 | BAQ | I Irving, Fourwinds, Woodburn Drive, Leyburn, DL8 5HU |
| G0 | BAR | Brickfields ARS c/o Douglas Fenna, 84 High Park Road, Ryde, PO33 1BX |
| G0 | BAU | S Craggs, 79 Silverdale Road, Cramlington, NE23 3LW |
| G0 | BAW | L Haynes, 9 Heather Gardens, Belton, Great Yarmouth, NR31 9PP |
| G0 | BAX | Alun Harrison, 10 Knoll Road, Sidcup, DA14 4QU |
| G0 | BAY | Stanley Parker, 85 Highfield Road, Glossop, SK13 8NZ |
| G0 | BBB | U Grunewald, Nuptown Orchard, Nuptown, Bracknell, RG42 6HU |
| GW0 | BBC | David Thomas, 88 Cefn Graig, Rhiwbina, Cardiff, CF14 6JZ |
| G0 | BBE | Lance Conlon, 4 Hill Crest Drive, Slack Head, Milnthorpe, LA7 7BB |
| G0 | BBJ | D Fry, 9 Brook Gardens, Emsworth, PO10 7JY |
| G0 | BBK | Philip Marshall, 35 Rosewood Close, Burnham-on-Sea, TA8 1HG |
| G0 | BBN | Kenneth Hendry, 23 Briscoe Way, Lakenheath, Brandon, IP27 9SA |
| G0 | BBO | J Stanbury, 6 Waterside Apartments, Weech Road, Dawlish, EX7 9FA |
| G0 | BBT | E Snape, 1 Stephen Crescent, Humberston, Grimsby, DN36 4DS |
| G0 | BBV | H Watts, 44 Laurel Road, Norwich, NR7 9LL |
| G0 | BCF | R Pickering, 14 Dalestorth Gardens, Skegby, Sutton-in-Ashfield, NG17 3FT |
| G0 | BCH | Peter Guppy, 37 Park Road, Aldershot, GU11 3PX |
| GD0 | BCJ | P Mcgrath, 69 Clagh Vane, Ballasalla, Isle of Man, IM9 2HF |
| GW0 | BCL | Raymond Johnson, 55 Maes yr Haf, Llanelli, SA15 3NF |
| GD0 | BCN | Hadyn Glaister, 42 Barrule Drive, Onchan, Douglas, Isle of Man, IM3 4NR |
| G0 | BCO | Alec Duffield, 32 Mount Close, Honiton, EX14 1QZ |
| G0 | BCP | Malcolm Jones, 44 Ainsdale Drive, Priorslee, Telford, TF2 9QJ |
| G0 | BCR | J Watts, 39 Turnberry Drive, Abergele, LL22 7UD |
| G0 | BCS | P Rose, 53 South St., Pennington, Lymington, SO41 8DY |
| G0 | BCT | R Sayer, Vignouse, Paimpont, France, 35380 |
| G0 | BCU | D Charnock, 44 Bramshill Close, Birchwood, Warrington, WA3 6TZ |
| G0 | BCW | David Grevett, 45 East Bridge Road, South Woodham Ferrers, Chelmsford, CM3 5SB |
| G0 | BCX | Graham Eden, 83 Windle Hall Drive, St. Helens, WA10 6QG |
| GM0 | BCY | Alexander Carslaw, 51 Stonefield Drive, Paisley, PA2 7QY |
| G0 | BCZ | R Johnson, 89 Arlington Gardens, Romford, RM3 0EB |
| G0 | BDB | P Stephens, 259 Beaumont Road, Plymouth, PL4 9EL |
| G0 | BDE | B Spencer, 2 Windmill Bank, Wigston, LE18 3SX |
| GU0 | BDI | D Prosser, Rustlings, Les Friquets, St. Andrew, Guernsey, GY6 8SJ |
| G0 | BDJ | K Dower, 35 Horton Way, Woolavington, Bridgwater, TA7 8JP |
| G0 | BDK | B Steele, 82 Margetts Road, Kempston, Bedford, MK42 8DT |
| G0 | BDM | D Briggs, 72 Showell Grove, Droitwich, WR9 8UD |
| G0 | BDN | A Slater, 1 Linglongs Avenue, Whaley Bridge, High Peak, SK23 7DT |
| G0 | BDP | R Wills, 8 Owlswood, [illegible], Salisbury, SP2 8DN |
| G0 | BDR | M Gee, 100 Plantation Hill, Worksop, S81 0QN |
| G0 | BDS | R Scroggs, 10 Lyon Close, Chelmsford, CM2 8NY |
| G0 | BDW | R Kelsall, 1+2 Tyn Rhos, Llanddona, Beaumaris, LL58 8YG |
| GI0 | BDZ | D Pavis, 269 Lower Braniel Road, Belfast, BT5 7NR |
| GI0 | BEB | T Magee, 10 Abernethy Park, Newtownabbey, BT36 6QQ |
| G0 | BEC | S Christmas, 4 The Close, Little Weighton, Cottingham, HU20 3XA |
| G0 | BEE | S Bloom, 8 South Riding, Bricket Wood, St. Albans, AL2 3ND |
| G0 | BEJ | Peter Mallett, Woodrow, Chalk Lane, Spalding, PE12 9YF |
| GM0 | BEL | G Lindsay, 6 Netherhouse Avenue, Lenzie, Glasgow, G66 5NG |
| G0 | BEN | D Pykett, 13 West Bank, Saxilby, Lincoln, LN1 2LU |
| G0 | BEP | D Boalch, 90 Belle Vue Road, Wivenhoe, Colchester, CO7 9EH |
| G0 | BES | S Illsley, 37 Pemerton Road, Weere, Winchester, SO22 6EU |
| G0 | BET | R Gough, 74 Merrick Road, Wolverhampton, WV11 3NU |
| G0 | BEU | M Williams, White Lodge, 21 Brook Lane, Felixstowe, IP11 7EG |
| G0 | BEV | M Hill, Windrush, Jesmond Gardens, Newcastle upon Tyne, NE2 2JN |
| G0 | BEX | A Penfold, 115 Applegarth Park, Seasalter Lane, Whitstable, CT5 4BZ |
| GI0 | BEY | Norman Turkington, 16 Glenshesk Park, Bangor, BT20 4US |
| G0 | BEZ | Victoria Stamps, Flat 85, Berglen Court, 7 Branch Road, London, E14 7JX |
| GI0 | BFA | R Mckersie, 4 Burnside Park, Belfast, BT8 6HU |

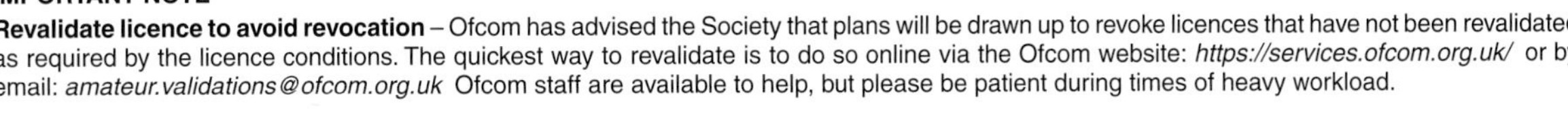

**IMPORTANT NOTE**

**Revalidate licence to avoid revocation** – Ofcom has advised the Society that plans will be drawn up to revoke licences that have not been revalidated as required by the licence conditions. The quickest way to revalidate is to do so online via the Ofcom website: *https://services.ofcom.org.uk/* or by email: *amateur.validations@ofcom.org.uk* Ofcom staff are available to help, but please be patient during times of heavy workload.

G0 BFC Clive Simkin, Flat 33, Weymouth House Balfour, Tamworth, B79 7BE
GI0 BFD Alwyn Magee, 70 Hillview Park, Enniskillen, BT74 6EU
G0 BFJ John Stocks, 96 North Street, Lockwood, Huddersfield, HD1 3SL
G0 BFK K James, Yew Tree Cottage, Whiston, Stafford, ST19 5QH
G0 BFM A Anderson, 16 Whitwell Road, Norwich, NR1 4HB
GD0 BFN J Kneale, 51 Maple Avenue, Onchan, Douglas, Isle of Man, IM3 3GA
GI0 BFO Robert Dixon, 16 Glenview Avenue, Belfast, BT5 7LZ
GM0 BFW A Mcclelland, 4 Walkerston Avenue, Largs, KA30 8ER
G0 BFZ A Nock, 57 Mushroom Green, Dudley, DY2 0EE
G0 BGA George Allan, 24 Leadbetter Drive, Bromsgrove, B61 7JG
G0 BGB M Davis, 48 The Longlands, Wombourne, Wolverhampton, WV5 0HQ
G0 BGH I Bamford, 60 Coast Drive, Greatstone, New Romney, TN28 8NX
G0 BGI S Knight, 14a Manor Road, Upton Lovel, Warminster, BA12 0JW
G0 BGR Roger Jones, Glenroy, 5 School Road, Blackpool, FY4 5DS
G0 BGV H Eastwood, 3 The Brambles, Thorpe Willoughby, Selby, YO8 9LL
G0 BGX Owen Pauley, 235 Roughton Road, Cromer, NR27 9LQ
G0 BGY James Moult, New Bungalow, Bar Bridge Lane, Swineshead, Boston, PE20 3PG
G0 BHA P White, 42 Abbey Road, Medstead, Alton, GU34 5PB
G0 BHH P Gregory, 20 Heyes Grove, Rainford, St. Helens, WA11 8BW
G0 BHK Edward Stiles, 16 Henry Avenue, Rustington, Littlehampton, BN16 2NY
G0 BHP Peter Fambely, 126 Ashton Lane, Sale, M33 5QJ
G0 BHR D Egan, 44 Wrights Lane, Warley, Cradley Heath, B64 6QX
G0 BHS D Angell, 12 Harrod Drive, Market Harborough, LE16 7EH
G0 BHT W Blake, Tylers Farm, Grantham Road, Grantham, NG33 5HG
G0 BHU M Jamieson, 3 Rowan Way, Bourne, PE10 9SB
G0 BIA R Armstrong, 2 West End Road, Laughton, Gainsborough, DN21 3GT
G0 BIE D Lucas, 9 Newborough Road, Alvaston, Derby, DE24 0LH
G0 BIN E Ashley, 2 School Road, Bulkington, Bedworth, CV12 9JB
G0 BIQ Roger Bellamy, Emathu, No. 48 28th April Street 1688, Xahhra, Malta, XRA1033
G0 BIR A Skinner, Halfway Lock Cottage, Upper Gambolds Lane, Bromsgrove, B60 3HB
G0 BIV J Young, 3 Kensington Close, Kings Sutton, Banbury, OX17 3XB
G0 BIW Martin Smith, 394 Longbridge Road, Barking, IG11 9EE
G0 BIX T Dansey, Woodlands, 19 Hill Chase, Chatham, ME5 9HE
G0 BJA G Vincent-Squibb, 7 Chudleigh Road, Henley Green, Coventry, CV2 1AF
G0 BJD S Wood, 55 Megdale, Wolds, Matlock, DE4 3TE
GI0 BJH T Hoey, 4 Bramble Avenue, Newtownabbey, BT37 0XL
G0 BJI Donald Peacock, 15 Farmfield Road, Banbury, OX16 9AP
G0 BJK D Thomas, 33 Chatsworth Road, Stretford, Manchester, M32 9QF
G0 BJL David Michael Oxley, 122 Windy House Lane, Sheffield, S2 1HE
G0 BJR G Oliver, 158 High Barn St., Royton, Oldham, OL2 6RW
G0 BJZ E Pritchard, 8 Stourmore Close, Willenhall, WV12 5RF
G0 BKA J Leedham, 27 St. Andrews Crescent, Stratford-upon-Avon, CV37 9QL
G0 BKB V Leedham, 27 St. Andrews Crescent, Stratford-upon-Avon, CV37 9QL
GM0 BKC Paul Glasper, 1 Lindertis Cottages, Kirriemuir, DD8 5NT
G0 BKD Barry Mawn, 13 Micklethwaite Grove, Moorends, Doncaster, DN8 4NU
G0 BKE Frederick Barker, 17 Walders Avenue, Sheffield, S6 4AY
G0 BKH B Minton, 8 Rosebank Walk, Barnton, Northwich, CW8 4PU
G0 BKJ M Glover, 4 New Hospital Villa, Hospital Road, Brecon, LD3 0DU
G0 BKL E Smith, 40 Grays End Close, Grays, RM17 5QR
G0 BKN Ian McIver, 3 Asgard Drive, Bedford, MK41 0UP
G0 BKP J Dadswell, 30 Barlow Road, Wendover, Aylesbury, HP22 6HS
G0 BKQ Paul Rowe, 45 Springhill Avenue, Wolverhampton, WV4 4ST
G0 BKR Gary Clark, 2 Roundheads End, Forty Green, Beaconsfield, HP9 1YB
G0 BKU Shaun Coles, 54 Cornwall Road, Shepton Mallet, BA4 5UR
G0 BKW Keith weston, 14 Auchinleck Court Burleigh Way, Crawley Down, Crawley, RH10 4UP
GM0 BKX T Stewart, 104 Barrhill Road, Cumnock, KA18 1PU
G0 BLB Richard Baker, Holmelea, Upper Bristol Road, Bristol, BS39 5RJ
G0 BLM P Mathews, 25 Shore Mount, Littleborough, OL15 8EN
G0 BLO Brian Osborn, 4 Highfield Close, Collier Row, Romford, RM5 3RX
G0 BLQ Henry Cooper, 26 Badgers Way, Buckingham, MK18 7EQ
G0 BLS A Wilkie, 7 Willow Drive, Droitwich, WR9 7QE
G0 BLT Gerald Blackmoor, 4 St. Godwalds Crescent, Aston Fields, Bromsgrove, B60 2EB
G0 BLU E Mustard, 108 Allandale, Hemel Hempstead, HP2 5AT
G0 BLV M Puncer, 17 St. Michaels Walk, Eye, Peterborough, PE6 7XG
G0 BLW A Crimlisk, 14 Long Lane, Aughton, Ormskirk, L39 5AT
G0 BMB Robert Morgan, 48 Newport Road, Gnosall, Stafford, ST20 0BN
G0 BMG Xmalcolm Green, 65 Rosamond Road, Bedford, MK40 3UG
G0 BMH James McLean, 24 Durham Drive, Oswaldtwistle, Accrington, BB5 3AT
G0 BML T Garvey, 162 Birchfields Road, Manchester, M14 6PE
G0 BMN K Hutchins, 23 Salisbury Road, Tunbridge Wells, TN4 9DJ
G0 BMP L Gyurgyak, 7 Lakeside Close, New Park, Newton Abbot, TQ13 9FE
G0 BMQ M Armstrong, 4 Medway Drive, Preston, Weymouth, DT3 6LF
G0 BMS I Cooper, 39 Doyle Road, Bolton, BL3 4SA
G0 BMT J Walkley, 10 Exton, Dunster Crescent, Weston-Super-Mare, BS24 9EH
G0 BMU T Drewitt, 6 Copse View Cottages, Ascot Road, Maidenhead, SL6 3JY
G0 BMZ B Lindgren, Berzeliigatan 26, Goteborg, Sweden, SE-41 53
G0 BNE A Knell, 13 Northumberland Road, Leamington Spa, CV32 6HE
G0 BNF Jeremy Holmes, 63 Grange Avenue, Street, BA16 9PF
G0 BNG Leonard Britton, 36 Frampton Court, Trowbridge, BA14 9HL
G0 BNJ B Northway, 3 Kingston Close, Kingskerswell, Newton Abbot, TQ12 5EW
G0 BNK Wearside Electronics at ARS c/o Ian Douglas, 13 Castlereagh Street, New Silksworth, Sunderland, SR3 1HJ
GW0 BNN J Bulpin, 27 Llys Dol, Morriston, Swansea, SA6 6LD
GW0 BNO D Lindley, 29 Belvedere Close, Kittle, Swansea, SA3 3LA
GM0 BNQ D Macdonald, Greenbrae Cottage, Auchterless, Turriff, AB53 8HD
G0 BNR Nigel Keightley, Wavendon, Daintree Road, Ramsey St. Marys, Huntingdon, PE26 2TF
G0 BNU D Wright, Dovetail Cottage, South Petherwin, Launceston, PL15 7LQ
G0 BNW W Wheeler, 201 Topsham Road, Exeter, EX2 6AN
G0 BNY C Lee, 29 Meadow Dale, Chilton, Ferryhill, DL17 0RW
G0 BNZ Jane Hewitt, 35 Birmingham Road, Alvechurch, Birmingham, B48 7TB
G0 BOC E Pitman, 35 Brackley Way, Totton, Southampton, SO40 3HP
G0 BOE Wojciech Piotrowski, 47 Saxon Way, Willingham, Cambridge, CB24 5UR
G0 BOF William Ardley, 353 Rush Green Road, Romford, RM7 0NJ
G0 BOH D Bowman, 6 Linksfield, Denton, Manchester, M34 3TE
G0 BOM Anthony SAMOUELLE, 4 Fox Road, Bourn, Cambridge, CB23 2TU
G0 BON Ivan Rogers, 26 Milton Road, Slough, SL2 1PF
G0 BOO Brian Gilbert, Silver Micha, Hunts Corner, Norwich, NR16 2HL
G0 BOQ P Macdonald, 77 Ousebank Way, Stony Stratford, Milton Keynes, MK11 1LD
G0 BOR John Goodall, 287 Newbrook Road, Atherton, Manchester, M46 9GZ
G0 BOT D Ashby, 36 Mersey Grove, Birmingham, B38 9LA
G0 BPA Bernard Lyford, 48 Wilverley Place, Blackfield, Southampton, SO45 1XW
GM0 BPF W Johnstone, Byre Cottage, Kilmichael Glassary, Lochgilphead, PA31 8QL
G0 BPK Nigel Ferguson, Royd Moor, Royd Moor Lane, Pontefract, WF9 1AZ
G0 BPL D Farnham, 24 Downham Road, Watlington, King's Lynn, PE33 0HS
G0 BPM D Coffey, 121 Worksop Road, Swallownest, Sheffield, S26 4WB
G0 BPQ J Jones, 16 Laurel Avenue, Darwen, BB3 3AG
G0 BPR S Whitnear, 55 Brier Crescent, Nelson, BB9 0QD
G0 BPS Richard Pascoe, 9 Horsley Close, Hawkinge, Folkestone, CT18 7FN
GM0 BPT A Murray, 1 Gordon Road, Edinburgh, EH12 6NB
G0 BPU Michael Johnson, 23 Camden Road, Ipswich, IP3 8JW
G0 BPX D Norridge, 125 Ferry Road, Marston, Oxford, OX3 0EX
G0 BPZ P Maple, Beech House, Derby Road, Ashbourne, DE6 1LZ
G0 BQB Pete Smith, 10 Denby Lane, Grange Moor, Wakefield, WF4 4ED
G0 BQC W scrivener, Folly Hall Farm, Kings Causeway, Nelson, BB9 0EZ
G0 BQE J Chown, 15 Hambleden Walk, Maidenhead, SL6 7UH
G0 BQG R Jefferies, 35 Lambrok Close, Trowbridge, BA14 9HH
G0 BQI M Chapman, 6a Rees Street, Islington, London, N1 7AR
G0 BQK M Hall, 31 Winchester Close, Stratton, Swindon, SN3 4HB
G0 BQO J Rawson, 3 Cooks Close, Kesgrave, Ipswich, IP5 2YT
G0 BQP J Simpson, 18 Southdean Drive, Hemlington, Middlesbrough, TS8 9HH
G0 BQQ I Laczko, 34 Airds Drive, Dumfries, DG1 4EW
G0 BQV M Ashdown, 42 Alpine Avenue, Tolworth, Surbiton, KT5 9RJ
G0 BQW S Robinson, 114 Hopefield Avenue, Sheffield, S12 4XE
GI0 BQX D Maguire, 16 Kilmacormick Drive, Enniskillen, BT74 6EP
G0 BQZ S Smith, 18 Stratford Drive, Eynsham, Witney, OX29 4QJ
G0 BRA Banbury Amateur R c/o Frank Humphris, 169 Bloxham Road, Banbury, OX16 9JU
G0 BRC BREDHURST RECEIVING & TRANSMITTING SOC. c/o Martin Pearson, 56 Parkwood Green, Parkwood, Gillingham, ME8 9PP
G0 BRH C Waters, 468 Buckfield Road, Leominster, HR6 8SD
GM0 BRJ D Wilson, Four Winds, High Barrwood Road, Glasgow, G65 0EE
G0 BRL M Bevan, 22 Spring Crescent, Brown Edge, Stoke-on-Trent, ST6 8QH
G0 BRM N Entwistle, Park Garden House, Park Garden, Bury St. Edmunds, IP28 8PB
GI0 BRO Raymond Wilson, 68 Kensington Road, Belfast, BT5 6NG
G0 BRQ A Morris, 108 Lytchett Drive, Broadstone, BH18 9NR
G0 BRS BORDER ARS c/o Landles Fairbairn, 4 The Avenue, Eyemouth, TD14 5DL
G0 BRW B Whatling, 6 Rock Road, Dursley, GL11 6LF
G0 BRX Stephen Shirras, 72 Sefton Avenue, Poulton-le-Fylde, FY6 8BL
G0 BRZ L Rimmer, 7 Dorfold Close, Sandbach, CW11 1EB
G0 BSA M Embling, 23 Sinodun Road, Didcot, OX11 8HP
G0 BSD David Tringham, 47 The Knoll, Palacefields, Runcorn, WA7 2UH
G0 BSF H Rolfe, 46 Great Gardens Road, Hornchurch, RM11 2BA
G0 BSH R Harman, 22 Ridgebrook Road, Kidbrooke, London, SE3 9QN
G0 BSJ E New, 6 Witchampton Close, Havant, PO9 5RY
G0 BSK K Bryant, 9 Cunningham Park, Mabe Burnthouse, Penryn, TR10 9HB
G0 BSN R Vincent-Squibb, 1 Alexander Avenue, Earl Shilton, Leicester, LE9 7AF
GD0 BSP R Snow, 73 Boxtree Road, Harrow Weald, Harrow, HA3 6TN
G0 BST J I'Anson-Holton, Lake View, Brookside Avenue, Telford, TF3 1LA
G0 BSX P Meiring, 18 Slayleigh Lane, Sheffield, S10 3RF
G0 BTA D Round, 21 Bunting Drive, Bradford, BD6 3XE
G0 BTB B Botham, The Cherries, Anglesey, LL74 8SR
G0 BTH P Clark, 180 Roselands Drive, Paignton, TQ4 7RW
GM0 BTK W Rossmann, Strathvale, Milton of Ogilvie, Forfar, DD8 1UN
G0 BTQ N Burge, 43 Bourn Rise, Pinhoe, Exeter, EX4 8QD
G0 BTT Eric Clay, 23 New Street, Sleaford, NG34 7HG
G0 BTU K Tupman, Magpies, 6 Larcombe Road, Petersfield, GU32 3LS
G0 BUB Michael Sharpe, New House, Highfield Farm, Grantham, NG32 3SJ
G0 BUC K Simpson, 5 Cedar Close, The Elms, Lincoln, LN1 2NH
G0 BUD John Spearing, 19 Elizabeth Square, London, SE16 5XN
GM0 BUE Alan Stark, 30 Kelvin Way, Kilsyth, Glasgow, G65 9UL
GM0 BUI G Williamson, 2 Laburnum Grove, Burntisland, KY3 9EU
G0 BUJ R Pelling, The Orchard, Shirwell, Barnstaple, EX31 4JR
G0 BUK I Mitchell, Cornerfield, Five Ash Down, Uckfield, TN22 3AP
G0 BUV John Howard, 130 Coventry Road, Coleshill, Birmingham, B46 3EH
G0 BUW P Martin, 39a The Grove, Bearsted, Maidstone, ME14 4JB
G0 BUX Robert Clinton, Appletrees, Alexandra Road, Mayfield, TN20 6UD
G0 BUZ E Piper, 44 Parsonage Estate, Rogate, Petersfield, GU31 5HJ
G0 BVA S Fawcett, 34 Wantsume Lees, Sandwich, CT13 9JF
G0 BVC G Broadhurst, Summerhayes, 10 Ottervale Close, Honiton, EX14 9TA
G0 BVD Philip Oakley, Elmsleigh House, New Street, Torrington, EX38 8BY
G0 BVG J Graham, Clintpark, Lockerbie, DG11 3JH
G0 BVK E Evers, 13 Grasmere Close, Penistone, Sheffield, S36 8HP
G0 BVM J Speers, 187 Worsley Road, Eccles, Manchester, M30 8BP
G0 BVO J Teasdale, 1 Newtown Bungalows, Newtown, Spennymoor, DL16 7QS
G0 BVQ Colin Hawkes, 12 Summer Hall Ing, Wyke, Bradford, BD12 8DN
G0 BVS M Arthur, 8 Hanbury Road, Bedworth, CV12 9BX
G0 BVT Anthony Tonge, 30 Cardigan Avenue, Morley, Leeds, LS27 0DP
G0 BVU D Davies, 30 Bullpit Road, Balderton, Newark, NG24 3LY
G0 BVV D Clench, 70 Shalbourne Crescent, Bracklesham Bay, Chichester, PO20 8RG
G0 BVW Q Curzon, 154 Clophill Road, Maulden, Bedford, MK45 2AE
G0 BWB John Chappell, 49 Midway, South Crosland, Huddersfield, HD4 7DA
G0 BWC Bolton Wireless Club c/o R Wilkinson, 84 Park Road, Bolton, BL1 4RQ
G0 BWE D Allen, 48 Castle Road, Crickhowell, NP8 1AP
G0 BWG J Raynes, 115 Deerlands Avenue, Sheffield, S5 7WU
G0 BWJ K Miller, 8 Horsham Gardens, Sunderland, SR3 1UJ
G0 BWK R Lee, 57 Hart Lane, Luton, LU2 0JF
G0 BWO D Wilkinson, 139 Grosvenor Road, Dalton, Huddersfield, HD5 9HX
G0 BWP B Passmore, 364 Franklin Road, Kings Norton, Birmingham, B30 1NG
G0 BWQ K Kitson, 278 Cowcliffe Hill Road, Huddersfield, HD2 2NE
G0 BWU R Smith, 9 Queen Elizabeth Drive, Castle Douglas, DG7 1HH
G0 BWV J Puttock, Sutton & Cheam Rs, 53 Alexandra Avenue, Sutton, SM1 2PA
G0 BWY G Fearnside, 16 Lee Court, Thwaites Brow, Keighley, BD21 4TL
G0 BXC P Hughes, 123 Garth Road, Morden, SM4 4LF
G0 BXD Anthony Wootton, Dower House, Hilton, Bridgnorth, WV15 5PB
G0 BXG R Dring, 22 Castle Street, Eastwood, Nottingham, NG16 3GW
G0 BXH B Hansell, 7 Fry Road, Stevenage, SG2 0QG
G0 BXJ A Peachey, 60 Upwell Road, March, PE15 9EA
G0 BXL Terence White, 79 Elmbridge, Harlow, CM17 0JY
G0 BXM Michael Smyth, 1 White Acres Road, Mytchett, Camberley, GU16 6EY
G0 BXP R Coleman, 16 Mouse Lane Rougham, Bury St. Edmunds, IP30 9JB
G0 BXS R Rennolds, 3 Woodlands Close, Milton-under-Wychwood, Chipping Norton, OX7 6LS
G0 BXU Alan Collick, 39 Beech Rise, Sleaford, NG34 8BJ
G0 BXV David Dixon, Flat 1, Norfolk House Oaklands Road, Havant, PO9 2RD
G0 BXZ L Jefferies, 24 Doctors Meadow, Ruyton XI Towns, Shrewsbury, SY4 1LX
G0 BYA Stanley Houlding, 90 Wordsworth Avenue, Stafford, ST17 9UE
G0 BYF Raymond Aucote, Meldon Bothy, Chagford, Newton Abbot, TQ13 8EJ
G0 BYH Geoffery Stokes, 23 Maynard Close, Bradwell, Milton Keynes, MK13 9HS
G0 BYK M Jackson, Stockswood, Stocks Lane, Southwold, IP18 6UJ
G0 BYL Jean Luxton, 2 Trinity Court, Westward Ho, Bideford, EX39 1LT
G0 BYQ S Ratcliffe, 173 Whinney Lane, New Ollerton, Newark, NG22 9TJ
G0 BYU F Cooper, 23 York Close, Gillow Heath, Stoke-on-Trent, ST8 6SE
G0 BYX Charles Pavier, 12a Friend Lane, Edwinstowe, Mansfield, NG21 9QZ
GW0 BYZ R Jones, 9 Dennithorne Close, Merthyr Tydfil, CF48 3HE
GW0 BZA K Duckfield, Ellesmere, 29 Esplanade Avenue, Porthcawl, CF36 3YS
G0 BZB Anthony Volpe, 31 Cleveland Gardens, Newcastle upon Tyne, NE7 7QE
G0 BZC Gordon Smith, 9 Blackett Avenue, Stockton-on-Tees, TS20 2EX
G0 BZF D Reid, Nicolaas Beetsstraat 29, Hengelo Ov, Netherlands, 7552HW
G0 BZH G Hodgson, 16 Dockroyd, Oakworth, Keighley, BD22 7RH
G0 BZJ J Newton, 21 Village Court, Penrhiw Avenue, Blackwood, NP12 0LU
GI0 BZM C Colhoun, 85 Whitehill Park, Limavady, BT49 0QF
G0 BZN C Johnson, 1 Cedar Close, Chesham, HP5 3LL
G0 BZP J Bates, 28 Westbourne Road, West Bromwich, B70 8LD
GM0 BZS Andrew Ince, Burnside, Braefield, Glen Urquhart, Inverness, IV63 6TN
G0 BZT R Targonski, 4 Woodville Gardens, Sedgley, Dudley, DY3 1LB
G0 BZU Keith Barlow, 105 Buller Street, Bury, BL8 2BQ
G0 BZV E Trett, 11 Langton Avenue, Bierley, Bradford, BD4 6BY
G0 BZW A Porter, 1 Bloomfield Terrace, Weston, Portland, DT5 2AB
G0 BZX P Smith, 8 Avalon Close, Orpington, BR6 9BS
GM0 CAD Dave Hewitt, Beechwood Lodge, Ardross, Alness, IV17 0XN
G0 CAE Keith Walsh, 13 Weston Park Homes, Weston Road, Portland, DT5 2DE
GD0 CAG David Talaber, 54 Southfield Park, North Harrow, Harrow, HA2 6HE
GI0 CAH Philip Leonard, 4 Clonurson Road, Enniskillen, BT92 3BU
G0 CAJ Tom Morgan, 12 Wodehouse Road, Southampton, SO19 2EQ
G0 CAK G Russell, 57 Silchester Road, Pamber Heath, Tadley, RG26 3ED
G0 CAL R Faulkner, 10 Fell Wilson St., Warsop, Mansfield, NG20 0PT
G0 CAM C Chislett, Woodview, Crofthandy, Redruth, TR16 5PT
G0 CAP J Burton, Bungalow, 22 Balance Hill, Uttoxeter, ST14 8HT
G0 CAS N Clarke, Lyme View, 3 East Cliff House, Dawlish, EX7 9JB
G0 CAX I Moore, 25 Granby Close, Winyates, Redditch, B98 0PJ
G0 CAY Kenneth Arnold, 14 Ford Close, St.Ive, Liskeard, PL14 3FN
GM0 CBA B Aitken, 48 Kenilworth Rise, Livingston, EH54 6JJ
G0 CBB M Phillips, 52 Rivington Drive, Burscough, Ormskirk, L40 7RP
G0 CBC J Johnstone, 12 Castle Acre, Ecclefechan, Lockerbie, DG11 3DU
G0 CBD G Roby, 40 Lulworth Drive, Hindley Green, Wigan, WN2 4QS
G0 CBI S McArthur, 36 Ingham Close, Bradshaw, Halifax, HX2 9PQ
G0 CBJ A Whittingham, 61 Hillcroft Road, Altrincham, WA14 4JE
G0 CBK Nigel Priestley, 13 Orchard Close, Charfield, Wotton-under-Edge, GL12 8TJ
GW0 CBL John Follant, 76 Wern Road, Skewen, Neath, SA10 6DL
G0 CBM Charles Wilkie, Bramcote, Grange Road, Sandilands, Mablethorpe, LN12 2RE
G0 CBN P Forster, 2 Rockingham Close, Birchwood, Warrington, WA3 6UY
G0 CBO Rodney Hoffman, 26 Penn Road, Taverham, Norwich, NR8 6NN
G0 CBP R McMahon, 34 Denmark Road, Poole, BH15 2DB
GM0 CBQ J Macleod, 61 Fox Covert Avenue, Edinburgh, EH12 6UH
G0 CBT S Peat, 64 Grange Road, Romford, RM3 7DX
G0 CBU W Drea, 146 Slewins Lane, Hornchurch, RM11 2BS
G0 CBW M Bentley, Nickers Hill Farm, Falhouse Lane, Dewsbury, WF12 0NL
G0 CCA G Coles, Walnut Lodge, Staunton Lane, Bristol, BS14 0QG
G0 CCB C Gill, 52 Southfield Road, Nailsea, Bristol, BS48 1JD
G0 CCC CAVERSHAM CON G c/o C Young, 18 Wincroft Road, Caversham, Reading, RG4 7HH
G0 CCF J Broome, 35 Claygate Road, Cannock, WS12 2RN
G0 CCG H Toh, 3 Priory Road, Dover, CT17 9RQ
G0 CCJ J Weinstock, 24 Hamilton Road, Tiddington, Stratford-upon-Avon, CV37 7DD
G0 CCL Cambridge Consultants ARC c/o Ludovic Laprade, 5 Greenacre Close, Godmanchester, Huntingdon, PE29 2UX
G0 CCM M Hurst, 2 Poplar Road, Oughtibridge, Sheffield, S35 0HR
G0 CCN Peter Hurst, 23 Cantilupe Crescent, Aston, Sheffield, S26 2AS
GW0 CCO L Holderness, 8 Jubilee Gardens, Templeton, Narberth, SA67 8ST
G0 CCQ J Smith, Dobie Lodge, Rochford, Tenbury Wells, WR15 8SR
G0 CCS John Zissler, 8 Norman Drive, Whittington, King's Lynn, PE33 9TQ
G0 CCT Derek Woolfall, 2 The Woodlands, Wirral, CH49 6NQ
G0 CCU L Whitelegg, 30 Chatsworth Road, Arnos Vale, Bristol, BS4 3EY
G0 CCV J Gaut, 18 First Avenue, Clipstone Village, Mansfield, NG21 9EA
G0 CDA M Ryder, 9 Lincoln Close, Woolston, Warrington, WA1 4LU
G0 CDB John May, 6 Hodson Close, Paignton, TQ3 3NU
GM0 CDC Frederick Shaw, 8 Morrison Street, Kirriemuir, DD8 5DB
GI0 CDM Joseph Neill, 16 Rathlin Street, Belfast, BT13 3DZ
G0 CDO Jeremy Faulkner-Court, Yew Tree Cottage, Avon Dassett, Southam, CV47 2AT
G0 CDP G Rudge, 18 Herbert Road, Warley, Smethwick, B67 5DD
G0 CDQ M Murphy, 10 Bayham Road, Sevenoaks, TN13 3XA
G0 CDR John Warwick, 29 Hay Brow Crescent, Scalby, Scarborough, YO13 0SG

G0 CDS CR Brind, 8 Pezenas Drive, Market Drayton, TF9 3UJ
G0 CDV Ray Evans, 0 Courthill Farm Cottages, Kelso, TD5 7RU
G0 CDY A Hulme, [illegible] Victoria Gardens, Ferndown, BH22 9JQ
G0 CDZ E Hicks-Arnold, Inglesides, Junction Road, [illegible], SP5 3AZ
GM0 CEA R Robertson, St. Madoes Cottage, St. Madoes, Perth, PH2 7NF
G0 CEB E Hutchinson, 58 Avon Rise, Retford, DN22 6QH
G0 CEC P Coenraats, 54 Falstaff Avenue, Earley, Reading, RG6 5TG
G0 CEF A Sheard, 8 Hazel Beck, Bingley, BD16 1LZ
G0 CEG P WORSDALE, 10 Manton Road, Lincoln, LN2 2JL
G0 CEI P Olliffe, Bitham Cottage, Oxford, OX12 8QW
G0 CEJ C Bayuley, 44 Royds Crescent, Rhodesia, Worksop, S80 3HG
G0 CEL Nigel Evely, 18 Teign View Place, Teignmouth, TQ14 8BX
G0 CEM T [illegible], 90 [illegible] Road, Hartcliffe, Bristol, BS13 9RB
[illegible] [illegible] F Field, 10 The Maples, Nailsea, Bristol, BS48 4RT
G0 CEO M Ohta, Suzuki-Apt202, 2-4-24 Shinden, Chiba, Japan, 292-0832
G0 CEP D Craker, Brynamlwg, Llanddewi Brefi, Tregaron, SY25 6PE
G0 CEQ D Griffiths, 297 Shurland Avenue, Barnet, EN4 8DQ
G0 CER D Harris, 9 Garden City, Tern Hill, Market Drayton, TF9 3QB
G0 CES M Smith, 8 Ridgeway Avenue, Marford, Wrexham, LL12 8ST
G0 CEU C Hawkins, 3 Offord Close, Tottenham, London, N17 0TE
G0 CEV K Towers, 7 Copeland Road, Hucknall, Nottingham, NG15 8EB
G0 CEW Peter Allanson, 16 Woodhouse Lane, Kirkhamgate, Wakefield, WF2 0SE
G0 CEY Gordon Cadey, 45 St. Mildreds Road, Westgate-on-Sea, CT8 8RJ
G0 CFB Richard Tyler, The Firs, Laundry Lane, Halesworth, IP19 0PY
GM0 CFC G Percival, 5 Murrell Terrace, Aberdour, Burntisland, KY3 0XH
G0 CFD Francis Dimmock, 78 Paxton Crescent, Shenley Lodge, Milton Keynes, MK5 7PY
G0 CFI R Deans, 23 Aldringham Park, Aldringham, Leiston, IP16 4QZ
GM0 CFK C Knight, 8 Ednam Drive, Glenrothes, KY6 1NA
G0 CFM B Robbins, 46 Purton Close, Kingswood, Bristol, BS15 9ZE
G0 CFN T Hastings, 1 Pottle Close, Botley, Oxford, OX2 9SN
G0 CFT Terry Feaviour, 9 Holbrook Crescent, Felixstowe, IP11 2NE
GM0 CFW K Moffat, 10 Greenfield Crescent, Balerno, EH14 7HD
G0 CGA W Roberts, 10 The Meadows, Station Road, Hodnet, TF9 3QF
G0 CGD Colin O'Neill, 1 Links Avenue, Little Sutton, Ellesmere Port, CH66 1QS
G0 CGE B Griffin, 75 Greenham Wood, Hanworth, Bracknell, RG12 7WH
G0 CGH R Hurley, 57 Cannons Close, Bishop's Stortford, CM23 2BQ
G0 CGI Brian Redfern, The Old Blacksmiths Forge, Old Newton, Craven Arms, SY7 9PG
G0 CGM A Cutcliffe, 9 Dixon Close, Paignton, TQ3 3NA
G0 CGQ Ray Hazlewood, 9 The Brambles, Haslington, Crewe, CW1 5RA
G0 CGS D Morris, 2 Willow Close, Brinsworth, Rotherham, S60 5JU
G0 CGT B Birch, 59 Shepperson Road, Sheffield, S6 4FG
G0 CGW G Ashcroft, 49 Tantallon, Birtley, Chester le Street, DH3 2JG
G0 CGZ G Allison, 24 Southfield Road, Scartho, Grimsby, DN33 2PL
G0 CHB B Small, 3 Katherine Crescent, Skegness, PE25 3LF
G0 CHC Martin Smith, 12 High Street, West Wickham, Cambridge, CB21 4RY
G0 CHE Kevin Piper, Flat 5, Marine Court, 4 Marine Drive West, Bognor Regis, PO21 2QA
G0 CHG Richard Moate, Garth House, Redbrook Street, Ashford, TN26 3QS
G0 CHJ R Poulter, 19 Homestead Way, Winscombe, BS25 1HL
G0 CHL K Dewhurst, Rock Cottage, 82 New Street, Stoke-on-Trent, ST8 7NW
GM0 CHM J Stephen, 26 Douglas Avenue, Elderslie, Johnstone, PA5 9NE
G0 CHN J Sutton, 40 Dane Valley Road, Margate, CT9 3RX
G0 CHO Clive Ousbey, 30 Hawthorn Way, Shipston-on-Stour, CV36 4FD
G0 CHP R Angus, 25 Jellicoe House, Capstan Road, Hull, HU6 7AS
GD0 CHQ J Pepper, 7 East Towers, Pinner, HA5 1TN
G0 CHR A Steward, 94 Whittington Avenue, Hayes, UB4 0AE
G0 CHV Thomas Sherriff, 5 Pembroke Avenue, Morecambe, LA4 6EJ
G0 CHY Patrick Moulton, 4 The Ridge, Withyham Road, Tunbridge Wells, TN3 9QU
G0 CIG Adrian Aldridge, 15 Doubletrees, St. Blazey, Par, PL24 2LD
GM0 CII W Rogers, 170 Boswall Parkway, Edinburgh, EH5 2JJ
G0 CIM S Dodd, 61 Church Road, Hove, BN3 2BP
G0 CIR R Jasper, 84 Rose Green Road, Bognor Regis, PO21 3EQ
G0 CIT G Davies, 86 Leatherhead Road, Ashtead, KT21 2SY
G0 CIX Robert Wilton, 30 Barrington Crescent, Birchington, CT7 9DF
G0 CJA L Harper, 23 Beech Avenue, Hazel Grove, Stockport, SK7 4QP
G0 CJC C Clode, 174 Kenn Road, Clevedon, BS21 6LH
G0 CJD K Spratley, 92 Plantation Hill, Worksop, S81 0QN
G0 CJG D Slatter, 13 Hill Burn, Henleaze, Bristol, BS9 4RH
G0 CJO R Nelson, Flat 1, 17 Ashburnham Road, Hastings, TN35 5JN
G0 CJQ R Saw, 33 Rectory Lane, Southoe, St. Neots, PE19 5YA
G0 CJV C Scholey, 11a Guildford Street, Grimsby, DN32 7PL
G0 CJX John Stott, Old Police House, Main Road, Saxmundham, IP17 1JY
G0 CJZ R Waller, 29 Valley Drive, Withdean, Brighton, BN1 5FA
G0 CKA G Waller, [illegible], 10 Old London Road, Brighton, BN1 8XQ
G0 CKD F Wall, 37 Newton Way, St. Osyth, Clacton-on-Sea, CO16 8RQ
G0 CKE J Centanni, 5 Mickle Meadow, Water Orton, Birmingham, B46 1SN
G0 CKF Bryan Woodward, 9 Shaftesbury Avenue, Hornsea, HU18 1LX
G0 CKH A Affolter, 12 Newfound Drive, Norwich, NR4 7RY
G0 CKI K Tarbett, 20 Leeholme, Houghton le Spring, DH5 8HR
GW0 CKK D Robinson, Island View Caravan Park, Beach Road, Penarth, CF64 5UG
GW0 CKL I Gulyas, 2 Eglwysilan Way, Abertridwr, Caerphilly, CF83 4EQ
G0 CKM C Gee, 6 Canterbury Close, Dukinfield, SK16 5RT
G0 CKU J Lewis, 34 Maple Crescent, Shaw, Newbury, RG14 1LL
G0 CKV O Lundberg, Rowan House, Cavendish Road, Weybridge, KT13 0JW
GW0 CKX D Matthews, 130 Parc Avenue, Morriston, Swansea, SA6 8HU
G0 CLC E Rogers, Room 21, Bodmeyrick Residential Home, Holsworthy, EX22 6HB
G0 CLD H Davey, 6 Cambridge Grove, Otley, LS21 1DH
G0 CLF R Moon, 22 Meagill Rise, Otley, LS21 2EJ
G0 CLG S Haynes, 9 Heather Gardens, Belton, Great Yarmouth, NR31 9PP
G0 CLH David Lingard, 17 Feltwell Road, Methwold Hythe, Thetford, IP26 4QJ
G0 CLJ S Banks, 15 Hunters Way, Saffron Walden, CB11 4DE
G0 CLM A Greig, 98 Appletree Lane, Redditch, B97 6TS
G0 CLR R Hunt, 197 North Walsham Road, Norwich, NR6 7QN
G0 CLT Philip Hodgson, 14 Catherine Howard Close, Thetford, IP24 1TQ
G0 CLV Stanley Barr, 11 Mallard Way, Wirral, CH46 7SJ
G0 CLX G Glotham, 6 Slim Road, Bentley, Walsall, WS2 0EG
G0 CMB R Burling, 28 Croydon Road, Arrington, Royston, SG8 0DJ
GW0 CMI K Voller, 5 Hillfield Road, Parcllyn, Cardigan, SA43 2DH
G0 CMK N Sherwood, The Orchards, Barton Road, Brigg, DN20 8SH
G0 CMM John Bull, Le Magnou, [illegible], France, [illegible]
GM0 CMO M Murray, 1 Gordon Road, Edinburgh, EH12 0ND
G0 CMP Vincent Warren, 129 Market Road, Thrapston, Kettering, NN14 4JT
G0 CMQ E Ward, 5 Ashley Grove, Hebden Bridge, HX7 5NF
G0 CMR R Melia, 14 Friar Park Road, Wednesbury, WS10 0TB
G0 CMT Egon Gadeberg, Hojmarksvej 35, Po Box 56, Horsens, Denmark, 8700
G0 CMU C Beecham, Moorview, 2 Endor Crescent, Ilkley, LS29 7QH
G0 CMW J Groom, Windyridge, High Road, Maidenhead, SL6 9JF
G0 CNA M Wyatt, 32 Stafford Road, Bridgwater, TA6 5PH
G0 CND I Sharrott, 31 The Fleet, Stoney Stanton, Leicester, LE9 4DZ
G0 [illegible] [illegible] Roberts, 70 Nairn Road, Walsall, WS3 3XB
[illegible] [illegible] [illegible] Edwards, 31 Eagleswell Road, Boverton, Llantwit Major, CF61 2UG
G0 CNK G Orchard, 8 Glyn Avenue, Prestatyn, LL19 9NN
G0 CNL C Rogers, 34 Martin Court, Werrington, Peterborough, PE4 6JS
GM0 CNP John Mullen, 46 Templars Crescent, Kinghorn, Burntisland, KY3 9XG
G0 CNU J Mckenzie, 4 Hadrian Court, Humshaugh, Hexham, NE46 4DE
G0 CNV Tony Mcmanus, 84 Beverley Road, Hessle, HU13 9BP
GM0 CNW H Taylor, 20 Woodhaven Avenue, Wormit, Newport-on-Tay, DD6 8LF
G0 COA G Cass, 1 Ash Brow, Flockton, Wakefield, WF4 4TE
G0 COC Robert Vallis, Bartley Villa, Southampton Road, Southampton, SO40 2NA
G0 COE E Coe, 20 Bitterne Way, Lymington, SO41 3PB
G0 COG K Stringer, 33 Brookes Road, Flitwick, Bedford, MK45 1BU
G0 COH J Rixon, Bro Afallon, Rhoslefain, Tywyn, LL36 9LY
G0 COI J Atkinson, 90 Priors Road, Cheltenham, GL52 5AN
G0 COJ B Ellery, 384 Sutton Way, Great Sutton, Ellesmere Port, CH66 3LL
G0 COL Colin Shepherd, 9 Wrea Head Close, Scalby, Scarborough, YO13 0RX
G0 COM A Redding, 28 Warwick Avenue, Bridgwater, TA6 5PF
G0 COQ W Kelsey-Stead, 65/4 Moo 2, Rawai, Phuket, Thailand, 83130
GW0 COU S Jones, 635 Clydach Road, Ynystawe, Swansea, SA6 5AX
G0 COX N Currey, 4 Saffron Drive, Oldham, OL4 2PU
G0 COY Terence Frearson, 31 Paradise Street, Rugby, CV21 3SZ
G0 COZ J Wayman, 1 Waterloo Close, Bredon, Tewkesbury, GL20 7WL
G0 CPA A Prichard, 1 Polton Dale, Swindon, SN3 5BN
G0 CPD G Forster, 2 Rockingham Close, Birchwood, Warrington, WA3 6UY
G0 CPF I Whyte, 36 Chestnut Road, Ashford, TW15 1DG
G0 CPJ P Walker, 46 Ribble Avenue, Freckleton, Preston, PR4 1RX
G0 CPN I Gurton, 28 Bloomfield Road, Harpenden, AL5 4DB
G0 CPO S NORM ALFRETON c/o G Childe, 20 Glenmore Drive, Stenson Fields, Derby, DE24 3HE
G0 CPP J Linfoot, Flat, 10 Pembroke Court, Oxford, OX4 1BY
G0 CPR John Stewart, 45 Dawn Crescent, Upper Beeding, Steyning, BN44 3WH
G0 CPT Alan Fielder, 46 Route de Pontivy, 22570 Plelauff, Plelauff, France, 22570
G0 CPU M Cracknell, 17 Windmill Fields, Harlow, CM17 0LQ
G0 CPV S Pocock, 14572 West, 152nd Place, Kansas, United States, 66062
G0 CPZ B Adams, 85 Copperfields, Lydd, Romney Marsh, TN29 9UU
G0 CQB Gerald Cant, 4 The Mount, Docking, King's Lynn, PE31 8LN
G0 CQC Nigel Wilding, Lyncombe Ridge, Lyncombe Vale Road, Bath, BA2 4LP
G0 CQD A Cooper, 10 Buckthorne Court, East Ardsley, Wakefield, WF3 2DD
G0 CQH T Neal, 34 Dene View, Ashington, NE63 8JF
G0 CQI G Baker, 26 Gardeners Road, Halstead, CO9 2TB
G0 CQJ J Kilmister, 9 Wheal Jane Meadows, Threemilestone, Truro, TR3 6EN
G0 CQK James Coombes, 22 Chollerford Close, Gosforth, Newcastle upon Tyne, NE3 4RN
GM0 CQL P Rudd, 41 Broadford Terrace, Broughty Ferry, Dundee, DD5 3EF
G0 CQO Kenneth Clark, Flat 9, Middlewood Hall, Doncaster Road, Barnsley, S73 9HQ
G0 CQP R Berrisford, 19 Moorlands Drive, Mayfield, Ashbourne, DE6 2LP
GM0 CQQ A Herring, Mountpleasant, 8 Linlithgow Road, Bo'ness, EH51 0DD
G0 CQR P Smith, 93 Nottingham Road, Long Eaton, Nottingham, NG10 2BY
G0 CQS S Faulkner, 96 Ashby Road East, Bretby, Burton-on-Trent, DE15 0PT
G0 CQT J Aulsebrook, 38 South Road, Beeston, Nottingham, NG9 1LY
G0 CQU Mark Dean, 3 Buxton Road West, Disley, Stockport, SK12 2AE
GM0 CQV B Hynes, 1 Hillside Court, Hillside Place, Peterculter, AB14 0TU
G0 CQY V Tharp, 14 Cumberland Road, Congleton, CW12 4PH
G0 CQZ N Gardner, 32 Holford Road, Witney, OX28 5NG
G0 CRB A Paddock, 15 Castle Close, Henley-in-Arden, B95 5LR
G0 CRD M Wallis, Quernmore, Hammer Lane, Hailsham, BN27 4JL
G0 CRE A Green, 59 London Road, Sleaford, NG34 7LQ
G0 CRF Terence Bailey, 65 Edge Lane, Chorlton cum Hardy, Manchester, M21 9JU
G0 CRG SUSSEX RAYNET c/o I Page, 127 Whyke Lane, Chichester, PO19 8AU
G0 CRJ Alan Reader, The Pastures, 8 Daddyhole Road, Torquay, TQ1 2ED
G0 CRK D Bowles, 9 The Meadows, Breachwood Green, Hitchin, SG4 8PR
G0 CRL K Moate, 32 Bournewood, Hamstreet, Ashford, TN26 2HL
G0 CRN B Hopper, 189 Western Road, Mickleover, Derby, DE3 9GT
G0 CRO Ronald Crow, 20 Victoria Grove, Wombourne, Wolverhampton, WV5 9AJ
G0 CRU Kwok Meng Tham, Flat 4, Warwick Court, 4 Lansdowne Road, London, SW20 8AP
G0 CRX K Pearson, Kalmia, Bishampton Road, Worcester, WR7 4DT
G0 CRY T Sparrey, 16 Rosemary Road, Parkstone, Poole, BH12 3HB
G0 CSK P senior, 13 St. Michaels Avenue, Swinton, Mexborough, S64 8NX
GM0 CSN R Trussler, 19 Royellen Avenue, Hamilton, ML3 8QH
G0 CSS H Shillitto, 25 Commonside, Selston, Nottingham, NG16 6FN
G0 CSU Terence Chadwick, 9 Ernest Street, Prestwich, Manchester, M25 3HZ
G0 CSV J Capindale, 2 Rivan Grove, Grimsby, DN33 3BL
G0 CSW William Hudson, 9 Nethergate, Dudley, DY3 1XW
G0 CSY W Dunstan, 5 Twemlow Lane, Holmes Chapel, Crewe, CW4 8DT
GM0 CSZ Frank Dinger, Shore Acre, Inver, Tain, IV20 1RX
G0 CTC Ernest Finnesey, 4 St. Catherines Close, Liverpool, L36 5RX
G0 CTD A Shipperley, 72 Hithercroft Road, Downley, High Wycombe, HP13 5RH
G0 CTF W Sargent, Likoma, 32 Seaton Down Road, Seaton, EX12 2SB
G0 CTG G Darrell, 10 Clatter Brune Estate, Presteigne, LD8 2LB
G0 CTH H Churchman, Westcroft, Church Road, Chester, CH4 9NG
GI0 CTI M Murphy, 97 Longfield Road, Mullaghbawn, Newry, BT35 9TX
G0 CTP James Smart, 4 Sycamore Close, Holmes Chapel, Crewe, CW4 7BT
G0 CTQ I Whiffin, 42 Canute Road, Birchington, CT7 9QH
G0 CTR Peter Martin, 32 Warkworth Court, Ellesmere Port, CH65 9EN
G0 CTS Rosalie Gumb, 17 Castle Lane, Bolsover, Chesterfield, S44 6PS
G0 CTZ F Gray, 18 Shepherds Court, Sheep House, Farnham, GU9 8LF
G0 CUA M Dissanayake, 9 Sweyn Place, Blackheath, London, SE3 0EZ
G0 CUB G Smith, 55 Countess Way, Euxton, Chorley, PR7 6PT
G0 CUH Steven Crane, 70 Highertown, Truro, TR1 3QD
G0 CUI L Hobson, 25 Bevan Close, Elsecar, Barnsley, S74 8DR
G0 CUL B Chilvers, 99 Links Avenue, Hellesdon, Norwich, NR6 5PQ
G0 CUN P Connolly, 21 Hartwood Green, Hartwood, Chorley, PR6 7BJ
G0 CUO J Hewitt, Peel House, Sacriston Lane, Durham, DH7 6TF
G0 CUX Robert Bedford, 26 Devon Avenue, Fleetwood, FY7 7EA
G0 CUZ C Morris, 12 Turners Hill Road, Lower Gornal, Dudley, DY3 [illegible]
G0 CVB K Albon, 65 Belmont Road, Kirkby in Ashfield, Nottingham, NG17 9DY
G0 CVC [illegible], 8 [illegible] Terrace, Halifax, HX1 4TA
GM0 [illegible] [illegible] Tigh, 7 Holmisdale, Isle of Skye, IV55 8WS
G0 CVH V Hughes, 27 Billy Lane, Clifton, Manchester, M27 8FS
G0 CVI Storm Khandro, 107 Castle Hill Gardens, Torrington, EX38 8LX
G0 CVM W Auty, 22 Hurley Close, Great Sankey, Warrington, WA5 1XG
G0 CVN I Howsham, West Street, North Kelsey, Lincoln, LN7 6EL
GM0 CVP O/B/O RHOSLLANNERCHRUGOG GROUP c/o C Phillips, Lonnie, Sanday, Orkney, KW17 2BA
GW0 CVY D Edwards, 26 Russell Terrace, Carmarthen, SA31 1SY
G0 CWA N Strong, 46 Malpas Drive, Great Sankey, Warrington, WA5 1HN
G0 CWD Ian Donachie, 72 Gresham Road, Norwich, NR3 2NQ
G0 CWF Michael Warr, 17 Moray Close, Peterlee, SR8 1DQ
G0 CWG A Evans, 8 Rhos Y Gad, Llanfairpwllgwyngyll, LL61 5JE
G0 CWH John Carden, 7 Rowland Close, Gillingham, ME7 3DJ
G0 CWK J Seymour, Mdina, Offwell, Honiton, EX14 9SA
G0 CWL D CHAPATON, 1 Rue De L'Angile, Lyon, France, 69005
G0 CWO L Gough, 7 Congreve Road, Stoke-on-Trent, ST3 2HA
G0 CWP K Stather, 5 St. Margarets Road, Bolton le Sands, Carnforth, LA5 8EN
G0 CWQ Alex Nesbitt, Amberley, Stokeinteignhead, Newton Abbot, TQ12 4QS
GM0 CWR William Pentland, Cambir, Faskally, Pitlochry, PH16 5LA
G0 CWS Alan Hattersley, The Bungalow, Top Lane, Buxton, SK17 8LP
G0 CWU F Humphreys, 31 Springfield Way, Oakham, LE15 6QA
G0 CWW R Gooden, 39 Heath Road, Ipswich, IP4 5RZ
G0 CWX M Shotter, Peverley, Newport Road, Cowes, PO31 8PE
G0 CWZ T Cattley, Yew Tree Cottage, Ifton Heath, Oswestry, SY11 3DH
G0 CXD T Moore, Highfields, Ashbourne Road, Belper, DE56 2LH
G0 CXF William Hamilton, 79 Ecroyd Park, Credenhill, Hereford, HR4 7EN
G0 CXJ A Beasley, 2 Ilmington Road, Blackwell, Shipston-on-Stour, CV36 4PG
GW0 CXK F Johns, Manteg, Penslade, Fishguard, SA65 9PB
G0 CXO S Wellings, 11 Matlock Road, Walsall, WS3 3QD
G0 CXU R Pitter, 57 Greenhill Way, Farnham, GU9 8TA
G0 CXV R Clark, Woodlands, Islet Road, Maidenhead, SL6 8HT
G0 CXW Thomas Pearce, 16 Beech Lodge, Rosewoodlane, Shoeburyness, SS3 9FA
G0 CXX James Jopling, 54 Redesdale Gardens, Gateshead, NE11 9XH
G0 CXY Colin Monteith, 46 Lochryan Street, Stranraer, DG9 7HR
G0 CYB Paul Kelsall, 7 Buttermere Crescent, Doncaster, DN4 5QF
G0 CYC R Curtis, 125 Handside Lane, Welwyn Garden City, AL8 6TA
G0 CYD W Snow, Willbeard Farm, Greenditch Street, Bristol, BS35 4HJ
GW0 CYG D Davies, 40a Furzeland Drive, Bryncoch, Neath, SA10 7UG
G0 CYI J Knight, Urbanizacao Quinta Da Torre, Edificio Perola, Armacao de Pera, Portugal, 8365-184
GW0 CYK S Radford, 11 South Parade, Maesteg, CF34 0AB
G0 CYL M Sefton, 8 Sandmoor Avenue, Leeds, LS17 7DW
G0 CYO G Beddow, 12 Wulfruna Gardens, Finchfield, Wolverhampton, WV3 9HZ
G0 CYR D Bramley, 10 Thirlmere Close, Huncoat, Accrington, BB5 6JQ
G0 CYU D Hooper, 6 Kingswell Avenue, Outwood, Wakefield, WF1 3DY
G0 CYX J Faulkner, 11 Valley View, South Elmsall, Pontefract, WF9 2DD
G0 CZD M Kinder, 12 Jessop Way, Haslington, Crewe, CW1 5FU
G0 CZL William Riley, 10 Hough, Halifax, HX3 7AP
GM0 CZM W Taylor, 49 Pentland Avenue, Port Glasgow, PA14 6LF
G0 CZR Keith Laws, 6 Crabbe's Close, Feltwell, Thetford, IP26 4BD
G0 CZU H Richardson, 14 Melbreak Close, Whitehaven, CA28 9TG
G0 CZY P Heath, The Cottage, Great Staughton Road, Bedford, MK44 2BA
G0 DAB D Buik, 54a Buckshaft Road, Cinderford, GL14 3DZ
G0 DAC D Cowley, 81 Ashtree Road, Walsall, WS3 4LS
G0 DAE C Tidwell, 86 Powerscourt Road, North End, Portsmouth, PO2 7JW
G0 DAF J Randall, 26 Marian Road, Boston, PE21 9HA
G0 DAG R Blackburn, Wyvernholme, 141 Whalley Road, Blackburn, BB1 9NE
G0 DAH B Smith, 1 High St., Over, Cambridge, CB4 5NB
G0 DAI D Isom, 36 Deerfold, Astley Village, Chorley, PR7 1UH
G0 DAL G Daley, 5 Linden Road, Forest Town, Mansfield, NG19 0EL
G0 DAM D Clayton, 171 Warning Tongue Lane, Cantley, Doncaster, DN4 6TU
G0 DAU Martyn Saunders, 22 Humphreys Close, St. Cleer, Liskeard, PL14 5DP
G0 DAV D Paine, Woodland View, St. Mellion, Saltash, PL12 6RH
G0 DAX David Burt, 19b Midhurst Road, Eastbourne, BN22 9HP
G0 DAY Keith Banks, 52 Hunter Avenue, Burntwood, WS7 9AQ
G0 DAZ Colin Mister, Woodbine Cottage, Hanley Childe, Tenbury Wells, WR15 8QY
G0 DBC J Hudson, 1 Linnet Way, Biddulph, Stoke-on-Trent, ST8 7UF
G0 DBD John West, Stonecroft, 4 Trevella Road, Bude, EX23 8NA
G0 DBE Lee Marsland, 154 Moss Lane, Litherland, Liverpool, L21 7NN
G0 DBI K Banks, 28 Warnes Lane, Burley, Ringwood, BH24 4EL
G0 DBJ G Willetts, Waterside, 48 Stourton Crescent, Stourbridge, DY7 6RR
GM0 DBK Duncan Kerr, 3 Glaisnock View, Cumnock, KA18 3GA
G0 DBM S Lovesey, 20 Ferry Gardens, Quedgeley, Gloucester, GL2 4PB
G0 DBS P Le Feuvre, 56 Greenfields Avenue, Alton, GU34 2EE
GM0 DBW Malcolm Bolton, 11 Covenanters Drive, Corston, Aberdeen, AB12 5AB
G0 DBX D Beale, 17 Rue Du Passolis, Montseret, France, 11200
G0 DBY Peter Dodge, 425 Sutton Road, Maidstone, ME15 8RA
G0 DCF R Green, Kenville, West Lane, Sheffield, S26 3XS
G0 DCG Martin Brown, 31 Victoria Road, Littlestone, New Romney, TN28 8NL
G0 DCI A Merrylees, 90 Grangehill Road, Eltham, London, SE9 1SE
G0 DCJ J Warren, The Old Barn, Scotgate Close, Thetford, IP24 1PF
G0 DCN G Cheeseman, 19 Hazelwood Grove, Leigh-on-Sea, SS9 4DE

UK Callsigns

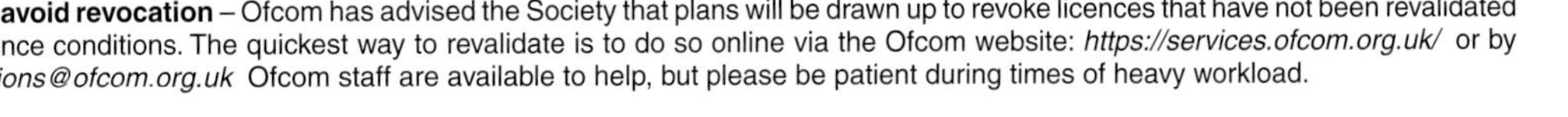
**IMPORTANT NOTE**

**Revalidate licence to avoid revocation** – Ofcom has advised the Society that plans will be drawn up to revoke licences that have not been revalidated as required by the licence conditions. The quickest way to revalidate is to do so online via the Ofcom website: *https://services.ofcom.org.uk/* or by email: *amateur.validations@ofcom.org.uk* Ofcom staff are available to help, but please be patient during times of heavy workload.

G0 DCO M Beirne, 14 Swiss Cottage, Bollinbrook Road, Macclesfield, SK10 3DJ
G0 DCP P Holdaway, Upper Flat, 9 Harecourt Road, London, N1 2LW
G0 DCR J Frost, 36 York Gardens, Braintree, CM7 9NF
G0 DCS P Ashton, 27 Dunsby Road, Luton, LU3 2UA
G0 DCU J Faithfull, 54 Cardiff Place, Bassingbourn, Royston, SG8 5LR
G0 DCW Aldo Corallini, 8 Britannia, Puckeridge, Ware, SG11 1TG
G0 DCZ Phillip Richards, 5 Pumphouse Close, Longford, Coventry, CV6 6RE
G0 DDA P White, 11 Elms Road, Fareham, PO16 0SQ
G0 DDE B Dignum, 16 Stirling Court Road, Burgess Hill, RH15 0PT
G0 DDF D Fairchild, 2 Linacre Road, Torquay, TQ2 8LE
G0 DDJ Anthony King, 4 Tyne Close, Wellingborough, NN8 5WT
GW0 DDK E Down, Silver Hill, Pen y Cwm, Haverfordwest, SA62 6JZ
GW0 DDL G Jones, 9 George Street, New Quay, SA45 9QR
G0 DDT James Barnes, 262 King Henrys Drive, New Addington, Croydon, CR0 0AA
G0 DDU James Norris, 15 Liverpool Road North, Burscough, Ormskirk, L40 5TN
G0 DDV N Hewett, 49 Harrow Way, Carpenders Park, Watford, WD19 5EH
G0 DDW M Hillier, 28 Meadow Walk, Bridgemary, Gosport, PO13 0YN
G0 DDY P Piper, 5 Goodwood Close, Midhurst, GU29 9JG
G0 DDZ M Eastman, 23 Haughgate Close, Woodbridge, IP12 1LQ
G0 DEB D Bannister, 60 St. Johns Avenue, Bridlington, YO16 4NL
G0 DEC D Willicombe, 26 Falkland Court, Braintree, CM7 9LL
G0 DEE Ralph Spilling, 20 Saxonfields, Poringland, Norwich, NR14 7JE
G0 DEF M Mutton, 39 Martin Road, Kettering, NN15 6HF
G0 DEH P Finbow, 6 Down Road, Teddington, TW11 9HA
G0 DEJ W Rutt, 24 Coopers Lane, Verwood, BH31 7PG
G0 DEK F Underwood, Hobletts, Fen Lane, Grays, RM16 3LT
G0 DEO W Batey, 13 Cassiobury Avenue, Feltham, TW14 9JE
G0 DEP D Jarrard, 26 Lingmell Court, Tolladine, Worcester, WR4 9YU
GM0 DEQ Robert Alexander, 9 Weston Place, Prestwick, KA9 2ED
G0 DER D Gillmore, 4 Holly Ridge, Fenns Lane, Woking, GU24 9QE
G0 DEU J Tournant, 47 High St., Linton, Cambridge, CB1 6HS
GM0 DEX A Goldie, 87 Ardrossan Road, Seamill, West Kilbride, KA23 9NF
G0 DEZ Dez Watson, 3 Brunel Drive, Biggleswade, SG18 8BT
G0 DFA D Allsopp, 10 Chalfont Close, Middleton-on-Sea, Bognor Regis, PO22 7SL
G0 DFC Leslie Cropley, The New Bungalow, Brundish Road, Eye, IP21 5LS
GI0 DFD R McAlister, 78 Cairn Road, Carrickfergus, BT38 9AP
G0 DFE J Stone, 12 Main Road, Hawkwell, Hockley, SS5 4JN
G0 DFF J Sidnell, 17 Barlings Road, Harpenden, AL5 2AL
G0 DFI D Oakley, 6 Staplehurst Gardens, Cliftonville, Margate, CT9 3JB
G0 DFO J Tomlinson, 33 Belgrave Street, Nelson, BB9 9HR
G0 DFT Joseph Maw, 10 Shamrock Close, Newcastle upon Tyne, NE15 8TW
G0 DFU E Spanner, 30 Lowtherville Road, Ventnor, PO38 1AP
G0 DFV R Cadd, 27 Grove Road, Brafield on the Green, Northampton, NN7 1BW
G0 DFY R Anthony, 2 Barrfield Road, Rhuddlan, Rhyl, LL18 2RY
G0 DGA D Gullick, Greenleas, Courthay Orchard, Langport, TA10 9AE
G0 DGB L Connell, 24 Finchale Road, Framwellgate Moor, Durham, DH1 5JN
G0 DGE J Gullick, Greenleas, Courthay Orchard, Langport, TA10 9AE
G0 DGF Malcolm Beakhust, 63 Chadacre Road, Stoneleigh Park, Epsom, KT17 2HD
G0 DGH G Daniels, 81 London Road, Clacton-on-Sea, CO15 3SR
GW0 DGJ C Riddle, 18 Windsor Mews, Adamsdown Square, Cardiff, CF24 0HS
GM0 DGK A Donaldson, 30 Jeanfield Crescent, Forfar, DD8 1JR
G0 DGQ G Dudley, 95 Alfreton Road, South Normanton, Alfreton, DE55 2BJ
G0 DGU P Brown, 17 Freyden Way, Frettenham, Norwich, NR12 7NB
G0 DGW D Barham, 64 Gorran Avenue, Rowner, Gosport, PO13 0NF
G0 DHA R Waller, 4 Rose Court, Ty Canol, Cwmbran, NP44 6JH
G0 DHB Ralph Evans, 20 Pulley Avenue, Eaton Bishop, Hereford, HR2 9QN
GM0 DHD A Lymer, 16 Gerson Park, Greendykes Road, Broxburn, EH52 6PL
G0 DHG B Ristic, 168 Heather Road, Newport, NP19 7QW
G0 DHI A Rutherford, 19 Briar Bank, Carlisle, CA3 9SN
G0 DHJ J Wraight, 59 Sandy Lane, Walton, Liverpool, L9 9AY
G0 DHL M Mason, 7 Clayhill Copse, Peatmoor, Swindon, SN5 5AL
G0 DHM D Moore, Stoke Hall Farm, Stoke on Tern, Market Drayton, TF9 2DU
G0 DHR H Rutt, 3 Russell Place, Highfield, Southampton, SO17 1NU
G0 DHS Alan Kitching, 1 Borrowdale, Albany, Washington, NE37 1QD
G0 DHT K Smith, Lower Carniggey Farm, Greenbottom, Truro, TR4 8QL
GI0 DHW Charles Cleland, 19 Sheskin Way, Belfast, BT6 0ER
GM0 DHZ H Johnsen, 7 Mure Place, Minishant, Maybole, KA19 8ES
G0 DIA Allan Holland, 156 Perry Rise, London, SE23 2QP
G0 DIB G Mcconnell, Paris House, 9 Bell Street, Brecon, LD3 0BP
G0 DIG P Johnson, 5 Brook Bank, Whitehaven, CA28 8PZ
G0 DIH Paul Strong, 2 Jasper Cottages, Cornworthy, Totnes, TQ9 7EY
G0 DIM A Latham, 49 Tithe Barn Road, Wootton, Bedford, MK43 9EZ
G0 DIP J Huggins, 7 Coniston Drive, Jarrow, NE32 4AE
GW0 DIQ M Smith, 7 Clos Gorsfawr, Grovesend, Swansea, SA4 4GZ
G0 DIR Stephen Caslake, Bishopwood Cottage, Wistow Common, Selby, YO8 3RD
G0 DIS P Pearce, 26 Milnthorpe Drive, Wakefield, WF2 7HU
G0 DIU T Rumble, 1 Victoria Street, Brighouse, HD6 1HH
GW0 DIV Rhys Griffiths, 5 Heol-y-Sarn, Llantrisant, Pontyclun, CF72 8DA
GW0 DIX Ronald Rees, 22 The Complex, Tan Y Bryn, Burry Port, SA16 0HP
G0 DIZ R Quaintance, 18 Queens Avenue, Ilfracombe, EX34 9LN
G0 DJA David Ackrill, 59 Moor Lane, Bolsover, Chesterfield, S44 6EW
G0 DJC Denis Collins, 71 Trench Road, Tonbridge, TN10 3HG
GM0 DJG J Walker, 5 Shiskine Drive, Kilmarnock, KA3 1PZ
G0 DJK D Keates, 13 Willow Rise, Witham, CM8 2LL
G0 DJL C Blount, 42 Penmere Drive, Newquay, TR7 1QQ
G0 DJM Michael Rawson, 16 Templar Gardens, Wetherby, LS22 7TG
G0 DJO A Robson, 19 Barnard Close, Bedlington, NE22 6NE
G0 DJQ R Salt, 1 Weaver Cottages, Rue Hill, Stoke-on-Trent, ST10 3HD
G0 DJS H Stemp, 5 Depot Road, Horsham, RH13 5HB
G0 DJT P Odegaard, Flat 4, 31 Birdhurst Road, South Croydon, CR2 7EF
G0 DJU David Lee, LD Le Champ Du Meslier, Romagne, France, 86700
G0 DJV B Joy, 1 Riverbourne Road, Salisbury, SP1 1NU
G0 DJX A Cullen, 30 Llys Cyncoed, Oakdale, Blackwood, NP12 0NQ
G0 DKF R Thomas, 48 Maryport Street, Usk, NP15 1AD
GW0 DKG Roy Malpas, 26 St. Davids Avenue, Whitland, SA34 0AF
GM0 DKK Arthur Mackenzie, 6 Princess Street, Bonnybridge, FK4 1BJ
G0 DKM S Daniels, 27 Willow Drive, Hutton, Weston-Super-Mare, BS24 9TJ
G0 DKN A McClelland, 12 Grove Heath North, Ripley, Woking, GU23 6EN
G0 DKO G Maskort, 133 Borstal Street, Rochester, ME1 3JU
G0 DKQ G Pool, 41 Radford St., Manton, Worksop, S80 2NQ
G0 DKR M Cole, 54 Ribble Road, Coventry, CV3 1AU
G0 DKS R Barnes, Pentwyn, Graeme Road, Yarmouth, PO41 0RX
G0 DKV E Peberdy, 18 Arden Road, Kenilworth, CV8 2DU
G0 DKX G Birk, 30 Maple Drive, Alvaston, Derby, DE24 0FT
G0 DKY J Bass, 8 Ann Close, Hassocks, BN6 8NB
G0 DKZ Brian Yates, 9 Cloister Walk, Whittington, Lichfield, WS14 9LN
GW0 DLA E Stuckey, 32 Stanley Road, Gelli, Pentre, CF41 7NJ
G0 DLB R Burdett, 11 Fisher Avenue, Rugby, CV22 5HN
G0 DLF A Keon, 72 Rochelle Way, Duston, Northampton, NN5 6YW
G0 DLK G Hulme, 15 Eastholme Drive, Rawcliffe Industrial Estate, York, YO30 5SU
G0 DLL W Jackson, 22 Cliff Gardens, Scunthorpe, DN15 7PJ
G0 DLP P Lee, 14 Downs Court Road, Purley, CR8 1BB
G0 DLQ G Orange, 27 Connery, Hucknall, Nottingham, NG15 7AH
G0 DLS D Sparey, 21 Buxton Road, Ashbourne, DE6 1EX
G0 DLT J Aizlewood, 7 Scott Avenue, Simonstone, Burnley, BB12 7HY
G0 DLW J Goldsmith, Woodlands, Candy, Oswestry, SY10 9AZ
G0 DMA J Slaney, 50 Laburnum Road, Langold, Worksop, S81 9RR
G0 DMB Colin Cade, 6 Court Close, Kirby Muxloe, Leicester, LE9 2DD
G0 DME J Hallam, 34 Danethorpe Vale, Nottingham, NG5 3DA
G0 DMH Bruce-Ernst Cox, 21 Shelthorpe Road, Loughborough, LE11 2PB
G0 DMJ Neil Beck, 36 Grove Street, Great Hale, Sleaford, NG34 9JZ
G0 DMK Dennis King, 94 Western Road, Mickleover, Derby, DE3 9GQ
G0 DMN S Langham, 43 Greenwood Drive, Kirkby-in-Ashfield, Nottingham, NG17 8JT
G0 DMP D Potter, 102 Normandy Avenue, Beverley, HU17 8PF
G0 DMS F Spencer, 35 Askew Grove, Repton, Derby, DE65 6GR
G0 DMU I Morison, 4 Arley Close, Macclesfield, SK11 8QP
G0 DMV Raymond Parrish, 89 Delamere Drive, Macclesfield, SK10 2PS
G0 DMW F Cosgrove, Denton Park Middle School, Linhope Road, Newcastle upon Tyne, NE5 2NW
G0 DND Neil Downs, 10 Oak Street, Northwich, CW9 5LJ
G0 DNF Dave Fisher, 3 Brookdale Close, Waterlooville, PO7 7NY
GM0 DNG G Wallace, 21 The Grange, Perceton, Irvine, KA11 2EU
GM0 DNH P Moore, 105 Fintry Drive, Dundee, DD4 9HQ
G0 DNI Gary Wann, Manor Barn, Shotatton, Shrewsbury, SY4 1JH
G0 DNQ Colin Anderton, 19 Berrylands Close, Wirral, CH46 7UT
G0 DNU R Hamblett, 49 Kneller Road, Twickenham, TW2 7DF
G0 DNV R Hammett, 47 Sowden Park, Barnstaple, EX32 8EJ
G0 DNY B Whysker, 21 Heyland Road, Manchester, M23 1HF
G0 DOA C Dale, 11 Roman Close, Newby, Scarborough, YO12 5RG
G0 DOB R Gray, 3 Perth Way, Immingham, DN40 1PW
G0 DOC H Kiff, Rock Cottage, The Cloudside, Congleton, CW12 3QG
G0 DOE T Purcell, 38a Moor Lane, Chessington, KT9 1BW
G0 DOG C Purcell, 76 Wensleydale Road, Great Barr, Birmingham, B42 1PL
G0 DOK Robert Cornish, 18 Rooksbury Croft, Havant, PO9 5HU
G0 DOM D Oskis, 10 Moultrie Way, Cranham, Upminster, RM14 1NB
G0 DOR P Davidson, 5 Derby Grove, Maghull, Liverpool, L31 5JJ
G0 DOU H Grandfield, 2 Bolshaw Road, Heald Green, Cheadle, SK8 3PJ
G0 DOW David Binns, Apple Tree Barn Highampton, Beaworthy, EX21 5LP
G0 DOZ John Rozday, 130 Walton Park, Pannal, Harrogate, HG3 1RJ
G0 DPC J Cook, 85 Newington Avenue, Southend-on-Sea, SS2 4SP
G0 DPE B Aldersey, 4 Salterbeck Terrace, Salterbeck, Workington, CA14 5HP
G0 DPG David Ganner, 58 Kilngate, Lostock Hall, Preston, PR5 5UW
G0 DPI D Barkley, 39 Fulbeck Avenue, Wigan, WN3 5QN
G0 DPJ K Ellis, 162 Wolverhampton Road, Dudley, DY3 1RF
G0 DPK P Mccaldon, 47 Merritt Road, Didcot, OX11 7DF
G0 DPO Kenneth Glazebrook, 86 Deveraux Drive, Wallasey, CH44 4DL
G0 DPQ A Henning, 24 Garfield Avenue, Draycott, Derby, DE72 3NP
G0 DPS J Fyrth, 2 Merton Gardens, Farsley, Pudsey, LS28 5DZ
G0 DPT M Smith, 32 Amesbury Drive, London, E4 7PZ
GI0 DPV James Mangan, Glenaulin House, Glen Road, Belfast, BT11 8BP
G0 DPX J Brown, 72 Whitcliffe Road, Cleckheaton, BD19 3BY
G0 DPY Andrew Garner, 7 Danes Court, Grimoldby, Louth, LN11 8TA
G0 DQB R White, 27 Windsor Walk, Scawsby, Doncaster, DN5 8NQ
GM0 DQC H Meikle, 20 Muirsland Place, Lesmahagow, Lanark, ML11 0FF
G0 DQH J Colwill, 18 Collingbourne Drive, Chandler's Ford, Eastleigh, SO53 4SW
G0 DQI D Harding, High Peak, Hillcrest Road, Deal, CT14 8EB
GI0 DQJ D Livingstone, 16 Stronge Court, Portadown, Craigavon, BT62 3QX
G0 DQM John Williams, Maes-yr-Awel, Suttonfield Road, Doncaster, DN6 9JX
G0 DQO W Cartwright, 50 Kings Road, Walsall, WS4 1JB
G0 DQQ S Power, 10 Beach Road, Hartford, Northwich, CW8 4BA
G0 DQS M Glen, 10 Field Lane, Dursley, GL11 6JE
G0 DQT A Duck, 15 Ambryn Road, New Inn, Pontypool, NP4 0NJ
GM0 DQV G George, 13 Balmoral Terrace, Elgin, IV30 4JH
G0 DQW Chris Hughes, 3 Crown Rise, Llanfrechfa, Cwmbran, NP44 8UG
G0 DQY G Hughes, 39 Thornhill Close, Upper Cwmbran, Cwmbran, NP44 5TQ
G0 DRA Derek Love, 4 St. Chads Road, Lichfield, WS13 7LZ
G0 DRC Dartmoor RC c/o Derek Dukes, 8 The Village, Jacobstowe, Okehampton, EX20 3RF
G0 DRD Sean Carvin, 43 Brackenstown Village, Dublin, Ireland
G0 DRE J Webster, 3 Badby Road West, Daventry, NN11 4HJ
G0 DRH B Harris, 9 Woodlands Close, Rayleigh, SS6 7RG
GW0 DRI J Smith, 7 Clos Gorsfawr, Grovesend, Swansea, SA4 4GZ
G0 DRJ June Connell, 24 Finchale Road, Framwellgate Moor, Durham, DH1 5JN
G0 DRL G Howarth, 15c Shaftesbury Road, Southsea, PO5 3JA
G0 DRM D Cookson, 70 Rope Lane, Wistaston, Crewe, CW2 6RD
G0 DRN M Jenkin, 22 Shelmore Way, Gnosall, Stafford, ST20 0DT
G0 DRO D Roberts, Flat 1, 129 Prestbury Road, Macclesfield, SK10 3DA
G0 DRQ R Hope, 3 Farm Crescent, Sittingbourne, ME10 4QD
G0 DRR Owen Perry, 9 Home Park Close, Bramley, Guildford, GU5 0JP
G0 DRS R Ford, 11 Lincoln Road, Ewloe, CH5 3RW
G0 DRT P Quested, Nethercroft, Southsea Avenue, Sheerness, ME12 2NH
GM0 DRU I Maclennan, 70 Kenneth Street, Stornoway, HS1 2DS
G0 DRV W Hoyle, 10 Picton Gardens, Rayleigh, SS6 7LB
G0 DRW Trevor Wright, 73 West Street, Ryde, PO33 2QQ
G0 DRX P Hill, 34 Church Road, Whitchurch, Bristol, BS14 0PP
G0 DSB T Winship, 32 Lytes Cary Road, Keynsham, Bristol, BS31 1XD
GI0 DSG William Mckeever, 17 The Hawthornes, Londonderry, BT48 8TH
G0 DSJ Edward Shipton, 51 Maes Stanley, Bodelwyddan, Rhyl, LL18 5TL
G0 DSK John Williams, Skelmorlie, 1 Dover Road, Sandwich, CT13 0BH
G0 DSN L Nash, Four Furlongs, Wells Road, Wells-next-The-Sea, NR23 1QE
G0 DSO R Calvert, Shortley Close, Robin Hoods Bay, Nr Whitby, YO22 4PB
G0 DSP M Lamb, 4 Hadfield Close, Connah's Quay, Deeside, CH5 4JP
G0 DSQ Khin Myint, 1 Belton Road, Camberley, GU15 2DE
G0 DSR D Jones, 20 Marsh Green, Wigan, WN5 0PU
G0 DSU G Fisher, 19 Orde Close, Crawley, RH10 3NG
G0 DSX N Hanking, 7 Clayside House, Kenton Court, South Shields, NE33 4HP
G0 DTC David Coxon, 13 Gate Farm Road, Shotley Gate, Ipswich, IP9 1QH
G0 DTI K Sumner, 7 Largs Road, Shadsworth, Blackburn, BB1 2JQ
G0 DTP L Quantrill, Innisfree, 1 Ironwell Lane, Hockley, SS5 4JY
G0 DTQ W Goldstraw, 5 Council Houses, Wantage Road, Hungerford, RG17 7DG
G0 DTT R Mycock, 14 Mount Pleasant, Tintwistle, Glossop, SK13 1LF
G0 DTW S Bolton, 40 Claytonwood Road, Stoke-on-Trent, ST4 6LD
G0 DUA S Linden, 4 Downing Drive, Great Barton, Bury St. Edmunds, IP31 2RP
G0 DUB Francis Mossop, 4 Brookdale Way, Waverton, Chester, CH3 7NT
G0 DUF R Hoare, 7 Springfield Close, Watlington, OX49 5RF
G0 DUG D Marsden, 127 Morley Crescent, Kelloe, Durham, DH6 4NP
G0 DUH P Murrell, 10 Irving Close, Braunton, EX33 1DH
G0 DUI Phillip Smith, 20 Deanscroft Way, Stoke-on-Trent, ST3 5XW
G0 DUK Keith Bennett, 78 Rectory Road, Upper Deal, Deal, CT14 9NB
G0 DUM D Gibbons, 17 Della Avenue, Barnsley, S70 6LG
G0 DUN David Wakeford, 2 Rooley House Cottage, Rooley Lane, Sowerby Bridge, HX6 1NS
GI0 DUP R Miskimmin, 15 Abbeydale Avenue, Newtownards, BT23 8RT
G0 DUQ R Wilkes, 47 Greenwood Park, Hednesford, Cannock, WS12 4DQ
G0 DUS M Marlow, 17 Ord Road, Fornham St. Martin, Bury St. Edmunds, IP31 1TB
GM0 DUX Robert McGowan, 35 Ochiltree, Dunblane, FK15 0DF
G0 DVB John Morris, 18 Ellingdon Road, Wroughton, Swindon, SN4 9HY
GD0 DVC P Robinson, 256 Victoria Road, Ruislip, HA4 0DW
G0 DVE S Hutchings, 5 Dales Close, Wimborne, BH21 2JU
G0 DVG T Callaghan, 27 Thealby Lane, Thealby, Scunthorpe, DN15 9AG
GM0 DVH N Mcnulty, 6 Main Road, Crookedholm, Kilmarnock, KA3 6JT
G0 DVJ Jonathan Mitchener, Cabins, Wenham Road, Ipswich, IP8 3EY
G0 DVL R Nash, 28 Squires Way, Wilmington, Dartford, DA2 7NW
GM0 DVO A GEMMELL, 23 Busby Road, Carmunnock, Glasgow, G76 9BN
G0 DVP G Gulliford, 29 Windsor Road, Seaham, SR7 8DG
G0 DVS E holding, 11 Dane Crescent, Ramsgate, CT11 7JU
G0 DVT John Brindle, 1 Holywell Close, Bury St. Edmunds, IP33 2LS
GI0 DVU James Henry, 3 Kirkwoods Park, Lisburn, BT28 3RR
G0 DVY R Ibbotson, Fern Lea, Alford, LN13 0JP
G0 DWB D Wathen-Blower, 61 Dykes End, Collingham, Newark, NG23 7LD
G0 DWC S Beadle, 18 The Shrubberies, Cliffe, Selby, YO8 6PW
G0 DWD John Hawes, Cherry Lodge, Woodwaye, Reading, RG5 3HA
G0 DWE Christopher Brown, 12 Forest Close, Newport, PO30 5SF
G0 DWF D Fouche, 17 Burlington Gardens, Rainham, Gillingham, ME8 8TA
GM0 DWH James Cobley, 15 Fintry Terrace, Bourtreehill South, Irvine, KA11 1JD
G0 DWJ N Hall, 10 Newnham Road, Leamington Spa, CV32 7SN
G0 DWM C Pearsons, 12 Abbey Way, Farnborough, GU14 7DA
GI0 DWN M Dougan, 97 Redrock Road, Collone, Armagh, BT60 2BN
G0 DWO P Labron, 22 Fourth Avenue, Morpeth, NE61 2HJ
GW0 DWQ Stephen Outen, 2 Heol Vaughan, Burry Port, SA16 0HF
G0 DWR P Bell, 5 Sages Lane, Privett, Alton, GU34 3NP
G0 DWS J Tubbs, 19 Greenhill Road, Northfleet, Gravesend, DA11 7EZ
G0 DWV Christopher Danby, Fir Trees, Hall Road, Norwich, NR10 3LX
GM0 DWY R Price, 80 Eastern Avenue, Largs, KA30 9EQ
G0 DWZ M Sharp, 50 Milton Drive, Southwick, Brighton, BN42 4NE
GM0 DXB W McGill, 112 West Main Street, Armadale, Bathgate, EH48 3JB
GM0 DXE Henry Antonine Quin, Flat/Edinbane Shop, Edinbane Shop, Portree, IV51 9PW
G0 DXF C Emmanuel, 29 Guillemot Way, Liverpool, L26 7WG
G0 DXG CWMBRAN CONTEST & DX GROUP c/o D Stoole, Brookside Farm, Baltic Terrace, Cwmbran, NP44 7AH
GM0 DXI H Lakhaney, 5 Snowberry Fields, Thankerton, Biggar, ML12 6RJ
G0 DXK Melvyn Bedford, 12 Winchester Drive, Mablethorpe, LN12 2AY
G0 DXO D Oates, 86 Queens Avenue, Maesgeirchen, Bangor, LL57 1NG
G0 DXT T Pearson, 7 Lathkill Close, Tibshelf, Alfreton, DE55 5LU
GU0 DXX P Guilbert, La Mignonette, Rue De La Foret, St Martin, Guernsey, GY4 6UB
GW0 DXZ Gloria Stephens, Ty Coch, Rhydwen Place, Clydach, SA6 5RN
GM0 DYD D Young, 4 Primrose Avenue, Rosyth, Dunfermline, KY11 2SS
GM0 DYF C O'Hare, 14 Pentland Road, Bellfield, Kilmarnock, KA1 3RS
G0 DYG D Doyle, 40 Howson St., Rock Ferry, Birkenhead, L42 2BR
G0 DYH S Fergusson, 37 Station Road, Old Colwyn, Colwyn Bay, LL29 9EL
G0 DYL Carol Jones, 63 Hockenhull Avenue, Tarvin, Chester, CH3 8LR
G0 DYM Mervyn Rivers, 5 Ann Carter Close, Hereford, HR2 7LS
GM0 DYU M Mcwhinnie, 35 Morrison Place Cruden Bay, Peterhead, AB423HZ
G0 DYW I Dowse, 57 Palmer Crescent, Leighton Buzzard, LU7 4HY
G0 DZA P Wentworth, 46 Woodside Avenue, Cinderford, GL14 2DW
G0 DZB Peter Onion, 56 New Park Street, Colchester, CO1 2NA
G0 DZC C Cosgrif, 53 Lower Manor Lane, Burnley, BB12 0EF
GD0 DZH D Gough, 20 Lawn Close, Ruislip, HA4 6ED
G0 DZI Charles Kuss, 20 Windermere Road, Haydock, St. Helens, WA11 0ES
GW0 DZL J Davies, 5 Talbot Street, Llanelli, SA15 1DG
G0 DZM Philip Gainey, Prencott, Harley Wood, Stroud, GL6 0LD
G0 DZQ B Stevens, 24 Waverley Crescent, Runwell, Wickford, SS11 7LN
G0 DZU P Barker, Peartree Cottage, Ash Hill Common, Romsey, SO51 6FU
G0 DZV K Rose, 57 Cheriton Road, Winchester, SO22 5AX
GM0 DZW R Webster, Tigh-Na-Darroch, Old Line Road, Ballater, AB35 5UT
G0 DZX John Huggins, 12 Willow Chase, North Anston, Sheffield, S25 4DQ
G0 DZY B Sparrow, 11b Croft Place, Mildenhall, Bury St. Edmunds, IP28 7LN
G0 DZZ John Hall, 4 Dorking Crescent, Clacton-on-Sea, CO16 8FQ
G0 EAE M Taylor, 26 St. Marys Road, Bozeat, Wellingborough, NN29 7JU
G0 EAG A Sammons, 15 Swallow Drive, Benfleet, SS7 5EN

| | | |
|---|---|---|
| GM0 | EAH | Alan McDougall, 18 Rutherwood Avenue, Glasgow, G13 2RJ |
| G0 | EAM | A Moore, 69 Renfrew Avenue, St. Helens, WA11 9RW |
| G0 | EAN | Colin Bibb, 51 Dorsett Road, Wednesbury, WS10 0JF |
| G0 | EAT | Stephen Anderson, 65 Sands Lane, Holme-on-Spalding-Moor, York, YO43 4HJ |
| G0 | EAU | Abdulla Surooprajally, 26 Walton Avenue, North Cheam, Sutton, SM3 9UB |
| G0 | EAW | J Humphries, 19 Tai Newydd, Llanfaelog, Ty Croes, LL63 5TW |
| G0 | EBD | M Element, 9 Longbridge Close, Shrewsbury, SY2 5YD |
| G0 | EBF | L Taylor, 18 Manifold Road, Eastbourne, BN22 8EH |
| G0 | EBG | H Fuller, 41 Burnham Road, Hullbridge, Hockley, SS5 6BG |
| G0 | EBI | J Speller, 43 Castle Hill Park, London Road, Clacton on Sea, CO16 9QP |
| G0 | EBK | Rodney Bickley, 8 East Woodhay Road, Winchester, SO22 6JH |
| G0 | EBL | Kevin Mayes, Stone House, [illegible] |
| G0 | EBP | A Bowmaker, 1 [illegible] Drive, Morecambe, LA4 4QD |
| G0 | EBQ | N Flatman, 2 Deben Valley Drive, Kesgrave, Ipswich, IP5 2FB |
| G0 | EBS | S Artus, 14 Bromley Road, East Peckham, Tonbridge, TN12 5BW |
| G0 | EBW | D Agar, 122 Salisbury Road, Moseley, Birmingham, B13 8JZ |
| G0 | EBZ | M Whitfield, Camp Farm, Elberton, Bristol, BS35 4AQ |
| G0 | ECB | M Hayhurst, 3 Burton Gardens, Brierfield, Nelson, BB9 5DR |
| G0 | ECG | Malcolm Higgin, 24 Tiverton Drive, Briercliffe, Burnley, BB10 2JT |
| G0 | ECI | R Matthews, 20a Main Street, Seaton, Oakham, LE15 9HU |
| G0 | ECJ | H Hughes, Asham, Walton Hill, Gloucester, GL19 4BT |
| G0 | ECK | P jenkins, 49 Ewell Park Way, Ewell, Epsom, KT17 2NW |
| G0 | ECL | K Clift, 28 Redgate Road, Girton, Cambridge, CB3 0PP |
| G0 | ECM | M Bell, 18 Linnet Close, Patchway, Bristol, BS34 5RN |
| G0 | ECN | Keith Watkinson, Sunnyview, Beacon Way, Skegness, PE25 1HL |
| G0 | ECQ | Charles Quinnin, 10 Willow Avenue, Blyth, NE24 1PG |
| G0 | ECS | G Westaby, 2 Goodwood, Bottesford, Scunthorpe, DN17 2TP |
| G0 | ECU | R Low, 56 George Street, Whithorn, Newton Stewart, DG8 8NZ |
| G0 | ECW | E Wilson, 20 Wivelsfield Road, Saltdean, Brighton, BN2 8FQ |
| G0 | ECX | Robert Mott, 2 Dennis Road, Weymouth, DT4 0NJ |
| G0 | ECZ | Brian Clarke, 4 Prospect Road, Langford, Biggleswade, SG18 9NY |
| G0 | EDC | B Hall, 7 Ferndale Close, Penyffordd, Chester, CH4 0NH |
| G0 | EDE | D Proctor, 7 Main Avenue, Westhill, Torquay, TQ1 4HZ |
| G0 | EDF | G Lunt, 45 Malvern Road, Liverpool, L6 6BN |
| G0 | EDH | S Boundy, Sha-Viste, Westground Way, Tintagel, PL34 0BH |
| GM0 | EDJ | P Temple, 23 Ramsay Place, Johnstone, PA5 0EX |
| G0 | EDK | Arnold Lee, 44 Lynn Road, North Shields, NE29 8HS |
| G0 | EDM | E Dennett, 123 Drift Road, Clanfield, Waterlooville, PO8 0PD |
| G0 | EDO | R Ormond, 75 Desford Road, Newbold Verdon, Leicester, LE9 9LG |
| GM0 | EDQ | J Shaw, 28 Drumcross Road, Bathgate, EH48 4HG |
| GM0 | EDR | J Bell, 52 Turnberry Road, Glasgow, G11 5AP |
| G0 | EDS | EASTBOURNE DISTRICT SCOUTS ARC c/o Anthony Seabrook, 63 St. Annes Road, Willingdon, Eastbourne, BN20 9NJ |
| G0 | EDT | J Hopwood, 53 St. Marys Road, Stratford-upon-Avon, CV37 6XG |
| G0 | EDY | Peter Gould, 53 Green Road, Kidlington, OX5 2EU |
| G0 | EEA | Graeme Reffell, 26 Barnwood Road, Gloucester, GL2 0RX |
| G0 | EEF | G Legg, 12 Churchill Road, Wimborne, BH21 2AU |
| GM0 | EEG | D McTaggart, 65 Oronsay Road, Airdrie, ML6 8FX |
| GM0 | EEH | J Hunter, 58 Crow Road, Lennoxtown, Glasgow, G66 7HU |
| G0 | EEJ | A Mitchell, 18 Burnards Court, Berrycombe Road, Bodmin, PL31 2NU |
| G0 | EEN | R Wright, 11 Newman Avenue, Lanesfield, Wolverhampton, WV4 6DA |
| GI0 | EEO | John Ryan, 11 Carnesure Heights, Comber, Newtownards, BT23 5RN |
| G0 | EES | S Henderson, 17 Bury Close, Warbstow, Launceston, PL15 8UZ |
| G0 | EET | Brian Henderson, Flat 7, John Wood House, Cathedral Views, Salisbury, SP2 7TW |
| GM0 | EEY | Derek Smith, Yeldavale, Harray, Orkney, KW17 2LE |
| G0 | EEZ | C Wright, 60 Grove Crescent, Hanworth, Feltham, TW13 6LZ |
| G0 | EFA | C Mansfield, 10 Priory Drive, Abbey Wood, London, SE2 0PP |
| GM0 | EFC | G Perry, 14b Meadowfoot Road, West Kilbride, KA23 9BX |
| GM0 | EFD | C Perry, 14b Meadowfoot Road, West Kilbride, KA23 9BX |
| G0 | EFG | B Balmer, 13 Chillingham Crescent, Ashington, NE63 8BQ |
| GM0 | EFH | Allan Buchan, Flat 7, 1 Castlebank Court, Glasgow, G13 2LA |
| G0 | EFI | V Newman, 14 Hilltop Close, Rayleigh, SS6 7TD |
| G0 | EFL | L Cohen, 17 Ashbury Drive, Blackwater, Camberley, GU17 9HH |
| G0 | EFN | H Dunne, 19 Bute Brae, Bletchley, Milton Keynes, MK3 7TA |
| G0 | EFO | Michael Shortland, 4 Hillier Road, Guildford, GU1 2JQ |
| G0 | EFP | Nigel Hughes, 43a Wellhouse Road, Beech, Alton, GU34 4AQ |
| GM0 | EFQ | Hendry Fisher, 1 Millhill Lane, Musselburgh, EH21 7RD |
| G0 | EFR | L Wheeler, 29 Belmont Park, Pensilva, Liskeard, PL14 5QT |
| G0 | EFS | P Hancock, 7 Carlton Avenue, Hayes, UB3 4AD |
| GM0 | EFT | R Neilson, 54 Macdonald Smith Drive, Carnoustie, DD7 7TB |
| GI0 | EFW | J O'Hara, 284 Foreglen Road, Dungiven, Londonderry, BT47 4PJ |
| G0 | EFY | H Warman, 177 Scotter Road, Scunthorpe, DN15 8AU |
| G0 | EFZ | Ian Gerrard, The Station House, Station Road, Market Rasen, LN7 6HZ |
| G0 | EGC | W Stormont, 3 Bridge Cottages, Greenham, Crewkerne, TA18 8QE |
| G0 | EGE | W Trotter, Bungalow, Stoupe Cross Farm, Whitby, YO22 4JU |
| G0 | EGG | David Jackson, 41 Colman Avenue, Wolverhampton, WV11 3RT |
| G0 | EGH | Daniel Taylor, 81 Goldfinch Close, Caldicot, NP26 5BW |
| GM0 | EGI | Brian Devlin, Borrodale, Main Street, Stirling, FK8 3PW |
| G0 | EGP | D Williamson, Wrybourne Lodge, 200 Bushbury Road, Wolverhampton, WV10 0NA |
| G0 | EGQ | G Peters, 10 Cedar Close, Buckley, CH7 2GE |
| G0 | EGR | C Davis, 49 Brackendale Road, Bournemouth, BH8 9HY |
| G0 | EGT | G Pitts, 119 Rusper Road, Crawley, RH11 0HW |
| G0 | EGW | V Readhead, White Lodge, Rendham Road, Saxmundham, IP17 2AA |
| G0 | EHA | I Pemberton, 26 Stanley Grove, Ruabon, Wrexham, LL14 6AH |
| G0 | EHE | Brian Evans, 27 Mulso Road, Finedon, Wellingborough, NN9 5DP |
| G0 | EHK | Gerard Cheetham, 172a Hesketh Lane, Tarleton, Preston, PR4 6AT |
| GM0 | EHL | B Armstrong, 31 Old Abbey Road, North Berwick, EH39 4BP |
| G0 | EHO | R Mellor, 1 The Square, Lybury Lane, St. Albans, AL3 7JB |
| G0 | EHQ | F Skinner, Halfway Lock Cottage, Upper Gambolds Lane, Bromsgrove, B60 3HB |
| G0 | EHR | Michael Clark, 60a Clatterford Road, Newport, PO30 1PA |
| G0 | EHS | F Fennah, 7 Y Ddol, Llanbrynmair, SY19 7DJ |
| G0 | EHT | J Tommey, The Birches, Upton Bishop, Ross-on-Wye, HR9 7UF |
| G0 | EHV | E Ashburner, 8 Shellbark, Houghton le Spring, DH4 7TD |
| G0 | EHW | Bernard Coles, 32 Victoria Street, Lostock Hall, Preston, PR5 5RA |
| G0 | EHX | Terence Elliott, 18 Hollinside Square, Sunderland, SR4 8AU |
| G0 | EIA | B Taylor, Namparra, Quarry Road, Liskeard, PL14 5NP |
| G0 | EIB | L Allen, 28 Clarence Place, Maltby, Rotherham, S66 7HA |
| G0 | EID | D Harbour, 20 Durkins Road, East Grinstead, RH19 2ER |
| G0 | EIF | P Massheder, 5 Hazel Close, Penwortham, Preston, PR1 0YE |
| G0 | EIG | M Holtham, 33 Daisy Meadow, Bamber Bridge, Preston, PR5 8DD |
| G0 | EIH | T Briley, 2 Usborne Close, Staplehurst, Tonbridge, TN12 0LD |
| G0 | EIM | M Heyes, 11 Beech Close, Isleham, Ely, CB7 5UU |
| G0 | EIQ | M Richards, 3 Derwent Crescent, Whetstone, London, N20 0QN |
| G0 | EIR | G Kemp, 27 Shady Grove, Alsager, Stoke-on-Trent, ST7 2NQ |
| GM0 | EIT | Alistair George, 7 Mid Street, Keith, AB55 5AG |
| G0 | EIY | S Pryce, 40 Gains Avenue, Bicton Heath, Shrewsbury, SY3 5AN |
| G0 | EIZ | William Kenyon, Flat 21 House 4, Copper Place, Manchester, M14 7FZ |
| G0 | EJD | M Moss, 1 Orchard Rise, Beckingham, Doncaster, DN10 4NG |
| GW0 | EJE | PEMBS RS c/o L [illegible], Haverfordwest, SA61 1BW |
| G0 | EJF | A Henderson, 63 Chosen Drive, Churchdown, Gloucester, GL3 2QS |
| G0 | EJI | L Garden, 227 Oak Crescent, Burlington, L7L 1H3 |
| G0 | EJO | G Nock, 20 Chigwell Road, Bournemouth, BH8 9HW |
| G0 | EJQ | Jim Stevenson, 18 Applegarth, Lincoln, LN5 0NW |
| G0 | EJR | A Love, 15 Mountain Ash, Weston Park, Bath, BA1 2UU |
| GI0 | EJT | S Rafferty, 81 Mullaghmore Drive, Omagh, BT79 7PQ |
| GI0 | EJU | V Hutchinson, 10 Golan Road, Knockmoyle, Omagh, BT79 7TJ |
| G0 | EJV | Les Hodges, 34 Wiseholme Road, Skellingthorpe, Lincoln, LN6 5TF |
| G0 | EKD | A Dickason, 15 Farsands, Oakley, Bedford, MK43 7SJ |
| G0 | EKH | K Harper, 2 Vale Road, Newton Abbot, TQ12 1DZ |
| G0 | EKK | Harvie Wright, 61d Clapgun Street, Castle Donington, Derby, DE74 2LF |
| GM0 | EKM | Cecil Duncan, Roadside Cottage, Hoswick, Shetland, ZE2 9HL |
| G0 | EKN | Christopher Nye, 6 Harding Road, Chadwell St. Mary, Grays, RM16 4XD |
| G0 | EKX | Malcolm Dodgson, 19 Selworthy Road, Southport, PR8 2NS |
| G0 | ELB | B Bray, 34 Newlands Drive, Forest Town, Mansfield, NG19 0HZ |
| G0 | ELC | Peter Hall, 28 Maria Drive, Fairfield, Stockton-on-Tees, TS19 7JL |
| G0 | ELG | R Zielinski, 154 Bensham Lane, Thornton Heath, CR7 7EN |
| G0 | ELJ | David Dawson, 19 Nightingale Avenue, Birmingham, B36 0RT |
| GM0 | ELL | Norman Elliot, Flat 1, Tarfside, Ascog, Isle of Bute, PA20 9EU |
| G0 | ELM | P Green, 14 Beech Avenue, Parbold, Wigan, WN8 7NS |
| G0 | ELN | S MacDonald, 61 Pavilion Road, Worthing, BN14 7EE |
| G0 | ELO | J Ellis, 21 Coxway, Clevedon, BS21 5AQ |
| GM0 | ELP | Douglas Maxwell, 29 Ambleside Rise, Hamilton, ML3 7HJ |
| G0 | ELU | K Kyriacou, Flat 4, Healey House, Beckenham, BR3 1RE |
| G0 | ELX | P Wilson, 39 Tintern Grove, Stockport, SK1 4DS |
| GD0 | ELY | J Brown, Cleckheaton, Ballaragh, Laxey, Isle of Man, IM4 7PW |
| G0 | ELZ | William Cross, 31 Joshua Close, Liverpool, L5 0TD |
| G0 | EMB | Hubert Blore, 17 Kendal Way, Wrexham, LL12 8AF |
| GM0 | EMC | K B McLaren, 3 Bracany Gardens, Fogwatt, Dalriada, Elgin, IV30 8SY |
| G0 | EMF | D Brown, 24 Parkside, Sacriston, Durham, DH7 6JU |
| G0 | EMK | Melvin Kendall, 88 Coldnailhurst Avenue, Braintree, CM7 5PY |
| G0 | EML | Raymond Bullock, 40 Little Harlescott Lane, Shrewsbury, SY1 3PY |
| G0 | EMM | Keith Dockray, 54 Kelsick Park, Seaton, Workington, CA14 1PY |
| GM0 | EMQ | John Vinton, 2 Luncarty Place, Turriff, AB53 4UD |
| G0 | EMR | Pamela Page, 144 Cody Road, Farnborough, GU14 0DD |
| G0 | EMS | Henry Brown, Priors Lea, Besford Road, Worcester, WR8 9AN |
| G0 | EMT | M Chapman, 102 Fangrove Park, Lyne, Chertsey, KT16 0BP |
| G0 | EMV | Walter Van Aswegen, 16 Spencer Road, Southampton, SO19 6QX |
| G0 | EMX | E Parr, 74 Stanley Road, Coventry, CV5 6FF |
| G0 | ENA | R Procter, 20 Costells Edge, Scaynes Hill, Haywards Heath, RH17 7PY |
| G0 | ENB | R Felton, 16 Quidenham Road, East Harling, Norwich, NR16 2JD |
| G0 | END | R McGarvie, Croftlands, Thornthwaite, Keswick, CA12 5SA |
| G0 | ENF | G Crawshaw, 51 Templeway West, Lydney, GL15 5JD |
| G0 | ENJ | D Buckingham, 208 Bannings Vale, Saltdean, Brighton, BN2 8DJ |
| G0 | ENM | D James, 70 Broadway West, Walsall, WS1 4DZ |
| G0 | ENN | Bernard Bowden, 49 Springfield Drive, Westcliff-on-Sea, SS0 0RA |
| G0 | ENO | Kevin Dempster, 48 Mount Pleasant Road, Pudsey, LS28 9AA |
| GM0 | ENQ | William Smith, 10 Woodlands Place, Inverbervie, Montrose, DD10 0SL |
| G0 | ENT | J Comerford, Bod Elen, Bontnewydd, Caernarfon, LL54 7YE |
| G0 | ENU | D Walters, 132 Tan Y Bryn, Valley, Holyhead, LL65 3ES |
| G0 | ENV | Alan Wood, 262 Egmanton Road, Meden Vale, Mansfield, NG20 9PY |
| G0 | ENW | G Fingerhut, Wild Rose, Behind Hayes, Templecombe, BA8 0BP |
| G0 | ENY | R Ford, 27 Albert Road, Millisons Wood, Coventry, CV5 9AS |
| G0 | ENZ | Trevor Trudgeon, 1 Bessy Beneath Cottages, Ruan High Lanes, Truro, TR2 5JX |
| G0 | EOF | G Potter, 88 Highlands Close, Kidderminster, DY11 6JU |
| G0 | EOG | J Jennings, 81 Newgate Street, Burntwood, WS7 8TX |
| G0 | EOH | D Bristow, Herniss Bungalow, Nr Penryn, TR10 9DT |
| G0 | EOI | Trevor Froggatt, 13 Leckford Road, Havant, PO9 5LH |
| G0 | EOJ | D Shore, 9 Hawthorn Close, Clowne, Chesterfield, S43 4SX |
| G0 | EOK | P Gibson, 66 Shevington Moor, Standish, Wigan, WN6 0SA |
| G0 | EOL | W Prater, 44 Alundale Road, Winsford, CW7 2QD |
| G0 | EOM | A Deakin, 36 The Ridgway, Romiley, Stockport, SK6 3EY |
| G0 | EON | F Cloke, 9 Mill Close, East Coker, Yeovil, BA22 9LF |
| G0 | EOP | L Herf, Old Chapel, Fore Street, South Molton, EX36 3HL |
| G0 | EOS | P Smith, 35 Tanglewood Close, Birmingham, B34 7QX |
| G0 | EOX | I Hunton, 123 Huddersfield Road, Diggle, Oldham, OL3 5NU |
| G0 | EOY | S Outterside, 21 Coquet Grove, Throckley, Newcastle upon Tyne, NE15 9JU |
| G0 | EOZ | P Korton, Mooraless, 11 North Filham Cot, Ivybridge, PL21 9D[illegible] |
| G0 | EPA | G Whitham, 55 Bisley Grove, Bransholme, Hull, HU7 4PY |
| G0 | EPE | D Cargill, 41 Grosvenor Road, Skegness, PE25 2DD |
| G0 | EPI | S Cooper, 14 Greenview Drive, Towcester, NN12 6DL |
| G0 | EPL | John Love, 191 High Street, Henley-in-Arden, B95 5BA |
| GM0 | EPO | J Shades, 15 Balminnoch Park, Doonfoot, Ayr, KA7 4EQ |
| G0 | EPP | A Przybyla, 18 Cherwell, Washington, NE37 3LA |
| G0 | EPR | P Wardale, 104 Rectory Place, Woolwich, London, SE18 5BY |
| G0 | EPU | Cyril Crosby, 37 Malwood Way, Maltby, Rotherham, S66 7HF |
| G0 | EPV | J Collins, 49 Alspath Road, Meriden, Coventry, CV7 7LU |
| G0 | EPY | Colin Hirst, 18 Lazenby Avenue, Fleetwood, FY7 8QH |
| G0 | EQC | I Williamson, Westholme, 4 Edgewell Road, Prudhoe, NE42 6JH |
| G0 | EQD | J Archer, 29 Easby Close, Bishop Auckland, DL14 0RW |
| G0 | EQE | David Cunningham, 2 Fairmead Road, Moreton, Wirral, CH46 8TX |
| G0 | EQH | E Shackleton, 2 Culcheth Avenue, Marple, Stockport, SK6 6NA |
| G0 | EQI | R Mallinson, 3 Captain Cooks Crescent, Whitby, YO22 4HL |
| GM0 | EQS | Stephen Palmer, Fintry Schoolhouse, Turriff, AB53 5RN |
| G0 | EQV | D Buckley, C/ Juan Ramon Jimenez 23, Formentera Del Segura, Alicante, Spain, 1379 |
| GM0 | EQW | Tygo Olsen, Ard Chuan, Taynuilt, PA35 1HY |
| GM0 | ERB | N Calder, 16 Camesky Road, Caol, Fort William, PH33 7EH |
| G0 | ERF | Ken Watkins, 29 Saddlers Close, Billingshurst, RH14 9QL |
| G0 | ERI | J Shaw, 1 Chestnut Way, Fareham, PO14 4LQ |
| G0 | ERL | W Marbus, Elm Farm House, Debenham Road, Stowmarket, IP14 5LP |
| G0 | ERS | R Smith, 79 Froxfield Road, Havant, PO9 5PW |
| GM0 | ERT | R Tannahill, 62 Caroline Park, Mid Calder, Livingston, EH53 0SJ |
| GM0 | ERV | S McLennan, 6 Mull Terrace, Oban, PA34 4YB |
| G0 | ERW | W Spencer, 40 Cairn Way, Bourne, PE10 9QY |
| G0 | ERY | R Saunders, 24322 Augustin Street, Mission Viejo, United States, 92691 |
| G0 | ESA | W Wilkinson, Les Mardeilles, Le Bourg, Chirac, France, 16150 |
| G0 | ESD | Thomas [illegible], Alicante, Spain, 3669 |
| G0 | ESH | G Cooper, The Dumbles, Shatterford, Bewdley, DY12 1TR |
| G0 | ESI | D Williams, Miramare, Egremont Road, St. Bees, CA27 0AS |
| G0 | ESK | C Williams, 3 Lodge Orchard, Mona Street, Amlwch, LL68 9RX |
| G0 | ESL | Matthew Musgrave, 11 Hillside Drive, Yealmpton, Plymouth, PL8 2NT |
| G0 | ESO | S Llewellyn, Eastfield Cottage, Mavis Enderby, Spilsby, PE23 4EJ |
| G0 | ESU | W Lee, 8 Bronheulog, Bodffordd, LL77 7SU |
| G0 | ESW | R Graham, 454 Lobley Hill Road, Lobley Hill, Gateshead, NE11 0BS |
| G0 | ESY | M Perrett, 9 Molyneaux Place, Plymouth, PL1 4RE |
| G0 | ETA | G Luhman, 31 Flexmore Way, Langford, Biggleswade, SG18 9PT |
| G0 | ETF | Stewart Rolfe, Tynlon, Minffordd, Bangor, LL57 4DR |
| G0 | ETI | Jeffrey Bedford, 41a Arden Road, Herne Bay, CT6 7UW |
| G0 | ETL | G Tatterson, 2 Eden Road, Leeds, LS4 2TT |
| G0 | ETM | J Davies, 1 Mount View, Plas Road, Blackwood, NP12 3RH |
| G0 | ETP | T Howe, 76 Birch Trees Road, Great Shelford, Cambridge, CB22 5AW |
| G0 | ETQ | J Curnow, 4 Penmere Court, Falmouth, TR11 2RN |
| G0 | ETU | E Cabban, Garmonfa, Capel Garmon, Llanrwst, LL26 0RG |
| G0 | ETV | L Mcquire, Springfield, Staynall Lane, Poulton-le-Fylde, FY6 9DR |
| G0 | ETZ | S Gurney, Crimond, The Common, Exmouth, EX8 5EE |
| G0 | EUC | R Burnet, 41 Douglas Crescent, Southampton, SO19 5JP |
| G0 | EUD | Charles Jackson, Red House Farm, Grange Road, King's Lynn, PE34 4HQ |
| GI0 | EUG | E Hagan, 34 Coolshinney Road, Magherafelt, BT45 5JF |
| G0 | EUJ | K Neville, 5 Coleville Avenue, Fawley, Southampton, SO45 1DA |
| G0 | EUL | Jacinto Estibeiro, The Joiners House, Preston, Duns, TD11 3TQ |
| GM0 | EUM | J Mackinnon, 60 Mount Stuart Drive, Wemyss Bay, PA18 6DX |
| G0 | EUN | James Nichol, 58 Benson Crescent, Doddington Park, Lincoln, LN6 3NU |
| G0 | EUP | M Rigg, 3 Fairway, Rochdale, OL11 3BU |
| G0 | EUR | B Barber, 1 Shore Place, Trowbridge, BA14 9TB |
| G0 | EUV | R Jones, Greens Cottage, Luton Road, Offley, Hitchin, SG5 3DR |
| G0 | EUZ | R Jones, 90 High Street, Cottenham, Cambridge, CB24 8SD |
| G0 | EVA | David Evans, 113 Denby Dale Road, Wakefield, WF2 8EB |
| G0 | EVD | J Noble, 22 Hadleigh, Letchworth Garden City, SG6 2LU |
| G0 | EVE | Valerie Berry, 40 Yerburgh Avenue, Colwyn Bay, LL29 7NB |
| G0 | EVF | John Mowbray, 44 Monkdale Avenue, Cowpen Estate, Blyth, NE24 4EB |
| G0 | EVG | Nigel Berry, 40 Yerburgh Avenue, Colwyn Bay, LL29 7NB |
| G0 | EVH | A Ferneyhough, 30 Bedford Drive, Sutton Coldfield, B75 6AU |
| G0 | EVI | S Monk, 310 Hinckley Road, Leicester, LE3 0TN |
| G0 | EVJ | Stephen Evans, 181 Curborough Road, Lichfield, WS13 7PW |
| G0 | EVM | Philip Stewardson, Hyland, St. Kenelms Road, Halesowen, B62 0NE |
| G0 | EVN | S Parkin, 13 Queens Drive, Nuthall, Nottingham, NG16 1EG |
| G0 | EVO | Freda Robinson, 42 The Paddock, York, YO26 6AW |
| G0 | EVP | K Griffiths, 44 Curzon Road, Poynton, Stockport, SK12 1YE |
| G0 | EVR | Anne Kittrick, 28 Jubilee Street, Hall Green, Wakefield, WF4 3JZ |
| G0 | EVS | H Angus, 111 Great Elms Road, Hemel Hempstead, HP3 9UQ |
| G0 | EVT | J Hoban, 3 Lake Lock Grove, Stanley, Wakefield, WF3 4JJ |
| G0 | EVU | Frank Samet, 4 Pembroke Grove, Glinton, Peterborough, PE6 7LG |
| G0 | EVV | David Stansfield, 22 Low Stobhill, Morpeth, NE61 2SG |
| G0 | EVW | Geoffrey Watts, 3 Maple Grove Knightsdale Road, Weymouth, DT4 0FE |
| G0 | EVX | A Cater, 2 Turkdean Road, Cheltenham, GL51 6AL |
| G0 | EVY | David Strobel, Dell Cottage, Copyholt Lane, Bromsgrove, B60 3AY |
| G0 | EVZ | S Males, 6 Lammas Path, Stevenage, SG2 9RN |
| G0 | EWD | P Kennedy, 262 Green Road, Springvale, Sheffield, S36 6BH |
| GI0 | EWE | P Holland, 35 Ashfield Road, Clogher, BT76 0HJ |
| GM0 | EWF | Derek Pettigrew, 112 South Street, Armadale, Bathgate, EH48 3JU |
| G0 | EWH | Richard Newton, 74 Walker Avenue, Stourbridge, DY9 9EL |
| G0 | EWI | John Daramy, 6 Boulton Close, Linacre Woods, Chesterfield, S40 4XJ |
| GI0 | EWP | John Hartin, 2 Berryhill Close, Dunamanagh, Strabane, BT82 0GZ |
| G0 | EWR | D Wale, 33 Westground Way, Tintagel, PL34 0BH |
| G0 | EWT | Anthony Coates, 11 Canterbury Road, Brotton, Saltburn-by-The-Sea, TS12 2XG |
| GM0 | EWU | C Craig, Knipoch Hotel, Knipoch, Oban, PA34 4QT |
| G0 | EWV | T Buckle, 15 Gleaves Avenue, Harwood, Bolton, BL2 [illegible] |
| GM0 | EWW | John Moore, 19 Mansefield, Tyndrum, Crianlarich, FK20 8RQ |
| GM0 | EWX | C Macpherson, 6 Borve, Skeabost Bridge, Portree, IV51 9PE |
| GW0 | EWY | W Woods, 40 Ger-y-Llan, Velindre, Llandysul, SA44 5YB |
| G0 | EWZ | I Mason, 56 Evenlode Crescent, Coventry, CV6 1BY |
| G0 | EXA | A Bendall, 3 St Michaels Gate, Brimfield, Ludlow, SY8 4NE |
| G0 | EXB | P Langdon, Dahlia Cottage, Kidderminster, DY14 9HP |
| G0 | EXD | Christopher Challinor, Bryn Tirion, Lower Frankton, Oswestry, SY11 4PA |
| G0 | EXN | John Eden, 4 Halescourt, Church Lane, Shifnal, TF11 [illegible] |
| G0 | EXU | Alun Davies-Jones, 10 Ponsford Road, Knowle, Bristol, BS4 2UP |
| G0 | EYA | J Morris, 31 Beldham Road, Farnham, GU9 8TW |
| G0 | EYE | Anthony Lunn, 45 St. Anthonys Avenue, Eastbourne, BN23 6LN |
| G0 | EYF | B Ford, 15 Derby Road, Barnstaple, EX32 7HW |
| G0 | EYG | C Tito, 13 Potter Way, Bedford, MK42 9RG |
| G0 | EYH | R Dawson, 74 Dylan Avenue, Cefn Fforest, Blackwood, NP12 3NG |
| G0 | EYL | William McAdam, 2 Manor Orchards, Knaresborough, HG5 0BW |
| G0 | EYM | R Rowlett, No 1 Bungalow, Main Road, Wisbech, PE14 9JR |
| G0 | EYO | Chris Pettitt, 12 Hennals Avenue, Redditch, B97 5RX |
| G0 | EYP | R Francis, 42 Carmarthen Road, Cheltenham, GL51 3LA |
| G0 | EYR | Peter Robinson, 35 Stoke Road, Taunton, TA1 3EH |
| G0 | EYT | M Fox, 49 Manor Drive, Esher, KT10 0AZ |
| G0 | EYU | C Talbot, 59 Heywood Avenue, Austerlands, Oldham, OL4 4AZ |

UK Callsigns

**IMPORTANT NOTE**

**Revalidate licence to avoid revocation** – Ofcom has advised the Society that plans will be drawn up to revoke licences that have not been revalidated as required by the licence conditions. The quickest way to revalidate is to do so online via the Ofcom website: *https://services.ofcom.org.uk/* or by email: *amateur.validations@ofcom.org.uk* Ofcom staff are available to help, but please be patient during times of heavy workload.

G0 EYW D Jones, 4 Granville Crest, Kidderminster, DY10 3QS
G0 EYX Derek Southey, 253 Sandon Road, Stafford, ST16 3HQ
G0 EYZ P Orchard, 5 Vicarage Road, Bletchley, Milton Keynes, MK2 2EZ
G0 EZB S Cowie, 37 Rockfield Drive, Llandudno, LL30 1PF
G0 EZI J Pitfield, 42 Spinney Green, Eccleston, St. Helens, WA10 5AH
G0 EZJ D Drake, 60 Jessopp Avenue, Bridport, DT6 4ES
G0 EZL Donald Hodgkinson, 16 Fortescue Avenue, Twickenham, TW2 5LS
GW0 EZQ LLANELLI A R S c/o Brian Davies, 2a Berwick Road, Bynea, Llanelli, SA14 9SS
GM0 EZR P Newton, 115 Napier Road, Glenrothes, KY6 1DU
G0 EZT E Humphries, 37 Grove Meadow, Cleobury Mortimer, Kidderminster, DY14 8AG
G0 EZU A Davies, 28 Sayer Road, Charing, Ashford, TN27 0JT
G0 EZX Colline Wood, 34 Rosemary Lane, Stourbridge, DY8 3EP
G0 EZY Terence Jeacock, 9 Parkwood Rise Barnby Dun, Doncaster, DN3 1LY
G0 FAA W Flindell, Pentley House, Marston Road, Sherborne, DT9 4BJ
G0 FAB H Kay, 51 Colin Crescent, Colindale, London, NW9 6EU
G0 FAD John Bowers, 22 Kilmiston Drive, Fareham, PO16 8DY
G0 FAE R Beer, 65 Bridgefield Road, Whitstable, CT5 2PH
G0 FAH Willam Wright, 46 Homestall Road, East Dulwich, London, SE22 0SB
G0 FAS George West, 33 Dorcis Avenue, Bexleyheath, DA7 4RL
G0 FAU Alan Doyle, 2 Ferndown Way, Weston, Crewe, CW2 5GS
G0 FAW Harold Jenkinson, Flat 2, 3 Kassima, Kissonerga/Paphos, Cyprus, 8574
G0 FBB MEOPHAM PARISH RC c/o I Burns, Little Delmar Farm, Leywood Road, Meopham, DA13 0UD
G0 FBC Colin Hyatt, 44 Barnes Lane, Sarisbury Green, Southampton, SO31 7BZ
G0 FBG Godfrey Hands, 74 Berrington Road, Nuneaton, CV10 0LB
G0 FBL B Sharman, 64 Collingwood Drive, Shiney Row, Houghton le Spring, DH4 7LP
G0 FBM D Lawson, 52 Ryefield Road, Eastfield, Scarborough, YO11 3DR
G0 FBO C Roberts, 86 Peake Road, Brownhills, Walsall, WS8 7BZ
G0 FBQ W Wootton, 94 Dyas Avenue, Great Barr, Birmingham, B42 1HF
G0 FBS George Nicolson, 34 Chalbury Close, Weymouth, DT3 6LE
G0 FBT C Hughes, Llanhennock Cheshire Home, Caerleon, Newport, NP18 1LT
G0 FBW A Armstrong, 1 Montfalcon Close, Peterlee, SR8 1DD
G0 FBX A Leigh, Roleystone, 7 Fieldfare, Gloucester, GL4 4WH
G0 FCA I Groom, 12 Billington Avenue, Rossendale, BB4 8UW
G0 FCB C Tatlow, Frenton Farm, Whitemoor, St. Austell, PL26 7XQ
G0 FCG P O'Neill, 36 Grantley Gardens, Mannamead, Plymouth, PL3 5BS
G0 FCH R Last, 39 Upham Road, Swindon, SN3 1DJ
GM0 FCI P Reid, 129 Mckinlay Crescent, Irvine, KA12 8DR
G0 FCJ Mark Lawson, 1a Hunters Close, Stroud, GL5 4UW
G0 FCM I Sircombe, 4 Long Eights, Northway, Tewkesbury, GL20 8QY
G0 FCO B Cooke, 49 Shepherds Croft, Slade, Stroud, GL5 1US
G0 FCQ D James, 15 Kington Gardens, Birmingham, B37 5HX
G0 FCT Ian Pawson, 3 Orion, Bracknell, RG12 7YX
G0 FCU S Kennedy, Laurel Cottage, Pond Lane, Guildford, GU5 9RS
G0 FCV A Woods, 8 Wareham Road, Lytchett Matravers, Poole, BH16 6DP
G0 FCX A Traynor, 2 Mansfield Road, Mossley, Ashton-under-Lyne, OL5 9JN
G0 FCZ Lee Standley, 26 Ullswater Drive, Middleton, Manchester, M24 5RL
G0 FDA R Dresser, 6 Acacia Avenue, Fencehouses, Houghton le Spring, DH4 6JG
G0 FDD Ciifford Shalley, 4 Almond Walk, Lydney, GL15 5LP
G0 FDE M Jackson, 2 Sunnybank, Watledge, Stroud, GL6 0AP
G0 FDH R Wishart, 15 Plumer Avenue, Tang Hall, York, YO31 0PX
G0 FDJ K Castley, 2 Thorpe Avenue, Moulton Chapel, Spalding, PE12 0XN
G0 FDP F Parradine, 87 Oakways, Eltham, London, SE9 2NZ
G0 FDS J Johnson, 150 Lenthall Avenue, Grays, RM17 5AB
G0 FDT M Flatman, 8 The Pines, Cringleford, Norwich, NR4 7LT
G0 FDV David Jardine, 11 Gorse Hill, Broad Oak, Heathfield, TN21 8TW
G0 FDX CENTRAL LANCS ARC c/o J Lawson, 14 Kentmere Avenue, Farington, Leyland, PR25 3UH
G0 FDZ Christopher Whitmarsh, 35 Dorchester Avenue, Bexley, DA5 3AH
G0 FEI Victor Ward, Romayne, 32 St. Johns Road, Great Yarmouth, NR31 9JT
G0 FEJ G Marshall, Birchlands, 3 Longbridge Close, Hook, RG27 0DQ
G0 FEK R Wilson, 34 Belfairs Drive, Chadwell Heath, Romford, RM6 4EB
G0 FEM G Felton, 10 Penbodeistedd, Llanfechell, LL68 0RE
G0 FEO A Quy, 17 Fircroft, Kingsbury, Tamworth, B78 2JU
G0 FEP Hector Maclean, 15 Keystone Cres, Kew East, Australia, 3102
G0 FEQ Keith Howard, 11 Station Road, Ulceby, DN39 6UQ
G0 FEU David Davies, Coedfryn, Halkyn, Holywell, CH8 8ES
G0 FEV Edward Kittrick, 28 Jubilee Street, Hall Green, Wakefield, WF4 3JZ
G0 FEZ K Wragg, 11a Fall Road, Heanor, DE75 7PQ
G0 FFA N Painter, 75 Chaytor Road, Polesworth, Tamworth, B78 1JS
G0 FFB Stephen Maughan, 17 Upper Dane, Desborough, Kettering, NN14 2LB
G0 FFF Richard Ford, 15 Holloway, Pershore, WR10 1HW
G0 FFK R McKenzie, 40 Fairway Avenue, West Drayton, UB7 7AN
G0 FFL Richard O'Keeffe, 40 Edinburgh Road, Maidenhead, SL6 7SH
G0 FFN E Fisher, 19 Keats Way, West Drayton, UB7 9DR
G0 FFQ R Baldock, 1a Thorneywood Road, Long Eaton, Nottingham, NG10 2DZ
G0 FGA A Walker, 2 Chelwood Drive, Sandhurst, GU47 8HT
G0 FGC J Biggs, 40 Packmore Street, Warwick, CV34 5BX
G0 FGE Basil Bolt, 106 Barley Farm Road, Exeter, EX4 1NJ
G0 FGG J Thompson, 17 Fryer Crescent, Darlington, DL1 2DX
GM0 FGI H Cromack, Pier View, Kilchattan Bay, Isle of Bute, PA20 9NW
G0 FGJ D Joyce, 44 St. Marys Close, Marston Moretaine, Bedford, MK43 0QZ
G0 FGK Brian Whitehouse, Flat 53, Peel House, Tamworth, B79 7BQ
G0 FGO W Waldron, 100 Porthmawr Road, Cwmbran, NP44 1NB
G0 FGP Richard Bradwell, Summer Fields School, Mayfield Road, Oxford, OX2 7EN
G0 FGR A Perkins, 111 Broadmead, Corsham, SN13 9AP
G0 FGS M Bosley, Crossroads Palmerston Road, Ross-on-Wye, HR9 5PN
G0 FGW Ronald Clements, 28 Willow Grove, Chippenham, SN15 1AR
G0 FGX Robert McCreadie, 45 Gwealhellis Warren, Helston, TR13 8PQ
G0 FGZ C Newton, Peartree Cottage, Little London, Longhope, GL17 0PH
G0 FHC Bryan Trimmer, Sydney Cottage, Salisbury Road, Romsey, SO51 6EE
GM0 FHD E Mottart, 1 Muirake Cottages, Cornhill, Banff, AB45 2BQ
G0 FHF Colin Ashdown, New Garth, Hallbankgate, Brampton, CA8 2NF
GM0 FHJ R Menzies, 105 Yoker Mill Road, Glasgow, G13 4HL
G0 FHK R Peart, 33 Fieldfare, Abbeydale, Gloucester, GL4 4WH
G0 FHL B Thomas, Plastirion, Padeswood Road, Buckley, CH7 2JL
G0 FHO G Taylor, 93 Fengate Mobile Home Park, Peterborough, PE1 5XE
GM0 FHS Simon Baird, Polfearn Cottage, Taynuilt, PA35 1JQ
G0 FHT Geoffrey Bate, 7 Albany Court, Redruth, TR15 2NY
G0 FHX A Hocking, 79 Cornish Crescent, Truro, TR1 3PE
G0 FHY J Hocking, 79 Cornish Crescent, Truro, TR1 3PE
G0 FIC Kenneth Tarry, 38 Tresithney Road, Carharrack, Redruth, TR16 5QZ
G0 FIG Alexander Trusler, 42 Mill Hill, Shoreham-by-Sea, BN43 5TH
G0 FIJ Christopher Jones, 46 Wilmington Close, Woodley, Reading, RG5 4LR
G0 FIN A Findlay, 4 Saxon Close, Rugby, CV22 7FJ
G0 FIP E Tugwell, 14 Martinique Way, Eastbourne, BN23 5TH
GM0 FIQ H Bohan, 12 Loch Way, Kemnay, Inverurie, AB51 5QZ
G0 FIT Konrad Menzel, 8 Higher Bockhampton, Dorchester, DT2 8QJ
G0 FIU R Edwards, Stanmore Cottage, Brockley Corner Cul, Bury St. Edmunds, IP28 6UA
G0 FIW I Osborne, Alacoo, Tan Lane, Clacton-on-Sea, CO16 9PS
G0 FJA Brian Samuels, 63 Mill Road, Okehampton, EX20 1PR
G0 FJB J Bailey, Powney Cottage, Powney Street, Ipswich, IP7 7AL
G0 FJD Derek Tyers, 18 Gardeners Close, Kidderminster, DY11 5DW
G0 FJE R Hill, 13 Maesglas Grove, Newport, NP20 3DJ
G0 FJH C Lonsdale, 6 Oak Tree Close, New Inn, Pontypool, NP4 0DG
G0 FJJ Adam Winkler, 10 Havenside, Shoreham-by-Sea, BN43 5LN
GW0 FJP David Stanley, 9 Haywain Court, Bridgend, CF31 2ED
GW0 FJQ A Marshall, 26 Avondale Road, Gelli, Pentre, CF41 7TW
G0 FJR D Paynter, 6 Blacksmiths Close, Ramsey Forty Foot, Huntingdon, PE26 2YW
G0 FJS Peter Copeland, 6 Waverley Road, Northampton, NN2 7DA
G0 FJZ Graham Bradley, 59 Main Road, Watnall, Nottingham, NG16 1HE
G0 FKF J Smith, Erw Las, The Cross Roads, Redruth, TR16 5PN
G0 FKG C Skillings, 11 Curtis Road, Norwich, NR6 6RB
G0 FKI A Brown, Panorama, Highway Lane, Redruth, TR15 1SE
G0 FKJ C Currey, 1 Newport Close, Portishead, Bristol, BS20 8DD
G0 FKK R Parrish, 11 Pitt Road, Maidstone, ME16 8PA
GM0 FKP James Tohill, 71 Campsie Road, Kilmarnock, KA1 3RY
G0 FKS K Stancliffe, 3 Upper Lambricks, Rayleigh, SS6 8BP
G0 FKW A Timms, 63 High Street, Astcote, Towcester, NN12 8NW
G0 FKX D Warren, 1 Ruby Terrace, Porkellis, Helston, TR13 0LD
G0 FKY J Merifield, 84 Wareham Road, Corfe Mullen, Wimborne, BH21 3LG
G0 FLA W Frewen, Tantalus, Main Street, Rye, TN31 6NB
G0 FLB I Gillingham, Elm Cottage, Reading Road, Didcot, OX11 0LU
G0 FLD K Pitman, Pump Cottage, High Street, Bridlington, YO15 1JT
G0 FLG B Parker, 120 Brooks Lane, Whitwick, Coalville, LE67 5DF
G0 FLI N Penistone, 114 Long Lane, Worrall, Sheffield, S35 0AF
G0 FLP J Hammond, Ashmond House, Queens Street, March, PE15 8SN
G0 FLQ R Cheetham, 65 Avondale Avenue, Hazel Grove, Stockport, SK7 4QE
G0 FLT A Auker-Howlett, 40 Shaftesbury Way, Royston, SG8 9DE
G0 FLU M Crane, 40 Dukes Way, Newquay, TR7 2RW
G0 FLV R Pennock, 4 Millers Way, Heckington, Sleaford, NG34 9JG
G0 FLW Louis Williams, 56 Meriden Avenue, Stourbridge, DY8 4QS
G0 FLX D Gray, Jaig House, Mill Lane, Selby, YO8 6QX
G0 FMB B Collins, 28 Marlborough Road, South Woodford, London, E18 1AP
G0 FMG J Voss, 4 Chaucer Avenue, Mablethorpe, LN12 1DA
G0 FMI R Friston, Savmar, 72 Bradenham Road, Thetford, IP25 7PJ
G0 FMJ Roy Bennett, 86 Westons Hill Drive, Emersons Green, Bristol, BS16 7DN
G0 FMN Alex McQuarrie, 47 Ramsons Avenue, Conniburrow, Milton Keynes, MK14 7BB
G0 FMO J Smethurst, 71 Altofts Lodge Drive, Normanton, WF6 2LB
G0 FMP I Hodgkiss, 41 Buckingham Rise, Worksop, S81 7ED
G0 FMT Dennis Unwin, 11 Carlton Rise, Melbourn, Royston, SG8 6BZ
G0 FMU A Turner, 17 The Dell, Great Warley, Brentwood, CM13 3AL
GM0 FMW D Enderby, 45 Oxgangs Park, Edinburgh, EH13 9LF
G0 FMX A Roberts, C/O Ellen Roberts, 42 Aec Haig Barracks, BFPO 30,
G0 FNA C Halliday, 5 Ovington Drive, Southport, PR8 6JW
G0 FNB S Mudd, 91 Chalkwell Avenue, Westcliff-on-Sea, SS0 8NL
G0 FND M Cousins, Wiccan Lodge, Mumbys Drove, Wisbech, PE14 9JT
GM0 FNE T Wilson, 20 Mace Court, Stirling, FK7 7XA
G0 FNF Ian Gilbert, 82 Abbotswood Road, Brockworth, Gloucester, GL3 4PF
G0 FNH James Toon, 2 Home Farm Park, Burton-on-Trent, DE13 9BJ
G0 FNJ David Earnshaw, 43 Bank Parade, Burnley, BB11 1UG
G0 FNM E Walker, 216 Milnrow Road, Rochdale, OL16 5BB
G0 FNP Peter Radcliffe, Hill View, Wilton, Pickering, YO18 7LE
G0 FNS J Brayshaw, 26 Ashfield Avenue, Malton, YO17 7LE
G0 FOB Neil Robinson, 15 Hollins Bank, Sowerby Bridge, HX6 2RU
G0 FOC M Clowes, 55 Landswood Close, Birmingham, B44 0LF
G0 FOE Peter Hume-Spry, 23 Appledore Avenue, Wollaton, Nottingham, NG8 2RE
G0 FOG C Singleton, 1 Abbey Close, Aslockton, Nottingham, NG13 9AF
G0 FOH M Holdsworth, Merrill, Ringwood Road, Southampton, SO40 7GY
G0 FOI John Wilde, 2 Bottoms Lane, Birkenshaw, Bradford, BD11 2NN
G0 FOK R Cox, 12 East Street, Thame, OX9 3JS
G0 FOT R Gibbons, 3 Fairfield, Gamlingay, Sandy, SG19 3LG
G0 FOU Garry Binns, 22 Carlyn Avenue, Sale, M33 2EA
G0 FOY G Grigg, 12 Townfield Road, Mobberley, Knutsford, WA16 7HF
G0 FPI J Spence, 60 Railey Road, Crawley, RH10 8BZ
G0 FPM Paul Fennell, 45 Badby Road West, Daventry, NN11 4HJ
G0 FPN D Waller, Apartment 23, 3 Woodbrooke Grove, Birmingham, B31 2FG
G0 FPO W Coulthard, 1 Lambton St., Eccles, Manchester, M30 8DD
G0 FPT Alan Pettigrew, 12 Pensford Close, Crowthorne, RG45 6QR
G0 FPU Michael Densham, 69 Mortimer Way, Leicester, LE3 1GR
G0 FPV I Marquis, 20 Hazelwood Grove, Leigh-on-Sea, SS9 4DE
G0 FPY J Bortowski, 4 Bryn Deiniol, Valley Road, Llanfairfechan, LL33 0SR
G0 FPZ Ritchie Wileman, 3 Primrose Way, Stamford, PE9 4BU
G0 FQA A Wallis, 14 Varley Close, Wellingborough, NN8 4UZ
G0 FQC P Wood, 26 Church Road, Bamber Bridge, Preston, PR5 6EP
G0 FQD R Harvey, 42 Groomsland Drive, Billingshurst, RH14 9HB
G0 FQF J Rae, Sunnybank House, Burnley Road East, Rossendale, BB4 9PX
G0 FQI D Breen, 53 Swift Close, Grange Park, Northampton, NN4 5AZ
G0 FQN John Procter, 168 St. Davids Road, Leyland, PR25 4UX
G0 FQO D Berry, The Bungalow, Basil Road, Kings Lynn, PE33 9RP
G0 FQP G Owens, 73 Edinburgh Road, Widnes, WA8 8BG
GM0 FQQ B Campbell, 93 Treeswoodhead Road, Kilmarnock, KA1 4PB
GM0 FQS A Smith, 60 Gordon Avenue, Bonnyrigg, EH19 2PQ
G0 FQU Alan Date, 1 Yew Tree Terrace, Gold Street, Milton Keynes, MK19 7LX
G0 FQV John Black, Solway View, Carlisle Road, Annan, DG12 6QX
G0 FQZ C Emblen, 12 Gefnan, Mynydd Llandygai, Bangor, LL57 4DJ
G0 FRB S Gresty, 4 Palace Road, Sale, M33 6WU
GM0 FRC Falkirk & District ARS c/o P Howson, 1 Howetown, Fishcross, Alloa, FK10 3AW
G0 FRD S Baldwin, 18 Derwent Road, Leighton Buzzard, LU7 2QW
G0 FRL R LOWE, 12 Cavenham Grove, Bolton, BL1 4UA
G0 FRM H White, 51 New Close, Knebworth, SG3 6NU
G0 FRN D Oakes, 56 Middle Way, Chinnor, OX39 4TP
G0 FRO Arthur Medcalf, 1 The Old School, School Lane, Didcot, OX11 0ES
G0 FRR FLIGHT REFUELLING ARS c/o A Baker, Highleaze, Deans Drove, Poole, BH16 6EQ
G0 FRS Farnborough Contest Group c/o Robert Konowicz, 12 Ambleside Crescent, Farnham, GU9 0RZ
G0 FRU C Osbourn, Bourn Bungalow, Back Lane, Newmarket, CB8 9NB
G0 FRV S Adams, 63 Turnbull Drive, Leicester, LE3 2JU
G0 FRX G Cowling, Laissez Faire, Reedness, Goole, DN14 8ET
G0 FRY John Walker, Wildersley, Wildersley Road, Belper, DE56 1PD
G0 FRZ Robert Swinney, 68 Pit House Lane, Leamside, Houghton le Spring, DH4 6QQ
G0 FSB Stewart Barton, 154 The Hill, Cromford, Matlock, DE4 3QU
G0 FSD Paul Robinson, 198 Westfield, Plymouth, PL7 2EJ
G0 FSF David Cawser, 26 Queen Street, Burton-on-Trent, DE14 3LR
G0 FSG C Easton, Dallmore, Lockwood Beck Road, Saltburn, TS12 3LE
G0 FSH W Mackenzie, Nambrac, Cairnmount, Jedburgh, TD8 6SA
G0 FSJ D Portnoy, 6 Birdsall Avenue, Nottingham, NG8 2EH
G0 FSL T Fleet, 51 The Crescent, Walsall, WS1 2DA
G0 FSM J Bent, 32 Cross Waters Close, Wootton, Northampton, NN4 6AL
G0 FSP J Pears, 19 Lichfield Close, Grantham, NG31 8RS
G0 FSR K Hamer, 18 Moor Hall Lane, Stourport-on-Severn, DY13 8RA
GM0 FSV J McGowan, 26 Wallace Gardens, Stirling, FK9 5LS
GM0 FSW N McAllister, 36 Kinneff Crescent, Dundee, DD3 9RG
GM0 FSY Donald Macdonald, 4a Brue, Isle of Lewis, HS2 0QW
GM0 FSZ Eric Sandilands, Eric Sandilands, 12 Kerr Court, Girvan, KA26 0BP
GM0 FTG Ron Hill, 9 Chambers Drive, Carron, Falkirk, FK2 8DX
GM0 FTH H Livingston, Monthouse, Parkhead Road, Linlithgow, EH49 7BS
G0 FTI R Hughes, 46 The Boundary, Oldbrook, Milton Keynes, MK6 2HT
G0 FTJ Norman Mitchinson, 85 Forest Road, Selkirk, TD7 5DD
G0 FTK W Kirk, 59 Silverbuthall Road, Hawick, TD9 7BH
G0 FTO Robert Morton, 44 Cromer Drive, Atherton, Manchester, M46 0QE
G0 FTR David Gill, 80 Bramwell Street, Sheffield, S3 7PB
G0 FTU Chris Jones, Surrey Assembly Hall, Brickhouse Lane, Godstone, RH9 8JW
GM0 FTX I Birkett, 25 Darnhall Crescent, Craigend, Perth, PH2 0HH
G0 FUE D Denton, 7 Uplands Avenue, East Ayton, Scarborough, YO13 9EU
G0 FUH R Douglas, 5 Portnalls Road, Coulsdon, CR5 3DD
G0 FUI Peter Abbott, 170 Hangleton Valley Drive, Hove, BN3 8FE
G0 FUN APAU CONTEST GROUP c/o W Somerville, Glendella, Wycombe Road, High Wycombe, HP14 3RP
G0 FUO Darrell Harrop, 7 Haythorne Way Swinton, Mexborough, S64 8SQ
G0 FUR Derek Buckle, 63 Ashley Drive South, Ashley Heath, Ringwood, BH24 2JP
G0 FUS Paul Fry, Flat 2, National Westminster Bank, North Street, Taunton, TA4 2JY
G0 FUU P Firmin, 25 The Heights, Hastings, TN35 5EP
G0 FUV J Mills, Smiths Hill, Petrockstow, Okehampton, EX20 3EZ
G0 FUW Stephen Hadley, 5 Sydenham Buildings, Bath, BA2 3BS
G0 FUY A Firth, 10 Holroyd Hill, Wibsey, Bradford, BD6 1PQ
G0 FVB G Stokes, 87a Wimborne Road, Southend on Sea, SS2 4JR
GW0 FVC J Newman, 57 Heritage Park, St. Mellons, Cardiff, CF3 0DQ
G0 FVD R Nichol, 32 Greenwood Avenue, Harworth, Doncaster, DN11 8HT
G0 FVF Adrian Howman, 32 Dereham Road, Pudding Norton, Fakenham, NR21 7NA
G0 FVH David Dolling, 41 Bullfinch Close, Poole, BH17 7UP
G0 FVI A Gilfillan, 57 Brant Road, Lincoln, LN5 8RX
GM0 FVJ L Martin, 17 King Street, Perth, PH2 8HR
G0 FVN P Johnston, Woburn House, 38 Chatsworth Avenue, Pontefract, WF8 2UP
G0 FVO J Leach, 19 Fyfe Crescent, Baildon, Shipley, BD17 6DR
G0 FVS J Jackson, 165 Hall St., Briston, Melton Constable, NR24 2LQ
G0 FVT David Lisney, 14 Clarendon Drive, Martham, Great Yarmouth, NR29 4TD
G0 FVU M Smith, 31 Cromford Road, Crich, Matlock, DE4 5DJ
G0 FVX A Barnes, 30 Holland Park, Barton under Needwood, Burton-on-Trent, DE13 8DU
G0 FWA P Waddington, 19 Olivers Drive, Witham, CM8 1QJ
G0 FWD J Adcock, 102 Richmond Way, Newport Pagnell, MK16 0LH
G0 FWF D Stanton, 53 Chester Road, London, N17 6EH
G0 FWP J Purvess, 389 Otley Old Road, Leeds, LS16 6BX
G0 FWU Philip Moreau, 24a Megacre, Wood Lane, Stoke-on-Trent, ST7 8PA
G0 FWX G Clarke, Norton House, Norton End, Saffron Walden, CB11 4JT
GM0 FWY Philip Gray, 21 Simpson Street, Glasgow, G20 6XZ
G0 FXC R Rowland, 60 Ombersley Road, Newport, NP20 3EE
G0 FXD D Harrison, 41 North End Lane, Malvern, WR14 2NG
G0 FXI R Oram, 4 Hardy Avenue, Bristol, BS3 2BP
G0 FXK J Paxton, 27 Holborn View, Leeds, LS6 2RD
G0 FXL D Garratt, 238 Hockley Road, Hockley, Tamworth, B77 5EY
G0 FXQ Ian Drury, 263 Waxwing Lane, Strasburg, United States, VA 22657
G0 FXR Roy Clark, 9 Kensington Avenue, Normanby, Middlesbrough, TS6 0QQ
G0 FXS Keith Rutter, 6 Chetney Close, Stafford, ST16 1XA
G0 FXT C Richardson, 47 Leighton Close, Crossgates, Scarborough, YO12 4LA
G0 FXY M Smith, 19 Legion St., South Milford, Leeds, LS25 5AY
GJ0 FYB C Le Jehan, Sundora, 4 St. Marys Village, St. Mary, Jersey, JE3 3BQ
G0 FYD I McCabe, 99 Edgeway Road, Blackpool, FY4 3NH

G0 FYE B Moss, 22 Battersby St., Ince, Wigan, WN2 2NA
G0 FYH R Butterworth, 12 Strickland Drive, Morecambe, LA4 6TB
G0 FYL Jean Samuels, 63 Mill Road, Okehampton, EX20 1PR
GW0 FYM A Williams, 100 [illegible] Avenue, Dunvant, Swansea, SA2 7XQ
G0 FYP G Holland, 44 Brightstowe Road, Burnham-on-Sea, TA8 2HP
G0 FYQ David Armstrong, 109 Kitchenman Apartments, Charlotte Close, Halifax, HX1 2NT
G0 FYU Timothy Liggins, 55 Kirkland Street, Pocklington, York, YO42 2BX
G0 FYW Brian Morrin, Flat 19 Elms Hall, Elms Rd, Morecambe, LA4 6DD
G0 FYX Stuart Swain, 40 Park Side, Havant, PO9 3PL
G0 FZA N Dickinson, 0 Coolidge Avenue, Lancaster, LA1 5FH
G0 FZB J Atherfold, 42 Mansell Road, Shoreham by Sea, [illegible]
G0 FZC G Palmer, 29 Lindsay Road, Leicester, LE3 2EJ
G0 FZD C Nichols, 7 Royston Avenue, Owlthorpe, Sheffield, [illegible]
G0 FZE B Gould, 2 Parkdale, Ibstock, LE67 6JW
G0 FZF C Quinan, 20 Putney Close, Oldham, OL1 [illegible]
G0 FZG Hilary Eddleston, Bridge End Farm, Threlkeld, Keswick, CA12 4SX
G0 FZH Derek Moore, 12 Newtown Park, Langport, TA10 9TF
G0 FZI D Hooper, 25 Jocelyn Drive, Wells, BA5 2ER
GM0 FZM K Blabey, 5 Grant Road, Prestonpans, EH32 9FE
G0 FZN Frank Rogers, 43 Beacon Road, Billinge, Wigan, WN5 7HF
G0 FZO C Lee, 21 Sandholme, Market Weighton, York, YO43 3ND
G0 FZP Christopher Price, 15 Gores Park, High Littleton, Bristol, BS39 6YG
GI0 FZT P Frend, 41 Brunswick Manor, Abbey Street, Bangor, BT20 4JD
G0 FZU J Tinsley, 21 Peckforton View, Kidsgrove, Stoke-on-Trent, ST7 4TA
GW0 FZY Justin Woolgar, Glan-yr-Afon, Cwmcerdinen, Felindre, Swansea, SA5 7PU
G0 FZZ John Foster, 23 Shrewsbury Street, Hartlepool, TS25 5RQ
G0 GAG M Cowley, 46 Mapletoft Avenue, Mansfield Woodhouse, Mansfield, NG19 8HT
G0 GAJ F Bunce, 45 Hailes Road, Gloucester, GL4 4RB
G0 GAL Eric Howells, 5 Bowland Close, Overdale, Telford, TF3 5HE
G0 GAP W Meecham, 38 Douglas Road West, Stafford, ST16 3NX
G0 GAQ Brian Johnson, 67 Nursery Lane, Northampton, NN2 7PT
G0 GAR A Robinson, 28 Briarsleigh, Wildwood, Stafford, ST17 4QP
GM0 GAT A Thomasson, 1 Eastside Green, Westhill, AB32 6XY
G0 GAW V Richards, Wayside, Penwithick Road, St. Austell, PL26 8UH
G0 GBC Pamela Dempster, 48 Mount Pleasant Road, Pudsey, LS28 9AA
G0 GBE C Thompson, 135 Stafford Road, Bloxwich, Walsall, WS3 3PG
G0 GBG B Wilkinson, 3 Friarage Avenue, Northallerton, DL6 1DZ
GM0 GBH P Young, 4 Primrose Avenue, Rosyth, Dunfermline, KY11 2SS
G0 GBI G Loake, 81 Duchess Road, Bedford, MK42 0SE
G0 GBL D Taylor, 38 Seward Road, Badsey, Evesham, WR11 7HQ
G0 GBN J Henshaw, 7 Gorsefield Close, Wirral, CH62 6BU
G0 GBP D Porter, 10 Broomholme, Shevington, Wigan, WN6 8DT
G0 GBQ H Whitfield, 81 Tenter Balk Lane, Adwick-le-Street, Doncaster, DN6 7EE
G0 GBR S Mattinson, 76 Fairway Avenue, Tilehurst, Reading, RG30 4QB
G0 GBS G Jarman, 212 High St., Clapham, Bedford, MK41 6BS
G0 GBU M McCallum, 65 Whalley Road, Altham West, Accrington, BB5 5DH
G0 GBV D Hackett, 50 Ship Road, Pakefield, Lowestoft, NR33 7DW
G0 GBW C Oswald, 3 Belvedere Drive, Bilton, Hull, HU11 4AX
G0 GBY M Francis, 50 Edinburgh Avenue, Leigh-on-Sea, SS9 3SG
G0 GCA M Collins, 22 Southfield Close, Woolavington, Bridgwater, TA7 8HJ
G0 GCJ Ronald Mellor, 2 Taxal View, Fernilee, High Peak, SK23 7HD
G0 GCK M Johnson, 7 Ash Grove, Northallerton, DL6 1RQ
GM0 GCO B Carson, 46 Tweed Drive, Bearsden, Glasgow, G61 1EJ
G0 GCQ John Rivers, Oakley, 19 Poplar Road, Tenterden, TN30 7NT
GM0 GDD Andrew Mcmillan, 10 Lyon Road, Erskine, PA8 6HG
GI0 GDF Ernest Hooks, 9 Curtis Walk, Lisburn, BT28 1HE
GW0 GDI N Erskine, 302 Caerphilly Road, Cardiff, CF14 4NS
G0 GDJ J Brooks, Bream, The Fitches, Saxmundham, IP17 1UX
G0 GDL M Rodgers, 22 Harewood Road, Allestree, Derby, DE22 2JN
G0 GDS J Hyde, 7 Wandells View, Brantingham, Brough, HU15 1QL
G0 GDT W Clapp, 21 Dronsfield Road, Fleetwood, FY7 7BW
G0 GDV John Strickland, 11 Chilworth Gardens, Waterlooville, PO8 0LD
G0 GEB Richard East, Bramleys, 34 Boyd Avenue, Dereham, NR19 1LU
GM0 GEE L Stapleton, 22 Ashie Road, Inverness, IV2 4EN
G0 GEF P Chipman, 2 Cornforth Close, Trinity Road, Tamworth, B78 2LA
G0 GEH Les Howarth, 12 Thomas St., Hemsworth, Pontefract, WF9 4AY
GW0 GEI Stephen Jones, Peterwell House, Maestir Road, Lampeter, SA48 7PA
G0 GEL J Pruden, 77 Browning Crescent, Ford Houses, Wolverhampton, WV10 6BQ
G0 GEP Graham Perry, 123 Green Lanes, Wylde Green, Sutton Coldfield, B73 5LT
G0 GEQ John Rogerson, 44 Romney Close, Clacton-on-Sea, CO16 8YE
G0 GER R Knighton, 262 Victoria Road West, Thornton-Cleveleys, FY5 3QB
G0 GEU S Holt, 22 Sulby Grove, Morecambe, LA4 6HD
G0 GEV T Hurst, Woodside, Parc Seymour, Caldicot, NP26 3AB
G0 GEZ Dennis Creek, The Coach House, Basketts Lane, Yarmouth, PO41 0PY
G0 GFA Christopher Armitage, 19 Park Road, Barlow, Selby, YO8 8ES
G0 GFC R Johnson, 3 Lance Drive, Chase Terrace, Burntwood, WS7 1FA
G0 GFD K Venn, 28 Streamleaze, Titchfield Common, Fareham, PO14 4NP
G0 GFE B Keechan, 24 Tolley Road, Kidderminster, DY11 7EW
G0 GFI C Burrows, 29 Hampden Road, Malvern Link, Malvern, WR14 1NB
G0 GFK J Denford, 10 Churchill Road, Bideford, EX39 4HG
GM0 GFL A Marriott, Parkview, Dunrossness, Shetland, ZE2 9JG
GW0 GFN D Anderson, Penrheol Farm, Meidrim, Carmarthen, SA33 5NX
G0 GFP A Fowler, 43 Eastbourne Heights, Oak Tree Lane, Eastbourne, BN23 8FB
G0 GFQ Keith Martin, 21 All Saints Close, Weybourne, Holt, NR25 7HH
G0 GFR B Holloway, 28 Elmsdale Road, Ledbury, HR8 2EG
GM0 GFV J Angiolini, 8 Park Way, Cumbernauld, Glasgow, G67 2BU
G0 GFY Conrad James, 180 Mitcham Road, Croydon, CR0 3JF
G0 GFZ P Taylor, 18 Redriff Road, Romford, RM7 8HD
G0 GGA Richard Parkins, 5 Ferndale Grove, Hinckley, LE10 0PH
G0 GGB A Norman, 13 Market Place, Great Yarmouth, NR30 1LY
G0 GGE P Burrow, 9 Minsmere Road, Belton, Great Yarmouth, NR31 9NX
G0 GGG N Rogers, 34 Broadway, Warminster, BA12 8EB
G0 GGH Frank Sell, 17 Auriel Avenue, Dagenham, RM10 8BS
G0 GGL D Ellins, 52 Littlewood, Stokenchurch, High Wycombe, HP14 3TF
G0 GGM Brian Fitzsimmons, 64 Belle Vue Road, Wivenhoe, Colchester, CO7 9LD
G0 GGN R Samways, 7 St. Michaels Way, Steeple Claydon, Buckingham, MK18 2QD
G0 GGQ G Redding, 50 Great Hill, Shefford, SG17 5EA
G0 GGT B Hartley, 25 Ridgeway, Barrowford, Nelson, BB9 8QP
GI0 GGY J Porter, [illegible], Londonderry, BT48 8AQ
G0 GHB J Graham, Caxtonian, [illegible] Road, Norwich, NR13 [illegible]
G0 GHD N Houghton, 67 Sough Road, South Normanton, Alfreton, DE55 2LD
G0 GHE A Goldspink, Danehaven, Themelthorpe, Dereham, NR20 5PS
GW0 GHF Brian Williams, 10 Pantycelyn Road, Llandough, Penarth, CF64 2PG
G0 GHG D Roberts, 16 Min Y Mor, Aberffraw, Ty Croes, LL63 5PQ
G0 GHH P Cross, Balls Farm Cottage, Musbury Road, Axminster, EX13 8TT
G0 GHI Timothy Reeves, Rowney Farm, Newcastle Road, Market Drayton, TF9 2QG
G0 GHM D Coxon, 7 Kingston Way, Nailsea, Bristol, BS48 4RA
GM0 GHN T Taylor, 10 Woodside Drive, Forres, IV36 2UF
G0 GHQ W Small, 24 Barrowdale Close, Exmouth, EX8 5PN
G0 GHT R Pearce, 52 Pearse Close, Hatherleigh, Okehampton, EX20 3GW
G0 GHW Geoffrey Whitehead, 27 Cheyney Walk, Westbury, BA13 3UH
G0 GIA R Keeley-Osgood, 2a St. Anns Crescent, Gosport, PO12 3JJ
GM0 GIB Martin Gibb, Am Fasgadh, [illegible], Glasgow, [illegible]
G0 GID G Renggli, 102 Stewart Road, Bournemouth, BH8 8NX
G0 GIE David Adams, 12 Milton Crescent, Dudley, DY3 3DR
G0 GIF James Catterson, St Sebastians, Gerald Road, Salford, M6 6DW
G0 GIH John Thomas, 41 The Uplands, Brecon, LD3 9HT
G0 GII B Jackson, 94 Nether Court, Halstead, CO9 2HF
G0 GIL J Carter, 112 Landor Road, Whitnash, Leamington Spa, CV31 2JZ
G0 GIN Keith Mills, 19 Thomas Bassett Drive, Colyford, Colyton, EX24 6PN
G0 GIT M Jinks, Caixa Postal 003, Campina Grande De Sul, Parana, Brazil
G0 GJA D Wright, 110 Mancroft Road, Caddington, Luton, LU1 4EN
G0 GJC M Hole, 10 Evans Grove, Marshalswick, St. Albans, AL4 9PJ
GW0 GJD R Radcliffe, 3 Bryncerdin Road, Newton, Swansea, SA3 4UB
G0 GJE R Cleverley, 22 The Tinings, Monkton Park, Chippenham, SN15 3LX
G0 GJG B Phillips, 6 Greenways, Penkridge, Stafford, ST19 5HD
G0 GJH G Hughes, 51 Kingsmead, Seaford, BN25 2HA
G0 GJL Sean Moyses, 78 Deerfield Road, March, PE15 9AG
G0 GJM J Rattigan, 4 Grosvenor Street, Barrow-in-Furness, LA14 4AH
G0 GJN Tony Roberts, 14 Glen Park Gardens, Bristol, BS5 7NE
G0 GJR A Beard, 7 Althorp Close, Tuffley, Gloucester, GL4 0XP
G0 GJV M Goodey, 62 Rose Hill, Binfield, Bracknell, RG42 5LG
G0 GJW Alan Sale, 5 Kingswood Road, Gillingham, ME7 1DZ
G0 GJX F Norton, 147 Wells Road, Glastonbury, BA6 9AN
GM0 GKF Fiona Stapleton, Ravenscourt, 23 Strathkinness High Road, St. Andrews, KY16 9UA
G0 GKH David Birch, 32 Union Street, Trowbridge, BA14 8RY
G0 GKI C Crawford, 70 Westmoreland Avenue, Welling, DA16 2QD
G0 GKK Anthony Pickering, 5 West Farm Court, Manor Road, Consett, DH8 6TL
G0 GKL M Stevens, 33 Langham Road, Hastings, TN34 2JE
G0 GKN J Dowse, 46 Nantwich Road, Middlewich, CW10 9HG
G0 GKO Godfrey Lumsden, Spencer Buildings, Front Street, Hartlepool, TS27 4RT
G0 GKP John Bonner, 40 Lyles Road, Cottenham, Cambridge, CB24 8QR
GM0 GKR John Stapleton, Ravenscourt, 23 Strathkinness High Road, St. Andrews, KY16 9UA
G0 GKY D Burton, 49 Aske Road, Redcar, TS10 2BP
G0 GLA M Hoppe, 67 Belmont Road, Maidenhead, SL6 6LG
G0 GLG Hamish Torunski, 12 Holm Way, Bicester, OX26 3YL
G0 GLH T Mitchell, 18 Park Avenue, Bedlington, NE22 7EJ
G0 GLI L Edwards, 23 Tan Y Bryn, Llanbedr Dc, Ruthin, LL15 1AQ
G0 GLJ Hamish Robertson, 13 St. Martins Road, Chatteris, PE16 6JF
G0 GLQ P Davenport, 1 Boobery, Sampford Peverell, Tiverton, EX16 7BS
G0 GLU Malcolm Fry, 1 Hebden Avenue, Warwick, CV34 5XD
G0 GLW Geoffrey White, 3 Kent Road, Gosport, PO13 0SP
G0 GLX M McFarland, Min Afon, Garreg Fawr Road, Caernarvon, LL54 7ED
G0 GLZ R Sugden, 7 Westbourne Grove, Goole, DN14 6NA
G0 GMA P Sweeting, 35 The Ridings, Market Rasen, LN8 3EE
G0 GMB M Baker, 25 Pentlands, Fullers Slade, Milton Keynes, MK11 2AF
G0 GMC C Cook, 45 Stonepound Road, Hassocks, BN6 8PR
GM0 GMD T Astbury, 8 Auchenlay Holdings, Auchenlay, Dunblane, FK15 9NA
G0 GME R Turner, 17 Wrose Brow Road, Shipley, BD18 2NT
GM0 GMJ J Bavin, Garvan, 10 Grampian Way, Glasgow, G78 2DH
G0 GMJ J Bowles, 38 Rydal Grove, Liversedge, WF15 7DN
G0 GML B Ogden, 14 Fermandy Lane, Crawley Down, Crawley, RH10 4UB
GM0 GMN James Bertram, 20 Kyles View, Largs, KA30 9ET
GM0 GMO G Wylie, 9 Friar Avenue, Bishopbriggs, Glasgow, G64 2HP
G0 GMS A Read, 7 Grange Court, Hixon, Stafford, ST18 0GQ
G0 GMY Paul Whitelock, 2 Shippards Road, Brighstone, Newport, PO30 4BG
G0 GNA J Weller, Pytchley, Chichester Close, Dorking, RH4 1LP
G0 GNE R Maddison, Tom Butt Cottage, Hope St, Godalming, GU8 6DE
G0 GNF Gervald Frykman, 8 Orchard Close, Bishops Itchington, Southam, CV47 2QS
G0 GNI M Anderson, 17 Orchard Road, Seaview, PO34 5JE
GM0 GNK INVERCLYDE AMATEUR RADIO GROUP c/o Andrew Givens, 5 Langhouse Place, Inverkip, Greenock, PA16 0EW
G0 GNP B Ackerman, 31 Melton Mill Lane, High Melton, Doncaster, DN5 7TE
G0 GNQ L West, 22 Lyndhurst Avenue, Margate, CT9 2PS
G0 GNU Jeffrey Norton, 7 Maudsley Street, Bradford, BD3 9JT
G0 GNV Mike Mundy, The Homestead, Homestead Lane, Burgess Hill, RH15 0RQ
G0 GNW G Goddard, 113 Linden Walk, Louth, LN11 9HT
GM0 GNY L Graupner, 5 Carden Close, Alves, Elgin, IV30 8FE
G0 GOB R Drage, 13 Manor Ride, Brent Knoll, Highbridge, TA9 4DY
G0 GOD John Perry, 3 Hurst Green Road, Minworth, Sutton Coldfield, B76 9AP
G0 GOE P Bozac, La Vigne, Motemboeuf, France, 16310
G0 GOH R Cornell, 81 Mercel Avenue, Armthorpe, Doncaster, DN3 3HS
G0 GOI J Macham, 9 Bankfield Grove, Scot Hay, Newcastle, ST5 6AR
GM0 GON A Harrison, Moneen, High Askomil, Campbeltown, PA28 6EN
G0 GOO Barrie Evans, 17 Clarence Place, Maltby, Rotherham, S66 7HA
G0 GOR W Jones, 42 The Park, Kingswood, Bristol, BS15 4BL
GM0 GOV Robert Dinning, South Brae, Aiket Road, Kilmarnock, KA3 4BP
G0 GOX F Goodger, 66 Selkirk Close, Wimborne, BH21 1TP
G0 GOZ Graham Reay, 53 Tithe Barn Road, Stafford, ST16 3PL
G0 GPB K Love, 16 Tenscore Avenue, Walsall, WS6 7BX
G0 GPE David Wells, Browtop, Old Lane, Crowborough, TN6 2AD
G0 GPF Peter Crowley, 45 Beeches Road, Birmingham, B42 2HJ
GI0 GPG Derek McKee, 38 Nursery Road, Armagh, BT60 4BL
G0 GPH Peter Butterworth, 1 The Avenue, Bury, BL9 5DQ
G0 GPK J Burrows, Applefields, Wilmingham Lane, Yarmouth, [illegible]
G0 GPO A Beer, 19 Island Road, Sturry, Canterbury, CT2 0EA
GW0 GPQ R Rees, Y Fedwen Arian, Salem, Llandeilo, SA19 7LY
G0 GPR R Wordsworth, 59 Highgate Lane, Goldthorpe, Rotherham, S63 9BA
G0 GPS I Jones, 107 Wolverley Road, Kidderminster, DY11 5JN
G0 GPV J Barnes, 24 Burleigh Place, Oakley, Bedford, MK43 7SG
G0 GPX M Wise, 440 Tuahiwi Road, RD1, Kaiapoi, New Zealand, 7961
G0 GPZ [illegible], Llandrindod Wells, LD1 5PU
GW0 GQC Brian Morgan, 5 Brynmawr, Bettws, Bridgend, CF32 8SD
GI0 GQG [illegible] Road, Omagh, BT79 0PH
G0 GQH [illegible], Milton Keynes, MK17 8QG
G0 GQI A Orton, 2 The Grove, Brampton Abbotts, Ross on Wye, HR9 7JH
G0 GQJ T Waters, 42 Tregundy Road, Perranporth, TR6 0EF
G0 GQK M Evans, St. Bega, Shay Lane, Newport, TF10 8DA
G0 GQO Stephen Taylor, 91 Severn Walk, Sutton Hill, Telford, TF7 4AS
G0 GQP D Jackson, 38 Chestnut Crescent, Bletchley, Milton Keynes, MK2 2LA
G0 GQT M Bishop, 52 Lingley Drive, Wainscott, Rochester, ME2 4NE
G0 GQV Paul Leech, Holly Tree Cottage, 1 Goat Lane, Norwich, NR13 4NF
G0 GQX M Chappell, 43 Aigburth Hall Avenue, Liverpool, L19 9EA
G0 GQY D Smith, 62 Beresford Road, Dorking, RH4 2DG
G0 GRB P Attew, 23 Kingsleigh Close, Trunch, North Walsham, NR28 0QU
G0 GRC GRANTHAM RAD CL c/o F Seddon, 13 Saltersford Road, Grantham, NG31 7HH
GM0 GRD R Kelly, 31 Cameron Crescent, Cumnock, KA18 3TA
G0 GRI Ian Carter, 12 Bobbin Lane, Westwood, Bradford-on-Avon, BA15 2DL
GM0 GRL D Moore, Parkview, Claredon Place, Dunblane, FK15 9HB
G0 GRM G Meanley, Hemplands, 8 Bennett Drive, Warwick, CV34 6QJ
G0 GRO B Phillips, 12 Fairview Avenue, Weston, Crewe, CW2 5LX
GW0 GRQ John Mitchell, 44 Crossways Street, Barry, CF63 4PQ
G0 GRR B Gray, Orchard End, 48 Smith House Lane, Brighouse, HD6 2LF
G0 GRS Gary Sawford, 17 Church Road, Pytchley, Kettering, NN14 1EL
G0 GRU R Foster, 30 Wimberley Way, South Witham, Grantham, NG33 5PU
G0 GRV Y Katoh, 4-6-3 Minami-Aoyama, Minato-Ku, Tokyo, Japan, 107-0062
GM0 GRW R Young, 2 Quarryhead Cottage, Trabboch, Mauchline, KA5 5JE
G0 GRX BOLTON RAYNET G c/o C Ashlin, 13 Brantfell Grove, Bolton, BL2 5LY
G0 GRZ F Pounder, 29 Read Avenue, Beeston, Nottingham, NG9 2FJ
G0 GSA Roland Hall, The Cottage, 47 Main Street, Rugby, CV23 0PE
G0 GSF Brian Austin, 110 Frankby Road, West Kirby, Wirral, CH48 9UX
GM0 GSG Hugh Cameron, 36 Lynn Crescent, Kirkwall, KW15 1FF
G0 GSH Christopher Summers, 18 Hays Lane, Hinckley, LE10 0LA
G0 GSJ David Howie, 22 Jason Street, Walney, Barrow-in-Furness, LA14 3EJ
G0 GSK W Blay, 50 Fir St., Cadishead, Manchester, M44 5AU
G0 GSL T Ritchie, 6 High Road East, Felixstowe, IP11 9JT
G0 GSM T Pritchard, 21 Newlyn Drive, Bredbury, Stockport, SK6 1EF
G0 GSN N Pope, 1 Knowsley Road West, Clayton le Dale, Blackburn, BB1 9PW
G0 GSR Frank Johnson, 9 Manor Close, Tavistock, PL19 0PN
G0 GST A Williams, 34 Gwydyr Road, Llandudno, LL30 1HQ
G0 GSX C Guerrero, 183 Edinburgh House, Queensway, Gibraltar
G0 GSY B Thomsen, 8 Richmond Road, Cleethorpes, DN35 8PD
G0 GSZ Peter Hunter, 28 Hanover Court, Canterbury Way, Thetford, IP24 1BZ
G0 GTC Derek Nicholls, 4 Nudger Close, Oldham, OL3 5AP
G0 GTI A Dickinson, 6 Church Lane, Bessacarr, Doncaster, DN4 6QB
GM0 GTL T Lorimer, 443 Delgatie Court, Pitteuchar, Glenrothes, KY7 4RW
G0 GTN John Bumford, 10 St. Alkmonds Square, Shrewsbury, SY1 1UH
G0 GTV S Moore, 2 Sheppards Close, Heighington, Lincoln, LN4 1TU
GW0 GUA John Waddell, 17 Heol Tydraw, Pyle, Bridgend, CF33 4AL
G0 GUC A Russon, 92 St. Georges Road, Dudley, DY2 8ER
G0 GUD D Law, 10 Derwent, Tamworth, B77 2LD
G0 GUF J Youde, 4 Greenacre, Hixon, Stafford, ST18 0QE
G0 GUG M Grove, 9 Foxcote Lane, Cradley Forge, Halesowen, B63 2JJ
GM0 GUJ J Cumming, 24 Parkhill Wynd, Leven, KY8 4LH
G0 GUN C Critchley, 3 Beaconsfield View, Robert Road, Slough, SL2 3XT
G0 GUO John Rosindale, Treworder Farm, Ruan Minor, Helston, TR12 7JL
G0 GUT N Reed, 30 Wrey Avenue, Liskeard, PL14 3HX
G0 GUV Roy Murkin, 59 Teagues Crescent, Trench, Telford, TF2 6RF
G0 GUW B Standen, 43 Westover Gardens, St. Peters, Broadstairs, CT10 3EY
GU0 GUX J scheffer, Route De Carteret, Cobo, Castel, Guernsey, GY5 7YS
GW0 GUY E Hollowell, 54 Portfield, Haverfordwest, SA61 1BW
G0 GVA J Lawson, 14 Kentmere Avenue, Farington, Leyland, PR25 3UH
G0 GVB Stuart Balme, 97 Backhold Drive, Halifax, HX3 9DT
G0 GVE S Thomas, 14 Goodley, Oakworth, Keighley, BD22 7PD
G0 GVF M Davidson, Cristina Bungalows, 26 Ave Del Pacific, Benelmadena Malaga, Spain
G0 GVN Colin Wackett, 7 Newell Road, Hemel Hempstead, HP3 9PD
G0 GVS N Clayton, 220 Milnrow Road, Rochdale, OL16 5BB
G0 GVT J Gibb, 37 St. James Road, Melton, North Ferriby, HU14 3HZ
G0 GVX B Sherwood, 363 Old Laira Road, Laira, Plymouth, PL3 6DH
G0 GVZ A Grimes, 37 Cavendish Avenue, Cambridge, CB1 7UH
G0 GWA C Browne, 109 Whirley Road, Macclesfield, SK10 3JL
G0 GWC D Roberts, 179 Southfield Avenue, Preston, Paignton, TQ3 1JX
G0 GWD A Brown, 6 Laing Square, Wingate, TS28 5JE
GW0 GWE R Edwards, 11 Trem Y Eglwys, Coed-Y-Glyn, Wrexham, LL13 7QE
G0 GWG C Partridge, 44 Pine Close, South Wonston, Winchester, SO21 3EB
G0 GWH Arthur Clampitt, 139 Smiths Rd, Emerald Beach, Australia, 2456
G0 GWI P Bell, 6 Dorchester Close, Hale, Altrincham, WA15 8PW
G0 GWL David Baker, Dunbar Lodge, 11 High Road, Bishop Auckland, DL14 8AE
G0 GWM Roy Sawkins, 48 South Road, Saffron Walden, CB11 3DN
G0 GWN J Faulconbridge, 32 Beridge Road, Halstead, CO9 1LB
G0 GWP A Horton, 184 Mount Pleasant, Keyworth, Nottingham, NG12 5ET
G0 GWS William Duff, Highfield, Maundown, Taunton, TA4 2BU
G0 GWY G BIRCH, 33 Kenilworth Road, Scunthorpe, DN16 1EY
G0 GXF P Hirst, 57 Etherington Drive, Hull, HU6 7JT
G0 GXH P Rogers, 20 Clayfield Close, Nottingham, NG6 8DG
G0 GXI Stephen Suter, Fair View, Station Road, York, YO62 4DG

**IMPORTANT NOTE**

**Revalidate licence to avoid revocation** – Ofcom has advised the Society that plans will be drawn up to revoke licences that have not been revalidated as required by the licence conditions. The quickest way to revalidate is to do so online via the Ofcom website: *https://services.ofcom.org.uk/* or by email: *amateur.validations@ofcom.org.uk* Ofcom staff are available to help, but please be patient during times of heavy workload.

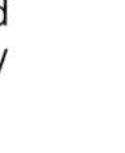

G0 GXO Henry Swaddle, 12 Belmont Gardens, Haydon Bridge, Hexham, NE47 6HG
G0 GXQ J Williams, 91 Mold Road, Buckley, CH7 2JA
G0 GXS R Wareham, 8 Meeting Lane Needingworth, St. Ives, PE27 4SN
G0 GXT Douglas Mellor, The Village Stores, Cleobury North, Bridgnorth, WV16 6RP
G0 GXU M Reeve, 25 Chiltern Road, Hitchin, SG4 9PJ
G0 GXX M Smallwood, 12 Rowan Walk, Hornsea, HU18 1TT
G0 GXZ Michael Butler, 1a Springhead Avenue, Hull, HU5 5HZ
G0 GYA C Taylor, 37 Manor Park Avenue, Allerton Bywater, Castleford, WF10 2DN
G0 GYH D Goodall, 5 Coach Drive, Eastwood, Nottingham, NG16 3DR
G0 GYI R Griffith, Wayside, Colchester Main Road, Colchester, CO7 8DH
G0 GYJ M Dawson, 11 Eastholme Drive, York, YO30 5SU
GM0 GYM T Quinn, 40 Drumry Road, Clydebank, G81 2LL
GM0 GYN P Gibson, 7 Rogerhill Drive, Kirkmuirhill, Lanark, ML11 9XS
G0 GYO Malcolm McPherson, 4 Highfield Place, Wideopen, Newcastle upon Tyne, NE13 7HW
G0 GYP C Fiedler, 51 Fleet Lane, Tockwith, York, YO26 7QD
GM0 GYQ Humphrey Garmany, Woodside Cottage, Shielhill, Dundee, DD4 0PW
GM0 GYT J Ritchie, 24 Kirkton Crescent, Knightswood, Glasgow, G13 3AQ
G0 GYU P Walters, Stonecroft, Main Street, Knaresborough, HG5 9LD
G0 GYY Peter Markham, 15 Victoria Road, Walton on The Naze, CO14 8BU
G0 GZB J Bartlett, 44 Beverington Road, Eastbourne, BN21 2SD
G0 GZE P WYLIE, 15 Semley Road, Hassocks, BN6 8PD
G0 GZF Andrew Pierce, 17b Alderford Street, Sible Hedingham, Halstead, CO9 3HX
G0 GZI R Jeffery, 7 Corfe Way, Winsford, CW7 1LU
G0 GZL Robert Daniels, 8 Dun Cow Close, Brinklow, Rugby, CV23 0NZ
G0 GZM Michael Johns, 3 Carew Close, Coulsdon, CR5 1QS
G0 GZN L Edmunds, 27a Sea View Road, Parkstone, Poole, BH12 3LP
G0 GZO Keith Walters, 18 Leabrooks Avenue, Sutton-in-Ashfield, NG17 5HU
G0 GZR Michael Heel, 27 Englefield Drive, Oakenholt, Flint, CH6 5SB
G0 GZT B Dixon, 36 Meadow Lane, Milehouse, Newcastle, ST5 9AJ
G0 GZU N Harris, 16 Gibbs Field, Bishop's Stortford, CM23 4EY
G0 GZV K Bailey, 35 Edgehill Road, Chislehurst, BR7 6LA
G0 GZW J Spoard, 29 Quarry Road, Alveston, Bristol, BS35 3JL
G0 HAD A Box, 21 St. Michaels Road, Melksham, SN12 6HN
G0 HAE R Isaac, 12 Abbeyfields Close, Netley Abbey, Southampton, SO31 5GR
G0 HAK P Owens, Flat 1, 45 The High Street, Enfield, EN3 4EF
G0 HAL A Paterson, 1 Birch Grove, Timperley, Altrincham, WA15 7YH
G0 HAS Adrian Jordan, 30 Spring Meadows, Trowbridge, BA14 0HD
G0 HAU E Hazell, Long Spring Cottage Gracious Lane, Sevenoaks, TN13 1TJ
G0 HAW C Blackmoor, 4 St. Godwalds Crescent, Aston Fields, Bromsgrove, B60 2EB
G0 HAY D Thomas, 29 Kimberley, Wilnecote, Tamworth, B77 5LD
G0 HBA R Curzon, 24 Edwards Drive, Wellingborough, NN8 3JJ
G0 HBB D Brown, 15 Osborne Avenue, Tuffley, Gloucester, GL4 0QN
G0 HBD A Kings, 13 Fairfield Road, Bulwark, Chepstow, NP16 5JP
GM0 HBF Charles Fraser, Rockside, Locheport, Isle of North Uist, HS6 5EU
G0 HBJ R Millett, 8 Sidestrand Road, Newbury, RG14 6HP
GM0 HBK Colin Robertson, 3 Sasaig, Teangue, Isle of Skye, IV44 8RD
G0 HBL Kevin Alderman, Ben, Wharf Road, Broxbourne, EN10 6HD
G0 HBN W Riley, 18 Leyland Close, Trawden, Colne, BB8 8TB
G0 HBO S Holmes, 17 Portland Gardens, Low Fell, Gateshead, NE9 6UX
G0 HBU Allan Dwyer, 10 Shaftway Close, Haydock, St. Helens, WA11 0YQ
G0 HBV T Sandilands, 8 Cheviot View, Lowick, Berwick-upon-Tweed, TD15 2TY
G0 HBW R Fitzgerald, 93 Finland Road, Brockley, London, SE4 2JQ
G0 HBX J Turbefield, 126 Preston Road, Chorley, PR6 7AU
G0 HBZ David Wright, Drws-Y-Nant, St. Asaph Avenue, Rhyl, LL18 5EY
GW0 HCB P Matheson, 34 Pentwyn, Radyr, Cardiff, CF15 8RE
G0 HCC HERTS COUNTY C/ c/o Trevor Groves, Emergency Planning, Hertfordshire County Council, Hertford, SG13 8DE
G0 HCD T Carroll, 2 West Durham Cottages, Roddymoor, Crook, DL15 9QX
G0 HCE Michael William Wilkinson, 98 Whitestiles, High Seaton, Workington, CA14 1LL
G0 HCI Michael Bodle, 46 Torrington Road, Portsmouth, PO2 0TW
GW0 HCK A Morgan, 3 Gelli Newydd, Golden Grove, Carmarthen, SA32 8LP
G0 HCN T Hayden, 7 Attlee Close, Garnlydan, Ebbw Vale, NP23 5ES
G0 HCO Mark Barnett, 26 Casewell Road, Kingswinford, DY6 9HB
G0 HCP David Goode, Wolfson College, Cambridge, CB3 9BB
GM0 HCQ M Gloistein, 27 Stormont Way, Scone, Perth, PH2 6SP
G0 HCR Stephen Turner, 71 Valley Road, Lillington, Leamington Spa, CV32 7RX
G0 HCT B Cooper, 8 Mitchell Way, Madeley, Telford, TF7 5SN
G0 HCX Michael Butler, 32 Harold Road, Deal, CT14 6QH
G0 HCY Graham Vine, 56 Colchester Road, St. Osyth, Clacton-on-Sea, CO16 8HB
G0 HDB Alan Davies, 11 Gravel Pits Close, Bredon, Tewkesbury, GL20 7QL
G0 HDC G Evans, 34 Coronation Drive, Donnington, Telford, TF2 8HY
G0 HDD A Adey, 37 Cranmere Avenue, Wolverhampton, WV6 8TR
G0 HDF R Bartlam, 34 Quarry Road, Selly Oak, Birmingham, B29 5NX
G0 HDG A Edwards, 29 Larch Road, Maltby, Rotherham, S66 8AZ
G0 HDH Colin Worsfold, 7 West View Road, Cowes, PO31 8NR
G0 HDI Brian Walker, Lantilla, Elmfield Lane, Southampton, SO45 1BJ
G0 HDJ A Douglas, Threave House, Blind Lane, Somerton, TA11 6BW
GI0 HDO M Deehan, 9 Farland Way, Londonderry, BT48 0RS
G0 HDP B Wainwright, 19 Durkar Low Lane, Durkar, Wakefield, WF4 3BL
G0 HDS R Hurt, 7 Atwood Close, Immingham, DN40 2DQ
G0 HDV K Coxon, 29 Chapel Road, Broughton, Brigg, DN20 0HW
G0 HDY R Finnis, 6 The Crescent, Cwmbran, NP44 7JG
G0 HDZ Ian Rose, 82 Little Brays, Harlow, CM18 6ES
G0 HEA B Griffin, 26 Hamer Street, Radcliffe, Manchester, M26 2RS
G0 HEE A Salt, 43 Waldorf Heights, Blackwater, Camberley, GU17 9JH
G0 HEF R Mckeever, 38 Brookfield Avenue, Runcorn, WA7 5RF
G0 HEJ E Lloyd, 43 Queensway, Upton, Chester, CH2 1PF
G0 HEL A Nicholls, 19b Dark Lane South, Steeple Ashton, Trowbridge, BA14 6EZ
G0 HEM G Gardner, New House, Birdbush Avenue, Saffron Walden, CB11 4DJ
G0 HEN G Henstock, 36 Cornwallis Road, Oxford, OX4 3NW
G0 HEQ Michael Farrant, 16 Camborne Grove, Yeovil, BA21 5DG
G0 HER K Kibblewhite, 53 Woodcote, Bedford, MK41 8EL
G0 HET Peter Nutkins, 31 Higher Spence Cottage, Bridport, DT6 6DF
G0 HEU Paul Stott, 70 Wansbeck, Washington, NE38 9EG
G0 HEV P Brindley, 6 Chaucer Close, Stowmarket, IP14 1GH
G0 HEW J Mchale, 21 Bonython Road, Newquay, TR7 3AW
G0 HEX D Cheetham, 4 Battersbay Grove, Hazel Grove, Stockport, SK7 4QW
G0 HFA J Lansdowne, 8 Nansloe Close, Helston, TR13 8BP
G0 HFC F Chadwick, 59 Beech Avenue, Greenfield, Oldham, OL3 7AW
G0 HFE T Cadman, 28 Denbigh Court, Ellesmere Port, CH65 5DX
G0 HFK R Fuller, Flat 13, Samuel Lewis Trust Dwellings, London, SW6 1BS
G0 HFL Nicholas Major, 26 Jubilee Drive, Earl Shilton, Leicester, LE9 7JF
G0 HFN R Stennett, West Willow, Milton Damerel, Holsworthy, EX22 7DL
G0 HFX C Parnell, 29 Southfield, Southwick, Trowbridge, BA14 9PW
G0 HGC B Roberts, Ty Clyd, Ffordd Mela, Pwllheli, LL53 5AP
GD0 HGG Stephen Entwisle, 30 Arden Mhor, Pinner, HA5 2HR
G0 HGH J Scott, 3 Westminster Drive, Spalding, PE11 2UW
G0 HGI B Boden, 71 Park Head Road, Sheffield, S11 9RA
G0 HGM J Jenkinson, 4 Greenways, Ilfracombe, EX34 8DT
G0 HGN Thomas Jones, 11 Lon Ogwen, Bangor, LL57 2UD
G0 HGO D Cady, 45 The Hill, Wheathampstead, St. Albans, AL4 8PR
GW0 HGP L Magfhogartai, Talaharian, Abergwili, Carmarthen, SA31 2JL
G0 HGV D Burgess, 243 Lichfield Avenue, Torquay, TQ2 8AJ
G0 HHA P Dixon, 68 Chelsea Road, Sheffield, S11 9BR
G0 HHC A Gorton, 7 Sterling Close, Colchester, CO3 9DP
G0 HHD G Williams, Gwastad Annas, Barmouth, LL42 1DX
GI0 HHE James Dowey, 19b The Bridges, Newtownabbey, BT37 0TD
G0 HHL Philip Scrivens, 1 Walnut Close, Milton, Derby, DE65 6WA
G0 HHN Philip Le-Brun, The Granary, Henton, Chinnor, OX39 4AE
GI0 HHV S Johnstone, 6 Gilbert Crescent, Bangor, BT20 4PE
G0 HHW Bryan Hayward, Tyn y Gerddi, Deiniolen, Caernarfon, LL55 3ND
GI0 HHZ R Fitzsimons, 9 Ingledene Park, Newtownards, BT23 8QT
G0 HIC J East, 35 Preachers Vale, Coleford, Radstock, BA3 5PT
G0 HID David Paine, 35 Marfield Close, Sutton Coldfield, B76 1YD
G0 HIF Alan Edmonds, 44 Blea Tarn Road, Kendal, LA9 7NA
GM0 HIG Christine Cameron, 36 Lynn Crescent, Kirkwall, KW15 1FF
G0 HIJ Wayne Roberts, 13 Roseacre Road, Elswick, Preston, PR4 3UD
G0 HIK N Gregory, Town End, Kirkby Road, Askam-in-Furness, LA16 7EY
GM0 HIM James Pert, 56 Lochiel Drive, Milton of Campsie, Glasgow, G66 8ET
G0 HIN T Hamilton, 2 Rowin Close, Hayling Island, PO11 9BT
G0 HIR A edwards, Flat 6, Oaktree Court, Fields Road, Cwmbran, NP44 3AZ
G0 HIW D Ross, 60 Kingsway, South Molton, EX36 4AL
G0 HIZ David Hughes, 39 Kestrel Drive, Crewe, CW1 3RY
G0 HJB F Little, 26 Hoghton Road, Longridge, Preston, PR3 3UA
G0 HJD J Ford, 42 The Grove, Walton-on-Thames, KT12 2HS
G0 HJK J Everett, 92 Thackeray Road, Ipswich, IP1 6JB
G0 HJL J Levesley, 96 Brookside Road, Bransgore, Christchurch, BH23 8NA
G0 HJM R Smith, 100 Braemar Road, Billingham, TS23 2AN
G0 HJR Robin Filby, 10 Malvern Avenue, Burton-on-Trent, DE15 9EB
GM0 HJU J Mackenzie, 17 Maple Avenue, Milton of Campsie, Glasgow, G66 8BB
GM0 HJV Craig McClure, 27 St. Andrews Drive, Gourock, PA19 1HY
G0 HJW Percy Webber, Palm Court Hotel, 1 Lansdowne Road, Falmouth, TR11 4BE
G0 HJX P Devine, 41 Carodoc Road, Wingate, TS28 5BT
G0 HJZ J Durrell, 8 Woodcote Cottages, Graffham, Petworth, GU28 0NY
G0 HKB R Connelly, Harford House, Wells Road, Radstock, BA3 4EX
G0 HKC Eric Chambers, Greenholme, 19 Marina Road, Salisbury, SP4 8DB
G0 HKE John Boyton, 3 Wenny Estate, Chatteris, PE16 6UX
G0 HKF Anthony Roberts, 149 Cannock Road, Burntwood, WS7 0BB
G0 HKI L Compton, 44 Overdown Rise, Portslade, Brighton, BN41 2YG
G0 HKK J Hall, 36 Sandfield Road, West Bromwich, B71 3NF
G0 HKN R Leigh, 19 Richmond Court, Worthing, BN11 4JB
G0 HKQ D Suddes, Villa Dobochet, Holway Road, Holywell, CH8 7DR
G0 HKW David Spencer, 48 Ingleborough Way, Leyland, PR25 4ZR
G0 HKZ S Liptrott, 8 Fox Bank Close, Widnes, WA8 9DP
G0 HLA T Barker, Peartree Cottage, Ash Hill Common, Romsey, SO51 6FU
G0 HLB Ronald Evans, 191 St. Leonards Road East, Lytham St. Annes, FY8 2HW
G0 HLI A Dudley, 7 St. Michaels Close, Willington, Derby, DE65 6EB
G0 HLJ I Barber, 17 Copley Crescent, Scawsby, Doncaster, DN5 8QW
GM0 HLK Morris Borthwick, 5/11 Dalgety Avenue, Edinburgh, EH7 5UF
G0 HLL R Steel, 43 Westfield Avenue, Wigston, LE18 1HY
G0 HLS S Deacon, 32 Westfield Way, Charlton, Wantage, OX12 7EW
G0 HLU A Mckay, 48 Holdenby Close, Retford, DN22 6UB
GM0 HLV Derek Gill, Old Post Office, Lairg, IV27 4PQ
G0 HLW Matthew Neale, 126 Brookvale Road, Solihull, B92 7JB
G0 HMD M Mills, 44 East Acridge, Barton-upon-Humber, DN18 5HH
G0 HME C Statham, 24 St. Johns Close, Heather, Coalville, LE67 2QL
G0 HMF J Tite, 40 Dingledery, Olney, MK46 5ES
G0 HMG S Finch, Combe Shorney Farm, Brompton Ralph, Taunton, TA4 2SB
G0 HMK T Angier, 29 Sunnyhill Road, Herne Bay, CT6 8LT
GM0 HMM F Robertson-Mudie, 18 Portnaguran, Isle of Lewis, HS2 0HD
G0 HMO E Cooke, 190 Newark Crescent, Nottingham, NG2 4NY
G0 HMS F Rafferty, 9 Mortlake Close, Hull, HU8 0QH
G0 HMX Maurice Wragg, 27 Rosedale, Worksop, S81 0TB
G0 HND A Simmonds, Fern Way, Purssells Meadow, High Wycombe, HP14 4SG
GW0 HNE K Luke, 19 Heol Y Gors, Cwmgors, Ammanford, SA18 1PE
G0 HNG B Horton, 3 Cromer Road, Finedon, Wellingborough, NN9 5LP
G0 HNI John Buckler, 181 Glen Road, Oadby, Leicester, LE2 4RJ
GM0 HNJ John Iain Cowan, 20a Harbour View, Invergordon, IV18 0EY
G0 HNL T Cannon, 10 Badger Close, Guildford, GU2 9PJ
G0 HNO T Callanan, 39 Greenlands Way, Henbury, Bristol, BS10 7PH
G0 HNP E Bottomley, 33 Duke Street, Coldstream, TD12 4BS
G0 HNQ Harry Brammeld, School House, Rosley, Wigton, CA7 8AU
GW0 HNS S Yates, 46 Y Berllan, Dunvant, Swansea, SA2 7RW
GW0 HNT W South, Kimberlee, 45 Dan y Bryn, Neath, SA11 3PJ
GM0 HNV P Mackenzie, 16 Cedar Drive, Milton of Campsie, Glasgow, G66 8AY
G0 HNW Paul Widger, Notre Revie, Cinderhills Road, Holmfirth, HD9 1EH
G0 HNZ Ronald Disney, 25 Davos Way, Skegness, PE25 1EL
G0 HOA T Goodwin, 63 Lynwood Drive, Merley, Wimborne, BH21 1UT
G0 HOB Jack Mc Glynn, 1 Primrose Crescent, Leeds, LS15 7QW
G0 HOC Mike Harris, Po Box 841, Hobart, Tasmania, Australia, 7001
G0 HOD N Colbourn, 7 Brookwood Road, Farnborough, GU14 7HH
G0 HOF Kenneth Barnett, 5 Morborne Road, Folksworth, Peterborough, PE7 3SS
G0 HOI A Mudd, 86 Manor Drive, Waltham, Grimsby, DN37 0NR
G0 HOJ J Hoyland, 67 Coronation Road, Wroughton, Swindon, SN4 9AT
G0 HOP G Evans, 12 Marlowe Close, Stevenage, SG2 0JJ
G0 HOQ M Piper, Applegarth, Milford Road, New Milton, BH25 5PW
G0 HOS B Tufnail, 197 Portland Road, Wyke Regis, Weymouth, DT4 9BH
G0 HOT M Reeds, 6 Leach Way, Riddlesden, Keighley, BD20 5DB
G0 HOV A Howes, 11 Stretton Road, Wolston, Coventry, CV8 3FR
G0 HOX I Atkins, 64 Twin Hill Lane, Stafford, Virginia, United States
G0 HPA G Smith, 3 Park Lane, Featherstone, Pontefract, WF7 6BL
GW0 HPC G Evans, 49 Pen Yr Ally Avenue, Neath, SA10 6DS
G0 HPG L Brookes, 177 Charnwood Close, Rubery, Birmingham, B45 0JY
G0 HPH P Brookes, 177 Charnwood Close, Rubery, Birmingham, B45 0JY
G0 HPK Andrew Gaston, 9 Lochans Mill Avenue, Stranraer, DG9 9BZ
G0 HPL T McCutcheon, 25 Millburn Court, Sheuchan Street, Stranraer, DG9 0DX
G0 HPM Richard Waddingham, 38 Station Road, Hockwold, Thetford, IP26 4HZ
G0 HPN Louis Denyer, 94 Wood Lane, Chippenham, SN15 3DZ
G0 HPQ P Delaney, 27 Grasmere Road, Redcar, TS10 1JA
G0 HPS G Bukin, Granary Cottage, Uffculme, Cullompton, EX15 3DN
G0 HPV Kenneth Hardy, 12 Liquorpond Street, Boston, PE21 8UF
GM0 HQF Eric Kolonko, West Manse, School Road, Kilbirnie, KA25 7LB
GM0 HQG George Flett, Stenadale, Orphir, Orkney, KW17 2RF
G0 HQK J Jones, 28 Clares Lane Close, The Rock, Telford, TF3 5DA
G0 HQN C Wilson, 34 Lyndhurst Street, Salford, M6 5YB
G0 HQT Bruce Bell, 4 Broadlee Bank, Tweedbank, Galashiels, TD1 3RF
G0 HQU S Broadhead, 39 Haw Avenue, Yeadon, Leeds, LS19 7XE
G0 HQX R Rea, 11 Wissage Lane, Lichfield, WS13 6DQ
G0 HRD L Lawrence, 119 Ladywood Road, Lane End, Dartford, DA2 7LP
G0 HRF B Barwick, 100 Westwood, Golcar, Huddersfield, HD7 4JY
G0 HRG Mold and District ARC c/o Stephen Studdart, 33 Linden Avenue, Connah's Quay, CH5 4SN
G0 HRH George Vaughton, Higher Woodhayne, Whitford, Axminster, EX13 7PB
G0 HRJ Christopher Mason, 22 Eskdale Avenue, Halifax, HX3 7NH
G0 HRK R Eselgroth, 15 Hedgerley Gardens, Greenford, UB6 9NT
G0 HRL G Ward, 67 Sebright Road, Wolverley, Kidderminster, DY11 5UA
G0 HRO John Bland, 5 King Street, Mansfield, NG18 2PX
G0 HRQ C Davies, 4 Reece Court, Barlow Road, Dukinfield, SK16 4AB
G0 HRR Kenneth Mott, 191 Joyners Field, Harlow, CM18 7QD
G0 HRS HILDERSTONE RS c/o R Marchant, Cascade, The Street, Canterbury, CT3 1LN
G0 HRT Robert Harwood, 24 Westerdale Drive, Banks, Southport, PR9 8DG
G0 HRW WIGAN & DIST AR c/o D Barkley, 39 Fulbeck Avenue, Wigan, WN3 5QN
G0 HRX Christopher Deakin, Wainscot, Lanreath, Looe, PL13 2NX
G0 HSA Andrew Bennett, 21 Hunstone Avenue, Norton, Sheffield, S8 8GE
GI0 HSB W Dickson, 8 Drumglass Avenue, Bangor, BT20 3HA
GM0 HSC Hugh Cumming, 29a/5 Summerside Place, Edinburgh, EH6 4NY
G0 HSD Andrew Shaw, 15 Austerby, Bourne, PE10 9JJ
G0 HSH D Denford, 20 Shaxton Crescent, New Addington, Croydon, CR0 0NU
G0 HSK Jeremy Hakes, 86 Station Crescent, Rayleigh, SS6 8AR
G0 HSN P Broadley, 8 Langley Gardens, Lowestoft, NR33 9JE
G0 HSR Huntingdonshire ARS c/o Philip Haylock, 25 Whitehouse Road, Sawtry, Huntingdon, PE28 5UA
G0 HSV S Jacob, 3 Broadfield Close, Gomeldon, Salisbury, SP4 6LX
G0 HSW B Statham, 11 Old Woods Hill, Torquay, TQ2 7NR
G0 HTD Henry Todd, 105 Brownhill Road, Blackburn, BB1 9QY
G0 HTF D Crucefix, 5 Gainsborough Mews, Kidderminster, DY11 6PZ
G0 HTG Peter Hague, 14 Camellia Close, Driffield, YO25 6QT
GM0 HTH John Grieve, Langamo, Harray, Orkney, KW17 2JU
G0 HTJ D Body, 53 Grove Road, Wollescote, Stourbridge, DY9 9AE
G0 HTK Trevor St John-Murphy, Sherbourne, Sherbourne Drive, Windsor, SL4 4AE
G0 HTL B Sargent, 25 Jordans Way, Bricket Wood, St. Albans, AL2 3SJ
G0 HTM R Bushell, 121 Rickmansworth Road, Watford, WD18 7JD
G0 HTO V Morris, 16 Wentworth Close, Longlevens, Gloucester, GL2 9RB
G0 HTS S ALDER, 56 Sparrowbill Way, Patchway, Bristol, BS34 5AU
GM0 HTT A Flett, Shannon, Dounby, Orkney, KW17 2HR
G0 HTX Anthony Coyle, 8 Farm Place, Kensington, London, W8 7SX
G0 HUD D Withers, 141 Broadway, Walsall, WS1 3HB
G0 HUF P Peterson, 13 Orchard Road, Smallfield, Horley, RH6 9QP
G0 HUG P Hemphill, Springhill, George Lane, Sudbury, CO10 7SB
G0 HUK G Roberts, 24 Cornwall Road, Bingley, BD16 4RN
G0 HUN M Roberts, 19 Bro Geirionydd, Trefriw, LL27 0JE
GM0 HUO A Pardoe, 59 Spottiswoode Gardens, St. Andrews, KY16 8SB
G0 HUQ R Monk, 43 Lichfield Drive, Bury, BL8 1BJ
G0 HUT T Hutton, 4 Victoria Gardens, Farnborough, GU14 9UH
G0 HUV R Fawkes, 9 Crickley Drive, Warndon, Worcester, WR4 9LB
G0 HUW A Dyson, 24 Newborough Close, Austrey, Atherstone, CV9 3EX
G0 HVA J Curtis, Glebe Farm, West Knighton, Dorchester, DT2 8PE
G0 HVB David Clark, 3 West Well Lane, Theale, Wedmore, BS28 4SW
G0 HVC W Hutchins, 25 Manor Road, Herne Bay, CT6 6RF
G0 HVH G Watson, 14 Burnholme Avenue, York, YO31 0LU
GI0 HVJ R Cunliffe, 82 Strabane Road, Newtownstewart, Omagh, BT78 4JZ
G0 HVN David Cottam, 14 Barnard Close, Rednal, Birmingham, B45 9SZ
G0 HVO R Minter, 15 Harold Close, Pevensey Bay, Pevensey, BN24 6SL
G0 HVP D Cooper, 11 Downland Court, Magazine Road, Ashford, TN24 8NF
G0 HVQ Darrell Moody, Old Chapel House, Malvern Road, Gloucester, GL19 3NZ
G0 HVS D Kearns, Ingleby, Sleegill, Richmond, DL10 4RH
G0 HVT A Harding, 17 St. Anns Road South, Heald Green, Cheadle, SK8 3DZ
G0 HVX R Mcquillan, 9 Sandpit Road, Welwyn Garden City, AL7 3TW
GD0 HWA Charles Howard, 5 Ballure Grove, Ramsey, Isle of Man, IM8 1NF
GM0 HWB Peter Lafferty, 18 Chesters Crescent, Motherwell, ML1 3QU
G0 HWC Paul Young, 14 Carisbrooke Avenue, Clacton-on-Sea, CO15 4RZ
G0 HWI Paul Mardle, The Pines, Franklin Road, Chelmsford, CM3 6NF
G0 HWK M Drew, 10 Marina Drive, Dunstable, LU6 2AH
GI0 HWO J Crawford-Baker, Georges Nest, 131 Gobbins Road, Larne, BT40 3TX
G0 HWP Anthony Wilson, 41 Bentley Drive, Walsall, WS2 8RX
GM0 HWQ P Wood, 25 Davidson Road, Edinburgh, EH4 2PE

G0 HWS P Dowsett, Furze Cottage, West Chiltington Road, Pulborough, RH20 2PR
G0 HWT P Littlechild, 13 Castle Hill, Daventry, NN11 4AQ
G0 HWU Adam Edwards, Higher Tregiddle Farm, Gunwalloe, Helston, TR12 7QW
G0 HWV B Lawton, 9 Byron Court, Kidsgrove, Stoke-on-Trent, ST7 4JF
G0 HWX S Unsworth, 26 Rupert Brooke Road, Rugby, CV22 6HQ
G0 HWY P Adams, 229 Upper Selsdon Road, South Croydon, CR2 0DZ
G0 HXC F Pegg, 20 Swaledale Avenue, Cowpen Estate, Blyth, NE24 4DT
G0 HXD P Halsall, 22 Northway, Northwich, CW8 4DF
G0 HXF A Etheridge, Wyngra, Haywards Heath Road, Haywards Heath, RH17 6NJ
GI0 HXH A McCaldin, 7 Mount Ida Road, Banbridge, BT32 4HF
G0 HXL Edwin Calthorpe, 49 Cross Coates Road, Grimsby, DN34 4QH
G0 HXM Ann [illegible], 11 Willoughby Road, Langley, Slough, SL3 8JH
G0 HXQ C Baron, 76 Princess Road, Rochdale, OL10 4DA
G0 HXR Geoffrey Martin, 34 Vicarage Farm, Parson Drove, Spalding, PE11 3QW
GW0 HXS T Williams, Bryndewi, Llanarth, SA47 0QN
G0 HXU Simon Dawson, Hamlet Cottage, West End, Tadcaster, LS24 9DL
G0 HYG R Taylor, 79 South Drive, Harwood, Bolton, BL2 3NS
GW0 HYH V Grayson, Willow Lodge, Croeslan, Llandysul, SA44 4SJ
GW0 HYL V Underwood, Rock Hill, Llanarthney, Carmarthen, SA32 8LJ
GD0 HYM Michael Dunning, 55 Station Park, Colby, Isle of Man, IM9 4NL
G0 HYN D Robertson, 5 Chandlers, Orton Brimbles, Peterborough, PE2 5YW
G0 HYP K Ferguson, 4 Honister Road, Whitehaven, CA28 8HS
G0 HYR R Deakin, 5 Burton Road, Oakthorpe, Swadlincote, DE12 7QU
G0 HYS S Hutchinson, 55 Richmond Avenue, Sheffield, S13 8TH
G0 HYT Patrick Gray, 2 Bryan Close, Sunbury-on-Thames, TW16 7UA
GW0 HYU K Richards, 7 Capel Edeyrn, Pontprennau, Cardiff, CF23 8XJ
GM0 HYW R Carmichael, 8 Winifred Street, Kirkcaldy, KY2 5SR
G0 HZA Melvyn Rolph, 60 Queen Street, Swaffham, PE37 7BT
G0 HZB M Roberts, 4 Elba Close, Paignton, TQ4 7LW
G0 HZC Graham Hutt, 21 Bentley Crescent, Fareham, PO16 7LU
G0 HZD W Brown, Longstone Cottage, Old Pound, St. Austell, PL26 7XS
G0 HZE R Howell, 161 Coneygree Road, Stanground, Peterborough, PE2 8LH
G0 HZG P Sturgess, 11 Laburnum Avenue, Newbold Verdon, Leicester, LE9 9LQ
GM0 HZI Neil McLaren, 10 Newton Avenue, Skinflats, Falkirk, FK2 8NP
G0 HZK R Muggleton, 136 Hurworth Avenue, Slough, SL3 7FQ
G0 HZL Martin Hawkins, 20 Salisbury Road, Blackpool, FY1 5QJ
GM0 HZO Thomas Leckie, 1 Dykehead, Port of Menteith, Stirling, FK8 3JY
G0 HZQ R Eldrett, 20 Hill Rise, Horspath, Oxford, OX33 1TJ
G0 HZY H Harding, 29 Brighton Avenue, Elson, Gosport, PO12 4BU
G0 IAA H Singer, Neptune Gap, Whitstable, CT5 1EL
G0 IAC Christopher Caspell, 28 Peel Terrace, Stafford, ST16 3HD
G0 IAD A Dobbyn, Roadways, Selwick Drive, Bridlington, YO15 1AP
G0 IAE D Yeo, 356 Radcliffe Road, Fleetwood, FY7 7NH
G0 IAG Anthony King, 2 Ebenezer Cottages, Thorney Road, Peterborough, PE6 7UB
G0 IAH T Preece, White House, Bredwardine, Hereford, HR3 6BY
G0 IAI P Dean, Down Farm, Lovaton, Yelverton, PL20 6PT
G0 IAK M Watts, Hainault, Coronation Avenue, Bradford-on-Avon, BA15 1AX
G0 IAL R Scott, 46 St. Albans Road, Hemel Hempstead, HP2 4BA
G0 IAP Peter Allen, 25 Wayside Avenue, Hornchurch, RM12 4LL
G0 IAS A Hickman, The Conifers, High Street, Retford, DN22 8AJ
G0 IAX Richard Bailey, 3 Charlotte Close, Birstall, Batley, WF17 9BX
G0 IAY Lee Wiltshire, 7 Burleaze, Chippenham, SN15 2AY
GI0 IBC RADIO AMATEUR INVALID AND BLIND CLUB c/o P Hallam, 95 Belfast Road, Carrickfergus, BT38 8BY
G0 IBE Richard Higgs, 60 Lichfield Avenue, Evesham, WR11 3EA
G0 IBG B Armstrong, 65 Asquith Road, Bentley, Doncaster, DN5 0NT
G0 IBI R Clarke, 14 Cranwell Avenue, Cranwell, Sleaford, NG34 8HG
G0 IBN Andrew KERSEY, 35 Sceptre Close, Tollesbury, Maldon, CM9 8XB
G0 IBR Stephen Dalley, 5 Anstey Mill Close, Alton, GU34 2QT
G0 IBS K Brown, Fernbank, Foreside Lane, Bradford, BD13 4EY
G0 IBT H Percy, 13 Cherry Tree Walk, Astley Cross, Stourport-on-Severn, DY13 0JT
G0 IBW D Jones, 5 Luccombe Close, Ingleby Barwick, Stockton-on-Tees, TS17 0NL
G0 IBY F Hubbard, 187 Standhill Road, Carlton, Nottingham, NG4 1LE
G0 IBZ M Hackford, Snuggles, Rockalls Road, Colchester, CO6 5AR
G0 ICB John Howell, 21 Peaslands Road, Saffron Walden, CB11 3ED
G0 ICC Richard Shirvington, 3 Main Street, Preston Bissett, Buckingham, MK18 4LH
G0 ICD D Matthews, 32 Stewards Avenue, Widnes, WA8 7BN
G0 ICE M Knott, 24 Walsingham Way, Ely, CB6 3AL
G0 ICG T Lees, Post Green, Lytchett Minster, Poole, BH16 6AP
G0 ICJ David Dawkes, 95 Houndsfield Lane, Wythall, Birmingham, B47 6LX
G0 ICK STH TOTTENHM AR c/o I Kraven, 55 Cranfield Crescent, Cuffley, Potters Bar, EN6 4DZ
G0 ICP J Ker, 1 The Willows, Ulcombe Road, Ashford, TN27 9QR
G0 ICT J Morrison, 52 Kimberley Close, Dover, CT16 2JW
G0 ICW M Bagnall, 6 Silver Fir Close, Hednesford, Cannock, WS12 4SU
G0 IDB Simon Jeffreys, 26 Meadway, Esher, KT10 9HF
G0 IDD D Clarke, 60 Dunedin Crescent, Burton-on-Trent, DE15 0EJ
G0 IDE David Dobson, 79 Wood Green, Leyland, PR25 2YL
G0 IDF D Fletcher, 12 Drayton Road, Dorchester-on-Thames, Wallingford, OX10 7PJ
G0 IDH M Frost, Kernyk, Trevanion Terrace, Redruth, TR15 3BP
G0 IDJ Janet Low, 56 George St., Whithorn, Newton Stewart, DG8 8NZ
G0 IDL M Bedell, 1 Pheasant Field Drive, Spondon, Derby, DE21 7LR
G0 IDP D Ford, 2 Bedelands Close, Burgess Hill, RH15 8BL
GD0 IDU N Bayliss, Teal, Tromode Road, Douglas, Isle of Man, IM2 5EH
GM0 IDV Alan Price, Upper Arsdale, Evie, Orkney, KW17 2NN
G0 IDZ L Bailey, 22 Coventry Grove, Wheatley, Doncaster, DN2 4QA
G0 IEB R Neal, 6 Wheatcroft Avenue, Scarborough, YO11 3BN
G0 IEE K Harris, 14 Dunstall Close, St. Marys Bay, Romney Marsh, TN29 0QX
G0 IEH T Lister, 6 Fordlands Crescent, Fulford, York, YO19 4QQ
G0 IEN R Wharton, 1407 Briar Bayou Dr, Houston, United States, 77077
G0 IEO R Woodberry, 35 Whybridge Close, Rainham, RM13 8BB
G0 IEQ G Kennedy, 19 Beech Grove, Whitby, Ellesmere Port, CH66 2PA
G0 IER Brian Smith, 73 Devon Street, Hull, HU4 6PL
G0 IES David Johnson, 32 Middleton Road, North Reddish, Stockport, SK5 6SH
GM0 IET M Shell, 15 Lundin View, Leven, KY8 5TL
G0 IEV M Rippington, 15 Queens Avenue, Canterbury, CT2 8AY
G0 IEW J Rose, 1 Nelson Place, Whiston, Prescot, L35 3PP
G0 IEY S Tribe, 6 Privett Road, Purbrook, Waterlooville, PO7 5HJ
G0 IFA K Keitch, 10 Sycamore Close, Willand, Cullompton, EX15 2SH
G0 IFC A Burnett, 2 Courtney Road, Tiverton, EX16 6EE
G0 IFD Thomas Street, Flat 18, 58 Mapledene Road, London, E8 3LE
G0 IFF Graham Spinney, 55 Highdale Avenue, Clevedon, BS21 7LU
G0 IFI Robert Finch, 6 Clover Way, Thetford, IP24 1LQ
G0 IFN Richard Moriarty, 46 Oak Avenue, Morecambe, LA4 6HS
G0 IFQ P Lait, Glenside, 8 Kingston Lane, Shoreham by Sea, BN43 6YB
G0 [illegible] Nigel Clayton, [illegible], Fair Lawn, Whitstable, CT5 3JZ
G0 [illegible] [illegible], Allerton, Bradford, BD15 9BD
GD0 IFU W Corkish, 17 [illegible] Drive, Onchan, Douglas, Isle of Man, IM3 4ND
G0 IFX B Galloway, 2 Summers Close, Knutsford, WA16 9AW
G0 IGA R Clark, 9 Windsor Close, Hove, BN3 6WQ
G0 IGB Vernon Elliott, 22 Kirkstead Road, Cheadle Hulme, Cheadle, SK8 7PZ
G0 IGC J Golightly, 123 Littlefield Lane, Grimsby, DN34 4PN
G0 IGH G Golightly, 123 Littlefield Lane, Grimsby, DN34 4PN
G0 IGJ J Dickson, Eilrig, Roberton, Hawick, TD9 7PR
G0 IGK G Knox, 33 St. Lukes Road, Aller, Newton Abbot, TQ12 4NE
G0 IGM M Manning, 31 Harcourt Crescent, Shrewsbury, SY2 5LQ
G0 IGT Harold Terry, 7 Marlow Drive, Irlam, Manchester, M44 6LR
G0 IGU Gary Morton, 23 Bembridge Road, Eastbourne, BN23 8DX
G0 IHA G Goodwin, Thumpers, 16 St. Catherines Road, Winchester, SO23 0PP
G0 IHC M Nash, 49 Oakfield Way, Sharpness, Berkeley, GL13 9UT
G0 IHE Q Reed, 3 Carre Gardens, Worle, Weston-Super-Mare, BS22 7YB
G0 IHF W Hough, 19 Farnham Close, Appleton, Warrington, WA4 3BG
G0 IHI Martin Eyers, 190 Greenhill Road, Herne Bay, CT6 7RS
G0 IHK Ian Tough, 15 Headlands Way Whittlesey, Peterborough, PE7 1RL
G0 IHO R Priestley, Le Haut Courtigne, Parcay Les Pins, Maine Et Loire, France, 49390
G0 IHU C Smith, 66 Bronte Farm Road, Shirley, Solihull, B90 3DF
G0 IIA J Stokes, The Beehive, Debenham Road, Ipswich, IP6 9TD
G0 IIB B Daniels, 18 Oakley Road, Chinnor, OX39 4HB
G0 IID James Ian Batley, 3 Folldon Avenue, Sunderland, SR6 9HP
G0 IIE John Colliton, PO Box 17, Ramsgate 2217, Sydney, Australia, 2219
G0 IIF John Green, 1 Huntley Grove, Sheffield, S11 7LX
G0 IIG G Fleming, 168 Blythway, Welwyn Garden City, AL7 1DU
G0 IIK N Ackland, 69 Great South West Road, Hounslow, TW4 7NH
G0 IIP R Chambers, 12 Dorchester Close, Maidenhead, SL6 6RX
G0 IIQ D Pykett, 35 Harneis Crescent, Laceby, Grimsby, DN37 7BA
GI0 IJB J Stevenson, 22 Knockfergus Park, Greenisland, Carrickfergus, BT38 8SN
GM0 IJD Peter Taggerty, 44 Ravenswood Drive, Glenrothes, KY6 2PA
G0 IJI A Muir, 21 Mansell Close, Towcester, NN12 7AY
G0 IJK B Puncher, Danbys Oast, Coldbridge Lane, Maidstone, ME17 2AX
G0 IJN A Slade, Skerries, Summerhill Althorne, Chelmsford, CM3 6BY
GM0 IJR I Ross, 14 Kilmundy Drive, Burntisland, KY3 0JW
G0 IJU P Elms, 12 Suffield Way, King's Lynn, PE30 3DE
G0 IJY I Roberts, 7 Heol Bradwen, Four Mile Bridge, Holyhead, LL65 2NF
G0 IJZ Marcus Walden, 181 Coleridge Road, Cambridge, CB1 3PW
G0 IKB K Brown, 1 Deerwood Close, Macclesfield, SK10 3RE
G0 IKC Roy Hole, 3 Holywell Park, Halwill, Beaworthy, EX21 5UD
G0 IKD A Galvin, 27 Hill Top Lane, Tingley, Wakefield, WF3 1HT
G0 IKE L Dodson, 1 Limmer Lane, Booker, High Wycombe, HP12 4QR
G0 IKI Michael Holbrough, 21 Malyns Close, Chinnor, OX39 4EW
G0 IKN S McDonald, 152 Stony Lane, Burton, Christchurch, BH23 7LD
G0 IKP T Clements, 72 Stanbridge Road, Haddenham, Aylesbury, HP17 8HN
G0 IKQ Anthony Cook, 17 Lothersdale, Wilnecote, Tamworth, B77 4HT
G0 IKR Michael Best, 21 Hubble Road, Corby, NN17 1JD
GM0 IKY A Diamond, 51 Marchlands Avenue, Bo'ness, EH51 9ER
G0 IKZ J Pugh, 1 The Lawns, Everton, Sandy, SG19 2LB
G0 ILA N Smith, 165 Upper Deacon Road, Southampton, SO19 5LN
GM0 ILB I Brown, Vadill, Brae, Shetland, ZE2 9QN
G0 ILC J Price, 30 Pottery Close, Whiston, Prescot, L35 3RW
G0 ILD B Harrison, 61 Foyle Road, Blackheath, London, SE3 7RQ
G0 ILH J Collins, 14 Balmain Crescent, Wolverhampton, WV11 1BG
G0 ILI G Pomroy, 17 Rock Close, Pengegon, Camborne, TR14 7TT
G0 ILK B Loram, 12 The Finches, Castleham, St. Leonards-on-Sea, TN38 9LQ
G0 ILN Richard Putnam, 95 Martyns Way, Bexhill-on-Sea, TN40 2SH
G0 ILO P Taylor, 16 Petrel Close, Herne Bay, CT6 6NT
GM0 ILQ I Ferguson, C/O Mr Jb Ferguson, 8 Cleveden Crescent, Glasgow, G12 0PR
G0 ILT T Drummond, 10 Delamere Road, Earley, Reading, RG6 1AP
G0 ILZ M Hanraads, 6 Oak Hill, Hollesley, Woodbridge, IP12 3JY
G0 IMA P Pearce, 7 Otter Road, Clevedon, BS21 6LQ
G0 IMB Robert Battersby, 4 Gorsey Brow, Urmston, Manchester, M41 9QE
G0 IMD R Clamp, 276 The Parade, Greatstone, New Romney, TN28 8UL
G0 IMG M Gotch, 44 Audley Road, Saffron Walden, CB11 3HD
GM0 IMH C Taylor, The Grange Smithy, Errol, Perth, PH2 7TB
G0 IMK N Sparrey, The Ashes, Mamble Road, Kidderminster, DY14 9HX
G0 IMP N Wheeldon, 28 Constance Avenue, Lincoln, LN6 8SN
G0 IMQ Neil Cummings, 14 Cemetery Road, Gloucester, GL4 6PB
G0 IMU David Spearman, 7 Farman Close, Salhouse, Norwich, NR13 6QD
G0 IMV Raymond Hill, Marclecote, Ledbury Road, Ross-on-Wye, HR9 7BE
GM0 IMW Andrew McBride, 4 Clova Place, Uddingston, Glasgow, G71 7BQ
G0 IMX C Lacey, 69 Field Lane, Pelsall, Walsall, WS4 1DQ
GM0 IMZ I McEwan, Granary Cottage, Vogrie Grange, Gorebridge, EH23 4NT
G0 INA E Cairns, 2 Stockhill Circus, Stockhill, Nottingham, NG6 0LS
G0 ING NTHANTS EXPD GR c/o Lionel Parker, 128 Northampton Road, Wellingborough, NN8 3PJ
G0 INJ P Shuttlewood, 10 Church Lane, Costock, Loughborough, LE12 6UZ
G0 INK S Taylor, 37 Crestfield Drive, Pye Nest, Halifax, HX2 7HG
GW0 INN P Garner, 44 Tycoch Road, Sketty, Swansea, SA2 9EQ
G0 INO D Elvin, 11a Lilyholt Road, Benwick, March, PE15 0XQ
G0 INQ C Hannell, Toat Lodge, Pulborough, RH20 1BZ
G0 INS P Yardley, 13 Blackthorn Road, Kenilworth, CV8 2DS
G0 INT J Lee, 2 York Terrace, Birchington, CT7 9AZ
G0 INZ R Bull, 24 Healey Close, Abingdon, OX14 5RL
GM0 IOA Samuel Aitken, 24 Laverock Avenue, Glenrothes, KY7 5HX
G0 IOE R Haynes, 6 Epping Walk, Furnace Green, Crawley, RH10 [illegible]
G0 IOF G Moore, 64 Walmer Road, Seaford, BN25 3TN
G0 IOI T Maunder, 19 St. Monica Road, Southampton, SO19 8FF
G0 IOK Gerald Markeson, 23 Chantry Lane, Tideswell, Buxton, SK17 8NP
GD0 IOM R Ferguson, Moaney Moar House, Corlea Road, Ballasalla, Isle of Man, IM9 3BA
G0 ION B Cooke, 5 Squires Gate, Rogerstone, Newport, NP10 0BP
G0 IOO A Willis, 46 Maryland Road, Thornton Heath, CR7 8DF
G0 IOP D Melville, Little Buckden, Milberry Lane, Chichester, PO18 9JJ
G0 IOQ DENTON PARK MIDDLE SCHOOL RC c/o F Cosgrove, Denton Park Middle School, Linhope Road, Newcastle upon Tyne, NE5 2NW
G0 IOR Michael [illegible]
GI0 IOT M Rabbett, 41 Richill Park, Londonderry, BT47 5QY
G0 IOU B Gleed, 19 Silver Birch Caravan Site, Walters Ash, High Wycombe, HP14 4UY
GM0 IOY E Wall, 43 Larkfield Road, Gourock, PA19 1YA
G0 IOZ Rob Southerington, 4 Reculver Close, Sunnyhill, Derby, DE23 1WN
G0 IPB L Collins, 32 Parkstone Road, Syston, Leicester, LE7 1LY
G0 IPC J Hilton, 99 Kirby Road, Stone, Dartford, DA2 6HD
G0 IPE Peter Gardner, 30 Wantage Close, Maidenbower, Crawley, RH10 7NU
G0 IPH D Sutton, 10 Normanhurst Road, Borough Green, Sevenoaks, TN15 8HT
GD0 IPK J Marsh, 44 Richmond Gardens, Harrow Weald, Harrow, HA3 6AJ
G0 IPN A Greenwood, 4 Roach Place, Rochdale, OL16 2DD
G0 IPO E Landor, Silverden, Silverden Lane, Cranbrook, TN18 5LX
G0 IPT D Griggs, 5 Collingwood Avenue, Muswell Hill, London, N10 3EH
GM0 IPV B Stephenson, 11 Bishop Forbes Crescent, Blackburn, Aberdeen, AB21 0TW
GM0 IPW R Dickie, Taynish, 11 Churchill Drive, Stornoway, HS1 2NP
G0 IPX FISTS/INTERNATIONAL MORSE PRE 'SOC' c/o John Griffin, 35 Cottage Street, Kingswinford, DY6 7QE
GI0 IQA S Ruff, 9 Cooleen Park, Newtownabbey, BT37 0RR
G0 IQC S Steed, 37 Greg Street, Stockport, SK5 7LB
GM0 IQD Donald Ross, 24 Eriskay Road, Inverness, IV2 3LX
G0 IQH F Manning, 180 Priestley Terrace, Wibsey, Bradford, BD6 1QU
G0 IQI Paul James, 14 Cairnmount, Jedburgh, TD8 6SA
G0 IQK W Chewter, 93 Eton Road, Ilford, IG1 2UF
G0 IQM S Atkinson, 26 Skipton Road, Trawden, Colne, BB8 8QS
G0 IQN C Jones, 189 Moor Lane, Cranham, Upminster, RM14 1HN
G0 IQP P Williams, 28 Cross Likey, Church Stoke, Montgomery, SY15 6AL
G0 IQQ G Bradley, Fairhaven, 29 Ashbourne Avenue, Cleckheaton, BD19 5JH
G0 IQZ W Williams, 31 Stad Ty Croes, Llanfairpwllgwyngyll, LL61 5JR
GW0 IRC D Knibbs, 24 Corbett Grove, Caerphilly, CF83 1SZ
G0 IRH D Pond, 31 Quintilis, Bracknell, RG12 7QQ
G0 IRI J Wade, 12 Kendal Road, Harlescott, Shrewsbury, SY1 4ER
G0 IRJ A Wilson, 42 Bacheler Street, Hull, HU3 2TZ
G0 IRK N Porter, 23 Calder Court, 7 Britannia Road, Surbiton, KT5 8TS
GD0 IRM N Chambers, 78 Durley Avenue, Pinner, HA5 1JH
GW0 IRP C Evans, 19 Bryn Terrace, Caerau, Maesteg, CF34 0UR
G0 IRQ Frederick Wagner, 4 Hamilton Close, South Walsham, Norwich, NR13 6DP
G0 IRT P Williams, 15 Rhymney Close, Rassau, Ebbw Vale, NP23 5TF
G0 IRY W Gallacher, 143 Acre Street, Huddersfield, HD3 3EJ
GM0 IRZ A Dick, 4 Westpark Gardens, Dundee, DD2 1NY
GM0 ISA Donald Arcari, 184 Fintry Drive, Fintry, Dundee, DD4 9LP
G0 ISC R Slyfield, 359 Ringwood Road, Poole, BH12 4LT
G0 ISE G Carter, 12 Somerford Road, Broughton, Chester, CH4 0SZ
G0 ISG R Hale, 5 Land Oak Drive, Kidderminster, DY10 2ST
G0 ISH Robert Pownall, Beechgrove, Field Road Whiteshill, Stroud, GL6 6AG
G0 ISI R Christopher, 33 Grange Park, Albrighton, Wolverhampton, WV7 3EN
G0 ISJ G Parkin, Mil-Rune, Marsh Gate, Camelford, PL32 9YN
G0 ISK M Glover, 50 Broadway, Brinsworth, Rotherham, S60 5ES
G0 ISL B Shersby, 4 Blenheim Gardens, Chichester, PO19 7XE
G0 ISM R Faversham, 19 Pelwood Road, Camber, Rye, TN31 7RU
G0 ISO M Edmunds, 27a Sea View Road, Parkstone, Poole, BH12 3LP
G0 ISP Michael Harrington, 32 South View Way, Prestbury, Cheltenham, GL52 5BN
GI0 ISQ David Christie, 8 Ballytober Road, Bushmills, BT57 8UX
GM0 IST J Purtell, 31 Daleally Crescent, Errol, Perth, PH2 7QA
G0 ISY J Davies, Winter Cottage, Church Road, St. Ives, TR26 3LE
GI0 ITJ D Linton, 4 Elmwood, Cullybackey, Ballymena, BT43 5PY
G0 ITL J Dunkley, 34 Sparrow Close, Ilkeston, DE7 4PW
G0 ITM J Dutton, 16 Briarfield, Washington, NE38 8RX
G0 ITO Alan O'shaughnessy, Southby, Buckland, Faringdon, SN7 8QR
G0 ITS R Wells, 35 Creswell Farm Drive, Stafford, ST16 1PG
G0 ITU M Wood, 78 Haycliffe Road, Bradford, BD5 9HB
G0 ITZ E Pieroni, 395 Huddersfield Road, Millbrook, Stalybridge, SK15 3HU
G0 IUA Stuart Ratcliffe, 20 Merrion Street, Farnworth, Bolton, BL4 7LG
G0 IUD Alan Wakeman, 133 Stanshawe Crescent, Yate, Bristol, BS37 4EG
G0 IUE Jonathan Wheeler, 27 Elley Green, Neston, Corsham, SN13 9TX
G0 IUH P Battershill, 15 Winkworth Road, Banstead, SM7 2QJ
G0 IUI K Jones, 12 Delamere Drive, Macclesfield, SK10 2PW
G0 IUK E Pickerill, 6 Pitmore Walk, Moston, Manchester, M40 0GB
G0 IUM J Mcculloch, 39 Beech Drive, St. Ives, PE27 6UB
G0 IUN C Stain, 6 Sutton Close, Sutton-in-Ashfield, NG17 3DP
GI0 IUP G Lyle, 40 Enagh Crescent, Maydown, Londonderry, BT47 6UG
G0 IUV K Knowles, 18 Croft Close, Pinxton, Nottingham, NG16 6RF
G0 IUW R Hill, 7 Berkeley Close, Cashe's Green, Stroud, GL5 4SA
G0 IUY J Tribe, 6 Privett Road, Purbrook, Waterlooville, PO7 5HJ
G0 IVB M Llewellyn, 28 North St., Clay Cross, Chesterfield, S45 9PL
G0 IVD P McGarvey, 125 Esme Road, Sparkhill, Birmingham, B11 4NJ
GW0 IVG William Jones, 30 Maesglas, Pontyates, Llanelli, SA15 5SG
G0 IVI Keith Edwards, 39 King Edward Street, Sandiacre, Nottingham, NG10 5BS
GI0 IVJ J McCausland, 17 Dunavon Heights, Dungannon, BT71 6TN
G0 IVO Brian Singleton, Langdales, Broadbury, Okehampton, EX20 4LL
G0 IVP S Parrish, Ambleside, Carlton Avenue, Hornsea, HU18 1JG

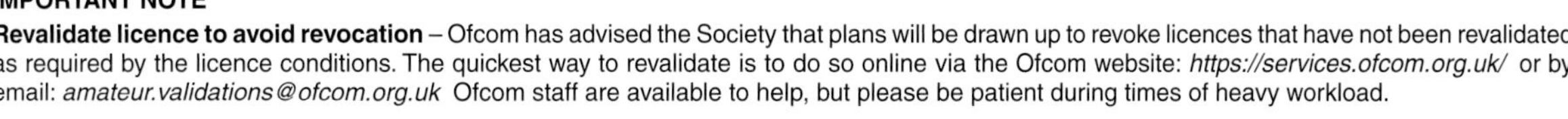

**IMPORTANT NOTE**

**Revalidate licence to avoid revocation** – Ofcom has advised the Society that plans will be drawn up to revoke licences that have not been revalidated as required by the licence conditions. The quickest way to revalidate is to do so online via the Ofcom website: *https://services.ofcom.org.uk/* or by email: *amateur.validations@ofcom.org.uk* Ofcom staff are available to help, but please be patient during times of heavy workload.

UK Callsigns

GM0 IVQ G McKinlay, 68 Hillend Road, Clarkston, Glasgow, G76 7XT
G0 IVR ITCHEN VALLEY ARC c/o P Baxter, Raka Lodge, Whitenap Lane, Romsey, SO51 5ST
G0 IVT T Jones, Heathbrook, Maesmawr Close, Brecon, LD3 7JF
G0 IVV James Sheldrake, 38 Andros Close, Ipswich, IP3 0SL
G0 IVX H Harrison, 8 North Leigh, Tanfield Lea, Stanley, DH9 9PA
G0 IVZ J Fisher, Farland, Rillaton, Callington, PL17 7PA
G0 IWB J Wells, 12 Church Road, Grafham, Huntingdon, PE28 0BB
G0 IWD A Aston, Ty Newydd, Y Ffor, Pwllheli, LL53 6UY
G0 IWF F Oakton, 180 Wragley Way, Stenson Fields, Derby, DE24 3DZ
G0 IWI David Ranson, 27 Kruses Road, North Warrandyte, Australia, VICTORIA 3113
G0 IWJ G Slingsby, 46 Bathurst Road, Staplehurst, Tonbridge, TN12 0LQ
G0 IWN D Greenhalgh, 8 Pleasant Road, Milton, Southsea, PO4 8JU
G0 IWW R Weedon, 50 Dryleaze Court, Wotton-under-Edge, GL12 7BL
GM0 IWX T Lorimer, 9 Orchard House, Orchard Grove, Leven, KY8 5XA
G0 IWZ A White, 13 Woodcote Drive, Poole, BH16 5RA
GW0 IXK R Griffiths, 71 Elder Grove, Llangunnor, Carmarthen, SA31 2LH
GW0 IXM Owen Williams, 11 Hafod Road, Tycroes, Ammanford, SA18 3QL
GM0 IXO Gerald Michie, 2 Moncur St., Townhill, Dunfermline, KY12 0HN
G0 IXP L Startup, 24 Parc Plas, Blackwood, NP12 1SJ
GW0 IXQ D Graham, 1 Maestir, Llanelli, SA15 3NS
G0 IXS Peter Turner-Hicks, 64 Sharpley Avenue, Coalville, LE67 4DT
G0 IXT G Lambert, Church House, Chapel Lane, Christchurch, BH23 6BE
G0 IXV P Biscombe, Keverine, Victoria Road, Folkestone, CT18 7JS
G0 IXZ D Fleetwood, Lynton House, Station Road, Chesterfield, S44 6BH
GM0 IYA W Brown, 29 Bankhead Crescent, Dennyloanhead, Bonnybridge, FK4 1RY
G0 IYD P Bates, 29 Juler Close, North Walsham, NR28 0SY
G0 IYE D Chalmers, 42 Thornbury Drive, Uphill, Weston-Super-Mare, BS23 4YH
G0 IYJ R Turley, 50 Hanscombe End Road, Shillington, Hitchin, SG5 3NB
G0 IYK M Harrison, 12 Richmond Rise, Reepham, Norwich, NR10 4LS
G0 IYM M Linton, 134 Eccles Old Road, Salford, M6 8QQ
G0 IYO Keith Stanmore, 48 St. Michaels Avenue, Bishops Cleeve, Cheltenham, GL52 8NX
GM0 IYP SUTHERLAND AND DISTRICT ARC c/o Clive Ohennessy, Savalbeg, Challenger Estate, Lairg, IV27 4ED
G0 IYQ N Mitchelson, Trevena, Winskill, Penrith, CA10 1PD
G0 IYS P Willis, 5 Binbrook Walk, Corby, NN18 9HH
G0 IYT C Savin, Jenalri, Union Lane, Preston, PR3 6SS
G0 IYU P Kirk, 38 Carleton Street, Morecambe, LA4 4NY
G0 IYV T Jones, 159 Cobden View Road, Crookes, Sheffield, S10 1HT
G0 IYW G Farndon, 28 Willow Close, Collycroft, Bedworth, CV12 8BE
G0 IYX Ian Roberts, 2 Samuel Fold, Pendlebury Lane, Wigan, WN2 1LT
G0 IYY M Lindsay, 5 Eatonhill, Norwich, NR4 7PY
G0 IZB H Smith, 69 Chapel View, South Croydon, CR2 7LJ
G0 IZC J Carver, 131 Rutland Avenue, High Wycombe, HP12 3JQ
G0 IZE R Bates, 36 Maple Crescent, Alveley, Bridgnorth, WV15 6LT
G0 IZI A Warwick, 58 Longworth Avenue, Tilehurst, Reading, RG31 5JY
G0 IZJ E Whittaker, 17 Packer Street, Bolton, BL1 3LD
G0 IZK Michael Dennehy, 45 Vine Road, Tiptree, Colchester, CO5 0LR
G0 IZL N Procter, 19 Manitoba Way, Selston, Nottingham, NG16 6FP
G0 IZN A Halliday, 12 Fowley Common Lane, Glazebury, Warrington, WA3 5JJ
G0 IZP R Mcdonald, 4 The Paddocks, Baunton, Cirencester, GL7 7DL
G0 IZQ J Bradnock, 5 Milverton Road, Knowle, Solihull, B93 0HX
G0 IZR Jason Bridson, 10 Clegg Street, Astley, Manchester, M29 7DB
G0 IZV D Gowers, 32 Silver Fox Crescent, Woodley, Reading, RG5 3JA
G0 JAA Tony Watson, 89 Addison Road, Wednesbury, WS10 0LW
G0 JAC M Hill, 9 Longacre, Woodthorpe, Nottingham, NG5 4JS
G0 JAF J Foy, 23 Lee Road, Nelson, BB9 8SD
G0 JAG R Glyn, 171 Bull Close Road, Norwich, NR3 1NY
G0 JAI Thomas Jones, 26 Treowain, Forge, Machynlleth, SY20 8EJ
G0 JAJ G West, 39 Court Farm Road, Eltham, London, SE9 4JL
G0 JAL W Bell, 244 Westbourne, Woodside, Telford, TF7 5QR
G0 JAM J Morrison, 107 Crown Meadow, Colnbrook, Slough, SL3 0LJ
G0 JAN J Ilston, 6 Dovedale, Canvey Island, SS8 8HX
G0 JAO L White, 56 Grange Road, Leigh-on-Sea, SS9 2HT
G0 JAP C Collins, 26 Nicholsons Wharf, Mather Road, Newark, NG24 1FN
G0 JAQ G Blackwood, 63 Illingworth Avenue, Halifax, HX2 9JH
G0 JAR R Offord, 116 Townsend Road, Snodland, ME6 5RL
G0 JBA P Boorman, Gladstone House, Marshborough Road, Sandwich, CT13 0PE
G0 JBC C Collins, South Peat Pitts Farm, Ogden, Halifax, HX2 9NR
GM0 JBE R McCart, Rovelea, Cumnock Road, Ayr, KA6 7PS
G0 JBH Ann Bingham, 84 Kathleen Road, Southampton, SO19 8LN
G0 JBJ A Tungate, 171 Marlborough Gardens, Faringdon, SN7 7DG
G0 JBM L Humphreys, 19 Clinch Green Avenue, Bexhill-on-Sea, TN39 5HN
G0 JBO B Lees, Preston Hall, Preston, Telford, TF6 6DH
G0 JBP R Lipscomb, Redmoor, Bickley Road, Bromley, BR1 2NF
G0 JBR Eric Thorley, 7 Drake Street, St. Helens, WA10 4JG
G0 JBS M Notman, 3 Pilling Avenue, Lytham St. Annes, FY8 3QF
G0 JBT David Roberts, 100 Sherwood Park Avenue, Sidcup, DA15 9JJ
G0 JBV A Mcintosh, 17 The Chase, Abbeydale, Gloucester, GL4 4WP
G0 JBY J Youd, 25 Hanson Road, Andover, SP10 3HL
G0 JBZ R Stevens, 51 Beacon Park Crescent, Upton, Poole, BH16 5PB
G0 JCA J Andress, 46 Bridwell Road, Plymouth, PL5 1AB
GW0 JCB Ceri Jones, 8 West Walk, Barry, CF62 8BY
G0 JCC Andrew Lancaster, Oak House, Sutton Hill Road, Bristol, BS39 5UT
G0 JCD P Busby, 7 Rimmer Green, Southport, PR8 5LP
GD0 JCF D Smith, 14 College Drive, Ruislip, HA4 8SB
G0 JCG A Butler, 8 The Furrows, Great Sutton, Ellesmere Port, CH66 2YL
G0 JCH J Head, 6 Rumbelow Road, Tiverton, EX16 6JT
G0 JCK W Pattinson, 2 The Green, Ticknall, Derby, DE73 7GY
G0 JCN M Lovatt, 37 Hartland Avenue, Bilston, WV14 9AN
G0 JCP D Grey, 7 Cemetery Lane, Tweedmouth, Berwick-upon-Tweed, TD15 2BS
G0 JCQ Brian Rimmer, 8a Mallee Avenue, Southport, PR9 8NL
GW0 JCT P Jones, Bronallt, Cenarth, Newcastle Emlyn, SA38 9JS
G0 JCY P Stuart, Skylark Corner, Seaborough Hill, Crewkerne, TA18 8PL
G0 JCZ M Martin, 2821 Bissonnet St, Houston, United States, 77005-4014
GM0 JDB James Ratter, Foulawick, Brae, Shetland, ZE2 9QS
G0 JDC John Gwynn, 117 Main Street, Goldthorpe, Rotherham, S63 9JW
G0 JDD I Douce, 67 Glenbervie Drive, Leigh-on-Sea, SS9 3JT
G0 JDE R Doyle, 61 St. Peters Road, West Mersea, Colchester, CO5 8LN
G0 JDG A Purseglove, 122 Chesterfield Road, Huthwaite, Sutton-in-Ashfield, NG17 2QF
G0 JDL J Clarke, 4 Swallowfields, Carlton Colville, Lowestoft, NR33 8TP
G0 JDM T Kewell, Appley Cottage, Iron Mill Lane Oldford, Frome, BA11 2NR
G0 JDO Trevor Thomas, 26 Corfe Crescent, Torquay, TQ2 7QX
G0 JDQ Michael Cozens, Lot 321, 1001 Starkey Rd, Florida, United States, 33771
GW0 JDS J Stonehouse, 49 Heol Y Gelynen, Upper Brynamman, Ammanford, SA18 1SB
GW0 JDW D Willis, 5 Dan Lan Rd, Llanelli, Dyfed, SA16 0NF
GW0 JDY M Lewis, 12 Fern Rise, Neyland, Milford Haven, SA73 1RA
G0 JEA R Kaye, 63 Coronation Drive, Birdwell, Barnsley, S70 5RL
G0 JEC D Naylor, 6 Front St., Kirk Merrington, Spennymoor, DL16 7HZ
G0 JEE Bernard Greer, Willow Farm, Main Road, Burton-on-Trent, DE13 9QD
GM0 JEF J Fysh, 7 Chestnut Place, Ellon, AB41 9HF
G0 JEH Steven Rosbottom, 26 Wellington Street, Preston, PR1 8TP
G0 JEK Chris Kelland, 11 The Meads, West Hanney, Wantage, OX12 0LJ
G0 JEQ Ronald Wicks, Gooseberry Cottage, Llangunllo, Knighton, LD7 1SW
G0 JEU Stuart Dunsmore, 33 Church Street, Messingham, Scunthorpe, DN17 3SB
GI0 JEV H Kernohan, 40 Lisnafillon Road, Gracehill, Ballymena, BT42 1JA
G0 JEW P Wells, 12 Shelley Drive, Lutterworth, LE17 4XF
G0 JEZ B Wells, 37 Elder Road, Denvilles, Havant, PO9 2UW
G0 JFA A Jones, Highercombe West, Highercombe, Dulverton, TA22 9PT
G0 JFC J Aithison, 14 Claymore Rise, Silsden, Keighley, BD20 0QQ
G0 JFD G Butcher, 15 Leander Drive, Gosport, PO12 4GG
G0 JFE P Sixsmith, 10 Wisbeck Road, Bolton, BL2 2TA
GI0 JFF William McBride, 5 Aylesbury Road, Newtownabbey, BT36 7YP
GM0 JFH J Mair, 43 Todhill Avenue, Onthank, Kilmarnock, KA3 2EQ
GM0 JFK C Harper, Glencoul Cottage, Cullicudden, Dingwall, IV7 8LL
GM0 JFL B Harper, The Police House, Broadford, Isle of Skye, IV49 9AB
G0 JFM Stephen Nicholls, Fieldway, The Street, Woodbridge, IP12 2QG
G0 JFP J Scott, 23 Botesworth Green, Milnrow, Rochdale, OL16 3PJ
GW0 JFQ N Williams, Flat 4, 234-237 Chapmans High Street, Swansea, SA1 1NZ
G0 JGB James Barnett, 43 Westsprink Crescent, Stoke-on-Trent, ST3 5JD
G0 JGF S Cox, 60 Leawood Road, Trent Vale, Stoke-on-Trent, ST4 6LA
G0 JGH A Loveridge, 7 Walnut Way, Northfield, Birmingham, B31 4ES
G0 JGI Christopher Davison, 27 Drake Close, Horsham, RH12 5UB
G0 JGM G Meares, 41 Tor Close, Worle, Weston-Super-Mare, BS22 6BZ
G0 JGV K Bewley, 38 Great Innings South, Watton at Stone, Hertford, SG14 3TF
GD0 JGX D Ginsberg, 26 Keeill Pharick Park, Glen Vine, Isle of Man, IM4 4EW
G0 JHC N Carr, 15 Westlands, Leyland, PR26 7XT
G0 JHD R Jennings, 54a Pensbury Street, Darlington, DL1 5LH
GM0 JHE Karen Hunter, 2/2 206 Skirsa Street, Glasgow, G23 5DJ
G0 JHG C Holmes, 17 Wenning Court, Morecambe, LA3 3SH
GW0 JHH Lyndon Ireland, 109 Dan-y-Cribyn, Ynysybwl, Pontypridd, CF37 3EU
G0 JHJ W Fowler, 1 East Orchard, Sileby, Loughborough, LE12 7SX
G0 JHK M Hedges, 24 Fletcher Avenue, St. Leonards-on-Sea, TN37 7QX
G0 JHL Richard Wilmot, 43 A 2, PÄÄSKYLÄNTIE, Jämsä, Finland, 42100
G0 JHQ Adrian Cotton, Coburg Cottage, Barton Estate, East Cowes, PO32 6NT
GI0 JHR H OXTOBY, 13 Castle Court, Cookstown, BT80 8QJ
G0 JHT D Talbot, Southways, Tichborne Down, Alresford, SO24 9PL
G0 JHU M Stephenson, 38 St. Helens Crescent, Low Fell, Gateshead, NE9 6DH
G0 JHW J Waterhouse, 81 Barkham Road, Wokingham, RG41 2RJ
G0 JIA Andrew Sharp, 11 Maple Grove, Mutley, Plymouth, PL4 6PZ
G0 JIB P Moran, 12 Sapphire Drive, Kirkby, Liverpool, L33 1UW
G0 JIF A Fennell, 12 Vale Road, Ramsgate, CT11 9LU
G0 JII Colin Davis, 10 Marnhull Road, Poole, BH15 2EX
G0 JIL G Hampson, 7 Merryfield Close, Bransgore, Christchurch, BH23 8BS
GD0 JIM J King, 4 Glenhurst Avenue, Ruislip, HA4 7LZ
G0 JIR A Potter, 8 Oaklands, The Street, Ashford, TN25 6NE
G0 JIS A Myland, 8 Burseldon Court, East Cliff Road, Dawlish, EX7 0BP
G0 JIT R Atherton, Frensham, Grange Lane, Northwich, CW8 2BQ
G0 JIV Francis Gurney-Smith, Levens, Surlingham Road, Norwich, NR14 7DN
G0 JIW Michelle Edwards, 98 Cornwallis Drive Eaton Socon, St. Neots, PE19 8TZ
G0 JJD G Thorne, 4 Barronwood Court, Tarleton, Preston, PR4 6TR
G0 JJE H Spratt, 9 Kennedy Close, Halesworth, IP19 8EG
G0 JJF I Mccormick, The Old Railway Inn, Station Road, Gilwern, NP7 0BY
G0 JJG J Butt, 24 Lowry Way, Stowmarket, IP14 1UF
G0 JJI P Forshaw, 73 Galloway Road, Hamworthy, Poole, BH15 4JS
G0 JJK T Atkins, 46 Fallowfield, Ampthill, Bedford, MK45 2TP
G0 JJM A Eves, 136 Thistle Grove, Welwyn Garden City, AL7 4AQ
G0 JJO M Wheatley, 25 Sheringham Drive, Etchinghill, Rugeley, WS15 2YG
G0 JJP roderick fowler, 5 Littlelawns Close, Walsall, WS8 7DH
G0 JJQ W Dillon, 49 Goring Way, Greenford, UB6 9NN
G0 JJR D Briggs, 37 Wainwright Avenue, Sheffield, S13 8EL
G0 JJS J Smith, 48 Oakleigh Gardens, Oldland Common, Bristol, BS30 6RH
G0 JJV S Hill, 26 Harborough Way, Sheffield, S2 1RG
G0 JJW John Walsh, Flints House, Coates, Peterborough, PE7 2DD
G0 JJY I Bowen, 169 Clopton Road, Stratford-upon-Avon, CV37 6TF
GD0 JKA A Brook, 9 Stonecrop Grove, Douglas, Isle of Man, IM2 7DX
G0 JKC K Clarke, 55 Compton Avenue, Aston-on-Trent, Derby, DE72 2AU
G0 JKE Stephen Ward, 27 Greenock Street, Sheffield, S6 4NB
G0 JKG Frank Shinn, 23 Cygnet Close, Brampton Bierlow, Rotherham, S63 6EY
G0 JKH D Hinson, 6 Nethergate, Stannington, Sheffield, S6 6DJ
G0 JKI M Edmunds, 2 Jubilee Road, Bungay, NR35 1RE
G0 JKJ Scott Marshall Marshall, 31 Postbridge Road, Coventry, CV3 5AG
G0 JKM P Renvoize, Flat 14 Block M, Peabody Buildings, London, EC1Y 8NL
G0 JKP A Whibley, 8 Ticehurst Road, Brighton, BN2 5PU
G0 JKU A Bowering, 137a Knole Lane, Brentry, Bristol, BS10 6JN
G0 JKY Brian Hadley, 60 Chapel St., Pensnett, Brierley Hill, DY5 4EF
G0 JKZ George Christofi, 19 Kingsland Avenue, Northampton, NN2 7PP
G0 JLE R Adams, 18 Dundridge Gardens, Bristol, BS5 8SZ
G0 JLF John Flowers, 77 Withies Park, Midsomer Norton, Radstock, BA3 2PB
G0 JLI T Davies, 31 Burnbush Close, Bristol, BS14 8LQ
GM0 JLJ E Mottart, 1 Muirake Cottages, Cornhill, Banff, AB45 2BQ
G0 JLL N Sheen, 26 Springvale Rise, Hemsworth, Pontefract, WF9 5HY
G0 JLP M Bray, 205 Woodlands Road, Gillingham, ME7 2SW
G0 JLS M Brown, 4 River Gardens, Shawbury, Shrewsbury, SY4 4LA
G0 JLU E Wood, 65 Walford Road, Rolleston-on-Dove, Burton-on-Trent, DE13 9AR
G0 JLV D Vaughan, 23 Beckmeadow Way, Mundesley, Norwich, NR11 8LP
G0 JLX Andrew Digby, 4 Paddock Close, South Wonston, Winchester, SO21 3EQ
G0 JMD J Davis, 62 Kingscote, Yate, Bristol, BS37 8YE
G0 JME J Pither, 74 Bucklands Road, Teddington, TW11 9QS
G0 JMI Michael Parkin, 17 Bolle Road, Alton, GU34 1PW
G0 JMJ W Williams, 16 Chapel Close, Elim Way, Blackwood, NP12 2AD
G0 JMK M Kemble, 74 Teg Down Meads, Winchester, SO22 5ND
G0 JML Mark Pivac, 72 Old Mill Road, Saffron Walden, CB11 3ER
G0 JMN H Marshall, 23 Cranbourne Drive, Chorley, PR6 0LJ
GM0 JMO John Bell, 5 Louisa Drive, Girvan, KA26 9AH
G0 JMR D Williams, 27 Grindlestone Hirst, Colne, BB8 8BF
G0 JMS M Standen, 11 Hazel Gardens, Sonning Common, Reading, RG4 9TF
G0 JMW M Williams, 51 Crackley Hill, Coventry Road, Kenilworth, CV8 2EE
G0 JMZ Peter Farrar, 2 Ancaster Avenue, Chapel St. Leonards, Skegness, PE24 5SL
G0 JNA R Janes, 37 Valley View, Market Drayton, TF9 1EA
G0 JNE T Royle, 35 Patrons Drive, Sandbach, CW11 3AS
G0 JNG A Stothard, Lanshaw Farm, Otley Road, Harrogate, HG3 1QX
G0 JNJ Alan Denny, 85 Delamere Drive, Macclesfield, SK10 2PS
G0 JNK A Powers, 42 Newbridge Road, Ambergate, Belper, DE56 2GS
G0 JNQ S MITCHELL, 78 Bellasize Park, Gilberdyke, Brough, HU15 2XU
G0 JNR S DOVETON, 2 Red Scar Drive, Scarborough, YO12 5RQ
G0 JNT Leslie Keeton, 66 Worlaby Road, Scartho Top, Grimsby, DN33 3JP
G0 JNV B O'Brien, 9 Tarbet Road, Dukinfield, SK16 4BE
G0 JNZ Terry Bray, 135 Fort Austin Avenue, Crownhill, Plymouth, PL6 5NR
G0 JOC J Lassemillante, 25 Rissington Walk, Thornaby, Stockton-on-Tees, TS17 9QJ
G0 JOD R Degg, 28 The Spinneys, Welton, Lincoln, LN2 3TU
G0 JOG Bernard Wilson, 41 Palmerston Close, Ramsbottom, Bury, BL0 9YN
GM0 JOL J Lincoln, 59 Obsdale Park, Alness, IV17 0TR
G0 JOM John Goddard, 65 New Street, North Wingfield, Chesterfield, S42 5JP
G0 JON J Swain, 1 Ganstead Way, Low Grange, Billingham, TS23 3SY
G0 JOP John Bryder, 110 Georgelands, Ripley, Woking, GU23 6DQ
G0 JOS E Christmas, 15 Norton Avenue, Surbiton, KT5 9DX
GM0 JOV A Farquhar, 57 Woodcroft Avenue, Bridge of Don, Aberdeen, AB22 8WY
G0 JOX D Sykes, 449 Westdale Lane, Mapperley, Nottingham, NG3 6DH
G0 JPC K Dale, 31 Cadshaw Close, Birchwood, Warrington, WA3 7LR
G0 JPE Peter Roberts, 61 Abbey Lane, Sheffield, S8 0BN
G0 JPF Stuart Lee, 15 Wilson Way, Earls Barton, Northampton, NN6 0NZ
GM0 JPG D Arnold, 1 Knockenhair Road, Dunbar, EH42 1BA
G0 JPI S Martin, 34 Parson Drove, West Pinchbeck, Spalding, PE11 3QW
G0 JPJ P Smith, S.V Kiwiroa, New Zealand Reg. Ship, New Zealand, ON-876019
G0 JPL D Smith, 16 Leander Close, Nottingham, NG11 7BE
G0 JPM J Mitchell, 8 Eldred Drive, Orpington, BR5 4PF
G0 JPQ C Barber, Charity Farm House, Mill Lane, Skegness, PE24 5NN
G0 JPU M Ferguson, 15 Squires Leaze, Thornbury, Bristol, BS35 1TB
G0 JPY D Polley, 42 Lindfield Road, Eastbourne, BN22 0AJ
G0 JPZ M Kirk, 5 The Paddock, Kirkby-in-Ashfield, Nottingham, NG17 8BT
G0 JQA HAMBLETON ARC c/o B Alderson, 43 Brompton Road, Northallerton, DL6 1ED
GM0 JQE I Templeton, 39 Cairngorm Court, Irvine, KA11 1PN
G0 JQK J Hughes, 18 Monmouth Road, Wallasey, CH44 3ED
G0 JQP G Bradley, 7 Copeland Row, Evenwood, Bishop Auckland, DL14 9PY
GI0 JQQ E Butler, 59 Ballinlea Road, Maghernahar, Ballycastle, BT54 6JL
G0 JQR D Allison, 7 Prospect Cottages, Bedlington, NE22 7AF
G0 JQS A Downing, 1 Raglans, Alphington, Exeter, EX2 8XN
GW0 JQT S Lloyd, 10 Park Crescent, Llanelli, SA15 3AE
G0 JQX C Ditchfield, 8 Meerbrook Way, Quedgeley, Gloucester, GL2 4QE
G0 JQZ S Chamberlain, 54 Henray Avenue, Glen Parva, Leicester, LE2 9QJ
G0 JRB J Reed, 28 Kealholme Road, Messingham, Scunthorpe, DN17 3ST
G0 JRC J Clayton, 49 Bramble Lane, Mansfield, NG18 3NP
GI0 JRD B Mcanespie, 23 Ashton Park, Belfast, BT10 0JQ
G0 JRE I Perks, 9 Atherton Close, Shalford, Guildford, GU4 8HZ
GW0 JRF F Rees, Caerleon, Picton Road, Tenby, SA70 7DP
G0 JRH P Scott-Dickinson, Ninicsu, 18 Pennington Drive, Weybridge, KT13 9RU
GI0 JRI K Murray, 67 Sicily Park, Belfast, BT10 0AN
G0 JRM C Brown, 8 The Elms, Horringer, Bury St. Edmunds, IP29 5SE
G0 JRN A Hansley, 238 Milton Road, Cowplain, Waterlooville, PO8 8SE
GM0 JRQ Brian Baker, The Ridge, Peat Inn, Cupar, KY15 5LH
G0 JRR C Curtis, 1 Westover Drive, Burton-upon-Stather, Scunthorpe, DN15 9HH
G0 JRT Thomas Hanratty, 12 Clarendon Street, Consett, DH8 5LS
G0 JRV John Broomfield, 14 Woodfen Crescent, Leominster, HR6 8SS
G0 JRX T Cave, 71 Cambo Drive, Cramlington, NE23 6TW
G0 JRY C Mulvany, 25 Redwing Close, Bicester, OX26 6SR
G0 JRZ T Brookes, Jemora, Littleham, Bideford, EX39 5HN
G0 JSA Russell Taylor, 11 Yeadon Close, Accrington, BB5 0FN
G0 JSC J Wheatley, 8 Winchester Close, Feniton, Honiton, EX14 3EX
G0 JSE J Edwards, 49 The Fleet, Stoney Stanton, Leicester, LE9 4DZ
G0 JSF B Halmshaw, 7 Gerrard Street, Rochdale, OL11 2EB
G0 JSG P Holden, 19 Briar Close, Lowestoft, NR32 4SU
G0 JSJ R Jackett, Bourne House, 105 Moor Road, Leyland, PR26 9HP
G0 JSL G Brown, 9 Western Drive, Leyland, PR25 1YB
G0 JSM J Brown, 9 Western Drive, Leyland, PR25 1YB
G0 JSO Rob Wynne, South Graceholme, High Lorton, Cockermouth, CA13 9UQ
G0 JSP P Fauchon, 114 Petersfield Avenue, Staines, TW18 1DJ
G0 JSR S Rickman, 35 Cedar Way, Basingstoke, RG23 8NG
G0 JST OBO JUBILEE SAILING TRUST (ARS) c/o J Wheatley, 8 Winchester Close, Feniton, Honiton, EX14 3EX
G0 JSU Andrew Collett, 5 Park View Drive, Lydiard Millicent, Swindon, SN5 3LX
G0 JSX R Davies, 8 Princes Park, Rhuddlan, Rhyl, LL18 5RW
GJ0 JSY Sydney Smith-Gauvin, 31 Le Jardin A Pommiers, La Rue de Patier, St. Saviour, Jersey, JE2 7LT
G0 JSZ J Crellin, 89 Wapshare Road, West Derby, Liverpool, L11 8LR
G0 JTA D Ellison, Blackburn Hall, Grinton, Richmond, DL11 6HH
G0 JTD R Lyne, 32 Davenwood, Upper Stratton, Swindon, SN2 7LL

GW0 JTE L Horne, 29 Station Terrace, Dowlais, Merthyr Tydfil, CF48 3PU
G0 JTF J Hosking, 14 School Terrace, Cwm, Ebbw Vale, NP23 7QY
GW0 JTJ T Watkins, Ty Unig, Forest Road, Treharris, CF46 5HG
G0 JTL J Hinchliffe, 19 The Terrace, Honley, Holmfirth, HD9 6DS
G0 JTM P Smith, 999 Manchester Road, Linthwaite, Huddersfield, HD7 5LS
G0 JTN C Smith, 35 Allendale Road, Earley, Reading, RG6 7PD
G0 JTP A Hill, 159 Sandford Road, Bradford, BD3 9NU
G0 JTR B Thatcher, 10 Harescombe, Yate, Bristol, BS37 8UA
G0 JTT S Hemsworth, 4 Spoonhill Road, Stannington, Sheffield, S6 5PA
G0 JTU Allan Lewis, 96 Roundhouse Close, Nantyglo, Ebbw Vale, NP23 4QY
G0 JUA J Hardcastle, 67 Caithness Road, Liverpool, L18 9SJ
G0 JUE T Cruse, Watch Tower House, The Ridgeway, London, NW7 1RS
G0 JUI E Gibson, 107 Church Avenue, Meanwood, Leeds, LS6 4JT
G0 JUK N Mayes, Cliffe Cottage, Rotherham Road, Barnsley, S71 [illegible]
G0 JUL George Hogben, 60 King Street, Yeadon, Leeds, [illegible]
G0 JUM J Barker, 11 Pennington Close, Copplestone, Crediton, EX17 5NA
G0 JUN R Softley, 14 Tupps Drive, Bedworth, CV12 0DE
G0 JUQ M Eborall, 26 Bishopton Lane, Stratford-upon-Avon, CV37 9JN
G0 JUR N Barnett, 60 Commercial Road, Spalding, PE11 2HE
G0 JUT D Horne, 24 Ringwood Drive, Leeds, LS14 1AP
G0 JUV Andrew Phillips, 2 New Zealand Terrace, Bridport, DT6 3PW
G0 JUW T Batchelor, 9 Somerset Close, Sittingbourne, ME10 1JU
G0 JUY Philip Barden, 38 Silver Close, Tonbridge, TN9 2UY
G0 JVB Michael Jackson, 114 Norman Street, Ilkeston, DE7 8NL
GM0 JVC M Grant, Monikie, Gryffe Road, Kilmacolm, PA13 4BB
G0 JVF David Cleaver, 4 Wyvern Close, Devizes, SN10 2UE
G0 JVH Arthur Puffett, 142 Cheltenham Road East, Gloucester, GL3 1AA
G0 JVI A Mawson, 38 Springbank Road, Gildersome, Leeds, LS27 7DJ
G0 JVK R Cook, 15 Hucklow Court, Mansfield, NG18 3QP
G0 JVL Alexander Sclater, 6 Balmoral Close, Alton, GU34 1QY
G0 JVN R Horne, 7 Alexander Road, Bentley, Walsall, WS2 0HJ
G0 JVT D Marjoram, 459 Norwich Road, Ipswich, IP1 5DR
G0 JVU N Pattinson, 4 Carlisle Road, Brampton, CA8 1SR
GM0 JVV J Stevenson, 52 Fernbrae Avenue, Rutherglen, Glasgow, G73 4AE
G0 JVW A Thornton, 1 Primula Close, Clifton, Nottingham, NG11 8SL
G0 JWD J Brambley, Trail View, Biggin, Buxton, SK17 0DH
G0 JWE Donald Cliffe, 5 Croft Close, Ockbrook, Derby, DE72 3RR
G0 JWF Brian Matthews, 25 Manor Park, Newbridge, Newport, NP11 4RS
G0 JWG J Gaunt, Fenton House, Church Road, King's Lynn, PE33 0HE
G0 JWJ Timothy Bridgland-Taylor, 1 Overbury Court, Hereford, HR1 1DG
G0 JWL K Lindsay, 11a Pyrford Close, Waterlooville, PO7 6BT
G0 JWO T Forbes, 63 Wardle Drive, Annitsford, Cramlington, NE23 7DE
GD0 JWR H Richardson, Pitcairn, Quarterbridge Road, Douglas, Isle of Man, IM2 3RQ
G0 JWV P Bryant, Crugsillick Cottage, Ruan High Lanes, Truro, TR2 5JP
G0 JWY J Curtis, Conway, 27 Southgate, Hornsea, HU18 1RE
G0 JXG M Price, 4 Vale View, Woodfieldside, Blackwood, NP12 0DB
G0 JXI M Brown, 11 Miles Close, Birchwood, Warrington, WA3 6QD
G0 JXJ D Copeland, 2b Rose Road, Canvey Island, SS8 0BP
G0 JXO R Smith, 24 Kirkstead Road, Carlisle, CA2 7RD
G0 JXP Kenneth Pile, 14 Semper Close, Knaphill, Woking, GU21 2NG
G0 JXQ D Davis, 17 Welbourne Close, Raunds, Wellingborough, NN9 6HE
G0 JXR P Keasley, 55 Hillside, Hoddesdon, EN11 8RW
G0 JXX Micheal Hoddy, 52 Hayling Rise, High Salvington, Worthing, BN13 3AG
G0 JYC P Hodgson, Uhland Str 4, Neunkirchen, Seelscheid, Germany, 53819
G0 JYD J Drinkwater, Springfield, Mudhurst Lane, Stockport, SK12 2BY
G0 JYE Peter Foss, 37 Ling Crescent, Ruddington, Nottingham, NG11 6GG
G0 JYF Simon Deakin, 25 Hungerford Avenue, Trowbridge, BA14 9ES
G0 JYH Henry Ryan, Fairview, Imperial Avenue, Sheerness, ME12 2HG
G0 JYI A Street, 11 Leigh Gardens, Leigh-on-Sea, SS9 2PX
G0 JYJ C Hodgson, 113 Roman Road, East Ham, London, E6 3RY
G0 JYK J Sharp, 22 Boat Lane, Irlam, Manchester, M44 6EN
G0 JYL J Bartram, 2 Reeves Piece, Bratton, Westbury, BA13 4TH
G0 JYN S Ashcroft, 90 Kestrel Close, Chipping Sodbury, Bristol, BS37 6XA
G0 JYQ M Gregory, 21 Jacaranda Close, Fareham, PO15 5LG
G0 JYS A Jepson, 45 Cotefield Road, Manchester, M22 1UR
G0 JYU D Smith, 104 Hanley Road, London, N4 3DW
G0 JYV J Rowlands, 67 Woodside, Gosport, PO13 0YX
G0 JYX T Cloke, 14 Bickley Close, Hanham Green, Bristol, BS15 3TB
G0 JYZ I Broomhall, 49 Funtley Hill, Fareham, PO16 7XA
G0 JZA Nigel Cox, Flat 2a, Arlington House, South View, Teignmouth, TQ14 8BJ
G0 JZE Alex Mcfadyen, 26 Lewis Road, Chipping Norton, OX7 5JS
G0 JZF N Rogers, 15 Templar Road, Yate, Bristol, BS37 5TF
G0 JZH R Morris, 96 Chandag Road, Keynsham, Bristol, BS31 1QE
G0 JZJ F Russell, 37 Overpool Road, Ellesmere Port, CH66 1JW
G0 JZL G Galley, 1 St. James Avenue, South Anston, Sheffield, S25 5DR
G0 JZS G Corbett, 359 London Road, Stoke-on-Trent, ST4 5AN
G0 JZT J Chappell, 2 Wayside, Knott End-on-Sea, Poulton-le-Fylde, FY6 0DD
G0 JZU W Etherington, 15 East Bank, North End Road, Arundel, BN18 0DJ
GM0 JZV John Warden, Westlea, Little Brechin, Brechin, DD9 6RQ
G0 JZW A Elford, 10 Meadowlands, Lymington, SO41 9LB
G0 JZY R Pocock, Flat 38, Brunel Court, 4 Harbour Road, Bristol, BS20 7JH
G0 KAB Allan Pilkington, Flat 17, Blackshaw House, Bolton, BL3 5NU
G0 KAK I Walker, 6 Granary Court, Northampton, NN4 0XX
G0 KAM G Jones, 14 Plantation Drive, Croesyceiliog, Cwmbran, NP44 2AN
G0 KAQ Edward Walker, 2 Newtown Road, Uppingham, Oakham, LE15 9TS
G0 KAS M Stevens, 20 Melton Place, Epsom, KT19 9EE
G0 KAT V Chapman, 20 St. Chad, Barrow-upon-Humber, DN19 7AU
G0 KAU R Crocker, 10 Westhall Close, Carlton-le-Moorland, Lincoln, LN5 9JD
GW0 KAX Petrie Owen, 13 Highland Close, Sarn, Bridgend, CF32 9SB
GM0 KAZ Adam White, 65 Orchard Street, Galston, KA4 8EJ
G0 KBA Irene Langtree, 243 Devonshire Road, Atherton, Manchester, M46 9QB
G0 KBJ E Burndred, 52 Everest Road, Kidsgrove, Stoke-on-Trent, ST7 4DY
G0 KBK R Sleigh, 14 Brook House Flats, Chetwynd End, Newport, TF10 7JD
G0 KBL Stan Rudcenko, 39 The Avenue, Sutton, SM2 7QA
G0 KBM D Manning, 2 Sluice Farm Cottages, Kirton, Ipswich, IP10 0QF
G0 KBN G Beech, The Old Hall, The Spinney, Ilkeston, DE7 4LZ
G0 KBO Vlad KRAVCHENKO, Flat 16, Birchfield House, London, E14 8EY
G0 KBP M Dearing, 1 Woodbine Villas, New Village Road, Cottingham, HU16 4NF
GM0 KBR John Mcfadyen, 8 Ramsay Crescent, Bathgate, EH48 1DD
G0 KBS L Kay, 2 Childwall Crescent, Childwall, Liverpool, L16 7PQ
G0 KBZ [illegible], Alford, LN13 0RN
G0 KCA I Walder-Davis, 60 Church St., St. Peters, Broadstairs, CT10 2TX
G0 KCB David Beckley, Fen Hill, Hall Road, Great Yarmouth, NR29 [illegible]
G0 KCC W Randolph, 13 Links Road, Poole, BH14 9QP
G0 KCD M Leech, 20 Walton Road, Frinton-on-Sea, CO13 0AQ
G0 KCE V Harding, 17 St. Anns Road South, Heald Green, Cheadle, SK8 3DZ
G0 KCF Christopher Fosbrook, 4a Yew Tree Road, Hayling Island, PO11 0QE
G0 KCG D Hall, 47 Sunningdale Road, Fareham, PO16 9PA
G0 KCH M Mccarthy, 75 Taynton Drive, Merstham, Redhill, RH1 3PX
GD0 KCL KINGS COLLEGE ARS c/o J Greenberg, 12 Broadhurst Avenue, Edgware, [illegible]
G0 KCN [illegible], Selkirk, TD7 5BB
GM0 KCY D Michael, 84 Dourtreehall, Girvan, KA26 0EL
G0 KCZ William Bowles, Willow Grove, Little Common North Bradley, Trowbridge, BA14 0TX
G0 KDA P Cooper, Rosebank, 17 Givendale Road, Scarborough, YO12 6LE
G0 KDB David Greenhalgh, Hillcroft, Colby, Appleby-in-Westmorland, CA16 6BD
GM0 KDC R Smith, 21 Glen View Crescent, Gorebridge, EH23 4BT
G0 KDD B Woodward, 58 Marine Drive, Bishopstone, Seaford, BN25 2RU
GM0 KDF R Thomson, 25 Cheviot Road, Silvertonhill, Hamilton, ML3 7HB
G0 KDG R Simpson, Hillfoot, Fernleigh Road, Grange-over-Sands, LA11 7HT
GI0 KDH ANDREW BROWN, 3 Gargrim Road, Fintona, Omagh, BT78 2EH
G0 KDI R Steans, 302 Walton Road, West Molesey, KT8 2HY
G0 KDL William Cooper, 24 Ambleside Road, Lightwater, GU18 5TA
GM0 KDO G Kirkland, 34 Langhouse Green, Crail, Anstruther, KY10 3UD
GM0 KDP Iain Dunbar, Mabruk, 25 Kinord Drive, Aboyne, AB34 5JZ
G0 KDQ M Ward, 2 Hollin Gate, Otley, LS21 2DP
G0 KDR R Lintott, Upper Grove Farm, Rendham, Saxmundham, IP17 2AS
G0 KDS S Lindsay, 27 Bagnell Road, Bristol, BS14 8PZ
G0 KDT P Cracknell, 54 Yannon Drive, Teignmouth, TQ14 9JP
G0 KDV Darenth Valley RS c/o M Wallace, 17 Leamington Avenue, Orpington, BR6 9QA
G0 KDW G Burrett, 10 Prospect Walk, Lower Burraton, Saltash, PL12 4RG
G0 KDX Brian Ashton, Squirrel Wood, Anderton Mill, Chorley, PR7 5PY
G0 KDY A Spry, Newlands Farm, Bradworthy, EX22 7RN
G0 KEB C Frost, 61 Selbourne Avenue, Surbiton, KT6 7NR
G0 KEC H Opitz, 26 Holme Court, Lower Warberry Road, Torquay, TQ1 1QR
G0 KED J Bower, Linwood, Stain Lane, Mablethorpe, LN12 1QB
G0 KEE Colin Simons, 51 Moorville Drive South, Carlisle, CA3 0AW
G0 KEI Derek Kennard, 63c Alton Gardens, Southend-on-Sea, SS2 6QU
G0 KEK B Curtis, Beggars Roost, Rea Barn Road, Brixham, TQ5 9EE
GD0 KEO A Birchenough, 20 St. Stephens Meadow, Sulby, Ramsey, Isle of Man, IM7 3DA
GM0 KEQ R Crawford, Glengarry, East Terrace, Kingussie, PH21 1JS
G0 KEV Kevin Gallagher, 8 Holme Grove, Burley in Wharfedale, Ilkley, LS29 7QB
G0 KEX Andrew O'Hara, 26 Thompson Avenue, Ainsworth, Bolton, BL2 5RJ
G0 KEY Steve Cole, 160 New Haw Road, Addlestone, KT15 2DN
G0 KFF Kathy Field, 50 Madrona, Amington, Tamworth, B77 4EJ
G0 KFG A Gauld, 1 Hirstead Road, Scarborough, YO12 6TW
GW0 KFL Robert Rees, 15 Taliesin Close, Pencoed, Bridgend, CF35 6JR
G0 KFM J Collins, 19 Brookside Park Homes, Waterloo Road, Wimborne, BH21 3SP
G0 KFO John Turner, 36 The Grove, Herne Bay, CT6 7QD
G0 KFQ Brian Wilson, 20 Peacock Way, Littleport, Ely, CB6 1AB
G0 KFS A Purcell, 33 Fishley Close, Bloxwich, Walsall, WS3 3QA
G0 KFT C Dickerson, 1 Park Farm Lane, Nuthampstead, Royston, SG8 8LT
G0 KFV M Evans, 72 Ambrose Street, York, YO10 4DR
G0 KFW W Cole, 5a Park Lane, Kemsing, Sevenoaks, TN15 6NU
G0 KFY P Elliot-West, 135 Tunstall Road, Sunderland, SR2 9BB
G0 KGA Andrew Danby, 104 Auchinleck Close, Driffield, YO25 9HE
G0 KGD V Zakharov, Ty-Brith, Cloddiau, Welshpool, SY21 9JE
G0 KGE H Johnson, 2 Thirlmere, Kennington, Ashford, TN24 9BD
G0 KGI J Coleman, 80 Ormston Avenue, Horwich, Bolton, BL6 7ED
G0 KGL G Lindsay, 66 Jubilee Crescent, Mangotsfield, Bristol, BS16 9AZ
G0 KGQ G Armatage, 39 Priors Grange, High Pittington, Durham, DH6 1DA
G0 KGR R Beadle, 2 Edward Cottages, Great Munden, Ware, SG11 1HT
G0 KGT B Williams, 8 Grimbald Road, Knaresborough, HG5 8HD
G0 KGY Maureen Mavin, 52 Bywell Road, Ashington, NE63 0LE
G0 KHA Keith Seddon, 17 Dunmail Drive, Kendal, LA9 7JG
G0 KHF P Witley, 18 Seagate Road, Hunstanton, PE36 5BD
G0 KHH A Rogers, 3 Ripley Drive, Wigan, WN3 6AJ
G0 KHJ John Warburton, 92 Worsley Road Farnworth, Bolton, BL4 9LX
G0 KHK P Shaw, 15 Greenfield Avenue, Marlbrook, Bromsgrove, B60 1HE
G0 KHQ Phillip Hughes, 4 Millards Close, Hilperton Marsh, Trowbridge, BA14 7UN
G0 KHR E Forsyth, 11 Brooklyn Road, Stockport, SK2 6BX
G0 KHY J Jenkins, 3 Gosslan Close, St. Ives, PE27 3YZ
G0 KHZ M Crane, Drewton House, Back Lane, Goole, DN14 7HD
G0 KIA Richard Harris, 19 Old Bath Road, Sonning, Reading, RG4 6SZ
G0 KIC B Hayward, 22 Waldron Street, Bishop Auckland, DL14 7DS
GW0 KIG Kevin O'Reilly, 14 Catherine Close, Abercanaid, Merthyr Tydfil, CF48 1YY
G0 KIK Steven Berry, 85 Lake View Close, West Park, Plymouth, PL5 4LT
G0 KIM J West, 242 Grane Road, Haslingden, Rossendale, BB4 4PB
G0 KIN T Harper, 72 School Road, Salford Priors, Evesham, WR11 8XN
G0 KIR K Jones, 1 Heyope Road, Heyope, Knighton, LD7 1PT
G0 KIU D Godfrey, 46 Park View Way, Mansfield, NG18 2RN
G0 KIY Norman Brook, 2 Back Regent Place, Harrogate, HG1 4QR
G0 KJC J Clayton, 49 Bramble Lane, Mansfield, NG18 3NP
G0 KJF Roderick Warner, Barley Hill Farm, Combe St. Nicholas, Chard, TA20 3HJ
G0 KJG B Bevington, 12 Buckingham Road, Rowley Regis, B65 9JN
G0 KJJ W Ritchie, 16 Avenue Mezidon Canon, Honiton, EX14 2TT
G0 KJK Keith Ranger, 144 Newton Street, Macclesfield, SK11 6RW
G0 KJM J RICHARDS, 14 Southwood Drive East, Bristol, BS9 2QP
G0 KJN J Windebank, 9 Townsend Place, St. Ippolyts, Hitchin, SG4 7RQ
G0 KJP D Scott, 19 The Fillybrooks, Aston, Stone, ST15 0DH
GW0 KJT Thomas Lewis, 2 Railway Terrace, Pontyberem, Llanelli, SA15 5HN
G0 KJU J Robertson, 28 Frith Road, Bognor Regis, PO21 5LL
GW0 KJZ J Jones, 64 Cleviston Park, Llangennech, Llanelli, SA14 9UP
G0 KKC D Browning, 81 Bishop Road, Bishopston, Bristol, BS7 8LU
G0 KKD Andrew Parker, Old Rectory, 1 Church Lane, Matlock, DE4 2GL
GM0 KKE I Coulson, 11 Redcliffe, Kingoodie, Dundee, DD2 5DL
G0 KKF B Hough, 54 Woodbourne Road, Sale, M33 3TN
G0 KKH J Henderson, 245 Crawley Close, Corringham, Stanford-le-Hope, SS17 7JU
G0 KKL Philip Mayer, Flat 7 Broomrigg, 5 Belle Vue Road, Poole, BH14 8UE
G0 KKO John Langan, 65 Armthorpe Drive, Little Sutton, Ellesmere Port, CH66 4NN
G0 KKR B Chapman, Millbrooke Cottage, Covenham St. Bartholomew, Louth, LN11 0PD
G0 KKS A Nance, 33 Oak Close, Copthorne, Crawley, RH10 3QT
G0 KKT Ian Osborne, 19 Lumber Leys, Walton on The Naze, CO14 8SS
G0 KKU J Howard, 111 Heath Road, Penketh, Warrington, WA5 [illegible]
G0 KKV M Lowe, 22 Stratford Avenue, Atherstone, CV9 [illegible]
G0 KLA Christopher Thompson, 3, 13 Evans St, Brooklyn, United States, NY 11201
G0 KLD Charles Wren, 38 Green Street, Hyde, SK14 1QX
G0 KLF N Anderton, 29 Cliftonville Drive, Swinton, Manchester, M27 5NA
G0 KLG R Hodds, 17 Oaklands Drive, Willerby, Hull, HU10 6BJ
G0 KLJ J Leader, 9 Southerwicks, Corsham, SN13 9NH
G0 KLK Arnold James, 14 Randle Drive, Sutton Coldfield, B75 5LH
GW0 KLN Christopher Ryalls, 6 Balmoral Close, Penycoedcae, Pontypridd, CF37 1XE
GM0 KLO C Grossart, 11 Woodlands Drive, Brightons, Falkirk, FK2 0TF
GM0 KLP J Pentland, 2 Glenniston Cottages, Auchtertool, Fife, KY5 0AX
G0 KLQ D Cross, 15 Fernside Road, West Moors, Ferndown, BH22 0EE
G0 KLT Dilwyn Rogers Jones, 20 Birchwood, Leyland, PR26 7QJ
G0 KLU P Fairhurst, 6 Audwick Close, Cheshunt, Waltham Cross, EN8 0RF
GW0 KLY R Jones, 134 Birchgrove Street, Porth, CF39 9UY
GM0 KMA Mervyn Rainey, 2 Shields Holdings, Lochwinnoch, PA12 4HL
G0 KMB K Bowdler, 18 Cavendish Street, Leigh, WN7 1SG
G0 KMC A Slaughter, 42 Goss Avenue, Waddesdon, Aylesbury, HP18 0LY
G0 KMF R Holmshaw, 142 Oakleigh Park Drive, Leigh-on-Sea, SS9 1RU
GM0 KMJ Paul Johnstone, 26 Lomond Crescent Stenhousemuir, Larbert, FK5 4LT
G0 KMK M Aslam, 38 Grey Street, Burnley, BB10 1BA
G0 KML Ben Hawes, 201 Ridgeway, Plympton, Plymouth, PL7 2HP
G0 KMN S Hepworth, 9 College View, Ackworth, Pontefract, WF7 7LA
G0 KMP G Aungiers, 6 Woodlands Crescent, Barton, Preston, PR3 5HB
G0 KMV H King, Que Lindo, Church Lane, Bristol, BS39 5UP
G0 KMW H Shepherd, Whydown, 3 White House Close, Abingdon, OX13 6LP
G0 KNH J Greenwood, 43 Townend Avenue, Ackworth, Pontefract, WF7 7HE
G0 KNJ R Bygrave, 69 Albert Gardens, Harlow, CM17 9QG
G0 KNL W Christlo, 6 Nether Ley Gardens, Chapeltown, Sheffield, S35 1AH
G0 KNM G Woodford, 81 Harrold Road, Rowley Regis, B65 0RL
G0 KNN M Gregg, 22 Mayfields, Spennymoor, DL16 6RN
GM0 KNT A Bates, Caberfeidh, Balnageith, Forres, IV36 2SG
G0 KNW C Wiles, Everest, Mile Road, Morpeth, NE61 5QW
G0 KNX Geoff Allen, 3 Ryton Close, Coventry, CV4 8HF
G0 KNY Keith Heaton, 66 Ashleigh Road, Exmouth, EX8 2JZ
G0 KOC Arthur Kinson, 6 Uplands Park, Broad Oak, Heathfield, TN21 8SJ
G0 KOE T Foxton, Dunbar, Dam Lane, Malton, YO17 9SJ
G0 KOF Donald Henretty, 13 Siskin Chase, Cullompton, EX15 1UD
G0 KOH J Dover, 11 Roman Road, Barton-le-Clay, Bedford, MK45 4QJ
G0 KOI M Cooper, 15 Woodleigh Avenue, Harborne, Birmingham, B17 0NW
G0 KOJ B Thomas, Torcottage, 12 The Dip, Newmarket, CB8 8AH
G0 KOK Peter Love, 2 Meadway, Dover, CT17 0PS
G0 KOO Alan McDowell, Fern Cottage, The Gride, Boston, PE22 9LS
G0 KOU B Arrowsmith, 25 Watchouse Road, Chelmsford, CM2 8PT
GI0 KOW Robert Cummings, 19 Bachelors Walk, Keady, Armagh, BT60 2NA
G0 KOY B Clues, 8 Acland Avenue, Colchester, CO3 3RS
G0 KPA June Ellison, 66 Fisher Street, Paignton, TQ4 5ES
GW0 KPD J james, 1 Pellau Road, Margam, Port Talbot, SA13 2LF
G0 KPE Carl McGowan, Belvedere, Tylers Lane, Reading, RG7 6TN
GI0 KPF Brendan McCausland, 5 Hollyfields, Dungannon, BT71 7BH
G0 KPG R Moore, 34 Fishponds Road, Kenilworth, CV8 1EZ
G0 KPH P Keighley, 15 Stuart Court, Warwick Terrace, Leamington Spa, CV32 5NU
GD0 KPN J McLoughlin, 8 Governors Hill, Douglas, Isle of Man, IM2 7AW
G0 KPQ J Morgan, 9 Consort Close, Plymouth, PL3 5TX
G0 KPU Robert Harper, 4 Gresford Road, Llay, Wrexham, LL12 0NW
G0 KPV G Owen, 2 Ffordd Beibio, Holyhead, LL65 2EF
G0 KPY Andrew Baker-Munton, 66 Stanway Road, Headington, Oxford, OX3 8HX
G0 KPZ David Portch, 148 Brixham Road, Welling, DA16 1EJ
G0 KQA Peter Davies, Silver Birches, Orchard Road, Basingstoke, RG22 6NU
GM0 KQB R Kemp, 1 Grendon Court, Stirling, FK8 2JX
GD0 KQE Kenneth Jordan, Engadine, Little Switzerland, Douglas, Isle of Man, IM2 6AG
G0 KQH Brian O'Donoghue, 30 Lake Drive, Hamworthy, Poole, BH15 4LT
G0 KQI L Painter, 185 Albion Street, Street Helens, WA10 2HA
G0 KQK T Chambers, Autumn, Water Lane, Grantham, NG33 4RT
G0 KQO G Elliott, 32 Chapel Street, Newport, PO30 1PZ
G0 KQP J Carroll, 5 Montagu View, Leeds, LS8 2RH
G0 KQR R Bundell, 24 Sylvan Avenue, East Cowes, PO32 6PS
G0 KQS Graeme Griffiths, Sherwood House Buggen Lane, Neston, CH64 0QD
G0 KQT A Holdway, 10 The Quantocks, Thatcham, RG19 3SF
GW0 KQU G Slatter, 6 Glannant St., Penygraig, Tonypandy, CF40 1JT
GW0 KQV D Clark, Martinique, Wolfscastle, Haverfordwest, SA62 5DY
G0 KQX William Cook, Fronoleu, Bryngwy, Rhayader, LD6 5BN
G0 KQY D Murrell, 11 Stokesay Drive, Cheadle, Stoke-on-Trent, ST10 1YU
G0 KRB Graham Phillips, 57 Hollytrees, Bar Hill, Cambridge, CB23 8SF
G0 KRC KIDDERMINSTER & DISTRICT ARS c/o P Harris, 22 Bramley Way, Bewdley, DY12 2PU
G0 KRD D Downes, 7 Sandy Lane, Fakenham, NR21 9ES
G0 KRG KEIGHLEY RAYNET GROUP c/o Lance Conlon, 4 Hill Crest Drive, Slack Head, Milnthorpe, LA7 7BB
G0 KRH C Tarrant, 91 Dunes Road, Greatstone, New Romney, TN28 8SW
G0 KRK Brian Cockfield, 47 Aston Road, Willenhall, WV13 3DG

**IMPORTANT NOTE**

**Revalidate licence to avoid revocation** – Ofcom has advised the Society that plans will be drawn up to revoke licences that have not been revalidated as required by the licence conditions. The quickest way to revalidate is to do so online via the Ofcom website: *https://services.ofcom.org.uk/* or by email: *amateur.validations@ofcom.org.uk* Ofcom staff are available to help, but please be patient during times of heavy workload.

UK Callsigns

| | | |
|---|---|---|
| G0 | KRL | Ian Capon, Pentre Garreg Bach, Marianglas, LL73 8PP |
| GW0 | KRQ | J Cartwright, 20 Castlefield Place, Cardiff, CF14 3DU |
| G0 | KRR | J Hough, 1 Rock Lane, Linslade, Leighton Buzzard, LU7 2QQ |
| G0 | KRS | KEIGHLEY ARS c/o Kathryn Conlon, 4 Hill Crest Drive, Slack Head, Milnthorpe, LA7 7BB |
| G0 | KRT | E Masters, 91 Mayfair Avenue, Worcester Park, KT4 7SJ |
| G0 | KRU | Alan Wright, Cherry Tree Cottage, High Road, Beighton, Norwich, NR13 3LA |
| G0 | KRX | P Ruder, 34 Chelmsford Road, South Woodford, London, E18 2PL |
| G0 | KRY | David Sanders, 149 Sutton Road, Walsall, WS5 3AW |
| G0 | KSC | Justin Johnson, Unit 1, Prout Industrial Estate, Canvey Island, SS8 7TJ |
| G0 | KSD | R Allgood, 7 The Chase, Blofield, Norwich, NR13 4LZ |
| G0 | KSJ | J Graves, 172 Hall Lane, Upminster, RM14 1AT |
| GD0 | KSL | R Torr, 68 East Towers, Pinner, HA5 1TL |
| G0 | KSN | H Dabhi, 23 Shalgrove Field, Fulwood, Preston, PR2 3SX |
| G0 | KSS | Keith Nobbs, 49 St. James Drive, Burton, Carnforth, LA6 1HY |
| G0 | KTC | C Ayres, 219 Ashingdon Road, Rochford, SS4 1RS |
| G0 | KTD | Andrew Bonney, 6 Mitchell Road, St. Austell, PL25 3AU |
| GW0 | KTE | P Bennett, 13 Thornbury Close, Baglan, Port Talbot, SA12 8EU |
| GM0 | KTH | D Wakefield, Millfield, Burray, Orkney, KW17 2SU |
| GM0 | KTJ | W Ward, 65 Durrockstock Road, Foxbar, Paisley, PA2 0AR |
| G0 | KTL | G Edmunds, 14 Avon Close, Bettws, Newport, NP20 7BZ |
| G0 | KTN | Trevor Smithers, 14 Georgian View, Bath, BA2 2LZ |
| GM0 | KTO | J Power, 0/2 20 Eastercraigs, Glasgow, G31 3LJ |
| G0 | KTP | R COCKBILL, 45 Mills Road, Melksham, SN12 7DT |
| G0 | KTR | R Farnley, 40 Barons Court, Failsworth, Manchester, M35 0LH |
| G0 | KTS | J Wright, 65 Groombridge Close, Welling, DA16 2BP |
| G0 | KTT | P Riley, 11 Pinewood Court, South Downs Road, Altrincham, WA14 3HY |
| G0 | KTU | Adrian Miller, 11 Blackbrook Avenue, Paignton, TQ4 7ND |
| G0 | KTV | John Edgington, 83 Woking Road, Guildford, GU1 1QL |
| G0 | KTW | J Moss, 1 Millers View, Windmill Way, Much Hadham, SG10 6BN |
| G0 | KTX | A Hornsby, 328 Pelham Road, Immingham, DN40 1PT |
| G0 | KTY | G Salisbury, Nythfa, 1 Stad-Y-Garnedd, Anglesey, L60 6BB |
| G0 | KUA | V Kathuria, 2 Bevan Road, Lovedean, Waterlooville, PO8 9QH |
| G0 | KUC | David Bloomfield, 14 Horsham Close, Luton, LU2 8JH |
| G0 | KUD | Philip Haith, 17 Lime Tree Avenue, Grimsby, DN33 2BB |
| G0 | KUE | P Webb, 119 Chipstead Valley Road, Coulsdon, CR5 3BP |
| G0 | KUF | N Buchanan, Meadowside, Jacobs Well Road, Guildford, GU4 7PD |
| GI0 | KUH | John McCabe, john mccabe, 121 Garvaghy Road, Craigavon, BT62 1EH |
| G0 | KUI | J Wood, 6 West Terrace, Stakeford, Choppington, NE62 5UL |
| GM0 | KUJ | J Mcgifford, 52 Gartons Road, Glasgow, G21 3HY |
| GM0 | KUP | F Mann, 21 Greenbank Court, Falkirk, FK1 5DS |
| G0 | KUQ | F Hills, 112 Boxfield Green, Stevenage, SG2 7DS |
| G0 | KUR | B Thompson, Bungal House, Main Road, Alford, LN13 0JP |
| G0 | KUU | F Gillham, 260 Summerhouse Drive, Wilmington, Dartford, DA2 7PB |
| G0 | KUW | Colin Bennett, 67 King St., Clowne, Chesterfield, S43 4BS |
| G0 | KUX | P Kay, 97 Avenue Road, London, N14 4DH |
| G0 | KUY | S Crane, 13 Kirkstone Drive, Royton, Oldham, OL2 6TP |
| G0 | KUZ | B Long, 81 Easthorpe Street, Ruddington, Nottingham, NG11 6LB |
| G0 | KVA | A Sargent, 25 Jordans Way, Bricket Wood, St. Albans, AL2 3SJ |
| G0 | KVB | R Edmonds, 87 Burton Road, Castle Gresley, Swadlincote, DE11 9EW |
| G0 | KVC | H Crouch, 21a Victoria Gardens, Horsforth, Leeds, LS18 4PJ |
| GM0 | KVD | C Mackay, 5 Cromer Gardens, Glasgow, G20 9JQ |
| GM0 | KVE | Adam Dickson, 17 Junction Road, Kinross, KY13 8TA |
| G0 | KVF | R Croucher, 26 Edith Avenue, Peacehaven, BN10 8JB |
| G0 | KVG | R Neal, 144 Netherton Road, Worksop, S80 2SB |
| G0 | KVJ | PETERLEE ARC c/o Andrew Pennell, 99 Westheath Avenue, Sunderland, SR2 9LQ |
| G0 | KVK | G Cooper, 33 Lawnswood Road, Wordsley, Stourbridge, DY8 5PH |
| G0 | KVM | W Gravenor, 3 Foxhill Grove, Queensbury, Bradford, BD13 2JN |
| G0 | KVO | D Pallister, 9 Curtis Hayward Drive, Quedgeley, Gloucester, GL2 4WJ |
| GI0 | KVQ | G Millar, 1 Mullybrannon Road, Dungannon, BT71 7ER |
| G0 | KVR | C Mayo, 118 Burden Road, Beverley, HU17 9LH |
| G0 | KVS | Mike Gill, 21 Priory Terrace, London, NW6 4DG |
| G0 | KVU | Helen Sheratte, Redcar Villa, 382 Buxton Road, Macclesfield, SK11 7ES |
| G0 | KVX | C Shearing, 52 Carrine Road, Truro, TR1 3XB |
| G0 | KVZ | S Adams, 34 Kingswood Chase, Leigh-on-Sea, SS9 3BD |
| GW0 | KWA | D Clark, 37 Rotherslade Road, Langland, Swansea, SA3 4QW |
| G0 | KWD | K Dyer, 79 Station Road, Woolton, Liverpool, L25 3PY |
| G0 | KWE | John Knowles, 10 Grove Hill, Hessle, HU13 0RT |
| G0 | KWF | W Taylor, 14 Rossiters Lane, St. George, Bristol, BS5 8TW |
| G0 | KWG | Stuart Lodge, 7 Primrose Drive, Milkwall, Coleford, GL16 7PU |
| GM0 | KWL | B Mulleady, 9 Elizabeth Crescent, Camelon, Falkirk, FK1 4JF |
| GD0 | KWM | Graham Brown, 10 Albert Street, Ramsey, Isle of Man, IM8 1JF |
| GW0 | KWO | K Williams, 39 Lewis Drive, Caerphilly, CF83 3FT |
| G0 | KWQ | Paul Bates, 46 Kingsley Avenue, Redditch, B98 8PL |
| GM0 | KWW | J Alexander, Shore Cottage, Girvan, KA26 9JH |
| G0 | KXD | H Worden, 3 Tower View, Darwen, BB3 3GZ |
| G0 | KXG | J Nicholls, 93 Swan Road, Hanworth, Feltham, TW13 6PE |
| G0 | KXL | Stephen Maton, 117 Woodchurch Road, Birkenhead, CH42 9LJ |
| G0 | KXV | Gary Meredith, 22 Kingfisher Road, Attleborough, NR17 2RL |
| G0 | KXW | A Fitzmaurice, 14 Welwyn Close, Thelwall, Warrington, WA4 2HE |
| G0 | KXY | P Wroe, 44 Hillberry Crescent, Warrington, WA4 6AF |
| G0 | KXZ | A Sockett, 35 Whernside Road, Woodthorpe, Nottingham, NG5 4LB |
| G0 | KYA | Steve Nichols, 20 Holly Blue Road, Wymondham, NR18 0XJ |
| G0 | KYB | M Kinder, 58 Longridge Avenue, Stalybridge, SK15 1HL |
| G0 | KYD | Alan Hull, 1 Occupation Lane, New Bolingbroke, Boston, PE22 7LW |
| G0 | KYE | L Landricombe, 19 Crackston Close, Eggbuckland, Plymouth, PL6 5SN |
| G0 | KYG | P Willetts, 197 Norwich Road, Fakenham, NR21 8LR |
| G0 | KYH | J Elliott, Tregerrick, Martinstown, Dorchester, DT2 9JN |
| G0 | KYJ | G Liversidge, 65 Rowelfield, Luton, LU2 9HL |
| G0 | KYK | R Beardsmore, 2 Fitzmaurice Road, Wednesfield, Wolverhampton, WV11 3EG |
| G0 | KYL | John Lawrence, 8 Murray Terrace, Dipton, Stanley, DH9 9HB |
| G0 | KYM | William Lowder, 24 Plantation Lane, Bearsted, Maidstone, ME14 4BH |
| G0 | KYN | R Markham, 25 Burndell Way, Hayes, UB4 9YF |
| G0 | KYR | David Cooper, 7 Kendal Rise, Bedlington, NE22 6PB |
| G0 | KYS | R Edgar, 45 Exeter Road, Dawlish, EX7 0AB |
| GM0 | KYU | J Robertson, 143 Rankin Court, Greenock, PA16 9AZ |
| G0 | KYX | Peter MORGAN, 4 Cromwell Mews, Burgess Hill, RH15 8QF |
| GW0 | KYY | Malcolm Warner, 76 Rhodfar Eos, Cwmrhydyceirw, Swansea, SA6 6SW |
| GJ0 | KYZ | Paul Mahrer, 2 Oakley, La Rue Parcqthee, St. Lawrence, Jersey, JE3 1FR |
| G0 | KZA | E Bishop, 21 Mandalay St., Basford, Nottingham, NG6 0BH |
| G0 | KZD | M Withey, 9 Marnhull Road, Longfleet, Poole, BH15 2EX |
| GW0 | KZE | Charles Dublon, Tyn-y-Waun, Dare Road, Aberdare, CF44 8UB |
| GM0 | KZG | A Adams, 157 High Road, Saltcoats, KA21 6JX |
| G0 | KZH | E Clayton, 220 Milnrow Road, Rochdale, OL16 5BB |
| G0 | KZI | J Williams, 18 St. Andrews Close, Holme Hale, Thetford, IP25 7EH |
| GM0 | KZJ | G Strang, 15 Almondbank Terrace, Edinburgh, EH11 1SS |
| GW0 | KZK | B Parsons, 31 Crymlyn Parc, Neath, SA10 6DG |
| G0 | KZM | D Egan, 56 Walker Avenue, Wollescote, Stourbridge, DY9 9EL |
| G0 | KZN | A Sargeant, 27 Sandygate Crescent, Old Leake, Boston, PE22 9RA |
| G0 | KZO | E Lomas, 2 Linney Road, Bramhall, Stockport, SK7 3JW |
| G0 | KZT | Andrew Briers, 33 Deans Walk, Coulsdon, CR5 1HR |
| G0 | KZW | W Jones, Tanglewood, 2 Bryntirion Avenue, Prestatyn, LL19 9PB |
| GM0 | KZX | Barry Spink, 9 St. Andrews Crescent, Dumbarton, G82 3ER |
| G0 | LAA | E Martin, 90 Grand Drive, Herne Bay, CT6 8LS |
| G0 | LAD | J Parfett, 65 Brompton Lane, Rochester, ME2 3BA |
| G0 | LAG | J Penney, 2a St John St., Wainfleet, Skegness, PE24 4DL |
| G0 | LAK | J Rogers, 186 Beavers Lane, Birleywood, Skelmersdale, WN8 9BP |
| GI0 | LAM | G Lamb, 31 Dromara Road, Ballyward, BT31 9SJ |
| G0 | LAN | Ann Taylor, 2 Gap Crescent, Hunmanby Gap, Filey, YO14 9QJ |
| G0 | LAU | J Barber, Jasmine Cottage, Spend Lane, Ashbourne, DE6 2AS |
| G0 | LAX | Anthony Duggan, 28 Higher Rads End, Eversholt, Milton Keynes, MK17 9ED |
| G0 | LAZ | J B A Pryer, Apesford Crossing Cottage, Apesford, Leek, ST13 7EX |
| G0 | LBA | A Hughes, Derwen Las, Valley Road, Llanfairfechan, LL33 0SS |
| G0 | LBB | Ronald Batty, 31 Spring Lane, Balderton, Newark, NG24 3NZ |
| G0 | LBE | Shaun Watkinson, 40 Wharfedale, Westhoughton, Bolton, BL5 3DP |
| GW0 | LBI | L Smart, Wordsley, Gwerthonor Road, Bargoed, CF81 8JS |
| GM0 | LBN | J Clark, 35 Jedburgh Avenue, Rutherglen, Glasgow, G73 3EN |
| G0 | LBO | J Ross, The Gables, Jack Lane, Northwich, CW9 8QA |
| G0 | LBQ | R Perks, 120 Cranes Park Road, Sheldon, Birmingham, B26 3ST |
| G0 | LBT | K Tromans, 7 Heathfield, Heath Charnock, Chorley, PR6 9LA |
| G0 | LBZ | P Cahill, 56 Dene Road, Headington, Oxford, OX3 7EE |
| G0 | LCB | A Cleaver, 19 Newlands Drive, Grove, Wantage, OX12 0NY |
| G0 | LCC | Howard Grinter, 9 Fieldway, Sandford, Winscombe, BS25 5PR |
| G0 | LCD | P Chinnock, 49 Wishings Road, Brixham, TQ5 9PB |
| G0 | LCE | Kenneth Robinson, 33 Mirlaw Road, Whitelea Chase, Cramlington, NE23 6UB |
| G0 | LCG | Steve Sorockyj, 8 Bowden Avenue, Bestwood Village, Nottingham, NG6 8XN |
| G0 | LCH | Martyn Nash, 22 Northleigh Close, Maidstone, ME15 9RP |
| G0 | LCJ | B Lucock, 15 Mayfield Road, Newquay, TR7 2DG |
| G0 | LCN | D Prendiville, 40 Caerleon Drive, Southampton, SO19 5LF |
| G0 | LCO | Robert Foster, 18 Stokesay Way, Sutton Hill, Telford, TF7 4QE |
| G0 | LCP | Brian Davey, 30 Long Down Gardens, Plymouth, PL6 8SB |
| G0 | LCR | LANCASHIRE COUNTY RAYNET c/o F Charnley, 30 Dunkirk Avenue, Fulwood, Preston, PR2 3RY |
| G0 | LCS | Kerry Rochester, 22 Langford Road, Cockfosters, Barnet, EN4 9DS |
| G0 | LCT | G Moss, 15 Coppice Avenue, Hatfield, Doncaster, DN7 6AH |
| G0 | LCU | B Walker, 70 King George Road, Loughborough, LE11 2PA |
| G0 | LCV | J Fidoe, 85 Sedgemoor Road, Bridgwater, TA6 5NS |
| G0 | LCX | David Weatherill, Northend Cottage, North End Road, Bristol, BS49 4AS |
| G0 | LDB | M Mallinson, 25 The Fairway, Banbury, OX16 0RR |
| GI0 | LDI | D Keys, 71 Madison Avenue, Eglinton, Londonderry, BT47 3PW |
| G0 | LDJ | Douglas Cansfield, 1 Brook Walk, Calmore, Southampton, SO40 2UY |
| G0 | LDO | R Summerfield, 64 Station Road, Broughton Astley, Leicester, LE9 6PT |
| G0 | LDP | Kevin Starkey, 13a Cardigan Road, Bedworth, CV12 0LY |
| GW0 | LDQ | G Kift, 23 Field Close, Morriston, Swansea, SA6 6QD |
| G0 | LDR | J Marlow, 21 Thames Rise, Kettering, NN16 9JL |
| G0 | LDU | K Allies, 6 Alston Close, Hazel Grove, Stockport, SK7 5LR |
| GM0 | LDX | K Strathdee, Keepers Cottage, Elginshill, Elgin, IV30 8NH |
| G0 | LDY | K Jenkinson, 2 Madeira Avenue, Codsall, Wolverhampton, WV8 2DS |
| GW0 | LDZ | B Garland, 20 Bryn Avenue, Upper Brynamman, Ammanford, SA18 1BD |
| G0 | LEA | Derek Gray, 63 Queen Street, Grange Villa, Chester le Street, DH2 3LU |
| GI0 | LEC | LOUGH ERNE ARC c/o Adrian Duffy, 81a Arney Road, Bellanaleck, Enniskillen, BT92 2DL |
| G0 | LEE | Robert Lee, 7 Long Meadow, Little Hoole, Preston, PR4 4RQ |
| G0 | LEF | T Bell, 16 North Seaton Road, Newbiggin-by-The-Sea, NE64 6XT |
| G0 | LEH | Graham Chatfield, 1a Sheringham Way, Orton Longueville, Peterborough, PE2 7AH |
| G0 | LEI | E hughes, 1 Leith Gardens, Tanfield Lea, Stanley, DH9 9LZ |
| G0 | LEJ | M Huggett, Rosslyn, Station Road, Brampton, CA8 1EX |
| G0 | LEL | F Sunley, 39 Winton Road, Northallerton, DL6 1QQ |
| G0 | LEN | WEST LINCS RY G c/o P Worsdale, Emergency Planning Department, Fire Brigade Headquarters, Lincoln, LN5 8EL |
| G0 | LEP | D Stewart, Buckskin, 16 Prescelly Close, Basingstoke, RG22 5DN |
| G0 | LES | E Simpson, Laneside, Cliff Lane, Bridlington, YO15 1JF |
| G0 | LEU | P Johnson, 5 The Hawthorns, Broadstairs, CT10 2NG |
| G0 | LEV | D Painter, Troutbeck, Mary Tavy, Tavistock, PL19 9PR |
| G0 | LEY | Robert Smith, 63 Hanna Street East, Windsor, Canada, ON N8X 2NI |
| G0 | LFA | N Swallow, 178 Barcroft Street, Cleethorpes, DN35 7DX |
| G0 | LFE | K Gray, 25 The Pastures, Blyth, NE24 3HA |
| G0 | LFF | R Hide, 74 Maple Drive, Burgess Hill, RH15 8DL |
| G0 | LFH | P Mustchin, 6 Spinney North, Pulborough, RH20 2AT |
| G0 | LFI | Frank Cotton, 49 Cornwall Road, Fratton, Portsmouth, PO1 5AR |
| G0 | LFM | V Sancto, Meadowbank, 15a Spratling Street, Ramsgate, CT12 5AW |
| G0 | LFN | S Southwell, Sullys, 12 Somerset Road, Southsea, PO5 2NL |
| G0 | LFP | Steven Courtney-Crowe, 28 Brymore Close, Prestbury, Cheltenham, GL52 3DY |
| G0 | LFQ | Paul Mason, Trevidavean, Antony, Torpoint, PL11 3AQ |
| G0 | LFV | P Fisher, Chevalier, Marks Corner, Newport, PO30 5UH |
| G0 | LFY | D Recardo, 1 Heronfield Close, Redditch, B98 8QL |
| G0 | LFZ | A Recardo, 1 Heronfield Close, Redditch, B98 8QL |
| G0 | LGA | R Letts, 28 Catlin Crescent, Shepperton, TW17 8EU |
| G0 | LGB | James Walker, 22 Temperance Field, Wyke, Bradford, BD12 9NR |
| G0 | LGC | Louis Culshaw, 15 Naunton Avenue, Leigh, WN7 4SX |
| G0 | LGE | Martin Brooman, 141 Northdown Park Road, Margate, CT9 3PX |
| G0 | LGF | Terry Evennett, 18 Fen Folgate, Shipdham, Thetford, IP25 7LT |
| G0 | LGG | N Challacombe, 17 Tanners Lane, Chalkhouse Green, Reading, RG4 9AD |
| G0 | LGJ | M Taylor, 6 Welden Road, Scarning, Dereham, NR19 2UB |
| G0 | LGK | E Wall, Shrubbery Cottage, Felderland Lane, Deal, CT14 0BT |
| G0 | LGO | A Turton, 58 Highfield Lane, Quinton, Birmingham, B32 1QT |
| GI0 | LGV | H Magill, 51 Ballybracken Road, Doagh, Ballyclare, BT39 0TQ |
| G0 | LGW | R Caine, 148 Dumpton Park Drive, Broadstairs, CT10 1RP |
| G0 | LGZ | Brian Grimes, Flat 12, Clyde House, Ventnor, PO38 1QL |
| G0 | LHB | Akira Okubo, 1427-9-608, Yamazaki-Cho Machida City, Tokyo, Japan, 195-0074 |
| G0 | LHD | R Caton, 13 Goss Barton, Nailsea, Bristol, BS48 2XD |
| G0 | LHE | R Wilmot, Elm Tres, Drayson Lane, Northampton, NN6 7SR |
| G0 | LHL | M Humphreys, 25 Dalestorth Close, Sutton-in-Ashfield, NG17 4EH |
| G0 | LHM | B Tuffrey, 53 Sheffield Road, Warmsworth, Doncaster, DN4 9QR |
| G0 | LHN | J Butterworth, 38 Stuart Avenue, Moreton, Wirral, CH46 9PF |
| G0 | LHR | L Robinson, 82 Grassholme, Wilnecote, Tamworth, B77 4BZ |
| G0 | LHU | J Lawton, 37 Southway, Horsforth, Leeds, LS18 5RN |
| G0 | LHV | R Kay, 10 Meadow Lane, Newport, Brough, HU15 2QN |
| G0 | LHX | Harold Passmore, 3 Cossington Lane, Woolavington, Bridgewater, TA7 8HL |
| G0 | LHZ | J Carter, 22 Orchard Coombe, Whitchurch Hill, Reading, RG8 7QL |
| G0 | LIA | R James, 77 Charlotte Close, Mount Hawke, Truro, TR4 8TT |
| G0 | LIB | R Weston, 38 Church Road, Peasedown St. John, Bath, BA2 8AF |
| G0 | LII | Steve Hodgson, 4 Nikolaou Michael Street, Dasaki Achnas, Cyprus, 5523 |
| GW0 | LIK | C Raymond, 23 Castle Pill Crescent, Steynton, Milford Haven, SA73 1HD |
| GM0 | LIM | J Duffy, 39 Kylerhea Road, Thornliebank, Glasgow, G46 8AB |
| G0 | LIN | C Smith, 2 Ha'penny Drive, Holbrook, Ipswich, IP9 2TT |
| G0 | LIQ | Joseph cunningham, 219 Alfreton Road, Underwood, Nottingham, NG16 5GX |
| GM0 | LIR | Philip Woods, 394 Glasgow Road, Wishaw, ML2 7SJ |
| G0 | LIS | A Wright, 8 Bryn Mor Terrace, Holyhead, LL65 1EU |
| G0 | LIW | P Hopkins, Stonelodge, St. Helens Avenue, York, YO42 2JF |
| GI0 | LIX | CARRICKFERGUS ARG c/o J Branagh, 17 Rathmoyle Park West, Carrickfergus, BT38 7NG |
| G0 | LIY | P Smit, 18 Owlwood Lane, Dunnington, York, YO19 5PH |
| G0 | LJB | P Williams, 44 Meadow Road, Mirehouse, Whitehaven, CA28 8EP |
| G0 | LJC | C Livesey, 14 Dene Drive, Longfield, DA3 7JR |
| G0 | LJD | B Howard, 15 Cambridge Road, Strood, Rochester, ME2 3HW |
| G0 | LJF | M Binks, 24 Mill Lane, Reddish, Stockport, SK5 6UU |
| G0 | LJG | D Green, The Archways, St. Georges Road, Trowbridge, BA14 6JQ |
| G0 | LJH | C Holmes, Manacor, 4 Dovefields, Uttoxeter, ST14 5LT |
| G0 | LJI | Graham Evans, 241 St. Johns Road, Newbold Moor, Chesterfield, S41 8PE |
| G0 | LJJ | D Mackenny, 21 Chilton Way, Hungerford, RG17 0JR |
| G0 | LJK | W Dancock, 11 St. Davids Close, Stourport-on-Severn, DY13 8RZ |
| G0 | LJM | D Roebuck, 8 Runnymede Court, Bradford, BD10 9JW |
| G0 | LJP | Paul McLeod, 4 Caple Avenue, Kings Caple, Hereford, HR1 4UL |
| G0 | LJS | H Sims, Pinewood Lodge, 276 Sandridge Lane, Chippenham, SN15 2JW |
| G0 | LJV | Stephen Swinbourne, 11 Stapleton Road, Warmsworth, Doncaster, DN4 9LA |
| G0 | LJW | D Goodwin, 25 Bevan Crescent, Blackwood, NP12 1EW |
| G0 | LKA | Carl Cruddas, 81 Church Walk, Atherstone, CV9 1PS |
| G0 | LKI | W Cockerell, 3 Churchford Road, Knowle, Braunton, EX33 2LT |
| G0 | LKJ | W Halliwell, 20 Llwynon Road, Oakdale, Blackwood, NP2 0LX |
| GM0 | LKS | Edward Mcgreevy, 47 Fairfield Drive, Renfrew, PA4 0EG |
| GM0 | LKT | A Ferris, 60 Appin Crescent, Kirkcaldy, KY2 6ES |
| G0 | LKY | G Civil, Whitehouse Farm, Magpie Lane, Brentwood, CM13 3DZ |
| G0 | LLB | R Smith, 34 Churchill Rise, Chelmsford, CM1 6FD |
| G0 | LLC | Malcolm Bridges, 7 Sun Road, Woodland, Bishop Auckland, DL13 5NF |
| GW0 | LLD | H Jones, Dolau Bran, Cynghordy, Llandovery, SA20 0LD |
| G0 | LLE | Paul Ferris, 116 Capel Road, Forest Gate, London, E7 0JS |
| G0 | LLG | Damien Davies, 4 Whitendale, Lancaster, LA1 5JD |
| GM0 | LLJ | B Borrows, 27 Craigdimas Grove, Dalgety Bay, Dunfermline, KY11 9XR |
| G0 | LLL | J Roberts, 69 Barnoldswick Road, Barrowford, Nelson, BB9 6BQ |
| G0 | LLP | L Proud, 26 Drayton Court, The Green, Nuneaton, CV10 0SL |
| G0 | LLX | A Bassett, 125 Stonyhill Avenue, South Shore, Blackpool, FY4 1PW |
| G0 | LMA | S Crooks, 10 Mere Close, Mountsorrel, Loughborough, LE12 7BP |
| G0 | LMD | M Butler, 44 East Stratton, Winchester, SO21 3DU |
| G0 | LMJ | E Garrott, Lynden, Clappers Lane, Chichester, PO20 7JJ |
| G0 | LMO | Adrian Everitt, 58 Eastwood Road, Aylestone, Leicester, LE2 8DB |
| GI0 | LMR | W Redmond, 6 Hazelwood Crescent, Craigywarren, Ballymena, BT43 6TA |
| G0 | LMX | V Denecker, Kernanderry, Faringdon Road, Abingdon, OX13 6QJ |
| G0 | LNA | R Henderson, 65 Rowarth Road, Manchester, M23 2UL |
| G0 | LNB | G Goodwin, 16 Hucklow Avenue, Newall Green, Manchester, M23 2YX |
| G0 | LNE | T Stokes, 22 Armada Close, Erdington, Birmingham, B23 7PB |
| G0 | LNI | J Stringer, 2 West End, Marston Magna, Yeovil, BA22 8BW |
| G0 | LNK | P Bower, 103 Henson Park, Chard, TA20 1NJ |
| GW0 | LNM | Paul Pentecost, Brynhyfryd, Maes y Bont Road, Llanelli, SA14 7NA |
| G0 | LNN | David Draycott, 3 Sycamore Gardens, Dymchurch, Romney Marsh, TN29 0LA |
| G0 | LNO | S Feeney, York House, York Drive, Altrincham, WA14 3HF |
| G0 | LNS | G Robinson, 9 Greenlands Court, Seaton Delaval, Whitley Bay, NE25 0BU |
| G0 | LNT | M Millward, 50 Barnsley Road, Moorends, Doncaster, DN8 4QT |
| G0 | LNV | T Appleyard, 78 Chelsea Road, Sheffield, S11 9BR |
| G0 | LNW | T Horabin, 69 Birchwood Avenue, North Gosforth, Newcastle upon Tyne, NE13 6QB |
| G0 | LNX | Ivan Davison, 20 Littlegreen Gardens Compton, Chichester, PO18 9NP |
| G0 | LOC | T Loraine, Fieldgate, Coltstaple Lane, Horsham, RH13 9BB |
| GM0 | LOD | G Collier, 64 Hadfast Road, Cousland, Dalkeith, EH22 2NZ |
| G0 | LOE | S Phillips, 26 Belvedere Drive, Dukinfield, SK16 5NW |
| G0 | LOF | F James, 6 Pinewood Close, East Preston, Littlehampton, BN16 1HF |
| G0 | LOH | K Dutson, 3 The Barracks, Wynford Eagle, Dorchester, DT2 0ER |
| GW0 | LOI | R Jones, 31 Three Arches Avenue, Cardiff, CF14 0NU |
| G0 | LOJ | C Budd, 10 Stanley Mead, Bradley Stoke, Bristol, BS32 0EG |
| GM0 | LOK | John Leggat, Ailach, St. Aethans Road, Elgin, IV30 2YR |
| G0 | LOL | Christopher Kidger, 25 Elmham Road Cantley, Doncaster, DN4 6LF |

| | | |
|---|---|---|
| GM0 | LOO | H Hunter, 25 Braehead Road, Kirkcaldy, KY2 6XP |
| G0 | LOP | G Tweedy, 8 Greencliffe Drive, York, YO30 6NA |
| GM0 | LOT | R Clasper, 32 Murieston Park, Livingston, EH54 9DT |
| G0 | LOW | THE SHORTWAVE SHOP c/o Duncan Kemp, 18 Fairmile Road, Christchurch, BH23 2LJ |
| G0 | LOZ | Ian Tomson, 13 Valley View, Bewdley, DY12 2JX |
| GM0 | LPB | J Gault, 25 Beech Brae, Bishopmill, Elgin, IV30 4NS |
| G0 | LPF | R Wilkins, 20 Fairholme Drive, Yapton, Arundel, BN18 0JH |
| G0 | LPG | B Gaunt, Po Box 50, Guildford, GU1 2FJ |
| G0 | LPN | B Alperowicz, 20 chemin du Cabanis, Meynes, France, 30840 |
| G0 | LPP | Charles Galea, 36 Godwit Close, Gosport, PO12 4JF |
| [illegible] | [illegible] | [illegible] 7 Oak Close, Whiston, Prescot, L35 3YG |
| G0 | LPT | G Webb, 23 Kerdane, Hull, HU9 9ED |
| G0 | LPU | A Newton, 10 Rowan Court, Greasby, Wirral, CH49 3QH |
| G0 | LPV | R Chappell, 40 Midway, South Crosland, Huddersfield, HD4 7DA |
| G0 | LPX | B Garbutt, 35 Westfield Road, Tockwith, York, YO26 7PY |
| G0 | LQC | D Briggs, 57 Charlton Drive, High Green, Sheffield, S35 3PA |
| G0 | LQD | P Valleley, 9 Lavender Road, Basingstoke, RG22 5NN |
| G0 | LQF | R Welch, 26 Stalls Road, Lower Gornal, Dudley, DY3 2UN |
| G0 | LQI | M Murphy, 133 Preston Road, Preston, Weymouth, DT3 6BG |
| G0 | LQK | Michael Hinchliffe, 2 Ash Grove, New Longton, Preston, PR4 4XJ |
| G0 | LQM | J Hipwell, 5 Dolphin Crescent, Paignton, TQ3 1AE |
| G0 | LQN | Leslie Fish, 44 Maycroft Avenue, Poulton-le-Fylde, FY6 7NE |
| G0 | LQO | R Taylor, 17 York Close, Clayton le Moors, Accrington, BB5 5RB |
| G0 | LQT | H Smith, 66 The Avenue, Clacton-on-Sea, CO15 4ND |
| G0 | LQU | P Fardell, 90 Beechwood Avenue, St. Albans, AL1 4XZ |
| G0 | LQV | Martyn Fordham, 24a Main Street, Prickwillow, Ely, CB7 4UN |
| G0 | LQW | Peter Macolive, 6 Pembroke Way, Hayes, UB3 1PZ |
| G0 | LQX | T Newstead, 31 Byron Road, Heysham, Morecambe, LA3 1UH |
| G0 | LQZ | Christopher Walkup, 1 Darley Hall, Luton, LU2 8PP |
| GM0 | LRA | LORN RADIO AMATEURS c/o Graham Henderson, Tigh An Drochaid, Kilchrenan, Argyll, PA35 1HD |
| G0 | LRE | Joseph Norman, 9a St. Johns Grove, Heysham, Morecambe, LA3 1ET |
| G0 | LRG | LEICESTERSHIRE REPEATER GROUP c/o G Dover, 31 Newbold Road, Kirkby Mallory, LE9 7QG |
| G0 | LRI | Stephen Kennedy, Colmans Farm, Elmstone-Hardwicke, Cheltenham, GL51 9TG |
| G0 | LRJ | P Daymond, 14 Philip Close, Plymouth, PL9 8QZ |
| G0 | LRK | Kevin Wall, 27 Broomfield Road, Fleetwood, FY7 7HA |
| G0 | LRM | A Littler, 365 Westhorne Avenue, London, SE12 9AB |
| G0 | LRO | D Watmough, 41 West Crayke, Bridlington, YO16 6XR |
| G0 | LRP | Peter Waters, Unit3, Site2 SandpitLane, Nr Beccles, NR34 7TH |
| G0 | LRR | RSSDALE RAYT GR c/o S Greenwood, Carter Place Farm, Hall Park, Rossendale, BB4 5BQ |
| G0 | LRU | F Alderson, Old School House, Tattersett, King's Lynn, PE31 8RS |
| G0 | LRW | M Simmons, 6 The Crescent, Bletchley, Milton Keynes, MK2 2QD |
| GI0 | LRZ | N Mitchell, 6 Brae Road, Newry, BT34 1NZ |
| G0 | LSE | Paul Harris, 7 Rowan Avenue, Egham, TW20 8AN |
| G0 | LSG | Ron Joy, 61 Grenville Drive, Tavistock, PL19 8DP |
| G0 | LSI | D Peachey, Thornbury Cottage, Ashmill, Beaworthy, EX21 5HA |
| G0 | LSJ | C Jones, 179 Blandford Road, Efford, Plymouth, PL3 6JZ |
| G0 | LSK | D Taylor, 22 Meon Road, Mickleton, Chipping Campden, GL55 6TD |
| G0 | LSP | L Pawlik, 2 Woodcock Close, Bamford, Rochdale, OL11 5QA |
| G0 | LSQ | D Williams, 41 Ravensgate Road, Charlton Kings, Cheltenham, GL53 8NS |
| G0 | LSU | Jeffrey Hair, 84 Oxford Street, Barrow-in-Furness, LA14 5QQ |
| G0 | LSX | D Barber, 179 Rye Hills, Bignall End, Stoke-on-Trent, ST7 8LP |
| G0 | LTB | A Veal, 64 Hither Bath Bridge, Bristol, BS4 5DJ |
| GW0 | LTC | G Edwards, 5 Lower Farm Court, Rhoose, Barry, CF62 3HQ |
| G0 | LTD | Brian Tugwell, 12 Anguilla Close, Eastbourne, BN23 5TS |
| G0 | LTE | D Prout, 8 Ferenberge Close, Farmborough, Bath, BA2 0DH |
| GI0 | LTF | Henry Irwin, 9 Edward Street, Armagh, BT61 7QU |
| G0 | LTO | R Summers, 18b Rose Road, Canvey Island, SS8 0BP |
| G0 | LTP | D Freeman, 71 Longleaze, Wootton Bassett, Swindon, SN4 8AS |
| G0 | LTR | TAMWORTH & LICHFIELD RAYNET c/o Robert Williams, 76 Quince, Amington, Tamworth, B77 4EU |
| GI0 | LTT | S Beattie, 28 Millers Lane, Newtownards, BT23 7AR |
| G0 | LTV | R Pearce, Talara, Kent Street, Battle, TN33 0SF |
| G0 | LTX | S Painting, Claytons, Inkpen, Newbury, RG17 9QE |
| G0 | LUA | David Jewell, 45 Church Walks, Llandudno, LL30 2HL |
| G0 | LUB | Anthony Nicholls, Flat 39, Bellair House, 41 East Street, Havant, PO9 1QS |
| G0 | LUC | P Dyke, 1102 Buckingham DR, Apt B, California, United States, 92626 |
| G0 | LUD | R Spacey, 18 Longdale Avenue, Ravenshead, Nottingham, NG15 9EA |
| GM0 | LUF | T Traill, 30 Strathesk Road, Penicuik, EH26 8EF |
| G0 | LUH | D Goodison, 33 Witham Road, Isleworth, TW7 4AJ |
| G0 | LUI | P Draper, 265 Nottingham Road, Ilkeston, DE7 5AT |
| G0 | LUK | D Palmer, Braidwood, Enborne Row, Newbury, RG20 0LY |
| G0 | LUL | P Clune, 50 St. Marks Road, Mitcham, CR4 2LF |
| G0 | LUM | Wesley Mitchell-Watson, 144 Shakespeare Crescent, Dronfield, S18 1ND |
| G0 | LUN | C Stayt, 100 Cromwell Way, Oddington, Kidlington, OX5 2LL |
| G0 | LUP | K Chambers, 9 Village Farm Road, Preston, Hull, HU12 8QH |
| G0 | LUQ | Terence Vale, Grange Farm Flat, Station Road, Bicester, OX26 5DX |
| G0 | LUU | K Blackburn, 63 Robsons Drive, Huddersfield, HD5 9JW |
| G0 | LUW | C Fells, 21 Garthorne Avenue, Darlington, DL3 9XL |
| G0 | LUY | Graham ONeill, 16 Aldam Street, Darlington, DL1 2HY |
| G0 | LVA | T Aanestad, 14 Overdale Gardens, Sheffield, S17 3HE |
| G0 | LVF | F Talmage, 54 Rodwell Avenue, Weymouth, DT4 8SG |
| G0 | LVG | Peter Nilan, 15 Broomhall Road, Pendlebury, Manchester, M27 8XP |
| GW0 | LVH | J Wimpenny, Gwili House, The Ropewalk, Milford Haven, SA73 3LW |
| GM0 | LVI | David Warburton, Lawvista, High Street, Perth, PH2 7QQ |
| G0 | LVJ | B Bradshaw, 18 Burley Avenue, Lowton, Warrington, WA3 2ES |
| GM0 | LVK | Leslie Alexander, 97 Land Street, Keith, AB55 5AP |
| GM0 | LVL | Colin McEwan, 42 Marionville Crescent, Edinburgh, EH7 6AU |
| G0 | LVN | David Byrne, 29 Holly Street, Tottington, Bury, BL8 3EZ |
| G0 | LVR | J Russell, 9 Pear Tree Lane, Rowledge, Farnham, GU10 4DW |
| G0 | LVT | B Thornber, 78 Skipton Road, Silsden, Keighley, BD20 9LL |
| G0 | LVX | T Burns, 52 Somerset Drive, Bury, BL9 9DQ |
| G0 | LVY | John Littler, 39 Wigan Road, Golborne, Warrington, WA3 3TZ |
| G0 | LWC | P Timlett, 10 Reynolds Gardens, Moulton, Spalding, PE12 6PT |
| GM0 | LWD | Lawrence McWilliams, 38 Churchill Street, Alloa, FK10 2JG |
| G0 | LWE | W Short, 6 Kensington Avenue, Normanby, Middlesbrough, TS6 0QQ |
| G0 | LWG | N Sheriden, 92 Station Road, Stallingborough, Grimsby, DN41 8AP |
| G0 | LWI | L Butler, [illegible] Road, Buckhurst Hill, IG9 0BY |
| G0 | LWL | D Spooner, 7 East Avenue, Althorne, Chelmsford, CM3 6DD |
| G0 | LWM | Anthony MOTHEW, 7 Ashfields, Loughton, IG10 1SB |
| G0 | LWN | D Parsons, 107 Larkswood Road, Chingford, London, E4 9DU |
| GI0 | LWO | S Magill, 40 Gardners Road, Lisburn, BT27 5PD |
| G0 | LWU | A Scarr, Kerrera, Chapel Lane, Morecambe, LA3 3JA |
| G0 | LXB | Christopher Tapping, 49 Coventry Gardens, Herne Bay, CT6 6SB |
| G0 | LXC | J Searle, 232 Park Lane, Frampton Cotterell, Bristol, BS36 2EN |
| G0 | LXF | R Turner, 5 Darenth Court, Quiller Road, [illegible], BH5 4NS |
| G0 | LXG | R Clulee, 25 Cloister Crofts, Leamington Spa, CV32 6QG |
| G0 | LXI | C Sidney, 25 John Mcguire Crescent, Binley, Coventry, CV3 2QG |
| G0 | LXL | G Truckel, 26 Elm Close, Chipping Sodbury, Bristol, BS37 6HL |
| GI0 | LXN | W Black, 14 Killyliss Road, Fintona, Omagh, BT78 2DL |
| G0 | LXP | R Cant, 25 Worcester Avenue, Garstang, Preston, PR3 1FJ |
| G0 | LXR | Barry Morgan, 208 Main Street, Burley In Wharfedal, Ilkey, LS29 7HS |
| G0 | LXV | M Lee, 23 Lyndale Road, Redhill, RH1 2HA |
| G0 | LXW | Alan Pollard, 16 Bellenger Way, Kidlington, OX5 1TR |
| G0 | LXX | J Mayfield, 9 Middlefell Way, Clifton, Nottingham, NG11 9JN |
| G0 | LYC | Paul Hindle, Three Ways, Oakwood, Hexham, NE46 4LE |
| GW0 | LYF | D Hobbs, 9 Llwynypia Terrace, Llwynypia, Tonypandy, CF40 2JD |
| G0 | LYG | F Aris, 5 Horley Road, Mottingham, London, SE9 4LF |
| GM0 | LYH | H Cochrane, 69 Balgray Avenue, Kilmarnock, KA1 4QT |
| G0 | LYI | W Stevens, 97 Kelvin Grove, Portchester, Fareham, PO16 8LF |
| G0 | LYJ | W Hughes, 26 Cambridge Cottages, Richmond, TW9 3AY |
| GW0 | LYK | S Parker, Maesycoed, Blaenycoed, Carmarthen, SA33 6ES |
| G0 | LYN | L Roper, 57 Burnt Hills, Cromer, NR27 9LW |
| GM0 | LYO | J Fletcher, 1 Silverwood Farm Cottage, Kilmarnock, KA3 6HJ |
| G0 | LYQ | L Selman, 156 Bradley Drive, Santa Cruz, United States, 95060 |
| G0 | LYR | P Spencer, 4 Jubilee Close, Duloe, Liskeard, PL14 4PA |
| GM0 | LYT | A Fegen, Acharn, Losset Road, Blairgowrie, PH11 8BU |
| G0 | LYX | D Brown, 67 Croft Road, Benfleet, SS7 5RL |
| G0 | LYZ | Clive Knaggs, 29 Wansford Road, Driffield, YO25 5NB |
| G0 | LZD | Robert Wood, 4 Burns Road, Royston, SG8 5PT |
| GM0 | LZE | Donald Morrison, 27B BENSIDE, Newmarket, Stornoway, HS2 0DZ |
| G0 | LZF | J Rivers, 211 Upper Wickham Lane, Welling, DA16 3AW |
| G0 | LZG | W Miles-Williams, 13 Cavendish Road, Chesham, HP5 1RW |
| G0 | LZI | M Tudor, 32 Ringwood, Oxton, Prenton, CH43 2LZ |
| G0 | LZJ | Colin Price, 4 Greenway Close, Helsby, Frodsham, WA6 0QX |
| G0 | LZL | D Bates, 92 Thirlmere Road, Partington, Manchester, M31 4PT |
| G0 | LZS | T Wright, 8 Glentham Close, Lincoln, LN6 8BX |
| G0 | LZV | C Nicholas, 19 Spring Crofts, Bushey, WD23 3AR |
| G0 | LZW | D Riddick, 289 Hatfield Road, St. Albans, AL4 0DH |
| G0 | LZX | Richard Knowles, 35 Poulton Road, Southport, PR9 7BE |
| G0 | LZY | R Smith, Pleasanton, Church Street, Halstead, CO9 3AZ |
| G0 | MAA | S Rawson, The Cabin, Tully, Four Mile House, Roscommon, Ireland |
| GM0 | MAC | A Macfarlane, Breadalbane, Ferrindonald, Isle of Skye, IV44 8RF |
| G0 | MAD | MAPPERLEY & DISTRICT ARC c/o Keith Moody, 5 Moore Road, Mapperley, Nottingham, NG3 6EF |
| G0 | MAF | Michael Plaskitt, 10 Grasby Crescent, Grimsby, DN37 9HE |
| G0 | MAH | G Humphrey, 57 Haig Avenue, Leyland, PR25 2DD |
| G0 | MAL | W Bowden, 43 Burlish Close, Stourport-on-Severn, DY13 8XW |
| GD0 | MAN | MANX STHRN DX G c/o Michael Dunning, 55 Station Park, Colby, Isle of Man, IM9 4NL |
| G0 | MAR | N Buchan, Acorns, 37 Forge Rise, Uckfield, TN22 5BU |
| G0 | MAS | Alan Sedgbeer, 102 Brewin Road, Tiverton, EX16 5DP |
| G0 | MAT | R Johnson, 30 Wheatlands, Titchfield Common, Fareham, PO14 4SL |
| GW0 | MAV | Richard Truran, 34 Coed y Brain Court, Llanbradach, Caerphilly, CF83 3JT |
| G0 | MAY | P Holmes, 11 Bingham Road, Cotgrave, Nottingham, NG12 3JS |
| G0 | MAZ | M Gurr, Elan, Sandown Road, Sandwich, CT13 9NY |
| G0 | MBA | Antony Horsman, 12 Hanwell Close, Clacton-on-Sea, CO16 7HF |
| G0 | MBB | Alan Cutter, Hundred House, Pink Road Lacey Green, Princes Risborough, HP27 0PG |
| G0 | MBD | J Curzon, 17 Bullfinch Way, Cottenham, Cambridge, CB24 8AW |
| G0 | MBG | B Drew, 59 Coventry Street, Kidderminster, DY10 2BZ |
| G0 | MBL | Andrew Plant, 99 Pegwell Road, Ramsgate, CT11 0ND |
| GW0 | MBN | P Salt, Acorns, Llwyncelyn, Cardigan, SA43 2PE |
| G0 | MBP | B Vanson, 25 St. Helens Road, Westcliff-on-Sea, SS0 7LA |
| G0 | MBQ | S Withers, 14 Rushes Road, Petersfield, GU32 3BW |
| G0 | MBR | MID-BEDS RAYNET GROUP c/o Ian McIver, 3 Asgard Drive, Bedford, MK41 0UP |
| G0 | MBS | B Sinclair, 48 East Crescent, Duckmanton, Chesterfield, S44 5ET |
| G0 | MBU | Gillian Buckwell, 24 Ings Road, Redcar, TS10 2DL |
| G0 | MBV | R Buckwell, 37 Scholars Gate, Guisborough, TS14 8LT |
| G0 | MBW | Malcolm Watkins, Llwyn Onn, Wainfelin Road, Pontypool, NP4 6DF |
| G0 | MBY | P Eaton, 3 Thirslet Drive, Heybridge, Maldon, CM9 4YN |
| G0 | MDZ | M Phillips, 14 Kingsclere Drive, Bishops Cleeve, Cheltenham, GL52 8TG |
| G0 | MCE | R Daw, Flat 11, Tong Court, Wolverhampton, WV1 1QQ |
| GM0 | MCJ | H Munro, Flat 6, Charlie Devine Court Bridge of Don, Aberdeen, AB22 8WG |
| G0 | MCM | M Sniezko-Blocki, 18 Westwoods Hollow, Burntwood, WS7 9AT |
| G0 | MCO | D Belcher, 7 Bower End, Chalgrove, Oxford, OX44 7YN |
| G0 | MCP | K Scott, 16 Hawton Road, Newark, NG24 4QD |
| G0 | MCQ | M Quicke, 53 Newfield Avenue, Farnborough, GU14 9PJ |
| G0 | MCT | Robert Craig, 109 Fordfield Road, Sunderland, SR4 0DD |
| G0 | MCV | S Morley, Mill Lane, Sileby, Loughborough, LE12 7UX |
| GM0 | MDD | John Clough, Obo Largs & District Ars, Redbank, Skelmorlie, PA17 5DX |
| G0 | MDJ | K Smithyes, Roseville, Childs Ercall, Market Drayton, TF9 2DG |
| G0 | MDK | C Hobson, 1 Martindale Avenue, Wimborne, BH21 2LE |
| G0 | MDM | R Robbins, 3 North Approach, Watford, WD25 0EH |
| G0 | MDN | B Millward, 5 Regency Close, Weddington, Nuneaton, CV10 0DF |
| G0 | MDO | Donald Ward, 10 Bircham Close, Bingley, BD16 3DY |
| G0 | MDQ | Paul Firmstone, 1 Holly Grange, Rhoswiel, Oswestry, SY10 7TU |
| G0 | MDR | F Lupton, 51 Bullens Green Lane, Colney Heath, St. Albans, AL4 0QR |
| G0 | MDV | Mark Bellas, 3 Elm Terrace, Penrith, CA11 7JY |
| GM0 | MDX | W Dempster, 124 Chatelherault Crescent, Hamilton, ML3 7PW |
| G0 | MEA | Philip Smith, 7 Prospect Drive, Keighley, BD22 6DD |
| G0 | MEC | M Spurgeon, 11 Homestead Road, Bodicote Chase, Banbury, OX16 9TW |
| G0 | MEE | B Shelton, 12 Meadowlands, Blundeston, Lowestoft, NR32 5AS |
| G0 | MEF | M Frear, 18 Boulsworth Road, Preston Grange, North Shields, NE29 9EN |
| G0 | MEN | A Fitzgerald, 39 rue Marcel Miquel, Issy-Les-Moulineaux, France, 92130 |
| G0 | MEO | R Davis, 17 Welbourne Close, Raunds, Wellingborough, NN9 6HE |
| G0 | MEQ | H Rigby, 33 Herne Rise, Ilminster, TA19 0HH |
| G0 | MEV | J Thorndyke, 23 Fordhams Close, Stanton, Bury St. Edmunds, IP31 2EE |
| G0 | MEW | L Whiteside, 9 Nutfield Gardens, Ilford, IG3 9TB |
| G0 | MEX | H Horne, 410 Bacup Road, Waterfoot, Rossendale, BB4 7JA |
| G0 | MEY | M Coulter, 52 Pine Close, Brant Road, Lincoln, LN5 9UT |
| G0 | [illegible] | [illegible] |
| G0 | MFD | J Keeley, Edge View Close, Grindleford, Hope Valley, S32 2[illegible] |
| G0 | MFH | P Lee, 50 [illegible] Road, Liverpool, L26 2HB |
| G0 | [illegible] | [illegible] Lostwithiel, PL[illegible] 0JD |
| G0 | MFR | Gareth Ayre, Emmotts Hay, Hartgrove, Shaftesbury, SP7 0LR |
| G0 | MFT | Graham Drake, The Bungalow, Church Lane, Wisbech, PE13 5LG |
| G0 | MFV | T Shelley, 4 Richens Drive, Carterton, OX18 3XT |
| G0 | MFY | L Choong, 14 School Close, Stevenage, SG2 9TY |
| G0 | MGC | G Clark, Holly Cottage, New Road, Bristol, BS35 4DX |
| G0 | MGH | A Strevens, 14 Larchfield Way, Horndean, Waterlooville, PO8 9HE |
| G0 | MGI | M Goodall, 2 Meadow Court, Littleport, Ely, CB6 1JW |
| G0 | MGJ | K Hancock, 12 Westmorland Close, Stoke-on-Trent, ST6 6UR |
| G0 | MGL | G Lees, 68 Green Lane, Oldham, OL8 3BA |
| G0 | MGM | R Dunne, 8 Telston Close, Bourne End, SL8 5TY |
| G0 | MGN | John Brand, 38 Canterbury Gardens, Hadleigh, Ipswich, IP7 5BS |
| GM0 | MGO | D Macdonald, 22 Christie Place, Elgin, IV30 4HX |
| GW0 | MGQ | Gordon Budge, 60 Bryntirion Road, Pontlliw, Swansea, SA4 9EB |
| G0 | MGT | D Carruthers, 19 Creek View Avenue, Hullbridge, Hockley, SS5 6LU |
| G0 | MGU | C Brown, The Cottage, Tylers Road, Harlow, CM19 5LJ |
| G0 | MGX | Mark Jones, 8 Stanton Avenue, Belper, DE56 1EE |
| G0 | MHA | P Bakrania, Bina, 31 South Priors Court, Northampton, NN3 8LD |
| GI0 | MHB | P Mckee, 168 Ballynamoney Road, Lurgan, Craigavon, BT66 6LD |
| G0 | MHC | GEORGE FORD, Thornley Road, Trimdon Station, TS29 6DA |
| GM0 | MHD | P Overton, Cluanie, Cairnballoch, Alford, AB33 8HQ |
| GM0 | MHE | B HYDE, The Cottage, Killiechronan, Isle of Mull, PA72 6JU |
| G0 | MHF | J Bisson, 14 Howbeck Drive, Oxton, Prenton, CH43 6UY |
| G0 | MHK | P Lee, 22 Bron Y Graig, Bodedern, Holyhead, LL65 3SY |
| G0 | MHN | David Tebay, 19 St. Johns Road, Newport, PO30 1LN |
| G0 | MHR | MID-HERTS RAY G c/o K Pollard, 5 Lodge Way, Stevenage, SG2 8DB |
| GM0 | MHS | D Rendall, Aranthrue, 17 Scapa Crescent, Kirkwall, KW15 1RL |
| G0 | MHY | Richard Jones, 29 Gray Road, RR #2, Ontario, Canada, K0K 2Y0 |
| G0 | MHZ | R Pardoe, 138 Fowler Road, Aylesbury, HP19 7QJ |
| G0 | MIA | C Murray, Brookfield, Collaroy Road, Thatcham, RG18 9PB |
| G0 | MIB | D Hussey, 15 The Ridings, Telscombe Cliffs, Peacehaven, BN10 7EF |
| G0 | MIC | M Constantine, 19 Elmcroft, Fairview Avenue, Woking, GU22 7NX |
| G0 | MID | R Jeffery, 3 New Road, Paddock Wood, Tonbridge, TN12 6HP |
| G0 | MIE | R Ebbs, 25 Foxtail Close, Gloucester, GL4 6DW |
| G0 | MIF | I Buckle, 28 Leybourne Road, Rochester, ME2 3QG |
| G0 | MIG | N May, Spring Barn, Eastwood Park, Wotton-under-Edge, GL12 8DA |
| G0 | MIH | Paul Swift, Midway, Norris Green, Callington, PL17 8DF |
| G0 | MIJ | C Nolan, 95 Strodes Crescent, Staines, TW18 1DG |
| G0 | MIK | Michael Ritson, 24 Chapel Road, Pawlett, Bridgwater, TA6 4SH |
| GM0 | MIS | S Buchanan, 7 Eilean Rise, Ellon, AB41 9NF |
| G0 | MIT | George Bromfield, 63 Herondale Road, Mossley Hill, Liverpool, L18 1JZ |
| GM0 | MIW | A Mcnicol, The Glebe House, Arbirlot, Arbroath, DD11 2NX |
| G0 | MIX | Malcolm Jones, 15 Quadrant Close, Murdishaw, Runcorn, WA7 6DW |
| G0 | MIZ | J Munro, Flat 3/4, 24 The Strand, Ryde, PO33 1JD |
| G0 | MJA | G Taylor, 32 Main Road, Great Holland, Frinton-on-Sea, CO13 0JL |
| G0 | MJB | R Daynes, 25 Redwood Close, Keighley, BD21 4YG |
| G0 | MJC | A Keeble, 5 Thistledown Road, Horsford, Norwich, NR10 3ST |
| G0 | MJF | M Weaver, 91 Mantle Street, Wellington, TA21 8BB |
| G0 | MJG | S Cartlidge, 19 Thornfield Road, Crosby, Liverpool, L23 9XY |
| G0 | MJJ | Leonard Challis, 30 London Road, Kirton, Boston, PE20 1JA |
| G0 | MJK | D Linnell, 19 Beech Avenue, Northampton, NN3 2HE |
| G0 | MJL | G Jennings, Ardwyn, Poolway Road, Coleford, GL16 7BE |
| G0 | MJO | Gerald Lucas, Flat 3, The Gateway, 2 Wilderton Road West, Poole, BH13 6EF |
| G0 | MJP | Richard Davies, 59 Gaunts Way, Letchworth Garden City, SG6 4PL |
| GM0 | MJR | E Baviello, 18 Glaskhill Terrace, Penicuik, EH26 0EL |
| G0 | MJT | M Tandy, 10 Palace Close, Rowley Regis, B65 9LG |
| G0 | MJU | A Moss, 67 Henacre Road, Kingsbridge, TQ7 1DP |
| G0 | MJV | A Williams, 16 Hoy Crescent, Seaham, SR7 0JT |
| G0 | MJX | J Harper, 109 Baxter Avenue, Kidderminster, DY10 2HB |
| G0 | MJY | D Gourley, 86 Upton Road, Kidderminster, DY10 2YB |
| G0 | MJZ | John Edwards, 42 Tenterfield Road, Ossett, WF5 0RU |
| G0 | MKA | T Chapman, 17 Trevor Road, Swinton, Manchester, M27 0YH |
| G0 | MKC | W Dunn, 25 Magdalene Court, Seaham, SR7 7DJ |
| G0 | MKD | Kenneth Davenport, 6 Malpas Road, Runcorn, WA7 4AD |
| G0 | MKE | J Attwell, 93 Manor Way, Peterlee, SR8 5RS |
| G0 | MKG | R Watson, Fairfield House, High Street, Boston, PE20 3LH |
| G0 | MKK | M Stockdale, First Floor, 22 Commercial Street, Harrogate, HG1 1TY |
| G0 | MKL | Robert Chell, 3 Elderberry Close, Stourport-on-Severn, DY13 8TF |
| G0 | MKN | K Brady, 17b Furzefield Road, Welwyn Garden City, AL7 3RL |
| G0 | MKP | N Crice, 13 Norbroom Drive, Newport, TF10 7RG |
| G0 | MKU | G Langford, 11 Nearhill Road, Kings Norton, Birmingham, B38 8LB |
| G0 | MKW | A Jones, 26 Clarendon Street, Bloxwich, Walsall, WS3 2HT |
| G0 | MKY | J Herries, Elmfold, Witney, OX8 6PZ |
| G0 | MKZ | T Pougher, 8 Wensleydale, Hull, HU7 6DE |
| G0 | MLB | Kenneth Walters, 14 Varo Terrace, Stockton-on-Tees, TS18 1JY |
| G0 | MLC | W Lowe, 54 St. Lesmo Road, Edgeley, Stockport, SK3 0TX |
| G0 | MLE | D Sabin, 1 West Nolands, Nolands Road, Calne, SN11 8YD |
| G0 | MLF | Paul Marshall, 31 Landseer Avenue, Northfleet, Gravesend, DA11 8NN |
| G0 | MLH | T Stilgoe, 37 Marlene Croft, Chelmsley Wood, Birmingham, B37 7JJ |
| G0 | MLJ | Mark Jones, 21 Fisherton Street, Salisbury, SP2 7SU |
| G0 | MLL | J Lyons, 40 Waddington Avenue, Burnley, BB10 4LB |
| G0 | MLM | T Leeman, 5 Serlby Rise, Nottingham, NG3 2LS |

**IMPORTANT NOTE**

**Revalidate licence to avoid revocation** – Ofcom has advised the Society that plans will be drawn up to revoke licences that have not been revalidated as required by the licence conditions. The quickest way to revalidate is to do so online via the Ofcom website: *https://services.ofcom.org.uk/* or by email: *amateur.validations@ofcom.org.uk* Ofcom staff are available to help, but please be patient during times of heavy workload.

GW0 MLN E Jones, 55 Blackoak Road, Cyncoed, Cardiff, CF23 6QU
G0 MLO K Packard, 19 Elm Drive, Rayleigh, SS6 8AB
G0 MMA K Plumridge, Flat 23, Barton Court, Tewkesbury, GL20 5RL
G0 MMB P Evans, 138 Pont Adam Crescent, Ruabon, Wrexham, LL14 6EG
G0 MMC J Cuthill, 17 Elmwood Drive, Keighley, BD22 7DN
G0 MMD G Shelton, 29 Viking Road, Northfleet, Gravesend, DA11 8ET
G0 MMH Peter Walker, 11 Flixton Drive, Crewe, CW2 8AP
G0 MMI C Underhill, 5 Grove Way, Waddesdon, Aylesbury, HP18 0LH
G0 MMJ D Wilkins, 18 Garendon Road, Loughborough, LE11 4QD
G0 MMO N Laud, 3 Woodlands, Wirksworth, Matlock, DE4 4PG
G0 MMQ Harry Dadak, 3 Cadogan Close, Holyport, Maidenhead, SL6 2JS
G0 MMT L Catherall, New Haven, Peckforton Hall Lane, Tarporley, CW6 9TF
G0 MMW L Roberts, 12 Deveron Close, Plymouth, PL7 2YF
G0 MMX David Hebden, 128 George Street, Shaw, Oldham, OL2 8DR
G0 MMY William Ellis, Broad Oak Cottage, Llyndir Lane, Burton, Wrexham, LL12 0AU
G0 MNA A Munir, 39 Gulberg V, Lahore, Pakistan
G0 MNC John Williams, 34 Brassington Street, Clay Cross, Chesterfield, S45 9NH
G0 MND T Rogers, 40 Rowell Way, Sawtry, Huntingdon, PE28 5WB
G0 MNH M Brown, 15 Hamilton Row, Waterhouses, Durham, DH7 9AU
G0 MNI Andy Carlile, Top Flat, 13b Mill Road, Cleethorpes, DN35 8HZ
G0 MNN Michael Franklin, 7 Auburn Close, Bridlington, YO16 7PN
G0 MNO N Bufton, 7 Laburnum Close, Rassau, Ebbw Vale, NP23 5TS
GW0 MNP Michael Butler, 1 Green Meadow, Cefn Cribwr, Bridgend, CF32 0BJ
GM0 MNV Raymond Gandy, 102 The Henge, Glenrothes, KY7 6XX
GM0 MNW K Carmichael, 8g Colonsay Terrace, Soroba, Oban, PA34 4YL
G0 MNY A Dagnall, 10 Rosebury Avenue, Leigh, WN7 1JZ
G0 MOF G Greenhalgh, 6 Clifton Grove, Rhyl, LL18 4AF
G0 MOH Robert Greaves, 2 Llys Dinas, Rhyl, LL18 4DZ
G0 MOI C Brigstocke, Pant-Y-Saer, Bwlch, Anglesey, LL74 8RG
G0 MOM S Kendall, 220 Marsh Street, Barrow in Furness, LA14 1BQ
GW0 MOQ N Brush, 25 Heol y Ffynnon, Efail Isaf, Pontypridd, CF38 1AU
G0 MOR S Morrey, 22 Wellpond Close, Sharnbrook, Bedford, MK44 1PL
G0 MOU R Clark, 20 Oakcroft Gardens, Littlehampton, BN17 6LT
G0 MOW R Harris, 25 Twynyffald Road, Blackwood, NP12 1HQ
GD0 MOX Daniel Gleek, 10 Castlereagh House, Lady Aylesford Avenue, Stanmore, HA7 4FP
G0 MPA Barry Coram, 18a Lake Green Road, Sandown, PO36 9HW
G0 MPI Stephen Sutcliffe, 142 Sandy Lane, Farnborough, GU14 9JQ
G0 MPJ B Osborne, 12 Arminers Close, Gosport, PO12 2HB
G0 MPK D Knights, 11 King Edward St., Kirton Lindsey, Gainsborough, DN21 4NF
G0 MPM John Claughton, 10 Witch Close, East Stour, Gillingham, SP8 5LB
G0 MPO A Neenan, 50 Middleton Road, Brownhills, Walsall, WS8 6JF
G0 MPP J Anderson, 93 Nab Wood Drive, Shipley, BD18 4EW
G0 MPQ J Wood, Garthmere, 4 Hunters Lane, Lincoln, LN4 4PB
G0 MPR Frederick Gibbons, 26 Tenbury Close, Bentley, Walsall, WS2 0NH
G0 MPT E Roddy, 701 Park Street, Stoughton, United States, 2072
G0 MPW J Woods, 26 Compton Road, Southport, PR8 4HA
G0 MPZ B Gifford, 20 Lewisham Road, Gloucester, GL1 5EL
G0 MQB J Lockyer, 19a Queens Avenue, Birchington, CT7 9QN
G0 MQC P Capewell, 191 Monyhull Hall Road, Birmingham, B30 3QN
G0 MQD Rebecca Field, 10 Somerville Close, Waddington, Lincoln, LN5 9QR
G0 MQE J Peirce, 600 Highland Avenue, Ottawa, Ontario, Canada, K2A 2K3
G0 MQH Derek Robinson, 2 Roman Close Castlefields, Runcorn, WA7 2JR
G0 MQI Robert Ingle, 16 Lintock Road, Norwich, NR3 3NU
G0 MQJ Peter Robinson, 92 Greasby Road, Greasby, Wirral, CH49 3NG
G0 MQK V Murton, 4 Cross Park Road, Wembury, Plymouth, PL9 0EU
G0 MQL Derek Franklin, 50 The Elms, Chatteris, PE16 6JN
G0 MQM M Hillier, 5 Sinodun Road, Didcot, OX11 8HP
GI0 MQN R Browning, 53 Caulside Park, Antrim, BT41 2DR
G0 MQR I Tuson, 6 Buffs Lane, Heswall, Wirral, CH60 2SG
G0 MQU Patrick Smyth, 19 North Avenue, Tredegar, NP22 3HE
G0 MQV N Cook, 17 Moorside Road, Richmond, DL10 5DJ
G0 MQW C Mcwhinnie, 32 Horse Close, Caversham, Reading, RG4 8TT
G0 MQX R Bowers, 54 Buxton Road, Dawley, Telford, TF4 2EW
G0 MRA E Southon, 20 Edinburgh Crescent, Kirton, Boston, PE20 1JT
G0 MRB R Broughton, 6 Lumley Place, Lincoln, LN5 7UT
G0 MRD D Gordon, 152 Oldham Road, Ashton-under-Lyne, OL7 9AN
G0 MRF David Bowman, 11 Crane Way, Twickenham, TW2 7NH
GM0 MRJ M Johnston, 27 Denholm Court, Glenrothes, KY6 1JP
G0 MRK J Kelly, 14 Arden Walk, Sale, M33 5NY
G0 MRL Laurence Bradshaw, 342 Manchester Road, Blackrod, Bolton, BL6 5BG
G0 MRM E Caligari, 209 Ormskirk Road, Upholland, Skelmersdale, WN8 0AA
G0 MRP Dave Pidgeon, 87 Suckling Green Lane, Codsall, Wolverhampton, WV8 2BY
G0 MRR Christopher Denton-Powell, 4 Korresia Walk, Bridgwater, TA5 2GT
G0 MRY Michael Hazzledine, 52 Springfield Road, Repton, Derby, DE65 6GP
G0 MRZ Brian Rowell, 73 Halsteads Road, Barton, Torquay, TQ2 8HB
G0 MSA A Hagland, 11 Coppice View, Heathfield, TN21 8YS
G0 MSF G Obey, 51 Chichester Close, Murdishaw, Runcorn, WA7 6DQ
GI0 MSG Thomas Mc Geown, 1 Drumcairn Road, Armagh, BT61 7SA
GI0 MSH Desmond McElroy, 81 Keady Road, Armagh, BT60 3AA
GI0 MSI ERIC NESBITT, 47 Mossfield, Glenanne, Armagh, BT60 2JF
GI0 MSK H Rattray, 20 Charlemont Gardens, Armagh, BT61 9BB
G0 MSO A Webb, 12 Forthlin Road, Allerton, Liverpool, L18 9TN
G0 MSR S Rutt, 3 Russell Place, Highfield, Southampton, SO17 1NU
G0 MSS J Taft, 8 Dresden Close, Mickleover, Derby, DE3 0RD
G0 MST John Scotter, 14 Tennyson Gardens, Fareham, PO16 7NW
G0 MSW E Goodwin, Tremayne, 11 Duchess Road, Monmouth, NP25 3HT
G0 MSY H Duggan, 41 Maesglas Road, Newport, NP20 3DE
G0 MSZ J Lycett, 24 Milbank Court, Darlington, DL3 9PF
G0 MTA F Bligh, 6 Woodlands Road, Bickley, Bromley, BR1 2AF
G0 MTB Paul Pearson, 53 Station Road, Dersingham, King's Lynn, PE31 6PR
G0 MTD S Topping, 7 Beckstone Close, Harrington, Workington, CA14 5QR
GI0 MTE P Robinson, 8 Annaboe Road, Kilmore, Armagh, BT61 8NP
G0 MTF G Sanders, 18 Impey Close, Thorpe Astley, Leicester, LE3 3SW
G0 MTI M White, 52 James St., Trethomas, Newport, NP1 8FY
G0 MTJ J Boothroyd, Quince Cottage, Church Lane, Ashford, TN26 1LS
G0 MTK I Chapman, 4312 Cedar Valley Drive, Plano, United States, TEXAS 75024
G0 MTN Lee Volante, Richmond House, Icknield Street, Birmingham, B38 0EP
G0 MTP Anthony Owen, 26 Gresham Street, Coventry, CV2 4EU
G0 MTQ John Baker, Moffat House, Church Road Broughton Moor, Maryport, CA15 7SS
G0 MTT R Williamson, 47 Ochre Dike Walk, Rotherham, S61 4DL
G0 MTV D Wright, Blakey Ridge, 2 Abbey Gardens, Wimborne, BH21 2EA
G0 MTW Francis Puffett, 4 Littlecote Close, Bishops Cleeve, Cheltenham, GL52 8TH
G0 MTY P Stunden, 3 Sunnybower Close, Blackburn, BB1 5QU
G0 MUC Simon Markwick, 5 Preston Road, Toddington, Dunstable, LU5 6EG
G0 MUD CHRISTCHURCH ARS c/o D Layne, 5 Howe Close, Christchurch, BH23 3JA
G0 MUH S Riley, 7 Crow Wood Avenue, Burnley, BB12 0JG
G0 MUJ Stewart Spink, 12 Chaucer Court, Ewelme, Wallingford, OX10 6HW
G0 MUK R Sanders, 36 Queens Road, Waterlooville, PO7 7SB
G0 MUN Adrien Collins, Flat 7, Haydon Court, Newton Abbot, TQ12 1GQ
G0 MUQ R Surrage, 80 Birch Road, Farncombe, Godalming, GU7 3NU
G0 MUR Robin Garrett, 27 Victoria Park Road, Buxton, SK17 7PU
G0 MUZ J Lockyer, 1 Rectory Cottage, The Common, Wallingford, OX10 6HP
G0 MVC C Neil, 208 Stonelow Road, Dronfield, S18 2ER
G0 MVE M Storkey, 9 Waterman Court, Acomb, York, YO24 3FB
G0 MVM A Frost, 15 Church Street Lacock, Chippenham, SN15 2LB
G0 MVP Martin Isted, 62 Chippers Road, Worthing, BN13 1DG
G0 MVR M Bentley, 1 Cotswold Road, Lupset, Wakefield, WF2 8EL
GW0 MVS D Clarke, 146 Clydach Road, Morriston, Swansea, SA6 6QB
G0 MVT W Brindley, 41 Boon Hill Road, Bignall End, Stoke-on-Trent, ST7 8LA
G0 MVU J Ellis, 41 Derby Road, Talke, Stoke-on-Trent, ST7 1SG
G0 MVV C Howes, 8 Alder Way, Hazel Slade, Cannock, WS12 0SX
G0 MVW W Barker, Fieldhead, School Lane, Thame, OX9 2NE
G0 MVX John Reeves, 5 Arrows Crescent, Boroughbridge, York, YO51 9LP
G0 MVY James Donnett, 51 Beaumont Avenue, St. Albans, AL1 4TT
G0 MWA D Sadler, Maritime Mobile, Ross Street, Opua, New Zealand
G0 MWE R Woodward, 22 Maryport Road, Dearham, Maryport, CA15 7EG
G0 MWH R Atkins, 24 Hill House Road, Norwich, NR1 4BE
GM0 MWJ D Robertson, 73 Kettilstoun Mains, Linlithgow, EH49 6SH
GD0 MWL Alan Crowther, 3 Lime Street, Port St. Mary, Isle of Man, IM9 5ED
G0 MWM E Bailey, 8 Blackthorn Close, Thornton-Cleveleys, FY5 2ZA
GW0 MWN David Harries, Rhydiau, Pencader, SA39 9BY
G0 MWS S Mcphee, 19 Lyttelton Road, Stourbridge, DY8 3RP
G0 MWT CHELMSFORD ARS c/o C Page, 1 The Leeway, Danbury, Chelmsford, CM3 4PS
G0 MWU B Pebody, 30 High St., Oakfield, Ryde, PO33 1EL
G0 MWV David Campbell, The Vicarage, High Street, Banbury, OX17 1NG
G0 MWW C Murt, 17 Drake Road, Padstow, PL28 8ES
G0 MWX C Hanks, 12 Pitsham Wood, Midhurst, GU29 9QZ
G0 MWY A Morgan, 31 Brindley Avenue, Latchford, Warrington, WA4 1RU
G0 MWZ R Rutherford, 26 St.Golder Road, Newlyn Coombe, Penzance, TR18 5QW
G0 MXB D Burnett, Cloonmaghaura, Williamstown, Ireland, 907
G0 MXD D P Wroe, Tylers House, Coton, Whitchurch, SY13 3LT
G0 MXE F Jennings, 2 Hickman Close, West Beckton, London, E16 3TA
GW0 MXG Pamela Taylor, 2 Pen-y-Dre, Caerphilly, CF83 2NZ
G0 MXH David Kay, 8 Meadowbrook Close, Lostock, Bolton, BL6 4HX
GM0 MXP R Park, The Loft, Front Street, Dunblane, FK15 9PX
GI0 MXT G Browne, 43 Northwood Drive, Belfast, BT15 3QP
G0 MXU Shaun Smith, 99 Greenwood, Bamber Bridge, Preston, PR5 8JX
G0 MXW D Houghton, 127 Melwood Drive, Liverpool, L12 8RN
G0 MXX B Clarke, Linden Cottage, School Lane, Nottingham, NG12 3FD
G0 MXY C Erratt, 60 Allen Court, Ridding Lane, Greenford, UB6 0JZ
GM0 MXZ T Goody, Lambs' Park, Forgandenny, Scotland, PH2 9HS
G0 MYA A Gray, 57 Dominie Cross Road, Retford, DN22 6NH
G0 MYC R Clifton, Heathwood, Thrigby Road, Great Yarmouth, NR29 3HJ
G0 MYD G Hughes, 292 Mount Pleasant, Redditch, B97 4JL
G0 MYH J Foster, 5 Jacobs Close, Glastonbury, BA6 8EJ
G0 MYJ D Sutherland, Devonia, Hamble Lane, Southampton, SO31 8EL
G0 MYL H Pigott, Brook Farm, Shobley, Ringwood, BH24 3HT
G0 MYM P Harrison, 16 Bodiham Hill, Garforth, Leeds, LS25 2LF
G0 MYN W Hughes, 60 Pineways, Appleton, Warrington, WA4 5EJ
G0 MYP D Wordsworth, 15 Tetbury Drive, Bolton, BL2 5NP
GM0 MYQ J Frati, 10 Benbecula Road, Aberdeen, AB16 6FU
G0 MYR Robin Worsley, Omaru, Higher Pennance, Redruth, TR16 5TQ
GM0 MYV I Reid, C/O Hamish Reid Tigh Au Rhuda, Luib, Isle of Skye, IV49 9AN
G0 MYX R Stephen, 97 Hunters Field Stanford in the Vale, Faringdon, SN7 8ND
GM0 MZD A Coutts, 37 West High Street, Bishopmill, Elgin, IV30 4DJ
G0 MZF B Nicholson, Flat 36, 154 Goswell Road, London, EC1V 7DX
GM0 MZH Robert Wallace, 80 Tourhill Road, Kilmarnock, KA3 2DA
G0 MZJ K Earp, 18 Main St., Kings Newton, Derby, DE73 1BX
G0 MZK R Little, Mythe House, 129 Slad Road, Stroud, GL5 1RD
G0 MZN J Nunn, 20 Somerton Gardens, Earley, Reading, RG6 5XG
G0 MZP R Kay, 7 Alderson Road, Worksop, S80 1UZ
G0 MZQ W Greed, 5 West View, Creech St. Michael, Taunton, TA3 5QP
G0 MZY Frank Woodhall, 13 Whitegate Drive, Clifton, Manchester, M27 8RE
G0 MZZ Anthony Benson, 1 Oxford Close, Gomersal, Cleckheaton, BD19 4RU
G0 NAA A Leake, Thorpe Garth, East Newton, Hull, HU11 4SD
GM0 NAE J Carlin, 24 Hillcrest Avenue, Paisley, PA2 8QW
GM0 NAI Jim Fisher, High Birches, Culbokie, Dingwall, IV7 8JS
G0 NAJ J Neary, 266 Yew Tree Lane, Dukinfield, SK16 5DN
G0 NAP P Howell, 29 South View Park, Plymouth, PL7 4JE
GM0 NAQ G Furmage, 25 Craigton Crescent, Alva, FK12 5DS
G0 NAR Christopher Fenton, Oakwell, Newtons Hill, Hartfield, TN7 4DH
G0 NAS L Aykroyd, 3 Bank Cottages, Orton Road, Penrith, CA10 3TW
G0 NAU M Best, 81 Maybury Road, Hull, HU9 3LB
G0 NAX D Edmonds, 1 Ashtree Close, Chelmsford, CM1 2RR
GM0 NAZ A Heggie, 75 Doon Walk, Craigshill, Livingston, EH54 5AD
GM0 NBA T Adam, 33 Lambie Crescent, Newton Mearns, Glasgow, G77 6JU
G0 NBB M Watkins, 9 Benacre Road, Whitstable, CT5 4NY
G0 NBC Leon Steenvoorden, 1 Thornbury Road, Immingham, DN40 1HH
G0 NBD Alfred Brown, 20 Sheen Road, Wallasey, CH45 1HA
G0 NBE R Allen, 5 Crompton Grove, Stoke-on-Trent, ST4 8UZ
GM0 NBG J G McVittie, 19 Beech Way, Girvan, KA26 0BX
G0 NBH John Goodwin, Hankelow Court, Hall Lane, Crewe, CW3 0JB
G0 NBI Graham Coomber, 2 Bracken Grove, Catshill, Bromsgrove, B61 0PB
G0 NBJ Neil Foster, 20a Pear Tree Road, Ashford, TW15 1PW
GM0 NBM John Hayes, 11 Nairn Way, Cumbernauld, Glasgow, G68 0HX
G0 NBP A Stevens, Gate Farm Barns, Earthcott Green, Bristol, BS35 3TA
G0 NBW B Nolan, 4 Shetland Road, Blackpool, FY1 6LP
GI0 NCA Robert Pinkerton, 9 Cloghole Road, Campsie, Londonderry, BT47 3JW
G0 NCE Owen Wheeler, 7 Aveling Close, Hoo, Rochester, ME3 9BZ
G0 NCH R Magri, 13 Roebuck Road, Chessington, KT9 1JY
G0 NCL R Harris, 5 Campbell Close, High Wycombe, HP13 5XY
G0 NCO D Murray, 11 Blandy Road, Henley-on-Thames, RG9 1PH
G0 NCQ M Hemmings, 2 Holly Walk, Nuneaton, CV11 6UU
G0 NCS Clarence Healey, 22 Stirling Road, St. Budeaux, Plymouth, PL5 1PD
G0 NCT F Norman, 101 Central Avenue, Canvey Island, SS8 9QP
GW0 NCU S David, 142 Robert Street, Manselton, Swansea, SA5 9NH
G0 NCW A Judge, 106 Bicknor Road, Park Wood, Maidstone, ME15 9PD
G0 NCX R Hughesdon, 3 Lyndhurst Road, Gosport, PO12 3QY
G0 NCY R Hartley, 23 Broomfield Road, Fleetwood, FY7 7HA
G0 NDA Stephen Frost, Fron Hyfryd, Nebo, Caernarfon, LL54 6EW
G0 NDB Rosemary Evans, 51532 Range Road 224, Sherwood Park, Canada, T8C 1H5
G0 NDC R Little, Maranatha, Higher Moresk Road, Truro, TR1 1BW
G0 NDD J Jackson, 26 Wadham Close, Peterlee, SR8 2NN
G0 NDF C Peters, 347 Mile Oak Road, Portslade, Brighton, BN41 2RD
G0 NDS NORTHAMPTON ARG c/o J Johnson, 30 Millside Close, Kingsley, Northampton, NN2 7TR
G0 NDU J Dykes, 33 Mill House Drive, Cheltenham, GL50 4RG
G0 NDV Paul Timmins, 34 Lumley Avenue, Skegness, PE25 2TH
G0 NDY R Wayne, Colkirk House, Manor House Street, Horncastle, LN9 5HF
GW0 NDZ Geraint Davies, 4 Crichton Street, Treorchy, CF42 6DF
G0 NEB John Dorning, 51 Osullivan Crescent, Blackbrook, St. Helens, WA11 9RE
G0 NEC Vincent Fletcher, 8 Plas Newydd, Llandudno, LL30 1GD
G0 NED Eric Dudley, 4 Lake Croft Drive, Stoke-on-Trent, ST3 7SS
G0 NEE Michael Stott, Wellview, 12 Castle View, Prudhoe, NE42 6AT
G0 NEF Alister Chapman, 13 Clayton Grove, Bracknell, RG12 2PT
G0 NEM Michael Purser, 17 Firecrest Road, Chelmsford, CM2 9SN
G0 NEN G Lewin, 19 Stafford Street, Brewood, Stafford, ST19 9DX
G0 NEO J Boland, 28 Vicarage Road, Orrell, Wigan, WN5 7AX
G0 NEP P Whitling, 17 Balcomb Crescent, Margate, CT9 3XJ
G0 NEQ Aldridge and Barr Beacon ARC c/o Leslie Horton, 4 Summer Lane, Walsall, WS4 1DS
G0 NER S Stones, 38 Mill View, Ferrybridge, Knottingley, WF11 8SR
G0 NES Donald Bryant, 35 Truemans Heath Lane, Hollywood, Birmingham, B47 5QE
GM0 NET AYRSHIRE RAYNET GROUP c/o T Stewart, 104 Barrhill Road, Cumnock, KA18 1PU
G0 NEU A Dawson, 182 Ladysmith Road, Enfield, EN1 3AE
G0 NEV Michael Carter, 1 Mill Lane, Preston, Weymouth, DT3 6DE
G0 NFA D Gilbert, 2 Greenfield Cottages, Bentley, Farnham, GU10 5HZ
G0 NFB L Raynor, 19 West View, Doncaster Road, Worksop, S81 9RA
G0 NFE R Ransome, 66 Spencer Way, Stowmarket, IP14 1UQ
G0 NFG D Hopper, 28 Western Avenue, Herne Bay, CT6 8TU
G0 NFH John Acton, 63 Bevington Close, Patchway, Bristol, BS34 5NP
G0 NFI P Edwards, 59 Treffry Road, Truro, TR1 1WL
G0 NFJ H Garland, 9 Field Close, Abridge, Romford, RM4 1DL
G0 NFL M George, 2 Jubilee Terrace, Isham, Kettering, NN14 1HG
GD0 NFN John Butler, 15 Church Close, Lonan, Laxey, Isle of Man, IM4 7JY
G0 NFO R Charteris, 7 Kennedy Close, Kidderminster, DY10 1LR
G0 NFR Roy Glynn, 118 Pelham Road, Birmingham, B8 2PD
G0 NFV A Hunt, 197 North Walsham Road, Norwich, NR6 7QN
G0 NFY A Cordwell, 71 Tadcaster Road, Norton Woodseats, Sheffield, S8 0RA
G0 NFZ R Hodges, 33 Harnall Close, Shirley, Solihull, B90 4QR
G0 NGA P Golder, 282 Noak Hill Road, Basildon, SS15 4DE
G0 NGD C Stenbacka, 11 Mount View, Billericay, CM11 1HB
G0 NGE William Clarke, 20 Langdale Road, Leyland, PR25 3AR
G0 NGG R Brown, 20 King Edward Road, Stanford-le-Hope, SS17 0EF
G0 NGH A Pemberton, 3 Parkside Drive, Exmouth, EX8 4LA
G0 NGI D Johnson, 15 St. Michaels Road, Sheerwater, Woking, GU21 5PY
GM0 NGJ A Caldwell, 7 Gladstone Terrace, New Deer, Turriff, AB53 6TE
G0 NGK Paul Walmsley, Valley View, Longworth Ave, Chorley, PR7 4PJ
G0 NGN Chris Jordan, Green Lane Cottage, Leintwardine, Craven Arms, SY7 0NB
G0 NGP Peter Wilson, Jansil, Worthing Road, Littlehampton, BN17 6JN
G0 NGQ J Gibbs, 3 Holts Green, Great Brickhill, Milton Keynes, MK17 9AJ
G0 NGW R Ramplin, 17 Cross St., Langold, Worksop, S81 9SL
G0 NHB W Stallard, 28 Wheatfield Road, Stanway, Colchester, CO3 0YJ
GU0 NHD K Benton, Keukenhof, Route de Carteret, Castel, Guernsey, GY5 7YS
GW0 NHE A Smith, 7 Clos Gorsfawr, Grovesend, Swansea, SA4 4GZ
G0 NHG J Godwin, 22 Stonebeck Avenue, Harrogate, HG1 2BW
G0 NHJ J Robson, 35 Melling Road, Cramlington, NE23 6AS
G0 NHK K Robson, 35 Melling Road, Cramlington, NE23 6AS
GM0 NHL Colin Waldron, 24/2 Vennel Street, Dalry, KA24 4AF
G0 NHM N Robertshaw, Cherry Tree House, Church Street, York, YO26 8DD
G0 NHO K Crookes, 64 Heron Drive, Audenshaw, Manchester, M34 5QX
G0 NHP Thomas Maguire, 1 Gosford St., Balsall Heath, Birmingham, B12 9ER
G0 NHR Nunsfield House Amateur Radio Group c/o K Clarke, 55 Compton Avenue, Aston-on-Trent, Derby, DE72 2AU
GM0 NHT H Cherrie, 6 Milking Hill, Tong, Isle of Lewis, HS2 0HU
G0 NHZ A Pollard, 2 Forest Close, Cowplain, Waterlooville, PO8 8JE
G0 NID E Page, Flat 29, Westfields, 212 Hall Lane, Manchester, M23 1LP
G0 NIF S Wilkins, 1 Chaucer Court, 75 Wendover Road, Staines, TW18 3DW
G0 NIG Nigel Smith, 45 The Gills, Otley, LS21 2BY
G0 NIK Philip Gell, 25 Westland, Martlesham Heath, Ipswich, IP5 3SU
G0 NIL D Woods, 30 Longridge Avenue, Stalybridge, SK15 1HG
G0 NIN N Beer, 2 Marcelle Court, School Road, Hindhead, GU26 6LR
G0 NIQ J Naylor, 6 Mallard Close, Christchurch, BH23 4DD
G0 NIX W Jones, 13 Kilrush Terrace, Woking, GU21 5EG
GW0 NIY R Milton, 49 Heol Y Deri, Rhiwbina, Cardiff, CF4 6HD
G0 NJD R Mallett, 71 Olivet Road, Woodseats, Sheffield, S8 8QR

G0 NJG K Binns, 31a Dellands, Overton, Basingstoke, RG25 3LD
G0 NJJ N Jones, 19 Foxhollow, Bar Hill, Cambridge, CB23 8EP
GM0 NJL R Watson, 24 Hillock Avenue, Redding, Falkirk, FK2 9UT
G0 NJO Joseph Howard, 37 Carisbrooke Road, St. Leonards-on-Sea, TN38 0JN
G0 NJP Mitsuru Murakami, 5 6 Takamidai, Takatsuki, Osaka, Japan, 5691020
G0 NJQ P Schlatter, Churchgate House, Sutton Road, Maidenhead, SL6 9SN
G0 NJS M Davie, 101 Upperfield Road, Welwyn Garden City, AL7 3LR
G0 NJT John Perkins, Flat 2block 122, Flat 2. Leachgreen Lane Rednal, Birmingham, B45 8EH
G0 NJZ T Dodds, 33 Westgate, Warley, Oldbury, B69 1BA
G0 NKC Michael Connolly, 12 Vincent Way, Martock, TA12 6DG
GW0 NKG Mike York, 9 Fox Hollows, Brackla, Bridgend, CF31 2NE
GW0 NKH R Hearne, Mora, Rhydypandy Road, Morriston, SA6 6NX
GW0 NKJ D Davies, 6 Dulais Fach Road, Tonna, Neath, SA11 3JW
G0 NKK Andrew Dipper, 4 Venn Close, Instow, Bideford, EX39 4LJ
G0 NKM Herbert Monks, 100 Crossfield Road, Cheadle Hulme, Cheadle, SK8 5PF
G0 NKQ V Nunns, 8 Trevithick Road, Tregurra, Truro, TR1 1RU
G0 NKU S Fowler, 58 Buxton Road, High Lane, Stockport, SK6 8BH
G0 NKZ K Everard, Woodside, Staple Lane, Taunton, TA4 4DE
G0 NLA R Bryan, 23 Quarry Lane, Halesowen, B63 4PB
GW0 NLB W Rees, 51 Heol Capel Itan, Pontyberem, Llanelli, SA15 5HF
G0 NLG R Chapman, 49 Walden Way, Frinton-on-Sea, CO13 0BH
G0 NLJ S Oliver, Chalk Lodge, Peters Lane, Princes Risborough, HP27 0LG
G0 NLL James Bowers, 35 Peveril Gardens, Newtown, Stockport, SK12 2RG
G0 NLM Colin Ridley, 20 Victoria Gardens, Ferndown, BH22 9JH
G0 NLN D Ashworth, 59a Normanby Road, London, NW10 1BU
G0 NLO G Wharton, Onanole, Clitheroe Road, Clitheroe, BB7 3DA
G0 NLQ P Dunn, 10 Endsleigh Close, Upton, Chester, CH2 1LX
G0 NLT D Wentworth, 7 Gilbeys Close, Stourbridge, DY8 4XU
GM0 NLU Neil Harvey, The Shieling, Tealing, Dundee, DD4 0QU
G0 NLV F Fairman, 26 Marina Gardens, Cheshunt, Waltham Cross, EN8 9QY
G0 NLX S Pearson, 18 New Road, Amersham, HP6 6LD
G0 NMB Arnold Haberman, 4 Allendale Drive, Copford, Colchester, CO6 1BP
G0 NMC T Neal, Sunnymead, Ewyas Harold, Hereford, HR2 0JA
G0 NMD Leslie Austin, 7 Kennedy Close, Chester, CH2 2PL
G0 NMH Brian Markey, 164 Bourn View Road, Netherton, Huddersfield, HD4 7JS
G0 NMJ John Denniss, 61 Checkstone Avenue, Bessacarr, Doncaster, DN4 7JY
G0 NMP Patrick Chapman, 1 Bader Close, Watton, Thetford, IP25 6FF
G0 NMS John Howes, 39 Pound Hill, Bacton, Stowmarket, IP14 4LP
G0 NMY Mark Longson, 54 Beresford Street, Shelton, Stoke-on-Trent, ST4 2EX
G0 NNE R Hart, River View, Kilkewydd, Welshpool, SY21 8RT
G0 NNF J Yates, 87 Princess Road, Warley, Oldbury, B68 9PW
G0 NNG Robin Westley, 91 Lincoln Way, Daventry, NN11 4SU
GI0 NNK John Cairns, 10 Dunsilly Road, Antrim, BT41 2JH
G0 NNN W Forbes, 29 High Street, Maryport, CA15 6BQ
G0 NNO M Shore, 12 Boscoppa Road, St. Austell, PL25 3DR
G0 NNR B Thomas, Creekside, Greenbank Road, Truro, TR3 6PQ
G0 NNS NORWICH NORTH SCOUTS FELLOWSHIP ARC c/o Colin Hendry, 109 Grove Avenue, New Costessey, Norwich, NR5 0HZ
G0 NNT Victor Martinelli, 62 St. Angelo Street, Sliema, Malta, SLM1334
G0 NNU L Payne, 147 Upper Marehay, Ripley, DE5 8JG
G0 NNZ J Belfield, 17 Burtondale Road, Crossgates, Scarborough, YO12 4JR
G0 NOB Lionel Leek, The Carriage House, Kiln Road, Paignton, TQ3 1SH
G0 NOH J Reed, 23 Morehall Avenue, Folkestone, CT19 4EQ
G0 NON T Behan, Maytree Cottage, Marley Lane, Haslemere, GU27 3RG
G0 NOO S Coburn, 54 Queensway, Hope, Wrexham, LL12 9PE
G0 NOP P Coburn, 54 Queensway, Hope, Wrexham, LL12 9PE
G0 NOU W Ayers, Peach Lodge, Foolow, Hope Valley, S32 5QB
G0 NOV Per Adams, 25 Appleton Avenue, Great Barr, Birmingham, B43 5LY
GI0 NOX Sean McAteer, 33 Knocknamuckly Lane, Portadown, Craigavon, BT63 5PF
G0 NPA R Stephens, 50 Windrush Way, Abingdon, OX14 3SX
G0 NPC G Hill, 1 Gleneagles Court, Edwalton, Nottingham, NG12 4DN
G0 NPE P Hulme, 92 London Road, Cowplain, Waterlooville, PO8 8EW
G0 NPF D Delacassa, The Tree House, Easthams Road, Crewkerne, TA18 7AQ
G0 NPG K Heaviside, 58 Arundel Drive, Ranskill, Retford, DN22 8PQ
G0 NPI J Podvoiskis, 3 Barnview Drive, Irlam, Manchester, M44 6WY
G0 NPJ L Jackson, 60 East Park Avenue, Darwen, BB3 2SQ
G0 NPK Douglas Goulbourne, Widowscroft Farm, Hollingworth, Hyde, SK14 8LE
G0 NPL Stuart Instone, 61 Llanfach Road, Abercarn, Newport, NP11 5LA
G0 NPM H Thomas, 34 Upland Road, Pontllanfraith, Blackwood, NP12 2ND
G0 NPN B Harris, 23 Pound Road, Highworth, Swindon, SN6 7LA
G0 NPO I Brown, Egremont, Arterial Road, Basildon, SS14 3JN
G0 NPP T Watson, 20 Ivanhoe View, Gateshead, NE9 7TR
G0 NPQ H Carruthers, 55 Inskip Terrace, Gateshead, NE8 4AJ
G0 NPU J Marshall, 28 The Dovecote, Horsley, Derby, DE21 5BS
G0 NPV B Ham, 37 Lower Moor, Barnstaple, EX32 8NW
G0 NPW M Ambach, Karwendelstrabe 7, Tyrol, Austria, A-6130
G0 NPY P Yates, 31 Wallpark Close, Brixham, TQ5 9UN
G0 NQA Alan Gurbutt, 16 Crabtree Lane, Sutton-on-Sea, Mablethorpe, LN12 2RT
GI0 NQC Adrian Dornford-Smith, 4 Rusheyhill Road, Lisburn, BT28 3TD
G0 NQE C Wilkinson, 8 Westfield Avenue, Knottingley, WF11 0JH
G0 NQG Stuart Latham, 27 Rockside Gardens, Frampton Cotterell, Bristol, BS36 2HL
G0 NQI J Shepherd, Yew Tree Cottage, The Stenders, Mitcheldean, GL17 0JF
G0 NQJ D Scaplehorn, 9 Stockwell Avenue, Mangotsfield, Bristol, BS16 9DR
G0 NQK R Edwards, Manor Cottage, Manor Road, Ipswich, IP7 6PN
G0 NQN T Fricker, 8 Folly View, Necton, Swaffham, PE37 8LU
G0 NQV D Litchfield, 37 Graeme Road, Enfield, EN1 3UU
G0 NQW D Marshall, 15 Whisby Court, Holton-le-Clay, Grimsby, DN36 5BG
G0 NQY S Seggar, 145 Mount View Road, Norton, Sheffield, S8 8PJ
G0 NRA Gerald Lowe, 25 Manor House Court, Kirkby-in-Ashfield, Nottingham, NG17 8LH
G0 NRB R Bellamy, 4 Wimbourne Walk, Corby, NN18 0BN
G0 NRF Gary Stilgoe, 47 Chesterton Close, Redditch, B97 5XS
G0 NRG NORTHAMPTON RAYNET c/o E Young, 1 Ashton Road, Roade, Northampton, NN7 2LF
G0 NRI William Hilton, 5 North Street, Williton, Taunton, TA4 4SL
G0 NRJ R Croucher, 17 Sundridge Road, Woking, GU22 9AU
G0 NRK James Butler, 14 Fairfield Road, Barnard Castle, DL12 8EB
G0 NRM Roy Stout, 7 Thornbridge Drive, Sheffield, S12 4YF
G0 NRN G Harrison, 14 Hardy Avenue, South Ruislip, Ruislip, HA4 0GX
GM0 NRT W Gardno, 52 Salisbury Terrace, Aberdeen, AB10 6QH
G0 NRW S Chadwick, 27 Retail Farm Estate, Four Mile Bridge, Holyhead, LL65 2EX
G0 NRX S Godbold, 13 Dawn Crescent, Upper Beeding, Steyning, BN44 3WH
G0 NRZ A Pill, 5 St. Leonards Close, Upton St. Leonards, Gloucester, GL4 8AL
G0 NSA T Brown, 5 St. Valentines Close, Kettering, NN15 5EG
G0 NSC W Thornton, 46 Lavender court, Croft road, Barnsley, S70 3FG
G0 NSG Paul Crespel, Via Leopardi 6, Cavaion Veronese, Italy, 37010
G0 NSH Bruce Piggott, 00 Lawrence Close, Hertford, SG14 2HH
G0 NSI J Muir, 10 Cheriton Close, Barnet, EN4 9TX
G0 NSK Andrew Pennell, 90 Westbourne Avenue, Sunderland, SR2 9LQ
G0 NSL C Russell, 100 Halton Road, Runcorn, WA7 5RJ
G0 NSO T Barfield, 91 Ollerton Road, New Southgate, London, N11 2JY
G0 NSP B Teasdale, 18 Valley Forge, Washington, NE38 7JN
G0 NSW B Byrne, 00 Archer Road, Redditch, B98 8DJ
G0 NSZ J Swinden, Meadow View, Llanfair Road, Abergele, LL22 8DH
GD0 NTA A Jarvis, Willowmead, Nugents Park, Pinner, HA5 4RA
GD0 NTB B Jarvis, Willowmead, Nugents Park, Pinner, HA5 4RA
GJ0 NTD G Blake, 29 Pied Du Cotil, St. Andrews Road, St. Helier, Jersey, JE2 3JF
G0 NTG Dominic Chawner, 49 St. Anns Road, Middlewich, CW10 9BY
G0 NTH Alan Gardner, Flat 6, Eastcliff Court, Shanklin, PO37 6EJ
GM0 NTI Lindsey Grieve, Elhanan, Myrtlefield Lane, Inverness, IV2 5UE
G0 NTJ A Williams, 23 Lancaster Gardens, Aylsham, Norwich, NR11 6LB
GM0 NTL Ron Fraser, Hopefield Cottage, Gladsmuir, Tranent, EH33 2AL
GM0 NTR James Harrison, 17b High Street, Oban, PA34 4BG
G0 NTT L Lloyd, 8 Coastal Rise Hest Bank, Lancaster, LA2 6HJ
GM0 NTY D Rankin, 11 Drummuir Foot, Girdle Toll, Irvine, KA11 1NW
G0 NUA Kathleen Franklin, 7 Auburn Close, Bridlington, YO16 7PN
G0 NUD B Bell, 74 Henderson Road, Carlisle, CA2 4PZ
G0 NUH M Darling, 1 Roman Way, Highworth, Swindon, SN6 7BU
GM0 NUI R Honeyman, 81 Glen Avenue, Largs, KA30 8RH
G0 NUL D Forster, 33 Wigeon Lane, Walton Cardiff, Tewkesbury, GL20 7RS
G0 NUN R Barker, 5 Wickridge Close, Uplands, Stroud, GL5 1ST
G0 NUO G Du Feu, 17 Oak Road, Tavistock, PL19 9LJ
G0 NUP K Prince, 59 Chantry Road, East Ayton, Scarborough, YO13 9ER
GM0 NUQ R Handyside, 113 Stockiemuir Avenue, Bearsden, Glasgow, G61 3LX
G0 NUR A Bushell, 121 Rickmansworth Road, Watford, WD18 7JD
G0 NUS G Dyer, 15 Park Road, Newbridge, Newport, NP11 4RE
G0 NUT David McKay, 43 Mordales Drive, Marske-by-the-Sea, Redcar, TS11 7HT
G0 NUU R Browne, Fox Cottage, Bukehorn Road, Peterborough, PE6 0QG
G0 NUZ L Wildman, 22 Berrys Wood, Newton Abbot, TQ12 1UP
G0 NVA F Stainsby, 11 Stonehouse Park, Thursby, Carlisle, CA5 6NS
G0 NVC D Hoppe, 354a Bourne Road, Pode Hole, Spalding, PE11 3LL
G0 NVD J Nothard, Ashmount, Fockerby, Scunthorpe, DN17 4RZ
G0 NVJ S Winter, Flat 32, Eyot House, London, SE16 4BN
G0 NVM J Chandler, 14 Highfield Road, Chelmsford, CM1 2NQ
G0 NVO P Oldham, 59 Wellspring Dale, Stapleford, Nottingham, NG9 7ET
G0 NVS M Fletcher, 51 Greasley St., Bulwell, Nottingham, NG6 8NG
G0 NVT P Boyle, 99 Heath Road, Penketh, Warrington, WA5 2BY
G0 NVV G Price, 58 Hollowfields Close, Redditch, B98 7NR
G0 NVX Christopher Watts, 41 Salter Street, Berkeley, GL13 9BU
G0 NVY Peter Hanson, 10 Parkfield Road, Ruskington, Sleaford, NG34 9HS
G0 NVZ J Welsh, 17 New House Park, St. Albans, AL1 1UA
G0 NWC H Cooper, 404a Clipsley Lane, Haydock, St. Helens, WA11 0SX
G0 NWE G Egan, 11 Shepherds Row, Castlefields, Runcorn, WA7 2LG
G0 NWF S Webster, 31 Park Estate, Shavington, Crewe, CW2 5AW
GI0 NWG A Williamson, 23 Iskymeadow Road, Armagh, BT60 3JS
G0 NWH North West Hampshire RAYNET c/o John Long, 1 Tangway, Chineham, Basingstoke, RG24 8SU
GM0 NWI Andrew Cunningham, 33 Broom Court, Stirling, FK7 7UL
G0 NWJ G Blomeley, 13 Edale Grove, Sale, M33 4RG
G0 NWL H Jordan, 33 Earlsbourne, Church Crookham, Fleet, GU52 8XG
G0 NWM TYNEMOUTH ARC c/o Graham Errington, 22 Willoughby Drive, Whitley Bay, NE26 3DY
GI0 NWN M Coyle, 67 Glen Road, Londonderry, BT48 0BY
G0 NWR N.W.R.R.C. c/o Edward Shipton, 51 Maes Stanley, Bodelwyddan, Rhyl, LL18 5TL
G0 NWS A Edwards, 130 Bedowan Meadows, Tretherras, Newquay, TR7 2TB
G0 NWT THE NORTH NORFOLK ARG c/o L Nash, Four Furlongs, Wells Road, Wells-next-The-Sea, NR23 1QE
G0 NWV D Brown, 65 Warstones Drive, Penn, Wolverhampton, WV4 4PF
G0 NWY I Peters, 62 Kingston Avenue, Seaham, SR7 8NL
G0 NXA Giles Herbert, Quarry Cottage, 1 Dingle Lane, Tewkesbury, GL20 0DW
G0 NXC R Sillito, 25 Naisbett Avenue, Peterlee, SR8 4BW
G0 NXD Peter Brazenall, Flat 80, Clent Court, Dudley, DY1 2AZ
G0 NXE Frank Rogers, 5 Station Gardens, Eckington, Pershore, WR10 3EZ
G0 NXF D Robinson, 5 Hazel Grove, Welton, Lincoln, LN2 3JX
G0 NXH P Cunningham, 2 The Park, Mistley, Manningtree, CO11 2AL
G0 NXI L Edgecumbe, 51 Aller Park Road, Newton Abbot, TQ12 4NH
G0 NXL Brian Ewald, Sto.Nino 2, Lower Casili, Cebu, Philippines, 6001
G0 NXM R Najman, 9 Bevin House, Alfred Street, London, E3 2BB
G0 NXN B Mitchell, 2 Mariners Court, Great Wakering, Southend-on-Sea, SS3 0DR
GM0 NXO George Fyall, 105 St. Kilda Crescent, Kirkcaldy, KY2 6DR
G0 NXQ W Love, 2 Longmead Cottages, Milborne St. Andrew, Blandford Forum, DT11 0HU
G0 NXS A Ellis, High Nentsberry, Alston, CA9 3LZ
G0 NXT S Platts, 15 Holywell Avenue, Smisby, Ashby-de-la-Zouch, LE65 2HL
GW0 NXW George Scanlin, 9 West Walk, Barry, CF62 8BY
G0 NXX J Lynch, 14 The Pastures, Cayton, Scarborough, YO11 3UU
G0 NXY Paul Martin, 48 Mill Lane, Fazeley, Tamworth, B78 3QD
G0 NYD Jonathan Burrow, 25 Hunting Hill Road, Carnforth, LA5 9JQ
G0 NYE J Dixon, 17 Marlowe Close, East Hunsbury, Northampton, NN4 0QQ
G0 NYH J Moseley, 42 Burford Road, Chipping Norton, OX7 5DZ
GI0 NYI John Benson, 18 Alexander Avenue, Armagh, BT61 7JD
G0 NYJ Sy Au, Flat 2, 1st Floor, Block C, Greenland Garden, Tuen Mun, N.T., Hong Kong
G0 NYK J Hoose, 91 Brevere Road, Hedon, Hull, HU12 8LX
G0 NYL B Jackson, 7 Ferriby Road, Scunthorpe, DN17 2EQ
G0 NYM Martin Borer, 37 Broadway, Ripley, DE5 3LJ
GM0 NYP N Purtell, 31 Daleally Crescent, Errol, Perth, PH2 7QA
G0 NYQ John Pape, 12a High Wiend, Appleby-in-Westmorland, CA16 6RD
G0 NYR R Cheetham, 8 Fairway, Huyton, Liverpool, L36 1UD
G0 NYS D Cox, 10 Calder Close, Bollington, Macclesfield, SK10 5LJ
G0 NYY S Nicholas, 39 Cubitts Close, Welwyn, AL6 0DZ
G0 NYZ S Maloney, 34 Keswick Road, Normanby, Middlesbrough, TS6 0BN
G0 NZA M Lowe, 25 Manor House Court, Kirkby-in-Ashfield, Nottingham, NG17 8LH
G0 NZE A Benfield, 12 St. Marys Court, Weald, Bampton, OX18 2HX
G0 NZI Carl Peake, 3 Marigold Walk, Bermuda Park, Nuneaton, CV10 7SW
G0 NZJ R [illegible], SS7 5HL
G0 NZN J Smith, 6 Cherry Grove, Croespenmaen, Newport, NP11 3DF
G0 NZR D Catterall, 86 Broomfield Road, Swanscombe, DA10 0LT
G0 NZT D Nash, 27 Sandling Avenue, Horfield, Bristol, BS7 0HS
G0 NZU Roy Blanning, 38 Northville Road, Northville, Bristol, BS7 0RG
GM0 OAA Michael Wigg, 1/1 2 Glencairn Drive, Glasgow, G41 4QN
G0 OAB D Griffith, 5 Upthorpe Drive, Wantage, OX12 7DF
G0 OAJ John Morrice, 184 Rowan Way, Newport, NP20 6JT
G0 OAP D Coulson, 76 Meadow Lane, Newhall, Swadlincote, DE11 0UW
G0 OAS N Goddard, 15 Canada Road, Cobham, KT11 2BB
G0 OAT R Petri, Tarnwood, Deneswav, Gravesend, DA13 0EA
G0 OAW W Waldron, Redstone Farm, Germans Week, Beaworthy, EX21 5BQ
G0 OAZ Jeffrey Fox, L'Auzisiere De St.Marsault, la Foret Sur Serve, France, 79380
G0 OBA K Barker, 4 Fort Widley Cottages, Southwick Hill Road, Portsmouth, PO6 3EU
G0 OBB W Evans, Brynawel, Cross Inn, Llanon, SY23 5NB
G0 OBE J Clarke, 16 Silver Birch Avenue, Bedworth, CV12 0AZ
G0 OBG A Cox, 8 Spence Avenue, Byfleet, West Byfleet, KT14 7TG
G0 OBH H Cox, Windrush, Malthouse Lane, Great Yarmouth, NR29 5QL
G0 OBJ G Pratley, 34 Druce Way, Thatcham, RG19 3PF
G0 OBK K Warnes, 3 Blue Bell Close, Underwood, Nottingham, NG16 5FN
G0 OBM W Noyce, 44 Appleton Road, Fareham, PO15 5QH
G0 OBN C Hodgson, 25 Pembroke Court, Sunderland, SR5 4DF
G0 OBO K Weeks, 11 Sandwich Road, Preston Grange, North Shields, NE29 9HT
G0 OBP K Hudson, 2 Colwill Walk, Plymouth, PL6 8XF
G0 OBQ G Lang, 63 Grosvenor Drive, Whitley Bay, NE26 2JR
G0 OBT S Fortt, 59 Coombe Dale, Sea Mills, Bristol, BS9 2JF
G0 OBV Martin Roberts, The Bourne, The Avenue, Reading, RG7 6NN
G0 OCB R Dingle, 29 Castle View Witton le Wear, Bishop Auckland, DL14 0DH
G0 OCC G Allen, 2 Haworth Drive, Bootle, L20 6EJ
G0 OCF R McCoye, 26 Hansby Close, Oldham, OL1 2UA
G0 OCK B Pilkington, 219 Brownhill Avenue, Burnley, BB10 4QH
G0 OCL F Pilkington, 219 Brownhill Avenue, Burnley, BB10 4QH
G0 OCR Franklin Batkin, 24 Sandyfields Road, Sedgley, Dudley, DY3 3LB
G0 OCS Mary Stabbins, Primrose Cottage, Carlidnack Lane, Falmouth, TR11 5HE
G0 OCT Lawrence Stabbins, Primrose Cottage, Carlidnack Lane, Falmouth, TR11 5HE
G0 OCW P Brazier, 1 Ravenshore Cottages, Holcombe Road, Rossendale, BB4 4AN
G0 OCY P Beeston, 100 Suffield Road, High Wycombe, HP11 2JL
G0 ODA Gerard Hill, 163 Parsonage Rd, Castle Hill, Australia, 2154
GM0 ODB J Kane, 21 Sersley Drive, Kilbirnie, KA25 6EY
G0 ODE D Williams, 212 Birchanger Lane, Birchanger, Bishop's Stortford, CM23 5QH
G0 ODH David Hibberd, 5 Victory Crescent, Cheadle, Stoke-on-Trent, ST10 1DN
G0 ODI R Sutton, 87 Downs Valley Road, Woodingdean, Brighton, BN2 6RG
G0 ODK W Everett, 120 Wantage Road, Reading, RG30 2SF
G0 ODM J Chomer, 14 Holly Park Gardens, Finchley, London, N3 3NJ
G0 ODN M Hall, 31 Meendhurst Road, Cinderford, GL14 2EF
G0 ODP P Warman, 107 Hillside Road, Corfe Mullen, Wimborne, BH21 3SB
G0 ODQ J Hall, Two Chestnuts, Emmington, Chinnor, OX39 4AA
G0 ODR Colin Hendry, 109 Grove Avenue, New Costessey, Norwich, NR5 0HZ
G0 ODS M Treacher, Barrons Cottage, Welburn, York, YO60 7DZ
G0 ODU K Petherick, 17 Castle Close, Totternhoe, Dunstable, LU6 1QJ
G0 ODX D Hughes, Balshult, Eriksmala, Sweden, 361 94
G0 ODY G Fleming, 27 Crawthorne Crescent, Huddersfield, HD2 1LB
G0 OEA A Moggridge, Outer Bailey, Kingsland, Leominster, HR6 9QN
G0 OEB Clive Donald, 8 Greenway, Walsall, WS9 8XE
G0 OED A Mardo, 10 Meadow View, Uffculme, Cullompton, EX15 3DS
GI0 OEH K Patterson, 8 Beechwood Gardens, Moira, Craigavon, BT67 0LB
G0 OEI M Hopkins, 30 Commonside, Brownhills, Walsall, WS8 7AY
G0 OEJ M Garbutt, 92 Owlet Road, Windhill, Shipley, BD18 2LT
G0 OEK D Spooner, 60 St. Pauls Road, Staines, TW18 0UH
G0 OEM George Cunningham, 7 Wykin Lane, Stoke Golding, Nuneaton, CV13 6HN
G0 OEQ J Blichfeldt, 2 Duck Cottages, Rolvenden Road, Cranbrook, TN17 4BT
G0 OER Paul Roberts, 1 Ardern Close, Bristol, BS9 2QT
G0 OES D Owen, 5 Vicarage Walk, Rosliston, Swadlincote, DE12 8LB
GW0 OET E Jacob, 5 Princess Louise Road, Llwynypia, Tonypandy, CF40 2LY
G0 OEW D Rooke, The Grange, 107 Wybunbury Road, Nantwich, CW5 7ER
G0 OEY Andrew Kerrison, 63 Stour Road, Harwich, CO12 3HS
G0 OFA R Dennis, 8 Newton Hall, Coach Road, Newton Abbot, TQ12 1ER
G0 OFB J Jones, 59 East Street, Long Buckby, Northampton, NN6 7RB
G0 OFD J Gilbert, 6 Mill Hill, Brancaster, King's Lynn, PE31 8AQ
G0 OFE James Smith, 38 Wilson Road, Bournemouth, BH1 4PH
G0 OFF G Hipkin, 62 Woodberry Way, Walton on The Naze, CO14 8EW
GW0 OFH S Williams, 5 Brynmelyn Avenue, Llanelli, SA15 3RU
GM0 OFL John Wilkie, Hope Cottage, 4 Main Street, Cupar, KY15 4SS
GM0 OFM J Park, Ardanach, Kingskettle, Cupar, KY15 7TN
G0 OFN Ian Clabon, 14 Melrose Avenue, Twickenham, TW2 7JE
G0 OFR G Borrowdale, 30 Barton View, Penrith, CA11 8AX
G0 OFT S Duncan, 10 Huntingdon Rise, Bradford-on-Avon, BA15 1RJ
G0 OFW P Lightfoot, 18 Fields Close, Alsager, Stoke-on-Trent, ST7 2ND
G0 OFX F Rawlins, 12 Arundel Road, Eastleigh, SO50 4PQ

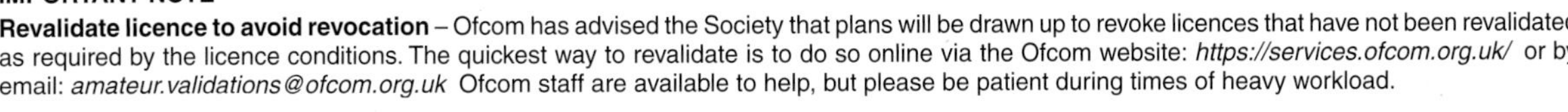
**IMPORTANT NOTE**

**Revalidate licence to avoid revocation** – Ofcom has advised the Society that plans will be drawn up to revoke licences that have not been revalidated as required by the licence conditions. The quickest way to revalidate is to do so online via the Ofcom website: *https://services.ofcom.org.uk/* or by email: *amateur.validations@ofcom.org.uk* Ofcom staff are available to help, but please be patient during times of heavy workload.

G0 OFY James Wane, 25 Holmdale Avenue, Crossens, Southport, PR9 8PS
G0 OFZ Alan Grace, 7 Sandringham Heights, St. Leonards-on-Sea, TN38 9UA
G0 OGB J Wallis, 10 Middlewood Road, Lanchester, Durham, DH7 0HL
G0 OGD G Nattrass, 24 Ritsons Road, Blackhill, Consett, DH8 0AW
G0 OGE Michael Free, 2 St Pauls Court, Princess Street, Maidenhead, SL6 1NX
G0 OGG J Brown, 1 Jackson Road, Houghton, Carlisle, CA3 0NW
G0 OGI D Keely, 15 Ffordd Cerrig Mawr, Caergeiliog, Holyhead, LL65 3LU
G0 OGJ K Marshall, 8 Porter Way, Northwich, CW9 7JA
G0 OGL S Glanville, Fron Haulog, Llanelian, Colwyn Bay, LL29 8UY
G0 OGM S Bowerman, 24 Bingham Close, Emerson Valley, Milton Keynes, MK4 2AU
GM0 OGN Richard Hall, 13 Cleat, Castlebay, Isle of Barra, HS9 5XX
G0 OGP Y Powell, 18 Carrington Road, Stockport, SK1 2QE
G0 OGS Stephen Malpass, 21 Tollhouse Way, Wombourne, Wolverhampton, WV5 8AF
G0 OGW C Douglas, 24 Avon Crescent, Romsey, SO51 5PY
G0 OGX J Pennington, 1 Chisel Close, Hereford, HR4 9XF
GM0 OGZ Richard Goodall, 3 Croftcrunie Cottages, Tore, Muir of Ord, IV6 7SB
G0 OHA A White, 11 Garden Close, Consett, DH8 5PA
G0 OHD Michael Holding, well house, Cartworth Bank Road, Holmfirth, HD9 2RF
G0 OHF F Wilson, 3 Foundry Mews, Burgh le Marsh, Skegness, PE24 5HQ
GI0 OHG E Bennett, 53 Condiere Avenue, Connor, Ballymena, BT42 3LD
G0 OHJ David Workman, 4 Rhuddlan Road, Buckley, CH7 3QA
G0 OHK Nigel King, 7 Fountains Close, Biddick Village, Washington, NE38 7TA
G0 OHR John Armitage, 17 Worsbrough Road, Birdwell, Barnsley, S70 5QR
GI0 OHT W Stanley, 95 Bangor Road, Newtownards, BT23 7BZ
GI0 OHU R King, 28 Moss Road, Waringstown, Craigavon, BT66 7QY
G0 OHW J Vasek, 20 West Hall Road, Richmond, TW9 4EE
G0 OHY A Mather, 15 Stanley Road, Worsley, Manchester, M28 3DT
G0 OID F Tett, 2 Church Park, Bradenstoke, Chippenham, SN15 4ER
G0 OIE M Gathergood, 54 Robin Lane, Bentham, Lancaster, LA2 7AG
G0 OIF Dawn Read, 17 Lumber Leys, Walton on The Naze, CO14 8SS
G0 OII Richard Pullen, 1 Ridings Court, 5 Crown Crescent, Scarborough, YO11 2BJ
G0 OIK P King, Chad Lane Farm, Chad Lane, St. Albans, AL3 8HW
G0 OIM B Turnbull, 56 Bryans Close Road, Calne, SN11 9AB
G0 OIN A Fairey, 1 Imbert Close, New Romney, TN28 8XP
G0 OIO J Fuller, 1 The Courtyard, Snape, Saxmundham, IP17 1FB
G0 OIQ Albert Welland, 6 Chartwell Road, Stafford, ST17 0AJ
G0 OIR G Wicks, 28 Old School Lane, Milton, Cambridge, CB24 6BS
G0 OIS T Smith, 33 Kites Nest Lane, Lightpill, Stroud, GL5 3PJ
G0 OIU A Smith Jones, 16 Armley Road, Liverpool, L4 2UN
G0 OIV A Sait, 124 Dicksons Drive, Newton, Chester, CH2 2BX
G0 OIW Mark Palmer, 28 Westfield Road, Caversham, Reading, RG4 8HH
G0 OIX David Wright, 132 Longmoor Lane, Liverpool, L9 9BZ
G0 OIY James Smith, 59 Charlecote Drive, Dudley, DY1 2GG
G0 OJB L Chadwick, 12 Brybank Road, Haverhill, CB9 7WD
G0 OJC Martin Crowly, 10 Green Lane, Ellesmere Port, CH65 0DJ
G0 OJF R Shireby, Norwood, Rookery Lane, Grantham, NG32 3RU
G0 OJG Joseph Omalley, 8 Hawks Court, Hallwood Park, Runcorn, WA7 2FR
G0 OJJ Anthony Green, 6 Goulds Close, Palgrave, Diss, IP22 1AR
G0 OJP R Melton, 4 Ashleigh Avenue, Maiden Newton, Dorchester, DT2 0BP
G0 OJR Barry Fox, 6 Bury Gardens, Elmdon, Saffron Walden, CB11 4LX
G0 OJS S John, 40 Elizabeth Avenue, Brixham, TQ5 0AY
G0 OJT W Jenkinson, 7 Moortown Road, Watford, WD19 6JH
G0 OJU M Goddard, Selsted Garage, Canterbury Road, Dover, CT15 7HJ
G0 OJW J Taylor, C/O 16 Caroline Place, Plymouth, PL1 3PS
G0 OJX P Williams, 3 Nutwell Cottages, Exmouth Road, Exmouth, EX8 5AP
G0 OJY Adrian Hurt, The Manse, High Oak Road, Ware, SG12 7PD
G0 OKA David Martin, 67 Mill Street, Torrington, EX38 8AL
G0 OKD Russell Bradley, 4 Paddocks Close, Pinxton, Nottingham, NG16 6JR
G0 OKF Sidney Bolam, 100 Bushfield Road, Scunthorpe, DN16 1NA
G0 OKI R Morris, 4 Greenway Gardens, Kings Norton, Birmingham, B38 9RY
GM0 OKJ J Fraser, 2 Barra Place, Stenhousemuir, Larbert, FK5 4UF
G0 OKK B Crowe-Haylett, 13 Lynton Close, Ely, CB6 1DJ
G0 OKL Janet Collins, 3 Burford Grove, Bristol, BS11 9RT
G0 OKN Russell Maloney, Rosewell, Jacobstow, Bude, EX23 0BN
G0 OKT C Parker, 1 Richmond Drive, Perton, Wolverhampton, WV6 7RR
G0 OKV K Cowell, 4 St. Georges Close, Colne, BB8 8DP
G0 OKX K Gardner, 13 Beanshaw, Eltham, London, SE9 3HL
G0 OKZ J Thorpe, Four Jays, 46a High Street, Doncaster, DN10 4BU
G0 OLD Duncan McLaren, 11 St. Matthew Close, Uxbridge, UB8 3SR
G0 OLE BOOTHFERRY ARS c/o Kenneth McCann, Treverven, Back Lane, Selby, YO8 6QP
GM0 OLF David Phillips, East Grange Steading, Inverarity, Forfar, DD8 2JN
G0 OLL E Platts, 38 Swanbourne Road, Sheffield, S5 7TL
G0 OLO D Collinson, 20 Carlisle Crescent, Penshaw, Houghton le Spring, DH4 7RD
G0 OLR L Roberts, Rose Cottage, Castle Hill, Leyburn, DL8 4QN
G0 OLS Tony Humphries, 23 Sycamore Drive, Lutterworth, LE17 4TR
G0 OLT L Tringale, 19 Lysander Road, Kings Hill, West Malling, ME19 4TT
G0 OLX Denis Stanton, 122 Foxon Lane, Caterham, CR3 5SD
G0 OLZ G Smith, 23 Gainsborough Close, Llantarnam, Cwmbran, NP44 3BX
G0 OMB Brian Walker, 15 Infirmary Road, Workington, CA14 2UG
GM0 OMC C Cook, Briarwood, 95 Old Edinburgh Road, Inverness, IV2 3HT
G0 OMD A Gilbert, 19 Farrs Avenue, Andover, SP10 2AH
G0 OMF Derek Hupton, 90 Warwick Road, Atherton, Manchester, M46 9PQ
G0 OMH P Burbeck, 5 Wouldham Terrace, Saxville Road, Orpington, BR5 3AT
G0 OMM S Adams, 13 Bells Drove, Sutton St. James, Spalding, PE12 0JG
G0 OMN G Charman, 4 Hornton Grove, Hatton Park, Warwick, CV35 7UA
G0 OMZ R Lomas, 7 Chaunterell Way, Abingdon, OX14 5PP
G0 ONA P Nicholls, 5 Dingle Road, Ashford, TW15 1HF
G0 ONB CSMT GROUP c/o Nicholas Shaxted, Heinrich-Heine-Allee 19, Dusseldorf, Germany, 40213
GI0 OND James Lappin, 46 Grange Road, Kilmore, Armagh, BT61 8NX
G0 ONF V Szendzielarz, 5 Granville Road, Urmston, Manchester, M41 0XY
G0 ONG J Mobbs, 5 Distaff Road, Poynton, Stockport, SK12 1HN
G0 ONH B FELLOWS, 2 White Harte Caravan Park, Kinver, Stourbridge, DY7 6HN
GM0 ONN I Barnetson, 38 Woodlands Drive, Lhanbryde, Elgin, IV30 8JU
G0 ONS J Chinnery, 31 Kingsway, Northampton, NN2 8HD
GW0 ONU D Harris, 2 Sheppard St., Pwllgwaun, Pontypridd, CF37 1HT
G0 ONW N Lees, 49 Flansham Park, Bognor Regis, PO22 6QH
GM0 ONX Leonard Paget, 40 Davaar Drive, Kilmarnock, KA3 2JG
G0 ONY John Edwards, 3 High Street, Bryngwran, Holyhead, LL65 3PL
G0 OOB D Walpole, 12 Damgate Lane, Acle, Norwich, NR13 3DH
G0 OOD Terence Chapman, 2 Links Close, Norwich, N6 5PJ
G0 OOF Reginald Williams, Dyffryn Coed, Union Road, Coleford, GL16 7QB
G0 OOI W Humphries, 76 Mortlake Road, Richmond, TW9 4AS
G0 OON Peter Healey, 10 Wroxham Road, Great Sankey, Warrington, WA5 3EE
G0 OOO SCARBOROUGH SEG c/o Roy Clayton, 9 Green Island, Irton, Scarborough, YO12 4RN
G0 OOR A Jex, 26 Springdale Crescent, Brundall, Norwich, NR13 5RA
G0 OOS Lenio Marobin, Flat 60, Tudor Court, London, N1 4NU
G0 OOU R Field, 34 Piltdown Close, Hastings, TN34 1UU
G0 OPA John Lee, Holly Lodge, Carrhouse Road, Doncaster, DN9 1PG
G0 OPC Mike Marriott, 188 Leverington Common, Leverington, Wisbech, PE13 5BP
G0 OPG C Knowlson, 28 Hill Drive, Handforth, Wilmslow, SK9 3AR
G0 OPI AC Bennett, 32 Gainsborough Road, Bournemouth, BH7 7BD
GM0 OPK Paul Thomson, Auchenlea, Neilston Road Barrhead, Glasgow, G78 1TY
G0 OPL W Cowell, 72a The Malting, Ramsey, Huntingdon, PE26 1LZ
G0 OPM G Melia, Sunnyside, Little Asby, Appleby-in-Westmorland, CA16 6QE
GW0 OPP R Owens, 62 Ty Llwyd Parc Estate, Quakers Yard, Treharris, CF46 5LB
G0 OPQ A Stocks, 3 Limestone Way, Burniston, Scarborough, YO13 0DQ
GM0 OPS John Dundas, 26 Balgray Road, Newton Mearns, Glasgow, G77 6PB
G0 OPT Peter Tennant, 128 Devonshire Street, Keighley, BD21 2QJ
G0 OPV R Heatley, 68 Jeckyll Road, Wymondham, NR18 0WQ
GM0 OPX D Mcferran, Ardlair, Milltimber, AB13 0ER
G0 OQE Frederick Porter, Kinross, 12 Brooklands Road, Milton Keynes, MK2 2RN
G0 OQI K Zak, 5 The Rookery, Sandy, SG19 2UR
G0 OQK Nick Garrod, 121 Totteridge Lane, High Wycombe, HP13 7PH
G0 OQP A Caton, 20 Lower Oxford Road, Newcastle, ST5 0PB
G0 OQQ Brett Wood, 52 Ashfield Avenue, Beeston, Nottingham, NG9 1PY
G0 OQR A Glen, 70 Moscow Road East, Stockport, SK3 9QL
G0 OQS N Dean, 13 St. Marys Avenue, Billinge, Wigan, WN5 7QL
GM0 OQV I Muir, 6 Dunard Court, Carluke, ML8 5RX
G0 OQX J East, 30 Auckland Road, Scunthorpe, DN15 7BT
G0 OQZ H Dawson, 6 Maer Top Way, Barnstaple, EX31 1RZ
G0 ORC Vincent Shirley, 160 Over Lane, Belper, DE56 0HN
G0 ORD E Chantler, Hilltop Gardens, High Beech Road, Ruardean, GL17 9UD
G0 ORE Nicholas Reddish, 15 Drakes Close, Redditch, B97 5NG
G0 ORG N Robertson, Clayhill Cottage, The Street, Ipswich, IP7 6NN
G0 ORH Kenneth Chandler, 4 Park Avenue, Thatcham, RG18 4NP
G0 ORJ J Bamford, 39 Skelldale View, Ripon, HG4 1UJ
G0 ORK S Humberstone, 4 Rowcroft Road, Paignton, TQ3 2RE
G0 ORL C Rowley, 31 Keepers Croft, East Goscote, Leicester, LE7 3ZJ
G0 ORM D Birch, 31 Grasmere Terrace, Maryport, CA15 7QN
G0 ORO Dennis Martin, 70 Moorlands Drive, Stainburn, Workington, CA14 4UJ
G0 ORP M Simpson, 3 Front Street, Barnby, Newark, NG24 2SA
G0 ORT D Leonard, Three Ashes Cottage, 442 Outwood Common Road, Billericay, CM11 1ET
G0 ORV V Wilton, Fairthorn Trotts Ln, Pooks Green, Southampton, SO4 4WQ
G0 ORX John Melton, 4 Charlwood Close, Copthorne, Crawley, RH10 3TG
G0 ORY A Moss, 10 Shakespeare Drive, Leicester, LE3 2SP
G0 OSA C Wilkinson, Les Mardeilles, Le Bourg, Chirac, France, 16150
GW0 OSB I Price, 16 Carmarthen Court, Caerphilly, CF83 2TX
G0 OSC G Mason, 18 Nithsdale Road, Liverpool, L15 5AX
G0 OSD G Alexander, 15 Brackley Way, Totton, Southampton, SO40 3HP
G0 OSG Robin Brazier, 9 Wheelers Walk, Blackfield, Southampton, SO45 1WX
G0 OSH WATCOMBE RC c/o H Davies, 33 Sandown Road, Ocean Heights, Paignton, TQ4 7RL
G0 OSI Kenneth Pallant, 7 Council Bungalows, Church Lane, Braintree, CM7 5SH
GM0 OSJ William Legge, The Manse, Muir of Fowlis, Alford, AB33 8JU
G0 OSK C Saggers, 49 Revels Road, Hertford, SG14 3JU
G0 OSO P Markham, Moor Farm, Moor Lane, Lincoln, LN3 4EG
G0 OSP P Leach, 17 The Wicket, Hythe, Southampton, SO45 5AU
G0 OSR H Middleton, Windermere, Stanway Green, Colchester, CO3 0QZ
G0 OSU John Collier, 27 Birdham Close, North Bersted, Bognor Regis, PO21 5TD
G0 OSV P Foster, 53 Garstang Road West, Poulton-le-Fylde, FY6 8AA
G0 OSW Roy Sainsbury, Bridge Farmhouse, Southampton Road, Salisbury, SP5 2ED
G0 OSX N Shackley, 20a Pear Tree Road, Ashford, TW15 1PW
GM0 OTB Raymond Pugh, 28 Pladda Road, Saltcoats, KA21 6AQ
GI0 OTC T Doherty, 37 Magheramenagh Drive, Portrush, BT56 8SP
G0 OTE E Bowell, 7 Bede House Bank, Bourne, PE10 9JX
G0 OTF Geoffrey George, 211 Bromford Road, Birmingham, B36 8HA
G0 OTH Robert Topliss, 12 Dorothy Avenue, Skegness, PE25 2BP
GM0 OTI John Grieve, Elhanan, Myrtlefield Lane, Inverness, IV2 5UE
G0 OTJ J Cummins, Tarr House, Lumb Lane, Matlock, DE4 2HP
GM0 OTS William Mcintosh, 14 East Road, Hopeman, Elgin, IV30 5SU
G0 OTT D W McDonald, 118 Torrington Avenue, Tile Hill, Coventry, CV4 9AA
GM0 OTU Alan king, 31 Pendreich Grove, Bonnyrigg, EH19 2EH
GW0 OTY W Cooper, 50 Tennyson Road, Penarth, CF64 2SA
GD0 OUD Stuart Hill, 54 Wybourn Drive, Onchan, Isle of Man, IM3 4AT
G0 OUG I Hole, 50 Westcroft Drive, Westfield, Sheffield, S20 8EF
GW0 OUH H Griffiths, 45 Jubilee Road, Godreaman, Aberdare, CF44 6DD
G0 OUJ R Walker, 270 Stourbridge Road, Halesowen, B63 3QR
G0 OUK J Hinton, Wayside, Beauchief Drive, Sheffield, S17 4RJ
GI0 OUM R Ferris, 3 Kingsland Drive, Belfast, BT5 7EY
G0 OUN Paul Bingham, 12 Sandpiper Road, Thorpe Hesley, Rotherham, S61 2UN
G0 OUO S Palk, Staverton, Western Lane, Minehead, TA24 8BZ
G0 OUR OPEN UNIVERSITY ARC c/o A Rawlings, Open University, Walton Hall, Milton Keynes, MK7 6AA
GW0 OUV M williams, 7 Heol Isaf, Nelson, Treharris, CF46 6NS
GI0 OUZ B Prunty, 16 Old Portadown Road, Lurgan, Craigavon, BT66 8RH
G0 OVA P Crake, 1 Ashdown Close, Bracknell, RG12 2SE
G0 OVC B Godfrey, 291 Collier Row Lane, Romford, RM5 3ND
GM0 OVD Robert Darroch, 36 Tweed Street, Dunfermline, KY11 4NA
G0 OVE K Mohammed, 63 Shirley Gardens, Barking, IG11 9XB
G0 OVK Roy Mansell, 2 Ambrose Close, Willenhall, WV13 3DQ
G0 OVQ A Bannister, 34 Morningside Drive, East Didsbury, Manchester, M20 5PL
G0 OVT B Navier, 12 Brooklyn Avenue, Brooklyn Street, Hull, HU5 1ND
G0 OVV M Bolton, 85 Oak Park Road, Wordsley, Stourbridge, DY8 5YJ
G0 OVY P Maggs, 85 Helsby Road, Sale, M33 2XF
G0 OWA John Wright, 10 Whalley Road, Heskin, Chorley, PR7 5NY
G0 OWC Peter Bush, 52 Asker Lane, Matlock, DE4 5LA
G0 OWE D Matthews, 54 The Wynding, Bedlington, NE22 6HW
G0 OWH J Dobbs, 9 Highlands, Littleborough, OL15 0DS
G0 OWI A Hawkridge, Thorntrees, 109 Allerton Road, Bradford, BD8 0AA
G0 OWJ A Cooper, 28 Belmont Road, Pensnett, Brierley Hill, DY5 4EX
G0 OWK S Searle, 14 Edison Gardens, Colchester, CO4 0AJ
GM0 OWM ORKNEY WRLSS MS c/o A Wright, Crosslea, Berstane Road, Kirkwall, KW15 1SZ
G0 OWP David Edwards, 9 Mark Road, Hightown, Liverpool, L38 0BG
G0 OWR C Howard, 75 Westbury Park, Wootton Bassett, Swindon, SN4 7DN
G0 OWU R Wilkes, 39 Hillside Road, Dudley, DY1 3LE
G0 OWV J Harbottle, 42 Littlemede, Eltham, London, SE9 3EB
G0 OXA G Landen-Turner, 59 Mill Road, Higher Bebington, Wirral, CH63 5PA
G0 OXB Philip Draycott, 41 Ashleigh Avenue, Bridgwater, TA6 6AX
G0 OXE THE MORSE CLUB c/o Christopher Tapping, 49 Coventry Gardens, Herne Bay, CT6 6SB
GI0 OXK Daniel Taggart, 106 Moorfield Road, Dromore, Omagh, BT78 3LR
G0 OXL Clifford Robinson, 33 Windsor Rd, WELLESLEY, Ma, United States, 2481
G0 OXP L Matthews, 6 Spotland Tops, Cutgate, Rochdale, OL12 7NX
GM0 OXS M Beith, 30 Raith Road, Fenwick, Kilmarnock, KA3 6DB
G0 OXT P Hutchinson, Rosebank Cottage, Marcombe Road, Torquay, TQ2 6LL
G0 OXV Keith Mahood, Brooklands Lodge, 1A Heskin Lane, Ormskirk, L39 1LR
G0 OXW V Soutter, 2 Hyde Barton, Churchill Way, Bideford, EX39 1NX
G0 OXX John Berridge, Bracklyn, St. Clare Road, Deal, CT14 7QB
G0 OXY Michael Gray, 142 Harrowden Road, Bedford, MK42 0SJ
G0 OXZ M Stone, 29 Chesterfield Road, Epsom, KT19 9QR
G0 OYA Michael Clapperton, 99 Bath Road, Bridgwater, TA6 4PN
G0 OYC K Saunders, 1 Chesham Way, Watford, WD18 6NX
G0 OYF S Harvey, 68 Stuart Road, Rowley Regis, B65 9HZ
G0 OYI G Holden, The House On The Green, Linstock, Carlisle, CA6 4PZ
G0 OYJ T Gonsalves, 30 Cunnington St., Chiswick, London, W4 5EN
G0 OYL W Waring, 1 Innerhaugh Mews, Haydon Bridge, Hexham, NE47 6DE
G0 OYM M Trahearn, 16 Grange Lane, Lichfield, WS13 7ED
G0 OYN Dave Hedley, 42 Liphook Road, Lindford, Bordon, GU35 0PP
G0 OYO D James, 7 Abbotts Road, Plymouth, PL3 4PD
G0 OYP B Barber, 3 Catherine Avenue, Mansfield Woodhouse, Mansfield, NG19 9AZ
G0 OYQ S Lowe, 14 Kensington Avenue, Kingswood, Hull, HU7 3AF
G0 OYR N Ashfield, 167 Greville Road, Warwick, CV34 5PU
G0 OYS D Temple, 4 Cameron Avenue, Abingdon, OX14 3SR
GM0 OYU M Chesters, Blackhill, Blackhill Road, Kirkwall, KW15 1FP
G0 OYX D Medley, 9 Northolme Crescent, Hessle, HU13 9HU
G0 OYY G Mantle, 26 Graham Road, Wordsley, Stourbridge, DY8 5PU
G0 OYZ Roy Bray, 10 Upwell Road, March, PE15 9DT
G0 OZB A Gardner, 28 Usk Court, Thornhill, Cwmbran, NP44 5UN
G0 OZG David Turner, 27 Aylesbury Avenue, Langney Point, Eastbourne, BN23 6AB
G0 OZJ G Gourley, 6a Longsight Lane, Cheadle Hulme, Cheadle, SK8 6PW
G0 OZL B Smith, 19 Fieldstone Court, Howick, Northpark, New Zealand, 1705
G0 OZM C Rapson, Kaloma, Northiam Road, Rye, TN31 6EP
G0 OZO Jamie Harris, 31 Grasby Road, Limber, Grimsby, DN37 8LB
G0 OZP B Salt, 9 Ashville Gardens, Pellon, Halifax, HX2 0PJ
GI0 OZQ D Gillespie, 81 Lisfannon Park, Londonderry, BT48 9DU
G0 OZR G Markham, 2 Edwin Avenue, Woodbridge, IP12 1JS
G0 OZS Ian Moffat, The Hatchets, The Street, Stowmarket, IP14 5PE
G0 PAB P Betts, 14 Saltergate Road, Messingham, Scunthorpe, DN17 3SZ
G0 PAD A Jacobs, 16 Clwyd Walk, Corby, NN17 2LN
G0 PAE C Hewitt, 28 Amersham Avenue, Langdon Hills, Basildon, SS16 6SJ
G0 PAG N Page, 54 Queensway, Old Dalby, Melton Mowbray, LE14 3QH
G0 PAI I Leitch, 70 Hanover Road, Rowley Regis, B65 9DZ
G0 PAN D Elkington, 45 Heathfield, Leeds, LS16 7AB
G0 PAO C Muddimer, 7 Tots Gardens, Acton, Sudbury, CO10 0DJ
G0 PAR D How, 25 Lovelace Road, West Dulwich, London, SE21 8JY
G0 PAS M Lord, 5 Wasdale Green, Cottingham, HU16 4HN
G0 PAU John Forsyth, 102 Langley Road, Watford, WD17 4PJ
G0 PAW P Weaving, 1 Hammy Close, Shoreham-by-Sea, BN43 6BL
G0 PAZ D Utley, 30 Station Road, Ackworth, Pontefract, WF7 7NA
G0 PBA K Garbutt, 92 Owlet Road, Windhill, Shipley, BD18 2LT
G0 PBB FOREST OF DEAN AMATEUR RADIO GROUP c/o William Bonser, 24 Meend Garden Terrace, Cinderford, GL14 2EB
G0 PBE D Yates, 101 Coach House Drive, Shevington, Wigan, WN6 8AU
G0 PBF James Brown, 71 Piccadilly Road, Swinton, Mexborough, S64 8LF
G0 PBH D Mason, 11 Bryony Close, Killamarsh, Sheffield, S21 1TF
G0 PBJ Leslie Wright, Cedar House, Old Aston Hill, Ewloe, CH5 3AL
G0 PBL P Davies, 85 Church Road, Byfleet, West Byfleet, KT14 7NG
G0 PBM A Razzell, 96 Weston Road, Aston-on-Trent, Derby, DE72 2BA
G0 PBN A Moulder, 10 Parsonage Road, Rainham, RM13 9LW
G0 PBO A Coleman, 16 Cowley Close, Swineshead, Boston, PE20 3ES
G0 PBP A Evans, 24 Oakleigh Avenue, Glen Parva, Leicester, LE2 9TH
G0 PBR R Clark, 4 Haigh Street, Cleethorpes, DN35 8QN
G0 PBS D Webber, 37 Woodhurst Drive, Denham, Uxbridge, UB9 5LL
G0 PBU Dennis Bradley, 2a Mitchell Street, Kettering, NN16 9HA
G0 PBV Nicholas Plumb, 35 Foamcourt Waye, Ferring, Worthing, BN12 5RD
G0 PBW R Brown, 26 Lynnes Close, Blidworth, Mansfield, NG21 0TU
G0 PBY Rodney Freer, 15 Fosse Close, Enderby, Leicester, LE19 2AW
G0 PCA K Godwin, 11 St. Lukes Way, Allhallows, Rochester, ME3 9PR
G0 PCB E Godwin, 11 St. Lukes Way, Allhallows, Rochester, ME3 9PR
G0 PCD S Farrow, 7 Bakewell Close, Hull, HU9 5LH
G0 PCE Robert Barnes, Flat 113, Queens Quay, 58 Upper Thames Street, London, EC4V 3EJ
G0 PCF B Foxall, 11 Cranley Gardens, Shoeburyness, Southend-on-Sea, SS3 9JP

| | | |
|---|---|---|
| GW0 | PCJ | C Watson, 4 Brookland Close, Maesycwmmer, Hengoed, CF82 7RH |
| G0 | PCK | A Lord, 47 Nottingham Road, Trowell, Nottingham, NG9 3PF |
| G0 | PCM | Ian Calvert, 16 Nab Wood Drive, Shipley, BD18 4EJ |
| G0 | PCP | H Baldock, 405 Court Lane, [illegible], Birmingham, B23 5LL |
| G0 | PCQ | [illegible], [illegible], Smithamn Hill, Bristol, BS10 6BZ |
| G0 | PCT | Douglas Hambly, Culver Park, Rattery, South Brent, TQ10 9LJ |
| GI0 | PCU | A Stewart, 1 Lislaynan, Ballycarry, Carrickfergus, BT38 9GZ |
| G0 | PCW | J Budden, Fieldgate, Durnstown, Lymington, SO41 6AL |
| G0 | PCY | J Radford, 93 Hook Road, Surbiton, KT6 5AF |
| G0 | PCZ | B Lody, 41 Galsworthy Road, Chertsey, KT16 8EP |
| G0 | PDA | William Cole, Y Marian, Bow Street, SY24 5BE |
| GW0 | PDB | G Griffiths, Dolcoed, Llandysul, SA44 4RJ |
| G0 | PDE | D Livingstone, 68 Brimley, Leonard Stanley, Stonehouse, GL10 3NA |
| G0 | PDH | D Hyde, 9 Empress Avenue, Marple, Stockport, SK6 7BG |
| GJ0 | PDJ | M Turner, 4 Le Clos Sara, St Lawrence, Jersey, JE3 1GT |
| G0 | PDK | W Marsden, 8 Albert Road, Eston, Middlesbrough, TS6 9QW |
| G0 | PDM | Michael Glover, 22 Fern Street, Sutton-in-Ashfield, NG17 2DW |
| GD0 | PDN | Dave Beedan, Ashmawr, Mount Rule Road, Douglas, Isle of Man, IM4 4QZ |
| G0 | PDP | A Farmer, 76 Wood Lane, Kingsnorth, Ashford, TN23 3AG |
| GM0 | PDQ | M Kusin, East Overhill Farm, Stewarton, Kilmarnock, KA3 5JT |
| G0 | PDV | R Netherway, 2 Avon Court, Lawn Road, Bristol, BS16 5BL |
| G0 | PDZ | Ian Lowe, 54 College Road, Margate, CT9 2SW |
| G0 | PEB | Robert Williams, 10 Barton Close, East Cowes, PO32 6LS |
| G0 | PEC | I Tutt, 1 Castle Road, Hadleigh, Ipswich, IP7 6JH |
| G0 | PEF | Ian Williams, 6 Newport Road, Godshill, Ventnor, PO38 3HR |
| G0 | PEG | J Jenner, Nuholm, Plain Road, Ashford, TN25 6RA |
| G0 | PEH | A Lifton, 70 Scrapsgate Road, Minster on Sea, Sheerness, ME12 2DJ |
| GM0 | PEI | A Pollock, 113 Gartmorn Road, Sauchie, Alloa, FK10 3PD |
| G0 | PEJ | Graham Ford, 5 Rosslyn Close, Hockley, SS5 5BP |
| G0 | PEK | Kevin Richardson, 35 Vidgeon Avenue, Hoo, Rochester, ME3 9DE |
| G0 | PEP | WATERS & STANTON ELECTRONICS c/o P Waters, 9 Tudor Way, Hawkwell, Hockley, SS5 4EY |
| G0 | PEQ | P Cook, 88 Sprowston Road, Norwich, NR3 4QW |
| G0 | PER | K Kreuchen, 211 Creek Road, March, PE15 8RY |
| G0 | PET | A Cunningham, 2 Orchard Close, Normandy, Guildford, GU3 2EU |
| G0 | PEV | Rodney Dawson, 6 Oxton Lane, Tadcaster, LS24 8AG |
| G0 | PEW | J Lyne, 157 Westwick Road, Sheffield, S8 7BW |
| GM0 | PEX | P Bendermacher, 1 Cedar Drive, Milton of Campsie, Glasgow, G66 8AY |
| G0 | PEY | Robert Pearson, 36 Masons Drive, Great Blakenham, Ipswich, IP6 0GE |
| G0 | PFA | M Sole, 44 Chestnut Avenue, Ewell, Epsom, KT19 0SZ |
| G0 | PFE | R Lees, Lyndric, 23 New Queen Street, Scarborough, YO12 7HL |
| G0 | PFH | Geoffrey Spurr, 31 Oakenshaw Court, Wyke, Bradford, BD12 9JE |
| G0 | PFI | Edward Ball, 57 Cherry Tree Road, Sheffield, S11 9AA |
| G0 | PFJ | F Poynter, 7 Howards Way, Cawston, Norwich, NR10 4AZ |
| GI0 | PFL | S McClean, 22 Whiteways, Newtownards, BT23 4UW |
| G0 | PFM | E Ashworth, 88 Hawthorn Avenue, Colchester, CO4 3JP |
| G0 | PFN | David Catchpole, 43 Welsford Road, Norwich, NR4 6QB |
| G0 | PFO | D Butler, 1901 Dean Avenue, Michigan, United States, 48842 |
| G0 | PFQ | S Streluk, 11 Ninefoot Lane, Belgrave, Tamworth, B77 2NA |
| G0 | PFT | M Farrell, Hobberley House, Hobberley Lane, Leeds, LS17 8LX |
| G0 | PFU | K Wignall, 4 Weavers Fold, Bretherton, Leyland, PR26 9AP |
| G0 | PFY | R Marshall, 66 Oakwood Hill, Loughton, IG10 3EP |
| G0 | PFZ | A Powell, Rich Lyn, Carmel Road, Holywell, CH8 7DF |
| G0 | PGA | Charles Smith, 5 Northfield Drive, Mansfield, NG18 3DD |
| G0 | PGB | C Hosking, 32 Queen Street, Penzance, TR18 4BH |
| GI0 | PGC | James Forsythe, 1 Coulson Avenue, Lisburn, BT28 1YJ |
| GM0 | PGD | A Paterson, 21 Kirkwood Avenue, Redding, Falkirk, FK2 9UF |
| G0 | PGI | David Beckly, Knighton, Buckland Monachorum, Yelverton, PL20 7LH |
| G0 | PGJ | G Smith, 16 Weeth Lane, Camborne, TR14 7JN |
| G0 | PGK | D Lawrence, 7 Richmond Road, Appledore, Bideford, EX39 1PE |
| G0 | PGL | Derek Blight, 73 Stoke Road, Taunton, TA1 3EL |
| G0 | PGQ | M Molloy, 20 The Lawn, Whittlesford, Cambridge, CB22 4NG |
| G0 | PGS | Phil Slater, 1 Greyhound Road, Glemsford, Sudbury, CO10 7SJ |
| G0 | PGT | J Newman, Sometimes (The Workshop), South Pew, Dorchester, DT2 9HZ |
| G0 | PGW | G Dunn, 6 Rosewood Avenue, Haslingden, Rossendale, BB4 5NG |
| G0 | PGX | S Thomas, Creekside, Greenbank Road, Truro, TR3 6PQ |
| G0 | PGY | J Underwood, 56 Bassenhally Road, Whittlesey, Peterborough, PE7 1RR |
| G0 | PGZ | Barry Hill, 48 Lackford Avenue, Totton, Southampton, SO40 9BT |
| G0 | PHC | Graham Woodhouse, 12 Matthew Street, Alvaston, Derby, DE24 0ER |
| G0 | PHD | C Whitehead, 27-28 Street Nicholas Street, Scarborough, YO11 2HF |
| G0 | PHE | P Long, 40d Curborough Road, Lichfield, WS13 7NQ |
| GM0 | PHG | D Mclaughlin, 96 Craighlaw Avenue, Eaglesham, Glasgow, G76 0HA |
| G0 | PHI | P Hirst, 4 Brook House, Brook House Lane, Huddersfield, HD8 8LX |
| G0 | PHP | K Green, 39 Fleetgate, Barton-upon-Humber, DN18 5QA |
| G0 | PHR | M Andrews, 8 Irving Road, Solihull, B92 9[illegible] |
| G0 | PHS | George Dodsworth, 9 South Street, Normanton, WF6 1EE |
| GM0 | PHW | Michael Whitehead, 185 Allanton Road, Allanton, Shotts, ML7 5AX |
| G0 | PHY | O Williams, 30 Franklin Road, Biggleswade, SG18 8DX |
| G0 | PIA | J Brown, 14 St. Georges Avenue, Hornchurch, RM11 3PD |
| G0 | PIB | K Simmonds, 9 Packman Drive, Ruddington, Nottingham, NG11 6GF |
| G0 | PID | B Thomas, 112 Pen Park Road, Bristol, BS10 6BP |
| G0 | PIK | A Clements, 37 Sun St., Isleham, Ely, CB7 5HU |
| G0 | PIL | T O'Brien, 45 Rossall Promenade, Thornton Cleveleys, FY5 1LP |
| G0 | PIN | A Pinnock, 1 Rutland Gardens, Ealing, London, W13 0ED |
| G0 | PIS | J Bird, 12 Beresford Gardens, Romford, RM6 6RX |
| G0 | PIT | A Freeman, 7 Palm Grove, Prenton, CH43 1TE |
| G0 | PIU | G Papadopoulos, 1 Darenth Road, London, N16 6EP |
| GM0 | PIV | M Black, Drumtochty, 37 Clepington Road, Dundee, DD4 7EL |
| G0 | PIY | Colin Pollock, Flat 5, 93 Priory Grove, London, SW8 2PD |
| G0 | PJA | Peter Baston, 27 Higher Common Road, Buckley, CH7 3NG |
| G0 | PJC | Adrian Jones, 35 Orchard Way, Letchworth Garden City, SG6 4RZ |
| G0 | PJF | B Ross, 25 Ty Hen, Rhostrehwfa, Llangefni, LL77 7EZ |
| G0 | PJG | J Geraghty, 61 Bridle Lane, Streetly, Sutton Coldfield, B74 3QE |
| GI0 | PJH | W Stewart, 23 Sandy Grove, Magherafelt, BT45 6PU |
| G0 | PJI | Peter Wood, 2 Central Crescent, Hethersett NR93EP, Norwich, NR9 3EP |
| G0 | PJM | M Hughes, The Cottage, Astley Burf, Stourport-on-Severn, DY13 0RX |
| G0 | PJO | M Waller, Olive Cottage, 6 Church Road, Ipswich, IP9 1HS |
| G0 | PJR | Paul Ruffle, 55 Nailers Drive, Burntwood, WS7 0ES |
| G0 | PJS | P Spicer, 86 Main St., Wilsford, Grantham, NG32 3NR |
| G0 | PJU | John Brown, 20 Stamford Avenue, Seaton Delaval, Whitley Bay, NE25 0PA |
| G0 | PJV | J Robertson, 18 Council Road, Ashington, NE63 8RZ |
| G0 | PJW | C Wormald, 22 Tulworth Road, Poynton, Stockport, SK12 1BL |
| G0 | PJY | P Graham, 11 Raby Court, Ellesmere Port, CH65 9DZ |
| G0 | PJZ | R Dorling, Aletheia, St. Marys Road, Colchester, CO7 8NN |
| G0 | PKE | C Tromble, 7 Allerton Grove, Birkenhead, CH42 5LR |
| GM0 | PKF | Peter French, 7 Knockothie Hill, Ellon, AB41 8BA |
| G0 | PKJ | D Stallon, 8 Hidcote Close, Eastcombe, Stroud, GL6 7EF |
| G0 | PKN | T Finneran, 23 Longdales Road, Lincoln, LN2 2JR |
| GM0 | PKP | W Carroll, 20 Pinewood Road, [illegible] |
| GM0 | PKQ | F Grant, Silverknowes, Arbeadie Road, Banchory, AB31 5XA |
| G0 | PKR | K Ritson, 14 Dunsdale Road, Holywell, Whitley Bay, NE25 0NG |
| G0 | PKT | [illegible] c/o Antony Hardman, [illegible] 7HF |
| GD0 | PKW | O Fairgrieve, 8 Aird, Point, Isle of Lewis, HS2 0EU |
| GM0 | PKX | Evan Michael, 8 Castlepark Grove, Kintore, Inverurie, AB51 0SN |
| G0 | PLA | Timothy Reddish, 72 Edgmond Close, Redditch, B98 0JQ |
| G0 | PLB | K Murray, Viamory, Wistanswick, Market Drayton, TF9 2BD |
| G0 | PLC | P Gosnell, 230 Rowley Gardens, London, N4 1HN |
| G0 | PLD | T Pogson, 64 New North Road, Slaithwaite, Huddersfield, HD7 5BW |
| G0 | PLG | W Hargreaves, 24 Moorside Road, Honley, Holmfirth, HD9 6ER |
| GM0 | PLH | W Chan, 5 Lansdowne Drive, Cumbernauld, Glasgow, G68 0JB |
| G0 | PLK | T Kennedy, Woodgreen, Williamstown, Ireland |
| G0 | PLK | T Kennedy, Woodgreen, Williamstown, Ireland |
| GW0 | PLP | Don Kirby, 7 Heol Enlli, Tanygroes, Cardigan, SA43 2JE |
| GD0 | PLQ | J Mitchell, 5 Westminster Drive, Douglas, Isle of Man, IM1 4EG |
| GD0 | PLR | William Smith, 1 High View Road, Douglas, Isle of Man, IM2 5BQ |
| G0 | PLS | I Wallis, 20 Gerard Avenue, Bishop's Stortford, CM23 4DU |
| G0 | PLX | J Parker, 24 Egmont Street, Salford, M6 7LA |
| G0 | PLZ | D Lindsay, 33 Varna Road, Bordon, GU35 0DG |
| G0 | PMB | G Banks, 10 Gregory Road, Glass Houghton, Castleford, WF10 4PH |
| G0 | PMF | G Dellbridge, 19 Cleeve Close, Astley Cross, Stourport-on-Severn, DY13 0NY |
| G0 | PMG | Robin Dellbridge, 24a Calder Road, Stourport-on-Severn, DY13 8QD |
| G0 | PMI | Robert Spencer, 4 Barstow Avenue, York, YO10 3HE |
| G0 | PMM | D Carrott, 5 Raeburn House, 42 Brighton Road, Sutton, SM2 5JH |
| GM0 | PMO | Andrew Fawcett, Glennairn, Stromness, KW16 3EX |
| G0 | PMP | Mark Overend, 58 Church Road, Liversedge, WF15 7LP |
| G0 | PMS | R Sweeney, 33 Traherne Close, Lugwardine, Hereford, HR1 4AF |
| G0 | PMU | R Nolson, 50 Shelf Hall Lane, Shelf, Halifax, HX3 7NA |
| G0 | PMW | Andrew Renwick, Bellevue House, Dornock, Annan, DG12 6SZ |
| G0 | PMX | J Garnham, 20 Deans Walk, Durham, DH1 1HA |
| G0 | PMY | C Rushton, 17 Sullom View, Garstang, Preston, PR3 1QF |
| G0 | PMZ | I Brydon, 12 Pearce Road, Maidenhead, SL6 7LF |
| G0 | PNA | Michael Cranwell, 21 Cockhaven Road, Bishopsteignton, Teignmouth, TQ14 9RF |
| G0 | PNB | R Hope, 7 Irwell Green, Taunton, TA1 2TA |
| GW0 | PNC | H Hartwell, Heulwen, Llanfair Clydogau, Lampeter, SA48 8LH |
| GW0 | PND | J Jones, 4 Lletai Avenue, Pencoed, Bridgend, CF35 5PW |
| G0 | PNE | D Hutson, Sandalwood, 60 Glyndwr Road, Colwyn Bay, LL29 8TA |
| G0 | PNF | W Warren, 38 Stoneyhurst Drive, Curry Rivel, Langport, TA10 0JH |
| G0 | PNG | G Buckley, 46 King Street, Portland, DT5 1NH |
| GW0 | PNI | D Pitkin, Charsfield Dental Practice, Priory Street, Cardigan, SA43 1BU |
| G0 | PNM | Peter Sobye, 2 Willowbank, Fraddon, St. Columb, TR9 6TW |
| G0 | PNN | R Waight, 41 Annalee Road, South Ockendon, RM15 5BZ |
| G0 | PNO | P Studdart, 656 Rayleigh Road, Hutton, Brentwood, CM13 1SJ |
| GI0 | PNP | Robert Pritchard, 79 Harbour Road, Ballyhalbert, Newtownards, BT22 1BW |
| G0 | PNQ | Andrew Varley, 37 Forest Road, Cambridge, CB1 9JA |
| G0 | PNR | G Mcllroy, 1 Belmont Walk, Worcester, WR3 7HY |
| G0 | PNS | RC of Pabay c/o Jeffery Harris, Cape Barn, Back Street Ash, Martock, TA12 6NY |
| G0 | PNT | S Poulter, 119 Aragon Road, Morden, SM4 4QG |
| GW0 | POA | M Hale, 5 Marchwood Close, Rumney, Cardiff, CF3 3LZ |
| GI0 | POB | G Eldridge, 59 Beechwood Gardens, Bangor, BT20 3JD |
| G0 | POC | P Elwood, 55 Madan Road, Westerham, TN16 1DX |
| GM0 | POD | W Mccallum, St Brides Way, Colyton, Ayr, KA6 6QG |
| G0 | POG | C Gavin, Hafod Wen, Bagillt Road, Bagillt, CH6 6JE |
| G0 | POK | D Quinnear, 5 Heath Drive, Chelmsford, CM2 9HA |
| G0 | POM | P Harris, 44 Boston Road, Heckington, Sleaford, NG34 9JE |
| G0 | POQ | D Kemp, 7 St. Nicholas Avenue, Hull, HU4 7AH |
| G0 | POT | Michael Sansom, 19 Baily Avenue, Thatcham, RG18 3EG |
| G0 | POU | J Crosby, 9 Hermitage Close, North Mundham, Chichester, PO20 1JZ |
| G0 | POY | A Eskeloon, 88 Charlton Crescent, Barking, IG11 0NL |
| GW0 | POZ | D Morgan, Coedybryn, Synod Inn, Dyfed, SA44 6JE |
| G0 | PPH | W Blythe, 4 Beresford Road, Stubbington, Fareham, PO14 2QX |
| G0 | PPI | Derek Chenery, 25 Aldreth Road, Haddenham, Ely, CB6 3PW |
| G0 | PPJ | P Johnson, 20 Bearmore Road, Warley, Cradley Heath, B64 6DU |
| G0 | PPK | W Gill, 21 Flockton Avenue, Standish Lower Ground, Wigan, WN6 8LH |
| G0 | PPL | Gerhard Lattka, 9 The Row, Sutton, Ely, CB6 2PD |
| G0 | PPM | K Powell, 88 Nortonwood, Forest Green, Stroud, GL6 0TR |
| G0 | PPQ | Peter Jackson, 24 Woodfield Park, Walton, Wakefield, WF2 6PL |
| G0 | PPR | G Whaling, 3 Bell Close, Little Snoring, Fakenham, NR21 0HX |
| G0 | PPS | PRUDENTIAL ARS c/o James Butler, 14 Fairfield Road, Barnard Castle, DL12 8EB |
| G0 | PPX | J Omara, 18 Tarrant Grove, Quinton, Birmingham, B32 2NW |
| G0 | PPY | N Turner, 31 Shamrock Avenue, Whitstable, CT5 4EL |
| G0 | PQA | A Heather, 4 Ridgeway Heights, Ridgeway Road, Torquay, TQ1 2ND |
| G0 | PQB | Stephen Slater, 118 Danziger Way, Borehamwood, WD6 5DG |
| G0 | PQD | Kenneth Skuse, 4 Barton Close, Berrow, Burnham-on-Sea, TA8 2NN |
| G0 | PQF | A Judge, 44 Thorley Lane, Bishop's Stortford, CM23 4AD |
| G0 | PQG | Alwyn Harper, 81 High Street, Great Houghton, Barnsley, S72 0AU |
| G0 | PQI | W Winters, 20 Y Berllan, Penmaenmawr, LL34 6HB |
| G0 | PQO | Kevin Martin, 8 Taylors Close, Meppershall, Shefford, SG17 5NH |
| G0 | PQR | C Wardle, P O Box N 3189, Nassau, Bahamas |
| GM0 | PQV | John Maguire, 64 High Street, Loanhead, EH20 9RR |
| G0 | PQW | P Bartholomew, 29 Beatrice Avenue, East Cowes, PO32 6HR |
| G0 | PQX | Sydney Shipley Shipley, 102 Jackson Street, Goole, DN14 6DH |
| G0 | PQY | A Langford, 53 Cambridge Avenue, Bottesford, Scunthorpe, DN16 3PH |
| G0 | PRF | John Goodwin, 146 Grimescar Road, Ainley Top, Huddersfield, HD2 2EB |
| GM0 | PRG | PERTH REPEATER GROUP c/o David Morris, Ash Cottage, Perth Road, Perth, PH2 0LW |
| G0 | PRH | Mario Grassi, Little Ash, Sleight Lane, Wimborne, BH21 3HL |
| G0 | PRI | L Ward, 20 The Green, Newby, Scarborough, YO12 5JA |
| G0 | PRM | Brian Goodier, 14 Meadowbank, Old Colwyn, Colwyn Bay, LL29 8EX |
| GM0 | PRO | P Greenway, 5 Java Place, Craignure, Isle of Mull, PA65 6BG |
| GM0 | PRQ | M Shield, Castleshield, Fiscavaig, Isle of Skye, IV47 8SN |
| G0 | PRS | POOLE RADIO SCOUTS (PRS) c/o Colin Baverstock, Butchers Coppice Scout Camp Site, Holloway Avenue, Bournemouth, BH11 9JW |
| G0 | PRT | James Lambrianou, 20 Woodhall Drive, Banbury, OX16 0TY |
| G0 | PRU | PRUDENTIAL ARS c/o D Dyer, 57 Garrison Lane, Felixstowe, IP11 7RR |
| G0 | PRV | D McNab, 19 Rainham Gardens, Alvaston, Derby, DE24 0DJ |
| G0 | PSD | P Hayward, 6 Greenock Close, Westlands, Newcastle, ST5 2LG |
| G0 | PSE | Thomas Taylor, 19 Derwent Grove, Taunton, TA1 2NJ |
| G0 | PSF | Peter Yeatman, 73 Roundway, Waterlooville, PO7 7QB |
| G0 | PSG | D Carvell, 26 Greenfield Avenue, Kettering, NN15 7LL |
| G0 | PSH | Aidan Goldstraw, 59 Lansbury Grove, Stoke-on-Trent, ST3 6JY |
| G0 | PSI | John Wood, 18 Kennedy Avenue, Long Eaton, Nottingham, NG10 3GF |
| G0 | PSJ | S Jacques, Torr Garth, 38 Cheyne Walk, Hornsea, HU18 1BX |
| G0 | PSK | G Hawkins, 8 Broughton Road, West Ayton, Scarborough, YO13 9JW |
| G0 | PSL | P Daddy, 52 Seafield Avenue, Hull, HU9 3JQ |
| G0 | PSO | Paul O'Nion, 11 Capitol Close, Swindon, SN3 4AB |
| GU0 | PSP | Michael Dowding, L'Ancrage, Les Marais, Guernsey, Guernsey, GY7 9LD |
| GW0 | PSV | G Wardman, 5 High Street, Trelewis, Treharris, CF46 6AB |
| G0 | PSY | Susan Brodie, Waterloo Cottage, Tanners Green, Norwich, NR9 4QS |
| G0 | PSZ | L Banaszak, 17 Stoney Piece Close, Bozeat, Wellingborough, NN29 7NS |
| G0 | PTA | R Attwood, 2 Elizabeth Road, Basingstoke, RG22 6AX |
| G0 | PTD | A Washington, 22 Elm Tree Drive, Bignall End, Stoke-on-Trent, ST7 8NG |
| G0 | PTE | Peter Davidson, 28 Daneswell Drive, Wirral, CH46 1QH |
| G0 | PTG | John Mattison, 21 Maynard House, Dunmow Road, Dunmow, CM6 2DL |
| G0 | PTI | H Aigeldinger, 14 Peregrine Avenue, Morley, Leeds, LS27 8TD |
| G0 | PTK | D Dunford, 25 Northfields Lane, Brixham, TQ5 8RS |
| G0 | PTL | D Caley, 5 Crosswood Close, Bransholme, Hull, HU7 5BU |
| G0 | PTM | A Baird, 65 Waterpump Court, Thorplands, Northampton, NN3 8UR |
| GI0 | PTQ | P Keenan, Drumbadreeuagh, Belleek, Enniskillen, BT93 3FT |
| G0 | PTR | Alan Ryland, Westholme, Blacksmiths Lane, Tewkesbury, GL20 7AH |
| G0 | PTT | Keith Caunce, 7 Trevanions Way, Totland Bay, PO39 0JL |
| G0 | PTU | J Davies, 8668 Ne Orchard Loop Road, Leland, United States, 28451 |
| GM0 | PTY | Alexander Higgins, 26 Kennerty Road, Bucksburn, Aberdeen, AB21 9HS |
| G0 | PUB | Peter Swynford, 6 The Rise, Cold Ash, Thatcham, RG18 9PD |
| G0 | PUD | D Shaw, 27 St. Davids Avenue, Romiley, Stockport, SK6 3JT |
| G0 | PUK | A Johnson, 3 Plantation Avenue, Swalwell, Newcastle upon Tyne, NE16 3JN |
| GW0 | PUM | D Jenkins, Gwalia House, 143a Priory Street, Carmarthen, SA31 1LR |
| GM0 | PUN | H Heritage, 6 Newton Place, Rosyth, Dunfermline, KY11 2LX |
| GW0 | PUP | George Brown, 17 High Street, Senghenydd, Caerphilly, CF83 4GG |
| G0 | PUQ | Hugh O'Hare, 39 Crichton Road, Carshalton, SM5 3LS |
| G0 | PUW | Graham Taylor, 33 Heol Aberwennol, Borth, SY24 5NP |
| G0 | PUY | C Duckworth, 121 Mill Gate, Newark, NG24 4UA |
| G0 | PVB | B Sketcher, 147 Moorside Road, Bradford, BD2 3HD |
| G0 | PVE | K Greaves, 10 Chatsworth Drive, Syston, Leicester, LE7 1HX |
| G0 | PVF | P Benson, 21 Farleigh Road, New Haw, Addlestone, KT15 3HS |
| GI0 | PVG | Thomas Lyons, 3 Clanbrassil Gardens, Portadown, Craigavon, BT63 5YD |
| G0 | PVI | R Loveland, 14 Ashmead, Bordon, GU35 0TL |
| G0 | PVJ | E Hewitt, 8 Embleton Road, Headley Down, Bordon, GU35 8AJ |
| G0 | PVN | C Fleet, 14 Fairwood Road, Penleigh, Westbury, BA13 4EA |
| G0 | PVO | L Hewitt, Sunny Nook, Grains Road, Oldham, OL2 8JF |
| G0 | PVP | C Duffy, 590 Chorley Old Road, Bolton, BL1 6AA |
| G0 | PVQ | P Fuller, 19 Greenwood Court, Webb Close, Crawley, RH11 9JH |
| G0 | PVR | J Davies, 13 The Close, Stalybridge, SK15 1HU |
| G0 | PVT | D henderson, 7 Love Avenue, Dudley, Cramlington, NE23 7BH |
| G0 | PVU | J Roberts, 31 Seaton Way, Marshside, Southport, PR9 9GJ |
| G0 | PVY | A Heward, 22 Ross Avenue, Leasowe, Wirral, CH46 2SB |
| G0 | PWA | D Williams, 31 Piper Hill Avenue, Manchester, M22 4DZ |
| G0 | PWC | B Dawe, 6 Ullswater Avenue, Stourport-on-Severn, DY13 8QP |
| G0 | PWH | Peter Hughes, 21A Erua Road, Waiheke Island, New Zealand, 1081 |
| G0 | PWK | S Alder, 1 Oakdene Terrace, Middlestone Moor, Spennymoor, DL16 7BA |
| G0 | PWL | S Wright, 33 Virginia Avenue, Lydiate, Liverpool, L31 2NN |
| G0 | PWO | A Boyes, 7 Thornwood Covert, Foxwood, York, YO24 3LF |
| G0 | PWQ | W Tonks, 295 Quinton Road West, Quinton, Birmingham, B32 1PG |
| GM0 | PWS | Neil Doherty, Cairdeas, Carrbridge, PH23 3AA |
| G0 | PWU | G Brown, 21 Armada Drive, Teignmouth, TQ14 9NF |
| G0 | PWV | H Chorley, 19 Cleeve Road, Priorswood, Taunton, TA2 8DX |
| G0 | PWW | Theodore Edwards, Kenneggy Lodge, Polperro Road, Looe, PL13 2JO |
| G0 | PWX | G Richards, 87 Woodlands Road, Ditton, Aylesford, ME20 6EF |
| G0 | PXA | G Petri, Mount Holly, Castledon Road, Billericay, CM11 1LH |
| G0 | PXB | C Marsh, 4 South Parade, South Kyme, Lincoln, LN4 4AQ |
| G0 | PXD | A Harrison, 25 Lansbury Avenue, New Rossington, Doncaster, DN11 0AA |
| G0 | PXE | Malcolm Cook, 16 Ascot Drive, Doncaster, DN5 8QA |
| G0 | PXF | K Linsley, 132 Bein Road, Tingley, Wakefield, WF3 1JB |
| G0 | PXG | M Hardman, 47 Oatlands Road, Manchester, M22 1AH |
| G0 | PXH | B Wilkinson, 22 Portree Crescent, Blackburn, BB1 2HB |
| G0 | PXI | P Rigby, 41 St. Huberts Road, Great Harwood, Blackburn, BB6 7AS |
| G0 | PXK | N Pratt, 23 Hall Lane, Whitwick, Coalville, LE67 5FD |
| G0 | PXL | D Martland, 6 Omega Way, Trentham, Stoke-on-Trent, ST4 8TF |
| G0 | PXM | G Kirby, Tralee, Brookside, Newport, PO30 4DJ |
| G0 | PXO | J Morgan, 5 Sealy Close, Wirral, CH63 9LP |
| G0 | PXP | T Cox, 60 Seven Oaks Crescent, Bramcote, Nottingham, NG9 3FP |
| G0 | PXQ | C Bell, 17 Jubilee Square, South Hetton, Durham, DH6 2TR |
| GM0 | PXR | F Coghill, Largiemore, Otter Ferry, Tighnabruaich, PA21 2DH |
| GI0 | PXS | J Madden, The Cottage, 53 Clarendon Street, Londonderry, BT48 7ER |
| G0 | PXT | E Denman, 15 Clare Way, Bexleyheath, DA7 5JU |
| GM0 | PXV | P Barclay, 15 Craigmount Avenue North, Edinburgh, EH12 8DH |
| G0 | PXX | E Mason, 36 Gattison Lane, New Rossington, Doncaster, DN11 0NQ |
| G0 | PXY | D Smith, 27 Hanbury Close, Cheshunt, Waltham Cross, EN8 9BZ |

UK Callsigns

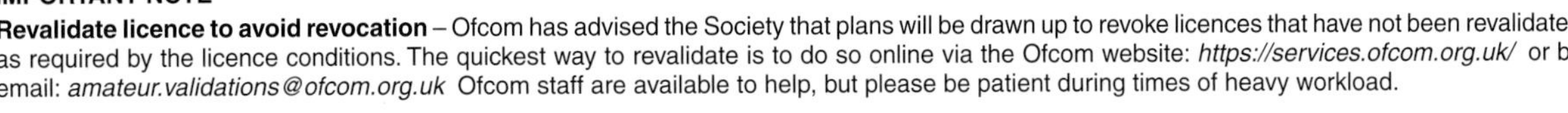

**IMPORTANT NOTE**

**Revalidate licence to avoid revocation** – Ofcom has advised the Society that plans will be drawn up to revoke licences that have not been revalidated as required by the licence conditions. The quickest way to revalidate is to do so online via the Ofcom website: *https://services.ofcom.org.uk/* or by email: *amateur.validations@ofcom.org.uk* Ofcom staff are available to help, but please be patient during times of heavy workload.

G0 PXZ G Walker, 54 Burnage Lane, Burnage, Manchester, M19 2NL
G0 PYD M Hague, Brookview, Milburn Grange, Kenilworth, CV8 2FE
G0 PYE Aidan Arnold, 2 Duck Lane, Haddenham, Ely, CB6 3UE
G0 PYF A De Buriatte, Tanglewood, East End, North Leigh, OX8 6PZ
G0 PYI George Bodaly, 41 Robert Street, Northampton, NN1 3BL
G0 PYJ J Swatton, 30 Squires Close, Crawley Down, Crawley, RH10 4JQ
GM0 PYM Paisley ARC c/o S McKinnon, 8 Rowanlea Avenue, Paisley, PA2 0RP
G0 PYS D Rose, 99 Blackfriars, Rushden, NN10 9PF
GW0 PYU H Clarke, 3 Tanyrallt Avenue, Bridgend, CF31 1PQ
G0 PYV M Hainesborough, 39 Princes Close, North Weald, Epping, CM16 6EW
G0 PYW A Haworth, 97 Challis Lane, Braintree, CM7 1AL
G0 PZB William Hattrick, 38 Nithsdale Road, Weston-Super-Mare, BS23 4JR
G0 PZC D Flitterman, Flat 7, 1 Rutland Gate, London, SW7 1BL
G0 PZD G Holmes, 6 Darleydale Drive, Eastham, Wirral, CH62 8EX
G0 PZF J O'Connell, Apartment 5, Roxboro House, Bailick Road, Co Cork, Ireland
G0 PZJ DUXFORD RS c/o R Pope, 95 Northolt Avenue, Bishop's Stortford, CM23 5DS
G0 PZM N Byron, 2 St. Aidans View, Boosbeck, Saltburn-by-The-Sea, TS12 3LS
G0 PZN A Jones, Edorene, Tincleton, Dorchester, DT2 8QR
G0 PZO Charles Jordan, 31 Rocky Bank Road, Birkenhead, CH42 7LB
G0 PZP William Rabbitt, 21 Barnfield Road, Woolston, Warrington, WA1 4NW
G0 PZR PENZANCE RC c/o Owen Prosser, 2 Caroline Close, Ventonleague, Hayle, TR27 4EX
G0 PZS T Edwards, 6 Cottage Home, Newborough, Llanfairpwllgwyngyll, LL61 6SY
G0 PZT E Allely, Dwyfor, Rhiw, Pwllheli, LL53 8AE
G0 PZU A Ward, 158 Mold Road, Mynydd Isa, Mold, CH7 6TF
G0 PZW J Birch, 32 Poplar Grove, Scotter, Gainsborough, DN21 3TZ
G0 PZX A Dennis, 44 Larksfield Road, Faversham, ME13 7ES
GW0 PZZ Mark Owen, 90 Shakespeare Avenue, Penarth, CF64 2RX
GW0 RAD John Lewis, 189 Heol y Gors, Cwmgors, Ammanford, SA18 1RF
G0 RAE R Walker, 12 Hill Drive, Ackworth, Pontefract, WF7 7LQ
G0 RAF RAF WADDINGTON ARC c/o Robert Pickles, The Pyewipe, Saxilby Road, Lincoln, LN1 2BG
GU0 RAG Keith De La Haye, Flat 4, Forest Lodge Flats, Forest Lane, St. Peter Port, Guernsey, GY1 1WJ
G0 RAL Philip Vallis, Gryphon, Dirtham Lane, Leatherhead, KT24 5SD
G0 RAM Maurice King, 65 Chepstow Close, Stevenage, SG1 5TT
G0 RAN M Jamil, 29 Harrow Close, Bury, BL9 9UD
GM0 RAO A Williamson, Cairn Cottage, Durris, Banchory, AB31 6DT
G0 RAR A Walton, Flat 25 Albert Weedall Centre, 23 Gravelly Hill North, Birmingham, B23 6BT
G0 RAS Victor Maddex, 3 The Vines, Shabbington, Aylesbury, HP18 9HH
G0 RAT W Barnes, 17 Saxon Way, Bradley Stoke, Bristol, BS32 9AR
G0 RAU D Woodnutt, 17 Hill Farm Road, Chalfont St. Peter, Gerrards Cross, SL9 0DD
G0 RAV R Ravenscroft, 4 The Paddock, Lidlington, Bedford, MK43 0RW
G0 RAX R Preston, 45 Long Meadow, Skipton, BD23 1BP
G0 RBA Edward Bannister, 59 Home Farm Park, Lea Green Lane, Nantwich, CW5 6ED
G0 RBB M Batchelor, 16 Clementi Avenue, Holmer Green, High Wycombe, HP15 6TN
GI0 RBC J Thompson, 3 Strandburn Park, Sydenham, Belfast, BT4 1ND
G0 RBD D Kiely, 45 Redland, Chippenham, SN14 0JB
G0 RBH R Hughes, 17 Pentrosfa Road, Llandrindod Wells, LD1 5NL
G0 RBI S Ward, Oaklands, Burtonwood Road, Warrington, WA5 3AN
G0 RBJ P Evans, Flat 7, 150 Booker Avenue, Liverpool, L18 9TB
G0 RBM C Boland, 13 Rushfield Crescent, Brookvale, Runcorn, WA7 6BN
GI0 RBO Jonathan Kernohan, 17 Tullygrawley Road, Teeshan, Ballymena, BT43 5NP
G0 RBQ R Gibbs, 13 Mulberry Gardens, Old Guildford Road, Horsham, RH12 3NH
GI0 RBS Robert Rainey, 4 Crossnadonnell Road, Limavady, BT49 0BD
G0 RBV P Brunton, Heathcote Kents Road, Torquay, TQ1 2NL
G0 RBW T Jones, 11 Coppice Close, Willaston, Nantwich, CW5 6NL
G0 RCF E Carrington, 4 Lancaster Drive, East Grinstead, RH19 3XF
GW0 RCG R Giddings, 28 Argrove, Port Talbot, SA12 8PP
G0 RCH Thomas Cullup, 13 London Street, Whittlesey, Peterborough, PE7 1BP
G0 RCI Alan Gibson, 1 Oakleigh Road, Grantham, NG31 7NN
G0 RCJ J Topham, 23 St. Nicholas View, West Boldon, East Boldon, NE36 0RF
G0 RCL Owen Baldwin, 23 Cherry Tree Walk, Tadcaster, LS24 9HS
G0 RCN N Allen, 78 Bargates, Christchurch, BH23 1QL
G0 RCP P Mellors, 64 Pinewood Way, North Colerne, Chippenham, SN14 8QU
G0 RCS Royal Signals ARS c/o Ian McGowan, Meld House, Hawthorn Road, Shrewsbury, SY3 7NB
G0 RCU R Thomas, 164 Kings Head Lane, Bristol, BS13 7BW
G0 RCW W CHESHIRE RAY c/o Francis Mossop, 4 Brookdale Way, Waverton, Chester, CH3 7NT
G0 RCX C Garbett, 27 Burghley Drive, West Bromwich, B71 3LX
G0 RCY M Crimes, 27 Dunmore Road, Little Sutton, Ellesmere Port, CH66 4PD
GM0 RDA G Adamson, 10 Rossend Terrace, Burntisland, KY3 0DQ
G0 RDB Christopher Fernie, 2 Hopkins Close, Cambridge, CB4 1FD
G0 RDC C Robinson, Greenhouse Farm, Lilliesleaf, Melrose, TD6 9EP
G0 RDD M Prendergast, 1 Olaman Walk, Peterlee, SR8 2EA
G0 RDF L Wolstenholme, The Hollies, Avondale Road, Chesterfield, S40 4TF
G0 RDG G Kowalski, 47 Graveney Place, Springfield, Milton Keynes, MK6 3LU
G0 RDH Brian Watson, 7 Branksome Drive, Morecambe, LA4 5UJ
GI0 RDJ I McMullan, 35 Howard Place, Lisburn, BT28 1EX
G0 RDK Christopher Wiseman, 42 Merlin Way, Kidsgrove, Stoke-on-Trent, ST7 4YL
GI0 RDM K Mcguckin, 20 Lisnahull Park, Dungannon, BT70 1UH
G0 RDN Gordon Johnston, 11 Granville Street, Deal, CT14 7EZ
G0 RDO J Snell, 5 Waverley Road, Newton Abbot, TQ12 2ND
G0 RDP David Peat, Jeswyn, Brookland Avenue, Mansfield, NG18 5NB
G0 RDR B Lowe, 18 Lowther Drive, Swillington, Leeds, LS26 8QG
G0 RDS A Williams, 30 Swan Close, Talke, Stoke-on-Trent, ST7 1TA
G0 RDT D Treen, 13 Peveril Road, Duston, Northampton, NN5 6JW
G0 RDU S Emms, 33 Whitworth Avenue, Stoke Aldermoor, Coventry, CV3 1EQ
G0 RDV L Davies, 3 Rydalside, Kettering, NN15 7DR
G0 RDX P Walker, Moze Cross Cottage, Beaumont Road, Harwich, CO12 5BQ
G0 RDY G Steel, Long Close, 82 Whatton Road, Derby, DE74 2DT
G0 RDZ S Smith, 12 Home Avenue, Duns, TD11 3HQ
G0 REA R James, Woodpeckers, Freshwater Lane, Truro, TR2 5AR
G0 REB C Salmon, 14 Surrey Drive, Congleton, CW12 1NU
GM0 RED EAST DUNBARTONSHIRE RAYNET GROUP c/o James Pert, 56 Lochiel Drive, Milton of Campsie, Glasgow, G66 8ET
G0 REE Dennis Jones, 120 Heathfield Road, Keston, BR2 6BF
G0 REF EPPING FOREST RAYNET GROUP c/o J Andrews, 85 Little Cattins, Harlow, CM19 5RN
G0 REL D Gaskell, 18 Woodcroft, Kennington, Oxford, OX1 5NH
G0 REN Chris Wienrich, 94 Sandling Lane, Penenden Heath, Maidstone, ME14 2EA
G0 REO Peter Hill, 16 Robins Way, Nuneaton, CV10 8PA
G0 REP Antony Blackburn, 2 Blackthorn Road, Stratford-upon-Avon, CV37 6TD
G0 REQ D HIBBERD, 11 Borrowdale, Brownsover, Rugby, CV21 1NH
G0 REU Tung Lam, 48 Lancaster Gate, Upper Cambourne, Cambridge, CB23 6AT
G0 REV A Bowmaker, Post Cottage, Ardley Road, Bicester, OX25 6LP
GM0 REW Julie Greig, 35 Sir Thomas Elder Way, Kirkcaldy, KY2 6ZR
GM0 REZ A Dailey, 82 Don Drive, Livingston, EH54 5LP
G0 RFA S Garczynski, 19 Thornhill Croft, Leeds, LS12 4JX
G0 RFE Alan Moore, 139 Argyle Street, Heywood, OL10 3RS
G0 RFF C Bourne, Essams, 11 The Grove, Hailsham, BN27 3HU
G0 RFG E Hyde, 63 Newlyn Drive, Sale, M33 3LH
G0 RFL T Pooley, 133 Hardie Road, Dagenham, RM10 7BT
G0 RFM John Copplestone, 25 Bruche Avenue, Paddington, Warrington, WA1 3HX
G0 RFN James Taylor, 121 Garesfield Gardens, Burnopfield, Newcastle upon Tyne, NE16 6LQ
G0 RFQ Geoffrey Buck, 3 Church View, Trawden, Colne, BB8 8SA
G0 RFS C Bradley, 71a Bagley Wood Road, Kennington, Oxford, OX1 5LY
G0 RFT R Lagar, 25 Neville Avenue, Warrington, WA2 9BQ
G0 RFX P Walford, Suite 184, 2 Old Brompton Road, London, SW7 3DQ
G0 RFY David Horton, 21 St. James Street, Waterfoot, Rossendale, BB4 7HN
G0 RGC J Bridge, Little House, Castle House Yard, Langport, TA10 9PR
G0 RGE M Jenkinson, 25 Porchester Close, Hucknall, Nottingham, NG15 7UB
G0 RGG John Hubbard, 4 Avondale, Ellesmere Port, CH65 6RW
G0 RGH HARIG c/o Jonathan Mitchener, Cabins, Wenham Road, Ipswich, IP8 3EY
G0 RGJ R Provins, 42 Forest View Road, Tuffley, Gloucester, GL4 0BX
G0 RGL D Edmondson, 64 Raleigh Avenue, Hayes, UB4 0EF
G0 RGM J Trice, 71 Deerswood Road, Crawley, RH11 7JP
G0 RGN B Woodhead, 16 Dow Street, Hyde, SK14 4BS
G0 RGO J Drummond, 14 Bulls Head Cottages, Turton, Bolton, BL7 0HS
G0 RGP A Gibbs, 17 Manor Bend, Galmpton, Brixham, TQ5 0PB
G0 RGU John O'Gorman, 141 Chesterfield Road, Huthwaite, Sutton-in-Ashfield, NG17 2QF
G0 RGW Richard Slatter, 1 Angells Meadow, Ashwell, Baldock, SG7 5QS
G0 RGX J Sandys, Tarn Cottage, The Maultway, Camberley, GU15 1PS
G0 RHB L Mulford, 55 Mill Farm Crescent, Hounslow, TW4 5PF
GW0 RHC K Dyer, 34 Lundy Drive, West Cross, Swansea, SA3 5QL
GW0 RHE Stephanie Williams, 5 Llys Yr Orsaf, Llanelli, SA15 2LB
G0 RHF P Ellwood, Coire Cas, Marsh Lane, Poulton-le-Fylde, FY6 9AW
G0 RHG D Stewart, 36 Monson Road, Redhill, RH1 2EZ
G0 RHH Dorothy Barnett, 26 Casewell Road, Kingswinford, DY6 9HB
G0 RHI B Dooks, 7 Manor Drive, Kirby Hill, York, YO51 9DY
G0 RHJ Richard Judson, 4 Hill Close, Cromer, NR27 0HX
G0 RHK P Ford, 19 Swan Bank, Hay-on-Wye, Hereford, HR3 5DW
G0 RHO John Belling, 77 Chantry Road, Marden, Tonbridge, TN12 9JD
GM0 RHP David Crooke, 2 Main Street, Carnock, Dunfermline, KY12 9JQ
G0 RHV J Parish, 83 Harold Road, Stubbington, Fareham, PO14 2QS
G0 RHY A Howarth, 12 Welbeck Road, Worsley, Manchester, M28 2SL
G0 RIC R Cannell, 284 Archway Road, Highgate, London, N6 5AU
G0 RIE D Reilly, 15 Shutewater Close, Bishops Hull, Taunton, TA1 5EH
G0 RIF Dean Barnes, 11 Back Lane, Whittington, Lichfield, WS14 9NH
G0 RII John Spacey, 43 Woodlands Road, Allestree, Derby, DE22 2HG
G0 RIJ W Sykes, Summerfield, Second Avenue, Ross-on-Wye, HR9 7HT
G0 RIK N Stockwell, 12 Weavers Mead, Great Cheverell, Devizes, SN10 5TP
G0 RIP J Austwick, 12 Dove Walk, Farnworth, Bolton, BL4 0RQ
G0 RIQ D Wisbey, 22 Rutland Drive, Hornchurch, RM11 3EN
G0 RIT A Nunneley, The Potter'S Wheel, Mullion Cove, Nr Helston, TR12 7ET
G0 RIU Peter Davis, 21 Newton Way, St. Osyth, Clacton-on-Sea, CO16 8RQ
GM0 RIV RAYNET INVERNES c/o Norman Baird, 23 Scorguie Avenue, Inverness, IV3 8SD
G0 RIX B Cook, 7 Rosewood Gardens, New Milton, BH25 5NA
G0 RIY G Watson, 1 Post Office Cottages, Loddiswell, Kingsbridge, TQ7 4QH
G0 RIZ B Body, 12a Elm Court Gardens, Truro, TR1 1DS
G0 RJA K Jones, 10 Dale Terrace, Lingdale, Saltburn-by-The-Sea, TS12 3EE
G0 RJC V Turner, 7 Highfield Crescent, Baildon, Shipley, BD17 5NR
G0 RJE R Enright, 17 Ripston Road, Ashford, TW15 1PQ
G0 RJG E Kelly, Durness, Newbridge, Dumfries, DG2 0QX
G0 RJI N Rapson, 21 Ashley Close, Penwithick, St. Austell, PL26 8UB
G0 RJJ E Foord, 65 Dane Court Gardens, Broadstairs, CT10 2SD
G0 RJL John Hilton, 177 Wilmot Road, Dartford, DA1 3BP
G0 RJN H Vicary, The Brambles, Wrotham Road, Gravesend, DA13 0QA
GI0 RJO L Douglas, 15 Bramhall Crescent, Londonderry, BT47 5HE
G0 RJT H Leak, 15 Sutherland Road, Tittensor, Stoke-on-Trent, ST12 9JQ
G0 RJV G Rogers, Maes Gwersyll, Garthmyl, Montgomery, SY15 6RS
G0 RJX Elizabeth Gaskell, 18 Woodcroft, Kennington, Oxford, OX1 5NH
G0 RKB D Roberts, 20 Beech Grove, Trowbridge, BA14 0HG
G0 RKC A Alecio, 18 Camperdown Terrace, Exmouth, EX8 1EH
G0 RKD S Gray, 613 London Road, Earley, Reading, RG6 1AT
G0 RKE Catherine Burgess, 12 Middleway, Grotton, Oldham, OL4 5SH
G0 RKG R Gaskell, 18 Woodcroft, Kennington, Oxford, OX1 5NH
G0 RKN H Burn, Ashleigh, Leek Road, Stoke-on-Trent, ST9 0DG
G0 RKP J Aubin, 46 Kenilworth Drive, Clitheroe, BB7 2QN
G0 RKQ Raymond Plumtree, 80 Dewsbury Avenue, Scunthorpe, DN15 8BP
G0 RKS Grant Goss, Little Ashcroft, Parkgate Road, Dorking, RH5 5DZ
G0 RKT D Dukesell, Mayfield, Ashbourne Road, Buxton, SK17 9RY
GM0 RKU P Craft, 2 Luke Place, Broughty Ferry, Dundee, DD5 3BN
G0 RKV Verdun Webley, 2 Octavian Drive, Bancroft, Milton Keynes, MK13 0PN
G0 RLA Paul Harvey, Rowlands Barn, Dunbridge Lane, Awbridge, Romsey, SO51 0GQ
G0 RLB B Stoneley, 44 Ilthorpe, Hull, HU6 9ER
G0 RLF George Low, 61 Fenwick Lane, Halton Lodge, Runcorn, WA7 5YU
G0 RLH E Miles, 31 Winnipeg Road, Bentley, Doncaster, DN5 0ED
G0 RLI J Thomas, 204 Watchouse Road, Galleywood, Chelmsford, CM2 8NF
G0 RLJ P Tyson, 44 Windmill Avenue, Kilburn, Belper, DE56 0PQ
G0 RLL T Dyson, 4 Lyspitt Common, Meppershall, Shefford, SG17 5GZ
G0 RLN K Taylor, 29 School Road, Pontefract, WF8 2AJ
G0 RLO Kathryn Conlon, 4 Hill Crest Drive, Slack Head, Milnthorpe, LA7 7BB
GW0 RLQ J Ellwood, 5 Smallwood Road, Baglan, Port Talbot, SA12 8AP
G0 RLS Philip Ashcroft, 38a Wood End, Bluntisham, Huntingdon, PE28 3LE
G0 RLT R Taylor, Flat 4, 3 St. Pauls Square, Southport, PR8 1NQ
G0 RLV Eric Jones, 16 Fisher Avenue, Rugby, CV22 5HN
G0 RLY J Karkoszka, 5 Wood Street, Haworth, Keighley, BD22 8BJ
GM0 RLZ Colin Brown, 9 Newton Crescent, Rosyth, Dunfermline, KY11 2QW
G0 RMB Stuart Ferris, Temorfa, Y Ffor, Pwllheli, LL53 6UB
G0 RMC M Charlton, 53 Dunstone View, Plymouth, PL9 8TW
G0 RMD P Calter, 8 Exeter Road, Scunthorpe, DN15 7AT
G0 RMG Roy Jones, 6 Wychwood Drive, Hunt End, Redditch, B97 5NW
G0 RMJ S Hogg, 38a High Street, Ventnor, PO38 1RZ
GM0 RML Arthur Smart, 6 Alton Bank, Nairn, IV12 5PJ
G0 RMN A Younger, 4 Esk Hause Close, West Bridgford, Nottingham, NG2 6SG
G0 RMO M Miller, 8 Pilton Walk, Newcastle upon Tyne, NE5 4PQ
G0 RMP R Seal, 2 Shaftesbury Road, Bridlington, YO15 3NP
G0 RMR Chris Rabey, 23 Thorn Lane, Four Marks, Alton, GU34 5BX
GM0 RMT Gordon Wilkie, 25 Barn Road, Stirling, FK8 1EP
G0 RMU R Clover, Teffont, 42 Warren Road, Addlestone, KT15 3UA
GM0 RMV M Verity, 19 Vivian Terrace, Edinburgh, EH4 5AW
G0 RMX Daniel Esdale, The Bell Inn, Central Lydbrook, Lydbrook, GL17 9SB
G0 RNA T Rawlinson, 330 Blackpool Old Road, Poulton-le-Fylde, FY6 7QY
G0 RNB NEIL BROOKS, 57 Mansel Crescent, Parson Cross, Sheffield, S5 9QR
G0 RNC A Pritchard, 27 Walkley Crescent Road, Walkley, Sheffield, S6 5BA
GD0 RNF I Hunnisett, 69 Cornwall Road, Ruislip, HA4 6AJ
G0 RNH M Ahmed, 75 Drove Road, Swindon, SN1 3AE
G0 RNI LUTON REP GRP c/o David Thorpe, 70 Willow Way, Ampthill, Bedford, MK45 2SP
GW0 RNK K Williams, 8 Trinity Place, Pontarddulais, Swansea, SA4 8RD
G0 RNM L Smith, 21 The Drive, Fareham, PO16 7NL
G0 RNN I Tutt, 1 Castle Road, Hadleigh, Ipswich, IP7 6JH
G0 RNO E Prothero, 5 Home Meadow Drive, Flackwell Heath, High Wycombe, HP10 9JY
G0 RNP D Eves, 64 Hillingdon Road, Gravesend, DA11 7LG
G0 RNQ B Willson, 4 Caldew Grove, Sittingbourne, ME10 4SL
G0 RNS John White, 24 Malines Avenue, Peacehaven, BN10 7PS
G0 RNV B Sherriff, 27 Magellan Way, Spalding, PE11 2FG
G0 RNX Sean Onions, 18 St. Cuthberts Crescent, Albrighton, Wolverhampton, WV7 3HW
G0 RNY A Attle, 2 Watson Park, Spennymoor, DL16 6NB
G0 ROA Henry Seidner, 401 East 80th Street, #33A, New York, United States, 10075
G0 ROB R Gilbert, 35 Lower Road, Malvern, WR14 4BX
G0 ROD C Reaney, 81a Bargate Road, Belper, DE56 1NE
G0 RON Ronald McNeil, 3 Thorncliffe Gardens, Auckley, Doncaster, DN9 3PE
G0 ROO Dover Construction Club c/o Ian Keyser, Rosemount, Church Whitfield Road, Dover, CT16 3HZ
G0 ROS R Kent, 40 Waxes Close, Abingdon, OX14 2NG
G0 ROT Michael Davis, 90 Belfield Road, Epsom, KT19 9SY
GM0 ROU Antony Butcher, 56 Glamis Road, Dundee, DD2 1TU
G0 ROV Arthur Hall, 21 Embry Close, Calne, SN11 8QS
G0 ROW Alan Gurnhill, 53 Millbrook Avenue, Denton, Manchester, M34 2DQ
G0 ROX D Lee, 131 Abbotsbury Road, Weymouth, DT4 0JX
G0 ROY R Biddle, 21 Kingsway West, Newton, Chester, CH2 2LA
G0 ROZ DORSET POLICE AR c/o Clive Hardy, 40 Beresford Road, Poole, BH12 2HE
G0 RPA I McAvoy, 5 Lytchett Way, Upton, Poole, BH16 5LS
G0 RPD J Barton, 183 Windy Arbor Road, Whiston, Prescot, L35 3SF
G0 RPF L Smith, 28 Chester Road, Stockton Heath, Warrington, WA4 2RX
G0 RPG J Riley, 1 Chatsworth Avenue, Culcheth, Warrington, WA3 4LD
G0 RPJ D Wesil, 8 Camber Way, Pevensey Bay, Pevensey, BN24 6RW
G0 RPL N Alison, 9 South Drive, Burgess Hill, RH15 9PY
G0 RPM Nigel Williams, 11 Berkeley Gardens Winchmore Hill, London, N21 2BE
G0 RPO R Dowd, Belgrano, 1 Watson Avenue, Warrington, WA3 3QX
G0 RPU J Symonds, La Cumbre, 35 Byward Drive, Scarborough, YO12 4JE
G0 RPV W Till, 97 Haslar Crescent, Waterlooville, PO7 6DD
G0 RPW Donald Wilson, 39 The Wintles, Bishops Castle, SY9 5ES
G0 RPX R Evans, 426 Hawthorn Crescent, Cosham, Portsmouth, PO6 2TX
G0 RPY C Button, 8 Heywood Road, Diss, IP22 4DJ
G0 RQF K Hales, 3 New Barnfields, Stretton Sugwas, Hereford, HR4 7AZ
G0 RQG J Gill, 24 Greenfields Court, Bridgnorth, WV16 4JS
G0 RQH D Hughes, 31 Sussex Drive, Pagham, Bognor Regis, PO21 4RN
G0 RQI Stephen Spragg, 4 Valley Road Arleston, Telford, TF1 2JP
G0 RQL Donald Roomes, View Field, Milton Damerel, Holsworthy, EX22 7NY
G0 RQN P Robertson, 1 Yaffle Mews, Great Cambourne, Cambridge, CB23 5HY
G0 RQO David Hillyer, 32a Belbroughton Road, Blakedown, Kidderminster, DY10 3JG
GW0 RQP Gerald Ashford, 26 Laura Street, Treforest, Pontypridd, CF37 1NW
G0 RQQ Keith Ballinger, 3 Cliff Court, Burton Road, Lincoln, LN1 3NN
G0 RQS L Pritchard, 86 Bryn Road, Markham, Blackwood, NP12 0QE
G0 RQX David Townend, 38 Kingston Drive, Shrewsbury, SY2 6SJ
G0 RQZ Richard Lawrence, 74 Principal Rise, 74 Principal Rise, York, YO24 1UF
G0 RRC R Smith, Lykkebo, The Street, Ipswich, IP8 3DN
G0 RRI Ian Burden, 97 Bevan Street West, Lowestoft, NR32 2AF
GM0 RRK Maurice Boyce, 93 Ledi Drive, Bearsden, Glasgow, G61 4JP
G0 RRM P Brumby, 69 Gilbert Walk, Nether Stowe, Lichfield, WS13 6AU
G0 RRO J Breingan, 44 Farmstead Road, Corby, NN18 0LG
G0 RRR J Marsden, 11 Firethorn Drive, Hyde, SK14 3SN
G0 RRV A Tomson, 5 Fordham Close, Ashwell, Baldock, SG7 5LJ
G0 RRX P Bethell, 7 Silverton Grove, Middleton, Manchester, M24 5JH

G0 RRZ R Carrington, 45 Crompton Road, Pleasley, Mansfield, NG19 7RG
G0 RSA John King, 39 Nursery Gardens, St. Ives, PE27 3NL
GM0 RSE Glenrothes & District ARC c/o Tam Brown, 11 Approach Row, East Wemyss, Kirkcaldy, KY1 4LB
G0 RSG 1st Ringmer Scout Group c/o Tim McConnell, 31 Langney Road, Eastbourne, BN21 3QD
GM0 RSI J Ritchie, 36 James Mitchell Place, Mintlaw, Peterhead, AB42 5ES
G0 RSL K White, 25 Curson Rise, Kendal, LA9 7PN
G0 RSR READING SCOUTS RADIO c/o S French, 22 Amity St., Newtown, Reading, RG1 3LP
G0 RSS R Simmonds, 4 Corys Close, Kirby Road, Norwich, NR14 7DP
G0 RSU G Weston, 2 Whitburn Road, Toton, Nottingham, NG9 6HP
G0 RSV W Webster, 21 Quince Tree Way, Hook, RG27 9SG
G0 RSW R Waters, 17 Wilson Road, Southend-on-Sea, SS1 1HG
G0 RSY A Gibbs, Orchard Court, Woodside Road, Wootton Bridge, Ryde, PO33 4JR
G0 RTA T Arakawa, 2-974-8-1502, Sayama, Osakasayama, Japan, 589-0005
G0 RTC T Chisholm, 316 Birchfield Road East, Northampton, NN3 2SY
G0 RTF I Slaney, 6 Little Shardeloes, High Street, Amersham, HP7 0EF
G0 RTH Alan Elcoate, 9 Parsonage Lane Laindon, Basildon, SS15 5YN
G0 RTI S Harriss, 6 Redland Road, Leamington Spa, CV31 2PB
G0 RTM P McKnight, 39 Dunmail Drive, Kendal, LA9 7JG
G0 RTN Gerry Lynch, Dean Chase, West Dean, Salisbury, SP5 1JJ
GW0 RTP Christopher Llewellyn, 16 Garth Street, Kenfig Hill, Bridgend, CF33 6EU
G0 RTQ David Lawrence, 75 Church Street, Ilkeston, DE7 8QP
GW0 RTR R Rees, 45 Sandy Road, Llanelli, SA15 4BR
G0 RTU Paul Kirkup, 337 Wheatley Lane Road, Fence, Burnley, BB12 9QA
GM0 RTY David Inns, 57 Craiglomond Gardens Balloch, Alexandria, G83 8RP
G0 RTZ G Hurst, 35a Trewsbury Road, London, SE26 5DP
GI0 RUC R Kerr, 194 Shore Road, Greenisland, Carrickfergus, BT38 8TX
GW0 RUD Peter Marriott, 16 Heol Morlais, Llannon, Llanelli, SA14 6BD
G0 RUF Neil Taylor, 30 Leonard Street, Hull, HU3 1SA
G0 RUH M Roberts, 82 Glover Road, Scunthorpe, DN17 1AS
G0 RUR Paul Simpson, Amber Lodge Nursing Home, 684-686 Osmaston Road, Derby, DE24 8GT
G0 RUS R Lenthall, 182 Chelmsford Avenue, Grimsby, DN34 5DB
G0 RUT Raymond Russell, 4 Hinton Road, Newport, PO30 5QZ
G0 RUV M Gent, 111 Portland St., Clowne, Chesterfield, S43 4SA
GM0 RUW J Coughtrie, 61 Bells Burn Avenue, Linlithgow, EH49 7LD
G0 RUX W Taylor, 21 Summerdale Road, Cudworth, Barnsley, S72 8XG
G0 RUY A Pritchard, 41 Borough Close, Kings Stanley, Stonehouse, GL10 3LJ
G0 RUZ C Farlow, 4 Nether Road, Silkstone, Barnsley, S75 4NN
G0 RVE A Pierce, 34 Church Close, Shawbury, Shrewsbury, SY4 4JX
G0 RVH K Dailey, 55 Chesterton Avenue, Harpenden, AL5 5SU
G0 RVI J Davis, 4 Stockbridge Close, Canford Heath, Poole, BH17 8SU
G0 RVK M Fogg, 15 Elm Grove, Bisley, Woking, GU24 9DG
G0 RVM A Gawthrope, 62 Meadow Way, Bradley Stoke, Bristol, BS32 8BP
GW0 RVR Robert Goodall, 8 Heol Penderyn, Brackla, Bridgend, CF31 2EA
G0 RVS Brian Roff, 1 Kennel Cottages, Arlington, Barnstaple, EX31 4LP
G0 RWA B Chorley, 19 Cleeve Road, Priorswood, Taunton, TA2 8DX
G0 RWI E Johns, 3 The Rowans, Portishead, Bristol, BS20 6SR
G0 RWJ D King, 78 Andersey Way, Abingdon, OX14 5NW
G0 RWL B Mackenzie, 73 Newstead Road, Weymouth, DT4 0AS
GW0 RWM R Martin, Cartref, Maenclochog, Clynderwen, SA66 7LB
GI0 RWO B Madden, 1 Skegoneill Drive, Belfast, BT15 3FY
G0 RWQ N Monument, C/Isla Cabrera 14.1.10, Regia Roig Blq 2, Orihuela, Spain, 3189
G0 RWS W Scott, Hunters Lodge, Broadmore Green, Worcester, WR2 5TE
G0 RWT R Pine, Rhodanna, Tennis Court Road, Bristol, BS39 7LU
G0 RWU J Ponton, Old Cottage Gardens, Legerwood, Earlston, TD4 6AS
G0 RWW M Barrass, 11 Flintham Court, Mansfield, NG18 4NB
G0 RWY D Ramsay, 2 Old Church Road, Colwall, Malvern, WR13 6ET
G0 RXA N Roscoe, 35 Kenilworth Road, Cheadle Heath, Stockport, SK3 0QL
G0 RXQ F Lockey, The Dormers, Cirencester Road, Tetbury, GL8 8HA
G0 RXU F Nethercott, 6 Laking Avenue, Broadstairs, CT10 3NE
GM0 RYA G Roberts, 8 Parkview, Lhanbryde, Elgin, IV30 8JZ
GM0 RYD J Van Dyke, 112 Alexander Avenue, Largs, KA30 9EX
GI0 RYK R White, 1 Woodland Park, Lisburn, BT28 1LD
G0 RYL Robert Hodges, 1a Clements Lane, Portland, DT5 1AS
G0 RYM S Goodwin, 14 Greenhill, Alveston, Bristol, BS35 2QX
G0 RYO John Webb, Little Johns Cottage, Milton Damerel, Holsworthy, EX22 7DL
G0 RYP C Martin, 7 St Lawrence Street, B'kara, Malta, BKR1521
G0 RYQ Peter Irwin, 11 Cowdale Cottages, Cowdale, Buxton, SK17 9SE
G0 RYR Timothy Ballinger, 9 Somerville Court, Cirencester, GL7 1TG
G0 RYS RICHMOND SCHOOL AMATEUR c/o M Vann, Richmond School, Darlington Road, Richmond, DL10 7BQ
GI0 RYU B Millar, 312 Churchill Park, Portadown, Craigavon, DT02 1EY
G0 RYV A Smith, Crown Cottage, Stone, Berkeley, GL13 9LE
G0 RYW S Preston, The Chapel, Robson Street, Shildon, DL4 1EB
G0 RZB D McDonnell, Glencoe, The Ridge, Salisbury, SP5 2LN
G0 RZG Richard Hayward, Old School Farm, Wickham Market, IP13 0HE
G0 RZI B Easdon, 20 Winder Gate, Frizington, CA26 3QS
G0 RZM R Judd, 24 Haywood Way, Reading, RG30 4QP
G0 SAC SUTTON AREA CONTEST GROUP c/o Alun Cross, 31 Mountcombe Close, Surbiton, KT6 6LJ
GW0 SAJ H Jones, 6 Westfa Road, Uplands, Swansea, SA2 0PR
G0 SAR SOUTH ANGLIA RAYNET c/o David Sparrow, 23 Tranmere Grove, Ipswich, IP1 6DU
G0 SAY C Thorpe, 78 Bowland Road, Wythenshawe, Manchester, M23 1JX
G0 SBA David Sant, Marjar, Marden, Hereford, HR1 3EP
G0 SBB David Barton, Manuka, West Hill, Worthing, BN13 3BZ
G0 SBC R Harris, 142 St. Nicolas Park Drive, Nuneaton, CV11 6EE
G0 SBH T Wernham, 2 Red House Farm Cottages, Station Road, Woodbridge, IP12 2DG
G0 SBK M Jenkins, 9 Tothill Road, Swaffham Prior, Cambridge, CB25 0JX
G0 SBM SOUTH DEVON RAYNET GROUP c/o Colin Coker, 46 Clarendon Road, Ipplepen, Newton Abbot, TQ12 5QS
G0 SBO E Hodgson, 21 Royd Avenue, Mapplewell, Barnsley, S75 6HH
G0 SBP Frederick Parkinson, 28 Gouldsmith Gardens, Darlington, DL1 2DU
G0 SBU Barry Wedgwood, 40 Ford Street, Consett, DH8 7AE
G0 SBV R Talbot, 11 Whitefield Road, Holbury, Southampton, SO45 2HP
G0 SBX E Barclay, 58 Stockton Road, Hartlepool, TS25 1RW
G0 SBY J Thompson, 4 Ridgemont, Fulwood, Preston, PR2 3FQ
G0 SBZ W Sandle, 507b Harrogate Road, Leeds, LS17 7DU
GM0 SCA S Edwards, The Old Police House, Broughton, Biggar, ML12 6HQ
G0 SCG A Leavey, 14 Cherry Close, Ealing, London, W5 4JW
G0 SCI Barry Young, 11 Gainsborough Avenue, Washington, NE38 7EF
G0 SCK D Britton, 31 Clay Bottom, Bristol, BS5 7EJ
G0 SCL S Lawrence, 4 Dale Park Rise, Leeds, LS16 7PP
G0 SCM F Binnington, 7 Webbs Close, Combs, Stowmarket, IP14 2NZ
G0 SCO SCOTTISH OFFICE ARC c/o J Lefever, 74 Ferneley Crescent, Melton Mowbray, LE13 1RZ
G0 SCQ D Brusch, 7 Tyrell Close, Stanford in the Vale, Faringdon, SN7 8EY
G0 SCR CATERHAM RADIO GROUP c/o Paul Lewis, 20 Annes Walk, Caterham, CR3 5EL
G0 SCT Roy Brickell, 82 Hills Road, Cobham Hills, Thetford, IP26 7EZ
G0 SCU F Taylor, 6 Shelley Close, Bolton le Sands, Carnforth, LA5 8HQ
G0 SCV G Belt, Flat 2, 3 King George Avenue, Leeds, LS7 4LH
GM0 SCW Robert Anderson, 10 Cyril Crescent, Paisley, PA1 1GT
G0 SCX Geoffrey Hobbs, 37 Winnards Park, Sarisbury Green, Southampton, SO31 7BX
G0 SCY W Dest, 61 Gainsborough, Hanworth, Bracknell, RG12 7WL
G0 SDC SOUTHERN DX CLUB c/o Peter Robinson, 11 The Avenue, Hambrook, Chichester, PO18 8TZ
G0 SDD C James, 4 Hill Top, Bream, Lydney, GL15 6JQ
G0 SDE B Jupp, 25 Briscoe Way, Lakenheath, Brandon, IP27 9SA
G0 SDF John Atkins, 30 Bransby Road, Chessington, KT9 2LA
G0 SDJ A McMullon, Carwood House, Hothersall Lane, Preston, PR3 2XB
G0 SDL J Wilson, Appletrees, Combeinteignhead, Newton Abbot, TQ12 4RE
G0 SDM Philip Robinson, The Swallows, Cock Fen Road, Lakes End, Wisbech, PE14 9QE
G0 SDQ L Talkowski, 83 Church Road, Peasedown St. John, Bath, BA2 8AB
GM0 SDS Brian Wills, 24 Hopes Avenue, Dalmellington, Ayr, KA6 7RN
G0 SDW M Pattman, 4 Branscombe Road, Bristol, BS9 1SN
G0 SDX WILLPOWER CONTEST GROUP c/o Jeremy Faulkner-Court, Yew Tree Cottage, Avon Dassett, Southam, CV47 2AT
G0 SDZ J Thomas, Flat 45, Kyrle Pope Court, Sudbury Avenue, Hereford, HR1 1XZ
G0 SEB J Shepherd, 25 Station Road, St. Helens, Ryde, PO33 1YF
G0 SEC J Curtis, 24 Brisbane Road, Weymouth, DT3 6RD
GM0 SEF David Stolting, 3 Eden Park, Cupar, KY15 4HS
GM0 SEI R Vennard, 4 Braehead, Girdle Toll, Irvine, KA11 1BD
G0 SEN S Thompson, Elmtree Cottage, Woodrow, Sturminster Newton, DT10 2AQ
GM0 SEP STRATHCLYDE EMERGENCY PLANNING UNIT c/o R Cowan, 85 Eastwoodmains Road, Clarkston, Glasgow, G76 7HG
G0 SET H Pearson, 55 Cowper Road, River, Dover, CT17 0PL
G0 SEU L Payas, C/O Jo-Anne Maclaren, 7801 Hibiscus Court , Gibraltar
G0 SEW K Green, 13 Knowle Road, Sheffield, S5 9GA
G0 SEY Edwin Russell, 60 Icknield Way, Tring, HP23 4HZ
G0 SFA B Hyde, 108 St. Bedes Crescent, Cambridge, CB1 3UB
G0 SFE K Spring, 18 Greenway, Woodmancote, Cheltenham, GL52 9HU
G0 SFG Tony Coneley, Flat 43, Crown Mews, 15 Clarence Road, Gosport, PO12 1DH
GD0 SFI B Hull, Uplands, Ballavitchel Road, Isle of Man, IM4 2DN
G0 SFJ Andrew Thomas, 21 Great Bowden Road, Market Harborough, LE16 7DE
G0 SFP B Rish, 15 Bryn Marl, Deganwy, Llandudno Junction, LL31 9BZ
GM0 SFQ John Stirling, 86 Obsdale Park, Alness, IV17 0TR
GI0 SFT P Mcdonald, 13 Heathfield, Culmore, Londonderry, BT48 8JD
G0 SFV D Burton, 3 Norwood, Carden Hill, Brighton, BN1 8AH
G0 SGF John Barrett, 12 Trent View Gardens, Radcliffe-on-Trent, Nottingham, NG12 1AY
GM0 SGH M Brown, 3 Arnott Road, Blackford, Auchterarder, PH4 1QE
G0 SGI J Sankey, 56 Gorsey Lane, Mawdesley, Ormskirk, L40 3TF
G0 SGM C Read, 2 Castle View, Gosport, PO12 4LS
G0 SGP A Danton, 3 Cliffe Close, Ruskington, Sleaford, NG34 9AT
G0 SGR S Rice, 94 St. Johns Avenue, Bridlington, YO16 4NL
G0 SGT David Huddleston, 162 Manor Road, Newton St. Faith, Norwich, NR10 3LG
G0 SGV John Allen, 57 Watford Road, Kings Langley, WD4 8DY
G0 SGX Frank Dingwall, 20 Whitehills Road, Loughton, IG10 1TS
G0 SHC Michael Lane, Cherry Tree House, Pipwell Gate, Spalding, PE12 8BA
GM0 SHD George Balfour, 6 Kirkden Street, Friockheim, Arbroath, DD11 4SX
G0 SHJ R Harrison, 22 East Anglian Way, Gorleston, Great Yarmouth, NR31 6QY
G0 SHM B Coates, 74 Colescliffe Road, Scarborough, YO12 6SB
G0 SHN Gerard Jacot, Boucle De L'Observatoire, Le Grand Revard, Pugny-Chatenod, France, 73100
G0 SHO B Lawrence, 70 Beacon Road, Rolleston-on-Dove, Burton-on-Trent, DE13 9EG
G0 SHP A Pratt, 4 Chestnut Close, Braunton, EX33 2EH
G0 SHT S Rossi, 21 Rattigan Gardens, Whiteley, Fareham, PO15 7EA
G0 SHU G Bennett, 57 Princess Way, Euxton, Chorley, PR7 6PL
G0 SHY M Bamber, 1 Penair Crescent, Truro, TR1 1YE
G0 SIA P Brooks, 3 Jamiesons Court, Kelso, TD5 7EU
G0 SIE A Swingler, 9 Princess Drive, Wistaston Green, Crewe, CW2 8HP
G0 SIG SIGNALLERS INTEREST GROUP c/o K Prince, 59 Chantry Road, East Ayton, Scarborough, YO13 9ER
G0 SII T Richards, 142 Princes Mews, Royston, SG8 9BN
G0 SIO G Goad, 2 Westholme, Orpington, BR6 0AN
G0 SIQ R Myerscough, Hamer, The Street, Holt, NR25 6NW
GW0 SIS Kirsten Barrett, Linroy, Stoney Lane, Bridgend, CF35 5AL
G0 SIU Bryan Durrant, 3 Parklands, Shoreham-by-Sea, BN43 6NN
G0 SIV Jack Wilcox, 31 Soane Street, Basildon, SS13 1QU
G0 SIW B Ellison, 6 Eskdale Road, Ashton-in-Makerfield, Wigan, WN4 8QT
G0 SIY A Hopkinson, 358 Edenfield Road, Rochdale, OL12 7NH
G0 SJB S Barraclough, 67 Sude Hill, New Mill, Holmfirth, HD9 7ER
G0 SJH S Harris, 19 Mundays Boro, Puttenham, Guildford, GU3 1AZ
G0 SJP Michael Windle, 8 Bylands, Danehill Road, Brighton, BN2 5PY
G0 SJR Rick Brand, Foxgrove, 19a Mill End Close, Dunstable, LU6 2FH
G0 SJS Dorothy Pugh, 31 Brocstedes Avenue, Ashton-in-Makerfield, Wigan, WN4 0NJ
G0 SJU T Bousfield, 8 Harpington View, Mordon, Stockton-on-Tees, TS21 2EZ
G0 SJV P Gostick, 25 Cashmere Lane, Cashmere, Queensland, Australia, 4500
G0 SKA Charles Mitchell, Old Tiles, Beaconsfield Road, Slough, SL2 3LZ
G0 SKC B Foote, Red Roofs, 5 Woodland Avenue, Colwyn Bay, LL29 9NL
G0 SKD Terry Ward, 4 Burrows Grove, Wombwell, Barnsley, S73 8PS
G0 SKI M Foy, 335 South Avenue, Southend-on-Sea, SS2 4HR
G0 SKJ K Cockburn, 11 Highlands Avenue, Barrow In Furness, LA13 0AU
G0 SKK David Chadwick, 386 Tamworth Road, Amington, Tamworth, B77 4AQ
G0 SKM M Tunstall, 24 Darbrook Avenue, Stoke on Trent, ST2 5UG
G0 SKN P Hartley, 9 Weston Road, Wimborne, BH21 2SF
G0 SKQ C HAINES, 29 Woodlands Close, Aston, Stone, ST15 0DX
G0 SKR J Goodall, 37 Woodfield Road, Bear Cross, Bournemouth, BH11 9EU
G0 SLB D Mullison, 11a Buckingham Drive, Colchester, CO4 3YH
GW0 SLC R Thomas, 9 [illegible] Close, Glyncoch, Pontypridd, CF37 3DF
G0 SLD Paul Westrupp, 2 Ridgeway, Horns Road, Cranbrook, TN18 4RA
G0 SLH Colin Shoesmith, 2 Caravelle Gardens, Northolt, UB5 6EU
G0 SLI T Day, 21 Mowbray Road, Ham, Richmond, TW10 7NQ
G0 SLJ D Pepper, 17 Cliffe House, Radnor Cliff, Folkestone, CT20 2TY
G0 SLK E Patterson, 45 Sandhurst Road, Rainhill, Prescot, L35 8NE
G0 SLL R Petrie, 116 Sherwood Avenue, Abingdon, OX14 3TU
G0 SLN P Grainger, 11 Smith Grove, Ryhope, Sunderland, SR2 0JU
G0 SLP M Coultas, 35 Monteigne Drive, Bowburn, Durham, DH6 5QB
G0 SLQ S Quinn, 48 Aldsworth Close, Springwell Village, Gateshead, NE9 7PG
G0 SLR Roy Lisle, 21 Porlock Close, Penketh, Warrington, WA5 2QE
G0 SLU C Barr, 17 Knighton Road, Otford, Sevenoaks, TN14 5LD
G0 SLV David Gregory, 6 Maxwell Grove, Blackpool, FY2 0QG
G0 SLW J Waite, 28 Overdown Rise, Portslade, Brighton, BN41 2YG
G0 SLY Carl Kratzer, 9900 Dale Ridge Ct, Vienna, United States, VA 22181
G0 SLZ A Matthews, 28 Sherwin Road, Stapleford, Nottingham, NG9 8PQ
G0 SMB Mark Browning, 76 Agincourt Road, Portsmouth, PO2 7AY
G0 SMH Barry Marchant, 20 Wrench Road, Norwich, NR5 8AS
G0 SMJ Michael Jackson, Sunny Cot, 44 Dulwich Road, Clacton-on-Sea, CO15 5NA
G0 SMM J O'Nion, 7 Ettington Close, Cheltenham, GL51 0NY
G0 SMN Anthony McKenzie, 311 Weston Road, Weston Coyney, Stoke-on-Trent, ST3 6HA
G0 SMO C Cash, 7 Park Lane, Park Village, Wolverhampton, WV10 9QE
G0 SMP S Pountain, 21 Hayfield Road, Chapel-en-le-Frith, High Peak, SK23 0JF
G0 SMR J Ballard, 7 Chapelcroft Court, Liverpool, L12 9GY
G0 SMS Diane Barnham, 35 Post Office Road, Frettenham, Norwich, NR12 7AB
GI0 SMU A Hanna, 39 Dalton Crescent, Comber, Newtownards, BT23 5HE
G0 SMZ R Clews, 99 Kilbury Drive, Worcester, WR5 2NG
G0 SNB William Bonser, 24 Meend Garden Terrace, Cinderford, GL14 2EB
G0 SNF J Culling, 4 Ash Road, Princes Risborough, HP27 0BQ
G0 SNG David Hill, 3 Morcar Road, Stamford Bridge, York, YO41 1PR
G0 SNK A Gill, Bradgate, Kings Lane, Lymington, SO41 6BQ
G0 SNM K Killick, 15 Popplechurch Drive, Swindon, SN3 5DE
G0 SNO D Lauder, 20 Sutherland Close, Barnet, EN5 2JL
G0 SNP J Du Heaume, 10 Water Lane, Pill, Bristol, BS20 0EQ
G0 SNQ Howard Davis, 44 Kenyon street, Ashton under Lyne, OL6 7DU
G0 SNS B Harrison, 8 Elm Park, Pontefract, WF8 4LG
GM0 SNT A Carpenter, 9 Glenbervie Road, Kirkcaldy, KY2 6HR
G0 SNU I Gray, 27 Meadow Close, Lavenham, Sudbury, CO10 9RU
G0 SNV Jack Worsnop, 1217 Thornton Road, Thornton, Bradford, BD13 3BE
G0 SNW P Rogers, 126 Bradford Road, Otley, LS21 3LE
G0 SNX N Johnson, 12 Bleach Mill Lane, Menston, Ilkley, LS29 6HE
G0 SNZ A Flood, 3 Ongar Walk, Blackley, Manchester, M9 8JD
G0 SOF G Hedley, 260e 100 Sts, Raymond, Alberta, Canada, TOK 250
G0 SOG Fred Mellings, 4 Kiln Lane, Horley, RH6 8JG
G0 SOK Peter Mayer, 248 Dimsdale Parade West, Wolstanton, Newcastle under Lyme, ST5 8EA
G0 SON R Peard, 28 New Road, Shoreham-by-Sea, BN43 6RA
G0 SOU Julie Pendleton, 17a Langley Drive, Kegworth, Derby, DE74 2DN
G0 SOV D Mason, 2 Lawrence Close, Charlton Kings, Cheltenham, GL52 6NN
G0 SOX P Chapman, 4 Churchill Close, Brightlingsea, Colchester, CO7 0RS
G0 SOY A Croft, 34 Fourth Avenue, Wolverhampton, WV10 9LZ
G0 SPA Paul Benson, 7 Crofton Close Attenborough, Nottingham, NG9 5HX
G0 SPB G Rusby, 12 Park Meadow, Princes Risborough, HP27 0EB
G0 SPC M Chawner, Timbertops, Churchfield Lane, Wallingford, OX10 6SH
G0 SPF T Hill, 19 Elmbank Road, Paignton, TQ4 5NG
G0 SPH K Brooks, 7 Mayfield Road, Northwich, CW9 7AS
G0 SPK D Neal, 490 Aureole Walk, Newmarket, CB8 7BQ
G0 SPL F Goodwin, 36 Grange Drive, Ryton, NE40 3LF
G0 SPQ I Wilson, 3 Caring Lane, Bearsted, Maidstone, ME14 4NJ
G0 SPS M Forder, 157 Kennington Road, Kennington, Oxford, OX1 5PE
G0 SPX J Sparks, 34 Green Park Avenue, Skircoat Green, Halifax, HX3 0SR
G0 SPY J Thorley, 8 Bryn Gwyn, Abergele, LL22 8JA
G0 SPZ M Rickard-Worth, 2 Harwood Road, Littlehampton, BN17 7AT
G0 SQE E Young, 14 Hanover Court, Gateshead, NE9 6TZ
G0 SQF J Bubez, 4 Southway, Burgess Hill, RH15 9ST
G0 SQH Derek Higbee, 12 Shelley Close, Ashley Heath, Ringwood, BH24 2JA
G0 SQI Nigel Blythe, 14 The Green, South Creake, Fakenham, NR21 9PD
G0 SQK M Stuckey, 212 Roughton Road, Cromer, NR27 9LQ
GD0 SQL R Bishop, 73 Broomgrove Gardens, Edgware, HA8 5RJ
G0 SQP F Ott, 16 Hornbeam Close, Chelmsford, CM2 9LW
G0 SQS M Hewitt, 1 Harpswell Hill Park, Hemswell, Gainsborough, DN21 5UT
GW0 SQT A Davis, 8 Cook Road, Barry, CF62 9HD
G0 SQX T Donley, 21 Elmridge, Leigh, WN7 1HN
GW0 SQY S Morgan, Oakfield House, Barry Road, Pontypridd, CF37 1HY
G0 SRC SOUTH DERBYSHIRE & ASHBY WOULDS ARG c/o Lewis Kirby, 41 Woodville Road, Overseal, Swadlincote, DE12 6LU
GW0 SRE D Price, 1 Rhas Cottages, Pontyates, Llanelli, SA15 5SF
GW0 SRF D Daniels, 73 Bethania Road, Upper Tumble, Llanelli, SA14 6DT
G0 SRG SUNDERLAND RAYNET GROUP c/o S Green, 133 Sevenoaks Drive, Sunderland, SR4 9NQ
GI0 SRL A Harbison, 26 Ballymartin Road, Templepatrick, Ballyclare, BT39 0BW
G0 SRO Iain Maclean, 7 Thirlestane Crescent, Lauder, TD2 6TT

**IMPORTANT NOTE**

**Revalidate licence to avoid revocation** – Ofcom has advised the Society that plans will be drawn up to revoke licences that have not been revalidated as required by the licence conditions. The quickest way to revalidate is to do so online via the Ofcom website: *https://services.ofcom.org.uk/* or by email: *amateur.validations@ofcom.org.uk* Ofcom staff are available to help, but please be patient during times of heavy workload.

| | | |
|---|---|---|
| GI0 | SRP | N Averill, 3 Edmund Court, Tobermore, Magherafelt, BT45 5QA |
| GM0 | SRQ | D Beacher, 17 Cairn Grove, Crossford, Dunfermline, KY12 8YD |
| G0 | SRR | M Baugh, 97 Wilson Avenue, Deal, CT14 9NJ |
| G0 | SRY | J Smart, 60 Blaze Park, Wall Heath, Kingswinford, DY6 0LN |
| G0 | SRZ | E Hart, 47 Northfield Crescent, Wells-next-The-Sea, NR23 1LR |
| GI0 | SSA | Jonathan Stevenson, 27 Kinnegar Rocks, Donaghadee, BT21 0EZ |
| G0 | SSC | George Mills, 49 Priestley Close, Doncaster, DN4 9DQ |
| G0 | SSE | N Morton, Cami Terrapico 54, Roquetes, Spain, 43520 |
| G0 | SSG | Reginald Andre, 54 Covertside, Wirral, CH48 9UL |
| G0 | SSJ | A Powell, 61 Albert Road, Grappenhall, Warrington, WA4 2PF |
| G0 | SSK | G Johnson, 503 Holden Road, Leigh, WN7 2JJ |
| G0 | SSL | A McEvoy, 12 Fountains Avenue, Haydock, St. Helens, WA11 0RS |
| G0 | SSN | C Ireland, 14 Castlefields, Istead Rise, Gravesend, DA13 9EJ |
| G0 | SSO | N Irwin, 1 Highfield Road, Barrow in Furness, LA14 5NZ |
| GM0 | SSQ | Alastair Winchester, 23 Craigmount, Avenue North, Edinburgh, EH12 8DL |
| G0 | SSV | Gary Kendall, 3 Brayton Avenue, Sale, M33 5HF |
| G0 | SSX | P Ellis, 4 Rostwold Way, Norwich, NR3 3NN |
| G0 | SSY | David Webb, 3 Cams Hill Lane, Hambledon, Waterlooville, PO7 4SP |
| G0 | SSZ | Roger Stanley, 219 Fartown, Pudsey, LS28 8NH |
| GM0 | STB | SCOTTISH TOURIST BOARD RADIO c/o R Aitkenhead, 11 Elm Court, Quarter, Hamilton, ML3 7FB |
| GI0 | STC | P Dellett, 14 Fox Park Drumnakilly, Omagh, BT79 0JX |
| G0 | STF | Tom Clements, 29 Bonchurch Drive, Wavertree, Liverpool, L15 4PW |
| G0 | STH | ST HELENS & DISTRICT ARC c/o Roy Vaughan, 6 Dellside Grove, St. Helens, WA9 5AR |
| G0 | STK | A Hardcastle, 25 Aire Crescent, Cross Hills, Keighley, BD20 7RW |
| GI0 | STM | Kevin Murray, 17 Glebe Court, Dungannon, BT70 3PU |
| GD0 | STR | W Shaw, 161 Springwood Crescent, Edgware, HA8 8SH |
| GI0 | STS | Ronnie Todd, 73 Lakeview Park, Drumgor, Craigavon, BT65 4AL |
| G0 | STW | C Kendrick, 18 Ainger Road, Upper Dovercourt, Harwich, CO12 4TS |
| G0 | SUA | A Edwards, 49 Griggs Meadow, Dunsfold, Godalming, GU8 4ND |
| G0 | SUB | G Thomas, 39 Seaway Road, Paignton, TQ3 2NX |
| GM0 | SUF | H Butcher, 14 Newbattle Road, Dalkeith, EH22 3DB |
| G0 | SUH | John Montgomery, Apt K, 220 Red Oak Drive West, Sunnyvale, United States, CA 94086 |
| G0 | SUI | G Scarlett, 7 Beamsley Grove, Gilstead, Bingley, BD16 3NE |
| G0 | SUL | J Ward, 8 Wilderness Road, Guildford, GU2 7QN |
| GU0 | SUP | Phil Cooper, 1 Clos Au Pre, La Route de la Hougue Du Pommier, Castel, Guernsey, GY5 7FQ |
| G0 | SUQ | Ian Johnson, 181 Broad Street, Bromsgrove, B61 8NQ |
| G0 | SUT | Jeffrey Burdett, 1 Main Street, Egginton, Derby, DE65 6HL |
| G0 | SUU | J Ogier, 37 Hill Park Road, Torquay, TQ1 4LD |
| GM0 | SUY | Colin Auld, 148 Echline Drive, South Queensferry, EH30 9XG |
| G0 | SVA | D Nock, 431 Locking Road, Weston-Super-Mare, BS22 8QN |
| G0 | SVB | P Herrmann, 5992 Royal Court, Lockport, United States |
| G0 | SVD | M Rhodes, 15 Manor Road, Hatfield, AL10 9LJ |
| G0 | SVH | E Wright, 26 Walmsley Close, Church, Accrington, BB5 4HL |
| G0 | SVJ | M Creswick, 5 Wheatlands Drive, Easington, Saltburn-by-The-Sea, TS13 4PB |
| G0 | SVK | John Hosfield, 29 Whitecroft, Gosforth, Seascale, CA20 1AY |
| G0 | SVN | N Savin, 138a Lillibrooke Crescent, Maidenhead, SL6 3XH |
| G0 | SVP | G Broadhurst, 9 Sharples Street, Accrington, BB5 0HQ |
| G0 | SVQ | A Haydock, 8 Corbridge Close, Blackpool, FY4 5EZ |
| GM0 | SVS | M Whiteley, 9 Pathfoot Avenue, Bridge of Allan, Stirling, FK9 4SA |
| G0 | SVU | R Morris, 57 Marten Drive, Huddersfield, HD4 7JX |
| G0 | SVX | I Roebuck, 3 Lodge Hill Drive, Kiveton Park, Sheffield, S26 5RU |
| G0 | SVZ | Martin Cooper, 20 Bankfield Road, Widnes, WA8 7UW |
| G0 | SWB | R Atkinson, 57 Jobling Avenue, Blaydon-on-Tyne, NE21 4RR |
| G0 | SWC | R Eeles, 50 Nightingale Road, Guildford, GU1 1EP |
| G0 | SWE | S Whitbourn, 50 Nightingale Road, Guildford, GU1 1EP |
| G0 | SWF | J Durrant, 16 Bugdens Lane, Verwood, BH31 6EY |
| G0 | SWH | B Fletcher, 58 Broomfield Avenue, Worthing, BN14 7SB |
| G0 | SWL | C Niles, 23 Randsfield Avenue, Brough, HU15 1BE |
| G0 | SWN | Gilwell Park Scout RC c/o Stuart Barber, Homedale, St. Monicas Road, Tadworth, KT20 6ET |
| G0 | SWO | Bernard Atkinson, 165 Alliance Avenue, Hull, HU3 6QY |
| G0 | SWS | Terry Stow, 38 The Strand, Mablethorpe, LN12 1BQ |
| G0 | SWU | Peter Broad, 78 Lenham Road, Sutton, SM1 4BG |
| G0 | SWW | M Stevens, Autumn Cottage, Silver Street, Horncastle, LN9 5NH |
| G0 | SWY | Michael Humphrey, 5 Ventnor Court, Southampton, SO16 3EB |
| G0 | SXA | William Daly, 85 Lordens Road, Huyton, Liverpool, L14 9PA |
| G0 | SXE | J Williams, 11 Courbet Drive, Connah's Quay, Deeside, CH5 4WP |
| G0 | SXG | Keith Stammers, 102 Eaton Road, Appleton, Abingdon, OX13 5JJ |
| G0 | SXK | A Dodd, 20 Braemar Avenue, Chelmsford, CM2 9PW |
| G0 | SXM | D Smith, Fernworthy, High Side, Wisbech, PE13 4LJ |
| G0 | SXN | B Watts, 9 Weavers Walk, Cullompton, EX15 1SS |
| GM0 | SXO | A Segar, 36 Kestrel Avenue, Dunfermline, KY11 8JL |
| GM0 | SXP | R Cliff, 32 Lochardil Road, Inverness, IV2 4LD |
| GW0 | SXS | Kevin Wheeler, The Glen, Dreenhill, Haverfordwest, SA62 3XH |
| G0 | SXU | C Bocock, 20 Court Avenue, Stoke Gifford, Bristol, BS34 8PJ |
| G0 | SXW | P Cressey, Skidby Hill Farm, Beverley Road, Cottingham, HU16 5TF |
| G0 | SXY | Phil Davies, 46 Wagstaff Way, Olney, MK46 5FB |
| GW0 | SXZ | J Green, 23 Litchard Park, Bridgend, CF31 1PF |
| G0 | SYF | Robert Pearce, 14 Shepherds Leaze, Wotton-under-Edge, GL12 7LQ |
| G0 | SYI | K Treasure, 30 Grace Park Road, Brislington, Bristol, BS4 5JA |
| G0 | SYP | Carsten Steinhoefel, 222 Stretford Road, Urmston, Manchester, M41 9NT |
| G0 | SYQ | M Wood, 4 Gordon Road, Hastings, TN34 3JN |
| G0 | SYR | Bryan Petifer, 14 Wood Lane, Caterham, CR3 5RT |
| G0 | SYS | D Hall, 72 Mansfield Road, Edwinstowe, Mansfield, NG21 9NH |
| G0 | SYT | D Firks, Bryn Garth Cottage, Hereford, HR2 8HJ |
| GM0 | SYV | James Kelly, 2 Ryeland Street, Strathaven, ML10 6DL |
| GM0 | SYY | J Hutchens, 2 Provost Gate, Larkhall, ML9 1DN |
| GM0 | SZA | Ian Stones, 8 Cloverfield Place, Bucksburn, Aberdeen, AB21 9RH |
| G0 | SZE | C Andrews, 33 Blackheath Road, Barnsley, S71 3RH |
| G0 | SZG | Joseph Greene, 308 Cedar Road, Nuneaton, CV10 9DY |
| GI0 | SZH | W McEwen, 234 Legahory Court, Legahory, Craigavon, BT65 5DH |
| G0 | SZJ | S Fletcher, Apartado 39, Ourique, Portugal, 7670 |
| G0 | SZK | L Joyce, 106a Victoria Drive, Bognor Regis, PO21 2EJ |
| G0 | SZN | M Lawrence, 1 Greenwood Cottages, Gelligroes, Blackwood, NP12 2JB |
| G0 | SZO | J Everard, Woodside, Staple Lane, Taunton, TA4 4DE |
| G0 | SZT | D Allibone, Virginia, North Street, Langport, TA10 9RH |
| GW0 | SZU | B Osborne, 163 Park Street, Bridgend, CF31 4BB |
| G0 | SZX | W Hunt, 8 Denmark Road, Exeter, EX1 1SL |
| G0 | TAA | W Chandler, 38 Falkland Road, Chandler's Ford, Eastleigh, SO53 3GD |
| GM0 | TAE | C Porter, 20 Baird Avenue, Kilwinning, KA13 7AR |
| G0 | TAG | Dave Beane, 12 Strafford Court, Pondcroft Road, Knebworth, SG3 6DF |
| G0 | TAH | K Aldus, 77 Springvale Road, Kings Worthy, Winchester, SO23 7ND |
| G0 | TAI | Ian Hawkins, 24 Flora Thompson Drive, Newport Pagnell, MK16 8ST |
| G0 | TAK | Roy Walker, 35 Romany Close, Letchworth Garden City, SG6 4LA |
| G0 | TAL | S Walsworth, 4 The Homestead, Heckmondwike, WF16 9JL |
| G0 | TAM | Alan Farrow, 18 The Green Trimingham, Norwich, NR11 8ED |
| G0 | TAN | Stephen Venner, 4 Cuckoo Lane, West End, Woking, GU24 9NG |
| G0 | TAO | R Lokuge, 11 Porchester Close, Southwater, Horsham, RH13 9XR |
| G0 | TAR | Brian Lucas, 8 Gilbert Close, Hempstead, Gillingham, ME7 3QQ |
| G0 | TAS | J Taylor, 19 Castle Close, Leconfield, Beverley, HU17 7NX |
| G0 | TAT | M Wilmot, Southview, Roman Road, Weston-Super-Mare, BS24 0AB |
| G0 | TAX | T Welch, 63 Vicarage Close, New Silksworth, Sunderland, SR3 1JF |
| GM0 | TAY | O.B.O. TAYSIDE RAYNET c/o I Strachan, 238 Coupar Angus Road, Muirhead, Dundee, DD2 5QN |
| G0 | TAZ | J Burgess, Combeside House, Symonsburrow, Cullompton, EX15 3XA |
| G0 | TBC | S Lawson, 27 Broadlands Avenue, Eastleigh, SO50 4PP |
| GM0 | TBH | James McMaster, 96 Cunningham Crescent, Ayr, KA7 3JB |
| G0 | TBI | Stuart McKinnon, 145 Enville Road, Kinver, Stourbridge, DY7 6BN |
| G0 | TBM | J Goulden, Wenffrwd Cottage, Llangollen Road, Llangollen, LL20 7UH |
| G0 | TBO | P Clark, 30 Alicia Avenue, Garlinge, Margate, CT9 5JZ |
| G0 | TBS | W Lucas, 12 Stuart Court, Priory Gate Road, Dover, CT17 9TX |
| G0 | TBU | R Brightwell, 40 Streete Court Road, Westgate-on-Sea, CT8 8BX |
| G0 | TBW | Gary DAVIS, 32 Medlock Close, Farnworth, Bolton, BL4 9QW |
| G0 | TCA | R Seaward, 13 Blythe Close, Catford, London, SE6 4UW |
| GM0 | TCC | Kenneth Walker, 174 South Seton Park, Port Seton, Prestonpans, EH32 0BP |
| G0 | TCE | P Taylor, 67 Rectory Park, South Croydon, CR2 9JR |
| G0 | TCF | Peter Loch, 400 Loughborough Road, West Bridgford, Nottingham, NG2 7FD |
| G0 | TCH | T Noszkay, 41 Brue Close, Weston-Super-Mare, BS23 3BX |
| G0 | TCI | B Hughes, 29 East View Close, Radwinter, Saffron Walden, CB10 2TZ |
| G0 | TCJ | T Worrall, 9 Barnstaple Close, Wigston, LE18 2QX |
| G0 | TCL | D Parsons, Aston Hall Res.Home, Lower Aston Hall Lane, Deeside, CH5 3EX |
| G0 | TCO | M Roper, 1 The Cottages, Norwich Road, Norwich, NR9 5BY |
| G0 | TCP | P Maynard, Seaton House, Lower Road, Ipswich, IP6 9AR |
| G0 | TCQ | A Cudlip, The Oaks, 16 Bilberry Close, Southampton, SO31 6XX |
| GM0 | TCU | C Pirie, 10 Annesley Park, Torphins, Banchory, AB31 4HG |
| G0 | TCV | A Thomas, 4 Gordon Terrace, King Street, Mold, CH7 1LD |
| G0 | TCW | A Furmston, 46 Twydall Lane, Gillingham, ME8 6JE |
| G0 | TCY | M Templeman, 44 Wisbeck Road, Bolton, BL2 2TA |
| GW0 | TDA | Peter Price, 1 Brynderi, Pontyates, Llanelli, SA15 5SU |
| G0 | TDC | S TINSLEY, 48 Smithy Lane, Croft, Warrington, WA3 7JG |
| G0 | TDE | R Barrett, Brookfield, Hobbacott Lane, Bude, EX23 0ES |
| G0 | TDG | N Hotson, Top Floor Flat, 36 Carew Road, Thornton Heath, CR7 7RE |
| G0 | TDJ | Stephen Smith, Westgarth Flat 4, 145-146 Marina, St.Leonards-on-Sea, TN38 0BT |
| G0 | TDM | J Sutton, 15 Lowther Street, Penrith, CA11 7UW |
| GI0 | TDP | J Driscoll, 67 Whinney Hill, Holywood, BT18 0HG |
| G0 | TDQ | P Luscombe, 33 Rea Barn Road, Brixham, TQ5 9ED |
| G0 | TDR | Trevor Round, 92 Church View Gardens, Kinver, Stourbridge, DY7 6EE |
| G0 | TDV | R King, 10 Lansdown Road, Kingswood, Bristol, BS15 1XB |
| G0 | TDX | Y Kimoto, 523 Yukinaga, Maizuru-City, Kyoto, Japan, 625-0052 |
| G0 | TDY | Andrew Frank, 15 Home Ground, Cricklade, Swindon, SN6 6JG |
| GM0 | TEA | A Cherry, 39 Clark Road, Edinburgh, EH5 3AR |
| G0 | TEB | C Sexton, Lapswater, Marsh, Honiton, EX14 9AL |
| G0 | TED | E Thane, 19 Churchill Road, Aston, Stone, ST15 0EB |
| G0 | TEE | T Speight, 1 Lyndene Avenue, Worsley, Manchester, M28 2RJ |
| G0 | TEI | R Sparks, Radio Licence Centre, Po Box 885, Bristol, BS99 5LG |
| G0 | TEL | Roger Bellenot, 22 Roderick Avenue, Peacehaven, BN10 8JT |
| G0 | TEM | A Merrix, 53 Pear Tree Road, Great Barr, Birmingham, B43 6HX |
| G0 | TEO | Leslie Baker, Flat 30, Francis Snary Lodge, 12 Chesterton Close, London, SW18 1SD |
| GD0 | TEP | A Kissack, 30 High View Road, Douglas, Isle of Man, IM2 5BH |
| G0 | TES | Roger Dicks, 10 Westfield Avenue, Raunds, Wellingborough, NN9 6DQ |
| G0 | TEV | John Mullin, 49 Leaders Way, Newmarket, CB8 0DP |
| G0 | TFB | WALLEN ANTENNAE RC c/o L Wallen, Lambda Works, 45a Whitehall Road, Ramsgate, CT12 6DE |
| G0 | TFC | A Stanley, 145 Tribune Drive, Houghton, Carlisle, CA3 0LF |
| G0 | TFD | D Eyre, 29 Old Acre Lane, Brocton, Stafford, ST17 0TW |
| GM0 | TFE | Charles Stewart, 185 Newbattle Abbey Crescent, Dalkeith, EH22 3LT |
| G0 | TFF | A McGhie, 16 Boyach Crescent, Isle of Whithorn, Newton Stewart, DG8 8LD |
| GD0 | TFG | J Dowling, 2 Ballabridson Park, Ballasalla, Isle of Man, IM9 2ES |
| G0 | TFI | D Roberts, 9 The Pound, Westoning, Bedford, MK45 5JN |
| G0 | TFK | J Sutcliffe, 4 Lancaster Gate, Nelson, BB9 0AP |
| G0 | TFL | J Williams, 107 Clay Lane, Rochdale, OL11 5QW |
| GD0 | TFO | J Bellis, 2 Cedar Walk, Douglas, Isle of Man, IM2 5NG |
| G0 | TFP | J Brett, 11 Manor Road, Astley, Manchester, M29 7PH |
| GM0 | TFQ | H Wignall, 7 Windyedge, Inverurie, AB51 3WJ |
| G0 | TFR | A Vining, The Cedars, Thorney Road, Peterborough, PE6 0LH |
| G0 | TFT | A Gibson, 28 Finchale Terrace, Jarrow, NE32 3TX |
| G0 | TFU | M Wilson, 6 The Chase, Calcot, Reading, RG31 7DN |
| G0 | TFV | T Carvell, 18 Park View, School Lane, Rye, TN31 6UR |
| G0 | TFX | S Roberts, 20 Beech Grove, Trowbridge, BA14 0HG |
| G0 | TGB | T Blore, Glendhoon, Laneham Street, Retford, DN22 0JX |
| GM0 | TGE | Ian Ross, Idlewild, Kintore, Inverurie, AB51 0XA |
| GM0 | TGG | L Mackenzie, 90 Tay Street, Newport on Tay, DD6 8AP |
| G0 | TGH | Stephen Leak, 12 Kentmere Approach, Leeds, LS14 1JP |
| G0 | TGK | P Lambert, Old Farm, Acomb, Hexham, NE46 4QF |
| G0 | TGM | S Fowler, 38 Meadowcroft, Aylesbury, HP19 9LN |
| G0 | TGP | W Ford, Flat 2, Hillyard Court, Wareham, BH20 4QX |
| G0 | TGQ | O Cubitt, 97 Sutton Lane, Langley, Slough, SL3 8AU |
| G0 | TGR | D Greenacre, 38 Toon Crescent, Bury, BL8 1JB |
| G0 | TGU | W Collier, 20 Wainers Croft, Greenleys, Milton Keynes, MK12 6AL |
| G0 | TGW | D HAY, 19 Munks Close, West Harnham, Salisbury, SP2 8PB |
| G0 | TGX | P Ireland, 2 The Ridings, Hull, HU5 5HW |
| G0 | THD | P Hart, 39 Barley Drive, Burgess Hill, RH15 9XG |
| G0 | THF | Keith Greatorex, 54 Lilac Grove, Glapwell, Chesterfield, S44 5NG |
| G0 | THH | H Hudders, 7 Cedar Crescent, Thame, OX9 2AX |
| G0 | THI | Walter Davey, 15 Park Avenue, Histon, Cambridge, CB24 9JU |
| G0 | THJ | John Stowell, 6 Westage Lane, Great Budworth, Northwich, CW9 6HJ |
| GI0 | THO | W Edmondson, 7 Rathcavan Drive, Ballymena, BT42 2QH |
| G0 | THQ | R King, 6 Mayon Green Crescent, Sennen, Penzance, TR19 7BS |
| GI0 | THR | J McDonald, 54 Bettys Hill Road, Newry, BT34 2ND |
| G0 | THS | Simon Graham, 25 South End Road, Ottringham, Hull, HU12 0DP |
| G0 | THV | J Coote, 8 St. Francis Chase, Bexhill-on-Sea, TN39 4HZ |
| G0 | THW | A Brentnall, Sandy Rise, Mill Lane, Woodbridge, IP12 3LL |
| G0 | THY | Martyn Preston, 15 Poplar Close, Kidlington, OX5 1HH |
| G0 | TIA | Paul Lodge, 79, Via Circonvallazione, 79, Milano, Italy, 20090 |
| G0 | TID | C Alexander, 25 Diamedes Avenue, Stanwell, Staines-upon-Thames, TW19 7JE |
| G0 | TIG | H Janes, 91 Thorpe Bay Gardens, Southend-on-Sea, SS1 3NW |
| G0 | TII | Stuart Graham, 4 Oakland Avenue, Ellenborough, Maryport, CA15 7BU |
| G0 | TIJ | D Page, 16 Arlington Close, Yeovil, BA21 3TB |
| G0 | TIL | J Parmenter, 48 Honey Way, Royston, SG8 7EU |
| G0 | TIP | G Thorne, 19 Lapwing Lane, Brinnington, Stockport, SK5 8JY |
| G0 | TIS | P Jones, 61 North Road, Bourne, PE10 9AU |
| G0 | TIW | Terry Parker, The Bungalow, 178 Green End Lane, Hemel Hempstead, HP1 2BQ |
| G0 | TIX | Paul Wilson, 2 Staley Close, Stalybridge, SK15 3HJ |
| G0 | TIZ | E Tometzki, 11 Southey Close, Enderby, Leicester, LE19 4QZ |
| G0 | TJC | Leonard Taylor, 35 Leafield Road, Darlington, DL1 5DF |
| G0 | TJD | A Wedgwood, 10 Milner Place, London, N1 1TN |
| G0 | TJE | S Sullivan, 7 Gosfield Road, Dagenham, RM8 1JY |
| G0 | TJG | M Spafford, Old School House, 11 Old School Lane, Chesterfield, S44 5UE |
| G0 | TJI | G Woods, 126 Luddenham Close, Ashford, TN23 5SA |
| GI0 | TJJ | A Hamilton, 60 Parkland Avenue, Lisburn, BT28 3JP |
| G0 | TJN | T Newland, 80 Burnway, Hornchurch, RM11 3SG |
| G0 | TJP | Alun Person, 78 Green Street, Ston Easton, Radstock, BA3 4BZ |
| G0 | TJQ | C Dowell, 19 Field Top, Bailiff Bridge, Brighouse, HD6 4EQ |
| G0 | TJR | L Ellams, 131 Broadway, Dunscroft, Doncaster, DN7 4HB |
| G0 | TJT | R Francis, 661 Osmaston Road, Derby, DE24 8NF |
| GI0 | TJV | A Gibson, 58 Cairnmore Park, Lisburn, BT28 2DN |
| G0 | TJY | T Herd, The Old Dairy, High Street, Ipswich, IP8 3AP |
| GM0 | TKB | Leslie Thomas, Greengeo, Scarfskerry, Thurso, KW14 8XN |
| GM0 | TKC | T Maxwell, 28 Cedar Grove, Dunfermline, KY11 8BH |
| GM0 | TKE | J McLuckie, 118 Lady Nairn Avenue, Kirkcaldy, KY1 2AT |
| G0 | TKF | W Stuart, 3 Rookery Vale, Deepcar, Sheffield, S36 2NP |
| G0 | TKG | D Mawson, 14 Windermere Close, Worksop, S81 7QE |
| G0 | TKJ | Terence Clayton, 40 Morrison Road, Darfield, Barnsley, S73 9ED |
| G0 | TKK | Geoffrey Eke, 27 Mercia Drive, Sheffield, S17 3QF |
| G0 | TKL | E Roberts, 43 Ashbourne Crescent, Sale, M33 3LQ |
| G0 | TKR | E Martin, 20 Easters Grove, Stoke-on-Trent, ST2 7PF |
| G0 | TKT | P Hamblett, 13 Ironside Close, Bewdley, DY12 2HX |
| G0 | TKU | O.B.O. TRIO-KENWOOD A.R.C. c/o D Wilkins, 8 Gainsborough Road, Ashley Heath, Ringwood, BH24 2HY |
| GW0 | TKX | Alan Mason, 101 Aneurin Bevan Avenue, Gelligaer, Hengoed, CF82 8ET |
| G0 | TKZ | R Hamblin, 5 Streatfield, Edenbridge, TN8 5DF |
| G0 | TLA | R Lythall, 1 Stotfold Drive, Thurnscoe, Rotherham, S63 0LZ |
| G0 | TLE | Peter Grigson, 60 Thrupp Close, Castlethorpe, Milton Keynes, MK19 7PL |
| G0 | TLI | C Vince, 8 Kent Road, Swindon, SN1 3NJ |
| GW0 | TLJ | Anthony Mathias, 75 Coombs Drive, Milford Haven, SA73 2NU |
| G0 | TLN | G Sim, 24 Fernmoor Drive, Irthlingborough, Wellingborough, NN9 5TL |
| G0 | TLP | A Whitwam, 9 Oak Apple Close, Stourport-on-Severn, DY13 0JR |
| G0 | TLQ | R Parsons, 10 Waterside, Isleham, Ely, CB7 5SH |
| G0 | TLS | Rodney Upton, 8 Talbot Gardens, Sparbrook House, Ellesmere, SY12 0QZ |
| G0 | TLT | D Davies, 14 Hammerwater Drive, Warsop, Mansfield, NG20 0DJ |
| G0 | TLU | P Thompson, Flat 4, 14 West End Way, Lancing, BN15 8RL |
| G0 | TLZ | J Trefry, 67 Axminster Close, Cramlington, NE23 2JE |
| G0 | TMA | Raymond Griffiths, Flat 1, Buckhurst Court, 29 Buckhurst Road, Bexhill-on-Sea, TN40 1QE |
| G0 | TME | Brian Park, West Lodge, Churchend, Stonehouse, GL10 3RX |
| G0 | TMF | M Fleetwood, 9 Reynolds Close, Swindon, Dudley, DY3 4NQ |
| G0 | TMH | Christopher Neary, 3 Wordsworth Close, Torquay, TQ2 6EA |
| G0 | TMJ | E Jones, 1 Ivel View, Sandy, SG19 1AU |
| G0 | TMK | W Howell, 41 Chestnut Avenue, Shavington, Crewe, CW2 5BJ |
| G0 | TML | A Rowley, 32 Spring Lane, Flore, Northampton, NN7 4LS |
| GI0 | TMS | M Smyth, 41 Coolaghy Road, Newtownstewart, Omagh, BT78 4LG |
| G0 | TMT | Mark Tuttle, 7 Mill Lane, Horsford, Norwich, NR10 3ES |
| GW0 | TMU | D Edwards, 68 Heol Y Meinciau, Pontyates, Llanelli, SA15 5RT |
| GW0 | TMV | T Vlismas, Maes Yr Awel, Newport Rd, Crymych, SA41 3RR |
| G0 | TMW | G Boswell, 7 Chestnut Avenue, Wootton, Northampton, NN4 6LA |
| G0 | TMZ | Norman Lilley, 18 Beechwood Avenue, New Milton, BH25 5NB |
| G0 | TNC | George Stephenson, 54 Rock Road, Sittingbourne, ME10 1JF |
| G0 | TNF | M Wills, Chapel View, Cliburn, Penrith, CA10 3AL |
| G0 | TNG | P Hayward, 63 Devereaux Crescent, Ebley, Stroud, GL5 4PX |
| G0 | TNH | L Crow, 181 Foxlydiate Crescent, Batchley Estate, Redditch, B97 6NS |
| GM0 | TNK | D Mackenzie, 32 Miller Gardens, Inverness, IV2 3DT |
| G0 | TNL | A Beattie, Pine Croft, Blitterlees, Wigton, CA7 4JJ |
| G0 | TNM | R Guppy, 12 Highfield Road, Caterham, CR3 6QX |
| G0 | TNO | C Billington, 5 Lamers Road, Luton, LU2 9BL |
| G0 | TNP | Peter Unstead, 91 Rising Parade, Albany, New Zealand, 632 |
| G0 | TNQ | A Lewis, 76 Sunnymede Avenue, Epsom, KT19 9TJ |
| G0 | TNS | H Hauton, 8 St. Catherines Crescent, Scunthorpe, DN16 3LQ |
| G0 | TNY | A Rose, Heronsgate, River Gardens, Maidenhead, SL6 2BJ |
| G0 | TOB | John Horsfall, 10 Derwent Close, Hebden Bridge, HX7 7ED |
| G0 | TOC | Marc Litchman, 26 Oak Tree Close, Loughton, IG10 2RE |
| G0 | TOD | T Northover, 13 Dagenham Avenue, Dagenham, RM9 6LD |
| G0 | TOE | Arthur Gallagher, 1a Wynsome Street, Southwick, Trowbridge, BA14 9RB |

| | | |
|---|---|---|
| GM0 | TOF | Martin Farnworth, 30e Roseangle, Dundee, DD1 4LY |
| G0 | TOK | Brian Dixon, 97 Sunny Blunts, Peterlee, SR8 1LN |
| GW0 | TOM | T Beedle, 2 Chestnut Grove, Maesteg, CF34 0NT |
| G0 | TOO | C Richmond, 116 Clarendon Road, Morecambe, LA3 1SD |
| G0 | TOQ | D Harrison, 2 Oak Lea Close, [illegible], Barnsley, S75 [illegible] |
| G0 | TOS | B Harper, 51 Cross Lane, Scarborough, YO12 0DQ |
| G0 | TOT | Richard Claxton, Purbeck View, 2 Hoburne Road, Swanage, BH19 2SL |
| GM0 | TOW | GB7SDX DX CLUSTER SUPPORT GROUP c/o R James, 4 Pentland Place, Bearsden, C/O Ray James, Glasgow, G61 4JU |
| G0 | TOX | C Wheeler, 190 Mount Pleasant, Redditch, B97 4JL |
| G0 | TOY | K Li, 16 Garthland Drive, Arkley, Barnet, EN5 3BB |
| G0 | TPA | A Taylor, The Grey House, Chipping Campden, GL55 6XP |
| G0 | TPB | D Smith, 84 Wheatlands, Fareham, PO14 4SL |
| G0 | TPD | J Gaythor, 33 Greenways, Winchcombe, Cheltenham, GL54 5LQ |
| G0 | TPE | Alwyn Davis, 200 Preston Old Road, Blackburn, BB2 2TX |
| G0 | TPG | D Taylor, 9 Stonepits Lane, Hunt End, Redditch, B97 5LX |
| G0 | TPH | Alan Horne, 54 Manor Road, Desford, Leicester, LE9 9JR |
| GM0 | TPI | D Harris, 42 Shira Terrace, East Kilbride, Glasgow, G74 2HU |
| G0 | TPJ | G Green, 11 Whitewell Drive, Upton, Wirral, CH49 4PF |
| G0 | TPK | Anthony Mitchell, 87 Bridgewater Road, Berkhamsted, HP4 1JN |
| GW0 | TPL | David Blundell, 30 Heol Ffynnon Wen, Cardiff, CF14 7TP |
| G0 | TPM | P Mercer, 10 Holmcliffe Avenue, Huddersfield, HD4 7RJ |
| G0 | TPN | M Padgett, 97 Larks Hill, Pontefract, WF8 4RP |
| G0 | TPO | Martin Cook, 11 Atherton Close, Shurdington, Cheltenham, GL51 4SB |
| G0 | TPP | P Pimblett, 5 Edgeside, Great Harwood, Blackburn, BB6 7JS |
| G0 | TPR | B Carter, Rhyd Y Mwyn, Cilgwyn Street, Llanerchymedd, LL71 8ED |
| G0 | TPY | J Brunt, 1 Dane Grove, Cheadle, Stoke-on-Trent, ST10 1QS |
| GM0 | TQB | Maarten De Vries, Old Post Office House, Knockbain, Munlochy, IV8 8PG |
| G0 | TQC | K Sharman, 7 Watkins Way, Paignton, TQ3 3JJ |
| GI0 | TQD | J Gough, 50 Culmore Point, Londonderry, BT48 8JW |
| G0 | TQG | D Houghton, 2 Hereford Avenue, Golborne, Warrington, WA3 3NA |
| G0 | TQJ | C Vernon, 99 High Street, Tarporley, CW6 0AB |
| G0 | TQP | J Smith, High Tree, Radford Lane, Wolverhampton, WV3 8JT |
| G0 | TQR | A Baker, 23 Trematon Drive, Ivybridge, PL21 0HT |
| G0 | TQS | C Forsyth, 8 Oriole Drive, Exeter, EX4 4SJ |
| G0 | TQT | John Joll, 16 Jephson Road, St. Judes, Plymouth, PL4 9ET |
| G0 | TQV | K Taylor, 1 Chapel Close, Reepham, Lincoln, LN3 4EJ |
| G0 | TQZ | M Emm, Highwood Cottage, Daggons Road, Fordingbridge, SP6 3DJ |
| G0 | TRB | Roger Betts, 15 Cleasby, Wilnecote, Tamworth, B77 4JL |
| G0 | TRC | TRAFFORD ARC c/o Peter Fambely, 126 Ashton Lane, Sale, M33 5QJ |
| G0 | TRD | Trond Thorman, Tir na Nog, Coombe Ridings, Kingston upon Thames, KT2 7JT |
| G0 | TRE | N Dimbleby, 4 Rossetti Place, Holmer Green, High Wycombe, HP15 6XA |
| G0 | TRG | Thames Amateur Radio Group c/o Norman Crampton, 7 Barneveld Avenue, Canvey Island, SS8 8NZ |
| G0 | TRH | D Brooke, 26 Ulverston Road, Hull, HU4 7HL |
| G0 | TRI | L Pritchard, The Granary, Greenways Farm, Ross-on-Wye, HR9 6DH |
| G0 | TRJ | R Makepeace, 55 Trethannas Gardens, Praze, Camborne, TR14 0LL |
| G0 | TRK | J Ogden, 65 Elm St., Middleton, Manchester, M24 2EQ |
| G0 | TRM | C Page, 1 The Leeway, Danbury, Chelmsford, CM3 4PS |
| G0 | TRN | J De Frece, 15 Kent Close, Highfield Road, Chesterfield, S41 7HA |
| G0 | TRP | M Williams, Trevean, Carpalla, St. Austell, PL26 7TY |
| G0 | TRU | Alexander Irvine, 21 Rutland Road, Partington, Manchester, M31 4NP |
| G0 | TRW | Ronald Cross, 7 The Island, Anthorn, Wigton, CA7 5AN |
| G0 | TRY | G Lyon, 33 Barlborough Road, Pemberton, Wigan, WN5 9HZ |
| GI0 | TSA | D Moore, 3 Knightsbridge, Londonderry, BT47 6FE |
| G0 | TSB | B Snell, 30 Queens Crescent, Brixham, TQ5 9PJ |
| GW0 | TSE | L Owen, Cartref, Llangain, Carmarthen, SA33 5AH |
| G0 | TSG | Paul ryder, 4 edgeway, Nottingham, NG86LY |
| G0 | TSH | Keith Hutt, Fenwick Crossing House, Fenwick Lane, Doncaster, DN6 0EZ |
| G0 | TSJ | S Ruud, 5 Wood Street, Haworth, Keighley, BD22 8BJ |
| G0 | TSK | Geoffrey Wilkins, 156 Buckingham Crescent, Bicester, OX26 4HB |
| GW0 | TSL | H Chapman, 302 Holton Road, Barry, CF63 4HW |
| G0 | TSM | Darren Collins, 6 Chalvington Road, Chandler's Ford, Eastleigh, SO53 3DX |
| G0 | TSN | O.B.O. HODDESDON RC c/o Donald Platt, 22 Charcroft Gardens, Enfield, EN3 7HA |
| G0 | TSQ | Christopher de Lacy, Monday Cottage, Hammerwood, East Grinstead, RH19 3QE |
| G0 | TSR | Nigel Depledge, 29 Scargill Drive, Spennymoor, DL16 6LY |
| GI0 | TSS | C Tait, 116 Aughnaskeagh Road, Dromara, BT25 2NT |
| G0 | TST | R Hawtree, 9 Stonechat Road, Billericay, CM11 2NX |
| G0 | TSU | D Michael, 5 Evelyn Close, Twickenham, TW2 7BL |
| G0 | TTE | Simon Sutherland, 11 Brentwood Road, Leicester, LE2 6AD |
| GW0 | TTF | W Price, Tramore, 67 High Street, Bridgend, CF32 0HL |
| G0 | TTG | M Warriner, 34 Cabrera Avenue, Virginia Water, GU25 4EZ |
| G0 | TTI | R Rayment, 145 Feeches Road, Southend-on-Sea, SS2 6TF |
| G0 | TTL | R Booth, Old School House, Old School Lane, Doncaster, DN11 9RW |
| G0 | TTM | Alan Radley, 16 Kingsley Lane, Thundersley, Benfleet, SS7 3TU |
| G0 | TTN | Paul Brzenczek, Flat 18, Ty Newydd House, Ty Newydd Court, Cwmbran, NP44 1LH |
| G0 | TTO | L chadwick, 2 Auden Place, Longton, Stoke-on-Trent, ST3 1SJ |
| G0 | TTQ | Brian Trivett, 712 East Myrtle Ave, Foley, United States, 36535 |
| G0 | TTR | A Bartle, 10 Holme Dene, Haxey, Doncaster, DN9 2JX |
| G0 | TTS | Paul Bulmer, 61 Middleham Avenue, York, YO31 9BD |
| G0 | TTW | K Yeates, Newlands, Ashby Lane, Lutterworth, LE17 4SQ |
| GM0 | TTY | W McBurney, 6 Hill Street, Tillicoultry, FK13 6HF |
| G0 | TUC | D Wilkes, 85 Moss Lane, Hesketh Bank, Preston, PR4 6AA |
| G0 | TUE | R Gilchrist, 13 Grammerscroft, Millom, LA18 5EQ |
| G0 | TUI | B Hudson, 5 Rylands Road, Southend-on-Sea, SS2 4LW |
| G0 | TUJ | Walter Spencer, 111 Rosmead Street, Hull, HU9 2TE |
| G0 | TUL | Roy Woollard, 68 Trunk Furlong, Aspley Guise, Milton Keynes, MK17 8HX |
| G0 | TUM | Barry Cooper, 13 Highbury Road, Leeds, LS6 4EX |
| G0 | TUN | A Powney, 16 Westbrook Way, Wombourne, Wolverhampton, WV5 0EA |
| G0 | TUO | Michael Brough, 24 St. Georges Road, Bletchley, Milton Keynes, MK3 5EN |
| G0 | TUP | N Callow, 3 Maunleigh, Forest Town, Mansfield, NG19 0PP |
| G0 | TUS | R Young, 50 Berrywell Drive, Duns, TD11 3HG |
| G0 | TUU | R Hudson, Norton House, 27 Torne View, Doncaster, DN9 3PQ |
| G0 | TUV | L Kennedy, 28 Murton Garth, Murton, York, YO19 5UL |
| G0 | TUW | G Woolfenden, 11 Chesshire Close, Areley Kings, Stourport-on-Severn, DY13 0EB |
| G0 | TUX | J Fossey, 12 Hitchin Road, Arlesey, SG15 6RP |
| G0 | TVB | Paul Rigg, 1 Stones Hey Gate, Widdop Road, Hebden Bridge, HX7 7HD |
| G0 | TVC | A Latham, 7 College Lane, Stratford upon Avon, CV37 6DD |
| G0 | TVD | Martin Smith, 9 Oak Green Way, Abbots Langley, WD5 0PJ |
| G0 | TVL | S Birkenshaw, 19a Vale Head Grove, Knottingley, WF11 8JL |
| G0 | TVM | Abdul Bashir, 70 Smith Lane, Bradford, BD9 6DQ |
| G0 | TVO | B Hewson, 6 Sanworth Street, Todmorden, OL14 5BU |
| G0 | TVR | C Binnell, 146 Hales Crescent, Warley, Smethwick, B67 6QX |
| G0 | TVS | C Saunders, 17 Bure Road, Friars Cliff, Christchurch, BH23 4ED |
| GM0 | TVT | Richard Hemmings, 2a Caversta, Isle of Lewis, HS2 9QE |
| G0 | TVU | M Binns, 49 Fairview, Pontefract, WF8 3NT |
| G0 | TVV | E Graham, 543 Lea Bridge Road, London, E10 7EB |
| G0 | TVW | David Rodman, 23 High Trees, Dore, Sheffield, S17 3GF |
| GW0 | TVX | R Ellis, 7 Manor Bal Gardens, St. Clears, Carmarthen, [illegible] |
| GM0 | TWB | I Lindsay, Talliady Cottage, Angus, DD8 2SP |
| G0 | TWD | [illegible] WIGAN DEANERY HIGH SCHOOL c/o J Drummond, 14 Bulls Head Cottages, Turton, Bolton, BL7 0HS |
| G0 | TWE | T Walker, The Rectory, Louth Road, Market Rasen, LN8 6BJ |
| GW0 | TWF | M Patterson, 6 Devonshire Place, Port Talbot, SA13 1SG |
| G0 | TWH | Martin Jewkes, 11 Maple Grove, Crewe, CW1 4DY |
| G0 | TWI | S Foote, Red Roofs, 5 Woodland Avenue, Colwyn Bay, LL29 9NL |
| G0 | TWL | Martin Lewis, 75 Lon Maesycoed, Newtown, SY16 1QQ |
| G0 | TWO | Peter Murdoch, 19 Penhelyg Road, Aberdovey, LL35 0PT |
| GW0 | TWR | Clive Harrison, 28 Brynau Wood, Cimla, Neath, SA11 3YQ |
| G0 | TWT | B Lee, Ridge Hill, Ledbury Road, Ledbury, HR8 1ND |
| G0 | TWV | K Stewart, 97 Chase Meadows, Blyth, NE24 4LB |
| GI0 | TWX | I Chin, 25 Meadowbrook, Islandmagee, Larne, BT40 3UG |
| G0 | TXA | Barbara Carter, 16 Hollow Street, Chislet, Canterbury, CT3 4DS |
| GM0 | TXJ | N Service, 13 Garden Terrace, Falkirk, FK1 1RL |
| G0 | TXL | Paul Elliott, 32 Crichton Avenue, Wallington, SM6 8HL |
| G0 | TXN | M Thoyts, 17 Solent Avenue, Lymington, SO41 3SD |
| GW0 | TXP | A Smith, Courtlands, Llysonnen Road, Carmarthen, SA33 5DR |
| G0 | TXU | G Redmond, 21 Grosvenor Gardens, Bognor Regis, PO21 3EZ |
| G0 | TXY | A Hicks, 29 Oak Tree Close, Strensall, York, YO32 5TE |
| G0 | TYC | W McGuire, 68 Boyd Avenue, Dereham, NR19 1ND |
| G0 | TYM | T Allison, 2 Westlands, Stokesley, Middlesbrough, TS9 5BU |
| G0 | TYN | M Holmes, 61 Maryland Lane, Wirral, CH46 7TS |
| G0 | TYQ | G Clark, 2 Keith Road, Swanton Morley, Dereham, NR20 4NQ |
| G0 | TYS | Stephen Allanson, 5 Kingsley Avenue, Crofton, Wakefield, WF4 1RN |
| G0 | TYZ | T Turley, 6 Rowan Grove, Oxford, OX4 7FD |
| G0 | TZC | P Sables, 45 Carr Head Lane, Bolton-upon-Dearne, Rotherham, S63 8DA |
| G0 | TZD | R Leah, 9 Trevia, Camelford, PL32 9UX |
| G0 | TZH | M Bushnell, Rose Cottage, Street Ashton, Rugby, CV23 0PH |
| G0 | TZM | K Winfield, Russettwalls, 58 Bretby Lane, Burton-on-Trent, DE15 0QW |
| G0 | TZO | R Nelson, 61 Broken Cross, Charminster, Dorchester, DT2 9QB |
| G0 | TZP | A Seals, 94a Snakes Lane, Southend-on-Sea, SS2 6UA |
| G0 | TZR | John Macknish, 21a Knoll Rise, Orpington, BR6 0EJ |
| G0 | TZV | G Bryant, 54 Drew Road, Stourbridge, DY9 0UP |
| G0 | TZY | A Davies, 20 Peel Park Close, Accrington, BB5 6PL |
| G0 | TZZ | C Soames, Stud Farm Bungalow, 3 The Street, Sporle, PE32 2EA |
| G0 | UAA | I Fallows, 47 Melrose Avenue, Burnley, BB11 4DN |
| G0 | UAC | C Martin, 17 Chambers Grove, Chapeltown, Sheffield, S35 2TD |
| G0 | UAD | Gordon Rowe, 8 Dove Close, Bolton-upon-Dearne, Rotherham, S63 8JL |
| GI0 | UAG | R Anderson, Derry Lodge, 2 Tullymally Road, Newtownards, BT22 1JX |
| G0 | UAI | S Marshall, 68 Parkfield Avenue, Hampden Park, Eastbourne, BN22 9SF |
| G0 | UAK | Ralph Thompson, 51 Rydal Avenue, Ramsgate, CT11 0PX |
| G0 | UAO | J Smith, 124 Parkside Avenue, Barnehurst, Bexleyheath, DA7 6NL |
| G0 | UAP | P Westbury, 5 Smithfield Place, Bournemouth, BH9 2QJ |
| G0 | UAS | A Smith, 3 Woodcourt Close, Sittingbourne, ME10 1QT |
| G0 | UAV | P Tristram, 26 St. Andrews Road, Paignton, TQ4 6HA |
| G0 | UAY | E Fletcher, 5 Butt Hill Court, Bury New Road, Manchester, M25 9NT |
| G0 | UBA | R Gibbs, 357 Downham Way, Bromley, BR1 5EW |
| G0 | UBG | David Endean, 6 Higher Westonfields, Totnes, TQ9 5QY |
| G0 | UBJ | K Beach, Taylor Cove, Harwich Road, Clacton-on-Sea, CO16 0AX |
| G0 | UBK | C Carvell, 6 Field Close, Whitby, YO21 3LR |
| G0 | UBL | M Stracey, 9 Boundary Drive, Hutton, Brentwood, CM13 1RH |
| G0 | UBM | J Horsfield, 8 St. Edmunds Road, Ipswich, IP1 3QZ |
| G0 | UBO | Ernest Martin, 9 The Valley Green, Welwyn Garden City, AL8 7DQ |
| G0 | UBX | Anthony Quince, 9 Biscay Close, Irchester, Wellingborough, NN29 7FD |
| G0 | UBY | Leigh Duffill, 14 Leonard Street, Hull, HU3 1SA |
| G0 | UCA | T Ogden, Tyn Llidiart, Brithdir, Dolgellau, LL40 2RP |
| G0 | UCC | M Sayegh, 1 Hoylake Road, East Acton, London, W3 7NP |
| G0 | UCD | Martin West, 12 Jenny Gill Crescent, Skipton, BD23 2RR |
| G0 | UCE | John Swartz, 15076 New Salem Bluff Road, Petersburg, United States, IL 62675 |
| G0 | UCH | Colin Hawes, 64 Whitmore Court, Whitmore Way, Basildon, SS14 2TN |
| G0 | UCI | R Jones, 29 Avon Dale, Newport, TF10 7LS |
| G0 | UCK | Barrington Linehan, 28, Hurlstone Grove, Milton Keynes, MK4 1EF |
| G0 | UCN | P Turner, 176 Crescent Road, Hadley, Telford, TF1 5LF |
| G0 | UCP | John Seager, 2 Waterford Road, Oxton, Prenton, CH43 6UT |
| G0 | UCS | W Dingley, 63 Hurst Green, Mawdesley, Ormskirk, L40 2QS |
| G0 | UCT | B Obrien, 59 Riddlesdown Avenue, Purley, CR8 1JL |
| G0 | UCX | T Taylor, 1 Cooke Gardens, Branksome, Poole, BH12 1QE |
| G0 | UDB | David Birch, 4 Godolphin Road, Helston, TR13 8PY |
| G0 | UDE | Clive Woodlock, 68 Morellement Jhuboo, Trou Aux Biches, Mauritius |
| G0 | UDG | K Deegan, 7 Oldcott Crescent, Kidsgrove, Stoke-on-Trent, ST7 4HF |
| G0 | UDH | R Pritchard, 1 Limetree Court, Station Road, Abergavenny, NP7 5JA |
| G0 | UDI | J Murphy, 8 Spencer Avenue, Wribbenhall, Bewdley, DY12 1DB |
| G0 | UDJ | D Jones, Bryn Awelon, Brynsannan, Holywell, CH8 8AX |
| GM0 | UDL | Andrew Cowan, House Of Shannon, Wester Templands, Fortrose, IV10 8RA |
| G0 | UDO | R Dodd, 33 Dogcroft Road, Stoke-on-Trent, ST6 6PE |
| G0 | UDP | Keith Fairbotham, 32 Northolme Drive, York, YO30 5RP |
| GM0 | UDY | A Hyslop, 9 Shalloch Square, Girvan, KA26 0EA |
| G0 | UDZ | Michael Baister, 7b Hepple Road, Spital Estate, Newbiggin-by-The-Sea, NE64 6ST |
| G0 | UEA | John Heald, 94 Haylings Road, Leiston, IP16 4DT |
| G0 | UEB | R Fisher, 14 Colindeep Lane, Sprowston, Norwich, NR7 8EG |
| G0 | UEC | J Bird, 56 Garsdale, Birtley, Chester le Street, DH3 2EY |
| G0 | UED | I Harkness, 2 Trevor Drive, Maidstone, ME16 0QP |
| GI0 | UEG | RADIO AMATEUR SPECIAL EVENT GROUP c/o Raymond Wilson, 68 Kensington Road, Belfast, BT5 6NG |
| G0 | UEH | R Hislop, 79 Norwood Avenue, Hasland, Chesterfield, S41 0NJ |
| G0 | UEK | S Payne, 55 Binstead Lodge Road, Binstead, Ryde, PO33 3TL |
| G0 | UEM | S Copley, 28 Kiln Close, Old Catton, Norwich, NR6 7HZ |
| G0 | UEO | J Ewan, 71 Kingston Drive, Connah's Quay, CH5 4TN |
| GM0 | UET | R Henderson, 22 Bowmont Place, East Kilbride, Glasgow, G75 8YG |
| G0 | UEU | A Jackson, 7 Rushfield, Sawbridgeworth, CM21 9NF |
| G0 | UEW | ST CHRISTOPHERS SCHOOL ARC c/o R Jones, Greens Cottage, Luton Road, Offley, Hitchin, SG5 3DR |
| G0 | UFB | P Short, 23 Barn Close, Hartford, Huntingdon, PE29 1XF |
| G0 | UFC | D Meacock, The Limes, Davids Lane, Boston, PE22 0DZ |
| G0 | UFD | Thomas Reynolds, 87 Clarendon Street, Rochdale, OL16 4UD |
| G0 | UFE | [illegible], 11 Ludlow Drive, [illegible], Telford, [illegible] 1EQ |
| G0 | UFF | Roy Reynolds, 5 Dymond Court, Bodmin, PL31 2FP |
| G0 | UFI | Gerald Brady, Thirn Grange, Thirn, Ripon, HG4 4AU |
| G0 | UFJ | Spencer Glover, 46 Merton Close, Oldbury, B68 8NG |
| G0 | UFL | T Reid, Menwith Hill Station, Po Box 905, Harrogate, HG3 2RF |
| G0 | UFN | Paul Dean, 80 Escallond Drive, Dalton-le-Dale, Seaham, SR7 8JZ |
| G0 | UFP | C Beesley-Reynolds, Kaos Roams, Palmerston Close, Leicester, LE8 0JJ |
| G0 | UFU | C Jameson, 35 Bilberry Grove, Taunton, TA1 3XN |
| G0 | UFV | P Shaw, 15 Moorfield Road, St. Giles-on-the-Heath, Launceston, PL15 9SY |
| G0 | UFW | J Marvill, 242 Hillmorton Road, Rugby, CV22 5BG |
| G0 | UFY | J Brown, 3 Slipper Mill, Slipper Road, Emsworth, PO10 8XD |
| G0 | UFZ | John Howard, 32 Eastgate, North Newbald, York, YO43 4SD |
| G0 | UGA | B Moody, 38 Bromwich Road, Willerby, Hull, HU10 6SF |
| G0 | UGD | Nigel Boyd, 16 Edensor Road, Eastbourne, BN20 7XR |
| GM0 | UGG | Alison Aird, 3 Graystones, Kilwinning, KA13 7DT |
| GM0 | UGH | M Westland, 142 Claremont, Alloa, FK10 2EG |
| G0 | UGI | H Argument, 9 Oxley Close, Shepshed, Loughborough, LE12 9LS |
| G0 | UGJ | D Basford, 91 Hollins Spring Avenue, Dronfield, S18 1RP |
| G0 | UGM | Eric Peasey, Greenfield Cottage, Greenfield Road, Colne, BB8 9PE |
| G0 | UGO | Angus Wilson, 253 South Avenue, Abingdon, OX14 1QT |
| G0 | UGQ | M Webb, 75 Bolingbroke Heights, Flint, CH6 5AN |
| G0 | UGR | M Chaloner, Barnhay House, Newport, Berkeley, GL13 9PY |
| G0 | UGS | F Taberner, 51 Canford View Drive, Colehill, Wimborne, BH21 2UW |
| G0 | UGW | Peter Grant, 37 Glenmore Park, Dundalk, County Louth, Ireland |
| G0 | UGX | J Kinsella, 85 Lowther Street, Coventry, CV2 4GL |
| G0 | UGY | S Jackson, 32 Sherwell Drive, Alcester, B49 5HA |
| GM0 | UHC | I Ropper, 2 Deerhill, Dechmont, Broxburn, EH52 6LY |
| G0 | UHD | K Hore, 15 Heriot Way, Great Totham, Maldon, CM9 8BW |
| G0 | UHF | R Darwent, 139 The Oval, Sheffield, S5 6SQ |
| G0 | UHG | R Warne, 60 Oldbury Orchard, Churchdown, Gloucester, GL3 2PU |
| G0 | UHI | D Norton, 52 Letchworth Road, Leicester, LE3 6FG |
| GW0 | UHJ | William Griffiths, 147 High Street, Tonyrefail, Porth, CF39 8PL |
| GD0 | UHK | M Peiperl, 45 High Street, Harrow, HA1 3HT |
| G0 | UHM | L Ruddock, 2 Cross Lane, Waterlooville, PO8 9TJ |
| GW0 | UHO | Peter Sage, 4 Gladstone Terrace, Miskin, Mountain Ash, CF45 3BS |
| G0 | UHQ | Michael Minihane, 60 Wolsey Drive, Walton-on-Thames, KT12 3BA |
| G0 | UHS | R Hatch, 99-101 Hornby Road, Blackpool, FY1 4QP |
| G0 | UHU | G Bloyce, 8 Stanwyn Avenue, Clacton-on-Sea, CO15 3AR |
| GW0 | UHX | S Jones, 64 Springfield Gardens, Hirwaun, Aberdare, CF44 9LY |
| G0 | UIB | Philip Pearson, 72 Marconi Road, Chelmsford, CM1 1QD |
| G0 | UID | John Parker, 1 Church End, Syresham, Brackley, NN13 5HU |
| G0 | UIF | G Fairbrass, 230 Kirkby Road, Barwell, Leicester, LE9 8FS |
| GM0 | UIG | C Cowan, 85 Eastwoodmains Road, Clarkston, Glasgow, G76 7HG |
| G0 | UIH | S Lawman, 44 Barnwell, Peterborough, PE8 5PS |
| G0 | UIL | Dean Brice, 4 Bishop Fox Drive, Taunton, TA1 3HQ |
| G0 | UIP | Niall Wallace, Tan y Bryn, Bryn Road, Flint, CH6 5HU |
| G0 | UIQ | William Furze, 2 Lynewood Road, Cromer, NR27 0EE |
| G0 | UIS | R Darby, 38 Abbotsbury, Great Hollands, Bracknell, RG12 8QU |
| G0 | UIW | Anthony Jones, 12 Hill Top Road, Harrogate, HG1 3AN |
| G0 | UIX | Andrew Lambert, 10 The Green, Cirencester, GL7 1AU |
| GW0 | UIZ | B Galsworthy, 30 Pen Yr Yrfa, Morriston, Swansea, SA6 6BA |
| G0 | UJD | M Bartle, 14 Litton Avenue, Skegby, Sutton-in-Ashfield, NG17 3AB |
| GI0 | UJG | R Stinson, 51 Cloncarrish Road, Portadown, Craigavon, BT62 1RN |
| G0 | UJI | D Sparkes, 31 Lockyers Drive, Ferndown, BH22 8AL |
| G0 | UJP | J Fleetwood, 11 Chichester Road, Southampton, SO18 6BB |
| G0 | UJU | D Barlow, 7 Prospect Terrace, Canal Road, Taunton, TA1 1PH |
| G0 | UJW | J Wintle, 23 Napier Walk, Andover, SP10 1PL |
| G0 | UKA | J Black, 8 Cornwood Close, Finchley, London, N2 0HP |
| G0 | UKB | Brian Jones, 47 Pine Crescent, Chandler's Ford, Eastleigh, SO53 1LN |
| GW0 | UKC | M Price, 50 Llangorse Road, Cwmbach, Aberdare, CF44 0HR |
| GM0 | UKD | Stuart Munro, 76 John Neilson Avenue, Paisley, PA1 2SX |
| G0 | UKF | J Griffith, Craig Artro, Llanbedr, LL45 2LU |
| G0 | UKG | A Powell, 80 St. Andrews Crescent, Abergavenny, NP7 6HN |
| G0 | UKJ | W Shrubsall, 55 Kenilworth Court, Sittingbourne, ME10 1TX |
| G0 | UKK | K Stanyer, 15 Wilbrahams Way, Alsager, Stoke-on-Trent, ST7 2NR |
| G0 | UKL | Paul Charlton, 73 King Street, Pinxton, Nottingham, NG16 6NL |
| G0 | UKM | M russell, 10 Baytree Grove, Ramsbottom, Bury, BL0 9UF |
| G0 | UKO | R Thealston, 130 Greenshaw Drive, Haxby, York, YO32 2DG |
| G0 | UKP | Brian Jopson, 21 Richmond Street, Southend-on-Sea, SS2 4NW |
| G0 | UKS | Roland Towler, 77 Glebe Road, Hull, HU7 0DU |
| G0 | UKT | W Waldron, Torfaen Scouts Arc, 23 Forest Close, Cwmbran, NP44 4TE |
| GM0 | UKZ | C Gibson, 45 Tiree Place, Newton Mearns, Glasgow, G77 6UJ |
| G0 | ULA | N Foster, 68 Brookfield Way, Lower Cambourne, Cambridge, CB23 5ED |
| GD0 | ULF | S Adams, c/o 55 Crosier Way, Ruislip, HA4 6HG |
| G0 | ULG | K Dewing, 124 Proctor Road, Norwich, NR6 7PH |
| G0 | ULH | Leslie Harris, 183a Painswick Road, Gloucester, GL4 4AG |
| G0 | ULI | M Kaliski, 132 Woodland Road, Hellesdon, Norwich, NR6 5RQ |
| G0 | ULL | Edward Williams, 50 Broad Lawn, New Eltham, London, SE9 3XD |
| G0 | ULM | Philip Wilson, 56 Highfield Road, Blacon, Chester, CH1 5AZ |
| G0 | ULN | L Fuller, 13 Higham Way, Brough, HU15 1NA |
| G0 | ULO | P Ravenscroft, 32 Tennyson Road, Wolverhampton, WV10 8NG |
| G0 | ULP | Leslie Parsons, 105 Victoria Road West, Prestatyn, LL19 7DS |
| G0 | ULQ | W Bucknell, 119 Fossway, York, YO31 8SQ |

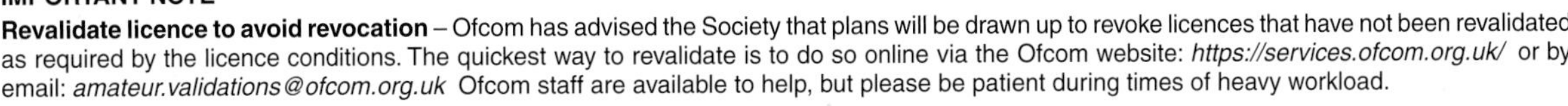

**IMPORTANT NOTE**

**Revalidate licence to avoid revocation** – Ofcom has advised the Society that plans will be drawn up to revoke licences that have not been revalidated as required by the licence conditions. The quickest way to revalidate is to do so online via the Ofcom website: *https://services.ofcom.org.uk/* or by email: *amateur.validations@ofcom.org.uk* Ofcom staff are available to help, but please be patient during times of heavy workload.

G0 ULS Frank Mortimer, 115 Dell Road, Lowestoft, NR33 9NX
G0 ULZ G Toop, 27 Lavington House, Jubilee Way, Blandford Forum, DT11 7UT
G0 UMD D Russell-Smith, 6 Braeside Close, Winchester, SO22 4JL
G0 UMI E Latter, 76 Crossway, Plympton, Plymouth, PL7 4HY
GM0 UMJ S Heerma Van Voss, Blarachaorachan, Fort William, PH33 6SZ
G0 UMK R Adams, 9 Chestnut Garth, Roos, Brooklands, Hull, HU12 0LE
G0 UML N Watling, 36 All Saints Walk, Mattishall, Dereham, NR20 3RF
G0 UMM N Roberson, 6 Long Lane, West Winch, King's Lynn, PE33 0PG
G0 UMP M Parker, 90 Cavendish Road, Patchway, Bristol, BS34 5HH
G0 UMS Robert Scott Petrie, 11 St. James's Close, Yeovil, BA21 3AH
G0 UMV P Johnson, 52 Evesham Road, Cookhill, Alcester, B49 5LJ
G0 UMY Christopher Lowe, 37 Parsonage Brow, Upholland, Skelmersdale, WN8 0JG
G0 UNB J Jeffers, 11 Polywell, Appledore, Bideford, EX39 1SG
G0 UNC George Hancock, 3c Richmond Street, Hull, HU5 3JY
G0 UND D Scargill, 10 Mendip Avenue, North Hykeham, Lincoln, LN6 9SZ
G0 UNE J Spencer, 6 Redcar Road, Sunderland, SR5 5QA
G0 UNF G Taylor, Flat 67a, Bramley Grange Hotel Flats, Guildford, GU5 0BL
G0 UNG Brian Massey, 40 East St., Ashton In Makerfield, Wigan, WN4 8ST
G0 UNK T Wright, 2 Regent Road, Church, Accrington, BB5 4AR
G0 UNW R Martin, 44 Threadneedle Cres, Willowdale, Ontario, Canada
G0 UNY M Lindley, 23 Townend Lane, Deepcar, Sheffield, S36 2TN
G0 UOB A Chilinski, 37 Swan Road, Dereham, NR19 1AG
G0 UOD E Sheather, Clare Cottage, North End, Shaftesbury, SP7 9HX
G0 UOI Robert Fox, 4 Mill House, School Lane, Polegate, BN26 6AU
G0 UOM N Taylor, 16 Josephine Road, Rotherham, S61 1BJ
G0 UOP O.B.O. UNIVERSITY OF PLYMOUTH A.R.S. c/o R Linford, Department Of Communication, And Electronic Engineering, Drake Circus Plymouth, PL4 8AA
G0 UOQ P Creissen, 143 Hawthorn Bank, Spalding, PE11 2UN
G0 UOS J Butterworth, 12 Ingswell Drive, Notton, Wakefield, WF4 2NF
GM0 UOU Colin Muir, 1/1 132 Falside Road, Paisley, PA2 6JT
G0 UOV S Porter, 20 Newbridge Road, Ambergate, Belper, DE56 2GR
G0 UOZ A Morris, 4 Pleasant Terrace, Lincoln, LN5 8DA
G0 UPB J Barker, Riverside House, Homecroft Drive, Cheltenham, GL51 9SN
G0 UPD Robert Brinkley, 70 Leopold Road, Felixstowe, IP11 7NR
GM0 UPE G Sutherland, Tigh Na Lachan, Stuckroech, Cairndow, PA27 8BZ
G0 UPG J Dunne, 40 Egmont Road, Poole, BH16 5BZ
G0 UPL H Summers, Flat 7, Drake House, 4 Victory Place, London, E14 8BG
G0 UPO M May, 53 Oakfield Road, Hadfield, Glossop, SK13 2BN
G0 UPS P Zimmermann, 85 Wimborne Road West, Wimborne, BH21 2DH
G0 UPV T Berrisford, 126 Star & Garter Road, Stoke-on-Trent, ST3 7HN
G0 UPY Martin Rogers, 12 Stephenson Way, Honeybourne, Evesham, WR11 7GH
G0 UQB D Brittain, 9 Highfield Road, Cookley, Kidderminster, DY10 3UB
G0 UQC R Birkett, 14 Greta Street, Keswick, CA12 4HS
G0 UQE Robert Weaver, 107 Patch Lane, Redditch, B98 7XE
G0 UQF G Merrils, 2 East St., Darfield, Barnsley, S73 9AE
GW0 UQH Sidney Provan, 12 Trewarren Road, St. Ishmaels, Haverfordwest, SA62 3SZ
G0 UQI L Snowden, 25 Brixham Road, Paignton, TQ4 7HG
G0 UQJ N Smith, 36 Maple Road, Sutton Coldfield, B72 1JP
GI0 UQK P Sinclair, 37 Willow Avenue, Banbridge, BT32 4RE
G0 UQO G Rekers, 18 The Ridge, Purley, CR8 3PE
G0 UQP Frederick Waters, 96 Stockley Road, Barmston Village, Washington, NE38 8DR
G0 UQQ Nigel Baskerville, 10 Park Avenue, Sprotbrough, Doncaster, DN5 7LW
G0 UQT S Nash, 26 Lyndhurst Crescent, Wembdon, Bridgwater, TA6 7QG
G0 UQU John Squires, 44 St. Marys Road, Doncaster, DN1 2NP
G0 UQV M Parkin, Chenet, 32 The Nooking, Doncaster, DN9 2JQ
G0 UQY P Cox, 17 Hyde Lane, Upper Beeding, Steyning, BN44 3WJ
G0 UQZ D Eyre, 41 Lindsay Road, Sheffield, S5 7WE
G0 URB J Morse, 14 Strathmore Road, Bournemouth, BH9 3NS
G0 URC P Ball, 12 Warren Way, Digswell, Welwyn, AL6 0DH
GM0 URD Ian Thomson, 33 Aytoun Grove, Dunfermline, KY12 9YA
G0 URF M Dodsworth, 359 Upper Town Street, Bramley, Leeds, LS13 3JX
GI0 URI S Robinson, 19 Pattonville, Dungannon, BT71 6DD
GI0 URN The Royal Naval (Ulster) ARC c/o Norman McKee, 54 Castlemore Park, Belfast, BT6 9RP
G0 URO D Fosh, 8 The Pines, Horsham, RH12 4UF
G0 URT Stuart Bradbury, 39 Grosvenor Road, Hyde, SK14 5AB
GM0 URU J Cartlidge, 14 Davidson Street, Broughty Ferry, Dundee, DD5 3AT
G0 USA L Civita, 53 Rockhurst Drive, Eastbourne, BN20 8XD
GI0 USC John Smith, 5 Old Turn, Carrickfergus, BT38 7EH
G0 USE J Davy-Jones, 1 Wensley Gardens, Emsworth, PO10 7RA
G0 USF Margaret James, 49 Church Street, Fontmell Magna, Shaftesbury, SP7 0NY
GM0 USI Alan Dimmick, 02 120 Shakespeare Street, Glasgow, G20 8LF
G0 USJ S Johnson, 36 Langar Woods, Langar, Nottingham, NG13 9HZ
G0 USK P Perera, 13 Dalcross Road, Hounslow, TW4 7RA
G0 USM B Parker, 24 Mayfield Road, Chorley, PR6 0DG
GI0 USQ P Fox-Roberts, 8 Lynwood Park, Holywood, BT18 9EU
GI0 USS K Chambers, 59 Ravenswood, Banbridge, BT32 3RD
GI0 USW Paul Mcdonald, 13 serpintine road, Newtownabbey, BT36 7HA
G0 UTB Philip Cordrey, Redline C.A Edif. Fannellis 5 Y 6, Av. Ricaurte, Cojedes, Venezuela, 2201
G0 UTC J Lomas, 5 Gwelfor, Rhos on Sea, Colwyn Bay, LL28 4AJ
GM0 UTD H Urquhart, The Cless, Peebles, EH45 8NU
GI0 UTE Desmond Auld, 37 Castlewellan Road, Rathfriland, Newry, BT34 5EL
G0 UTM Bernard Watson, 6 Shakespeare Avenue, Scunthorpe, DN17 1SA
G0 UTN S Harrison, 22 Dales Avenue, Sutton-in-Ashfield, NG17 4BY
G0 UTP Ronald Walker, 1 Farmcote Court, Hemlington, Middlesbrough, TS8 9LJ
G0 UTR David Coleman, 1 Kerstin Close, Cheltenham, GL50 4SA
G0 UTT DENGIE HUNDRED ARS c/o A Slade, Skerries, Summerhill Althorne, Chelmsford, CM3 6BY
G0 UTU Peter Humphreys, 47 Crescent Road, Locks Heath, Southampton, SO31 6PE
GI0 UTV I Ross, 312 Castlereagh Road, Belfast, BT5 6AD
G0 UTZ R Davies, 36 Woodside Avenue, Brown Edge, Stoke-on-Trent, ST6 8RX
G0 UUA W Hutchings, Bitternsdale Farm, Bustomley Lane, Stoke-on-Trent, ST10 4PE
GM0 UUB G Matthews, 88 Nevis Crescent, Alloa, FK10 2BN
G0 UUC W Slater, 44 Hope Street, Chesterfield, S40 1DG
G0 UUF S Errington, 23 Pinewood Drive, Bletchley, Milton Keynes, MK2 2HT
G0 UUG M Harris, 82 Wyckham Road, Castle Bromwich, Birmingham, B36 0HS
G0 UUI Raymond Newport, 17 College Crescent, Oakley, Aylesbury, HP18 9QZ
G0 UUM J Lewis, Burley Cottage, Shortwood, Stafford, ST21 6RG
G0 UUN J Lewis, Burley Cottage, Shortwood, Stafford, ST21 6RG
G0 UUP M Stevens, 24 Oakroyd Close, Burgess Hill, RH15 0QN
G0 UUR C Branch, 10 Queens Road, Thame, OX9 3NQ
G0 UUS G Hanson, 34 Old Garden Close, Locks Heath, Southampton, SO31 6RN
G0 UUT Elan Paim, 10 West Road, Norwich, NR5 0NE
G0 UUU P Earnshaw, 7 Hampton Road, Scarborough, YO12 5PU
G0 UUX W Ellis, 23 Cliff Boulevard, Kimberley, Nottingham, NG16 2JJ
G0 UUZ A Breeze, 1 Park Road West, Wollaston, Stourbridge, DY8 3NG
GI0 UVD William Bustard, 66 Hertford Crescent, Lisburn, BT28 1SQ
G0 UVE Ian Reynolds, Five Bars, Hereford Road Weobley, Hereford, HR4 8SW
G0 UVG P Ellis, 9 Matilda Gardens, Shenley Church End, Milton Keynes, MK5 6HT
GU0 UVH T Bosher, Highlea, Les Mouriaux, Alderney, Guernsey, GY9 3UY
G0 UVL L Cartwright, 14 Norley Hall Avenue, Wigan, WN5 9TG
G0 UVN B Goolding, 10 Oakwell Close, Stevenage, SG2 8UG
G0 UVP R Jermy, 4 Ramsdean Avenue, Wigston, LE18 1DX
G0 UVQ A Davidson, 51 Wentworth Crescent, Beggarwood, Basingstoke, RG22 4WU
G0 UVR G Akse, Drive Cottage, Ebberston, Scarborough, YO13 9PA
G0 UVT J Goldie, The Coachmans Cottage, Pulford Lane, Chester, CH4 9NN
G0 UVX Geoffrey Valentine, 15 Kingsway, Stotfold, Hitchin, SG5 4EL
G0 UWA E Shanklin, The Coachmans Cottage, Pulford Lane, Chester, CH4 9NN
G0 UWB Ralph Moyle, Aitchill House, Lower Brailes, Banbury, OX15 5AP
G0 UWD J Matthews, 42 Wexham Street, Beaumaris, LL58 8HW
G0 UWF Steven Bowles, 37 Manor Road, Paignton, TQ3 2HZ
G0 UWH J Hartigan, 83 Parsons Road, Redditch, B98 7EJ
G0 UWI H Chipper, 36 Newbridge Way, Truro, TR1 3LX
G0 UWK I Goodier, 2 Chatterley Drive, Kidsgrove, Stoke-on-Trent, ST7 4HW
G0 UWO N Winfield, Oaklea, Chyvogue Meadow, Truro, TR3 7JP
G0 UWS Andrew Sharman, 3 Deben Crescent, Swindon, SN25 3QB
G0 UWU Mike Claridge, 105 Barnwood Avenue, Gloucester, GL4 3AG
G0 UWV John Reynolds, The Paddock, Leswalt, Stranraer, DG9 0LJ
G0 UWW Sidney Fisk, 45 Stow Road, Wiggenhall St. Mary Magdalen, King's Lynn, PE34 3BX
G0 UWX D Pollard, 191 High Road, Halton, Lancaster, LA2 6QB
GI0 UXD D Burns, 207 Rathfriland Road, Dromara, Dromore, BT25 2EQ
G0 UXF C Whittaker, 3a Oak Avenue, Horwich, Bolton, BL6 6JE
G0 UXG Patrick Blunt, 17 Offens Drive, Staplehurst, Tonbridge, TN12 0LR
G0 UXH A Mawson, 39 Glenridding Drive, Barrow-in-Furness, LA14 4PA
G0 UXI M Whitehead, 45 Riverton Road, Manchester, M20 5QH
GW0 UXJ Ashley Burns, 34 Lakeside Gardens, Merthyr Tydfil, CF48 1EN
G0 UXN T Cahill, 22 Lornes Close, Southend-on-Sea, SS2 4PX
G0 UXO B Hillman, 2 Holmes Chapel Road, Congleton, CW12 4NE
G0 UXR T Gilmore, 48 Ash Lane, Hale, Altrincham, WA15 8PD
GW0 UXX D Cumiskey, 16 Delapoer Drive, Haverfordwest, SA61 1HJ
G0 UYA S Chamberlin, 15 Bull Close, East Tuddenham, Dereham, NR20 3LX
G0 UYC Doug Rolph, 3 Bell Close, Bawdeswell, Dereham, NR20 4SL
G0 UYE A Colton, 9 Pineway, Bridgnorth, WV15 5DS
G0 UYF P Moss, 11 The Meadows, Carlton, Goole, DN14 9QZ
G0 UYG A Forster, 9 Pinewood Walk, Stokesley, Middlesbrough, TS9 5HU
G0 UYH A Smart, Nine Hamelin Street, Pye Green Road, Cannock, WS11 2SE
G0 UYM OBO THAMES AMATEUR RADIO c/o Jonathan Hall, Pump Farm Cottage, Pump Lane South, Marlow, SL7 3RB
G0 UYP R Arkell, 33 Chatsworth Crescent, Rushall, Walsall, WS4 1QU
G0 UYQ Michael Melbourne, 32 Lake Farm Road, Rainworth, Mansfield, NG21 0ED
GM0 UYS Simon Jeffrey, 33 The Rowans, Insch, AB52 6ZD
G0 UYT John Hemming, 62 Beaumont Road, Birmingham, B30 2DY
G0 UYV B Smith, 80a Bramcote Lane, Beeston, Nottingham, NG9 4ES
GI0 UYY L O'Flaherty, 1 Ravensdale Villas, Newry, BT34 2PG
GM0 UYZ John Wheeler, 54 Wittet Drive, Elgin, IV30 1TB
G0 UZD A Gisby, 2 Lakeside Brighton Road, Lancing, BN15 8LN
G0 UZE Karl Brookes, Kohima, Spout Lane, Stoke-on-Trent, ST2 7LR
G0 UZF I Riley, 102 Harrison Road, Chorley, PR7 3HS
G0 UZJ Kevin Hogg, 30 Dale Park Avenue, Winterton, Scunthorpe, DN15 9UY
GW0 UZK A Rushton, 29 Queen Street, Blaengarw, Bridgend, CF32 8AH
GM0 UZV Winston Cargill, 23 Ceres Road, Craigrothie, Cupar, KY15 5QB
G0 UZW N Donald, 20 Parkhill Road, Barnby Dun, Doncaster, DN3 1DP
GW0 UZX F Thompson, 15 Aneurin Crescent, Twynyrodyn, Merthyr Tydfil, CF47 0TB
G0 UZY J Shotter, 41 Summerheath Road, Hailsham, BN27 3DR
GI0 VAB Philip Moore, 59 Belmont Road, Belfast, BT4 3DE
G0 VAD J Whitehall, 29 Melrose Terrace, Newbiggin-by-The-Sea, NE64 6XN
G0 VAE Michael Clarke, 21 Sycamore Road, Greenstead Estate, Colchester, CO4 3NF
G0 VAG P Mason, 8 Westbourne Park, Scarborough, YO12 4AT
G0 VAH Alan Heyworth, Caldecott Farm, Caldecott, Chester, CH3 6PE
G0 VAI Robert Frow, Valley Yard, Skeete Road, Folkestone, CT18 8DS
G0 VAJ S Hobden, 10 Turton Close, Brighton, BN2 5DA
G0 VAL D Wyatt, 96 Woodlands Close, Clacton-on-Sea, CO15 4RU
G0 VAM A Witter, 44 Regent Street, Newton le Willows, WA12 9LS
G0 VAP G Withers, 5 Whorlton Close, Kemplah Park, Guisborough, TS14 8LW
G0 VAR C Wilson, 15 Biddenden Close, Bearsted, Maidstone, ME15 8JP
G0 VAS V Ikonomou, 18 Canhams Road, Great Cornard, Sudbury, CO10 0EP
G0 VAU T James, 114 Broadway, Loughborough, LE11 2JG
G0 VAV J Farrington, 6 The Gravel, Mere Brow, Preston, PR4 6JX
G0 VAX Brian Bowers, 31 Gresford Avenue, Wirral, CH48 6DA
G0 VAY G Gower, 14 Beck Garth, Hedon, Hull, HU12 8LH
G0 VAZ R Barker, 56 Southend Road, Weston-Super-Mare, BS23 4JZ
GM0 VBE B Higton, The Straith, Priestland, Darvel, KA17 0LP
G0 VBG W Chandler, 6 St. Thomas Close, Hinton Waldrist, Faringdon, SN7 8RP
G0 VBI Panikos Christodoulou, Flat 27, Citizen House, London, N7 7ND
G0 VBK M Blackmore, Flat 20, Hill View House, Bristol, BS15 1TA
G0 VBM R Caddy, 4 Londesborough Park, Seamer, Scarborough, YO12 4QT
G0 VBN J Cressey, 32 Ballifield Road, Sheffield, S13 9HX
G0 VBP F Madely, 138 Coningsby Drive, Kidderminster, DY11 5LZ
G0 VBQ F Webster, 14 Redbank Avenue, Erdington, Birmingham, B23 7JR
G0 VBR Mark Dunstan, 3 Coronation Road, Rawmarsh, Rotherham, S62 5LW
G0 VBT G Dawson, 22 High St., Tean, Stoke-on-Trent, ST10 4DZ
G0 VBX John Collingwood, 36 Bishop Street, Alfreton, DE55 7EF
G0 VBZ Dennis Stinton, 43 The Meadows, Bidford-on-Avon, Alcester, B50 4AP
G0 VCA P Wilson, The Elms, Manor Lane, Cheltenham, GL52 9QX
G0 VCB D McCormick, 2 Littleworth Cottage, Duncote, Towcester, NN12 8AL
G0 VCD Leslie Browne, 7 Mornington Drive, Cheltenham, GL53 0BH
G0 VCJ K Emblen, 17 Larch Close, Bognor Regis, PO22 9LA
GM0 VCN W Long, 52 Stirling Road, Milnathort, Kinross, KY13 9XG
G0 VCV John Partridge, 27 Leigh Road, Penhill, Swindon, SN2 5DE
G0 VCW R Evans, 27 Webb Road, Raunds, Wellingborough, NN9 6HH
G0 VCY P Clark, 62 Ashburnham Road, Southend-on-Sea, SS1 1QE
G0 VDE William Rothwell, 30 Wellbrook Way, Girton, Cambridge, CB3 0GP
G0 VDJ E Webster, 33 Cherry Gardens, Bitton, Bristol, BS30 6JA
G0 VDN B Logan, 22 Chiltern House, Hillcrest Road, London, W5 1HL
G0 VDO Cecil Pond, 316 St. Faiths Road, Old Catton, Norwich, NR6 7BL
G0 VDP K Hutley, Three Ways, 1 Walden House Road, Maldon, CM9 8PJ
G0 VDQ A Ramsden, 7 Florence Road, Pakefield, Lowestoft, NR33 7BX
G0 VDR L Goffin, The Hollies, Belaugh Green Lane, Norwich, NR12 7AJ
G0 VDT P Clark, 21 Uplands, Welwyn Garden City, AL8 7EN
G0 VDU J Newman, Shangri-La, Treverbyn Road, St. Austell, PL26 8TL
G0 VDV D Steele, Tanglewood, Beckingham Street, Maldon, CM9 8LL
G0 VDZ Nigel Newby, 167 Watersplash Road, Shepperton, TW17 0EN
G0 VEB Robert Verrall, 28 Higher Holcombe Road, Teignmouth, TQ14 8RJ
G0 VEC B Hoare, Kilclare, Carrick On Shannon, Drumaleague House, Co. Leitrim, Ireland
G0 VEH John Mulye, 83 Forest Drive East, London, E11 1JX
G0 VEI Barry Davison, Pond House, Moores Lane, Colchester, CO7 6RF
GM0 VEK Peter Davie, 62 Monkland Avenue, Kirkintilloch, Glasgow, G66 3BW
GW0 VEM A Mccleverty, 3 The Fold, Upper Thornton, Milford Haven, SA73 3UE
G0 VEP Paul Steed, Unit 32, Regents Trade Park, Gosport, PO13 0EQ
G0 VET P Burnand, 5 Northgate, Hornsea, HU18 1ES
G0 VEU G Chantry, Summerfield, The Avenue, Oswestry, SY11 4LF
GW0 VEW D Davies, 27 Twyniago, Pontarddulais, Swansea, SA4 8HX
G0 VEX P Dickinson, 34 Marshall Avenue, Bridlington, YO15 2DS
G0 VFB D Banks, 6 Kirkstall Close, Walsall, WS3 2SS
GM0 VFD A Adam, 10 Greenmount Road North, Burntisland, KY3 9JQ
G0 VFE W Vernon-Jones, 13 Erleigh Dene, Newbury, RG14 6JG
GW0 VFF Stephen Emanuel, 98 Moorland Road, Neath, SA11 1JL
G0 VFL K Gardner, Str Matei Vasilescu 82, Drobeth Turnu Sevferin 1500, Mehedinti, Romania
G0 VFM A Whitlock, 50 Greenfield Avenue, Northampton, NN3 2AF
G0 VFS R Bailey, 13 Whiteland Rise, Westbury, BA13 3HP
G0 VFU Francis Cokayne, 101 Neston Drive, Bulwell, Nottingham, NG6 8QY
G0 VFV G Marley, 41 Scalby Road, Burniston, Scarborough, YO13 0HN
G0 VFW Terence Thirlwell, 58 Chesham Road, Bovingdon, Hemel Hempstead, HP3 0EA
GM0 VFY Gordon Stuart, Easter Ardoe Cottage, Ardoe, Aberdeen, AB12 5XT
G0 VFZ R Barnes, 18 Battle Road, Tewkesbury, GL20 5TZ
G0 VGB D London, 113 Westbrooke Avenue, Hartlepool, TS25 5HZ
G0 VGC B Jenkins, 6 Stuart Drive, Thetford, IP24 3GA
G0 VGD J Constance, 4 Hopgarden Road, Tonbridge, TN10 4QS
GM0 VGI G Anderson, 21 Bydand Gardens, Inverurie, AB51 4FL
G0 VGJ D Graham, Parkburn, Colby, Appleby-in-Westmorland, CA16 6BD
G0 VGK P jenkinson, 11 St. Marys Road, Little Haywood, Stafford, ST18 0NJ
GI0 VGL George Warnock, 98 Skerriff Road, Altnamachin, Newry, BT35 0PJ
G0 VGN A Fawcett, 26 Merlin Court, Oswaldtwistle, Accrington, BB5 3TA
G0 VGP B Davies, 18 Wyresdale Gardens, Lancaster, LA1 3FA
G0 VGR P Tamplin, 74 Asquith Road, Gillingham, ME8 0JD
G0 VGT A Trent, 18 Castle Drive, Reigate, RH2 8DQ
GI0 VGV A Wright, 29 Wynfort Lodge, Moira, Craigavon, BT67 0QT
G0 VGY P Keen, 89 Raleigh Avenue, Hayes, UB4 0EF
G0 VHB I Herron, 1 Durham Court, Sacriston, Durham, DH7 6UZ
G0 VHF Colchester Contest Group c/o J Lemay, Carlton House, White Hart Lane, Colchester, CO6 3DB
GI0 VHG P Hughes, 17 Cardinal Dalton Park, Keady, Armagh, BT60 3TS
G0 VHH G Mann, 10 Earsham Drive, King's Lynn, PE30 3UZ
G0 VHI W Stainforth, 2 Grangefield Terrace, New Rossington, Doncaster, DN11 0LT
G0 VHJ G Holland, 29 Plantation Drive, Ellesmere Port, CH66 1JT
G0 VHK D Godding, 46 Brockley Close, Tilehurst, Reading, RG30 4YP
G0 VHL P Jones, 28 Helena Road, Capel-le-Ferne, Folkestone, CT18 7LQ
G0 VHO Donovan Ross, Rosehill, Leyland Lane, Leyland, PR26 8LB
G0 VHQ A Harding, Po Box 10620 Apo, Grand Cayman, Cayman Islands
G0 VHR Colin Cook, 1-2 Hardenmains Farm Cottages, Jedburgh, TD8 6RB
G0 VHT Paul Morrison, 64 Wolf Lane, Windsor, SL4 4YZ
G0 VHY R Wilson, 7 Scarratt Close, Forsbrook, Stoke-on-Trent, ST11 9AP
G0 VIA C Cannon, 2 Garden Cottages, Southside, Driffield, YO25 4ST
GI0 VIB A Smith, 42 Ballycullen Road, Moy, Dungannon, BT71 7HT
G0 VID D Morris, 117 Lonsdale Avenue, Doncaster, DN2 6HF
G0 VIE David Brooks, 10 Meadow Close, Whaley Bridge, High Peak, SK23 7BD
GI0 VIF D Canning, 10 Cotswold Close, Saintfield, Ballynahinch, BT24 7FQ
G0 VIG L Merrick, 4 Berryfield Glade, Churchdown, Gloucester, GL3 2BT
G0 VII G Boundey, 25 Ivy Place, Tantobie, Stanley, DH9 9PT
G0 VIJ J Johnson, 1 Rosa Vella Drive, Dereham, NR20 3SB
GD0 VIK Daniel Wood, The Hawthorns, Port Erin, Isle of Man, IM9 6EL
G0 VIM M Rivers, Snagsmount, Lambden Road, Ashford, TN27 0RB
G0 VIQ E Sully, 10 The Paddock, Pound Hill, Crawley, RH10 7RQ
GM0 VIT W Henderson, Strone View, Bridge of Cally, Blairgowrie, PH10 7JL
GM0 VIV J Dunn, 50 Duddingston Road, Edinburgh, EH15 1SG
G0 VIX M Rutland, Roselea, 28 Eastfield Avenue, Fareham, PO14 1EG
GM0 VIY J Oates, 14 Craighlaw Avenue, Eaglesham, Glasgow, G76 0EU
G0 VJB B Vaughan, 43 Bankfield Road, Shipley, BD18 4AW
G0 VJC C Smith, 19 Wiscombe Avenue, Penkridge, Stafford, ST19 5EH

GI0 VJE C Hannigan, 4 Silverhill Road, Strabane, BT82 0AE
G0 VJH Jeffrey Herrington, 84 Glenn Road, Poringland, Norwich, NR14 7LU
G0 VJI R Clay, 38 Hubbards Road, Chorleywood, Rickmansworth, WD3 5JJ
G0 VJJ D Gould, 87 Wentworth Drive, Bedford, MK41 8QD
G0 VJK M Stanton, 11 Eldean Road, Duston, Northampton, NN5 6PF
G0 VJM Alan Howell, 16 Blueberry Way, Worle, Weston-Super-Mare, BS22 6SF
G0 VJN P Andres, Flat 80, Northfield House, Bristol, BS3 1XB
GJ0 VJP Nigel Collier Webb, Chatolot, Les Marais Avenue, La Route de la Haule, Jersey, JE3 1LE
G0 VJR R Henshall, 9 Murrayfield Drive, Willaston, Nantwich, CW5 6QF
GW0 VJS OBO MID GLAM AMATEUR RADIO GROUP c/o M Carey, 47 Heol Ty Gwyn, Maesteg, CF34 0DD
G0 VJY R Welbourn, 30 Bower Road, Swinton, Mexborough, S64 8NU
G0 VKC Kelvin Cunningham, Claerwen Cottage, Three Ashes, Hereford, HR2 8NA
G0 VKE A Willis, 5 Robin Hill, Choppenhangers Road, Maidenhead, SL6 2QZ
G0 VKF M George, 3 Oak Crescent, Cherry Willingham, Lincoln, LN3 4AX
GM0 VKG OBO LARGS & DISTRICT ARS c/o John Clough, Obo Largs & District Ars, Redbank, Skelmorlie, PA17 5DX
G0 VKH H Venus, 45 St. Albans Road, Seven Kings, Ilford, IG3 8NN
G0 VKI A Hopkinson, 34 Welby St., Fenton, Stoke-on-Trent, ST4 4PL
G0 VKL R Butler, 34 Commissioners Road, Strood, Rochester, ME2 4EB
G0 VKS Harry Klein, 30 Odenwaldstrasse, Frankfurt, Germany, 60528
G0 VKX R Smith, 16 Southbrook Road, Langstone, Havant, PO9 1RN
G0 VKY D Russell, 1 Debden Close, Ernesettle, Plymouth, PL5 2DB
G0 VLC Andrew Betts, Edgewood, Hoe Court, Lancing, BN15 0QX
GI0 VLE William Dalton, 16 Junction Road, Randalstown, Antrim, BT41 4NP
G0 VLF Raymond McAleer, 44 Harvey Avenue, Durham, DH1 5ZG
G0 VLI Diana Taylor, Greys Mead Lodge, Thame Park Road, Thame, OX9 3PL
G0 VLJ G Heald, 2 Holyrood Terrace, Weymouth, DT4 0BE
G0 VLK A Palfreeman, 29 Boulby Road, Redcar, TS10 5EB
G0 VLQ B Close, 74 Heston Avenue, Hounslow, TW5 9EX
G0 VLR LEICESTER RAYNET GROUP c/o Andrew Holmes, 5 Launde Park, Market Harborough, LE16 8BH
G0 VLV D Hardman, 15 Wordsworth Avenue, Bolton le Sands, Carnforth, LA5 8HJ
G0 VMA Glyn Skupski, 57 Three Nooks, Bamber Bridge, Preston, PR5 8EN
G0 VMC Jeffrey Williams, 3 Woodbury Lane, Salisbury, SP2 8FE
GW0 VMD K Peacey, 2 Robin Close, Cardiff, CF23 7HN
G0 VME M Macdonald, 44 Hillesden Avenue, Elstow, Bedford, MK42 9YX
G0 VMF P Quirk, 70 Sands Lane, Oulton, Lowestoft, NR32 3HS
G0 VMK Peter Nicholls, 11 Minsmere Road, Belton, Great Yarmouth, NR31 9NX
G0 VMN David Turton, 68 Bartholomew Street, Wombwell, Barnsley, S73 8LD
G0 VMP J Roze, 9 Ralfland View, Shap, Penrith, CA10 3PF
G0 VMQ P Ellis, 104 Gravesend Road, Strood, Rochester, ME2 3PN
G0 VMR Patrick Smith, Bron Awel, Brynisa Road, Wrexham, LL11 6NS
GW0 VMS L Beedle, 2 Chestnut Grove, Maesteg, CF34 0NT
G0 VMT T Moorhead, 53 Childwall Priory Road, Liverpool, L16 7PA
GM0 VMV E Kennedy, 1 Greenbank Gardens, Edinburgh, EH10 5SL
G0 VMW R Price, 8 Tanllwyfan, Old Colwyn, Colwyn Bay, LL29 9LQ
GW0 VMZ John Davies, 34 Penlan View, Ynysfach, Merthyr Tydfil, CF47 8NJ
G0 VNA G Kent, Plum Tree Cottage, Colesbrook Lane, Gillingham, SP8 4HH
G0 VNB M Brain, Coldharbour, Low Road, Devizes, SN10 4JY
GW0 VND M Goodridge, 17 Charles St., Neyland, Milford Haven, SA73 1SA
G0 VNE H Stokes, 9 Causeway Glade, Dore, Sheffield, S17 3EZ
G0 VNH C Langdon, 652 Hotham Road South, Hull, HU5 5LE
G0 VNI S Williams, The Croft, Ringwood Road, Southampton, SO40 7LA
G0 VNJ M Guy, 38 Sandy Lane, Charlton Kings, Cheltenham, GL53 9DQ
G0 VNK J Harders, Kalckreuthweg 17, Hamburg, Germany, 22607
G0 VNO D Johns, 8 Hill Fold, Dawley Bank, Telford, TF4 2QE
G0 VNQ W Askam, 10 Staunton Close, Castle Donington, Derby, DE74 2XA
G0 VNV E Troughton, 31 Leyburn Road, Blackburn, BB2 4NQ
G0 VNW J Carrington, 101 Papplewick Lane, Hucknall, Nottingham, NG15 8BG
G0 VNY John Crawford, 2c Papillon House, Balkerne Gardens, Colchester, CO1 1PR
G0 VOB K Fuller, 28 Bradshaw Way, Irchester, Wellingborough, NN29 7DP
G0 VOE C Iles, 2 Mountview Terrace, Pawlett, Bridgwater, TA6 4SL
G0 VOF Mark Walmsley, 121 Roe Lee Park, Blackburn, BB1 9SA
G0 VOG D Roberts, 16 Pentre Isaf, Old Colwyn, Colwyn Bay, LL29 8UT
G0 VOJ S Williams, 6 Oak Road, Clanfield, Waterlooville, PO8 0LJ
G0 VOK Niall Reilly, 22 Lee Drive, Northwich, CW8 1BW
GM0 VOL V Nerurkar, 26 Fothringham Drive, Monifieth, Dundee, DD5 4SW
GM0 VOU P Scott, 30 Main St., Newmills, Dunfermline, KY12 8SS
G0 VOV D Hartwell, 57 Wyatts Covert, Denham, Uxbridge, UB9 5DJ
GU0 VPA Rudie Peeters, 17 Grosse Hougue, Saltpans Road, St. Sampson, Guernsey, GY2 4NS
G0 VPC M Davies, 27 Bedford Road, Orpington, BR6 0QJ
G0 VPE READING AND WEST BERKSHIRE RAYNET c/o Denis Plowman, 20 Marathon Close, Woodley, Reading, RG5 4UN
GM0 VPG N Thackrey, 23 Scorguie Avenue, Inverness, IV3 8SD
G0 VPH M Austin, 107 Spicer Close, London, SW9 7UE
G0 VPJ John Stacey, 16 Crane Drive, Verwood, BH31 6QB
G0 VPO Alan Robinson, 2 Fort Street, Sandown, PO36 8BA
G0 VPS D Bennett, 3 Sivilla Road, Kilnhurst, Mexborough, S64 5TY
G0 VPT L Smith, Hillside, Kings Mill Lane, Stroud, GL6 6SA
G0 VPU M Burbidge, 3 Kirklands, Hest Bank, Lancaster, LA2 6EH
G0 VPV G Pesarini, 53 Llanvanor Road, London, NW2 2AR
G0 VPW M Rhodes, 102 Malvern Crescent, Little Dawley, Telford, TF4 3JF
G0 VPX M Worsfold, 9 Montacute Road, Lewes, BN7 1EN
G0 VPY Eric Weston, 33 William Street, Tunbridge Wells, TN4 9RP
G0 VPZ Jane Greenfield, 36 Barttelot Road, Horsham, RH12 1DQ
G0 VQA J Groves, 24 Gimble Way, Pembury, Tunbridge Wells, TN2 4BX
G0 VQB Michael Grainger, 174 Woodlands Road, Gillingham, ME7 2SX
G0 VQD P Newberry, 3 Fieldview, Horley, RH6 9DX
G0 VQG M Thomas, 10 Finchale Crescent, Darlington, DL3 9SA
G0 VQJ J Metcalfe, 158 Barrowford Road, Colne, BB8 9QR
G0 VQK S Haigh, 2 Locker Avenue, Warrington, WA2 9PS
G0 VQL Geoff Guild, 15 Canalside Cottages, Chester Road, Runcorn, WA7 3AQ
G0 VQM Charles Reid, 54 Montacute Road, New Addington, Croydon, CR0 0JE
G0 VQO Nicholas Haynes, 139 Hull Road, Anlaby, Hull, HU10 6ST
G0 VQR Thomas Cannon, 35 Loddon Bridge Road, Woodley, Reading, RG5 4AP
G0 VQS P Beck, 6 Holly Close, Little Bealings, Woodbridge, IP13 6PL
G0 VQT F Seabourne, 11 Heyford Avenue, London, SW20 9JT
G0 VQW A Jack, 1 Dockle Way, Upper Stratton, Swindon, SN2 7LQ
G0 VQX D Thomas, 11 Lordwells Drive, Bracknell, RG12 9YT
G0 VQY P Woodling, 31 Douglas Avenue, Brixham, TQ5 9EL
GW0 VQZ N Jones, 59 Woodfield Terrace, Penrhiwceiber, Mountain Ash, CF45 3YA
G0 VRE A Challis, 12 Moorland Crescent, Boultham Moor, Lincoln, LN6 7NL
G0 VRF Martin Waples, 42 Butts Road, Wellingborough, NN8 2PU
G0 VRH D Bower, 29 Worksop Road, Mastin Moor, Chesterfield, S43 3DH
G0 VRK C Seabridge, 13 Hillside Avenue Forsbrook, Stoke-on-Trent, ST11 9BH
GW0 VRL C Saunders, 14 Portway, Bishopston, Swansea, SA3 3JR
G0 VRM Andrew Russell, 3 St. Nicholas Close, North Newbald, York, YO43 4TT
GM0 VRP R Phillips, 18 Broomridge Road, St. Ninians, Stirling, FK7 0DT
G0 VRQ Keith Gillon, 33 Norwich Close, Ashington, NE63 9RY
G0 VRT M Ritson, 14 Dunsdale Road, Holywell, Whitley Bay, NE25 0NG
G0 VRU B Gee, 11 Spitfire Avenue, Grimoldby, Louth, LN11 8UJ
G0 VRV S Philipps, 24 Acres End, Amersham, HP7 9DZ
G0 VRW P Wadhams, Brickwood, Bigbury Road, Canterbury, CT4 7ND
G0 VRX Christopher Jenkins, 31 Ashbrook Crescent, Rochdale, OL12 9AJ
G0 VRY P Maccormick, 7 Poplar Close, Horsford, Norwich, NR10 3SE
G0 VRZ K Tuer, Broad Ing, Penrith, CA10 2LL
G0 VSB A Gibbs, 18 Grange Close, Ludham, Great Yarmouth, NR29 5PZ
G0 VSG I Smyth, 97 Leas Drive, Iver, SL0 9RB
G0 VSH M Wright, 3 St Anthony, Melleiha, Malta
G0 VSJ P Taylor, 133 Beech Drive, Shifnal, TF11 8HZ
G0 VSK A Thomas, 3 Barnfield Way, Cannock, WS12 0PR
G0 VSL K Hill, 18 Baker Close, Chasetown, Burntwood, WS7 4GU
G0 VSM Trevor Day, 46 Beatrice Avenue, Saltash, PL12 4NG
G0 VSO Philip Burchell, 5 Brentwood Place, Ebbw Vale, NP23 6JR
G0 VSP Allan Sheldon, 28a Mount Road, Tettenhall Wood, Wolverhampton, WV6 8HT
G0 VSS R King, 19 Greenhayes, Cheddar, BS27 3HZ
GW0 VST P Sandham, 6 Skomer Close, Nottage, Porthcawl, CF36 3QH
GW0 VSW J Mason, 2 Golwg-Y-Bryn, Off Woodland Road, Neath, SA10 6SP
G0 VSY G Forster, 3 Foxmoor, Bishops Cleeve, Cheltenham, GL52 8SS
G0 VSZ D Forster, 3 Foxmoor, Bishops Cleeve, Cheltenham, GL52 8SS
G0 VTA R Collins, 30 Upham Road, Swindon, SN3 1DN
GD0 VTC L King, 4 Glenhurst Avenue, Ruislip, HA4 7LZ
G0 VTD Susan Towler, 77 Glebe Road, Hull, HU7 0DU
G0 VTI Terence Ibbitson, 36 Knoll Park, East Ardsley, Wakefield, WF3 2AX
G0 VTJ A Jameson, 9 White Laithe Green, Leeds, LS14 2EP
G0 VTL H Bradfield, Glebe Farm, Wharf Street, Leicester, LE4 8AY
G0 VTM J Pearson, 23 Glebe Avenue, Mitcham, CR4 3DZ
GI0 VTS R Boyle, 13 Sherwood Road, Ballymacormick, Bangor, BT19 6DJ
G0 VTU David Stocks, 215 Watling Street, Wilnecote, Tamworth, B77 5BB
G0 VTV K Groom, 2 Ruins Barn Road, Tunstall, Sittingbourne, ME10 4HS
G0 VUC C Boughton, 79 Hawfield Lane, Burton-on-Trent, DE15 0BY
G0 VUH A France, 1 Narrow Lane North Anston, Sheffield, S25 4BJ
G0 VUL Andy Hardie, Tana-Merah, Church Lane, Retford, DN22 9NQ
G0 VUM P Cooper, 16 Mortomley Close, High Green, Sheffield, S35 3HZ
G0 VUN T Wooding, 101 Park Farm Road, Ryarsh, West Malling, ME19 5JX
G0 VUT Christopher Turner, 28 Reading Street, Broadstairs, CT10 3AZ
G0 VUX BOLTON SCHOOL ARC c/o C Walker, Bolton School Ltd, Chorley New Road, Bolton, BL1 4PA
GM0 VUY Richard Turnbull, 6 Letham Gait, Dalgety Bay, Dunfermline, KY11 9GT
G0 VVA M Newbold, 10 Shaw Street, Derby, DE22 3AS
GI0 VVC J Serridge, 21 Lassara Heights, Warrenpoint, Newry, BT34 3PG
G0 VVF David Turner, 25 Princess Close, Woodville, Swadlincote, DE11 7EG
G0 VVG A Lomas, 14 Ingledene Caravan Site, Lawsons Road, Thornton-Cleveleys, FY5 4DL
G0 VVK P Neale, 15 Kenmore Walk, Wibsey, Bradford, BD6 3JQ
G0 VVP R Hatton, Plot 2, Sutton Road, Hull, HU7 5YY
G0 VVQ J Mahon, 2 Rob Lane, Newton-le-Willows, WA12 0DR
G0 VVR C Leigh, 4 Corrick Close, Draycott, Cheddar, BS27 3UB
G0 VVT Edward Murphy, 21 Standard Street, Stoke-on-Trent, ST4 4NG
G0 VVX CROHAM CALLERS c/o Mark Samuel, 71 Brighton Road, South Croydon, CR2 6EE
G0 VVY M Soper, 16 Queen Elizabeth Drive, Crediton, EX17 2EJ
G0 VVZ David Matthiae, 7 Bustlers Rise, Duxford, Cambridge, CB22 4QU
G0 VWB M Davies, 23 Star Lane, Folkestone, CT19 4QH
GW0 VWD Allan Ball, 20 Hillcrest, Brynna, Pontyclun, CF72 9SJ
G0 VWE P Whitfield, 55 Greenways, Sutton, Woodbridge, IP12 3TP
G0 VWF C Howell, 24 Bellring Close, Belvedere, DA17 6LP
G0 VWH M Wright, 27 Ellesmere Close, Hucclecote, Gloucester, GL3 3DH
G0 VWP T Sayner, 59 Horner Street, York, YO30 6DZ
G0 VWQ D Coulburn, Spindleberry, Weare Giffard, Bideford, EX39 4QR
G0 VWT T Holland, 135 Newcastle Road, Stone, ST15 8LF
GI0 VWU A McCabe, 40 Rydalmere Street, Belfast, BT12 6GF
G0 VWV R Giles, 9 Bower Green, Lords Wood, Chatham, ME5 8TN
G0 VWW T Robson, 58 Burton Road, Lincoln, LN1 3LB
G0 VWX E Collinson, 40 Cock Robin Lane, Catterall, Preston, PR3 1YL
GM0 VWZ Samuel Lawrie, 4 Glenavon Drive, Airdrie, ML6 8QG
GM0 VXA P Marriott, Craiglea Cottage, Omoa Road, Motherwell, ML1 5LQ
G0 VXB S cockburn, Jacks House, Welcombe, Bideford, EX39 6HE
G0 VXC Martin Coles, 133 Highthorn Road, Kilnhurst, Mexborough, S64 5UU
G0 VXD G Clayton, 27 Simpsons Lane, Knottingley, WF11 0HG
G0 VXE D Herbert, 50 St. Leonards Crescent, Scarborough, YO12 6SP
G0 VXG R Wilkinson, 139 Church Road, Jackfield, Telford, TF8 7ND
G0 VXJ M Finch, 73 Cordingley Way, Donnington, Telford, TF2 7LJ
G0 VXK J Davies, 26 Beverley Close, Wylde Green, Sutton Coldfield, B72 1YF
G0 VXM G Simmons, 90 Pollards Fields, Knottingley, WF11 8TD
GM0 VXQ P Mulheron, 78 South Commonhead Avenue, Airdrie, ML6 6PA
G0 VXS C Newman, 89 Goose Cote Lane, Oakworth, Keighley, BD22 7NQ
G0 VXV P Nicholls, 11 Westfield Mount, Yeadon, Leeds, LS19 7NL
G0 VXW M Rendall, 633 Moston Lane, Manchester, M40 5QD
G0 VXX A Fry, 96 Westcroft Gardens, Morden, SM4 4DL
G0 VXY G Forde, 22 Alloa Road, London, SE8 5AJ
GM0 VXZ I Terris, 1 Linden Avenue, Wishaw, ML2 8SE
G0 VYB Tom Cleghorn, 9 Bridge of Aldouran, Leswalt, Stranraer, DG9 0LW
G0 VYC M Dawson, 11 Owls Retreat, Colchester, CO4 3FE
GW0 VYF Phillip Simmons, 16 Frederick Street Neyland, Milford Haven, SA73 1TG
G0 VYG GB/ADX CSG/CYMRU CONTEST GROUP c/o T Jones, Penrhiw Bach, Bryngwran, Holyhead, LL65 3RD
G0 VYK R Hobson, 41 Ridgeway, Pembury, Tunbridge Wells, TN2 4ER
GM0 VYL P Maver, 69 Mayfield Crescent, Musselburgh, EH21 6EX
G0 VYN M Guest, 53 Ringwood Close, Furnace Green, Crawley, RH10 6HQ
G0 VYP W Southworth, 58 Moyse Avenue, Walshaw, Bury, BL8 3BL
G0 VYQ T Mcinerney, 41 Newton Way, Tongham, Farnham, GU10 1BY
G0 VYT Michael Garry, 14 Adrians Close, Mansfield, NG18 4HG
G0 VYU R Raynor-Smith, 10 Marsh Road, Trowbridge, BA14 7PR
G0 VYV E Smith, Magnolia House, Main Road, Highbridge, TA9 3QZ
G0 VYX P Rumsam, 24 Hamer Avenue, Rossendale, BB4 0QH
GM0 VYY D Macconnell, Mine Cottage, Tashieburn Road, Lanark, ML11 9SS
G0 VZA B Howlett, 156 Larnach Road, Waverley, Dunedin, New Zealand
G0 VZD Robert Ferris, Polmarth House, Carnmenellis, Redruth, TR16 6NT
G0 VZE Edward Cook, 8 Calvert Grove, Newcastle, ST5 8QA
G0 VZH R Morris, 16 Stephens Close, Luton, LU2 9AN
G0 VZI P Robinson, 5 Coppice Close, Haxby, York, YO32 3RR
G0 VZK C Sampson, 1 Warton Lane, Austrey, Atherstone, CV9 3EJ
G0 VZL E Levring, 3 Evelyn Croft, Wylde Green, Sutton Coldfield, B73 5LF
G0 VZN K Vollor, 20 Browns Lane, Uckfield, TN22 1RY
G0 VZO J Koops, 64 Winchester Avenue, Nuneaton, CV10 0DW
G0 VZT S Hopton, 5 Wellington Close, Marske-by-the-Sea, Redcar, TS11 6NW
G0 VZV D A'Bear, 7 Meadow Close, Bembridge, PO35 5YJ
G0 VZX P Dart, 208 Elburton Road, Plymouth, PL9 8HU
G0 VZZ D Taylor, 16 Ambleside Place, Canterbury, CT2 7UJ
G0 WAB William Neale, 5 Gibbs Court, Dane Close, Wirral, CH61 3XS
G0 WAC Keith Wells, 42 Eggesford Road, Derby, DE24 3BH
G0 WAD C Dyson, 21 Highmoor, Kirkhill, Morpeth, NE61 2AS
G0 WAE D Phillips, 14 Seymour Road, Newton Abbot, TQ12 2PU
GI0 WAH W Hutchman, 35 Carlingford Park, Newry, BT34 2NY
G0 WAL Walter Reed, 10 Ashmeads Way, Wimborne, BH21 2NZ
G0 WAM S Stevens, 25 Busticle Lane, Sompting, Lancing, BN15 0DJ
G0 WAN M Gill, 23 Walmers Avenue, Higham, Rochester, ME3 7EH
G0 WAS Alan Smith, 9 Moor Lane, Maulden, Bedford, MK45 2DJ
G0 WAT P Brice-Stevens, 31 Lodgefield, Welwyn Garden City, AL7 1SD
G0 WAW Irene Oura, The Quoins, Gloucester Road, Bath, BA1 8AD
G0 WAX L Merrin, 10 Hawkesbury Road, Canvey Island, SS8 0EX
G0 WAY R Slimmon, Rhosygorlan, Devils Bridge, Aberystwyth, SY23 4RF
G0 WBA K Fradgley, 84 Church Road, Wordsley, Stourbridge, DY8 5AU
G0 WBC Christopher Mortimer, 6 Honeycomb Close, Narborough, Leicester, LE19 3PS
G0 WBL Steven Hughes, 43 The Cloisters, Rickmansworth, WD3 1HL
G0 WBR T Johnson, 7 Southover Way, Hunston, Chichester, PO20 1NY
G0 WBS MONMOUTH SCHOOL AMATEUR SCHOOL SOCIETY c/o P Wentworth, 46 Woodside Avenue, Cinderford, GL14 2DW
G0 WBT J Woodcock, 54 Longworth Road, Horwich, Bolton, BL6 7BE
G0 WBV R Burchell, 1 Broad Forstal Farm Cottages, Tilden Lane, Tonbridge, TN12 9AX
G0 WCB A Bathurst, 81 Heatherstone Avenue, Dibden Purlieu, Southampton, SO45 4LE
GI0 WCE D Cromie, 11 Cherryvalley Park West, Belfast, BT5 6PU
G0 WCH B Prestage, 17 Moorgate Road, Hindringham, Fakenham, NR21 0PT
G0 WCI Mark Hand, 24 Fozdar Crescent, Bilston, WV14 9UH
G0 WCJ J Slater, 25 Croft Lane, Diss, IP22 4NA
G0 WCK D Hornby, 7 Milton Street, West Bromwich, B71 1NJ
G0 WCO Peter Anness, 18 Plaisir Place, Thurston Road, Lowestoft, NR32 1RY
G0 WCR M Knott, 76 New Barns Avenue, Mitcham, CR4 1LF
G0 WCS S Ormerod, 6 New Wokingham Road, Crowthorne, RG45 7NR
G0 WCU R Roberts, 9 Stones Close, Hogsthorpe, Skegness, PE24 5NZ
G0 WCZ Graeme Sutherland, 9 Old Court Close, Brighton, BN1 8HF
G0 WDC D Clark, Meadowcroft Bungalow, Ugthorpe, Whitby, YO21 2BL
GM0 WDF J Dunlop, West Dougliehill Farm, Dougliehill Road, Port Glasgow, PA14 5XF
G0 WDG L Morgan, Flat 30, Raleigh Court, Sherborne, DT9 3EQ
G0 WDK D Hackett, 131 Station Road South, Walpole St. Andrew, Wisbech, PE14 7LZ
G0 WDQ S Collins, 30 Upham Road, Swindon, SN3 1DN
G0 WDT R Emery, 10 Penarth Place, Newcastle, ST5 2JL
G0 WDU Martin Round, 2 Bell Mead, Studley, B80 7SH
GW0 WDV S Steddy, 1 Springfield Close, Wenvoe, Cardiff, CF5 6DA
G0 WDW W Williamson, Monfa, Walford Heath, Shrewsbury, SY4 2HT
G0 WDX WORLD DX RC c/o R Morgan, 56 The Meadows, Hull, HU7 6EE
G0 WEA Richard Foster, 60 Sandford Close, Bransholme, Hull, HU7 4HN
G0 WEB R Webb, 25 Greenfield Croft, Bilston, WV14 8XD
GM0 WED Edmund Holt, Ashwell, Cannigall, Kirkwall, KW15 1QX
G0 WEF Nicholas Kenworthy, 19 Parkside Close, Radcliffe, Manchester, M26 2QS
G0 WEK A Robinson, 43 Fremantle Road, High Wycombe, HP13 7PQ
G0 WEO D Waterfield, 21 Cedar Road, Mickleton, Chipping Campden, GL55 6SZ
G0 WER P Moran, 1 Plas Issa, Brook Street, Wrexham, LL14 3EE
G0 WET Robert Wright, P.o. Box 7, P.o. Box 7, Valencia, Philippines, 6215
G0 WEV S McMullen, 70 Sylvan Avenue, Timperley, Altrincham, WA15 6AB
G0 WEX Stephen Elliott, 1 The Gables, Oddfellows Road, Hope Valley, S32 1DU
G0 WEY J Doorea, 14 Parc Tyddyn, Red Wharf Bay, Pentraeth, LL75 8NQ
GM0 WEZ Peter Ewing, Kildonan House, Caerlaverock Farm, Crieff, PH5 2BD
GM0 WFA Timothy Clark, 23 Letham Place, St. Andrews, KY16 8RB
GM0 WFB John Keenan, 53 Clermiston Crescent, Edinburgh, EH4 7DF
G0 WFD Michael Farrell, 24 Lindsay Street, Stalybridge, SK15 2LT
G0 WFE S Dean, 42 North Street, Wareham, BH20 4AQ
G0 WFF C Whelan, 50 Garrick Close, Ings Road Estate, Hull, HU8 0ST
G0 WFG Anthony Vinters, 106 Halifax Road, Ripponden, Sowerby Bridge, HX6 4AG
G0 WFH C Gresswell, 121 Granby Court, Bletchley, Milton Keynes, MK1 1NG
G0 WFK H Bamford, Upper Twynings Farm, Pumphouse Lane, Droitwich, WR9 7EB
G0 WFL M Street, 262 Carter Knowle Road, Sheffield, S7 2EB
G0 WFM P Wallace, 8 Wemmick Close, Rochester, ME1 2DL

**IMPORTANT NOTE**

**Revalidate licence to avoid revocation** – Ofcom has advised the Society that plans will be drawn up to revoke licences that have not been revalidated as required by the licence conditions. The quickest way to revalidate is to do so online via the Ofcom website: *https://services.ofcom.org.uk/* or by email: *amateur.validations@ofcom.org.uk* Ofcom staff are available to help, but please be patient during times of heavy workload.

G0 WFO D Tarry, 2 Kestrel Heights, Codnor Park, Nottingham, NG16 5PW
G0 WFP W Painz, 8 Warminger Court, Ber Street, Norwich, NR1 3ED
G0 WFQ Malcolm Troy, 22 Jackie Wigg Gardens, Totton, Southampton, SO40 9LZ
G0 WFT D Saunders, 14 Shelton Avenue, Toddington, Dunstable, LU5 6EL
G0 WFV A Corbett, Lan Y Llyn, 10 Rutland Avenue, Waddington, LN5 9FW
G0 WGA M Merrin, 10 Hawkesbury Road, Canvey Island, SS8 0EX
G0 WGB J Atkins, 6 Carolina Gardens, Plymouth, PL2 2ER
GW0 WGE D Thomas, 85 Tan Y Bryn, Burry Port, SA16 0LD
G0 WGH J Edwards, 21 Ridgeway, Ottery St. Mary, EX11 1DT
G0 WGI K Sherwin, 1 Nursery Close, Wroughton, Swindon, SN4 9DR
G0 WGJ Gerald Jarvis, 38 West Cliff Road, Dawlish, EX7 9DY
G0 WGL P Furness, 6 Westbrook Square, West Gorton, Manchester, M12 5PU
G0 WGM Trevor Jones, 26 HEOL MAENOFFEREN, Blaenau Ffestiniog, LL41 3DL
GW0 WGN J Lyons, 4 London Road, Pembroke Dock, SA72 6DU
G0 WGP Ronald Glover, 5 The Vineries, Burgess Hill, RH15 0ND
G0 WGV I French, Penmellyn, Tarrandean Lane, Truro, TR3 7NW
GW0 WGW Brian James, 72 Park Place, Bargoed, CF81 8NB
G0 WHC William Chesterton, 61 Butt Lane, Blackfordby, Swadlincote, DE11 8BG
G0 WHD G Hassall, 3 Sunny Bank, Cark in Cartmel, Grange-over-Sands, LA11 7PF
G0 WHL E Barnes, 10 Cranbourne Road, Rochdale, OL11 5JD
G0 WHN J Edwardes, Pippins, 1 Horse Lane Orchard, Ledbury, HR8 1PP
G0 WHO R Clutson, 151 Stepney Road, Scarborough, YO12 5NJ
G0 WHQ K Wilson, 11 Harbour Close, Murdishaw, Runcorn, WA7 6EH
GD0 WHV R Bessell, 6 Bayford Lodge, Wellington Road, Pinner, HA5 4NJ
G0 WHY Paul White, 12 Broadleas Crescent, Devizes, SN10 5DH
G0 WHZ T Tipping, Flat 24, The Moorings, Kingsbridge, TQ7 1LP
GM0 WIB M Mcdermott, Upper Outseat, Banff, AB45 3RB
G0 WIC E Clark, 8 Rose Cottages, Shotton Colliery, Durham, DH6 2NF
G0 WID P Gill, 1 Harbour View, Truro, TR1 1XJ
G0 WIE P Swan, The Old Rectory, Sampford Brett, Taunton, TA4 4LA
G0 WIG P Ellison, 28 Kingscote Road East, Cheltenham, GL51 6JS
G0 WIH DACORUM AR TRANSMITTING SOC c/o Anthony Mitchell, 87 Bridgewater Road, Berkhamsted, HP4 1JN
G0 WIL M Williams, The Croft, Ringwood Road, Southampton, SO40 7LA
G0 WIS Geoffrey Woodbury, 4 Henley Drive, Droitwich, WR9 7RX
G0 WIT M Perry, 52 Somerset Avenue, Chessington, KT9 1PN
G0 WIW N Appleby, 2 Stella Farm, Narborough, King's Lynn, PE32 1HY
G0 WIX A Beeching, 174 Grove Road, Rayleigh, SS6 8UA
G0 WIY W King, 3 Grove Court, The Waterloo, Cirencester, GL7 2PZ
G0 WIZ I Waugh, 2 Gilloch Avenue, Dumfries, DG1 4DN
G0 WJA L Coleman, 1 Kerstin Close, Cheltenham, GL50 4SA
G0 WJC M Briscoe, 34 Winterton Drive, Low Moor, Bradford, BD12 0UX
G0 WJD Derek Jackson, 10 Wisdoms Green, Coggeshall, Colchester, CO6 1SG
G0 WJH John Kemp, 85 St. Andrews Way, Church Aston, Newport, TF10 9JQ
GI0 WJI R Bicker, 62 Spa Road, Ballynahinch, BT24 8PT
G0 WJJ D Mullaney, 62 Darby Road, Wednesbury, WS10 0PN
G0 WJK A Cobb, 37 Nordham, North Cave, Brough, HU15 2LT
G0 WJN P Howland, 73 Timberdine Avenue, Worcester, WR5 2BG
G0 WJS David Mellings, 36 Hillwood Drive, Glossop, SK13 8RJ
G0 WJU Roy Crewe, 11 Osbert Close, Norwich, NR1 2NL
G0 WJV C Page, 73 Two Saints Close, Hoveton, Norwich, NR12 8QR
G0 WJX Ronald Davies, 84 Hob Hey Lane, Culcheth, Warrington, WA3 4NW
G0 WJZ Ian Donachie, 72 Gresham Road, Norwich, NR3 2NQ
G0 WKA M Drever, 66 Milton Road, Branton, Doncaster, DN3 3PB
G0 WKH Mike Thomas, 63 Corfe Way, Broadstone, BH18 9ND
G0 WKI H Yearl, 191 Tamworth Road, Kettlebrook, Tamworth, B77 1BT
G0 WKJ C Towle, 32 Charlwood Road, Luton, LU4 0BU
G0 WKL R Martin, 1 Collins Lane, West Harting, Petersfield, GU31 5NZ
G0 WKM N Purchon, Mitchells Elm House, Wanstrow, Shepton Mallet, BA4 4SN
G0 WKN K Whitmore, Amosford, Lutton Gowts, Spalding, PE12 9LQ
G0 WKQ South Tyneside ARS c/o David Harbron, 48 Sheridan Road, Biddick Hall, South Shields, NE34 9JJ
G0 WKT Mark Pugh, 28 Marsh Road, Wilmcote, Stratford-upon-Avon, CV37 9XR
G0 WKU P Yea, 89 Laxton Road, Taunton, TA1 2XF
G0 WKW V Pleshkevich, c/o E.Kvarnstrom, Skarbacksvagen 13, Vitaby, Sweden, 27736
G0 WKZ B Pybus, 29 Howick Park, Monkwearmouth Shore, Sunderland, SR6 0AQ
G0 WLC B Cole, 499 Lightwood Road, Stoke-on-Trent, ST3 7EN
G0 WLD M Russell, 23a St Ann'S Road, Barnes, London, SW13 9LH
G0 WLF W Storace-Rutter, 46 Norbury Court Road, London, SW16 4HT
G0 WLG Paul Blizzard, 12 Hilton Road, Malvern, WR14 3NP
GW0 WLI J Ruddle, Flat 58, Thomas Court, Cardiff, CF23 5EZ
G0 WLJ J Lees, 30 Clowes Avenue, Alsager, Stoke-on-Trent, ST7 2RL
GW0 WLN C Purcell, 31 Brookdale Court, Church Village, Pontypridd, CF38 1RP
GW0 WLQ R Evans, 4 Llantrisant Road, Tonyrefail, Porth, CF39 8PP
G0 WLR Mark Pettigrew, 26 Victoria Court, Sheffield, S11 9DR
G0 WLS B Chamberlain, 2 Smiths Walk, Lowestoft, NR33 8QN
GI0 WLW H Quinn, The Parochial House, 35 St. Patricks Street, Armagh, BT60 3TQ
G0 WLX Alan Suckling, 54 Lake Hill, Sandown, PO36 9HF
G0 WMB A Houghton, 2 Beanhill Crescent, Alveston, Bristol, BS35 3JG
G0 WMC P Norman, 36 Saddlers Park, Eynsford, Dartford, DA4 0HA
G0 WMD M De Silva, 31 Rosemary Avenue, Hounslow, TW4 7JQ
G0 WMG G Studd, 34 The Broadway, Lancing, BN15 8NY
G0 WMH Anthony Hughes, 7 Eden Grove, Kirkpatrick Fleming, Lockerbie, DG11 3AT
G0 WMJ J Walker, 7 Chesterfield Road, Liverpool, L23 9XL
G0 WMN W McNab, 74 Parkhouse Road, Minehead, TA24 8AF
G0 WMQ Edward Williams, 16 Birch Grove, Wallasey, CH45 1JG
GW0 WMT Stuart Robinson, 23 Thornhill Road, Cardiff, CF14 6PE
G0 WMU Ray Davis, 49 Goodway Road, Great Barr, Birmingham, B44 8RL
G0 WMW D Roberts, 17 Edgecombe Avenue, Weston-Super-Mare, BS22 9AY
G0 WMX G Giuliani, 32 Davison Street, Newburn, Newcastle upon Tyne, NE15 8NB
G0 WMY P Hanmer, 182 Tollemache Road, Prenton, CH43 7SE
G0 WMZ R Duckworth, 79 Werneth Road, Woodley, Stockport, SK6 1HR
G0 WNB S Blumson, 15 Bryn Colwyn, Colwyn Bay, LL29 9LJ
G0 WND T Sunouchi, 200-20 Morooka-Cho, Kouhoku-Ku, Yokohama, Japan, 222-0002
G0 WNF E Clark, Flat B, 145 Church Street, Whitby, YO22 4DE
G0 WNJ M Bottomley, 21 Priory Park, Grosmont, Whitby, YO22 5QQ
GM0 WNR A Campbell, 17 Arran Road, Motherwell, ML1 3NA
GM0 WNS Ian Calder, Cut the Wind Cottage, Arbroath, DD11 4RH
G0 WOA M Lyon, 182 Wilmslow Road, Heald Green, Cheadle, SK8 3BG
G0 WOC J Doxey, 41 Shady Grove, Hilton, Derby, DE65 5FX
G0 WOI AVON SCOUTS ARC c/o Rex Laney, 7 Downfield Close, Alveston, Bristol, BS35 3NJ
G0 WOM T Renshaw, 1 Basford View, Cheddleton, Leek, ST13 7HJ
G0 WON M Kelly, 9 Thetford Road, Brandon, IP27 0BS
G0 WOP P Dabell, 21 Tatton Road North, Heaton Moor, Stockport, SK4 4RL
G0 WOU David Atkinson, 596 Wolseley Road, St. Budeaux, Plymouth, PL5 1UX
G0 WPC Nigel Crossley, 58 Woodhouse Gardens, Sheffield, S13 7LS
G0 WPF J Towle, 36 Old Orchard, Haxby, York, YO32 3DT
G0 WPH P Howard, 162 High St., Hanham, Bristol, BS15 3HH
G0 WPL J Wozniak, 40 Cockhill, Trowbridge, BA14 9BQ
G0 WPM N Gilboy, 6 Talisman Close, Sherburn Village, Durham, DH6 1RJ
G0 WPO Neil Griffiths, 14 Howarth Cross Street, Rochdale, OL16 2PB
GM0 WPU K Faloon, Moss-Side Croft, 6 Rothiemay, Huntly, AB54 5NY
GI0 WPV Owen Price, 18 Hill Crest Walk, Bangor, BT20 4DF
GM0 WPW Terence Halligan, 4 Trainers Brae, North Berwick, EH39 4NR
G0 WPX WPX CONTEST GROUP c/o Donald Beattie, Hares Cottage, Woolston, Church Stretton, SY6 6QD
G0 WQA J Ward, 49 Woodhall Drive, Lincoln, LN2 2AE
G0 WQC J Keeling, 31 Tudor Drive, Otford, Sevenoaks, TN14 5QP
G0 WQQ D Bennett, Shrove Furlong, Longwick Road, Princes Risborough, HP27 9HE
G0 WQW J Thompson, 78 Lowestoft Road, Worlingham, Beccles, NR34 7RD
G0 WQY L Mansfield, 25 Carlton Road, Derby, DE23 6HB
G0 WRC WYTHALL CONTEST GROUP c/o Lee Volante, Richmond House, Icknield Street, Birmingham, B38 0EP
G0 WRE Paul Scarratt, 339 Utting Avenue East, Norris Green, Liverpool, L11 1DF
GM0 WRH E Castle, 38 Davieland Road, Giffnock, Glasgow, G46 7LU
GW0 WRI Richard Lewis, 26 Bryn Road, Upper Brynamman, Ammanford, SA18 1AU
G0 WRK P Taylor, Old Acres, Priory Road, Yeovil, BA22 8NY
G0 WRL Derek Webb, 16 Burrowfield, Bruton, BA10 0HR
G0 WRM LICHFIELD RAYNET GROUP c/o M White, 62 Dalebrook Road, Burton-on-Trent, DE15 0AD
G0 WRN J Hodges, 48 Beach Road, Severn Beach, Bristol, BS35 4PF
G0 WRQ O Baxter, 5 Church Path, Bridgwater, TA6 7AJ
GM0 WRR John Scott, 70 Montford Avenue, Glasgow, G44 4PA
G0 WRS WARRINGTON ARC c/o Michael Isherwood, 32 Franklin Close, Old Hall, Warrington, WA5 8QL
G0 WRT Paul Winfield, 150 Tinshill Road, Leeds, LS16 7PN
GM0 WRU Robert Holmes, 3 Buchan Street, Wishaw, ML2 7HG
GM0 WRV Roderick Spink, 44 West Park, Inverbervie, Montrose, DD10 0TT
G0 WSA Jeffrey Proctor, Manor Farm Bungalow, Wendlebury, Bicester, OX25 2PS
G0 WSB M Brimley, 42 Grange Road, Netley Abbey, Southampton, SO31 5FE
G0 WSC Robert Connett, 15 Channels Lane, Horton, Ilminster, TA19 9QL
G0 WSD G Swann, 1 Beaver Close, Chichester, PO19 3QU
G0 WSH Robert Munt, Box 166, Jarfalla, Sweden, SE-177 23
G0 WSI W Warburton, 4 Haynes Way, Dibden Purlieu, Southampton, SO45 5QQ
G0 WSP Phil Croft, 82 Granby Road, Buxton, SK17 7TJ
GM0 WSR STRATHCLYDE REGIONAL RAYNET GROUPS c/o D Mackinnon, 60 Mount Stuart Drive, Wemyss Bay, PA18 6DX
G0 WSY J Adams, 53 Princess Road, Rochdale, OL16 4AY
G0 WTA N Midworth, 1 Highfields Drive, Loughborough, LE11 3JS
G0 WTB C Hicks, 59 Elmsfield Avenue, Norden, Rochdale, OL11 5XW
G0 WTC W Clark, 41 Brook Close, Jarvis Brook, Crowborough, TN6 2ET
G0 WTD T Stokes, 33 Talbot Drive, Euxton, Chorley, PR7 6PD
G0 WTF A Wright, 4 Wilshere Road, Welwyn, AL6 9PX
G0 WTG D Way, 4 Pergin Crescent, Poole, BH17 7AJ
G0 WTI Gavin Lightfoot, Dovedale, Woodville Road, Bude, EX23 9JA
G0 WTK Robert Jenkins, 11 Westfield Drive, Worksop, S81 0JS
G0 WTL G Fisher, 6 Totternhoe Road, Dunstable, LU6 2AG
G0 WTM David Sutton, 32 Queensway, Euxton, Chorley, PR7 6PW
G0 WTO E Birch, 4 Kynnesworth Gardens, Higham Ferrers, Rushden, NN10 8NH
G0 WTR M Robertson, Clayhill Cottage, The Street, Ipswich, IP7 6NN
G0 WTW Grayson Thompson, Lyngrove, Seaton Lane, Seaham, SR7 0LP
G0 WUA P Harris, 1 Newlands, Landkey, Barnstaple, EX32 0NJ
G0 WUG G Bishop, 8 Bulstrode Place, Kegworth, Derby, DE74 2DS
G0 WUH H White, 111 Beacon Glade, South Shields, NE34 7QU
G0 WUI W Cassidy, 17 Catcheside Close, Whickham, Newcastle upon Tyne, NE16 5RX
GW0 WUL C Minard, 3 Riverside Close, Aberfan, Merthyr Tydfil, CF48 4RN
G0 WUM E Roobottom, Puffin View, Abergwyngregyn, Llanfairfechan, LL33 0LL
G0 WUO Ronald Berkeley, 4 Raleigh Road, Leasowe, Wirral, CH46 2QZ
GM0 WUP William Steele, 1 James Street, Bannockburn, Whins of Milton, Stirling, FK7 0NQ
GM0 WUQ J Steele, 35 Devlin Court, Whins of Milton, Stirling, FK7 0NP
GM0 WUR G Steele, 1 James Street, Bannockburn, Whins of Milton, Stirling, FK7 0NQ
G0 WUU John Purcell, 6 Templeman Drive, Carlby, Stamford, PE9 4NQ
G0 WUV John Dickinson, 112 Stoneleigh Avenue, Longbenton, Newcastle upon Tyne, NE12 8XQ
G0 WUW V Claridge, 105 Barnwood Avenue, Gloucester, GL4 3AG
GM0 WUX J Donnan, 41 Annick Drive, Dreghorn, Irvine, KA11 4ER
G0 WUY A Williamson, Millfield Lodge, 151 Hull Road, York, YO10 3JX
G0 WUZ Ronald Rous, 8 Avon Close, Macclesfield, SK10 3DB
G0 WVA D Townsend, 21 Honeysuckle Close, Margate, CT9 5UN
G0 WVD Terence Lishman, 13 Meadow Way, Sandown, PO36 8QE
G0 WVE B Richardson, 12 Stoney Lane, Barrow, Bury St. Edmunds, IP29 5DD
G0 WVL Mark Jones, 80 Mount Pleasant Estate, Brynithel, Abertillery, NP13 2HU
G0 WVM M Brooker, 4 May Close, Sidlesham, Chichester, PO20 7RR
G0 WVT W Carwood, 57 Upton Road, Kidderminster, DY10 2YB
G0 WVV D Jones, 4 Back Bower Lane, Gee Cross, Hyde, SK14 5NS
G0 WVW David Smith, 16 Browning Drive Great Sutton, Ellesmere Port, CH65 7BW
G0 WVY F Holt, 22 First Avenue, Tottington, Bury, BL8 3JA
G0 WWA David Nicholas, 41 Grayling Road, Stourbridge, DY9 7AZ
G0 WWD D Weston, 25 Ley Lane, Kingsteignton, Newton Abbot, TQ12 3JE
G0 WWE S Fitton, 7 Blackburn Way, West Wick, Weston-Super-Mare, BS24 7GT
G0 WWF P Baron, 55 Church View, Brompton, Northallerton, DL6 2RD
G0 WWH Alan Houghton, 3 Billinge Close, Bolton, BL1 2JP
G0 WWL A Babbage, 10 Heather Close, Honiton, EX14 2YP
G0 WWM Toshinobu Yukawa, 5349 Route 12, Birch Hill, Richmond, Canada, C0B 1Y0
G0 WWO Craig Sawyer, 4 Padley Close, Ripley, DE5 3FG
G0 WWP S Griffith, 12 Beech Grove, Cliffsend, Ramsgate, CT12 5LD
G0 WWQ H Burton, Vale View, Porth y Waen, Denbigh, LL16 4BU
G0 WWR Roger Prior, 20 Churchfield Road, Walton on The Naze, CO14 8BL
G0 WWT James Smith, 54 Greenfield Avenue, Kettering, NN15 7LL
G0 WWU R Pratt, 4 King John Avenue, Gaywood, King's Lynn, PE30 4QA
GM0 WWX R Johnstone, 8 Harris Court, Alloa, FK10 1DD
G0 WWZ M Charlesworth, Po Box 841, 9506, Korondal City, South Cotabato, Philippines
G0 WXA S Everitt, 125 Victoria Road, Warley, Oldbury, B68 9UL
G0 WXC R Needham, 1 The Leas, Sedgefield, Stockton-on-Tees, TS21 2DS
G0 WXD P Atkinson, 27 Ranworth Road, Bramley, Rotherham, S66 2SP
G0 WXE A Kelleher, 144 Alibon Road, Dagenham, RM10 8DE
G0 WXF P Wright, 60 Farnborough Road, Clifton, Nottingham, NG11 8GF
G0 WXG J Lamb, 10 Lenhurst Avenue, Leeds, LS12 2RE
G0 WXH S Croot, 58 Dixie Street, Jacksdale, Nottingham, NG16 5JZ
G0 WXJ Peter Badham, 75 Newtown Road, Worcester, WR5 1HH
G0 WXL E Harper, 20 Bar Meadows, Malpas, Truro, TR1 1SS
G0 WXN Malcolm Thomas, 1 Veiga, Monforte de Lemos, Spain, 27416
GW0 WXO Anthony Davies, 81 Brynifor, Mountain Ash, CF45 3AB
G0 WXW GRESFORD & DISTRICT ARS c/o Thomas Rogers, The Willows, 48 Hillock Lane, Wrexham, LL12 8YL
GM0 WXX F Mackay, Shalimar, Alligin, Achnasheen, IV22 2HB
G0 WXZ David Milne, 1 Windsor Road, Christchurch, BH23 2EE
G0 WYA N Taylor, 61 Oldbury Road, Rowley Regis, B65 0NP
GI0 WYB J Simpson, 66 Ballyportery Road, Dunloy, Ballymena, BT44 9BN
G0 WYF Chris Rudge, 1 Mill Lane, Alfington, Ottery St. Mary, EX11 1PF
G0 WYG David Biginton, 67 Capstone Road, Bromley, BR1 5NA
G0 WYI Charles Sharpe, 35 Fairmead Close, Nottingham, NG3 3EQ
GI0 WYK William Carress, 12 Ashbourne Park, Newtownards, BT23 7RE
G0 WYM A Shields, 8 Thames Drive, Melton Mowbray, LE13 0DS
G0 WYN Anthony Godwin, 27 Melbourne Avenue, Dronfield Woodhouse, Dronfield, S18 8YW
GI0 WYO Ronald Kilgore, 3 Summer Meadows Manor, Londonderry, BT47 6SE
G0 WYP S Barker, C/O, 14 Cordova Avenue, Manchester, M34 2WP
G0 WYQ J Richardson, 24 Brockhall Road, Kingsley, Northampton, NN2 7RY
G0 WYR Christopher Fox, 36 Haig Drive, Slough, SL1 9HB
G0 WYS Keith Matthew, 3 Marconi Close, Helston, TR13 8PD
G0 WYT I Seabright, 78 Lazy Hill, Birmingham, B38 9PA
G0 WYU E Martin, 2 Bospowis, St. Martins Crescent, Camborne, TR14 7HN
G0 WYV P Little, 24 Pickwick Crescent, Rochester, ME1 2HZ
G0 WYY R Bell, 41 Ravenshill Road, West Denton, Newcastle upon Tyne, NE5 5EA
G0 WYZ Susan Gillen, 33 Norwich Close, Ashington, NE63 9RY
G0 WZA D Wood, 2 Buckingham Mews, Flitwick, Bedford, MK45 1TB
G0 WZB Brian Burdis, 10 Johnston Avenue, Hebburn, NE31 2LJ
G0 WZC S Cooper, 27 Polweath Road, Penzance, TR18 3PW
G0 WZD J Paloschi, 1 Lander Close, Milton, Cambridge, CB24 6EB
G0 WZG K Norris, 15 East View, Choppington, NE62 5UF
G0 WZH S Frankum, 39 Leighton Road, Wingrave, Aylesbury, HP22 4PA
G0 WZJ J Pickering, 11 Smithson Court, Malton, YO17 7BQ
G0 WZK Nicholas Collis Bird, 1 Drayton Park, London, N5 1NU
G0 WZL John Rushton, 18 Landless Street, Brierfield, Nelson, BB9 5LA
G0 WZM I kitchen, 25 Christchurch Mount, Epsom, KT19 8LU
GM0 WZO I Finlayson, 10b Flesherin, Isle of Lewis, HS2 0HE
G0 WZV K Aston, 10 Browning Close, Lexden, Colchester, CO3 4JJ
GI0 WZW R Pollock, 5 Brooke Grove, Banbridge, BT32 3YA
G0 WZX B Stevens, 172 Gordon Avenue, Camberley, GU15 2NT
G0 WZY Michael Davies, 45 Elm Grove, Swainswick, Bath, BA1 7BA
G0 WZZ Malcolm Bobby, Hafan, Church Street, Penycae, LL14 2RL
G0 XAA ANSTY CONTEST CLUB c/o Keith Evans, Littlefield House, Bolney Road, Haywards Heath, RH17 5AW
G0 XAB R Dawson, 28 Calf Close, Haxby, York, YO32 3NS
GI0 XAC A Chin, 25 Meadowbrook, Islandmagee, Larne, BT40 3UG
G0 XAD B Wright, 96 Ellenborough Close, Thorley, Bishop's Stortford, CM23 4HU
G0 XAE M Buckland, The Homestead, Bollow, Westbury-on-Severn, GL14 1QX
G0 XAF Paul Coles, 10 Springfield Close, Mangotsfield, Bristol, BS16 9BZ
G0 XAH David Tecklenberg, 1 Proby Close, Yaxley, Peterborough, PE7 3ZF
G0 XAI S Bennett, Flat 1-4, 208 High Street, Lewes, BN7 2NS
G0 XAK Stephen Curtis, 389 Portway, Shirehampton, Bristol, BS11 9UF
G0 XAM L Camber, Sundown, Strawberry Gardens, Newick, BN8 4QX
G0 XAN Gary Aylward, 53 Overdown Rise, Portslade, Brighton, BN41 2YF
G0 XAO P Cole, 18 Mundys Field, Ruan Minor, Helston, TR12 7LF
G0 XAP B Blake, 39 Heol Sant Gattwg, Llanspyddid, Brecon, LD3 8PD
G0 XAR Stephen Farthing, 21 Cavell Close Swardeston, Norwich, NR14 8DH
G0 XAS J Gavin, 22 Rotherwick Way, Cambridge, CB1 8RX
G0 XAT R Wallbank, 32 Truro Place, Heath Hayes, Cannock, WS12 3YJ
G0 XAU C Martin, 7 Mayfair Close, Dukinfield, SK16 5HR
GM0 XAV A Main, 1 Border Avenue, Saltcoats, KA21 5NH
G0 XAW A Santillo, 34 Wearde Road, Saltash, PL12 4PP
G0 XAY Richard Elford, Prospects, Tormarton Road, Badminton, GL9 1HP
G0 XAZ H Simmons, 96 Porlock Road, Southampton, SO16 9JF
G0 XBA A Hill, 13 Sycamore Way, Winklebury, Basingstoke, RG23 8AD
G0 XBC M Soane, 24 Nurseries Road, Wheathampstead, St. Albans, AL4 8TP
G0 XBG A Marinho, 13 Pipkin Way, Oxford, OX4 4AR
G0 XBL H Watts, 7 Hartwood Road, Liverpool, L32 7QH
G0 XBO E Smith, 8 Nene Road, Hunstanton, PE36 5BZ
G0 XBQ Chris Levingston, 44 Lewis Road, South Australia, Glynde, Australia, 5070
G0 XBV A Nottage, 99 Fermor Way, Crowborough, TN6 3BH

| | | |
|---|---|---|
| G0 | XCF | Colin Foley, 12 Cross Street, Northam, Bideford, EX39 1BS |
| G0 | XDI | M Cabban, 2 Sandycroft Road, Amersham, HP6 6QL |
| G0 | XDL | Gareth Edwards, 82 Regent Drive, Skipton, BD23 1BB |
| G0 | XFG | D Riches, 21 Brinkley Way, Felixstowe, IP11 9TX |
| GM0 | XFK | John Neary, 17 Harkins Avenue, Blantyre, Glasgow, G72 0RQ |
| G0 | XGL | G Lawrence, 20 Branewick Close, Fareham, PO15 5RS |
| G0 | XGM | Ronald Burgess, 22 Lee Close, Kidlington, OX5 2XZ |
| G0 | XIT | B Davis, Westfield House, Wood End Road, Bedford, MK43 9BB |
| G0 | XJS | James Sliman, 4 Coleridge Close, Exmouth, EX8 5SP |
| G0 | XKK | K Keenan, 25 Harris Road, Harpur Hill, Buxton, SK17 9JS |
| G0 | XOX | P Thorndike, 56 Durham Road, Southend-on-Sea, SS2 4LU |
| G0 | XPD | Antony Sutton, Karena, Gweek, Helston, TR12 6UP |
| G0 | XRC | EXMOUTH ARC c/o R Maynard, Clarnard, 7 Phillipps Avenue, Exmouth, EX8 3HY |
| G0 | XTA | [illegible] |
| G0 | XTL | T [illegible], 11 Crossfield, Fernhurst, Haslemere, GU27 3JI |
| G0 | XTM | N Marshall, 3 The Green, Dorking Road, Tadworth, KT20 5SQ |
| G0 | XVC | Richard Nixon, 31 Ashfield, Shotley Bridge, Consett, DH8 0RF |
| G0 | XVI | D Veale, 5 Heathfield Close, Dronfield, S18 1RJ |
| G0 | XVS | J Thompson, 22 Kendal Drive, Dronfield Woodhouse, Dronfield, S18 8NA |
| G0 | XXX | 5XX GROUP DAVENTRY c/o D Knowler, Apartedo 1009, 8670 - 999, Aljezur, Portugal |
| G0 | XYL | J Hockley, 44 Brookfields, Crickhowell, NP8 1DJ |
| GI0 | XYZ | Co-Antrim A.R.S DX Group c/o T Hoey, 4 Bramble Avenue, Newtownabbey, BT37 0XL |
| G0 | XZT | Jeremy Gilbert, 17 Phoenix Way, Portishead, Bristol, BS20 7FG |
| G0 | YAE | Douglas Phillips, 28 Prince Of Wales Road, Great Totham, Maldon, CM9 8PX |
| G0 | YBU | Daniel Cuff, 7 Parsons Pool, Shaftesbury, SP7 8AL |
| G0 | YDX | Robert Bates, Apartment 608, 465 West Dominion Drive, Wood Dale, United States, 60191-2309 |
| G0 | YKC | G Clampin, Inverkeris, Tydd Road, Spalding, PE12 0HP |
| G0 | YKK | Kevin Kelsall, 56 Glenwood Avenue, Baildon, Shipley, BD17 5RS |
| G0 | YLO | WINCANTON LADIES CONTEST GROUP c/o C Monksummers, 29 Cloverfields, Peacemarsh, Gillingham, SP8 4UP |
| G0 | YOU | C Haye, 15 Byron Close, Yateley, GU46 6YW |
| G0 | YRT | B Howarth, 23 Yew Tree Road, Denton, Manchester, M34 6JY |
| G0 | YSS | A Jones, 122 Slater St., Latchford, Warrington, WA4 1DW |
| G0 | YYY | Robert Konowicz, 12 Ambleside Crescent, Farnham, GU9 0RZ |
| GI0 | ZAK | Fredrick Tanner, 4 Leopold Gardens, Belfast, BT13 3XN |
| GM0 | ZAM | J Glennon, 68 Carronshore Road, Carron, Falkirk, FK2 8EE |
| G0 | ZAP | Robert Crozier, 930-932 Burnley Road, Loveclough, Rossendale, BB4 8QL |
| G0 | ZAT | William Skipper, 18 Central Avenue, South Shields, NE34 6AZ |
| GW0 | ZDL | D Lister, Driftwood, Sunnyside, Haverfordwest, SA62 3BG |
| G0 | ZEE | C Monksummers, 29 Cloverfields, Peacemarsh, Gillingham, SP8 4UP |
| G0 | ZEP | Richard Carter, 12 Glebe Close, Abbotsbury, Weymouth, DT3 4LD |
| GI0 | ZER | Eddie Robinson, 281 Pomeroy Road, Pomeroy, Dungannon, BT70 3DT |
| G0 | ZHP | THE POLISH SCOUT ARC - LONDON c/o Krzysztof Jasinski, 35 Friars Place Lane, London, W3 7AQ |
| G0 | ZIG | B Lennox, 66 Albacore Crescent, London, SE13 7HP |
| G0 | ZIP | F Marston, 1 Weaver Road, Leicester, LE5 2RL |
| G0 | ZMC | M Conlon, 3 Selside, Brownsover, Rugby, CV21 1PG |
| G0 | ZMH | M Howell, Orchard House, Blennerhasset, Wigton, CA7 3QX |
| G0 | ZPV | J Roberts, Long Meadow, Kidderton Lane, Nantwich, CW5 8JD |
| G0 | ZZZ | Philip Measom, 44 Heath Side, Petts Wood, Orpington, BR5 1EY |

## G*1

| | | |
|---|---|---|
| G1 | AAC | P Watterson, 25 Church Lane, Mablethorpe, LN12 2NU |
| G1 | AAD | N Youd, 8 Forest Road, Piddington, Northampton, NN7 2DA |
| G1 | AAG | A Withers, 23 Fernie Road, Guisborough, TS14 7LZ |
| G1 | AAH | P Worledge, 8 Forest Edge Road, Sandford, Wareham, BH20 7BX |
| G1 | AAK | P Webster, 30 Belvoir Road, Widnes, WA8 6HR |
| G1 | AAL | A Woodward, 40 Berwood Farm Road, Wylde Green, Sutton Coldfield, B72 1AG |
| G1 | AAP | M Wilson, 18 Briars Close, Southwood, Farnborough, GU14 0PB |
| G1 | AAQ | Evan Writer, 78 Henley Way, Ely, CB7 4YJ |
| G1 | AAR | M West, The Lair, Lewes Road, Newhaven, BN9 9AH |
| G1 | ABA | D Lambert, 72 Johnson Drive, Barrs Court, Bristol, BS30 7BS |
| G1 | ABJ | A Norton, 29 Long Lane, Chapel-en-le-Frith, High Peak, SK23 0TA |
| G1 | ABM | I Macdonald, 23 Cymberline Way, Warwick Gates, Warwick, CV34 6FQ |
| G1 | ABQ | F Macdonald, Boothlands Farm, Newdigate, Dorking, RH5 5BS |
| G1 | ABW | B Webber, 5 Sheldon Way, Berkhamsted, HP4 1FG |
| GW1 | ABX | C Passey, 38 Bendrick Road, Barry, CF63 3RE |
| G1 | ACA | J Garner, Cobwebs, Lewes Road, Haywards Heath, RH17 7PG |
| G1 | ACD | C Clifford, 12 Green Park, Brinkley, Newmarket, CB8 0SQ |
| G1 | ACD | Peter Hughes, 7 Dellfield Lane, Liverpool, L31 6AS |
| GI1 | ACN | T Hutton, 23 Enniscrone Park, Portadown, Craigavon, BT63 5DQ |
| G1 | ACV | W Harrison, Top Flat, Craiglwyd Hall, Penmaenmawr, LL34 6ER |
| G1 | ACY | M Holley, 95 Lyes Green, Corsley, Warminster, BA12 7PA |
| G1 | ADB | M Hunt, 46 Cunningham Drive, Bury, BL9 8PD |
| G1 | ADE | H Kirk, 34 Tilworth Road, Hull, HU8 9BN |
| GM1 | ADI | R Stevens, 10 Tiel Path, Glenrothes, KY7 5AX |
| GW1 | ADY | L Rees, 40 High St., Abergwili, Carmarthen, SA31 2JB |
| G1 | AEA | A Head, 16 Western Close, Penton Park, Chertsey, KT16 8QB |
| G1 | AEB | J Savage, 30 Green Meadows, Caravan Park, Cheltenham, GL51 6SN |
| G1 | AEF | R Smith, 130 Winchester Road, Ford Houses, Wolverhampton, WV10 6EZ |
| G1 | AEI | P Stevenson, 2 Lazonby Hall Cottages, Lazonby, Penrith, CA10 1BA |
| G1 | AEJ | L Smith, 4 Penhale Road, Braunstone, Leicester, LE3 2UU |
| G1 | AEQ | Derek Lewis, 10 Addington Road, Bolton, BL3 4QZ |
| G1 | AET | F Lawson, 10 Avebury Close, Tuffley, Gloucester, GL4 0TS |
| G1 | AEU | A Lott, 27 Queens Crescent, Brixham, TQ5 9PJ |
| G1 | AEX | G Muggeridge, Gribble House, Wey Street, Ashford, TN26 2QH |
| G1 | AFI | D Mills, 25 Lower Park Crescent, Poynton, Stockport, SK12 1EF |
| G1 | AFJ | P Keyte, 11 Woodward Road, Pershore, WR10 1LW |
| G1 | AFK | B Kelsey, 18 Parkside Avenue, Littlehampton, BN17 6BG |
| G1 | AFW | Robert Harris, 15 Rodmer Close, Minster on Sea, Sheerness, ME12 2BS |
| G1 | AGA | Adrian Gee, 74 New Street, Milnsbridge, Huddersfield, HD3 4LD |
| G1 | AGB | Denise Gee, 74 New Street, Milnsbridge, Huddersfield, HD3 4LD |
| G1 | AGK | D Woodruffe, 7 Orchard Close, Melton, Woodbridge, IP12 1LD |
| G1 | AGM | G Williams, 22 Moor Tarn Lane, Walney, Barrow-in-Furness, LA14 3LP |
| G1 | AGW | D Yeatman, 1 Apple Tree Close, Deeping St James, Peterborough, PE6 8RF |
| GM1 | AHF | A Mccormack, 10 Harris Court, North Muirton, Perth, PH1 3DD |
| GM1 | AHG | N Mccormack, 18 Harris Court, North Muirton, Perth, PH1 3DD |
| G1 | AHM | A MARTLAND, Knowleswood, Wrennals Lane, Chorley, PR7 5PW |
| G1 | AHQ | D Mobley, 17 Butts Close, Aynho, Banbury, OX17 3AE |
| G1 | AHS | Alan Mitchell, 7 Cross Park St., Horbury, Wakefield, WF4 6AE |
| G1 | AHT | Christopher Wager-Bradley, Swallowdale, Longhoughton Road, Alnwick, NE66 3AI |
| G1 | AHW | T Rodden, Greenwells, Worsall Road, Yarm, TS15 9EF |
| G1 | AIA | F Oliver, 9 Taylor Terrace, West Allotment, Newcastle upon Tyne, NE27 0EF |
| G1 | AIB | E Robinson, 12 Belroyal Avenue, Bristol, BS4 4RT |
| G1 | AIF | R Robinson, 9 Prospect Terrace, New Kyo, Stanley, DH9 7TR |
| G1 | AIG | S Rimell, 1 Pullin Court, Warmley, Bristol, BS30 8YL |
| G1 | AII | James Rushton, 8 Keats Road, Stonebroom, Alfreton, DE55 6JG |
| G1 | AIO | A Peters, 28 [illegible] Avenue, Didcot, OX11 0AD |
| G1 | AJC | T Howe, 4 Willow Crescent, Broughton Gifford, Melksham, SN12 8NB |
| G1 | AJD | M Jacobsen, 3 Green Lane, Tickton, Beverley, HU17 9RH |
| G1 | AJE | M Chambers, 1 Green Lane, Tickton, Beverley, HU17 9RH |
| G1 | AJK | Dennis Neale, 161 Antrobus Road, Birmingham, B21 9NU |
| G1 | AJN | George Overy, Flat2 37 Magdalene rd, St Leonards on Sea, TN376ET |
| G1 | AJS | N Parker, 7 The Hollies, Clee Hill, Ludlow, SY8 3NZ |
| G1 | AJT | C Pickering, 16 Ashworth Way, Newport, TF10 7EG |
| G1 | AJU | J Parsons, 40 Tynings Close, Kidderminster, DY11 5JP |
| G1 | AJV | E Roberts, 4 Willow Way, Redditch, B97 6PH |
| G1 | AJY | A Young, 4 Woodlea, Leybourne, West Malling, ME19 5QY |
| G1 | AJZ | P Rayson, 1 Grange Gardens, Taunton, TA2 7EN |
| G1 | AKA | Harold Rowley, 44 Tytherington Drive, Macclesfield, SK10 2HJ |
| G1 | AKB | Steve Ryder, Corner Cottage, Crown Lane, Worcester, WR8 9BE |
| G1 | AKD | A Rideout, 7 Beech Road, Martock, TA12 6DT |
| G1 | AKE | Paul Rowland, 7 Maxwell Street, Bury, BL9 7QA |
| G1 | AKV | T Alexander, 8 Greenway, Eastbourne, BN20 8UG |
| G1 | ALA | J Bird, 17 Sherrards Way, Barnet, EN5 2BW |
| G1 | ALD | Alan Brown, 40 Sutherland Road, Edmonton, London, N9 7QG |
| G1 | ALK | Robert Batchelor, 20 Mallard Way, Lower Stoke, Rochester, ME3 9ST |
| G1 | ALL | Michael Bond, 8 Alfred Street, Irchester, Wellingborough, NN29 7DR |
| G1 | ALR | Anthony Stafford, 24 Bourne Street, Croydon, CR0 1XL |
| G1 | ALU | K Spiers, Robins Post, North Heath Lane, Horsham, RH12 5PJ |
| GW1 | ALV | Alan Shaw, Derlwyn, Efailwen, Clynderwen, SA66 7JP |
| G1 | AML | Iain Tidey, April Cottage, Cansiron Lane, East Grinstead, RH19 3SE |
| G1 | AMN | N Webb, 3 Allens Lane, Norwich, NR2 2JB |
| G1 | AMS | J Winter, Flat 23, Knightlow Lodge Knightlow Avenue, Coventry, CV3 3HH |
| G1 | ANA | James Coles, 10 Westgate Hill Street, Bradford, BD4 0SJ |
| G1 | ANF | T Crowe, 15 Lambert Road, Kendray, Barnsley, S70 3AA |
| GI1 | ANG | Gerard Carvill, 18 Hospital Road, Newry, BT35 8PW |
| G1 | ANI | M Cooper, 33 Park View, Royston, Barnsley, S71 4AA |
| G1 | ANK | R Bond, 21 Dinglebank Close, Lymm, WA13 0QH |
| G1 | ANQ | Paul Fincher, 11 Verney Mews, Reading, RG30 2NT |
| G1 | ANS | K Elvin, 97 Jeans Way, Dunstable, LU5 4PR |
| G1 | ANV | R Edgar, 19 Butt Hedge, Long Marston, York, YO26 7LW |
| GW1 | ANW | David Evans, 6 Oakfield Terrace, Ammanford, SA18 2NG |
| G1 | ANZ | P Thirst, The Haywain, Thirsts Farm, Norwich, NR12 0RU |
| G1 | AOC | Richard Cecil Doughty, 7 High Street, Garlinge, Margate, CT9 5LN |
| G1 | AOE | J Darley, 16 Ivydene, Knaphill, Woking, GU21 2TA |
| G1 | AOF | R Dean, 10 Livingstone Road, Ellesmere Port, CH65 2BE |
| G1 | AOQ | I Gorsuch, Elmstone Farm, Fosten Green, Ashford, TN27 8ER |
| G1 | AOR | Peter Grayshon, 90 Park Lea, Bradley, Huddersfield, HD2 1QP |
| G1 | AOZ | James Henderson, The Bungalow, Appleby Grammar School, Appleby-in-Westmorland, CA16 6XU |
| G1 | APA | C Hibbert, 19 Fern Road, Maia, Dunedin, New Zealand, 9022 |
| G1 | APL | R Hopkins, 15 Oak Meadow, Shipdham, IP25 7FD |
| G1 | APQ | Alan Howells, 16 Oakley Wood Road, Bishops Tachbrook, Leamington Spa, CV33 9RN |
| G1 | APU | Keith Pierson, The Brambles, Rockwell Lane, Oswestry, SY10 9QR |
| G1 | AQF | A Matthews, 44 Essex Close, Dines Green, Worcester, WR2 5RW |
| G1 | AQI | J Aldersey, 36 Walls Road, Salterbeck, Workington, CA14 5JA |
| G1 | AQP | A Bell, 1 Purbeck Drive, Lostock, Bolton, BL6 4JF |
| G1 | AQV | J Blackman, 30 Parklands Way, Penrith, CA11 8SD |
| G1 | AQX | K Armstrong, 30 Cobholm Place, Cambridge, CB4 2UN |
| GD1 | AQY | Stephen Broad, Ballaberragh, Farm, Ramsey, Isle of Man, IM73EB |
| G1 | ARD | G Ashton, 101 Wickenby Garth, Bransholme, Hull, HU7 4RF |
| G1 | ARF | J Makin, 6 Cambridge House, Courtfield Gardens, London, W13 0HP |
| G1 | ARH | Rodney Lupton, 10 Avenue Close, Harrogate, HG2 7LJ |
| G1 | ARL | C Mandall, 11a Hazel Road, Park Street, St. Albans, AL2 2AH |
| G1 | ARU | A Judge, 44 Thorley Lane, Bishop's Stortford, CM23 4AD |
| GM1 | ASA | H Kilpatrick, 80 Livingstone Terrace, Irvine, KA12 9DN |
| G1 | ASD | K Knight, 80 Winton Road, Reading, RG2 8HJ |
| G1 | ASG | T Stokes, 21 Guildford View, Sheffield, S2 2NZ |
| G1 | ASN | James Stansfield, Flat 12, Harty House, Church Street, Manchester, M30 0LT |
| G1 | ASR | David Thornton, 8 Chestnut Court, Toft Hill, Bishop Auckland, DL14 0TG |
| G1 | ASU | Andrew Clews, 47 Gaydon Road, Solihull, B92 9BJ |
| GM1 | ASY | John Christie, 99 Meadow Crescent, Elgin, IV30 6EP |
| G1 | ATA | D Cotton, 24 Stirling Rise, Stretton, Burton-on-Trent, DE13 0JP |
| G1 | ATC | AIR CADET ARS c/o Victor Tuff, 8 Millcroft Court, Blyth, NE24 3JG |
| G1 | ATG | C Clarke, 29 Huyton Hey Road, Huyton, Liverpool, L36 5SF |
| G1 | ATL | K Bishop, 8 Sandbanks Grove, Hailsham, BN27 3LS |
| G1 | ATQ | H Rogers, 74 Front Road, Murrow, Wisbech, PE13 4HU |
| GM1 | ATW | E King, Marionville, Donibristle, Cowdenbeath, KY4 8EU |
| G1 | ATY | B Jarratt, 7 Kings Road, Aylesbury, HP21 7RR |
| G1 | ATZ | G Morris, 18 Grosvenor Road, Shotton, Deeside, CH5 1NU |
| G1 | AUH | Richard Holland, 53 Manor Farm Way, Middleton, Leeds, LS10 3RB |
| G1 | AUI | Chris Hayes, 37 St. Radigunds Street, Canterbury, CT1 2AA |
| G1 | AUM | G Gumbrell, 24 Tattershall Drive, Market Deeping, Peterborough, PE6 8BS |
| G1 | AUQ | Richard Green, 2 Fleets Lane Cottage, Fleets Road, Lincoln, LN1 2DN |
| G1 | AUR | H Graham, 4 Eisenhower Road, Basildon, SS15 6JR |
| GW1 | AUT | Neil Gordon, 21 Picket Mead Road, Newton, Swansea, SA3 4SA |
| G1 | AUU | S Goan, 36 Wellers Grove, Cheshunt, Waltham Cross, EN7 6HU |
| G1 | AUY | R Curzon, 24 Edwards Drive, Wellingborough, NN8 3JJ |
| G1 | AVA | D Carter, 10 Opal Street, Keighley, BD22 7BP |
| G1 | AVB | Rodney Davidson, 3 Eastridge Drive, Bishopsworth, Bristol, BS13 8HQ |
| G1 | AVC | J Dean, 27 Ionic Road, Liverpool, L13 3DU |
| G1 | AVF | B Eastick, 30 Farmcote Road, Aldermans Green, Coventry, CV2 1SA |
| G1 | AVW | H Sterry, 23 Eardisley Close, Matchborough, Redditch, B98 0BX |
| G1 | AVZ | J Toolan, 12 Stillington Road, Huby, York, YO61 1HW |
| G1 | AWD | T Wells, 2 Stephens Close, Mortimer Common, Reading, RG7 3TL |
| G1 | AWF | A Wyspianski, 60 Alington Crescent, Kingsbury, London, NW9 8JJ |
| GW1 | AWH | Ian Robinson, 13 Clas Tywern, Cardiff, CF14 4SB |
| G1 | AWJ | John Moyle, Amberley, Cotswold Close, Shipston-on-Stour, CV36 4ND |
| G1 | AWK | [illegible] |
| G1 | AWU | Alan Pickles, 40 Hermitage Street, Crewkerne, TA18 8ET |
| GM1 | AXI | T Ball, 3 Kerland Place, Glengarnock, Beith, KA14 3BQ |
| GW1 | AXU | J Cook, 22 Northlands Park, Bishopston, Swansea, SA3 3JW |
| G1 | AXW | G Dann, Bramley House, School Lane, Gloucester, GL2 4YG |
| G1 | AYH | Mark Newell, 15 The Grove, Luton, LU1 5PE |
| G1 | AYI | T Smith, 19 Leaf Road, Houghton Regis, Dunstable, LU5 5JG |
| G1 | AYP | Keith Lawton, Meadowbank, Sutton St. Nicholas, Hereford, HR1 3BJ |
| GM1 | AYT | J McAllister, Redwood Lodge, Coach Road, Glasgow, G65 0PR |
| G1 | AYU | C Chambers, 1 Saunders Close, Pound Hill, Crawley, RH10 7AE |
| G1 | AZA | J Cook, 72 Valebridge Road, Burgess Hill, RH15 0RP |
| G1 | AZC | Malcolm Cannings, 7 Whinlatter Place, Newton Aycliffe, DL5 7DR |
| G1 | AZD | A Drage, 51 Greenbank Avenue, Kettering, NN15 7EF |
| G1 | AZE | Brian Davies, 22 Hillside Road, Four Oaks, Sutton Coldfield, B74 4DQ |
| G1 | AZJ | Robert Endersby, 3 Woodside, Vigo, Gravesend, DA13 0SU |
| G1 | AZZ | G Taylor, 31 Ashfurlong Crescent, Sutton Coldfield, B75 6EN |
| G1 | BAA | F Whittaker, 14 Mill Lane, Wombourne, Wolverhampton, WV5 0LG |
| G1 | BAB | R Wilson, Barley Corners, Arundel Road, Seaford, BN25 4LZ |
| GW1 | BAI | K Lowe, 18 Aberdovey Close, Dinas Powys, CF64 4PS |
| G1 | BAL | James Mahoney, 61 Wood Street, Barnsley, S70 1NA |
| GM1 | BAN | Donald Morrison, 4 West Murkle, Murkle, Thurso, KW14 8YT |
| G1 | BAQ | A Miller, 44 Spring Gardens, Newport Pagnell, MK16 0EE |
| G1 | BAR | B Norris, 1 Earleswood, Benfleet, SS7 1DN |
| G1 | BAX | J Peters, Ferndale Cottage, Brea, Camborne, TR14 9AT |
| G1 | BBA | T Tilley, 10 Elmhurst Close, Haverhill, CB9 8EG |
| G1 | BBC | J Duxbury, 7 Osprey Close, Blackburn, BB1 8LP |
| G1 | BBH | J Sharkey, 33 Ffordd Morfa, Llandudno, LL30 1ES |
| G1 | BBI | K Ford, 3 Crossfell, Wildridings, Bracknell, RG12 7RX |
| G1 | BBK | B Artingstall, 19 Town Lane, Denton, Manchester, M34 6AF |
| G1 | BBT | J Dackham, 4 Overbury Close, Weymouth, DT4 9UE |
| G1 | BBY | Walter Brown, 53 Drummonds Close, Longhorsley, Morpeth, NE65 8UR |
| G1 | BCB | A Blake, 58 Greenacres, Bath, BA1 4NR |
| G1 | BCE | B Brough, 6 Higgs Road, Wednesfield, Wolverhampton, WV11 2PD |
| GW1 | BCI | A Gray, 69 Tyn Y Parc Road, Cardiff, CF14 6BJ |
| G1 | BCN | A Hopkinson, 104 Everill Gate Lane, Wombwell, Barnsley, S73 0YJ |
| G1 | BCU | Richard Tagg, 38 Salhouse Road, Rackheath, Norwich, NR13 6QH |
| GW1 | BDF | K Jones, 10 Trinity Road, Tonypandy, CF40 1DQ |
| GW1 | BDG | O Jones, 10 Trinity Road, Tonypandy, CF40 1DQ |
| G1 | BDH | Brian Jones, 8 Walton Crescent, Llandudno Junction, LL31 9ER |
| G1 | BDI | B Jones, 56 Mount Grace Road, Luton, LU2 8EP |
| G1 | BDP | M Broad, 7 Steventon Road, Drayton, Abingdon, OX14 4JX |
| G1 | BDQ | Andrew Bell, Doddington Mill, Mill Lane, Nantwich, CW5 7NN |
| G1 | BDU | A Bradshaw, Lyndale, 145 Alder Lane, Wigan, WN2 4ET |
| G1 | BDY | Martin Bragg, 33 Mosley Street, Barnoldswick, BB18 5BS |
| G1 | BEB | R Cocking, 22 Dunbeath Avenue, Rainhill, Prescot, L35 0QH |
| G1 | BEC | R Collins, 29 Brook Drive, Verwood, BH31 6DH |
| G1 | BEG | J Ellsmore, 15 Greenbush Drive, Halesowen, B63 3TJ |
| G1 | BEJ | I Dixon, 60a Woodlands Road, Allestree, Derby, DE22 2HF |
| G1 | BEK | G Death, 105 Belvedere Road, Ipswich, IP4 4AD |
| G1 | BES | James Gibbard, 2 Almond Court, Liverpool, L19 2QZ |
| G1 | BET | A Waters, 12 Anvil Court, Whittonstall, Consett, DH8 9JU |
| GI1 | BEU | Leonard Gough, 76-78 Culmore Point, Londonderry, BT48 8JW |
| GW1 | BFB | Anthony Frayne, 20 Springfield Avenue, Upper Killay, Swansea, SA2 7HW |
| G1 | BFF | T Fishlock, 62 Red Barn Road, Brightlingsea, Colchester, CO7 0SJ |
| G1 | BFG | Andrew Harper, 144 Ashfield Road, Blackpool, FY2 0EN |
| G1 | BFK | Roger Else, 134 Market Street, South Normanton, Alfreton, DE55 2EJ |
| G1 | BFS | P Sainsbury, 103a Pilot Road, Hastings, TN34 2AU |
| G1 | BGC | S Javes, 95 West Way, Lancing, BN15 8LZ |
| G1 | BGF | D Mclachlan, 1 North Holme Court, Northampton, NN3 8UX |
| G1 | BGH | P Nicholls, 40 Sedgefield Close, Worth, Crawley, RH10 7XG |
| G1 | BGJ | E Morgan, 79 Mayland Avenue, Canvey Island, SS8 0BU |
| G1 | BGK | O Lang, Delford Barn, Ashburton, Newton Abbot, TQ13 7HT |
| G1 | BGM | P Leslie, 5 Maple Croft, Netherton, Huddersfield, HD4 7HS |
| G1 | BGO | D Lee, 36 Westwick, Hedon, Hull, HU12 8HQ |
| G1 | BGQ | P Sampson, 23 Westfield Road, Mirfield, WF14 9PW |
| G1 | BHB | S Matthews, 222 Widney Lane, Solihull, B91 3JY |
| G1 | BHF | S Oldfield, The Sycamores, Fulford Road, Stoke-on-Trent, ST11 9QT |
| G1 | BHG | B Osborne, 12 Arminers Close, Gosport, PO12 2HB |
| G1 | BHO | B Palmer, 28 Westminster Crescent, Burn Bridge, Harrogate, HG3 1LY |
| G1 | BHQ | R Pritchard, 41 Greenland Avenue, Maltby, Rotherham, S66 7EU |
| G1 | BHR | I Habbitt, 66 Parkfield Avenue, Delapre, Northampton, NN4 8QB |
| G1 | BHS | M Rumbelow, The Chase, Knott Park, Leatherhead, KT22 0HR |
| G1 | BHV | R Green, 20 Haygate Drive, Wellington, Telford, TF1 2BY |
| G1 | BHW | C Vaughan, 11 Fremantle Road, Aylesbury, HP21 8EH |
| G1 | BIA | Harold Wilmshurst, Langholm Lodge, Raydaleside, Darlington, DL3 7SJ |
| G1 | BIF | T Whelan, 43 Martin Avenue, Little Lever, Bolton, BL3 1NX |
| G1 | BIM | R Ross, 46 Arbour Close, Rugby, CV22 6EH |
| G1 | BIN | B Talbot, 27 Shuttleworth Road, Clifton upon Dunsmore, Rugby, CV23 0DB |
| G1 | BIU | Clive Utting, 22 Yew Tree Drive, Bromsgrove, B60 1AL |
| G1 | BJE | A Stimpson, 2 Church Avenue, Kings Sutton, Banbury, OX17 3RJ |
| G1 | BJK | P Lough, 87 Finchley Road, Kingstanding, Birmingham, B44 0LB |
| G1 | BJN | C Stroud, 6 Church Road, Ideford, Newton Abbot, TQ13 0BB |

UK Callsigns

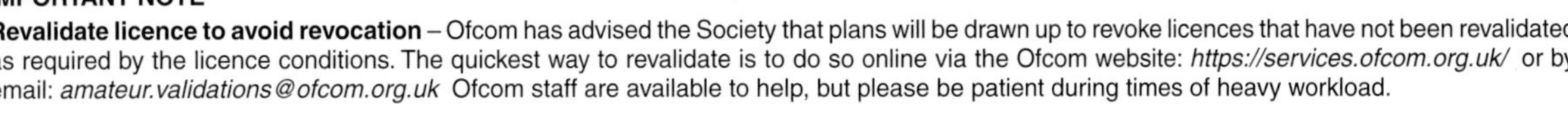

**IMPORTANT NOTE**

**Revalidate licence to avoid revocation** – Ofcom has advised the Society that plans will be drawn up to revoke licences that have not been revalidated as required by the licence conditions. The quickest way to revalidate is to do so online via the Ofcom website: *https://services.ofcom.org.uk/* or by email: *amateur.validations@ofcom.org.uk* Ofcom staff are available to help, but please be patient during times of heavy workload.

G1 BJZ Gavin Garlick, 1 Shannon Way, Burton Latimer, Kettering, NN15 5SX
G1 BKB S Stokes, 52 Brantley Avenue, Wolverhampton, WV3 9AR
G1 BKI Michael York, 3 Cam Close, Corby, NN17 2LJ
G1 BKJ J Prior, 6 Emfield Grove, Grimsby, DN33 3BS
G1 BKL D Ambler, Corrig, 4 Old Main Road, Bridgwater, TA6 4RY
GM1 BKR J Rankin, 3 Spalding Drive, Largs, KA30 9BZ
G1 BKZ W Moore, 10 Progress Street, Darwen, BB3 2DT
G1 BLB P Thurman, 11 Copperfield Drive, Langley, Maidstone, ME17 1SX
G1 BLJ S Lovell, 12 The Holloway, Swindon, Dudley, DY3 4NT
G1 BLK C Ridley, 14 Painswick Road, Hall Green, Birmingham, B28 0HH
G1 BLO Norman Swan, 8 Tyrrells Court, Bransgore, Christchurch, BH23 8BU
G1 BLQ P Lloyd, 35 Westfield Road, Hertford, SG14 3DL
G1 BLV S Davies, Swallow Cottage, The Chantry, Leyburn, DL8 4NA
GM1 BLX I Dewar, 11 Abbotshall Road, Kirkcaldy, KY2 5PH
G1 BMB G Fallows, 66 Ulverston Road, Swarthmoor, Ulverston, LA12 0JF
G1 BMN N Lamb, 106 St. Davids Road, Leyland, PR25 4XY
G1 BMP S Norminton, 63 Candish Drive, Plymouth, PL9 8DB
G1 BMT P Tuthill, 12 Herbert Road, Salisbury, SP2 9LF
G1 BMW P Winterton, 35 Paynesfield Road, Tatsfield, Westerham, TN16 2AT
GM1 BNA R Main, 14 Hunters Grove, East Kilbride, G74 3HZ
G1 BNE Andrew Perkins, 10 Roman Gardens, Houghton Regis, Dunstable, LU5 5QR
G1 BNG S Marsh, 28 Orcheston Road, Bournemouth, BH8 8SR
G1 BNN Stuart Tilly, 24 Whinham Way, Morpeth, NE61 2TF
GM1 BNP R Holt, Whitlam Farmhouse, Newmachar, Aberdeen, AB21 0RS
GM1 BNS W Graham, 6 Braemar View, Clydebank, G81 3RR
G1 BNV S Henderson, 7 Havering, Castlehaven Road, London, NW1 8TH
G1 BNX R Huxley, 83 Gleneagles Road, Wyken, Coventry, CV2 3BH
G1 BOB C Balsdon, 4 Queens Hayes, Willey Lane, Okehampton, EX20 2NG
G1 BOO F Crompton, 24 Alcester Road, Sale, M33 3QP
GM1 BOT G Ogg, 65 Downie Park, Dundee, DD3 8JW
G1 BOX M Platten, 48 Brier Road, Sittingbourne, ME10 1YL
G1 BPD Ian Attridge, 12 Ascot Road, Orpington, BR5 2JF
G1 BPE M Barwick, 32 St. Georges Road, Harrogate, HG2 9BS
G1 BPS Martin Rowell, 1 Willow Street, Haslingden, Rossendale, BB4 5NA
G1 BPU L Staal, 5 Hunt Court, 236 Chase Side, London, N14 4PG
G1 BPV A Stalker, 68 Finch Drive, Seaford, NG34 7US
G1 BQG A Bright, 15 Cross Road, Maldon, CM9 5EE
G1 BQH John Murphy, 34 Knights Hill, Walsall, WS9 0TG
G1 BQI G Smith, 92 Lime Road, Accrington, BB5 6BJ
GM1 BQP William Macrobbie, 44 Moray Park Terrace, Culloden, Inverness, IV2 7RW
G1 BQQ J Sowerbutts, 22 Worsley Street, Accrington, BB5 2PA
G1 BQR M Spinks, 26 Church Hill, Royston, Barnsley, S71 4NH
G1 BQV Michael Sunderland, 36 Moorlands Avenue, Yeadon, Leeds, LS19 6AD
G1 BRB Adrian Patton, 72 Sanctuary Way, Grimsby, DN37 9RZ
G1 BRD J Smith, 127 Wolverhampton Road, Cannock, WS11 1AR
G1 BRF P Williams, 292 Hagley Road, Hasbury, Halesowen, B63 4QG
G1 BRP K Crawley, 19 Park Mount, Harpenden, AL5 3AS
G1 BRS BOURNEMOUTH RS c/o M Stevens, 16 Golf Links Road, Ferndown, BH22 8BY
GM1 BSG John Ross, 16 Myreton Drive, Bannockburn, Stirling, FK7 8PX
GI1 BSJ J Cunningham, 4 Garvaghy Road, Portglenone, Ballymena, BT44 8EF
G1 BSY K Morris, 3 Moravian Close, Dukinfield, SK16 4EW
G1 BSZ R Nash, Roann, Bedmond Road, Hemel Hempstead, HP3 8SH
G1 BTF Alvin Hardy, 14 Parsonage Road, Rainham, RM13 9LW
G1 BTI K Forrest, 61 Woodbury Road, Bridgwater, TA6 7LJ
GM1 BTL William Erskine, 30 Market Road, Kirkintilloch, Glasgow, G66 3JL
G1 BTN I Evans, 37 Lyndale Avenue, Lostock Hall, Preston, PR5 5UU
G1 BTV David Holt, 2 London Heights, Dudley, DY1 2QZ
G1 BUJ A Bates, 17 Walkers Heath Road, Kings Norton, Birmingham, B38 0AB
G1 BUQ J Spink, 38 Hemlingford Road, Sutton Coldfield, B76 1JQ
G1 BUV M Osborne, 5 Wells Road, Riseley, Bedford, MK44 1DY
GM1 BUY C Rose, 14 Spoutwells Drive, Scone, Perth, PH2 6RR
GM1 BVA L Nieto, 10 Gardrum Gardens, Shieldhill, Falkirk, FK1 2TB
GM1 BVT W Pettett, 15 Strude Howe, Alva, FK12 5JU
G1 BVV G Pemberton, 8 Hotchin Road, Sutton-on-Sea, Mablethorpe, LN12 2NP
G1 BWG R Dodd, 5 Halesworth Road, Wolverhampton, WV9 5PH
G1 BWH Peter Eaton, 8 Chester Road, Barnwood, Gloucester, GL4 3AX
G1 BWI J Eastham, 81 Park Lee Road, Blackburn, BB2 3NZ
G1 BWP W Webb, The New Bungalow, Tram Road, Coleford, GL16 8DN
GU1 BWW D Ash, Trigale House, La Trigale, St Anne, Guernsey, GY9 3TX
G1 BWZ Nigel Fieldsend, 47 Hollycroft, Barmston, Driffield, YO25 8PP
GM1 BXI R Clark, 5 Abbotsfield Terrace, Auchterarder, PH3 1DD
G1 BXL Nigel Mann, 332 Stonehouse Lane, Quinton, Birmingham, B32 3AL
G1 BXQ Jonathan Smith, Noonhill Farm, Grove Road, Coventry, CV7 9JE
G1 BXT I Thomas, Myrtle Cottage, Swan Lane, Ashford, TN25 6EB
G1 BXX Adrian Gillard, 10 Parc Pendre, Brecon, LD3 9ES
G1 BYI J Moore, 11 Sherborne Road, Wallasey, CH44 2EY
G1 BYJ G Noon, 48 Sanderling Close, Letchworth Garden City, SG6 4HY
G1 BYO S Slaughter, 96 Adys Road, London, SE15 4DZ
G1 BYP P Snitch, 79 Albion Avenue, York, YO26 5QZ
G1 BYQ D Hatton, 34 Avocet Way, Bicester, OX26 6YP
G1 BYS A Kempton, 14 Lower Gravel Road, Bromley, BR2 8LT
G1 BYT N Kinselley, 5 Helford Close, Bedford, MK41 7TU
G1 BZD P Ward, 8 Dingle Drive, Droylsden, Manchester, M43 7NR
G1 BZE C Weeds, 2 Kendray Close, Belper, DE56 0EY
G1 BZM P Endean, 11 Forrester Drive, Brackley, NN13 6NE
G1 BZR D Cameron, 14 Queen Street, Castle Douglas, DG7 1HX
GI1 BZT Leslie Gornall, 14 Ballymoghan Lane, Magherafelt, BT45 6HW
G1 BZU Royal Naval ARS c/o Joe Kirk, 111 Stockbridge Road, Chichester, PO19 8QR
G1 BZW R Kimber, 38 Greenmere, Brightwell-cum-Sotwell, Wallingford, OX10 0QG
GI1 CAI J McBride, Dunaree, 20 Oldcastle Road, Omagh, BT78 4HX
G1 CAN P Shonfield, 242 Chickerell Road, Weymouth, DT4 0QY
G1 CAY D Shea, 38 Ranworth Avenue, Hoddesdon, EN11 9NR
G1 CBB L Walker, The Biel, Furze Vale Road, Bordon, GU35 8EP
G1 CBK Stephen Mole, 53 Parkfield Road, Rainham, Gillingham, ME8 7TA
G1 CBL L Mason, Reflow, Unit 2 Spring Lane North, Malvern, WR14 1BU
G1 CBS J Hatt, 77 Pentland Close, Basingstoke, RG22 5BQ
G1 CBY Sarah Fountaine, 142 Elvaston Road, North Wingfield, Chesterfield, S42 5GA
GM1 CCI C Watson, 11 Ladybridge Houses, Banff, AB45 2JR
G1 CCM T McMillan, 47 Sandsend Road, Eston, Middlesbrough, TS6 8AF
GM1 CCN Cleland Orr, Easter Cowden Farm, Dalkeith, EH22 2NS
G1 CCW F Haselden, 7 Chestnut Avenue, Gosfield, Halstead, CO9 1TD
G1 CCX P Kennedy, 30 Firs Drive, Harrogate, HG2 9HB
G1 CDH D Davies, 10 Bryn Castell, Abergele, LL22 8QA
G1 CDN D Wormall, 20 Greenfield Road, Hemsworth, Pontefract, WF9 4RL
G1 CDO R Wormall, 17 Newstead Grove, Fitzwilliam, Pontefract, WF9 5DS
G1 CDQ A Sturman, 3 Windward House, 73, Lytham St. Annes, FY8 1LZ
G1 CDY D Forth, 20 West Common Crescent, Scunthorpe, DN17 1DJ
G1 CEI Peter Hirons, Flat 4, 423 Winchester Road, Southampton, SO16 7DE
GM1 CEJ R Stout, 16 Ardoch Park, Balgeddie, Glenrothes, KY6 3PJ
G1 CEO R Day, 17 Barry Avenue, Bicester, OX26 2DZ
GI1 CET Jim Barr, 2 Willowvale Close, Islandmagee, Larne, BT40 3SD
G1 CEU John Clarke, 31 Station Road, Ormesby, Great Yarmouth, NR29 3NH
G1 CFA P Middleton, 5 Fieldhouse, Holmfirth, HD9 1EN
G1 CFB K Rhodes, 34 Bannister Drive, Hull, HU9 1EJ
G1 CFE Alan Wyatt, 32 Wensleydale Avenue, Blackpool, FY3 7RS
G1 CFG Christopher Rigby, 4 Humber Street, Longridge, Preston, PR3 3WD
G1 CFJ G Gardner, 165 Brookhouse Road, Brookhouse, Lancaster, LA2 9NY
G1 CFK D Bowles, Fiddlers Nook, Thurston Road, Bury St. Edmunds, IP31 2PL
G1 CFZ John Bottomley-Mason, 13 Santa Monica Road, Bradford, BD10 8QX
G1 CGD S Oliver, 21 Hillside Court, Holywell, CH8 7PJ
G1 CGH A Rawlins, 4 Low Park, West Woodburn, Hexham, NE48 2SQ
G1 CGJ M Davies, Kuling, Bridgwater Road, Winscombe, BS25 1NB
G1 CGP A Gillard, 28 Moor Tarn Lane, Walney, Barrow-in-Furness, LA14 3LP
G1 CGU G Fitzpatrick, 24 Spring Gardens, Alcombe, Minehead, TA24 6BH
G1 CHE David OGarr, 10 Ellastone Grove, Stoke-on-Trent, ST4 5EE
G1 CHM C Milburn, Field House, Copper Hill, Hayle, TR27 4LY
G1 CHN A James, The Red House, Gandish Road, Colchester, CO7 6TP
G1 CHQ M Wells, 7 Vint Rise, Idle, Bradford, BD10 8PU
GW1 CHS John Hughes, Reservoir House, St. Lythan's, Cardiff, CF5 6BQ
GM1 CHT A Hyde, 19 Drum Brae Gardens, Edinburgh, EH12 8SY
G1 CHV C Compton, 21 Vange Riverview Centre, Vange, Basildon, SS16 4NE
G1 CIA M Ferentiuk, 74 Fallowfield Drive, Rochdale, OL12 6LZ
G1 CIM S Hancock, Monrad, Back Street, Gainsborough, DN21 3DL
G1 CIT M WHALLEY, 6 Rookery Walk, Clifton, Shefford, SG17 5HW
G1 CIV D Owen, 23 Mannings Drive, Hinckley, LE10 0LG
GW1 CIY Gareth Evans, Maes yr Haf, Beulah, Newcastle Emlyn, SA38 9QB
G1 CJC L Gilbert, Holmefield Cottage, Oker, Matlock, DE4 2JJ
G1 CJI Paul Arnold, 36 Gopsall Road, Hinckley, LE10 0DY
G1 CJJ P Williams, 6 Parc Ffynnon, Llysfaen, Colwyn Bay, LL29 8SA
G1 CJK Robert Baiey, 318 Plumstead Common Road, London, SE18 2RT
G1 CKF Julie Darby, 97 Littlehaven Lane, Roffey, Horsham, RH12 4JE
G1 CKJ T Martin, 201 Gloucester Road, Kidsgrove, Stoke-on-Trent, ST7 4DQ
G1 CKR T Miller, 21 Lighton Close, Hereford, HR1 1UH
G1 CKT R Nelson, 11 Meadow Way, Plymouth, PL7 4JB
GI1 CKU T Gardiner, 17 Grange Valley Gardens, Ballyclare, BT39 9HE
G1 CKV N Derbyshire, 54 Windy Arbor Road, Whiston, Prescot, L35 3SG
G1 CKY P Turner, 16 Pendragon Way, Leicester Forest East, Leicester, LE3 3EY
G1 CLD S Patterson, Dunedin, Little Ness, Shrewsbury, SY4 2LG
G1 CLJ P Kennedy, 12 Newbroke Road, Rowner, Gosport, PO13 9UJ
G1 CLT R Bokor, 75 Attlee Road, Middlesbrough, TS6 7NA
G1 CMC Andrew Lickley, 19 Sandy Rise, Selby, YO8 9DW
GM1 CMF Peter Carnegie, 29 Dalgetty Court, Muirhead, Dundee, DD2 5QJ
G1 CMH J Norwood, Flat 28 The Manor, Church Road, Gloucester, GL3 2HT
G1 CMZ Stephen Lewkowicz, 7 The Mart, Locking Road, Weston-Super-Mare, BS23 3DE
GM1 CNH Norman Stewart, 160 Carrick Knowe Drive, Edinburgh, EH12 7EW
G1 CNI S Dwyer, Po Box 44, Tahmoor, Australia, 2573
G1 CNN Philip Beeson, Flat 6, Oxford Court, London, W3 0HH
G1 CNV T Thornton, Orchard Walk, 23 Crookham Road, Fleet, GU51 5DP
G1 CNZ Robert Reid, 34 Wellesley Street, Taunton, TA2 7DT
GM1 COF Paul McGowan, 38 Mckenzie Crescent, Lochgelly, KY5 9LT
G1 COV COVENTRY RAYNET GROUP c/o D Green, 67 Coombe Park Road, Binley, Coventry, CV3 2NW
G1 COW R Penfold, 1 Padworth Road, Burghfield Common, Reading, RG7 3QE
G1 COX A Berkeley, 42 Ringley Drive, Whitefield, Manchester, M45 7LR
G1 COY C Robson, 43 Longdyke Drive, Carlisle, CA1 3HT
G1 CPA P Nairne, 137 Barden Road, Tonbridge, TN9 1UX
G1 CPC Jeremy Arthur, St. Aubin, Plomer Green Lane, High Wycombe, HP13 5XN
G1 CPD G Ghetti, 7 Rue De Provence, Paris, France, 75009
G1 CPM Edward Rose, 26 Lavender Way, Bourne, PE10 9TT
G1 CPO Colin Haygarth, 3 Rew Close, Ventnor, PO38 1BH
G1 CPU Gregory Milligan, 163 Grindle Road, Longford, Coventry, CV6 6DS
G1 CPX Ian Clarke, 19 Welbeck, Bracknell, RG12 8UQ
G1 CQA Robert Chaney, 55 Bartlow Road, Linton, Cambridge, CB21 4LY
GM1 CQC H Smith, 601 Ferry Road, Edinburgh, EH4 2TT
G1 CQF P Simms, 141 Brays Road, Sheldon, Birmingham, B26 2UL
G1 CQG David Perry, 5 Beech Hill, Wellington, TA21 8ER
G1 CQR D Fuller, 26 Longfields, Ely, CB6 3DN
G1 CQT P Turley, 35 Alwinton Avenue, Stockport, SK4 3PU
G1 CRN W Murray, 91 Chaucer Avenue, Hounslow, TW4 6NA
G1 CRT W Cambridge, 36 Selwyn Avenue, Richmond, TW9 2HA
G1 CSA J Walton, 23 Keighley Avenue, Sunderland, SR5 4BU
G1 CSN R Beasley, 26 Retford Close, Harold Hill, Romford, RM3 9NA
G1 CSO John Dent, 18 New Street, St Neots, PE19 1AE
G1 CSR CIVIL SERVICE ARS c/o Neil Sanderson, 54 Kelvedon Close, Chelmsford, CM1 4DG
G1 CSS M Wilson, 210 London Road, Worcester, WR5 2JT
G1 CSY E Hartley, 2 Lamberts Close, Weasenham, King's Lynn, PE32 2TE
G1 CTO David Powell, 88 Church View, Chirk, Wrexham, LL14 5PF
G1 CTQ Nicolo Losardo, 14 Arnside Close, Clayton le Moors, Accrington, BB5 5GG
G1 CUC H Mattinson, 11 Riverside Park, Canonbie, DG14 0UY
G1 CUG DA Laughton, 2 Stamford Road, Careby, Stamford, PE9 4EB
G1 CUH Brian Reid, 32 Arlington Drive, Alvaston, Derby, DE24 0AU
G1 CUM J Rawlings, Castle House, Barrow Haven, Barrow-upon-Humber, DN19 7EY
GW1 CUQ Nigel Paull, 6 Llys Caradog, Creigiau, Cardiff, CF15 9JP
G1 CUZ S Seal, Crantock, Bellingdon, Chesham, HP5 2XW
G1 CVR D Paine, 34 Brockenhurst Street, Burnley, BB10 4ET
G1 CWD P bullough, 11 Druids View, Bingley, BD16 2DY
G1 CWI M Kemp, Casa Lucia, Vale Formosilho, S.Marcos Da Serra, Portugal, P8375
G1 CWJ A Burton, Avenida Rio Ebro, 48. Plot 101, Alicante, Spain, 03170 ROJALES
G1 CWQ Brian Wyatt, 3 Shipley Close, Blackpool, FY3 7UJ
G1 CWW G Stone, 37 Canterbury Drive, Ashby-de-la-Zouch, LE65 2QQ
G1 CWZ D Penrose, 7 Two Ashes, Bayston Hill, Shrewsbury, SY3 0QF
G1 CXQ C Roberts, 11 Adel Wood Drive, Leeds, LS16 8JQ
G1 CYQ B Wheeldon, 27 Lawrence Walk, Newport Pagnell, MK16 8RF
G1 CYY T Brien, 54 Central Avenue, Fartown, Huddersfield, HD2 1DA
G1 CZH P Harkins, 27 Hillfoot Green, Liverpool, L25 7UH
G1 CZN P Burrows, 7 Eton Terrace, Ince, Wigan, WN3 4NS
G1 CZU M Abram, 28 Langport Drive, Vicars Cross, Chester, CH3 5LY
G1 CZW R Silcocks, 69 Kennaway Road, Clevedon, BS21 6JJ
G1 DAE I Rusby, 12 Park Meadow, Princes Risborough, HP27 0EB
G1 DAK S Felton, 8 Clancutt Lane, Coppull, Chorley, PR7 4NS
G1 DAT P Burnett, 14 Hollywalk Drive, Middlesbrough, TS6 0PL
G1 DAU G Speirs, 19 Poets Way, Winchester, SO22 5BX
G1 DAV D Forsey, 3 Northwood Drive, Newbury, RG14 2HB
G1 DAX P Costigan, 10 The Paddock, Clevedon, BS21 6JU
G1 DAZ S Burchell, 31 Thornton Road, Girton, Cambridge, CB3 0NP
G1 DBH B Cobb, 28 Sandringham Road, Newton Abbot, TQ12 4HA
G1 DBI G Doig, 78 Plane Tree Drive, Crewe, CW1 4ES
G1 DBL L Owen, 27 Coniston Drive, Holmes Chapel, Crewe, CW4 7LA
G1 DBR D Ross, 113 Nun House Drive, Winsford, CW7 3LE
G1 DBZ R Cooper, 31 Erskine Crescent, Sheffield, S2 3LQ
GM1 DCB Marian Senior, The Raw, Bridgend, Isle of Islay, PA44 7PZ
G1 DCI Justin Griffith, My Home, Highworth Road, Swindon, SN3 4SF
G1 DCU P Gardner, 40 Wynall Lane South, Stourbridge, DY9 9AH
G1 DCX M Race, 76 Lonsdale Road, Stamford, PE9 2SG
G1 DCY S Richmond, 1042 Evesham Road, Astwood Bank, Redditch, B96 6ED
G1 DCZ J Sandall, 7 St. Road, Compton Dundon, Somerton, TA11 6PX
G1 DDA Frederick Wood, 7 Yew Tree Park, The Rowe, Newcastle, ST5 4EN
G1 DDF Francis Griffin, 77 Widmore Drive, Hemel Hempstead, HP2 5JL
G1 DDI C Cadman, 32 Breedon Hill Road, Derby, DE23 6TG
G1 DDK M Abraham, Skywave Marine Services, Unit 1, The Arcade, Falmouth, TR11 2TD
G1 DDR R Oakley, 20 Halton Lane, Wendover, Aylesbury, HP22 6AR
G1 DDS D Seccombe, 14 Millfield, Bedlington, NE22 5DZ
G1 DEN P Edinburgh, 77 Westerley Lane, Shelley, Huddersfield, HD8 8HP
G1 DEO Brian Davies, 12 Woodbine Close, Newport, PO30 1AF
G1 DEP J Dunhill, 8 Brentwood Avenue, Thornton-Cleveleys, FY5 3QR
G1 DEQ D Gilbey, 7 Victory Way, Cottenham, Cambridge, CB24 8TG
G1 DER J Hacker, 4 Foxglove Close, Bamber Bridge, Preston, PR5 6XR
GD1 DES D Smith, 14 College Drive, Ruislip, HA4 8SB
G1 DEU D Hagger, 7 Pecockes Close, Great Cornard, Sudbury, CO10 0NQ
G1 DEV E Hardwick, 5 Seaview, Oakmere Park, Little Neston, CH64 0XB
G1 DEX H Irvin, 15 Rock Edge, Knowler Hill, Liversedge, WF15 6DY
G1 DEY Christopher Jacob, 10 Wynchgate, Southgate, London, N14 6RR
G1 DEZ P Baxter, 27 Manor Crescent, Brinsworth, Rotherham, S60 5HG
G1 DFF M Smith, 22 Cedars Avenue, Wombourne, Wolverhampton, WV5 0JX
G1 DFI J Swift-Hook, 12 Warwick Drive, Newbury, RG14 7TT
G1 DFM A Westlake, 47 Quarry Road, Kingswood, Bristol, BS15 8NZ
G1 DFN Frederick Wright, 1 Old Engine Houses, Brusselton, Shildon, DL4 1QA
G1 DFP G Fielding, 35 Amos Avenue, Litherland, Liverpool, L21 7QH
G1 DFR Alaister Gemmill, 32 Larks Rise, Cleobury Mortimer, Kidderminster, DY14 8JJ
G1 DFT Ian Hampson, 293 Sandbrook Road, Southport, PR8 3RP
G1 DFW D Hoare, 51 Hartington Road, Dronfield, S18 2LE
G1 DFZ R Jobbins, 8 Newark Road, Hartlepool, TS25 2LA
G1 DGL Richard Simpson, 51 Ramleaze Drive, Salisbury, SP2 9PA
G1 DGW I Johnson, 24 York Road, Maghull, Liverpool, L31 5NL
G1 DGY A Koch, 65 Collier Lane, Ockbrook, Derby, DE72 3RP
G1 DHB Robin Fagence, 5 Balmoral Close, Billericay, CM11 2LL
G1 DHM G Miller, 32 Belbroughton Close, Lodge Park, Redditch, B98 7NH
G1 DHQ D Palmer, 60 Heathcote Drive, Sileby, Loughborough, LE12 7ND
G1 DHY N Roe, 41 Highfield Lane, Chaddesden, Derby, DE21 6PH
G1 DIA Peter Rowe, 5 Bramble Close, Great Boughton, Chester, CH3 5XN
G1 DIF Devon Data Group c/o Donald Roomes, View Field, Milton Damerel, Holsworthy, EX22 7NY
G1 DIG S Cadman, 71 Gayfield Avenue, Withymoor, Brierley Hill, DY5 2BU
G1 DIK Alexis Smith, Windycross, Newbourne Road, Woodbridge, IP12 4PT
G1 DIM C Smith, 37 Ivory Close, Tuffley, Gloucester, GL4 0QY
G1 DIO Annette Heath-Anderson, 12 The Medway, Daventry, NN11 4QU
G1 DIR A Wilkinson, 15 St. Margarets Grove, Leeds, LS8 1RZ
G1 DJI John Short, 7 Bushfields, Loughton, IG10 3JT
G1 DJQ N Lofthouse, Cambridge Park, 8 Abbott Clough Avenue, Blackburn, BB1 3LP
G1 DJU C Whitby, 7 Wentworth Way, Stoke Bruerne, Towcester, NN12 7SA
G1 DKE Malcolm Spry, 71 High Street, Topsham, Exeter, EX3 0DY
G1 DKI Machiel Lindenbergh, 26 Manston Drive, Perton, Wolverhampton, WV6 7LX
G1 DKK J Price, Maen Dylan, Pontllyfni, Caernarfon, LL54 5EF
G1 DKV G Charlton, 20 Bailey Crescent, South Elmsall, Pontefract, WF9 2TL
G1 DKX A Thomas, 92 Singleton Crescent, Goring-by-Sea, Worthing, BN12 5DJ
G1 DKY J Miller, 40 Central Avenue, Herne Bay, CT6 8RX
G1 DLA R Deacon, 22 Islip Gardens, Northolt, UB5 5BX
G1 DLB J Desborough, 106 Grand Avenue, Lancing, BN15 9QD
G1 DLH M Ogle, 22 Warwick Street, Daventry, NN11 4AL
G1 DLJ G Hope, 3 Farm Crescent, Sittingbourne, ME10 4QD
G1 DLP W Jones, 160 Christchurch Road, Newport, NP19 7SA
GM1 DLS Craig Barry, 32 Prospect Drive, Ashgill, Larkhall, ML9 3AJ
G1 DMH Leonard Lees, 3 Ockbrook Court, Muskham Avenue, Ilkeston, DE7 8EY

G1 DMN P Snow, 14 Beechwood Avenue, Darlington, DL3 7HP
G1 DMR Richard Manser, 53 Downs Barn Boulevard, Downs Barn, Milton Keynes, MK14 7LL
G1 DMS D Segal, Flat 1, Masons House, London, NW9 0NG
G1 DMW F Latham, Higher Lane, Parbold, Wigan, WN8 7HA
G1 DNA John Birkmyre, Swarland, Morpeth, NE65 9JW
G1 DNI W Darling, 2 Strathaird Avenue, Walney, Barrow-in-Furness, LA14 3DE
G1 DNK B Cunningham, 14 Leeson Drive, Ferndown, BH22 9QQ
G1 DNO Christopher Birtchnell, Linnetts Roost End, Sturmer, Haverhill, CB9 7XW
G1 DNP R Collins, 12 Bean Oak Road, Wokingham, RG40 1RL
G1 DNT Mark Cole, 52 Lower Meadow, Quedgeley, Gloucester, GL2 4YY
G1 DNY R Clay, 38 Hubbards Road, Chorleywood, Rickmansworth, WD3 5JJ
G1 DNZ Graham Clarke, 150 Minver Crescent, Nottingham, NG8 5PN
G1 DOA K Chappell, 17 Linton Close, Winyates, Redditch, B98 0NA
G1 DOG I [illegible], 118 Uttoxeter Road, Blythe Bridge, Stoke-on-Trent, ST11 9NT
G1 DOJ T Brodrick, 16 Wallenge Drive, Paulton, Bristol, BS39 7PX
G1 DOL Robert Breakspear, 7 Woodside, North Leigh, Witney, OX29 6SQ
G1 DON Donald [illegible], 50 MacDonald Street, Orrell, Wigan, WN5 0AJ
G1 DOT David Barker, 80 Rolvenden Road, Wainscott, Rochester, ME2 4PG
G1 DOX John Acton, 63 Bevington Close, Patchway, Bristol, BS34 5NP
G1 DPI Anthony Barratt, 23 Wilberforce Road, South Anston, Sheffield, S25 5EG
G1 DPJ C Beasley, 12 East Leys Court, Moulton, Northampton, NN3 7TX
GW1 DPL Martyn Beer, 67 Killan Road, Dunvant, Swansea, SA2 7TH
G1 DPN D Bettany, 10 Redbrook Crescent, Melton Mowbray, LE13 0EU
G1 DPT Thomas Cairney, Dolgoch, Hall Lane, Lutterworth, LE17 5RP
GW1 DPV Miles Carter, 14 Cerdin Avenue, Pontyclun, CF72 9ER
G1 DPW S Cmoch, 25 Monro Place, Epsom, KT19 7LD
G1 DPX R Colley, 12 Glenfield Road, Banstead, SM7 2DG
G1 DQD C Anderson, 11 Swallowfield Drive, Hull, HU4 6UG
G1 DQF P Buckmaster, 7 Yew Tree Close, New Ollerton, Newark, NG22 9UP
G1 DQL R Bradley, 1 Audley Place, Sutton, SM2 6RW
GD1 DQQ D Dwight, 19 The Highway, Stanmore, HA7 3PL
G1 DQU M Elliott, 52 Wellfield Road, Alrewas, Burton-on-Trent, DE13 7EZ
GW1 DQV Brian Emary, 2 The Paddocks, Penarth, CF64 5BW
G1 DRG Gareth Foster, 19 Asquith Avenue, Burnholme, York, YO31 0PZ
G1 DRI K Gill, 33 Hazel Croft, Werrington, Peterborough, PE4 5BJ
G1 DRP I Hand, 28 Chartwell Close, Werrington, Stoke-on-Trent, ST9 0PQ
G1 DRR N Hamilton, 8 Sudbury Court, Mansfield, NG18 3RZ
G1 DRW David Hart, 71 Breinton Road, Hereford, HR4 0JY
G1 DRY S Cox, 25 Church Close, Stoke St. Gregory, Taunton, TA3 6HA
G1 DSA J Critchley, 3 Beaconsfield View, Robert Road, Slough, SL2 3XT
G1 DSB J Darling, 145 Hartlands, Bedlington, NE22 6JJ
G1 DSF A Daw, 19 Rowan Close, Yarnfield, Stone, ST15 0EP
G1 DSG M Degerdon, 25 Rosslyn Road, Billericay, CM12 9JN
G1 DSJ P Morgan, 29 Brisbane Road, Reading, RG30 2PE
GM1 DSK David Keay, Parkhill, Cromwell Park, Perth, PH1 3LW
G1 DSM A Hicks, 5 Restwell Avenue, Cranleigh, GU6 8PQ
G1 DSP SPALDING & DISTRICT ARC c/o J Hill, The New Gatehouse, Gubboles Drove, Spalding, PE11 4AX
G1 DSZ J Phillips, 20 The Meadows, Broomfield, Herne Bay, CT6 7XF
GW1 DTA M Pilot, 92 Llanllienwen Road, Cwmrhydyceirw, Swansea, SA6 6LU
G1 DTE W Merz, 38 Lime Avenue, Colchester, CO4 3NL
G1 DTF A Middleton, 2 Beccles Way, Bramley, Rotherham, S66 2SJ
G1 DTS E Kier, 9 Newbridge Way, Truro, TR1 3LX
G1 DUI P Norman, 3 Church View, Witchford, Ely, CB6 2HH
G1 DUJ B Oakley, 6 Staplehurst Gardens, Cliftonville, Margate, CT9 3JB
G1 DUO P Richards, 114 Northleach Close, Redditch, B98 8RD
G1 DUS D Roberts, Westpark, 296 Westleigh Lane, Leigh, WN7 5PW
G1 DUT J Robertson, Everslea, Congleton Road, Congleton, CW12 2LL
G1 DVA Paul Middlehurst, 7 Statham Drive, Lymm, WA13 9NW
G1 DVD B Marshall, 1 Anglers Way, Chesterton, Cambridge, CB4 1TZ
G1 DVH J Knighton, 90 Sherwood Crescent, Market Drayton, TF9 1NP
G1 DVO C Hepworth, 20 Station Avenue, Duns, TD11 3HW
G1 DVU J Green, 788 The Ridge, St. Leonards-on-Sea, TN37 7PS
G1 DWC Robert Everett, 73 Fordwych Road, London, NW2 3TL
GU1 DWO A Smith, La Cambrette, La Rue Des Reines, Forest, Guernsey, GY8 0JB
G1 DWT J Dwight, 59 Highfield Road, Bramley, Leeds, LS13 2BX
G1 DWU A Swales, 90 Earlswood Road, Dorridge, Solihull, B93 8RN
G1 DXD P Darke, 18 Colchester Close, Southend-on-Sea, SS2 6HR
G1 DXH R Crissell, 1 Medlar Drive, South Ockendon, RM15 6TS
G1 DXM David Walling, 37 Ulverston Road, Swarthmoor, Ulverston, LA12 0JB
G1 DXQ Paul Postle, 20 Courtenay Close, Norwich, NR5 9LB
G1 DYC D Winkley, Southall Cottage, Hadley, Droitwich, WR9 0AU
G1 DYL K Tysoe, 5 Greenbank, Droitwich, WR9 7QS
G1 DYN Jeffery Snowling, 5 Verbena Close, Beechwood, Runcorn, WA7 3JA
G1 DYQ N Prosser, 35 Holmfirth Close, Belmont, Hereford, HR2 7UG
G1 DYR R Munday, 12 Glisson Road, Hillingdon, Uxbridge, UB10 0HH
G1 DYT P Allen, 8 Ashmead Crescent, Birstall, Leicester, LE4 4GS
G1 DZB N Babbage, 248 Molesey Avenue, West Molesey, KT8 2ET
G1 DZD L Ball, 14 St. Wilfrids Road, Burgess Hill, RH15 8BD
G1 DZY K Bricknall, 21 Uplands Way, Springwell Village, Gateshead, NE9 7NQ
G1 DZZ K Bridle, 8 Hardy Avenue, Dorchester, DT1 1LL
G1 EAB A Bolton, 5 Willow Crescent, Gedling, Nottingham, NG4 4BL
G1 EAE A Broughton, Cobwebs, The Fleet, Pulborough, RH20 1HS
GM1 EAH W Buchanan, 38 Kenmount Place, Kennoway, Leven, KY8 5LT
G1 EAJ M Bunting, 22 Ling Close, Coltishall, Norwich, NR12 7HZ
G1 EAM A Bush, 15 Pelton Avenue, Sutton, SM2 5NN
G1 EAN A Butler, Ty Ni, Hall Lane, Leamington Spa, CV33 9HG
G1 EAV S Davies, Laburnum House, Guilsfield, Welshpool, SY21 9PX
G1 EAX R Dawkins, 17 Dacer Close, Stirchley, Birmingham, B30 3BZ
G1 EBB A Di Duca, 15 Moray Close, Halesowen, B62 9PP
G1 EBP C Jermany, 5 Lexington Close, Hemsby, Great Yarmouth, NR29 4ES
G1 EBT Alistair Jones, 56 Copthorne Drive, Shrewsbury, SY3 8RX
G1 EBV S Dawswell, 66 Priory Walk, Leicester Forest East, Leicester, LE3 3PP
G1 EBW P Challen, 20 Drummond Road, Cawston, Rugby, CV22 7TN
G1 EBX Steven Challen, 35 Hazel Crescent, Towcester, NN12 6UQ
G1 EBZ R Charlton, Meer Booth Rd, Boston, PE22 7AB
G1 ECC D Chippendale, 19 East Park Avenue, Darwen, BB3 2SQ
G1 ECE B Clark, 9 Conigree, Chinnor, OX9 4JY
G1 ECI J Christy, 1 Edinburgh Drive, Hindley Green, Wigan, WN2 4HL
G1 ECK Nessa Preval, 63 Dudley Avenue, Leicester, LE5 2EF
G1 ECS C Frettsome, 16 Botany Avenue, Mansfield, NG18 5NG
G1 ECV J Gardener, 32 Beckington Crescent, Chard, TA20 2BU
G1 ECY G Giles, 74 The Larches, Uxbridge, UB10 0DN
G1 EDA R Goff, 21 Findon Road, Elson, Gosport, PO12 4EP
G1 EDE David Graydon, 16 Long Row, Port Mulgrave, Saltburn by The Sea, TS13 5LF
G1 EDH T Hacker, 179a Churchill Avenue, Chatham, ME5 0DQ
G1 EDK P Hammond, 160 Westlands Caravan Park, Herne Bay, CT6 7LE
G1 EDM J hargreaves, 5 Nuttall Avenue, Little Lever, Bolton, BL3 1PW
G1 EDP M Hazell, 15 Lords Hill, Coleford, GL16 8BG
G1 EDT W Hewitt, 99 Derrydown Road, Perry Barr, Birmingham, B42 1RY
G1 EDU A Hicks, 20 Victoria Road, Meltham, Holmfirth, HD9 5NL
G1 EDX Michael Holtam, 18 Cowley Close, Cheltenham, GL51 6NP
G1 EEA S Howcroft, 23 Alderley Avenue, Blackpool, FY4 1QG
G1 EEO M Kirby, Church Cottage, Burrington, Umberleigh, EX37 9JG
G1 EEZ J Lewis, 51b Wellsway, Bath, BA2 2UD
G1 EFF A Marriott, 75 St Johns Road, Cudworth, Barnsley, S72 [illegible]
G1 EFG Christopher Mcara, 6 Winniford Close, Chideock, Bridport, DT6 6SA
G1 EFK G Means, Ferry Farm, Witham Bank, Lincoln, LN4 4QA
G1 EFL Martyn Medcalf, 47 Paddock Drive, Chelmsford, CM1 6UX
G1 EFO Paul Hyde, 24 Grassam Close, Preston, Hull, HU12 8XF
G1 EFP A Jarrett, 4 Langstone Close, Horwich, Bolton, BL6 5SZ
G1 EFS S Newell, 7 Edward Road West, Walton Park, Clevedon, BS21 7DY
G1 EFT Peter Nicholson, 20 Rowley Road, Torquay, TQ1 4PX
G1 EFU Alan Nixon, 14 Carlton Road, Lowton, Warrington, WA3 2EP
G1 EFX C Nutkins, Higher Spence, Bridport, DT6 6DF
G1 EGB G Page, 23 Maskelyne Close, Battersea, London, SW11 4AA
G1 EGE K Pay, 34 Vaudrey Crescent, Congleton, CW12 3HP
G1 EGI Donald Phillips, Bethune, Rame Cross, Penryn, TR10 9DZ
G1 EGK J Preece, The Grange, Harewood Road, Wetherby, LS22 5BL
G1 EGL Robert Preston, 45 Gaynor Close, Wymondham, NR18 0EA
G1 EGZ Andrew Adams, Radnor, Shorts Road, Carshalton, SM5 2PB
G1 EHB P Allcock, 4 Eden Court, Ticehurst, Wadhurst, TN5 7AF
G1 EHE M Appleton, Flat 2, Black Swan Buildings, Winchester, SO23 9DT
G1 EHF D Austen, Tudorlands, Silchester Road, Tadley, RG26 5DG
G1 EHI R Davies, 1 Mount View, Plas Road, Blackwood, NP12 3RH
G1 EHM Paul Bird, 4 Parkside Avenue, Tilbury, RM18 8DT
G1 EHS B Brodribb, 1 Ponswood Road, St. Leonards-on-Sea, TN38 9BU
G1 EHU M Hostekens, 1 Ponswood Road, St. Leonards-on-Sea, TN38 9BU
G1 EHX C Cameron, Rose Cottage, Orchard Way, Coleford, GL16 7AQ
G1 EIB N Purkins, 16 Nunburnholme Avenue, North Ferriby, HU14 3AN
G1 EIG J Ryan, 71b Gunterstone Road, London, W14 9BS
G1 EIH D Samber, 102 Midsummer Avenue, Hounslow, TW4 5BB
G1 EIO Brian Smith, 43 Oak Avenue, Hindley Green, Wigan, WN2 4LZ
G1 EIP G Smith, 52 Penhill Crescent, St Johns, Worcester, WR2 5PX
G1 EIR R Smith, 29 Windmill Lane, Henbury, Bristol, BS10 7XE
G1 EIV S Stanley, 11 Mandeen Grove, Mansfield, NG18 4FA
G1 EIX H Stephens, 16 Addison Drive, Stratford-upon-Avon, CV37 7PL
G1 EIZ M Stewart, 2 Patmore Link Road, Hemel Hempstead, HP2 4PX
G1 EJA R Stone, 1 Poplar Close, Ashford, TN23 3DY
G1 EJK G Tomlinson, 4 Werneth Close, Denton, Manchester, M34 6LR
G1 EJQ A Walker, Gymru Fach, 51 The Crescent, Consett, DH8 5JF
G1 EKC M Davis, Minffordd, Oakeley Square, Blaenau Ffestiniog, LL41 3PU
G1 EKM Seppi Evans, 25 Church Street, Hatfield, AL9 5AS
G1 EKU Neil Kernahan, 15 Howgill Lane, Sedbergh, LA10 5DE
G1 ELE R Watt, 88 Graham Crescent, Portslade, Brighton, BN41 2YB
G1 ELJ R Williams, 53 Springhill Park, Wolverhampton, WV4 4TR
G1 ELK P Wilson, Barkstan Lodge, Quadring Bank, Spalding, PE11 4RF
GI1 ELP David Allen, 12 Scriggan Road, Limavady, BT49 0DH
G1 ELQ B Bentley, Sandy Ridge, Church Street, Stoke-on-Trent, ST7 4RS
G1 ELX Andrew Challinor, 24 West End Rise, Horsforth, Leeds, LS18 5JL
G1 ELZ M Cook, Well Cottage, Salisbury, SP5 3AR
G1 EME OBO THE WORCESTER MOONBOUNCE SOCIETY c/o W Day, 4 Queenswood Drive, Worcester, WR5 3SZ
G1 EML Hilary Hill, 5 Wentworth Gardens, Alton, GU34 2BJ
G1 EMM K Hill, 40 Dearbourne Avenue, Toronto, Canada, M4K 1M7
G1 EMW R Dearsley, Prince William Farm, Lynn Road, King's Lynn, PE33 9BD
G1 ENA G Edwards, 22 Whalley Lane, Uplyme, Lyme Regis, DT7 3UR
G1 END Mark Friedman, Flat 28, Hertford Mews, Potters Bar, EN6 1XW
GW1 ENG P Gibson, The Nook, Trimsaran Road, Kidwelly, SA17 4EB
G1 ENP P Hewett, Tarn Hows, 7 Lakeside Drive, Presteigne, LD8 2EG
G1 ENR A Deakin, The Farmhouse, New House Farm, Tenbury Wells, WR15 8TW
G1 EOA J Minaudo, Meadowside, Newbridge, Dumfries, DG2 0QX
G1 EOH S Draybrooke, 6 Tubbenden Lane, Orpington, BR6 9PN
GW1 EOI G John, 31 Gellifawr Road, Morriston, Swansea, SA6 7PN
G1 EOJ M Kay, 107 St. Martins Road, Blackpool, FY4 2DZ
G1 EOK E Keeble, 17 Moat Avenue, Green Lane, Coventry, CV3 6BT
G1 EOM H Kinghorn, 29 Meadowview Road, Sompting, Lancing, BN15 0HU
GI1 EOS P Leitch, 212 Belfast Road, Muckamore, Antrim, BT41 2EY
G1 EPD Derek Hathaway, 46 Blackwell Avenue, Newcastle upon Tyne, NE6 4DR
G1 EPF L Marshall, 75 Acacia Crescent, Wigan, WN6 8NJ
G1 EPL R O'Callaghan, 78 Marine Parade, Fleetwood, FY7 8RD
G1 EPO J Pragnell, Sundale, Northampton Road, Brackley, NN13 7TY
G1 EPR R Rees, 16 Railway View, Caldicot, NP26 5GB
G1 EPS Michael Rhodes, 155 High Park Road, Southport, PR9 7BY
G1 EQF P Uttridge, Springers Rest, Beck Lane, Hull, HU12 9RG
G1 EQJ M Whittle, Churchfield Cottage, West Road, Wareham, BH20 5RY
G1 EQL P Wootton, 20 Oakhill Road, Dronfield, S18 2EJ
G1 EQM Raymond Agacy, 23 Highgate Lane, Bolton-upon-Dearne, Rotherham, S63 8HR
G1 EQU Robert Percival, 23 Plumtree Road, Thorngumbald, Hull, HU12 9QG
GW1 ERA A Price, 2 Ger Y Coed, Brackla, Bridgend, CF31 2LA
G1 ERF S Rogers, 31 Morgan Road, Southsea, PO4 8JS
G1 ERM D Salter, 94 Clifton Street, Swindon, SN1 3QA
G1 ERQ R Stevens, 172 Branksome Avenue, Stanford-le-Hope, SS17 8DE
G1 ERS D Strange, 16 Cheltenham Gardens, London, E6 3DH
G1 ERU S Mole, 17a Marlborough, Seaham, SR7 7SA
G1 ERY A Moseley, 15 Gillsway, Northampton, NN2 8HT
G1 ERZ Allen Moules, 5 Hill Road, Borstal, Rochester, ME1 3NJ
G1 ESC C Mountain, 16 Temples Court, Helpston, Peterborough, PE6 7EU
G1 ESW K Laughton, 30 Tiverton Close, Radcliffe, Manchester, M26 3UJ
G1 ESX K Love, 63 Buxton Road, Spixworth, Norwich, NR10 3PP
G1 ETD J Newland, 84 Waterman Way, London, E1W 2QW
G1 ETQ Brian Sweeney, 51 Tristram Avenue, Hartlepool, TS25 5PA
G1 ETZ Chris Walker, 1 Shepherds Close, Shepshed, Loughborough, LE12 9SQ
G1 EUA B Wall, 3 Grenville Avenue, Teignmouth, TQ14 9NJ
G1 EUD D Wiles, 62 Taylor Street, Tunbridge Wells, TN4 0DX
G1 EUF R Wilson, Street Farm, Henny Street, Sudbury, CO10 7LS
G1 EUG D Wolfe, 48 Wilby Lane, Great Doddington, Wellingborough, NN29 7TP
G1 EUH J Woods, 1 Mill Lane, Burscough, Ormskirk, L40 5TJ
G1 EUI Martin Wright, 71 Oakridge Road, High Wycombe, HP11 2PL
G1 EUM S Poole, [illegible], South Hanningfield Road, Wickford, SS11 [illegible]
G1 EUN A Friend, 43 [illegible], Peterborough, PE4 6QY
G1 EUQ A Gammon, 5 Sommerville Close, Faversham, ME13 8HP
G1 EUT M Coupe, 10 Wenlock Drive, Grassmoor, Chesterfield, S42 5BH
G1 EUU Mark Gibson, 1 [illegible] Road, Grantham, NG31 [illegible]
G1 EUZ M Hannam, 98 Angelica Road, Lincoln, LN1 1AY
G1 EVA Mark Hattersley, 190 Elmton Road, Creswell, Worksop, S80 4DY
G1 EVI B Green, 49 Brockman Crescent, Dymchurch, Romney Marsh, TN29 0UA
G1 EVR P Lowe, 155 Long Lane, Bolton, BL2 6EU
G1 EVV C Mylchreest, 21 Bexhill Gardens, St. Helens, WA9 5FQ
G1 EWC A Webster, 49 Uplands Croft, Werrington, Stoke-on-Trent, ST9 0LF
G1 EWE T Williams, 20 Sandringham Drive, Dartford, DA2 7WB
G1 EWH S Bell, The Haven, Low Street, Retford, DN22 0LN
G1 EWM Thomas Conlin, 5 Morland Drive, Rochester, ME2 3LW
GW1 EWW K Edwards, 25 Woodland Road, Neath, SA11 3AL
G1 EWY Bob Owen, Llys Helen, Croesor, Penrhyndeudraeth, LL48 6SR
GJ1 EXC M Green, 1 Peel Terrace, La Route Du Fort, St. Helier, Jersey, JE2 4PA
G1 EXG J Hare, 1 The Copse, 50-52 Princes Road, Brighton, BN2 3RH
G1 EXK S Bussey, 7 Ilderton Crescent, Seaton Delaval, Whitley Bay, NE25 0FH
G1 EXM Christopher Bussey, 6 Ray Court, Wimblington, March, PE15 0FE
G1 EXR W Cosgrove, 62 Twyford Avenue, Great Wakering, Southend-on-Sea, SS3 0EX
G1 EXU R Cloke, 24 Cornflower Drive, Chelmsford, CM1 6XY
G1 EXV D Cooper, 75 Merevale Avenue, Nuneaton, CV11 5LU
G1 EYD K Rutter, 12 Berwick Terrace, North Shields, NE29 7AW
G1 EYG G Shipperley, 72 Hithercroft Road, Downley, High Wycombe, HP13 5RH
G1 EYJ Ian Smith, 21 Gorsehill Road, Wallasey, CH45 9JA
G1 EYS Pedro Jorquera, 21 Highlands Road, Orpington, BR5 4JP
G1 EYT P Vickers, 21 Blackwood Drive, Sutton Coldfield, B74 3QP
G1 EYY Dennis Whincup, 172 Appleton Road, Hull, HU5 4PF
G1 EYZ S White, 22 Silverley Way, Ashley, Newmarket, CB8 9DY
G1 EZF Michael Allmark, 11 Potternewton Crescent, Leeds, LS7 2DY
G1 EZI S Armstrong, 5 Dashpers, Brixham, TQ5 9LJ
G1 EZJ C Barker, 52 Spode Street, Stoke-on-Trent, ST4 4DY
G1 EZU D Harpham, 16 Scotts Way, Kirkby-in-Ashfield, Nottingham, NG17 9DN
G1 FAA Steven Jeffery, 35 Lynton Avenue, Orpington, BR5 2EH
G1 FAD Thomas Kenney, 7 Hickin Close, Charlton, London, SE7 8SH
GM1 FAF J Marshall, Drummorlie, Wallyford Toll, Musselburgh, EH21 8JT
GM1 FAI A Miller, 21 Merker Terrace, Linlithgow, EH49 6DD
G1 FBE D Telford, 9 Central Avenue, Carlisle, CA1 3QB
G1 FBI J Hughes, 47 Stambourne Way, Upper Norwood, London, SE19 2PY
GW1 FBL T Boorman, 43 Ffordd Taliesin, Killay, Swansea, SA2 7DF
GM1 FBM Bradley Borland, Beechwood Cottage, Muirhall Road, Perth, PH2 7LL
G1 FBQ S Fraser, Walnut Tree Cottage, Main Road, Abingdon, OX13 5LN
GD1 FBS P Gracie, 16 Antoneys Close, Pinner, HA5 3LP
G1 FBU K Goodchild, 115 Gloucester Road, Newbury, RG14 5JJ
G1 FBW Terence Howchen, 1 Ash Road, Canvey Island, SS8 7EA
G1 FBZ William Hicks, 7 Meadow Close, Thundersley, Benfleet, SS7 3RJ
G1 FCU Simon Reed, 20 Mead Crescent, Bookham, Leatherhead, KT23 3DU
G1 FCW Essex CW Club c/o S Cocks, 1 Church Road, Laindon, Basildon, SS15 4EH
G1 FDD Neil Groeber, 113 Kings Road, Kings Heath, Birmingham, B14 6TN
G1 FDL Vernon Kelk, 7 Rowan Place, Garforth, Leeds, LS25 2JR
G1 FDN W Kenyon, 22 Barons Way, Lower Darwen, Darwen, BB3 0RG
G1 FDO T King, 32 Bagnall Avenue, Arnold, Nottingham, NG5 6FT
G1 FEF C Smith, 23 Main Road, Naphill, High Wycombe, HP14 4QD
G1 FEJ J Saveall, Nascott Bungalow, Beach Road, Woolacombe, EX34 7BT
GM1 FEM I Smith, 13 Newmills Grove, Balerno, EH14 5SY
G1 FEO Gillian Jones, 8 Kenilworth Road, Lighthorne Heath, Leamington Spa, CV33 9TH
G1 FEP D Twidale, 18 Kinnaird Road, Wallasey, CH45 5HN
G1 FET P Taylor, 29 Dunstall Road, Halesowen, B63 1BB
G1 FEV D Watson, 112 Keith Way, Southend-on-Sea, SS2 [illegible]
G1 FEX John Allsop, 15 Woodland Grove, Mansfield Woodhouse, Mansfield, NG19 8AZ
G1 FFH T Firth, 126 Tombridge Crescent, Kinsley, Pontefract, WF9 5HE
G1 FFO P Thomas, 9 Awefield Crescent, Smethwick, Smethwick, B67 6PR
G1 FFR S Tucker, 28 Peregrine Road, Offerton, Stockport, SK2 5UR
G1 FFU D Woolmer, 2 Muccleshell Close, Havant, PO9 2HR
G1 FGC P Chalkley, 10 Preston Gardens, Luton, LU2 7NL
G1 FGI Paul Escreet, Little Worsall, Husthwaite, York, YO61 4PX
G1 FGK W Grint, 15 Ivythorn Road, Street, BA16 0TE
GM1 FGN Reginald Hussey, 21 Maidenfield, Mossbank, Shetland, ZE2 0TD
G1 FHH P Johnson, 30 Copplestone Grove, Longton, Stoke-on-Trent, ST3 5UD
G1 FHI S Murray, 51 Huddersfield Road, Newhey, Rochdale, OL16 3QZ
G1 FHK C Peart, 33 Fieldfare, Abbeydale, Gloucester, GL4 4WH
G1 FHO B Rivers, 211 Upper Wickham Lane, Welling, DA16 3AW
G1 FHR Reggie Roots, 14 Sussex Close, London Road, Sevenoaks, TN15 6BB
G1 FHY S Wise, 6 Honeysuckle Close, Eastbourne, BN23 8DA
G1 FIM P McDonnell, 55 Lodge Hall, Harlow, CM18 7SY
G1 FIP S Rowlandson, 48 Greville Road, Warwick, CV34 5PB
G1 FIZ J Pottinger, 16 Valley Drive, Yarm, TS15 9JQ
G1 FJD A Wood, 2 Towning Close, Deeping St. James, Peterborough, PE6 8HR
G1 FJF A Bawden, 67 Silo Drive, Farncombe, Godalming, GU7 3NZ
G1 FJH P Bruce, 26 Queens Road, Wilbarston, Market Harborough, LE16 8QJ

UK Callsigns

**IMPORTANT NOTE**

**Revalidate licence to avoid revocation** – Ofcom has advised the Society that plans will be drawn up to revoke licences that have not been revalidated as required by the licence conditions. The quickest way to revalidate is to do so online via the Ofcom website: *https://services.ofcom.org.uk/* or by email: *amateur.validations@ofcom.org.uk* Ofcom staff are available to help, but please be patient during times of heavy workload.

G1 FJI R Bullock, 32 Tinmans Green, Redbrook, Monmouth, NP25 4NB
G1 FJJ M Cammish, 20 Chantry Avenue, Hartley, Longfield, DA3 8DD
G1 FJS K Davis, 7 Hebden Road, Lower Westwood, Bradford-on-Avon, BA15 2BX
G1 FKJ George Portlock, 1 Windmill Cottage, Oxford Street, Marlborough, SN8 2DH
GW1 FKL Gareth Howells, 70 Meadow Street, Treforest, Pontypridd, CF37 1SS
G1 FKM Jon Kendall, 6 Wellington Street, Allerton, Bradford, BD15 7QZ
G1 FKP D Le Vine, Anglecroft, Borough Road, Westerham, TN16 2LA
G1 FKS C MacLiesh, 33 River Way, Twickenham, TW2 5JP
G1 FKT J Mansley, 2 Beech Tree Close, Cuerden Residential Park, Leyland, PR25 5PA
GW1 FKY Kenneth Eaton, 21 Westminster Way, Bridgend, CF31 4QX
G1 FLI Charles Franklin, 3 Park Road, Rugby, CV21 2QU
G1 FLL P Gray, 31 Rudgard Avenue, Cherry Willingham, Lincoln, LN3 4JQ
GM1 FLQ Russell Monahan, 35a Wilson Street, Beith, KA15 2BE
G1 FLV M Parr, 5 Suffolk Grove, Leigh, WN7 4TA
G1 FLW P Partridge, 18 Chaucers Drive, St Peters Field, Nuneaton, CV10 9SD
G1 FLX S Pickstone, 48 Oak Tree Drive, London, N20 8QH
G1 FLY A Rosier, 19 Imber Drive, Highcliffe, Christchurch, BH23 5BE
G1 FMA A Robinson, 5 Alford Road, Heaton Chapel, Stockport, SK4 5AW
G1 FMC Richard Rose, 44 Linden Road, Newport, PO30 1RJ
G1 FMT C Hardy, 110 Jubilee Road, Waterlooville, PO7 7RG
G1 FMU Keith Harris, 8 Trelawney Rise, Callington, PL17 7PT
G1 FMV George Hind, 2 Jamiesons Court, Kelso, TD5 7EU
G1 FMW D Lewis, 4 Westwood Grove, Solihull, B91 1QB
G1 FMX W McCandlish, Lingdowey, Stoneykirk Road, Stranraer, DG9 7BX
G1 FNA L Phillips, 2 Stratton Green, Bedgrove, Aylesbury, HP21 7EP
G1 FND N Stephens, 7 Quarry Road, Alveston, Bristol, BS35 3JL
G1 FNF Brian Walker, 47 Coppice Avenue, Eastbourne, BN20 9QJ
G1 FNN William Ball, 56 Howard Avenue, Bexley, DA5 3BE
G1 FNP Mark Bellas, 3 Elm Terrace, Penrith, CA11 7JY
G1 FNS B Cutts, 7 Lych Gate Close, Sandhurst, GU47 8JH
G1 FNU J Dodd, 38 The Quadrant, North Shields, NE29 7HP
GM1 FNX P Ewing, Arisaig, Priestland, Darvel, KA17 0LP
G1 FOA Peter Franklin, Flat 2, The Willows, Chelmsford, CM1 1TN
G1 FOE P Holt, 27 Sandown Close, Blackwater, Camberley, GU17 0EN
G1 FOF M James, 9 Denham Crescent, Mitcham, CR4 4LZ
G1 FOM P Lyttle, 3 Woodlands, East Ardsley, Wakefield, WF3 2JG
G1 FON Michael Mangan, Schuetzenstr. 54, Kaiserslautern, Germany, 67659
G1 FOW K Worsley, 102 Cabul Close, Warrington, WA2 7SE
G1 FPC A Sands, 7 Kimberley Avenue, Seymour Street, Hull, HU3 5PP
GM1 FPD D King, 18 Ford Spence Court, Benderloch, Oban, PA37 1PY
G1 FPK R Kerridge, 80 Melton Road, Wymondham, NR18 0DE
G1 FPP Nevil Sirkett, Flat 2, Barfield House, Ryde, PO33 2JP
G1 FPY A Watson, 9 Linthurst Newtown, Blackwell, Bromsgrove, B60 1BP
G1 FPZ A Wilcox, 2 Dawkins Road, Poole, BH15 4JD
G1 FQD S Eldredge, 2 Chelmsford Drive, Worcester, WR5 1QX
G1 FQI D Jackson, 273 James Reckitt Avenue, Hull, HU8 8LQ
G1 FQX J Lamb, 205 Springfield Road, Sutton Coldfield, B76 2SY
G1 FRD G Bennett, 14 Thessaly Road, Stratton, Cirencester, GL7 2NG
G1 FRL G Mott, 191 Joyners Field, Harlow, CM18 7QD
G1 FRM R doughty, 4 Trinity Road, Wisbech, PE13 3UN
G1 FSE K Stokes, 33 The Crescent, Burntwood, WS7 2PA
G1 FSF J Williams, 52b Pensford Drive, Eastbourne, BN23 7NY
GI1 FSJ N Colgan, 11 St. Johns Park, Moira, Craigavon, BT67 0NL
GM1 FSU Iain Menzies, 33 Lochside Drive, Bridge of Don, Aberdeen, AB23 8EH
G1 FSW R Consolante, 19 Chestnut Gardens, Stamford, PE9 2JY
G1 FSX John Nash, 21 St. Marys Close, Peterborough, PE1 4DR
GM1 FSZ K Hall, 12 Brockhill Rise, Inverurie, AB51 5RH
G1 FTD J Hobbs, Fetchalls, The Green, Bury St. Edmunds, IP30 9AF
GM1 FTG Mark Lonnen, Hunt Hall, Glendevon, Dollar, FK14 7JZ
G1 FTH A Marston, 92 Sorrell Road, Nuneaton, CV10 7AW
G1 FTK N Apps, 71 De Cham Road, St. Leonards-on-Sea, TN37 6HF
G1 FTU John Pearson, Largo, Hemming Green, Chesterfield, S42 7JQ
G1 FTV A Bryant, 42 Kemerton Walk, Swindon, SN3 2EA
GM1 FTZ Henry Simpson, Clachan Farm Cottages, Rosneath, Helensburgh, G84 0QR
G1 FUG Anthony Sagar, 28 Rangoon Road, Solihull, B92 9DB
G1 FUJ B Jones, 4 Caddick Close, Kingswood, Bristol, BS15 4RT
G1 FVA K Irons, 14 Beech Grove, Houghton, Carlisle, CA3 0NU
G1 FVC I Batten, 17 Cornfield Road, Birmingham, B31 2EB
G1 FVE D Ward, 36 Croxby Avenue, Scartho, Grimsby, DN33 2NW
G1 FVH M Jordan, 160 Beta Road, Farnborough, GU14 8PH
G1 FVP R Slone, 55 Chilton Way, Hungerford, RG17 0JR
G1 FVS Paul Dean, 10 Moor Terrace, Bradford, BD2 4SG
G1 FVU K Bloomfield, 14 Manners Road, Fornham St. Martin, Bury St. Edmunds, IP31 1TE
G1 FWC W Williams, Llwynprenteg, Llanafan, Ceredigion, SY23 4BQ
G1 FWE J Duggan-Keen, Bodlondeb, Chapel Street, Mold, CH7 5AE
G1 FWF H Morgan, 2 Mayfield Park South, Fishponds, Bristol, BS16 3NG
G1 FWR P Harman, 35 Point Clear Road, St. Osyth, Clacton-on-Sea, CO16 8EP
G1 FWS F Swaine, 2 Norwich Close, Stevenage, SG1 4NU
G1 FWU T Norbury, 19 Charles Cope Road, Orton Waterville, Peterborough, PE2 5ER
G1 FWY Christopher Pitt, 31 D'Arcy Way, Tolleshunt D'Arcy, Maldon, CM9 8UD
G1 FWZ B Lakey, 3 Simons Close, Worle, Weston-Super-Mare, BS22 6DJ
G1 FXB A Turquand, 63 Sundown Avenue, Dunstable, LU5 4AL
G1 FXC R Lambourne, Rosedale, Townsend, Bicester, OX27 0EY
G1 FXD O Rogers, The Barn, Millways, Bodmin, PL30 5PJ
G1 FXL Simon Annetts, Hengwm, Rhayader, LD6 5LD
G1 FXM Terence Young, Elm Field House, Tollesbury Road, Maldon, CM9 8UA
G1 FXS P Striplin, 64 Ebrington Road, Malvern, WR14 4NL
G1 FXT R Robinson, 5 Lilac Close, Newton Longville, Milton Keynes, MK17 0DQ
G1 FXX Robert Tunbridge, 35 Coworth Close, Ascot, SL5 0NR
G1 FYE C Sterland, 103 Main St., Distington, Workington, CA14 5UJ
G1 FYF C Bradley, 7 Coltsfoot, Biggleswade, SG18 8SR
G1 FYQ PONTEFRACT AND DISTRICT ARS c/o K Taylor, 29 School Road, Pontefract, WF8 2AJ
G1 FYS Kevin Boothroyd, 16 Kelvin Avenue, Dalton, Huddersfield, HD5 9HG
G1 FYU Michael Blockley, 56 St. Michaels Avenue, Gedling, Nottingham, NG4 3PE
G1 FZL P Dyer, 18 Christopher Close, Yeovil, BA20 2EH
G1 FZR Roderick Delve, 18 Thame Road, Piddington, Bicester, OX25 1PX
G1 FZS R Fleet, 17 Crown Road, Portslade, Brighton, BN41 1SJ
G1 FZV A Ogden, 5 Lower Bristol Road, Clutton, Bristol, BS39 5PB
G1 GAD F Mcloughlin, 21 Darwin Crescent, Newcastle upon Tyne, NE3 4TT
G1 GAN Peter Cartwright, 1 Railway Cottages, Sutton Bingham, Yeovil, BA22 9QW
G1 GAR M Kipping, 46 Old Hardenwaye, Totteridge, High Wycombe, HP13 6TJ
G1 GAS C Kelley, 5 Russell Way, Wootton, Bedford, MK43 9EX
G1 GAT R Fernihough, 3 Sandpiper Close, Quedgeley, Gloucester, GL2 4LZ
G1 GBC W Boucher, 12 Highfield Terrace, Ilfracombe, EX34 9LG
G1 GBF J Delaney, 31 Roose Road, Barrow-in-Furness, LA13 9RG
G1 GBH Stephen Parkins, 34 Rudgrave Square, Wallasey, CH44 0EL
G1 GBI M Pearson, 34 Downside Road, Sutton, SM2 5HP
G1 GBR A Ziemacki, 3 Wheatcroft Road, Rawmarsh, Rotherham, S62 5JR
G1 GBV D Evans, 59 Watlington Road, Benfleet, SS7 5DT
G1 GBX D Vanbeck, 101 Upper St., Islington, London, N1 1QN
GM1 GCB Stuart Raisey-Skeats, 20 Gordon Street, Boddam, Peterhead, AB42 3AY
G1 GCF F Clough, 2 Hudson Close, Tadcaster, LS24 8JD
G1 GCJ G Morris, 21 Orchard Place, Deer Park, Ledbury, HR8 2XD
G1 GCY G Gowland, 7 Canewdon Hall Close, Canewdon, Rochford, SS4 3PY
G1 GDA Michael Austin, 10 Simon Place, Wideopen, (Division Of Tyneside Motor Sport, Newcastle upon Tyne, NE13 7HT
G1 GDB Derek Thwaytes, 1 Old Chapel Close, Bothel, Wigton, CA7 2HJ
G1 GDJ C Godward, 3 Court Close, Brighton, BN1 8YG
G1 GDM James Coombes, 22 Chollerford Close, Gosforth, Newcastle upon Tyne, NE3 4RN
GM1 GDO J Morton, 6 Deanpark Place, Balerno, EH14 7ED
G1 GDR B Smith, 4 Planetree Close, Bromsgrove, B60 1AW
G1 GDS John Hunt, 47 Wyche Road, Malvern, WR14 4EF
G1 GDT C Groom, Woodstock Cottage, 2 Woodstock Terrace, Dursley, GL11 5SW
GM1 GEQ T Menzies, 239 Eskhill, Penicuik, EH26 8DF
G1 GER A Goodings, 2 Mulberry Grove, Bradwell, Great Yarmouth, NR31 8QJ
G1 GES W Barbour, 27 Drove Road, Langholm, DG13 0JW
G1 GET F Wood, The Square, Skillington, Grantham, NG33 5HB
G1 GEV Kevin Argyle, 62 Yew Tree Drive, Leicester, LE3 6PL
G1 GEY D Stoker, 20 The Rowans, Gateshead, NE9 7BN
G1 GFA N Pitt, 37 Shelley Drive, Four Oaks, Sutton Coldfield, B74 4YD
G1 GFC S Bradley, 75 New Road, Sawston, Cambridge, CB22 3BN
G1 GFD A Crook, 54 Somerset Way, Paulton, Bristol, BS39 7YX
G1 GFF Vernon Thomas, Flat 38, Lazonby Court, St. Leonards-on-Sea, TN38 0QP
G1 GFW J Jacobs, 11 Delamere Close, Castle Bromwich, Birmingham, B36 9TW
G1 GFZ R Pelling, Spring Cottage, Main Road, Hastings, TN35 4SL
G1 GGB R Phillips, 58 Baranscraig Avenue, Patcham, Brighton, BN1 8RE
G1 GGI Steven Hargreaves, 10 Reedham Crescent, Cliffe Woods, Rochester, ME3 8HT
G1 GGK F Coldham, 5 Church Lane, Towersey, Thame, OX9 3QL
G1 GGN R Barkley, 9 Eagle Close, Erpingham, Norwich, NR11 7AW
G1 GGT R Sharp, Arosa, The Park, Didcot, OX11 0HB
G1 GHG Keith Knibbs, 8 Ferguson Way, Huntington, York, YO32 9YG
GD1 GHK William Corlett, 9 Kerrocruin, Kirk Michael, Isle of Man, IM6 1AF
G1 GHU T Smith, 19 Higher Holcombe Road, Teignmouth, TQ14 8RJ
G1 GHY A Laszkiewicz, 38 Langley Lane, Ifield, Crawley, RH11 0NA
GM1 GHZ BACKPACKERS RADIO ACTIVITY GROUP c/o Paul Thompson, 31 St. Marys Drive, Perth, PH2 7BY
G1 GIA I Sinclair, 40 Holders Hill Gardens, London, NW4 1NP
G1 GID G Watt, 48 Southdown Road, Portslade, Brighton, BN41 2HN
G1 GIE R Buckley, 6280 Hawkes Bluff Avenue, Davie, Fort Lauderdale, United States, 33331-3419
G1 GIJ D Hadjidakis, 19 Eastfield Road, Royston, SG8 7ED
G1 GJD Stephen Corson, 9 St. Marys Way, Weedon, Northampton, NN7 4QL
G1 GJT P Jackson, 10 Claremont Road, Nottingham, NG5 1BH
G1 GKA R Mason, 32 Linden Drive, Evington, Leicester, LE5 6AH
G1 GKF R Mann, Little Chysauster, Penzance, TR20 8XA
G1 GKH George Hinds, 35 Lime Grove, Burntwood, WS7 0HA
GI1 GKI Thomas Campbell, 69 Limehill Road, Lisburn, BT27 5LR
G1 GKK S Brooke, 14 Saxton Avenue, Heanor, DE75 7PZ
G1 GKN Andrew Tyler, 41 Beadle Way, Great Leighs, Chelmsford, CM3 1RT
G1 GKR A Barlow, 454 Shaw Road, Royton, Oldham, OL2 6PG
G1 GKV Peter Jones, Pen y Galchen Farm, Pwlldu, Pontypool, NP4 9SS
G1 GKW Edward Perryman, Flat 4, Garth House, Bognor Regis, PO21 1HQ
G1 GLG Stanley Padgham, 4 Hollamby Park, Hailsham, BN27 2LX
G1 GLN David Beasley, 40 Susannah Street, London, E14 6LS
G1 GLS Ronald Lees, 41 Banks Court, Greenwood Close, Altrincham, WA15 7HH
G1 GLZ B Start, 9 Front Street, Corbridge, NE45 5AP
GI1 GME A McIlwee, 62 Grange Road, Ballymena, BT42 2DU
G1 GMF R Sievert, 5 Sandmoor Road, New Marske, Redcar, TS11 8BP
G1 GMG S Gainswin, 1 Buckfast Road, Buckfast, Buckfastleigh, TQ11 0EA
G1 GMH D Peat, 23 Hill Bottom Close, Whitchurch Hill, Reading, RG8 7PX
G1 GMM Henry Barczynski, 64 Kings Acre, Coggeshall, Colchester, CO6 1NY
G1 GMV A Brewer, 25 Ackerman Road, Dorchester, DT1 1NZ
G1 GMX J Mold, The Brambles, 10 Snowberry Avenue, Belper, DE56 1RE
G1 GNP D Roe, 9 The Orchard, Fairfield Road, Ilkeston, DE7 6DD
G1 GNX Gillian Leonard, Ye Olde Noggin Cottage, School Lane, Warrington, WA4 4QB
G1 GOP A Abbott, 5 Heathcote Gardens, Rudheath, Northwich, CW9 7JB
G1 GOQ S Abbott, 5 Heathcote Gardens, Rudheath, Northwich, CW9 7JB
G1 GOY Peter Miles, 37 Central Avenue, Northampton, NN2 8EA
G1 GPE David Murray, 8 Tweed Crescent, Rushden, NN10 0GS
G1 GPM M Stevenson, 6 Charnock Crescent, Sheffield, S12 3HB
G1 GQB Jonathan Bagshaw, 21 Havelock Square, Thornton, Bradford, BD13 3EZ
G1 GQJ Catherine Clark, 9 Conigre, Chinnor, OX39 4JY
G1 GQQ M Rowbotham, 37 Crawford Rise, Arnold, Nottingham, NG5 8QF
G1 GQY A Armstrong, 18 Flaxfield Way, Kirkham, Preston, PR4 2AY
G1 GQZ C Armstrong, 18 Flaxfield Way, Kirkham, Preston, PR4 2AY
G1 GRB Roy Arnold, 26 Pinehurst Park, West Moors, Ferndown, BH22 0BW
G1 GRM Brenda Sinclair, 97 Lear Drive, Wistaston Green, Crewe, CW2 8DS
G1 GRN L Skorupinski, 49 Pool Lane, Winterley, Sandbach, CW11 4RZ
G1 GRP P Slater, 32 Winthorpe Avenue, Morecambe, LA4 4RE
G1 GRT G Thomas, 39 Seaway Road, Paignton, TQ3 2NX
G1 GRZ W Baker, 26 Gardeners Road, Halstead, CO9 2TB
G1 GSB P Standley, Bligh House, 1 Norwich Road, Norwich, NR16 1DJ
G1 GSG M Taylor, 7 Marshall Road, Cropwell Bishop, Nottingham, NG12 3DP
G1 GSJ W McLaren, Ingleside, Waterloo, Whitchurch, SY13 2PX
G1 GSK M Baylis, 45 Florence Avenue, Hove, BN3 7GX
G1 GSN P Bradfield, 118 East Road, Langford, Biggleswade, SG18 9QP
G1 GST J Thomas, 59 Cross Lane, Dudley, DY3 1PD
GW1 GSW C Tinker, Coach House Cottage, Tegryn, Llanfyrnach, SA35 0BD
G1 GSY Gary Bridle, 43 Cornflower Close, Locks Heath, Southampton, SO31 6SP
G1 GTA Paul Butler, 25 Harringdale Road, High Harrington, Workington, CA14 4NU
G1 GTF E Chilton, 33 Kersall Court, Nottingham, NG6 9DT
G1 GTH Colin Clark, 24 Daisy Royd, Huddersfield, HD4 6RA
G1 GTK David Towers, 50 Westbeech Road, Pattingham, Wolverhampton, WV6 7AQ
G1 GTM Andrew Turner, Carnbrae, Woodhouse Hill, Lyme Regis, DT7 3SL
G1 GTP B Warnaby, 69 Caledonian Road, Hartlepool, TS25 5LB
G1 GTQ Anthony Clarke, 18 Waterloo Road, Brighouse, HD6 2AT
G1 GTR Paul Clarke, 13 Mitchell Street, Brighouse, HD6 2AY
G1 GTS R Clarke, 47 Peartree Road, Enfield, EN1 3DE
G1 GUI Simon Watts, 16 Northampton Close, Bracknell, RG12 9EF
G1 GVJ Patricia Allen, 17 Winfield Road, Sedbergh, LA10 5AZ
G1 GVM D Gale, 27 Portal Road, Southampton, SO19 8LD
G1 GVP J Gibbon, 18 Eagle Street, Penn Fields, Wolverhampton, WV3 7DN
G1 GWE A Friel, 10 Marvejols Park, Cockermouth, CA13 0QR
G1 GWF B Maxwell, Hillcrest, Castle View, Egremont, CA22 2NA
G1 GWJ A Gillon, 94 Pelham Street, Ashton under Lyne, OL7 0DU
G1 GWO J Green, 32 Elizabeth Close, Highwoods, Colchester, CO4 9YU
G1 GWS J Mossop, 14 Websters Lane, Great Sutton, Ellesmere Port, CH66 2LH
G1 GWX R Patrick, 9 Brant Avenue, Illingworth, Halifax, HX2 8DL
G1 GXB K Ray, 4 Elm Road, Bishops Waltham, Southampton, SO32 1JR
GD1 GXC S Ray, 75 The Meads, Edgware, HA8 9HE
G1 GXF T Scott, 9 Walker Drive, Leigh-on-Sea, SS9 3QS
GM1 GXH I Sinclair, Clan Sinclair House, Nosshead Lighthouse, Caithness, KW1 4QT
G1 GXQ R Tulk, Home Farm Lodge, Pen y Lan, Wrexham, LL14 6HS
G1 GXW B Woodhouse, 8 Firs Drive, Rugby, CV22 7AQ
G1 GXX Alan Mayes, 31 Holsey Lane, Bletchley, Milton Keynes, MK2 3FH
G1 GYC Martin Hallsworth, 87 Talbot Street, Hazel Grove, Stockport, SK7 4BJ
G1 GYF D Harvey, 264 Rangefield Road, Bromley, BR1 4QY
G1 GYH J Hay, 23 Manor Close, Wilmslow, SK9 5PX
G1 GYJ Frank Mallows, 31 Booth Road, Hartford, Northwich, CW8 1RD
G1 GYM P McEwen, 26 Walton Avenue, North Shields, NE29 9BS
G1 GYQ A Hayward, 1 Cleveland Road, Basildon, SS14 1NF
G1 GYT Timothy Down, 1 Park View, East Tytherley Road, Romsey, SO51 0LW
G1 GZG M Newport, 9 Highbury Park, Exmouth, EX8 3EJ
G1 GZI A Farmar, Hawkes Place, Horslett Hill, Holsworthy, EX22 6RS
G1 GZK K Feay, 19 Dorset Avenue, Diggle, Oldham, OL3 5PL
G1 GZM Ian Ford, 97 Green Rock Lane, Walsall, WS3 1NQ
G1 GZZ Nigel Pugh, 3 Chapel Crescent, Darliston, Whitchurch, SY13 2AR
G1 HAB H Birkmyre, Swarland, Morpeth, NE65 9JW
G1 HAC J Hilton, 32 Dowry St., Fitton Hill, Oldham, OL8 2LP
G1 HAH Brian Hodgson, 40 Trentham Drive, Bridlington, YO16 6ES
G1 HAX N Bevan, Mountain View, Whip Lane, Oswestry, SY10 8HU
G1 HBC T Hopkins, 58 Broom Grove, Knebworth, SG3 6BQ
G1 HBD A Hornby, 2 Maple Close, Winnersh, Wokingham, RG41 5PE
G1 HBE Andrew Howlett, 43 Cheetham Hill Road, Dukinfield, SK16 5JL
G1 HBF M Hughes, 2 Chaldon Road, Canford Heath, Poole, BH17 8DB
G1 HBR K Jackaman, 186 Charlton Road, Charlton, London, SE7 7DW
G1 HBV E Jones, 37 Sluice Road, Denver, Downham Market, PE38 0DY
G1 HBW F Jones, 184 Harwich Road, Little Clacton, Clacton-on-Sea, CO16 9PU
G1 HCC E Kent, 100 Waskerley Road, Washington, NE38 8DS
G1 HCI Richard De Ste Croix, 49 Oxford Street, Grimsby, DN32 7JE
G1 HCJ G De Ste Croix, 49 Oxford Street, Grimsby, DN32 7JE
G1 HCM Frederic Dawson, 33 Oakwood Road, Ryde, PO33 3JU
G1 HCU Gordon Gratton, 5 Nursery Avenue, Ovenden, Halifax, HX3 5SZ
G1 HDG P Greed, 12 Bailey Close, Windsor, SL4 3RD
G1 HDK William Akhurst, 20 Newton Road, Faversham, ME13 8DZ
G1 HDO A Appleton, Flat 9, 19-21 West Cliff Road, Bournemouth, BH4 8AT
G1 HDR Robert Stanford, 1 South End, Bassingbourn, Royston, SG8 5NG
G1 HDX Mary Robertson, 12 James Park Homes, Egremont, CA22 2QQ
G1 HEA A Steele, 92 Eelholme View Street, Keighley, BD20 6AY
G1 HEJ J Alexander, 1 Locarno Road, Swanage, BH19 1HY
G1 HEN D Coates, 2 Penfold Drive, Countesthorpe, Leicester, LE8 5TP
G1 HEP C Heptonstall, Badger Cottage, 27 Bolster Moor Road, Huddersfield, HD7 4JU
G1 HEQ Keith Tucker, 507 New North Road, Ilford, IG6 3TF
G1 HER G Dasilva-Hill, 12 St. Stephens Crescent, Thornton Heath, CR7 7NP
G1 HEU G Tybora, 37 Nunsfield Drive, Alvaston, Derby, DE24 0GH
GW1 HEV Dilwyn Thomas, 3 Oaklands Terrace, Wiston, Haverfordwest, SA62 4PR
G1 HEW P Travers, 49 West Bank Drive, South Anston, Sheffield, S25 5JG
G1 HEX J Dunn, 8 Ettrick Terrace North, Craghead, Stanley, DH9 6BE
G1 HEY J Todd, Atlast, 7 Marine Avenue West, Mablethorpe, LN12 2TX
G1 HEZ C Tindale, 4 Station Road, Beamish, Stanley, DH9 0QU
G1 HFA R WINDER, 176 Ambleside Road, Lancaster, LA1 3ND
G1 HFE Sally Wood, 18 Rosemellin, Camborne, TR14 8QF
G1 HFH D Ward, 10 Fulshaw Avenue, Wilmslow, SK9 5JA
G1 HFK W Willoughby, 27 Foxwood Grove, Sheffield, S12 2FN
G1 HFS K Burgess, 32 Hendon Street, Leigh, WN7 1TS
G1 HFT Ernest Bennett, 20 Cromford Road, Clay Cross, Chesterfield, S45 9RE
G1 HFW F Hewins, Hillcrest, Wern Road, Mold, CH7 6PY
G1 HFY B Watson, 20 St. Marys Gardens, Hilperton Marsh, Trowbridge, BA14 7PG
G1 HGA Kevan Yates, 3 Flaxland Crescent, Sileby, Loughborough, LE12 7SB
G1 HGB J Neville, 44 Thorpe House Avenue, Sheffield, S8 9NG
G1 HGC Brian Newton, 42 Heath Road, Widnes, WA8 7NQ
G1 HGD M Newell, 189 Humber Road, Coventry, CV3 1NZ
G1 HGF T Standring, 52 Beechcroft Road, Ipswich, IP1 6BD

| | | |
|---|---|---|
| G1 | HGT | A Berkerey, 36 Erlesmere Gardens, London, W13 9TY |
| G1 | HGY | Peter Parry, Forge House, Church Road, Wellingborough, NN9 6BQ |
| G1 | HHB | Christopher Brown, 12 Forest Close, Newport, PO30 5SF |
| G1 | HHC | D Bolt, C/O Old School Hse, Maristow Roborough, Plymouth, PL6 7BY |
| G1 | HHD | E Bolt, Old School House, Roborough, Plymouth, PL6 7BY |
| G1 | HHH | HASTINGS ELEC RD c/o T Ransom, 9 Lyndhurst Avenue, Hastings, TN34 2BD |
| G1 | HHM | K Roberts, Gwenallt, Lon Crecrist, Holyhead, LL65 2AZ |
| G1 | HHO | Gordon Reeve, 10 Badgers Copse, New Milton, BH25 5PE |
| G1 | HHQ | Jamie Brown, 14 The Green, Winscombe, BS25 1AL |
| G1 | HHS | A Burrows, 1 Browns Avenue, Runwell, Wickford, SS11 7PT |
| G1 | HHT | A Benstock, 10 Wike Ridge Avenue, Leeds, LS17 9NL |
| G1 | HHU | Norman Ball, 140 Albert Avenue, Prestwich, Manchester, M25 0HE |
| G1 | HHW | W Curtis, Rio Taibilla 11, San Miguel de Salinas, Alicante, Spain, 3193 |
| GD1 | HIA | P Smith, 48 Silverburn Crescent, Ballasalla, Isle of Man, [illegible] |
| G1 | HIB | M Standing, 7 Oxcliffe New Farm Caravan Park, Oxcliffe Road, Morecambe, LA3 3EF |
| G1 | HIG | I Ravelini, 15 Clarendon Green, Orpington, BR5 2NY |
| G1 | HII | R Spencer, 2 Windmill Bank, Wigston, LE18 3SX |
| G1 | HIJ | R Dimmock, 67 Meadway, Dunstable, LU6 3JT |
| G1 | HIN | Ronald Ellwood-Thompson, 15 Skinner Street, Aberystwyth, SY23 2JU |
| G1 | HIO | Moreen Horsfield, 59 Queens Drive, Newton-le-Willows, WA12 0LY |
| G1 | HIP | Keith Horsfield, 59 Queens Drive, Newton-le-Willows, WA12 0LY |
| G1 | HIU | J Clarke, The New Vicarage, Low Road, Saxmundham, IP17 1LJ |
| G1 | HJD | B Taylor, 111 High St., Warboys, Huntingdon, PE17 2TB |
| G1 | HJL | Ian Copland, 6 Victoria Street, Cullingworth, Bradford, BD13 5AE |
| G1 | HJO | James Cornall, Fern Holme, Taylors Lane, Preston, PR3 6AB |
| G1 | HJP | Trevor Carter, 84 Colvile Road, Wisbech, PE13 2EL |
| G1 | HJS | Michael Todd, 38 The Churchlands, New Romney, TN28 8LB |
| GM1 | HJX | Mark Williams, 33/9 Marlborough Street, Edinburgh, EH15 2BD |
| G1 | HKF | C Maclennan, 72 Sandsfield Lane, Gainsborough, DN21 1DD |
| G1 | HKM | Frank Woods, 275 Scotter Road, Scunthorpe, DN15 7EH |
| G1 | HKP | P Wooldridge, 10 Woodfield Road, Coventry, CV5 6AL |
| G1 | HKR | T Whittam, 27 Dimples Lane, Garstang, Preston, PR3 1RD |
| G1 | HKS | D Wilson, 34 Belfairs Drive, Chadwell Heath, Romford, RM6 4EB |
| G1 | HKT | R Woodley, 1 Melton Drive, Didcot, OX11 7JP |
| G1 | HKU | C Weatherley, 7 Rochester Way, Croxley Green, Rickmansworth, WD3 3NE |
| G1 | HLP | D Plant, 15 Heathcombe Road, Bridgwater, TA6 7PD |
| G1 | HLQ | Maureen Edwards, 11 Lightwood, Crown Wood, Bracknell, RG12 0TR |
| G1 | HLS | W Etherton-Scott, 62 Spencer Road, Walthamstow, London, E17 4BD |
| G1 | HLT | Ian Fay, 7 Oakridge Close, Forest Town, Mansfield, NG19 0EY |
| G1 | HLV | J Lee, Deighton Manor, Deighton, Northallerton, DL6 2SN |
| G1 | HLY | R Powell, 55 Lumley Road, Horley, RH6 7JF |
| G1 | HMI | T Rock, 4 Hunters Gate, Much Wenlock, TF13 6BW |
| G1 | HML | H Willard, M V Irma, Paglesham Boatyard, Paglesham Eastend, SS4 2ER |
| G1 | HMT | Geoffrey Gray, Home Farm, Furlong Drove, Ely, CB6 2EQ |
| G1 | HMW | W Gardner, 25 Prospect Place, Wing, Leighton Buzzard, LU7 0NT |
| G1 | HMY | P Batty, 14 Woodville Road, Penwortham, Preston, PR1 9DR |
| G1 | HMZ | K Breedon, 17 Emmanuel Avenue, Arnold, Nottingham, NG5 9QN |
| G1 | HND | M Burling, 28 Croydon Road, Arrington, Royston, SG8 0DJ |
| G1 | HNF | N Bufton, 7 Laburnum Close, Rassau, Ebbw Vale, NP23 5TS |
| GW1 | HNG | P Beesley, 12 Bryngolwg, Aberdare, CF44 0ER |
| G1 | HNH | P Bannister, 38 Regent Street, Stowmarket, IP14 1RH |
| G1 | HNN | R Cockman, 31 Kensington Road, Southend-on-Sea, SS1 2SX |
| G1 | HNU | M Gray, 28 The Close, Bradwell, Great Yarmouth, NR31 8DR |
| GM1 | HNZ | Alexander Simmers, Loanside, Crossroads, Keith, AB55 6LP |
| G1 | HOD | Andrew Smith, 14 Bridge Street, Shepshed, Loughborough, LE12 9AD |
| G1 | HOI | Nigel Ballard, 45, 185 NW Harwood, Prineville, United States, 97754 |
| G1 | HOJ | B Baylis, 118 Eastgate, Deeping St. James, Peterborough, PE6 8RD |
| G1 | HOL | S Cook, 40 Fairfax Road, Birmingham, B31 3ST |
| G1 | HOP | R Chaston, 132 Jackers Road, Coventry, CV2 1PF |
| G1 | HOU | A Kirby, 185 High Street, Dunsville, Doncaster, DN7 4BU |
| G1 | HPB | Richard Wearing, 163 Birmingham Road, Stratford-upon-Avon, CV37 0AP |
| G1 | HPS | Trefor Jones, 25 Foxcotte Road, Charlton, Andover, SP10 4AR |
| G1 | HPU | P James, 8 Pipers Wood Cottages, Little Missenden, Amersham, HP7 0RQ |
| G1 | HPV | T Jones, 175 New Road, Great Wakering, Southend-on-Sea, SS3 0AR |
| G1 | HPZ | I Russell, 368 Whitehall Road, St. George, Bristol, BS5 7BT |
| G1 | HQE | C Close, 3 Hay Green, Therfield, Royston, SG8 9QL |
| G1 | HQG | A Coley, 5 Arundel Way, Highcliffe, BH23 5DX |
| G1 | HQH | I Church, 26 Usk Way, Didcot, OX11 7SQ |
| G1 | HQJ | D Robinson, 16 Green Lane, Platts Heath, Maidstone, ME17 2NS |
| G1 | HQK | I Richardson, 1 Cedar Drive, Lowestoft, NR33 9HA |
| G1 | HQN | J Rattenbury, Compton Lodge, High Ham, Langport, TA10 9DH |
| G1 | HQO | G Spaven, Spout House Farm, Macclesfield Road, High Peak, SK23 7QU |
| G1 | HQQ | F Jensen, 79 The Drakes, Shoeburyness, Southend-on-Sea, SS3 9NY |
| G1 | HQW | J Kiernan, 26 Popes Lane, Gorefield Road, Wisbech, PE13 5BD |
| G1 | HRA | D Lloyd, 35 Charles Close, Abbotts Barton, Winchester, SO23 7HT |
| G1 | HRD | Vince Allen, 20 Harbut Road, Battersea, London, SW11 2RA |
| G1 | HRH | Melvyn Gregory, 45 Larksfield Avenue, Bournemouth, BH9 3LW |
| G1 | HRJ | P Deakes, 108 Glaisdale Drive East, Nottingham, NG8 4LZ |
| G1 | HRL | E Dillow, 18 Laburnum Grove, Warwick, CV34 5TG |
| G1 | HRM | T Davenport, 36 Rydale Road, Nottingham, NG5 3GS |
| G1 | HRQ | Brian Madore, 66a West Street, Ryde, PO33 2QF |
| G1 | HRU | Stephen Hill, 26 Crescent Road, Netherton, Dudley, DY2 0NW |
| G1 | HRV | K Higgins, 22 Thatchers Lane, Cliffe, Rochester, ME3 7TN |
| GM1 | HRY | Allen Davis, 3 High Shore, Banff, AB45 1DB |
| G1 | HSA | S Arnold, 30 Pine Avenue, Newton-le-Willows, WA12 8JE |
| G1 | HSF | M Evans, Fernlea, Pit Hill Lane, Bridgwater, TA7 9BT |
| G1 | HSG | Neil Evans, 25 Chetwyn Avenue, Bromley Cross, Bolton, BL7 9BN |
| G1 | HSH | M Ellerby, 3 Gilwern Court, Ingleby Barwick, Stockton-on-Tees, TS17 5DJ |
| G1 | HSI | M Glazier, 19 West Place, Brookland, Romney Marsh, TN29 9RG |
| G1 | HSJ | L Godden, The Conifers, 14 Pirehill Lane, Stone, ST15 0JN |
| G1 | HSL | J Girt, 22 Medway Road, Ipswich, IP3 0QH |
| G1 | HSM | Leon Heller, 1 Princes Road, St. Leonards-on-Sea, TN37 6EL |
| G1 | HSO | Mark Hoey, 37 Newhouse Road, Blackpool, FY4 4JJ |
| G1 | HSP | C Hunt, 700 Western Boulevard, Nottingham, NG8 5FH |
| G1 | HSX | P Kimber, 16 Sycamore Close, Lydd, Romney Marsh, TN29 9LF |
| G1 | HTF | H Heron, 43 Cheetham Hill Road, Dukinfield, SK16 5JL |
| G1 | HTL | J Foster, 25 Hunter Road, Arnold, Nottingham, NG5 6QZ |
| G1 | HTM | J Froggatt, 17 Queensway, Saxilby, Lincoln, LN1 2QR |
| G1 | HTN | P Farrow, 10 St. Thomas Close, Chilworth, Guildford, GU4 8LQ |
| G1 | HTO | Richard Fortescue, 7 Dodkin Lane, Weymouth, DT3 [illegible] |
| G1 | HTT | L Gray, 87 London Road, Coventry, CV1 2JQ |
| GU1 | HTY | B Ayres, Rousay, Bailiffs Cross Road, St. Andrew, Guernsey, GY6 8RY |
| G1 | HUM | R BEECH, 6 Law Cliff Road, Birmingham, B42 1LP |
| GD1 | HVL | P Howarth, 5 Carrick Mews, Bay View Road, Port St. Mary, Isle of Man, IM9 5AN |
| G1 | HWA | K Harris, 27 Middle Field Road, Rotherham, S60 3JJ |
| G1 | HWH | Graham Jackson, Pathways, Down Barton Road, Birchington, CT7 0PY |
| G1 | HWJ | P Milner, 3 Larne Avenue, Cheadle Heath, Stockport, SK3 0[illegible] |
| G1 | HWK | T McCarthy, 25 Henley Avenue, North Cheam, Sutton, SM3 9SG |
| G1 | HW[illegible] | [illegible], 77 Richmond Way, Oadby, Leicester, LE2 5TR |
| G1 | HWP | R Davies, 7 The Nook, Tupsley, Hereford, HR1 1NH |
| G1 | HWR | L Mills, 54 Petters Road, Ashtead, KT21 1NE |
| G1 | HWY | M Jupp, 54 Shooting Field, Steyning, BN44 3RQ |
| G1 | HXN | G King, [illegible], Norwich, NR12 7[illegible] |
| G1 | HXP | J Lennard, 10 Orston Road East, West Bridgford, Nottingham, NG2 5FU |
| G1 | HXR | R Davis, 6 Fairway Drive, Northmoor, Wareham, BH20 4SG |
| G1 | HXT | G Eden, Heathend Cottage, Cromhall, Wotton under Edge, GL12 8AS |
| G1 | HXZ | C Cave, 20 Meadow View, Banbury, OX16 9SR |
| G1 | HYA | N Cramp, 3 Sowood Court, Ossett, WF5 0TJ |
| G1 | HYC | Dale Curson, 25 Colbert Park, Swindon, SN25 4YJ |
| G1 | HYG | B Crowther, 104 John St., Beamish, Stanley, DH9 0QP |
| GU1 | HYN | B Bolderston, 12 Hartlebury Estate, Steam Mill Lane, St. Martin, Guernsey, GY4 6NH |
| G1 | HYO | Michael Green, 26 Hunters Field, Stanford in the Vale, Faringdon, SN7 8LR |
| G1 | HYQ | D Chenoweth, 20 Churchlands Road, Bedminster, Bristol, BS3 3PW |
| G1 | HYT | E Cook, 129 Days Lane, Sidcup, DA15 8JT |
| G1 | HYU | Kenneth Church, 31 Riversway, King's Lynn, PE30 2ED |
| G1 | HYX | A Chance, 18 Egdon Glen, Crossways, Dorchester, DT2 8BQ |
| G1 | HZD | C Legate, 101 Butchers Lane, Walton on The Naze, CO14 8UD |
| G1 | HZI | Ian Dodd, Garden Cottage, Sandhow, Hexham, NE46 4LW |
| G1 | HZJ | Michael Devine, Fern House, 7 South Parade, Seascale, CA20 1PZ |
| G1 | HZL | M Donaldson, 54 Glebe Street, Burnley, BB11 3LH |
| G1 | HZN | Chris Dadd, 60 Rosaire Place, Scartho, Grimsby, DN33 2JS |
| G1 | HZR | K Farrar, 8 Ascot Avenue, Cantley, Doncaster, DN4 6HE |
| G1 | IAB | S Matthews, 66 West End Road, Epworth, Doncaster, DN9 1LB |
| G1 | IAD | David Morton, 12 Pennygate Drive, Lowestoft, NR33 9HL |
| G1 | IAG | P Morris, 18 Greenway Close, London, NW9 5AZ |
| G1 | IAL | A Plant, 148 Chatsworth Road, Halesowen, B62 8TH |
| G1 | IAQ | D Memory, 43 Welford Road, Blaby, Leicester, LE8 4FT |
| G1 | IAV | P Costello, Newgrange, Poplar Road, New Milton, BH25 5XP |
| G1 | IAW | Ian Woodward, Corlander, Middle Road, Wrexham, LL11 3TW |
| G1 | IBF | Geoffrey Tullock, 16 Ward Lea, Nafferton, Driffield, YO25 4JZ |
| G1 | IBJ | C Diaper, 163 Edwin Road, Gillingham, ME8 0AQ |
| G1 | IBO | D Buss, Rlc, Po Box 885, Bristol, BS99 5LG |
| G1 | IBP | A Heaysman, 325 Broomfield Road, Chelmsford, CM1 4DU |
| G1 | IBS | W Chadwick, 102 Feltham Road, Ashford, TW15 1DP |
| G1 | IBX | Charles Drayton, The Lindens, Main Street, York, YO19 6RG |
| G1 | ICA | D Keable, 90 King Edward Road, Rugby, CV21 2TE |
| G1 | ICH | M Adams, 61 Monks Park Road, Northampton, NN1 4LU |
| G1 | ICI | Michael Thorpe, 18 Sherrier Way, Lutterworth, LE17 4NW |
| G1 | ICK | D Winton, 16 Lord Avenue, Clayhall, Ilford, IG5 0HP |
| G1 | ICQ | A Orgee, 54 Riverview Close, Hallow, Worcester, WR2 6DA |
| G1 | ICX | D Palmer, 18 Newfields, Sporle, King's Lynn, PE32 2UA |
| G1 | IDE | Stuart Roy, 28 Kingston Rise, New Haw, Addlestone, KT15 3EY |
| G1 | IDF | A Edwards, 51 Redrock Road, Rotherham, S60 3JN |
| G1 | IDJ | G Perry, 61 Ollands Road, Reepham, Norwich, NR10 4EL |
| G1 | IDQ | G Arnold, 36 Market Street, Rugeley, WS15 2JL |
| G1 | IDV | George Bain, 99 Longford Lane, Gloucester, GL2 9HB |
| G1 | IDZ | D Young, 70 North Malvern Road, Malvern, WR14 4LX |
| G1 | IEB | L Tatham, Hebron Stores, Llangwnadl, Pwllheli, LL53 8NW |
| G1 | IEC | P Walton, 2 Albert Road, Bromsgrove, B61 7BE |
| GM1 | IEL | J Bruce, 24 South Green Drive, Airth, Falkirk, FK2 8JP |
| G1 | IEO | J Turner, 2 Hilberry Road, Canvey Island, SS8 7EL |
| G1 | IEP | G Tomkins, The Close, Broomfield, Bradford, BD14 6PJ |
| G1 | IEX | D Broughton, 33 Queens Park Flats, Queens Park Close, Mablethorpe, LN12 2AS |
| G1 | IEY | S Reigate, 9 Effingham Road, Croydon, CR0 3NF |
| G1 | IFF | A Rose, 15 Elderwood Way, Tuffley, Gloucester, GL4 0RA |
| G1 | IFH | Graham Reading, 36 Radford Close, Ravenfield, Rotherham, S65 4LD |
| G1 | IFV | Nicholas Ginger, Barnlea, Fairwarp, Uckfield, TN22 3DT |
| G1 | IFW | W Gain, 14 Clarence Road, St. Leonards-on-Sea, TN37 6SD |
| G1 | IFX | D Garratt, 3 Fort Road, Mountsorrel, Loughborough, LE12 7HB |
| G1 | IGA | C Brooks, 78 Leggatts Wood Avenue, Watford, WD24 6RP |
| G1 | IGC | S Brookes, 52 Larch Grove, Kendal, LA9 6AU |
| G1 | IGE | S Bates, 6 The Green, Swanwick, Alfreton, DE55 1BL |
| G1 | IGN | Gary Scroggs, 52 Eastern Road, Burnham-on Crouch, CM0 8BT |
| G1 | IGP | G Spinks, 89 Uplands Road, Oadby, Leicester, LE2 4NT |
| G1 | IGW | D Cliff, 23 Grey Towers Drive, Nunthorpe, Middlesbrough, TS7 0LT |
| G1 | IHA | R Stevens, 75 Telford Road, London, N11 2RL |
| G1 | IHB | P Spashett, Llys-Y-Coed, Trefriw, Gwynedd, LL27 0QA |
| G1 | IHE | John Smith, 65 Woods Avenue, Hatfield, AL10 8QF |
| G1 | IHI | Michael Godsave, 35 Furlong Close, Midsomer Norton, Radstock, BA3 2PR |
| G1 | IHJ | A Homer, 6 Ensall Drive, Wordsley, Stourbridge, DY8 4XX |
| G1 | IHL | Stephens Hopkins, 98 Court Road, Kingswood, Bristol, BS15 9QP |
| G1 | IHS | C Currie, 33 Ashridge Drive, Bricket Wood, St. Albans, AL2 3SR |
| G1 | IHY | J Shilson, 3 Hereford Close, Desborough, Kettering, NN14 2XA |
| G1 | III | Christopher Smith, 199a Richardshaw Lane, Stanningley, Pudsey, LS28 6AA |
| G1 | IIO | B Thornton, 21 Valley Road, Banbury, OX16 9BQ |
| GU1 | IIW | R Loveridge, Shamley, Route de Portinfer, Vale, Guernsey, GY6 8LN |
| G1 | IIX | B Lee, 58 Shaw Avenue, Normanton, WF6 2TT |
| G1 | IIY | D Turner, 154 Rowlett Road, Corby, NN17 2BS |
| GW1 | IIZ | J Underwood, Rock Hill, Llanarthney, Carmarthen, SA32 8LJ |
| G1 | IJC | Michael Williamson, 2 Lancaster Close, Fakenham, NR21 8DW |
| G1 | IJJ | J Lainchbury, 17 Pearmain Avenue, Wellingborough, NN8 4SF |
| G1 | IJM | M Shoosmith, 18 Pottery Close, Aylesbury, HP19 7FY |
| G1 | IJQ | D Martin, 27 St. Andrews Road, Stratton, Bude, EX23 9AG |
| G1 | IJY | Richard Wise, Flat 1, Canelia House, 18-21 West Parade, Worthing, BN11 3RB |
| G1 | IKF | R Blakemore, 31 Millstone Rise, Liversedge, WF15 7BW |
| G1 | IKG | P McGahon, 520 Chessington Road, West Ewell, Epsom, KT19 9HH |
| G1 | IKH | Ian Mclaughlin, 28 Jarvis Avenue, Nottingham, NG3 7BH |
| G1 | IKL | P O'Sullivan, Japonica Cottage, Ardens Grafton, Alcester, B49 6DR |
| G1 | IKT | S Elliott, Manor House, Bewholme, Driffield, YO25 8DX |
| G1 | IKV | B Austin, 16 Heathlands, Westfield, Hastings, TN35 4QZ |
| G1 | ILC | S colley, 8 Tennyson Road, Maltby, Rotherham, S66 7LU |
| G1 | ILF | Paul Ellis, 36 Poplar Road, Skellow, Doncaster, DN6 8AH |
| G1 | ILG | Dilan Evans, 12 The Mead, [illegible] |
| G1 | ILH | G Tarr, 18 The Loont, Winsford, CW7 1FU |
| G1 | ILJ | C Wood, Eastburn Warren Farmhouse, Garton Road, Driffield, YO25 9DR |
| G1 | ILO | R Dell, 80 West Avenue, Lightcliffe, Halifax, HX3 8TJ |
| G1 | ILY | C Sims, 226 Exeter Road, Exmouth, EX8 3NB |
| G1 | IMD | M Hall, 6 Poplar Avenue, New Mills, High Peak, SK22 4HH |
| G1 | IME | N Hopkins, 22 Cornfield Way, Ashton-under-Hill, Evesham, WR11 7TA |
| G1 | IMI | C Foreman, Thornham Farm, Wansford, Driffield, YO25 8JJ |
| G1 | IMM | A Gee, 24 Granhams Close, Great Shelford, Cambridge, CB22 5LG |
| G1 | IMS | I Stewart, 34 Newgate Street Village, Hertford, SG13 8RB |
| G1 | IMY | R Laycock, 24 Farmcroft Road, Mansfield Woodhouse, Mansfield, NG19 8QT |
| G1 | INA | A Lowe, 47 Springfield Park Road, Chelmsford, CM2 6EB |
| G1 | INB | P Laker, 207 Columbia Road, Ensbury Park, Bournemouth, BH10 4EE |
| G1 | IND | V Lowe, 35 Elm Place, Armthorpe, Doncaster, DN3 2DE |
| G1 | INI | B Ginsburg, 27 Park Crescent, Elstree, Borehamwood, WD6 3PT |
| G1 | INJ | R Ginsburg, 3 Basing Hill, London, NW11 8TE |
| G1 | INK | Stephen Green, 8 Granby Road, Fairfield, Buxton, SK17 7TW |
| GM1 | INS | B Skakle, 190 West Road, Fraserburgh, AB43 9NL |
| G1 | INU | Mark Sweet, 67 Swinley Road, Wigan, WN1 2DL |
| GD1 | IOM | IOM ARS (IOMARS) c/o Andrew Morgan, Thal'loo Glass, Nassau Road, The Dog Mills, Mwyllyn Moddey, Isle of Man, IM7 4AQ |
| G1 | IOO | D Camac, 6 Wisbeck Road, Tonge Fold, Bolton, BL2 2TA |
| G1 | IOP | A Cheer, 15 Stibbs Way, Bransgore, Christchurch, BH23 8HG |
| G1 | IOR | R Hebdige, 17 Tunstall Way, Chesterfield, S40 2RH |
| G1 | IOT | Hydren Harrison, 2 Hendre, Newtown, Ebbw Vale, NP23 5FE |
| G1 | IPD | D Mobbs, 64 Cranford Road, Kingsley, Northampton, NN2 7QX |
| G1 | IPE | A Medcalf, 23 Allesborough Drive, Pershore, WR10 1JH |
| G1 | IPI | Roger Taylor, Lower Manaton, South Hill Road, Callington, PL17 7LW |
| G1 | IPP | M Allen, 23 Waterloo Crescent, Countesthorpe, Leicester, LE8 5SU |
| G1 | IPU | George Coote, 20 Weeley Road, Little Clacton, Clacton-on-Sea, CO16 9EN |
| G1 | IPY | J Rowlands, 2 Wellfield, Longton, Preston, PR4 5BX |
| G1 | IQA | G Adkins, 117 Connolly Drive, Rothwell, Kettering, NN14 6TN |
| G1 | IQE | F Angwin, 171 Windsor Road, Wellingborough, NN8 2LZ |
| G1 | IQF | Michael Ames, 7 Northgate, Leyland, PR25 3NR |
| G1 | IQG | A Bloodworth, 79 Hands Road, Heanor, DE75 7HB |
| G1 | IQK | B Shaw, 23 Lodge Drive, Culcheth, Warrington, WA3 4ES |
| G1 | IQN | J Spicer, 5 Berries Mount, Bude, EX23 8AP |
| GW1 | IQS | I Jones, 7 The Oaks, Quakers Yard, Treharris, CF46 5HQ |
| G1 | IQU | D Jolley, 212 Eastern Esplanade, Southend-on-Sea, SS1 3AD |
| G1 | IRG | Simon Manning, 11 Broomhill Crescent, Southfields, Northampton, NN3 5BH |
| G1 | IRQ | N Tansley, 11 Juniper Close, Lutterworth, LE17 4US |
| G1 | IRX | Marion Weir, 17 Pasteur Drive, Apley, Telford, TF1 6PQ |
| G1 | ISK | F Davies, Hendref, Red Wharf Bay, Pentraeth, LL75 8YG |
| G1 | ISN | T Evans, 22 Malthouse Lane, Ashover, Chesterfield, S45 0AL |
| G1 | ISP | B Etherington, 24 Broomcroft Road, Ossett, WF5 8LH |
| G1 | ISR | Peter Kenington, Trap Farm, Devauden, Chepstow, NP16 6PE |
| G1 | ISS | Brian Lyons, 51 Wade Reach, Walton on The Naze, CO14 8RE |
| G1 | ISX | C Hall, 9 Moneyhill Court, Dellwood, Rickmansworth, WD3 7DY |
| G1 | ISY | Norman Morris, 15 Turners Close, Highnam, Gloucester, GL2 8EH |
| G1 | ITE | P Hayler, 27 Birch Way, Heathfield, TN21 8BB |
| G1 | ITJ | Kenneth Edmett, Solstice, Youngs Paddock, Salisbury, SP5 1RS |
| G1 | ITL | D Gilbey, 34 Farnhurst Road, Barnham, Bognor Regis, PO22 0JN |
| G1 | ITS | T Williams, 86 Hillcrest Road, Rochdale, OL11 2QB |
| G1 | ITV | K Ward, 5 BROOMFIELD ROAD, DEANE, Bolton, BL3 4DB |
| G1 | IUA | Nigel Harris, Sunnyside Lodge, Mongeham Road, Deal, CT14 8JW |
| G1 | IUD | C Sermons, 17 Wellside, Marks Tey, Colchester, CO6 1XG |
| G1 | IUF | M Kilkenny, 138 Stanbury Road, Hull, HU6 7BW |
| G1 | IUL | Bryan Jackson, 10 Wood End Croft, Coventry, CV4 9RN |
| G1 | IUT | T Christmas, 3 Mount Road, Cosby, Leicester, LE9 1SX |
| G1 | IUW | G Diaper, 89 East Street, Sudbury, CO10 2TP |
| G1 | IUZ | STEPHEN GROVES, 135 Ring Road, Crossgates, Leeds, LS15 7QE |
| G1 | IVF | David Lowe, 21 Farndon Road, Market Harborough, LE16 9NW |
| G1 | IVG | Colin Lowe, 22 Ryelands Close, Market Harborough, LE16 7XE |
| G1 | IVI | M Grey, 7 Cemetery Lane, Tweedmouth, Berwick-upon-Tweed, TD15 2BS |
| G1 | IVK | T Garnham, 45 Buscot Drive, Abingdon, OX14 2BL |
| G1 | IVL | Richard Hudson, 80 Drake Avenue, Worcester, WR2 5RR |
| G1 | IVO | Lesley Ladner, 7 Polventon Close, Heamoor, Penzance, TR18 3LD |
| G1 | IVP | J Lamb, 5 Honeycroft, Loughton, IG10 3PR |
| G1 | IVV | G Merrington, Cartref, Ball Lane, Frodsham, WA6 8HP |
| G1 | IWE | Terance Coombs, 114 Talbot Street, Whitwick, Coalville, LE67 5AZ |
| G1 | IWH | Graham Dean, 33 Birdwell Common, Birdwell, Barnsley, S70 5TL |
| G1 | IWT | Reginald Moore, 9 Rowland St., Allenton, Derby, DE24 9BT |
| G1 | IXE | V Green, 50 Alcove Road, Fishponds, Bristol, BS16 3DR |
| G1 | IXF | Ivor Green, 50 Alcove Road, Bristol, BS16 3DR |
| G1 | IXV | Colin Haver, 18 Church Lane, Edenham, Bourne, PE10 0LS |
| G1 | IYA | Alfred Greenwood, 172 Bradford Road, Shipley, BD18 3DE |
| G1 | IYB | B Haines, 66 North Drive, Grove, Wantage, OX12 7PN |
| G1 | IYE | I Hawes, 129 Manor Road, Ash, Aldershot, GU12 6QB |
| G1 | IYO | Ronald Reed, 482 Baring Road, London, SE12 0EG |
| G1 | IZA | D Lamb, 33 Cherston Road, Loughton, IG10 3PL |
| G1 | IZB | Francis Smith, 6 Mill Close, Marshchapel, Grimsby, DN36 5TP |



**IMPORTANT NOTE**

**Revalidate licence to avoid revocation** – Ofcom has advised the Society that plans will be drawn up to revoke licences that have not been revalidated as required by the licence conditions. The quickest way to revalidate is to do so online via the Ofcom website: *https://services.ofcom.org.uk/* or by email: *amateur.validations@ofcom.org.uk* Ofcom staff are available to help, but please be patient during times of heavy workload.

G1 IZD C Stubbs, 3 Cartmel Close, Macclesfield, SK10 3PE
G1 IZH Anthony Trueman, Fairacre, Worrall Hill, Lydbrook, GL17 9QD
G1 IZN R Mitchell, 45 Kent Close, Mitcham, CR4 1XN
G1 JAA R Lees-Oakes, 6 Tabley Street, Mossley, Ashton-under-Lyne, OL5 9PD
G1 JAB J Burke, 48 Medina Road, Portsmouth, PO6 3HD
G1 JAG K Powell, 89 Avenue Road, Leicester, LE2 3EA
G1 JAH J Hagues, 7 Eastern Green Park Two, Eastern Green, Penzance, TR18 3BA
G1 JAL P Westbury, 6 Bradford Road, Rode, Frome, BA11 6PR
G1 JBB L Richards, 14 St. Julitta, Luxulyan, Bodmin, PL30 5ED
G1 JBE Derek Blackburn, 24 Reservoir Street, Darwen, BB3 1LQ
G1 JBG James Beacon, 14 Siskin Close, Bishops Waltham, Southampton, SO32 1RQ
G1 JBJ S Bartlett, Decoy Cottage, Decoy Road, Peterborough, PE6 7QD
G1 JBM Martyn Clark, 28 Compton Crescent, West Moors, Ferndown, BH22 0BZ
G1 JBT E Davey, 19 Northfield Road, Swaffham, PE37 7JB
G1 JBW B Ellison, 931 Burnley Road, Todmorden, OL14 7ET
G1 JBZ I Halsey, Cowtrott, Back Lane, Great Yarmouth, NR29 5ED
G1 JCC Ian Jefferson, 125 Telscombe Way, Luton, LU2 8QP
G1 JCL M Munn, 11 Foxley Road, Queenborough, ME11 5AW
G1 JCP J Pasfield, Fairlands, White Lodge Crescent, Clacton-on-Sea, CO16 0HT
G1 JCT S Farrant, The Bungalow, Brewery Yard, Stroud, GL5 4JW
G1 JCW Anthony Duffy, 14 Garden Street, Padiham, Burnley, BB12 8NP
G1 JDE O Graffham, 106 Barford Road, Edgbaston, Birmingham, B16 0EF
G1 JDF D Gray, 11 Field Close, Bollington, Macclesfield, SK10 5JG
GM1 JDJ L Mcleman, 14 Flures Place, Erskine, PA8 7DH
G1 JDO P Oliver, 67 High St., Great Houghton, Barnsley, S72 0AU
G1 JDP M Overton, 28 Broadviews, Great Lumley, Chester le Street, DH3 4HN
G1 JDQ P Paterson, Oak Lea, 11a Fletsand Road, Wilmslow, SK9 2AD
G1 JDT Graham Palmer, 5 Dunstar Avenue, Audenshaw, Manchester, M34 5LJ
G1 JDV J Vasey, 22 Rickleton Village Centre, Washington, NE38 9ET
G1 JEA David Hart, 10 Marina Drive, March, PE15 0AU
G1 JEH Kenneth Schneider, 65 Alpha Road, Birchington, CT7 9ED
G1 JER John Johnson, 5 Hunters Ride, Appleton Wiske, Northallerton, DL6 2BD
G1 JEZ S Taylor, 72 Molyneux Drive, Wallasey, CH45 1JT
G1 JFF A Weddell, 10 High Street, Eyemouth, TD14 5EU
G1 JFL M Woolridge, 23 Marina Drive, May Bank, Newcastle, ST5 9NL
G1 JFQ B Acheson, 32 Lords Lane, Brighouse, HD6 3RF
G1 JFT R Beaugie, 32 Court Gardens, Rogerstone, Newport, NP10 9FU
G1 JFU D Bryant, 22 Highfield Park, Heaton Mersey, Stockport, SK4 3HD
G1 JGD N Cullis, 39 Gilbert Drive, Langdon Hills, Basildon, SS16 6SP
G1 JGE Maurice Colley, 118 Devon Crescent, Birtley, Chester le Street, DH3 1HP
G1 JGF L Cox, 7 Timberdine Avenue, Worcester, WR5 2BD
G1 JGM B Easey, 4 Ash Trees, East Brent, Highbridge, TA9 4DQ
G1 JGR C Fortnum, 11 Ayr Close, Stamford, PE9 2TS
G1 JGS M Garland, 4a St Andrews Way, Freshwater, PO40 9NH
G1 JGT John Giller, 9 Alberta Crescent, Huntingdon, PE29 1TL
G1 JGY Howard Ketley, 24 Farmcroft Road, Mansfield Woodhouse, Mansfield, NG19 8QT
G1 JHB C Lawrence, Flat 22, St. Pauls Court, Salford, M7 3NZ
G1 JHD Martin Plant, 13 Willoughby Close, Alveston, Bristol, BS35 3RW
G1 JHG Tamsin McCormick, 33 Bryanston Road, Aigburth, Liverpool, L17 7AL
G1 JHL M Hubball, 6 Cypress Close, Woolston, Warrington, WA1 4EB
G1 JHM A Harding, 11 Mallard Close, Basingstoke, RG22 5JP
G1 JHN Francis Harris, 4 Parc an Ithan, The Lizard, Helston, TR12 7PA
G1 JHP H Hamer, 126 Mellor Brow, Mellor, Blackburn, BB2 7PN
GI1 JHQ Joe Selfridge, 6 Belmont Place, Coleraine, BT52 1QH
GM1 JHU J Adams, 3a Glenpatrick Road, Elderslie, Johnstone, PA5 9BH
G1 JHX A Bennett, 32 Park Road, Stretford, Manchester, M32 8DQ
G1 JHY A Potter, 25 Robinsons Meadow, Ledbury, HR8 1SU
G1 JHZ S Potter, 25 Robinsons Meadow, Ledbury, HR8 1SU
G1 JIE Keith Robertson, Tyn y Pwll, Fachwen, Caernarfon, LL55 3HD
G1 JIG Stuart Ridgard, 5 Orchard Way, Luton, LU4 9LT
G1 JIH E Rowell, 80 Kings Delph, Whittlesey, Peterborough, PE7 2PD
G1 JIJ J Passfield, 2 Parker Road, Chelmsford, CM2 0ES
G1 JIR D Robinson, 4 Mayorlowe Avenue, Stockport, SK5 8DB
G1 JIW Peter Spooner, 62 Chester Crescent, Newcastle, ST5 3RW
G1 JJA S Bessent, 407 Evesham Road, Crabbs Cross, Redditch, B97 5JA
G1 JJE N Banks, 6 Bylands Place, Newcastle, ST5 3PQ
G1 JJK P Naylor, 14 Wrockwardine Road, Wellington, Telford, TF1 3DB
G1 JJQ J Schulz, 4 Collinge Close, East Malling, West Malling, ME19 6QS
G1 JJR V Smith, 99 Dewhurst Road, Fartown, Huddersfield, HD2 1BN
G1 JJT N Stubbs, 8 West Avenue, Hilton, Derby, DE65 5FY
G1 JKE Nigel Knapton, 4 Crabmill Lane, Easingwold, York, YO61 3DE
G1 JKF N Leaney, 31 Saxon Way, Willingham, Cambridge, CB24 5UR
GM1 JKJ A Britton, 15 Glenbrook, Balerno, EH14 7JE
G1 JKL A Crouch, 107 Waterleat Avenue, Paignton, TQ3 3UD
G1 JKN P Cooper, 5 Lower Leys Way, Leominster, HR6 0SS
G1 JKO Gareth Cooper, Hazeldene, Fleet Coy, Spalding, PE12 0RU
G1 JKP R Coleman, 18 London Road, Thatcham, RG18 4LQ
G1 JKV T Whittaker, Flat 2, Paul Vanson Court, New Berry Lane, Walton-on-Thames, KT12 4HQ
G1 JKX J West, 4 Coronation Terrace, Longhorsley, Morpeth, NE65 8UN
G1 JLB M Denison, 9 Derwent Road, Harrogate, HG1 4SG
G1 JLE David Freeborough, 12 Meadow Lane, Newmarket, CB8 8FZ
G1 JLG B Giddings, 71 Tyrone Road, Southend-on-Sea, SS1 3HD
G1 JLM Steve Brosnan, 4 Black Rod Close, Hayes, UB3 4QJ
G1 JLP K Robson, 13 Woodstock Avenue, Galashiels, TD1 2EE
G1 JLQ John Yarnall, 4 Parklands, Evesham, WR11 2QJ
G1 JLX Nicholas Povey, 33 Church Lane, Fradley, Lichfield, WS13 8NJ
G1 JMC Ben Harris, 55 Valiant Way, Melton Mowbray, LE13 0GE
G1 JMD Peter Hall, 64 Synehurst Crescent, Badsey, Evesham, WR11 7XX
G1 JMF A Hooper, 5 Nine Elms Road, Longlevens, Gloucester, GL2 0HA
G1 JMH J Hickey, 36 Station Road, Alderholt, Fordingbridge, SP6 3RB
G1 JMK M Justice, 6 Stanley Terrace, Devizes, SN10 5AJ
G1 JMN R Andrews, Mount View, Park Lane, Worcester, WR2 6PQ
G1 JMP R Ainsworth, 95 Heysham Close, Murdishaw, Runcorn, WA7 6DT
G1 JMS J Stoddart, Apartment 11, 251 Wigan Road, Wigan, WN1 2RF
G1 JMV Peter Slark, 11 Hillfield Walk, Bolton, BL2 2UR
G1 JMW W Smith, 37 Peake Road, Brownhills, Walsall, WS8 7BZ
G1 JMY F Taylor, Wold Lodge, Pocklington Road, York, YO42 1YJ
GM1 JNC A Campbell, 17 Moulin Circus, Cardonald, Glasgow, G52 3JY
G1 JNG G Eccles, 1 Bridge Place, Amersham, HP6 6JF
G1 JNI A Fennah, 7 Y Ddol, Llanbrynmair, SY19 7DJ
G1 JNQ P Auld, 80 Milestone Road, Stone, Dartford, DA2 6DN
G1 JNR Pamela Adcock, Bleak House, Cefn Coch, Welshpool, SY21 0AE
GM1 JNS Gerald Bartram, 6 Craigewan Crescent, Peterhead, AB42 1HL
G1 JNX S Langston, 67 Oak Tree Drive, Hassocks, BN6 8YA
G1 JNY A Larkin, 16 Thetford Close, Corby, NN18 9PH
G1 JOA Barry Marsh, 90 Ellingham Road, Hemel Hempstead, HP2 5LL
G1 JOD R Norton, Middleton House, Bleathwood, Ludlow, SY8 4LX
G1 JOJ B Smith, 146 Battram Road, Ellistown, Coalville, LE67 1GB
G1 JOL B Shane, 7 Oakwood Glade, Holbeach, Spalding, PE12 7JS
G1 JON John Storey, Hampstead House, Fairfax Road, Birmingham, B31 3QY
G1 JOO R Seymour, 5 Clifton Place, Easton, Bristol, BS5 0SE
G1 JOR John Ormsby-Rymer, 109 Goldcrest Road, Chipping Sodbury, Bristol, BS37 6XJ
GW1 JOV Timothy Moore, Dan Bryn Coch, Llandyfan, Ammanford, SA18 2TY
G1 JOW Christopher Oates-Miller, 32 Burnlee Road, Holmfirth, HD9 2PS
G1 JPC Martin Ward, Four Winds, 13 Westfield Avenue, Wellingborough, NN9 6DQ
G1 JPF D Barton, 31 Belgrave Road, Fairbourne, LL38 2AZ
G1 JPI Melvyn Taylor, 26 Monks Close, Lancing, BN15 9DD
GM1 JPJ R Jamieson, 6a Mary Street, Stonehaven, AB39 2AD
G1 JPK T Jefferies, 98 St. Johns Road, Frome, BA11 2BD
G1 JPP A Hawes, 25 Folly Close, Fleet, GU52 7LN
G1 JPS Graham Chilton, 5 Horner Court, London, E11 4HY
G1 JPT B Gleave, 1 Fearnley Way, Newton-le-Willows, WA12 8SQ
G1 JQH Dennis Spalding, 171 Minster Road, Minster on Sea, Sheerness, ME12 3LH
G1 JQK S Gibbs, 43 Reddish Vale Road, Stockport, SK5 7EU
GI1 JQP John Innes, 22 Ashley Lodge, Dunmurry, Belfast, BT17 0AF
G1 JQR D Sell, 17 Auriel Avenue, Dagenham, RM10 8BS
G1 JRD John Barber, 4 Holyrood, Harwich, CO12 4UH
G1 JRF D Bruckshaw, 18 Old Moat Drive, Northfield, Birmingham, B31 2LY
G1 JRL H Cook, 31 Butley Road, Felixstowe, IP11 2NY
G1 JRP C Davis, 5 Redwing Avenue, Chippenham, SN14 6XJ
G1 JRR R Chalker, 182 Bridge Road, Chessington, KT9 2EY
G1 JRU D Evans, 63 Malwood Road West, Hythe, Southampton, SO45 5DL
G1 JRW D Gilchrist, 204 Great West Road, Heston, Hounslow, TW5 9AW
G1 JRX M Girgis, Rozel, Wilson Road, Kidderminster, DY11 7XU
G1 JRZ G Hobbs, 3 Glebe Cottage, Bremhill, Calne, SN11 9LD
G1 JSK P Lees, 2 Russet Close, Braintree, CM7 1DR
G1 JST P Johnson, 156 Norby Estate, Norby, Thirsk, YO7 1BQ
G1 JTC S Baskerville, Shalimar, Grove Lane, Leeds, LS6 2AP
GM1 JTK A Doig, 18 Gotterstone Drive, Broughty Ferry, Dundee, DD5 1QW
G1 JTM M Ferris, 22 Route De Matha, Aigre, France, 16140
G1 JTX L Sharman, 14 Northlands Avenue, Orpington, BR6 9LY
G1 JTZ B Linn, 35 New St., Carcroft, Doncaster, DN6 8EH
G1 JUD R Robinson, 24 Affleck Avenue, Radcliffe, Manchester, M26 1HN
G1 JUI M Lister, Beaconfield, Middle Road, Poole, BH16 6HJ
G1 JUO I Penney, 11 Eclipse Drive, Sittingbourne, ME10 2HR
G1 JUP M Baker, 9a, Manor Road, Alton, GU34 2PF
GW1 JVB Gerald Evans, 2 Old Village Road, Barry, CF62 6RA
G1 JVF Geoffrey Hannan, 20 Arlington Drive, Stockport, SK2 7EB
G1 JVH C Kirkman, The Nant, Nantmawr, Oswestry, SY10 9HN
G1 JVL J Leggett, 10 Home Park Road, Nuneaton, CV11 5UB
G1 JVM David Turton, 8 Lightwoods Road, Warley, Smethwick, B67 5AY
G1 JVN F Ursell, 110 Watt Lane, Sheffield, S10 5RE
G1 JVO Christine Ursell, 110 Watt Lane, Sheffield, S10 5RE
G1 JVU R Aitken, 81 Rashgill, Locharbriggs, Dumfries, DG1 1QN
G1 JVY A Smith, Woodlands, Old School Lane, Biggleswade, SG18 9JL
G1 JWD R Osborne, 24 Brockington Road, Bodenham, Hereford, HR1 3LR
G1 JWG David McKay, 15 Wellington Crescent, Baughurst, Tadley, RG26 5PJ
GM1 JWJ Ron Male, 13 Briar Grove, Forfar, DD8 1DQ
G1 JWL M Warren, Flat 15, Fulton Lodge, Harrogate, HG3 2UT
G1 JWO A Stone, 32 Berrynarbor Park, Sterridge Valley, Ilfracombe, EX34 9TA
G1 JWY Terry Yorke, 12 Shanklin Drive, Weddington, Nuneaton, CV10 0BA
G1 JXA R Whatley, 7 Okefield Road, Crediton, EX17 2DN
GI1 JXE G Murray, Ashgrove, 61 Monteith Road, Banbridge, BT32 5RD
G1 JXG Alan McMillan, 183 Forest Road, Clipstone Village, Mansfield, NG21 9DS
G1 JXL Mark Phillips, 71 Juniper Square, Havant, PO9 1HZ
G1 JXP Ian Wilson, Meadow Lodge, Kilhallon, Par, PL24 2RL
G1 JXS C Bunkum, 7 Goose Green Close, Wolvercote, Oxford, OX2 8QT
G1 JXX Haydn Williams, 24 Vaughan Close, Four Oaks, Sutton Coldfield, B74 4XR
G1 JYB Barrie Cartledge, Oysterber Farm, Burton Road, Lancaster, LA2 7ET
G1 JYH M Cherry, 36 Meads Avenue, Hove, BN3 8EE
G1 JYK S Langdale, Bramley House, Dishforth Road, Ripon, HG4 5BU
G1 JYR D Fraley, 1334 Warwick Road, Knowle, Solihull, B93 9LQ
G1 JYZ Alan Philpott, 2 Ocean View Road, Ventnor, PO38 1AA
G1 JZG M Wilmshurst, Chalklands, Main Street, Horncastle, LN9 5PT
G1 JZK Andrew Austin, 44 Mendip Crescent, Bedford, MK41 9EP
G1 JZL S Beckett, 15 Peaks Avenue, New Waltham, Grimsby, DN36 4LJ
GM1 JZM D Johnstone, 7 Gleneagles Avenue, Glenrothes, KY6 2QA
G1 JZN John Jacklin, 26 Rockmill End, Willingham, Cambridge, CB24 5HY
G1 JZT R Everitt, 55 Risborough Road, Bedford, MK41 9QR
G1 JZU D Goulden, 26 Derwent Walk, Greenacres, Oldham, OL4 2DJ
G1 JZX John Hesketh, 735 Manchester Road, Over Hulton, Bolton, BL5 1BA
G1 JZY T Mitchell, 22 Grundy Avenue, Prestwich, Manchester, M25 9TG
G1 JZZ D Porter, 2 Flour Mill Close, Burscough, Ormskirk, L40 5TL
G1 KAG A Watson, 60 Beresford Avenue, Surbiton, KT5 9LJ
G1 KAK C Buttery, Yew Tree Cottage, Chapel Lane, Newport, PO30 3DD
G1 KAO A Hawxby, 73 Anthea Drive, Huntington, York, YO31 9DB
G1 KAR SOUTHDOWN A R S c/o P Appleby, Flat 14, Maryan Court, Hailsham, BN27 3DJ
G1 KAS M Hughes, 43 Coach Road, Baildon, Shipley, BD17 5HS
G1 KAT Clive Lawrence, 23 Brutus Drive, Coleshill, Birmingham, B46 1UF
G1 KBC S Barrington, Fawley Cottage, Butt Lane, Loughborough, LE12 5EE
G1 KBE Thomas Bradley, 32 Laurel Gardens , Marlowe Road, Hartlepool, TS25 4NZ
G1 KBF C Halls, 16 Stoats Close, South Molton, EX36 4JU
G1 KBG D Arthur, Durnaford, Callington, PL17 7HP
G1 KBJ A Wagstaff, 124 Langwith Road, Langwith Junction, Mansfield, NG20 9RP
G1 KBL M Rack, 212 Willingham Street, Grimsby, DN32 9PY
GM1 KBZ S Crockford, 6 Kinmundy Gardens, Peterhead, AB42 2HW
GJ1 KCB A Du Chemin, Les Jardins des Sablons, La Grande route des Sablons, Fuschia, Grouville, Jersey, JE3 9HS
GM1 KCH W Curran, 10 The Laurels, Dundee, DD4 0AD
G1 KCR J Smith, 125 De Montfort Way, Coventry, CV4 7DU
G1 KCS A Scrutton, Ashleigh, Butt Hill, Southam, CV47 8NE
G1 KCU J Warrington, 204 High Street, Feltham, TW13 4HX
G1 KCW H Elleray, 34 Trent Close, Plymouth, PL3 6PB
G1 KDO D Cattell, 22 Budmouth Avenue, Weymouth, DT3 6JW
GI1 KDS K Lewis, 763 Antrim Road, Belfast, BT15 4EP
G1 KEB R Farmer, 72 Bradleys Lane, Wallbrook, Bilston, WV14 8YW
G1 KEI P Smith, 47 Bostock Road, Abingdon, OX14 1DW
G1 KEP J Houghton, Glenwood, Raby Road, Wirral, CH63 4JS
G1 KEV H Denton, 27 Melrose Gardens, Hersham, Walton-on-Thames, KT12 5HF
G1 KFB Timothy Jesson, The Hawthorns, The Outwoods, Hinckley, LE10 2UD
G1 KFG Judith Feay, 19 Dorset Avenue, Diggle, Oldham, OL3 5PL
G1 KFH J Richmond, 11 Elm Avenue, Pennington, Lymington, SO41 8BD
G1 KFQ P King, 10 Hockley Lane, Eastern Green, Coventry, CV5 7FR
G1 KGA O Himmo, 227b Caterham Drive, Coulsdon, CR5 1JS
G1 KGC Peter Simpson, 100 Pitchford Avenue, Maddington, Australia, WA 6109
G1 KGE M Phelps, Windermere, Wyson, Ludlow, SY8 4NQ
G1 KGL S Dorrington, Rosewood, 74 Hangleton Way, Hove, BN3 8EQ
G1 KGO S Coben, 106 Fleming Mead, Mitcham, CR4 3LW
G1 KGQ C Buxton, 6 Stoney Lane, Selston, Nottingham, NG16 6ET
G1 KGU J Amos, Mite View, Ravenglass, CA18 1SW
G1 KGV Carol Ashcroft, Wood End House, Wood End, Huntington, PE28 3LE
GI1 KGZ Peter Knott, 47 Pretoria Street, Belfast, BT9 5AQ
GI1 KHF Robert Maternaghan, 1 Pinegrove Crescent, Ballymena, BT43 6TL
G1 KHH Harold Mosley, 17 Cadwgan Road, Old Colwyn, Colwyn Bay, LL29 9PY
G1 KHM K Morgan, 157 Headlands, Fenstanton, Huntingdon, PE28 9LP
G1 KHS David Tucker, 5 Uplands Close, Hawkwell, Hockley, SS5 4DN
GM1 KHU Christopher Wall, 27 Golf Terrace, Insch, AB52 6JY
G1 KIB J Martin, The Old Bakehouse, Shotteswell, Banbury, OX17 1JA
G1 KII D Beale, 88 Long Innage, Cradley Forge, Halesowen, B63 2UY
G1 KIJ David Marsters, 21 Cow Lane, Rampton, Cambridge, CB24 8QG
G1 KIT John Fisher, 63 Rogers Avenue, Creswell, Worksop, S80 4JR
G1 KIW J Moss, 42 Chantry Lane, Necton, Swaffham, PE37 8ET
G1 KIZ T Head, 36a Ashacre Lane, Worthing, BN13 2DH
G1 KJG C Robinson, 24 Hackamore, Benfleet, SS7 3DU
G1 KJH Joseph Beach, 8 Harvey Lane, Norwich, NR7 0BQ
G1 KJQ Michael Haymes, 174 Queens Promenade, Blackpool, FY2 9JN
G1 KJX B Hobbs, 33 Hawthorn Drive, Heswall, Wirral, CH61 6UP
G1 KKA P MONTGOMERY, 7 Birchwood Close, Tavistock, PL19 8DR
G1 KKD T Elcock, Little Grange, 33 Cromford Drive, Derby, DE3 9JT
G1 KKE D Rose, 6 The Holt, Mollington, Banbury, OX17 1BE
G1 KKF A Lawes, 15 Leybourne Road, Brighton, BN2 4LT
G1 KKG M White, 7 Tyneham Close, Sandford, Wareham, BH20 7BE
G1 KKH P Cunliffe, 37 Rectory Road, Worthing, BN14 7PE
GM1 KKI K Johnston, Innisfree, Gulberwick, Shetland, ZE2 9JX
G1 KKJ K Taylor, 23 Vardre Avenue, Deganwy, Conwy, LL31 9UT
G1 KKS I Gott, Tayman House, The Street, Badminton, GL9 1HH
G1 KLI Mark Smith, Sycamore Farm, 6 Station Road, Spalding, PE12 0NP
G1 KLK A Dutton, 111 St. Michaels Road, Crosby, Liverpool, L23 7UL
G1 KLP Andy Jackson, 81 Suffield Way, King's Lynn, PE30 3DX
G1 KLW P Golding, 80 Birdbrook Road, London, SE3 9QP
G1 KLZ Doug Ellershaw, 38 Lakeber Avenue, Bentham, Lancaster, LA2 7JN
G1 KMJ J Couzins, 30 Camden Road, St. Peters, Broadstairs, CT10 3DR
G1 KMN Norman Thompson, 10 Belmont Crescent, Swindon, SN1 4EY
G1 KMS I Millar, The Grange, 105 High Street, Northampton, NN3 3JX
G1 KNA J Burdett, Glencorse, 13 Fairfax Avenue, Selby, YO8 4AZ
G1 KNI S Martin, 6 Prinsted Walk, Fareham, PO14 3AD
G1 KNK G Mellors, 21 Church Close, Stoke St. Gregory, Taunton, TA3 6HA
G1 KNQ G Roberts, 22 Grove St., New Balderton, Newark, NG24 3AZ
G1 KNU P Sharp, Purbrook Cottage, Lyme Road, Axminster, EX13 5BL
G1 KNX ST Jones, 31 Church Street, Tewkesbury, GL20 5PD
G1 KNZ Jeremy Washby, 2 Olivier Court, Council Avenue, Hull, HU4 6RW
G1 KOD John Rodgers, 5 Bridge Avenue, Latchford, Warrington, WA4 1RJ
G1 KOG E Beir, 17 Deansway, Hemel Hempstead, HP3 9UE
G1 KOH G Hall, 14 Evington Parks Road, Leicester, LE2 1PR
G1 KON L Mccoy, 56 Curate Road, Anfield, Liverpool, L6 0BZ
G1 KOP Liverpool Raynet Group c/o James Gibbard, 2 Almond Court, Liverpool, L19 2QZ
G1 KOR A Waddoups, 20 Stevenson Way, Wickford, SS12 9DY
G1 KOT Michael Lynn, 52 Vowler Road, Langdon Hills, Basildon, SS16 6AQ
G1 KOX T Williams, 24 Elm Tree Close, Northolt, UB5 6AR
G1 KPI T Houghton, 24a Studley Road, Torquay, TQ1 3JN
G1 KPU D Skilton, 137 Coast Drive, Lydd on Sea, Romney Marsh, TN29 9NS
G1 KPV A Rayner, 147 Ramuz Drive, Westcliff-on-Sea, SS0 9JN
G1 KPZ M Gillott, 2 Firthwood Avenue, Coal Aston, Dronfield, S18 3BQ
G1 KQE H Sweet, 5 Dence Close, Herne Bay, CT6 6BH
G1 KQH Steve Wigg, 45 Cambrian Lane, Rugeley, WS15 2XH
G1 KQN A Bowyer, 80 Holme Fen, Holme, Peterborough, PE7 3PR
G1 KQP Simon Price, Whitton Paddocks, Pulley, Shrewsbury, SY3 0AG
G1 KQU George Gray, 10a Albert Close, Rayleigh, SS6 8HP
GW1 KQV C Caudy, 43 Graham Avenue, Pen-y-Fai, Bridgend, CF31 4NR
GW1 KQY B Francis, 4 Heol Tir Coch, Efail Isaf, Pontypridd, CF38 1BW
G1 KQZ Antony Quinn, 581 Chorley Old Road, Bolton, BL1 6BL
G1 KRU Arthur Graves, 49 Robin Lane, Edgmond, Newport, TF10 8JL
G1 KRX B Piper, 26 Hare Law Gardens, Stanley, DH9 8DG
G1 KSC S Harrison, 44 Rosslyn Road, Whitwick, Coalville, LE67 5PT
G1 KSE A Robinson, 2 Frome Close, Marchwood, Southampton, SO40 4SL

G1 KSH P Sherwood, 3 Otham Close, Canterbury, CT2 7QX
G1 KSI S Bennington, The Oaks, Lynn Road, Wisbech, PE14 7DF
G1 KSK F Mullin, 26 Fearnhead Lane, Fearnhead, Warrington, WA2 0BE
G1 KSN Victor Tankard, 55 Filching Road, Eastbourne, BN20 8SD
G1 KST Jack Almond, 49 The Promenade, Withernsea, HU19 2DW
G1 KSW Enrico Giacani, 21 Barton Terrace, Leeds, LS11 8TP
G1 KTF Dominic Webb, Fairway, Barkham Road, Wokingham, RG41 4DH
G1 KTS G Goodridge, 110 Quarrendon Road, Amersham, HP7 9EP
G1 KTW C Thomas, 10 Acrefield Avenue, Guilsfield, Welshpool, SY21 9PN
G1 KTY R Rider, Rose Cottage, Coley Road, Bristol, BS40 6AP
G1 KTZ David Robins, Aynlo, Higher Road, Liskeard, PL14 5NQ
G1 KUG Lloyd Pinnoo, 1 Warwick Close, Pocklington, York, YO42 2FQ
GM1 KUI G Smith, 38 Crown Cottages, Stuartfield, Peterhead, AB42 5HH
G1 KUN T Sexton, 41 St. Bedes Gardens, Cambridge, CB1 3UF
G1 KUQ P Andrews, 37 Lonsdale Road, Harborne, Birmingham, B17 9QX
G1 KVC J Norris, 3 St. Pauls Close, Adlington, Chorley, PR6 9RS
G1 KVI Derek Bedson, Crud yr Awel, Old Llanfair Road, Harlech, LL46 2SS
G1 KVO M Haydon, 1 Glencrofts, Hockley, SS5 4GN
G1 KVP B Froggatt, 59 Queens Road, Rushall, Walsall, WS4 1HP
G1 KVQ Anthony Mallin, 88 Highbridge Road, Burnham-on-Sea, TA8 1LN
G1 KVR M Wood, Kviabol, Sheep Pen Lane, Seaford, BN25 4QR
G1 KVW I Wilson, 45 Meadway, Halstead, Sevenoaks, TN14 7EY
G1 KWA Bruce Simpson, Fassifern, Minto, Hawick, TD9 8SG
G1 KWF Stephen Watson, 1 Owl Way, Hartford, Huntingdon, PE29 1YZ
GM1 KWG J Winterbourne, Birkenbush, Clochan, Buckie, AB56 5AL
G1 KWK M Duerden, 12 Masefield Avenue, Bradford, BD9 6EX
GM1 KWX Charles Carter, Annachmor House, Clynder, Helensburgh, G84 0QD
G1 KXJ Anthony Poole, 17 Adelaide Street, Stonehouse, Plymouth, PL1 3JF
G1 KXP Stephen Lindsay-Smith, 47 Shaftesbury Avenue, Timperley, Altrincham, WA15 7NP
G1 KXQ M Bloxham, 34 Northcote Road, Farnborough, GU14 9EA
G1 KXX Andy Shaw, 70 Field Gardens, Steventon, Abingdon, OX13 6TF
G1 KXZ A Moore, 103 Park Grove, Barnsley, S70 1QE
G1 KYK M Partridge, Flat 2, Lymington Court Station Road, Sutton Coldfield, B73 5JY
G1 KYN D Jackson, 84 Devon Crescent, Birtley, Chester le Street, DH3 1HP
G1 KYV S Youngs, Glenlovat, Oakley Road, Cheltenham, GL52 6NZ
G1 KZA E Kilner, 3 Ruskin Close, Wath-upon-Dearne, Rotherham, S63 6NU
G1 KZD T Asker, 34 Post Office Road, Frettenham, Norwich, NR12 7AB
GM1 KZG John Southworth, 2 School Street, New Pitsligo, Fraserburgh, AB43 6NE
G1 KZI N Lansley, 126 The Promenade, Peacehaven, BN10 7JA
G1 LAN D Roberts, 192 Boothferry Road, Hull, HU4 6EW
G1 LAO J Flower, 12 Balmoral Close, Alton, GU34 1QY
G1 LAP J Beecham, Newholme, Whitemill Lane, Stone, ST15 0EG
G1 LAR L Rogers, 37 Gilbey Road, Tooting, London, SW17 0QQ
G1 LAT Stephany Kirkwood, 1 Nether View, Wennington, Lancaster, LA2 8NP
G1 LAW E Boyce, 214 London Road, Benfleet, SS7 5SJ
G1 LBH Peter Tomkins, 64 Glenwood Gardens, Bedworth, CV12 8DA
GI1 LBI H Budina, 46 Dunderg Road, Macosquin, Coleraine, BT51 4NE
G1 LBK M Kirk, Badgers Rise, Dunley Gardens, Stourport-on-Severn, DY13 0LL
G1 LBM Clifford Taylor, 28 Grace Court, Dial Lane, Bristol, BS16 5UP
G1 LBU Raymond Parry, 84 Sulgrave Road, Washington, NE37 3BZ
G1 LCC J Edwards, 1 Herons Way, Runcorn, WA7 1UH
G1 LCE S Turner, Rainow Villa, Under Rainow Road, Congleton, CW12 3PL
G1 LCN G Clegg, 16 The Pastures, Lower Westwood, Bradford-on-Avon, BA15 2BH
G1 LCR LEICESTER RAYNET GRP c/o Derek Harrison, 7 Shirley Close, Castle Donington, Derby, DE74 2XB
G1 LCS J Burrows, 1 Browns Avenue, Runwell, Wickford, SS11 7PT
G1 LCY R Brown, 3 Penmare Close, Hayle, TR27 4PJ
G1 LDC P Gibson, 1 Binney Road, Northwich, CW9 5PZ
G1 LDN J Burnet, 41 Douglas Crescent, Southampton, SO19 5JP
G1 LDY A Smith, 7 Gladstone Road, Broughton, Chester, CH4 0RN
G1 LED K Pinkard, 2 Lonsdale Court, Lache Lane, Chester, CH4 7LZ
G1 LEH M Nicholson, 7 Ellerbeck Close, Workington, CA14 4HY
G1 LEL A rowland, 66 Viking Way, Connah's Quay, Deeside, CH5 4JW
G1 LEN Leonard Taylor, 35 Leafield Road, Darlington, DL1 5DF
G1 LEO J Brittain, 26 Saxby Close, Eastbourne, BN23 7BH
G1 LES John Buckle, 14 Alder Close, Mapplewell, Barnsley, S75 6JA
G1 LEX J Blything, Blythwood, 319 Great Brickkiln Street, Wolverhampton, WV3 0PY
G1 LFD V Middleton, 5 Fieldhouse, Holmfirth, HD9 1EN
G1 LFI G Hepworth, 3 College View, Ackworth, Pontefract, WF7 7LA
G1 LFM David Mawdsley, 3 Chapel Lane, Cronton, Widnes, WA8 4NT
GW1 LFN D Rees, 17 Cwm Mwyn, Gorslas, Llanelli, SA14 7HY
G1 LFR D Love, 17 Longstaff Avenue, Rawnsley, Cannock, WS12 0QE
G1 LFX M Lamb, 4 Hadfield Close, Connah's Quay, Deeside, CH5 4JP
G1 LGB Graeme Gundry, 181 Stoneleigh Avenue, Worcester Park, KT4 8YA
G1 LGJ S Stephens, 41 Elliotts Lane, Codsall, Wolverhampton, WV8 1PG
GI1 LGM C Dowdall, 24 Glasmullen Road, Glenariffe, Ballymena, BT44 0QZ
G1 LGQ P White, 25 Witton Road, Ferryhill, DL17 8QE
G1 LGW D Hamilton, 4 Bannister Drive, Hull, HU9 1EJ
G1 LGY R Bishop, 26 Robin Gardens, Totton, Southampton, SO40 8US
G1 LHD M Goodes, 17 Ashmead Close, Lords Wood, Chatham, ME5 0NY
G1 LHE G Courtney, 24 Greenwood Close, Bognor Regis, PO22 9DG
G1 LHL Barry Mayson, 19 Hudson Close, Sturry, Canterbury, CT2 0HX
G1 LHQ Stephen King, 24 Cagney Drive, Swindon, SN25 4YR
G1 LHV J O'Nions, Pant-Glas, Gegin Lane, Wrexham, LL11 3YT
G1 LIG M Coker, 5 Penling Close, Cookham, Maidenhead, SL6 9NF
G1 LIK S Church, 24 Lovel End, Chalfont St. Peter, Gerrards Cross, SL9 9PA
G1 LJL B Duncan, 13 Westwick Grove, Sheffield, S8 7DP
GM1 LKD James Craib, 10 Cameron Road, Bridge of Don, Aberdeen, AB23 8QN
GW1 LKG R Cannon, 43 Plas St. Pol De Leon, Portway Marina, Penarth, CF64 1TR
G1 LKH R Hilton, 8 Hogshill Lane, Cobham, KT11 2AQ
G1 LKJ Philip Manning, 1 Waverley Gardens, Ash Vale, Aldershot, GU12 5JP
G1 LKK Colin Wardle, 16 Tedworth Avenue, Stenson Fields, Derby, DE24 3BS
G1 LKL J Wilkinson, 160 Staines Road, Feltham, TW14 9ED
G1 LLA C Davis, Fourwinds, Ringwood Road, Christchurch, BH23 7BE
G1 LLQ P Dollimore, 96 Hayes Bridge Ct, Uxbridge Road, Hayes, UB4 0JH
G1 LLU I Olver, 10 Celtic Road, Deal, CT14 9EE
G1 LLW D Rogers, 36 Guessens Road, Welwyn Garden City, AL8 6RH
G1 LLZ M Jackson, 146 Sea Road, Chapel St. Leonards, Skegness, PE24 5RY
G1 LMC J Trainer, 86 Plessey Road, Blyth, NE24 3LX
G1 LMI J Sanders, 35 Halkingcroft, Slough, SL3 7BB
G1 LML J Thornton, 7 Queens Road, Vicars Cross, Chester, CH3 5HA
G1 LMN S Froggatt, 255 Rushton Road, Desborough, Kettering, NN14 2QB
G1 LMQ Graham Donachie, 31 Eastfields, Narborough, King's Lynn, PE32 1SS
G1 LMS jonathan honeyball, 3 Mill End Close, Warboys, Warboys, PE28 2FP
G1 LMT Steve Longson, 7 John Street, Knutton, Newcastle, ST5 6DT
G1 LMU Stephen Hodgetts, 4 Sedgefield Walk, Catshill, Bromsgrove, B61 0SE
G1 LMW D Partridge, 44 Trumpet Terrace, Cleator, CA23 3DY
G1 LMZ C Clennell, 69 Seventh Row, Ashington, NE63 8HX
G1 LNA Dave Wood, 18 Rosemellin, Camborne, TR14 8QF
G1 LNQ P Dury, 2 Manor Rise, Thornton in Craven, Skipton, BD23 3TP
G1 LNR L Button, 37 Abbots Way, Preston Farm, North Shields, NE29 8LU
G1 LOE Kevan Gosling, 485b Blandford Road, Plymouth, PL3 6JF
G1 LOK Paul Blyth, 12 Beulah Street, King's Lynn, PE30 4DN
G1 LOL G Davidson, 18 Gotham Lane, Bunny, Nottingham, NG11 6QJ
G1 LOU L Vaisey, Esperanza, 5 Hudson Close, Ringwood, BH24 1XL
G1 LOV Paul Foster, 6 Croxteth Road, Bootle, L20 5EA
G1 LOW Phyllis Collick, 39 Beech Rise, Sleaford, NG34 8BJ
G1 LPQ D Thorpe, 167 Southwell Road West, Mansfield, NG18 4HD
G1 LPS Terry Roxby, 3 Coulton Terrace, Kirk Merrington, Spennymoor, DL16 7HN
G1 LQB L Clarke, 10 Stonecliff Park, Prebend Lane, Lincoln, LN2 3JS
G1 LQC P Chambers, 26 Drummond Gardens, Christ Church Mount, Epsom, KT19 8RP
G1 LQH M Lewis, 8 Wetherdown, Herne Farm, Petersfield, GU31 4PN
G1 LQM C Costello, The Coach House, The Green, Norwich, NR12 9PZ
G1 LQT Colin Matthews, 37 Arundel Way, Newquay, TR7 3AG
GD1 LQV G Austin, 10 St. Peters Close, Ruislip, HA4 9JT
G1 LQX R Saunders, 3 Lancaster Road, Cressex Industrial Estat, High Wycombe, HP12 3NN
G1 LRK C Bassett, 11 Redcastle Road, Thetford, IP24 3NF
G1 LRM R Cartmell, 16 Churchfield Drive, Wigginton, York, YO32 2FL
GW1 LRN Robert Metcalfe, 99 Dol y Dderwen, Ammanford, SA18 2GE
G1 LRR Laurence Pinto, 43 Napoleon Road, Twickenham, TW1 3EW
G1 LRU R Jones, 8 Mullen Avenue, Downs Barn, Milton Keynes, MK14 7LU
G1 LSB P Brockett, 146 Winsover Road, Spalding, PE11 1HQ
G1 LSK Ian Wiseman, Flat 4, Granville House, Hunstanton, PE36 6BS
G1 LSN M Felton, 1 Barnwell Close, Wistaston, Crewe, CW2 6TG
G1 LSX J Humphreys, 16 Church Croft, Madley, Hereford, HR2 9LT
G1 LSZ P Lambert, 22 Cullingworth Avenue, Hull, HU6 7DD
G1 LTC E Rodd, 32 Cranfield, Plympton, Plymouth, PL7 4PF
G1 LTE G Eades, 117 Booths Farm Road, Birmingham, B42 2NU
G1 LTG Graham Perry, 123 Green Lanes, Wylde Green, Sutton Coldfield, B73 5LT
G1 LTH D Williams, 5 West Road, Ormesby, Great Yarmouth, NR29 3RJ
G1 LTI P Smith, 65 Oatlands, Gossops Green, Crawley, RH11 8EH
G1 LTK Adrian Sturgess, 11 Keats Close, Earl Shilton, Goonhilly, Leicester, LE9 7DU
G1 LTL D Gardner, New House, Birdbush Avenue, Saffron Walden, CB11 4DJ
GM1 LTM Ursula Wallace, No 1 Holding Wester, Kincardie, Tayside, PH7 3RP
G1 LTV Stephen Eames, 4 Dabey Close, Markfield, LE67 9UJ
G1 LUC N Spring, The Orchard, Copyhold Lane, Dorchester, DT2 9LT
G1 LUF J Wright, 298 Field Road, Bloxwich, Walsall, WS3 3NB
G1 LUN P Baker, 651D Puketona Road, Paihia, New Zealand, 204
G1 LUX C Deacon, 12 Russet Way, Burnham-on-Crouch, CM0 8RB
GM1 LUZ C Campbell, 18 Parkview Avenue, Falkirk, FK1 5JX
G1 LVH D Barnett, Steep Holme, Front Street, Newark, NG23 7AA
G1 LVR Peter Cousins, 38 Braunston Drive, Hayes, UB4 9RB
G1 LVV A Critchlow, 51b West Road, Buxton, SK17 6HQ
G1 LVW P Parker, 43 Meadow Close, Farmoor, Oxford, OX2 9PA
GD1 LVY J Dorman, 1 Sprucewood Rise, Foxdale, Douglas, Isle of Man, IM4 3JP
G1 LVZ F West, 14 Ashley Drive, Twickenham, TW2 6HW
G1 LWE J Etchells, 34 Link Avenue, Urmston, Manchester, M41 9NJ
G1 LWF T Finch, 9 Halstead Road, Southampton, SO18 2PQ
G1 LWH B Whitehouse, 105 Quarry Road, Birmingham, B29 5LE
G1 LWL M Patrick, 39 Poplar Road, Healing, Grimsby, DN41 7RE
G1 LWX Michael Berry, 133 Rectory Road, Ashton-in-Makerfield, Wigan, WN4 0QF
G1 LWY K Slater, 56 Berners Road, Sheffield, S2 2GB
GM1 LXA Stephen Sellick, 1 Hatton Home Farm Cottages, Turriff, AB53 8ED
GM1 LXM Sandra McIntier, G/L 42 Boyd Street, Largs, KA30 8LE
G1 LZF B Dickinson, 178 Wycliffe Gardens, Shipley, BD18 3JB
G1 LZH P Taylor, 10 Pickenham Road, Birmingham, B14 4TG
G1 LZL Derek Atkinson, 74 Ruskin Close, High Harrington, Workington, CA14 4LS
G1 LZS Stephen Stallworthy, 6 Kenwood Close, Hastings, TN34 2AT
G1 LZZ D Dennett, Redhill Cottage, 14 Main Road, Wareham, BH20 5RN
G1 MAC Malcolm MacBeth, 58 The Common, Abberley, Worcester, WR6 6AY
G1 MAD Moorlands And District ARS c/o Christopher Beesley, 15 Byron Close, Cheadle, Stoke-on-Trent, ST10 1XB
G1 MAL D Simmons, 47 Lower Street, Haslemere, GU27 2NY
G1 MAR MIDLAND ARS c/o Norman Gutteridge, 68 Max Road, Quinton, Birmingham, B32 1LB
G1 MAV Gerard Seaman, 22a Mount Road, Bexleyheath, DA6 8JS
GW1 MAX Robert Pullen, 8 Carmarthen Court, Caerphilly, CF83 2TX
G1 MBE C Batty, 32 The Warings, Heskin, Chorley, PR7 5NZ
G1 MBG J Brown, The Meadows, Canworthy Water, Launceston, PL15 8UB
G1 MBM M Kitson, 54 Hollins Lane, Sowerby Bridge, HX6 2RP
G1 MBN P Doyle, 11 Clifford Avenue, Longton, Preston, PR4 5BH
G1 MBR MORECAMBE BAY ARS c/o C Edgar, 61 Winchester Avenue, Lancaster, LA1 4HX
GW1 MBV J Follett, 136 Westbourne Road, Penarth, CF64 3HH
G1 MBW M Smith, 32 Amesbury Drive, London, E4 7PZ
G1 MCG D Minton, 8 Rosebank Walk, Barnton, Northwich, CW8 4PU
G1 MCI A Rawson, 43 County Road North, Hull, HU5 4HN
GM1 MCN C Stanley, Nazareth House, 34 Claremont Street, Aberdeen, AB10 6RA
G1 MCR MERSEYSIDE CTY c/o E Hampson, Raynet Group, 21 Marlowe Road, Wallasey, CH44 3DA
G1 MCT G Williams, 29 Coleridge Road, Barnby Dun, Doncaster, DN3 1AN
G1 MCW W Higgins, 15 Redburn Close, Liverpool, L8 4XR
G1 MCY C Toogood, 16 Penlea Avenue, Bridgwater, TA6 6JU
G1 MDC D Tommey, 99 Fairfield Park Road, Bath, BA1 6JR
G1 MDF D Harrison, 10 Oakwood Road East, Rotherham, S60 3ER
G1 MDG CHESHAM&DIST AR c/o W Ray, 54 Gladstone Road, Chesham, HP5 3AD
G1 MDJ J Murch, Downings, Prinsted Lane, Emsworth, PO10 8HS
GM1 MDO J Stewart, 104 Barrhill Road, Cumnock, KA18 1PU
G1 MDQ Malcolm Mitchell, 310 Parlaunt Road, Slough, SL3 8AX
G1 MDS M Lewis, Westbank, 46 Weyside Road, Guildford, GU1 1HX
G1 MET R Heywood, 22 Catterall Close, Blackpool, FY1 3RB
G1 MFK M Morris, 1 Fruitlands, Malvern, WR14 4AH
G1 MGF A Hall, 172 Aldershot Road, Guildford, GU2 8BL
G1 MGI L Long, 18 Pentre Poeth Road, Bassaleg, Newport, NP10 8LL
G1 MGN T Jones, 35 Manta Road, Dosthill, Tamworth, B77 1PE
G1 MGU J Studd, 64 Moggs Mead, Herne Farm, Petersfield, GU31 4NX
G1 MGZ M Brophy, 78 Foley Road West, Streetly, Sutton Coldfield, B74 3NP
G1 MHA C Corner, 100 Monkseaton Drive, Whitley Bay, NE26 3DJ
G1 MHB Gareth Patten, 31 Sea View Road, Skegness, PE25 1BN
G1 MHF G Fleming, 25 Waverton Avenue, Prenton, CH43 0XB
G1 MHM WJ Taylor, Corsend Farmhouse, Corsend Road, Gloucester, GL19 3BP
G1 MHN Michael Dawson, Fairview, Wilkins Road, Wisbech, PE14 8DQ
G1 MHP David Gilbert, 10 Pigeon Grove, Bracknell, RG12 8AP
G1 MHZ John Richens, 168 Frobisher Drive, Swindon, SN3 3EW
G1 MIE Keith Martin, 21 All Saints Close, Weybourne, Holt, NR25 7HH
G1 MIL P Thompson, 19 Charleston Road, Penrhyn Bay, Llandudno, LL30 3HB
GD1 MIP Andrew Morgan, Thal'loo Glass, Nassau Road, The Dog Mills, Mwyljyn Moddey, Isle of Man, IM7 4AQ
G1 MIY R Lunnon, 9 Hennerton Way, High Wycombe, HP13 7UE
G1 MJA Peter Griffin, 147 Kingshayes Road, Aldridge, Walsall, WS9 8SN
G1 MJI Dennis Fry, West Tennis Court Cottage, Saunton, Braunton, EX33 1LL
GI1 MJJ F Gilliland, 48 Malone Heights, Belfast, BT9 5PG
G1 MJN Michael Newbold, Sawubona, Vicarage Lane, Skegness, PE24 4JJ
G1 MJO A Hemming, 64 Haslucks Green Road, Shirley, Solihull, B90 2EJ
G1 MJT R Naylor, 6 Alden Close, Morley, Leeds, LS27 0SG
G1 MJV N Liddiard, Orchard End, Dunwich Lane, Saxmundham, IP17 2JP
GM1 MKC C Strong, Little Coucher Cairn Crofts, St. Katherines, Inverurie, AB51 8TQ
G1 MKE R Cox, Carlingford, Brimley Drive, Teignmouth, TQ14 8LE
G1 MKP M Pauley, 12 Coxs End, Over, Cambridge, CB24 5TZ
G1 MKR MILTON KEYNES RAYNET GROUP c/o Donald Kirkwood, Greensand Lodge, High Street, Lidlington, Bedford, MK43 0QR
G1 MKS Paul Healy, 43 Brockley View, London, SE23 1SL
G1 MKV A Coe, Woodbrook Cottage, Paddock Row, Clwyd, LL14 6DD
G1 MLC H Taha, 53 Barbers Hill, Werrington, Peterborough, PE4 5ED
G1 MLK Andrew Bull, Field View House, 38 Sandles Road, Droitwich, WR9 8RA
GM1 MLS A Pearson, 1 Bridgefield, Inverbervie, Montrose, DD10 0SR
G1 MLV P Kelsey, 573 Stannington Road, Stannington, Sheffield, S6 6AB
GM1 MLW Colin Lindsay, 51 Perrays Crescent, Dumbarton, G82 5HP
GM1 MLY Michael Bull, 20 Grenitote, Lochmaddy, Isle of North Uist, HS6 5BP
G1 MMA S Butcher, 155 Crow Lane East, Newton-le-Willows, WA12 9UD
G1 MMD Keith Morris, 9 Hexham Close, Mansfield, NG18 3GR
G1 MME Alan Morgan, 6 Hartington Court, Hartington Grove, Cambridge, CB1 7TZ
G1 MMI T Ryder, 117 Cotefield Drive, Leighton Buzzard, LU7 3DN
GM1 MMK Kenneth Cupples, 16 Glebe Crescent, Airdrie, ML6 7DH
G1 MMN C Lovett, 49 Tame St. East, Walsall, WS1 3LB
G1 MMT S Quick, 4 David Way, Poole, BH15 4QX
G1 MMZ A Roffey, 32 Hertford Road, Digswell, Welwyn, AL6 0DB
GW1 MNC A Dykes, 16 The Mercies, Porthcawl, CF36 5HN
G1 MNP Howard Gray, Frankfurter Strasse 15, Backnang, Germany, D-71522
GW1 MNU I Jukes, 14 Beechwood Place, Narberth, SA67 7EE
G1 MNX P Lane, 1 St. Davids Close, Lower Willingdon, Eastbourne, BN22 0UZ
G1 MNY Richard Smith, 3 Florence Farm Mobile Home Park, London Road, Sevenoaks, TN15 6BP
G1 MOB N Booth, Greencotes, Warden, Hexham, NE46 4SS
G1 MOK Terry Dommett, 5 Bay View Gardens, Burnham-on-Sea, TA8 1LD
GM1 MON J McQueen, Rowan Cottage Redcastle, Lunanbay, Arbroath, DD11 5SS
G1 MOS H Whitbread, Foresters, Main Road, Woodbridge, IP12 4SL
G1 MOV M Ballantyne, 248 Calshot Road, Great Barr, Birmingham, B42 2BX
G1 MOW D Mallin, 60 Arundel Road, Littlehampton, BN17 7DF
G1 MOZ James Nicholson, 11 West View Rise, Huddersfield, HD1 4UR
G1 MPC M Human, 28 Lincoln Drive, Croxley Green, Rickmansworth, WD3 3NH
G1 MPD Melvin Champion, 31 West Street, Earl Shilton, Leicester, LE9 7EL
G1 MPG Colin Alefs, 27 Millfields, Beckermet, CA21 2YY
G1 MPI Marc Hillman, 28 Murndal Dr, Donvale, Australia, 3111
G1 MPL C Hitcham, 27 Kirby Cane Walk, Lowestoft, NR32 3EL
G1 MPP A Baxter, 94 Abbeyfield Drive, Fareham, PO15 5PF
G1 MPR G Roberts, Derwynno, Heol Offa, Wrexham, LL11 3EN
G1 MPT D Bowden, 4 Cornmill Close, Bardsey, Leeds, LS17 9EG
G1 MPU P Brolan, 2 Mount Road, Barnet, EN4 9RL
G1 MPW Stephen Cooke, 21 Woaldon Close, Southwater, Horsham, RH13 9HP
GM1 MQA Robin Cooke, Taigh na Greine, Lower Bayble, Isle of Lewis, HS2 0QB
G1 MQB P Barker, 7 Verbena Close, Nottingham, NG3 4PZ
G1 MQC DEVON COUNTY COUNCIL EMERGENCY PLNG c/o M Newport, 9 Highbury Park, Exmouth, EX8 3EJ
GM1 MQE Andrew Norris, 1 Bellard Road, West Kilbride, KA23 9JT
G1 MQQ G Stainton, 168 Slades Road, Golcar, Huddersfield, HD7 4JR
G1 MRE P Langford, 33 Briscoe Road, Hoddesdon, EN11 9DG
G1 MRI P Gardiner, 12 Weston Road, Aston Clinton, Aylesbury, HP22 5EG
G1 MRP H Powell, 5 Carter Road, Great Barr, Birmingham, B43 6JR
GM1 MRS Leslie Alexander, 97 Land Street, Keith, AB55 5AP
G1 MRX P Young, 30 Badminton Road, Maidenhead, SL6 4QT
GM1 MRY W Andrew, 11 Eddington Gardens, Chryston, Glasgow, G69 0JW
G1 MSA Alan Taylor, 8 Hartland Avenue, Coventry, CV2 3EQ
G1 MSB E Turner, 14 Lauderdale Gardens, Bushbury, Wolverhampton, WV10 8AY
G1 MSD Raymond Newsome, 6 Woodhouse Grove, Fartown, Huddersfield, HD2 1AS
G1 MSG M White, 11 Beck Way, Loddon, Norwich, NR14 6UZ

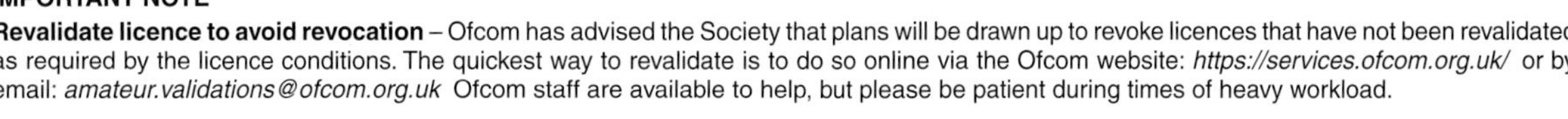

**IMPORTANT NOTE**

**Revalidate licence to avoid revocation** – Ofcom has advised the Society that plans will be drawn up to revoke licences that have not been revalidated as required by the licence conditions. The quickest way to revalidate is to do so online via the Ofcom website: *https://services.ofcom.org.uk/* or by email: *amateur.validations@ofcom.org.uk* Ofcom staff are available to help, but please be patient during times of heavy workload.

G1 MSK James Riley, Hillcrest, Norwich Road, Norwich, NR14 6BQ
GM1 MSN W Clark, 66 Winstanley Wynd, Kilwinning, KA13 6EB
GM1 MSO David Clark, 50 Dalry Road, Kilwinning, KA13 7HE
G1 MSR J Cockroft, 8 Harris Road, Standish, Wigan, WN6 0QR
GM1 MSS I Coates, 55 Whitehaugh Park, Peebles, EH45 9DB
G1 MSY K Ludgate, 138 Halton Road, Runcorn, WA7 5RW
G1 MTA Robin Grey, 23 Chapel Rise, Ringwood, BH24 2BL
G1 MTB Stephen White, 12 Morgan Close, Bexhill-on-Sea, TN39 5EQ
GW1 MTH Huw Williams, Flat 1, 74a Pontmorlais, Merthyr Tydfil, CF47 8UN
G1 MTJ S Tickle, 14 Rothesay Drive, Crosby, Liverpool, L23 0RF
G1 MTP N Mcneil, 3 Grosvenor Court, Water Lane, York, YO30 6PX
G1 MTU D Ward, 27 Penzer Street, Kingswinford, DY6 7AA
G1 MUC C Shingles, 20 Spencer Close, Lingwood, Norwich, NR13 4BB
G1 MUM H Phillips, 6 Peaks Down, Peatmoor, Swindon, SN5 5BH
GU1 MUP P Rudd, Val Des Arquets, Les Arquets, St. Pierre Du Bois, Guernsey, GY7 9HE
G1 MUQ P Hillier, Bythan, Avenbury Lane, Bromyard, HR7 4LB
G1 MUT P Lawrence, 2 Chapel Terrace, Station Street, Ashbourne, DE6 1DF
GM1 MUY Leslie Coxon, 40 Hamilton Street, Broughty Ferry, Dundee, DD5 2RE
G1 MVE P Tither, 32 Manor Avenue, Marston, Northwich, CW9 6DS
G1 MVF Stig Rasmussen, 10 The Pightle, Grafham, Huntingdon, PE28 0UU
G1 MVG F Marshall, Hartwell, Newgrounds, Fordingbridge, SP6 2LJ
G1 MVI P Bowe, 197 Gloucester Avenue, Chelmsford, CM2 9DX
G1 MVQ S Hamilton-Cooper, 4 Wren Close, Appleby Magna, Swadlincote, DE12 7BD
G1 MVT D Driscoll, 25 Broom Close, Dawlish, EX7 0RP
GW1 MVZ D Petrie, 48a Lower Quay Road, Hook, Haverfordwest, SA62 4LR
GM1 MWK James Grieve, 10 Jubilee Court, Kirkwall, KW15 1XR
G1 MWS Roger Bell, 92 Dean Drive, Wilmslow, SK9 2EY
G1 MWT M Clancy, 34 High Meadows, Greetland, Halifax, HX4 8QF
G1 MXC K Sampson, 40 Crisp Road, Lewes, BN7 2TX
G1 MXD G Waldron, 55 Sheringham Road, Poole, BH12 1NS
GM1 MXE G Schafers, Dahlsteven, Orkney, KW17 2RD
G1 MXM I Hunt, Four Seasons, Westmarsh, Canterbury, CT3 2LP
G1 MXO C Ratcliffe, 28 Vicarage Lane, Wilpshire, Blackburn, BB1 9HX
GM1 MYF Richard Jones, 46B Forest Road, Aberdeen, AB15 4BP
GD1 MYM E Emons, 18 Haig Road, Stanmore, HA7 4EP
G1 MYO F Ross, 2 Mount Pleasant, Steeple Claydon, Buckingham, MK18 2QS
G1 MYQ P England, Moonstones, Down Ampney, Cirencester, GL7 5QS
GM1 MYR R Cook, 95 Old Edinburgh Road, Inverness, IV2 3HT
G1 MZD Dave Barlow, 34 Mays Way, Potterspury, Towcester, NN12 7PP
G1 MZG S Banks, 18 Sheerstock, Haddenham, Aylesbury, HP17 8EU
G1 MZH E Schamp, 21 Beechwood Avenue, Melton Mowbray, LE13 1RT
GD1 MZJ J Sutherland, Archallagan Park, Marown, Isle of Man, IO9 9SU
G1 MZM M Bignell, 53 Rosebay Avenue, Birmingham, B38 9QT
G1 MZP A Froggatt, 16 Bridgwater Close, Walsall, WS9 9PL
G1 MZT Terence Tipper, 114 Paddock Lane, Redditch, B98 7XT
G1 MZW Ian Gillson, 13 Beech Green, Southcourt, Aylesbury, HP21 8JG
GM1 MZZ John Frearson, 27 Miltonbank Crescent, Guardbridge, St. Andrews, KY16 0XE
G1 NAA M Crabtree, 23 Ava Crescent, Richmond Hill, Ontario, Canada, L4B 2X1
G1 NAB G Rainy Brown, Old Stores Cottage, Newbury, RG20 8SE
G1 NAN A Gateley, 2 Langmere Road, Watton, Thetford, IP25 6LG
G1 NAP J Hopkins, 3 De Havilland Road, Upper Rissington, Cheltenham, GL54 2NZ
G1 NAQ Anthony Ashton, 6 Lansdowne Crescent, Darton, Barnsley, S75 5PW
G1 NAT D Mcgowan, 7 Eccles Close, Henley Green, Coventry, CV2 1EF
G1 NAU Jonathan Cawsey, 134 Goddard Avenue, Swindon, SN1 4HX
G1 NBK W Winning, Plump House, Terrington, York, YO60 6QB
G1 NBO R Mewis, 52 Princess Street, Burton-on-Trent, DE14 2NP
G1 NBP P Allan, 16 Farmstead Close, Grove, Wantage, OX12 0BD
G1 NBT Angela Bradley, 59 Main Road, Watnall, Nottingham, NG16 1HE
G1 NBU L Wellbeloved, 8 Orchard Close, South Wonston, Winchester, SO21 3EY
GW1 NBW Wayne Morris, 17 Fairway, Port Talbot, SA12 7HG
G1 NBY R Roeschlaub, 20 Pannatt Hill, Millom, LA18 5DB
G1 NCD P Moss, 20 Baristow Close, Chester, CH2 2EA
G1 NCG Kenneth Powell, 32 South View Avenue, Swindon, SN3 1EA
G1 NCK D Bates, 71 Nicholas Crescent, Fareham, PO15 5AJ
G1 NCL C Holmes, 16 Industrial St., Pelton, Chester le Street, DH2 1NR
G1 NCM Angus Urquhart, 75 Springvale Road, Winchester, SO23 7ND
G1 NCN H Jones, 8 Warren Close, Old Catton, Norwich, NR6 7NL
G1 NCO Peter Robinson, 12 Mountain Ash Avenue, Leigh-on-Sea, SS9 4SZ
G1 NCR NORTH CHESHIRE RC c/o G Gourley, 6a Longsight Lane, Cheadle Hulme, Cheadle, SK8 6PW
G1 NDK Keith Dunn, Marylands, Maidstone Road, Tonbridge, TN12 0RH
G1 NDL K Harrison, 20 Springfield Avenue, Ashbourne, DE6 1BJ
G1 NDQ M Taylor, 29 Harewood Avenue, Halifax, HX2 0LU
G1 NDV R Airey, 30 White Horse Crescent, Grove, Wantage, OX12 0PY
G1 NEB A Dearman, 232 Birmingham Road, Redditch, B97 6EL
GW1 NED H Anderson, Penrheol Farm, Meidrim, Carmarthen, SA33 5NX
G1 NEG Pat McGarry, 10 Douglas Avenue, Soothill, Batley, WF17 6HG
G1 NEN Anthony Hefford, 31 High Street, Rushton, Kettering, NN14 1RQ
GM1 NET STRATHCLYDE R G c/o R Campbell, 32 Harvie Avenue, Newton Mearns, Glasgow, G77 6LQ
G1 NEV David Dawson, 4 Hawksworth Lane, Guiseley, Leeds, LS20 8HA
GM1 NEW J Kerins, 30 Beech Avenue, Newton Mearns, Glasgow, G77 5PP
G1 NEZ Brian Stiff, 4 Timberlaine Road, Pevensey Bay, Pevensey, BN24 6DE
G1 NFB D Bagley, 38 Ashchurch Road, Tewkesbury, GL20 8BT
G1 NFE C Duffy, 25 Redcar Avenue, Thornton-Cleveleys, FY5 2LG
G1 NFN R Hammond, 124 Maney Hill Road, Sutton Coldfield, B72 1JU
G1 NFO B Webb, 63 Rother Road, Rotherham, S60 2UZ
G1 NFQ R Vivian, Flat 1, 26 Beer Road, Seaton, EX12 2PD
G1 NGE Roy Nelson, Woodlands, Norwich Road, Hevingham, NR10 5QX
G1 NGL Kevin Roberts, 535 Kbel Yliniemi Lane, Independence, United States, 97351
G1 NGN P Johansson, 63 Grange Road, Rhyl, LL18 4AD
G1 NGR Roderick Sharman, Flat 1, 11 Sherbourne Road, Blackpool, FY1 2PW
G1 NHG M Hausler, 5 Balaton Place, Snailwell Road, Newmarket, CB8 7YP
G1 NHX Peter Severn, 310 Worlds End Lane, Birmingham, B32 2SB
G1 NIC Nicholas James, Hillandale, Seaway Lane, Torquay, TQ2 6PN
G1 NIT Michael Virtue, 50 Borthwick Park, Orton Waterville, Peterborough, PE2 6YY
G1 NIV John Young, The Estate Office, Granary Court, Chadwell Heath, RM6 6PY
G1 NJG B Asker, 34 Post Office Road, Frettenham, Norwich, NR12 7AB
G1 NJI R Foster, 35 Colin Road, Barnwood, Gloucester, GL4 3JL
G1 NJV Michael Wing, 27 Hill Street, Hunstanton, PE36 5BS
G1 NKF I Spindler, 1 Spring Cottages, Brewery Lane, Stroud, GL5 2EA
G1 NKN A Mason, 43 Rosebank, Epsom, KT18 7RS
G1 NKT Peter Grant, 37 Glenmore Park, Dundalk, County Louth, Ireland
G1 NKV Francis Smith, 18-26 Hendon Rise, Thorneywood, Nottingham, NG3 3AN
G1 NLQ D Arter, 18 Essex Road, Westgate-on-Sea, CT8 8AP
G1 NLS Geoffrey Borrett, 45 Yarwells Headland, Whittlesey, Peterborough, PE7 1RF
G1 NLZ R Brown, 15 Johnson Road, Great Baddow, Chelmsford, CM2 7JL
G1 NML J Hyde, 6 Crown Green, Coventry, CV6 6FA
G1 NMN M Durey, 71 Orchard Road, Maldon, CM9 6EW
G1 NMP N Fenner, 22 Gowers Field, Aylesbury, HP20 2QT
G1 NMQ L Gibbs, 45 Woolavington Hill, Woolavington, Bridgwater, TA7 8HQ
G1 NMR M Kelly, 18 Fitzmaurice Road, Christchurch, BH23 2DY
G1 NMW Alex Stone, 5 Bridge Street, Cheltenham, GL51 9DQ
G1 NNA B Lloyd, 17 Brooklands Road, Brantham, Manningtree, CO11 1RN
G1 NNB G Lloyd, 9 Hornbeam Walk, Witham, CM8 2SZ
G1 NNF R Kenny, 35 Broom Leys Road, Coalville, LE67 4DD
G1 NNN S Moore, 104 Gloucester Avenue, Chelmsford, CM2 9LF
G1 NNR David Newman, 78 Vale Road, Poole, BH14 9AU
G1 NNU Michael Lowe, 1 Warecroft Road Kingsteignton, Newton Abbot, TQ12 3DN
G1 NOO Ian Allgood, 53 The Avenue, Leighton Bromswold, Huntingdon, PE28 5AW
G1 NOR G Galbraith, 44 Parker Road, Grays, RM17 5YN
G1 NOS K Brookes, 20 School Avenue, Guide Post, Choppington, NE62 5DN
G1 NPA P Allan, 214 Westwood Road, Sutton Coldfield, B73 6UQ
G1 NPC G Hammond, 31 Earlsway, Macclesfield, SK11 8RJ
G1 NPI C Badcock, 7 Heathfield Road, Chandler's Ford, Eastleigh, SO53 5RP
G1 NPJ Keith Hyslop, gallions point marina, royal docks, London, E16 2QY
G1 NPN Kenneth McDougall, 51 Argyll Avenue, Wirral, CH62 8EB
G1 NPP Peter Pritchard, South View, 16 Ashby Road, Leicester, LE7 4WF
G1 NQB Chris Carpenter, 19 Hambrook Lane, Stoke Gifford, Bristol, BS34 8QB
G1 NQH Marlene Cook, Brooksdie Cottage, Brook Lane, Market Harborough, LE16 8SJ
G1 NQN K Benfold, 56 Cornwall Avenue, Blackpool, FY2 9QW
G1 NQO James Jacques, 65 Daggers Hall Lane, Marton, Blackpool, FY4 4AX
G1 NQU M Peacock, 19 Ashfield Terrace, Haworth, Keighley, BD22 8PL
G1 NRE Derek Paul, 99 Wilkinson Road, Bedford, MK42 7FR
G1 NRF Michael Halloway, 41 Trenoweth Estate, North Country, Redruth, TR16 4AQ
G1 NRG Northants Raynet Group c/o Simon Manning, 11 Broomhill Crescent, Southfields, Northampton, NN3 5BH
G1 NRK Peter Slater, 12a Apsley Close, Bishop's Stortford, CM23 3PX
G1 NRM A Harrison, 34 Marsh Lane, Mill Hill, London, NW7 4QP
G1 NRN Kenneth Johnson, 98 Wroxham Road, Great Sankey, Warrington, WA5 3NU
G1 NRS NEWPORT ARS c/o Margaret Hill, 13 Maesglas Grove, Newport, NP20 3DJ
G1 NRX P Hill, 13 Onslow Road, Newent, GL18 1TL
G1 NRY I Leach, 36 Harrowden, Bradville, Milton Keynes, MK13 7DA
G1 NSB Roy Winterburn, Flat 15, Elms Farm, Mather Avenue, Manchester, M45 8NT
G1 NSD D Young, 13 Crawshaw Park, Pudsey, LS28 7EP
G1 NSG H Goodwin, 91 Grange Lane, Sutton Coldfield, B75 5LD
G1 NSQ C Joyce, 70 Campbell Road, Twickenham, TW2 5BY
G1 NST Sally Dodd, Chedburgh House, Hall Lane, North Walsham, NR28 0RZ
G1 NSV David Kemplen, 2 Vicarage Close, Menheniot, Liskeard, PL14 3QG
G1 NTI B Longstaff, 23 CHESTER ROAD ESTATE, Stanley, DH9 0QD
G1 NTK P Hull, Hazelwood, 2 Cheats Road, Taunton, TA3 5JW
G1 NTL J McShane, 12 Virginia Gardens, Middlesbrough, TS5 8BT
G1 NTN A Edwards, 9 Lincoln Close, Woodley, Romsey, SO51 7TJ
G1 NTP B Haden, 72 Charlton Road, Blackheath, London, SE3 8TT
G1 NTR Steven Mayer, 41 Lowe Street, Macclesfield, SK11 7NJ
G1 NTV D Merry, 18 Tremabe Park, Dobwalls, Liskeard, PL14 6JS
G1 NTX S Taylor, 5 Collingwood, Farnborough, GU14 6LX
G1 NUH R Hardwick, 5 Seaview, Oakmere Park, Little Neston, CH64 0XP
G1 NUN S Lomas, 62 Springfield Road, Etwall, Derby, DE65 6LA
G1 NUO Edward Oram, 31 Nathaniel Walk, Tring, HP23 5DG
G1 NUS John Thornley, 270 Hurdsfield Road, Macclesfield, SK10 2PN
G1 NVE V Wood, 175 Windleshaw Road, Dentons Green, St. Helens, WA10 6TP
G1 NVL G Officer, Flat 5, 8 Charlton Drive, Sale, M33 2BJ
G1 NVN A Parkin, 1 Dunelm Walk, Leadgate, Consett, DH8 7QT
G1 NVO B Kimber, 27 Court Road, Brockworth, Gloucester, GL3 4ES
G1 NVS R Pennington, 5 Park Close, Northway, Tewkesbury, GL20 8RB
G1 NVV Michael Timlett, 111 Poynters Road, Dunstable, LU5 4SQ
G1 NVY K Peers, 47 Walpole Avenue, Whiston, Prescot, L35 2XX
G1 NWA Colin Rickerby, 113 Cliftonville Road, Woolston, Warrington, WA1 4BJ
GW1 NWF R Gray, 36 Heol Pentre Felen, Morriston, Swansea, SA6 6BY
G1 NWG C Scates, 17 Trecastle Way, Carleton Road, London, N7 0EL
G1 NWH H Walker, 24 Castleton Avenue, Riddings, Alfreton, DE55 4AG
G1 NWM James Westwood, 9 Landbeach Road, Milton, Cambridge, CB24 6DA
G1 NWO J Sharp, 3 Inkerman Road, Eton Wick, Windsor, SL4 6LE
G1 NWT B Worviell, Cliddesden, The Street, Shaftesbury, SP7 9PF
G1 NWZ Michael Spacey, 4 Hickman Court, Copenhagen Close, Luton, LU3 3TW
G1 NXB G Fardoe, 3 Park Avenue, Wallasey, CH44 9DZ
G1 NXI Colin Foster, 18 Ellsworth Rise, Nottingham, NG5 5LT
G1 NXR Veronica Barrett, 4 Alexandra Street, Heywood, OL10 2AU
G1 NXS B Martlew, 15 Dunscar Close, Birchwood, Warrington, WA3 7LS
G1 NXT K Rushton, 9 Laburnum Avenue, Woolston, Warrington, WA1 4NY
G1 NXV R Roycroft, Roadside House, Knutsford Road, Macclesfield, SK11 9AS
G1 NYI T Rozier, 26 Watersmeet Way, London, SE28 8PU
G1 NYJ David Hopton, 32 Braemar Avenue, Urmston, Manchester, M41 6HP
G1 NYN D Stilgoe, 35 Pildacre Lane, Chickenley, Dewsbury, WF12 8NR
G1 NYP M Fleet, 152 Bridge Road, Chessington, KT9 2EY
G1 NYS R Parkhurst, 26 Valebridge Road, Burgess Hill, RH15 0QY
G1 NYZ Derek Robinson, 4 Rushden Drive, Reading, RG2 8LJ
G1 NZD P Bates, 132 Brownings Avenue, Chelmsford, CM1 4HJ
GW1 NZF R Stuckey, 8 Gelli Crossing, Gelli, Pentre, CF41 7UD
G1 NZH Jack Edgecock, 13 Holmsdale Close, Durgates, Wadhurst, TN5 6UT
G1 NZK A Davis, 73a Milton Road, Taunton, TA1 2JQ
G1 NZL Colin Elliott, 7 Elizabeth Diamond Gardens, South Shields, NE33 5HX
G1 NZN James Rolley, 12 Ravenscroft Drive, Chaddesden, Derby, DE21 6NX
G1 NZP E Elliott, 7 Red House Road, Hebburn, NE31 2XS
G1 NZQ Richard Finch, 12 Simcox Street, Hednesford, Cannock, WS12 1BG
G1 NZZ Richard Nicol, 37 Thicknall Drive, Stourbridge, DY9 0YH
G1 OAE R Steel, 7 Derwent Bank, Seaton, Workington, CA14 1EE
G1 OAM Trevor Wilson, 15 Whipperley Way, Luton, LU1 5LB
G1 OAR P Wallace, 7 Trinity View, Ketley Bank, Telford, TF2 0DX
G1 OAU V English, 29e High Street, Eye, Peterborough, PE6 7UP
G1 OAW J Smith, 62 Elson Lane, Elson, Gosport, PO12 4EU
G1 OAX Leroy Pugh, 15 Didcott Way, Appleby Magna, Swadlincote, DE12 7AS
G1 OAZ Graham Sugden, 247 Yorkland Avenue, Welling, DA16 2LH
G1 OBA I Bpophy, 78 Foley Road West, Streetly, Sutton Coldfield, B74 3NP
G1 OBC I Evans, 6 Park End, Lichfield, WS14 9US
G1 OBM J Miller, 2 Penpethy Close, Brixham, TQ5 8NP
G1 OCH C Hillman, Mayjon, Crow, Ringwood, BH24 3ER
G1 OCK R Liepziger, 21 Third Avenue, Woodside Park, Poulton-le-Fylde, FY6 0PW
G1 OCL D Potter, 9 Beachcroft Place, Lancing, BN15 8JN
G1 OCR Neil Law, April Cottage, Station Road, Lymington, SO41 6AB
G1 OCS A Heap, 56 Moorside Road, Eccleshill, Bradford, BD2 3RB
G1 OCY K Minihane, 60 Wolsey Drive, Walton-on-Thames, KT12 3BA
G1 ODB D Price, 21 Orchard Road, Nailsea, Bristol, BS48 2DZ
G1 ODD B Westlake, 47 Quarry Road, Kingswood, Bristol, BS15 8NZ
G1 ODE Godfrey Wheeler, 14 Mina Road, Bristol, BS2 9TB
G1 ODJ Trevor Boycott, 17 Brook Street, Whitley Bay, NE26 1AF
G1 ODK G Cole, 87 Chichester Road, Ramsgate, CT12 6NZ
G1 ODN Julian Thompson, 14 Redwing Close, Horsham, RH13 5PE
G1 ODQ C Bond, 6 Copse Close, Hugglescote, Coalville, LE67 2GL
G1 ODT Alan Pargeter, The Acres, Longhorsley, Morpeth, NE65 8QH
G1 ODZ G Fountain, Tinkers Lane, Wigginton, Tring, HP23 6JB
G1 OEB A Orchard, Flat No 3, 35 The High Street, Hemel Hempstead, HP3 0HG
G1 OEF E Earland, 7 Paxford House Square, Ottery St. Mary, EX11 1BX
G1 OEM Terence Webb, 95 Devereaux Crescent, Ebley, Stroud, GL5 4PX
G1 OEP J Harris, 109 Hook Rise South, Surbiton, KT6 7NA
G1 OEQ D Tribute, 'Pathey', Lower Polstain Road, Truro, TR3 6BQ
G1 OER A Parkin, 4 Waverley Road, Farnborough, GU14 7EY
G1 OET Keith Doswell, 15a Queen Street, Desborough, Kettering, NN14 2RE
G1 OFG P Howard, Cork Farm, Ruthern Bridge, Bodmin, PL30 5LU
G1 OFL Rahim Hall-Osman, 67 Livingstone Road, Gravesend, DA12 5DN
G1 OFW Steven Gadsby, 30 Woodside Close, Knaphill, Woking, GU21 2DD
G1 OFX W Coates, 3 Graysmead, Sible Hedingham, Halstead, CO9 3NX
G1 OFY P Hendy, 4 The Pack, Burgh-by-Sands, Carlisle, CA5 6BE
G1 OGB Peter Campion, 16 Hurrell Court, Efford Lane, Plymouth, PL3 6LT
G1 OGC J Kelly, 18 Mount Pleasant, Riddings, Alfreton, DE55 4BL
G1 OGE A Wilson, Moor Cottage, Ellastone Road, Stoke-on-Trent, ST10 3ER
G1 OGH K Atkinson, 62 Fines Park, Stanley, DH9 8QY
G1 OGR R Welch, 2 Broadlands Avenue, Waterlooville, PO7 7JE
G1 OGV R Bicknell-Thompson, 4 Linden Court, Greenfrith Drive, Tonbridge, TN10 3LW
G1 OGY D Gilligan, 21 Daen Ingas, Danbury, Chelmsford, CM3 4DB
GM1 OGZ D Madden, Flat 2/2, 24 Collier Street, Johnstone, PA5 8AR
G1 OHD B Beswick, 2 Ferndale Road, Peak Dale, Buxton, SK17 8AY
G1 OHH S Griffin, 6 Raygill Place, Lancaster, LA1 2UQ
G1 OHL W Morden, Apartment 3, Fernside House 49 Hollington Park Road, St. Leonards-on-Sea, TN38 0SE
G1 OHU John Freeman, 81 West Hill, Kimberworth, Rotherham, S61 2EX
G1 OHV Roger De Havilland, 11 Morvale Street, Stourbridge, DY9 8DE
G1 OHX V Pears, 10 Fremantle Road, South Shields, NE34 7RF
G1 OIB R Jones, 10 Ferndale Crescent, Gobowen, Oswestry, SY11 3PJ
GW1 OII S Lloyd, 10 Park Crescent, Llanelli, SA15 3AE
G1 OIK William Jaggard, 2 Aled Drive, Rhos on Sea, Colwyn Bay, LL28 4UU
G1 OIO W Benton, 2 Regents Close, Seaford, BN25 2EB
G1 OIS N Ellis, 1a Northcote Road, Croydon, CR0 2HX
G1 OIZ R Young, 1 Croft Walk, Whitwell, Worksop, S80 4UD
G1 OJB P Critchley, 4 Shandon Avenue, Northenden, Manchester, M22 4DP
G1 OJD D Facer, 7 Lowry Close, Bedworth, CV12 8DG
G1 OJL Stanley Evenden, 11 Chapel Street, Tavistock, PL19 8DX
G1 OJO D Pearson, 16 Hilldown Road, Hayes, Bromley, BR2 7HX
G1 OJQ Peter Brooks, 7 Ashbourne Road, Underwood, Nottingham, NG16 5EH
G1 OJS Alan Robinson, 1 The Heights, Fareham, PO16 8TL
G1 OJT R Barnish, 64 Braithwell Road, Maltby, Rotherham, S66 8JU
G1 OKB A Ibbotson, 62 Crag View Crescent, Oughtibridge, Sheffield, S35 0GD
G1 OKD R Penfold, 12 Kings Avenue, Chippenham, SN14 0UJ
G1 OKF D Cannon, 44 Grange Bottom, Royston, SG8 9UQ
G1 OKI D Hyde, 108 St. Bedes Crescent, Cambridge, CB1 3UB
G1 OKK Leslie Railton, 41 Signal Hayes Road, Walmley, Sutton Coldfield, B76 2RP
G1 OKP Robert Lloyd, 52 Roman Way, Ross-on-Wye, HR9 5RL
G1 OKV G Hendricks, 105 Hillcrest Park, Wilbury Hills Road, Letchworth Garden City, SG6 4LF
G1 OKY J Cottrell, Bryn Dewi, Llanallgo, Moelfre, LL72 8HB
G1 OLE D Bates, 10 Upton Gardens, Worthing, BN13 1DA
G1 OLM J Hesketh, 87 Condor Grove, Blackpool, FY1 5NA
G1 OLQ William Dacey, 69 Freshfield Gardens, Allerton, Bradford, BD15 7PR
G1 OLT Craig Sawyer, 4 Padley Close, Ripley, DE5 3FG
G1 OLY Oliver Brooks, 34 Jubilee Road, Stokenchurch, High Wycombe, HP14 3SJ
GI1 OMD Terry McQuaid, 5 Edenamohill, Drumkeen, Enniskillen, BT93 0FQ
G1 OMI John Canning, 130 Main Road, Duston, Northampton, NN5 6RA
G1 OMX P Cumiskey, 1 York Terrace, Gateshead, NE10 9NB
G1 OMY D Ainscough, 11 Tressel Drive, Sutton Manor, St. Helens, WA9 4BS

G1 OMZ E Rogers, 5 Brooking Way, Saltash, PL12 4TJ
G1 ONC P Maitland, 7 Spinners Court, Stalham, Norwich, NR12 9EQ
G1 OMR D Hodtry, 44 Kilsyth Close, Fearnhead, Warrington, WA2 0SQ
G1 ONE Bolton Wireless Club c/o Derek Lewis, 10 Addington Road, Bolton, BL3 4QZ
G1 ONH F Slater, 32 Winthorpe Avenue, Morecambe, LA4 4RE
G1 ONJ D Tennant, 128 Devonshire Street, Keighley, BD21 2QJ
G1 ONK L Boston, Lissa Park Sitesi No 16 (86/R), 965 Sokak Mustafa Kemal Bulvari, Calis, Turkey, FETHIYE 48300
G1 ONQ Frederick Pearce, 1b Council Street, Bozeat, Wellingborough, NN29 7LS
G1 ONV Robert Bonar, 6 Harepark, Allerford, Minehead, TA24 8HI
G1 OOB J Clark, The Bungalow, Sutton Lane, Street, BA16 9RJ
G1 OOG A Cooper, Riverfield, Creek View Avenue, Hockley, SS5 6LU
G1 OOJ M Watson, 15 Ellis Park, Gt. Glemham, Saxmundham, IP17 2DA
G1 OOM Barry Woodcock, 27 Main Street, Cosby, Leicester, LE9 1UW
G1 OOS E Rowthorn, 4 Woburn Court, Rushden, NN10 9HL
G1 OOU J Pearce, 19 Cawthorne Road, Kettlethorpe, Wakefield, WF2 7HW
G1 OOW D Strooter, 78 Stockfield Road, Acocks Green, Birmingham, B27 6BB
G1 OOZ D Parsons, 1 Carlyle Road, Rowley Regis, B65 9BQ
G1 OPA D Smith, 15 Billington Close, Coventry, CV2 5NQ
G1 OPD P Elsom, 7b Church Lane, Keelby, Grimsby, DN41 8ED
G1 OPG Ken Chappell, 21 Victoria Street, Long Eaton, Nottingham, NG10 3EW
G1 OPJ Alex Golding Brown, 17 Main Street, Withybrook, Coventry, CV7 9LT
GM1 OPO George Askew, 49 Kittlegairy Road, Peebles, EH45 9LX
G1 OPT P Harvey, 4 Linden Grove, Teddington, TW11 8LT
G1 OPV P Drew, 20 Russell Street, Accrington, BB5 2NF
G1 OPW F Cox, 44 Mountain Wood, Bathford, Bath, BA1 7SB
G1 OPZ Fred Anderson, Fansana, 2 Monmouth Paddock, Bath, BA2 7LA
G1 OQB R Tams, 7 Hermitage Road, Abingdon, OX14 5RN
G1 OQF C Caines, 6 Abel Smith Gardens, Branston, Lincoln, LN4 1NN
G1 OQG David Fryer, 16 elston place, Aldershot, GU124HY
G1 OQI C Kill, 169 Spring Road, Southampton, SO19 2NU
G1 OQM M Palmer, 250 Kinson Road, East Howe, Bournemouth, BH10 5EP
G1 OQO M Williamson, Greenfields Farm, Plumley Moor Road, Knutsford, WA16 9SB
GM1 OQT James Watson, 64 Anstruther Street, Law, Carluke, ML8 5JG
G1 OQU A Windsor, 8 Tresawla Court, Tolvaddon, Camborne, TR14 0HF
G1 OQV John Connor, 28 Church Street, Hungerford, RG17 0JE
G1 OQW A Boot, 63 Hunters Way, Stoke-on-Trent, ST4 5EF
G1 OQX Margaret Dunham, 5 King Street, Wimblington, March, PE15 0QF
G1 ORB David Tomsett, 20 North Avenue, Bognor Regis, PO22 6HG
G1 ORC OLDHAM AM RAD C c/o G Oliver, 158 High Barn St., Royton, Oldham, OL2 6RW
G1 ORG David Stanley, 25 Kingsley Crescent, Bulkington, Bedworth, CV12 9PS
G1 ORK D Spicer, 35 Strood Road, St. Leonards-on-Sea, TN37 6PN
G1 ORL C Barlow, 16 Fosseway South, Midsomer Norton, Radstock, BA3 4AN
G1 ORN F Daniels, 6 Middlemead, Stratton-on-the-Fosse, Radstock, BA3 4QH
G1 ORP J Marlow, West Bulthy, Bulthy, Welshpool, SY21 8ER
G1 ORS Bruce Williams, 3 Welton Close, Wilmslow, SK9 6HD
G1 ORT Simon Smale, The Old Vicarage, 68 Cardigan Road, Bridlington, YO15 3JT
G1 OSA A Sleigh, 2 Rock Terrace, Buxton, SK17 6HN
G1 OSE R Howlett, 37 Waveney Drive, Hoveton, Norwich, NR12 8DP
G1 OSG W Beilby, 119 Beaconfield, Withernsea, HU19 2EW
G1 OSH G Slater, 12a Apsley Close, Bishop's Stortford, CM23 3PX
G1 OSI John Nicholson, 117 Lower Meadow, Harlow, CM18 7RF
G1 OSJ M Howard, 8 Abbotts Crescent, St. Ives, Huntingdon, PE17 6YB
G1 OSL Gordon Hunter, 2 Dilloway Street, St. Helens, WA10 4LN
G1 OSO Andrew Durbridge, 16 Nightingale Drive, Mytchett, Camberley, GU16 6BZ
G1 OSP Jonathan Woollons, 28 Columbus Ravine, Scarborough, YO12 7JT
GM1 OST Thomas Ferguson, 40 Dallowie Road, Patna, Ayr, KA6 7ND
G1 OTA D Lee, 25 Elm View, Steeton, Keighley, BD20 6SZ
G1 OTI Kenneth Nickson, 25 Burntwood Road, Buckley, CH7 3EL
G1 OTN R Weight, 1 Crowland Road, Thornton Heath, CR7 8RP
G1 OTZ J Macdonald, 42 Lion Lane, Haslemere, GU27 1JD
G1 OUA H Saunders, 8 Norfolk Road, Luton, LU2 0RE
G1 OUG Kevin Anderson, 9 Bradford Park Drive, Bolton, BL2 1PA
GW1 OUP D George, 24 Ty Fry Close, Brynmenyn, Bridgend, CF32 8YB
G1 OUX P Bolderson, 113 Kirkdale Crescent, Leeds, LS12 6AY
G1 OUY T Willans, 3 Highfield, Hatton Park, Warwick, CV35 7TQ
G1 OVG T Powell, 11 Wymering Lane, Portsmouth, PO6 3QT
G1 OVH Neil Waud, 21 Olivia Road, Brampton, Huntingdon, PE28 4RP
GM1 OVJ W Forsyth, Aldernaig, Braehead, Avoch, IV9 8QL
G1 OVO R Cox, 60 Prospect Crescent, Whitton, Twickenham, TW2 7EA
GM1 OVW R Hetherington, 37 Brockwood Avenue, Penicuik, EH26 9AN
G1 OVY P McClelland, 30 Bowyer Road, Abingdon, OX14 2EF
G1 OWD M Forsyth, 13 Hillside Close, Paulton, Bristol, BS39 7PN
G1 OWI M Branch, 38 Kynaston Road, Didcot, OX11 8HD
G1 OWJ A Copsey, 13 Monro Avenue, Crownhill, Milton Keynes, MK8 0BB
G1 OWK George Garner, 8 Lansdowne Road, Swadlincote, DE11 9DZ
G1 OWM G Woodley, 16 Albert St., St. Barnabas, Oxford, OX2 6AY
G1 OWZ John Scott, 10 Beethoven Close, Old Farm Park, Milton Keynes, MK7 8PL
G1 OXB D Owen, 8 Kingsdown Road, Epsom, KT17 3PU
G1 OXF R Lewis, 0 Wethordown, Herne Farm, Petersfield, GU31 4PN
G1 OXH A Price, 10 Low Meadow, Whaley Bridge, High Peak, SK23 7AY
G1 OXJ Ian Jones, 15 Victoria Road, Penygroes, Caernarfon, LL54 6HD
G1 OXO J Ilston, 6 Dovedale, Canvey Island, SS8 8HX
G1 OXQ I McKune, 16 Queensberry Court, Dumfries, DG1 1BT
G1 OXT Roy Faulkner, 10 Fell Wilson Street, Warsop, Mansfield, NG20 0PT
G1 OYF V Shirley, 18 Crotch Crescent, Marston, Oxford, OX3 0JJ
G1 OYG D Crowe, 18 Bengairn Avenue, Patcham, Brighton, BN1 8RH
G1 OYH Howard Grinter, 9 Fieldway, Sandford, Winscombe, BS25 5PR
G1 OYM A Ingram, 78 Kenwood Gardens, Gants Hill, Ilford, IG2 6YG
G1 OYU Beryl Toon, 2 Marstonlane Park, Rolleston On Dove, Staffs, DE13 9BJ
G1 OYZ P Smith, 189 Rolleston Road, Burton-on-Trent, DE13 0LD
G1 OZB A Saunders, Firtrees, Wheatcroft Avenue, Bewdley, DY12 1DD
G1 OZD J Anderson, 179 Rolleston Road, Burton-on-Trent, DE13 0LD
G1 OZR I Thomson, 2 Casaubon Close, Dereham, NR19 1EG
G1 OZV Malcolm Jolly, The Oaks, 6 Gwealhellis Warren, Helston, TR13 8PQ
GW1 OZW B Donovan, Honyegol, Drope Road, Cardiff, CF5 6EP
G1 PAD D Reeves, Isla Blanca, 21 Falmouth Road, Congleton, CW12 3BH
G1 PAF P Foster, 9 The Greenway, Ickenham, Uxbridge, UB10 8LS
G1 PAK C Parker, 6 Chilham Close, Hemel Hempstead, HP2 4UU
G1 PAT P Chapman, 24 Broad Lane, Moulton, Spalding, PE12 6PN
GW1 PAV Philip Grey, 4 Lon Carreg Bica, Birchgrove, Swansea, SA7 9QH
G1 PBF D Pearce, 32 Marshall Road, Willenhall, WV13 3PB
G1 PBX E Churchill, 87 Bradley Crescent, Shirehampton, Bristol, BS11 9SR
G1 PBY A Parrott, 54 Dockin Hill Road, Doncaster, DN1 2QU
G1 PCA R Blandford, 16b Sherwood Road, Keynsham, Bristol, BS31 1DB
G1 PCD Paul Dicken, Rosgol, Deiniolen, Caernarfon, LL55 3LU
G1 PCG J Dunwell, 8 Violet Grove, Thatcham, RG18 4DQ
G1 PCN E Benzie, 6 Priors Park, Emerson Valley, Milton Keynes, MK4 2BT
G1 PCU Andrew Popplewell, 00 Wellington Street, Hull, HU6 9QQ
G1 PCR Christopher Carter, 80 Cranbrook Drive, Maidenhead, SL6 6SS
G1 PCU C Mather, 5 Knolles Road, Cowley, Oxford, OX4 3HT
G1 PDA E Evans, 11 Turrot Road, Wallasey, CH45 5HE
G1 PDS S Cliffe, 24 Dalehurst Road, Bexhill-on-Sea, TN39 4BN
G1 PEE S JACKSON, 71 Slyne Road, Bolton le Sands, Carnforth, LA5 8AQ
G1 PEK M Sutton, 17 Barton Close, Witchford, Ely, CB6 2HS
GM1 PEL B Taynton, 32 Broomhall Road, Edinburgh, EH12 7PD
G1 PER M Payne, 14 Linacres Drive, Chellaston, Derby, DE73 6XH
G1 PEU Graham GIBBONS, 43 Buckland Avenue, Basingstoke, RG22 6JA
GW1 PFK A Romano, The Glen, Glen Road, Swansea, SA3 5QJ
GM1 PFU C Hewlett, 6 Glenturret Terrace, Perth, PH2 0AR
G1 PFY J Gold, 6 Woodland Avenue, Bournemouth, BH5 2DJ
G1 PFZ W Gill, 2 Rufford Court, Rufford Avenue, Leeds, LS19 7ED
G1 PGD K Biggs, 30 Elder Lane, Burntwood, WS7 9BT
G1 PGH J Baker, 10 Forest Way, Ashtead, KT21 1JL
G1 PGI Colin Smith, The Laurels, Maple Court Rodmersham, Sittingbourne, ME9 0LR
G1 PGJ D Castle, 8 Woodhall Court, Welwyn Garden City, AL7 3TD
G1 PGN D Brooks, 10 Elmtree Road, Ruskington, Sleaford, NG34 9BT
G1 PGQ R Pennycook, 28 Marine Court, Southsea, PO4 9QU
G1 PGS D Hands, 45 Croft Avenue, West Wickham, BR4 0QH
G1 PGV Jo Davidson, 5 Hanover Parc, Indian Queens, St. Columb, TR9 6ER
G1 PGX Michael Bingham, 6 Bittern Close, Hull, HU4 6SQ
G1 PHA G Day, 102 Meadlands Drive, Ham, Richmond, TW10 7ED
GM1 PHD N Muir, 25 Drylaw House Gardens, Edinburgh, EH4 2UE
G1 PHJ H Johnson, 27 Ridgeway Avenue, Gravesend, DA12 5BD
G1 PHK R Baines, 319 Pontefract Road, Featherstone, Pontefract, WF7 5AB
G1 PHN M Morley, 8 The Becks, Alvechurch, Birmingham, B48 7NE
G1 PHS P Street, 12 Ledston Avenue, Garforth, Leeds, LS25 2BP
G1 PHU B Burton, Natson, Tedburn St. Mary, Exeter, EX6 6ET
G1 PHV C Marsh, 85 Cromwell Crescent, Market Harborough, LE16 9JW
G1 PIF D Rogers, 20 Chapel Close, Acomb, Hexham, NE46 4RX
G1 PIH H Owen, Llys Gwynedd, Bethel, Caernarfon, LL55 1YB
G1 PII G Cooper, 39 Church Road, Harlington, Dunstable, LU5 6LE
G1 PIX R Bibby, 40 Morval Crescent, Runcorn, WA7 2QS
G1 PIY F Cholerton, 17 Stringer Crescent, Warrington, WA4 1QN
G1 PJB Thomas Henderson, 18 Roundhaye Road, Bournemouth, BH11 9JB
G1 PJC Graham Siarey, 23 Celsus Grove, Swindon, SN1 4GE
G1 PJI S Scott, 11a Lodge Crescent, Orpington, BR6 0QE
G1 PJK J Foster, 15 Parklands Way, Liverpool, L22 3YX
G1 PJL D Brookfield, The Barns, Milwr, Holywell, CH8 8HE
G1 PJM P Mitchell, 11 Wingle Tye Road, Burgess Hill, RH15 9HR
G1 PJO Kenneth Thompson, 13 Kirby Walk, Peterborough, PE3 9UD
G1 PJP Phillip Probert, 7 Albany Road, Blackwood, NP12 1DZ
G1 PJR J Garrett, 2 Wantsume Lees, Sandwich, CT13 9JF
G1 PJT R Wood, Lynwood, Halley Road, Heathfield, TN21 8TG
G1 PJV N Saunders, 24 Gateland Close, Haxby, York, YO32 2ZZ
G1 PJZ J Rogers, 55 York Road, Driffield, YO25 5AY
G1 PKG A Stockton, 190 Sommerfield Road, Woodgate, Birmingham, B32 3TA
GW1 PKM P Miller, Ddaugae Farm, Gwrhyd Road, Swansea, SA9 2RY
GM1 PKN D Gillies, 10 Killeonan, Campbeltown, PA28 6PL
G1 PKO C Jones, 52 The Drive, Bury, BL9 5DL
G1 PKP M Lloyd, 243 Stand Lane, Radcliffe, Manchester, M26 1JA
G1 PKR D Bannister, 11 Keats Drive, Swadlincote, DE11 0DS
G1 PKS Victor Johnston, 14 Auckland Close, London, SE19 2DA
G1 PKV John Sennitt, 44 Pear Tree Avenue, Newhall, Swadlincote, DE11 0NB
G1 PKW Patrick Kingsley-Williams, Banhadlen Uchaf, Back Road, Mold, CH7 4QD
G1 PLE C Griffith, 5 Park Close, Yaxley, Peterborough, PE7 3JW
G1 PLJ John Morgan, Holly Cottage, Old Racecourse, Oswestry, SY10 7PQ
G1 PLU S Goy, 352 Chanterlands Avenue, Hull, HU5 4ED
G1 PLV R Cilia, 18 London Fields House, Kensington Road, Crawley, RH11 9NS
G1 PMA David Harding, 37 Junction Cottages, London Road, Pulborough, RH20 1LA
G1 PMF P Felton, 13 New Street, Sudbury, CO10 6JB
G1 PMJ D Loon, 18 Stourcliffe Road, Wallasey, CH44 3AF
G1 PMK Andrew Mills, 12 Sydney Street, Kimberley, Nottingham, NG16 2LQ
G1 PNB Steven Carwell, 8 Shakespeare Road, Prestwich, Manchester, M25 9GW
G1 PNC I Collins, 1 Artillery Place, Hollyhedge Road, Manchester, M22 4GG
GW1 PND K Hassall, 110 Waterloo Road, Hakin, Milford Haven, SA73 3PF
G1 PNL D Johnson, 27 Ridgeway Avenue, Gravesend, DA12 5BD
G1 PNX D Staples, 2 Bulcote Road, Clifton, Nottingham, NG11 8FD
GM1 POA J Jamieson, 11 Binns Road, Glasgow, G33 5HU
G1 POC C Elsom, 8 King Avenue, Maltby, Rotherham, S66 7HX
G1 POD E Mitchell, 11 Wingle Tye Road, Burgess Hill, RH15 9HR
G1 POJ Keith Phillips, 10 Weale Court, Vyne Road, Basingstoke, RG21 5NN
G1 POK D Harris, 3 Middle Close, Camberley, GU15 1NZ
G1 POM E Baker, 19 Ramsey Road, Thornton Heath, CR7 6BX
G1 POR A Porter, 1125 Yardley Wood Road, Warstock, Birmingham, B14 4LS
G1 POV R Smith, 18 Hornby Avenue, Westcliff-on-Sea, SS0 0LE
G1 PPB T Roots, 11 Windermere Avenue, Eastern Green, Coventry, CV5 7GP
G1 PPD A Shons, 108 Southdown Road, Catherington, Waterlooville, PO8 0NF
G1 PPG G Joyner, Valle De Los Nogales 609, Fraccionamiento Real Del Valle, Nuevo Leon, Mexico, ZP 66350
G1 PPK Joseph Knight, 183 Northumberland Avenue, Thornton-Cleveleys, FY5 2JS
G1 PPO A Wade, 40 Throxenby Lane, Scarborough, YO12 5HW
G1 PPQ S Walker, 22 Ward Close, Aylestone, Leicester, LE2 8NJ
G1 PPU D Walker, 115 Kilby Road, Fleckney, Leicester, LE8 8BP
G1 PPX G Nicholls, 2 Leybrook Croft, Hemsworth, Pontefract, WF9 4JA
G1 PPZ A Porton, 17 Causey Farm Road, Halesowen, B63 1EQ
G1 PQJ Paul Wilson, Orchard Cottage, Rectory Road, Norwich, NR14 8TT
G1 PQK P Harrison, 20 Priory Road, Stanford-le-Hope, SS17 7EW
G1 PQO Sally Stevens, 49 The Beeches, Upton-upon-Severn, Worcester, WR8 0QQ
G1 PQT J Mayes, 44 Foxwarren, Claygate, Esher, KT10 0JZ
G1 PQX Robert Moat, 27 Pioneer Road, Dover, CT16 2AR
G1 PQY G Jones, 25 Myvod Road, Wednesbury, WS10 9BT
G1 PRE K Moore, Dinas No.10, Treriever, Downderry, Torpoint, PL11 3LY
G1 PRF Steven Haden, 33 Poplar Avenue, Chelmsley Wood, Birmingham, B37 7RD
G1 PRU M Watkins, 9 Adrian Road, Whitstable, CT5 4NY
G1 PRL R Williams, 51 Windways, Little Sutton, Ellesmere Port, CH66 1JF
G1 PRM Ann Webber, 60 Trowley Hill Road, Flamstead, St. Albans, AL3 8EE
G1 PRP A Moore, Wyndrush, Northend Lane, Southampton, SO32 3QN
G1 PRS C Ladley, 25 Laburnum Crescent, Louth, LN11 8SG
G1 PRW A Swift, 38 Knightsbridge Way, Stretton, Burton-on-Trent, DE13 0WJ
G1 PRZ B Saich, 65 Orchard Rise West, Sidcup, DA15 8TA
G1 PSH L Walton, 2 Church Road, Colmworth, Bedford, MK44 2JX
G1 PSL Christopher Sparrow, 112 Hill Cot Road, Bolton, BL1 8RW
G1 PSS A Laszkiewicz, 13 Darwall Drive, Ascot, SL5 8NB
GM1 PST P Stanhope, The Roundal, Alva, FK12 5HU
GM1 PSU I Manson, 25 Etna Court, Armadale, Bathgate, EH48 2TD
G1 PSW A Bunting, 11 Lon Y Gaer, Deganwy, Conwy, LL31 9RG
GM1 PSZ D Liddle, 9 Rullion Road, Penicuik, EH26 9HS
G1 PUK David Jones, 100 Cop Lane, Penwortham, Preston, PR1 0UR
G1 PUO David West, 30 Farm Avenue, Swanley, BR8 7JA
G1 PUQ C Tripp, Kingshill House, Church Street, Somerton, TA11 6ER
GM1 PUR Daniel Wood, 50 Riverside Road, Eaglesham, Glasgow, G76 0DG
G1 PUU Steven Pantall, 27 Woodlands Drive, Foston, Derby, DE65 5DL
G1 PUV Christopher Wiseman, 42 Merlin Way, Kidsgrove, Stoke-on-Trent, ST7 4YL
G1 PUY J De Renzi, Bankside, South Newington, Banbury, OX15 4JE
G1 PUZ S Box, 103 Ilkeston Road, Bramcote, Nottingham, NG9 3JT
G1 PVA Graham Kemp, 5 Gosselin Street, Whitstable, CT5 4LA
G1 PVD Hugh Murray, 35 Elliot Rise, Hedge End, Southampton, SO30 2RU
GI1 PVE R Law, 147 Lone Moor Road, Londonderry, BT48 9LA
G1 PVN Phil Jones, 72 Lon Maesycoed, Newtown, SY16 1QQ
G1 PVR R Kelly, Hogbrook Farm, Banbury Road, Leamington Spa, CV33 9QL
G1 PVT D Broad, 14 Albion Road, Westcliff-on-Sea, SS0 7DR
G1 PVU J Allen, 23 Beech Road, Street, BA16 0RY
G1 PVZ J Vincent, Brookfield, North Street, Crewkerne, TA18 7AX
G1 PWF D King, 79 Wootton Drive, Hemel Hempstead, HP2 6LA
GM1 PWL K Robertson, 7 Meadows Crescent, Lochgilphead, PA31 8AG
G1 PWM J Bulman, 3 South View, Littlethorpe, Ripon, HG4 3LL
G1 PWO P Owens, Flat 1, 45 The High Street, Enfield, EN3 4EF
G1 PWS R Manson, Smavollen 11, Stavanger, Norway, 4017
G1 PWU D Dwyer, 24 Alder Way, Melksham, SN12 6UL
G1 PWY P Gardner, 2 South Road, Morecambe, LA4 5RA
G1 PXH C Daily, 1 Railway Cottages, Wendens Ambo, Saffron Walden, CB11 4LA
G1 PXM R Blakeway, Ty Nantglyn, Glascwm, Llandrindod Wells, LD1 5SE
G1 PXQ A Scivetti, 10 Rippleside, Basildon, SS14 1UA
G1 PXW P Hollands, 16 Hazel Crescent, Thornbury, Bristol, BS35 2LX
G1 PYJ P Powell, 10 Selsdon Close, Wythall, Birmingham, B47 6HP
G1 PYY A Thomas, 22 Sea Road, Abergele, LL22 7BU
G1 PZA W Grech-Cini, Byram Garnge, Great North Road, Byram, WF11 9PA
G1 PZD Mark Pattison, 13 Mixes Hill Road, Luton, LU2 7TX
G1 PZP C Collett, Yaffle, 26 Hertford Road, Hoddesdon, EN11 9JR
GM1 PZT G Hardacre, 242 Sutherland Way, Knightsridge, Livingston, EH54 8JB
GI1 RAA Thomas Hourican, 43 Burren Road, Warrenpoint, Newry, BT34 3SA
G1 RAE Margret Buckley-Brown, 2 Brothertoft Road, Boston, PE21 8HD
G1 RAF Royal Air Force Halton RS c/o Alfred Mockford, 58 Wendover Heights, Old Tring Road, Aylesbury, HP22 6PH
G1 RAG James Jones, 10 Huntington Close, Winyates, Redditch, B98 0NF
G1 RAO K Hughes, 20 Pickering Close, Bury, BL8 1UE
G1 RAP R Prosser, 27 Dorset Gardens, Rochford, SS4 3AH
G1 RAX D Bendall, Brambles, 17 Berryfield Road, Lymington, SO41 0HQ
G1 RBA A Wheatley, 3 Woodsbank Terrace, Wednesbury, WS10 7RQ
G1 RBH Keith Dunn, 65 Lime Street, Sutton-in-Ashfield, NG17 4GA
GI1 RBI W McKeown, 15 Laragh Lee, Ballycassidy, Enniskillen, BT94 2JA
G1 RBX R Baker, 3 Hazelton Close, Solihull, B91 3GA
G1 RBY P Hurp, 55 Brooklyn Grove, Coseley, Bilston, WV14 8YH
G1 RBZ G Norris, 26 Westwood Road, Leyland, PR25 3NS
G1 RCD DARTMOOR RC c/o D Bolt, C/O Old School Hse, Maristow Roborough, Plymouth, PL6 7BY
G1 RCE Anthony Hills, 12 Heathway, Chaldon, Caterham, CR3 5DL
G1 RCI P Mason, 34 Central Park Avenue, Wallasey, CH44 0AQ
G1 RCN Peter Wilson, 146 Wilkinson Street, Nottingham, NG8 5FJ
G1 RCV CRAY VALLEY RS c/o Adrian Styles, 6 Hill Brow, Crayford, Dartford, DA1 0NX
G1 RCW J Chetwynd, 35 Cordelia Close, Dibden, Southampton, SO45 5UD
G1 RCX Paul Twency, 0 Dovehouse Close, Eynsham, Witney, OX29 4EW
GM1 RDG J Horsburgh, Donvilla, 66 Harlaw Road, Inverurie, AB51 4TB
G1 RDJ Christopher Price, 44 Poplar Road, Stourbridge, DY8 3BD
G1 RDU P Fanning, 9 Fishermans Walk, Shoreham-by-Sea, BN43 5LW
G1 RDX S Newbold, 7 Rookery Meadow, Holmer Green, High Wycombe, HP15 6XF
G1 REL G Hawker, 46 Southfield Drive, North Ferriby, HU14 3DX
G1 REO J Telford, 85 Medway, Great Lumley, Chester le Street, DH3 4HU
G1 RET Roger Taylor, 3 Solent View, Calshot, Southampton, SO45 1BH
G1 RFB R Palmer, 55 Edinburgh Road, Freshwater, PO40 9DL
G1 RFC R Armitage, 15 Northolmby St., Howden, Goole, DN14 7JL
G1 RFH FOREST HEATH RAYNET c/o John Slater, 47 Broom Road, Lakenheath, Brandon, IP27 9EZ
G1 RFI Anthony Venables, 7 Kempsey Close, Redditch, B98 7TL



**IMPORTANT NOTE**

**Revalidate licence to avoid revocation** – Ofcom has advised the Society that plans will be drawn up to revoke licences that have not been revalidated as required by the licence conditions. The quickest way to revalidate is to do so online via the Ofcom website: *https://services.ofcom.org.uk/* or by email: *amateur.validations@ofcom.org.uk* Ofcom staff are available to help, but please be patient during times of heavy workload.

G1 RFQ D Jenks, Tre-Vorgan, Courtenay Road, Tavistock, PL19 0EE
G1 RFS T Niner, 281 Nightingale Road, Edmonton, London, N9 8QL
G1 RFX K Tonner, Millstream Cottage, Golden Valley, Malvern, WR13 6AA
G1 RGG P King, 124 Henley Grove Road, Rotherham, S61 1RY
GM1 RGM D Keddie, 26 Daleally Crescent, Errol, Perth, PH2 7QA
GD1 RGT G Williams, 76 Eastern Avenue, Pinner, HA5 1NJ
G1 RHB K Sears, 21 Sandhurst Road Bulwell, Nottingham, NG6 8DL
G1 RHE Walter Berry, The Bungalow, Basil Road, Kings Lynn, PE33 9RP
GD1 RHT E Moore, 73 Reayrt y Chrink, Port Erin, Port Erin, Isle of Man, IM9 6DL
G1 RHW Terence Ibbitson, 36 Knoll Park, East Ardsley, Wakefield, WF3 2AX
GI1 RIB B Rafferty, 81 Mullaghmore Drive, Omagh, BT79 7PQ
GM1 RIG I Mcgowan, Feddal Lodge, Braco, Dunblane, FK15 9RA
G1 RIR L Shears, 7 Lower Furlongs, Brading, Sandown, PO36 0DX
G1 RIV T Dyson, 4 Lyspitt Common, Meppershall, Shefford, SG17 5GZ
G1 RIX Iain Steele, 10 Manor Road, Irby, Wirral, CH61 4UA
G1 RJA M Johnson, 28 Bittles Green, Motcombe, Shaftesbury, SP7 9NX
G1 RJD W Blower, 129 Kingsway, Kirkby-in-Ashfield, Nottingham, NG17 7FH
G1 RJN S Hall, Little Dene, Eastwick Road, Leatherhead, KT23 4BJ
GM1 RJS J Hambrook, 32 Blackdales Avenue, Largs, KA30 8HU
G1 RJW R Wicks, 32 Shelley Close, Northcourt, Abingdon, OX14 1PR
G1 RKD B Allport, 1 Percy Drive, Swarland, Morpeth, NE65 9JN
GM1 RKI A Morris, 39 Old Town, Peebles, EH45 8JE
G1 RKJ Robert Wilson, 107 Hamilton Avenue, Uttoxeter, ST14 7FE
G1 RKR R Luker, Grantham Cottage, Haywards Heath Road, Lewes, BN8 4DS
G1 RLA Rory Hall, Bliss Lodge, Worcester Road, Chipping Norton, OX7 5XS
G1 RLB R Lawson, 28 Hallett Way, Bude, EX23 8PG
G1 RLD D Beeton, 50 Hanson Avenue, Shipston-on-Stour, CV36 4HS
G1 RLF R Walter, 10 Birch Meadow, Clehonger, Hereford, HR2 9RH
G1 RLI Paul Webb, 40 Links Road, Penn, Wolverhampton, WV4 5RF
G1 RLK Bradshaw Ian, 35 Stanley Road, Heysham, Morecambe, LA3 1UR
G1 RLR P Pedley, 24 Appledore Road, Walsall, WS5 3DT
G1 RLT Peter Vipond, The Old Forge, Nentsbury, Alston, CA9 3LH
GM1 RLV Douglas Mackay, Burnlea, Harrapool, Isle of Skye, IV49 9AQ
G1 RMC SW LONDON RAYNE c/o Ian Jackson, 5 Vivien Close, Chessington, KT9 2DE
G1 RMN M Richards, 20 Tas Combe Way, Willingdon, Eastbourne, BN20 9JA
G1 RNL M Kinsella, The Nook, Eaudyke Road, Boston, PE22 8RU
G1 RNV Brian Purse, 28 Holford Road, Guildford, GU1 2QF
G1 RNZ Graham Saville, 4 Shannon Court, Downs Barn, Milton Keynes, MK14 7PP
G1 ROD Irene Lupton, 19 Avenue Close, Harrogate, HG2 7LJ
G1 ROE I Roe, Tyddyn Berth, Chwilog, Pwllheli, LL53 6RQ
G1 ROH D Gower, 68 Wood Common, Hatfield, AL10 0UB
G1 ROK Paul Court, Hamara, Shortlands Grove, Bromley, BR2 0LS
GM1 ROM D O'Connor, 2 Latch Farm Cottages, Kirknewton, EH27 8DQ
G1 RON F Donnachie, 2 The Mall, Patrington Haven Leisure Park, Patrington, HU12 0PT
GM1 ROX Alastair Donald, South Sandlaw House, Alvah, Banff, AB45 3UD
G1 RPE Ian Smith, 66 Aire Road, C/O Beech Croft, Wetherby, LS22 7UE
G1 RPO Chris Reed, Colins, Throcking Road, Buntingford, SG9 9RA
G1 RPP C Broadbent, 7 Wharfe Park, Addingham, Ilkley, LS29 0QZ
G1 RPT M Tribe, 11 Heathlands Close, Crossways, Dorchester, DT2 8TS
G1 RPV R Hardiman, 27 Staithe Road, Martham, Great Yarmouth, NR29 4PT
GM1 RQD D Marwick, 17 Laverock Road, Kirkwall, KW15 1EE
G1 RQI P Quirk, 75 Harcourt Road, Folkestone, CT19 4AF
GW1 RQM M Aquilina, 3 Aldergrove Close, Port Talbot, SA12 8EY
G1 RRE J Eckersley, 88 New Heys Way, Bradshaw, Bolton, BL2 4AQ
G1 RRG H Johnstone, 16 Riverside Crescent, Otley, LS21 2RS
GM1 RRJ B Elliott, 195 Braehead Road, Cumbernauld, Glasgow, G67 2BL
G1 RRR K Bareham, 19 Northfield Road, Ringwood, BH24 1LS
G1 RRU Paul Atkinson, 19 Haggar Street, Wolverhampton, WV2 3ET
G1 RRW A Henderson, 39 Stowell Crescent, Wareham, BH20 4PT
G1 RSC S Hall, 17 Nevill Road, Rottingdean, Brighton, BN2 7HH
G1 RSE J Rod, 42 Westwood Avenue, Ferndown, BH22 9HN
G1 RSF A Dean, Les Monneries, Combieres, Charente, France
G1 RSK C Broughton, 65 Manby Road, Immingham, DN40 2SG
GI1 RSR C Fogarty, 96 Killycomain Drive, Portadown, Craigavon, BT63 5JS
G1 RTW G White, 101 London Road, Hailsham, BN27 3AH
G1 RTX I Poole, 8 Bates Close, Higham Ferrers, Rushden, NN10 8HF
G1 RUG A Kay, Pear Tree Cottage, Hale House Lane, Farnham, GU10 2JG
G1 RUL Anthony Cake, 8 Carrick Close, Dorchester, DT1 2SB
G1 RUZ E Byrne, 25 South Road, Grassendale Park, Liverpool, L19 0LS
G1 RVC Russell Baker, 24 Nant Road, Connah's Quay, Deeside, CH5 4AL
G1 RVF B Dempster, 5 Church Walk, Bozeat, Wellingborough, NN29 7ND
G1 RVH C Stancer, 20 Overton Avenue, Willerby, Hull, HU10 6AR
G1 RVK Ian Brooks, 10 Foxgloves, Deeping St. James, Peterborough, PE6 8SH
G1 RVP P Scott, 22 Powys Road, Llandudno, LL30 1HZ
G1 RVT B Kavanagh, 73 Esh Wood View, Ushaw Moor, Durham, DH7 7FD
G1 RWR S Lee, 7 Ridge Way Close, Rotherham, S65 3NH
G1 RWT K Whitton, 11 Dursley Road, Shirehampton, Bristol, BS11 9XB
G1 RWX K Ikin, 15 Broadway, Farnworth, Bolton, BL4 0HQ
GI1 RXL Christopher O'Connell, 15 Grange Road, Coleraine, BT52 1NG
GI1 RXM A Murphy, 3 Church Lane, Crossgar, Downpatrick, BT30 9PX
G1 RXV Stephen Mugele, 19 Ambassador, Bracknell, RG12 8XP
GD1 RYF M O'Callaghan, 79 Kingsfield Avenue, Harrow, HA2 6AQ
G1 RYM R Knox, 13 Grosvenor Road, Billingham, TS22 5HA
G1 RYQ Simon Marshall, 15 Moulton Close, Belper, DE56 0EA
G1 RYS I McCulloch, 5 Knighthead Point, The Quarterdeck, London, E14 8SR
G1 RYY Robert Howse, 37 Great Eastern Road, Hockley, SS5 4BX
GW1 RZE P Morgan, Flat 24, Ynysderw House, Swansea Road, Swansea, SA8 4AA
G1 RZJ Lawrence Ward, 19 Spring Close View, Sheffield, S14 1RJ
G1 RZZ Raymond Wood, 40 Ashville Gardens, Pellon, Halifax, HX2 0PL
G1 SAJ R White, 61 Bournemead Avenue, Northolt, UB5 6PX
G1 SAK M Weatherley, 95 Cambalt Road, Putney, London, SW15 6EX
G1 SAM A Hodgkinson, 64 Rhodfa Wen, Llysfaen, Colwyn Bay, LL29 8LE
G1 SAN SANDWELL ARS c/o A Hollyoake, Sandwell Arc, Broadway, Warley, B68 9DP
G1 SAR SOUTH ANGLIA RAYNET c/o Keith Gaunt, 21 Abbey Close, Rendlesham, Woodbridge, IP12 2UD
G1 SAT INMARSAT ARC c/o R Smith, 32 Wolseley Gardens, London, W4 3LR
GM1 SBD Michael Angiolini, Innis Chonain, Mill Road, Stirling, FK7 9LP
G1 SBK John Spink, Highfields, Church Lane, Leeds, LS16 8DE
G1 SBN J Davison, 29 Glenfield Avenue, Wetherby, LS22 6RN
G1 SBW I Case, 70 Heathfield Park, Widnes, WA8 9WX
G1 SBZ T Lumley, 32 Downland Road, Woodingdean, Brighton, BN2 6DJ
G1 SCA S Allen, 28 Neville Road, Luton, LU3 2JJ
G1 SCB Kevin Gray, Donkleywood House, Donkleywood, Hexham, NE48 1AQ
G1 SCL N Stackhouse, 16 Tintern Avenue, Urmston, Manchester, M41 6FJ
G1 SCN Christopher Taylor, 4 Tunnel Road, Beaminster, DT8 3BQ
G1 SCO R Brown, 52 Challenger Drive, Sprotbrough, Doncaster, DN5 7RY
G1 SCQ Kevin Kent, 5 Jubilee Road, Heacham, King's Lynn, PE31 7AR
G1 SCR Shropshire Raynet c/o Mark Jones, 8 Sunfield Gardens, Bayston Hill, Shrewsbury, SY3 0LA
G1 SCT Michael Lane, Cherry Tree House, Pipwell Gate, Spalding, PE12 8BA
G1 SCV Alan Faulkner, Northwood, Cranham, Gloucester, GL4 8HB
G1 SCY Frederick West, Kensey, Haye Road, Callington, PL17 7JJ
G1 SDH J Shepherd, 140 The Broadway, Herne Bay, CT6 8HY
G1 SDJ Charles Cooper, Tapshays Cottage, Burton Street, Sturminster Newton, DT10 1PS
G1 SDK Soderis Karpasitis, Riverdene, Blythe Road, Hoddesdon, EN11 0BB
G1 SDN Robert Mordue, 29 Sycamore Close, Witham, CM8 2PE
G1 SDX G Taylor, 14 Maple Road, Brixham, TQ5 0DG
G1 SED P Hedicker, Hilvista, 26 Malden Road, Sidmouth, EX10 9LS
G1 SEF A Fearnley, 1 Dover Road, London, E12 5DZ
G1 SEO Martin Lines, 158 Nine Mile Ride, Finchampstead, Wokingham, RG40 4JA
G1 SES M Halden, 19 Fenwick Lane, Halton Lodge, Runcorn, WA7 5YU
G1 SEW Anthony Goatman, 150 Merlin Park, Portishead, Bristol, BS20 8RW
G1 SFU D Coffey, 14 Shawbridge, Harlow, CM19 4NJ
G1 SGA R Blunt, 8 The Crescent, Wolverhampton, WV6 8LA
GW1 SGE A Morgan, 3 Gelli Newydd, Golden Grove, Carmarthen, SA32 8LP
G1 SGG Mervyn Jones, 48 Maes Alltwen, Dwygyfylchi, Penmaenmawr, LL34 6UA
GW1 SGH R Thorne, 6 Cromwell Avenue, Rhyddings, Neath, SA10 8DW
G1 SGM R Parkin, Craigside, 15 Holly Drive, Leeds, LS16 6EF
G1 SGP N Barnes, 44 Cromford Road, Wirksworth, Matlock, DE4 4FR
G1 SGR N Marsh, 16 Daytona Quay, Eastbourne, BN23 5BN
G1 SGS Barrie Kimber, 4 Nautilus Drive, Minster on Sea, Sheerness, ME12 3NJ
G1 SGZ P Gamble, 9 Windmill Close, Ockbrook, Derby, DE72 3TE
G1 SHH A Compton, Fairlight, 25 Framfield Road, Uckfield, TN22 5AH
G1 SHI M Kuik, 196 Prestbury Road, Macclesfield, SK10 3BS
G1 SHN G Richardson, 12 Northenhay Walk, Morden, SM4 4BS
G1 SHQ I Woodford, 81 Harrold Road, Rowley Regis, B65 0RL
G1 SHT D Wooster, 34 New Road, Penn, High Wycombe, HP10 8DL
G1 SHU N Carr, 7 Pear Tree Drive, Sedgeberrow, Evesham, WR11 7GQ
G1 SID C Siddons, 423 London Road, Grays, RM20 4AB
G1 SIG Brian Scholte, 266 Hednesford Road, Heath Hayes, Cannock, WS12 3DS
G1 SIM Simon Hallam, 46 Holte Road, Atherstone, CV9 1HN
G1 SIO J Robson, Ealands, The Stanners, Corbridge, NE45 5BA
G1 SIP R Webb, 54 Ashby Avenue, Chessington, KT9 2BU
G1 SIU A Bingley, 51 Kirkby Folly Road, Sutton-in-Ashfield, NG17 5HP
G1 SIX Robert Foster, Gorsethorpe Cottage, Gorsethorpe Cottages, Mansfield, NG21 9HJ
G1 SJB C Thompson, 135 Stafford Road, Bloxwich, Walsall, WS3 3PG
G1 SJG E Boydon, 56 Oliver Leese Court, Ten Butts Crescent, Stafford, ST17 9HP
G1 SJO A Banthorpe, 32 Long Close, Station Road, Henlow, SG16 6JS
G1 SJT A Long, 23 Beech Road, Sutton Weaver, Runcorn, WA7 3ER
G1 SJU D Maciver, 176 Burges Road, London, E6 2BS
G1 SJZ P Randall, 7 Eastbourne Avenue, Featherstone, Pontefract, WF7 6LQ
G1 SKE Davina turner, 26 Carlton Way, Cleckheaton, BD19 3DG
G1 SKI K Weaver, 26 Southdown Close, Haywards Heath, RH16 4JR
G1 SKQ J Forster, 4 Rydal Road, Lemington, Newcastle upon Tyne, NE15 7LR
G1 SKR Edwin Smith, 22 Fleece Road, Long Ditton, Surbiton, KT6 5JN
G1 SKV Barry Abell, 39 Etty Avenue, York, YO10 3TJ
G1 SKW R Sidwell, 81 Oakengates Road, Donnington, Telford, TF2 7LQ
G1 SLA C Baker, 17 Dawlish Avenue, Chadderton, Oldham, OL9 0RF
G1 SLE R Drabble, 68 St. Lawrence Avenue, Bolsover, Chesterfield, S44 6HT
G1 SLG Michael Butcher, 1 Mushroom Field Road, Northampton, NN3 5AD
G1 SLI Brian Elliott, 51 Allerhope, Hall Close Grange, Cramlington, NE23 6SX
G1 SLO K Turner, 44b Foxgrove Road, Beckenham, BR3 5DB
G1 SLP Eric Methven, 31 Ghyll Field Road, Durham, DH1 5HT
G1 SLU K Ford, 123 Stockwood Lane, Bristol, BS14 8SZ
G1 SMB M Chitty, Timbercroft, Faris Lane, Addlestone, KT15 3DL
G1 SMC Steven Mccloy, 28 Ferndown Drive, Godmanchester, Huntingdon, PE29 2LU
G1 SMJ F Beavan, Uplands, Bronllys, Brecon, LD3 0HN
G1 SMP Paul Devlin, 52 Manor Rise, Lichfield, WS14 9RF
GW1 SMT C Manning, Unit 235 209 City Road, Roath, Cardiff, CF24 3JD
G1 SMY S Fisher, 19 Sandown Road, Sandown, PO36 9JL
G1 SNI P Smith, 27 Briar Close, Gillingham, SP8 4SS
G1 SNO DR Tubb, 42 Hill Farm Road, Marlow, SL7 3LU
G1 SNQ I Boss, 11 Penkridge Road, Church Gresley, Swadlincote, DE11 9FH
G1 SNU D Tunbridge, 12 Burnham Road, Latchingdon, Chelmsford, CM3 6EU
G1 SOB T Yetton, 7 Warwick Close, Canvey Island, SS8 9YB
G1 SOG R Stearn, 18 Kings Avenue, Chippenham, SN14 0UJ
G1 SOX William Stennett, 26 Moorfield Road, St. Giles-on-the-Heath, Launceston, PL15 9SY
G1 SOY J Moseley, 72 Wisden Road, Stevenage, SG1 5JA
G1 SPA Derek Harrison, 7 Shirley Close, Castle Donington, Derby, DE74 2XB
G1 SPJ A Gibbs, 86 Broadmark Road, Slough, SL2 5PN
G1 SPM A Hughes, 8 Rigby Grove, Little Hulton, Manchester, M38 0FQ
G1 SPQ D Mccauley, 7 Bonnets Lane, Wareham, BH20 4HA
G1 SPT Steve Tatem, 55 Chelwood Road, Chellaston, Derby, DE73 5SJ
G1 SPU Alistair Burnett, 16 Shielding Way, Stafford, ST16 3WG
GW1 SPW B Saunders, Bishops Mill, Llanwrda, SA19 8AD
G1 SPX K Shires, 19 Prince Charles Avenue, Sittingbourne, ME10 4NA
G1 SQA John Yates, 60 Leamington Road, Branston, Burton-on-Trent, DE14 3HX
G1 SQC Ken Boote, 51 Sunnyfield Oval, Stoke-on-Trent, ST2 7PA
G1 SQG P Dennis, Fuchsia House, 18 West View Road, Yelverton, PL20 7DD
G1 SQI J Bewley, 21 Duloe Gardens, Pennycross, Plymouth, PL2 3RS
G1 SQW R Rawson, 43 County Road North, Hull, HU5 4HN
GM1 SQZ Gerald Pocock, 1 Pitcairn Grove, East Kilbride, Glasgow, G75 8TN
G1 SRA T Binns, Crossfarm Cottage, Keighley, BD22 9LE
GW1 SRB Kenneth Gough, 2 Church Road, Abertridwr, Caerphilly, CF83 4DL
G1 SRD P Foulds, 7 Bridge Road, Little Sutton, Spalding, PE12 9EG
GM1 SRP John McCulloch, Wester Curr Cottage, Dulnain Bridge, Grantown-on-Spey, PH26 3LX
GM1 SRR Michael Christmas, Lindens, Smithy Loan, Dunblane, FK15 0HQ
G1 SSL M Belcher, 52 Kynaston Road, Didcot, OX11 8HD
G1 SSS John Banfield, 2 Laleham Close, Eastbourne, BN21 2LQ
G1 SSZ Francis Bowhill, 78 East Gomeldon Road, Gomeldon, Salisbury, SP4 6NB
G1 STK A Goddard, 65 Langley Hall Road, Solihull, B92 7HE
G1 STO William Bamford, 7 Poole Avenue, Stoke-on-Trent, ST2 7JJ
G1 STP V Trolan, Hilbre, East Taphouse, Liskeard, PL14 4NJ
G1 STQ Lewis Taylor, 15 Woodbridge Road, Newcastle, ST5 4LA
GM1 STW Richard Gerrard, 1 Craigmar Court, Mains of Concraig, Aberdeen, AB15 8RL
G1 SUH John Speakman, 130 Dicconson Street, Wigan, WN1 2BA
G1 SUK A Dimmock, Gwyndy, Llandegfan, Menai Bridge, LL59 5PW
G1 SUM T Cull, 25 Queensway, Ponteland, Newcastle upon Tyne, NE20 9RZ
G1 SUP Tim Stone, Le Moulin, Puylagarde, Tarn Et Garonne, France, 82160
G1 SVD R Roper, 57 Burnt Hills, Cromer, NR27 9LW
G1 SVI T Metcalfe, 38 Station Road, Branston, Lincoln, LN4 1LH
G1 SVJ Christopher Murphy, 13 Northfield Road, Ringwood, BH24 1LS
G1 SVL K Stocker, 22 Hadlow Down Close, Luton, LU3 2PY
G1 SVN M Churchman, Westcroft, Church Road, Chester, CH4 9NG
G1 SVP B Howard, 15 Four Acres, Bideford, EX39 3RW
G1 SVQ C Pringle, 12 Atkinson Road, Dumfries, DG2 7DH
G1 SVR SEVERN VALLEY RD c/o E Churchyard, 11 Greenfields Drive, Bridgnorth, WV16 4JW
GW1 SVV D Osborne, 22 Springfield Gardens, Hirwaun, Aberdare, CF44 9LY
G1 SWE M Foster, 15 Parklands Way, Liverpool, L22 3YX
GD1 SWF S Roberts, 43 Lawn Close, Ruislip, HA4 6ED
G1 SWH G School, 5 Canal Row, Haigh, Wigan, WN2 1NA
G1 SWI B Gillett, 18 Rookery Close, Fenny Drayton, Nuneaton, CV13 6BB
G1 SWK Timothy King, 10 Berkeley Close, Ipswich, IP4 2TS
G1 SWR Warwickshire Avon Raynet Group c/o Clive Ousbey, 30 Hawthorn Way, Shipston-on-Stour, CV36 4FD
G1 SWU G Denham, 36 Redstone Farm Road, Hall Green, Birmingham, B28 9NT
G1 SWX Christopher Harrap, 68 Stafford Road, Weston-Super-Mare, BS23 3BS
G1 SWZ R Wright, 61 Quarry Road, Hurtmore, Godalming, GU7 2RW
G1 SXB G Whetstone, 60 Worple Road, Staines, TW18 1EE
G1 SXJ Patrick Duckles, 8 Railway Cottages, Skillings Lane, Brough, HU15 1EN
G1 SXT M Kerry, 40 Oaklands Road, Sebastopol, Pontypool, NP4 5BZ
G1 SXU P Janes, 19 Fair View, Chepstow, NP16 5BX
GM1 SXX A Copland, 74 Whitehaugh Avenue, Paisley, PA1 3SR
G1 SXY Y Entwistle, Park Garden House, Park Garden, Bury St. Edmunds, IP28 8PB
GM1 SYC W Graham, 7 Brunt Place, Dunbar, EH42 1RT
GI1 SYM G Thompson, 57 Rosepark, Donaghadee, BT21 0BN
G1 SYP Mark Thornton, 41 Godley Street, Royston, Barnsley, S71 4DH
G1 SYU M Oubridge, 54 Cantle Avenue, Downs Barn, Milton Keynes, MK14 7QS
G1 SYV B Goodier, 42 Manchester Road, Clifton, Manchester, M27 6WY
G1 SYZ D Setterfield, 3 Waldon Close, Plympton, Plymouth, PL7 2ZA
GI1 SZC D Mcmanus, 38 Deanfield, Bangor, BT19 6NX
G1 SZD T Trengove, 8 Kemp Close, Truro, TR1 1EF
G1 SZK Robert Frost, 68 Wessex Road, Didcot, OX11 8BP
GM1 SZM W Robertson, 28 Dewars Avenue, Kelty, KY4 0BG
G1 SZT W Dowkes, Woodlea, Gillamoor Road, York, YO62 6EL
G1 TAI Philip Gabel, 4 Blacksmiths Green, Shutlanger, Towcester, NN12 7RS
G1 TAR C Anderton, 5 Leyland Avenue, Hindley, Wigan, WN2 3SB
G1 TAU J Drewry, 10 Drayton Drive, Heald Green, Cheadle, SK8 3LF
G1 TAY A Shearer, 101 Millside, Stalham, Norwich, NR12 9PB
G1 TAZ Frederick Perkin, 5 Highgrove, Trevadlock Hall Park, Launceston, PL15 7PW
G1 TBE John Kelday, 20 Lowfield, Eastfield, Scarborough, YO11 3LQ
G1 TBI W Monk, Brook House, River View, Buxton, SK17 8SW
G1 TBK D Watson, 72 Dawes Avenue, West Bromwich, B70 7LS
G1 TBN A Malhi, 5 Beech Avenue, Warton, Preston, PR4 1BY
G1 TBT David Taylor, Top Wath Laer, Top Wath Road, Harrogate, HG3 5PG
GM1 TBW A Napier, Jehrada Cottage, Longhaven, Peterhead, AB42 0NY
G1 TBX G Williams, 16 Coppice Road, Talke, Stoke-on-Trent, ST7 1UB
G1 TCK Dorothy Adams, 28 Greenside, Stoke Prior, Bromsgrove, B60 4EB
GM1 TCN Thomas Curran, Whinstone Cottage, Wester Galcantray, Nairn, IV12 5XX
GM1 TCP J Campbell, 16 Barony Road, Auchinleck, Cumnock, KA18 2LL
G1 TDL Mike Mundy, The Homestead, Homestead Lane, Burgess Hill, RH15 0RQ
G1 TDN T Collinson, 26 Westway Avenue, Hull, HU6 9SA
G1 TDO F SANCHEZ-GARCI, 74 Gorthorpe, Hull, HU6 9EZ
G1 TDP J Spry, 21 Christchurch Gardens, Waterlooville, PO7 5BT
GM1 TDT D Robertson, 131 Foxbar Road, Paisley, PA2 0BD
GM1 TDU J Rooney, Rob Roy, Kinnell, Montrose, DD10 0TG
G1 TDV A Potts, 4 Bloomfield Close, Newport, NP19 9ET
G1 TEX N Swann, 9 Alexandra Road, Parkstone, Poole, BH14 9EL
G1 TFB A Hughes, 38 Llys Dyffryn, St. Asaph, LL17 0SX
GI1 TFC R McMaster, 43 Craigs Road, Carrickfergus, BT38 9RL
GW1 TFL J Robson, 16 Dunraven Road, Sketty, Swansea, SA2 9LG
G1 TFM J Bishop, 93 The Vale, Feltham, TW14 0JY
G1 TFY D Mackinnon, 20 Saxon Grange, Sheep Street, Chipping Campden, GL55 6BY
GM1 TFZ Susanne Bavin, Garvan, 10 Grampian Way, Glasgow, G78 2DH
GM1 TGY Charles Christie, Firlands, Spey Valley Drive, Aberlour, AB38 9NU
G1 TGZ R Mclintock, 10 The Close, Riverhead, Sevenoaks, TN13 2HE
G1 THA Victor Collins, Flat 2, Marsh Mead, Glebe Road, Petersfield, GU31 5SB
G1 THD A Simmons, 22 Willow Way, Princes Risborough, HP27 9AY
G1 THF THE HAM FELLOWSHIP c/o Alan Nixon, 14 Carlton Road, Lowton, Warrington, WA3 2EP
G1 THG D Moore, East View, The Common, Gillingham, SP8 5NB

G1 THJ M Solley, 34 Firbank Road, Dawlish, EX7 0NW
GM1 THR Nicola Harrison, 127 Bruntsfield Place, Flat 1F3, Edinburgh, EH10 4EQ
GM1 THS G Slessor, 27 Scurdie Ness, Aberdeen, AB12 3NG
G1 THW A Taylor, 5 Brookside, Beare Green, Dorking, RH5 4QH
G1 TIF J Batey, The Hemmel, Barrasford, Hexham, NE48 4BD
G1 TIH F Bell, 143 Peter Street, Blackpool, FY1 3NN
G1 TIJ P Seaman, 18 Earlsford Road, Mellis, Eye, IP23 8DY
G1 TIK Denyse Waters, Gardeners Cottage, Sandhoe, Hexham, NE46 4LU
G1 TIQ S Crellin, 89 Wapshare Road, West Derby, Liverpool, L11 8LR
G1 TJH C Bartoot, 6 Maldon Close, Bishopstoke, Eastleigh, SO50 6RD
GW1 TJK A Evans, Maes Yr Onnen, 134 Waterloo Road, Ammanford, SA18 3RY
GJ1 TJP J Poole, Jardin Du Puits, La Longue Rue, St. Martin, Jersey, JE3 6ED
G1 TJR Richard Bromley, 33 Bromley Road, Lytham St Annes, FY8 1PQ
G1 TJT W Adam, [illegible], DN15 6YE
G1 TJW B Smith, 25 The Forge, Tetbury, GL8 8JE
G1 TKE Allan Gibson, 9 Fishers Mead, Dulverton, TA22 9EN
G1 TKQ I Burdon, 72 Greenway Road, Taunton, TA2 6LE
G1 TKY P Draper, 4 [illegible] Croft, Hatton, Derby, DE65 [illegible]
G1 TLA B Pattenden, Inshallah, Abbeystrewery, Skibbereen, Ireland
G1 TLC A Myers, 7 Hillside, Chelveston, Wellingborough, NN9 6AQ
G1 TLE Malcolm Rowe, 11 Parkfield, Stillington, York, YO61 1JR
G1 TLH D Koopman, High View, Graffham, Petworth, GU28 0QE
G1 TLW A Lloyd, 96 Fairdene Road, Coulsdon, CR5 1RF
G1 TMF J Firth, 29 Curzon Street, Newcastle, ST5 0PD
GD1 TMW A Caspersz, 25 Cheltenham Place, Harrow, HA3 9NB
G1 TNK J Flattley, 53 The Drive, Bredbury, Stockport, SK6 2ED
G1 TNP A Karande, Flat 1, 6 St. Dominics Close, Torquay, TQ1 4UN
G1 TNR D Jordan, 21 Rosewood Park, Walsall, WS6 7HD
G1 TOB T Hall, 3 Saville Road, Twickenham, TW1 4BQ
G1 TOL S Talbot, 8 Thornford Drive, Swindon, SN5 7BB
G1 TPA David Mercer, 84 New North Road, Reigate, RH2 8NA
G1 TPC M Bellamy, 2 Nelson Drive, Rothwell, Kettering, NN14 6DZ
G1 TPN R Whateley, 14 Eastfield Road, Delapre, Northampton, NN4 8PE
G1 TPO Robert Steel, 15 Thornbury Avenue, Seghill, Cramlington, NE23 7RT
G1 TPV D Juett, 10 Leys Road, Cambridge, CB4 2AU
G1 TQH K Scroggins, 44 Hillcroft Road, Herne, Herne Bay, CT6 7EW
G1 TQN A Hobbs, The Sail Loft, 604 Blandford Road, Poole, BH16 5EQ
G1 TQR J Harris, 48 Beech Close, Corby, NN17 2AF
G1 TQT Mark Costello, Flat 24, Wisley House, London, SW1V 2QS
G1 TQU C Chambers, 8 Dagtail Lane, Redditch, B97 5QT
G1 TQY P Bishop, 2 Spruce Avenue, Whitehill, Bordon, GU35 9TA
G1 TRF K Nicholson, 11 Latton Close, Chilton, Didcot, OX11 0SU
G1 TRI Christopher Curtis, 24 Oakwood Road, Horley, RH6 7BU
G1 TRL J Deacon, 28 Dollicott, Haddenham, Aylesbury, HP17 8JG
GI1 TRZ W Hamilton-Sturdy, 319 Old Glenarm Road, Larne, BT40 1TU
G1 TST P Jackman, 1 Palmer Road, Trowbridge, BA14 8QP
G1 TSV S Corrigan, 163 Blackburn Road, Heapey, Chorley, PR6 8EJ
G1 TTB G Birkby, 44 Lady Bay Road, West Bridgford, Nottingham, NG2 5DS
G1 TTC K Howard, 73 Challacombe, Furzton, Milton Keynes, MK4 1DP
G1 TTG J Brickwood, Datemachi House #301, 3-33-5, Tokyo, Japan, 150
G1 TTH Edmond Roughton, 18 Church Close, Braybrooke, Market Harborough, LE16 8LD
G1 TTK G Lewis, 57 Edgecumbe Road, Roche, St. Austell, PL26 8JH
G1 TTL Francis Moss, 64 Birch Avenue, Cuerden Residential Park, Leyland, PR25 5PD
G1 TTX A Docherty, Sunnybrae, Wickhurst Road, Sevenoaks, TN14 6LY
G1 TUI J Tracy, 18 Preston St., Kirkham, Preston, PR4 2ZA
G1 TUL S Neale, 28 Needham Drive, Sutton St. James, Spalding, PE12 0EG
G1 TUS R Rodley, Meadow Cottage, Cold Ashby Road, Northampton, NN6 8QP
G1 TUU M Dixon, 118 Kings Ash Road, Paignton, TQ3 3TU
G1 TUZ P Nelson, 42 York Avenue, East Cowes, PO32 6RU
G1 TVW J Halliday, 24 Duncan Avenue, Otley, LS21 3LN
G1 TWF L Barker, Flat 2, The Limes, Colchester, CO3 3SJ
G1 TWH S Greenfield, Byways, Brightlingsea Road, Colchester, CO7 8JH
G1 TWS M Dench, 110 Eastwood Road, Rayleigh, SS6 7JR
G1 TWT A Scott, 21 Hexham, Oxclose, Washington, NE38 0NR
G1 TWW R Reeves, 40 Kennett Road, Romsey, SO51 5PQ
G1 TWY B Panton, Lavers, Preston Road, Sudbury, CO10 9QD
G1 TXO M Riches, 32 Wyncham Avenue, Sidcup, DA15 8ER
G1 TYP E Summers, 262 Huddersfield Road, Stalybridge, SK15 3DZ
G1 TYU R Ward, 1 Kirkcroft Close, Thorpe Hesley, Rotherham, S61 2UH
G1 TZC Neil Cheese, 219 Cordell Road, Long Melford, Sudbury, CO10 9ET
G1 TZZ Michael Foster, 7 Orion Way, Braintree, CM7 9UR
G1 UAF Adam Webster, 7 Castlehythe, Ely, CB7 4BU
G1 UAL C Bagwell, 1 Waldegrave Court, Movers Lane, Barking, IG11 7UW
G1 UAY Keith Varnals, Regent Studio, Skidden Hill, St Ives, TR26 2DU
G1 UAZ C Musson, 14 Alfreton Road, South Normanton, Alfreton, DE55 2AS
G1 UBC J Pudley, 24 Appledore Road, Walsall, WS5 3DT
G1 UBH M Howell, 4 Chattisham Close, Stowmarket, IP14 2RE
G1 UBL A Camm, 24 Croyde Avenue, Greenford, UB6 9LS
G1 UBN Stuart Knight, 46 Hollybank Road, Hythe, Southampton, SO45 5FQ
G1 UBT E Dunn, 118 James Turner Street, Birmingham, B18 4NE
G1 UBV N Brigden, 176 Hulverston Close, Sutton, SM2 6UA
G1 UCC D Boot, 13 Westland Street, Stoke-on-Trent, ST4 7HE
G1 UCG Wayne Roberts, 13 Roseacre Road, Elswick, Preston, PR4 3UD
G1 UCI D Roberts, 56 Meadow Lane, Ainsdale, Southport, PR8 3RS
G1 UCN M Davis, 5 Coniston Close, Bridlington, YO16 6HQ
G1 UCO D Ralph, 55 St. Marys Road, Warley, Smethwick, B67 5DH
G1 UCR Richard Irving, 52 Holsworthy Square, London, WC1X 0BG
G1 UCT D Pye, 95 Lansdowne Way, High Wycombe, HP11 1UB
G1 UCZ D Dewar, 224 Seaside Road, Aldbrough, Hull, HU11 4RY
G1 UDB G Wardle, 53 Braine Road, Wetherby, LS22 6NP
G1 UDE D Grayson, 28 Chesterfield Road, Swallownest, Sheffield, S26 4TL
G1 UDR John Scott, 47 Corinthian road, Eastleigh, so53 2ay
G1 UDS P Spence, 17 Springvale Rise, Parkside, Stafford, ST16 1TE
G1 UDT M Boydon, 56 Oliver Leese Court, Ten Butts Crescent, Stafford, ST17 9HP
G1 UDW D Upton, 14 Ipley Way, Hythe, Southampton, SO45 3LJ
G1 UDX M Stasuik, 30 Ramsey Drive, Arnold, Nottingham, NG5 6QL

G1 UEA D Mills, 56 Canterbury Road, Birchington, CT7 9AS
G1 UEO J Aldred, 1e North End Road, Steeple Claydon, Buckingham, MK18 2PF
G1 UEQ J Sims, 345 Blandford Road, Hamworthy, Poole, BH15 4HP
G1 UEV Stephen Brown, 10 Walkers Lane South, Blackfield, Southampton, SO45 1YN
G1 UFA B Bailey, 10 Milton Road, Waterlooville, PO7 6AA
G1 UFH J Coyne, 20 Hawthorn Hill, Letchworth Garden City, SG6 4HG
G1 UFJ R Davis, 27 Tolsford Close, Etchinghill, Folkestone, CT18 8BU
G1 UFL K Whistance, 20 Solent Way, Milford on Sea, Lymington, SO41 0TE
G1 UFM S Pugh, 18 Styal Avenue, Stretford, Manchester, M32 9SJ
G1 UFS N Chandler, 7 Sherlock Avenue, Parklands, Chichester, PO19 3AE
G1 UFT P Zara, 69 Castleton Road, Wigston, LE18 1FQ
G1 UFX A Hern, Flat 9, Block P, Peabody Estate, London, SE1 8DU
G1 UGB Joy Slater, 57 Freshbrook Road, Lancing, BN15 8DF
G1 UGG [illegible], Blackpool, FY1 3AQ
G1 UGH T Chaplin, 21 Shilltoe Close, Bury St. Edmunds, IP33 3DU
G1 UGJ Barry Lowe, 5 Kingfisher Drive, Necton, Swaffham, PE37 8NN
G1 UGL P Beardshaw, 12 Halesworth Close, Chesterfield, S40 3LW
G1 UGO [illegible], 70 Halsbury Road, Westbury Park, Bristol, BS6 7SU
G1 UGV D Bodman, 34 Churchlands, North Bradley, Trowbridge, BA14 0TD
G1 UHB S tomkins, The Close, Broomfield, Bradford, BD14 6PJ
GW1 UHF Michael Perrett, Penlan, Whitland, SA34 0QX
G1 UHO H Court, 10 Dorset Road, West Kirby, Wirral, CH48 6DJ
G1 UIB R Moxon, 16 Kielder Oval, Harrogate, HG2 7HQ
G1 UID H Kentfield, 1 Torquay Avenue, Gosport, PO12 4NS
G1 UIO M McVittie, 19 Alder Crescent, Poole, BH12 4BD
GM1 UIR A Perks, The Lodge, Cemetery Drive, Dumbarton, G82 5HD
G1 UJX J Huddlestone, 8 Wilmot Avenue, Chaddesden, Derby, DE21 6PL
G1 UKA M Wilkie, 1 Celandine Close, Billericay, CM12 0SU
G1 UKG C Brown, 42 Mandeville Close, Vanbrugh Park, London, SE3 7AH
G1 UKH R Rodgers, 88 Norwich Road, Watton, Thetford, IP25 6DW
G1 UKS Dennis Simmons, 191 Lancaster Road, Morecambe, LA4 5QR
G1 UKW C Crane, 3 Hawkshead Drive, Royton, Oldham, OL2 6TW
G1 UKZ Philip Hall, 2 Broadway Close, Urmston, Manchester, M41 7NR
G1 ULB Graham Walker, 7 Tuscany View, Salford, M7 3TX
G1 ULG David Porter, 44 Weir Road, London, SW12 0NA
G1 ULP T Garner, Wolhara, 32 Gorefield Road, Wisbech, PE13 5AS
G1 ULQ P Rowsell, Thomley Hall Farm, Worminghall, Aylesbury, HP18 9JZ
G1 ULR D Ross, 2 Jemmetts Close, Dorchester-on-Thames, Wallingford, OX10 7RA
G1 UMS J Preece, White House, Bredwardine, Hereford, HR3 6BY
G1 UMY R Wiltshire, Danesboro, Stonehill Road, Chertsey, KT16 0ER
G1 UNB P Durrant, Wagtails, 138 Tilt Road, Cobham, KT11 3HR
G1 UNG R Spencer, 93 Waveney Road, St. Ives, PE27 3FN
G1 UNN Philip Coldicott, 45 Orrian Close, Stratford-upon-Avon, CV37 0TT
G1 UNQ Roger Davis, Lowlands, Station Road, Evesham, WR11 7QG
G1 UNU Keith Etwell, Hawthorn Cottage, Old Worcester Road, Kidderminster, DY11 7XS
G1 UOD Peter Farmer, 22 Nortune Close, Birmingham, B38 8AJ
G1 UOR W Rodgers, 9 Hillcrest, Skelmersdale, WN8 9JZ
GW1 UOV Adrian Ham, Gwen y Wawr, New Inn, Pencader, SA39 9AZ
G1 UOY William Bagley, Glan Severn, Trefeglwys Road, Llanidloes, SY18 6HZ
G1 UPP Michael Maude, 6 Malin Parade, Portishead, Bristol, BS20 7FW
G1 UPT J Ravelini, 15 Clarendon Green, Orpington, BR5 2NY
G1 UPX T Giles, 108 Queensway, Didcot, OX11 8SW
G1 UQC David Rusbridge, 1 Ray Bond Way, Aylsham, Norwich, NR11 6UT
G1 UQF R Davies, 33 Melbourne House, Melbourne Road, Northampton, NN5 5LW
G1 UQK Neal Lewis, Woodstock, Bridge Street, Pershore, WR10 2PL
G1 UQT S Alliott, 32 Broad Gates, Silkstone, Barnsley, S75 4HD
G1 URD R Ameson, 5 Godre'r Gaer, Llwyngwril, LL37 2JZ
G1 URF A Jones, Hafandeg, Southgate, Aberystwyth, SY23 1RY
G1 URG SOUTH KENT RAYNET c/o Anthony Phillpott, Southways, Stombers Lane, Folkestone, CT18 7AP
G1 URJ Nick Capon, 13 West Croft, Berinsfield, Wallingford, OX10 7NL
G1 URQ David Traynor, 6f Thornwythe Grove, Great Sutton, Ellesmere Port, CH66 2PZ
G1 URR C Gough, 67 Pickmere Lane, Wincham, Northwich, CW9 6EB
G1 URW J Carman, 5 Melbourne Road, Blacon, Chester, CH1 5JQ
G1 URZ T Bennett, 16 Montgomery Avenue, Hemel Hempstead, HP2 4HE
G1 USF P Napp, 23 Harriot Drive, Newcastle upon Tyne, NE12 7EU
GD1 USI A Tawney, Croym dty Chione, The Howe, Isle of Man, IM9 5PR
G1 USK Robert Firth, 40 Ashfield Road, Chippenham, SN15 1QQ
GM1 USN J Challis, Bay Villa, Strachur, Cairndow, PA27 8DE
G1 USV R York, 10 Severn View Road, Thornbury, Bristol, BS35 1AY
G1 USW Philip Daniels, 29 Station Road, Wickwar, Wotton-under-Edge, GL12 8NB
G1 USZ Michael [illegible], 10 Oldends Lane, Stonehouse, GL10 2DG
G1 UTC C Thomas, Hazel Mount, Lockhams Road, Southampton, SO32 2BD
G1 UTF R Beer, 8 Littlestone Court, Grand Parade, New Romney, TN28 8NF
G1 UTJ A Lees, Timbercroft, Elliotts Orchard, Warwick, CV35 8LD
G1 UTM P Yearsley, 25 Dinmor Road, Manchester, M22 1NN
G1 UTN C King, 7 Hillcrest, Hyde, SK14 5LJ
G1 UTP Stephen Darlington, 17 Eleanor Road, Royton, Oldham, OL2 6BH
G1 UTS Graham May, 95 Moorfield Avenue, Denton, Manchester, M34 7TX
G1 UTZ A Peters, 10 Hill View Close, Grantham, NG31 7PH
G1 UUF I Grounsell, 23 Loughbrow Park, Hexham, NE46 2QD
G1 UUJ D Eyre, 29 Old Acre Lane, Brocton, Stafford, ST17 0TW
G1 UUK G Perry, 6 Morgan Close, Arley, Coventry, CV7 8PR
G1 UUL S Brough, 9 Beech Tree Lane, Cannock, WS11 1AZ
G1 UUP A Robins, 38 Eastbourne Road, Willingdon, Eastbourne, BN20 9NS
G1 UUS M Carter, 4 Asbury Road, Balsall Common, Coventry, CV7 7QN
G1 UUT R Bygate, 91 Woodcote Avenue, Kenilworth, CV8 1BE
G1 UUV R Fairholm, 63 Rugby Road, Clifton upon Dunsmore, Rugby, CV23 0DE
G1 UUZ M Davis, 20 Pavilion Avenue, Warley, Smethwick, B67 6LA
G1 UVD R Rothery, 2 Highcroft, Mount Pleasant, Batley, WF17 7NT
G1 UVE A Shackleton, 31 Ashton Street, Leeds, LS8 5BY
G1 UVI M Stanley, 35 Moorgate Road, Kippax, Leeds, LS25 7ET
G1 UVJ D Rogers, 11 Beech Crescent, Mexborough, S64 9EH
G1 UVK Andy Linfoot, 19 Vicarage Close, Bubwith, Selby, YO8 6LN

G1 UVN John Jones, Silversprings, Llanelly Church Road, Abergavenny, NP7 0EL
G1 UWD D Peachey, 28 Broad Street, Truro, TR1 1JD
GM1 UWE M Peachey, 8/7 Durar Drive, Edinburgh, EH4 7HN
G1 UWQ J Naughton, 80 Brookford Avenue, Coventry, CV6 2GQ
G1 UWV B Dixon, Terridene, Park Road, New Milton, BH25 6QE
G1 UXW Martin Lloyd Lewis, Gorse Cottage, Graig Road, Cwmbran, NP44 5AS
G1 UYT S Tomkinson, 3 Heysham Close, Weston Coyney, Stoke on Trent, ST3 6RG
G1 UYW K Wallis, Cartref Newydd, Llanarmon D.C., LL20 X
G1 UYZ D Payne, 8 Gascoigne Drive, Spondon, Derby, DE21 7GL
G1 UZC P Jarrett, 17 Wolmers Hey, Great Waltham, Chelmsford, CM3 1DA
G1 UZD J Yates, 8 Holt Drive, Wickham Bishops, Witham, CM8 3JR
G1 UZO E Legg, 11a Edale Avenue, Mickleover, Derby, DE3 9FY
G1 UZW N Boag, 60 Harebell, Amington, Tamworth, B77 4NA
G1 VAA N Willis-Browne, 2 Parkfield Road, Aigburth, Liverpool, L17 8UH
G1 VAB [illegible], 54 Palmerston Street, Derby, DE23 6PF
G1 VAC David Allsebrook, 12 Portman Chase, Stenson Fields, Derby, DE24 3DQ
GM1 VAD Stephen Scanlain, Crossraquel, Old Newton, Nairn, IV12 5RA
G1 VAG A Grant, 26 Fountains Avenue, Boston Spa, Wetherby, LS23 6PX
G1 VAJ A Thain, 8 Eskdale, Brownsover, Rugby, CV21 1NJ
G1 VAL M Pearson, 5 Craven Court, Warwick Drive, Barnoldswick, BD10 0WA
G1 VAN P Van Falier, 572 Stafford Road, Ford Houses, Wolverhampton, WV10 6NN
G1 VAO Arthur Andrews, 12 Kings Lea, Ossett, WF5 8RY
GW1 VAW B Williams, 10 Tynybedw Terrace, Treorchy, CF42 6RL
G1 VAY D Hearn, 90 Princes Drive, Valley Dip, Seaford, BN25 2TX
GI1 VAZ G Richardson, 6 Cedarhurst Rise, Belfast, BT8 7RJ
G1 VBA Patricia Buckle, 63 Ashley Drive South, Ashley Heath, Ringwood, BH24 2JP
G1 VBB R Bedwell, 36 Spindleberry Grove, Nailsea, Bristol, BS48 1QF
GM1 VBD R Scott, Enzie Slackhead, Buckie, Moray, AB5 2BJ
GM1 VBE Stuart Clink, Southsyde, Woodhead Avenue, Glasgow, G71 8AR
G1 VBL R Kelsall, 11 Manor Road, Ducklington, Witney, OX29 7YD
G1 VBO R Piper, Tatnam Farm, St. Mary's Road, Romney Marsh, TN29 0PW
G1 VBP P Smith, The Rufford Care Centre, Room 1, Gateford Road, S81 7BH
G1 VBQ P Wright, 81 High St., Syston, Leicester, LE7 1GQ
G1 VBY K Moreton, 29 Tiber Drive, Chesterton, Newcastle, ST5 7QD
G1 VCU Allan Smithson, 118 High Street, Linton, Cambridge, CB21 4JT
G1 VCZ A Hewes, 89 Fronks Road, Dovercourt, Harwich, CO12 4EQ
G1 VDC M Jeffers, 20 Three Crowns Road, Colchester, CO4 5AD
G1 VDE D Taylor, 21 Munday Close, Bussage, Stroud, GL6 8DG
G1 VDO S Preston, 50 Milton Avenue, Malton, YO17 7LB
G1 VDP Christopher Colclough, 52 Alexandra Street, Nuneaton, CV11 5RL
G1 VDT P Whatley, Fairacre, Bishton Lane, Chepstow, NP16 7LG
G1 VDW L Marquardt, 2 Pembroke Terrace, Varteg, Pontypool, NP4 7UJ
GM1 VDZ G Neil, 7 Dalfarson Avenue, Dalmellington, Ayr, KA6 7TX
GM1 VFQ J Murdoch, 8 Primpton Avenue, Dalrymple, Ayr, KA6 6EL
GM1 VFR Henry McDonald, 43 Southfield Road, Cumbernauld, Glasgow, G68 9DZ
G1 VFW I Beeby, 32 Edditch Grove, Bolton, BL2 6BJ
G1 VGA K Crane, 92 Dimond Road, Southampton, SO18 1JS
G1 VGI K Waterson, 20 Cadogan Road, Bury St. Edmunds, IP33 3QJ
G1 VGK Mark Whittington, 253 Kings Drive, Eastbourne, BN21 2UR
G1 VGM Simon Ball, 8 Ringwood Road, Ryde, PO33 3NX
G1 VGO D Holdsworth, 28 Moorbank Close, Wombwell, Barnsley, S73 8RX
G1 VGP G Anderson, 1 White Rose Mead, Garforth, Leeds, LS25 2EG
G1 VHC Peter Ward, 13 St. Laurence Road Winslow, Buckingham, MK18 3BD
G1 VHN D Barnes, 46 Lawnswood Avenue, Wordsley, Stourbridge, DY8 5LR
G1 VHW W Killeen, 12 Meriden Grove, Lostock, Bolton, BL6 4RQ
G1 VHY D symonds, 79 Kingsway, Kirkby-in-Ashfield, Nottingham, NG17 7EH
G1 VID T Howe, The Old School House, 3 Bairds Hill, Broadstairs, CT10 3AA
G1 VIF Carl Morphett, 138 Healds Road, Dewsbury, WF13 4HT
G1 VIG D Carrick, 5 Kings Close, Market Overton, Oakham, LE15 7PS
G1 VII N Pooley, 64 Lynwood Grove, Orpington, BR6 0BH
G1 VIN C Kirkland, 29 Shelley Road, Enderby, Leicester, LE19 4QX
GD1 VIO A Eades, 41 Woodhall Gate, Pinner, HA5 4TX
G1 VIP M Walker, Bessbrook, 43 Wimborne Road, Wimborne, BH21 3DS
G1 VIR Ian Berry, 1-2 Pottery Cottages, Trefonen, Oswestry, SY10 9DF
G1 VIS Sheila Grimes, 73 Ryston Road, Denver, Downham Market, PE38 0DP
G1 VIT Noel Cliff, 12 New Church Road, Wellington, Telford, TF1 1JH
G1 VIW R Paterson, 27 Copt Heath Drive, Knowle, Solihull, B93 9PA
G1 VIY T Depledge, 13 Peel Drive, Astbury, Congleton, CW12 4RF
G1 VIZ A Hemenway, 69 Sixth Avenue, Heworth, York, YO31 0UR
G1 VJB B Flounders, 76 Springfield Road, Sebastopol, Pontypool, NP4 5BX
GM1 VJD R Terras, 8 Haddington Gardens, Lawthorn, Irvine, KA11 2EB
G1 VJE Mark Sayers, 21 Chesham Road North, Weston-Super-Mare, BS22 8AD
G1 VJG C Rhenius, 8 Rutland Close, Cambridge, CB4 2HT
G1 VJJ C Silvey, 106 Southchurch Avenue, Southend-on-Sea, SS1 2RP
G1 VJN B Jones, 73 Tonge Road, Murston, Sittingbourne, ME10 3NR
G1 VJQ T Monaghan, 15 Mulgrave Road, Worsley, Manchester, M28 2RW
G1 VJY Andrew Birch, Maximillien, Les Ecovets, Chesieres, Switzerland, 1885
G1 VKB A Morris, 140 Astwood Road, Worcester, WR3 8EZ
G1 VKC D Close, 60 Lead Lane, Ripon, HG4 2LN
G1 VKG J Howarth, 10 Poplar Place, Penrith, CA11 9HN
G1 VKJ B Duffy, 26 Cherry Blossom Close, Billing, Northampton, NN3 9DN
G1 VKN Stanley Matthews, Matilda - Kings Marina, Mather Road, Newark, NG24 1FW
G1 VKT Eric Dale, 29 Hulme Road, Leigh, WN7 5BT
G1 VLA A Lee, Sandana, Kirkpatrick Fleming, Lockerbie, DG11 3BA
G1 VLD A Burkitt, 6 Chewells Close, Haddenham, Ely, CB6 3XE
G1 VLS Gareth Turner, 19 Hector Road, Darwen, BB3 0AY
G1 VLU David Green, 5 New Mill Road, Holmfirth, HD9 7SG
GW1 VMA M Hills, Ty Newydd, Rhos, Llandyssul, SA44 5HE
GI1 VMF J Oliver, 29 Callan Bridge Park, Armagh, BT60 4BU
G1 VMX Ernest Driver, 39 Witham Road, Woodhall Spa, LN10 6RW
G1 VNE S Nocera, Strada Provinciale, Mulazzano 88, Parma, Italy, 43010
G1 VNH V Palmer, 57 Old Tiverton Road, Exeter, EX4 6NG
G1 VNL A Smith, 112 Manor Lane, Charfield, Wotton-under-Edge, GL12 8TN
G1 VNM Susan Eyers, 190 Greenhill Road, Herne Bay, CT6 7RS
G1 VNS K Baum, 11 St. Ives Road, Wigston, LE18 2JB

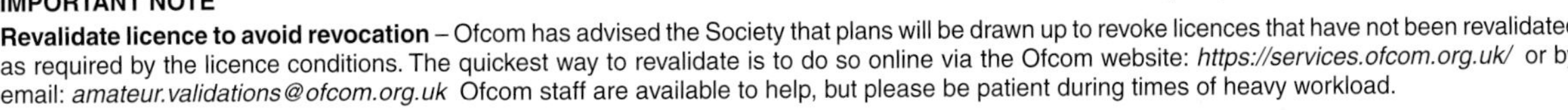

**IMPORTANT NOTE**

**Revalidate licence to avoid revocation** – Ofcom has advised the Society that plans will be drawn up to revoke licences that have not been revalidated as required by the licence conditions. The quickest way to revalidate is to do so online via the Ofcom website: *https://services.ofcom.org.uk/* or by email: *amateur.validations@ofcom.org.uk* Ofcom staff are available to help, but please be patient during times of heavy workload.

G1 VNU S Exell, 22 Woodside Road, Beare Green, Dorking, RH5 4RH
G1 VNV Melvyn Gold, 14 Brewers Lane, Badsey, Evesham, WR11 7EU
G1 VNZ A Atkins, 28 Third Avenue, Pebsham, Bexhill-on-Sea, TN40 2PG
G1 VOB M Prior, 36 Bassnage Road, Halesowen, B63 4HQ
G1 VOC A James, 70 Martin Croft, Silkstone, Barnsley, S75 4JS
G1 VOJ A Duce, 16 Gillmans Road, Orpington, BR5 4LA
G1 VON S Howard, 95 Greenbarn Way, Blackrod, Bolton, BL6 5TE
G1 VOP D Hettiarratchi, 2 Carham Close, Gosforth, Newcastle upon Tyne, NE3 5DX
G1 VOQ Philip Tandy, Old Channel Hill Farm, North End, Fordingbridge, SP6 3HA
G1 VOY R Ward, Overdale, Egton, Whitby, YO21 1UE
GI1 VPA D Smythe, 57 Ballymacormick Avenue, Bangor, BT19 6AY
G1 VPE Richard Wilcockson, 58 Didcot Close, Chesterfield, S40 2UF
G1 VPH S Currie, 36 Stonecrop Close, Birchwood, Warrington, WA3 7PD
G1 VPS Michael Wright, 25 St. Matthews Road, Kettering, NN15 5HE
G1 VQB G Doughty, 95 Buxton Road, Chaddesden, Derby, DE21 4JL
G1 VQG Gary Hannaford, 8 Porthmellon Gardens, Callington, PL17 7QL
G1 VQH Glenn King, Mill Cottage, 48 Mill Street, Swadlincote, DE12 8ES
G1 VQI Craig Coates, 10 Woodcote House Queen Street, Hitchin, SG4 9TL
G1 VQK E Wright, 94 Bachelor Gardens, Harrogate, HG1 3EA
G1 VQV S Gent, 66 Apperley Way, Cradley Forge, Halesowen, B63 2PY
G1 VRA E Jones, Bramley Lodge, Back Lane, Royston, SG8 6DD
G1 VRC Trevor Nicholson, 8 East Street, High Spen, Rowlands Gill, NE39 2HD
G1 VRJ John Ager, 20 Kirktonhill Road, Westlea, Swindon, SN5 7AF
G1 VRP M Carter, 9 Westward Road, Malvern, WR14 1JX
GW1 VRR J Williams, 31 Syr Davids Avenue, Cardiff, CF5 1GH
GW1 VRW Colin King, 27 Gadlys Road West, Barry, CF62 7HX
G1 VSD W Bennett, 90 Garwood Road, Yardley, Birmingham, B26 2AW
G1 VSE D Herridge, 63 High Road, Orsett, Grays, RM16 3HB
G1 VSH G Pettit, 7 Dunster Crescent, Hornchurch, RM11 3QD
G1 VSK A Heyes, 53 Gilderdale Close, Birchwood, Warrington, WA3 6TH
G1 VSM Derek Pratt, 17 Worcester Gardens, Greenford, UB6 0BH
G1 VSO Christopher Champ, 31 Nobles Close, Oxford, OX2 9DN
GM1 VSR A Bain, 4 Bawdley Head, Fraserburgh, AB4 5SE
G1 VTE M Joynson, 90 Fairhope Avenue, Bare, Morecambe, LA4 6LA
G1 VTN Corrinne Peacock, 1 Furnace Lane, Madeley, Crewe, CW3 9EU
G1 VTO B Kemp, 193 Cavalry Park, March, PE15 9DL
G1 VTP Joseph Bridgehouse, 12 Castle Hall Close, Stalybridge, SK15 2HR
G1 VTQ B Gorman, 40 Maudland Bank, Preston, PR1 2YL
G1 VTS J Smith, 9 Birchway, Hayes, UB3 3PA
G1 VTU R Arnold, The Ridge, Hickman Road, Nuneaton, CV10 9NG
G1 VUG M Cooper, Osborne House, Main Street, Louth, LN11 0XF
G1 VUK S Hunt, 1 Lucknow Cottages, Northbridge Street, Robertsbridge, TN32 5NP
G1 VUP Andrew Cheeseman, Flat 5 Dubarry House, Hove Park Villas, Hove, BN3 6HP
G1 VUY D White, 14 Waggoners Way, Bugbrooke, Northampton, NN7 3QT
G1 VVB S Patrick, 9 Brant Avenue, Illingworth, Halifax, HX2 8DL
G1 VVE G Bindon, 74 Cashford Gate, Taunton, TA2 8QB
G1 VVF M Clews, 137 Wyatt Road, Sutton Coldfield, B75 7ND
G1 VVH Peter Fisher, Flat 46 Morris House, Fairchild Close, London, SW11 2SU
G1 VVL Albert Proctor, 448 Tuttle Hill, Nuneaton, CV10 0HR
G1 VVM G Brett, 47 Windermere Avenue, Huncoat, Accrington, BB5 6JG
G1 VVT Howard Mawson, 118 Byron Street, Loughborough, LE11 5JW
G1 VVU John Stephens, 34 King Street, Seahouses, NE68 7XR
G1 VVX I Andronov, 53 Broad Street, Ludlow, SY8 1NH
G1 VVY John Waters, 2 Tea Caddy Cottages, Worthing Road, Horsham, RH13 8LG
GM1 VWA J Gilruth, 88 Fintry Crescent, Dundee, DD4 9EX
G1 VWL Richard Knowles, 35 Poulton Road, Southport, PR9 7BE
G1 VWP Simon Goodwin, Keep House, 1 Deal Castle Road, Deal, CT14 7BB
G1 VWU Alan Driver, Grace Barn, Pencarrow, Advent, Camelford, PL32 9RZ
G1 VWZ Robert Huntley, 49 Main Street, Wetwang, Driffield, YO25 9XL
GM1 VXE Alistair Rennie, 5 Barbieston Cottage, Drongan, Ayr, KA6 7EF
G1 VXS Steven Shirvington, 22 Bassett Road, Northleach, Cheltenham, GL54 3QJ
G1 VXX A Powell, 9a Shaftesbury Close, Bracknell, RG12 9PX
G1 VXY Terence Tomkins, 40 Diksmuide Drive, Ellesmere, SY12 9QA
G1 VYA H Seddon, 15 Westfield Grove, Wigan, WN1 2QJ
G1 VYB T Smith, 11 River Terrace, Wisbech, PE13 1PZ
GM1 VYF P Letters, 23 East Lennox Drive, Helensburgh, G84 9JD
GM1 VYG J McDonald, 4 Braeside, Bowfield Road, Johnstone, PA9 1BP
G1 VYM D Eveleigh, 7 Malin Road, Littlehampton, BN17 6NN
G1 VYS A Walker, 49 Selworthy Road, Stoke-on-Trent, ST6 8PL
G1 VZB C Lillis, 6 Whitelake View, Urmston, Manchester, M41 8UT
GM1 VZG T Gilmour, Fiold, Rope Walk, Kirkwall, KW15 1XJ
G1 VZT Barry Rayner, 44 Foxhall Fields, East Bergholt, Colchester, CO7 6QY
G1 VZW B Hitchen, 40 Methuen Avenue, Fulwood, Preston, PR2 9QX
G1 WAB DERBYS WORKD ALL BRITAIN G.ARC c/o J Wainwright, 8 Common Lane, Cutthorpe, Chesterfield, S42 7AN
G1 WAC WYTHALL RAD CLUB c/o David Dawkes, 95 Houndsfield Lane, Wythall, Birmingham, B47 6LX
G1 WAE Charles Rogers, 63 Greenwell Road, Haydock, St. Helens, WA11 0SQ
G1 WAP Brian Stott, 35 Sheridan Road, Laneshawbridge, Colne, BB8 7HW
G1 WAS P Delaney, 61 Lyndale Avenue, Eastham, Wirral, CH62 8DG
G1 WAW WESSEX AW CLUB c/o Paul Mitchell, 3 High Howe Close, Bournemouth, BH11 8NN
G1 WCY RIPON & DISTRICT ARS c/o John Reeves, 5 Arrows Crescent, Boroughbridge, York, YO51 9LP
G1 WDQ T Hutton, 23 Dines Close, Wilstead, Bedford, MK45 3BU
G1 WEF R Cornwell, 13 Milford Road, Thurrock, Grays, RM16 2QL
G1 WEV Ian Henderson, 1 Prestbury Road, Pennywell, Sunderland, SR4 9DW
G1 WFA C Ward, 7 Coates Close, Heybridge, Maldon, CM9 4PB
G1 WFG S Lake, 42 Haling Park Road, South Croydon, CR2 6NE
G1 WFJ A Woodhouse, 5 Dudley Road, Kingswinford, DY6 8BT
G1 WFO P Burden, 110 Westbury Leigh, Westbury, BA13 3SH
G1 WFS Stephen Rogers, 19 Stoke Street, Hull, HU2 9BL
G1 WFU R Dickson, 49 Ashgrove, Peasedown St. John, Bath, BA2 8EF
GI1 WGK W Steele, 19 William Street, Donaghadee, BT21 0HL
G1 WGM K Perry, 25 Hillary Drive, Crowthorne, RG45 6QF
G1 WGO Andrew Smith, 72 Barnsley Road, South Kirkby, Pontefract, WF9 3QE
GW1 WGR WEST GLAMORGAN RAYNET c/o Michael Rowles, 7 Gelli Deg, Bryncoch, Neath, SA10 7PL
G1 WHT Martyn Watts, 11 Bywood Place, Grimsby, DN37 9RH
G1 WHU David Baines-Jones, 22-24 Grove Royd, Halifax, HX3 5QU
G1 WHY J Lowe, 23 Hoylake Drive, Tividale, Oldbury, B69 1QA
G1 WID J Hartridge, 15 Hundred Acres, Wickham, Fareham, PO17 6JB
G1 WIS D Mountain, 45 Westway Gardens, Redhill, RH1 2JB
G1 WIW R Dowdeswell, 5 Croft Close, Barwell, Leicester, LE9 8EW
GU1 WJA W Ayres, Rousay, Bailiffs Cross Road, St. Andrew, Guernsey, GY6 8RY
G1 WJG J Coates, The Old Timbers, 23 Yoells Lane, Waterlooville, PO8 9SG
G1 WJK G Richards, 3 Pleasant Close, Kingswinford, DY6 9TQ
G1 WJO A Blackwell, 1 Gladstone Terrace, Hinckley, LE10 1HE
G1 WJR William Rollins, 5 Little Clacton Road, Great Holland, Frinton-on-Sea, CO13 0ET
GM1 WKH N Graves, The Lythe, Tree Road, Ellon, AB41 7JY
G1 WKK James Arnott, 27 Main Road, Tadley, RG26 3NJ
G1 WKO Reuben Reichmann, 9 Rue Du Croteau, la Neuville Les Wasigny, France, 8270
G1 WKS WEST KENT ARS c/o Leslie Featherstone, The Bungalow, Prices Wood, Tonbridge, TN11 8HP
G1 WKZ G Evans, 4 The Mallards, Fareham, PO16 7XR
G1 WLD Sally Evans, 4 The Mallards, Fareham, PO16 7XR
GI1 WLJ George McCutcheon, 34 British Road, Aldergrove, Crumlin, BT29 4DJ
G1 WLN N Roskruge, 2 Lanner Green Terrace, Lanner, Redruth, TR16 6DQ
G1 WLO M Head, 20 Clock Tower Court, Park Avenue, Bexhill-on-Sea, TN39 3HP
G1 WLU Steve Brookes, Ivy Cottage, Haselor, Alcester, B49 6LX
G1 WLW A Fordyce, 41 Benscliffe Drive, Loughborough, LE11 3JP
G1 WLX J Goacher, 41 Clay Hill, Two Mile Ash, Milton Keynes, MK8 8AY
G1 WMJ Michael Hudson, 70 High Street, Wingham, Canterbury, CT3 1BJ
G1 WMK John Mayo, Mavis House, Rectory Lane, Bicester, OX27 8DX
G1 WMN K Harvey, 61 Westfield Road, Northchurch, Berkhamsted, HP4 3PW
G1 WMS R Cadwallader, Rambla Grande, Los Reyes, Urcal, Spain, 4691
GM1 WMU Stephen Webster, 15 Forrest Place, Armadale, Bathgate, EH48 2GZ
G1 WMV B Catchpoole, 8 Buckland Avenue, Basingstoke, RG22 6JL
G1 WNL F Thomas, 38 Partridge Avenue, Yateley, GU46 6PB
G1 WNZ Giuseppe Sollazzo, 24a-24c Palace Gates Road, London, N22 7BN
G1 WOR WORTHING & DIST ARC c/o Phil Godbold, 13 Dawn Crescent, Upper Beeding, Steyning, BN44 3WH
G1 WPG B Groome, The Old Smithy, High Street, Dorchester, DT2 8JW
G1 WPH L Eden, 23 Elm Green Close, Worcester, WR5 3HD
G1 WPL R Balkwell, 2 Franklyn Road, Droylsden, Manchester, M43 6DS
G1 WPR Terry Bromley, 7 Brookside, Desborough, Kettering, NN14 2UD
G1 WQC R Pratt, 11 Park Road, Ryde, PO33 2BG
G1 WQH R Booth, 66 Fairburn Crescent, Pelsall, Walsall, WS3 4PU
G1 WQL Peter Kenyon-Brodie, 415 Cottingham Road, Hull, HU5 4AA
G1 WQN S Mangan, 48 Emblett Drive, Newton Abbot, TQ12 1YJ
G1 WQU T Gregg, 27 Somerleaze Close, Wells, BA5 1UD
G1 WQY K Webber, 2 Henniker Road, Ipswich, IP1 5HD
G1 WRC Wisbech Amateur Radio & Electronics Club c/o James Balls, 7 Rowan Close, Holbeach, Spalding, PE12 7BT
G1 WRD Mark Simpson, Orchard House, Todwick Grange, Sheffield, S26 1JQ
G1 WRE R Constantine, 35 Heckler Lane, Ripon, HG4 1PU
G1 WRF Alan Jolly, 27 Murrayfield Drive, Brandon, Durham, DH7 8TG
G1 WRH G Marshall, 1 Portland Close, Braintree, CM7 9NJ
G1 WRN NORTH WARKS RAYNET GROUP c/o Terry Yorke, 12 Shanklin Drive, Weddington, Nuneaton, CV10 0BA
G1 WRO M Smith, 8 Milldale Road, Farnsfield, Newark, NG22 8DQ
G1 WRS Wakefield and District RS c/o Darryl Burden, 16 Milnthorpe Lane, Wakefield, WF2 7DE
G1 WRU John Jinks, 27 Taryn Drive, Darlaston, Wednesbury, WS10 8XY
GW1 WRV R Marston, 34 Kevin Ryan Court, Georgetown, Merthyr Tydfil, CF48 1EE
G1 WRY Ann White, 34 Pain's Way, Amesbury, Salisbury, SP4 7RG
G1 WSC Johathan Bland, 9 Earl Street, Grimsby, DN31 2NB
G1 WSD David Garratt, 87 Garden Road, Eastwood, Nottingham, NG16 3FY
G1 WSE J Frizell, 17 St. Johns Terrace, Lewes, BN7 2DL
G1 WSF D Pettican, 52 Shepherds Way, Saffron Walden, CB10 2AH
G1 WSM M Franklin, 6 Norbury Close, Lancing, BN15 0QL
G1 WSN J Spillett, Mockbeggar Cottage, Mockbeggar, Ringwood, BH24 3NQ
G1 WSW Malcolm Flewitt, 38 Laburnum Avenue, Newbold Verdon, Leicester, LE9 9LQ
G1 WTB E Musson, 110 Marples Avenue, Mansfield Woodhouse, Mansfield, NG19 9DW
G1 WTH C Piddock, 118 Howley Grange Road, Halesowen, B62 0HU
G1 WTL John Beachey, 16 Morgan Street, Blaenavon, Pontypool, NP4 9ER
G1 WTN R Wroe, 13 Silverdale Drive, Barnsley, S71 2PP
G1 WTS M Roper, 19 Normay Rise, Newbury, RG14 6RY
G1 WTW Craig Underwood, 4 Hawthorn Road, Godalming, GU7 2NE
G1 WTX P Egan, 13 Beechcroft Drive, Guildford, GU2 7SA
G1 WTY B Parkes, 11 Hampton Grove, Cheadle Hulme, Cheadle, SK8 6DG
GW1 WTZ Christopher Green, 11 Brookfield Close, Gorseinon, Swansea, SA4 4GW
G1 WUC M Garner, 40 Studley Road, Harrogate, HG1 5JU
G1 WUH Kenneth Carr, 41 Surrey Road, Dagenham, RM10 8ES
G1 WUM R Miles, Haseley Lodge, Birmingham Road, Warwick, CV35 7HF
G1 WUU John Neate, 23 Crossley Moor Road, Kingsteignton, Newton Abbot, TQ12 3LE
G1 WUY J Wilkins, 21 Stocks Loke, Cawston, Norwich, NR10 4BS
G1 WVD M Shrago, 12 Oakwood Road, Bricket Wood, St. Albans, AL2 3PU
G1 WVK Jeremy Power, 45 Grace Gardens, Cheltenham, GL51 6QE
G1 WVM R Vowles, 47 Tyndale Avenue, Yate, Bristol, BS37 5EX
G1 WVO D Willey, 17 Bridge Place, Saxilby, Lincoln, LN1 2QA
G1 WVR WELLAND VALLEY ARS c/o David Lowe, 21 Farndon Road, Market Harborough, LE16 9NW
G1 WVS P Gibson, 14 Nene Side Close, Badby, Daventry, NN11 3AD
G1 WVV Robert Sutton, 28 Shrubbery Gardens, Wem, Shrewsbury, SY4 5BX
G1 WVW R Jones, 8 Downing Avenue, Newcastle, ST5 0JY
G1 WVZ Richard McCutcheon, 13 The Beeches, Rugeley, WS15 2QY
G1 WWA R Williams, Coombe Farm Cottage, Stottesdon, Kidderminster, DY14 8LS
G1 WWB R Eeles, 23 Elgin Avenue, Ashford, TW15 1QE
GW1 WWE J Peake, Winley, 70 Higher Lane, Swansea, SA3 4PD
G1 WWH Alan Benn, Burneston, Bedale, DL8 2HT
G1 WWI M Dronfield, White Lodge Farm, High Bradfield, Sheffield, S6 6LJ
G1 WWP J Sharpe, 10 Stocking Green Close, Hanslope, Milton Keynes, MK19 7NH
G1 WWR S Cockshoot, 23 Dane Mount, Margate, CT9 3SA
G1 WWW William Edmondson, Glaslyn, Penysarn, LL69 9YB
G1 WWY L Donald, 53 Andrews Way, Raunds, Wellingborough, NN9 6RD
G1 WXC David Blackman-Wells, 15 Purbeck Place, Littlehampton, BN17 5DP
G1 WXF J Pearson, 17 Hebden Avenue, Woodloes Park, Warwick, CV34 5XD
G1 WXK M Bell, 151 Towngate, Ossett, WF5 0PP
G1 WXS P Springall, 31 The Orchards, Epping, CM16 7BB
G1 WXT M Moorecroft, 4 St. Davids Road, Locks Heath, Southampton, SO31 6EP
G1 WXU G Parsons, 73 Worthing Avenue, Elson, Gosport, PO12 4DB
G1 WXW P Prescott, 13 The Boltons, Waterlooville, PO7 5QR
G1 WYA Richard Webb, Calle Pavo Real, (Argenta 2), Las Chismosas 1 puerta 7, 03189 Orihuela Costa, Spain, ALACANTE
G1 WYB E Coupe, Killidina, 28 Wellington Road, Blackburn, BB2 2NQ
G1 WYC S Smith, 82 Wignals Gate, Holbeach, Spalding, PE12 7HR
G1 WYD A Darlington, 15 Kestrel Close, Carterton, OX18 3LS
G1 WYG David Biginton, 67 Capstone Road, Bromley, BR1 5NA
G1 WYM Manoranjan Cheema, Home Farm, Back Lane, Market Harborough, LE16 9SE
G1 WYP H Milsom, 1 Wyld Court, Blunsdon, Swindon, SN25 2EE
GM1 WYV A Henderson, 30 Pentland Crescent, Larkhall, ML9 1UR
GI1 WYZ R Kennedy, 3 St. Annes Crescent, Newtownabbey, BT36 5JZ
G1 WZB I Ross, 46 Fordbridge Road, Ashford, TW15 2SJ
G1 WZG J Endicott, 16 Packs Close, Harbertonford, Totnes, TQ9 7TL
GW1 WZI J Cartwright, 20 Castlefield Place, Cardiff, CF14 3DU
G1 WZK D Heaton, 39 Bridgwater Road, Romford, RM3 7UB
G1 WZM Clive Turner, 2 Martins Mews, Haverhill, CB9 7FU
G1 WZO Leslie Leach, Leyland, The Street, Gloucester, GL2 7ED
G1 WZQ Adrian Utting, 20 Davenport Road, Leicester, LE5 6SA
G1 XAA Hilary Jacklin, 26 Rockmill End, Willingham, Cambridge, CB24 5HY
G1 XAJ J Franklin, 16 Mountbatten Drive, Colchester, CO2 8BH
G1 XAL A Dangerfield, Brookside, High Street, Gloucester, GL2 7LW
G1 XAM J Bryant, 12 Dale Tree Road, Barrow, Bury St. Edmunds, IP29 5AD
G1 XAP P Whittingham, 28 Wedge Avenue, Haydock, St. Helens, WA11 0DY
G1 XAS J Hume, 2 Llain Wen, Pentrefelin, Amlwch, LL68 9PD
G1 XBE T Beecher, 77 Grime Lane, Sharlston Common, Wakefield, WF4 1EH
GW1 XBG P Smith, 35 Terrace Road, Swansea, SA1 6HN
GM1 XBK Kenneth McClure, 9 Cumnock Road, Mauchline, KA5 5AE
G1 XBL B Darby, Pippins, Green Street, Worcester, WR5 3QB
G1 XBR S Loney, 4 Mendip Road, Southampton, SO16 4BN
G1 XBX A Manning, 9 Seymour Caravan Park, Liverton, Newton Abbot, TQ12 6HA
G1 XCB P Davies, 91 Station Road, Hadfield, Glossop, SK13 1AR
G1 XCC Michael Lockwood, 5 Marsh Street, Cleckheaton, BD19 5BW
G1 XCK William Potter, Flat, 1 Tabernacle Walk, Blandford Forum, DT11 7DL
G1 XCY Elizabeth Knott, 24 Walsingham Way, Ely, CB6 3AL
G1 XDJ H Opitz, 26 Holme Court, Lower Warberry Road, Torquay, TQ1 1QR
G1 XDK David Rouse, 23 Montgomery Way, King's Lynn, PE30 4YH
G1 XDS J Fyson, One Redlands Estate, Ibstock, LE6 1HT
G1 XDV T Gale, 1a Aldridge Road, Streetly, Sutton Coldfield, B74 3TU
GM1 XEA P Thomson, 13 Westwood Drive, Westhill, AB32 6WW
GM1 XEB M McCulloch, 6 Learmont Place, Milngavie, Glasgow, G62 7DT
G1 XEH R Burton, 18 Churchfield, Harpenden, AL5 1LL
G1 XEP L Lambert, 124 Frankland Road, Croxley Green, Rickmansworth, WD3 3AU
G1 XES E Turner, 1104 Wimborne Road, Bournemouth, BH10 7AA
G1 XET Christopher Burton, 14 Fotherley Road, Mill End, Rickmansworth, WD3 8QG
GW1 XFB D Evans, Bwthyn Bach, 2 Old Village Road, Barry, CF62 6RA
G1 XFE J King, 40 Galway Avenue, Chaddesden, Derby, DE21 6TR
G1 XFL K Lanham, 22 Ascot Close, Ladywood, Birmingham, B16 9EY
G1 XFM Jane Shaw, 33 Park Farm Close, Horsham, RH12 5EU
G1 XFO M Price, 67 Broadway, Oldbury, B68 9DP
G1 XFR B Bance, 36 Leafy Oak Road, Grove Park, London, SE12 9RS
G1 XFX A Harvey, 26 Kidborough Road, Gossops Green, Crawley, RH11 8HW
G1 XGE Daren Hutchinson, 32 The Causeway, Kingswood, Hull, HU7 3AL
G1 XGM Michael De-Wynter, 8 Eldon Place, Cutler Heights, Bradford, BD4 9JH
G1 XGN Mark Brady, 42 Arden Avenue, Middleton, Manchester, M24 1PN
G1 XGP S Blinkhorn, 12 Eloura Lane, New South Wales, Australia, 2577
G1 XGW A Gray, 12 Peak Close, Oldham, OL4 2TH
G1 XGZ David Richards, Flat 8, Thackeray Court, London, SW3 3LB
G1 XHA J De Bank, 5 Horn Hill View, Beaminster, DT8 3PJ
G1 XHO T Williams, 145 Bulwell Lane, Old Basford, Nottingham, NG6 0BS
G1 XHR E Goodwin, Hankelow Court, Hall Lane, Crewe, CW3 0JB
GM1 XHZ Thomas Valentine, 4 Angus Cottages, Friockheim, Arbroath, DD11 4SR
GI1 XIB J Wilkinson, 67 Glenwood, Ahoghill, Ballymena, BT42 1GW
G1 XIC CALLINGTON COMMUNITY COLLEGE RC c/o Keith Harris, 8 Trelawney Rise, Callington, PL17 7PT
G1 XIE R Dyer, 21 Allden Avenue, Aldershot, GU12 4AG
G1 XIH Lee Taylor, 76 Sidney Road, Blackley, Manchester, M9 8AT
G1 XII T Lovatt, 5 Acre Rise, Willenhall, WV12 4SL
GM1 XIN W Allan, Corse Farm, Kininmonth, Peterhead, AB42 4JU
G1 XIO A Faram, 4 Wellington Road, Gillingham, ME7 4NN
G1 XIV G Reynolds, The Thatched Cottage, St. Thomas Drive, Bognor Regis, PO21 4TN
G1 XIY Matthew Nottingham, 11 Taverners Drive, Ramsey, Huntingdon, PE26 1SF
GM1 XJE J Hopkins, 9 Pathfoot Avenue, Bridge of Allan, Stirling, FK9 4SA
GW1 XJJ H Worgan, 29 Mayfield Avenue, Laleston, Bridgend, CF32 0LH
G1 XJK F Tilley, 37b Fant Lane, Maidstone, ME16 8NP
G1 XJM Robert Davidson, 17 Willows Avenue, Alfreton, DE55 7ER
G1 XJN D Jones, 12 Brockhill Close, Kettering, NN15 7DS
G1 XJO Nigel Snowden, 11 Marion Drive, Shipley, BD18 2EY
G1 XJT A Nichols, Pentre, Trelawney Road, Truro, TR3 7EN
G1 XJZ D Layton, The Nook, 7 Turton Street, Kidderminster, DY10 2TH
G1 XKD G Lawton, 23 Fiske Court, Cavendish Road, Sutton, SM2 5ER

G1 XKJ K Higlett, 3 Clover Way, Killinghall, Harrogate, HG3 2WE
G1 XKL Brian Smith, 1 Hirsts Cottages, Spa Lane, Ormskirk, L40 6JG
G1 XKN A Chambers, 34 Haunchwood Drive, Sutton Coldfield, B76 1JR
G1 XKQ B Neale, 30 Berry Avenue, Paignton, TQ3 3QN
G1 XKY E Marsh, 15 Beacon Close, Rubery, Birmingham, B45 9DA
G1 XKZ A Bard, 9 Linden Road, Cullompton, EX15 1TE
G1 XLE Paul Bryan, 187 Wolverhampton Road, Pelsall, Walsall, WS3 4AW
G1 XLG C Proctor, 24 Orchard Way, Southam, CV47 1EX
GM1 XLH C Cullingworth, Lochmoss, Ythanwells, Huntly, AB54 6HA
G1 XLL Patrick Greetham, Flat 10, Hillman House, Coventry, CV1 1FZ
G1 XLN J Danko, 216 Hanley Road, Sneyd Green, Stoke-on-Trent, ST1 6DD
G1 XLT R Perrat, 18 Pells Hill, Northolt, UB5 4NI
G1 XLW A Harrington, 44 Fairburn Crescent, Pelsall, Walsall, WS3 4PU
GD1 XMA M Haley, Yn Croit, Ballamanagh Road, Sulby, Isle of Man, IM7 2HB
G1 XMH Robert Fisher, White House, Slough Road, Manningtree, CO11 1NS
G1 XMI J Brown, 78 Park Way, St. Austell, PL25 4HR
G1 XMP C Brookes, 58 Brookwood Drive, Stoke on Trent, ST3 6HY
G1 XNC Robert Gearing, 6 Boughton Close, Gillingham, ME8 6ND
G1 XNG Christopher Whitehead, 6 Welbeck Street, Sutton-in-Ashfield, NG17 4AY
G1 XNI N Dingle, 29 Castle View, Witton le Wear, Bishop Auckland, DL14 0DH
G1 XNK R Rafter, 8 Bishops Walk, Ilchester, Yeovil, BA22 8NS
G1 XNN R Harding, Highview, High Road, Wallingford, OX10 0QT
G1 XNX Martyn Cowell, 105 Belgrave Road, Darwen, BB3 2SF
G1 XOG I Wilson, 131 Weatherly Road, Torbay, Auckland, New Zealand, 630
GM1 XOI Mid Lanark ARS c/o Carrie Welsh, 28 Peacock Wynd, Motherwell, ML1 4ZL
G1 XOT Bryan Blake, Lafiteau, Benque, Aurignac, France, 31420
G1 XOW Steven Wragge, Tree Tops, Priory Road, Nottingham, NG14 7GW
G1 XOZ Clive Harding, 24 Bryer Close, Bridgwater, TA6 6UR
G1 XPB L Rohrlach, 19 Moatfield Road, Bushey, WD23 3BP
G1 XPD L Wheatley, 25 Hobbis House, Redditch Road, Birmingham, B38 8LS
GM1 XPE J Graham, Lodge, Stronsay, Orkney, KW17 2AN
G1 XPF A Duell, 3 Jail Lane, Biggin Hill, Westerham, TN16 3SA
G1 XPW M King, 104 Green Lane, Vicars Cross, Chester, CH3 5LE
G1 XQI Keith Bates, 2, 28 Synge Street, Portobello, Ireland, DUBLIN 8
G1 XQP J Jackson, Greengable Upcott, Bishops Hull, Taunton, TA4 1AQ
G1 XRE S Staton, 6 Greenhowsyke Lane, Northallerton, DL6 1HP
G1 XRF Brian Rogers, 5 Springfield Road, Ruskington, Sleaford, NG34 9HG
G1 XRJ Steve Hancock, 8 Elanor Road, Sandbach, CW11 3FZ
G1 XRM Michael Baybrook, 12 Bossington Close, Rownhams, Southampton, SO16 8DW
G1 XRO C Frost, 110 Spring Hill, Weston-Super-Mare, BS22 9BD
G1 XRQ J Koenig, 216 Bretch Hill, Banbury, OX16 0LU
G1 XRT R Taylor, 24 Hoestock Road, Sawbridgeworth, CM21 0DZ
G1 XSA S Cattle, 5 Highworth Drive, Newcastle upon Tyne, NE7 7FB
G1 XST Suzanne Davies, 64 Oakfields, Worth, Crawley, RH10 7FL
G1 XSV Melvyn Scarr, 15 Biddesden Lane, Ludgershall, Andover, SP11 9PG
G1 XTA S Hodson, Flagstones, 12 Duns Tew, Bicester, OX6 4JR
G1 XTD I Clark, Flat, Redhill Farm, Penrith, CA11 0DT
GI1 XTK B Braniff, 5 Cintons Park, Downpatrick, BT30 6NS
GW1 XUD R Andrews, 270 Barry Road, Barry, CF62 8BJ
G1 XUE G Henne, 57 Heaf Gardens, Bentley Close, Aylesford, ME20 7SF
G1 XUH M Thornton, 46 Lavender court, Croft road, Barnsley, S70 3FG
G1 XUU Simon Bishop, 22 John St., Brightlingsea, Colchester, CO7 0NA
G1 XUW D Austin, 17 Patricia Avenue, Horstead, Norwich, NR12 7EW
G1 XVC S Ward, Beech Cottage, Saron Road, Goytre, NP4 0BN
G1 XVD C Snow, 77 Oxford Drive, Hadleigh, Ipswich, IP7 6AW
G1 XVF J Pottage, 18 Pennine Close, Huthwaite, Sutton-in-Ashfield, NG17 2QD
G1 XVL B Perry, 152 Stanborough Avenue, Borehamwood, WD6 5LR
G1 XVM J Duggan, 112 Gaer Park Drive, Newport, NP20 3NR
G1 XVR D Briggs, 17 The Lonnen, South Shields, NE34 8EJ
G1 XVT P Hannan, 151 Star Road, Peterborough, PE1 5HG
G1 XVW Anton Pogorzelski, 28c Mosslea Road, London, SE20 7BW
G1 XVY R Sacharewicz, 15 Milford Close, Walkwood, Redditch, B97 5PZ
G1 XWD A Rhodes, 2 Kent Avenue, Theddlethorpe, Mablethorpe, LN12 1QE
G1 XWK A Rich, The Court House, Wadborough Road, Worcester, WR7 4RF
G1 XWM M Cox, 17 Tybalt Close, Heathcote, Warwick, CV34 6XB
G1 XWN G Andrews, 22 Arnhem Grove, Braintree, CM7 5UQ
G1 XWO F Williams, 15 Hartsbourne Way, Stafford, ST17 4NR
G1 XWS H Seatory, Ivydene, The Street, Woodbridge, IP12 3QU
G1 XWX D Brunton, 29 Norfolk Road, Wangford, Beccles, NR34 8RE
G1 XWZ F Millbank, Room 216 Kate House, Pitchill House Nursing Home, Evesham, WR11 8SN
G1 XXE Peter Yeates, 9 Arlington Road, St. Annes, Bristol, BS4 4AF
G1 XXF J Ellison, 68 Rocket Way, Forest Hall, Newcastle upon Tyne, NE12 9RL
G1 XXH Richard Tapp, 26a Main Road, Grendon, Northampton, NN7 1JW
GW1 XXL R Flynn, 9 Railway Terrace, Blaenclydach, Tonypandy, CF40 2DA
G1 XXR Sharon Austin, 5 Mercia Road, Baldock, SG7 6RZ
G1 XXV T Blackmore, 50 Fraser Close, Shoeburyness, Southend-on-Sea, SS3 9YS
G1 XXW Peter King, 2 Ebenezer Cottages, Thorney Road, Peterborough, PE6 7UB
G1 XYD A Bates, 29 Juler Close, North Walsham, NR28 0SY
G1 XYF C Hudson, 8 College Road, Bredon, Tewkesbury, GL20 7EH
G1 XYG Richard Butler, 15 Bracknell Crescent, Nottingham, NG8 5EU
G1 XYN I Pritchard, 8 Hoon Avenue, Newcastle, ST5 9NY
G1 XYO Steven Glazzard, 109 Highfields Road, Chasetown, Burntwood, WS7 4QS
G1 XYR D Searle, 33 Claypool Road, Kingswood, Bristol, BS15 9QJ
G1 XYS Allen Brown, 5 Somersby Drive, Kenton, Newcastle upon Tyne, NE3 3TN
G1 XYV M Minshull, 12 Dunnett Close, Attleborough, NR17 2NG
G1 XYZ KINGS LYNN ARC c/o Edward Haskett, 23 Gloucester Road, Gaywood, King's Lynn, PE30 4AB
G1 XZA Ian Rawlingson, 1 Wadham St., Penkhull, Stoke-on-Trent, ST4 7HF
G1 XZB J Rawlingson, 1 Wadham St., Penkhull, Stoke-on-Trent, ST4 7HF
G1 XZG D Collins, 5 Elmwood Close, Lincoln, LN6 0LZ
GW1 XZI Roy Magwood, 13 Inverness Place, Cardiff, CF24 4RU
G1 XZQ R Carville, 66 Ludlow Road, Paulsgrove, Portsmouth, PO6 4AE
G1 XZV John Heys, 2 Oakenhill Walk, Bristol, BS4 4LP
G1 XZW Roy Hudson, 7 Grange Avenue, Luton, LU4 9AS
G1 XZX B Strutt, 10 Park Cottages, Lower Somersham, Ipswich, IP8 4PP
G1 YAB Clive ROGERS, 221 Dales Road, Ipswich, IP1 4JY
G1 YAE R Thompson, 1 Littlehoughton Farm Cottages, Littlehoughton, Alnwick, NE66 3JZ
G1 YAF A Tyler, 16 Harridge Road, Leigh-on-Sea, SS9 4HA
G1 YAH J Mcsoley, 88 Rodings Avenue, Stanford-le-Hope, SS17 8DT
G1 YAS D Elliott, 48a Great Lane, Reach, Cambridge, CB5 0JF
G1 YBA I Hardaker, 5 Nursery Road, Arnold, Nottingham, NG5 7ET
G1 YBB S Clements, 46 Brampton Road, Newton Farm, Hereford, HR2 7DF
GW1 YBF L Ward, 11 Verlands Way, Pencoed, Bridgend, CF35 6TY
G1 YBG J Ballance, Orchid Bank, Woolhope, Hereford, HR1 4RQ
G1 YBI A Jones, 43 Oakleigh Road, Droitwich, WR9 0RP
G1 YBK J Fyson, 1 Rodlands Estate, Ibstock, Leicester, LE6 1HT
G1 YBM John Pedley, 92 Ashfield Drive, Moira, Swadlincote, DE12 6HQ
G1 YBT J Bagshaw, 2 Boulton Court, Robin Hood Road, Skegness, PE25 3QU
G1 YCK M Travis, 10 Victoria Road, Kearsley, Bolton, BL4 8NR
G1 YCM A Laughlan, 33 Park House, Gorseyfields, Manchester, M43 6DX
G1 YCN D Lewis, 81 Ashton Avenue, Rainhill, Prescot, L35 0QR
G1 YCR R Lawrence, 82 Moseley Street, Southend on Sea, SS2 4NN
G1 YDA M Davies, 10 Rue Alphonse Delaveau, Pouzauges, France, 85700
G1 YDD A Forster, 56 Tantobie Road, Denton Burn, Newcastle upon Tyne, NE15 7DQ
G1 YDG A Miles, Yew Tree House, Main Street, Wantage, OX12 0HT
G1 YDI C Lambeth, 159 Barns Road, Oxford, OX4 3RB
G1 YDJ T Polley, 9 Otter Road, Clevedon, BS21 6LQ
G1 YDQ John Carpenter, 34b Carey Park, Killigarth, Looe, PL13 2JP
GI1 YEA L O'Flaherty, 1 Ravensdale Villas, Newry, BT34 2PG
G1 YED W Ross-fraser, 47 Lichford Road, Sheffield, S2 3LB
G1 YEP F Russell, 7 Glenmore Avenue, Liverpool, L18 4QE
G1 YES B Underhay, 24 Rutland Road, Southall, UB1 2UP
G1 YEU M Eales, 32 Selston Drive, Nottingham, NG8 1DE
G1 YEV D Martin, 73 Summerfields Way, Ilkeston, DE7 9HE
G1 YEW R Jane Hill, 10 The Moorings, Littlehampton, BN17 6RG
G1 YEZ A Lord, 66 Salcombe Drive, Glenfield, Leicester, LE3 8AF
G1 YFA J Rymsza, 24 Green Lane, Studley, B80 7HD
G1 YFC Paul Neades, 19 Perrystone Lane, Hereford, HR1 1QY
G1 YFD S Lycett, 27 Ropewalk, Alcester, B49 5DD
G1 YFE J Dent, 90 Eastwood, Chatteris, PE16 6RX
G1 YFI J Simmonds, 94 Gravel Hill, Tile Hill, Coventry, CV4 9JH
GM1 YFO P Mirtle, 11 Humbie Road, Kirkliston, EH29 9AN
GW1 YFP M Hearne, Mora, Rhydypandy Road, Morriston, SA6 6NX
G1 YFQ D Scothern, 5 Wilkinson Drive, Middle Rasen, Market Rasen, LN8 3LD
G1 YGP S Jarman, 55 The Meadows, Todwick, Sheffield, S26 1JG
GM1 YGV Robert Johnstone, 10 Lundy Road, Inverlochy, Fort William, PH33 6NX
GM1 YGW G Craib, 55 Kingfisher Place, Dunfermline, KY11 8JN
G1 YGY C Weaver, 11 Thirlmere, Swindon, SN3 6LA
GW1 YHA P George, 24 Ty Fry Close, Brynmenyn, Bridgend, CF32 8YB
G1 YHB John Moggeridge, 22 St. Michaels Court, Faircross Avenue, Weymouth, DT4 0DS
G1 YHE D Coate, 74 Wimborne Road, Poole, BH15 2BZ
G1 YHG M Kennedy, Milestones, Blandford Forum, DT11 9DW
G1 YHI K Davies, 2 Orchard Close, Lytchett Minster, Poole, BH16 6JH
G1 YHJ G Williams, 2 Cotton Close, Broadstone, BH18 9AJ
GW1 YHL D Crawshaw, 12 Glanmor Crescent, Uplands, Swansea, SA2 0PJ
G1 YHN S Rhodes, 221 Ormonds Close, Bradley Stoke, Bristol, BS32 0DW
G1 YHP J Hogg, 1 Deepdale, Guisborough, TS14 8JY
G1 YHV Simon Briscoe, 8b Corfe View Road, Corfe Mullen, Wimborne, BH21 3LZ
G1 YIL Anthony Kaye, 2 Church Place 135 Edward Road, Balsall Heath, Birmingham, B12 9JQ
G1 YIQ J Stafford, 6 Gardners Drive, Hullavington, Chippenham, SN14 6EL
G1 YJB Glenda Evans, 16 Kynaston Drive, Wem, Shrewsbury, SY4 5DE
G1 YJH Martin Blackman, 39 Lyndhurst Avenue, Mill Hill, London, NW7 2AD
G1 YJI P Kay, 97 Avenue Road, London, N14 4DH
G1 YJJ R Colman, 197 Coppins Road, Clacton-on-Sea, CO15 3LA
G1 YJL William Pond, The Wheatlands, Calais Street, Sudbury, CO10 5JA
G1 YJQ John Duffy, Flat 14, The Sycamores, Newcastle upon Tyne, NE4 7ER
G1 YJR J Davies, 70 Ash Road, Sandiway, Northwich, CW8 2PB
G1 YJW Samuel Bates, 21 St. Peters Close, Clayton le Dale, Blackburn, BB1 9HH
G1 YJY Paul Sengupta, 48 Badger Close, Guildford, GU2 9WA
GM1 YKE J Campbell, 16 Barony Road, Auchinleck, Cumnock, KA18 2LL
G1 YKI J Heathfield, 82 Auriel Avenue, Dagenham, RM10 8BT
G1 YKK Susan O'Connor, 32 Whitfield Cross, Glossop, SK13 8NW
G1 YKL J Brackenridge, 21 St. Mark Road, Deepcar, Sheffield, S36 2TF
GW1 YKT W John, 4 Heol Y Bryn, Rhiwbina, Cardiff, CF4 6HY
G1 YKX M Rouse, 105 Great Spenders, Basildon, SS14 2NS
G1 YKY S Jones, 14 Plantation Drive, Croesyceiliog, Cwmbran, NP44 2AN
G1 YKZ Richard Burt, 11 Long Common, Heybridge, Maldon, CM9 4US
G1 YLB Steven Doyle, 2 The Greenways, Paddock Wood, Tonbridge, TN12 6LS
G1 YLE Paul Adams, 25 Main Road, Kesgrave, Ipswich, IP5 1AQ
G1 YLG A Hodkin, 18 Habershon Drive, Chapeltown, Sheffield, S35 2ZT
G1 YLJ A Hunt, 14 Sandalwood Close, Willenhall, WV12 5YJ
G1 YLM Elaine Bradshaw, 30 Whitoford Drive, Kettering, NN15 6HH
G1 YLN M Swetman, 11 Outer Circle, Taunton, TA1 2BS
G1 YLQ C Thomas, Apartment 231, Bournville Gardens Village, Birmingham, B31 2FS
G1 YLV M Bayliss, 2 Plattens Court, Wroxham, Norwich, NR12 8SQ
G1 YMA William Scoles, 26 The Close, Brancaster Staithe, King's Lynn, PE31 8BS
G1 YMC S Fitzpatrick, 19 Claremont Falls, Killigarth, Looe, PL13 2HT
GM1 YME J Hein, 78 Montgomery Street, Edinburgh, EH7 5JA
G1 YMH M Homer, 86 Victoria Road, Brierley Hill, DY5 1DB
G1 YMJ A Smith, 6 Norton Crescent, Towcester, NN12 6DN
G1 YMN G Brooks, 16 Cameron Close, Lillington, Leamington Spa, CV32 7DZ
G1 YMP S Ellis, 6 Newark Road, Hindley, Wigan, WN2 3HR
G1 YMR P Webster, 117 Warley Road, Blackpool, FY1 2RW
G1 YMV H Johnson, 2 Greenbank Avenue, Storth, Milnthorpe, LA7 7JP
G1 YNH Christopher Arundel, 54 Broadmead, Castleford, WF10 4SE
G1 YNJ M Rocke, Orchard House, 55 Tarvin Road, Chester, CH3 5DY
G1 YNO Bernard Surtees, 5 Haweswater Grove, West Auckland, Bishop Auckland, DL14 9LQ
G1 YNQ J Crow, 71 Stockshill Road, Ashby, Scunthorpe, DN16 2LQ
G1 YOA G Hawkins, 11523 Sun Ray Court, San Diego, United States, 92131
G1 YOF A Lockwood, 9 Hartley Road, Exmouth, EX8 2SG
G1 YOS M Avenell, Limo House, Worlds End, Newbury, RG20 8SD
GJ1 YOT N Paisnel, 11 Bon Air Apartments, La Grande Route de la Cote, St. Clement, Jersey, JE2 6SE
G1 YOU Stephen Nicholls, Fieldway, The Street, Woodbridge, IP12 2QG
G1 YPH A Roberts, 18 Surtees Grove, Stoke-on-Trent, ST4 3HH
GM1 YPJ I Davies, 24 Ardgour Road, Caol, Fort William, PH33 7PQ
G1 YPM R Northcott, 32 Lichfield Road, Exwick, Exeter, EX4 2EU
G1 YPQ Karl Hobson, 7 Heather Croft, Sharlston Common, Wakefield, WF4 1TJ
G1 YPR G Dearden, 125 Campsall Field Road, Wath-upon-Dearne, Rotherham, S63 7ST
G1 YPT Geoffrey Hartshorn, 11 Lime Avenue, Ripley, DE5 3HD
G1 YPU E Howberry, 69 Alpha Terrace, Trumpington, Cambridge, CB2 9HS
G1 YPZ Leslie Mitchell, Flat 75, Dalehead, London, NW1 2JL
G1 YQI P Bennett, 7 Woburn Avenue, Firwood Ind Est, Bolton, BL2 3AY
G1 YQL C Stagg, 559 Dividy Road, Stoke-on-Trent, ST2 0BX
G1 YQM Richard Evans, Maesyronnen, Sarnau, Llanymynech, SY22 6QL
G1 YQN Patrick Meehan, Heathercot, Cross Drive, Maidstone, ME17 3NP
G1 YQP M West, 27 Nidderdale Road, The Meadows, Wigston, LE18 3XW
G1 YQU J Stapleford, 7 Garfield Road, Hugglescote, Coalville, LE67 2HU
G1 YQY William Oakes, 2 Hillcrest, Scotton, Catterick Garrison, DL9 3NJ
G1 YRC YORK ARC c/o Arthur Palfrey, 51 The Village Wigginton, York, YO32 2PR
G1 YRE S Piper, Willow End, The Street, Ipswich, IP6 9HG
G1 YRF David Buggs, 2 Archway Cottages Valley Road, Leiston, IP16 4AR
G1 YRJ Matthew Stott, 8 Kingfisher Way, Stowmarket, IP14 5BB
G1 YRM Charles Mead, 32 Sandy Road, Potton, Sandy, SG19 2QQ
G1 YRQ Richard Nock, 43 Delph Drive, Brierley Hill, DY5 2LQ
G1 YRR W Fry, 227 London Road North, Merstham, Redhill, RH1 3BN
G1 YRY S Roberts, 23 Deal Court, Haldane Road, Southall, UB1 3NT
G1 YSA M Crick, 85 Ashurst Road, London, N12 9AU
GI1 YSG Patrick Kennedy, Oakwood, 29 Barnfield Road, Lisburn, BT28 3TQ
G1 YSR Peter Shaw, The Hawksmoor Suite, Colesdane, Stede Hill, Maidstone, ME17 1NP
G1 YSX D Taylor, 103 Southend, Garsington, Oxford, OX44 9DL
G1 YTG Alan Thackray, 32 Sumerlin Drive, Clevedon, BS21 6YW
G1 YTL H Vyvyan, 13 Shearwater Close, Peel Common, Gosport, PO13 0RB
G1 YTO K Wenman, 2 Hythe Road, Sittingbourne, ME10 2LR
G1 YTV S Fisher, Farlands, Lower Rillaton, Callington, PL17 7PF
G1 YTX Terence Clayton, 40 Morrison Road, Darfield, Barnsley, S73 9ED
G1 YUB B HARRISON, 15 Helmington Terrace, Hunwick, Crook, DL15 0LQ
G1 YUL James Robson, 28 Eastfield Street, Sunderland, SR4 7SA
G1 YUN Keith Hetherington, 29 Broomridge Avenue, Newcastle upon Tyne, NE15 6QN
G1 YUS A Lees, 692 Walmersley Road, Bury, BL9 6RN
G1 YUU A Bagworth, 127 Barnsley Road, Darfield, Barnsley, S73 9PE
G1 YUX J Garnett, 21 Vicarage Close, Mossley Hill, Liverpool, L18 7HU
G1 YVI Keith Biddlecombe, 29 Stone Close, Worthing, BN13 2AU
G1 YVS Christopher Newby-Robson, 1 Bramley Drive, Offord D'Arcy, St. Neots, PE19 5SF
G1 YVV I Coleman, 69 Glebelands, West Molesey, KT8 2PY
G1 YVZ A Bell, 159 Hounslow Road, Hanworth, Feltham, TW13 6PX
G1 YWI A Williams, 26 Matlock Road, Bloxwich, Walsall, WS3 3QD
G1 YWN A Whitworth, 183 Logan St., Bulwell, Nottingham, NG6 9FX
G1 YWY Mark Jones, 16 Cumnock Road, Castle Cary, BA7 7FE
G1 YXA B Dixon, 16 Dyrham Parade, Patchway, Bristol, BS34 6EF
G1 YXG B Davidson, 2 Tennyson View, Elm Lane, Newport, PO30 4JS
G1 YXH Raymond Harrison, 14 St. Leonards Avenue, Chatham, ME4 6HL
G1 YXJ A Palmer, 14 Garibaldi Road, Redhill, RH1 6PB
GW1 YXR N Williams, 31 Syr Davids Avenue, Cardiff, CF5 1GH
G1 YXT C Wise, 28 Southlands, East Grinstead, RH19 4BZ
G1 YXY L White, The Garden Flat, 47 Hamerton Road, Gravesend, DA11 9DX
G1 YYD E Brown, 25 Cork Road, Lancaster, LA1 4BD
G1 YYH J Heaton, 85 Morris Green Lane, Bolton, BL3 3JD
G1 YYP M Illston, 4 The Sett, Oxhill, Warwick, CV35 0RE
G1 YYU William Fludgate, 7 The Slade Newton Longville, Milton Keynes, MK17 0DR
G1 YYY BRAINTREE RAYNET GROUP c/o D Willicombe, 26 Falkland Court, Braintree, CM7 9LL
G1 YZF D Edwards, 240 Berthin, Greenmeadow, Cwmbran, NP44 4LB
G1 YZH R Baxter, 4 Kendal Gardens, Woodley, Stockport, SK6 1BL
G1 YZI P Rennison, Foxhall Cottage, Kelshall, Royston, SG8 9SE
G1 YZJ R Rennison, Foxhall Cottage, Kelshall, Royston, SG8 9SE
G1 YZT E Fenlon, 17 Hawes Avenue, Ramsgate, CT11 0RN
G1 ZAG I Shaw-Ashton, Delamere, Westland Avenue, Darwen, BB3 2ST
G1 ZAK A Powney, 16 Westbrook Way, Wombourne, Wolverhampton, WV5 0EA
G1 ZAR S Tyler, 38 Wordsworth Road, Loughborough, LE11 4LQ
G1 ZAW Mark Smoker, 41 Queens Gardens, Darford, DA2 6HZ
G1 ZAY A Robinson, 31 Heathview Road, Socketts Heath, Grays, RM16 2RS
G1 ZBA P Leeman, 449 Montagu Road, Edmonton, London, N9 0HR
G1 ZBB S Brown, 6 Pathfinder Way, Ramsey, Huntingdon, PE26 1LX
G1 ZBG M Bacon, 52 Wroslyn Road, Freeland, Witney, OX29 8HH
G1 ZBH G Brock, 148 Lonsdale Drive, Rainham, Gillingham, ME8 9HX
G1 ZBJ A Manning, 12 Clifford Drive, Heathfield, Newton Abbot, TQ12 6GX
G1 ZBL N Marsh, 16 Laurel Close, North Warnborough, Hook, RG29 1BH
G1 ZBO E McLucky, 11 Ripon Road, Killinghall, Harrogate, HG3 2DG
G1 ZBP A Whipp, 114 Lower Manor Lane, Burnley, BB12 0EF
G1 ZBU G Newby, 77 Darby Road, Garston, Liverpool, L19 9AN
G1 ZBW William Baker, 103 New Lane, Bolton, BL2 5BY
G1 ZBY S Parker, 8 Greenbank Drive, Lincoln, LN6 7LQ
G1 ZCC Clive Feather, 10 Thruffle Way, Bar Hill, Cambridge, CB23 8TR
GM1 ZCD Philip Hughes, 26 Kingseat Road, Dunfermline, KY12 0DD
G1 ZCS E Davis, 10 Fairfield Drive, Lowestoft, NR33 8QG
G1 ZDF D Fleetwood, 9 Reynolds Close, Swindon, Dudley, DY3 4NQ
G1 ZDG P Whittaker, 48 Elgin Gardens, Guildford, GU1 1UB
G1 ZDR J Angus, 8 Gravel Road, Bromley, BR2 8PF
G1 ZDT A Gregory, 9 Fordbridge Road, Ashford, TW15 2TD
G1 ZDU B Rowles, 4 Milton Road, Aston Clinton, Aylesbury, HP22 5LA
G1 ZDX M Bodecott, The Old Parsonage, 3 Church Walk, Peterborough, PE6 7HZ

**IMPORTANT NOTE**

**Revalidate licence to avoid revocation** – Ofcom has advised the Society that plans will be drawn up to revoke licences that have not been revalidated as required by the licence conditions. The quickest way to revalidate is to do so online via the Ofcom website: *https://services.ofcom.org.uk/* or by email: *amateur.validations@ofcom.org.uk* Ofcom staff are available to help, but please be patient during times of heavy workload.

JK Callsigns

G1 ZDY C Tipp, 77 Kimberley Road, Croydon, CR0 2PZ
G1 ZEA P Jones, 21 Hill Top Rise, Harrogate, HG1 3BW
G1 ZEC Gordon Stevens, 25 Avenue Road, New Milton, BH25 5JP
G1 ZED B Haworth, 139 Manchester Road, Accrington, BB5 2NY
G1 ZEI John Wyatt, Ciampia, 10 St. Georges Hill, Perranporth, TR6 0DZ
G1 ZEK David Ault, 68 Moira Dale, Castle Donington, Derby, DE74 2PJ
G1 ZEU S Aspey, 9 Colwell Court, Newton Aycliffe, DL5 7PS
G1 ZEW David Pentin, 35 Seafore Close, Liverpool, L31 2JS
G1 ZEX R Davies, 71 Higher Croft Road, Lower Darwen, Darwen, BB3 0QT
G1 ZFB K Barton, Hameau de Cassoulet, Miribel Lanchatre, France, 38450
G1 ZFD J Davies, 71 Higher Croft Road, Lower Darwen, Darwen, BB3 0QT
G1 ZFF T Voisey, 26 Gorlands Road, Chipping Sodbury, Bristol, BS37 6LA
G1 ZFG J Stephenson, 5 Hunstrete, Pensford, Bristol, BS39 4NT
G1 ZFS N Woolard, 159 Medway Road, Worcester, WR5 1LL
G1 ZFX J Milosevic, 38 Thornhill Close, Upper Cwmbran, Cwmbran, NP44 5TQ
G1 ZGF R Jackson, 37 Carisbrooke Road, Harpenden, AL5 5QS
G1 ZGH J Sharpe, 204a Featherstone Lane, Featherstone, Pontefract, WF7 6AH
G1 ZHD Adam Gilmore, Ashfields, Naseby Road, Market Harborough, LE16 9RZ
G1 ZHI C Owens, 7 Frondeg, Southsea, Wrexham, LL11 6RH
G1 ZHL Moshin Dharas, 225 Redmile Walk, Peterborough, PE1 4UR
G1 ZHM P Hunt, 32 Tudor Road, Leigh-on-Sea, SS9 5AX
G1 ZHN M Griffiths, 2 Muirway, Benfleet, SS7 4LS
G1 ZHZ J Hall, 27 Quarry Hill Road, Ilkeston, DE7 4DA
G1 ZIM Simon Agnew, 48 Wasdale, Millom, LA18 4JJ
G1 ZIV J Large, 9 Maitland Terrace, Kildrochat, Stranraer, DG9 9EX
G1 ZJK David Ellard, 35 Edgehill Drive, Daventry, NN11 0GR
G1 ZJP Robert Offer, Chapel Yard Cottage, Quadring Eaudyke, Spalding, PE11 4QB
G1 ZJQ Derek Smith, 44 Yarmouth Drive, Cramlington, NE23 1TS
G1 ZKE M Grindle, 57 Islwyn Street, Cwmfelinfach, Newport, NP11 7HY
G1 ZKN Raymond Ogden, Plas Yn Bonwm Farm, Holyhead Road, Corwen, LL21 9EG
G1 ZKZ Gary Kenealy, 20 Penny Lane, Haydock, St. Helens, WA11 0QS
G1 ZLA M Roberts, 18 Craster Drive, Nottingham, NG6 7FJ
G1 ZLB J Kynaston, Smithy Cottage, Main Street, Nottingham, NG12 5PY
G1 ZLC P Ashby, 12 Treeford Close, Solihull, B91 3PW
G1 ZLD Michael Bignell, 57 Ramsey Road, Halstead, CO9 1AS
GW1 ZLL D Ball, 38 Heol Sirhwi, Barry, CF62 7TG
G1 ZLY Nick Cooper, 56 Kingfisher Road, Mansfield, NG19 6EG
G1 ZME Roy Coatman, Ivy Cottage, St. Buryan, Penzance, TR19 6DT
G1 ZMG R Hoad, Broad Lea, Amsbury Road, Maidstone, ME17 4DN
G1 ZMJ D Roberts, 31 Seaton Way, Marshside, Southport, PR9 9GJ
G1 ZMS MID SUSSEX ARS c/o C Cook, 45 Stonepound Road, Hassocks, BN6 8PR
G1 ZMW C Tubey, 2 Rowley Close, Swadlincote, DE11 8LX
GW1 ZNC S Elworthy, 70 Maple Drive, Brackla, Bridgend, CF31 2PF
G1 ZND A Soble, 6 The Glebe, Hildersley, Ross-on-Wye, HR9 5BL
G1 ZNK A Edwards, 68 Middlemarch Road, Coventry, CV6 3GF
GM1 ZNR V Roberts, 4 Ladieside, Brae, Shetland, ZE2 9SX
G1 ZNT W Barton, 27 Hornby Crescent, Clock Face, St. Helens, WA9 4RY
G1 ZNV R Charteris, 7 Kennedy Close, Kidderminster, DY10 1LR
G1 ZNX R Agnew, 156 Goswell End Road, Harlington, Dunstable, LU5 6NT
G1 ZNZ A Steele, 72 Park Lane, Knypersley, Stoke-on-Trent, ST8 7AS
G1 ZOB R Brown, 28 Albertus Road, Hayle, TR27 4JQ
G1 ZOQ A Mather, 330 Lever Street, Radcliffe, Manchester, M26 4PT
G1 ZOS Colin Wood, Wurzerstr. 180, Bonn, Germany, 53175
GM1 ZOX Neil Senior, 36 Lathro Park, Kinross, KY13 8RU
G1 ZOY M Knowles, 17 Stainmore Close, Birchwood, Warrington, WA3 6TP
G1 ZPA Norman Johanssen, 10 Waverley Court, Verulam Place, St. Leonards-on-Sea, TN37 6QR
G1 ZPC C Rule, 1 Park en Venton, Mullion, Helston, TR12 7JH
G1 ZPJ P Read, 58 Godolphin Road, Helston, TR13 8QJ
G1 ZPO A Brookes, 212 Pontefract Road, Featherstone, Pontefract, WF7 5AG
G1 ZPQ J Pitchford, 7 Firecrest Drive, Leegomery, Telford, TF1 6FZ
G1 ZPU Robert Compton, 18 Drove Road, Gamlingay, SG193NY
G1 ZQE D Marsden, 94 Blackford Road, Shirley, Solihull, B90 4BX
GM1 ZQF George Milne, 6 Alexandra Street, Alyth, Blairgowrie, PH11 8AS
G1 ZQG Paul Huntley, 5 Beacon Avenue, Barton-upon-Humber, DN18 5DP
G1 ZQN Paul Gibson, 7 Greenfields Road, Horley, RH6 8HW
G1 ZQO C Stokes, 21 Deerswood Lane, Bexhill-on-Sea, TN39 4LT
G1 ZQR Roger Twyman, Farmend, Halls Lane, Reading, RG10 0JB
G1 ZQV Michael Lowe, 34 Woodbank Road, Groby, Leicester, LE6 0BN
G1 ZRE R Ellis, 7 Bromley Close, Blackpool, FY2 0SD
G1 ZRP M Crook, 21 Treyew Road, Truro, TR1 2BY
G1 ZRQ D Barrett, 10 Trelawny Road, Menheniot, Liskeard, PL14 3TS
G1 ZRR Nigel Youngman-Smith, 12 Timber Way, Chinnor, OX39 4EU
G1 ZRS J Cantwell, 10 Cathedral Drive, Fairfield, Stockton-on-Tees, TS19 7JT
G1 ZRT C Guymer, 74 West Common Lane, Scunthorpe, DN17 1DU
G1 ZSE Kelvin Crocker, 32 Godmanston Close, Poole, BH17 8BU
G1 ZSF A Lewis, 76 Reading Road, Finchampstead, Wokingham, RG40 4RA
G1 ZSG Christopher Bell, 41a Handel Road, Canvey Island, SS8 7HL
G1 ZSK T Adams, 26 Hillside Avenue, Plymouth, PL4 6PR
G1 ZSR M Salzman, 89 Delamere Road, Bedworth, CV12 8SG
G1 ZST L Sherwood, 50 Thornton Road, Manchester, M14 7WT
G1 ZSV Roger Mercer, 23 Larne Road, Bilton Grange, Hull, HU9 4UE
G1 ZSY A Hughes, 37 Brisbane Road, Reading, RG30 2PE
G1 ZSZ Horst Hossle, Flat 91, Castlemeads Court, 143 Westgate Street, Gloucester, GL1 2PB
G1 ZTB W Bell, 77 Bongate, Jedburgh, TD8 6DU
G1 ZTG Paul Wilsdon, 64 Chestnut Avenue, Euxton, Chorley, PR7 6BS
G1 ZTJ G Bailey, 34 Newton Road, Bideford, EX39 2LL
G1 ZTK W Causer, 47 Sandringham Road, Wombourne, Wolverhampton, WV5 8EF
G1 ZTM Andrew Corp, 158 Somerton Road, Street, BA16 0SA
G1 ZTN Paul Harvey, 64 Privett Road, Gosport, PO12 3SX
G1 ZUB P Thompson, Berry Brow, Wetherby Road, Leeds, LS14 3AU
G1 ZUC R Johnson, Mile House, Lansdown Road, Bath, BA1 5SY
G1 ZUH M Pomroy, 21 Nook Farm Avenue, Syke, Rochdale, OL12 0SH
G1 ZUS N Watts, The Vista, Churchill Way, Bideford, EX39 1PA
G1 ZUU AVON VALLEY ARS c/o Steve Brookes, Ivy Cottage, Haselor, Alcester, B49 6LX
G1 ZUZ Jon Rowling, 11 Barncroft, Norton, Runcorn, WA7 6RJ
G1 ZVC Steven Beith, 18 Avenue Road, New Milton, BH25 5JP
G1 ZVE H Barugh, Westwinds, 40 Ruden Way, Epsom, KT17 3LN
GM1 ZVJ John Hilton, 25 Alford Way, Dunfermline, KY11 8BF
G1 ZVO W Scott, 16 Sweetbriar Lane, Holcombe, Dawlish, EX7 0JZ
G1 ZVZ Kevin Fish, 11 Little Meadow Way, Bideford, EX39 3QZ
G1 ZWB B Ward, 13 St. Laurence Road, Winslow, Buckingham, MK18 3BD
G1 ZWH T Sanders, Sandalwood, 41 Tinney Drive, Truro, TR1 1AT
G1 ZWQ J Bowker, 9 Scarthwood Close, Bolton, BL2 4DU
G1 ZYJ A Ainger, 16 Hillside Road, Harpenden, AL5 4BT
G1 ZYN N Suffolk, 2 Tamerton Road, Leicester, LE2 9DD
G1 ZYS F Woodland, 12 Toll House Way, Chard, TA20 1FH
G1 ZZA Paul Harvey, 10 Barnfield Close, Wirral, CH47 7DA
G1 ZZC P Golds, 70 Meredith Road, Worthing, BN14 8EE
G1 ZZG Brian Thomas, 4 Gilbert Close, Torquay, TQ2 6BS
G1 ZZL Michael Low, 23 Larch Crescent, Tonbridge, TN10 3NN

## G*2

GW2 ABJ G Edwards, 2 Heol Y Glo, Tonna, Neath, SA11 3NJ
G2 ABR C Mayman, Greenacre, Stones Green Road, Harwich, CO12 5BS
G2 AIW M Lambeth, 11 Ellerman Avenue, Twickenham, TW2 6AA
G2 AJW J Jack, Malindella, Main Road, Dumfries, DG1 1RZ
G2 ALM R Wilkins, 36 Offington Gardens, Worthing, BN14 9AU
G2 ALN Lee Taylor, 76 Sidney Road, Blackley, Manchester, M9 8AT
G2 AMG H Mitchell, Stone Cottage, Yeovil Road, Yeovil, BA22 9RR
G2 ANC J Bromiley, 28 Clive Road, Westhoughton, Bolton, BL5 2HR
G2 API H Batty, 64 North St., Scalby, Scarborough, YO13 0RU
G2 AQJ R Collins, 33 Elm Close, Laverstock, Salisbury, SP1 1SA
G2 ART F Cawson, 43 Trafalgar Road, Southport, PR8 2HF
G2 ARU R Loveland, Apartment 8, Royal Bay Court, 86a Barrack Lane, Bognor Regis, PO21 4DY
G2 ARV Robert Bennett, 16 Emily Street, St. Helens, WA9 5LZ
G2 ARY George Lee, 16 Phoenix Chase, North Shields, NE29 8SS
G2 AS Sheffield HF DX Group c/o Peter Day, Sheffield HF DX Group, 38 Broomhill Road, Chesterfield, S41 9DA
G2 AXO William Purser, Bethel and Bethesda Residential Home, Equity Road East, Leicester, LE9 7FY
G2 AXQ J Walker, 11 Burrett Gardens, Wisbech, PE13 3RP
G2 AZM E Oakley, 67 South Road, Northfield, Birmingham, B31 2QZ
G2 BAR Barrington Hill, 38 Westons Brake, Emersons Green, Bristol, BS16 7BP
G2 BBC ARIEL RAD GROUP c/o David Pick, 178 Alcester Road South, Kings Heath, Birmingham, B14 6DE
G2 BBI L Steel, 1b Trinity Avenue, Westcliff-on-Sea, SS0 7PU
G2 BGG J Garner, Barbon, Aigburth Hall Road, Liverpool, L19 9DG
G2 BHG G Harrison, 13 High View Park, Cromer, NR27 0HQ
G2 BHY A Bonner, 57 Downsview, Heathfield, TN21 8PF
G2 BJK G Brown, 25 The Cloisters, South Street, Wells, BA5 1SA
G2 BKZ Robert McTait, 20 Rowland Road, Stevenage, SG1 1TE
G2 BOF D Harris, 3 Middle Close, Camberley, GU15 1NZ
G2 BQP P Gully, 23 Lawrence Grove, Henleaze, Bristol, BS9 4EL
G2 BQY TROWBRIDGE + DISTRICT ARC c/o Ian Carter, 12 Bobbin Lane, Westwood, Bradford-on-Avon, BA15 2DL
G2 BRS BOURNEMOUTH RS c/o M Stevens, 16 Golf Links Road, Ferndown, BH22 8BY
G2 BSJ R Biltcliffe, 3 Church View, Steeple Claydon, Buckingham, MK18 2QR
G2 BSW R Ward, Serendipity, 17 Marlpit Lane, Seaton, EX12 2HH
G2 BTZ E Moreman, 5 Sheridan Way, Longwell Green, Bristol, BS30 9UE
G2 BUJ S Greenwood, 11 Clifton Street, Swindon, SN1 3PY
G2 BUP H Gibson, 8 Springfield, Norton St. Philip, Bath, BA2 7NR
G2 BWW Andrew Barrett, 38 Haw Lane, Bledlow Ridge, High Wycombe, HP14 4JJ
GI2 BX CITY OF BELFAST RADIO AMATEUR SOCIETY c/o Frank Hunter, 2 Wandsworth Court, Belfast, BT4 3GD
G2 BXP Martin Prestidge, 48 Parkfield Road, Warley, Oldbury, B68 8PT
G2 BZR R Bassford, 59 Watling St., Dordon, Tamworth, B78 1SY
G2 CD R Matthews, 7 Coolgardie Avenue, Chigwell, IG7 5AU
G2 CFC G Fretwell, 17 Cross Lane, Stocksbridge, Sheffield, S36 1AY
GW2 CGF S Griffiths, 1 Nicholl Court, Mumbles, Swansea, SA3 4LZ
G2 CHI W Bailey, 25 Lenham Road East, Saltdean, Brighton, BN2 8AF
G2 CIW J Moseley, 33 Cathedral Court, London Road, Gloucester, GL1 3QE
G2 CJK A Clarkson, 6 Mather Avenue, Accrington, BB5 5AU
G2 CKR M Garfitt, 90 Wedderburn Road, Malvern, WR14 2DQ
G2 CNN Simon Ball, 8 Ringwood Road, Ryde, PO33 3NX
G2 CO F Cooknell, 65 Coombe Valley Road, Preston, Weymouth, DT3 6NL
G2 CP Scarborough ARC c/o D Herbert, 50 St. Leonards Crescent, Scarborough, YO12 6SP
G2 CQX Victor Pugh, 8 Beech Close, Hanwood, Shrewsbury, SY5 8RA
G2 DAN S Whiteley, 142 Brisbane Road, Mickleover, Derby, DE3 9JW
G2 DBH George Dodd, St Nicholas Cottage, 14 Bury Fields, Guildford, GU2 4AZ
G2 DD L James, Pinecroft, Green Drive, Wokingham, RG40 2HT
G2 DGB A Short, 12 Grosvenor Crescent, Dorchester, DT1 2BA
GW2 DHM W Andrews, 69 Fairwater Grove West, Cardiff, CF5 2JN
G2 DJ DERBY & DISTRICT ARS c/o R Buckby, 20 Eden Bank, Ambergate, Belper, DE56 2GG
G2 DJM Neil Chilton, 38 Kingswood Avenue, Newcastle upon Tyne, NE2 3NS
G2 DLK Gwilym Williams, Adre, Ffordd Caergybi, Llanfairpwllgwyngyll, LL61 5YX
G2 DLX Douglas Mitchell, 1 Denstroude Cottages, Denstroude Lane, Canterbury, CT2 9JX
G2 DML John Crossfield, Forest Lodge, Chopwell Wood, Rowlands Gill, NE39 1LT
G2 DNJ N Brierley, Minera, 6 Trinity Crescent, Llandudno, LL30 2PQ
G2 DPA Mario Brashill, 42 Bannister Street, Withernsea, HU19 2DT
G2 DPL P Smith, Obo Bury Radio Soc, Moses Yth Comm Cntr, Bury, BL9 0BS
G2 DPQ K Amos, 1 Byron Close, Upper Caldecote, Biggleswade, SG18 9DF
G2 DPY Des Silverson, 63 Downside, Shoreham-by-Sea, BN43 6HF
G2 DSY J Dale, 17 Lansdowne Gardens, Romsey, SO51 8FN
G2 DWB N Webster, 1 Gratton Dale, Carlton Colville, Lowestoft, NR33 8WP
G2 DX FARNBOROUGH AND DISTRICT RS c/o Graham Roff, 47 Penshurst Rise, Frimley, Camberley, GU16 8XX
G2 DZH N Talbot, 105 Westwood Lane, Welling, DA16 2HJ
G2 FA FOLKESTONE & DISTRICT ARS c/o D Pepper, 17 Cliffe House, Radnor Cliff, Folkestone, CT20 2TY
G2 FCP F Varley, 39 Nettleton Road, Mirfield, WF14 9AW
G2 FFD D Skipworth, Melrose, West End Road, Boston, PE22 0BU
G2 FGT R Rogers, 67 Kingswell Avenue, Arnold, Nottingham, NG5 6SY
G2 FHF Jon Illsley, 55 Avalanche Road, Portland, DT5 2DJ
G2 FJA MARTS c/o Kevin Earl, 210 Churchill Avenue, Chatham, ME5 0JS
G2 FKO Appledore and District ARC c/o John Lovell, Kowloon, Slade, Bideford, EX39 3LZ
G2 FKZ RS of Great Britain c/o Graham Coomber, 2 Bracken Grove, Catshill, Bromsgrove, B61 0PB
G2 FLW Morris Clarkson, Causeway Cottage, Sawley Road, Clitheroe, BB7 4RS
G2 FM FLAXTON MOOR CONTEST GROUP c/o C Quarton, Flaxton Gatehouse, Flaxton, York, YO60 7QT
G2 FMW E Baker, 86 Osborne Gardens, Herne Bay, CT6 6SE
GW2 FOF Rhondda ARS c/o John Howells, Bronllys, Vicarage Road, Rhondda-Cynon-Taff, CF40 1HR
G2 FQZ Robin Day, Resting Oak Cottage, Resting Oak Hill, Lewes, BN8 4PS
G2 FSH B Weeden, 24 Berkeley Close, Rochester, ME1 2UA
G2 FSJ K Levitt, 1 Charnwood Close, Andover, SP10 2RB
G2 FSR J Hunt, 4 Warmdene Road, Brighton, BN1 8NL
G2 FT D Blake, Kandy, 5 Mill Road, Cromer, NR27 0BG
G2 FUU T Knight, Homefield, Back Lane, Waltham Abbey, EN9 2DD
G2 FVL Leslie Carrick-Smith, Highfields House, Sheffield Road, Clowne, Chesterfield, S43 4AP
G2 FXJ Stephen Moisy, 15 Charles Street, Redditch, B97 5AA
G2 FXQ Stanley Saddington, South Ridding, Sibson Road, Atherstone, CV9 3RE
G2 FXV M Middleton, Dolphin View Nursing And, Residential Home, Harbour Road, Morpeth, NE65 0AP
G2 FXZ J Hodgetts, 59 Woodland Road, Halesowen, B62 8JS
G2 FYO H Terraneau, 2653 Nutmeg Circle, Simi Valley, United States, 93065-1327
G2 HBA C Spencer, 12 Sunny Bank, London, SE25 4TQ
G2 HCA L Sanders, 2 Cae Neuadd, Penybontfawr, Oswestry, SY10 0NS
G2 HCG Bertram Sykes, Flat 7, Solent Pines, Whitby Road, Lymington, SO41 0UX
G2 HDF MIDLAND CONTEST GROUP c/o M Waldron, 32 Windmill Street, Upper Gornal, Dudley, DY3 2DQ
G2 HFP Stan Trudgill, 55 Orchard Road, Lytham St. Annes, FY8 1PG
G2 HFR Jonathan Kelly, Arosfa, Westminster Road, Wrexham, LL11 6DN
G2 HHH T Bayliss, 55 Foxlydiate Crescent, Redditch, B97 6NJ
G2 HIX David Craig, Pear Tree Cottage, Cripps Corner Road, Robertsbridge, TN32 5QS
G2 HKG R Lowson, Moss House, Penton, Carlisle, CA6 5RT
G2 HKQ A Knight, 17 Moorland Crescent, Upton, Poole, BH16 5LA
G2 HKS R Udall, Longfield, 20 Upper Way, Rugeley, WS15 1QA
G2 HKU E Trowell, 316 Minster Road, Minster on Sea, Sheerness, ME12 3NR
G2 HLB C Maltby, The Willows Farm, Stallingborough Road, Grimsby, DN40 1NR
G2 HLP D Hearsum, 1225 Duckview Court, Centerville, United States, 45458-2784
G2 HMK T Brown, 99 Brinkburn Drive, Darlington, DL3 0JY
G2 HNA J Weaver, 7 Cramer Street, Stafford, ST17 4BX
G2 HNI L Hewitt, 60 Shaftesbury Avenue, Southampton, SO17 1SD
G2 HS Peter Hale, 13 Seaton Drive, Ashford, TW15 3ET
G2 HW South Manchester RC c/o Ronald Smith, 16 Coniston Avenue, Sale, M33 3GT
G2 HX Gloucester Amateur Radio & Electronics Society c/o Leslie Harris, 183a Painswick Road, Gloucester, GL4 4AG
G2 IF W Setterfield, 54 Hallam Road, Nelson, BB9 8AB
G2 JL Terence Mortimer, 10 Harold Road, Hayling Island, PO11 9LT
G2 KF Trevor Harris, Summerfield, Coombe Road, Lanjeth, St. Austell, PL26 7TL
G2 KG C Hill, 47 Belswains Lane, Hemel Hempstead, HP3 9PW
G2 KQ Brian Hawes, 3 Orchard Close, Cassington, Witney, OX29 4BU
G2 LL OBO HASTINGS ELECTRONICS AND RC c/o T Ransom, 9 Lyndhurst Avenue, Hastings, TN34 2BD
G2 LO Ariel radio 2L0 c/o Ian Jefferson, 7 Bluebell Close, Rugby, CV23 0UH
G2 LW Crystal Palace Radio & Electronics Club c/o Robert Burns, 84 Portnalls Road, Coulsdon, CR5 3DE
GM2 MP NORTH OF SCOTLAND CONTEST GROUP c/o K Kerr, East Loanhead, Auchnagatt, Ellon, AB41 8YH
G2 NF A Canning, 261 Loddon Bridge Road, Woodley, Reading, RG5 4BL
G2 NM AMBERLEY MUSEUM ARC c/o A Leigh, 50 Colbourne Road, Hove, BN3 1TB
G2 OA SOUTHPORT & DISTRICT ARC c/o Brian Rimmer, 8a Mallee Avenue, Southport, PR9 8NL
G2 OG J Hogg, Bwthyn Y Briallu, Ynys Ferw Bach, Gaerwen, LL60 6NW
GW2 OP Pembrokeshire Contest Group c/o Martin Shelley, Sunray, Pendine, Carmarthen, SA33 4PD
G2 OT RAOTA c/o K Jones, Field House, Wragby Road, Lincoln, LN2 2QU
G2 OU FARMORS SCHOOL RC c/o D Tatlow, Mulberry House, Bettys Grave, Cirencester, GL7 5ST
G2 PA P Dyke, 1102 Buckingham DR, Apt B, California, United States, 92626
G2 PB P Baron, 55 Church View, Brompton, Northallerton, DL6 2RD
G2 PK J Ellison, Jowsers, Northfield Lane, Wells-next-The-Sea, NR23 1JZ
GU2 RS Richard Robilliard, Moss Bank, La Mare, St Andrew, Guernsey, GY6 8XX
G2 RSA P King, 32 Millstream Way, Leegomery, Telford, TF1 6QR
G2 SH J Shearme, Chevin, Penn Street, Amersham, HP7 0PY
G2 SR Surrey Raynet c/o Timothy Dabbs, 4 Caverleigh, Cadogan Road, Surbiton, KT6 4DH
G2 SU NORTHERN HGTS RD c/o Allan Robinson, 9 Illingworth Close, Illingworth, Halifax, HX2 9JQ
G2 SZ David Goyder, 8 Bloomsbury Walk, Southampton, SO19 9GB
G2 TO BURY ST EDMUNDS A.R.C. c/o G Woods, Bamburgh House, Hunston, Bury St. Edmunds, IP31 3EN
GD2 TV BAIRD MUSEUM CLUB c/o D Mclean, 24 Montgomery Road, Edgware, HA8 6NT
G2 UG Halifax and District ARS c/o John Wake, 60 Cloverville Approach, Odsal, Bradford, BD6 1ET
G2 UH D Hayward, Hope, Churchwell Street, Sherborne, DT9 6RG
G2 UT K Reid, 4 Harles Acres, Hickling, Melton Mowbray, LE14 3AF
G2 VS Roy Barrett, 76 Westgate Park, Sleaford, NG34 7QP

G2 XG E Davie, 7 Cranworth Crescent, Chingford, London, E4 7HN
G2 XP SUTTON & CHEAM c/o J Puttock, Sutton & Cheam Rs, 53 Alexandra Avenue, Sutton, SM1 2PA
G2 XQ I Marshall, 6 Shrubbery Lane, Wyke Regis, Weymouth, DT4 9LU
G2 XV Cambridge & District RC c/o John Donner, 10 Lyles Road, Cottenham, Cambridge, CB24 8QR
G2 YC Robert McKnight, Gortadrohid, Reengaroga, Co. Cork, Ireland
G2 YL Sally Quarton, Flaxton Gatehouse, Flaxton, York, YO60 7QT
G2 YT P Fox, Hillside House, Almshoebury, Hitchin, SG4 7NT

## G*3

G3 AAF K Avery, 4 Whiphill Close, Doncaster, DN4 6DY
G3 AAS M Glynn, 39 Moor Allerton Drive, Leeds, LS17 6RY
G3 AAZ G Gibbs, Windward, 7 The Grove, Huntingdon, PE29 1YD
G3 AB A Chadwick, 5 Thorpe Chase, Ripon, HG4 1UA
G3 ACQ H Harmsworth, 43 Cornelian Avenue, Scarborough, YO11 3AN
G3 ADZ Keith Gaunt, 21 Abbey Close, Rendlesham, Woodbridge, IP12 2UD
GM3 AEI J Rousselle, The Raw, Bridgend, Isle of Islay, PA44 7PZ
G3 AER G Wright, 70 Gunton Drive, Lowestoft, NR32 4QB
G3 AFB D Tait, 34 Mount Street, Dorking, RH4 3HX
G3 AFT THE GRAFTON ARS c/o Brian Bond, 86 Agar Grove, Camden Town, London, NW1 9TL
G3 AGC W Curphey, 8 Emily Davison Avenue, Morpeth, NE61 2PL
G3 AGF Ray Edginton, 9 Churchill Road, Seaford, BN25 2UL
G3 AH Riviera ARC c/o Alan Wyatt, 78 Marldon Road, Torquay, TQ2 7EH
G3 AHE R James, 40 Barrack Road, Hounslow, TW4 6AG
GM3 AHR A Thomson, Meadowrise, 4 Law View Gardens, Leven, KY8 5SW
G3 AIK K Watkins, Bow House, Hurst, Martock, TA12 6JU
G3 AIO S Fenwick, 28 Gimble Way, Pembury, Tunbridge Wells, TN2 4BX
G3 AJD T Moore, 6 Old Parsonage Court, Otterbourne, Winchester, SO21 2EP
G3 AJK Robert Earland, 7 Trews Weir Court, Exeter, EX2 4JS
G3 AJW C Eatwell, 15 Broadlawn, Woolavington, Bridgwater, TA7 8EP
G3 AKF Reading & District ARC c/o Vincent Robinson, 4 Hilltop Road, Caversham, Reading, RG4 7HR
G3 AKI F Knowles, 1 Mayfield Close, Bishops Cleeve, Cheltenham, GL52 8NA
G3 AKJ A Wheele, 4 Mannings Way, Barnstaple, EX31 1QF
G3 AKN Robert Milne, 19 Musgrave Road, Chinnor, OX39 4PL
G3 ALG G Starling, 207 Shirley Road, Croydon, CR0 8SB
G3 ALK E Holmes, 7 Castle Drive, Ilford, IG4 5AE
GM3 ALZ Fred Gordon, Croft of Torrancroy, Strathdon, AB36 8US
GJ3 AME P Landor, Lauge, Rue Des Raisies, St. Martin, Jersey, JE3 6AT
G3 AMF K Thompson, 11 Ten Bell Lane, Soham, Ely, CB7 5BJ
G3 AMH H Green, 9 Robert Avenue, Cundy Cross, Barnsley, S71 5RB
G3 AMK B Littleproud, 25 Fern Avenue, Lowestoft, NR32 3JF
G3 AMW Hull & District ARS c/o Bryan Aslin-Smith, 161 Sharp St., Newland Avenue, Hull, HU5 2AE
GI3 AMY J Collett, 10 Cronstown Road, Newtownards, BT23 8QS
G3 ANG J Emmott, 6 Meadowcroft, Euxton, Chorley, PR7 6BU
G3 APL J Russon, 59 Ridge Road, Kingswinford, DY6 9RE
G3 APS Leslie Shergold, 8 The Moors, Lydiard Millicent, Swindon, SN5 3LE
G3 APU J Andrews, 44 Eastridge View, East The Water, Bideford, EX39 4RS
G3 AQB W Stephenson, 20 Chapel Court, Chapel Row, Seahouses, NE68 7TD
G3 AQF Anthony Kearns, 8 Pennyfathers Lane, Welwyn, AL6 0EN
G3 ARE F Chubb, 2 Brook Close, Plympton, Plymouth, PL7 1JR
GW3 ARS J Sagar, 75 Hookland Road, Newton, Porthcawl, CF36 5SG
G3 ASE Harold King, 7 Needingworth Road, St. Ives, PE27 5JN
G3 ASG Raymond Fautley, 7 Kingfisher Road, Downham Market, PE38 9RQ
GD3 ASR EDGWARE & DISTRICT RS c/o H Haria, 34 Larkfield Avenue, Harrow, HA3 8NF
G3 AST J Plowman, 17 Orchardleigh, East Chinnock, Yeovil, BA22 9EN
G3 ASV G Pope, 5 Penn Crescent, Haywards Heath, RH16 3HW
GW3 ASW Aberdare ARS c/o Barry Werrell, 26 Glynhafod Street, Cwmaman, Aberdare, CF44 6LD
G3 ASX D Paine, 43 Wilton Road, Muswell Hill, London, N10 1LX
G3 ATC OBO AIR CADET RS c/o William Green, 2 Irkdale Avenue, Enfield, EN1 4BD
G3 ATI Alan Williams, 74 Broadfield Road, Bristol, BS4 2UW
G3 ATX A Perry, The Cottage, The Green, Bristol, BS48 3BG
G3 ATZ G Morris, 18 Grosvenor Road, Shotton, Deeside, CH5 1NU
G3 AUE A McGhie, 1 Boyach Crescent, Isle of Whithorn, Newton Stewart, DG8 8LD
G3 AVE F Flanner, 1 Ludford Close, Sutton Coldfield, B75 6DW
G3 AVL R Reynolds, 12 Eastham Rake, Wirral, CH62 9AA
G3 AVN P Parker, Flat 15, The Rise Care Home, Dawlish, EX7 0QL
G3 AWK N Gough, 20 Earlsfield, Branston, Lincoln, LN4 1NP
G3 AWP F Gifford, 21 Dongal Road, Bournemouth, BH9 2ND
GM3 AXX A Fraser, 58 Rigghead, Stewarton, Kilmarnock, KA3 3DQ
G3 AZI A McCann, 105 Todd Lane North, Lostock Hall, Preston, PR5 5UP
G3 AZW A Bates, 68 Hill St., Hilperton, Trowbridge, BA14 7RS
G3 BAC R Bastow, 2a New Road, Meopham, Gravesend, DA13 0LS
G3 BBK J Orrin, Greenacres, Church Street, Heathfield, TN21 9AL
G3 BBX D Holloway, 10 Spencers Orchard, Bradford-on-Avon, BA15 1TJ
G3 BCE D Nichols, Marsh Farm, Camp Road, Templecombe, BA8 0TH
G3 BDQ J Heys, White Friars, Friars Hill, Hastings, TN35 4EP
G3 BDT A Searle, 30 Hawthorne Grove, Poulton-le-Fylde, FY6 7PN
G3 BEX W Short, Highland Light, 26 Howard Crescent, Beaconsfield, HP9 2XP
G3 BFL H Siebert, 3 Greenlands Road, Kingsclere, Newbury, RG20 5RJ
G3 BGF R Winkworth, 1 Collingwood Drive, Mundesley, Norwich, NR11 8JB
G3 BHA N Taylor, 8 Aragon Way, Bournemouth, BH9 3SB
G3 BHF E Hasted, 54 Plaxtol Road, Erith, DA8 1NL
G3 BHM H Kempson, 8 Hounds Way, Hayes, Wimborne, BH21 2LD
G3 BII A Clark, 19 Lakes Lane, Beaconsfield, HP9 2LA
G3 BIK Edwin Chicken, Ivy Thorn Cottage, Morpeth, NE61 6LQ
G3 BJ Donald Beattie, Hares Cottage, Woolston, Church Stretton, SY6 6QD
G3 BJC Reginald Sparry, Everstone House, Peterstow, Ross-on-Wye, HR9 6LH
G3 BJD John Maxwell, 10 Castle View, Egremont, CA22 2NA
G3 BKJ Harold Alderson, 31 Rumbold Road, Edgerton, Huddersfield, HD3 3DB
G3 BLS David Walker, 32 South St., Osney, Oxford, OX2 0BE
G3 BMI A Bolton, 20 Bullen Close, Cambridge, CB1 8YU
G3 BMO H Speed, 45 Willow Glade, Huntington, York, YO32 9NJ
G3 BMQ G Humphrey, 56a Park Lane, Wallington, SM6 0TN
G3 BNE G Alderman, 35 Lynswood Drive, Sidcup, DA14 6JQ
G3 BNF A Embleton, 34 Riverdale Park, Dent Lane, Chesterfield, S43 3UH
G3 BNP H Gowing, 11 Curbridge Road, Witney, OX28 5JT
G3 BNW J Bailey, 13 Heywood Road, Alderley Edge, SK9 7PN
G3 BOK Sarah Rutt, Granthorpe, Hull Road, Hull, HU11 5RN
G3 BPF Arthur Painter, Cold Green, Rochford, Tenbury Wells, WR15 8SP
GD3 BPG J Richards, 64 Crescent Gardens, Ruislip, HA4 0TA
G3 BPK DOUGLAS VALL AR c/o D Snape, 30 Culcross Avenue, Wigan, WN3 6AA
G3 BPP R Hampton, 11 Greenlands, Hutton Rudby, Yarm, TS15 0JQ
G3 BPQ E Smith, 99 The Ladysmith, Ashton-under-Lyne, OL6 9AP
GD3 BQE R Fussey, 9 Alicia Gardens, Harrow, HA3 0JD
G3 BQT L Hulme, 21 Brookside Crescent, Greenmount, Bury, BL8 4BG
G3 BRQ K Tackley, 1 Greenways, Fleet, GU52 7UG
G3 BRS BURY RADIO SOC c/o P Smith, Old Bury Radio Soc, Moses Yth Comm Cntr, Bury, BL9 0RS
G3 BSA West Manchester RC c/o T Speight, 1 Lyndene Avenue, Worsley, Manchester, M28 2RJ
G3 BSN P D Stanley, 1 Thames View, Cliffe Woods, Rochester, ME3 8LR
GM3 BSQ ABERDEEN ARS c/o I Munro, 57 Craigiebuckler Avenue, Aberdeen, AB15 8SF
GM3 BST John Tuke, 2/23 Hawthorn Gardens, Loanhead, EH20 9EE
G3 BV Quentin Cruse, Cerrig Mawr, Talybont, Aberystwyth, SY24 5DJ
G3 BVA E Digman, 75 Ramsden Road, Orpington, BR5 4LU
G3 BVB D Adair, 3 Belmont Close, Shaftesbury, SP7 8NF
G3 BWI W Timms, 22 Padway, Penwortham, Preston, PR1 9EL
G3 BWV F Springate, 150 Mackenzie Road, Beckenham, BR3 4SD
G3 BXS A Stacey, 22 Montagu Road, Datchet, Slough, SL3 9DJ
G3 BYG N Williams, Chappel Lake Farm, Beaworthy, EX21 5UF
G3 BZB R Cunliffe, 5 Silk Mill Lane, Tutbury, Burton-on-Trent, DE13 9LE
G3 BZU Royal Naval ARS c/o Joe Kirk, 111 Stockbridge Road, Chichester, PO19 8QR
G3 CAJ R Prince, 52 Mafeking Road, Southsea, PO4 9BG
G3 CAZ J Shaw, 128 Perth Road, Ilford, IG2 6AS
GW3 CBA Henry Kellaway, 34 Winston Road, Barry, CF62 9SW
G3 CCL Gordon Ireland, 20 St. Chads Road, Withington, Manchester, M20 4WH
G3 CCO D Williams, 27 Dene Way, Upper Caldecote, Biggleswade, SG18 9DL
G3 CCX Peter Craw, 117 Sea Lane, Rustington, Littlehampton, BN16 2RU
G3 CDM I Gardner, 30 Pierremont Crescent, Darlington, DL3 9PB
G3 CEI C Brown, Downlands, Off Hackwood Lane, Basingstoke, RG25 2NH
GI3 CFH NTH WEST IRELAN c/o D Fulton, 120 Dunnalong Road, Bready, Strabane, BT82 0DP
G3 CFR J Jowett, Ashleigh, Kilmington, Axminster, EX13 7ST
G3 CFV F Gay, 61 Abbey Road, Yeovil, BA21 3EY
G3 CGD J Yeend, 30 St. Lukes Road, Cheltenham, GL53 7JJ
G3 CGE R Gardner, 62 Rosewall Road, Southampton, SO16 5DW
G3 CIK H Romer, 96 Mortlake Road, Richmond, TW9 4AS
G3 CIL Michael Holley, 6586 196th Street, Langley, Canada, BC V2Y 1R3
G3 CIM S Denney, 52a Intwood Road, Cringleford, Norwich, NR4 6AA
G3 CIO ROYAL SIGNALS ARS c/o Bryan Downes, 6 Greenland Crescent, Beeston, Nottingham, NG9 5LB
G3 CJD L Allen, 21 Inghead Road, Slaithwaite, Huddersfield, HD7 5DS
G3 CKE S Mason, 46 Frankton Close, Redditch, B98 0HJ
G3 CLW L Hutton, 46 Penwill Way, Paignton, TQ4 5JQ
G3 CMH YEOVIL ARC c/o George Davis, Broadview, East Lanes, Yeovil, BA21 5SP
G3 CMU H Meyers, Cornerways, 2 Old Mill Lane, Polegate, BN26 5NS
G3 CNO FORT PURBROOK ARC c/o Michael Ponsford, 83 Grant Road, Portsmouth, PO6 1DU
G3 CNX GRIMSBY ARS c/o G Smith, 6 Fenby Close, Grimsby, DN37 9QJ
G3 CO Colchester Radio Amateurs Club c/o Herbert Yeldham, 19 Wade Reach, Walton on The Naze, CO14 8RG
GM3 COB J Paterson, 10 Hathaway Drive, Giffnock, Glasgow, G46 7AE
G3 CON L Crabbe, 6 Node Hill, Studley, B80 7RR
GM3 COQ D Oswald, 8 Redfield Road, Montrose, DD10 8TW
G3 CPC R Charlton, 7 St. Margarets Drive, Twickenham, TW1 1QL
G3 CPG L Damon, 18 Scafell Court, Dewsbury, WF12 7PD
G3 CPN M Stevens, 16 Golf Links Road, Ferndown, BH22 8BY
G3 CPT D Capp, 46 Stoke Road, Bletchley, Milton Keynes, MK2 3AD
G3 CQL M Clarke, 26 Lingfield Drive, Rochford, SS4 1EA
G3 CQU K Raffield, 113 Waddington Avenue, Coulsdon, CR5 1QP
GW3 CR Raymond Richards, 77 Church Road, Llanstadwell, Milford Haven, SA73 1EA
G3 CRH Hubert Sanders, Little Orchard, 68a Park Road, Burton-on-Trent, DE13 7AJ
G3 CRS Royal Naval ARS c/o Joe Kirk, 111 Stockbridge Road, Chichester, PO19 8QR
G3 CSA ELLESMRE P&D AR c/o Thomas Saggerson, 18 Ploughmans Way, Great Sutton, Ellesmere Port, CH66 2YJ
G3 CSC S Roddan, 18 Muncaster Drive, Rainford, St. Helens, WA11 8NR
G3 CSR CIVIL SERVICE ARS c/o Neil Sanderson, 54 Kelvedon Close, Chelmsford, CM1 4DG
G3 CSY K Hill, 30 Hestham Avenue, Morecambe, LA4 4PZ
G3 CTP J Swift, 20 Leighlands, Crawley, RH10 3DW
G3 CTQ H Westwell, 224 Dickson Road, Blackpool, FY1 2JS
G3 CTZ Alfred Jones, 17 Oaklea Way, Old Tupton, Chesterfield, S42 6JD
G3 CUF Harry Ashworth, 97 Winchcombe Road, Sedgeberrow, Evesham, WR11 7UZ
G3 CUR R Collette, 8a Woolwich Road, Belvedere, DA17 5EW
G3 CUY E Paul, 91 Windmill Drive, Brighton, BN1 5HH
G3 CVI B Thwaites, 118 Baddow Hall Crescent, Chelmsford, CM2 7BU
G3 CVK P Bolton, 50 Meadow Road, West Malvern, Malvern, WR14 2SD
G3 CWH R Rogers, 107 Rotherham Road, Coventry, CV6 4FH
G3 CWI Richard Newstead, 89 Victoria Road, Macclesfield, SK10 3JA
G3 CWT F Vale, 40 Ferry St., Stapenhill, Burton-on-Trent, DE15 9EY
G3 CXP R Gill, 45 Biggin Lane, Ramsey, Huntingdon, PE26 1NB
G3 CYL Geoffrey Bennett, 16 Coxheath Road, Church Crookham, Fleet, GU52 6QJ
G3 CYU J Wilson, 1 Beeches Farm Road, Crowborough, TN6 2NY
G3 CYX P Lambert, 11 Marlborough Close, Musbury, Axminster, EX13 8AP
G3 CZL R Buckman, Heathfield, Hang Hill Road, Lydney, GL15 6LQ
G3 CZU DORKING & DISTRICT R.S c/o W Blanchard, The Trundle, Tower Hill, Dorking, RH4 2AN
G3 DAF C Bland, 84 Milton Road, Grimsby, DN33 1DE
G3 DAO R Braithwaite, 32 Rupert Crescent, Queniborough, Leicester, LE7 3TU
G3 DAV John Waller, 17 Spencer Close, Marske-by-the-Sea, Redcar, TS11 6BD
G3 DBJ David Buggs, 2 Archway Cottages Valley Road, Leiston, IP16 4AR
G3 DBV S Hedges, 25 Rudland Close, Thatcham, RG19 3XW
G3 DCE F Humphries, Little Hayes, 1 Meadway, Sidmouth, EX10 9JA
G3 DCO Brian Coyne, 58 Osborne Road, New Milton, BH25 6AB
G3 DCV A Watson, 93 St. Dunstans Drive, Gravesend, DA12 4BJ
G3 DCZ R McDonald, 60 Dudley Drive, Morden, SM4 4RJ
GM3 DDL J Jackson, 74 Cairngorm Crescent, Paisley, PA2 8AW
G3 DEJ T Wiseman, 70 Dove House Lane, Solihull, B91 2EA
G3 DEN Richard Lea, 17 The Owell Pakenham, Bury St. Edmunds, IP31 2LE
G3 DEY L Ford, 177 Latters Orchard, Old Road, Maidstone, ME18 5PR
G3 DFY N Devine, 46 Tytton Lane West, Wyberton, Boston, PE21 7HL
G3 DOH D Hardcastle, 829 Carrol, Harlingen, United States, 78550
G3 DID J Doyle, 16 Park Hall Crescent, Birmingham, B36 9SN
GM3 DIE T Dickson, 91 Milton Road West, Edinburgh, EH15 1RA
GM3 DIN A Clark, 11 Regent Park Square, Glasgow, G41 2AF
G3 DIT PRTSMTH & DARS c/o Terence Mortimer, 10 Harold Road, Hayling Island, PO11 9LT
G3 DMO C Earnshaw, 35 Rogersfield, Langho, Blackburn, BB6 8HB
G3 DNH J Spicer, 291 Green Lane, Coventry, CV3 6EH
G3 DNN G Saville, 2 Gaskell Close, Littleborough, OL15 8EB
G3 DNS N King, 31 Great Norwood Street, Cheltenham, GL50 2AW
GM3 DOD A Murray, 50 Castlepark Drive, Fairlie, Largs, KA29 0DG
G3 DOV D Dove, 3 Walnut Grove, Watton, Thetford, IP25 6EY
G3 DPM D Cooknell, 23 The Hyde, Winchcombe, Cheltenham, GL54 5QR
G3 DQG Gerald Mortimer, 122a Roker Lane, Pudsey, LS28 9ND
G3 DQQ D Winterburn, 47 Hilda Avenue, Tottington, Bury, BL8 3JE
G3 DQT J Ayres, 8 Cornfield Road, Seaford, BN25 1SW
G3 DQW PETERBOROUGH RADIO AND ELECTRONIC SOC. c/o B Vaughan, 7 Oundle Road, Chesterton, Peterborough, PE7 3UA
G3 DQY John Vaughan, Flat 9, St. Leonards Court, St. Leonards-on-Sea, TN38 0PS
G3 DRN E Allen, 30 Bodnant Gardens, Wimbledon, London, SW20 0UD
GW3 DRV O Jones, 4 Chalybeate Gardens, Aberaeron, SA46 0DL
G3 DSZ A Kent, 23 Pagehall Close, Scartho, Grimsby, DN33 2HF
G3 DT L Boorman, 2 Bull Lane Cottages, Bull Lane, Ashford, TN26 3HA
G3 DTP Andrew Jackson, Flat 6, St. Albans Court, Rochdale, OL11 4HW
G3 DTU C Prior, 36 Bassnage Road, Halesowen, B63 4HQ
G3 DTX I Duck, Chenies, Loudhams Wood Lane, Chalfont St. Giles, HP8 4AR
G3 DUW R Hodgson, The Shealing, Forest Moor Drive, Knaresborough, HG5 8JT
GJ3 DVC The Jersey ARS c/o R Taylor, 21 Samares Avenue, La Grande Route De St. Clement, St. Clement, Jersey, JE2 6NY
G3 DVF G Cain, 23 Wiltshire Avenue, Crowthorne, RG45 6NR
G3 DWI G Lusty, Sundial House, High Street, Chipping Campden, GL55 6AG
G3 DXD D Rolph, 2 Victoria Court, Victoria Road, Marlow, SL7 1DR
G3 DXZ Charles Fletcher, 12 Park Crescent, Retford, DN22 6UF
G3 DYO N Alder, Greenwoods, Eastfield Road, Ross on Wye, HR9 5JY
G3 DZJ F Pardy, 5 Y Bryn, Glan Conwy, Colwyn Bay, LL28 5NJ
G3 DZS Harold Fudge, 12 Rosemoor Road, Torrington, EX38 7NB
G3 EAE G Billington, 75 Mount Vernon Road, Barnsley, S70 4DW
G3 EAO Victor Donald Bullett, 38a Sidney Road, Walton-on-Thames, KT12 2LY
G3 EBP Michael Courcoux, 10 Baskerfield Grove, Woughton on the Green, Milton Keynes, MK6 3EN
G3 EBV S Squire, Leafield, 4 Little Green Lane, Rickmansworth, WD3 3JQ
GJ3 ECC R Taylor, 21 Samares Avenue, La Grande Route De St. Clement, St. Clement, Jersey, JE2 6NY
G3 ECM P Bowles, 29 Coleman Avenue, Hove, BN3 5ND
G3 ECP J Brown, Manor Cottage, 2 The Maltings, Huntingdon, PE28 4DZ
GI3 ECQ G McGarry, 18 Marna Brae Park, Lisburn, BT28 3PD
G3 EDD Brian Armstrong, 39 Angle End, Great Wilbraham, Cambridge, CB21 5JG
G3 EEH J Watkinson, The Moorings, 63 Ruffa Lane, Pickering, YO18 7HN
G3 EEZ UK Microwave Group c/o John Worsnop, 20 Lode Avenue, Waterbeach, Cambridge, CB25 9PX
GD3 EFD M Thompson, Whitehouse Cottage, St. Marks, Ballasalla, Isle of Man, IM9 3AH
GM3 EFH G Syme, 9a Maitland Street, Tayport, DD6 9DL
G3 EFL W Preston, 8 Pencraig View, Greytree, Ross-on-Wye, HR9 7JR
G3 EFS W Borland, 25 Broadoaks Way, Bromley, BR2 0UA
GD3 EFX RAD SOC HARROW c/o C Friel, 102a Sharps Lane, Ruislip, HA4 7JB
G3 EGC J Hoban, 13 Druids Close, Egerton, Bolton, BL7 9RF
G3 EGF T Kellett, Braville, St. Ives Road, Consett, DH8 7SJ
G3 EGV R Staniforth, 26 Winslow Road, Preston, Weymouth, DT3 6NE
G3 EHQ H Bone, 2 Waterville Gardens, Orton Waterville, Peterborough, PE2 5LG
G3 EHW J Watkins, 19 Barrow Grove, Sittingbourne, ME10 1LB
GM3 EIY Michael Joyner, 1 Spey Walk, Holytown, Motherwell, ML1 4ST
G3 EIZ C Lyon, Ardraeth, The Drive, Bodorgan, LL62 5AW
G3 EJH W Peatman, 110 Cator Lane, Beeston, Nottingham, NG9 4BB
GW3 EJR John Armstrong, Mirianog, 1 Bryn Bedw, Cardigan, SA43 2NY
G3 EKE L Stockley, C/O Glebe Cottage, Baylham, Ipswich, IP6 8JS
G3 EKJ Harry Mallacks, Fieldfaro, Eastbourne Road, Lewes, BN8 6PS
G3 EKT OBO ROYAL AIR FORCE ARS CO DURHAM c/o B Burke, 12 Evesham Grove, Hurworth, Darlington, DL2 2YE
G3 EKW ARC OF NOTTM c/o S Williams, Haywood Community Centre, 46 Haywood Road, Nottingham, NG3 6AD
G3 ELS Bernard Rudd, Orchard Bungalow, 14 Walmer Close, Colchester, CO7 0PE
G3 ELV Royal Air Force Henlow Radio & Electronics Club c/o Roy Walker, 35 Romany Close, Letchworth Garden City, SG6 4LA
G3 ENI A Pegler, Brook House, Forest Close, Leatherhead, KT24 5BU
G3 ENO R Green, 8b The Beck, Elford, Tamworth, B79 9BP
GD3 ENV Peter Poole, 2c The Avenue, Hatch End, Pinner, HA5 4EP
GM3 EOB C Merrilees, 6 Spoutwells Drive, Scone, Perth, PH2 6RR
G3 EOO J Hamlett, 23 Riddings Road, Timperley, Altrincham, WA15 6BW

**IMPORTANT NOTE**

**Revalidate licence to avoid revocation** – Ofcom has advised the Society that plans will be drawn up to revoke licences that have not been revalidated as required by the licence conditions. The quickest way to revalidate is to do so online via the Ofcom website: *https://services.ofcom.org.uk/* or by email: *amateur.validations@ofcom.org.uk* Ofcom staff are available to help, but please be patient during times of heavy workload.

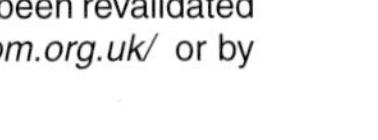

G3 EPO K Procter, 11 Boucher Road, Budleigh Salterton, EX9 6JF
G3 EQM J Theobald, 2a Retreat Road, Topsham, Exeter, EX3 0LF
G3 ERD DERBY DIST ARS c/o Jack Anthony, 77 Brayfield Road, Littleover, Derby, DE23 6GT
G3 ESK Louis Potter, 2 Linden Drive, Chatteris, PE16 6DZ
G3 ESP Wakefield & District RS c/o Stuart Adaway, 20 Foundry Street, Barnsley, S70 1PL
G3 ESY P Jones, 13 Blenheim Close, Hereford, HR1 2TY
G3 ETP Peter Woodyard, 65 Raglan Street, Lowestoft, NR32 2JS
G3 EUE E Jones, White Lodge, The Street, Steyning, BN44 3WE
G3 EUG EDDYSTONE USER GROUP c/o William Moore, 4b Colville Road, Newton, Wisbech, PE13 5HH
G3 EVT R Mutton, Summer Hayes, Mill Lane, Alcester, B49 6LF
G3 EWF A Harris, 5 Wickham Court, Stapleton, Bristol, BS16 1DQ
G3 EWM P Green, 23 Tilton Road, Borough Green, Sevenoaks, TN15 8RS
G3 EWT C Tamkin, 4 Stanmer Villas, Brighton, BN1 7HP
G3 EXL David Derham, 3 Riverbank Cottages, Old Ferry Road, Saltash, PL12 6BJ
G3 EZB J Rackett, Little Vectis, Folgate Lane, Norwich, NR8 5DP
G3 FBT Stephen Briggs, 20 Bluebell Close, Newton Aycliffe, DL5 7LN
G3 FBU W Brown, 79 Mill Hill, Deal, CT14 9EW
G3 FCM A Cowley, 13 Steward Close, Stuntney, Ely, CB7 5TW
GM3 FDN J Petrie, 4 Cruachan Place, Grangemouth, FK3 0BU
G3 FDW A Gibbings, 16 Turnberry Avenue, Eaglescliffe, Stockton-on-Tees, TS16 9EH
G3 FDZ D Whitehead, Tyddyn Bach, Dyffryn Ardudwy, LL44 2RQ
G3 FET L Rawlings, Flat 5, Francis Court, 47 Church Street, Littlehampton, BN17 5PY
G3 FEW Edward Rule, 15 Norwich Road, Lenwade, Norwich, NR9 5SH
GI3 FFF BALLYMENA ARC c/o Jeffrey Clarke, 154 Galgorm Road, Ballymena, BT42 1DE
G3 FFR W Darbyshire, Flat 29, Poplar Court, Lytham St. Annes, FY8 1NZ
G3 FGP R Brooks, 10 The Oval, Longfield, DA3 7HD
G3 FHG M Hopkins, Hylton Cottage, Grafton, Tewkesbury, GL20 7AT
G3 FHL G Bagley, 49 Green Lane, Malvern Wells, Malvern, WR14 4HT
G3 FHN E Aldworth, Glenaire, 15 Heather Way, Hastings, TN35 4BL
G3 FHT Anthony Lewis, 8 Lutyens Fold, Milton Abbot, Tavistock, PL19 0NR
G3 FIA A Lowden, 3 Boscobel Road, Great Barr, Birmingham, B43 6BB
G3 FIC J Glover, 53 Swanpool Lane, Aughton, Ormskirk, L39 5AY
G3 FIR B Farrow, Gardencourt, 135 Tally Ho Road, Ashford, TN26 1HW
GM3 FJA W Sleat, 9 Doocot Road, St. Andrews, KY16 8QP
G3 FJE SHEFFORD & DISTRICT ARS c/o D Ross, 3 Little Lane, Clophill, Bedford, MK45 4BG
G3 FJI E Jones, 8 Merllyn Road, Rhyl, LL18 4HH
G3 FJL J Hall, 250 Scraptoft Lane, Leicester, LE5 1PA
G3 FJO A Ellefsen, 121 The Furlongs, Ingatestone, CM4 0AL
GI3 FJX J Davidson, 7 Keel Point, Dundrum, Newcastle, BT33 0NQ
GD3 FKI Eric Lambert, 6 Abercorn Gardens, Kenton, Harrow, HA3 0PB
G3 FKY J Parker, 36 North Avenue, Leek, ST13 8DP
GD3 FLH IOM ARS c/o Arthur Sinclair, 1, Marathon Drive, Douglas, Isle of Man, IM4 2BP
G3 FLV L Keighley, 24 St. Annes Road, Headingley, Leeds, LS6 3NX
G3 FMO G Elliott, Oatlands, Southend Road, Chelmsford, CM2 7TD
G3 FMR C Dwyer, Ystrad, 29 The Oval, Llandudno, LL30 2BU
G3 FMU D McDiarmid, 102 Shalloak Road, Broad Oak, Canterbury, CT2 0QH
G3 FMW J Stockley, 22 Manor Gardens, Killinghall, Harrogate, HG3 2DS
G3 FNL R Grubb, 7762 Brockway Drive, Boulder, United States, 80303
G3 FNO G Morgan, 27 Kestrel Close, Downley, High Wycombe, HP13 5JN
G3 FNZ J Lambert, 49 Rede Court Road, Strood, Rochester, ME2 3SP
G3 FOQ D Delanoy, Martindale, Halls Lane, Bury St. Edmunds, IP31 3LG
G3 FP B Arnold, 5 Salcott Road, Beddington, Croydon, CR0 4PS
G3 FPH J Hayes, 4 St. Marys Drive Northop Hall, Mold, CH7 6JF
G3 FPQ D Courtier-Dutton, Markham Oak Cottage, Dockenfield Road, Farnham, GU10 4LP
G3 FPY J Dew, 62 Monks Park Avenue, Horfield, Bristol, BS7 0UH
G3 FRE W Frith, 56 Ringleas, Cotgrave, Nottingham, NG12 3NE
GM3 FRU D Wark, Flat 37a, Northwood House, Edinburgh, EH9 2EL
G3 FRV Ronald Vaughan, 1 Langstone Close, Maidenbower, Crawley, RH10 7JR
G3 FSA A Davis, Willow Cottage, Hedging, Bridgwater, TA7 0DE
GW3 FSP Leighton Davies, Glanmor, Brynna Road, Bridgend, CF35 6PD
G3 FSX R Ellis, Laura House, 79 Sunte Avenue, Haywards Heath, RH16 2AB
G3 FTH John Hale, 136 Bush Road, Cuxton, Rochester, ME2 1HB
G3 FTK L Gray, 109 Foxholes Road, Poole, BH15 3NE
GI3 FTT W Brennan, 10 Dunhugh Park, Londonderry, BT47 2NL
G3 FUJ W Scott, 10 Pavilion Road, Littleover, Derby, DE23 6XL
G3 FVA STH MANCHESTR RD c/o David Armitage, 12 Loughborough Close, Sale, M33 5UF
G3 FVR R Bannister, 22 Manton Road, Hitchin, SG4 9NW
G3 FWD B Purchase, 126 Renton Road, Wolverhampton, WV10 6XH
G3 FWI William Sutton, Pendle, 6 Cuperham Close, Romsey, SO51 7LH
G3 FWU L Richardson, Belmont Cottage, Christys Lane, Shaftesbury, SP7 8NQ
G3 FXI P CARDWELL, 3 Old Talbot, Llanwnog, Caersws, SY17 5JG
GD3 FXN A Radcliffe, 3 Cronk Drine, Union Mills, Douglas, Isle of Man, IM4 4NG
G3 FYF P Acke, Kinghurst Farm, Holne, Newton Abbot, TQ13 7RU
G3 FYQ PONTEFRACT DIS c/o C Wilkinson, 8 Westfield Avenue, Knottingley, WF11 0JH
G3 FYX R Emery, 30 Station Road, Winterbourne Down, Bristol, BS36 1EP
GW3 FZV R Lewis, 1 Victoria Avenue, Penarth, CF64 3EN
G3 GAA W Jeans, 36 Pimms Grove, High Wycombe, HP13 7EF
G3 GAF Colin Dollery, 101 Corringham Road, London, NW11 7DL
G3 GAH D Johnson, 31 Coniston Avenue, Penketh, Warrington, WA5 2QY
G3 GAQ D Bottomley, 24 Midhope Road, Woking, GU22 7UE
G3 GBD S Hancock, 53 Friary Grange Park, Winterbourne, Bristol, BS36 1NA
GD3 GBG A Moore, 114 Ballabrooie Drive, Douglas, Isle of Man, IM1 4HQ
G3 GBN S Feldman, Flat 5 Maitland Joseph House, 35 Marlowes, Hemel Hempstead, HP1 1LB
G3 GBS M Sandoz, Edelweiss, Broad Lane, Solihull, B94 5DP
G3 GBU STOKE ON TRENT ARS c/o Albert Allen, 3 Wayfield Grove, Harpfields, Stoke-on-Trent, ST46DB
GM3 GBZ OBO STRATHMORE ARC c/o George Balfour, 6 Kirkden Street, Friockheim, Arbroath, DD11 4SX
GD3 GCE P Gordon, Dormer House, Walpole Drive, Ramsey, Isle of Man, IM8 1NA
G3 GCW B Jones, 44 Winner Hill Road, Paignton, TQ3 3BT
G3 GDB G Bird, 16 Simnel Road, London, SE12 9BG
G3 GDH D Silveston, 192 Rosemary Avenue, Minster on Sea, Sheerness, ME12 3HX
G3 GEF J Andrews, 45 Sandes Court, Sandes Avenue, Kendal, LA9 4LN
G3 GEG E Cooper, Ciren, 19 Ventnor Road, Sandown, PO36 0JT
G3 GEI SOLUHULL ARS c/o Roger Hancock, 80 Ulleries Road, Solihull, B92 8EE
G3 GEJ L Airey, 32 Brookside Close, Bedale, DL8 2DR
G3 GEV Stephen Hollingshurst, 9 Watersedge Court, 1 Wharfside Close, Erith, DA8 1QW
G3 GEX Peter Burton, 18 Tankerfield Place, Romeland Hill, St. Albans, AL3 4HH
GM3 GG BANFF AND BUCHAN ARC c/o S Roberts, 10 Mill Place, Tarland, Aboyne, AB34 4YG
G3 GGG R Bishop, 31 Blenheim Close, Didcot, OX11 7JQ
G3 GGH P Horn, Darfield, 50 Barrack Road, Bexhill-on-Sea, TN40 2AZ
G3 GGI A Laurence, 70 Firs Avenue, London, N11 3NQ
G3 GGK P Simpson, 109 Highfields Road, Highfields Caldecote, Cambridge, CB23 7NX
G3 GGL D Wormald, 160 Sutton Park Road, Kidderminster, DY11 6LF
G3 GGN David Shute, 100 Wick Street, Wick, Littlehampton, BN17 7JS
G3 GGR J Sykes, 49 Chapel St., Pelsall, Walsall, WS3 4LW
G3 GGS W Waring, 51 Church Road, Leyland, PR25 3AA
G3 GGU G Smith, Greenacres, Top Road, Chesterfield, S44 5AE
G3 GHN CLIFTON ARS c/o Stephen Fletcher, 90 Westcombe Park Road, Blackheath, London, SE3 7QS
G3 GHS J Holland, Tanglewood, Portheast Way, St. Austell, PL26 6JA
G3 GIB A Wake, 42 Charles Avenue, Watton, Thetford, IP25 6BZ
G3 GIH J Bird, The Old Stackyard, Daisy Green, Bury St. Edmunds, IP31 3HX
G3 GIZ Chester & District RS c/o Paul Holland, Chatterton, Chapel Lane, Malpas, SY14 7AX
G3 GJA Clive Reynolds, 49 Westborough Way, Anlaby Common, Hull, HU4 7SW
G3 GJJ P Watson, 5 High Garth, Winston, Darlington, DL2 3RY
G3 GJL Worcester Radio Amateur Association c/o Richard Moles, 14 Dorsett Road, Stourport-on-Severn, DY13 8EL
G3 GJQ R Handley, Flat 2, 11 Trinity Square, Llandudno, LL30 2RA
G3 GJW T Lundegard, Saxby, Botsom Lane, Sevenoaks, TN15 6BL
G3 GKC I Rosevear, 20 Christchurch Road, Bradford-on-Avon, BA15 1TB
G3 GKG G Horsfall, 183 Chester Road, Macclesfield, SK11 8QA
GM3 GKJ A Gordon, The Paddock, Greenhead, Tranent, EH34 5EH
G3 GKS R Christian, 27 Howey Rise, Frodsham, WA6 6DN
G3 GLA B Mase, 18 Norton Drive, Norwich, NR4 6JD
G3 GLB J Lacey, 50 Petersham Avenue, Byfleet, West Byfleet, KT14 7HY
G3 GLL T Green, 6 Woodrolfe Road, Tollesbury, Maldon, CM9 8SB
G3 GLW P Willis, 26 Snellgrove Close, Calmore, Southampton, SO40 2WD
G3 GLX J Simmonds, 99 Foljambe Avenue, Chesterfield, S40 3EY
G3 GMC P McVey, 18 Worlebury Hill Road, Weston-Super-Mare, BS22 9SP
G3 GML F Murray, 3 Rosemary Close, Tiptree, Colchester, CO5 0QD
G3 GMM Eric McFarland, 60 Sutton Oaks, London Road, Crewe, CW4 7AS
G3 GMS M Thayne, 14 Tyndale Avenue, Monkseaton, Whitley Bay, NE26 3BA
G3 GMW L Nichols, 5 Middle Pasture, Peterborough, PE4 5AU
G3 GMY F Green, 5 Silvercliffe Gardens, New Barnet, Barnet, EN4 9QT
G3 GNA D Macmillan, Brook Farm, Broadwas, Worcester, WR6 5NE
G3 GNR Robert Short, 270 Exeter Road, Exmouth, EX8 3NL
G3 GOS P Peach, The Firs, Goldsmith Lane, Axminster, EX13 7LU
G3 GOT Bryan Le Grys, 8 Kitchener Way, Shotley Gate, Ipswich, IP9 1RW
G3 GQC OBO MANFIELD AM RS c/o D Riley, 9 Century Avenue, Mansfield, NG18 5EE
G3 GQK John Wall, P.O Box 631, Nambucca Heads, Australia, 2448
GM3 GRG D Rollo, 25 Beaufort Drive, Kirkintilloch, Glasgow, G66 1AX
G3 GRL Simon Houlton, 97 Mansfield Road, Alfreton, DE55 7JP
G3 GRQ C Hebden, 129 Millers Way, Honiton, EX14 1JB
G3 GRS GRAVESEND ARS c/o D Lawley, 1515 High Road, London, N20 9PJ
G3 GRV George Halse, 10 Charnock Close, Hordle, Lymington, SO41 0GU
G3 GRY F Wiseman, 14 Parkway, Crowthorne, RG45 6EN
G3 GSL A Rennison, 23 Windermere Drive, Adlington, Chorley, PR6 9PD
G3 GTA J Shute, 32 Woodborough Drive, Winscombe, BS25 1HB
G3 GTF B Harris, 25 Rother View, Burwash, Etchingham, TN19 7BN
GM3 GTQ A Mcphedran, 3 Argyll Road, Bearsden, Glasgow, G61 3JX
GI3 GTR Roy McKinty, 3 Rhanbuoy Road, Craigavad, Holywood, BT18 0DY
G3 GUE A Dowling, Church Cottage, Frittenden, Cranbrook, TN17 2DD
G3 GUR J Scully, 1 Wyde Feld, Bognor Regis, PO21 3DH
G3 GUX John Brimecombe, Llwyn Onn, Llangoed, Beaumaris, LL58 8PH
G3 GVM F Robins, 59 Titchfield Road, Stubbington, Fareham, PO14 2JF
G3 GWB United Trades Club c/o John Cockrill, 28 Northampton Road, Harpole, Northampton, NN7 4DD
G3 GWC E Ramsdale, 8 May Cottages, Monkswell Lane, Coulsdon, CR5 3SX
G3 GWE A Daum, 100 Shawbridge, Harlow, CM19 4NW
G3 GXG Colin Lee, 61 Roberts Road, Barton Stacey, Winchester, SO21 3RU
G3 GXI Eccles and District ARS c/o Christopher Harrison, 11 Ringley Park, Whitefield, Manchester, M45 7NT
G3 GXQ Walter Roberts, 24 Leeds Road, Barwick in Elmet, Leeds, LS15 4JD
G3 GYC P Ingram, 28 John Amery Drive, Stafford, ST17 9NA
G3 GYE Peter Pitts, Westmoors House, Trezelah, Penzance, TR20 8XD
G3 GYQ Christopher Spackman, 7 Hatchers Crescent, Blunsdon, Swindon, SN26 7AQ
G3 GZT Reg Moores, 117 Horton Road, Brighton, BN1 7EG
G3 GZX A Bladon, 6 Quarry Bank, Mold Road, Denbigh, LL16 4DT
G3 GZZ A Bevan, 14 Parsonage Road, Berrow, Burnham-on-Sea, TA8 2NL
G3 HAA J Morgan, 10 Bamber Gardens, Southport, PR9 7PQ
G3 HAL Ronald Parrott, 3 Ash Grove, Chard, TA20 1BZ
GM3 HAM LOTHIANS RS c/o P Bates, 10 Swanston Avenue, Edinburgh, EH10 7BU
G3 HAN M Hitchman, 12 Briar Walk, Oadby, Leicester, LE2 5UE
G3 HBW A Mynett, Po Box 493, Puerto Del Rosario, Canary Islands, Spain
G3 HCO G Errock, 307 Main Road, Emsworth, PO10 8JG
G3 HCQ S Gabriel, Millbrook House, 3 Mill Drove, Bourne, PE10 9BX
G3 HCS H Stratton, 26 Marjorie Road, Chaddesden, Derby, DE21 4HQ
G3 HCT J Bazley, C/O Mr P Chadwick, Three Oaks, Swindon, SN5 0AD
G3 HCZ B Edmondson, 1 Harbour Lane, Turton, Bolton, BL7 0PA
G3 HDF K Groves, 6 Overleigh Drive, Buckley, CH7 2PA
G3 HDM S Campbell, Carrer De Ses Sevines, 1 Bajo, Mallorca, Spain
G3 HDT J Graham, 18 Wheatriggs Avenue, Milfield, Wooler, NE71 6HU
G3 HDX C Massey, 33 Ash Hill Gardens, Leeds, LS17 8JW
G3 HEE J Fancourt, 35 St.Paul`s Street, Stamford, PE9 2BH
G3 HEH Edmund Parker, 7 Sandfield Court, Wrenbury, Nantwich, CW5 8HW
G3 HEJ Derek Stanners, Tanglewood, Samarkand Close, Camberley, GU15 1DG
GM3 HEN A White, Byeways, Whiting Bay, Isle of Arran, KA27 8QH
G3 HEU D Rickers, 4 St. Marks Terrace, Wrexham, LL13 0PQ
GD3 HFC F Arrowsmith, The Evergreens, South Cape, Laxey, Isle of Man, IM4 7JB
G3 HFM A Vickers, Foxcroft, 4 Woodlands End, Macclesfield, SK11 9BF
GU3 HFN THE GUERNSEY AR c/o Phil Cooper, 1 Clos Au Pre, La Route de la Hougue Du Pommier, Castel, Guernsey, GY5 7FQ
G3 HFZ J Yardley, Conifers, 30 Ulwell Road, Swanage, BH19 1LL
GM3 HGA J McCall, 1 Pinewood Place, Aberdeen, AB15 8LT
G3 HGD V Best, 3 Old Auction Mart, Kirkby Lonsdale, Carnforth, LA6 2AF
G3 HGE T Withers, Woodpeckers, West Stow, Bury St. Edmunds, IP28 6ER
G3 HGI I Soars, 5 Ferndale Road, Church Crookham, Fleet, GU52 6LJ
G3 HGL B Clark, 97 Rhos Road, Rhos on Sea, Colwyn Bay, LL28 4TT
G3 HHD T Hayward, Skirt Bank, Nether Silton, Thirsk, YO7 2LL
G3 HHU J Ickringill, 28 Deena Close, Queens Drive, London, W3 0HR
G3 HIF A Reid, 205 Mortimer Road, South Shields, NE34 0RT
G3 HIU Milton Keynes ARS c/o David White, 1 Whaddon Road, Shenley Brook End, Milton Keynes, MK5 7AF
GI3 HJH R McBurney, 8 Main Road, Ballymartin, Newry, BT34 4NU
G3 HJK B Mitchell, 98 Queensway, Heald Green, Cheadle, SK8 3ET
G3 HJP G Cooper, 25 Plantation Avenue, Shadwell, Leeds, LS17 8TB
G3 HJS R Woodford, 19 Fairlie Road, Littlemore, Oxford, OX4 3SW
G3 HKA C Booth, 88 Green Drive, Thornton-Cleveleys, FY5 1JD
G3 HKD D Money, 125 Wroxham Road, Norwich, NR7 8AD
G3 HKF B Ferris, 5 Guildway, Todwick, Sheffield, S26 1JN
G3 HKH M Harrison, 3 Stert Street, Abingdon, OX14 3JF
G3 HKN F Shakespeare, Fairways, 53 Ashby Road East, Burton-on-Trent, DE15 0PS
G3 HKO D Wood, 28 Hillcrest Avenue, Scarborough, YO12 6RQ
G3 HKT A Partner, 10 The Tanners, Titchfield Common, Fareham, PO14 4BH
G3 HLG D Johnson, Robins, 4 Station Road, Newark, NG23 7RA
G3 HLI Malcolm Bradford, 101 Oxendon Way, Binley, Coventry, CV3 2HA
G3 HLN Patrick Woods, 145 Hollybush Lane, Welwyn Garden City, AL7 4JT
G3 HMB I Elliot, Grange House, Manningtree Road, Ipswich, IP9 2SW
G3 HMG A Macgregor, 14 Quantock Grove, Williton, Taunton, TA4 4PD
G3 HMO J Osborne, 141 Chadwick Road, London, SE15 4PY
G3 HMQ J Robson, 32 St. Stephens Road, Cold Norton, Chelmsford, CM3 6JE
G3 HMR Guy Moser, 30 Blackhall Croft, Blackhall Road, Kendal, LA9 4UU
G3 HMV N Bolton, 2 Selborne Villas, Clayton, Bradford, BD14 6JZ
G3 HNC B Dyer, 30 Smithson Avenue, Castleford, WF10 3HN
GM3 HNE G Campbell, 17 Roseburn Terrace, Edinburgh, EH12 5NG
GI3 HNM C Davies, 121 Comber Road, Toye, Downpatrick, BT30 9PD
GM3 HOM J Reilly, 30 Park Crescent, Bishopbriggs, Glasgow, G64 2NS
G3 HPB F Tooley, 70 Langbury Lane, Ferring, Worthing, BN12 6QA
G3 HPC W Stonehouse, 3 Clifton Close, Plymouth, PL7 4BL
G3 HPD Fred Dews, 341 Crossley Lane, Mirfield, WF14 0NR
G3 HQG G Atkins, 20 Mansfield Road, Killamarsh, Sheffield, S21 2BX
G3 HQS C Baker, Roffensis, 16 Boulderside Close, Norwich, NR7 0JJ
G3 HQT P Ball, 68 Brook Lane, Warsash, Southampton, SO31 9FG
G3 HQX John Brodzky, 3 Ropewalk House, Hyde Abbey Road, Winchester, SO23 7XH
G3 HRE F Watson, 54 Tavistock Road, Cambridge, CB4 3ND
G3 HRH R Hills, 2 The Dell, Otterbourne Road, Winchester, SO21 2DE
G3 HRK D Willies, 17 Campion Way, Sheringham, NR26 8UN
G3 HRU G Senior, 27 Leconfield Court, Wetherby, LS22 6TY
G3 HRX J Hilling, 24 Gloucester Road, Gaywood, King's Lynn, PE30 4AB
G3 HST Geoff Allen, Moor Farm Cottage, Salcombe, TQ8 8PW
G3 HSU K Richards, 25 Weir Road, Hemingford Grey, Huntingdon, PE28 9EH
G3 HSV D Alesbury, 23 Cullerne Road, Swindon, SN3 4HU
G3 HTA John Forward, Sunrays, Barnstaple Cross, Crediton, EX17 2EP
G3 HTB Malcolm Squance, Church Lane, Cubbington, Leamington Spa, CV32 7JT
G3 HTC C Storey, 12 Vereker Drive, Sunbury-on-Thames, TW16 6HF
G3 HTF Les Barclay, Appletrees, Studley Road, Ripon, HG4 2QJ
G3 HTJ William Walker, 53 Wolfridge Ride, Alveston, Bristol, BS35 3PR
G3 HTO R Dolton, 43 Jubilee Meadow, St. Austell, PL25 3EX
G3 HTT W Cheeseworth, 10 Barton Mill Court, Station Road West, Canterbury, CT2 7JZ
G3 HTX Walter Hipwell, 33 Dolphin Court Road, Paignton, TQ3 1AG
G3 HUA R Holloway, Sturrow Cottage, Lewes Road, Haywards Heath, RH16 2LQ
G3 HUB M Harrison, Rolling Hills, Brandy Lane, Lostwithiel, PL22 0QH
G3 HUD Margeret Brown, 10 Park House Mews, Congleton Road, Sandbach, CW11 4SP
G3 HUK M Morrissey, 1 Hamilton Road, Church Crookham, Fleet, GU52 6AS
G3 HUL D Mallett, 45 Crown Road, New Costessey, Norwich, NR5 0ES
GM3 HUN W Hunter, 4 Baird Grove, Edinburgh, EH12 5RP
G3 HUO K Young, 80 Darbys Lane, Oakdale, Poole, BH15 3ET
G3 HUR D Brough, 18 Lark Hall Road, Macclesfield, SK10 1QP
G3 HUX J Matthews, 4 Berrington Grove, Ashton-in-Makerfield, Wigan, WN4 9LD
G3 HVA D Pinnock, 2 Oak Close, Oakley, Basingstoke, RG23 7DD
G3 HVJ A Chappell, 22206 Del Valle St, Woodland Hills, United States, 91364-1515
GM3 HVK J Craig, 147 Avon Road, Larkhall, ML9 1RA
G3 HWF South and West Yorkshire Wing ATC c/o David Taylor, 76 Heworth Village, York, YO31 1AL
G3 HWM John Cowling, 19 The Drive, Hullbridge, Hockley, SS5 6LZ
G3 HWS G Marshall, 39 Kew Road, Southport, PR8 4HH
G3 HWW York ARS c/o Christopher Rouse, 86 Melton Avenue, Clifton, York, YO30 5QG
G3 HWX B Whitty, Fourways Morris Lane, Halsall, Ormskirk, L39 8SX

G3 HXK P Nethercot, Ronhill, Stoodleigh, Tiverton, EX16 9PJ
G3 HXN Jonathan Crisp, 371 Stroud Road, Tuffley, Gloucester, GL4 0DA
G3 HYG Douglas Topping, Bentley, Middle Street, Waltham Abbey, EN9 2LB
G3 HYH C Hay, 37 Acres Road, Leicester Forest East, Leicester, LE3 3HB
GM3 HYX Charles Rattray, 58 Aberdour Road, Dunfermline, KY11 4PE
G3 HZP H James, 10 Playsted Lane, Cambourne, Cambridge, CB3 6GA
G3 HZR B Harris, 1 Newbridge Cottages, Midgehole, Hebden Bridge, HX7 7AL
G3 HZT P Fraser, 45 The Martlet, Hove, BN3 6NT
G3 HZW D Mainhood, Gunrod, Holcombe Lane, Bridgwater, TA7 9BX
G3 IAR M Crowther-Watson, The Snicket, 14 The Avenue, Sevenoaks, TN13 3LA
G3 IAZ Alfred Wickham, Apartment 6, Panama Beach, Eastbourne, BN23 5PL
G3 IBI Peter Scott, 62 Old Street, Fareham, PO14 3HW
G3 IBQ K Hill, 61 Milford Avenue, Nepean, Ontario, Canada, K2G 1C1
GM3 IBU A Wright, Crosslea, Berstane Road, Kirkwall, KW15 1SZ
G3 IBY I Wilmshurst, 4 Eastern Road, West End, Southampton, SO30 3EQ
G3 ICA G Adams, Sue Marey, Selsley Hill, Stroud, GL5 5JS
G3 ICB A Bull, 91 Lower Way, Thatcham, RG19 3HS
G3 ICG K Mcfarlane, Clifton, 18 Needham Road, Harleston, IP20 9JY
G3 ICU George Davis, Broadview, East Lane, Yeovil, BA21 5DP
G3 ICZ W Clowes, 144 Norton Lane, Norton-in-the-Moors, Stoke-on-Trent, ST6 8BZ
G3 IDB A Brooks, 45 Northfield Road, Townhill Park, Southampton, SO18 2QE
G3 IDW R Reynolds, 6 Church Way, Stratton, Swindon, SN3 4NF
G3 IDY R Robson, 66 Tilstock Crescent, Shrewsbury, SY2 6HQ
G3 IEJ Stewart Watson, 6 Hope Street, Lytham St. Annes, FY8 3SL
G3 IFB F Bliss, Coppulex, North Road East, Cheltenham, GL51 6RE
G3 IFX A Cooke, 9 Lee Crescent, Ilkeston, DE7 5EF
G3 IGC A Garforth, 110 Foxdenton Lane, Chadderton, Oldham, OL9 9QR
G3 IGH H Sanders, 8 Caldicot Close, Aylesbury, HP21 9UF
G3 IGQ University of Surrey EARS c/o Laurence Stant, EARS, University of Surrey Student's Union, Guildford, GU2 7XH
G3 IGU K Coates, 76 Copley Crescent, Scawsby, Doncaster, DN5 8QP
G3 IGV J Birkbeck, 4 Tregullan View, Bodmin, PL31 1BH
G3 IGZ D Bruce, 22 Brownspring Drive, New Eltham, London, SE9 3JX
G3 IHX N Bond, 333 Hillandale Drive, Charlotte, United States, 28270
G3 IIN Michael Griffin, Michaelmas, Southdown Road, Freshwater, PO40 9UA
G3 IIO D Harriott, 23 Hamsey Crescent, Lewes, BN7 1NP
G3 IIV A Davies, Paarl, 129 Cotwall End Road, Dudley, DY3 3YQ
G3 IIW M Sands, Beech Lea, St. Marks Road, Tunbridge Wells, TN2 5LU
G3 IJA J Allan, 5 Terrington Court, Strensall, York, YO32 5PA
G3 IJL A Sephton, 16 Bloemfontein Avenue, Shepherds Bush, London, W12 7BL
G3 IJS J Stratfull, 55 Craigweil Lane, Aldwick, Bognor Regis, PO21 4XN
G3 IJU E Briggs, 32 Lethbridge Road, Wells, BA5 2FN
G3 IJV R Harvey, 16 Gatesgarth Close, Hartlepool, TS24 8RB
G3 IJW G Garrett, 226 Rydal Drive, Bexleyheath, DA7 5DG
G3 IKB D Giddens, 89 Pollards Oak Road, Oxted, RH8 0JE
G3 IKL R Craxton, 103 Clifton Road, Rugby, CV21 3QH
G3 IKN V Stagg, 7 The Hermitage, Warfield Street, Bracknell, RG42 6AS
G3 IKQ R Chilton, 80 Plantation Road, Hextable, Swanley, BR8 7SB
G3 IKX R Hall, Reen Manor Cottage, Perranporth, TR6 0AJ
G3 ILE E Marsh, 63 Willows Lane, Accrington, BB5 0SQ
G3 ILO S Spencer, 9 Vaisey Field, Whitminster, GL2 7pt
G3 IMW Samuel Whitfield, 7 Sir Alex Walk, Topsham, Exeter, EX3 0LG
G3 IMX E Jolliffe, 96 Cowes Road, Newport, PO30 5TP
G3 INP G Stanway, Bramble Edge Cottage, Bates Lane, Frodsham, WA6 9LL
G3 INQ Brian Podmore, 6 Alfred Court, Furlong Road, Bourne End, SL8 5AZ
G3 INR P Buchan, 79 Cavendish Avenue, Cambridge, CB1 7UR
G3 INU R Appleby, 14 Truro Court, Canterbury Way, Stevenage, SG1 4LF
G3 INY E Tudor, Mowhills House, 133 High Street, Bedford, MK43 7ED
G3 INZ J Tournier, Avalon, 13 Greenlands, High Wycombe, HP10 9PL
G3 IOB P Revell, 54 Lytham Road, Perton, Wolverhampton, WV6 7YY
G3 IOI N Pascoe, 118 London Road, Wickford, SS12 0AR
G3 IOJ B Rixon, 1 Carde Close, Hertford, SG14 2EU
G3 IOM R Chidzey, 8 Dormans Close, Dormansland, Lingfield, RH7 6RL
G3 IOR Patrick Gowen, 17 Heath Crescent, Norwich, NR6 6XD
G3 IPD C Oakley, 4 Cross Keys Lane, Low Fell, Gateshead, NE9 6DA
G3 IPG George Phipps, 12 Mill Close, Pulham Market, Diss, IP21 4TQ
G3 IPL R Winters, 43 Manor Close, Harpole, Northampton, NN7 4BX
G3 IPP M Dance, Golf Cottage, 8 St. Johns Road, Crawley, RH11 7BD
G3 IQF R Fowler, 49 Westhorpe Park, Westhorpe, Marlow, SL7 3RH
G3 IQX E Popplewell, 71 Thornbury Road, Southbourne, Bournemouth, BH6 4HU
G3 IQY Arthur Rees, 59 Hillside Gardens, Barnet, EN5 2NQ
G3 IRA John Wren, 29 Carisbrook Terrace, Chiseldon, Swindon, SN4 0LW
G3 IRQ P Rackham, Upyonda, Otley Bottom, Ipswich, IP6 9NG
G3 ISB Colin Brock, 24 Glebelands Road, Knutsford, WA16 9DZ
G3 ISD Edward Hatch, 147 Borden Lane, Sittingbourne, ME10 1BY
G3 IST S Turner, 8001 Bayshore Drive, Seminole, United States, 34646
G3 ISX Clifford Leal, 61 Light Oaks Avenue, Light Oaks, Stoke-on-Trent, ST2 7NF
GJ3 IT Jersey ARS c/o M Turner, 4 Le Clos Sara, St. Lawrence, Jersey, JE3 1GT
G3 ITB Thomas Bartlett, 19 Hardley Street, Hardley, Norwich, NR14 6BY
G3 ITF B Freeman, 47 Gorham Avenue, Rottingdean, Brighton, BN2 7DP
G3 ITH R Franklin, 2 Berkeley Drive, Kingswinford, DY6 9DX
G3 ITL J Humpelota, 76 Marlborough Road, Braintree, CM7 9LR
GM3 ITN L Hamilton, Halls Land, Cochno Road, Clydebank, G81 6NR
G3 ITT J Cairns, 2 Ffordd Tirion, Sychdyn, Mold, CH7 6DY
G3 IUB BIRMINGHAM A R S c/o David Cottam, 14 Barnard Close, Rednal, Birmingham, B45 9SZ
G3 IUC R Mcmillan, East Orchard, Almeley, Hereford, HR3 6LF
G3 IUE M Newell, 35 Ingleside Crescent, Lancing, BN15 8EN
G3 IUJ R Rogerson, 19 Martins Road, Shortlands, Bromley, BR2 0EE
G3 IUO G Allen, 157 Lynton Road, Bedminster, Bristol, BS3 5LN
G3 IUV G Loveday, 2 St. Aldwyns Close, Horfield, Bristol, BS7 0UQ
G3 IUW Laurie Pritchard, Green Horizons, Send Hill, Woking, GU23 7HR
G3 IUY J Presland, 6 PIPPIN CLOSE, Sutton Ely, CB6 2RX
G3 IUZ H Davis, 6 St. Thomas Terrace, Wells, BA5 2XG
G3 IVC A Sycamore, Fir Tree Cottage, Compton Valence, Dorchester, DT2 9ES
G3 IVF H Smith, Greenacres, The Green, Ashbourne, DE6 4NN
G3 IVH E Younge, 11 Charlottes, Washbrook, Ipswich, IP8 3HZ
G3 IVK D Evans, 11 Hill View, Bryn-y-Baal, Mold, CH7 6SL
G3 IVP A Page, 153 Westfield, Plymouth, PL7 2EQ
G3 IVQ E Brickstock, 3 Vauxhall Road, Stourbridge, DY8 1EX
GW3 IVR Norman Bromley, 176 Westbourne Road, Penarth, CF64 5BR
G3 IW BRITISH AEROSPACE ARS c/o W Wilkie, 14 Horseshoe Close, Northwood, Cowes, PO31 8PZ
G3 IWE A Wyse, 29 Tregainlands Park, Washaway, Bodmin, PL30 3AU
G3 IWH Ivor Hall, 46 Bushmead Road, Luton, LU2 7EU
G3 IWM M Holland, 7 Willans Court, Willans Drive, Newtown, SY16 4DB
G3 IWR James Parker, 47F 474 Castle Lane West, Bournemouth, BH8 9UD
G3 IWW Roland Hopkins, 34 Shelley Close, Abingdon, OX14 1PR
GM3 IWX W Ritchie, 8 Cheviot Place, Grangemouth, FK3 0DE
G3 IXI K Landon, 1 The Laurels, Leedons Park, Broadway, WR12 7HB
G3 IXN M Lovejoy, 75 Stoneham Lane, Swaythling, Southampton, SO16 2NZ
G3 IXZ Robert Rowden, 41 Brockington Road, Bodenham, Hereford, HR1 3LP
G3 IYI D Baker, Long Haul, 3 Chapel Lane, Lincoln, LN6 0EX
G3 IZA D Allison, 71 South Hill Road, Bromley, BR2 0RW
G3 IZD Ivan Davies, 13 Thurlow Way, Barrow-in-Furness, LA14 5XP
G3 IZF T Irvine, 24 Woodville Avenue, Crosby, Liverpool, L23 3BZ
G3 IZG J Wells, 23 Gainsborough Road, Blackpool, FY1 4DZ
G3 IZM J Harper Bill, 1 Shepherds Close, Staple Hill, Bristol, BS16 5LE
G3 IZQ H Hyman, 19 Black Horse Drive, Acton, United States, 1720
G3 IZW D Weaver, 7 Rolfe Drive, Burgess Hill, RH15 0LA
G3 JAF A Trigell, 15 Danecrest Road, Hordle, Lymington, SO41 0HZ
G3 JAL Roger Taylor, 304 Brigstock Road, Thornton Heath, CR7 7JE
G3 JAU C Davies, 107 Talbot Road, Bournemouth, BH9 2JE
G3 JBF L Brown, Ladygate, St. Michaels Road, Stafford, ST19 5AH
G3 JBJ F Mathers, 17 Penlon, Menai Bridge, LL59 5LR
GW3 JBZ J Brace, 12 Heol Gwili, Gorseinon, Swansea, SA4 4GE
G3 JCJ C Antrobus, 10 Rodger Road, Woodhouse, Sheffield, S13 7RH
G3 JCK F Chilvers, 5 Low Common Close, Foulsham, Dereham, NR20 5TW
G3 JCM David Bolwell, 3 Mildmays, Danbury, Chelmsford, CM3 4DP
G3 JCR K Smith, 20 Manor House Gardens, Abbots Langley, WD5 0DH
G3 JDD R Dobson, 16 Howden Road, Fulham, Australia, 5024
G3 JDM P Wright, 10 Hillcrest, Stafford, ST17 9YA
G3 JDO H Martin, 7 Nairn Street, Jarrow, NE32 4HX
G3 JDT B Read, Glenside, 4 Hatton Lane, Warrington, WA4 4BY
G3 JDY OBO ROYAL AIR FORCE ARS E RIDING AREA c/o Bernard Atkinson, 165 Alliance Avenue, Hull, HU3 6QY
G3 JFC B Stone, 12 Robertson Avenue, Leasingham, Sleaford, NG34 8NJ
G3 JFD B Brown, 130 Ashland Road West, Sutton-in-Ashfield, NG17 2HS
GM3 JFG STORNOWAY REPEATER GROUP c/o Jon Hague, Ocean View, 36b Lower Bayble, Stornoway, HS2 0QB
G3 JFR N Cottrell, 28 Colley Wood, Kennington, Oxford, OX1 5NF
G3 JFS Peter Cole, 25 Wardlow Gardens, Plymouth, PL6 5PU
G3 JFT B Dare, 128 Sancroft Road, Spondon, Derby, DE21 7ES
G3 JFW P Beevers, Hill Farm Granary, Lower Somersham, Ipswich, IP8 4PU
G3 JGA J Lawrence, 40 Aberconway Road, Prestatyn, LL19 9HL
G3 JGE V Owen, 17 Knowles Avenue, Prestatyn, LL19 8SG
G3 JGP Edward Robinson, 16 Shaw Green, Storth, Milnthorpe, LA7 7JB
G3 JHH S Burgess, 34 Redcliffe Road, London, SW10 9NJ
G3 JHI R Hathaway, 30 Berkeley Drive, Hornchurch, RM11 3PY
G3 JHP E Allen, 11 Newlands Close, Horley, RH6 8JR
G3 JHS Alan Wilson, Bardon, 7 Main Street, Loughborough, LE12 6RH
G3 JHU C Pavey, 3 Field Close, Chatham, ME5 9TD
G3 JIE Donald Youngs, 12 Fox Grove, East Harling, Norwich, NR16 2PS
GM3 JIG Kenneth Hodge, 66 Ardrossan Road, Seamill, West Kilbride, KA23 9LX
GM3 JIJ Jon Hague, Ocean View, 36b Lower Bayble, Stornoway, HS2 0QB
G3 JIP J Hill, Calle El Palomar 13, El Romeral, Malaga, Spain, 29130
G3 JIR J Hardcastle, 8 Norwood Grove, Rainford, St. Helens, WA11 8AT
G3 JIS R Heaton, 20 Tewkesbury Avenue, Urmston, Manchester, M41 0RJ
GD3 JIU Michael Thompson, 3 Close Cam, Port Erin, Isle of Man, IM9 6NB
G3 JIX K Smith, Staple Farm House, Durlock Road, Canterbury, CT3 1JX
GM3 JJQ D Millar, 51 Tiree Crescent, Polmont, Falkirk, FK2 0UX
G3 JJR J Rickwood, 44a The Bridle Path, Madeley, Crewe, CW3 9EL
G3 JJT C Kempson, 8 Arle Gardens, Cheltenham, GL51 8HR
G3 JKB D Simmonds, Parsonage Farm House, Binbrook, Market Rasen, LN8 6BN
GM3 JKC C Cooper, 28 Kippford Street, Glasgow, G32 9BW
G3 JKD Jack Davison, 6 Eden Close, Heighington Village, Newton Aycliffe, DL5 6RU
G3 JKE G Thomas, 13 Essex Drive, Taunton, TA1 4JX
G3 JKF Kenneth Franklin, 4 Princes Close, Seaford, BN25 2EW
G3 JKL John Lovell, Kowloon, Slade, Bideford, EX39 3LZ
G3 JKM Dennis Buckland, 102 St. Peters Road, Wiggenhall St. Peter, King's Lynn, PE34 3HF
G3 JKS F Claytonsmith, 16 Templand, Crossmichael, Castle Douglas, DG7 3BF
G3 JKV W Blanchard, The Trundle, Tower Hill, Dorking, RH4 2AN
G3 JKX Michael Street, 12 Ullswater Close, Priorslee, Telford, TF2 9RB
G3 JLK C Jeffery, Flat 23, Alexandra House, Weston Super Mare, BS23 3SH
G3 JLN F Blain, High Ridge, Howgate Lane, Bembridge, PO35 5QW
G3 JLQ B Thomas, 34 Barton Road, Market Bosworth, Nuneaton, CV13 0LQ
G3 JLZ V Ludlow, 6 Raleigh Crescent, Stevenage, SG2 0EQ
G3 JME M Watson, Kimberley, Chilsworthy, Holsworthy, EX22 7DQ
G3 JMJ Donald Nunn, Oak Lea, Crouch House Road, Edenbridge, TN8 5EL
G3 JMK D Hurrell, 3 Garfield Close, Bishops Waltham, Southampton, SO32 1AQ
GM3 JMM J Murdoch, 4 Cedar Drive, Milton of Campsie, Glasgow, G66 8AY
G3 JMX P Hayward, 31 Pinewood Avenue, Lowestoft, NR33 9AQ
G3 JMZ J Hilton, Windsor House, Preston Road, Chorley, PR7 5HH
G3 JNB Victor Brand, 8 Greenway, Campton, Shefford, SG17 5BN
G3 JNI J Pitcher, 63 The Chase, Holland-on-Sea, Clacton-on-Sea, CO15 5PZ
G3 JNJ Donald Platt, 22 Charcroft Gardens, Enfield, EN3 7HA
G3 JNM T Whittaker, 16 Acresdale, Lostock, Bolton, BL6 4PJ
G3 JNP J Halman, 131 Reservoir Road, Gloucester, GL4 6SX
GM3 JOB G Bryce, 3 West Bowhouse Way, Girdle Toll, Irvine, KA11 1NJ
G3 JOE Joseph Brown, 10 Park House Mews, Congleton Road, Sandbach, CW11 4SP
G3 JOR V Capell, Endways, 15 Copse Road, Bexhill-on-Sea, TN39 3UA
G3 JOT F Whatley, 1 Mill Close, Wroughton, Swindon, SN4 9AR
G3 JOX A Greaves, Jacobs Well, Woodhill Road, Chelmsford, CM2 7SF
GI3 JOZ J Williamson, 26 Avonbrook Gardens, Coleraine, BT52 1SS
G3 JPB Charles Noden, Brownhills Cottage Farm, Brownhills, Market Drayton, TF9 4BE
G3 JPQ D Parker, 6 Cambridge Road, Chingford, London, E4 7DP
G3 JPJ J Peerless, 101 Greenside, Borehamwood, WD6 4JD
G3 JPM B Grainge, 4 Maltings Close, Chevington, Bury St. Edmunds, IP29 5RP
G3 JPO M Fielding, 68 Mitford Road, South Shields, NE34 0EQ
G3 JPT C Reynolds, Beacon View, Bronwylfa Road, Welshpool, SY21 7RD
G3 JPU D Plant, Briarfields, Raby Crescent, Shrewsbury, SY3 7JN
G3 JPZ I Denney, 5 Howard Close, Harleston, IP20 9HY
G3 JQ A Webster, 8 Brookside Court, 142 Prestbury Road, Macclesfield, SK10 3BR
G3 JQG G Hawksworth, 16 Birkhead Street, Heckmondwike, WF16 0DE
G3 JQK E Jones, Appleton Thorne, Lower Broad Lane, Redruth, TR15 3HJ
G3 JQJ J Haggan, 22 Alnwick Road, Newton Hall, Durham, DH1 5NL
G3 JQS J Guttridge, Victoria Cottage, The Common, Cambridge, CB21 5LR
G3 JRD R Dancy, 1 Ladds Corner, Eastcourt Lane, Gillingham, ME7 2UW
G3 JRF F Thomas, 99 Eastfield Road, Wollaston, Wellingborough, NN29 7RS
G3 JRH P Horne, The Annexe, Burntwood Farm, Winchester, SO21 1AF
G3 JRK J Knight, 10 Lynton Drive, Burnage, Manchester, M19 2LQ
G3 JRL Frederick Armstrong, 4 Medway Drive, Preston, Weymouth, DT3 6LF
G3 JRM PYE LOWESTOFT RD c/o John Elsdon, 15 Union Road, Lowestoft, NR32 2BZ
G3 JRS Anthony Kidd, 35 Hollands Way, Kegworth, Derby, DE74 2GQ
G3 JRY Selwyn Auty, 3 Rochford Crescent, Boston, PE21 9AE
G3 JSA D Wilcox, 13 Richards Drive, Dartmouth, Canada, NS B3A 2P1
G3 JSG John Gunn, Flat 18, Bro Llewelyn, Penrhyndeudraeth, LL48 6AL
G3 JSJ D Pritchard, 19 Thorne Close, Verwood, BH31 6QG
G3 JSK D Dean, 8 Bradford Road, Corsham, SN13 0QR
G3 JSR P Chapman, 27 Ilfracombe Gardens, Romford, RM6 4RL
G3 JSU L Sampson, 107 South Street, Lancing, BN15 8AS
G3 JSV D Holmes, Fair Oaks, Berriew, Welshpool, SY21 8AU
G3 JTJ J Jones, Westerland, 1 Roborough Close, Plymouth, PL6 6AH
G3 JTK G Allen, 119 Haymoor Road, Poole, BH15 3NR
G3 JTO F Gell, 93 Pasture Road, Stapleford, Nottingham, NG9 8HR
G3 JTQ R Griffiths, 7 Dever Way, Oakley, Basingstoke, RG23 7AQ
G3 JTT Peter Thompson, 30 Farnol Road, Birmingham, B26 2AF
G3 JUL G Voller, 56 Marlborough Road, Ashford, TW15 3QA
G3 JUU D Adams, 23 Arlington Gardens, Attleborough, NR17 2NH
G3 JUW C Lovell, 5 Montpelier Road, Ilfracombe, EX34 9HP
G3 JUX J McFarlane, 141 Tyler Grove, Aston, Stone, ST15 0JA
G3 JVC J Cleeve, 44 Ditton Hill Road, Long Ditton, Surbiton, KT6 5JD
G3 JVL M Walters, 26 Fernhurst Close, Hayling Island, PO11 0DT
GD3 JVM R Medcraft, 134 Dulverton Road, Ruislip, HA4 9AG
G3 JVN D Keen, 14 Penina Avenue, Newquay, TR7 2LE
G3 JVP V Purdy, 99 Belmont Road, Uxbridge, UB8 1QX
G3 JVR D Nokes, 16 Salisbury Grove, Giffard Park, Milton Keynes, MK14 5QA
G3 JWH J Harber, 7 Balmoral Road, Hornchurch, RM12 4NR
G3 JWI R Page-Jones, 34 Edwards Way, Hutton, Brentwood, CM13 1BT
G3 JWN F Walker, 2 Croft Place, Brighouse, HD6 4AP
G3 JWQ B Maycock, Hill House, Bullock Lane, Alfreton, DE55 4BP
G3 JXC C Gregory, 51 Calgary Avenue, Blackburn, BB2 7DS
G3 JYG J Kirby, 14 Grovelands Road, Hailsham, BN27 3BZ
G3 JYS R Finch, 8 Chalfont Close, Allesley, Coventry, CV5 9HL
G3 JZF John Smith, 17A, Sutton Coldfield, B74 2QA
G3 JZL W Montford, 3 The Close, Main Street, Coventry, CV8 3JF
G3 JZP J Hodgkins, Bridge House, Hunton, Bedale, DL8 1PX
G3 JZT R Cheetham, 7 Parkway, Stockport, SK3 0PX
G3 KAC UNIVERSITY OF BRISTOL ARC c/o Kenneth Stevens, 20 Coberley, Bristol, BS15 8ES
G3 KAE John Rowley, 41 Main Street, East Ayton, Scarborough, YO13 9HL
G3 KAF J France, Cobham Cottage, 34 Ladythorn Road, Stockport, SK7 2ER
G3 KAG Anthony Parker, Hillside, Roston, Ashbourne, DE6 2EH
G3 KAJ D Jagger, Briarside, Gorn Road, Llanidloes, SY18 6DQ
G3 KAK J Hunter, 14 Curlew Rise, Gretna, DG16 5LB
G3 KAN A Shrewsbury, 36 Winchester Road, Delapre, Northampton, NN4 8AY
G3 KAP R Taylor, 2 Brenchley Mews, Charing, Ashford, TN27 0JQ
G3 KAR D Hammond, Christen Mares, Willersey Hill, Broadway, WR12 7PF
G3 KAU L Laszkiewicz, 38 Langley Lane, Ifield, Crawley, RH11 0NA
GW3 KAX Geoffrey MacKrell, Preseli, Newchapel, Boncath, SA37 0EH
G3 KBH M Hughes, Northdean, Brimstone Lane, Gravesend, DA13 0BW
G3 KBI T Waller, 12 Skelton Road, Brotton, Saltburn-by-The-Sea, TS12 2TJ
GM3 KBP A Kerr, 47 Hillpark Avenue, Edinburgh, EH4 7AH
G3 KBR R Huntsman, 23 Worts Causeway, Cambridge, CB1 8RJ
G3 KBS N Coupe, 19 Hillary Drive, Crowthorne, RG45 6QF
GM3 KC MONTROSE AMATEUR RADIO STATION c/o B Murray, Sherwood Cottage, Farnell, Brechin, DD9 6UH
G3 KCB B Green, 18 Kenilworth Road, Sale, M33 5FB
G3 KCD P Bedwell, 3 Roman Way, Dibden Purlieu, Southampton, SO45 4RP
G3 KCF Rodney Kent, 8 Woodfield Lane, Stowmarket, IP14 1BN
G3 KCG D Tyerman, 20 Grace Gardens, Bishop's Stortford, CM23 3EX
GW3 KCQ John Williams, Y-Fedw, Cwmann, Lampeter, SA48 8DT
G3 KCT D Blythe, 6 Penn House, Mallory Street, London, NW8 8SX
G3 KCV J Saunders, 7 Stone Lane, Yeovil, BA21 4NN
GM3 KCY G Buchanan, 30 Gilmour Avenue, Clydebank, G81 6AW
G3 KCZ W Siertsema, 21 Rowles Close, Kennington, Oxford, OX1 5LX
GW3 KDB P Miles, Y Gorlan, Cross Inn, Llandysul, SA44 6NP
G3 KDD V Barrett, 2 Carlisle Close, Sandy, SG19 1TX
G3 KDE P Cheesman, Laroch Post Office, Lane North Mundham, Chichester, PO20 1JY
G3 KDP A Bounds, 32 Tregwary Road, St. Ives, TR26 1BL
G3 KDU M crawford, 95 Victoria Avenue, Princes Avenue, Hull, HU5 3DW
G3 KDW H Turnbull, 13 Linden Court, Wessex Road, Southampton, SO18 3RB
G3 KDY Roy Folgate, Stile Cottage, Wilkinson Drive, Market Rasen, LN8 3LD
G3 KEG C Rogers, 100 Sparth Road, Clayton le Moors, Accrington, BB5 5QD
G3 KEK G Carr, 88 Woodrow Crescent, Knowle, Solihull, B93 9EQ
G3 KEL R Bray, Croft House, Blencogo, Wigton, CA7 0BZ

UK Callsigns

**IMPORTANT NOTE**

**Revalidate licence to avoid revocation** – Ofcom has advised the Society that plans will be drawn up to revoke licences that have not been revalidated as required by the licence conditions. The quickest way to revalidate is to do so online via the Ofcom website: *https://services.ofcom.org.uk/* or by email: *amateur.validations@ofcom.org.uk* Ofcom staff are available to help, but please be patient during times of heavy workload.

G3 KEP David Pratt, 11 Moorleigh Close, Kippax, Leeds, LS25 7PB
G3 KEQ J Mitchell, Chellow Dene, Viewlands Avenue, Westerham, TN16 2JE
G3 KEV M Hamilton, 2 Wordsworth Close, Scalby, Scarborough, YO13 0SN
GM3 KEZ J Little, 33 Manor Court, Forfar, DD8 1AD
G3 KFB N Parkinson, 16 Collinson Avenue, Scunthorpe, DN15 8AB
G3 KFG H Taylor, 17 Rose Acre Road, Littlebourne, Canterbury, CT3 1SY
G3 KFP Alan Olds, 43 Fourth Avenue, Teignmouth, TQ14 9DT
G3 KFS D Preston, Ashfield, Tanworth Lane, Redditch, B98 9EH
G3 KFU P Barry, 21 Old Pasture Road, Frimley, Camberley, GU16 8SA
G3 KGA A Morrison, 30 Sullington Gardens, Worthing, BN14 0HR
GW3 KGI Malcolm Bowen, 24 Parklands View, Sketty, Swansea, SA2 8LX
G3 KGM D Maclennan, 12 Windermere Close, Aylesbury, HP21 7HP
G3 KGN A Edwards, 48 Fillebrook Avenue, Leigh-on-Sea, SS9 3NT
G3 KGP M Palmer, Fairways, 8 Gwealdues, Helston, TR13 8JZ
G3 KGT J Nicolson, 24 Pottersfield Road, Woodmancote, Cheltenham, GL52 9PY
G3 KGV Kenneth Bates, 5 Ffordd Nant Goch, Llangadfan, Welshpool, SY21 0PW
GM3 KHH W Cecil, Innes House, Oran, Buckie, AB56 5EP
G3 KHK D Connolly, Jonquil, 6 Sanderson Mews, Colchester, CO5 7HF
G3 KHQ A Langley, 58 Dumbarton Road, Brixton Hill, London, SW2 5LU
G3 KHR J Fox, 25 Langdale Crescent, Bexleyheath, DA7 5DZ
G3 KHU Ralph Gabbitas, 12 Thornyville Drive, Oreston, Plymouth, PL9 7LF
G3 KHZ D Cox, 18 Station Road, Castle Bytham, Grantham, NG33 4SB
G3 KII G Lively, 9 Wilson Road, Shurdington, Cheltenham, GL51 4SN
G3 KIJ E Lugmayer, 17 Borough End, Beccles, NR34 9YW
G3 KIL R Messer, The Shambles, Swinbrook Road, Oxford, OX18 1DX
G3 KIP K Grover, 1 Powdermill Close, Tunbridge Wells, TN4 9DR
G3 KIQ J Elliot, 2 Pennine Close, Blackley, Manchester, M9 6HR
G3 KIW Geoffrey Jenner, Pogles Wood Cottage, Paradise Lane, Reading, RG7 6NU
G3 KJC R Church, Three Birches, Sandy Close, Thatcham, RG18 9QP
GM3 KJE J Scott, 5 Garthdee Terrace, Aberdeen, AB10 7JE
GM3 KJI E Pollard, 265 Eldon Street, Greenock, PA16 7QE
G3 KJK L Wilkes, Parc Crane, Penmenner Road, Helston, TR12 7NN
G3 KJN I Winter, 5 Uwch Y Nant, Mynydd Isa, Mold, CH7 6YP
G3 KJO HUDDERSFIELD TECHNICAL COLLEGE ARS c/o Mark Lupton, 18 The Paddock, Kirkheaton, Huddersfield, HD5 0ER
G3 KJS William Smith, 32 Lumley Road, Chester, CH2 2AQ
G3 KJW P Allely, Dwyfor, Rhiw, Pwllheli, LL53 8AE
G3 KJX B Alderson, 43 Brompton Road, Northallerton, DL6 1ED
G3 KJY John York, 48 Browhead Court, Shackleton Street, Burnley, BB10 3DS
GM3 KJZ G Paterson, 3 Ferry Barns Court, North Queensferry, Inverkeithing, KY11 1ET
G3 KKC A Rumbelow, 7 Hoof Close, Littleport, Ely, CB6 1HU
G3 KKD I Waters, 39 Stow Road, Stow-cum-Quy, Cambridge, CB25 9AD
G3 KKJ Alexander Shannon, 23 Glebeland Close, West Stafford, Dorchester, DT2 8AE
G3 KKP J Burgess, Moorend, Main Street, Leeds, LS20 8NX
G3 KKZ P Champion, The Lodge, Tydcombe Road, Warlingham, CR6 9LU
G3 KLC J Bennett, Koivula, Station Road, Boston, PE20 3QT
G3 KLD R Russell, 43 Ingestre Road, Hall Green, Birmingham, B28 9EQ
G3 KLF I Crowther, 3 Glenelg, Fareham, PO15 6JU
G3 KLH David Alexander, 20 Deans Court, Milford on Sea, Lymington, SO41 0SG
G3 KLK B Page, 7 Marconi Way, Southall, UB1 3JP
G3 KLN N Whittaker, 111 Burnley Road, Colne, BB8 8DT
G3 KLP James Young, Woodglades, 34 The Demesne, Ashington, NE63 9TP
G3 KLT Ivan Eamus, Hampden House, Cottesmore Road, Oakham, LE15 7LJ
G3 KLV Gordon Vine, 4 Tollgate Close, Northampton, NN2 6RP
G3 KLY M Lloyd, 26 Oaklands Avenue, Newcastle, ST5 0EX
G3 KLZ D Enoch, 7a Mount Road, Evesham, WR11 3HE
G3 KMA R Balister, La Quinta, Mimbridge, Woking, GU24 8AR
G3 KMD J Bass, 3 Tennyson Avenue, Grays, RM17 5RG
G3 KME L Pennell, 182 Northampton Road, Wellingborough, NN8 3PJ
GM3 KMF G J Robbins, 39 Locheil Gardens, Glenrothes, KY7 6YL
G3 KMG D Plumridge, Rose Cottage, Castleside, Consett, DH8 9AP
G3 KMI Southampton University Wireless Society c/o Philip Crump, 99a Mayfield Road, Southampton, SO17 3SY
G3 KML Robert Whitfield, 42 Greenwood, Tweedmouth, Berwick-upon-Tweed, TD15 2EB
G3 KMM John Crowther, 15 Chemin De Bausses, Villelongue D'Aude, Limoux, France, 11300
G3 KMO Michael Birch, 12 The Heath, Hevingham, Norwich, NR10 5QW
G3 KMQ R Heslop, Fairways, Meadow Drive, Bude, EX23 8HZ
G3 KMS D Swain, 3 Nevy Fold Avenue, Horwich, Bolton, BL6 6QG
G3 KMV R Birchall, Willow Tree House, Poole, Nantwich, CW5 6AL
G3 KMY Dennis Radford, Glenburn, Brassington Lane, Chesterfield, S42 6LB
G3 KND John Hardy, Vogelenzang, 1b Roberts Road, Aldershot, GU12 4RD
G3 KNG A Embrey, 59 Oaken Lanes, Codsall, Wolverhampton, WV8 2AW
G3 KNJ John Otter, 7 Longacre Road, Dronfield, S18 1UQ
G3 KNU P Jackson, 7 Ferriby Road, Scunthorpe, DN17 2EQ
G3 KNZ A Eccles, 78 Uplands Avenue, Connah's Quay, Deeside, CH5 4LG
G3 KOA T Robinson, 32 Campbell Crescent, East Grinstead, RH19 1JR
G3 KOB R Goodman, 8 Decouttere Close, Church Crookham, Fleet, GU52 0UR
G3 KOD P Kay, 7 St. Regis Close, London, N10 2DE
G3 KOJ R Ezra, 39 Buckland Close, Waterlooville, PO7 6ED
G3 KOM Francis Foulkes, Bankside Bungalow, Maidstone Road, Tonbridge, TN12 7HA
G3 KOQ B Parker, 9 Yewdale Avenue, Heysham, Morecambe, LA3 2LR
G3 KOS B Faithfull, 68 Lampton Road, Long Ashton, Bristol, BS41 9AQ
G3 KOX N Waite, 7 Lanercost Close, Welwyn, AL6 0RW
G3 KOZ W Henderson, 9 Chiselbury Grove, Salisbury, SP2 8EP
G3 KPO Rosemary Amateur Radio Group c/o A Thornton, Little Gables, Rosemary Lane, Ryde, PO33 2UX
G3 KPU Eric Prince, 9 Alwyn Road, Thorne, Doncaster, DN8 5JG
G3 KPV J Killeen, 10 Den Brook Close, Torquay, TQ1 3TP
G3 KQF Jack Anthony, 77 Brayfield Road, Littleover, Derby, DE23 6GT
G3 KQG E James, The Meadows, Kennford, Exeter, EX6 7TZ
G3 KQQ C Mattacks, 68 Middlesex Drive, Bletchley, Milton Keynes, MK3 7EU
G3 KQR D Foster, 56 Elmbridge Avenue, Tolworth, Surbiton, KT5 9HA
G3 KQV J Ryley, 30 St. Helens Drive, Leicester, LE4 0GS
G3 KQY R Disley, 6 St. Margarets Road, Farington, Leyland, PR25 4XT
GD3 KRT G Hodges, 102 Torrington Road, Ruislip, HA4 0AU
G3 KRW K Whelan, Killiney, Longsplatt, Corsham, SN13 8DF
G3 KRX W Addy, 14 Cresttor Road, Liverpool, L25 6DW
G3 KRZ John Greenwood, 36 Minster Court, Bracebridge Heath, Lincoln, LN4 2TS
G3 KSF R Harper, 21 Howard Oliver House, Harvey Gardens, Southampton, SO45 3LS
G3 KSP P Hooper, 1 Victoria Mews, Morecambe, LA4 5QD
G3 KTA P Munt, 130 Chipstead Way, Woodmansterne, Banstead, SM7 3JR
G3 KTC R Tucker, 7 Knights Close, Ashby-de-la-Zouch, LE65 2JQ
G3 KTH M Darkin, 3 Adrian Close, Shell, Droitwich, WR9 7AY
G3 KTI M Rees, Blue Pillars, 6 Grove Crescent, Coleford, GL16 8AZ
G3 KTJ Gerald Rigby, 30a Pimbo Lane, Upholland, Skelmersdale, WN8 9QQ
G3 KTM R Atthill, Heronsway, West Street, Salisbury, SP3 4AH
G3 KTP Derek West, 84 High Street, Castle Donington, Derby, DE74 2PQ
G3 KTT M Gallon, 41 Dene Gardens, Newcastle upon Tyne, NE15 8RL
G3 KTU J Ault, 63 Baring Road, Bournemouth, BH6 4DT
G3 KTZ C Lindsay, 38 St. Vincents Close, Littlebourne, Canterbury, CT3 1TZ
G3 KUD J Duncan, 9 Springhill Close, Westlea, Swindon, SN5 7BG
G3 KUE PRESTON ARS c/o A McPhail, 300 Fletcher Road, Preston, PR1 5HJ
G3 KUL D Stephenson, 16 Chard Court, Fortfield Road, Bristol, BS14 9NL
GI3 KVD D Jones, 5 Whitehill Park, Limavady, BT49 0QF
G3 KVE T Wright, 24 Stuart Road, Bootle, L20 9ET
G3 KVG J Charles, 87 Lees Hall Road, Sheffield, S8 9JL
G3 KVJ S Tomlinson, 31 The Quarry, Alwoodley, Leeds, LS17 7NH
G3 KVP D Kitchen, Folkingham Place, Market Place, Sleaford, NG34 0SE
G3 KVR S Davis, 3 Coronation Road, Banwell, BS29 6AZ
G3 KVT A Smith, Winston House, Felthorpe Road, Norwich, NR9 5TF
G3 KVX R Pattinson, 15 Maes Yr Eglwys, Llansantffraid, SY22 6BE
G3 KWB R Neville, 35 Beechcroft Road, Newport, NP19 8AG
G3 KWJ N Valentine, The White House, Dene Road, Ashtead, KT21 1EB
G3 KWK R Nolan, 6 Plymouth Close, Redditch, B97 4NP
G3 KWN W Nicoll, Yonder, Milldown Road, Blandford Forum, DT11 7DE
G3 KWO K Dawson, 44 Avondale Road, Darwen, BB3 1NS
G3 KWW R Wilkinson, 83 Palewell Park, London, SW14 8JJ
G3 KWY A Swain, 17 Anson Road, Shepshed, Loughborough, LE12 9LA
G3 KXB D Pantony, 71 South Street, Whitstable, CT5 3EJ
G3 KXE E Bettles, 15 St. Francis Avenue, Southampton, SO18 5QL
G3 KXF Donald Wallis, 17 Upper Belgrave Road, Seaford, BN25 3AD
G3 KXI D Keeler, 16 Honeysuckle Way, Witham, CM8 2XG
GM3 KXQ S Floyd, 3 Crarae Place, Newton Mearns, Glasgow, G77 6XX
G3 KXS H Perry, 688 Durham Road, Madison, United States, 6443
G3 KXV V Johnston, 9 Holbeck Avenue, Middlesbrough, TS5 8DR
GW3 KXX R Weaver, 59 Broad St., Leckwith, Cardiff, CF1 8BZ
G3 KYE J Orr, 102 Manor House Lane, Yardley, Birmingham, B26 1PR
G3 KYF K Sullivan, 14 Wigston Road, Blaby, Leicester, LE8 4FU
G3 KYM H Stamper, The Bungalow, School Hill, St Austell, PL26 7TP
GI3 KYP A Patterson, 24 Cyprus Avenue, Belfast, BT5 5NT
G3 KYZ D Clarke, Primrose Mount, Old Neighbourhood, Stroud, GL6 8AA
G3 KZB Murray Ward, Flat 14, Meadrow Court, Godalming, GU7 3HG
G3 KZC R Harknett, 28 Woodyleaze Drive, Hanham, Bristol, BS15 3BY
G3 KZE J Davies, 45 Dahn Drive, Ludlow, SY8 1XZ
G3 KZG Arthur Bills, Brooklands, 2 The Acre, Stourbridge, DY7 6HW
G3 KZN D Blakeley, 338 Thong Lane, Gravesend, DA12 4LQ
GW3 KZO M Dennis, Esgair Newydd, Tresaith, Cardigan, SA43 2JG
G3 KZR I Davies, Lusty Hill Farm, Lusty Gardens, Bruton, BA10 0BS
G3 KZT A James, 143 Gaer Park Drive, Newport, NP20 3NS
G3 KZU Michael Dolan, 15 Ringwood Road, Headington, Oxford, OX3 8JB
G3 KZX L Loveland, 21 Roseland Close, Keyworth, Nottingham, NG12 5LQ
G3 KZZ D Forster, 281 Mortimer Road, South Shields, NE34 0DR
G3 LAA Anthony Sedman, 69 Beechwood Avenue, Locking, Weston-Super-Mare, BS24 8DS
G3 LAG H Gow, 43 Ringstead Crescent, Weymouth, DT3 6PT
G3 LAI Graham Livingston, 24 Duncannon Drive, Falmouth, TR11 4AQ
G3 LAS John Butcher, The Stables, Priory Farm, Lincoln, LN4 3SL
G3 LAU F Adkin, 5 Cosway Mansions, Shroton Street, London, NW1 6UE
GM3 LAW W Walker, 45 Watts Gardens, Cupar, KY15 4UG
G3 LAZ R Gerrard, RNIB Wavertree House, 211 Somerhill Road, Hove, BN3 1RN
G3 LBM A Mulcahy, 8 Old Barn Close, Winkleigh, EX19 8JX
G3 LBS G Cleeton, 24 Severn Drive, Newcastle, ST5 4BH
G3 LCF P Baldwin, 49 King George Vi Mansions, Court Farm Road, Hove, BN3 7QX
G3 LCH M Pharaoh, 1 Madeira Road, Mitcham, CR4 4HD
G3 LCI H Young, 23 Willow Grove, Wirral, CH46 0TU
G3 LCL A Baylis, Queen Oak Inn, Bourton, Gillingham, SP8 5AL
G3 LCQ M Williams, Dwyros, 12 Penrhos Avenue, Llandudno Junction, LL31 9EL
G3 LCS D Shepherd, 35 The Crescent, Haversham, Milton Keynes, MK19 7AN
G3 LCY J Tamlin, 53 Hele Gardens, Plympton, Plymouth, PL7 1JY
G3 LDC John Phillips, 9 Trelawny Close, Usk, NP15 1SP
G3 LDG B Gee, Daisy Bank, Carlton Road, Bedford, MK43 7JL
G3 LDI Roger Cooke, The Old Nursery, The Drift, Norwich, NR14 8LQ
G3 LDJ K Day, 45 Thick Hollins Drive, Meltham, Holmfirth, HD9 4DR
G3 LDY R Taylor, Flat 2, 2a Moorlands Road, Budleigh Salterton, EX9 6AG
GI3 LEG D Wilson, 189 Cregagh Street, Belfast, BT6 8NL
G3 LEK L Kitching, Woodsyde Lower Road, Harmer Hill, Shrewsbury, SY4 3QX
G3 LEO G Brigham, The Manor, Carthorpe, Bedale, DL8 2LF
G3 LEQ Gordon Adams, 2 Ash Grove, Knutsford, WA16 8BB
G3 LET Peter Hobbs, Middle House, Tilgate Forest Lodge, Crawley, RH11 9AF
GW3 LEW G Weale, Winfield, Templeton, Narberth, SA67 8SP
GD3 LFD R Widders, 82 Azalea Walk, Eastcote, Pinner, HA5 2EH
G3 LFE Raymond Clark, Flat 44, Townsend Court Green Lane, Leominster, HR6 8TD
GJ3 LFJ H Mesny, La Trigale, Route de L'Eglise, St. Lawrence, Jersey, JE3 1LA
G3 LFR M Everett, Thrupp Wharf, Cosgrove, Milton Keynes, MK19 7BE
G3 LFV R Manser, 39 Long Meadow, Markyate, St. Albans, AL3 8JN
G3 LFX D Pedder, 37 Hersham Road, Walton-on-Thames, KT12 1LE
G3 LGA M Hayward, Brindle, Romsey Road, Stockbridge, SO20 8DB
G3 LGF G Falding, 10 Angel Court, Shaftesbury, SP7 8HX
G3 LGK B Sandall, Amber Croft, Main Road, Alfreton, DE55 6EH
GD3 LGQ P Marsden, 49 Southfield Park, North Harrow, Harrow, HA2 6HF
G3 LGR M Hooles, 114 Cassiobury Drive, Watford, WD17 3AQ
G3 LGT J Tate, Pine Holt, 34 Queens Road, Fleet, GU52 7LE
GM3 LGU Robert Pryde, 5 Dobson's Walk, Haddington, EH41 4RU
G3 LGV R Hardman, 80 Green Lane, Oldham, OL8 3BA
G3 LGW D Spencer, Paladin, 89 Watling Street, Tamworth, B78 3DE
G3 LHB W Blanchard, 46 King George Avenue, Loughborough, LE11 2NU
G3 LHG E Smith, 3 Meadow Avenue, Wetley Rocks, Stoke-on-Trent, ST9 0BD
G3 LHI Alan Smith, 110 Broadstairs Road, Broadstairs, CT10 2RU
G3 LHJ Derrick Webber, 43 Lime Tree Walk, Milber, Newton Abbot, TQ12 4LF
GW3 LHK G Griffiths, Glyndwr, Lampeter Road, Aberaeron, SA46 0ED
G3 LHN R Muir, 19 Eastwick Drive, Bookham, Leatherhead, KT23 3PY
G3 LHS L Matthews, 14 Mayforth Gardens, Ramsgate, CT11 0LL
G3 LHU Michael Dixon, 20 Dunster Gardens, Cheltenham, GL51 0QT
G3 LHZ M Underhill, Hatchgate, Tandridge Lane, Lingfield, RH7 6LL
G3 LIK Michael Puttick, 21 Sandyfield Crescent, Cowplain, Waterlooville, PO8 8SQ
G3 LIO J Gibbs, 13 Bromley Road, Macclesfield, SK10 3LN
G3 LIV John Melvin, 2 Salters Court, Newcastle upon Tyne, NE3 5BH
GM3 LIW A Wood, 2 Palmer Place, Birkhill, Dundee, DD2 5RB
G3 LJD J Davies, 57 Madeira Court, Knightstone Road, Weston Super Mare, BS23 2BH
G3 LJR Timothy Saxton, 38 William Street, Dalbeattie, DG5 4EN
GW3 LJS T Bloxam, 15 Cleveland Avenue, Mumbles, Swansea, SA3 4JD
G3 LKV Daniel Locke, 4 Glebe Close, Doveridge, Ashbourne, DE6 5NY
G3 LKW D Wiltshire, 71 Ferndale, Waterlooville, PO7 7PG
GM3 LKY Philip Cohen, Tigh Ruadh, Tain, IV19 1NF
G3 LLD S Collier, 64 Slonk Hill Road, Shoreham-by-Sea, BN43 6HY
G3 LLE K Webster, 25 Carlin Gate, Blackpool, FY2 9QT
G3 LLG R Loveday, 42 Bridle Path, Woodcote, Reading, RG8 0SE
G3 LLJ A Hodgkinson, 46 Lamberts Lane, Congleton, CW12 3AU
G3 LLK J Gale, 66 Burys Bank Road, Crookham Common, Thatcham, RG19 8DD
GM3 LLP Bernard Watson, 4 Caldwell Road, West Kilbride, KA23 9LE
G3 LLV J Mcelvenney, 10 Bignor Place, Sheffield, S6 1JE
G3 LLZ Dennis Goacher, 27 Glevum Road, Swindon, SN3 4AA
G3 LME K Taylor, 7 Bowen Close, Cheltenham, GL52 5EG
G3 LMH Robert Wellbeloved, 8 Orchard Close, South Wonston, Winchester, SO21 3EY
G3 LMQ James Hamer, 7 Arundel Road, Coventry, CV3 5JT
G3 LMR John Eley, 25 Peckleton View, Desford, Leicester, LE9 9QF
G3 LMX T Mitchell, 27 Hanmer Road, Simpson, Milton Keynes, MK6 3AY
G3 LNL P Lovelady, 14 Maunders Court, Liverpool, L23 9YU
G3 LNM Raymond Scrivens, 6 Highland Close, Cantley, Norwich, NR13 3SW
G3 LNN John Symes, 19 Boundary Close, Kirkby-in-Ashfield, Nottingham, NG17 8RS
G3 LNP Anthony Preedy, 2, Worthen, Shrewsbury, SY5 9JQ
GW3 LNR A Gwynne, 77 Edward St., Pant, Merthyr Tydfil, CF48 2BB
G3 LNS G Beasley, PO Box 1344, Paphos, Cyprus
G3 LNW J McGuire, 5 Primrose Way, Trevadlock Hall Park, Launceston, PL15 7PW
G3 LOD D Rowse, 48 Oatlands Avenue, Bar Hill, Cambridge, CB23 8EQ
G3 LOE W Roberts, 13 Brean Road, Stafford, ST17 0PA
G3 LOF G Peskett, 13 Warneford Road, Oxford, OX4 1LT
G3 LOV M Francis, Cherry Tree Cottage, Atlantic Close, Tintagel, PL34 0EL
G3 LOX B Johnson, Rivendel Cottage, Manor Farm, Dorchester, DT2 0JJ
G3 LPC Weald ARS c/o R Evans, 30 Chandler Close, Bampton, OX18 2NW
G3 LPL P Sherdley, 2 Stable Yard, Taylors Lane, Preston, PR3 6AP
G3 LPN J Hunt, 28 Robins Bow, Camberley, GU15 3NR
G3 LPS E Pickering, 7 Hob Green, Mellor, Blackburn, BB2 7EP
G3 LPT G Woods, Bamburgh House, Hunston, Bury St. Edmunds, IP31 3EN
G3 LPU E Burrell, 27 Blandford Avenue, Oxford, OX2 8EA
GU3 LPV A Catts, PO Box 1029, 7 Little Street, Alderney, Guernsey, GY9 3JD
G3 LPY Richard Nye, Beech Cottage, Gorelands Lane, Chalfont St. Giles, HP8 4HQ
G3 LQB Kenneth Bishop, Friedrich-Ebert Strasse 23, Osterode, Germany, 37520
G3 LQC R Evans, 30 Chandler Close, Bampton, OX18 2NW
GW3 LQE A Ernest, 6 Kymin Terrace, Penarth, CF64 1WW
G3 LQJ R Cox, 12a Kelling Close, Holt, NR25 6RU
G3 LQO E Harris, 10 Girdle Road, Walsworth, Hitchin, SG4 0AN
G3 LQP R Brown, 262 Fir Tree Road, Epsom, KT17 3NL
G3 LQR S Freeman, West Farm, Cransford, Woodbridge, IP13 9PQ
G3 LQW K Wallace, 55 Lamborne Road, Leicester, LE2 6HQ
G3 LQX M Nicholls, 6 Lyme Bay Road, Teignmouth, TQ14 8RS
GI3 LQY John Stronach, 20 Monaville Drive, Lisburn, BT28 2DR
G3 LRA C Eley, 1 Hilldale View, Gaisgill, Penrith, CA10 3UE
GM3 LRG James Gray, 47 South Street, Greenock, PA16 8QG
G3 LRH G Frampton, 1 Ludlow Road, Church Stretton, SY6 6DD
G3 LRI J Blakey, 10 Wilson Terrace, Newcastle upon Tyne, NE12 7JP
G3 LRL R Bowell, 16 Margarite Way, Wickford, SS12 0ER
G3 LRP P Ackley, Camelot, Greenside, Wakefield, WF4 2BG
G3 LRQ M Humphries, 2 South View Close, Twyford, Reading, RG10 9AY
G3 LRS OBO LEICESTER RS c/o Roger Talbott, 33 Highfield Street, Anstey, Leicester, LE7 7DU
G3 LRU J Miller, 57 Clarendon Villas, Hove, BN3 3RE
G3 LRX R Durell, Middleton Farm, Hubbards Hill, Lenham, ME17 2EJ
G3 LSA D Moore, 5 Seahaven Springs Estate, Seaholme Road, Mablethorpe, LN12 2QS
GD3 LSF Edward Ellis, Ballahams, 3 Glen Road, Laxey, Isle of Man, IM4 7AP
G3 LSJ C Gerrard, 6 Bridle Close, Sleaford, NG34 7TD
G3 LSQ P Aitchison, Upper Weston House, Cot Lane, Chichester, PO18 8SU
G3 LST Peter Clarke, Half Moon House, Church Street, Colchester, CO6 4QH
G3 LSX G Townsend, 21 Grange Avenue, East Barnet, Barnet, EN4 8NJ
G3 LTF Peter Blair, Woodleigh, Upper Wyke, Andover, SP11 6EA
G3 LTK Christopher Kenny, 4 Midhurst Close, Ferring, Worthing, BN12 5BP
G3 LTM B Moyler, 1 Bay Walk, Aldwick, Bognor Regis, PO21 4ET
G3 LTX Robert Savage, Plas Gwyntog, Rhoslefain, Tywyn, LL36 9ND
G3 LUA Alan Knowles, 73 Kingslea Road, Solihull, B91 1TJ
G3 LUC E Bate, 5 Elm Road, Shildon, DL4 1BH
G3 LUH K Reader, 21 Broadwater Avenue, Poole, BH14 8QY

G3 LUK 59 SQUADRON AIR c/o Ian Halliwell, 61 Cliffe Road, Shepley, Huddersfield, HD8 8AG
G0 LUN SPECIAL COMMUNICATIONS (TA) ASSOCIATION c/o D Smith, The Old Forge, High Street, Newmarket, CB8 0SP
G3 LUO Christopher Evans, High Haycote, Gawthrop, Sedbergh, LA10 5QH
G3 LUW B Whittaker, Woodlands, Newton Down, Lifton, PL16 0AS
G3 LUZ Fred Machin, 70 Poors Lane, Hadleigh, Benfleet, SS7 2LN
GM3 LVA D Simpson, Larchwood, Tomatin, Inverness, IV13 7YR
G0 LVB G Brooks, 11 Prince Charles Way, Seaton, EX12 2[illegible]
G3 LVL Brian Ash, 5 Church Close, Wickham Bishops, Witham, CM8 3LN
G3 LVP K Eastty, 7 The Grange, The Reddings ROAD, Cheltenham, GL51 6RL
G3 LVW R Smith, 40 Highwoods Drive, Marlow Bottom, Marlow, SL7 3PY
G3 LWD P Stone, Bramley, Stone Street, Hythe, CT21 4JP
G3 LWE L Franklin, 32 Neeld Crescent, Chippenham, SN14 0HT
G3 LWJ C Way, 8 Stratford Place, Eaton Socon, St. Neots, PE19 8HY
G3 LWM Jeffery Harris, Cape Barn, Back Street Ash, Martock, TA12 6NY
G3 LWR Julian Evans, 18 Mandeville Road, Isleworth, TW7 6AD
G3 LWT Peter Buck, 17 Sanden Close, Hungerford, RG17 0LA
G3 LWU G Brisbar, 97 Chambers Lane, Mynydd Isa, Mold, CH7 6UZ
G3 LXB S Jones, 43 New St., Chase Terrace, Walsall, WS7 8BT
G3 LXE J Boden, Plas Heulwen, Llanfair Road, Newtown, SY16 3JY
G3 LXJ F Fisher, 6 Napier Close, Hornchurch, RM11 1FL
G3 LXQ D Gallop, 4 Volunteer Road, Theale, Reading, RG7 5DN
GU3 LYC T De Putron, Shieling Cottage, La Rue Marquand, St. Andrew, Guernsey, GY6 8RB
G3 LYD E Henderson, The Homestead, High Street, Ventnor, PO38 3HZ
G3 LYG A Macgregor, 10 Balroy Court, Forest Hall, Newcastle upon Tyne, NE12 9AW
G3 LYP M Scott, The Magnolias, Marlow Road, High Wycombe, HP14 3JW
G3 LYU David Price, Llansilin, Oswestry, SY10 7QB
G3 LYW John Weston, Chenery Lodge, 44 Old Newbridge Hill, Bath, BA1 3LU
G3 LYZ Brian Currey, 42 Westfield Avenue, Goole, DN14 6JX
G3 LZC A Stirland, 98 Aldreds Lane, Heanor, DE75 7HG
G3 LZI J Oates, Cherry Tree Cottage, Green Moor, Sheffield, S35 7DQ
G3 LZM M Bush, 5 Quay Close, Hereford, HR1 2RQ
G3 LZN G Ellison, Little Flushing, St. Peters Road, Falmouth, TR11 5TJ
G3 LZO Peter Thomas, 20 Bleasdale Court, Longridge, Preston, PR3 3TX
G3 LZR E Speller, 78 Chelmsford Road, Holland-on-Sea, Clacton-on-Sea, CO15 5DJ
G3 LZZ Andrew Pomfret, Flat 2, Ingwell House, Grange-over-Sands, LA11 6DP
G3 MAE Anthony Wilson, 8 The Paddock Appleton Wiske, Northallerton, DL6 2BE
G3 MAH B Bailey, Herbs And Honey, Summer Hill, Althorne, CM3 6BX
G3 MAI R Stevens, 138 Grange Drive, Stratton, Swindon, SN3 4LA
G3 MAJ E Holden, 10 Rowan Tree Close, Greasby, Wirral, CH49 3AW
G3 MAR MIDLAND ARS c/o Norman Gutteridge, 68 Max Road, Quinton, Birmingham, B32 1LB
GM3 MAS A Pringle, 1 Falloch Road, Milngavie, Glasgow, G62 7RR
G3 MAU J Wardle, 17 Frederick Neal Avenue, Coventry, CV5 7EH
G3 MAV J Bradley, 17 Talboys Walk, Tetbury, GL8 8YU
G3 MAZ H Bell, 35 Elm Trees, Long Crendon, Aylesbury, HP18 9DG
GI3 MBB A McMurtry, 20 Towerview Crescent, Bangor, BT19 6BA
GD3 MBC R Wernham, Fair Isle, Lhoobs Road, Douglas, Isle of Man, IM4 3JB
G3 MBD Herbert Dannatt-Brader, 20 Shire Place, Northampton, NN3 8DE
G3 MBK D Underdown, 26 Birch Road, Farncombe, Godalming, GU7 3NT
G3 MBM J Masters, 8 Purbeck Terrace Road, Swanage, BH19 2DE
G3 MBN B Gibbs, 15 Moor Barton, Neston, Corsham, SN13 9SH
G3 MBO Brian Aspinwall, 33 Clipstone Crescent, Leighton Buzzard, LU7 3LU
G3 MBU M Standige, 7 Hill Crest Avenue, Burnley, BB10 4JA
G3 MCA D Owen, 1 Mosslea Road, Orpington, BR6 8HP
G3 MCB Anthony Williams, 1 Wyvern Road, Sutton Coldfield, B74 2PS
G3 MCC K Worrall, 21 Northwood Avenue, Middlewich, CW10 0HR
G3 MCD K Holland, Ravendale, St. Lawrence, Bodmin, PL30 5JL
G3 MCE L Lee, 34 Westby Way, Poulton-le-Fylde, FY6 8AD
G3 MCK Gerald Stancey, 22 Peterborough Avenue, Oakham, LE15 6EB
G3 MCL C Simpkins, 6 Compton Way, Olivers Battery, Winchester, SO22 4EY
G3 MCP P Goadby, 535 Welford Road, Leicester, LE2 6FN
G3 MCV B Vaughan, 17 Richmond Close, West Town, Hayling Island, PO11 0ER
G3 MCX W Kennedy, 22 Croham Park Avenue, South Croydon, CR2 7HH
G3 MD Timothy Drew, 10 Sparkey Close, Witham, CM8 1QR
G3 MDD Brian Mudge, 9 Crossmead, Woolavington, Bridgwater, TA7 8ER
G3 MDG CHESHAM & DISTRICT A.R.S. c/o W Ray, 54 Gladstone Road, Chesham, HP5 3AD
G3 MDI M Plummer, Kembali, 14 Turnberry Drive, Woodhall Spa, LN10 6UE
G3 MDK Raymond Jones, Woodcote, 37 Coed Pella Road, Colwyn Bay, LL29 7BB
G3 MDM Gordon McGee, 2 Ilynton Avenue, Firsdown, Salisbury, SP5 1SH
G3 MDR M Hallet, 33 Latimer Street, Romsey, SO51 8DF
G3 MEC J Pearce, 86 Sopers Lane, Poole, BH17 7EU
G3 MED Frank Griffiths, 105 Hillcroft Crescent, South Oxhey, Watford, WD19 4PA
G3 MEH R Piper, 8 Osborne Way, Wigginton, Tring, HP23 6EN
G3 MEV C Cory, Tokelex, Chapel Lane, Thatcham, RG19 8BE
G3 MEY Jack Lawrence, 16 Waverley Court, Corsham, SN13 9NN
G3 MFG D Close, 27 High St., Collyweston, Stamford, PE9 3PW
G3 MFH Geoffrey Dale, 20 Blythe Avenue, Stoke-on-Trent, ST3 7JY
G3 MFJ G Firth, 13 Wynmore Drive, Bramhope, Leeds, LS16 9DQ
G3 MFK M Camp, 82 Leicester Road, Hinckley, LE10 1LT
G3 MFL Alan Hussell, Pear Tree Cottage, Savernake Road, Marlborough, SN8 3AS
G3 MFO P Elliot, 3 Shickle Place, Hopton, Diss, IP22 2QR
G3 MFW Harry Woodhouse, 143 Bodmin Road, Truro, TR1 1RA
GW3 MFY W Lee, Wallas Farm House, Wick Rd, Bridgend, CF35 5AE
G3 MGH P Clegg, 1 Frogmore Close, Hughenden Valley, High Wycombe, HP14 4LN
G3 MGJ P Robinson, 4 St. Johns Court, Stratford-upon-Avon, CV37 9AD
G3 MGL A Davis, 22 Yarmouth Close, Crawley, RH10 6TH
G3 MGS C Stephens, 12 Berkshire Road, Bishopston, Bristol, BS7 8EX
G3 MGU A Dodson, 53 Simons Lane, Wokingham, RG41 3HG
G3 MGW R Wheeler, 51 Seaview Road, Brightlingsea, Colchester, CO7 0PR
G3 MGX J Tomlinson, 34 Bentley Road, Tacolneston, Norwich, NR16 1DL
G3 MHD A Williams, 1259 Huon Road, Neika, Tasmania, Australia, 7005
G3 MHF M Ockenden, 11 Ratton Road, Eastbourne, BN21 2LU
G3 MHR W Lee, 6 Highfield Road, Swanwick, Alfreton, DE55 1BW
G3 MHT Edwin Landon, 14 The Blackthorns, Broughton, Brigg, DN20 0BB
G3 MHV Terence Langdon, 58 Upper Marsh Road, Warminster, BA12 9PN
G3 MHX M Tate, 40 Crossgates, Bedwell Plash, Stevenage, SG1 1LS
G3 MHY R Morris, Manfield, 3 Wrockwardine Road, Telford, TF1 3DA
G3 MIP S Heilbron, 8 Beechwood Drive, Formby, Liverpool, L37 2DG
G3 MIQ M Akehurst, 70 Corda Road, New Eltham, London, SE9 3SJ
G3 MJM Anthony Marshall, 2 Westwood Way, Beverley, HU17 8GE
G3 MJN L Harvey, 755a London Road, Westcliff on Sea, SS0 9[illegible]
G3 MJW C Edmunds, 51 Whiston Road, Kingsley, Northampton, NN2 7RR
G3 MKE W Smith, 12 Benscliffe Drive, Loughborough, LE11 3JP
GW3 MKT M Hooks, 1 Llwyn Castan, Pentwyn, Cardiff, CF23 7DA
G3 MKU A Bower, 82 Anson Road, Shepshed, Loughborough, LE12 9PU
G3 MKV C Curtis, 24 Rodney Road, Hartford, Huntingdon, PE29 1RZ
G3 MLO P Weatherall, Woodside, 6 Mile Garage, Stone St, Canterbury, CT4 6DN
G3 MLQ S Blundell, 12 Brookfield Street, Melton Mowbray, LE13 0NB
G3 MLS D Nappin, New Edge Farm, Heptonstall, Hebden Bridge, HX7 7PG
G3 MMA D Mayes, 12 Parsons Court, St. Andrews Road, Halstead, CO9 2TF
G3 MME Peter Whitford, Throo Pieces, Vernon Lane, Kelstedge, Chesterfield, S45 0EA
GI3 MMF William McAleer, 90 Gortin Park, Belfast, BT5 7EQ
GI3 MMG David Noon, 34 Rodney Park, Bangor, BT19 6FN
G3 MMJ Graham browne, 39a Cromwell Road, Canterbury, CT1 3LD
G3 MMN Basil Newman, 101 Tally Ho Road, Shadoxhurst, Ashford, TN26 1HW
G3 MMS G Whiting, 25 Obthorpe Lane, Thurlby, Bourne, PE10 0ES
G3 MMX Eric Lawley, 3 Barnicott Close, Newton Ferrers, Plymouth, PL8 1BP
G3 MNB Howard Benjamin, Flat 12, Caldecote, Kingston upon Thames, KT1 3EH
G3 MNJ James Yates, 23 Kingsfield Gardens, Bursledon, Southampton, SO31 8AY
G3 MNS I Swan, Flat 3, Mason Court, Alford Road, Mablethorpe, LN12 2GY
G3 MNV P Darragh, 48 Goodwood Park Road, Northam, Bideford, EX39 2RR
G3 MOA James Ruff, 17 Harts Close, Teignmouth, TQ14 9HG
G3 MOE J Moxey, 11 Westbury Road, Cheltenham, GL53 9EN
G3 MOL J Lixenberg, Orchard House, 77a Pembroke Crescent, Hove, BN3 5DF
G3 MON D Gent, 12 Field Road, Billinghay, Lincoln, LN4 4EA
GM3 MOR R Webster, Meric, 7 Woodmuir Crescent, Newport-on-Tay, DD6 8HL
G3 MOT Joshua Lambert Hurley, 19 Hill Close, West Bridgford, Nottingham, NG2 6GQ
GW3 MOV Colin Smith, 38 Plas Taliesin, Penarth, CF64 1TN
G3 MPB A Smith, 10 Goodwood Road, Redhill, RH1 2HH
G3 MPD OBO POLDHU ARC c/o Leslie Jones, Treharne Cottage, Meaver Road, Helston, TR12 7DN
G3 MPF C Smith, 29 Cloisters, Tarleton, Preston, PR4 6UL
G3 MPN David Johnson, 54 Norwich Road, Wymondham, NR18 0NT
G3 MPP G Price, 17 Celtic Close, Undy, Caldicot, NP26 3PB
G3 MPW A Walker, 14 St. Joans Drive, Scawby, Brigg, DN20 9BE
G3 MQD P Greed, The Bungalow, Townsend, Devizes, SN10 4RR
G3 MQI MARCH ARMY CADET FORCE c/o R Gill, 45 Biggin Lane, Ramsey, Huntingdon, PE26 1NB
GM3 MQO G Olesen, 8 Rowallan Crescent, Prestwick, KA9 2HE
G3 MQR D Robinson, 32 Bullock Wood Close, Colchester, CO4 0HX
G3 MRC B Poole, 18 Grosvenor Avenue, Kidderminster, DY10 1SS
G3 MRQ D Byne, Storm Bay, Church Street, Daventry, NN11 3YT
G3 MRT R Strafford, Chy Lowarth, Sparnock, Truro, TR3 6EB
G3 MRV G Carrick, 4 Kingfisher Lane, Gretna, DG16 5JS
G3 MRX P Robinson, 9 Barton Close, Cambridge, CB3 9LQ
G3 MRZ M Crutchley, 40 Ulfton Crescent, Shirley, Solihull, B90 3SA
G3 MSL Robert Ives, 11 Coombe Drive, Fleet, GU51 3DY
G3 MSM G Grieve, 1a Coombe Valley Road, Preston, Weymouth, DT3 6NH
G3 MSO E Tunstall, 11 The Broadway, Charlton on Otmoor, Kidlington, OX5 2UB
G3 MSW K Ashcroft, Fendley Corner, Common Lane, Harpenden, AL5 5DW
G3 MTD Barrie Kissack, 13 Church Street, The Old Saddlers, Braunton, EX33 2EL
G3 MTG Richard Prior, 35 Hanson Drive, Fowey, PL23 1ET
G3 MTJ R Skoyles, 2 Hay Close, Great Oakley, Corby, NN18 8HX
G3 MTP P Gadsden, Rose Cottage, Salwayash, Bridport, DT6 5HX
G3 MTR B Wolfe, 24 Marchbank Drive, Cheadle, SK8 1QY
GM3 MTW C Wolstencroft, 29 Fasach, Glendale, Isle of Skye, IV55 8WP
GM3 MUA P Lawlor, Woodside, North Kessock, Inverness, IV1 3XG
G3 MUJ H Morris, 47 Waldron Thorns, Heathfield, TN21 0AD
G3 MUO G Gott, 10 Churchill Crescent, Marple, Stockport, SK6 6HJ
G3 MUX C Benson, Orchard Croft, 85 Runcorn Road, Warrington, WA4 6UA
G3 MVM P Pierson, 7 Beehive Road, Goffs Oak, Waltham Cross, EN7 5NL
G3 MVU A Adkins, 80 Marine Parade, Leigh-on-Sea, SS9 2NL
G3 MVV N Miller, Avon, Gardiners Lane North, Billericay, CM11 2XA
G3 MVX J Burke, 120 Seabourne Road, Bexhill-on-Sea, TN40 2SD
G3 MVZ Frank Garrett, 18 Wolfe Close, Chichester, PO19 6BY
G3 MWM Donald Murden, Po Box 06, Curitiba, Parana, Brazil, 80011 970
G3 MWO D A Beales, 2 Wood Close, Tostock, Bury St. Edmunds, IP30 9PX
G3 MWQ P Groves, 326 Alcester Road, Burcot, Bromsgrove, B60 1BH
GM3 MWX A Winton, 2 Castlehill Cottages, Brisbane Glen Road, Largs, KA30 8SN
G3 MWZ J Casling, 19 Orchard Close, Tavistock, PL19 8HA
G3 MXA B Collins, 64 Park Avenue, Sittingbourne, ME10 1QY
G3 MXF P Cutler, 14 Verulam Road, Poole, BH14 0PP
G3 MXH T Downing, 8 Auction Yard, Haughley, Stowmarket, IP14 3GA
G3 MXJ Dennis Andrews, Coupelle, Levignac de Guyenne, France, 47120
G3 MXK D Price, 10 Laburnum Grove, Banbury, OX16 9DP
GM3 MXN T Sorbie, 9 Lynn Court, Larkhall, ML9 1QT
G3 MXP J Palfrey, Caprice, 4 Laverstock Park West, Salisbury, SP1 1QL
G3 MXV H Pierson, 65 Station Road, Countesthorpe, Leicester, LE8 5TB
G3 MXZ R Wesson, 58 Rochester Road, Middlesbrough, TS5 6QE
G3 MYA A Martindale, 16 Charles Miller Court, Leiston, IP16 4BY
G3 MYC C Cheatle, 56 Ashfurlong Crescent, Sutton Coldfield, B75 6EN
G3 MYG R Inman, 60 Abercorn Road, Mill Hill, London, NW7 1JL
G3 MYI John Lewis, 50 Robin Gardens, Waterlooville, PO8 9XF
G3 MYM R Micklewright, 5 Sandringham Road, Yeovil, BA21 5JE
G3 MYY S Boston, Beavers, Mill Lane, Ipswich, IP8 4AU
G3 MYZ Peter Nicholson, 3 Welborn Court, Main Street, Scarborough, YO11 3XA
G3 MZA E Hamblen, 64 Tollers Lane, Coulsdon, CR5 1BB
G3 MZB G Ward, 7 Wells Drive, Market Rasen, LN8 3EF
G3 MZC C Sutcliffe, 1 Tollgate Road, Culham, Abingdon, OX14 4NL
G3 MZI J Hood, 89 Freemens Way, Deal, CT14 9DQ
G3 MZN R Lightfoot, 28 Wheal Gorland Road, St. Day, Redruth, TR16 5LT
G3 MZV Gordon Brown, 52a Rivelands Road, Swindon Village, Cheltenham, GL51 9RF
G3 MZX M Pedreschi, Clary Lodge, Corse of Clary, Newton Stewart, DG8 6BH
G3 MZY James Last, Orchard House, Gwyllt Road, Llanfairfechan, LL33 0EG
G3 MZZ A Kightley, 29 The Parkway, Gosport, PO13 0PT
G3 NAE C Richardson, 10 Fielders Way, East Wellow, Romsey, SO51 6EX
G0 NAI R Norman, 50 Birchfield Street, Oldbury, B69 0QH
G3 NAK G Mallinson, 145 Huddersfield Road, Meltham, Holmfirth, HD9 4AJ
G3 NAN Robert Henderson, 48 Cartwright Crescent, Brackley, NN13 6HA
G3 NAP B Sowter, 56 Alderminster Road, Coventry, CV5 7JU
G3 NAQ G Grayer, Bagatelle, 2 Southend, Newbury, RG20 7BE
G3 NAT LONDON RAYNET G c/o A Brooker, 18 Honeybourne Way, Petts Wood, Orpington, BR5 1EZ
G0 NAV Edward Cook, Edward D Cook, 162 O St. Spc.51, Lincoln, United States, 95648
G3 NAW J Ryan, 4 Ferry Lane, Bath, BA2 4HS
G3 NAY S Whithorn, 53 Torbay Road, Allesley, Coventry, CV5 9JY
G3 NBL J Larson, Nyhem, Whitton Village, Stockton on Tees, TS21 1LQ
G3 NBN R Weaving, 27 Beech Drive, Nailsea, Bristol, BS48 1QA
G3 NBQ Peter Burt, 3335 Mountain Highway, North Vancouver, Canada, V7K 2H4
G3 NBS A Bairstow, 27 Williams Way, Longwick, Princes Risborough, HP27 9RP
G3 NBY Howard Murray, 36 Sterndale Road, Davenport, Stockport, SK3 8QU
G3 NBZ K Thorne, 3 Cherry Gardens, Abstacle Hill, Tring, HP23 4EA
G3 NCB H Bourner, Woodpeckers, 11 Richborough Road, Sandwich, CT13 9JE
G3 NCN John Ellerton, 7 Cotterell Close, Bracknell, RG42 2HL
GM3 NCO Alexander Mustard, Tigh Ard, Knockhouse Hill, Dunfermline, KY12 8PT
GW3 NCT R Lord, 8 Llys Steffan, Llantwit Major, CF61 2UF
GW3 NDB G Wyatt, 3 Creidiol Road, Mayhill, Swansea, SA1 6TZ
GD3 NDC C Deamer, Gatehouse, Warren Lane, Stanmore, HA7 4LD
G3 NDI C Fry, 7 Thornbury Close, Crowthorne, RG45 6PE
G3 NDK R Webb, 142 Penrose Avenue, Carpenders Park, Watford, WD19 5AA
G3 NDN D Newey, 15 Clent View Road, Stourbridge, DY8 3JE
G3 NDO P Sorab, Woodgaston Cottage, Woodgaston Lane, Hayling Island, PO11 0RL
G3 NDS R Oliver, Flat 21, Exeter Court, 52 Wharncliffe Road, Christchurch, BH23 5DF
G3 NEH J Isles, 3 Drovers Croft, Greenleys, Milton Keynes, MK12 6AN
G3 NEO P bagshaw, 48 Kiveton Lane, Todwick, Sheffield, S26 1HL
G3 NEP Christopher Wager-Bradley, Swallowdale, Longhoughton Road, Alnwick, NE66 3AT
GM3 NEQ A Finlay, 19 Fraser Avenue, Newton Mearns, Glasgow, G77 6HP
G3 NFB J Leviston, 9 Barnes Avenue, Fearnhead, Warrington, WA2 0BL
G3 NFC Burton & District RS c/o Geoffrey Newstead, 97 Hawthorn Crescent, Burton-on-Trent, DE15 9QN
G3 NFJ M Coward, High Bank, 51 High Street, Warminster, BA12 7AP
GI3 NFM K McElhatton, 2a Orpheus Drive, Dungannon, BT71 6DR
G3 NFP L Beckwith, 5 Hunts Mead, Bromham, Chippenham, SN15 2JP
G3 NFV R Sykes, 16 The Ridgeway, Fetcham, Leatherhead, KT22 9AZ
G3 NFW J Carroll, White Lodge, Hunston, Chichester, PO20 1PA
G3 NFY Barry Twist, 11 Church Street, Minehead, TA24 5JU
G3 NGJ William Epton, 2 Eastcliff Road, Lincoln, LN2 5RU
G3 NGK D Chapman, 6 Pickhurst Green, Hayes, Bromley, BR2 7QT
GM3 NGW W Webb, 5 Thornlea Drive, Giffnock, Glasgow, G46 6DB
G3 NGX Harry Hogg, Crossways, Ferry Road, Reading, RG8 0JL
G3 NGZ Pelican Radio Group Little Rissington c/o Michael Grierson, 1 Blenheim Close, Upper Rissington, Cheltenham, GL54 2QX
G3 NHB David Bowyer, 41a High Street, Trumpington, Cambridge, CB2 9HR
G3 NHE M Dann, 61 Alms Hill Road, Parkhead, Sheffield, S11 9RR
G3 NHF Joseph Noble, 27 Chestnut Avenue, Donington, Spalding, PE11 4XH
G3 NHL C Lewis, Chy An Mor, Greatwood, Falmouth, TR11 5SR
G3 NHP G Peacock, Hallowsgate House, Flat Lane, Tarporley, CW6 0PU
GM3 NHQ Thomas Harrison, 7 Cults Gardens, Broughty Ferry, Dundee, DD5 1QT
G3 NHR H Rogers, Aughavore, Church Walk, Louth, LN11 8LJ
G3 NHS J Carp, 82 Imperial Avenue, Victorian Road, London, N16 8HW
G3 NHV D Hare, White Lodge, Mount Gabriel, Schull, Ireland
G3 NHX G Quarterman, 2 Milton Avenue, Sutton, SM1 3QB
G3 NIC Kenneth Plant, Rose Cottage, Lincoln Road, Lincoln, LN2 2NE
G3 NID I Douglas, 6 Ansley Road, Houghton, Huntingdon, PE28 2DQ
G3 NIE Charles Bell, 85 Waterbeach Road, Landbeach, Cambridge, CB25 9FA
GM3 NIG Dennis Cram, 61 Gailes Road, Troon, KA10 6TB
G3 NII R Porter, 6 Clifton Road, Shefford, SG17 5AA
G3 NIJ Brian Barker, 4 Glantlees, West Denton, Newcastle upon Tyne, NE5 2PJ
G3 NIL Graham Munden, 126 Stanley Green Road, Poole, BH15 3AQ
G3 NIN James Brogan, 38 Graig Park Circle, Newport, NP20 6HE
G3 NIQ R Gorton, 2 Clyde Court, Clyde Close, Redhill, RH1 4AY
G3 NIR G Miles, 15 Elizabeth Way, Herne Bay, CT6 6ES
G3 NIW P Ives, Allt Na Criche, Ockham Road South, Leatherhead, KT24 6QJ
G3 NJA TORBAY ARS c/o Derrick Webber, 43 Lime Tree Walk, Milber, Newton Abbot, TQ12 4LF
G3 NJB WACRAL (World Association of Christian Radio Amate c/o Peter Jackson, 24 Woodfield Park, Walton, Wakefield, WF2 6PL
G3 NJG T George, 8 Lanehays Road, Hythe, Southampton, SO45 5ER
G3 NJV P Randall, Mycroyke, Ruan Minor, Helston, TR12 7LU
G3 NJX R Beeson, The Grove, Main Road, Ripley, DE5 3RE
G3 NJY M Bibby, 47 Whitney Tavern Road, Weston, United States, 2493
G3 NKC D Sharred, 4 Rufford Close, Wistaston, Crewe, CW2 6XP
GM3 NKG A Campbell, 22 Saltire Crescent, Larkhall, ML9 2LG
G3 NKH R Dowling, Orchard House, Oughtrington Lane, Lymm, WA13 0RD
G3 NKJ R Gill, 45 Biggin Lane, Ramsey, Huntingdon, PE26 1NB
G3 NKL R Jones, 5 Apsley Fold, Longridge, Preston, PR3 3TY
GW3 NKM C Jones, 77 Margam Road, Port Talbot, SA13 2LB
G3 NKQ C Burchell, 4 Bakers Way, Perry, Huntingdon, PE28 0BS
G3 NKS D Thom, 78 Farmfield Road, Cheltenham, GL51 3RA
G3 NKW H White, 16 Turnberry Close, Lymm, WA13 9LY
GM3 NLB F Inglis, 3 Fleming Road, Bishopton, PA7 5HW
G3 NLY R Smethers, 46 Church Road, Burntwood, WS7 9EA

**IMPORTANT NOTE**

**Revalidate licence to avoid revocation** – Ofcom has advised the Society that plans will be drawn up to revoke licences that have not been revalidated as required by the licence conditions. The quickest way to revalidate is to do so online via the Ofcom website: *https://services.ofcom.org.uk/* or by email: *amateur.validations@ofcom.org.uk* Ofcom staff are available to help, but please be patient during times of heavy workload.

G3 NMD Houghton ARC c/o Ian Laidler, 5 South St., West Rainton, Houghton le Spring, DH4 6PA
G3 NMH Hal Perkins, 31 Dorchester Road, Weybridge, KT13 8PE
G3 NMJ G Knapp, 4 Venture Close, Bexhill-on-Sea, TN40 1TU
G3 NML M Slater, 46 Ladywood, Eastleigh, SO50 4RW
GM3 NMN R Dunlop, 39 Braid Drive, Glenrothes, KY7 4ES
G3 NMT R Fernandez, 52 Windermere Road, Noctorum, Prenton, CH43 9SW
G3 NMW Tony Whateley, 285 Harborne Road, Birmingham, B15 3JB
G3 NMX Derek Wills, 4002 Amy Circle, Austin, United States, 78759
G3 NMZ George Bath, 11 Heron Way, Hickling, Norwich, NR12 0YQ
G3 NN C Bolt, 147 Swan Avenue, Bingley, BD16 3PL
G3 NNA M Codd, 71a Higher Road, Longridge, Preston, PR3 3SY
G3 NNB R Evans, Cemlyn, Ffordd Dewi Sant, Pwllheli, LL53 6EG
G3 NNG Colin Desborough, 22 Westland Road, Faringdon, SN7 7EY
G3 NNN P Mason, 1 Morley Lane, Stanley, Ilkeston, DE7 6EZ
G3 NNO M George-Powell, Old Church Lane Cottage, Pateley Bridge, Harrogate, HG3 5LY
G3 NNT S Pilkington, The Quarries, Quarry Drive, Ormskirk, L39 5BG
G3 NNV P Swanson, 11 Grassmoor Close, Wirral, CH62 7JY
G3 NNW K Taylor, 34 Shore Road, Warsash, Southampton, SO31 9FU
G3 NNY John Bolter, 32 Rotherham Avenue, Luton, LU1 5PN
GM3 NNZ B East, 26 Hyndford Road, Lanark, ML11 9AE
G3 NOA Peter Reynolds, Brook Bushes, Bramshaw, Lyndhurst, SO43 7JB
G3 NOC A Waldie, Gwyn Lyn, 85 Park Road, Coleford, GL16 7AG
G3 NOI Raymond Cumming, 21 Britannia Way, Woodmancote, Cheltenham, GL52 9QP
G3 NOP David Peacock, Robin Hill, Cottingham, HU16 5JG
G3 NOX Jeremy Royle, Keepers Cottage, Duddenhoe End, Saffron Walden, CB11 4UU
G3 NPA G Anderson, 46 Bearcroft, Weobley, Hereford, HR4 8TA
G3 NPC J Swanson, 23 Oatlands Road, Tadworth, KT20 6BS
G3 NPI G Suggate, 26 Highlands Road, Buckingham, MK18 1PL
G3 NPJ J Jones, 47 Rhodesway, Wirral, CH60 2UA
G3 NPL E Matthews, 20 Stockwell Furlong, Haddenham, Aylesbury, HP17 8HD
G3 NPM A Macdonald, 5 Arlington Close, Swindon, SN3 3NB
GI3 NPP Ronnie Gibson, 109 Bush Road, Dungannon, BT71 6QG
G3 NPS B Harrad, 32 Woodfield Avenue, Northfleet, Gravesend, DA11 7QG
G3 NPT G Bell, 1 Fountain Close, Hessle, HU13 0LB
G3 NPY J Joslin, 150 Roman Bank, Skegness, PE25 1SE
G3 NPZ T Griffiths, 18 Lulworth Road, Lee-on-The-Solent, PO13 9HU
G3 NQA S Hall, 76 Cheltenham Drive, Bromford, Birmingham, B36 8QG
G3 NQF R Fenton, Harmins Green, France Lynch, Stroud, GL6 8LZ
G3 NQK J Beddows, 17 Rue Francois Mitterand, Pleven, France, 22130
G3 NQN M Hartung, 31 Ellenbrook Road, Hatfield, AL10 9RW
G3 NQT R Levi, 24 Stanmore Way, Loughton, IG10 2SA
G3 NQX William Brown, 73 Church Avenue, Preston, PR1 4UD
G3 NQZ G Lockhart, 179 Poolbrook Road, Malvern, WR14 3JZ
G3 NR Andrew Birt, 190 Epsom Road, Guildford, GU1 2RR
G3 NRD John Packer, Butts Bank Farm, Gulval, Penzance, TR18 3BB
G3 NRH B Perrin, Hanami, Back Street, Gillingham, SP8 5JY
G3 NRM M Moore, 127 Adel Lane, Leeds, LS16 8BL
G3 NRQ C Higgins, Billdoro, Mill Lane, Louth, LN11 7SA
G3 NRU D Brook-Foster, 246 St. Margarets Banks, High Street, Rochester, ME1 1HY
G3 NRW Ian Wade, 7 Daubeney Close, Harlington, Dunstable, LU5 6NF
G3 NRX R Murphy, 3 Lady Leasow, Shrewsbury, SY3 6AB
G3 NRZ Christopher Hogg, 7 Elm Grove, Erith, DA8 3BL
G3 NSD Brian Styles, York House, Bluntisham Road, Huntingdon, PE28 3LY
G3 NSF Thomas Simpson, 41 Benyon Grove, Orton Malborne, Peterborough, PE2 5XS
G3 NSI N Ashman, 11 Burdon Place, Peterlee, SR8 5QZ
G3 NSL I Whitter, The Old Hall, Hall Lane, Lincoln, LN3 4HT
G3 NSO G Brookes, 27 Pineside Avenue, Cannock Wood, Rugeley, WS15 4RG
G3 NSP J Lennox, Kestrel, School Lane, Bicester, OX25 4AW
G3 NSS T Spain, Manor View, Shotatton, Shrewsbury, SY4 1JD
GI3 NSV M Donnelly, Tirgarve, 7 Blackwatertown Road, Armagh, BT61 8EZ
G3 NSW R Kay, 7 Lea Drive, Blackley, Manchester, M9 7AR
G3 NTA G Couzens, 47 Holmstead Avenue, Whitby, YO21 1NA
G3 NTD A Marsden, 15 Northfield Way, Retford, DN22 7LJ
G3 NTF I Neary, 65 Vicarage Road, Ashton-under-Lyne, OL7 9QY
G3 NTI R Blain, 11 Mill Bank, Ness, Neston, CH64 4BJ
G3 NTM William Brown, 18 Georgian Close, Staines, TW18 4NR
G3 NUA J Hogg, 16 Moorston Close, Naisberry Park, Hartlepool, TS26 0PJ
G3 NUB M Bursnall, Panorama, Church Lane, Bishops Castle, SY9 5AF
G3 NUG E Cheadle, Lower Withers Barns, Middleton on the Hill, Leominster, HR6 0HY
G3 NUL V Johnston, 119 High St., Cheveley, Newmarket, CB8 9DG
G3 NUN A Langford-Brown, 9 Orchard Lane, Corfe Mullen, Wimborne, BH21 3SU
GW3 NUO P Williams, Crud Y Gwynt, 27 Mynydd Garnllwyd Road, Swansea, SA6 7PB
G3 NUQ I Macarthur, 2 Bramley Close, Bramhall, Stockport, SK7 2DT
GM3 NUU John Reid, Rochelle, Findon, Aberdeen, AB12 3RL
G3 NVB Allan Bryant, 1 Downlands Road, Winchester, SO22 4ET
G3 NVJ Geoffrey Hubber, 3 Antron Way, Mabe Burnthouse, Penryn, TR10 9HS
G3 NVL R Allen, 692 Hitchin Road, Luton, LU2 7UH
G3 NVM D Arigho, 7 Burgess Close, Odiham, Hook, RG29 1PG
G3 NVP B Mapp, 33 Cotswold Drive, Redcar, TS10 4AG
GM3 NVQ G Martin, 39 St. Johns Drive, Dunfermline, KY12 7TB
GI3 NVW William Pollock, 155 Doogary Road, Omagh, County Tyrone, BT79 0HF
G3 NVX R Davison, 76 Poplars Way, Beverley, HU17 8PU
G3 NWH A Collis, C/O 510 Lowther Road, Dunstable, LU6 3LJ
G3 NWL A Lock, 7 Heather Close, St. Leonards, Ringwood, BH24 2QJ
G3 NWR Wirral ARS c/o Gordon Hunter, 151 Norwich Drive, Wirral, CH49 4GD
G3 NWS Francis Clare, Glen View, Newport Road, Caldicot, NP26 3BZ
G3 NWW Malcolm Wakely, Chyandour, 3 Ganges Close, Mylor Harbour, Falmouth, TR11 5UG
G3 NWX K Morgan, 97 Elmwood, Sawbridgeworth, CM21 9NN
G3 NWY D Forster, 79 Westbrooke Avenue, Hartlepool, TS25 5HX
G3 NXC Anthony Plant, 178 Clay Lane, Yardley, Birmingham, B26 1DY
G3 NXK O Diplock, North Lodge, Messing Park, Colchester, CO5 9TD
G3 NXL P Lamming, 25 Leconfield Garth, Follifoot, Harrogate, HG3 1NF
G3 NXN F Wickens, 32 Kenilworth Avenue, Wimbledon, London, SW19 7LW
G3 NXO A Watt, Little Owls, Singleton, Chichester, PO18 0EX
GW3 NXR T Miles, West Uplands Lodge, Upland Arms, Dyfed, SA32 8DX
G3 NXS Frederick Shaw, 69 Finedon Road, Irthlingborough, Wellingborough, NN9 5TY
G3 NXT W Fletcher, 3 Orchard Close, Metheringham, Lincoln, LN4 3DT
G3 NXV R Jennings, Edensor, Grendon, Atherstone, CV9 3DP
G3 NXX Ian Miller, 11 Lynton Drive, High Lane, Stockport, SK6 8JE
G3 NXZ J Howe, 18 Laburnum Grove, Conisbrough, Doncaster, DN12 2JW
G3 NYB W Bingham, 7 Bolton Hill Road, Doncaster, DN4 6DQ
G3 NYD D Coles, 113 Berrow Road, Burnham-on-Sea, TA8 2PH
G3 NYE A Taylor, 25 Burnside Road, Gatley, Cheadle, SK8 4NA
GM3 NYG Joan Fish, 31 Oaklands Avenue, Irvine, KA12 0SE
GI3 NYJ S Currie, 122 Belfast Road, Comber, Newtownards, BT23 5QP
G3 NYK A Melia, 67a Deben Avenue, Martlesham Heath, Ipswich, IP5 3QR
G3 NYR D Rayner, 42 Canford Drive, Allerton, Bradford, BD15 7AU
G3 NYS C Whiteley, 30 Lynch Hill Park, Whitchurch, RG28 7NF
G3 NYX J Heaviside, Grisedale, 110 Cuckfield Road, Hassocks, BN6 9RZ
G3 NYZ A Stafford, Blakefield, Jawbone Lane, Derby, DE73 1BW
G3 NZK P Robson, Glenross, Whiteshoot, Salisbury, SP5 2PR
G3 NZL Howard Chapman, 57 Athelstan Road, Southampton, SO19 4DE
G3 NZO Graham Kidder, Oxspring, Chalvington, Hailsham, BN27 3TG
G3 NZP Malcolm Harman, 19 Hill House Close, Turners Hill, Crawley, RH10 4YY
G3 NZR William Young, 5 Grasmere Grove, Frindsbury, Rochester, ME2 4PN
G3 NZS H Parkes, 35 Dovey Road, Tividale, Oldbury, B69 1NT
G3 NZV A Park, Waterside Cottage, Bowden Lane, High Peak, SK23 0QF
G3 NZW S James, Beresford, Latchmoor Ave, Gerrards Cross, SL9 8LJ
G3 NZY R Shelley, 4 Fairview Court, St. Martins Avenue, Scarborough, YO11 2DA
G3 OAD T Haydu Jones, 1 Beggars Roost, Golf Course Road, Stroud, GL6 6TJ
G3 OAF W Jeffs, Silver Jay, Colehill Lane, Wimborne, BH21 7AN
G3 OAH P Whittlestone, Turangi, 14a Croft Bank, Malvern, WR14 4DW
G3 OAJ Cecil Davies, 11 BEAUMARIS WAY, Grove Park, Blackwood Gwent, NP12 1DF
G3 OAL E Lincoln, Lynholme, Millbank, Newton Aycliffe, DL5 6RF
G3 OAR G Greenwood, 1 Maltkiln Lane, Castleford, WF10 4LF
G3 OAX I Anderson, 102 Hopefold Drive, Worsley, Manchester, M28 3PW
G3 OAY R Graham, 5 The Langlands, Hampton Lucy, Warwick, CV35 8BN
G3 OAZ J Randall, 243 Paddock Road, Basingstoke, RG22 6QP
GM3 OBC Richard Thomson, 1 Knowehead, Star, Glenrothes, KY7 6LA
GM3 OBG P Bridges, 29 Kirkbank, Auchmithie, Arbroath, DD11 5SY
G3 OBL J Tyrrell, 2 Briar Close, Yeovil, BA21 5XA
GI3 OBO D Waugh, 16 Seaview Avenue, Millisle, Newtownards, BT22 2BN
G3 OBV P Harris, 15 Ratliffe Road, Rugby, CV22 6HB
G3 OBZ Michael Birkett, Hazelwood, Cromwell Ave, Woodhall Spa, LN10 6TH
G3 OCA K Frankcom, 1 Chesterton Road, Spondon, Derby, DE21 7EN
G3 OCB Clive Bowden, Tregwyn, Tregonning Road, Truro, TR3 7FG
G3 OCH J Hulett, 21 Exmoor Avenue, Leicester, LE4 0BJ
G3 OCI D Hayter, 31a High View Rise, Crays Hill, Billericay, CM11 2XU
G3 OCP D Wallace, 11 Station Road, Haddenham, Aylesbury, HP17 8AN
G3 OCR Stuart Nutt, 23a Hesketh Drive, Southport, PR9 7JX
G3 OCW F Cubberley, The Cedars, Worcester, WR2 4TE
G3 ODB Alan Pritchard, 41 Maes Cantaba, Ruthin, LL15 1YP
G3 ODC Donald Martin, 7 Seaview Avenue, Eastham, Wirral, CH62 0BD
G3 ODD E Stables, Manor Croft, Water Lane, Selby, YO8 6QL
G3 ODO W Buckett, 1659 Laurel Ridge Lane, Lawrenceville, Georgia
GM3 ODP T Salvesen, Easter Catter, Croftamie, Glasgow, G63 0EX
G3 ODX Stephen Clarke, 18 Dunedin Drive, Caterham, CR3 6BA
G3 OEB R Downs, 23 Old London Road, Benson, Wallingford, OX10 6RR
GD3 OEC C Isham, 2 Lime Grove, Ruislip, HA4 8RY
G3 OEF R Maule, 6 Deepwater Bay Rd, House 1, Hong Kong, China, ZZ6 9DE
G3 OEQ D BUNN, Heliophilia Gardens 1, Agias Annis St. 17, Kato Pafos, Cyprus, 8036
G3 OFI B Bisley, 132-1919 St Andrews Place, Courtenay, British Columbia, Canada, V9N 9J4
G3 OFP G Cunnah, 225 Springwell Lane, Balby, Doncaster, DN4 9AJ
G3 OFT Peter Bower, An Cluain, Ballplay Road, Moffat, DG10 9JU
G3 OFW Herbert Blake, 19 Segsbury Grove, Harmans Water, Bracknell, RG12 9JL
G3 OFX R Welch, 112 Copsewood Road, Bitterne, Southampton, SO18 1QR
G3 OGE J Rose, 1 Westgate House, 22 Westgate, Hornsea, HU18 1BP
G3 OGH A Brooker-Carey, 29 Byron St., Amble, Morpeth, NE65 0ER
G3 OGK Gerald Kennedy, Thayers, Edwyn Ralph, Bromyard, HR7 4LY
G3 OGP R Powell, Garlands Farm, The Haven, Billingshurst, RH14 9BH
G3 OGQ G Fare, 6 Rockfield Mews, Alexandra Road, Warrington, WA4 2AE
G3 OGX J Allsop, 17 Hambro Hill, Rayleigh, SS6 8BN
G3 OGZ M Beer, 24 Byron Court, Beech Grove, Harrogate, HG2 0LL
G3 OHC Graham Badger, 3 Hesketh Close, Cranleigh, GU6 7JB
G3 OHH R Hargreaves, 46 Castle Road, Mow Cop, Stoke-on-Trent, ST7 3PH
G3 OHL Douglas White, Holme Fell Cottage, Hallbankgate, Brampton, CA8 2NJ
G3 OHM SOUTH BIRMINGHAM RS c/o John Storey, Hampstead House, Fairfax Road, Birmingham, B31 3QY
G3 OHN K Whitehouse, 27a Howdles Lane, Brownhills, Walsall, WS8 7PL
G3 OHP Michael Winter, 9 Higham Road, Cliffe, Rochester, ME3 7SH
G3 OHS J Perry, 517 Longbridge Road, Barking, IG11 9DD
G3 OHX I Jackson, Brattle House, Manor Road, Beaconsfield, HP9 2QU
GM3 OIB K Younger, 183 Main Street, Pathhead, EH37 5SQ
G3 OIC I Croxford, 16 Chesterwood, Hollywood, Birmingham, B47 5EN
G3 OIF P Squires, 191 Station Road, Knowle, Solihull, B93 0PT
G3 OIH Bernard Shields, 24 Churchfield, Fulwood, Preston, PR2 8GT
G3 OIL Michael Wills, 23 Falcons Way, Salisbury, SP2 8NR
G3 OIN J Nicholas, 28 Hardy Avenue, Rhyl, LL18 3BG
G3 OIP P Holker, Drovers Cottage, 44 West Street, Tetbury, GL8 8DR
GM3 OIV Walter Anderson, 6 Winchburgh Road, Winchburgh, Broxburn, EH52 6QB
G3 OJ J Hobin, 14 St. Martins Green, Trimley St. Martin, Felixstowe, IP11 0UU
G3 OJG P Gale, Garden Cottage, Sacombe Green, Ware, SG12 0JQ
G3 OJI J Sleight, Orchard House, School Hill, Southam, CV47 8NN
G3 OJK J Bates, 8 Spaxton Road, Durleigh, Bridgwater, TA5 2AP
G3 OJL Malcolm Plaster, Combe House, Milton Lane, Wells, BA5 1DG
G3 OJQ E Wilson, 2 Wharf Road, Stamford, PE9 2DU
G3 OJS H Braham, 10 Glebe Way, Frinton-on-Sea, CO13 9HR
G3 OJV P Waters, 9 Tudor Way, Hawkwell, Hockley, SS5 4EY
G3 OJX A Hobbs, 65 Spurfield, West Molesey, KT8 1RR
G3 OJZ Brian Todd-White, 3 Alexandra Road, Capel-le-Ferne, Folkestone, CT18 7LB
G3 OKA John Share, 82 Birkenhead Road, Meols, Wirral, CH47 0LB
G3 OKB Michael Ireson, 15 Digby Drive, North Luffenham, Oakham, LE15 8JS
G3 OKD Z Nilski, The Poplars, Wistanswick, Market Drayton, TF9 2BA
G3 OKH G Hillman, 504 Chester Road, Kingshurst, Birmingham, B36 0LG
G3 OKS S Smithies, Moorcroft, Fernhill ROAD, Horley, RH6 9SY
G3 OKT J Thompson, The Old Place, Old Racecourse, Oswestry, SY10 7HL
G3 OKU M Cross, 6 Brackendale Close, Camberley, GU15 1HP
G3 OKY D Vincent, 10 Leaveland Close, Beckenham, BR3 3PL
G3 OLB T Boucher, Hedgerows, Sheldon, Honiton, EX14 4QS
G3 OLH A Remsbury, Nodali, 16 Little Green Lane, Chertsey, KT16 9PH
G3 OLP B Wadsworth, 5 Birch Avenue, Todmorden, OL14 5NX
G3 OLU John Saunders, APARTAMENTO 6306, FORUM MARE NOSTRUM, CAMINO DE PINXO 2, Alfaz Del Pi, Spain, 3580
G3 OLW J Burnett, Wenrisc, Chapel Lane, Tewkesbury, GL20 8HS
G3 OLX J Parker, Palfreys, Picquets Way, Banstead, SM7 1AJ
G3 OMA Stan Kay, 5 Chevalier Close, Middleleaze, Swindon, SN5 5TS
G3 OMB R Spurgeon, 57 Laburnum Crescent, Kirby Cross, Frinton-on-Sea, CO13 0QH
G3 OMD A Callegari, Danebridge Nursery, Much Hadham, SG10 6JG
G3 OMJ P Judkins, 18 St. Johns Square, Wakefield, WF1 2RA
G3 OMK T Kirk, 54 Highfields Drive, Loughborough, LE11 3JT
GW3 OMN May Jenkins, 25StepneyRoad, Swansea, SA2 0FZ
G3 OMR M Russoff, Flat 3 Hartsbourne Court, Hartsbourne Road, Bushey Heath, WD23 1PZ
G3 OMS R Simpson, 23 Larkhill, Rushden, NN10 6BG
G3 OMT A Russell, 5 Little Close, Swadlincote, DE11 0EB
G3 OMY D Hancock, 17 Forestlake Avenue, Ringwood, BH24 1QU
G3 OMZ David Lee, Wellow Green Cottage, Wellow Top Road, Yarmouth, PO41 0TA
G3 OND John Denman, 167 Minnis Road, Birchington, CT7 9QD
GI3 ONF Robert Sinton, 35 The Rose Garden, Tandragee, Craigavon, BT62 2NJ
G3 ONI D Woods, Flat 26, Chapel Court, Wilmslow, SK9 5EN
GU3 ONJ A Richmond, The Cedars Holly Drive, 3 Braye Road, Vale, Guernsey, GY3 5PQ
G3 ONL P Brodribb, 18 Ipswich Road, Debenham, Stowmarket, IP14 6LB
G3 ONQ R Goodall, Hazelmere, 21 Church Lane, Halifax, HX2 0EF
G3 ONR Barry Reynolds, 17 Cresswells Mead, Holyport, Maidenhead, SL6 2YP
G3 ONU D Barry, 2 Catherine Close, Shrivenham, Swindon, SN6 8ER
G3 ONV J Verity, Tall Pine, Station Road, Leicester, LE9 2EN
G3 OOH Gerald Lander, 132 Chemin De Saule, Bernex, Switzerland, 1233
G3 OOK J Plenderleith, 3D Deluxe Court, Jalan Pahlawan Kepayan, Kota Kinabalu, Malaysia, 88200
G3 OOL J Hatch, 628-707 Esquimalt Road, Victoria, Canada, BC V9A 3L7
G3 OOP BRYAN HAVENHAND, 15 Sandiway, Chesterfield, S40 3HG
G3 OOU Robert Burns, 84 Portnalls Road, Coulsdon, CR5 3DE
G3 OOW Melvyn Docker, Apartment 219, Clarence Park, 415 Worcester Road, Malvern, WR14 1FU
G3 OPB M Bues, 7A Alice Parkins Close, Hadleigh, IP7 6FE
GW3 OPC Norman Ward, 17 Heol Nant, Llanelli, SA14 8EL
G3 OPE C Urwin, Clifton House, 1 Nelson Terrace, Newcastle upon Tyne, NE17 7JR
G3 OPG Roy Tingay, 18 Grove Road, Newbury, RG14 1UH
G3 OPH R Atkinson, 1 Lake Walk, Adderbury, Banbury, OX17 3PF
G3 OPJ C Harrisson, 129 Granville Way, Sherborne, DT9 4AT
G3 OPW J Cook, Upwood Park, Black Moor Road, Keighley, BD22 9SS
G3 OPX Clive Green, Gothic, Plymouth Road, Totnes, TQ9 5LH
G3 OQC J Woods, 1 Dean Road, Cosham, Portsmouth, PO6 3DG
G3 OQD Martin Emmerson, 6 Mounthurst Road, Hayes, Bromley, BR2 7QN
G3 OQF R Kay, 7 Chemin Des Grands-Champs, Bogis-Bossey, Switzerland, CH 1279
GM3 OQI J Ramsay, 150 City Road, Dundee, DD2 2PW
GW3 OQK Andrew Fairgrieve, 3 Pleasant Road, Gorseinon, Swansea, SA4 9WH
G3 OQO David Henley, 36 Main Street, Newbold, Rugby, CV21 1HW
GI3 OQR Dick Gibson, 93 Cavan Road, Dungannon, BT71 6QN
G3 OQT R Mclachlan, Trotters Ash, Hollington Lane, Ashbourne, DE6 3AE
G3 ORD D Munton, Burlands Bungalow, Burlands Lane, York, YO26 6PS
G3 ORG I Taylor, 10 Westfield Road, Henlow, SG16 6BN
G3 ORI J Vickers, 45 Willow Park Drive, Stourbridge, DY8 2HL
G3 ORK R Talbot, 9 Bracebridge Drive, Southport, PR8 6XH
G3 ORL Douglas Williams, 14 Seymour Avenue, Parc Seymour, Caldicot, NP26 3AG
G3 ORN W Thomas, 20 Vinnicombes Road, Stoke Canon, Exeter, EX5 4BB
G3 ORP P Pickering, 21 Palmar Road, Maidstone, ME16 0DL
G3 ORV M Saunders, 40 Archfield Road, Cotham, Bristol, BS6 6BE
G3 ORX Arthur Rumbold, 31 Springfield Close, Hawthorn, Corsham, SN13 0JR
G3 ORY R Titterington, Wyclif House, St. Marys Road, Lutterworth, LE17 4PS
G3 OS A Boor, 2 The Orchard, Hayton, Retford, DN22 9LJ
G3 OSI D Swanson, 48 Moscow Drive, Liverpool, L13 7DJ
G3 OSP S Plumtree, Flat 18, Oliver Leese Court, Stafford, ST17 9HP
G3 OSQ D Beakhust, Nonsuch Lodge, Morgans Vale Road, Salisbury, SP5 2HU
G3 OSR Paul Hughes, Glan Y Mor, 14 Stockdove Way, Thornton Cleveleys, FY5 2AR
G3 OST D Wilson, Chemin D'Arques, Ambrumesnil, Offranville, France, 76550
G3 OTD W Spilman, 73 Wainfleet Road, Skegness, PE25 3RZ
G3 OTH Charles Cook, Swiss Cottage, Netherton Lane, Bedlington, NE22 6DR
G3 OTK R Harris, 10 South Street, South Petherton, TA13 5AD
G3 OTN P Seaman, 5 Berkeley Close, Maidenhead, SL6 5JP
G3 OTR M Beckley, Mallards, Albury Road, Ware, SG11 2DN
GI3 OTU A Burge, 38 Bayview Road, Bangor, BT19 6AR
G3 OTV Paul O'Kane, 36 Coolkill, Sandyford, Dublin, Ireland, 18
G3 OTW W Miller, 418 Old Chester Road, Birkenhead, CH42 4PD
G3 OTY Robin Cogzell, 242 Clydesdale Tower, Holloway Head, Birmingham, B1 1UJ

G3 OUA David Tarr, 17 Allendale Avenue, Findon Valley, Worthing, BN14 0AH
G3 OUC P Painting, 15 Turnpike Road, Shaw, Newbury, RG14 2ND
G3 OUI I Dickinson, 94a Richmond Street, College Park, Adelaide, Australia, 5069
G3 OUT A Walker, High Beacon Farm, Fulletby, Horncastle, LN9 6LB
GM3 OUU G Rennie, 60 Woodend Place, Aberdeen, AB15 6AN
G3 OUV Philip Perkins, 47 Priory Avenue, High Wycombe, HP13 6SN
G3 OVD Robert Baker, Up Yonder, Pant Yr Hesg Rd, Newport, NP1 4TB
G3 OVE Malcolm Brown, 25 Carpenters Lane, West Kirby, Wirral, CH48 7EY
G3 OVH A Abbey, 1 The Fairway, Kirby Muxloe, Leicester, LE9 2EU
G3 OVK John Frearson, 5 Caxton Close, New Whittington, Chesterfield, S43 2EA
G3 OVW M Hubbard, 7 Creake Road, Syderstone, King's Lynn, PE31 8SF
G3 OVX H Hammett, 27 Courtman Road, Tottenham, London, N17 7HT
G3 OWB J Holland Carter, 37 Highfield Avenue, Cambridge, CB4 2AJ
G3 OWE D Saunders, 4a Ullswater Crescent, Radipole, Weymouth, DT3 5HE
G3 OWJ P Jarvis, 44 Torrin Drive, Shrewsbury, SY3 6AW
G3 OWQ J Clarke, 29 Long Brackland, Bury St. Edmunds, IP33 1JH
GM3 OWU Victor Stewart, 9 Baberton Park, Juniper Green, EH14 5DW
G3 OWX John Greany, flat 3 Crete Hill House, Cote House Lane, Bristol, BS9 3UW
GM3 OXA Alan Fosters, 16 Reid Crescent Milnathort, Kinross, KY13 9TB
G3 OXG David Thompson, 31 Sandy Road, Potton, Sandy, SG19 2QQ
G3 OXK J Carson, 23 Whinny Rig, Heathhall, Dumfries, DG1 3RJ
G3 OXL D Westbury, Rose Cottage, Cruise Hill, Redditch, B97 5UA
G3 OXN D Swainson, 4 Grasmere Avenue, Spondon, Derby, DE21 7JZ
G3 OXR Paul Garthwaite, The Grey Horse, Barnsley, S75 2TX
G3 OXS N Rivett, 42 Sunningvale Avenue, Biggin Hill, Westerham, TN16 3BX
GM3 OXX G Burt, Clunie Lodge, Netherdale, Turriff, AB53 4GN
G3 OYB W Waters, 4 Calartha Road, Pendeen, Penzance, TR19 7DZ
GI3 OYG J Semple, 5 Tullaghgore Road, Ballymoney, BT53 6QF
G3 OYL D Gilbert, 348 Willington Road, Kirton End, Boston, PE20 1NU
G3 OYN G Saunders, 17 Chester St., Caversham, Reading, RG4 8JH
G3 OYT G Clinton, 2 Greenways, Abbots Langley, WD5 0EU
G3 OYU Brian Davies, Red Roofs, Crowhurst Road Crowhurst, Lingfield, RH7 6DG
G3 OYX Michael Rignall, Ashdown, Nupend, Stroud, GL6 0PY
GM3 OZB Allan McKay, 2 Osprey Drive, Kilmarnock, KA1 3LQ
G3 OZC J Holstead, 72 Woodlands Avenue, Feniscowles, Blackburn, BB2 5NN
G3 OZD P Cross, 5 Lings Lane, Hatfield, Doncaster, DN7 6AB
G3 OZE J Grainger, 6 Fulford Cross, Fulford, York, YO10 4PB
GM3 OZJ I Morgan, 43 Dalgety Gardens, Dalgety Bay, Dunfermline, KY11 9LF
G3 OZK Michael James, 36 Iver Lane, Iver, SL0 9LF
G3 OZL A Jeavons, Wadsley Grove, Worrall Road, Sheffield, S6 4BE
G3 OZN E Badger, 20 Tennyson Drive, Worksop, S81 0EE
G3 OZP P Smith, 39 Sherborne Avenue, North Shields, NE29 8NT
G3 OZT R German, 10 Beverley Road, Dibden Purlieu, Southampton, SO45 4HS
GI3 OZW P Dynes, 1 Rossin View, Donaghmore, Dungannon, BT70 1SZ
G3 OZZ J Ramsay, Fairview, Briar Close, Hastings, TN35 4DP
G3 PAG J Davies, Cedar Croft, School Lane, West Malling, ME19 5EH
G3 PAI J Rabson, 60 Bixley Road, Ipswich, IP3 8PG
GM3 PAK M Senior, The Raw, Bridgend, Isle of Islay, PA44 7PZ
G3 PAQ J Davis, 76 Allfarthing Lane, London, SW18 2AJ
G3 PAX J Barker, 2 Barons Hall Lane, Fakenham, NR21 8HB
G3 PBF J Orford, 63 Flowerhill Way, Istead Rise, Gravesend, DA13 9DS
G3 PBI A Davies, 69 Sycamore Road, Chalfont St. Giles, HP8 4LG
G3 PBR A Green, 6 Shipley Close, Woodley, Reading, RG5 4RT
G3 PBT R Hilsley, 1 Chelmerton Avenue, Great Baddow, Chelmsford, CM2 9RE
G3 PCG Dave Askew, Lapthorne, Adsborough, Taunton, TA2 8RP
G3 PCJ T Walford, Upton Bridge Farm, Long Sutton, Langport, TA10 9NJ
G3 PCL Donald Shaw, 3 Randolph Close, Cheltenham, GL53 7RT
G3 PCT P Hurst, Anchorage House, Upper Wood Lane, Dartmouth, TQ6 0DQ
G3 PCW M Watling, 8 Preetz Way, Blandford Forum, DT11 7XG
G3 PCX B Dodge, 34 Downs Road, Penenden Heath, Maidstone, ME14 2JN
G3 PCY John Wilson, Pantawel, Cilmery, Builth Wells, LD2 3PB
G3 PDC R Curwen, 53 Karslake Road, Liverpool, L18 1EY
G3 PDD J Dolby, Oaklea, School Lane, Belper, DE56 2AL
G3 PDE D Paterson, 19 Allison House, Westward Road, Southampton, SO3 4NR
G3 PDH Malcolm Prestwood, Salatiga, Bell Lane, Norwich, NR13 6RR
GI3 PDN R Harbison, 26 Ballymartin Road, Templepatrick, Ballyclare, BT39 0BW
G3 PDP A Ralls, 12 Oakhill Close, Bursledon, Southampton, SO31 1AP
GM3 PDX J Barker, 44 Priory Road, Linlithgow, EH49 6BS
G3 PEJ P Watson, 37 Chestnut Bank, Scarborough, YO12 5QJ
G3 PEK B Simpson, 20 Monterey Street, St Ives, Australia, NSW 2075
G3 PEM C Thomson, 109 Hillside Grove, Chelmsford, CM2 9DD
G3 PEN David Penny, 79 Grove Avenue, New Costessey, Norwich, NR5 0JA
G3 PET A Widdowson, 34 Highfields Road, Chasetown, Burntwood, WS7 4QU
G3 PEW John Hudson, 68 Lower Street, Stansted, CM24 8LR
GW3 PEX Leslie France, 8 Conway Drive, Cwmbach, Aberdare, CF44 0LL
G3 PEZ John Gutteridge, 66 Croft Drive, Moreton, Wirral, CH46 0QT
G3 PFE G Spriggs, Brookbank Cottage, Newcastle Road, Nantwich, CW5 7EJ
G3 PFH M Blunden, 24 Mill View Close, Woodbridge, IP12 4HR
G3 PFJ J Harris, 3 Chimney Mills, West Stow, Bury St. Edmunds, IP28 6ES
G3 PFM A Baker, Highleaze, Deans Drove, Poole, BH16 6EQ
G3 PFO C Barr, Riders Way, Collum Green Road, Slough, SL2 4AX
G3 PFT A Heeley, 108 Valley Lane, Lichfield, WS13 6ST
G3 PFV Keith Robbins, 1 Rhiw Parc Road, Abertillery, NP13 1BS
G3 PFX C Small, Overlangs, Kingston, Kingsbridge, TQ7 4PF
G3 PGA A Hammond, 23 St. Andrews Road, Fremington, Barnstaple, EX31 3BS
G3 PGC Ralph Armstrong, 6 Barnstaple Road, North Shields, NE29 8QA
GW3 PGJ R Bashford, Bwlch Y Fforest, Llandovery, SA20 0US
G3 PGK Colline Pearless, 26 Church Road, Preston, Weymouth, DT3 6RP
G3 PGN H Buckenham, Tweed Cottage, Tilbury Road, Halstead, CO9 4JG
G3 PGQ David Yates, 159 Church Road, Kessingland, Lowestoft, NR33 7SQ
GM3 PGY Anthony Mc Ewen, 4 Reef Terrace, Crossapol, Isle of Tiree, PA77 6UT
G3 PHD I Gardiner, 189 Brennan Road, Tilbury, RM18 8BA
G3 PHG Alan Gibbs, 223 Crimea Street, Noranda, Australia, WA 6062
G3 PHJ J Johnston, 9 Appleby Glade, Haxby, York, YO32 3YW
G3 PHL B Davies, 17 Linksway, Leigh-on-Sea, SS9 4QY
G3 PHO Peter Day, Sheffield HF DX Group, 38 Broomhill Road, Chesterfield, S41 9DA
G3 PIA HARWELL ARS c/o Colin Desborough, 22 Westland Road, Faringdon, SN7 7EY
G3 PID P Chandler, 528 Goffs Lane, Goffs Oak, Waltham Cross, EN7 5EW
G3 PIJ Peter Mellett, 16 Tutton Hill, Colerne, Chippenham, SN14 8DN
GM3 PIL Raymond Munro, 20 County Cottages, Piperhill, Nairn, IV12 5SE
G3 PIN J Patten, 8 Leacroft Road, Penkridge, Stafford, ST19 5BX
G3 PIO C Owen, 13 Brynffynnon, Star, Gaerwen, LL60 6BA
G3 PIY Charles Isaacs, Holme View, Brick Lane, Christchurch, BH23 8DU
G3 PIZ Timothy Watts, 28 Woodger Close, Guildford, GU4 7XR
G3 PJB Peter Bailey, 34 Pinks Hill, Swanley, BR8 8NJ
G3 PJC C Arnold, 47 Peartree Lane, Danbury, Chelmsford, CM3 4LS
G3 PJQ Anthony Aldridge, 1 Mary Grove, Highnam, Gloucester, GL2 8NH
G3 PJT Robert Whelan, 99 Green End, Comberton, Cambridge, CB23 7DY
G3 PJV P Walsh, 23 Moss Fold Road, Darwen, BB3 0AQ
G3 PJW R Unsworth, 6 Coleridge Road, Billinge, Wigan, WN5 7EB
G3 PJY Raymond Millman, 103 The Crescent, Walsall, WS1 2DA
G3 PKC J Tinker, 72 Jackson Avenue, Leeds, LS8 1NB
G3 PKD R Sharples, 40 Greetham Road, Cottesmore, Oakham, LE15 7DR
G3 PKL C Fox, 2 Mill Cottages, Wareham Road, Poole, BH16 6FT
G3 PKQ J Holmes, 36 Hillside Gardens, Walthamstow, London, E17 3RJ
G3 PKR K Parker, 263 High Street, Hayes, UB3 5ET
GM3 PKV H Thornton, 8 Ferry Row, Fairlie, Largs, KA29 0AJ
G3 PKY P Okelly, The Ravel, School Lane, Drogheda, Ireland
GW3 PLB Ronald Howe, Brooklands, Caeffynnon, Kidwelly, SA17 5EJ
G3 PLE David Barlow, Pine, Churchtown, Helston, TR12 7BW
G3 PLJ P Fairnington, 30 Orchard Street, Weston Super Mare, BS23 1RQ
GI3 PLL R Moore, 818 Seacoast Road, Castlerock, Coleraine, BT51 4SD
G3 PLN Graham Smith, 7 Coniston Avenue, Grimsby, DN33 3EE
GM3 PLO J gray, Norland, South End, Stromness, KW16 3DJ
G3 PLP Robert Cox, 30 Brooks Road, Sutton Coldfield, B72 1HP
G3 PLR D Skye, 16 Lulworth Avenue, Poole, BH15 4DQ
G3 PLT Gordon Lawes, 7 Tormynton Road, Weston-Super-Mare, BS22 9HU
G3 PLW John Norton, 32 Fismes Way, Wem, Shrewsbury, SY4 5YD
G3 PLX John Martinez, High Blakebank Farm, Underbarrow, Kendal, LA8 8HP
G3 PLY George McNeil, 168 Chobham Road, Ascot, SL5 0HU
GM3 PMB W Miller, Whiteleys Farm, Ayr, KA7 4EG
G3 PMH March And District Radio Amateur Society c/o E Campbell, 2 Russell Avenue, March, PE15 8EL
G3 PMJ S Revell, 11 Mere Fold, Worsley, Manchester, M28 0SX
GM3 PML David Smith, East Neuk, Netherley, Stonehaven, AB39 3RB
G3 PMO A Spencer, 297 Liverpool Road, Walmer Bridge, Preston, PR4 5QD
G3 PMR Alan Jubb, Psathi Village, Pafos, Cyprus, 8749
G3 PMV A Feist, 1 Lowry Drive, Marple Bridge, Stockport, SK6 5BR
G3 PMW K Dews, 14 Baddow Place Avenue, Great Baddow, Chelmsford, CM2 7JN
G3 PND S Appleyard, Plumtree House, Mill Lane, Cromer, NR27 9PH
G3 PNO I Hawkins, Victoria House, Victoria Street, Totnes, TQ9 5EF
G3 PNP J Ward, 90 Monarch Road, Eaton Socon, St. Neots, PE19 8DF
G3 PNQ A Floyd, 27 Beechfield, Parbold, Wigan, WN8 7AR
G3 PNT C Durell, 17 Ryders Avenue, Westgate-on-Sea, CT8 8LW
G3 PNU E Clark, 1 Station Road, Drigg, Holmrook, CA19 1XH
G3 POG David Mawdsley, 20 Cable Street, Formby, Liverpool, L37 3LX
GM3 POI C Penna, North Windbreck, Deerness, Orkney, KW17 2QL
G3 POM Guy Morgan, 7 Quantock Grove, Williton, Taunton, TA4 4PD
G3 POQ P Hayes, 16 Melton Drive, Storrington, Pulborough, RH20 4LU
GI3 POS A Smyth, 91a Gilford Road, Lurgan, Craigavon, BT66 7EB
GM3 POT J Walford, Chorcaill, Reay, Thurso, KW14 7RG
GW3 PPB J Perkins, 12 The Promenade, Swansea, SA1 6EN
G3 PPC Douglas Taylerson, 18 The Grove, Teddington, TW11 8AS
G3 PPE M Eccles, Newtonlees Bungalow, Kelso, TD5 7SZ
G3 PPO L Hook, 79 Whiteley Crescent, Bletchley, Milton Keynes, MK3 5DQ
GW3 PPQ N Mackinnon, Fairybank, Grondre, Dyfed, SA66 7HH
G3 PPR J Beavon, 24 Cromer Road, Mundesley, Norwich, NR11 8BE
G3 PPT Lionel Sear, 4 Mount Pleasant Road, Threemilestone, Truro, TR3 6BB
G3 PPU P Smith, 56 Alphington Avenue, Frimley, Camberley, GU16 8LR
G3 PQA J Rogers, Dromore, Strande Lane, Maidenhead, SL6 9DN
G3 PQB Sidney Harbour, 43 Warbon Avenue, Peterborough, PE1 3DS
G3 PQC P Turk, 13 The Crescent, Farnborough, GU14 7AR
G3 PQD Derek St John, 26 Henry Street, Rainham, Gillingham, ME8 8HE
G3 PQF D Dell, 7 Blunden Road, Cove, Farnborough, GU14 8QJ
G3 PQJ Brian Cole, 17 Coburg Court, East Cowes, PO32 6SS
G3 PQM M Thorp, Cecil, Thorrington Road, Clacton-on-Sea, CO16 9ES
G3 PQP Tom Foster, 136 Sladepool Farm Road, Kings Heath, Birmingham, B14 5EF
G3 PQY J Lawrence, 2a Hall Road, Hull, HU6 8SA
G3 PRC PLYMOUTH RADIO COMMUNITY c/o Peter Connor, 20 Longfield, Lutton, Ivybridge, PL21 9SN
G3 PRE W Armstrong, 24 Newbury Street, South Shields, NE33 4UF
G3 PRF Richard Findlay, 8 Sycamore Court, Moor Street, Derby, DE21 7EA
G3 PRH M Coward, 51 Farleigh Road, Backwell, Bristol, BS48 3PB
G3 PRI David Quigley, 1a Elizabeth Road, Bishop's Stortford, CM23 3RJ
G3 PRK Ahmet Yilmaz, 26 The Crest, London, N13 5JT
G3 PRL David Snow, Rhwngyddwy Dre, Brynsiencyn, Llanfairpwllgwyngyll, LL61 6TZ
G3 PRO R Evans, West Winds, Main Street, York, YO23 7DA
G3 PRQ E Woodon, Mullins, Windsor Road, Alton, GU34 5EF
G3 PRR Ian Partridge, 4 Thames Street, Louth, LN11 7AD
G3 PRU John Nicholas-Letch, Honey Trees, Furlong Drove, King's Lynn, PE33 9SX
G3 PRW John Dolan, 24 Pen Derwydd, Llangefni, LL77 7QE
G3 PSA A McCann, 5 Arrowsmith Drive, Hoghton, Preston, PR5 0DT
G3 PSC J Holton, 1204 Greenford Road, Greenford, UB6 0HQ
G3 PSG NORTH RIDING RAFARS c/o R Armstrong, 2 West End Road, Laughton, Gainsborough, DN21 3GT
G3 PSM Colin Thomas, 16 Fordlands, Thorpe Willoughby, Selby, YO8 9PD
GM3 PSP Alan Masson, 20 Frogston Avenue, Edinburgh, EH10 7AQ
GI3 PSQ C Bristow, 58 Bristow Park, Belfast, BT9 6TJ
G3 PSR M Gibbs, 62 Abinger Drive, Chatham, ME5 8UL
G3 PSS M Kent, 99 London Road, Newington, Sittingbourne, ME9 7RH
G3 PSU P Martin, 15 St. Lukes Close, Cannock, WS11 1BB
G3 PSV David Park, 18 Widworthy Drive, Broadstone, BH18 9BD
G3 PSW P Spencer, 32 Harfield Road, Sunbury-on-Thames, TW16 5PT
G3 PSZ Kenneth Jones, 24 Station Road, Okehampton, EX20 1EA
G3 PTB A Tomalin, Chapel Street, Barford, Norwich, NR9 4AB
G3 PTG R Quaily, 11 Wrestfield, Eaton Bray, Dunstable, LU6 2JQ
G3 PTI K Atter, 60 Hough Road, Barkston, Grantham, NG32 2NS
G3 PTQ Terry Chapman, 5 Maple Close, Bottisham, Cambridge, CB25 9DQ
G3 PTS G Holt, 7 Beech Close, Olivers Battery, Winchester, SO22 4JY
G3 PTX L Buckley, 188 Compstall Road, Romiley, Stockport, SK6 4JF
G3 PTZ A Bensley, 13 Lime Grove, Cherry Willingham, Lincoln, LN3 4BE
G3 PUO L Rooks, 17 The Close, Clayton le Moors, Accrington, BB5 5RJ
G3 PUQ N Semmens, 4 South Park, Redruth, TR15 3AW
G3 PUR Robert Tarr, 37 Warwick Avenue, Coventry, CV5 6DJ
G3 PUX J Champion, Mill Bungalow, Billingshurst, RH14 9DY
GM3 PUY Iain Forsyth, 68 Drumover Drive, Glasgow, G31 5RP
G3 PUZ D Hogan, 17 Buckingham Mansions, Bath Road, Bournemouth, BH1 2PG
G3 PVB L Santer, 3 Barn Park, Liverton, Newton Abbot, TQ12 6HE
G3 PVG J Bennett, 11 Enderby Road, Thurlaston, Leicester, LE9 7TF
G3 PVH D Sumner, 20 Woodlands Way, Southwater, Horsham, RH13 9HZ
G3 PVJ H Coltman, 68 Cressex Road, High Wycombe, HP10 4TY
G3 PVU J Hunt, 28 Harris Road, Lincoln, LN6 7PN
G3 PWB I Dufour, 3 Western Close, Rushmere St. Andrew, Ipswich, IP4 5UU
G3 PWJ Robert Fisher, 34 Doctors Hill, Stourbridge, DY9 0YE
G3 PWK Jack Braithwaite, 50 Crossways Drive, Harrogate, HG2 7DH
G3 PWN G Grimshaw, 1 Sandsacre Drive, Bridlington, YO16 6UA
G3 PWS Richard Dalton, 23 Muswell Road, Mackworth, Derby, DE22 4HN
G3 PWY David Gresswell, 10 Cherrywood Gardens, Flackwell Heath, High Wycombe, HP10 9AX
G3 PXF A Petts, 88 Northfield Lane, Horbury, Wakefield, WF4 5JF
G3 PXH M Bartlett, 3 Jessopp Avenue, Bridport, DT6 4AN
G3 PXI Anthony Evans, Apartado 286, Luz Lagos, Portugal, 8601-929 LUZ GS
G3 PXL Albert Hickin, Buzon 24, Caseta 1, Cuevas Del Almanzora, Spain, 4610
G3 PXU G Grove, 11 Croft Close, Warwick, CV34 6QY
G3 PXV Robert Wiseman, 3 Springfield Road, Ruskington, Sleaford, NG34 9HG
G3 PXX John Phillips, 18 Rockfarm Drive, Little Neston, Neston, CH64 4DZ
GW3 PYD D Stephens, 1 Awelfryn Terrace, Merthyr Tydfil, CF47 9YP
G3 PYE Cambridgeshire Repeater Group c/o Phillip Nice, 5 Walden Close, Doddington, March, PE15 0TW
G3 PYF J Green, 68 Magdalen Lane, Wingfield, Trowbridge, BA14 9LQ
G3 PYH Arnold Broadbent, 52 Norman Street, Failsworth, Manchester, M35 9EJ
G3 PYI David Coy, 26 Hardy Road, Bishops Cleeve, Cheltenham, GL52 8BN
G3 PYO J Dann, 1 Ffinch Close, Ditton, Aylesford, ME20 6ET
G3 PYP G Hibberd, Flat 15, The Croft, Meadow Drive, Devizes, SN10 3BJ
G3 PYW A Speight, Glebe Cottage, Hollow Lane, Woodbridge, IP13 8LZ
GW3 PYX J Chetcuti, 3 Beechwood Drive, Penarth, CF64 3RB
G3 PYZ Royal Signals ARS c/o John West, 9 Bainbridge Court, St. Helen Auckland, Bishop Auckland, DL14 9EJ
G3 PZB Alan Ash, 34 Coronation Avenue, Cowes, PO31 8PN
G3 PZE C Burkitt, The Old Wheelwright, 17 Oxford Road, Hitchin, SG4 8NP
G3 PZF G Dale, 16 Palfrey Close, St. Albans, AL3 5RE
G3 PZL P Brown, Little Langford, Newlands Lane, Henley-on-Thames, RG9 5PS
G3 PZN C Wood, 24 Talveneth, Pendeen, Penzance, TR19 7UT
G3 PZU B Brown, 138 First Avenue, Sudbury, CO10 1YU
G3 PZV P Greed, The Bungalow, Townsend, Devizes, SN10 4RR
G3 PZX A Ward, 20 Tower Close, Costessey, Norwich, NR8 5AU
G3 PZZ P Smith, 38 Leasway, Wickford, SS12 0HE
G3 QI Flaxton Moor Contest Group c/o C Quarton, Flaxton Gatehouse, Flaxton, York, YO60 7QT
G3 RAC THALES ARC c/o D Waterworth, 116 Reading Road, Woodley, Reading, RG5 3AD
G3 RAF ROYAL AIR FORCE ARS c/o M Garrett, 489 Dorchester Road, Weymouth, DT3 5BP
G3 RAL Loughborough and Disttrict ARC c/o Andrew Harrison, 44 Rosslyn Road, Whitwick, Coalville, LE67 5PT
G3 RAM Christopher Langmaid, Flat 4, Woodlawn High Street, Partridge Green, Horsham, RH13 8HR
G3 RAR E Hodgson, 12 Cornmoor Road, Whickham, Newcastle upon Tyne, NE16 4PU
G3 RAU Derek Moffatt, Mill House, Middle Street, Glentworth, Gainsborough, DN21 5BZ
G3 RBD F Hanson, 207 Grant Road, Liverpool, L14 0LG
G3 RBJ A Payne, Laurel Bank, Sand Road, Wedmore, BS28 4BZ
G3 RBP R Parsons, 120 Cavendish Road, Matlock, DE4 3HE
G3 RBY A Stagles, 8 Goodwood Close, Cowplain, Waterlooville, PO8 8BG
G3 RCB N Kingsley, 1 Wensleydale Gardens, Hampton, TW12 2LU
G3 RCD Christopher Brockbank, 31 Park Hill, Church Crookham, Fleet, GU52 6PW
G3 RCE Robert Albright, 50 Portsdown Road, Portsmouth, PO6 4QH
G3 RCM SHEFFIELD ARC c/o Andrew Bennett, 21 Hunstone Avenue, Norton, Sheffield, S8 8GF
G3 RCQ David Cole, Amber Lights, Market Lane, Wisbech, PE14 7LT
G3 RCV CRAY VALLEY RS c/o Adrian Styles, 6 Hill Brow, Crayford, Dartford, DA1 3NX
G3 RCW WORKSOP ARS c/o Roy Henson, 2 Byron Street, Shirebrook, Mansfield, NG20 8PJ
G3 RCX I Gibson, 7 Heycroft Road, Eastwood, Leigh-on-Sea, SS9 5SW
G3 RCZ G Thompson, 22 Warton Avenue, Heysham, Morecambe, LA3 2LX
G3 RDA S Whitehead, 98 Oak Road, Fareham, PO15 5HP
GW3 RDB HOOVER ARC c/o Thomas George, 80 Yew Street, Troedyrhiw, Merthyr Tydfil, CF48 4EE
G3 RDC A Wood, Orchard View, Sutton Road, Hereford, HR1 3NL
G3 RDF J Jeffrey, Old Church Cottage, Ipsden, Wallingford, OX10 6AE
G3 RDG K Michaelson, 40 The Vale, Golders Green, London, NW11 8SG
G3 RDH James Barnes, 4 Deepdene Drive, Dorking, RH5 4AD
G3 RDN E Sharp, 11 Forge Way, Billingshurst, RH14 9LJ
G3 RDP H Cutts, 50 Cropton Road, Hull, HU5 4LP
G3 RDQ David Griffiths, Upcote Cottage, Chilbolton, Stockbridge, SO20 6BA
G3 RDR P Rudwick, 29 Fuller Street, London, NW4 4RR
G3 RDW A Kendrick, Long Drive, Roman Road, Sutton Coldfield, B74 3AA

**IMPORTANT NOTE**

**Revalidate licence to avoid revocation** – Ofcom has advised the Society that plans will be drawn up to revoke licences that have not been revalidated as required by the licence conditions. The quickest way to revalidate is to do so online via the Ofcom website: *https://services.ofcom.org.uk/* or by email: *amateur.validations@ofcom.org.uk* Ofcom staff are available to help, but please be patient during times of heavy workload.

G3 RDZ J Walker, 94 Keys Park, Parnwell Way, Peterborough, PE1 4SN
G3 RE Stuart Dixon, 33 Medhurst Crescent, Gravesend, DA12 4HJ
G3 REB Roger Cole, Lyndale, Brimscombe Lane, Stroud, GL5 2RF
G3 RED David Sylvester, 10 Ivy Grove, Gunthorpe, Peterborough, PE4 7TW
G3 REH Henry Neale, Thornlea, Fishergate, Spalding, PE12 0EZ
G3 REL B Woodfield, 49 Oakfield Road, Blackwater, Camberley, GU17 9DZ
G3 REM M Austin, 20 Dimple Park, Egerton, Bolton, BL7 9QE
G3 REP R Parkes, 2 Saxon Road, Steyning, BN44 3FP
G3 REU G Hearn, 70 Cranmer St., Long Eaton, Nottingham, NG10 1NL
G3 REV Raymond Pulling, 410 Leach Lane, Sutton Leach, St. Helens, WA9 4NA
G3 REW D Morris, 5 Eden Park, Brixham, TQ5 9LS
GM3 RFA D Garrington, 3 Sutherland Avenue, Fort William, PH33 6JS
G3 RFH Kenneth Randall, 25 Kingsway, Thornton-Cleveleys, FY5 1DL
GD3 RFK Douglas Dodd, Ellan Geay, Ballayockey Lane, Regaby, Ramsey, Isle of Man, IM7 3HP
G3 RFN G Wild, 17 New Church Close, Clayton le Moors, Accrington, BB5 5GH
G3 RGB A Moon, 6 Troon Close, Saltersgill, Middlesbrough, TS4 3HX
G3 RGC T Matthews, 38 Foxhill, Grimsby, DN37 9QL
G3 RGD R Dobdinson, 73 Watwood Road, Hall Green, Birmingham, B28 0TW
G3 RGE K King, BLANDFORD GARRISON A.R.C., COLE BLOCK, Blandford Forum, DT11 8RH
G3 RGJ Richard Weston, 43 Pearce Avenue, Parkstone, Poole, BH14 8EG
G3 RGM D Mullins, Flat 23, Kennington Palace Court, London, SE11 5UL
G3 RGN L Binns, Leamar, 707 Halifax Road, Cleckheaton, BD19 6LJ
G3 RGP R Pratt, 1 Colebrooke Avenue, Ealing, London, W13 8JZ
G3 RGQ T Harding, 73 Balland Field, Willingham, Cambridge, CB24 5JT
G3 RGS D Thomson, Skippers Down, Old Coach Road, Sevenoaks, TN15 7NR
GM3 RGU Joseph Connelly, 9 Glenhead Crescent, Hardgate, Clydebank, G81 6LW
G3 RHP John Garrett, Shrubbery Farm, Otley, Ipswich, IP6 9PD
G3 RHQ Keith Vickers, Hillview, Barton-upon-Humber, DN18 5DZ
G3 RHR K Drinkwater, Brearton Lodge, Brearton, Harrogate, HG3 3BX
G3 RHU M Stanbridge, 183 Charlton Park, Midsomer Norton, Radstock, BA3 4BR
G3 RHW C Cushion, 3 The Copse, Bridgwater, TA6 4DW
G3 RHZ A Wilkinson, 18 Tansey Crescent, Stoney Stanton, Leicester, LE9 4BT
G3 RIB W Huxley, No2 Bungalow, Nant Mawr Rd, Clwyd, CH7 2BS
G3 RID David Nancarrow, 6 Trythogga Road, Gulval, Penzance, TR18 3NA
GW3 RIH Wesley Elton, 15 Main Avenue, Peterston-super-Ely, Cardiff, CF5 6LQ
G3 RIK Dave Carden, 9 Wood Hey Grove, Rochdale, OL12 9TY
G3 RIM T Emeney, 10 Kilnside, Claygate, Esher, KT10 0HS
G3 RIR Neil Ackerley, 24 Macaulay Road, Lutterworth, LE17 4XB
G3 RIX Martin Tetley, 87 Main Street, Irton, Scarborough, YO12 4RJ
GW3 RIY A Chapman, 14 Birch Walk, Danygraig, Porthcawl, CF36 5AN
G3 RJE J Hunt, 33 Rainhill Road, Rainhill, Prescot, L35 4PA
G3 RJF I Walker, 28 Norrington Road, Maidstone, ME15 9RA
G3 RJH R Harding, High Trees, Arrowsmith Road, Wimborne, BH21 3BG
G3 RJI Alan PAUL, 3 Brunswick Avenue, Upminster, RM14 1NA
G3 RJM R Cutts, 60 Holmpton Road, Withernsea, HU19 2QD
G3 RJS P Barry, 235 Manor Way, Aldwick, Bognor Regis, PO21 4HT
G3 RJT C Garland, 48 Underbank End Road, Holmfirth, HD9 1ES
G3 RJV G Dobbs, 9 Highlands, Littleborough, OL15 0DS
G3 RKF Terence Roeves, 33 York Crescent, Wilmslow, SK9 2BB
G3 RKH John Marshall, 166 Calton Road, Gloucester, GL1 5ER
G3 RKJ Neil Summers, 126 Kestrel House, 1 Alma Road, Enfield, EN3 4QE
G3 RKK A Shepherd, 59 Lime Avenue, Camberley, GU15 2BH
G3 RKL A Whitaker, 160 Derbyshire Lane, Sheffield, S8 8SE
G3 RKM J Meaker, 11 Woodend View, Mossley, Ashton-under-Lyne, OL5 0SN
G3 RKP Don Evetts, 35 Wood End Road, Kempston, Bedford, MK43 9BB
G3 RKQ A Balmforth, Leam Brink, Lutton Gowts, Spalding, PE12 9LQ
GW3 RKV Richard Volck, Maes Y Bryn, Rosebush, Dyfed West Wales, SA66 7QS
G3 RKZ Brian Tibbert, 99a Main Street, Horsley Woodhouse, Ilkeston, DE7 6AW
G3 RLA Colin Phillips, Bella Vista, The Moorings, Wirral, CH60 9JT
G3 RLD R Ramshaw, 132 Main Road, Duston, Northampton, NN5 6RA
G3 RLE B Turner, 56 Bamford Way, Rochdale, OL11 5NB
G3 RLF D Price, 8 Newland Road, Droitwich, WR9 7AF
G3 RLJ John Harper, East View, 35 John Street, Sutton in Ashfield, NG17 4EN
G3 RLL D Grindell, 23 Park Hall Avenue, Walton, Chesterfield, S42 7LR
G3 RLO D Cadman, 32 Breedon Hill Road, Derby, DE23 6TG
G3 RLT W Stewart, 4 Denmark Road, Kingston upon Thames, KT1 2RU
G3 RLV M Vann, 3 Mile Planting, Richmond, DL10 5DB
G3 RMD Francis Regan, 7 Hilltop Road, Cheltenham, GL50 4NW
G3 RMF Brendan Magill, 14 Barry Street, Worcester, WR1 1NR
G3 RMJ Philip Jennings, Myrtle Cottage, Harpers Lane, Presteigne, LD8 2AN
G3 RMK R Ruaux, Park View, Wallage Lane, Crawley, RH10 4NG
G3 RMN M Smith, 121 Shirley Way, Croydon, CR0 8PN
G3 RMQ J Ingham, High Croft, 1 Layton Crescent, Leeds, LS19 6RJ
G3 RMX W Hall, 52 Barley Gate, Leven, Beverley, HU17 5NU
G3 RMY J Andrews, 12 Gerald Close, Burgess Hill, RH15 0NB
G3 RMZ A pink, 37 Shute Park Road, Plymouth, PL9 8RB
GI3 RNO P Greenan, 9 Ashville Park, Antrim, BT41 1HH
G3 RNP J Price, 125 Oakfield Road, Malvern, WR14 1DT
G3 RNV C Galloway, 105 Dumbarton Road, Stockport, SK5 7EX
G3 RNX W Walker, 44 South Road, Weston-Super-Mare, BS23 2HE
G3 ROC R Collins, Thorn Acacia, Rye Road, Rye, TN31 6NJ
G3 ROD R Davenport, 7 Nether Close, Duffield, Belper, DE56 4DR
G3 ROG Geoffrey Morgan, 22 Monks Road, Winchester, SO23 7EQ
G3 ROM B Sweetman, Cortijasa Los Perez 220, Benajarafe, Malaga, Spain, 29790
G3 ROO Ian Keyser, Rosemount, Church Whitfield Road, Dover, CT16 3HZ
G3 ROP Maurice Goodrick, 18 Milford Street, Cambridge, CB1 2LP
G3 ROQ R Gill, 45 Biggin Lane, Ramsey, Huntingdon, PE26 1NB
G3 ROS H Williams, Roslyn, Whalley Road, Burnley, BB12 7HT
G3 ROW S Smith, 810 West 6th Street, Silver City, United States, 88061
G3 RPA J Knowles, Springhill, Gilpins Ride, Berkhamstead, HP4 2PD
G3 RPB Keith Spicer, Grove Cottage, Dallinghoo, Woodbridge, IP13 0LR
G3 RPD G Clinch, 2 Storrs Close, Bovey Tracey, Newton Abbot, TQ13 9HR
G3 RPL T Neyland, 22 Pax Hill, Bedford, MK41 8BT
GM3 RPM J McAvoy, 120 Donaldswood Road, Paisley, PA2 8EB
G3 RPO F Seddon, 23 Countessway, Euxton, Chorley, PR7 6PT
G3 RPV T Venn, 22 Eaton Close, Hartford, Huntingdon, PE29 1SR
G3 RPZ H Trunley, Bijou House, 75 Belgrave Road, Leigh-on-Sea, SS9 5EL
G3 RQF Duncan Keith, 108 Lower Northam Road, Hedge End, Southampton, SO30 4FT
GM3 RQQ Hugh Robertson, 102 Orchy Crescent, Bearsden, Glasgow, G61 1RE
G3 RQR N Kirtley, 14 Byron Avenue, Winchester, SO22 5AT
G3 RQS R Rimmer, 25 Haig Court, Chesterton, Cambridge, CB4 1TT
GI3 RQU S Laverty, 572 Antrim Road, Belfast, BT15 5GL
G3 RQX P Lewis, 20 Osborne Road, Penn, Wolverhampton, WV4 4AY
G3 RQZ Peter Madagan, 40 Lagham Park, South Godstone, Godstone, RH9 8ER
G3 RRG P Taylor, 44 Leegate Road, Stockport, SK4 4AX
G3 RRI Richard Wilmot, Swinside, Branthwaite Lane, Workington, CA14 1HE
G3 RRM J Hughes, 41 Highfield Avenue, Great Sankey, Warrington, WA5 2TW
G3 RRN K Jones, Field House, Wragby Road, Lincoln, LN2 2QU
G3 RRP R Pine, 21 Hatherden Avenue, Poole, BH14 0PJ
G3 RRS Rutherford Appleton Laboratory ARC c/o J Wright, 2 Barnfield, Charney Bassett, Wantage, OX12 0HA
G3 RRW J Francis, 5 Central Park Avenue, Plymouth, PL4 6NW
G3 RSB Raymond Scaife, 7 Woodgates Close, North Ferriby, HU14 3JS
G3 RSC Sutton Coldfield RS c/o Barry Adkins, 4 Orion Close, Ward End, Birmingham, B8 2AU
G3 RSD J Reynolds, 6 Fairfield Court, Cleethorpes, DN35 0QW
G3 RSE C Cheney, 35 Metcalfe Road, Cambridge, CB4 2DB
G3 RSF Alan Notschild, 8 Hillpark, BUCKLAND BREWER, Bideford, EX39 5HY
G3 RSI Finlay McKeracher, Wickets, 1 Marshal Close, Alton, GU34 1RA
G3 RSM Frederick Burnett, 4 Woodlands Drive Fulwood, Preston, PR2 9SQ
G3 RSP A Pampling, 47 Altham Grove, Harlow, CM20 2PQ
G3 RST R Southern, 30 Barnfield, Crowborough, TN6 2RY
G3 RSU D Bindon, Forth House, Water Street, Langport, TA10 0HH
G3 RSV R Dowsett, 23 South Wootton Lane, King's Lynn, PE30 3BS
G3 RSW W Mullarkey, The Barn, Skull House Lane, Wigan, WN6 9DJ
GW3 RTA W Lewis, 5 Galon Uchaf Road, Merthyr Tydfil, CF47 9TP
G3 RTB R Bell, 14 Wacker Field Road, Rendlesham, Woodbridge, IP12 2UT
G3 RTD J Gailer, Shelleys, King Stag, Sturminster Newton, DT10 2BE
G3 RTE G Kellaway, 55 Ladbrooke Drive, Potters Bar, EN6 1QW
GM3 RTJ Graham Henderson, Tigh An Drochaid, Kilchrenan, Argyll, PA35 1HD
G3 RTM Alan Chaddock, 2 Willow Lane, Milton, Abingdon, OX14 4EG
G3 RTO N Pratt, 87a Dovecote Road, Newthorpe, Nottingham, NG16 3QL
G3 RTP J Pennington, Brambling, Forest Road, Waterlooville, PO7 6UE
G3 RTR R Rowse, Polvarth, Park Lane, Chester, CH4 0HN
G3 RTY H Meers, 10 Lawnswood Avenue, Chasetown, Burntwood, WS7 4YD
G3 RUD E Workman, Sunset, 2 Burnham Drive, Weston-Super-Mare, BS24 9LW
G3 RUE E Edwards, Ceris, Ruthin Road, Denbigh, LL16 3EU
G3 RUG G Twiss, 9 Brae Head, Eaglescliffe, Stockton-on-Tees, TS16 9HP
G3 RUH J Miller, 3 Bennys Way, Coton, Cambridge, CB23 7PS
GM3 RUI Roy Furness, 43 Glebe Street, Leven, KY8 4QN
G3 RUJ Roger Powell, 13 Bridges Drive, Bristol, BS16 2UB
G3 RUO W Williamson, 84 Atfield Drive, Whetstone, Leicester, LE8 3NE
GM3 RUP C Morton, 295 Byres Road, Hillhead, Glasgow, G12 8TL
G3 RUV A James, 4 The Chestnuts, Aylesbeare, Exeter, EX5 2BY
G3 RUZ D Martin, 4 Chilbolton Mews, 19 Chilbolton Avenue, Winchester, SO22 5HU
G3 RVA R Crowe, 37 Huccaby Close, Brixham, TQ5 0RJ
G3 RVC Peter Cochrane, Willow Barn, East Lane, Woodbridge, IP13 6EB
GW3 RVG sidney sedgebeer, 50 Minffrwd Road Pencoed, Bridgend, CF35 6SD
G3 RVI J Walch, 52 Marsh House Road, Sheffield, S11 9SP
GM3 RVL Harry Brash, 5 Hillview Drive, Edinburgh, EH12 8QW
G3 RVM C Trusson, 27a Roman Way, Thatcham, RG18 3BP
G3 RVX J Colegate, 1 Oldmere Cottages, High Street, Bath, BA1 7TJ
G3 RVY P Colegate, 65 Forest Road, Melksham, SN12 7AB
G3 RWE T Yates, 3 Sycamore Crescent, Macclesfield, SK11 8LL
G3 RWF Peter Henwood, Conifers, Church Road, Canterbury, CT3 1UA
G3 RWI P Cross, Home Farm House, Icomb, Cheltenham, GL54 1JD
G3 RWL R Limebear, 60 Willow Road, Enfield, EN1 3NQ
G3 RWP Craig Bell, 2 The Pastures, Long Bennington, Newark, NG23 5EG
G3 RWQ Colin Bray, 313 Droitwich Road, Claines, Worcester, WR3 7SR
G3 RWV Michael Sanders, 7 Netherby Close, Tring, HP23 5PJ
G3 RWW G Southern, 27 Eldred Road, Liverpool, L16 8NZ
GW3 RWX David Thomas, 88 Cefn Graig, Rhiwbina, Cardiff, CF14 6JZ
G3 RXA J Thomas, Blair House, Market Place, Norwich, NR16 2AN
G3 RXD G Llewelyn, Bayano, 33 Awelfryn, Amlwch, LL68 9DG
G3 RXG R Burgess, 11 Beech Road, Shipham, Winscombe, BS25 1SA
G3 RXI E Blundell, 29 Garden Close, Hook, RG27 9QZ
G3 RXM J Hope, 1 Drummond Close, Bracknell, RG12 2QG
G3 RXO Roger Brown, Lower Dicker, Hailsham, BN27 4BG
G3 RXP David Mason, 5 Spa Top, Caistor, Market Rasen, LN7 6RB
G3 RXS W Scarlett, 14 Warren Drive, Bingley, BD16 3BX
G3 RXU I Macpherson, 1 Broomie Dell, Earlston, TD4 6BN
GI3 RXV N Graham, 3 Shilgrove Place, Castledawson, Magherafelt, BT45 8AL
GM3 RXZ R Marshall, 52 Lumsdaine Drive, Dalgety Bay, Dunfermline, KY11 9YU
GW3 RYE J Harris, Treweryn, Llwyndafydd Road, Llandysul, SA44 6BT
G3 RYH J Brodie, Huntlands Farm, Gaines Road, Worcester, WR6 5RD
G3 RYK I Grayson, 156 Little Brays, Harlow, CM18 6EY
G3 RYP D Craggs, New House, Dacre Banks, Harrogate, HG3 4EW
G3 RYQ J Eason, 16 Laburnum Close, Cheshunt, Waltham Cross, EN8 8NB
GW3 RYR Colin Morgan, 33 West Grove, Merthyr Tydfil, CF47 8HJ
G3 RYW D Wardlaw, 21 Tormey Street, Balwyn North, Australia, VIC 3104
G3 RYY Norman Penketh, Greenways, Bolton Road, Chorley, PR7 4AJ
G3 RYZ M Byrne, 16 Downham Gardens, Tamerton Foliot, Plymouth, PL5 4QE
G3 RZC Roy Pellett, La Biochere, Aizenay, France, 85190
G3 RZF David Horton, 26 The Crescent, Slough, SL1 2LQ
G3 RZG Michael Box, 18 Stottingway Street, Weymouth, DT3 5QA
G3 RZI M Moss, 1082 Evesham Road, Astwood Bank, Redditch, B96 6ED
G3 RZJ G Hall, 185 Dialstone Lane, Stockport, SK2 7LQ
G3 RZP Peter Chadwick, Three Oaks, Braydon, Swindon, SN5 0AD
G3 RZV A Lawrance, 97 Dorchester Road, Oakdale, Poole, BH15 3QZ
G3 RZY C Abrey, 31 Yew Tree Lane, Leeds, LS15 9JD
G3 SAD Stevenage & District A.R.S c/o Robert McTait, 20 Rowland Road, Stevenage, SG1 1TE
GM3 SAE R McMillan, 54 Birchwood, Invergordon, IV18 0BG
G3 SAH R Matthews, 14 Parmington Close, Callow Hill, Redditch, B97 5YL
GM3 SAN Simpson Weir, 19 Ellismuir Road, Baillieston, Glasgow, G69 7HW
G3 SAO John Midgley, 3 Chipping Fold, Milnrow, Rochdale, OL16 4YD
G3 SAQ A Coles, 6 Eden Park, Lancaster, LA1 4SJ
G3 SAR R Warner, Cubs Wood, Rycroft Lane, Sevenoaks, TN14 6HT
G3 SB T Bryant, Trenewydd Residential Home, Ger Y Tarell, Brecon, LD3 8DE
G3 SBA Richard Marshall, 30 Ox Lane, Harpenden, AL5 4HE
GM3 SBC E Murphy, 65 Silverknowes Crescent, Edinburgh, EH4 5JA
G3 SBF B Eames, 4 Dabey Close, Markfield, LE67 9UJ
G3 SBI C Horrabin, Ivydene, Barker Lane, Blackburn, BB2 7ED
G3 SBL STAFFORD & DISTRICTS ARS c/o Graham Reay, 53 Tithe Barn Road, Stafford, ST16 3PL
G3 SBM D Turner, 50 Hardings, Chalgrove, Oxford, OX44 7TJ
G3 SBP R Gynn, Tremonde, Kelly, Lifton, PL16 0HH
G3 SBT W Turnbull, 15 Marshallsay Road, Chickerell, Weymouth, DT3 4BB
G3 SCD David Dunn, Littledale, West Road, Horncastle, LN9 6QP
G3 SCJ David Power, 47 Marlborough Street, Gainsborough, DN21 1BT
G3 SCL Roger Houghton, Hans-Miederer-Str. 10b, Schliersee, Germany, 83727
GI3 SCM Thomas McCullough, 16 Mccormack Gardens, Lurgan, Craigavon, BT66 8LE
G3 SCT Thurrock Sea Cadet Corp - Tilbury c/o Nicholas Wilkinson, 12 Woodlands Close, Grays, RM16 2GB
G3 SCV G Stanton, 8 Kennett Close, Norwich, NR4 7JA
GW3 SCX J Naylor, 32 Graig Y Tewgoed, Cwmavon, Port Talbot, SA12 9YE
G3 SCY Clive Seldon, 97 Gunners Road, Shoeburyness, Southend-on-Sea, SS3 9SB
G3 SCZ R Brown, 22 Lordswood, Silchester, Reading, RG7 2PZ
G3 SDC DE MONTFORT UNIVERSITY ARS c/o R Titterington, Wyclif House, St. Marys Road, Lutterworth, LE17 4PS
G3 SDG J Bottom, 48 Chesterton Avenue, Harpenden, AL5 5SU
G3 SDH P Kelly, Martyndale, The Street, Bristol, BS40 6JE
G3 SDL David Court, Connogue, River Lane, Shankill, Ireland, DUBLIN 18
G3 SDO K Heathfield, 2 Georgian Close, Broadway, Weymouth, DT3 5PF
G3 SDS SOUTH DORSET RS c/o Geoffrey Watts, 3 Maple Grove Knightsdale Road, Weymouth, DT4 0FE
G3 SDT D Allen, Chelsea Cottage, The Turnpike, Norwich, NR16 1RS
G3 SDW Kenneth Underwood, 43 Belmont Road, Kirkby-in-Ashfield, Nottingham, NG17 9DY
G3 SDY Gerald Edinburgh, 77 Westerley Lane, Shelley, Huddersfield, HD8 8HP
G3 SEA Paul Perretta, 1511 Punahou St., Apt 208, Honolulu, United States, 96822
G3 SED E Devereux, 191 Botley Road, Burridge, Southampton, SO31 1BJ
G3 SEF R Frew, Sawley House, 82 Wormholt Road, London, W12 0LP
G3 SEG W Gordon, 55 Trajan Avenue, South Shields, NE33 2AN
G3 SEJ E John, Obo St. Dunstans Ars, 52 Broadway Avenue, Wallasey, CH45 6TD
G3 SEK I White, 2 Appleby Cottages, Whithorn, Newton Stewart, DG8 8DQ
G3 SEM P Cort-Wright, Redlands, 11-13 Hardingham Street, Norwich, NR9 4JB
G3 SEN Ronald Dawes, 18 Sutherland Road, Nottingham, NG3 7AP
G3 SEQ John Crossfield, Forest Lodge, Chopwell Wood, Rowlands Gill, NE39 1LT
GM3 SER H Bremner, 2 Rowan Crescent, Lenzie, Glasgow, G66 4RE
G3 SES Philip Stevens, 20 Abbots Park, Chester, CH1 4AN
G3 SET G Aram, 5 Lancaster Green, Hemswell Cliff, Gainsborough, DN21 5TQ
G3 SEY R Mackey, The Tudor, 44 South Street, Ossett, WF5 8LF
G3 SEZ P Luft, Swan House, Livesey Road, Ludlow, SY8 1EY
G3 SFB C Hale, 16 Windmill Court, East Wittering, Chichester, PO20 8RJ
GW3 SFC Alwyn Richards, 30 Well Place, Aberdare, CF44 0PB
G3 SFE Pierre Everett, 58 Greenwood Avenue, Bognor Regis, PO22 9EX
G3 SFG SOUTHGATE ARC c/o Donald Berry, 4 Holly Hill, London, N21 1NP
G3 SFK P Kerry, 251 Upper Rainham Road, Hornchurch, RM12 4EY
G3 SFM A Bailey, 12 Nelson Road, Rainham, RM13 8AL
GW3 SFQ R Mugford, 27 Highfield Close, Dinas Powys, CF64 4LR
G3 SFU P Woodfield, 49 Oakfield Road, Blackwater, Camberley, GU17 9DZ
G3 SFV E Meachen, 46 Rainsborough Gardens, Market Harborough, LE16 9LW
GI3 SG Jack Patty, 3c Finwood Park, Belfast, BT9 6QR
G3 SGA A Jones, Po Box 355, Crestholme 3652, Natal, South Africa
G3 SGC Graham Morris, Norcrest, Beach Road, Norwich, NR12 0AL
G3 SGF P Casemore, 9 Wellcroft Cottages, Church Lane, Hassocks, BN6 9BZ
G3 SGK B King, 41 Ridgeway Avenue, Newport, NP20 5AH
G3 SGL Ana Isaacs, Holme View, Brick Lane, Christchurch, BH23 8DU
G3 SGR John Craig, Ferndown, Tilley Lane, Hailsham, BN27 4UT
G3 SGS J Clements, 147 Luton Road, Dunstable, LU5 4LP
G3 SGV John Fallon, Ryelands, Carkeel, Saltash, PL12 6PH
G3 SGX R Bona, 1 Maxwell Road, Broadstone, BH18 9JG
G3 SGY Arnold Nesbitt, 28 Fairfax Road, Middleton St. George, Darlington, DL2 1HF
G3 SGZ T Chapple, 39 Maynards Park, Bere Alston, Yelverton, PL20 7AR
G3 SHD L Dray, 1 Chalfont Close, Bradville, Milton Keynes, MK13 7HS
G3 SHF Bernard Naylor, 47 Chester Road, Poynton, Stockport, SK12 1HA
G3 SHK Richard Pett, 5 Kingford Close, Woodfalls, Salisbury, SP5 2NQ
G3 SHL John Harlow, 19, 5th St, Section L, Fairview Park, Yuen Long, Hong Kong, NT
GM3 SHR J Coster, 17 Glamis Place, Dalgety Bay, Dunfermline, KY11 9UA
G3 SHX R West, 15 St. Andrews Close, Margate, CT9 4HA
G3 SHY Richard Cottrell, 157 Ridge Lane, Watford, WD17 4SU
G3 SHZ J Whittington, Twyford Manor, Bicester Road, Buckingham, MK18 4EL
G3 SIA B Keyte, 9 Swanns Meadow, Bookham, Leatherhead, KT23 4JX
G3 SIG Royal Signals ARS c/o Bryan Downes, 6 Greenland Crescent, Beeston, Nottingham, NG9 5LB
GW3 SIK Keith Pugh, Tanybanc, Blaenporth, Cardigan, SA43 2BD
G3 SIO J Beardsmore, 67 Beachwood Avenue, Kingswinford, DY6 0HL
G3 SIR D Durham, 29 Waverley Road, Stratton St Margaret, Swindon, SN3 4AY
G3 SIT R Kressman, 12 School Lane, Fenstanton, Huntingdon, PE28 9JR
G3 SIU P Hearson, 14 Osgood Gardens, Orpington, BR6 6JU
GW3 SIY G Steele, 5 Golden Close, West Cross, Swansea, SA3 5PE
G3 SJH C Eyles, 9 St. Peters Road, Harborne, Birmingham, B17 0AT
G3 SJI M Batt, 9 Grange Park, Westbury-on-Trym, Bristol, BS9 4BU
G3 SJJ J Burbanks, 16 Cotgrave Road, Plumtree, Nottingham, NG12 5NX
G3 SJK Stephen Cherry, 4 West Hill Road South, South Wonston, Winchester, SO21 3HP
G3 SJR William Tynan, 22 Belchmire Lane, Gosberton, Spalding, PE11 4HG

G3 SJW S Haigh, 8 Spuldhurst Road, Hackney, London, E9 7EH
G3 SJX P Hart, The Willows, Paice Lane, Alton, GU34 5PR
GM0 CJY C Lawrenson, Hollyburn, West Port, Cupar, KY15 7DW
GD0 CKH C Black, 33 Claughbane Drive, Ramsey, Isle of Man, IM8 2BH
G3 SKI Ronald Bravery, 19 Lindum Road, Worthing, BN13 1LX
G3 SKN Denis Naylor, 52 RUE DU PORT, Pontorson, France, 50170
G3 SKR A Gold, 30 Argyll Road, London, W8 7DO
G3 SKV S Hobday, 31 Sackville Crescent, Harold Wood, Romford, RM3 0EJ
G3 SKY IOW RAD SOC c/o Alan Ash, 34 Coronation Avenue, Cowes, PO31 8PN
GD3 SKZ K Manktelow, Tramman House, Ballabeg, Isle of Man, IM9 4HA
G3 SLI A Osborne, 18a Cumnor Road, Boars Hill, Oxford, OX1 5JP
G3 SLJ [illegible], Germany, 85084
G3 SLK R Pickering, 147 Windermere Avenue, Nuneaton, CV11 6HN
G3 SLL Harry Tyrcman, 37 Lavinia Street, Seven Hills, Australia, NSW 2147
G3 SLS A Hancock, 6 The Fairway, Mablethorpe, LN12 1LL
G3 SLT David Ormerod, 21 Vallotta Close, Chelmsford, CM1 2PT
G3 SLX J Smith, 256 Stone Road, Stoke-on-Trent, ST4 8NJ
G3 SMD R Turner, 7 Paddocks Lane, Cheltenham, GL50 4NU
G3 SMF I Hamill, 74 Lampits Hill, Corringham, Stanford-le-Hope, SS17 9AJ
G3 SMM W Furness, 18 Riddings Court, Timperley, Altrincham, WA15 6BG
G3 SMN R Forster, 28 Springbridge Road, Manchester, M16 8PW
G3 SMT P Torry, Pen-Y-Rhos, Oswestry, SY10 7HP
G3 SMV J Smith, 18 Hounslow Road, Mackworth, Derby, DE22 4BW
G3 SMZ Raymond Hill, 68 Chestnut Street, Chadderton, Oldham, OL9 8HH
G3 SNA Stuart Andrew, Berry Brow House, Berry Brow, Oldham, OL3 7EJ
GJ3 SND B Walster, Le Ponterrin Cottage, La Rue Du Ponterrin, St. Saviour, Jersey, JE2 7HP
G3 SNG A Ambler, 12 Oakdene Road, Marple, Stockport, SK6 6PJ
G3 SNH W Harrison, 44 Briar Road, Thornton-Cleveleys, FY5 4NB
G3 SNN A Woolford, 39 Apple Orchard, Prestbury, Cheltenham, GL52 3EH
G3 SNO Graham Smith, Stonycroft, Godsons Lane, Southam, CV47 8LX
G3 SNP M Pitcher, Sandycot, Cadsden Road, Princes Risborough, HP27 0NB
G3 SNR G Morgan, Eaton House, Eaton Bank, Belper, DE56 4BH
G3 SNT R Dixon, Copper Beeches, Witton Gilbert, Durham, DH7 6TW
G3 SNU Kenneth Selleck, Westphalia, Dartington, Totnes, TQ9 6DJ
G3 SOA W Mccartney, Lychgate House, Uffington, Shrewsbury, SY4 4SN
G3 SOE R Jennings, 31 Copper Beech Drive, Wombourne, Wolverhampton, WV5 0LH
GI3 SOO M Foley, 5 Woodland Drive, Cookstown, BT80 8PL
G3 SOU Southampton ARC c/o Malcolm Troy, 22 Jackie Wigg Gardens, Totton, Southampton, SO40 9LZ
GW3 SPA R Alban, 73 Plymouth Road, Penarth, CF64 3DD
G3 SPI I Dawe, 10 Selsden Close, Elburton, Plymouth, PL9 8UR
G3 SPJ C Wooff, 55 Bostall Hill, Abbey Wood, London, SE2 0QX
G3 SPL P Lee, 10 Antony Gardner Crescent, Whitnash, Leamington Spa, CV31 2TQ
G3 SPN Neil Collins, Flat 1, Vista Mare West, 44 West Parade, Worthing, BN11 5EF
G3 SPO Patrick Oneill, Recreation Cottage, Slad Road, Stroud, GL6 7QA
G3 SPP A Minett, 45 Patterdale Drive, Worcester, WR4 9HS
GM3 SPT G McKay, 152 Inveresk St., Greenfield, Glasgow, G32 6TA
G3 SPV K Richardson, Brookfield Grove, The Dukes Drive, Bakewell, DE45 1QQ
G3 SPY GPT(COVENTRY)ARS c/o Roger Harris, Clevelands, Tamworth Road, Coventry, CV7 8JJ
G3 SQA Peter Moss, The Maples, South Road, Horncastle, LN9 6QB
G3 SQD Russell Tollerfield, 11 Spencer Court, Merton Road, Southsea, PO5 2AJ
G3 SQN J Grant, 8 Thornhill Way, Mannamead, Plymouth, PL3 5NP
G3 SQO David Best, Tanglewood, Showley Road, Blackburn, BB1 9DP
G3 SQQ J Franks, 11 Thoresby Avenue, Kirkby-in-Ashfield, Nottingham, NG17 7LY
G3 SQU Christopher Clarke, 14 Woodlea Gardens, Newcastle upon Tyne, NE3 5BY
G3 SQW T Hutchinson, Box 196, Gilgil, Kenya, 20116
G3 SQX Edwin Taylor, 115 St. Albans Avenue, London, W4 5JS
G3 SRA SILVERTHORN RC CONTEST GROUP c/o A Mowbray, 33 Shepherds Walk, Benfleet, SS7 2LP
G3 SRC SURREY RA CON C c/o Maurice Fagg, 113 Bute Road, Wallington, SM6 8AE
GW3 SRF D Woolen, Rose Cottage, Newcastle Hill, Bridgend, CF31 4EY
GW3 SRG A Peake, 70 Higher Lane, Langland, Swansea, SA3 4PD
G3 SRJ G Carlisle, 3 Grimms Meadow, Walters Ash, High Wycombe, HP14 4UH
G3 SRM S Hulme, 64 Salem Street, Amlwch, LL68 9BT
GD3 SRN Geoffrey Bray, 4 Ledway Drive, Wembley, HA9 9TQ
G3 SRQ Raymond Bisseker, 24 Millgate, High Wycombe, HP11 1GL
G3 SRR Melvyn Rees, 83 Salisbury Road, Farnborough, GU14 7AE
G3 SRT SALOP A.R.S. c/o Richard Golding, 7 Belvidere Avenue, Shrewsbury, SY2 5PF
GM3 SRV R Tatton, 17 Paties Road, Edinburgh, EH14 1EF
G3 SRX N Down, 23 Christopher Close, Heckington, Sleaford, NG34 9SA
GW3 SSK J Williams, 5a Derllwyn Close, Tondu, Bridgend, CF32 9DH
G3 SSN J Brand, 133 Hatfield Avenue, Fleetwood, FY7 7DU
G3 SSW S Frents, 50 Blandy Avenue, Southmoor, Abingdon, OX13 5DB
G3 SSZ L Lavelle, 49 Jones Road, Goffs Oak, Waltham Cross, EN7 5JT
G3 STF Peter Sandiford, 11 Calle Menendez Pelayo, San Javier, Spain, 30730
G3 STG Geoffrey Griffiths, 14 Mansion House Gardens, Melton Mowbray, LE13 1LE
G3 STJ Brian Riley, 2 Watson Street, Swinton, Manchester, M27 6AQ
G3 STP Philip La Pierre, 42 Berry Vale, South Woodham Ferrers, Chelmsford, CM3 5GY
G3 STT W Haynes, 37 Hawthorn Grove, Southport, PR9 7AA
G3 STZ Christopher Thorn, 4 Riveredge, Framilode, Gloucester, GL2 7LH
G3 SUA E Winstanley, Padua, 1 Drews Court, Gloucester, GL3 2LD
G3 SUG J Jarvis, 56 Upper Churnside, The Beeches, Cirencester, GL7 1AP
GW3 SUH K Hughes, 2 Graig Terrace, Ferndale, CF43 4EU
G3 SUI J Burrows, 68 Grosvenor Road, Sale, M33 6NW
G3 SUK M Baker, 8 Wynton Rise, Stowmarket, IP14 2AB
G3 SUL D Waller, 66 Wallace Drive, Dunstable, LU6 2DF
GI3 SUM J Gould, 13 Maralin Avenue, Bangor, BT20 4RQ
G3 SUN Grahame Hodgkinson, Cfone Communications, 9 Adler Industrial Estate, Betam Road, Hayes, UB3 1ST
G3 SUS Stuart Jacobs, 16 Mayfield Park, Thorley, Bishop's Stortford, CM23 4JL
G3 SUV D Ashby, 1 Cedar Court, Lexden Road, Colchester, CO6 3BT
G3 SUX D Bradshaw, 25 Meare Close, Tadworth, KT20 5RZ
G3 SUY [illegible], Chelmsford, CM3 4RG
GM3 SUZ Daniel McLean, Whitecroft Farm, Barrs Brae, Port Glasgow, PA14 5QG
G3 SVC SPEN VALLEY ARS c/o John Wilde, 2 Bottoms Lane, Birkenshaw, Bradford, BD11 2NN
G3 SVD A Hewitt, Redwood House, Adbury Holt, Newbury, RG20 9BW
G3 SVI David Davis, 188 Eastwood Old Road, Leigh-on-Sea, SS9 4RY
G3 SVJ LUTON VHF GROUP c/o Andy Barter, 300 Northdown Road, Margate CT9 3HD
G3 SVK Fred Curtis, 32 Elgin Avenue, Harold Wood, Romford, RM3 0YT
G3 SVQ [illegible], Bedford, MK43 7JX
G3 SVR SEVERN VALLEY RD c/o E Churchyard, 11 Greenfields Drive, Bridgnorth, WV16 4JW
G3 SVW Ronald Smith, 16 Coniston Avenue, Sale, M33 3GT
G3 SVZ Paul Laxton, 62 Reddington Road, Plymouth, PL3 6PT
G3 SWC B Tinton, Farthings, 1 Bridge Road, Horsham, RH12 3HD
G3 SWH P Whitchurch, 21 Dickensons Grove, Congresbury, Bristol, BS49 5HQ
GM3 SWK Alexander Shearer, 12 Coolin Drive, Portree, IV51 9DN
G3 SWU T Heeley, 34 Worlaby Road, Scartho Top, Grimsby, DN33 3JT
G3 SWW H Cooper, 9 Fortyfoot, Bridlington, YO16 7SA
G3 SXA J Croft, 14 Stanstead Road, Forest Hill, London, SE23 1BW
G3 SXC A Critchley, 39 Westcliffe, Great Harwood, Blackburn, BB6 7PH
G3 SXE Lester Lethbridge, 24 Furze Road, High Salvington, Worthing, BN13 3BH
G3 SXH Andrew Henderson, 50 Sylvan Road, Exeter, EX4 6EY
G3 SXI D Ashmore, Flat 8, Carter Bench House, Clarence Road, Macclesfield, SK10 5JZ
G3 SXP John Redford, Woodgates Harling Road, Gt Hockham, IP24 1NP
G3 SXQ E Rockett, Devoran, The Causeway, Highbridge, TA9 4QT
G3 SXR A Read, Readymoney Cove, Fowey, PL23 1JH
G3 SXT Paul Sweeny, 11a Newfield Place, Sheffield, S17 3ER
G3 SXV B Vincent, 18 Rowanhayes Close, Ipswich, IP2 9SX
G3 SXW R Western, 7 Field Close, Chessington, KT9 2QD
G3 SYA D Ashworth, 31 Belmont Avenue, Ribbleton, Preston, PR2 6DH
G3 SYB Herbert Barker, Azul Avion, Main Road, Alford, LN13 0JP
G3 SYD Sydney Beauchamp, 1 Gosden Close, Furnace Green, Crawley, RH10 6SE
G3 SYG J Marsh, 17 Fish Lane, Aldwick, Bognor Regis, PO21 3AH
GW3 SYL R Price, 49 Pant Hirwaun, Heol-y-Cyw, Bridgend, CF35 6HH
G3 SYM D Coltart, Tregerry Farm, Treneglos, Launceston, PL15 8UF
G3 SYS Darrel Emerson, 4269 N.Soldirer Trail, Tucson, United States, AZ 85749
G3 SYZ A Rogers, Draycott, Primrose Hill, Hastings, TN35 4DN
G3 SZE James Evrall, 38 Eastlang Road, Fillongley, Coventry, CV7 8ER
G3 SZF R Frost, 24 Mount Pleasant, Hertford Heath, Hertford, SG13 7QU
G3 SZG John Wright, 39a Sion Hill, Kidderminster, DY10 2XT
G3 SZJ M Shardlow, 19 Portreath Drive, Allestree, Derby, DE22 2BJ
G3 SZM John Wuille, 45 Keymer Crescent, Goring-by-Sea, Worthing, BN12 4LD
G3 SZR C Davis, 148 Birkbeck Road, Beckenham, BR3 4SS
G3 SZS Ronald Bee, 19 Hazelcroft, Churchdown, Gloucester, GL3 2DS
G3 SZU Keith Radford, 30 Whitendale Drive, Bolton le Sands, Carnforth, LA5 8LY
G3 SZV B Ward, 138 County Road, Ormskirk, L39 1NN
G3 SZY Gary Douglas, 169 High Street, Cheveley, Newmarket, CB8 9DG
G3 TA C Lambert, Stonecroft, Notch Road, Cirencester, GL7 7JU
G3 TAA Keith Jessop, 15 Courtenay Gardens, Newton Abbot, TQ12 1HS
GI3 TAC D Campbell, 23 Belmont Avenue, Bangor, BT19 1NG
G3 TAF D Cassere, 9 St. Marys Garth, East Keswick, Leeds, LS17 9ER
G3 TAG Ronald Gouldstone, 11 School Lane, Toft, Cambridge, CB23 2RE
G3 TAI C Ward, 50 Lakeside, Bracknell, RG42 2LE
G3 TAJ R Marchant, Cascade, The Street, Canterbury, CT3 1LN
GM3 TAL Malcom Hamilton, 3 Charles Court, Limekilns, Dunfermline, KY11 3LG
G3 TAO William Eaton, 8e St. Aubyns Road, London, SE19 3AD
G3 TAQ N Bullock, 29 St. Marys Road, Stowmarket, IP14 1LP
G3 TAW C Wood, 22 Habberley Road, Kidderminster, DY11 6AA
G3 TAX J Boydell, 13 Lynch Road, Farnham, GU9 8BZ
G3 TAY Alan Yarker, 6 Moor Top Road, Halifax, HX2 0NP
G3 TAZ R Davies, 69 Stopsley Way, Luton, LU2 7UU
G3 TBF Harvey Wilkins, 17 Bathleaze, Kings Stanley, Stonehouse, GL10 3JN
G3 TBG G Goulbourn, 41 Rutland Road, Stamford, PE9 1UP
G3 TBJ C Webster, 20 Piggotts Road, Caversham, Reading, RG4 8EN
G3 TBK John Cree, 24 Old Lincoln Road, Caythorpe, Grantham, NG32 3EJ
G3 TBL Raymond Ashman, Cartref, Salisbury Road, Salisbury, SP4 9QZ
G3 TBW T Westbury, 6 Ellerdene Close, Redditch, B98 7PW
G3 TCG M Trundle, 20 Denehurst Gardens, Hastings, TN35 4PB
G3 TCI Alan Bye, 7 Larkfield Avenue, Gillingham, ME7 2LN
G3 TCL Michael Dawson, 11 St. Georges Close, Brampton, Huntingdon, PE28 4US
G3 TCO A Preece, 12 South Dene, Bristol, BS9 2BW
G3 TCQ RAF SOUTH YORKSHIRE AREA-ARC c/o R Clayton, 171 Warning Tongue Lane, Cantley, Doncaster, DN4 6TU
G3 TCT Graham Kimbell, Eastfield Farmhouse, Fair Place, Somerton, TA11 7DN
G3 TCU Phil Guttridge, 33 Franklyn Road, Godalming, GU7 2LD
G3 TCV J Edwards, Pen y Maes, Trehelig, Welshpool, SY21 8SG
GM3 TCW J Kelly, 144a Manse Road, Newmains, Wishaw, ML2 9BL
G3 TCY J Lewis, 10 Sheringham Drive, Etchinghill, Rugeley, WS15 2YG
G3 TCZ R Freeman, 3 Hoffmann Gardens, South Croydon, CR2 7GE
G3 TDC J Yates, Ferncliffe, Stourbridge, DY7 6DX
G3 TDF P Farley, Rainbows End, Tom Lane, High Peak, SK23 9UN
G3 TDH R Stevens, 19 Canberra Road, Bramhall, Stockport, SK7 1LG
G3 TDL R Davis, 105 Eldred Avenue, Brighton, BN1 5EL
G3 TDM Roger Mason, 19 Lawrence Road, St. Agnes, TR5 0XQ
G3 TDT I Hollingsbee, 89 Swift Road, Abbeydale, Gloucester, GL4 4XJ
G3 TDX E Ingram, 36 Kenwood Road, Knighton, Leicester, LE2 3PJ
G3 TEB G Addis, 34 Ryhill Way, Lower Earley, Reading, RG6 4AZ
G3 TEC T Rutherford, 17 Rosedale Avenue, Sunderland, SR6 8BD
G3 TEE F Stork, 20 Gay Meadows, Stockton on the Forest, York, YO32 9UJ
G3 TEH Alan Storey, 23 Foster Street, Barnsley, S70 3EW
G3 TEI Thorn EMI ARC c/o John Gaffney, 77 South Street, Pennington, Lymington, SO41 8DY
G3 TEL P McPherson, 2 Osborne Place, Lower Street, Merriott, TA16 5NP
G3 TEP Brian Atkinson, 20 King Street, Seahouses, NE68 7XP
G3 TEU A Sherer, 35 Beverley Road, Willerby, Hull, HU10 6AW
G3 TEV Michael Mills, Shepton, 3 Tylers Way, Stroud, GL6 8ND
G3 TEX Phillip Painter, Linden House, Barkleys Hill, Bristol, BS16 1FH
G3 TFA G Whenham, Hogs Hollow, Welsh Road East, Southam, CV47 1NE
G3 TFF G Fuller, 99 Stanbury, Keighley, BD22 0HA
G3 TFI G Rogers, 19 Manor Road, Henley-on-Thames, RG9 1LT
G3 TFO J Auty, 64 Ainley Road, Huddersfield, HD3 3QX
G3 TFR J Hardstone, 17 Whitefield, Stockport, SK4 2PF
G3 TFV E Tokley, 14 Maple Way, Earl Shilton, Leicester, LE9 7HW
G3 TFX Richard Fusniak, 70 Cromwell Road, Cambridge, CB1 3EG
GM3 TFY D Guest, 31 Newmills Crescent, Balerno, EH14 5SX
G3 TGB [illegible], GM7 2PE
G3 TGD Michael Allenson, 4 The Orchard, Powick, Worcester, WR2 4SL
G3 TGE Derek Cahill, Hillcrest, 1 The Greenyard, Northampton, NN7 1EQ
G3 TGF C Bonner, 57 Downsview, Heathfield, TN21 8PF
G3 TGG T Gratton, 23 Culhorn Road, Stranraer, DG9 8DB
G3 TGL A Fantham, 52 Calverley Road, Kings Norton, Birmingham, B38 8PW
G3 TGN P Collar, 5 Oak Tree Lane, Tavistock, PL19 0DA
G3 TGO B Vaughan, 7 Oundle Road, Chesterton, Peterborough, PE7 3UA
G3 THC David Stimson, 94 Casterton Road, Stamford, PE9 2UB
G3 THF B McHugh, 283 Coppice Road, Poynton, Stockport, SK12 1SP
G3 THG A Kent, The Coach House, Dipford, Taunton, TA3 7NR
GM3 THI R Harkess, Friarton Bank, Rhynd Road, Perth, PH2 8PT
G3 THM L Best, 3a Chipstead Lane, Sevenoaks, TN13 2AH
G3 THQ Brian Greenaway, 5 Lansdowne Grove, Neasden, London, NW10 1PL
G3 THS P Last, 4 Hillside, Marham, King's Lynn, PE33 9JJ
G3 THT Ernest Bennett, 20 Cromford Road, Clay Cross, Chesterfield, S45 9RE
G3 THV G Swindells, 4 Fitzhenry Mews, Norwich, NR5 9BH
G3 THW P Walters, 22 Windmill Rise, Woodhouse Eaves, Loughborough, LE12 8SG
G3 TIE A Dutton, 130 Wades Hill, Winchmore Hill, London, N21 1EH
G3 TIG P Turner, 11 Manor Court Road, Hanwell, London, W7 3EJ
GI3 TIJ Frederick Eccles, 31 Ballydawley Road, Moneymore, Magherafelt, BT45 7NU
G3 TIK D French, 37 Warner Road, Ware, SG12 9JN
G3 TIN B Taylor, Perry House, 188 Walstead Road, Walsall, WS5 4DN
G3 TIR D Stewart, Apt 347, 8601-902, Luz-Lagos, Algarve, Portugal
G3 TIX R Hardy, 522 Halifax Road, Bradford, BD6 2LP
G3 TJA R Street, 11 Royal Close, Rugeley, WS15 2DD
G3 TJC E Ross, 20 Briar Wood, Shipley, BD18 1NB
G3 TJE Peter Smith, 7 Tower Walk, Weston-Super-Mare, BS23 2JR
G3 TJH W Bickham, 22 Ash Crescent, Galmington, Taunton, TA1 5PW
G3 TJI Graham Roff, 47 Penshurst Rise, Frimley, Camberley, GU16 8XX
GI3 TJJ J Boyce, 19 Dunvale Park, Londonderry, BT48 0AU
GI3 TJM Richard Miller, 47a Newtownards Road, Donaghadee, BT21 0PY
G3 TJP D Lankshear, 28 Monmouth Place, Newcastle, ST5 3DF
G3 TJS P Goodenough, Llys Aderyn, 11 Guildford Road, Lightwater, GU18 5RZ
G3 TJU L Grant, 4 Berry Close, Purdis Farm, Ipswich, IP3 8SP
G3 TJX Geoffery Tillson, 95 Kelverlow Street, Oldham, OL4 1LX
G3 TKA P Duncan, 18 Pickering Road, Hull, HU4 6TL
G3 TKB John Foster, 4 Hookergate Lane, Rowlands Gill, NE39 2AD
G3 TKD A Cooper, 10 Lon Cerys, Denbigh, LL16 5UY
G3 TKF R Thompson, 179 Newbridge Hill, Bath, BA1 3PY
GW3 TKG D Locke, 201 Tyn Y Tower, Baglan, Port Talbot, SA12 8YE
GW3 TKH K Winnard, 208 Heol Hir, Thornhill, Cardiff, CF14 9LA
G3 TKK Peter Doughty, Mallows, Ballam Road, Preston, PR4 3PN
G3 TKN Vincent Lear, 53 Chaplains Avenue, Cowplain, Waterlooville, PO8 8QH
G3 TKS J Sanderson, 28 Finmere, Hanworth, Bracknell, RG12 7WF
GM3 TLA D Pearson, 23 Binghill Road West, Milltimber, AB13 0JB
G3 TLD M Selwyn, 50 Tufthorn Avenue, Coleford, GL16 8PT
G3 TLH Ian Brown, 15 Juniper Close, Exeter, EX4 9JT
G3 TLK Andrew Endacott, Redacres Market Garden, St. Marychurch Road, Newton Abbot, TQ12 4SB
G3 TLP I Jones, Tyddyn Brith, Gaerwen, Gwynedd, LL60 6HD
G3 TLU John Serlin, 9 Woodside Grove, London, N12 8QT
G3 TLV G Wynes, Hill View, Wrenbury Wood, Nantwich, CW5 8HH
G3 TLY S Alexander, Pinetrees, Wilmslow Avenue, Woodbridge, IP12 4HW
G3 TMA I Buffham, 62/70 Soi Sukhumuit 13, Sukhumvit Rd, Klongtoey Nua, Bangkok, Thailand, 10110
G3 TMB J Baker, 29 Garstang Road, Southport, PR9 9XW
G3 TMD Edward Parsons, 22 Colins Walk, Scotter, Gainsborough, DN21 3SR
GI3 TME Robert Hargan, 13 Drumlerry, Londonderry, BT48 8GQ
GW3 TMJ A Taylor, 24 Emroch St., Goytre, Port Talbot, SA13 2YE
G3 TMP John Jones, Haulfryn, Stryt-Cae-Rhedyn, Mold, CH7 4SS
G3 TMQ R Harrison, 57 Rue Des Bouviers, Mansle, France, 16230
G3 TMR David Emmett, 22 Syljon, 114 Villiers Road, Walmer, South Africa, 6070
GW3 TMS D Smith, 2 Glan Yr Afon Gardens, Sketty, Swansea, SA2 9HY
G3 TMU C Neale, 63 Rosemary Gardens, Blackwater, Camberley, GU17 0NJ
G3 TMX S Bennett, 12 Angel Lane, Bury St. Edmunds, IP33 1RF
G3 TNE C Wantling, 28 Moss Green, Welwyn Garden City, AL7 3TE
G3 TNI J Clingan, 41 Cranham Close, Headless Cross, Redditch, B97 5AY
GI3 TNK Stanley Dornan, 9 Clonallon Gardens, Belfast, BT4 2BY
G3 TNQ C Davis, 963 Manchester Road, Bury, BL9 8DN
GD3 TNS Arthur Sinclair, 1, Marathon Drive, Douglas, Isle of Man, IM4 2BP
G3 TNX V Allison, 24 Colston Gate, Cotgrave, Nottingham, NG12 3JY
G3 TNY K Spooner, Penncroft, 16 Booton Road, Norwich, NR10 4AH
G3 TOA B Otter, Po Box 34554, Lusaka, Zambia
GW3 TOB Allan Coughlin, 37 Parc Y Felin, Creigiau, Cardiff, CF15 9PB
G3 TOF R Brown, 177 Radburn Close, Harlow, CM18 7EH
G3 TOJ G Steel, 10 Rossmere Avenue, Rochdale, OL11 4BT
G3 TON A Fentham, 106 Elm Road, New Malden, KT3 3HP
G3 TOP A Peperell, 47 Glade Rd, Marlow, SL7 1DQ
G3 TOQ N Taylor, Ministro Raul Fernandes, 180 Apt 1805, Bothfogo, Rio de Janeiro, Brazil, 22260-040
G3 TOV G Miles, 200 Ladybank Road, Mickleover, Derby, DE3 0RR
G3 TOW Anthony Hirst, 4 Parc Issa, Bryn-y-Baal, Mold, CH7 6NH
G3 TOY R Wright, High View Cottage, Tatenhill Common, Burton-on-Trent, DE13 9RT

**IMPORTANT NOTE**

**Revalidate licence to avoid revocation** – Ofcom has advised the Society that plans will be drawn up to revoke licences that have not been revalidated as required by the licence conditions. The quickest way to revalidate is to do so online via the Ofcom website: *https://services.ofcom.org.uk/* or by email: *amateur.validations@ofcom.org.uk* Ofcom staff are available to help, but please be patient during times of heavy workload.

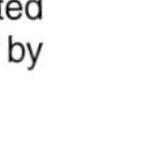

G3 TOZ Gary Parkhurst, Fruit Farm Kittles Lane Marsham, Norwich, NR10 5QE
G3 TPB J Knight, 2120 North Pantops Drive, Charlottesville, United States, 22901
G3 TPH R Henville, 67 Salisbury Road, Blandford Forum, DT11 7LW
G3 TPI Ted Wager, 1 Sundown Close, New Mills, High Peak, SK22 3DH
G3 TPJ Oliver TILLETT, 27 Cranbrook Drive, Gidea Park, Romford, RM2 6AP
G3 TPO C Ockendon, 29 Garlies Road, Forest Hill, London, SE23 2RU
G3 TPP B Eyre, 56a Ounsdale Road, Wombourne, Wolverhampton, WV5 8BH
G3 TPQ Geoffrey Harris, 12 Highridge Close, Purton, Swindon, SN5 4BS
G3 TPV F Robinson, 28 Homer Park, West Common, Soton, SO45 1XP
G3 TPW S Webb, 1 The Green, Swinton, Malton, YO17 6SY
G3 TQA Allan Robinson, 9 Illingworth Close, Illingworth, Halifax, HX2 9JQ
G3 TQC J Sunderland, 7 Beavers Close, Guildford, GU3 3BX
G3 TQD R Avery, 42 Lineholt Close, Redditch, B98 7YU
G3 TQF G Findon, 3 The Paddock Newton, Rugby, CV23 0EE
G3 TQL J Jones, 9 Stonehouse Avenue, Willenhall, WV13 1AP
G3 TQQ John David Bottomley, 32 Ruffa Lane, Pickering, YO18 7HN
G3 TQX George Grimshaw, 50 Rembrandt Way, Bury St. Edmunds, IP33 2LT
G3 TQY M Knights, Springside Farm, Tismans Common, Horsham, RH12 3DU
G3 TQZ Roger Allan, Longfield, Upper Wick Lane, Worcester, WR2 5SU
G3 TRB Terence Barber, 48 Newland Road, Droitwich, WR9 7AZ
G3 TRC R Collins, 8 Sylvan Way, Redhill, RH1 4DE
G3 TRD J Bellamy, 9 Croutel Road, Felixstowe, IP11 7EF
G3 TRE K Rigby, 9 Batcliffe Drive, Leeds, LS6 3QB
G3 TRG R Green, 2 Ragley Walk, Rowley Regis, B65 9NT
G3 TRH R Farrance, 63 Salisbury Close, Rayleigh, SS6 9UH
G3 TRK Donald Kitson, 11 Deerstone Road, Nelson, BB9 9LN
G3 TRL Anthony Green, Rembrandt Stud, Clotton Common, Tarporley, CW6 0HQ
G3 TRR Anthony Mills, 207 Sutherland Drive, Wirral, CH62 8EQ
G3 TRV M Smith, 161 Batley Road, Kirkhamgate, Wakefield, WF2 0SP
G3 TRX C Bailey, 15 Seymour Avenue, Margate, CT9 5HT
G3 TRY Mid-Thames Radio Direction-Finding Club c/o W Pechey, Jays Lodge, Crays Pond, Reading, RG8 7QG
G3 TSA J Denby, 107 Station Road, Fenay Bridge, Huddersfield, HD8 0DE
G3 TSC TRINITY SCHOOL RC c/o R Evans, 7 Westland Drive, Hayes, Bromley, BR2 7HE
G3 TSE D Brealy, 1 Lydford Close, Ivybridge, PL21 0YW
G3 TSF E Glasscott, 26 Columbus Circle, Bluffton, United States, 29909
G3 TSM V Mallows, 13 Greatfield Way, Rowland's Castle, PO9 6AG
G3 TSO Michael Grierson, 1 Blenheim Close, Upper Rissington, Cheltenham, GL54 2QX
GW3 TSQ J Bowen, 33 Parklands View, Sketty, Swansea, SA2 8LT
G3 TSR P Reader, 42 Chiltley Way, Liphook, GU30 7HG
G3 TSS C Waters, 1 Chantry Estate, Corbridge, NE45 5JH
G3 TSV T Clay, 132 Underdale Road, Shrewsbury, SY2 5EF
G3 TSZ A MacWalter, 142 Altrincham Road, Wilmslow, SK9 5NQ
G3 TTB P Clegg, 6 Ricketts Drive, Billericay, CM12 0HH
G3 TTC K Orchard, 32 Myton Crescent, Warwick, CV34 6QA
G3 TTG Victor Batchelor, 31BSupakarn Condo, 1057 Charoen Nakorn Road, Bangkok, Thailand, 10600
G3 TTI L Meikle, 3 Hillcrest, West Woodburn, Hexham, NE48 2RZ
G3 TTJ James Barber, 33 Midford Lane, Limpley Stoke, Bath, BA2 7GR
G3 TTP B Horsey, Nethercotts, Gurney Street, Bridgwater, TA5 2HW
G3 TTU Ronald Holt, 50 Alverley Lane, Doncaster, DN4 9AR
G3 TTY B Field, Greenleaves, 2 Duke Street, Southport, PR8 1RS
G3 TUF F Long, 37 St. Catherines Road, Bitterne, Southampton, SO18 1LS
G3 TUL J Copson, 16 Fothergill Way, Wem, Shrewsbury, SY4 5NX
G3 TUU Christopher Keeble, 86 Kirby Road, Walton on The Naze, CO14 8RL
G3 TUW Peter Moore, 54 Herbert Ave, Palmerston North, New Zealand, 4412
GU3 TUX C Rees, 2 Rue De la Saline, Alderney, Guernsey, GY9 3XD
G3 TUY M Bruce, 3 Redlands Place, Wokingham, RG41 4ED
G3 TVC L Rice, Beechwood, 11 Barnoldby Road, Grimsby, DN37 0JR
G3 TVD J Shersby, 29 Vale Square, Ramsgate, CT11 9DE
G3 TVH J Harknett, 60 Windmill Drive, Croxley Green, Rickmansworth, WD3 3FE
G3 TVI R Stevens, 64 Ferndale, Waterlooville, PO7 7PB
G3 TVL P Hunt, 14 Walnut Close, Epsom, KT18 5JL
G3 TVM H Fletcher, 20 Westfield Road, Great Shelford, Cambridge, CB22 5JW
G3 TVN R Williams, 23a Acacia Avenue, Liverpool, L36 5TN
G3 TVR E Churchyard, 11 Greenfields Drive, Bridgnorth, WV16 4JW
G3 TVS THAMES VALL ART c/o A Pegler, Brook House, Forest Close, Leatherhead, KT24 5BU
G3 TVT I Fraser, 18 Savick Avenue, Bolton, BL2 6JJ
G3 TVU I Brown, 63 Peak View Drive, Ashbourne, DE6 1BR
G3 TVV A Coates, 35 Mogg St., St. Werburghs, Bristol, BS2 9UB
G3 TVW H Davison, 31 Box Hill, Scarborough, YO12 5NQ
G3 TVX D Ashwood, 2 Elm Road, Congleton, CW12 4PR
G3 TVY J Sutton, 3 Sunrise Avenue, Nottingham, NG5 1NH
G3 TWB R Ballard, 31 South Devon Avenue, Nottingham, NG3 6FT
G3 TWJ M Roach, 104 Old Lodge Lane, Purley, CR8 4DH
G3 TWN F Mason, Awel Mon, Bodffordd, Llangefni, LL77 7LJ
G3 TWX David Woodhouse, 38 Jenny Road, Spixworth, Norwich, NR10 3QW
G3 TWY G Mills, 11 Milton Street, Narborough, LE9 4FX
G3 TXC N Harris, April Cottage, Sheepcote Green, Saffron Walden, CB11 4SJ
G3 TXE A Parker, 7 St. Peters Court, Claydon, Ipswich, IP6 0HZ
G3 TXF N Cawthorne, Falcons, St. Georges Avenue, Weybridge, KT13 0BS
G3 TXH B Levett, 18 Forge Road, Little Sutton, Ellesmere Port, CH66 3SQ
G3 TXK Christopher Moss, 2 Sutton Lane, Adlington, Chorley, PR6 9PA
G3 TXL Angus Graham, Woodtown, Sampford Spiney, Yelverton, PL20 6LJ
G3 TXQ Stephen Hunt, 21a Green Street, Milton Malsor, Northampton, NN7 3AT
G3 TXX B Tiffany, 18 Fairfax Road, Bingley, BD16 4DR
G3 TXZ C Tucker, Flat 35, Martlets Court, Crowborough, TN6 1JF
G3 TYA J Grant, Tanyanga, Wheal Leisure, Perranporth, TR6 0EY
G3 TYG B Winslow, 10 Almond Walk, Hazlemere, High Wycombe, HP15 7RE
GW3 TYI David West, 44 Glanmor Park Road, Sketty, Swansea, SA2 0QE
G3 TYO J Stringer, 1 Hazel Road, Tavistock, PL19 9DN
G3 TYP I Jackson, 20 Daventry Road, Barby, Rugby, CV23 8TR
GM3 TYS I Drysdale, 82 Kirk Brae, Cults, Aberdeen, AB15 9QQ
G3 TZA J Riley, Wheal Eliza, Compton Street, Somerton, TA11 6PS
GI3 TZB W J M McKinney, 33 Heatherstone Road, Bangor, BT19 6AE
G3 TZD R Mansell, 354 Allen Road, Salt Point, United States, 12578
G3 TZE R Armitage, 3 Holst Mead, Stowmarket, IP14 1TD
G3 TZG J Glanville, 3 Seneschal Road, Cheylesmore, Coventry, CV3 5LF
G3 TZL P Bowen, White House, Durleigh Marsh, Petersfield, GU31 5AX
G3 TZM William Mahoney, 61 Starbold Crescent, Knowle, Solihull, B93 9LA
G3 TZO Paul Holland, Chatterton, Chapel Lane, Malpas, SY14 7AX
G3 TZQ Stan Ridgway, 8 Almeria Court, Plympton, Plymouth, PL7 1TX
GW3 TZT M Mead, 23 Murlande Way, Rhoose, Barry, CF62 3HL
G3 TZU J Harding, 5 Salisbury Road, Whitchurch, SY13 1RQ
GI3 TZX William Nesbitt, 101 Belfast Road, Bangor, BT20 3PP
GM3 UA A Pairman, Seabank, Largiebeg, Brodick, KA27 8RL
G3 UAA D Ramsey, The Orchard, Carmen Grove, Leicester, LE6 0BA
G3 UAE J Gill, 22 Maddever Crescent, Liskeard, PL14 3PT
G3 UAF Malcolm Smith, 138 Market St., Clay Cross, Chesterfield, S45 9LY
GM3 UAG J Davidson, Cairntoul, Ellon, AB41 8QS
G3 UAP Peter Parker, Avenue Kersbeek 116, 1190 Brussels, Belgium
GD3 UAS Trevor Morgan, 2 Park View, Hatch End, Pinner, HA5 4LN
G3 UAX R Stansfield, 22 Reeds Avenue, Earley, Reading, RG6 5SR
GW3 UAY C Butters, 39 Parc y Ffynnon, Ferryside, SA17 5TQ
G3 UAZ E Sweetman, Flat 39, White Lion Courtyard, Ringwood, BH24 1AJ
GI3 UBA R Reid, 21 Ballymaconnell Road, Bangor, BT20 5PN
G3 UBB D Fill, 2 Brook Close, Packington, Ashby-de-la-Zouch, LE65 1WA
G3 UBC William Guilfoyle, St. Davids Nursing Home Priory Road, Ascot, SL5 8RS
G3 UBD G Higgins, Lower Laithe Farm, Providence Lane, Keighley, BD22 7QS
G3 UBH J Pugh, 5 Pen Y Maes, Llanfechain, SY22 6XL
G3 UBI M Fisher, Bank Top Farm, Cropton, Pickering, YO18 8HH
GM3 UBJ W Hossack, Kincrig, 39 Skene Street, Macduff, AB44 1RP
G3 UBL C Ledger, Kinrara, Sandhills Road, Salcombe, TQ8 8JP
G3 UBP Colin Riches, 28 Saxondale Avenue, Burnham-on-Sea, TA8 2PS
G3 UBS B Speakman, Merrydown, Burley Lane, Derby, DE6 4JS
G3 UBV D Roberts, 8 Churnet Close, Bedford, MK41 7ST
G3 UBX Peter Burden, 68 Coalway Road, Wolverhampton, WV3 7LZ
G3 UBY A Clark, Sans Souci, Fairmead Road, Saltash, PL12 4JH
G3 UCA Peter Sinclair, 32 Barn Meadow, Bamber Bridge, Preston, PR5 8DU
G3 UCD R Pescod, 7 Brian Close, Chelmsford, CM2 9DZ
G3 UCF Alan Passmore, 16 Chaffinch Close, Basingstoke, RG22 5QD
GM3 UCH W Wright, 460 Main Street, Stenhousemuir, Larbert, FK5 3JU
GM3 UCI Gordon McCallum, 15 Quarry Road, Law, Carluke, ML8 5HB
GW3 UCJ Martin Evans, 31 Cilmaengwyn Road, Pontardawe, Swansea, SA8 4QL
G3 UCK G Downs, 2 Dyehouse, Wilsden, Bradford, BD15 0BE
G3 UCL University College London ARS c/o George Smart, Old Queens Head, Ipswich Road, Diss, IP21 4XP
GM3 UCN F Hetherington, 4 Rosebery Place, Livingston, EH54 6RP
G3 UCQ J Farrar, 2 Marsh Lane, Hayle, TR27 4PS
G3 UCT M Taylor, Orchard House, Leigh, Sherborne, DT9 6HL
G3 UCW M Pettit, 3c Clive Court, Grand Parade, Eastbourne, BN21 3DD
G3 UD G Bloor, 26 Leveson Road, Hanford, Stoke-on-Trent, ST4 4QP
G3 UDA Kenneth Linney, Sunnybank, Oak Lane, Shrewsbury, SY3 5BW
G3 UDC K Bran, 9 Carnoustie Grove, Bletchley, Milton Keynes, MK3 7RP
G3 UDD Stephen Chandler, Malt House, Box, Stroud, GL6 9HF
G3 UDH Patrick Butcher, 55 Offington Lane, Worthing, BN14 9RJ
G3 UDI R Butcher, Temple Lodge, Six Mile Bottom Road, Cambridge, CB21 5LD
G3 UDN MID WARWICKSHIRE A.R.S. c/o D Darkes, 70 Braemar Road, Lillington, Leamington Spa, CV32 7EY
G3 UDP M Brown, 4 Boyfields, Quadring, Spalding, PE11 4QQ
G3 UDV P Lindsley, Oak Lodge, Cromer Road, Cromer, NR27 9QT
G3 UED J Jones, 12 Francis Groves Close, Bedford, MK41 7DH
G3 UEE D Diamond, 36 Darbys Lane, Oakdale, Poole, BH15 3ET
G3 UEG D Gould, 2 Mayfield Close, Harlow, CM17 0LH
G3 UEK John Whitehouse, P.O. Box 583, Laceys Spring, Alabama, United States, 35754
G3 UEN R Stiles, Outgate, Southsea Road, Bridlington, YO15 1AE
GW3 UEP Roger Plimmer, Fronhaul, Llandysul, SA44 4NA
G3 UEQ Andrew Hearn, 53 Twyford Gardens, Salvington, Worthing, BN13 2NT
G3 UES Echelford ARS c/o Stuart Roy, 28 Kingston Rise, New Haw, Addlestone, KT15 3EY
G3 UEU John Holmes, 23 School Lane Berry Brow, Huddersfield, HD4 7RA
GI3 UEX Raymond Thompson, 94 Orangefield Crescent, Belfast, BT6 9GJ
G3 UEY D Browning, 13 Beechcombe Close, Pershore, WR10 1PW
G3 UEZ R Gilbert, 18 Peckham Avenue, New Milton, BH25 6SL
G3 UFB Neil Brinkworth, 11 Haycroft Road, Stevenage, SG1 3JL
G3 UFF P Rodway, 37 Neville Avenue, Portchester, Fareham, PO16 9NR
G3 UFI P Conway, 1 The Woodlands, Hastings, TN34 2SF
G3 UFJ L Symons, 31 Springfield Way, Threemilestone, Truro, TR3 6BJ
G3 UFQ D Eckley, 27 Apsley Grove, Dorridge, Solihull, B93 8QP
G3 UFS C Smith, 50 Grand Avenue, Lancing, BN15 9PZ
G3 UFV P Crawshaw, 35 Bishopton Avenue, Stockton-on-Tees, TS19 0RA
G3 UFX H Julian, Brigantine, Lower Market Street, Penryn, TR10 8BH
G3 UFY S Knowles, 77 Bensham Manor Road, Thornton Heath, CR7 7AF
G3 UGC J Smethurst, 81 Springside Road, Bury, BL9 5JG
G3 UGF Richard Constantine, 18 Hillbeck, Halifax, HX3 5LU
G3 UGJ D Smith, 33 Rippington Drive, Marston, Oxford, OX3 0RJ
G3 UGX R Heaton, Flat 5, 73 Belsize Park Gardens, London, NW3 4JP
G3 UHF STH MANCHESTER c/o C Ward, 2 Arlington Drive, Stockport, SK2 7EB
G3 UHJ R Gordon, 77 Alwyn Road, Darlington, DL3 0AH
G3 UHK J Baldwin, 19 Lutyens Close, Stapleton, Bristol, BS16 1WL
G3 UHN Peter Neale, 98 Meadway, Harpenden, AL5 1JQ
G3 UHS C Houltby, 9 Bayard Street, Gainsborough, DN21 2JZ
G3 UHT W Garner, Sarkshields Cottage, Eaglesfield, Lockerbie, DG11 3AE
G3 UHU D Hampton, 9 Portwey Close, Weymouth, DT4 8RF
G3 UHV C Sutton, Braehead, Old Lane, Stoke-on-Trent, ST6 8TG
G3 UHW H Tomlinson, 32 Manor Road, Farnborough, GU14 7EU
G3 UHX A Thorpe, 12 Newnham Lane, Ryde, PO33 4ED
G3 UI L Cobb, 27 Moorlands Crescent, Halifax, HX2 8AA
G3 UIB C Hearn, 8 The Poles, Upchurch, Sittingbourne, ME9 7EX
G3 UID Kevin Baldock, 284 Rocky Mountain High, Camano Island, United States, WA 98282
G3 UIF G Thorne, Flagstaff House, Main Street, Hull, HU12 0RY
GI3 UIH W Aylward, 37 Stewartstown Avenue, Belfast, BT11 9GF
G3 UIJ M Peake, 73 Jamieson House, 4 Edgar Road, Hounslow, TW4 5QH
G3 UIK J Young, Shirley Lodge, 45 Graham Road, Malvern, WR14 2HU
G3 UIS A Stone, West Lodge, The Downs, Poulton-le-Fylde, FY6 7EG
G3 UIT Brian Seedle, 54 Normoss Road, Blackpool, FY3 0AL
G3 UJA B Mcclory, 12 The Crescent, Mottram St. Andrew, Macclesfield, SK10 4QW
G3 UJB Brian Davis, 2 Rawden Close, Harwich, CO12 4BW
G3 UJE Brian Gale, Tall Trees Farm, Noah's Ark Lane, Great Warford, WA16 7AX
G3 UJG J Seal, Aspenden Cottage, Coltsfoot Green, Newmarket, CB8 8UW
G3 UJI S Turner, 51 Hilton Road, Stoke-on-Trent, ST4 6QZ
G3 UJK J Burnham, 86 Icknield Court, Berryfield Road, Princes Risborough, HP27 0HE
G3 UJO Peter Bradley, 60 Weyland Road, Headington, Oxford, OX3 8PD
G3 UJU M Haslam, Updene, The Dene, Salisbury, SP3 6EE
G3 UJV Robert Heath, 26 Lancaster Avenue, Hadley Wood, Barnet, EN4 0EX
G3 UJZ J Mcnaught, Ryton House, Lechlade, GL7 3AR
G3 UK John Whittaker, 48 Baunton, Cirencester, GL7 7BB
G3 UKB R Cowdery, 80 Caxton End, Eltisley, St. Neots, PE19 6TJ
G3 UKC University of Kent c/o Frederick Barnes, 4 Pound Close, Ducklington, Witney, OX29 7TH
G3 UKD A Golding, 40 Unicorn Lane, Eastern Green, Coventry, CV5 7LJ
G3 UKE P Adams, 34 Mount Pleasant Close, Lightwater, GU18 5TP
GM3 UKG G Grant, 35 Inward Road, Buckie, AB56 1DD
G3 UKH P Hopwood, 58 Bolbec Road, Newcastle upon Tyne, NE4 9EP
G3 UKI Barry Curnow, 25 Manchester Square, London, W1U 3PY
G3 UKL M Bennett, Shireley, Munns Lane, Sittingbourne, ME9 7SY
G3 UKM M Leighton, 85 Kemps Green Road, Balsall Common, Coventry, CV7 7QF
G3 UKV M Vincent, 9 Sleapford, Long Lane, Telford, TF6 6HQ
G3 UKW M Newton, 11 Chestnut Close, Rushmere St. Andrew, Ipswich, IP5 1ED
G3 ULD G Cawkwell, 50 Station Road, Patrington, Hull, HU12 0NE
G3 ULL D Walker, Little Chapple, Skilgate, Taunton, TA4 2DP
G3 ULN M Hibbitt, 123 Stanborough Road, Plymstock, Plymouth, PL9 8PJ
G3 ULO I Spencer, Fichtenweg loc, Much, Germany, 53084
GM3 ULP Gordon Hunter, 12 Airbles Drive, Motherwell, ML1 3AS
G3 ULT READING & DISTRICT ARC c/o J Carter, 22 Orchard Coombe, Whitchurch Hill, Reading, RG8 7QL
GW3 UMD N Maxwell, 1 Nant Fawr Crescent, Cardiff, CF2 6JN
G3 UMF Alan Simpson, Forest Farm, Old Road, Shotover Hill, Oxford, OX3 8TA
G3 UML L Margolis, 52 Park View Gardens, Hendon, London, NW4 2PN
G3 UMM P Hudson, 105 Southlands, Weston, Bath, BA1 4DZ
G3 UMT Brian Turvey, 90 Jenkinson Road, Towcester, NN12 6AW
G3 UMV P Johnson, 52 Evesham Road, Cookhill, Alcester, B49 5LJ
GU3 UMX D Ozanne, Eturs Lodge, Les Eturs, Castel, Guernsey, GY5 7DT
G3 UNA David Cutter, 34 Greengate Lane, Knaresborough, HG5 9EL
G3 UNI T Wood, 4 Musgrave Road, Chinnor, OX39 4PL
G3 UNM A Matthews, Winsford, The Common, Stoke-on-Trent, ST10 2PA
G3 UNS T Mills, 22 The Dingle, Crawley, RH11 7JE
G3 UOA University of Aston RS c/o Peter Best, 21 Greening Drive, Edgbaston, Birmingham, B15 2XA
G3 UOC D Brown, Rexfield, Alcester Road, Henley-in-Arden, B95 6BH
G3 UOD Michael Spencer, Cleeve House, Melton Road, Melton Mowbray, LE14 3QG
G3 UOI J Firby, 19 Cliffe Avenue, Harden, Bingley, BD16 1LN
G3 UOJ J Hartwell, Fulling Mill Oast, Caring Lane, Maidstone, ME17 1TJ
G3 UOM D Horsburgh, 11 Delamare Way, Oxford, OX2 9HZ
G3 UON D Geere, TINOS Premier Marinas Ltd, Western Concourse, Brighton, BN2 5UP
G3 UOO D Rogers, Green Tops, Megs Lane, Buckley, CH7 2AG
GU3 UOQ P Le Boutillier, Vue Du Pre, Route de St. Andre, St. Andrew, Guernsey, GY6 8TU
G3 UOS A Whitaker, Univer Of Sheffield, Dept Of Elec Eng, Sheffield, S1 3JD
G3 UPA Michael Foden, 10 Maud Road, Water Orton, Birmingham, B46 1PD
G3 UPD Howard Del Monte, 5 Scotts Close, Colden Common, Winchester, SO21 1US
GI3 UPG R McKimm, 227 Millisle Road, Donaghadee, BT21 0LN
G3 UPI T Codling, 21 Willow Close, Saxilby, Lincoln, LN1 2QL
G3 UPJ D Trainer, 153 High Street, Cherry Hinton, Cambridge, CB1 9LN
G3 UPM T Burke, 12 Worthing Road, Laindon, Basildon, SS15 6AL
G3 UPN K Snape, Delamere, Ryston End, Downham Market, PE38 9AX
G3 UPS Richard Keyte, Low Farm Cottage, New Road, Great Yarmouth, NR31 9HT
G3 UPW P Smith, 9 Ash Road, Shepperton, TW17 0DN
G3 UPY D Houghton, 119 Welsby Road, Leyland, PR25 1JD
G3 UPZ Howard James, 32 John Bunyan Close, Whiteley, Fareham, PO15 7LE
G3 UQD R Whittington, 65 King Edward Avenue, Worthing, BN14 8DG
G3 UQL Micheal Baker, 10 Catchpole Close, Greenleys, Milton Keynes, MK12 6LR
G3 UQR D Robinson, 3 Marriott Close, Irthlingborough, Wellingborough, NN9 5RB
GM3 UQU J Birtwistle, 3 Schoolhill Terrace, Lossiemouth, IV31 6JZ
G3 UQW Alan Ball, 3 Orchard Lea, Sherfield-on-Loddon, Hook, RG27 0ES
G3 URA R Whittering, 3 Church Cottages, Rye Road, Cranbrook, TN18 5PN
G3 URE John Thexton, 78 Greenfield Road, Newcastle upon Tyne, NE3 5TQ
G3 URI NEWBURY VINTAGE WIRELESS SOCIETY c/o M Franks, 13 Fifth Road, Newbury, RG14 6DN
G3 URJ A Moss, 17 Surrey Drive, Finchfield, Wolverhampton, WV3 9LW
G3 URK I Campbell, 27 Lewis Close, Adlington, Chorley, PR7 4JU
G3 URL C Adams, 25 Avon Road, Cannock, WS11 1LJ
G3 URN M Jolley, 34469 N Circle Drive, Round Lake, United States, 60073
G3 URQ J Letts, Bridgeways, Snows Lane, Leicester, LE7 9JS
G3 URU R Edworthy, 44 Middleton Avenue, Littleover, Derby, DE23 6DL
G3 URV F Stevens, 38 Endhill Road, Birmingham, B44 9RR
G3 URX J Speake, 211 Milton Road, Cambridge, CB4 1XG
G3 URZ B Ewen-Smith, 1 Kinnersley, Severn Stoke, Worcester, WR8 9JR
G3 USA C Taylor, 39 School Road, Great Alne, Alcester, B49 6HQ
G3 USC M Hall, Redthorn Bungalow, Upton, Langport, TA10 9NL
G3 USD David Mason, 2a Devon Road, Bedford, MK40 3DF
G3 USE Stephen Down, 1 Dove Close, Honiton, EX14 2GP
G3 USF M Harrison, 1 Church Fields, Keele, Newcastle, ST5 5HP
GI3 USK H Kernaghan, 1 Elizabeth Road, Holywood, BT18 0PL
GM3 USL CUNNINGHAME & DISTRICT AMT RC c/o J Walker, 5 Shiskine Drive,

Kilmarnock, KA3 1P7

G3 USO C Walker, St. Jude, Stoney Lane, Wilmslow, SK9 6LG

G3 USR G Holland, The Lodge, 3a Reeves Lane, Oakham, LE15 8SD

G3 UST John Turner, Flat 4, Barnett Janner House, Leicester, LE1 0UR

G3 USW W Clough, 32 Jackson Crescent, Rawmarsh, Rotherham, S62 7EN

G3 USX Michael Robertson, 1 Lindvale, Horsell Rise, Woking, GU21 4BG

G3 UTA K Smyth, 154 Scrub Lane, Benfleet, SS7 2JP

G3 UTC G Farr, 26 Burstead Drive, Billericay, CM11 2QN

G3 UTE N Wright-Williams, Trinco, 9 Orpine Close, [illegible]

G3 UTG Alan Antley, 12 Fairfield Avenue, Rhyl, LL18 3EE

G3 UTL R Darker, 51 Rockfield Drive, Llandudno, LL30 1PF

GM3 UTO I Halloch, 28 Brentwood Drive, Glasgow, G53 7LU

G3 UTS T Belshaw, 20 Greencroft Road, Delves Lane Industrial Estate, Consett, DH8 7DY

G3 UUB N Bateman, 10 Telford Crescent, Woodley, Reading, RG5 4QT

G3 UUC J Nurse, 25 Dobson Road, Crawley, RH11 7UH

G3 UUF J Hansom, 12 Torquay Avenue, Hartlepool, TS25 3DP

G3 [illegible] [illegible], Marden, Tonbridge, TN12 9TQ

G3 UUI M Mapson, 253 Central Avenue, Southend-on-Sea, [illegible]

G3 UUL T Jones, 32 Oakwood Drive, Hucclecote, Gloucester, GL3 3JF

G3 UUM A Page, 22 Tower Road, Feniscowles, Blackburn, BB2 5LE

G3 UUQ A Clelland, Rieschbogen 7, Hohenkirchen, Germany, 85635

G3 UUR David Gordon-Smith, The Chalet, Bell Road, Attleborough, NR17 1UL

G3 UUT John Wilson, 20b High Green, Great Shelford, Cambridge, CB22 5EG

G3 UUU L Newman, Eastholme, Mill Street, Newton Abbot, TQ13 8AR

G3 UUV Richard FROST, 27 Bowles Court, Westmead Lane, Chippenham, SN15 3GU

G3 UUY D Wright, St. Julians, 55 Old Road, Harlow, CM17 0HD

G3 UUZ H Bluer, 20 Trewellard Road, Pendeen, Penzance, TR19 7ST

G3 UVA D Knowles, The Clappers, Spon Green, Buckley, CH7 3BL

G3 UVB D Barnes, 27 Royal Court, Worksop, S80 2DL

G3 UVC SOUTHAMPTON INSTITUTE ARC c/o J Mcleod, 91 Gorselands Way, Rowner, Gosport, PO13 0DG

G3 UVM M Simpson, 36 Rectory Close, Newbury, RG14 6DD

G3 UVQ Neil Mercer, 19 Sycamore Road, Brookhouse, Lancaster, LA2 9PB

G3 UVR Denis Jones, 39 Pensby Road Heswall, Wirral, CH60 7RA

G3 UVU J Curry, Clonlea, New Ridley, Stocksfield, NE43 7RQ

G3 UVW COVENTRY TECH ARC c/o Roger Harris, Clevelands, Tamworth Road, Coventry, CV7 8JJ

G3 UVY L Parkin, 8 Smithfield Close, Ripon, HG4 2PG

G3 UWE R Simpson, 30 Heath Lawns, Fareham, PO15 5QB

G3 UWH J Endicott, The Mill House, Halse, Taunton, TA4 3AQ

GW3 UWL Sir David Grant, The Court, 19 Marine Parade, Penarth, CF64 3BE

G3 UWM P Marchant, 12 Laurel Way, Ickleford, Hitchin, SG5 3UP

G3 UWP Robin Pickering, 41 Maiden Greve, Malton, YO17 7BE

G3 UWR C Bonsall, Parkside, Lodge Road, Doncaster, DN6 8EB

GW3 UWS UWS RS c/o T Davies, 44 Carnglas Road, Sketty, Swansea, SA2 9BW

G3 UWT P Myers, 22 High St., Barnby Dun, Doncaster, DN3 1DS

GM3 UWX J Stirling, 25 Maxwell Road, Bishopton, PA7 5HE

G3 UWZ Maurice Newman, 26 Highbank, Westdene, Brighton, BN1 5GB

G3 UXH P Carey, 44 Monteney Gardens, Sheffield, S5 9DY

G3 UXJ B Donders, Apple Tree Cottage, Bryn-y-Gwenin, Abergavenny, NP7 8AA

G3 UXM J Greaves, 23 Woodhouse Road, Intake, Sheffield, S12 2AY

G3 UXO A Eardley, 43 Cranbourne Drive, Harpenden, AL5 1RJ

G3 UXQ M Smart, 12 Bromwich Lane, Stourbridge, DY9 0QZ

G3 UXR N Goddard, 1 Aston Mead, St. Catherine's Hill, Christchurch, BH23 2SP

G3 UXY Arthur Baker, 1 Napier Road, Maidenhead, SL6 5AR

G3 UYB Michael Shaw, Beech Farm Cottage, Hawkhurst Road, Battle, TN33 0QS

G3 UYC J Peirson, Ashfield Farm, Ulting, Maldon, CM9 6QP

G3 UYD E Clarke, 65 Oakmount Road, Chandler's Ford, Eastleigh, SO53 2LJ

G3 UYE Michelle Richer, 1 Station Road, Surfleet, Spalding, PE11 4DA

G3 UYG J Clegg, 11 South Park Road, Gatley, Cheadle, SK8 4AL

G3 UYK P Kemble, 74 Teg Down Meads, Winchester, SO22 5ND

G3 UYL David Knott, 22 Linden Close, Prestbury, Cheltenham, GL52 3DU

G3 UYN Clifford Malcolm, Glen Mor, Trenance, Helston, TR12 6QL

GM3 UYR P Gamble, 21 St. Marys Drive, Perth, PH2 7BY

G3 UYX J Ball, 7 Moorfield Road, Woodbridge, IP12 4JN

G3 UYY Edward Bradley, 3 Windrush, Wargrave Road, Henley-on-Thames, RG9 2LX

G3 UZB J Shewan, 42 Stirling Road, Redcar, TS10 2JZ

G3 UZD F Bilke, 2 Walcott Avenue, Christchurch, BH23 2NG

G3 UZF Adrian Green, 75 Higher Woolbrook Park, Sidmouth, EX10 9ED

G3 UZI I Wollen, Courtyard Cottage, Brownston Street, Ivybridge, PL21 0RQ

GI3 UZJ D Singleton, 38a Cloughey Road, Portaferry, Newtownards, BT22 1NQ

G3 UZM C Haddock, 26 Featherbed Lane, Exmouth, EX8 3NE

GW3 UZS J Diplock, Cartref, 98 Pendwyallt Road, Cardiff, CF14 7EH

G3 UZW R Andrews, 10 Hilltop Rise, Bookham, Leatherhead, KT23 4DB

G3 UZX F Mitchell, 158 Cobham Road, Fetcham, Leatherhead, KT22 9JR

GI3 VAF Robert Best, 6 Knightsbridge Court, Bangor, BT19 6SD

G3 VAJ Ian Gray, 1 Greenside Avenue, Berwick-upon-Tweed, TD15 1B7

G3 VAK M Sutcliffe, 26 Weald Road, Burgess Hill, RH15 9SP

G3 VAL George Talbot, Calton Cottage, Solkirk, TD7 5LS

G3 VAO Michael Farmer, Horton Brook Cottage, Horton, Shrewsbury, SY4 5NB

GM3 VAP C Weston, 18 Kirkbrae Mews, Cults, Aberdeen, AB15 9QF

G3 VAS G Jones, 294 Halling Hill, Harlow, CM20 3JU

GI3 VAW R Sherrard, 39 Shanreagh Park, Limavady, BT49 0SF

G3 VBA K Hatton, 38 Doric Avenue, Frodsham, WA6 6QQ

G3 VBE F Miles, 65 Montgomery Street, Hove, BN3 5BE

G3 VBG B Morris, 80 Newcastle Road, Leek, ST13 7AA

G3 VBI H Christopher, 19 Spa Hill, Kirton Lindsey, Gainsborough, DN21 4BA

G3 VBL Christopher Pedder, Thorncliffe, 5 Royalty Lane, Preston, PR4 4JD

G3 VBQ D Wright, 5 Padin Close, Chalford, Stroud, GL6 8FB

GM3 VBT T Logan, 137 Buccleuch St., Garnethill, Glasgow, G3 6QN

G3 VBU John Lynch, 11 Rosenthorpe Road, London, SE15 3EG

G3 VBV Stea Boyce, 58 Woodbury Road, Halesowen, B62 9AW

GM3 VBY F Hindley, The White House, 17 Main Road, Elgin, IV30 8UR

G3 VC Martin Bridge, Boundary House, Waste Green Lane, Boston, PE20 2AT

G3 VCA Robert Pickles, The Pyewipe, Saxilby Road, Lincoln, LN1 2BG

G3 VCG D Wilks, 36 Greenways, Chelmsford, CM1 4EF

G3 VCH Roy Philpott, Ernst Batzer Str. 2, Offenburg, Germany, 77652

GI3 VCI M McFadden, 121 Greystown Avenue, Belfast, BT9 6UH

G3 VCK J Fenwick, 78 Loveridge Road, London, NW6 2DT

G3 VCL D Clark, 60 Somerset Avenue, Harefield, Southampton, SO18 5FS

G3 VCM I Anderson-Moehrle, 10214 Hunt Club Lane, Palm Beach Gardens, United States, 33418

G3 VCN Paul Kalas, 110a Underlane, Plympton, Plymouth, PL7 1QZ

G3 VCQ Colin Wilson, 82 Lennox Road, Sheffield, S6 4FN

G3 VCR C Rooney, 103 Hart Plain Avenue, Cowplain, Waterlooville, PO8 8PN

G3 VCT H Hemmings, Wood View, Cryers Hill Road, High Wycombe, HP15 6JR

G3 VCV D Prout, 7 Chemin Des Estimeurs Nord, Plan De La Dame, Valreas, France, 84600

G3 VCX D Bridgen, 22 Maple Grove, Immingham, DN40 2JH

G3 VCY C Clayton, Wildfield, West Flexford Lane, Guildford, GU3 2JW

G3 VDB J Evans, 7 Durncroft Close, Chelford, Macclesfield, SK11 9SW

G3 VDC J Sellors, Blacktoft Grange, Blacktoft Grange Road, Brough, HU15 2ZU

G3 VDF H Gregory, 44 Mowlands Close, Sutton in Ashfield, NG17 [illegible]

G3 VDH R Godwin, Hopworthy Moor Cottage, Pyworthy, Holsworthy, EX22 6XX

G3 [illegible] [illegible], 8 Minnie Street, Keighley, BD21 1HY

G3 VDL J St Leger, Warmbrook, Throwleigh, Okehampton, [illegible]

G3 VDO I Hacking, 1 Pine Crescent, Poulton-le-Fylde, FY6 8EB

G3 VDS Roy Higham, 91 Welch Road, Hyde, SK14 4DJ

G3 VDU P Bennett, 56 Winchester Avenue, Weddington, Nuneaton, CV10 0DW

G3 VDV N Brinnen, 134 Victoria Road, Mablethorpe, LN12 2AJ

G3 VDZ A Richardson, 18 Spencer Road, Ryde, PO33 2NY

G3 VEB R Bridson, 14 Zig Zag Road, Wallasey, CH45 7NZ

G3 VEF Fareham & District ARC c/o Derek Clarkson, Fareham Sailing & Motorboat Club, Lower Quay, Fareham, PO16 0RA

G3 VEH Chris Morcom, 15 Markson Road, South Wonston, Winchester, SO21 3EZ

GM3 VEI I Sheffield, 37 Bellevue Court, Queens Road, Dunbar, EH42 1YR

G3 VEK Steve Holden, 113 Lower Camden, Chislehurst, BR7 5JD

G3 VER VERULAM ARC c/o Robert Heath, 26 Lancaster Avenue, Hadley Wood, Barnet, EN4 0EX

G3 VES H Martin, 1 Houghton Park Cottages, Hazelwood Lane, Bedford, MK45 2EY

G3 VET Michael Langwade, 19 South Wootton Lane, King's Lynn, PE30 3BS

G3 VEV R Butterfield, Flat 4, The Green, Lincoln, LN3 5TY

GW3 VEW K Godfrey, Kimberley, Ludchurch, Narberth, SA67 8JE

GM3 VEY Findlay Baxter, 8 Northcote Park, Aberdeen, AB15 7SX

G3 VFB A Matthews, Chicks Barn, Fitzhead, Taunton, TA4 3LA

G3 VFC Terry Chipperfield, 5 Lullingstone Close, Hempstead, Gillingham, ME7 3TS

G3 VFD C Westwood, Uplands, The Hillside, Orpington, BR6 7SD

G3 VFF D Hine, Whirlwind, Chesboule Lane, Spalding, PE11 4EU

G3 VFH L Moore, 15 Elmete Drive, Roundhay, Leeds, LS8 2LA

G3 VFL Albert Lightly, 9 The Kymin, Monmouth, NP25 3SD

G3 VFO T Hart, The Hawthorns, 163 Hastings Road, Battle, TN33 0TP

G3 VFU S Street, Po Box 107, Chiang Nai Post Office, Chiang Mai, Thailand, 50000

GD3 VFX D Davison, 28 Treve Avenue, Harrow, HA1 4AJ

G3 VFZ M Hughes, Cefn Dinas, Bangor, LL57 4DP

G3 VG J Wood, 7 Sherring Close, Bracknell, RG42 2LD

G3 VGD David Jones, 31 Meadow Road, Windermere, LA23 2EU

G3 VGE M Hickman, 75 Carlton Road, Redhill, RH1 2BZ

G3 VGG BROMSGROVE & DISTRICT ARC c/o C Margetts, 16 Lahn Drive, Droitwich, WR9 8TQ

G3 VGH B Hutchinson, 78 Strensall Road, Huntington, York, YO32 9SH

G3 VGK Kevin Blackburn, 57 Hope Street, Leigh, WN7 1NB

G3 VGR D Aldridge, 62 Roding View, Buckhurst Hill, IG9 6AQ

G3 VGW R Buckby, 20 Eden Bank, Ambergate, Belper, DE56 2GG

G3 VGX Richard Orton, 15 Middleton Close, Cambridge, CB4 1DG

G3 VGY R Ricketts, 30 Water Lane, Tiverton, EX16 6RB

G3 VGZ Brian Duffell, 7 Potto Close, Yarm, TS15 9RZ

G3 VHE R Evans, 23 Hardwell Close, Grove, Wantage, OX12 0BN

G3 VHF Miguel Eavis, 61 Hitchmead Road, Biggleswade, SG18 0NL

G3 VHH John Delves, 11 Willoughby Road, Langley, Slough, SL3 8JH

G3 VHI G Boultbee, 6 Laxton Close, Heckington, Sleaford, NG34 9TS

GD3 VHK J Robinson, 8 Lorraine Park, Harrow, HA3 6BX

G3 VHL H Buttress, 132 Elan Avenue, Stourport-on-Severn, DY13 8LR

GI3 VHM V Addidle, 23 Church Lodge, Moneyrea, Newtownards, BT23 6ES

G3 VHN J Burge, 14 Robinson Place, Brant Broughton, Lincoln, LN5 0SJ

G3 VHS J Cobb, Middle Cottage, Abingdon, OX13 5LR

G3 VHW N Humphrey, 10 Pembroke Close, Eastleigh, SO50 4QY

G3 VHZ Bryan Neary, 30 Laneham Close, Doncaster, DN4 7HU

G3 VIC D Barney, 5 Station Road, Sheringham, NR26 8RE

G3 VID Tyrone Howe, 80 Devon Gardens, Birchington, CT7 9SR

G3 VIP G Wood, 47 Church Lane, Holton-le-Clay, Grimsby, DN36 5AQ

G3 VIR Roland Brade, 9 Magness Road, Deal, CT14 9JF

G3 VIX Thomas Stevens, 97 Broad Acres, Hatfield, AL10 9LE

G3 VIY H Vasper, 31 Oakland Road, Forest Town, Mansfield, NG19 0EJ

G3 VJE H Cole, 3 Canberra Crescent, Grantham, NG31 9RD

G3 VJG M Deutsch, 80 Windermere Road, Kettering, NN16 8UF

G3 VJI J Steel, Oakdene, 4 Broom Close, Kendal, LA9 6BN

G3 VJJ F Smedley, 13 Justice Avenue, Saltford, Bristol, BS31 3DR

G3 VJM A Wood, Danehill, Brookhill Road, Crawley, RH10 3PS

G3 VJN Adrian Ryan, 13c Spyros Kiprianou, Limassol, Cyprus, 4717

G3 VJR J Longstaff, 23 Harlington Road, Adwick-upon-Dearne, Mexborough, S64 0NL

G3 VJV C Hartley, 16 Cyril Bell Close, Lymm, WA13 0JS

G3 VJX M Gill, Upper Bean Hall, Church Road, Redditch, B69 6RN

GM3 VJY J Evans, 64 Craigmount Avenue North, Edinburgh, EH12 8DL

G3 VKB J Orr, 18 Randall Close, Langley, Slough, SL3 8RJ

G3 VKF K Kelly, 2 Longden Lane, Macclesfield, SK11 7EN

G3 VKI F Turner-Smith, 26 Ash Church Road, Ash, Aldershot, GU12 6LX

G3 VKK CHESTERFIELD & DISTRICT ARS c/o John Otter, 7 Longacre Road, Dronfield, S18 1UQ

GW3 VKL Barry ARS c/o Philip King, 11 Lord Street, Penarth, CF64 1DD

G3 VKM Roger Basford, Newgate, Thorpe Road Haddiscoe, Norwich, NR14 6PP

GM3 VKN Philip Mansell, Broad Meadows, Fort Augustus, PH32 4DW

G3 VKQ Colin McEwen, 37 Malvern Way, Twyford, Reading, RG10 9PY

G3 VKT R Smith, 32 Wolseley Gardens, London, W4 3LR

G3 VKU D Hollingsworth, 4 Cairn View, Longframlington, Morpeth, NE65 8JT

G3 VKV G Jones, 32 The Grove, Hales Road, Cheltenham, GL52 6SX

G3 VKW Keith Evans, Littlefield House, Bolney Road, Haywards Heath, RH17 5AW

G3 VLB A Saunders, 6 Douglas Crescent, Kelso, TD5 8BB

G3 VLC C Hawkins, 2 Benett Drive, Hove, BN3 0PL

G3 VLD T Denney, Spindrift-East-, Terrace, Walton on Naze, CO14 8PX

G3 VLF T Desmond, Park View, Middle Lane, Matlock, DE4 5EG

G3 VLG Hinckley Amateur Radio and Electronics Society c/o Vincent Hopkins, 109 Smith Street, Coventry, CV6 [illegible]

G3 VLH J Longhurst, 13 Hophurst Drive, Crawley Down, Crawley, RH10 4XA

G3 VLJ A Hansen, 1829 Francisco St, Berkeley, United States, CA 94703

G3 VLL G Gauntlett, 7 Riverside Drive, Sprotbrough, Doncaster, DN5 7LH

G3 VLN John Allin, 57 Burleigh Road, West Bridgford, Nottingham, NG2 6FQ

G3 VLO John Owen Jones, Spread Eagle, Rishworth, Sowerby Bridge, HX6 4RA

G3 VLR B Rippin, 37 Ferry Road, South Cave, Brough, HU15 2JG

G3 VLW P Martin, Orchard Rise, Littlemoor Road, Highbridge, TA9 4NG

G3 VM Daniel Grace, 107 Bush Avenue, Little Stoke, Bristol, BS34 8NG

G3 [illegible] [illegible], March, PE15 9EQ

G3 VMJ J Peters, 93 Baddow Hall Crescent, Chelmsford, [illegible]

G3 VMK Neville Chadwick, 2 Orchard Close, Gunthorpe, Nottingham, NG14 7FE

G3 VMP B Mills, Highlands Cottage, Crow Lane, Clacton-on-Sea, CO16 9AN

G3 VMQ P Tory, 29 King Edwards Avenue, Gloucester, GL1 5DD

G3 VMR R Redding, 53 Cadwell Drive, Maidenhead, SL6 3YS

G3 VMT T Poole, 64 Humber Close, Thatcham, RG18 3DT

G3 VMU C Davis, 23 Vernon Walk, Northampton, NN1 5ST

G3 VMV Colin Whiting, 5 Carlton, Elloughton, Brough, HU15 1FF

G3 VMW Stephen Wilson, 3 Crag Gardens, Bramham, Wetherby, LS23 6RP

G3 VMY Edward Searle, 203 Church Road, Earley, Reading, RG6 1HW

G3 VMZ D Nicholls, 26 Highfield Close, Semington, Trowbridge, BA14 6JZ

G3 VNB R Thomas, 7 Lane Gardens, Bushey Heath, Bushey, WD23 1PE

G3 VNG D Hind, 4 Thornyville Villas, Plymouth, PL9 7LA

G3 VNH Peter Hardy, LAMBDA HOUSE, SEANOR LANE, Chesterfield, S45 8DH

G3 VNI Simon Cammies, 5 Sheringham Close, Allington, Maidstone, ME16 0NF

G3 VNP P Dowles, 78 Wantz Road, Maldon, CM9 5DE

G3 VNQ M Pritchard, 9 Tamarack Drive, Cortlandt Manor, United States, 10567

G3 VNT Lindsay Pearson, Hatherly, The Street, Stowmarket, IP14 6LX

G3 VNU J Finch, 286 Sea Front, Hayling Island, PO11 0AZ

GM3 VNW J Macphee, 24 Bourtreehall, Girvan, KA26 9EL

G3 VNY I Walker, 45 Terry Drive, Walmley, Sutton Coldfield, B76 2PT

GW3 VNZ D Jacklin, 40 Westbourne Road, Penarth, CF64 3HF

G3 VOB David Vivian, Belle View Cottage, Blandford Road North, Poole, BH16 5PP

G3 VOF Martin Foster, 1 Clavering Court, Lincombe Drive, Torquay, TQ1 2HH

GW3 VOL J Phillips, 96 Maes y Sarn, Pentyrch, Cardiff, CF15 9QR

G3 VOM D Lane, 2 Eden Close, Wilmslow, SK9 6BG

G3 VOO Michael Barnett-Bone, 7 Dorchester Hill, Milborne St. Andrew, Blandford Forum, DT11 0JG

G3 VOS R Cottrell, Larkhill, 47 Bullsland Lane, Rickmansworth, WD3 5BD

G3 VOT G Webster, Red House Farm, Ashford Lane, Bakewell, DE45 1NJ

G3 VOU J Barlow, 68 Willow Avenue, Cheadle Hulme, Cheadle, SK8 6AX

G3 VOV M Lane, 56 Main Street, Bushby, Leicester, LE7 9PP

G3 VOW M Fereday, Spindlewood, Stoney Lane, Thatcham, RG18 9HQ

G3 VPA M Rose, 59 Park Drive, Sittingbourne, ME10 1RD

G3 VPE H Pinchin, 61 Cole Bank Road, Hall Green, Birmingham, B28 8EZ

G3 VPF E Harland, 5 Bramdon Lane, Portesham, Weymouth, DT3 4HG

G3 VPG C Jacob, 18 Compton Way, Olivers Battery, Winchester, SO22 4HS

G3 VPH J Mayall, 10 Manor Close, Droitwich, WR9 8HG

G3 VPK W Mcclintock, 8 Fort Victoria Cottages, Westhill Lane, Yarmouth, PO41 0SA

GM3 VPN J Gardner, Taringa, Edentown, Cupar, KY15 7UH

G3 VPQ Ivor Westwood, Flat1, 43 Augustus Road, London, SW19 6LW

G3 VPR R Harrison, 512 Broadgate, Weston Hills, Spalding, PE12 6DA

G3 VPS P Lennard, 5 Parkside, East Grinstead, RH19 1JG

G3 VPT Paul Burgess, 26 William Peck Road, Spixworth, Norwich, NR10 3QB

GI3 VPV R Aughey, 30 Glen Road, Hillsborough, BT26 6ES

G3 VPW J Wright, 2 Barnfield, Charney Bassett, Wantage, OX12 0HA

G3 VPX I Sumner, 132 Barrs Road, Cradley Heath, B64 7EZ

G3 VQF J Moorhouse, 185 Aldermoor Road, Southampton, SO16 5NQ

G3 VQG R Beadle, 5 Badgeney Road, March, PE15 9AP

G3 VQM D Harrington, 27 Bayview Road, Peacehaven, BN10 8QD

G3 VQO L Allwood, 9 Gorse End, Horsham, RH12 5XW

GM3 VQQ M Hall, 3 Shierlaw Gardens, Airth, Falkirk, FK2 8RB

G3 VQR Adrian Henshaw, 3 Lewens Close, Wimborne, BH21 1JJ

G3 VQS Ronald Kirby, 197 Longfield, Falmouth, TR11 4SR

G3 VQW Barry Fawkes, 6 Oak Avenue, Worcester, WR4 9UG

G3 VQY J Cumming, Camelot, Cheltenham Road, Hockley, SS5 5HJ

G3 VRB James Nias, 49 St. Margarets Road, Bishopstoke, Eastleigh, SO50 6DG

G3 VRF John Charlton, 57 Victoria Road, Bidford on-Avon, Alcester, B50 4AR

G3 VRU P Ford, 15 Doles Lane, Whitwell, Worksop, S80 4SN

G3 VRV Michael Huish, Beckets, Woodbury, Exeter, EX5 1JD

G3 VRW P Lamb, 5 The Templars, Bridge End, Warwick, CV34 6PF

G3 VRY J Pitt, 30 Hillcroft Road, Chesham, HP5 3DJ

G3 VSB G Jones, Braemar, Alton Road, Uttoxeter, ST14 5DH

G3 VSC K Thompson, 3 Parkside, Morecambe, LA4 4TJ

G3 VSH D Freedman, Rivermeade, Irwell Vale, Bury, BL0 0QA

G3 VSI N Prince, 96 Foxglove Way, Springfield, Chelmsford, CM1 6QR

G3 VSJ David Chaloner, 38 Barnfield Close, Hoddesdon, EN11 9EP

G3 VSK Terence McCurry, 148 Moorgate Road, Rotherham, S60 3AZ

G3 VSL J Aroott, 122 Woodlands Road, Ashurst, Southampton, SO40 7AL

G3 VSQ R West, 10 Hawkshill Drive, Hemel Hempstead, HP3 0BS

G3 VSR Thomas Barraclough, 27 Kestrel Park, Skelmersdale, WN8 6TA

G3 VST Frank Moore, Causeway House, Risbury, Leominster, HR6 0NG

G3 VSU A Moore, The Saint Lawrence Tavern, High Street, Ramsgate, CT11 0QP

G3 VSV D Middleton, 8 Fulmar Close, Bradwell, Great Yarmouth, NR31 8JG

GM3 VTB V Budas, 20 Oak Avenue, Bearsden, Glasgow, G61 3HD

G3 VTD R Price, 36 Hadleigh Rise, Pontefract, WF8 4SJ

G3 VTE R Swetmore, 18 Tideswell Road, Stoke-on-Trent, ST3 5EG

**IMPORTANT NOTE**

**Revalidate licence to avoid revocation** – Ofcom has advised the Society that plans will be drawn up to revoke licences that have not been revalidated as required by the licence conditions. The quickest way to revalidate is to do so online via the Ofcom website: *https://services.ofcom.org.uk/* or by email: *amateur.validations@ofcom.org.uk* Ofcom staff are available to help, but please be patient during times of heavy workload.

GM3 VTH D Coutts, 29 Barons Hill Avenue, Linlithgow, EH49 7JU
G3 VTL J Levett, 402 Cromwell Lane, Burton Green, Kenilworth, CV8 1PL
G3 VTO M Coombs, 10 Horseshoe Walk, Widcombe, Bath, BA2 6DE
G3 VTR A Davis, Fieldings, Bury Road, Bury St. Edmunds, IP29 4PL
G3 VTS C Walker, 2 Georgian Close, Abbeydale, Gloucester, GL4 5DG
G3 VTT Colin Turner, 182 Station Road, Rainham, Gillingham, ME8 7PR
G3 VUD P Bentley, 12 West Terrace, Seaton Sluice, Whitley Bay, NE26 4RE
G3 VUE T Mowbray, Elmhirst House, Lincoln Road, Horncastle, LN9 5AW
G3 VUH Mike Blackwell, 31 Britannia Walk, Market Harborough, LE16 8BF
G3 VUI Michael Harris, Box 226, Port Stanley, South Atlantic, Falkland Islands
G3 VUK Ronald Knight, 8 Narromine Drive, Calcot, Reading, RG31 7ZL
G3 VUL J Lotz, 29 Burton Manor Road, Stafford, ST17 9QJ
G3 VUN G Ackerley, The Paddock, Stoney Lane, Tarporley, CW6 0SX
G3 VUO J Mills, 9 Sandpiper Walk, Chelmsford, CM2 8XJ
G3 VUR Keith Evans, 68 Downs Road, Hastings, TN34 2DZ
G3 VUS David Latimer, 44 Lyndale Avenue, Barrow-in-Furness, LA13 9AR
G3 VUY David Bradley, 4 Felthorpe Close, Upton, Wirral, CH49 4GY
G3 VVC John Parry, 19 Lon Hedydd, Llanfairpwllgwyngyll, LL61 5JY
G3 VVE H Robinson, 4 Cross Street, Mansfield Woodhouse, Mansfield, NG19 9NA
GM3 VVF A Ross, 17 Tarvit Green, Glenrothes, KY7 4SJ
G3 VVG B Salt, Cruets, The Village, Yelverton, PL20 7NA
G3 VVL Kenneth Lax, 17 Malt Rise, Crew Green, Shrewsbury, SY5 9EU
G3 VVR J Grace, Woodside, Easthorpe, Malton, YO17 6QX
G3 VVT Robert Wilkinson, 18 Green Road, Kendal, LA9 4QR
G3 VVW J Forrest, Little Orchard, 21 Bell Lane, Leatherhead, KT22 9ND
G3 VWA C Marflow, 13 Walthew Green, Roby Mill, Skelmersdale, WN8 0QT
G3 VWC A Marriott, 28 Horseshoe Walk, Bath, BA2 6DF
G3 VWD C Bean, 11 Nightingale Lane, Coventry, CV5 6AY
G3 VWH B Wilde, 34 Grangefields Road, Shrewsbury, SY3 9DB
G3 VWJ G Westwood, 133 Torrisholme Road, Lancaster, LA1 2TZ
G3 VWK Albert Hammett, (Hammett), Ladock, Truro, TR2 4PQ
G3 VWQ P Forster, 59 Woodland View, Stratton Strawless, Norwich, NR10 5LT
G3 VWX E Perks, The Oaklands, Bromfield Road, Ludlow, SY8 1DW
GM3 VWY I Malcolm, 2 Morton Crescent, St. Andrews, KY16 8RA
G3 VXA M Harrold, 26 Leys Close, Harefield, Uxbridge, UB9 6QB
G3 VXE G Brindle, 8 Peckover Drive, Pudsey, LS28 8EF
G3 VXF B Ellis, Whitmore Lodge, Ridgemoor, Hindhead, GU26 6QX
G3 VXH Roger Huffadine, 19 Cumberland Street, Worcester, WR1 1QE
G3 VXJ R Rylatt, 16 First Avenue, Worthing, BN14 9NJ
G3 VXK R Porter, 16 Millcroft, Crosby, Liverpool, L23 9XJ
G3 VXM D Clemens, 66 Mayles Road, Milton, Southsea, PO4 8NP
G3 VXS D Peach, Flat 35, Homeshire House, 36 Sandbach Road South, Stoke-on-Trent, ST7 2LP
G3 VXY B Cotton, 12 Tower Gardens, Bassett, Southampton, SO16 7EL
G3 VYA B Atkiss, 47 Russell Road, Partington, Manchester, M31 4DY
G3 VYD J Bourne, Tyndalls, 8 Kelvedon Road, Witham, CM8 3LZ
G3 VYE J Doswell, 20 Little Pittern, Kineton, Warwick, CV35 0LU
G3 VYF M Lee, 11 Sturrocks, Vange, Basildon, SS16 4PQ
G3 VYG Roger Walpole, 1 Woodfarm Cottage, Reymerston, Norwich, NR9 4QZ
G3 VYI Michael Franklin, 6 Tor Road, Farnham, GU9 7BX
GM3 VYJ T Jameson, 8 River View, Dalgety Bay, Dunfermline, KY11 9YE
G3 VYK P Frost, 164 Newthorpe Common, Newthorpe, Nottingham, NG16 2EN
G3 VYN M Turner, Plumtree Cottage, Spring Lane, Norwich, NR15 2NT
G3 VYS A Nell, Teaselwood, Parkham, Bideford, EX39 5PL
G3 VYU Robert Chamberlain, 1 Thornemead, Werrington Meadows, Peterborough, PE4 7ZD
G3 VYW D Carter, Oak Hall Cottage, Bartholomew Street, Hythe, CT21 5BT
G3 VYX Christopher Burr, The Old Rectory House, Radipole Lane, Weymouth, DT4 9RN
GI3 VYY B Hamilton, 4 Castleton Court, 16 Osborne Park, Belfast, BT9 6HA
G3 VYZ L Thompson, 44 Tillmouth Avenue, Holywell, Whitley Bay, NE25 0NP
G3 VZE David Kennedy, 79 High Street, Dunsville, Doncaster, DN7 4BS
G3 VZF John Adams, Chilterns, Bellingdon, Chesham, HP5 2XL
G3 VZG Richard Golding, 7 Belvidere Avenue, Shrewsbury, SY2 5PF
G3 VZH C Doran, 16 Wordsworth Road, Penge, London, SE20 7JG
G3 VZJ A Clemmetsen, The Danes, Shellbridge Road, Arundel, BN18 0ND
G3 VZL R Newman, 20 Glapthorn Road Oundle, Peterborough, PE8 4JQ
G3 VZM F Houghton, 14 Windfield Gardens Little Sutton, Ellesmere Port, CH66 1JJ
G3 VZO V Hartshorn, 61 Fulmerton Crescent, Redcar, TS10 4NJ
G3 VZR E Thompson, Meadowside, Bromsberrow Heath, Ledbury, HR8 1NX
G3 VZT R Johnson, The Hollies, Belaugh Green Lane, Norwich, NR12 7AJ
G3 VZU W Mooney, 538 Liverpool Road, Great Sankey, Warrington, WA5 3LU
G3 VZV G Shirville, Birdwood, Heath Lane, Milton Keynes, MK17 8TN
G3 WAB P Harrison, 8 Buxtons Lane, Guilden Morden, Royston, SG8 0JU
G3 WAE I Harris, Orchard Cottage, The Street, Devizes, SN10 2LD
G3 WAG D Gillett, 20 Redcar Avenue, Hereford, HR4 9TJ
G3 WAH N Hodgson, 42 Tofts Grove, Rastrick, Brighouse, HD6 3NP
G3 WAL J Barker, 76 Halebrose Court, Seafield Road, Bournemouth, BH6 3DU
G3 WAM M Taplin, 111a Coupe Lane, Old Tupton, Chesterfield, S42 6HD
GM3 WAP A Philp, Philp House, High Street, Blairgowrie, PH11 8DW
G3 WAS LICHFIELD A.R.S c/o R Smethers, 46 Church Road, Burntwood, WS7 9EA
G3 WBA I Currell, 47 Highdale Avenue, Clevedon, BS21 7LU
G3 WBB E Avery, 2 Blythe Avenue, Thornton-Cleveleys, FY5 2LL
G3 WBC Roger Bryant, 12 Laburnum Grove, Luton, LU3 2DW
G3 WBG H Hindle, 6 Windsor Road, Conisbrough, Doncaster, DN12 3DF
G3 WBI Peter Lewis, 15 Norwood Road, Lytham St. Annes, FY8 2QN
G3 WBK Paul Tofts, 48 Rugby Road, Brighton, BN1 6EB
G3 WBL Keith Weller, Charbury House, Bayton, Kidderminster, DY14 9LJ
G3 WBN A Thurlow, Chesnet House, Croydon, CR0 5BA
G3 WBP J Broadley, 13 Portland Close, Bedford, MK41 9NE
G3 WBQ Trevor Brook, 22 Downside Road, Guildford, GU4 8PH
GI3 WBR R Mccrea, 1 Killynoogan Terrace, Killynoogan, Enniskillen, BT93 8DF
G3 WBS Derek Thomson, 2a The Landway, Kemsing, Sevenoaks, TN15 6TG
GW3 WBU B Vodden, 22 Heath Avenue, Penarth, CF64 2QZ
GW3 WCA P Dunbar, Pengwern Fach, Penrherber, Newcastle Emlyn, SA38 9RL
G3 WCB D JOHN, 41a Chequers Orchard, Iver, SL0 9NJ
G3 WCD Christopher Dillon, 63 High Street, Toseland, St. Neots, PE19 6RX
G3 WCE B Edwards, Elder Cottage, Norwich, NR10 5BB
G3 WCJ P Hackett, Lot49, Bodeguero Way, Wooroloo, Australia, 6558
G3 WCL J Croker, 29 Alexandra Road, Bedminster Down, Bristol, BS13 7DF
G3 WCM F Chidlow, 64 Mitchell Avenue, Northside, Workington, CA14 1AA
G3 WCO John Rollason, Pavitts Cottage, Keers Green, Dunmow, CM6 1PQ
G3 WCQ Richard Bailey, 43 Earlsdon Avenue South, Coventry, CV5 6DR
G3 WCU J Pealing, 93 Fernside Road, Poole, BH15 2JQ
GW3 WCV David Howell, 6 Douglas Close, Cardiff, CF5 2QT
G3 WCY Brian Smith, 26 Sandhurst Lane, Blackwater, Camberley, GU17 0DH
G3 WDD T Horrobin, 29 Ambleside Road, Maghull, Liverpool, L31 6BY
G3 WDE Peter Ford, 11 Brook Lane, Felixstowe, IP11 7EG
G3 WDG C Suckling, 314a Newton Road, Rushden, NN10 0SY
G3 WDI Terry Weatherley, 16 Beverley Court, Carlton Colville, Lowestoft, NR33 8JZ
G3 WDL John Boast, 118 Barnsley Road, Moorends, Doncaster, DN8 4QR
G3 WDM C Care, 127 Brooklands Crescent, Fulwood, Sheffield, S10 4GF
G3 WDN Edward Fielding, The Birches, 3 Sneath Road, Norwich, NR15 2DS
G3 WDS D Spooner, 45 Otterburn Avenue, Whitley Bay, NE25 9QR
G3 WDU I Peterkin, 3 Mill Lane, Thorp Arch, Wetherby, LS23 7DZ
G3 WDX P Hickey, 16 Cross Road, South Oxhey, Watford, WD19 4DH
G3 WEA A Cross, 34 Pinewood Drive, Potters Bar, EN6 2BD
G3 WEB G Gardiner, 11 Langdale Avenue, Ramsgate, CT11 0PQ
GM3 WED A Rose, Craiglea, Schoolcroft, Dingwall, IV7 8LB
G3 WEF A Beazley, 24 Tealsbrook, Covingham, Swindon, SN3 5AU
G3 WEG Philip Webster, 22 Whincroft Drive, Ferndown, BH22 9LJ
G3 WEI D Turner, Birchwood, Heath Top, Market Drayton, TF9 4QR
G3 WEJ SB Bradshaw, 11 Meadow Park, Dawlish, EX7 9BS
GI3 WEL Raymond Knox, 91 Banbridge Road, Waringstown, Craigavon, BT66 7RU
GI3 WEM V Gracey, 23 Cascum Road, Banbridge, BT32 4LF
G3 WEQ C Collins, 21 Bron Wern, Llanddulas, Abergele, LL22 8JD
G3 WEU Keith Gregory, 67 Clowne Road, Barlborough, Chesterfield, S43 4EH
G3 WEW Roger Wood, 8305 El Matador Drive, Gilroy, United States, 95020
G3 WEZ J Lawrence, 3 Siskin Crescent, Rogiet, Caldicot, NP26 3UW
G3 WF D Cockings, Elettra, 207a Birchfield Road, Redditch, B97 4LX
G3 WFF Barry Tew, 96 Mill Lane, Sawston, Cambridge, CB22 3HZ
G3 WFH D Morris, 27 Albert Square, Bowdon, Altrincham, WA14 2ND
GM3 WFJ Robert Andrew, The Old Manse, Kirkmichael, Blairgowrie, PH10 7NY
G3 WFL Ian Drake, 34 The Holt, Bishops Cleeve, Cheltenham, GL52 8NQ
G3 WFM J Crabbe, 47 Torrington Drive, Potters Bar, EN6 5HU
GI3 WFP P McAlpine, 20 Gransha Road South, Bangor, BT19 7QB
G3 WFT D Holland, 32 Woodville Drive, Sale, M33 6NF
G3 WFW K Hampson, 11 Gladstone Grove, Stockport, SK4 4BX
G3 WGC WELWYN-HATFIELD c/o K Pollard, 5 Lodge Way, Stevenage, SG2 8DB
G3 WGE E Law, 4 Brograve Close, Galleywood, Chelmsford, CM2 8YA
G3 WGH M Reeve, 18-20 Radford Road, Nottingham, NG7 5FS
G3 WGK Bernard Wormwell, 26 Windsor Avenue, Longridge, Preston, PR3 3EL
G3 WGN David Aslin, Oakstone, Spreyton, Crediton, EX17 5AL
G3 WGQ John Hartley, 2 Hall Street, Cockbrook, Ashton-under-Lyne, OL6 6SD
G3 WGU Steve Williamson, 120 Warbreck Hill Road, Blackpool, FY2 0TR
G3 WGV John Linford, Pennine View, Sleagill, Penrith, CA10 3HD
G3 WGY Howard Ashford, 56 Guarlford Road, Malvern, WR14 3QP
G3 WGZ G Sowden, Villa Clare, The Lizard, Helston, TR12 7NU
GI3 WHA L Hanna, Igrangeville Park, Newtownards, BT23 8TE
G3 WHB Stephen Christie, 4 Dairy Court, Holyport, Maidenhead, SL6 2US
G3 WHG M Key, 12 Great Melton Road, Hethersett, Norwich, NR9 3AB
G3 WHJ A Johnson, 49 Tennyson Drive, Malvern, WR14 2UL
GM3 WHT M Smith, Vakterlee, Cumliewick, Shetland, ZE2 9HH
G3 WHU G Parrott, Flat 5, 33 Princes Drive, Colwyn Bay, LL29 8PD
G3 WI Mick Hulme, 44 Thirlmere Avenue, Ashton-under-Lyne, OL7 9HN
G3 WIA R Ottley, 15 Orchard Way, Thrapston, Kettering, NN14 4RE
G3 WII F Clarke, 8 Tristram Close, Chandler's Ford, Eastleigh, SO53 4TT
GM3 WIJ Norman Mackenzie, 57 Countesswells Terrace, Aberdeen, AB15 8LQ
G3 WIK Michael Shorland, Baxhill Bungalow, Upper Colwall, WR13 6DL
GM3 WIL D Cossar, 52 Bentfield Drive, Prestwick, KA9 1TT
G3 WIM WIMBLEDON & DISTRICT ARS c/o Jim Gale, Barn End, Highampton, Beaworthy, EX21 5LT
G3 WIN Windscale ARS c/o Richard Wood, Abbots Croft, Abbey Road, St. Bees, CA27 0EG
G3 WIO E Obrien, Tanglewood, Anthonys Way, Wirral, CH60 0BP
G3 WIP G Bulger, Flat C/21, Herbal Hill Gardens, 9 Herbal Hill, London, EC1R 5XB
G3 WIS B Day, 54 South Avenue, Hope Carr, Leigh, WN7 3BU
G3 WIU W Bekenn, 35 Blackdown Avenue, Rushmere St. Andrew, Ipswich, IP5 1AY
G3 WIW A Leach, 199 Braemor Road, Calne, SN11 9EA
GM3 WJE J Thom, 64a Hawick Drive, Dundee, DD4 0TA
G3 WJG G Lean, 54 Blacketts Wood Drive, Chorleywood, Rickmansworth, WD3 5QH
G3 WJH W Wilkinson, Chiriqui, 15 Camerton Road, Workington, CA14 1LP
G3 WJI P White, Linden House, Willisham, Ipswich, IP8 4SP
G3 WJJ David Finnemore, 4 Purbeck Gardens, Felton Road, Poole, BH14 0QS
G3 WJM B Schoth, 3 Solent Drive, Hythe, Southampton, SO45 5FP
G3 WJN R Hassell Bennett, 30 Greenlands Avenue, Redditch, B98 7QA
G3 WJP J Parnell, 40 Dolcoath Road, Camborne, TR14 8RW
G3 WJS J Starling, 16 Queenscliffe Road, Ipswich, IP2 9AS
G3 WKA Neil Bardell, 4 Church End, Arlesey, SG15 6UY
GM3 WKB David Topham, Dairy Cottage, Kilmany, Cupar, KY15 4PT
G3 WKE S Braidwood, 48 Inwood Avenue, Hounslow, TW3 1XG
G3 WKF Maurice Richards, Wayside Cottage, Penwithick Road, St. Austell, PL26 8UH
G3 WKH Robert Martin, 14029 23rd Place NE, Seattle, United States, WA 98125
G3 WKI M Hewins, 37 Ringwood Close, Furnace Green, Crawley, RH10 6HQ
G3 WKL J Gould, 116 Wolverton Road, Newport Pagnell, MK16 8JG
G3 WKP Peter King, Nirvana, Comprigney Hill, Truro, TR1 3TX
G3 WKR M Goodwin, 6 Hobbs Hill, Rothwell, Kettering, NN14 6YG
G3 WKS West Kent ARS c/o David Green, St. Annes, Poundfield Road, Crowborough, TN6 2BG
G3 WKW Robert Thornton, 26 Florence Road, Fleet, GU52 6LQ
G3 WKX MAIDENHEAD & DISTRICT ARC c/o Mark Palmer, 28 Westfield Road, Caversham, Reading, RG4 8HH
G3 WLA Arthur Macpherson, 15a Monkstone Drive, Berrow, Burnham-on-Sea, TA8 2NW
G3 WLD J Hall, 22 Haverhill Road, Stapleford, Cambridge, CB22 5BX
G3 WLG M Griffiths, The Oaklands, Hollybush Lane, Worcester, WR6 6HQ
G3 WLH C Pell, 1 Glenville Gardens, Hindhead, GU26 6SX
G3 WLM R Joyce, 20 Barking Close, Luton, LU4 9HG
GW3 WLN J Pritchard, 13 Cefn Graig, Rhiwbina, Cardiff, CF14 6SW
G3 WLO E Denton, 11 Highland Road, Amersham, HP7 9AU
G3 WLT D Firth, 3 School Lane, Shaldon, Teignmouth, TQ14 0DG
G3 WLV J Bushby, 14 Clayton Drive, Thurnscoe, Rotherham, S63 0RZ
G3 WLW R Millar, 1229 Leeds Road, Bradley, Huddersfield, HD2 1UY
G3 WLY John Harwood, 12 Longwood Avenue, Cowplain, Waterlooville, PO8 8HX
G3 WMA W Shepperd, 28 Tyne Road, Oakham, LE15 6SJ
G3 WMD J Whomes, 44 Russell Close, Steeple Morden, Royston, SG8 0NE
G3 WME M Groom, 409 Finchampstead Road, Finchampstead, Wokingham, RG40 3RL
G3 WMJ G Jillings, 16 Greenwich Road, Diep River, South Africa, 7800
GW3 WMP J Hopton, 9 Bryneuraidd, Ammanford, SA18 3TG
G3 WMQ M Watson, Chant House, Dark Lane, Stroud, GL6 0DR
G3 WMS I Vance, Larkfield, Debden Road, Saffron Walden, CB11 3RU
G3 WMT R Dowling, 80 Elmfield Way, Sanderstead, South Croydon, CR2 0EF
G3 WMX C Knott, 154 Park View, Crewkerne, TA18 8JJ
G3 WMY S Smith, Five Oaks, Sandy Lane, Henfield, BN5 9UX
GM3 WNB J Ohare, 208 Gilmartin Road, Linwood, Paisley, PA3 3ST
G3 WNC R Todd, 17 Tudor Road, West Bridgford, Nottingham, NG2 6EB
G3 WND Robyn Aston, 16 St. Johns Road, Mortimer Common, Reading, RG7 3TR
G3 WNP T Baker, 54 Hamilton Road, Reading, RG1 5RD
G3 WNQ E Lingard, Tedulf Rottenrow, Theddlethorpe, Mablethorpe, LN12 1NX
G3 WNR Kenneth Grey, 15 Woodbourne Avenue, Leeds, LS17 5PQ
G3 WNS A Willson, Hilltop, Cryers Hill Road, High Wycombe, HP15 6LJ
G3 WNV D Field, Rackhay, Prescott, Cullompton, EX15 3BA
G3 WNW David Bailey, 31 Antrobus Street, Congleton, CW12 1HE
G3 WOA John Goodman, 15 Highway, East Taphouse, Liskeard, PL14 4NW
G3 WOD J Welford, 303 Scalby Road, Scarborough, YO12 6TF
G3 WOE Michael White, 76 Birch Row, Bromley, BR2 8FG
G3 WOH E Grossmith, 4 Lincoln Way, Rainhill, Prescot, L35 6PJ
GM3 WOJ Chris Tran, Achnacoille, Lamington, Invergordon, IV18 0PE
G3 WOK D Clifton, 59 Grantham Road, Bracebridge Heath, Lincoln, LN4 2LE
G3 WOM M Muir, 6 Broadstairs Court, Sunderland, SR4 8NP
G3 WOO R Brace, 11 Cedar Close, Sawbridgeworth, CM21 9NT
G3 WOR Worthing and District Amateur Club c/o Andrew Cheeseman, Flat 5 Dubarry House, Hove Park Villas, Hove, BN3 6HP
G3 WOS Christopher Gare, Old White Lodge, 183 Sycamore Road, Farnborough, GU14 6RF
G3 WOT M Meads, 12 Burlington Way, Hemingford Grey, Huntingdon, PE28 9BS
G3 WOV G MacNaught, 30 West End Falls, Nafferton, Driffield, YO25 4QA
GU3 WOW Peter Hancock, La Breloque, Les Grandes Rues, Les Buttes, Guernsey, GY79EL
GM3 WPA S Hutchinson, 4 Wiston Place, Dundee, DD2 3JR
G3 WPB Paul Smith, 180 Victoria Road, Ferndown, BH22 9JE
G3 WPD A Smith, 118 Bois Moor Road, Chesham, HP5 1SS
G3 WPF Reginald Unsworth, Spurs Lodge, Sagars Road, Wilmslow, SK9 4HE
G3 WPG D Dye, 10 Headington Close, Bradwell, Great Yarmouth, NR31 8DN
G3 WPH M Chamberlain, 10 Clifton Rise, Wargrave, Reading, RG10 8BN
G3 WPN Vyvyan Bennellick, The Wolverns, Castle Frome, Ledbury, HR8 1HG
G3 WPP David Minett, Melrose Cottage, South Road, Truro, TR3 7AD
G3 WPQ M Kaye, Pucknell Lodge, Hollywell Lane, Kidderminster, DY14 9NR
G3 WPR Christopher Richmond, 1 Grangeway Gardens, Ilford, IG4 5HN
G3 WPT R Brown, 65 Staining Rise, Staining, Blackpool, FY3 0BU
G3 WPV D Lamont, 6B Route De Mailhac, Bize Minervois, France, 11120
G3 WQ P Jackson, 5 Rhodes Terrace, Osgodby, Selby, YO8 5HF
G3 WQG David Chalmers, 25 Willow Close, Flackwell Heath, High Wycombe, HP10 9LH
G3 WQK SOUTHDOWN ARS c/o John Vaughan, Flat 9, St. Leonards Court, St. Leonards-on-Sea, TN38 0PS
G3 WQL Alan Conway, 17 Mountcastle Road, Leicester, LE3 2BW
G3 WQU P Mckay, C/O Unifil, Po Box 5852, New York, United States, 10163-5852
G3 WQY T Codrai, Sealand, Coast Road, Norwich, NR12 0PD
G3 WRA Stuart Powell, Rowan House, Lugwardine, Hereford, HR1 4AE
G3 WRD R Richardson, Common Crest, Drapery Common, Sudbury, CO10 7RW
GW3 WRE B Jones, 6 Pentyla, Maesteg, CF34 0BB
G3 WRI Paul Brown, 30 Applerigg, Kendal, LA9 6EA
G3 WRJ R Bacon, The Gyffen, 3 Gosmore Road, Hitchin, SG4 9AN
G3 WRK G Oakes, 13 Tidnock Avenue, Congleton, CW12 2HN
G3 WRL E Northwood, 8 Derwood Grove, Werrington, Peterborough, PE4 5DD
G3 WRO K Haynes, 34 Pear Tree Mead, Harlow, CM18 7BY
G3 WRR Quin Collier, 19 Grangecliffe Gardens, South Norwood, London, SE25 6SY
G3 WRS Wakefield & District RS c/o Stuart Adaway, 20 Foundry Street, Barnsley, S70 1PL
G3 WRT Ian Dilworth, Ashpound Cottage, Pound Lane, Ipswich, IP9 2JB
G3 WSB K Band, 11 Denewood Close, Watford, WD17 4SZ
G3 WSC Crawley ARC c/o John Pitty, 12 St. Leonards Road, Horsham, RH13 6EJ
G3 WSD Alf Fisher, 63 Spencer Close, Potton, Sandy, SG19 2QR
G3 WSM B Storry, 508 Arleston Lane, Stenson Fields, Derby, DE24 3AA
GM3 WSR Vernon Clark, 6 Parkhill Circle, Dyce, Aberdeen, AB21 7FN
GW3 WSU Colin Beynon, 16 Hardy Close, Barry, CF62 9HJ
G3 WSV James Lawson, 3 Pearmains, Great Leighs, Chelmsford, CM3 1QS
G3 WSW John Holmes, 37 Redwood Avenue, Leyland, PR25 1RN
G3 WSZ P Gilson, 9 Little Preston Hall Park, Hall Road, Little Preston, LS26 8UW
G3 WTB R Baxter, 10 Windsor Court, Oxford Road, Southport, PR8 2JJ
G3 WTD J Davis, 71 Broughton Road, Croft, Leicester, LE9 3EB
G3 WTN R Limehouse, 56 Lincoln Way, Daventry, NN11 4SX
G3 WTO J Spencer, 76 Durranhill Road, Carlisle, CA1 2SZ
G3 WTP Bedford and District ARC c/o Robert Leask, 80 Mill Road, Sharnbrook, Bedford, MK44 1NP
G3 WTQ Peter Angold, 10 Hartford Avenue, Wilmslow, SK9 6LP
G3 WTR D Wright, 8 Calverley Park, Tunbridge Wells, TN1 2SH

G3 WTS J Smith, Windycross, Newbourne Road, Woodbridge, IP12 4PT
G3 WTT Colin Goodwin, 1 School Lane Canwick, Lincoln, LN4 2RP
G3 WTV K Baker, 33 Reading Road, Woodley, Reading, RG5 3DA
G3 WTY P Hodgkiss, 28 Beaumont Rise, Worksop, S80 1YA
G3 WTZ M Jones, 55 Rowan Way, Malpas, Newport, NP20 6JN
G3 WUA B Lindop, 56 Marina Court, 9-19 Mount Wise, Newquay, TR7 2EJ
G3 WUB P Rice, 23 Christchurch Square, Homerton, London, E9 7HU
G3 WUG Ian Elvins, 6 Bay Road, Unit 23, Newmarket, United States, NH 03857
G3 WUH W Dufton, 22 Windsor Road, Bexhill-on-Sea, TN39 3PB
G3 WUI G Snink, 60 Woodhouse Hill, Huddersfield, HD2 1DH
G3 WUK Jaques Spencer Chapman, Apartado De Correos 156, Mojacar 04638, Almeria, 222 31P
G3 WUL Roy Whillier, 9 Tudor Drive, Yateley, GU46 6BX
G3 WUN D Holden, 99 Sheerstock, Haddenham, Aylesbury, HP17 8EY
GI3 WUO I Waring, 16 Belfast Road, Holywood, BT18 9EL
G3 WUW A Panworth, 3570 Corey Road, Malabar, United States, 32950
GM3 WUX Terry Robinson, 82 Albert Road, Glasgow, G42 8DH
G3 WUZ P Brown, The Briers, Brent Road, Burnham-on-Sea, TA8 2JT
G3 WVG Kenneth Pritchard, 9 Golf Close, Pyrford, Woking, GU22 8PE
G3 WVL J Jones, 17 Leamore Crescent, Shrewsbury, SY3 7QB
G3 WVM Christopher Loosemore, 24 Myrtlebury Way Hill Barton, Exeter, EX1 3GA
G3 WVQ J Barratt, 26 Johnstone Road, Newent, GL18 1PZ
G3 WVR John Green, Honeysuckle Cottage, New Green, Braintree, CM7 5EG
GW3 WVV R Barker, Henllys, Maenygroes, New Quay, SA45 9RL
G3 WWG J Ross, 24 Raby Road, Stockton-on-Tees, TS18 4JA
GW3 WWH Reginald Taylor, Brynderi Farm, Blaenawun, Whitland, SA34 0JD
G3 WWI R Oxley, 1 Elm Grove, Maidstone, ME15 7RT
G3 WWL B Tipper, 271 Blackberry Lane, Four Oaks, Sutton Coldfield, B74 4JS
G3 WWS M Southall, 61 Grange Close, Horam, Heathfield, TN21 0EF
G3 WWT J Teed, 47 West Cliff Road, Dawlish, EX7 9DZ
GI3 WWY M Anderson, 8 Loughbrickland Road, Gilford, Craigavon, BT63 6BH
GW3 WXA J GOUGH, Traleen, Rhydlewis, Llandysul, SA44 5PN
G3 WXC Peter Brooker, 28 Uplands Road, Northwood, Cowes, PO31 8AL
G3 WXD C Zammit, 9 Sandbanks Drive, Basingstoke, RG22 4UL
G3 WXG I Habens, 48 Carden Avenue, Brighton, BN1 8NE
G3 WXH J Arnold, 6 The Spinney, Weston-Super-Mare, BS24 9LH
G3 WXM M Smith, Linden, 86 Grove Road, Tring, HP23 5PB
G3 WXN Lawrence McKown, Flat D, 310 Oldham Road, Oldham, OL2 5AS
G3 WXU S Allbutt, 8 Langton Close, Vinters Park, Maidstone, ME14 5PG
G3 WXW Cecil Traveller, 13 Cosy Corner, North Walsham, NR28 0EN
G3 WYB A Tring, 1 Crownbourne Court, St. Nicholas Way, Sutton, SM1 1JE
G3 WYD P Patmore, 141 Cannons Close, Bishop's Stortford, CM23 2BL
G3 WYH R Hutton, 6 The Sidings, Ruskington, Sleaford, NG34 9GA
G3 WYK P Bysshe, Orchard House, High Road, Maidenhead, SL6 9JT
GM3 WYL A Ritchie, 83 Larkfield Road, Lenzie, Glasgow, G66 3AS
G3 WYN John GIBSON, Four Oaks, Tylers Green, Haywards Heath, RH17 5DZ
G3 WYP Derek Allan, 283 Cliffe Lane, Gomersal, Cleckheaton, BD19 4SB
G3 WYT Martin Edwards, 23 Burnside, Waterlooville, PO7 7QQ
G3 WYW P Bigwood, 18 The Martins, Thatcham, RG19 4FD
G3 WZA W Burnet, 56 Sleaford Road, Boston, PE21 8EU
G3 WZE Peter Cleary, 531 Diamond Street, San Francisco, United States, CA94114-3223
G3 WZG P Murtha, 16 Approach Road, Margate, CT9 2AN
G3 WZH N Ghani, 52b Stormore, Dilton Marsh, Westbury, BA13 4BH
G3 WZI K Reeves, 9 Tibberton Close, Solihull, B91 3UD
G3 WZJ A Watt, Manor Farm, Eagle Hall, Lincoln, LN6 9HZ
G3 WZK Stephen Beal, 16 Clovelly Avenue, Warlingham, CR6 9HZ
G3 WZO John Kyriakides, 16 Wise Lane, London, NW7 2RE
G3 WZP G Budden, 7 Ashburton Gardens, Ensbury Park, Bournemouth, BH10 4HP
G3 WZR R Wright, 2 Jackson Close, Devizes, SN10 3AP
G3 WZS H Williams, 7 Munn Drive, Po Box 276, Tobermory, Canada, N0H 2R0
G3 WZT John Matthews, 46 Park Lane, West Grinstead, Horsham, RH13 8LT
GM3 WZV Stewart Hunter, Balnagown Cottage, Muir of Ord, IV6 7RS
G3 WZW G Laycock, 1 Campsall Cottage, Churchfield Road, Doncaster, DN6 9BY
G3 WZZ Andrew Huddleston, Willow Bank Cottage, Willow Bank, Keighley, BD20 5AN
G3 XAB D Whittaker, 2 Stone Edge, Halifax Road, Burnley, BB10 3QH
G3 XAC Christopher Whitehead, 10 Berkeley Drive, Read, Burnley, BB12 7QG
G3 XAG J Gibbon, The Bungalow, Manless Terrace, Saltburn-by-The-Sea, TS12 2DQ
G3 XAN W Forrester, 34 Keble Drive, Liverpool, L10 3LD
G3 XAP A Ashton, 2 Wickham Road, Thwaite, Eye, IP23 7EE
G3 XAQ Alan Ibbetson, Katalin, Town Lane Chartham Hatch, Canterbury, CT1 7NN
G3 XAS Colin Riggs, 1 Alms Walk, Wimborne St. Giles, Wimborne, BH21 5LZ
G3 XAU T Woodward, 33 Common Road, Hemsby, Great Yarmouth, NR29 4LT
G3 XAW M Chouings, 32 Nunney Close, Keynsham, Bristol, BS31 1XG
G3 XAX A Paley, 19 Arbour Lane, Wickham Bishops, Witham, CM8 3NS
G3 XAZ R Stoppard, 4 Beaumaris Drive, Beeston, Nottingham, NG9 5PB
G3 XBE H Walton, 12 Le Page Court, Nottingham, NG8 3ES
G3 XBF BARKING RADIO AND ELECTRONICS SOCIETY c/o S Peat, 64 Grange Road, Romford, RM3 7DX
G3 XBH G Thompson, 25a Copleston Road, London, SE15 4AN
G3 XBI Peter Boast, Laurel Bank, 19 Main Road, Highbridge, TA9 3QU
G3 XBM Roger Lapthorn, 7 Mill Close, Burwell, Cambridge, CB25 0HL
G3 XBN F Chamberlain, 43 Old Mill Close, Patcham, Brighton, BN1 8WE
G3 XBQ A Weseley, Loves House, Goudhurst Road, Tonbridge, TN12 9NB
G3 XBW M Wells, 15 Rivers Reach, Frome, BA11 1AQ
G3 XBX DW Harris, 119 Stanlake Road, London, W12 7HQ
G3 XBY D Harvey, 38 School Road, Shirley, Solihull, B90 2BB
G3 XBZ P Ciotti, 6 Bascott Road, Bournemouth, BH11 8RH
G3 XCD H Martin, 17 Vyner Road, Wallasey, CH45 6TE
G3 XCE Ernest Wells, 23 Briarfield Road, Poulton-le-Fylde, FY6 7PW
G3 XCJ W Burden, 44 Spekehill, Eltham, London, SE9 3BW
G3 XCK J Pegrum, 14 The Leys, Langford, Biggleswade, SG18 9RS
G3 XCO D Meldrum, 34 Graham Road, Ipswich, IP1 3QF
G3 XCS C Squires, 5 Frith Road, Saltash, PL12 6EL
G3 XCT D Dade, 40 Compton Avenue, Brighton, BN1 3PS
G3 XCW G Winter, 14 Drakes Lea, Evesham, WR11 3DJ
G3 XCY Keith Bristow, 34 Stanier Road, Preston, Weymouth, DT3 6PD
GI3 XCZ G Martin, 100 Drumconnolly Road, Gortaclare, Omagh, BT79 0XS
G3 XDA R Holderness, 16 Helmsley Way, Spalding, PE12 6BG
GI3 XDD J Crampton, 105a Ballymena Road, Doagh, Ballyclare, BT39 0TN
G3 XDK Andrew Maris, 140 Edward Street, Brighton, BN2 0JL
G3 XDL A Long, 2a Hawthorndene Road, Bromley, BR2 7DY
G3 XDM Alan Benson, 31 Oakhill Drive, Welwyn, AL6 9NW
G3 XDP C Wilkinson, 509 Warrington Road, Culcheth, Warrington, WA3 5QY
G3 XDS P Wilde, 5 Ruddington Court, Mansfield, NG18 4QD
G3 XDU K Whitbread, 27 Duckmill Crescent, Duckmill Lane, Bedford, MK42 0AE
GI3 XDX G Mcdowell, 13 Bedford Road, Cullybackey, Ballymena, BT43 5PR
G3 XDY J Quarmby, 12 Chestnut Close, Rushmere St. Andrew, Ipswich, IP5 1ED
G3 XDZ Zygmunt Skrobanski, 1035 Pine Grove Pointe Drive, Roswell, United States, GA 30075-2704
G3 XEC Geoffrey Grundy, Route de L'Angouiniere, la Roche Sur Yon, France, 00000
G3 XED C Masters, 79 Kings Head Lane, Bristol, BS13 7DB
G3 XEF Michael Fleetwood, Hemmet, 235 Shingle Hill Way, Gundaroo, Australia, NSW 2617
G3 XEI J Hooper, Long Barn House, Bolney Road, Horsham, RH13 8AZ
G3 XEJ Contest Cymru c/o Stephen Cole, 101 Allt-yr-yn Road, Newport, NP20 5EF
G3 XEN P Mullineaux, 27 Ashfield Avenue, Lancaster, LA1 5EB
G3 XEP WHITE ROSE ARS c/o E Hannaby, 34 Woodlea Lane, Meanwood, Leeds, LS6 4SX
GI3 XEQ J Bailie, 25 Upper Knockbreda Road, Belfast, BT6 0NA
G3 XER Damien Mannix, 34 Ashby Road, Ticknall, Derby, DE73 7JJ
G3 XEV John Cooper, 34 Arcal Street, Dudley, DY3 1TG
G3 XEW G Childs, 115 Summerhouse Drive, Bexley, DA5 2ER
G3 XEY A Robinson, 48 Colton Road, Shrivenham, Swindon, SN6 8AZ
G3 XFB David Jewson, 8 Johnsgate, Brewood, Stafford, ST19 9HZ
G3 XFD Rob Mannion, Flat 1, 1 Spencer Road, Bournemouth, BH1 3TE
G3 XFF E Tuddenham, 42 Garrison Lane, Felixstowe, IP11 7RP
G3 XFL J Harding, Whispers, 27 Northfield Drive, Truro, TR1 2BS
G3 XFN G Coffin, 45 Egerton Road, Streetly, Sutton Coldfield, B74 3PG
G3 XFU I Hasman, Fleetway, The Spinney, Newark, NG24 2NT
G3 XG BRAINTREE & DISTRICT ARS c/o Melvin Kendall, 88 Coldnailhurst Avenue, Braintree, CM7 5PY
G3 XGC G Cottrell, 36 Davenant Road, Oxford, OX2 8BY
G3 XGD G Watson, 6 The Avenue, Lyneal, Ellesmere, SY12 0QJ
G3 XGE P Greenhalgh, 13 Primrose Avenue Urmston, Manchester, M41 0TY
G3 XGH W Jamison, Horseshoe Cottage, Town Fold, Stockport, SK6 5BT
G3 XGK C Langley, Clarence Cottage, Commodore Road, Lowestoft, NR32 3NF
G3 XGU Keith Hill, 42 Greenleafe Avenue, Doncaster, DN2 5RF
G3 XGV G Fowles, Ruby House, Broad Marston, Stratford-upon-Avon, CV37 8XY
G3 XGW Keith Yates, Tibblestone Lodge, Ashton Road, Tewkesbury, GL20 7AU
G3 XGY B Harris, 36 Holland Way, Blandford Forum, DT11 7RU
G3 XGZ Malcolm Lambert, 3 Holyoake Avenue, Blackpool, FY2 0QL
G3 XHB Grahame Bentley, 2 Conan Drive, Richmond, DL10 4PQ
GW3 XHD B Walters, 16 Broomhill, Port Talbot, SA13 2US
GW3 XHG David Griffiths, 7 Canning Street Ton Pentre, Pentre, CF41 7HF
GI3 XHL W Rawson, 79 Loopland Drive, Castlereagh, Belfast, BT6 9DW
G3 XHM A Lewis, Bradley Villa, 41 West Street, Ryde, PO33 2UH
G3 XHP G Biddulph, 12 Cobbs Lane, Hough, Crewe, CW2 5JN
G3 XHW J Morris, 2 The Corniche, Sandgate, Folkestone, CT20 3TA
G3 XHZ John Farrer, Woodside Cottage, Catmere End, Saffron Walden, CB11 4XG
G3 XIA Peter Bates, 28 Salvington Hill, Worthing, BN13 3AT
G3 XIB Brian Johnson, 30 Tamar Way, Gunnislake, PL18 9DH
G3 XIG C Graham, 1 Arnhem Green, Dorchester, DT1 2PS
G3 XIH W Dixon, 17 Chestnut Bank, Scarborough, YO12 5QJ
G3 XII F Harrison, 78 Lancaster Lane, Leyland, PR25 5SP
G3 XIP D Aspinall, 53 Springhill Road, Fen Drayton, Cambridge, CB24 4SR
G3 XIQ K FINCH, 32 Main Street, Pymoor, Ely, CB6 2ED
GW3 XIS R Belcher, 8 Bishops Grove, Sketty, Swansea, SA2 8BE
G3 XIU William Byers, 29 Newtown, Frizington, CA26 3QQ
G3 XIV G Bulleyment, 30 Brackley Avenue, Fair Oak, Eastleigh, SO50 8FL
G3 XIX J Hobin, 14 St. Martins Green, Trimley St. Martin, Felixstowe, IP11 0UU
G3 XIY R Hall, 22 Cumbria Close, Thornbury, Bristol, BS35 2YE
G3 XIZ Christopher Osborn, 116 Holme Court Avenue, Biggleswade, SG18 8PB
G3 XJA D Williams, 5 Coed Eithen Street, Blaenavon, NP4 9LQ
GW3 XJC Bill Luke, 33 Maiden Street, Cwmfelin, Maesteg, CF34 9HP
G3 XJF P Duffett-Smith, 4b The Avenue, Godmanchester, Huntingdon, PE29 2AF
G3 XJI W Wilkinson, 1 Scafell Drive, Kendal, LA9 7PE
G3 XJM J Sawdy, 41 Ashbarn Crescent, Winchester, SO22 4QH
G3 XJN H Duncombe, La Rochelle, Thakeham Copse, Pulborough, RH20 3JW
G3 XJP Peter Rhodes, Danvers House, Wigmore, Leominster, HR6 9UF
GW3 XJQ Martin Shelley, Sunray, Pendine, Carmarthen, SA33 4PD
G3 XJR A Dickinson, 1 Pearl Bank, Apartment #33-04, Singapore, Singapore, 169016
G3 XJS Peter Barville, Felucca, Pinesfield Lane, West Malling, ME19 5EN
G3 XJW L Rix, 63 Edendale Road, Melton Mowbray, LE13 0EW
G3 XJY J Jardine, Field Head Farm, Highwood, Uttoxeter, ST14 8PU
G3 XJZ Colin Sykes, 15 Morpeth Close, Wirral, CH46 6HQ
G3 XKB K Bevan, Renhold, 25 Bryn Gannock, Conwy, LL31 9UG
G3 XKD M King, 15 Glebe Road, Prestbury, Cheltenham, GL52 3DG
G3 XKE C Evans, 8 Blakelands Avenue, Sydenham, Leamington Spa, CV31 1RJ
G3 XKG R Stanton, 16 Ashwood Park, Fetcham, Leatherhead, KT22 9NT
G3 XKH B Ward, 12 Pagets Road, Bishops Cleeve, Cheltenham, GL52 8AG
G3 XKL C Gill, Little Acre, Plough Lane, Devizes, SN10 5SR
G3 XKS H Grattan, Grattan Grange, St. Breward, Bodmin, PL30 3PN
G3 XKU M Rose, 71 Maryon Road, Ipswich, IP3 9NJ
G3 XKV R Stratton, 60 Lateward Road, Brentford, TW8 0PL
G3 XKX D Wills, 70 Hidcote Road, Oadby, Leicester, LE2 5PF
G3 XKY G Schrager, Flat 4, 3 The Park, London, N6 4EU
G3 XLB Michael Giddings, FLAT 1 165 ROXBURGH STREET, Bootle, L20 9NH
G3 XLE J Vaughan, Eastwood Lodge, Main Road, Boston, PE22 7JU
G3 XLG Raymond Spreadbury, Lings Farm, Blacksmiths Lane Forward Green, Stowmarket, IP14 5ET
G3 XLI P Holland, 38 Marlin Square, Abbots Langley, WD5 0EG
GI3 XLK W Magee, 6 Cherryvalley Park West, Belfast, BT5 6PU
G3 XLL John Lockwood, 22 Egremont Road, Diss, IP22 4NF
G3 XLN D Russell, 29 Gold St., Hanslope, Milton Keynes, MK19 7LU
G3 XLP I Richardson, Brockwood, Grove Road, Ryde, PO33 3LH
G3 XLR A Bunyan, 87 Seymer Road, Romford, RM1 4LA
G3 XLS T Williams, 5 Greenwood Drive, Bolton le Sands, Carnforth, LA5 8AP
G3 XLW David Powell, Broomhill, Aveton Gifford, Kingsbridge, TQ7 4NE
G3 XLX Roger Littlewood, Brewery Farm, Old Coach Road, Axbridge, BS26 2EH
G3 XLZ J Tozer, 54 Ganges Road, Plymouth, PL2 3AZ
G3 XMB Robert Richardson, 42 King Edwards Road, South Woodham Ferrers, Chelmsford, CM3 5PQ
G3 XMC David Brain, Orchardleigh, Bristol Road, Bristol, BS10 0HF
G3 XMG Mike Graham, 30 Moorlands Road, Thornton, Liverpool, L23 1US
G3 XMK Arthur Flather, 10 Oakleigh Court, Stone, ST15 8LA
G3 XMM Thomas Morgan, 32 Grasmere Road, Longlevens, Gloucester, GL2 0NQ
G3 XMP A Brasier, The Bears, Moor End Lane, Fakenham, NR21 0EJ
G3 XMQ Peter Eggleton, 8 Hammond Green, Wellesbourne, Warwick, CV35 9LY
GM3 XMY D Hobden, 1 Larch Avenue, Glenrothes, KY7 5TE
G3 XNE A Smyth, 3 Carteret Road, Bude, EX23 8DD
G3 XNG Robert Leask, 80 Mill Road, Sharnbrook, Bedford, MK44 1NP
G3 XNK D Kidd, 48 Layton Lane, Shaftesbury, SP7 8EY
G3 XNN R Jephcott, 3 Chatsworth Park, Thornbury, Bristol, BS35 1JF
G3 XNP Kenneth Arnold, 22 Silverbirch Court, Friends Avenue, Waltham Cross, EN8 8LZ
GD3 XNU J Craine, Mwyllin Squeen, Station Road, Ballaugh, Isle of Man, IM7 5AH
G3 XNX Derek Chivers, 51 Alma Road, Brixham, TQ5 8QR
G3 XOB David Ellacott, 39 Canford Lane, Bristol, BS9 3DQ
G3 XOC M Cooley, 21 Castle Road, Newport, PO30 1DT
G3 XOD R Horsman, 65 Pendennis Park, Brislington, Bristol, BS4 4JL
G3 XOI Alan Gordon, 20 Hawkins Crescent, Shoreham-by-Sea, BN43 6TP
GJ3 XOJ D Gray, La Brecque, La Grande Route de la Cote, St. Clement, Jersey, JE2 6FP
G3 XOK Robert Kearney, 32 Springfield Road, Lower Somersham, Ipswich, IP8 4PQ
G3 XOP Paul Featherstone, Airwave Services Ltd, 2 Firs Close, Aylesbury, HP22 4LH
GM3 XOQ Peter Weller, Mither Tap, Bridge Road, Inverurie, AB51 5QT
G3 XOU David Wright, Cross Cottage, 208 Whitchurch Road, Tavistock, PL19 9DQ
G3 XOV R Johnson, 29 Hungary Hill, Stourbridge, DY9 7PS
G3 XPA R Bevan, Sitio Do Laranjeiro 616f, Moncarapacho, Olhao, Portugal, 8700 077
G3 XPC Roy Chapman, 22 Windsor Ride, Finchampstead, Wokingham, RG40 3LG
G3 XPD D Smith, 5 Peel Street, Stafford, ST16 2DZ
G3 XPI B Hallows, 3 Southdown Close, Rochdale, OL11 4PP
G3 XPJ K George, 34 Third Avenue, Northville, Bristol, BS7 0RT
G3 XPK John Dore, Henfaes Isaf, Llangurig, Llanidloes, SY18 6SN
G3 XPM R Tinson, 29 Appleby Crescent, Knaresborough, HG5 9LS
G3 XPQ G Black, 24 Mount Drive, Leyburn, DL8 5JQ
G3 XPR I Bassett-Smith, Grey Gables, Southam Road, Cheltenham, GL52 3BB
G3 XPT Gordon Symonds, 45 Westfield Road, Dereham, NR19 1JB
G3 XPU Clive Woodley, 170 Rugby Road Burbage, Hinckley, LE10 2ND
G3 XPW Christopher Moller, 344 High Street, Cottenham, Cambridge, CB24 8TX
G3 XPZ J Appleton, 66 Bolton Road West, Ramsbottom, Bury, BL0 9ND
G3 XQE K Brown, 54 Wendover Rise, Allesley, Coventry, CV5 9JU
G3 XQJ G Wren, 2 Netheredge Close, Knaresborough, HG5 9BZ
G3 XQM Anthony Finch, 15 The Brooks, Burgess Hill, RH15 8TR
G3 XQO Philip Salomon, 28 Ansell Road, Wrexham, LL13 9NQ
GM3 XQP A French, 83/16 Hopetoun Street, Edinburgh, EH7 4NJ
G3 XQZ P Simpson, Orchard House, Church Road, Oakham, LE15 8AD
G3 XRC D Carlsen, 57 Chignal Road, Chelmsford, CM1 2JA
G3 XRD G Knight, 57b Oliver Road, Kirk Hallam, Ilkeston, DE7 4JY
G3 XRI Peter Williams, 2 Sycamore Avenue, Newton-le-Willows, WA12 8LT
G3 XRJ Sidney Chappell, Lyonesse, Trebehor, Penzance, TR19 6LX
G3 XRK D Griffin, 12 Charles Road, Whittlesey, Peterborough, PE7 2RG
G3 XRM David Dunn, 9 Mill Bank Estate, Llandegfan, Menai Bridge, LL59 5RD
G3 XRN ROYAL NAVAL AUXILIARY SERVICE ARS c/o Geoffrey Axford, 24 Jack Branch Court, Wash Lane, Clacton-on-Sea, CO15 1EJ
GI3 XRQ Bangor and District ARS c/o Richard White, 28 Lord Warden's Parade, Bangor, BT19 1YU
G3 XSC Keith Southgate, 10 Cott Road, Lostwithiel, PL22 0ET
G3 XSD Graham King, 1 Spring Gardens, North Baddesley, Southampton, SO52 9JG
G3 XSI T Haslam, 25 Backmoor Road, Norton, Sheffield, S8 8LP
G3 XSN B DONN, 7 Thurne Way, Liverpool, L25 4SQ
G3 XSR A Sutton, 3 Pendre Walk, Tywyn, LL36 0AY
G3 XSV A Hydes, Woodcroft, Bath Road, Bristol, BS40 5ED
G3 XSZ F Mundy, 5 The Paddocks, New Haw, Addlestone, KT15 3LX
G3 XTC P Borrett, 21 Kenley Walk, Cheam, Sutton, SM3 8ES
G3 XTH G King, 73 Grand Avenue, Hassocks, BN6 8DD
G3 XTI J Jarvie, 11 Guild Road, Aston Cantlow, Henley-in-Arden, B95 6JA
G3 XTN R Hough, Millfield, 9 Heads Drive, Grange-over-Sands, LA11 7DY
G3 XTP K Lloyd, 10 The Verneys, Cheltenham, GL53 7DB
G3 XTQ M Draycott, 119a High Street North, Stewkley, Leighton Buzzard, LU7 0EX
G3 XTR Bruce Dunn, 12 Campbell Road, Westville, South Africa, 3629
G3 XTT D Field, 105 Shiplake Bottom, Peppard Common, Henley-on-Thames, RG9 5HJ
G3 XTZ Graham Phillips, 27 Stanley Road, Ashford, TW15 2LP
G3 XUB F Boardman, Woodside, Burtonwood Road, Warrington, WA5 3AN
G3 XUC A Keohane, 6 Birchwood Fields, Tuffley, Gloucester, GL4 0AL
G3 XUD Paul Kirby, 7 Lillywhite Close, Burgess Hill, RH15 8TF
G3 XUF A Warner, 79 Kelvin Grove, Portchester, Fareham, PO16 8LF
G3 XUH R Pearson, 8 St. Benets Close, Walton-le-Dale, Preston, PR5 4UT
G3 XUM Joseph Moran, 30 Elsie Street, Farnworth Bolton, BL4 9HT
G3 XUP G Everest, 13 Noel Rise, Burgess Hill, RH15 8BW
GM3 XUW R Johnston, 123 Craigmount Brae, Edinburgh, EH12 8XW
G3 XUX E Fitzgerald, 4 Southwick Road, Wickham, Fareham, PO17 6HS

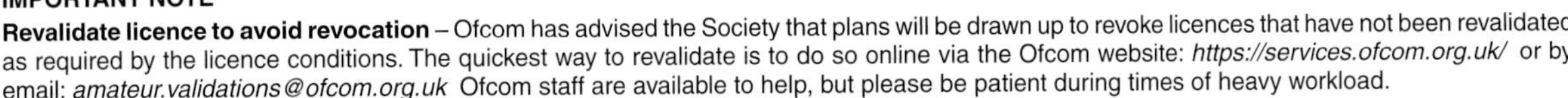

**IMPORTANT NOTE**

**Revalidate licence to avoid revocation** – Ofcom has advised the Society that plans will be drawn up to revoke licences that have not been revalidated as required by the licence conditions. The quickest way to revalidate is to do so online via the Ofcom website: *https://services.ofcom.org.uk/* or by email: *amateur.validations@ofcom.org.uk* Ofcom staff are available to help, but please be patient during times of heavy workload.

G3 XVA D Pickles, 26 Padstow Avenue, Fishermead, Milton Keynes, MK6 2ES
G3 XVB A Vizoso, Stones Farm House, Faringdon Road, Faringdon, SN7 8NP
G3 XVC M Collopy, New Gardens, Burnt House Lane, Dartford, DA2 7SP
G3 XVG T Barraclough, 52 Denshaw Avenue, Denton, Manchester, M34 3NX
G3 XVH S Franklin, 337 Hendon Way, London, NW4 3NB
G3 XVL Christopher McCarthy, 31 Philip Road, Ipswich, IP2 8BQ
G3 XVN Philip Norris, 18 Primrose Avenue, Downham Market, PE38 9EU
G3 XVP P Pimblott, 40 Richmondfield Lane, Barwick in Elmet, Leeds, LS15 4EZ
G3 XVQ F Jinks, 28 St. Anns Road, Bonnie View, Blackwood, NP12 3PG
G3 XVR D Higgins, Lyndenhurst, The Shrave, Alton, GU34 5BJ
G3 XVS R Kitching, Ashby Down, Streetway Road, Grateley, SP11 7EH
G3 XVV Malcolm Salmon, 54 Church Road, Rivenhall, Witham, CM8 3PH
G3 XVW K Ball, 39 Spinney Close, Northfield, Birmingham, B31 2JG
G3 XVY P Coull, 40 Wear Bay Crescent, Folkestone, CT19 6BA
G3 XWA J Ennis, 30 Hillcrest Avenue, Carlisle, CA1 2QJ
G3 XWB C Cadogan, 8 Horncliffe Close, Rawtenstall, Rossendale, BB4 6EE
G3 XWD Derek Watts, 40 Outlands Drive, Hinckley, LE10 0TW
G3 XWH Richard Horton, 23 Back Lane, Whixley, York, YO26 8BG
G3 XWK W Pinnell, 4 Rue Du Petit Caladou, le Poujol Sur Orb, France, 34600
G3 XWL J Cripps, 3 Queens Court, Queens Road, Cranbrook, TN18 4JE
G3 XWM C Harvey, 51 Sandyfield Crescent, Waterlooville, PO8 8SG
G3 XWN G Laycock, 48 Marina Terrace, Golcar, Huddersfield, HD7 4RA
G3 XWO T Lowe, 688 East J Street, Chula Vista, United States, 92010
G3 XWU Robert Pearson, 10 Eastleigh Close, Boldon Colliery, NE35 9NG
G3 XWV Carol Hamilton, The Stables Foinavon, Newsham, Richmond, DL11 7RD
GW3 XXB A Evans, 74 Celyn Avenue, Cardiff, CF23 6EQ
G3 XXC K Rigelsford, 14 Glebelands Avenue, South Woodford, London, E18 2AB
G3 XXE P Williams, 4 Plantation Close, Aller, Newton Abbot, TQ12 4NS
G3 XXF C Vine, 14 Hamilton Road, St. Albans, AL1 4PZ
G3 XXG P Sharpen, 3 Western Road, Urmston, Manchester, M41 6LE
G3 XXH S Watts, 58 Cambridge Avenue, New Malden, KT3 4LE
G3 XXM D Richards, 229 Battle Road, St. Leonards-on-Sea, TN37 7AJ
G3 XXN Fred Pickersgill, 3 Church St., Langold, Worksop, S81 9NW
G3 XXO E Birks, 46 Curzon Drive, Worksop, S81 0LP
G3 XXQ Leonard Dixon, 24 Angerton Gardens, Newcastle upon Tyne, NE5 2JB
G3 XXR Roger Higton, 13 Wilton Avenue, Bradley, Huddersfield, HD2 1RN
G3 XXX S Bradnam, 39 Pelham Way, Cottenham, Cambridge, CB24 8TQ
G3 XYA S Charles, 48 Elms Farm Road, Elm Park, Hornchurch, RM12 5RD
G3 XYB Graham Maitland, 7 Battery Road, Cowes, PO31 8DP
G3 XYC P Crust, 16 London Lane, Wymeswold, Loughborough, LE12 6UB
G3 XYD W Gordon-Laycock, 51 Overbrook, Swindon, SN3 6AR
G3 XYE J Clifton, Romford Cottage, Romford, Verwood, BH31 7LE
G3 XYF John Wresdell, Bracey Bridge Farm, Harpham, Driffield, YO25 4DE
G3 XYG M George, 30 Northfield Park, Barnstaple, EX31 1QA
G3 XYH J Hill, 35 Windmill Avenue, Marshalswick, St. Albans, AL4 9SJ
G3 XYI D Fearnley, 14 Salforal Close, Rettendon Common, Chelmsford, CM3 8EL
G3 XYJ R Walker, 27 Archer Close, Kings Langley, WD4 9HF
G3 XYO S Line, Cottles House, Cottles Lane, Exeter, EX5 1EE
G3 XYV I Cooper, 118 Stagsden Road, Bromham, Bedford, MK43 8QJ
GW3 XYW D Jones, 22 Alltiago Road, Pontarddulais, Swansea, SA4 8HU
G3 XYZ KINGS LYNN ARC c/o Edward Haskett, 23 Gloucester Road, Gaywood, King's Lynn, PE30 4AB
G3 XZB N Edwards, 14 Churchill Close, Cowes, PO31 8HQ
GJ3 XZE R Allenet, Les Sablons, La Grande Route de la Cote, St. Clement, Jersey, JE2 6FW
G3 XZG J Browne, 82 Cresswell Road, Chesham, HP5 1TA
G3 XZJ Graham Maynard, 6 Brimblecombe Close, Emmbrook, Wokingham, RG41 1QH
G3 XZK D Gething, 31 Lower Lodge Lane, Hazlemere, High Wycombe, HP15 7AT
GI3 XZM D Vance, The Eaves, Reagh Island, Newtownards, BT23 6EN
G3 XZO Martin Rhodes, 21 Halford Road, Ettington, Stratford-upon-Avon, CV37 7TL
G3 XZP David Holburn, 19 Whitwell Way, Coton, Cambridge, CB23 7PW
G3 XZV John Sonley, Ravenscliffe, Lands Lane, Knaresborough, HG5 9DE
G3 XZX J Lowe, 24 Candish Drive, Plymouth, PL9 8DB
G3 XZY T Garner, 122 Wainfleet Road, Skegness, PE25 3RX
GM3 YAC Philip Howarth, Kilbeg House Teangue, Isle of Skye, IV44 8RQ
G3 YAD M Goodrich, 2 Highworth Crescent, Yate, Bristol, BS37 4EY
GW3 YAF T Davies, Rhyd Wen, Llangyndeyrn, Kidwelly, SA17 5EN
G3 YAG W Thompson, 2 Fern Close, Frimley, Camberley, GU16 9QU
G3 YAI T Mills, 16 Hunts Hill, Glemsford, Sudbury, CO10 7RL
G3 YAJ D Sellen, Prospect House, Wignall Street, Manningtree, CO11 2HX
GM3 YAO F Offler, Ar Dachaidh, 3 King David Drive, Montrose, DD10 0SW
G3 YAR Ian Gildersleve, 7 Oak Park Road, Newton Abbot, TQ12 1RQ
G3 YAS P Ellis, 152 Cambridge Road, Hounslow, TW4 7BH
G3 YBA Ernest Cooper, Flat, 23 Westminster Crescent, Sheffield, S10 4EU
G3 YBE E Gilbert, 2 Church Field, Stanford, Ashford, TN25 6UA
G3 YBG J Rabjohns, Quarries Bungalow, Barley Lane, Exeter, EX4 1TA
G3 YBH Phil Storey, PO Box 47060, Denman Street Postal Outlet, Vancouver, Canada, V6G 3E1
G3 YBK R Donno, 6 Mincinglake Road, Exeter, EX4 7EA
G3 YBM R Mitchell, 98 Marlborough Drive, Burgess Hill, RH15 0EU
GW3 YBN C DAVIES, 31 Park Prospect, Graigwen, Pontypridd, CF37 2HF
G3 YBO Roger Baines, 327 Langer Lane, Wingerworth, Chesterfield, S42 6TY
G3 YBP A Young, Hillcrest, Graynfylde Drive, Bideford, EX39 4AP
GM3 YBQ K Horne, 10 Blair Place, Kirkcaldy, KY2 5SQ
G3 YBR Stuart Cook, 6 Essex Court, MARLTON, New Jersey, United States, 8053
G3 YBS Robert Lindsay-Smith, 58 Chalgrove Road, London, N17 0JD
G3 YBU Brian Whittle, Holmlea, Main Road, Hull, HU12 9NG
G3 YBY Ian McCarthy, 76 High Street, Purton, Swindon, SN5 4AD
GI3 YBZ J Mccann, 61 Glengawna Road, Glengawna, Omagh, BT79 7WJ
GM3 YCB S Riddell, 16 Lewis Drive, Old Kilpatrick, Glasgow, G60 5LE
G3 YCE J Peck, 7 Paddock Close, Radcliffe-on-Trent, Nottingham, NG12 2BX
G3 YCH J Sharman, 7 Watkins Way, Paignton, TQ3 3JJ
G3 YCJ P Sheard, 52 Victoria Road, Elland, HX5 0QA
G3 YCN W Kent, Old Cottage, Hermitage Lane, Maidstone, ME14 3HP
G3 YCO R Lewis, 115 Chester Road, Whitby, Ellesmere Port, CH65 6SB
G3 YCV J Hibbert, 5 Cliff View Road, Cliffsend, Ramsgate, CT12 5ED
G3 YCX Arthur Cain, 42 Wood Lane, Prescot, L34 1LW
G3 YCY R Barrett, 47 Marshals Drive, St. Albans, AL1 4RD
G3 YDD Hereford ARS c/o Timothy Bridgland-Taylor, 1 Overbury Court, Hereford, HR1 1DG
G3 YDE William Bagwell, 93 Broadley Drive, Torquay, TQ2 6UT
GI3 YDH M McIntyre, 36 Beechgrove Park, Belfast, BT6 0NR
G3 YDL James Thornton, 29 Farrar Avenue, Mirfield, WF14 9ED
GI3 YDM J Dunlop, 34 Ballybentragh Road, Dunadry, Antrim, BT41 2HJ
GM3 YDN Dennis Nutt, Little Craigfin, Kilkerran, Maybole, KA19 8LR
G3 YDO William McNally, 2607 The Highlands Dr., SUGAR LAND, Texas, United States, 77478
G3 YDT William Patterson, 32 The Pagoda, Maidenhead, SL6 8EU
G3 YDX Ron Stone, Taranaki, Four Crosses, Llanymynech, SY22 6RJ
G3 YDY P Selwood, 43 Keene Way, Galleywood, Chelmsford, CM2 8NT
G3 YEC R Edmondson, 16 Orchard Close, Copford, Colchester, CO6 1DB
G3 YED R Nettleton, 4 Sycamore Close, Stratford-upon-Avon, CV37 0DZ
G3 YEG N Sears, 17 Walls Road, Bembridge, PO35 5RA
G3 YEK R Johnston, Flat 906, Orchard Plaza, Poole, BH15 1EH
GD3 YEO R Rimmer, 27 Manor Lane, Farmhill, Douglas, Isle of Man, IM2 2NP
G3 YEP Richard Wakeley, Fir Bank, Fell Lane, Penrith, CA11 8BJ
G3 YEQ Leslie Miller, 28 Arthur Road, Margate, CT9 2EN
G3 YER D Lowe, Flat 3, Southdowns, The Avenue, Bristol, BS8 3GE
G3 YEU Barry Short, 83 Rowanfield Road, Cheltenham, GL51 8AF
GM3 YEW David Morris, Ash Cottage, Perth Road, Perth, PH2 9LW
G3 YFD W Hewitt, 22 Derby Road, Stockport, SK4 4NE
G3 YFE J Shaw, 57 London Road, Amesbury, Salisbury, SP4 7EE
G3 YFG Stephen Westell, 2 Whiteacre Lane, Barrow, Clitheroe, BB7 9BJ
G3 YFK Patrick McAlister, Lower Winnington Farm, Winnington, Shrewsbury, SY5 9DJ
G3 YFL C Wright, Holmwood, Brackley Avenue, Hook, RG27 8QX
G3 YFM M Smyth, 21 The Paddock, Longworth, Abingdon, OX13 5BX
G3 YFO Mark Bunce, 36 Burlington Road, Burnham, Slough, SL1 7BQ
G3 YFP John Bottomley, 42 Birkdale Court, Fornham St. Martin, Bury St. Edmunds, IP28 6XF
G3 YFU E Tomlin, Indalo, Magna Mile, Market Rasen, LN8 6AJ
G3 YFV P I'Anson, Flat 297, Latymer Court, London, W6 7LD
G3 YFW P Rosen, 2 Hadley Heights, Hadley Road, Barnet, EN5 5QH
G3 YGA E Warwick-Oliver, St. Madron, Throwleigh, Okehampton, EX20 2HX
G3 YGB John Coleman, 18 Chester Street, Coventry, CV1 4DJ
G3 YGC E Elliott, 18 Bear St., Lowerhouse, Burnley, BB12 6NQ
G3 YGD D Brown, Cappers Farm, Well Head Road, Burnley, BB12 9LR
G3 YGE J Okas, Dipley Springs, Dipley Common, Hook, RG27 8JS
G3 YGF J Gannaway, Highview House, Winchester Road, Eastleigh, SO50 7HB
G3 YGG John Kelly, 79 West Hill Avenue, Epsom, KT19 8JX
GW3 YGH A Hughes, 68 Higher Lane, Langland, Swansea, SA3 4PD
G3 YGJ D Brierley, 4 Waterloo Terrace, Bideford, EX39 3DJ
G3 YGL Frank Smith, 34 Bridle Road, Eastham, Wirral, CH62 8BR
G3 YGM M Osborne, Cheriton, Alexandra Road, St. Ives, TR26 1ER
G3 YGR C Thomas, Oakdene, School Lane, Reading, RG7 3ES
G3 YGZ A Walsh, Green Royd, Saddleworth Road, Halifax, HX4 8NU
G3 YHB James Baker, 109 Bermuda Road, Wirral, CH46 6AX
G3 YHC W Hermes, 22 Mallinson Crescent, Harrogate, HG2 9HP
G3 YHD T Holroyd, 27 Lady Acre Close, Lymm, WA13 0SN
G3 YHF C Skelcher, 51 Blenheim Road, Moseley, Birmingham, B13 9TY
G3 YHG D Harding, 17 Summerfield Close, Wokingham, RG41 1PH
G3 YHH J Froud, Summer Park, New Road, Teignmouth, TQ14 8UF
G3 YHI R Vale, High House, Mounts Lane, Daventry, NN11 3ES
G3 YHK J Clemence, Aroha, 67 Tomline Road, Felixstowe, IP11 7NX
G3 YHM R Harvey, 26 Birkdale Road, Worthing, BN13 2QY
G3 YHN C Pedley, 25 Fallowfield Road, Walsall, WS5 3DH
G3 YHO R Yaxley, Ashness, Swaffham Road, Dereham, NR19 2LX
G3 YHQ D Mercer, 19 Kingsfield Drive, Didsbury, Manchester, M20 6JA
GJ3 YHU D Robinson, 16 SEAGROVE COURT, LA RUE DE LA CORBIERE, St. Brelade, Jersey, JE3 8HN
G3 YHV Colin Chidgey, 46 Station Road, Shirehampton, Bristol, BS11 9TX
G3 YIA M Harris, 100 Chapel Lane, Wymondham, NR18 0DN
G3 YIB Michael Knowler, 2 Vulcan Street, Southport, PR9 0TW
G3 YIC V Sedgley, 25 Avenue Road, Weymouth, DT4 7JH
G3 YIE E Lusty, Stanley End Farm, Bell Lane, Stroud, GL5 5JY
G3 YIF J Weiner, 1 Chippendayle Drive, Harrietsham, Maidstone, ME17 1AD
G3 YIH F Cobb, Mon Reve, Rhodfa Nant, Abergele, LL22 9ND
G3 YII N Smith, Creekside, Nursery Lane, King's Lynn, PE30 3NA
G3 YIK John Morgan, Cedars, Springhill, Abingdon, OX13 5HL
G3 YIN E Ellery, 14 Four Acres, Bideford, EX39 3RW
G3 YIQ R Jones, The Old Vicarage, Upper South Wraxall, Bradford-on-Avon, BA15 2SB
G3 YIR ANGLO AMERICN RD c/o A Ward, 20 Tower Close, Costessey, Norwich, NR8 5AU
G3 YIW J Gallop, 55 Somervell Drive, Fareham, PO16 7QW
G3 YIY R Ingram, 11 Bank Terrace, Mevagissey, St. Austell, PL26 6QZ
G3 YJA B Porter, 49 Beverley Road, Leamington Spa, CV32 6PW
G3 YJD J Davies, 25 Harkness Close, Bletchley, Milton Keynes, MK2 3NB
G3 YJE P Merriman, The Old Croft, Brimpsfield, Gloucester, GL4 8LD
G3 YJG G Mason, 8 Leighton Road, Sunderland, SR2 9HQ
G3 YJN R Hodge, 36 Binswood End, Harbury, Leamington Spa, CV33 9LN
G3 YJP Derek Benham, 19 Benham Road, Otis, United States, 1253
G3 YJQ F Bourne, 78 Normandy Way, Plymouth, PL5 1SR
G3 YJS M Roche, 8 Northdown Close, Penenden Heath, Maidstone, ME14 2ER
G3 YJW R Whitehouse, White Court, Camp Hill, Tonbridge, TN11 8LE
G3 YJZ A Mitchell, 89 Queen Annes Grove, Enfield, EN1 2JU
GM3 YKA John Wiewiorka, 47 Albert Avenue, Grangemouth, FK3 9AT
G3 YKB J Hodgson, 18 Nascot Place, Watford, WD17 4QT
G3 YKC D Fayers, 1 Tismeads Crescent, Swindon, SN1 4DP
G3 YKI K Vickers, The Bungalow, Cleeve Road, Evesham, WR11 8JU
G3 YKK Chris Donne, The Hideaway, Lease Lane, Immingham, DN40 3PT
G3 YKO D Darwood, Briarwood Cottage, Packhorse Lane, Birmingham, B38 0DN
GM3 YKP W Sutton, Old Police Station, Dunbeath, KW6 6EA
G3 YKS R Butlin, 48 Roman Way, Market Harborough, LE16 7PQ
G3 YKW R Walker, 17 Ballantyne South, Montreal West, Canada, QC H4X 2BI
G3 YKZ Michael Biddiscombe, 20 Arlington Close, Malpas, Newport, NP20 6QF
G3 YLA J Bacon, 37 Burgh Lane, Mattishall, Dereham, NR20 3QP
GM3 YLD J Frew, Queens Cottage, 87 Queen Street, Dunoon, PA23 8AX
GJ3 YLI A Morrissey, Flat 4, 1 Springfield Crescent, St. Helier, Jersey, JE2 4GL
G3 YLJ Gill Whitehead, 23 Coulsons Road, Bristol, BS14 0NN
G3 YLL R Brooks, 10 Leopold Close, Bognor Regis, PO22 8JJ
GJ3 YLN John Speller, Lindau, Gorey Village Main Road, Jersey, JE3 9EP
G3 YLU W Taylor, 4 Newbie Barns, Newbie, Annan, DG12 5QL
G3 YLV P Jones, Oaklea, Cadney Lane, Whitchurch, SY13 2LW
G3 YLW P Lascelles, 4 Meadowsweet Close, Snettisham, King's Lynn, PE31 7UG
G3 YLY B Hawes, 15 Bridge Lane, Wimblington, March, PE15 0RR
G3 YMC David Sergeant, 8 Toll Gardens, Bracknell, RG12 9EX
G3 YMD Dover ARC c/o Peter Love, 2 Meadway, Dover, CT17 0PS
G3 YMH R Wainwright, The Olives, High Street, Uckfield, TN22 4LB
G3 YMM T Campbell Davis, 9 Cloister Road, Acton, London, W3 0DE
G3 YMN Jeremy Rhys, 33 Esher Place Avenue, Esher, KT10 8PU
GI3 YMT M Higgins, 1 Cairnshill Park, Belfast, BT8 6RG
G3 YMU John Hibberd, Barn Cottage, School Lane, Crewe, CW3 0BA
G3 YMV I MacHardie, 20 Arley Close, Swindon, SN25 4TP
G3 YMW D Sapsworth, 16 Laxton Avenue, Hardwick, Cambridge, CB23 7XL
GM3 YMX David Ferguson, 21 Pentland Drive, Edinburgh, EH10 6PU
GI3 YMY N Newell, 18 Kilmaine Avenue, Bangor, BT19 6DU
G3 YMZ Allan Plummer, 6 Shelley Place, KALLAROO, Perth, Australia, 6025
G3 YNC C Adams, 18 Glenavon Road, Highcliffe, Christchurch, BH23 5PN
GM3 YND I Simpson, 3 Ravenscraig Terrace, Steelend, Dunfermline, KY12 9LU
G3 YNF B Turner, 48a High Street, Great Houghton, Northampton, NN4 7AF
G3 YNG F Higgins, 27 School Road, Thornton-Cleveleys, FY5 5AW
G3 YNJ C Powell, 38 Braeside Road, St. Leonards, Ringwood, BH24 2PH
G3 YNK D Evans, Michigan Villa, Wall Road, Hayle, TR27 5HA
G3 YNN QRZ AMATEUR RADIO GROUP OF SUSSEX c/o S Constable, 9 Ridgeway Close, Heathfield, TN21 8NS
G3 YNO Mike Booth, 28 Humber Road, North Ferriby, HU14 3DW
G3 YNU I Stevenson, 18 Sittingbourne Road, Wigan, WN1 2RR
G3 YOA A Adams, The Gables, Chapel Road, North Walsham, NR28 0QG
G3 YOC R Moore, 38 Sandygate, Wath-upon-Dearne, Rotherham, S63 7LR
GM3 YOI Kenneth Falconer, Lumbo Farmhouse, St. Andrews, KY16 8NS
G3 YOL Stephen Cole, Halebrook, Bridgwater Road, Winscombe, BS25 1NH
G3 YOM K Beddoe, 30 Tamella Road, Botley, Southampton, SO30 2NY
G3 YON F Webster, 16 Pembroke Road, Dronfield, S18 1WH
G3 YOO J Webster, 7 Hardwick Avenue, Allestree, Derby, DE22 2LN
G3 YOQ L Spinks, 2 Ventnor Avenue, Grantham, NG31 7EA
GM3 YOR Andrew Givens, 5 Langhouse Place, Inverkip, Greenock, PA16 0EW
G3 YOV Trevor Gammage, 23 Artizan Road, Northampton, NN1 4HU
G3 YOY Henry Clark, 2 Chestnut Close, Peakirk, Peterborough, PE6 7NW
G3 YPD P Chester, 44 Richmond Drive, Lichfield, WS14 9SZ
G3 YPE M Greenwood, 21 Dobb Top Road, Holmbridge, Holmfirth, HD9 2PQ
G3 YPK Stephen Wallis, Mas La Floride, 250 Chemin De Magarnaud, Sommieres, France, 30250
G3 YPL D Gray, 19 Westbury Gardens, Higher Odcombe, Yeovil, BA22 8UR
G3 YPM R Moore, 20 Ebrington Road, Malvern, WR14 4NL
G3 YPS STUART ATKINSON, 13 Charles Street, Gainsborough, DN21 2JA
G3 YPT P Tomes, 86 Hurn Road, Christchurch, BH23 2RP
G3 YPU P Koker, 123 Lower Howsell Road, Malvern, WR14 1DH
G3 YPW Peter Willingham, 49 Creek Road, Hayling Island, PO11 9RA
G3 YPY A Head, 32 Weald View Road, Tonbridge, TN9 2NQ
G3 YPZ John Petters, 218 New House Farm, Hospital Drove, Spalding, PE12 9EN
G3 YQ S Hunt, 6 Adlard Grove, Humberston, Grimsby, DN36 4JZ
G3 YQA Frank Wilson, 26 Humber Road, North Ferriby, HU14 3DW
G3 YQB D Rankin, 105 Sparrowhawk Way, Hartford, Huntingdon, PE29 1XY
G3 YQC John Wood, 14 Little Paradise, Marden, Hereford, HR1 3DR
G3 YQF R Linford, Department Of Communication, And Electronic Engineering, Drake Circus Plymouth, PL4 8AA
G3 YQG N Waylett, Po Box 4148, Santa Clara, United States, 95056
GW3 YQH Robert Evens, 41 Heol Gwys, Upper Cwmtwrch, Swansea, SA9 2XQ
G3 YQJ P Burnet, 166 Oundle Road, Thrapston, Kettering, NN14 4PQ
GM3 YQK J Dillon, Steelbank Cottage, Dalgraven, Kilwinning, KA13 6PL
G3 YQL Peter Murfitt, 53a Codnor Denby Lane, Codnor, Ripley, DE5 9SP
G3 YQM D Wynford-Thomas, Coach House, The Hollies, Pentrepoeth Road, Newport, NP10 8RT
G3 YQN Robert Trott, 5 Farrant Close, Orpington, BR6 6AY
G3 YQO Donald Kirkwood, Greensand Lodge, High Street, Lidlington, Bedford, MK43 0QR
G3 YQP C Hardie, 3 Berth Glyd, Gyffin, Conwy, LL32 8NP
G3 YQV C Railton, 14 Copford Lane, Long Ashton, Bristol, BS41 9NF
G3 YQW M Funnell, 15 Mcindoe Road, East Grinstead, RH19 2DD
G3 YQZ M Johnson, 36 Coventry Road, Bulkington, Bedworth, CV12 9ND
G3 YRC YARMOUTH RC c/o Peter Nicholls, 11 Minsmere Road, Belton, Great Yarmouth, NR31 9NX
G3 YRH Brian Dodds, 1 Croft View, Killingworth, Newcastle upon Tyne, NE12 6BT
GI3 YRL J Branagh, 17 Rathmoyle Park West, Carrickfergus, BT38 7NG
G3 YRP I Dudley, Tynewydd, Llansantffraid, SY22 6TW
G3 YRQ Ian Parkinson, 61 Cinnamon Lane, Fearnhead, Warrington, WA2 0AG
G3 YRU P Wilby, Green Farm, Main Street, York, YO42 1YQ
G3 YRX I Elston, 11 Knowle Drive, Exwick, Exeter, EX4 2DF
G3 YSD J Murdoch, 32 Scalegill Road, Moor Row, CA24 3JL
G3 YSG M Taylor, 4 Yew Tree Court, Botley Road, Southampton, SO31 1EA
G3 YSK A Button, 13 Taplings Road, Weere, Winchester, SO22 6HE
G3 YSM Michael Davidson, 14 Fuchsia Walk, Wirral, CH49 3AG
G3 YSN Henry Smith, Carnview, Tregender Lane, Penzance, TR20 8DJ
G3 YSQ A Pratt, 7 The Croft, West Hanney, Wantage, OX12 0LD
G3 YSR Christine Beattie, Mayerin, Churchway, Aylesbury, HP17 8RG
G3 YSW Nigel Thrower, 8 Upton Gardens, Worthing, BN13 1DA
G3 YSX Stewart Bryant, 154 London Road North, Merstham, Redhill, RH1 3AA
G3 YTC George Rutherford, 29 Sisial y Mor, Rhosneigr, LL64 5XB
GD3 YTE P Gill, Hollybank, Sulby Bridge, Sulby, Isle of Man, IM7 2AY
G3 YTG Tom Blair, 56 Rue Du Golfe De Barbareu, Etaules, France, 17750
G3 YTI S Cooper, 24 Cambridge Street, Darwen, BB3 3JH
G3 YTL C Lewis, 26 Ffordd Gwenllian, Llay, Wrexham, LL12 0UW
G3 YTN R Hill, 35 Coxwold View, Wetherby, LS22 7PU

G3 YTQ C Kidd, 118 Segensworth Road, Fareham, PO15 5EQ
G3 YTR M Hall, 1 Larches Way, Crawley Down, Crawley, RH10 4UJ
GM3 YTS R Ferguson, 19 Leighton Avenue, Dunblane, FK15 0EB
G3 YTT W Taylor, 10 Mickleton, Wilnecote, Tamworth, B77 4QY
G3 YTU Christopher Coward, 27 Rothley Chase, Haywards Heath, RH16 3PE
G3 YTW Graham Clarke, 117 Bermuda Village, Nuneaton, CV10 7PW
G3 YTX G Clamp, 9 Furse Close, Camberley, GU15 1BF
G3 YTY M Edib, 84 Connaught Gardens, London, N13 5DT
G3 YTZ Mark Readman, Flat 1-8, 365 Wilmslow Road, Manchester, M14 6AH
GW3 YUC D Davies, 30 Wern Isaf, Dowlais, Merthyr Tydfil, CF48 3NY
G3 YUD Peter Hawkins, 37 Alexandra Road, Dorchester, DT1 2LZ
G3 YUH R Ayling, 25 Nash Court Road, Margate, CT9 4DH
G3 YUJ R Stead, 63 Colchester Road, Ipswich, IP4 0DT
G3 YUQ F Isley, 26 Elmsdale Road, Wootton, Bedford, MK43 9JW
G3 YUU Christopher Lord, Green Sleeves, Marley Road, Maidstone, ME17 1BS
G3 YUX D Moore, 60 Ivatt, Tamworth, B77 2HQ
G3 YUZ I Wilson, 23 Alyth Road, Bournemouth, BH3 7DQ
G3 YVA Bernard Edwards, 774, Calle 21 S.O., Puerto Rico, Puerto Rico, 921
G3 YVH A Boyne, 18 Crow Lane West, Newton-le-Willows, WA12 9YG
G3 YVI R Gilbert, Mayville, New Copse, Alton, GU34 5NP
G3 YVK Henry Tabberer, 101 Broadclyst Gardens, Southend-on-Sea, SS1 3QU
GW3 YVN N Little, Brynhyfryd, Llansteffan, Carmarthen, SA33 5HA
GU3 YVV Richard Outhwaite, Le Courtillet, La Route de Sausmarez, Guernsey, Guernsey, GY4 6SF
G3 YVW Brian Blackwell, 19 Tokely Road, Frating, Colchester, CO7 7GA
GM3 YVX D Coupar, 32 Gillies Place, Broughty Ferry, Dundee, DD5 3LE
G3 YVY DENNIS HANLEY, 5 Hallcroft Close, Billingham, TS23 1QN
G3 YVZ Thomas Gardner, 200b Heaton Road, St Teresas Men Club, Newcastle upon Tyne, NE6 5HP
G3 YWA E Pepper, 30 Westfield Drive, Harpenden, AL5 4LP
G3 YWF P Smith, 18 David Avenue, Cliftonville, Margate, CT9 3DU
G3 YWG Anthony Emery, 11 Highmoor Road, Bournemouth, BH11 8RT
G3 YWH F Hill, 24 Mount St. James, Guide, Blackburn, BB1 2DR
G3 YWL Clifford Coverdale, 2 Lillypilly Lane, Cooranbong, Australia, 2265
G3 YWM P Hubert, 575 Bramford Lane, Ipswich, IP1 5JX
G3 YWO J Tripp, 15 Lime Grove, Bottesford, Nottingham, NG13 0BH
G3 YWS J Smith, 16 Woodlands, Winthorpe, Newark, NG24 2NL
G3 YWT P Smith, Beechwood, Clarendon Road, Salisbury, SP5 3AT
G3 YWU Stephen Fisher, Arkle, 31 Frith Avenue, Northwich, CW8 2JB
G3 YWW A Carpenter, 17 Victoria Avenue, Upwey, Weymouth, DT3 5NG
G3 YWX I Poole, 17 Glebe Road, Dorking, RH4 3DS
G3 YXH Simon Marshall, The Bunglaow, Llamedos Stables, Fieldhead Lane, Bradford, BD11 1JL
GM3 YXJ D Begg, 48 Eskhill, Penicuik, EH26 8DG
G3 YXM David Pick, 178 Alcester Road South, Kings Heath, Birmingham, B14 6DE
G3 YXN Paul Whalley, Elms View, Newmarket, CB8 8TZ
G3 YXO Des Watson, Norton, Gote Lane, Lewes, BN8 5HX
G3 YXQ R Ireland, 31 St James Street, Sackville, New Brunswick, Canada, E4L 4L7
G3 YXS D Naylor, 7 Ruthven Court, Litherland, Liverpool, L21 2PE
G3 YXW P Dunford, 22 Summerdown Road, Eastbourne, BN20 8DT
GM3 YXY A Thomson, Roselea, Lanark, ML11 7SE
G3 YYC G Sharples, 3a Green Lane Park Homes, Breinton, Hereford, HR4 7PN
G3 YYE H Lawrence, 14 Manor Lane, Dinnington, Sheffield, S25 2SW
G3 YYG J Bolton, 3 Fyne Drive, Linslade, Leighton Buzzard, LU7 2YG
G3 YYK Robin North, 13 Nicholson Way, Havant, PO9 3AZ
G3 YYN Peter Spurr, Windmill Farm Barn, High Street, Milton Keynes, MK14 5AX
G3 YYQ L Balls, 2 Bure Close, Great Yarmouth, NR30 1QU
G3 YYR A Parry, 17 May Pole Knap, Somerton, TA11 6HP
G3 YYW G Wills, 137 Aldermans Drive, Peterborough, PE3 6BB
G3 YYZ J Cuthbert, Fulbeck, 48 Mayes Lane, Harwich, CO12 5EJ
G3 YZB Samuel Bassford, 27 Millpit Furlong, Littleport, Ely, CB6 1HT
G3 YZK G West, 6 Lammas Close, Cowes, PO31 8DT
G3 YZN R Awbery, Dashwood, Beacon View Road, Godalming, GU8 6DU
G3 YZO Robert Wilson, Kellers, Duck Street, Saffron Walden, CB11 4JU
G3 YZQ Paul WILLIAMS, 41 Church Street, St. Georges, Telford, TF2 9JZ
G3 YZR John Porter, Birklands, 16 The Oval, Scarborough, YO11 3AP
G3 YZT Anthony Slaney, 1 Marlborough Way, Goring-by-Sea, Worthing, BN12 4HG
G3 YZV Ernest Woollard, 24 Griffin Close, Twyford, Banbury, OX17 3HR
G3 YZW A Armstrong, 423 Bideford Green, Linslade, Leighton Buzzard, LU7 2TY
G3 YZY H Brindle, Currabawn, Stakes Hill Road, Waterlooville, PO7 7BD
G3 YZZ K Deverstock, 16 Chaucer Close, Emmer Green, Reading, RG4 8PA
G3 ZAE S Benstead, 15 Les Congenies, Bouniagues, France, 24560
G3 ZAG B Taylor, 27 Ridgeway, Wellingborough, NN8 4RU
G3 ZAJ D Sutton, Deer Wood, Canterbury Road, Ashford, TN25 4DF
G3 ZAL L Huggett, 24 Hockers Lane, Detling, Maidstone, ME14 3JN
G3 ZAU Eric Lord, 41 Daven Road, Congleton, CW12 3RB
G3 ZAV R Mills, 45 Hugh Price Close, Murston, Sittingbourne, ME10 3AS
G3 ZAW K Bird, Leys House, Main Street, Northampton, NN7 4SH
G3 ZAY M Atherton, 41 Enniskillen Road, Cambridge, CB4 1SQ
G3 ZBB G Jackson, Old Fir Tree Inn, Peacemarsh, Gillingham, SP8 4EU
G3 ZBF M Matthews, 10 Burgill Crescent, Ramsgate, CT12 6EZ
G3 ZBG G Moorfield, 43 Broadmark Lane, Rustington, Littlehampton, BN16 2HH
G3 ZBI NUNSFIELD HOUSE c/o K Frankcom, 1 Chesterton Road, Spondon, Derby, DE21 7EN
G3 ZBM W Worthington, 32 Princess Drive, Wistaston Green, Crewe, CW2 8HS
G3 ZBP M Baker, 82 Folkestone Road, Copnor, Portsmouth, PO3 6LR
GM3 ZBR Kenneth Melvin, 1 Charleston Village, Charleston, Forfar, DD8 1UF
G3 ZBS J Mccall, 5 Sundew Close, Wokingham, RG40 5YB
G3 ZBU Alister Watt, 5 Brambling Road, Horsham, RH13 6AX
G3 ZBZ B Cross, 176 Outwood Road, Heald Green, Cheadle, SK8 3LL
G3 ZCA D Lake, 9 Grafton Close, King's Lynn, PE30 3EZ
G3 ZCD R FOGG, 24 Edinburgh Gardens, Windsor, SL4 2AN
G3 ZCG K Young, Flat 36, Brookhurst Court, Leamington Spa, CV32 6PB
G3 ZCH D Hill, 11 Chapeltown Road, Radcliffe, Manchester, M26 1YF
G3 ZCI R O'Brien, 9 Holmwood Garth, Hightown, Ringwood, BH24 3DT
G3 ZCJ M Allerton Austin, 13 Kilpin Green, North Crawley, Newport Pagnell, MK16 9LZ
GI3 ZCK Gerard Ward, 6b Stranmillis Road, Belfast, BT9 5AA
G3 ZCL G Hammersley, 74 Cammel Road, West Parley, Ferndown, BH22 8SB
G3 ZCT James Boehler, 17 [illegible] Road, Leicester, LE3 0SA
G3 ZCV N Harper, 9 The Orchard, Market Deeping, Peterborough, PE6 8JS
G3 ZCX P Fort, 1 Lowther Lane, Foulridge, Colne, BB8 7JY
G3 ZCY R Hogg, Green Pastures, Darley, Harrogate, HG3 2QF
G3 ZCZ Joseph Kasser, 13-03, 150 Clementi Rd, Singapore, Singapore, 129789
G3 ZDF Joe Kirk, 111 Stockbridge Road, Chichester, PO19 8QR
G3 ZDG Steven Cole, 1 The Copse, Exmouth, EX8 4EY
GM3 ZDH Robert Dixon, 2 Maidens, East Kilbride, Glasgow, G74 4RS
G3 ZDK P Given, 4 Elvoden Drive, Ilkeston, DE7 9JW
G3 ZDM Richard Muriel, 13 York Road, Sale, M33 6EZ
G3 ZDQ I Flemming, Rudderhams Cottage, Blandford Hill, Blandford Forum, DT11 0AA
G3 ZDT P Morrison, Saddlers, Upper Green Road, Tonbridge, TN11 9PL
G3 ZDU C Marshall, Bassington, Hulne Park, Alnwick, NE66 3JE
G3 ZDW R Hyde, 25 The Pastures, Cottesmore, Oakham, LE15 7DZ
G3 ZDY D Palmer, Flat 3, 60 Millbank, London, SW1P 4RW
G3 ZEB B Robinson, The Gables, The Street, Great Yarmouth, NR29 4EA
G3 ZED A Rothwell, Brandon, Manor Brow, Keswick, CA12 4AP
G3 ZEF R Stephen, 97 Hunters Field Stanford in the Vale, Faringdon, SN7 8ND
G3 ZEJ R Smith, Kilderkins, 7 Birch Grove, Ottery St. Mary, EX11 1XP
G3 ZEK M Bailey, 12 Bridgers Mill, Haywards Heath, RH16 1TF
G3 ZEM Robert Henderson, Po Box 62155, Pafos, Cyprus, 8061
G3 ZEN A Glaser, 155 Little Breach, Chichester, PO19 5UA
G3 ZEO S Wilders, Old Farm Barn, Silkstead Lane, Winchester, SO21 2LG
G3 ZEQ M Brenig-Jones, Orchard House, Larters Lane, Stowmarket, IP14 5HB
G3 ZER I Mercer, 28 West Way, Rickmansworth, WD3 7EN
G3 ZES A Downing, Oaktree Bungalow, The Endway, Chelmsford, CM3 6DU
GM3 ZET LERWICK RC c/o T Goodlad, 72 North Lochside, Lerwick, Shetland, ZE1 0PJ
GM3 ZEU C Clarkson, 8 Moor Place, Portlethen, Aberdeen, AB12 4TF
GD3 ZEX C Douglas, Sea Villa, The Promenade, Laxey, Isle of Man, IM4 7DF
G3 ZEZ G Coleman, 16 Kestrel Way, Clacton-on-Sea, CO15 4JE
G3 ZFC C Davis, 3 Cross Road, Haslington, Crewe, CW1 5SY
G3 ZFF R Hornbuckle, 54 Gladys Avenue, Cowplain, Waterlooville, PO8 8HS
G3 ZFP Ray Penberthy, 10 Lancot Avenue, Dunstable, LU6 2AW
G3 ZFR Roger Harris, Clevelands, Tamworth Road, Coventry, CV7 8JJ
G3 ZFT A Magnus, Woodland Cottage, Linkside East, Hindhead, GU26 6NY
G3 ZFV J Watts, Riverside, St. Georges Road, Barnstaple, EX32 7AS
G3 ZFX B Cornwall, Hoadswood, Battle Road, St. Leonards-on-Sea, TN37 7BS
G3 ZFZ G Gibson, 174 Roose Road, Barrow-in-Furness, LA13 0EE
G3 ZGA J Hart, Peter-Vischer-Str.9, Marktredwitz, Germany, 95615
G3 ZGC Richard Jolliffe, 54 Glendale Avenue, Wash Common, Newbury, RG14 6RU
G3 ZGG Philip Storey, 8-9 Chadley Lane, Godmanchester, Huntingdon, PE29 2AL
GM3 ZGH R Yeoman, 162 Jamphlars Road, Cardenden, Lochgelly, KY5 0ND
G3 ZGI Terry O'Neill, Braeside, Cookbury, Holsworthy, EX22 7YG
G3 ZGN P Swarbrick, 1 Hill View, Charminster, Dorchester, DT2 9QX
G3 ZGP Richard Cridland, 13 Clarendon Avenue, Redlands, Weymouth, DT3 5BG
G3 ZGQ L Mead, 12 Ferniefields, High Wycombe, HP12 4SP
G3 ZGT B Bruce, 25 Boothgate Drive, Howden, Goole, DN14 7EN
G3 ZGU K Richens, The Marsh, Marsh Lane, Market Drayton, TF9 2SF
G3 ZGY G Paddock, 56 Clee View Road, Wombourne, Wolverhampton, WV5 0BD
G3 ZGZ David Woodhall, 15 Cherrywood Avenue, Thornton-Cleveleys, FY5 1SU
G3 ZHA G Gillam, 58 Downhall Road, Rayleigh, SS6 9LY
G3 ZHB Alan Stuart, 207 Saunders Lane, Mayford, Woking, GU22 0NT
G3 ZHC N Willmot, 2 Athlone Road, Walsall, WS5 3QX
G3 ZHE A Heyes, 20 Walsingham Road, Penketh, Warrington, WA5 2AQ
G3 ZHJ P Moss, 20 Leamington Road, Luton, LU3 3XQ
G3 ZHK T Kellow, Glenvale, St. Dominick, Saltash, PL12 6TD
G3 ZHL J Morgan, Cedars, Springhill, Abingdon, OX13 5HL
G3 ZHO Ronald Wilton, 23 Moorland View, Liskeard, PL14 3TQ
G3 ZHP D Marsden, 8 Ogden View Close, Halifax, HX2 9LY
G3 ZHS R Ray, 37 Doxey Fields, Stafford, ST16 1HJ
G3 ZHT B Lundean, 13 Isis Close, Lympne, Hythe, CT21 4JQ
G3 ZHU G Clark, Kenzie, Canterbury Road, Dover, CT15 7HR
G3 ZHV Robert Bowler, 21 Pine Close, South Wonston, Winchester, SO21 3EB
G3 ZHY M Irving, 75 Southbourne Coast Road, Bournemouth, BH6 4DX
G3 ZHZ A Macfadyen, 19 Oldfield Road, Lower Willingdon, Eastbourne, BN20 9QD
G3 ZIB D Tye, 11 Townsend Close, Bracknell, RG12 0XE
G3 ZIC John Viney, 121 Eleanor Road, Prenton, CH43 7QP
G3 ZID Anthony Greathead, 20 Westland, Martlesham Heath, Ipswich, IP5 3SU
G3 ZIE Edric Brown, 21 Newbridge Way, Pennington, Lymington, SO41 8BG
G3 ZIF H Wilson, 6 Hicborrow Close, Etwall, Derby, DE65 6HY
G3 ZIG Roy Reed, Oak Cottage, Dereham Road, Dereham, NR20 4AA
G3 ZII M Rathbone, 25 Halsall Road, Southport, PR8 3DB
G3 ZIJ J Stables, 9 Milbanke Close, Ouston, Chester le Street, DH2 1JJ
G3 ZIK A Mather, 15 Claughton Avenue, Bolton, BL2 6US
G3 ZIL G Griffiths, 14 Bassett Close, Southampton, SO16 7PE
G3 ZIM R Wolton, 12 Well Lane, Liverpool, L16 5ET
G3 ZIN G Spencer, 16 West Lawn, Ipswich, IP4 3LJ
G3 ZIO C Harvey, Ham Cottage, 1a Elstan Way, Croydon, CR0 7PR
G3 ZIV K Nolan, West End Cottage, Woodhall, Selby, YO8 6TG
G3 ZIY Richard Drinkwater, 15 Woodside Crescent, Smallfield, Horley, RH6 9NA
G3 ZJF P Broughton, 7 Old Mill Way, Wells, BA5 2JU
G3 ZJG John Garner, 50 Thorndale, Ibstock, LE67 6JT
G3 ZJJ Michael Peet, 31 White Street, West Lavington, Devizes, SN10 4LP
G3 ZJK C Milner, The Everglades, Sawbridge Road, Rugby, CV23 8DN
G3 ZJO Edward Bennett, 44 Central Avenue, White Hills, Northampton, NN2 8DZ
G3 ZJP W Fenton, 50 Orion Road, Rochester, ME1 2UH
G3 ZJQ R Walker, 2 Chelwood Drive, Sandhurst, GU47 8HT
GW3 ZJS John Smith, Llainfran, New Quay, SA45 9RR
G3 ZJT R Clayton, 42 Hibernia Street, Scarborough, YO12 7DH
G3 ZJV M Firth, 63 Sycamore Road, East Leake, Loughborough, LE12 6PP
G3 ZJW B McCombe, 208 Thorpe Road, Peterborough, PE3 6LB
G3 ZJX B Castle, 10 Oakley Drive, Bromley, BR2 8PP
G3 ZJY J Greenwood, 91 Keyhaven Road, Milford on Sea, Lymington, SO41 0TF
G3 ZJZ John Mason, 35 Broad Way, Hockley, SS5 5EL
G3 ZKD W Ball, 6 Coronation Drive, Penketh, Warrington, WA5 2DD
G3 ZKE Brian Bond, 66 Agar Grove, Camden Town, London, NW1 9TL
G3 ZKG J Riley, 41 Church Avenue, West Sleekburn, Choppington, NE62 5XF
G3 ZKH I Bateman, Jaysville, The Strand, Pershore, WR10 3JZ
G3 ZKI A Williams, 38 Seneca Street St. George, Bristol, BS5 8DX
G3 ZKN Donald Morgan, 9 Summerland Drive, Churchdown, Gloucester, GL3 2LZ
G3 ZKO P Lee, 20 Little Haseley, Oxford, OX44 7LH
G3 ZKQ A Walton, 3 Fox Hill Close, Selly Oak, Birmingham, B29 4AH
G3 ZKZ John Shaw, 2 Castle Close, Felixstowe, IP11 9NN
G3 ZLD J Barker, 57 Broom Park, Teddington, TW11 9RS
G3 ZLE D Ward, 11 Spruce Grove Avenue, Baden, Ontario, Canada, N3A 3P7
G3 ZLF R Nelson, 225 Walton Road, Chesterfield, S40 3BT
G3 ZLJ E Dalton, 29 Windmill Lane, Castlecroft, Wolverhampton, WV3 8HJ
G3 ZLM Roger Hook, 55 Parkwood Crescent, Hucclecote, Gloucester, GL3 3JH
G3 ZLO Michael Adams, 41 Primrose Lane, Yeovil, BA21 5SH
G3 ZLR A Ridley, 28 Riverbank, Laleham Road, Staines, TW18 2QE
G3 ZLS S Craske, Brooklands, Daddon Hill, Bideford, EX39 3PW
G3 ZLX F Jones, 94 Westbrook End, Newton Longville, Milton Keynes, MK17 0BY
GM3 ZMA James Butler, 11 Quartalehouse, Stuartfield, Peterhead, AB42 5DE
G3 ZMD J Reed, The Firs, Gloucester Road, Newent, GL18 1EH
G3 ZME TELFORD&DIST AR c/o M Vincent, 9 Sleapford, Long Lane, Telford, TF6 6HQ
G3 ZMG John Maughan, 83 Oak Road, Peterlee, SR8 3HU
G3 ZMH D McAuslan, Golden Sedge, Street End, Bristol, BS40 7TL
G3 ZMK J Askew, PO BOX 4487, Linstead, St. Catherine, Jamaica
G3 ZML M Owen, 3 Gordon Road, Mount Waverley, Victoria, Australia, 3149
G3 ZMM R Hodgkinson, 39 Oxford Road, Carlton-in-Lindrick, Worksop, S81 9BD
G3 ZMN M Turner, 12 Purley Bury Avenue, Purley, CR8 1JB
G3 ZMO J Callum, Dales Barn Top, Town Head, Hawes, DL8 3RH
G3 ZMS MID SUSSEX ARS c/o C Cook, 45 Stonepound Road, Hassocks, BN6 8PR
G3 ZMX Frank Jackson, 86 Longfield Grove, Todmorden, OL14 6NP
G3 ZNB Peter Hannam, Bogg Hall, Oulston, York, YO61 3RE
GM3 ZNC J Mulheron, 10 Devonview Place, Airdrie, ML6 9DF
G3 ZND James Bartlett, Chickamauga, Tinnahinch, Clonaslee, Ireland, CO.LAOIS
G3 ZNE N Ingle, 81 Redmoor Close, Tavistock, PL19 0ER
G3 ZNG S Birt, 82 Aintree Road, Thornton-Cleveleys, FY5 5HP
G3 ZNH Robert Coombes, 133 Masefield Road, Warminster, BA12 8HY
G3 ZNK David Hainsworth, 48 Greenacre Park, Rawdon, Leeds, LS19 6AR
G3 ZNR D Bailey, 12 St. Philips Drive, Burley in Wharfedale, Ilkley, LS29 7EN
G3 ZNT R Brown, 282 Luton Road, Dunstable, LU5 4LF
G3 ZNU M Appleby, 6 Mandeville Road, Prestwood, Great Missenden, HP16 9DS
G3 ZNV J Green, Rickwood House, Meadhurst Road, Chertsey, KT16 8HU
G3 ZNW R Blasdell, 32 Fulham Close, Broadfield, Crawley, RH11 9NY
G3 ZOC John Cunliffe, 12 Lords Avenue, Lostock Hall, Preston, PR5 5HH
G3 ZOG A Elliott, 15 Braemar Gardens, East Herrington, Sunderland, SR3 3PX
G3 ZOH B George, 14 Pondfield Road, Orpington, BR6 8HJ
G3 ZOI D Deane, 10 Stephens Road, Mortimer Common, Reading, RG7 3TU
G3 ZOL J Powell, 156 Avon Way, Colchester, CO4 3YP
GU3 ZOM D Pearson, Tequesta, York Avenue, Port Soif, Vale, Guernsey, GY6 8HS
G3 ZON K Lacy, 9 Rhodes Way, Tilgate, Crawley, RH10 5DQ
GM3 ZOT J Hewitt, 145 Queens Road, Fraserburgh, AB43 9PU
G3 ZOU R France, 25a Greenfield Road, Middleton on the Wolds, Driffield, YO25 9UL
G3 ZOW K Clamp, 12 Cowlishaw Close, Shardlow, Derby, DE72 2GS
G3 ZOX Philip Durham, 18 Maldon Road, Great Totham, Maldon, CM9 8PR
G3 ZOY A Alldrick, 23 Coxs Close, Nuneaton, CV10 7ET
G3 ZPA David White, 1 Whaddon Road, Shenley Brook End, Milton Keynes, MK5 7AF
G3 ZPB Peter Burton, 202 Coulsdon Road, Coulsdon, CR5 2LF
G3 ZPI G Braund, 184 Faversham Road, Kennington, Ashford, TN24 9AE
G3 ZPJ Michael Symons, 11 Tudor Lodge Park, Truthwall, Penzance, TR20 9BW
G3 ZPK H Willis, 12 Combe View, Hungerford, RG17 0BZ
G3 ZPL Neil Richardson, 501 Forest Avenue, Palo Alto, United States, CA 94301
G3 ZPM A Stormont, The Hawthorns, Church Lane, Louth, LN11 7JR
G3 ZPR David Mason, 26 Upton Road, Fleetsbridge, Poole, BH17 7AH
G3 ZPS Steven Shorey, 47 Stanham Road, Dartford, DA1 3AN
G3 ZPU Anthony Nightingale, 42 Spilsby Road, Horncastle, LN9 6AW
G3 ZPW Michael Brown, 6 Castle Court, Praa Sands, Penzance, TR20 9SX
G3 ZQB Anthony Seabrook, 63 St. Annes Road, Willingdon, Eastbourne, BN20 9NJ
G3 ZQC J Smith, 22 Harley Road, Condover, Shrewsbury, SY5 7AX
G3 ZQF S Carpenter, 4 Mount Court, West Wickham, BR4 9AH
G3 ZQH D Barrett, Linden House, Clifton Lane, Nottingham, NG11 6AA
G3 ZQI Brian Downer, 159 Howlands, Welwyn Garden City, AL7 4RL
G3 ZQJ B Stagg, 1 Hamilton Way, Leckhampton, Cheltenham, GL53 7DQ
G3 ZQL Stephen Murray, P9, Atico 3C, Tigaiga 3, Parque de la Reina, Spain, 38632
G3 ZQM The TyneSide ARS c/o Thomas Gardner, 200b Heaton Road, St Teresas Men Club, Newcastle upon Tyne, NE6 5HP
G3 ZQQ James Peden, 51a Bewdley Road, Kidderminster, DY11 6RL
G3 ZQR B Downton, Lickwith Cottage, Monkokehampton, Winkleigh, EX19 8SL
G3 ZQS INT. MORSE PRESERVATION SOCIETY c/o Robert Walker, P.O.Box 6743, Tipton, DY4 4AU
G3 ZQT Joseph Yu, Hunterscombe, Dorking Road, Leatherhead, KT22 8JT
G3 ZQU Martin Goodrum, Cedars, Church Lane, Stowmarket, IP14 5JL
G3 ZQV Ralph Hague, 111 Mount Vernon Road, Worsbrough, Barnsley, S70 4HH
G3 ZQW B Barrington, Pinlands Cottage, Bines Road, Horsham, RH13 8EQ
G3 ZQY N Clark, Chelsworth, Heaton Grange Road, Romford, RM2 5PP
G3 ZRA R Elliot, 1321 East Bailey Road, Naperville, United States, 60565
G3 ZRB D Hill, 872 Oldham Road, Rochdale, OL11 2BN
G3 ZRE P Ottewell, 30 Cumberland Avenue, Leyland, PR25 1BH
G3 ZRG I Steward, Keeper's Cottage, Banville Lane, Cromer, NR27 9RN
G3 ZRH Anthony Stokes, 34 Shenfield Crescent, Brentwood, CM15 8BW
G3 ZRJ Anthony Butler-Roskilly, 2 North Road, Pennymoor, Tiverton, EX16 8LQ
G3 ZRL Allan McWatters, 38 Sutherland Mount, Leeds, LS9 6DP
G3 ZRM M Payne, 3 Waterside Close, Bordon, GU35 0HB
G3 ZRN David Catherwood, 14 Hatton Lane, Hatton, Warrington, WA4 4BY



**IMPORTANT NOTE**

**Revalidate licence to avoid revocation** – Ofcom has advised the Society that plans will be drawn up to revoke licences that have not been revalidated as required by the licence conditions. The quickest way to revalidate is to do so online via the Ofcom website: *https://services.ofcom.org.uk/* or by email: *amateur.validations@ofcom.org.uk* Ofcom staff are available to help, but please be patient during times of heavy workload.

G3 ZRP R Perry, Little London Cottage, Little London, Market Rasen, LN7 6JP
G3 ZRQ D Maxfield, 40 Fegg Hayes Road, Stoke-on-Trent, ST6 6RA
G3 ZRR Mark Samuel, 71 Brighton Road, South Croydon, CR2 6EE
G3 ZRS P Rodmell, 2 Meadow Way, Walkington, Beverley, HU17 8SD
GM3 ZRT W Strachan, Glennoch, Glencoe, Argyll, PH49 4HN
G3 ZRX THE RAYNET ASSOCIATION c/o T Lundegard, Saxby, Botsom Lane, Sevenoaks, TN15 6BL
G3 ZRY G Stott, 8 Willow Road, Chinnor, OX39 4RA
G3 ZSB V Poore, 216 Powder Mill Lane, Twickenham, TW2 6EJ
G3 ZSF A Houltby, 2 Sinderson Road, Humberston, Grimsby, DN36 4UF
GM3 ZSH John Donaldson, 53 Izatt Avenue, Dunfermline, KY11 3BL
G3 ZSJ Ronald Troughton, 4 Owletts, Worth, Crawley, RH10 7SQ
G3 ZSK MERLIN SKELETON CLUB c/o John Tysiorowski, 52 Meadow Croft, Penrith, CA11 8EH
G3 ZSQ R Dunham, 42 Marsdale Drive, Stockingford, Nuneaton, CV10 7DE
G3 ZSS Peter Bacon, 18 Monks Well, Farnham, GU10 1RH
G3 ZST T Surgey, The Littlewood, Windmill Lane, Southam, CV47 2BN
G3 ZSU Shaun Scannell, 20 Queens Road, Wilbarston, Market Harborough, LE16 8QJ
G3 ZSX K Craig, 20 Alexander Close, Abingdon, OX14 1XA
G3 ZSZ R James, 283 High St., New Whittington, Chesterfield, S43 2AP
G3 ZTB Richard Ranson, 107 West End Lane, Horsforth, Leeds, LS18 5ES
GW3 ZTH J Ludlow, 44 Fox Hollows, Brackla, Bridgend, CF31 2NG
G3 ZTI K Marshall, 2 Keepers Mill, Woodmancote, Cheltenham, GL52 9QS
G3 ZTJ C Morgan, The Villa, The Green, Wallsend, NE28 7PH
GI3 ZTL F Convery, 2 Coolagh Road, Maghera, BT46 5JR
GM3 ZTP S Elwell-Sutton, 30 Orchard Court, Dundee, DD4 9DB
G3 ZTR D Lockwood, 25 Thorntondale Drive, Bridlington, YO16 6GW
G3 ZTT MID CHESHIRE AR c/o D Bevan, 46 Park Lane, Hartford, Northwich, CW8 1PY
G3 ZTU J Mace, Well Cottage, Tismans Common, Horsham, RH12 3DU
G3 ZTV P Webster, 3 Templemere, Norwich, NR3 4EF
G3 ZTX Peter Angell, Star Hill House, Star Hill, Forest Green, Stroud, GL6 0NJ
G3 ZTY John Yale, 15 Rectory Avenue, Corfe Mullen, Wimborne, BH21 3EZ
G3 ZTZ P Howell, 1 Jasmine Close, Littlehampton, BN17 6UP
G3 ZUB M Davison, 10 Springfield Close, Loughborough, LE11 3PT
G3 ZUC Doris Cardell, 22 Millview Road, Heckington, Sleaford, NG34 9JP
G3 ZUE A Nicholas, Verriotts Lane, Morecombelake, Bridport, DT6 6DU
G3 ZUI Malcolm Johnson, Greentiles, Alford Road, Alford, LN13 0JW
G3 ZUK Roger Whitehead, Church View, Church End, Cambridge, CB21 5PE
G3 ZUL Brian Kennedy, 21 The Croft, Kidderminster, DY11 6LX
G3 ZUM Brian Lonnon, 5 Mickle Meadow, Water Orton, Birmingham, B46 1SN
G3 ZUN Dannatt Sharpe, 12 Belgrave Crescent, Seaford, BN25 3AX
G3 ZUO David Ham, Upton Manor Lodge, Upton Manor Road, Brixham, TQ5 9QZ
G3 ZUS Nicholas Ewer, Ashtree Farm, Buckland, Faringdon, SN7 8PX
G3 ZUT M Thorne, Rossendale, Main Road, Bristol, BS35 5RE
G3 ZVC B Comer, Corner House, High Street, Fairford, GL7 4EQ
G3 ZVH David Bedford, 64 St. James Avenue, Upton, Chester, CH2 1NL
G3 ZVI P Longhurst, 18 Austen Close, Exeter, EX4 8HB
G3 ZVK J Simons, 120 Bond Way, Hednesford, Cannock, WS12 4SN
G3 ZVM A Greenbank, Grahamsley, Westburn, Ryton, NE40 4EU
G3 ZVN G Peck, 4 Koonowla Close, Biggin Hill, Westerham, TN16 3BJ
G3 ZVQ J Bridge, 8 Highfield Grove, Lostock Hall, Preston, PR5 5YB
G3 ZVS E Park, Waterside Cottage, Bowden Lane, High Peak, SK23 0QF
G3 ZVT G Rabstaff, Evesham, 26 Lichfield Drive, Manchester, M25 0HX
G3 ZVV J Gellatly, 11 Arches Drive, Bilsthorpe, NG22 8QF
G3 ZVW Stephen White, Moorcroft, Crewkerne Road, Axminster, EX13 5SY
GI3 ZVZ D Nicholls, 2 Printshop Road, Templepatrick, Ballyclare, BT39 0HZ
G3 ZWD Paul Flicos, 35 The Broadway, Northbourne, Bournemouth, BH10 7EU
G3 ZWF P Harpur, 16 Lime Avenue, Upminster, RM14 2HY
GM3 ZWG W Frame, 13 Conon Avenue, Bearsden, Glasgow, G61 1EN
G3 ZWK D Raimbach, 21 Waldorf Heights, Blackwater, Camberley, GU17 9JQ
G3 ZWM I Morrison, The Vicarage, 1a Church Road, Sandy, SG19 2JY
G3 ZWN A Slingsby, 10d Stanbury Road, Thruxton, Andover, SP11 8NS
G3 ZWP Richard Davies, Flat 24, Hurley Court, Bracknell, RG12 9QH
G3 ZWR N Hay, 19 Logan Road, Walkerville, Newcastle upon Tyne, NE6 4SY
G3 ZWW Michael Quee, Goslings Barn, Sheepcotes Lane, Braintree, CM77 8ER
G3 ZWY D Ansell, 1 Carslake Close, Sidmouth, EX10 9FJ
G3 ZXA A Wenham, 28 Pine Wood, Sunbury-on-Thames, TW16 6SG
GM3 ZXB Allan Robertson, 77 Cobden Street, Dundee, DD3 6DD
G3 ZXD Maurice French, PO Box 217, Leigh, New Zealand, 947
G3 ZXF David Corner, 122 Stortford Hall Park, Bishop's Stortford, CM23 5AP
GM3 ZXG John Higgins, 9 Waverley Street, Greenock, PA16 9DH
G3 ZXM M Brown, Curraghmore, Tullogher, Co. Kilkenny, Ireland, X91 R642
G3 ZXO J Burnie, 1 Chapel Meadow, Buckland Monachorum, Yelverton, PL20 7LR
G3 ZXV P Veale, 13 Lawford Gardens, Kenley, CR8 5JJ
G3 ZXW J Midmore, 9 Whiteways, Wimborne, BH21 2PQ
G3 ZXY Peter Holtham, 21 Sherborne Place, Chapel Hill, Australia, QLD 4069
G3 ZXZ Martin Stokes, 3 Priory Way, Mirfield, WF14 9QS
G3 ZYC M Sneap, Farm Close, Pentrich, Derby, DE5 3RR
G3 ZYD R Alton, 23 Cemetery Road, Belper, DE56 1EJ
G3 ZYE Robin Bellerby, Glenamour, Newton Stewart, DG8 7AE
G3 ZYL Gordon Bowhay, Windwhispers, Lewannick, Launceston, PL15 7QD
G3 ZYN Roy Gledhill, 25 Sandpiper Way, Torquay, TQ2 7GJ
G3 ZYP A Matheson, 1 St. Edmunds Close, Bromeswell, Woodbridge, IP12 2PL
G3 ZYQ A Robinson, 6 The Crescent, Minster on Sea, Sheerness, ME12 3BQ
G3 ZYR A Booer, Lower Farm Barn, Duck End Lane, Witney, OX29 5RH
G3 ZYV D Ferigan, 191 Gillingham Road, Gillingham, ME7 4EP
G3 ZYX Richard Offord, 10 Barberry Way, Blackwater, Camberley, GU17 9DX
G3 ZYY Trevor Day, 46 Beatrice Avenue, Saltash, PL12 4NG
G3 ZYZ M Joiner, 22 Sanderstead Court, Addington Road, South Croydon, CR2 8RA
GM3 ZZA P Rose, 4 Heatherfield Glade, Livingston, EH54 9JE
G3 ZZD Stephen Ireland, PO Box 55, Glen Forrest, Glen Forest, Australia, 6071
GD3 ZZF D Woolley, 21 Lulworth Avenue, Wembley, HA9 8TP
G3 ZZH M Battersby, 25 Lowther Close, Emmbrook, Wokingham, RG41 1JE
G3 ZZI Gregory Smith, 34 Huria Lane, Woodend, New Zealand, 7610
G3 ZZL S Keightley, Flat 4, Petworth Court, Rustington, BN16 2LF
G3 ZZM M Robinson, 7400 Old Bunch Road, Wendell, United States, 27591
GD3 ZZN Martyn Rickward, Dunsandle, 12 Fairway Close, Port Erin, Isle of Man, IM9 6LS
G3 ZZP D Matthews, 7 Boulsworth Avenue, Hull, HU6 7DZ
G3 ZZQ Richard Ludwell, Church Lodge, Thetford Road, Thetford, IP24 2QX
G3 ZZS Richard Wills, 21 Woodford Road, Glenholt Park, Plymouth, PL6 7HX
G3 ZZU C Waldron, 22 Windermere Road, Patchway, Bristol, BS34 5PW
G3 ZZV G Evans, 20 Creekside View, Tresillian, Truro, TR2 4BS
G3 ZZW P Chimber, 202 Wintersdale Road, Leicester, LE5 2GP
G3 ZZX A Evans, Ashlea, Aston Munslow, Craven Arms, SY7 9ER
G3 ZZZ J Gibbs, 10 Waverley Road, Margate, CT9 5QB

## G*4

GI4 4OKU Thomas Patton, 29 Greystone Park, Limavady, BT49 0EQ
GM4 AAF Dundee ARC c/o Paul Glasper, 1 Lindertis Cottages, Kirriemuir, DD8 5NT
G4 AAH K Lawson, 233 Southwell Road West, Mansfield, NG18 4HF
G4 AAL J Layton, Meadow View, Martley, Worcester, WR6 6QA
G4 AAQ Philip Butterfield, 29 Aire Street, Knottingley, WF11 9AT
G4 AAR ASHFORD ARC c/o John Wellard, 19 Southment Motto, Kingsnorth, Ashford, TN23 3NJ
G4 AAU W Bowen, 126 Westfield Lane, Kippax, Leeds, LS25 7HU
G4 AAW P Miller, 14 Blackmanstone Way, Maidstone, ME16 0NT
G4 AAX NORTHUMBRIA ARC c/o G Emmerson, 72 The Gables, Widdrington, Morpeth, NE61 5RB
G4 ABC Thornbury and South Gloucestershire ARC c/o Peter Cabban, Ivydene, Upper Tockington Road, Bristol, BS32 4LQ
G4 ABE J Ellis, 4 Hazelmount Crescent, Warton, Carnforth, LA5 9HS
G4 ABI Donald Radley, P.O. Box 61251, Paphos, Cyprus, 8132
G4 ABL A Howell, 46 Acre Court, Andover, SP10 1HH
G4 ABM Cyril Todd, 7 Sandown Road, Billingham, TS23 2BQ
G4 ABN T Atkins, 55 Havenbrook Blvd, Willowdale, Ontario, Canada
G4 ABQ Jonathan Hudson, 46 High Street, Odell, Bedford, MK43 7BB
G4 ABT Phil Oliver, 12 Croft Road, Edwalton, Nottingham, NG12 4BW
G4 ABW L Willey, 7 Oaklands Road, Four Oaks, Sutton Coldfield, B74 2TB
G4 ABX B Macaulay, The Old Chapel, Chapel Street, Lutterworth, LE17 6AZ
G4 ABY D Green, 2 Foss Field, Winstone, Cirencester, GL7 7JY
G4 ACF ACF/CCF INT R N c/o Robin Cogzell, 242 Clydesdale Tower, Holloway Head, Birmingham, B1 1UJ
G4 ACI J Blackburn, 40 Carlton Avenue, Upholland, Skelmersdale, WN8 0AE
G4 ACJ H Reeve, 11 Heather Drive, Ferndown, BH22 9SD
G4 ACL D Atkinson, 38 Hornbeam Road, Theydon Bois, Epping, CM16 7JX
GM4 ACM A Miller, 38 Randolph Road, Broomhill, Glasgow, G11 7LG
G4 ACP John Scherrer, 26 Grange Way, Willington, Bedford, MK44 3QW
G4 ACS G Weale, 11 Heather Drive, Kinver, Stourbridge, DY7 6DR
G4 ACU M Levy, 34428 Yucaipa Blvd, E346, Yucaipa, United States
G4 ACW N Roe, 10 Ramsdean Road, Stroud, Petersfield, GU32 3PJ
G4 ACY Robert Ratcliffe, 173 Montague Road, Bilton Hill, Rugby, CV22 6LG
G4 ACZ R Mutton, Summer Hayes, Mill Lane, Alcester, B49 6LF
G4 ADD W Ricalton, 4 South Road, Longhorsley, Morpeth, NE65 8UW
G4 ADE Michael Woollin, 14 St. Nicholas Drive, Hornsea, HU18 1EW
G4 ADG Peter West, 2 Quilp Drive, Chelmsford, CM1 4YA
G4 ADJ Peter Hampton, 45 Mortlake Avenue, Redhill, Worcester, WR5 1QB
G4 ADK R CUTBUSH, 60 Culver Way, Sandown, PO36 8QL
GW4 ADL T Davies, 44 Carnglas Road, Sketty, Swansea, SA2 9BW
G4 ADM A Maish, 73 Edenfield Gardens, Worcester Park, KT4 7DX
G4 ADP Peter McCurrie, Lakefields, Drake Close, Southampton, SO40 4XB
G4 ADR N Ayres, Glendevon, 68 Queens Road, Thame, OX9 3NQ
G4 ADS John Chisman, 115 St. Lukes Avenue, Ramsgate, CT11 7HT
G4 ADV NEWQUAY & DISTRICT ARS c/o K Francks, 63 Parc Godrevy, Pentire, Newquay, TR7 1TY
G4 AEB Thomas Baines, 6 Brading Avenue, Clacton-on-Sea, CO15 4PA
G4 AED B Cator, 9 Saham Road, Watton, Thetford, IP25 6EA
G4 AEE Michael Bedford, 4 Holme House Lane, Oakworth, Keighley, BD22 0QY
G4 AEG I Kemp, 21 Rednal Road, Kings Norton, Birmingham, B38 8DT
G4 AEH James Lee, 44 Howard Road, Nuneaton, CV10 7ES
G4 AEI Geoffrey Prater, Heathfield, Wyndham Lane, Salisbury, SP4 0BY
GM4 AEK P Boswell, Coralach, Dunvegan, Isle of Skye, IV55 8WF
G4 AEL R Cox, 34 Ratcliffe Drive, Stoke Gifford, Bristol, BS34 8UD
G4 AEM P Ellis, 96 Whitelands Avenue, Chorleywood, Rickmansworth, WD3 5RG
G4 AEO Peter Hunt, 93 Park Road, Coalville, LE67 3AF
G4 AEP W Thomas, 58 Bearwood Road, Wokingham, RG41 4SY
G4 AER R Sobey, 5 Fairway Close, Liphook, GU30 7XD
G4 AES K Walker, 3 Glen View, Stile, Sowerby Bridge, HX6 1NL
G4 AEV R Anderson, Trinafour, Abingdon Road, Abingdon, OX13 6NU
G4 AEY Robert Twist, 1 Birchwood Drive, Rushmere St. Andrew, Ipswich, IP5 1EB
G4 AEZ B Oughton, 176 South Lodge Drive, Southgate, London, N14 4XN
G4 AFA Nigel Porter, 13 Charfield Close, Winchester, SO22 4PZ
G4 AFE Jonathan Tallentire, 4 Alston Road, Middleton-in-Teesdale, Barnard Castle, DL12 0UU
GM4 AFF S Cooper, Ambleside, Lauriston, Montrose, DD10 0DJ
GI4 AFH G Phillips, 38 Sketrick Ind Park, Newtownards, BT23 3BN
G4 AFI A Cheetham, 39 Burns Avenue, Church Crookham, Fleet, GU52 6BN
G4 AFJ G Dover, 31 Newbold Road, Kirkby Mallory, LE9 7QG
G4 AFQ David Warner, Treeside, School Lane, Norwich, NR11 8HJ
G4 AFR F Nicholson, 5 Friars Terrace, Barrow-in-Furness, LA13 0BX
G4 AFS T Bucknell, 7 Alexander Court, Sandbeds, Keighley, BD20 5NW
G4 AFT David Randles, Long Reach, Westerns Lane, Harrogate, HG3 3PB
G4 AFU Paul Rollin, Farthings, Burneston, Bedale, DL8 2JE
G4 AFX Adrian Moore, Garden Wing, Copinger Hall, Stowmarket, IP14 3DJ
G4 AFY R Perrin, 8 Granville Crest, Kidderminster, DY10 3QS
G4 AFZ Vivian Bott, 25 Finkle Street, Hensall, Goole, DN14 0QY
G4 AGC C Wortham, 57 Cranleigh Drive, Swanley, BR8 8NZ
G4 AGE Raymond Evans, Mansfield, 1 Horsehead Lane, Chesterfield, S44 6HU
GM4 AGG W SCOTLAND ARS c/o A Stewart, Three Acres, Cochno Road, Clydebank, G81 6PX
G4 AGH S Pearson, 75 Gloucester Road, Thornbury, Bristol, BS35 1JH
GM4 AGL William Ferguson, 72 High Parksail, Erskine, PA8 7HX
GD4 AGM Robert Williams, Flat 32, St. Johns Court 59 Murray Road, Northwood, HA6 2FY
G4 AGN John Porter, 109 Heacham Drive, Leicester, LE4 0LL
G4 AGQ J Billingham, 14 St. Matthews Court, Sutherland Road, Brighton, BN2 2EX
G4 AGY G Rippengill, 5 Bridge Farm Drive, Liverpool, L31 9AL
G4 AHC T O'Neill, 10 Dalehurst Close, Wallasey, CH44 8AE
GI4 AHD Frederick Elder, 44 Learmount Road, Claudy, Londonderry, BT47 4AQ
G4 AHG SHIREHAMPTON ARC c/o Colin Chidgey, 46 Station Road, Shirehampton, Bristol, BS11 9TX
G4 AHJ Michael Downey, 11 Woodlands Drive, Lepton, Huddersfield, HD8 0JB
G4 AHK B Palin, 11 Ashgrove Close, Marlbrook, Bromsgrove, B60 1HW
G4 AHM J Stratton, 10 Brownshill, Maulden, Bedford, MK45 2BT
G4 AHN D Lax, 1 Gardeners Hill Road, Wrecclesham, Farnham, GU10 4RL
G4 AHO K Jones, 13 Upland Grove, Bromsgrove, B61 0EL
GI4 AHP T Sloan, 13 Mount Royal, Lisburn, BT27 5BF
G4 AHT M Niven, 16 Treewall Gardens, Bromley, BR1 5BT
G4 AHW A Thomson, 392 Glen Ross Road, Quinte West Ontario, Canada, K0K 2C0
G4 AHZ John Kynaston, 19 Sharples Drive Wrea Green, Preston, PR4 2EL
G4 AIB P Holt, 41 Garden Avenue, Ilkeston, DE7 4DF
G4 AIE W Mackie, 23 College Park, Horncastle, LN9 6RE
G4 AIJ R Jones, Sycamores, The Sheet, Ludlow, SY8 4JT
GI4 AIO R Lindsay, 67 Halfpenny Gate Road, Moira, Craigavon, BT67 0HP
G4 AIR D Bieber, Tonkins Quay House, Lanteglos-by-Fowey, PL23 1NB
G4 AIU Eugene Morgan, 12 Kitts, Wellington, TA21 9AX
G4 AIW A Scarsbrook, 16 Greenbank Avenue, Uppermill, Oldham, OL3 6EB
G4 AJA Christoper Hoare, 16 Shrivenham Road, Highworth, Swindon, SN6 7BZ
G4 AJE Paul Brown, 33a March Road, Wimblington, March, PE15 0RW
G4 AJG Peter Perera, 43 Hillside Avenue, Woodford Green, IG8 7QU
G4 AJJ G Smith, 39 Hornbeam Way, Kirkby-in-Ashfield, Nottingham, NG17 8RL
G4 AJO R Finch, 48 Allens Lane, Sprowston, Norwich, NR7 8EJ
G4 AJQ Nigel Johnson, 503-97 Lawton Blvd, TORONTO, Ontario, Canada, M4V 1Z6
GM4 AJR Lewis Donaldson, 11 Highgate Gardens, Aberdeen, AB11 7TZ
G4 AJU I Aldridge, 28 Robert St., Williton, Taunton, TA4 4PG
GM4 AJV Malcolm MacKinnon, 55 Fairbae, Edinburgh, EH11 3GZ
G4 AJW A Wade, 139 Gilbert Road, Cambridge, CB4 3PA
G4 AJY David Ellis, 26 Drake Close, Benfleet, SS7 3YL
G4 AKA Michael Diprose, 4a Russet Close, Staines, TW19 6AX
G4 AKB Melvin Court, Escorredor 138, Dolores, Spain, 3150
G4 AKC D Starkie, 5 Kidbrooke Avenue, Blackpool, FY4 1QR
G4 AKD I Alexander, 46 Pettitts Lane, Dry Drayton, Cambridge, CB23 8BT
G4 AKE C Gent, 27 Walnut Avenue, Alvaston, Derby, DE24 0PP
G4 AKG P Fry, 11 Park Road, Burgess Hill, RH15 8EU
G4 AKQ Maurice Bernard, 16 Mountbatten Avenue, Chatham, ME5 0JX
G4 AKR G Slack, 16 East Carr, Cayton, Scarborough, YO11 3TS
G4 AKW Geoffrey Robinson, 2 Hasketon Road, Woodbridge, IP12 4JR
G4 AKY David Hayes, Trenley, 68 London Road, Sevenoaks, TN13 2UG
G4 AL John Wood, 18 Kennedy Avenue, Long Eaton, Nottingham, NG10 3GF
G4 ALA John Hardwick, 455 Hatton Road, Feltham, TW14 9QP
G4 ALB N Castledine, 1 Johns Close, Burbage, Hinckley, LE10 2LY
G4 ALC J Balls, 48 Collingwood Road, Great Yarmouth, NR30 4LR
G4 ALD Francis Donovan, 4 Rembrandt Drive, Northfleet, Gravesend, DA11 8NQ
G4 ALE Addiscombe ARC c/o Michael Franklin, 6 Tor Road, Farnham, GU9 7BX
G4 ALF Kenneth Law, 93 Measham Drive, Stainforth, Doncaster, DN7 5TQ
G4 ALR Michael Down, 5 Juniper Mead, Stotfold, Hitchin, SG5 4RU
G4 ALT Anthony Taylor, 21 Gould Avenue West, Kidderminster, DY11 7HD
G4 ALY Ralph Bird, 6 The Cross, St. Dominick, Saltash, PL12 6SP
G4 ALZ R Bridgland, 20 Newling Way, Worthing, BN13 3DG
G4 AMD Chris Heavens, 14905 31st Avenue SE, Mill Creek, United States, 98012
G4 AMF Jack Cresswell, 7 Glinton Avenue, Blackwell, Alfreton, DE55 5HD
G4 AMI M Hearn, 63 Greswolde Road, Solihull, B91 1DX
G4 AMJ David Evans, 330 Weld County Road 16 1/2, Longmont Co, United States, 80504-9467
G4 AMN Christopher Wainwright, 60 Main Street, Hoby, Melton Mowbray, LE14 3DT
G4 AMP Brian Flack, Ave Des Hospitaliers De, St Jean 7, Waterloo, Belgium, 1410
G4 AMT Terry George, Sea Call, Sennen Cove, Penzance, TR19 7BT
G4 AMX Joseph Barrett, Flat 5, Rhos Abbey, Rhos Promenade, Colwyn Bay, LL28 4QA
G4 AMY Roger Briggs, Nickey Nook View, Lancaster New Road, Preston, PR3 1NL
G4 ANB J Morris, 4111 Eve Road, Simi Valley, United States, 93063
G4 AND J King, Chetwynd, Henfield Road, Steyning, BN44 3TF
G4 ANE H Leach, 30 Taywood Road, Thornton-Cleveleys, FY5 2RT
GW4 ANK R Davenport, 14 Milward Road, Barry, CF63 3QD
G4 ANN Richard Hadfield, 45 Erica Way, Copthorne, Crawley, RH10 3XG
G4 ANP M Valentine, 10 Thellusson Avenue, Scawsby, Doncaster, DN5 8QN
G4 ANQ P Clayton, 90 Littleheath Road, South Croydon, CR2 7SD
G4 ANT EAST ANGLIAN CONTEST GROUP c/o Roy Reed, Oak Cottage, Dereham Road, Dereham, NR20 4AA
G4 ANU C Columbine, 5 Thornbury Drive, Mansfield, NG19 6NB
G4 ANV P Hudson, 3 Rowan Drive, Kilburn, Belper, DE56 0PG
G4 ANW T Slack, 16 Woodside Avenue, Alverstone Garden Village, Sandown, PO36 0JD
G4 ANY D Stephens, Croeso Cottage, 31 Coton, Whitchurch, SY13 2RA
G4 ANZ B Warren, 46 Old Mill Gardens, Berkhamsted, HP4 2NZ
G4 AOA H Mason, 9 Chatsworth Drive, Little Eaton, Derby, DE21 5AP
G4 AOJ R Horton, 31 Furze Lane, Purley, CR8 3EJ
G4 AOK Timothy Winter, 6 Cunliffe Drive, Brooklands, Sale, M33 3WS
G4 AOL D Harmer, 4 Somerton Gardens, Earley, Reading, RG6 5XG
G4 AOP David Hibbin, 95a Thorpe Acre Road, Loughborough, LE11 4LF
G4 AOQ David Ward, 60 New Road, High Wycombe, HP12 4LG
GM4 AOR Kenneth Henderson, 97 Granton Road, Edinburgh, EH5 3NH
G4 AOS J West, Horsley House, Rochester, Newcastle upon Tyne, NE19 1TA
G4 AP J Rooke, 12 Hellings Gardens, Broadclyst, Exeter, EX5 3DX
G4 APB K May, 53 Shearwood Crescent, Crayford, Dartford, DA1 4SU
G4 APD RUGBY AMATEUR TRANSMITTING SOCIETY c/o P Wells, 12 Shelley Drive, Lutterworth, LE17 4XF
GW4 APF M Richards, 9 Bank Road, Llangennech, Llanelli, SA14 8UB
G4 APG Michael Pellatt, 21 Carlisle Road, Dartford, DA1 1XB

GM4 API D Hebenton, Craigmill Cottage, 3 Craigmill Road, Dundee, DD0 0PH
G4 APJ K Punshon, 24 Newcombe Road, Ramsbottom, Bury, BL0 9UT
G4 APL Paul Lowis, 20 Annes Walk, Caterham, CR3 5EL
G4 APO R Hirst, 21 Manor Farm Court, Thrybergh, Rotherham, S65 4NZ
G4 APP W Grogan, 92 School Road, Thornton Cleveleys, FY5 5AP
G4 APS D Fiander, 2 Snowshill Close, Nuneaton, CV11 4XQ
G4 AQA P Hall, 39 Mill Lane, Kirk Ella, Hull, HU10 7JF
G4 AQB Stephen Macdonald, 58a Torbot Drive, Bolton, BL2 6LT
G4 AQE P Saunders, Orchard Cottage, Vale Road, Broadstairs, CT10 2JG
G4 AQG UNIVERSITY OF SUSSEX ARS c/o Andrew Maris, 140 Edward Street, Brighton, BN2 0JL
G4 AQJ K Gordon, 96 Pear Tree Crescent, Shirley, Solihull, B90 1LF
G4 AQK D Davis, 23 Matley Moor, Liden, Swindon, SN3 6NL
G4 AQR Ian Cordingley, Orchard Cottage, Compton, Paignton, TQ3 1IA
G4 AQS M Bliss, 53 Rowallan Drive, Bedford, MK41 8AS
G4 AQT John Rowbotham, 18 Weldbank Close, Beeston, Nottingham, NG9 6FU
G4 AQZ Geoffrey Axford, 24 Jack Branch Court, Vitch Lane, Clacton on Sea, CO15 1EJ
G4 ARC Rhyl District Amateur Radio c/o Alan Evans, 4 Elm Grove, Rhyl, LL18 3PE
G4 ARE EXETER ARS c/o A Jordan, 21 Madison Avenue, Exeter, EX1 3AH
G4 ARF FURNESS A.R.S. c/o David Latimer, 44 Lyndale Avenue, Barrow-in-Furness, LA13 9AR
G4 ARI T Raven, 15 Preston Close, Stanton under Bardon, Markfield, LE67 9TX
GM4 ARJ J Ferguson, 26 Cleuch Avenue, Tullibody, Alloa, FK10 2RX
G4 ARN NORFOLK ARC c/o Anthony Hall, 122 Norwich Road, New Costessey, Norwich, NR5 0EH
G4 ARO T Covey, 68 Wellington Close, Walton-on-Thames, KT12 1BB
G4 ARS CARLISLE & DIS. ARS c/o C Wolf, 35a Moorhouse Road, Carlisle, CA2 7LU
GM4 ARU J McIntyre, 12 Johnstone Lane, Carluke, ML8 4NR
G4 ARX B Curley, 22 Churchill Crescent, Sheringham, NR26 8NQ
G4 ARY A Langford, 33 Briscoe Road, Hoddesdon, EN11 9DG
G4 ASF R Mccurrach, Isa Coed, Bowden LANE, Bude, EX23 9BJ
G4 ASG Philip Bayley, 9 Westbrook Green, Bromham, Half Acre, Chippenham, SN15 2EF
G4 ASH I Roberts, 20 Queensway, Burton Latimer, Kettering, NN15 5QW
G4 ASI F Emery, Room 10, Building 448, Westerham, TN16 3BN
G4 ASK E Rayland, 40 Sycamore Close, Taunton, TA1 2QJ
G4 ASL S Ayling, Kitnocks, 89 Queens Road, Alton, GU34 1JA
G4 ASM Alan Murphy, Apartment 28, Trinity Gardens, 1 Kingsmead Road South, Prenton, CH43 6TA
GU4 ASO R Ayres, Langaller, Rue Colin, Vale, Guernsey, GY6 8LA
G4 ASP J Holding, Old Pearmain, Eardisland, Leominster, HR6 9DN
G4 ASQ Michael Jordan, 4 Marchfont Close, Nuneaton, CV11 6GA
G4 ASR D Butler, Yew Tree Cottage, Lower Maescoed, Hereford, HR2 0HP
G4 ASW M Yorke, 8 St John Place, Port Washington, New York, United States, 11050
G4 ASX O Perry, 60 Malines Avenue, Peacehaven, BN10 7RS
G4 ASY D Yeaman, Paddock View, Hurstbourne Priors, Whitchurch, RG28 7SE
G4 ASZ M Hurst, 21 Bankside, Dunton Green, Sevenoaks, TN13 2UA
G4 ATA John Hotchin, 151 Winchester Road, Grantham, NG31 8RX
G4 ATB R Shapland, 14 Charney Court, Grange-over-Sands, LA11 6DL
G4 ATG BARTG c/o Andrew Thomas, The Stone Barn, 1 Home Farm Close, Bicester, OX26 1TZ
G4 ATL David Bloomfield, 26 Preston Crowmarsh, Wallingford, OX10 6SL
G4 ATQ Geoffrey Hawkins, 18 Brook Street, Leighton Buzzard, LU7 3LH
G4 ATR J Rogers, 24 Treza Road, Porthleven, Helston, TR13 9NB
G4 ATU S Brown, Mullins View, 1D Turnpike Road, Ormskirk, L39 3LD
G4 AUB A Smith, 56 Longrood Road, Rugby, CV22 7RE
G4 AUC S Baugh, 70 Madingley, Bracknell, RG12 7TF
G4 AUD Anthony Lacy, Llanoris, Llanerfyl, Welshpool, SY21 0EP
G4 AUE Andrew Rose, 18 Highview Gardens, St. Albans, AL4 9JX
GD4 AUF C Friel, 102a Sharps Lane, Ruislip, HA4 7JB
G4 AUG R Mortimer, 19 St. Monance Way, Colchester, CO4 0PJ
G4 AUL Graham Mitchell, 10 Wealden Close, Hildenborough, Tonbridge, TN11 9HB
G4 AUN R Collett, 70 Clifton Road, Darlington, DL1 5DX
GM4 AUP I Suart, 37 Meldrum Mains, Glenmavis, Airdrie, ML6 0QQ
G4 AUQ F Barker, 90 Hall Road, Hull, HU6 8SB
G4 AUR Jeffrey McBurney, 4 Fownhope Road, Sale, M33 4RF
G4 AUS James Anderson, 72 Saffron, Amington, Tamworth, B77 4EP
G4 AUV G Wing, 105 Moore Avenue, Norwich, NR6 7LG
G4 AUY R Sherwood, 43 Kingsland, Arleston, Telford, TF1 2LE
G4 AVC David Bowers, 31 Clarence Road, Wrexham, LL11 2EU
G4 AVE L Cates, 45 Smoke Lane, Reigate, RH2 7HJ
G4 AVF Alan Fletcher, 11 Little Oak Close, Lees, Oldham, OL4 3LW
G4 AVJ Geoffrey Pople, 3 Leighton Drive, Creech St. Michael, Taunton, TA3 5DW
G4 AVK Steve Ripley, 62 Palewell Park, London, SW14 8JH
G4 AVL P Newby, 238a Wherstead Road, Ipswich, IP2 8JZ
G4 AVN Thomas Thompson, 146 Hawthorn road, Ashington, NE63 9BG
G4 AVS R Wilson, Aerial House, 1 The Fields, Woodbridge, IP12 2HZ
G4 AVV G Cluer, 12 Bingham Road, Addiscombe, Croydon, CR0 7ED
G4 AVX A Newman, 101 Washbrook Road, Portsmouth, PO6 3SD
GM4 AWA R Payne, Conifers, 10 Lambourn, Wolfhill, PH2 6TQ
GM4 AWB Roderick Macduff, 17 Larchfield Road, Bearsden, Glasgow, G61 1AP
G4 AWF D Wilson, Barnside, Straight Road, Stowmarket, IP14 2LZ
G4 AWG G Higgs, Firtree House, Perry Wood, Faversham, ME13 9SF
G4 AWJ Gordon Thomas, 9 Highcroft Crescent, Heathfield, TN21 8HE
G4 AWK Malcolm Roberts, 2 Thurlby Close, Washingborough, Lincoln, LN4 1HG
G4 AWM D Norfolk, 13 Oakwood Crescent, Greenford, UB6 0RF
G4 AWO R Gray, 10 Stone Park, Broadsands, Paignton, TQ4 6HT
G4 AWU R Lane, 8 Town Street, Lound, Retford, DN22 8RS
G4 AWW Neil Shepherd, Jo Kebi, Stonehall Road, Dover, CT15 7JS
G4 AWY R Mekka, 57 St. Johns Road, Caversham, Reading, RG4 5AL
G4 AWZ P Matthews, 22 Rydens Road, Walton-on-Thames, KT12 3DA
G4 AXA N Pope, Silver Hill, Norwich Road, Great Yarmouth, NR29 5PB
G4 AXC Cyril Burden, Cedar Croft, Hengar Lane, Bodmin, PL30 3PH
G4 AXD Graham Edy, 44 Roseholme, Maidstone, ME16 8DR
G4 AXF J Jacques, 30 Centurian Way, Bedlington, NE22 6LD
G4 AXI C Gerrard, 22 Kelso Drive, The Priorys, North Shields, NE29 9NS
G4 AXO J Wills, 48 Fairfield Road, Winchester, SO22 6SG
GM4 AXS P Wilberforce, 8 Ferryfield Road, Connel, Oban, PA37 1SR
G4 AXU Geoffrey Parr, Chesil Coppice, West Bexington, Dorchester, DT2 9DD
GI4 AXV J Doherty, 172 Dunmore Road, Ballynahinch, BT24 8QQ
G4 AXW S Jones, 4b Dewdley Court, Evesham, WR11 4AH
G4 AXX M Marsden, 38 Lambert Cross, Saffron Walden, CB10 2DP
G4 AXY A Mort, 88 Longfield Road, Winnall, Winchester, SO23 0NU
G4 AYB A Kelle, Urb.Sorries 10, la Massana, Andorra, AD.400
G4 AYD T Hodgetts, 15 Wiltons, Wrington, Bristol, BS40 5LS
G4 AYH Geoffrey Monks, 7 Town Street, Rawdon, Leeds, LS19 6PU
G4 AYK MID SEVERN VALL c/o P Perrins, 9 Merrick Close, Hayley Green Halesowen, B63 1JY
G4 AYL E Lambert, 41 Brand Hill Drive, Crofton, Wakefield, WF4 1PJ
G4 AYM GLOS AR & ELECTRONICS SOCIETY c/o Anne Reed, 32 Hollis Garden, Cheltenham, GL51 6JQ
G4 AYO M Hewitt, 10 Blacka Moor View, Sheffield, S17 3GZ
G4 AYQ J Durrans, 87 The Links, Trevethin, Pontypool, NP4 8DQ
G4 AYR T Greenwood, 30 Ringwood Road, Headington, Oxford, OX3 8JA
G4 AYS A Crook, 153 Shortheath, Shortheath, Swadlincote, DE12 6BL
G4 AYU Norman Kenyon, 74 Albert Road, Leyland, PR25 4YJ
G4 AZA Roger Winkworth, 13 Bagley Close, Kennington, Oxford, OX1 5LS
G4 AZC Paul Martin, Stoneovers, Wellow Top Road Ningwood, Yarmouth, PO41 0TL
G4 AZD A Edgecock, Sunnydene, Station Road, Colchester, CO7 8JA
G4 AZG G Macdonald, Pilgrims Cottage, Church Lane, Canterbury, CT4 6HX
G4 AZH M Bushnell, Rose Cottage, Street Ashton, Rugby, CV23 0PH
GW4 AZI D Thomas, Sunnydale, Scurlage, Swansea, SA3 1BA
GD4 AZJ R Troughton, Flat 2, Waterfront Apartments, Mooragh Promenade, Ramsey, Isle of Man, IM8 3AN
GD4 AZL P Justin, Garth, Park View Road, Pinner, HA5 3YF
G4 AZM Colin Wilson, 17719 phil c peters road, Winter Garden, United States, 34787
G4 AZQ M Sex, 81 Woodlands Way, Southwater, Horsham, RH13 9TF
G4 AZS A Bayling, 55 Shelton Road, Shrewsbury, SY3 8SU
G4 AZT Terence Barker, 1 Links Road, Kennington, Oxford, OX1 5RX
G4 AZU J Tiller, 21 Portal Road, Winchester, SO23 0PX
G4 AZX J Robinson, 19 Sunnycroft Gardens, Cranham, Upminster, RM14 1HP
G4 BAN P Godfrey, 5 Parkway, Southgate, London, N14 6QU
G4 BAO John Worsnop, 20 Lode Avenue, Waterbeach, Cambridge, CB25 9PX
GM4 BAP Alastair Beaton, 21 Airyhall Terrace, Aberdeen, AB15 7QN
G4 BAQ Richard Chambers, 17 Exmoor Close, Worthing, BN13 2PW
G4 BAS CLUB OF FRIENDSHIP c/o Howard Ketley, 24 Farmcroft Road, Mansfield Woodhouse, Mansfield, NG19 8QT
G4 BAU Richard Russell, 228 Broomhill, Downham Market, PE38 9QY
G4 BAV J GEE, 11 Charlton Avenue, Ipswich, IP1 6BH
G4 BBA Peter Chilcott, 321 Eastfield Road, Peterborough, PE1 4RA
G4 BBD M Tooley, 4 Shelley Road, Bath, BA2 4RJ
GI4 BBE R Bolton, Ohmvilla, 69 Newcastle Street, Newry, BT34 4AQ
G4 BBH R Ferryman, 25 Winant Way, Dover, CT16 2AX
G4 BBI Paul Nixon, 8 White Edge Close, Chesterfield, S40 4LE
G4 BBJ R Ramsay, 1 Sapho Park, Gravesend, DA12 4NA
G4 BBL A Thackery, 19 Pyne Point, Clevedon, BS21 7RL
G4 BBQ D King, 62 Ansley Road, Nuneaton, CV10 8NU
G4 BBT Roger Hancock, 80 Ulleries Road, Solihull, B92 8EE
G4 BBU P Whittle, 20 Marlbrook Lane, Marlbrook, Bromsgrove, B60 1HN
G4 BBY R Edwards, 27 Provis Mead, Chippenham, SN15 3UA
G4 BBZ S Ball, 1 Brindlegate, Pocklington, York, YO42 2HB
G4 BCA D Tunnicliffe, Old Golf House, Pirton Lane, Gloucester, GL3 2QE
G4 BCB K Johnston, 92C McDowalls Road, Yugar, Australia, 4520
GJ4 BCC R Davies, 2 Manor View Close, La Grande Route de St. Pierre S, St. Peter, Jersey, JE3 7AZ
GW4 BCF R Newman, 138 Newton Nottage Road, Porthcawl, CF36 5EE
G4 BCG G W Wale, 2 The Jordans, Coventry, CV5 9JT
G4 BCH P Burgess, Tretawn, Kite Hill, Ryde, PO33 4LG
G4 BCP L Graves, The Beach Hut, 6 Hauxley Links, Morpeth, NE65 0JS
G4 BCS J Buckingham, 2c Main Road, Biddenham, Bedford, MK40 4BA
G4 BCT A Gordon, 4 Victoria Road West, Thornton-Cleveleys, FY5 1BU
G4 BCV Essex Raynet c/o Neil Smith, Clare Cottage, White Ash Green, Halstead, CO9 1PD
G4 BCX Anthony Helm, 38 Blandford Road, Lower Compton, Plymouth, PL3 5DU
G4 BCZ Jeremy White, 13 Stokes Court, Ponthir, Newport, NP18 1RY
G4 BDC K Collerton, 3 Ness Lane, Preston, Hull, HU12 8SG
G4 BDJ R Mccartney, Calmdhu, Walter Street, Langholm, DG13 0AX
GI4 BDL Victor Simpson, 25 Waringstown Road, Lurgan, Craigavon, BT66 7HH
G4 BDQ Peter Harris, 53 Fairfax Court, Southampton, SO19 6FU
GI4 BDR Noel Evans, 87A Oldtown Road, Castledawson, Magherafelt, BT45 8BZ
G4 BDW James Bagley, 19 Low Road, Keswick, Norwich, NR4 6TZ
G4 BDX Michael Horoszko, 1 Woodgarth Cottages, Reedness, Goole, DN14 8EX
G4 BEB R Browning, 11 Ragmans Close, Marlow, SL7 3QW
G4 BEI James Palmer, 124a High Street, Wyke Regis, Weymouth, DT4 9NU
G4 BEL Roger Taylor, 12 The Rampart, Haddenham, Ely, CB6 3ST
G4 BEM Stanley Ford, 3 Hill View, Stoke-on-Trent, ST2 7AR
G4 BEO B Hailstone, 6 Larkswood Rise, St. Albans, AL4 9JU
G4 BEQ G Hotchkiss, Flat 54, Sanderling Lodge, Gosport, PO12 1EN
G4 BER Andrew Trend, Mr Ag Trend, 23 Bosley View, Congleton, CW12 3TU
G4 BEU J Small, 20 Hastings Road, Birkdale, Southport, PR8 2LW
G4 BEV R Taylor, 6 Churchill Crescent, Marple, Stockport, SK6 6HJ
G4 BEZ J Phillipson, 3 Montrose Close, New Hartley, Whitley Bay, NE25 0TA
G4 BFC A Riddell, 12 Sunrise, Malvern, WR14 2NJ
G4 BFR D Baldwin, 112 Moorland View Road, Chesterfield, S40 3DF
G4 BFS Terry Sargent, 15 Pound Lane, Blofield, Norwich, NR13 4NB
G4 BFT C Johnson, 32 Nightingale Close, Daventry, NN11 0GU
G4 BFV D Sinclair, 46 Church Lane, Mablethorpe, LN12 2NU
GM4 BFX A Milne, 65 Lord Hays Grove, Aberdeen, AB24 1WT
G4 BG A Duckworth, Ambergate, 2 Ashleigh Drive, Teignmouth, TQ14 8QX
GI4 BGB P Kelly, 30 Cahore Road, Draperstown, Magherafelt, BT45 7LY
GW4 BGD R Williams, Tan Y Bryn, Rhigos, Aberdare, CF44 9DJ
G4 BGH A Ruddell, 9 Parsonage Close, Charlton, Wantage, OX12 7HP
G4 BGM C Zeal, 20 Hurst Park, Midhurst, GU29 0BP
G4 BGP Clifford Barker, 45 Cuerdale Lane, Walton-le-Dale, Preston, PR5 4BP
GM4 BGS Sam Liddell, 49 Inchbrae Road, Cardonald, Glasgow, G52 3HA
G4 BGT M Staton, 30 Shaftesbury Avenue, Chandler's Ford, Eastleigh, SO53 3BS
G4 BGW I Wilson, Whitethorn, Sandhurst Lane, Gloucester, GL2 9NW
G4 BHC F Stevens, 11 Hen Wythva, Camborne, TR14 7XN
G4 BHD Trevor Goldsworthy, Trevarth, Atlantic Terrace, Camborne, TR14 7AW
G4 BHE B Macklin, 4 Bramdown Heights, Basingstoke, RG22 4UB
G4 BHJ M Fochtmann, 1 Chapmans Way, Over, Cambridge, CB24 5PZ
G4 BHL J Firth, 10 Ridgway Avenue, Darfield, Barnsley, S73 9DU
G4 BHP G Benwell, The Bid, Tunley, Bath, BA2 0DZ
G4 BHT Michael Hulands, 100 Avenue Road, Rushden, NN10 0SJ
GM4 BHU D Aitkenhead, 37/3 Cavalry Park Drive, Edinburgh, EH15 3QG
G4 BIA H Hood, 8 Fayre Meadow, Robertsbridge, TN32 5AU
G4 BIC J Clegg, 75 Patch Lane, Bramhall, Stockport, SK7 1HR
G4 BID W Boyd, 2 The Ramblers, Poringland, NR14 7QN
G4 BII David Williams, 3 Main Street, Pounden, Bicester, OX27 9AZ
G4 BIK P Mellor, 10 Greenfields, Earith, Huntingdon, PE28 3QH
G4 BIM P Bentley, Blakes Hill, Limerstone Road, Newport, PO30 4AE
G4 BIN N Long, Homedale, Bayford Hill, Wincanton, BA9 9LS
GW4 BIS A Davies, 12 Church St., Troedyrhiw, Merthyr Tydfil, CF48 4HD
GM4 BIT R Wilson, 5 Collins Drive, Loans, Troon, KA10 7HA
G4 BIX David Price, 34 Vanda Crescent, St. Albans, AL1 5EX
G4 BIY Michael Corbett, 6 Windgap Lane, Haughley, Stowmarket, IP14 3PA
G4 BIZ A Paxton, Cleveland House, Bartley Road, Southampton, SO40 7GP
G4 BJB Christopher Hurst, 28 Hengistbury Road, Barton-on-Sea, BH25 7LU
G4 BJC INTERNATIONAL SHORTWAVE LEAGUE c/o Arthur Kinson, 6 Uplands Park, Broad Oak, Heathfield, TN21 8SJ
G4 BJD Gary Overton, 14 Aylestone Drive, Hereford, HR1 1HT
G4 BJF B Marshall, 23 Sandgate Avenue, Birstall, Leicester, LE4 3HQ
G4 BJG P Smith, 11 Chatsworth Avenue, Clowne, Chesterfield, S43 4SR
G4 BJJ H Tickell, 26 Shear Brow, Blackburn, BB1 7EX
GI4 BJK Kenny Patterson, 1a Demesne Gate, Saintfield, Ballynahinch, BT24 7BE
G4 BJN D Harvey, 23 Lapwing Close, Hemel Hempstead, HP2 6DS
G4 BJO B Greeves, 65 Stowupland Road, Stowmarket, IP14 5AN
G4 BJP Simon Popek, 42 Victoria Road, Polegate, BN26 6DA
G4 BJS J Loose, Flat 30 Highbury Court, Howard Road East, Birmingham, B13 0RQ
G4 BJT M Ware, 20 Bath Road, Buxton, SK17 6HH
G4 BJX W Whatmore, 51 The Fairways, Sherford, Taunton, TA1 3PA
G4 BKA A Neaves, 8 East House Drive, Hurley, Atherstone, CV9 2HB
G4 BKB G Jessup, 68 Danes Road, Bicester, OX26 2LR
G4 BKE David Wright, 4 Wynne Close, Broadstone, BH18 9HQ
G4 BKF Thomas Howarth, 71 Ford Road, Wirral, CH49 0TD
GW4 BKG Stephen Emlyn-Jones, 26 Lime Tree Way, Porthcawl, CF36 5AU
G4 BKH Arthur Chorley, 354 Denton Lane, Chadderton, Oldham, OL9 8QD
G4 BKI Paul Evans, 15 Watch Knob Lane, Swannanoa Nc, United States, 28778
G4 BKO J Francis, 22 Earlswood Drive, Mickleover, Derby, DE3 9LN
G4 BKQ R Gubbins, 29 Meadow End, Gotham, Nottingham, NG11 0HP
G4 BKR W Taggart, Calle Zarauz 61, Urb. San Luis, Alicante, Spain, 3180
G4 BKS P Erkiert, 129 Cannock Road, Aylesbury, HP20 2AS
G4 BLD C Croucher, 13 Magnolia Way, Pilgrims Hatch, Brentwood, CM15 9QS
G4 BLE Stephen Cole, 101 Allt-yr-yn Road, Newport, NP20 5EF
G4 BLH Michael Crawshaw, 50 Kibble Grove, Brierfield, Nelson, BB9 5EW
G4 BLL P Burnett, 4 Park Avenue, Hyde, SK14 4JS
GM4 BLO George Milne, 65 Millburn Avenue, Clydebank, G81 1ER
G4 BLS P Appleby, Flat 14, Maryan Court, Hailsham, BN27 3DJ
G4 BLT R Sterry, 9 Finch Avenue, Wakefield, WF2 6SE
G4 BM Tom Searle, 2 Woolfall Terrace, Seaforth, Liverpool, L21 4PJ
G4 BMC D Barrell, 26 Yerville Gardens, Hordle, Lymington, SO41 0UL
G4 BMD Michael Hayes, Apartment 6 The Court, Dunboyne Castle, Dunboyne, Ireland, CO MEATH
G4 BMH Nicholas Brooks, Bridge Cottage, Glenside North, Spalding, PE11 3NJ
G4 BMK M Kerry, 2 Beacon Close, Seaford, BN25 2JZ
G4 BMM Paul Knight, 75 Ashcroft Road, Luton, LU2 9AX
G4 BMO D Cloke, Church Cottage, East Coker, Yeovil, BA22 9LY
G4 BMP R Sadler, 9 Meade King Grove, Woodmancote, Cheltenham, GL52 9UD
G4 BMQ David Harrop, 1 Edgecombe Crescent, Rowner, Gosport, PO13 9RD
G4 BMR D Riddle, 26 Avonfield Avenue, Bradford-on-Avon, BA15 1JE
G4 BMU Stephen East, 2 Linscott House, 64d Russell Road, Buckhurst Hill, IG9 5QE
G4 BMW C Pescod, 7 Brian Close, Chelmsford, CM2 9DZ
G4 BNB R Wynn, 48 Darnley Road, Woodford Green, IG8 9HY
G4 BNE R Herring, 96 St. Fabians Drive, Chelmsford, CM1 2PR
GW4 BNJ D Williams, 48 St. Hilary Drive, Killay, Swansea, SA2 7EH
G4 BNK W Wright, 27 St. Johns Road, Farnborough, GU14 9RL
G4 BNL R Morley, 63 Holt Park Crescent, Holt Park, Leeds, LS16 7SL
G4 BNM S Homans, 3 Hilton Mews, Bramhope, Leeds, LS16 9LF
G4 BNO M Ayling, 68 Littledown Avenue, Queens Park, Bournemouth, BH7 7AS
G4 BNP J Burgess, 11 Winters Lane, Ottery St. Mary, EX11 1AR
G4 BNS Alan Collinson, 30 Thornton Road, Pickering, YO18 7HZ
G4 BNT Gordon Moore, 3 Manor Park, Sheriff Hutton Road, Strensall, YO32 5TL
G4 BNW M Knight, 18 Friary Road, Abbeymead, Gloucester, GL4 5FD
G4 BNX I Middleton, 9 Bronthall Court, Kirk Close, Nottingham, NG9 5EZ
GM4 BOA Eric Margetts, 17 Farndale, East Kilbride, Glasgow, G74 4QS
G4 BOB A Wallwork, 4 Woodland View, School Lane, Chorley, PR6 8PJ
G4 BOF Peter Harry, 5 St. Michaels Avenue, Kingsland, Leominster, HR6 9QR
G4 BOH Christopher Cummings, Castle View, Childs Lane, Congleton, CW12 4TQ
G4 BOJ N Greenstreet, 223 Upperthorpe, Sheffield, S6 3NG
G4 BOL Ronald Fineman, 4 Sherbourne Avenue, Bradley Stoke, Bristol, BS32 8BB
G4 BON J Strutt, 163 Scalby Road, Scarborough, YO12 6TB
G4 BOO D Rumens, 3 Flecker Close, Thatcham, RG18 3BA
G4 BOP Paul Berwick, Beech Croft, West Hill, Ottery St. Mary, EX11 1UY
G4 BOQ John Hall, 15 Main Street, Greetham, Oakham, LE15 7NJ
G4 BOU J Chance-Read, 15 Garrard Way, Wheathampstead, St. Albans, AL4 8PE
G4 BOV A Horton, Martletts, 52 Lower Cookham Road, Maidenhead, SL6 8JZ
G4 BOZ A Brock, 1 Carpenter Drive, St. Leonards-on-Sea, TN38 9RX

**IMPORTANT NOTE**

**Revalidate licence to avoid revocation** – Ofcom has advised the Society that plans will be drawn up to revoke licences that have not been revalidated as required by the licence conditions. The quickest way to revalidate is to do so online via the Ofcom website: *https://services.ofcom.org.uk/* or by email: *amateur.validations@ofcom.org.uk* Ofcom staff are available to help, but please be patient during times of heavy workload.

G4 BP SCARBOROUGH ARS c/o M Day, 33 Ryndle Walk, Scarborough, YO12 6JT
G4 BPE A Evans, Fairfield Main St, Claypole, Newark, NG23 5BA
G4 BPJ Brian Stone, 12 Forbes Road, Newlyn, Penzance, TR18 5DQ
G4 BPN N Kerstein, 40 Davidson Close, Hythe, Southampton, SO45 6JT
G4 BPO PO RESEARCH CNT c/o Christopher Hoare, 16 Shrivenham Road, Highworth, Swindon, SN6 7BZ
G4 BPR Phillip Truran, 29 Queens Crescent, Marshalswick, St. Albans, AL4 9QG
G4 BPV Peter Barker, 2 Oriole Drive, Exeter, EX4 4SJ
G4 BQA M Morley, 8 The Ridings, Seaford, BN25 3HW
G4 BQB John Crocker, The Railway House, Stogumber, Taunton, TA4 3TR
G4 BQC B Makeham, 64 Benomley Road, Almondbury, Huddersfield, HD5 8LS
GM4 BQD R Muir, 9 Craigs Court, Torphichen, Bathgate, EH48 4NU
G4 BQF M Duce, 28 Thompson Avenue, Canvey Island, SS8 7TS
G4 BQH D Livsey, 18 Tollards Road, Countess Wear, Exeter, EX2 6JJ
GI4 BQI William Mccullough, 16 Ballylisk Lane, Portadown, Craigavon, BT62 3RN
G4 BQJ A Hill, 3 Cambrai Avenue, Warrington, WA4 6QU
G4 BQN Norman Marsden, 32 Chard Road, Drimpton, Beaminster, DT8 3RF
G4 BQR W Carmichael, 47 Neath Drive, Ipswich, IP2 9TA
G4 BQS B Prichard, The Gables, Wootton Lane, Canterbury, CT4 6RT
G4 BQV R Mullard, 46 Green Lane, Clanfield, Waterlooville, PO8 0JX
G4 BQW W Glover, Swallows Meadow Court, 33 Swallows Meadow, Solihull, B90 4PH
G4 BQY P Aburrow, 25 Hill Crescent, Worcester Park, KT4 8NB
G4 BRA BRACKNELL ARC c/o M Goodey, 62 Rose Hill, Binfield, Bracknell, RG42 5LG
G4 BRB A G Stewart, 121 William Street, Dalbeattie, DG5 4EE
G4 BRC KENT RAYNET GROUP c/o T Lundegard, Saxby, Botsom Lane, Sevenoaks, TN15 6BL
G4 BRF Ronald Mickleburgh, 85 Carey Park, Killigarth, Looe, PL13 2JP
G4 BRH J Sniadowski, 42 Milesmere, Two Mile Ash, Milton Keynes, MK8 8DP
G4 BRK Neil Whiting, Forge End, Garford, Abingdon, OX13 5PF
G4 BRL A Moore, 14 Heath Road, Ipswich, IP4 5SA
GM4 BRM A Long, 34 Thornly Park Drive, Paisley, PA2 7RP
GM4 BRN Kingdom ARS c/o Peter Merckel, 1 Mortimer Court, Dalgety Bay, Dunfermline, KY11 9UQ
GW4 BRS Barry ARS c/o Steven Trahearn, 148 Gladstone Road, Barry, CF62 8ND
G4 BRW M Gordon, 57 Taunton Road, Bridgwater, TA6 3LP
G4 BSA M Draper, The Wallow, Mount Road, Bury St. Edmunds, IP31 2QU
G4 BSC John Wells, Tredworth, Sunnyfield Lane, Cheltenham, GL51 6JE
G4 BSD David Hoose, Leonard Cheshire He, Oaklands, Garstang, PR3 1RD
G4 BSK M Rhind-Tutt, Oldfield, Moor Road, Bridgwater, TA7 9AR
G4 BSM S Grove, 31 Sheppard Way, Minchinhampton, Stroud, GL6 9BZ
G4 BSS John Spence, 4 Langford Lane, Burley in Wharfedale, Ilkley, LS29 7NR
G4 BSV Alan Cox, 175 Hillcrest, Weybridge, KT13 8AS
G4 BSW Nigel Hadley, 323 Canterbury Road, Margate, CT9 5JA
G4 BTE M Smith, 24 Lea Bank, Wolverhampton, WV3 9HN
GI4 BTG Brian Davidson, 106 Tudor Park, Newtownabbey, BT36 4WL
G4 BTI D Case, 8 Fawley Road, Reading, RG30 3EN
G4 BTK A Whitehouse, 690 Kingstanding Road, Kingstanding, Birmingham, B44 9SS
G4 BTN Christopher Brion, Passaford House, Hatherleigh, Okehampton, EX20 3LU
G4 BTS Mexborough & District ARS c/o Darrell Harrop, 7 Haythorne Way Swinton, Mexborough, S64 8SQ
G4 BTW Ian Jolly, 1 Llewelyn Drive, Bryn-y-Baal, Mold, CH7 6SW
G4 BTX N Monument, 6 Manor Road, Martlesham Heath, Ipswich, IP5 3SY
GM4 BUA Thomas Shepherd, 1 Spruce Gardens, Cupar Muir, Cupar, KY15 5WN
G4 BUB Peter Cox, 53 Boleyn Avenue, Enfield, EN1 4HR
G4 BUD B Underwood, 3 Jackson Close, Hampton Magna, Warwick, CV35 8SZ
G4 BUE Christopher Page, Highcroft Farmhouse, Gay Street, Pulborough, RH20 2HJ
G4 BUF G Jolley, 70 Hempstead Road, Holt, NR25 6DG
G4 BUH Michael Banahan, 18 Lynn Road, Ely, CB6 1DA
G4 BUI John Simpson, 19 Greenacres, Wetheral, Carlisle, CA4 8LD
GI4 BUJ J Sander, 696 Doagh Road, Newtownabbey, BT36 4TP
G4 BUL D Brudenell, 84 Porthcawl Green, Tattenhoe, Milton Keynes, MK4 3AL
G4 BUO D Lawley, 1515 High Road, London, N20 9PJ
G4 BUP P Moss, Amalrie, Franklin Road, Chelmsford, CM3 6NF
G4 BUS R Prosser, 18 Danes Way, Leighton Buzzard, LU7 3LS
G4 BUW Kevin Lamb, 35 Snode Hill, Beech, Alton, GU34 4AX
G4 BUX BUXTON RADIO AMATEURS c/o D CARSON, 21 Harris Road, Harpur Hill, Buxton, SK17 9JS
GW4 BUZ John Howells, Bronllys, Vicarage Road, Rhondda-Cynon-Taff, CF40 1HR
G4 BVB R Pridham, Victoria House, Chilsworthy, Cornwall, PL18 9PB
GM4 BVD A Sampson, 47 Muirend Road, Perth, PH1 1JD
G4 BVE John Clifford, Dippers Barn, Pool Quay, Welshpool, SY21 9JY
G4 BVF M Sinclair, 28 Roker Park Avenue, Ickenham, Uxbridge, UB10 8ED
G4 BVG A Young, 90 Pine Ridge, Carshalton, SM5 4QH
G4 BVH P Reed, 20 Greenfield Crescent, Brighton, BN1 8HJ
G4 BVI Glen Chenery, 44 Belstead Road, Ipswich, IP2 8AZ
GW4 BVJ R Mortimore, 76 Cwmfferws Road, Tycroes, Ammanford, SA18 3UA
G4 BVK Kenneth Stevens, 20 Coberley, Bristol, BS15 8ES
G4 BVM Charles Newman, 19 Clare Road, Peterborough, PE1 3DT
G4 BVP M Noble, Harbet, Shipley Road, Horsham, RH13 9BG
G4 BVQ P Kennedy, 18 Rushmere Avenue, Levenshulme, Manchester, M19 3EH
G4 BVS S Overend, Monticello, 73 Court Road, Plymouth, PL8 1BZ
G4 BVT R Osborne, Plas-Y-Bryn, 1 Belle Vue Gardens, Brecon, LD3 7NY
GM4 BVU N Macdonald, 3 Townhill Road, Hamilton, ML3 9UX
G4 BVV P Goben, 1 Petal Close, Maltby, Rotherham, S66 7HJ
G4 BVW A Reilly, 4 Moreton Drive, Poulton-le-Fylde, FY6 8ED
G4 BVY I Dixon, 5 The Howsells, Lower Howsell, Malvern, WR14 1AD
GM4 BVZ J Davidson, Rosemount, Whiting Bay, Isle of Arran, KA27 8PR
G4 BWB Robert Andrews, 2 Church Road, Frampton Cotterell, Bristol, BS36 2NA
G4 BWC BRADLEY WOOD SCOUT RADIO GROUP c/o M Bray, 2 Camborne Drive, Fixby, Huddersfield, HD2 2NF
G4 BWE S Price, 9 Spurcroft Road, Thatcham, RG19 3XX
G4 BWF Rodney Johnson, 27 Oakfield Avenue, Markfield, LE67 9WH
G4 BWG Stephen Marsh, 26 Station Road, Whyteleafe, CR3 0EP
G4 BWJ B Cook, 34 Boscombe Crescent, Bristol, BS16 6QR
G4 BWL Howard Morris, 2 Brickwall Lane, Curry Rivel, Langport, TA10 0NX
GI4 BWM J McCullagh, 2 Holestone Road, Doagh, Ballyclare, BT39 0SB
G4 BWN P Funnell, 6 Bolero Close, Wollaton, Nottingham, NG8 2BZ
G4 BWO D Tyler, 5 Brentry Avenue, Bristol, BS5 0DL
G4 BWP Frederick Handscombe, Sandholm, Bridge End Road Red Lodge, Bury St. Edmunds, IP28 8LQ
G4 BWR M Hildich, 7 Claverham Park, Claverham, Bristol, BS49 4LS
G4 BWV A Burchmore, 49 School Lane, Horton Kirby, Dartford, DA4 9DQ
G4 BWX S Egerton, 15 Hyde Road, Torrisholme, Morecambe, LA4 6NU
G4 BWY P Willcocks, 27 Manor Road, Barnet, EN5 2LE
GI4 BXB R Brown, Apartment 10, Anchor Watch, Donaghadee, BT21 0GA
G4 BXC Harold Pearce, 32 Marshall Road, Willenhall, WV13 3PB
G4 BXD Bernard Nock, 47 Oakfield Road, Kidderminster, DY11 6PL
G4 BXH D Hardy, Box 52831, Dubai, United Arab Emirates
G4 BXI C Godden, 84 Crescent Road, Ramsgate, CT11 9QZ
G4 BXQ A Pressley, 22 Springbank Avenue, Farsley, Pudsey, LS28 5LW
G4 BXS John Morris, Church Terrace Cottage, Dunkeswell, Honiton, EX14 4QZ
G4 BXY H Barker, 31 Briants Avenue, Caversham, Reading, RG4 5AY
G4 BXZ J Howell, 3 Gate Farm Road, Shotley Gate, Ipswich, IP9 1QH
GW4 BYA P Braham, 23 Gilfach y Gog, Penygroes, Llanelli, SA14 7RJ
G4 BYB R Penman, 9 Southall Avenue, Worcester, WR3 7LR
G4 BYD A Atkinson, 5 Ashfield Avenue, Skelmanthorpe, Huddersfield, HD8 9BW
G4 BYE T Miller, 4 Jessop Road, Stevenage, SG1 5NF
GM4 BYF P Bates, 10 Swanston Avenue, Edinburgh, EH10 7BU
G4 BYG Victor Lindgren, 143 Hull Road, Anlaby, Hull, HU10 6ST
G4 BYI Albert Wilson, 223 Waingaro Road, RD 1, Ngaruawahia, New Zealand, 3793
G4 BYL B Smith, 27 Thorneyholme Road, Accrington, BB5 6BD
G4 BYM B Buzzing, 1 Westmead Close, Droitwich, WR9 9LG
G4 BYO W Tee, 87 Higher Blandford Road, Broadstone, BH18 9AE
G4 BYR Ian Maslen, 9 Church Lane, Marsworth, Tring, HP23 4LX
G4 BYS Graham Warren, 96 Parkside Drive, Cassiobury, Watford, WD17 3BB
GM4 BYT R Cook, 132 Clachtoll, Lochinver, Lairg, IV27 4JD
G4 BYW J Lekesys, 4 Gleneagles Way, Fixby, Huddersfield, HD2 2NH
G4 BYY K Plumridge, 32 Hawkhurst Close, Southampton, SO19 9AW
G4 BYZ Christopher Mills, North Lodge, Margery Wood Lane, Tadworth, KT20 7BA
G4 BZA J Tyblewski, Field Farm Bungalow, Belchford, Horncastle, LN9 6LF
G4 BZB D Parsons, 27 St. Leodegars Way, Hunston, Chichester, PO20 1PE
G4 BZE P Bradley, Woodlands, Longdown, Exeter, EX6 7SR
G4 BZF Martin Reed, 1 The Cottages, Farm Lane, Plymouth, PL6 5RJ
G4 BZG R Smith, 17 Styrrup Road, Harworth, Doncaster, DN11 8LL
G4 BZI R Bracey, 7 Park Estate, Shavington, Crewe, CW2 5AW
G4 BZJ A Mitchell, 18 Malham Fell, Bracknell, RG12 7DU
G4 BZL D Simpson, Ivy Cottage, Princess Street, Leeds, LS19 6BS
G4 BZM Mike Edwards, 13 Lechmere Crescent, Malvern, WR14 1TY
G4 BZP F Partington, 21 East Road, Wymeswold, Loughborough, LE12 6ST
G4 BZR F Jordan, 16 Elterwater Crescent, Barrow-in-Furness, LA14 4PH
G4 BZS M Pasek, 10 Prospect Place, Norwood Green, Halifax, HX3 8QF
G4 BZU B Beaven, 7 Glamorgan Road, Up Hatherley, Cheltenham, GL51 3JF
G4 BZV N Barton, 147 Whinney Lane, New Ollerton, Newark, NG22 9TJ
G4 CAA NATS & CAA RS c/o S Rossi, 21 Rattigan Gardens, Whiteley, Fareham, PO15 7EA
GM4 CAB Stephen Reynolds, 39 Panmure St., Broughty Ferry, Dundee, DD5 2EU
G4 CAF David Hogg, Fairview, Dordale Road, Bromsgrove, B61 9JT
G4 CAJ Michael Farr, 23 Waterfall Way, Barwell, Leicester, LE9 8EH
G4 CAK Michael Scarlett, 44 Derwent Road, Linslade, Leighton Buzzard, LU7 2QW
GM4 CAM Donald Hamilton, 7 High Langside Holding, Craigie, Kilmarnock, KA1 5ND
GM4 CAQ Robert Miles, 15 Clark Avenue, Linlithgow, EH49 7AP
G4 CAT Nigel Schofield, Maen Llwyd-Tan Yr Allt, Llanllyfni, Caernarfon, LL54 6RT
GM4 CAU T Wratten, 89 Hilton Road, Aberdeen, AB24 4HX
G4 CAX D Borley, 95 Meadow Lane, Moulton, Northwich, CW9 8QQ
G4 CAY C Parker, 25 Meadow Dale, Chilton, Ferryhill, DL17 0RW
G4 CAZ J Lefever, 74 Ferneley Crescent, Melton Mowbray, LE13 1RZ
G4 CBA S Mulligan, 49 Springhead Avenue, Hull, HU5 5HZ
G4 CBD J Swanson, 9 Park House Gardens, Twickenham, TW1 2DF
GI4 CBG R Smyth, 58 Gilnahirk Road, Belfast, BT5 7DH
G4 CBL P Tomlinson, 55 Reldene Drive, Hull, HU5 5HS
G4 CBM G Blakeley, Stowe House, Preston Gubbals Road, Shrewsbury, SY4 3LY
G4 CBO D Aiken, 16 Woodland Gardens, North Wootton, King's Lynn, PE30 3PX
GJ4 CBQ Philip Daniells, Le Belon, 2 Clos Vallios, St. Lawrence, Jersey, JE3 1GP
G4 CBS N Swain, Hill Cottage, Camerton Hill, Bath, BA2 0PS
G4 CBT H Wall, 54 Little Harlescott Lane, Shrewsbury, SY1 3PZ
G4 CBW Anthony Horsfall, 60 Talke Road, Red Street, Newcastle, ST5 7AH
G4 CBY T Cooper, Lincolnshire House, Brumby Wood Lane, Scunthorpe, DN17 1AF
G4 CBZ A Mepham, 1a Grand Crescent, Rottingdean, Brighton, BN2 7GL
GW4 CC SWANSEA ARS c/o Roger Williams, 114 West Cross Lane, West Cross, Swansea, SA3 5NQ
G4 CCA Michael Fadil, 25 North Parade, Horsham, RH12 2DA
G4 CCB A Brown, 14 Holly Rise, New Ollerton, Newark, NG22 9UZ
G4 CCC C Young, 18 Wincroft Road, Caversham, Reading, RG4 7HH
G4 CCE R Angell, 214 Lower Higham Road, Chalk, Gravesend, DA12 2NN
G4 CCF Royal Signals ARS c/o John West, 9 Bainbridge Court, St. Helen Auckland, Bishop Auckland, DL14 9EJ
G4 CCH H Ling, 8 Spa Hill, Kirton Lindsey, Gainsborough, DN21 4NE
G4 CCI J Chapman, 7 Ravensthorpe Drive, Loughborough, LE11 4PU
GM4 CCN T Keats, Tigh na Luch, Skye of Curr Road Dulnain Bridge, Grantown-on-Spey, PH26 3PA
G4 CCQ M Stanton, 84 Forest Hill, Maidstone, ME15 6TH
G4 CCT S Hyman, 49 Southover, Woodside Park, London, N12 7JG
G4 CCY P Fagg, 113 Bute Road, Wallington, SM6 8AE
G4 CCZ P Simons, Westwood, Faris Lane, Addlestone, KT15 3DJ
G4 CDC E Morton, 6 Norfolk Avenue, Burton-upon-Stather, Scunthorpe, DN15 9EW
G4 CDD Denby Dale And District ARS c/o John Chappell, 49 Midway, South Crosland, Huddersfield, HD4 7DA
G4 CDF M Naylor, 6 Holsworthy Close, Lower Earley, Reading, RG6 3AH
G4 CDG A Davidson, Po Box Hm150, Hamilton Hmax, Bermuda
G4 CDH J Brade, 11 Old Farm Place, Ash Vale, Aldershot, GU12 5SF
G4 CDI G Boardman, 9 Byron Road, Weston-Super-Mare, BS23 3XQ
G4 CDJ Peter Jarrett, 36 Ferndale Road, Teignmouth, TQ14 8NH
G4 CDL F Mepham, Avenida Robleda 16/22, San Luis, Torrevieja, Spain, 3180
G4 CDN Richard Banester, Fairfield, Church Road, Norwich, NR12 9SA
G4 CDR C Winstanley, 3 Peter Street, Blackburn, BB1 5HQ
G4 CDW G Trickey, 3 Fairleigh Rise, Kington Langley, Chippenham, SN15 5QF
G4 CDX P Wheeler, 69 Waterside Road, Slyfield Green, Guildford, GU1 1RQ
G4 CDY Terry Giles, 37 Smitham Downs Road, Purley, CR8 4NG
G4 CDZ J Boden, 2 The Coppice, Whaley Bridge, High Peak, SK23 7LH
GM4 CEA R Mccracken, 6 Binnie Street, Gourock, PA19 1JS
G4 CEC P Knight, 26 Meadway, Harrold, Bedford, MK43 7DR
G4 CEI M Baker, 17 Whitehills Green, Goring, Reading, RG8 0EB
G4 CEJ Raymond Moore, 17 Somme Avenue, Flookburgh, Grange-over-Sands, LA11 7LJ
G4 CEK J Bird, 140 Meadvale Road, London, W5 1LS
G4 CEL S Hudson, Frekes Cottage, Moorside, Sturminster Newton, DT10 1HQ
G4 CEN D Davies, 35 Ruthellen Road, Chelmsford, United States, 1824
G4 CEP G Morris, 7 Manor Road, Sandy, SG19 1DT
G4 CES ROYAL AIR FORCE COSFORD ARC c/o Michael Farmer, Horton Brook Cottage, Horton, Shrewsbury, SY4 5NB
G4 CEU D Jarvis, Flat 1, Gunnery House, 2 Chapel Road, Southend-on-Sea, SS3 9SL
G4 CEX C Durant, 63 Ulleries Road, Solihull, B92 8DX
G4 CEY J Ball, 68 Swallows Court, Pool Close, Spalding, PE11 1GZ
G4 CFB K Henry, 80 Fernwood Rise, Westdene, Brighton, BN1 5EP
G4 CFC Llyr Gruffydd, 45 Maes Yr Hafod, Menai Bridge, LL59 5NB
G4 CFG P Arnold, 14 George Birch Close, Brinklow, Rugby, CV23 0NN
G4 CFH J Hill, 10 Albert Clarke Drive, Willenhall, WV12 5AU
G4 CFK Leigh Smith, Apartment 624, 12 Leftbank, Manchester, M3 3AG
G4 CFP William Bones, 22 Rotherhead Close, Horwich, Bolton, BL6 5UG
GI4 CFQ J Mcsweeney, 109 Twaddell Avenue, Belfast, BT13 3LG
G4 CFS Glyn Dodwell, 14 Cricklewood Close, Bishops Waltham, Southampton, SO32 1SJ
G4 CFV Rolande Hall, Pinewood Lodge, 16 Tullyvarraga Hill, Co Clare, Ireland
G4 CFW R Raven, 9 Southwood Close, Ferndown, BH22 9HW
G4 CFY A Nailer, 12 Weatherbury Way, Dorchester, DT1 2EF
G4 CFZ M Stevens, 3 Rip Croft, Portland, DT5 2EE
G4 CGA D Sellwood, 47 Waterhall Avenue, London, E4 6NA
G4 CGB D Tromans, 68 Brook St., Wall Heath, Kingswinford, DY6 0JG
G4 CGD A Richardson, 24 West House Close, Wimbledon, London, SW19 6QU
GW4 CGE W Dore, Sea Vista, Gelliswick, Milford Haven, SA73 3RS
G4 CGF Wladmiar Badz, Bottom Flat, 36 Luckington Road, Bristol, BS7 0US
G4 CGG Richard l'Anson, 87 Tranby Lane, Anlaby, Hull, HU10 7DT
G4 CGH Martin Davies, 2 Manor Close, Berrow, Burnham-on-Sea, TA8 2LN
G4 CGL J Miller, 29 Springhill Road, Wednesfield, Wolverhampton, WV11 3AW
G4 CGM M Duff, Clittaford Club, Moses Close, Plymouth, PL6 6JP
G4 CGO James Pollock, 71 Stevenson Street, Kew, Australia, 3101
G4 CGP P Wright, 4 Avill Way, Wickersley, Rotherham, S66 1DL
G4 CGR K Davies, High View, Alcester Road, Henley-in-Arden, B95 6BH
G4 CGU R Taylor, 23 Ridgacre Lane, Birmingham, B32 1EL
G4 CGV Colin Manklow, 37 Brittons Crescent, Barrow, Bury St. Edmunds, IP29 5AG
G4 CGW J Dunglinson, Blenheim, Willow Lane, Camberley, GU17 9DL
GW4 CGZ David Newman, 138 Twyn Carmel, Merthyr Tydfil, CF48 1PH
G4 CHD T Adams, 1 Francis Drive, Westward Ho, Bideford, EX39 1XE
G4 CHG P Ashton, 7 Conway Grove, Cheadle, Stoke-on-Trent, ST10 1QG
G4 CHH J Heathershaw, 29 Tranmere Park, Hessle, HU18 1QZ
G4 CHI P Robinson, Longcroft House, Longcroft Lane, Burton-on-Trent, DE13 8NT
G4 CHJ A Williams, 10 Olde Hall Road, Featherstone, Wolverhampton, WV10 7BB
G4 CHL P Howe, 6 Rue du Doyen Guyon, Aix en Provence, France, 13090
G4 CHM R Mcewan, Fifth Acre, Carr Lane, Alfreton, DE55 2DN
GM4 CHX James Kyle, 7 Fasaich, Strath, Gairloch, IV21 2DH
GU4 CHY R Allisette, Lilyvale House, Rue Des Houmets, Castel, Guernsey, GY5 7XZ
G4 CIA W Cooper, 20 Planton Way, Brightlingsea, Colchester, CO7 0LB
G4 CIB Brian Woodcock, The Larches, Poolhay Close, Gloucester, GL19 4NY
G4 CIC S Edmondson, 7 Browns Road, Bradley Fold, Bolton, BL2 6RQ
G4 CID R McClements, Eskdail, Newtown St. Boswells, Melrose, TD6 0RY
G4 CIG A Lincoln, 6 The Old Common, Chalford, Stroud, GL6 8JN
G4 CIJ J Chennells, 10 Lower Cippenham Lane, Slough, SL1 5DF
G4 CIO M Phillips, Chapel House, The Cross, Stonehouse, GL10 3TU
G4 CIZ A Wallbank, 1 Pollards Cottages, Clanville, Andover, SP11 9JD
G4 CJJ Michael Viner, 15 St. Anthonys Drive, Hedon, Hull, HU12 8NT
G4 CJK V Roney, 76 Hilton Lane, Great Wyrley, Walsall, WS6 6DT
G4 CJM J Alcock, 1 Alma St., Fenton, Stoke-on-Trent, ST4 4PH
G4 CJO A Mountifield, 6 Sawyers Close, Teg Down, Winchester, SO22 5JX
G4 CJP V Duffy, 2 Moor View Close, High Harrington, Workington, CA14 4NX
G4 CJR Colin Crick, 19 The Drive, Coulsdon, CR5 2BL
G4 CJT K Hughes, 4 Epsom Place, Cranleigh, GU6 7ET
G4 CJV A Kerton, 8 Fabian Drive, Stoke Gifford, Bristol, BS34 8XN
G4 CJY B Payne, 78 Carver Hill Road, High Wycombe, HP11 2UA
G4 CK Rowland Stellig, Geenstone, The Street, Shaftesbury, SP7 9PE
G4 CKB J Banester, Fairfield, Church Road, Norwich, NR12 9SA
G4 CKH G JACKSON, 86 Lloyds Avenue, Kessingland, Lowestoft, NR33 7TR
G4 CKK P Atkins, 60 Wentworth Way, Harborne, Birmingham, B32 2UX
G4 CKQ A Horne, 1 Upper Halliford Road, Shepperton, TW17 8RX
G4 CKS D Fitzgerald, 36 Vardens Road, London, SW11 1RH
G4 CKT R Gwynne, 17 Dorrington Close, Stoke-on-Trent, ST2 7BZ
G4 CKW Paul Hawes, 181 Eastfield Road, Southsea, PO4 9EL
G4 CKX Stephen Taylor, 5 Chiltern Avenue, Bishops Cleeve, Cheltenham, GL52 8XP
G4 CLA P Lindsay, The Barn, Main Street, Lutterworth, LE17 5HY
G4 CLB Chris Brown, 66 Denham Lane, Chalfont St. Peter, Gerrards Cross, SL9 0ES
G4 CLC D Lewis, Brandywine, Westbury, BA13 4NY

G4 CLD G Beavor, The Gables, Reading Road, Reading, RG7 3BU
G4 CLE Trevor Baker, 18 Prescott Avenue, Hufford, Ormskirk, L40 1TT
G4 CLF James Bryant, Hillhead Cottage, Calshot Road, Southampton, SO45 1BH
G4 CLG Stephen Whittingham, 18 Northcroft, Shenley Lodge, Milton Keynes, MK5 7AJ
G4 CLI David Lockwood, 75 Foxroyd Lane Estate, Dewsbury, WF12 0BB
G4 CLJ P Eccles, Inghams, The Town, Dewsbury, WF12 0QX
G4 CLK R Goodchild, Grey Stone House, Owmby Cliff Road, Market Rasen, LN8 2HL
G4 CLN P Redfern, 12 Wilbarn Road, Paignton, TQ3 2BN
G4 CLP J Harrison, 47 Mason Way, Padbury Wa, Australia, 6025
G4 CLR I Hewer, 23 Thoresby Avenue, Tuffley, Gloucester, GL4 0TD
G4 CLY Nigel Thompson, 6 Miena Way, Ashtead, KT21 2HU
G4 CMC P Carr, 73 Mendip Road, Yatton, Bristol, BS49 4HP
G4 CMG Thomas Milne, Lynwood, Clovelly Road, Hindhead, GU26 6RP
G4 CMH D Spendlove, 22 Green Bank, Harwood, Bolton, BL2 3NG
GM4 CMI Robert Campbell, 1 Gibraltar Terrace, Dalkeith, EH22 1EE
G4 CMK Richard Harker, 140 Victoria Road, Beverley, HU17 8PJ
G4 CML J Livesley, Rivendell, 71b Hilltop Road, Hitchin, SG5 3NS
G4 CMM C Pope, Silver Hill, Norwich Road, Great Yarmouth, NR29 5PB
G4 CMP P Lennon, 53 Rycot Road, Speke, Liverpool, L24 3TH
G4 CMQ David Stephens, 42 Radcliffe Drive, Ipswich, IP2 9QZ
G4 CMR E Beckett, Clifton, 3 Spring Grove, Leatherhead, KT22 9NN
G4 CMT RAYWELL PARK SCOUT ARS c/o Andrew Russell, 3 St. Nicholas Close, North Newbald, York, YO43 4TT
G4 CMU George Brind, 9 Becket Wood, Newdigate, Dorking, RH5 5AQ
G4 CMX Paul Rossiter, 48 Park Drive, Hucknall, NG15 7LU
G4 CMY Anthony Mann, 13 Rosedale Avenue, Stonehouse, GL10 2QH
G4 CMZ K Archer, 24 Willson Road, Littleover, Derby, DE23 1BZ
G4 CNH Les Carpenter, 166 Abbey View, Garsmouth Way, Watford, WD25 9DZ
G4 CNI P Geiger, Lloyd Mount, Howard Drive, Altrincham, WA15 0LT
GW4 CNL Gordon Goodfield, 10 Lewis Street, Church Village, Pontypridd, CF38 1BY
G4 CNW R Bold, 10 Haddon Close, Stanground, Peterborough, PE2 8LS
G4 CNX Frederick Ramsey, 75 Smithy Lane, Tingley, Wakefield, WF3 1QB
G4 CNZ D Allen, 344 Coventry Road, Hinckley, LE10 0NH
G4 COE D Smith, 54 Warrington Road, Leigh, WN7 3EB
G4 COJ C Roberts, 8 Oaklands Park Drive, Rhiwderin, Newport, NP10 8RB
G4 COL I Braithwaite, 28 Oxford Avenue, St. Albans, AL1 5NS
G4 COM J Compton, Aysgarth, Durley Brook Road, Southampton, SO32 2AR
G4 COQ J Holland, 19 Calder Avenue, Freckleton, Preston, PR4 1DN
G4 COR I Harvey, 50 Callow Hill Way, Littleover, Derby, DE23 3RL
G4 COS J Hansell, 87 Garratts Way, High Wycombe, HP13 5XT
G4 COT S Brett, 8 Pinewood Grove, Hull, HU5 5YY
G4 COV C Cardwell, 11 Manor Cottages, Heronsgate Road, Rickmansworth, WD3 5BJ
GM4 COX John Hood, 4 Murray Road, Law, Carluke, ML8 5HR
G4 CPA G Hanson, 11 Churchill Way, Cross Hills, Keighley, BD20 7DN
G4 CPC S Wright, 20 Stillwell Grove, Wakefield, WF2 6RN
G4 CPD G Knox, Glencairn, 6 Aldborough Road, York, YO51 9EA
G4 CPE A Turner, 7 Slate Hall, Sundon, Luton, LU3 3PY
G4 CPG M Howkins, 16 Beckett Court, Gedling, Nottingham, NG4 4GS
G4 CPI John Housden, 9 Mill Lane Close, Hogsthorpe, Skegness, PE24 5NJ
G4 CPL C Mcgee, Zafra, The Knapp, Bromyard, HR7 4BD
G4 CPM A Fielding, 95 Hillcrest, Weybridge, KT13 8AS
G4 CPN J Bird, The Vicarage, Exwick Hill, Exeter, EX4 2AQ
G4 CPQ N Scrogie, 46a Stoneleigh Broadway, Epsom, KT17 2HS
G4 CPV Robert Fisk, 16 Sterry Drive, Thames Ditton, KT7 0YN
G4 CPW P Wilson, 5 Pebble Close, Lowestoft, NR32 4DR
G4 CPY N Grassby, 11 Eider Close, Whetstone, Leicester, LE8 6YB
G4 CQA George Angell, 55 Golden Riddy, Leighton Buzzard, LU7 2RH
G4 CQH J Sperry, 50 Lochinver, Hanworth, Bracknell, RG12 7LD
G4 CQI A Lanfear, 120 Charlton Road, Kingswood, Bristol, BS15 1HF
GI4 CQL N Kyle, 197 Aghafad Road, Clogher, BT76 0XE
G4 CQM Derek Hilleard, Hazeldene, Bridgerule, Holsworthy, EX22 7EW
G4 CQN M Mawby, 61 Carter Drive, Beverley, HU17 9GL
G4 CQO S Burgess, Tretawn, Kite Hill, Ryde, PO33 4LG
G4 CQQ R Taylor, 8 Park Avenue, Markfield, LE67 9WA
G4 CQR David Wood, 49 Wolsey Crescent, Morden, SM4 4TD
G4 CQS Anthony Rowsby, 10 Echells Close, Bromsgrove, B61 7EB
G4 CQT D Price, Vine Cottage, Garth Road, Cwmbran, NP44 7AB
G4 CQV P Baldwin, 26 Ashford Road, Fulshaw Park, Wilmslow, SK9 1QE
G4 CQW A Lane, 21 Winterbourne Road, Poole, BH15 2ES
G4 CQX L Palfrey, C/O Po Box 314, Cyprus, XX99 1AA
G4 CQZ M Doig, Helenfa, Ystrad Road, Denbigh, LL16 3HE
G4 CRB W Oxley, Cloudshill, Well Lane, Cheltenham, GL51 1DD
G4 CRC CORNISH ARC c/o Kenneth Tarry, 38 Tresithney Road, Carharrack, Redruth, TR16 5QZ
G4 CRE D Rush, 8 Sheaf Place, Worksop, S81 7LE
G4 CRG K Burgin, The Pike Lock House, Eastington, Stonehouse, GL10 3RT
G4 CRH M Worvill, The Berwyns, Domgay Road, Llanymynech, SY22 6SL
G4 CRK R Sellman, 43 Mount Avenue, Stone, ST15 8LW
G4 CRM J Lennon, 107 Andrew Crescent, Waterlooville, PO7 6BG
G4 CRN A Hall, Woothill, Bear Lane, Tewkesbury, GL20 6BB
G4 CRP K Tyler, Pinfold House, 3 Pinfold Lane, Leeds, LS25 1HE
G4 CRS E.C A.R.C. c/o Alistair Mackay, 2 Highcliffe Grove, New Marske, Redcar, TS11 8DU
G4 CRT Lewis Kirby, 41 Woodville Road, Overseal, Swadlincote, DE12 6LU
G4 CRW A Holmes, 4 Castle Avenue, Datchet, Slough, SL3 9BA
G4 CSD P Hyde, 25 Merton Road, Basingstoke, RG21 5UA
G4 CSE M Lewis, 10 Kenmore Drive, Bristol, BS7 0TT
G4 CSI C Osborn-Jones, Hudnall House, Hudnall Lane, Berkhamsted, HP4 1QQ
G4 CSM D Chaplin, 35 Lanes End, Totland Bay, PO39 0AL
GI4 CSO J McCormack, 12 Glengoland Crescent, Dunmurry, Belfast, BT17 0JG
GI4 CSP G Robinson, 23 Old Moy Road, Dungannon, BT71 6PS
G4 CSV J Jackson, 43 Ambleside, Boundary Court, Stockport, SK8 1BA
GW4 CSY B Vickery, 6 Duffryn Close, St. Nicholas, Cardiff, CF5 6SS
G4 CSZ M Riley, 5 Dunstarn Gardens, Leeds, LS16 8EJ
G4 CTA Alfred Raymond Clewer, 6 Frensham Close, Stanway, Colchester, CO3 0HP
G4 CTC T Cann, Noahs Rough, Old Coach Road, Sevenoaks, TN15 7NR
G4 CTD C Vernon, 50 Copthall Road West, Ickenham, Uxbridge, UB10 8HS
G4 CTE Patrick Bradshaw, 13 Hill Top Road, Grenoside, Sheffield, S35 8PE
G4 CTI F Ashcroft, 7 Kings Ripton Road, Sapley, Huntingdon, PE28 2NU
G4 CTM R Barrett, 3 Bramshott Close, Hitchin, SG4 9EP
G4 CTT Timothy Thirst, Thirsts Farm, Happisburgh Road, Happisburgh, NR12 0RU
G4 CTU B Hitchins, 12 Parkland Avenue, Kidderminster, DY11 6BX
G4 CTV G Moss, Cysgod Y Graig, Cwmsymlog, [illegible], SY23 3EZ
G4 CTY Alexander Lightbody, 3 Elphicks Place, Tunbridge Wells, TN2 5NB
G4 CTZ Ian Cage, 334 Stockton Lane, York, YO31 1JW
G4 CUE W Pechey, Jays Lodge, Crays Pond, Reading, RG8 7QG
G4 CUG D Worsell, 9 Waterworks Cottages, Old Willingdon Road, Eastbourne, BN20 0AS
G4 CUI G Cook, 1 St. Albans Road, Fulwood, Sheffield, S10 4DN
G4 CUQ D Hughes, 30 Fuller Road, Dagenham, RM8 2TU
G4 CUR Philip Burton, 70 Station Road, Cholsey, Wallingford, OX10 9QB
GI4 CUV N Atkins, 38 Rosscoole Park, Belfast, BT14 8JX
GM1 CUX G Winchester, 33 Craigmount Avenue North, Edinburgh, EH12 8DL
G4 CVA John Wardle, 27 First Avenue, Bridlington, YO15 2JW
G4 CVC J Everist, 11 Redding Close, Dartford, DA2 6NB
G4 CVD P Petty, 41 Hensley Road, Bath, BA2 2DR
G4 CVF B Sheppard, 21 Lambourne Court, St. Johns Close, Uxbridge, UB8 2UL
G4 CVG W Bullock, 14 Saxon Drive, Rillington, Malton, YO17 8LZ
G4 CVK OLDSWINFORD HOSPITAL ARC c/o James Kimpton, 28 Clifton Street, Stourbridge, DY8 3XT
G4 CVM Robert Watson, 36 Abbots Close, Knowle, Solihull, B93 9PP
G4 CVN D Williams, Lyngarth, Leatherhead Road, Leatherhead, KT23 4RR
G4 CVO W Wyer, 11 Nether Close, Wingerworth, Chesterfield, S42 6UR
G4 CVS B Pearson, 8 The Pastures, Edlesborough, Dunstable, LU6 2HL
G4 CVU J Swingewood, 5 Blaze Park, Wall Heath, Kingswinford, DY6 0LL
G4 CVW Edward Law, 6 Blossom Hill, Erdington, Birmingham, B24 9DN
G4 CVX Russell Sims, 345 Blandford Road, Hamworthy, Poole, BH15 4HP
G4 CW NORTH KENT RS c/o F Connor, 134 Summerhouse Drive, Bexley, DA5 2ES
G4 CWA William Burton, 23 Purok 5, San Pedro Ii, Pampanga, Philippines
G4 CWB D Andrews, 100 Duchy Road, Harrogate, HG1 2HA
G4 CWC R BARRETT, Lumina, Bridegate Lane, Melton Mowbray, LE14 3QA
G4 CWE A Humm, 32 Layton Road, Hounslow, TW3 1YH
G4 CWG G Crossland, 32 Long Bridge Street, Llanidloes, SY18 6AR
G4 CWH C Smithers, 10 Grange Park, Bishops Stortford, CM23 2HX
G4 CWL Brian Fletcher, 5 Rue Des Charmes, Champagne le Sec, France, 86510
G4 CWM J Pickles, 111 Linden Avenue, Prestbury, Cheltenham, GL52 3DT
G4 CWP W Pevy, Brambletye, Ashstead Lane, Godalming, GU7 1SY
G4 CWU Barry Heppenstall, Gwelfor, Llanrhyddlad, Holyhead, LL65 4BG
G4 CWV F Parr, 5 Benenden Road, Wainscott, Rochester, ME2 4NU
G4 CWY A Sharp, Flat 3, The Manor, 6 Stourwood Avenue, Bournemouth, BH6 3PN
G4 CXE P Bolton, 93 Westfields, Narborough, King's Lynn, PE32 1SY
GM4 CXF J Thomson, 31 Teviot Place, Troon, KA10 7EE
G4 CXJ B Mussell, Stars End, 18 Orion Avenue, Gosport, PO12 4GL
G4 CXK R Evans, 74 Alexandra Street, Ebbw Vale, NP23 6JF
G4 CXL R Menday, Huf House, Horseshoe Ridge, Weybridge, KT13 0NR
GM4 CXM R James, 4 Pentland Place, Bearsden, C/O Ray James, Glasgow, G61 4JU
G4 CXP Derrick Dance, 18 Masons Court, Kelso, TD5 7NJ
G4 CXQ David Dyer, 26 Locking Road, Weston Super Mare, BS23 3DF
G4 CXT Malcolm Bell, Quebec Cottage, Curlew Green, Saxmundham, IP17 2RA
G4 CXW Graham Spencer, 17 Rockland Road, Bristol, BS16 2SW
G4 CXZ Andrew Thompson, 6 Ducks Walk, Twickenham, TW1 2DD
G4 CYA J Otley, 13 Cruise Road, Sheffield, S11 7EE
G4 CYB Frank Burnett, Herons Siege, Blundies Lane, Stourbridge, DY7 5HU
G4 CYC K Green, Ao-Te-Aroa, 12 Hill Road, Fareham, PO16 8LB
G4 CYF C Tully, Harmony, The Crescent, Clacton-on-Sea, CO16 0EP
G4 CYG D Darkes, 70 Braemar Road, Lillington, Leamington Spa, CV32 7EY
G4 CYI John Palfrey, Lower Trewince Farm, Newquay, TR8 4AW
G4 CYO Keith Robinson, 3 The Woodhouses, Patshull Road, Wolverhampton, WV6 7DU
G4 CYR Stephen Allen, 96a Churchway, Haddenham, Aylesbury, HP17 8DT
GI4 CYU H McIlroy, 28 Currans Brae, Moy, Dungannon, BT71 7SY
G4 CYY Clive Lewis, 54 Whelpley Hill Park, Whelpley Hill, Chesham, HP5 3RJ
G4 CYZ Lawrence Large, Captains Farmhouse, Streat Lane, Hassocks, BN6 8SB
G4 CZA K Newman, 2 Skys Wood Road, St. Albans, AL4 9NZ
G4 CZB John Cockrill, 28 Northampton Road, Harpole, Northampton, NN7 4DD
G4 CZH Eric Brindley, 150a Woods Lane, Derby, DE22 3UE
G4 CZK A Mercer, Penhale, Glen Gardens, Bideford, EX39 3PH
GI4 CZO G Mccomb, 1 Magheraboy Drive, Portrush, BT56 8GP
G4 CZP Richard Crossley, 2 Jewel View, 31 Downside, Ventnor, PO38 1AL
G4 CZR Clive Redfern, 6 PONT CROIX, Mellionnec, France, 22110
G4 CZU Philip Hadler, 30 Hillview Road, Whitstable, CT5 4HX
GI4 CZW C Corderoy, 3 The Limes, Drumlyon, Enniskillen, BT74 5NQ
G4 CZX I Godden, 163 Ringmor Road, Worthing, BN13 1DZ
G4 CZZ R Aggus, 68 Conifer Walk, Stevenage, SG2 7QS
G4 DAC D Squires, 91 Croham Valley Road, South Croydon, CR2 7JJ
G4 DAF G Walker, 56 Goodwin Road, Croydon, CR0 4EG
G4 DAM R Denoo, 32 Hayeswood Road, Stanley Common, Ilkeston, DE7 6GB
G4 DAP C Ison, 19 Grays Close, Chalgrove, Oxford, OX44 7TN
G4 DAQ W Silvester, 2 Tudor Close, Barton-le-Clay, Bedford, MK45 4NE
G4 DAT R Davidson, 5 St. Lucians Lane, Wallingford, OX10 9ER
GI4 DAV D Hart, 31 Downshire Road, Carrickfergus, BT38 7QD
G4 DAX David Smith, Red Roof, Goathland, Whitby, YO22 5AN
G4 DAY David Sawyer, 2 Blunts Wood Road, Haywards Heath, RH16 1NB
G4 DBD A Borland, 39 Green Lane, Willaston, Nantwich, CW5 7HY
G4 DBE J Clark, 20 Sandy Lane, Irby, Wirral, CH61 0HD
G4 DBF D Freeston, 20 Coningham Road, Whitley Wood, Reading, RG2 8QP
G4 DBG F Kneale, 60 Summertrees Road, Great Sutton, Ellesmere Port, CH66 2BJ
G4 DBM B Mcgennity, 46 St. Andrews Road, Boreham, Chelmsford, CM3 3BY
G4 DBN Neil Smith, Birch Tree House, Asselby, Goole, DN14 7HE
G4 DBP J Anderson, 9 Hopgrove Lane North, York, YO32 9TF
G4 DBQ B Roberts, 7 North Square, London, NW11 7AA
G4 DBR Chris Ewing, 130 Uttoxeter Road, Hill Ridware, Rugeley, WS15 3QX
G4 DBW Robert Hammond, 51 Poplar Drive, Herne Bay, CT6 7PY
G4 DBX Leslie Stubbs, The Cottage, Middlewich Road, Crewe, CW1 4DA
G4 DBY P Walker, 48 Whitefields Drive, Richmond, DL10 7DL
G4 DBZ D Martin, 12 South Park, Redruth, TR16 0AW
GI4 DCC W Chesney, 52 Taylorstown Road, Toomebridge, Antrim, BT41 3RT
G4 DCD Chris Stephenson, 8 Livingstone Close, Rothwell, Kettering, NN14 6HT
G1 DCE J Sketchley, 48 Coverdale, Whitwick, Coalville, LE67 5BP
G4 DCF M Booth, 15 Nether Royd View, Silkstone Common, Barnsley, S75 4QQ
G4 DCH Christopher Tucker, Georgian House, Greenhill, Sherborne, DT9 4EP
G4 DCI Philip HOPEWELL, 3 Hunts Orchard, Hathern, Loughborough, LE12 5HQ
G4 DCJ David Jarrett, 15 Groveside, East Rudham, King's Lynn, PE31 8RL
G4 DCK Michael Holliday, 1a Fairmile Gardens, Longford, Gloucester, GL2 9ED
GM4 DCL T Main, 1b Polton Road, Lasswade, EH18 1AB
G4 DCM P Rhodes, Parcela 1352, Calle De Zurbaran 8, Alicante, Spain, 3170
G4 DCP P Hull, Seymour Cottage, Forest Road, Waterlooville, PO7 6UA
G4 DCW D Walker, 70 High Street, Cranfield, Bedford, MK43 0DF
G4 DCX E Trickey, 53 Hollyguest Road, Hanham Green, Bristol, BS15 9NN
G4 DCY W Bradshaw, Flat 6, Heath Mount Hall, Ilkley, LS29 8JH
G4 DDB Enrico Connor, 8 Russell Street, Dover, CT16 1PX
G4 DDC DUNSTABLE DOWNS RC c/o P Seaford, 14 Nevis Close, Leighton Buzzard, LU7 2XD
G4 DDD R John, 32 Hundred Acre Road, Streetly, Sutton Coldfield, B74 2LA
G4 DDE L Rooke, 57 Lichfield Road, Brownhills, Walsall, WS8 6HR
G4 DDH R Hodges, 11 Inlands Rise, Daventry, NN11 4DQ
G4 DDI Colin Guy, 7 Herrick Court, Clinton Park, Lincoln, LN4 4QU
G4 DDK S Jewell, Blenheim Cottage, Falkenham, Ipswich, IP10 0QU
G4 DDL Michael Pemberton, 37 Woodmancott Close, Forest Park, Bracknell, RG12 0XU
G4 DDM R Finch, 1 Cherry Tree Cottage, Church Road, High Wycombe, HP10 8LN
G4 DDN G Leonard, 65 Qualitas, Bracknell, RG12 7QG
G4 DDP Richard Clark, 41 Avenue Road, Bexleyheath, DA7 4EP
G4 DDS A Dalton-Kirby, 45 Sutton Road, Kirk Sandall, Doncaster, DN3 1NY
G4 DDT Alan Ives, 24 Johnson Crescent, Heacham, King's Lynn, PE31 7LQ
G4 DDV Russell Bowman, 13 Wellington Road, St. Albans, AL1 5NJ
G4 DDX Ronald Pratt, 16 Thurlow Close, Stevenage, SG1 4SD
G4 DDY Maurice Fagg, 113 Bute Road, Wallington, SM6 8AE
G4 DDZ N Turner, Rose Cottage, Catshill Cross, Stafford, ST21 6LT
G4 DEA P Dunning, Cold Harbour, Bishop Burton, Beverley, HU17 8QA
GM4 DEK I Sturrock, 7 Knowetop Place, Roslin, EH25 9NR
G4 DEM D Walker, The Horseshoe Inn, 1 Horseshoe Court, Bristol, BS36 2FD
G4 DEN A Leigh, 6 Navenby Road, Wigan, WN3 5QJ
G4 DEO A Wallis, 5 Nancevallon, Higher Brea, Camborne, TR14 9DE
G4 DEP David Dabinett, Pentre Isaf, Llangyniew, Powys, SY21 0JT
G4 DEQ A Derrick, 4 Hillside Cottages, Barrow Street, Bristol, BS48 3RX
G4 DEU Anthony Fuge, 6 Haythorne Court, Staple Hill, Bristol, BS16 5QS
G4 DEV Stanley Newport, 18 Chacewater Crescent, Worcester, WR3 7AN
G4 DEW J Males, 49 Gunthorpe Road, Peterborough, PE4 7TN
GM4 DEX J Sharp, 72 Broom Road, Rimbleton, Glenrothes, KY6 2BQ
G4 DF B Slatter, 87 Windy Arbour, Kenilworth, CV8 2BJ
G4 DFA Tom Ellinor, 53 Hillside, Banstead, SM7 1HG
G4 DFB Donald Berry, 4 Holly Hill, London, N21 1NP
G4 DFC C Goldingay, 71 Kingham Close, Redditch, B98 0SB
G4 DFD K Bailey, 6 Stebbings Close, Hollesley, Woodbridge, IP12 3QY
G4 DFE William Raybould, 33 Roberts Green Road, Dudley, DY3 2BB
G4 DFG P Gibbs, 41 Mill Lane, Tettenhall Wood, Wolverhampton, WV6 8EZ
G4 DFI O Cross, 28 Garden Avenue, Bexleyheath, DA7 4LF
G4 DFJ Jeremy Klein, 5 Cranley Gardens, London, N10 3AA
G4 DFN S Widdett, Reabrook House, Mill Pool Lane, Kidderminster, DY14 8EZ
G4 DFO S Wainwright, 39 Ascot Road, Birmingham, B13 9EN
G4 DFP A Morecroft, 4 Arran Close, Bolton, BL3 4PP
G4 DFQ N Dear, Hollybush Cottage, Candy, Oswestry, SY10 9BA
G4 DFS Stephen Booth, 13 Milner Avenue, Penistone, Sheffield, S36 9DB
G4 DFT R Perrin, 131 Acacia Avenue, Ottawa, Ontario, Canada, K1M 0R2
G4 DFU Frank Skillington, 4 Haywood Court, Rainworth, Mansfield, NG21 0GX
G4 DFV Duncan Walters, 11 King George V Avenue, Mansfield, NG18 4ER
G4 DFX J Taylor, 26 Courthope Road, London, NW3 2LD
G4 DFY R Dedman, 2 Forest Villas, Long Mill Lane, Sevenoaks, TN15 8LQ
G4 DFZ Kenneth Knight, 61 Westbourne Road, Sutton-in-Ashfield, NG17 2FB
G4 DGB T Crute, 26 Runcorn, Sunderland, SR2 0BP
G4 DGF A Matthews, 14 Hardy Green, Wellington College, Crowthorne, RG45 7QR
GI4 DGI D Coyle, 16 Northland Avenue, Londonderry, BT48 7JN
G4 DGL E Mills, C/O Ac Clarke, 243 Barton Road, Cambridge, CB23 7BU
G4 DGM J Meddings, 100 Goldthorn Hill, Wolverhampton, WV2 3HU
G4 DGQ J Dussart, Seagarth, Cresswell, NE61 5JU
GM4 DGT William Stirling, 16 Shire Way, Alloa, FK10 1NQ
G4 DGW A Gagnon, 60 Woodruff Avenue, Hove, BN3 6PJ
G4 DHF David Johnson, Dean Cottage, Dowsby Fen, Bourne, PE10 0TU
G4 DHK Roger Stanleigh, Shallow Pool Bungalow, Looe, PL13 2ND
G4 DHL C Durnall, 143 Green Lane, Wolverhampton, WV6 9HB
GM4 DHN N Macleod, 54 Drum Brae South, Edinburgh, EH12 8TB
G4 DHT R Haverson, Kerri, Sunton, Marlborough, SN8 3DZ
G4 DHU Derek Spender, Rose Cottage, Huntsmans Lane, Sudbury, CO10 7JX
G4 DHV C Jones, 49 Newport Pagnell Road, Hardingstone, Northampton, NN4 6ER
G4 DHW P Mcelroy, 2 Donohue Lane, Manchester, United States, 6040
G4 DHY D Bird, 50 Cambridge Road, Crowthorne, RG45 7ER
G4 DIA B Powell, 1 The Heights, Market Harborough, LE16 8BQ
G4 DIC Richard Phipps, 4 Mill Court, Wells-next-The-Sea, NR23 1HF
G4 DIE I Dredge, 60 Springfield Close, Rudloe, Corsham, SN13 0JR
G4 DIG D Hine, 5 Star Road, Uxbridge, UB10 0QH
G4 DIH R Coates, 57 Dalebrook Road, Burton-on-Trent, DE15 0AB
G4 DII Alva Excell, 7 Kingslake Villas, Taunton Road, Bridgwater, TA6 6BW
GM4 DIJ J Howie, 36 Clermiston Road, Edinburgh, EH12 6XB
GM4 DIN Norman Burns, 24 Garioch Road, Inverurie, AB51 4RQ
G4 DIP B Chapman, 83 Courtenay Road, Great Barr, Birmingham, B44 8JB

UK Callsigns

**IMPORTANT NOTE**

**Revalidate licence to avoid revocation** – Ofcom has advised the Society that plans will be drawn up to revoke licences that have not been revalidated as required by the licence conditions. The quickest way to revalidate is to do so online via the Ofcom website: *https://services.ofcom.org.uk/* or by email: *amateur.validations@ofcom.org.uk* Ofcom staff are available to help, but please be patient during times of heavy workload.

| | | |
|---|---|---|
| G4 | DIS | Kenneth Mills, 7 Montgomery Close, Colchester, CO2 8SJ |
| G4 | DIT | R Siddall, 79 The Knoll, Palacefields, Runcorn, WA7 2UH |
| G4 | DIU | A Walker, 26 Sketchley Court, Nottingham, NG6 7DL |
| G4 | DIV | L Day, 86 Copperfield Road, Southampton, SO16 3NY |
| G4 | DIY | R Bennett, 17 Truro Close, St. Helens, WA11 9EL |
| GM4 | DIZ | H Lydall, 27 Calder Road, Edinburgh, EH11 3PF |
| G4 | DJB | Peter Roberts, 10 Tintagel Drive, Frimley, Camberley, GU16 8XQ |
| G4 | DJC | Richard Baker, 42 Rushleydale, Springfield, Chelmsford, CM1 6JX |
| G4 | DJD | D Carter, 43 Sturton Road, Sheffield, S4 7DE |
| G4 | DJJ | C Callicott, Clare House, Hepscott, Morpeth, NE61 6LT |
| G4 | DJK | D Corkill, 1a Hardie Crescent, Braunstone, Leicester, LE3 3DQ |
| G4 | DJP | John Chivers, 33 Hazelwood Road, Duffield, Belper, DE56 4DP |
| G4 | DJX | Alan Gray, 5 Meadow Close, St. Albans, AL4 9TG |
| G4 | DJY | Charles Steeden, Parklands, Chapel Road, Blackpool, FY4 5HT |
| G4 | DJZ | A Petrie, 3 Sharma Leas, Peterborough, PE4 6ZH |
| G4 | DKB | Paul Hubbard, 67 Knightbridge Walk, Billericay, CM12 0HL |
| G4 | DKC | Terence Smith, 9a Rubens Street, London, SE6 4DH |
| G4 | DKD | Edward Pascoe, 48 Bull Baulk, Middleton Cheney, Banbury, OX17 2QQ |
| G4 | DKH | K Hastie, 3 The Woodlands, Kings Worthy, Winchester, SO23 7QQ |
| G4 | DKM | R Lacken, 378 Wallisdown Road, Wallisdown, Bournemouth, BH11 8PS |
| G4 | DKP | J Etheridge, 14a Areley Court, Stourport-on-Severn, DY13 0AR |
| G4 | DKQ | J Loughlin, 17 Saffron Hill, Letchworth Garden City, SG6 4DB |
| G4 | DKV | M Pipes, Lower Hough Park, Hulland Ward, Ashbourne, DE6 3EN |
| G4 | DKX | Norman Cartwright, Little Bulmer Farm, Wiston Road, Colchester, CO6 4LT |
| G4 | DKZ | D Dodd, 8 Lindal Close, Dalton-in-Furness, LA15 8NL |
| G4 | DLA | L Turner, 160 Sandbach Road, Church Lawton, Stoke-on-Trent, ST7 3RB |
| G4 | DLC | B Bourne, 15 Rhos Fawr, Morfa, Abergele, LL22 9YH |
| G4 | DLD | M Garwood, Flat D, 6 St. Pauls Terrace, Northampton, NN2 6ET |
| G4 | DLG | Robert Bower, Tigh na Bruaich, Port Logan, Stranraer, DG9 9NE |
| G4 | DLP | R Stoddon, 34 Cromwell Road, Lancaster, LA1 5BD |
| G4 | DLT | R Hill, Rose Lodge, 35 Colne Fields, Huntingdon, PE28 3DL |
| GM4 | DLU | A Mccudden, 9 Dryburgh Lane, East Kilbride, Glasgow, G74 1BQ |
| G4 | DLY | P Collister, Flat 1, 57 South Parade, Wirral, CH48 0QQ |
| G4 | DLZ | W Taylor, 8 Park Avenue, Markfield, LE67 9WA |
| G4 | DMB | W Green, 38 Greenlands Way West, Sheringham, NR26 8XP |
| G4 | DMC | R Cleverley, 13 The Close, Melksham, SN12 6AG |
| G4 | DMF | John Wright, 10 Thorpes Road, Heanor, DE75 7GQ |
| G4 | DMG | D Griffiths, 1 Shepherds Down, Alresford, SO24 9PP |
| G4 | DMH | M Horton, 47 Checkstone Avenue, Doncaster, DN4 7JY |
| GM4 | DMI | C Armistead, 45 Swanston Gardens, Edinburgh, EH10 7DF |
| G4 | DML | xGraham Moore, Calvers Farm, Norwich Road, Thelveton, IP21 4NG |
| G4 | DMM | Ian Stinchcombe, 16 Reynell Road, Ogwell, Newton Abbot, TQ12 6YA |
| G4 | DMP | David Pratt, 11 Moorleigh Close, Kippax, Leeds, LS25 7PB |
| GM4 | DMQ | J Pritchard, 36 Craigleith Hill Crescent, Edinburgh, EH4 2JU |
| G4 | DMR | D Bevan, 3 Trem Y Foryd, Kinmel Bay, Rhyl, LL18 5JE |
| G4 | DMS | P Freeman, 1 Littleworth, Towcester, Towcester, NN12 8AL |
| G4 | DMT | J Southall, 4 Tye Lane, Willisham, Ipswich, IP8 4SR |
| G4 | DNA | S Green, 119 Oxford Road, Abingdon, OX14 2AB |
| G4 | DND | J Kennedy, Tor View Cottage, Postbridge, Yelverton, PL20 6SY |
| G4 | DNE | George Swaysland, 35 Keyhaven Road, Milford on Sea, Lymington, SO41 0QW |
| G4 | DNG | M Whitaker, 332 Milton Road, Cambridge, CB4 1LW |
| G4 | DNH | J Easteal, The Chalkers, Ermin Street, Hungerford, RG17 7TS |
| G4 | DNI | P Weldon, 7 Coverdale, Heelands, Milton Keynes, MK13 7LZ |
| G4 | DNJ | A Grisley, 7 Arnhill Road, Gretton, Corby, NN17 3DN |
| G4 | DNK | Mark Lelliott, Well Lane Corner, Lower Froyle, Alton, GU34 4LJ |
| GD4 | DNP | Robin Travis, 14 Elmstead Avenue, Wembley, HA9 8NX |
| GI4 | DNW | Matthew Getty, 34 Magheralave Park East, Lisburn, BT28 3BT |
| G4 | DNX | D Dyer, 57 Garrison Lane, Felixstowe, IP11 7RR |
| G4 | DOA | Anthony Mead, 11 Yarnton Close, Nine Elms, Swindon, SN5 5UQ |
| G4 | DOC | D James, 76 Grove Road, Harpenden, AL5 1HF |
| G4 | DOE | J Alford, 26 Edmunds Avenue, St. Pauls Cray, Orpington, BR5 3LF |
| GM4 | DOF | R Davidson, 3 Hillcrest Avenue, Kirkcaldy, KY2 5TU |
| G4 | DOL | P Atkins, 28 Victoria Place, Easton, Portland, DT5 2AA |
| GI4 | DOM | D Cafolla, 87 Stockmans Lane, Belfast, BT9 7JD |
| GW4 | DOO | A Kenyon, 6 Abbey Road, Port Talbot, SA13 1HA |
| G4 | DOQ | J Willis, Kilncroft, Broadlayings, Newbury, RG20 9TS |
| GM4 | DOZ | T Findlay, 37 Adamton Road North, Prestwick, KA9 2HY |
| G4 | DPA | G Austin, 16 Courtenay Road, Wantage, OX12 7DN |
| GM4 | DPC | T Wilson, School House, Boarhills, Fife, KY16 8PP |
| G4 | DPD | Colin Richardson, Domaine De Calcat, Route De Cates, la Sauvetat Sur Lede, France, 47150 |
| G4 | DPF | Iain Ross, 12 Manor Close, Buckden, St. Neots, PE19 5XR |
| G4 | DPH | Graham Jones, 7 The Avenue, Yatton, Bristol, BS49 4DA |
| G4 | DPJ | D Wear, 84 Hulham Road, Exmouth, EX8 3LA |
| GD4 | DPK | F Quayle, 1 Birch Hill Gardens, Onchan, Douglas, Isle of Man, IM3 4ET |
| G4 | DPO | A Nixon, 174 Davidson Road, Croydon, CR0 6DE |
| G4 | DPP | P Slade, Derlee House, East Lane, Abbots Langley, WD5 0QG |
| G4 | DPT | Timothy Upstone, 12 Glentworth Road West, Morecambe, LA4 4SZ |
| G4 | DPU | J Pilling, 223 Manchester Road, Accrington, BB5 2PF |
| G4 | DPV | Stanley Ford, 3 Hill View, Stoke-on-Trent, ST2 7AR |
| G4 | DPW | P Leslie-Reed, 43 Milehouse Lane, Newcastle, ST5 9JZ |
| G4 | DPZ | David Johnson, 96 Summerfields Avenue, Halesowen, B62 9NR |
| G4 | DQA | David Macken, 17 Culvercroft, Binfield, Bracknell, RG42 4DF |
| G4 | DQB | Geoff Wallis, 18 Kenilworth Grove, Newcastle, ST5 0LE |
| G4 | DQG | A Riley, 378 Hungerford Road, Crewe, CW1 6HD |
| GM4 | DQJ | R Grant, 31 Stormont Park, Scone, Perth, PH2 6SD |
| G4 | DQL | N Hall, 5 Brooklyn Crescent, Cheadle, SK8 1DX |
| G4 | DQN | G Spenceley, 168 Robin Hood Lane, Walderslade, Chatham, ME5 9LA |
| G4 | DQP | Vincent Lewis, Four Winds Cottage, Main Street, Brough, HU15 1RJ |
| G4 | DQQ | W Thomas, 64 West End, Silverstone, Towcester, NN12 8UY |
| G4 | DQT | S Taaffe, 56 San Marino, Muirhevna, Dundalk Eire, Ireland |
| G4 | DQW | J Krzymuski, 3079 Aberdeen Ct, Marietta, United States, 30062 |
| G4 | DQZ | W Ellis, Gillhams House, Gillhams Lane, Haslemere, GU27 3ND |
| G4 | DR | D Urquhart, 7 Padwell Lane, Bushby, Leicester, LE7 9PQ |
| G4 | DRA | S Chester, 63 Hawkshead Street, Southport, PR9 9BT |
| G4 | DRI | I Selby, 2 Ashley Close, Welwyn Garden City, AL8 7LH |
| GD4 | DRO | Thomas Brosnan, 168 Abbots Road, Edgware, HA8 0SA |
| G4 | DRR | G Spencer, Tyn Cae, Llanfwrog, Anglesey, LL65 4YL |
| G4 | DRS | J Wayman, Oak Tree Lodge, Redbridge Road, Dorchester, DT2 8BG |
| G4 | DRU | B plastow, 185 Allesley Old Road, Coventry, CV5 8FL |
| G4 | DRV | J Harris, Flat 36, Colonel Stevens Court, 10a Granville Road, Eastbourne, BN20 7HD |
| G4 | DRX | Dave Mckone, 12 Hawkshead Road, Knott End-on-Sea, Poulton-le-Fylde, FY6 0QE |
| G4 | DRZ | G Carney, 94 Combe Avenue, Portishead, Bristol, BS20 6JX |
| G4 | DSA | Gary Kemp, 4 Chapter Way, Monk Bretton, Barnsley, S71 2HP |
| G4 | DSC | O Boniface, 11 Holmefield Road, Ripon, HG4 1RZ |
| G4 | DSD | Ronald Woodman, 89a Western Way, Ponteland, Newcastle upon Tyne, NE20 9AW |
| G4 | DSE | Peter Zollman, 92 Well Lane, Curbridge, Witney, OX29 7PA |
| G4 | DSF | S Jones, 11 Alba Close, Middleleaze, Swindon, SN5 5TL |
| G4 | DSI | I Mcandrew, South Winds, Outrigg, St. Bees, CA27 0AN |
| G4 | DSN | John Dryden, 33 Old Station Road, Newmarket, CB8 8DT |
| G4 | DSO | Terry Hughes, 15 Boreland Road, Kirkcudbright, DG6 4HL |
| G4 | DSP | SPALDING & DISTRICT ARS c/o David Hall, 4 Burns Close, Peterborough, PE1 3JJ |
| G4 | DSQ | Roger Coombe, 150 Tean Road, Cheadle, Stoke-on-Trent, ST10 1LW |
| G4 | DSR | B Irwin, 97 Offerton Lane, Stockport, SK2 5BS |
| G4 | DSY | R Miller, 21 Woodstock Avenue, Sutton, SM3 9EG |
| G4 | DTB | M Bryan, 127 Ledbury Road, Hereford, HR1 1RQ |
| G4 | DTC | R Howgego, 39 Harestone Valley Road, Caterham, CR3 6HN |
| GM4 | DTH | P Dick, Napier House, 8 Colinton Road, Edinburgh, EH10 5DS |
| GM4 | DTJ | R Henderson, 2 Burdiehouse Avenue, Edinburgh, EH17 8AW |
| G4 | DTL | W Young, 56 Lincoln Road, Washingborough, Lincoln, LN4 1EG |
| G4 | DTM | P Marrable, 6 Piccadilly Close, Roselands, Northampton, NN4 8RU |
| G4 | DTP | D Pells, 6 Clarence St., Stonebroom, Alfreton, DE55 6JW |
| G4 | DTQ | D Gibbon, 90 Grosvenor Road, Prestatyn, LL19 7TS |
| G4 | DTT | W Brooks, 11 Lowther Grove, Garforth, Leeds, LS25 1EN |
| G4 | DTU | A Roberts, Brynlludw, Van, Llanidloes, SY18 6NP |
| G4 | DTW | S Parsons, 54 Furze Cap, Kingsteignton, Newton Abbot, TQ12 3TE |
| G4 | DTZ | R Holland, 1 Station Road, Castle Donington, Derby, DE74 2NJ |
| G4 | DUA | R Bearne, 35 Harnwood Road, Salisbury, SP2 8DD |
| G4 | DUB | Richard Harden, 23 Old Road, Barlaston, Stoke-on-Trent, ST12 9EQ |
| G4 | DUE | A Parker, 5 Geddes Close, Hawkinge, Folkestone, CT18 7QL |
| G4 | DUF | B Phillips, Woody Nook, Petworth Road, Godalming, GU8 5TU |
| G4 | DUI | P Wilson, 6 Hereford Road, Colne, BB8 8JX |
| G4 | DUJ | T Morley, 34 Bickerton Point, South Woodham Ferrers, Chelmsford, CM3 5YG |
| G4 | DUL | M Coburn, 16 Chapel Close, Toddington, Dunstable, LU5 6AZ |
| G4 | DUM | Victor Long, 2 B Pinnacle Hill, Bexleyheath, DA7 6AF |
| G4 | DUO | F Taylor, 7 Osterley Lodge, Church Road, Isleworth, TW7 4PQ |
| G4 | DUQ | P Keane, 45 Bramblewood Road, Worle, Weston-Super-Mare, BS22 9LW |
| G4 | DUT | D Elliott, 17 Baverstock Close, Chellaston, Derby, DE73 6ST |
| G4 | DUV | T Hynes, 2 Deynes Road, Debden, Saffron Walden, CB11 3LH |
| G4 | DUW | J Goldbey, Waylands Gate, St. Johns Road, New Milton, BH25 5SD |
| G4 | DUX | K Hampson, 9 North Crescent, Garlieston, Newton Stewart, DG8 8BA |
| G4 | DVA | T Stanway, 24 Fellbrook Lane, Bucknall, Stoke-on-Trent, ST2 8AQ |
| GW4 | DVB | B Price, 156 Parc Bryn Derwen, Llanharan, Pontyclun, CF72 9TX |
| G4 | DVG | J Douglas, 367 Wightman Road, London, N8 0NA |
| G4 | DVI | M Small, 15 Cannock Drive, Stockport, SK4 3JB |
| G4 | DVJ | R Hall, 12 Britannia Gardens, Westcliff-on-Sea, SS0 8BN |
| G4 | DVK | M Lang, 52 Gloucester Road, Burnham-on-Sea, TA8 1JA |
| G4 | DVM | M Cartwright, Seacue, 8 Adelaide Avenue, West Bromwich, B70 0SL |
| G4 | DVN | S Whalley, 1 Radley Way, Werrington, Stoke-on-Trent, ST9 0JN |
| G4 | DVP | P Hicks, 7912 Eagle Trail, Dallas, United States, TX 75238 |
| G4 | DVV | J Thomas, 57 Bourton Avenue, Patchway, Bristol, BS34 6EB |
| G4 | DVX | Roy Farr, 1 Dimple Lea, Warner Beach, South Africa, 4126 |
| G4 | DVZ | T Beaumont, 39 Meadow Road, Garforth, Leeds, LS25 2EN |
| G4 | DWC | D Cannings, 5 Rowan Close, Brackley, NN13 6PB |
| G4 | DWF | D Faulkner, 1 Westland, Martlesham Heath, Ipswich, IP5 3SU |
| G4 | DWM | Thomas Hunt, 25 Pike Purse Lane, Richmond, DL10 4PS |
| GW4 | DWN | Huw Richards, 112 Shelone Road, Neath, SA11 2NG |
| G4 | DWO | W Ingham, Westfield Villa, Westfield Villas, Wakefield, WF4 6EQ |
| G4 | DWR | Mary Molloy, 153 Palmdale Drive, Scarborough, Canada, MIT 1P2 |
| G4 | DWU | J Blowers, 28 Keld Close, Scarborough, YO12 6UF |
| G4 | DWX | Michael Smith, Tonn Marr, Bronybuckley, Welshpool, SY21 7NQ |
| G4 | DWZ | Peter Tucker, 8 Garland Close, Hemel Hempstead, HP2 5HU |
| G4 | DXB | Brian Chester, 147 Sanctuary Way, Grimsby, DN37 9RX |
| GI4 | DXK | W Gordon, 17 Ballyheather Road, Ballymagorry, Strabane, BT82 0BD |
| G4 | DXN | G Williams, 6 Nightingale Court, Leam Terrace, Leamington Spa, CV31 1DQ |
| G4 | DXO | Peter Jones, 40 Furze Road, Worthing, BN13 3BH |
| G4 | DXP | C Howells, 11 West Garth, Carlton, Stockton-on-Tees, TS21 1DZ |
| G4 | DXT | T Shaman, 3 Padshall Park, Bideford, EX39 3NE |
| G4 | DXW | Ronald Smith, 29 George Street, Peterborough, PE2 9PD |
| G4 | DXY | John Spendlove, 15 Grammer Street, Denby Village, Ripley, DE5 8PQ |
| G4 | DYC | Michael Cooke, 4 Geddes Way, Mattishall, Dereham, NR20 3RE |
| GI4 | DYE | Eamonn MacIntyre, 115 Bell Doo, Strabane, BT82 9QL |
| G4 | DYG | Peter Rich, 392 Doncaster Road, Stairfoot, Barnsley, S70 3RH |
| G4 | DYH | G Dunn, Croft Cottage, Innocence Lane, Ipswich, IP10 0PL |
| G4 | DYI | C Titheridge, 41 Church Walk, Worthing, BN11 2LT |
| G4 | DYJ | J Cope, Brookfield, Willoughby Drive, Spilsby, PE23 5EX |
| G4 | DYK | John Williams, 11 Heath Farm Road, Stourbridge, DY8 3AX |
| G4 | DYM | E Auty, Jesla, 5 Silverstone Way, Bristol, BS49 5ES |
| G4 | DYO | B McCartney, 123 Reading Road, Finchampstead, Wokingham, RG40 4RD |
| G4 | DYR | Roy Page, 28291 Misty Morning Lane, BELOIT, Ohio, United States, 44609 |
| G4 | DYT | D March, 86 Henley Avenue, Norton, Sheffield, S8 8JJ |
| G4 | DYU | B Hazlewood, 62 Cooper Avenue, Brierley Hill, DY5 3PE |
| G4 | DYV | B Whiting, Fourwinds, Buttercake Lane, Boston, PE22 9QX |
| G4 | DYY | R Mander, Meadowlands, Severn Lane, Welshpool, SY21 7BB |
| G4 | DZC | M Bayes, 25 Welby Gardens, Grantham, NG31 8BN |
| G4 | DZH | H Davies, 33 Sandown Road, Ocean Heights, Paignton, TQ4 7RL |
| G4 | DZJ | Duncan Paul, Chy Leweth, Vorvas, St. Ives, TR26 3HL |
| G4 | DZK | G Stocker, 8 Brook Drive, Astley, Manchester, M29 7HS |
| GM4 | DZM | Ian Shewan, Springbank, Distillery Road, Inverurie, AB51 0ES |
| G4 | DZS | A D Watson, 59 Merdon Avenue, Chandler's Ford, Eastleigh, SO53 1GD |
| G4 | DZU | Douglas Parker, 50 Rein Road, Tingley, Wakefield, WF3 1HZ |
| GM4 | DZX | Robert MacLeod, Skalvik, Auckengill, Wick, KW1 4XP |
| G4 | EAB | J Blackburn, 9 Pitchford Road, Albrighton, Wolverhampton, WV7 3LS |
| GM4 | EAF | PERTH & DIST AR c/o Alexander Hutton, 4 Linn Road, Stanley, Perth, PH1 4QS |
| G4 | EAG | Simon Ruffle, 39 Nightingale Avenue, Cambridge, CB1 8SG |
| G4 | EAJ | B Wignall, 119 Ounsdale Road, Wombourne, Wolverhampton, WV5 8BL |
| G4 | EAK | M Betts, 19 Maracas Cove, Western Australia, Australia, 6028 |
| G4 | EAN | Ian Brothwell, 56 Arnot Hill Road, Arnold, Nottingham, NG5 6LQ |
| G4 | EAQ | Andrew Churchley, 46 Birchdale Road, Appleton, Warrington, WA4 5AW |
| G4 | EAS | Christopher Ellery, 17 Wessex Way, Dorchester, DT1 2NR |
| GM4 | EAU | C Murray, 43 Malleny Avenue, Balerno, EH14 7EJ |
| GM4 | EAW | J Mathers, 36 Alexander Street, Dunoon, PA23 7EW |
| G4 | EAX | J GELL, 21 Maylands Avenue Breaston, Derby, DE72 3EE |
| G4 | EAZ | P Holliman, 17 Arundel Road, Tewkesbury, GL20 8AT |
| GD4 | EBA | David Kinrade, 16 Snaefell Road, Douglas, Isle of Man, IM2 6LX |
| G4 | EBE | George Harcourt, Hard Farm, Little Marsh Lane, Holt, NR25 7LL |
| G4 | EBF | G Reason, 37 Park End, Croughton, Brackley, NN13 5LX |
| G4 | EBG | Barry Meredith, 20 Kestrel Avenue, Thorpe Hesley, Rotherham, S61 2TT |
| G4 | EBI | A Hamm, 166 Sylvan Road, London, SE19 2SA |
| G4 | EBK | G Smith, 6 Fenby Close, Grimsby, DN37 9QJ |
| G4 | EBL | Ralph Whitwell, 14 Green Lane, Yarpole, Leominster, HR6 0BG |
| G4 | EBN | Michele Valente, Glenville, Abbey Road, Durham, DH1 5DQ |
| G4 | EBO | W Gibbs, 25 Belvedere Road, Exmouth, EX8 1QN |
| G4 | EBQ | N Talbot, 59 Heywood Avenue, Austerlands, Oldham, OL4 4AZ |
| GI4 | EBS | J Mcnerlin, 5 Rosendale Avenue, Limavady, BT49 0AE |
| G4 | EBT | David Taylor, 3 Crofters Drive, Cottingham, HU16 4SD |
| GM4 | EBX | P Hopkinson, The Coaches, Kingussie, PH21 1NY |
| G4 | EBY | G Head, 7 Partridge Way, High Wycombe, HP13 5JX |
| G4 | ECE | John Martin, 38 Parklands, Mablethorpe, LN12 1BY |
| G4 | ECF | G Penney, 8 Drake Park, Bognor Regis, PO22 7QG |
| G4 | ECO | B Palmer, Small Pine, Hedgerow Lane, Leicester, LE9 2BN |
| G4 | ECS | Anthony Wisbey, 12 Livingstone Road, Caterham, CR3 5TG |
| G4 | ECT | CHESHUNT DIS AR c/o J Crabbe, 47 Torrington Drive, Potters Bar, EN6 5HU |
| G4 | EDC | Robert Vane-Stobbs, 2 Wood Cottage, Walford Heath, Shrewsbury, SY4 3AZ |
| G4 | EDD | J Fletcher, 5 Hayeswood Road, Stanley Common, Ilkeston, DE7 6GB |
| G4 | EDG | Steven Taylor, 80 Nadder Park Road, Exeter, EX4 1NX |
| G4 | EDH | George Rose, 15 Pendennis Close, Winklebury, Basingstoke, RG23 8JD |
| G4 | EDK | A Ball, Tinten House, 2 Tinten Lane, Dorchester, DT1 3WP |
| G4 | EDM | W Concannon, 155 Walton Road, Sale, M33 4FS |
| G4 | EDN | Kelvin Currie, 37 Golden Ridge, Freshwater, PO40 9LF |
| G4 | EDQ | R Gulliver, 32 Lavender Close, Thornbury, Bristol, BS35 1UL |
| G4 | EDR | David Mappin, 13 Willow Close, Filey, YO14 9NY |
| G4 | EDW | P Eaton, Orchard House, Oxford Road, Winchester, SO21 3JG |
| G4 | EDX | John Fletcher, 69 Thackerays Lane, Woodthorpe, Nottingham, NG5 4HU |
| G4 | EDY | M Grindrod, 20 Castle Mead, Kings Stanley, Stonehouse, GL10 3LD |
| G4 | EDZ | W Russell, Saltwood House Cottage, Rectory Lane, Hythe, CT21 4QA |
| G4 | EEE | Alan Wood, Pezula, Brimpton, Reading, RG7 4TR |
| G4 | EEF | S Foster, 6 Webster Close, Hornchurch, RM12 6TF |
| G4 | EEH | Donald Greer, 5 Potto Close, Yarm, TS15 9RZ |
| G4 | EEJ | Robin Arak, 76 Halifax Road, Brighouse, HD6 2EP |
| G4 | EEL | Alan Cheshire, 5 Spion Kop, Oadby, Leicester, LE2 4QN |
| G4 | EEQ | F Robinson, 26 Winstanley Road, Little Neston, Neston, CH64 0UZ |
| G4 | EES | P Smith, Forge House & Stables, Whistley Road, Devizes, SN10 5TD |
| G4 | EET | Steven Greep, 5 Berkswell Close, Solihull, B91 2EH |
| G4 | EEV | D Warwick, Orchard Cottage, Colber Lane, Harrogate, HG3 3JR |
| G4 | EEZ | Martin Bath, 146 North Road, Hertford, SG14 2BZ |
| G4 | EFB | Clive McCloud, 34 St. Stephens Road, Portsmouth, PO2 7PG |
| G4 | EFD | D Stubbs, 63 Moss Lane, Wardley, Manchester, M27 9RD |
| G4 | EFE | Martin Peters, 11 Filbert Drive, Tilehurst, Reading, RG31 5DZ |
| G4 | EFG | D Watton, 247 Bloxwich Road, Walsall, WS2 7BB |
| G4 | EFH | A Johnson, Winchcombe, 3 Merrivale Lane, Ross-on-Wye, HR9 5JL |
| G4 | EFO | Michael Senior, 16 Cherry Tree Close, Billingshurst, RH14 9NG |
| GM4 | EFR | James Moar, Hansel, Stangergill Cres, Caithness, KW14 8UT |
| G4 | EFS | P Buzzing, 39 Kendlewood Road, Kidderminster, DY10 2XG |
| G4 | EFX | A Levitt, Strines Clough Farm, Blackshaw Head, Hebden Bridge, HX7 7JA |
| G4 | EFY | J Hurst, 12 Dukes Mead, Fleet, GU51 4HA |
| G4 | EGB | J Fletcher, 114 Scholes Park Road, Scarborough, YO12 6RA |
| GM4 | EGD | I Brownlie, 16 Border Street, Greenock, PA15 2EE |
| G4 | EGG | W Higginson, 7 Arundale, Westhoughton, Bolton, BL5 3YB |
| G4 | EGM | R Webster, 230 Huyton Lane, Huyton, Liverpool, L36 1TH |
| G4 | EGN | V Coles, 205 Farmers Close, Witney, OX28 1NS |
| G4 | EGQ | Peter Pennington, 6 Highland Close, Folkestone, CT20 3SA |
| G4 | EGR | David Barwood, 41 Wingfield Road, Bristol, BS3 5EG |
| GW4 | EGS | M Price, 19 Pencaerfenni Park, Crofty, Swansea, SA4 3SE |
| G4 | EGU | Philip Wolfe, 69 Alderney Road, Slade Green, Erith, DA8 2JH |
| GM4 | EGX | Roger Howard, 22 Kirkbrae Drive, Cults, Aberdeen, AB15 9RH |
| G4 | EGY | S Liptrott, 40 Mapperley Orchard, Arnold, Nottingham, NG5 8AG |
| GM4 | EHB | Edward Brockie, Ach na Shee, Breadalbane Street, Isle of Mull, PA75 6PX |
| G4 | EHD | W Tait, 51 Broadley Crescent, Halifax, HX2 0RL |
| G4 | EHG | C Bryan, 9 Brandy Hole Lane, Chichester, PO19 5RL |
| G4 | EHJ | Eric Wilby, 5 Matuku Street, Heretaunga, Wellington, New Zealand, 5018 |
| G4 | EHK | D Goulbourne, 6 Grovewood Drive, Appley Bridge, Wigan, WN6 9JF |
| G4 | EHN | J Axe, 5 Hillgate Place, London, W8 7SL |
| GM4 | EHP | I Petrie, Ugie Cottage, Victoria Road, Peterhead, AB42 4NL |
| G4 | EHQ | M Holley, 1 Willow Grange, Tilley Close, Rochester, ME3 9HS |
| G4 | EHR | D Kirton, 16 Silver Innage, Halesowen, B63 2PP |
| G4 | EHT | Wiliam Watson, 7 Darwin Close, Lichfield, WS13 7ET |
| G4 | EHW | Peterborough & District ARC c/o Ronald Smith, 29 George Street, Peterborough, PE2 9PD |
| G4 | EHX | Jeremy Fearn, 37 Bourne Square, Breaston, Derby, DE72 3DZ |
| G4 | EHY | F Greenough, 7 Carnforth Avenue, Hindley Green, Wigan, WN2 4LD |

G4 EIA M Wallis, 34 St. Aidans Close, Bristol, BS5 8RH
G4 EIC E Calvert, 163 Milner Road, Heswall, Wirral, CH60 5RY
G4 EIE D Francis, 10 Isgoed, Beaumaris, LL58 8HH
G4 EIG J Vickerstaff, 5 Luddington Road, Solihull, B92 0QH
G4 EII A Cunliffe, 35 Coultshead Avenue, Billinge, Wigan, WN5 7HT
G4 EIK Robert Currell, Brookside, Treworga, Truro, TR2 5NP
G4 EIL G Oughtibridge, 1 Lincoln Drive, Liversedge, WF15 7NJ
G4 EIM J Beaumont, 132 Hull Road, Woodmansey, Beverley, HU17 0TH
G4 EIN D Jones, Vine Tree Cottage, Mill Lane, Abergavenny, NP7 9SA
GD4 EIP C Baillie-Searle, 2 Marguerite Place, Foxdale, Douglas, Isle of Man, IM4 3HE
G4 EIV J Sondhis, 47 Emlyn Road, Horley, RH6 8RX
GM1 EIW J Dunnington, 4 Woodburn Way, Cumbernauld, Glasgow, G68 9BJ
G4 EIX D Whalley, 1 Lees Farm Drive, Madeley, Telford, TF7 5SU
G4 EIY D Thomas, 8 Whitehill Road, Barton-le-Clay, Bedford, MK45 4PF
GI4 EIZ W Stewart, 66 Ballyoillan Park, Belfast, BT14 8HD
G4 EJD C Bourne, 5 Brempton Croft, Hilderstone, Stone, ST15 8XL
G4 EJE J Brown, 2 Coriander Gardens, Littleover, Derby, DE23 2UB
G4 EJG Ian Adams, 9 Copper Hall Close, Rustington, Littlehampton, BN16 3RZ
G4 EJH Keith Middleton, 92 South Road, Portishead, Bristol, BS20 7DY
GM4 EJI G Lucas, 20 Myreside Gardens, Kennoway, Leven, KY8 5TR
G4 EJK David Reardon, 65 Blenheim Road, Caversham, Reading, RG4 7RP
G4 EJM M West, 19 Park Drive, Trentham, Stoke-on-Trent, ST4 8AB
G4 EJP P Sheppard, 220 Beckfield Lane, York, YO26 5QS
G4 EJQ R Clare-Noon, 91 Budshead Road, Higher St. Budeaux, Plymouth, PL5 2PJ
G4 EJS E Stainer, 66 Herbert Road, Bath, BA2 3PP
G4 EJU R Hands, 19 Orwell Road, Walsall, WS1 2PJ
G4 EJW N Perkins, 231 Burnham Road, Burnham-on-Sea, TA8 1LT
GM4 EJX Alastair Murray, 67 Carronvale Road, Larbert, FK5 3LH
G4 EKB Davd Epton, 61 Cartmel Drive, Dunstable, LU6 3PT
GM4 EKC J Mackinnon, 185 Deeside Gardens, Aberdeen, AB15 7QA
G4 EKD P Spelman, 68 Hardwick St., Tibshelf, Alfreton, DE55 5QH
G4 EKF S Sinclair, 29 Lealands, Lesbury, Alnwick, NE66 3QN
G4 EKG Michael Tittensor, 16 Durcott Road, Evesham, WR11 1EQ
GM4 EKI Grahame Marsh, Sorak, Swordale Road, Dingwall, IV16 9UZ
G4 EKJ C Shaw, 10 St. Helens Road, Harrogate, HG2 8LB
G4 EKL J Lorton, 2 Charlotte Close, Kirton, Newark, NG22 9LW
G4 EKM S Green, 133 Sevenoaks Drive, Sunderland, SR4 9NQ
G4 EKQ D Baylis, 2 Greswolde Park Road, Acocks Green, Birmingham, B27 6QD
G4 EKS R Holtham, 27 Peyton Close, Eastbourne, BN23 6AF
G4 EKT Hornsea ARC c/o Richard I'Anson, 87 Tranby Lane, Anlaby, Hull, HU10 7DT
G4 EKV M Lobb, 52 Ridge Park Avenue, Mutley, Plymouth, PL4 6QA
G4 EKW Michael Shaw, 50 White Road, Nottingham, NG5 1JR
G4 EKZ D Saul, 78 Ingleton Drive, Lancaster, LA1 4QZ
G4 ELA Robert Dawson, 2 Bertram Drive, Wirral, CH47 0LQ
G4 ELC G Keay, 9 Buchanan Avenue, Bournemouth, BH7 7AA
G4 ELG D Campbell, 12 Newton Close, Newton Solney, Burton-on-Trent, DE15 0SL
G4 ELI S Brown, Helford Lodge, The Fairway, Falmouth, TR11 5LR
G4 ELJ Duncan Clark, 24b Heatherdale Road, Camberley, GU15 2LT
G4 ELK A Lewis, 1 Springcroft, Parkgate, Neston, CH64 6SF
G4 ELL Robert James-Robertson, 8 Whittington Road, Worcester, WR5 2JU
G4 ELM Edward Jewell, 12 Patricks Copse Road, Liss, GU33 7DL
G4 ELP David Stockley, C/O Icom UK Ltd, Blacksole House, Herne Bay, CT6 6GZ
GI4 ELQ J Cushnahan, 34 Cornakinnegar Road, Lurgan, Craigavon, BT67 9JN
GM4 ELV D Dhuglas, 1 Micklehouse Road, Baillieston, Glasgow, G69 6TG
G4 ELW Ian Bontoft, 5 Kings Drive, Westonzoyland, Bridgwater, TA7 0HJ
G4 ELY R Panting, 124 Loddon Bridge Road, Woodley, Reading, RG5 4AW
G4 ELZ Jeffrey Pascoe, 3 Aller Brake Road, Aller, Newton Abbot, TQ12 4NJ
G4 EMA Ian Welburn, 33 Bowland Way, Clifton, York, YO30 5PZ
G4 EMB N Lockett, 18 Seagers, Great Totham, Maldon, CM9 8PB
G4 EMD R Edge, 4 Mortimer Hill Cleobury Mortimer, Kidderminster, DY14 8QQ
G4 EMH G Parsley, 7 Rowan Road, Martham, Great Yarmouth, NR29 4RY
G4 EMK G Parker, 6 Cedar Drive, Bourne, PE10 9SQ
G4 EML Colin Durbridge, 2 Send Villas, Sandy Lane, Send, Woking, GU23 7AP
G4 EMQ J Purchon, 19 Warburton, Emley, Huddersfield, HD8 9QP
G4 EMT J Tawn, 11 Wyke Road, Prescot, L35 5HL
G4 EMV P Johnson, 4 Chapel Lane, Blackwater, Camberley, GU17 9ET
G4 EMW E Warrington, 3 Long Meadow, Wigston, LE18 3TY
GM4 EMX Christopher Hall, 18 Sumburgh Crescent, Aberdeen, AB16 6WF
G4 ENA P Asquith, Well Cottage, The Green, Stroud, GL5 5LN
G4 ENB C Asquith, 36 Sunningdale, Luton, LU2 7TE
G4 ENC J Fenton-Coopland, 14 Chevril Court, Wickersley, Rotherham, S66 2BN
GM4 ENF A Fyffe, 39 Watts Gardens, Cupar, KY15 4UG
G4 ENJ Keith Hunter, 1 Markers Park, Payhembury, Honiton, EX14 3NL
G4 ENK P Kelly, 14 Manville Road, Wallasey, CH45 5AY
G4 ENL P Jewitt, Colenco Power Engineering, Taferstrasse 26, Baden, Switzerland, CH5402
GM4 ENN Alan Rae, 183 Campsie Street, Glasgow, G21 4XY
GM4 ENP John Johnston, 4 Lawhoad Road West, St. Andrews, KY16 9NE
G4 ENR K Brook, 154 Ridge Nether Moor, Swindon, SN3 6NF
G4 ENS Alan Morris, 6 Barrowby Gate, Grantham, NG31 7LT
G4 ENW Alan Gilbert, 47 Rayleigh Avenue, Eastwood, Leigh-on-Sea, SS9 5DN
G4 ENZ M Church, 2b Meadow Way, Churchdown, Gloucester, GL3 2AU
G4 EOA Timothy Strickland, 22a Branksome Road, St. Leonards-on-Sea, TN38 0UA
G4 EOB G Lawrance, 77 Bigland Drive, Ulverston, LA12 9PD
G4 EOC Richard Grunwald, 20 Sunny Bank Road, Batley, WF17 0LJ
G4 EOD D King, 87 Staverton Road, Werrington, Peterborough, PE4 6LY
G4 EOE R Everest, 2 Burley Road, Parkstone, Poole, BH12 3DA
G4 EOF S Lawrence, 34 Stanley Drive, Leicester, LE5 1EA
G4 EOG Alan Heritage, Badgers Sett, Draynes, Liskeard, PL14 6RY
G4 EOJ T Ilott, 45 Parkside, Snettisham, King's Lynn, PE31 7QF
GU4 EON M Allisette, Les Amballes Lodge, Les Amballes, St. Peter Port, Guernsey, GY1 1WU
G4 EOR P Stevens, 3 Ash Grove, Great Bromley, Colchester, CO7 7UQ
G4 EOT D Bussell, 26 Norbreck Crescent, Wigan, WN6 7RF
GM4 EOU John Smith, 6 Rodger Street, Cellardyke, Anstruther, KY10 3HU
G4 EOW P Baxter, Raka Lodge, Whitenap Lane, Romsey, SO51 5ST
G4 EOX Nigel Davenport, 26 Prairie Crescent, Burnley, BB10 1FU
G4 EPA J Pepper, 52 King Style Close, Orick, Northampton, NN6 7ST
G4 EPC J Stevens, 16 Brindles Field, Tonbridge, TN9 2YS
G4 EPD R Heeley, 263 Barnsley Road, Cudworth, Barnsley, S72 8JP
GW4 EPF J Pile, 5 Western Close, Mumbles, Swansea, SA3 4HF
G4 EPH John Splaine, 765 Wells Road, Bristol, BS14 0PR
GI4 EPK L Coyle, 14 Colby Avenue, Culmore, Londonderry, BT48 8PF
G4 EPL L Ward, 49 Edgewood Drive, Hucknall, Nottingham, NG15 6HY
G4 EPM N Lewis, 97 Orsett Road, Grays, RM17 5HA
G4 EPN A Wright, 34 Wallis Way, Stoney Stanton, Leicester, LE9 1DW
G4 EPU Michael Gray, 33 Claremont Drive, Pitsea, Basildon, SS16 4TL
G4 EPW L Goulding, 24 Lancaster Drive, Lydney, GL15 5SL
G4 EPX D Chater-Lea, 7 Greenside, Crowthorne, RG45 6EY
GI4 EQA E Mooney, 33 Piney Hill, Magherafelt, BT45 6PY
G4 EQC B Smith, 11 Tean Close, Burntwood, WS7 9JS
G4 EQD N Smith, 1 Park View, Messingham, Scunthorpe, DN17 3TT
G4 EQE D Smith, 7 Demesne Gardens, Martlesham Heath, Ipswich, IP5 3UA
G4 EQJ John Lee, 12 Gainsborough Close, Folkestone, CT19 5NH
G4 EQK M Hale, 9 Cramer Gutter, Oreton, Kidderminster, DY14 0UA
G4 EQL Michael Townsend, 39 Main Street, Fleckney, Leicester, LE8 8AP
G4 EQM W Evans, 9 Edwin Road, Didcot, OX11 8LG
GI4 EQN C Cupples, 20 Westland Avenue, Ballywalter, Newtownards, BT22 2TR
G4 EQP A York, 8 Granville Close, Hanham, Bristol, BS15 3TJ
G4 EQR David Evans, 103 Rhea Hall Estate, Highley, Bridgnorth, WV16 6JY
G4 EQS Kevin Dowson, Fyling Hall Lodge, Fylingdales, Whitby, YO22 4QN
G4 EQX J Mcilroy, 17 Brownsfield Road, Yardley Gobion, Towcester, NN12 7TY
GM4 EQY J Hately, 10 Crags Road, Paisley, PA2 6RA
G4 EQZ Keith Faulkner, 18 Milton Crescent, Talke, Stoke-on-Trent, ST7 1PF
GW4 ERB B SKIDMORE, Milton Oak House, Oxland Lane, Milford Haven, SA73 1LG
G4 ERD Alastair Hamilton, 2905 Nancy Creek Road Nw, Atlanta, United States, 30327
G4 ERF P Jones, Woodview, 10 Barrow Hill, Bury St. Edmunds, IP29 5DX
G4 ERH J Perry, C/O Wagons/Lits Apt168, San Antonio, Baleric Isles, Spain, ZZ9 9CO
G4 ERL E Lawley, 23 Briar Rigg, Keswick, CA12 4NN
GI4 ERM K Bones, 54 Derryvolgie Park, Lisburn, BT27 4DA
G4 ERN E Newsham, 31 Christopher Close, Yeovil, BA20 2EH
G4 ERO Colin Leonard, 24 Lower Road, Stuntney, Ely, CB7 5TN
G4 ERP Richard Marshall, 40 Evesham Road, Bishops Cleeve, Cheltenham, GL52 8SA
G4 ERQ Tony Birchall, 10 Avon Court, Alsager, Stoke-on-Trent, ST7 2BA
G4 ERR John Cummins, 5 Blenheim Orchard, Shurdington, Cheltenham, GL51 4TG
G4 ERS J Gamblen, Illfield, High Wych, Sawbridgeworth, CM21 0HX
G4 ERT H Marriott, 108 Leicester Road, Quorn, Loughborough, LE12 8BB
G4 ERV W Coombes, 33 Clarence Park Road, Boscombe East, Bournemouth, BH7 6LF
G4 ERW D Lurcook, 7 Bournes Place, Woodchurch, Ashford, TN26 3PD
G4 ERX R Elliott, No 6 Aphrodities Rock Village, 19 Lykourisson Street, Paphos, Cyprus, 8852
G4 ERY D Tyson, 12 Melbury Road, Woodthorpe, Nottingham, NG5 4PG
G4 ERZ Alan Wells, 38 Sextant Road, Hull, HU6 7BA
G4 ESG Derek Neal, 2 St. Margarets Avenue, Ashford, TW15 1DR
GI4 ESI S Mcclean, 14 Bamber Park, Ballymena, BT43 5HE
G4 ESL P Edwards, 14 Northfield Close, Caerleon, Newport, NP18 3EZ
G4 EST Charles Cartmel, 46 Lathom Drive, Rainford, St. Helens, WA11 8JR
G4 ESU Christopher Rouse, 86 Melton Avenue, Clifton, York, YO30 5QG
G4 ESY D Jackson, 16 Melrose Park, Beverley, HU17 8JL
G4 ETC Peter Keeble, 18 Shrubland Drive, Rushmere St. Andrew, Ipswich, IP4 5SX
G4 ETD Anthony Firth, 1 Wee Cottage, Crook, Kendal, LA8 8LH
G4 ETG D Humphries, 7 Shakespeare Drive, Upper Caldecote, Biggleswade, SG18 9DD
G4 ETI J Shaw, 20 Castleton Grove, Jesmond, Newcastle upon Tyne, NE2 2HD
G4 ETK C Bourne, 12 Sheepcoat Close, Shenley Church End, Milton Keynes, MK5 6JL
G4 ETM John Taylor, 2 Gap Crescent, Hunmanby Gap, Filey, YO14 9QJ
G4 ETN B Smith, Cleeve Valley Farm, Chipstable, Taunton, TA4 2QF
G4 ETO J Roach, 33 Pound Lane, Topsham, Exeter, EX3 0NA
G4 ETP T Pinch, 1 Fernhill Close, Ivybridge, PL21 9JE
G4 ETS Julian Forsey, 3 Orchard Leaze, Dursley, GL11 6HY
G4 ETW WILLENHALL & DISTRICT ARS c/o Malcolm Gibbons, 117 Ettingshall Road, Bilston, WV14 9XF
G4 ETX David Ludlow, 26 Bastion House, East Walls, Chichester, PO19 1QZ
G4 ETZ F Webb, 166 Glastonbury Road, Yardley Wood, Birmingham, B14 4DS
GW4 EUA G Smith, 13 Lapwing Close, Penarth, CF64 5GA
G4 EUC George Mendoza, 32 The Circuit, Cheadle Hulme, Cheadle, SK8 7LG
G4 EUF George Mayo, 28 Ring Fence, Shepshed, Loughborough, LE12 9HY
G4 EUG G Payno, 28 Pollards Drive, Horsham, RH13 5HH
G4 EUJ Robert Whiteley, 35 Wood Lane, Gedling, Nottingham, NG4 4AD
G4 EUK G Adcock, 2 Erringham Road, Shoreham-by-Sea, BN43 5NQ
G4 EUR Michael Tout, 25 Booth Rise, Northampton, NN3 6HP
G4 EUW B Keeling, 41 Regent Road, Brightlingsea, Colchester, CO7 0NN
G4 EUZ Durham and District ARS c/o Raymond McAleer, 44 Harvey Avenue, Durham, DH1 5ZG
G4 EVA Christopher Roberts, Rosemary, York Road, West Byfleet, KT14 7HX
G4 EVC K Chadwick, 1 Parklands, Southport, PR9 7HX
G4 EVD E Parry, 60 Hunters Forstal Road, Herne Bay, CT6 7DW
G4 EVE P Webster, The Old School, School Hill, Cirencester, GL7 2LS
G4 EVI J Howard, 127 Goldcroft, Yeovil, BA21 4DD
GW4 EVJ George Watson, 19 Kelvin Road, Clydach, Swansea, SA6 5JP
G4 EVK Ian Shepherd, 12 Watsons Lane, Harby, Melton Mowbray, LE14 4DD
GW4 EVL T Hopkins, 39 Glen Road, West Cross, Swansea, SA3 5PR
G4 EVN S Garrett, Church Farmhouse, The Street, Stowmarket, IP14 6LX
G4 EVP C McPartland, 55 Elliotts Lane, Codsall, Wolverhampton, WV8 1PG
G4 EVR Anthony Davies, 8 Half Acres, Bishop's Stortford, CM23 2QP
GM4 EVS D Johnstone, Sycamore House, Kirk Loan, Perth, PH2 6TD
G4 EVW G Sutton, 2 Orchard Close, Uttoxeter, ST14 7DZ
G4 EVX Ronald Price, 19 New Brighton Road, Sychdyn, Mold, CH7 6EF
G4 EVZ M Powle, 31 The Grove, Billericay, CM11 1AU
G4 EWE D Overton, Hawthorn Cottage, Chale Green, Ventnor, PO38 2JN
G4 EWI F Warner, 10 Brookfield Road, Walsall, WS9 8JF
G4 EWJ Brian Jordan, 42 Ben Nevis Road, Birkenhead, CH42 6QY
G4 EWK David Mellor, 18 Briar Close, Newhall, Swadlincote, DE11 0RX
GM4 EWL R Macleod, 22 Firthview, Dingwall, IV15 9PF
GM4 EWM Edward McLean, 21 Milnefield Avenue, Elgin, IV30 6EJ
G4 EWT S Mason, 15 Northfield Close, Bishops Waltham, Southampton, SO32 1FW
G4 EWV Ian Alexander McPherson, 12 Victoria Crescent, Ashford, TN23 7HL
G4 EWW T James, 2 The Green, Bottom Street, Southam, CV47 2EJ
G4 EWZ J Halford, The Anchor Inn, Chesterfield Road, Alfreton, DE55 7LP
G4 EXD Ian Marsh, 8 South Esk, Culgaith, Penrith, CA10 1QR
G4 EXE B Hope, Oriel, Moelfre, LL72 8HN
G4 EXF A Grindrod, Mullions, Church Street, Stonehouse, GL10 3HX
GI4 EXI Garry Crothers, 46 Culmore Point, Londonderry, BT48 8JW
G4 EXK T Bradbury, 18 Allsprings Drive, Great Harwood, Blackburn, BB6 7DN
G4 EXN L Dolman, 46 Norfolk Street, Norwich, NR2 2SN
G4 EXT J Corben, 65 Oatley Park Avenue, Oatley, New South Wales, Australia, 2223
G4 EXU D Fisher, 1 Francolin Close, Woodhaven, Natal South Africa, ZZ1 9FR
G4 EXZ R Fidler, 55 Sunnyvale Drive, Longwell Green, Bristol, BS30 9YQ
G4 EYA C Evans, 64 Boyd Avenue, Toftwood, Dereham, NR19 1ND
G4 EYB Dennis Fernie, Shepherds Close, Reigate Road, Leatherhead, KT22 8RD
G4 EYE A Free, Homeric, Harwich Road, Harwich, CO12 5JF
G4 EYJ DA Davies, 21 Russell Drive, Malvern, WR14 2LE
G4 EYM J Shardlow, 19 Portreath Drive, Allestree, Derby, DE22 2BJ
G4 EYN K Wright, 61 Albert Road, Chaddesden, Derby, DE21 6SH
G4 EYO C Carver, 8 Overlea Drive, Hawarden, Deeside, CH5 3HS
G4 EYR S Phillips, Millbank Cottage, Mill Half, Hereford, HR3 6HY
G4 EYT C Williams, 35 Heath Crescent, Norwich, NR6 6XF
G4 EYV P Skolar, Apartment 12, Fircroft, Ascot, SL5 9GF
G4 EYX P Davies, 14 Saville Road, Blackpool, FY1 6JP
G4 EZC H Spencer, 5 Carlyn Drive, Chandler's Ford, Eastleigh, SO53 2DJ
G4 EZE Jeremy Hinton, 7 The Glebelands, Crowborough, TN6 1TF
G4 EZF W Logan, 27 Shaw St., Mottram, Hyde, SK14 6LE
G4 EZG Martin Mac Gregor Of Stirling, Flat 4, Grange Lodge, The Grange, London, SW19 4PR
GM4 EZJ K Glendinning, 14 Craiglockhart Avenue, Edinburgh, EH14 1HW
G4 EZM E Green, 6 Downham Place, Blackpool, FY4 1QS
G4 EZN James Keeler, 67 Perne Avenue, Cambridge, CB1 3RY
G4 EZP I Melville, 3 Crescent Road, Benfleet, SS7 1JL
G4 EZQ C Doman, 6 Churnet Close, Bedford, MK41 7ST
G4 EZU W Peterson, 22 Weston Close, Potters Bar, EN6 2BQ
G4 EZW NEWPORT ARS c/o Margaret Hill, 13 Maesglas Grove, Newport, NP20 3DJ
G4 EZX D Titheridge, 2 The Oaks, Wilsden, Bradford, BD15 0HH
G4 EZZ D Andrews, 38 Rue Du Moulin Du Gue, St. Malo, France, 35400
G4 FAA Lawrence Atkinson, 56 The Spinney, Sidcup, DA14 5NF
G4 FAB S Fox, 16 The Teasels, Bingham, Nottingham, NG13 8TY
G4 FAD Richard Langford, Foxholes, Parsonage Farm, Hereford, HR4 8AJ
G4 FAE S Hodgetts, 79 Field Lane, Alvaston, Derby, DE24 0GQ
G4 FAH D Jones, 41 Sorrel Walk, Brierley Hill, DY5 2QG
G4 FAI Anthony Smith, 13 Old Library Mews, Norwich, NR1 1ET
G4 FAJ R Sadler, East Lynne, 202 Shire Oak, Walsall, WS9 9PD
G4 FAL Nicholas Totterdell, Moscar Cross House, Hollow Meadows, Sheffield, S6 6GL
G4 FAP R Painting, 15 Surrey Walk, Aldridge, Walsall, WS9 8JG
G4 FAQ David Jones, 7 Camrose Gardens, Pendeford, Wolverhampton, WV9 5RN
G4 FAS Geoffrey Royle, 56 Branksome Drive, Heald Green, Cheadle, SK8 3AJ
G4 FAT Nick Trollope, 2 Elm Villa Holly Green, Upton-upon-Severn, Worcester, WR8 0PD
GM4 FAU John Walker, West Lodge, Otterstone, Dunfermline, KY11 7HZ
G4 FAV A Bevan, 330 Stourbridge Road, Halesowen, B63 3QR
G4 FAW David Cutts, 51 Brook Lane, Felixstowe, IP11 7LG
G4 FAX Robert Macfie, 97 Chesford Road, Luton, LU2 8DP
G4 FAZ George Brownett, Apartment 264, The Crescent, Bristol, BS1 5JR
G4 FBB D Ellis, 17 Victoria Avenue, Yeadon, Leeds, LS19 7AS
G4 FBG D Shone, 6 Windlehurst Road, High Lane, Stockport, SK6 8AB
G4 FBI E Creasy, 16 Birchwood Close, Horley, RH6 9TX
G4 FBK M Kipp, 55 Hollybrook Mews, Yate, Bristol, BS37 4GB
G4 FBN B Neale, Badgers Sett, Harbertonford, Totnes, TQ9 7PU
G4 FBO W Fish, Chestnut House, 4 Jenkins Close, York, YO42 2PA
GM4 FBR I Dean, 33 Woodstock Road, Aberdeen, AB15 5EX
G4 FBQ J Hobley, 31 The Pippins, Glemsford, Sudbury, CO10 7PQ
G4 FBS Horndean and District ARC c/o Christopher Jacobs, Flat 33, The Lodge, Waterlooville, PO7 8BX
GM4 FBU H Macdougall, 17 Prospecthill Street, Greenock, PA15 4HH
G4 FBV D Vaughan, 6 Swallow Close, Felixstowe, IP11 9LR
G4 FBY B Surgey, Courtlands, Monks Corner, Saffron Walden, CB10 2RW
G4 FBZ William Kitching, 3 Prince Charles Crescent, Telford, TF3 2JX
G4 FCA J Haddon, 8 Oaklands, Cradley, Malvern, WR13 5LA
G4 FCB N Edwards, 40 Camden Street, Walsall, WS1 4HF
G4 FCC G Freeman, 12 The Haven, Beadnell, Chathill, NE67 5AW
G4 FCD R Girling, Heath Farmhouse, Cottisford, Brackley, NN13 5SN
G4 FCF W Wade, 11 St. Marys Road, Bluntisham, Huntingdon, PE28 3XA
G4 FCI Andrew Cullup, 201 Elm Low Road, Elm, Wisbech, PE14 0DF
G4 FCL P Lawson, 1 Beehive Cottages, Wickham Road, Fareham, PO16 7JF
G4 FCN Colin Coker, 46 Clarendon Road, Ipplepen, Newton Abbot, TQ12 5QS
G4 FCT J Gunn, 8 College Gardens, Hornsea, HU18 1EF
G4 FCU David Restall, 7 Medway Close, Skelton-in-Cleveland, Saltburn-by-The-Sea, TS12 2JZ
G4 FCV Robert JONES, 2 Pen-y-Cwarel Road, Wyllie, Blackwood, NP12 2HP
G4 FCX B Pearl, 66 Benfleet Road, Benfleet, SS7 1QB
G4 FCY I Smith, 17031 Los Cerritos, Los Gatos, United States, 95030
G4 FCZ Miles Thomas, The Old School, Church Lane, Lowestoft, NR32 5LL

UK Callsigns

**IMPORTANT NOTE**

**Revalidate licence to avoid revocation** – Ofcom has advised the Society that plans will be drawn up to revoke licences that have not been revalidated as required by the licence conditions. The quickest way to revalidate is to do so online via the Ofcom website: *https://services.ofcom.org.uk/* or by email: *amateur.validations@ofcom.org.uk* Ofcom staff are available to help, but please be patient during times of heavy workload.

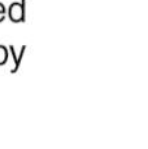

G4 FDA W Chapman, 34 Saxon Close, Oake, Taunton, TA4 1JA
G4 FDD J Livingston, 26 Dikelands Lane, Upper Poppleton, York, YO26 6JB
G4 FDF V Cunningham, 9 Lacon Road, Bramford, Ipswich, IP8 4HD
G4 FDG R Taylor, Porters Barn, Stewley LANE, Ilminster, TA19 9NJ
G4 FDI Simon Giles, Conifers, Kington Magna, Gillingham, SP8 5EW
G4 FDK R Palmer, 26 Silverstone Way, Congresbury, Bristol, BS49 5ES
GM4 FDM T Wylie, 3 Kings Crescent, Elderslie, Johnstone, PA5 9AD
G4 FDN P G McGuinness, 9 Farmdale Road, Carshalton, SM5 3NG
G4 FDP Robert Miller, 65 West Road, Oakham, LE15 6LT
G4 FDS John Ingram, NINE BARROW DOWN, SWANAGE, Swanage, BH19
GM4 FDT R Kerr, Rosskeen Bridge, Invergordon, IV18 0PR
G4 FDU R Mckinlay, 54 Barn Meadow Lane, Bookham, Leatherhead, KT23 3EY
G4 FDX I Offer, Southease, Balmer Lawn Road, Brockenhurst, SO42 7TT
G4 FEA C Beezley, 19 Beech Avenue, Claverton Down, Bath, BA2 7BA
G4 FEB D Emery, 424 Clement Avenue, Charlotte, North Carolina, United States, 8204
GM4 FEI A Marsden, 63 Carlogie Road, Carnoustie, DD7 6EX
G4 FEJ B Fawcett, 75 Ark Royal, Bilton, Hull, HU11 4BN
G4 FEM P Greatorex, 2 Briar Briggs Road, Bolsover, Chesterfield, S44 6SE
GM4 FEO John Gaughan, 12 Fernbank Avenue, Windygates, Leven, KY8 5FA
G4 FEQ Henry Stogdale, 14 Main Street, Ledston, Castleford, WF10 2AA
G4 FET T Ransom, 9 Lyndhurst Avenue, Hastings, TN34 2BD
G4 FEU T Southwell, 12 Chequer Lane, Upholland, Skelmersdale, WN8 0DE
G4 FEV David Whitty, 146 Avenue Road, Rushden, NN10 0SW
GD4 FFA R Harris, 98 Evelyn Avenue, Ruislip, HA4 8AJ
G4 FFC Martin Packer, Ricmaes Cottage, Chadwell End, Bedford, MK44 2AU
G4 FFE Leon Marriott, 94 Lyndhurst Road, Worthing, BN11 2DW
GM4 FFF GM Flora And Fauna c/o Jurij Phunkner, 7 Plenshin Court, Glasgow, G53 6QW
GI4 FFL James Finnegan, 15 Mossgreen, Richhill, Armagh, BT61 9JX
G4 FFM D Bailey, 10 Manor Road, Stutton, Tadcaster, LS24 9BR
G4 FFN C Baker, 78 Station Road, Whittlesey, Peterborough, PE7 1UE
GM4 FFP I Campbell, 35 Radernie Place, St. Andrews, KY16 8QR
G4 FFS D Hodge, 15 Buckland Close, Peterborough, PE3 9UH
G4 FFU C Forster, 48 Woolsington Gardens, Woolsington, Newcastle upon Tyne, NE13 8AR
G4 FFV I Forster, 48 Woolsington Gardens, Woolsington, Newcastle upon Tyne, NE13 8AR
G4 FFW M Betts, 56 Kingswood Road, Fallowfield, Manchester, M14 6RX
G4 FFX R Clear, 33 Cedars Road, Beddington, Croydon, CR0 4PU
G4 FFY Ray Howells, 16 Handel Walk, Tonbridge, TN10 4DG
G4 FGC D Cutts, 62 Forest Drive, Broughton, Chester, CH4 0QJ
G4 FGF J Drakeley, 31 Goldstar Way, Birmingham, B33 0YP
GI4 FGH William Tweedy, 11 Beechgrove Rise, Belfast, BT6 0NH
G4 FGJ G Mcgowan, 6 Caldecote Green, Upper Caldecote, Biggleswade, SG18 9BX
GM4 FGL Gerald Williams, 1 Fife Street, Keith, AB55 5EH
G4 FGO Alfred Oliver, Beaver Lodge, Dale Road, Brough, HU15 1HY
G4 FGP F Preece, 44 Broadmeadow, Aldridge, Walsall, WS9 8JA
G4 FGR S Porter, 138 Broad Lane, Essington, Wolverhampton, WV11 2RQ
GM4 FGS I Douglas, 47 Meadowpark, Ayr, KA7 2LW
G4 FGW C Hall, 49 Chilton Road, Richmond, TW9 4JB
G4 FGY J Maltby, Ingle Nook, Lenton Road, Grantham, NG33 4HA
GM4 FH Alexander Hamilton, 10/3 Fox Street, Edinburgh, EH6 7HN
GI4 FHB W McFaul, 9 Durham Park, Londonderry, BT47 5YD
G4 FHF John Walker, Roseberry, Owmby Road, Barnetby, DN38 6BD
G4 FHK T Knight, 3 Eaton Close, Rainworth, Mansfield, NG21 0AR
G4 FHN Robert Lovell, 16 North View, Staple Hill, Bristol, BS16 5RU
G4 FHQ Michael Hardy, 6 Apple Tree Close, Bromyard, HR7 4UL
G4 FHU Robert Pearson, 13 Mill Drove, Bourne, PE10 9BX
G4 FI N Seath, 6 Harvester Way, Sibsey, Boston, PE22 0YD
G4 FIA M Mucklow, 7 Burns Close, Newport Pagnell, MK16 8PL
G4 FIC D Pearson, Hope Cottage, Parkhouse, Monmouth, NP25 4QD
G4 FIE P Groom, 2a The Chestnuts, Countesthorpe, Leicester, LE8 5TL
G4 FIF David Cherrington, 4 Bloomfield Close, Wombourne, Wolverhampton, WV5 8HQ
G4 FIG B Callaway, 44 Grover Avenue, Lancing, BN15 9RQ
G4 FIH E Fernandes, 2c Northampton Park, London, N1 2PJ
G4 FIN K Mullaney, 5 St. Peters Way, Cogenhoe, Northampton, NN7 1NU
G4 FIQ G Clegg, 6 Vergette Court, Towngate West, Market Deeping, Peterborough, PE6 8DJ
G4 FIT J Chapman, 83 High St., Sutton, Ely, CB6 2NW
G4 FIV P Morley, Ash House, Germansweek, Beaworthy, EX21 5BP
GM4 FIZ A Murray, Woodhouse, Mount High, Balblair, IV7 8LH
G4 FJB J Dodd, 7 Hornbrook Grove, Solihull, B92 7HH
G4 FJC Richard De La Rue, Linden Lea, Balls Chase, Halstead, CO9 1NY
G4 FJF Michael Thacker, Kings Walden, 25 London Road, Dover, CT17 0SF
G4 FJH D Powell, 18 Exley Close, Warmley, Bristol, BS30 8YD
GD4 FJI R Allison, 20 Droghadfayle Park, Port Erin, Isle of Man, IM9 6EP
G4 FJJ D Bayliss, 20 Midhill Drive, Rowley Regis, B65 9SD
G4 FJK Timothy Hugill, Swandhams House, Sampford Peverell, Tiverton, EX16 7ED
G4 FJP J Perry, 108 Elm Road, New Malden, KT3 3HP
G4 FJT C Cuthbert, 44 Towse Close, Clacton-on-Sea, CO16 8US
G4 FJW C Hook, 11 Battlesmere Road, Cliffe Woods, Rochester, ME3 8TR
G4 FJX I Perera, 1 Francis Road, Perivale, Greenford, UB6 7AD
G4 FKA G Plucknett, 9 Oakwood Gardens, Coalpit Heath, Bristol, BS36 2NB
GM4 FKD D Smillie, Muir House, Daviot, Inverness, IV2 5ER
G4 FKE C White, 111 Waterbeach Road, Slough, SL1 3JU
G4 FKG G Kirk, 124 Star Road, Peterborough, PE1 5HF
G4 FKH Gwyn Williams, 21 Borda Close, Chelmsford, CM1 4JY
G4 FKI David Thorpe, 70 Willow Way, Ampthill, Bedford, MK45 2SP
G4 FKP B Tarry, 6 Beech Gardens, Rainford, St. Helens, WA11 8DJ
G4 FKQ M Barnwell, 77 Elmfield Road, Peterborough, PE1 4HA
G4 FKR ROBERT HAMMOND, 3 Rutledge Drive, Littleton, Winchester, SO22 6FE
G4 FKU K Salter, Alton, 12 Perinville Road, Torquay, TQ1 3NZ
G4 FKX Roger Abel, 23 Edward Gardens, Wickford, SS11 7EH
G4 FKY R Sharpe, 1 Park Copse, Horsforth, Leeds, LS18 5UN
GI4 FLG K Mayne, 8 Grandmere Park, Bangor, BT20 5RF
G4 FLM F Crofts, 43 Broadlands Drive, East Ayton, Scarborough, YO13 9ET
GM4 FLP I Strachan, 238 Coupar Angus Road, Muirhead, Dundee, DD2 5QN
G4 FLR Daniel Tanner, 4 Duckpitts Cottages, Bramling, Canterbury, CT3 1LY
G4 FLS A Snow, 1a Park Avenue, Longlevens, Gloucester, GL2 0DZ
G4 FLW F White, Delamain, Bracken Rise, Paignton, TQ4 6JU
GM4 FLX A Lovegreen, 16 Grahams Avenue, Lochwinnoch, PA12 4EG
G4 FLY Garry Haynes, 39 Zinzan Street, Reading, RG1 7UG
G4 FMA K Fraser, Broomfields, Courtlands, Uckfield, TN22 3LS
GD4 FMB David Taylor, Burnt Mill House, Mount William, Isle of Man, IM2 4PE
G4 FMC M Constable, 2 The Banks, Long Buckby, Northampton, NN6 7QQ
G4 FMI F Connor, 29 Parkdale Road, Paddington, Warrington, WA1 3EN
G4 FMJ L Cooke, 23 Widecombe Road, Stoke-on-Trent, ST1 6SL
G4 FMM T Walsh, 106 Westgate, Elland, HX5 0BB
G4 FMO Colin Palmer, 29 Paget Rise, Abbots Bromley, Rugeley, WS15 3EF
G4 FMQ H Charlesworth, 195 Fylde Road, Southport, PR9 9XZ
G4 FMY D Larsen, C/O Salbu (Pty) Ltd, Private Bag X 2352, Wingate Park, South Africa, 153
G4 FNC L Harper, Three Oaks, Braydon, Swindon, SN5 0AD
G4 FND D Yeates, 61 Martins Hill Lane, Burton, Christchurch, BH23 7NW
G4 FNG R Walker, South Moor Farm, Langdale End, Scarborough, YO13 0LW
G4 FNI K Nichols, 11 Tregonwell Road, Bournemouth, BH2 5NR
G4 FNJ Paul Fuller, 4 Whitworth Road, Minehead, TA24 8EB
G4 FNK Andrew Jackson, 6 Blandys Hill, Kintbury, Hungerford, RG17 9UE
G4 FNL Graham Bubloz, 42 Hillcrest, Westdene, Brighton, BN1 5FN
G4 FNO G Lloyd, 15 Budden Crescent, Caldicot, NP26 4PP
G4 FNP J Guite, 15 Marlborough Avenue, Falmouth, TR11 2RW
G4 FNQ C Wedgbury, 32 Cloverdale, Stoke Prior, Bromsgrove, B60 4NF
G4 FNR D Rabone, 6 Cranwell Grove, Kesgrave, Ipswich, IP5 2YN
GI4 FNU M McDowell, 50 Dunraven Parade, Belfast, BT5 6BT
G4 FNZ D Bannister, 7 Sudeley Close, Malvern, WR14 1LP
G4 FOB John Samuels, 25 Oxenpill, Meare, Glastonbury, BA6 9TQ
G4 FOC FIRST CLASS C.W OPERATORS CLUB c/o Michael Puttick, 21 Sandyfield Crescent, Cowplain, Waterlooville, PO8 8SQ
G4 FOD T Yeomans, 15 Turner Road, Woodfield, Dursley, GL11 6LT
G4 FOH Stephen Foote, 14 High Street, Chrishall, Royston, SG8 8RP
GW4 FOI John Doyle, 54 Bryncatwg, Cadoxton, Neath, SA10 8BG
G4 FOL J Bell, 60 Queens Close, West Moors, Ferndown, BH22 0HN
GW4 FOM R Rowles, 37 Vincent Court, Vincent Road, Cardiff, CF5 5AQ
G4 FON R Goff, Spring Fields, Bayswater Road, Oxford, OX3 9RZ
G4 FOR D Hawkes, 19 Taj Court, Ottawa, Canada, K1G 5K7
G4 FOS I Gilmore, 4 Borton Road, Blofield, Norwich, NR13 4RU
G4 FOT H Exley, 16 Croft Street, Horncastle, LN9 6BE
G4 FOW R Strangeway, 88 Old Manor Way, Portsmouth, PO6 2NL
G4 FOX Melton Mowbray ARS c/o G Mason, 120 Scalford Road, Melton Mowbray, LE13 1JZ
G4 FOY K Scott, 20 Tower Street, Alton, GU34 1NU
GM4 FOZ David Moodie, 1 Lageonan Road, Grandtully, Aberfeldy, PH15 2QY
G4 FPA J Shorthouse, 20 Boxgrove Road, Sale, M33 6QW
G4 FPB C Roper, 3 Dorset Road, Wallasey, CH45 5DB
G4 FPE G Butterfield, 13 Windsor Walk, Batley, WF17 0JL
G4 FPG R Hawke, Basque Close, Hastingleigh, Ashford, TN25 5JB
G4 FPI B Wood, 193 Robin Way, Chipping Sodbury, Bristol, BS37 6JU
G4 FPM E Keeler, 18 Clyde Road, Worthing, BN13 3LG
G4 FPO K Wilson, 14 Stuart Grove, Eggborough, Goole, DN14 0JX
G4 FPV S Perkins, 6 Delamere Road, Malvern, WR14 2BQ
G4 FPY K Jones, 58 Woodlands Road, Allestree, Derby, DE22 2HF
GM4 FPZ Michael Lisle, Cromalt, 50 Lade Braes, St. Andrews, KY16 9DA
GM4 FQE Eric Thirkell, 20 The Glebe, Crail, Anstruther, KY10 3UJ
G4 FQF P Herring, 34 Woodlands Road, Romford, RM1 4HD
GM4 FQG R McLaren, Lethendry, North Road, Dunbar, EH42 1AY
G4 FQH Bea Nelmes, Birchgrove, 17 Woodfield Road, Dursley, GL11 6HB
G4 FQI Malcolm Smith, Wilson Hall Farm, Slade Lane, Wilson, Derby, DE73 8AG
G4 FQM Samuel Morris, 23 Ellesmere Way, Carlisle, CA2 6LZ
G4 FQO Gerard kelly, 15 Dartmouth Drive, Windle, St. Helens, WA10 6BP
G4 FQP C Bamford, 12 Lincoln Drive, Caistor, Market Rasen, LN7 6PA
G4 FQR D Jones, 56 Grebe Crescent, Horsham, RH13 6ED
G4 FQT Ronald Gregory, 24 Tilton Road, Borough Green, Sevenoaks, TN15 8RS
G4 FQU Ieuan Jones, Rhandirmwyn, Llandygai, Bangor, LL57 4LD
G4 FQV D Gray, 8 Foxglove Close, Wyke, Gillingham, SP8 4TW
G4 FQW B Dunn, 17 Duke St., Clayton Le Moors, Accrington, BB5 5NQ
G4 FQZ D Simms, 1 Old Barn Close, Little Eaton, Derby, DE21 5AX
G4 FRA V Lane, 3 Lawnway, York, YO31 1JD
G4 FRB G Morse, Riversdale, High Street, Salisbury, SP3 5JL
G4 FRD John Walton, 2 Billy Mill Avenue, North Shields, NE29 0QX
G4 FRF K Johnson, 7 Bridge Croft, Clayton le Moors, Accrington, BB5 5XP
G4 FRH R Dawkins, 22 Derwen Fawr, Crickhowell, NP8 1DQ
G4 FRI G Bird, Holmwood, 101 Brookfield Road, Gloucester, GL3 2PN
G4 FRK John Rodway, 9 York Avenue, Thornton-Cleveleys, FY5 2UG
G4 FRL Nigel Ambridge, 41 Lower Icknield Way, Chinnor, OX39 4DZ
G4 FRM P Hill, 8 Davenport Park Road, Stockport, SK2 6JS
G4 FRO G Orford, 29 Church Road, Stoke Bishop, Bristol, BS9 1QP
G4 FRR Kenneth Redford, 2 Cressington Cottages, Westend, Stonehouse, GL10 3SN
G4 FRV R Vincent, Little Poulner, White Horse Lane, Bideford, EX39 1NW
G4 FRW Mark Nicholson, 10 Beechfield Road, Cheadle Hulme, Cheadle, SK8 7DS
G4 FRX J Nelson, Bank Cottage, Crew Green, Shrewsbury, SY5 9AS
G4 FRZ Adrian Jarrett, 73 Abbots Road, Abbots Langley, WD5 0BJ
GM4 FSB G Millar, 30 Albert Crescent, Newport-on-Tay, DD6 8DT
G4 FSD John Creasey, 144 Belthorn Road, Belthorn, Blackburn, BB1 2NN
G4 FSE P Biner, 295 Daws Heath Road, Rayleigh, SS6 7NS
GM4 FSF K Horne, 80 New Row, Dunfermline, KY12 7EF
G4 FSG P Murchie, 42 Catherine Road, Woodbridge, IP12 4JP
G4 FSH J Bagnall, Rainow, Under Rainow Road, Congleton, CW12 3PL
G4 FSJ David Whittaker, 23 Havelock Road, Penwortham, Preston, PR1 9RH
G4 FSK P Vallow, 1 Carolbrook Road, Ipswich, IP2 9JF
G4 FSN Eric Walton, 68 Mary Street West, Horwich, Bolton, BL6 7JU
G4 FSQ J Morley, 65 Longfield Avenue, Golcar, Huddersfield, HD7 4BT
G4 FSS D Hocking, 10 Garfit Road, Kirby Muxloe, Leicester, LE9 2DE
G4 FSU Ian Greenshields, 3 Lovers Walk, Wells, BA5 2QL
G4 FTA Robert Earle, 10 Crosslands, Fringford, Bicester, OX27 8DF
G4 FTG Charles Norton, 2 Heathlands Drive, Maidenhead, SL6 4NF
G4 FTI A Bowhill, 9 West Park Drive East, Roundhay, Leeds, LS8 2EE
G4 FTK Nigel Cridland, Alfriston, 105 Elvetham Road, Fleet, GU51 4HN
G4 FTL G King, 59 Rookery Lane, Northampton, NN2 8BX
G4 FTP Eugene Kraft, 6 The Nook, Wivenhoe, Colchester, CO7 9NH
G4 FTQ Peter Clutterbuck, 19 Warwick Close, Dorking, RH5 4NN
G4 FTW Michael ROWLAND, 16 Hayter Close, West Wratting, Cambridge, CB21 5LY
G4 FTX G Knock, 31 Northmead, Ledbury, HR8 1BE
G4 FTY R Page, Mercury House, 19 Green Lane, Birmingham, B46 3NE
G4 FTZ Brian Togwell, 26 Garraways, Royal Wootton Bassett, Swindon, SN4 8LL
G4 FUA Graham Cheater, 34 Robbins Close, Bradley Stoke, Bristol, BS32 8AS
GI4 FUE C Morrison, 60 Windslow Drive, Carrickfergus, BT38 9BB
G4 FUG P Clark, 42 Shooters Hill Road, Blackheath, London, SE3 7BG
G4 FUH SCUNTHORPE STEEL ARC c/o K Turner, Clifton, High Street, Doncaster, DN9 1JS
G4 FUI Martin Rigby, 16 Juniper Way, Penrith, CA11 8UF
G4 FUJ Graham Wright, 35 Langdale Road, Cheltenham, GL51 3LX
GI4 FUM William Hutchinson, 40 Oldstone Hill, Muckamore, Antrim, BT41 4SB
G4 FUO J Nowell, Crofters Cottage, Back Lane, York, YO23 3SH
G4 FUP N Braeman, Flat 37, Hamble Court, East Cliff Manor, Bournemouth, BH1 3PH
G4 FUR Coulsdon Amateur Transmitting Society c/o Andrew Briers, 33 Deans Walk, Coulsdon, CR5 1HR
G4 FUU M Pothecary, 61 Inglewood, Pixton Way, Croydon, CR0 9LN
G4 FUY Peter Bonson, 26 Shreen Way, Gillingham, SP8 4EL
G4 FUZ Arnold Mallows, Kilmurry House, Kilmurry Estate, Kilmurry Fermoy Cork, Ireland, 00 00
G4 FVA P Catling, The Heights, 10 Adams Road, Cambridge, CB25 0JU
G4 FVB I Clabon, 2 Farm View Lodge, Lodge Hill Lane, Rochester, ME3 8NE
G4 FVK D Sewell, 11 Haddon Close, Stanground, Peterborough, PE2 8LS
G4 FVL G Rankin, 25 The Chase, Coulsdon, CR5 2EJ
G4 FVM James Edgar, 7 Welltower Park, Ayton, Eyemouth, TD14 5RR
GM4 FVO Clifford Evans, East Cottage, Mount Melville, St. Andrews, KY16 8NT
G4 FVP Clive Davies, 28 Neville Road, Darlington, DL3 8HY
GM4 FVQ Alan Dimmick, 02 120 Shakespeare Street, Glasgow, G20 8LF
GM4 FVS G Cusiter, 4 Elphin Hill, Ellon, AB41 8BH
G4 FVU A Sweetapple, Bent Oak, Axminster Road, Axminster, EX13 8AQ
G4 FVV B Vincent, 27 Naseby Walk, Leeds, LS9 7SY
G4 FVW D Hooper, 8 Barn Close, Crewkerne, TA18 8BL
G4 FVX R Johns, 42 Lansdown Road, Redland, Bristol, BS6 6NS
G4 FVZ P Gould, Evergreen, Benty Heath Lane, Neston, CH64 1RZ
G4 FWA John Beckett, 9 Gleneagles Drive, Ipswich, IP4 5SD
G4 FWK P Mooney, 57 Johnstown Road, Co Dublin, Dun Laoghaire, Ireland
G4 FWM C Webb, 6 Chatsworth Avenue, Fleetwood, FY7 8EG
G4 FWN N May, Sandock Nurseries, Middle Dimson, Gunnislake, PL18 9NG
GD4 FWQ C MATTHEWMAN, 26 King Orry Road, Glen Vine, Douglas, Isle of Man, IM4 4ES
G4 FWR A Johnson, 86 Meadow Close, Thatcham, RG19 3RL
G4 FWT S O'Shanohun, The Corner Stone, Treskinnick Cross, Bude, EX23 0DT
G4 FXA V Arnold, 435 Manchester Road, Clifton, Manchester, M27 6WH
G4 FXE Robert Lott, 83 Manor Road, Dover, CT17 9LQ
GW4 FXF G Swan, Long Acre, New Road, Neath, SA10 8HT
G4 FXI P Overell, 48 Bedgrove, Aylesbury, HP21 7BD
GM4 FXL A Docherty, 10 Dumyat Road, Menstrie, FK11 7DG
G4 FXM G Farnie, Barn End, Rughill, Wedmore, BS28 4HL
G4 FXQ C Williams, 92 Beechwood Gardens, Ilford, IG5 0AQ
G4 FXR William Wunderlich, 31 College Road, Bromley, BR1 3PU
G4 FXT N Burkitt, 31 Loxwood, Earley, Reading, RG6 5QZ
G4 FXU R Napper, 12 Brumell Drive, Lancaster Park, Morpeth, NE61 3RB
G4 FXY philip staton, 52 School Road, Newborough, Peterborough, PE6 7RG
G4 FYB S Carter, 48 Gorse Bank Road, Hale Barns, Altrincham, WA15 0AS
G4 FYE G Coggon, 45 Ansten Crescent, Cantley, Doncaster, DN4 6EZ
G4 FYG M Newlands, 72 Town Acres, Tonbridge, TN10 4NG
GM4 FYH Catherine Waddington, Wester Lathallan, Leven, KY8 5QP
G4 FYI T Fallick, 44 Cypress Road, Newport, PO30 1HA
G4 FYJ J Lemon, 30 Iveagh Court, Farm Hill, Exeter, EX4 2LR
G4 FYM David Wiggs, 8 Bulbery, Abbotts Ann, Andover, SP11 7BN
G4 FYO Terence Foley, 16 Buckingham Road, Winslow, Buckingham, MK18 3DY
G4 FYQ M Robins, 36 Wolverley Avenue, Wollaston, Stourbridge, DY8 3PJ
G4 FYT D Lawrence, 23 Parkmead Road, Wyke Regis, Weymouth, DT4 9AL
G4 FYY P Head, 97 Malthouse Road, Southgate, Crawley, RH10 6BJ
G4 FZA John Bladen, 4 St. James Close, Hanslope, Milton Keynes, MK19 7LF
G4 FZC Alan Chapman, Avda De Jerez 278, Urb Torrenueva, Malaga, Spain, 29649
GI4 FZD Paul Menown, 34 Cairnburn Road, Belfast, BT4 2HS
G4 FZF Anthony Grinling, 32 Maybridge Square, Goring-by-Sea, Worthing, BN12 6HR
G4 FZG B Sirignano, Eversholt, 22 Cleevelands Drive, Cheltenham, GL50 4QB
G4 FZH Clive Smith, Ravenstone House, Whithorn, Newton Stewart, DG8 8DU
G4 FZL L Povoas, 9 Masons Drive, Necton, Swaffham, PE37 8EE
GW4 FZM M Davey, Penywaen, Capel Isaac, Llandeilo, SA19 7UL
G4 FZP A Drury, 31 Brook Drive, Whitefield, Manchester, M45 8FR
G4 FZR K Dally, Ealand Grange, Ealand, Scunthorpe, DN17 4DG
G4 FZS H Bulmer, 10 Southfield Lodge, South End Villas, Crook, DL15 8NN
G4 FZV P Redall, 106 Stowey Road, Yatton, Bristol, BS49 4EB
G4 FZY J Turner, 26 Clydesdale Gardens, Richmond, TW10 5EF
G4 FZZ D Holmes, 12 Chestnut Close, Rushmere St. Andrew, Ipswich, IP5 1ED
G4 GAB Ron Padbury, 8 Osbourne Drive, Holton-le-Clay, Grimsby, DN36 5DS
G4 GAF Albert McCann, Lower Fiddlers Green, Felindre, Knighton, LD7 1YT
G4 GAI K Taylor, 31 Stonehill Drive, Rochdale, OL12 7JN
G4 GAK M Sykes, 21 Croft Walk, Broxbourne, EN10 6LD
G4 GAP Humphrey Fitzherbert, 36 Westover Road, Broadstairs, CT10 3ES
G4 GAT B Denton, 2 Seacroft Road, Broadstairs, CT10 1TL
G4 GBA Charles Brookson, Orchard View, The Street., Stonham Aspal, Stowmarket, IP14 6AJ
G4 GBC Francis Orchard, 39b Breach Road, Marlpool, Heanor, DE75 7NJ

| | | |
|---|---|---|
| G4 | GBE | R Blacker, 20 Claremont Park, Lincoln Road, Sleaford, NG34 8AE |
| G4 | GBI | A Edwards, 96 Bathurst Road, Winnersh, Wokingham, RG41 5JF |
| G4 | GBK | C Appleton, 249 Devonshire Road, Atherton, Manchester, M46 9QB |
| G4 | GBP | Colin North, Somerholme, Forest Road, [illegible] |
| G4 | GBT | Ian Coleman, 12 Headington Close, Bradwell, Great Yarmouth, NR31 8DN |
| G4 | GBW | J Wilcox, 533 Upper Brentwood Road, Gidea Park, Romford, RM2 6LD |
| G4 | GBX | W Greed, 18 Nursteed Park, Devizes, SN10 3AN |
| G4 | GBY | John Robson, 35 Hankin Avenue, Dovercourt, Harwich, CO12 5HE |
| G4 | GCI | Neville Palmer, 14 Cambria Drive, Dibden, Southampton, SO45 5UW |
| G4 | GCJ | F Fuller, 7 Prestwick Close, Bletchley, Milton Keynes, MK3 7RQ |
| G4 | GCL | John Tyler, 1 Mansefield Road, Tweedmouth, Berwick-upon-Tweed, TD15 2DX |
| GI4 | GCN | R Booth, 12 Priory Drive, Carrickfergus, BT38 8HZ |
| G4 | GCQ | R Thomas, 7 Sandringham Close, Rushden, NN10 9ER |
| G4 | GCT | North Bristol ARC c/o Richard Ufford, Prospects, Tormarton Road, Badminton, GL9 1HP |
| G4 | GCU | Zygmunt Kowalczyk, 6 St Georges Crescent, Redcar, TS11 8BT |
| G4 | GDC | S Wilcox, Conifers, Aisthorpe, Lincoln, LN1 2SG |
| G4 | GDF | J Cain, 24 Agnew Crescent, Wigtown, Newton Stewart, DG8 9DT |
| G4 | GDG | Robert Smith, 47 Windsor Road, Levenshulme, Manchester, M19 2FA |
| G4 | GDL | M Ellis, 32 Pegholme Drive, Otley, LS21 3NZ |
| G4 | GDM | J Owens, Yr Hafan I Maes Gyn, An Llanarmon-Yn-Ial, Clwyd, CH7 4PY |
| G4 | GDO | F Lamb, 13-336 Queen St. South, Mississauga, Ontario, Canada, L5M 1M2 |
| G4 | GDP | J O'Shea, 30 Sue Ryder Homes, Owning, Brighton, BN2 7HA |
| G4 | GDR | Adrian Heath, 227 Windrush, Highworth, Swindon, SN6 7EB |
| G4 | GDS | D Jones, 3 Kingfisher Drive, Benfleet, SS7 5ES |
| G4 | GDT | D Wood, 20 Varndean Gardens, Brighton, BN1 6WL |
| G4 | GDU | I Hoskin, 14 Trevingey Parc, Redruth, TR15 3BZ |
| G4 | GDX | I Smith, 25 Windrush Avenue, Brickhill, Bedford, MK41 7BS |
| G4 | GDY | M Edwards, 9 Earls Walk, Binley Woods, Coventry, CV3 2AJ |
| G4 | GED | David Richardson, 68 Beech Tree Road, Holmer Green, High Wycombe, HP15 6UT |
| G4 | GEE | Robert Nash, 135 Farren Road, Coventry, CV2 5EH |
| GI4 | GEL | R Penn, 9 Milltown Road, Donaghcloney, Craigavon, BT66 7NE |
| G4 | GEN | Alan Morriss, Pipinford Park, Millbrook Hill Nutley, Nutley, TN22 3HX |
| G4 | GEO | C Tomkinson, 13 Mainwaring Road, Over Peover, Knutsford, WA16 8TR |
| G4 | GEP | Victor Peake, 24 Holyoke Grove, Leamington Spa, CV31 2RB |
| G4 | GET | Ivor Jordan, 70 Hungerhill Road, Kimberworth, Rotherham, S61 3NP |
| G4 | GEW | Peter Lee, 190 Chaldon Way, Coulsdon, CR5 1DH |
| G4 | GEY | J Carter, 30 Braemar Road, Hazel Grove, Stockport, SK7 4QG |
| G4 | GEZ | R Evans, 2 Greyfriars Lane, West Common, Harpenden, AL5 2QJ |
| G4 | GFC | S Wright, 163 Croham Valley Road, South Croydon, CR2 7RE |
| G4 | GFD | A Gilman, 10 Hanwell Close, Leigh, WN7 3NU |
| G4 | GFE | D Foulds, 12 Royal Beach Court, North Promenade, Lytham St. Annes, FY8 2LT |
| G4 | GFI | M Broadway, 91 Tattenham Grove, Epsom, KT18 5QT |
| G4 | GFJ | L Frankham, 47 St. Marys Gardens, Hilperton Marsh, Trowbridge, BA14 7PH |
| G4 | GFL | ABERGAVENNY RS c/o A Hopkins, 30 Wavell Drive, Newport, NP20 6QN |
| G4 | GFM | David Hessom, 89 Pond Close, Overton, Basingstoke, RG25 3LZ |
| G4 | GFN | Simon Dabbs, 52 Hayling Rise, Worthing, BN13 3AG |
| G4 | GFT | V Leach, Lakehead Cottage, Wellow Lane, Newark, NG22 9DG |
| G4 | GFV | J Simpson, 19 Hollinside Close, Whickham, Newcastle upon Tyne, NE16 5QZ |
| G4 | GFY | P King, 78 Gweal Wartha, Helston, TR13 0SN |
| G4 | GFZ | S Dunkerley, Po Box Hm 2215, Hamilton, Bermuda, ZZ9 9PO |
| G4 | GGC | Michael Marsh, 21 Stour Gardens, Great Cornard, Sudbury, CO10 0JN |
| G4 | GGE | D Nicholson, 41 Thurstons Barton, Bristol, BS5 7BQ |
| GM4 | GGF | V Mason, 19 Sherwood Crescent, Bonnyrigg, EH19 3LQ |
| G4 | GGH | Paul Ledbury, 12 Sandfield Close, Lichfield, WS13 6BF |
| G4 | GGI | Roger Williamson, Burwood, Wych Hill Lane, Woking, GU22 0AA |
| G4 | GGL | T Grainger, 34 Maple Avenue, Ripley, DE5 3PY |
| G4 | GGR | F Gemmell, 89 Coach Road, Guiseley, Leeds, LS20 8AY |
| G4 | GGT | Martin Masterson, 44 Highstone Avenue, London, E11 2PP |
| G4 | GGX | S Randall, 66 Park Court, Harlow, CM20 2PZ |
| G4 | GGZ | J Birch, 13 Alison Way, Aldershot, GU11 3JX |
| G4 | GHA | J Cleaton, 1 Avon Drive, Northmoor, Wareham, BH20 4EL |
| G4 | GHB | Bill Kitchen, 73 Birch Street, Ashton-under-Lyne, OL7 0JD |
| G4 | GHI | Richard Crabb, 29 Horsecastles Lane, Sherborne, DT9 6BU |
| G4 | GHK | J Donovan, 6 Manor Place, Church, Accrington, BB5 4DX |
| G4 | GHL | M Ward, 9 Woodshears Drive, Malvern, WR14 3EA |
| G4 | GHM | J Mills, 2 Old Vicarage Close, Chilton Polden, Bridgwater, TA7 9DY |
| G4 | GHO | Stephen Webb, 9 Fifth Avenue, Chelmsford, CM1 4HB |
| G4 | GHQ | P Fisher, 95 Slaithwaite Road, Thornhill Lees, Dewsbury, WF12 9DN |
| G4 | GHR | D Humphreys, 64 Holne Chase, Plymouth, PL6 7UB |
| G4 | GHT | M Skyner, 15 Dart Close, Alsager, Stoke-on-Trent, ST7 2HY |
| G4 | GHZ | P Collins, 18 Linksway, Hendon, London, NW4 1JR |
| GI4 | GID | Joe Heasley, 36 Collinbridge Gardens, Newtownabbey, BT36 7SU |
| G4 | GIG | Jane Mullany, Flat 3 Michelle Close, Hollybank Road, Birmingham, B13 0PH |
| G4 | GIM | B Waters, 60 Whitewood Way, Whittington, Worcester, WR5 2LN |
| GM4 | GIO | R Marshall, 15 Craigleith Hill, Edinburgh, EH4 2EF |
| G4 | GIR | Ian Frith, 50 Rowallan Drive, Bedford, MK41 8AS |
| G4 | GIS | John Darbyshire, 7 Sandle Road, Bishop's Stortford, CM23 5HY |
| G4 | GIV | R Howarth, 91 Armadale Road, Bolton, BL3 4PB |
| G4 | GIX | Terence Kearns, 7 Flitwick Grange, Milford, Godalming, GU8 5DN |
| G4 | GIY | R Harris, 303 Northgate, Cottingham, HU16 5RL |
| GW4 | GJA | K Austen, 6 Caernarvon Grove, Merthyr Tydfil, CF48 1JS |
| G4 | GJE | D Davis, 6 Regina Drive, Walsall, WS4 2HB |
| G4 | GJI | R Whitley, 22 Pen Y Bryn Road, Colwyn Bay, LL29 6AF |
| G4 | GJO | Donald Blampied, 113 Green Street, Enfield, EN3 7JF |
| G4 | GJR | Terry Aldridge, 2 Shelley Close, Newport Pagnell, MK16 8JB |
| G4 | GJS | W Owens, Dorfstrasse 49, Effeld, Germany, D-41849 |
| G4 | GJU | P Moxham, 233 Walsall Road, Aldridge, Walsall, WS9 0QA |
| G4 | GJV | Allan Horne, 22 Hedingham Close, London, N1 8UA |
| G4 | GJY | S Simmonds, 14 Lindsey Crescent, Kenilworth, CV8 1FL |
| G4 | GKC | C Willoughby, 79 Liskeard Road, Walsall, WS5 3ES |
| GM4 | GKH | G Duke, 3 Woodlands Grove, Westhill, Inverness, IV2 5DU |
| G4 | GKK | A Hawkins, 101 Tobyfield Road, Bishops Cleeve, Cheltenham, GL52 8NZ |
| G4 | [illegible] | [illegible], Manchester, M29 7[illegible] |
| G4 | GKU | J Cooper, 44 Belvedere Road, Bridlington, YO15 3NA |
| G4 | GKX | John Trevett, 12 Churchill Road, Blandford Forum, DT11 7HH |
| G4 | GKY | C Williams, 12a Parc An Dix Lane, Phillack, Hayle, TR27 5AB |
| G4 | GKZ | Richard Revill, 102 Hurst Drive, Stretton, Burton-on-Trent, DE13 0EE |
| G4 | GLC | D Hamilton, Rome Lea, 4 Lane Ends, Settle, BD24 0AG |
| G4 | GLG | C Edwards, 56 Stonehall Road, Lydden, Dover, CT15 7JY |
| G4 | GLH | D Bennett, Flat 3, Falcon Crag, Cowan Head, Kendal, LA8 9HL |
| G4 | GLI | M King, 28 Topcliffe Way, Cambridge, CB1 8SH |
| GD4 | GLM | Godfrey Manning, 63 The Drive, Edgware, HA8 8PS |
| G4 | GLN | A Bellfield, 50 Highfield Road, Biggin Hill, Westerham, TN16 3UU |
| G4 | GLP | Dennis Dale Green, 31 Robins Bow, Camberley, GU15 3NP |
| G4 | GLQ | John Tysiorowski, 52 Meadow Croft, Penrith, CA11 8EH |
| GW4 | GLU | Mark Norbury, 16 Pont Aur, Ynyscedwyn Road, Flat, Swansea, SA9 1BP |
| G4 | GLV | Alan Burgess, 12 Middleway, Grotton, Oldham, OL4 5EH |
| G4 | GLW | C Redmayne, 20 Kings Road, Accrington, BB5 6BS |
| G4 | GMD | D Hitchins, 21 Colwell Court, Newton Aycliffe, DL5 7PS |
| G4 | GMI | J Seddon, 8 Upper Elms Road, Aldershot, GU11 3ET |
| G4 | GMK | M North, 10 Long Lane, Pott Shrigley, Macclesfield, SK10 5SD |
| G4 | GMN | R Caswell, 15 Murtwell Drive, Chigwell, IG7 5ED |
| G4 | GMS | Leslie Hicks, 108 Northorpe, Thurlby, Bourne, PE10 0HZ |
| G4 | GMT | A Aedy, 35 Ashlea Avenue, Brighouse, HD6 3SR |
| G4 | GMW | M Weaver, 22 Greenhill Road, Alveston, Bristol, BS35 3LZ |
| G4 | GMZ | J Alder, 104 Park Lane, Congleton, CW12 3DE |
| G4 | GNA | D Townend, 442 Blackmoorfoot Road, Crosland Moor, Huddersfield, HD4 5NS |
| G4 | GND | R Culpan, 23 Aldreth Road, Haddenham, Ely, CB6 3PP |
| G4 | GNG | C Pemberton, 2 Henthorn St., Shaw, Oldham, OL2 7AY |
| GD4 | GNH | R Ferguson, Moaney Moar House, Corlea Road, Ballasalla, Isle of Man, IM9 3BA |
| G4 | GNK | P Jones, 7 Cromwell Road, Ware, SG12 7JS |
| G4 | GNO | JP Callaghan, Evergreen, Seale Lane, Farnham, GU10 1LE |
| G4 | GNP | Stephen McGrory, The Paddocks, High Street, Goole, DN14 5NY |
| G4 | GNQ | Geoffrey Sims, 85 Surrey Street, Glossop, SK13 7AJ |
| GM4 | GNR | W Thow, 11 St. Marys Place, Ellon, AB41 8QW |
| G4 | GNS | Stephen Henry, 28 Marion Avenue, Shepperton, TW17 8AY |
| GI4 | GNT | Joseph Taggart, Windy Brae, 5 Glasvey Drive, Limavady, BT49 9HQ |
| G4 | GNU | Andrew Cross, 15 Louise Road, Rayleigh, SS6 8LW |
| G4 | GNV | S Jones, 12 Yew Tree Close, Yeovil, BA20 2PD |
| G4 | GNW | T Hennigan, 128 Dimsdale View West, Newcastle, ST5 8EL |
| G4 | GNX | Alan Baker, 11 Fairfield Close, Shoreham-by-Sea, BN43 6BH |
| G4 | GNY | Martin DAVIES, Laburnum House, Guilsfield, Welshpool, SY21 9PX |
| G4 | GOA | J Harris, 28 Campion Drive, Bradley Stoke, Bristol, BS32 0BH |
| G4 | GOG | T Densham, 37 Bovingdon Park, Roman Road, Hereford, HR4 7SW |
| G4 | GOJ | G Porter, Ye Olde Homestede, High Street, Grimsby, DN36 5PL |
| GI4 | GOL | Gerald Brennan, 69 Kashmir Road, Belfast, BT13 2SB |
| G4 | GOM | F Smith, 11 Reed Field, Bamber Bridge, Preston, PR5 8HT |
| G4 | GON | J Guest, 6 The Tyning, Bath, BA2 6AL |
| G4 | GOO | M Kimmitt, Old Oaks, Tilston Road, Malpas, SY14 7DB |
| G4 | GOP | D Benn, 36 Church Avenue, Horsforth, Leeds, LS18 5LD |
| G4 | GOR | J Cross, 57f Grasmere Road, Blackpool, FY1 5HP |
| GI4 | GOS | H Sinclair, 43 Edgcumbe Gardens, Belfast, BT4 2EH |
| G4 | GOT | R Bradbury-Harrison, 11 Derwent Drive, Goring-by-Sea, Worthing, BN12 6LA |
| G4 | GOU | M Wilson, 32 Jordans Way, Bricket Wood, St. Albans, AL2 3SL |
| GI4 | GOV | Philip Barr, 5 Rosewood Park, Belfast, BT6 9RX |
| GM4 | GOW | R Armstrong, Lera Cottage, Charleston Village, Forfar, DD8 1UF |
| G4 | GOX | R Pearson, 33 Livedge Hall Lane, Liversedge, Liversedge, WF15 7DP |
| G4 | GOZ | E Cockerill, 6 Richmond Avenue, Barnoldswick, BB18 5JB |
| GI4 | GPA | William Otterson, 34 Ashbourne Park, Coleraine, BT51 3RE |
| G4 | GPB | R Cooper, 17 Cavendish Drive, Claygate, Esher, KT10 0QE |
| GI4 | GPC | J Ferguson, 7 Lairds Road, Katesbridge, Banbridge, BT32 5NN |
| G4 | GPD | William Horn, 9 Springwell View, Love Lane, Bodmin, PL31 2QP |
| G4 | GPF | Howard Winwood, 16 Brook Lane Hackenthorpe, Sheffield, S12 4LF |
| G4 | GPJ | N Bailey, 12 Carmarthen Close, Callands, Warrington, WA5 9UU |
| G4 | GPL | A Fish, 32 Deacons Hill Road, Elstree, Borehamwood, WD6 3LH |
| GM4 | GPP | C Auty, Valsgarth, Haroldswick, Shetland, ZE2 9EF |
| G4 | GPQ | T Stockill, 26 Hunters Close, Chatteris, PE16 6BD |
| G4 | GPR | A Mills, 116 Mays Lane, Barnet, EN5 2LS |
| G4 | GPV | A Brown, 12 Winstone Gardens, Cirencester, GL7 1GJ |
| G4 | GPW | B Ainsworth, 23 Cokeham Road, Sompting, Lancing, BN15 0AE |
| G4 | GPY | S Edwards, 71 St. Leonards Road Molescroft, Beverley, HU17 7HP |
| G4 | GPZ | S Culpan, 32 Riverside Drive, Hambleton, Poulton-le-Fylde, FY6 9EB |
| G4 | GQA | J Chmielewski, 2 Wolverton Avenue, Kingston upon Thames, KT2 7QD |
| G4 | GQE | Nicholas Harris, Mere Farmhouse, Matlasko Road, Norwich, NR11 7BE |
| GM4 | GQM | Gerry Firmin, Prestegaard, Uyeasound, Shetlandisle, ZE2 9DL |
| G4 | GQP | R Foote, 8 Hippings Vale, Oswaldtwistle, Accrington, BB5 3LH |
| G4 | GQR | BRIGHTON & DIST c/o P Thompson, Flat 4, 14 West End Way, Lancing, BN15 8RL |
| G4 | GQS | B Bentley, 25 Edinburgh Drive, North Anston, Sheffield, S25 4HB |
| G4 | GQV | Jack Barrett, 13 Church Bank, Church, Accrington, BB5 4JQ |
| G4 | GQY | C Loo, 74 Ilkeston Road, Trowell, Nottingham, NG9 3PX |
| G4 | GQZ | Dennis Tweedie, 39 Frenchfield Way, Penrith, CA11 8TW |
| GM4 | GRC | Glenrothes & District ARC c/o Dave Francis, 2 Morlich Crescent, Dalgety Bay, Dunfermline, KY11 9UW |
| G4 | GRG | Grajon Radio Group c/o Graham Badger, 3 Hesketh Close, Cranleigh, GU6 7JB |
| G4 | GRJ | D Gower, 2 Norview Road, Whitstable, CT5 4DN |
| G4 | GRK | P England, 2a Firs Close, Cowes, PO31 7NF |
| G4 | GRM | Leslie Horton, 4 Summer Lane, Walsall, WS4 1DS |
| G4 | GRN | Terence Griffiths, 75 Central Avenue, Waltham Cross, EN8 7JJ |
| G4 | GRP | George Gardiner, 2a Walkers Close Airmyn, Goole, DN14 8LR |
| G4 | GRR | G Searle, Hucking Court Barn, Church Road, Maidstone, ME17 1QT |
| G4 | GRS | Martin Williams, Flat 2, High Point, London, N6 4BA |
| G4 | GRT | D Mounter, 36 Norwich Road, Watton, Thetford, IP25 6DB |
| G4 | GRU | David jones, 36 Moor Lane, Woodford, Stockport, SK7 1PP |
| G4 | GRZ | R Marsh, 54 Waverton Road, Bentilee, Stoke-on-Trent, ST2 0QY |
| G4 | GSA | Peter Milsom, 214 Ormonds Close, Bradley Stoke, Bristol, BS32 0DZ |
| G4 | GSB | M Hall, 35 Bunns Lane, Dudley, DY2 7RA |
| G4 | GSC | J Osborne, 3 Temple Gardens, Staines, TW10 0NQ |
| G4 | GSD | Andrew Wallin, 11 [illegible] |
| G4 | GSG | E Warner, 99 St. Peters Park, Northop, Mold, CH7 6YU |
| GW4 | GSH | Margaret Beynon, 16 Hardy Close, Barry, CF62 9HJ |
| G4 | GSK | P Barnett, Dunelm, Barley Hill, Romsey, SO51 0LF |
| G4 | GSL | J Foster, 14 Braemar Grove, Heywood, OL10 0RR |
| G4 | GSM | J Oconnor, 11 Dalston Grove, Winstanley, Wigan, WN3 6EN |
| G4 | GSO | H Elliott, 40 Dene House Road, Seaham, SR7 7BQ |
| G4 | GSR | D Roberts, The Mead, Beaconsfield Road, Liverpool, L25 6EJ |
| G4 | GSS | Ray Bennett, Penrhiw Old Rd, Bwlchgwyn, Clwyd, [illegible] |
| GI4 | GST | Wendell Johnston, 3 Glenview, Comber, Newtownards, BT23 5HR |
| G4 | GSY | M Bainbridge, 21 Cockey Moor Road, Bury, BL8 2HD |
| G4 | GSZ | Kathleen Court, Vereda Escorredor 138, Alicante, Spain, 0100 |
| G4 | GTC | COLEG MENAI RC c/o D Davies, Rhoeyr, Llanfair Pg, LL61 5JR |
| G4 | GTD | R Ford, 2 Jersey Avenue, St. Annes, Bristol, BS4 4RA |
| G4 | GTE | David Evans, Glendale, Mount Pleasant Road, Buckley, CH7 3ET |
| G4 | GTH | M Linda, 16 Woodlinken Close, Verwood, BH31 6DD |
| G4 | GTN | P Reeve, 2 Court Road, Tunbridge Wells, TN4 8ED |
| G4 | GTS | D Fairhurst, 61 Wetherby Way, Stratford-upon-Avon, CV37 9LU |
| G4 | GTU | Steven Pocock, Popes Cottage, Main Street, York, YO20 9RQ |
| GM4 | GTV | Norman Mackenzie, 57 Countesswells Terrace, Aberdeen, AB15 8LQ |
| G4 | GTX | W Craigen, 19 Nilverton Avenue, Sunderland, SR2 7TS |
| GI4 | GTY | LAGAN VALLEY AR c/o James Henry, 3 Kirkwoods Park, Lisburn, BT28 3RR |
| G4 | GTZ | M Phillips, 12 Reydon Avenue, Wanstead, London, E11 2JD |
| G4 | GUA | J Overton, 1 Pigeon House Farm, Pigeon House Lane, Waterlooville, PO7 5SF |
| G4 | GUC | D Bailey, 12 Westbeck, Ruskington, Sleaford, NG34 9GU |
| G4 | GUE | I Pope, P O Box 662, Durbanville, South Africa, 7551 |
| G4 | GUG | Michael Meadows, 8 Beeches Park Hampton Fields, Minchinhampton, Stroud, GL6 9BA |
| GI4 | GUH | J Clarke, 1 Rathview, Banbridge, BT32 4PY |
| G4 | GUJ | R Merrell, 40 Fanton Walk, Wickford, SS11 8QT |
| G4 | GUK | K Scott-Green, 1 Pickwick, Corsham, SN13 0JD |
| GM4 | GUL | S Macdonald, 5 Lower Glebe, Aberdour, Burntisland, KY3 0XJ |
| G4 | GUN | G Le Good, 45 Kingsfield Crescent, Witney, OX28 2JB |
| G4 | GUO | Charles Brain, 7 Elverlands Close, Ferring, Worthing, BN12 5PL |
| G4 | GUQ | Ewan Crawford, 28 McCullogn Drive, Erin, Ontario, Canada, N0B 1T0 |
| G4 | GUS | J Firmin, Warren Cottage, Hill House Road, Norwich, NR14 7EE |
| G4 | GUV | Joseph Aindow, 2 Cutlers Close, Sydling St. Nicholas, Dorchester, DT2 9RG |
| G4 | GUW | G Baggott, 105 The Crescent, Walsall, WS1 2DA |
| G4 | GUX | John Kuipers, 27 Shirley Street, Hove, BN3 3WJ |
| G4 | GUY | Thomas Eaves, 3 Barons Road, Dousland, Yelverton, PL20 6NG |
| G4 | GVE | J Hawkings, 2 Balfour Drive, Biddulph, Stoke-on-Trent, ST8 7SZ |
| G4 | GVG | V Gormley, 24 Beech Road, Garstang, Preston, PR3 1FS |
| G4 | GVI | Brian Spencer, 5 Dios Polieos, House No 3, Paphos, Cyprus, 8820 |
| GM4 | GVJ | G Marshall, Drummorlie, Wallyford Toll, Musselburgh, EH21 8JT |
| GM4 | GVK | I Munro, 57 Craigiebuckler Avenue, Aberdeen, AB15 8SF |
| G4 | GVN | D Barrott, 18 Church Street, Somersham, Huntingdon, PE28 3EG |
| G4 | GVQ | Simon Eatough, 48 Mount Marua Way, Upper Hutt, New Zealand, 5018 |
| G4 | GVR | Roger Mason, 3 Coronation Close, Hellesdon, Norwich, NR6 5HF |
| GI4 | GVS | P Hallam, 95 Belfast Road, Carrickfergus, BT38 8BY |
| G4 | GVV | S Fox, Flat 3, Woodford House, Aldershot, GU11 3EL |
| G4 | GVW | P Gillen, 86 Meadowlands, Kirton, Ipswich, IP10 0PP |
| G4 | GVZ | D Morris, 40e Lansdown Crescent, Cheltenham, GL50 2NG |
| G4 | GWB | I Gibbs, 9 The Square, Choppington, NE62 5DA |
| G4 | GWE | J Martin, 57 Crescent Road East, Palm Beach, Auckland, New Zealand, 1001 |
| G4 | GWF | Harry Haden, 3 Colwyn Grove, Atherton, Manchester, M46 9XE |
| G4 | GWG | D Snape, 30 Culcross Avenue, Wigan, WN3 6AA |
| G4 | GWH | M Steventon, Pantpurlais Cefnllys, Llandrindod Wells, LD1 5PD |
| G4 | GWI | James Sheehan, 1 Osierground Cottages, Agester Lane, Canterbury, CT4 6NP |
| G4 | GWJ | John Butcher, Mount Pleasant, Trampers Lane, Fareham, PO17 6DG |
| G4 | GWP | B Langford, Dulce Verano, 29m San Jaime, Alicante, Spain, 3720 |
| GD4 | GWQ | A MATTHEWMAN, 26 King Orry Road, Glen Vine, Douglas, Isle of Man, IM4 4ES |
| G4 | GWR | Andrew Scott-Green, 58a High Street, Sutton Benger, Chippenham, SN15 4RL |
| G4 | GWT | A Kittle, 28 Clare Crescent, Towcester, NN12 6QQ |
| G4 | GWU | T Chapman, 11 Ash Court, Brampton, Huntingdon, PE28 4FH |
| G4 | GWV | R Hookham, 50 Billy Mill Avenue, North Shields, NE29 0QN |
| G4 | GWX | Jeffrey Travis, 4 Merrial Close, Bakewell, DE45 1JB |
| G4 | GWZ | R Whitehead, 14 Southgate Crescent, Rodborough, Stroud, GL5 3TS |
| G4 | GXB | Philip Butcher, 52 Chandos Road, Rodborough, Stroud, GL5 3QZ |
| G4 | GXD | D Travis, 1 Hawthorn Close, Whixall, Whitchurch, SY13 2ND |
| G4 | GXI | Peter Pearson, 58 Winchester Road, Grantham, NG31 8AD |
| G4 | GXK | SALTASH DIST AR c/o Kevin Hale, 58 St. Stephens Road, Saltash, PL12 4BJ |
| G4 | GXL | Stephen Fletcher, 43 Philip Rudd Court, Pott Row, King's Lynn, PE32 1WA |
| G4 | GXM | Roger Corr, 16 Waterdell Lane, St. Ippolyts, Hitchin, SG4 7RA |
| G4 | GXN | Michael Wright, 5 Woodview Park, The Donahies, Dublin, Ireland, DUBLIN 13 |
| G4 | GXO | R Taylor, 16 Chestnut Close, Culgaith, Penrith, CA10 1QX |
| G4 | GXP | KIDDERMINSTER & DARS c/o B Hitchins, 12 Parkland Avenue, Kidderminster, DY11 6BX |
| G4 | GXQ | Paul Bernard William Swain, 39 Coniston Drive, Handforth, Wilmslow, SK9 3NN |
| GM4 | GXR | John Higginbotham, Woodlands, Gairlochy, Spean Bridge, PH34 4EQ |
| G4 | GXW | G Cahill, 21 Moresby Close, Westlea, Swindon, SN5 7BX |
| G4 | GXY | Edward Dowlman, 4 Beald Way, Ely, CB6 3DA |
| G4 | GXZ | A Warrilow, Gyse Lodge, Gussage All Saints, Wimborne, BH21 5ET |
| G4 | GYA | Roy Williscroft, 91 Parkfield Crescent, Tamworth, B77 1HB |
| G4 | GYF | G Hiscoe, 1 Greendale Close, Fleetwood, FY7 8BQ |

UK Callsigns

**IMPORTANT NOTE**

**Revalidate licence to avoid revocation** – Ofcom has advised the Society that plans will be drawn up to revoke licences that have not been revalidated as required by the licence conditions. The quickest way to revalidate is to do so online via the Ofcom website: *https://services.ofcom.org.uk/* or by email: *amateur.validations@ofcom.org.uk* Ofcom staff are available to help, but please be patient during times of heavy workload.

| Prefix | Suffix | Details |
|---|---|---|
| G4 | GYI | P Ward, 23 Ropewalk, Alcester, B49 5DD |
| G4 | GYJ | R Littlefield, 7 Carron Mead, South Woodham Ferrers, Chelmsford, CM3 5GH |
| G4 | GYL | M Denby, 13 Hunger Hills Avenue, Horsforth, Leeds, LS18 5JS |
| G4 | GYN | Robert Colson, 46 Westwood Drive, Amersham, HP6 6RJ |
| G4 | GYO | G Humpston, 10 Winwood Drive, Quainton, Aylesbury, HP22 4AZ |
| G4 | GYP | L Ratcliff, 15 Spring Close, Biggleswade, SG18 0HL |
| G4 | GYS | J Plested, 24 Farm Way, Bushey, WD23 3SS |
| G4 | GZ | GRIMSBY ARS c/o G Smith, 6 Fenby Close, Grimsby, DN37 9QJ |
| G4 | GZA | D Ayris, 16 Chapel Lane, Northorpe, Gainsborough, DN21 4AF |
| G4 | GZB | K Turner, Clifton, High Street, Doncaster, DN9 1JS |
| G4 | GZC | Paul Teanby, 34 High Street, Belton, Doncaster, DN9 1LR |
| GM4 | GZD | G Smith, Ardvourlie, Loaneckheim, Beauly, IV4 7JQ |
| G4 | GZG | Lawrence Stringer, 2 Lion Cottages, Toot Hill Road, Ongar, CM5 9QL |
| G4 | GZH | D Andrew, Little Stone House, The Crescent, Steyning, BN44 3GD |
| G4 | GZK | H Dalton, 24 Church Lane, Coven, Wolverhampton, WV9 5DE |
| G4 | GZL | D Barker, 79 South Parade, Boston, PE21 7PN |
| G4 | GZM | A Mcmillan, 21 Clifford Avenue, Kingsteignton, Newton Abbot, TQ12 3NZ |
| G4 | GZN | K Andreang, 62 Castleton Avenue, Barnehurst, Bexleyheath, DA7 6QU |
| G4 | GZO | A Thurbon, 37 Lealand Road, Drayton, Portsmouth, PO6 1LZ |
| GM4 | GZQ | John McGinty, 77 Crawford Road, Houston, Johnstone, PA6 7DA |
| G4 | GZS | Keith Wallace, 11 Orson Leys, Rugby, CV22 5RG |
| G4 | GZT | P Jensen, 7 Union Street, Mosman, New South Wales, Australia, 2088 |
| G4 | GZU | R Woodcock, 143 Berry Hill Road, Mansfield, NG18 4RT |
| GM4 | GZW | E Simon, 100 Findhorn Place, Edinburgh, EH9 2NZ |
| GW4 | GZX | J Hunter, 245 Heathwood Road, Heath, Cardiff, CF14 4HS |
| G4 | HAA | DUMFRIES AND GALLOWAY RAYNET c/o Richard Hopkins, 15 Station Drive, Dalbeattie, DG5 4FA |
| GD4 | HAB | R Rubins, 28 Dudley Road, South Harrow, Harrow, HA2 0PR |
| G4 | HAC | C Denscombe, High Holme, 4 Kendricks Bank, Shrewsbury, SY3 0EX |
| G4 | HAG | Jack Long, 9 Denbrook Avenue, Bradford, BD4 0QH |
| G4 | HAI | Peter Levitt, 23 Castello Drive, Birmingham, B36 9TB |
| G4 | HAJ | D Magee, 2 Holt Park Vale, Holt Park, Leeds, LS16 7QX |
| G4 | HAK | P Torrance, 1 Clifton Lawn, Ramsgate, CT11 9PB |
| GM4 | HAO | R Mackean, 10a Dick Place, Edinburgh, EH9 2JL |
| G4 | HAP | H Lavin, 30 Greenslate Road, Billinge, Wigan, WN5 7BG |
| G4 | HAS | D Buck, 4687 Bracknell Road, BURLINGTON, Ontario, Canada, L7M 0E5 |
| GW4 | HAT | Philip Jones, 68 Pastoral Way, Sketty, Swansea, SA2 9LY |
| G4 | HAY | C Baker, 13 Pines Ridge, Horsham, RH12 1PZ |
| G4 | HBA | Roger Horne, 51 Welland Court, Higham, Barnsley, S75 1PZ |
| G4 | HBD | P Trepess, 3 Lawford Rise, Wimborne Road, Bournemouth, BH9 2BZ |
| G4 | HBI | F Cassidy, 55 High Bank Road, Droylsden, Manchester, M43 6FS |
| G4 | HBK | Daniel Lewis, 23 Gelligroes Road, Pontllanfraith, Blackwood, NP12 2JU |
| G4 | HBL | G Hardy, The Mill House, Thearne, Beverley, HU17 0RU |
| GM4 | HBQ | A Taylor, 6 Bowling Green St., Methil, Leven, KY8 3DH |
| G4 | HBR | J Mcgee, 3 Hedgelea Road, East Rainton, Houghton le Spring, DH5 9RR |
| G4 | HBS | S Illidge, 24 Maes Briallen, Llandudno, LL30 1JJ |
| G4 | HBT | Mike Foreman, 27 Winsbury Way, Bradley Stoke, Bristol, BS32 9BF |
| G4 | HBV | A Martin, 21 Ashwood Way, Hucclecote, Gloucester, GL3 3JE |
| G4 | HBY | M Cotton, Esterith, 113 Belvedere Road, Burton-on-Trent, DE13 0RF |
| G4 | HBZ | Brian Clowes, 7 Dukesfield Drive, Buckley, CH7 3HN |
| G4 | HCB | J Harrison, 36 Elmlea Avenue, Bristol, BS9 3UU |
| G4 | HCC | Michael Hodgkinson, 34 Pennine Way, Brierfield, Nelson, BB9 5DT |
| G4 | HCD | A Reed, 28 Russell Street, Sutton in Ashfield, NG17 4BE |
| GM4 | HCE | K Kirkland, 11 Marchfield Park Lane, Edinburgh, EH4 5BF |
| G4 | HCG | R Gordon, Middle House, 9 Fotheringhay Road, Peterborough, PE8 5HP |
| G4 | HCI | M Foreman, 39 Artists View Drive, Calgary, Alberta, Canada, T3Z 3N4 |
| G4 | HCK | Nicholas Wilkinson, 12 Woodlands Close, Grays, RM16 2GB |
| GI4 | HCN | Jeffrey Clarke, 154 Galgorm Road, Ballymena, BT42 1DE |
| GM4 | HCO | V Kusin, East Overhill Farm, Stewarton, Kilmarnock, KA3 5JT |
| GI4 | HCX | I Magill, 205 Whitechurch Road, Ballywalter, Newtownards, BT22 2LA |
| G4 | HCY | Michael Stokes, 14 Shillitoe Avenue, Potters Bar, EN6 3HG |
| G4 | HCZ | L Fellows, 19 Grosvenor Road, Lower Gornal, Dudley, DY3 2PS |
| GW4 | HDB | M Greatrex, 4 Lee St., St. Thomas, Swansea, SA1 8HQ |
| G4 | HDD | S Rose, 14 Highgate West Hill, London, N6 6JR |
| G4 | HDE | S Green, 6 Poveys Mead, Kingsclere, Newbury, RG20 5ER |
| GW4 | HDF | V Hill, 9 Cae Pant, Caerphilly, CF83 2UW |
| GI4 | HDJ | B McGarry, 43 Umrycam Road, Feeny, Londonderry, BT47 4TJ |
| G4 | HDL | N Sedgwick, Flat 3, Hartford Court, 33 Filey Road, Scarborough, YO11 2TP |
| G4 | HDO | A Kirkland, 4 Laurelwood Road, Droitwich, WR9 7SE |
| G4 | HDR | Alan Evans, 4 Elm Grove, Rhyl, LL18 3PE |
| G4 | HDS | Paul Unwin, Mycroft, Rochester, Newcastle upon Tyne, NE19 1RH |
| G4 | HDU | Barry Keal, 46 Eastway, Maghull, L31 6BS |
| G4 | HDY | G Burgess, 44 Clifton Road, Winchester, SO22 5BU |
| GW4 | HDZ | D Birch, 16 Llanharry Road, Brynsadler, Pontyclun, CF72 9DB |
| G4 | HEB | P Tuffs, 48 Mackie Drive, Guisborough, TS14 6DJ |
| G4 | HEC | P Stracey, 14 Portfield Road, Christchurch, BH23 2AG |
| G4 | HEE | W Dallas, 21 Jubilee Avenue, Asfordby, Melton Mowbray, LE14 3RY |
| G4 | HEJ | W Reid, Comphurst, Comphurst Lane, Hailsham, BN27 4TX |
| GM4 | HEL | HELENSBURGH ARC c/o Barry Spink, 9 St. Andrews Crescent, Dumbarton, G82 3ER |
| G4 | HER | S Rogers, Green Tops, Megs Lane, Buckley, CH7 2AG |
| G4 | HES | W Ray, 54 Gladstone Road, Chesham, HP5 3AD |
| G4 | HEV | Gordon Cass, 18 Rawcliffe Drive, York, YO30 6PE |
| G4 | HEW | G Hancock, 12122-244th Street, Maple Ridge, Canada, BC V4R 1I1 |
| G4 | HFG | Graham Eckersall, 65 Lowside Drive, Roundthorn, Oldham, OL4 1AS |
| G4 | HFI | Malcom Roberts, The Willows, Riverside, Hayle, TR27 5JD |
| G4 | HFO | Martin Blythe, Trethullan Farmhouse, Sticker, Saint Austell, PL26 7EH |
| G4 | HFQ | G Freeth, 9 South Avenue, New Milton, BH25 6EY |
| G4 | HFS | M Davies, The Granary, Chequers Lane, High Wycombe, HP14 3PH |
| G4 | HFU | Philip Spooner, The Birches, Wingrave Road, Aylesbury, HP22 4LT |
| G4 | HFZ | S Mccann, 211 East Common Lane, Scunthorpe, DN16 1HL |
| G4 | HGB | D France, 28 Arlbury Road, Northampton, NN3 8QJ |
| G4 | HGH | Anthony Selmes, 35 Windmill Rise, Hundon, Sudbury, CO10 8EQ |
| GW4 | HGJ | G Carruthers, Henllys Farm, Cardigan, SA43 2RH |
| G4 | HGK | John Davis, Hurstbourne, Westdown Road, Bexhill-on-Sea, TN39 4DY |
| G4 | HGL | J Buckley, Sandringham, Neston Road, Neston, CH64 4AT |
| G4 | HGM | Martyn Gregory, 30 Tanner Way, Bridgton, United States, ME04009 |
| G4 | HGN | David Hoyle, Pharmacy Cottage, Queen Street, Buxton, SK17 8JT |
| G4 | HGR | M Baker, 39 The Cherry Orchard, Hadlow, Tonbridge, TN11 0HU |
| GW4 | HGS | Grayham Passmore, 127 High Street, Neyland, Milford Haven, SA73 1TR |
| G4 | HGT | J Wilkinson, 7 Hilton Grange, Bramhope, Leeds, LS16 9LE |
| G4 | HGV | M Leach, 15 Beech Lea, Blunsdon, Swindon, SN26 7DE |
| G4 | HHA | K Stalley, The Forge, Woodbridge Road, Woodbridge, IP12 2JE |
| G4 | HHH | P Walker, East Rigg, Fylingdales, Whitby, YO22 4QG |
| G4 | HHJ | D Thomas, 64 Marconi Way, St. Albans, AL4 0JG |
| G4 | HHL | V Gorny, 22 Park Road, Shirehampton, Bristol, BS11 0EF |
| G4 | HHM | David Ryder, 96 Huttoft Road, Sutton-on-Sea, Mablethorpe, LN12 2QZ |
| G4 | HHO | C Buckley, Curraghmore, Model Farm Road, Co. Cork, Ireland |
| G4 | HHS | L May, 20 Crescent Road, Marland, Rochdale, OL11 3LF |
| G4 | HHX | Richard Edmonds, 14 Singledge Lane, Whitfield, Dover, CT16 3EJ |
| G4 | HHY | C Goode, Tall Trees, Woodbury Salterton, Exeter, EX5 1QB |
| G4 | HHZ | A Harwood, 55 Nichol Road, Chandler's Ford, Eastleigh, SO53 5AX |
| G4 | HIA | M Nicholls, 12 Bents Drive, Sheffield, S11 9RP |
| G4 | HIC | M Maddison, 34 Maple Avenue, Sandiacre, Nottingham, NG10 5EF |
| G4 | HIE | M Hammond, 53 Chiltern Road, Baldock, SG7 6LT |
| G4 | HIF | D Mallet, 41 Kiln Close, Calvert, Buckingham, MK18 2FD |
| G4 | HIH | R Wilson, 4 Dinmont Place, Hall Close Grange, Cramlington, NE23 6DN |
| G4 | HIJ | R Woolley, 29 Belle Vue Road, Ashbourne, DE6 1AT |
| G4 | HIN | R Twiggs, 31 Westlands Avenue, Slough, SL1 6AH |
| G4 | HIQ | A Sturman, 22 St. Crispins Avenue, Wellingborough, NN8 2HT |
| G4 | HIV | B Milne, 11 Station Road, Thorpe-on-the-Hill, Lincoln, LN6 9BS |
| G4 | HIW | C Vernon, 2 Standing Butts Close, Walton-on-Trent, Swadlincote, DE12 8NJ |
| G4 | HIX | P Duncan, 89 Felstead Crescent, Sunderland, SR4 0AE |
| G4 | HIY | B Burke, 12 Evesham Grove, Hurworth, Darlington, DL2 2YE |
| G4 | HIZ | J Easdown, 38 North Street, Barming, Maidstone, ME16 9HF |
| G4 | HJB | C Hall, 10 Porlock Court, Cramlington, NE23 3TT |
| G4 | HJD | A Goy, 352 Chanterlands Avenue, Hull, HU5 4ED |
| G4 | HJE | S Small, 102 Crestway, Chatham, ME5 0BH |
| G4 | HJF | W Dredge, 10 Lime Close, Locking, Weston-Super-Mare, BS24 8BH |
| G4 | HJH | Mark Hardaker, PO Box 82267, Budaiya, Bahrain |
| G4 | HJI | Jonathan Bright, 43 Rue Vauban, Village Neuf, France, 68128 |
| GM4 | HJK | R Mitchell, 9 Pine Way, Perth, PH1 1DT |
| G4 | HJL | Marc Zarattini, The Pippins, Orchard Street, Derby, DE3 0DF |
| GM4 | HJO | Marek Mozolowski, The Auld Manse, 8 Sandport, Kinross, KY13 8DN |
| GM4 | HJQ | D Mackenzie, 58 High Street, East Linton, EH40 3BH |
| G4 | HJS | Phillip Tempest, 15 Charles Avenue, Leeds, LS9 0AE |
| G4 | HJT | David Lloyd, 39 High Street, 40 Bertrand Drive, Princeton, United States, 8540 |
| G4 | HJW | Bernard Wright, 39 High Street, Little Wilbraham, Cambridge, CB21 5JY |
| G4 | HJY | M Black, 28 Cricketers Close, Chessington, KT9 1NL |
| G4 | HKB | Patricia Turner, 1 Longridge, Colchester, CO4 3FD |
| G4 | HKC | I Butson, 60 Churnwood Road, Parsons Heath, Colchester, CO4 3EY |
| G4 | HKO | Thurrock Acorns ARC c/o Nicholas Wilkinson, 12 Woodlands Close, Grays, RM16 2GB |
| G4 | HKP | C Turner, 150 Shingara Sands, Petroy Drive, Four Ways, South Africa, 2191 |
| G4 | HKQ | Christopher Marsh, 33 Southview Road, Hockley, SS5 5DY |
| G4 | HKR | A Reed, 85 Ringway, Garforth, Leeds, LS25 1BZ |
| G4 | HKS | Martin Lynch, Wessex House, Drake Avenue, Staines-upon-Thames, TW18 2AP |
| GM4 | HKV | J Henderson, 7 Lumsden Crescent, St. Andrews, KY16 9NQ |
| G4 | HKX | R Rowlands, 4 Glascoed, Hermon, Bodorgan, LL62 5LF |
| G4 | HKY | L Bower, 1 Elmfield Drive, Skelmanthorpe, Huddersfield, HD8 9BT |
| G4 | HKZ | Julie Butcher, Mount Pleasant, Trampers Lane, Fareham, PO17 6DG |
| G4 | HLA | J Sullivan, 1 Godley Hill Road, Hyde, SK14 3BW |
| G4 | HLB | R Hallam, 16 Hall Road, Haconby, Bourne, PE10 0UY |
| G4 | HLF | Paul Westwell, 11 Cheshire Park, Warfield, Bracknell, RG42 3XA |
| G4 | HLI | J Friend, 62 St. Catherines Hill, Bramley, Leeds, LS13 2LE |
| G4 | HLL | WALSALL ARC c/o C Willoughby, 79 Liskeard Road, Walsall, WS5 3ES |
| G4 | HLN | Lawrence Bennett, 26 Winchester Road, Burnham-on-Sea, TA8 1HY |
| G4 | HLO | W Davies, Erw Deg, 11 Madoc Street, Porthmadog, LL49 9BU |
| G4 | HLT | M Eckhoff, 6 Ramsbury Drive, Earley, Reading, RG6 7RT |
| G4 | HLW | K Turnell, 31 Greenbank Terrace, Ringstead, Kettering, NN14 4DD |
| G4 | HLX | N Taylor, 7 Badgers Gardens, Charlton Road, Wantage, OX12 8FE |
| G4 | HLZ | Mark Wood, 52 Priory Lane, Grange-over-Sands, LA11 7BJ |
| G4 | HMA | M Smith, 8a Duke Street, Cullompton, EX15 1DW |
| G4 | HMC | James Oliver, Chalk Lodge, Peters Lane, Princes Risborough, HP27 0LG |
| GD4 | HMD | H Drury, 11 Batchworth Lane, Northwood, HA6 3AU |
| G4 | HME | L Bailey, 47 Millers Park, Wellingborough, NN8 2NQ |
| GM4 | HML | S Mcluckie, 12 Croft Place, Eliburn, Livingston, EH54 6RJ |
| G4 | HMM | B Dearing, 44 Woodlands Way, Southwater, Horsham, RH13 9HZ |
| GM4 | HMN | A Cumming, 18 South Covesea Terrace, Lossiemouth, IV31 6NA |
| G4 | HMR | D Morris, Hafodty Cottage, Tregarth, Bangor, LL57 4NS |
| G4 | HMX | J Halliday, 16 Ennerdale Drive, Congleton, CW12 4FR |
| G4 | HND | A Course, 5 Conway Drive, Burton Latimer, Kettering, NN15 5TA |
| G4 | HNF | D Waterworth, 116 Reading Road, Woodley, Reading, RG5 3AD |
| G4 | HNG | G Poulton, The Leas, Higher Sea Lane, Bridport, DT6 6BB |
| G4 | HNJ | George Wheatley, 67 Moorlands Road, Verwood, BH31 7PD |
| GM4 | HNK | Colin Ferguson, Leckuary, Kilmichael Glassary, Lochgilphead, PA31 8QL |
| G4 | HNO | S Wilson, 8 Hillcrest Avenue, Stockport, SK4 3JS |
| G4 | HNQ | J Bryden, 32 Jerusalem Road, Skellingthorpe, Lincoln, LN6 5TW |
| G4 | HNU | Peter Vaughan, 26 Canterbury Road, Worthing, BN13 1AE |
| G4 | HNW | S Walls, 11 Copperfield Close, Malton, YO17 7YN |
| G4 | HNX | E Beal, 49 Ambersham Crescent, East Preston, Littlehampton, BN16 1AJ |
| G4 | HNZ | S Bannister, 14 Amery Close, Worcester, WR5 2HL |
| G4 | HOC | Mark Oliver, 34 Manderley Close, Coventry, CV5 7NR |
| G4 | HOD | M Gunby, 128 Heath Road, Runcorn, WA7 4XL |
| G4 | HOF | Patrick Warrener, 3 Ashby Close, Holton-le-Clay, Grimsby, DN36 5BD |
| G4 | HOI | Walter Skeels, 141 Woodward Road, Dagenham, RM9 4ST |
| G4 | HOJ | Philip Hobson, High Rising, 4 Dovecote Lane, Lincoln, N5 0AD |
| G4 | HOK | J McKay, 2 Bransghyll Terrace, Horton-in-Ribblesdale, Settle, BD24 0HG |
| G4 | HOL | Michael Capstick, Avda. Jardines del Almanzora No 62, La Alfoquia de Zurgena, Zurgena, Spain, 4661 |
| G4 | HOM | Frederick Garratt, 90 Brushfield Road, Birmingham, B42 2QJ |
| G4 | HON | C Ward, 2 Arlington Drive, Stockport, SK2 7EB |
| G4 | HOP | S Fordham, 61 Cemetery Road, Dronfield, S18 1XX |
| G4 | HOU | L Anstead, 21 Tickenor Drive, Finchampstead, Wokingham, RG40 4UD |
| G4 | HOW | Nigel Cleaver, 18 Old Cleeve, Minehead, TA24 6HJ |
| G4 | HOY | John Fennell, Bajamar House, Belton Road, Doncaster, DN9 1JL |
| GD4 | HOZ | David Osborn, Kionlough House, Kionlough Lane, Ramsey, Isle of Man, IM7 4AG |
| G4 | HPD | Barry Constable, Dukes Pleasure Long Headland, Ombersley, Droitwich, WR9 0DX |
| G4 | HPE | Steven Richards, 6 Heathfield, Royston, SG8 5BW |
| G4 | HPH | J Littler, 363 Atherton Road, Hindley, Wigan, WN2 3XD |
| GM4 | HPK | David Moore, Rashfield Farm By Kilmun, Dunoon, PA23 8QT |
| GD4 | HPN | Richard Baker, Clea Ghlass, Ballaragh Road, Laxey, Isle of Man, IM4 7PG |
| G4 | HPS | P Barker, 11 Dipton Gardens, Sunderland, SR3 1AN |
| G4 | HPT | D Oliver, Ashdell, Newlands Lane, Birmingham, B37 7EE |
| G4 | HPX | J Trotter, 29 Broad Park Road, Bere Alston, Yelverton, PL20 7AH |
| G4 | HPY | R Spragg, 3 Truro Gardens, Luton, LU3 2AP |
| G4 | HQA | Richard Knowles, 22 Thornley Road, Ribbleton, Preston, PR2 6EY |
| G4 | HQB | Philip Sandell, 1 St. Margaret Road, Ludlow, SY8 1XN |
| G4 | HQC | C Wilcox, 42 Kentmere Close, Cheltenham, GL51 3PD |
| G4 | HQD | R Bagley, 8 Bishop Ruzar Furrugia Street, Xaghra, Xra, Malta, 103 |
| GM4 | HQF | D Lindsay, 39 Seamount Court, Aberdeen, AB25 1DQ |
| G4 | HQH | Samuel Parker, 20 Swaddale Avenue, Chesterfield, S41 0SU |
| G4 | HQM | D Waspe, 28 Wilman Way, Salisbury, SP2 8QS |
| GM4 | HQU | N Gent, 4 Eskview Villas, Eskbank, Dalkeith, EH22 3BN |
| G4 | HQX | Peter Morys, 41 Salter Street, Berkeley, GL13 9BU |
| GM4 | HQZ | A Morrison, Block 19, 2 Sandpiper Road, Edinburgh, EH6 4TR |
| G4 | HRB | D Taylor, 8 Fambridge Close, Maldon, CM9 6DJ |
| G4 | HRC | HAVERING & DISTRICT ARC c/o David Nuttall, 92 Long Road, Lowestoft, NR33 9DH |
| G4 | HRE | D Hollow, 8 Vermont Woods, Finchampstead, Wokingham, RG40 4PF |
| G4 | HRG | R Denley, 50 Cranmere Avenue, Wergs, Wolverhampton, WV6 8TS |
| G4 | HRH | A Allen, The Hollies, Sedgeford, Whitchurch, SY13 1EX |
| GM4 | HRJ | J Mcniff, East Cove Cottage, Main Road, Port Glasgow, PA14 6XP |
| GM4 | HRL | Anthony Sergeant, 24 Academy Road, Bo'ness, EH51 9QD |
| G4 | HRS | HORSHAM ARC c/o John Matthews, 46 Park Lane, West Grinstead, Horsham, RH13 8LT |
| G4 | HRU | R Profitt, 10 Taunton Vale, Hunters Hill, Guisborough, TS14 7NB |
| G4 | HRY | D Farn, 14 Corfe Close, Coventry, CV2 2JG |
| G4 | HS | S Hopper, 16 Stanford Avenue, Hassocks, BN6 8JL |
| G4 | HSB | P Rovardi, 8 Cambridge Road, Linthorpe, Middlesbrough, TS5 5NQ |
| G4 | HSC | Harry Hughes, 16 Dalton Drive, Goose Green, Wigan, WN3 6TQ |
| G4 | HSD | R Smithers, 16 Derby Road, Sutton, SM1 2BL |
| GW4 | HSH | Roger Williams, 114 West Cross Lane, West Cross, Swansea, SA3 5NQ |
| G4 | HSK | S Glass, 36 Pickwick Avenue, Chelmsford, CM1 4UN |
| G4 | HSM | R Hurrell, 97 Dovercliffe Close Se, Calgary, Alberta, Canada, T2B 1W4 |
| G4 | HSN | A Chorley, Leycot, Cornells Lane, Saffron Walden, CB11 3SP |
| G4 | HSO | P Baker, South Lodge, Kimpton, Hitchin, SG4 8ER |
| GM4 | HSR | D Gillies, 56 Forehill Road, Ayr, KA7 3DT |
| G4 | HSS | P Forshaw, 54 The Park, Penketh, Warrington, WA5 2SG |
| GJ4 | HSW | F Le Quesne, Brookhill House, Princes Tower Road, St. Saviour, Jersey, JE2 7UD |
| G4 | HSX | F Cole, 3 Wadsworth Avenue, Todmorden, OL14 7NF |
| G4 | HSZ | P Thacker, 23 Lulworth Avenue, Leeds, LS15 8LW |
| G4 | HTB | T Rance, 2 Glenavon Gardens, Slough, SL3 7HN |
| G4 | HTD | Laurence Mason, Forest Farm, Folly Drove, Stewley, Ilminster, TA19 9NW |
| G4 | HTE | E Sergeant, 13 Morven Close, Potters Bar, EN6 5HE |
| G4 | HTG | Alan Brunton, 409 Outwood Common Road, Billericay, CM11 1ET |
| G4 | HTH | R Herringshaw, 35 Oxley Close, Shepshed, Loughborough, LE12 9LS |
| G4 | HTL | A Mcculloch, 91 Broadhurst, Farnborough, GU14 9XA |
| G4 | HTO | Ian Myford, 33 Station Road, Edingley, Newark, NG22 8BX |
| GM4 | HTU | Anthony Langton, 71 Gray Street, Aberdeen, AB10 6JD |
| G4 | HTV | ITV WEST RC c/o R Thompson, 179 Newbridge Hill, Bath, BA1 3PY |
| G4 | HTW | P McVeigh, The Dale, Bowns Hill, Matlock, DE4 5DG |
| G4 | HTX | Richard Houghton, Elmtrees, Church End, Bedford, MK44 2RP |
| G4 | HTY | D Stokes, Flat 6, 35-37 Gratton Road, London, W14 0JX |
| G4 | HTZ | Stephen Barrett, 266 Wakering Road, Shoeburyness, Southend-on-Sea, SS3 9TP |
| G4 | HUA | T Ellam, 3115 Carleton Street SW, Calgary, Canada, AB T2T3L5 |
| G4 | HUD | J Bramall, 55 Wood Lane, Louth, LN11 8RY |
| G4 | HUE | A Nehan, Danisway, Queens Road, Bedford, MK44 2LA |
| G4 | HUF | P Baguley, 16 Churchill Road, Broadheath, Altrincham, WA14 5LT |
| G4 | HUG | William Daniels, 48 Mellanear Road, Hayle, TR27 4QT |
| G4 | HUH | P Chapman, 1291 Los Amigos Avenue, California, United States, 93065 |
| GM4 | HUL | W Savory, 20 Broomfield, Carradale East, Campbeltown, PA28 6RZ |
| G4 | HUM | D Hazzard, 69a Beaconsfield Avenue, Portsmouth, PO6 2PS |
| G4 | HUN | N Whiteside, 2 Reed Cottages, Great Cambourne, Cambridge, CB23 6GR |
| G4 | HUO | M Bennett, 9 Lavender Avenue, Blythe Bridge, Stoke-on-Trent, ST11 9RN |
| G4 | HUP | D Powis, Fircroft, Pound Lane, Woodbridge, IP13 0LN |
| G4 | HUQ | M Crake, 12 Bosburn Drive, Mellor Brook, Blackburn, BB2 7PA |
| G4 | HUT | David Consitt, Saxtorpsvagen 210, Landskrona, Sweden, 26194 |
| G4 | HUW | S Faulkner, Vaarveien 8, Oslo, Norway, 1182 |
| GM4 | HUX | Ron Lindsay, 32a James Street, Alva, FK12 5AL |
| GU4 | HUY | Roger Sarre, Le Clercs, Clos Du Murier, Rue de Bas, St. Sampson, Guernsey, GY2 4HJ |
| G4 | HVC | A Kiddle, 19 Old Lincoln Road, Caythorpe, Grantham, NG32 3DF |
| G4 | HVD | T Barnett, East View, Squires Road, Lydbrook, GL17 9QL |
| G4 | HVF | Christopher Bracewell, 14 Woodlane, Falmouth, TR11 4RF |
| G4 | HVG | J Phipps, 5 Akeman Close, St. Albans, AL3 4NJ |
| GI4 | HVI | A Hamilton, 11 Norwell Park, Castlerock, Coleraine, BT51 4TS |
| GM4 | HVM | A Douglas, 24 Plane Grove, Dunfermline, KY11 8RA |
| G4 | HVO | J Fitzwater, The Olde Cottage, Babylon Lane, Tadworth, KT20 6XE |
| G4 | HVR | G Southwell, 4a Neve Avenue, Wolverhampton, WV10 9BU |
| GM4 | HVS | R Teperek, 8 Forest Park, Stonehaven, AB39 2GF |
| G4 | HVT | N Wilkinson, Breidablikkbakken 15, Porsgrunn, Norway, 3911 |
| G4 | HVV | HAVEN VALLEY CONTEST CLUB c/o Chris Goadby, Heligan, 12 School Road, Newmarket, CB8 9RX |
| G4 | HVW | F MOODY, 87 Whitegate Walk, Rotherham, S61 4LP |

G4 HWA Bernard Morton, Yew Tree House, 14 Baker Street, Northampton, NN7 3EZ
G4 HWC [illegible] Marske By The Sea, Redcar, TS11 7LP
G4 HWF R Rudd, 41 Chester Terrace, Brighton, BN1 6GB
G4 HWH A Jandrell, 21 Wildacres, Stourbridge, DY8 0PH
G4 HWI Michael Allin, 50 Swallow Rise, Knaphill, Woking, GU21 2LH
G4 HWJ M Dawson, Mulberry Cottage, The Hamlet, Ely, CB6 1SB
G4 HWK Fred Pilling, Shrublands, Bradfield Road, Manningtree, CO11 2SL
G4 HWM D Jeffery, 11 Beechwood Crescent, Chandler's Ford, Eastleigh, SO53 5PA
G4 HWN R Heath, Flat 172, Hagley Road Retirement Village, 330 Hagley Road, Birmingham, B17 8BN
GM4 HWO C Wright, 3 Stanedykehead, Edinburgh, EH16 6YE
GM4 HWS O Roberts, 10 Mill Place, Tarland, Aboyne, AB34 4YG
G4 HWV T Wiles, Manor Farm, Manor Close, Middlesbrough, TS8 9AG
G4 HWW R Scott, Flat 57 Tatton Court, 35 Derby Rd, Stockport, SK4 4NI
G4 HXC D Edwards, 170 Pallott Drive, Nuneaton, CV11 0JA
G4 HXE Alan Tilbee, 61 Pacific Close, Southampton, SO14 3TY
G4 HXH R Pope, 99 Northolt Avenue, Bishop's Stortford, CM23 5DS
G4 HXK F Rendell, 64 Rivermead, Stalham, Norwich, NR12 9PJ
G4 HXL Laurence Manderson, 16 Archery Avenue, Foulridge, Colne, BB8 7NH
G4 HXN D Kelly, 27 Keswick Road, Bookham, Leatherhead, KT23 4DQ
GW4 HXO Michael Probert, 1 Ynys Dawel, Solva, Haverfordwest, SA62 6UA
G4 HXQ G Burlington, Podgwell Cottage, Seven Leaze Lane, Stroud, GL6 6NJ
G4 HXU D Mcdermott, 6 Chiltern Grove, Thame, OX9 3NH
G4 HXX CROSSWAYS CONTEST GROUP c/o Colin Dollery, 101 Corringham Road, London, NW11 7DL
G4 HXY S Simmons, 48 Copland Road, Stanford-le-Hope, SS17 0DF
G4 HYD Anthony Oakley, 2 Manor Close, Beverley, HU17 7BP
GM4 HYF G Allan, Flat 65, Stonelaw Court, Glasgow, G73 2PH
G4 HYG Chris Moulding, 106 Barton Road Farnworth, Bolton, BL4 9PT
G4 HYI E Towers, Belway, Beaconsfield Road, Haywards Heath, RH17 7JU
GM4 HYR M Bond, 1 Saughtonhall Crescent, Edinburgh, EH12 5RF
G4 HYT Philip Kurian, 22a Lindisfarne Avenue, Blackburn, BB2 3EH
G4 HYW Andrew Wilkes, Efford Park, Milford Road, Lymington, SO41 0JD
G4 HYY Thomas Jackson, 86 Lascelles Avenue, Withernsea, HU19 2EB
G4 HYZ B Green, 28 Sunnybank Road, Griffithstown, Pontypool, NP4 5LT
G4 HZE E Hill, 14 Station Road, Saltash, PL12 4DY
G4 HZF R Scarlett, 1 St. Martins Crescent, Grimsby, DN33 1BG
G4 HZG M White, 62 Dalebrook Road, Burton-on-Trent, DE15 0AD
GW4 HZH Daniel Doherty, 3 Llys Penpant, Morriston, Swansea, SA6 6DA
G4 HZI W Backhouse, 191 Wigmore Road, Gillingham, ME8 0TL
G4 HZJ Leslie Jackson, 1 Belvedere Avenue, Atherton, Manchester, M46 9LQ
GW4 HZM John Styles, 5 Heol-y-Berth, Caerphilly, CF83 1SP
G4 HZN T Lockwood, 8 St. Nicholas Road, Thorne, Doncaster, DN8 5BS
G4 HZP A Charlton, The Crook, Rowelton, Carlisle, CA6 6LH
G4 HZR D Saunders, 4 Furzedene, Furze Hill, Hove, BN3 1PP
G4 HZT T Morton, 3 Grandstand Road, Hereford, HR4 9NE
G4 HZV R Bagwell, 30 Christmas Pie Avenue Normandy, Guildford, GU3 2EN
G4 HZW A Usher, 14 Bucklow Avenue, Mobberley, Knutsford, WA16 7ET
G4 HZX N Squibb, 127 Copers Cope Road, Beckenham, BR3 1NY
G4 IAB A Bell, 10 Long Acre, Weaverham, Northwich, CW8 3PT
G4 IAD D Crompton, The Beeches, 6 St. Johns Wood, Bolton, BL6 4FA
G4 IAG Terry Court, Woodview, Breach Oak Lane, Coventry, CV7 8AU
G4 IAJ Timothy Jefferson, Flat 3, 4 Esplanade Gardens, Scarborough, YO11 2AW
G4 IAL John Heywood, 46 The Close Wyre Vale Park, Garstang, Preston, PR3 1PL
G4 IAO A Robertson, 7 Big Back Lane, Chedgrave, Norwich, NR14 6BH
G4 IAQ Judith Brooks, 28 Avon Vale Road, Loughborough, LE11 2AA
G4 IAR David Brooks, 28 Avon Vale Road, Loughborough, LE11 2AA
G4 IAT B Smith, 69 Birch Hall Avenue, Darwen, BB3 0JW
G4 IAU D Lilley, 65 Peel St., Horbury, Wakefield, WF4 5AN
G4 IAY F Whittaker, 91 Oakdale, Worsbrough, Barnsley, S70 5NR
G4 IBC Radio Amateur Invalid and Blind Club c/o Kelvin Marsh, Highgrove, Creech Heathfield, Taunton, TA3 5EW
G4 IBH D Dockery, 20 Saffron Way, Sittingbourne, ME10 2EY
G4 IBI W Mitchell, Wychwood, The Ridgeway, Cranleigh, GU6 7HR
G4 IBM C Murphy, 15 Loders Close, Poole, BH17 9BF
G4 IBN K Pointon, 1 Deans Court, Pontefract, WF8 1NH
G4 IBS Geoff Baxendale, Sarno, Granville Road, Darwen, BB3 2SS
GI4 IBV S Johnston, 61 Ravenhill Park, Belfast, BT6 0DG
G4 IBW R Ropinski, 38 The Leys, Little Eaton, Derby, DE21 5AR
G4 IBZ P Richardson, 10 Mosgrove Close, Gateford, Worksop, S81 8TD
G4 ICB B Clarke, 59 Baden Powell Crescent, Pontefract, WF8 3QD
G4 ICC Michael Gater, 17 Douglas Road, Northampton, NN5 6XX
G4 ICE A Mitchell, 11 Poplar Lane, Cannock, WS11 1NQ
G4 ICF A Denison, 40 Leysholme Drive, Leeds, LS12 4HQ
G4 ICH Chris Wickenden, Chalfont, Little Whelnetham, Bury St. Edmunds, IP30 0DG
G4 ICI Roger Perks, Drayton Lodge, Drayton Manor Drive, Tamworth, B78 3TJ
G4 ICM ICOM (UK) AR c/o David Stockley, C/O Icom UK Ltd, Blacksole House, Herne Bay, CT6 6GZ
G4 ICP Richard Witney, 36 Dapifer Drive, Braintree, CM7 3LG
G4 ICU Anthony Jones, 15 High Street, Sedgley, Dudley, DY3 1RL
G4 ICZ B Greatrix, 12 Swainsfield Road, Yoxall, Burton-on-Trent, DE13 8PT
G4 IDC Michael Rudge, 8 Penrallt Estate, Llanystumdwy, Criccieth, LL52 0SR
G4 IDD D Dockar, 49 Dixon Lane, Wortley, Leeds, LS12 4RR
G4 IDF David Hobro, 60 Linksview Crescent, Worcester, WR5 1JJ
G4 IDG Graham Tonge, 6 Bickford Close, Lapley, Stafford, ST19 9JZ
G4 IDH I Harris, 47D Tower 2 Queens Terrace, 1 Queen Street, Sheung Wan, Hong Kong, 12345
G4 IDJ D MacGregor, 29 Terrington Hill, Marlow, SL7 2RE
G4 IDL T Wade, 47 Rig Drive, Swinton, Mexborough, S64 8UL
G4 IDR D Redman, 13 Halifax Road, Golcar, Huddersfield, HD7 4NS
G4 IDT F Heywood, 62 Southleigh Road, Leeds, LS11 5SG
G4 IDU K Kniveton, 32 Minster Avenue, Bude, EX23 8RY
G4 IDV P Brown, 3 Lon Llewelyn, Abergele, LL22 7DG
G4 IDW A Compton, Aysgarth, Durley Brook Road, Southampton, SO32 2AR
G4 IEB C Williamson, 72 Granville Drive, Kingswinford, DY6 8LL
G4 IEC A Everard, 2 Oak Wood Road, Wetherby, LS22 7QY
GM4 IEF A Hancock, Pitlair House Nursing Home, Cupar, KY15 5RF
G4 IEG [illegible] Close, Uxbridge, [illegible] 9HX
G4 IEH S Lindell, 60 Lakenheath, Oakwood, London, N14 4RP
G4 IES W Pitt, 1 Windy Ridge, James Street, Stourbridge, DY7 0ED
G4 IET John French, 10 Sunridge Avenue, Luton, LU2 7JL
G4 IEU W Griffiths, 3 Garreglwyd Park, Holyhead, LL65 1NW
G4 IEV P Gill, 48 Meeting House Lane, Balsall Common, Coventry, CV7 7FX
G4 IEZ R Senior, 5 Cwm Arthur, Denbigh, LL16 4BD
GW4 IFE A Strachan, 1 Cornelius Close, South Cornelly, Bridgend, CF33 4RQ
G4 IFI C Loftus, C/O, 15 Chappell Road, Manchester, M43 7UQ
G4 IFJ M Daniels, 8 Hathersage Drive, Glossop, SK13 8RG
G4 IFM Stanley Petraitis, 16 Brookbank Road, Dudley, DY3 2RX
G4 IFR P Hanson, 42 Oak Avenue, Newport, TF10 7EE
G4 IFT D Howorth, 11a Norwood Drive, Torrisholme, Morecambe, LA4 6LT
G4 IFU P Griffin, 8 Kelsey Close, St. Helens, WA10 4GY
G4 IFX Christopher Deacon, Spring Valley, Churt Road, Farnham, GU10 2QU
G4 IGE Leslie Hall, 57 Station Hill, Swannington, Coalville, LE67 8RJ
G4 IGF Peter Higgs, Oulton, Daisy Lane, Parkside, Wrexham, LL12 8BP
G4 IGG N Bennett, 1 Burnham Avenue, Oxley, Wolverhampton, WV10 6DX
G4 IGK M Wickham, 43 Bishopstone, Aylesbury, HP17 8SH
G4 IGL R Coombes, 9 Beechwood Close, Evington, Leicester, LE5 6SY
GM4 IGS R Chapman, 65 Lochgreen Avenue, Troon, KA10 6UP
G4 IGT R Roberts, Clydfan, Lon Ganol, Menai Bridge, LL59 5TH
G4 IGU K Blackett, 46 Lansdown, Yate, Bristol, BS37 4LR
G4 IGZ D Pellowe, 191 Preston New Road, Blackpool, FY3 9TN
GD4 IHC R Furness, Breryk, Windsor Road, Ramsey, Isle of Man, IM8 3EB
G4 IHI P Ferrari, Maggie, Back Road, Halesworth, IP19 9DY
G4 IHM I Wingfield, Keyhaven, 2 Belmont Close, Abergavenny, NP7 5HW
G4 IHO D CARSON, 21 Harris Road, Harpur Hill, Buxton, SK17 9JS
G4 IHR N Allen, 8 Shoulbard, Fleckney, Leicester, LE8 8TX
G4 IHS Gerald Donn, Flat 31, Rex Cohen Court, Liverpool, L17 1AB
G4 IHT R Riddington, Beech House, Tetbury, GL8 8SN
G4 IHX P Honour, 21 Castle St., Marsh Gibbon, Bicester, OX27 0HJ
G4 IHY R Clarkson, 11 Seymour Road, Newton Abbot, TQ12 2PT
G4 IHZ M Hyde, 23 Northumberland Way, Barnsley, S71 5DH
G4 IIA Michael Stamford, The Old Wheelwrights, East Street, Leominster, HR6 9HB
G4 IIB K Marshall, Alderbaran, Ruckcroft, Carlisle, CA4 9QR
G4 IIC C Clifford, 11 Halfcot Avenue, Stourbridge, DY9 0YB
G4 IID Colin Eastland, 40 Hillside Road, Bushey, WD23 2HA
G4 IIH P Henson, 70 Mell Road, Tollesbury, Maldon, CM9 8SR
G4 III P Godwin, Holgate, Selby Road, Goole, DN14 0LN
G4 IIK C Lodge, 35 Beaumont Cottages, Kelsale, Saxmundham, IP17 2NW
G4 IIN N Evans, 56 Homerton Road, Middlesbrough, TS3 8LX
G4 IIO Philip Howe, 59 Days Road, Samford Valley, Australia, 4520
GM4 IIR Andrew Nelson, 5 Scarletmuir, Lanark, ML11 7PS
G4 IIX Christopher Wherrett, 14 Sails Drive, York, YO10 3LR
G4 IIY Ian Fugler, Lees Hill Farm, Lees Hill, Brampton, CA8 2BB
G4 IJA B Barnes, 28 Oaklands Park, Roughton Moor, Woodhall Spa, LN10 6UU
G4 IJB R Butterworth, 3 Derriman Glen, Sheffield, S11 9LQ
G4 IJD J Seddon, 38 Kemple View, Clitheroe, BB7 2QD
GU4 IJF Nigel Roberts, Maison Du Cotil, Alderney, Guernsey, GY9 3YZ
G4 IJI M Walker, 19 Highbury Place, Headingley, Leeds, LS6 4HD
G4 IJJ Alan Spratt, 8 Pheasant Rise, Copdock, Ipswich, IP8 3LF
G4 IJM Ian Arnold, 44 Elwick Avenue, Acklam, Middlesbrough, TS5 8NT
G4 IJO G Gaunt, 7 Marine Parade, Saltburn-by-The-Sea, TS12 1DP
G4 IJR Brian Moyse, 1703 Twin Pond Circle, College Station, Texas, United States, 77845-3051
G4 IJU J Coles, 84 Mansfield Lane, Calverton, Nottingham, NG14 6HL
G4 IJV B Dowling, Box Cottage, Box, Stroud, GL6 9HB
GI4 IKF T Black, 147 Old Westland Road, Belfast, BT14 6TE
G4 IKI Paul Gabriel, 4 Four Cottages, Whippingham Road, East Cowes, PO32 6NG
G4 IKJ P Edwards, 34 Albion Road, Malvern Link, Malvern, WR14 1PU
G4 IKL R Hibbin, 2 Phoenix Close, West Wickham, BR4 0TA
G4 IKQ R Kitchener, 43 Haven Close, Swanley, BR8 7JY
GM4 IKT R Purves, 5 Forth Court, Port Seton, Prestonpans, EH32 0TN
G4 IKX D Thomas, 18a Stockwell Lane, Aylburton, Lydney, GL15 6DN
G4 IKY D Sillars, 34 Sandown Road, Stevenage, SG1 5SF
G4 ILA William McKae, 3 Grantham Close, Wirral, CH61 8SU
GM4 ILE J Smy, 2 Dungavel Gardens, Hamilton, ML3 7PE
G4 ILF A Hyde, 68 Broxburn Road, Warminster, BA12 8EZ
G4 ILH J Acott, 2 Park Hill Road, Sidcup, DA15 7NL
G4 ILI Grant Cratchley, 2 The Maples, The Reddings, Cheltenham, GL51 6RW
G4 ILL J Beynon, 28 Princes Meadow, Newcastle upon Tyne, NE3 4RZ
G4 ILM M Turnbull, Southlea, Newbury, Gillingham, SP8 4QJ
G4 ILN G Fitt, 15 Sidegate Avenue, Ipswich, IP4 4JJ
G4 ILP C Borkowski, 25 Stroud Road, Wimbledon, London, SW19 8DQ
G4 ILQ R Manton, 18 Barnetts Close, Kidderminster, DY10 3DG
G4 ILR C Howett, Meadow Cottage, Church Close, Norwich, NR12 7DL
GM4 ILS R Adam, 1 Woodlands Crescent, Bishopmill, Elgin, IV30 4LY
G4 ILT Gary Barnacle, 58 Cotley Road, Leicester, LE4 2LH
G4 ILW James Dingwall, Flat 3, Baltic House, London, SW2 1NQ
G4 ILX S Sliwinski, 9 Oakhill Road, Sheffield, S7 1SJ
GI4 ILZ W Sharpe, 22 Tweskard Park, Belfast, BT4 2JZ
G4 IMB P Gascoigne, 108 Blandford Avenue, Castle Bromwich, Birmingham, B36 9JD
GW4 IMC Trevor Waters, 34 Woodlands Park, Betws, Ammanford, SA18 2HF
G4 IMH V Tatman, 271 London Road, Bedford, MK42 0PX
G4 IML M Giles-Holmes, 8 Drakes Close, Plymouth, PL6 5XL
G4 IMP Anthony Phillpott, 15 Southways, Stombers Lane, Folkestone, CT18 7AP
G4 IMS John Roe, 5 Lawford Lane, Writtle, Chelmsford, CM1 3EA
G4 IMU Keith Holley, 18 Sandford Avenue, Loughton, IG10 2AJ
G4 IMV J Mollart, 8 Harrison Street, Newcastle, ST5 1NH
G4 INA Philip Grice, 48 Repington Road, Tamworth, B77 4AA
G4 INB B Dupree, 3 Hillary Road, Cheltenham, GL53 9LB
G4 INF B Walpole, Bridge Farm, Stony Lane, Exeter, EX5 1PP
G4 ING J Hartley, 50 Waverley Road, Hyde, SK14 5AU
G4 INI J Church, Belle Vue, Gas Lane, Torrington, EX38 7BE
G4 INU Frank Haighton, 2028 Cheviot Court, BURLINGTON, on, Canada, L7P 1W8
G4 INX Arthur Harris, [illegible] Close, Milborne Port, Sherborne, DT9 5ED
G4 IOA P Hill, 2 Salisbury Avenue, Ramsgate, CT11 7LH
GM4 IOB R Smith, Hestivald, Downies Lane, Stromness, KW16 3EP
G4 IOD W Marshall, 92 High Street, Ossett, WF5 9RQ
G4 IOE F Stevenson, Raakollveien 20A, Rolvsoy, Norway, N-1663
G4 IOG J Blackett, 70 Church Lane, Newington, Sittingbourne, ME9 7JU
G4 IOJ M Fielding, 35 Windmill Grove, Fareham, PO16 0HP
G4 IOK Christopher Marshall, 100 Hailey Road, Witney, OX28 1HQ
GD4 IOM Isle of Man ARS c/o Michael Webb, Coastguard House, 1 Mount Morrison, Peel, Isle of Man, IM5 1PN
G4 ION IONSPHERIC P GR c/o E Warrington, Dept Of Engineering, University of Leicester, Leicester, LE1 7RH
GI4 IOO R Chambers, 32 Victoria Road, Sydenham, Belfast, BT4 1QU
G4 IOQ A White, Wddyn Cottage, Treflach, Oswestry, SY10 9HQ
G4 IOR Brian Rowell, 2 The Willows, Burton-on-the-Wolds, Loughborough, LE12 5AF
G4 IOV Peter Emmerton, 5 Portsmouth Wood Close, Lindfield, Haywards Heath, RH16 2DQ
G4 IPB P Hodgkinson, Woodedge, Snaisgill Road, Barnard Castle, DL12 0RP
G4 IPF L Horseman, 55 Sackville Avenue, Hayes, Bromley, BR2 7JS
G4 IPH R Bass, 292 Thornhills Lane, Clifton, Brighouse, HD6 4JQ
G4 IPI D Foster, 1 Thorn Court, Four Marks, Alton, GU34 5BY
G4 IPJ C Jeans, 20 Parkfield Road, Ickenham, Uxbridge, UB10 8LN
GM4 IPK Andrew Steven, Pangdene, Virkie, Shetland, ZE3 9JS
G4 IPL L Winters, 58 Larkhall Lane, Harpole, Northampton, NN7 4DP
G4 IPM Nick Terry, 15 Baldwins Close, Bourn, Cambridge, CB23 2TH
G4 IPN W Flindall, 3 Meadow Drive, Gressenhall, Dereham, NR20 4LR
G4 IPR Tony Jones, 130 Turkey Street, Enfield, EN1 4PS
G4 IPV G Mayne, 228 Tutbury Road, Burton-on-Trent, DE13 0NY
G4 IPY A White, 3 Guarlford Road, Malvern, WR14 3QW
G4 IQA Reginald Lloyd, Llwyn Celyn, Pandy, NP7 8DN
G4 IQB D Fuller, 9 Llwyn Onn, Croesyceiliog, Cwmbran, NP44 2AL
GD4 IQD N Sivapragasam, 1 Treve Avenue, Harrow, HA1 4AL
G4 IQF S Wilkinson, 18 Tansey Crescent, Stoney Stanton, Leicester, LE9 4BT
G4 IQJ P Brannon, 90 Jacksmere Lane, Scarisbrick, Ormskirk, L40 9RS
G4 IQK G Evans, 14 Beach Priory Gardens, Southport, PR8 1RT
G4 IQO Christopher Britton, 271 Havant Road, Farlington, Portsmouth, PO6 1DB
G4 IQQ R Phillips, Moonraker, 2 The Close, Dartford, DA2 7ES
G4 IQR Nick Troop, 8 Fox Green, Great Bradley, Newmarket, CB8 9NR
G4 IQV G Menzies, 40 Epsom Lane North, Epsom, KT18 5PY
G4 IQW Adrian Langford, 42 Amis Way, Stratford-upon-Avon, CV37 7JF
G4 IQZ J Long, 51 Bratton Road, Westbury, BA13 3ES
G4 IRB John Heath, 19 Anson Road, Swinton, Manchester, M27 5GZ
G4 IRC IPSWICH RAD CLUB c/o J GEE, 11 Charlton Avenue, Ipswich, IP1 6BH
G4 IRD R Richards, 39 North Holme Court, Northampton, NN3 8UX
G4 IRG E Turner, 9 Wallingford Road, Handforth, Wilmslow, SK9 3JT
G4 IRH Trevor Pendleton, 17a Langley Drive, Kegworth, Derby, DE74 2DN
GD4 IRP Frank Boocock, 109 Northumberland Road, Harrow, HA2 7RB
G4 IRS R Ball, 1 Mount Hindrance C, Chard, TA20 1DZ
G4 IRU N Ashcroft, Oaklands, 11 Greenway, Wilmslow, SK9 1LU
G4 IRV J Hastie, 13 Thornlands, Easingwold, York, YO61 3QQ
G4 IRY R Gladden, 145a Hampton Road, South Fremantle, Australia, WA 6162
GI4 ISH M Fearis, 205 Dunluce Avenue, Belfast, BT9 7AX
G4 ISJ Peter Martin, 11 Winchester Way, Cheltenham, GL51 3EZ
G4 ISK David Brighton, 39 Les Forets, Glenac, France, 56200
GM4 ISM M Hughes, 6 Hawthorn Gardens, Larkhall, ML9 2TD
G4 ISN Andrew Holmes, 5 Launde Park, Market Harborough, LE16 8BH
G4 ISQ B Jones, 7 Timbertree Road, Warley, Cradley Heath, B64 7LE
GI4 ISR C Mcclurg, 4 Gracefield Lodge, Dollingstown, Craigavon, BT66 7UA
G4 ISS J Proudfoot, Laburnum Cottage, Corby Hill, Carlisle, CA4 8PL
G4 ISU N Whittingham, The Lilacs, 4 Ridgedale Mount, Pontefract, WF8 1SB
G4 ITB James Stone, 35 Landseer Avenue, Chapel St. Leonards, Skegness, PE24 5QZ
G4 ITC Christopher Claydon, 69 Abingdon Road, Dorchester-on-Thames, Wallingford, OX10 7LB
G4 ITG B Davey, 31 Somervell Drive, Fareham, PO16 7QL
G4 ITJ C Hard, 3 Longbridge, Ponthir, Newport, NP18 1GT
G4 ITP C Owen, 334 Beaumont Leys Lane, Leicester, LE4 2BJ
G4 ITQ B Lindley, 3 Orchard Way, Fontwell, Arundel, BN18 0SH
G4 ITR K Fisher, 51 Edge Hill, Ponteland, Newcastle upon Tyne, NE20 9RR
G4 ITV B Dingle, 74 Fenay Lane, Almondbury, Huddersfield, HD5 8UJ
G4 ITX M Payne, 34 Thales Drive, Arnold, Nottingham, NG5 7NF
G4 ITY D Hardie, 42 Lagoon Road, Pagham, Bognor Regis, PO21 4TJ
G4 IUA Jeff Campbell, 61 Telegraph Lane, Claygate, Esher, KT10 0DT
G4 IUF M Parker, 23 Pannal Avenue, Pannal, Harrogate, HG3 1JR
G4 IUH R Pye, 7 Meadow View, Potterspury, Towcester, NN12 7PH
G4 IUJ J Wroe, 25 Yew Tree Lane, Poynton, Stockport, SK12 1PU
GW4 IUK H Morley, 63 Lewis Road, Neath, SA11 1DJ
GW4 IUL D Pullin, 32 Clinton Road, Penarth, CF64 3JD
GD4 IUM G Adams-Spink, 55 Hawthorn Drive, Harrow, HA2 7NU
GW4 IUN R Janes, 3 Greenway Avenue, Rumney, Cardiff, CF3 3HQ
G4 IUP R Limbort, 21 Staincliffe Drive, Keighley, BD22 6FF
GM4 IUS N Bethune, 9 Links Gardens, Leith, Edinburgh, EH6 7JH
G4 IVB Ray Wollaston, 35 Main Road, Bilton, Hull, HU11 4AP
G4 IVC F Wood, 20a Lynwood Avenue, Felixstowe, IP11 9HS
G4 IVD A James, Church House, Cornfield Drive, Gloucester, GL2 4QJ
GI4 IVI A Kerr, 29 The Rose Garden, Tandragee, Craigavon, BT62 2NJ
G4 IVL Timothy King, Flat 1, 159 Cheriton Road, Folkestone, CT19 5HG
G4 IVO R Hargreaves, 23 Bracken Road, Long Eaton, Nottingham, NG10 4DA
G4 IVT G Coleman, 111 Woodland Drive, Watford, WD17 3DA
G4 IVU A Dixon, 66 Longacre, Chelmsford, CM1 3BJ
G4 IVZ G Harper, 12 Bletchley Road, Stewkley, Leighton Buzzard, LU7 0ER
G4 IWA John Arrowsmith, 16 Mancetter Road, Mancetter, Atherstone, CV9 1NZ
G4 IWD G Craig, 83 Pearl Road, Walthamstow, London, E17 4QY
G4 IWF G Mason, 51 Egerton Road, Streetly, Sutton Coldfield, B74 3PG

UK Callsigns

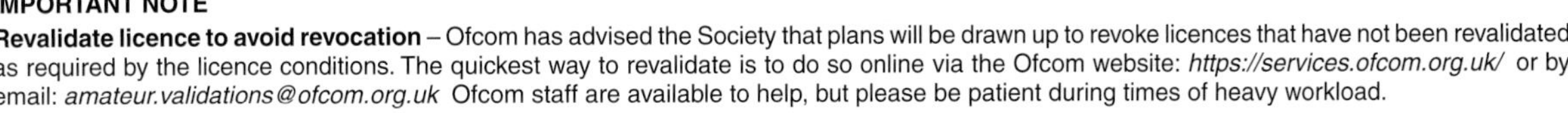

**IMPORTANT NOTE**

**Revalidate licence to avoid revocation** – Ofcom has advised the Society that plans will be drawn up to revoke licences that have not been revalidated as required by the licence conditions. The quickest way to revalidate is to do so online via the Ofcom website: *https://services.ofcom.org.uk/* or by email: *amateur.validations@ofcom.org.uk* Ofcom staff are available to help, but please be patient during times of heavy workload.

G4 IWI John Stocking, Bildersbrook, Grove Road, Melton Constable, NR24 2DE
G4 IWN J Andrews, 5 Chapman Avenue, Maidstone, ME15 8EG
G4 IWO Nicholas Bradley, 29 Raphaels, Basildon, SS15 5EA
GI4 IWP E Maclaine, 105 Bencran Road, Sixmilecross, Omagh, BT79 9QA
G4 IWQ D Cannon, 57 Halswell Road, Clevedon, BS21 6LE
G4 IWR S Berry, 40 Warrendale, Barton-upon-Humber, DN18 5NH
G4 IWS C Caine, 10 Goodwood Close, Burghfield Common, Reading, RG7 3EZ
G4 IWU John Scrivens, 7 Normandy Way, Fordingbridge, SP6 1NW
G4 IWV I Parker, 43 Longdown Road, Congleton, CW12 4QH
G4 IXB C Tuvey, 1 Dorset Way, Heston, Hounslow, TW5 0NF
G4 IXD I Palgrave Brown, The Abbey House, The Street, King's Lynn, PE33 9HP
G4 IXE G Walmsley, Warwick Farm House, Cracknore Hard Lane, Southampton, SO40 4UT
G4 IXF D Toon, 26 Reddish Avenue, Whaley Bridge, High Peak, SK23 7DP
GM4 IXH J Finlayson, 7 Abbotshall Road, Cults, Aberdeen, AB15 9JX
GD4 IXL THE 120 RC c/o Andrew Hosking, 30 Edrick Road, Edgware, HA8 9JD
G4 IXQ A Constable, Oakside, The Street, Bury St. Edmunds, IP31 1NG
G4 IXT Ian Jefferson, 7 Bluebell Close, Rugby, CV23 0UH
G4 IXW Geoffrey Hampson, 38 Draycott Road, Southmoor, Abingdon, OX13 5BZ
G4 IXY P Beardsmore, 2 Spencer Place, Sandridge, St. Albans, AL4 9DW
G4 IYA M Adams, 8 Boltons Close, Brackley, NN13 6ND
G4 IYC B Couchman, 48 Eastfields, Blewbury, Didcot, OX11 9NS
G4 IYE Ray Smith, 72 Worthing Road, Patchway, Bristol, BS34 5HX
G4 IYK Stuart Dixon, 33 Medhurst Crescent, Gravesend, DA12 4HJ
GI4 IYO K Burnside, 4 Cuttles Road, Comber, Newtownards, BT23 5YX
G4 IYP F Dearden, 22 Claremont Road, Chorley, PR7 3NH
G4 IYS D Burgess, 15 Prince George Avenue, Oakham, LE15 6GE
GM4 IYZ James Potts, Eastwood Court, 1 Eastwoodmains Road, Glasgow, G46 6QB
G4 IZA David Howard, 276 Dahlia Court, Bradenton, United States, FL 34212
GI4 IZF M Weller, 58 Manse Road, Ballycarry, Carrickfergus, BT38 9LF
G4 IZH P Robinson, 24 Haveroid Way, Crigglestone, Wakefield, WF4 3PG
G4 IZI C Thomas, 25 Loverock Crescent, Rugby, CV21 4AJ
G4 IZJ P Rennick, 41 Church Road, Pontnewydd, Cwmbran, NP44 1AT
GD4 IZL Garry Brookes, 44 Magherchirrym, Port Erin, Port Erin, Isle of Man, IM9 6DB
G4 IZQ A Scarth, 1 Beechwood Avenue, Whitley Bay, NE25 8EP
G4 IZS R Sexton, 31 Fosters Lane, Woodley, Reading, RG5 4HH
G4 IZU D Byers, 16 Tealby Court, Georges Road, London, N7 8HY
G4 IZX P Beards, 3 Elm Drive, Brightlingsea, Colchester, CO7 0LA
G4 IZZ Michael Eggleton, 49 Gretton Road, Gotherington, Cheltenham, GL52 9QU
G4 JA P Stenning, 20 Galba Road, Caistor, Market Rasen, LN7 6GN
G4 JAA P Hawkins, 38 Davidson Close, Great Cornard, Sudbury, CO10 0YU
G4 JAC Roy Emeny, 28 Manor House Way, Brightlingsea, Colchester, CO7 0QR
GM4 JAE I Miller, Moorgate, 5 Heathcote Gardens, Inverness, IV2 4AZ
G4 JAJ Brian Noble, 19 Ayrton Avenue, Blackpool, FY4 2BW
G4 JAQ M Crofts, 43 Broadlands Drive, East Ayton, Scarborough, YO13 9ET
G4 JAR HADRABS CONT GR c/o I Melville, 3 Crescent Road, Benfleet, SS7 1JL
G4 JAV W Bird, 10 Beech Avenue, Tamworth, B77 2BE
G4 JAX A Lunn, 11 Dibden Lodge Close, Hythe, Southampton, SO45 6AY
G4 JBA J Alderman, 38 Greenacres, Shoreham-by-Sea, BN43 5WY
G4 JBD Graham Laming, 72 Fildyke Road, Meppershall, Shefford, SG17 5LU
G4 JBE D Lacey, 16 Abbots Way, Monks Risborough, Princes Risborough, HP27 9JZ
G4 JBF G Lester, Lufflands, Yettington, Budleigh Salterton, EX9 7BP
G4 JBH A Dening, 42 Grove Avenue, Yeovil, BA20 2BD
G4 JBK A Maude, 5 Darrowby Close, Thirsk, YO7 1FJ
G4 JBL C White, Pegasus, Gotts Corner, Sturminster Newton, DT10 1DD
G4 JBQ Julian Cleak, Dantre, Newport Road, Cwmbran, NP44 3AE
G4 JBR P Dixon, Hardwick House, New Road, South Molton, EX36 4BH
G4 JBW D Barber, 3 Vestry Road, Street, BA16 0HY
G4 JBY G Bowden, 78 Lynwood Avenue, Darwen, BB3 0HZ
G4 JCA Christopher How, 9 Chanctonbury Walk, Storrington, Pulborough, RH20 4LT
G4 JCF G Hoey, Foehrer Strasse 8, Muenster, Germany, 64839
G4 JCG P Chapman, 4 Chester Close, Garstang, Preston, PR3 1LH
G4 JCH Brian Hercombe, 13 Dovecote, Shepshed, Loughborough, LE12 9RW
G4 JCJ C Newman, 4 Winchilsea Drive, Gretton, Corby, NN17 3BT
GW4 JCK N Warnock, 2 Sheepcourt Cottages, Bonvilston, Cardiff, CF5 6TN
G4 JCL D bryan, 3 New Lane, Skelmanthorpe, Huddersfield, HD8 9EH
GM4 JCM Alan Glashan, 35a Lochinver Crescent, Gourdie, Dundee, DD2 4UA
G4 JCP Philip Cadman, 21 Scotts Green Close, Dudley, DY1 2DX
G4 JCS J Stevenson, Highfields Farm, Saltburn by The Sea, TS13 4UG
G4 JCX Christopher Gallacher, 345 London Road, Clanfield, Waterlooville, PO8 0PJ
G4 JCY R Thornton, Binesfield, Bines Green, Horsham, RH13 8EH
G4 JCZ Anthony Clifton, 87 Aubrey Road, Quinton, Birmingham, B32 2BA
G4 JDC Leslie Boddington, Flat 33, Sorrel House, Birmingham, B24 0TQ
G4 JDE Gerwyn Evans, 32 Radstock Court, Abergavenny, NP7 5BQ
G4 JDF Peter Scovell, Flat 23, Derek Horn Court, Camberley, GU15 3JL
G4 JDG C Aitchison, Upper Weston House, Cot Lane, Chichester, PO18 8SU
G4 JDH R Purbrick, 2 Oyster Cottages, Tinnocks Lane, Southminster, CM0 7NF
GM4 JDK Maureen Hopkinson, The Coaches, Kingussie, PH21 1NY
G4 JDO R Tew, 4 Chetwode Close, Allesley, Coventry, CV5 9NA
G4 JDP S Pallett, 6 Lancaster Close, Coalville, LE67 4TG
G4 JDS L Radley, 34 Queens Road, Chelmsford, CM2 6HA
G4 JDT Harvey Lexton, 11 Mulberry Close, Romford, RM2 6DX
G4 JDW L Nelson-Jones, 15 Gainsborough Road, Bournemouth, BH7 7BD
GW4 JDZ D Samuel, 61 Bolgoed Road, Pontarddulais, Swansea, SA4 8JF
G4 JED Keith Bird, 25 Knowsley Way, Hildenborough, Tonbridge, TN11 9LG
G4 JEF Damian Wood, Little Burgate Farm, Markwick Lane, Godalming, GU8 4BD
G4 JEI N Osborne, 17 Rogate Close, Sompting, Lancing, BN15 0DY
GM4 JEJ Michael Thomson, Ravenside, Mill Road, Carnoustie, DD7 7SQ
GM4 JEM W Redpath, 69 Ulster Crescent, Edinburgh, EH8 7JL
G4 JEO F Kemp, 42 Baker Road, Abingdon, OX14 5LW
G4 JES M Wells, 12 Fulbeck House, Turner Avenue, Lincoln, LN6 7NQ
G4 JEY R Furness, Mermaid Lodge, 68/70 Brighton Lodge, Lancing, BN15 8LW
G4 JFC G Hainsworth, The Annexe, 16 Rowlandson Close, Northampton, NN3 3PB
G4 JFD David Featherstone, 6 Claremont Gardens, Tunbridge Wells, TN2 5DD
G4 JFF C Webb, 68 Higgs Field Crescent, Warley, Cradley Heath, B64 6RB
G4 JFG J May, Midsummers Eve, Third Cliff Walk, Bridport, DT6 4HX
GM4 JFH S Draycott, Kinmount, Whiting Bay, Isle of Arran, KA27 8QH
G4 JFN R Hudson, 15 Fellows Road, Farnborough, GU14 6NU
GI4 JFP D Goodman, 60 Castlewood Avenue, Coleraine, BT52 1EW
G4 JFS J FitzSimons, 27 Brese Avenue, Warwick, CV34 5TS
G4 JFV R Oldroyd, 197 Inner Promenade, Lytham St. Annes, FY8 1DW
G4 JFX Bernard Mount, 4 Maplestone Road, Whitchurch, Bristol, BS14 0HH
G4 JGF J Fitzgerald, 21 St. Aidans Avenue, Darwen, BB3 2BS
G4 JGG J Pether, 7 Celina Close, Bletchley, Milton Keynes, MK2 3LS
G4 JGH A Allchin, 9 Ashfield Road, Kings Heath, Birmingham, B14 7AS
G4 JGQ J Bevan, 10 Streamdale, Abbey Wood, London, SE2 0PD
G4 JGS S Harding, 9 Lightsfield, Oakley, Basingstoke, RG23 7BL
GW4 JGU A Green, 9 Westbourne Grove, Sketty, Swansea, SA2 9DT
G4 JGV S Sharred, Flat 2, 63 Severne Road, Birmingham, B27 7HJ
GW4 JGW K Simpson, 59 Midland Place, Llansamlet, Swansea, SA7 9QX
G4 JGX Robert Calver, Meadowbank, Lowertown, Helston, TR13 0BY
G4 JHA R Thomas, 2 Woodlands Road, Astley, Manchester, M29 7BH
GU4 JHH R Harvey, Courtil Masse, Les Landes, Vale, Guernsey, GY3 5JD
G4 JHI David Miller, 10 Fairview, Horsham, RH12 2PY
G4 JHN J Unwin, 28 Wallett Avenue, Beeston, Nottingham, NG9 2QR
G4 JHP J Hawes, 13 Broadmead Road, Colchester, CO4 3HB
G4 JHQ C Kear, 60 Haywoods Lane, Somerset, Tasmania, Australia, 7322
G4 JHS P Hey, 47 Hillcrest Road, Thornton, Bradford, BD13 3PQ
G4 JHU Norman Fineman, Deansway, 2 The Drive, Rickmansworth, WD3 4EB
G4 JHW D Morrison, Flat 1, 118 Anerley Park, London, SE20 8NU
GI4 JIC P Mcauley, 68 Ballylenaghan Heights, Belfast, BT8 6WL
G4 JIE Ronald Newman, 20 Marshmoor Park, Wallow Lane, Ipswich, IP7 7BZ
G4 JIG Ernest White, 12a Partridge Close, Great Oakley, Harwich, CO12 5DH
G4 JIH K Adams, 12 Hawkewood Avenue, Waterlooville, PO7 6EB
G4 JII R Green, Kingswood, Red House Lane, Doncaster, DN6 7EA
G4 JIJ I Kraven, 55 Cranfield Crescent, Cuffley, Potters Bar, EN6 4DZ
G4 JIK D Bird, 6 Wyebank, Bakewell, DE45 1BH
G4 JIN Virginia Gledhill, 25 Sandpiper Way, Torquay, TQ2 7GJ
G4 JIO K Mason, 5 Davenport Avenue, Hessle, HU13 0RL
G4 JIQ W Barker, 69 Britten Road, Brighton Hill, Basingstoke, RG22 4HN
G4 JIR A Rixon, 12 Vancouver Street, Darlington, DL3 6HN
G4 JIU Ian McGarrigle, 58 Langland Close, Corringham, Stanford-le-Hope, SS17 7LB
G4 JIV C DAVIES, 6 Valerie Avenue, Baulkham Hills, Australia, 2153
GI4 JIW J Ferrin, 38 Dalewood, Newtownabbey, BT36 5WR
G4 JIX James Bentley, 33 Lime Road, Ferryhill, DL17 8DL
GI4 JJF K A McLlroy, 69 Morston Park, Bangor, BT20 3ER
G4 JJH J Herbert, 8 Falmouth Road, Springfield, Chelmsford, CM1 6HY
GM4 JJJ D Anderson, Braeside, Urquhart, Fife, KY12 8QL
G4 JJM M Allison, 19 Ash Grove, Kirklevington, Yarm, TS15 9NQ
G4 JJP Richard Thomas, 71b St. Thomas Street, Wells, BA5 2UY
G4 JJQ J Wheway, 25 Mount View Avenue, Scarborough, YO12 4EW
G4 JJR L James, 65 Fflorens Road, Newbridge, Newport, NP11 3DW
G4 JJS Simon Harrison, Seacroft Grange Care Village, The Green, Leeds, LS14 6JL
GW4 JJV M Bell, 6 Owain Close, Cyncoed, Cardiff, CF23 6HN
GW4 JJW A Bell, 6 Owain Close, Cyncoed, Cardiff, CF23 6HN
G4 JJX Michael Grange, 6 Draysfield, Wormshill, Sittingbourne, ME9 0TY
G4 JJY J Carline, 101 Cemetery Road, Scunthorpe, DN16 1EB
G4 JKA J Ewen-Smith, 1 Kinnersley, Severn Stoke, Worcester, WR8 9JR
G4 JKB J Barnes, Capricorn, 13 Marchhill Drive, Dumfries, DG1 1PP
G4 JKC P Howard, 72 Marlowe Way, Lexden, Colchester, CO3 4JP
G4 JKE Derek King, Flat 21, Anchor Court, 2 Carey Place, London, SW1V 2RT
G4 JKF B Hodges, Gramaur, Mucklestone Wood Lane, Market Drayton, TF9 4ED
G4 JKH J Phillips, 57 New Sturton Lane, Garforth, Leeds, LS25 2NW
GW4 JKK A Bexley, Pennar Fach Farm Plwmp, Llandysul, SA44 6ES
G4 JKQ Ted Bowen, 40 Grange Road, Ibstock, LE67 6LF
G4 JKR David Wilson, 94 Lon Hedydd, Llanfairpwllgwyngyll, LL61 5JY
G4 JKS M Claytonsmith, Hares Cottage, Woolston, Church Stretton, SY6 6QD
GM4 JKT O Thores, 5 Havens Edge, Limekilns, Dunfermline, KY11 3LJ
G4 JKV M Rackham, 31 Severn Road, Pontllanfraith, Blackwood, NP12 2GA
G4 JKY Elizabeth Lennox, Lowfield House, Low Street, York, YO61 4QA
G4 JKZ K Leggett, Barton Lodge, Corfe, Taunton, TA3 7AQ
GM4 JLD P Woods, 12 Dalriada Place, Kilmichael Glassary, Lochgilphead, PA31 8QA
GI4 JLF R Russell, 1 Belmont Drive, Belfast, BT4 2BL
G4 JLG D Yorke, 40 Edge Fold Road, Worsley, Manchester, M28 7QF
G4 JLJ L Bailey, 3 Eden Close, Hutton Rudby, Yarm, TS15 0HT
G4 JLO H Dyson, 15 Swallow Grove, Netherton, Huddersfield, HD4 7SR
G4 JLP Derek Clarkson, Fareham Sailing & Motorboat Club, Lower Quay, Fareham, PO16 0RA
G4 JLV Julian Brower, 37 High Street, Steventon, Abingdon, OX13 6RZ
G4 JLX Harry Braggs, 47 Manor Road, Sandown, PO36 9JA
GM4 JLZ E Philip, 1 Pitstruan Terrace, Aberdeen, AB10 6QW
G4 JMB P Weaver, Flat 27, Regatta Point, 38 Kew Bridge Road, Brentford, TW8 0EB
G4 JMC J Trickett, 86 School Road, Thurcroft, Rotherham, S66 9DL
G4 JMF David Ollerhead, 15 Kingsley Road, Chester, CH3 5RR
GD4 JMG J Gorton, 12 Apsley Close, Harrow, HA2 6AP
G4 JMM John McFadyen, 33 Weymouth Bay Avenue, Weymouth, DT3 5AE
G4 JMO A Oakley, The Laithe, Coal Pit Lane, Colne, BB8 8NR
G4 JMP J Kelk, 10 Burton Fields, Herne Bay, CT6 6JU
G4 JMT M Firth, 6 Eastfield Drive, Woodlesford, Leeds, LS26 8SQ
GM4 JMU K Maxted, 18 Castleton Avenue, Newton Mearns, Glasgow, G77 5NF
G4 JMY D Liversidge, 6 Yardley Way, Grimsby, DN34 5UQ
GM4 JNB Norman Baird, 23 Scorguie Avenue, Inverness, IV3 8SD
G4 JNE C Houghton, 22 Rainow Road, Macclesfield, SK10 2PF
G4 JNH R Barker, 171 Leicester Road, New Packington, Ashby-de-la-Zouch, LE65 1TR
G4 JNK N Kendall, 35 Lodgefield Park, Stafford, ST17 0YE
G4 JNL P Senior, 9 Seely Close, Heighington, Lincoln, LN4 1TT
G4 JNQ E Allison, 7 Abbey Road, Flitcham, King's Lynn, PE31 6BT
G4 JNS David Hughes, Flat 23, Margaret Hill House, 77 Middle Lane, London, N8 8NX
G4 JNT Andrew Talbot, 15 Noble Road, Hedge End, Southampton, SO30 0PH
G4 JNU P Smith, 248a Kidmore Road, Caversham, Reading, RG4 7NE
G4 JNW L norton, 27 Guildford Square Lynemouth, Morpeth, NE61 5XP
G4 JNX N Whyborn, Kimberlin, Southwood Road, Norwich, NR13 3AB
G4 JNZ Christopher Barron, The Marling Pitts, Coughton, Ross-on-Wye, HR9 5ST
G4 JOA K Wood, Flat 98, Harbour Tower, Gosport, PO12 1HE
G4 JOB J Barker, 6 Larkswood Close, Tilehurst, Reading, RG31 6NP
G4 JOD F Rawlings, 14 Haddon Way, Carlyon Bay, St. Austell, PL25 3QG
GW4 JOG P Truberg, 106 Johnston Road, Llanishen, Cardiff, CF14 5HJ
G4 JOI R Tidnam, 21 Manor Lane, Lewisham, London, SE13 5QW
GD4 JOO Christopher Harman, 46 Chandos Crescent, Edgware, HA8 6HL
GI4 JOR J Farrell, 36 Cumber Park, Drumaness, Ballynahinch, BT24 8GA
G4 JOT S Carfoot, 24 Marble Church Grove, Bodelwyddan, Rhyl, LL18 5UP
G4 JOU R Bowden, Flat 7, 39 Anstey Road, Alton, GU34 2RD
G4 JOV John Wedderburn, 12 Victoria Avenue, Market Harborough, LE16 7BQ
G4 JOW Jonathan Butler, Pickstock Manor, Pickstock, Newport, TF10 8AH
G4 JPA Robert Jarvis, 2135 Oak Beach Blvd, 2135 Oak Beach Blvd, Sebring Fl, United States, 33875
G4 JPB J Beaumont, 9 Warren Bridge, Oundle, Peterborough, PE8 4DQ
GW4 JPC Gareth Woods, 178 Saron Road, Saron, Ammanford, SA18 3LN
G4 JPE B Hatley, 9 Somerstown Court, Tilehurst Road, Reading, RG1 7TY
G4 JPG I Wilson, 11 Ellwyn Terrace, Galashiels, TD1 2BA
GW4 JPJ H Genon, Dolau Cwerchyr, Penrhiwllan, Llandysul, SA44 5NZ
G4 JPK Jonpaul Pymm, Larkfield, Goxhill Road, Barrow-upon-Humber, DN19 7EE
G4 JPP E Jones, 1 Awel Y Mor, Cambrian Road, Tywyn, LL36 0AG
G4 JPQ J Hopewell, 2 Pyes Meadow, Elmswell, Bury St. Edmunds, IP30 9UF
G4 JPS BRISTOL RAYNET c/o A Williams, 38 Seneca Street St. George, Bristol, BS5 8DX
G4 JPX Ian Harrison, 61 Charles Street, Golborne, Warrington, WA3 3DF
GM4 JPZ C Hall, 42 Torridon Road, Broughty Ferry, Dundee, DD5 3JG
G4 JQB R Wickham, 35 Ashley Road, Bathford, Bath, BA1 7TT
G4 JQF M Key, 14 Ascot Road, Wigginton, York, YO32 2QE
G4 JQJ R Field, 12 Granson Way, Washingborough, Lincoln, LN4 1EY
G4 JQK S Casey, 14 Harrison Close, Emersons Green, Bristol, BS16 7HB
G4 JQL S Wayman, Oaktree Lodge, Redbridge Road, Dorchester, DT2 8BG
G4 JQN R Ward, 1 Dursley Road, Heywood, Westbury, BA13 4LG
GW4 JQQ R Henry, 1 Afan Valley Road, Neath, SA11 3SS
G4 JQS C Boulton, Manor Cottage, Stratton, Dorchester, DT2 9RY
G4 JQU Z Pokusinski, 362 Long Banks, Harlow, CM18 7PG
G4 JQV C Mee, 26 De Lisle Court, Loughborough, Leicester, LE11 4PP
G4 JQW Frank Lobban, 20 Evering Avenue, Poole, BH12 4JQ
G4 JQX Charles Riley, 1 Coulston, Westbury, BA13 4NX
G4 JR Andrew Anderson, 232 Annan Road, Dumfries, DG1 3HE
GI4 JRA J Harrigan, 124 Drones Road, Pharis, Ballymoney, BT53 8JT
G4 JRB M Hahn, 21 Stanley Road South, Rainham, RM13 8AJ
G4 JRD R De Muth, 66 Perkins Road, Ilford, IG2 7NQ
GM4 JRF H Hamilton, 8 Ardlui Gardens, Milngavie, Glasgow, G62 7RL
G4 JRJ S North, 2 Robey Drive, Eastwood, Nottingham, NG16 3DP
GW4 JRK R Kentish, Eryl, Ponthirwaun, Cardigan, SA43 2RJ
G4 JRW K Burton, 93 Truncliffe, Bradford, BD5 8NX
G4 JRY T Wislocki, 30 Kingston Road, Scunthorpe, DN16 2BE
G4 JS Darwen ARC c/o William Kenyon, Flat 21 House 4, Copper Place, Manchester, M14 7FZ
G4 JSD J Hamilton, 89 The Paddocks, Old Catton, Norwich, NR6 7HE
G4 JSE R Salaman, 39 Arthur Street, Unley, Australia, SA 5061
G4 JSK Lewis Welch, 3 Sunnyfield Avenue, Cliviger, Burnley, BB10 4TE
G4 JSM P Hart, 112 Shelton Avenue, Hucknall, Nottingham, NG15 7QA
G4 JSP Christopher Perkins, The Laurels, Higher Heath, Whitchurch, SY13 2HZ
G4 JSQ D Piper, 102 Redhouse Lane, Walsall, WS9 0DB
G4 JSS V Waddington, 1 Bridle Lane, Netherton, Wakefield, WF4 4HN
G4 JST F Ogden, 11 Stocklands Close, Cuckfield, Haywards Heath, RH17 5HH
G4 JSV N Hingley, 29 Mayfield Road, Hurst Green, Halesowen, B62 9QW
G4 JSW H Butler, 73 Bunbury Way, Epsom, KT17 4JP
G4 JSX Martin Owen, Thatched Cottage, Main Street, Rugby, CV23 0JA
G4 JSZ D Fry, The Stocks, Lyth Bank, Shrewsbury, SY3 0BE
GM4 JTA B Elliott, 14 Thornlea Drive, Giffnock, Glasgow, G46 6BZ
G4 JTC John Bautista, 47 Valiant House, Varyl Begg Estate, Gibraltar, GX11 1AA
G4 JTE P Djali, 177 Mount Pleasant, Kingswinford, DY6 9SS
GI4 JTF E H Squance, 11 Ballymenoch Road, Holywood, BT18 0HH
G4 JTK J Lee, 57 Capenhurst Lane, Whitby, Ellesmere Port, CH65 7AQ
G4 JTM J Llewellyn, Pier Road, Enniscrone, Co Sligo, Ireland
G4 JTO H Young, 72 Perrinsfield, Lechlade, GL7 3SD
G4 JTP G Parker, 14 Maplewood, Ashurst, Skelmersdale, WN8 6RJ
G4 JTR Vincent Robinson, 4 Hilltop Road, Caversham, Reading, RG4 7HR
GI4 JTS Robin MacRory, 8 Manse Road, Newtownards, BT23 4TP
G4 JTX P Simon, Dutch Cottage, Gainsborough Road, Gainsborough, DN21 5PE
GW4 JUC Hugh Woodward, 11 Pant-yr-Odyn, Sketty, Swansea, SA2 9GR
G4 JUD F Loach, 39 Park Road West, Wolverhampton, WV1 4PL
GM4 JUE J Cormack, 16 Shore Lane, Wick, KW1 4NT
G4 JUH R Wilkinson, 3 Anglesey Road, Dronfield, S18 1UZ
G4 JUI D Draper, Bryn Erin, Llangoed, Beaumaris, LL58 8SU
G4 JUK Michael Neville, 103 Walsall Road, Great Wyrley, Walsall, WS6 6LD
G4 JUM B Buller, 36 Grove Road, Ashtead, KT21 1BE
G4 JUN Victor Winton, Ty Cerrig, Rhosesmor Road, Holywell, CH8 8DL
G4 JUR R Harrod, 6 Carnforth Road, Barnsley, S71 2RA
G4 JUV C Bauers, 21 Nethergate Street, Bungay, NR35 1HE
G4 JUW W Cole, The Mill House, Maesbrook, Oswestry, SY10 8QH
G4 JUZ N Gabriel, 156 Clarence Avenue, New Malden, KT3 3DY
G4 JVA G Butler, 37 Turmore Dale, Welwyn Garden City, AL8 6HT
G4 JVC I Jones, 4 Grove Crescent South, Boston Spa, Wetherby, LS23 6AY
G4 JVD P Hainsworth, 74 Ravensbourne Drive, Woodley, Reading, RG5 4LJ
G4 JVH G Onions, 3 Tower Rise, Tividale, Oldbury, B69 1NP
G4 JVJ R Ashman, 44 Conan Doyle Walk, Swindon, SN3 6JB
G4 JVM F Pearson, Coach House, The Park, Manningtree, CO11 2AL
GJ4 JVP John ARTHUR, 13 Les Quennevais Park, St. Brelade, Jersey, JE3 8GB

G4 JVT G Howell, 25 Thornhill Road, Hednesford, Cannock, WS12 4LR
G4 JVV R Greaves, 15 Beech Ride, Sandhurst, GU47 8PR
G4 JVX D Powell, 6 Curlew Close, Winsford, CW7 1SW
G4 JVZ M Glennon, 5 Whiteley Croft Rise, Otley, LS21 3NR
G4 JWA D Naylor, 19 Dindbarrow, Burton Bradstock, Bridport, DT6 4RG
G4 JWK L Ball, Tree Tops, Bodiam, Robertsbridge, TN32 5UG
G4 JWL P Woodward, Le Rosey, Rolle, Switzerland, 1180
G4 JWV N Lyons, 114 Spring Hill, Weston-Super-Mare, BS22 9BD
GI4 JWW T Martin, 57 Oneill Road, Newtownabbey, BT36 6UN
G4 JXC R Butler, 6 Woodland Avenue, Dursley, GL11 4EW
G4 JXE P King, 21 Compton Way, Olivers Battery, Winchester, SO22 4HS
G4 JXH P McGivern, 83 Birdhill Avenue, Reading, RG2 7JU
G4 JXI [illegible], 12 Coronation Drive, Leigh, WN7 [illegible]
G4 JXJ Clive Blewitt, 12 Salton Street, Secret Harbour, Australia, 6173
G4 JXK David Bonfield, 14 Springdale Close, Brixham, TQ5 9HL
GI4 JXM J Welsh, 20 Bryansburn Brae, Doagh, Ballyclare, BT39 0RJ
G4 JXN Gareth Roberts, 4 Fronddeg, Ffordd Penmynydd, Llanfairpwllgwyngyll, LL61 5AX
GM4 [illegible] S Brown, 10 Barclay Park, Aboyne, AB34 5JE
G4 JXR G Wilde, 26 Fleetham Grove, Hartburn, Stockton-on-Tees, TS18 5LH
G4 JXU K Lee, 34 Evergreen Way, Wokingham, RG41 4BX
G4 JXZ I Terrell, 10 Red Lion Close, Cranfield, Bedford, MK43 0JA
GM4 JYB B Sparks, Windhaven, Brough, Thurso, KW14 8YE
G4 JYE David Sargent, 15 Wilton Road, Balsall Common, Coventry, CV7 7QW
G4 JYF C Golley, 10 New Molinnis, Bugle, St. Austell, PL26 8QL
G4 JYG Trevor Watson, 59 Vincent Road, Norwich, NR1 4HQ
G4 JYH A Curtis, 19 Donnelly Road, Bournemouth, BH6 5NW
GI4 JYJ Geoff McMaw, 26 Watch Hill Road, Ballyclare, BT39 9QW
G4 JYK P Leather, 35 Somerset Close, Congleton, CW12 1SE
G4 JYL Gwendalyn Thomas, 16 Fordlands, Thorpe Willoughby, Selby, YO8 9PD
G4 JYN WATERSIDE ARS c/o Timothy Williams, 31 Manor Road, Holbury, Southampton, SO45 2NQ
G4 JYP N Shelley, 25 Threeways, Cuddington, Northwich, CW8 2XJ
G4 JYQ J Tierney, 39 Daneway, Southport, PR8 2QW
G4 JYT R Armstrong, 38 Watson Avenue, Market Harborough, LE16 9NA
G4 JYU Nigel Bourner, 11 Richborough Road, Sandwich, CT13 9JE
G4 JYW David Proctor, 36 Westlands, Pickering, YO18 7HJ
G4 JZA S Geary, Bella Vista, The Square, Truro, TR2 4DS
GM4 JZB David Gardner, 7 Croft Road, Auchterarder, PH3 1EW
G4 JZC A Leigh, 50 Colbourne Road, Hove, BN3 1TB
G4 JZF G Taylor, 1 Threshers Drive, Willenhall, WV12 4AN
G4 JZL Jeremy Adams, 1 Powell Close, Creech St. Michael, Taunton, TA3 5TE
G4 JZQ M Noakes, 333 St. Neots Road, Hardwick, Cambridge, CB23 7QL
G4 JZR E Williams, 7 Laurel Drive, Willaston, Neston, CH64 1TN
GD4 JZS David Dix, 1 Highfield Crescent, Northwood, HA6 1EZ
G4 JZV R Bellamy, 16 Cedar Close, Grafham, Huntingdon, PE28 0DZ
G4 JZY Jonathan Price, 18 Woodland Drive, Bassaleg, Newport, NP10 8PA
G4 JZZ C Gadd, 40 Stanley Mount, Sale, M33 4AE
GD4 KAB L Rose, 2 Westglade Court, Woodgrange Close, Harrow, HA3 0XQ
G4 KAE D Wood, 7 Mead Close, Cheddar, BS27 3XN
G4 KAL B Thompson, 23 South Street, Keelby, Grimsby, DN41 8HE
G4 KAM Steve Greenwood, Little Oaks, Green Lane, Axminster, EX13 5TD
G4 KAR R Jeffries, 22 Ingrams Way, Hailsham, BN27 3NP
G4 KAT M Westwater, 3 Burns Way, Harrogate, HG1 3NA
G4 KAU T Mansfield, 2 Stratford Crescent, Cringleford, Norwich, NR4 7SF
GM4 KAV F Bowles, 40 Craigbarnet Road, Milngavie, Glasgow, G62 7RA
G4 KAX M Haswell, 5 Westcombe Avenue, Leeds, LS8 2BS
G4 KAZ Brian Davies, 2 Glan Llyn Terrace, Bethel, Caernarfon, LL55 1YL
G4 KBA Kenneth Boucher, 22 Emery Close, Walsall, WS1 3AL
G4 KBB Brian Bristow, 13 Princes Street, Piddington, High Wycombe, HP14 3BN
G4 KBH R Hodgson, 29456 Trailway Lane, Agoura Hills, United States, 91301
G4 KBI C Wainman, 9 Willson Drive, Riddings, Alfreton, DE55 4AF
G4 KBK R Fisher, 80/72 Kangan Drive, Berwick, Australia, 3806
GJ4 KBM B Nelson, 1 La Genetiere, La Route Orange, St. Brelade, Jersey, JE3 8GP
G4 KBP Michael Ford, Micalma, Anslow Lane, Burton-on-Trent, DE13 9DS
G4 KBQ J Haslam, C/O Po Box 269, C111 Seeb Airport, Muscat, Oman
G4 KBS N Benton, 69 Stonehill Rise, Doncaster, DN5 9HD
GI4 KBW Peter Henderson, 7 Clonaslea, Newtownabbey, BT37 0UL
G4 KBX C Chapple, Woodend, Hebron, Morpeth, NE61 3LA
G4 KCC H Holmden, 29 Cambridge Road West, Farnborough, GU14 6QA
G4 KCD B Dean, 3 Marchant Court, Gunthorpe Road, Marlow, SL7 1UW
G4 KCF Kenneth Sanderson, 39 Kirkland Street, Pocklington, York, YO42 2BX
G4 KCM C Sanders, 13 Meadow Court, Whiteparish, Salisbury, SP5 2SE
G4 KCN David Salmon, The Pines, 5a Westfield Avenue, Harpenden, AL5 4HN
GI4 KCO Karen Wright, 72 Elm Corner, Dunmurry, Belfast, BT17 9PY
G4 KCP D Appleton, 28 Edgewood, Shevington, Wigan, WN6 8HR
GW4 KCQ P Evans, 2 Cwmnantllwyd Road, Gellinudd, Swansea, SA8 3DT
G4 KCR S Dunn, 4 St. Ronans Road, Harrogate, HG2 8LE
G4 KCT Barry Firth, 8 Lyndale Avenue, Osbaldwick, York, YO10 3QB
G4 KCU Dennis Greatbatch, 56 Riverside, Repps With Bastwick, Great Yarmouth, NR29 5JY
G4 KCV R Murray-Shelley, 5 Dan y Wern, Pwllgloyw, Brecon, LD3 9PW
G4 KCX P Hicks, 14 Oakwood, Flackwell Heath, High Wycombe, HP10 9DW
GW4 KCY P Trimmer, 15 Cypress Court, Landare, Aberdare, CF44 8YB
G4 KCZ C Conduit, 21 Shadybrook Lane, Weaverham, Northwich, CW8 3PN
G4 KDE A Lamont, 50 Hockley Road, Rayleigh, SS6 8EB
G4 KDH K Howe, Woodlands, St. Peters Road, Hockley, SS5 6AA
G4 KDI R Stanton, 33 Brook Road, Shotton, CH5 1HH
G4 KDK J Riggs, 28 Long Hill, Mere, Warminster, BA12 6LR
G4 KDL A Seago, 50 Kimberley Road, Lowestoft, NR33 0TZ
G4 KDM J Pearson, 75 Broomfield Road, Marsh, Huddersfield, HD1 4QF
G4 KDN J Phaff, Abefield, Fingest Lane, High Wycombe, HP14 3LS
G4 KDR I Wassell, 21 Speedwell Way, Horsham, RH12 5WA
G4 KDS Colin Lafferty, 31 Oran Court, Oranmore, Ireland
G4 KDU Graham Baldwin, 31 Kilnhurst Road, Todmorden, OL14 6AX
G4 KDW I Davidson, 24 Queenswood Drive, Hitchin, SG4 0LG
G4 KDX Christopher Arundel, 25 Howard Street, York, YO10 4BQ
G4 KEB L Bright, 49 Fellows Avenue, Wall Heath, Kingswinford, DY6 9ET
G4 KEC Robert Cookson, 4 Wellington Gardens, Selsey, Chichester, PO20 0EE
G4 KEE V Tomkins, 58 Chancellors Way, Beacon Hill, Exeter, EX4 9DY
G4 KEG D Fryer, 28 Hudson Road, Eastwood, Leigh-on-Sea, SS9 5NX
[illegible] Christopher Easton, Heyword, Marolingdy Lane, Heathfield, TN21 0EY
G4 KEL S Kell, 8 Thornhill Close, Shildon, DL4 1FL
G4 KEN K Smith, 32 St. Clements Road, Harrogate, HG2 8LX
GD4 KEP H Haria, 34 Larkfield Avenue, Harrow, HA3 8NF
GI4 KEQ D McMahon, 36 Ballycraigy Road, Newtownabbey, BT36 5ST
G4 KES B Bloomer, 2 Magor Hill Cottages, Magor Hill, Camborne, TR14 0JF
GD4 KEW R Marshall, 80 Drake Road, Harrow, HA2 9EA
G4 KEX B Hubbard, 16 Shelf Moor, Halifax, HX3 7PW
G4 KEY David Turner, 22 Westhawe, Bretton, Peterborough, PE3 8BA
G4 KEZ Brian Archer, 86 York Road, Swindon, SN1 2JU
G4 KFA T Bearpark, 19a Humber Lane, Patrington, Hull, HU12 0PJ
G4 KFB M Bird, 84 Penwill Way, Paignton, TQ4 5JQ
G4 KFC A Scandrett, 72 Hesketh Road, Yardley Gobion, Towcester, NN12 7TX
GW4 KFD R Wilson, 1a Treetops, Llanelli, SA14 8DN
G4 KFF R Hewson, 2 Ribchester Way, Brierfield, Nelson, [illegible]
G4 KFH E Dickinson, 21 Old Hall Park, Longthorpe, York, YO51 9BZ
G4 KFI D Bromfield, 3 Warwick Road, Brynithel, Ebbw Vale, NP23 4AH
G4 KFJ C Baker, 5 Holly Bank Rise, Dukinfield, SK16 5EG
G4 KFL R Rowney, 50 Wychdell, Stevenage, SG2 8JD
G4 KFP J Marshall, 92 High Street, Ossett, WF5 9RQ
G4 KFS T Wood, 47 Marsh View, Beccles, NR34 9RT
G4 KFT M Rothwell, 3 Chiltern Road, Prestbury, Cheltenham, GL52 5JQ
G4 KFY J Edwards, 15 The Meadows, Llandudno Junction, LL31 9LP
G4 KFZ R Stanton, 50 Plymstock Road, Plymouth, PL9 7NU
G4 KGA M Hattam, 11 Dukes Wood Avenue, Gerrards Cross, SL9 7LA
G4 KGC P Suckling, 314a Newton Road, Rushden, NN10 0SY
G4 KGE J Baldwin, 30 Petters Road, Ashtead, KT21 1NE
G4 KGF G Brooks, 1 Highfield Close, Pembury, Tunbridge Wells, TN2 4HG
GM4 KGK Norman Munro, Windyridge, Lower Bayble, Isle of Lewis, HS2 0QB
G4 KGL M Lees, 15 Blacklock, Chelmsford, CM2 6QL
G4 KGN D Mitchinson, 49 Baildon Way, Skelmanthorpe, Huddersfield, HD8 9GY
G4 KGO R Matthews, 191 Valley Road, Ipswich, IP4 3AH
G4 KGP K Pollard, 5 Lodge Way, Stevenage, SG2 8DB
G4 KGT J Hughes, 74 Fairacres, Prestwood, Great Missenden, HP16 0LF
G4 KGU B Thomas, 12 Link Road, Sale, M33 4HP
G4 KGX W Green, 3 Amos Road, Leicester, LE3 6NA
G4 KGY T Lawford, 32 Holm Oak Close, Canterbury, CT1 3JL
GM4 KGZ S Low, Gartwood, 14 Dundas Avenue, North Berwick, EH39 4PS
GM4 KHE G Phanco, 28 Park Road, Clydebank, G81 3LH
G4 KHG E Scholes, 19 Castle Hill, Newton-le-Willows, WA12 0DU
GM4 KHI Thomas Ferrie, 17 Bargarron Drive, Paisley, PA3 4LL
G4 KHJ Brian Geeson, 24 Rydal Avenue, Poulton-le-Fylde, FY6 7DJ
G4 KHK P Martin, 24 Heddington Close, Trowbridge, BA14 0LH
G4 KHM J Whitington, 18 Somerset Road, Ferring, Worthing, BN12 5QA
G4 KHQ J Woodland, 7 Lighthouse Park, St. Brides Wentlooge, Newport, NP10 8SL
G4 KHR R North, 21 St. Augustine Grove, Bridlington, YO16 7DB
G4 KHS KELSO ARS c/o G Chalmers, 38 Grove Hill, Kelso, TD5 7AS
G4 KHT A Lord, 5 Wasdale Green, Cottingham, HU16 4HN
G4 KHU P Hawkins, Temple View, High Street, Templecombe, BA8 0JG
G4 KHY M Edwards, Nafferton, Killerton Road, Bude, EX23 8EN
G4 KIF A Sansom, 1881-9 Avenue S E, Salmon Arm, Canada, BC V1E 2J6
G4 KIH W Bartlett, 48 Barrymore Walk, Rayleigh, SS6 8YF
G4 KIK D Whyborn, 33 Church Road, Trull, Taunton, TA3 7LG
G4 KIM P Newman, 16 Oakwood Road, Westlea, Swindon, SN5 7EF
G4 KIN Philip Taylor, 22 Windermere Drive, Rainford, St. Helens, WA11 7LD
G4 KIP J Ball, Moss Nook Farm, Moss Nook Lane ROAD, St. Helens, WA11 8AG
G4 KIQ A Brooks, 10 St. James Avenue East, Stanford-le-Hope, SS17 7BQ
G4 KIR Kenneth Chattenton, 29 Wand Hill Gardens, Boosbeck, Saltburn-by-The-Sea, TS12 3AP
G4 KIT E Sandaver, 33 North Farm Road, Lancing, BN15 9BT
G4 KIU Nigel Peacock, 47b De la Warr Road, East Grinstead, RH19 3BS
GI4 KIX D Gilmore, The Overlook, 29 Ballymaconaghy Road, Belfast, BT8 6SB
G4 KIY K Hancock, 5 St. Andrews Place, Whittlesey, Peterborough, PE7 1BX
G4 KIZ D Holmes, Lancaster House, Magna Mile, Market Rasen, LN8 6AD
G4 KJA B Preston, 24 Nursery Close, Hucknall, Nottingham, NG15 6DQ
GI4 KJC N Quinn, 54 Moyle Road, Newtownstewart, Omagh, BT78 4JT
G4 KJD I Pitkin, Clover Cottage, Kenny, Ilminster, TA19 9NH
G4 KJJ J Smith, 30 Rookery Close, St. Ives, PE27 5FX
G4 KJK David Oliver, 15 Brixham Avenue, Cheadle Hulme, Cheadle, SK8 6JG
G4 KJP L Jordan, Cami De Fuster No 1, Marxuquera Alta, Valencia, Spain, 46700
G4 KJS Allen Gregory, 1-3 Nargate Street, Littlebourne, Canterbury, CT3 1UH
G4 KJU R Fisher, 85 Larkway, Brickhill, Bedford, MK41 7JP
G4 KJV J Densem, Cotswold, Startley, Chippenham, SN15 5HG
G4 KKB Keith Blamey, 123 St. Edmunds Walk, Wootton Bridge, Ryde, PO33 4JJ
G4 KKG J Taylor, 12 Glenthorne Avenue, Yeovil, BA21 4PG
G4 KKJ H Perryman, 15 Queen Mary Crescent, Kirk Sandall, Doncaster, DN3 1JU
G4 KKN P Roberts, 2 Samuals Fold, Pendlebury Lane, Wigan, WN2 1LT
G4 KKO J Walton, 17 Wychperry Road, Haywards Heath, RH16 1HJ
G4 KKR R Page, 156 High Road, Newton, Wisbech, PE13 5ET
G4 KKS Albert Morris, 4 Woodville Gardens, Wigston, LE18 1JZ
G4 KKT J Mahoney, 18 Park Avenue, London, N22 7EX
G4 KKU A Imianowski, 97 Bloomfield Road, Brislington, Bristol, BS4 3QP
GM4 KKV P Rucklidge, 8 Stanehead Park, Biggar, ML12 6PU
G4 KKZ K Robinson, 13 Race Hill, Launceston, PL15 9BB
G4 KLA John Nelson, 67 Swarthmore Road, Birmingham, B29 4NH
G4 KLB Colin Watts, 42 Truscott Avenue, Bournemouth, BH9 1DB
G4 KLD C Dewhurst, 56 Collett Way, Priorslee, Telford, TF2 9SL
G4 KLE Mervyn Foster, 7 Church Street, Fenstanton, Huntingdon, PE28 9JL
G4 KLF Anthony Selmes, 82 Beaufort Court, Beaufort Road, St. Leonards-on-Sea, TN37 6PF
G4 KLJ D Wellings, 41 Wroxham Drive, Wollaton, Nottingham, NG8 2QR
G4 KLM Peter Raven, Wedgewood, Green Lane West, Norwich, NR13 6LT
GM4 KLN I Moore, 7 Greenside Avenue, Rosemarkie, Fortrose, IV10 8XA
GM4 KLO M Mistofsky, 18 Troon Place, Newton Mearns, Glasgow, G77 5TQ
G4 KLT L Jones, 52 The Drive, Bury, BL9 5DL
G4 KLX J Naylor, 12 Hayman Close, Mansfield Woodhouse, NG19 8BP
G4 KMB A Griggs, Barleycombe, Banwell Road, Axbridge, BS26 2XZ
G4 KME John Henry Horley, 50 Hillswood Drive, Endon, Stoke-on-Trent, ST9 0DW
G4 KMF E Collier, 31 Mosyer Drive, Orpington, [illegible]
G4 KMH Stephen Cottis, 61 Oaken Grove, Maidenhead, SL6 6HN
G4 KMJ D Edwards, 210 Hillside Road, Hastings, TN34 2QT
G4 KMK Robert Blower, 133 Almondbury Bank, Huddersfield, HD5 8EX
G4 KMM Philip Northmore, 96 The Avenue, London, N17 6TG
G4 KMP Glen Ramsey, 21 Goldsmith Road, Eastleigh, SO50 5EN
G4 KMW R Greenhough, 36 Churchbalk Lane, Pontefract, WF8 2QQ
G4 KMX R Cope, 41 Hall Lane, Witherley, Atherstone, CV9 3LT
G4 KNI David Rickard, 12 Dabryn Way, St. Stephen, St. Austell, PL26 7PF
G4 KNN A Leggott, 3 Hayes Mead, Holbury, Southampton, SO45 2JZ
G4 KNO Andrew Summers, Broxwood, Bury Road, Bury St. Edmunds, IP29 4PH
G4 KNQ H Smith, Grey Gables, Humphrey Gate, Buxton, SK17 9TS
G4 KNR G Mason, 6 Brompton Close, Bridlington, YO16 7LL
G4 KNS J Wallett, 40 Aldreth Road, Haddenham, Ely, CB6 3PW
G4 KNT I Morton, 65 Manton Road, Hitchin, SG4 9NP
G4 KNV D Wilkinson, Woodview, Old Byland, York, YO62 5LG
G4 KNX A Bennett, 4 Chelmarsh Close, Redditch, B98 8SQ
G4 KNZ S Davies, 17 Haywood, Bracknell, RG12 7WG
GW4 KOE R Lines, 19 Magnolia Close, Cardiff, CF23 7HQ
GM4 KOI S Milne, 24 St. Ternans Road, Newtonhill, Stonehaven, AB39 3PF
G4 KOJ J Wilson, 54 Devonshire Drive, Mickleover, Derby, DE3 9HB
G4 KOK Jonathan Stockley, Clee View, Leys Lane, Leominster, HR6 0AZ
G4 KON L Butt, 16a Kestrel Crescent, Oxford, OX4 6DX
G4 KOO Sidney Cawthorne, 8 Captains Brae, Twynholm, Kirkcudbright, DG6 4PE
G4 KOQ G Birkhead, 103 Roselawn Road, Castleknock, Dublin, Ireland, 15
G4 KOR A Hughes, 55 Welford Road, Shirley, Solihull, B90 3HX
G4 KOT G Lindsay, 10 Northamptonshire Drive, Durham, DH1 2DF
G4 KOU G Martin, 68 Golden Avenue, East Preston, Littlehampton, BN16 1QU
G4 KOV H Wright, Sandpiper Cottage, Standard Road, Wells-next-The-Sea, NR23 1JY
G4 KOW Donald McLachlan, 48 Nursery Avenue, Bexleyheath, DA7 4JZ
G4 KOY R Gill, 87 Penkett Road, Wallasey, CH45 7QQ
G4 KPD A Grant, Chandlers, Welsh Street, Chepstow, NP16 5LU
G4 KPE Paul Griggs, 6 Nightingale Way, Sutton Bridge, Spalding, PE12 9RG
G4 KPF T Hart, 15 Whitefriars Meadow, Sandwich, CT13 9AS
G4 KPG W Lam, 2 Wistaria Road, Flat 3a, Kowloon, Hong Kong
G4 KPH D Lewis, 4 Raymond Court, Pembroke Road, London, N10 2HS
G4 KPI J Lorton, 14 Provis Mead, Chippenham, SN15 3UA
G4 KPL M Young, 8 Tweed Close, Worcester, WR5 1SD
G4 KPM M Pitt, 20 Little Halt, Portishead, Bristol, BS20 8JQ
G4 KPP C Kelly, 115 Kingsdown Crescent, Dawlish, EX7 0HB
G4 KPS Christopher Saunders, 26 Henley Fields, St. Michaels, Tenterden, TN30 6EL
G4 KPT Castles and Stately Homes On The Air c/o Chris Darlington, 24 West Ridge, Northampton, NN2 7RA
G4 KPU G Taylor, 179 Bradway Road, Bradway, Sheffield, S17 4PF
G4 KPV F Dunn, 12 Streete Court, Westgate-on-Sea, CT8 8BT
G4 KPX R Burton, 28 Mulberry Way, Ely, CB7 4TH
G4 KPZ Vernon Cracknell, 106 High St., Upwood, Huntingdon, PE26 2QE
GW4 KQ D Phillips, 37 Saint Margarets Park, Lower Ely, Cardiff, CF5 4AP
GI4 KQA T Moffitt, 36 Greenview, Parkgate, Ballyclare, BT39 0JP
G4 KQC G Leatherbarrow, 6 Queens Walk, Thornton-Cleveleys, FY5 1JW
G4 KQD Alison Down, 5 Juniper Mead, Stotfold, Hitchin, SG5 4RU
G4 KQE A Mead, 9 Abraham Drive, Silver End, Witham, CM8 3SP
G4 KQH D Howes, 14 Manitoba Way, Eydon, Daventry, NN11 3PR
G4 KQK Charles Barnes, Glebe Farmhouse, Stafford, ST18 9DQ
G4 KQL A Daulman, 2 Trentham Road, Hartshill, Nuneaton, CV10 0SN
G4 KQO R Ferguson, 8 Rutland Gardens, Croydon, CR0 5ST
G4 KQP S Jones, 114 Portland Road, Toton, Nottingham, NG9 6EW
G4 KQQ R Jones, 2 Bubwith Walk, Wells, BA5 2EN
GM4 KQS Alexander Smith, 9 Woodmill, Kilwinning, KA13 7PT
G4 KQT F Ullman, 20 Deepdale Drive, Leasingham, Sleaford, NG34 8LR
G4 KQV C Hands, 41 Coverdale Road, Solihull, B92 7NU
G4 KQY M Pearce, 51 Grove Avenue, New Costessey, Norwich, NR5 0JB
G4 KQZ T Thorne, 17 Pine St. South, Bury, BL9 7BU
G4 KRD M Khalaf, 508 London Road, Thornton Heath, CR7 7HQ
G4 KRF Richard Moore, Flat 15, Nelson Court, 130 Rowson Street, Wallasey, CH45 2LZ
G4 KRH R Cane, 24 South End, Longhoughton, Alnwick, NE66 3AW
G4 KRJ E Gaffney, 54 Dockham Road, Cinderford, GL14 2BH
G4 KRN Alan Troy, 1b Lidderdale Road, Liverpool, L15 3JG
G4 KRO Bryan Hay, 2 Lapworth Oaks, Lapworth, Solihull, B94 6LE
G4 KRT Michael Davis, 35 Mullion Croft, Kings Norton, Birmingham, B38 8PH
G4 KRW H Waterman, 170 Station Road, Mickleover, Derby, DE3 9FJ
G4 KSA D Mountain, 178 Wragby Road, Lincoln, LN2 4PT
G4 KSG R Ralph, 62 Northdown Road, Solihull, B91 3ND
GI4 KSH H Morrow, 2 Carnhill Grove, Newtownabbey, BT36 6LS
G4 KSK R Benyon, C/ Grecia 17, Villalbilla, Madrid, Spain, 28819
GI4 KSO D Mawhinney, 233 Ballynahinch Road, Annahilt, Hillsborough, BT26 6BH
G4 KSQ B Morris, 22 Burdell Avenue, Headington, Oxford, OX3 8ED
G4 KSR Sylvie Norris, 17 Montroy Close, Henleaze, Bristol, BS9 4RS
G4 KST T Hughes, 42 Western Drive, Hanslope, Milton Keynes, MK19 7LD
G4 KSU Keith Prettyjohns, 99 Richmond Street, Sheerness, ME12 2QS
G4 KSY Alvey Street, 43 Ridgedale Road, Bolsover, Chesterfield, S44 6TX
G4 KTB T Cottham, 4 Talisman Close, Tiptree, Colchester, CO5 0DT
G4 KTG H Wilson, 24 Clumber Avenue, Newark, NG24 4DT
G4 KTP D Dimambro, 16 Southam Road, Radford Semele, Leamington Spa, CV31 1TA
G4 KTQ Gordon Davies, 56 Ffordd Cynan, Bangor, LL57 2NS
G4 KTR D Burrell, 67 Newfield Drive, Nelson, BB9 9RR
GW4 KTT Paul Valerio, The Brackens, Reynoldston, Swansea, SA3 1AE
G4 KTU Kenneth White, 22 Ridyard Street, Wigan, WN5 9PA
G4 KTW E Dale, The Woodlands, Cotheridge, Worcester, WR6 5LZ
G4 KTX J Goldsmith, The Maltings, Flacks Green, Chelmsford, CM3 2QS
G4 KTZ P Cullen, 5 Swaledale Gardens, Fleet, GU51 2TE

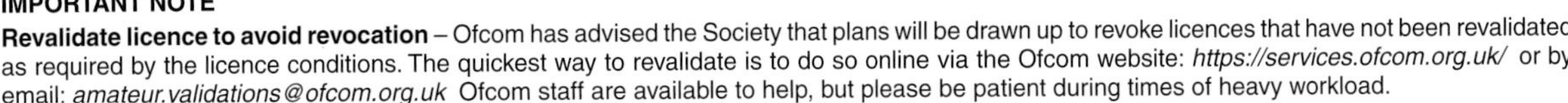
**IMPORTANT NOTE**

**Revalidate licence to avoid revocation** – Ofcom has advised the Society that plans will be drawn up to revoke licences that have not been revalidated as required by the licence conditions. The quickest way to revalidate is to do so online via the Ofcom website: *https://services.ofcom.org.uk/* or by email: *amateur.validations@ofcom.org.uk* Ofcom staff are available to help, but please be patient during times of heavy workload.

G4 KUC J Goodier, 20 Poleacre Lane, Woodley, Stockport, SK6 1PG
G4 KUD Barry Whittles, 12 Locksley Gardens, Birdwell, Barnsley, S70 5SU
G4 KUE Christopher Raspin, 35 Allesley Hall Drive, Coventry, CV5 9NS
G4 KUF A Redman, 42 Gallows Hill Lane, Abbots Langley, WD5 0DA
G4 KUJ Trevor Groves, 31 Tunnel Wood Close, Watford, WD17 4SW
G4 KUK B Tayler, 8 St. Martins Field, Otley, LS21 2FN
G4 KUL D Hepplestone, 104 Albert Road, London, DA5 1NW
GI4 KUM William Glenn, 1 Meadowside, Antrim, BT41 4HD
G4 KUQ P Goodfellow, 10 St. Agnes Walk, Knowle, Bristol, BS4 2DL
G4 KUR Stuart Hammonds, 22 The Croft, Meriden, Coventry, CV7 7NQ
GW4 KUS Herbert Hemmens, 44 Cecil Street, Manselton, Swansea, SA5 8QN
G4 KUX Nicholas Peckett, Four Winds, Woodland, Bishop Auckland, DL13 5RH
G4 KUY M Hill, 203 Biggleswade Road, Upper Caldecote, Biggleswade, SG18 9BJ
GI4 KUZ W Hamill, 47 Gracefield, Gracehill, Ballymena, BT42 2RP
G4 KVC R Mitchell, 6 Green Street, Smethwick, B67 7BX
G4 KVD James McMahon, 5 Victoria Walk, Wokingham, RG40 5YL
G4 KVI Christopher Dunn, 71 Redfield Road, Midsomer Norton, Radstock, BA3 2JH
G4 KVK P Park, 2 Leyburn Drive, High Heaton, Newcastle upon Tyne, NE7 7AP
G4 KVL B Tharme, 4 Longcroft Avenue, Liverpool, L19 4TB
G4 KVP A Woodland, 45 Walsingham Road, Wallasey, CH44 9DX
G4 KVQ R Scott, 20 Forest Hill, Carlisle, CA1 3HF
G4 KVR P Bell, 24 Onslow Gardens, Ongar, CM5 9BG
G4 KVT Jason Fairfax, 382 Wells Road, Bristol, BS4 2QP
G4 KVU M Shearer, Appleacre, Mill Road, Haverhill, CB9 7NN
G4 KVX P Bleiker, Waterside, 31 North Shore Road, Hayling Island, PO11 0HL
G4 KWE Tim Peel, Herongate, Derwent Lane, Hope Valley, S32 1AS
G4 KWF E Pickup, 36 Werneth Road, Glossop, SK13 6NF
G4 KWH Christopher Meadows, 16 Dart Road, Bedford, MK41 7BT
G4 KWJ J Hakes, Commonbank Cottage, Lancaster, LA2 9AN
G4 KWK K Hakes, Commonbank Cottage, Lancaster, LA2 9AN
G4 KWM P Deville, Bexton Doncaster Road, Mexborough, S64 0JD
G4 KWO G Phillips, 20 Eastfield Drive, Solihull, B92 9ND
G4 KWQ Andrew Soltysik, 24 Cottage Close, Hednesford, Cannock, WS12 1BS
G4 KWT Denis Pibworth, 20 Marathon Close, Woodley, Reading, RG5 4UN
G4 KWW J Ilott, 23 Keats Gardens, Stroud, GL5 4DJ
G4 KWX B Cox, Sylvan House, Alton Road, Farnham, GU10 5EL
G4 KWY D Gasser, 49 Pennycress, Locks Heath, Southampton, SO31 6SY
G4 KWZ Geoffrey Harris, Windmill House, Ripon Road, Kirby Hill, York, YO51 9DP
G4 KXF J Farrar, New Cottage, Messing Park, Colchester, CO5 9TD
G4 KXG Ken Jackson, 17 Copperfield Close, Kettering, NN16 9EW
G4 KXK J Ward, 38 Stonechat Avenue, Abbeydale, Gloucester, GL4 4XD
G4 KXL J Redman, 488 Blair Road, Georgia, United States, 30563
G4 KXO M Reynolds, Ilex House, Redwick Road, Bristol, BS12 3LQ
G4 KXP John Brockett, 17 Swan Drive, Droitwich, WR9 8WA
G4 KXQ M Wogden, 28 Magnolia Close, Barnstaple, EX32 8QH
G4 KXR A Tipper, 10 Tithebarn Copse, Exeter, EX1 3XP
G4 KXU Graham Robinson, 25 Stable Way, Kingswood, Hull, HU7 3FA
G4 KXV D Rigby, 145 Knightlow Road, Harborne, Birmingham, B17 8PY
G4 KXW G Redhead, 18 Paddock Way, Dronfield, S18 2FF
G4 KYE T Carhart, C6 Tamar Park, Coxpark, Gunnislake, PL18 9BD
G4 KYH A Waddilove, 2 Gwel Trencrom, Hayle, TR27 6PJ
G4 KYI R Shipton, 3 Fiery Lane, Uley, Dursley, GL11 5DA
G4 KYK J Jones, 7 Frankwell Street, Tywyn, LL36 9EP
G4 KYO Godfrey Barber, 25 Queensway, Hayle, TR27 4NJ
GW4 KYT D Thomas, 3 New Road, Trebanos, Swansea, SA8 4DL
G4 KYU R Ringrose, Melford House, George Street, Ipswich, IP8 3NH
G4 KYX D Gee, 13 Dart Road, Brickhill, Bedford, MK41 7BT
G4 KYY P Day, 46 Beatrice Avenue, Saltash, PL12 4NG
G4 KZB P Hazelwood, 12 Ryecroft, Stourbridge, DY9 9EH
G4 KZD John Young, 30 Crofton Way, Enfield, EN2 8HS
G4 KZI B Clark, 21 Church Road, Binstead, Ryde, PO33 3TA
G4 KZK R Smith, 15 St. Anthonys Way, Brandon, IP27 0DN
G4 KZO Andrew Keir, Kingfisher House, Nantwich Road, Tarporley, CW6 9JT
G4 KZQ Robert Bennett, 16 Emily Street, St. Helens, WA9 5LZ
G4 KZT B Ashdown, 19 Station Road, Helmdon, Brackley, NN13 5QT
G4 KZU N Rathbone, 7 Foreland Way, Keresley, Coventry, CV6 2NN
G4 KZV J Parkin, 18 Bradnock Close, Birmingham, B13 0DL
G4 KZW Stuart Haydock, 60 Tong Street, Bradford, BD4 9LX
G4 KZX A Still, 17 Arundel Road, Newhaven, BN9 0ND
G4 KZZ Nigel Roberts, 13 Rosemoor Close, Hunmanby, Filey, YO14 0NB
G4 LAB LEICESTERSHIRE WORKED ALL BRITTAIN GROUP c/o David Brooks, 28 Avon Vale Road, Loughborough, LE11 2AA
G4 LAD LEEDS&DIST ARC c/o M Howes, Yarnbury Rufc, Brownberrie Lane, Leeds, LS18 5HB
G4 LAE Christopher Wordley, Whispering Winds, 7 Fulcher Avenue, Chelmsford, CM2 6QN
G4 LAF R Brodrick, 16 Wallenge Drive, Paulton, Bristol, BS39 7PX
G4 LAI C Fone, 12 Chiltern Rise, Ashby-de-la-Zouch, LE65 1EU
G4 LAJ R Hackett, 4 Ryton Grove, Birmingham, B34 7RS
G4 LAK R Procter, 83 Twickenham Road, Newton Abbot, TQ12 4JG
G4 LAM R Lamberton, 28a Newtown Road, Raunds, Wellingborough, NN9 6LX
G4 LAN P Conway, 14 Leahall Lane, Rugeley, WS15 1JE
GM4 LAO Archie Waddell, 13 Auchenglen Road, Braidwood, Carluke, ML8 5PH
G4 LAP M Winter-Kaines, The Coach House, 15 Maltravers Street, Arundel, BN18 9AP
G4 LAU Charles Stevens, 9 Newbury Avenue, Melton Mowbray, LE13 0SR
G4 LAW F Craven, 2 Barn Owl Way, Stoke Gifford, Bristol, BS34 8RZ
G4 LAY G Dobbs, Chaka, Grimsby Road, Market Rasen, LN8 6DH
GM4 LBE Arthur Tait, 12 Greenwell, Gott, Shetland, ZE2 9UL
G4 LBH Richard Giles, Lodge N48, Merley House, Wimborne, BH21 3AA
G4 LBJ L Gurney, 3 Lathom House, Lathom Park, Ormskirk, L40 5UP
G4 LBM South London RAYNET c/o Ian Jackson, 5 Vivien Close, Chessington, KT9 2DE
GM4 LBN B Armistead, 45 Swanston Gardens, Edinburgh, EH10 7DF
G4 LBQ J Philipson, Clifton Farm House, Pullover Road, King's Lynn, PE34 3LS
G4 LBS BORDEN GRAM SCH c/o K Groom, 2 Ruins Barn Road, Tunstall, Sittingbourne, ME10 4HS
G4 LBT Richard Harmer-Knight, 3 Grendon Drive, Sutton Coldfield, B73 6QA
G4 LBU E Kersey, 98 Campbell Road, Ipswich, IP3 9RE
G4 LBX William Stopforth, 10 Cedar Crescent, Ormskirk, L39 3NT
G4 LBY S Wright, 22 Crown Street, Mansfield, NG18 3JL
G4 LCB M Goldman, 19 Myddelton Park, Whetstone, London, N20 0HT
G4 LCE N Watson, 14 Mill Lane, Whittlesford, Cambridge, CB22 4NE
G4 LCF Graham Williams, 2 The Paddocks, Lodge Hill, Newport, NP18 3BZ
G4 LCH Mark Gregory, 113 Masons Way, Solihull, B92 7JF
GM4 LCJ C Gove, 16 Aboyne Gardens, Kirkcaldy, KY2 6EL
G4 LCL E Beardmore, Kilaguni, The Avenue, Stoke-on-Trent, ST9 9LW
G4 LCM P Allsopp, 32 Linden Close, Cheltenham, GL52 3DU
GM4 LCP J Staruszkiewicz, 1 Nether Kirkton Way, Neilston, Glasgow, G78 3PZ
G4 LCU M Brownlow, The Croft, 1 Byne Close, Pulborough, RH20 4BS
G4 LDA R Lawrence, Tintern, Chepstow, NP16 6SE
G4 LDB T Kendall, 86 Rockford Close, Redditch, B98 7YL
G4 LDC A Wallis, 4 Trevose Close, Chandler's Ford, Eastleigh, SO53 3EB
G4 LDD P Harling, Pimlico House, Gisburn Road, Clitheroe, BB7 4ES
G4 LDJ F Gabell, 25 Woodland Way, Crowborough, TN6 3BQ
G4 LDL Anthony Bettley, 1 Dovetrees, Covingham, Swindon, SN3 5AX
GI4 LDN S McQuaid, Mullaghrodden, Dungannon, Co Tyrone, BT70 3LU
GW4 LDP I Dobby, 43 Chestnut Avenue, West Cross, Swansea, SA3 5NL
G4 LDR Neil Underwood, Blandings, Yarmley Lane, Salisbury, SP5 1RB
G4 LDS Chris Baker, 14 Clarendon Road, Morecambe, LA4 4HS
G4 LDT D Holland, 42 Front Street, Sunniside, Bishop Auckland, DL13 4LW
G4 LDW M Morris, 14 Batavia Road, Sunbury-on-Thames, TW16 5NB
GM4 LDX M McForsyth, Haltoun, Eddleston, Peebles, EH45 8PW
G4 LEA F Veale, 6 Grantson Close, Brislington, Bristol, BS4 4NA
G4 LED Adrian Wood, 67 Bay View Road, Duporth, St. Austell, PL26 6BN
G4 LEG Peter Brent, 14 Stagelands, Crawley, RH11 7PE
G4 LEM Edward Goodman, 83 Avondale Road, Kettering, NN16 8PL
G4 LEN Alan Kendall, 3 Australia Court, Cambridge, CB3 0JA
G4 LEP C Jacobs, 16 Woodyard Close, Mulbarton, Norwich, NR14 8AS
G4 LEQ LEQTRONICS R C c/o Gordon Adams, 2 Ash Grove, Knutsford, WA16 8BB
GM4 LER T Goodlad, 72 North Lochside, Lerwick, Shetland, ZE1 0PJ
G4 LES L Macvean, 27 Babs Field, Bentley, Farnham, GU10 5LS
G4 LEV C Veitch, 108 Racecourse Road, RD2, Otane, New Zealand, 4277
G4 LEX Gerald Train, 29 Waggoners Way, Morton, Bourne, PE10 0XR
G4 LFA Gordon Cooper, 22 Lilli Pilli Close, Laurieton, Australia, NSW 2443
G4 LFB H White, 184 Thistle Grove, Welwyn Garden City, AL7 4AJ
G4 LFE R Broom, Lakeside House, Thurlby Moor, Lincoln, LN6 9QF
GW4 LFF John Devonshire, 19 Voss Park Drive, Llantwit Major, CF61 1YD
G4 LFG M Davis, 53 St. Georges Avenue, South Shields, NE33 3EH
GM4 LFK Leslie McLean, Lower Hatton Cottage, Dunkeld, PH8 0ET
GM4 LFL John Rennie, 1 The Banks, Brechin, DD9 6JD
G4 LFQ J Holloway, Flat 1, 66 Unthank Road, Norwich, NR2 2RN
G4 LFS R Draycott, 25 Flat Lane, Whiston, Rotherham, S60 4EF
G4 LFT Anthony Busby, 44 Scrub Rise, Billericay, CM12 9PG
GW4 LFV B crow, Lindisfarne, Pen y Waun, Cardiff, CF15 9SJ
GW4 LFW Terence Cross, 50 Ty-Newydd, Whitchurch, Cardiff, CF14 1NQ
G4 LGB B Graham, 19 Cannerby Lane, Sprowston, Norwich, NR7 8NQ
G4 LGH K Garside, 191 Kenton Road, Newcastle upon Tyne, NE3 4NR
G4 LGK Bernard Kates, 28 Palm Drive, East Albury, Australia, NSW 2640
GM4 LGM John McGregor, 26 Engels Street, Alexandria, G83 0RZ
G4 LGO N Thomas, 24 Victoria Road, Teddington, TW11 0BG
GI4 LGP Steven McCracken, 29 Norwood Gardens, Belfast, BT4 2DX
G4 LGU W Hills, Alperton, Rowhill Road, Dartford, DA2 7QQ
G4 LGX J Hall, 30 Chatsworth Road, Harrogate, HG1 5HS
G4 LGY P Harber, 28 Regent Road, Epping, CM16 5DL
G4 LHA Grant Reoch, 16339 Granite Park Ct, Cypress, United States, 77429
G4 LHE J Lee, 41a Orchard Road, Seer Green, Beaconsfield, HP9 2XH
G4 LHF S Moffat, 14 Churchill Rise, Burstwick, Hull, HU12 9HP
G4 LHI Peter Rosamond, 13 Newnham Close, Hartford, Huntingdon, PE29 1RP
GM4 LHJ Joe Campbell, 23 Napier Avenue, Bathgate, EH48 1DF
G4 LHL M Edwards, 16 Maes Crugiau, Rhydyfelin, Aberystwyth, SY23 4PP
GM4 LHQ W Herron, 21 Southfield Avenue, Paisley, PA2 8BY
G4 LHR E Williams, 10 Eastbourne Close, Ingol, Preston, PR2 3YR
GD4 LHT James McSorley, 117 Park Avenue, Ruislip, HA4 7UL
GM4 LHW S Burnett, 17 Crusader Drive, Roslin, EH25 9NP
G4 LHY E Coventon, Mansa, 21 Trerieve Estate, Torpoint, PL11 3LY
G4 LIA J Gordon, 36 Warbeck Close, Newcastle upon Tyne, NE3 2FG
G4 LIC J Graham, 14 Ashbourne Grove, London, W4 2JH
GI4 LIF R Goligher, Mountjoy East, County Tyrone, BT79 7JJ
G4 LIG Paul Hesketh, 5 Beeches Close, Ixworth, Bury St. Edmunds, IP31 2EW
G4 LIJ R Nutt, 4 Mercers Drive, Bradville, Milton Keynes, MK13 7AY
G4 LIL C Brown, Sandysike Cottage, Sandysike, Carlisle, CA6 5SS
G4 LIM Geoff Moody, 37 Pine Street, Stockton-on-Tees, TS20 2SP
G4 LIO J Marshman, 12 Neelands Grove, Cosham, Portsmouth, PO6 4QL
G4 LIQ P Williams, 54 High St., Yelling, St. Neots, PE19 6SD
G4 LIR P Taylor, 64 Walford Road, Rolleston-on-Dove, Burton-on-Trent, DE13 9AR
GM4 LIS D Wilkes, 11 Trinity Crescent, Beith, KA15 2HG
G4 LIX Geoffrey Greenwood, 11 James Street, Holywell Green, Halifax, HX4 9AS
G4 LIY C Ware, 4 Highfield Terrace, Lower Bentham, Lancaster, LA2 7EP
G4 LJB J Wild, 6 Chestnut End, Headley, Bordon, GU35 8NA
GU4 LJC B Le Lievre, Calabar Forest Road, Forest, Guernsey, GY8 0AB
G4 LJF I Shepherd, Hutts Farm, Blagrove Lane, Wokingham, RG41 4AX
G4 LJG D Seabrook, 16 Blinco Road, Rushden, NN10 0DT
G4 LJI R Pellatt, Milverton House Nursing Home, 99 Ditton Road, Surbiton, KT6 6RJ
G4 LJK R T Mckee, 5 Moorcroft, Ossett, WF5 9JL
G4 LJN R Bartlett, 37 Church Road, Ferndown, BH22 9ES
G4 LJR G Garden, 87 Cootes Avenue, Horsham, RH12 2AF
GW4 LJS Paul Harding, Harbour Light, Five Roads, Llanelli, SA15 5AQ
G4 LJT W Hayward, 155 The Fairway, Dymchurch, Romney Marsh, TN29 0QF
G4 LJU Colin Howell, 43 Copsleigh Close, Salfords, Redhill, RH1 5BJ
G4 LJW J Jenkins, 73 Greville Road, Southville, Bristol, BS3 1LE
G4 LJY J Warren, Clifden Farm, Quenchwell ROAD, Truro, TR3 6LN
G4 LKD J Spurgeon, Whitgift House, Whitgift, Goole, DN14 8HL
GW4 LKE R Sabido, 29 Taff Vale Estate, Edwardsville, Treharris, CF46 5NJ
G4 LKF B Thompson, 21 Birling Place, Corby, NN18 0LZ
GI4 LKG V Tait, 30 Corby Drive, Lisburn, BT28 3HG
G4 LKM G Clarke, 33 Mulberry Avenue, Penwortham, Preston, PR1 0LL
G4 LKP K Craven, 8 Melander Close, York, YO26 5RP
GW4 LKS W Evans, Dan Y Craig, Craig Road, Swansea, SA7 9HS
G4 LKT P Goodman, 34 Fullers Road, South Woodford, London, E18 2QA
G4 LKU Dennis Hill, 8 Lingfield Walk, Corby, NN18 9JS
G4 LKW Peter Head, 36a Ashacre Lane, Worthing, BN13 2DH
G4 LKX D Hepworth, 2 Granby Crescent, Doncaster, DN2 6AN
G4 LKZ C Shuttleworth, 17 Stirling Close, Clitheroe, BB7 2QW
G4 LLG P West, 5 Stonehill Close, Appleton, Warrington, WA4 5QD
G4 LLI G Matthews, 101 Trafalgar Road, Horsham, RH12 2QL
G4 LLL N Rudgewick-Brown, 8 The Avenue, Wheatley, Oxford, OX33 1YL
G4 LLM Thomas Barnes, 20 Mayes Close, Warlingham, CR6 9LB
G4 LLN R Connolly, Newfane, Temple Way, Slough, SL2 3HE
G4 LLQ A Leeming, 52 Kingfisher Drive, Pickering, YO18 8TA
G4 LLW A Saunders, 59 Fleming Road, Quinton, Birmingham, B32 1NB
G4 LLZ A Barr, 28 Roundway, Honley, Holmfirth, HD9 6DD
G4 LMA J Baylis, 41 Ailesbury Way, Burbage, Marlborough, SN8 3TD
G4 LMF R Harrison, 18 Gunners Lane, Studley, B80 7LX
GM4 LMG D Heasman, 41 Honeyberry Drive, Rattray, Blairgowrie, PH10 7RB
G4 LMK J Morris, 17 Overbrook Grange, Nuneaton, CV11 6BQ
G4 LML W Turner, 11 Field View Close, Exhall, Coventry, CV7 9BJ
G4 LMM P Stears, 127 Hughenden Avenue, High Wycombe, HP13 5SS
G4 LMN David Piper, 68 The Derings, Lydd, Romney Marsh, TN29 9BN
G4 LMR British Railways ARS c/o Geoffrey Sims, 85 Surrey Street, Glossop, SK13 7AJ
G4 LMV W Loxley, 92 Needlers End Lane, Balsall Common, Coventry, CV7 7AB
G4 LMW Rob Thomson, Shire Jee Neevas, Cold Ash Hill, Thatcham, RG18 9PH
G4 LMX S Crosson Smith, The Old Pump House, Engine Road, Downham Market, PE38 0EN
G4 LMY J Piggott, 30 Farleigh Fields, Orton Wistow, Peterborough, PE2 6YB
G4 LNC A Friis, 22 Garthwaite Crescent, Shenley Brook End, Milton Keynes, MK5 7AX
G4 LNE N Howorth, 42 Fairfield Avenue, Rossendale, BB4 9TQ
G4 LNG F Hollis, 97 Manor Road, Chesterfield, S40 1HZ
G4 LNM D Brown, 26 The Brucks, Wateringbury, Maidstone, ME18 5PX
GM4 LNN Christine Foden, 4 Coastguard Houses, Cromwell Road, Kirkwall, KW15 1LN
GW4 LNP Bridgend & District ARC c/o Thomas Hulmes, 12 Blackmill Road, Bryncethin, Bridgend, CF32 9YW
G4 LNQ K Marshall, 44 Rosemary Drive, Alvaston, Derby, DE24 0TA
G4 LNR L Miles, 130 Well Lane, Willerby, Hull, HU10 6HS
G4 LNT B Thompson, 113 Gordon Road, Stanford-le-Hope, SS17 7QZ
G4 LNY A Thurgood, 19 Froment Way, Milton, Cambridge, CB24 6DT
G4 LNZ G Langford, 15 Ambleside Drive, Hereford, HR4 0LP
G4 LOA M Rudd, Williford House, Hermitage, Dorchester, DT2 7BB
G4 LOB A Major, 33 Borough Road, Bridlington, YO16 4HN
G4 LOD D Parrott, 39 Groves Road, Newport, NP20 3SP
G4 LOE Gary Tuppeny, 5 Ashlawn Crescent, Solihull, B91 1PR
G4 LOF Maurice Adams, 7 Finningley Drive, Allestree, Derby, DE22 2XP
G4 LOG R Farley, Linden Lea, Close Hill, Redruth, TR15 1EW
G4 LOH T Fern, South Boderwennack Farm, Trevenen Bal, Helston, TR13 0PR
G4 LOI B Howell, 13 Westfield, Plympton, Plymouth, PL7 2DY
G4 LOJ C Black, Charisma, Church Road, Norwich, NR14 7PB
G4 LOM John Boult, Findon Lodge, Hartside, Durham, DH1 5RJ
G4 LON J Berg, 25 Larch Close, Billinge, Wigan, WN5 7PX
G4 LOO D Ross, 3 Little Lane, Clophill, Bedford, MK45 4BG
G4 LOP Christopher Hannah, 5 Gunby Road, Orby, Skegness, PE24 5HT
G4 LOR W Mooney, 21 Windsor Court, Poulton-le-Fylde, FY6 7UX
G4 LOV J Sutherland, 31 Kensington Road, Sandiacre, Nottingham, NG10 5PD
G4 LOX D Morton, 9 Metford Grove, Bristol, BS6 7LG
G4 LOY B Carr, Spring House, Station Road, Grimsby, DN36 5QS
G4 LPA D Williams, 6 St. Annes Close, Bexhill-on-Sea, TN40 2EL
G4 LPD R Mills, 3 Whitfield Close, Wilford, Nottingham, NG11 7AU
G4 LPF S Swain, 9 Brickyard, Stanley Common, Ilkeston, DE7 6FR
GM4 LPG W Maslen, Broomie Knowe, Skye of Curr Road, Grantown-on-Spey, PH26 3PA
G4 LPJ G Kolbe, Riccarton Farm, Newcastleton, TD9 0SN
G4 LPL I Davis, P K 215, Bodrum, Turkey, 48400 TR
G4 LPO J Hampson, Ivy Lodge, Tor Side, Rossendale, BB4 4AJ
G4 LPP P Holt, 119a Hertford Road, Stevenage, SG2 8SH
G4 LPS T Stansfield, 174 Perry Street, Billericay, CM12 0NX
G4 LPT J Hopkins, 19 Cairnport Road, Stranraer, DG9 8BQ
G4 LPU J Jones, 9 Aelybryn, Ceinws, Machynlleth, SY20 9EZ
G4 LPW Christopher Clarke, 5 The Cottages, Low Road, Dereham, NR20 3DG
G4 LPY J Carter, 147 Maidenway Road, Paignton, TQ3 2PT
G4 LPZ R Doran, 1 Maple Drive, Chellaston, Derby, DE73 6RD
G4 LQD Tony Alderman, 8 Melrose Road, Weybridge, KT13 8UP
G4 LQE N Bishop, 33 Pollards Green, Chelmsford, CM2 6UH
G4 LQF N Field, 14 Regent Road, Harborne, Birmingham, B17 9JU
G4 LQG C Richardson, 25 Hookstone Drive, Harrogate, HG2 8PR
G4 LQH R Sharpe, Owl Cottage, Royal Oak Lane, Lincoln, LN5 9DT
G4 LQI Erwin David, 105 Kingsdown Park, Whitstable, CT5 2DH
G4 LQJ John Spridgen, 4 Cheveril Lane, Bury, Huntingdon, PE26 2NH
G4 LQL D Lander, 1 Colby Close, Forest Town, Mansfield, NG19 0LS
G4 LQM T McCrimmon, Helm Bar, Lazonby, Penrith, CA10 1BX
GM4 LQR J Reid, 80 Bellside Road, Cleland, Motherwell, ML1 5NU
GI4 LQU R Mckinney, 12 Old Coach Gardens, Belfast, BT9 5PQ
G4 LQW Derek McNiel, C/O G Mcniel, Stable Cottage, Alresford, SO24 0HP
G4 LQX R Coleman, 35 Meadowside Road, Upminster, RM14 3YT
G4 LQZ Peter Dolling, Apartado 229, Alicante, Spain, 3340
G4 LRB Kenneth Geen, 34 Kensington Road, Ipswich, IP1 4LD
G4 LRD D Holt, 241 New Hey Road, Oakes, Huddersfield, HD3 4GH
G4 LRG John West, 9 Bainbridge Court, St. Helen Auckland, Bishop Auckland, DL14 9EJ
G4 LRH G Obermaier, 9 Milton Park Avenue, Southsea, PO4 8JG
G4 LRL P Wilkins, 12 Chadcote Way, Catshill, Bromsgrove, B61 0JT

G4 LRN Neil Barker, 3 Silesbourne Close, Birmingham, B36 9ST
G4 LRO Roger Talbott, 33 Highfield Street, Anstey, Leicester, LE7 7DU
G4 LRP Adrian Boyd, 5 Walmer Close, Southwater, Horsham, RH13 9XY
G4 LRQ R Collett, 5 Miles Drive, Grove, Wantage, OX12 7JA
G4 LRT S Berry, Hillview, Stanford Close, Northampton, NN6 0LW
GM4 LRU Thomas Hood, 20 Thomson Crescent, Port Seton, Prestonpans, EH32 0AN
G4 LRV N Bundle, Crosses Farm, Pleck, Sturminster Newton, DT10 1NY
G4 LRY R Ratcliffe, 12 Palmer Close, Nine Mile Ride, Wokingham, RG40 3EB
G4 LSA J Bell, Ryanna Cottage, Sturbridge, Stafford, ST21 6LF
GM4 LSD Gavin Newton, 2 Cochrane Court, Milngavie, Glasgow, G62 6QT
G4 LSE K Darton, 18 Hightield Avenue, Bishop's Stortford, CM23 5LS
G4 LSG Stanley Smith, 1 Parkland Crescent, Norwich, NR6 7RQ
G4 LSK A Soto, 3 Lynns Hall Close, Great Waldingfield, Sudbury, CO10 0FH
G4 LSL B Lawrence, 42 Cross St., Crowle, Scunthorpe, DN17 4LH
G4 LSQ P Elmer, 6 Elmore Lane, Kesgrave, Ipswich, IP5 2GW
G4 LSU A Burnett, 72 Ightham Road, Erith, DA8 1LU
G4 LSV C Herrell, 61 Mansfield Road, Alfreton, DE55 7JN
G4 LSX G Pearce, 2 Clark Close, Wraxall, Bristol, BS48 1JL
G4 LTG J Dimont, 16 Riverside Walk, Isleworth, TW7 6HW
G4 LTH John Allan, 13 Vincent Close, Corringham, Stanford-le-Hope, SS17 7QL
G4 LTI M Coverdale, 1a Halton Chase, Westhead, Ormskirk, L40 6JR
G4 LTK P Hinks, 1 Richard Joy Close, Holbrooks, Coventry, CV6 4EY
GM4 LTL N Hyde, 18 Mansefield, Methlick, Ellon, AB41 7DF
G4 LTM G Hudsmith, 17 Greenside Close, Dukinfield, SK16 5HS
G4 LTR Jack Bryant, The Whistlers, Frogham, Fordingbridge, SP6 2HT
G4 LTS B Packington, 83 Fitzroy Road, Whitstable, CT5 2LE
G4 LTT Alan Willetts, 43 Galloway Avenue, Birmingham, B34 6JL
G4 LTY S Richards, 39 Trenowah Road, St. Austell, PL25 3EB
G4 LTZ J Peake, 8 Surrey Drive, Congleton, CW12 1NU
G4 LUA R Gathergood, 37 Hawkley Drive, Tadley, RG26 3YH
G4 LUB D Waldron, 1 Galbraithe Close, Bilston, WV14 8HX
GM4 LUD R Bannerman, 20 Post Box Road, Birkhill, Dundee, DD2 5PX
G4 LUE Ernest Bailey, 8 Hild Avenue, Cudworth, Barnsley, S72 8RN
G4 LUF R Irish, 15 Tenter Hill, Wooler, NE71 6DB
G4 LUN A Stickland, 3 Kivernell Road, Milford on Sea, Lymington, SO41 0PP
G4 LUO C Morgans, Merlewood, Maidstone Road, Sittingbourne, ME9 7QA
G4 LUQ M Tust, 28 Osprey Close, Beechwood, Runcorn, WA7 3JH
GM4 LUS S Smith, 80 Deanburn Park, Linlithgow, EH49 6HA
G4 LUT W Terry, Morston, 121 Lodge Lane, Grays, RM17 5SF
G4 LUW E Johnson, 29 Watering Lane, Collingtree, Northampton, NN4 0NJ
G4 LUX Peter Biddle, 14 Crossways Park, Howey, Llandrindod Wells, LD1 5RD
G4 LUY Dennis Chubb, 11 Pelham Close, Bembridge, PO35 5TS
G4 LVA A Lucas, 4 Hewell Close, Kingswinford, DY6 7RQ
GI4 LVC J Chapman, 21 Coolshinney Close, Magherafelt, BT45 5DR
G4 LVD B Durrant, 140 Fletcher Road, Ipswich, IP3 0LA
G4 LVG D Halls, 7 Raeburn Road, Ipswich, IP3 0EW
G4 LVI Andrew Entwistle, 68 Sandy Lane, Stretford, Manchester, M32 9BX
G4 LVK A Kelly, 8 Green Slade Crescent, Marlbrook, Bromsgrove, B60 1DS
G4 LVN K Robinson, Marston House, Churchthorpe, Louth, LN11 0XL
G4 LVO V Stretch, 5 Ledwych Road, Droitwich, WR9 9LA
G4 LVR B Roe, 7 Abbey Fields, Crewe, CW2 8HJ
G4 LVV Alan Hanson, 1 Church Street, Kempsey, Worcester, WR5 3JG
GM4 LVW M Bowman, 5 Whinfield Gardens, Prestwick, KA9 2PW
G4 LVY W Hughes, 7 Cambridge Avenue, Marton-in-Cleveland, Middlesbrough, TS7 8EH
G4 LWB Philip Smith, 2a Kirby Lane, Kirby Lodge, Melton Mowbray, LE13 0BY
G4 LWC L Collins, 44 Hollybush Lane, Penn, Wolverhampton, WV4 4JJ
GW4 LWD W Chandler, 19 Cilhaul Terrace, Mountain Ash, CF45 3ND
G4 LWF P Green, 1 Haddon Croft, Hayley Green, Halesowen, B63 1JQ
G4 LWG Ian Lambert, 21 East View Terrace, Barnoldswick, BB18 5NW
GW4 LWL K Edwards, 25 Gareth Close, Thornhill, Cardiff, CF14 9AF
G4 LWN Raymond Nock, 83 Coles Lane, West Bromwich, B71 2QW
G4 LWQ G Simpson, Floral Cottage, 11 Summerwood Road, Street, BA16 0RL
G4 LWU R Moore, 45 Lime Kiln Way, Salisbury, SP2 8RN
G4 LWV E Videan, 40 Guessens Grove, Welwyn Garden City, AL8 6RF
G4 LWY J Bryce, 6a Cawley Avenue, Culcheth, Warrington, WA3 4DF
G4 LWZ CHEPSTOW + DISTRICT A.R.S c/o Stephen Trott, 6 Mounton Drive, Chepstow, NP16 5EH
G4 LXA D Gibson, 10 Church Close, Braybrooke, Market Harborough, LE16 8LD
G4 LXC P Johnson, 148 Broadmead, Tunbridge Wells, TN2 5NN
G4 LXD F De Bass, Hawthorns, 10 Melville Road, Thetford, IP24 1NG
G4 LXH D Jones, 34 Alpha Grove, Isle of Dogs, London, E14 8LH
G4 LXJ C Phillips, Bangla, Silchester Road, Tadley, RG26 5EP
GM4 LXM Graham Low, 23 Bellfield Road, North Kessock, Inverness, IV1 3XU
GW4 LXO Jonathan Eastment, 211 Pantbach Road, Rhiwbina, Cardiff, CF14 6AE
G4 LXR R Hooper, 88 Ninehams Road, Caterham, CR3 5LJ
G4 LXU Chris Lennox, Lowfield House, Low Street, York, YO61 4QA
G4 LXV A Rose, 40 Wilson Drive, Outwood, Wakefield, WF1 3DN
G4 LXW A Trousdale, 65 Low Moor Side, New Farnley, Leeds, LS12 5EA
G4 LXX Ian Taylor, Heatherlands, Felixstowe Road, Ipswich, IP10 0DE
G4 LXY David Millin, Flat G12a, Elizabeth Court, Bournemouth, BH1 3DX
G4 LYB C Kitchener, 4 Cramswell Close, Haverhill, CB9 0QL
G4 LYC P Collett, 7 Saxon Rise, Earls Barton, Northampton, NN6 0NY
G4 LYD D Palmer, 123 Bucklesham Road, Kirton, Ipswich, IP10 0PF
G4 LYE Norman Pilling, 22 Templar Way, Selby, YO8 9XH
G4 LYF Anthony Webb, 11 Crowland Road, Haverhill, CB9 9LE
G4 LYG J Lavis, Briar Cottage, Turleigh, Bradford-on-Avon, BA15 2HG
G4 LYL Helen Bonnor, 1 Christchurch Road, Winchester, SO23 9SR
G4 LYM Geoffrey Schiffeldrin, 68 The Fairway, Alwoodley, Leeds, LS17 7PD
G4 LYU Brian Gauntlett, 4 Sandbanks Gardens, Hailsham, BN27 3TL
GM4 LYV W Hattie, 47 Border Way, Kirkintilloch, Glasgow, G66 2BD
G4 LYW PRIOR PARK COLL c/o John Weston, Chenery Lodge, 44 Old Newbridge Hill, Bath, BA1 3LU
G4 LYX John Wylie, 15 Semley Road, Hassocks, BN6 8PD
G4 LYY J Schoolar, 140 Slades Road, Golcar, Huddersfield, HD7 4JR
G4 LZE C Lugard, 5 Woodland Gardens, South Croydon, CR2 8PH
G4 LZF C Chapman, 101 Stoneygate Road, Luton, LU4 9TL
G4 LZJ P Garnett, Drewen Garth, Church St, Hull, HU11 4RN
G4 LZK R Broughton, 18 Elim Court Gardens, Crowborough, TN6 1BS
GM4 LZO G McDonald, Ellrigg, Ballencrieff Toll, Bathgate, EH48 4LD
G4 LZP Robert Smith, 4 Glan Ysgethin, Talybont, LL43 2BB
G4 LZQ G Williams, 1 Portland Place, The Green, Gloucester, GL2 7ET
GI4 LZR W Turner, 61 Thiepval Avenue, Belfast, BT6 8JF
GI4 LZS J Smyth, 12 Cleland Park Central, Bangor, BT20 3EP
G4 LZT R Brown, 40 Pegholme Drive, Otley, LS21 3NZ
G4 LZU Edward Hayden, Firbank, 1 Watery Lane, Taunton, TA3 5BX
G4 LZV K Brazington, 38 Tamworth Road, Amington, Tamworth, B77 3BT
G4 LZZ A Siemieniago, 3 Skye Close, Highworth, Swindon, SN6 7HR
G4 MAB Michael Barry, 19 Trinity Road, Northampton, NN3 3FA
GI4 MAC M McKinney, 29b Saintfield Road, Ballygowan, Newtownards, BT23 6HB
G4 MAG D Lucas, 23 Rectory Close, Wistaston Green, Crewe, CW2 8HG
GM4 MAI Angus McWilliam, Lochaber, Braehead, Avoch, IV9 8QL
GI4 MAJ William Mcclintock, 37 Belfast Road, Larne, BT40 2PH
G4 MAK J Gregg, 63 Low Common, Methley, Leeds, LS26 9AF
G4 MAN M Mansell, 251 Long Road, Canvey Island, SS8 0JG
G4 MAR C Rowe, 20 Lucknow Road, Willenhall, WV12 4QF
G4 MAS Christopher Day, 59 Hoe Lane, Ware, SG12 9LS
G4 MAU D Birchall, 6 Hillmorton Road, Knowle, Solihull, B93 9JL
GI4 MAZ Phelim Canny, 7 Drumachose Park, Limavady, BT49 0NY
G4 MB J Bowes, 20 Broomfield Road, Bexleyheath, DA6 7PA
G4 MBA Anne Cowsill, 21 Manor Close, Bromham, Bedford, MK43 8JA
G4 MBC MID-BEDS CONTEST ASSOC c/o Frederick Handscombe, Sandholm, Bridge End Road Red Lodge, Bury St. Edmunds, IP28 8LQ
G4 MBD I Moth, 145 Carisbrooke Road, Newport, PO30 1DG
G4 MBE R Scargill, 17 Springfield Lane, Morley, Leeds, LS27 9PL
GM4 MBG I SIMPSON, 1 Knockhall Road, Newburgh, Ellon, AB41 6BJ
G4 MBH Alec Embleton, 4 Daventry Close, Mickleover, Derby, DE3 0QT
G4 MBJ R Hyett, 18 Escley Drive, Hereford, HR2 7LU
G4 MBK J Broadbent, Buttercross Cottage, Low Road, Gainsborough, DN21 4ER
G4 MBL S Elmore, Eirianfron, Llangoed, Beaumaris, LL58 8PG
GI4 MBM Samuel McConnell, 8 Carnesure Drive, Comber, Newtownards, BT23 5LP
GI4 MBQ Colin Black, 5 Woodbrook Park, Warrenpoint, Newry, BT34 3HL
G4 MBZ P Taylor, 227 Woodland Walk, Aldershot, GU12 4FQ
G4 MCA J Hanton, 5 St. Davids Drive, Thorpe End, Norwich, NR13 5HR
G4 MCE A Audcent, 9 Woodlands, Axbridge, BS26 2AX
G4 MCF C Begg, 11 Lilac Road, Normanby, Middlesbrough, TS6 0BS
GI4 MCH N Crymble, 60 Princes Drive, Newtownabbey, BT37 0AZ
G4 MCM D Hadaway, 66 St. Annes Close, Winchester, SO22 4LQ
G4 MCQ Stephen Bailey, 50 Quantock Close, Warmley, Bristol, BS30 8UT
GD4 MCR Derek Cannon, 44 Derby Road, Peel, Peel, Isle of Man, IM5 1HP
G4 MCU G Stow, 15 Hawthorne Gardens, Hockley, SS5 4SW
GM4 MCV A Patterson, 10a Hermitage Place, Edinburgh, EH6 8AF
GI4 MCW G Edgar, 51 Kempe Stones Road, Newtownards, BT23 4SQ
G4 MD Paul Howett, 2, Parkfield Road, Stourbridge, DY8 1HD
G4 MDB R Tokley, 9 Peel Road, Springfield, Chelmsford, CM2 6AQ
G4 MDC J Divall, 2 Brockswood Lane, Welwyn Garden City, AL8 7BG
GI4 MDD I Gibson, 4 Ilford Avenue, Belfast, BT6 9SF
G4 MDE Kenneth Hodkinson, 13 Clovelly Road, Edenthorpe, Doncaster, DN3 2PE
G4 MDF MID-THAMES RDF c/o B Bristow, Club Jays Lodge, Reading, RG8 7QG
G4 MDG J Baily, 13 Longleigh Lane, Bexleyheath, DA7 5SL
G4 MDH G Feary, 76 Parsons Way, Royal Wootton Bassett, Swindon, SN4 8DJ
G4 MDJ A Smith, 48 Milltown Way, Leek, ST13 5SZ
G4 MDK G Winfield, 328 Stone Road, Stoke-on-Trent, ST4 8NJ
G4 MDM P Brassington, 42 Dartmouth Avenue, Newcastle, ST5 3NY
G4 MDN David Fowler, The Dees, Cross Lanes, Gerrards Cross, SL9 0LR
GI4 MDO S Hewitt, 23 Drumard Road, Portadown, Craigavon, BT62 4HP
G4 MDR Alan Farmer, 42 Sunridge Close, Newport Pagnell, MK16 0LT
G4 MDT G Fitton, 29 Okus Grove, Upper Stratton, Swindon, SN2 7QA
G4 MDU J Gudgeon, Shillingsworth Cottage, Leckhampstead Road, Milton Keynes, MK19 6BY
GD4 MDY S Keenan, Fenella Villa, Peveril Road, Peel, Isle of Man, IM5 1PJ
G4 MDZ Shaun Cline, 24 Petrel Way, Hawkinge, Folkestone, CT18 7GZ
G4 MEA R Hutchings, 16 Le Marchant Road, Frimley, Camberley, GU16 8RW
G4 MEB P Green, 1 Haddon Croft, Hayley Green, Halesowen, B63 1JQ
G4 MEE D Mobbs, 39 Bramwell Road, Freckleton, Preston, PR4 1SS
G4 MEF B Rudkin, 18 Beechfield Avenue, Birstall, Leicester, LE4 4DA
G4 MEH M Hughes, 8 Elm Beds Road, Poynton, Stockport, SK12 1TG
G4 MEI I Williams, 27 Y Glyn, Caernarfon, LL55 1HF
G4 MEK C Chappell, 6 Brayside Avenue, Cowcliffe, Huddersfield, HD2 2PQ
G4 MEM Mark David, Ridgeway, Grainbeck Lane, Harrogate, HG3 2AA
G4 MEO B Elliott, 13 Spring Grove, Sandy, SG19 1EU
GI4 MEQ M Kelly, 6 Beechdene Gardens, Lisburn, BT28 3JH
G4 MES J Willis, 1 Cedar Crescent, Royston, SG8 5BP
G4 MET E Robinson, 60 Huntsmans Drive, Hereford, HR4 0PN
GD4 MEX M Care, 12 Hallowell Road, Northwood, HA6 1DW
GM4 MFB C Bowden, West Reidford, Drumoak, Banchory, AB31 5AU
G4 MFD W Dean, Bowerdene, Staplehay Trull, Taunton, TA3 7HH
G4 MFE R Kaiser, Blackcoombe Farm, Henwood, Liskeard, PL14 5BW
G4 MFI D Roberts, 88 Woodhouse Road, Urmston, Manchester, M41 7WX
G4 MFK G Dennick, 29 Vale Cottage, Old Post Office Lane, Evesham, WR11 5UF
GM4 MFL EASTER ROSS RC c/o R Kerr, Rosskeen Bridge, Invergordon, IV18 0PR
G4 MFN M Jones, 67 Dosthill Road, Two Gates, Tamworth, B77 1JD
GM4 MFO M Mackenzie, Ash Lodge, 8 Brookend Brae, Helensburgh, G84 0QZ
G4 MFP W Woollon, Greensward, Townsend, Didcot, OX11 0DX
G4 MFQ R Dunstan, 100 Trevithick Road, St. Austell, PL25 4RJ
G4 MFR C Shanks, 225 Freshfield Road, Brighton, BN2 9YE
G4 MFS M Smith, 23 Sutcliffe Avenue, Weymouth, DT4 9SA
G4 MFV J Marshall, 278 Derby Road, Bramcote, Nottingham, NG9 3JN
G4 MFW Barry Fletcher, 53 Onslow Gardens, London, SW7 3QF
G4 MFX Graham Davis, Orchard House Peartree Avenue, Martham, Great Yarmouth, NR29 4RJ
G4 MGB H Mayor, Lock House, Canal Bank, Preston, PR4 6HD
G4 MGG S Esposito, 21a Spencefield Lane, Leicester, LE5 6PT
G4 MGH Roger Hampson, 30 Witts Lane, Purton, Swindon, SN5 4EX
G4 MGI P Kemmis, 14 Rochester Road, London, NW1 9JH
G4 MGK W Brown, 2 Ashworth Park, Knutsford, WA16 9DE
G4 MGN M Norman, 7 Kingsway, Seaford, BN25 2NE
G4 MGO J Newman, 20 Marshmoor Mobile Home Park, Wallow Lane, Ipswich, IP7 7BZ
G4 MGP H Buddy, Greyholme, West Lane, Scarborough, YO13 9AR
G4 MGQ [illegible]
G4 MGR WIRRAL & DIST RD c/o Denis Jones, 39 Pensby Road Heswall, Wirral, CH60 7RA
G4 MGV Ronald Pass, 6 High Street, Hanslope, Milton Keynes, MK19 7LQ
G4 MGW Alan Lodge, 92 Cheltenham Road, Gloucester, GL2 0LX
G4 MGX J Freeman, 5a Beech Avenue, Briar Bank Park, Bedford, MK45 3WE
G4 MGY J Gibbs, 27 East Hill Park, Knatts Valley, Sevenoaks, TN15 6YF
G4 MHA Peter Wright, 2 Windsor Mews, Stanley, DH9 8UH
G4 MHC MALVERN HILLS RADIO AMATEURS CLUB c/o David Hobro, 60 Linksview Crescent, Worcester, WR5 1JJ
GI4 MHD G Quaite, 21 Broomhill Road, Ballynahinch, BT24 8QD
G4 MHE Ralph Musto, BOHINJSKA BELA 76, Bohinjska Bela, Slovenia, 4263
G4 MHF J Marchall, Chaseborough House, Village Hall Lane, Wimborne, BH21 6SG
G4 MHJ Robert Hewitt, 38 Eastry Road, Erith, DA8 1NN
G4 MHK T Fougere, 40 Longland Road, Eastbourne, BN20 8HY
G4 MHQ Alec Bell, 22 Ryde Place, Lee on The Solent, PO13 0AU
G4 MHR C Norman, 31 Cae Braenar, Holyhead, LL65 2PN
G4 MHS N Naish, 85 Wear Bay Road, Folkestone, CT19 6PR
G4 MHX B Smith, 6 Howbeck Crescent, Wybunbury, Nantwich, CW5 7NX
G4 MIA G O'Keeffe-Wilson, 20 South Drive, Upton, Wirral, CH49 6LA
G4 MIB David Senior, Court Barton, Bull Street, Taunton, TA3 5PW
G4 MID Edward Pratt, 65 Barton Road, Thurston, Bury St. Edmunds, IP31 3PD
G4 MIE A Weeden, Jacaranda, Sheringham Road, Sheringham, NR26 8TG
GM4 MIG Iain Giffen, 57 Glengarry Crescent, Falkirk, FK1 5UE
G4 MIH David Fenton, Waverley, Warrington Road, Northwich, CW8 2LW
GW4 MII Philip Jenkins, 2 Gwynfi Street, Treboeth, Swansea, SA5 7DW
G4 MIJ R Hunt, 21 Springwell, Ingleton, Darlington, DL2 3JJ
G4 MIK M Bull, Toad In The Hole, 25 Prospect Park, Tunbridge Wells, TN4 0EQ
GM4 MIM Iain Morrison, 53 Eastcroft Drive, Polmont, Falkirk, FK2 0SU
G4 MIO Peter Davies, 9 Place Albert 1e, la Hulpe, Belgium, 1310
GW4 MIP P Phillips, Woodreefe, Amroth, Narberth, SA67 8NR
G4 MIS N Allen, 17 Winfield Road, Sedbergh, LA10 5AZ
G4 MIT G Hurst, The Hollies, Derby, DE6 6NB
G4 MIV Kenneth GIBSON, 179 Ratcliffe Road, Sileby, Loughborough, LE12 7PX
G4 MIX Michael Howland, 1 Swanton Farm Cottages, Lydden, Dover, CT15 7JN
G4 MJA M Swift, 4 Embleton Drive, Chester le Street, DH2 3JS
G4 MJC Flemming Jul-christensen, 66 Bushey Lodge Cottages, Firle, Lewes, BN8 6LS
GI4 MJD M Dunne, 26 Duncreggan Road, Londonderry, BT48 0AD
G4 MJF M Hill, 42 Oaklands Drive, Northampton, NN3 3JL
G4 MJI T Sturmey, 35 Lane Court, Boscobel Crescent, Wolverhampton, WV1 1QH
G4 MJT F Harrison, 98 The Stray, South Cave, Brough, HU15 2AL
G4 MJU E Smith, 256 Stone Road, Stoke-on-Trent, ST4 8NJ
G4 MJW Stephen Carey, 50 Renals Way, Calverton, Nottingham, NG14 6PH
G4 MJX G Fisher, 87 Ethersall Road, Nelson, BB9 0RP
G4 MKD W Kendal, 9 The Glade, Furnace Green, Crawley, RH10 6JS
G4 MKE A Plaice, 10 Stockhill Road, Chilcompton, Radstock, BA3 4JL
G4 MKF M Franks, 13 Fifth Road, Newbury, RG14 6DN
G4 MKG P Opie, Timbers, Stockers Hill Road, Sittingbourne, ME9 0PL
G4 MKI Duncan Bray, 180 Greenhill Road, Herne Bay, CT6 7RS
G4 MKP T Burbidge, 11 The Drift, Little Gransden, Sandy, SG19 3DX
G4 MKQ Kevin Barnes, 2 Zeus Lane, Waterlooville, PO7 8AG
G4 MKR R Byford, 7 Sutton Mill Road, Potton, Sandy, SG19 2QB
G4 MKT Barry Jackson, Meadow Top Farm, Edgeside Lane, Rossendale, BB4 9SD
GM4 MKU James Flett, 40 Commerce Street, Lossiemouth, IV31 6QH
G4 MKW P Bowden, 12 Honeywood, Roffey, Horsham, RH13 6AE
G4 MKX C Gericke, Pear Tree House, Water Street, Bristol, BS40 6AD
G4 MLB Roger Padmore, 3 Uldale Close, Nelson, BB9 0ST
G4 MLG A Denyer, 94 Wood Lane, Chippenham, SN15 3DZ
G4 MLI B Mitchell, 16 Perhaver Park, Gorran Haven, St. Austell, PL26 6NZ
G4 MLO G Rae, 62 Brunel Drive, Upton, Northampton, NN5 4AJ
G4 MLQ J Lamont, 9 Deepdale Croft, Barugh Green, Barnsley, S75 1QG
G4 MLR J Norris, Freshfield House, Freshfield Lane, Haywards Heath, RH17 7HE
G4 MLV L Gaunt, 31 Moat Hill, Birstall, Batley, WF17 0DX
G4 MLW Ian Jones, 114 Tennent Road, York, YO24 3HG
G4 MLY I Vincent, 40 Treetops Close, London, SE2 0DN
G4 MM BOX 25 CONTEST GROUP c/o J Clayton, 217 Prestbury Road, Cheltenham, GL52 3ES
GD4 MMA K Barnard, 89 Kings Road, Harrow, HA2 9LD
G4 MMG A Beecher, 27 Normandale, Bexhill-on-Sea, TN39 3LU
G4 MMH M Evans, Corners, Howbourne Lane, Uckfield, TN22 4QB
G4 MMI R Hodge, 20 Linden Grove, Roydon, Diss, IP22 4GJ
GI4 MMJ L Kirk, 26 Wallace Hill Road, Downpatrick, BT30 9BU
G4 MMT A Haley, 25 Sutherland Avenue, Broadstone, BH18 9EB
G4 MNA L Meale, 57 Chestnut Drive, Newton Abbot, TQ12 4JZ
G4 MNB Ronald sharp, 77 Cloche Way, Swindon, SN2 7JN
G4 MNE J White, 25 Fulwith Drive, Harrogate, HG2 8HW
GI4 MNF N Foote, 4 Bushfield Road, Moira, Craigavon, BT67 0JB
G4 MNI W Loucks, 155 Brentwood Rd N, Toronto, Ontario, Canada, M8X 2C8
GI4 MNN R Barr, 13 Fairhill Walk, Belfast, BT15 4GR
G4 MNP M Ward, 14 Grayling Mead, Fishlake Meadows, Romsey, SO51 7RU
G4 MNT R Calkin, 5 Bergen Court, Maldon, CM9 6UH
G4 MOC Paul Fawkes, 118 Rhoon Road, Terrington St. Clement, King's Lynn, PE34 4HZ
G4 MOE S Noke, 48 Hoadley Green, Salisbury, SP1 3HS
GW4 MOG C Tombs, 14 Heol Merioneth, Boverton, Llantwit Major, CF61 2GS
G4 MOH Kenneth Moran, 1 Temple Close, Barnwood, Gloucester, GL4 3ER
G4 MOI D Bone, 69 Pick Hill, Waltham Abbey, EN9 3LD
G4 MOK V Cashmore, 31 Maes y Dyffryn, Greenfield, Holywell, CH8 7QR
G4 MOL K Highley, 3 West Hill Drive, Hythe, Southampton, SO45 6DL
G4 MOP Harry Williams, 15 Hiawatha, Wellingborough, NN8 3SH
G4 MOT K Watson, 12 Regency Park Grove, Pudsey, LS28 8QD

UK Callsigns

**IMPORTANT NOTE**

**Revalidate licence to avoid revocation** – Ofcom has advised the Society that plans will be drawn up to revoke licences that have not been revalidated as required by the licence conditions. The quickest way to revalidate is to do so online via the Ofcom website: *https://services.ofcom.org.uk/* or by email: *amateur.validations@ofcom.org.uk* Ofcom staff are available to help, but please be patient during times of heavy workload.

G4 MOV E Durey, 71 Orchard Road, Maldon, CM9 6EW
GW4 MOZ John Upstone, 31 Broadway, Llanblethian, Cowbridge, CF71 7EX
G4 MPA G Squibb, 36 Frognal Gardens, Teynham, Sittingbourne, ME9 9HU
GM4 MPC D Smith, 13 Fernie Place, Dunfermline, KY12 9BX
G4 MPG P Grace, 3 Warwick Grange, Solihull, B91 1DD
G4 MPH Nigel Simmonds, 77 Main Street, Long Whatton, Loughborough, LE12 5DF
G4 MPI William Sharples, Foxgrove, Bonchurch Shute, Ventnor, PO38 1NX
G4 MPJ Malcolm Whitfield, 26 Kingsclere Drive, Bishops Cleeve, Cheltenham, GL52 8TG
G4 MPK S Foster, 25 Thorne Crescent, Bexhill-on-Sea, TN39 5JH
G4 MPL T Grimbleby, 109 Downfield Avenue, Hull, HU6 7XE
G4 MPO C Duffy, 87 Allenby Drive, Leeds, LS11 5RX
G4 MPQ K Clark, 33 Landers Reach, Lytchett Matravers, Poole, BH16 6NB
GM4 MPR Donald Miller, Old School, Ackergill, Wick, KW1 4RG
G4 MPT D Abbott, 42 Rosebery Avenue, Blackpool, FY4 1LB
G4 MPW M Corbett, Braemar, Heath Mill Lane, Guildford, GU3 3PR
GM4 MPY Alastair Bell, 48 Greenlaw Crescent, Paisley, PA1 3RT
GI4 MQA Matthew Bradley, 28 Church Road, Moneyrea, Newtownards, BT23 6BB
G4 MQB Martin Pill, 33 Queens Road, Cheltenham, GL50 2LX
G4 MQF A Ramsey, 51 Queens Road, Warmley, Bristol, BS30 8EJ
G4 MQG C Winters, 45 Blackbush Spring, Harlow, CM20 3DY
G4 MQK C Cubitt, Ostend Place Chalets, Flat 2, Norwich, NR12 0NJ
G4 MQL R Cuddington, 20 Kingscourt Lane, Rodborough, Stroud, GL5 3QR
G4 MQM D Cressey, 8 Parklands Drive, Harlaxton, Grantham, NG32 1HX
G4 MQP P Crowe, 22 Ringsbury Close, Purton, Swindon, SN5 4DE
G4 MQQ Simon Jones, 30 Lindford Chase, Lindford, Bordon, GU35 0TB
G4 MQR Gareth Blower, 30 The Glebe, Cumnor, Oxford, OX2 9QA
G4 MQS P Elliott, 153 Glenhills Boulevard, Leicester, LE2 8UH
G4 MQV James Sanderson, 5 Babbacombe Drive, Ferryhill, DL17 8DA
G4 MQW R Richardson, Radbourne Cottage Farm, Upper Radbourne, Southam, CV47 1NQ
G4 MRB John Feeley, 177 Rock Street, Sheffield, S3 9JF
G4 MRD Christopher Scrase, 67 Old Barrack Road, Woodbridge, IP12 4ED
G4 MRK J Veitch, 14 Dunmore Avenue, Sunderland, SR6 8ET
G4 MRL Nigel Thomas, 31 Gloucester Street, London, SW1V 2DB
GI4 MRN J McCrea, 14 Fairfield Park, Bangor, BT20 4TX
G4 MRQ R Marchington, 78 Buxton Road, Dove Holes, Buxton, SK17 8DW
G4 MRS MARTLESHAM RS c/o A Cook, The Old Vicarage, High Road, Ipswich, IP6 9LP
G4 MRU Peter Sables, 76 Sandyfields View, Carcroft, Doncaster, DN6 8JQ
G4 MRW R Westmeckett, 3 Alton Grove, Portchester, Fareham, PO16 9NJ
G4 MRX J Cooch, 6 Blackthorn Close, Newton, Preston, PR4 3TU
GI4 MRZ E Smith, 114 Bloomfield Road South, Bangor, BT19 7HR
G4 MSA David Price, Fregatwal 4, Leiden, Netherlands, 2317GN
GD4 MSE J Sivapragasam, 1 Treve Avenue, Harrow, HA1 4AL
G4 MSG W Turton, 22 Meadow Road, Ripley, DE5 3EP
G4 MSI Philip Needham, Cil y Sarn, Llanegryn, Tywyn, LL36 9SB
G4 MSJ T Moan, 23 Laurel Grove, Sunderland, SR2 9EE
G4 MSK W Wilkinson, 24 Greenway, Bromley, BR2 8EY
GM4 MSL George Wallace, 21 Rosslyn Court, Rosslyn Avenue, Perth, PH2 0GY
G4 MSN R Slator, 27 The Drive, Alwoodley, Leeds, LS17 7QB
G4 MSP Paul Shaw, 972a New Hey Road, Huddersfield, HD3 3FE
G4 MSQ Frank Watson, Syne Hurst Cottage, Kimbolton Road, Bedford, MK44 2EW
G4 MSV S O'Donnell, Men A Vaur, Wall Rd Gwinear, Hayle, TR27 5HA
G4 MSW L Morgan, 22 Stonelea Road, Hemel Hempstead, HP3 9JY
G4 MSY R Naylor, 89 Pelham Avenue, Grimsby, DN33 3NG
GW4 MTD H McMurray, 14 Hopkin St., Brynhyfryd, Swansea, SA5 9HN
GW4 MTE Richard Smith, 6 Lavender Court, Brackla, Bridgend, CF31 2ND
G4 MTF G Williams, 8 Blythe Close, Newport Pagnell, MK16 9DN
G4 MTG John Sillitoe, 42 Marsham Road, Kings Heath, Birmingham, B14 5HD
G4 MTH A Smith, 25 Lindsay Close, Stanwell, Staines, TW19 7LF
GM4 MTI D Spence, Royal Fern, Dunollie Road, Oban, PA34 5JQ
G4 MTP John Theodorson, 7 Kingfisher Court, Overstone Lakes, Ecton Lane, Northampton, NN6 0BD
G4 MTR Donald Coulter, 15 Woodville Way, Whitehaven, CA28 9LT
G4 MTW F Cook, 2 Burford Gardens, Sunderland, SR3 1LX
GI4 MTZ G Downs, 19 Mullaghboy Road, Islandmagee, Larne, BT40 3TT
G4 MUA A Kingdon, 4 Castlemead Walk, Northwich, CW9 8GP
GI4 MUE J Gwilt, 207 Clandeboye Road, Bangor, BT19 1AA
G4 MUI D Brown, 114 Telford Way, High Wycombe, HP13 5TA
G4 MUJ Roger Parker, Llanedw, Rhulen, Builth Wells, LD2 3UU
G4 MUL D Mayo, 6 Leigh Avenue, Marple, Stockport, SK6 6DF
GI4 MUN A Lennon, 5 The Drumlins, Ballynahinch, BT24 8HW
G4 MUP G Rouse, 43 Oakwood Drive, Prenton, CH43 7NX
G4 MUQ C Ashlin, 13 Brantfell Grove, Bolton, BL2 5LY
G4 MUS David Elwell, 18 Padgetts Way, Hullbridge, Hockley, SS5 6LR
G4 MUT Terence Hackwill, 6 Ramsbury Drive, Earley, Reading, RG6 7RT
G4 MUU F Westall, 4 Francesca Lodge, Somerford Way, Christchurch, BH23 3QN
G4 MUV Stephen Nicolle, 17 Allensmore Close, Matchborough, Redditch, B98 0AS
G4 MUW G Weaver, 60 Crispin Road, Winchcombe, Cheltenham, GL54 5JX
GM4 MUZ H Angus, South Grange Care Centre, Grange Road, Dundee, DD5 4HT
G4 MVA Glynn Burhouse, 40 High Park, Hawarden, CH5 3EF
G4 MVB A Berrow, 657 Old Lode Lane, Solihull, B92 8NB
G4 MVE D Casey, 18 Sandholme Drive, Ossett, WF5 8QP
G4 MVP J Paskins, 190 Gore Road, New Milton, BH25 5NQ
GI4 MVQ D McCluney, 49 Upper Cairncastle Road, Larne, BT40 2EG
G4 MVS G Mellett, Weston, Byways, Chichester, PO20 0HY
G4 MVX M Gardiner, 206 Caulfield Road, East Ham, London, E6 2DQ
GW4 MVY D Davies, 48 Bryn Eglur Road, Morriston, Swansea, SA6 7PQ
G4 MVZ R Pepper, 4 Marine Avenue, Skegness, PE25 3ER
GI4 MWA Fred Ruddell, 16 Beechfield Manor, Aghalee, Craigavon, BT67 0GB
G4 MWC WEST MANCHESTER RC c/o T Speight, 1 Lyndene Avenue, Worsley, Manchester, M28 2RJ
G4 MWD Ian Shaw, 33 Park Farm Close, Horsham, RH12 5EU
G4 MWF P Wilkinson, 14 Grasmere Close, Penistone, Sheffield, S36 8HP
G4 MWG I Grant, 25 Wilton Drive, Romford, RM5 3TJ
G4 MWH R Blythe, 4 Ashlea Close, Selby, YO8 4NY
G4 MWJ R Featherstone, 3 Fairfield, Coningsby, Lincoln, LN4 4SP
G4 MWL A Keyworth, 14 Robinson Road, Sheffield, S2 5QW
G4 MWO Paul Gaskell, 131 Greenfield Road, Dentons Green, St Helens, WA10 6SH
G4 MWP Terry Underhill, 5 Lyndhurst Croft, Eastern Green, Coventry, CV5 7QE
G4 MWQ C Weir, 1 Ashfield Place, Ilkley Road, Otley, LS21 3PN
G4 MWS Macclesfield & District ARS c/o Alan Denny, 85 Delamere Drive, Macclesfield, SK10 2PS
G4 MWW J Mcleod, 91 Gorselands Way, Rowner, Gosport, PO13 0DG
G4 MWX L Payne, 40 Westmorland Drive, Costhorpe, Worksop, S81 9JT
G4 MXE R SWINNERTON, 8 Maple Close, Brereton, Sandbach, CW11 1SQ
G4 MXF R Wallace, 161 Alma Avenue, Hornchurch, RM12 6AT
G4 MXI Keith Bruntlett, Cronk-Ny-Mona, King Street, Louth, LN11 0PN
G4 MXM W Hawkridge, 7 Langdale Gardens, Leeds, LS6 3HB
G4 MXP Derek Aunger, 2 Leyland Lodge, Morval, Looe, PL13 1PN
G4 MXQ Ben Acres, 7 Haywain Close, Weavering, Maidstone, ME14 5UX
GI4 MXV Matthew McKee, 4 Loran Parade, Larne, BT40 2DF
GI4 MXW D McKinney, 13 Lynden Gate, Portadown, Craigavon, BT63 5YH
G4 MXX T Fenwick, Lower Rill Farm, Chillerton, Newport, PO30 3HQ
G4 MXY B Sowerby, 3 Goughs Lane, Bracknell, RG12 2JR
G4 MYB C Barham, 10 Little Brook Road, Sale, M33 4WG
G4 MYE B Chase, 9 Claremont Drive, Taunton, TA1 4JE
G4 MYN J Thomas, 42 Allington Drive, High Grange, Billingham, TS23 3UA
G4 MYQ Graham Pettican, 6 Baltic Wharf, Norwich, NR1 1QA
G4 MYS Andrew Sillence, 74 Atherley Road, Southampton, SO15 5DS
GI4 MYT W Stewart, 11 Fairway Gardens, Castlereagh, Belfast, BT5 7PS
G4 MYU A Summers, 6 Rothesay Road, Brierfield, Nelson, BB9 5RS
G4 MYW B Mallinson, 7 Barnes Wallis Way, Churchdown, Gloucester, GL3 2TR
G4 MYY E Ball, 59 Queen Anne Gardens, Falmouth, TR11 4SW
G4 MYZ R Roscoe, 4 Cobham Villas, Longden, Shrewsbury, SY5 8EP
G4 MZB R Mills, Heather Lea, Oaklands, Welshpool, SY21 8HL
G4 MZC B Horsman, 53 Meadow Drive, Bembridge, PO35 5XU
G4 MZF D Earnshaw, 21050 Summit Road, Los Gatos, United States, 95033-8501
G4 MZI G Dunn, 20 The Grange, Wombourne, Wolverhampton, WV5 9HX
G4 MZK P Ansell, 46 Rochford Way, Croydon, CR0 3AD
G4 MZL E Ailsby, 15 Norman Close, Bridport, DT6 4ET
G4 MZM P Harper, 9 The Orchard, Market Deeping, Peterborough, PE6 8JS
G4 MZN M Huntsman, Pear Tree Cottage, Hildersham, Cambridge, CB21 6BU
G4 MZQ Roy Bickley, 12 Cemetery Road, Market Drayton, TF9 3BD
G4 MZU C Harrison, 6 Woodlands Close, Chandler's Ford, Eastleigh, SO53 5AT
G4 MZV R Privett, 2 Stevenson Court, Eaton Ford, St. Neots, PE19 7LF
G4 MZY David Plater, 58 Mead Fields, Bridport, DT6 5RF
G4 MZZ J Powell, 40 Kent Road, Formby, Liverpool, L37 6BQ
G4 NAC David Bosworth, 13 Burns Road, Kettering, NN16 9LA
GI4 NAE D Jackson, 1 Cloughey Road, Portaferry, Newtownards, BT22 1ND
G4 NAJ N Ashdown, Cobwebs, Wilderness Lane, Uckfield, TN22 4HT
G4 NAK R Morey, 1 Bradfield Cottages, Queens Road, Freshwater, PO40 9HB
G4 NAQ C Maby, 13 Dingle Road, Bristol, BS9 2LN
G4 NAV G Quayle, 10 Abbotsford Grove, Timperley, Altrincham, WA14 5AZ
G4 NBC Michael Hoare, Chy Noweth, Seworgan, Falmouth, TR11 5QN
G4 NBF Anthony Topsfield, Wild Willow Cottage, Hancock Lane, Truro, TR2 5DD
G4 NBG C Budd, 24 Ash Road, Horfield, Bristol, BS7 8RN
G4 NBH W Cockshaw, 14 Shropshire Road, Leicester, LE2 8HW
G4 NBI Leslie Everton, 18 Markham Road, Sutton Coldfield, B73 6QR
G4 NBM M Jones, Rhos Eithin, Brynsiencyn, Llanfairpwllgwyngyll, LL61 6TZ
G4 NBN A Bergman, River House, Suggs Lane, Ilminster, TA19 9RJ
GI4 NBO G Barr, 51 Hillhead Road, Dundonald, Belfast, BT16 1XD
G4 NBP David Coggins, 28 Main Road, Marlesford, Woodbridge, IP13 0AD
G4 NBQ Raymond Hassell, 2 Regnum Close, Eastbourne, BN22 0XH
G4 NBR J Hill, The New Gatehouse, Gubboles Drove, Spalding, PE11 4AX
G4 NBS Anthony Collett, 10 Quince Road, Hardwick, Cambridge, CB23 7XJ
G4 NBW J Alford, 86 Grindleford Road, Great Barr, Birmingham, B42 2SQ
GW4 NBY Keith Barrett, Linroy, Stoney Lane, Bridgend, CF35 5AL
G4 NCA P Cook, 38 Oak Road, Kettering, NN15 7AP
G4 NCB Kenneth Woolfindin, Viewlands, Milford Road, Leeds, LS25 6AF
G4 NCD K Morey, Iona, Colwell Road, Totland Bay, PO39 0AH
G4 NCF W Prickett, 105 High St., Wootton Bassett, Swindon, SN4 7AU
G4 NCI Robert Smith, 25 Sisters Way, Birkenhead, CH41 4FF
G4 NCJ J Short, Allybere, Marhamchurch, Bude, EX23 0HY
G4 NCK Patricia Shapero, 3 Princess Court, Leeds, LS17 8BY
G4 NCP M Carver, 45 Harvester Way, Crowland, Peterborough, PE6 0DG
G4 NCS C Angove, 9a Wanstead Road, Bromley, BR1 3BL
G4 NCU M Hewitt, Hillcrest Bungalow, Middle Street, Crewkerne, TA18 8LY
G4 NCV L Hollingworth, 55 Glenfield Avenue, Nuneaton, CV10 0DZ
G4 NCY I Woomans, 223 Umberslade Road, Selly Oak, Birmingham, B29 7SG
G4 NCZ J Ramsay, 79 Humphrey Lane, Urmston, Manchester, M41 9PT
G4 NDC Derek Mason, 133 Bath Road, Atworth, Melksham, SN12 8LA
G4 NDD J Lloyd, 72 Thornyville Villas, Plymouth, PL9 7LD
G4 NDL R Davies, 2 Torrhill Cottages, Godwell Lane, Ivybridge, PL21 0LT
G4 NDM R Carter, 49 Cambridge Road, West Bridgford, Nottingham, NG2 5NA
G4 NDP J Burnett, 42 Wentworth Drive, South Kirkby, Pontefract, WF9 3RY
G4 NDT R Bramley, 8 Ivy Bank Park, Bath, BA2 5NF
G4 NDU A Bramley, 8 Ivy Bank Park, Bath, BA2 5NF
GM4 NDV W Rattray, 17 Brownside Road, Cambuslang, Glasgow, G72 8NL
G4 NEA S Rice, 13 Wigram Way, Stevenage, SG2 9TP
G4 NEE D Foreman, 1 Stour Valley Close, Upstreet, Canterbury, CT3 4DB
G4 NEG William Glew Glew, Carinya, Beltoft Belton, DN9 1MB
G4 NEH J Hankin, The Brewhouse, High Swinside Farm, Cockermouth, CA13 9UA
G4 NEI K Hodge, Bryn Hyfryd, 16 Mold Road, Mold, CH7 6TD
G4 NEJ Kevin Jackson, 13 Jellicoe Avenue, Gosport, PO12 2PA
G4 NEL NEL RYNT L/B WT c/o D Bird, 154 Cherrydown Avenue, Chingford, London, E4 8DZ
G4 NEO Chris Digby, 7 Dagnall Road, Olney, MK46 5BJ
G4 NEQ R Welsh, Holme View, Farleton, Lancaster, LA2 9LF
G4 NER P Lidbetter, 1 Moor Lane, Westfield, Hastings, TN35 4QU
G4 NEY Jonathan Jarvis, 116 Balland Field, Willingham, Cambridge, CB24 5JU
G4 NFA F Austin, 45 Southdown Crescent, Cheadle Hulme, Cheadle, SK8 6EQ
G4 NFE J Edwards, 5 Windmill Rise, York, YO26 4TU
G4 NFF S Frisby, 19 Woodcock Close, Norwich, NR3 3TB
GI4 NFH Raymond Jennings, 117 Belsize Road, Lisburn, BT27 4BS
GM4 NFI David Leckie, 6 Galloway Place, Fort William, PH33 6UH
G4 NFL C Peel, The Ferns, Park Wood Drive, Newcastle, ST5 5EU
G4 NFO D Porter, 10 Blakes Crescent, Highbridge, TA9 3LE
G4 NFP Michael Saunby, Teachmore, Jacobstowe, Okehampton, EX20 3AJ
G4 NFR R Tyson, 49 Strathaird Avenue, Walney, Barrow-in-Furness, LA14 3DE
G4 NFS Norman Sharples, 47 New Fosseway Road, Bristol, BS14 9LW
G4 NFT G McAvoy, 5 Lytchett Way, Upton, Poole, BH16 5LS
G4 NFV James Clark, 12 Ogle Avenue, Morpeth, NE61 2PN
GI4 NFW John Hegarty, 1 Cookstown Road, Moneymore, Magherafelt, BT45 7QF
G4 NFY A Clarke, Ravenswood, Gull Road, Wisbech, PE13 4ER
G4 NGB Bill Gliddon, Apartment 1, Idle Shores, Woolacombe, EX34 7BX
G4 NGD Michael Hadnum, 79 Mowbray Road, Bedford, MK42 9UX
G4 NGF M Chapman, Millway, Dunton Lane, Lutterworth, LE17 5HX
GM4 NGJ C Bridges, Highfield, Ballinluig, Pitlochry, PH9 0LG
G4 NGL John Gass, 2a Orchard Way, Breachwood Green, Hitchin, SG4 8NT
GI4 NGP D Paul, 4 Draperstown Road, Tobermore, Magherafelt, BT45 5QG
G4 NGR C Webber, 107 Northfields Lane, Brixham, TQ5 8RN
G4 NGS G Towler, 77 Worrin Road, Shenfield, Brentwood, CM15 8JL
G4 NGV Anthony Whittaker, 31 Rydal Road, Haslingden, Rossendale, BB4 4EE
G4 NHA P Hastilow, 18 Broadway Avenue, Croydon, CR0 2LP
GW4 NHB P Boyce, 28 Alfreda Road, Cardiff, CF14 2EH
G4 NHC F Armstrong, 18 Mowbray Road, South Shields, NE33 3AU
G4 NHD M Brightman, Patriot'S Arms, 6 New Road, Swindon, SN4 0LU
G4 NHE C Evans, 4 Fryers Copse, Wimborne, BH21 2HR
G4 NHF A Jones, 51 Wiclif Way, Nuneaton, CV10 8NH
GW4 NHH M Buck, 23 Velindre Road, Cardiff, CF14 2TE
GM4 NHI J Cramond, Robson'S Croft, Dunecht, Westhill, AB32 7EQ
G4 NHL T Dixon, 30 Green Lane, Stamford, PE9 1HF
G4 NHN C Rowsell, 31 Kingsfield Gardens, Bursledon, Southampton, SO31 8AY
G4 NHO J Pattemore, 2 Edes Cottages, Ottways Lane, Ashtead, KT21 2PG
G4 NHP S Porter, 5 Clearheart Lane, Kings Hill, West Malling, ME19 4GT
G4 NHQ A Mcmackin, 33 Poynder Place, Hilmarton, Calne, SN11 8SQ
G4 NHR I Turner, 74 Diban Avenue, Elm Park Estate, Hornchurch, RM12 4YF
G4 NHT MOORLANDS&DIS A c/o Christopher Beesley, 15 Byron Close, Cheadle, Stoke-on-Trent, ST10 1XB
G4 NHW D Collins, 22 Stalyhill Drive, Stalybridge, SK15 2TR
GM4 NHX G Brooks, The Old Post Office, Scotscalder, Caithness, KW12 6XJ
G4 NIA James Houghton, Kingsway, North Road, Lifton, PL16 0EH
G4 NID Graham Bromley, 46 Independent Hill, Alfreton, DE55 7DG
G4 NIF Dennis Lee, 22 Woodland Rise, Parkend, Lydney, GL15 4JX
G4 NIJ K Sheldon, Whitehaven, May Tree Road, Pershore, WR10 2NY
G4 NIL R Henshall, St. Anns, Staplehay, Taunton, TA3 7HB
G4 NIP DR Lewis, 76 Reading Road, Finchampstead, Wokingham, RG40 4RA
G4 NIV N Rowcroft, 110 Linceslade Grove, Loughton, Milton Keynes, MK5 8BL
G4 NIX G Cooke, 7 Lime Grove, Royston, SG8 7DJ
G4 NIY S Cooke, 5 Honey Way, Royston, SG8 7ES
G4 NIZ R King, 20 Woodside East, Thurlby, Bourne, PE10 0HT
G4 NJA R Hewson, 6 Talisman Drive, Bottesford, Scunthorpe, DN16 3SW
G4 NJB K Hepke, 25 Victoria Avenue, Willerby, Hull, HU10 6DD
G4 NJI A Corker, 59 Foljambe Road, Rotherham, S65 2UA
G4 NJJ Peter Cousins, 28 Church Road, Clenchwarton, King's Lynn, PE34 4EA
G4 NJK Richard Elliott, Fatherford Farm, Okehampton, EX20 1QQ
G4 NJN A Bowman, 18 Essex Avenue, Isleworth, TW7 6LF
GI4 NJQ Peter Igo, 84 Glebetown Drive, Downpatrick, BT30 6PZ
G4 NJR Michael Doe, 2 Summerfields, Yarnfield, Stone, ST15 0RH
G4 NJT R Smith, 2131 - 21st Ne, Salmon Arm, British Columbia, Canada, V1E 2DN
G4 NJW George Lowes, 26 Huttoft Road, Sutton-on-Sea, Mablethorpe, LN12 2QY
GI4 NKB Frank Hunter, 2 Wandsworth Court, Belfast, BT4 3GD
G4 NKC M Jones, Racecourse Farm, Church Lane, Bridgnorth, WV16 4NW
G4 NKI John Shaw, 1a Greenways, The Stookes, Chesterfield, S40 3HF
GI4 NKK K Planck, 4 Westland Drive, Ballywalter, Newtownards, BT22 2TH
G4 NKP David Mellin, 99 The Borough, Downton, Salisbury, SP5 3LX
G4 NKR G Lewis, The Old, Post Office Cottage, Brecon, LD3 0UR
G4 NKU David Dunn, 37 Ridgemead, Calne, SN11 9EW
G4 NKW Graham Hilton, 8 Sandwich Close, St. Ives, PE27 3DQ
G4 NKX Peter Digby, Primrose Cottage, Avenue Road, Hayling Island, PO11 0LX
GI4 NKY W Campbell, 68 Richmond Court, Lisburn, BT27 4QX
G4 NLA G Goodrich, 29 Cresswells, Corsham, SN13 9NJ
G4 NLB B Horne, 77 Surrey Hills Residential Park, Boxhill Road, Tadworth, KT20 7LZ
G4 NLC Peter Hedison, 85 Moorhouse Lane, Whiston, Rotherham, S60 4NH
G4 NLD Paul Frost, 1 Chester Street, Rhyl, LL18 3ER
G4 NLG Frank Askew, 9 The Hall Spinney, Howden, Goole, DN14 7FD
G4 NLH David Haydon, 50 Ward Close, Stratton, Bude, EX23 9BB
G4 NLI R Scott, 10 Middle St., North Perrott, Crewkerne, TA18 7SG
G4 NLJ John Martin, Whitehill Foot Farm, Kelso, TD5 8LB
G4 NLK R Southall, 7 The Willows, Brereton, Rugeley, WS15 1EP
G4 NLL D Bell, 5 Byron Court, Dalton on Tees, Darlington, DL2 2PX
G4 NLO Keith Butcher, 18 Windmill Street, Whittlesey, Peterborough, PE7 1HJ
GI4 NLQ P Burns, 41 Lambeg Road, Lambeg, Lisburn, BT27 4QA
G4 NLU R Williams, 50 Hemerdon Heights, Plympton, Plymouth, PL7 2EY
G4 NLW J Wilding, 8 Millbrook Way, Brierley Hill, DY5 3YY
G4 NMA Alan Townsend, 12 Fieldfare, Stevenage, SG2 9NJ
G4 NMC D Willis, 41 Chadbrook Crest, Richmond Hill Road, Birmingham, B15 3RL
G4 NMD Graham Smith, 18 Poynings Place, Old Portsmouth, Portsmouth, PO1 2PB
G4 NME R Smith, 47 Cinder Road, Dudley, DY3 2RH
G4 NMF W Squire, 7 Essex Crescent, Seaham, SR7 8DZ
G4 NMK K Petre, 24 Harrogate Terrace, Murton, Seaham, SR7 9PQ
G4 NMP B Dudhill, WESTFIELD, Mablethorpe, L N12
G4 NMS Stephen Burgess, Borne House, Romsey Road, Stockbridge, SO20 6PR
G4 NMT A Godsiff, 59 Vinson Close, Orpington, BR6 0EQ
G4 NMU Richard Jones, Flat 4, Russell Court, Southport, PR9 8NY
G4 NMV J Baxter, 16 Avon Close, Weston-Super-Mare, BS23 4QS
G4 NMY P Bennett, 45 Ravenbank Road, Luton, LU2 8EJ
G4 NNB G Mackie, 8 The Avenue, Biggleswade, SG18 0PS
GM4 NNH J Barber, 156 Jamphlars Road, Cardenden, Lochgelly, KY5 0ND

G4 NNI Edward Bailey, 213 Ashby Road, Hinckley, LE10 1SJ
G4 NNJ A Forster, 3 Willow Heights, Lydney, GL15 5LR
GM4 NNK [illegible], 22 [illegible] Crescent, Blackburn, Aberdeen, AB21 0JZ
G4 NNL M Jones, Jodanare, 72b Princes Drive, Colwyn Bay, LL29 8PW
GI4 NNM J McGillian, 48 Millfield, Ballymena, BT43 6PB
G4 NNN C Winterflood, 12 Bourne Road, Colchester, CO2 7LQ
G4 NNO T Hadley, Staplins Bungalow, Tewkesbury Road, Gloucester, GL19 4AH
G4 NNP [illegible], Paignton, TQ4 [illegible]
G4 NNS Brian Coleman, Woodlands, Redenham, Andover, SP11 9AN
G4 NNX D Ward, 18 Honders, Stony Stratford, Milton Keynes, MK11 1RB
G4 NNY D Ransford, 52 Loughborough Road, Bunny, Nottingham, NG11 6QD
G4 NNZ G Martorano, 81 Sapcote Drive, Melton Mowbray, LE13 1HG
G4 NOB R Durbeck, 20 St Johns Road, Smalley, Ilkeston, DE7 6EG
G4 NOC N Black, 7 Woodland Crescent, Bracknell, RG42 2LH
G4 NOE M Hollinghurst, 30 Hall Road, Cheltenham, GL53 0HE
G4 NOK North Wakefield RC c/o Robert Rawson, 68 Broom Nook, Leeds, LS10 3LR
G4 NOL Bill Robinson, [illegible] Belmont Road, Sale, M33 6HY
GW4 NOO Michael Oliver, 6 Flemish Close, St. Florence, Tenby, SA70 8LT
G4 NOP Richard Simpson, 14 Cumberland Way, Dibden, Southampton, SO45 5TW
G4 NOR Clive Heaps, 12 Oak Tree Close, Mansfield, NG18 3EN
GW4 NOS R Hopkins, 8 Shady Road, Gelli, Pentre, CF41 7UG
G4 NOT Christopher Sturgeon, Windyridge, Linkside West, Hindhead, GU26 6PA
G4 NOU R Wigmore, Little Landguard, Whitecross Lane, Shanklin, PO37 7EJ
G4 NOX K Campbell, 4 Orchard Close, Rowlands Gill, NE39 1EQ
G4 NOY J Mills, 103 Irby Road, Wirral, CH61 6UZ
G4 NPA Andrew Abbot, 4 Nursery Drive, Birmingham, B30 1DR
G4 NPB S Abbot, 4 Nursery Drive, Birmingham, B30 1DR
GW4 NPC R Blayney, 42 Ty Draw, Church Village, Pontypridd, CF38 1UF
G4 NPD G Chamberlain, 13 Mayford Close, Beckenham, BR3 4XS
G4 NPE E Dench, 30 Gravel Walk, Emberton, Olney, MK46 5JA
G4 NPG P Duffy, 15 The Glade, Sheldon, Birmingham, B26 3PW
G4 NPH John Arnold, 2 Duck Lane, Haddenham, Ely, CB6 3UE
G4 NPN David Westwood, 1 Elmfield Road, Hartlebury, Kidderminster, DY11 7LA
G4 NPS J Woolliss, Wharfdale, 245 Scartho Road, Grimsby, DN33 2EA
G4 NPT R Williamson, 1 Pygall Avenue, Gotham, Nottingham, NG11 0JW
G4 NPU P Williams, 122 Longhurst Lane, Mellor, Stockport, SK6 5PG
G4 NPY C Read, 3 Wyrley Close, Lichfield, WS14 9DA
G4 NQC J Neal, 60 Balsdean Road, Brighton, BN2 6PF
G4 NQI R Atterbury, Lomas de San Jose, Malaga, Spain, 29712
G4 NQJ Christopher Brown, Kingsdown Cottage, Fron, Montgomery, SY15 6SB
G4 NQL M Churms, 123 Urb Buenavista, Guadamar Del Segura, Alicante, Spain, 3140
G4 NQM M Drohan, 23 Lindholme Drive, Rossington, Doncaster, DN11 0UP
G4 NQQ N Hemmings, 3 Church Close, Shapwick, Bridgwater, TA7 9LS
G4 NQS R Young, 79 Cradge Bank, Spalding, PE11 3AF
G4 NQW P Perrins, 9 Merrick Close, Hayley Green, Halesowen, B63 1JY
G4 NQZ R Riley, 103 St. Nicolas Park Drive, Nuneaton, CV11 6DZ
G4 NRA John Tisdale, 12 Digby Road, Kingswinford, DY6 7RP
G4 NRC RADIO AMATEURS EMERGENCY NETWORK c/o Geoffrey Griffiths, 14 Mansion House Gardens, Melton Mowbray, LE13 1LE
G4 NRD A Lindsay, 21 Willow Road, Four Pools Industrial Estate, Evesham, WR11 1YW
G4 NRE William Ward, 88 Central Road, Cromer, NR27 9BW
G4 NRF Harold Westwood, 67 Bedford Close, Featherstone, Pontefract, WF7 5LH
G4 NRG Roger Greengrass, 19 Worcester Way, Attleborough, NR17 1QU
G4 NRH D Whitehead, 50 Southey Lane, Kingskerswell, Newton Abbot, TQ12 5JG
G4 NRI Mohinder Dhami, 118 Havelock Drive, Brampton, Ontario, Canada, L6W 4E3
G4 NRK Newborough Radio Klub c/o philip staton, 52 School Road, Newborough, Peterborough, PE6 7RG
G4 NRP Jeremy Fish, Kennels Cottage, Manor Road, Redditch, B97 5TB
GI4 NRQ COLERAINE & DIS c/o James Hamill, 67 Windsor Avenue, Coleraine, BT52 2DR
G4 NRR Nigel Astbury-Rollason, Fern Lodge, The Parade, Newton Abbot, TQ13 0JH
G4 NRT D Bondy, 19 Harriet Drive, Rochester, ME1 1DY
G4 NRV Anthony Diplock, 24 Billings Hill Shaw, Hartley, Longfield, DA3 8EU
G4 NRW Dennis Reeve, Trincoe, Mill Road, Wells-next-The-Sea, NR23 1BZ
G4 NRX S Mantell, 50 Coleshill St., Fazeley, Tamworth, B78 3RA
G4 NRY I Mantell, 34 Piccadilly, Tamworth, B78 2EP
G4 NRZ Keith Moody, 5 Moore Road, Mapperley, Nottingham, NG3 6EF
G4 NS J Hudson, 22 Essex Gardens, Marsden, South Shields, NE34 7JQ
G4 NSA D Morgan, 12 Rosalind Avenue, Bebington, Wirral, CH63 5JR
G4 NSB John Winterburn, South Farm Cottage, Ypres Road, Swindon, SN4 0JF
G4 NSC W Weatherspoon, 12 Greenacres Close, Crawcrook, Ryton, NE40 4TD
G4 NSD Ian Mitchell, Greenway Cottage, Greenway, Westerham, TN16 2BT
G4 NSE J Rank, 18 Coldyhill Lane, Scarborough, YO12 6SF
G4 NSH Sam Robinson, Upper Hambleton Hill Farm, Wainstalls, Halifax, HX2 7TX
G4 NSJ G Heffer, 35 Henty Road, Worthing, BN14 7HF
GM4 NSL George Greenlees, 22 Hunters Grove, Hunters Quay, Dunoon, PA23 8LQ
G4 NSM A BRUCE, 5 Orchard Croft, Epworth, Doncaster, DN9 1LL
G4 NSN Michael Craft, 8 Juniper Road, Farnborough, GU14 9XU
G4 NSO S AUCKLAND, 6 Lombard Crescent, Darfield, Barnsley, S73 9PP
GI4 NSS Lawrence Robinson, 18 Lord Warden's Avenue, Bangor, BT19 1YE
G4 NST S Thorpe, 11 Grove Road, Hethersett, Norwich, NR9 3JP
GI4 NSV A Donnelly, Tirgarve, 7 Blackwatertown Road, Armagh, BT61 8EZ
G4 NSW F Lacey, 27 Howards Gardens, New Balderton, Newark, NG24 3FJ
G4 NSZ M Stanway, 72 Sheldons Court, Winchcombe Street, Cheltenham, GL52 2NR
G4 NTA P Allan, 2 Park View, Queensbury, Bradford, BD13 1PL
G4 NTC David Henderson, 4 Vincent Street, Bolton, BL1 4SA
G4 NTG J Williamson, 5 Rochester Close, Headless Cross, Redditch, B97 5FP
G4 NTJ A Rick, 9 Sheldon Close, Loughborough, LE11 5EZ
G4 NTL James Mc Dermott, Milking Green Gate, Elkock, Sanquhar, DG4 6LD
GD4 NTR Godfrey Kelly, 5 Tynwald Close, Peel, Peel, Isle of Man, IM5 1JJ
G4 NTT P Dunthorne, 277 Westleigh Lane, Leigh, WN7 5PW
G4 NTV A Wilkes, 34 Tideswell Road, Great Barr, Birmingham, B42 2DT
G4 NTW W Douglas, 6 Ripon Court, Sacriston, Durham, DH7 6XF
GM4 NTX K Elliott, Northfield Cottage, Denny, FK6 6RB
G4 NTY [illegible] Avenue, Worsley, Manchester, M28 [illegible]
G4 NUA E Williams, 45 Bayford Place, Cambridge, CB4 2UF
G4 NUB M Tyler, 27 Shakespeare Drive, Dinnington, Sheffield, S25 2RP
G4 NUF Colin Muller, 118 Park Lane, Northampton, NN5 6PZ
G4 NUG R Needs, 13 Greenway, Great Horwood, Milton Keynes, MK17 0QR
G4 NUJ Kenneth Symonds, The Beeches, 30 Fairlea Crescent, Bideford, EX39 1BD
G4 NUK J Brown, 33 Balmoral Drive, Leicester, LE3 3AD
GM4 NUN Gordon MacKenzie, 1 Walnut Grove, Blairgowrie, PH10 6TH
G4 NUO A MacKenzie, Ivy House, 116 High Street, Redcar, TS11 6JX
G4 NUS A Layland, 16 Park Road, Quarry Bank, Brierley Hill, DY5 2DA
GM4 NUU Hamish Park, [illegible], Upper [illegible], Dunfermline, KY12 9LP
G4 NUX C Smith, Wealdon Cottage, Dunsfold Road, Billingshurst, RH14 0PJ
G4 NUY E Whorton, Vandling, Well Bank, Bedale, DL8 2QF
G4 NUZ A Davies, 22 Meadow Road, Cornmeadow Green, Worcester, WR3 7PP
G4 NVA J Dyke, 2 Brooklands Drive, Goostrey, Crewe, CW4 8JB
G4 NVH Graeme Boull, 80 Ascot Road, Baswich, Stafford, ST17 0AQ
GM4 NVI D Chapman, 9 Baillieswells Terrace, Bieldside, Aberdeen, AB15 9AR
G4 NVL T Copeman, 1 Chestnut Avenue, Welney, Wisbech, PE14 9RG
G4 NVM J Duddridge, 19 Ridgeway, Hurst Green, Etchingham, TN19 7PJ
G4 NVN George Harper, Pathways, 41 Somerset Close, Congleton, CW12 1SE
G4 NVP K Kett, 24 Deancourt Drive, New Duston, Northampton, NN5 6PY
G4 NVQ D Shirley, 93 Alfred Road, Hastings, TN35 5HZ
G4 NVT Michael Musgrave, 49 Vowler Road, Langdon Hills, Basildon, SS16 6AQ
G4 NVV Roger English, 124 Hillside Road, Portishead, Bristol, BS20 8LG
G4 NVY Michael Juffs, 32 Brooklands Park, Longlevens, Gloucester, GL2 0DP
G4 NWJ J Chick, Moonrakers, Allington, Salisbury, SP4 0BX
GM4 NWK T Hill, 3 Swift Crescent, Glasgow, G13 4QN
G4 NWM M Talbott, 44 Tamworth Road, Amington, Tamworth, B77 3BT
G4 NWN W Talbott, 44 Tamworth Road, Amington, Tamworth, B77 3BT
G4 NWO K Chaplin, 1 Beechwood Crescent, Amington, Tamworth, B77 3JH
G4 NWR NTH WILTS RAYNE c/o H Woolrych, 20 Meadow Drive, Devizes, SN10 3BJ
G4 NWS A Wheeler, 11 Barley Way, Rothley, Leicester, LE7 7RL
G4 NWW J Edgeley, 12 The Glade, Horsham, RH13 6DD
G4 NXA Donald Kirk, Glebelands, Old Vicarage Gardens, Vicarage Lane, Southam, CV47 7RT
G4 NXB J Evans, 11 Columbia Close, Selston, Nottingham, NG16 6GP
G4 NXD R Johns, 12 Woodfield Road, New Inn, Pontypool, NP4 0PT
G4 NXG A Birch, 6 Crescent Road, Wallasey, CH44 0BQ
G4 NXI R Light, 72 Badger Rise, Portishead, Bristol, BS20 8AX
GI4 NXJ M McFall, 4 Riverdale Close, Ballyclare, BT39 9WE
G4 NXL M Street, 41 Shaw Lane, Holbrook, Belper, DE56 0TG
G4 NXO SHEPPEY WESTERN c/o Anthony Collett, 10 Quince Road, Hardwick, Cambridge, CB23 7XJ
G4 NXP Derek Taylor, The Acreage, View Farm, 11 Malvern Road, Worcester, WR2 4SF
G4 NXR Peter Rose, Cuptree Cottage, 17 The Cross, Colchester, CO7 9QQ
G4 NXS M Davis, 446 Upper Wortley Road, Scholes, Rotherham, S61 2SS
GM4 NXT W Davidson, 7a South Street, Aberchirder, Huntly, AB54 7XR
G4 NXV David Gadsden, 37 Cambridge Street, Wymington, Rushden, NN10 9LG
G4 NXW K Chesters, 4 Kirton Crescent, Lytham St. Annes, FY8 4BJ
G4 NYA R Hyams, 6 Gaudick Road, Eastbourne, BN20 7QE
G4 NYB R Knighton, Merry Ways, The Green, Derby, DE72 2BJ
G4 NYD I Watling, 5 Claylands Court, Bishops Waltham, Southampton, SO32 1JS
G4 NYG D Routledge, 7 Littleton Croft, Solihull, B91 3XR
G4 NYJ C Webb, 65 Littlebeck Drive, Darlington, DL1 2TU
G4 NYK R Williams, 67A Sea Mills Lane, Stoke Bishop, Bristol, BS9 1DR
G4 NYL S Brown, 2 Windsor Close, Read, Burnley, BB12 7QH
GU4 NYT Nigel Le Page, Heathwick, Les Martins, Guernsey, Guernsey, GY4 6QJ
G4 NYV Michael Katzmann, 6654 Barnaby Street Nw, Washington Dc, United States, 20015-2357
G4 NYW Roger Schoales, Ventura, Towngate, Hope Valley, S33 9JX
G4 NYY Peter Ernster, 36 Forest End, Fleet, GU52 7XE
G4 NYZ J Battle-Welch, 25 Moorcroft Close, Callow Hill, Redditch, B97 5WB
G4 NZB P Neville, 66 Oak Lodge Avenue, Chigwell, IG7 5HZ
G4 NZC B Manchett, 12 Old Parsonage Court, Otterbourne, Winchester, SO21 2EP
G4 NZE E Third, 1 Roman Road, Corby, NN18 8FZ
G4 NZG Simon Parsons, 2 The Pentelows, Covington, Huntingdon, PE28 0RY
G4 NZK B Laniosh, 47 Barley Mow Lane, Catshill, Bromsgrove, B61 0LU
G4 NZN Keith Lowe, Springwood, Priesthorpe Road, Pudsey, LS28 5RE
G4 NZO Gilbert Frederick, 98 Coleridge Park Drive, Winnipeg, Canada, MB R3K 0B5
G4 NZQ P Brooks, 7 Lindford Drive, Eaton, Norwich, NR4 6LT
G4 NZU R Wilson, 9 Greythorn Drive, West Bridgford, Nottingham, NG2 7GG
G4 NZX D Cooper, 6 Swinburne Close, Barnby Dun, Doncaster, DN3 1BS
G4 NZY David Coles, 91 Grove Lane, Harborne, Birmingham, B17 0QT
G4 NZZ B Coulson, 19 St. Lukes Close, Kettering, NN15 5HD
G4 OAB M Clutton, 8 Ash Grove, Runcorn, WA7 5LR
G4 OAE D Crisp, 20 Crawford Close, Earley, Reading, RG6 7PE
G4 OAG A Dymott, St. Huberts, Ashey Road, Ryde, PO33 4BB
G4 OAI H Richter, 84 Roehampton Drive, Wigston, LE18 1HU
G4 OAK Stephen Richards, 1 Stocks Mead, Washington, Pulborough, RH20 4AU
G4 OAN L Wild, The Drey, Penny Lane, Bedlington, NE22 6HD
G4 OAR N Mclaren, 596 Woodchurch Road, Prenton, CH43 0TT
GM4 OAS Gordon Liddle, Ashbeck Morar, Mallaig, PH40 4PD
G4 OAU G Austin, 38 Willow Crescent, Hatfield Peverel, Chelmsford, CM3 2LJ
G4 OAV Stanley Ames, 21 Common Lane, Harpenden, AL5 5BT
G4 OAX W Joiner, 15 Laxton Grove, Great Holland, Frinton-on-Sea, CO13 0SF
G4 OBA R Jarvis, 10 Daddlebrook Road, Alveley, Bridgnorth, WV15 6NU
G4 OBB D Kaylor, 10 Teal Close, Oxford, OX4 7GU
G4 OBC M Taylor, 5 Hawford Avenue, Kidderminster, DY10 3BH
GM4 OBD Graham Sangster, 36 St. Marys Drive, Ellon, AB41 9LW
G4 OBE R Snary, 12 Borden Avenue, Enfield, EN1 2BZ
G4 OBK Philip Catterall, Westlands Whitby Road, Pickering, YO18 7HJ
G4 OBN Stewart Harding, 21 Abbey Road, Medstead, Alton, GU34 5PB
G4 OBT Norman Day, Tanhouse Farm, Rusper Road, Dorking, RH5 5BX
G4 OBV E Day, 42 Grosvenor Close, Thorley, Bishop's Stortford, CM23 4JP
G4 OBX R Dobson, 40 Dipton Gardens, Sunderland, SR3 1AN
GM4 OCA P Windsor, Hillside Overbrae, Fisherie, Turriff, AB53 [illegible]
G4 OOO John Johnston, 6182 Doctor Blair Crescent, North Gower, Ontario, Canada, K0A 2T0
G4 OCF Douglas Clayton, 8 Wains Close, Clevedon, BS21 6NZ
G4 OCH K Dickens, The Old Post Office, Cleobury Road, Ground Floor Flat, Bewdley, DY12 2QG
G4 OCJ E Mullock, 4 Ashdene, Healey Dell, Rochdale, OL12 6DJ
GI4 OCK J Mackay, 12 Lynne Road, Bangor, BT19 1NT
GI4 OCL R McCurry, 82 Cumberland Road, Dundonald, Belfast, BT16 2RR
G4 OCN E Crotton, 83 Open Hearth Close, Griffithstown, Pontypool, NP4 5LU
G4 OCQ I Blackman, 69 Thorntons Close, Pelton, Chester le Street, DH2 1QH
G4 OCR M Butler, 41 Neale Road, Chorlton cum Hardy, Manchester, M21 [illegible]
G4 OCS Ivan Boxon, 236 Westdale Lane, Carlton, Nottingham, NG4 4FL
G4 OCU D Gipp, 9 Yew Tree Road, Elkesley, Retford, DN22 8AY
G4 OCX L Jewell, Ashbrook Farm, Mill Hill, Bedford, MK44 2HP
G4 OCZ C Richardson, 148 Old Fort Road, Shoreham-by-Sea, BN43 5HL
G4 ODA B Tatnall, Poplar House, Delgate Bank, Spalding, PE12 6DH
G4 ODD Michael Mathers, Rose Cottage, Kirton Road, Newark, NG22 0HF
G4 ODE N Dovaston, 53 Elmway, Chester le Street, DH2 2LX
G4 ODF D Faulkner, Amber Croft, Dale Close, Mansfield, NG20 9EB
G4 ODG Vincent Cawthron, 8 Clay Hill Road, New Quarrington, Sleaford, NG34 7TF
G4 ODI Alan Dyer, Knoll View, Biddisham, Axbridge, BS26 2RE
G4 ODM Clive Mott-Gotobed, 14 Copse Road, New Milton, BH25 6ES
GW4 ODN A Whitticombe, 160 Haven Drive, Hakin, Milford Haven, SA73 3HN
G4 ODR Enfield Contest Group c/o John Young, 30 Crofton Way, Enfield, EN2 8HS
G4 ODS W Allan, 50 Byland Road, Harrogate, HG1 4ET
GI4 ODT W Barker, 96 Highlands Road, Limavady, BT49 9LY
G4 ODV John Coyne, 19, Karouli Demetriou, Larnaca, Cyprus, 7577
GM4 ODW David Maclean, Gramaiche, Donavourd, Pitlochry, PH16 5JS
GJ4 ODX S Langlois, L'Amarrage, La Route Orange, St. Brelade, Jersey, JE3 8GP
GD4 OEA C Gerrard, 6 Rheast Lane, Peel, Isle of Man, IM5 1BE
G4 OEB Paul Grant, 24 Dowlands Road, Bournemouth, BH10 5LG
G4 OEC Ernest McPheat, OLD SCHOOL, Holford, Bridgwater, TA5 1SF
G4 OED G Perry, 12 Boydell Close, Shaw, Swindon, SN5 5QT
G4 OEF D Bayliss, 38 Yarborough Crescent, Lincoln, LN1 3LU
G4 OEH N Rumble, 24 Firle Road, North Lancing, Lancing, BN15 0NZ
GW4 OEJ A England, 9 Priory Road, Milford Haven, SA73 2DS
G4 OEK Roy Jones, 17b Plumpton Park Road, Doncaster, DN4 6SQ
G4 OEM G Hooker, 42a Nether Hall Road, Doncaster, DN1 2PZ
G4 OEP A Smith, 15 Dyrham Close, Henleaze, Bristol, BS9 4TF
G4 OEQ C Thomas, 69 Quakers Road, Downend, Bristol, BS16 6JG
G4 OER David Warner, 28 Jameson Bridge Street, Market Rasen, LN8 3EW
GW4 OES A Pickard, 89 Ael Y Bryn, Llanedeyrn, Cardiff, CF3 7LL
G4 OEU Q Campbell, 8 Cookson Close, Corbridge, NE45 5HB
G4 OEX G Jones, 3 Kings Mews, Bedford Street, Warrington, WA4 6GY
G4 OEY T Sanderson, Backershagenlaan 32, Wassenaar, Netherlands, 2243 AD
GM4 OEZ W Taylor, 2 Jubilee Terrace, Findochty, Buckie, AB56 4QA
G4 OFA B Millican, 24 Wellington Terrace, Bramley, Leeds, LS13 2LH
GM4 OFC J Snelgrove, Mill House, South Bridgend, Crieff, PH7 4DH
GM4 OFI John Robertson, Springwell House, Auckengill, Wick, KW1 4XP
G4 OFN Peter Edmonds, 170 Halton Road, Sutton Coldfield, B73 6NZ
G4 OFO N Baynes, 62 Cromford Way, New Malden, KT3 3BA
G4 OFP J Knowles, 6 Seaway Gardens, Paignton, TQ3 2PE
G4 OFU Roland Burdess, 33 The Green, Dartford, DA2 6JS
G4 OGB L Elliott, Elm Lodge, 2 Hood Croft, Doncaster, DN9 2FB
G4 OGG J West, 2 Sanders Close, Kempston, Bedford, MK42 8RX
G4 OGL John O'Brien, 5 Highfields Mead, East Hanningfield, Chelmsford, CM3 8XA
GM4 OGM S Mather, Pentland View, Limekiln Road, West Linton, EH46 7BA
G4 OGO S Williams, 31 Kensington Park, Magor, Caldicot, NP26 3QG
GI4 OGQ W Kernohan, 3 Camphill Park, Ballymena, BT42 2DQ
G4 OGW D Thomas, Handley Cross Cottage, Harewood End, Hereford, HR2 8JT
G4 OGZ Michael Walker, 52 Derwent Road, Harpenden, AL5 3NX
GW4 OH A David, 45 Amanwy, Llanelli, SA14 9AH
GI4 OHA Laurence Lux, Hyde Brae, Hyde Hill, Stroud, GL6 8NY
G4 OHB Paul Taylor, Flat 11, Romsley Hill Grange, Farley Lane, Halesowen, B62 0LN
G4 OHC R Poore, 8 Ainsdale Close, Worthing, BN13 2QX
G4 OHF G Bean, 141 Narborough Road, Leicester, LE3 0PB
GI4 OHH Derrick Cox, 32 Kilmaconnell Road, Castleroe, Coleraine, BT51 3QZ
G4 OHJ Jeffrey Porter, 77 Westholme Road, Bidford-on-Avon, Alcester, B50 4AN
G4 OHM 5TH BHAM RNT GR c/o J Parkin, 18 Bradnock Close, Birmingham, B13 0DL
G4 OHP L Sharps, 68 Vicarage Lane, Elworth, Sandbach, CW11 3BU
G4 OHQ J Reade, 7 Wilmar Close, Hayes, UB4 8ET
GM4 OHT Trevor Mitchell, 12 Dalriada Place, Kilmichael Glassary, Lochgilphead, PA31 8QA
G4 OHV Colin Addison-Lees, 18 Langley Avenue, Somercotes, Alfreton, DE55 4LT
GI4 OHW N Bell, Rocklyn, 16 Dromore Road, Omagh, BT78 1QZ
GM4 OHY R Cameron, 88 Little Vennel, Cromarty, IV11 8XF
G4 OIA Simon Hartgroves, Wayside, Stamford Hill, Budo, EX23 9AZ
G4 OID John Storry, 99 Swineshead Road, Wyberton Fen, Boston, PE21 7JG
G4 OIE Roger Neale, Field House, Recreation Road, Mansfield, NG19 8TL
G4 OIG Gerald Peck, 45 Bentley Close, Wootton, Northampton, NN3 5JS
G4 OII M Morley, Padagi, Town Road, Grimsby, DN36 5JE
GM4 OIJ B Robertson, Woodside House, Feabuie, Inverness, IV2 5EQ
G4 OIK J Price, 4 Housman Walk, Kidderminster, DY10 3XL
G4 OIL J Price, 26 Hales Park, Bewdley, DY12 2HT
G4 OIM P Marchant, 29 Hilldrop Road, Bromley, BR1 4DB
G4 OIN A Reeley, Gibraltar House, 53 Pegasus Gardens, Gloucester, GL2 4NP
G4 OIQ P Storey, 4 Sorrel Close, Wootton Bassett, Swindon, SN4 7JG
G4 OIR Neil Robinson, 14 Glendale, Orton Wistow, Peterborough, PE2 6YL
G4 OIS G Reid, 65 Rowelfield, Luton, LU2 9HL
G4 OIV Donald Mavin, 52 Bywell Road, Ashington, NE63 0LE

UK Callsigns

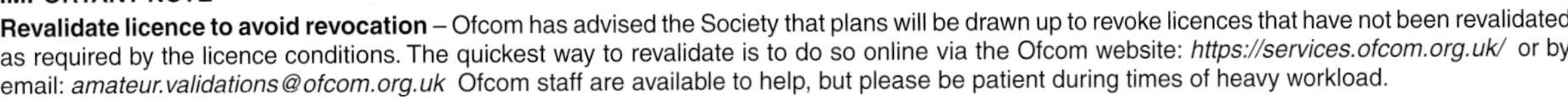

**IMPORTANT NOTE**

**Revalidate licence to avoid revocation** – Ofcom has advised the Society that plans will be drawn up to revoke licences that have not been revalidated as required by the licence conditions. The quickest way to revalidate is to do so online via the Ofcom website: *https://services.ofcom.org.uk/* or by email: *amateur.validations@ofcom.org.uk* Ofcom staff are available to help, but please be patient during times of heavy workload.

G4 OIW Charles Pine, 2 Grange Drive, Stokesley, Middlesbrough, TS9 5PQ
G4 OJB William Tatum, 51 Priory Close, Tavistock, PL19 9DG
G4 OJD M Guy, 73 Penn Meadows, Brixham, TQ5 9PF
G4 OJF Gerald Ball, 12 Kelstern Close, Northwich, CW9 5QR
G4 OJG J Glass, 70 Canterbury Road, Lydden, Dover, CT15 7ES
G4 OJI David Schofield, 18 Berrow Walk, Bristol, BS3 5ES
G4 OJJ V Holyoake, 281 Causeway, Green Road, Oldbury, B68 8LT
G4 OJK H Baxendale, 72c De Villiers Avenue, Liverpool, L23 2XF
G4 OJL John Bevan, 5 Selsdon Close, Wythall, Birmingham, B47 6HP
G4 OJN A Semark, 11 Fir Tree Close, Thorpe Willoughby, Selby, YO8 9PF
G4 OJP Raymond Prosser, 17 Lloyd Street, Hereford, HR1 2HB
G4 OJQ Alan Rowland, Mole Cottage, Chapel Close, Morwenstow, EX23 9JR
G4 OJR A Stone, 96 Reading Road, Farnborough, GU14 6NP
G4 OJS John Rowlands, 70 Braces Lane, Marlbrook, Bromsgrove, B60 1DY
G4 OJU Christopher Lock, 11 Stockton Close, Bristol, BS14 0DS
G4 OJV A Picton, 5 Tuttles Lane East, Wymondham, NR18 0EN
G4 OJW Louis Szondy, 6 Stanhope Gardens, London, N6 5TS
G4 OJY A Wright, 2 Wards End Cottages, Tow Law, Bishop Auckland, DL13 4JS
G4 OKA N Crook, 3 College Road, Reading, RG6 1QE
G4 OKB Barrie Bloomfield, 2 Walstead Manor Cottages, Scaynes Hill Road, Haywards Heath, RH16 2QG
G4 OKC A Gardner, 19 Lower Rea Road, Brixham, TQ5 9UD
G4 OKD A Forryan, 21 Blakesley Road, The Meadows, Wigston, LE18 3WD
G4 OKE M Gould, 10 Canterbury Close, Pelsall, Walsall, WS3 4PB
GW4 OKF Paul Granby, 104 Priory Road, Milford Haven, SA73 2ED
G4 OKH Martin Fisher, Witham Lodge, Fen Road, Wisbech, PE13 5HT
G4 OKM Mike Smith, 7 Russley Green, Wokingham, RG40 3HT
G4 OKO D Rycroft, 1 Littlefields Avenue, Banwell, BS29 6BE
G4 OKS Michael Porter, Penlee, 11 Penwithick Road, St. Austell, PL26 8UQ
GI4 OKU Thomas Patton, 29 Greystone Park, Limavady, BT49 0EQ
G4 OKW C Trayner, 2 Herisson Close, Pickering, YO18 7HB
G4 OKY R Wilkinson, 10 Mildenhall Road, Loughborough, LE11 4SN
G4 OKZ D Wilson, 20 The Square, Worsthorne, Burnley, BB10 3NG
G4 OLA R McCubbin, 20 Wellesley Park, Wellington, TA21 8PY
G4 OLF A Thomson, 7 Bloomstiles, Salthouse, Holt, NR25 7XJ
GM4 OLH I Walsh, 43 Sandyhill Road, Tayport, DD6 9NX
G4 OLK Alistair Mackay, 2 Highcliffe Grove, New Marske, Redcar, TS11 8DU
G4 OLL David Wilson, 10 Winchester Drive, Stourbridge, DY8 2LH
G4 OLO G Spencer, 5 Pitchcroft Lane, Church Aston, Newport, TF10 9AQ
G4 OLP R Parker, 2 Laurel Road, Norwich, NR7 9LL
G4 OLS John Lloyd, 16 Gilbanks Road, Stourbridge, DY8 4RN
G4 OLU D Steward, 30 Rifhams Drive, Great Baddow, Chelmsford, CM2 7DD
G4 OLY C Morgan, 316 Middle Road, Southampton, SO19 8NT
G4 OLZ I Sirley, The Barn, Littleworth Lane, Horsham, RH13 8JF
G4 OMD D Wilson, 1 Witchford Close, Lincoln, LN6 0SS
G4 OMG C Prescott, 15 Sarabeth Drive, Tunley, Bath, BA2 0EA
G4 OMI A Proudler, 17 Sunnyside Road, Ketley Bank, Telford, TF2 0DT
GI4 OMK P Murphy, 11 Danesfort Apartments, Villa 1, Belfast, BT9 5QL
G4 OMN D Thorndike, 48 Cressingham Road, Reading, RG2 7JR
G4 OMP M Nyman, 26 Silverstone Court, River Brook Drive, Birmingham, B30 2SH
G4 OMS R Reynolds, 90 Manchester Road, Blackpool, FY3 8DP
G4 OMT M Taylor, 320 Duffield Road, Derby, DE22 1EQ
G4 OMV D Marlow, 53 The Lawns, Corby, NN18 0TA
G4 OMZ Linda Welding, 67 Sunningdale Close, Burtonwood, Warrington, WA5 4NS
G4 ONC E Westcott, 8 Portal Place, Ivybridge, PL21 9BT
G4 ONF Paul Sergent, 6 Gurney Close, New Costessey, Norwich, NR5 0HB
G4 ONG W Lowe, 34 Ridgeway, Lowton, Warrington, WA3 2QL
G4 ONH C Weller, Mole End, Brent Hall Road, Braintree, CM7 4JZ
GW4 ONI David Mears, 7 Tanydarren, Cilmaengwyn, Swansea, SA8 4QT
G4 ONJ A Lightfoot, 13 Midhurst Close, Ifield, Crawley, RH11 0BS
G4 ONP LOUGHTON & EPPING FOREST ARS c/o John Mulye, 83 Forest Drive East, London, E11 1JX
G4 ONS M Slade, 5 Pedder Road, Clevedon, BS21 5HB
G4 ONV G Parker, 49 Newlands, Dawlish, EX7 0EA
G4 ONZ P Tebbutt, 115 Whitfield Mill, Meadow Road, Bradford, BD10 0LP
G4 OO THE G4OO RADIO MEMORIAL GROUP c/o J Hill, The New Gatehouse, Gubboles Drove, Spalding, PE11 4AX
G4 OOB J WESTERMAN, 7 Gascoigne Court, Barwick in Elmet, Leeds, LS15 4NY
G4 OOC B Simister, 3 Beech Tree Road, Featherstone, Pontefract, WF7 5EB
G4 OOE Adrian Langmead, 10 The Copse, Scarborough, YO12 5HG
G4 OOH S Parker, Flat 4, 72 Springfield Mount, Leeds, LS18 5QE
G4 OOI C Parker, 18 Langdale Gardens, Leeds, LS6 3HB
G4 OOJ C Rose, 3 Harley Drive, Leeds, LS13 4QY
G4 OOK S Stobbs, 78 Hershall Drive, Town Farm, Middlesbrough, TS3 8NX
G4 OOL J Yeandel, Fairfield Farm, Penhallow, Truro, TR4 9LT
G4 OOQ N Ward, 8 Meadow View Road, Kempston, Bedford, MK42 7BE
GM4 OOU John Forsyth, 10 Rowallan Crescent, Prestwick, KA9 2HE
G4 OOX L Harvey, 27 Guernsey Drive, Birmingham, B36 0PB
G4 OOY D Bird, 13 Kilvington Road, Arnold, Nottingham, NG5 7HQ
G4 OPB N Hydes, 2 Stable Court, Martlesham Heath, Ipswich, IP5 3UQ
G4 OPD A Blissett, 26 Cherry Orchard, Holt Heath, Worcester, WR6 6ND
G4 OPE M Hodges, 40 Ennersdale Road, Coleshill, Birmingham, B46 1EP
GI4 OPH T Crawford, 50 Thornleigh Gardens, Bangor, BT20 4NP
G4 OPI Anthony Easom, 1 Station Close, West Ayton, Scarborough, YO13 9JQ
G4 OPK D Carrett, 80 Rotherfield Way, Emmer Green, Reading, RG4 8PL
G4 OPL Tony Bayliss, 4 Sycamore Close, Polgooth, St. Austell, PL26 7BW
G4 OPN W broxup, 9 Kingsway, Hapton, Burnley, BB11 5RB
G4 OPO Clive Haddrell, 9 Counterpool Road, Kingswood, Bristol, BS15 8DQ
G4 OPP Ian Horsefield, 61 Lewis Court Drive, Boughton Monchelsea, Maidstone, ME17 4LG
G4 OPR R Hayward, Sunnyfields, Lighthouse Road, Dover, CT15 6EJ
G4 OPT A Kemsley, Newfield Lodge Rest Home, 93-99 St. Andrews Road South, Lytham St. Annes, FY8 1PU
GM4 OPU J Houston, 26 Clerk Drive, Corpach, Fort William, PH33 7LE
G4 OPV J Jackson, 15 Jackson Crescent, Stourport-on-Severn, DY13 0EW
GW4 OPW B Jones, 12 Ashbourne Court, Aberdare, CF44 8HA
G4 OPY S Balmer, 101 Marsh Lane, Shepley, Huddersfield, HD8 8AP
G4 OQ G Lidstone, 76 Thames Drive, Leigh-on-Sea, SS9 2XD
GW4 OQB A Greatrex, Clydfan, Dinas Cross, Newport, SA42 0XS
G4 OQG M Ayres, 3 Wicks Drive, Chippenham, SN15 3EL
G4 OQH H Callaghan, 5 Manor Park View, Manor Park Road, Glossop, SK13 7TL
G4 OQJ I James, 4 Lancaster Gardens, Earley, Reading, RG6 7PA
G4 OQK R Alderton, 1 Comfrey Way, Thetford, IP24 2UU
G4 OQL Christopher Bowley, Plum Tree House, Walk Close, Derby, DE72 3PN
G4 OQN M Barr, 30 Hounslow Road, Twickenham, TW2 7EX
G4 OQP J Gerrity, 14 Lostock Avenue, Hazel Grove, Stockport, SK7 5JN
G4 OQR Malcolm Huddart, Grange Coach House, High Street, Lincoln, LN6 9LU
G4 OQU J Davenport, 1 Lowfields, Staveley, Chesterfield, S43 3QB
G4 OQV R Beecham, 7 Crummock Close, Coventry, CV6 6GY
G4 OQX G Cooper, 61 Fallowfield Road, Hasbury, Halesowen, B63 1BZ
GI4 OQY J K Chambers, 44 Ballywillin Road, Portrush, BT56 8JN
G4 OQZ Brian Dawson, Isca, 12 Lestock Way, Fleet, GU51 3EB
G4 ORB G Busby, The Terrace, Terrace Road South, Bracknell, RG42 4DS
G4 ORC OLDHAM AM RC c/o G Oliver, 158 High Barn St., Royton, Oldham, OL2 6RW
G4 ORE Alan Charles, 14 Chorleywood Bottom, Chorleywood, Rickmansworth, WD3 5JD
GI4 ORI James Hamill, 67 Windsor Avenue, Coleraine, BT52 2DR
G4 ORJ A Jones, Fairview, Frenchs Road, Wisbech, PE14 7JF
G4 ORP M Parsons, 15 Sherbourne Road, Hangleton, Hove, BN3 8BA
G4 ORQ A Walker, 4a Winston Drive, Eston, Middlesbrough, TS6 9LY
G4 ORS William Ragg, 14 Mocatta Way, Burgess Hill, RH15 8UR
G4 ORU G Wadwell, 7 Barkhart Drive, Wokingham, RG40 1TW
G4 ORV D Whatmough, Flat 3, 170 Buxton Road, Stockport, SK2 6HA
G4 ORW A Atherley, 2 Haydock Close, Dosthill, Tamworth, B77 1QR
G4 ORX P Baggett, 33 Foxglove Way, Thatcham, RG18 4DL
G4 ORY C Bates, 335 Clarence Road, Four Oaks, Sutton Coldfield, B74 4LU
G4 OSB Tom Arris, 7 Rowan Road, North Hykeham, Lincoln, LN6 8LY
GI4 OSG D Robinson, 17 Dalton Glen, Comber, Newtownards, BT23 5RJ
G4 OSH A Nevison, 10 Birchway, Tunbridge Wells, TN2 3DA
G4 OSI David Whitehouse, 10 Felstead Street, Stoke-on-Trent, ST2 7HJ
G4 OSJ P Brewer, 2 Mill Close, Wing, Oakham, LE15 8RH
G4 OSK K Hall, 21 Eardulph Avenue, Chester le Street, DH3 3PR
G4 OSO E Binns, Fieldview, 5 Moorside, Cleckheaton, BD19 6JH
G4 OSP M Binns, Fieldview, 5 Moorside, Cleckheaton, BD19 6JH
G4 OSR S Roberts, 1 Lakeside Crescent, Long Eaton, Nottingham, NG10 3GH
GM4 OSS S Campbell, 14 Hillhouse Place, Stewarton, Kilmarnock, KA3 3HT
G4 OST Peter Cabban, Ivydene, Upper Tockington Road, Bristol, BS32 4LQ
G4 OSU M Dixey, 50 Sandon Road, Ford Houses, Wolverhampton, WV10 6EN
GM4 OSV C Dunn, 66 Glen Doll Road, Neilston, Glasgow, G78 3QP
G4 OSX Jonathan Griffiths, 5015 Mattos CT, Fremont, United States, CA 94536
G4 OSY David Gascoigne, 2 Thorncliffes, Chapel Lane, Pontefract, WF9 3NJ
G4 OTB N Hailes, Yew Tree Cottage, The Hollies Common, Stafford, ST20 0JD
G4 OTC Peter Gagen, 16 Melbourne Road, Bromsgrove, B61 8PE
G4 OTD C Taylor, 2 Kismet Avenue, Highbury, Australia, 5089
G4 OTE M Grayson, One Elm, 58 Kaye Lane, Huddersfield, HD5 8XU
GI4 OTG A McNeice, 148 Doagh Road, Newtownabbey, BT36 6BA
G4 OTI P Stockbridge, 11 Fairways, Frodsham, WA6 7RU
G4 OTJ J Witchell, Pennyquick Cottage, Broomhill Lane, Bristol, BS39 5SA
G4 OTL M Baker, 5 Lumb Lane, Liversedge, WF15 7QH
G4 OTS G Eccleston, 24 Orton Lane, Wombourne, Wolverhampton, WV5 9AW
G4 OTU David Fagan, 3 Oxenham Green, Torquay, TQ2 6DX
G4 OTV David Green, St. Annes, Poundfield Road, Crowborough, TN6 2BG
G4 OTX G Hibberd, 2 Carr Bank, Oakamoor, Stoke-on-Trent, ST10 3EA
G4 OUB John Whetstone, 70 Heanor Road, Smalley, Ilkeston, DE7 6DX
G4 OUG Christopher Beesley, 15 Byron Close, Cheadle, Stoke-on-Trent, ST10 1XB
G4 OUH J Brockway, 9 Warren Close, Princes End, Tipton, DY4 9PQ
G4 OUI M Brockway, 9 Warren Close, Princes End, Tipton, DY4 9PQ
G4 OUJ S Carrigan, 1 Milford Crescent, Littleborough, OL15 9EF
G4 OUK S Bradshaw, 14 Sheringham Drive, Crewe, CW1 3XJ
G4 OUM T Bolton, 25 Woodfield Drive, Lichfield, WS14 9HH
GI4 OUN D Fulton, 120 Dunnalong Road, Bready, Strabane, BT82 0DP
GI4 OUP S Henderson, 47 Donaghedy Road, Bready, Strabane, BT82 0DB
G4 OUS D Bean, 18 Witham Court, Higham, Barnsley, S75 1PX
G4 OUT I Cornes, 17 Chilwell Avenue, Little Haywood, Stafford, ST18 0QZ
GW4 OUU D Grace, The Laurels, Gwbert Road, Cardigan, SA43 1AF
G4 OUZ L Day, 3 Harris Way, Lee Mill Bridge, Ivybridge, PL21 9EU
G4 OVD G Rugen, 24 Highgate Road, Lydiate, Liverpool, L31 0DA
GI4 OVE J McElvanna, 26 Lissummon Road, Newry, BT35 6NA
G4 OVF P Sampson, 34 Solway Road, Moresby Parks, Whitehaven, CA28 8XJ
G4 OVG J Thompson, 29 Arun, East Tilbury, Tilbury, RM18 8SX
G4 OVH H Owen, 5 Arllwyn Cefn Road, Bwlchgwyn, Wrexham, LL11 5YF
G4 OVI Ronald Rideout, 8 Wiltshire Close, Gillingham, SP8 4LZ
G4 OVJ R Read, 29 Imber Road, Shaftesbury, SP7 8RX
G4 OVL M Allen, 18 Philip Garth, Wakefield, WF1 2LS
G4 OVM C Barnes, 13 Waterworks Road, Farlington, Portsmouth, PO6 1NG
GI4 OVN S Dawson, 2 Glencraig Park, Craigavad, Holywood, BT18 0BZ
G4 OVO J Featherstone, Garden Close, Greenway Road, Torquay, TQ1 4NJ
G4 OVR D Fillingham, 6 Kings Chase, Rothwell, Leeds, LS26 0HL
G4 OVS F Goddard, 4 St. Peters Close, Barnburgh, Doncaster, DN5 7EN
G4 OVT JOSEPH GRANT, 3 Craggwood Close, Horsforth, Leeds, LS18 4RL
G4 OVV N Howard, 25 Whitecroft Road, Wigan, WN3 5PS
G4 OVW B Jempson, Flat 1, 3 Dacre, Scarborough, YO11 2SP
G4 OVX Mark Kennett, Toms Cottage, Kendal Lane, York, YO26 7QN
GI4 OWA G Elliott, 4 Fernbrae Gardens, Londonderry, BT47 5XS
GI4 OWB J Fallows, 22 Meadow Way, Ballygowan, Newtownards, BT23 5TQ
G4 OWH Geoffrey Gregor, 41 Stonebridge Drive, Frome, BA11 2TN
G4 OWK T Pearsall, 6 Vernon Close, Martley, Worcester, WR6 6QX
G4 OWL Christopher Payne, 7 Ellis Avenue, Onslow Village, Guildford, GU2 7SR
G4 OWN A Turner, 1 Milton Road, Flitwick, Bedford, MK45 1QA
G4 OWQ Donald Scott, Isyfoel, Morfa Bychan, Porthmadog, LL49 9YD
G4 OWS N Cunliffe, 44 Shore Road, Hesketh Bank, Preston, PR4 6RB
G4 OWT S Harwood, 24 Firle Crescent, Lewes, BN7 1QG
G4 OWY Raymond Howes, 202 Abbotsbury Road, Weymouth, DT4 0NA
G4 OXD Terence Rose, 41 Keats Way, Hitchin, SG4 0DP
G4 OXG Nicholas Wood, 1 Firth Gardens, London, SW6 6QB
G4 OXK R Waygood, 2 Brookside Close, Bransgore, Christchurch, BH23 8BT
GW4 OXL W Smith, 11 Connacht Way, Pembroke Dock, SA72 6FB
GI4 OXO W Fitzsimons, 83 Boghill Road, Newtownabbey, BT36 4QT
G4 OXR C Mortimer, Meadow Bank, Dowlish Wake, Ilminster, TA19 0NZ
G4 OXU D Whan, 1 Hillclose Avenue, Darlington, DL3 8BH
G4 OYC J Cook, 31 Vicarage Hill, Marldon, Paignton, TQ3 1NH
GI4 OYG Mervyn Black, 38 Town Park, Carrickfergus, BT38 8FG
G4 OYH A Bowyer, 37 St. Mark Drive, Colchester, CO4 0LP
GI4 OYI D Chambers, 238 Donaghadee Road, Newtownards, BT23 7QP
GI4 OYL John Cuthbert, 19 Antrim Road, Ballymena, BT42 2BJ
GI4 OYM William ELLIOTT, 23 Castle View Park, Portrush, BT56 8AS
G4 OYN A Fuller, 17 Brington Drive, Barton Seagrave, Kettering, NN15 6UW
G4 OYO A Bramley, 13 Moorland Avenue, Stapleford, Nottingham, NG9 7FY
G4 OYP G Savin, 19 Hailey Avenue, Loughborough, LE11 4QW
G4 OYR Nigel Lee, Silverstone, Alverstone Road, Sandown, PO36 0LH
G4 OYT H Moss, 101 Barnford Crescent, Oldbury, B68 8PR
G4 OYX David Porter, 8 Stanton Drive, Ludlow, SY8 2PH
G4 OYZ Malcolm Spencer, 29 Kliffen Place, Halifax, HX3 0AL
G4 OZC W Smith, 15 Henbury Drive, Woodley, Stockport, SK6 1PY
G4 OZD R Woolley, 7 Geveze Way, Broughton Astley, Leicester, LE9 6HJ
G4 OZG Edward Haskett, 23 Gloucester Road, Gaywood, King's Lynn, PE30 4AB
GI4 OZI A Kinghan, 14 Sunningdale Park North, Belfast, BT14 6RZ
GI4 OZJ G Allen, 6 Lougherne Road, Annahilt, Hillsborough, BT26 6BX
G4 OZL P Ingram, Rosehill Cottage, Mount Carmel Road, Andover, SP11 7ER
G4 OZM D Bradberry, 6 The Close, Easton on the Hill, Stamford, PE9 3NA
G4 OZN Philip Donaldson-Badger, Heckdyke Cottage, Heckdyke, Doncaster, DN10 4BE
G4 OZQ G Fisher, 9 Shrubbery Road, Drakes Broughton, Pershore, WR10 2AX
GW4 OZU P Hyams, Tricklewood, Pembroke Dyfed, SA71 5HY
G4 OZX C Goble, 12 Longfield Road, Emsworth, PO10 7TR
G4 OZY A Osborn, 35 Griston Road, Watton, Thetford, IP25 6DN
G4 PAA Philip Booth, 22 Charters Lane, Brandesburton, Driffield, YO25 8QJ
G4 PAC Peter Caldwell, 3 Orchard Mead, Broadwindsor, Beaminster, DT8 3RA
GW4 PAF J Thomas, 2 Tudor Way, Llantwit Fardre, Pontypridd, CF38 2NH
G4 PAH M Rollason, Ash Ridge, Clint, Harrogate, HG3 3DS
G4 PAI P Wells, 15 Apple Tree Grove, Ferndown, BH22 9LA
G4 PAS P Searles, 63 Whalley Road, Ramsbottom, Bury, BL0 0DP
G4 PAT J Thirsk, 47 Chestnut Avenue, Euxton, Chorley, PR7 6BP
G4 PAV G Brutnall, 57 Wollaston Road, Irchester, Wellingborough, NN29 7DA
G4 PBC John Kilroy, 119 Station Road, Brimington, Chesterfield, S43 1LJ
G4 PBD R Hughes, 8 Frinton Court, The Esplanade, Frinton-on-Sea, CO13 9DW
G4 PBE L Hewett, 19 Parkers Place, Martlesham Heath, Ipswich, IP5 3UX
G4 PBF P Baker, 23 Orde Close, Crawley, RH10 3NG
G4 PBJ Brian Oakley, 6 Windmill Way, Haxby, York, YO32 3NL
G4 PBN John Vivian, 3 Station Road, Gunnislake, PL18 9DX
G4 PBO D Smith, 21 Sydney Road, Benfleet, SS7 5RD
G4 PBP Russel Stewart, 424 Wood End Road, Wolverhampton, WV11 1YD
G4 PBR Dennis Suttenwood, White Lodge, Mersea Road, Colchester, CO5 7LJ
GI4 PBS T Wilson, 39 Woburn Road, Millisle, Newtownards, BT22 2HY
GI4 PBT T Wilson, Brambly Hedge, 39 Woburn Road, Newtownards, BT22 2HY
G4 PBY Brian Jones, 13 Albert Street, Cheltenham, GL50 4HS
G4 PBZ T Ashton, 90 Secker Avenue, Warrington, WA4 2RE
G4 PCB Alan Cox, 12 Merrymeet, Whitestone, Exeter, EX4 2JP
G4 PCD M Dally, 11 Wrightson Terrace, Doncaster, DN5 9ST
G4 PCE R Collins, 389 Lode Lane, Solihull, B92 8NN
G4 PCF P Goodson, 46 Southwold, Bracknell, RG12 8XY
GW4 PCJ Roderick Belcher, Parciau, Bronwydd Arms, Carmarthen, SA33 6BN
G4 PCK Barrie James, Rivendell, Kingsgate Close, Torquay, TQ2 8QA
G4 PCL Barry Walker, 22 Peveril Road, Tibshelf, Alfreton, DE55 5LQ
G4 PCN C Lambert, 23 Palmars Cross Hill, Rough Common, Canterbury, CT2 9BL
GW4 PCO P Mogford, 27 Ynysymaerdy Road, Briton Ferry, Neath, SA11 2TE
G4 PCP C Shelton, 18 Beaconsfield Drive, Coddington, Newark, NG24 2RX
GI4 PCQ J Quinn, 86 Knocknacarry Road, Cushendun, Ballymena, BT44 0NS
G4 PCR J O'Hara, 12 Ray Avenue, Nantwich, CW5 6HJ
GM4 PCT A Gordon, 2 Duchray St., Riddrie, Glasgow, G33 2DD
G4 PCW A Tucker, 3 Eston Close, Mabe Burnthouse, Penryn, TR10 9JW
GW4 PCX R Price, 2 Grassholm Place, Broadway, Haverfordwest, SA62 3HX
G4 PCZ D St Quintin, 16 Cromwell Road, Sprowston, Norwich, NR7 8XH
G4 PDD F Bibby, 14 St. Clare Terrace, Chorley New Road, Bolton, BL6 4AZ
G4 PDE R Bradshaw, 44 Hawthorn Street, Derby, DE24 8BD
G4 PDG B Hillard, Farmlea, Hele Lane, South Petherton, TA13 5AP
G4 PDI Bryan Kenzie, 9 Goodliffe Avenue, Balsham, Cambridge, CB21 4AD
G4 PDK Roy Davies, 11 Tamar Green, Corby, NN17 2LA
G4 PDM David Mills, 31 Claremont Drive, Hartlepool, TS26 9PD
G4 PDQ J Clayton, 217 Prestbury Road, Cheltenham, GL52 3ES
G4 PDR D Hughes, 19 Burnsall Close, Farnborough, GU14 8NN
G4 PDU C Carrington, 3 Jeake Drive, Rye, TN31 7FH
G4 PDY John Brandhuber, 3 Brigham Place, Felpham, Bognor Regis, PO22 7NW
G4 PEA R Flanders, 51 Rookwood Court, Guildford, GU2 4EL
G4 PED J Hinde, 12a Station Parade, Ockham Road South, Leatherhead, KT24 6QN
G4 PEF W Ingram, 141 Churchill Road, Willesden Green, London, NW2 5EH
G4 PEK Les Dymond, 1 Pinslow Cross, St. Giles-on-the-Heath, Launceston, PL15 9SB
G4 PEL W Threapleton, Cobbs Nook Farm, Newstead Lane, Stamford, PE9 4JJ
G4 PEM Simon Rodda, 1-2 Penrose Terrace, Penzance, TR18 2HH
G4 PEN R Potts, 18 Parkside Road, Hoyland, Barnsley, S74 0AL
G4 PEO John Pitty, 12 St. Leonards Road, Horsham, RH13 6EJ
GI4 PES N Robinson, 3 Moorland Drive, Lisburn, BT28 2XU
G4 PET J Smith, Pasturefields House, Pasturefields Lane, Stafford, ST18 0RD
G4 PEU K Smith, Pasturefields House, Pasturefields Lane, Stafford, ST18 0RD
G4 PEW Richard Wood, Abbots Croft, Abbey Road, St. Bees, CA27 0EG
GW4 PEX W Williams, 168 Mumbles Road, West Cross, Swansea, SA3 5AN
G4 PEY R Wilmot, 1 Retreat Cottages, Church Lane, Horsham, RH12 3ND
G4 PFA P Wheeler, 21 Browns Road, Holmer Green, High Wycombe, HP15 6SL
G4 PFE J Laverick, 5 York Crescent, Newton Hall Estate, Durham, DH1 5PU
G4 PFF John Potter, 5 Poplar Close, Carlton, Nottingham, NG4 1HF
G4 PFG M Spooner, 6 Cross Road, Starston, Harleston, IP20 9NQ

G4 PFJ James Backus, 2 Southview Villas, Dunmow Road, Bishop's Stortford, CM22 6SW
G4 PFK G Gifford, 184 Chantrey Crescent, Great Barr, Birmingham, B43 7PQ
G4 PFL P Freestone, 6 Harcourt Road, Llandudno, [illegible]
G4 PFO J Grogory, 22 Tower View Road, Great Wyrley, Walsall, WS6 6HE
G4 PFQ NORTHWEST DURHAM RAYNET c/o Thomas Hanratty, 12 Clarendon Street, Consett, DH8 5LS
G4 PFR J Harding, 19 Carrington Crescent, Wendover, Aylesbury, HP22 6AW
G4 PFT James Harris, 45 Redehall Road, Smallfield, Horley, RH6 9QA
G4 PFU D Blunt, 12 Mallard Place, East Grinstead, RH19 4TF
G4 PFW Howard Palmer, 5 Hurst Close, Crawley, RH11 8LQ
G4 PFX D Palmer, 14 Garibaldi Road, Redhill, RH1 6PB
G4 PFY J O'Hagan, 13 Chapel Road, Stanford in the Vale, Faringdon, SN7 8LE
G4 PFZ John Aspland, 6 Trilithon Close, Hellesdon, Norwich, NR6 5EP
G4 PGA B Gage, 2 Wellsbourne Road, Stone Cross, Pevensey, BN24 5QX
G4 PGB P Hayward, 22 Falconers Park, Sawbridgeworth, CM21 0AU
G4 PGD H Hall, 10 Park View, Truro, TR1 2BW
G4 PGG P Beesley, 15 Byron Close, Cheadle, Stoke-on-Trent, ST10 1XB
GI4 PGH J Crawford, 2 Holywood Road, Newtownards, BT23 4TQ
G4 PGJ D Ward, 48 Moat Bank, Bretby, Burton-on-Trent, DE15 0QJ
GM4 PGM P Brash, 4 Union Street, Lossiemouth, IV31 6BA
GI4 PGN J Bailie, 4 Quarry Road, Greyabbey, Newtownards, BT22 2QF
G4 PGO David Fernant, 2 Lonsboro Road, Wallasey, CH44 9BR
G4 PGQ D Harrison, 77 Leigh Road, Hindley Green, Wigan, WN2 4SZ
G4 PGS Peter Clark, 23 Nova Mews, Sutton, SM3 9HY
GM4 PGV P Lawless, 37 Oaklands Avenue, Irvine, KA12 0SE
G4 PGW Nigel Puttick, 33 Alder Hill Drive, Totton, Southampton, SO40 8JB
G4 PGX Martin Williams, 114 Ferry Street, Burton-on-Trent, DE15 9EY
G4 PGY R White, Beech Hill, Northampton, NN7 4LL
G4 PHB W Vickers, Creigiau, Penrhyndeudraeth, LL48 6LS
G4 PHC Geoffery Stearn, 31, Regents Way, Minehead, TA24 5HS
G4 PHK P Colbeck, 76 Church Road, Winterbourne Down, Bristol, BS36 1BY
G4 PHL P Green, 6 Yews Close, Worrall, Sheffield, S35 0BB
G4 PHP David Foster, 120 Green Lane, Cookridge, Leeds, LS16 7HF
G4 PHR T Clough, 37 Park Avenue, Mirfield, WF14 9PB
GW4 PHT D Dalling, 308 Townhill Road, Mayhill, Swansea, SA1 6PD
G4 PHV Gary Bennison, 35 Ermine Street, Thundridge, Ware, SG12 0SY
G4 PIA L Roberts, 18 Turret Grove, London, SW4 0ET
GI4 PID Brian Little, 8 Ballynoe Road, Antrim, BT41 2QT
G4 PIE David Tyler, 12 Bernards Way, Flackwell Heath, High Wycombe, HP10 9EQ
G4 PIJ John Goodman, 4 Maloren Way, West Moors, Ferndown, BH22 0BQ
G4 PIP C Bottoms, Treboro House, Ullenhall, Henley-in-Arden, B95 5NN
G4 PIQ A Cook, The Old Vicarage, High Road, Ipswich, IP6 9LP
G4 PIR John Child, 12 Beachill Road, Havercroft, Wakefield, WF4 2EJ
G4 PJD H Hoare, Farvardale, The Street, Bishop's Stortford, CM22 7LT
G4 PJE Raymond Kershaw, 13 Silver Hill, Milnrow, Rochdale, OL16 3UJ
G4 PJJ N Garbutt, Tudor Cottage, Main Road, Gloucester, GL2 8JP
G4 PJK R Mosedale, 21 Druids Avenue, Aldridge, Walsall, WS9 8LA
G4 PJL R Bailey, 5 Braemar Road, Doncaster, DN2 5HN
G4 PJP M Clay, 24 Begonia Drive, Burbage, Hinckley, LE10 2SW
GM4 PJR N Yarrow, 10 Coxburn Brae, Bridge of Allan, Stirling, FK9 4PS
G4 PJS Peter Shields, 81 Flaxton, Skelmersdale, WN8 6PE
G4 PJT S Schofield, 18 Ascot Close, Mexborough, S64 0JG
G4 PJY G Taylor, 23 Welland Way, Oakham, LE15 6SL
G4 PJZ J Towle, 46 Querneby Road, Nottingham, NG3 5HY
G4 PKE R Badham, Caedman, Terrace Road North, Bracknell, RG42 5JG
G4 PKF E Wood, 68 Baswich Crest, Stafford, ST17 0HJ
GM4 PKJ D Smith, Haremuir Bungalow, Benholm, Montrose, DD10 0HX
G4 PKK S Juden, 17a Astonville St., Southfield, London, SW18 5AN
G4 PKM J Derrick, 37 Admiralty Street, Keyham, Plymouth, PL2 2BR
G4 PKO D French, 14 Linden Close, Prestbury, Cheltenham, GL52 3DU
G4 PKP John Jones, Jason Photographic, New Moss Farm, Liverpool, L37 0AH
G4 PKT D Lewin, 14a Warwick New Road, Leamington Spa, CV32 5JG
GD4 PKV D Griffiths, 61 The Drive, North Harrow, Harrow, HA2 7EJ
G4 PKW C Gerard, 7 Parkwood Road, Sidemoor, Bromsgrove, B61 8UA
G4 PKX F Gallimore, 3 Wilson Crescent, Lostock Gralam, Northwich, CW9 7QH
G4 PKZ R Richardson, Manor Farm, Manor Lane, Oakham, LE15 7JL
G4 PLH R Hughes, 73 Upland Road, Sutton, SM2 5JA
GM4 PLI J Nellis, 64 Kirkwood Avenue, Clydebank, G81 2ST
G4 PLK S Lewis, 189 Ashburton Road, Hugglescote, Coalville, LE67 2HE
G4 PLL I Thomas, 15 Wakefield Road, Fitzwilliam, Pontefract, WF9 5AJ
G4 PLS A Haigh, White Horse Cottage, Maypole, Canterbury, CT3 4LN
G4 PLT Peter Teather, 11 Huntingdon Way, Burgess Hill, RH15 0NR
G4 PLU J Bates, 63 Sunny Blunts, Peterlee, SR8 1LP
G4 PLV M Seton, 12 Chatsworth St., Roundthorn, Oldham, OL4 5LF
G4 PLW Peter Walker, Willow Corner, Bendish, Hitchin, SG4 8JH
G4 PLX Anthony Salata, 64 Wildwood Road, London, NW11 6UP
G4 PLY Vivian Morris, 21 Cranhill Road, Street, BA16 0BY
G4 PLZ Peter Connors, Manor Cottage, Mill Road Banningham, Norwich, NR11 7DT
G4 PMA David Pearson, 42 Church Street, Stapleford, Nottingham, NG9 8DJ
G4 PMB Francis Thompson, 38 Hanson Park, Northam, Bideford, EX39 3SB
G4 PMG Martin Green, Huntley, Chesham Road, Tring, HP23 6HH
GM4 PMH S Dunn, 4 Mid Street, Rosehearty, Fraserburgh, AB43 7JS
G4 PMJ Allan Santos, 17 Elm Garth, Roos, Baytree, Hull, HU12 0HH
GM4 PMK R Blackwell, Willowbank, Pennyghael, Isle of Mull, PA70 6HB
G4 PMM R Williams, 2 Keepers Close, Bestwood Village, Nottingham, NG6 8XE
GI4 PMP H Smith, 1a Taylor Park, Limavady, BT49 0NT
G4 PMS P Steele, 107 Lower Shelton Road, Marston Moretaine, Bedford, MK43 0LW
GM4 PMT Alastair Ross, 6 Burnside Street, Findochty, Buckie, AB56 4QW
G4 PMV K Grime, 13 Runnymede Court, Jackson Street, Bolton, BL3 5HX
G4 PMW Roy Dunn, 117 All Saints Way, West Bromwich, B71 1RU
G4 PMY George Bell, Linden Lea, Crewe Road, Sandbach, CW11 4RE
G4 PMZ P Butcher, 9 Little Platt, Guildford, GU2 8JU
G4 PNB Alan Bathurst, 64 Oakfields, Guildford, GU3 3AU
G4 PNC D HOOD, 32 Bishops Wood, Nantwich, CW5 7QD
G4 PND A Daniel, 10 Tamarisk Close, Hatch Warren, Basingstoke, RG22 4UX
G4 PNH G Aungiers, 17 Broadwood Drive, Fulwood, Preston, PR2 9SS
G4 PNI R Bishop, 40 Auburn Grove, Blackpool, FY1 5NJ
G4 PNK T Copeland, Park Farm House, Park Road, Bedford, MK43 7QF
G4 PNL A Cox, 118 Harborough Road, Desborough, Kettering, NN14 2QY
G4 PNM A Wixon, Riverview Cottage, Melrose, TD6 9JB
G4 PNP B Deak, 57 Arundel Road, Peacehaven, BN10 8RP
G4 PNQ R Dhami, 3327 Smoke Tree Road, Mississauga, Ontario, Canada, L5N 7M6
G4 PNT A Hellewell, 41 Woodlea Grove, Armthorpe, Doncaster, DN3 2HN
G4 PNV G Powrie, 2 Glyn Garth Court, Menai Bridge, LL59 5PD
G4 PNX David Painter, 93 Oxclose Lane, Arnold, Nottingham, NG5 6FN
G4 POB T Hutchings, 9 Little Dell, Welwyn Garden City, AL8 7HZ
GI4 POC Robert Drain, 5 Ravelstone Avenue, Bangor, BT19 1EQ
G4 POD J Gould, 30 Weymouth Avenue, Middlesbrough, TS8 9AD
G4 POF John Hart, 1 Meadow Court, Fordingbridge, SP6 1LW
G4 POG W Evans, 3 Coastline Village, Ostend Road, Norwich, NR12 0NF
G4 POI D Lambert, 8 Stretton Road, Barnsley, S71 1XQ
G4 POL B Robertson, 12 Green Lane, Woodstock, OX20 1JY
G4 POP Terence Genge, 28 Hillside Road, Burnham on Crouch, CM0 8EY
G4 POR B Banks, 30 Hospital Road, Burntwood, WS7 0LD
G4 POT David Girling, 20 Fore Street, Praze, Camborne, TR14 0JX
G4 POU P Dyer, 36 Margate Road, Ipswich, IP3 9DF
G4 POW A Owen, 60 Brighton Avenue, Elson, Gosport, PO12 4BX
G4 POY Robert Kent, Talstr 4, Eriskirch, Germany, 88097
G4 PPB E Marshall, 75 Acacia Crescent, Wigan, WN6 8NJ
G4 PPC B Lowe, 19 Wolverhampton Road, Bloxwich, Walsall, WS3 2EZ
G4 PPD V Dann, 37 St. Brelades Avenue, Poole, BH12 4JR
G4 PPE Malcolm Bell, 55 Park Road, Hampton Hill, Hampton, TW12 1HX
G4 PPG J O'Sullivan, 40 Sheldon Avenue, Standish, Wigan, WN6 0LW
G4 PPH A Betts, 41 Long Lane, Shirebrook, Mansfield, NG20 8AZ
G4 PPJ S Bone, 6 Manor Road, Folksworth, Peterborough, PE7 3SU
G4 PPK C Everley, 5 Firs Close, Hazlemere, High Wycombe, HP15 7TF
G4 PPL Patrick Fisher, 10 Magdalene Court, Seaham, SR7 7DJ
G4 PPN David Chapman, 22 Horsley Drive, Kingston upon Thames, KT2 5GG
G4 PPP R Bailey, 10 Epping Close, Walsall, WS3 1TT
G4 PPR M Spencer, 67 Holmley Lane, Dronfield, S18 2HQ
G4 PPS D Herbert, 4 Sadler Drive, Marton-in-Cleveland, Middlesbrough, TS7 8HJ
GM4 PPT R Hodge, 34 Craig View, Coylton, Ayr, KA6 6LB
G4 PPU M Roy, 17 Elgar Avenue, Tolworth, Surbiton, KT5 9JH
G4 PPV Graham Jarrett, 3 Carisbrooke Road, Strood, Rochester, ME2 3SN
G4 PPW A Keech, 2 Mountfield Road, Irthlingborough, Wellingborough, NN9 5SY
G4 PPZ J Young, Nonsuch, Oxbridge, Bridport, DT6 3UB
G4 PQB D Mathers, Dovedale Lodge, Bourton on the Hill, Moreton-in-Marsh, GL56 9TE
G4 PQI J Raybould, 2 Woodland Avenue, Brierley Hill, DY5 1EQ
G4 PQM E James, 59 Queensway, Euxton, Chorley, PR7 6PN
G4 PQP Philip Malme, Newhaven, Mill Lane East Runton, Cromer, NR27 9PH
G4 PQS W Tedbury, Tyting House, Exeter Road, Honiton, EX14 1AX
G4 PQU A Harwood, 108 Tudor Green, Jaywick, Clacton-on-Sea, CO15 2PE
GI4 PQV T Pollock, 33 Seahill, Donaghadee, BT21 0SH
G4 PQW M Davis, 478 Eastern Avenue, Gants Hill, Ilford, IG2 6EQ
G4 PQX Derrick Taylor, 28 Main Street, Broadmayne, Dorchester, DT2 8EB
G4 PQY A Williams, 7 Bower Hall Drive, Steeple Bumpstead, Haverhill, CB9 7ED
G4 PRB Peter Ball, 21 Doonamana Road, Dun Laoghaire, Ireland, CO DUBLIN
G4 PRD P Dakin, 12 Spinney Close, Kidderminster, DY11 6DQ
G4 PRF S Brown, 27 The Court, Anderby Creek, Skegness, PE24 5YQ
GI4 PRH D Simpson, 31 Beech Green, Doagh, Ballyclare, BT39 0QB
G4 PRJ M Worsfold, 5 Turner Close, Langney, Eastbourne, BN23 7PF
G4 PRL Roy Hunt, 13 Westlake Rise, Heybrook Bay, Plymouth, PL9 0DS
GM4 PRO T Oneil, 187 Main St., Chapelhall, Airdrie, ML6 8SF
GW4 PRP S Lane, 12 Carlos Street, Port Talbot, SA13 1YD
G4 PRQ Ivan Hooper, Flat 5, Wenlock House, 41 Stanstead Road, London, SE23 1HG
G4 PRS POOLE RAS c/o David Mason, 26 Upton Road, Fleetsbridge, Poole, BH17 7AH
G4 PRW philip Whitten, 2 Eastmead, Woking, GU21 3BP
G4 PSE M Grime, 10 East Park Avenue, Darwen, BB3 2SQ
G4 PSH Terry Owen, Touchwood, The Street, Norwich, NR12 9RF
G4 PSI C Franks, 11 Orchard Close, Crook, DL15 8QU
GM4 PSJ R Stroud, 24 Cullen St., Portsoy, Banff, AB45 2PJ
GM4 PSL Terence Grice, 35 Approach Row, East Wemyss, Kirkcaldy, KY1 4LB
G4 PSO A Little, 20 Vicarage Close, Shillington, Hitchin, SG5 3LS
G4 PSP S Gardner, 191 Charlton Park, Midsomer Norton, Radstock, BA3 4BR
G4 PSR Colin Sartorius, 39 Althorne Gardens, London, E18 2DA
G4 PSS Stephen Black, 71 Bellerby Drive, Ouston, Chester le Street, DH2 1UF
G4 PST D Turner, Hurdletree Bank Farm, Hurdletree Bank, Spalding, PE12 8QQ
G4 PSU Alan Davidson, 5 Hanover Parc, Indian Queens, St. Columb, TR9 6ER
G4 PTE K Lown, Maurice House, Callis Court Road, Broadstairs, CT10 3AH
G4 PTF C Keeping, 12 St. Francis Avenue, Southampton, SO18 5QJ
G4 PTK G Mason, 120 Scalford Road, Melton Mowbray, LE13 1JZ
G4 PTM J Fyrth, 2 Merton Gardens, Farsley, Pudsey, LS28 5DZ
GM4 PTQ S Bennie, 13a Scotland Street, Stornoway, HS1 2JN
G4 PTU Blandford Garrison CW and DX Group c/o Richard Carter, 12 Glebe Close, Abbotsbury, Weymouth, DT3 4LD
GD4 PTV B Brough, 4 The Bretney, Jurby, Isle of Man, IM7 3BL
G4 PTW A Brunning, 6 Newstead Road, Barnwood, Gloucester, GL4 3TQ
G4 PTZ John Topley, 27 Inveraray Close, Sinfin, Derby, DE24 3JA
G4 PUB BASILDON DIST RD c/o S Wensley, 7 Bradshaw Close, Windsor, SL4 5PS
GW4 PUC R Rees, 16 Brynheulog, Llanelli, SA14 8AE
G4 PUD B Langdon, 80 Glen Rise, Birmingham, B13 0EJ
G4 PUM R Brookes, Broadeaves, Bridgemere Lane, Nantwich, CW5 7PN
G4 PUO Wojciech Stumpf, 18 Saxhorn Rd, High Wycombe, HP14 3JN
G4 PUP B Philipp, 2 Red Lion Park, Denbigh Road, Battle, TN33 9ET
G4 PUQ P Mcewen, Southerly, Church Road, Halesworth, IP19 0EA
GM4 PUS J Murray, Ose Farm House, Isle of Skye, IV56 8FJ
G4 PUX Brian Gayther, Coed Park, Penisarwaun, Caernarfon, LL55 3PW
G4 PUZ Nick Barrington, 18 Brambleside, Thrapston, Kettering, NN14 4PY
G4 PVC Alexander Smith, 21, Darwin Crescent, Morley, Australia, 6062
G4 PVM P Tittensor, 47 St. Johns Road, Chelmsford, CM2 0TY
G4 PVN A Taylor, 3 Mond Crescent, Billingham, TS23 1DL
G4 PVP P Painter, 80 Willowsbrook Road, Hurst Green, Halesowen, B62 9RF
GM4 PVQ D Ross, 11 Edinview Gardens, Stonehaven, AB39 2HG
G4 PVS J Maude, Anthony Fold Farm, Bury Old Rd, Ramsbottom Lancs, BL0 0RW
G4 PVU A Wilkinson, 1 Langley Close, Penrhyn Bay, Llandudno, LL30 3LN
G4 PVX A Daws, 9 Wellow Mead, Peasedown St. John, Bath, BA2 8SA
G4 PVY Royston Limb, 3 Canford Heights, Western Road, Poole, BH13 7BE
G4 PVZ G Loach, 39 Park Road West, Wolverhampton, WV1 4PL
G4 PWA P Dane, Oakhill Lodge, Hewelsfield, Lydney, GL15 6UN
G4 PWB G Smith, 9a Lansdowne Drive, Rayleigh, SS6 9AL
G4 PWD M Mchale, 41 Sheringham Drive, Etchinghill, Rugeley, WS15 2YG
G4 PWE J Veness, 59 St. Helens Down, Hastings, TN34 2BG
G4 PWF Richard Harris, The Mill, Mill Lane, Market Rasen, LN8 3LF
G4 PWG J Hubbard, 2 Carlton Road, Portchester, Fareham, PO16 8JW
G4 PWI P Heredge, 110 Oxford Crescent, Didcot, OX11 7AX
G4 PWM David East, 39 Chapel Lane, Navenby, Lincoln, LN5 0ER
G4 PWP D Blackwell, 58 Bleadon Hill, Weston Super Mare, BS24 0JW
GM4 PWQ J Foster, 185 Sea Road, Methil, Leven, KY8 2EQ
GM4 PWR Alan [illegible], 91 Lundy Road, Inverlochy, Fort William, PH33 6NY
G4 PWS S Keen, 34 Unwin Road, Isleworth, TW7 6HX
G4 PWV D Howton, 4 Stonepine Close, Wildwood, Stafford, ST17 4QS
G4 PWY Patrick Hennessy, 5 Smedley Court, Egginton, Derby, DE65 6HD
GW4 PWZ W Evans, Windyridge Bungalow, Mount View, Merthyr Tydfil, CF47 0UX
GM4 PXB M Gale, 83 King Street, Peterhead, AB42 1UQ
G4 PXC A Lord, 16 Lark Valley Drive, Fornham St. Martin, Bury St. Edmunds, IP28 6UG
G4 PXE A Brend, 42 West Garth Road, Exeter, EX4 5AJ
G4 PXF M Harries, 63 Oakhill Road, Dronfield, S18 2EL
GM4 PXG T Worthington, 5 Clairmont Place, Lerwick, Shetland, ZE1 0BR
G4 PXH E Southwell, 60 Solent Breezes, Hook Lane, Southampton, SO31 9HG
G4 PXJ J Peet, 66 Barry Road, Northampton, NN1 5JS
GI4 PXM William Nelson, 111 Rutherglen Street, Belfast, BT13 3LR
G4 PXN C Sissons, 9 Mount Pleasant, Goldenbank, Falmouth, TR11 5BW
G4 PXR T Geldart, Langdale, Coast Road, Ulverston, LA12 9QZ
G4 PXX P Styles, 23 Mereweather Avenue, Frankstow, Victoria, Australia, 3199
GD4 PXY G Bloomfield, 15 Beaulieu Drive, Pinner, HA5 1NB
G4 PYA A Ledger, 32 St. Augustines Crescent, Whitstable, CT5 2NW
G4 PYD C Johnson, 51 Newstead Avenue, Holton-le-Clay, Grimsby, DN36 5BQ
G4 PYG Ivor Bennett, Collins Green, School Road, Colchester, CO5 9TH
G4 PYH D Jackson, 9 Stour Close, Altrincham, WA14 4UE
G4 PYI B Burman, 53 Field Avenue, Hatton, Derby, DE65 5ER
GM4 PYJ James Balfour, 36 Causewayhead Road, Stirling, FK9 5EU
G4 PYQ Albert Hill, 37 Rock St., Gee Cross, Hyde, SK14 5JX
G4 PYR R Maskill, 21 Clayton, Orton Goldhay, Peterborough, PE2 5SB
G4 PYS E Devereux, 15 Severn Close, Paulsgrove, Portsmouth, PO6 4BB
G4 PYU S Harding, Los Huertos, Apartado de Correos 42, Malaga, Spain, 29754
G4 PYV Stephen Parkin, 17 Magellan Drive, Worksop, S80 3QZ
G4 PYW M Zubrzycki, 4 Falklands Court, Easington, Hull, HU12 0QE
G4 PZJ C Christopher, 15 Inman Road, Earlsfield, London, SW18 3BB
G4 PZL Reg Pain, Hegerston, Long Road West, Colchester, CO7 6ES
G4 PZN Christopher Poulson, 14 Castlegate, Penrith, CA11 7HZ
G4 PZU F Day, 27 Prince Charles Road, Lewes, BN7 2HY
G4 PZV A Terry, 6 Seaton Close, Stubbington, Fareham, PO14 2PX
G4 PZW R Proctor, 6 North Street, Burwell, Cambridge, CB25 0BA
G4 PZX A Tracey, Well 'N' Garden, Abberton Road, Colchester, CO5 7AS
G4 QA R Brown, The Lilacs, Scar Lane, Whitby, YO21 3SD
G4 RAA Michael Brown, 12 Mead Way, Burnham, Slough, SL1 6HD
G4 RAB Dorothy Ellis, 1 Showering Close, Bristol, BS14 8DY
G4 RAC J Cooper, 134 Jordan Avenue, Stretton, Burton-on-Trent, DE13 0JD
G4 RAE Rex Laney, 7 Downfield Close, Alveston, Bristol, BS35 3NJ
G4 RAF RAF SEALAND ARC c/o A Ward, 158 Mold Road, Mynydd Isa, Mold, CH7 6TF
GD4 RAG J Martin, Tradewinds, Mount Gawne Road, Port St. Mary, Isle of Man, IM9 5LX
GM4 RAH P Robertson, 32 Crosswood Crescent, Balerno, EH14 7HS
GM4 RAI Robert Shand, 12 Bexley Terrace, Wick, KW1 5HQ
G4 RAJ J Shaw, 31 Dartmouth Avenue, Almondbury, Huddersfield, HD5 8UP
G4 RAK J Hornby, 21 West Wools, Portland, DT5 2EA
G4 RAP V Seaman, 15 Dukes Orchard Nicholas Close, Writtle, Chelmsford, CM1 3JZ
G4 RAR P Clemens, 18 Ladylea Road, Horsley, Derby, DE21 5BN
G4 RAV P Evans, 27 Terence Airey Court, Harleston, IP20 9JP
G4 RAY G Pickering, 1 Warrens Yard, Wells-next-The-Sea, NR23 1PA
GM4 RAZ B Smith, 8 Moss Side Drive, Portlethen, Aberdeen, AB12 4NY
G4 RBC C Hawkridge, 2 Windward Close, Littlehampton, BN17 6QX
G4 RBH T Farmer, 12 Rose Avenue, Mitcham, CR4 3JS
G4 RBP Rowena Purdy, 4 York Road, Brookenby, Market Rasen, LN8 6EX
G4 RBQ D Love, Highridge, South Road, Haywards Heath, RH17 7QS
G4 RBR Chris Randall, 38 Kilmorey Gardens, St. Margaret's, Twickenham, TW1 1PY
G4 RBU T Crookes, 167 Willow Drive, Handsworth Hill, Sheffield, S9 4AU
G4 RBZ Christopher Dervin, 24 Willow Park Way, Aston-on-Trent, Derby, DE72 2DF
G4 RCB Derek Thorp, West View, West Lane, Winkleigh, EX19 8QU
G4 RCC CARAVAN & CAMPING ARC c/o A Wright, 34 Webbs Way, Stoney Stanton, Leicester, LE9 4BW
G4 RCD M Capstick, 186 Forest Lane, Harrogate, HG2 7EE
G4 RCE M , Gladbachstrasse 19, Zurich, Switzerland, 8006
G4 RCF J O'Dell, 5 Further Ends Road, Freckleton, Preston, PR4 1RL
G4 RCG John Muzyka, 2 Engine Fold, Wrenthorpe, Wakefield, WF2 0PP
G4 RCH S Thompson, 2 Allenby Drive, Leeds, LS11 5RP
G4 RCJ Derek Underwood, Kilnhurst, Kilnhurst Road, Todmorden, OL14 6AX
GI4 RCK W McMillen, 26 Maymount Street, Belfast, BT6 8BH
G4 RCM W Williams, 154 Llysfaen Road, Old Colwyn, Colwyn Bay, LL29 9HP
GM4 RCN J Young, 13 Craig Crescent, Causewayhead, Stirling, FK9 5LR
G4 RCP C Marriott, 19 Beechey Close, Denver, Downham Market, PE38 0DH
G4 RCR P Starley, Oak House, Birmingham Road, Warwick, CV35 7DX
G4 RCY A Beglin, 3 The Mead, Shipham, Winscombe, BS25 1TR



**IMPORTANT NOTE**

**Revalidate licence to avoid revocation** – Ofcom has advised the Society that plans will be drawn up to revoke licences that have not been revalidated as required by the licence conditions. The quickest way to revalidate is to do so online via the Ofcom website: *https://services.ofcom.org.uk/* or by email: *amateur.validations@ofcom.org.uk* Ofcom staff are available to help, but please be patient during times of heavy workload.

G4 RCZ I Dempster, 54 Fashoda Road, Selly Park, Birmingham, B29 7QJ
G4 RDA U Harris, Preston Lodge, Kentisbury, Barnstaple, EX31 4NH
G4 RDC J Gumb, 16 Raggleswood Close, Earley, Reading, RG6 7LH
G4 RDG G Murray, 176 Golfwood Drive, Hamilton, Canada, L9C 7B8
G4 RDH M Wirthner, 51 College Road, Upper Beeding, Steyning, BN44 3TB
GM4 RDI J White, 1 Banknowe Road, Tayport, DD6 9LG
G4 RDL LEEDS RAYNET GROUP c/o G Belt, Flat 2, 3 King George Avenue, Leeds, LS7 4LH
G4 RDM P Rouget, 7 Palmer Close, Wellingborough, NN8 5NX
G4 RDS B Wood, 100 Lower Road, Hullbridge, Hockley, SS5 6DD
GW4 RDW Leuan Jones, 6, Norton Terrace, Glyncorrwg, SA13 3AN
G4 RDY L Harrison, 48 Bleasdale Avenue, Thornton-Cleveleys, FY5 3RQ
G4 REC A Marrows, 7 Victoria Close, Yeadon, Leeds, LS19 7AU
G4 REE P Miller, 2 The Pavilions End, Camberley, GU15 2LD
GM4 REF W McLean, 159 Castlemilk Road, Glasgow, G44 4NA
G4 REG A Boocock, 2 Vine Garth, Clifton, Brighouse, HD6 4JZ
G4 REH R England, 5 Weir Road, Congresbury, Bristol, BS49 5HL
GW4 REI David May, 19 Sycamore Street, Pembroke Dock, SA72 6QN
G4 REK J Tylee, 40 Luna Road, Thornton Heath, CR7 8NY
GM4 REN Brian Strathdee, 85 Weavers Knowe Crescent, Currie, EH14 5PP
G4 REU J Taylor, 18 Fackley Way, Stanton Hill, Sutton-in-Ashfield, NG17 3HT
G4 REX Philip Hassmann, 11 Oak Close, Bulwark, Chepstow, NP16 5RL
G4 RFA M Tew, 25 Broad Oak Lane, Penwortham, Preston, PR1 0UX
G4 RFC Stephen Fletcher, 90 Westcombe Park Road, Blackheath, London, SE3 7QS
G4 RFF Tariq Mundiya, Apt 5, 35 Mercer Street, New York, United States, 10013
GI4 RFH T Robinson, 21 Carnhill Road, Newtownabbey, BT36 6LA
GD4 RFI R Linden, 24 Hartland Drive, Edgware, HA8 8RH
G4 RFJ I Mccann, Maythorne, Rosslyn Avenue, Poulton-le-Fylde, FY6 0HE
GD4 RFK Margaret Dodd, Ellan Geay, Ballayockey Lane, Ramsey, Isle of Man, IM7 3HP
G4 RFN A Robey, 54 Jarrett Avenue, Wainscott, Rochester, ME2 4NL
G4 RFO B Wood, 11 Oakdale Avenue, Wibsey, Bradford, BD6 1RP
G4 RFP A Goodall, 10 Beacon Close, Everton, Lymington, SO41 0LQ
G4 RFR Flight Refuelling ARS c/o Julian Smith, 157 Churchill Road, Poole, BH12 2JB
G4 RFU D Abbott, 21 Leckhampton Road, Cheltenham, GL53 0AZ
G4 RFV Brian Adams, 14 Foxcroft Drive, Wimborne, BH21 2JZ
G4 RGA J Dunnett, 43 Oakfield Park, Wellington, TA21 8EX
G4 RGB Mark Rogers, 4 Hill Corner, Ledgemoor, Hereford, HR4 8QG
G4 RGE N Lovely, Dolphin Cottage, Upper Green Road, Ryde, PO33 1XE
G4 RGF P McCall, 11 Elworthy Drive, Wellington, TA21 9AT
G4 RGH D Mclaughlin, 90 Broom Lane, Rotherham, S60 3EW
GW4 RGI William Baker, 4 Connaught Place, Pembroke Dock, SA72 6EZ
G4 RGM Greater Manchester Raynet c/o Neikolas Czernuszka, 12 Durham Drive, Ashton-under-Lyne, OL6 8BP
G4 RGO John Crocker, 8 Oakwood Avenue, Havant, PO9 3RA
G4 RGP E Hall, 93 Sthbourne Coast, Bournemouth, BH6 4DX
GD4 RGR K Grattan, 41 Carrick Park, Sulby, Ramsey, Isle of Man, IM7 2EY
GM4 RGS Ramsay Smith, 8 Mosside Drive, Portlethen, Aberdeen, AB12 4NY
GM4 RGU D Nicolson, Silver Birches, Blebo Craigs, Cupar, KY15 5UF
G4 RGY Burnham ARC c/o Brian Mudge, 9 Crossmead, Woolavington, Bridgwater, TA7 8ER
G4 RHB M Bailey, 10 Greenwood Avenue, Bolton le Sands, Carnforth, LA5 8AW
G4 RHC M Kellett, 1 Spa Cottages, Gilsland, Brampton, CA8 7AL
G4 RHJ P Vickers, 2 Firbank Drive, Woking, GU21 7QT
G4 RHK L Woodcock, 2 Poolhay Close, Corse Lawn, Gloucester, GL19 4NY
G4 RHL Richard Langdon, 15 St. Cuthberts Way, Sherburn Village, Durham, DH6 1RH
G4 RHR K Backhouse, 113 Bucklesham Road, Kirton, Ipswich, IP10 0PF
G4 RHX A Moore, 1 St. Andrews Road, New Marske, Redcar, TS11 8AU
G4 RHY M Pratt, The Bays, Back Lane, Doncaster, DN9 3AJ
G4 RHZ B Coupe, 9 School Lane, Auckley, Doncaster, DN9 3JR
G4 RIB D Stoole, Brookside Farm, Baltic Terrace, Cwmbran, NP44 7AH
G4 RIE D Littler, 16 Lee Bank, Westhoughton, Bolton, BL5 3HQ
G4 RIH Neil Thorne, 61 Horsham Avenue, London, N12 9BG
G4 RIK R Kirkwood, 42 Porters Hill, Harpenden, AL5 5HR
G4 RIM A Day, 3 Harris Way, Lee Mill Bridge, Ivybridge, PL21 9EU
G4 RIO S Williams, 18 The Leas, Barkston, Grantham, NG32 2PD
G4 RIP John Creaseyy, 8 Church Street, Billingborough, Sleaford, NG34 0QG
G4 RIQ Leslie Rushforth, 90 Brearley Avenue, New Whittington, Chesterfield, S43 2DZ
G4 RIS B Didmom, 45 Millstrood Road, Whitstable, CT5 1QF
G4 RIU R Jones, 67 Plover Road, Larkfield, Aylesford, ME20 6LA
G4 RIV WIGTOWNSHIRE ARC c/o Andrew Gaston, 9 Lochans Mill Avenue, Stranraer, DG9 9BZ
G4 RJA Ivor Wilkinson, 24 Isis Way, Hilton, Derby, DE65 5LP
G4 RJD Kenneth Ward, 3 Levetts Hollow, Hednesford, Cannock, WS12 2AW
G4 RJF J Weatherer, 20 Gilloch Crescent, Dumfries, DG1 4DW
G4 RJG I Toon, 18 Barrowfield Road, Stroud, GL5 4DF
G4 RJM G Heward, 4 Hillside Drive, Little Haywood, Stafford, ST18 0NN
G4 RJO B Robertson, 28 Heath Lane, Blackfordby, Swadlincote, DE11 8AA
GM4 RJX James Hatton, 64 Abercromby Crescent, Helensburgh, G84 9DN
G4 RJY C Sidney, 10 Colville Close, Bampton, OX18 2NN
G4 RJZ Anne Phillpott, Southways, Stombers Lane, Folkestone, CT18 7AP
G4 RKB J Conlon, 24 Goldcrest Close, Colchester, CO4 3FN
GI4 RKC S Jennings, 34 Palmer Avenue, Lisburn, BT28 3QB
G4 RKD T Clarke, 19 Ratby Lane, Markfield, LE67 9RJ
G4 RKF B Peart, 7a Grundy Close, Abingdon, OX14 3SD
G4 RKG J Whiting, 19 Watermore Close, Frampton Cotterell, Bristol, BS36 2NQ
GM4 RKH T Llewellyn, The Shepherds Cottage, Buckies Farm, Thurso, KW14 7XH
GW4 RKI K Perryman, 52 Darwin Road, Port Talbot, SA12 6BS
G4 RKK I Welford, Mistletoe House, Watton Road, Thetford, IP24 1PB
G4 RKL W Welford, Bowling Green House, Griffin Lane, Attleborough, NR17 2AD
GM4 RKM Thomas Cassidy, 34 Torr-na-Faire, Lochaline, Oban, PA80 5XS
G4 RKN Peter Wells, 24 Common Close, West Winch, King's Lynn, PE33 0LB
G4 RKO Barry Cooper, 20 The Paddock, Alconbury, Huntingdon, PE28 4WS
G4 RKP Raymond Groom, Tryst, Rackhams Corner, Lowestoft, NR32 5LB
G4 RKR D Geddes, 9 Rosenella Close, Northampton, NN4 8RX
G4 RKU B Bamber, 14 Ellesmere Avenue, Thornton-Cleveleys, FY5 5JD
G4 RKV L Adams, 50 Selsea Avenue, Herne Bay, CT6 8SD
GW4 RKX G Cook, 22 Northlands Park, Bishopston, Swansea, SA3 3JW
GW4 RKZ R Cleverley, 33 Tylchawen Crescent, Tonyrefail, Porth, CF39 8AL
G4 RLA Colin Butcher, 7 Lascelles Hall Road, Kirkheaton, Huddersfield, HD5 0AT
G4 RLC T Isom, 64 Cuffling Drive, Leicester, LE3 6NF
G4 RLF Martyn Wright, 24 Wessex Road, Wilton, Salisbury, SP2 0LW
G4 RLL J Woods, 4 Wheatfield Drive, Burton Latimer, Kettering, NN15 5YL
G4 RLM J Hiscock, 62 East Borough, Wimborne, BH21 1PL
G4 RLN D Rosevear, 37 Sharaman Close, St. Austell, PL25 3DH
G4 RLO G Seal, 160 Conway Drive, Shrewsbury, SY2 5UG
G4 RLP T Varney, 29 Ffordd Eryri, Caernarfon, LL55 2UR
G4 RLR James Hogan, 13 Strawberry Fields, Great Barford, Bedford, MK44 3BQ
G4 RLS John Elsdon, 15 Union Road, Lowestoft, NR32 2BZ
G4 RLT R Langford, 43 Oldminster Road, Sharpness, Berkeley, GL13 9US
G4 RLU P Quickfall, Quickfall, 14 Lade Fort Crescent, Romney Marsh, TN29 9YF
GM4 RLV W Duguid, Villach, 7 Hawthorn Place, Ballater, AB35 5QH
G4 RLX H Cave, 3 Grace Meadow, Whitfield, Dover, CT16 3HA
GI4 RMA L McCullough, Down Lodge, Downpatrick, BT30 7LY
G4 RMC D Marsden, 67 Fourth Avenue, Watford, WD25 9QH
G4 RMD J Cobley, 4 Briars Close, Hatfield, AL10 8DQ
G4 RMG Eric Guy, 9 Longshore Apartments, Dane Road, Newquay, TR7 1EN
G4 RMJ A Cattani, Rozel, Marsh Road, Spalding, PE12 9PJ
GW4 RML D Davies, 101 Westlands, Port Talbot, SA12 7DE
G4 RMN M Hogan, 16 Freshfield Close, West Earlham, Norwich, NR5 8RA
G4 RMQ J Dudley, 2 Heathcote Grove, London, E4 6RT
G4 RMS RUSHEY MEAD SCHOOL AR CLUB c/o G Dover, 31 Newbold Road, Kirkby Mallory, LE9 7QG
G4 RMT Paul Johnson, 4 High Beech, Lowestoft, NR32 2RY
G4 RMV M Buckle, 3 Tilesford Park, Tilesford, Pershore, WR10 2LA
G4 RMX J Phelps, 33 Thirlmere Drive, North Anston, Sheffield, S25 4JP
G4 RNA P Dronfield, Grange Farm, Castleton, Hope Valley, S33 8WB
G4 RNC C Blezard, 26 Welford Avenue, Lowton, Warrington, WA3 2RN
G4 RND C Hawkins, 57 Links Road, Knott End-on-Sea, Poulton-le-Fylde, FY6 0DF
G4 RNF John Handley, Flat 31, Croft Manor Mason Close, Freckleton, Preston, PR4 1RG
G4 RNI George Tuck, 10 Redberry Way, South Shields, NE34 0BQ
G4 RNK Robert Dodson, 22 Southgate Crescent, Rodborough, Stroud, GL5 3TS
GI4 RNP Victor McFarland, 13 Railway Street, Derriaghy, Belfast, BT17 9EU
G4 RNR J Maunder, 56 Conery Lane, Enderby, Leicester, LE19 4AB
G4 RNT D Thorpe, 10 Stoke Road, Taunton, TA1 3EJ
G4 RNW M Stewart, 29 Elstree Road, Bushey Heath, Bushey, WD23 4GH
G4 RNX Antony Walker, 5 Christchurch Road, Malvern, WR14 3BH
G4 RNZ K Page, 51 Bournville Road, Weston-Super-Mare, BS23 3RR
G4 ROA A Chamberlain, 16 Okehampton Road, Stivichall, Coventry, CV3 5AU
G4 ROB Robert Taylor, 18 Spruce Avenue, Selston, Nottingham, NG16 6DX
G4 ROC Lawrence Odell, 30 St. Hybalds Grove, Scawby, Brigg, DN20 9DG
G4 ROH W Smith, 10 Playford Road, Rushmere St. Andrew, Ipswich, IP4 5RH
G4 ROI S Kiernan, 60 Riverview Road, Epsom, KT19 0LB
G4 ROJ R Stafford, 21 Kittiwake Drive, Kidderminster, DY10 4RS
G4 ROK G Sambrook, 73 Hayes Drive, Barnton, Northwich, CW8 4JX
G4 ROM M Ellis, Field Cottage, Hole Lane, Farnham, GU10 5LP
G4 ROP C White, 18 Ashton Gardens, Old Tupton, Chesterfield, S42 6JF
G4 ROR D Harrison, 55 Hudson Close, Worcester, WR2 4DP
G4 ROS F Sweetingham, 38 Pippins Green Avenue, Kirkhamgate, Wakefield, WF2 0RU
G4 ROU W Maudsley, 42 Crawford St., Clock Face, St. Helens, WA9 4XH
G4 ROV Phillip Weaver, 24 Montclaire Avenue, Blackwood, NP12 1EE
G4 ROX A Capel, 33 Romney Avenue, Bristol, BS7 9ST
G4 RPA D Court, 4 Rucrofts Close, Aldwick, Bognor Regis, PO21 3SL
G4 RPC R Cassling, 14 Canada Way, Lower Wick, Worcester, WR2 4DJ
G4 RPD A Else, 77 Sherwood Street, Mansfield Woodhouse, Mansfield, NG19 9NB
GM4 RPE J McCabe, 109 Weirwood Avenue, Baillieston, Glasgow, G69 6LQ
G4 RPF P Osborne, 27 Orange Tree Close, Chelmsford, CM2 9ND
G4 RPI C Parrish, 16 Charter Close, Boston, PE21 9PD
G4 RPJ Irene Flaherty, 10 Highfield Park, Heaton Mersey, Stockport, SK4 3HD
G4 RPK J Kaine, 74 Camden Mews, London, NW1 9BX
G4 RPL D Ingham, Auchengray, 51 Helena Street, Mexborough, S64 9PF
G4 RPO T Gemmell, 30 Goldie Crescent, Lochside, Dumfries, DG2 0AJ
G4 RPP G Kyte, 25 Brasted Close, Bexleyheath, DA6 8HU
G4 RPT Martin Edis, 28 High Street, Broughton, Kettering, NN14 1NG
G4 RPV K Baker, 153 Long Nuke Road, Birmingham, B31 1DX
G4 RPW F Charnley, 30 Dunkirk Avenue, Fulwood, Preston, PR2 3RY
G4 RQA T Mills, 14 Oxford Close, Padiham, Burnley, BB12 7DB
G4 RQF R Langer, Elms Bungalow, Queens Road, Sheffield, S20 1AW
G4 RQG Steve Baggaley, 35 Hayner Grove, Weston Coyney, Stoke-on-Trent, ST3 6PQ
G4 RQI David Warr, 5 Monckton Drive, Castleford, WF10 3HT
G4 RQJ Robert Hannan, 87 Plymouth Street, Walney, Barrow-in-Furness, LA14 3AN
G4 RQK Andrew Johnson, 14 Highfield, Duddington, Stamford, PE9 3QD
G4 RQL M Wilson, The Old Chapel, Poulshot Road, Devizes, SN10 1RW
G4 RQO J Pulford, 68 York Avenue, Droitwich, WR9 7DQ
G4 RQP G Hallett, 9 Dolcroft Road, Rookley, Ventnor, PO38 3NT
G4 RQQ Terfel Jones, 19 Penlon, Menai Bridge, LL59 5LR
G4 RQS J Leighton, 12 Morley Avenue, Connah's Quay, Deeside, CH5 4RE
G4 RQU D Young, 9 Mercedes Avenue, Hunstanton, PE36 5EJ
G4 RQW A McEwen, Apartment 9, 10 Lismore Place, Carlisle, CA1 1LX
G4 RRA Paul Pasquet, Honey Blossom Cottage Spreyton, Crediton, EX17 5AL
G4 RRD Ian Reynolds, 4 Chappel Hill, Fakenham, NR21 9HW
G4 RRH Jonathan Green, 83a High Street, Ramsey, Huntingdon, PE26 1BZ
G4 RRL John WILLIAMS, Cartref, Capel Garmon, Llanrwst, LL26 0RG
G4 RRM Peter Walker, 11 Flixton Drive, Crewe, CW2 8AP
G4 RRN C Harrold, Boundary Farm, Felbrigg Road, Norwich, NR11 8PD
GM4 RRP R Morris, The Gables, Highfield, Muir of Ord, IV6 7XN
G4 RRQ Ian Wright, 5 Parc an Gate, Mousehole, Penzance, TR19 6TT
G4 RRR T Bunce, Pear Tree House, Greaves Lane, Malpas, SY14 7AR
G4 RRU R Crooks, 6 Whylands Avenue, Worthing, BN13 3HG
G4 RRX R Saxton, 7 Huxley Road, Old Lakenham, Norwich, NR1 2JR
G4 RSC Reading School ARC c/o Thomas Walter, Main House, Erleigh Road, Reading, RG1 5LW
G4 RSD David Bones, Flint Cottage, Ipswich Road, Woodbridge, IP13 7PP
G4 RSE SOUTH ESSEX ARS c/o David Speechley, 15 Southwick Road, Canvey Island, SS8 0EP
G4 RSF M Booth, 45 Park Avenue, Thackley, Bradford, BD10 0RJ
G4 RSG R Booth, 45 Park Avenue, Thackley, Bradford, BD10 0RJ
GI4 RSI K Allen, 25 Knockgreenan Avenue, Omagh, BT79 0EB
G4 RSL Christopher Bagley, 47 Meadow View Road, Weymouth, DT3 5PB
G4 RSN R Burman, Woodlands Vale, Calthorpe Road, Ryde, PO33 1PR
G4 RSP D Sandy, The Chestnuts, Dumbs Lane, Norwich, NR10 3BH
G4 RSS J Upton, 24 Heritage Drive, Clowne, Chesterfield, S43 4ST
G4 RST D Martin, 111 Arkwright Road, Irchester, Wellingborough, NN29 7EE
G4 RSU P Winnett, 148 Green Lanes, Epsom, KT19 9UL
G4 RSW A Bairstow, 63 Barnes Road, Stafford, ST17 9RL
G4 RSX M Dean, 117 Waltham Way, Chingford, London, E4 8HD
G4 RTA K Mellor, 2 Clune St., Clowne, Chesterfield, S43 4NJ
G4 RTC X Iona, 13 Vicars Close, Enfield, EN1 3DW
G4 RTH R Hamstead, 1a The Close, North Walsham, NR28 9HS
G4 RTI E Handy, 80 Watwood Road, Shirley, Solihull, B90 2HY
G4 RTJ C Howe, 113 Fatfield Park, Washington, NE38 8BP
GM4 RTN T Morton, 15 Craig Crescent, Causewayhead, Stirling, FK9 5LR
G4 RTO Gregg Calkin, 54 Pattermead Crescent, Ottawa, Ontario, Canada, K1V 0G2
G4 RTP A Shattock, The Stone House, Westport Road, Co. Galway, Ireland
G4 RTQ I Whitehead, 3 Botany Close, Thatcham, RG19 4GJ
G4 RTS W Bateson, 10 Priestfield Avenue, Colne, BB8 9QJ
G4 RTV C Tucker, 4 Kelsey Park Road, Beckenham, BR3 6LJ
G4 RTW Glyn Rolf, Flat 4, Hawksworth House, 73 St. Johns Road, Sandown, PO36 8HE
G4 RTX G Kingdon, Flat 12, Courtfields, Lancing, BN15 8PA
G4 RTY R Hayward, Alverstone, 28 Chatsworth Avenue, Shanklin, PO37 7NZ
G4 RUA R Medcalf, 21 Greenbank, Falmouth, TR11 2SW
G4 RUE Ian Worsdale, 10 Manton Road, Lincoln, LN2 2JL
G4 RUI A Keeble, 9 Horsley Avenue, Shiremoor, Newcastle upon Tyne, NE27 0UF
G4 RUJ P Evans, 706 St. Johns Road, Clacton-on-Sea, CO16 8BN
GU4 RUK A Jefferys, 2 Les Douze Maisons, Collings Road, St. Peter Port, Guernsey, GY1 1FQ
G4 RUL A Turner, 42 Brassey Avenue, Hampden Park, Eastbourne, BN22 9QG
G4 RUN M Beesley, 60 Ainsbury Road, Canley Gardens, Coventry, CV5 6BB
GM4 RUP J Campbell, 96 Boghead Road, Lenzie, Glasgow, G66 4EN
G4 RUR Mark Baker, 26 Irlam Road, Ipswich, IP2 9QR
G4 RUS J Childs, 38 Carlton Road, Wilbarston, Market Harborough, LE16 8QD
G4 RUT Hugh Edwards, 19 Cameron Road, BURPENGARY, Queensland, Australia, 4505
G4 RUW R Daniel, 4 Gloucester Road, Newbury, RG14 5JP
GW4 RUX A Jones, Forest Lodge, Glynhafod Street, Aberdare, CF44 6LD
G4 RUZ A Athawes, Holly Croft, Faringdon, SN7 7NG
GW4 RVA T Nicholas, 15 Maes Llewelyn, Carmarthen, SA31 1JJ
G4 RVE C Andrews, 29 Dell Drive, Angmering, Littlehampton, BN16 4HE
GI4 RVF J Burke, 45 Shorelands, Greenisland, Carrickfergus, BT38 8FB
G4 RVG I Binding, 40 Parklands, South Molton, EX36 4EW
G4 RVH D Hird, 27 Red Beck Park, Cleator Moor, CA25 5EU
G4 RVJ D Jones, 6 Priory Close, Pilton, Barnstaple, EX31 1QX
G4 RVK David Bentley, 106 Pargeter Street, Walsall, WS2 8RR
G4 RVL D Gentle, 1 Sunny Hill, Milford, Belper, DE56 0QR
G4 RVO T Collinson, 8 Brownberrie Drive, Horsforth, Leeds, LS18 5PP
G4 RVP S O'Donnell, Men A Vaur, Wall Rd Gwinear, Hayle, TR27 5HA
GD4 RVQ Jon Wornham, 64 Seafield Close, Onchan, Isle of Man, IM3 3BU
G4 RVS Andrew Rodgers, 278 Norton Lane, Norton, Sheffield, S8 8HE
GI4 RVT R Jenkins, 11 Willowvale Crescent, Islandmagee, Larne, BT40 3SQ
G4 RVU P Wigley, 7 Cavendish Close, Duffield, Belper, DE56 4DF
G4 RVV Martin Stoneham, Hafnia, 139 Hever Avenue, Sevenoaks, TN15 6DT
G4 RVY G Richardson, 69 O'Neill Drive, Peterlee, SR8 5UD
G4 RVZ G Smith, Dormie House, 61 Cable Road, Wirral, CH47 2AZ
G4 RWA N Van Stigt, 93 Park Road, Teddington, TW11 0AW
G4 RWD K Cheetham, 71 Westmead Road, Barton under Needwood, Burton-on-Trent, DE13 8JR
GM4 RWE D Brown, Willow Crook, Turin, Forfar, DD8 2UZ
G4 RWF Mark Piecha, Oaklea, Gordons Close, Taunton, TA1 3DA
G4 RWG R Guest, 67 Hanbury Road, Dorridge, Solihull, B93 8DN
G4 RWH R Williams, 1 The Meadows, Newhall Green, Coventry, CV7 8BF
G4 RWI Nigel Spear, 79a Lower Icknield Way, Chinnor, OX39 4EA
G4 RWK W Tolman, Pulland Cottage, West Down, Ilfracombe, EX34 8NH
G4 RWM P Titherington, 5 Hayland Green, Hailsham, BN27 1SR
G4 RWN F Rowan, 1 Massey Walk, Wythenshawe, Manchester, M22 5JY
G4 RWQ Brian Wilkes, 3 Alsop Crest, Acton Trussell, Stafford, ST17 0SJ
G4 RWR Rhys Thomas, Ystrad Isa, Ystrad, Denbigh, LL16 4RL
G4 RWS Steven Valentine, 65 Holland Street, Bolton, BL1 8PA
G4 RWV P Paling, 15 Longfellow Road, Banbury, OX16 9LB
G4 RWW P Glaisher, The Firs, 279 Addiscombe Road, Croydon, CR0 7HY
G4 RWY Andrew Jones, 81 Barston Road, Warley, Oldbury, B68 0PU
G4 RXB K Hawkings, 2 Balfour Grove, Biddulph, Stoke-on-Trent, ST8 7SZ
GM4 RXD Royston Gasken, 3 Hameravirin, Glendale, Isle of Skye, IV55 8WL
G4 RXF Grantley Bence, 10 Valley Road, Mangotsfield, Bristol, BS16 9HN
G4 RXG A Mumford, Upper Cross Farm, Thornton Lane, Sandwich, CT13 0EU
G4 RXH H Fowler, 164 Rectory Road, Deal, CT14 9NP
G4 RXK P McMullan, 6 Deepdale Road, Blackpool, FY4 4UD
GI4 RXM T Stitt, 51 Lakeland Road, Hillsborough, BT26 6PW
GW4 RXO P Alexander, 19 Saron Road Bynea, Llanelli, SA14 9LT
G4 RXQ R Buck, 2 Talbot Cottages, Birtley, Chester le Street, DH3 1AR
G4 RXR R Raine, 47 Buckingham Road, Peterlee, SR8 2DT
GI4 RXS R Burnside, 19 Hilton Park, Portglenone, Ballymena, BT44 8HH
GM4 RXW N Webster, Meric, 7 Woodmuir Crescent, Newport-on-Tay, DD6 8HL
GI4 RXX S Tweedie, 12 Glencraig Close, Newtownabbey, BT36 5GZ
G4 RYB J Baldwin, 31 Beech Road, Branston, Lincoln, LN4 1PG
G4 RYE David Cocker, 34 Beechfield, New Farnley, Leeds, LS12 5QS
G4 RYH F Appleby, 10 Buckingham Orchard, Chudleigh Knighton, Newton Abbot, TQ13 0EW

G4 RYI D Ashcroft, 9 Aldermere Crescent, Urmston, Manchester, M41 8UE
G4 RYJ A Salisbury, Heddwch, Ash Grove, Flint, CH6 5RX
G4 RYK A Richards, Castell Forwyn, Abermule, Powys, SY15 [illegible]
GI4 RYL M Mccallan, 16 Abbey Crescent, Newtownabbey, BT37 9PD
G4 RYM Ian Spalding, 2 Briery Lands, Heath End, Stratford-upon-Avon, CV37 0PP
G4 RYO Peter Allan, 7 Homelands Place, Kingsbridge, TQ7 1QU
GI4 RYP J Ferguson, Drumbee-More, Armagh, BT60 1HP
G4 RYQ K Edwards, 10 Bala Drive, Rogerstone, Newport, NP10 9HN
G4 RYS Norman Black, 12 Stonegate Way, Leeds, LS17 6FD
G4 RYT J Pickup, 274 Mauldeth Road West, Chorlton cum Hardy, Manchester, M21 7TQ
G4 RYV D Rumbold, 15 Lodge Grove, Yateley, GU46 7AD
G4 RZC Lars Ingerslev, 20 Stoney Run Lane, Marlton, United States, [illegible] 1019
G4 RZD L Bradley, 138 Templeton Road, Birmingham, B44 0BY
G4 RZE A Taylor, 5 Wyebank Rise, Tutshill, Chepstow, NP16 7DS
G4 RZF G May, 5 The Burlongs, Glebe Road, Swindon, SN4 7DR
G4 RZI Mike Nagle, The Woodlands, Pilton West, Barnstaple, EX31 4JQ
G4 RZM B Williams, Warren Cottage, Polyphant, Launceston, PL15 7PU
G4 RZN R Leeds, 1 Arbor Hill, Cromer, NR27 9DN
G4 RZQ K RUSSELL, Courtiles, Main Road, Ventnor, PO38 3NH
G4 RZR Roger Tooth, Orchard Leigh, Sampford Brett, Taunton, TA4 4JZ
GM4 RZW D Taylor, 42 Craiglockhart Road, Edinburgh, EH14 1HG
G4 RZY L Baker, The Novers Park Community Centre, Rear of 122-124, Bristol, BS4 1RN
G4 RZZ I Griffin, 15 Hesselyn Drive, Rainham, RM13 7EJ
G4 SAB C Leat, 8 White Point Court, Whitby, YO21 3UR
G4 SAC S Collings, 4 Glamis Close, Waterlooville, PO7 8JN
G4 SAJ Christopher Green, 76 Dibleys, Blewbury, Didcot, OX11 9PU
GI4 SAM Sam Noble, 19 New Line, Dundonald, Belfast, BT16 1UU
G4 SAS R Jones, 42 Fastmoor Oval, Birmingham, B33 0NR
G4 SAT INMARSAT ARC c/o D JOHN, 41a Chequers Orchard, Iver, SL0 9NJ
G4 SAV Frederick Hepworth, 5 Snydale Avenue, Normanton, WF6 1SS
G4 SAW M Arbon, 106 The Tideway, Rochester, ME1 2NN
GI4 SBA K Branagh, 17 Rathmoyle Park West, Carrickfergus, BT38 7NG
G4 SBB C Fay, Driftwood, Middle Road, Lymington, SO41 6BB
G4 SBD G Bax, 8 Hockeredge Gardens, Westgate-on-Sea, CT8 8AN
G4 SBE Kenneth Bowden, 14 Pool Hey Lane, Scarisbrick, Southport, PR8 5HS
G4 SBF P Fry, 54 Studley Avenue, Holbury, Southampton, SO45 2PP
G4 SBG Harry Crossland, 107 Fairway, Normanton, WF6 1SN
G4 SBM R Harding, 12 Keswick Avenue, Loughborough, LE11 3RL
G4 SBN J Ayers, 3 Sovereign Way, Ryde, PO33 3DL
GM4 SBP G Allan, 31 Jubilee Grove, Glenrothes, KY6 1HW
G4 SBQ Michael Rushton, 14 Acorn Close, Leyland, PR25 3AF
G4 SBS R Phillips, 4 Cumberland Drive, Fazeley, Tamworth, B78 3YA
G4 SBU Brian Gundry, 37 Stoneham Park, Petersfield, GU32 3BT
G4 SBW T Carberry, 10 Honeymeade Close, Stanton, Bury St. Edmunds, IP31 2EF
G4 SCB M Sargent, 19 Pine Tree Close, Cowes, PO31 8DX
G4 SCE Paul Whitehead, Carrick View, Bank End, Carlisle, CA5 6QW
GD4 SCG C Curson, 3 Cranmer Road, Edgware, HA8 8UA
G4 SCJ David Meakins, 19 Booth Lane North, Northampton, NN3 6JQ
GW4 SCK R Hancock, 6 Alexandra Terrace, Abernant, Aberdare, CF44 0RG
G4 SCL M Starkey, Cutlers Forth Farm, Radley Road, Newark, NG22 8AP
G4 SCM John Claxton, Camino Del Perpen 25, Catral, Spain, 3158
G4 SCO N Drury, 3 Northam Close, Marshside, Southport, PR9 9GA
G4 SCS Andrew Fletcher, stonehouse stores, West rd, Ormesby, NR29 3RJ
G4 SCV I Gammon, The Haven, Craddock, Cullompton, EX15 3LH
G4 SCY P Bradbury, 52 Moss Park Avenue, Werrington, Stoke-on-Trent, ST9 0EP
G4 SDI L Footring, 26 Ernest Road, Wivenhoe, Colchester, CO7 9LG
G4 SDJ R Freeman, Flat 2, Russett Court, 15 Kirtleton Avenue, Weymouth, DT4 7PS
G4 SDL B Dorricott, 6 Knowsley Avenue, Urmston, Manchester, M41 7BT
G4 SDO D Phillips, Trem y Fammau, Tri Thy, Tir y Fron Lane, Mold, CH7 4TU
G4 SDT Steven Lansdown, 11 Redbrook Road, Newport, NP20 5AA
G4 SDU P Smart, 6 Nobold Close, Baschurch, Shrewsbury, SY4 2EH
G4 SDX G Townend, 9 Warren Park Close, Hove Edge, Brighouse, HD6 2RU
G4 SDZ M Gayler, 39 Holmefield Av Ws, Leicester Forest Es, Leicester, LE3 3FF
G4 SEA Roy Seabridge, 7 Heritage Avenue, Frankston South, Melbourne, Australia
G4 SEF Roger Jenkinson, 4 Apple Croft Skidby, Cottingham, HU16 5UG
G4 SEG A Clayton, 448 Gisburn Road, Blacko, Nelson, BB9 6LZ
G4 SEJ B Vane, 3 Charnwood, Chestfield, Whitstable, CT5 3QD
G4 SEK K Aylwin, 9 Hockeredge Gardens, Westgate-on-Sea, CT8 8AN
G4 SEL G Wilkes, 49 Charlemont Road, Walsall, WS5 3NQ
G4 SEN N Whitham, The Cottage, Castle Gate, Penzance, TR20 8BQ
G4 SEP C Turner, Saxavord, Humberston Road, Grimsby, DN36 5NJ
G4 SEQ D Vickers, 48 Bromley Road, Hanging Heaton, Batley, WF17 6EH
G4 SET Clive Hall, 8 Sharps Court, Exmouth, EX8 1DT
G4 SEU Jeremy Russell, 9 Batten Close, Christchurch, BH23 3BJ
G4 SEV N Le Gresley, 32 Churchill Road, Welton, NN11 2JH
G4 SEW K Weir, Les Ginestes Appt 231, 28 Av Dr Gerhardt, Peymeinade, France, 6530
G4 SEZ C Greenland, 21 Penleigh Close, Corsham, SN13 9LE
GM4 SFA A Keenan, Darwin, Coalhall, Ayr, KA6 6ND
G4 SFB D Knowler, Apartedo 1009, 8670 - 999, Aljezur, Portugal
G4 SFD D Birks, Flat, 1 Pewsham House, Chippenham, SN15 3RX
GI4 SFE J McCullough, 12 Bramble Grange, Newtownabbey, BT37 0XH
G4 SFG Peter O'Connor, 36 Heron Road, Oldbury, B68 8AQ
G4 SFH N Richardson, 22 Bramshott Drive, Hook, RG27 9EY
G4 SFJ S Stott, 20 Lingfield Crescent, Wigan, WN6 8QA
G4 SFN J Tetlow, 14 Fountains Crescent, Hebburn, NE31 2HT
G4 SFP John Nash, 259 Weald Drive, Furnace Green, Crawley, RH10 6PN
G4 SFQ S Fletcher, The Bakery, Keswick Road, Norwich, NR12 0HF
G4 SFS Peter Grosjean, Garden House, West Horrington, Wells, BA5 3ED
GM4 SFT Daniel McAlonan, Glenonan House Cromlech St, Dunoon, PA23 8PQ
GM4 SFW J Stuart, Tigh Na Coille, Bishop Kinkell, Dingwall, IV7 8AW
G4 SFY Raymond Baker, 15 Northfield House 46 High Street, Mundesley, Norwich, NR11 8JW
GI4 SFZ Charles Hought, 37 Oldpark Avenue, Ballymena, BT42 1AX
G4 SGA G Barnes, 3 Blandford Avenue, Castle Bromwich, Birmingham, B36 9HX
G4 SGD Steven Simpson, 17 Astley Way, Ashby-de-la-Zouch, LE65 1LY
G4 SGE Barry Hughes, 86 Lewis Avenue, Wolverhampton, WV1 2AR
G4 SGF Kenneth Ruiz, Flat 12, The Woodlands, 39 Shore Lane, Sheffield, S10 3DU
G4 SGG David Earp, 88 Linton Rise, Nottingham, NG3 7BY
G4 SGI Simon Collings, 46 St. Michaels Road, Cheltenham, GL51 3RR
G4 SGJ John Campbell, 9 Blackdown Close, Dibden Purlieu, Southampton, SO45 5QS
G4 SGN P Playle, 6 Walnut Tree Close, Cheshunt, Waltham Cross, EN8 0NH
G4 SGQ Peter Hruza, 18 Withy Avenue, Fordon, Welshpool, SY21 8NJ
GW4 SGR S.GLAM RYNT GRP c/o Roy Magwood, 13 Inverness Place, Cardiff, CF24 4RU
G4 SGU G Gilbertson, 6 The Stray, South Cave, Brough, HU15 2AL
G4 SGV K Jones, 228 Evesham Road, Headless Cross, Redditch, B97 5EP
G4 SGW W Prouse, 1 Springfield Cottages, Bishops Tawton, Barnstaple, EX32 0DF
G4 SGX Iain Haywood, 5 Pump Corner, Marsham, Norwich, NR10 5PW
G4 SGY J Willis, 94 Wharncliffe Road, Loughborough, LE11 1SN
G4 SHA M Webb, 9 Steele Close, Devizes, SN10 3SL
G4 SHB P Sheridan, 17 Boakes Drive, Stonehouse, GL10 3QW
G4 SHC R Bentham, 12 Tanners Way, Nantwich, CW5 7FL
G4 SHF Stephen Purser, 80 John Bold Avenue Stoney Stanton, Leicester, LE9 4DN
G4 SHH P Brooking, 49 Binstead Lodge Road, Ryde, PO33 3TL
G4 SHJ N Douglas, 87 Hutton Avenue, Hartlepool, TS26 9PR
G4 SHK George Clifford, Turnpike Road, Blunsdon, Swindon, SN26 7EA
G4 SHM M Baker, 92 Moy Avenue, Eastbourne, BN22 8UQ
G4 SHN John Pitts, 4 Tannery Close, Dagenham, RM10 7EX
G4 SHO F Dibden, 127 Mayola Road, Clapton, London, E5 0RG
G4 SHY Lewis Afford, 44 Shepperton Court, Coventry Road, Nuneaton, CV11 4NP
GM4 SID Sidney Will, 53 Bishop Forbes Crescent, Blackburn, Aberdeen, AB21 0TW
G4 SIE R Mason, 35 Princes Gardens, Blyth, NE24 5HL
G4 SIF Richard Rowsell, 61 Barrack Road, Bexhill-on-Sea, TN40 2AZ
G4 SII P Garston, 85 Wood Lane, Hawarden, Deeside, CH5 3JG
G4 SIJ B Hammond, 10 Grampian Close, Sleaford, NG34 7WA
G4 SIL E Tubman, 54 Summerfield Avenue, Whitstable, CT5 1NS
GI4 SIP J McKavanagh, 28 Thompsons Grange, Carryduff, Belfast, BT8 8TG
G4 SIS R Keefe, 28 Burstead Drive, South Green, Billericay, CM11 2QN
GI4 SIW ANTRIM & DIS AR c/o William Hutchinson, 40 Oldstone Hill, Muckamore, Antrim, BT41 4SB
GI4 SIZ Trevor Thompson, 135 Glenhead Road, Limavady, BT49 9LR
GI4 SJB J Bruce, 36 Ringbuoy Cove, Cloughey, Newtownards, BT22 1LL
G4 SJD Stanley Davis, 33 Pollard Close, Plymstock, Plymouth, PL9 9RR
G4 SJG G Upton, 18 Cranthorne Drive, Bakersfield, Nottingham, NG3 7HD
G4 SJH B Lewis, 23 Lightwater Meadow, Lightwater, GU18 5XH
G4 SJI Anthony Harris, 10 Egroms Lane, Withernsea, HU19 2LZ
G4 SJJ Henry MATTHEWS, 30 Rosewood Avenue, Burnham-on-Sea, TA8 1HE
G4 SJL Steven Thomson, 11 Beverley Road, London, W4 2LL
G4 SJM John Reeves, 5 Arrows Crescent, Boroughbridge, York, YO51 9LP
G4 SJN B Hunt, Tralee, Oakridge Lynch, Stroud, GL6 7NY
GW4 SJO M Edwards, Aelwyd Y Don, Tresaith, Cardigan, SA43 2JH
G4 SJP Stephen Prior, East Brantwood, Manor Road, Barnstaple, EX32 0JN
GI4 SJQ G Frazer, 20 Old Rectory Park, Portadown, Craigavon, BT62 3QH
G4 SJT W Young, Kiplings, Graynfylde Drive, Bideford, EX39 4AP
G4 SJU S Browne, 38 Aldrin Road, Pennsylvania, Exeter, EX4 5DN
G4 SJV D Chapman, 24 Broad Lane, Moulton, Spalding, PE12 6PN
G4 SJW Bridges RC Hampshire c/o Malcolm Troy, 22 Jackie Wigg Gardens, Totton, Southampton, SO40 9LZ
GW4 SKA John Barber, 49 Blackmill Road, Bryncethin, Bridgend, CF32 9YN
GM4 SKB M Whyatt, Backburn Cottage, Castleton Road, Auchterarder, PH3 1JS
G4 SKM Maltby and District ARS c/o R Cochrane, 134 Moor Lane South, Ravenfield, Rotherham, S65 4QR
G4 SKN K Romang, 11 Moor Barton, Neston, Corsham, SN13 9SH
G4 SKO Mark Brooke, 6 Sun Street, Stanningley, Pudsey, LS28 6DJ
GW4 SKP A Clark, 27 Heol Sant Bridget, St. Brides Major, Bridgend, CF32 0SL
G4 SKU John Blain, 53 Edward Arnold Court, Emmerton Road, Bedford, MK42 8QS
G4 SLG K Oliver, 42 Minster Drive, Cherry Willingham, Lincoln, LN3 4NA
GW4 SLI John Plumley, 34 Graigwen Crescent, Abertridwr, Caerphilly, CF83 4BN
G4 SLL John Buckland, 245 Saunders Lane, Mayford, Woking, GU22 0NU
GI4 SLQ K Boyd, 29 Benburb Road, Moy, Dungannon, BT71 7SQ
GM4 SLV John Pumford-Green, Greenmeadow, Clousta, Shetland, ZE2 9LX
G4 SLW Dominic Dudkowski, 7 White Street, Brighton, BN2 0JH
GM4 SLY John Bell, 13 Corrie Place, Troon, KA10 6TZ
G4 SMA M Goode, Meadowgreen, Batch Valley, Church Stretton, SY6 6JW
G4 SMB Michael Briggs, 70 Auchinleck Close, Kellythorpe, Driffield, YO25 9HE
G4 SMD D Blackwell, 2 Courtry Cottages, Bridgehampton, Yeovil, BA22 8HF
G4 SME SKMRSDLE & D AR c/o J Rogers, 186 Beavers Lane, Birleywood, Skelmersdale, WN8 9BP
G4 SMK K Wilson, 15 Woodside Avenue, Cottingley, Bingley, BD16 1HR
GW4 SML T Morgan, 3 Beddgelert Field, Bridgend, CF31 5FH
G4 SMM M Manley, Rolleston, Parkgate Road, Chester, CH1 6JS
G4 SMQ D McDonald, 3 Cloverton Drive, Bridgwater, TA6 4HQ
G4 SMT J Short, 42 Alt Road, Formby, Liverpool, L37 6DF
G4 SMX J Wilson, 168 Elms Vale Road, Dover, CT17 9PN
GI4 SNA D Ross, 127 Pond Park Road, Lisburn, BT28 3RE
G4 SND M Newey, 148 High Street, Pensford, Bristol, BS39 4BH
G4 SNI H Farley, 11 College Lane, Hatfield, AL10 9PB
G4 SNJ C Frenzel, Butlers Hall, Butlers Hall Lane, Bishop's Stortford, CM23 4BL
G4 SNL I Dunworth, 51 The Crest, Sawbridgeworth, CM21 0ER
G4 SNN N Booth-Isherwood, 65 Burnaby St., Alvaston, Derby, DE24 8RN
G4 SNO A Duggins, Dowles Bungalow, Dowles Road, Bewdley, DY12 3AA
G4 SNQ T Wadsworth, 20 Rook Wood Way, Little Kingshill, Great Missenden, HP16 0DF
G4 SNR E Meekers, 5 Frobisher Close, Mudeford, Christchurch, BH23 3SN
G4 SNU R Hardie, 12 Hopland Close, Longwell Green, Bristol, BS30 9XB
G4 SNV G Eastgate, 103 Western Road, Leigh-on-Sea, SS9 2PB
G4 SNW Mark Donnelly, 29 Chain Terrace, Creetown, Newton Stewart, DG8 7HN
GM4 SNZ D J McCrandles, 21 Wall St., Camelon, Falkirk, FK1 4QB
G4 SOA P Instone, 19 Dickenson Road, Swindon, SN25 1WQ
G4 SOB W Hammond, 27 Gainsborough Road, Colchester, CO3 4QN
G4 SOC V Shaw, 130 Aberthaw Road, Ringland, Newport, NP19 9QS
G4 SOF Jeffrey Blight, Lowbell, Handy Cross, Bideford, EX39 3ET
G4 SOH Martyn Spence, 5 St. Helens Avenue, Benson, Wallingford, OX10 6RY
G4 SOI Mike Wray, The Old Croft, Top Street, Retford, DN22 0LG
G4 SOK R Hollow, The Beeches, Grove Lane, Penzance, TR20 9HN
G4 SOL I Bale, 3b Holes Lane, Knottingley, WF11 0LH
G4 SOM J Spiteri, 27 Hillcote Close, Sheffield, S10 3PT
G4 SOP K Lillingstone, 4 Old Bakery Court, Coltishall, Norwich, NR12 7DQ
G4 SOQ B Lyons, 16 Ashdale Close, Sawtry, Huntingdon, PE28 5SN
G4 SOR A Collins, 19 Cavendish Road, Skegness, PE25 2QZ
G4 SOT Dennis Goodwin, Route De Samatan, Frontignan Saves, France, 31230
GI4 SOY M Dornan, 2 Beechgrove, Ballynahinch, BT24 8NQ
G4 SOZ D Herd, Capricorn Cottage, Low Common, Norwich, NR14 7DU
G4 SPA Buxton ARS c/o R Marchington, 78 Buxton Road, Dove Holes, Buxton, SK17 8HW
G4 SPC Brian Escreet, 190 Front Street, Sowerby, Thirsk, YO7 1JN
G4 SPD Neville Jarvis, 25 St. Augustines Close, Aldershot, GU12 4SF
G4 SPE G Callaghan, 1 Wessex Close, Semington, Trowbridge, BA14 6SA
GW4 SPL P Ace, 116 Gellionen Road, Clydach, Swansea, SA6 5HF
G4 SPR F Rattray, 4 Winton Manor Court, Winton, Kirkby Stephen, CA17 4HR
G4 SPS N Beggs, 11 Orion Close, Fareham, PO14 2SQ
G4 SPT H Golding, Barany Uyca 2, Hodmezovasarhely, Hungary, 6800
GI4 SPU Norma Alcock, 22 Chippendale Avenue, Bangor, BT20 4PT
G4 SPV A Adamson, 520 York Road, Stevenage, SG1 4EP
G4 SPW T Devlin, 32 Kestrel Drive, Dalton-in-Furness, LA15 8QA
G4 SPY A Kay, 36 Yorks Wood Drive, Birmingham, B37 6DL
G4 SPZ P Harris, 22 Bramley Way, Bewdley, DY12 2PU
G4 SQA D Yeoman, 12 Melrose Drive, Old Fletton, Peterborough, PE2 9DN
G4 SQG R Nicholson, 7 Half Mile Gardens, Leeds, LS13 1BL
G4 SQI G Gulliford, 18 Purslane, Abingdon, OX14 3TR
G4 SQJ G Turner, 21 Barlow Road, Chichester, PO19 3LD
G4 SQK G Middleton, 212 East Markham Avenue, Durham, United States, 27701
GI4 SQL S Adrain, 10 Highgate Drive, Newtownabbey, BT36 4WQ
GM4 SQM D Anderson, 34 Culzean Crescent, Kilmarnock, KA3 7DT
GM4 SQO R Riddiough, 1 Cedar Road, Ayr, KA7 3PE
G4 SQQ C Hollister, Rosemead, 326 Passage Road, Bristol, BS10 7TE
G4 SQV J Hart, 88 Breckhill Road, Woodthorpe, Nottingham, NG5 4GQ
G4 SRD Ronald Sealy, 10 Mallard Close, Bowerhill, Melksham, SN12 6TQ
GW4 SRE John Reeves, 5 Lon Y Bryn, Glynneath, Neath, SA11 5BG
G4 SRF C Radcliffe, 85 Brian Avenue, Cleethorpes, DN35 9DE
GW4 SRI A Gray, 24 Waunarlwydd Road, Cockett, Swansea, SA2 0GB
GM4 SRL R Cowan, 85 Eastwoodmains Road, Clarkston, Glasgow, G76 7HG
G4 SRP Gerald Brierley, 6 Yeo Drive, Appledore, Bideford, EX39 1RD
GI4 SRQ William McHugh, 47 Main St., Hamiltonsbawn, Armagh, BT60 1LP
G4 SRV D Darby, 28 Coleshill Close, Hunt End, Redditch, B97 5UN
G4 SRX George Sampson, 47 Netherthorpe Way, North Anston, Sheffield, S25 4FL
GM4 SSA Hans Hassel, Sumra, Eshaness, Shetland, ZE2 9RS
G4 SSC A Taylor, 38 Summershades Lane, Grasscroft, Oldham, OL4 4ED
G4 SSD South Devon RC c/o John May, 6 Hodson Close, Paignton, TQ3 3NU
GI4 SSF Stephen Craig, 6 Kingswood Park, Belfast, BT5 7EZ
G4 SSH Roy Clayton, 9 Green Island, Irton, Scarborough, YO12 4RN
G4 SSJ R Bolton, 83 Sandicroft Close, Birchwood, Warrington, WA3 7LY
G4 SSL W Lancashire, 111 Pentire Avenue, Newquay, TR7 1PF
G4 SSO Alan Mcmillan, Marl Bank, St. Marys Road, Ramsey St. Marys, Huntingdon, PE26 2SN
G4 SSP Keith Waghorne, 23 Bramley Hill, Mere, Warminster, BA12 6JX
G4 SSV S smith, 1 Buckfast Road, Lincoln, LN1 3JS
G4 SSW J Walker, 15 Hillfield Road, Bilton, Rugby, CV22 7EW
G4 SSZ D Fox, 49 Turnpike Hill, Hythe, CT21 4SE
G4 STB P Lock, 82 Rownhams Road, Throop, Bournemouth, BH8 0NL
G4 STD B West, 13 Tanglewood Close, Upper Shirley, Croydon, CR0 5HX
G4 STE Stephen Farr, 37a Bromsgrove Road, Studley, B80 7PG
G4 STH Gerald Timbrell, Crossing Cottage, Lamyatt, Shepton Mallet, BA4 6NG
G4 STI M Pinder, 36 West Ridge, Allesley, Coventry, CV5 9LN
G4 STK E Brown, 16 Springfield, Ovington, Prudhoe, NE42 6EH
G4 STO P Rose, Pinchbeck Farmhouse, Mill Lane, Sturton by Stow, Lincoln, LN1 2AS
G4 STP Tony Mangles, 46 Cedar Crescent, Willington, Crook, DL15 0DA
G4 STV Hadley Wood Contest Group c/o Simone Wilson, 21 Plumian Way, Balsham, Cambridge, CB21 4EG
G4 STW A Buckley, 7 Lobelia Drive, Bottesford, Scunthorpe, DN17 2GE
G4 STZ Thomas Bromsgrove, 11 Moelwyn Drive, Ellesmere Port, CH66 1TY
G4 SUA B Common, 59 Thornbera Gardens, Bishop's Stortford, CM23 3NP
GW4 SUD K Jones, 111 Ewenny Road, Bridgend, CF31 3LN
G4 SUE Margaret Hill, 13 Maesglas Grove, Newport, NP20 3DJ
GM4 SUF Phillip Gane, Ardmore Lodge, Station Road, Tain, IV19 1LA
G4 SUK M Kent, 304 Reculver Road, Herne Bay, CT6 6SR
G4 SUM K Summerhill, 23 Ellis Close, Cottenham, Cambridge, CB24 8UN
GW4 SUN Bourke Le Carpentier, 43 Aber-Nant Road, Aberdare, CF44 0PY
G4 SUO Peter Barwick, Brook Cottage, 94 Ambleside Road, Lightwater, GU18 5UJ
GM4 SUR R Aitken, 2 Eskdale Drive, Bonnyrigg, EH19 2LD
G4 SUS S Morgan, Hollaway, Northbourne Road, Deal, CT14 0LA
G4 SUU M Nairn, 13 Hanover Court, Lacey Street, Ipswich, IP4 2PJ
G4 SUX R Payne, 9 Shelburne Way, Derry Hill, Calne, SN11 9PA
G4 SVA Jaqueline Appleton, 66 Bolton Road West, Ramsbottom, Bury, BL0 9ND
G4 SVB A Gatrell, Sunnyside, Muddles Green, Lewes, BN8 6HW
G4 SVC Timothy Axtell, 146 Olivers Battery Road South, Winchester, SO22 4LF
G4 SVE John Hewett, 84 Dunsgreen, Ponteland, Newcastle upon Tyne, NE20 9EJ
G4 SVG S Wensley, 7 Bradshaw Close, Windsor, SL4 5PS
G4 SVI Christopher Ames, Heatherdene, 58 The Street, Norwich, NR8 6AB
G4 SVL I Hewitt, 35 Birmingham Road, Alvechurch, Birmingham, B48 7TB
GM4 SVM G Hudson, 17 Drylaw Crescent, Edinburgh, EH4 2AU
G4 SVQ P Haffenden, 113 Pavilion Road, Worthing, BN14 7EG
G4 SVR W Baddeley, 12 Stockport Road, Altrincham, WA15 8ET

**IMPORTANT NOTE**

**Revalidate licence to avoid revocation** – Ofcom has advised the Society that plans will be drawn up to revoke licences that have not been revalidated as required by the licence conditions. The quickest way to revalidate is to do so online via the Ofcom website: *https://services.ofcom.org.uk/* or by email: *amateur.validations@ofcom.org.uk* Ofcom staff are available to help, but please be patient during times of heavy workload.

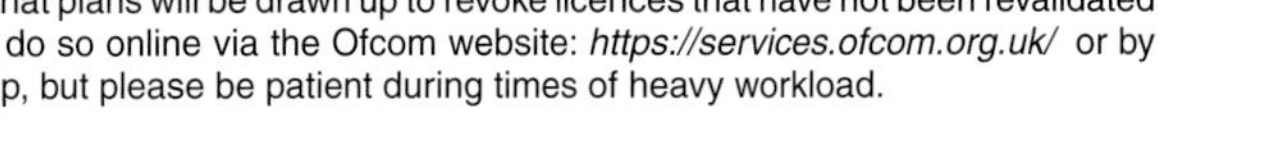

G4 SVS D Bush, 8 Oldbury Chase, Bristol, BS30 6DY
G4 SVV David Lees, 1, Davies Ave, Cheadle, SK8 3PF
GM4 SVW Bill McLaren, 4 Firth Crescent, Gourock, PA19 1EW
G4 SVY J Perez, 9 Rectory Road, Shanklin, PO37 6NX
G4 SWA S Authers, 9 Conway Avenue, Birmingham, B32 1DR
G4 SWH R Jones, Tangmere, 40 Lordship Lane, Letchworth Garden City, SG6 2BL
G4 SWM D Crosby-Clarke, 55 Robin Lane, Sandhurst, GU47 9AU
G4 SWN S Griffiths, New House, Thompsons Hill, Malmesbury, SN16 0PZ
G4 SWO S Griffiths, 25 Hanks Close, Malmesbury, SN16 9UA
G4 SWQ R Torence-Smith, Birch Lodge, Flax Lane, Sudbury, CO10 7RS
G4 SWR I Hill, 7 Cosford Close, Matchborough, Redditch, B98 0BH
G4 SWY D Turner, 15 Brooke Close, Bushey, WD23 1FB
GW4 SXA Paul Ap-Dafydd, 11 Cemetery Road, Trecynon, Aberdare, CF44 8HL
G4 SXD D Jones, Edorene, Tincleton, Dorchester, DT2 8QR
G4 SXE B Holden, 76 The Lawns, Rolleston-on-Dove, Burton-on-Trent, DE13 9DE
G4 SXG D Fraser, 63 Vicars Hall Gardens, Worsley, Manchester, M28 1HW
G4 SXH L Fletcher, 2 Hillside Close, Heddington, Calne, SN11 0PZ
GM4 SXJ R Malcolm, 43 Kinghorne St., Hospitalfield, Arbroath, DD11 2LZ
G4 SXK Alan Leighs, 16 Spode Close, Cheadle, Stoke-on-Trent, ST10 1DT
GU4 SXM P Bannier, 10 Le Bouet, Longstore, St Peter Port, Guernsey, GY1 2BA
G4 SXQ D Lempriere, Harewarren Lodge, Salisbury, SP2 0NF
G4 SXR Colin Bracher, 29 Bungalow Park, Holders Road, Amesbury, Salisbury, SP4 7PJ
G4 SXT David Ayers, 22 Holders Road, Amesbury, Salisbury, SP4 7PP
GI4 SXV Eric Barker, 39 Birchwood, Omagh, BT79 7RA
G4 SXX J Key, 29 Scots Court, Hook, RG27 9QJ
G4 SXY Gerald Haines, 25 Hook Hill, South Croydon, CR2 0LB
G4 SXZ Richard Hiles, 19 Station Road, Kirton Lindsey, Gainsborough, DN21 4BB
G4 SYA C Allen, 11 Chandos Court, Martlesham, Woodbridge, IP12 4SU
G4 SYB P Loveland, 25 White Acres Road, Mytchett, Camberley, GU16 6JJ
G4 SYC G Lomas, 2 Linney Road, Bramhall, Stockport, SK7 3JW
G4 SYD H Cook, 24 Front St., Sherburn Hill, Durham, DH6 1PA
G4 SYE M Wilson, The Old Post Office, Knowl Hill, Reading, RG10 9YD
G4 SYF N Wallace, 2 Mansefield, Leitholm, Coldstream, TD12 4JQ
G4 SYG P Tattersall, Anchor Cottage, The Street, Ipswich, IP10 0EU
GD4 SYI Lucien Wilder, Dental Surgery, 5 Dovercourt Gardens, Stanmore, HA7 4SJ
G4 SYL Janet Frost, 68 Wessex Road, Didcot, OX11 8BP
GI4 SYM W Donaldson, 44 Drumman Hill, Armagh, BT61 8RW
GW4 SYO H Thomas, 1 Cambrian Terrace, Llwynypia, Tonypandy, CF40 2HN
GU4 SYQ Lesley Le Page, Heathwick, Les Martins, Guernsey, Guernsey, GY4 6QJ
G4 SYR S Field, 4 Lyndale, Kelvedon Hatch, Brentwood, CM15 0BQ
G4 SYT D Chambers, 26 Drummond Gardens, Christ Church Mount, Epsom, KT19 8RP
G4 SYV R Ackroyd, 14 Isis Avenue, Bicester, OX26 2GS
G4 SYW A Hargreaves, 13 Linworth Road, Bishops Cleeve, Cheltenham, GL52 8PA
G4 SZA I Donaldson, 25 Alwyn Road, Maidenhead, SL6 5EG
G4 SZB E Flannigan, 14 Westbank Avenue, Blackpool, FY4 5BT
GM4 SZG J Freeland, 48 Elgin Place, Shawhead, Coatbridge, ML5 4JQ
G4 SZI Allan Parry, 18 Spinney Lane, Rabley Heath, Welwyn, AL6 9TF
G4 SZO Kevin Painter, Delbrueckstrasse 14, Berlin, Germany, 14193
GI4 SZP N Hughes, 32 Kinedale Park, Ballynahinch, BT24 8YS
GI4 SZQ K Murphy, 8 Rosscolban Meadows, Kesh, Enniskillen, BT93 1UH
G4 SZS Trevor Harber, 27 Yarlington Close, Norton Fitzwarren, Taunton, TA2 6RR
GI4 SZU John McCurry, 30 Carrowdoon Road, Dunloy, Ballymena, BT44 9DL
GW4 SZV ABERPORTH YMCA ARC c/o G Carruthers, Henllys Farm, Cardigan, SA43 2HR
GI4 SZW Michael Keenan, 30 Ballynabee Road Camlough, Newry, BT35 7HD
G4 SZX M Stockton, 7 The Croft, Thorne, Doncaster, DN8 5TL
GI4 SZY Robert Wilson, 57 Mill Green, Doagh, Ballyclare, BT39 0PH
G4 TAD Mervyn Wooltorton, 4 Hall Lane, Oulton, Lowestoft, NR32 5DJ
G4 TAG T Gammage, 31 Kennet Drive, Congleton, CW12 3RH
G4 TAH I Conibear, 5 Brunswick Street, Barton Hill, Bristol, BS5 9QN
GI4 TAJ J Bingham, 35 Rathmena Drive, Ballyclare, BT39 9HZ
G4 TAK John Hancock, 29 Convent Close, Aughton, Ormskirk, L39 4XP
GM4 TAL A Blyth, 73 Glassel Park Road, Longniddry, EH32 0TA
G4 TAM B Chambers, 93 Main Road, Hoo, Rochester, ME3 9EU
G4 TAO S Rogers, Gaythorpe, Blacketts Wood Drive, Rickmansworth, WD3 5QQ
GI4 TAP S Mccabe, 27 Baronscourt Road, Carryduff, Belfast, BT8 8BQ
G4 TAT D Ruth, 1 Derwent Drive, Appley, Ryde, PO33 1NT
G4 TAU E Davies, Gwelydon, Lon Brynteg, Ynys Mon, LL59 5UA
GI4 TAV M Doherty, 20 Drumcairn Close, Belfast, BT8 8HQ
G4 TAW N Perrott, 56 Park Avenue, Chatsworth, Nsw, Australia, 2067
G4 TAY D Sommerfield, 51 Spindletree Drive, Oakwood, Derby, DE21 2DG
G4 TAZ Francis Rewaj, 5 Spring Gardens, Grizebeck, Kirkby-in-Furness, LA17 7XJ
G4 TBF E Popham, 741 Landbase Australia, Locked Bag 25, Gosford, Australia, 2250
G4 TBG Denis Smith, 34 Grays Lane, Downley, High Wycombe, HP13 5TZ
G4 TBI P Cornell, 22 Ravine Road, Bournemouth, BH5 2DU
G4 TBJ R Smith, 184 Solihull Road, Shirley, Solihull, B90 3LG
G4 TBK D Nix, 75 Mayfield Road, Chaddesden, Derby, DE21 6FX
G4 TBM M Tester, 6 Harvard Close, Lewes, BN7 2EJ
G4 TBN Colin Anderson, 10 School Lane, Everdon, Daventry, NN11 3BW
G4 TBO Christopher Harris, 33 Brent Street, Brent Knoll, Highbridge, TA9 4DT
G4 TCA R Hawthorn, Weirsmeet Bungalow, Mill Lane, Derby, DE74 2EJ
G4 TCB P Holland, 9 Garmont Road, Leeds, LS7 3LY
G4 TCC Andrew Hawkins, 29 Warwick Road, Oldbury, B68 0NE
G4 TCE P Wade, 356 Shirehall Road, Sheffield, S5 0JP
G4 TCG R Tams, 4 Langdale Close, Fryston, Castleford, WF10 2RB
G4 TCI Michael Soars, 8 Nomis Park, Congresbury, Bristol, BS49 5HB
G4 TCK N Nicholls, Flat 17, Link House, Nelson Road, Bideford, EX39 1HS
G4 TCM W Smith, 65 Larchwood Road, Yew Tree Estate, Walsall, WS5 4HE
G4 TCO David Preece, Tyning House, Westrip, Stroud, GL6 6EY
G4 TCP J Caddick, 3 Church Walk, Avonwick, South Brent, TQ10 9EJ
GI4 TCR Andrew Jackson, 9 Cove Crescent, Groomsport, Bangor, BT19 6HW
GI4 TCS W Jackson, Shantara, 21 Carnreagh, Hillsborough, BT26 6LJ
G4 TCT Philip Johnson, 5 Moorside Drive, Drighlington, Bradford, BD11 1HE
G4 TCX D Shingler, 39 Oaklands, Bridgnorth, WV15 5DU
G4 TDB D Waterhouse, 19 Finsbury Drive, Brierley Hill, DY5 3NY
G4 TDC L Wilson, Eastwood, Common Road, Tadcaster, LS24 9PQ
G4 TDF R Copsey, 7 Musson Close, Marston Green, Solihull, B37 7HS
G4 TDG Robert Dowson, 14 The Warren, Tuffley, Gloucester, GL4 0TT
G4 TDI D Day, 1 Kings Paddock, Ossett, WF5 8EN
G4 TDO Brian Fereday, 16 Glentworth Gardens, Wolverhampton, WV6 0SF
G4 TDP S Bowden, 62 Manor House Road, Wednesbury, WS10 9PH
G4 TDQ L Ashton, Bacton Wood Mill, Spa Common, North Walsham, NR28 9SH
G4 TDR C Jay, Hill House, Badgers Orchard, Pershore, WR10 3HJ
G4 TDU B Brandon, 8 Moor Park Avenue, Castleton, Rochdale, OL11 3JG
G4 TDV R Stokes, Sunnybank, The Arch, Exeter, EX5 1LL
G4 TDW R Maskew, 23 Daventry Road, Rochdale, OL11 2LN
G4 TDZ Alec Jones, Riverbank, Main Street, Newark, NG22 0PP
G4 TEB P Larbalestier, 54 Churchill Avenue, Halstead, CO9 2BE
GI4 TED K Doherty, 77 Drumflugh Road, Benburb, Dungannon, BT71 7QF
GM4 TEF A Chalmers, Mayfield, Malcolm Road, Peterculter, AB14 0NX
G4 TEK J Churchill, Yew Tree Cottage, Grove Lane, Salisbury, SP5 2NR
GD4 TEM J Freestone, C/O Kololi Holiday Services, Po Box 335, Douglas, Isle of Man, IM99 2QF
G4 TEN John Burch, 1 South Farm Close, Tarrant Hinton, Blandford Forum, DT11 8JY
G4 TEP L Kennedy, 69 Drayton Road, Borehamwood, WD6 2DA
G4 TEQ John Mattocks, Brynheulog, Bronygarth Road, Oswestry, SY10 7RQ
G4 TEU J Burr, 4 The Fleet, Royston, SG8 5BB
G4 TEW R Cooper, 4 Battle Close, Boroughbridge, York, YO51 9GN
G4 TEZ John Lever, 30 Radcliffe Road, Winsford, CW7 1RE
G4 TFB B Gray, 29 Verity Walk, Stourbridge, DY8 4XS
G4 TFC D Gray, 29 Verity Walk, Stourbridge, DY8 4XS
G4 TFD B Pickard, 3 Lodore Road, Bradford, BD2 4HY
G4 TFF P Ayre, 1 Spring Gardens, Broadmayne, Dorchester, DT2 8PP
G4 TFH P Mackinven, 43 Stringers Lane, Aston, Stevenage, SG2 7EF
G4 TFI W White, 9 Saffron Way, Tiptree, Colchester, CO5 0AY
GM4 TFJ E Wallace, Lochiel Villa, Corpach, Fort William, PH33 7LR
G4 TFM A Davis, 51 Gungrog Hill, Welshpool, SY21 7UL
G4 TFO A Coaton, 138 Ratby Road, Groby, Leicester, LE6 0BT
G4 TFP Patrick Herman Cranmer, 38 Barbrook Lane, Tiptree, Colchester, CO5 0EF
GW4 TFS A Jones, 6 Gower View, Llanelli, SA15 3SN
G4 TFT C Greenwood, 3 Moorfield Drive, Oakworth, Keighley, BD22 7EX
G4 TFU A Gerrard, 2 Dudley Road, Timperley, Altrincham, WA15 6UE
G4 TFV F Kelly, 2 Victoria Terrace, Kirkstall, Leeds, LS5 3HX
G4 TFW H Guy, 24 The Mall, Binstead, Ryde, PO33 3SF
GW4 TFX B James, 9 Brangwyn Close, Morriston, Swansea, SA6 6AS
G4 TFZ D Jarvis, 21 Ashurst Place, Stannington, Sheffield, S6 5LN
GW4 TGA S Marvelley, 36 Muirfield Drive, Mayals, Swansea, SA3 5HS
G4 TGB David Meadows, 39 Sylvester Street, Mansfield, NG18 5QS
G4 TGE John Pullen, 71 Barrow Road, Barton-upon-Humber, DN18 6AE
G4 TGG G Sifford, 25 Kingsley Court, Fraddon, St. Columb, TR9 6PD
G4 TGJ R Tomlinson, 25 Beverley Rise, Ilkley, LS29 9DB
G4 TGK W Wimble, 87 Rolfe Lane, New Romney, TN28 8JL
GW4 TGL W Protheroe-Thomas, Golwg Y Llan, Carmarthen, SA32 8PR
G4 TGM A Sherratt, Anlyn, Norbury Drive, Brierley Hill, DY5 3DP
G4 TGP R barling, Maranello House, Pay Street, Folkestone, CT18 7DZ
G4 TGQ P Creighton, 6 Kirkwall Drive, Penketh, Warrington, WA5 2HX
GI4 TGR T Greer, Ticino, 82 Purdysburn Hill, Belfast, BT8 8JZ
G4 TGS D Swarbrook, 6 Westview Close, Leek, ST13 8ES
GW4 TGT T Threlfall, 14 Clarence Street, Pembroke Dock, SA72 6JP
G4 TGV I Soaft, 55 The Close, Thurleigh, Bedford, MK44 2DT
G4 TGW D Starmer, 20 Garners Way, Harpole, Northampton, NN7 4DN
G4 THA Mark Crook, 28 Porter Street, Preston, PR1 6QN
G4 THC M Arnison, 57 Heywood Road, Cinderford, GL14 2QU
G4 THF Brian Smith, 63 Hitchin Road, Stotfold, Hitchin, SG5 4HT
G4 THG J Thompson, 1 Berkley Close, Chippenham, SN14 0PS
G4 THI A Robson, 5 Wetton Lane, Tibshelf, Alfreton, DE55 5NA
GW4 THK William Moore, 2 Heol Cae Glas, Sarn, Bridgend, CF32 9UG
G4 THN M Anthony, Middlewood House, Blacksmiths Lane, Stowmarket, IP14 5ET
GM4 THP D Last, 38 Lumsden Way, Balmedie, Aberdeen, AB23 8TS
G4 THU J Read, 35 Maytree Hill, Droitwich, WR9 7QU
G4 THV Roger Biddlecombe, 24 West Avenue, Althorne, Chelmsford, CM3 6DF
G4 THX G Donoughue, 1 Kings Mount, Leeds, LS17 5NS
G4 THY K Cornes, 6 Haywood Heights, Little Haywood, Stafford, ST18 0UR
G4 TIA S Jarvis, 1 Wakenslade Cottages, School House, Chard, TA20 4PJ
G4 TIC Basil Helman, The Dingle, Redway, Minehead, TA24 8QF
G4 TID D Hall, 6 St. Augustine Drive, Droitwich, WR9 8QR
G4 TIF M Jones, 5 Congreve Close, Warwick, CV34 5RQ
G4 TIG William Pope, Greenlands, Dunkeswell, Honiton, EX14 4RE
G4 TIH C Kay, 26 Clare Close, Elstree, Borehamwood, WD6 3NJ
G4 TIM G Clementson, 74 Pentley Park, Welwyn Garden City, AL8 7SG
G4 TIQ D Edwards, Rosedene, Trewint Estate, Liskeard, PL14 3RL
G4 TIV C Roper, 12 Canford Drive, Allerton, Bradford, BD15 7AU
G4 TIW David Wood, 159 Liswerry Road, Newport, NP19 9QR
G4 TIX H Woolrych, 20 Meadow Drive, Devizes, SN10 3BJ
G4 TIZ P Wyles, The Lawns, Halkyn Road, Holywell, CH8 7SJ
G4 TJA Peter Prosser, 47 Devereux Drive, Watford, WD17 3DD
G4 TJC Simon Melhuish, 22 Mayflower Close, Glossop, SK13 8UD
GM4 TJD M Maclennan, 10 Ruilick, Beauly, Inverness, IV4 7EY
G4 TJI R Slim, 58 Fairways Drive, Harrogate, HG2 7ER
G4 TJK M Porter, 17 Lancaster Avenue, Guildford, GU1 3JR
GM4 TJL J Hebborn, Elysian Fields, Spean Bridge, PH34 4EX
G4 TJM E Mordas, 6 Walsingham Gate, School Close, High Wycombe, HP11 1PA
G4 TJN George Smallwood, Rushbrit, The Spinney, Old Road, Wrexham, LL11 5UF
GW4 TJQ J Wallis, 17 Pantbach Place, Whitchurch, Cardiff, CF14 1UN
G4 TJS R Dent, 111 New Road, Bromsgrove, B60 2LJ
G4 TJU F Richards, 9 Dales Grove, Worsley, Manchester, M28 7JW
G4 TJY Lawrence Barker, 75 Hills Road, Saham Hills, Thetford, IP25 7EW
G4 TKF P Tuck, 178 St. Ediths Marsh, Bromham, Chippenham, SN15 2DJ
G4 TKO John Sharman, 102 Commercial Road, Skelmanthorpe, Huddersfield, HD8 9DS
G4 TKP R Peel, 3 Martins Hill Lane, Burton, Christchurch, BH23 7NJ
G4 TKS J Clancy, 22 Audrey Needham House, Victoria Grove, Newbury, RG14 7RB
G4 TKW D Hamilton, 38 Gosport Road, Lee-on-The-Solent, PO13 9EN
G4 TLE H Kennard, Chestnut Cottage, Main Street, Rye, TN31 6UL
G4 TLL J Jones, Amusement Depot, Station Road, Cullompton, EX15 1BQ
G4 TLM B Jennings, 33 Queen Margarets Drive, Brotherton, Knottingley, WF11 9HR
G4 TLO P Johnson, 19d Leigh Road, Havant, PO9 2ET
G4 TLR Brian Richards, 37 Salisbury Grove, Sutton Coldfield, B72 1YE
G4 TLS J Norton, 2 Hill Rise, Great Rollright, Chipping Norton, OX7 5SW
G4 TLT A Price, 127 Millfield Avenue, London, E17 5HN
G4 TLW H Allen, 425 Broadway, Chadderton, Oldham, OL9 8AP
G4 TLY Edgar Holmes, 36 Corn Gastons, Malmesbury, SN16 0DR
G4 TMA P Fielding, 8 Lime Grove, Poulton-le-Fylde, FY6 8EL
GI4 TMB Michael Beggs, 15 Marsham Court, Cotswold Drive, Bangor, BT20 4RS
G4 TMC P Barnett, 8 Parsonage Road, Horsham, RH12 4AR
G4 TMD L Downes, 357 Stone Road, Stafford, ST16 1LD
G4 TMF Peter Aisthorpe-Buckley, 17 Pine View Road, Verwood, BH31 6LQ
G4 TMG C Sherwood, 14 Amberley Road, Rustington, Littlehampton, BN16 2EF
G4 TMI Philip Johnson, 3a Railway Street, Tow Law, Bishop Auckland, DL13 4DU
G4 TML B Parr, 5 Ashes Lane, Almondbury, Huddersfield, HD4 6TE
G4 TMQ J Martin, 22 Wansbeck Court, Front Street East, Bedlington, NE22 5BU
G4 TMR S Lacy, 26 Sterndale Drive, Newcastle, ST5 4HS
G4 TMV E Gale, 4 Waingap Crescent, Whitworth, Rochdale, OL12 8PX
G4 TMX Wilfred Armstrong, 121 Bede Street, Sunderland, SR6 0NT
G4 TMY N Hounslow, 18 Crompton Place, Blackburn, BB2 6LW
G4 TMZ David Gillott, 132 Racecommon Road, Barnsley, S70 6JY
G4 TNA K Pope, 305 Hulton Lane, Bolton, BL3 4LF
G4 TNB Peter Dollery, 22 Barley Mead, Danbury, Chelmsford, CM3 4RP
G4 TND D Jenkins, 27 Glebelands, Chudleigh, Newton Abbot, TQ13 0GB
G4 TNE Donald Horsman, 33 Chanters Hill, Barnstaple, EX32 8DN
G4 TNF T Jones, 9 Hinsley Drive, Wrexham, LL13 9QH
G4 TNJ R Milenkovic, 10 Loganbarns Road, Dumfries, DG1 4BU
GM4 TNP J BURKE, 25 Duncan Road, Auchmuty, Glenrothes, KY7 4HS
G4 TNU Andrew Scott, 79 Westwood Drive, Amersham, HP6 6RR
G4 TNY David Womack, 58 Nelson Road North, Great Yarmouth, NR30 2AT
GM4 TOE Barry Horning, Cemetery Lodge, Colleonard Road, Banff, AB45 1DZ
G4 TOG Barry Grainger, 23 Heath Road, Hordle, Lymington, SO41 0GG
G4 TOH R Russell, 23 Milfoil Avenue, Conniburrow, Milton Keynes, MK14 7DY
G4 TOI Peter Andrews, 12 Cedarwood Grove, Sunderland, SR2 9EJ
G4 TOM T Turbert, 200 Salisbury Terrace, York, YO26 4XP
G4 TOO M Final, 3 Borda Close, Chelmsford, CM1 4JY
GM4 TOQ A Stewart, Three Acres, Cochno Road, Clydebank, G81 6PX
GI4 TOR Aubrey Kincaid, 63 Carolhill Park, Ballymena, BT42 2DG
G4 TOT R James, Brantholme, Hasty Brow Road, Lancaster, LA2 6AG
G4 TOX J Glover, Burrows Farm, Toot Hill Road, Ongar, CM5 9QW
G4 TOY R Handstock, 38 Watson Close, Upavon, Pewsey, SN9 6AE
G4 TOZ Simon Marchini, Flat 9, Crofthill Court, Rochdale, OL12 9UX
GW4 TPG M Evans, 14 Heol Dewi, Hengoed, CF82 7NP
G4 TPH T Brockman, Pin Mill, High Street, Hungerford, RG17 0NE
GI4 TPI G Anderson, 13 Ashley Park, Bangor, BT20 5RQ
G4 TPJ R Mepham, 36 Bramble Close, Hildenborough, Tonbridge, TN11 9HQ
G4 TPK Paul Phillips, 83 Arundel Road, Benfleet, SS7 4EE
GD4 TPM A Malcher, 68 Maryatt Avenue, South Harrow, Harrow, HA2 0SX
G4 TPO Stephen McCulloch, 125 Comptons Lane, Horsham, RH13 5NZ
GM4 TPQ William Milligan, 7 Girvan Road, Turnberry, Girvan, KA26 9LP
G4 TPR J Mitchell, 65 Robb Place, Castle Douglas, DG7 1LW
G4 TPS P Stinton, 37 Harriet Close, Sutton Bridge, Spalding, PE12 9QU
G4 TPV George Jones, 397 Fishponds Road, Eastville, Bristol, BS5 6RJ
G4 TPW H Igglesden, Treeways, Littleworth Lane, Horsham, RH13 8ER
GM4 TPX K Gerard, 9 Overdale Crescent, Prestwick, KA9 2DB
GI4 TPY K Boag, 12 Plantation Road, Bangor, BT19 6AF
G4 TQB P Grannell, 6 Fermain Close, Seabridge, Newcastle, ST5 3EF
G4 TQC Mike Anson, 15 Clover Ridge, Cheslyn Hay, Walsall, WS6 7DP
GW4 TQD J Gulley, Lawnfields, Brynhoffnant, Llandysul, SA44 6EA
G4 TQE J Burton, 43 Perry Road, Gobowen, Oswestry, SY10 7BX
G4 TQL K Prince, Room 36, Millfield, Bury New Road, Heywood, OL10 4RQ
G4 TQO P Fowler, 7 Ormonde Avenue, Orpington, BR6 8JP
G4 TQR R Wilkes, 47 Richmond Drive, Glen Parva, Leicester, LE2 9TJ
G4 TQS A Wallis, 27 Church Farm Road, Upchurch, Sittingbourne, ME9 7AG
G4 TQT I Waller, 25 Livingstone Road, Chapeltown, Sheffield, S35 2UG
G4 TQY M Addison, 6 Hanley Orchard, Hanley Swan, Worcester, WR8 0DS
G4 TQZ P Foster, 18 Stokesay Way, Sutton Hill, Telford, TF7 4QE
G4 TRA S Redway, Hill House, Rodbourne, Malmesbury, SN16 0ES
G4 TRD R Dafter, 49 Balmoral Road, Salisbury, SP1 3PZ
G4 TRE B Boon, Orchards, School Lane, Woodbridge, IP13 0ES
G4 TRF Ian Boon, 327 Broomfield Road, Chelmsford, CM1 4DU
G4 TRG I Willmer, 30 Portland Road, East Grinstead, RH19 4EA
GM4 TRH Allan Macdonald, 1 Struan Road, Portree, Isle of Skye, IV51 9EG
G4 TRI Susan March, 23 Pebworth Close, Church Hill North, Redditch, B98 9JX
G4 TRM S Burgess, Muston Farm, Winterborne Muston, Blandford Forum, DT11 9BU
G4 TRN J Everingham, 17 Collingwood Road, Redland, Bristol, BS6 6PD
G4 TRP Mervyn Fell, 5 Sandown Close, Goring-by-Sea, Worthing, BN12 4QA
G4 TRR Peter Rose, St. Margarets, Westrop, Swindon, SN6 7HJ
GM4 TRS A Pierce, Mains Of Auchreddie, New Deer, Turriff, AB53 6SL
G4 TRU A Thompson, 47 The Signals, Feniton, Honiton, EX14 3UP
G4 TRV R Pears, 24 Westbourne Drive, St. Austell, PL25 5EA
G4 TRW K Prior, 14 Bincombe Rise, Weymouth, DT3 6AS
G4 TRY Anthony Moriarty, 7 Meadow Drive, Bolton le Sands, Carnforth, LA5 8HA
GM4 TRZ T McLeod, 1 Lochside Cottages, Otterston, Burntisland Fife, KY3 0RZ
G4 TSA R TURLEY, 39 Lawnswood Park Road, Swinton, Manchester, M27 5NJ
G4 TSB R Cooper, 8 Hollyfield Drive, Barnt Green, Birmingham, B45 8HP
G4 TSD R Edwards, 8 Boney Hay Road, Burntwood, WS7 9AB
G4 TSF P Scholefield, 10 Gainsborough Avenue, Leeds, LS16 7PG
G4 TSG John WILLIAMS, Cartref, Capel Garmon, Llanrwst, LL26 0RG

G4 TSH Justin Snow, 104 Redfern Avenue, Hounslow, TW4 5LZ
GI4 TSK James Skillen, 3 Copeland Drive, Comber, Newtownards, BT23 5JJ
G4 TSN Jonathan Lee, 46 Little Lane, Huthwaite, Sutton-in-Ashfield, NG17 2NA
GD4 TSO Peter Oliver, BAL ABBRIG, 3 Cronk-y-Thatcher, Colby, Isle of Man, IM9 4LN
G4 TSQ M Levett, 5 Park Road, Yapton, Arundel, BN18 0JE
G4 TST D Richardson, L'Ancresse, Uplands Road, Waterlooville, PO7 6HE
G4 TSV J Robinson, 2 Bridge Mill Road, Nelson, BB9 7BD
G4 TSW TIVERTON SOUTH WEST ARC c/o A Burnett, 2 Courtney Road, Tiverton, EX16 6EE
G4 TTA DRAGON ARC c/o John Parry, 19 [illegible], Llanfairpwllgwyngyll, LL61 5JY
G4 TTB Alex Gordon, Flat 3, Woodlands, Ipswich, IP8 3RR
GM4 TTC F Howes, 43 Tanzieknowe Road, Cambuslang, Glasgow, G72 0ND
GM4 TTD N Loughrey, 47 Obsdale Road, Alness, IV17 0TU
G4 TTF BISHOP AUCKLAND AR CLUB c/o T Bevan, 6 Buttermere Grove, West Auckland, Bishop Auckland, DL14 9LG
G4 TTG H Bryant, 141 Shakespeare Road, Fleetwood, FY7 7HH
G4 TTJ J Lee, Flat 15, Whitwell Hatch, Haslemere, GU27 3AW
GI4 TTL Michael Corcoran, 20 Ringbuoy Cove, Cloughey, Newtownards, BT22 1LL
G4 TTM J Betts, The Cottage, Meaford, Stone, ST15 0PX
G4 TTN G Hedgewell, 121 Gubbins Lane, Harold Wood, Romford, RM3 0DL
G4 TTQ R Philpot, 68 Brocksparkwood, Hutton, Brentwood, CM13 2TJ
G4 TTS Christopher Harrison, The Mobile Home, Langar Airfield, Nottingham, NG13 9HY
G4 TTX R Smith, 405 Windmill Avenue, Kettering, NN15 6PS
G4 TTY E Macdonald, 7 Alder Close, Crawley Down, Crawley, RH10 4UL
G4 TTZ R Margolis, 12a Wyndham Close, Yateley, GU46 7TT
G4 TUA Thomas Higgs, Town Head Farmhouse, Ravenstonedale, Kirkby Stephen, CA17 4NQ
G4 TUD Iwan Williams, 52 Bridge Street, Aberystwyth, SY23 1QB
G4 TUF Anthony Wilday, 12 Duke Street, Bamber Bridge, Preston, PR5 6FT
G4 TUH S Elsdon, 22 The Swallows, Welwyn Garden City, AL7 1BY
GI4 TUJ W Konos, 27 Hillhead Road, Ballynahinch, BT24 8LB
G4 TUK R Scarfe, Freshfields, Great Melton Road, Norwich, NR9 3NR
G4 TUM J Speakman, 33 Leyburn Avenue, Bispham, Blackpool, FY2 9AQ
G4 TUO E Whitworth, 129a Broomhill, Downham Market, PE38 9QU
G4 TUP D Norris, 26 Freckleton Road, Southport, PR9 9XE
GI4 TUV R Bailie, 26 Moatview Park, Dundonald, Belfast, BT16 2BE
G4 TUX J Baines, 12b Tall Trees Park, Old Mill Lane, Mansfield, NG19 0JP
GM4 TVB Colin Burnet, 138 Hilton Drive, Aberdeen, AB24 4NH
G4 TVC J Darby, 58 Cloverlands, Crawley, RH10 8EH
G4 TVD Gordon Hector, 6 Benford Close, Bristol, BS16 2UD
GW4 TVE S Edwards, Aelwyd Y Don, Tresaith, Cardigan, SA43 2JH
G4 TVJ A Johns, 5 Oakfields, Loddon, Norwich, NR14 6UT
G4 TVN Bryan Yates, 39 Moss Lane, Garstang, Preston, PR3 1PD
GW4 TVQ R Thomas, 3 Tor Y Mynydd, Baglan, Port Talbot, SA12 8LE
G4 TVT G Spencer, 322 Colchester Road, Ipswich, IP4 4QN
GW4 TVU V Sedgebeer, 40 Pen Y Bryn, Cymmer, Port Talbot, SA13 3SD
G4 TVW R stone, 51 Elaine Avenue, Rochester, ME2 2YW
G4 TVX R LAMB, 27 The Ridgeway, Braintree, CM7 1EB
G4 TWB F Openshaw, 9 The Dale, Abergele, LL22 7DS
G4 TWC D Powell, 8 Cranbrook Drive, Sittingbourne, ME10 1RF
G4 TWG S Greenwood, Carter Place Farm, Hall Park, Rossendale, BB4 5BQ
G4 TWH G Wood-Hill, 26 Bramerton Road, Hockley, SS5 4PJ
G4 TWK H Hart, 20 Cowdray Drive, Goring-by-Sea, Worthing, BN12 4LH
G4 TWL T Lee, 19a Imperial Avenue, Mayland, Chelmsford, CM3 6AQ
G4 TWP L Miles, 23 Milland Road, Hailsham, BN27 1TQ
G4 TWS S Holmes, 7 Parkland Crescent, Old Catton, Norwich, NR6 7RQ
G4 TWT H Holmes, 7 Parkland Crescent, Old Catton, Norwich, NR6 7RQ
G4 TWW T Bevan, 98 Heage Road, Ripley, DE5 3GH
G4 TXA D Mccartney, 3 Cwmcarn, Emmer Green, Reading, RG4 8LE
G4 TXD Michael Robbins, 2 Tolview Terrace, Hayle, TR27 4AG
G4 TXE A Goode, Tudor House, Chenhalls Road, Hayle, TR27 6HJ
G4 TXF C White, 7 Woodward Road, Pershore, WR10 1LW
G4 TXG N Hamilton, North View, Chawston Lane, Bedford, MK44 3BH
G4 TXK T Stanley, 35 Moorgate Road, Kippax, Leeds, LS25 7ET
G4 TXL Alan Stevenson, Szabadság u. 32., Veresegyház, Hungary, 2112
G4 TXM G Porter, 20 Fitzwilliam Drive, Barton Seagrave, Kettering, NN15 6RG
GM4 TXN A Newlands, 21 Castle Crescent, Inverbervie, Montrose, DD10 0SB
G4 TXO John Middleton, 8 Cullen Close, Newark, NG24 1DF
G4 TXV A Turner, 11 Holmcroft Road, Kidderminster, DY10 3AQ
G4 TYA C Carter, 12 Grove Street, Leamington Spa, CV32 5AJ
G4 TYD Anthony Kelly, Brook House, Tremar Coombe, Liskeard, PL14 5EN
G4 TYH R Roberts, Bryn Gwyfan, Hiraddug Road, Rhyl, LL18 6HS
G4 TYN D Bell, 12 Parker Gardens, Stapleford, Nottingham, NG9 8QG
G4 TYO G Lilley, 100 Trentham Drive, Nottingham, NG8 3NE
G4 TYP K Ward, 9 Porlock Close, Long Eaton, Nottingham, NG10 4NZ
G4 TYR C Miles, 23 Redacre Road, Sutton Coldfield, B73 5EA
G4 TYT A Hunt, 141 Pickhurst Lane, Bromley, BR2 7HF
G4 TYW R Wilson, 95 Longfield Road, Todmorden, OL14 6ND
G4 TYY John Worley, 37 Fall Road, Heanor, DE75 7PQ
G4 TZA Christopher Read, 58 Somerset Road, Chiswick, London, W4 5DN
G4 TZF C Toby, 32 Swallow Road, Langley Green, Crawley, RH11 7RF
G4 TZG S Mellor, 52 Tetbury Drive, Bolton, BL2 5NS
G4 TZK G Prater, 297 Highfield Road North, Chorley, PR7 1PH
G4 TZL K Rogers, 7 Buckleigh Road, Wath-upon-Dearne, Rotherham, S63 7JB
G4 TZM I Paterson, 21 Beech Grove, Little Oakley, Harwich, CO12 5NN
G4 TZO P Pledger, Mas Trabuch, Brunyola, Gerona, Spain, 17441
G4 TZQ D Rouse, 14 Kestrel Close, Downley, High Wycombe, HP13 5JN
G4 TZR R Stringfellow, 18 Cline Court, Crownhill, Milton Keynes, MK8 0DB
G4 TZT K Horton, 1 Mulberry Close, Woolston, Warrington, WA1 4ED
G4 TZX Gordon Everest, 20 Seaway Road, St. Marys Bay, Romney Marsh, TN29 0RU
G4 UAA John Gaffney, 77 South Street, Pennington, Lymington, SO41 8DY
G4 UAF J Higgins, 124 Cromwell Road, South Kensington, London, SW7 4ET
G4 UAI P Cockman, 29 Kensington Road, Southend-on-Sea, SS1 2SX
GW4 UAJ Eric Allwood, 10 Fairfield Close, Aberdare, CF44 0PF
G4 UAL J Guffogg, 31 Pavillion Gardens, Lincoln, LN6 8BD
G4 UAM A Gould, 3 Clarkson Road, Lingwood, Norwich, NR13 4BA
G4 UAQ Ian Weston, 53 Dickens Road, Maidstone, ME14 2QR
G4 UAT A Thomas, 10 Brisco Avenue, Loughborough, LE11 5HB
G4 UAU J Parish, 50 Far Hey Close, Radcliffe, Manchester, M26 3GL
G4 UAV A Waltham, 100 Middleton Road, London, E8 4LN
G4 UAW M Spillett, 17 Washbourne Close, Brixham, TQ5 9TG
G4 UAY D Grant, 115 Clayton Road, Newcastle, ST5 3EW
G4 UBB John Brown, 21 Coulsdon Road, Sidmouth, EX10 9JJ
G4 UBC K Durrant, 26 Dozule Close, Leonard Stanley, Stonehouse, GL10 3NL
GM4 UBF A Pontiero, 1 Dalmeny Road, Hamilton, ML3 6PP
G4 UBI A Priddy, 44 Frys Hill, Kingswood, Bristol, BS15 4QJ
GM4 UBJ William Tracey, 65 Kirkland Street, Motherwell, ML1 3JW
G4 UBK K Martin, 19 Rosevale Gardens, Luxulyan, Bodmin, PL30 5EP
G4 UDM D Bryant, Plover, Hareby Road, Spilsby, PE23 4JB
G4 UBQ C Powles, 14 Willow Close, Four Crosses, Llanymynech, SY22 6NF
G4 UBH Peter Richardson, 18 New Road, Heage, Belper, DE56 2BA
G4 UBT K Stone, 63 Banks Road, Pound Hill, Crawley, RH10 7BS
G4 UCC W Trinder, 351 Livesey Branch, Road, Blackburn, BB2 4QJ
G4 UCE B Davies, 9 Paisley Avenue, Eastham, Wirral, CH62 8DL
G4 UCJ Sean Gilbert, [illegible] Crescent, Buckingham, MK18 1LZ
GW4 UCK Graham Jones, Frondeg, Chapel Rd, Swansea, SA4 3PU
G4 UCL A Fallows, 72 Soutergate, Ulverston, LA12 7ES
G4 UCT A Cooke, 7 School Lane, Warmingham, Sandbach, CW11 3QL
G4 UCU S Hebel, 2 Sandringham Close, Barrowford, Nelson, BB9 6PT
G4 UCV Ian Hynes, BRYN, LLANDDONA, Beaumaris, LL58 8UE
G4 UCX P Johnson, 38 Bristol Road, Ipswich, IP4 4LP
G4 UCY Cyril Laird, 31 Foxlease, Bedford, MK41 8AP
G4 UCZ M Kirk, 2 Denton Gardens, East Cowes, PO32 6EJ
G4 UDB C Fay, 36 Shooters Hill Close, Southampton, SO19 1FW
G4 UDD S Chapman, 2 Birds Croft, Great Livermere, Bury St. Edmunds, IP31 1JJ
G4 UDE M Ellis, Coed Cottage, Llansilin, Oswestry, SY10 9BS
G4 UDF I Fox, 8 Priestfields, Leigh, WN7 2RG
G4 UDG C Fawkes, 24 Moreton Close, Kidsgrove, Stoke-on-Trent, ST7 4HP
G4 UDH P Harley, 6 Huntsbank Drive, Newcastle, ST5 7TB
GI4 UDI J McCullagh, 53 Fernagh Road, Omagh, BT79 0PL
G4 UDK R Wood, 102 Ombersley Close, Redditch, B98 7UT
G4 UDN C Peake, 279 Mansfield Road, Skegby, Sutton-in-Ashfield, NG17 3AP
GD4 UDT Yves Remedios, 44 Kingsway, Wembley, HA9 7QR
G4 UDU Phil Godbold, 13 Dawn Crescent, Upper Beeding, Steyning, BN44 3WH
G4 UDV R Green, 8 Briarbeck, Shelfield, Walsall, WS4 1XA
G4 UDW P Hersey, Cranbrook, Waghorns Lane, Uckfield, TN22 4JA
G4 UDY Beverley Moorecroft, 4 St. Davids Road, Locks Heath, Southampton, SO31 6EP
G4 UDZ S Tyler, 28 Rushen Drive, Hertford Heath, Hertford, SG13 7RB
G4 UEA P Robinson, 46 Waidshouse Road, Nelson, BB9 0SB
G4 UED Gary Henstridge, 21 John Gay Road, Amesbury, Salisbury, SP4 7NN
G4 UEF K Dalton, 64 Coppice Road, Kingsclere, Newbury, RG20 5RT
GM4 UEH Alan Ford, 14 Corsankell Wynd, Saltcoats, KA21 6HY
GW4 UEJ M Charman, Noddfa, Dwrbach, Fishguard, SA65 9RL
G4 UEL G Hollebon, Flat 24, Providence Place, Farnham, GU9 7RQ
G4 UEN K Foskett, 2 Ambleside Gardens, Southampton, SO19 8EY
G4 UEO David Stewart, The Paddock, Allendale, Hexham, NE47 9EL
GW4 UEP John Morgan, Arosfa, Upper Bridge Street, Newport, SA42 0PL
G4 UET J rolfe, 56 Elmhurst Road, Thatcham, RG18 3DH
G4 UEV A Gray, Fairhaven, 11 Cook Road, Thetford, IP25 7DJ
G4 UFC Peter Fretwell, 5 Main St, Brinsley, NG16 5BG
GM4 UFD Robert Gall, 49a Ugie Street, Peterhead, AB42 1NX
G4 UFG A Johnson, 27 Walden Avenue, Oldham, OL4 2PW
G4 UFJ N Taylor, The Olde Barn, 369a Leymoor Road, Huddersfield, HD7 4QQ
G4 UFK Alan Watts, 23 St. Marys Close, Torrington, EX38 8AS
G4 UFL Neil Wood, 244 Leymoor Road, Golcar, Huddersfield, HD7 4QP
G4 UFP C Ross, The Old Cottages, Middlestead, Selkirk, TD7 5EY
G4 UFQ B Jackson, Bryn Tirion, Maes y Waen, Bala, LL23 7SF
G4 UFR E Horsfield, 13 St. Leonards Way, Ardsley, Barnsley, S71 5BS
G4 UFS David Pearson, 48 Nuneham Grove, Westcroft, Milton Keynes, MK4 4DH
G4 UFU B Steen, 30 Shady Grove, Alsager, Stoke-on-Trent, ST7 2NH
G4 UFX D Blackwell, Rosegarth, 31 Main Street, Ilkeston, DE7 6AU
G4 UFZ R Greenwood, 128 Towngate, Netherthong, Holmfirth, HD9 3XZ
G4 UGB Richard Bracegirdle, 3 Westover Grove, Warton, Carnforth, LA5 9QR
G4 UGD Ian Clover, West Lodge, Oulton Park, Tarporley, CW6 9BN
GM4 UGF David Duff, Felcanty, Monikie, Broughty Ferry, Dundee, DD5 3QN
GW4 UGI R Crowley, 15 Rudry Street, Penarth, CF64 2TZ
G4 UGK C Cattrall, 57 Stonebridge, Orton Malborne, Peterborough, PE2 5NT
G4 UGM D Wade, 28 Hazel Road, Altrincham, WA14 1JL
GM4 UGN D Duckworth, 16 Kennedy Court, Caol, Fort William, PH33 7PF
G4 UGO T Farmer, York House, Old Gloucester Road, Bristol, BS35 3LQ
G4 UGQ John Davies, 42 Boxworth Road, Elsworth, Cambridge, CB23 4JQ
G4 UGR T Burke, 10 Goodwood Road, Lancaster, LA1 4LZ
G4 UGT David Hockin, 18 Lower Down Road, Portishead, Bristol, BS20 6PF
G4 UGU R Hall, 19 Buckingham Place, Downend, Bristol, BS16 5TN
G4 UGV M Hurrell, 74 Southcote Road, Bournemouth, BH1 3SS
G4 UGW Philip Jones, 46 Sergeants Lane, Whitefield, Manchester, M45 7TS
GI4 UHA J Maguire, 4 Lawnakilla Park, Enniskillen, BT74 7JN
GD4 UHB J Parslow, Iraie Vane, Lhergy Dhoo, Peel, Isle of Man, IM5 2AE
G4 UHI D Westby, 55 Tarn Road, Thornton-Cleveleys, FY5 5AY
G4 UHJ David Lee, 1 West View Cottage, The Level, Lydney, GL15 4QD
G4 UHK J Newell, 8 Belgrave Road, Abergavenny, NP7 7AL
G4 UHM Stephen Parsons, 248 Filey Road, Scarborough, YO11 3AQ
G4 UHQ B Carr, 23 Belford Drive, Bramley, Rotherham, S66 3YW
G4 UHR D Reekie, 37 Harvey Way, Saffron Walden, CB10 2AP
G4 UHS R Rowlands, 18 Green Crescent, Rowner, Gosport, PO13 0DP
G4 UHT W O'Reilly, 12 Singledge Avenue, Whitfield, Dover, CT16 3LQ
G4 UHU M Rumens, 18 De Legh Grove, West Allington, Bridport, DT6 5QY
G4 UHW D Allsopp, Karveden, 7 Dowthorpe Hill, Northampton, NN6 0PB
G4 UHZ D Goulsbra, Delfour, 3 Chapel Street, Market Rasen, LN8 3AG
G4 UIA Derek Johnson, 56 Warkworth Drive, Chester le Street, DH2 3TH
GW4 UIE S Williams, 17 Brettenham Street, Llanelli, SA15 3ED
G4 UIF Anthony Marston, Stormsfield, Station Road, Limerick, Ireland
G4 UIH Sue Woolgar, 8 Bowerland Avenue, Torquay, TQ2 8QH
G4 UII D Woolgar, 8 Bowerland Avenue, Torquay, TQ2 8QH
G4 UIL R Lewis, Tal Engan, Tudweiliog, Pwllheli, LL53 8ND
G4 UIO A Cochrane, 136 Osward, Courtwood Lane, Croydon, CR0 9HE
G4 UIQ Joseph Greenhough, 58 Gorsey Bank, Wirksworth, Matlock, DE4 4AD
G4 UIR John Patterson, Fairhaven, Tai Terfyn, Caerwys Road, Cwm Dyserth, Rhyl, LL18 6HT
G4 UIT D Geraghty, 114 Little Oaks, Penryn, TR10 8QF
G4 UIW P Wood, 61 Stoke Road, Bromsgrove, B60 3EP
G4 UIY S Hamilton, 89 The Paddocks, Old Catton, Norwich, NR6 7HE
G4 UJF M Finnigan, 3 Frances Avenue, Rhyl, LL18 2LW
G4 UJI E Cowperthwaite, Woodlands, Garstang Road, Lancaster, LA2 0EG
G4 UJJ David Bodman, 66a Martins Road, Keevil, Trowbridge, BA14 6NA
G4 UJL B Poole, 1 Hungerford Piece, Studley, Calne, SN11 9LR
G4 UJO M Greer, The Pines, 5a Leek Road, Congleton, CW12 3HE
G4 UJP B Smith, 17 Thornley Road, Wirral, CH46 6HR
G4 UJS Robert Harrison, Green Lane House, Whixall, Whitchurch, SY13 2PT
G4 UJV J Fenn, 10 Mildenhall Road, Fordham, Ely, CB7 5NR
G4 UJW Charles Elliott, 52 Wellfield Road, Alrewas, Burton-on-Trent, DE13 7EZ
G4 UKA C Hawkridge, 57 Wilkes Wood, Creswell, Stafford, ST18 9QH
GD4 UKD B Gibson, 101 Torbay Road, Harrow, HA2 9QF
GM4 UKG M Manokshaw, 32 Inchcolm Drive, North Queensferry, Inverkeithing KY11 1LD
G4 UKO N Hill, The Hollies, 40a Hampden Road, Ashford, TN23 6JL
G4 UKP D Ford, 43 Chapel Lane, Rode Heath, ST7 3SE
G4 UKU Peter Jones, 8 Tyn y Pwll Estate, Llanbedrog, Pwllheli, LL53 7PG
G4 UKV Ian Leonard, Ye Olde Noggin Cottage, School Lane, Warrington, WA4 4QB
G4 UKW K Wevill, 6 Henacre Wood Court, Queensbury, Bradford, BD13 2LJ
G4 UKX R Miller, 31 Gladstone Road, Corton, Lowestoft, NR32 5HJ
G4 UKZ R Rounce, Field House Farm, Blakeney Road, Fakenham, NR21 0BU
G4 ULD R Todd, 1700 E. Lakeside Drive #21, Gilbert, United States, 85234
G4 ULG Janet Rawlings, Welwyn, Church Walk, Lydney, GL15 4NY
G4 ULI B Long, 2 The Limes, Castor, Peterborough, PE5 7BH
G4 ULM J Martin, 25 Mcnish Court, Grenville Way, St. Neots, PE19 8PE
G4 ULN Ian Purdy, 76 Lea Road, Dronfield, S18 1SD
G4 ULP David Pritchard, 1 Sandstone Cottages, Walton-in-Gordano, Clevedon, BS21 7AJ
G4 ULQ Gordon Judd, 1 Mayfield Way, Ferndown, BH22 9HP
G4 ULT Leslie Walker, Oaklea, 13 Manor Road, Sandown, PO36 9JA
G4 ULV D Woodman, 66 Southfield Avenue, Kingswood, Bristol, BS15 4BQ
G4 ULZ R Ottway, 9 Grove Road, Burgess Hill, RH15 8LE
GM4 UMA S MacLennan, 10 Ruilick, Beauly, Inverness, IV4 7EY
G4 UMB P Howard, 63 West Bradford Road, Waddington, Clitheroe, BB7 3JD
G4 UME Hugh Park, 11a Morecambe Road, Morecambe, LA3 3AA
G4 UMG M Brassington, 42 Dartmouth Avenue, Newcastle, ST5 3NY
G4 UMJ R Carslake, 38 Loppets Road, Tilgate, Crawley, RH10 5DW
G4 UMM A Curran, 103 Highlander Drive, Donnington, Telford, TF2 8JU
G4 UMP Gary Daisley, 10 Arundel Road, Benfleet, SS7 4EF
G4 UMS Michael Kinger, 5 Fore Street, Gunnislake, PL18 9BN
G4 UMT G Elliott, 10 Farningham Close, Spondon, Derby, DE21 7DZ
G4 UMV P Johnson, 52 Evesham Road, Cookhill, Alcester, B49 5LJ
G4 UMW R Browning, 28 Mowbray Close, Bromham, Bedford, MK43 8LF
G4 UMY M Strong, 92 Cobham Road, Halesowen, B63 3JX
G4 UNB David Williams, 12 Saxon Street, Radcliffe, Manchester, M26 3TB
G4 UNE Stephen Sharples, 1 Garners End, Chalfont St. Peter, Gerrards Cross, SL9 0HE
G4 UNF K East, 39 Chapel Lane, Navenby, Lincoln, LN5 0ER
G4 UNG S Patterson, 44 Stoke Farthing, Broad Chalke, Salisbury, SP5 5ED
G4 UNH A Pyne, 414 Beacon Road, Bradford, BD6 3DJ
G4 UNI T Hepple, 18 King Charles Walk, London, SW19 6JA
G4 UNJ B Walters, 17 Oakway, Birkenshaw, Bradford, BD11 2PG
G4 UNL R Charlesworth, PO Box 841, 9506 Koronadal City, South Cotobato, Philippines
G4 UNM R Bushell, 12 Sandham Close, Sandown, PO36 9DS
G4 UNO John Dobson, 27 Darkfield Way, Woolavington, Bridgwater, TA7 8JB
G4 UNS David Brown, 8 Gaynes Court, Upminster, RM14 2JH
G4 UNW P Everard, The Bungalow, Toynton Fenside Road, Spilsby, PE23 5DB
G4 UNX Jonathan Fry, 4 South Road, Brighton, BN1 6SB
GM4 UOD L Drake-Brockman, 59 Sunnyside, Culloden Moor, Inverness, IV2 5ES
G4 UOI R Butterfield, 33 Orchard Square, Wormley, Broxbourne, EN10 6JA
G4 UON P Prowse, 9 Fairway, Carlyon Bay, St. Austell, PL25 3QE
G4 UOO John Bleaney, 58 Jeans Way, Dunstable, LU5 4PW
G4 UOR C Bourke, 36 The Drive, Fareham, PO16 7NL
G4 UOS G Newton, 5 Southend Gardens, Highbridge, TA9 3LD
G4 UOW J Cosgrove, 8 Wandsworth Road, Newcastle upon Tyne, NE6 5AD
G4 UOZ E Ball, 50 Keldgate, Beverley, HU17 8HY
G4 UPA J Poxon, 22 Sandhills Road, Bolsover, Chesterfield, S44 6EY
G4 UPB Grahame Read, 9 Askrigg Avenue, Little Sutton, Ellesmere Port, CH66 4TW
GI4 UPC W Millar, 121 Ballypollard Road, Magheramorne, Larne, BT40 3JG
G4 UPD Michael Parks, 240 Stainbeck Road, Leeds, LS7 2NN
GW4 UPG V Jackson, 24 Bishop Road, Ammanford, SA18 3HA
G4 UPI J Green, St. Annes, Poundfield Road, Crowborough, TN6 2BG
G4 UPK D Thompson, 112 Lexton Drive, Churchtown, Southport, PR9 8QW
GM4 UPN Paul Ingram, 4/6 Balmwell Avenue, Edinburgh, EH16 6HF
G4 UPO Dave Ince, 4 Roadside Court, Lowton, Warrington, WA3 2DU
G4 UPR John Dickson, 33 Ringwood Grove, Weston-Super-Mare, BS23 2UA
G4 UPS E Collins, 27 Parklands, Hemyock, Cullompton, EX15 3RY
G4 UPU R Ainsworth, 14 Edge Fold Crescent, Worsley, Manchester, M28 7EX
G4 UPX Ian Wilson, 18 High Street, Jedburgh, TD8 6AG
G4 UPY A Hodge, Brambledown, 116a Broad Road, Eastbourne, BN20 9RD
G4 UQA Michael Goodman, Randoms, Holt Road, Holt, NR25 7UA
GM4 UQD Alistair Boyd, 86 Ravenswood Rise, Livingston, EH54 6PG
G4 UQE Tim FitzGerald, 11 Hillcrest Road, Camberley, GU15 1LF
G4 UQF M Sole, 17 Hyholmes, Bretton, Peterborough, PE3 8LG
GM4 UQG R Aitkenhead, 11 Elm Court, Quarter, Hamilton, ML3 7FB
G4 UQI A Lether, 16 The Dingle, Fulwood, Preston, PR2 3EX
G4 UQK J Roberts, 24 Minster Crescent, Darwen, BB3 3PY
G4 UQM D Grainger, 25 Westwood Heath Road, Leek, ST13 8LN
G4 UQN K Stockley, 19 The Lawns, Wisbech, PE13 1SW
GD4 UQO P Parker, 46 Ballaquane Park, Peel, Isle of Man, IM5 1PX

UK Callsigns

**IMPORTANT NOTE**

**Revalidate licence to avoid revocation** – Ofcom has advised the Society that plans will be drawn up to revoke licences that have not been revalidated as required by the licence conditions. The quickest way to revalidate is to do so online via the Ofcom website: *https://services.ofcom.org.uk/* or by email: *amateur.validations@ofcom.org.uk* Ofcom staff are available to help, but please be patient during times of heavy workload.

G4 UQR John Gibbs, 3 Holts Green, Great Brickhill, Milton Keynes, MK17 9AJ
G4 UQS J Blundell, 68 Alton Road, Leicester, LE2 8QA
G4 UQU D Smith, 2 Niton Road, Weddington, Nuneaton, CV10 0BX
G4 UQW David Beckett, 433 New St., Biddulph Moor, Stoke-on-Trent, ST8 7NG
G4 UQY M Regan, 36 Moor Park Gardens, Leigh-on-Sea, SS9 4PY
G4 URA A Haynes, Thorrington Nurseries, Tenpenny Hill, Colchester, CO7 8JB
GW4 URB R Teesdale, 22 Cwmgelli Drive, Treboeth, Swansea, SA5 9BS
G4 URD R Caira, 12 West Hill Road, Herne Bay, CT6 8HG
G4 URG S Richardson, 19 South Avenue, Thornton-Cleveleys, FY5 1JY
G4 URM P Butler, Tanglewood, Elms Lane, Wolverhampton, WV10 7JS
G4 URN M Turvey, 106 Foxwell St., Worcxester, Worcester, WR5 2ET
G4 URP R Powell, 57 Bartons Drive, Yateley, GU46 6DW
G4 URS J Osborne, 64 Old Warren, Taverham, Norwich, NR8 6GA
G4 URV W Peel, 34 Carlyn Avenue, Sale, M33 2EA
G4 URW J Allison, 17 Gordon Terrace, Stakeford, Choppington, NE62 5UE
G4 URX T Robinson, 26 Keeble Drive, Washingborough, Lincoln, LN4 1DZ
GM4 URZ HELENSBURGH ARC c/o Barry Spink, 9 St. Andrews Crescent, Dumbarton, G82 3ER
G4 USC Kevin Appleton, 5 Hart Hill Crescent, Full Sutton, York, YO41 1LX
G4 USD BOMCHIL CASTRO GOODRICH CLARO AROSMENA c/o D Brill, 25 Boulevard Barbes, Paris, France, 75018
G4 USK B Finlay, 4 Henden Mews, Maidenhead, SL6 4GY
G4 USN Mark Havard, 61 Northwood End Road, Haynes, Bedford, MK45 3QB
G4 USP Simon Hall, 22 Leam Road, Lighthorne Heath, Leamington Spa, CV33 9TE
G4 USQ T Hodgetts, 14 St. Peters Road, Portishead, Bristol, BS20 6QY
G4 UST A Forbes, Field Cottages, 44 Walkmills, Church Stretton, SY6 6NJ
G4 USW William Jenkins, 5 Seatoller Place, Barrow-in-Furness, LA14 4NH
G4 USX E Pritchard, 18 New Ridd Rise, Hyde, SK14 5DD
G4 UTC D Abraham, 42 Lower Greenfield, Ingol, Preston, PR2 3ZT
G4 UTE M Chaudhry, 613 Service Road, G 10/4, Islamabad, Pakistan, ZZ7 9PO
G4 UTF A Cockman, 31 Kensington Road, Southend-on-Sea, SS1 2SX
G4 UTG F Collins, 31 Mount Pleasant Road, Poole, BH15 1TU
G4 UTJ J Gorton, 4 Weavers Close, Colchester, CO3 4LT
GM4 UTK D James, Baltic House, Baltic Street, Montrose, DD10 8EX
G4 UTM Brian Dennis, Thistledown Yallands Hill, Monkton Heathfield, Taunton, TA2 8NA
G4 UTN Graham Bromley, 46 Independent Hill, Alfreton, DE55 7DG
G4 UTQ M adamson, 13 Towers Close, Bedlington, NE22 5ER
G4 UTR M Alder, 342 Church St., Edmonton, London, N9 9HP
G4 UTS E Bracey, 3 Dyffryn Road, Waunlwyd, Ebbw Vale, NP23 6UA
G4 UTV Arthur Cockerill, 90 Stockton Road, Middlesbrough, TS5 4AJ
G4 UTX Graham Eagle, 1 Kestrel Way, Plymouth, PL6 7SY
G4 UTY G Fooks, 10 Bincombe Drive, Crewkerne, TA18 7BE
G4 UUA M Robinson, 2 Bridge Mill Road, Nelson, BB9 7BD
G4 UUB M Lemin, Mill House, Lingwood Road, Norwich, NR13 4AH
GI4 UUC W Thompson, Maranne, 21 Watch Hill Road, Ballyclare, BT39 9QW
G4 UUE Liam McGrogan, 239 Haslingden Road, Rossendale, BB4 6RX
G4 UUF N Kelly, 3 The Terrace, Gawcott, Buckingham, MK18 4HL
G4 UUG D Payne, 23 Laburnum Avenue, Newbold Verdon, Leicester, LE9 9LQ
G4 UUH S Rogers, 31 Coleridge Road, Ottery St. Mary, EX11 1TD
G4 UUI S Rooker, 67 Hawks Way, Ashford, TN23 5UW
G4 UUJ E Jeffery, 11 Furze Hill Road, Shanklin, PO37 7PA
G4 UUM Peter SKIVINGTON, 52 Downlands, Stevenage, SG2 7BH
G4 UUQ T Tallis, 1 High Peak Cottages, High Peak Junction, Matlock, DE4 5HN
G4 UUT A Turner, 30 Wheatlands Road, Paignton, TQ4 5HU
G4 UUU Christopher Clayton, 17 Meadow Dene, East Ayton, Scarborough, YO13 9EL
G4 UUW David Williams, 12 Springfield Road, Exmouth, EX8 3JX
G4 UVA P Money, Meadow View, Podmore Lane, Dereham, NR19 2NS
G4 UVB Paul Gibson, Rivendell, 9 Mallard Close, Ormskirk, L39 5QJ
GW4 UVC J Stephens, 105 Wern Street, Tonypandy, CF40 2DH
G4 UVD D Asquith, 516 Old Bedford Road, Luton, LU2 7BY
G4 UVF J Taylor, 6 The Stray, South Cave, Brough, HU15 2AL
G4 UVG D Stewart, 4 Towles Pastures, Castle Donington, Derby, DE74 2RX
G4 UVJ David Speechley, 15 Southwick Road, Canvey Island, SS8 0EP
GW4 UVN Jeffrey Travers, 40 Birchgrove Road, Birchgrove, Swansea, SA7 9JR
G4 UVV D Pike, 22 Stable Court, Gatchell Oaks, Taunton, TA3 7EG
G4 UVW E Underhill, 5 Lyndhurst Croft, Eastern Green, Coventry, CV5 7QE
G4 UVX P Lee, 51 Ashford Road, Faversham, ME13 8XN
G4 UVZ A Whatmore, Hollybank, Sellicks Green, Taunton, TA3 7SD
G4 UWF M Kebbell, 56 King Edward Avenue, Hastings, TN34 2NQ
G4 UWG N Dunn, 4 Swyneghyll, Temple Sowerby, Penrith, CA10 2AW
G4 UWM James McKenna, 18 Frobisher Close, Goring-by-Sea, Worthing, BN12 6EY
GM4 UWN R Kane, 39 Tollohill Drive, Aberdeen, AB12 5DQ
G4 UWP L Flynn, 20 Heather Lea Place, Sheffield, S17 3DN
G4 UWR V Thomas, 6 Hillside Court, Pontnewydd, Cwmbran, NP44 1LS
G4 UWS A Walker, 53 Parkstone Avenue, Parkstone, Poole, BH14 9LW
G4 UWW Anthony Prior, 62 Gainsborough Drive, Lawford, Manningtree, CO11 2JU
GM4 UWX John Rennie, 19 Harbour Place, Portknockie, Buckie, AB56 4NR
G4 UXB R Ball, 144 Broad Lane, Hampton, TW12 3BW
G4 UXC Michael Butler, Field Farm Bungalow, Longdon Hill, Evesham, WR11 7RP
G4 UXD Derek Brandon, 1 Woodlands Road, Saltney, Chester, CH4 8LB
G4 UXG Jonathan Dew, 8 Silverbeck Way, Stanwell Moor, Staines-upon-Thames, TW19 6BT
G4 UXH Colin Wilkinson, 14 Ryleyfield Road, Milnthorpe, LA7 7PT
G4 UXJ T Ager, 5 Matthews Close, Bedhampton, Havant, PO9 3NJ
G4 UXL K Jones, 10 Whetstone Hey, Great Sutton, Ellesmere Port, CH66 3PH
G4 UXO N Emson, 9 Sands Close, Pattishall, Towcester, NN12 8LU
G4 UXP M Huxham, 34 The Close, Brixham, TQ5 8RF
G4 UXV C Osborn, 19 Maple Drive, Huntingdon, PE29 7JE
GM4 UXX Anne Hood, 4 Murray Road, Law, Carluke, ML8 5HR
G4 UXY Christopher Boulter, 18a Red Lion Street, Chesham, HP5 1EZ
GM4 UYE Hugh Martin, 11 Ewing Court, Broomridge, Stirling, FK7 0QP
G4 UYF L Aldhous, 5 Banks Lane, Heckington, Sleaford, NG34 9QY
G4 UYI R Heselwood, 1 Well Garth Mount, Leeds, LS15 7LF
G4 UYJ B Crow, 690 Walmersley Road, Bury, BL9 6RN
GM4 UYK James Caddis, 30 Newlands Drive, Kilmarnock, KA3 2DW
G4 UYM F Roberts, 5 Manor Farm Close, Broughton, Kettering, NN14 1SL
GM4 UYP J Smith, 10 Witchknowe Avenue, Caprington, Kilmarnock, KA1 4LQ
G4 UYR Raymond Noble, Fallowfield, Chandler Road, Norwich, NR14 8RG
GW4 UYT R Jenkins, 1 Lon Y Bryn, Glynneath, Neath, SA11 5BG
GM4 UYZ Robert Glasgow, 7 Castle Terrace, Port Seton, Prestonpans, EH32 0EE
G4 UZC D Ralph, Tal-Y-Maes, Llanbedr, Powys, NP8 1SY
GD4 UZE C Mason, 145 Park Avenue, Ruislip, HA4 7UN
G4 UZF B MATTHEWS, 12 School Road, Thurston, Bury St. Edmunds, IP31 3SP
G4 UZG Gordon Price, 28 Leewood Close, Brampton Bierlow, Rotherham, S63 6ET
G4 UZN A Quest, 445 St. Lane, Leeds, LS17 6HQ
G4 UZO K Richards, Trigg Court, Trewetha, Port Isaac, PL29 3RU
GM4 UZR J Low, 4 Smith Avenue, Inverness, IV3 5ES
G4 VAF N Dorrington, Im Buergel 4, Woertham, Germany, 63939
G4 VAG H Green, 2 Whitchurch Road, Bangor On Dee, Wrexham, LL13 0AY
G4 VAH P Hudson, 15 Fellows Road, Farnborough, GU14 6NU
G4 VAL V Pellowe, 191 Preston New Road, Blackpool, FY3 9TN
G4 VAM Paul Harrison, 2 Allington Close, Bainton, Stamford, PE9 3AG
G4 VAO M Jordan, Petalouda, Low Street, Wymondham, NR18 9RY
G4 VAP I Kenyon, 4 Well Lane, Warton, Carnforth, LA5 9QZ
G4 VAS E Cooper, 39 Violet Road, Southampton, SO16 3GZ
G4 VAV A Brooks, 17 Grosvenor Avenue, Carshalton, SM5 3EJ
G4 VAX S Cope, 24 Metcalf Road, Newthorpe, Nottingham, NG16 3NL
GM4 VAY A Newlands, 7 Muir Close, Stewarton, Kilmarnock, KA3 3HG
GD4 VBA Rob Harrison, 108 Ballacriy Park, Colby, Isle of Man, IM9 4NB
G4 VBD S Mcadam, 92 Armstrong Close Birchwood, Warrington, WA3 6DJ
GM4 VBE R Fairholm, 28 Queensberry Avenue, Clarkston, Glasgow, G76 7DU
G4 VBI R Harte, 32 Kingsgate Avenue, Kingsgate, Broadstairs, CT10 3QP
G4 VBJ B Kay, 19 Langham Grove, Timperley, Altrincham, WA15 6DY
G4 VBK R Deeprose, 70 Hollington Old Lane, St. Leonards on Sea, TN38 9DP
G4 VBM C Leighton, 12 Morley Avenue, Connah's Quay, CH5 4RE
G4 VBO J Mattock, 1633 Dufferin Cres, Nanaimo, Canada, V9S 5T4
G4 VBP B Patchett, 48 Parsley Hay Road, Sheffield, S13 8NJ
G4 VBQ Stefan Largent, 31 Penzance Road, Kesgrave, Ipswich, IP5 1LU
G4 VBS P Chapman, 10 School Cottages, Hargrave, Bury St. Edmunds, IP29 5HR
GW4 VBV R Beckers, 13 Taplow Terrace, Pentrechwyth, Swansea, SA1 7AD
G4 VBX A Currell, 33 The Oval, Saham Toney, Thetford, IP25 7HW
G4 VCA G Mason, Lilac Cottage, Tremar Coombe, Liskeard, PL14 5EL
G4 VCB C Melvin, 3304 Woodglen Drive, Mckinney, United States, 750 71
G4 VCE S Sewell, Medway, The Rosery, Norwich, NR14 8AL
G4 VCJ Robert Percival, 6 Bulmer Place, Hartlepool, TS24 9BQ
G4 VCL Robert Parry, 2 Campanula Drive, Rogerstone, Newport, NP10 9JG
G4 VCN Andy Soars, 118 Braddon Road, Loughborough, LE11 5YZ
G4 VCO D Seddon, Zante, 31 Pembridge Road, Hemel Hempstead, HP3 0QN
G4 VCP C Smith, 24 Watling Way, Whiston, Prescot, L35 7NG
G4 VCQ R Hogan, 10 Lisle Place, Wotton-under-Edge, GL12 7BJ
G4 VCX M Beaumont, 71 Lime Tree Avenue, Coventry, CV4 9EZ
G4 VCY J Eley, 94 Raphael Close, Whoberley, Coventry, CV5 8LS
GI4 VCZ Peter Donnelly, 64 Aghnagar Road, Garvaghy, Dungannon, BT70 2EL
G4 VDB D Brocklehurst, 73 Ridgeway, Clowne, Chesterfield, S43 4BD
G4 VDF A Palmer, 11 Cedar Avenue, Blackwater, Camberley, GU17 0JE
GM4 VDG James Rankin, 64 Forrest Walk, Uphall, Broxburn, EH52 5PN
G4 VDH W Peak, 10 The Oval, Scarborough, YO11 3AP
G4 VDJ Brian Lee, 61 Pendleway, Pendlebury, Manchester, M27 8QS
G4 VDX Joseph Menguy, 6 Laurel Grove, Lowton, Warrington, WA3 2EE
G4 VDZ F Williamson, 6 Lord Tedder Court, Rothbury, Morpeth, NE65 7TU
G4 VEA M Piercy, 14 Maybury Mews, London, N6 5YT
GW4 VEB David Lintern, 108 Pontygwindy Road, Caerphilly, CF83 3HF
G4 VEC Michael Elliott, 20 Haysel, Sittingbourne, ME10 4QE
G4 VEG S Hall, 15 Hurst Close, Staplehurst, Tonbridge, TN12 0BX
G4 VEH L Skinner, 15 Ridge Close, Portishead, Bristol, BS20 8RQ
GW4 VEI C Barrett, 30 Brecon Road, Hirwaun, Aberdare, CF44 9ND
G4 VEL John Smith, 43 Ash Close, Thetford, IP24 3HQ
G4 VEO G Barratt, Charnwood, Great North Road, Retford, DN22 8NL
G4 VEQ T Jones, Penrhiw Bach, Bryngwran, Holyhead, LL65 3RD
G4 VER Robert Heath, 26 Lancaster Avenue, Hadley Wood, Barnet, EN4 0EX
G4 VET N Greaves, Forty Shilling Cottage, Oghill, Co. Kildare, Ireland
G4 VEW G Davies, 40 Derby Road, Talke, Stoke-on-Trent, ST7 1SG
G4 VEY F Havard, Whitehalgh Farm, Whitehalgh Lane, Blackburn, BB6 8ET
G4 VFC Deirdre Monnery, 8 Reeds Lane, Southwater, Horsham, RH13 9DQ
G4 VFE C Davies, 3 Bryn Onnen, Flint, CH6 5QB
G4 VFG Peter Lewis, 18 Bittaford Wood, Bittaford, Ivybridge, PL21 0ET
G4 VFH G Fuller, 61 The Underwood, London, SE9 3EP
G4 VFJ S Shenton, 36 Walleys Drive, Newcastle, ST5 0NG
G4 VFK C Archer, 118 Cator Lane, Beeston, Nottingham, NG9 4BB
G4 VFL Andrew Holland, 12 Riverside Drive, Egremont, CA22 2EH
G4 VFQ D Nuttall, The Mount, Church Lane, Maidstone, ME14 4EF
G4 VFR K Hackwell, 15 Standish Avenue, Billinge, Wigan, WN5 7TF
G4 VFU Carl White, Saltfleet House, Pump Lane, Louth, LN11 7RL
G4 VFX Christopher Perkins, 32 Empshott Road, Southsea, PO4 8AU
GW4 VGB R Harper, 114 Pantbach Road, Cardiff, CF14 1UE
G4 VGF Brian Ramsden, 2 Brookfield Cottages, Mill Lane, Northwich, CW8 2TA
G4 VGL Stefan Luckhaus, Wingertstrasse 5, Kleinwallstadt, Germany, 63839
G4 VGM E Grindel, Bischofsheimer, Platz 24, Frankfurt, Germany, D 60326
G4 VGN V Havran, Kurt-Schumacher, Ring 31, Dreieich, Germany, D-63303
GM4 VGR James Buchanan, 114 Glasgow Road, Whins of Milton, Stirling, FK7 0LJ
GM4 VGU A Lyttle, 23 Heathfield Drive, Kirkmuirhill, Lanark, ML11 9SR
G4 VGY D Cline, 68 Frenchgate, Richmond, DL10 7AG
G4 VHB C Averill-Elias, 12 Bubwith Close, Chard, TA20 2BL
G4 VHE R Haase, 674 Valley View Lane, Strafford, United States, 19087
G4 VHG J Fowler, 6 Cridlake, Axminster, EX13 5BS
G4 VHH F Atkin, 18 East Lane, Corringham, Gainsborough, DN21 5QU
G4 VHI M Sawyers, 20 Fairways, Ferndown, BH22 8BA
G4 VHJ J Taylor, 29 Meadow Walk, Ewell, Epsom, KT17 2EF
G4 VHK Ronald Leslie, Tranquil, Rectory Lane Kingston, Cambridge, CB23 2NL
G4 VHL T Langford, 11 The Grove, Blackawton, Totnes, TQ9 7BA
G4 VHM Michael Hindley, 12 Tremayne Avenue, Brough, HU15 1BL
GI4 VHO David Calderwood, 43 Rathview Park, Mullybritt, Enniskillen, BT94 5EW
G4 VHP S Murdoch, 55 Pendre Avenue, Prestatyn, LL19 9SH
G4 VHQ E Smith, Borodino, Manor Road, Bridgwater, TA7 9HB
G4 VHS E Williams, Newhaven, Kinmel Way, Abergele, LL22 9NE
G4 VHV A Sims, 38 Giffard Drive, Welland, Malvern, WR13 6SE
G4 VHX J Allen, 18 Horsley Close, Chesterfield, S40 4XD
GM4 VHZ N Brown, 7 Mid Road, Beith, KA15 2AJ
G4 VIA J McSherry, 1 Station Houses, Corkickle, Whitehaven, CA28 7XG
G4 VIF Roberts Watts, 116 Hassall Road, Sandbach, CW11 4HL
G4 VII J Lawrence, 25 Sylvia Crescent, Totton, Southampton, SO40 3LP
GM4 VIK T Irwin, 6 Inverarnie Park, Inverarnie, Inverness, IV2 6AX
G4 VIL J Fielding, 35 Amos Avenue, Litherland, Liverpool, L21 7QH
G4 VIM B Pulfrey, 21 Emfield Road, Grimsby, DN33 3BW
G4 VIO Arthur Greenbank, 3 Cooperative Terrace, Stanley, Crook, DL15 9SE
GI4 VIP THE STH BELFAST c/o P Murphy, 11 Danesfort Apartments, Villa 1, Belfast, BT9 5QL
G4 VIQ P Brushwood, 2 High Trees, Waterlooville, PO7 7XP
G4 VIS H Cameron, 14 Queen Street, Castle Douglas, DG7 1HX
G4 VIT M Wood, 42 Buckingham Drive, Willenhall, WV12 5TD
G4 VIW William Watts, 11 Meadow Way, Upminster, RM14 3AA
G4 VIX THE EAST COAST VHF GROUP c/o D Bartlett, 80 Burnway, Hornchurch, RM11 3SG
GI4 VIZ M Jamieson, 59 Curragh Road, Coleraine, BT51 3RZ
G4 VJB V Bloor, 22 Regency Close, Talke Pits, Stoke-on-Trent, ST7 1RH
G4 VJI C Lindsay, 31 Barnes Close, Blandford Forum, DT11 7NG
G4 VJL J Oldfield, 62 Chipperfield Drive, Bristol, BS15 4DR
G4 VJN Stephen Matthews, 30 Broadgate Lane, Deeping St. James, Peterborough, PE6 8NW
G4 VJT K Farmer, 61 Queens Road, Beckenham, BR3 4JJ
GI4 VJZ Thomas Wilson, Wilden, 18 Ahoghill Road, Antrim, BT41 3BJ
G4 VKC David Lawrence, 6 Hollycombe Close, Liphook, GU30 7HR
G4 VKE R Pearce, 1a Green Lane, Dalton-in-Furness, LA15 8LZ
GW4 VKG W Weston, 3 Factory Terrace, Aberkenfig, Bridgend, CF32 9AF
GM4 VKI M Kavanagh, 4 Old Auchans View, Dundonald, Kilmarnock, KA2 9EX
G4 VKJ TP Grant, 81 Hillworth Road, Devizes, SN10 5HD
G4 VKO J Whittock, 18 Westons Brake, Emersons Green, Bristol, BS16 7BP
GI4 VKS A Mccallion, 3 Lisky Road, Strabane, BT82 8NW
G4 VKV T Linacre, 69 Elizabeth Road, Fazakerley, Liverpool, L10 4XL
G4 VKX I Wade, 59 St. Annes Road, Kettering, NN15 5EQ
G4 VKY Newsham Community Hall c/o D Hamby, 13 Trident Drive, Blyth, NE24 3RL
G4 VLA R Trudgill, The Retreat, Kiln Lane, Bedford, MK45 4DA
G4 VLF John Wingfield, 56 Manor Fields, Liphook, GU30 7BS
G4 VLH B Pash, Dales Stores, Station Road, Cheltenham, GL54 4HP
G4 VLI R Allen, 39 Deerpark, Co Meath, Ireland
G4 VLK D HASLEHURST, 23 Yew Tree Drive, Shirebrook, Mansfield, NG20 8QH
G4 VLL C Denham, 3 Glenmore Close, Flackwell Heath, High Wycombe, HP10 9DF
G4 VLN M Evans, 2a Moreton Road, Worcester Park, KT4 8EZ
G4 VLP H Knatchbull, 19 Riverside Road, West Moors, Ferndown, BH22 0LG
G4 VLS P Turnham, 71 Theobald Road, Norwich, NR1 2NX
G4 VLT C Tunna, 52 Shaftoe Road, Springwell, Sunderland, SR3 4EZ
G4 VLU M Hatwood, Calgary, Denbigh Circle, Rhyl, LL18 5HW
G4 VLV Andrew Flint, 4 Churchill Way, Painswick, Stroud, GL6 6RQ
G4 VLW Roger Davey, 35 The Pines, Faringdon, SN7 8AT
GM4 VLX J Brown, 33 Gartmore Road, Paisley, PA1 3NG
G4 VLZ Malcolm Nettleship, 141 Hollybank Drive, Sheffield, S12 2BU
G4 VMA Michael Anderson, Vanburgh House, Ainderby Quernhow, Thirsk, YO7 4HX
G4 VMB D Stoddart, 16 Market Place, Long Buckby, Northampton, NN6 7RR
G4 VMC P Coates, 20 The Flashes, Gnosall, Stafford, ST20 0HL
G4 VMD C Hackney, Mar Azul 9, Apt.20 2 Fase, Alicante, Spain, 03710 CALPE
G4 VME Richard Youell, 1 Greenside, Waterbeach, Cambridge, CB25 9HW
G4 VMF Stuart Schofield, 55 St. Botolphs Green, Leominster, HR6 8ER
G4 VMG John Holmes, 10 Chapel Road, Morley St. Botolph, Wymondham, NR18 9TF
G4 VMI M Pickworth, 39 Pollard Road, Weston-Super-Mare, BS24 7BZ
G4 VMM S Tidmarsh, 4 The Grange, Earl Shilton, Leicester, LE9 7GT
G4 VMQ J Harris, 23 Brookvale Grove, Solihull, B92 7JH
G4 VMR J Watkins, One Ash, Frogshall Lane, Ware, SG11 1JH
GW4 VMT G Williams, 12 Heol Johnson, Talbot Green, Pontyclun, CF72 8HR
G4 VMU E Gordon, 11 Apperley, West Denton, Newcastle upon Tyne, NE5 2JS
G4 VMW J Curtis, 18 Carlidnack Close, Mawnan Smith, Falmouth, TR11 5HF
G4 VMX Anthony Ritchie, 24 Swift Close, Newport Pagnell, MK16 8PP
G4 VMY A Cooper, 3 Marina Way, Ripon, HG4 2LJ
G4 VMZ A Jones, 16 Hawe Farm Way, Broomfield, Herne Bay, CT6 7UD
G4 VNA Richard Bell, Long Meadows, Haddockstones, Harrogate, HG3 3LA
G4 VNC M Leak, Papermill House, Nordham, Brough, HU15 2LT
G4 VNE D Hunt, 233 Kingsley Road, Kingswinford, DY6 9RP
G4 VNG Robert McCallum, 9 Hardwick Close, Blackwell, Alfreton, DE55 5LL
GW4 VNK Leslie Smith, Blaenlluest, Cilcennin, Lampeter, SA48 8RP
G4 VNM S Frost, 32 Hunters Lodge, Fareham, PO15 5NE
G4 VNR Roger Sharp, 22 Sunderton Lane, Clanfield, Waterlooville, PO8 0NU
G4 VNS R Sims, 61 Constable Drive, Newport, NP19 7QB
G4 VNX W Wood, 11 Walbert Avenue, Thurnscoe, Rotherham, S63 0TN
G4 VOB G Sunter, 16 Waindale Close, Mount Tabor, Halifax, HX2 0UL
G4 VOG A Hepworth, 9 Linden Grove, Kirkby-in-Ashfield, Nottingham, NG17 8JJ
G4 VOJ A Tennant, 2 Chapel Hill Farm Cottage, Lower Lane, Preston, PR3 3SL
G4 VOK Peter Dresser, 6 Acacia Avenue, Fencehouses, Houghton le Spring, DH4 6JG
G4 VOT R Bullimore, 30 Podsbrook House, Guithavon Street, Witham, CM8 1DR
G4 VOU A Pinkney, 1 Hester Gardens, New Hartley, Whitley Bay, NE25 0SH
G4 VOV David Hallsworth, 6 Eastholme, Luddington Road, Scunthorpe, DN17 4PP
G4 VOW B Pluckrose, 104 Edward Road, West Bridgford, Nottingham, NG2 5GB
G4 VOY R Powell, Old School House, Broxwood, Leominster, HR6 9JQ
G4 VOZ J Jennings, Mill Side, Mill Road, Lutterworth, LE17 5DE
G4 VPA J Martindale, The Old School House, Ipswich Way, Stowmarket, IP14 6DJ
G4 VPC E Ikin, 30 Kelsborrow Way, Kelsall, Tarporley, CW6 0NL
G4 VPD M Pugh, 44 Simms Lane, Hollywood, Birmingham, B47 5HY
G4 VPE R Derricott, Birches, The Paddock, Stourbridge, DY9 0RE
G4 VPF O Davies, 16 Central Way, Horninglow, Burton-on-Trent, DE13 0UU
G4 VPI R Riley, 161 Botany Road, Kingsgate, Broadstairs, CT10 3SD

G4 VPJ D Bridgnell, Penvale, 1 Tretherras Road, Newquay, TR7 2RB
G4 VPL J Villena Bota, Santa Ana 74, Estartit, Gerona, Spain, 17258
G4 VPM A [illegible], 900 Sparrow Branch Circle, Jacksonville, United States, 32259
G4 VPS B Lewis, Juno, Ingleton, Carnforth, LA6 3AN
G4 VPU John Armstrong, 12 Firtree Gardens, Whitley Bay, NE25 8XB
G4 VPW Paul Wilcock, 12 Napier Road, Eccles, Manchester, M30 8AG
GW4 VPX Allan Jones, Maes Y Llyn, Maesycrugiau, Pencader, SA39 9DH
G4 VPZ A Hill, 36 Narrow Lane, Halesowen, B62 9NQ
G4 VQE R Spencer, [illegible]
G4 VQF P White, 4 Barnett Lane, Wonersh, Guildford, GU5 0SA
G4 VQH M Clutton, Cumberwell, Cumberland Lane, Whitchurch, SY13 2NJ
G4 VQI P Hughes, 92 Freshwater Drive, Paignton, TQ4 7SD
G4 VQJ D Northwood, 5 Beech Grange, Landford, Salisbury, SP5 2AL
GI4 VQK Alphonsus Ward, 50 Derry Road, Strabane, BT82 8LD
G4 VQL G Dymond, 28 The Green, Exmouth, EX8 2QR
G4 VQP Chris Smith, Mount Elland, Carnmenellis, Redruth, TR16 6PB
G4 VQR E Rood, 139 Potovens Lane, Outwood, Wakefield, WF1 2LF
G4 VQS Hwfa Jones, 47 Penkett Road, Wallasey, CH45 7QG
G4 VQT [illegible]
GM4 VQY G Leiper, 76 Martin Drive, Stonehaven, AB39 2LU
G4 VQZ J Oakley, 152 Little Breach, Chichester, PO19 5UA
G4 VRB Kevin Raine, 30 St. Andrews Gardens, Shepherdswell, Dover, CT15 7LP
G4 VRC R Doran, 28 Buckingham Road, Petersfield, GU32 3AZ
GM4 VRE P Henderson, 134 Gray Street, Aberdeen, AB1 6JU
GI4 VRF F Macdonald, 5 Glenview Crescent, Castlereagh, Belfast, BT5 7LX
G4 VRG F Margrave, Templars, Peerley Road, Chichester, PO20 8DW
G4 VRJ R Clifft, 11 Hambleton Street, Wakefield, WF1 3NW
G4 VRM A Berry, 148 Maple Way, Gillingham, SP8 4RR
G4 VRN M Blewett, 32 Miltons Crescent, Godalming, GU7 2NT
GW4 VRO Peter Parsons, 9 Military Road, Pennar, Pembroke Dock, SA72 6SH
G4 VRP R Porter, 47 Milford Avenue, Wick, Bristol, BS30 5PP
G4 VRS AYLESBURY VALE RS c/o Victor Gerhardi, 24 Putnams Drive, Aston Clinton, Aylesbury, HP22 5HH
G4 VRT Barry Barker, The Nook, Park Lane, Sheffield, S36 8WW
G4 VRU B Stephenson, 12 Claremont Terrace, York, YO31 7EJ
G4 VRW K Newbould, 188 Brooklands Avenue, Leeds, LS14 6RH
G4 VRX G Brown, 1 Dog Kennel Lane, Oldbury, B68 9LU
G4 VSB A Brown, 6 The Firs, Rushbrooke Lane, Bury St. Edmunds, IP33 2SY
G4 VSD P Samuels, 6 Miriam Close Caister-on-Sea, Great Yarmouth, NR30 5PH
GW4 VSE M Carey, 47 Heol Ty Gwyn, Maesteg, CF34 0BD
G4 VSI A Stone, 29 Nottingham Road, Belper, DE56 1JG
G4 VSJ K Drakeford, Sunnyside, Frolesworth Lane, Lutterworth, LE17 5AS
G4 VSK B Skelton, 8 Dunelm Drive, West Boldon, East Boldon, NE36 0HJ
G4 VSL T Watkins, One Ash, Frogshall Lane, Ware, SG11 1JH
G4 VSO R Carter, Mundens, Horsley Road, Stroud, GL6 0JR
G4 VSQ Alastair Bolton, 17 Lomond Avenue, Caversham, Reading, RG4 6PL
G4 VSR S Alston, 21 Hilltop Road, Wingerworth, Chesterfield, S42 6RX
G4 VSS Michael Isherwood, 32 Franklin Close, Old Hall, Warrington, WA5 8QL
G4 VSV G Ingham, Courthaven, South Duffield Road, Selby, YO8 5HP
G4 VSW M Taylor, 2 Bickerton Drive, Hazel Grove, Stockport, SK7 5QY
G4 VSX Peter Reilly, 40 Bollin Drive, Lymm, WA13 9QA
G4 VSY K Hutton, 7 Roseveare Drive, Roseveare Park, Gothers, St. Austell, PL26 8GY
G4 VTA J Taylor, 219 Mandarin Way, Cheltenham, GL50 4SB
GM4 VTB M Budas, 20 Oak Avenue, Bearsden, Glasgow, G61 3HD
G4 VTC Anthony Croydon, Harvesters, Newdigate Road, Dorking, RH5 4QB
G4 VTD I Daniels, 24 Ockley Lane, Keymer, Hassocks, BN6 8BB
GW4 VTG E Smith, 21 St. Davids Road, Pembroke, SA71 5JH
G4 VTM J Hicks, Cory House, Kilworth Road, Lutterworth, LE17 6JW
G4 VTN J Rodda, 22 Balmoral Drive, Felling, Gateshead, NE10 9TZ
G4 VTO P Tanner, Beechcroft, Station Hill, Newton Abbot, TQ13 0EE
G4 VTQ D Rainer, 11 Ryhill Way, Lower Earley, Reading, RG6 4AZ
G4 VTU R George, 19 Apthorpe Street, Fulbourn, Cambridge, CB21 5EY
G4 VUA Alan Burton, 26 Woffindin Close, Great Gonerby, Grantham, NG31 8LP
G4 VUD John Head, 21 Reynell Avenue, Newton Abbot, TQ12 4HE
G4 VUF C Tugman, 41 Chatsworth Road, Hunstanton, PE36 5DJ
GM4 VUG C Green, 4 Gallowhill Gardens, Kinross, KY13 8RT
G4 VUH J Washington, Flat 9, Llys Canol, Holywell, CH8 7XG
G4 VUI P Sweeney, 15 Alford Road, West Bridgford, Nottingham, NG2 6GJ
G4 VUK L Wolfson, 7 Gilmore Drive, Prestwich, Manchester, M25 1NB
G4 VUM D Stocks, 78 Moor Street, Mansfield, NG18 5SQ
G4 VUN P Norris, Thirn Farm, Thirn, Ripon, HG4 4AU
G4 VUP G Newton, 6 Yardley Way, Grimsby, DN34 5UQ
G4 VUR I Daniels, 20a Stalham Road, Hoveton, Norwich, NR12 8DG
G4 VUV S Valori, 7 Upton Close, Norwich, NR4 7PD
G4 VUW D Kemp, 35 Rushett Drive, Dorking, RH4 2NR
G4 VVD P Taylor, 8 High St., Clive, Shrewsbury, SY4 3JL
G4 VVE E Macmanus, 41 Oldfield Crescent, Stainforth, Doncaster, DN7 5PE
G4 VVF Nicholas Allen, Lilac Cottage, Roddhurst, Presteigne, LD8 2LH
G4 VVK Bernard Murray, La Casa, 30 Middlegate Green, Rossendale, BB4 8PY
G4 VVL Kenneth Arrowsmith, 1 Northcroft Road, Gosport, PO12 3DR
G4 VVM A Bennett, 28 Kinglake Drive, Taunton, TA1 3RR
G4 VVP Basil Gillard, Charmaine, Broadway, Radstock, BA3 4GT
G4 VVQ Fred Shead, 7 White Cottages, Fuller Street, Chelmsford, CM3 2AY
G4 VVS J Blanchard, 41 Deane Drive, Galmington, Taunton, TA1 5PQ
G4 VVT A Moss, 9 Summerfield Drive, Middleton, Manchester, M24 2TQ
GM4 VVX Clive Ohennessy, Savalbeg, Challenger Estate, Lairg, IV27 4ED
GM4 VVY Duncan Davis, 3 High Shore, Banff, AB45 1DB
G4 VVZ C Wilson, 2 Bainton Close, Bradford-on-Avon, BA15 1SE
G4 VWA S Ward, 88 Little Barn Lane, Mansfield, NG18 3JJ
G4 VWB R Osgathorpe, 1 Medway Drive, Allestree, Derby, DE22 2UB
GI4 VWC G Christie, The Brambles, 9 Burnet Park, Newtownabbey, BT37 0XY
G4 VWE M Courteney, 36 Nursery Close, Hellesdon, Norwich, NR6 5SJ
G4 VWF Reginald Hawkins, 43 The Courtyard, Taylor Avenue, Northampton, NN3 2DD
G4 VWG Stuart VanKassel, 9 Tarragon Close, Swindon, SN2 2SG
G4 VWI D Hatton, 9 Gregory Close, Thurmaston, Leicester, LE4 8BP
G4 VWL John Owen, 7 Linear Park, Wirral, CH46 6FL
G4 VWO J Bloodworth, Sibrwd yr Awel, Penrhyndeudraeth, LL48 6AY
G4 VWP R Bloodworth, Sibrwd yr Awel, Penrhyndeudraeth, LL48 6AY
G4 VWS C Davies, Essex House, 42 Boxworth Road, Cambridge, CB23 4JQ
G4 VWT B Padgett, 105 Middlecroft Road, Staveley, Chesterfield, S43 3XH
GM4 VWV Robert Mcewan, 12 Valleyfield Drive, Cumbernauld, Glasgow, G68 9NW
G4 VWX A Shone, 27, Pump Street, Malvern, WR14 4LU
GW4 VWY G Whiteway, 4 Nicholas Court, Gorseinon, Swansea, SA4 4PR
GM4 VXA Paul Williams, 21 St. Clair Way, Ardrishaig, Lochgilphead, PA30 8FB
G4 VXB M Ellis, 16 Fielding Street, Faversham, ME13 7JZ
G4 VXD Russell King, 1 Emmas Crescent, Stanstead Abbotts, Ware, SG12 8AZ
G4 VXE Timothy Kirby, Willowside, Bow Bank, Longworth, OX13 5ER
G4 VXG Geoffrey Buxton, 12 Jute Road, York, YO26 5EN
G4 VXH A Rean, 17 Mount Pleasant Road, Dawlish Warren, Dawlish, EX7 0NA
GM4 VXM I Munro, 12 Greenstone Place, Dundee, DD2 1XD
G4 VXN C Bennett, 49 Keats Avenue, Redhill, RH1 1AF
G4 VXP R Ward, 19 Canterbury Road, Leyton, London, E10 6EE
G4 VXU J Haig, 3 Hartland Court, Gaping Lane, Hitchin, SG5 2JH
G4 VXV R Boulton, 23 Stamford Bridge West, Stamford Bridge, York, YO41 1AQ
G4 VXW R Seddon, 255 Westleigh Lane, Leigh, WN7 5PN
G4 VXX S Parkes, 6 Wychwood Park, Weston, Crewe, CW2 5GP
G4 VYA J Jacobs, 17 Cotswold Drive, Albrighton, Wolverhampton, WV7 3DQ
G4 VYC Victor Packman, 241 Gurnard Pines Cockleton Lane, Cowes, PO31 8RL
G4 VYE John Harris, 23 Balmoral Drive, Hednesford, Cannock, WS12 4LT
G4 VYF K Thomas, 15 Meakers Way, Huttoft, Alford, LN13 9TR
G4 VYG B Roberts, 52 School Lane, Toft, Cambridge, CB23 2RE
G4 VYH N Baker, Roffensis, 16 Boulderside Close, Norwich, NR7 0JJ
G4 VYI Melvyn Dalley, 195 Marlcliffe Road, Sheffield, S6 4AH
G4 VYJ J O'Sullivan, 30 Highbank Road, Kingsley, Frodsham, WA6 8AE
G4 VYK R Vaughan, 73b Westward Drive, Pill, Bristol, BS20 0JR
G4 VYL R Reilly, 4 Moreton Drive, Poulton-le-Fylde, FY6 8ED
G4 VYN J Lawrence, 1 Naples Close, Hopton, Great Yarmouth, NR31 9SB
G4 VYP Dorothy Rimmer, 8a Mallee Avenue, Southport, PR9 8NL
GM4 VYQ W Harvey, 32 Upper Glenfyne Park, Ardrishaig, Lochgilphead, PA30 8HH
G4 VYR G McCartney, 12 Timway Drive, West Derby, Liverpool, L12 4YR
G4 VYU Ogilvie Jackson, Cossarshill Farm, Selkirk, TD7 5JB
G4 VZB I Ray, Jagarbacken 5, Danderyd, Sweden, SE 182-39
G4 VZC P Stokes-Herbst, 72 Devonshire Road, Middlesbrough, TS5 6DP
G4 VZH A Hobkirk, 216 Northwick Road, Worcester, WR3 7EH
GM4 VZI William Lawrie, Sandana Old Woodhouselea, 2, Roslin, EH25 9QJ
G4 VZK D Perkins, 10 The Foxes, Sutton Hill, Telford, TF7 4NH
G4 VZL J Caddick, 5 Great Hay Drive Sutton Hill, Telford, TF7 4DT
G4 VZR Douglas Cormack, Lukes Orchard, Far Green, Dursley, GL11 5EL
G4 VZS L Andrew, 15 Farndale Road, Knaresborough, HG5 0NY
G4 VZT P Green, 61 Gravel Hill, Wimborne, BH21 3BJ
G4 VZW Kenneth Rankine, Calle Alfredo L. Jones 34, Building Perez Rocha, Las Palmas, Spain, 35008
GM4 VZY D Deans, 17 Montrose Way, Dunblane, FK15 9JL
G4 WAB THE WORKED ALL c/o G Darby, 5 Lumsden Terrace, Catchgate, Stanley, DH9 8EQ
G4 WAC WYTHALL RAD CLUB c/o David Dawkes, 95 Houndsfield Lane, Wythall, Birmingham, B47 6LX
G4 WAF Anthony Fewkes, 21 Tong Road, Bishops Wood, Stafford, ST19 9AB
G4 WAG Terence Morley, 84 Cliff Lane, Ipswich, IP3 0PJ
GI4 WAH J Keenan, 24 Leode Road, Hilltown, Newry, BT34 5TJ
G4 WAK Neil Rumbol, 66 The Avenue, Hadleigh, Benfleet, SS7 2DL
G4 WAL P Walton, 6 Gorse Grove, Longton, Preston, PR4 5NP
G4 WAM M Lockley, 37 Farmside Lane, Biddulph Moor, Stoke-on-Trent, ST8 7LY
G4 WAO James Kimpton, 28 Clifton Street, Stourbridge, DY8 3XT
G4 WAP R Southern, 31 Burnsall Road, Brighouse, HD6 3JS
G4 WAS K Atack, 29 High Hill, Essington, Wolverhampton, WV11 2DW
G4 WAV Anthony Medland, Trevilla, 93 Pengelly Road, Delabole, PL33 9AT
G4 WAW South Bristol ARC c/o Andrew Jenner, 24 The Willows, Nailsea, Bristol, BS48 1JQ
G4 WAX Jeffrey Moon, 25 Shotley Gardens, Gateshead, NE9 5DP
G4 WAY Roger Holyoake, 105 Onslow Road, Croydon, CR0 3NZ
G4 WAZ N MacKinnon, 49 Balmoral Way, Worle, Weston-Super-Mare, BS22 9AL
G4 WBA Brian Westbrook, 14 Pickering Street, Maidstone, ME15 9RS
G4 WBC WEST BROMWICH RC c/o I Leitch, 70 Hanover Road, Rowley Regis, B65 9DZ
G4 WBF Paul Finney, New Rivernook Farm, 10 Kinnersley Manor, Reigate, RH2 8QJ
G4 WBG R Dunn, 13 Horton Gate, Giffard Park, Milton Keynes, MK14 5JQ
G4 WBH P Jackson, 15 Bankside, Retford, DN22 7UW
G4 WBI Steven Haydon, 58 Deanfield Road, Henley-on-Thames, RG9 1UU
G4 WBO Keith Johnson, 23 Rotherhead Close, Horwich, Bolton, BL6 5UG
G4 WBP D Hatfield, 29 Awbridge Road, Netherton, Dudley, DY2 0HZ
G4 WBT S Clifton, 15 Cae Clyd, Craig-y-Don, Llandudno, LL30 1BL
G4 WBV Adrian Fry, 128 Sylvan Way, Sea Mills, Bristol, BS9 2LU
G4 WBW K Odlum, 17 Glebe Street, Talke, Stoke-on-Trent, ST7 1NP
GD4 WBY Michael Jerrome-Jones, Fairfield, Jurby Road, Ramsey, Isle of Man, IM7 2EB
G4 WCD D Longstaff, 83 Spring Gardens, Anlaby Common, Hull, HU4 7QG
GM4 WCE P Kirsop, 295 Gilmerton Road, Edinburgh, EH16 5UL
G4 WCK Colin Baverstock, Butchers Coppice Scout Camp Site, Holloway Avenue, Bournemouth, BH11 9JW
G4 WCO Donald Foy, 37 Gorsey Croft, Eccleston Park, Prescot, L34 2RS
G4 WCP Stuart Richardson, 25 Kenmure Avenue, Patcham, Brighton, BN1 8SH
G4 WCY H Bottomley, 8 Leyburn Place, Filey, YO14 0DQ
G4 WDA John Curtis, 8 King Street, Wilton, Salisbury, SP2 0AX
G4 WDC G Cooke, 106 Wirral Drive, Winstanley, Wigan, WN3 6LD
G4 WDH Brian Cowley, 12 St. Annes Road, Southampton, SO19 9FF
G4 WDO A Douglas, 3 North Lodge Cottages, Ladykirk, Berwick-upon-Tweed, TD15 1SU
G4 WDP David Preston, 77 Wensley Road, Woodthorpe, Nottingham, NG5 4JX
G4 WDR West Devon Raynet c/o I Harley, 302 Tavy House, Duke Street, Plymouth, PL1 4HL
G4 WDS Robert Silvera, 10 White Hill, Kinver, Stourbridge, DY7 6AD
G4 WDZ K Bennett, Lilac Cottage, St. Neots Road, Bedford, MK44 2ER
G4 WEC G Ieesley, Marsh View, Main Road, Alford, LN13 0JP
G4 WED N Tipping, 18 Collingworth Rise, Park Gate, Southampton, SO31 1DA
G4 WEE S Leech, 9 Parkside Drive, Old Catton, Norwich, NR6 7DP
G4 WEH M Pepper, 56 Meadow Lane, Burgess Hill, RH15 9JA
G4 WEL J Bolton, 110 Vale Road, Ash Vale, Aldershot, GU12 5HS
G4 WEM A Penney, 110 Vale Road, Ash Vale, Aldershot, GU12 5HS
G4 WEN K Porter, Thornfield, Mount Carmel Road, Andover, SP11 7EE
G4 WEP W Hewitt, 101 Sunnyside Avenue, Ball Green, Stoke-on-Trent, ST6 8DZ
G4 WET TRIPLE B.C.G c/o Michael Butler, Field Farm Bungalow, Longdon Hill, Evesham, WR11 7RP
G4 WEV A Russell, 6 Bartlemy Road, Newbury, RG14 6JX
GM4 WEW Christine Brown, Glencraig, Ballantrae, Girvan, KA26 0PA
G4 WEY B Bush, 49 Mimosa Avenue, Wimborne, BH21 1TU
G4 WEZ K Westley, 29 The Limes, Sawston, Cambridge, CB22 3DH
G4 WFC Michael Morris, Fieldhead Farm, Denholme, Bradford, BD13 4LZ
G4 WFE J Langton, Grobdale of Girthon, Laurieston, Castle Douglas, DG7 2PZ
G4 WFF C McGuire, 17 Victoria Road, Emsworth, PO10 7NU
G4 WFK E Seddon, 20 Pinfold Lane, Bottesford, Nottingham, NG13 0AR
G4 WFL P Ford, 24 Tonstall Road, Epsom, KT19 9DP
GW4 WFM Geoffrey Moller, 31 Wyngarth, Winch Wen, Swansea, SA1 7EF
G4 WFR R Cooper, 53 Saturn Close, Southampton, SO16 8HF
G4 WFT Timothy Kearsley, 142 Avenue Road, Rushden, NN10 0SW
GM4 WFV S Duguid, 29 Park Circus, Ayr, KA7 2DJ
G4 WFW P Edwards, 1 Radley Avenue, Wickersley, Rotherham, S66 2HZ
G4 WFZ P Marsh, Columbia, 28 Orcheston Road, Bournemouth, BH8 8SR
G4 WGA R Hall, Hillside, Potten End Hill, Hemel Hempstead, HP1 3BN
G4 WGB Roy Vaughan, 6 Dellside Grove, St. Helens, WA9 5AR
GM4 WGC P Naughton, 16 Holton Crescent, Sauchie, Alloa, FK10 3DZ
G4 WGD C Pearse, 77a Nutfield Road, Merstham, Redhill, RH1 3ER
G4 WGE Alun Cross, 31 Mountcombe Close, Surbiton, KT6 6LJ
G4 WGF G Fairhurst, 42 Chorley Road, Standish, Wigan, WN1 2SS
G4 WGJ M Collins, 185 Church Road, Haydock, St. Helens, WA11 0NB
G4 WGK G Kemp, 38 Merlin Way, Leckhampton, Cheltenham, GL53 0LU
G4 WGN Kenneth Wilson, 102 Waddicar Lane, Melling, Liverpool, L31 1DY
G4 WGR Robert Gibson, 52 Broomfields, Denton, Manchester, M34 3TH
G4 WGT William Taylor, 27 Netherley Road, Coppull, Chorley, PR7 5EH
G4 WGU G Tarry, 8 Wareham Road, Blaby, Leicester, LE8 4BE
G4 WGX Brian Rivers, Maybank, Athelney Bridge, Bridgwater, TA7 0SB
G4 WGZ A Brooker, 18 Honeybourne Way, Petts Wood, Orpington, BR5 1EZ
G4 WHA Geoffrey Harper, 15 Seaforth Park, Annan, DG12 6HX
G4 WHF K Wilson, 111 Marple Road, Stockport, SK2 5EP
G4 WHK R Cann, 39 Grafton Road, Harwich, CO12 3BD
G4 WHL P Callaghan, 8 Abbey Road, Edwinstowe, Mansfield, NG21 9LQ
G4 WHM P Callaghan, 8 Abbey Road, Edwinstowe, Mansfield, NG21 9LQ
G4 WHN C Walker, 14 Collingwood Road, Long Eaton, Nottingham, NG10 1DR
G4 WHO N Foot, Oakfield Farm, Horton Way, Verwood, BH31 6JJ
G4 WHT W Tattersall, 45 Russell Avenue, Alsager, Stoke-on-Trent, ST7 2BN
G4 WHV Mady Langdon, 58 Upper Marsh Road, Warminster, BA12 9PN
G4 WHY M Foot, Oakfield Farm, Horton Way, Verwood, BH31 6JJ
G4 WHZ D Cater, 104 St. Johns Road, Clacton-on-Sea, CO16 8DB
G4 WIA Ivan Whitmore, Sunny Bank, Commercial Road, Helston, TR12 6LY
G4 WIG P Lees, 107 Balmoral Road, Wordsley, Stourbridge, DY8 5JJ
G4 WIL J Wilkinson, 147 Alder Lane, Hindley Green, Wigan, WN2 4ET
G4 WIM Timothy Forrester, Dow Brook House, Brades Lane Freckleton, Preston, PR4 1HG
G4 WIP Alan Crickett, 40 Ousden Close Cheshunt, Waltham Cross, EN8 9RQ
G4 WIR I Page, 127 Whyke Lane, Chichester, PO19 8AU
G4 WIS Valerie Gleek, Fieldgate, The Warren, Radlett, WD7 7DU
G4 WIY A Clark, Applecroft Care Home, Sanctuary Close, Chilton Way, Dover, CT17 0ER
G4 WIZ D Burleigh, 39 Neville Close, Basingstoke, RG21 3HG
GM4 WJA J Fraser, Cherrybrae Croft, Aultmore, Keith, AB55 6QU
G4 WJB R Barratt, 37 Cemetery Road, Whittlesey, Peterborough, PE7 1RT
G4 WJE M Fenelon, 72 Fieldside, Epworth, Doncaster, DN9 1DP
G4 WJG David Nicolson, 142 Shireburn Caravan Park, Edisford Road, Clitheroe, BB7 3LB
G4 WJH Paul Mathews, 1 Erith Road, Belvedere, DA17 6HB
G4 WJJ Peter Short, 60 Town Park, Crediton, EX17 3JN
G4 WJM W Cooper, 32 High St., Thurlby, Bourne, PE10 0EE
G4 WJO Ronald Thomas, 6 Ty Mawr Estate, Holyhead, LL65 2DN
G4 WJQ Bruce Blain, 31 Crest Road, Bromley, BR2 7JA
G4 WJR J Singleton, 5 Cavan Drive, Haydock, St. Helens, WA11 0GN
G4 WJS W Somerville, Glendella, Wycombe Road, High Wycombe, HP14 3RP
G4 WJV John Forrest, 3 Martindale Park, Houghton le Spring, DH5 8EX
G4 WJW T Murphy, 7 The Knapp, Templecombe, BA8 0JP
G4 WJX Martin Kessel, 4 Harington Drive, Stoke-on-Trent, ST3 5ST
G4 WJZ A Kerr, Braemoray, Dalditch Lane, Budleigh Salterton, EX9 7AS
G4 WKB H Poulton, 1 Marnhull Close, Coventry, CV2 2JS
G4 WKD K Dunstan, 41 Gravel Lane, Wilmslow, SK9 6LS
G4 WKG C Dawley, Glassonby Lodge, Glassonby, Penrith, CA10 1DT
G4 WKQ Lorna Jones, 8 Tyn y Pwll Estate, Llanbedrog, Pwllheli, LL53 7PG
G4 WKT Norman Bleek, 49 Lowhills Road, Peterlee, SR8 2DJ
G4 WKW Richard Benton, The Bungalow, Hugus Road, Threemilestone, Truro, TR3 6DY
G4 WKY Nigel Lee, 68 Andlers Ash Road, Liss, GU33 7LR
G4 WKZ G Fitch, Tides Reach, 39 Hen Gel Llechi, Y Felinheli, LL56 4PD
G4 WLA D Dell, Bushmead, 12 Penfield Gardens, Dawlish, EX7 9NQ
G4 WLE N Slater, 17 Hall Park Drive, Lytham St. Annes, FY8 4QR
G4 WLG K Dunwell, 8 Violet Grove, Thatcham, RG18 4DQ
G4 WLI P Nutt, 40 Parkfield Drive, Middleton, Manchester, M24 4ED
G4 WLJ N Bell, 16 Amersham Close, Urmston, Manchester, M41 7WH
G4 WLK M Morgan, 6 Blakeley Heath Drive, Wombourne, Wolverhampton, WV5 0HW
G4 WLP S McCombe, Willow Dene, Lower Broad Oak Road, Ottery St. Mary, EX11 1XH
G4 WLS C Smith, 15 Bearsdown Close, Plymouth, PL6 5TX
G4 WLT J Williams, 7 Tynewydd, Nantybwch, Tredegar, NP22 3SG
G4 WLV D Gladwin, Dorset House, St. Annes Road, Eastbourne, BN21 2HR
G4 WMA P Haslam, 55 Jesmond Park West, Newcastle upon Tyne, NE7 7BX
G4 WMB W Bell, 33 King Harry Lane, St. Albans, AL3 4AS
GW4 WMD William David, Sirmione, Lawrenny Road, Kilgetty, SA68 0SY
GI4 WME F Hull, 44 Killynether Walk, Belfast, BT8 7DB

**IMPORTANT NOTE**

**Revalidate licence to avoid revocation** – Ofcom has advised the Society that plans will be drawn up to revoke licences that have not been revalidated as required by the licence conditions. The quickest way to revalidate is to do so online via the Ofcom website: *https://services.ofcom.org.uk/* or by email: *amateur.validations@ofcom.org.uk* Ofcom staff are available to help, but please be patient during times of heavy workload.

G4 WMF G Blake, Flat 5, 46 Marlborough Road, Ipswich, IP4 5AX
GU4 WMG J Gallienne, Westward, Rue Des Marettes, St. Martin, Guernsey, GY4 6JW
G4 WMH Warwick Hall, 45 Dorchester Road, Solihull, B91 1LN
GW4 WMK D Turner, 20 Long Mains, Monkton, Pembroke, SA71 4NB
GM4 WMM W McMillan, Rennabreck, Rendall, Orkney, KW17 2EZ
G4 WMN John Robb, 3 Silver Dell, Watford, WD24 5LT
G4 WMO P Stainton, Fairview, Mareham on the Hill, Horncastle, LN9 6PQ
G4 WMP Melvyn Bangle, 21 Oakhill Road, Addlestone, KT15 1DH
G4 WMQ A Richardson, 1 Silverton Terrace, Rothbury, Morpeth, NE65 7QS
G4 WMV R Bridge, 11 Wheatfield Close, Bredbury, Stockport, SK6 1EW
G4 WMY G Kay, High Trees, Stockland Bristol, Bridgwater, TA5 2PZ
G4 WMZ K Law, 50 Main Street, Little Downham, CB6 2ST
G4 WNA H Williams, 37 Mickledales Drive, Marske-by-the-Sea, Redcar, TS11 6DF
G4 WND R Banks, Dingle Cottage, Church Stoke, Montgomery, SY15 6TJ
G4 WNF F Rhodes, 248 Woolwich Road, London, SE2 0DW
G4 WNG T Furness, 129 North Ridge, Bedlington, NE22 6DF
GI4 WNH E Loughran, 6 Oaklea Road, Magherafelt, BT45 6NH
G4 WNI J Howarth, 80 John F Kennedy Estate, Washington, NE38 7AL
G4 WNP R Tant, 34 Manor Road, Wheathampstead, St. Albans, AL4 8JD
GM4 WNQ P Ramsey, 1 Skye Place, Stevenston, KA20 3DG
G4 WNU John Smith, George Bungalow, The Street, Axminster, EX13 7RW
G4 WNV S Robinson, 18 Headley Grove, Tadworth, KT20 5JF
G4 WNW T Almond, Maranatha, Lumber Lane, Warrington, WA5 4AX
G4 WNZ Mark Watson, 17 Hatherton Road, Shanklin, PO37 7NA
G4 WOB Joe Bazyk, Heyford Cedars, Watling Street, Northampton, NN7 4SB
G4 WOD J Sheppard, 37 Oakfield Road, Kingswood, Bristol, BS15 8NT
G4 WOE David Hudson, 2 Muirfield Rise, St. Leonards-on-Sea, TN38 0XL
G4 WOH P Thwaytes, 1 Sunningdale, Waltham, Grimsby, DN37 0UA
G4 WOI Roger Allen, 115 Trerice Drive, Newquay, TR7 2TE
G4 WOL R Tenwolde, 376 Buxton Road, Macclesfield, SK11 7ES
G4 WOQ Anthony Leach, Wendover, Park Road, Carlisle, CA4 8AT
G4 WOS D Flello, 1 St. Andrews Way, Tilmanstone, Deal, CT14 0JH
GW4 WOV Kevin Williams, 16 Eiddwen Road, Penlan, Swansea, SA5 7EN
GD4 WOW J Jones, Ballagarrow, Glen Auldyn, Ramsey, Isle of Man, IM7 2AF
GW4 WPA T Leary, 21 Gelli Glas Road, Morriston, Swansea, SA6 7PS
G4 WPB Peter Bruce, Seascape, Surf Crescent, Sheerness, ME12 4JU
G4 WPE M Bland, 18 Hill Street, Newhall, Swadlincote, DE11 0JR
G4 WPG L Hatton, 51 Castner Avenue, Weston Point, Runcorn, WA7 4EH
G4 WPH Stephen Valentine, Unit 21, Industrial Estate, Bala, LL23 7NL
G4 WPI J Fuller, 42 Kitchener Road, Amesbury, Salisbury, SP4 7AD
G4 WPO D Bevan, 32 Thorley Park Road, Bishop's Stortford, CM23 3NQ
G4 WPR Dominic Trotman, 71 Bexley Street, Windsor, SL4 5BX
G4 WPT David Jackman, 41 Wannock Lane, Eastbourne, BN20 9SD
G4 WPW Roy Frettsome, 7 Wheatfield Crescent, Mansfield Woodhouse, Mansfield, NG19 9HH
G4 WQB Keith Hamlyn, 1 Elm Tree Cottages The Common, Winchmore Hill, Amersham, HP7 0PN
GW4 WQC D Williams, 149 Rhiwr Ddar, Taffs Well, Cardiff, CF4 7PD
G4 WQD J Jocys, 28 Vaudrey Drive, Timperley, Altrincham, WA15 6HQ
GM4 WQH John Naughton, 124 Churchill Street, Alloa, FK10 2JU
G4 WQL Michael Bender, Ivy Chimney Villa, Skinners Bottom, Redruth, TR16 5DT
G4 WQO P Truitt, 2a Queens Gate Place, London, SW7 5NS
G4 WQS N Reading, 30 Clifton Rise, Windsor, SL4 5TD
G4 WQT Rita Watson, 158 Kingsfold Drive, Penwortham, Preston, PR1 9EQ
G4 WQU P Barrett, 9 Mabena Close, St. Mabyn, Bodmin, PL30 3BS
G4 WQZ John Wiles, 12a Ashling Gardens, Denmead, Waterlooville, PO7 6PR
G4 WRA WORDSLEY ARC c/o Susan Sands, 33 High Street, Kinver, Stourbridge, DY7 6HF
G4 WRB Keith Beech, 40 Star Street, Wolverhampton, WV3 9BL
G4 WRC WANLIP RAD CLUB c/o A Wheeler, 11 Barley Way, Rothley, Leicester, LE7 7RL
G4 WRD Nigel Underwood, 44 East View, Barnet, EN5 5TN
GI4 WRJ R Jennings, 12 Garnerville Gardens, Belfast, BT4 2PA
G4 WRK I Edwards, 9 Long Lane, Wellington, TF6 6HH
GU4 WRP D Fletcher, Celicia, 5 La Neuve Rue Estate, St. Peter Port, Guernsey, GY1 1SF
G4 WRQ D Wring, 8a Rectory Road, Easton-in-Gordano, Bristol, BS20 0QB
GJ4 WRR F Leighton, 4 Victoria Village, Estate Trinity, Jersey, JE4 9VI
G4 WRX D Cherrington, 4 Bloomfield Close, Wombourne, Wolverhampton, WV5 8HQ
G4 WSB Arthur Bowditch, 28 Selby Crescent, Freshbrook, Swindon, SN5 8PE
G4 WSE Thomas Saggerson, 18 Ploughmans Way, Great Sutton, Ellesmere Port, CH66 2YJ
G4 WSF R Smith, 37 Lyngford Road, Taunton, TA2 7EF
G4 WSH FRANCIS BLAXLAND, Wenman, The Lizard, Helston, TR12 7NZ
G4 WSI A Brown, 81 Ipswich Crescent, Great Barr, Birmingham, B42 1LY
G4 WSL Roy Cable, 4a Ermine Close, St. Albans, AL3 4JZ
G4 WSM WESTON SUPER MARE RS c/o Alastair Nussey, 9 Brent Street, Brent Knoll, Highbridge, TA9 4DU
G4 WTA M Jones, 57 Mountway Road, Bishops Hull, Taunton, TA1 5DS
G4 WTD C Flatman, 36 Skoner Road, Bowthorpe Industrial Estate, Norwich, NR5 9AX
G4 WTE M Rye, 33a Darnley Street, Gravesend, DA11 0PH
GM4 WTK R Fortune, Stewarton Lodge, Eddleston, Peebles, EH45 8PP
GU4 WTN A Hamon, 22 Mount Row, St. Peter Port, Guernsey, GY1 1NT
G4 WTQ Neil Harvey, 5 Harvey Gardens, Loughton, IG10 2AD
GM4 WTS William Stevenson, 11 West Drive, Airdrie, ML6 8BL
GI4 WTT T McDonnell, 52 Moira Road, Glenavy, Crumlin, BT29 4JL
G4 WTU D Pay, Longmeadow House, Dunsford, Exeter, EX6 7AD
G4 WTX V Hansford, Whitehouse Farm, Hardway, Bruton, BA10 0RJ
G4 WTZ J Hall, 23 St. James Avenue, Congleton, CW12 4DY
G4 WUA Geoffrey Brown, 13 Francis Avenue, Moreton, Wirral, CH46 6DH
G4 WUB D Farr, 10 Yeomanside Close, Whitchurch, Bristol, BS14 0PZ
G4 WUG K Medley, 3 Beck Lane, Horsham St. Faith, Norwich, NR10 3LD
G4 WUH Ian Hopkins, 1 Beauchamp Villas, Kempley Green, Dymock, GL18 2BW
G4 WUI John Marr, 11 Morley Crescent, Kelloe, Durham, DH6 4NN
G4 WUJ N Plant, 73 Robert Burns Avenue, Cheltenham, GL51 6NX
G4 WUK D Dyer, 64 Churchill Close, Sturminster Marshall, Wimborne, BH21 4BH
G4 WUM Frank Amos, 53 Valley View, Jarrow, NE32 5QT
G4 WUO P Bullock, 4 Yarmouth Road, Blofield, Norwich, NR13 4JS
G4 WUQ P Harding, Flat D, 106 Bushey Hill Road, London, SE5 8QQ
GM4 WUR C Phillips, Lonnie, Sanday, Orkney, KW17 2BA
G4 WUS William Bingham, 67 Coronation Street, Carlin How, Saltburn-by-The-Sea, TS13 4DW
G4 WUU P Williamson, The Laurels, Norwich Road, Cawston, NR10 4HA
G4 WUV C Baker, 48 Hazell Road, Farnham, GU9 7BP
G4 WUW M Baker, 48 Hazell Road, Farnham, GU9 7BP
G4 WUX Philip Bourne, 6 Blythe Mount Park, Blythe Bridge, Stoke-on-Trent, ST11 9PP
G4 WVB J Williams, Arfryn, Windsor Road, Wrexham, LL14 1ST
G4 WVC M Jones, 24 Whitford Road, Birkenhead, CH42 7JA
G4 WVD M Bundy, 5 Dawe Crescent, Bodmin, PL31 1PY
G4 WVF C Farley, 8 Church Road, Mellor, Stockport, SK6 5PR
G4 WVH T Hathaway, 24 Oxford Meadow, Sible Hedingham, Halstead, CO9 3QN
G4 WVK David Davies, 20 Broadlands Way, Oswestry, SY11 2YD
G4 WVM M Keating, 13 Conway Road, Paignton, TQ4 5LF
GI4 WVN H Gilbody, 5 The Plateau, Piney Hills, Belfast, BT9 5QP
GW4 WVO Wenvoe ARC c/o Petrie Owen, 13 Highland Close, Sarn, Bridgend, CF32 9SB
G4 WVP F Russell, 56 Gatesgarth Road, Middleton, Manchester, M24 4JJ
G4 WVR WELLAND VALLEY ARS c/o S Day, 14 The Crescent, Market Harborough, LE16 7JJ
G4 WVT J Stageman, Sunray, Kennford, Exeter, EX6 7XS
GD4 WVW J Lane, 41 Ravenswood Crescent, Harrow, HA2 9JL
G4 WVY J Joynt, Gurtymadden, Loughrea, County Galway, Ireland
G4 WWB William Bennett, 61 L Mansion Drive, Liverpool, L11 9DP
GI4 WWF V Fails, 38 Fortsandel Avenue, Coleraine, BT52 1TL
G4 WWG A Brown, 44 Earlswood, Skelmersdale, WN8 6AT
G4 WWH Philip Pavelin, 7A Castletown, Portland, DT5 1BD
G4 WWL I Rowe, 19 Poplar Avenue, Wetherby, LS22 7RA
GW4 WWN Michael Rowles, 7 Gelli Deg, Bryncoch, Neath, SA10 7PL
G4 WWP D Barry, Plough End, 8 Dell Lane, Bishop's Stortford, CM22 7SJ
G4 WWR THREE COUNTIES ARC c/o D Kamm, Delabole Head, Week St. Mary, Holsworthy, EX22 6UU
GM4 WWU R Steel, 19 St. Brides Road, Newlands, Glasgow, G43 2DU
G4 WWY P Brown, White Cottage, Woodside, Epping, CM16 6LF
G4 WXC Steve Vaughan, Norman House, 58 Norman Avenue, Abingdon, OX14 2HL
G4 WXF Rikki Logan, 66 Friars Orchard, Gloucester, GL1 1GF
G4 WXG C Lees, 152 Birmingham Road, Redditch, B97 6EN
G4 WXI F McKeown, 1 Thirlmere Road, Preston, PR1 5TR
G4 WXJ Robert Harnett, 41 Stepney Road, Scarborough, YO12 5BT
G4 WXK G Sears, 36 Cedars Road, Exhall, Coventry, CV7 9NJ
G4 WXO John Pemberton, Dunkirk Cottage, Dunkirk Lane, Chester, CH1 6LU
GM4 WXQ William Goudie, 5 North Lochside, Lerwick, Shetland, ZE1 0PA
G4 WXR B Hayes, 363 Watnall Road, Hucknall, Nottingham, NG15 6EP
G4 WXT G Shead, 37 Shalford Road, Rayne, Braintree, CM77 6BY
G4 WXX J Charnock, 20 Clifton Road, Ashton-in-Makerfield, Wigan, WN4 0AZ
G4 WXZ J Roach, 63 Truro Close, St. Helens, WA11 9EL
G4 WYC S Powell, 10 Foresters Square, Bracknell, RG12 9ES
GI4 WYE P Doran, 143 Gransha Road, Bangor, BT19 7RB
G4 WYF C Ellison, 31 Dudley Avenue, Blackpool, FY2 0TU
G4 WYH A McPhail, 300 Fletcher Road, Preston, PR1 5HJ
G4 WYI A Huff, 4 Greding Walk, Hutton, Brentwood, CM13 2UF
G4 WYL C Levett, 5 Park Road, Yapton, Arundel, BN18 0JE
G4 WYN D Harries, 1 St. Michaels Close, Ashby-de-la-Zouch, LE65 1ES
G4 WYO K Brewer, 14 Poplar Road, Kensworth, Dunstable, LU6 3RS
G4 WYW M Fisher, 147 Outer Circle, Southampton, SO16 5HB
GW4 WYX W Thomas, 24 Pontneathvaughan Road, Glynneath, Neath, SA11 5NT
G4 WYZ M Prescott, Rathgael, 44 Glamis Drive, Chorley, PR7 1LX
G4 WZA A Nokes, 24 Braces Lane, Marlbrook, Bromsgrove, B60 1DY
G4 WZB H Worley, 22 Cross Road, Wellingborough, NN8 4AT
GM4 WZD J Nicholl, 7 Holmisdale, Glendale, Isle of Skye, IV55 8WS
GM4 WZG Bernard McIntosh, 14 River View, Dalgety Bay, Dunfermline, KY11 9YE
G4 WZH A Le Couteur Bisson, 36 Gibson Way, Porthleven, Helston, TR13 9AW
G4 WZI Richard Hilton, 19 Oxford Street, New Rossington, Doncaster, DN11 0TD
G4 WZJ M Wiblin, 60 Shepherds Lane, Bracknell, RG42 2BT
GM4 WZL John Scott, 5 Barnwood Gate, Galston, KA4 8NA
G4 WZM S Johnston, Burn Moor End Farm, Wheathead Lane, Nelson, BB9 6LD
G4 WZN B Turner, 15 Smardon Avenue, Brixham, TQ5 8JN
GM4 WZP J Gentles, Culra, 11 Corbiehill Avenue, Edinburgh, EH4 5DT
G4 WZQ I Smith, 24 Sea View Road, Herne Bay, CT6 6JA
G4 WZS A Glynn, Cartref, Llanfachraeth, Holyhead, LL65 4UY
G4 WZT Nigel Thomas, Flat 18, Livability, Hereford, HR4 9HP
G4 WZU L Thompson, 12 Long St., Great Gonerby, Grantham, NG31 8LN
G4 WZZ Brenda Hunt, 5 Osprey Close, Whitstable, CT5 4DT
GI4 XAA C D McCann, Drumbally Hue House, Rock, Tyrone, BT70 3JY
G4 XAB R Hunt, 3 Osprey Close, Whitstable, CT5 4DT
G4 XAE A Cole, 13 Centre Close, Beccles, NR34 9JJ
G4 XAG B Mahany, 3 Portland Road, Frome, BA11 4JA
G4 XAH D Mahany, 3 Portland Road, Frome, BA11 4JA
G4 XAL Peter Lawrence, 4 Monkshood Close, Wokingham, RG40 5YE
G4 XAN C Goddard, 75 Downs Road, Slough, SL3 7DA
GI4 XAP J Seaman, 109 Belvoir Drive, Belfast, BT8 7DN
G4 XAR Michael Kearns, 16 Fieldton Road, Liverpool, L11 9AG
G4 XAT R Evans, 7 Westland Drive, Hayes, Bromley, BR2 7HE
G4 XAU J Rutkowski, 7 Beach Road, Holyhead, LL65 1ES
GM4 XAV J Stevens, 10 Tiel Path, Glenrothes, KY7 5AX
G4 XAW P Nelson, Croc Ard, Botany Street, Newton Stewart, DG8 9JG
GW4 XAZ Ian Mitchell, 18-19 Hendre-Wen Road, Blaencwm, Treorchy, CF42 5DR
G4 XBC A Turner, 19 Trelawney Road, St. Austell, PL25 4JA
G4 XBD G Nash, 36 Lynton Avenue, Arlesey, SG15 6TS
G4 XBE John Easey, Bojangles, Hackmans Lane, Chelmsford, CM3 6RE
G4 XBF M Ray, Willow Mead House, Willow Mead, Godalming, GU8 5NR
G4 XBG Kevin Murphy, 34 Hawkenbury Way, Lewes, BN7 1LT
G4 XBI D Parslow, 1 Willington Close, Harlescott, Shrewsbury, SY1 3RH
G4 XBJ P Kemp, 9 Moorfield Way, Wilberfoss, York, YO41 5PL
G4 XBS C Smith, 1 Langley Court, St. Ives, PE27 5WX
G4 XBU J Atkinson, 8 Woodcock Road, Flamborough, Bridlington, YO15 1LJ
G4 XBW Robert Head, The White House, School Lane, St Austell, PL25 3TJ
G4 XBX A Wallman, 8 Oakbank Avenue, Manchester, M9 4EX
G4 XBZ A Roberts, Apartment 17, Fleur de Lis Duttons Road, Romsey, SO51 8LH
G4 XCB Ken Rook, 232 Wick Road, Brislington, Bristol, BS4 4HN
G4 XCE A Tamplin, Browtop, Old Lane, Crowborough, TN6 2AD
G4 XCK Stephen Boden, 14 Potters Way, Ilkeston, DE7 5EX
G4 XCM J Tavener, The Cube, North Drive, Wirral, L60 0BD
G4 XCQ L Prescott, 58 Blenheim Road, Ashton-in-Makerfield, Wigan, WN4 9JN
G4 XCR G Wood, The Old Corner Smithey, New Road, Dereham, NR20 5TA
G4 XCV R Barnett, 10 Boscaswell Terrace, Pendeen, Penzance, TR19 7DS
G4 XCX C Clarke, 33 James Road, Kidderminster, DY10 2TP
G4 XCY F Mills, 14 Seagram Close, Aintree, Liverpool, L9 0NA
G4 XDB A Parry, 189 Kimbolton Road, Bedford, MK41 8DR
G4 XDC M Taylor, 7 Malt Fallows, Crew Green, Shrewsbury, SY5 9AT
G4 XDE S Crosskey, 25 Meadow Gardens, Baddesley Ensor, Atherstone, CV9 2DA
G4 XDG D Humphreys, 129a Chester Road, Northwich, CW8 4AA
G4 XDJ Brian Fields, 64 Collins Street, Waikouaiti, New Zealand, 9510
G4 XDK Nigel Heasman, 66 Waterford Road, Ipswich, IP1 5NN
G4 XDL M Norman, 52 Turkdean Road, Cheltenham, GL51 6AL
G4 XDM S Comis, 178 Lordswood Road, Birmingham, B17 8QH
G4 XDP Peter Knowles, 18 Brookside, Pill, Bristol, BS20 0JX
G4 XDR A Walker, Stanley Cottage, Station Road, Wrexham, LL13 0LJ
G4 XDT Leslie Booth, 40 St. Georges Road, New Mills, High Peak, SK22 4JT
G4 XDU David Chislett, Hilltops, 2a St. Marks Road, Maidenhead, SL6 6DA
G4 XDV R Hall, 9 Stone Court, South Hiendley, Barnsley, S72 9DL
G4 XDW A Chidwick, 4 Burgess Close, Whitfield, Dover, CT16 3NP
G4 XDX S Garbett, 2 Redruth Court, Launceston Road, Wigston, LE18 2FU
GU4 XEA P Carre, La Petite Miellette, La Miellette Lane, Vale, Guernsey, GY3 5EN
G4 XED S Cowdell, 6 Pearl St., Bedminster, Bristol, BS3 3EA
G4 XEE Derek Bate, 15 Martins Drive, Ferndown, BH22 9SG
G4 XEF B Passmore, 16 Epworth Road, Rhyl, LL18 2NU
G4 XEI David Mcloughlin, 23 St. Marys Court, Clayton le Moors, Accrington, BB5 5LA
G4 XEJ A Allen, 86 Grayswood Park Road, Quinton, Birmingham, B32 1HE
G4 XEL S Evans, 72 Sandown Road, Toton, Nottingham, NG9 6JW
G4 XEO David Holmes, 32 Eastcheap, Rayleigh, SS6 9JZ
GW4 XES D Johns, 16 Maes yr Haf, Llansamlet, Swansea, SA7 9ST
G4 XET J Rawlinson, Hollydene, Newbiggin, Penrith, CA10 1TA
GD4 XEW I Rosenberg, 11 Parkside Drive, Edgware, HA8 8JU
G4 XEX Peter Rivers, 34 Coales Gardens, Market Harborough, LE16 7NY
G4 XEZ A Smith, 3 The Fold, Wolverhampton, WV4 5QY
G4 XFC John Fulton, 8 Park View, Legsby, Market Rasen, LN8 3QP
GI4 XFE A Calvin, 20 Orangefield Crescent, Armagh, BT60 1DS
G4 XFF J Holdsworth, 37 Harewood Crescent, Old Tupton, Chesterfield, S42 6HS
G4 XFG Nigel Goodman, Lindum, Holton Road, Grimsby, DN36 5LS
G4 XFM D Steer, 24 Manor Drive, Ivybridge, PL21 9BD
GI4 XFN G Smith, 4 Conway Court, Belfast, BT13 2DR
GI4 XFS C Gilbody, 5 The Plateau, Piney Hills, Belfast, BT9 5QP
G4 XFT J Tranter, 275 Bosty Lane, Aldridge, Walsall, WS9 0QE
GM4 XFU William Davidson, 31 Glenmuir Crescent, Logan, Cumnock, KA18 3EY
G4 XFV Andrew McKechnie, 2 Batts Pond Lane, Dropping Holms, Henfield, BN5 9YU
GI4 XFX R Reid, 6 Sperrin Heights, Townhill Road, Ballymena, BT44 8AD
GI4 XFY E Townley, 27 Windmill Road, Kilkeel, Newry, BT34 4LP
G4 XFZ R Griffin, 53 St. Johns Avenue, Warley, Brentwood, CM14 5DG
GU4 XGB Andrew Bichard, The Swallows, 17 Clos Du Murier, Guernsey, Guernsey, GY2 4HJ
G4 XGD P Harman, 25 Pitts Road, Slough, SL1 3XG
G4 XGI R Hales, 239 Charlton Road, Shepperton, TW17 0SH
G4 XGN P Riggott, 1 Mill Lane, Queensbury, Bradford, BD13 1LP
GI4 XGO George Armstrong, 45 Rathmena Drive, Ballyclare, BT39 9HZ
G4 XGP M Kelly, Birkby Lodge, Brickley Park Road, Bromley, BR1 2AT
GI4 XGQ T Devine, 141 Longland Road, Dunamanagh, Strabane, BT82 0PP
G4 XGR S Clark, 1 Holcroft, Orton Malborne, Peterborough, PE2 5SL
G4 XGT John Dawson, 21 Church Street, Needingworth, St. Ives, PE27 4TB
GM4 XGY Gerald Smith, 1/2 1 Seres Court, Clarkston, Glasgow, G76 7PL
G4 XHC F Jackson, 34 High Street, Blyton, Gainsborough, DN21 3JY
G4 XHE R Cook, 7 New Road, Worthing, BN13 3JG
G4 XHF P Chamberlain, 114 Grattons Drive, Crawley, RH10 3JP
GM4 XHH R franklin, 8 Hawdene, Broughton, Biggar, ML12 6FW
G4 XHK L Soutter, 2 Hyde Barton, Churchill Way, Bideford, EX39 1NX
GI4 XHO F Orr, 29a Mccraes Brae, Whitehead, Carrickfergus, BT38 9NZ
G4 XHP D Daniels, 40 Rennie Street, Dean Bank, Ferryhill, DL17 8NG
GM4 XHQ G Mcinnes, 14 East Croft, Ratho, Newbridge, EH28 8PD
G4 XHT E Wilkinson, 83 Gleneagles Road, Urmston, Manchester, M41 8SB
GM4 XHV G Horsburgh, 3 Dumyat Road, Alva, FK12 5NN
G4 XHX Mandy Powers, 16 Roman Avenue North, Stamford Bridge, York, YO41 1DP
G4 XHZ F Jolley, 30 Oban Drive, Shadsworth, Blackburn, BB1 2HY
G4 XIE R Shard, 76 Clipsley Lane, Haydock, St. Helens, WA11 0UB
G4 XIL B Hurst, 25 Hoadly Road, Cambridge, CB3 0HX
G4 XIM R Bradfield, 118 East Road, Langford, Biggleswade, SG18 9QP
G4 XIN A Henstock, 16 The Coppice, Enfield, EN2 7BY
G4 XIP J Baker, 11 London Road, Old Basing, Basingstoke, RG24 7JE
G4 XIQ J Lainchbury, 33 Ennersdale Road, Coleshill, Birmingham, B46 1EP
GI4 XIR W Bird, 198 Ashmount Gardens, Lisburn, BT27 5DB
GU4 XIT Richard Bird, Redroof, La Mare de Carteret, Castel, Guernsey, GY5 7XD
G4 XIU M Kelleway, Greenhills, Newport Road, Ventnor, PO38 2QW
G4 XIW S Mackenzie, 6 Bridge Farm Close, Grove, Wantage, OX12 7QF
G4 XIX M Purnell, The Olde Cottage, Lewdown, Okehampton, EX20 4DQ
G4 XIZ Roland Heath, 9 Woodside Lane, Leek, ST13 7AN
GI4 XJC John Colley, 243 Jordanstown Road, Newtownabbey, BT37 0LX
GI4 XJD J Doherty, 75 Drumflugh Road, Benburb, Dungannon, BT71 7QF
G4 XJE Douglas Brawn, 16 Mansel Close Cosgrove, Milton Keynes, MK19 7JQ
GM4 XJF J Park, The Stables, Whiting Bay, Isle of Arran, KA27 8QH
G4 XJG R Hirst, 47a Rowley Lane, Fenay Bridge, Huddersfield, HD8 0JG

GI4 XJJ W Mccaughey, 60 Ballynure Road, Newtownabbey, BT36 5SJ
GW4 XJK R King, 30 Railway Terrace, Llanelli, SA15 2RH
G4 XJL D Rogers, Gabwell House, Stokeinteignhead, Newton Abbot, TQ12 4QS
G4 XJN J Williamson, 12 Honeysuckle Road, Widmer End, High Wycombe, HP15 6BW
G4 XJS J Smith, 84 Oakwood Drive, St. Albans, AL4 0XA
GM4 XJY D McMinn, Crestholme, East Bay, Mallaig, PH41 4QF
G4 XKA P Adams, 19 Thistledown, Tilehurst, Reading, RG31 5WE
G4 XKC A Sieroslawski, 8 Poot Hall, Dewhirst Road, Rochdale, OL12 0AS
G4 XKD K Dixon, 23 Dorking Walk, Corby, NN18 9JJ
GW4 XKE Dennis Egan, 19 Sycamore Close, Dinas Powys, CF64 4TQ
G4 XKF Derick Browne, 67 Benfield Way, Portslade, Brighton, BN41 2DN
GI4 XKI J O'Neill, 225 Dungannon Road, Killeshill, Dungannon, BT70 1TH
G4 XKK K Hurston, Lan Singala, Hope Corner Lane, Taunton, TA2 7PR
G4 XKL R Whetton, 117 Tutbury Road, Burton-on-Trent, DE13 0NU
G4 XKM Paul Andrews, 88 Connegar Leys, Blisworth, Northampton, NN7 3DF
GM4 XKP K Macgillivray, 87 Castle Street, Forfar, DD8 3AG
G4 XKR Russell Coward, 10 Market Street, Hambleton, Poulton-le-Fylde, FY6 9AP
G4 XKV G Pigott, 67 Mayplace Road West, Bexleyheath, DA7 4JL
G4 XKZ Melvyn Collins, 30 Denver Road, Dartford, DA1 3JU
G4 XLA I Carruthers, Flat 43, House 119, St. Petersberg, Russian Federation, 193024
GI4 XLB SUNSPOTS RAC c/o Gordon Curry, 87 Burren Road, Ballynahinch, BT24 8LF
G4 XLC E Metcalfe, 18 Kirkstone Drive, Morecambe, LA4 5XP
G4 XLG M Rollings, 39 Summerleys, Edlesborough, Dunstable, LU6 2HR
G4 XLM J Todd, Dorothy House, 127 Dorothy Avenue North, Peacehaven, BN10 8DS
GM4 XLN J Durrand, 9 Breadalbane Crescent, Wick, KW1 5AS
G4 XLO Kevin Tatlow, 54 Moorpark Road, Birmingham, B31 4HD
GM4 XLU E Wallace, 10 Gean Court, Cumbernauld, Glasgow, G67 3LU
G4 XLY E Grint, 15 Ivythorn Road, Street, BA16 0TE
G4 XMA B Easton, 8 Church View Road, Camborne, TR14 8RQ
GM4 XMD William McDicken, 4 Baillie Drive, Logan, Cumnock, KA18 3HS
G4 XME L Shone, 3 Ascot Drive, Dudley, DY1 2SN
G4 XMJ Geoff Wiggins, Cherry Trees Thorney Road, Emsworth, PO10 8BN
G4 XML P Glydon, 24 Imperial Road, Knowle, Bristol, BS14 9ED
G4 XMO S Ashfield, 28 Long Grove, Baughurst, Tadley, RG26 5NY
G4 XMP C Balderston, 57 Puttenham Road, Chineham, Basingstoke, RG24 8RB
G4 XMQ Terrance Cooling, 17 Hawthorn Avenue Cherry Willingham, Lincoln, LN3 4JS
G4 XMR Mark Richardson, 1 Cross Tree Crescent, Kempsford, Fairford, GL7 4EX
G4 XMS C Munton, 86 Amsbury Road, Hunton, Maidstone, ME15 0QH
GW4 XMU D Jones, 30 Sorrell Drive, Penpedairheol, Hengoed, CF82 8LA
GW4 XMV D Palmer, Hazelgrove, Mwtswr Lane, Cardigan, SA43 3HZ
G4 XMX Leslie Johnson, 6 Hurst Court, Bunbury, Tarporley, CW6 9QX
G4 XMY J Colson, 718 East Buckingham Drive, Lecanto, United States, 34461
G4 XMZ Peter Toms, Longfleet, Shipton Lane, Bridport, DT6 4NQ
G4 XNA A Willis, 20 Oakenbrow, Sway, Lymington, SO41 6DY
GM4 XND W Clark, 173 Dunnikier Road, Kirkcaldy, KY2 5AD
G4 XNE F Handy, 429 Penn Road, Penn, Wolverhampton, WV4 5LN
G4 XNF J Cameron, 23 Farley Crescent, Oakworth, Keighley, BD22 7SH
G4 XNK H Johnson, 74 Kirkham Gardens, Bromyard, HR7 4EA
G4 XNO Michael Goodearl, Glenhurst, Wood Lane, Dartmouth, TQ6 0DP
G4 XNP D Rayner, 69 Saracen Road, Hellesdon, Norwich, NR6 6PB
GM4 XNQ D Muir, 28 Grange Crescent, Edinburgh, EH9 2EH
G4 XNR P Morris, The Overlands, Church Minshull, Nantwich, CW5 6DX
G4 XNS N Speak, Le Bois Trainard, Lizant, France, 86400
G4 XNV D Owen, 39 Smithford Walk, Tarbock Green, Prescot, L35 1SF
G4 XNW J Simmonds, 19 Red Admiral Apartments, Worcester Street, Stourbridge, DY8 1AJ
GD4 XOD W Jones, Ballanard Road, Onchan, Douglas, Isle of Man, IM4 5EA
G4 XOE Martyn Swaby, 16 Daimler Avenue, Herne Bay, CT6 8AE
G4 XOG C Wood, 9 Tamar Close, Walsall, WS8 7LH
G4 XOH D Blackwell, 10 High Oaks Gardens, Bournemouth, BH11 9LJ
GM4 XOI A McGill, 37 Barlae Avenue, Eaglesham, Glasgow, G76 0DA
G4 XOJ N Wade, 6 Aisthorpe, Capel St. Mary, Ipswich, IP9 2HT
G4 XOL Mark Osborne, 27 Silverdale Road, Newton-le-Willows, WA12 0JT
G4 XOM R Egan, 56 Walker Avenue, Stourbridge, DY9 9EL
G4 XOP T Cooper, 55 Meadway, St. Austell, PL25 4HT
G4 XOU Ronald Hague, 37 Vernon Drive, Nuthall, Nottingham, NG16 1AR
G4 XOW D Lomas, Galmpton, Cannon Lane, Maidenhead, SL6 3NR
G4 XPI Peter O'Dea, 5 Matthews Court, Blackpool, FY4 2BT
G4 XPJ A Gridley, 13 Brockwell, Oakley, Bedford, MK43 7TD
G4 XPP J Davies-Bolton, 39 Newholme Estate, Station Road, Wingate, TS28 5EJ
G4 XPT A Fernandez, 2 Silverston Ave, Bognor Regis, PO21 2RB
G4 XPU M Bennett, 7 Woburn Avenue, Firwood Ind Est, Bolton, BL2 3AY
G4 XPV P Maisey, 155 Parkfield Drive, Birmingham, B36 9TY
G4 XPY David Fuller, 51 Evenlode, Banbury, OX16 1PQ
G4 XQA Kenneth James, 6 Holly Grove, Paddington, Warrington, WA1 3HB
G4 XQB T Rowe, 1 Mere Road, Marston, Northwich, CW9 6DR
G4 XQD D Cast, 100 Priory Court, Priory Park, Ipswich, IP10 0JX
G4 XQE A Turner, 4 Taylor Road, Ashtead, KT21 2HY
G4 XQF N Clacher, 3 Annan Crescent, Marton, Blackpool, FY4 4RQ
G4 XQG K Icke, 15 Shannon Way, Evesham, WR11 3FF
G4 XQH J Davies, 109 Croesonnen Parc, Abergavenny, NP7 6PF
GM4 XQJ Brian Waddell, 3a Polmont Road, Laurieston, Falkirk, FK2 9QQ
G4 XQQ J Freeman, 5a Beech Avenue, Briar Bank Park, Bedford, MK45 3WE
G4 XQV T Sismey, Southland House, 1b West End Lane, Leeds, LS18 5JP
G4 XQW Robert Stacey, 2 Valley Way, Fakenham, NR21 8PH
G4 XQX D Oliver, 6 Kensington Road, Gosport, PO12 1QY
G4 XQZ J Fisher, 6 Castle Way, Havant, PO9 2RZ
G4 XRA R Avery, 64 Burnmill Road, Market Harborough, LE16 7JF
G4 XRB J Gagg, 20 Stanstead Avenue, Tollerton, Nottingham, NG12 4EA
G4 XRD G Pope, 16 Catchpole Close, Corby, NN18 8DE
G4 XRG E Godlieb, 4 Tytherington Park Road, Macclesfield, SK10 2EL
G4 XRJ Jean Mills, Aquila, 4 Westhill Road South, Winchester, SO21 3HP
G4 XRK J Lord, 16 Lark Valley Drive, Fornham St. Martin, Bury St. Edmunds, IP28 6UG
G4 XRM B Foster, 5 Jacobs Close, Glastonbury, BA6 8EJ
G4 XRO S Hill, 32 Hunters Croft, Haxey, Doncaster, DN9 2NX
GM4 XRP J Porter, 1 Loney Crescent, Denny, FK6 5EG
G4 XRR M Willgoss, 9b Elwell Manor Gardens, Weymouth, DT4 8HJ
GM4 XRT Mark Taylor, The Old Croft, Forse, Lybster, KW3 6BX
G4 XRV Rupert Bullock, Putnams, Hawridge, Chesham, HP5 2UQ
G4 XRW R Wood, Bwlcyn, Eifl Road, Caernarfon, LL54 5HG
G4 XRX Roger Headland, 18 Blucher Street, Liverpool, L22 8QB
GM4 XRY Alastair Rimmer, 16 Johnston Drive, Barassie, Troon, KA10 6SD
G4 XSA A Boniface, 33 Caraway Place, Wallington, SM6 7AG
G4 XSD W Bridgen, 11 Furness Grove, Thurnscoe, Rotherham, S63 0TY
G4 XSC Geoffrey Trim, 731 Dorchester Road, Weymouth, DT3 5LF
GI4 XSF Michael Stevenson, 69 Portaferry Road, Cloughey, Newtownards, BT22 1HP
G4 XSG Stanley Stuart, 102 Mitton Road, Whalley, Clitheroe, BB7 9JN
G4 XSI J Saunders, Top Hill Farm, Woodside Green, Maidstone, ME17 2ET
G4 XSM G Davey, 49 Maltward Avenue, Bury St. Edmunds, IP33 3XQ
G4 XST Paul Cheeseman, 10 Limden Close, Stonegate, Wadhurst, TN5 7EG
GW4 XSX M Tovey, Frondewi, Aberaeron, SA46 0JG
G4 XTA Paul Godolphin, Shepherds Cottage, Flakebridge, Appleby in Westmorland, CA16 6JZ
GI4 XTC W Armstrong, 8 Killowen Crescent, Lisburn, BT28 3DS
G4 XTE J Johnson, Winterwood, West Lodge Crescent, Huddersfield, HD2 2EH
G4 XTF N Hancocks, Wesley House, Allensmore, Hereford, HR2 9BE
G4 XTG I Brown, 453 Blackburn Road, Turton, Bolton, BL7 0PW
G4 XTK A Kurnatowski, 24 Eversleigh Rise, Darley Bridge, Matlock, DE4 2JW
G4 XTO W Reade, 106 Wellington Road, Bollington, Macclesfield, SK10 5HT
G4 XTR N Hearn, Horsebrook Farm, South Brent, TQ10 9EU
G4 XTS John Strutt, Woodland, Gardiners Lane North, Billericay, CM11 2XE
GD4 XTT W Brown, Cleckheaton, Ballaragh, Laxey, Isle of Man, IM4 7PW
G4 XTU J Jones, 3 Blackstope Lane, Retford, DN22 6NW
G4 XTW A Bowes, 3 Cameron Mews, Mill Street, Bury Street Edmunds, IP28 7DP
G4 XTX Chris Cooper, 31 Beacon Park Drive, Skegness, PE25 1HE
G4 XTZ Alan Taylor, 36 Bodmin Avenue, Slough, SL2 1SL
G4 XUA R Walton, 275 Ridgacre Road, Quinton, Birmingham, B32 1EG
GW4 XUE D Thomas, 5 Parcydelyn, Carmarthen, SA31 1TS
G4 XUG G Patterson, Jurys, Fore Street, South Molton, EX36 3HL
G4 XUI John Gordon, 54 Guibal Road, London, SE12 9LX
G4 XUJ K Traill, 57 Ashfield Drive, Dumfries, DG2 9BP
G4 XUM Martin Platt, 10 Chesterton Way, Weston, Crewe, CW2 5NZ
G4 XUQ S Winters, 16 Rushton Grove, Harlow, CM17 9PR
G4 XUR D Smith, 47 Laburnum Street, Taunton, TA1 1LB
GM4 XUS G Smith, 80 Deanburn Park, Linlithgow, EH49 6HA
G4 XUV D Bevan, 46 Park Lane, Hartford, Northwich, CW8 1PY
G4 XUW David Hudson, 54 Montfitchet Walk, Stevenage, SG2 7DT
G4 XUZ Ray Chandler, 43 The Drive, Shoreham-by-Sea, BN43 5GD
G4 XVE J Francis, Pintail Cottage, St. Helena, Saxmundham, IP17 3ED
G4 XVF A Henk, 3 Well Road, Tweedmouth, Berwick-upon-Tweed, TD15 2BB
G4 XVH D Van Haaren, 7 Middle Boy, Abridge, Romford, RM4 1DT
G4 XVI Janet Ames, 58 The Street, Ringland, Norwich, NR8 6AB
G4 XVM Michael Brett, 25a First Avenue, Galley Hill, Waltham Abbey, EN9 2AL
G4 XVO Stephen Bate, High Trees, Much Hadham, SG10 6AX
G4 XVP P Hart, 4 Kings Ride, Penn, High Wycombe, HP10 8BL
G4 XVR B Hughes, Annies Cottage, Gravel Walk, Malpas, SY14 8JQ
G4 XVS K Hughes, Annies Cottage, Gravel Walk, Malpas, SY14 8JQ
G4 XVV E Davies, 11 Herons Close, Fareham, PO14 2HA
G4 XVW Ian Dobson, Pine View, Forest Dale Road, Marlborough, SN8 2AS
G4 XVY D Bastin, 94 Clyfton Close, Broxbourne, EN10 6NY
G4 XWA Malcolm Cohen, 7 Northdale Park, Swanland, North Ferriby, HU14 3RH
GW4 XWC W Crooks, 52 St. Catherines Road, Baglan, Port Talbot, SA12 8AS
G4 XWD Jmes Cookson, 24 St. Johns Avenue, Kidderminster, DY11 6AU
G4 XWE L Perrett, 1 Churchill Close, Wells, BA5 3HY
GD4 XWF J Harrison, 33 Tynwald Close, St. Johns, Douglas, Isle of Man, IM4 3LZ
GM4 XWL S Gaw, 10 Scotstoun Park, South Queensferry, EH30 9PQ
G4 XWM F Walton, 36 Cranfield Road, Wavendon, Milton Keynes, MK17 8AS
G4 XWN H Walker, 46 Golden Grove, Rhyl, LL18 2RS
G4 XWP C Boyce, 41 Furlong Close, Buckfast, Buckfastleigh, TQ11 0ER
G4 XWQ D Cottle, The Brambles, Landkey Road, Barnstaple, EX32 9BW
G4 XWR P Grainger, 26 Beattie Street, South Shields, NE34 0NJ
GM4 XWS D Munro, Eriskay, 4 Boswell Crescent, Inverness, IV2 3ET
G4 XWT F Donachie, 57 Avon Road West, Christchurch, BH23 2DF
G4 XWW G Winyard, 76 West Elloe Avenue, Spalding, PE11 2BJ
G4 XWZ D Lerner, 6 Willow Road, Kings Stanley, Stonehouse, GL10 3HS
G4 XXA F Mills, 66 Beeches Road, Charlton Kings, Cheltenham, GL53 8NQ
G4 XXB E Mills, 66 Beeches Road, Charlton Kings, Cheltenham, GL53 8NQ
G4 XXD S White, 15 Spurway Road, Canal Hill, Tiverton, EX16 4ER
G4 XXF B Morris, 62 Gerllan, Tywyn, LL36 9DE
G4 XXG STOCKTON&DIS AR c/o D Burton, 49 Aske Road, Redcar, TS10 2BP
G4 XXH Raymond Miles, Lone Oak, Clappers Farm Road, Reading, RG7 2LH
G4 XXI G Lee, 5 Morton, Tadworth, KT20 5UA
G4 XXJ Nicholas Jones, Hillesley, Montpellier Park, Llandrindod Wells, LD1 5LW
G4 XXK David Hart, 52 Scalwell Lane, Seaton, EX12 2DJ
G4 XXM David Frederick, 16 Phoenix Drive, Eastbourne, BN23 5PG
GM4 XXO Ian Carbry, 24 Craigenhill Road, Kilncadzow, Carluke, ML8 4QT
G4 XXP J Bancroft, 101 Meliden Road, Prestatyn, LL19 8LU
G4 XXS G Cooper, 44 Nursery Close, Hucknall, Nottingham, NG15 6DQ
G4 XXT J Cassidy, 137 Heath Park Road, Gidea Park, Romford, RM2 5XJ
G4 XXW J Groeger, Waldweg 11, Schneverdingen, Germany, D-29640
G4 XXX F Pullen, 35 Berrycroft, Willingham, Cambridge, CB24 5JX
G4 XXZ D Palfreman, 43 Southfield Close, Scraptoft, Leicester, LE7 9UR
G4 XYB M Kingdon, Watersmeet Cottage, Brewers Lane, Calne, SN11 8EZ
G4 XYC A Reynolds, 90 Windfield, Leatherhead, KT22 8UJ
G4 XYD R YOUNG, 5 Edge Hill, Chellaston, Derby, DE73 6RP
G4 XYG Stephen Randall, 129 Ryeland Way, Andover, SP11 6RH
G4 XYH W Stock, The Cottage, Hollow Road, Winscombe, BS25 1TG
G4 XYI P Coombs, 28 Cae Braenar, Holyhead, LL65 2PN
G4 XYK Peter Mitchell, 19 Ashbourne Avenue, Whetstone, London, N20 0AL
G4 XYM J Hoskins, 37 Green Close, Didcot, OX11 8TE
G4 XYN R Savin, 7 Bannard Road, Maidenhead, SL6 4NG
G4 XYP Robert Jobes, 11 Eglinton Street North, Sunderland, SR5 1DY
G4 XYR W Clarkson, 53 Acre Drive, Eccleshill, Bradford, BD2 2LS
G4 XYS J Mundy, 19 Brickfield Grove, Halifax, HX2 9AZ
G4 XYW Andrew Pevy, 2 Oaktree Way, Sandhurst, GU47 8QS
G4 XYY J England, 2 Clifford Road, Bramham, Wetherby, LS23 6RN
G4 XZA E Wardle, 57 Brook View Drive, Keyworth, Nottingham, NG12 5RA
G4 XZC P Gass, 7 Chipperfield Close, Upminster, RM14 3EA
G4 XZF B Irwin, Rockside, Frog Lane, Braunton, EX33 1BB
G4 XZG P Lawton, 5 Belvedere Gardens, Leeds, LS17 8BS
G4 XZI G Hall, 22 Templenewsam View, Leeds, LS15 0LW
G4 XZJ H Jones, Hafan, 7 Tan y Bryn Street, Tywyn, LL36 9UY
G4 XZK Stephen Bond, 12 Richmond Road, Farsley, Pudsey, LS28 5DY
G4 XZM K Pickles, 79 Mill Lane, Hanging Heaton, Batley, WF17 6DZ
GM4 XZN J MacDonald, 16 Muir Wood Drive, Currie, EH14 5EZ
G4 XZP E Wood, Bwlcyn, Eifl Road, Caernarfon, LL54 5HG
G4 XZS R Wooton, 2 Gill Park, Efford, Plymouth, PL3 6LX
GM4 YAA J Sinclair, 3 Ben More Drive, Paisley, PA2 7NU
G4 YAA Stephen Barnwell, 4 Railway Cottages, Reforne, Portland, DT5 2AP
G4 YAB J Livesley, 79 Mellor Road, New Mills, High Peak, SK22 4DP
G4 YAC N Howarth, 2 Sudeley Grove, Hardwick, Cambridge, CB23 7XG
G4 YAF A Trudgen, 14 Park An Pyth, Pendeen, Penzance, TR19 7ET
G4 YAH G Warnes, 20 Clivedon Way, Halesowen, B62 8TB
G4 YAJ S Woodhead, 804 Huddersfield Road, Dewsbury, WF13 3LZ
G4 YAK C Dobinson, 37 Ladram Road, Thorpe Bay, Southend-on-Sea, SS1 3PX
G4 YAL S Tuffin, 21 Garraways, Wootton Bassett, Swindon, SN4 8NQ
G4 YAM C Atkin, 23 Brewster Avenue, Immingham, DN40 1DW
G4 YAN R Page, 26 Colne Road, High Wycombe, HP13 7XN
G4 YAP G South, 76 Lilac Crescent, Hoyland, Barnsley, S74 9PW
G4 YAQ B Setter, Briarwood, Alexandra Road, Crediton, EX17 2DH
G4 YAR P Read, 45 Beaconsfield Road, Epsom, KT18 6HY
G4 YAS E Lucas-Davis, 27 Cadbury Road, Sunbury-on-Thames, TW16 7NA
GM4 YAT Thomas Turner, 5 Braemore Place, Fort William, PH33 6HX
GM4 YAU W Scott, Garden House, Fetternear, Inverurie, AB51 5LY
G4 YAV R Dixon, 91 Bondicar Terrace, Blyth, NE24 2JR
G4 YAW Jim Lawton, Cathedral View, Aberystwyth, SY23 1HH
G4 YAX D Diss, 130 Beridge Road, Halstead, CO9 1JU
G4 YAZ H Sher, Sea Echo, 53 Leonard Road, New Romney, TN28 8RX
G4 YBA G Collis, 13 Westbrook Close, Horsforth, Leeds, LS18 5RQ
G4 YBB M Coombs, Rose Cottage, Horns Cross, Bideford, EX39 5DJ
G4 YBD Graham Reed, 6 Tentergate Close, Knaresborough, HG5 9BJ
G4 YBG A White, Northdale, Goughs Lane, Bracknell, RG12 2RA
G4 YBH Brian Hawkins, Andorra, Haw Lane, High Wycombe, HP14 4JG
G4 YBJ J Davies, 8 Randle Meadow Court, Great Sutton, Ellesmere Port, CH66 2BL
GJ4 YBM Anthony Alexandre, Merryvale Cottage, La Vallee de St. Pierre, St. Lawrence, Jersey, JE3 1EZ
G4 YBN I Ansell, 9 Sewell Harris Close, Harlow, CM20 3HB
G4 YBP P Darcy, 20 Turner Croft, Fradley, Lichfield, WS13 8SA
G4 YBS Morecambe Bay ARS c/o C Edgar, 61 Winchester Avenue, Lancaster, LA1 4HX
G4 YBT E Tracey, 100 Booth Close, Kingswinford, DY6 8SP
GU4 YBW Paul Wadley, Gironde, Lorier Lane, Vale, Guernsey, GY3 5JG
G4 YBX E Scleparis, 8 Devonshire Park, Reading, RG2 7DX
G4 YCD M Lowe, Crossley Farm, Bristol, BS17 1RH
G4 YCE L Ball, 16 Kelston View, Whiteway, Bath, BA2 1NW
G4 YCG C Beeston, 74 Liss Road, Southsea, PO4 8AS
G4 YCJ A Clift-Jones, Coed Tew Mill, Nant Glas, Llandrindod Wells, LD1 6PD
GW4 YCO D Gill, 19 Rowling St., Williamstown, Tonypandy, CF40 1QY
G4 YCP G Newman, 2 Grange Road, Eldwick, Bingley, BD16 3DH
G4 YCS I Carby, Springfield, Main Street, York, YO30 1AA
GW4 YCT Carmarthen ARS c/o Roderick Belcher, Parciau, Bronwydd Arms, Carmarthen, SA33 6BN
G4 YCV Clive Vickery, 7 Higher Redgate, Tiverton, EX16 6RJ
G4 YCW C Croxford, Bodley Cottage, Parracombe, Barnstaple, EX31 4PR
GI4 YCZ J Rainey, 40 Cranny Lane, Portadown, Craigavon, BT63 5SW
G4 YDB F Heald, Brightling House, Alexandra Road, Heathfield, TN21 8ED
GM4 YDC S Hunt, 5 Highland Road, Crieff, PH7 4LE
G4 YDD William Paul Davidson, 171 Ramsey Road, St. Ives, PE27 3TZ
G4 YDE E Metcalf, Beech Lee, Vicarage Lane, Alresford, SO24 0DU
G4 YDH G Innes, Stonehaven, Holwell Road, Hitchin, SG5 3SL
G4 YDI R Benbow, 54 Park Lea, Bradley Grange, Huddersfield, HD2 1QH
G4 YDM John Allsopp, 30 Manor Park, Concord, Washington, NE37 2BT
G4 YDO Paul Boaler, 21 Harts Close, Birmingham, B17 9LE
GI4 YDP George Moore, 12 Irish Green Street, Limavady, BT49 9AD
G4 YDQ D Hannant, 36 Coslany Street, Norwich, NR3 3DT
G4 YDR David Reed, 14 Glenholt Road, Plymouth, PL6 7JA
G4 YDT John Craddock, 1 Brookes Road, Broseley, TF12 5SB
G4 YDW Patricia Grant, 3 Craggwood Close, Horsforth, Leeds, LS18 4RL
G4 YDX K Gill, 16 Hafodarthen Road, Llanhilleth, Abertillery, NP13 2RY
G4 YDZ M Massen, The Old School, Horning Rd, Norwich, NR12 8JH
G4 YEB D Whitton, 61 Greenacre Park, Gilberdyke, Brough, HU15 2TY
G4 YEE P Hall, Barnlea, Knapp Lane, Romsey, SO51 9BT
G4 YEF B Ronnor, 95 Roids Piece, Purton, Swindon, SN5 4BA
G4 YEG R Paganuzzi, 3 St. Johns Close, Hook, RG27 9HW
G4 YEI Simon Masterman, 5 Leggs Lane, Heyshott, Midhurst, GU29 0DJ
G4 YEJ A Ayton, 3 Links Avenue, Norwich, NR6 5PE
G4 YEK S Clack, 23 Cameron Grove, York, YO23 1LE
G4 YEO Louis Gillain, 1 Willow Avenue, Denham, UB94AG
G4 YEQ GALA & DIST ARS c/o John Campbell, 50 Glebe Place, Galashiels, TD1 3JW
G4 YER David Davies, 248 WEST STREET, HOYLAND NR, Barnsley, S749EE
G4 YES J Thompson, 3 Newport Mount, Headingley, Leeds, LS6 3DB
G4 YET Robert Littlewood, 2 High Street, Scotton, Gainsborough, DN21 3QZ
G4 YEX J Kennedy, 60 Burnway, Albany, Washington, NE37 1QQ
G4 YFB Stephen Coleman, 60 Wilson Road, Reading, RG30 2RN
G4 YFC T Hill, 11 Paget Cottages, Munden Road, Ware, SG12 0NL
G4 YFD F Fish, 25 Bannockburn Way, Billingham, TS23 3QN
G4 YFF Ernest Reynolds, 4 Underwood Close, Stafford, ST16 1TB
G4 YFI R Larter, 12 Ashby Road, Hinckley, LE10 1SL



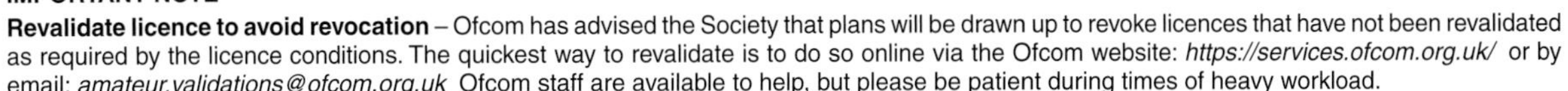
**IMPORTANT NOTE**

**Revalidate licence to avoid revocation** – Ofcom has advised the Society that plans will be drawn up to revoke licences that have not been revalidated as required by the licence conditions. The quickest way to revalidate is to do so online via the Ofcom website: *https://services.ofcom.org.uk/* or by email: *amateur.validations@ofcom.org.uk* Ofcom staff are available to help, but please be patient during times of heavy workload.

G4 YFJ N Von Fircks, 4 Park Street, Salisbury, SP1 3AU
G4 YFK M Aitchison, 21 St. Pauls Road West, Dorking, RH4 2HT
G4 YFO George Wardy, 7 Yew Tree Road, Crewe, CW2 8BN
G4 YFS S Van Praag, 12 Derby Street, Darlington, DL3 0NW
G4 YFT B Page, 7 Muirfield Crescent, Tividale, Oldbury, B69 1PW
G4 YFU Michael Parker, 85 Elston Road, Aldershot, GU12 4HZ
G4 YFV Nolan Banham, Lodge Bungalow, Norwich Road, Diss, IP21 4EE
G4 YFX P Johnson, 119 Riverstone Way, Northampton, NN4 9QW
G4 YFZ B Jones, Jetza, Rose Avenue, Burton-on-Trent, DE13 0DQ
G4 YGA Stephen Howarth, 44 Church Street, Old Catton, Norwich, NR6 7DR
G4 YGB A Graph, 39 Poets Corner, Margate, CT9 1TR
G4 YGD G Aldred, 212 Reepham Road, Norwich, NR6 5SW
G4 YGE C Oakley, 9 Fitzroy Road, Landport, Lewes, BN7 2UB
G4 YGH D Hart, 30 Dartford Avenue, Edmonton, London, N9 8HD
G4 YGJ L Rozentals, 67 Linden Close, Eastbourne, BN22 0TT
G4 YGL G Reece, 34 Priestley Gardens, Romford, RM6 4SL
GD4 YGM P Reynolds, 34 Thurlstone Road, Ruislip, HA4 0BT
GM4 YGN F Albers, 71 Old Evanton Road, Dingwall, IV15 9RB
G4 YGP John Vinson, 11 Ripon Way, Carlton Miniott, Thirsk, YO7 4LR
G4 YGQ M Tate, Stone Cottage, The Street, Norwich, NR13 3PL
GM4 YGS G Wells, Squaredoch, Deskford, Buckie, AB56 5YD
G4 YGT W Waldron, 16 Barke St., Highley, Bridgnorth, WV16 6LQ
G4 YGU John Hetherington, 1 Downs Wood, Vigo, Gravesend, DA13 0SQ
G4 YGV Roger Barnes, 7 Maycroft Close, Ipswich, IP1 6RG
G4 YGW Washington ARC c/o Frederick Waters, 96 Stockley Road, Barmston Village, Washington, NE38 8DR
G4 YGY R Fawke, 25 Derwent Drive, Tewkesbury, GL20 8BA
G4 YGZ K Ghillyer, 54 Longmeadow Road, Saltash, PL12 6DR
G4 YHG J Hubner, 30 Orchard Rise, Olveston, Bristol, BS35 4DZ
G4 YHK H Jackman, 8 The Buchan, Camberley, GU15 3XB
G4 YHN A Gehammar, The Lodge, 119 Ashdon Road, Saffron Walden, CB10 2AJ
G4 YHP Cliff Jobling, Joycliff, 20a Poplar Road, Grimsby, DN41 7RD
G4 YHQ D Webb, 11 Chant Close, Wherwell, Andover, SP11 7JA
GM4 YHS S Grant, 16 Netherton Place, Westmuir, Kirriemuir, DD8 5LD
G4 YIA B Robinson, 196 Bristol Avenue, Farington, Leyland, PR25 4QZ
G4 YIC A Pitt, 16 Highlands Drive, North Nibley, Dursley, GL11 6DX
GW4 YID Malcolm James, 14 Carmel Road, Winch Wen, Swansea, SA1 7JY
G4 YIE C Kelley, Dunkeld, Bridge Street, Southam, CV47 2XY
G4 YIF Stephen Lumbard, 26 Waverleigh Road, Cranleigh, GU6 8BZ
G4 YIG J Cluley, 24 Avon Green, Wyre Piddle, Pershore, WR10 2JE
G4 YIH W Maycey, 21 Brook Drive, Wickford, SS12 9EQ
G4 YIM J Cameron, 20 Fellmead, East Peckham, Tonbridge, TN12 5EQ
G4 YIS J O'Farrell, Shannon, Gunville Road, Salisbury, SP5 1PP
G4 YIT I Toon, 56 Cockbank, Turves, Peterborough, PE7 2HN
G4 YIV B Whyle, 4 Britannia Gardens, Rowley Regis, B65 8DT
G4 YIZ A Mansfield, 90 Vicarage Road, Mickleover, Derby, DE3 0EE
G4 YJA B Lambert, Firbank, East Street, Horsham, RH12 4RE
G4 YJB Ronald Briggs, 32 Waterside, Evesham, WR11 1BU
G4 YJC Graham Thornton, 115 High Street, Studley, B80 7HN
G4 YJD J Donin, 2 Crablands, Selsey, Chichester, PO20 9AX
G4 YJH John Hockey, 11 Amulet Way, Shepton Mallet, BA4 4TL
GW4 YJI T Tilley, 47 Stratton Way, Neath Abbey, Neath, SA10 7BU
G4 YJK P Duke, 4 Doggets Lane, Fulbourn, Cambridge, CB21 5BT
G4 YJM M Leonard, 7 Moorside Parade, Drighlington, Bradford, BD11 1HR
G4 YJN Michael Griggs, Tudor Rose Cottage, Malting Green, Colchester, CO2 0JE
G4 YJP S Stanton, 10 Jenner Crescent, Northampton, NN2 8NB
G4 YJQ D Cutts, 36 Lodge Road, Little Oakley, Harwich, CO12 5EE
G4 YJS Barrie Parsons, 65 Foster Street, Widnes, WA8 6ET
G4 YJT Christopher Roberts, 59a Wharncliffe Road, Loughborough, LE11 1SL
G4 YJU Christopher Mount, 6 Almond Close, Countesthorpe, Leicester, LE8 5TG
G4 YJW G Grundy, 47 Northiam Road, Eastbourne, BN20 8LP
G4 YJX John Boyes, 1 Fuller Close, Shepton Mallet, BA4 5PX
G4 YJY W Marsden, 33 Kilton Crescent, Worksop, S81 0AX
G4 YK B Morrissey, 50 Fingringhoe Road, Langenhoe, Colchester, CO5 7LB
G4 YKB H Ramsden, 23 Nandywell, Little Lever, Bolton, BL3 1JU
G4 YKE Keith Howell, 25 Shelleycotes Road, Brixworth, Northampton, NN6 9NE
G4 YKG J Pether, Stead Farm, Quarry Land Lane, Nr Axbridge, BS26 2QW
G4 YKH C Littler, 11 Richards Road, Stoke D'Abernon, Cobham, KT11 2SX
G4 YKK Ann Babbage, 247-248 Molesey Avenue, West Molesey, KT8 2ET
GW4 YKM K Martich, 25 Pentwyn Isaf, Energlyn, Caerphilly, CF83 2NR
G4 YKQ P Welford, 11 Ridgeside, Bledlow Ridge, High Wycombe, HP14 4JN
G4 YKR K Ramsdale, 770 Warrington Road, Risley, Warrington, WA3 6AQ
G4 YKV K Wood, 257 Church Road, Haydock, St. Helens, WA11 0LY
G4 YKW A Hopkins, 30 Wavell Drive, Newport, NP20 6QN
G4 YKX M Blakeley, Stowe House, Preston Gubbals Road, Shrewsbury, SY4 3YL
G4 YKZ R Harvey, Richlyn House, Cedar Road, Norwich, NR9 3JY
GW4 YLF John Dixon, 73 Brynhyfryd Terrace, Ferndale, CF43 4HT
G4 YLG E Jacquemai, 26 The Crescent, Brighton, BN2 4TD
G4 YLI Russell Barnes, The Cottage, Hutton Row, Penrith, CA11 9TR
G4 YLK D Adams, 16 Centurion Close, Birchwood, Warrington, WA3 6NE
GM4 YLN C Grierson, 174 Baberton Mains Drive, Edinburgh, EH14 3DZ
G4 YLO H Timbrell, Crossing Cottage, Lamyatt, Shepton Mallet, BA4 6NG
GJ4 YLP Christopher Landor, L'Auge, La Rue Du Puchot, Jersey, JE3 6AS
G4 YLQ R May, 53 Oakfield Road, Hadfield, Glossop, SK13 2BN
G4 YLT J Smith, 43 Pagitt Street, Chatham, ME4 6RE
G4 YLW P Yaxley, 10 Leybourne Close, Brighton, BN2 4LU
G4 YMB M Brass, 11 Lealholm Way, Guisborough, TS14 8LN
G4 YMC John McCallum, 7 Jasmine Terrace, Birtley, Chester le Street, DH3 1RW
GM4 YMD William MacDiarmid, 167 Glasgow Road, Whins of Milton, Stirling, FK7 0LH
G4 YME M Elsey, Trekeek Farm, Camelford, PL32 9UB
G4 YMF J Cracklow, 4 The Lawns, Sidcup, DA14 4ET
G4 YMG Peter Chorley, 6 Conference Close, Warminster, BA12 8TF
G4 YMH M Harney, 14 Druce Way, Thatcham, RG19 3PF
GM4 YMI Anthony Dodds, 9 Lower Wellheads, Dunfermline, KY11 3JG
GW4 YMJ W Fitzgerald, 20 Yr Hendre, Kenfig Hill, Bridgend, CF33 6EG
G4 YML E Jones, 15 St. Joseph Place, Llantarnam, Cwmbran, NP44 3HH
GM4 YMM Christine Dons, 37 Ashley Drive, Edinburgh, EH11 1RP
G4 YMQ A Kimm, 9 Tennis Street, Burnley, BB10 3AG
G4 YMT Malcolm Taylor, 64 Elmdene Road, Kenilworth, CV8 2BX
GJ4 YMX D Warncken, Flat 2, 3 Norfolk Terrace, St. Helier, Jersey, JE2 3ZB
G4 YMY R Nicol, 32 Mayfair Drive, Newbury, RG14 6EE
G4 YMZ John May, Glanrhyd Uchaf, Llangeitho, Tregaron, SY25 6QU
G4 YNC C Philpott, 4 Footways, Wootton Bridge, Ryde, PO33 4NQ
G4 YNG M Garlick, Church View, School Lane, Kettering, NN14 3LQ
G4 YNH S Payas, 36 Tintern Close, Popley, Basingstoke, RG24 9HE
G4 YNI H Beckman, 16 Wilton Road, Crumpsall, Manchester, M8 4WQ
G4 YNK N Fletcher, 11 Parkgate Drive, Bolton, BL1 8SD
G4 YNL THE A TEAM CONTEST GROUP c/o R Banks, Dingle Cottage, Church Stoke, Montgomery, SY15 6TJ
G4 YNM Benedict Spencer, 33 New King Street, Bath, BA1 2BL
G4 YNO P Barnes, 69 Southborne, Overcliff Drive, Bournemouth, BH6 3NN
G4 YNS S Shakeshaft, Belvedere Cottage, Wrexham Road, Chester, CH4 9DG
G4 YNT Martin Yallop, 39 Warrenne Keep, Stamford, PE9 2NX
G4 YNU J Scriven, 1 Holgate Road, Pontefract, WF8 4ND
G4 YNV H Snaden, 92 Avon Way, Portishead, Bristol, BS20 6LU
G4 YNX B Bassford, 12 Little Brum, Grendon, Atherstone, CV9 2ET
G4 YOA F Harvey, 137 Epping New Road, Buckhurst Hill, IG9 5TZ
G4 YOC D Gully, 46 Shellards Road, Longwell Green, Bristol, BS30 9DU
G4 YOF G Coomber, 6 Birch Way, Birch, Colchester, CO2 0NQ
G4 YOR R Greenwood, 26 Littlefield Walk, Bradford, BD6 1UU
G4 YOS Dawn Corallini, 8 Britannia, Puckeridge, Ware, SG11 1TG
G4 YOT L Zalicks, 3 Retford Path, Harold Hill, Romford, RM3 9NL
G4 YOV John Metcalfe, 3 Castle Close, Stockton-on-Tees, TS19 0SL
GU4 YOX R Beebe, San Grato, Les Hougettes, Castel, Guernsey, GY5 7DZ
G4 YOZ B Starkey, 52 Bermuda Road, Nuneaton, CV10 7HP
G4 YPA J Aisher, 44 Cranleigh Road, Portchester, Fareham, PO16 9DN
G4 YPC P Croucher, 66 Loop Road, Kingfield, Woking, GU22 9BQ
G4 YPE N Hanney, 62 Avonfield Avenue, Bradford-on-Avon, BA15 1JF
G4 YPF Wesley Taylor, 3 Westcroft, Leominster, HR6 8HE
G4 YPG Reuben Mason, Flat 4, Lysander House, Washington Road, Pulborough, RH20 4RF
G4 YPH D Rothwell, 37 Eamont Avenue, Crossens, Southport, PR9 9YX
G4 YPI A Maires, 26 Dunmow Road, Thelwall, Warrington, WA4 2HQ
G4 YPK P Knowles, 6 Dorchester Close, Basingstoke, RG23 8EX
GM4 YPL R Thompson, Lochview West, 2 St. Ninians Avenue, Linlithgow, EH49 7BP
G4 YPQ K Simpson, 5 Plover Fields, Madeley, Crewe, CW3 9EG
GI4 YPR William Swail, 30 The Gables, Ballyphilip Road, Newtownards, BT22 1RB
GI4 YPS Albert Bradley, 11 Ness Gardens, Londonderry, BT47 2HH
G4 YPV David Ramsden, 76 Brigg Lane, Camblesforth, Selby, YO8 8HD
G4 YQA Michael Lawson, 2 Low Lane, Embsay, Skipton, BD23 6SD
G4 YQC Paul Whiting, 77 Melford Way, Felixstowe, IP11 2UH
G4 YQD T Mayfield, 14 Wheatley Grange, Coleshill, Birmingham, B46 3LZ
G4 YQG B Hodgetts, 15 Wiltons, Wrington, Bristol, BS40 5LS
G4 YQH J Frampton, 161 Longmead Avenue, Bristol, BS7 8QG
G4 YQJ F Collie, 58 Waarem Avenue, Canvey Island, SS8 9DZ
G4 YQK K Taylor, 22 Anderson, Dunholme, Lincoln, LN2 3SR
G4 YQL R Silvey, 9 Kempe Road, Finchingfield, Braintree, CM7 4LE
G4 YQP Michael Simmens, 1 Meaver Cottages, Meaver Road, Helston, TR12 7DN
G4 YQQ Allen Booth, 656 Southmead Road, Filton, Bristol, BS34 7RD
G4 YQS T White, Rosewall Bungalow, Towednack Road, St. Ives, TR26 3AL
G4 YQW K Lawton, 52 Gamble Lane, Leeds, LS12 5LP
G4 YRA J Hills, 27 Wellington Road, Denton, Newhaven, BN9 0RD
G4 YRC YORK ARC c/o A Williamson, Millfield Lodge, 151 Hull Road, York, YO10 3JX
GM4 YRE E Marcus, 11 Parkview, Fettercairn, Laurencekirk, AB30 1XZ
G4 YRF K Amos, 1 Byron Close, Upper Caldecote, Biggleswade, SG18 9DF
G4 YRH H Turner, Riding Hill House, 3 Riding Hill, Halifax, HX3 7TS
G4 YRM R Maynard, Clarnard, 7 Phillipps Avenue, Exmouth, EX8 3HY
GM4 YRO W Patterson, 11 Almond Place, Comrie, Crieff, PH6 2BB
GI4 YRP T Hutchinson, 47 Ballylough Road, Donaghcloney, Craigavon, BT66 7PQ
G4 YRR R Rushton, 53 Crossfield Avenue, Blythe Bridge, Stoke-on-Trent, ST11 9PL
G4 YRT G Cogger, 40 The Crescent, Southwick, Brighton, BN42 4LA
G4 YRV Michael White, 34 Pain's Way, Amesbury, Salisbury, SP4 7RG
G4 YRX J Hilliard, 44 Lerwick Way, Corby, NN17 2DZ
G4 YRY M Holloway, 70 Baring Road, Southbourne, Bournemouth, BH6 4DT
G4 YRZ Robert Denton, 33 Godfreys Court, Worksop, S80 1TT
G4 YSB J Leary, 18a Chestnut Avenue, Andover, SP10 2HE
G4 YSE George Ring, 31 Studland Park, Westbury, BA13 3HQ
G4 YSF J Stimpson, 12 Fairhaven Court, Pittville Circus Road Roa, Cheltenham, GL52 2QR
G4 YSG Alan Cooper, 85 Mansfield Road, Aston, Sheffield, S26 2BR
G4 YSH C Bowers, Cornbury, Seymour Plain, Marlow, SL7 3BZ
G4 YSJ Peter Rowland, 17 Hemel Hempstead Road, Redbourn, St. Albans, AL3 7NL
G4 YSN I Brown, Redland House, Westruther, Gordon, TD3 6NF
G4 YSO M Nixon, 32 Gilbert Sutcliffe Court, Cleethorpes, DN35 0SF
G4 YSP K Metcalf, 34 Framland Drive, Melton Mowbray, LE13 1HY
G4 YSQ T Rogers, Lodge 20, Benson Waterfront, Riverside Park, Oxon, OX10 6SJ
G4 YSS John Earnshaw, Dunelm, Ayton Road, Scarborough, YO12 4RQ
G4 YSZ R Painton, 17 Brookside, Pill, Bristol, BS20 0JX
G4 YTA Michael Owen, 3 Canford View Drive, Wimborne, BH21 2UW
G4 YTB Dave Slade, 10 Larch Close, Weaverham, Northwich, CW8 3ED
G4 YTC Michael Tyson, 1 Warwick Road, Bude, EX23 8EU
G4 YTD Tim Booth, 12 St. Quintin Field, Nafferton, Driffield, YO25 4PD
G4 YTF Peter Godber, 3 Chalvington Close, Evington, Leicester, LE5 6XT
G4 YTG Anthony Gilbey, 83 Chignal Road, Chelmsford, CM1 2JA
G4 YTH T Handford, 20 Minehead Road, Knowle, Bristol, BS4 1BN
G4 YTI Nicola Terry, 2 Crosley House, Crosley Wood Road, Bingley, BD16 4QD
G4 YTJ J Pagett, 26 Rednal Hill Lane, Rednal, Birmingham, B45 9LR
G4 YTK S Hopley, 35 Norton Grange, Norton Canes, Cannock, WS11 9QZ
G4 YTL David Hilton-Jones, Home Farm, Lillingstone Lovell, Buckingham, MK18 5BJ
G4 YTM Ian Pettinger, 266 West Street, Hoyland, Barnsley, S74 9EQ
G4 YTN L Thomas, 31 Claude Avenue, Oldfield Park, Bath, BA2 1AE
G4 YTO M Yeomans, 6 Badsey Close, Northfield, Birmingham, B31 2EJ
G4 YTQ David Richardson, Holmlea, Town Street, Immingham, DN40 3DA
G4 YTT Michael Curran, 40 Barnpark Road, Teignmouth, TQ14 8PN
G4 YTU K Maskell, 2 Birkhall Close, Chatham, ME5 7QD
G4 YTV Richard Guttridge, Ivy House, Rise Road, Hull, HU11 5BH
G4 YTY Alan Dodd, 109 Perinville Road, Babbacombe, Torquay, TQ1 3PD
G4 YUA M Rowland, 27 Wilmot Close, Witney, OX28 5NL
G4 YUF C Sharon, 7 Waverley Gardens, Barkingside, Ilford, IG6 1PJ
G4 YUG Clive ROGERS, 221 Dales Road, Ipswich, IP1 4JY
G4 YUI R Smith, 27 Laburnum Road, Bournville, Birmingham, B30 2BA
G4 YUK A Ince, 3 Craycroft Road, Westwoodside, Doncaster, DN9 2DG
G4 YUL R Hudson, 5 Common Lane, Hemingford Abbots, Huntingdon, PE28 9AN
G4 YUN Matthew Fox, 10 Alderhay Lane, Rookery, Stoke-on-Trent, ST7 4RQ
G4 YUO E Hodges, 2 Joeys Field, Bishops Nympton, South Molton, EX36 4PX
G4 YUZ Ian Parker, 7 Cherry Tree Road, Hoddesdon, EN11 9JS
G4 YVA J Guest, 57 Park Road, Quarry Bank, Brierley Hill, DY5 2HT
G4 YVB J Finch, Hillside House, Chapel Street, Camelford, PL32 9UP
G4 YVD Peter Challinor, Las Ciguenas, 33 Hill Road, Telford, TF2 8NA
G4 YVE C Herwig, 29 New Road, Cupernham, Romsey, SO51 7LL
G4 YVF F James, 70 Broadway West, Walsall, WS1 4DZ
G4 YVI P Rimmer, 1 Pear Tree Close, Weaverham, Northwich, CW8 3HD
G4 YVJ Linda Beal, 17 Park Street, Cleethorpes, DN35 7NG
G4 YVK J Felgate, 31 Melbourne Road, Ipswich, IP4 5PP
G4 YVM David Perry, 11 St. Lawrence Close, Stratford sub Castle, Salisbury, SP1 3LW
GW4 YVN G Bertos, Farallon, Beach Road, Pembroke Dock, SA72 6TP
G4 YVO G Howarth, 79 Eden Avenue, Edenfield, Bury, BL0 0LD
G4 YVQ R Hargreaves, Lawnswood, Lee Road, Blackpool, FY4 4QS
G4 YVV P Leetham, 26 Petersham Drive, Alvaston, Derby, DE24 0JU
G4 YVW P Speed, 52 Hunter Avenue, Shenfield, Brentwood, CM15 8PF
G4 YVX J Rafferty, 130 Hurst Road, Coseley, Bilston, WV14 9EU
G4 YVY Timothy Williams, 31 Manor Road, Holbury, Southampton, SO45 2NQ
G4 YWA P Cartwright, Danae, Glen Road, Deal, CT14 8DD
G4 YWD W Davies, 104 Bromborough Village Road, Wirral, CH62 7EX
G4 YWG D Fowler, 22 Larchwood Crescent, Leyland, PR25 1RJ
GM4 YWI Thomas Ross, 40 North Gyle Loan, Edinburgh, EH12 8JH
G4 YWJ P Ganley, 60 Bole Hill Road, Sheffield, S6 5DD
GW4 YWM W Matthews, Cornerways, William Street, Swansea, SA9 1AT
G4 YWN G Morris, Rivendell, The Street, Braintree, CM7 5HN
G4 YWR N Lill, 15 Gloucester Road, Bingley, BD16 4RW
GM4 YWS G McKay, Reay House, St. Vigeans, Arbroath, DD11 4RA
GI4 YWT John Crichton, 10 Bann Drive, Londonderry, BT47 2HW
GM4 YWU J Bledowski, 13 Riggend Road, Arbroath, DD11 2DR
GM4 YWV W Watson, 21 Cameron Way, Bridge of Don, Aberdeen, AB23 8QD
G4 YWX Allan Bell, 43 Bigsby Road, Retford, DN22 6SF
G4 YWZ C Winning, 14 Fairmead Way, Totton, Southampton, SO40 7JH
G4 YXB A Utley, 3 Dene Grove, Silsden, Keighley, BD20 9NR
GM4 YXI K Kerr, East Loanhead, Auchnagatt, Ellon, AB41 8YH
G4 YXJ T Trethewey, 8 Sunningdale Road, Saltash, PL12 4BN
G4 YXR P Waygood, 89 George Street, Wellington, TA21 8HZ
G4 YXS Francis Lake, 77a Wood Lane, Chapmanslade, Westbury, BA13 4AT
G4 YXU S Rawcliffe, 4 Rue Des Contamines, Gex, France, 1170
G4 YXX N Varnes, Kelneath, West Hill, Wincanton, BA9 9BZ
G4 YYB Ernest Holme, 29 Sandy Lane, Hindley, Wigan, WN2 4EJ
G4 YYC D Craig, 48 Fairholme, Bedford, MK41 9DD
G4 YYD Alan Birtwistle, 6 Solness Street, Bury, BL9 6PP
G4 YYE K Smith, 27 Fairleas, Branston, Lincoln, LN4 1NW
G4 YYF S Collings, Marsden, Heugh Road, Stranraer, DG9 8TD
G4 YYG Gerald Plant, 74 Elmwood Drive, Blythe Bridge, Stoke-on-Trent, ST11 9NX
G4 YYH R Blemings, 1 Trethern Close, Troon, Camborne, TR14 9ER
G4 YYI S Tear, 18 The Chase, Sinfin, Derby, DE24 9PD
G4 YYL T Anderton, 12 Oaklands Close, Halvergate, Norwich, NR13 3PP
G4 YYM M Remnant, Redwood Court, Tolcarne Road, Camborne, TR14 9AA
G4 YYO P Sutcliffe, Rosemead, Cheadle Road, Stoke-on-Trent, ST10 4BH
G4 YYP Patricia Plant, 74 Elmwood Drive, Blythe Bridge, Stoke-on-Trent, ST11 9NX
G4 YYR Stanley Gibbs, 14 Castle Mead, Kings Stanley, Stonehouse, GL10 3LD
G4 YZA Donald Brown, 104 Kineton Green Road, Solihull, B92 7EE
G4 YZC E Smith, Willow Cottage, Mill Road, Louth, LN11 9TF
G4 YZD F Benstead, 19 Davis Court, Eastland Road, Bristol, BS35 1DP
G4 YZF Brian Alston-Pottinger, 16 Vincent Close, Great Yarmouth, NR31 0HR
G4 YZH B Calvert-Toulmin, Brandesby House, 31 West End, Scunthorpe, DN15 9NR
G4 YZK R Marsh, Hazatree, 71 Station Road, Worcester, WR3 7UP
G4 YZL G Woollams, Pebbles, Pebsham Lane, Bexhill-on-Sea, TN40 2NT
G4 YZM S Green, 4 Countess Drive, Walsall, WS4 1HT
G4 YZN K Chapman, 10 Beck Lane, Collingham, Wetherby, LS22 5BW
G4 YZO J Badger, 87 Blackberry Lane, Four Oaks, Sutton Coldfield, B74 4JF
G4 YZP J McMahon, 15 Chatteris Park, Runcorn, WA7 1XE
G4 YZR M Baker, 62 Court Farm Road, Whitchurch, Bristol, BS14 0EG
GM4 YZT J Simpson, Flat 1/B, Isla Court, Perth, PH2 7HJ
G4 ZA G Anderson, 46 Bearcroft, Weobley, Hereford, HR4 8TA
GD4 ZAB Lynda Taylor, Burnt Mill House, 1, 2 Mount William, Douglas, Isle of Man, IM2 4PE
G4 ZAC M Lewis, Oak Fruit Farm, Devils Highway, Reading, RG7 1XS
G4 ZAD Gary Kitson, 50 Willis Pearson Avenue, Bilston, WV14 8DA
G4 ZAG George Woodworth, 136 Wepre Park, Connah's Quay, CH5 4HW
GI4 ZAH Frederick Anderson, Flat 6, 25 Main Street, Coleraine, BT51 4RA
G4 ZAI A Stevenson, 11 Alexandra Road, Malvern, WR14 1HA
G4 ZAL N Head, 11 Crowden Crescent, Tiverton, EX16 4ET
G4 ZAM Robert Lowe, The Chimes, 4 Broadway, Swindon, SN25 3BT
G4 ZAO D Holmes, 17 Green Lane, Scarborough, YO12 6HL
G4 ZAP A1 Contest Group c/o C Wilson, 2 Bainton Close, Bradford-on-Avon, BA15 1SE
G4 ZAQ R Burke, 66 Vineyard Road, Newport, TF10 7RU
G4 ZAR D Flanagan, 13 Bryn Awelon, Flint, CH6 5QA
G4 ZAS J Searle, 21 Chetwynd Drive, Southampton, SO16 3HY
GW4 ZAW J Aspinall, 66 Lake Road East, Cardiff, CF23 5NN
G4 ZAX S Jones, 71 Milford Road, Pennington, Lymington, SO41 8DN

G4 ZAY J Tench, 20 Waterfield Meadows, North Walsham, NR28 9LD
G4 ZBC L Brazier, 65 Cadman Crescent, Fallings Park, Wolverhampton, WV10 0UH
G4 ZBE T Anderson, 38 Redwood Drive, Chase Terrace, Burntwood, WS7 2AC
G4 ZBF Graham Starkey, 45 Chalcot Drive, Hednesford, Cannock, WS12 4SF
G4 ZBG Terence Green TWO, 1 Conduit Road, Stamford, PE9 1QQ
G4 ZBH A Holder, 47 Church Road, Gurnard, Cowes, PO31 8JP
G4 ZBK J Olsen, Klintevej 218, Hjertebjerg, Stege, Denmark, DK 4780
G4 ZBL D Bateman, 20 Nags Head Hill, St George, Bristol, BS5 8LN
GW4 ZBN L Connery, 37 Thomas St., Abertridwr, Caerphilly, CF8 4AU
G4 ZBO R Parker, 34 Sandgate, Kendal, LA9 6HT
G4 ZBQ J Thomas, 113 Southwood Drive, Coombe Dingle, Bristol, BS9 2QR
G4 ZBS A Appleyard, 5 Rowan Close, Puriton, Bridgwater, TA7 0AL
GW4 ZBU J Johns, 10 Maes yr Haf, Llansamlet, Swansea, SA7 9ST
G4 ZBW John Sccal, South Low, Lyth, Kendal, LA8 8DJ
G4 ZBZ Ann Cooper, 26 Burlington Road, Skegness, PE25 2EW
G4 ZCA R McCormick, 22 Eric Road, Wallasey, CH44 5RQ
G4 ZCD Ian Wilson, 65 Cupernham Lane, Romsey, SO51 7LE
G4 ZCG Anthony Ashworth, 210 Liverpool Road, Hutton, Preston, PR4 5HD
G4 ZCJ H Cresswell, 34 Kingsgate Avenue, Birstall, Leicester, LE4 3HB
GW4 ZCL P Jones, 14 Fonmon Road, Rhoose, Barry, CF62 3DZ
G4 ZCM K Frowd, 290 Pilton Vale, Newport, NP20 6LS
G4 ZCN Barry Grylls, 22 Aldeburgh Close, Hartlepool, TS25 2RG
G4 ZCP B Roberts, 70 Coombe Park Road, Coventry, CV3 2PE
G4 ZCR C Glenn, 61 Ansell Road, Erdington, Birmingham, B24 8LX
G4 ZCS Christopher Saunders, Garlands, Malthouse Lane, Burgess Hill, RH15 9XA
G4 ZCT Christopher Thomas, 3 Poldice Terrace, Poldice, Redruth, TR16 5QA
GM4 ZCV E Prietzel, 4 Cherry Lane, Cupar, KY15 5DA
G4 ZCW David Reed, 4 Allwood Drive, Carlton, Nottingham, NG4 3EH
G4 ZCY F Barwell, Galahad, Penisarwaun, Caernarfon, LL55 3BN
G4 ZDD B Brookfield, 17 St. Stephens Drive, Aston, Sheffield, S26 2EP
G4 ZDE R Boss, 11 Penkridge Road, Church Gresley, Swadlincote, DE11 9FH
G4 ZDF T Langham, 1 Chatsworth Avenue, Radcliffe-on-Trent, Nottingham, NG12 1DG
G4 ZDG Richard Mather, 27 Bridgeacre Gardens, Coventry, CV3 2NQ
GM4 ZDH David Hepworth, The Boskage, West Linton, EH46 7BH
G4 ZDN G Titterington, 2 South Road, Sandy, SG19 1HE
G4 ZDP S Williams, 18 Croft Road, Newbury, RG14 7AL
G4 ZDQ A Siddons, 18 Earlswood Road, Evington, Leicester, LE5 6JB
G4 ZDR A Perrett, 99 Welsford Avenue, Wells, BA5 2HZ
G4 ZDT David Brodie, Waterloo Cottage, Tanners Green, Norwich, NR9 4QS
G4 ZDU S Kirkwood, The Acre, Church Road, Hereford, HR2 9SE
G4 ZDX A Staniforth, 2 Park View, Mapperley, Nottingham, NG3 5FD
G4 ZDY Roy Haining, 2 Keswick Close, Kirby Cross, Frinton-on-Sea, CO13 0TG
GW4 ZEA Edwards Hawkins, 12 Marine Drive, Ogmore-by-Sea, Bridgend, CF32 0PJ
G4 ZEB A Richardson, 117 Polgrean Place, St. Blazey, Par, PL24 2LH
G4 ZEG E Cross, 15 Carisbrooke Crescent, Barrow-in-Furness, LA13 0HU
G4 ZEJ Robert Coombes, 20 Gaskyns Close, Rudgwick, Horsham, RH12 3HE
G4 ZEL D Rampton, Chalemar, Eddeys Lane, Bordon, GU35 8HU
G4 ZEN G Gardner, 10 Chestnut Close, Ventnor, PO38 1DQ
G4 ZES Richard Mills, 5 Summerlands Road, Marshalswick, St. Albans, AL4 9XB
GM4 ZET William Connolly, 4 Lomond Bank, Glenfarg, Perth, PH2 9PF
G4 ZEU W Dix, 16 Shakespeare Road, Radstock, BA3 3XL
G4 ZEW Doug Adams, 4 St. Georges Close, Brampton, Huntingdon, PE28 4US
GM4 ZEX G Duncan, Mansewood, Woodhead, Turriff, AB53 8LT
G4 ZEY E Gough, 41 Matlock Green, Matlock, DE4 3BT
G4 ZEZ J Curwen, 1 Oak Drive, Halton, Lancaster, LA2 6QJ
G4 ZFC R Marks, 14 Carnation Road, Rochester, ME2 2YE
G4 ZFD D Roper, Lunesdale, Halifax Road, Nelson, BB9 0EG
G4 ZFE R Everitt, 6 Ormathwaites Corner, Warfield, Bracknell, RG42 3XX
G4 ZFJ C Roberts, 122 Lower Road, Hullbridge, Hockley, SS5 6BH
G4 ZFP Peter Lewis, 12 St. James Park, Tunbridge Wells, TN1 2LH
G4 ZFQ Alan Reeves, 41 Nodes Road, Cowes, PO31 8AD
G4 ZFR FELIXSTOWE DARS c/o Paul Whiting, 77 Melford Way, Felixstowe, IP11 2UH
GM4 ZFS S Graham, 8 Kirkton Crescent, Dundee, DD3 0BN
G4 ZFT Mary Nurse, 67 Grasleigh Way, Allerton, Bradford, BD15 9BD
G4 ZFV D Green, 56 Southfields Road, Littlehampton, BN17 6PA
G4 ZFX J Blades, 3 Briery Croft, Stainburn, Workington, CA14 1XJ
G4 ZFY J Bowes, 8 Coxford Drove, Southampton, SO16 5FD
G4 ZGC P Crouch, 85 Stomp Road, Burnham, Slough, SL1 7NA
G4 ZGE T Stocks, 1 Church Street Messingham, Scunthorpe, DN17 3SB
G4 ZGG B Storey, 8-9 Chadley Lane, Godmanchester, Huntingdon, PE29 2AL
G4 ZGM C Macdonald, 3 Shaftesbury Avenue, Doncaster, DN2 6DT
G4 ZGP Geoff Pritchard, 26 Anglesey Drive, Poynton, Stockport, SK12 1BU
G4 ZGQ D Richardson, 55 Barton Road, Central Treviscoe, St. Austell, PL26 7PT
GM4 ZGU R Camley, 16 Ferness Oval, Balornock, Glasgow, G21 3SQ
GM4 ZGV D Davidson, 12 The Paddock, Peterculter, AB14 0UE
G4 ZGZ Simon Harris, 39 Trevithick Avenue, Torpoint, PL11 2PX
G4 ZHA D Raine, 160 Aragon Road, Morden, SM4 4QN
G4 ZHD P Crofts, 8 Sandown Avenue, Mickleover, Derby, DE3 0QQ
G4 ZHE D Cannon, 69 Hayfield Road, Oxford, OX2 6TX
G4 ZHG John Nevin, 26 Beech Avenue, Newark, NG24 4DY
G4 ZHI John Howell-Pryce, Bwlch Teulu, Tynygraig, Ystrad Meurig, SY25 6AJ
G4 ZHK D Lennard, 24 Southdown Road, Shoreham-by-Sea, BN43 5AN
GM4 ZHL G Graham, 18 Hamarsgarth, Mossbank, Shetland, ZE2 9TH
G4 ZHN D Young, The Gables, Aerodrome Road, Canterbury, CT4 5EX
G4 ZHT C Strevens, 11 Kenley Road, London, SW19 3JJ
G4 ZHX J Burton, 23 Dorchester Close, Dartford, DA1 1ND
G4 ZHY R Nicholls, 9 Roberts Close, Stretton on Dunsmore, Rugby, CV23 9EZ
G4 ZHZ Adrian Nash, 10 Broome Close, Yateley, GU46 7SY
G4 ZIB Anthony Roberts, 25 Sebright Road, Wolverley, Kidderminster, DY11 5TZ
G4 ZID L Chapman, 6 Barholm Avenue, Lutton, Spalding, PE12 9HS
G4 ZIF M Taylor, Holly House, Faussett Hill, Canterbury, CT4 7AH
G4 ZIH R West, 51 Glen Avenue, Herne Bay, CT6 6HU
G4 ZII A Taylor, 50 Long Hill Rise, Hucknall, Nottingham, NG15 6GN
G4 ZIL A Brown, Skellies Knowes East, Leswalt, Stranraer, DG9 0RY
G4 ZIS R Beech, 131 Bounces Road, Lower Edmonton, London, N9 8LJ
GM4 ZIT J Brown, 51 Braeside Park, Balloch, Inverness, IV2 7HN
G4 ZIU M O'Connell, 5 Beckwith Close, Harrogate, HG2 0BJ
G4 ZIW M Hutchings, 31 Newtown Road, Little Irchester, Wellingborough, NN8 2DX
G4 ZIY Mal Haddon, 1 Victoria Place, Weston, Portland, DT5 2AA
G4 ZIZ R Barrett, Willow Lodge, Links Road, Leicester, LE9 2BP
G4 ZJC P Berry, 3 Village Farm Road, Preston, Hull, HU12 8UH
G4 ZJD L Taylor, 14 Spring Grove, Chiswick, London, W4 3NH
G4 ZJE K Feighney, 67 Moorside Road, Brookhouse, Lancaster, LA2 9PJ
G4 ZJH I Tickle, 7 Ashfords Close, Saxmundham, IP17 1WB
GM4 ZJI Christopher Claydon, 33 Craigievar Drive, Glenrothes, KY7 4PH
G4 ZJK Ronald White, 29 Princes Road, Clacton-on-Sea, CO15 5LA
G4 ZJL Derek Wood, 29 Oakville Road, Heysham, Morecambe, LA3 2TB
G4 ZJO Harry Docherty, 4 Prospect Drive, Hest Bank, Lancaster, LA2 6HX
G4 ZJP M Ward, Laurels, Eastergate Lane, Chichester, PO20 3SJ
G4 ZJR E Knibb, The Cottage, Cold Newton Road, Leicester, LE7 9DA
G4 ZKA J Watson, 158 Kingsfold Drive, Penwortham, Preston, PR1 9EQ
G4 ZKD J Moule, Silver Dale, Callow Hill Rock, Kidderminster, DY14 9DP
G4 ZKE M Chapman, 18 The Winter Knoll, Littlehampton, BN17 6ND
G4 ZKG J Corfield, 5 Beasley Close, Great Sutton, Ellesmere Port, CH66 2SX
G4 ZKH Monty Curtis, 11 Pentreath Terrace, Lanner, Redruth, TR16 6HP
G4 ZKI M Day, 76 Freeman Road, Didcot, OX11 7DB
G4 ZKJ W Applebee, 9 The Glade, Bucks Horn Oak, Farnham, GU10 4LU
G4 ZKM W Ingram, 39 Ainsdale Drive, Peterborough, PE4 6RL
G4 ZKN P Robinson, 1 Stennack, Troon, Camborne, TR14 9JT
G4 ZKQ J Garner, Craythorne, Amberstone, Hailsham, BN27 1PJ
G4 ZKR J Olbrien, 14 Ryecroft Close, Middlewich, CW10 0PJ
G4 ZKS A Howland, Hollydene, Station Road, Colchester, CO7 8LJ
G4 ZKT A Hale, 32 Russell Road, Northolt, UB5 4QS
G4 ZKW T Davies, 7 Medway, Sturton by Stow, Lincoln, LN1 2DY
GI4 ZLD G Breslin, 85 Whitehouse Park, Londonderry, BT48 0QA
G4 ZLF L Forde, 3 Heather Way, Rosudgeon, Penzance, TR20 9PT
G4 ZLI T Schofield, 25 Kingsfield, Ringwood, BH24 1PH
G4 ZLJ P Aspinall, 20 Carr Lane, New Hall Hey, Rossendale, BB4 6BE
G4 ZLK S Pike, 32 Rosewood Road, Dudley, DY1 4DZ
G4 ZLN Brian Phillips, 2 Oriole Grove, Kidderminster, DY10 4HG
G4 ZLP N Crook, 10 Shuttle Close, Rossington, Doncaster, DN11 0FR
G4 ZLT R Winkup, 92 Barnes Crescent, Bournemouth, BH10 5AW
G4 ZLU T Clark, Thaw House, Brunswick Street, Nelson, BB9 0HZ
G4 ZLX A Whillock, 74 Chettell Way, Blandford St. Mary, Blandford Forum, DT11 9PH
G4 ZMA J Smith, 7a The Green, East Leake, Loughborough, LE12 6LD
G4 ZMB D Sharples, 11 Lina Street, Accrington, BB5 1SL
G4 ZMH Gerald Robinson, 8 Fenlands Crescent, Lowestoft, NR33 9AW
GM4 ZMK Richard Coyle, 216 Faifley Road, Clydebank, G81 5EG
G4 ZML D Coupe, 14 Maltby Road, Thornton, Middlesbrough, TS8 9BU
G4 ZMM Jonathan Roberts, Vernann House, Gayton, Stafford, ST18 0HJ
G4 ZMN P Shepherd, 315 Daws Heath Road, Benfleet, SS7 2TY
G4 ZMP D Butler, 42 Coombe Farm Avenue, Fareham, PO16 0TR
G4 ZMR M Reynolds, 21 Jubilee Gardens, Nantwich, CW5 7BS
G4 ZMS E Rennie, 26 Kingshill Avenue, Collier Row, Romford, RM5 2SD
G4 ZMU Anthony Vernon, 35 Cornworthy, Shoeburyness, Southend-on-Sea, SS3 8AN
G4 ZMW M Williams, 3 Holly Lodge Close, Bristol, BS5 7XG
G4 ZMY A Wakely, 177 St.Hermans Estate, Hayling Island, PO11 9NE
GM4 ZNC William Findlay, 46 Rowallan Drive, Kilmarnock, KA3 1TU
G4 ZNI A Wragg, 11a Fall Road, Heanor, DE75 7PQ
G4 ZNK R Walsh, 16 Pinewood Grove, Midsomer Norton, Radstock, BA3 2RH
GM4 ZNS John Callaghan, 31 Hillview Road, Darvel, KA17 0DQ
GM4 ZNX D Stockton, 13 Dunvegan Court, Crossford, Dunfermline, KY12 8YL
G4 ZNY K Brown, 1 Normandy Way, Bletchley, Milton Keynes, MK3 7UN
G4 ZNZ D Neilson, 11 Craigs Way, Thirsk, YO7 1UD
GM4 ZOA S Mcgregor, 35 Pentland Gardens, Edinburgh, EH10 6NN
G4 ZOB P Harris, 47 North Park Grove, Roundhay, Leeds, LS8 1EW
G4 ZOC J Lawton, Grenehurst, Pinewood Road, High Wycombe, HP12 4DD
G4 ZOF A Hughes, Kemble Motors, Unit 9, Coundon, BISHOP AUCKLAND
G4 ZOG D Andrews, 3 St. Davids Road, Thornbury, Bristol, BS35 2JE
G4 ZOH Colin Hetherington, 10 Westway, Cowes, PO31 8QP
G4 ZOI Dene Hunsdale, 15 Ash Street, Bury, BL9 7BT
G4 ZOK K Mills, 75 Caistor Lane, Caistor St. Edmund, Norwich, NR14 8RB
G4 ZON A Edwards, 50 Wyatts Green Lane, Wyatts Green, Brentwood, CM15 0PY
G4 ZOQ John Dennis, 44 The Drive, Uckfield, TN22 1BZ
G4 ZOR Eric Wand, 1 Firs Chase, West Mersea, Colchester, CO5 8ND
GI4 ZOS W Boyd, 51 South Sperrin, Knock, Belfast, BT5 7HW
G4 ZOU D Nuthall, 29 Bloxham Crescent, Hampton, TW12 2QG
G4 ZOX C Moore, Spion Cop, Blacksmiths Lane, Lincoln, LN5 9SW
G4 ZOY D Elliott, 6 Linden Close, Stakeford, Choppington, NE62 5LD
G4 ZPA B Watling, 10 Nutbourne Road, Farlington, Portsmouth, PO6 1NR
G4 ZPB R Alexander, 1 Locarno Road, Swanage, BH19 1HY
G4 ZPC P Collier, 7 Cavendish Close, Bicton Heath, Shrewsbury, SY3 5PG
G4 ZPH F Machniak, 18 Wyatt Road, Kempston, Bedford, MK42 7EN
G4 ZPI David Madden, 17 Canberra Gardens, Birmingham, B34 7LP
G4 ZPJ C Marks, 85 Madrona, Tamworth, B77 4EJ
G4 ZPL Colin Barwell, Galaghad, Penisarwaun, Caernarfon, LL55 3BN
G4 ZPM D Thompson, 1 West Kinmel Street, Rhyl, LL18 1DA
G4 ZPN M Brown, 47 Threlfall Road, Blackpool, FY1 6NW
G4 ZPO G Belt, 45,Prospect Road, Dorchester, DT1 2PF
G4 ZPP B Hope, 19 Seaview Court, Hillfield Road, Chichester, PO20 0JS
G4 ZPQ S Drury, 24 Mollison Road, Hull, HU4 7HB
GD4 ZPR R Wilson, 1 Larkfield Avenue, Harrow, HA3 8NQ
G4 ZPW D McKie, 16 Guys Close, Addison Square, Ringwood, BH24 1PQ
G4 ZPZ Ian Macpherson, 18 Mountbatten Avenue, Dukinfield, SK16 5BU
G4 ZQC R Alderson, Old School House, Tattersett, King's Lynn, PE31 8RS
G4 ZQF Roger Worth, 2 Orchard Drive, Otterton, Budleigh Salterton, EX9 7JL
GM4 ZQH J Howell, 26 Bonaly Crescent, Colinton, Edinburgh, EH13 0EW
G4 ZQJ A Mayes, 103 Lionel Road, Canvey Island, SS8 9DJ
G4 ZQL N Higgins, 7 Staveley Close, Middleton, Manchester, M24 4RU
G4 ZQM J Neary, 29 Willow Avenue, Torquay, TQ2 8DH
G4 ZQS L Brown, 17 Chaucer Walk, Langney, Eastbourne, BN23 7QT
G4 ZQT Jeffrey Wright, 85 Kingfisher Drive, Beacon Park Home Village, Skegness, PE25 1TQ
G4 ZQV Ian Bradford, The Meadows, Penyrheol, Pontypool, NP4 5XS
GW4 ZQY M Couch, 37 Heol Rhosyn, Morriston, Swansea, SA6 6ER
G4 ZRA G Moffatt, 30 Rose Walk, St. Albans, AL4 9AF
G4 ZRB W Gerrard, 9 St. Marys Gardens, Bagshot, GU19 5JX
G4 ZRC M Cole, 25 Holly Gardens, West Drayton, UB7 9PF
G4 ZRD B Rosewarn, 16 Stoke Park Close, Bishops Cleeve, Cheltenham, GL52 8HH
G4 ZRF T Emery, 23 Richmondfield Way, Barwick in Elmet, Leeds, LS15 4HJ
GM4 ZRH Alexander Hutton, 4 Linn Road, Stanley, Perth, PH1 4QS
GW4 ZRK H Stephens, Ty Coch, Rhydwen Place, Clydach, SA6 5RN
G4 ZRM D Line, 20 Wykeham Road, Higham Ferrers, Rushden, NN10 8HU
GM4 ZRR I Watt, 21 Glenwood Way, Edinburgh, EH12 8QA
G4 ZRT Mark Johnson, 5 Donigers Dell, Swanmore, Southampton, SO32 2TJ
G4 ZRV Tom Russell, 59 Durban Road, Grimsby, DN32 8BA
GW4 ZRW Thomas George, 80 Yew Street, Troedyrhiw, Merthyr Tydfil, CF48 4EE
GM4 ZRX J Lindsay, 6 Netherhouse Avenue, Lenzie, Glasgow, G66 5NG
G4 ZRY A Clemons, 2 Cherry Tree Road, Rainham, Gillingham, ME8 8JU
G4 ZRZ G Baker, 33 Twycross Road, Wokingham, RG40 5PE
G4 ZSA A Smith, 16 Burley Close, South Milford, Leeds, LS25 5BT
G4 ZSC A Kent, 9 Tolmers Gardens, Cuffley, Potters Bar, EN6 4JE
G4 ZSD W Guy, 102 Bonington Road, Mansfield, NG19 6QQ
G4 ZSG J Savegar, 39 Little Lane, Roundfield, Reading, RG7 6RA
G4 ZSM G Driver, 216 Court Lane, Birmingham, B23 5RH
G4 ZSO Nicolas Dakin, 3 Tavistock Road, West Bridgford, Nottingham, NG2 6FH
G4 ZSP G Marriott, 6 The Pastures, Barrow upon Soar, Loughborough, LE12 8LA
G4 ZSR D Woollams, Pebbles, Pebsham Lane, Bexhill-on-Sea, TN40 2NT
G4 ZSS S Simpson, 8 Hallfield Avenue, Micklefield, Leeds, LS25 4AU
G4 ZST David Nuttall, 92 Long Road, Lowestoft, NR33 9DH
G4 ZSV John Broughton, 55 Webbs Close, Wolvercote, Oxford, OX2 8PX
G4 ZSW P Withall, 19 Highfield Drive, Ewell, Epsom, KT19 0AU
G4 ZSX P Johnson, 56 Sycamore Avenue, Lowestoft, NR33 9PJ
G4 ZSY C Ingram, Woodwind, Kingstone, Hereford, HR2 9HD
G4 ZSZ B Watts, 74 Westfield Road, Caversham, Reading, RG4 8HJ
G4 ZTA J Reed, Easton Villa, Grangemoor Road, Morpeth, NE61 5PU
G4 ZTC T Cleghorn, 12 Rennington Close, Stobhill Gate, Morpeth, NE61 2TQ
G4 ZTD Kelvin Wright, 63 Dudley Avenue, Leicester, LE5 2EF
G4 ZTF John Scott, Kemsley Street Cottage, Kemsley Street, Gillingham, ME7 3LS
G4 ZTG K Ford, Tan y Bryn, Llanbedr, LL45 2ND
G4 ZTM N Rohsler, 107 Quinton Lane, Quinton, Birmingham, B32 2TT
GM4 ZTO John Mcllwraith, 54 Foreland, Ballantrae, Girvan, KA26 0NQ
G4 ZTQ Sidney Chappell, 67 Swanfield Drive, Chichester, PO19 6GL
G4 ZTR J Lemay, Carlton House, White Hart Lane, Colchester, CO6 3DB
G4 ZTS C Thorne, High Trees, Bradford on Tone, Taunton, TA4 1EX
GI4 ZTU Hugh Morgan, 42 Ardmore Road, Holywood, BT18 0PJ
G4 ZTW C Gallagher, 9 St. Marys Road, New Romney, TN28 8JB
G4 ZTY D Dalton, 22 Fernleigh Avenue, Mapperley, Nottingham, NG3 6FL
G4 ZTZ Keith Taylor, 107 Trelowarren Street, Camborne, TR14 8AW
GW4 ZUA W Webb, 5 Shop Houses, Llwydcoed, Aberdare, CF44 0TH
G4 ZUC G Allin, 1 Brookhill Court, Sutton-in-Ashfield, NG17 1EP
G4 ZUD Brian Perry, Preswylfa, Carno, Caersws, SY17 5JP
G4 ZUE Reginald Hopkins, 259 Croft Road, Nuneaton, CV10 7EE
G4 ZUH J Rowles, The Haven, 5 Honey Lane, Chatteris, PE16 6LG
G4 ZUI P Bevington, 40 Carnarthen Street, Camborne, TR14 8UP
GW4 ZUJ Bernard Willis, 5 Park Street, Penrhiwceiber, Mountain Ash, CF45 3YW
GM4 ZUK Allan Duncan, Barrhill House, Peterculter, AB14 0LN
G4 ZUL S Cocks, 1 Church Road, Laindon, Basildon, SS15 4EH
G4 ZUN C Gee, 100 Plantation Hill, Worksop, S81 0QN
G4 ZUP R Angel, The Olde Cheese House, Upton Lovell, Warminster, BA12 0JW
G4 ZUS Gordon Smith, 5 Adbolton Lodge, Carlton, Nottingham, NG4 1DR
GD4 ZUU Paul Chambers, 15 Ballagorry Drive, Ramsey, Isle of Man, IM7 1HE
G4 ZUW A Hockley, 44 Brookfields, Crickhowell, NP8 1DJ
G4 ZUX N Sparks, 36 Tormynton Road, Worle, Weston-Super-Mare, BS22 9HT
G4 ZVA T Webster, 42 The Meadow, Mount Pleasant Residential Park, Crewe, CW4 8JU
G4 ZVB G Mantovani, 74 Barnsley Road, South Kirkby, Pontefract, WF9 3QE
G4 ZVD J Birse, 3 Main Road, Rathmell, Settle, BD24 0LH
GM4 ZVF M Sheriff, Schoolhouse Dunmore, Kilberry Road, Tarbert, PA29 6XY
G4 ZVK J Taylor, 123 Lancaster Road, Hindley, Wigan, WN2 4JA
G4 ZVL Michael Higgins, 8 Clos Rheidol, Caldicot, NP26 4JD
G4 ZVN Paul Baxter, 20 Thorpe Street, Thorpe Hesley, Rotherham, S61 2RP
GW4 ZVO B Froley, 20 Sandymeers, Porthcawl, CF36 5LP
G4 ZVP B Rhodes, 13 Amanda Road, Harworth, Doncaster, DN11 8HP
G4 ZVQ Stephen Evans, 6 Eastfield Way, Caerleon, Newport, NP18 3EU
G4 ZVS C Ford, 19 Listowel Road, Kings Heath, Birmingham, B14 6HH
G4 ZVU T Chadwick, 102 Feltham Road, Ashford, TW15 1DP
GW4 ZVV B Raby, 4 Tyfica Road, Pontypridd, CF37 2DA
G4 ZVW D Ilsley, 3 Peel Yard, Martlesham Heath, Ipswich, IP5 3UL
G4 ZVX M Russell, 67 Dugard Road, Cleethorpes, DN35 7SD
G4 ZVZ S Josko, 69 Newborough Road, Shirley, Solihull, B90 2HB
G4 ZWA G Johnson, The Cottage, Mareham on the Hill, Horncastle, LN9 6PQ
G4 ZWB R Ayers, 94 Carr Avenue, Leiston, IP16 4AT
G4 ZWD Carol Spicer, 27 Carden Crescent, Patcham, Brighton, BN1 8TQ
G4 ZWE Peter Burtenshaw, 9 Winchester Way, Eastbourne, BN22 0JP
G4 ZWI Fred Cooper, 29 Mayfair Avenue, Mansfield, NG18 4EQ
GM4 ZWJ S Macfarlane, 6 Edward Drive, Helensburgh, G84 9QP
G4 ZWM H Hoy, The Meadows Cottage, Stow Heath Road, North Walsham, NR28 0LR
G4 ZWN T Anziani, 42 Tyn Rhos Estate, Penysarn, LL69 9BZ
G4 ZWO Adrian Green, Bryn Y Coed, Llanfair Road, Abergele, LL22 8DH
G4 ZWQ P Smith, 16 Church St., Owston Ferry, Doncaster, DN9 1RG
G4 ZWR D Edwards, 2 Mason Close, Headless Cross, Redditch, B97 5DF
G4 ZWX Perceval Harrison, Brocas Street, Eton, Windsor, SL4 6BW
G4 ZWY Steven Icke, 11 Church Lane, Bromyard, HR7 4DZ
G4 ZXA R Smith, 1 Hall Lane, Wolvey, Hinckley, LE10 3LF
G4 ZXB A Tomlins, 44 Newlands, Balcombe, Haywards Heath, RH17 6JA

UK Callsigns

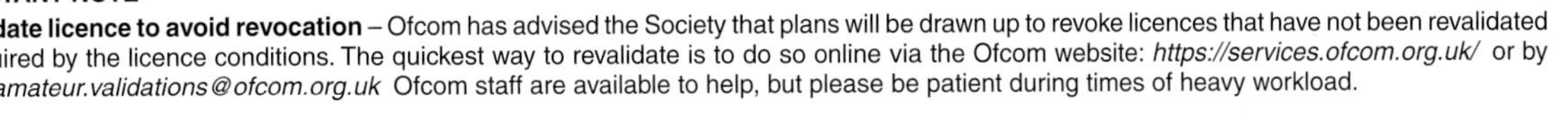

**IMPORTANT NOTE**

**Revalidate licence to avoid revocation** – Ofcom has advised the Society that plans will be drawn up to revoke licences that have not been revalidated as required by the licence conditions. The quickest way to revalidate is to do so online via the Ofcom website: *https://services.ofcom.org.uk/* or by email: *amateur.validations@ofcom.org.uk* Ofcom staff are available to help, but please be patient during times of heavy workload.

G4 ZXF Kevin Kimber, Low Harland Cottage, Farndale, York, YO62 7JX
GW4 ZXG L Thomas, Roughton, Corntown Road, Bridgend, CF35 5BH
G4 ZXI Nicholas Parnell, 4 Forge Lane, Headcorn, Ashford, TN27 9QQ
G4 ZXJ J Burns, 7 Johns Road, Eyemouth, TD14 5DX
G4 ZXN M Ward, 25 Margeson Close, Coventry, CV2 5NU
G4 ZXO P Horbaczewskyj, 27 Sheddingdean Close, Burgess Hill, RH15 8JQ
G4 ZXP V Legge, 26 Goldcroft Avenue, Weymouth, DT4 0ET
G4 ZXQ Andrew Mainwaring, The Old Smoke House, Lodge Farm Barns, Hereford, HR4 8NN
G4 ZXS E Loach, 99 Gorse Lane, Clacton-on-Sea, CO15 4RJ
G4 ZXT M Twigg, 30 Valley Drive, Yarm, TS15 9JQ
G4 ZXV W Bailey, 225 Holburne Road, Kidbrooke, London, SE3 8HF
G4 ZXZ W Johnson, 12a Kings Road, Spalding, PE11 1QB
G4 ZYH M Timms, 5 Lytchett Way, Nythe, Swindon, SN3 3PJ
G4 ZYL J Anderson, 44 Overhill Road, Burntwood, WS7 4SU
G4 ZYM W Williams, Plum Tree Farm, Commonwood Road, Wrexham, LL13 9TA
G4 ZYN M Sherlock, Flat 1, 34 Duke Street, Southport, PR8 1JA
G4 ZYO B Lawrance, 14 Warren Close, Porthleven, Helston, TR13 9BL
G4 ZYR H Webber, 6 Barn Ground, Highnam, Gloucester, GL2 8LJ
GW4 ZYV J Raymond, 23 Castle Pill Crescent, Steynton, Milford Haven, SA73 1HD
G4 ZYY G Fildes, 62 Higher Days Road, Swanage, BH19 2LB
G4 ZYZ T Seymour, 21 Chainhouse Road, Needham Market, Ipswich, IP6 8ER
G4 ZZD A Hellier, 64 Penlee Park, Torpoint, PL11 2PZ
GM4 ZZH H Firth, Edan, Berstane Road, Kirkwall, KW15 1NA
G4 ZZK B Tutt, 15 Alexandria Drive, Herne Bay, CT6 8HX
G4 ZZL R Lyons, 15 Winston Avenue, Tiptree, Colchester, CO5 0JU
GD4 ZZN A Rickward, 14 Ballakneale Avenue, Port Erin, Isle of Man, IM9 6ND
G4 ZZP K Lock, 5 Copthorne Crest, Shrewsbury, SY3 8RU
G4 ZZS D Bamber, 3 Abbotts View, Sompting, Lancing, BN15 0NG
G4 ZZV M Singh-Gill, 30 King Edwards Gardens, London, W3 9RQ
GM4 ZZW R Watts, 1c Kirklands, 100 Greenock Road, Largs, KA30 8PG
G4 ZZY T Watts, Carne Grey Cottage, Trethurgy, St. Austell, PL26 8YE
G4 ZZZ T Smith, Lower Carniggey Farm, Greenbottom, Truro, TR4 8QL

## G*5

G5 AU Andrew Albinson, 86 Pelican Parade, Ballajura, Australia, WA 6066
G5 BBL Jan Verduyn, 14 Ragleth Grove, Trowbridge, BA14 7LE
G5 BCO Patrick Gautier-Lynham, 95 Oxford Road, Marlow, SL7 2PL
GM5 BDW Robert Watson, 37a Muirfield Crescent, Dundee, DD3 8PY
G5 BH M Coleman, Flat A, 53 De Parys Avenue, Bedford, MK40 2TR
G5 BK CHELTENHAM A.R.A c/o A Woolford, 39 Apple Orchard, Prestbury, Cheltenham, GL52 3EH
G5 BW W Waugh, 67 Cragside, Whitley Bay, NE26 3EF
G5 CDC Joel Kornreich, 35 Charlotte Drive, Spring Valley, United States
GM5 CGA Leonard Landers, 33 Newmanswalls Avenue, Montrose, DD10 9DD
GM5 CX R Ferguson, 19 Leighton Avenue, Dunblane, FK15 0EB
G5 DJW Arthur Osmond, 10 Deerhurst Park, Forest Row, RH18 5GD
G5 EDQ S Hahn, 26 Watling Street, Gillingham, ME7 2YH
G5 FM Martin Wheeler, 114 Boundary Way, Glastonbury, BA6 9PH
G5 FZ LINCOLN SHORT WAVE CLUB c/o P Rose, Pinchbeck Farmhouse, Mill Lane, Sturton by Stow, Lincoln, LN1 2AS
G5 GX John Smith, 4 Townend Villas, Humbleton, Hull, HU11 4NR
G5 HI Robin Birch, 15 Chester Street, Cirencester, GL7 1HF
G5 HY D Wilkins, 8 Gainsborough Road, Ashley Heath, Ringwood, BH24 2HY
G5 JJ Taunton and District ARC c/o D Rosewarn, 16 Charles Crescent, Taunton, TA1 2XN
G5 KC C Quarton, Flaxton Gatehouse, Flaxton, York, YO60 7QT
G5 KN Kettering & District ARS c/o Keith Doswell, 15a Queen Street, Desborough, Kettering, NN14 2RE
G5 KW THE UK SIX METRE GROUP c/o D Toombs, 1 Chalgrove, Welwyn Garden City, AL7 2QJ
G5 LK Reigate Amateur Transmitting Society c/o Peter Tribe, The Paddock, Wix Hill, Leatherhead, KT24 6ED
G5 LP Lionel Parker, 128 Northampton Road, Wellingborough, NN8 3PJ
G5 MS MANCHESTER & DISTRICT ARS c/o Kev Hudson, 20 Claude Street, Crumpsall, Manchester, M8 5AW
G5 MUN H Van Driel, 20 Links Avenue, Little Sutton, Ellesmere Port, CH66 1QT
G5 MW MEDWAY A.R.T. c/o John Hale, 136 Bush Road, Cuxton, Rochester, ME2 1HB
G5 MY H Mee, 268 Victoria Rd East, Leicester, LE5 0LF
G5 NB N Brown, 3 Mulberry Tree Close, Filby, Great Yarmouth, NR29 3HD
G5 NF Roger Ward, Lower Ton-y-Felin Farm, Croespenmaen, Newport, NP11 3BE
G5 OW W Wigg, 7 Brendon Way, Long Eaton, Nottingham, NG10 4JS
G5 PI PHILIPS TELE LT c/o John Wilson, 20b High Green, Great Shelford, Cambridge, CB22 5EG
G5 QK SOUTHEND&DIS AR c/o Alan Radley, 16 Kingsley Lane, Thundersley, Benfleet, SS7 3TU
G5 RP VOWHARS c/o I White, 2 Appleby Cottages, Whithorn, Newton Stewart, DG8 8DQ
G5 RR HUCKNALL ROLLS ROYCE A.R.C c/o Steve Sorockyj, 8 Bowden Avenue, Bestwood Village, Nottingham, NG6 8XN
G5 RS GUILDFORD CONTEST GROUP c/o P Croucher, 66 Loop Road, Kingfield, Woking, GU22 9BQ
G5 RV MID SUSSEX ARS c/o Gavin Keegan, 12 Allington Road, Newick, Lewes, BN8 4NA
G5 TO Sheffield & District Wireless Society c/o Peter Day, Sheffield HF DX Group, 38 Broomhill Road, Chesterfield, S41 9DA
G5 UI Roger Perkis, 11 Epping Road, Corby, NN18 8GS
G5 UM LEICESTER RS CONTEST GROUP c/o D Wills, 70 Hidcote Road, Oadby, Leicester, LE2 5PF
GM5 VG OBO WINDY YETT CONTEST GROUP c/o W Miller, Whiteleys Farm, Ayr, KA7 4EG
G5 VH P Chapman, 112 Sharpland, Leicester, LE2 8UP
G5 VO N Clarke, Brimham Lodge Farm, Brimham Rocks Road, Harrogate, HG3 3HE
G5 VZ Christopher Pearson, 4 Brentwood Close, Thorpe Audlin, Pontefract, WF8 3ES
G5 WQ Ian Williams, Alma Cottage, South Marston, Swindon, SN3 4SN
G5 XV Newbury & District ARS c/o Michael Sansom, 19 Baily Avenue, Thatcham, RG18 3EG
G5 XX Ariel Radio Group c/o P Richmond, 57 The Fairway, Daventry, NN11 4NW
G5 YC ICARS c/o S Bunting, 17 Sunnydene Avenue, Highams Park, London, E4 9RE
G5 ZG BISHOPS STORTFORD AR SOCIETY c/o A Judge, 44 Thorley Lane, Bishop's Stortford, CM23 4AD

## G*6

G6 AAB T Sloane, 42 Ashbury Drive, Blackwater, Camberley, GU17 9HH
G6 AAC Paul McGoldrick, 23 Green Acre, Trebullett, Launceston, PL15 9QL
GW6 AAG F Steadman, 10 Oaktree Avenue, Sketty, Swansea, SA2 8LL
GM6 AAJ Graham Scattergood, 14 Market Street, Forfar, DD8 3EY
G6 AAK James Smith, 16 Cross Keys, Ossett, WF5 9SJ
G6 AAR David Bolingford, Cobb Gate, School Lane, Pulborough, RH20 4LL
G6 AAZ K Woodward, 19 Hazel Grove, Winchester, SO22 4PQ
G6 ABA P Dobson, 16 Glenair Avenue, Parkstone, Poole, BH14 8AD
G6 ABG Denis Coldbeck, 101 Westlands Road, Hull, HU5 5NX
G6 ABJ M Claydon, 4 Sandringham Gardens, London, N12 0NX
G6 ABM Angela Chick, The Rowans, Bourne View, Salisbury, SP4 0AA
G6 ABO R Campbell, 207 Seabank Road, Wallasey, CH45 1HD
G6 ABP C Cave, 31 Mill Road, Rearsby, Leicester, LE7 4YN
G6 ABU M Dale, 2 Ward Avenue, Mapperley, Nottingham, NG3 6EQ
G6 ACJ D Frampton, 28 Horsham Road, Owlsmoor, Sandhurst, GU47 0YY
G6 ADD T Hallam, 98 Keppel Road, Sheffield, S5 0TY
G6 ADG M Kennedy, 96 Kingsway, Boston, PE21 0AU
G6 ADO S Nicholas, Greenbank, Chester High Road, Neston, CH64 7TR
G6 AEB S Neil, 55 Colne Road, Brightlingsea, Colchester, CO7 0DU
G6 AEC D Nicholls, 22 Yeo Way, Clevedon, BS21 7UP
GM6 AES M Clark, 12 Achaphubil, Fort William, PH33 7AL
G6 AFA Philip Paskin, 36 Lewarne Road, Newquay, TR7 3JT
GD6 AFB Nigel Bazley, Newhaven, Mill Road, Ballasalla, Isle of Man, IM9 2EG
G6 AFE R Plested, 33 Hartbury Close, Cheltenham, GL51 0NZ
G6 AFG A Afford, 2 Holly Court, Sandiway, Northwich, CW8 2PP
G6 AFK J Adams, 6 Austen Road, Guildford, GU1 3NP
G6 AFL P Blay, Treetops, Mount Pleasant, Crewkerne, TA18 7AH
G6 AFS Denis Bell, 7 Chichester Drive, Cotgrave, Nottingham, NG12 3JJ
G6 AFT A Carr, 4 Tansor Close, Corby, NN17 2QP
G6 AFX Alan Crickett, 40 Ousden Close Cheshunt, Waltham Cross, EN8 9RQ
G6 AFZ M Tipper, 10 Lowdale Avenue, Scarborough, YO12 6JW
G6 AGA G Clark, 2 Whitton Manor Road, Isleworth, TW7 7NL
G6 AGN David Darby, 12 Laburnum Close, Clacton-on-Sea, CO15 2DD
G6 AGO B Bean, 19 Coleshill Road, Sutton Coldfield, B75 7AA
G6 AGP A Patterson, 10 Pear Tree Close, Wirral, CH60 1YD
G6 AGR M Taylor, 8 Clifford Close, Long Eaton, Nottingham, NG10 3BT
GW6 AGS R Thomas, 4 Duffryn Avenue, Cardiff, CF23 6LF
G6 AGT R Thomas, 9 Bayswater Close, Runcorn, WA7 1NY
G6 AGY A Smith, 103 Station Road, Seaham, SR7 0BD
G6 AGZ George Smith, 103 Station Road, Seaham, SR7 0BD
G6 AHC T Snook, 116 Rosemary Road, Poole, BH12 3HE
G6 AHD M Sumner, Jaggen, Maldon Road, Chelmsford, CM3 6LF
G6 AHE P Young, 8 The Slype, Wheathampstead, St. Albans, AL4 8RY
G6 AHF C Waterworth, 16 Fountains Walk, Lowton, Warrington, WA3 1EU
G6 AHH Christopher Walden, The Briers, Scures Hill, Hook, RG27 9JS
G6 AHK C Wallwork, Baileys Farm Cottages, 40-44 Henwood Green Road, Tunbridge Wells, TN2 4LF
G6 AHN S Reynolds, 12 Lowlands Crescent, Great Kingshill, High Wycombe, HP15 6EG
G6 AHO A Oakes, 12 Bridge Mill Court, Chorley, PR6 9DU
G6 AHR Richard Redpath, 11 View Terrace, The Platt, Lingfield, RH7 6QX
G6 AHV J Spriggs, Kreuzstr. 18, Tuerkenfeld, Germany, D-82299
G6 AHX S Evans, 18 Hillview Lane, Twyning, Tewkesbury, GL20 6JW
G6 AIB G Farline, Willow Cottage, Manor View Road, Scarborough, YO11 3PB
G6 AIG Hugh Gibson, 10 Trafalgar Street, Cambridge, CB4 1ET
G6 AII Robin George, Timbers, Lake Lane, Bognor Regis, PO22 0AD
G6 AIK J Gill, Millside, Mill Road, Steyning, BN44 3LN
G6 AIO Philip Hillier, 20 Firtree Road, Norwich, NR7 9LG
G6 AIQ M Homer, 29 Holmefield Avenue, Fareham, PO14 1EF
G6 AIU L Harland, 16 Burford Close, Dagenham, RM8 3ST
G6 AIZ M Holmes, 15 Anderton Way, Garstang, Preston, PR3 1RF
G6 AJ BARNSLEY & DISTRICT ARC c/o Paul Garthwaite, The Grey Horse, Barnsley, S75 2TX
GM6 AJA M Hunt, Gaoith The Saorsa, 30 Kanachrine Place, Ullapool, IV26 2TX
G6 AJC I Hodgkins, 2 Seagrave Road, Coventry, CV1 2AA
G6 AJG T Jenkins, 134 Frankland Road, Croxley Green, Rickmansworth, WD3 3AU
G6 AJK P Jones, Falkland House, Mountfields, Wrexham, LL13 0BZ
G6 AJS A Sharp, 17 Beechwood Avenue, Flanshaw, Wakefield, WF2 9JZ
G6 AJT B Kenneally, 5 Havengore, Pitsea, Basildon, SS13 1JU
G6 AJV Garry Larcombe, 2 Balmoral Drive, Hednesford, Cannock, WS12 4LT
G6 AJW David Lucas, 42 Falcon Way, Ashford, TN23 5UR
G6 AJX S Lampard, 111 Whitworth Way, Wilstead, Bedford, MK45 3EF
G6 AK J Brister, 49 Tiverton Road, Loughborough, LE11 2RU
G6 AKG Richard Ayley, 1 Ballam Close, Upton, Poole, BH16 5QT
G6 AKK P Archer, 26 Freshfield Drive, Macclesfield, SK10 2TU
G6 AKN M Bentley, 9 Tinkers Castle Road, Seisdon, Wolverhampton, WV5 7HF
G6 AKS F Barwell, Galahad, Penisarwaun, Caernarfon, LL55 3BN
G6 AKX Tony Blackburn, 42 Thames Drive, Biddulph, Stoke-on-Trent, ST8 7HL
G6 ALB A Burge, 32 High Street, Swaffham Prior, Cambridge, CB25 0LD
G6 ALG N Cutmore, 3 Linden Close, Tadworth, KT20 5UT
G6 ALJ Trevor Collins, 11 Sutton Road, Maidstone, ME15 9AE
G6 ALN G Colclough, 20 Pembroke Drive, Whitby, Ellesmere Port, CH65 6TD
G6 ALR R Delamare, 14 Brandreth Road, Plymouth, PL3 5HQ
G6 ALU S Drury, 25 Crosslands, Stantonbury, Milton Keynes, MK14 6AY
G6 ALW B Darby, 96 Bassnage Road, Halesowen, B63 4HG
G6 ALZ A Davis, 2 Wolverhampton Road, Essington, Wolverhampton, WV11 2DB
G6 AMF Brian Elliott, 41 Henwick Lane, Thatcham, RG18 3BN
GW6 AMK William Needham, Cnwc y Rhedyn, Aberporth, Cardigan, SA43 2DA
G6 AML J Newcombe, 75 Gargrave Road, Skipton, BD23 1QN
G6 AMV Lowry McConnell, 12 Marlborough Drive, Weston-Super-Mare, BS22 6DQ
G6 AMW C Williams, 133 Devon Drive, Chandler's Ford, Eastleigh, SO53 3GJ
G6 AMX Peter Helm, 90 Horne Street, Bury, BL9 9HS
G6 ANA Phillip Miller, Flat 907, De Montfort House, Leicester, LE1 5XR
GI6 ANC A Murphy, 53 Whitehouse Park, Newtownabbey, BT37 9SH
G6 ANI J Baverstock, Meadow View, Newbridge, Cadnam, Southampton, SO40 2NW
G6 ANJ C Perrott, 15 Chestnut Drive, Claverham, Bristol, BS49 4LN
G6 ANO L Goodwin, Gallifrey, The Village, Chelmsford, CM3 1AS
G6 ANR S Garfirth, 19 Ingleside Drive, Stevenage, SG1 4RN
G6 ANV B Gulliford, 12 Hawthorn Road, Eynsham, Witney, OX29 4NT
G6 AOA J Hewes, 15 Emmanuel Drive, Bottesford, Scunthorpe, DN16 3PE
G6 AOB A O'Brien, 25 Sands Road, Paignton, TQ4 6EG
G6 AOF J Henshaw, 88 Lower Blandford Road, Broadstone, BH18 8ND
G6 AOH R Hoblin, 4 Portiswood Close, Pamber Heath, Tadley, RG26 3UQ
GM6 AOJ William Hay, 11 Lovat Road, Glenrothes, KY7 4RU
GM6 AOR George Robertson, 32 The Square, Ellon, AB41 9JB
G6 AOS S Pilbeam, 74 Southbank Avenue, Marton Moss, Blackpool, FY4 5BX
G6 AOV C White, 20a The Beacon, Ilminster, TA19 9AH
G6 APB Geoffrey Taylor, 1 Haigh Street, Greetland, Halifax, HX4 8JF
G6 APD L Sawford, 16 Queens Close, Lee-on-The-Solent, PO13 9NA
G6 APE Adam Schofield, 38 Edinburgh Road, Broseley, TF12 5PE
G6 APH Colin Shiradski, 69 Masefield Avenue, Borehamwood, WD6 2HG
G6 APJ Graham Smith, 18 Poynings Place, Old Portsmouth, Portsmouth, PO1 2PB
GW6 APK George Sinclair, 4 Nant y Mynydd, Seven Sisters, Neath, SA10 9BU
G6 APQ F Hill, 12 Woodbine Walk, Chelmsley Wood, Birmingham, B37 6SB
G6 APW Trevor Harvey, 79 Vine Road, Stoke Poges, Slough, SL2 4DQ
G6 APX Stephen Handley, 8 Nabb Close, St. Georges, Telford, TF2 9PT
GM6 AQB A Riddell, 16 Lewis Drive, Old Kilpatrick, Glasgow, G60 5LE
G6 AQI T Smith, 1 St. Jude Gardens, Colchester, CO4 0QJ
G6 AQL A Ryan, 6 Cumloden Court, Newton Stewart, DG8 6AB
GM6 AQR H Wynne, 103 New City Road, Glasgow, G4 9JX
G6 AQW N Wiltshire, 66 Neville Road, Shirley, Solihull, B90 2QW
G6 ARC ANDOVER RADIO AMATEURS CLUB c/o C Keens, Toad Hall, 69 Lillywhite Crescent, Andover, SP10 5NA
G6 ARM N Kett, High View House, 1 Parnell Close, Northampton, NN6 7GJ
G6 ARO I Kendall, 65 Olive Grove, Swindon, SN25 3DB
G6 ARR Stephen Kimber, 3 Gloucester Way, Glossop, SK13 8RZ
G6 ART Stephen Langton, Corner Cottage, Harlow Road, Bishop's Stortford, CM22 7NB
G6 ASA N Lipman, Meadowcroft, Cotswold Road, Oxford, OX2 9JG
G6 ASH N Ash, 16 St. Marys Road, Sawston, Cambridge, CB22 3SP
G6 ASJ Michael Bradley, Flat 20, Crown Court, Portsmouth, PO1 1QN
G6 ASK J Matthews, Moor View, Oldways End, Tiverton, EX16 9JQ
GI6 ATD G Rodgers, 23 Rathmore Park, Bangor, BT19 1DQ
G6 ATK K Austin, 13 North End Grove, Portsmouth, PO2 8NF
G6 ATS Douglas Bowen, Flat 9, Moorfields Court, Silver Street, Bristol, BS48 2AG
GW6 ATT M Bryan, 10 Woodlands Road, Barry, CF63 4EF
G6 ATW R Czajkowski, 37 The Great Court, Royal Naval Hospital, Great Yarmouth, NR30 3JU
GI6 ATZ Gordon Curry, 87 Burren Road, Ballynahinch, BT24 8LF
G6 AUC H Whitfield, Apartment 88, Rishworth Palace, Rishworth Mill Lane, Sowerby Bridge, HX6 4RZ
G6 AUD S Challis, 22 Allens Road, Ramsden Heath, Billericay, CM11 1JF
G6 AUE Gary Cosham, 85 Capsey Road, Ifield, Crawley, RH11 0UF
GI6 AUI D Doherty, 42 Silverbrook Park, Newbuildings, Londonderry, BT47 2RD
G6 AUK G Evans, 36 Walton Road, Folkestone, CT19 5QS
G6 AUO M Graffham, 106 Barford Road, Edgbaston, Birmingham, B16 0EF
G6 AUP Barry Goodyear, 13 Moorland Avenue, Barnsley, S70 6PQ
G6 AUR B Golding, 67 Milford Avenue, Wick, Bristol, BS30 5PP
G6 AUS P Humby, 126 Middleton Road, Oswestry, SY11 2XA
G6 AUW Raymond Howes, 202 Abbotsbury Road, Weymouth, DT4 0NA
G6 AUY D Hawley, The Old Dairy, Edgefield Hall Barns, Edgefield, NR24 2RD
G6 AVI R Tucker, Foxhall Cottage, Dukes Lane, Attleborough, NR17 1BL
G6 AVK Colin Thomson, 160 Down Hall Road, Rayleigh, SS6 9PD
G6 AVL H Thompson, 6 Alexandra Chase, Cramlington, NE23 6AA
G6 AVN D Shaw, 33 The Fairway, Halifax, HX2 9PZ
G6 AVP A Rowe, 79 Shelvers Way, Tadworth, KT20 5QQ
G6 AVS J Russell, 13 Stonebridge Lea, Orton Malborne, Peterborough, PE2 5LY
G6 AVT G Stanhope, 39 Denham Close, Stubbington, Fareham, PO14 2BQ
G6 AVY D Lane, 230 Raeburn Avenue, Eastham, Wirral, CH62 8BB
G6 AWF David Miller, 33 Springfield Park, Twyford, Reading, RG10 9JG
G6 AWM C Montgomery, 70 Campbell Road, Twickenham, TW2 5BY
G6 AWO R Mansel, Ashcroft House, Ashfield Road, Bury St. Edmunds, IP30 9HJ
G6 AWP A McHardy, The Haven, Hull Road, Hull, HU12 0TE
G6 AWY N Armstrong, 16 Clay Hill Road, Sleaford, NG34 7TF
G6 AWZ Philip Ashdown, 1 Wheelers Patch, Emersons Green, Bristol, BS16 7JL
G6 AXC R Beaumont, The New Hall, Fletchergate, Hull, HU12 8ET
G6 AXE G Broad, 14 Albion Road, Westcliff-on-Sea, SS0 7DR
G6 AXH P Brothers, 101 Bridgewater Drive, Northampton, NN3 3AF
G6 AXK Peter Butler, 25 Orrishmere Road, Cheadle Hulme, Cheadle, SK8 5HP
G6 AXO A Bell, 24 Onslow Gardens, Ongar, CM5 9BG
G6 AXY P Coombes, Harleyford, Lower Wokingham Road, Crowthorne, RG45 6BT
GM6 AXZ K Cocks, 60a Palmerston Place, Edinburgh, EH12 5AY
G6 AY G Kellaway, 55 Ladbrooke Drive, Potters Bar, EN6 1QW
G6 AYD D Chorley, Sunnylands, Sandpitts Hill, Langport, TA10 0NG
G6 AYE Stephen Cotterill, Arcadia, Leicester Lane, Leicester, LE9 9JJ
G6 AYH K Cooke, 28 Curland Place, Longton, Stoke-on-Trent, ST3 5JL
GW6 AYR R Shearing, Woodstock, 6 Fair View Estate, Merthyr Tydfil, CF48 1HW
G6 AYS T Ramsden, 1a Fox Grove, Walton-on-Thames, KT12 2AT
G6 AYU P Rice, 4 Council St., Walton, Peterborough, PE4 6AQ
G6 AYX D Robinson, 3 King Edward Avenue, Wickham Market, Woodbridge, IP13 0SL
G6 AYY T Rumbold, 23 Montague Road, Saltford, Bristol, BS31 3LA
G6 AZE A Roberts, 9 Littlemoor Lane, Newton, Alfreton, DE55 5TY
G6 AZG Stephen Pearless, 22 Fieldridge, Newbury, RG14 2QD
G6 AZL P Tarmey, 20 Merlin Crescent, Branston, Burton-on-Trent, DE14 3JF

| | | |
|---|---|---|
| G6 | AZP | D Glover, 16 Cardigan Grove, Trentham, Stoke-on-Trent, ST4 8XY |
| G6 | AZR | A Granshaw, 38 Tudor Gardens, Stony Stratford, Milton Keynes, MK11 1HX |
| G6 | AZX | R Hughes, 4 Brittania Terrace, Porthmadog, LL49 9NB |
| G6 | BAH | G Davis, 2 New House, Ponthir Road, Newport, NP6 1PE |
| G6 | BAL | David De La Haye, Terzay, Oiron, FRANCE, 79100 |
| G6 | BAM | J Draper, 42 Pitt Street, Broadwaters, Kidderminster, DY10 2UN |
| GM6 | BAO | Anthony Devine, 12 Auchengate, Barassie, Troon, KA10 0UQ |
| G6 | BAT | D Falstein, 3 Gracefields, 121 The Avenue, Fareham, PO14 3AA |
| G6 | BAY | C Howes, 1 Wharrage Road, Alcester, B49 6QY |
| G6 | BBD | R Hancock, 16 Buttermere, Wellingborough, NN8 3ZA |
| G6 | BBG | A Harland, 23 Shelley Drive, Stratford sub Castle, Salisbury, SP1 3JZ |
| G6 | BBH | N Burton, 63 Salcombe Drive, Glenfield, Leicester, LE3 8AG |
| G6 | BBI | P Ward, 63 Salcombe Drive, Glenfield, Leicester, LE3 8AG |
| G6 | BBK | S Nelson, 10 Wragg Drive, Newmarket, CB8 7SD |
| G6 | BBM | G Tremain, 25 Hurst Road, East Molesey, KT8 9AQ |
| G6 | BBN | Jeffery Temple-Heald, Shires, 28 West End, Cambridge, CB22 4LX |
| G6 | BBR | M Thomas, 17 Rectory Park Avenue, Sutton Coldfield, B75 7BL |
| G6 | BBW | J Witts, 35 Warton Road, Basingstoke, RG21 [illegible] |
| G6 | BCG | R Whitehouse, 5 Parkland Drive, Darlington, DL3 9DT |
| G6 | BCL | Norman Miller, BC House, East Hanningfield Road, Chelmsford, CM3 8EW |
| G6 | BCM | S Ward, 33 All Saints Way, Aston, Sheffield, S26 2FJ |
| G6 | BD | Martin Farmer, Tara Cottage, 16 Beckside, Lincoln, LN2 2PH |
| G6 | BDH | J Kennard, 52 Lavender Lane, Stourbridge, DY8 3EF |
| GI6 | BDI | A King, 43 Orby Gardens, Belfast, BT5 5HS |
| G6 | BDM | Christopher Parker, Jasmine Cottage, Apperley, Gloucester, GL19 4DE |
| GI6 | BDN | R Larke, 11 Ballymaconnell Road South, Bangor, BT19 6DG |
| G6 | BDW | Andrew Sibley, 25 Vesta Avenue, St. Albans, AL1 2PG |
| G6 | BDY | R Southern, 208 Puxton Drive, Kidderminster, DY11 5HJ |
| G6 | BEB | J Lines, Karen House, 11 Hill Street, Brierley Hill, DY5 2AY |
| G6 | BEH | K Penaluna, 5 Holkham Close, Rushmere St. Andrew, Ipswich, IP4 5DW |
| G6 | BEL | Steven Fairweather, 65 Ambleside Avenue, Hornchurch, RM12 5EU |
| G6 | BEN | A Burke, 24 Wentworth Close, Farnham, GU9 9HJ |
| G6 | BER | S Boote, The Shippen, Downgate, Callington, PL17 8JX |
| GM6 | BEY | Michael Craig, 7 Hallyards Cottages, Kirkliston, EH29 9DZ |
| G6 | BFM | A Green, 117 Acanthus Road, Liverpool, L13 3DY |
| G6 | BFP | L Humphrey, Four Gables, 2 Gilletts Lane, High Wycombe, HP12 4BB |
| G6 | BGA | K Turvey, St. Vincents Cottage, St. Vincents Lane, West Malling, ME19 5BW |
| G6 | BGH | I MacDiarmid, 73 Stadium Avenue, Blackpool, FY4 3QA |
| GM6 | BGJ | I Maclennan, 70 Kenneth Street, Stornoway, HS1 2DS |
| GM6 | BGL | K Maclean, Gramaiche, Donavourd, Pitlochry, PH16 5JS |
| GM6 | BGQ | D Small, 17 Toll Court, Lundin Links, Leven, KY8 6HH |
| G6 | BGY | J Meek, Flat 26, Wickham Court, Clevedon, BS21 7TN |
| G6 | BHA | R Smart, 67 Corkland Road, Chorlton cum Hardy, Manchester, M21 8XT |
| G6 | BHB | J Seager, 58 Lone Valley, Widley, Waterlooville, PO7 5EB |
| G6 | BHE | N Rogers, 66 East Beach Park, 66 East Beach Park, Shoeburyness, SS3 9SG |
| G6 | BHH | D Palmer, Firdene, Abbey Road, Alton, GU34 5PB |
| G6 | BHI | A Palmer, Firdene, Abbey Road, Alton, GU34 5PB |
| G6 | BHQ | K Williams, 19 Narberth Crescent, Llanyravon, Cwmbran, NP44 8RJ |
| GM6 | BHR | R Warbrick, 8 Bathurst Drive, Alloway, Ayr, KA7 4QN |
| G6 | BHS | J Watson, 58 St. Georges Drive, Cheltenham, GL51 8NX |
| G6 | BHX | Chrisropher Walker, 19 Springfield Grove, Corby, NN17 1EN |
| G6 | BHY | R Vicarage, 10 Fleming Avenue, Sidford, Sidmouth, EX10 9NY |
| G6 | BIA | R Thompson, 39 Grotto Road, South Shields, NE34 7AQ |
| GM6 | BIG | David Anderson, 20 Greenrig Road, Hawksland, Lanark, ML11 9QA |
| G6 | BIM | J Bowers, 6 Fairview Park, Hetton-le-Hole, Houghton le Spring, DH5 0SE |
| G6 | BIT | D Crossley, 25 Newhaven Close, Bury, BL8 1XX |
| G6 | BIU | David Carter, 23 First Street, Low Moor, Bradford, BD12 0JQ |
| G6 | BIX | E Donbavand, 6 Springmeadow, Charlesworth, Glossop, SK13 5HP |
| G6 | BJB | A Forsyth, 14 Highgrove Road, Lancaster, LA1 5FS |
| G6 | BJG | I Hancock, 64 Swanswell Road, Solihull, B92 7EY |
| G6 | BJJ | I Harley, 302 Tavy House, Duke Street, Plymouth, PL1 4HL |
| G6 | BJL | R Harding, 12 Aller Vale Close, Exeter, EX2 5NH |
| G6 | BJO | P McTaggart, 33 Manor Farm Close, Barton-le-Clay, Bedford, MK45 4TB |
| G6 | BJQ | R Hanrahan, 53 Main St., Walton, Street, BA16 9QQ |
| G6 | BJR | K Hulbert, 15 St. Germans Road, Forest Hill, London, SE23 1RH |
| G6 | BJY | D Vivash, 16 Whitchurch Close, Maidenhead, SL6 7TZ |
| G6 | BK | Blackwood Contest Group c/o Robert JONES, 2 Pen-y-Cwarel Road, Wyllie, Blackwood, NP12 2HP |
| G6 | BKD | Julie Scotney, 30 Trinity Road, Rothwell, Kettering, NN14 6HY |
| G6 | BKL | P Metcalfe, 65 Saville Road, Whiston, Rotherham, S60 4DZ |
| G6 | BKY | Nigel Arkwright, 1 Penrith Avenue, Heysham, Morecambe, LA3 2DJ |
| G6 | BLA | S Woodford, The Lord Nelson, 1 Hale Road, Thetford, IP25 7RA |
| G6 | BLC | B Conway, 29 Mandeville Road, Southgate, London, N14 7NJ |
| G6 | BLK | A Johnson, Edelweiss, Boxley Road, Chatham, ME5 9JG |
| G6 | BLU | B Nicholls, 29 Wittmead Road, Mytchett, Camberley, GU16 6ER |
| G6 | BME | D Gibb, 46 School Road, Charing, Ashford, TN27 0JN |
| G6 | BMG | J Hind, 80 Forge Fields, Sandbach, CW11 3RD |
| GM6 | BML | A Ramsay, 15 Dunalistair Gardens, Broughty Ferry, Dundee, DD5 2RJ |
| G6 | BMP | A Roberts, 16a High Street, Llangefni, LL77 7NA |
| GW6 | BMR | S Roberts, 3 West Grove, Merthyr Tydfil, CF47 8HJ |
| G6 | BMY | R Satterthwaite, 47 Aberford Road, Baguley, Manchester, M23 1JY |
| G6 | BMZ | Michael Williams, 194 Hucknall Road, Nottingham, NG5 1FB |
| GI6 | BNI | D Mawhinney, 14 Cayman Avenue, Bangor, BT19 6XG |
| G6 | BNJ | J Bonnett, 87 Well Road, Otford, Sevenoaks, TN14 5PT |
| G6 | BNO | David Dallaway, 17 Bantams Close, Birmingham, B33 0YL |
| GM6 | BNS | Sean Lewis, Eyin Helga, Evie, Orkney, KW17 2PJ |
| G6 | BNW | J Garcia-Rodriguez, St. Albans, Mill Lane, Dover, CT15 4HR |
| G6 | BOF | G Hollidge, Clifton Close, Boundstone, Farnham, GU10 4TP |
| G6 | BOK | Peter King, 10 Heath Hey, Woolton, Liverpool, L25 4TJ |
| G6 | BOP | Alec Reid, 115 Robingoodfellows Lane, March, PE15 8JH |
| G6 | BOQ | Elizabeth Parker, Jasmine Cottage, Apperley, Gloucester, GL19 4DE |
| G6 | BOX | Simone Wilson, 21 Plumian Way, Balsham, Cambridge, CB21 4EG |
| G6 | BPH | F Bennewitz, 1 Millfield Avenue, Saxilby, Lincoln, LN1 2QN |
| G6 | BPK | S Cook, 50 Bath Road, Swindon, SN1 4AY |
| G6 | BPN | Robert Edmondson, 91 Lewin Road, London, SW16 6JX |
| G6 | BPY | W Roe, 39 Marlborough Road, Southwold, IP18 6LR |
| G6 | BQC | M Stuart, 207 Saunders Lane, Mayford, Woking, GU22 0NT |
| G6 | BQE | P Tilley, 22 Meadowsweet, Waterlooville, PO7 8RS |
| G6 | BQM | P Bentley, Sandy Ridge, Church Street, Stoke on Trent, ST7 4RS |
| G6 | BQQ | M Barnes, Drovers, Crampshaw Lane, Ashtead, KT21 2UF |
| G6 | BRA | Bracknell ARC c/o Ian Pawson, 3 Orion, Bracknell, RG12 7YX |
| GW6 | BRC | Barry ARS c/o Steven Trahearn, 148 Gladstone Road, Barry, CF62 8ND |
| G6 | BRD | W Hammond, 215 Broadoak Road, Ashton-under-Lyne, OL6 8RP |
| G6 | BRP | Philip Walter, 2 Hallams Lane, Beeston, Nottingham, NG9 5FH |
| G6 | BRS | BURY RADIO SOC c/o P Smith, Sbs Bury Radio Soc, Moses Yth Comm Cntr, Bury, BL9 0BS |
| GM6 | BRU | J Steele, 54 Myrtle Crescent, Bilston, Roslin, EH25 9SB |
| G6 | BRV | R Shelford, Wellbeech House, High Street, Uckfield, TN22 5JU |
| G6 | BRW | S Sumner, 7 St. Marys Close, Pirton, Hitchin, SG5 3RG |
| G6 | BRY | Christopher Thomas, 52 Derwent Road, Burton-on-Trent, DE15 9FR |
| G6 | BSP | S Guest, 2 Tanyard, Evershot, Dorchester, DT2 0JX |
| G6 | BSS | G Higgs, 68 Otterfield Road, West Drayton, UB7 8PF |
| G6 | BTB | C Pringle, 38 Priory Road, Littlemore, Oxford, OX4 4NE |
| G6 | BTO | A Layton, 17 Maplehurst, Leatherhead, KT22 9NB |
| G6 | BTP | E Beswarick, 2 Hurst, Beaminster, DT8 3ES |
| G6 | BTR | M Challis, 18 Castlefield Close, Eastleaze, Swindon, SN5 7EG |
| G6 | BTX | K Holmes, 313 Havering Road, Romford, RM1 4BZ |
| G6 | BUH | Joseph Walsh, 13 Byam Street, London, SW6 2RB |
| G6 | BUP | Christopher Tung-Lam Chan, 11 The Paddocks, Welwyn Garden City, AL7 2BW |
| G6 | BUT | HARLOW & DISTRICT ARS c/o Mike Simkins, 37 St. Andrews Meadow, Harlow, CM18 6BL |
| G6 | BUU | Raymond Costello, 6 Qua Fen Common, Soham, Ely, CB7 5DH |
| G6 | BUV | A Cutts, Highthorns Cottage, North Frodingham, Driffield, YO25 8LS |
| G6 | BUW | I Davies, Garthewyn, Caernarfon, LL55 2RL |
| G6 | BUY | R Gingell, 23 Woodfarm Road, Malvern Wells, Malvern, WR14 4PL |
| G6 | BVF | Nigel Linge, 21 Pennant Drive, Prestwich, Manchester, M25 3BT |
| GI6 | BVQ | T Finlay, 4 Station Road, Eglinton, Londonderry, BT47 3PR |
| G6 | BVR | R Gammage, 12 The Butts, Warwick, CV34 4SS |
| G6 | BVS | J Hayman, 22 Princess Street, Abertillery, NP3 1AR |
| G6 | BWA | C Clarke, 11 Eastmoor Villas, Epworth Road, Doncaster, DN9 2LH |
| G6 | BWE | K Edwards, 289 Monks Walk, Buntingford, SG9 9DZ |
| G6 | BWJ | J Richards, 44 Swain Street, Watchet, TA23 0AG |
| G6 | BWK | T Wallis, 17 Alderbank, Wardle, Rochdale, OL12 9NH |
| G6 | BWM | Robert Smith, 49 Aubourn Avenue, Lincoln, LN2 2JW |
| G6 | BWN | J Stewart, 101 West Way, Lancing, BN15 8LZ |
| G6 | BWO | Jack Taberner, 20 Stevenson Drive, Wirral, CH63 9AH |
| G6 | BWP | D Weaver, 8 Strathmore Close, Worthing, BN13 1PQ |
| G6 | BWT | A Bajjon, 35a Blackford Road, Shirley, Solihull, B90 4BU |
| G6 | BXO | C Blackwell, 20 Southworth Avenue, Blackpool, FY4 3LH |
| G6 | BXR | R Calvert, 3a Panxworth Road, South Walsham, Norwich, NR13 6DY |
| G6 | BXS | D Ellison, Riverside, Old Mill Drive, Colne, BB8 0TX |
| G6 | BXT | Martin Fry, 61 Swift Road, Abbeydale, Gloucester, GL4 4XH |
| G6 | BXU | Edward Hatherall, 101 Park Crescent, Abergavenny, NP7 5TL |
| G6 | BXV | D Willis, Rivendell, Shirnall Hill, Alton, GU34 3EJ |
| G6 | BYF | C Gomez, The Gazebo, Military Road, Rye, TN31 7NY |
| G6 | BYK | Jim Parkes, 65 Ferrier Road, Stevenage, SG2 0NZ |
| G6 | BYL | D Lycett, 1 Saredon Close, Pelsall, Walsall, WS3 4DH |
| G6 | BZE | M James, 9 Wyke Mark, Winchester, SO22 5DJ |
| G6 | BZG | L Green, 37 Park Road, Northville, Bristol, BS7 0RH |
| G6 | BZL | M Adams, The Vicarage, Intake Lane, Ormskirk, L39 0HW |
| G6 | BZQ | G Doubleday, 1 St. Johns Avenue, Chelmsford, CM2 0UA |
| G6 | BZW | Eric Butt, 97 Hawthorn Crescent, Yatton, Bristol, BS49 4RG |
| G6 | CAC | J Hallett, 16 Streche Road, Swanage, BH19 1NF |
| GI6 | CAG | W Millar, 9 Lynnehurst Drive, Comber, Newtownards, BT23 5LN |
| G6 | CAR | Anthony Baldwin, Rathlin, Dromnea, Kilcrohane, Bantry, Ireland, P75 Y300 |
| G6 | CBB | D Beddow, 34 Loweswater Road, Stourport-on-Severn, DY13 8LP |
| G6 | CBL | D Leslie, 8 The Avenue, Swarland, Morpeth, NE65 9JL |
| G6 | CBP | A Pidgeon, 106 Winchester Avenue, St Johns, Worcester, WR2 4JQ |
| G6 | CBY | M Jeeves, 52 Castlefields, Istead Rise, Gravesend, DA13 9EJ |
| G6 | CCB | A Stonehouse, 105 Humberston Avenue, Humberston, Grimsby, DN36 4ST |
| G6 | CCN | L Armstrong, 19 Barton Close, North Shields, NE30 2TG |
| G6 | CCQ | R Powell, Manuela, Jack Haye Lane, Stoke-on-Trent, ST2 7NG |
| G6 | CDT | G Henshaw, 18 Queens Avenue, Ilkeston, DE7 4DL |
| G6 | CDU | G Keeble, 4 Bardfield Way, Frinton-on-Sea, CO13 0AN |
| G6 | CDV | A Morling, 33 Russell Court, Chesham, HP5 3JH |
| G6 | CDW | N Miller, 3 Upwood Gorse, Tupwood Lane, Caterham, CR3 6DQ |
| G6 | CEM | Evan Weir, 10 St. Georges Crescent, Whitley Bay, NE25 8BJ |
| G6 | CEP | Alan Kneebone, 34 Henver Road, Newquay, TR7 3BN |
| G6 | CEZ | R Brand, 17 Park Road, Fordingbridge, SP6 1EQ |
| G6 | CFA | J Carrick Smith, 15 The Vale, Oakley, Basingstoke, RG23 7LB |
| G6 | CFC | G Purchon, 33 Lancaster Avenue, Hitchin, SG5 1PA |
| G6 | CFU | N Shaw, The Gables, Camp Lane, Banbury, OX17 1DH |
| G6 | CGC | R Sheppard, 51 Marks Road, Wokingham, RG41 1NR |
| G6 | CGI | Martin Rowat, 154 Hollingwood Lane, Bradford, BD7 4DB |
| G6 | CGO | E Parr, 18 Arundel Close, Macclesfield, SK10 2NS |
| G6 | CGQ | R Hatch, 4 Springfield Crescent, Parkstone, Poole, BH14 0LL |
| G6 | CGY | Robert Percival, 6 Bulmer Place, Hartlepool, TS24 9BQ |
| G6 | CHA | Ernest Povey, Hillcroft, Schoolfields, Henley-on-Thames, RG9 4DH |
| G6 | CHC | V Appleton, 15 Pinewood Crescent, Ramsbottom, Bury, BL0 9XE |
| G6 | CHD | Paul Bridle, 11a Romsey Grove, Wigan, WN3 6JQ |
| G6 | CHI | A Bowley, Plum Tree House, Walk Close, Derby, DE72 3PN |
| G6 | CHJ | M Carter, 14 North Star Drive, Leighton Buzzard, LU7 3DP |
| G6 | CHT | M Hall, 31 Meendhurst Road, Cinderford, GL14 2EF |
| G6 | CHX | P Holland, High Lea Cottage, Witchampton Lane, Wimborne, BH21 5AF |
| G6 | CIA | Thomas Kenyon, 31 Marble Hill Gardens, Twickenham, TW1 3AU |
| G6 | CIE | R Townsend, 3 Cranfield View, Darwen, BB3 2HP |
| G6 | CIF | D Taylor, 8 Russell Drive, Wollaton, Nottingham, NG8 2BH |
| G6 | CIO | J Robinson, 31 Church Road, Banks, Southport, PR9 8ET |
| G6 | CIP | P Ralston, Laund House, 9 College Avenue, Liverpool, L37 3JL |
| G6 | CIT | R Young, 143 Rodmell Avenue, Saltdean, Brighton, BN2 8PH |
| G6 | CJB | P White, 8 Kingswood Court, Maidenhead, SL6 1DD |
| GW6 | CJJ | J Alexander, Awel Ingli, Cilgwyn, Newport, SA42 0QS |
| G6 | CJR | Stuart Barber, Homedale, St. Monicas Road, Tadworth, KT20 6ET |
| G6 | CJT | B Bradshaw, 28 Park House Walk, Low Moor, Bradford, BD12 0PL |
| G6 | CKD | L Newbury, 97 Johns Avenue, Hendon, London, NW4 4EN |
| G6 | CKE | C Evans, 21 Snowdrop Close, Crawley, RH11 9EQ |
| G6 | CKH | J Muir, 150 Thorntree Road, Thornaby, Stockton-on-Tees, TS17 8LX |
| G6 | CKJ | David Morris, 255 Lichfield Road, Wolverhampton, WV11 3EW |
| G6 | CKK | R Martin, 1 Rosemount Court, Rochester, ME2 3NF |
| G6 | CKL | I Martin, 24 Heddington Close, Trowbridge, BA14 0LH |
| G6 | CKM | D Langdon, 17 Forest Grove, Eccleston Park, Prescot, L34 2RY |
| G6 | CKN | R Morrison, 38 Burnfoot Road, Hawick, TD9 8EN |
| G6 | CKW | R Beattie, 11 Pine Grove, Bricket Wood, St. Albans, AL2 3ST |
| G6 | CKY | M Gray, 20 Ravenstone Street, London, SW12 9SS |
| G6 | CKZ | Paul Berwick, 4 Brewer Road, Crawley, RH10 6BP |
| G6 | CLA | Graham Blacksell, 152 Hawthorn Avenue, Colchester, CO4 3YA |
| G6 | CLD | Geraldine Coker, 46 Clarendon Road, Ipplepen, Newton Abbot, TQ12 5QS |
| G6 | CLK | Peter Carter, 19 Felix Road, Walton-on-Thames, KT12 2LB |
| G6 | CLP | John Miller, 7 Malvern Crescent, Ashby de la Zouch, LE65 2JZ |
| G6 | CLU | [illegible], 8 High Beech, Chalet Park, Battle Road, St Leonards on Sea, TN37 7BS |
| G6 | CLW | B Lloyd, 243 Stand Lane, Radcliffe, Manchester, M26 1JA |
| G6 | CLX | David Lloyd, 506 Manchester Road, Bury, BL9 9NZ |
| GI6 | CMA | R Dawson, 31 Clonmore Manor, Lisburn, BT27 4EW |
| G6 | CMB | Ian Dalton, 10 St. Vincents Villas, Temple Hill, Dartford, DA1 5HT |
| G6 | CMD | C Driver, 23 Mercers Row, St. Albans, AL1 2QS |
| G6 | CMF | A Daborn, 49 Crescent Road, Locks Heath, Southampton, SO31 6PE |
| G6 | CML | John Sykes, 20 Woodend Road, Bournemouth, BH9 2JQ |
| G6 | CMN | A Shaw, 14 Delph Crescent, Clayton, Bradford, BD14 6RY |
| GM6 | CMQ | Daniel Robson, 35 Lady Nairne Road, Dunfermline, KY12 9YD |
| G6 | CMS | Mark Robertson, 13 Orchard Cottages, Main Road, Chelmsford, CM3 3AD |
| G6 | CMV | D Palmer, Spidrift, Landsdown Road, Malvern, WR14 1HX |
| G6 | CMX | James Pell, 33 Low Street, Winterton, Scunthorpe, DN15 9RT |
| G6 | CND | John Oliver, 67 High Street, Great Houghton, Barnsley, S72 0AU |
| G6 | CNF | J Payne, 71 Waarden Road, Canvey Island, SS8 9AB |
| G6 | CNK | R Freshwater, 82 Sandford Road, Chelmsford, CM2 6DH |
| G6 | CNL | P Farnell, 40 Thorney Lane, Luddendenfoot, Halifax, HX2 6UX |
| G6 | CNQ | Terence Genes, 28 Hillside Road, Burnham on Crouch, CM0 8EY |
| GW6 | CNS | Jeffrey Graham, 23 Somerset Road, Barry, CF62 8BL |
| G6 | CNW | J Gibson, Penrose Cottage, Carne, St. Austell, PL26 8DB |
| G6 | CNX | J Goodwin, 10 Abingdon View, Worksop, S81 7RT |
| G6 | COB | John Hodkinson, 3 Cypress Close, Market Drayton, TF9 3HJ |
| G6 | COG | D Holdsworth, Middle Pasture, Heath Lane, Halifax, HX3 0AG |
| G6 | COL | LINCOLN SHORTWAVE c/o P Rose, Pinchbeck Farmhouse, Mill Lane, Sturton by Stow, Lincoln, LN1 2AS |
| G6 | COZ | R Turner, 73 Digby Court, Nottingham, NG7 1RG |
| G6 | CP | David Cutter, 34 Greengate Lane, Knaresborough, HG5 9EL |
| G6 | CPE | K Stanley, 35 St. Blaize Road, Romsey, SO51 7JY |
| G6 | CPF | J Stephenson, 16 Greenways, Driffield, YO25 5HX |
| G6 | CPO | N Wysocki, 6 Rose Dene, Stourport-on-Severn, DY13 8SU |
| G6 | CPS | A Yates, 12 Graham Drive, Middleton, King's Lynn, PE32 1RL |
| G6 | CPX | Mark Waples, 24 Constable Drive, Wellingborough, NN8 4UX |
| G6 | CPY | E Whitham, 72 Bole Hill, Treeton, Rotherham, S60 5RE |
| G6 | CQB | M Wilson, 23 Claydown Way, Slip End, Luton, LU1 4DU |
| G6 | CQC | Alan Varty, Wisteria, Hillcrest, Durham, DH7 0BQ |
| G6 | CQG | L Constantine, 18 Hillbeck, Halifax, HX3 5LU |
| G6 | CQH | J Abbishaw, Hastings House Farm, Littletown, Durham, DH6 1QB |
| G6 | CQR | C Bailey, 32 Ryland Road, Moulton, Northampton, NN3 7RE |
| G6 | CRC | CHESHUNT & DISTRICT A.R.S c/o Robert Gray, 51 Wyatt Close, Ickleford, Hitchin, SG5 3XY |
| G6 | CRD | S Brown, 5 Keepside Close, Ludlow, SY8 1BQ |
| G6 | CRF | Terence Bailey, 65 Edge Lane, Chorlton cum Hardy, Manchester, M21 9JU |
| G6 | CRG | B Bowes, 1 Rockall Close, Southampton, SO16 8EH |
| G6 | CRR | R Solomons, 32 Church Road, Pembury, Tunbridge Wells, TN2 4BT |
| GM6 | CRX | F McLeod-Stangroom, 6 Leonach, Strathlachlan, Cairndow, PA27 8DB |
| G6 | CSC | William Skidmore, 29 The Meadows, Grisedale Road, Bakewell, DE45 1TP |
| G6 | CSK | A Beal, 115 Maldon Road, Witham, CM8 1HR |
| G6 | CSL | Christopher Redding, 20 Bromley Street, Workington, CA14 2TP |
| G6 | CSN | Geoffrey Chadwick, 25 Passmonds Crescent, Rochdale, OL11 5AW |
| G6 | CSR | H Calloway, 6 Franchise Gardens, Wednesbury, WS10 9RQ |
| G6 | CTA | J Davidson, 12 Hanbury Close, Dronfield, S18 1RF |
| G6 | CTC | COVTRY TECH ARC c/o James Witt, 67 Dillotford Avenue, Coventry, CV3 5DS |
| G6 | CTE | Leigh Duffill, 14 Leonard Street, Hull, HU3 1SA |
| G6 | CTH | E Dunne, 16 Ulleswater Close, Little Lever, Bolton, BL3 1UD |
| G6 | CTP | H Wakefield, 32 Mandene Gardens, Great Gransden, Sandy, SG19 3AP |
| G6 | CTV | F Foggs, 62 Laurel Manor, 18 Devonshire Road, Sutton, SM2 5EJ |
| G6 | CTY | C Edwards, 54 Thoroughgood Road, Clacton-on-Sea, CO15 6DP |
| G6 | CUA | H Erridge, 15 Maurice Road, Southsea, PO4 8HH |
| G6 | CUE | J Frampton, 54 Hudson Road, Bexleyheath, DA7 4PG |
| G6 | CUK | Andrew Fisher, 14 Whitefields Drive, Richmond, DL10 7DQ |
| G6 | CUQ | Neil Wedgbury, 12 The Ridgeway, Astwood Bank, Redditch, B96 6LT |
| GW6 | CUR | Stephen Williams, 371 Coed Y Gores, Llanedeyrn, Cardiff, CF23 9NR |
| G6 | CUT | J Whitehurst, Serendipity, 97 Noke Common, Newport, PO30 5TY |
| G6 | CUV | K Wyeth, 3 West Palace Gardens, Weybridge, KT13 8PU |
| G6 | CUY | J Wildsmith, Lingmoor, 7 Lambert Road, Uttoxeter, ST14 7QG |
| G6 | CVB | J Taylor, 12 Fairview Drive, Westcliff-on-Sea, SS0 0NY |
| G6 | CVD | C Thornley, Sylvastone House, Herne Street, Herne, CT6 7HG |
| G6 | CVE | R Tanfield, 8 Rede Close, Bedford, MK41 7UH |
| G6 | CVP | David Wilkins, 74 Wood Lodge Lane, West Wickham, BR4 9NA |
| G6 | CVV | Mark Gumbrell, 13 Crowfields, Deeping St. James, Peterborough, PE6 8NY |
| G6 | CVW | W Griffiths, 6 Stanway Close, Middleton, Manchester, M24 1HE |
| G6 | CVY | H Gibbons, 13 Woburn Avenue, Bolton, BL2 3AY |
| G6 | CW | ARC OF NOTTINGHAM c/o Michael Shaw, 50 White Road, Nottingham, NG5 1JR |

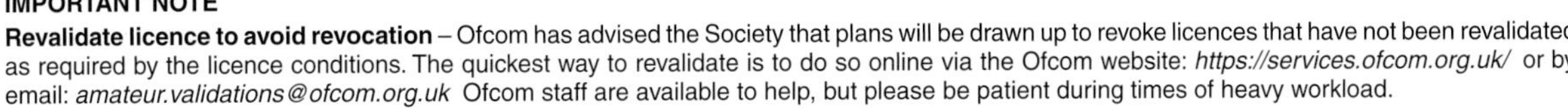

**IMPORTANT NOTE**

**Revalidate licence to avoid revocation** – Ofcom has advised the Society that plans will be drawn up to revoke licences that have not been revalidated as required by the licence conditions. The quickest way to revalidate is to do so online via the Ofcom website: *https://services.ofcom.org.uk/* or by email: *amateur.validations@ofcom.org.uk* Ofcom staff are available to help, but please be patient during times of heavy workload.

G6 CWF C Hazell, 18 Cleeve Hill, Downend, Bristol, BS16 6HN
G6 CWH S Harwood, 24 Firle Crescent, Lewes, BN7 1QG
G6 CWW Victor Holbrook, 84 Haddon Street, Derby, DE23 6NQ
G6 CWZ D McCallum, Glan Alaw, Llanddeusant, Holyhead, LL65 4AG
GI6 CXD R McWhirter, 200 Townhill Road, Portglenone, Ballymena, BT44 8AR
G6 CXI A Long, 35 Heath Court, Grampian Way, Derby, DE24 9NG
G6 CXM G Lees, Timbercroft, Elliotts Orchard, Warwick, CV35 8ED
G6 CXN K Lankshear, 28 Monmouth Place, Newcastle, ST5 3DF
G6 CXO D Lloyd, 16 Kingsley Road, Brighton, BN1 5NH
G6 CXV Kevin Phillips, Stockings Barn, Whitbourne, Worcester, WR6 5SR
G6 CXY R Revan, 50 Woodland Rise, Welwyn Garden City, AL8 7LF
G6 CYA Roger King, 55 Coppins Road, Clacton-on-Sea, CO15 3HS
G6 CYE A Read, 36 West St., Tollesbury, Maldon, CM9 8RJ
G6 CYF David Richards, 433-435 Cronton Road, Widnes, WA8 5QG
G6 CYH I Roberts, 32 Priory Drive, Plymouth, PL7 1PU
G6 CYO Ian Jarvis, The Garden House, Walkley Wood, Stroud, GL6 0RT
G6 CYR Ronald Jenkins, 29 Ebnal Close, Barons Cross, Leominster, HR6 8SL
G6 CYT R Kempton, 14 Bloxam Gardens, Rugby, CV22 7AP
G6 CYU M Kendrick, 157 Pinar De Gariata, La Nucia, Alicante, Spain, 3530
G6 CYV P Kirkham, 9 Bluebell Close, Biddulph, Stoke-on-Trent, ST8 6TJ
G6 CZB R Poffley, 3 Bowerhill Road, Salisbury, SP1 3DN
G6 CZD Michael Swanwick, 45 Coach Way, Willington, Derby, DE65 6ES
GW6 CZE C Peacock, 8 Heol Ewenny, Pencoed, Bridgend, CF35 5QA
GM6 CZM I McAulay, 9 Randolph Cliff, Edinburgh, EH3 7TZ
G6 CZO David Mcghie, 54 School Road, Newborough, Peterborough, PE6 7RG
G6 CZS C Moore, 4 Woodhurst Road, Peterborough, PE2 8PF
G6 CZX W Aitchison, 18 Kerensa Green, Falmouth, TR11 2HE
G6 CZZ J Abram, 3 Frenchies View, Denmead, Waterlooville, PO7 6SH
G6 DAC Martin Bounds, 5 Ingleby Close, Heacham, King's Lynn, PE31 7SA
G6 DAD D Blagburn, 10 Tottington Avenue, Springhead, Oldham, OL4 4RY
G6 DAH Donald Budd, 81 Bohemia Chase, Leigh-on-Sea, SS9 4PW
G6 DAI N Brickwood, 4 Vale Cottages, Shillingstone, Blandford Forum, DT11 0SS
G6 DAN Brian Daniels, 113 Orchard Way, Wymondham, NR18 0NZ
G6 DAO G Bradbury, 3 Westfield Bank, Barlborough, Chesterfield, S43 4EG
G6 DAP J Balmford, Upper Brook Farm House, The Avenue, Aylesbury, HP18 9LD
G6 DAQ A Boonham, 1 Oakleigh Drive, Sedgley, Dudley, DY3 3LH
G6 DAU Philip Bidwell, 156 Elstree Park, Barnet Lane, Borehamwood, WD6 2RP
G6 DAY M Pemberton, 37 Bardsley Close, Croydon, CR0 5PS
G6 DBC A Norfolk, 18 Middle Lane, Amcotts, Scunthorpe, DN17 4AT
G6 DBJ J Fairhurst, 4 Glenmoor Road, Buxton, SK17 7DD
G6 DBP J Firmstone, 1 Holly Grange, Rhoswiel, Oswestry, SY10 7TU
G6 DBQ D Fryer, Norwood, 105 Chester Road, Stockport, SK7 6HG
G6 DBU R Gambles, 5 College Way, Horspath, Oxford, OX33 1SQ
G6 DBX A Grover, 44 Stirling Court Road, Burgess Hill, RH15 0PT
G6 DBY P Gould, Derna, Surrey Lane, Colchester, CO5 0QT
G6 DBZ S Griffin, 50 Cherrybrook Drive, Broseley, TF12 5SH
G6 DCH J Molyneux, 18 Bay Close, Horley, RH6 8LF
G6 DCS Gary Norris, 1 Pear Tree Avenue, Newhall, Swadlincote, DE11 0LZ
G6 DCT David Littlewood, 50 Industry Road, Sheffield, S9 5FQ
GI6 DCX E Lyons, Creevy Tennant Lodg, 17 Brae Road, Ballynahinch, BT24 8UN
G6 DDA Andrew Moss, 20 Black-A-Tree Court, Black-A-Tree Road, Nuneaton, CV10 8BD
G6 DDC J Leatherbarrow, 10 Henley Drive, Southport, PR9 7JU
G6 DDF John Morris, 45 Ffordd Pentre Mynach, Barmouth, LL42 1EN
G6 DDJ Stuart Pillinger, Calle Sileno 10, Mailbox (buzon) 470, Fortuna, Spain, MURCIA 30620
G6 DDO R Owen, 36 Foley Road, Stourbridge, DY9 0RT
G6 DDP R Oakden, 38 Brookfield Avenue, Hucknall, Nottingham, NG15 6FF
G6 DDR Leonard Horne, 8 Kingsway Avenue, Broughton, Preston, PR3 5JN
G6 DDU G Goddard, 30 Western Avenue, Holbeach, Spalding, PE12 7QD
G6 DEA Geoffrey Goss, 1 Willow Bank Close, Throckmorton, Pershore, WR10 2JW
G6 DEG T Hampson, 6 Rushmere Drive, Bury, BL8 1DW
G6 DEP M Harris, 11 Lower Rawlinson Terrace, Tredegar, NP2 4JD
G6 DER K Hewitt, 6 Church Grove, Monk Bretton, Barnsley, S71 2EY
G6 DET Marc Heighton, 3 Warner Road, Codsall, Wolverhampton, WV8 1SA
G6 DEV D Harris, 15 Millwood Road, Orpington, BR5 3LG
GI6 DEY Frank Hunter, 2 Wandsworth Court, Belfast, BT4 3GD
G6 DFA C Willies, 17 Campion Way, Sheringham, NR26 8UN
G6 DFB C Smith, 83 Sledmore Road, Dudley, DY2 8DY
G6 DFC P Johnson, 3 Lance Drive, Chase Terrace, Burntwood, WS7 1FA
G6 DFH John Roberts, 155 Langley Hall Road, Solihull, B92 7HB
G6 DFM John Phelps, Windy Dene, Green Lane, Chessington, KT9 2DT
G6 DFR Trevor Parfitt, 4 Back Street, Lakenheath, Brandon, IP27 9HF
G6 DFV Allan Parker, 13 Hartley Street, Colne, BB8 9DF
GW6 DFX D James, 5 Lon Y Parc, Cardiff, CF14 6DF
G6 DFY G Joly, 116 Hind Grove, London, E14 6HP
G6 DFZ M Jones, 28 Winston Avenue, Colchester, CO3 4NQ
G6 DGK Gavin Keegan, 12 Allington Road, Newick, Lewes, BN8 4NA
G6 DGQ J Baines, 2 Moor Close, Radcliffe, Manchester, M26 4QF
G6 DGR N Bean, 19 Coleshill Road, Sutton Coldfield, B75 7AA
GW6 DGU Roy Britton, LLWYNON, 95 North Road, Cardigan, SA43 1LT
G6 DGV C Brock, 37 Ashington Drive, Bury, BL8 2TS
G6 DGW E Ball, 59 Queen Anne Gardens, Falmouth, TR11 4SW
G6 DGX J Raby, Cedar House, Coppenhall, Stafford, ST18 9DA
G6 DHD A Rollason, Fern Lodge, The Parade, Newton Abbot, TQ13 0JH
G6 DHI David Kennedy, 1 Lynton Road, Hindley, Wigan, WN2 4EH
G6 DHT P Chace, 3 Nightingale Way, Sutton Bridge, Spalding, PE12 9RG
G6 DHU Michael Chace, 34 Shortill Farms Road, Buxton, United States, 4093
G6 DHW I Clayton, 15 Ashbourne Drive, Desborough, Kettering, NN14 2XG
G6 DIC Philip Dickinson, 28 Chaucer Crescent, Newbury, RG14 1TR
G6 DID John Davis, 38 Dover Close, Southwater, Horsham, RH13 9XX
G6 DIE G Drohan, 23 Lindholme Drive, Rossington, Doncaster, DN11 0UP
G6 DIF V Van Den Bergh, St. Francis C of E Church, Masefield Drive, Tamworth, B79 8JB
G6 DIM T Eves, Banks Farm, Manor Road, Romford, RM4 1NH
G6 DIO R Everson, Eversons Farm, Bardfield Road, Braintree, CM7 5HU
G6 DIQ J Wilkins, 6709 Western Avenue, Darien, United States, 60561
G6 DIR Martin Wray, 18 Cleveland Street, Loftus, Saltburn-by-The-Sea, TS13 4JB
G6 DIZ Denise Feeley, 177 Rock Street, Sheffield, S3 9JF
G6 DJH D Harvey, 23 Sprules Road, Brockley, London, SE4 2NL
G6 DJQ G Tomlinson, 10 Ashbourne Road, Underwood, Nottingham, NG16 5EH
G6 DJS D Sojkowski, 7 Spenlow Drive, Chelmsford, CM1 4UQ
G6 DJY W Telford, The Walnuts, Main Road, Boston, PE20 2LQ
G6 DKE E Reynolds, 11 New Street, Sudbury, CO10 1JB
G6 DKF L Marsh, 18 Northgate, Hornsea, HU18 1ES
G6 DKI R Tew, 11 Huson Road, Warfield, Bracknell, RG42 2QX
G6 DKK S Simes, 53 Waterford Lane, Cherry Willingham, Lincoln, LN3 4AN
G6 DKM L Sandford, 150 Tipton Road, Woodsetton, Dudley, DY3 1AL
GI6 DKQ Brian McKeen, 27 Old Grange Drive, Carrickfergus, BT38 7HG
G6 DKS R Saverton, Flat 8, 2 Christ Church Road, Surbiton, KT5 8JJ
G6 DLJ P Bridges, 30 New Road, Hythe, Southampton, SO45 6BP
G6 DLM Q Borthwick, 106 Westpole Avenue, Cockfosters, Barnet, EN4 0BB
G6 DLZ Paul Bosanquet-Bryant, Flat 21, Westcliff Court, Clacton-on-Sea, CO15 1LA
G6 DMF D Wilkins, 58 High Road, Wormley, Broxbourne, EN10 6JN
G6 DMG Stephen Wellon, 71 Toftdale Green, Lyppard Bourne, Worcester, WR4 0PE
G6 DMM K Webster, 27 Glendale Close, Horsham, RH12 4GR
G6 DNA T Cattermole, 24 Cromwell Road, Colchester, CO2 7EN
G6 DNH M Carvell, 12 Liskeard Drive, Allestree, Derby, DE22 2GW
GI6 DNI D Chapman, 3 Brustin Lee, Ballygally, Larne, BT40 2QA
G6 DNL K Snellin, 3 Turnberry, Bracknell, RG12 8ZJ
G6 DNV G Taylor, 10 Scott Close, Hexham, NE46 2QB
GW6 DOC Robert Yarnold, 47 Small Meadow Court, Caerphilly, CF83 3RT
G6 DOD Mark Wheeler, 105 High Street, Wootton Bridge, Ryde, PO33 4LU
G6 DOF C Wankling, 60 Castle Road, Rayleigh, SS6 7QF
G6 DOI Clive Wigginton, 4 Copes Haven, Shenley Brook End, Milton Keynes, MK5 7HA
G6 DOK C Williams, Caermai, Stad Pen y Berth, Llanfairpwllgwyngyll, LL61 5YT
G6 DON J Walsh, 7 Unicorn Place, Ball Green, Stoke-on-Trent, ST6 6LX
G6 DOQ H Davies, 76 Brook Lane, Timperley, Altrincham, WA15 6RS
G6 DOR D Durrant, 22 St. Martinsfield, Martinstown, Dorchester, DT2 9JU
G6 DOV L Dunn, 24 Mynchen Road, Beaconsfield, HP9 2BA
G6 DOW A Deacon, 1 Connaught Gardens, Crawley, RH10 8NB
G6 DOX D Dodd, 5 Orchard Garth, Wreay, Carlisle, CA4 0RN
G6 DOZ Leslie Dell, 205 Thelwall Lane, Warrington, WA4 1NF
G6 DPE D Evans, 631 Chatsworth Road, Chesterfield, S40 3NT
G6 DPH Barrie Flinn, 65 Marina Avenue, Great Sankey, Warrington, WA5 1JH
G6 DPL L Green, 76 Dibleys, Blewbury, Didcot, OX11 9PU
G6 DPS M Harrison, 33 Campion Park, Up Hatherley, Cheltenham, GL51 3WA
G6 DPW D Waghorne, 5 Freelands Drive, Church Crookham, Fleet, GU52 0TE
G6 DQA Matthew Morgan, 125 Holymoor Road, Holymoorside, Chesterfield, S42 7DR
G6 DQB John Mitchell, Y Graigwen, Cadnant Road, Menai Bridge, LL59 5NG
G6 DQH D Moore, 71 Woodlands Avenue, Talgarth, Brecon, LD3 0AT
G6 DQK P McBride, 34 Arundel Close, Carrbrook, Stalybridge, SK15 3LS
G6 DQO Iain Martin, 6 Hollow Oak Lane, Cuddington, Northwich, CW8 2XN
G6 DQT W Lasbury, Sonserra Flats, Flat 4 Fekruna St, St Pauls Bay, Malta
G6 DQU L Mullin, 16 Springfield, Sowerby Bridge, HX6 1AD
G6 DQY J Orrells, Perry Willows, Yeaton, Shrewsbury, SY4 2HY
G6 DQZ N Perry, 10 Carlyle Avenue, Kidderminster, DY10 3QZ
G6 DRC D Cooper, Linden House, Greenhill Park Road, Evesham, WR11 4NL
G6 DRG Terry Place, 73 Williams Street, Langold, Worksop, S81 9NX
G6 DRH D Hickton, 27 Vanguard Road, Long Eaton, Nottingham, NG10 1DX
GI6 DRK Ian Humes, 160 North Road, Belfast, BT4 3DJ
G6 DRN P HAYLOR, 76 Beauchamp Road, Billesley Common, Birmingham, B13 0NR
G6 DRP D Hemmins, 18 Burn Walk, Burnham, Slough, SL1 7EW
G6 DSA R Jeffery, 7 Corfe Way, Winsford, CW7 1LU
G6 DSD R Jones, 20 Bibsworth Avenue, Moseley, Birmingham, B13 0BA
G6 DSG N Austin, 184 Tunstall Road, Knypersley, Stoke-on-Trent, ST8 7AH
G6 DSP C Addis, 1 Newchurch Lane, Culcheth, Warrington, WA3 5RW
G6 DTH A Allnutt, The Squirrels, Nutcombe Lane, Dorking, RH4 3DZ
G6 DTN David Crake, Kentolop, Holyhead Road, Shrewsbury, SY4 1EE
G6 DTT A Campbell, Eden Park, Den Cross, Edenbridge, TN8 5PW
G6 DTW Alvin Challen, Links Corner Cottage, Links Road, Ashtead, KT21 2EG
G6 DUC A Rowlands, Hill House, Bridge Road, Bristol, BS8 3PE
G6 DUH David Crewe, 71 Ladybalk Lane, Pontefract, WF8 1LA
G6 DUI Ian Castle, 26 Lonsdale Drive, Sittingbourne, ME10 1TS
G6 DUN R Burrows, 32 Frenchs Farm Road, Poole, BH16 5RT
G6 DUT Malcolm Bluck, 26 Mayfield Avenue, Scarborough, YO12 6DF
G6 DVE A Redshaw, 417 Marston Road, Marston, Oxford, OX3 0JG
G6 DVP R Vickers, 51 Charlecote Close, Redditch, B98 0TQ
G6 DWM Gunam Sohal, 15 Icknield Road, Luton, LU3 2NY
G6 DWO R Smart, 52 Devonshire Avenue, Southsea, PO4 9EF
G6 DWS N Shearer, 64 Balsall Heath Road, Edgbaston, Birmingham, B5 7NE
G6 DXC Clive Ellis, 43 Epsom Walk, Hereford, HR4 9NJ
G6 DXD A Edwards, 35 Eldon Road, Cheltenham, GL52 6TX
G6 DXP M Gentry, Maeldune, Orsett Road, Stanford-le-Hope, SS17 8NS
G6 DYK Simon Hicks, 53 Hillfield Road, Oundle, Peterborough, PE8 4QR
G6 DYM Gary Hudgell, 18 Fellowes Lane, Colney Heath, St Albans, AL4 0QA
G6 DYR David Bettie, 54 Grendon Road, Polesworth, Tamworth, B78 1NU
G6 DYU L Horn, 9 Musson Close, Irthlingborough, Wellingborough, NN9 5XW
G6 DYW Chris Hughes, 3 Crown Rise, Llanfrechfa, Cwmbran, NP44 8UG
G6 DZI Charles Kuss, 20 Windermere Road, Haydock, St. Helens, WA11 0ES
G6 DZJ Stephen Kitchener, 101 Highfield Road, Tring, HP23 4DS
G6 DZT D Anstock, 12 Raymoor Avenue, St. Marys Bay, Romney Marsh, TN29 0RD
G6 DZX J Beardmore, 6 Essex Close, Congleton, CW12 1SH
G6 EAH R Carrington, 45 Crompton Road, Pleasley, Mansfield, NG19 7RG
G6 EAM James Calder, 50 High Street, Shrewsbury, SY1 1ST
G6 EAR P Dowler, 21a Wash Lane, Clacton-on-Sea, CO15 1UW
G6 EAX Stephen Hufschmied, 99 Leverstock Green Road, Hemel Hempstead, HP3 8PR
G6 EAZ R Hildebrand, Meadow View, Cunningham Place, Bakewell, DE45 1DD
G6 EBL Michael Brundle, 36 Campion Street, Derby, DE22 3EF
G6 EBO B Beckers, 6 Patmore Way, Collier Row, Romford, RM5 2HF
GI6 EBX S Bird, 70 Greencastle Road, Kilkeel, Newry, BT34 4JJ
G6 ECN Bernard Clay, 3 Sandy Close, Bollington, Macclesfield, SK10 5DT
G6 ECS P Buckingham, Thrimley House, Thrimley Lane, Bishop's Stortford, CM23 1HX
G6 ECT Robert Close, 208 Northampton Road, Wellingborough, NN8 3PW
G6 EDC Frank Davis, 28 Western Drive, Claybrooke Parva, Lutterworth, LE17 5AG
G6 EDD S Donald, 5 Windsor Road, Royston, SG8 9JF
G6 EDF D Evans, 107 Bradbury Road, Solihull, B92 8AL
G6 EDJ S Edwards, 10 Ermin Close, Baydon, Marlborough, SN8 2JQ
G6 EDM D Evans, Caithness, Greenlands Road, Sevenoaks, TN15 6PG
G6 EDR R Fletcher, 31 Snowdrop Close, Broadfield, Crawley, RH11 9EG
G6 EDT M Fletcher, Chusan, 32a Mill Road, Bedford, MK44 1NX
G6 EDU M Firth, Kasamily, 73 Lions Lane, Ringwood, BH24 2HH
G6 EEB William Moodie, 141 Wood Lane, Handsworth, Birmingham, B20 2AQ
G6 EED N Mockridge, 6 Dunkerton Rise, Norton Fitzwarren, Taunton, TA2 6TF
G6 EEE A Mead, 17 Beadle Way, Great Leighs, Chelmsford, CM3 1RT
G6 EEF D Malekout, 59 Glebelands Avenue, Ilford, IG2 7DL
GI6 EEH S Mccullagh, 18 Village Walk, Portadown, Craigavon, BT63 5TL
G6 EER G Middleton, 37 Hamdon Close, Stoke-Sub-Hamdon, TA14 6QN
G6 EES Phillip Morris, 8 Millfield, Lambourn, Hungerford, RG17 8YQ
G6 EET David Monk, 311 Birmingham Road, Lickey End, Bromsgrove, B61 0ER
G6 EEU M Meredith, 55 New Barn Lane, Cheltenham, GL52 3LB
GU6 EFB K Le Boutillier, Tiverton, Bailiffs Cross Road, St. Andrew, Guernsey, GY6 8RT
G6 EFE S Weiss, 7 Tennyson Avenue, Grays, RM17 5RG
G6 EFO K Goodchild, 2 Westfield Close, Norden, Rochdale, OL11 5XB
GI6 EGE R Hadden, 28 Belfast Road, Comber, Newtownards, BT23 5EW
GI6 EGJ J Potts, 217 Donaghanie Road, Beragh, Omagh, BT79 0RZ
G6 EGO D Pink, 31 The Fairway, Daventry, NN11 4NW
G6 EGU B Nixon, 87 Field Avenue, Canterbury, CT1 1TS
G6 EGY I Niven, Keepers Cottage, Sulby, Northampton, NN6 6EZ
G6 EHE William Ward, 88 Central Road, Cromer, NR27 9BW
G6 EHG R Quiney, 59 Malham Road, Stourport-on-Severn, DY13 8NT
G6 EHJ John Parker, 14 Southland Road, Leicester, LE2 3RJ
G6 EHL D Partington, 6 Celandine Avenue, Cowplain, Waterlooville, PO8 9BE
G6 EIH Robert McCracken, 16 Station Road, Rolleston-on-Dove, Burton-on-Trent, DE13 9AA
G6 EIO A Mitchell, 85 Farriers Green, Monkton Heathfield, Taunton, TA2 8PP
GI6 EIR D Mullan, 5 Mountfield Drive, Coleraine, BT52 1TW
G6 EIZ J Austin, 17 New Road, Ascot, SL5 8QB
G6 EJD David Bird, 59 Speedwell Close, Melksham, SN12 7TE
G6 EJF R Amos, 89 Stanstrete Field, Great Notley, Braintree, CM77 7JW
G6 EJH J Bradley, 66 Belmont Road, Parkstone, Poole, BH14 0DB
G6 EJI Paul Barrett, 91 Victoria Street, Shaw, Oldham, OL2 7AA
G6 EJM T Burrows, 11 Louis Close, Old Catton, Norwich, NR6 7BG
G6 EJT John Bibby, 19 Richmond Crescent, Mossley, Ashton-under-Lyne, OL5 9LQ
G6 EJU Christopher Biddles, 129 Hallam Crescent East, Leicester, LE3 1FG
GI6 EJW William McCormick, 46 Gortlane Drive, Greenisland, Carrickfergus, BT38 8SY
G6 EKM Richard Perks, The Stables, Bishops Offley, Stafford, ST21 6EX
G6 EKS A Stelfox, 6 Surrey Street, Glossop, SK13 7AH
G6 EKT HORNSEA ARC c/o R Guttridge, Ivy House, Rise Road, Hull, HU11 5BH
G6 ELG M Wright, 6 Tregalister Gardens, St. Germans, Saltash, PL12 5NQ
G6 EMB G Collins, 33 West Hay Grove, Kemble, Cirencester, GL7 6BE
G6 ENA M Pyrah, 53 St. Georges Road, Ramsgate, CT11 7EF
G6 ENN David Gordon, 38 Deer Park Road, Langtoft, Peterborough, PE6 9RB
G6 ENO B Garrett, 226 Rydal Drive, Bexleyheath, DA7 5DG
GJ6 ENP J Gready, Avon Cottage, La Rue D'Elysee, St. Peter, Jersey, JE3 7DT
G6 ENR Stuart Grant, 104 Front Street, Lockington, Driffield, YO25 9SH
G6 ENS S Gordon, 20 Hawkins Crescent, Shoreham-by-Sea, BN43 6TP
G6 ENT D Gordon, 29 The Mannings, Surry Street, Shoreham-by-Sea, BN43 6RP
G6 ENU I Gordon, 9 Park Road, Camberley, GU15 2SP
G6 ENY Noel Graham, Millers Croft, Queens Road, Freshwater, PO40 9ES
G6 ENZ G Holmes, 10 Birch Road, Stamford, PE9 2FB
G6 EOA G Hill, 14 Oak Drive, Colwall, Malvern, WR13 6RA
G6 EOK R Lewis, 12 Station Road, Wimborne, BH21 1RG
G6 EON P Martin, 40 Carnarthen Street, Camborne, TR14 8UP
G6 EOO R Machin, 236 Tamworth Road, Kettlebrook, Tamworth, B77 1BY
G6 EOR W Power, 31 Darbys Hill Road, Tividale, Oldbury, B69 1SE
G6 EPL Richard Jonas, 49 Clarendon Road, Aylesham, Canterbury, CT3 3AQ
G6 EPN P Knight, Hawkwind, Elcot Lane, Marlborough, SN8 2AZ
G6 EPU B Sherman, 240 Annan Road, Dumfries, DG1 3HE
G6 EPX P Shuttleworth, 12 Oak Avenue, Penwortham, Preston, PR1 0XQ
G6 EQB John Singleton, 48, Pennine Way, Ashby-de-la-Zouch, LE65 1EW
G6 EQD B Stirk, 31 Edinburgh Drive, Walsall, WS4 1HS
G6 EQF R Skinner, 23 Woodstock Road, Worcester, WR2 5ND
G6 EQI R Smith, 3 Mendip Edge, Weston-Super-Mare, BS24 9JF
G6 EQL S Thomas, 64 Victoria Road, Aigburth, Liverpool, L17 0DP
G6 EQP T Thompson, 7 West Bank, Dorking, RH4 3BZ
G6 EQS J Hastings, 2 Coltsfoot Road, Rushden, NN10 0GE
G6 EQT J Aston, 3 Valley Road, Darley, Harrogate, HG3 2QE
G6 EQZ R Bracken, 72 Brampton Way, Portishead, Bristol, BS20 6YT
G6 ERI R Couch, 54 Hill Park Road, Gosport, PO12 3EB
G6 ERJ A Croucher, 73 Loxley Close, Church Hill North, Redditch, B98 9JH
G6 ERK A Cunliffe, 28 Rosebank Close, Ainsworth, Bolton, BL2 5QU
G6 ERZ Clifford Jones, 46 Ryedale Way, Tingley, Wakefield, WF3 1AJ
G6 ESJ P Wookey, 16 Danvers Way, Westbury, BA13 3UE
G6 ESK D Whittle, 2 Wilcox Leys, Moreton Morrell, Warwick, CV35 9BG
G6 ESM D Tankaria, 23 Oakwood Avenue, Southall, UB1 3QD
G6 ESQ P Baker, 12 College Close, Coltishall, Norwich, NR12 7DT
G6 ETC J Brown, 44 Perowne Way, Sandown, PO36 9BX
G6 ETL Peter Cooke, 55 Priory Road, Portbury, Bristol, BS20 7TQ
G6 ETP J Cookson, Barker Fold Farm, Tockholes Road, Darwen, BB3 0LU
G6 ETX M Carter, 22 John Morgan Close, Hook, RG27 9RP
G6 ETZ C Chalmers, 8 Westbury Close, Crowthorne, RG45 6NL
GM6 EUC D Cruickshank, 61 Woodside Road, Banchory, AB31 4EN
G6 EUF Paul Raynor, 29 Kilvin Drive, Beverley, HU17 9PG
G6 EUG C Slater, 70 Windsor Avenue, Ashton-on-Ribble, Preston, PR2 1JD

G6 EUI C Shaw, 19 Church Road, Teversham, Cambridge, CB1 9AZ
G6 EUO John Slater, 47 Broom Road, Lakenheath, Brandon, IP27 9EZ
G6 EUR Philip Williams, Llwyn, Manaton, Welshpool, SY21 8DJ
G6 EUT Alan Williams, Brynfield, Kingswood Forden, Welshpool, SY21 8TS
G6 EUU Anthony Wilson, Flat 5, Shelley House, London, E2 0HE
G6 EUW A Sheridan, 6 Mill Road, Burnham-on-Crouch, CM0 8PZ
G6 EUY W Shadwell, 2 POPPY CLOSE, YAXLEY, Peterborough, PE7 3FA
G6 EVC C Sleight, Orchard House, School Hill, Southam, CV47 8NN
G6 EVX A Wood, 54 Wilton Park Road, Shanklin, PO37 7DU
G6 EVY H Woolrych, 20 Meadow Drive, Devizes, SN10 3BJ
G6 EWH E Turton, 27 Langdale Avenue, Hesketh Bank, Preston, PR4 6TD
G6 EWJ D Taylor, Holmehurst, Church Street, Horsham, RH12 3ET
G6 EWK David Mason, 15 Windmill Gardens, Prenton, CH43 7YQ
GI6 EWO B Davis, 49 The Roddens, Larne, BT40 1QL
G6 EWP D Davy, 22 Scott Gardens, Lincoln, LN2 4LX
G6 EWQ C Dormer, 39 Eastmoor Road, Newport, NP19 4NX
G6 EWX Nicholas Evans, 1 Tan y Fedwen, Llandderfel, Bala, LL23 7PT
G6 EXC Paul Gibson, Rivendell, 9 Mallard Close, Ormskirk, L00 5QJ
G6 EXE M Graham, 11 Robert Moffat, High Legh, Knutsford, WA16 6PS
G6 EXG M Gee, 100 Plantation Hill, Worksop, S81 0QN
G6 EXN E Hall, 9 Valance Avenue, Chingford, London, E4 6DR
G6 EXU A Jobber, Church Hill, Kings North, Ashford, TN23 3EG
G6 EXX Brian Kent, 4 Bedmond Road, Pimlico, Hemel Hempstead, HP3 8SH
G6 EXZ A Kent, 166 Louth Road, Scartho, Grimsby, DN33 2LG
G6 EYA P Kershaw, On Y Va Devizes Marina Village, Horton Avenue, Devizes, SN10 2RH
G6 EYD A Mott, 2 Woodside Close, Chesterfield, S40 4PW
G6 EYI Charles Moore, Glen View, Fosseway, Radstock, BA3 4BB
G6 EYJ D Morton, 27 Beechfield Way, Hazlemere, High Wycombe, HP15 7TP
G6 EYS Andrew Morne, 16 Warmden Avenue, Baxenden, Accrington, BB5 2PR
G6 EZG I Prince, 31 Gillshill Road, Hull, HU8 0JG
G6 EZH Steven Pepper, 149 The Hill, Glapwell, Chesterfield, S44 5LU
G6 EZI James Oldroyd, 357 Commercial St #704, Boston, United States, MA 02109-1240
G6 EZM D Winters, 13a St. Catherines Road, Bournemouth, BH6 4AE
G6 EZR Francis Thompson, 38 Hanson Park, Northam, Bideford, EX39 3SB
G6 EZY David Powell, 82 Belmont Street, Southport, PR8 1JH
G6 FAF C Narroway, 26 Fern Way, Watford, WD25 0HG
G6 FAH K Lawrence, 54 Sheldrake Road, Christchurch, BH23 4BP
G6 FAL R Stoneman, 9 Winchester Road, Northampton, NN4 8AZ
G6 FAX Peter Brooks, Flat 4, 8 Glencathara Road, Bognor Regis, PO21 2SF
G6 FBA J Butters, 21 Erleigh Road, Reading, RG1 5LR
G6 FBB R Chidgey, 14 Drury Road, Colchester, CO2 7UX
G6 FBH Gary Davis, Westbury House, 3 Windermere, Tamworth, B77 5TD
G6 FBJ John Endicott, 1 Elm Tree Park, Yealmpton, Plymouth, PL8 2ED
GW6 FBV T Howell, 19 Uwchgwendraeth, Drefach, Llanelli, SA14 7AR
G6 FCI C Mcmahon, 130 Newton Drive, Blackpool, FY3 8JA
G6 FCJ P Magnus-Watson, 95 Sutton Lane, Slough, SL3 8AU
G6 FCL Jim Mahoney, Winton Dene, The Street, Sudbury, CO10 8JP
G6 FCS David Lane, 10 Whylands Close, Worthing, BN13 3HB
G6 FDD R Pinchin, 10 Epping Drive, Melton Mowbray, LE13 1UH
G6 FDG Ian Rivers, 35 Cloverville Approach, Bradford, BD6 1ET
G6 FDI B Raymer, 19 Caithness Drive, Crosby, Liverpool, L23 0RG
G6 FDK S Maskrey, The Hayloft, Stamford Lane, Chester, CH3 7QD
G6 FDO I Moody, 54 Lansdowne Road, Studley, B80 7RD
G6 FDP Stuart Litobarski, 7 Exeter Road, Southsea, PO4 9PZ
GM6 FDQ G Allan, Corse Farm, Kininmonth, Peterhead, AB42 4JU
G6 FDS D Allen, 21 Kelvin Road, Thornton-Cleveleys, FY5 3AF
G6 FDU R Butterworth, 49 Swandene, Pagham, Bognor Regis, PO21 4UR
G6 FDX Christopher Bicknell, Flat 25, Madderfields Court, London, N11 2JL
G6 FED D Corsi, 4 Horsley Drive, Wrexham, LL12 8BE
G6 FEI D Harris, 53 Welwyn Drive, Salford, M6 7PQ
G6 FEJ R Hawkes, 1 The Fairway, Wellingborough, NN9 5YS
G6 FEM Anne Harris, April Cottage, Sheepcote Green, Saffron Walden, CB11 4SJ
GI6 FEN Paul Irwin, 6 Cairnburn Avenue, Belfast, BT4 2HT
G6 FEQ Philip Jolly, 22 Wellhouse Road, Barnoldswick, BB18 6DD
G6 FES S Jones, 12 Meadow Croft, Cross Lanes, Wrexham, LL13 0UJ
G6 FEX J Sandford, 23 South Lawn, Locking, Weston-Super-Mare, BS24 8AD
G6 FFB D Meaker, 181 Dovecote, Yate, Bristol, BS37 4PF
G6 FFH T Sallis, 54 West Way, Hove, BN3 8LQ
G6 FFL T Short, Freemans Farm, Itchington, Bristol, BS35 3TL
G6 FFQ F Bilton, 50 Coldwell Road, Crossgates, Leeds, LS15 7HA
G6 FFR B Berry, 7 Barlow Close, Telford, TF3 2NQ
G6 FFU Philip Coogan, 24 High Street, Kingsley, Stoke-on-Trent, ST10 2AE
G6 FGA Richard Holyhead, 42 Dockham Road, Cinderford, GL14 2BH
G6 FGC Chris Hawkins, 80 Duston Wildes, Northampton, NN5 6ND
G6 FGJ Christopher Tandy, 7 The Swallows, Patrons Way West, Uxbridge, UB9 5PB
G6 FGL T Toulson, 30 Old Park Avenue, Sheffield, S8 7DR
G6 FGV John Webber, 5 Leda Mews, Achilles Close, Hemel Hempstead, HP2 5WR
G6 FGW Ray Weekes, 84 Vera Road, Yardley, Birmingham, B26 1TT
G6 FGY Eric Westbrook, 66 Nelson Close, Croydon, CR0 3SW
GI6 FHD A McPartland, 4 Clanbrassil Gardens, Portadown, Craigavon, BT63 5YD
G6 FHK C Leonard, 138 Sundridge Drive, Chatham, ME5 8JD
G6 FHM Donald Sunderland, 1 Allfield Cottages, Condover, Shrewsbury, SY5 7AP
G6 FHR Roger Plant, 32 Buckland Road, Pen Mill Trading Estate, Yeovil, BA21 5HA
G6 FIB T Wicks, 123 The Crescent, Andover, SP10 3BN
GM6 FIK D Stevenson, 2 Melbourne Court, Braidpark Drive, Glasgow, G46 6LA
G6 FIL D Smith, 323 Colchester Road, Ipswich, IP4 4SF
G6 FIN A Stevens, 16 Tremlett Grove, Ipplepen, Newton Abbot, TQ12 5BZ
G6 FIO J Slater, 154 Ralph Road, Shirley, Solihull, B90 3JZ
G6 FIP I Tebboth, 20 Glebe Road, Stratford-upon-Avon, CV37 9JU
G6 FIT Anthony Lewis, 81 Ashton Avenue, Rainhill, Prescot, L35 0QR
G6 FIV M Leitch, 95 Malvern Road, North Shields, NE29 9ET
G6 FJA Frederick Aunger, 2, Lowick, Woodthorpe, York, YO24 2RF
G6 FJE L Plewa, 174 Dorset Avenue, Chelmsford, CM2 8YY
G6 FJG N Pinkney, 4 St. Hughs Road, Buckden, St. Neots, PE19 5UB
G6 FJI M Richards, 1 Ashenden Close, Abingdon, OX14 1QE
G6 FJL Philip Standen, 17 Canberra Road, Worthing, BN13 3HH
G6 FJO D Turner, 14 The Poplars, Launton, Bicester, OX26 5DW
G6 FJP Robert Wild, 15 Cartridge Street, Heywood, OL10 3AF
G6 FKA Geoffrey Valler, 12 Charlotte Drive, Gosport, PO12 4GS
G6 FKB E Taylor, 19 Chester Road, Saltney Ferry, Chester, CH4 0AQ
G6 FKE K Redmond, 8 George Street, Morecambe, LA4 5SU
G6 FKL L Taylor, 127 Dundee Close, Fearnhead, Warrington, WA2 0UJ
G6 FKN M Lee, 55 Woodland Avenue, Guildford, GU2 4LA
G6 FKP S Moore, 25 Overdale Avenue, Mynydd Isa, Mold, CH7 6US
G6 FKR D Roberts, 10 Woodville Terrace, Darwen, BB3 2JH
G6 FKS S Robinson, 4 Grayling Close, Cambridge, CB4 1NP
G6 FKW Leslie Palmer, 17 Rushlake Green, Birmingham, B34 6TL
G6 FKY I Norris, The Old Post Office, Arundel Road, Arundel, BN18 0SD
G6 FLE N Scott, 1 Lakeside, Fareham, PO17 5EP
G6 FLH J Smith, 25 Seafield Close, Seaford, BN25 3JR
G6 FLK Joseph Walton, 168 Park Road, Stanley, DH9 7AJ
GM6 FLL A Simpson-Fraser, 430 Millcroft Road, Cumbernauld, Glasgow, G67 2QW
G6 FLQ Geoffrey Smith, Top of the Hill, Hockerton Road, Newark, NG22 8PB
G6 FLR John Smith, 33 HOP POLE GREEN, LEIGH SINTON, Malvern, WR13 5DP
G6 FLU C Mock, Homelea, Royal Oak Hill, Newport, NP18 1JF
G6 FLW C Thompson, 27 Queensland Drive, Colchester, CO2 8UD
G6 FLY H Lee, 26 Ratcliffe Avenue, Branston, Burton-on-Trent, DE14 3DA
G6 FMF D Timson, 40 Rockwood Road, Calverley, Pudsey, LS28 5AA
G6 FMN P Rogers, 12 St. Peters Rise, Headley Park, Bristol, BS13 7LY
G6 FMS M Peers, 46 Lowndes Park, Driffield, YO25 5BG
GW6 FNB Dennis Morris, 11 Ffordd-y-Mynach, Pyle, Bridgend, CF33 6HT
G6 FNJ Robert Oglesby, 1 High Street, Littleton Panell, Devizes, SN10 4EL
G6 FNQ A Smith, 16 Hazel Way, Barwell, Leicester, LE9 8GP
G6 FNY T Strand, 129 Malmesbury Road, Chippenham, SN15 1PZ
G6 FOF C Morris, Flat IV, Brummel Court, Worcester Road, Droitwich, WR9 0DF
G6 FOH C McLean, Trefuge, Coads Green, Launceston, PL15 7NB
G6 FOI A Regan, 153 Acre Lane, Cheadle Hulme, Cheadle, SK8 7PB
GI6 FOR Nicholas Lane, 117a Hillhall Road, Lisburn, BT27 5BT
GM6 FOT T Armour, 69 Hillend Road, Clarkston, Glasgow, G76 7XT
G6 FOV D Aldridge, 17 Priory Close, Tavistock, PL19 9DJ
G6 FOW Paul Atkinson, 30 Spital Terrace, Gainsborough, DN21 2HQ
G6 FOX WEST MIDLAND RDF c/o Thomas Ray, 1 Providence Lane, Leamore, Walsall, WS3 2AQ
G6 FPC J Body, The Durhams, Church Street, Salisbury, SP5 5BH
G6 FPF G Barnes, Rockleigh, 17 Savile Park, Halifax, HX1 3EA
G6 FPH M Cole, 45 Gainsborough Road, Tilgate, Crawley, RH10 5LD
G6 FPK N Cooper, 53 Stanway Road, Benhall, Cheltenham, GL51 6BU
G6 FPN Y Dunn, 117 All Saints Way, West Bromwich, B71 1RU
G6 FPO Graham Faulkner, Flat 2, 1265 Melton Road, Syston, Leicester, LE7 2EN
G6 FPP A Ford, 8 Merganser Drive, Bicester, OX26 6UQ
G6 FPQ C Groves, Wyandell Hailsham Road, Heathfield, TN21 8AS
G6 FPX S Hill, 7 Meadowcroft Court, Runcorn, WA7 2NS
G6 FQL Robin Heath, Flat 2, Portland View, 62 High Street, Great Yarmouth, NR31 6RQ
G6 FQP B Kneebone, 1 Chapel Terrace, Carnkie, Helston, TR13 0DT
GI6 FQT D McConville, 28 Derrycor Lane, Derryadd, Craigavon, BT66 6QW
G6 FRB John Pearce, 25 Boughton Street, St. Johns, Worcester, WR2 4HE
G6 FRS FARNBOROUGH & DISTRICT c/o Michael Hearsey, Halycon, Lawday Link, Farnham, GU9 0BS
G6 FS Leiston ARC c/o Martin Danfer, The Nook, Mill Common, Woodbridge, IP12 2ED
G6 FSG William Chamberlain, Cwmmelyn, Kings Road, Newton Stewart, DG8 8PP
G6 FSK David Fisher, 6 Small Holdings Road Clenchwarton, King's Lynn, PE34 4DY
G6 FSU Martyn Apperly, Chaundlers, Church Lane, Cambridge, CB23 2NG
G6 FTA M Everall, 17 Golden Park Avenue, Torquay, TQ2 8LR
G6 FTB F Jackson, 27 Prairie Crescent, Burnley, BB10 1EU
G6 FTE P Bondar, Keepers Cottage, Islebeck, Thirsk, YO7 3AN
G6 FTH D Clark, 43 Glenfield Crescent, Chesterfield, S41 8SF
G6 FTJ P Carter, 145 Wakefield Road, Dewsbury, WF12 8AJ
G6 FTL Paul Dixon, 37 Carlton Close, Parkgate, Neston, CH64 6RB
GI6 FTM J Dynes, 30 Breagh Road, Portadown, Craigavon, BT63 5LT
G6 FTY R Miles, 60 Aylesham Way, Yateley, GU46 6NT
G6 FUD David Robins, Ivy Cottage, Lyme Road, Stockport, SK12 1TH
G6 FUT I Donn, 3 The Willows, Amersham, HP6 5NT
G6 FUY R Fishwick, 13 Wethersfield Road, Prenton, CH43 9UN
G6 FVB R Baker, 23 Disraeli Road, London, W5 5HS
G6 FVD J Henville, 5 Station Road, Hemyock, Cullompton, EX15 3SE
G6 FVF P Fenn, 38 Harwood Close, Welwyn Garden City, AL8 7SN
G6 FVJ A James, 82 Sandringham Drive, Spondon, Derby, DE21 7QA
G6 FVL R Young, 134 Harport Road, Redditch, B98 7PD
G6 FVM A Williamson, 31 Poulton Road, Southport, PR9 7BE
G6 FVZ Robert Munns, 2 The Tyleshades, Romsey, SO51 5RJ
G6 FWK D Booth, 54 Shaw Drive, Knutsford, WA16 8JR
G6 FWO Paul Dover, 92 The Roundway, Claygate, Esher, KT10 0DW
G6 FXE C Walton, 49 Blandford Drive, Walsgrave, Coventry, CV2 2JD
G6 FXH K Bradley, 1a Redlands Close, Exeter, EX4 8BE
G6 FXR D Ainslie, Brackendene, 17 Sandhurst Road, Crowthorne, RG45 7HR
GI6 FXY E Connolly, 21 Clanrye Avenue, Newry, BT35 6EH
G6 FXZ A Carnall, 3 Main St., Glenluce, Newton Stewart, DG8 0PN
G6 FYA Clement Collins, 29, Seaford, BN25 1SP
G6 FYC J Cowee, 26 Arundel Road, Heatherside, Camberley, GU15 1DL
G6 FYE C Das Neves Pedro, The Leas, Bearwood, Leominster, HR6 9EE
G6 FYL N Harris, 104 Blandford Drive, Walsgrave, Coventry, CV2 2NE
G6 FYR Alan Johnson, 14 Norman Road, Newhaven, BN9 9LJ
G6 FYU J Walker, 33 Erica Way, Horsham, RH12 5XL
G6 FZC D Hall, 1 Westfall, Wearhead, Bishop Auckland, DL13 1JD
G6 FZV W Day, 4 Queenswood Drive, Worcester, WR5 3SZ
G6 FZW A Eaves, 3 Station Cottages, Station Road, Leighton Buzzard, LU7 0SQ
G6 GA R Ainge, Happy Cottage, Majors Barn, Stoke-on-Trent, ST10 1PY
G6 GAB W Honey, 20 Pennor Drive, St. Austell, PL25 4UW
G6 GAC Gordon Jenkins, 75 Rectory Road, Coltishall, Norwich, NR12 7HW
G6 GAF A Franklin, C/O 4 Princes Close, Seaford, BN25 2EW
GI6 GAG N Orr, 405 Enniskeen, Drumgor, Craigavon, BT65 4AB
G6 GAK M Tyrrell, 189 Runcorn Road, Barnton, Northwich, CW8 4HR
GI6 GAQ Henry Warke, 5 Meadow View, Ballymoney, BT53 7AH
G6 GAW D Peters, 46 Sheridan Road, Worthing, BN14 8ET
GI6 GBK John Anderson, 1 Claragh Hill Drive, Kilrea, Coleraine, BT51 5YR
G6 GBL A Abbott, 164 Bath Road, Reading, RG30 2HA
G6 GBT I Cole, 21 Quincey Drive, Erdington, Birmingham, B24 9LX
G6 GBU A Dixon, 7 Dragwell, Kegworth, Derby, DE74 2EL
G6 GCI C Burnett, 88 Mill Lane, Romsey, SO51 8EQ
G6 GCJ J Burnett, 44 Bourne Vale, Hungerford, RG17 0LL
G6 GCK J Cook, St. Davids, Chepstow Road, Newport, NP18 2JR
G6 GCM T Collings, 55 Flora Thompson Drive, Newport Pagnell, MK16 8SH
G6 GCO D Davies, 79a Spenser Road, Bedford, MK40 2BE
G6 GCW I Wiltshire, 4 Nether Close, Eastwood, Nottingham, NG16 3DL
G6 GCY John Robinson, 84 Hereford Way, Middleton, Manchester, M24 2NN
G6 GDD Richard Forster-Pearson, 202A Far Laund, Belper, DE56 1FP
G6 GDI Victor Gerhardt, 24 Putnams Drive, Aston Clinton, Aylesbury, HP22 5HH
G6 GDN O Price Gore, 32 Oakham Close, Desborough, Kettering, NN14 2FH
G6 GEK A Elliott, Karwia House, Rooke Road, Leatherhead, KT22 0BW
G6 GEL K Inman, 15 Waterbridge Court, Appleton, Warrington, WA4 3BJ
G6 GEN R Ainsworth, 18 Washington Drive, Slough, SL1 5RE
G6 GEP Anthony Holdup, Tunnel Farm, Tunnel Rd, Imbil (Po 155), Australia, 4570
G6 GES R Haywood, 16 The St., Kingston, Canterbury, CT4 6JB
G6 GEV D Ashton, 12 Little Lees, Charlbury, Chipping Norton, OX7 3HB
G6 GEX C Farley, 1 Wesley Cottages, Mutley, Plymouth, PL3 4RB
G6 GFA P Arscott, 122 Woodlands Road, Ashurst, Southampton, SO40 7AL
G6 GFC Jim Burrows, 4 Cavendish Crescent, Alsager, Stoke-on-Trent, ST7 2EF
G6 GFG P Cook, 109 Crosthwaite Avenue, Wirral, CH62 9DF
G6 GFJ H Goozee, 45 Brighton Road, Purley, CR8 2LR
GM6 GFL David Begg, 12 Broomhill Road, Penicuik, EH26 9EE
G6 GFO N Atrill, 22 Lester Close, Plymouth, PL3 6PX
GM6 GFQ C Barnard, 122 Union Grove, Aberdeen, AB10 6SB
G6 GFR T Crook, 21 Cleveland Close, Maidenhead, SL6 1XE
G6 GGN Michael Hoskin, 7 Worrall Mews, Clifton, Bristol, BS8 2HF
G6 GGT Malcolm Huntley, 81-82 The Avenue, Sunderland, SR2 7EZ
G6 GGV Brian Hollingworth, 62 Illingworth Avenue, Halifax, HX2 9JD
G6 GGW N Gautrey, 11 Bracken Close, Crawley, RH10 8JR
G6 GGY L Fitzwater, The Old Cottage, Babylon Lane, Tadworth, KT20 6XE
G6 GGZ Michael Ferne, 24 Essex Gardens, Leigh-on-Sea, SS9 4HG
G6 GHE A Rawdon, 44 Southgate, Hornsea, HU18 1AL
G6 GHP R Vansittart, 74 Westbrook Avenue, Margate, CT9 5HD
G6 GHU R Wood, 12 Roundhead Drive, Thame, OX9 3DG
GI6 GIE John Pinkerton, 40 Seacon Park, Seacon, Ballymoney, BT53 6QB
G6 GIF Michael Oram, 25 Jerome Close, Marlow, SL7 1TX
G6 GIU A Stephens, 4 Falcon House, Gurnell Grove, London, W13 0AE
G6 GJD C Harper, Flat 2, Dove Tree Court, Blackpool, FY4 4NA
G6 GJN T Biggs, 3 Pentathlon Way, Cheltenham, GL50 4SE
G6 GJV William Willis, 24 Old Hall Lane, Walton on The Naze, CO14 8LE
GM6 GJW J Leith, Appiehouse, Stenness, Stromness, KW16 3LB
G6 GJY S Smith, 36 Greenfields, Earith, Huntingdon, PE28 3QH
G6 GKG W Hodson, 27 Belvedere Grove, London, SW19 7RQ
G6 GKK A Barton, Orchard Bungalow, Westfield Road, Retford, DN22 7BT
G6 GKL Matthew Borrow, 189 Crofton Road, Orpington, BR6 8JB
G6 GKP J Coyne, 44 Brompton Avenue, Rhos on Sea, Colwyn Bay, LL28 4TF
G6 GKT Ian Houldridge, 57 Heads Lane, Hessle, HU13 0JH
G6 GLB G Walker, 141 DEERLEAP, Bretton, Peterborough, PE3 9YD
G6 GLH P Burfield, 33 St. Ediths Road, Kemsing, Sevenoaks, TN15 6PT
G6 GLO GLOUCESTERSHIRE COUNTY RAYNET c/o Jerry Pallister, 13 Dock Road, Sharpness, Berkeley, GL13 9UA
G6 GLR GT LUMLEY ARS c/o B Corker, 46 Danelaw, Great Lumley, Chester le Street, DH3 4LU
G6 GLT R Bennett, 11 Powys Close, Haslingden, Rossendale, BB4 6TH
G6 GLW J Fisher, 4 Chancery Close, Lincoln, LN6 8SD
G6 GLZ D Clews, 11 Roping Road, Yeovil, BA21 4BD
G6 GMF Michael Inness, 6 Denning Road, Wrexham, LL12 7UG
G6 GMH Gordon Russell, East Gate, Baslow, Bakewell, DE45 1SG
G6 GMR Greater Manchester Raynet c/o Neikolas Czernuszka, 12 Durham Drive, Ashton-under-Lyne, OL6 8BP
GM6 GMZ N Saunders, 6 Haughs Of Clinterty, Kinellar, Aberdeen, AB21 0TZ
GI6 GNA H Wright, 2 Duncans Road, Lisburn, BT28 3LP
G6 GNC J Thornber, 7 Buckland Close, Peterborough, PE3 9UH
G6 GND R Lambert, 10 Ambleside, Rugby, CV21 1JB
G6 GNE James Sugden, 3 Castle Keep, Hibaldstow, Brigg, DN20 9JG
G6 GNO B Cooper, 8 Stanley Road, Doncaster, DN5 8RR
G6 GOG A Kerr, 10 Hillcrest Road, Crosby, Liverpool, L23 9XS
G6 GOS Michael Jones, 56 Newton Road, Lewes, BN7 2SH
G6 GOV B Wood, 8 Chichester Drive, Chelmsford, CM1 7RY
G6 GOW H Wheeler, 2 Heather Close, Brereton, Rugeley, WS15 1BB
G6 GOX Leslie Timbrell, 3 Rushden Road, Wymington, Rushden, NN10 9LN
G6 GPF J Woodhouse, 130 Hangleton Way, Hove, BN3 8EQ
GM6 GPH James Robertson, 41 Balgarvie Crescent, Cupar, KY15 4EF
G6 GPR David Jefferies, 96 Broad Street, Wood Street Village, Guildford, GU3 3BE
G6 GPV Kenneth Patching, 36 Cleveland Drive, Fareham, PO14 1SW
G6 GQF G Martin, 9 Clarkes Avenue, Kenilworth, CV8 1HX
G6 GQG I Moston, 19 Wegnalls Way, Leominster, HR6 8TQ
G6 GQI R Swann, 3 Elizabeth Avenue, Newmarket, CB8 0DJ
G6 GQJ J Davies, 1 Woodland Road, Halesowen, B62 8JS
GI6 GRV J Barnett, 2 Donegall Park, Whitehead, Carrickfergus, BT38 9ND
G6 GS GUILDFORD & D RD c/o Andrew Pevy, 2 Oaktree Way, Sandhurst, GU47 8QS
G6 GSF Keith Edwards, Whitehaven, High Street, Uckfield, TN22 4JU
G6 GSG Robert Grimley, 11 Sewell Wontner Close, Kesgrave, Ipswich, IP5 2GB
G6 GSI David Millington, 9 Roxburgh Croft, Leamington Spa, CV32 7HT
GW6 GSR SOUTH GLAMORGAN RAYNET GROUP c/o P Williams, 5 Whitewell Drive, Llantwit Major, CF61 1TA
G6 GTB J Tracey, 100 Booth Close, Kingswinford, DY6 8SP
G6 GTC Phil Willson, 37 The Grove, Sidcup, DA14 5NG
G6 GTH S Leonard, 231 Hale Road, Hale, Altrincham, WA15 8DN

**IMPORTANT NOTE**

**Revalidate licence to avoid revocation** – Ofcom has advised the Society that plans will be drawn up to revoke licences that have not been revalidated as required by the licence conditions. The quickest way to revalidate is to do so online via the Ofcom website: *https://services.ofcom.org.uk/* or by email: *amateur.validations@ofcom.org.uk* Ofcom staff are available to help, but please be patient during times of heavy workload.

G6 GTJ Louise Baldwin, The Fields, Pinewood Road, Market Drayton, TF9 4QE
G6 GTS P Dudman, Chapel House, Pen y Bryn, Wrexham, LL14 1UA
G6 GTZ P Wilson, 162 Bowerdean Road, High Wycombe, HP13 6XW
G6 GUC D Ellis, Field End, Northwood Green, Westbury-on-Severn, GL14 1NB
G6 GUD M Everley, 5 Firs Close, Hazlemere, High Wycombe, HP15 7TF
G6 GUH Peter Bonds, The Gables, Crosslane Head, Bridgnorth, WV16 4SJ
G6 GUT B Turley, 12 Legh Drive, Woodley, Stockport, SK6 1PT
G6 GVF Keith Waters, 25 Edwin Road, Twickenham, TW2 6SP
G6 GVH J Marks, 124 Stowey Road, Yatton, Bristol, BS49 4EB
G6 GVI R Wilkinson, 84 Park Road, Bolton, BL1 4RQ
G6 GVL M Longley, 78 Priory Road, Eastbourne, BN23 7BE
G6 GVM P Martin, 21 Baldwin Avenue, Eastbourne, BN21 1UJ
G6 GVO M Pearce, 64 Goongarrie Drive, Wa, Australia, 6169
G6 GVR Gerry Whittle, 5 Chantry Close, Westhoughton, Bolton, BL5 2LY
G6 GVS R Wood, 6 Timberlaine Road, Pevensey Bay, Pevensey, BN24 6DE
G6 GVU S Wood, 30 Ramsay Way, Eastbourne, BN23 6AL
G6 GVZ E Rigby, 12 Sorrel Avenue, Tean, Stoke-on-Trent, ST10 4LY
G6 GW BLACKWOOD & DISTRICT AMATEUR RADIO SOC. c/o Daniel Lewis, 23 Gelligroes Road, Pontllanfraith, Blackwood, NP12 2JU
G6 GWE M Ranger, 13 Springfield Close, Crowborough, TN6 2BN
G6 GWP John Briggs, Wood Lea, Bawtry Road, Doncaster, DN10 5BS
G6 GWU P Hopkinson, 59 Mulberry Drive, Upton-upon-Severn, Worcester, WR8 0ET
G6 GWX Christopher Hore, 45 Medrose Street, Delabole, PL33 9BN
G6 GWY K Dodd, 1 Nansen St., Bulwell, Nottingham, NG6 9JE
G6 GXE L Jordan, 20 Coniston Road, Folkestone, CT19 5JF
G6 GXG David Ridden, 6 Maple Drive, Witham, CM8 2LH
G6 GXK David Wrigley, 45 Norford Way, Rochdale, OL11 5QS
G6 GXS D McLean, Quartier Les Tourres, Pourcieux, France, 83470
G6 GXY M White, 8 Browning Avenue, Droylsden, Manchester, M43 6QG
G6 GXZ B Vaslet, Heatherlea, Adbury Holt, Newbury, RG20 9BN
G6 GYC D Oultram, 61 Bolton Road, Westhoughton, Bolton, BL5 3DN
G6 GYF M Marshman, 12 Neelands Grove, Cosham, Portsmouth, PO6 4QL
G6 GYG D Langridge, 4 The Puddledocks, Puddledock Lane, Weymouth, DT3 6LZ
G6 GYM D Popely, 24 Lawson Avenue, Stanground, Peterborough, PE2 8PL
G6 GYN P Price, 67 Bennetts Road, Keresley End, Coventry, CV7 8HY
G6 GYV Brian Thompson, 17 Avenue Road, Askern, Doncaster, DN6 0AR
G6 GZS P Empringham, Paulzanne, Bank End, Louth, LN11 7LN
G6 GZZ A Hammond, 23 St. Andrews Road, Fremington, Barnstaple, EX31 3BS
G6 HAA A Johns, Glan Y Nant, Murcot, Broadway, WR12 7HS
G6 HAT P Calpin, 36 Chatsworth Grove, Harrogate, HG1 2AS
G6 HBF A Bischtschuk, 30 Livingstone Road, Wirral, CH46 2QR
G6 HBJ T Charman, 1 Bowler Lea, Downley, High Wycombe, HP13 5UD
G6 HBQ A Ford, 1 Hem Heath Cottage, Longton Road, Stoke-on-Trent, ST4 8HP
G6 HBZ S Jenkinson, Field End, Castleton, Hope Valley, S33 8WB
GD6 HCB A Kennaugh, 36 Seafield Close, Onchan, Douglas, Isle of Man, IM3 3BU
G6 HCF L Carter, Hattersbrick Farm, Lancaster Road, Preston, PR3 6BN
G6 HCH Christopher Back, The Gift Shop, 3 Albion Villas, Main Road, Wareham, BH20 5RQ
G6 HCI B Byrne, 13 Tittensor Road, Newcastle, ST5 3BS
G6 HCQ Stephen Crawford, 71 Harewood Road, Bedford, MK42 9TH
G6 HCT The Home Counties ATV Group c/o Thomas Grady, 63 Bridport Close, Lower Earley, Reading, RG6 3DG
G6 HCW D Fieldsend, 3 Rosehall Close, Redditch, B98 7YD
G6 HDD P Ingham, 8 Eagle Street, Schenectady, New York, United States, 12307
G6 HDF S Kelly, 4 Franklyn Close, Perton, Wolverhampton, WV6 7SB
G6 HEB P Ballance, 6 Coronation Terrace, Knaresborough, HG5 8JN
G6 HEE A Grimmett, 3 Tydd Low Road, Long Sutton, Spalding, PE12 9AR
G6 HEF David Bailey, 2 Hodgson Gardens, Millom, LA18 5LE
G6 HEJ Graham Stewart, 3 Harvest Crescent, Carterton, OX18 1FF
G6 HFB Ann Wedgwood, 2 Hodgson Gardens, Millom, LA18 5LE
G6 HFF Glenn Bates, 9 Parkdene Close, Harwood, Bolton, BL2 3LH
GM6 HFH I Baker, 31 Strathaven Road, Stonehouse, Larkhall, ML9 3EN
G6 HFK L Dutton, 5 Beaver Close, Stoke-on-Trent, ST4 6PR
G6 HFS Brian Shaw, 43 Egremont Road, Hardwick, Cambridge, CB23 7XR
G6 HFW J Graham, 142 Shakerley Lane, Atherton, Manchester, M46 9TZ
G6 HFZ S Homer, 31 Shaftmoor Lane, Acocks Green, Birmingham, B27 7RU
G6 HGD R Waller, Mauray, 22 Nightingale Lane, Thetford, IP26 4AR
G6 HGE D Heale, 3 Evans Wharf, Hemel Hempstead, HP3 9WU
GM6 HGF David Hulin, 18 Bartonholm Gardens, Irvine, KA12 8TD
G6 HGG R Ireson, 6 Walker Square, Wellingborough, NN8 5PQ
G6 HGI Peter Johnston, 566 Woodchurch Road, Prenton, CH43 0TT
G6 HGK Eric Kesterton, 24 Alexandra Road, Illogan, Redruth, TR16 4DY
G6 HGM Robert Buckle, Bissom House, Parrotts Lane, Tring, HP23 6NE
G6 HGR K Potts, 31 Sparnon Close, Redruth, TR15 2RJ
G6 HGU P Saunders, 19 Sharpfield Avenue, Rawmarsh, Rotherham, S62 7QF
GM6 HGW Colin Topping, 26 Crathes Close, Glenrothes, KY7 4SS
G6 HGX B Waterloo, 55 Solent Road, Hill Head, Fareham, PO14 3LB
G6 HH HASTINGS ELECTRONICS & RC c/o T Ransom, 9 Lyndhurst Avenue, Hastings, TN34 2BD
G6 HHE J Avern, 8 Napier Crescent, Fareham, PO15 5BL
G6 HHH Graham Dowse, 60 Lower Mortimer Road, Southampton, SO19 2HF
G6 HHK D Birkbeck, Plantation Cottage, Saltburn, TS12 3JZ
G6 HIA A Cook, Woodlands House, Hempstead Road, Hemel Hempstead, HP3 0DS
G6 HIB Christopher Craven, 24 Links Drive, Bexhill-on-Sea, TN40 1TE
G6 HIE Brian Edwards, 4 Hythe Crescent, Seaford, BN25 3TU
G6 HIG G Edmonds, Wellwood End, Waterworks Lane, Dover, CT15 5JW
G6 HIO D Ollerton, 91 Church Road, Bickerstaffe, Ormskirk, L39 0EB
G6 HIQ C Lavis, Glen Orchard, East Lydford, Somerton, TA11 7HD
G6 HIU N Lasher, 21 Longfield Avenue, London, NW7 2EH
G6 HIV J Martin, 1 Marsh Street, Strood, Rochester, ME2 4BB
G6 HIX J O'Hagan, Brubell, 13 Chapel Road, Faringdon, SN7 8LE
G6 HJU J Binns, 2 Gawsworth Close, Poynton, Stockport, SK12 1XB
G6 HJV J Evill, 54 Copsey Grove, Farlington, Portsmouth, PO6 1NB
GI6 HKE W Leitch, 16 Cabin Hill Gardens, Belfast, BT5 7AP
G6 HKF Roger Mew, Tehig, La Mustais, Sion Les Mines, France, 44590
G6 HKH P Randell, 38 Hanover Drive, Brackley, NN13 6JS
G6 HKL D Martin, 9 Twinberrow Lane, Woodmancote, Dursley, GL11 4AP
G6 HKN W McCue, 2 Downham Avenue, Culcheth, Warrington, WA3 5RU
G6 HKP D Merrington, 31 North Road, Wellington, Telford, TF1 3ED
G6 HKS R Mason, 11 Dyers Mews, Neath Hill, Milton Keynes, MK14 6ER
G6 HKY W Metcalfe, 81 Westminster Drive, Bromborough, Wirral, CH62 6AN
G6 HKZ R Moses, 80 Edgeworth, Yate, Bristol, BS37 8YW
G6 HL Club Hut c/o William Ward, 88 Central Road, Cromer, NR27 9BW
G6 HLL B Allman, 38 Whinchat Drive, Birchwood, Warrington, WA3 6PB
G6 HLR G Marshall, 118 Heather Road, Small Heath, Birmingham, B10 9TB
GM6 HLT J Melville, Shirva, 6 Dixon Avenue, Dunoon, PA23 8NA
G6 HLU T Miller, 23 Manchester Road, Altrincham, WA14 4RQ
G6 HMA Patrick Matthews, 47 Slyne Road, Morecambe, LA4 6PD
G6 HMF Roger Venison, Brookland, Sharnbrook Road, Bedford, MK44 1EX
G6 HMG R Trowsdale, 422 Leatherhead Road, Chessington, KT9 2NN
GW6 HMJ A Upcott, 67 Hunters Ridge, Brackla, Bridgend, CF31 2LJ
G6 HMN R Sunter, 15 Wellhead, Winewall, Colne, BB8 8BW
G6 HMS E Veall, 24 Meadow Drive, Tickhill, Doncaster, DN11 9ET
G6 HMV R Tilley, 41 Rookery Road, Knowle, Bristol, BS4 2DX
G6 HMX D Tucker, 2 Chardonnay Crescent, Thornton-Cleveleys, FY5 3UH
G6 HNI D Baker, 16 Warners Bridge Chase, Rochford, SS4 1JE
G6 HNJ Ian Bennett, Homefield, Winchester Road, Southampton, SO32 2DH
G6 HNN M Bugg, 39 Glencoe Road, Ipswich, IP4 3PP
G6 HNP P Beever, 33 Masterton Road, Stamford, PE9 1SN
G6 HNQ Kevin Blackburn, 57 Hope Street, Leigh, WN7 1NB
G6 HNR J Ball, 94 Marshall Lake Road, Shirley, Solihull, B90 4PN
G6 HNS R Ball, 139 Bedford Road, Sutton Coldfield, B75 6DB
G6 HOB David Brebner, 2 Oldborough Drive, Loxley, Warwick, CV35 9HQ
G6 HOC A Bird, 95 Hundred Acre Road, Streetly, Sutton Coldfield, B74 2BS
G6 HOR D Padfield, 9 Dunster Close, Minehead, TA24 6BY
G6 HOS C Playford, 6 Nutberry Close, Teynham, Sittingbourne, ME9 9SP
G6 HPE Paul Simms, Link Communications, 61 Ipswich Crescent, Great Barr, B42 1LY
G6 HPK D Scott, 8 Lynton Road, Chesham, HP5 2BU
G6 HPL R Seddon, 255 Westleigh Lane, Leigh, WN7 5PN
G6 HPT D Sumner, 34 Japonica Close, Bicester, OX26 3YB
G6 HQ Peter Bushell, The Fairway, Well Lane, Neston, CH64 4AN
G6 HQX A Cook, 90 Ramsbury Walk, Trowbridge, BA14 0UX
G6 HRA J Chesterman, Danby, 69 Heath Lane, Woodstock, OX20 1RZ
GW6 HRL H Winter, 48 The Moorings, Glanteifion, St. Dogmaels, SA43 3LH
G6 HRX T Whitehead, 14 Somerset Road, Willenhall, WV13 2RY
G6 HSC V Williamson, 28 Mill Park Drive, Braintree, CM7 1XF
G6 HSD R Willmott, 85 Malthouse Lane, Earlswood, Solihull, B94 5RZ
G6 HSG A Walsh, 14 The Rydings, Langho, Blackburn, BB6 8BQ
G6 HSI S Wallace, 26 Parsons Drive, Glen Parva, Leicester, LE2 9NS
G6 HSR Julian Hague, 33 East Rise, Royal Sutton Coldfield, B75 7TH
G6 HSW Leslie Hagger, 48 Little Meadow, Bar Hill, Cambridge, CB23 8TD
G6 HTA P Hartas, 6 Newton Street, Whitby, YO21 1QX
G6 HTB E Brodie, 116 Pagham Road, Pagham, Bognor Regis, PO21 4NN
G6 HTH C Hall, Oakenhill, North Pole Road, Maidstone, ME16 9HH
G6 HTS R Hooper, 11 Joy Wood, Boughton Monchelsea, Maidstone, ME17 4JY
G6 HTT G Reece, 9 Lambert Close, Framlingham, Woodbridge, IP13 9TE
G6 HTY Allyson Rollitt, St. Peters, 29 High Street, Lincoln, LN5 0EE
G6 HTZ A Rogers, 11 Avebury Close, Curzon Park, Calne, SN11 0EP
GW6 HUD R Rees, 5 Rhydyffynnon, Pontyates, Llanelli, SA15 5UG
G6 HUH S Ross, Candletrees, Hyde Heath, Amersham, HP6 5RW
G6 HUI B Tanner, 2 Doveys Cottage, Kington Langley, Chippenham, SN15 5NT
G6 HUN Alan Thompson, Hillier Garden Centre, Priors Court Road, Thatcham, RG18 9TG
G6 HUO J Thompson, Belmoor Lodge, Pilton Lane, Exeter, EX1 3RA
G6 HUP Michael Thompson, 2 Cotman Road, Lincoln, LN6 7PA
G6 HUR Nick Thursfield, Highmoor, Llanymynech, SY22 6HB
G6 HV D Bradley, Thelbridge Hall, Witheridge, Tiverton, EX16 8NZ
G6 HVA Martin Vernon, 33 Ffordd Morfa, Llandudno, LL30 1ES
G6 HVE N Martin, 12 Coppice Side, Hull, HU4 6XJ
G6 HVQ L Dodson, 24 Ashcombe Terrace, Tadworth, KT20 5EW
G6 HVX David Gladwish, 36 All Saints Street, Hastings, TN34 3BJ
GM6 HVY Rupert Goodwins, 3 Westmost Close, Edinburgh, EH6 4TE
G6 HWA F Glover, 11 Esk Valley, Grosmont, Whitby, YO22 5BG
G6 HWI Brian Wilson, 102 Woodlands Road, Woodlands, Doncaster, DN6 7JZ
G6 HWR Malcolm Fern, 8 Hackney Road, Hackney, Matlock, DE4 2PW
G6 HWT Clifford Freeman, Mill House, Great Bricett, Ipswich, IP7 7DE
G6 HXB Mark Aston, Flat 51, Acton House, 253 Horn Lane, London, W3 9EJ
G6 HXL Derek Latham, 89 Kestrel Park, Skelmersdale, WN8 6TA
G6 HXR D Lawrence, 31 Taylor Road, Snodland, ME6 5HJ
G6 HXU E Loader, 13 Vale Road, Hartford, Northwich, CW8 1PL
G6 HXW Lionel Leighton, 177 Terringes Avenue, Worthing, BN13 1JS
G6 HXZ Philip Lovett, Betula, High Halden, Ashford, TN26 3LY
G6 HYF C IRONMONGER, 77 Boston Road, Spilsby, PE23 5HH
G6 HYI P Ingle, 8 Slayleigh Delph, Sheffield, S10 3RZ
G6 HYJ Edwin Sykes, 5 Farm Close, High Wycombe, HP13 7YA
G6 HYP N Jones, 7 Church Terrace, Church Road, Norwich, NR16 2NA
G6 HZG K Purcer, 6 Parkway, Ryde, PO33 3UX
G6 HZH Stephen Prosser, 53 Broadlands Rise, Lichfield, WS14 9SF
G6 HZI L Ashford, 24 Buchanan Road, Walsall, WS4 2EN
G6 HZJ D Pemberton, 7 Riddings Lane, Hartford, Northwich, CW8 1NB
G6 HZK George Partridge, 53 Acres Road, Brierley Hill, DY5 2XY
G6 HZX R Purdy, 49 Mansfield Road, Eastwood, Nottingham, NG16 3DY
G6 IAJ W Saunders, 61 Nether Court, Halstead, CO9 2HN
G6 IAN Ian Brooks, 10 Windermere Close, Dunstable, LU6 3DD
G6 IAT T Bruce, 17 Blaydon Road, Luton, LU2 0RP
G6 IBD D Bowles, 23 Broughton Way, Rickmansworth, WD3 8GW
GI6 IBL Mike Barr, 4 Sandelwood Avenue, Coleraine, BT52 1JW
G6 IBN M Bodill, 24 Dalbeattie Close, Arnold, Nottingham, NG5 8QX
G6 IBP G Burnett, 314 Highcliffe, Spittal, Berwick-upon-Tweed, TD15 2JN
G6 IBU M Squance, 83 Earls Mill Road, Plympton, Plymouth, PL7 2BX
G6 IBW R Savigar, 95 Hillier Road, Devizes, Devizes, SN10 2FB
G6 ICC I Campbell, 273 Crystal Palace Road, London, SE22 9JH
G6 ICH R Brothwood, Amberley, Coombe Cross, Newton Abbot, TQ13 9EP
GD6 ICR Michael Webb, Coastguard House, 1 Mount Morrison, Peel, Isle of Man, IM5 1PN
G6 ICV Ian Whittaker, 20 Manor Drive, Leicester, LE4 1BL
G6 ICZ Richard Waller, 48 Lambfield Way, Ingleby Barwick, Stockton-on-Tees, TS17 5BG
GM6 IDF H Stinton, Lower Inchlumpie, Strathrusdale, Alness, IV17 0YQ
G6 IDG C Stringer, Meadowbank, Station Road, Sidmouth, EX10 0ER
G6 IDL Michael Waud, 7 Chalkpit Lane, Candlesby, Spilsby, PE23 5SE
G6 IDO A Davies, 8 Alexandra Road, Wednesbury, WS10 9LH
G6 IDU I Rose, 144 Overton Road, Benfleet, SS7 4DT
G6 IDW T Roberts, 9 Dixons Road, Market Deeping, Peterborough, PE6 8AG
G6 IEE M Elsley, 25 Elmsdale Road, Wootton, Bedford, MK43 9JW
G6 IEI P Williams, Peanjays, 4 Cutbush Close, Reading, RG6 4XA
GI6 IES C Hagan, 15d Ballygalget Road, Portaferry, Newtownards, BT22 1NE
G6 IFE P Holland, 3 Manor Villas, Chilton Road, Aylesbury, HP18 0DN
G6 IFH J Rimington, 8 Harvesters, Tolleshunt D'Arcy, Maldon, CM9 8UF
G6 IFN L Rouse, 69 Shackerdale Road, Wigston, LE18 1BR
G6 IFQ S howcroft, Warwick Cottage, 5 Ecclesgate Road, Blackpool, FY4 5DW
G6 IFR G Horwood, 25 Briar Road, Shepperton, TW17 0JB
G6 IFS N Hollinshead, 35 Parkside Drive, May Bank, Newcastle, ST5 0NL
G6 IFV John Hunt, 77 Scott Street, Burnley, BB12 6NJ
G6 IGK L Glasscock, 37 Huntingfield Road, Bury St. Edmunds, IP33 2JA
G6 IGO D Gospel, Sunnydene Farm, The Common, Fakenham, NR21 9JB
G6 IGU Andrew Greenleaf, The Lindens, Frating Road, Colchester, CO7 7SY
G6 IGV David Gregson, 8 Lennox Gate, Blackpool, FY4 3JQ
G6 IGW N Gutten, 8 Chalfield Close, Crewe, CW2 6TJ
G6 IGY J Mead, 1 Tudor Court, Fagl Lane, Wrexham, LL12 9PJ
G6 IHB F Norton, 62 Moorlands Drive, Shirley, Solihull, B90 3RE
G6 IHD S Maxwell, 17 Gerard Road, Wallasey, CH45 6UQ
G6 IHG H Mitchell, 17 Burners Close, Burgess Hill, RH15 0QA
GI6 IHM Ronald McDowell, 3 Lord Warden's Park, Bangor, BT19 1YG
G6 IHO V Marks, 176 Middlemarch Road, Coventry, CV6 3GL
G6 IHU J Measom, 41 Glebe Road, Thringstone, Coalville, LE67 8NU
G6 IHW A Mackinlay, 26 Anderson Road, Erdington, Birmingham, B23 6NN
G6 IIA A Stansfield, 22 Low Stobhill, Morpeth, NE61 2SG
G6 IIF Robert Sharpe, 14 Dansie Court, Compton Road, Colchester, CO4 0EA
G6 IIK D Gill, 79 Heather Walk, Bolton-upon-Dearne, Rotherham, S63 8BZ
G6 IIM P Jones, 2 Farmers Heath, Great Sutton, Ellesmere Port, CH66 2GX
G6 IIN P Currigan, 5 Gayton Avenue, Wallasey, CH45 9LJ
G6 IIP J Clarke, 19 Kensington Road, Gaywood, King's Lynn, PE30 4AT
G6 IIU R Cooper, 69 Vicarage Lane, Elworth, Sandbach, CW11 3BU
G6 IIZ J Clark, Brooklyn Cottage, Milton Combe, Yelverton, PL20 6HP
G6 IJK A Clayphon, 71 Blagrove Drive, Wokingham, RG41 4BD
G6 IJQ W Cartwright, 3 Masefield Rise, Halesowen, B62 8SH
G6 IJW Gerrit Stoelwinder, Hampt Cottage, Middle Hampt, Callington, PL17 8NR
G6 IKC S Saunders, 16 Hill Close, Pennsylvania, Exeter, EX4 6HG
G6 IKE S Lynch, 21 Mill Lane, Bolton le Sands, Carnforth, LA5 8HR
G6 IKH Anthony Simpson, 7 Hartington Street, Newcastle, ST5 8DR
G6 IKM D Tarbuck, 23 Kingsway, Newton-le-Willows, WA12 8LZ
GM6 IKN K Towns, Pluscarden Abbey, Pluscarden, Elgin, IV30 8UA
G6 IKU J Stockton, 183 Eskdale, Tanhouse, Skelmersdale, WN8 6ED
G6 ILC G Sword, Garden Cottage, Kiln Road, Salisbury, SP5 2HT
G6 ILD Paul Southgate, Flat 1, The Old Yard, Mill Road, Holsworthy, EX22 7RT
G6 ILH Antony Davies, 27 Whitecroft Lane, Mellor, Blackburn, BB2 7HA
G6 ILN J Dodge, 5 Moat Way, Queenborough, ME11 5BU
G6 ILT Eric Elliston, 117 Willbye Avenue, Diss, IP22 4NP
G6 ILU E Edmonds, 1 Chalkhole Cottages, Flete Road, Margate, CT9 4LL
G6 ILX B Edward, 27 Barford Close, Ainsdale, Southport, PR8 2RS
G6 ILY William Evans, Treetops, Whitchurch Road, Wrexham, LL13 0BL
G6 ILZ I Fullerton, 54 Brashland Drive, Northampton, NN4 0SS
G6 IMH R Firth, Kasamily, 73 Lions Lane, Ringwood, BH24 2HH
G6 IMJ K Walker, 37 Willingdon Road, Liverpool, L16 3NE
G6 IML I Walsh, 7 Winchester Avenue, Hartshead Green, Ashton-under-Lyne, OL6 8BU
G6 IMN Keith Wetherell, 12 Parc Stephney, Budock Water, Falmouth, TR11 5EJ
G6 IMQ J Wild, 20 Sandy Lane, Cholsey, Wallingford, OX10 9PY
G6 IMS Thomas Vernalls, 5 Min y Traeth, Minffordd, Penrhyndeudraeth, LL48 6EG
G6 IMW T Rogers, Lodge 20, Benson Waterfront, Riverside Park, Oxon, OX10 6SJ
G6 INA Philip Reidy, Famagusta Avenue 45, House 2, Sotira, Cyprus, 5390
G6 INF J Markham, 4 Ty Arfon, Tywyn, LL36 0TA
G6 ING S Meigh, 75 Botteslow St., Hanley, Stoke-on-Trent, ST1 3NE
G6 INI C Mahony, 20 Kenchester, Bancroft, Milton Keynes, MK13 0QP
G6 INK Eric McGlen, 22 Stratford Avenue, Sunderland, SR2 8RX
G6 INM Kevin O'Reilly, 1 Parkfield, Crewe, CW1 4TT
G6 INO R Ottolini, 154 Barwick Road, Leeds, LS15 8SW
G6 INU D Port, 8 Betterton Drive, Sidcup, DA14 4PS
G6 INV David Pratley, 2 Haseldine Meadows, Hatfield, AL10 8HE
G6 INW Martin Purnell, 14 Cranleigh Close, Bournemouth, BH6 5LD
G6 INX G Pryke, 38 Colne Drive, Walton-on-Thames, KT12 3SQ
GW6 IOA C Crow, Lindisfarne, Pen y Waun, Cardiff, CF15 9SJ
G6 IOB P Coghlan, 96 Cambridge Road, Ely, CB7 4HU
G6 IOE G Crawford, 4 Beverley Gardens, Gedling, Nottingham, NG4 3LF
G6 IOM M Cunningham, 16 Cherry Waye, Eythorne, Dover, CT15 4BT
G6 ION R Civil, 7 Sunnybanks, Hatt, Saltash, PL12 6SA
GI6 IOU M Pollock, 33 Seahill, Donaghadee, BT21 0SH
G6 IOV Paul Phelps, 57 Southbrook Road, Havant, PO9 1RL
G6 IOW D Peachey, 4 Windermere Drive, Great Notley, Braintree, CM77 7UA
G6 IOX A Pearce, 49 Bishopswood Road, Tadley, RG26 4HF
G6 IPB B Doyle, Ashwell Croft, Brunthwaite LANE, Keighley, BD20 0ND
G6 IPC I Downes, 21 Caldbeck Court, Beeston, Nottingham, NG9 5NH
G6 IPH R Distin, The Martletts, Broad Oak, Rye, TN31 6DN
G6 IPN M Davies, 54 Helmside Road, Oxenholme, Kendal, LA9 7HA
G6 IPQ R Deakin, 55 Pendeen Crescent, Plymouth, PL6 6RE
GW6 IPR Peter Drew, 6 Clos Cae'r Wern, Caerphilly, CF83 1SQ
G6 IPW S Featherstone, 36 Denton Avenue, Grantham, NG31 7JL
G6 IQC S Leak, 5 Stelfox Lane, Audenshaw, Manchester, M34 5HE
G6 IQF Richard Harris, Greve De Lec, 14a All Saints Lane, Clevedon, BS21 6AY
GM6 IQH Richard Wickenden, 2 Buail-Bhan, Ballinluig, Pitlochry, PH9 0NH
G6 IQI P White, 16 Charnwood Close, Chandler's Ford, Eastleigh, SO53 5QP
G6 IQM M Wooding, 5 Ware Orchard, Barby, Rugby, CV23 8UF
G6 IQP D Marriott, 10 Springfield Drive, Nottingham, NG6 8WD

| | | |
|---|---|---|
| G6 | IQY | J Price, 67 Broadway, Oldbury, B68 9DP |
| G6 | IRE | R Aynge, 9 Sedgebrook Road, Blackheath, London, SE3 8LR |
| G6 | IRF | S Atwell, 10 Belding Avenue, Manchester, M40 3SF |
| G6 | IRG | M Andrews, 1 Garrick Close, Dudley, DY1 3DF |
| G6 | IRJ | G Andronov, 90 Overbury Close, Northfield, Birmingham, B31 2HD |
| GI6 | IRL | J Agnew, 23 Berwick Heights, Moira, Craigavon, BT67 0SZ |
| G6 | IRP | C Gardner, 7 Lesley Close, Bexley, DA5 1LX |
| G6 | IRU | B Green, 23 Freemantle Avenue, Blackpool, FY4 1SX |
| G6 | IRW | K Holmes, Gable Cottage, Low Hill, Helmsby, Warrington, WA6 0NW |
| G6 | IRX | C Holt, 1 Vale View, Common Road, Wincanton, BA9 0RB |
| G6 | IRY | R Hobbs, 120 Misbourne Road, Hillingdon, Uxbridge, UB10 0HP |
| G6 | IRZ | Wayne Hughes, 3 Wattles Lane, Acton Trussell, Stafford, ST17 0RE |
| G6 | ISA | John Frederick Hutchins, 18 Derby Road, Sale, M33 5PR |
| G6 | ISB | Alan Hunt, 10 Sturton Street, Forest Fields, Nottingham, NG7 6HU |
| G6 | ISG | P Hancock, 2 Culloden Road, Leamington Spa, CV32 5LU |
| G6 | ISM | John Hancock, 7 Hollies Close, Houghton-on-the-Hill, Leicester, LE7 9GW |
| G6 | ISY | L Hill, 21 Liddiards Way, Purbrook, Waterlooville, PO7 5QW |
| GW6 | ITD | [illegible], 118 Heol [illegible], Rhiwbina, Cardiff, CF14 6SS |
| G6 | ITJ | M Jones, 4 Plastirion Avenue, Prestatyn, LL19 9DU |
| G6 | ITM | D Jupp, 26 Audley Avenue, Gillingham, ME7 3AY |
| G6 | ITO | P Kelly, 39 Copeland Avenue, Tittensor, Stoke-on-Trent, ST12 9JA |
| G6 | ITU | M Bunn, 45 Red Cat Lane, Burscough, Ormskirk, L40 0RA |
| G6 | ITV | L Parker, 15 Savile Place, Mirfield, WF14 0AJ |
| G6 | ITW | M Blundell, 68 Alton Road, Leicester, LE2 8QA |
| G6 | IUD | Ron Bracey, 50 Harrow Way, Watford, WD19 5ET |
| G6 | IUF | R Brittain, Sarenchel, St Cross, Harleston, IP20 0NY |
| G6 | IUK | R Bastable, Gwynfryn, Ffordd Caergybi, Llanfairpwllgwyngyll, LL61 5SZ |
| G6 | IUQ | M Gaylard, 66 Runnymede Road, Yeovil, BA21 5SU |
| G6 | IUS | B Gilbert, 22 Oaklands Way, Hildenborough, Tonbridge, TN11 9DA |
| G6 | IVB | Alan Gornall, 28 Woodward Close, Winnersh, Wokingham, RG41 5NW |
| G6 | IVC | Martyn Griffiths, 25 Lethbridge Road, Southport, PR8 6JA |
| G6 | IVD | P Guy, 11 Ludlow Crescent, Redcar, TS10 2LQ |
| G6 | IVE | D Barnes, 30 Royds Avenue, Accrington, BB5 2LE |
| GI6 | IVJ | Robert Brown, 157 Newtownards Road, Bangor, BT20 4HS |
| G6 | IVP | J Burton, 22 Pear Tree Lane, Hempstead, Gillingham, ME7 3PT |
| G6 | IVR | ITCHEN VALLEY ARC c/o P Baxter, Raka Lodge, Whitenap Lane, Romsey, SO51 5ST |
| G6 | IVW | R Balderson, 15 Woodrush Way, Moulton, Northampton, NN3 7HU |
| G6 | IVY | Clare Somerville, 1 Glyn Isaf, Llandudno Junction, LL31 9HF |
| G6 | IWC | Mark Saunders, Rose Cottage, Pleasant Lane, Wrexham, LL11 5DH |
| G6 | IWD | Linda Sherratt, Anlyn, Norbury Drive, Brierley Hill, DY5 3DP |
| G6 | IWK | Ronald Skingley, 1 Lower Parc Estate, Gweek, Helston, TR12 7AG |
| G6 | IWT | M Rea, Osmary, Station Road, Bishop's Stortford, CM22 6LG |
| G6 | IWU | Jeremy Rodwell, 20 Nelson Street, King's Lynn, PE30 5DY |
| G6 | IWZ | David Jefferys, 22 Cleveland Gardens, Cricklewood, London, NW2 1DY |
| G6 | IXE | Mike James, 58 Spitfire Way, Hamble, Southampton, SO31 4RT |
| G6 | IXH | D Hodges, 5 Greenlands, Leighton Buzzard, LU7 3UJ |
| G6 | IXM | Graham Hares, 30 Copped Hall Drive, Camberley, GU15 1NP |
| G6 | IXN | G Hewitt, 66 Portland Drive, Forsbrook, Stoke-on-Trent, ST11 9AU |
| G6 | IXS | S Henson, 4 Monaco Place, Westlands, Newcastle, ST5 2QT |
| G6 | IYA | H Woodnutt, Latitude, Gannock Park West, Conwy, LL31 9HQ |
| G6 | IYD | D Noakes, 117 Kingsmead Park, Allhallows, Rochester, ME3 9TA |
| GM6 | IYJ | M Plested, Cregneash, Platcock Wynd, Fortrose, IV10 8SQ |
| G6 | IYP | R Parry, Glengarriff, Rhyl Road, Rhyl, LL18 2TP |
| G6 | IYS | Ian Porter, Wold View, Park Lane, Louth, LN11 8US |
| G6 | IYY | Richard Adamek, The Old Engine House, Top Road, Norwich, NR12 8XB |
| G6 | IZA | I Alderton, 6 Hurford Drive, Thatcham, RG19 4WA |
| G6 | IZK | Anthony Collier, 2 The Hollies, Trerise Road, Camborne, TR14 7HB |
| G6 | IZQ | James Coad, 17 Dilly Lane, Barton on Sea, New Milton, BH25 7DQ |
| G6 | IZU | K Frame, 102 High Street, Galashiels, TD1 1SQ |
| GW6 | IZZ | C Evans, 20 Glanhafan, Llangwm, Haverfordwest, SA62 4JB |
| G6 | JAC | P Dalley, 32 Albert Road, Erdington, Birmingham, B23 7LT |
| G6 | JAF | Ian Evans, 19 Grange Road, Stone, ST15 8PR |
| G6 | JAK | S Deacon, 3 Blenheim Grove, Offord D'Arcy, St. Neots, PE19 5RD |
| G6 | JAL | Anthony Dixon, 23 Appleby Drive, Barrowford, Nelson, BB9 6EX |
| G6 | JAM | Malcolm Dainty, 5 Woodman Close, Wednesbury, WS10 9UA |
| G6 | JAP | G Denison, 28 Long Street, Thirsk, YO7 1AP |
| G6 | JAR | M Drake, 7 Orient Road, Paignton, TQ3 2PB |
| G6 | JAS | A Balding, 8 Winston Way, Farcet, Peterborough, PE7 3BU |
| G6 | JAY | R Luckett, 20 Leicester Villas, Hove, BN3 5SQ |
| GM6 | JBF | D Mardlin, 35 Uist Road, Aberdeen, AB16 6FN |
| G6 | JBL | G Moore, 2 Meadow Lea, Worksop, S80 3QJ |
| G6 | JBN | Roger Thomas, Post Office, Llanbedr, LL45 2HH |
| G6 | JBQ | G Taylor, 54 Bowershott, Letchworth Garden City, SG6 2EU |
| G6 | JBY | J Bibby, 24 Assarts Lane, Malvern, WR14 4JR |
| G6 | JCI | W Henson, 1 Bonser Close, Carlton, Nottingham, NG4 1DP |
| G6 | JCK | Allan Harvey, Cornerways, Mount Pleasant, Holyhead, LL65 1SN |
| G6 | JCM | J Harrold, Turner Close, Huntington, York, YO31 4PF |
| G6 | JCT | H Haslehurst, Westlands, Stinting Lane, Mansfield, NG20 8EQ |
| G6 | JCV | A Haslehurst, Westlands, Stinting Lane, Mansfield, NG20 8EQ |
| G6 | JCX | Fred Hewitt, 12 Woodside, North Walsham, NR28 0XA |
| G6 | JCY | B Hedge, Birchwood Lodge, Barnards Road, Norwich, NR28 9RG |
| G6 | JDC | T Kemp, 85 Roschill Road, Rawmarsh, Rotherham, S62 7BX |
| G6 | JDF | G Walker, Lluesty, Bryn Hyfryd Road, Tywyn, LL36 9HG |
| G6 | JDH | G Webster, 153 Frogmore Lane, Waterlooville, PO8 9RD |
| G6 | JDO | Nicholas Wright, 99 School Road, Saxon Street, Newmarket, CB8 9RY |
| G6 | JDP | J Mott-Gotobed, 14 Copse Road, New Milton, BH25 6ES |
| G6 | JDW | T Scarfe, Freshfields, Great Melton Road, Norwich, NR9 3NR |
| G6 | JEB | J Bailey, 22 Hainfield Drive, Solihull, B91 2PL |
| G6 | JEF | Stephen Wardley, 5 Swindon Street, Bridlington, YO16 4JD |
| GM6 | JEP | Frank Cassidy, 9 Spey Road, Troon, KA10 7DY |
| G6 | JEU | P Chrysostomou, 45 Leyborne Avenue, Ealing, London, W13 9RA |
| G6 | JEY | M Cooper, Flat 3, 32 Lansdowne Road, Worthing, BN11 5HB |
| G6 | JFL | D Barsby, 42 New Terrace, Pleasley, Mansfield, NG19 7PY |
| G6 | JFN | Peter Brazier, Stud House, Mentmore, Leighton Buzzard, LU7 0QE |
| GM6 | JFP | D Brown, 15 Eliots Park, Peebles, EH45 8HB |
| G6 | JFU | A Mayman, Lingmell, Cedar Grove, Hull, HU11 4QH |
| GW6 | JFV | T Morris, 29 Heol Croes Faen, Nottage, Porthcawl, CF36 3SW |
| GI6 | JGB | M McNinch, 5 Bangor Road, Groomsport, Bangor, BT19 6JF |
| G6 | JGF | A Morgan, 8 Shaftesbury Road, Watford, WD17 2RQ |
| GM6 | JGH | Jonathan Masshedar, 2b Pentland Park, Loanhead, EH20 9PA |
| G6 | JGP | A Lawrence, Columbine Cottage, Ford, Salisbury, SP4 6DJ |
| G6 | JGR | J Richardson, 30 Shaftesbury Avenue, Chandler's Ford, Eastleigh, SO53 3BS |
| G6 | JGT | A Davis, 7 Kennedy Crescent, Gosport, PO12 2NL |
| G6 | JHG | J Grieve, 65 Royal Lane, Uxbridge, UB8 3QU |
| GM6 | JHH | W Gunn, Tullochard, Scouriemore, Lairg, IV27 4TG |
| G6 | JIF | J [illegible], 24 [illegible] Road, Houghton Conquest, Bedford, MK45 3LU |
| GM6 | JIL | Ronald Mcmillan, 17b Kingston Road, Neilston, Glasgow, G78 3JA |
| GD6 | JIM | J King, 4 Glenhurst Avenue, Ruislip, HA4 7LZ |
| G6 | JIR | N Brown, 11 Tudor Green, Jaywick, Clacton-on-Sea, CO15 2PA |
| G6 | JIY | R Bamber, Hermes, 84 Paynesfield Road, Westerham, TN16 2BQ |
| G6 | JJA | N Billingham, 33 Scalwell Park, Seaton, EX12 2DD |
| G6 | JJB | B Banks, 16 Park Road, Burntwood, WS7 0EE |
| G6 | JJF | C Byrne, 104 Ripon Hall Avenue, Ramsbottom, Bury, BL0 9RE |
| G6 | JJG | K Breakwell, 91 Lynton Avenue, Claregate, Wolverhampton, WV6 9NQ |
| G6 | JJI | A Bromfield, 11 Blackthorn Croft, Clayton le Woods, Chorley, PR6 7TZ |
| G6 | JJK | J Bourne, 91 Burwell Road, Exning, Newmarket, CB8 7DU |
| GM6 | JJN | R Berry, Sylvan House, Glenmoriston, Inverness, IV63 7YJ |
| G6 | JJP | J Pinson, 10 Kenelm Close, Clifton-on-Teme, Worcester, WR6 6EB |
| GI6 | JJR | Nick Loughrey, 12 Billys Road, Newry, BT34 2NA |
| G6 | JJT | Elizabeth Ferguson, Willowmead, Church End, Bedford, MK44 2RP |
| G6 | JJX | R Price, Arfryn, Trecastle, Brecon, LD3 8UP |
| G6 | JKF | Michael Lowe, 4 Virginia Avenue, Stafford, ST17 4YA |
| G6 | JKK | G Orchard, 189 Sopwith Crescent, Wimborne, BH21 1SR |
| GM6 | JKU | George Henderson, 34 Soutar Crescent, Perth, PH1 1QB |
| G6 | JKV | Richard Henneman, Lydbury, Wistanswick, Market Drayton, TF9 2BB |
| G6 | JKY | Brian Hadley, 60 Chapel St., Pensnett, Brierley Hill, DY5 4EF |
| G6 | JLI | P Dixey, Rose Cottage, 197 Raikes Lane, Batley, WF17 9QF |
| G6 | JLL | S Douglas, 1030 Shields Road, Walkerville, Newcastle upon Tyne, NE6 4SR |
| G6 | JLU | A Millar, 8 Eisenhower Road, Shefford, SG17 5UP |
| G6 | JMB | J Mountain, Thurlow House, Aldenham Avenue, Radlett, WD7 8HJ |
| G6 | JMC | D Miller, Maes Hyfryd, Llanfynydd, Wrexham, LL11 5HH |
| GI6 | JMD | J Moller, 29 Carlton Heights, Bangor, BT19 6ZB |
| G6 | JME | Ray Powell, 39 Compton Drive, Eastbourne, BN20 8DA |
| G6 | JMG | Philip Parton, 12 Duchess Drive, Bridgnorth, WV16 4JD |
| G6 | JMJ | K Renton, 87 Shirley Gardens, Barking, IG11 9XB |
| G6 | JMO | J Page, 9 Ascot Close, Elstree, Borehamwood, WD6 3JH |
| G6 | JMX | B Wendon, 89 Palewell Park, London, SW14 8JJ |
| GM6 | JNJ | D Anderson, 34 Culzean Crescent, Kilmarnock, KA3 7DT |
| GM6 | JNQ | I Cox, 8 Traill St., Castletown, Thurso, KW14 8UG |
| G6 | JNS | P Crosland, Sprackets Orchard, Curry Rivel, Langport, TA10 0PP |
| G6 | JNV | M Carter, 17 Mcwilliam Road, Woodingdean, Brighton, BN2 6BE |
| G6 | JNW | S Carter, 84 Barnett Road, Brighton, BN1 7GH |
| G6 | JNZ | William Caine, 116b Hill Street, Hednesford, Cannock, WS12 2DR |
| GM6 | JOA | A White, Brodiescroft, Banff, AB45 3BR |
| GM6 | JOD | T Lawless, 2 Lawers Place, Bourtreehill North, Irvine, KA11 1LR |
| G6 | JOL | Robert Young, 53-55 Kirby Road, Leicester, LE3 6BD |
| GI6 | JOP | A Wallace, 61 Locksley Park, Belfast, BT10 0AS |
| G6 | JOR | David Webb, 1 Corelli Road, Basingstoke, RG22 4NB |
| GM6 | JOS | Kevin Arrowsmith, The Cairn House, Smoo, Durness, IV27 4QA |
| G6 | JOV | John Ashworth, 33 Edgar Road, West Drayton, UB7 8HN |
| G6 | JPC | Clive Jenkins, 10 Marsh Court, Abergavenny, NP7 5HQ |
| G6 | JPE | A Jephcott, 12 Clos Du Beauvoir, Rue Cohu, Castel, Guernsey, |
| G6 | JPG | J Gilliver, 5 Yew Tree Park Homes, Charing, Ashford, TN27 0DD |
| G6 | JPM | Stephen Green, 7 Brook View, Totnes, TQ9 5FH |
| GI6 | JPO | Herbie Graham, Cavancarragh, Lisbellaw, Co Fermanagh, BT94 5GL |
| G6 | JPQ | John Gould, 108 Newton Road, Burton-on-Trent, DE15 0TT |
| G6 | JPR | Mark Glover, 3 Sammons Way, Coventry, CV4 9TD |
| G6 | JPS | Jack Skertchly, 132 Derby Road, Spondon, Derby, DE21 7LX |
| G6 | JPT | D Gleave, 1 Fearnley Way, Newton-le-Willows, WA12 8SQ |
| G6 | JQD | R Skinner, 23 Hardy Road, Greatstone, New Romney, TN28 8SF |
| G6 | JQE | Gary Tannahill, 15 Bowfell Avenue, Newcastle upon Tyne, NE5 3XB |
| GU6 | JQF | M Trenchard, Mont Gibel, 3 Clifton Stairs, St. Peter Port, Guernsey, GY1 2PL |
| G6 | JQH | J williams, 18 The Leas, Barkston, Grantham, NG32 2PD |
| G6 | JQX | John Wingfield, 56 Manor Fields, Liphook, GU30 7BS |
| G6 | JRE | Stuart Stanton, 6 Trevor Road Beeston, Nottingham, NG9 1GR |
| G6 | JRI | Ian Wright, 3 Sykes Court, Wheldrake, York, YO19 6GE |
| G6 | JRL | M Bernard, 36 Garth Drive, Hambleton, Selby, YO8 9QD |
| G6 | JRM | H Bottomley, Nerefield, Aylesbury Road, Aylesbury, HP18 0BL |
| G6 | JRS | A Cuthbertson, 72 Bulford Road, Durrington, Salisbury, SP4 8DJ |
| GM6 | JRX | D Fraser, Kylerhea, Harbour Road, Thurso, KW14 8TG |
| GI6 | JRY | Hedmond Getty, 6 Rocheville, Cookstown, BT80 8QE |
| G6 | JRZ | S Gunn, 55 Station Road, West Byfleet, KT14 6DT |
| G6 | JSF | Michael Hayward, 1 Station Road, Grateley, Andover, SP11 8LG |
| G6 | JSI | A haswell, 66 White Hart Lane, Fareham, PO16 9BQ |
| GW6 | JSJ | David James, 6 Chave Terrace, Maesycwmmer, Hengoed, CF82 7RZ |
| G6 | JSN | Alfred Sym, 1 Beech Close, Spetisbury, Blandford Forum, DT11 9HG |
| GW6 | JSO | Andy Hubbard, Pant-y-Meillion, Velindre, Penboyr, Llandysul, SA44 5JA |
| G6 | JSR | Ann Mason, 5 Birch Road, Kippax, Leeds, LS25 7DY |
| G6 | JTC | Martin Whiteley, 9 Fernie Close, Barton Seagrave, Kettering, NN15 6RE |
| G6 | JTD | Derek Walker, 100 Clifton Road, Kingston upon Thames, KT2 6PN |
| G6 | JTI | H Martin, 80 Topcliffe Road, Sowerby, Thirsk, YO7 1RT |
| G6 | JTK | R Nokes, 99 Harmers Hay Road, Hailsham, BN27 1TW |
| G6 | JTO | John Parfrey, 97 Gordon Road, Camberley, GU15 2JQ |
| G6 | JTT | J Trett, 1 Moorland Way, Bridgwater, TA6 4JL |
| G6 | JTV | Ron Allen, 65 Atherstone Road, Measham, Swadlincote, DE12 7EG |
| G6 | JTW | H Marshall, Mardachrob, 4 Howell Road, Sleaford, NG34 9RX |
| G6 | JTX | E Bielawski, Rowanlea, Quarry Brow, Wrexham, LL12 8SJ |
| GM6 | JUA | D Brown, 10 Culmore Place, Falkirk, FK1 2RP |
| G6 | JUI | K Dare, One Bee, 1 Gloucester Road, Reading, RG30 2TH |
| G6 | JUP | J Sutton, 252 Rawling Road, Gateshead, NE8 4UH |
| G6 | JUQ | Gary Williams, 21 Arden Close, Southport, PR8 2RR |
| G6 | JUT | Jonathan Whiting, Castle Marina Road, Nottingham, NG7 1TN |
| G6 | JVA | J Greevy, 11a Norman Road, Walsall, WS5 3QJ |
| G6 | JVB | R Griffiths, 26 Brynglas, Gilwern, Abergavenny, NP7 0BP |
| G6 | JVK | M Jeffery, 14 Rosemary Avenue, Earley, Reading, RG6 5YQ |
| G6 | JVO | M Kidd, 99 Ferry Road West, Scunthorpe, DN15 8UG |
| G6 | JVP | P Cole, 100 Regents Park Road, Southampton, SO16 8NY |
| G6 | JVT | Colin Santer, 51 Limbrick Lane Goring-by-Sea, Worthing, BN12 6AB |
| G6 | JVX | H Schofield, 15 Deerfield Road, March, PE15 9AH |
| G6 | JWD | James Davies, Welfare, High Street, Borth, SY24 5JD |
| GM6 | JWF | Alistair Paul, 20 Upper Bridge Street, Alexandria, G83 0LL |
| GM6 | JWH | D Taylor, 3 Abbotsgrange Road, Grangemouth, FK3 9JD |
| G6 | JWL | Harold Roberts, Pen yr Erw, Graigfechan, Ruthin, LL15 2EY |
| G6 | JWM | David Le Grove, Apartment 3, Beechwood, Ilkley, LS29 8AH |
| G6 | JWO | A Legg, 2 Alkington Farm Lane Cottage, Heathfield, Berkeley, GL13 9PL |
| G6 | JXC | Peter Chadbund, 20 Northlands Road, Adstock, Buckingham, MK18 2JH |
| GI6 | JXG | W Collins, 33 New Row, Kilrea, Coleraine, BT51 5TA |
| G6 | JXS | W Hughes, 37 Winchester Close, Ashington, NE63 9QJ |
| G6 | JYB | Murray Niman, 55 Harrow Way, Chelmsford, CM2 7AU |
| G6 | JYN | D Watkinson, 18 Romsey Road, Lyndhurst, SO43 7AA |
| G6 | JYO | C Allen F B S, 20 Hollywood Lane, Hollywood, Birmingham, B47 5PX |
| G6 | JYR | Damon Benton, 221 Prestwood Road, Wolverhampton, WV11 1HF |
| G6 | JYX | R Drew, Derwent House, Landing Lane, Selby, YO8 0RA |
| G6 | JZE | P Graham, 14 Carlaw Road, Birkenhead, CH42 8QA |
| G6 | JZN | A Ogden, 21 Glenmore Road, Brixham, TQ5 9BT |
| G6 | JZV | K Lummis, 161 Stowupland Road, Stowmarket, IP14 5AP |
| G6 | JZW | Colin Muller, 5 Ash Close, Flitwick, Bedford, MK45 1JY |
| G6 | KAE | John Bailey, 54 Dimsdale Road, Northfield, Birmingham, B31 5RD |
| G6 | KAI | Michael Brighton, 11 West Close, Norwich, NR5 0NH |
| GM6 | KAM | A Drummond, Flat 4f, Crossfolds Crescent, Peterhead, AB42 1RD |
| G6 | KAV | H Hughes, Hendre Bach, Cerrigydrudion, Corwen, LL21 9TB |
| G6 | KAW | G Instone, 19 Dickenson Road, Swindon, SN25 1WG |
| GM6 | KAY | C Bates, 10 Swanston Avenue, Edinburgh, EH10 7BU |
| G6 | KBC | C Philpot, 17 Jervis Court, Ilkeston, DE7 8PX |
| G6 | KBD | D Potts, 11 Walmer Road, Newport, NP19 8NU |
| G6 | KBQ | T Williams, 2 Hazelwood, Greasby, Wirral, CH49 2RQ |
| G6 | KBS | J Musgrave, 57 Chiltern Road, Baldock, SG7 6LT |
| GI6 | KBX | J Turner, 45 Gloonan Hill, Ahoghill, Ballymena, BT42 1PU |
| G6 | KBZ | Norman Wilkins, Norjen, Thorpe Market Road, South Repps, NR11 8NG |
| G6 | KCG | P Sharpe, 46 Beaumont Road, New Costessey, Norwich, NR5 0HG |
| G6 | KCJ | David Wynters, 11 Heritage Lane, Ascott-under-Wychwood, Chipping Norton, OX7 6AD |
| G6 | KCV | P Willmott, C/O Oil Management Serv. Ltd., P.O. Box Hm 1751, Hamilton Hm Gx, Bermuda |
| GI6 | KCX | J Madden, 17 Avondale, Antrim, BT41 2AT |
| GM6 | KDB | Kevin Lee, West Skares, Glens of Foudland, Huntly, AB54 6AT |
| GM6 | KDD | D Scobbie, 17 Roselea Drive, Brightons, Falkirk, FK2 0TJ |
| GI6 | KDN | Kenn Mcinnes, 39 St. Johns Place, Belfast, BT7 3HA |
| G6 | KDU | Malcolm Thorley, 5 Burland Road, Newcastle, ST5 7ST |
| G6 | KDY | Andrew Perkins, 3 Greenway Close, Radcliffe-on-Trent, Nottingham, NG12 2BU |
| G6 | KEH | John Golightly, 1 Pannier Mews, Castle Street, Torrington, EX38 8EE |
| G6 | KEN | Kenneth DaSilva-Hill, 5 Station Road, Charing, Ashford, TN27 0JA |
| GM6 | KEV | David Smith, Mandala, Belhaven Road, Dunbar, EH42 1NW |
| G6 | KEZ | P Pattison, 18 Broadgate Lane, Deeping St. James, Peterborough, PE6 8NW |
| G6 | KFD | P Stockwell, 62 Golden Cross Road, Ashingdon, Rochford, SS4 3DQ |
| GM6 | KFO | G Gordon, 31 Stoneyhill Avenue, Musselburgh, EH21 6SB |
| G6 | KFR | David Jones, 2 The Orchard Mill Lane, Kings Sutton, Banbury, OX17 3RG |
| G6 | KGA | L Coleman, Lilac Cottage, Coley Lane, Stafford, ST18 0XB |
| G6 | KGG | W Duffner, 23 Grange Close, Misterton, Doncaster, DN10 4EN |
| G6 | KGK | Geoffrey Gudgin, 7 Merchant Place, Middleton, Milton Keynes, MK10 9JL |
| G6 | KGL | Graham Gudgin, 890 W Iowa Ave, Sunnyvale, United States, 94086 |
| G6 | KGR | Martin Buck, Upper Glaisfer, Llangynidr, Crickhowell, NP8 1LN |
| G6 | KGU | David Craig, Pear Tree Cottage, Cripps Corner Road, Robertsbridge, TN32 5QS |
| G6 | KHA | T Hyde, 14 Wyley Road, Coventry, CV6 1NW |
| G6 | KHD | K Bierton, 44 Stalmine Hall Park, Hall Gate Lane, Poulton-le-Fylde, FY6 0LD |
| G6 | KHG | R Champion, 25 Congreve Road, Worthing, BN14 8EL |
| G6 | KHM | L Edwards, 71 Gleneagles Road, Yardley, Birmingham, B26 2HT |
| G6 | KHN | S Harvey, 53 Winleigh Road, Handsworth Wood, Birmingham, B20 2HN |
| G6 | KHW | I Bultitude, 48 Forty Acres Road, Devizes, SN10 3DG |
| G6 | KIA | C Duckles, 8 Railway Cottages, Skillings Lane, Brough, HU15 1EN |
| G6 | KIB | Paul Duesbury, The Bungalow, Robins Lane, Cambridge, CB23 8HH |
| G6 | KIE | D Banks, 145 Compton Crescent, Chessington, KT9 2HG |
| G6 | KIH | P Ball, Fernley Green Road, Knottingley, WF11 8DH |
| G6 | KIV | S Blythe, 17 Ashlea Road, Wirral, CH61 5UG |
| G6 | KIW | M Dennis, 16 Gwel Y Llan, Llandegfan, LL59 5YH |
| G6 | KIZ | M Griffiths, 70 Towcester Road, Far Cotton, Northampton, NN4 8LQ |
| G6 | KJA | Patrick Hoyle, Sunset Cottage, Water End Lane, Ayot St. Peter, Welwyn, AL6 9BB |
| GI6 | KJC | W Abram, 1 The Driggs, Groomsport, Bangor, BT19 6HY |
| G6 | KJE | Alan Dolby, 27 Tucker Road, Ottershaw, Chertsey, KT16 0HD |
| G6 | KJH | P Horobin, 12 Laurel Road, Blaby, Leicester, LE8 4DL |
| G6 | KJK | J Chappell, 15 Edmund Avenue, Stafford, ST17 9FT |
| G6 | KJM | J Mirams, 29 Martello Court, Jevington Gardens, Eastbourne, BN21 4HR |
| G6 | KJT | Stephen Brabbins, 8 Park Drive, Bingley, BD16 3DF |
| G6 | KJY | L Cartwright, 18 High Causeway, Much Wenlock, TF13 6BZ |
| G6 | KKA | John Edmondson, 6 Park Lea, Bradley, Huddersfield, HD2 1QH |
| G6 | KKN | P Clowes, 14 Derek Drive, Sneyd Green, Stoke-on-Trent, ST1 6BY |
| GM6 | KKP | A Dick, 4 Westpark Gardens, Dundee, DD2 1NY |
| G6 | KKW | Robert Rogers, 31 Westgate Bay Avenue, Westgate on Sea, CT88AH |
| G6 | KLC | Anthony Morris, Bodvel Hall, Llannor, Pwllheli, LL53 6DW |
| G6 | KLF | A Lythaby, 25 Greenhill Road, Otford, Sevenoaks, TN14 5RR |
| G6 | KLH | R Taylor, 57 Walnut Tree Road, Shepperton, TW17 0RP |
| G6 | KLQ | Jeffrey Laing, Penyboncyn, Pen-y-Garnedd, Oswestry, SY10 0AN |
| G6 | KMG | I Turnbull, 45 Elton Road, Darlington, DL3 8HU |
| GM6 | KMK | S Windsor, Hillside Overbrae, Fisherie, Turriff, AB53 5QP |
| G6 | KMQ | C Meadows, 47 Widney Lane, Solihull, B91 3LL |
| G6 | KNE | J Wright, Rhumbles, Station Approach, Dorking, RH5 5HT |
| G6 | KNK | J Solomon, 11 Angle Close, Hillingdon, Uxbridge, UB10 0BS |

JK Callsigns

**IMPORTANT NOTE**

**Revalidate licence to avoid revocation** – Ofcom has advised the Society that plans will be drawn up to revoke licences that have not been revalidated as required by the licence conditions. The quickest way to revalidate is to do so online via the Ofcom website: *https://services.ofcom.org.uk/* or by email: *amateur.validations@ofcom.org.uk* Ofcom staff are available to help, but please be patient during times of heavy workload.

G6 KNM R Suttenwood, 51 High St., Rowhedge, Colchester, CO5 7ET
G6 KNU Hop Wing Man, 115 Northdown Park Road, Margate, CT9 3PX
G6 KOB Jan Sobanski, 10 Robert Avenue, Barnsley, S71 5RB
G6 KOE A Reilly, 14 Carleton Gardens, Carleton, Poulton-le-Fylde, FY6 7PB
GM6 KON T Wilkins, The Farmhouse, Freswick, Wick, KW1 4XX
GM6 KOR K Osborne, 42 India Street, Edinburgh, EH3 6HB
G6 KPD J Perrett, 12 Horne Close, Stratton-on-the-Fosse, Radstock, BA3 4SS
G6 KPJ P Vaughan, The Views, Bedford Road West, Northampton, NN7 1HB
GM6 KPL A Wilson, 1 Union Street, Newmilns, KA16 9BJ
G6 KPT Neil Woodley, 20 St. Edwards Road, Cheddleton, Leek, ST13 7JP
G6 KPW S Taylor, 17 Crays Hill, Leabrooks, Alfreton, DE55 1LN
G6 KPX A Thorne, 31 Oak Farm Close, Sutton Coldfield, B76 1PJ
G6 KQ Keith Spicer, Grove Cottage, Dallinghoo, Woodbridge, IP13 0LR
G6 KQD Gary Morris, 20 Victoria Way, Stafford, ST17 0NU
G6 KQJ Helen Moon, 14 Elmwood Road, Eaglescliffe, Stockton-on-Tees, TS16 0AQ
G6 KQK C Sanders, The Bungalow, North Pill, Saltash, PL12 6LJ
G6 KQN Graeme Robertson, 24 Begonia Avenue, Farnworth, Bolton, BL4 0DS
G6 KQS John Newton, 122 Dane Valley Road, Margate, CT9 3RY
G6 KQZ B Wiseman, 307 Kempshott Lane, Basingstoke, RG22 5LY
G6 KRC KIDDERMINSTER RD c/o Anthony Hartland, 11 Linden Avenue, Kidderminster, DY10 3AA
G6 KRG Jonathan Freeman, 18a Five Bells, Watchet, TA23 0HZ
GW6 KRK E Karklins, Lonlas House, Lonlas, Neath, SA10 6SD
G6 KRN M Everitt, Mallory, St. Johns Road, Ventnor, PO38 3EF
G6 KRS Nigel Ashall, 21 Buxton Lane Droylsden, Manchester, M43 6HL
G6 KRY C Pieters, 32 Olde Farm Drive, Blackwater, Camberley, GU17 0DU
G6 KSK Anthony Hodgson, 33 Higham Road, Wainscott, Rochester, ME3 8BE
G6 KSO P Lash, 7 Park Road, Stockport, SK4 4PY
G6 KSR F Patman, Northcote, 31 Church Road, Lincoln, LN6 5UW
G6 KSV A Sayers, 145 Campkin Road, Cambridge, CB4 2NP
G6 KTB W Curtis, Innsbruck, Trevingey Crescent, Redruth, TR15 3DF
G6 KTC Wayne Bramwell, 15 Chadswell Heights, Lichfield, WS13 6BH
G6 KTE D Brunt, 31 The Green, Kingsley, Stoke-on-Trent, ST10 2AG
G6 KTG D Clements, 3 Tilefields, Hollingbourne, Maidstone, ME17 1TZ
G6 KTK C Handley, Torcroft, 40 New Village Road, Little Weighton, HU20 3XH
G6 KTN Michael Lamerick, 12 Woodhouse Close, Birchwood, Warrington, WA3 6QP
G6 KTO J Martyn Clark, 127 Blackpool Road North, Lytham St. Annes, FY8 3DB
GM6 KTP K Morrison, 8 St. Helena Crescent, Hardgate, Clydebank, G81 5PD
G6 KTR Graham Birch, 11 Orsino Walk, Colchester, CO4 3LU
G6 KTX A King, 2 Longstaff Gardens, Fareham, PO16 7RR
G6 KUI Peter Walker, 23 Denstone Drive, Alvaston, Derby, DE24 0HZ
G6 KUJ F Moulding, 28 Woodbine Road, Bolton, BL3 3JH
G6 KVA R Dresser, 6 Acacia Avenue, Fencehouses, Houghton le Spring, DH4 6JG
G6 KVE C Payne, 4 George Street, Helpringham, Sleaford, NG34 0RS
G6 KVG S Reid, 223 The Greenway, Epsom, KT18 7JE
G6 KVI B Gosling, 15 Cherry Chase, Tiptree, Colchester, CO5 0AE
G6 KVK G Howell, 19 Constable Avenue, Eaton Ford, St. Neots, PE19 7RH
G6 KVR Peter Roberts, 30 Baldwins Lane, Hall Green, Birmingham, B28 0QX
GI6 KVS H Porter, 30 Twinburn Road, Newtownabbey, BT37 0EL
G6 KVY Stephen Trotter, 61 Trinity Road, Billericay, CM11 2RY
G6 KWA D King, 20 Trinity Close, Haslingfield, Cambridge, CB23 1LS
G6 KWH Jonathan Dixon, 16 Forest Lane, Martlesham Heath, Ipswich, IP5 3ST
G6 KWU R Adamson, 46 Taliesin Avenue, Shotton, Deeside, CH5 1HY
G6 KWZ J Manning, 280 Ledbury Road, Hereford, HR1 1QL
G6 KXB R Linzey, 29 Arkle Court, Alnwick, NE66 1BS
G6 KXD M Parker, Hazel House, Talkin, Brampton, CA8 1LE
G6 KXN Roger Perry, 6 Morgan Close, Arley, Coventry, CV7 8PR
G6 KXP D Flanagan, Ryan Mar, Stair Drive, Stranraer, DG9 8EY
G6 KXW A Blair, 35 South Court Avenue, Dorchester, DT1 2BY
G6 KYE M Davis, 86 Upper Shaftesbury Avenue, Southampton, SO17 3RT
G6 KZI R Gregory, 75 Station Road South, Belton, Great Yarmouth, NR31 9LZ
G6 LAE John Clifton, 6 Chester Close, Newbury, RG14 7RR
G6 LAU D Tanswell, Highstead Farmhouse, Bradford, Holsworthy, EX22 7AA
G6 LAW Chris Rudge, 1 Mill Lane, Alfington, Ottery St. Mary, EX11 1PF
G6 LBE John Massey, 10 Rapley Avenue, Storrington, Pulborough, RH20 4QL
G6 LBG Nick Orgill, 32 Upland Avenue, Chesham, HP5 2EB
G6 LBJ Peter Shadbolt, 39 Ringstead Crescent, Weymouth, DT3 6PT
G6 LBL Jeremy Scarr, 6 Farmanby Close, Thornton Dale, Pickering, YO18 7TD
G6 LBO K Batty, 19 Breckland Close, Stalybridge, SK15 2QQ
G6 LBQ A Hunter, 22 Lynthorpe Avenue, Cadishead, Manchester, M44 5JQ
G6 LBR Anthony Ledger, 9 Fox Wood, Westlea, Swindon, SN5 7AW
G6 LCL T Mallett, 11 Caragh Road, Chester le Street, DH2 3EA
G6 LCP P Muzyka, 2 Engine Fold, Wrenthorpe, Wakefield, WF2 0PP
G6 LCS John McNeill, 2 Greenwood Close, Weaverham, Northwich, CW8 3RH
G6 LCU J Retter, 12 Palmerston Road, Grays, RM20 4YR
G6 LCX Michael Pomfret, 17 Lovers Lane, Atherton, Manchester, M46 0PG
G6 LD Denby Dale ARS c/o John Stocks, 96 North Street, Lockwood, Huddersfield, HD1 3SL
G6 LDA J Round, 53 Furlong Lane, Halesowen, B63 2TB
GM6 LDG Paul Clements, Dhualton Cottage, Kirtomy, Thurso, KW14 7TB
G6 LDJ R Wilkinson, 2 Conway Avenue, Billingham, TS23 2HX
G6 LDM D Shippen, 11a Pear Tree Drive, Wincham, Northwich, CW9 6EZ
G6 LDO C Seeney, 91 Dovehouse Close, Eynsham, Witney, OX29 4EW
G6 LDP D Scott, 7 Greenfield Mount, Wrenthorpe, Wakefield, WF2 0TJ
G6 LDW J Tottle, 327a Edificio, Calle Miguel Machado, Son Caliv, Calvia, Spain, 07181,
G6 LDY J Seddon, 11 Hilda Street, Leigh, WN7 5DG
G6 LEB Tim Leader-Chew, 10 Hawmead, Crawley Down, Crawley, RH10 4XY
G6 LEI Stephen Meadwell, 25 Redland Road, Oakham, LE15 6PH
G6 LEK R Mason, 23 Fulmodeston Road, Stibbard, Fakenham, NR21 0LT
G6 LEU David Last, Hillview, New Road, Bridport, DT6 4NY
G6 LEY D Miller, 44 Long Lane, Ickenham, Uxbridge, UB10 8TA
GM6 LEZ John Mcdermott, 12a Margaret Street, Greenock, PA16 8AS
G6 LFA A Machin, 10 Buttlehide, Maple Cross, Rickmansworth, WD3 9TZ
G6 LFC J McHale, Glen Elg, The Green, Skipton, BD23 4LB
GD6 LFD J Corderoy, 1 Alandale Drive, Pinner, HA5 3UP
G6 LFG John Bradbury, 281 Peter Street, Macclesfield, SK11 8EX
G6 LFJ Jonathan Aslan, 16 Guildford Street, Brighton, BN1 3LS
G6 LFQ R Cross, 84a Cranborne Avenue, Surbiton, KT6 7JT
G6 LFT G Cooke, 37 Hertford Close, Woolston, Warrington, WA1 4EZ
G6 LFW J Ford, 24 Tonstall Road, Epsom, KT19 9DP
G6 LGM I Rogers, Tremayne Cottage, Calloose Lane, Hayle, TR27 5ET
G6 LGR Alan Picot, 14 Ringshall Road, St. Pauls Cray, Orpington, BR5 2LZ
G6 LGW Anthony Pollard, 75 Ridgeway Avenue, Dunstable, LU5 4QL
G6 LHA F Priestnall, 56 Badger Gate, Threshfield, Skipton, BD23 5EN
GW6 LHF Danny Owen, Grasmere, Pontycleifion, Cardigan, SA43 1DW
G6 LHG B O'Shea, 37 Gardeners Road, Halstead, CO9 2TA
G6 LHQ R Harber, 7 Hamilton Avenue, Cobham, KT11 1AU
G6 LI LINCOLNSHIRE POACHERS CONTEST GROUP c/o David Johnson, Dean Cottage, Dowsby Fen, Bourne, PE10 0TU
G6 LIB J Baker, 5 Larkspur Close, Bishop's Stortford, CM23 4LL
G6 LIK Leslie Clark, 56 Rembrandt Avenue, South Shields, NE34 8RU
GM6 LIN Joe Quinn, 8 Cluny Drive, Newton Mearns, Glasgow, G77 6YG
G6 LJC Gary Weston, 11 Friars Road, Abbey Hulton, Stoke-on-Trent, ST2 8DQ
G6 LJE Robin Waitt, Orchard Cottage, Canonbie, DG14 0RZ
G6 LJF J White, 8 Well Side, Marks Tey, Colchester, CO6 1XG
G6 LJH M Wilson, Hillside, Chapel Lane, Kettering, NN14 4EA
G6 LJR David Twyman, 77 Essex Road, Maldon, CM9 6JH
G6 LJU John Whitehouse, The Paddock, Westmancote, Tewkesbury, GL20 7EP
G6 LJX Simon Williams, 187 London Road, Northwich, CW9 8AR
G6 LKA B Woolnough, 57 Cranborne Road, Potters Bar, EN6 3AB
G6 LKB David Warburton, 36 Bigland Drive, Ulverston, LA12 9PD
G6 LKG R Milne, 9 Brunstath Close, Wirral, CH60 1UH
G6 LKH C Dunlop, 32 Court Way, Twickenham, TW2 7SN
G6 LKJ John Depledge, 37 Higher Bents Lane, Bredbury, Stockport, SK6 1EE
GM6 LKQ E Fry, Rosehall, Beauly, IV4 7AW
G6 LKV G Ashbee, 6 The Green, Wimbledon, London, SW19 5AZ
G6 LKW Thomas Ashbee, Plough Heights, Main Road, Itchen Abbas, Winchester, SO21 1BQ
G6 LKZ D Bentley, 4 Highway, Crowthorne, RG45 6HE
G6 LLD George Bell, 4 Dallymore Drive, Bowburn, Durham, DH6 5ES
G6 LLF Philip Bennett, 1 The Briars, Newcastle, ST5 9PU
G6 LLG M Broadway, 69 The Brambles, Crowthorne, RG45 6EF
G6 LLL D Burrows, 32 Whitfield Cross, Glossop, SK13 8NW
G6 LLP Robert Farey, 38 Trent Close, Yeovil, BA21 5XQ
G6 LLU D Setterfield, 10 Birch Walk, Bride Street, Todmorden, OL14 5ET
G6 LMB P Steadman, 41 The Linkway, Brighton, BN1 7EJ
G6 LMC P Webb, 63 Trinity Road, Halstead, CO9 1ED
G6 LMI J Evans, 91 Queens Avenue, Flint, CH6 5JP
G6 LMJ Geoffrey Eardley, 45 Little Moss Lane, Scholar Green, Stoke-on-Trent, ST7 3BL
G6 LMR Kenneth Fisher, 26 Manila Street, Sunderland, SR2 8RS
G6 LNF Nicholas Clare, 4 Arlington, Weymouth, DT4 9SG
G6 LNL I Dobson, 4a North Terrace, Seaham, SR7 7EU
G6 LNS James Duxbury, Woodlands, Wallace Lane, Preston, PR3 0BB
G6 LNU J Durban, 62 Westfield Way, Charlton, Wantage, OX12 7EP
G6 LNV John Cunliffe, 142 Hall Road, Hull, HU6 8SB
G6 LOC T Stirrup, 23 Round Wood, Penwortham, Preston, PR1 0BN
G6 LOJ N Pettit, 10 Broom Road, Lakenheath, Brandon, IP27 9ES
G6 LOR E Wood, 57d Halesowen Street, Rowley Regis, B65 0HF
G6 LPB R Steele, 27 Beasley Grove, Great Barr, Birmingham, B43 7HG
G6 LPC Alan Samways, 61 Cooper Road, Rye, TN31 7BG
G6 LPD A Tucker, 63 Oakes Road, Bury St. Edmunds, IP32 6PU
G6 LPG Steven Taylor, 76 Queensdown Gardens, Bristol, BS4 3JF
G6 LPS T Biddle, Flat 8, Fountain House, Torquay, TQ1 1EQ
G6 LPT Robert Bruckner, 41 Uphill Grove, London, NW7 4NH
G6 LPV David Blackmore, 2 Witten Gardens, Northam, Bideford, EX39 3RE
G6 LPX P Brown, 5 Fairview Close, Amington, Tamworth, B77 3LA
G6 LQE D Byrom, 206 Didsbury Road, Stockport, SK4 2AA
G6 LQG Derith Bate, 45 Arlington Way, Stoke-on-Trent, ST3 7WH
G6 LQI Nigel Bird, 2 Manor Valley, Weston-Super-Mare, BS23 2SY
G6 LQM G Barker, 99 Sheffield Road, Wymondham, NR18 0HS
G6 LQP D Brown, Hillingswood, Acton, Newcastle, ST5 4EG
G6 LQR Brett Walker, 5-6 Cochrane Terrace, Willington, Crook, DL15 0HN
G6 LRT Cola Johnson, 52 Evesham Road, Cookhill, Alcester, B49 5LJ
G6 LRU R Jones, 53 Wavertree Road, Blacon, Chester, CH1 5AF
G6 LRY Chris Kelland, 11 The Meads, West Hanney, Wantage, OX12 0LJ
G6 LSB Nigel Key, 6 The Hollow, Stanwick, Wellingborough, NN9 6PY
G6 LSC M Kitchener, 5 Whinbush Grove, Hitchin, SG5 1PT
G6 LSD P Kerry, 35 Victoria Drive Blackwell, Alfreton, DE55 5JL
GW6 LSL S Wood, 2 Radyr Road, Llandaff North, Cardiff, CF14 2FU
G6 LSO C Wolf, 35a Moorhouse Road, Carlisle, CA2 7LU
G6 LST D Rhodes, 1 Tanpit Cottages, Winstanley, Wigan, WN3 6JY
G6 LSW A Stevenson, 37 Hillside Road East, Bungay, NR35 1JU
G6 LTB Peter Townrow, 64 Millham Road, Bishops Cleeve, Cheltenham, GL52 8BG
G6 LTK Kevin Wilson, 44 Campbell Road, Caterham, CR3 5JN
G6 LTN A Wanford, 4 Willows Close, Tydd St. Mary, Wisbech, PE13 5QR
G6 LTR J Warner, 32 Rolleston Road, Wigston, LE18 2EP
G6 LUD A Ryan, 34 Mead Road, Gravesend, DA11 7PP
G6 LUE T Yates, 5 Manor Garth, Kellington, Goole, DN14 0NW
G6 LUF Anthony Yates, 59 Worden Lane, Leyland, PR25 3BD
G6 LUJ R Perry, Thornaby, Queen Street, Colyton, EX24 6JU
G6 LUK Jeremy Russell, 9 Batten Close, Christchurch, BH23 3BJ
G6 LUM J Papworth, 339 Gayfield Avenue, Brierley Hill, DY5 3JE
G6 LUO B Maynard, 11 Denham Road, Canvey Island, SS8 9HB
G6 LUU Alan Marshall, 13 Barn Owl Close, Northampton, NN4 0RQ
G6 LUY Nevil Mattey, 2 A Spring Grove, Gravesend, DA12 1LL
G6 LVB H Long, Flat 10, Delahay House, 15 Chelsea Embankment, London, SW3 4LA
G6 LVC J Shergold, 35 Orchard Grove, New Milton, BH25 6NZ
G6 LVG S Normandale, 5 The Beacon, Ilminster, TA19 9AH
G6 LVI P Hickey, 36 Station Road, Alderholt, Fordingbridge, SP6 3RB
G6 LVJ R Hickey, Mallorin, Blackfield Road, Southampton, SO45 1EG
G6 LVM C Holderness, 7 Oakfield Avenue, Clayton le Moors, Accrington, BB5 5XG
G6 LVN Robert Hope, 35 Pinewood Gardens, North Cove, Beccles, NR34 7PQ
G6 LVS Derek Mallalieu Howard, Flat 17, The London Well Street, Ryde, PO33 2SS
G6 LVT Christopher Harvey, 619 West Street, Crewe, CW2 8SH
G6 LWA C Hall, 147 Gordon Avenue, Camberley, GU15 2NR
G6 LWC Reginald Hardman, 4 Alverstone Road, Wallasey, CH44 9AA
G6 LWD Peter Hurley, 18 Pear Tree Lane, Wolverhampton, WV11 1BD
G6 LWK M Horsfield, 13 St. Leonards Way, Ardsley, Barnsley, S71 5BS
G6 LWT J Mills, Smiths Hill, Petrockstow, Okehampton, EX20 3EZ
G6 LWZ L Miles, 1 Wyndham Wood Close, Fradley, Lichfield, WS13 8UZ
G6 LXE Terry Doyle, Leal House, Sparrow Pit, Buxton, SK17 8ET
G6 LXF P Duley, 4 Brean Road, Stafford, ST17 0PA
G6 LXL J Ellis, Goosters Green, Hope Bagot, Ludlow, SY8 3AE
G6 LXP D English, 14 Elm Close, Ryde, PO33 1ED
G6 LXU S Westall, 4 South View, Great Harwood, Blackburn, BB6 7NL
G6 LXV David Woods, 110 Sandy Lane, Warrington, WA2 9JA
G6 LXW John Weigh, 167 Farm View Road, Rotherham, S61 2BL
G6 LYA Peter Whysall, 1 Greenlees Close, Fareham, PO17 5GS
G6 LYD Colin Weaver, Linton Lodge, Pluckley Road, Ashford, TN27 0AQ
G6 LYE G Whiles, 7 Thorndale Street, Hellifield, Skipton, BD23 4JE
G6 LYJ J young, 1 Stevenson Place, Annan, DG12 6BU
G6 LYM D Miller, 9 High Mead, Hockley, SS5 4QG
G6 LZB P Adams, 464 Whippendell Road, Watford, WD18 7PT
G6 LZM G Beddington, Konrei, Tower Hill, Norwich, NR8 5AX
G6 LZX B Broad, 1 Sussex Close, Laindon, Basildon, SS15 6PR
G6 LZZ Jim Bolton, Huds House, Cowgill, Sedbergh, LA10 5TQ
G6 MAA G Bishop, Oyston Lodge, Lynstone Road, Bude, EX23 8LR
GW6 MAB John Barwick, 13 Greenfields Avenue, Bridgend, CF31 4SR
G6 MAC B McDonnell, 68 Chaigley Road, Longridge, Preston, PR3 3TQ
G6 MAD Terence Morley, 84 Cliff Lane, Ipswich, IP3 0PJ
G6 MAJ A Mulvaney, 38 Ramwells Brow, Bromley Cross, Bolton, BL7 9LL
G6 MAM Kelvin Wright, 63 Dudley Avenue, Leicester, LE5 2EF
G6 MAR Graham Wratten, 10 The Boundary, Seaford, BN25 1DG
G6 MAT Anthony Valentine, 21 Naseby Road, Congleton, CW12 4QX
G6 MAW Rex Underwood, 35 Greenfields, Langley Mill, Nottingham, NG16 4GJ
G6 MAY D Pool, 6 Rivett Close, Clothall Common, Baldock, SG7 6TW
G6 MBD J Durston-Wyatt, 62 St. Johns Road, Epping, CM16 5DP
G6 MBF David Surgey, 4 Down Lane, Bathampton, Bath, BA2 6UE
G6 MBH W Stiling, 11 Carrol Grove, Cheltenham, GL51 0PP
G6 MBI D Stainton, Tilton House, 39 Redland Grove, Nottingham, NG4 3ET
G6 MBL M Snow, 32 Orchard Avenue, Worthing, BN14 7PY
G6 MBR MID BEDS R/NET c/o Ian McIver, 3 Asgard Drive, Bedford, MK41 0UP
G6 MBV C Sutcliffe, 652 Newchurch Road, Newchurch, Rossendale, BB4 9HG
G6 MC Brimham Contest Group c/o Susan Clarke, Brimham Lodge Fm, Harrogate, HG3 3HE
G6 MCB M Baldry, 10 Kingfisher Court, Lowestoft, NR33 8PJ
G6 MCC L Crompton, 6 Moss Avenue, Ashton-on-Ribble, Preston, PR2 1SH
G6 MCE P Garde, 21 Leicester Avenue, Timperley, Altrincham, WA15 6HR
G6 MCG C Garnham, 1 Ennerdale Close, Felixstowe, IP11 9SS
G6 MCN Andrew Gillespie, Elm Tree Cottage, Chilbolton, Stockbridge, SO20 6BA
G6 MCQ J Grane, 15 Pinelands Way, Osbaldwick, York, YO10 3QJ
G6 MCV J McVicar, 2 Lilliardsedge Par, Mr Ancrum, Jedburgh, TD8 6TZ
G6 MCX Paul Garland, 6 Barn Piece, Chandler's Ford, Eastleigh, SO53 4HP
G6 MCY Michael Goddard, 65 Langley Hall Road, Solihull, B92 7HE
GM6 MD CLYDE COAST CONTEST CLUB c/o A Dunn, Shankston Farm, Patna, Ayr, KA6 7LD
G6 MDC M Green, 9 Greencroft Avenue, Northowram, Halifax, HX3 7EP
G6 MDF M Pethen, 2 Saltdean Way, Bexhill-on-Sea, TN39 3SS
G6 MDG James Tyson, 1102 Rochdale Road, Blackley, Manchester, M9 7EQ
G6 MDM Wendy Smyth, 4 Dereham Road, Pudding Norton, Fakenham, NR21 7NA
G6 MDN G Tillett, 43 Chippenham Road, Harold Hill, Romford, RM3 8HJ
G6 MDR I Stanley, 6 Kennedy Avenue, Long Eaton, Nottingham, NG10 3GF
G6 MDS A Scott, 3 Majestic Road, Hatch Warren, Basingstoke, RG22 4XD
G6 MED Peter Cartmell, 16 Churchfield Drive, Wigginton, York, YO32 2FL
G6 MEH John Turner, 44b Foxgrove Road, Beckenham, BR3 5DB
G6 MEI Carl Thacker, 30 Fairview Drive, Adlington, Chorley, PR6 9SB
G6 MEW Anthony Hunt, 7 Wood Lodge, Calmore, Southampton, SO40 2UP
G6 MFB R Hodges, 21a Preston Lane, Lyneham, Chippenham, SN15 4AR
G6 MFM D Hills, 14 Curzon Road, Thornton Heath, CR7 6BR
GD6 MFR David Hunt, 67 Dorchester Avenue, North Harrow, Harrow, HA2 7AX
G6 MFU N Cowley, 126 Racecourse Road, Swinton, Mexborough, S64 8DS
G6 MGA Stephen Cooper, 27 Huntsmans Gate, Bretton, Peterborough, PE3 9AU
G6 MGH M Cooper, 3 Marina Way, Ripon, HG4 2LJ
G6 MGN D Richardson, 14 Wingfield Avenue, Lakenheath, Brandon, IP27 9HS
G6 MGQ A Reddish, Wheelwrights House, Luckeys Corner, Ipswich, IP7 7LR
G6 MGZ James Wilfred Middleton, 187 Balcombe Road, Horley, RH6 9EA
GM6 MHC Ralph McNaught, 5 Bonnymuir Crescent, Bonnybridge, FK4 1GD
G6 MHF Adrian Marshall, Thistledome, First Avenue, Watford, WD25 9PS
G6 MHO I Pomfret, 20 Sandown Road, Bury, BL9 8HN
G6 MHR J Castelow, 7 Langford Close, Burley in Wharfedale, Ilkley, LS29 7NP
GW6 MHV B Cooke, 51 Celyn Avenue, Cardiff, CF23 6EJ
G6 MIC M Clayden, 121 North Lane, East Preston, Littlehampton, BN16 1HB
G6 MID P Croft, Exchange Buildings, Exchange Street, Normanton, WF6 2AA
G6 MIF David Cooper, 42 St. James Court, Harpur Hill, Buxton, SK17 9RE
GW6 MIH M Cleverley, 33 Tylchawen Crescent, Tonyrefail, Porth, CF39 8AL
G6 MIS S Ransom, 1 Bilberry Road, Clifton, Shefford, SG17 5HB
G6 MIU William Livesey, 20 West Way, Little Hulton, Manchester, M38 9GL
G6 MJA Michael Addison, Berrymead, Oxford Street, Great Missenden, HP16 9JH
G6 MJB D Lloyd, Rangelands, Old Guildford Road, Camberley, GU16 6PH
G6 MJM S Parker, 19 Sundour Crescent, Wednesfield, Wolverhampton, WV11 1AP
G6 MJQ Adrian Peake, 1 The Common, Evington, Leicester, LE5 6EA
G6 MJW G Davis, 30 Bonny Wood Road, Hassocks, BN6 8HR
GM6 MJY C Donald, 126 Newburgh Circle, Bridge of Don, Aberdeen, AB22 8XB
G6 MKD Michael Douglass, 20 Cadshaw Close Birchwood, Warrington, WA3 7LR
G6 MKJ N Ellis, 140 Wollaston Road, Irchester, Wellingborough, NN29 7DH
G6 MKL Roger Ellis, 4a Elmdale Road, Earl Shilton, Leicester, LE9 7HQ
G6 MKO S Everett, 11 Chepstow Road, Felixstowe, IP11 9BU
G6 MKQ Geoffrey Evans, 31 Queen Elizabeth Crescent, Accrington, BB5 2AS
GW6 MKR Christopher Foster, Pentwyn House, Delfryn, Aberdare, CF44 0TU

G6 MKZ Paul Fisher, 2 Leas Drive, Iver, SL0 9RD
G6 MLH Gillian Marshall, Fern House, Church Road, Newark, NG23 7ED
G6 MLI D Morgan, Northwood Hotel, 47 Rhos Road, Colwyn Bay, LL28 4RS
GW6 MLL Bernard Murphy, 99 Deepglade Close, St. Thomas, Swansea, SA1 8EJ
G6 MLU Trevor Abson, 177 Meadowhall Road, Kimberworth, Rotherham, S61 2JW
GD6 MLV K Barker, 8 Shelley Gardens, Wembley, HA0 3QG
G6 MMA Martin Barlow, 56 Pasturegreen Way, Irlam, Manchester, M44 6TE
G6 MMB Jeremy Bulbrook, 33 Stonecross Way, March, PE15 9DH
G6 MMD S Burrows, 2 Luscombe Farm Cottages, Heath End, Stratford-upon-Avon, CV37 0PP
G6 MMG D Brown, 28 Bishop Drive, Whiston, Prescot, L35 3JL
G6 MMJ P Bromley, 3 Georgia Avenue, Broadwater, Worthing, BN14 8AZ
G6 MML V Bates, 9 Parkdene Close, Harwood, Bolton, BL2 3LH
GW6 MMM J Bowen, 18 Admirals Walk, Sketty, Swansea, SA2 8LQ
G6 MMR A Salter, 143 Eastwood Road North, Leigh-on-Sea, SS9 4NB
G6 MMS J Young, 45 Eaves Lane, Chadderton, Oldham, OL9 8RG
G6 MMT J Ward, 64 Gladstone Road, Ipswich, IP3 8AT
G6 MNB M Bulmer, Highfield, 7 Fountain Avenue, Altrincham, WA15 8LY
GW6 MNC William Turner, 37 Dan-y-Bryn Avenue, Radyr, Cardiff, CF15 8DD
G6 MNI Rachel Andrews, 10 Summerfield Close, Mevagissey, St. Austell, PL26 6TZ
G6 MNJ P Andrews, 10 Summerfield Close, Mevagissey, St. Austell, PL26 6TZ
G6 MNL Anthony Butler, 45 Roewood Close, Holbury, Southampton, SO45 2JT
G6 MNN M Broad, 18 Ramptons Meadow, Tadley, RG26 3UR
G6 MOD P Boden, 54 Avill, Hockley, Tamworth, B77 5QF
G6 MOI Andrew Bates, 44 Chalfont Drive, Sileby, Loughborough, LE12 7RQ
G6 MOT S Kilmister, 9 Wheal Jane Meadows, Threemilestone, Truro, TR3 6EN
G6 MOZ P Sealey, 45 Haydon Way, Coughton, Alcester, B49 5HY
G6 MPE John Simmons, 282 Bishopton Road West, Stockton-on-Tees, TS19 7LY
G6 MPJ W Smith, Boleyn Service Station, 77 River Road, Barking, IG11 0DS
G6 MPK T Smith, 87 Swanland Road, Hessle, HU13 0NS
G6 MPN A Shalders, 29 Princess Drive, Sandbach, CW11 1BS
G6 MPT P Pritchard, 5 Charlemont Road, Stone Cross, West Bromwich, B71 3HX
G6 MPX David Prince, 40 Ffordd Gryffydd, Llay, Wrexham, LL12 0RT
G6 MQD Paul Mullins, 29 Windmill Way, Tring, HP23 4HH
G6 MQG C Northrop, Mayfield, 47b Hardhorn Road, Poulton-le-Fylde, FY6 7SR
G6 MQI G Pointon, 448 Stockport Road, Thelwall, Warrington, WA4 2TR
G6 MQJ P Racher, 2 Heron Way, Horsham, RH13 6DG
G6 MQK I Stinton, 57 Wildfields Road, Clenchwarton, King's Lynn, PE34 4DE
G6 MQN R Wyatt, 1 Ivy Villa, Kings Hill, Haverhill, CB9 7NA
G6 MQP Nicholas Roberts, 81 Broad Lane, Coventry, CV5 7AH
G6 MQU B Plumtree, Sunnyside, Station Road, Skegness, PE24 5ES
G6 MQY Peter Wilson, Laurel Cottage, 43 Newnham Road, Ryde, PO33 3TE
G6 MQZ T Wilson, Orchard House, Whitmoor Lane, Guildford, GU4 7QB
G6 MRN N Parr, 24 Park Avenue, Awsworth, Nottingham, NG16 2RA
G6 MRO J Reddaway, Voltaire House, Ffordd Uchaf, Wrexham, LL11 5UN
G6 MRP K Playford, 1 Cherwell Close, Abingdon, OX14 3TD
G6 MRW P Grant, 117 Hazel Avenue, Farnborough, GU14 0DW
G6 MRY C Guy, 78 Park Road, Bolton, BL1 4RQ
G6 MSC T Glover, 70 Sandown Road, Toton, Nottingham, NG9 6JW
G6 MTB D Hesketh, flat 8 Sefton Court, Bedford road, Torquay, TQ1 3LJ
G6 MTE S Heath, 12 The Medway, Daventry, NN11 4QU
G6 MTF T Horn, 9 Gipton Wood Avenue, Leeds, LS8 2TA
G6 MTG David Knight, 119 Bracebridge Street, Nuneaton, CV11 5PD
GI6 MTL M McCutcheon, 10 Chestnut Brae, Gilford, Craigavon, BT63 6FA
G6 MTY Margaret Matthews, 30 Broadgate Lane, Deeping St. James, Peterborough, PE6 8NW
G6 MUJ C James, 5 Arbroath Road, Luton, LU3 3LA
G6 MUP W Jones, Pen-Y-Berth, Pen-Y-Garth, Gwynedd, LL55 1EY
G6 MUQ Terry Jones, 21 Lynwood Gardens, Croydon, CR0 4QH
G6 MUW B Kent, 105 Church Road, Dordon, Tamworth, B78 1RN
G6 MUX A Kelly, 9 Cotswold close, Dibden Purlieu, Southampton, SO45 5QW
GM6 MUZ Charlie Duncan, 12 Juniper Park Road, Juniper Green, EH14 5DX
G6 MVD P Sweeting, 35 The Ridings, Market Rasen, LN8 3EE
G6 MVF Christopher Stokes, 7 St. Nicholas Close, Arnold, Nottingham, NG5 6GU
G6 MVN Roger Suckling, 21 Warren Court, Meadowside, Dartford, DA1 2RZ
G6 MVQ James Simmonds, Overbeck South, Stokesley Road, Guisborough, TS14 8DL
G6 MVR Derek Scott, 20 Belmont View, Harwood, Bolton, BL2 3QN
G6 MVS M Sandler, 21 Kenilworth Gardens, Hornchurch, RM12 4SE
G6 MVW E Sayer, 27 Glenmere Park Avenue, Benfleet, SS7 1SS
G6 MWB Tim Gordon, 101 Dorset Road, Bexhill-on-Sea, TN40 2HU
G6 MWD Christopher Goodhand, 22 Somin Court, Doncaster, DN4 8TN
G6 MWM Brian Hall, 132 Lon Penrhyn, Benllech, Tyn-Y-Gongl, LL74 8RW
G6 MWS Anthony Hueck, 9 Corden Avenue, Mickleover, Derby, DE3 9AQ
G6 MXE Robert Peeling, 13 Greenview Crescent, Hildenborough, Tonbridge, TN11 9DR
G6 MXL C Rodwood, 53 Woodpecker Drive, Poole, BH17 7SB
G6 MXV A Poupard, Woodlea, Crouch House Road, Edenbridge, TN8 5EN
G6 MYH Charles Mallory, 11 Baymead Meadow, North Petherton, Bridgwater, TA6 6QW
G6 MYL Carol Jones, 63 Hockenhull Avenue, Tarvin, Chester, CH3 8LR
G6 MYO B Johnson, 2 Plumtree Cottages, Hill Street, Swadlincote, DE12 7PW
G6 MYT Clive King, 3 Huntingdon Gardens, Christchurch, BH23 2TW
G6 MYY David Davies, Penrallt, Abercaseg Road, Gerlan, Bangor, LL57 3SP
G6 MYZ S Doorey, 11 Langley Gardens, Petts Wood, Orpington, BR5 1AD
G6 MZF Lesley Cromar, 17 Ipley Way, Hythe, Southampton, SO45 3LG
G6 MZN M Esser, 10 Van Diemans Road, Wombourne, Wolverhampton, WV5 0BQ
G6 MZT Brian Fereday, 16 Glentworth Gardens, Wolverhampton, WV6 0SF
G6 MZV R Titmuss, 70 Mallards Rise, Church Langley, Harlow, CM17 9PL
G6 MZW D Speak, 42 Penn Lea Road, Twerton, Bath, BA1 3RB
G6 NAD Michael McDermott, 91 Hargwyne Street, London, SW9 9RH
G6 NAG D Lang, 8 Church Hill, Cheddington, Leighton Buzzard, LU7 0SY
G6 NAH P Proudlove, 14 Heath Avenue, Rode Heath, Stoke-on-Trent, ST7 3RY
G6 NAJ T Leyland, The Rectory, Alrewas Road, Burton-on-Trent, DE13 7HP
G6 NAL R Pain, The Hornbeams, Boxley Road, Chatham, ME5 9JG
G6 NAP Edward Lester, 178 Newtown Road, Malvern, WR14 1PJ
GI6 NAQ S McCullagh, 7 Clanbrassil Gardens, Portadown, Craigavon, BT63 5YD
G6 NAV Robin Martin, 110 Binscombe, Godalming, GU7 3QJ
G6 NAX S Moring, 1 Burrows Cottages, Toot Hill Road, Ongar, CM5 9QN
G6 NBE P Bethell, 7 Silverton Grove, Middleton, Manchester, M24 5JH
G6 NBF Jeffrey Baron, 9 Milton Avenue, Doncaster, DN5 8ER
G6 NBI John Pratt, 107 Cranbrook Drive, Maidenhead, SL6 6RY
G6 NBK R Perry, 11 Elliotts Lane, Codsall, Wolverhampton, WV8 1PG
G6 NBL R Barnett, 5 Overbrook, Evesham, WR11 1DE
G6 NBM G Bryant, 37 Broad Leas Court, Broad Leas, St Ives, PE27 5XG
G6 NBP Peter Blease, 15 Shadybrook Lane, Weaverham, Northwich, CW8 3PN
G6 NCE M Craig, 194 Elm Grove, Brighton, BN2 3DA
G6 NCL R Pickstone, 33 Shore Mount, Littleborough, OL15 8EN
GU6 NCZ P Wild, Honfleur, La Rue Du Le Hurel, Vale, Guernsey, GY3 5AF
G6 NDA Christopher Venn, Stantor, High Road, Templecombe, BA8 0DN
G6 NDH P Walker, 37 Cromwell Road, Grimsby, DN31 2DN
G6 NDJ A Wilson, 23 Claydown Way, Slip End, Luton, LU1 4DU
GI6 NDM G O'Boyle, 27a Draperfield Road, Cookstown, BT80 8RS
G6 NEA J Dean, 15 Park Close, Sonning Common, Reading, RG4 9RY
G6 NEK P Diss, 130 Beridge Road, Halstead, CO9 1JU
G6 NEZ R Emerson, 4 Freeford Gardens, Lichfield, WS14 9RJ
G6 NFB Timothy Wright, 182 Lansdowne Road, Oxton, Prenton, CH43 7SQ
G6 NFC Andrew Young, 1 Rapley Green, Bracknell, RG12 7PS
G6 NFE Richard White, 36 Normandy Way, Ashford, TN23 5LN
G6 NFJ John Eden, 23 Elm Green Close, Worcester, WR5 3HD
GI6 NFK Stephen Ferguson, 20 Old Road, Loughgall, Armagh, BT61 8JD
G6 NFR R Foden, 1a Garden Cottages, Eaton Road, Liverpool, L12 3HQ
G6 NGA J LAMBLE, 17 Willowvale, Lowestoft, NR32 4UB
G6 NGF M Tatlow, 20 Windmill Way, Tysoe, Warwick, CV35 0SB
G6 NGM S Cross, 7 April Place, Buckhurst Road, Bexhill-on-Sea, TN40 1UE
G6 NGN D Simpson, The Hawthorns, Slacken Lane, Stoke-on-Trent, ST7 1NQ
G6 NGR Peter Thornton, 99 Hollingworth Road, Littleborough, OL15 0AZ
G6 NGV Wayne Taylor, 29 Holmdale Road, Syston, Leicester, LE7 2JN
G6 NHA M Malyon, 16 Tintern Road, Gosport, PO12 3QN
GW6 NHB Graham Mahoney, 684 Beechley Drive, Pentrebane, Cardiff, CF5 3SS
G6 NHG Stuart Marshall, 25 Carlcroft, Wilnecote, Tamworth, B77 4DL
G6 NHK Nicholas Martin, Stonea House, Middle Road, March, PE15 0AJ
G6 NHL A McCallum, Trosgol, Deiniolen, Caernarfon, LL55 3LU
G6 NHO Ron Smith, 20 Dryden Way, Higham Ferrers, Rushden, NN10 8DH
G6 NHU Keith Maton, 41 Bemerton Gardens, Kirby Cross, Frinton-on-Sea, CO13 0LQ
G6 NHV David Meakins, 19 Booth Lane North, Northampton, NN3 6JQ
G6 NHW P Minchin, 122 Mildenhall Road, Great Barr, Birmingham, B42 2PQ
G6 NHY Keith Marriott, 1 Holbeck Road, Hucknall, Nottingham, NG15 7SR
GM6 NIA D McCall, 11 Craiglockhart Dell Road, Edinburgh, EH14 1JW
GM6 NIC J McAulay, 9 Randolph Cliff, Edinburgh, EH3 7TZ
G6 NID J Matthew, 1 West View, Haslingden, Rossendale, BB4 5DA
G6 NIO Mike Smith, 39 Seliot Close, Poole, BH15 2HQ
G6 NIW F Shaw, 43 Egremont Road, Hardwick, Cambridge, CB23 7XR
G6 NIX E Samuels, 156 Fulmer Close, Hampton, TW12 3YN
G6 NIZ A Scott, The Conifers, Back Lane, York, YO30 2DF
G6 NJE John Smith, 414 Sparrowhawk Drive, Willow Grove Park, Poulton-le-Fylde, FY6 0RS
G6 NJJ Mervyn Swift, Spa Cottage, Spa Lane, Ormskirk, L40 6JQ
G6 NJL M Spittle, Stablecleugh, Ewes, Langholm, DG13 0HJ
G6 NJO P Askham, 1 Park House Cottage, Carr Lane, Thirsk, YO7 3PF
G6 NJR P Nikolic, 18 Harper Grove, Tipton, DY4 9SR
G6 NJT F Neill, 7 Bellevue Terrace, Southampton, SO14 0LB
G6 NKI K Brown, 73 Church Avenue, Preston, PR1 4UD
G6 NKL Derek Baldock, Deeside, Platts Lane, Woodhall Spa, LN10 5DY
G6 NKS G Pearn, 59 London Road, Kessingland, NR33 7PN
G6 NLC C Rabe, 79 Rectory Avenue, Corfe Mullen, Wimborne, BH21 3EZ
G6 NLD Arthur Reed, 6 Brancaster Close, Nottingham, NG6 8SL
G6 NLE H Roberts, 3 Short Avenue, Allestree, Derby, DE22 2EH
G6 NLG M Ritchie, Bruern Abbey, Bruern, Chipping Norton, OX7 6QA
G6 NLN J Burdass, Cedarwood, High Road, Wallingford, OX10 0PT
GW6 NLO Timothy Bott, 37 Kent Row, Pembroke Dock, SA72 6DF
G6 NLP Michael Bryant, The Nook, Llanarmon Road, Wrexham, LL11 5YP
G6 NLQ T Fradley, 40 Higher Green, Poulton-le-Fylde, FY6 7BL
G6 NLS Doris Budd, Valhalla, 81 Bohemia Chase, Leigh-on-Sea, SS9 4PW
G6 NLU S Buxton, 22 Maurice Drive, Nottingham, NG3 5GF
G6 NLW I Rylett, 3 Martin Street, Brighouse, HD6 1DA
G6 NLX George Richardson, 22 Clarkfield, Mill End, Rickmansworth, WD3 8FH
G6 NLZ M Reynolds, 24 Mill Road, Lydd, Romney Marsh, TN29 9EJ
G6 NMA P Ayers-Hunt, 15 Kelvin Road, Leamington Spa, CV32 7TF
G6 NMK Malvyn Grimes, 73 Ryston Road, Denver, Downham Market, PE38 0DP
G6 NMQ G Goodyer, Flat, 54 Wyndham Road, Petworth, GU28 0EQ
G6 NMU Joseph Greenhough, 58 Gorsey Bank, Wirksworth, Matlock, DE4 4AD
G6 NNA Adrian Liggins, 97 Vessel Crescent, Scarborough, Ontario, Canada, M1C 6K5
G6 NNB David Owen, Pen Y Bont, Ty Nant, Corwen, LL21 0PE
G6 NNK P Frampton, 118 Ramnoth Road, Wisbech, PE13 2JD
G6 NNO J Evans, 74 Trejon Road, Cradley Heath, B64 7HJ
GI6 NNP V Hagan, 7 Emania Terrace, Armagh, BT60 4AS
G6 NNS J Hunt, 35 Wallisdean Avenue, Portsmouth, PO3 6HA
G6 NOI J Whelan, 8 Welland Road, Higher Bebington, Wirral, CH63 2JU
G6 NOL J Weir, 17 Pasteur Drive, Apley, Telford, TF1 6PQ
GM6 NOO C Wood, 5 Damhead Steading, Kinloss, Kinloss, IV36 3UA
G6 NOW R Beck, 26 Cheshire Court, Ravenall Close, Birmingham, B34 6PZ
G6 NPC R Carlson, 45 Firs Road, Milnthorpe, LA7 7QF
G6 NPE A Coates, 2 Kipling Avenue, Burntwood, WS7 2HS
G6 NPJ J Copeland, Little Cophall, Dowlands Lane, Crawley, RH10 3HX
G6 NPP Raymond Willis, Moorhaven, Boyton, Launceston, PL15 9RL
G6 NPW S Caine, 19 Turner Drive, Tingley, Wakefield, WF3 1UD
G6 NPZ P Chard, 29 Nettle Gap Close, Wootton, Northampton, NN4 6AH
G6 NQB S Clements, 39 Redland Close, Marlbrook, Bromsgrove, B60 1DZ
G6 NQL John Wilkinson, The Old Joinery, Garsdale, Sedbergh, LA10 5PJ
G6 NQM K Whitchurch, 65 Honey Hill Road, Kingswood, Bristol, BS15 4HN
G6 NQO D Hawken, The Old House, Hophurst Place, Hophurst Lane, Crawley, RH10 4LN
G6 NQQ Andrew Wilson, 17 Rook Way, Horsham, RH12 5FR
G6 NQU James Hoy, 39 Blackbird Road, Caldicot, NP26 5RE
G6 NQY R Heath, 222 Congleton Road, Talke, Stoke-on-Trent, ST7 1LW
G6 NRH D Halifax, 22 Wendover Way, Welling, DA16 2BN
G6 NRK Arthur Hunt, 39 Circular Road West, Liverpool, L11 1AY
G6 NRL D Hargreaves, Viridis, Rufford Road, Rufford, DN22 0DW
G6 NRM Simon Hanscombe, 21 St. Marks Drive, Wellington, Telford, TF1 0QA
G6 NSG J Jones, 26 Spring Road, Wrexham, LL11 2LU
G6 NSK A Jones, 4 Park View, Llanddew, Brecon, LD3 9RL
G6 NSQ Peter James, 9 Smallholding, Tutbury Road, Burton-on-Trent, DE13 0AL
G6 NSU Peter Lewis, 18 Bittaford Wood, Bittaford, Ivybridge, PL21 0ET
G6 NSZ D Lawton, 48 Woodlands Road, Woodlands, Doncaster, DN6 7JZ
G6 NTE J Lyons, 8 Anstie Close, Devizes, SN10 2EN
G6 NTM E Murphy, 25 Warrington Road, Ashton-in-Makerfield, Wigan, WN4 9PJ
GI6 NTP David McAlpine, 35 Carnamena Avenue, Castlereagh, Belfast, BT6 9PJ
G6 NTQ R Morgan, 1 Hillmeads Drive, Dudley, DY2 7TS
G6 NTW K Gooboo, 64 Connaught Gardens, Palmers Green, London, N10 5DG
G6 NTY Brian Griffiths, 18 Julius Drive, Coleshill, Birmingham, B46 1HL
G6 NUI D Chamberlain, 44 Parsonage Chase, Minster on Sea, Sheerness, ME12 3JX
GM6 NUJ R Crawford, Glengarry, East Terrace, Kingussie, PH21 1JS
G6 NUS A Croft, 15 St. Marys Road, Bozeat, Wellingborough, NN29 7JU
G6 NUX S Clack, Flat 9, The Grange, Emsworth, PO10 7QP
G6 NUZ A Charlton, 26 Saundergate Lane, Wyberton, Boston, PE21 7BZ
G6 NVC K Castley, 2 Thorpe Avenue, Moulton Chapel, Spalding, PE12 0XN
G6 NVD C Webb, 2 Rykhill, Chadwell St. Mary, Grays, RM16 4RR
G6 NVF David John Mcglasson Mcglasson, 19 Kennedy Street, Ulverston, LA12 9EA
G6 NVH M Morris, 19 Gowy Close, Alsager, Stoke-on-Trent, ST7 2HX
G6 NVI S Mindel, Longwood House, Arkley Lane, Barnet, EN5 3JR
G6 NVJ C Marlow, 27 Sandy Way, Connah's Quay, Deeside, CH5 4SH
G6 NVS P Harrison, 41 Chestnut Close, Handsacre, Rugeley, WS15 4TH
G6 NVT Mark Harrison, 244 Providence Road, Sheffield, S6 5BH
G6 NVU Michael Haynes, 10 Cypress Grove, Denton, Manchester, M34 6EA
G6 NVW Paul Higgins JP, 49 Milton Road, Hoyland, Barnsley, S74 9AX
G6 NVY A Hilbourne, 24 Tamarisk Avenue, Reading, RG2 8JB
G6 NWC Peter Holley, Wotton Farm, Buckfastleigh, TQ11 0HB
G6 NWK M Parker, 31 Sandholme Drive, Burley in Wharfedale, Ilkley, LS29 7RG
G6 NWN I Poyser, 24 Overstone Close, Sutton-in-Ashfield, NG17 4NL
G6 NWS Jeffrey Hildreth, 69 Mason Street, Sutton-in-Ashfield, NG17 4HQ
G6 NWT W Taylor, 33 Lancaster Avenue, Dawley, Telford, TF4 2HS
GM6 NX STIRLING AND DISTRICT ARS c/o Hugh Martin, 11 Ewing Court, Broomridge, Stirling, FK7 0QP
GW6 NXH W REES, 1 St. Marys Close, Briton Ferry, Neath, SA11 2JU
G6 NXL T Rees, Ty Goleu, Llwyngwril, LL37 2UZ
G6 NXM R Rixon, 11 The Ridings, Waltham Chase, Southampton, SO32 2TR
GM6 NXN William Roy, 1 Belts, Strachan, Banchory, AB31 6NL
G6 NXP S Rafferty, 11 Gilbert Road, Peterlee, SR8 2AN
G6 NXV Margaret Shannon, 129 Hampton Lane, Blackfield, Southampton, SO45 1WF
G6 NXW K Sykes, 68 Newtown Avenue, Cudworth, Barnsley, S72 8DY
G6 NYC R Ashman, 44 Conan Doyle Walk, Swindon, SN3 6JB
G6 NYF J Aylward, 7 Manygates Lane, Wakefield, WF1 5NT
G6 NYG Roy Adams, 28 Greenside, Stoke Prior, Bromsgrove, B60 4EB
G6 NYH Gary Austin, 21 St. Georges Place, Northampton, NN2 6EP
G6 NYL P Baylis, 118 Eastgate, Deeping St. James, Peterborough, PE6 8RD
G6 NYR Allan Davis, 9 Taliesin Street, Llandudno, LL30 2YE
GM6 NYT J Danton, 12 Laburnum Road, Methil, Leven, KY8 2HA
G6 NZ M Newnham, 75 Liss Road, Southsea, PO4 8AS
G6 NZA M Davenport, 8 Cedar Avenue, Chesterfield, S40 4ES
G6 NZG S Edson, Inglewood, Camelot Gardens, Mablethorpe, LN12 2HP
G6 NZL P Fletcher, 43 Merlin Way, Woodville, Swadlincote, DE11 7QU
G6 NZN Glenville Fowler, 10 Ullswater Road, Wimborne, BH21 1QT
G6 NZO M Finney, 49 Ashcroft Drive, Old Whittington, Chesterfield, S41 9PA
G6 NZW D Smith, 90 Endhill Road, Kingstanding, Birmingham, B44 9RP
G6 NZY Barry Sparke, 1 Norwich Street, Mundesley, Norwich, NR11 8DN
G6 OAI S Baverstock, 43 Tatnam Road, Longfleet, Poole, BH15 2DW
G6 OAN C Bryan, 113 Hoe View Road, Cropwell Bishop, Nottingham, NG12 3DJ
G6 OAS Anthony Inglis, 59 Chapel Street, Forsbrook, Stoke-on-Trent, ST11 9DA
G6 OAU M Jones, 17 Puddingmoor, Beccles, NR34 9PL
G6 OAV C Jones, 1 Stonehill Close, Leigh-on-Sea, SS9 4AZ
G6 OAW Thomas Jones, Marvor, Madyn Road, Amlwch, LL68 9DL
G6 OBA M Kaznowski, 85 St. Albans Road, Kingston upon Thames, KT2 5HH
G6 OBB P Kerr, Burrow Farm, Burrowbridge, Bridgwater, TA7 0RH
G6 OBD T Keeling, 1 New Lane, Brown Edge, Stoke-on-Trent, ST6 8TQ
G6 OBE S Kimblin, 4 Horsham Close, Westhoughton, Bolton, BL5 2GR
G6 OBG J Kay, 12 Williams Avenue, Newton-le-Willows, WA12 0NN
G6 OBJ Graham Webster, Flat 3 Charlotte Broadwood, Vicarage Lane, Dorking, RH5 5LL
G6 OBO L Weiss, 7 Tennyson Avenue, Grays, RM17 5RG
G6 OBT Harry Wilson, 11 Palmerston Close, Haslington, Crewe, CW1 5QE
G6 OBU S Wright, 21 Poplars Close, Watford, WD25 7EW
G6 OCA M Barnes, 451 Hough Fold Way, Harwood, Bolton, BL2 3PU
G6 OCB D Byers, 11 Heath Road, Ashton-in-Makerfield, Wigan, WN4 9DY
GI6 OCC K Brennan, 1 Ballyscullion Lane, Bellaghy, Magherafelt, BT45 8NQ
G6 OCE Clive Stapleton, 7 Hawksdale Close, Chellaston, Derby, DE73 6PS
G6 OCF R Wallis, 187 Langtons Meadow, Farnham Common, Slough, SL2 3NT
G6 OCM Eric Bunyan, 1 Talman Close, Ifield, Crawley, RH11 0RB
G6 OCO A Bruce, 26 Regency Close, Wigmore, Gillingham, ME8 0LA
G6 ODA A Bardy, 67 Chase Side, Enfield, EN2 6NQ
G6 ODE B Aylward, 49 The Ruffetts, South Croydon, CR2 7LT
G6 ODF D Adamson, 19 Cliveden Avenue, Walsall, WS9 8HG
G6 ODT K Lamford, 41 Drayton Road, Irthlingborough, Wellingborough, NN9 5TA
G6 ODU Robert Leong, 55 Liverpool Road, Aughton, Ormskirk, L39 5AP
G6 ODW L Liffchak, 6 Ashmore Grove, Welling, DA16 2RU
G6 OEI T Blest, 2 Gayton Avenue, Littleover, Derby, DE23 1GA
G6 OEJ A Barnard, 36 St. Pauls Road, Walton Highway, Wisbech, PE14 7DN
G6 OEM J Bolland, 18 Ward Avenue, Formby, Liverpool, L37 2JD
G6 OER J Topping, 3 Dean Road, Handforth, Wilmslow, SK9 3AF
G6 OES M Smith, 36 Wendover Rise, Coventry, CV5 9JU
G6 OET M Telford, 11 Twyford Close, Swinton, Mexborough, S64 8UH

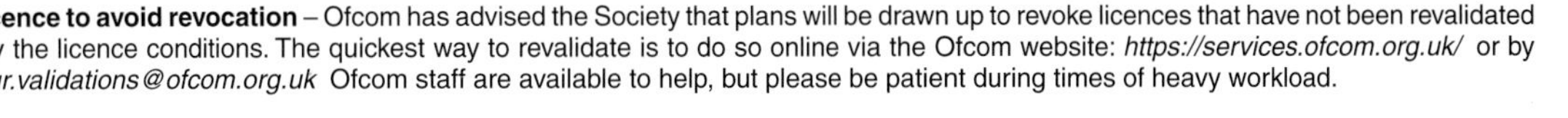
**IMPORTANT NOTE**

**Revalidate licence to avoid revocation** – Ofcom has advised the Society that plans will be drawn up to revoke licences that have not been revalidated as required by the licence conditions. The quickest way to revalidate is to do so online via the Ofcom website: *https://services.ofcom.org.uk/* or by email: *amateur.validations@ofcom.org.uk* Ofcom staff are available to help, but please be patient during times of heavy workload.

G6 OEW C Thorn, 20 Kiln Road, Shaw, Newbury, RG14 2HA
GM6 OFB James McArdle, 40 Rodney Drive, Girvan, KA26 9DZ
G6 OFD L Morrison, 29 Mead Hatchgate, Hook, RG27 9PU
G6 OFM John Cordial, 146 Freshwater Drive, Poole, BH15 4JF
GM6 OFO M Clark, 38 Dunsinane Drive, Perth, PH1 2DU
G6 OFV S Crossland, 16 Holland Road, High Green, Sheffield, S35 4HF
G6 OFZ J Roberts, 8 Woodgate Close, Market Harborough, LE16 8EX
GW6 OGD S Dawber, 7 Heol Y Pentir, Rhoose, CF62 3LQ
GM6 OGN A Sives, 4 Fir Grove, Craigshill, Livingston, EH54 5JP
G6 OGT A Scholes, 45 Howden Road, Blackley, Manchester, M9 0RQ
G6 OGZ J Mitchell, 17 Spring Close, Lutterworth, LE17 4DD
GM6 OHF D Macliver, 20 Lancaster Avenue, Beith, KA15 1AR
G6 OHK Peter Butler, 219 Ridge Avenue, Burnley, BB10 3JF
G6 OHM Andy Dunham, 28 Kingfisher Close, Chatteris, PE16 6TP
G6 OHQ Graham Finch, 77 Furnivall Crescent, Lichfield, WS13 6DB
G6 OHR R Edwards, 11 Littlington Court, Surrey Road, Seaford, BN25 2NZ
G6 OIA P Rattenbury, 1 Holmewood, Furzton, Milton Keynes, MK4 1AR
G6 OIB John Riley, 11 Sutton Road, Mepal, Ely, CB6 2AQ
G6 OIF A POSTANS, 62 Elm Grove, Bromsgrove, B61 0DX
G6 OIH RICHARD PHILLIPS, 1 Forge Close, Ashendon, Aylesbury, HP18 0HJ
G6 OIO M Thomas, 13 Chestnut Grove, The Bryn, Blackwood, NP12 2PU
G6 OIX J Roberts, 9 Tower Close, North Weald, Epping, CM16 6HA
G6 OIY J Roberts, 6 Weavers Close, Braintree, CM7 2WB
GI6 OJC O Okane, 39 Harberton Park, Ballymena, BT43 6NF
G6 OJK Alan Mayall, 7 Cortay Park, Llanyre, Llandrindod Wells, LD1 6DT
G6 OJN K Michael, 59 Mattock Lane, Ealing, London, W13 9LA
G6 OJV James Greenley, 22 Langley Drive, Norton, Malton, YO17 9AR
G6 OJX E Grayson, Manor Gate, 2 Polsue Way, Truro, TR2 4BE
G6 OJZ P Anstock, 12 Raymoor Avenue, St. Marys Bay, Romney Marsh, TN29 0RD
G6 OKA C Glover, 16 Woodfield Road, Radlett, WD7 8JD
G6 OKB Roy Gilham, Wren Cottage, Wayborough Hill, Ramsgate, CT12 4HR
G6 OKC M Gerrard, 29 Forest Drive, Broughton, Chester, CH4 0QT
G6 OKU M Law, 23 Yeldersley Close, Chesterfield, S40 4LG
G6 OLJ D Hill, 33 Cleveland Close, Thornbury, Bristol, BS35 2YD
GM6 OLM R Hendry, 43 Barone Road, Rothesay, Isle of Bute, PA20 0DY
G6 OLU P Hobson, 220 Station Road, Burton Latimer, Kettering, NN15 5NT
G6 OLV A White, 20 Wyles Street, Gillingham, ME7 1ND
G6 OLY A Williams, 1 Leslie Drive, Leigh-on-Sea, SS9 5NW
G6 OMH B Staddon, 311 Cheney Manor Road, Swindon, SN2 2PE
G6 OMN G Rogers, 7 Flordon, Birch Green, Skelmersdale, WN8 6PA
GW6 OMV Paul Rea, 159 Mill View Estate, Maesteg, CF34 0DP
G6 ONE Dave Williams, 16 Church St., Owston Ferry, Doncaster, DN9 1RG
G6 ONI Bernard Steponitis, Flat 7, 6, Second Avenue, Hove, BN3 2LH
G6 ONV Michael Johnson, 16 Gardner Close, Raunds, Wellingborough, NN9 6HN
G6 ONW S Jackson, 256 Perry Road, Sherwood Rise, Nottingham, NG5 1GP
G6 ONZ P Kinsey, Glyn Elwy, Allt Goch, St. Asaph, LL17 0BP
GM6 OOA Derek King, Marionville, Donibristle, Cowdenbeath, KY4 8EU
G6 OOH John Stone, 13 Winchester Close, Newport, PO30 1DR
G6 OOK M Stewart, Toll Bar Gardens, Lower Bentham, Lancaster, LA2 7DD
G6 OOT C Atkins, 11 Brambledown, West Mersea, Colchester, CO5 8RY
G6 OPD L Middleton, 24 Townshend Road, Worle, Weston-Super-Mare, BS22 7FW
G6 OPK M Lee, 23 Camford Close, Beggarwood, Basingstoke, RG22 4UJ
G6 OPV Edward Starkey, 71 Elwick Drive, Liverpool, L11 4UW
G6 OPY Roger van Cleak, 19 Hanbury Road, Stoke Heath, Bromsgrove, B60 4LS
G6 OQJ W Castle, 2 Wellington Close, Mundesley, Norwich, NR11 8JF
GI6 OQL J Craig, 8 Muckamore View, Muckamore, Antrim, BT41 2EU
GM6 OQN R Campbell, 32 Harvie Avenue, Newton Mearns, Glasgow, G77 6LQ
G6 OQO J Douthwaite, 38 Burnside Road, Newcastle upon Tyne, NE3 2DU
G6 OQV B Everitt, The Hermitage, The Rookery, Nuneaton, CV10 9PB
G6 ORE R Trangmar, Ffynnon Bach Isaf, Tregarth, Bangor, LL57 4PA
G6 ORH D Wright, 23 Oakenhall Avenue, Hucknall, Nottingham, NG15 7TF
G6 ORJ A Weller, 104 Medina Avenue, Newport, PO30 1HG
G6 ORL David Woodhouse, 5 Swallow Wood, Fareham, PO16 8UF
G6 ORM Stephen Whiley, 34 Oulton Close, Kidderminster, DY11 5DY
G6 ORO Glen Walsh, 36 Westminster Street, Newtown, Wigan, WN5 9BH
G6 ORS Anthony Bennett, 39 West View, Parbold, Wigan, WN8 7NT
G6 ORT Lewis Bailey, 19 Winckley Road, Preston, PR1 8EL
G6 OSH Douglas Ridley, 37 Harewood Close, Whickham, Newcastle upon Tyne, NE16 5SZ
G6 OSJ J Roberts, 1 Ollerdale Close, Allerton, Bradford, BD15 9BT
G6 OSK Edward Robinson, 32 Ardeen Road, Intake, Doncaster, DN2 5EU
G6 OSO D Parr, Lordings, Station Road, Pulborough, RH20 1AH
G6 OSR Paul Price, 20 Froglands Way, Cheddar, BS27 3NY
G6 OSV Ian Woodward, 20 Boyle Avenue, Warrington, WA2 0EZ
GM6 OSZ B Williams, 29 St. Ternans Road, Newtonhill, Stonehaven, AB39 3PF
GW6 OTD Paul Sizer, Gambos End, Reynoldston, Swansea, SA3 1BR
G6 OTE D Shaw, 19 Upper Moors, Great Waltham, Chelmsford, CM3 1RB
G6 OTL G Blades, 11 Willard Grove, Stanhope, Bishop Auckland, DL13 2XY
G6 OTP Michael Rainbow, 38 Moselle Drive, Churchdown, Gloucester, GL3 2RY
G6 OTQ Timothy Roddy, 26 Chapeltown Road, Radcliffe, Manchester, M26 1YF
G6 OTS W Peck, 5 Stirling Crescent, Horsforth, Leeds, LS18 5SJ
G6 OTV A Ricalton, 33 Tintagel Close, Cramlington, NE23 1NZ
G6 OTW A Ricalton, 84 Wansdyke, Morpeth, NE61 3RA
G6 OTZ Andrew Shaw, 4 Jones Lane, Burntwood, WS7 9DS
G6 OUA Robert Saunders, 8 Norfolk Road, Luton, LU2 0RE
G6 OUI M Burnell, 49 Ashfield Road, Carterton, OX18 3QZ
G6 OUJ B Bozman, 33 Maple Road, Loughborough, LE11 2JL
GM6 OUL N Bowry, 18 Mortonhall, Park Gardens, Edinburgh, EH17 8SR
G6 OUM Breaden Breaden, 6 Breydon Road, Sprowston, Norwich, NR7 8EE
G6 OUO P Burgess, 232 Hightown Road, Luton, LU2 0DN
G6 OUT David Andrew, 14 Westfield Grove, Morecambe, LA4 4LQ
G6 OUX S Smith, 71 Rockford Close, Redditch, B98 7SZ
G6 OVA N Styne, 2 Greenway, Burton-on-Trent, DE15 0AR
G6 OVC Barry Thurlow, 1 Sheffield Way, Earls Barton, Northampton, NN6 0PF
GW6 OVD Maldwyn Clee, 42 Station Road, Hirwaun, Aberdare, CF44 9TA
G6 OVX Roger Hadfield, 2 Bridge Street, Shaw, Oldham, OL2 8BG
G6 OWB Edward Hill, 213a Leicester Road, Markfield, LE67 9RF
G6 OWI P Haworth, 2 Heys Court, Blackburn, BB2 4PQ
G6 OWS Ken Jones, 1 Chadlow Road, Liverpool, L32 7QR
G6 OWT G Kelly, Brook House, Liskeard, PL14 5EY
GD6 OXG John Williams, Brookfield, Douglas Road, Ballabeg, Isle of Man, IM9 4EF
G6 OXI C Webb, 50 Ridgeway, Eynesbury, St. Neots, PE19 2QY
G6 OXJ D Webb, 10 Nuns Meadow, Gosfield, Halstead, CO9 1UB
GM6 OXL Ian Wilkins, 133 Gavin Street, Motherwell, ML1 2RL
G6 OXN Ian Walker, 66 Wood Street, Kettering, NN16 9SB
G6 OXQ A Day, 7 Seagers, Great Totham, Maldon, CM9 8PB
G6 OXZ Michael Charlton, 20 Bailey Crescent, South Elmsall, Pontefract, WF9 2TL
G6 OYF M Matthews, 14 Ralph Court, Stafford, ST17 9FR
G6 OYU R Spinner, 9 Lindholme Road, Lincoln, LN6 3RQ
G6 OYV T Silvers, 15 Stanford Way, Walton, Chesterfield, S42 7NH
G6 OZH D Martin, 2 Farm View Road, Kirkby-in-Ashfield, Nottingham, NG17 7HF
G6 OZT P White, 3 South View, Whitwell, Worksop, S80 4NP
G6 OZU A Wilson, 67 Sandpits, Leominster, HR6 8HT
G6 PAA James Brownsett, 10 Great Aldens, Bedford, MK41 8JS
G6 PAE R Hillum, 48 Lydiard Way, Trowbridge, BA14 0UJ
G6 PAJ Richard Green, 1 Knightsbridge Road, Messingham, Scunthorpe, DN17 3RA
G6 PAO M Hale, 7 Craigside, Biddulph, Stoke-on-Trent, ST8 6BP
G6 PAP S Hale, 19 Nailers Drive, Burntwood, WS7 0ES
G6 PAR P Rhodes, 1 Killinghall Avenue, Bradford, BD2 4SA
GI6 PAZ W Mcconnell, 17 Beech Green, Doagh, Ballyclare, BT39 0QB
G6 PBG Norman Munnery, 3 Monnington Lane, Poundbury, Dorchester, DT1 3RJ
G6 PBI K Partington, 38 Queensgate Drive, Royton, Oldham, OL2 5SD
G6 PBN David Talbot, 10 Chapel Lane, Wirksworth, Matlock, DE4 4FF
G6 PBO J Tobin, 5 Ashley Close, Ringwood, BH24 1QX
G6 PBQ Frank Watson, 9 Brennand Street, Clitheroe, BB7 2HG
G6 PBW J Wainwright, 8 Common Lane, Cutthorpe, Chesterfield, S42 7AN
G6 PBZ P Wright, 75 Preston Road, Abingdon, OX14 5NG
G6 PCC R Slade, 2 South Lodge Drive, Fornham St. Genevieve, Bury St. Edmunds, IP28 6TQ
G6 PCE R Stamford, 30 Craft Way, Steeple Morden, Royston, SG8 0PF
G6 PCN J Bailey, 3 Maple Drive, Hucknall, Nottingham, NG15 6GG
G6 PCP J Brown, 1 Whitehouse Cottages, Woodham Walter, Maldon, CM9 6LR
GM6 PCW Peter Boyd, 144 Brown Street, Paisley, PA1 2JE
G6 PCX J Beresford, 1 Russell Place, Maltby, Rotherham, S66 7HB
G6 PDE Christopher Irish, 128 Rushmere Road, Ipswich, IP4 4JX
G6 PDJ Stephen Jubb, 4 Manor End, Worsbrough, Barnsley, S70 5JB
G6 PDM S Procter, 1b York Villas, York Street, Colne, BB8 0ND
GW6 PDR S Riggs, 3 Lawrence Terrace, Llanelli, SA15 1SW
G6 PEA B Terry, 389 Sutton Road, Maidstone, ME15 9BU
G6 PEG C Price, 42 Kipling Road, Kettering, NN16 9JZ
G6 PEH A Rands, 20 Riby Road, Keelby, Grimsby, DN41 8ER
G6 PEP J Morris, 22 St. Amand Drive, Abingdon, OX14 5RQ
G6 PFF A Willis, Kilncroft, Broadlayings, Newbury, RG20 9TS
G6 PFJ G Gott, 21 Hamilton Avenue, Dumfries, DG2 7LW
GW6 PFK L Griffiths, 36 Lonydd Glass, Llanilid, Pontyclun, CF72 9FZ
G6 PFN A Hewitt, 29 Brabazon Road, Oadby, Leicester, LE2 5HF
G6 PFP S Hill, 10 Honeycrft Drive, St. Albans, AL4 0GE
G6 PFX I Harris, 4 Hopton Close, Bartestree, Hereford, HR1 4DQ
G6 PFZ Anthony Holroyd, 59 Southern Parade, Preston, PR1 4NJ
G6 PGG D Jones, Bramble Cottage, Newtown, Nantwich, CW5 8BG
G6 PGJ J Kyle, 1a Lynmouth Gardens, Greenford, UB6 7HR
G6 PGM D Kaye, The Pantiles, Bildeston Road, Stowmarket, IP14 2JT
G6 PGN C King, 18812 Thornwood Circle, Huntington Beach, California, United States
G6 PGO B Key, 65 Ravenhurst Road, Birmingham, B17 9TB
G6 PGP S Kinton, 7 Ferndale Drive, Ratby, Leicester, LE6 0LH
G6 PGQ M Karaszy-Kulin, Ridgeway, Salisbury Road, Horsham, RH13 0AL
G6 PGT John Chapman, 43 Balas Drive, Sittingbourne, ME10 5AS
GM6 PGV N Phillips, 4 High Street, Pitlessie, Cupar, KY15 7ST
G6 PHC Phillip Dewick, Corner House, High Street, Gainsborough, DN21 4SW
G6 PHF M Dent, 23 Spruce Avenue, Lancaster, LA1 5LB
G6 PHH Philip Dickens, 2 Millfield Avenue, Marsh Gibbon, Bicester, OX27 0HP
G6 PHJ Peter Daines, The Hollies, Main Street Thurlaston, Leicester, LE9 7TP
G6 PHM Paul Durbin, 27 Kenneth Road, Bristol, BS4 5AE
G6 PHT Simon Fitzhugh, 25 Bridge Meadow, Denton, Northampton, NN7 1DA
G6 PHU David Ford, 6 School Lane, Stewartby, Bedford, MK43 9NG
G6 PHX Craig Johnson, 37 Oakfield Drive, Mirfield, WF14 8PX
G6 PHZ P Maddox, 7 Keats Road, Flitwick, Bedford, MK45 1QD
G6 PIB M Livingston, 22 Oak Avenue, Elloughton, Brough, HU15 1LA
G6 PII D Simpson, 20 Belvoir Place, Balderton, Newark, NG24 3HH
G6 PIM P Lawford, 44 Clarendon Road, Broadstone, BH18 9HY
G6 PJC P Brown, 13 Hillside Close, Biddulph Moor, Stoke-on-Trent, ST8 7PF
G6 PJD M Belshaw, Tara Cottage, 11 Hectors Way, Blandford Forum, DT11 9QP
G6 PJE D Bull, 2 School Road, St. Johns Fen End, Wisbech, PE14 8JR
G6 PJP L Bealing, 18 Avon Road, Oakley, Basingstoke, RG23 7DJ
G6 PJT John Coates, 139 Berrow Road, Burnham-on-Sea, TA8 2PN
G6 PKG Robert Davis, 72 Windyridge, Bisley, Stroud, GL6 7DA
G6 PKM John Allen, 27 Grafton Road, Whitley Bay, NE26 2NR
GM6 PKP J Allardyce, 17 Hallglen Terrace, Glen Village, Falkirk, FK1 2AP
G6 PKS B Bean, 46 Grand Drive, Herne Bay, CT6 8JS
G6 PKV Gary Branagan, 434 Manchester Road West, Little Hulton, Manchester, M38 9XU
G6 PKY R Bush, 3 Charnwood Avenue, Keyworth, Nottingham, NG12 5JX
G6 PLF J Smoker, 9 Anson Way, Bicester, OX26 4UH
GM6 PLG P Sloan, 24 Hythe View, Lossiemouth, IV31 6TP
GI6 PLO I Bell, 3 Stratford Drive, Bangor, BT19 6ZW
GD6 PLR L Chandless, 16 Crest Gardens, Ruislip, HA4 9HD
G6 PLT E Cheetham, 172a Hesketh Lane, Tarleton, Preston, PR4 6UD
G6 PLU A Chenery, 43 Wessex Estate, Ringwood, BH24 1XD
G6 PMC Raymond Evans, 16 Monmouth Grove, Prestatyn, LL19 8TS
G6 PMD Clifford Eagling, 96 Regent Road Brightlingsea, Colchester, CO7 0NZ
G6 PMF H Langsley, 39 Lavender Road, Basingstoke, RG22 5NN
G6 PMJ S Murphy, 1 Orchard Cottage, Golden Valley, Newent, GL18 1HN
G6 PMO I Parker, 27 St. Audries Road, Worcester, WR5 2AL
G6 PMR Philip Shaw, 52 Belvedere Parade, Bramley, Rotherham, S66 3WA
G6 PMW G Goodier, 3 The Paddock, Beckingham, Lincoln, LN5 0FD
G6 PNG P Hill, 28 Somerton Grove, Thatcham, RG19 3XE
GM6 PNJ S Hammond, 51st (Scottish) Brigade, G6 Regional Radio Controller, Forthside, FK7 7RR
G6 PNO P Hill, 33 The Pastures, South Beach, Blyth, NE24 3HA
G6 POC R Kinrade, 23 Crofthill Road, Slough, SL2 1HG
G6 POE John Knott, 3 Lords Wood, Welwyn Garden City, AL7 2HF
G6 POI John Wright, Chez Mon, Burton Road, Carnforth, LA6 1QN
G6 POJ I Worthy, 7 The Paddocks, Pilsley, Chesterfield, S45 8ET
G6 POO R Smallwood, 12 Oak Close, Connah's Quay, CH5 4GG
G6 POP Graham Starkey, 45 Chalcot Drive, Hednesford, Cannock, WS12 4SF
G6 POV Martin Walker, 232 Bideford Green, Leighton Buzzard, LU7 2TS
G6 POZ R Farrall, 7 The Meadows, South Cave, Brough, HU15 2HR
G6 PPA T Farmer, 35 Ascot Drive, Dudley, DY1 2SN
G6 PPD Andrew Morgan, 14b Hawthorn Gardens, Ryton, NE40 3ED
G6 PPU H Chappell, Oanley, East Lyng, Taunton, TA3 5AU
G6 PPV P Caswell, 94 Dewsbury Road, Luton, LU3 2HJ
G6 PPY Stephen Carter, 3 Mary Street, Burnley, BB10 4AJ
G6 PQI J Finch, The Croft, Dalby Road, Melton Mowbray, LE14 3EX
G6 PQP Paul Brooks, Cherating, Hanning Road Horton, Ilminster, TA19 9QH
G6 PRA J Whittaker, 6 Bradley Gardens, Burnley, BB12 6JT
G6 PRE E Snell, 156 Brookdale Avenue South, Greasby, Wirral, CH49 1SS
G6 PRL Dennis Brown, 63a Great Northern Street, Huntingdon, PE29 7HJ
G6 PRP Walter Barker, 297 Williamthorpe Road North Wingfield, Chesterfield, S42 5NT
G6 PSA N Turnham, 153 Canterbury Road, Urmston, Manchester, M41 0PY
G6 PSC M Horn, 3 Church Cottages, Pound Lane, Beccles, NR34 0EX
G6 PSO I Russell, 24 Standard Avenue, Tile Hill, Coventry, CV4 9BW
G6 PSQ R Langton, Dawn Cliffe, Goodwin Road, Dover, CT15 6ED
G6 PSZ T Shackleton, 27 Court Crescent, Kingswinford, DY6 9RJ
G6 PTF J Wilson, Belle Vue House, Common Side, Workington, CA14 4PU
G6 PTI L Gooch, 27 Stanstead Way, Frinton on Sea, CO13 0BG
G6 PTT Peter Hedley, 7 Midhill Close, Langley Park, Durham, DH7 9YA
G6 PTX G Gane, 28 Queens Croft, Kelso, TD5 7NN
G6 PUE M MacKmin, 89 Wellingborough Road, Rushden, NN10 9YJ
G6 PUR James Howells, 66 Rochester Avenue, Burntwood, WS7 2DL
G6 PUV W Holding, 20 Lingfield Crescent, Wigan, WN6 8QA
G6 PVA M Green, 97 Langley Hall Road, Solihull, B92 7HD
G6 PVC P Coates, Jacaranda, Cotswold Close, Staines, TW18 2DD
G6 PVK G Jones, 18 Mountain Close, Hope, Wrexham, LL12 9SE
G6 PVT L Adamson, 58 Halesworth Road, Wolverhampton, WV9 5PJ
G6 PVU A Appleyard, 18 Earl Street, Nelson, BB9 9JA
G6 PVV C Burt, 1 Chapter Court, Vicarage Road, Egham, TW20 8NL
G6 PVW George Bayliffe, 179 Breedon Street, Long Eaton, Nottingham, NG10 4EW
G6 PWF S Choules, 43 Ashbrook Road, Old Windsor, Windsor, SL4 2LT
G6 PWJ Jonathan Chiddick, Unioninkatu 45A10, Helsinki, Finland, 170
G6 PWL Richard Cloutman, 35 Camlet Way, St. Albans, AL3 4TL
G6 PWQ David Dick, 140 Chatham Street, Stockport, SK3 9JU
G6 PWS R Fuller, 18 St. Leonards Crescent, Sandridge, St. Albans, AL4 9EH
G6 PXJ A Harrison, Nirvana Cottage, 42 Bell Lane, Goole, DN14 8RP
G6 PXN R Bee, 80 Hospital Road, Burntwood, WS7 0EQ
G6 PXQ R Boyce, 3 Castleton Cottages, Westhide, Hereford, HR1 3RF
GW6 PXW Ellis Davies, 6 Park View, Cwmaman, Aberdare, CF44 6PP
G6 PXX L Dempsey, 24 James Street, Great Harwood, Blackburn, BB6 7JE
GM6 PYD Allan Dunnett, 11 Silverknowes View, Edinburgh, EH4 5PY
G6 PYE Cambridge Reapeater Group c/o Phillip Nice, 5 Walden Close, Doddington, March, PE15 0TW
G6 PYF David Hills, 9 Brook Gardens, Devizes, SN10 2FX
G6 PYI Mick Jones, 15 Rowan Rise, Kingswinford, DY6 8EE
G6 PYL P Hatter, 14 Morland Avenue, Bromborough, Wirral, CH62 6BE
G6 PYM A Hedges, 25 The Lanes, Cheltenham, GL53 0PU
GI6 PYP Alan Gault, 134 Leighan Road, Randalshough, Enniskillen, BT93 7DN
G6 PYR H Adams, Hill Sixty, Happisburgh, Norwich, NR12 0RB
G6 PZ Paul Beecham, Cornerstones, Eastertown, Weston-Super-Mare, BS24 0HY
G6 PZE D Jefferson, 48 Neston Road, Walshaw, Bury, BL8 3DB
G6 PZF P James, 33 Headley Chase, Warley, Brentwood, CM14 5BN
G6 PZH Brian Hickman, 7 Nina Close, Stourport-on-Severn, DY13 9RZ
G6 PZN Michael McLoughlin, 13 Old Manor Gardens, Wymondham, Melton Mowbray, LE14 2AN
G6 PZS D Carr, 5 Church Meadow, Hyde, SK14 4RT
G6 QA L Jopson, 68 Greenmount Park, Kearsley, Bolton, BL4 8NT
G6 QM Southgate ARC c/o Donald Berry, 4 Holly Hill, London, N21 1NP
G6 QN T Blakeman, 6 Rutland Close, Ashtead, KT21 1PY
G6 RAF ROYAL AIR FORCE ARS c/o M Garrett, 489 Dorchester Road, Weymouth, DT3 5BP
G6 RAH Roger Hammond, 126 Otley Drive, Ilford, IG2 6QY
GM6 RAK D Brown, 14 Newton Crescent, Carnoustie, DD7 6HW
G6 RAO D Griffiths, 35 Greystones Crescent, Mardy, Abergavenny, NP7 6JY
G6 RAQ S Hayter, 2 Shelsley Drive, Langdon Hills, Basildon, SS16 6NA
G6 RAV W Keeley, 93 Park Crescent, Abergavenny, NP7 5TL
G6 RAZ J Paton, 16 Homefield, Thornbury, Bristol, BS35 2EW
GI6 RBD K Brady, 26 Kilbroney Road, Rostrevor, Newry, BT34 3BJ
G6 RBM R Jeffery, 15 Greenway, Hulland Ward, Ashbourne, DE6 3FE
G6 RBO W Bennett, 44 Wood Lane, Streetly, Sutton Coldfield, B74 3LR
G6 RBP R Pearsey, 21 Ashwood Drive, Newbury, RG14 2PN
G6 RBR M Allen, 1 Allens Yard, Chatteris, PE16 6QE
G6 RBZ Robert Coombes, 53a South Road, Oakfield, Cwmbran, NP44 3EL
G6 RC CRAWLEY ARC c/o Richard Hadfield, 45 Erica Way, Copthorne, Crawley, RH10 3XG
G6 RCD P Clark, 166 Attenborough Lane, Attenborough, Nottingham, NG9 6AB
GW6 RCK H Fray, 17 Homelands Road, Cardiff, CF14 1UH
G6 RCT T Stellar, 27 Blackmore Chase, Wincanton, BA9 9SB
G6 RCX Denis Smithies, 26 Coed Mor, Penyffordd, Holywell, CH8 9HY
G6 RCY David Reed, 11 Grenville Close, Corby, NN17 2RP
G6 RDD I Senter, 33 King Coel Road, Colchester, CO3 9AQ
G6 RDO A Shaw, 1 Chapel Lane, Clifford, Wetherby, LS23 6HU

GW6 RDV B Clarke, 47 Oakway, Pentrebane, Cardiff, CF5 3EH
G6 REA Johnston Gilpin, 28 Ivel Road, Sandy, SG19 1AX
G6 REC M Hart, 7 Ullswater Avenue, South Wootton, King's Lynn, PE30 3NJ
GW6 REF D Jones, 16 New Road, Llandovery, SA20 0ED
G6 REG Andrew Joyce, Ashdene, Highworth Road, Swindon, SN? 4SE
G6 REH J Staplehurst, 12 Trotter Way, Epsom, KT19 7EW
G6 REM I Atkinson, 537753, 7 Headquarter Squadron, BFPO 36,
G6 REQ W Vize, Cefn Rhos, Bethel, Caernarfon, LL55 1YB
G6 REV Jonathan Yam, 4537 Mossburg Court, Marietta, United States, 30066
G6 REW G Seymour-Smith, Glencarne, Bridgerule, Holsworthy, EX22 7ED
G6 REY R McMinn, 1c Bickley Avenue, Sutton Coldfield, B74 4DY
G6 RFH David Ruck, 50 Kiel Walk, Corby, NN18 9DE
G6 RFJ L Waite, The Towers, Castle Street, Nottingham, NG2 4AE
G6 RFL Richard Rothery, 12 Reevy Crescent, Bradford, BD6 2BT
G6 RFM A Warren, 20 Wolverhampton Road, Stafford, ST17 4BP
G6 RFR Alan Brammer, 5 The Green, Reepham, Lincoln, LN3 4DH
G6 RFS A Brown, 22 Mount Wise Crescent, Plymouth, PL1 4QQ
G6 RFU P Csapo, 87 Latchmere Road, Kingston upon Thames, KT2 5TU
G6 RGA J Ewen, 28 Court Road, Eastbourne, BN22 9F?
GM6 RGD T Murray, 2 The Glebe, Edzell, Brechin, DD9 7SZ
G6 RGI Andrew Shead, 95 Sea Front, Hayling Island, PO11 0AW
G6 RGN W Stockley, 10 Swan Road, Timperley, Altrincham, WA15 6BX
G6 RGT M Morgan, 17 Dunstable Road, Newport, NP19 9NE
GM6 RGY W Hardie, 96 Carmuirs Avenue, Camelon, Falkirk, FK1 4PB
G6 RHA Stephen Howes, 46 High Street, Upwood, Huntingdon, PE26 2QE
G6 RHB Robert Howlett, 106 Ewell by Pass, Epsom, KT17 2PP
G6 RHJ M Swain, 17 Sponnes Road, Towcester, NN12 6ED
G6 RHK C Spencer, 83 Oulton Road, Ipswich, IP3 0QE
G6 RHL P West, 6 Iveldale Drive, Shefford, SG17 5AD
G6 RHN D Morris, Rowley Farm, Rowley Lane, Borehamwood, WD6 5PE
G6 RHV I Smith, 126a High Street, Teddington, TW11 8JB
G6 RIC Michael Ellis, 28 High Meadows, Romiley, Stockport, SK6 4PT
G6 RIG N Golding, Coppice View, 16 Littlewood Road, Walsall, WS6 7EU
G6 RII M Dodson, Tree Tops, Badgeworth Lane, Cheltenham, GL51 4UW
G6 RIJ R Fletcher, 33 Littlewood Lane, Cheslyn Hay, Walsall, WS6 7EJ
G6 RIM C Berry, 258 Lowerhouse Lane, Burnley, BB12 6NG
G6 RIQ G Dunn, 11 Ellesmere Rise, Grimsby, DN34 5PE
G6 RIY A Wilkinson, 34 Coppice Lane, Hellifield, Skipton, BD23 4JW
G6 RIZ Barry Jones, 96 Somerset Road, Farnborough, GU14 6DS
G6 RJH J Proffitt, 38 Hockley Road, Poynton, Stockport, SK12 1RW
G6 RJW Roger Woodgate, Roger Woodgate, c/o P.D.C. Main Road, Turakina, New Zealand, 5152
G6 RKF R Taylor, 20 Scraley Road, Heybridge, Maldon, CM9 4BL
G6 RKG J Walters, The Gables, Lavenham Road, Sudbury, CO10 0RN
G6 RKJ R Butland, 4 Park Close, Sonning Common, Reading, RG4 9RY
G6 RKS R Domville, 12 Craig Terrace, Peterlee, SR8 3AJ
G6 RLG J Kerr, 3 Lime Kiln Way, Salisbury, SP2 8RN
G6 RLM R Maddison, Tom Butt Cottage, Hope St, Godalming, GU8 6DE
G6 RMA D Glover, 14 Fitzgerald Avenue, Herne Bay, CT6 8LN
G6 RMJ William Clements, 3 May Street, Durham, DH1 4EN
GI6 RMO D Johnston, Olanda, Lisreagh, Co Fermanagh, BT94 5BX
G6 RMV Richard Sharp, The Old School, Bridge Street, Bridport, DT6 5LS
G6 RNA Trevor Lovell, 4 Maes Rathbone, Waen, St. Asaph, LL17 0AD
G6 RNF R Smith, 12 East View, West Bridgford, Nottingham, NG2 7QN
G6 RNR Sean Allison, The Lodge, Hospital Road, Richmond, DL10 6DX
G6 RNT G Kingdon, Wymering, Copley Drive, Barnstaple, EX31 2BH
G6 RNV C Brewster, 35 Ffordd Las, Sychdyn, Mold, CH7 6DU
GI6 ROI J Polson, 6 Castlemara Drive, Carrickfergus, BT38 7RB
G6 ROS J Warwick, 12 Oak Street, Sutton in Ashfield, NG17 3FF
G6 RPD R Montford, 394 Selbourne Road, Luton, LU4 8NU
G6 RPK Nicholas Waterton, 1270 Killaby Drive, MISSISSAUGA, Ontario, Canada, L5V 1B1
G6 RPW S Andrews, 1 Swynford Close, Kempsford, Fairford, GL7 4HN
G6 RQA D Nicholls, 15 Poplar Avenue, Heacham, King's Lynn, PE31 7EB
G6 RQJ R Sherlock, 34 St. Cecilias Road, Belle Vue, Doncaster, DN4 5EG
GM6 RQU B Hamilton, 51 Grange Road, Grange, Edinburgh, EH9 1UF
GM6 RQW Thomas Christie, 4 Glebe Park, Bressay, Shetland, ZE2 9ER
G6 RQZ B Cripps, 3 Sabre Court, Aldershot, GU11 1YY
G6 RRJ R Urwin, 43 Sykefield Avenue, Leicester, LE3 0LD
G6 RRS D Rotgans, 18 Minter Avenue, Densole, Folkestone, CT18 7DS
G6 RRV R Weston, The Old Dairy, Slate Cross, Bridgwater, TA7 8QR
G6 RRY M Dunbar, 42 Wickham Way, Shepton Mallet, BA4 5YG
G6 RSE SOUTH ESSEX ARS c/o David Speechley, 15 Southwick Road, Canvey Island, SS8 0EP
G6 RSI L Hart, 25 Murcroft Road, Stourbridge, DY9 9HT
G6 RST Horndean And District ARC c/o John Wiles, 12a Ashling Gardens, Denmead, Waterlooville, PO7 6PR
G6 RSU Roy Anstee, 12 Ashmore Avenue, Stockport, SK3 0QY
G6 RTD G Small, 6 Mary St., Longridge, Preston, PR3 3WN
G6 RTC Jolyon Menhinick, Coburg, Barton Estate, East Cowes, PO32 6NT
G6 RTG John Sutton, 29 Victory Avenue, Darlaston, Wednesbury, WS10 7RR
G6 RTN Kenneth Bone, Makerstoun School House, Kelso, TD5 7PB
G6 RTY D Johnson, Penvern, Nacton, Ipswich, IP10 0EW
G6 RUE J Newey, 8 Llandaff Row, Penpentre, Brecon, LD3 8DH
G6 RUM Ian Nice, 6 Malden Road, Sidmouth, EX10 9LS
G6 RUO J Griffiths, 4 Lon Elan, Meliden, Prestatyn, LL19 8LP
G6 RUP J Horner, 43 Birch Close, Patchway, Bristol, BS34 5SA
G6 RUY J Heaney, 15 Perth Street, Nelson, BB9 8EE
G6 RVH R Jamieson, 3 Waterpark Road, Prenton Park, Birkenhead, CH42 9NZ
G6 RVP Christopher Pung, 73 John Mace Road, Colchester, CO2 8WW
G6 RVS Raymond Sohst, 2 Shaftesbury Drive, Maidstone, ME16 0JS
G6 RVZ D Carruthers, 168a Wanstead Park Avenue, London, E12 5EF
GU6 RWD Simon Hancock, L'Hirondel, Hubits de Bas, Guernsey, Guernsey, GY4 6NB
GW6 RWJ D Silcox, Troedyrhiw, Penparc, Cardigan, SA43 2AE
G6 RXD Martin Yirrell, 40a St. Albans Hill, Hemel Hempstead, HP3 9NG
G6 RXF G Priestley, 7 Affleck Avenue, Radcliffe, Manchester, M26 1HN
G6 RXK A Orchard, Kilimani, Cuilfail, Lewes, BN7 2BE
G6 RXP Stephen Dwyer, 10 Swan Street, Darwen, BB3 2LW
G6 RXQ Austin Gordon, 2 Merse Avenue, Kirkcudbright, DG6 4RN
G6 RXV Kevin Keen, 26 Brogden Close, Botley, Oxford, OX2 9DS
G6 RXY David Ferguson, Aneataprint Four Ltd 3-5 Lord Street, Watford, WD17 2LN
G6 RYM H Wogg, 47 Highfield Road, Birkenhead, CH43 0DU
G6 RYW V Smith, 9 Pinewood Drive, Mansfield, NG18 4PG
G6 RZG W Dewhurst, Top Townhead, Grindleton, Clitheroe, BB7 4QT
G6 RZJ O Somers, Houghwood House, Red Barn Road, Wigan, WN5 7UA
G6 RZR Nigel Stevens, 151 Ferme Park Road, London, N8 9BP
G6 RZS R Wood, 115 Anchorway Road, Green Lane, Coventry, CV3 6JH
G6 RZY N Harper, 15 Epsom Close, Boothill, Tamworth, B77 1QT
G6 SAQ Geoffrey Hines, 11 Montagu Gardens, Wallington, SM6 8EP
G6 SBD G Davies, 2 Ffordd Aled, Wrexham, LL12 7PP
G6 SBG A Lubrani, Ranscombe Manor, Sherford, Kingsbridge, TQ7 2DP
G6 SBI D Smith, 8 Corunna Drive, Horsham, RH13 5HG
G6 SBN Malcolm Searl, 130 Chatham Street, Reading, RG1 7HT
GI6 SBW Arthur Alcock, 22 Chippendale Avenue, Bangor, BT20 4PT
G6 SCG M Lockwood, 33 Elmtree Road, Calverton, Nottingham, NG14 6QA
G6 SCM J Webb, Oakdene, 22 Meeting House Lane, Coventry, CV7 7FX
G6 SDC Bernard Simmons, Wootton Leas, 35 Benenden Green, Alresford, SO24 9PE
G6 SDE A Curley, 21 Trinity Rise, Penton Mewsey, Andover, SP11 0RE
G6 SDG N Knowles, Waterloo West, Waterloo Road, Grantham, NG32 3DX
G6 SDI P Hall, 28 Grangeway, Rushden, NN10 9EZ
GM6 SDV James More, 51 Hilton Drive, Aberdeen, AB24 4NJ
G6 SDW Mark Rowan, 1 Yanleigh Close, Bristol, BS13 8AQ
G6 SDY R Beecroft, 28 Hall Garth Lane, West Ayton, Scarborough, YO13 9JA
G6 SEE Z Feast, 2 Dyrham Close, Burnham-on-Sea, TA8 2TT
G6 SEF J Feast, 2 Dyrham Close, Burnham-on-Sea, TA8 2TT
G6 SEK P Ovey, 35 Lower Fairfield, St. Germans, Saltash, PL12 5NH
GM6 SEV Ian Carr, 36a Broomieknowe, Lasswade, EH18 1LN
G6 SFC T Foulds, Deer Park, Detling Avenue, Broadstairs, CT10 1SR
G6 SFE S Al-Kattan, 8 Little John Drive, Rainworth, Mansfield, NG21 0JJ
G6 SFF C Hewes, 15 Heathfield Road, Nottingham, NG5 1NL
G6 SFH P Barber, 17 Wheelwright Avenue, Leeds, LS12 4UW
GI6 SFO Gordon Miskimmin, 332 Rathfriland Road, Dromara, Dromore, BT25 2HN
G6 SFR FLT REFUELNG ARS c/o A Baker, Highleaze, Deans Drove, Poole, BH16 6EQ
G6 SFW P Dodd, 9 Rudge Croft, Kitts Green, Birmingham, B33 9NZ
G6 SFY G Barker, 18 Penryn Close, Nuneaton, CV11 6FF
G6 SGA S Thornber, 18 Lichfield Road, Talke, Stoke-on-Trent, ST7 1SQ
G6 SGD David Carding, Mill House, Walcot, Telford, TF6 5ER
G6 SGE P Bayliss, 36 Slingates Road, Stratford-upon-Avon, CV37 6ST
G6 SGM C Macey, 29 Burleigh Road, Sutton, SM3 9NE
G6 SGV Matthew Durkin, Selsdon House, 23 Jameson Road, Bexhill-on-Sea, TN40 1EG
G6 SGW John Miller, 6 Saunders Mews, Southsea, PO4 9XZ
G6 SGY Brian Sales, 2 Highview, Hurley, Atherstone, CV9 2RP
G6 SGZ J Smith, 1 Markby Close, Moorside, Sunderland, SR3 2RG
G6 SHD G McBrien, 26 Lumb Carr Avenue, Ramsbottom, Bury, BL0 9QG
G6 SHF M Trolan, Hilbre, East Taphouse, Liskeard, PL14 4NJ
G6 SHQ M Brown, 6 Snell Hatch, West Green, Crawley, RH11 7JB
G6 SHS K Eldridge, 44 Merley Gardens, Merley, Wimborne, BH21 1TB
G6 SIG Gerald Timbrell, Crossing Cottage, Lamyatt, Shepton Mallet, BA4 6NG
G6 SIM John Simarpi, 12b Westfield Road, Wells, BA5 2HS
G6 SIQ W Whitcombe, 11 The Elms, Deerton Street, Sittingbourne, ME9 9LH
G6 SIX Peter Macmillen, Spinney Cottage, Sychnant Pass Road, Conwy, LL32 8NS
G6 SJA W Barnes, 17 Greenhill Road, Long Buckby, Northampton, NN6 7PU
G6 SJG T Hurton, 4 Athlone Close, Enham Alamein, Andover, SP11 6JY
G6 SKF L Hopson, 39a Fenside Road, Boston, PE21 8HY
G6 SKK T Parkin, 8 Horsley Crescent, Holbrook, Belper, DE56 0UB
G6 SKM W Taylor, 5 Gadbury Avenue, Atherton, Manchester, M46 0LQ
G6 SKP A Whitgreave, 2 Oaklea Avenue, Hoole, Chester, CH2 3RE
G6 SKR G Walker, 81 Normanshire Drive, Chingford, London, E4 9HE
G6 SKS C Learoyd, Leofric House, 31 Leofric Avenue, Bourne, PE10 9QT
G6 SKT W Learoyd, Leofric House, 31 Leofric Avenue, Bourne, PE10 9QT
G6 SL Chris Pettitt, 12 Hennals Avenue, Redditch, B97 5RX
G6 SLG A Berry, Emberley Leys, Ratley, Banbury, OX15 6DS
G6 SLH Annette Bland, 18 Hill Street, Newhall, Swadlincote, DE11 0JR
G6 SLO P Charnley, 9 Bryn Crescent, Rhuddlan, Rhyl, LL18 5RF
G6 SLY D Lewis, 30 Printers Park, Hollingworth, Hyde, SK14 8QH
G6 SLZ Jacqueline Mackenzie, 11 Upper Heyshott, Petersfield, GU31 4QA
G6 SMI G Langstaff, Flat 22 Benwell Close, Benwell Grange, Newcastle-upon-Tyne, NE15 6RZ
G6 SMJ M Pitts, 30 Sandhurst Avenue, Surbiton, KT5 9BS
G6 SMW Geoffrey Harper, 15 Seaforth Park, Annan, DG12 6HX
G6 SNA Donald Heather, 65 Fairview Road, Headley Down, Bordon, GU35 8HQ
G6 SND W Jarvis, Owlpen, 20 Park Road, Devizes, SN10 4ED
G6 SNI G Platt, 15 Mount Close, Nantwich, CW5 6JJ
G6 SNN D Ramsey, 11 Pendle Close, Basildon, SS14 3NA
GJ6 SNQ A Leighton, 2 Le Petit Menage, Fountain Lane, St. Saviour, Jersey, JE2 7RL
G6 SNV R Smith, 3 Payne Road, Wootton, Bedford, MK43 9JL
G6 SOA Richard Wilday, 4 Kenelm Road, Clifton-on-Teme, Worcester, WR6 6DW
G6 SOX John Gerrard Bradley, 68 Rosedale Avenue, Alvaston, Derby, DE24 0FJ
G6 SOY P Berry, Bundys Cottage, Colwood Lane, Haywards Heath, RH17 5QQ
G6 SOZ A Byrne, Holly Cottage, Deacons Lane, Thatcham, RG18 9RJ
G6 SPB D Corder, 140 Edward Road, Somerford, Christchurch, BH23 3EW
G6 SPG P Cesnavicius, 52 Boundary Road, Irlam, Manchester, M44 6HD
G6 SPH J Crookbain, 11 Champlain Avenue, Canvey Island, SS8 9QL
GD6 SPI S Carwood, 34 Flemming Avenue, Ruislip, HA4 9LF
G6 SPN R Barton, 82 Buckingham Road, South Woodford, London, E18 2NJ
G6 SPQ Wallace Holmes, 37 Barmpton Lane, Darlington, DL1 3HH
G6 SQL Andrew Lowthian, 38 Arthur Street, Ryde, PO33 3BU
G6 SQS F Sivyer, 22 Boxley Road, Walderslade, Chatham, ME5 9LF
G6 SQT C Wall, 151 Bisley Road, Stroud, GL5 1HS
G6 SRE B Stone, Reindene, Faversham Road, Ashford, TN25 4PQ
G6 SRJ A Waring, 2 Wroxton Close, Thornton-Cleveleys, FY5 3EY
G6 SRS Stourbridge & District ARS c/o David Scott, Hyde Bungalow, The Hyde, Stourbridge, DY7 6LS
G6 SRT D Armstrong, 102 Victoria Road, Oxford, OX2 7QG
G6 SRU C Alford, 6 Meadow Rise, Bournville, Birmingham, B30 1UZ
G6 SRV R Andrews, 11 Holly Grove, Verwood, BH31 6XA
G6 SRY S Aram, 63 Newbury Street, Wantage, OX12 8DJ
G6 SRZ W Baxter, 19 Westbury Road, Nottingham, NG5 1EP
G6 SSH Victor Bates, 382 Hindley Road, Westhoughton, Bolton, BL5 2DT
G6 SSM O Burgess, 14 Leys Close, Harefield, Uxbridge, UB9 6QB
G6 SSN Kevin Burton, 2 Council House, Stainfield Road, Bourne, PE10 0SG
G6 SSQ Doreen Bolton, Huds House, Cowgill, Sedbergh, LA10 5TQ
GD6 SSV J Cott, 50 Parkside Way, North Harrow, Harrow, HA2 6DG
G6 STD David Macey, Affric House, New Street, Banbury, OX15 0SR
G6 STE B Stevens, 16 Fowey Close, Wellingborough, NN8 5WW
G6 STF Graham Smith, Sunny Patch, Western Backway, Kingsbridge, TQ7 1QB
G6 STI H Staddon, 45 Saxony Parade, Hayes, UB3 2TQ
G6 STJ S Smith, 9 Beadon Drive, Salcombe, TQ8 8NU
G6 STK R Sweet, 9 Seafield Road, Colwyn Bay, LL29 7HB
G6 STS G Jones, The Bungalow, Castle Street, Pontypool, NP4 9QL
G6 SUK B Trevor, 30 Clayton Crescent, Brentford, TW8 9PT
G6 SUR P Thornsby, 25 Kipling Way, Stowmarket, IP14 1TS
G6 SUV J Barlow, 3 Shaw Brook Close, Rishton, Blackburn, BB1 4ES
G6 SVH Kevin Henderson, 42 Chartwell Avenue, Wingerworth, Chesterfield, S42 6SP
G6 SVJ S Harvey, 148 Smithfield Road, Uttoxeter, ST14 7LB
G6 SVL Simon Harris, 34 Butterfly Crescent, Nash Mills Wharf, Hemel Hempstead, HP3 9GS
G6 SVV R Gray, Willett, 28 Hoe Lane, Romford, RM4 1AX
G6 SWD A Gibbings, 3 Bonville Crescent, Tiverton, EX16 4BN
G6 SWJ J Askey, The Maltings, Brewery Yard, Kettering, NN14 3BT
G6 SWO Michael Fincher, Brickyard Farm, Lincoln Road, Horncastle, LN9 5NW
G6 SWT S French, 47 Horn Lane, Woodford Green, IG8 9AA
G6 SWW S Ellin, 7 Crawshaw Avenue, Beauchief, Sheffield, S8 7DZ
G6 SWZ M Davy, 22 Scott Gardens, Lincoln, LN2 4LX
G6 SXB Leslie Dunham, 5 King Street, Wimblington, March, PE15 0QF
G6 SXC J Mallichan, 17 Napier Road, Gillingham, ME7 4HB
G6 SXD E Drinkwater, 57 Ludlow Road, Bridgnorth, WV16 5AH
G6 SXN G Dixon, 4 Yarborough Road, Keelby, Grimsby, DN41 8HG
G6 SYA M Mills, 6 Bower Road, Hextable, Swanley, BR8 7SE
GD6 SYB John Malcom, 62 Linden Avenue, Ruislip, HA4 8UA
G6 SYI Peter Somerfield, 27 Ormerod Street, Worsthorne, Burnley, BB10 3NU
G6 SYW B Bauly, Poplar Farm Mendlesham, Stowmarket, IP14 5SN
G6 SYX G Brookes, 47 Lucas Avenue, York, YO30 6HL
G6 SZB J Barton, 76 Elvaston Road, North Wingfield, Chesterfield, S42 5HH
GM6 SZJ M Burke, 5 Braeview Crescent, Star, Glenrothes, KY7 6LZ
G6 SZP N Carter, 23 Sandhall Drive, Highroad Well Moor, Halifax, HX2 0DL
G6 SZS Richard Crook, 26 Chapel Street, Rishton, Blackburn, BB1 4NP
G6 TAF P Penny, 79 Grove Avenue, New Costessey, Norwich, NR5 0JA
G6 TAH D Palmer, 17 Atyeo Close, Burnham-on-Sea, TA8 2EJ
G6 TAI James Peel, 9 Hillspring Road, Springhead, Oldham, OL4 4SJ
G6 TAK P Reay, 26 Clifton Court, Workington, CA14 3HR
G6 TAN Michael Shread, 21 The Strand, Mablethorpe, LN12 1BQ
G6 TAP David Squire, Green Valley, Raleigh Road, Barnstaple, EX31 4HY
G6 TAS Roger Wroe, 11 Malvern Close, Banbury, OX16 9EL
GI6 TBC V Loughran, 10 Oakwood, Armagh, BT60 1QR
G6 TBE P Lowrie, 11 Berrymoss Court, Kelso, TD5 7NP
G6 TBJ J Larssen, 228a Barnsole Road, Gillingham, ME7 4JB
G6 TBT G Davey, 55 Lion Lane, Haslemere, GU27 1JF
G6 TBV K Cheers, 112 Rickerscote Road, Stafford, ST17 4HB
G6 TCD R Gleeson, 47 Shore Avenue, Shaw, Oldham, OL2 8DA
G6 TCV J Halliday, 42 New Road, Bignall End, Stoke-on-Trent, ST7 8QF
G6 TDG K Hodges, 18 Leycester Close, Birmingham, B31 4SS
G6 TDJ G Ward, 65 Birtles Road, Macclesfield, SK10 3JG
G6 TDR Robert McDermott, 6 John Gilmour Way, Burley in Wharfedale, Ilkley, LS29 7SR
G6 TDW J Mead, 12 Coltsfoot Close, Ixworth, Bury St. Edmunds, IP31 2NJ
GD6 TDX D Yarrow, 193 Ladygate Lane, Ruislip, HA4 7RD
G6 TEB A Varga, 2 Yew Tree Lane, Malvern, WR14 4LJ
G6 TEL Stephen Mayer, 453 Wimborne Road, Poole, BH15 3EE
GW6 TEO G Smith, 11 Sandy Leys, Castlemartin, Pembroke, SA71 5HJ
G6 TEQ I Stuckey, Rancliffe, Higher Downs Road, Torquay, TQ1 3LD
G6 TER R Monksummers, 29 Cloverfields, Peacemarsh, Gillingham, SP8 4UP
G6 TET B Smith, 8 Devon Street, Leigh, WN7 2NG
G6 TEX T Speak, 40 Orchard Close, Bolsover, Chesterfield, S44 6DY
G6 TFE D Standen, 54 Park Way, Hastings, TN34 2PJ
GI6 TFF R Symington, 8 Thorndene Park, Carrickfergus, BT38 9EA
G6 TFJ R Smith, 251-258 Valley View, Broad Street, Icklesham, TN36 4AS
G6 TFP Christopher Mann, Woodford, Listowel, Ireland
G6 TFV D Owen, 18 Prescott Avenue, Atherton, Manchester, M46 9LN
G6 TGB A Pennells, 56 Wilverley Place, Blackfield, Southampton, SO45 1XW
G6 TGE G Holman, 5 Ingleton Road, Newsome, Huddersfield, HD4 6QX
G6 TGJ J Hirons, Furlong House, Racecourse Lane, Shrewsbury, SY3 5BJ
G6 TGM W Howard, 2 Heather Drive, Rise Park, Romford, RM1 4SP
G6 TGQ Sidney Houghton, 259b St. Faiths Road, Norwich, NR6 7BB
G6 TGR Gwyn Jones, 3 Tan y Buarth Estate, Bethel, Caernarfon, LL55 1UP
G6 TGW R Jones, 49 Sycamore Drive, Huntingdon, PE29 7JA
G6 THC M Johnson, Happy Valley, Highfield Road, Westerham, TN16 3UX
G6 THM Colin Smith, 8 Terry Close, Stoke-on-Trent, ST3 6NS
G6 THP L Curwen, 12 Garden Close, St. Ives, PE27 3XZ
GM6 TIB I Campbell, 35 Thornwood Avenue, Lenzie, Glasgow, G66 4FI
G6 TID M Coleman, 1 Burdon Drive, Bartestree, Hereford, HR1 4DL
G6 TIQ A Price, Flat 2, South Elms, 69 Silverdale Road, Eastbourne, BN20 7EU
G6 TIU Mark Rainer, 101 Gwydir Street, Cambridge, CB1 2LG
G6 TIW D Reeve, 12 Lambourne Road, Birstall, Leicester, LE4 4FU
G6 TJC S Deville, 39 Acre Close, Maltby, Rotherham, S66 8BL
GM6 TJD J Doull, 52 Howburn Road, Thurso, KW14 7ND
G6 TJE R Dawson, 10 St. Julien Close, New Duston, Northampton, NN5 6QX
G6 TJJ R Downham, 11 Churchills Rise, High Street, Cullompton, EX15 3AU
G6 TJK A Dowell, 54 Station Street, Castle Gresley, Burton-on-Trent, DE14 1BS



**IMPORTANT NOTE**

**Revalidate licence to avoid revocation** – Ofcom has advised the Society that plans will be drawn up to revoke licences that have not been revalidated as required by the licence conditions. The quickest way to revalidate is to do so online via the Ofcom website: *https://services.ofcom.org.uk/* or by email: *amateur.validations@ofcom.org.uk* Ofcom staff are available to help, but please be patient during times of heavy workload.

G6 TJY I Randle, 12 Cuckoo Avenue, Hanwell, London, W7 1BT
G6 TJZ Peter Rendell, 6 The Park, Bradley Stoke, Bristol, BS32 0AP
G6 TKB K Slaughter, 652 Newchurch Road, Newchurch, Rossendale, BB4 9HG
GU6 TKE C Wild, Honfleur, La Rue Du Le Hurel, Vale, Guernsey, GY3 5AF
G6 TKH J Torring, Ivy Cottage, Royal Oak, Filey, YO14 9QE
G6 TKR E Tratt, 162 Stoddens Road, Burnham-on-Sea, TA8 2EL
G6 TKV T Tyrer, 85 Swann Lane, Cheadle Hulme, Cheadle, SK8 7HU
G6 TKW P Tomlinson, 158 Seamore Avenue, Benfleet, SS7 4LA
G6 TKY E Caligari, 209 Ormskirk Road, Upholland, Skelmersdale, WN8 0AA
G6 TLA Paul Curran-Bilbie, 198 Birchwood Lane, Somercotes, Alfreton, DE55 4NF
G6 TLB P Curran, 422 Carlton Road, Worksop, S81 7QW
G6 TLN D Allen, 35 Fortescue Chase, Thorpe Bay, Southend-on-Sea, SS1 3SS
G6 TLP David Attree, 36 Furze Road, Norwich, NR7 0AS
G6 TLS J Balding, 7 Mouse Lane, Rougham, Bury St. Edmunds, IP30 9JB
G6 TLX C Bull, 35 Manor Road, Wokingham, RG41 4AR
G6 TM CONWAY VALL ARC c/o Raymond Jones, Woodcote, 37 Coed Pella Road, Colwyn Bay, LL29 7BB
GM6 TMH D Bell, 11 Shebster Court, Thurso, KW14 7ES
G6 TMN H Bryan, 2 Ashbrook Close, Hesketh Bank, Preston, PR4 6LY
G6 TMQ L Saagi, 17 Broughton Close, Anstey, Leicester, LE7 7EU
G6 TNA C Walton, 6 Gorse Grove, Longton, Preston, PR4 5NP
G6 TNE G Skulski, 30 Eastfield Road, Laindon, Basildon, SS15 4JE
G6 TNI B Telford, 18 Kirkstall Close, South Anston, Sheffield, S25 5BA
G6 TNK J Turnbull, 34 Bridge Avenue, Hanwell, London, W7 3DJ
G6 TNQ N Bosanquet-Bryant, 2 Dupont Close, Clacton-on-Sea, CO16 8YD
G6 TNR D Blackman, 115 Ringwood, Bracknell, RG12 8XU
G6 TNW Ian Webb, Cornerways, Orchard Road, St. Neots, PE19 7AN
G6 TOC W Forster, 35 Beaconsfield Road, Burton-on-Trent, DE13 0NT
G6 TOI Andrew Edgcombe, 2 Providence Place, Fore Street, Kingsbridge, TQ7 4QP
G6 TOX B Taylor, Swn-Y-Don, Beaumaris, LL58 8RW
G6 TOY D Williams, Hollybank, Royston Road, Taunton, TA3 7RE
GJ6 TPD Evelyn Langlois, L'Amarrage, La Route Orange, Jersey, JE3 8GP
G6 TPE Michael Long, 28 Wentworth Drive, Wirral, CH63 0JA
G6 TPG John Littler, 39 Wigan Road, Golborne, Warrington, WA3 3TZ
G6 TPI C Bryan, 3 Hales Place, Stoke-on-Trent, ST3 4NF
G6 TPO D Bathe, Moel Tryfan, Grange Lane, Harlow, CM19 5HG
G6 TQC W George, 28 Melbourn Close, Duffield, Belper, DE56 4FX
G6 TQF P Game, 15 Nightingale Close, Gosport, PO12 3EU
G6 TQH G Giudice, 31 Woodfield Cross, Tredegar, NP22 4JG
G6 TQL T Law, 27 Grove Road, London, N12 9EB
G6 TQZ Richard Andrews, 10 Fourth Avenue, Havant, PO9 2QX
G6 TRA D Andrew, The Willows, Nordelph, Downham Market, PE38 0BY
G6 TRG TODMORDEN RYT G c/o Paul Rigg, 1 Stones Hey Gate, Widdop Road, Hebden Bridge, HX7 7HD
G6 TRM Michael Bryant, 104 Manor Road, Dover, CT17 9JZ
G6 TRN D Best, 64 New Hey Road, Cheadle, SK8 2AQ
G6 TRO B Wilcox, 1 Parklands, Stanwick, Wellingborough, NN9 6QX
G6 TRQ J Wright, 9 Willow Close, Broadmeadows, Alfreton, DE55 3AP
G6 TRW A Toas, 116 Rownhams Road, North Baddesley, Southampton, SO52 9EU
G6 TRX J Sugrue, 124 Hall Lane, Upminster, RM14 1AL
G6 TRY Gordon Smillie, 5 Fleckers Drive, Up Hatherley, Cheltenham, GL51 3BB
G6 TSC V Simmons, 88 Wellcome Avenue, Dartford, DA1 5JW
G6 TSE Alan Sullivan, 20 Crockerne Drive, Pill, Bristol, BS20 0LF
G6 TSF Paul Shayler, 38 Maryside, Slough, SL3 7ET
G6 TSJ P Hannam, 7 Bodenham Close, Buckingham, MK18 7HR
G6 TSL Raymond Hill, Marclecote, Ledbury Road, Ross-on-Wye, HR9 7BE
G6 TSM W Hirst, 8 Moss Road, Alderley Edge, SK9 7HZ
G6 TSP Kenneth Hendry, 23 Briscoe Way, Lakenheath, Brandon, IP27 9SA
G6 TSX Charles Heater, 8 Whitethorn Close Ash, Aldershot, GU12 6NZ
G6 TSZ D Hall, 282 Dereham Road, Norwich, NR2 3TL
G6 TTD R Hewitt, 11614 Waesche Drive, Mitchellville, United States, 20271
G6 TTX W Kenyon, 13 Baskerfield Grove, Woughton on the Green, Milton Keynes, MK6 3ES
G6 TUD M Prosser, 18 Thornhill Way, Rogerstone, Newport, NP10 9FT
GM6 TUE P Mclaren, Dalriada, Fogwatt, Moray, IV30 8SY
G6 TUG Ian Metcalfe, 8 Oakbrook, Woodfield Road, Crawley, RH10 8GD
G6 TUS R Page, 53 The Brambles, Bar Hill, Cambridge, CB23 8SZ
G6 TVA D Biram, 124 Keresforth Hill Road, Red Gables, Barnsley, S70 6RG
G6 TVB Richard Steele, 98 Obelisk Rise, Boughton Green, Northampton, NN2 8QU
G6 TVC D Spooner, Thorny How, Canon Pyon, Hereford, HR4 8NT
G6 TVD Eric Sims, Hafan, Engedi, Holyhead, LL65 3RR
G6 TVE John Tyreman, 12 Richmond Close, Rochdale, OL16 4RJ
G6 TVI G Sculthorpe, 50 Station Road, Dersingham, King's Lynn, PE31 6PR
G6 TVJ I Bennett, 47 Bakers Ground, Stoke Gifford, Bristol, BS34 8GD
G6 TVK Andrew Baker, 22 Benbow Close, Daventry, NN11 4JR
G6 TVP Stephen Burke, 17 The Crescent, Wragby, Market Rasen, LN8 5RF
G6 TVR John Black, Solway View, Carlisle Road, Annan, DG12 6QX
G6 TVX Michael Collins, Coburg Cottage Mount Road, East Cowes, PO32 6NT
G6 TW SOUTH CHESHIRE ARS c/o Peter Walker, 11 Flixton Drive, Crewe, CW2 8AP
G6 TWA A Woollard, 30 John Grinter Way, Wellington, TA21 9AR
G6 TWB South Cheshire ARS c/o Barrie Rigby, 76 Woodland Road, Rode Heath, Stoke-on-Trent, ST7 3TL
G6 TWD W West, Lectric, Alexandra Road, Crediton, EX17 2DH
GD6 TWF C Wood, Deep Water, Glen Rushen Road, Peel, Isle of Man, IM5 3BA
G6 TWR Keith Simmonds, Hazeldene, Barnstaple Road, South Molton, EX36 3RD
G6 TWX Alan Tatterton, 28 Kinloch Drive, Bolton, BL1 4LZ
G6 TXB Bernard Thompson, 12 Albion Road, Chatham, ME5 8SR
G6 TXH Robert Webb, 90 Queens Road, Tunbridge Wells, TN4 9JU
G6 TXP Alan Cronk, 65 Russell Court, Chatham, ME4 5LE
G6 TXQ V Covell-London, 15 St. Nicholas Way, Potter Heigham, Great Yarmouth, NR29 5LG
G6 TXV Stephen Calver, 144a Smugglers Club Ground, Bridgemarsh Lane, Chelmsford, CM3 6DQ
G6 TXY B Coulstock, 32 Climping Park, Bognor Road, Littlehampton, BN17 5DW
G6 TYB J Cooke, 106 Wirral Drive, Winstanley, Wigan, WN3 6LD
G6 TYF R Duley, Denova, Hornsby Lane, Grays, RM16 3AU
GW6 TYO B Young, Apartment 64, St. Margarets Court, Swansea, SA1 1RZ
G6 TYT S White, 8 Bilford Avenue, Worcester, WR3 8PJ
GM6 TYX Christopher MacLeod, Morven, Marybank, Isle of Lewis, HS2 0DD
G6 TZE Jon Richards, 17 Chaffers Mead, Ashtead, KT21 1NA
G6 TZO N Roberts, 11 Mallard Road, Barrow upon Soar, Loughborough, LE12 8BF
G6 TZT T Rogers, 36 Goodacre Road, Ullesthorpe, Lutterworth, LE17 5DL
G6 UAJ P Longstaff, 27 Feather Wood, Westlea, Swindon, SN5 7AG
G6 UAN Michael Moseley, 5 Solent View, Calshot, Southampton, SO45 1BH
G6 UAP E Magnuszewski, 49 Elvaston Road, Nottingham, NG8 1JU
GW6 UAS J Mcmurray, 30 St. Martins Crescent, Llanishen, Cardiff, CF14 5QA
G6 UAW Martyn Egerton, 46 Badgers Copse, Camberley, GU15 1HW
G6 UBH Mike Faithfull, 99 Bramble Road, Hatfield, AL10 9SB
G6 UCI G Miller, 11 Friars Avenue, Great Sankey, Warrington, WA5 2AR
GM6 UCN William Mckenzie, 14 Bridgend, Dunblane, FK15 9ES
G6 UCO G Bee, 80 Hospital Road, Burntwood, WS7 0EQ
G6 UCQ M Burgess, 20 Norfolk Road, Luton, LU2 0RE
G6 UCT P Bowron, 52 Eastcotes, Tile Hill, Coventry, CV4 9AU
G6 UCW Andrew Brookes, 8 Cedar Close, Ruskington, Sleaford, NG34 9FH
G6 UCY K Porter, 60 Spitfire Road, Wallington, SM6 9GL
G6 UDA R Pyrah, Whispering Waves, The Shore, Poulton-le-Fylde, FY6 9EA
G6 UDB F Scott, 15 Sunningdale Close, Kirkby-in-Ashfield, Nottingham, NG17 8NW
G6 UDF P Phelps, 14 The Warren, Hazlemere, High Wycombe, HP15 7ED
G6 UDG A Price, Brook House, Drury Lane, Shrewsbury, SY4 1DT
G6 UDI S Phillips, 79 Selwyn St., Stoke, Stoke-on-Trent, ST4 1ED
G6 UDX Brian Oldford, 16 Ludlow Drive, Stirchley, Telford, TF3 1EG
G6 UED Tim Lloyd, 18 Coleville Road, Minworth, Sutton Coldfield, B76 1XR
G6 UEG Ian Norman, 33 High Street, Puckeridge, Ware, SG11 1RN
G6 UEH E Naylor, 18 Mackenzie Crescent, Cheadle, Stoke-on-Trent, ST10 1LU
G6 UEI P Norman, 20 Meadow Close, Budleigh Salterton, EX9 6JN
G6 UEQ R Rix, Patterdale, Roe Downs Road, Alton, GU34 5LG
G6 UER V Rice, 24 Harewood Close, Tuffley, Gloucester, GL4 0SR
G6 UEU G Reid, 9 King George Place, Walsall, WS4 1EQ
G6 UEV M Piper, 26 Hare Law Gardens, Stanley, DH9 8DG
GW6 UFH S Fry, 10 Heaseland Place, Killay, Swansea, SA2 7EQ
G6 UFI P Fanning Ba Bd Cf, 45 Roman Road, Wattisham Airfield, Ipswich, IP7 7RW
G6 UFL S Duckett, 35 Fowlmere Road, Foxton, Cambridge, CB22 6RT
GI6 UFO John Fitzgerald, 15 Bunnahesco Road, Bunnahesco, Enniskillen, BT94 5HJ
GI6 UFU J Campbell, 22 Sheridan Drive, Helens Bay, Bangor, BT19 1LB
G6 UFV Carl Spencer, 18 Coatsby Road, Kimberley, Nottingham, NG16 2TH
G6 UFZ K Chamba, 63 Patricia Avenue, Wolverhampton, WV4 5AQ
G6 UGA M Pinkney, 169 Sandringham Road, Perry Barr, Birmingham, B42 1PZ
GW6 UGC G Phillips, 83 Heol Y Llwynau, Trebanos, Swansea, SA8 4DB
G6 UGE C Power, 29 Alderley, Digmoor, Skelmersdale, WN8 9NB
G6 UGG A Panton, 35 Long Water Drive, Gosport, PO12 2UP
G6 UGS M Allison, 6 Eden Road, Beverley, HU17 7HD
G6 UGT T Aherne, 21 Burbage Place, Alvaston, Derby, DE24 8NP
G6 UGW Malcolm Bell, 61 Oldbury Orchard, Churchdown, Gloucester, GL3 2PU
G6 UGZ A Scott, 23 Wingfield Road, Great Barr, Birmingham, B42 2QB
GM6 UHC A Stewart, 3 Pinkie Road, Newmachar, Aberdeen, AB21 0RG
G6 UHD Brian Scott, The Cottage, Piece, Redruth, TR16 6SF
GM6 UHE Alison Wilson, Lochend, Beith, KA15 2LN
G6 UHL B Ritchie, 65 Ransome Avenue, Worcester, WR5 3AL
G6 UHS A Coe, 22 St. Annes Way, Spalding, PE11 3PN
GW6 UHY Leslie Crompton, 6 Morgans Terrace, Pontrhydyfen, Port Talbot, SA12 9TP
G6 UIF I Clark, 41 Brook Close, Jarvis Brook, Crowborough, TN6 2ET
G6 UIM Stephen Daniels, 46 Freshwater Drive, Paignton, TQ4 7SD
G6 UIT W Dillon, 49 Goring Way, Greenford, UB6 9NN
G6 UJC C Dukes, 4 Westfields, West Woodhay, Newbury, RG20 0BW
GM6 UJG Victor Simpson, 43 Fortingall Place, Perth, PH1 2NF
G6 UJI B Staton, 99 Linden Avenue, Prestbury, Cheltenham, GL52 3DT
G6 UJR Mark Severs, 125 Hawthorne Way, Shelley, Huddersfield, HD8 8QF
G6 UKC Michael Brown, 33 Stonegate, Cowbit, Spalding, PE12 6AH
G6 UKM Stephen Brown, 22 Asquith Close, Biddulph, Stoke-on-Trent, ST8 7LN
G6 UKN W Bailey, 35 Elton Lane, Winterley, Sandbach, CW11 4TN
G6 UKO Colin Barwell, Galaghad, Penisarwaun, Caernarfon, LL55 3BN
G6 UKQ John Riley, 56 Church St., Bignall End, Stoke-on-Trent, ST7 8PE
G6 ULD R Humphrys, 10 St. Andrews Road, Bexhill-on-Sea, TN40 2BQ
G6 ULJ Richard Green, Branford House, Valley Road, Tasburgh, Norwich, NR15 1NG
G6 ULS P Kent-Woolsey, 32 Yaxham Road, Dereham, NR19 1AJ
G6 UMH E Harding, 49 Compass Close, Murdishaw, Runcorn, WA7 6DL
G6 UML T Reader, 76 West View Road, Dartford, DA1 1TR
G6 UMN L Gibson, 27 Farm Street, Barrow-in-Furness, LA14 2RX
G6 UMT Andrew Handcocks, Woodpeckers, Chapel Lane, Southampton, SO45 1YX
G6 UMU Anthony Haigh, Nant Fawr, Corwen, LL21 9AA
G6 UMX J Hibbert, 125 Chase Hill Road, Arlesey, SG15 6UF
G6 UNA Win Stoneman, 5 Creaton Road, Hollowell, Northampton, NN6 8RP
GM6 UNL J Leslie, 7 Charters Street, Stirling, FK7 0QE
G6 UNN R Lewis, 5 Popham Close, Bridgwater, TA6 4LD
GM6 UNQ E Leask, 2/7 Barnton Avenue West, Edinburgh, EH4 6EB
G6 UNR Raymond Rogers, 97 Sutherland Avenue, Biggin Hill, Westerham, TN16 3HH
G6 UNU Anthony Lunn, 45 St. Anthonys Avenue, Eastbourne, BN23 6LN
G6 UNX Barry Lockhart, Tan Yr Onnen, Llangernyw, Abergele, LL22 8PP
G6 UOH I Thacker, 3 Webster Way, Gonerby Hill Foot, Grantham, NG31 8GH
G6 UOO D Wilde, 3 Canal Cottages, Buxworth, High Peak, SK23 7NF
G6 UOX Michael Walker, 94 Lambert Road, Uttoxeter, ST14 7QY
G6 UPA D Wiseman, 22 Queens Crescent, Clapham, Bedford, MK41 6DA
G6 UPH M Hackney, Mar Azul 9, Apt.20 2 Fase, Alicante, Spain, 03710 CALPE
G6 UPI B Hurrell, 33 Meadow Way, Hellesdon, Norwich, NR6 5NN
G6 UPL T Hayhurst, The Paddock, Crooklands, Milnthorpe, LA7 7NL
G6 UPM Andrew Hayhurst, The Paddock, Crooklands, Milnthorpe, LA7 7NL
G6 UPQ David Holloway, 48 Wenrisc Drive, Minster Lovell, Witney, OX29 0RQ
G6 UPR B Hingston, Hazelwood Farm, Paignton, TQ3 1SQ
G6 UQ STOCKPORT RS c/o Bernard Naylor, 47 Chester Road, Poynton, Stockport, SK12 1HA
G6 UQA S Duckles, 8 Railway Cottages, Skillings Lane, Brough, HU15 1EN
G6 UQC Alan King, 73 Higher Hillgate, Stockport, SK1 3HD
G6 UQI E Payne, 4 Richmond Crescent, Barons Cross, Leominster, HR6 8RX
G6 UQO D Oliver, 37 Milford Avenue, Elsecar, Barnsley, S74 8DT
G6 UQZ Andrew Parkhurst, 14 Church Street, Clare, Sudbury, CO10 8PD
G6 URF A Hartley, 18 Smithy Close, Cronton, Widnes, WA8 5BT
G6 URK J Jennings, 354 Williamthorpe Road, North Wingfield, Chesterfield, S42 5NS
G6 URM Brett Johnson, 6 Winston Avenue, Plymouth, PL4 6AZ
GM6 URP Ria Gray-Jones, Flat C, 7 Nelson Street, Aberdeen, AB24 5EP
G6 URR I Kirk, 12 Edinbane Close, Rise Park, Nottingham, NG5 5DU
G6 URT C Kapoutsis, 7a East Lane, Morton, Bourne, PE10 0NW
G6 USA Peter Love, 2 Meadway, Dover, CT17 0PS
G6 USD M Matthews, 213 Hucclecote Road, Gloucester, GL3 3TZ
G6 USG Gary Comer, 27 Peckforton View, Kidsgrove, Stoke-on-Trent, ST7 4TA
G6 USL B Cowell, 46 Gattison Lane, New Rossington, Doncaster, DN11 0NQ
G6 USO Paul Chamings, 52 Crown Street, Redbourn, St. Albans, AL3 7PF
G6 USR Mark Davis, Sunny Bank, Headcorn Road, Maidstone, ME17 2AN
G6 UST P Drury, 5 Bede Place, Peterborough, PE1 4EE
G6 USU T Derbyshire, 32 Hardie Avenue, Wirral, CH46 6BJ
G6 USX W Dennison, 41 Tarbert Walk, Stepney, London, E1 0EE
G6 USZ David Deverell, 23 Frankmarsh Park, Barnstaple, EX32 7HN
G6 UTF David Foster, 11 Dingle Road, Leeswood, Mold, CH7 4SN
G6 UTK G Fisher, 16 Somerset Lane, Lansdown, Bath, BA1 5SW
G6 UTL S Foulser, 32 Langhorn Road, Southampton, SO16 3TN
G6 UTT P Sheppard, Round Corners, 7 First Avenue, Bognor Regis, PO22 6ED
GI6 UUC J Thompson, 21 Watch Hill Road, Ballyclare, BT39 9QW
G6 UUQ L Sheward, 7 Harlington Avenue, Grove, Wantage, OX12 7NQ
G6 UUR Samuel Whitehead, 94 Cranmore Boulevard, Shirley, Solihull, B90 4RU
GI6 UUT William Page, 4 Glebe Manor, Hillsborough, BT26 6NS
G6 UVB C Wilson, Rustleigh, 7 Stagbury Close, Coulsdon, CR5 3PH
G6 UVN M Henman, 4 Lyne Walk, Hackleton, Northampton, NN7 2BW
G6 UVO C Heritage, 29 Hill Head, Glastonbury, BA6 8AW
G6 UVS P Hannington, 21 Little Gate, Westhoughton, Bolton, BL5 2SD
G6 UVU J Handy, 77 Abbeyfield Road, Wolverhampton, WV10 8TH
G6 UW CAMBRIDGE UWS c/o James Keeler, 67 Perne Avenue, Cambridge, CB1 3RY
GM6 UWF J Allan, 87 Needless Road, Perth, PH2 0LD
G6 UWI N Bradshaw, 26 Suffolk Gardens, South Shields, NE34 7JF
G6 UWK J Barden, 2 Pond Hall Cottages, Bradfield Road, Manningtree, CO11 2SP
G6 UWO D Bullock, 1 Selby Close, Beeston, Nottingham, NG9 6HS
G6 UWS M Byles, 108 Kingsway, Wellingborough, NN8 2EN
G6 UWW M Williams-Davies, Plas Penrhos, Llwyngwril, LL37 2QB
G6 UX Ian Truslove, 44 John Street, Hinckley, LE10 1UY
G6 UXE Robert Wheeler, 36 Kimbolton Crescent, Stevenage, SG2 8RJ
G6 UXF K Young, 8 Magnolia Close, Worcester, WR5 3SJ
G6 UXG A Webb, 35 Hill House Drive, Minster, Ramsgate, CT12 4BE
G6 UXK David Wookey, 3 Westland Close, Boscombe Down, Salisbury, SP4 7QS
G6 UXM S Vinnicombe, 8a Cross Road, Cholsey, Wallingford, OX10 9PE
G6 UXU C Stanley, 494 Blackburn Road, Darwen, BB3 0AJ
G6 UXW P Mundy, 25 Lonsdale Avenue, Cosham, Portsmouth, PO6 2PU
G6 UXX P Leese, 4 Harefield, Harlow, CM20 3EF
G6 UXY A Lightly, 8 Smithville Close, St. Briavels, Lydney, GL15 6TN
G6 UYJ Alan Page, 32 Aotea Terrace, Christchurch, New Zealand, 8022
G6 UYK Philip Russell, 1 Larch Grove, Kendal, LA9 6AU
G6 UYM David Richards, 25-27 Burnivale, Malmesbury, SN16 0BL
G6 UYN A Rumney, Church House Farm Cottage, Cheltenham, GL51 0TW
G6 UZA A Kotowicz, 47 Portree Drive, Rise Park, Nottingham, NG5 5DT
G6 UZG Phil Ashby, 26 Van Diemens Lane, Bath, BA1 5TW
G6 UZJ James Austen, 13 Coverdale, Whitwick, Coalville, LE67 5BP
G6 UZL Paul Bunn, 6 St. Benedicts Drive, Little Haywood, Stafford, ST18 0QH
G6 UZM Simon Byford, 21 Clarke Drive, Shaw, Swindon, SN5 5SH
G6 UZO M Brunsdon, 7 Oldberg Gardens, Brighton Hill, Basingstoke, RG22 4NP
G6 UZR A Brown, Badgers Way, Holton, Wincanton, BA9 8AL
G6 UZT D Brown, 29 School Lane, Upton-upon-Severn, Worcester, WR8 0LQ
G6 UZY M Owen, 49 Southdale Drive, Carlton, Nottingham, NG4 1DA
G6 VAA G Perks, 55 Andrew Road, Tipton, DY4 0AJ
G6 VAD P Purdy, 4 Hethersett Road, East Carleton, Norwich, NR14 8HX
G6 VAE T White, 117a Western Road, Southall, UB2 5HN
G6 VAL A Oughton, 176 South Lodge Drive, Southgate, London, N14 4XN
G6 VAR John Smith, 6 Hollams Road, Tewkesbury, GL20 5DG
G6 VAW Carol Soars, 118 Braddon Road, Loughborough, LE11 5YZ
G6 VAX R Saunders, 93 Oaks Avenue, Worcester Park, KT4 8XG
G6 VAZ D Thomas, 25 Lime Close, Mildenhall, Bury St. Edmunds, IP28 7PR
G6 VBA D Townend, 25 De Trafford Street, Huddersfield, HD4 5DR
G6 VBD J Savage, 2 Alvecote Cottages, Alvecote Lane, Tamworth, B79 0DJ
G6 VBE R Ransom, 1 Bilberry Road, Clifton, Shefford, SG17 5HB
G6 VBJ Peter Tasker, Chenar House, Mile Path, Woking, GU22 0JL
G6 VBK Desmond Hatton, 24 Langdale Road, Leyland, PR25 3AR
GW6 VBN Martin Hunt, 23 Swansea Road, Pontardawe, Swansea, SA8 4AL
G6 VBQ A Haddock, 1 Heron Way, St. Ives, PE27 6SS
GW6 VBR L Cowley, 3 Pleasant Villas, Pontarddulais, Swansea, SA4 8QF
G6 VCF Peter Brackstone, 3 Wentworth Close, Beverley, HU17 8XB
GI6 VCG J Brownlees, 8 Cairnbeg Park, Larne, BT40 1UB
GI6 VCL K Cunningham, 4 Garvaghy Road, Portglenone, Ballymena, BT44 8EF
G6 VCR H Eden, 142 Ringway, Thornton-Cleveleys, FY5 2NW
GM6 VCV William Ferguson, Benview Lodge, Benview, Alloa, FK10 3AP
G6 VCY Patricia Eley, 94 Raphael Close, Whoberley, Coventry, CV5 8LS
G6 VDA A Sutton, 3 Cornflower Close, Willand, Cullompton, EX15 2TT
G6 VDK P Lutas, 616 Queens Drive, Swindon, SN3 1AZ
G6 VDW R Olliver, 39 Nutshalling Avenue, Rownhams, Southampton, SO16 8AY
G6 VDX Iain Ogilvie, 8 Devonshire Road, Prenton, CH43 4UL
G6 VDY H Jeffery-Wright, 55 Burland Avenue, Wolverhampton, WV6 9JJ
G6 VED Robert Straughan, 1 Crossroads, Gilwern, Abergavenny, NP7 0DX

| Prefix | Suffix | Name and address |
|---|---|---|
| G6 | VEG | T Gray, 8 Holystone Grange, Holystone, Newcastle upon Tyne, NE27 0UX |
| G6 | VEI | D Pierce, 3 Druids Close, Gorsedd, Holywell, CH8 8QY |
| G6 | VEJ | Francis Stone, 51 The Glen, Yate, Bristol, BS37 5PJ |
| G6 | VEN | A Rose, 4 Llys Clwyd, Kinmel Bay, Rhyl, LL18 5EW |
| GW6 | VET | J Goodson, 22 Pant Gwyn, Bridgend, CF31 5BA |
| G6 | VEY | Ian Haver, 4 Campion Way, Bourne, PE10 0QE |
| G6 | VEZ | Geoffrey Helm, 31 Faringdon Avenue, South Shore, Helmsman Electronics Ltd, Blackpool, FY4 3QQ |
| G6 | VF | S Illman, 66 Frieth Road, Marlow, SL7 2QU |
| G6 | VFA | M Hine, Tall Trees, Lime Lane, Derby, DE21 4RF |
| G6 | VFB | W Hogan, 279 Halliwell Road, Bolton, BL1 3PE |
| G6 | VFC | David Hooton, 80 Portland Road, Rushden, NN10 0DJ |
| GW6 | VFH | R Jenkins, 29 Pemberton Street, Llanelli, SA15 2RB |
| G6 | VFI | Adrian Jones, 10 Woodside Gardens, Chineham, Basingstoke, RG24 8EU |
| G6 | VFO | J Stokes, 109 Hollyhedge Road, West Bromwich, B71 3BT |
| G6 | VGA | C Mccall, 18A Dovetree House, Little Park, Wadhurst, TN5 6DL |
| G6 | VGC | R Woolley, 82 Pennycroft Road, Uttoxeter, ST14 7ET |
| G6 | VGH | I Allen, 14 Bellridge Place, Wellesbourne, Warwick, [illegible] |
| G6 | VGN | K Ratcliffe, 173 Whinney Lane, New Ollerton, Newark, NG22 9TJ |
| G6 | VGO | M Barrett, 1 Walterstead Cottage, Ladykirk, Berwick-upon-Tweed, TD15 1XW |
| G6 | VGS | Martyn Bradbury, 55 Crowthorp Road, Northampton, NN3 5EY |
| G6 | VGT | David Bowlas, 38 Senneleys Park Road, Northfield, Birmingham, B31 1AL |
| G6 | VGV | I Craig, 1 Whitton Drive, Chester, CH2 1HF |
| G6 | VGZ | D Cheriton, 5 Cornwall Close, Warwick, CV34 5HX |
| GM6 | VHA | Michael Deverill, Flannan House, Aird Uig Timsgarry, Isle of Lewis, HS2 9JA |
| G6 | VHE | M Entwistle, 34 Webbs Court, Lyneham, Chippenham, SN15 4TR |
| G6 | VHG | Anthony Foster, 35 Gloucester Place, Peterlee, SR8 2HB |
| G6 | VIC | Victor Jones, Gwel y Mor, Porth y Felin Road, Holyhead, LL65 1BG |
| G6 | VIF | B Morris, 21 Loxley Gardens, Southdown, Bath, BA2 1HS |
| G6 | VIK | I King, 11 Cockhall Close, Litlington, Royston, SG8 0RB |
| G6 | VIN | J Walker, 44 Albany Road, Kilnhurst, Mexborough, S64 5UG |
| GD6 | VIO | M Wingrove, 46 Clifford Road, Wembley, HA0 1AE |
| G6 | VIQ | M Watson, Salt Pie Farm, Birdsedge, Huddersfield, HD8 8XP |
| G6 | VIY | A Wood, 12 Bishops Meadow, Sutton Coldfield, B75 5PQ |
| G6 | VJA | I Taylor, 97 George Street, Cleethorpes, DN35 8PL |
| G6 | VJC | John Taylor, 17 Aintree Way, Castle Security Systems And Networking, Dudley, DY1 2SL |
| G6 | VJK | Albert Maslin, 2 Clarks Cottages, White Horse Road, Colchester, CO7 6TX |
| G6 | VJM | D Lynch, 30 Whitecroft View, Baxenden, Accrington, BB5 2QP |
| G6 | VJP | David Pilkington, 45 High Meadows, Midsomer Norton, Radstock, BA3 2RZ |
| G6 | VJR | Derek Reading, 23 Elwy Circle, Ash Green, Coventry, CV7 9AU |
| G6 | VJU | Steven Wyles, 14 Drovers Way, Bracknell, RG12 9EY |
| G6 | VKA | C Thompson, Fourwinds, Walton Hill, Gloucester, GL19 4BT |
| G6 | VKI | C Richardson, Tany Y Bryn, Garndolbenmaen, LL51 9UQ |
| G6 | VKL | David Mayers, 15 Oakfield Road, Poynton, Stockport, SK12 1AR |
| G6 | VKP | J Littlewood, 5 Laburnum Grove, Harrogate, HG1 4EH |
| G6 | VKS | I Morgan, Leigh House, 64 Widney Road, Solihull, B93 9AW |
| G6 | VKX | Michael Webber, 23 Ramsey Close, Horley, RH6 8RE |
| GW6 | VKY | A White, 86 Derlwyn, Dunvant, Swansea, SA2 7QE |
| G6 | VLC | S Paxton, 11 Synderford Close, Didcot, OX11 7UT |
| G6 | VLK | Simon Thompson, 89 Henley Crescent, Solihull, B91 2JH |
| G6 | VLT | John Higgins, 190 Little Glen Road, Glen Parva, Leicester, LE2 9TT |
| G6 | VLV | D Colman, 22 Peerley Close, East Wittering, Chichester, PO20 8PB |
| GI6 | VLY | J Earle, 25 Carnesure Park, Comber, Newtownards, BT23 5LT |
| G6 | VMB | C Gibson, 103 Lydalls Road, Didcot, OX11 7DT |
| G6 | VMF | R Hope, 30 Greendale Gardens, Hetton-le-Hole, Houghton le Spring, DH5 0EF |
| G6 | VMI | D Milne, 22 Eastnor Road, Reigate, RH2 8NE |
| G6 | VMR | Martin Adams, 122 Green Lane, Castle Bromwich, Birmingham, B36 0BX |
| G6 | VMV | S Brocklehurst, Bank View, Reades Lane, Congleton, CW12 3LL |
| G6 | VNC | Robert Davies, 27 Smiths Way, Water Orton, Birmingham, B46 1TW |
| G6 | VNI | Gillian Duggan, 28 Higher Rads End, Eversholt, Milton Keynes, MK17 9ED |
| G6 | VNO | N Hanson, 100 Bassett Green Road, Southampton, SO16 3EF |
| G6 | VNW | Barry Major, 3 Tithebarn Grove, Wavertree, Liverpool, L15 6TG |
| G6 | VOE | D Simpkins, 34 Rose Avenue, Weldon, Corby, NN17 3HB |
| G6 | VOV | R Leavold, 8 Wilkinson Way, North Walsham, NR28 9BB |
| G6 | VPH | R Gorton, 3 Pickford Avenue, Little Lever, Bolton, BL3 1PN |
| G6 | VPJ | G Hall, 54 Townfields, Sandbach, CW11 4PQ |
| G6 | VPK | K Higbee, 5 Davoren Walk, Bury St. Edmunds, IP32 6QA |
| G6 | VPL | J Hopkinson, 4 Marwood Croft, Streetly, Sutton Coldfield, B74 3JU |
| G6 | VPN | B Jameson, 42 Eastgate, Fleet, Spalding, PE12 8NA |
| G6 | VPU | J Mulligan, [illegible] St. Helens Road, Leigh, WN7 [illegible] |
| G6 | VPV | J Wake, 15 Deepdale Way, Darlington, DL1 2TA |
| G6 | VPW | R Stoate, 19 Jean Road, Brislington, Bristol, BS4 4JT |
| G6 | VQC | Anthony Read, Huenibachstrasse 75, Huenibach, Switzerland, CH-3626 |
| G6 | VQN | Andrew Morris, 67 Broad Oak Way, Cheltenham, GL51 3LL |
| G6 | VQV | S Shenfield, 3 Blackberry Grove, Bradwell-on-Sea, Southminster, CM0 7QE |
| G6 | VQW | R Seaton, Wisteria Cottage, Welsh Road, Leamington Spa, CV33 9AQ |
| GM6 | VRC | James Brown, Inchbcag Cottage, Inchcoonans, Perth, PH2 7RB |
| G6 | VRF | B Crowther, 10 Askrigg Close, Marton Moss, Blackpool, FY4 5RE |
| G6 | VRI | Gerald Eccleshare, 22 Barley Close, Herne Bay, CT6 7XG |
| GW6 | VRN | Maldwyn Jones, 66 Brondog, Heolgerrig, Merthyr Tydfil, CF48 1TP |
| G6 | VRU | G Giles, 42 Owls Road, Verwood, BH31 6HJ |
| G6 | VSE | A Bansal, Fernley, 2 Seaview Cotts, Chideock, DT6 6JE |
| G6 | VSG | D Harris, 9 Garden City, Tern Hill, Market Drayton, TF9 3QB |
| G6 | VSQ | Frank Whitehurst, Roselands, Clarke Lane, Macclesfield, SK10 5AH |
| G6 | VSY | C Sheppard, 42 Freeman Road, Didcot, OX11 7DD |
| G6 | VTA | Melanie Fisher, 46 Hedgerow Walk, Andover, SP11 6FD |
| G6 | VTE | S Chambers, 1 Tatling Grove, Walnut Tree, Milton Keynes, MK7 7EG |
| G6 | VTH | Robert Carney, 29 Hayton Close, Sunderland, SR5 2BU |
| G6 | VTN | P Green, 79 The Spinney, Bar Hill, Cambridge, CB23 8SU |
| G6 | VTR | Andrew Holmes, 11 Deerness Road, Bishop Auckland, DL14 6UB |
| G6 | VTX | M Brindley, 53b Seabridge Road, Newcastle, ST5 2HU |
| GW6 | VTZ | Donald Campbell, [illegible] |
| G6 | VUL | D Dutton, [illegible] |
| G6 | VUF | F Caulfield Kerney, 47 Freemans Close, Stoke Poges, Slough, SL2 4ER |
| G6 | VUG | R Collins, 37 Warwick Road, Twickenham, TW2 6SW |
| G6 | VUJ | J Davis, 69 Bryanston Road, Solihull, B91 1BS |
| G6 | VUN | M Hodson, 17 Marshfield Close, Redditch, B98 8RW |
| G6 | VUX | S Challoner, Grosvenor Farm, Holme Street, Chester, CH3 8EQ |
| G6 | VVE | Sally Banks, 29 Froxmere Close, Crowle, Worcester, WR7 4AP |
| G6 | VVG | G Caldwell, 10 Craigmath, Dalbeattie, DG5 4EB |
| G6 | VVL | K Hotchen, 6 Nourse Close, Leckhampton, Cheltenham, GL53 0NQ |
| G6 | VVS | H Jackson, 46 Ashford Road, Maidstone, ME14 5DH |
| GM6 | VVX | N Dunnachie, 12 Blackhill View, Law, Carluke, ML8 5JZ |
| G6 | VVZ | T Butler, 103 Spring Gardens, Anlaby Common, Hull, HU4 7QH |
| G6 | VWF | J Holbrook, 1 Segrave Grove, Hull, HU5 5DJ |
| G6 | VWI | C Kowcun, 27 Mill Crescent, Kingsbury, Tamworth, B78 2LX |
| GI6 | VWS | J Quigg, 9 Springhill Terrace, Limavady, BT49 9BS |
| [illegible] | [illegible] | [illegible], Nottingham, NG6 8PU |
| G6 | VXE | D Hilton, 35 Sandringham Road, Northolt, UB5 5HN |
| G6 | VXL | T Buck, 170 Rover Drive, Castle Bromwich, Birmingham, B36 [illegible] |
| G6 | VXR | H Metcalf, Beech Lee, Vicarage Lane, Alresford, SO24 0DU |
| G6 | VXZ | A Sorab, Woodgaston Cottage, Woodgaston Lane, Hayling Island, PO11 0RL |
| G6 | VYC | A Handley, 4 Southwood Drive, Thorne, Doncaster, DN8 5QS |
| G6 | VYK | E Williams, 20 Borough Road, Bridlington, YO16 4HL |
| GM6 | VYY | Albert McMinn, Siarardh, Mallaig, PH41 4QY |
| GM6 | VYZ | W McMinn, Glengyle, East Bay, Mallaig, PH41 4QF |
| G6 | VZB | M Lennox, 17 Coed y Fron, Holywell, CH8 7UJ |
| G6 | VZF | Antony Dawes, 1a Lower Olland Street, Bungay, NR35 1BY |
| G6 | VZG | M Frosdick, 48 Woodfield, Briston, Melton Constable, NR24 2JY |
| G6 | VZM | John Johnson, 62 Julien Road, Ealing, London, W5 4XA |
| G6 | VZS | D Goodall, 94 Camp Mount, Pontefract, WF8 4BX |
| G6 | VZU | C Hunt, 23 Beccles Road, Gorleston, Great Yarmouth, NR31 0PW |
| G6 | VZZ | J Hackett, 18 Brow Edge, Rossendale, BB4 7TT |
| G6 | WAG | D Jones, Bradford House, The Square, Corwen, LL21 0DL |
| G6 | WAN | R Rayner, 12 Weedon Way, King's Lynn, PE30 4YY |
| G6 | WAO | N Austin, 6 Riches Close, Tasburgh, Norwich, NR15 1NX |
| G6 | WAS | D Carpenter, 2 Milton Road, Little Irchester, Wellingborough, NN8 2DY |
| G6 | WAU | Sean Connor, 1 Tallis Walk, Grange Park, Swindon, SN5 6BQ |
| G6 | WAY | J Randall, 3 Steins Lane, Humberstone, Leicester, LE5 1ED |
| G6 | WAZ | Andrew Ronnie, 7 Beechwood Avenue, Stranraer, DG9 0AU |
| G6 | WBG | P Smith, Juniper Cottage, Palestine, SP11 7ER |
| G6 | WBT | I Thorp, Pinelodge, Carleton Green, Pontefract, WF8 3NJ |
| G6 | WBX | Patrick Yorke, 27 Luard Court, Havant, PO9 2TN |
| G6 | WCI | M Richards, 72 Carlton Avenue, Westcliff-on-Sea, SS0 0QL |
| G6 | WCX | Desmond Mardle, 22 Wayfield Link, Avery Hill, London, SE9 2LP |
| G6 | WDC | L Baldwin, 26a Cheney Hill, Heacham, King's Lynn, PE31 7BS |
| G6 | WDR | A Tett, 19 Park Road, Shoreham-by-Sea, BN43 6PF |
| G6 | WDS | F Deravi, Jennison Building, Canterbury, CT2 7NT |
| G6 | WEH | R Burrows, 6 Frensham Drive, Hitchin, SG4 0QP |
| G6 | WEI | G Cockcroft, 31 Holt Road, Kintbury, Hungerford, RG17 9UY |
| G6 | WEL | J Reynolds, 4 Rosewood Drive, Winsford, CW7 2UW |
| GW6 | WEU | K Turner, 115 Newton Road, Newton, Swansea, SA3 4SW |
| G6 | WEW | J Fitzsimons, 63 School Lane, Chapel House, Skelmersdale, WN8 8EN |
| G6 | WFF | G Solkow, 12a Manor Court, Penkhull, Staffs, ST4 5DW |
| G6 | WFM | K Farr, 3 Sheppard Drive, Chelmsford, CM2 6QE |
| G6 | WFS | G Quantrill, 47 Lambeth Road, Leigh-on-Sea, SS9 5XR |
| G6 | WFW | A Humphreys, 45 Cwm Place, Llandudno, LL30 1LP |
| GI6 | WFX | R Johnston, 51 Kennedy Drive, Lisburn, BT27 4JA |
| G6 | WGA | A Swift, 56 Birch Hall Avenue, Darwen, BB3 0JB |
| G6 | WGE | N Riding, 15 Church Lane, Dewsbury Moor, Dewsbury, WF13 4EN |
| G6 | WGM | Malcolm Reilly, Flat 5, 57 Cheriton Road, Folkestone, CT20 1DF |
| G6 | WGY | Raymond Clague, 11 Trebor Avenue, Bryntirion Park, Bagillt, CH6 6DP |
| G6 | WGZ | David Collier, 133 Woodstock Road, Moston, Manchester, M40 0DG |
| G6 | WHH | Stephen Martin, 19 Old Manor Road, Rustington, Littlehampton, BN16 3QU |
| G6 | WHS | N Read, 296 Westdale Lane, Mapperley, Nottingham, NG3 6EU |
| G6 | WHT | Keith Willard, 5 Waltham Way, Frinton-on-Sea, CO13 9JE |
| G6 | WHY | K Daniels, Greenacre, 71 Little Yeldham Road, Halstead, CO9 4LN |
| GI6 | WHZ | R Freeburn, 6 Killycurragh Road, Cookstown, BT80 9LB |
| G6 | WIG | G Crowton, 64 Atlantic Road, Birmingham, B44 8LQ |
| G6 | WIL | Gabriel Wilden, 39 The Laurels, Morris Avenue, Jaywick, CO15 2JN |
| G6 | WIO | B Mchugh, 63 Three Butt Lane, Liverpool, L12 7HE |
| G6 | WIT | G Anderton, 12 Oaklands Close, Halvergate, Norwich, NR13 3PP |
| G6 | WJD | John Dobson, 13 Elgin Close, Bedlington, NE22 5HJ |
| G6 | WJJ | A Kendal, LE TABUTAIS, LA ROCHE CHEVREUX, Prissac, France, 36370 |
| G6 | WJM | I Smith, Freshwinds, 4 Lakeside Estate, Rhosneigr, LL64 5JW |
| G6 | WJW | H Hutton, Cassiobury, The Street, Diss, IP22 2PS |
| G6 | WJX | E Jackson, Melford, 36 Ickleton Road, Cambridge, CB22 4RT |
| G6 | WKI | R Lewis, 42 Launceston Close, Romford, RM3 8HQ |
| G6 | WKN | Chris Reed, 14 Fletcher Drive, Wickford, SS12 9FA |
| G6 | WKO | J Richards, 8 Westminster Crescent, Burn Bridge, Harrogate, HG3 1LY |
| G6 | WKQ | Philip Rowe, 131 Cambridge Road, Great Shelford, Cambridge, CB22 5JJ |
| G6 | WKU | W Walker, 18 Parc Sychnant, Conwy, LL32 8SB |
| G6 | WKZ | Matthew Jaques, 3 The Rowans, Baldock, SG7 6HJ |
| G6 | WLA | K Jarratt, Belvoir, Rose, Truro, TR4 9PF |
| G6 | WLE | R Bailey, The Malt House, Great Shefford, Hungerford, RG17 7ED |
| GM6 | WLJ | D Milne, 30 Bruceland Road, Elgin, IV30 1SF |
| G6 | WLM | S Simmonds, 3 Robert Cramb Avenue, Tile Hill, Coventry, CV4 9LA |
| G6 | WLP | Gordon Smith, High Croft, 91 Rannerdale Drive, Whitehaven, CA28 6JZ |
| G6 | WLQ | M Smith, 10 Riffams Court, Riffams Drive, Basildon, SS13 1BQ |
| G6 | WLX | A Davey, Highdale, 82 Silver Street, Nailsea, BS48 2DS |
| GM6 | WMA | D Elam, Achnacree, 38 Hunter Avenue, Loanhead, EH20 9SN |
| G6 | WME | B Gray, 33 Long Barrow Drive, North Walsham, NR28 9YA |
| G6 | WMG | D Hastings, Westering, Norwich, NR13 6RQ |
| G6 | WML | J Barrasford, 34 Barnard Avenue, Ludworth, Durham, DH6 1LS |
| G6 | WMR | WEST MIDLANDS COUNTY RAYNET c/o John Barnett, 11 Ridge Street, Stourbridge, DY8 4QF |
| G6 | WMT | [illegible], TD6 9TY |
| G6 | WMU | D Pearce, 247 Wigston Lane, Aylestone, Leicester, LE2 8DJ |
| GJ6 | WMZ | M L'Amy, Tamarind, Le Mont de St. Anastase, St. Peter, Jersey, JE3 7ES |
| G6 | WNB | G Bennett, 6 Danescroft, Bridlington, YO16 7PZ |
| G6 | WNG | J Haines, The Westlands, Wilcott, Shrewsbury, SY4 1BJ |
| G6 | WNX | D Mitchell, 65 Robb Place, Castle Douglas, DG7 1LW |
| GW6 | WOB | H Stevens, Parc Y Dilfa, Talley, Llandeilo, SA19 7YT |
| GM6 | WOF | A Firth, Edan, Berstane Road, Kirkwall, KW15 1NA |
| G6 | WOI | C Flint, 782 College Road, Birmingham, B44 0AL |
| G6 | WOT | David Fishlock, 93 Shackstead Lane, Godalming, GU7 1RL |
| G6 | WPE | S Mason, 46 Frankton Close, Redditch, B98 0HJ |
| G6 | WPJ | Matthew Phillips, Woodside, Bures, CO8 5BN |
| G6 | WPK | John Puttock, 8 Millfield, St. Margarets-at-Cliffe, Dover, CT15 6JL |
| G6 | WPL | S Lawson, 33 Country Meadows, Market Drayton, TF9 3LP |
| G6 | WPO | A Brislin, Greengage, Plough Road, Droitwich, WR9 7NL |
| G6 | WPR | D Fleetwood, 31a Upper Highway, Hunton Bridge, Kings Langley, WD4 8PP |
| G6 | WQH | J Wilson, 15 Hampstead Court, Hull, HU3 1UP |
| G6 | WQJ | A Tidswell, [illegible] Avenue, Holywell, CH8 7UG |
| G6 | WQN | W Convery, 20 Grove Road, Hethersett, Norwich, NR9 3JP |
| G6 | WRB | Mark Ashby, 15 Snowford Close, Luton, LU3 3XU |
| G6 | WRC | Warrington ARC c/o Paul Middlehurst, 7 Statham Drive, Lymm, WA13 9NW |
| G6 | WRY | G Smith, 41 Glebe Place, Galashiels, TD1 3JW |
| G6 | WSF | M Strickland, 25 Coniston Drive, Aylesham, Canterbury, CT3 3HZ |
| G6 | WSN | David Westgate, 72 Bosworth Street, Leicester, LE3 5RA |
| G6 | WSX | W Carter, 49 The Oval, Holmfirth, HD9 3ET |
| G6 | WSZ | J O'Hara, 4 Lower Mill Close, Goldthorpe, Rotherham, S63 9BY |
| G6 | WTD | R Kenward, The Bungalow, 20 Church Road, Coventry, CV8 3ET |
| GM6 | WTH | T Callaghan, 18 Kingswell Avenue, Onthank, Kilmarnock, KA3 2EZ |
| GW6 | WTK | B Wiegold, 16 Hafan Werdd, Caerphilly, CF83 3BU |
| G6 | WTM | M Higlett, 3 Clover Way, Killinghall, Harrogate, HG3 2WE |
| GM6 | WTP | Nicholas Sanders, 8 Danube Street, Edinburgh, EH4 1NT |
| GM6 | WTT | David Brian Anderson, Hallmoss Farm, Inverugie, Peterhead, AB42 3BP |
| G6 | WUD | Raymond Green, 10 Torwood Court, Cramlington, NE23 2BZ |
| G6 | WUR | T Price, 54 Medeway, Lake, Sandown, PO36 9HQ |
| G6 | WVD | J Williams, 48 Belvedere Drive, Plas Coch, Wrexham, LL11 2BG |
| G6 | WVL | J Parr, 114 Ashton Road, Golborne, Warrington, WA3 3UX |
| G6 | WVM | D Harrison, 22 Oswin Grove, Coventry, CV2 5GJ |
| G6 | WVO | Christopher Hunt, Greenfield House, Heapham, Gainsborough, DN21 5PT |
| G6 | WVR | Stephen Worner, 1 Tynedale, Hull, HU7 6EL |
| G6 | WVS | P Child, 36 Crosslands, Caddington, Luton, LU1 4ER |
| G6 | WWA | T Banham, 28 Norwood Avenue, High Lane, Stockport, SK6 8BJ |
| G6 | WWM | J Slade, 10 Larch Close, Weaverham, Northwich, CW8 3ED |
| G6 | WWR | THREE COUNTIES ARC c/o D Kamm, Delabole Head, Week St. Mary, Holsworthy, EX22 6UU |
| G6 | WWS | B Smith, 17 Thornley Road, Wirral, CH46 6HB |
| G6 | WWV | P Mann, 9 Holcombe Road, Blackpool, FY2 0SR |
| G6 | WWY | George Miller, Silvermine, Cooks Lane, Axminster, EX13 5SQ |
| G6 | WXI | G Ball, Ciss Green Farm, Watery Lane, Congleton, CW12 4RS |
| G6 | WXJ | L Bagnall, 15 Ypres Road, Allestree, Derby, DE22 2NA |
| G6 | WXK | Ian Buckie, 156 Greenfield Crescent, Horndean, Waterlooville, PO8 9EW |
| G6 | WXM | Anthony Burt, 17 Western Road, Wolverton, Milton Keynes, MK12 5AY |
| G6 | WXN | C Bennett, 12 Sherwood Road, Winnersh, Wokingham, RG41 5NJ |
| G6 | WXS | R Archer, 37 Caroline Street, Preston, PR1 5UY |
| G6 | WXZ | A Collier, 2 Viceroy Court, Gordon Road, Stanford-le-Hope, SS17 8NL |
| G6 | WYD | S Chambers, 52 Chapel Lane, Spondon, Derby, DE21 7JW |
| G6 | WYE | A Clack, 2 St. Christophers Close, Cranwell Village, Sleaford, NG34 8XB |
| G6 | WYF | R Cook, Arnel Ltd, Arnel House, 1 Peerglow Centre, Ware, SG12 9QL |
| G6 | WYH | N Daniels, 2 Homelye Lane, Dunmow, CM6 3AW |
| G6 | WYL | J Williamson, 5 Frensham Close, Stanway, Colchester, CO3 0HP |
| G6 | WYQ | S Quade, 60 Carlton Mews, Birmingham, B36 0AD |
| G6 | WYS | W Patching, 7 Bursledon Road, Hedge End, Southampton, SO30 0BP |
| G6 | WZA | D Wickens, Auchensail, 3 Bews Lane, Chard, TA20 1JU |
| G6 | WZC | D Slatter, 5 Opendale Road, Burnham, Slough, SL1 7LY |
| G6 | WZD | C Sillence, 104 Coleford Bridge Road, Mytchett, Camberley, GU16 6DT |
| G6 | WZE | P Robinson, 108 Station Road, Mickleover, Derby, DE3 9FP |
| G6 | WZL | Barry Walker, 81 Stacey Avenue, Wolverton, Milton Keynes, MK12 5DN |
| G6 | WZM | H Collinson, 28 Tadcaster Avenue, Leicester, LE2 9GA |
| G6 | WZN | M Hodges, 2 Coral Avenue, Westward Ho, Bideford, EX39 1UW |
| G6 | WZP | D Rogers, 39 Fore Street, Seaton, EX12 2AD |
| G6 | WZY | A Gerrard, 51 Sheringham Drive, Crewe, CW1 3XJ |
| G6 | WZZ | B Gibson, 55 Ledward Street, Winsford, CW7 3EN |
| G6 | XAG | A Higgs, Neatsfold, Hilton, Blandford Forum, DT11 0DQ |
| G6 | XAK | C Harding, 15 The Stampers, Tovil, Maidstone, ME15 6FF |
| G6 | XAN | Stephen Harding, 20 Wey Barton, Byfleet, West Byfleet, KT14 7EF |
| G6 | XAR | L Hall, 170 Macers Lane, Wormley, Broxbourne, EN10 6EE |
| G6 | XAT | D Armstrong, 69 Station Crescent, Rayleigh, SS6 8AR |
| G6 | XAV | G Lawrence, 5 Longwood View, Furnace Green, Crawley, RH10 6PB |
| G6 | XAW | D Lawrence, 3826 Se 1st Place, Cape Coral, United States, FL33 904 |
| G6 | XBD | Sidney Meakin, 25 Derby Road, London, E18 2PZ |
| G6 | XBG | J Lines, 6 Hawthorn Road, Denmead, Waterlooville, PO7 6LJ |
| G6 | XBS | John Newman, 21 Stains Close, Cheshunt, Waltham Cross, EN8 9JJ |
| G6 | XBV | Kenneth Simpson, 6 New Market Street, Usk, NP15 1AT |
| G6 | XCC | J Sayer, 19 Arras Boulevard, Hampton Magna, Warwick, CV35 8TY |
| G6 | XCD | E Ashworth, 232 Clifton Road, Darlington, DL1 5EA |
| G6 | XCK | S Bishop, 1 Walsh Close, Hitchin, SG5 2HP |
| GU6 | XCM | Mason Paul, La Tourelle, La Houte Des Blanches, St. Martin, Guernsey, GY4 6AF |
| G6 | XCO | R Piper, 3 The Haven, Langley Park, Durham, DH7 9UW |
| G6 | XCU | R Willis, 10 Nayling Road, Braintree, CM7 2RZ |
| G6 | XCV | K Williams, 23 Finchdean Road, Rowland's Castle, PO9 6DA |
| G6 | XD | John Taylor, 14 Woodway Close, Teignmouth, TQ14 8QG |
| G6 | XDB | J Woodnutt, 17 Hill Farm Road, Chalfont St. Peter, Gerrards Cross, SL9 0DD |
| G6 | XDG | J Page, 18 Winifred Road, Dagenham, RM8 1PP |

UK Callsigns

**IMPORTANT NOTE**

**Revalidate licence to avoid revocation** – Ofcom has advised the Society that plans will be drawn up to revoke licences that have not been revalidated as required by the licence conditions. The quickest way to revalidate is to do so online via the Ofcom website: *https://services.ofcom.org.uk/* or by email: *amateur.validations@ofcom.org.uk* Ofcom staff are available to help, but please be patient during times of heavy workload.

G6 XDI C Packman, 4 Angel Lane, Hayes, UB3 2QX
G6 XDK Valerie Oag, Parkside, Stratton Park, Biggleswade, SG18 8QS
G6 XDN D Lindop, 44 Young Road, London, E16 3RR
G6 XDY K Gibson-Ford, 123 Hawthorn Crescent, Cosham, Portsmouth, PO6 2TJ
G6 XDZ N Glover, 21a Jason Close, Bridlington, YO16 6JA
G6 XEB D Green, 47 Siston Common, Bristol, BS15 4PA
G6 XEF P Hammond, 31 Honey Way, Royston, SG8 7ES
G6 XEL D Hawkins, Travellers Lodge, Bere Road, Wareham, BH20 7PA
G6 XEN R Hill, 114 Moorside Crescent, Sinfin, Derby, DE24 9PT
G6 XEX A Croft, Exchange Buildings, Exchange Street, Normanton, WF6 2AA
G6 XFB B Roe, Po Box 836, Lincoln, LN4 4WR
G6 XFR F Fielder, 103 Acworth Court, Acworth Crescent, Luton, LU4 9JE
G6 XFU Anthony Edge, 1 Newquay Drive, Macclesfield, SK10 3NQ
GW6 XGA David Collins, 12 Penybedd, Pembrey, Burry Port, SA16 0HJ
G6 XGF D Cadman, 32 Breedon Hill Road, Derby, DE23 6TG
G6 XGJ J Davis, 446 Upper Wortley Road, Scholes, Rotherham, S61 2SS
G6 XGK M Drinkall, 11 Rossefield Gardens, Bramley, Leeds, LS13 3RQ
G6 XGT M Thornsby, 2 Shelley Way, Bacton, Stowmarket, IP14 4TP
G6 XGV M Valenti, 545 Gander Green Lane, North Cheam, Sutton, SM3 9RF
G6 XHF Simon Richards, 58 Holm Lane, Oxton, Prenton, CH43 2HS
GD6 XHG Ed Rixon, 65 Friary Park, Ballabeg, Ballabeg, Isle of Man, IM9 4EP
G6 XHI K Ridgwell, 4 Prykes Drive, Chelmsford, CM1 1TP
G6 XHJ Paul Raxworthy, 32 St. Marys Avenue, Alverstoke, Gosport, PO12 2HX
G6 XHK Ian Roper, 109 BIRSTALL PK CT, Birstall, WF17 9DL
G6 XID Stephen Mann, Station House, 1 Station Road, Bristol, BS49 4AJ
G6 XIF C Milton, 31 Morley Road, Tiptree, Colchester, CO5 0AA
G6 XII J Miller, The Oast House, Houghton Green Lane, Rye, TN31 7PJ
G6 XIR M Bennett, Ravenswood, The Shires, Southampton, SO3 4BA
G6 XJB D Wratten, 42 North Road, Petersfield, GU32 2AX
G6 XJC Leslie Whitehead, Flat 2, Masons Court, Clacton-on-Sea, CO15 3SE
G6 XJD D Whysall, Christ Church Vicarage, 587 Nuthall Road, Nottingham, NG8 6AD
G6 XJE J Whysall, Christ Church Vicarage, 587 Nuthall Road, Nottingham, NG8 6AD
G6 XJF A Webb, 255 Bambury Street, Stoke-on-Trent, ST3 5QY
G6 XJI D Wiblin, 98 Pemberton Road, Slough, SL2 2JY
G6 XJJ S McKay, 11 Brough Meadows, Catterick, Richmond, DL10 7LQ
G6 XJN G Valenti, 31 Stratton Court, Bognor Regis, PO22 8DP
G6 XJT David Ramsden, 76 Brigg Lane, Camblesforth, Selby, YO8 8HD
G6 XKE H Papworth, 339 Gayfield Avenue, Brierley Hill, DY5 3JE
G6 XKF A Parfitt, 242 Hook Road, Chessington, KT9 1PL
G6 XKJ I Pinkard, 10 Westminster Green, Handbridge, Chester, CH4 7LE
G6 XKK Hazel Parrott, 3 Fox Gardens, Lymm, WA13 9EY
G6 XKO Robert McLellan, 74 Mount Ambrose, Redruth, TR15 1QR
G6 XKV D Bodenham, 75 Rosedale Avenue, Stonehouse, GL10 2QH
G6 XKX R Newell, 57 Evendene Road, Evesham, WR11 2QA
G6 XKY G Ogden, 10 Hartington Drive, Standish, Wigan, WN6 0UA
G6 XLB R Morris, 10 Danetre Drive, Daventry, NN11 4GY
G6 XLC J Mills, 6 Borrowdale Road, Halfway, Sheffield, S20 4HL
G6 XLG P Pulley, 7 St. Peters Close, Pirton, Worcester, WR8 9EH
G6 XLL Laurence Segal, Flat 1, Masons House, London, NW9 9NG
G6 XLR Stephen Nightingale, Cefn Ydfa, Bartwood Lane, Ross-on-Wye, HR9 5TA
G6 XMA Susan Butler, 45 Roewood Close, Holbury, Southampton, SO45 2JT
G6 XMB T Betts, 3 Burns Avenue, Mansfield Woodhouse, Mansfield, NG19 9JR
G6 XML W Barnes, 49 Sunningdale Road, Haydon Wick, Swindon, SN25 3AZ
G6 XMM T Bugg, Gravel Hill, Nayland, Colchester, CO6 4BJ
G6 XMT Miriam Samson, 115a Far Gosford Street, Coventry, CV1 5EA
G6 XMU G Smith, 71 Mount Pleasant Road, Wisbech, PE13 3NQ
G6 XN Wey Valley Amateur Radio Group c/o Andrew Vine, Hilden, Woodland Avenue, Cranleigh, GU6 7HZ
G6 XND P Smith, 6 Nuthatch, Longfield, DA3 7NS
G6 XNI A Taylor, 20 Mythop Road, Marton, Blackpool, FY4 4UZ
G6 XNJ John Taylor, 21 Greystone Avenue, Elland, HX5 0QH
G6 XNK J Theedom, 5 Rodbridge Drive, Southend-on-Sea, SS1 3DF
G6 XNN Eric Townsend, 10 Little Oak Avenue, Kirkby-in-Ashfield, Nottingham, NG17 9BG
G6 XNP A Trett, 236 Avondale, Ash Vale, Aldershot, GU12 5NQ
G6 XNQ Robert Taylor, 53 Hutton Park, Hutton Moor Lane, Weston-Super-Mare, BS24 8RZ
G6 XNU V Williams, 24 Sunny Bank Avenue, Blackpool, FY2 9EQ
G6 XOD Edward Whitby, 1 Gloucester Avenue, Beeston, Nottingham, NG9 1HE
G6 XOE F Whitby, 1 Gloucester Avenue, Beeston, Nottingham, NG9 1HE
G6 XOG C Wells, Troutbeck, Arthington Lane, Otley, LS21 1JZ
G6 XOR David Winfield, 1 Underhill Close, Derby, DE23 1RH
G6 XOU Herbert Yeldham, 19 Wade Reach, Walton on The Naze, CO14 8RG
G6 XOX A Patrick, 22 Falcon Way, Dinnington, Sheffield, S25 2NY
G6 XPB R Partner, 22 Moordale Avenue, Priestwood, Bracknell, RG42 1RT
G6 XPF Graham Love, 8 Scotts Way, Tunbridge Wells, TN2 5RG
G6 XPO Anthony Parsons, Tops House, Llansilin, Oswestry, SY10 7QB
G6 XPY R Chappell, 17 Redcar Avenue, Hereford, HR4 9TJ
G6 XPZ A Carter, 28 Smithwell Lane, Heptonstall, Hebden Bridge, HX7 7NX
G6 XQB R Carter, 56 Main Road Naphill, High Wycombe, HP14 4QB
G6 XQO P Gait, 6 Martindale Road, Churchdown, Gloucester, GL3 2DW
G6 XQP G Garner, Tredore, Haugh Road, Norwich, NR16 2DE
G6 XQR F Gizzi, 19 Kings Field, Bursledon, Southampton, SO31 8EN
G6 XQT N Godwin, 9 Broadway, Barnsley, S70 6QQ
G6 XQX S Gray, 21 Sixth Avenue, Flint, CH6 5ND
G6 XQY J Griffin, 6 Heathfield, Royston, SG8 5BW
G6 XRE B Helsdon, 23 Kintore Drive, Great Sankey, Warrington, WA5 3NW
G6 XRF John Hicks, Flat 213, Enterprise House, 112 Kings Head Hill, London, E4 7ND
G6 XRH James Hoare, 8 Sunnyheath, Havant, PO9 3BW
G6 XRI S Hobbs, 19 Ashfield Road, Kenilworth, CV8 2BE
G6 XRK M Huggins, Black Firs, Pinewood Road, Iver, SL0 0NJ
G6 XRL J Hunt, 11 Vicarage Lane, Poynton, Stockport, SK12 1BG
G6 XRS OBO LEICESTER RS c/o Roger Talbott, 33 Highfield Street, Anstey, Leicester, LE7 7DU
G6 XRY G Kobiela, 61 Earith Road, Willingham, Cambridge, CB24 5LS
G6 XSB M Dower, 19 Fullwell Court, Fullwell Avenue, Ilford, IG5 0RZ
G6 XSC I Denison, 5 Hazelwood Close, Cheltenham, GL51 5RX
G6 XSK E Firth, 2 Gladstone Close, Littlemoor, Weymouth, DT3 6RH
G6 XSL Charlie Franklin, Troy Cottage, Hyde Heath, Amersham, HP6 5RW
G6 XSS B Gell, 27 Park Road, Barnstone, Nottingham, NG13 9JF
G6 XSY J Goodey, 62 Rose Hill, Binfield, Bracknell, RG42 5LG
G6 XSZ D Graham, 127 Shephall View, Stevenage, SG1 1RP
G6 XTC A Tripp, 3 Ash Close, Oathills, Malpas, SY14 8JB
G6 XTD R Hallsworth, 27 Westfield Avenue, Heanor, DE75 7BN
G6 XTG B Haynes, 6 Epping Walk, Furnace Green, Crawley, RH10 6LX
G6 XTJ Raymond Harris, 88 Earles Meadow, Horsham, RH12 4HR
G6 XTK Douglas Harris, Claws Cottage, Crablands, Chichester, PO20 9AY
G6 XTT R Holgate, 5 Exley Gardens, Halifax, HX3 9EE
G6 XTZ T Jarvis, 1 Whitehall Avenue, Mirfield, WF14 0AQ
GD6 XUD S Justin, Garth, Park View Road, Pinner, HA5 3YF
G6 XUJ G Collins, Little Orchard, Hemp Lane, Tring, HP23 6HF
G6 XUV Dave Lee, 188 Manstone Avenue, Sidmouth, EX10 9TJ
G6 XUX R Mettam, 12 School Lane, Marsh Lane, Sheffield, S21 5RS
G6 XVQ P Braybrooke, 6 Tubbenden Lane, Orpington, BR6 9PN
G6 XVY T Barker, 2 The Beeches, Chapel Lane, Morecambe, LA3 3HU
G6 XVZ A Barker, 33 Willoughby Avenue, Kenilworth, CV8 1DG
GM6 XW A Winton, 2 Castlehill Cottages, Brisbane Glen Road, Largs, KA30 8SN
G6 XWD C Breckons, Low Wood Farm, Lamonby, Penrith, CA11 9SS
G6 XWK S Branton, 30 Warren Lane, Martlesham Heath, Ipswich, IP5 3SH
G6 XWM R King, 52 Ford Road, Tiverton, EX16 4BE
G6 XWY D Clarke, 10 Dorchester End, Colchester, CO2 8AR
G6 XWZ T Cooper, Tamarisk, Exbury Road, Southampton, SO45 1XD
G6 XXB D Cook, Stepping Stones, 31 Vicarage Hill, Paignton, TQ3 1NH
G6 XXE S Crowther, 17 Carr Gate Crescent, Carr Gate, Wakefield, WF2 0QR
G6 XXJ David Clubley, 37 Appleton Road, Beeston, Nottingham, NG9 1NE
G6 XXL Greg Carter, Rivendell, North Reston, Louth, LN11 8JD
G6 XXN Anthony Clarke, 138 High Street, Barwell, Leicester, LE9 8DR
G6 XXQ B Dodds, 21 Lynton Drive, Lords Wood, Chatham, ME5 8QA
G6 XXY K Dobson, 152 Foryd Road, Kinmel Bay, Rhyl, LL18 5LS
G6 XYD K Elsworth, 88 Mungo Park Way, Orpington, BR5 4EQ
G6 XYF R Ediss, 5 Stirling Crescent, Totton, Southampton, SO40 3BN
G6 XYL Jean Luxton, 2 Trinity Court, Westward Ho, Bideford, EX39 1LT
G6 XYO J Fazey, 90 Beecher Road, Halesowen, B63 2DW
G6 XYR J Scothern, 24 Cavendish Crescent, Kirkby-in-Ashfield, Nottingham, NG17 9BN
G6 XYS A Searle, 22 Crowther Close, Southampton, SO19 1BX
G6 XYU J Stanton, Waters & Stanton Plc, 22 Main Road, Hockley, SS5 4QS
G6 XYV E Strode, 26 Churchill Close, Congleton, CW12 4QU
G6 XYX B Slater, 47 Broom Road, Lakenheath, Brandon, IP27 9EZ
G6 XZA M Scott, 28 Penwarden Way, Bosham, Chichester, PO18 8LF
G6 XZC Christopher Shaw, 17 South Street, Pilsley, Chesterfield, S45 8BQ
G6 XZM Christopher Smith, 104 Warren Road, Banstead, SM7 1LB
G6 XZO Stephen Sandilands, 5 Sallywood Close, Stenson Fields, Derby, DE24 3ES
G6 XZP Richard Sammons, 42 Woodcote Avenue, Wallington, SM6 0QY
G6 XZS J Thorn, 20 Kiln Road, Shaw, Newbury, RG14 2HA
G6 YAH C Wheeler, 11 Brooklands Way, Redhill, RH1 2BN
G6 YAI I Wilson, 2 Kingswood Close, Owlthorpe, Sheffield, S20 6SD
G6 YAK Paul Willetts, 49 Summervale Road, Hagley, Stourbridge, DY9 0LX
G6 YAQ F Barker, 13 Ashbourne Road, Eccles, Manchester, M30 0HW
G6 YAR R Porteus, 22 North View, Meadowfield, Durham, DH7 8SQ
G6 YAS Mario Brashill, 42 Bannister Street, Withernsea, HU19 2DT
G6 YB CITY BRISTOL GR c/o D Bailey, 41 Lippiatt Lane, Timsbury, Bath, BA2 0JF
G6 YBC D Anderson, 142 Tyldesley Road, Atherton, Manchester, M46 9AB
G6 YBH A White, 85 Goddard Way, Saffron Walden, CB10 2EB
G6 YBN A West, 82 Mount Pleasant Road, Alton, GU34 2RS
G6 YBV S Hunt, 33 Rutland Street, Ashton-under-Lyne, OL6 6TX
G6 YCE A Brooke, 14 Counting House Road, Disley, Stockport, SK12 2DB
G6 YCF M Bartlett, 206 Victoria Road, Romford, RM1 2NP
G6 YCG A Bennett, 5 Fifth Avenue, Northville, Bristol, BS7 0LP
G6 YCI M Buck, 178 Rover Drive, Castle Bromwich, Birmingham, B36 9LL
G6 YCL M Banner, 7 Lowdham Road, Gedling, Nottingham, NG4 4JP
G6 YCM Robert Brookes, 52 Larch Grove, Kendal, LA9 6AU
G6 YCN R Brassington, Above Park Farm, Leek Road, Stoke-on-Trent, ST10 2PT
G6 YCO J Baddeley, 52 Stephens Way, Bignall End, Stoke-on-Trent, ST7 8PL
GW6 YCT M Le Ves Conte, 74 Glan Road, Aberdare, CF44 8BW
G6 YCV T Leach, 2 Selkirk Gardens, Cheltenham, GL52 5LX
G6 YCW B Lancaster, 1 Belgrave Close, Dodleston, Chester, CH4 9NU
G6 YCZ J Massey, 10 Rapley Avenue, Storrington, Pulborough, RH20 4QL
G6 YDN John Mountain, 15 Eldon Close, Chapel-en-le-Frith, High Peak, SK23 0PX
G6 YDO Fraser Mirams, 10 Ravenoak Park Road, Cheadle Hulme, Cheadle, SK8 7EH
G6 YDP Arthur Gallagher, 1a Wynsome Street, Southwick, Trowbridge, BA14 9RB
G6 YDT E Gittins, 40 Melyd Avenue, Prestatyn, LL19 8RN
G6 YEA N Guy, 43 Hereford Road, Bolton, BL1 4NJ
G6 YEK Desmond Heard, 103 Moorland Road, Weston-Super-Mare, BS23 4HU
G6 YEY R Hope, 26 Chaucer Avenue, Andover, SP10 3DS
G6 YFF Grace Hunter, 57 The Cedars, Hailsham, BN27 1TU
G6 YFG S Lles, 3 Petersway Gardens, St. George, Bristol, BS5 8TA
G6 YFH R Ingle, 48 Barlborough Road, Clowne, Chesterfield, S43 4RF
G6 YFL H Jones, 15 Bonchurch Walk, Manchester, M18 8BP
G6 YFY I Pitfield, 27 Winchester Crescent, Fulwood, Sheffield, S10 4ED
G6 YFZ D Paul, Enfield, Gunton Road, Wymondham, NR18 0QP
G6 YGB Robert Preston, 188 Dumers Lane, Radcliffe, Manchester, M26 2GF
G6 YGH Alan Richardson, 9 Webbers Way, Puriton, Bridgwater, TA7 8AS
G6 YGI Brian Rogers, Fronucha, Rhewl, Oswestry, SY10 7AS
G6 YGJ R ROBINSON, 128 Norman Avenue, Bradford, BD2 2NE
G6 YGP D Lee, 4 Blythe Cottages, Blythe Lane, Ormskirk, L40 5UA
G6 YGV M Lane, Harewood Villa, Harewood Place, Halifax, HX2 7PN
GM6 YGW Brian Finch, Anchor Cottage, Lybster, KW3 6AS
G6 YHE G MANDER, 70 Copthall Way, New Haw, Addlestone, KT15 3TU
G6 YHF R Marchant, 12 Poplar Close, Huntingdon, PE29 7BP
G6 YHK T Miller, 6 Captains Walk, Falmouth, TR11 4HR
G6 YHL C Miller, 5 Lodge Lane, Bewsey, Warrington, WA5 0AG
G6 YHP C Molyneux, 23 Kemp Close, Chatham, ME5 9SP
G6 YHW Graham Murly, Vinge Redonde, 24360, Champniers et Reilhac, Dordogne, France, .
G6 YIE Stewart Forbes, 8 Nutmeg Close, Earley, Reading, RG6 5GX
G6 YII K Everington, 1 Norfolk Road, Wigston, LE18 4WH
G6 YIJ J Elford, 7 Cunliffe Road, Stoneleigh, Epsom, KT19 0RJ
G6 YIK Neil Drury, 444 Upper Shoreham Road, Shoreham-by-Sea, BN43 5NE
G6 YIO T Chapman, 17 Trevor Road, Swinton, Manchester, M27 0YH
G6 YIP Andrew Cohen, 9 Terrace Rd, 9 Terrace Rd, Plymouth Meeting, United States, 19462
G6 YIQ J Dixon, 8 East View, St. Ippolyts, Hitchin, SG4 7PD
G6 YIS Robert Chell, 3 Elderberry Close, Stourport-on-Severn, DY13 8TF
G6 YIU P Dawson, Ivy Dene, Middle Lane, Wolverhampton, WV8 2BE
G6 YIW W Gilroy, Little Harewood Farm, Clamgoose Lane, Stoke-on-Trent, ST10 2EG
G6 YJD J Govier, 111 Pearson Crescent, Wombwell, Barnsley, S73 8SF
G6 YJH Alf Haills, The Cherries, Main Road, Chelmsford, CM3 1NR
G6 YJJ P Hambly, 22c Windsor Road, London, W5 5PD
G6 YJO David Arscott, 20 Orchid Vale, Kingsteignton, Newton Abbot, TQ12 3YS
G6 YJR J Angel, 33 Grovewood Close, Chorleywood, Rickmansworth, WD3 5PX
G6 YLA John Howard, 11 Lightwood, Crown Wood, Bracknell, RG12 0TR
G6 YLB G Howse, 1 Sutherland Close, Woodloes Park, Warwick, CV34 5UJ
G6 YLD G Hope, 17 Church Road, Sutton at Hone, Dartford, DA4 9EX
G6 YLN M Hobbs, 22 Swan Place, Reading, RG1 6QD
G6 YLO P Hizzey, Borde Neuve, Maurens, France, 31540
G6 YLQ D Harrop, C/Mariano Aguilo 2a, Edificio Formentera 1, Mallorca, Spain, 7181
G6 YLR K Harris, 20 Rose Walk, Wicken Green Village, Fakenham, NR21 7QE
G6 YLV James Cromack, 45 Chelsea Road, Aylesbury, HP19 7BG
G6 YLW T Cannon, 36 St. Margarets Drive, Wigmore, Gillingham, ME8 0NR
G6 YLX Anthony John Crabtree, 15 Richmond Gardens, Redhill, Nottingham, NG5 8JS
G6 YLZ P Cornes, 46 Newland Avenue, Stafford, ST16 1NL
GI6 YM THE CITY OF BELFAST YMCA RC c/o William McAleer, 90 Gortin Park, Belfast, BT5 7EQ
G6 YMA Nicholas Clark, 2 Barleycroft, Stevenage, SG2 9NP
G6 YMD Michael Cooke, 22 Durham Close, Grantham, NG31 8RL
G6 YMH C Hughes, 85 Benson Gardens, Wortley, Leeds, LS12 4LA
G6 YMI Anthony Harris, 10 Egroms Lane, Withernsea, HU19 2LZ
G6 YMS Peter Humphreys, Tyn Llan Bodffordd, Llangefni, LL77 7DZ
G6 YMU D Hutchings, 2 Burghley Avenue, Bishop's Stortford, CM23 4PD
G6 YMY P Jacques, Caprius, The Parks, Evesham, WR11 8JP
G6 YNA A Johnston, 70 Queendown Avenue, Gillingham, ME8 9NZ
G6 YNL Robin Perry, Straight Mile Cottage, Gloucester Road, Bristol, BS35 3SB
G6 YNT Stuart Pentecost, 3 Delamare Road, Cheshunt, Waltham Cross, EN8 9AP
G6 YNV S Raddy, 32 Berry Park, Saltash, PL12 6EN
G6 YNW M Reeves, 17 Newark Avenue, Putnoe, Bedford, MK41 8NX
G6 YOG Mike Rutt, 31 Succombs Place, Southview Road, Warlingham, CR6 9JQ
G6 YOP P Harding, 54 Manor Road, Stretford, Manchester, M32 9JB
G6 YOR J Gillott, 132 Racecommon Road, Barnsley, S70 6JY
G6 YOZ J Addison, 20 Wychwood Rise, Great Missenden, HP16 0HB
GW6 YPA Michael Attfield, 16 Rhodfar Eos, Cwmrhydyceirw, Swansea, SA6 6TF
G6 YPF J Armstrong, 14 Rickwood Park, Horsham Road, Dorking, RH5 4PP
G6 YPJ J Brown, 9 Avenue Road, Walkford, Christchurch, BH23 5QH
G6 YPK Antony Bradbury, 20 Arden Close, Warwick, CV34 5SN
G6 YPM J Willats, 17 Purcell Road, Crawley, RH11 8XJ
G6 YPY S Davis, 30 Bonny Wood Road, Hassocks, BN6 8HR
GM6 YQA C Davies, 2 Henry Street, Thurso, KW14 7NW
G6 YQI E Fletcher-Cowen, 18 Buckingham Avenue, Horwich, Bolton, BL6 6NR
G6 YQJ David Fisher, 86 Parsons Lane, Littleport, Ely, CB6 1JS
G6 YQN N Fox, 32 Westmorland Avenue, Kidsgrove, Stoke-on-Trent, ST7 1AT
G6 YQT S Forbes, 11 Henfield View, Warborough, Wallingford, OX10 7DB
G6 YQU R Fuller, The New House, Main Street, Lutterworth, LE17 6NT
G6 YQW John Taylor, 7 Caddick Road, Birmingham, B42 2RL
G6 YRB J Stewart, 107 Turnberry, Skelmersdale, WN8 8EG
G6 YRC A Smith, 4 Wesley Grove, Burnley, BB12 0JJ
GM6 YRH A Smith, Robsland, Strathaven Road, Lanark, ML11 0HY
G6 YRI S Sizmur, 38 Longbourne Way, Chertsey, KT16 9ED
G6 YRJ T Simmons, 3 West Hill Place, Brighton, BN1 3RU
G6 YRK Stephen Wright, 24 Green Lane, Glossop, SK13 6XY
GM6 YRN Alexander Stewart, 6 Lawers Place, Aberfeldy, PH15 2BE
G6 YRV D Bedford, 28 Durfold Drive, Reigate, RH2 0QA
G6 YRY R Bearchell, 81 Leaves Green Road, Keston, BR2 6DG
G6 YSB J Bates, 16 Harewood Avenue, Great Barr, Birmingham, B43 6QE
G6 YSL S Watts, 15 Churchill Way, Northam, Bideford, EX39 1DF
G6 YSN K Ward, 8 Hinckley Road, St. Helens, WA11 9HU
G6 YSO Philip Wayer, 4 Chatburn Avenue, Waterlooville, PO8 8UB
G6 YSQ S Tricker, 1 Drewitt Court, 75 Godstow Road, Oxford, OX2 8PE
G6 YSZ P Tonge, 1 View Hill, Stalybridge, SK15 2TH
G6 YTB R Watts, 41 Watford Road, Crick, Northampton, NN6 7TT
G6 YTO Richard Cassidy, 9 Langham Way, Ely, CB6 1DZ
G6 YTR R Broughton, Brookside, Blagdon Terrace, Newcastle upon Tyne, NE13 6EY
G6 YTV A Black, Redholme, The Street, Thetford, IP25 6NL
G6 YTW A Bennett, 29 Kennington Road, Kennington, Oxford, OX1 5NZ
G6 YTX R Burnett, 46 Dorset Waye, Heston, Hounslow, TW5 0ND
G6 YTY Anthony Bournes, 115 Abbotts Ann Down, Andover, SP11 7BX
G6 YTZ William Van-Den-Bergh, 12 Bishops Drive, Langport, TA10 9HW
GW6 YUC E Brooksbank, 22 King Street, Carmarthen, SA31 1BS
G6 YUX Brian Clough, Ashby Powerboat School, 31 Countess Road, Salisbury, SP4 7AS
G6 YUY Frederick Crockford, 41 Coram Green, Hutton, Brentwood, CM13 1LW
G6 YVD G Wood, Tethers End, Angarrack Lane, Hayle, TR27 5JF
G6 YVJ Colin Ward, 416a Portsmouth Road, Southampton, SO19 9AT
G6 YVS J Wilson, 36 North Warren Road, Gainsborough, DN21 2TU
G6 YWL A Griffiths, 45 Clarence Road, Bilston, WV14 6NZ

G6 YWN Peter Groom, 2 Alms Road, Doveridge, Ashbourne, DE6 5JZ
G6 YWU Derek Harding, 20 D'Arcy Road, Tiptree, Colchester, CO5 0RP
G6 YWV M Harrison, Barn Cottage, Parwich, Ashbourne, DE6 1QH
G6 YWY [illegible] Oakford, Tiverton, EX16 9DW
G6 YXB D Hewson, Woodwells, 52 Elmham Road, Dereham, NR20 4BW
G6 YXO S Fisher, 37 Elmlands Grove, York, YO31 1ED
G6 YXT Brian Evans, c/o Angel 18, Xerta Tarragona, Spain, 43592
G6 YXV K Faulkner, 5 Tregarrick, West Looe, Looe, PL13 2SD
G6 YXW P Foulkes, 23 Callowbrook Lane, Rubery, Birmingham, B45 9HW
G6 YXX N Frederick, 72 Cheltenham Street, Barrow-in-Furness, LA14 5HW
G6 YXY C Edwards, Seymore, Greenhill Park Road, Evesham, WR11 4NL
G6 YYN Kenneth McCann, Treverven, Back Lane, Selby, YO8 6QP
G6 YYQ N Munro, 19 Lowndes Court, Queens Road, Bromley, BR1 3EA
G6 YYU A Mutimer, 52 Sycamore Avenue, Wymondham, NR18 0HX
G6 YZB L Nunn, 103 Bladindon Drive, Bexley, DA5 3BT
G6 YZF S Alston-Pottinger, 86 Main Street, Walton, Street, BA16 9QN
G6 YZH M Smith, 5 Derwent Close, North Anston, Sheffield, S25 4GD
G6 YZU L Nixon, 87 Field Avenue, Canterbury, CT1 1TS
G6 ZAA John Wellard, 19 South Motto, Kingsnorth, Ashford, TN23 3NJ
G6 ZAC A Wilson, 21 Lakes Close, Chilworth, Guildford, GU4 8LL
G6 ZAF D Walker, 27 Daltons Close, Langley Mill, Nottingham, NG16 4GP
GM6 ZAK Andrew Sutton, 22 St. Michaels Drive, Cupar, KY15 5BS
G6 ZAL S Ward, 125 Heys Lane, Blackburn, BB2 4NG
G6 ZAM Phillip Waldron, 15 The Pastures, High Wycombe, HP13 5LZ
G6 ZAX R Hollick, 7 Grenfell Road, Bournemouth, BH9 2UD
G6 ZAY R Hope, 129 Lunedale Road, Dartford, DA2 6JX
G6 ZBO M Julians, 29 Trentdale Road, Carlton, Nottingham, NG4 1BU
G6 ZBT David Green, 6 Garth Villas, Rimswell, Withernsea, HU19 2DB
G6 ZBV A Higham, 12 Lakenheath Drive, Sharples, Bolton, BL1 7RJ
G6 ZCI James Anderson, 72 Saffron, Amington, Tamworth, B77 4EP
G6 ZCR John Phillips, 39 Bryn Glas, Rhosllanerchrugog, Wrexham, LL14 2EA
G6 ZCS V Priamo, 58 Ffordd Glyn, Coed-y-Glyn, Wrexham, LL13 7QW
GM6 ZCX M Rochester, Eadar Da' Sloc, Achmelvich, Lairg, IV27 4JB
GM6 ZCY M Rochester, Eadar Da' Sloc, Achmelvich, Lairg, IV27 4JB
G6 ZDB Gene Reddington, 2 South St., Newton, Alfreton, DE55 5TT
G6 ZDE D Ellingworth, 3 Leighton Park West, Westbury, BA13 3RW
G6 ZDH R Roberts, All-Y-Coed, Sychnant Pass Road, Conwy, LL32 8EU
G6 ZDP K Baum, 25 Lakers Meadow, Billingshurst, RH14 9NP
G6 ZDS Raymond Baldwin, 25 Jacey Road, Birmingham, B16 0LL
G6 ZDV Adrian Beales, Broomhill Bungalow, Mappleton Road, Hull, HU11 4UW
G6 ZEM John Hollerbach, 119 Mead End, Biggleswade, SG18 8JU
G6 ZEN J Homan, 55 Ark Royal, Bilton, Hull, HU11 4BN
G6 ZEQ Robert Hubert, 107 Kingsmead Park, Allhallows, Rochester, ME3 9QS
G6 ZET David Jackson, 19 Shelley Close, Bolton le Sands, Carnforth, LA5 8HQ
G6 ZEW M Jennings, 6 Broomroyd, Worsbrough, Barnsley, S70 5DU
G6 ZEY Kenneth Johnson, 66 Godwin Way, Cambridge, CB1 8QR
G6 ZEZ C Jones, 709 Bath Road, Taplow, Maidenhead, SL6 0PB
G6 ZFA J Justice, 6 Stanley Terrace, Devizes, SN10 5AJ
G6 ZFG Paul Wood, 31 Larches Lane, Tettenhall, Wolverhampton, WV3 9PX
G6 ZFI D Smith, 39 High Croft, Kelso, TD5 7NB
G6 ZFK E Toohey, No6 Block E, Peabody Avenue, London, SW1V 4AS
G6 ZFO D Tate, 73 Sparth Avenue, Clayton le Moors, Accrington, BB5 5QH
G6 ZFU C Stephen, 12 Beaufort Close, Leegomery, Telford, TF1 6XU
G6 ZFV T Wynne-Jones, 37 Oakleigh Drive, Croxley Green, Rickmansworth, WD3 3EE
G6 ZFX P senior, 13 St. Michaels Avenue, Swinton, Mexborough, S64 8NX
G6 ZFZ M Turner, 461 Bushbury Lane, Bushbury, Wolverhampton, WV10 8JX
G6 ZG GORLESTON ARS c/o Brian Alston-Pottinger, 16 Vincent Close, Great Yarmouth, NR31 0HR
G6 ZGA M Smith, 6 Norton Crescent, Towcester, NN12 6DN
G6 ZGB C Salmon, 20 Lime Close, Sandbach, CW11 1BZ
G6 ZGC Mark Tann, 11 St. Margarets Grove, Redcar, TS10 2HW
G6 ZGF Douglas Simpson, 18 Croft House View, Morley, Leeds, LS27 8NS
G6 ZGH Richard York, 44 Denmark Street, Lancaster, LA1 5LY
G6 ZGI Christine Butler, 8 Douglas Walk, Chelmsford, CM2 9XQ
G6 ZGK G Weston, 2 Gill Park, Efford, Plymouth, PL3 6LX
G6 ZGO N Carr, 15 Westlands, Leyland, PR26 7XT
G6 ZGU Jeffrey Brown, 10 Cherry Trail, Coldwater, Ontario, Canada, L0K 1E0
GW6 ZGY Raymond Bennett, 88 Coychurch Road, Pencoed, Bridgend, CF35 5NA
G6 ZHB John Booth, 9 New Street, Abingdon, OX14 3PE
G6 ZHF S Bailey, Silverthorne House, North Piddle, Worcester, WR7 4PR
G6 ZHJ G Butler, 23 Roman Meadow, Downton, Salisbury, SP5 3LB
G6 ZHL Malcolm Leack, 68 Dale Street, Lancaster, LA1 3AW
GW6 ZHM R Lannon, 16 Heol Mabon, Rhiwbina, Cardiff, CF14 6RL
G6 ZHO G Lattin, 5 Seymour Road, Broadfield, Crawley, RH11 9ES
G6 ZHS K Lupton, Oak Tree Cottage, Post Office Lane, Frodsham, WA6 8JJ
G6 ZHU P Lightfoot, 7 Fearns Avenue, Newcastle, ST5 8ND
G6 ZHY A Mayers, 2 Wyndham Gardens, Wrexham, LL13 9LY
G6 ZIC Joseph McComb, Bridge End Cottage, Bridge End, Hexham, NE48 2RY
G6 ZIO N Dessau, 20 Coventry Circle, Mahopac, United States, 10541
GI6 ZIR Adrian Duffy, 01a Arney Road, Bellanaleck, Enniskillen, BT92 2DL
G6 ZIY A Fairhurst, 16 Waverley Road, Hindley, Wigan, WN2 3DN
G6 ZJD E Fensome, 77 Church Green Road, Bletchley, Milton Keynes, MK3 6BY
G6 ZJI A Washby, 57 Cromwell Road, Hedon, Hull, HU12 8GF
G6 ZJK F Webster, 1 Fir Tree Cottages, Lower Ansford, Castle Cary, BA7 7JY
G6 ZJM D Leese, 22 Elm Road, Abram, Wigan, WN2 5XG
G6 ZJN R Williams, 220 Euston Grove, Morecambe, LA4 5LJ
G6 ZJS L Wright, 17 Drayton St., Alumwell Estate, Walsall, WS2 9QB
G6 ZJV A Winterbottom, 38 Heaton Avenue, Dewsbury, WF12 8AQ
G6 ZKC David Usher, 26 Meneth, Gweek, Helston, TR12 6UW
G6 ZKM J Cornell, 10 Craneswater Park, Southsea, PO4 0NT
G6 ZKS M Staniland, 2 Epsom Road, Cantley, Doncaster, DN4 6HX
G6 ZKU B Sawyers, 36 Frome Road, Bath, BA2 2QB
G6 ZKX R Smith, Smith Farms, Herne Lane, Dereham, NR19 1QE
G6 ZKY E Stebbings, 1 Coupland Road, Wootton, Abingdon, OX13 6DU
G6 ZKZ V Smith, 40 Princess Gardens, Blackburn, BB2 5EJ
G6 ZLD P Bent, 7 Bandon Rise, Wallington, SM6 8PT
G6 ZLJ Mark Adams, 62 Woodlands Road, Holmcroft, Stafford, ST16 1QP
G6 ZLS G Ashbee, 34 Manorgate Road, Kingston upon Thames, KT2 7AL
G6 ZLY Donald Brasenell, 18 Whitelaw Avenue, Castle Douglas, DG7 1GB
G6 ZMD S Roberts, 96 Hill Crescent, Dudleston Heath, Ellesmere, SY12 9NA
G6 ZME TELFORD DIST AR c/o James Walsnell, 15 Cuckoo Oak Green Madeley, Telford, TF7 4HT
G6 ZMG G Mills, 37 Holborough Road, [illegible]
GW6 ZMN W McDowall, 36 Adenfield Way, Rhoose, Barry, CF62 3EA
G6 ZMO James Mooney, 2 Madford Lane, Launceston, PL15 9EB
G6 ZMX Alan O'shaughnessy, Southby, Buckland, Faringdon, SN7 8QR
G6 ZNJ A Reeve, 188 Dorset Avenue, Great Baddow, Chelmsford, CM2 8YY
G6 ZNO I Martin, 21 Baldwin Avenue, Eastbourne, BN21 1UJ
G6 ZNT I Cross, 5 Upper Crescent, Minster Lovell, Witney, OX29 0RT
G6 ZNW Stan Cascino, 3 Connaught Road, Folkestone, CT20 1DA
G6 ZOB A Crowther, 16 Linden Avenue, Tuxford, Newark, NG22 0JR
G6 ZOE C English, 124 Hillside Road, Portishead, Bristol, BS20 8LG
G6 ZOJ Adrian Buchan, 5 Copythorne Close, Brixham, TQ5 8QG
G6 ZOL P Lancaster, 134 Wigan Road, Euxton, Chorley, PR7 6JW
G6 ZOT J Leary, 24 Howard Drive, Old Whittington, Chesterfield, S41 9JU
G6 ZPL Philip Manning, 21 Whitethorn Way, Oxford, OX4 6ER
G6 ZPR Alexander Morris, 32 New Road, Wonersh, Guildford, GU5 0SE
G6 ZPV N Mansfield, 2 Little Halt, Portishead, Bristol, BS20 8JQ
G6 ZQA C Nolan, 94 St. Andrews Road, Burgess Hill, RH15 0PH
G6 ZQJ A Doughty, 42 Thornton Road, Ilford, IG1 2ER
G6 ZQS Mark Charlton, 104 Foundry Street, Horncastle, LN9 6AF
G6 ZQU D Crook, Bedford Road, Sherington, Newport Pagnell, MK16 9NQ
G6 ZRO B Stoner, Montrose, Wesley Road, Whitby, YO22 4RW
G6 ZRS B Starr, 121 Pretoria Road, Patchway, Bristol, BS34 5PY
G6 ZRV Peter Stainton, Corpusty Lodge West, Heydon, Norwich, NR11 6RX
G6 ZSF D Neely, 3 Sidestrand Road, Newbury, RG14 6HP
G6 ZSG L Onions, 8 Prince Charles Close, Rubery, Birmingham, B45 0NB
G6 ZSH P Owen, 288 Chervil Rise, Wolverhampton, WV10 0HR
G6 ZSQ John Pepper, 39 Marina Court, 9-19 Mount Wise, Newquay, TR7 2EJ
G6 ZSU P Payton, 11 Hexham Way, Dudley, DY1 2UN
G6 ZTD Brian Robinson, 23 Croft Drive, Millhouse Green, Sheffield, S36 9NE
G6 ZTF A Bowler, 12 Wrenbury Drive, Coventry, CV6 6JZ
G6 ZTH Malcolm Richardson, Better View, Back Lane, Sutton-in-Ashfield, NG17 2LL
G6 ZTL Bernard Rogers, 24 Marmion Road, Coningsby, Lincoln, LN4 4RG
G6 ZTM D Redmill, 38 Whitland Road, Carshalton, SM5 1QT
G6 ZTP G Down, 8a Abbeville Close, Exeter, EX2 4SJ
G6 ZTR S Davies, 111 Southend, Garsington, Oxford, OX44 9DL
G6 ZTT O.B.O. MID-CHESHIRE CONTEST GROUP c/o M Baguley, 2 Kensington Way, Northwich, CW9 8GG
G6 ZTZ S French, 22 Amity St., Newtown, Reading, RG1 3LP
G6 ZUE William Edwards, 31 Cumberland Avenue, Benfleet, SS7 5NU
G6 ZUO D Gibson, 14 Lowfield Road, Dewsbury Moor, Dewsbury, WF13 3SR
GW6 ZUS John Gray, 36 Heol Pentre Felen, Llangyfelach, Swansea, SA6 6BY
G6 ZUV John Griffin, 35 Cottage Street, Kingswinford, DY6 7QE
G6 ZUZ J Hampshire, 14 Fellows Road, Cowes, PO31 7JN
G6 ZVB K Harris, 14 Dunstall Close, St. Marys Bay, Romney Marsh, TN29 0QX
G6 ZVD Malcolm Hicken, 60 St. Denys Crescent, Ibstock, LE67 6NX
G6 ZVL Martin Hoskins, 22 Rosedale Gardens, Thatcham, RG19 3LE
G6 ZVO K Howarth, 79 Eden Avenue, Edenfield, Bury, BL0 0LD
G6 ZVU Stewart Hughes, 50 Albany Road, Dalton, Huddersfield, HD5 9UW
G6 ZVV Nigel Hull, 3 Goring House, Charter Way, Colchester, CO4 5JL
G6 ZWC C Brown, 16 Old Croft Close, Good Easter, Chelmsford, CM1 4SJ
G6 ZWI A Angove, 22 Bramble Close, Newquay, TR7 2SU
G6 ZWL C Wright, 19 Redwood Glen, Chapeltown, Sheffield, S35 1EA
G6 ZWM Robert Wade, 104 Brookehowse Road, London, SE6 3TW
G6 ZWZ J Sexton, 31 Hurst Green, Mawdesley, Ormskirk, L40 2QS
G6 ZXN Ian Carter, 12 Bobbin Lane, Westwood, Bradford-on-Avon, BA15 2DL
G6 ZXO J Crowe, 15 Lambert Road, Kendray, Barnsley, S70 3AA
GW6 ZYI Brian Jones, 10 Hughes Street, Penygraig, Nr Tonypandy, CF40 1LX
G6 ZYM K Keeble, Hall Cottage, Hardwick Road, Harleston, IP20 9PU
G6 ZYX Graham Spruce, 158 Wolverhampton Street, Wednesbury, WS10 8UB
G6 ZYZ Patrick Skerritt, 39 Bache Street, West Bromwich, B70 7EW
G6 ZZE P Read, 11 Fairview Avenue, Whetstone, Leicester, LE8 6JQ
G6 ZZF P Thomas, 42 Wyndham Road, Abergavenny, NP7 6AF
G6 ZZR D Wiltshire, 19 Heron Way, Basingstoke, RG22 5QF
G6 ZZS D Watts, 176 Blatchcombe Road, Paignton, TQ3 2JP

## G*7

G7 AAI Anthony Hickey, 11 Barker Road, Wirral, CH61 3XH
GM7 AAJ Peter McManus, 59 Mauchline Road, Hurlford, Kilmarnock, KA1 5AB
G7 AAR Peter Comben, Vicarage Farmhouse, Church Way, Aylesbury, HP17 8RG
G7 AAS D Hawkins, 8 Braybrook Street, East Acton, London, W12 0AP
G7 AAU Helen Studdart, 33 Linden Avenue, Connah's Quay, Deeside, CH5 4SN
G7 AAV Stephen Studdart, 33 Linden Avenue, Connah's Quay, Deeside, CH5 4SN
G7 AAY Karrl Richardson, 514 Obelisk Rise, Northampton, NN2 8SX
G7 ABE C Duberley, 2 The Grove, Greenford, UB6 9BY
G7 ABF Keith Austin, 6 Boothey Close, Biggleswade, SG18 0DG
G7 ABQ D Ferns, 18 Sandelswood End, Beaconsfield, HP9 2AE
G7 ABR A Clark, Brookside, Milford, Bakewell, DE45 1DX
G7 ABT David Hepworth, 1 Greengate Crescent Epworth, Doncaster, DN9 1HA
G7 ABZ M Bromage, 14 Rhuddlan Way, Kidderminster, DY10 1YH
G7 ACA B Pearce, 39 Fairholme Park, Ollerton, Newark, NG22 9AS
G7 ACD Richard Cariss, 6 Granville Avenue, Newport, TF10 7DX
G7 ACG Jack Baker, 10 Green Lane, Rugeley, WS15 2AR
G7 ACJ G Mantle, 6 North Green, Wolverhampton, WV4 4RQ
G7 ACK George Bromfield, 63 Herondale Road, Mossley Hill, Liverpool, L18 1JZ
G7 ACM S Pinkney, 39 Butterley Drive, Loughborough, LE11 4PX
G7 ACN C Hayes, 9 Grenville Way, Thetford, IP24 2JH
G7 ACO Roy Horton, 1 Stonehill Rise, Doncaster, DN5 9HD
G7 ACR Peter Blakemore, 50 Longley Farm View, Sheffield, S5 7JX
G7 ADF I Bradbury, 11 St. Stephens Avenue, Wigan, WN1 3UQ
G7 ADH Gordon Williams, 18 Luther Road, Bournemouth, BH9 1LH
G7 ADP C Baker, 17 Coronation Road, Illogan, Redruth, TR16 4SG
G7 ADS Adrian Cresswell, 31 New Street, Doddington, March, PE15 0SP
GM7 ADU L Morrison, 22 Lodge Park, Kilmacolm, PA13 4PY
G7 ADW G Laycock, 18 Montague Crescent, Garforth, Leeds, LS25 2EP
GM7 ADY M Morrison, 22 Lodge Park, Kilmacolm, PA13 4PY
G7 AEA Gloucestershire County Raynet c/o Richard Large, 6 Jasmine Close, Abbeydale, Gloucester, GL4 6FJ
G7 AEC CHELTENHAM RYNT c/o Paul Kent, 82 Despenser Road, Tewkesbury GL20 5TW
G7 AEE TEWKESBURY RYNT c/o C Davis, 38 Courtney Close, Tewkesbury, GL20 5FB
G7 AEF FRST OF DEAN RY c/o Graham Harden, 13 Greenfield Road, Coleford, GL16 8BY
G7 AEH COTSWOLD RAYNET c/o G Hayter, 22 Golden Farm Road, Beeches Estate, Cirencester, GL7 1DX
G7 AEQ R Murphy, 17 Valley View Road, Paulton, Bristol, BS39 7QB
G7 AES Peter Crook, 40 St. Aubins Avenue Brislington, Bristol, BS4 4NX
G7 AEY D Martin, 12 The Willows, Kemsley, Sittingbourne, ME10 2TE
GW7 AFC Mark Grant, 11 Golwg yr Eglwys, Pontarddulais, Swansea, SA4 8EE
GM7 AFE A Erwood, Lunna House, Lunna, Shetland, ZE2 9QF
G7 AFL A Fountaine, 19 Metcalfe Grove, Blakelands, Milton Keynes, MK14 5JY
G7 AFO K Hollingsworth, 34 Marconi Drive, Yaxley, Peterborough, PE7 3ZR
G7 AFQ K Marlow, Computer Science, Po Box 363, Edgbaston Birmingha, B15 2TT
G7 AFS Malcolm Sheldon, 5 Runnymede Mews, Faversham, ME13 8RU
G7 AFT K Brazier, 4 Conifer Close, Hythe, Southampton, SO45 5EL
G7 AFV M Weston, 10 Flete Avenue, Newton Abbot, TQ12 4EH
G7 AFW M Towers, 44 Ravenscroft Drive, Chaddesden, Derby, DE21 6NX
G7 AFZ D Payea, 10 Royal Drive, Seaford, BN25 2XW
G7 AGA K Askew, 3a Craven Drive, Broadheath, Altrincham, WA14 5JF
G7 AGB P Rothwell, 20 Henbury Close, Corfe Mullen, Wimborne, BH21 3TF
G7 AGC M Collis, 2 Westwood Avenue, Urmston, Manchester, M41 9NG
G7 AGG R Ricketts, 2 Brynystwyth, Penparcau, Aberystwyth, SY23 1SS
G7 AGI David De Silva, 22 Bishop Road, Bristol, BS7 8LT
G7 AGO J Lee, 188 Manstone Avenue, Sidmouth, EX10 9TJ
G7 AGR AYLESBURY RYN GROUP c/o R Clark, 9 Conigre, Chinnor, OX39 4JY
GM7 AHA V Turnbull, 18 Easterfield Court, Livingston Village, Livingston, EH54 7BZ
G7 AHB Tim Green, 12 Springfields, Ambrosden, Bicester, OX25 2AH
G7 AHO Peter Russell, 27 Main Street, Haconby, Bourne, PE10 0UR
G7 AHP Stephen Crask, 14 Southfield Road, Paignton, TQ3 2SW
G7 AHR M Brett, 106 Trinity Avenue, Llandudno, LL30 2YQ
G7 AHT Declan Smith, 14 Ashmead Green, Dursley, GL11 5EW
G7 AIB Norman Denton, 41 Monks Dale, Yeovil, BA21 3JB
G7 AIC V Newman, 35 Netherton Road, Yeovil, BA21 5NY
G7 AIF D Grevatt, 17 Foxdale Drive, Angmering, Littlehampton, BN16 4HF
G7 AIH R Whitenstall, 4 Monksmead, Borehamwood, WD6 2LQ
G7 AIR William Moore, 4b Colville Road, Newton, Wisbech, PE13 5HH
G7 AIY V Lamb, 19 Pemba Drive, Buckley, CH7 2HQ
G7 AJE F Salt, 6 Bodycoats Road, Chandler's Ford, Eastleigh, SO53 2GX
G7 AJG T Ellis, 29 St. Annes Road, Clacton-on-Sea, CO15 3NF
G7 AJJ A Hammond, 5 Durness Close, Kettering, NN15 5BN
G7 AJK Alan Knowler, 385 Capstone Road, Gillingham, ME7 3JE
G7 AJN H Lister, 68 Spring Avenue, Gildersome, Leeds, LS27 7BT
G7 AJP Brian Staniforth, 1 Pylon Cottages, Donington-on-Bain, Louth, LN11 9RQ
G7 AJR S Godfrey, 7 Laburnum Close, North Baddesley, Southampton, SO52 9JT
G7 AJS T Green, 34 Thorn Close, Kettering, NN16 9BU
G7 AJT M Carter, 17 Ash Crescent, Higham, Rochester, ME3 7BA
G7 AJX R King, 31 Lambert Road, Sprowston, Norwich, NR7 8AA
G7 AKI P Shambrook, 7 The Close, Cheltenham, GL53 0PQ
G7 AKJ W Wrench, 2 Maunders Place, Otterton, Budleigh Salterton, EX9 7JE
G7 AKM David Pearson, 8 Walnut Way, Swanley, BR8 7TW
G7 AKP Yvonne Branch, 38 Kynaston Road, Didcot, OX11 8HD
G7 AKV E Grantham, 18 Fen End Lane, Spalding, PE12 6AD
G7 ALC L Civita, 53 Rockhurst Drive, Eastbourne, BN20 8XD
GI7 ALH Joseph Bailie, 42d John Street Lane, Newtownards, BT23 4LY
GM7 ALI A Crighton, Tighnacreag, Pacemuir Road, Kilmacolm, PA13 4JJ
G7 ALR P Goodman, 85 Rantree Fold, Basildon, SS16 5TW
G7 AMD Gerald Blakemore, 6 Pine Tree Close, Hednesford, Cannock, WS12 4JT
G7 AMQ J Morstatt, 32 Elwy Circle, Ash Green, Coventry, CV7 9AU
G7 AMS A Steel, Royal Oak Mews, 74 Caradoc Road, Prestatyn, LL19 7PF
G7 AMW P Jackson, 4 Abbotsbury, Orton Malborne, Peterborough, PE2 5PS
G7 ANA Robin Jackson, 5 Home Farm Court, Hooton Pagnell, Doncaster, DN5 7BL
G7 ANB D Ball, 16 Kelston View, Whiteway, Bath, BA2 1NW
GM7 ANE William Jamieson, 90 Highpark Avenue, New Cumnock, Cumnock, KA18 4HH
G7 ANG Angelo Santagata, 25 Swyncombe Avenue, London, W5 4DR
G7 ANH F Pattinson, 4 Carlisle Road, Brampton, CA8 1SR
G7 ANK Colin Bryan, Beau Rivage, Seaholme Road, Mablethorpe, LN12 2DF
G7 ANO T Hyde, 10 Castleton Avenue, Riddings, Alfreton, DE55 4AG
G7 ANQ J Hedges, 31 Meadow Road, Hartshill, Nuneaton, CV10 0NL
G7 ANV S O'Malley, 140 Allerburn Lea, Alnwick, NE66 2QP
G7 ANX A Smith, Rose Lane Cottage, Huxham, Exeter, EX5 4EP
G7 ANY A King, 31 Springhill, Pennycross, Plymouth, PL2 3QZ
G7 AOA P Gash, 2 Betjeman Walk, Yateley, GU46 6YP
GW7 AOE Guy Williams, 10 Strawberry Place, Morriston, Swansea, SA6 7AG
GJ7 AOG C Eve, 2 The Elms, La Rue Des Cosnets, St. Ouen, Jersey, JE3 2BJ
G7 AOK R Swynford-Lain, 19 Kenton Road, Earley, Reading, RG6 7LQ
GM7 AOM John Curr, 56 Drygate Street, Larkhall, ML9 2DA
G7 AOQ J Johnson, 32 Bradlea Rise, Rawmarsh, Rotherham, S62 5QJ
GU7 APA P Ash, Trigale House, Alderney, Alderney, Guernsey, GY9 3TZ
G7 APD RUGBY AM TRA SO c/o Stephen Tompsett, 9 Ashlawn Road, Rugby, CV22 5ET
G7 API S Garlick, 37 Edith Road, Kettering, NN16 0QB
G7 APL S Bonham, 4 St. Martins Avenue, Studley, B80 7JJ
G7 APM Clive Mockford, 21 Nantwich Road, Middlewich, CW10 9HE
G7 APO Colin Monckton, 52 Chalkpit Lane, Dorking, RH4 1EY
G7 APP Ian Capon, Pentre Garreg Bach, Marianglas, LL73 8PP
G7 APQ A Jones, 6 Heatherbreea Gardens, Rushden, NN10 6EH
G7 APS S Ellison, 16 Beechtree Road, Walsall, WS9 9LS
G7 APU L Young, 7 Tudor Rose, 28 Northgate, Hunstanton, PE36 6AP
G7 AQA M Hawkshaw, 5 Carr Hill Road, Calverley, Pudsey, LS28 5PZ

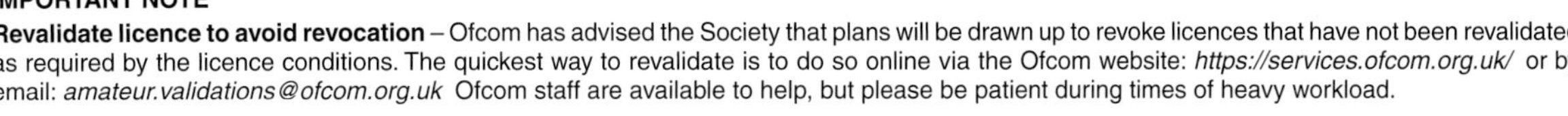
**IMPORTANT NOTE**

**Revalidate licence to avoid revocation** – Ofcom has advised the Society that plans will be drawn up to revoke licences that have not been revalidated as required by the licence conditions. The quickest way to revalidate is to do so online via the Ofcom website: *https://services.ofcom.org.uk/* or by email: *amateur.validations@ofcom.org.uk* Ofcom staff are available to help, but please be patient during times of heavy workload.

G7 AQD W Williams, 2 Lightfoot Lane, Fulwood, Preston, PR2 3LP
G7 AQF A Gregory, 13 Combe Avenue, Portishead, Bristol, BS20 6JR
G7 AQK N McGrath, 48 Willersley Avenue, Orpington, BR6 9RS
G7 AQL B Burbage, 9 Westerdale Drive, Frimley, Camberley, GU16 9RB
G7 AQN I Cooper, Ceylon, 70 Carshalton Park Road, Carshalton, SM5 3SW
GI7 AQO R Todd, 14 Glencroft Road, Newtownabbey, BT36 5GD
G7 AQV A Russell, 73 Seymour Road, Newton Abbot, TQ12 2PX
G7 ARF S Wright, 63 Cambridge Road, St. Albans, AL1 5LF
G7 ARJ J Baber, 130 Lumsden Road, Southsea, PO4 9LR
G7 ARK Alan Wright, 3 Wyborn Close, Hayling Island, PO11 9HY
G7 ARP R Orchard, 31 Chiswick House, Bell Barn Road, Birmingham, B15 2AA
GD7 ARS W Wrigley, 20 Fairy Hill Close, Ballafesson, Port Erin, Isle of Man, IM9 6TJ
G7 ART M Blake, 20 Triumphal Crescent, Plymouth, PL7 4RW
G7 ASF COVENTRY ARS c/o John Beech, 124 Belgrave Road, Coventry, CV2 5BH
GW7 ASL P Jones, 27 Hawthorn Road East, Llandaff North, Cardiff, CF14 2LR
G7 ASY P Matkin, 31 Southgate End, Cannock, WS11 1PS
G7 ASZ N Blair, 19 Church Street, Bourn, Cambridge, CB23 2SJ
G7 ATJ Robert Williams, 10 Barton Close, East Cowes, PO32 6LS
G7 ATW G Johnson, 63 Laurel Drive, Bradwell, Great Yarmouth, NR31 8PB
G7 AUE B Oubridge, 54 Cantle Avenue, Downs Barn, Milton Keynes, MK14 7QS
G7 AUF K Gebhardt, 15 Jubilee Road, Corfe Mullen, Wimborne, BH21 3NH
G7 AUP A White, Tioram, Garthends Lane, Selby, YO8 6QW
GW7 AUQ Nick Smith, 7 Lili Mai, Barry, CF63 1DW
G7 AUR S Davis, 7 Kennedy Crescent, Gosport, PO12 2NL
G7 AUU R Selwood, 33 Chandlers, Sherborne, DT9 3RT
GM7 AUW J Milne, 24 Lorne Street, Edinburgh, EH6 8QP
GM7 AUX E Ramsay, Tighnduin, 2 Queen Street, Dundee, DD5 4HG
GI7 AUY I Potts, 46 Richmond Park, Omagh, BT79 7SJ
G7 AVB D Daymond, 15 Constance Street, Newport, NP19 7DB
G7 AVF P Honeybone, 30 The Hordens, Barns Green, Horsham, RH13 0PJ
G7 AVU Robert Fisk, 25 Cromwell Street, Gainsborough, DN21 1DH
G7 AVZ Laurence Collings, 37 Armstrong Road, Mansfield, NG19 6HZ
G7 AWG V Watson, 3 Anderton Rise, Millbrook, Torpoint, PL10 1DA
G7 AWK David Easton, 86 Dryburn Road, Kelloholm, Sanquhar, DG4 6SN
G7 AWW M Gynane, 164 Stockbridge Lane, Huyton, Liverpool, L36 8EH
GI7 AXB W Scott, 89 Henryville Manor, Ballyclare, BT39 9FP
G7 AXL A Marchington, 30 Warwick Avenue, Golcar, Huddersfield, HD7 4BX
G7 AXM J Smith, 7 Ainsworth Court, Cameron Close, Freshwater, PO40 9JH
G7 AXN R Moxham, 8 Dunroyal Close, Helperby, York, YO61 2NH
G7 AXW A Parkes, 96 Oakham Road, Dudley, DY2 7TQ
G7 AYA G Jessup, 25 Harrier Green, Holbury, Southampton, SO45 2EY
G7 AYB R Upton, 35 Weston Street, Swadlincote, DE11 9AT
G7 AYE R Phipps, 39 Perrinsfield, Lechlade, GL7 3SD
G7 AYI L Faragher, 4 Kirloe Avenue, Leicester Forest East, Leicester, LE3 3LA
G7 AYL Pat Thompson, Flat 11, Old Gaol, 16 Grove Street, Bath, BA2 6PJ
G7 AYO Linda Hutt, Fenwick Crossing House, Fenwick Lane, Doncaster, DN6 0EZ
G7 AYP A Gregory, 9 Fordbridge Road, Ashford, TW15 2TD
G7 AYQ A Tregay, 53 Haverscroft Close, Taverham, Norwich, NR8 6LT
G7 AYS D Webster, 5 Eastfield Road, Princes Risborough, HP27 0JA
GM7 AYW J Hunter, 1 Mitchell Drive, Rutherglen, Glasgow, G73 3QP
G7 AZA J Cash, 89 Peacocks, Harlow, CM19 5NZ
G7 AZC A Rogers, Yoke Farm, Upper Hill, Leominster, HR6 0JZ
G7 AZH V Barber, 8 Hollyberry Close, Redditch, B98 0QT
G7 AZJ B Sayers, 13 Mulberry Close, Cambridge, CB4 2AS
G7 AZM W Knowler, 33 Cherry Tree Road, Rainham, Gillingham, ME8 8JY
G7 AZT Vincent Mills, 18 Muir Road, Maidstone, ME15 6PX
G7 AZV R Quick, Halfway House, Upton Scudamore, Warminster, BA12 0AE
G7 AZW M Ney, 4 Rathen Road, Withington, Manchester, M20 4GH
G7 BAB W Foden, 209 Lord Lane, Failsworth, Manchester, M35 0PX
G7 BAC M Gohl, 23 Maplewood Avenue, Hull, HU5 5YE
G7 BAE T Searle, Haven Orchard, Exwick Lane, Exeter, EX4 2AP
GM7 BAS W Hunter, 2 Wallace Cottages, Southend, Campbeltown, PA28 6RX
G7 BAV A Roberts, 4 Rocky Park Road, Plymouth, PL9 7DQ
G7 BBC BBC CLB ARIEL RD c/o Gareth Rowlands, 59 Barlee Crescent, Uxbridge, UB8 2EX
G7 BBD C Hartley, 102a Bedford Road, Cranfield, Bedford, MK43 0HA
G7 BBJ B Jenkinson, 14 Sandhill Way, Harrogate, HG1 4JN
G7 BBN E Crookall, 17 Dundee St., Moorlands, Lancaster, LA1 3DS
G7 BBU C Sharp, Dept. Of Astronomy, University Of Arizona, Tuscon, United States, 85721
G7 BBY M Jones, 20 Kingston Close, Droitwich, WR9 7RY
GM7 BCC Robert Sutherland, Tigh - Na - Coille, Mill Road, Nairn, IV12 5EW
G7 BCI V White, 6 Laburnum Close, South Anston, Sheffield, S25 5GL
G7 BCK N Gillies, 5 Pickmere Terrace, Dukinfield, SK16 4JJ
G7 BCO A Adey, 8 Spinners Court, Telford, TF5 0PG
G7 BCS K Wood, 3 Penrose Way, Four Marks, Alton, GU34 5BG
G7 BCW H Fitches, 29 Harrington Way, Oakham, LE15 6SE
GM7 BDD Alan Mankin, 10 Kerloch Crescent, Banchory, AB31 5ZF
G7 BDK P Blackett, 32 Woodstock Road, Carshalton, SM5 3DZ
G7 BDR A Davis, 5 Ludlow Close, Loughborough, LE11 3TB
G7 BDS J Mills, 42 Maple Grove, Welwyn Garden City, AL7 1NL
G7 BEJ Wayne Young, 24 Peebles Road, Newark, NG24 4RW
G7 BEP M Van Der Steeg, Horsebrook Farm, South Brent, TQ10 9EU
GI7 BET R Griffin, 19 Jubilee Park, Cookstown, BT80 8LJ
G7 BFH Nigel Lambert, 2 Sandhills Close, White Hills, Northampton, NN2 8EB
G7 BGM D Allison, 57 Algarth Road, Pocklington, York, YO42 2HJ
G7 BGO P Toll, 83 Pepper Street, Lymm, WA13 0JT
G7 BGT Royston Pykett, 20 Bolton Street, Swanwick, Alfreton, DE55 1BU
G7 BGY M Bellaby, 21 Sprydon Walk, Nottingham, NG11 9ET
G7 BGZ M Hickman, 45 St. Martins Avenue, Doncaster, DN5 8HZ
G7 BHE Paul Gledhill, 36 Tylers Ride, South Woodham Ferrers, Chelmsford, CM3 5ZT
G7 BHG R Gill, 24 Larkfield Crescent, Rawdon, Leeds, LS19 6EH
G7 BHH Leigh Kemp, Flat 1, 85 Red Lion Road, Surbiton, KT6 7QG
G7 BHR D Coombes, 2 Ormesby Drive, Potters Bar, EN6 3DZ
G7 BHU T Mayfield, 184 Wharf Road, Pinxton, Nottingham, NG16 6LQ
G7 BHW John Wilson, 61 New Lane, Hilcote, Alfreton, DE55 5HT
G7 BHY P Bailey, 21 Westhall Road, Mickleover, Derby, DE3 0PA
G7 BIK D Clark, 33 Landers Reach, Lytchett Matravers, Poole, BH16 6NB
G7 BIL W Knox, 9 Harrow Close, Caerleon, Newport, NP18 3EF
G7 BIM Stephen Beazley, 10 Barnecut Close, St. Cleer, Liskeard, PL14 5RU
G7 BIP G Roffey, 31 Saxville Road, Orpington, BR5 3AN
G7 BIQ K LLoyd, 9 Hornbeam Walk, Witham, CM8 2SZ
G7 BIV Richard Hudson, 12 Magnus Drive, Colchester, CO4 9WQ
G7 BIX T Houlihane, 1 Pepper Close, Bassingbourn, Royston, SG8 5HX
G7 BIY Brian Ansell, 26 Stubby Lane, Wolverhampton, WV11 3NL
G7 BJB Charles Foster, 10 Handel Street, Derby, DE24 8AZ
G7 BJC M Wiggins, 158 Prince Charles, Avenue, Derby, DE3 4LQ
G7 BJD Robert Ward, 12 Meadow Lea, Worksop, S80 3QJ
G7 BJE Ian Godlington, 46 Bren Way, Hilton, Derby, DE65 5HP
G7 BJG I Harvey, 20 Hawke Road, Stafford, ST16 1PZ
G7 BJN J Barlow, 38 St. Pauls Road, Newcastle, ST5 2PQ
G7 BJR G Mitchell, 8 Addison Road, Mexborough, S64 0DJ
G7 BKJ Christopher Watney, 23 The Wad, West Wittering, Chichester, PO20 8AH
G7 BKL Brian Moulton, 70 St. Georges Avenue, Westhoughton, Bolton, BL5 2EU
G7 BKN R Hatcher, 61 Holland Road, Oxted, RH8 9AU
G7 BLD Colin Holden, 26 Valebridge Drive, Burgess Hill, RH15 0RW
G7 BLJ R Maytum, 62 Coronation Close, Great Wakering, Southend-on-Sea, SS3 0JG
G7 BLK T Dicks, 4 Nicholas Drive, Reydon, Southwold, IP18 6RE
G7 BLT William Manthorp, 49 Cassell Road, Bristol, BS16 5DE
G7 BLX M Rowe, 31 Thornhill Avenue, Thornhill, Southampton, SO19 6PS
G7 BMC P Hollis, 2 Falcon Drive, Whittington, Lichfield, WS14 9PF
G7 BME D Watts, 9 Filwood Drive, Kingswood, Bristol, BS15 4HT
G7 BMM S Hallam, 125 Charnwood Road, Shepshed, Loughborough, LE12 9NL
G7 BMP N Gough, 27 Oak Place, Stoke-on-Trent, ST3 5PN
G7 BMT S Hodges, 15 Middlewich Street, Crewe, CW1 4BS
G7 BMY J Moore, Fairview, 28 Bulmer Lane, Great Yarmouth, NR29 4AF
G7 BNB Stephen Reynolds, Dowgill Head House, North Stainmore, Kirkby Stephen, CA17 4EX
G7 BNC John Beadle, The Coach House, Cwmdauddwr, Rhayader, LD6 5HA
G7 BND B Welthy, 8 Du Cane Place, Witham, CM8 2UQ
GM7 BNF Mike Harrington, Mount Pleasant House, North Road, Wick, KW1 4DN
G7 BNI N Pope, 21 Moultrie Road, Rugby, CV21 3BD
G7 BNK D Wood, 16 Church Road, Pelsall, Walsall, WS3 4QN
G7 BNL A Creek, Westmoor House, Wisbech Road, Ely, CB6 1RQ
G7 BNM A Ison, 32 Station Road, Lode, Cambridge, CB25 9HB
G7 BNN Karen Sheppard, Woodlands Bungalow, Gunby Road, Skegness, PE24 5HT
G7 BNO Christopher Sheppard, 11 Mosscar Close, Spion Kop, Mansfield, NG20 0BW
G7 BNS T Healey, 5 St. Johns Crescent, Huddersfield, HD1 5DY
G7 BNW Graham Ingmire, 93 Havelock Road, Luton, LU2 7PP
G7 BNZ Huw Williams, 14 Mahonia Drive, Langdon Hills, Basildon, SS16 6SD
G7 BOB R Kin, The Poplars, Kingsmead Road, High Wycombe, HP11 1JL
G7 BOH P Craig, 1 Liddel Way, Chandler's Ford, Eastleigh, SO53 4QF
GD7 BOJ Michael Rodgers, 1 Kings Court, Ramsey, Ramsey, Isle of Man, IM8 1LJ
GM7 BOW R King, 118 Boswell Road, Inverness, IV2 3EW
G7 BOY B Hodgkinson, 16 Swain Avenue, Buckley, CH7 3BR
GM7 BOZ A Bowie, 374a High Street, Leslie, Glenrothes, KY6 3AX
G7 BPF D Rose, 8 Ambrose Avenue, Hatfield, Doncaster, DN7 6QQ
G7 BPG David Dixon, 5 Denbigh Close, Newcastle under Lyme, ST5 3DL
G7 BPI Christopher Thompson, 13 Wentworth Avenue, Luton, LU4 9EN
G7 BPM Neil Hemingway, 24 Ealees Road, Littleborough, OL15 0HQ
G7 BPN K Pentecost, 46 Austen Way, Crook, DL15 9UT
G7 BPO R Ashman, 44 Conan Doyle Walk, Swindon, SN3 6JB
G7 BPQ Michael Meadows, 4 The Grove, Wharncliffe Side, Sheffield, S35 0EA
G7 BPR S Winlove-Smith, 7 Maughan Street, Shildon, DL4 1AP
G7 BPX Andrew Gifford, 56 Seymour Road, Gloucester, GL1 5QD
G7 BPZ Jean May-Golding, 9 St. Barts Road, Sandwich, CT13 0BG
G7 BQA Robert Golding, 9 St. Barts Road, Sandwich, CT13 0BG
G7 BQM Edward Turk, Sunny Meadow, Three Bridges Road, Long Buckby Wharf, Northampton, NN6 7PP
G7 BQS Malcolm Rodgers, 14 North Street, Rawmarsh, Rotherham, S62 5NH
G7 BQT B Miller, 22b Avondale Road, Fleet, GU51 3BS
G7 BQU G Daynes, 25 Redwood Close, Keighley, BD21 4YG
G7 BQY Arthur Brighton, 22 Langport Drive, Vicars Cross, Chester, CH3 5LY
G7 BRA J Riley, 132 Barrs Road, Warley, Cradley Heath, B64 7EZ
G7 BRB J Marsh, 39 Palace Gate, Odiham, Hook, RG29 1JZ
G7 BRC BREDHURST RECEIVING & TRANSMITTING SCY c/o Martin Pearson, 56 Parkwood Green, Parkwood, Gillingham, ME8 9PP
G7 BRF D Oliver, 36 Baker Avenue, Stratford-upon-Avon, CV37 9PN
G7 BRJ J Mills, 70 Crescent Road, Rochdale, OL11 3LG
GM7 BRL D O'Donnell, 188 Warriston Street, Glasgow, G33 2LD
G7 BRM George Jeffery, 48 Minnis Lane, River, Dover, CT17 0PR
G7 BRP J Biggs, 5 Churchgate St., Soham, Ely, CB7 5DS
G7 BRS J Rushton, 391 Rossendale Road, Burnley, BB11 5HP
G7 BRU J Rhodes, Little Meadow, Roke, Wareham, BH20 7JF
G7 BRX R Bell, 10 Old Mill Way, Weston-Super-Mare, BS24 7AS
G7 BRZ C Gaunt, 30 Sonja Crest, Immingham, DN40 2EQ
G7 BSC R Snelling, 91 Oakfield Road, Newport, NP20 4LP
G7 BSF Ann Lewis, 20 Annes Walk, Caterham, CR3 5EL
G7 BSG BEMERTON SC GRP c/o A Carter, 28 Springfield Road, Wellington, TA21 8LG
G7 BSK Jason Ingram, 15-47 Northoaks, 20 Woodlands Crescent, Singapore, Singapore, 738081
G7 BSL P Bedford, 9 Woodland Avenue, Kirkthorpe, Wakefield, WF1 5TD
G7 BSO Andrew NOBLE, 42 Upper Street, Salisbury, SP2 8LY
G7 BSP Sybil Farmer, Horton Brook Cottage, Horton, Shrewsbury, SY4 5NB
GW7 BTC CARDIFF AND DISTRICT ARC c/o Roy Magwood, 13 Inverness Place, Cardiff, CF24 4RU
G7 BTI MADLEY AMATEUR RADIO GROUP c/o N Prosser, 35 Holmfirth Close, Belmont, Hereford, HR2 7UG
G7 BTP P Jensen, 16 Hawthorn Avenue, Immingham, DN40 1AR
G7 BUF Paul Parkin, 2 The Knoll, Dronfield, S18 2EH
G7 BUK D Brinnen, 134 Victoria Road, Mablethorpe, LN12 2AJ
G7 BUL Ed Williamson-Brown, Meadowside, Green Lane, Stowmarket, IP14 5DS
G7 BUN H Ellis, Home Cottage, Drury Square, King's Lynn, PE32 2NA
G7 BUR G Robinson, 228 Bradford Road, Riddlesden, Keighley, BD20 5JT
G7 BUS G Payne, 155 Camping Hill, Stiffkey, Wells-next-The-Sea, NR23 1QL
G7 BVH M Gurr, Elan, Sandown Road, Sandwich, CT13 9NY
G7 BVL K Lambert, 1 Langton Road, Chichester, PO19 3LY
G7 BVS N Hill, 16 Bittern Avenue, Abbeydale, Gloucester, GL4 4WA
G7 BVZ B Allen, 84 Holland Road, Little Clacton, Clacton-on-Sea, CO16 9RS
G7 BWE E Gibbons, 19 Queens Park Road, Caterham, CR3 5RB
G7 BWF A Gray, 147 Kirby Road, Walton on The Naze, CO14 8RL
G7 BWI J White, 11 Rowdown, Upper Lambourn, Hungerford, RG17 8RF
G7 BWO D Morgan, 26 Lyndhurst Road, Exmouth, EX8 3DT
G7 BWV R Fosbraey, 122 East Street, Sittingbourne, ME10 4RX
G7 BWW Michael Widdows, 27 Market Close, Barnham, Bognor Regis, PO22 0LH
G7 BXA P Austin, 24 Fairfield Terrace, Bramley, Leeds, LS13 3DH
G7 BXG F Clarke, 95 Kirklington Road, Rainworth, Mansfield, NG21 0JZ
G7 BXJ P Turner, 260 New Lane, Huntington, York, YO32 9LY
G7 BXL Michael Bickerton, 54 Swinnel Brook Park, Grane Road, Rossendale, BB4 4FN
G7 BXS R Wake, 55 Bearsdown Road, Eggbuckland, Plymouth, PL6 5TR
G7 BXU Stephen Welton, 18 Coningham Road, Reading, RG2 8QP
GM7 BYB A McIntyre, 18 Seal Craig Gardens, Altens, Aberdeen, AB12 3SH
G7 BYE J Hersom, 10 Young St., Gilesgate, Durham, DH1 2JU
G7 BYG A Found, 18 Mead Fields, Bridport, DT6 5RF
G7 BYI M Baldry, 10 Kingfisher Court, Lowestoft, NR33 8PJ
G7 BYK Ronald Barry, 14 Home Mead Creswicke Road, Bristol, BS4 1UQ
G7 BYN D Bendrey, 73 Kestrel Close, Chipping Sodbury, Bristol, BS37 6XB
G7 BYS J Pollard, 25 Bridgemere Close, Radcliffe, Manchester, M26 4FS
G7 BYU W Mcguffie, 1 Norbury Drive, Marple, Stockport, SK6 6LL
G7 BYV Terry Connolly, Windways, Brooke Road, Ashford, TN24 8HN
G7 BYW C Stone, 60 Staddon Park Road, Plymouth, PL9 9HJ
G7 BZC Roy Pickett, 1 The Cottage, Hospital Road, Wingland, Spalding, PE12 9YR
G7 BZD P Yates, Kingsomborne, The Broadway, Totland Bay, PO39 0BL
G7 BZE W Gillott, 14 Oakham Place, Barnsley, S75 2ND
G7 BZM R Brown, 2 Hay Green Close, Bournville, Birmingham, B30 1RQ
G7 BZQ G Reynolds, 187 Steelhouse Lane, Wolverhampton, WV2 2AU
G7 BZR James Shurmer, 126 Gaerwen Uchaf Estate, Gaerwen, LL60 6JW
G7 BZU A Mccoll, 105 North East Road, Southampton, SO19 8AF
G7 BZY Peter McFarland, 1 Felin Graig, Llangefni, LL77 7RL
G7 CAA N Royle, 62 Cynthia Close, Poole, BH12 3JW
G7 CAF S Bond, 38 Hampsfell Drive, Morecambe, LA4 4TU
G7 CAG A Malpass, 48 Geoffrey Barbour Road, Abingdon, OX14 2ES
G7 CAH Mark Astley, 30 Lon Ceirios, Newtown, SY16 1PR
G7 CAS W Welburn, 31 West Bank, Scarborough, YO12 4DX
G7 CBI Michael Higgins, 2 Walden Road, Keynsham, Bristol, BS31 1QW
G7 CBR A Phoenix, 1 Conway Road, Redcar, TS10 2EN
G7 CBU J Griffiths, 143 Brynglas, Hollybush, Cwmbran, NP44 7LL
G7 CBW S Duffield, 129 Badsey Road, Oldbury, B69 1BU
G7 CBY Leslie Cotton, 6 Blacksmith Row, Lytham St. Annes, FY8 4UE
G7 CBZ S Cotton, 6 Blacksmith Row, Lytham St. Annes, FY8 4UE
G7 CCH D Woods, 26 Compton Road, Southport, PR8 4HA
G7 CCL S Cullingworth, 66 Meadow Road, Garforth, Leeds, LS25 2EN
G7 CCR FLINTSHIRE RAYNET c/o M Ellett, 14 Canon Drive, Bagillt, CH6 6LS
G7 CCS G Oliver, 17 Jack Stephens Estate, Penzance, TR18 2QE
G7 CCV P UPTON, 73 Allington Close, Taunton, TA1 2NA
G7 CDI P Matthews, Finingley House, 128 Thealby Gardens, Doncaster, DN4 7EG
G7 CDO A Corps, 6a Salisbury Road, Leigh-on-Sea, SS9 2JX
G7 CDU G Prince, 75 Queens Avenue, Bromley Cross, Bolton, BL7 9BJ
G7 CEA David Smith, 62 Cheshire View, Brymbo, Wrexham, LL11 5AW
G7 CEB James Redpath, 41 Sandford Green, Banbury, OX16 0SB
G7 CEC John Evans, 33 Viscount Bridgeman Court, Queen Elizabeth Drive, Oswestry, SY11 2UF
G7 CED K Barnett, 126 Oldham St., Latchford, Warrington, WA4 1EX
G7 CEN T Martin, 10 Hardy Close, Galley Common, Nuneaton, CV10 9SG
G7 CEQ W Jones, 26 Cwm Silyn, The Park, Caernarfon, LL55 2AG
G7 CER C Riley-Moxon, 51 Nuttall Lane, Ramsbottom, Bury, BL0 9JX
G7 CEW D Everard, 6 Leith Hill Green, St. Pauls Cray, Orpington, BR5 2SB
G7 CEY Peter Copeland, 7 Stuart Avenue, Draycott, Stoke-on-Trent, ST11 9AA
G7 CFC Andrew Chamberlain, 11 Woden Crescent, Wolverhampton, WV11 1PR
G7 CFS David Halliday, 33 Brentnall Close, Great Sankey, Warrington, WA5 1XN
G7 CFT G Taylor, 48 Westwood Heath Road, Leek, ST13 8LL
G7 CFW K Morrison, 7 Turnstone End, Yateley, GU46 6PE
G7 CFX A Newell, 4 Dexter Square, Cricketers Way, Andover, SP10 5DB
G7 CGB M Biddulph, 26 Bramwell Close, Upper Stratton, Swindon, SN2 7SN
G7 CGC P Oliver, 32 Pearmain Way, Stanway, Colchester, CO3 0NP
G7 CGN S Hodkinson, 17 Thorn Well, Westhoughton, Bolton, BL5 2PJ
G7 CGT A Day, 8 The Garth, Ash, Aldershot, GU12 6QN
G7 CHB Roy Montague, 71 Middlethorpe Road, Cleethorpes, DN35 9PP
G7 CHC J Anderson, Tre Noce Cottage, 75 St. Michaels Road, Paignton, TQ4 5NA
G7 CIA Jonathan Adams, 14 Kedleston Close, Northampton, NN4 0WF
G7 CIH Charles Hayes, 2 Castleford House, Castle Road, Okehampton, EX20 1HZ
G7 CIK Michael Kielthy, 35 Alexandra Road, Sheringham, NR26 8HU
G7 CIQ A Wiseman, 61 Hilton Avenue, Horwich, Bolton, BL6 5RH
G7 CIT Thomas McGuigan, 18 Flexbury Gardens, Harlow Green, Gateshead, NE9 7TH
G7 CIV Barry Perrin, 8 Station Road, Gretton, Corby, NN17 3BU
G7 CIY Keith Gaunt, 21 Abbey Close, Rendlesham, Woodbridge, IP12 2UD
G7 CJC J Hughes, Hollies Bungalow, Valeswood, Shrewsbury, SY4 2LH
G7 CJD C Dyer, 6 Witcombe, Yate, Bristol, BS37 8SA
G7 CJG Graham Stringer, 13 Garfield Street, Kettering, NN15 7HX
G7 CJO Jason Graves, 7 Matthews Chase, Binfield, Bracknell, RG42 4UR
G7 CJS David Evans, 16 Cruden Road, Gravesend, DA12 4HD
G7 CJW J Wardle, 9 Leefield Road, Chapel-en-le-Frith, High Peak, SK23 0LF
G7 CKG Richard Varley, 23 Manor Court, Bingley, BD16 1QD
G7 CKL B Taylor, 32 Marples Avenue Mansfield Woodhouse, Mansfield, NG19 9HA

G7 CKP John Surman, 122 Burwell Meadow, Witney, OX28 5JQ
G7 CKQ Kenneth Hartley, [illegible] Raleigh Road, Sunderland, SR5 5RD
G7 CKS David Davies, 2 London Road, Battle, TN33 0EU
[illegible]
G7 CLH D Smith, 7 Dunlin Close, Norton, Stockton on Tees, TS20 1SJ
G7 CLM Janet Harrington, 9 High House Estate, Sheering Road, Harlow, CM17 0LL
G7 CLO C Gaukroger, Meadow View, Lansallos, Looe, PL13 2PU
G7 CLR F Charnley, 30 Dunkirk Avenue, Fulwood, Preston, PR2 3RY
G7 CLX K Marsden, 3 Lane Head, Heptonstall, Hebden Bridge, HX7 7PB
G7 CLY J Hill, 55 The Oval, Welton, Brough, HU15 1DA
G7 CMB David Kennett, 6 the Holdings, Hatfield, AL95HQ
GI7 CMC N Moore, 164 Ardenlee Avenue, Belfast, BT6 0AE
G7 CMF R Jones, Afallon, Rhostrehwfa, Llangefni, LL77 7YP
GU7 CMH G Simon, 3 Mahaut Villas, Collings Road, St. Peter Port, Guernsey, GY1 1FP
G7 CMI B Birch, 4 [illegible] Gardens, Higham Ferrers, Rushden, NN10 8NH
G7 CMM G Jones, 13 Palace Close, Flint, CH6 5YE
G7 CMN CHESHIRE COUNTY RAYNET c/o Bruce Williams, [illegible] Wolton Close, Wilmslow, SK9 6HD
G7 CMP B Fielding, The Copse, Charmouth Road, Axminster, EX13 5SZ
G7 CNC Dennis Gray, Flat 57, Wesley Court, 1 Millbay Road, Plymouth, PL1 3LB
G7 CND Joseph Bell, 2 Rake Lane, Milford, Godalming, GU8 5AB
GU7 CNI M Elliston, La Guillard Lane, St. Andrew, Guernsey, GY6 8YJ
G7 CNP H Weatherhead, 39 Meadow Park, Dawlish, EX7 9BU
GM7 CNW G Dryburgh, 86 Normand Road, Dysart, Kirkcaldy, KY1 2XP
G7 CNX P Hammond, 23 Peppers Close, Weeting, Brandon, IP27 0PU
G7 CNZ Kenneth Ford, 8 Blakedon Road, Wednesbury, WS10 7HY
G7 COA R Johnson, 7 West Parade, Warminster, BA12 8LY
G7 COB S Coburn, 54 Queensway, Hope, Wrexham, LL12 9PE
G7 COC N Whelan, 54 Boroughbridge Road, Northallerton, DL7 8BN
G7 COD Andrew Kitchen, Newton Hall Farm, Bank Newton, Skipton, BD23 3NT
G7 COG Jason Lister, 6 Fordlands Crescent, Fulford, York, YO19 4QQ
G7 COP P Payton, 3 Astor Close, Winnersh, Wokingham, RG41 5JZ
G7 COQ K Raxworthy, 9 Harrow Drive, Edmonton, London, N9 9EQ
G7 COU R Stormes, 1 Meadowbank, Belton, Doncaster, DN9 1NW
GM7 CPJ G Currie, The Old Post Office, Cuminestown, AB53 5TQ
GM7 CPL C Scott, 115 Tarvit Terrace, Springfield, Cupar, KY15 5SE
G7 CPN S Burgess, 59 Back Lane, Congleton, CW12 4PY
G7 CPQ C Ambrose, 47 Whitton Close, Swavesey, Cambridge, CB24 4RT
GM7 CPR J Wright, 9 Meadowhead Road, Plains, Airdrie, ML6 7JF
GM7 CPY Shirley Leggat, Ailach, St. Aethans Road, Elgin, IV30 2YR
G7 CQA Roy Ginn, 91 High St., Shoeburyness, Southend on Sea, SS3 9AR
G7 CQB D Locock, Bank House, Selattyn, Oswestry, SY10 7DX
G7 CQG Robert Bradshaw, 11 Glebe Avenue, Orton Waterville, Peterborough, PE2 5EN
G7 CQK I Phillips, Goldsworthy Farm, Stony Lane, Gunnislake, PL18 9BL
GU7 CQN Jeremy Gardner, The Ferns, Rue De La Girouette, St. Saviour, Guernsey, GY7 9NN
GM7 CQQ A Donaldson, 36 Rothes Park, Leslie, Glenrothes, KY6 3LH
G7 CQW A Riley, 4 Birtle Drive, Astley, Manchester, M29 7RE
G7 CQX D Read, 31 Grace Gardens, Bishop's Stortford, CM23 3EU
G7 CQZ E Courtnell, 12 Bray Street, Birkenhead, L41 8BX
G7 CRA I Brelsford, 78 Borough Road, Redcar, TS10 2EQ
G7 CRG CENTRAL CHESHIRE RAYNET GROUP c/o Peter Fox, 5 Llandovery Close, Winsford, CW7 1NA
G7 CRK S Halbertsma, 65 Gareth Grove, Bromley, BR1 5EG
G7 CRM D Stump, 9 Shipton Grove, Swindon, SN3 1BZ
G7 CRN R Phin, 35 Parkland Close, Newquay, TR7 3EB
G7 CRQ Adam Haywood, Flat 21, Atholl House, 178 Woodcote Road, Wallington, SM6 0PB
G7 CRR J Wheeler, 8 Slimbridge Close, Worcester, WR5 3SH
G7 CRS MAXPAK c/o M Hall, 35 Bunns Lane, Dudley, DY2 7RA
G7 CRU Paul Wallis, 6 Lancelott Court, Pershore, WR10 1RE
G7 CRV Ben Burnside, 4 Manor View Oswaldkirk, York, YO62 5YJ
G7 CRY James Bain, 7 Wrights Lane, Sutton Bridge, Spalding, PE12 9RH
G7 CSF J Plowright, 11 Coddington Street, Newport, United States
G7 CSI M Morris, 20 Bracken Way, Chobham, Woking, GU24 8PR
G7 CSJ K Chapman, 19 St. Johns Rise, Woking, GU21 7PN
GW7 CSK D Winter, 25 Pembroke St., Thomastown, Porth, CF39 8DU
G7 CSL R Dent, 9 Dovey Close, St. Ives, Huntingdon, PE27 6HW
G7 CSM Robert Knight, 32 Linnet Close, Abbeydale, Gloucester, GL4 4UA
G7 CSS M Budd, 2 Orpwood Close, Hampton, TW12 3YE
G7 CST Adrian Clarke, 86 Roebuck Road, Walsall, WS3 1AL
G7 CSV SPEN VALLEY ARS c/o T Clough, 37 Park Avenue, Mirfield, WF14 9PB
G7 CSX A Keen, 20 Horam Park Close, Horam, Heathfield, TN21 0HW
G7 CTE Nigel Farmer, 8 Mill Hill Lane, Sandbach, CW11 4PN
G7 CTG E Ives, 15 Northlands, Adwick-le-Street, Doncaster, DN6 7AX
G7 CTI S Lock, 17 Elgar Crescent, Droitwich, WR9 7SP
GM7 CTV A Smith, 13 Park Terrace, Markinch, Glenrothes, KY7 6BN
GI7 CTW E Regan, 4 Lecumpher Road, Desertmartin, Magherafelt, BT45 5LY
G7 CUA Robert Cookson, Briarswood, Snow Hill Lane, Preston, PR3 1BA
G7 CUB D Price, Summer Fields, Mayfield Road, Oxford, OX2 7EN
G7 CUD J Richardson, 68 Place Farm Way, Monks Risborough, Princes Risborough, HP27 9JY
G7 CUF Gregory Adrian, Flat 43, Farriers House, London, EC1Y 8TB
G7 CUL R Bourn, 7 Clitheroes Lane, Freckleton, Preston, PR4 1SD
G7 CUO John Folland, 41 Rydal Avenue, Billingham, TS23 1HX
G7 CUP Phillip Ingle, 8 Burnstone Gardens, Moulton, Spalding, PE12 6PS
G7 CUU David Stainforth-Small, 10 Balland Park, Ashburton, Newton Abbot, TQ13 7BS
G7 CUW B Minish, Raheens, Castlebar, Ireland
G7 CUY Kevin Martin, 8 Taylors Close, Meppershall, Shefford, SG17 5NH
G7 CVA E Curnow, 9 Moreton Bay, Bilton, Hull, HU11 4ER
G7 CVC Maurice Porter, 19 Thistle Downs, Northway, Tewkesbury, GL20 8RE
G7 CVF S Evans, 34 Kent Road, Southport, PR8 4BJ
G7 CVM D Illman, 27 Blackborough Road, Reigate, RH2 7BS
G7 CVY K Helgesen, 13 Mara Court, White Road, Chatham, ME4 5TW
G7 CVZ Andrew Bevins, 12 Wheatstone Road, Formby, Liverpool, L37 6BF

G7 CWE D Ishmael, 38 Greenford Close, Orrell, Wigan, WN5 8RH
G7 CWI I Green, 25 Riley Avenue, Lytham St. Annes, FY8 1HZ
G7 CWM James Denton, 48 Seas End Road, Surfleet, Spalding, PE11 4DQ
[illegible] 2 Main Road, Shortwood, Mangotsfield, BS16 9NH
G7 CWO Daren Ward, 107 Oundle Road, Birmingham, B44 8EH
G7 CWT Michael Joy, Cheddon Corner, Cheddon Fitzpaine, Taunton, TA2 8LD
G7 CXB T Mcinnes, 7 Hilary Drive, Merry Hill, Wolverhampton, WV3 7NJ
G7 CXM D Lindsay, Avda De Las Delicias, 41, 46183 La Eliana, Valencia, Spain
G7 CXO V Cassar, 51 Aylesford Avenue, Beckenham, BR3 3SR
G7 CXT S Haywood, 7 Stamford Gardens, Dagenham, RM9 4ET
G7 CXU S Power, 8 Green Lane, Chislehurst, BR7 6AG
G7 CYD Alan Jenkins, 15 Tilstone Avenue, Eton Wick, Windsor, SL4 6NF
G7 CYF C Wardill, 85 Station Road, Chellaston, Derby, DE73 5SU
G7 CYN Calvin Casper, 22 Eshton Road, Gargrave, Fell View, Skipton, BD23 3SE
G7 CYQ Edward Hornby, 14 Essex Road, Stevenage, SG1 3EZ
GW7 CYT David Phillips, 15 Herbert Street, Treorchy, CF42 6AW
GM7 CZC R Johnson, 3 Hamilton Gardens, Edinburgh, EH15 1NH
G7 CZF Jonathan Jenkins, 18 High Beeches, Gerrards Cross, SL9 7HX
G7 CZI Graham Miles, 7 Dobbin Close, Rawtenstall, Rossendale, BB4 7TH
GM7 CZU J McLaughlan, 2 Donaldson Drive, Irvine, KA12 0QG
G7 DAB Ian Weeks, 19 St. Michaels Road, Tunbridge Wells, TN4 9JG
G7 DAH C Moore, 168 Church End Lane, Runwell, Wickford, SS11 7DN
GM7 DAJ Robert Hepburn, 44 MacIndoe Crescent, Kirkcaldy, KY1 2JG
G7 DAL J Ratigan, 81 Cunningham Drive, Unsworth, Bury, BL9 8PD
GM7 DAP Alan Lord, 5 Windsor Terrace, Brechin, DD9 6SD
G7 DAR Richard Charlton, 13 Hollywood Avenue, Walkerville, Newcastle upon Tyne, NE6 4TN
G7 DAZ James Watson, Flat 1, 53 Castle Street, Bolton, BL2 1AD
G7 DBN Edward Turner, Rectory Cottage, Little Marsh, Marsh Gibbon, Bicester, OX27 0AP
G7 DBO I Leaver, 2 Marnhull Close, Coventry, CV2 2JS
G7 DBT Ron Claridge, 3 Wentworth Avenue, Leagrave, Luton, LU4 9EN
G7 DBV D Ritson, C/O 12 Tudor Grange, Easington Village, Peterlee, SR8 3DF
GI7 DBZ W Hollinger, 51 Collin Road, Ballyclare, BT39 9JS
G7 DCF Alan McLennan, Flat 2, 86 Chatsworth Road, Croydon, CR0 1HB
G7 DCJ S Warner, 96 Walter Nash Road East, Kidderminster, DY11 7BY
G7 DCM Lawson Bant, 58 Severn Way, Cressage, Shrewsbury, SY5 6DS
G7 DCT A Horsfall, 2 Temple Walk, Halton, Leeds, LS15 7SQ
G7 DDD P Scott, 18 Westhaven Court, Station Road, Nuneaton, CV13 0PR
G7 DDF Phil Johnson, 6 Rugby Road, Lilbourne, Rugby, CV23 0SP
G7 DDN Christopher Rolinson, 534 Haslucks Green Road, Shirley, Solihull, B90 1DS
G7 DDQ W Roberts, 12 Camberley Drive, Penn, Wolverhampton, WV4 5RP
G7 DDR R Murray, 8 Church Lane, Kirk Langley, Ashbourne, DE6 4NG
G7 DDV Christopher James, 15 Willow Crescent, Warminster, BA12 9LH
G7 DEC A Grundy, 647 Preston Old Road, Feniscowles, Blackburn, BB2 5ER
G7 DEE K Denniss, 4 Aysgarth Road, Sheffield, S6 1HU
G7 DEG R Williams, 6 Ralfland View, Shap, Penrith, CA10 3PF
G7 DEH H Carpenter, 44 Bowbridge Road, Newark, NG24 4BZ
G7 DEI Steven Courtney-Crowe, 28 Brymore Close, Prestbury, Cheltenham, GL52 3DY
G7 DEU G Renton, 58 St. Christophers Road, Humberston, Grimsby, DN36 4EA
G7 DEY P Knowles, 35 Raby Park Road, Neston, CH64 9SW
G7 DFC Jack Worsnop, 1217 Thornton Road, Thornton, Bradford, BD13 3BE
GM7 DFI K Trinder, 29 Woodside Road, Brookfield, Johnstone, PA5 8UB
G7 DFP J Fitzpatrick, 22 Ferry Road, Surlingham, Norwich, NR14 7AR
G7 DFV G Jelley, 28 Blanches Road, Partridge Green, Horsham, RH13 8HZ
G7 DFW A Crisp, 17 Gaitskell House, Howard Drive, Borehamwood, WD6 2PB
G7 DFX G Allan, Kent House, 106 Kent Road, Sheffield, S8 9RL
G7 DGC M Lewis, 21 Woodlands Road, Ashton-under-Lyne, OL6 9DU
G7 DGD Brian Walton, 28 Durham Terrace, Durham, DH1 5EH
G7 DGE Albert Biggin, 14 Coultas Avenue, Deepcar, Sheffield, S36 2PT
G7 DGF David Coupe, 22 West Street, South Normanton, Alfreton, DE55 2AJ
G7 DGP B Barrass, 7 The Crescent, Easton on the Hill, Stamford, PE9 3LZ
GM7 DHA Kevin Pugh, 1 Barrington Gardens, Beith, KA15 2BA
G7 DHD D Dyson, 5 Warwick Street, Church, Accrington, BB5 4AL
G7 DHG A Gardner, 28 Usk Court, Thornhill, Cwmbran, NP44 5UN
G7 DHJ Michael Harding, 79 Beaumont Walk, Leicester, LE4 0PP
G7 DHM W Holt, 20 Lingfield Mount, Leeds, LS17 7EP
G7 DHQ G Davies, 17 Remington Road, Walsall, WS2 7EJ
G7 DHW David Neale, Greyhills Farm, Diptford, Totnes, TQ9 7NQ
G7 DIB A Finon, Radford House, Hall Lane, Wymondham, NR18 9TB
G7 DIE Stephen Salmon, 35 Westgate Road, Lytham St. Annes, FY8 2SG
G7 DIG Randolph Dee, 10 Sanderson Street, Coxhoe, Durham, DH6 4DG
G7 DIL P Davis, The Willows, Park View, Cwmbran, NP44 1RB
G7 DIO Christopher Harper, 132 Park Avenue Eas, Dallas, United States, GA 30157
G7 DIR A Brinton, 136 Efford Road, Plymouth, PL3 6NQ
GI7 DIT D Roberts, 6 Plantation Road, Bangor, BT19 6AF
G7 DIU David Leech, 4 Rydal Close, Huntingdon, PE29 6UF
G7 DIW A Saul, 18 Elm Bank Close, Cubbington, Leamington Spa, CV32 6LR
G7 DIZ M Beatrup, 34 Springfield Drive, Halesowen, B62 8EU
G7 DJI T Davis, 127 Lon Glanyrafon, Newtown, SY16 1QT
G7 DJN A Coates, 19 Bretton Avenue, Bolsover, Chesterfield, S44 6XN
G7 DJT D Tinley, 2 Rosemount Close, Loose, Maidstone, ME15 0AJ
G7 DKB D Simons, 65 Dolphin Court Road, Paignton, TQ3 1AB
G7 DKY Graham Rowntree, 59 Greenside, Euxton, Chorley, PR7 6AS
G7 DKZ J Stearn, Half Acre, Hatch Green, Taunton, TA3 6TN
G7 DLD R Hilton, 9 Waterloo Fields, Kingswood, Welshpool, SY21 8LF
G7 DLE J Hodges, 70 Chestnut Drive, Brixham, TQ5 0DD
GM7 DLY J Whitcomb, 1/1 30 Highburgh Road, Glasgow, G12 9DZ
G7 DME Robert Gornall, 29 Park View, Wetherden, Stowmarket, IP14 3JT
G7 DMG William Hetherington, 32 Almond Way, Lutterworth, LE17 4XJ
G7 DMH Stuart Hetherington, 8 Kings Lane Yelvertoft, Northampton, NN6 6LX
G7 DMK P Drage, Cranford House, 167 Rockingham Road, Kettering, NN16 9JA
GM7 DMN Yvonne Benting, Suthainn, Askernish, Isle of South Uist, HS8 5SY
G7 DMP J Barnes, 26 Fairthorn Road, Sheffield, S5 6LX
G7 DMQ Simon Rafferty, 22 Hengist Close, Horsham, RH12 1SB
G7 DMS M Horsfall, 8 Greenbrook Road, Burnley, BB12 6NZ

G7 DMX N Taylor, 5 Miranda Road, Preston, Paignton, TQ3 1LE
G7 DMZ B Knight, Conchardrin, Tibberton, Gloucester, GL2 8EB
G7 DNF Tom Haye, Woodside, Sutton Wood Lane, Alresford, SO24 9SG
G7 DNG I Casey, 38 Wordsworth Road, Salisbury, SP1 3BH
[illegible]
GJ7 DNJ I Meade, Etape De Base, La Rue Des Platons, Trinity, Jersey, JE3 5AA
G7 DNM E Ashworth, 10 Wisteria Drive, Lower Darwen, Darwen, BB3 0QY
GD7 DNP Atul Patel, 33 Christchurch Avenue, Harrow, HA3 8ND
G7 DNQ S Howard, 5 Grummock Avenue, Ramsgate, CT11 0RR
G7 DNR K Crookes, 64 Heron Drive, Audenshaw, Manchester, M34 5QX
G7 DNT Keith Handscombe, 8 Fletcher Road, Ipswich, IP3 0LF
G7 DNV L Carter, 3 Cleviscroft, Stevenage, SG1 1UJ
G7 DNX M Morris, 4 Meadow Brook Road, Northfield, Birmingham, B31 1NE
G7 DOA David Morris, Flat 3, 748 Melton Road, Leicester, LE4 8BD
G7 DOE Ron Mount, 4 Hermitage Road, Abingdon, OX14 5RN
G7 DOF K Mount, 4 Hermitage Road, Abingdon, OX14 5RN
G7 DOL Royal Naval ARS c/o Joe Kirk, 111 Stockbridge Road, Chichester, PO19 8QR
G7 DOR Dorking & District RS c/o George Brind, 9 Becket Wood, Newdigate, Dorking, RH5 5AQ
G7 DOS B Smith, 7 School Walk, Chase Terrace, Burntwood, WS7 1NQ
G7 DOW G Smith, 59 Radipole Lane, Weymouth, DT4 9RN
G7 DOY J Baddeley, 22 Scott Road, Denton, Manchester, M34 6FT
G7 DPF G Brightman, 5 Meadow Rise, Lacey Green, Princes Risborough, HP27 0QY
GD7 DPG J Wrigley, 20 Fairy Hill Close, Ballafesson, Port Erin, Isle of Man, IM9 6TJ
GM7 DPI Philip Blacklaw, Flat 12, Servite House, Tayport, DD6 9JZ
G7 DPR F Overbury, 47 The Maltings, Dunmow, CM6 1BY
G7 DPU David Reynolds, 19 Clipstone Close, Wigston, LE18 3QS
G7 DPV A Marks, 7 Saunby Close, Arnold, Nottingham, NG5 7LA
G7 DPW A Butterworth, 3 Fir Tree Avenue, Worsley, Manchester, M28 1LP
G7 DPZ E Curd, 11 Ashkirk Close, Waldridge, Chester le Street, DH2 3HY
G7 DQA J Hallin, 12 Church Park Road, Plymouth, PL6 7SA
G7 DQC Peter Anderson, 11 Kelly Close, St. Budeaux, Plymouth, PL5 1DS
G7 DQE M Meerman, University Of Surrey, Dept Elec. Eng., Guildford, GU2 5XH
G7 DQL P Perkins, 29 Parkhill, Middleton, King's Lynn, PE32 1RJ
G7 DQQ Michael Lampett, 130 Clopton Road, Birmingham, B33 0RL
G7 DQZ D Keen, 5 London Road, Uckfield, TN22 1HU
G7 DRD T Platt, 30 Great Ellshams, Banstead, SM7 2BA
G7 DRG R Moseley, 307 Archer Road, Stevenage, SG1 5HF
G7 DRO W Webber, Springfield Lodge, Broadway, Winscombe, BS25 1UE
G7 DRR David Bellinger, Holly Cottage, Deacons Lane, Thatcham, RG18 9RJ
G7 DRT M Dickinson, 1 Tregaron Avenue, Cosham, Portsmouth, PO6 2JU
G7 DRU A Tink, 13 The Wicketts, Bristol, BS7 0SR
G7 DRW V Wynn, 12 Holly Road, Orpington, BR6 6BE
G7 DRX J Stelmasiak, Golden Rocks, Coed Lane, Montgomery, SY15 6AB
GM7 DRY S Graham, 15 Stone Crescent, Mayfield, Dalkeith, EH22 5DT
G7 DSA S Jeffery, 7 Corfe Way, Winsford, CW7 1LU
GU7 DSB P Blampied, 9 Rue Des Grons Estate, St. Martin, Guernsey, GY4 6JT
G7 DSO P Yarnold, 162 Harwill Crescent, Aspley, Nottingham, NG8 5LF
G7 DSQ R Roberts, 16 Poplar Drive, Pucklechurch, Bristol, BS16 9QF
G7 DST Derek Thomson, 96 Rydal Avenue, Loughborough, LE11 3RX
G7 DSU Christopher Tong, 24 James Road, Cuxton, Rochester, ME2 1DJ
G7 DSV Dennis Crinson, 12 Atherton Street, Stockport, SK3 9JN
GJ7 DTA A Lange, Les Bois, La Rue de la Pointe, St. Peter, Jersey, JE3 7AQ
GW7 DTB R Dixon, 19 The Burrows, Porthcawl, CF36 5AJ
GM7 DTC J Arthur, 15 St. Andrews Place, Beith, KA15 1JE
G7 DTG S Le Poer Trench Brown, 71 Studland Park, Westbury, BA13 3HN
G7 DTK T Timms, 16 Claverdon Road, Coventry, CV5 7HP
G7 DTR Martin Penny, 79 Grove Avenue, New Costessey, Norwich, NR5 0JA
G7 DTS A Lowe, 33 Dandies Chase, Eastwood, Leigh-on-Sea, SS9 5RF
G7 DTT Alan Reeves, Pendeen, 13 Maple Way, Leavenheath, Colchester, CO6 4PQ
G7 DTV H Partridge, 11 Elm Close, Stourbridge, DY8 3JH
G7 DUB P Spencer, 11 Two Trees Estate, Wadebridge, PL27 7PG
G7 DUC B Tonkin, 9 Penhallick Road, Carn Brea, Redruth, TR15 3YJ
G7 DUE W Davis, 6 Bushy Mead, Waterlooville, PO7 5DY
G7 DUI C Teague, 20 Herriot Grove, Bircotes, Doncaster, DN11 8ET
G7 DUK M Clarke, 12 Moat Court, Shaw Close, Chertsey, KT16 0PH
G7 DUY Raymond Jones, 14 Lunsford Road, Liverpool, L14 0NU
GD7 DUZ S Kelly, 44 Westhill Avenue, Castletown, Isle of Man, IM9 1HY
GW7 DVJ D Maxted, 33 Bryn Dryslwyn, Bridgend, CF31 5BT
G7 DVO Terry Spearing, 139 Holt Road, Hellesdon, Norwich, NR6 6UA
GI7 DWF J Murphy, 19 Kilburn Park, Armagh, BT61 9HA
G7 DWH E Shaw, 5 Charlock Grove, Cannock, WS11 7FR
G7 DWI Andrew Davies, 23 Holly Place, Eastbourne, BN22 0UT
G7 DWM C Hadjigeorgiou, 26 Priory Gardens, Hampton, TW12 2PZ
G7 DWN A Keen, 29 Churchman Close Melton, Woodbridge, IP12 1RN
G7 DWO S Prisk, 86 Wycliffe Grove, Werrington, Peterborough, PE4 5DF
G7 DWU Michael Stapleton, 15 Haviland Way, Cambridge, CB4 2HA
G7 DWV K Webster, 9 King John Avenue, Fareham, PO16 9AP
G7 DWX M Twells, Camels, Annscroft, Shrewsbury, SY5 8AN
G7 DWY Ian Barraclough, Maru, 25 Blaithroyd Lane, Halifax, HX3 9PS
G7 DXB Lynne Watson, 40 Dalesman Drive, Carlisle, CA1 3TH
G7 DXC J Maxwell, Tysties, Tile Barn, Newbury, RG20 9UY
GM7 DXE A Todd, Waterfurrows, Breakachy, Beauly, IV4 7AE
G7 DXN Patrick Chapman, 1 Bader Close, Watton, Thetford, IP25 6FF
G7 DXQ K Salt, 4 Burghwood Road, Ormesby, Great Yarmouth, NR29 3LT
GM7 DXT J MacLeod, 59 Fife Street, Keith, AB55 5EG
G7 DXV Peter Shepherd, 25 Tomkins Close, Stanford-le-Hope, SS17 8QU
G7 DXX J Walker, 121 Park Drive, Upminster, RM14 3AU
G7 DYB A Rudling, 1 St. Anthonys Close, Ottery St. Mary, EX11 1EN
G7 DYD Michael Green, 59 Brand End Road, Butterwick, Boston, PE22 0JD
G7 DZD Kevin Quinlan, Polidoris Cottage, Polidoris Lane, High Wycombe, HP15 6XD
GI7 DZE S Thompson, 19 Windsor Heights, Larne, BT40 1UL
GM7 DZK J Malone, 8 St. Margarets Crescent, Polmont, Falkirk, FK2 0UP
G7 DZR G Shand, 5 Bromyard Drive, Chellaston, Derby, DE73 6PF
G7 DZY B Daniel, Tamar Bay Road, Freshwater Bay, PO40 9QS

UK Callsigns

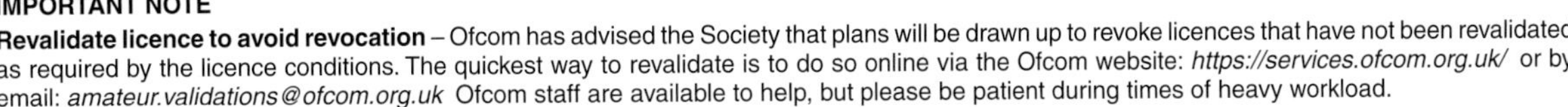
**IMPORTANT NOTE**

**Revalidate licence to avoid revocation** – Ofcom has advised the Society that plans will be drawn up to revoke licences that have not been revalidated as required by the licence conditions. The quickest way to revalidate is to do so online via the Ofcom website: *https://services.ofcom.org.uk/* or by email: *amateur.validations@ofcom.org.uk* Ofcom staff are available to help, but please be patient during times of heavy workload.

G7 EAH P Stewart, 34 North Park Road, Bramhall, Stockport, SK7 3JS
G7 EAQ D Potten, 151 Sherborne Road, Yeovil, BA21 4HF
G7 EAR ECHELFORD ARS c/o Patrick Gray, 2 Bryan Close, Sunbury-on-Thames, TW16 7UA
G7 EAT J Hatfield, 22 Blackhorse Crescent, Amersham, HP6 6HP
G7 EBF Paul Langfield, 88 Counthill Road, Oldham, OL4 2PE
G7 EBI Michael Evans, 114 Penwill Way, Paignton, TQ4 5JW
G7 EBL Grayhame Orlebar, 21 Field Lane, Willersey, Broadway, WR12 7QB
G7 EBR M'HD & E.BERK RD c/o Robert Mclachlan, Heathersett, Lightlands Lane, Maidenhead, SL6 9DH
G7 EBX W Parkin, 40 Cliffe Avenue, Carlin How, Saltburn-by-The-Sea, TS13 4DT
G7 ECA Caleb Price, 18 Armley Park Road, Leeds, LS12 2PG
G7 ECE S Holmes, 31 Brightside Avenue, Staines, TW18 1NE
G7 ECG K thompson, 32 The Crescent, Pattishall, Towcester, NN12 8NA
G7 ECQ N Murray, East End House, Oak Lane, Sheerness, ME12 3QR
G7 ECU EASTBOURNE & WEALDEN RAYNET c/o P Appleby, Flat 14, Maryan Court, Hailsham, BN27 3DJ
G7 EDA R Woodcock, 5 Walton Road, Sidcup, DA14 4LJ
G7 EDF D Hall, 29 Airedale Avenue, Tickhill, Doncaster, DN11 9UH
G7 EDK I Barkley, 39 Fulbeck Avenue, Wigan, WN3 5QN
G7 EDZ Mark Hedges, 17 Fairview Road, Dudley, DY1 2RT
G7 EED N Winter, 3 Princes Boulevard, Bebington, Wirral, CH63 5LH
G7 EEE Anthony MOTHEW, 7 Ashfields, Loughton, IG10 1SB
G7 EEG T Cole, Walden, The Common, Lydney, GL15 6NT
G7 EEJ D Swift, 63 Guiness Trust Buildings, Fulham Palace Road, London, W6 8BD
G7 EEN S Paget, 2 Willow Cottages, Huxley Lane, Chester, CH3 9BE
GM7 EEY Peter Webster, 6 Hillcrest Avenue, Kirkcaldy, KY2 5TU
G7 EFG M Marston, Apt 12b La Palmeras De Benavista, Calle Virgo, Malaga, Spain, 29680
G7 EFL Andy Crooks, 7 The Cleave, Harpenden, AL5 5SJ
G7 EFV W Waterton, 23 Mill Drive, Leven, Beverley, HU17 5NR
G7 EGQ Ian Harrop, 35 Langdale Crescent, Dalton-in-Furness, LA15 8NR
G7 EGU Peter Adams, 12 The Birches, Benfleet, SS7 4NT
G7 EGX Michael Miller, 12 Leighfields Avenue, Leigh-on-Sea, SS9 5NN
GW7 EHD D Cotton, 135 Main Road, Bryncoch, Neath, SA10 7TW
GM7 EHN J Baird, 26 Bearside Road, Stirling, FK7 9BY
G7 EHR R Watson, 7 Glatton Road, Sawtry, Huntingdon, PE28 5SY
G7 EHS Ian Tooley, L'Eree, Burnham Road, Chelmsford, CM3 6DP
G7 EHU B Walley, 52 Main St., Rosliston, Swadlincote, DE12 8JW
G7 EHY David Robinson, 130 Magnolia Drive, Colchester, CO4 3LX
G7 EIA S Ralph, 70 Mickleburgh Hill, Herne Bay, CT6 6DX
G7 EIE Timothy Jacobs, 43 Winfields, Pitsea, Basildon, SS13 1HA
G7 EIK G Edlin, 2 Ashby Road, Donisthorpe, Swadlincote, DE12 7QG
G7 EIS E Shaddick, 6 Haylands, Portland, DT5 2JZ
G7 EJH J Tyerman, 7 Veronica Close, Branston, Lincoln, LN4 1PU
G7 EJK John Turner, 50 The Plain, Brailsford, Ashbourne, DE6 3BZ
G7 EJN Graham Prosser, 23 Guiting Road, Selly Oak, Birmingham, B29 4RD
G7 EJO I Oxley, 29 The Gables, Newhall, Swadlincote, DE11 0TG
G7 EKC Andrew Taylor, 106 Raeburn Avenue, Surbiton, KT5 9EA
G7 EKD P Kneebone, 25 Rookery Close, Fenny Drayton, Nuneaton, CV13 6BB
G7 EKG P Ainscow, 10 Rectory Road, Felling, Gateshead, NE10 9DH
G7 EKJ S Cox, 137 Perry Walk, Blackrock Road, Birmingham, B23 7XL
G7 EKL Nicholas Denker, 103 Springhill Road, Burntwood, WS7 4UJ
G7 EKM K McGeough, 57 Stonehouse Park, Thursby, Carlisle, CA5 6NS
G7 EKT G Aucott, 21 Riverway, Wednesbury, WS10 0DN
G7 EKW Stuart Foxall, 25 Western Road, Sutton Coldfield, B73 5SP
G7 ELA M Dunn, 45 Chaddock Lane, Worsley, Manchester, M28 1DE
G7 ELC Neil Fountain, The Venture, Green Lane, Huntingdon, PE28 5YE
G7 ELE Leslie Dean, 2 Ellar Ghyll Cottage, Ellar Ghyll, Otley, LS21 3DN
GD7 ELF Chris Dennis, 1 Ballanoa Meadow, Santon, Isle of Man, IM4 1HQ
G7 ELG Alan Scarisbrick, 106 Edward Street, Grantham, NG31 6JG
G7 ELH Keith Graham, 44a Adelaide Drive, Colchester, CO2 8UB
G7 ELS A Bartram, 47 Temple Gate Crescent, Leeds, LS15 0EZ
G7 ELV Peter Lockwood, 61 Beverley Road, Whitley Bay, NE25 8JQ
G7 ELX John Northfield, 48 Gleanings Drive, Halifax, HX2 0PA
G7 ELZ T Leahy, 31 Harrow Road, Carshalton, SM5 3QH
G7 EME WORCS MNBOUNCE c/o D Palmer, Spidrift, Landsdown Road, Malvern, WR14 1HX
G7 EMH A Adem, 38 Cantley Gardens, Ilford, IG2 6QA
GW7 EMO David Brough, 19 Cameron Street, Cardiff, CF24 2NW
GW7 EMV M O'Reilly, 40 St. Anthony Road, Heath, Cardiff, CF14 4DJ
G7 EMZ Ian Mellor, 124 Ryknield Road, Kilburn, Belper, DE56 0PF
G7 ENA D Neal, 33 Swallow Drive, Louth, LN11 0DN
G7 ENC D Flatters, 7 Cornwall Crescent, Diggle, Oldham, OL3 5PW
G7 ENM S Yates, 60 Leamington Road, Branston, Burton-on-Trent, DE14 3HX
G7 ENQ W Donald, 15 Kingsland Parade, Portobello, Dublin, Ireland
G7 ENR Clive Davies, 138 Cannons Gate, Clevedon, BS21 5HN
G7 ENS N Swift, 19 Carlton Road, Caversham, Reading, RG4 7NT
G7 ENT Steven Alexander, 18 Southbourne Grove, Hockley, SS5 5EE
G7 EOA G Harrold, Birches, 2 Mill Hill Lane, Wymondham, NR18 9DD
G7 EOC C Hopkins, 16 Maypole Road, Gravesend, DA12 2LP
G7 EOE Esther David, 105 Kingsdown Park, Whitstable, CT5 2DH
G7 EOG Carl Flynn, 2 Trafalgar Avenue, Grimsby, DN34 5RE
G7 EOH Grahame Newnham, 22 Warren Place, Calmore, Southampton, SO40 2SD
G7 EOK P Wilson, 18 Riversdale Road, Halton, Runcorn, WA7 2AP
G7 EPE D Chamberlain, 33 Drake Close, New Milton, BH25 5JG
G7 EPL L Wong, 320 Wilbraham Road, Chorlton cum Hardy, Manchester, M21 0UX
G7 EPM John Enever, 21 Waldegrave Way, Lawford, Manningtree, CO11 2DT
G7 EPN Jonathan Webb, 36 Westfield Drive, Knutsford, WA16 0BN
G7 EPR R Jeeves, 78 Willowhale Green, Bognor Regis, PO21 4LW
G7 EPX B Coffin, 2 Pound Farm Close, Hilperton Marsh, Trowbridge, BA14 7PZ
G7 EPY W Wright, 53 Forshaw Avenue, Grange Park Estate, Blackpool, FY3 7PW
G7 EQG J Sturman, 41 Jay Close, Haverhill, CB9 0JR
G7 EQK Alan Grime, 23 Claremont Road, Milnrow, Rochdale, OL16 4EZ
G7 EQO M Ryan, 13 Normandy Road, Heavitree, Exeter, EX1 2SR
G7 EQR D Gammans, 35 Chute Avenue, High Salvington, Worthing, BN13 3DS
G7 EQX Paul Wickers, 4 Little Foxburrows, Colchester, CO2 7UG
G7 ERC HASTINGS & ROTHER RAYNET GRP c/o G Hodge, 8 Stainsby Street, St. Leonards-on-Sea, TN37 6LA
G7 ERH V Houghton, 8 Wheatfield Road, Cronton, Widnes, WA8 5BU
G7 ERI A Brown, 3 Wyebank View, Tutshill, Chepstow, NP16 7DR
G7 ERQ George Prowse, 1 Woodview, Penryn, TR10 8QA
G7 ERS J Hauton, 15 Bourne Close, Lincoln, LN6 7DR
G7 ESE Chas Tully, 19 Glyn Place East Melbury, Shaftesbury, SP7 0DP
GW7 ESF T Ford, 14 Hillsnook Road, Ely, Cardiff, CF5 5DD
G7 ESI S Gregory, 73 Princess Way, The Walshes, Stourport-on-Severn, DY13 0EL
GM7 ESM J Grundey, 8 Fraser Avenue, Blairgowrie, PH10 6QJ
G7 ESO K Reynolds, 20 Wentwood Gardens, Plymouth, PL6 8TD
GD7 ESU C Ellis, Ballahams, 3 Glen Road, Laxey, Isle of Man, IM4 7AP
G7 ESX Stephen Bowman, 39 Pearson Street, Spennymoor, DL16 6HP
G7 ESY Ian Bowman, 22 Bryan Street, Spennymoor, DL16 6DW
G7 ETC Simon Abel, 121 Angela Road, Horsford, Norwich, NR10 3HF
G7 ETK Stewart Yates, 14 Wright Street, Horwich, Bolton, BL6 7HZ
G7 ETM Andrew Sherratt, 22 Lane Green Avenue, Codsall, Wolverhampton, WV8 2JT
G7 ETS S Hambleton, 6 Wylde Green Road, Sutton Coldfield, B72 1HB
G7 EUB Richard Tabor, 10 Lone Pines Close, Matchams Lane, Christchurch, BH23 6LP
G7 EUF G Rhodes, 54 Chell Green Avenue, Stoke-on-Trent, ST6 7JY
G7 EUG William Moore, 4b Colville Road, Newton, Wisbech, PE13 5HH
GW7 EUL Michael Prince, Courthope, Cardigan, SA43 2LA
G7 EUT D Richards, Orchard Cottage, Ashbourne Road, Ashbourne, DE6 4NJ
G7 EVC Phillip Stone, 2 The Russets, Lees Close, Ashford, TN25 6RW
G7 EVF G Keene, 28 Little Hoddington, Upton Grey, Basingstoke, RG25 2RN
G7 EVG Garry Nicholas, Room 2 Middle Corridor, Abergele Hospital, Abergele, LL22 8DP
G7 EVI John Galbraith, 13 Simeons Walk, Quarry Bank, Brierley Hill, DY5 2EL
G7 EVK A Wade, 3 Ashendene Grove, Stoke-on-Trent, ST4 8NW
G7 EVP M Pritchard, 155 Elliott Road, March, PE15 8HF
G7 EVQ M Jordan, 139 Camping Hill, Stiffkey, Wells-next-The-Sea, NR23 1QL
G7 EVR Stephen Sprint, 14 Monterey Drive, Allerton, Bradford, BD15 9LP
G7 EVT C Mills, 16 Broom Close, Wath-upon-Dearne, Rotherham, S63 7JU
G7 EVY Graham Lawton, 35 Liverpool Road, Rufford, Ormskirk, L40 1SA
G7 EWA A Hampton, 5 Willow Crescent, Worthing, BN13 2SU
G7 EWD D Siviter, Cilgeraint Farm, St. Anns Bethesda, Bangor, LL57 4AX
G7 EWH D Hughes, Balshult, Eriksmala, Sweden, 361 94
G7 EWK V Thomas, 4 The Common, Whissonsett, Dereham, NR20 5SZ
G7 EWL Shane Hogarth, 34 High Street, Irthlingborough, Wellingborough, NN9 5TN
G7 EWS P Breck, 209 Eden Park Avenue, Beckenham, BR3 3JW
G7 EWX Neil Price, 245 Anchor Road, Longton, Stoke-on-Trent, ST3 5DX
G7 EWY Paul Kember, 2 Sandhills Crescent, Wool, Wareham, BH20 6HB
G7 EXD R Fletcher, 160 Barnsley Road, Denby Dale, Huddersfield, HD8 8QW
G7 EXH Barrie Mee, Anncott, Hylas Lane, Rhyl, LL18 5AG
G7 EXO K Brown, 15 Gloucester Road, Aldershot, GU11 3SL
GW7 EXQ Richard Morris, Llainlan, Pontyberem, SA15 5HP
G7 EXT H Jarvis, Dovecote Farm, Patmans Lane, Boston, PE22 8QJ
G7 EXX E Gwilliam, 15 Sheppard Way, Minchinhampton, Stroud, GL6 9BZ
G7 EXZ S Hobbs, 15 The Valley, Salisbury, SP2 9EJ
G7 EYE S Finnegan, 25 Westcliff Gardens, Margate, CT9 5DT
G7 EYL M Dixon, 19 Stanford Way, Broadbridge Heath, Horsham, RH12 3LH
G7 EYM M Parkyn, Brookfield, Clee St. Margaret, Craven Arms, SY7 9DX
GW7 EYP Martyn Jenkins, 23 Guenever Close, Thornhill, Cardiff, CF14 9AH
G7 EYR P Wiles, 16 Churchill Road, Broadheath, Altrincham, WA14 5LT
G7 EYS C Chance, 19 White Beam Rise, Clanfield, Waterlooville, PO8 0LQ
G7 EYV A Little, 444 Dunsbury Way, Leigh Park, Havant, PO9 5BJ
G7 EZE Adrian Byrne, 23 The Deansway, Kidderminster, DY10 2RH
G7 EZH Paul Higginson, 93 Oakfield Road, Wollescote, Stourbridge, DY9 9DE
G7 FAD V Ritson, 24 Chapel Road, Pawlett, Bridgwater, TA6 4SH
G7 FAQ E Cloude, 39 Wentworth Close, Weybourne, Farnham, GU9 9HJ
G7 FAR RAF Waddington ARC c/o Martin Farmer, Tara Cottage, 16 Beckside, Lincoln, LN2 2PH
G7 FAS Alasdair Warnock, 7 Diglis Road, Worcester, WR5 3BW
G7 FAZ Andrew Mould, 8, Foxwood, Brierley Hill, DY5 2PH
G7 FBE E Dare, 17 Montgomery Drive, Spencers Wood, Reading, RG7 1BQ
G7 FBT E Woodhouse, 7 Cow Heys, Dalton, Huddersfield, HD5 9RG
GW7 FBV S Hathaway, 9 Mirehouse Place, Angle, Pembroke, SA71 5BD
G7 FBY R Furniss, 7 Elizabeth Road, Sutton Coldfield, B73 5AR
G7 FCC David Bent, 21 Loughborough Avenue, Nottingham, NG2 4LN
G7 FCJ P Honeywell, 38 Kipling Avenue, Tilbury, RM18 8HF
G7 FCL Derek Hames, 10 Downs Close, East Studdal, Dover, CT15 5BY
GI7 FCM Stephen Fleming, 15 Castle Green, Ballynure, Ballyclare, BT39 9GN
GI7 FCP J McCormick, 14 Ballyoran Park, Portadown, Craigavon, BT62 1JN
G7 FCU Frank Smith, 3 Downside Close, Findon Valley, Worthing, BN14 0EZ
GI7 FCW P Quinn, 53 Dernanaught Road, Dungannon, BT70 3BU
G7 FDD W Cooper, 33 Elm Drive, Cherry Burton, Beverley, HU17 7RJ
GM7 FDS Thomas Whitehead, Four Winds, Papigoe, Wick, KW1 4RD
G7 FDW Christopher Milburn, 9 Woodhall Avenue, Bradford, BD3 7BY
G7 FEA M Codling, 18 Ash Grove, Pinehurst, Swindon, SN2 1RX
G7 FED E Wells, 1a Brocklewood Avenue, Poulton-le-Fylde, FY6 8BZ
G7 FEE F Harvey, 39 Simonside Terrace, Heaton, Newcastle upon Tyne, NE6 5JY
G7 FEF J Pipkin, 46 Charles Avenue, Albrighton, Wolverhampton, WV7 3LF
G7 FEG Ian Kilkenny, 23 Hazelhurst Road, Stalybridge, SK15 1HD
G7 FEL J Slater, 313 Southend Road, Stanford-le-Hope, SS17 8HL
G7 FEP Darren Birt, 3 South Brent Close, Brent Knoll, Highbridge, TA9 4BS
G7 FEQ B Shipton, 4 School Close, Kilcott Road, Wotton-under-Edge, GL12 7RH
G7 FFB D Egleton, 16a Frescade Crescent, Basingstoke, RG21 3NF
G7 FFC V Prall, 20 Marlowe Close, Basingstoke, RG24 9DD
G7 FFI Keith Harding, 21 Doulton Way, Ashingdon, Rochford, SS4 3BX
G7 FFK A James, 66 Rydal Crescent, Worsley, Manchester, M28 7JD
G7 FFM P Malpass, 30 Countisbury Road, Norton, Stockton-on-Tees, TS20 1PZ
G7 FFR J Rutherford, 270 Milburn Road, Ashington, NE63 0PL
G7 FFS Andrew Pike, 63 Mill Lane, Bentley Heath, Solihull, B93 8NN
G7 FFV I Miller, 5 Avenue Terrace, Sunderland, SR2 7HB
G7 FFW Stephen Lonsdale, 16 Hinkler Street, Cleethorpes, DN35 8PR
G7 FFZ Jacqueline Humphries, 23 Sycamore Drive, Lutterworth, LE17 4TR
G7 FGA David Moreland, 179 Carr Lane, York, YO26 5HQ
G7 FGD J Brown, St. Winnolls House, St. Winnolls, Torpoint, PL11 3DX
GM7 FGH John Bartolo, 84 Calderbraes Avenue, Uddingston, Glasgow, G71 6ED
GI7 FGQ Peter Faulkner, 40 Glenariff Drive, Comber, Newtownards, BT23 5HA
G7 FGR Grahame Cluley, 1 Shepherds Lane, Greetham, Oakham, LE15 7NX
GJ7 FGS Joseph Bette- Bennett, 2 Aspley Villas, Bagatelle Road, Jersey, Jersey, JE2 7TA
G7 FGZ D Mitchell, 55 Halewick Lane, Sompting, Lancing, BN15 0ND
G7 FHA C Brailsford, 65 Cherry Orchard, Codford, Warminster, BA12 0PW
GI7 FHU Ian Davies, 14 Stramore Terrace, Gilford, Craigavon, BT63 6EU
G7 FHV T Beeching, 11 Kents Road, Haywards Heath, RH16 4HL
GI7 FHZ E Mccrystal, 33 Richmond Park, Omagh, BT79 7SJ
G7 FIA P Page, 67 Teesdale Road, Dartford, DA2 6LB
G7 FIJ B Parsons, 20 High Park Road, Halesowen, B63 2JA
G7 FIK Sean Regan, 92-94 Lytham Road, Blackpool, FY1 6DZ
GM7 FIS J Russell, 15 Glen View, Cumbernauld, Glasgow, G67 2DA
G7 FJC Paul Fellingham, 152a Freshfield Road, Brighton, BN2 9YD
G7 FJK T Brown, 22 Moss Road, Congleton, CW12 3BN
G7 FJU Christopher Ovenden, 2 Firemans Cottage, Fortis Green, London, N10 3PB
GI7 FJY Nigel Gamble, 21 Anderson Crescent, Waterside, Londonderry, BT47 2BY
G7 FJZ P Selley, 2 Coronation Street, Barnstaple, EX32 7AY
G7 FKF R Phillips, 24 Harris Lane, Wistow, Huntingdon, PE28 2QG
G7 FKJ C Holloway, 23 Ryecroft Road, Stretford, Manchester, M32 9BS
G7 FKP D Henderson, 2 Beverley Court, Beverley Road, York, YO43 3NB
G7 FKS H Ellis, 10 Gardens Quay, Pitwines Close, Poole, BH15 1XL
G7 FKX C Wood, 2 Plain Cottages, Plain Road, Tonbridge, TN12 9LS
G7 FKZ R Bowden, 35 Glebelands, Biddenden, Ashford, TN27 8EA
GM7 FLG D Pegg, 11 Glenward Avenue, Lennoxtown, Glasgow, G66 7EP
G7 FLI John Moyse, 2 Kestle Drive, Truro, TR1 3PT
G7 FLS Kevin Lawson, 7 Church Lane, Castle Donington, Derby, DE74 2LG
G7 FLX B Timms, 74 Park Gwyn, St. Stephen, St. Austell, PL26 7PN
GM7 FLZ E Chesters, Blackhill, Blackhill Road, Kirkwall, KW15 1FP
G7 FMB R Burns, 43 Gibson St., Bickershaw, Wigan, WN2 5TF
G7 FMF Peter Jennings, Millerdale Cottage, Gorst Hill, Kidderminster, DY14 9YR
G7 FMI M Kensall, 40 Eskdale Avenue, Ramsgate, CT11 0PB
G7 FMJ Grant Mitchell, 7 Buxton Close, Whetstone, Leicester, LE8 6NT
G7 FML P Clarke, 93 Commercial Road, Spalding, PE11 2YU
G7 FMQ J Sutton, 10 Cathcart Road, Stourbridge, DY8 3UZ
G7 FMV D Sweet, 50 Mereside, Soham, Ely, CB7 5XE
G7 FMW Paul Olson, 23 Dennett Close, Liverpool, L31 5PD
G7 FND Edward Millership, 16 Bramble Way, Wirral, CH46 7UP
G7 FNM J Walmsley, Bank Field, 5 Dimples Lane, Preston, PR3 1RD
G7 FNN Anthony Morley, 87 Epsom Drive, Ipswich, IP1 6SS
GI7 FNP H Massey, 156 Killaughey Road, Donaghadee, BT21 0BQ
G7 FNQ David Jones, Noddfa, High Street, Bodorgan, LL62 5AS
G7 FNU R Beaumont, 49 Vincent Close, Broadstairs, CT10 2ND
GI7 FOD Paul McCollam, 32 Robinson Way, Bangor, BT19 6NR
G7 FOT Walter Sanger, Tregonning Lea, Laddenvean, Helston, TR12 6QD
G7 FOX Melton Mowbray ARS c/o G Mason, 120 Scalford Road, Melton Mowbray, LE13 1JZ
G7 FPJ J Marks, 591 London Road, Stoke-on-Trent, ST4 5AZ
GM7 FPN Alistair McPherson, 3 Fulmar Road, Elgin, IV30 4HL
G7 FPR Gerard Flanagan, Flat 9, West Cliff Court, 25 Portarlington Road, Bournemouth, BH4 8BX
G7 FPS S Harrison, 3 Smallbridge Close, Worsley, Manchester, M28 7XS
G7 FPU C Wissun, 111 Berkeley Vale Park, Berkeley, GL13 9TQ
G7 FPW P Howard, 3 Hollies Close, Shepton Mallet, BA4 5LG
G7 FPZ D Foster, 29 Harrowfield Road, Stechford, Birmingham, B33 9BU
G7 FQE J Whiffen, 5 Sharpthorpe Close, Lower Earley, Reading, RG6 4DB
G7 FQP S Earle, 12 Ray Lea Close, Maidenhead, SL6 8QN
G7 FQY G Spark, Park Lodge, Cheltenham Drive, Sale, M33 2DQ
GM7 FRC FIFE RAYNET GROUP c/o J BURKE, 25 Duncan Road, Auchmuty, Glenrothes, KY7 4HS
G7 FRH A Russ, 21 Francis Road, St. Pauls Cray, Orpington, BR5 3LY
G7 FRW K McCaffery, 34 Ringwood Road, Luton, LU2 7BG
G7 FSA R Colclough, 8 Parker Jervis Road, Stoke-on-Trent, ST3 5RP
G7 FSC Keith Davis, 26 Mendip Drive, Nuneaton, CV10 8PT
G7 FSD Anthony Holles, 20 Sapcote Road, Burbage, Hinckley, LE10 2AU
G7 FSH F Pearson, 15 York Road, Driffield, YO25 5AT
G7 FSI Barry Williams, 68 Beeston Road, Sheringham, NR26 8EJ
GD7 FSJ Patrick O'Connor, 149 Roxeth Green Avenue, Harrow, HA2 0QJ
G7 FSR A Wyard, 85 Swaledale, Bracknell, RG12 7ET
G7 FTA Michael Collins, 15 Trevethan Rise, Falmouth, TR11 2DX
G7 FTD K Taber, 110 Uplands, Peterborough, PE4 5AF
G7 FTF B Lee, 43 Longview Road, Saltash, PL12 6EF
GM7 FTK Michael Long, 52 Stirling Road, Milnathort, Kinross, KY13 9XG
G7 FTM S Clayton, 22 Orchard Avenue, North Anston, Sheffield, S25 4BW
G7 FTS O Whiteside, 9 Beech Grove House, Beech Grove, Harrogate, HG2 0ES
G7 FUM N Seath, 6 Harvester Way, Sibsey, Boston, PE22 0YD
G7 FUQ Anne Vincent, 12 Spelman Road, Norwich, NR2 3NJ
G7 FUV Ian Marsh, 56b Oliver Crescent, Farningham, Dartford, DA4 0BE
G7 FUW J Birch, 15 Adstone Grove, Birmingham, B31 4AU
G7 FVA Richard Hunter, 26 Templer Road, Preston, Paignton, TQ3 1EL
G7 FVH R Barrick, Orchard Bungalow, Pasture Lane, Middlesbrough, TS6 8EH
G7 FVR C Heywood, 29 Smallwood Mews, Wirral, CH60 6TE
GM7 FWA Alexander Pratt, 129 Brodie Court, Glenrothes, KY7 4UE
G7 FWD Bruce Nicholls, 15 Canal Way, Devizes, SN10 2UB
G7 FWE A Smith, 12 Northgate, Beccles, NR34 9AS
G7 FXO Peter Werba, 47 Ulwell Road, Swanage, BH19 1LG
G7 FXW Philip Whitworth, 67 Staddiscombe Road, Staddiscombe, Plymouth, PL9 9LU
GW7 FXX B Harries, 12 Panteg, Llanelli, SA15 3TF
G7 FXY P Hallett, 30 Summerdown Walk, Trowbridge, BA14 0LJ
G7 FXZ G Hodgetts, 2 Friars Gorse, Stourton, Stourbridge, DY7 6SP
GM7 FYB D Wemyss, 24 Brucklay Court, Peterhead, AB42 2UF
G7 FYG Christopher Wright, 12 Bryn Teg, Arddleen, Llanymynech, SY22 6PZ

G7 FZB Graeme Ridgeway, 15 Blenheim Court, Alsager, Stoke-on-Trent, ST7 2BY
G7 FZJ M Whatley, Woodside West, Wood Lane, Halifax, HX3 8HB
G7 FZN P Du Plessis, 42 La Providence, Rochester, ME1 1NB
[illegible]
GM7 GAE Ian Mackenzie, 62/5 Craighall Road, Edinburgh, EH6 4RU
G7 GAG J O'Neill, 24 Lily Lane, Bamfurlong, Wigan, WN2 5JN
G7 GAH Robert Dore, Maespoeth Cottage, Corris, Powys, SY20 9RD
G7 GAK Donald Garbutt, 8 Yorkshire Road, Partington, Manchester, M31 4GW
G7 GAP J Cartwright, 109 Kneller Road, Twickenham, TW2 7DT
G7 GAZ B Kerrison, 45 Bramley Crescent, Bearsted, Maidstone, ME15 8JZ
GM7 GBD G Macgregor, 6 Kincaidfield, Milton of Campsie, Glasgow, G66 8ER
G7 GBE Stephen Burgoine, 47 Squirrel Close, Hounslow, TW4 7NU
G7 GBJ J Kaczmarek, 2 Westgate Terrace, London, SW10 9BJ
G7 GBN P Baird, 168 Plumberow Avenue, Hockley, SS5 5AT
G7 GBZ P Leach, 21 Abbess Close, Chelmsford, CM1 2SE
G7 GCB R Bishop, 67 East Road, West Mersea, Colchester, CO5 8HB
G7 GCD Stephen Lee, Flat 1, 2 Granby Road, Harrogate, HG1 4ST
G7 GCF P Kell, 56a Central Parade, New Addington, Croydon, CR0 0JL
G7 GCI M Collett, 54 Dalkeith Road, Harpenden, AL5 5PW
G7 GCU M Edge, 2 Yew Tree Place, Walsall, WS3 3DG
G7 GCW Piers Andrew, 3 Grayway Close, Highfields Caldecote, Cambridge, CB23 7UZ
G7 GDA Trevor Wootton, 1 Lingfield Drive, Walsall, WS6 6LS
G7 GDC A Gosden, 10 Radcliffe Way, Northolt, UB5 6HP
GM7 GDE Andrew Hood, 26 Annan Avenue, East Kilbride, Glasgow, G75 8XT
G7 GDV A Porteous, 73 Dowgate Close, Tonbridge, TN9 2EJ
G7 GEA James Broadfoot, 65a Swan Meadow, Pewsey, SN9 5HP
G7 GEE J Gee, 51 Hattons Lane, Childwall, Liverpool, L16 7QR
G7 GEF G Duthie, 15 Wagtail Close, Twyford, Reading, RG10 9ED
G7 GEI Mark Arliss, 55 Hartland Crescent, Edenthorpe, Doncaster, DN3 2PQ
G7 GEL J MANSELL, 8 Himley Gardens, The Straits, Dudley, DY3 3AS
G7 GEP Christopher Danks, Ashmore Nurseries, Radford Lane, Wolverhampton, WV4 4XP
G7 GES B Norcott, 5 The Shrubbery, Upminster, RM14 3AH
G7 GEU V Bruntnell, 4 Cypress Avenue, Dudley, DY3 2JF
G7 GEX N Potter, 4 Eastleigh Drive, Mickleover, Derby, DE3 9HZ
G7 GFC David Mullock, 18 Tewkesbury Close, Upton, Chester, CH2 1NF
G7 GFH W Baker, 41 Kenwood Park Road, Sheffield, S7 1NE
G7 GFK K Percival, 1608 Scant Row, Chorley Old Road, Bolton, BL6 6PZ
G7 GFM Jeremy Hunt, 43 Felton Close, Redditch, B98 0AG
G7 GFP I Bishop, 115 Burman Road, Shirley, Solihull, B90 2BQ
G7 GFQ M Charlwood, 60 Alfred Road, Feltham, TW13 5DJ
G7 GFR B Clifford, 8 Caldbeck Place, North Anston, Sheffield, S25 4JY
G7 GFX Peter Everard, 56 Hawkins Crescent, Shoreham-by-Sea, BN43 6TP
G7 GGA N Dawson, 4 Bathurst Close, Staplehurst, Tonbridge, TN12 0NA
G7 GGF Craig Martin, 3 Jasmine Close, Lutterworth, LE17 4GR
G7 GGG G Richardson, 11 Queensway, Forest Town, Mansfield, NG19 0BX
G7 GGH G hurrell, 13 Hinton Road, Newport, PO30 5QZ
G7 GGJ A Edwards, 45 Chilton Grove, Yeovil, BA21 4AW
G7 GGM D Thomalla, 14 Walkers Lane, Penketh, Warrington, WA5 2PA
G7 GGN J Williams, 41 Cote Green Lane, Marple Bridge, Stockport, SK6 5EB
G7 GGT S Mullins, 549 Bromford Lane, Washwood Heath, Birmingham, B8 2EA
GI7 GHC Thomas Lyons, 3 Clanbrassil Gardens, Portadown, Craigavon, BT63 5YD
G7 GHE David Pearson, Warren Cottage Pontfadog, Llangollen, LL20 7AT
G7 GHH I Wraith, 7 Bowman Close, Sheffield, S12 3LR
G7 GHI W Stainforth, 2 Grangefield Terrace, New Rossington, Doncaster, DN11 0LT
G7 GHP G Fellows, 34 The Ridings, Bexhill-on-Sea, TN39 5HU
GM7 GIF Kenneth Juner, 56 Queens Gardens, East Calder, Livingston, EH53 0EG
G7 GIG Richard Vincent, 14 Trevenson Street, Camborne, TR14 8JB
G7 GIJ J Barnett, 20 Springford Gardens, Southampton, SO16 5SW
GM7 GIO W Mackinnon, 31 Kirk Bauk, Symington, Biggar, ML12 6LB
G7 GIS M Glendinning, 148 Gala Park, Galashiels, TD1 1HD
G7 GJA Peter Cockayne, 7a Wrekin Drive, Bradmore, Wolverhampton, WV3 7HZ
G7 GJI D Sager, 29 Station Road, Mickleover, Derby, DE3 9GH
G7 GJM C Unsworth, 42 Whitemill Lane, Stone, ST15 0EG
G7 GJN A Khachaturian, 377 Watford Road, St. Albans, AL2 3DD
G7 GJO P Morris, 117 Lonsdale Avenue, Doncaster, DN2 6HF
G7 GJS Norman Cheesewright, 5 Duberly Close, Perry, Huntingdon, PE28 0BP
G7 GJT W Everton, Fencott, Fen Road, Lincoln, LN4 1AE
G7 GJU G Darby, 5 Lumsden Terrace, Catchgate, Stanley, DH9 8EQ
G7 GJV Anne Gordon, 1 Surrey Street, Hetton-le-Hole, Houghton le Spring, DH5 9LX
GI7 GJX N Simmons, 116 Killyglen Road, Larne, BT40 2HX
G7 GJY J Chapman, 77a Carnforth Gardens, Elm Park, Hornchurch, RM12 5DR
G7 GJZ Chris Brown, 73 Ringstone, West Huntspill, Highbridge, TA9 3RF
GI7 GKC Ian Boyd, 21 Fulmar Avenue, Lisburn, BT28 3HS
G7 GKD L Tryhorn, 46 Mill Green Road, Amesbury, Salisbury, SP4 7RE
GW7 GKN [illegible], 9 Oakridge Acres, Tenby, SA70 8DR
G7 GKQ L Measures, 163 Huddersfield Road, Meltham, Holmfirth, HD9 4AJ
GM7 GKT R Smith, 27 Elm Lane, Foresters Lodge, Glenrothes, KY7 5TD
G7 GKX J Rough, 10 Beaconsfield Road, Shotton, Deeside, CH5 1EZ
G7 GLA Jim Mitchinson, 93 Hinckley Road, Leicester Forest East, Leicester, LE3 3GN
G7 GLH J Birch, 29 West Road, Dibden Purlieu, Southampton, SO45 4RH
GM7 GLJ A Potter, 42 Pendor Gardens, Rumford, Falkirk, FK2 0BJ
G7 GLL Leonard Carlile, 26 The Bungalows, Stonebroom, Alfreton, DE55 6LH
G7 GLQ Dennis Cottrell, 2 Foss Court, Summerhill Road, Bristol, BS5 8HF
G7 GLR GRTR LNDN RAYNE c/o Ian Jackson, 5 Vivien Close, Chessington, KT9 2DE
G7 GLS J Pinna, 31 Bowness Road, Little Lever, Bolton, BL3 1UB
G7 GLW R Cains, 58 Sunnydale Road, Lee, London, SE12 8JN
G7 GLZ R Hourston, 12 The Warren, Chesham, HP5 2RY
G7 GMB John Craig, 1 Eldon Road, Eastbourne, BN21 1UD
G7 GMD Max Ollerton, 1 Hammy Way, Shoreham-by-Sea, BN43 6GH
G7 GMQ D Smith, 65 St. Anthonys Road, Kettering, NN15 5JB
G7 GMR B Golland, 15 Turpin Close, Gainsborough, DN21 1PA
G7 GMU G Lamb, Parisfield, Headcorn Road, Tonbridge, TN12 0BT
G7 GMZ W Newton, 7 Moss Close, Bridgwater, TA6 4NA
G7 GNA Leonard Smith, 13 Eagle Avenue, Waterlooville, PO8 9UB
GM7 GNO Nicholas Goodall, 26 Greenbank Loan, Edinburgh, EH10 5SJ
G7 GN[illegible] J Olive, [illegible] Road, Chipping Sodbury, Bristol, BS37 6DQ
G7 GNU R Brayshaw, 38 Chilfrome Close, Canford Heath, Poole, BH17 9WE
G7 GOA S Constable, 18 Salvington Gardens, Worthing, [illegible]
GM7 GOE M Doig, 18 Gotterstone Drive, Broughty Ferry, Dundee, DD5 1QW
G7 GOK Neil Breckell, 4 Folly Lane North, Farnham, GU9 0HX
G7 GOV Michael Hill, 31 Brocklesby Avenue, Immingham, DN40 2AS
G7 GPG A Jakowuik, 167 Magdala Terrace, Galashiels, TD1 2HZ
G7 GPI A Baily, 13 Longleigh Lane, Bexleyheath, DA7 5SL
G7 GPJ Ray Banks, Highview, New Road, Sturminster Newton, DT10 2HF
G7 GPL Nick Giles, 6 Bridgewater Mews, London Road, Warrington, WA4 6LF
G7 GPU A Sharman, 9 Silver Close, Minety, Malmesbury, SN16 9QT
G7 GQA Andrew Doswell, 14 Carisbrooke Drive, Charlton Kings, Cheltenham, GL52 6YA
G7 GQB M Woodhouse, 18 Soame Close, Aylsham, Norwich, NR11 6JF
G7 GQC G Beckingham, 20 Baptist Close, Abbeymead, Gloucester, GL4 5GD
G7 GQD David Pearce, 2 Mell Avenue, Hoyland, Barnsley, S74 9HF
G7 GQH R Hannemann, 112 Northern Road, Aylesbury, HP19 9QY
G7 GQL J Sutton, 15 Lowther Street, Penrith, CA11 7UW
G7 GQM E Sutton, 15 Lowther Street, Penrith, CA11 7UW
G7 GQO R Harman, The Briars, Brambleberry Lane, Skegness, PE24 5DQ
G7 GQW D Williams, 28 Mill Lane, Great Sutton, Ellesmere Port, CH66 3PF
G7 GQX Derek Howard, 6 Draycote Close, Solihull, B92 9PT
G7 GRB Henry Ewing, 100 Warren Road, Dartford, DA1 1PL
G7 GRC Grantham ARC c/o Kevin Burton, 2 Council House, Stainfield Road, Bourne, PE10 0SG
G7 GRH N Hardie, 38 Sentry Knowe, Selkirk, TD7 4BG
G7 GRJ R Sedge, 25 Furfield Close, Maidstone, ME15 9JR
G7 GRO D Stimpson, 193 Weaver Street, Winsford, CW7 4AJ
G7 GRQ A Gray, 79 Brougham Terrace, Hartlepool, TS24 8EU
G7 GRR Shaun Everett, 4 Ilkley Place, Newcastle, ST5 6QP
G7 GRU M Lucas, 22 Ferny Brow Road, Wirral, CH49 8EE
GI7 GRY S Gordon, 138 Mullalelish Road, Richhill, Armagh, BT61 9LT
GI7 GSB Alan Wiese, 105 Milltown Avenue, Lisburn, BT28 3TR
G7 GSC N Godden, 23 Rapsons Road, Willingdon, Eastbourne, BN20 9RJ
G7 GSD Kevin Osborn, 2a Sullington Gardens, Worthing, BN14 0HR
G7 GSF Simon Blandford, Flat 18, Avro House, 5 Boulevard Drive, London, NW9 5HF
G7 GSR John Shrubsall, 54 Park Avenue, Sittingbourne, ME10 1QY
G7 GSX Charles Penfold, 149 Shuttlewood Road, Bolsover, Chesterfield, S44 6NX
G7 GTG A Hyndman, Norman House, Railway Terrace, Kings Langley, WD4 8JE
G7 GTH A Marriott, Norman House, Railway Terrace, Kings Langley, WD4 8JE
GM7 GTS C Richman, 18 Nigel Rise, Livingston, EH54 6LT
G7 GTU G Sharples, 24 Kelboro Avenue, Audenshaw, Manchester, M34 5UH
GM7 GTX M Kaye, 146 Newlands Road, Grangemouth, FK3 8NZ
G7 GUB A Alderton, 7 Bigland Drive, Ulverston, LA12 9NU
G7 GUG G Wales, 7 Montgomery Avenue, Hampton-on-the-Hill, Warwick, CV35 8QP
G7 GUK D Nelson, 101 Gledhow Lane, Roundhay, Leeds, LS8 1NE
GM7 GUL C Jordan, 3 Birch Avenue, Rosemount, Blairgowrie, PH10 6XE
G7 GUO S Falconer, 6 Ogilvie Road, High Wycombe, HP12 3DS
GI7 GUT D Watt, 51 Rashee Road, Ballyclare, BT39 9HT
GM7 GVD D Innes, 6 Mamore Terrace, Inverness, IV3 8PF
GI7 GVI T Henderson, 7 Legaloy Road, Ballyclare, BT39 9PS
G7 GVJ S Fletcher, Fernleigh, Ash Lane, Gloucester, GL2 9PS
G7 GVP Colin Price, 16 Woodlands Drive, Warton, Preston, PR4 1UQ
G7 GWA A Jakins, 29 Burchnall Close, Deeping St. James, Peterborough, PE6 8QJ
GW7 GWO S Evans, Hazelbrook, Felin Ban Farm Estate, Cardigan, SA43 1PG
G7 GWT Graham Taylor, 33 Heol Aberwennol, Borth, SY24 5NP
GM7 GWW S Gardiner, Kyendigaet, Whiteness, Shetland, ZE2 9GJ
G7 GXE P Kitson, 15 Louvain Road, Derby, DE23 6DA
GM7 GXI G Cowan, 15 Waterhaughs Grove, Glasgow, G33 1RS
G7 GXR B Clewes, 19 Church Mews, Denton, Manchester, M34 3GL
GI7 GXZ Stanley Dornan, 3 Hampton Lane, Bangor, BT19 7GB
G7 GYN Colin Barlow, 2/5 Hospital Steps, Gibraltar, GX11 1AA
G7 GYR K Wade, Eccleston Hall, Lydiate Lane, Chorley, PR7 6LY
G7 GZB Chris Davies, 84 Hob Hey Lane, Culcheth, Warrington, WA3 4NW
G7 GZC David Coles, 19 Somerville House, 1 Rodney Road, Twickenham, TW2 7AL
G7 GZJ Kevin Oliver, 155 Old Road, East Cowes, PO32 6AX
G7 GZK I Croft, 34 Laburnum Drive, Armthorpe, Doncaster, DN3 3HE
G7 GZU Steven Selwyn, 50 Tufthorn Avenue, Coleford, GL16 8PT
G7 GZV Harold Houldershaw, The First Bungalow, Fen Road, Boston, PE22 8EX
G7 GZZ E Gaffney, 1 White Hart Lane, Wistaston Green, Crewe, CW2 8EX
G7 HAE C M Davies, Afallon, 3 Penygraig, Aberystwyth, SY23 2JA
G7 HAF William Hunton, 60a Bondgate, Helmsley, York, YO62 5EZ
G7 HAR B Ferris, 5 Guildway, Todwick, Sheffield, S26 1JN
G7 HAS A Newton, [illegible], Victoria Road, Malvern, WR14 [illegible]
G7 HBN P Osborne, 11 Galston Road, Luton, LU3 3JZ
G7 HBO R Cornell, 18 Holland Park Avenue, Ilford, IG3 8JR
G7 HBU T Hickling, 6 Harrold Road, Bozeat, Wellingborough, NN29 7LP
G7 HBV C Heard, 42 Hallowell Down, South Woodham Ferrers, Chelmsford, CM3 5FS
G7 HCB D Atterbury, 7 Ross Court, Stevenage, SG2 0HD
G7 HCC Dennis Jones, 120 Heathfield Road, Keston, BR2 6BF
G7 HCJ A Parr, 52b Trent Boulevard, West Bridgford, Nottingham, NG2 5BD
G7 HCL Peter Good, 80 Meredith Road, Stevenage, SG1 5QS
G7 HCN Alexander Jones, 179 Blandford Road, Efford, Plymouth, PL3 6JZ
G7 HCO N Lambert, Bradfields Farm, Burnmills Road, Wickford, SS12 9JX
G7 HCQ D Browne, 293 St. Albans Road, Hemel Hempstead, HP2 4RP
G7 HCR Graham Richardson, The Homestead Washway Road, Holbeach, Spalding, PE12 7PP
G7 HCT K Moore, 8 Lilac Close, Toftwood, Dereham, NR19 1JY
G7 HDC A Rowe, 5 Church Street, Knighton, LD7 1AG
G7 HDR D Horder, 77 Grove Avenue, Harpenden, AL5 1EZ
GW7 HDS Simon Barker, 11 Prosser Street, Treharris, CF46 5LN
G7 HDU J Tombs, Mariedown, Bustards Lane, Wisbech, PE14 7PQ
G7 HDW Jennifer Bigger, 128 Trueway Drive South, Shepshed, Loughborough, LE12 9DY
G7 HDZ A Rowell, 105 Hedgehope Road, Newbiggin Hall, Newcastle upon Tyne, NE5 4LB
[illegible] DY1 [illegible]
G7 HEK Andrew Owen, 57 Melrose Avenue, Vicars Cross, Chester, CH3 5JB
G7 HEN Michael Priestley, 29 Birchlands Avenue, Wilsden, Bradford, BD15 0HB
G7 HEP Arthur Ellis, Eikly Tregada, Launceston, PL15 9NA
G7 HEY L Morrell-Cross, Delta Lodge, 14 Rushton Crescent, Bournemouth, BH3 7AF
G7 HEZ James French, 2 Pepper Hill, Stourbridge, DY8 1BJ
G7 HFE S Hitches, 7 Church Close, Chedgrave, Norwich, NR14 6NH
G7 HFL C Elphick, 2 Vine Way, Brentwood, CM14 4UU
G7 HFP C Castle, 2 Wellington Close, Mundesley, Norwich, NR11 8JF
G7 HFS Ian Harling, 114 Latimer Road, Eastbourne, BN22 7DR
G7 HFW A Wood, 76 Russet Road, Weaverham, Northwich, CW8 3HZ
GW7 HFZ A Strachan, 16 Clos y Wiwer, Llantwit Major, CF61 2SG
G7 HGB Joan Dunn, 10 Endsleigh Close, Upton, Chester, CH2 1LX
G7 HGD Philip Allott, 1 Abbey Court, Abbey Road, Knaresborough, HG5 8HX
G7 HGF I Simpson, Honeysuckle Cottage, 39 Chewton Street, Nottingham, NG16 3GY
G7 HGI R Roberts, 13 Tudor Way, Wickford, SS12 0HS
G7 HGQ D Horwood, 12 Curtis Close, Mill End, Rickmansworth, WD3 8QA
G7 HGT P Stimpson, 93 Chaucer Road, Farnborough, GU14 8SR
GW7 HGU M Howard, 64 Lawrenny St., Neyland, Milford Haven, SA73 1TB
GM7 HHB John Brown, 133 Meadowbank Road, Kirknewton, EH27 8BH
G7 HHI Simon Curry, Barnabus Communications, Barnabus Cottage, Egley Road, Mayford, Woking, GU22 0NQ
G7 HHK R Johnson, Honeysuckle Cottage, Front Street, Northallerton, DL6 2AA
G7 HHL R Garner, 3 Cozens Hardy Road, Sprowston, Norwich, NR7 8QE
G7 HHM L Dring, 22 Castle Street, Eastwood, Nottingham, NG16 3GW
G7 HHN K Glover, 7 Mill Lane, Cressing, Braintree, CM77 8HW
G7 HHQ R Saunders, The Grange, High Road, Wisbech, PE13 4RG
G7 HHT M Gotts, 23 Beechcroft Avenue, Croxley Green, Rickmansworth, WD3 3EG
G7 HHU A Edwards, 3 Simonside Close, Morpeth, NE61 2XY
G7 HHW Gregory Phillips, 14 Orchard Close, Plymouth, PL7 2GT
G7 HHZ J whelan, 1 Chevin Road, Milford, Belper, DE56 0QH
G7 HIC K Bow, 16 Brook Road, Ivybridge, PL21 0AX
G7 HID Michael Burgess, 63 Chalvey Park, Slough, SL1 2HX
GD7 HIH R pedro, 65 Glebe Crescent, Harrow, HA3 9LB
G7 HII D Lloyd, No.5 The Close, Burton Gardens, Hereford, HR4 8RQ
G7 HIJ J Gunia, 21 Campbell Avenue, Leek, ST13 5RR
G7 HIK J Doherty, 101 Padacre Road, Torquay, TQ2 8QQ
G7 HIN P Riddell, 4 Pear Tree Road, Addlestone, KT15 1SR
G7 HIO W Austin, 53 Giantswood Lane, Congleton, CW12 2HQ
G7 HIQ Jonathan Hickey, 53 Norwood Avenue, Hasland, Chesterfield, S41 0NN
GM7 HIR Ann Pert, 56 Lochiel Drive, Milton of Campsie, Glasgow, G66 8ET
G7 HIT P Chambers, 7 Redland Close, Beeston, Nottingham, NG9 5LA
G7 HIU Robert Hurst, 33 Northern Road, Aylesbury, HP19 9QT
G7 HIX R Gray, 12 St. Francis Close, Deal, CT14 9LS
G7 HIY T Jefford, 7 Bellevue Street, Folkestone, CT20 1HY
G7 HJD G Holland, 11 Swanton Drive, Dereham, NR20 4DW
G7 HJG Robert Blewitt, 9 Durlston Close, Amington, Tamworth, B77 3QG
G7 HJJ H Holman, 62 The Ridge, Kennington, Ashford, TN24 9EU
G7 HJK Richard Kearnes, 25 Epsom Close, Clacton-on-Sea, CO16 8FE
GW7 HJN Stuart Tweed, 257 Penybanc Road, Ammanford, SA18 3QW
G7 HJQ Michael Erber, 75 St. Andrews Road North, Lytham St. Annes, FY8 2JF
G7 HJR T Rudderham, 24 Casswell House, Grimsby, DN32 7SB
G7 HJT Tony Reynard, 12 Acorn Close, Selsey, Chichester, PO20 9HL
G7 HJX D Raybould, 63 Rochester Avenue, Burntwood, WS7 2DL
G7 HKN P Walsh, 2 Elm Road, Winwick, Warrington, WA2 9TW
G7 HKQ I Tideswell, 2 Pangbourne Avenue, Urmston, Manchester, M41 0GF
G7 HKT Colin Fowle, 9 Haffenden Meadow, Charing, Ashford, TN27 0JR
G7 HKU J Turner, 7 Highfield Crescent, Baildon, Shipley, BD17 5NR
G7 HKZ T Allen, 15 Manning Road, Cotford St. Luke, Taunton, TA4 1NY
G7 HLD C Woolley, 12 Heathfield Road, Stroud, GL5 4DQ
G7 HLG B Morrell-Tourle, 77 Mallard Road, Bournemouth, BH8 9PJ
G7 HLP K Baldock, 66 Port Road, New Duston, Northampton, NN5 6NL
G7 HLU V Meads, May Tree Barn, Upper Main Street, Peterborough, PE8 5AN
G7 HLV J Jordan, Woodroyd, 66 Spring Ave, Keighley, BD21 4TA
G7 HLW G Burn, 4 Goston Gardens, Thornton Heath, CR7 7NQ
G7 HLZ Raymond Davies, Parclands, Raglan, Usk, NP15 2BX
G7 HMA Ian Smith, 4 Stour Road, Grays, RM16 4BS
G7 HMB G Bull, 48 Spragg House Lane, Stoke-on-Trent, ST6 8DX
G7 HMF E Last, 134 New Queens Road, Sudbury, CO10 1PJ
G7 HMI Richard Shelford, 3 Browning Chase, Littleport, Ely, CB6 1FH
G7 HMK A Baldwin, B M Box 6902, London, WC1N 3XX
G7 HMN C Boutell, 6 Willow Way, Harwich, CO12 4HR
G7 HMO B Boult, 20 Perry Road, Long Ashton, Bristol, BS41 9FE
G7 HMU J Stratton, 22 Tufton Gardens, West Molesey, KT8 1TE
G7 HMV Matthew Wood, 26 Parkfield Crescent, Kimpton, Hitchin, SG4 8EQ
G7 HMW W Knight, 30 Stretford Road, Urmston, Manchester, M41 9JZ
G7 HMZ A Murfin, 21 Kings Road, St. Neots, PE19 1LD
G7 HNF J Baldwin, 71 Norfolk Road, Littlehampton, BN17 5HE
G7 HNG Andrew Lord, 726 Derby Road, Wingerworth, Chesterfield, S42 6LZ
G7 HNL Martin Carter, 1 Corner Cottages, White-Ladies-Aston, Worcester, WR7 4QJ
G7 HNM Gerald Greatrix, West Cottage, Main Road, Boston, PE20 3PZ
G7 HNN Kenneth Chandler, 4 Park Avenue, Thatcham, RG18 4NP
G7 HNR A Ball, 39 Deepdale Avenue, Birmingham, B26 3EL
GM7 HNU G Banks, 9b Powis Crescent, Aberdeen, AB2 3YS
G7 HOA WIDNES & RUNCORN ARC c/o D Wilson, 12 New Street, Elworth, Sandbach, CW11 3JF
GW7 HOC Darren Warburton, 71 Richards Terrace, Cardiff, CF24 1RW
G7 HOE P Goode, 23 Byworth Road, Farnham, GU9 7BT
G7 HOK P Kellingley, 290 Calmore Road, Calmore, Southampton, SO40 2RF
G7 HOL D Martin, Alken, The Covert, Orpington, BR6 0BT
GW7 HOM Vivian Cole, 77 Parc Castell Y Mynach, Creigiau, Cardiff, CF15 9NZ

**IMPORTANT NOTE**

**Revalidate licence to avoid revocation** – Ofcom has advised the Society that plans will be drawn up to revoke licences that have not been revalidated as required by the licence conditions. The quickest way to revalidate is to do so online via the Ofcom website: *https://services.ofcom.org.uk/* or by email: *amateur.validations@ofcom.org.uk* Ofcom staff are available to help, but please be patient during times of heavy workload.

G7 HON Stephen Martin, Broad Oak, Pheasant Lane, Maidstone, ME15 9QR
G7 HOT J Scott, 16 Hawton Road, Newark, NG24 4QB
G7 HOV J Bertram, 21 Mayfair Avenue, Twickenham, TW2 7JG
G7 HPI C Vance, 64 Caulfield Road, Swindon, SN2 8BT
G7 HQC I Sorrell, 67 Northfield Drive, Pontefract, WF8 2DJ
G7 HQF P Smith, 17 Beverley Avenue, Canvey Island, SS8 0DN
G7 HQH M Croxford, 34 Brington Road, Long Buckby, Northampton, NN6 7RW
G7 HQJ Adrian Baker, 34 Clare Street, Stoke-on-Trent, ST4 6ED
GW7 HQL Jon Caswell, 31 Pontalun Close, Barry, CF63 1QJ
G7 HQP J Woods, 1 Dean Road, Cosham, Portsmouth, PO6 3DG
GM7 HQW Brian Currie, Fawn House, Abriachan, Inverness, IV3 8LB
G7 HQY K Walton, Springfield, Green Lane, Uckfield, TN22 5LA
G7 HRF S Crutchley, 8 Cloverland Drive, Hemsby, Great Yarmouth, NR29 4JY
G7 HRH R Conway, 9 Whitworth Lane, Loughton, Milton Keynes, MK5 8EB
G7 HRJ T Jackson, 6 The Ridge, Letchworth Garden City, SG6 1PP
G7 HRL T Turner, 21 Spurgate, Hutton, Brentwood, CM13 2LA
G7 HRM S Baker, 8 Vista Avenue, Salem, United States, 1970
G7 HRP I Booth, 16 Sandstone Drive, Leeds, LS12 5SU
G7 HRQ D Gorse, 4 St. Michaels Close, Southport, PR9 9QY
G7 HRR HUCKNALL ROLLS ROYCE ARC c/o Steve Sorockyj, 8 Bowden Avenue, Bestwood Village, Nottingham, NG6 8XN
G7 HRZ S Haynes, 10 Cypress Grove, Denton, Manchester, M34 6EA
G7 HSA A Cramp, 7 St Margarets Road, Ludlow, SY8 1XN
G7 HSB A Green, Moss View, Southport Road, Ormskirk, L39 7JU
G7 HSL Timothy Reddish, 72 Edgmond Close, Redditch, B98 0JQ
G7 HSN J Calder, Grassington, Station Road, Bedale, DL8 1SX
GD7 HSO D Hardinges, 4 The Close, Eastcote, Pinner, HA5 1PH
G7 HSS J East, 102 Westfield Lane, Wyke, Bradford, BD12 9LS
G7 HSW I Edgington, 108 Ger Y Llan, Penrhyncoch, Aberystwyth, SY23 3TR
G7 HSY Brian Stanton, 107 Beaconside, South Shields, NE34 7PT
GD7 HTG Stuart Hill, 54 Wybourn Drive, Onchan, Isle of Man, IM3 4AT
G7 HTN Peter Seitz, 6 Meadow Rise, Iwade, Sittingbourne, ME9 8SB
G7 HTU P Jenkins, 28 King Edward Road, Brynmawr, Ebbw Vale, NP23 4SD
GJ7 HTV A Mourant, Little Mead, Claremont Road, St. Saviour, Jersey, JE2 7RT
G7 HUC M Fiorentini, 22 Pytchley Crescent, Upper Norwood, London, SE19 3QT
GM7 HUD Andrew Sinclair, 5 Murieston Wood, Livingston, EH54 9EE
G7 HUG Nathan Markley, 79 Peake Close, Peterborough, PE2 9JE
G7 HUJ Stephen Telford, 44 Northcote Crescent, Leeds, LS11 6NN
G7 HUK P Hart, 104 St. Austell Drive, Wilford, Nottingham, NG11 7BQ
G7 HUO Carol Terry, Sandfields, Long Lane, Newbury, RG14 2TH
G7 HUP Mark Terry, Sandfields, Long Lane, Newbury, RG14 2TH
GW7 HVA Norman Callan, 24a Ynysmeurig Road, Abercynon, Mountain Ash, CF45 4SY
GI7 HVC T Kennedy, 19 Orchard Avenue, Newtownards, BT23 7AF
G7 HVF Stuart Garlick, 4 Oakfield Avenue, Kingswinford, DY6 8HH
G7 HVL Charles Spires, 15 Staple Hill Road, Bristol, BS16 5AA
G7 HVN M Templeman, 28 Kewstoke Road, Kewstoke, Weston-Super-Mare, BS22 9YD
G7 HVO R Gerrard, 12 Goldrill Gardens, Bolton, BL2 5NL
G7 HWM Adrian Brookes, 8 Peppersgate, Lower Beeding, Horsham, RH13 6ND
G7 HXI Ian Duffin, Mirabella, Bush Drive, Bush Estate, Norwich, NR12 0SF
G7 HXW C Rolph, The Hollies, Back Lane, Eastgate, Norwich, NR10 4HL
G7 HYG R Barber, 180 Beechfield, Hoddesdon, EN11 9QN
G7 HYM Nicholas Singer, 11 Langley Road, Beckenham, BR3 4AE
G7 HYS D Germaney, 22 Westbrook Road, Weston-Super-Mare, BS22 8JX
GI7 HYU J Adams, 2 Dorset Close, Galgorm, Ballymena, BT42 1QP
G7 HYZ M Thompson, 23 Hare Park Lane, Crofton, Wakefield, WF4 1HS
G7 HZQ S Breen, 20 Goodwood Close, Clophill, Bedford, MK45 4FE
G7 HZS G West, West Cottage, Eaudyke Road, Boston, PE22 8RU
G7 HZU A Bateman, 4 Fair Meadows, High Street, Rugeley, WS15 3LD
G7 HZZ Alan Clayton, 6 Albert Road, Bunny, Nottingham, NG11 6QE
G7 IAE R Lindley, 23 Quadrant Close, Murdishaw, Runcorn, WA7 6DW
G7 IAK Chris Hughes, 16 Morgraig Avenue, Newport, NP10 8UP
G7 IAM Malcolm Chrzanowski, 53 Lamb Street, Kidsgrove, Stoke-on-Trent, ST7 4AL
G7 IAS Roger Bell, 3 Haywards Heath Road, Balcombe, Haywards Heath, RH17 6NG
G7 IAT Martin Howells, 34 Cobden Street, Cross Keys, Newport, NP11 7PF
G7 IAU C Penney, 9 Elm Lane, Minster on Sea, Sheerness, ME12 3SQ
G7 IAW C Walsh, 4 Musbury Crescent, Rossendale, BB4 6AY
G7 IBD Steven Beaumont, 29 Tiln Lane, Retford, DN22 6RT
G7 IBF R Waller, 6 Pitchcombe, Yate, Bristol, BS37 4JX
G7 IBH K Ashton, 13 Laceys Avenue, Leverton, Boston, PE22 0BG
G7 IBL Mike Whatley, 2 Thompsons Hill, Sherston, Malmesbury, SN16 0PZ
GM7 IBM M Robertson, Woodside House, Feabuie, Inverness, IV2 5EQ
G7 IBN F Goodes, 17 Ashmead Close, Lords Wood, Chatham, ME5 8NY
GW7 IBT A Earp, 42 Tudor Gardens, Neath, SA10 7RX
G7 IBU D Nicholls, 19 Kimmeridge, Crown Wood, Bracknell, RG12 0UD
G7 IBX V Finlayson, 92 Herlington, Orton Malborne, Peterborough, PE2 5PR
G7 ICD James Mcdowall, 19 Plaistow Court, Hallwood Park, Runcorn, WA7 2GR
G7 ICE Robert Catlow, 137 Haven Lane, Oldham, OL4 2QQ
G7 ICV S Hardes, 21 Chevening Close, Chatham, ME5 7PZ
G7 IDH Marc Newton, 5 Granville Avenue, Newcastle, ST5 1JH
G7 IEB R Emberton, 10 Lodway Close, Pill, Bristol, BS20 0DE
G7 IED R Martin, 45 Quail Holme Road, Knott End-on-Sea, Poulton-le-Fylde, FY6 0BT
GD7 IEF D Roadnight, 14 Newquay Crescent, Harrow, HA2 9LJ
GD7 IEH M Blackburn, 63 Westbourne Drive, Douglas, Isle of Man, IM1 4BB
G7 IEO A Cook, 26 Worcester Road, Stourport-on-Severn, DY13 9PB
G7 IER Arthur Bent, THREE GABLES, CRAGGS HILL, Carnforth, LA6 1DJ
G7 IET A Dunlop, High View, Milton Avenue, Sevenoaks, TN14 7AU
GM7 IEU A Steele, 20 Stewart Way, Alford, AB33 8UB
G7 IEY Derek Chenery, 25 Aldreth Road, Haddenham, Ely, CB6 3PW
GI7 IEZ Thomas Mc Geown, 1 Drumcairn Road, Armagh, BT61 7SA
G7 IFB Simon Thompson, 3 Boudicca Walk, Wivenhoe, Colchester, CO7 9JB
G7 IFD Robin Hodder, 1 Stitches Farm House, Manea Road, March, PE15 0PE
G7 IFI B Jones, 25 Milton Drive, Wistaston Green, Crewe, CW2 8BT
G7 IFJ Richard Page, 68 The Ridgeway, St. Albans, AL4 9PS
G7 IFL Peter King, 1 Rue Du Canelots, Saint Frajou, France, 31230
G7 IFM J Hewitt, 9 Alford Fold, Fulwood, Preston, PR2 3UU
G7 IFO N Rigby, 2 Mill Lane, Sutton Manor, St. Helens, WA9 4HW
G7 IFR Terry Chibnell - Smith, Nursery Cottage, Whitney-on-Wye, Hereford, HR3 6HT
G7 IFU M Mccartney, 1 Tollemache Close, Manston, Ramsgate, CT12 5LX
GI7 IFW S Boskett, 314 Shore Crescent, Belfast, BT15 4JU
GM7 IFX J Barnett, 72 Cameron Toll Gardens, Edinburgh, EH16 4TG
G7 IGF I Fields, Boarzell Cottage, London Road, Etchingham, TN19 7QY
G7 IGR C Crowhurst, 143 Drayton High Road, Drayton, Norwich, NR8 6BD
G7 IGU L Evans, 58 Westminster Drive, Bromborough, Wirral, CH62 6AW
G7 IGV W Ballard, 9 Fife Close, Stamford, PE9 2YX
G7 IHD James Bolsover, 2 Kintyre Way, Heysham, Morecambe, LA3 2YF
G7 IHE H Robinson, 16 Coniston Avenue, Ashton-in-Makerfield, Wigan, WN4 8AY
GM7 IHJ M Alexander, 38 The Wynd, Dalgety Bay, Dunfermline, KY11 9SJ
G7 IHL W Humphries, 47 Fouracre Crescent, Downend, Bristol, BS16 6PT
G7 IHN Stuart Harvey, Gabled Cottage, Shipton Oliffe, Cheltenham, GL54 4HZ
G7 IHP Kevin Weston, 2 Beech Grove, Somerton, TA11 6LG
GM7 IHR R Brodie, Midgeloch Cottage, Arbuthnott, Scotland, AB3 1NX
G7 IHV Geoffrey Havell, Flat 13, Waldron House, London, SW2 1PA
G7 IHX James Allan, 60 Godfrey Road, Halifax, HX3 0SU
GM7 IHZ G Hayes, Flat 6, 87 London Road, Edinburgh, EH7 5TT
G7 IIB P Shields, 3 Hawthorn Road, Tavistock, PL19 9DL
G7 IIC S Oliphant, Homeside, Compton, Paignton, TQ3 1TD
G7 IID Edward Caunt, 5 Littledale, Pickering, YO18 8PS
G7 IIF Andrew Barrett, 188 Birkenhead Road Meols, Wirral, CH47 0NF
G7 IIH J Bond, 2 Kent Road, Fleet, GU51 3AH
G7 III I Young, 71 Sherbourne Crescent, Coventry, CV5 8LG
GM7 IIL A Ferguson, 33 West Park Road, Newport-on-Tay, DD6 8NP
G7 IIN Martyn Hewitt, 2 Hill View, Worstead, North Walsham, NR28 9SD
G7 IIO Brian Bellamy, 71 High Road, Benfleet, SS7 5LH
G7 IIQ Jennifer Davis, 21 Newton Way, St. Osyth, Clacton-on-Sea, CO16 8RQ
G7 IIS C Beatrup, Bon Air, 34 Springfield Drive, Halesowen, B62 8EU
G7 IIZ Graham Dooley, 93 Springfields, Walsall, WS4 1JX
G7 IJC David Wells, 21 Kings Road, Barnetby, DN38 6HF
G7 IJD A Carter, Harmony Cottage, 26 The Green, Fakenham, NR21 7LG
G7 IJI D Gibbs, 40 Arcot Road, Birmingham, B28 8LZ
G7 IJL D McClew, 135 Parkview Terrace, Hermitage Street, Blackburn, BB1 4ND
G7 IJW B Rushton, Cherrydene, New Road, Windermere, LA23 2LA
G7 IJY B Evans, 51 Katrina Grove, Featherstone, Pontefract, WF7 5LW
GM7 IKB Gideon Riddell, Lawhead Croft, Tarbrax, West Calder, EH55 8LW
G7 IKG A Thynne, 1 Earlston Way, Birmingham, B43 5JR
G7 IKM W Willan, 31 St. Oswalds Lane, Bootle, L30 5QD
G7 IKS A Raistrick, 10 Orchard Way, Chinnor, OX39 4UD
G7 ILA T Brennan, 9 Mill Lane, Felixstowe, IP11 7RL
G7 ILD P Brown, 15a Barton Court Avenue, Barton on Sea, New Milton, BH25 7EP
G7 ILG Ian Glossop, 1 Harborough Hill Cottages, Birmingham Road, Kidderminster, DY10 3LH
G7 ILI A Page, 29 Lambourne Close, Fareham, PO14 1SL
G7 ILJ BANBURY RAY GRP c/o B Thornton, 21 Valley Road, Banbury, OX16 9BQ
G7 ILL Peter Atherton, Findern Lane, Willington, Derby, DE65 6DW
G7 ILP K Naylor, 3 Windrush Close, Bicester, OX26 2AR
G7 ILS I Warrilow, 84 Marple Road, Stockport, SK2 5RN
G7 ILX R Voges, 43 Eastgate, Fulwood, Preston, PR2 3HS
G7 ILY D Barber, 2 St. Jamess Mews, Church, Accrington, BB5 4JR
G7 IMB S Jeffcoate, 25a Northampton Road, Lavendon, Olney, MK46 4EY
G7 IMD A Spittlehouse, 7 fernbank, Battle Green, Doncaster, DN9 1LJ
G7 IMH M Fortescue, 98 Campbell Road, Florence Park, Oxford, OX4 3NU
G7 IMQ P Bannister, 222 Haslucks Green Road, Shirley, Solihull, B90 2LN
G7 IMR Mark Taft, 44 Langcomb Road, Shirley, Solihull, B90 2PR
G7 IMT D Gerard, 15 Nyetimber Lane, Bognor Regis, PO21 3HQ
GI7 IMU Andy Reid, 18 Orby Grove, Belfast, BT5 6AL
G7 IMV R King, Old Orchard, South Milton, Kingsbridge, TQ7 3JZ
G7 IMY S Kemp, 16 Douglas Road, Aylesbury, HP20 1HW
G7 IMZ G Smith, 19 Parker Road, Humberston, Grimsby, DN36 4TT
G7 INC George Bacon, 36 Warnadene Road, Sutton-in-Ashfield, NG17 5BD
G7 ING Michael Darbyshire, 50 Gaythorne Avenue, Preston, PR1 5TA
GI7 INR A Greer, 6 Ashley Gardens, Banbridge, BT32 4BN
G7 INY J Coady, Sunset, Station Road Wisbech St. Mary, Wisbech, PE13 4RT
G7 IOB C Knowlson, 28 Hill Drive, Handforth, Wilmslow, SK9 3AR
G7 IOC R Mitchell, 2 Corbar Road, Stockport, SK2 6EP
G7 IOF K Roebuck, 1 Hollingthorpe Road, Hall Green, Wakefield, WF4 3NH
G7 IOI N Telford, 18 Kirkstall Close, South Anston, Sheffield, S25 5BA
G7 ION M Tennant, 64 Aldenham Road, Kemplah Park, Guisborough, TS14 8LD
G7 IOO P Horton, 408 Woodcrest Way, Forney, United States, 75126
G7 IPA M Clements, 23 Pudding Lane, Gadebridge, Hemel Hempstead, HP1 3JU
G7 IPH P baker, 6 Firework Close, Kingswood, Bristol, BS15 4LT
G7 IPI P Crane, 64 Bridge Avenue, Cheslyn Hay, Walsall, WS6 7EP
GI7 IPO H Stokes, 32 Islay Street, Antrim, BT41 2TS
GW7 IPS S Hamlyn, 6 New Road, Newcastle Emlyn, SA38 9BA
G7 IPX C Bowden, 36 Aspin Drive, Knaresborough, HG5 8HQ
G7 IQD Robert Cook, I.T.tomaree Lodge Little Lakes Leisure, Lye Head, Bewdley, DY12 2UZ
G7 IQM P Jaggs, 218 New Road, London, E4 9SJ
G7 IQO David Flatters, 87 Albert Promenade, Loughborough, LE11 1RD
G7 IQZ Russell Norman, 87 Edenfield Gardens, Worcester Park, KT4 7DX
G7 IRD T Jones, 3 Woodlands Close, St. Arvans, Chepstow, NP16 6EF
G7 IRF A Saunders, 61 Southlands Drive, Timsbury, Bath, BA2 0HB
G7 IRG G Wisbey, 4 Avenue Road, Streatham, London, SW16 4HL
G7 IRH Wendy Moth, 145 Carisbrooke Road, Newport, PO30 1DG
GM7 IRI K Baxter, Flat 2, 29b Corbiehill Road, Edinburgh, EH4 5BQ
GI7 IRJ Patrick Mcateer, 36 Ballyquillan Road, Aldergrove, Crumlin, BT29 4RH
G7 IRK D Deacon, 14 Dukes Road, Braintree, CM7 5UE
G7 IRN Mark Scott, 3 Summerhill, Ticehurst, Wadhurst, TN5 7JA
G7 IRP D Williams, 66 Gover Road, Hanham, Bristol, BS15 3JZ
G7 IRS D Mellor, 31 High Street, Swinderby, Lincoln, LN6 9LW
G7 IRU Carlo Hosegood, 4 The Orchard, Sixpenny Handley, Salisbury, SP5 5QL
G7 IRW T Reynolds, Hilbre, Kingsway Lane, Ruardean, GL17 9XT
G7 ISD C Rizzo, Downside, Downs Road, Chichester, PO18 9LS
G7 ISE Gavin Walters, 12 Portstone Close, Northampton, NN5 6QP
G7 ISR G Lines, 11a Gloucester Road North, Bristol, BS7 0SG
GI7 ISX S Butler, 25 Chippendale Avenue, Bangor, BT20 4PX
GM7 ITG R Young, 3 Collieston Path, Bridge of Don, Aberdeen, AB22 8LY
G7 ITM G Clarkson, 40 Wharf Road, Ash Vale, Aldershot, GU12 5AY
G7 ITO Michael De Banks, 56 Blackwater Drive, Aylesbury, HP21 9RX
G7 ITS Mark Fasham, 29 Granville Avenue, Ramsgate, CT12 6DX
G7 ITT Seamus Import, The Old Rectory, Dufton, Appleby-in-Westmorland, CA16 6DA
G7 ITU Stewart Marlow, 329 Norcot Road, Tilehurst, Reading, RG30 6AG
G7 ITW D Fennelly, 23 Trent View Gardens, Radcliffe-on-Trent, Nottingham, NG12 1AY
G7 ITX Timothy Emblem-English, 4 Mark Avenue, London, E4 7NR
G7 ITZ Barbara Calvert, Turners Hill Road, Crawley Down, Shepherds Farm, Crawley, RH10 4HQ
G7 IUB Leigh Porter, 324-326 Lillie Road, London, SW6 7PP
G7 IUE G Clem, 25 Alexander Close, Waterlooville, PO7 5TB
GM7 IUF Wilson Howie, 24 Newfield Drive, Dundonald, Kilmarnock, KA2 9EW
G7 IUI W Fagan, 1135 Melton Road, Syston, Leicester, LE7 2JS
G7 IVF C Flux, 28 Lodden Avenue, Berinsfield, Wallingford, OX10 7QB
G7 IVG K Graham, 10 Summerfields, Dalston, Carlisle, CA5 7NW
G7 IVN P Jagdev, 10 St. Johns Road, Southall, UB2 5AN
G7 IVU R Walker, 16 Norman Drive, Stilton, Peterborough, PE7 3RS
G7 IVW Peter Hester, 24 Halton Fenside, Halton Holegate, Spilsby, PE23 5BD
GI7 IVX R Connolly, 21 Eleastan Park, Kilkeel, Newry, BT34 4DA
G7 IWA A Maunder, 2 Downhouse Road, Waterlooville, PO8 0TX
G7 IWK G Blackburn, 10 Lodge Close, Redhill, Nottingham, NG5 8NZ
G7 IWU H Judge, 8 Fontenoy Road, Balham, London, SW12 9LU
G7 IWV Adrian Godley, 177 Cheriton Road, Folkestone, CT19 5HG
G7 IWW R Gibbs, 32 Beswick Avenue, Ensbury Park, Bournemouth, BH10 4EY
G7 IWZ R Murray, 92 North Lane, East Preston, Littlehampton, BN16 1HE
G7 IXC R Barkley, 39 Fulbeck Avenue, Wigan, WN3 5QN
G7 IXG Dieter Doermann, 19 Jackman Close, Fradley, Lichfield, WS13 8PW
G7 IXH Peter Lawton, 207 Eachelhurst Road, Sutton Coldfield, B76 1EA
G7 IXK G Smith, 129 Chiltern Way, Duston, Northampton, NN5 6BW
G7 IXM M Hooks, 299 Cotton End Road, Wilstead, Bedford, MK45 3DT
G7 IXP P Hammersley, 30 Bonner Grove, Aldridge, Walsall, WS9 0DU
G7 IYA B Whittock, 12 Hillside Crescent, Midsomer Norton, Radstock, BA3 2NB
G7 IYF Pieter Van Klinkenberg, 59 Watlington Street, Reading, RG1 4RF
G7 IYG Nicholas Hobbs, 7 Maygoods Lane, Uxbridge, UB8 3TE
G7 IYH L Hobbs, 7 Maygoods Lane, Cowley, Uxbridge, UB8 3TE
G7 IYI B Goddard, 3 Spring Gardens, Quenington, Cirencester, GL7 5BG
G7 IYM Timothy Mann, 34 Crows Grove, Bradley Stoke, BS32 0DA
G7 IYN A Attack, 44 Globe Road, Hornchurch, RM11 1BW
G7 IYQ K Fulcher, Derventio, High Street, Gainsborough, DN21 5LY
G7 IYX R Dodds, 33 Westgate, Warley, Oldbury, B69 1BA
G7 IZA G Griffiths, Newcastle Court, Evancoyd, Presteigne, LD8 2PA
G7 IZC R Phillips, 25 Carlile Hill, Hemlington, Middlesbrough, TS8 9SL
G7 IZE Keith Palmer, 6 Parklands Close, Arnold, Nottingham, NG5 9QU
G7 IZM F Lucas, Ivella, Recreation Street, Dudley, DY2 9EU
G7 IZN M Dunn, 39 Gainsbrook Crescent, Norton Canes, Cannock, WS11 9TN
G7 IZU A Smith, 38 Chaucer Road, Tavistock, PL19 9AJ
G7 IZV C Funnell, 61 Blackwatch Road, Coventry, CV6 3GS
G7 IZW F Chilton, 127 Nicholls Field, Harlow, CM18 6EB
G7 JAE C Pritchard, 11 Willow Green, Needingworth, St. Ives, PE27 4SW
G7 JAF Andrew Lambert, 56 Marlborough Road, Sheffield, S10 1DB
GI7 JAM K Gibson, 4 Ilford Avenue, Belfast, BT6 9SF
G7 JAN J Martyn, Aspiration, Queens Road, Crowborough, TN6 1QQ
G7 JAO C King, 39 West Street, Huntingdon, PE29 1WT
G7 JAQ R Adam, 8 Lexington Court, Purley, CR8 1JA
G7 JAS David Harris, 68 Tomlinson Avenue, Luton, LU4 0QW
G7 JAV D Wilkins, 6 Crown Lane, Rothwell, Kettering, NN14 6LR
G7 JBD Christopher Storrie, 3 Stocken Hall Mews, Stretton, Oakham, LE15 7RL
G7 JBW P Hoath, 1 Red Lodge Drive, Bilton, Rugby, CV22 7TT
G7 JBZ Richard Cone, 6 Renault Drive, Bracebridge Heath, Lincoln, LN4 2QG
G7 JCD Michael Jones, 4 Bell Street, Tipton, DY4 8HZ
G7 JCF S Beamish, The Old Vicarage, Vicarage Road, Woodbridge, IP13 8DT
G7 JCQ G Blunt, 24 Maxton Road, Liverpool, L6 6BJ
G7 JCX J Price, 37 The Court, Anderby Creek, Skegness, PE24 5YQ
G7 JDA Andrew Roberts, 5 Colmar Close, Daventry, NN11 9BT
G7 JDB John Blackburn, 2 Heath Drive, Sutton, SM2 5RP
G7 JDE G Dickie, 49 Longbridge Close, Tring, HP23 5HG
G7 JDF J Hope, 48 Holbeck, Bracknell, RG12 8XE
G7 JDH A Nevill, 47 Tranquil Walk, New Rossington, Doncaster, DN11 0RY
G7 JDI D Carslake, 21 Kestrel Drive, Bingham, Nottingham, NG13 8QD
G7 JDK R Rothwell, 11 St Marks Road, Stourbridge, DY9 7DT
G7 JDN M Collins, 8 Newfield Road, Marlow, SL7 1JW
G7 JDQ M Softley, 7 Dale End, Brancaster Staithe, King's Lynn, PE31 8DA
G7 JDR Clive Bennett, The Old Cottage, Waterside Road, Southminster, CM0 7QT
GM7 JDS Brian Reid, 10 Badenoch Road, Kirkintilloch, Glasgow, G66 3NX
GW7 JDX M Ghassempoory, 102 Colchester Avenue, Penylan, Cardiff, CF23 9AZ
GI7 JEB M Gibson, 1 Downshire Park, Bangor, BT20 3TP
GM7 JED I MacDonald, 3 Anderson Road, Stornoway, HS1 2PG
G7 JEJ T Hyder, 83 Beam Hill Road, Burton-on-Trent, DE13 0AD
GI7 JEM David Branagh, 146 Craigs Road, Carrickfergus, BT38 9XA
G7 JFI Terry Steeper, 16 High Street, Eagle, Lincoln, LN6 9DH
G7 JFM S Smith, 26 Broadsands Avenue, Paignton, TQ4 6JN
GM7 JFN K Maclean, 10b Knockaird, Port of Ness, Isle of Lewis, HS2 0XF
G7 JFU R Evison, 26 Mill Pond Road, Windlesham, GU20 6JT
G7 JGE Christopher Hobson, 28 Withering Road, Swindon, SN1 4GU
G7 JGF Timothy Froggatt, 20 The Boulevard, Hollingworth, Hyde, SK14 8PL
GM7 JGH A Bruce, 20 Weir Crescent, Milton, Wick, KW1 5SS
G7 JGI John Dilks, Handleys Farm Bungalow, The Clays, Lincoln, LN5 0RN
G7 JGQ A Greenland, 19 The Ridgeway, Potton, Sandy, SG19 2PS
GM7 JGR John Howie, 29, Coates Gardens, Edinburgh, EH12 5LG
GI7 JGT Martin Mc Namee, 22 St. Patricks Park, Rosslea, Enniskillen, BT92 7QY
G7 JGW W Holroyd, 8 Carr Dene Court, Preston Street, Preston, PR4 2XA

G7 JGY P Smith, 174 Willerby Road, Hull, HU5 5JW
G7 JGZ R Brooks, 8 Chichester Place, Tiverton, EX16 4BW
[illegible]
G7 JHE G Beckett, Royston, 2a Cadewell Lane, Torquay, TQ2 7AU
GW7 JHK P Brettle, 27 Neath Road, Resolven, Neath, SA11 4AA
G7 JHM J McCollin, 17 Lamsey Road, Hemel Hempstead, HP3 9HB
G7 JHV D Gervais, Seven Gables Lodge, Buckingham Road, Buckingham, MK18 3NA
G7 JHW Robert Johnson, 30 Thorpe Downs Road, Church Gresley, Swadlincote, DE11 9FB
G7 JHX J Williams, 40 Tythe Barn Lane, Shirley, Solihull, B90 1RW
G7 JHZ D Randles, 20 Felix Road, Ealing, London, W13 0NT
G7 JIB Luke Evans, Polvellan, School Hill, St. Austell, PL26 6TG
G7 JIF Stuart Ruffell, 2 Beulah Cottage, Church Street, Gillingham, SP8 5RL
G7 JIM W Barton, 4 Hawthorn Flats, Hawthorn Road, Dorchester, DT1 2PE
G7 JIN C Willis, 9 Avington Close, Sedgley, Dudley, DY3 3LN
G7 JJC Patrick Gerrard, 6 Ellabank Road, Heanor, DE75 7HF
G7 JJD A Harrington, 38 Pilgrims Road, Halling, Rochester, ME2 1HW
G7 JJG K Watts, 68 Kentwood Hill, Tilehurst, Reading, RG31 6DE
G7 JJJ Clive Marshall, Gladstan House, 70 Chester Road, Runcorn, WA7 3DY
G7 JJP L Towler, 8 Stowehill Road, Peterborough, PE4 7PY
G7 JJW S Coffin, 5 Colt Close, Streetly, Sutton Coldfield, B74 2EA
G7 JJX R Wallace, 31 Salts Road, West Walton, Wisbech, PE14 7EJ
GI7 JKA J McCullagh, 2 Holestone Road, Doagh, Ballyclare, BT39 0SB
G7 JKD M Coward, The Hollies, Fenton Lane End, Brampton, CA8 9LE
G7 JKH C Hyde, 42 Fern Road, Whitby, Ellesmere Port, CH65 6PB
G7 JKK J Mossman, 13 Tynrhos Estate, Caergeiliog, Holyhead, LL65 3HS
GI7 JKM S Glendinning, 2 Scotts Road, Moneymore, Magherafelt, BT45 7TW
G7 JKW S Avery, Wilding Farm Cottage, Cinder Hill, Lewes, BN8 4HP
G7 JKY S Smith, Oak Cottage, South Street, Alresford, SO24 0DY
G7 JLC A Edwards, 34 Albion Road, Malvern Link, Malvern, WR14 1PU
GI7 JLD J Hunter, 29 Mullaghacall Road, Portstewart, BT55 7EG
G7 JLF Ryan Pike, 63 Bishopstone, Aylesbury, HP17 8SH
G7 JLG A Williams, 2 Nant Y Berllan, Llanfairfechan, LL33 0SN
G7 JLK R Elliott, 39 Amanda Way, Pensilva, Liskeard, PL14 5RA
G7 JLO N Townend, 124 Rylands Road, Southend-on-Sea, SS2 4LJ
G7 JLS D Bryant, Knowle Barns, Broadhempston, Totnes, TQ9 6DA
G7 JLT K Bryant, 18 Loundyes Close, Thatcham, RG18 3EB
G7 JMB J Baker, Green Lane Farmhouse, Rugeley, WS15 2AR
G7 JME P Good, 11 Moorland Road, Didsbury, Manchester, M20 6BB
G7 JMQ Marcus Tidmarsh, 16 Castleton Road, Mitcham, CR4 1NY
G7 JMU Darren Butterworth, 27 Royds Avenue, Linthwaite, Huddersfield, HD7 5QU
G7 JMW A Weaver, 116 Maldon Road, Tiptree, Colchester, CO5 0BN
G7 JMZ J Bache, 62 Whittingham Road, Halesowen, B63 3TP
G7 JNM A White, 6 Greenbank, Hadfield, Glossop, SK13 1PD
G7 JNS S McLennan, 179 King John Avenue, Bear Wood, Bournemouth, BH11 9SJ
G7 JOA RSC OF CHESHIRE c/o Colin Rickerby, 113 Cliftonville Road, Woolston, Warrington, WA1 4BJ
G7 JOW John Ashbee, 49 Sandwich Road, Whitfield, Dover, CT16 3LT
G7 JPN Michael Bateman, 22 Bowling Green Lane, Albrighton, Wolverhampton, WV7 3HL
G7 JQF W Booth, 8 Park Crescent, Bacup, OL13 9RL
G7 JQT E Barry, 8 Astley Crescent, Scotter, Gainsborough, DN21 3SL
G7 JQW H Derrick, 28 Great Parks, Holt, Trowbridge, BA14 6QP
G7 JQZ D Beadle, 4 Harlaxton Drive, Lincoln, LN6 3NR
G7 JRC David Smith, 10 Kibroyd Drive, Darton, Barnsley, S75 5DF
G7 JRD Tristan Alwyn-Clark, 1 Blackfriars Road, Lincoln, LN2 4WS
GI7 JRG Alaister McNerlin, 27 Roeview Park, Limavady, BT49 9BQ
G7 JRJ Catherine Wainwright, 31 Queens Road, Leytonstone, London, E11 1BA
G7 JRK Peter Dixon, 7 Pincey Mead, Basildon, SS13 3EW
G7 JRM Christopher Hinton, 65 South Street, Tarring, Worthing, BN14 7NE
G7 JRP T Pratley, 28 Charles Avenue, Watton, Thetford, IP25 6BZ
G7 JRT J Tonge, Bracken Brae, Gwalchmai, Holyhead, LL65 4SL
G7 JRU A Martin, 36 Saxon Road, Lowestoft, NR33 7BT
G7 JSB J Buxton, 38 Maulden Road, Flitwick, Bedford, MK45 5BW
G7 JSC Roger Brotherton, 167 Pershore Road, Hampton, Evesham, WR11 2NB
G7 JSE S Almond, 5 Coronation Road, Rawmarsh, Rotherham, S62 5LW
G7 JSH James Field, Dan-y-Coed, North Beach Road, Aberystwyth, SY23Â £DT
G7 JSQ P Domachowski, 39 Wycliffe Road West, Coventry, CV2 3DX
G7 JSS Colin Watson, 26 Jupiter Gate, Stevenage, SG2 7ST
G7 JST JUBILEE SAILING TRUST(ARS) c/o J Wheatley, 8 Winchester Close, Feniton, Honiton, EX14 3EX
G7 JSV W McAreavey, 3 Hall Farm Cottage, East Heckington, Boston, PE20 3QG
G7 JSW R Steward, 2 Glonister House, 238 Avondale Drive, Hayes, UB3 3PP
G7 JTB Ronald Pluck, The Garden House, St. Leonards Avenue, Blandford Forum, DT11 7PA
G7 JTD D Lockett, 10 Cornwall Drive, Bayston Hill, Shrewsbury, SY3 0ED
G7 JTF A Harvey, Rose House, Rose Grove, Doncaster, DN3 3AJ
G7 JTH John Carter, 30 Swift Way, Sandal, Wakefield, WF2 6SR
G7 JTI Gerard Cuskin, 57 Aln Street, Hebburn, NE31 1XT
G7 JTK S Bell, 6 Broom Wood Court, Prudhoe, NE42 6HB
G7 JTR D Lock, Pelican House, Chilton Candover, Alresford, SO24 9TX
G7 JTV J Caswell, 3 Birch Road, Finchampstead, Wokingham, RG40 3LB
G7 JTZ Robert Smith, 17 Julian Road, Spixworth, Norwich, NR10 3QA
GW7 JUB Thelma Jones, Glyn Coch Farm, Ffynnongain, Carmarthen, SA33 4AR
G7 JUC K Marsh, 21 Edward Road, Eynesbury, St. Neots, PE19 2QF
G7 JUD H Aviss, 249 Kings Drive, Eastbourne, BN21 2UR
GI7 JUH T Cox, 13 Shrewsbury Gardens, Belfast, BT9 6PJ
G7 JUJ P Moss, 23 Lees Row, Padfield, Glossop, SK13 1EN
G7 JUL Albert Whitcher, 12 Battersby Street, Bury, BL9 7SG
G7 JUN Mike Steadman, 26 Walkers Green, Marden, Hereford, HR1 3DU
G7 JUP J Beckingham, 20 Baptist Close, Abbeymead, Gloucester, GL4 5GD
G7 JUR P Lock, 1 Carters Walk, Farnham, GU9 9AY
G7 JUV C Broadbent, Unit 4, The Old Slate Quarry, Aberllefenni, SY20 9RU
GM7 JUX Win Dyer, 24 Southfield Road, Cumbernauld, Glasgow, G68 9DZ
G7 JUZ R Shams-Nia, 1090 Eastern Avenue, Ilford, IG2 7SF
G7 JVB P Wade, 41 Prospect Avenue, Stanford-le-Hope, SS17 0NH
G7 JVC M Hewitt, 1 Harpswell Hill Park, Hemswell, Gainsborough, DN21 5UT
G7 JVE Nigel Cook, 35 Glanville Road, Hadleigh, Ipswich, IP7 5SQ
G7 JVF S Mobley, 2 Lingham Close, Solihull, B92 9NW
[illegible]
G7 JVJ Edward Peacock, Octon Lodge, Langtoft, Driffield, YO25 3BJ
G7 JVK R Hardie, 12 Hopland Close, Longwell Green, Bristol, BS30 9XB
G7 JVN D Greywolf, 3 Denham Close, St. Leonards-on-Sea, TN38 9RS
G7 JVO K Saxby, 184 Brodrick Road, Eastbourne, BN22 9RH
G7 JVQ F Sparks, 36 High View Road, Guildford, GU2 7RT
G7 JWD T Place, 34 Holcroft, Orton Malborne, Peterborough, PE2 5SL
G7 JWE A Liddell, 4 Russet Court, Kingswood, Wotton-under-Edge, GL12 8SG
G7 JWH A Butler, 43 Severn Way, Cressage, Shrewsbury, SY5 6DS
G7 JWI G Harrison, 58 Hollywall Lane, Stoke-on-Trent, ST6 5PP
G7 JWJ E Hickman, Eriska, 33 Romany Way, Stourbridge, DY8 3JR
G7 JWL E Oakes, 30 Linden Avenue, Stourport-on-Severn, DY13 0EQ
G7 JWO R Allcock, 44 Newmount Road, Stoke-on-Trent, ST4 3HQ
G7 JWQ B Priestley, Priorswood Cottage, Tyndale Road, Gloucester, GL2 7DJ
G7 JWV R Ebbetts, Markway House, Blackbush Road, Lymington, SO41 0PB
G7 JWW Simon Charters, Beechgrove, Haselor Lane Hinton-on-the-Green, Evesham, WR11 2QZ
G7 JWX B Maley, 47 Kemp Road, Whitstable, CT5 2PY
G7 JXB K Cox, 16 Henty Close, Walberton, Arundel, BN18 0PW
G7 JXD J Pritchard, 22 Osborne Way, Haslingden, Rossendale, BB4 4DZ
G7 JXF M Forknell, 24 Sherbourne Avenue, Nuneaton, CV10 9JH
G7 JXJ C Smith, 30 Rookery Close, St. Ives, PE27 5FX
G7 JXL D KERRIDGE, 7 Haslers Place, Haslers Lane, Dunmow, CM6 1AJ
G7 JXQ D Greenwood, 70 Manor Rise, Lichfield, WS14 9RF
G7 JXR Garry Wiseman, 7 Barton Road, Woodbridge, IP12 1JQ
G7 JXT I Ballantyne, 49 Cricketers Way, Chatteris, PE16 6UR
G7 JXU M Barker, 103 Friarswood Road, Newcastle, ST5 2EF
G7 JXX Ian Thaiss, 4a Union Street, Market Rasen, LN8 3AA
G7 JXY I Guffick, 13 Alderwood Close, Hartlepool, TS27 3QR
G7 JYD G Leggett, Burghfield, Junction Road, Cold Norton, Chelmsford, CM3 6HU
G7 JYG H Odd, Verona, Harrow Road, Sevenoaks, TN14 7JU
G7 JYJ Tim Gittoes, 14 Ithon Close, Llandrindod Wells, LD1 6BD
GI7 JYK Peter Lowrie, 15 Elderburn, Newtownabbey, BT36 5NF
G7 JYL J Sage, 2 Grandsire Gardens, Hoo, Rochester, ME3 9LH
G7 JYQ Timothy Dabbs, 4 Caverleigh, Cadogan Road, Surbiton, KT6 4DH
GM7 JYW P Lawrence, Gateside Smithy, Munlochy, IV8 8PA
G7 JYY M Penn, 5 Angus Close, Kenilworth, CV8 2XH
G7 JYZ S Turley, 22 Powlers Close, Stourbridge, DY9 9HH
G7 JZC A Upchurch, 68 Lindleys Lane, Kirkby-in-Ashfield, Nottingham, NG17 8AD
G7 JZI W Hilton, 8 Ashfield Avenue, Hindley Green, Wigan, WN2 4RG
G7 JZJ Mike Doyle, 133a Pope Lane, Penwortham, Preston, PR1 9DD
G7 JZK W Hancox, 30 Barlows Close, Liverpool, L9 9HH
G7 JZM G Priestley, 24 Saxton Avenue, Bradford, BD6 3SW
G7 JZS Mark Budd, Blacksmiths Cottage, Catfoss Road, Driffield, YO25 8DX
G7 JZY Kevan Long, Manor Farm, 27 Church Street, Hull, HU11 4RN
G7 KAK Ian Clewley, 31 Kenilworth Road, Basingstoke, RG23 8JF
GD7 KAM Andrew Swearman, 56 Garth Avenue, Surby, Port Erin, Isle of Man, IM9 6QU
G7 KAO D Clarke, 2 Wilmot Road, Dartford, DA1 3BA
G7 KAT P Hulse, 56 The Platters, Rainham, Gillingham, ME8 0DJ
G7 KAV N Stemp, 3 Loxwood, East Preston, Littlehampton, BN16 1DT
G7 KAX R Jones, 13 Tir Estyn, Deganwy, Conwy, LL31 9PY
G7 KBD A Carlton, 32 Culver Road, Bradford-on-Avon, BA15 1HZ
G7 KBE B Mcintyre, Flat 24, Napier Court, Sherborne, DT9 6BG
G7 KBH K Wainwright, 25 Tithebarn Road, Rugeley, WS15 2QW
G7 KBI G Dreiling, Picton Farm, Holywell, CH8 9JQ
GM7 KBK Ernest Pratt, 46 Sheddocksley Drive, Aberdeen, AB16 6NX
G7 KBR Paul Phillips, 10 Byron Grove, East Grinstead, RH19 1SG
G7 KBZ S Hutchinson, 32 Uppleby, Easingwold, York, YO61 3BB
G7 KCC J Durdin, 55 Cedric Road, Bath, BA1 3PE
G7 KCE Jeremy Hannaford, 22 Barn Park, Stoke Gabriel, Totnes, TQ9 6SR
G7 KCN B Elcoate, 9 Parsonage Lane, Laindon, Basildon, SS15 5YN
G7 KDG Paul Edmondson, 20 Mill Road, Impington, Cambridge, CB24 9PE
G7 KDH D Edmondson, 4 Elm View, Steeton, Keighley, BD20 6SZ
G7 KDI P Stevenson, Nant Fach Cerrigydrudion, Corwen, LL21 0SB
G7 KDJ A Chadwick, 2 Auden Place, Longton, Stoke-on-Trent, ST3 1SJ
G7 KDM Colin Campbell, 21 Sellwood Drive, Carterton, OX18 3AZ
G7 KDN A Thomas, 49 Tristan Close, Calshot, Southampton, SO45 1BN
G7 KDQ K Roan, 133 Woodhouse Lane, Beighton, Sheffield, S20 1AD
G7 KDR B Hopkins, 14 Falkenham Rise, Basildon, SS14 2JQ
GW7 KDU Mark Lewis, 111 Willowbrook Gardens, St. Mellons, Cardiff, CF3 0BY
G7 KDX R Bell, 5 Byron Avenue, Blyth, NE24 5RN
G7 KEA R Chapman, Flat 3, Goda Court, Littlehampton, BN17 6AS
GI7 KEC J Stafford, 31 Shimna Close, Belfast, BT6 0DZ
G7 KEE B Daw, 19 Rowan Close, Yarnfield, Stone, ST15 0EP
G7 KEI B Edgley, 2 Queens Close, Hyde, SK14 5RE
G7 KEK R HORSFALL, 7 Lytham Close, Doncaster, DN4 6UT
G7 KEP A Reeve, 27 Mendip Vale, Coleford, Radstock, BA3 5PP
G7 KFM I Hasman, Fleetway, The Spinney, Newark, NG24 2NT
G7 KFN Christine Hasman, Fleetway, The Spinney, Newark, NG24 2NT
G7 KFQ N Camp, 1 Higher Tresillian Cottages, Tresillian, Newquay, TR8 4PL
GM7 KFS Anthony Wood, Seaward, Toward, Dunoon, PA23 7UA
G7 KFZ Robert May, 5 Lenham Walk, Manchester, M22 1GE
G7 KGD H Wrighton, 43 Bryn Celyn, Colwyn Bay, LL29 6DH
G7 KGH M Forder, 157 Kennington Road, Kennington, Oxford, OX1 5PE
G7 KGI E Gould, 53 Green Road, Kidlington, OX5 2EU
G7 KGP J chisholm, 162 Ardington Road, Northampton, NN1 5LT
G7 KGR Christopher Saunders, 10 Hillyfields, Dunstable, LU6 3NS
G7 KGV Ian Lewis, Whitehill Lodge, Hextalls Lane, Redhill, RH1 4QT
GM7 KHA S Grant, 2 Clayton Avenue, Irvine, KA12 0TR
G7 KHE M Knowlson, 23 Hawthorne Avenue, Shipley, BD18 2JB
G7 KHF S Bates, 282 Wennington Road, Rainham, RM13 9UU
G7 KHL S Smith, 287 Campkin Road, Cambridge, CB4 2LD
GI7 KHR W Smyth, 35 Davarr Avenue, Dundonald, Belfast, BT16 2NT
G7 KHT Andrew Haw, 16 Sunnybank Crescent, Yeadon, Leeds, LS19 7TE
G7 KHV Richard Irvine, 1 Nutana Avenue, Hornsea, HU18 1JU
G7 KHW D Nook, 112 Helmsley Close, Bewsey, Warrington, WA5 0GB
G7 KHZ Russell Hobbs, 3 Duncombe Close, Bridgwater, TA6 4UT
[illegible]
G7 KIE N Kirkman, 4 Woodhall Crescent, Saxilby, Lincoln, LN1 2HZ
G7 KIF C Davis, 91 Station Road, Barton under Needwood, Burton-on-Trent, DE13 8DS
G7 KII M Chilcott, 16 Mount Gould Avenue, St. Judes, Plymouth, PL4 9EZ
G7 KIL C Hunt, 39 Withdean Crescent, Brighton, BN1 6WG
G7 KIN B Kinsella, 8 Sherwood Park Road, Sutton, SM1 2SQ
G7 KIO G Hawthorn-Slater, Ty Croes, Garndolbenmaen, LL51 9UJ
G7 KIQ P Hyde, 10 Highfield Crescent, Taunton, TA1 5JH
G7 KIS B Latta, 17 Park Lane, Holywell, CH8 7UR
G7 KIT D Hogg, 26 Grenville Drive, Church Crookham, Fleet, GU51 5NR
G7 KIV R Gadney, 6 Dan Yr Eppynt, Tirabad, Llangammarch Wells, LD4 4DR
G7 KIW R Henery, 117 Marlborough Road, Swindon, SN3 1NJ
GM7 KIY J Webster, 40 Greenhill Terrace, Knockentiber, Kilmarnock, KA2 0BZ
G7 KJA R Early, 11 Wenlock Drive, Newport, TF10 7HH
G7 KJD John Smallwood, 6 Thatchers Croft, Copmanthorpe, York, YO23 3YD
G7 KJE A Wilkes, 51 Shrewsbury Drive, Newcastle, ST5 7RQ
G7 KJO Michael Wray, Dinas Bran, Ceidio, Pwllheli, LL53 8UG
G7 KJP R McMahon, 8 Meadow Close, Holburn Estate, Ryton, NE40 3RU
G7 KJR C Baxter, 6 Merrington Close, Kirk Merrington, Spennymoor, DL16 7HU
G7 KJT S Mills, 49 Temple Gate Crescent, Leeds, LS15 0EZ
G7 KJV Marc Litchman, 26 Oak Tree Close, Loughton, IG10 2RE
G7 KJW Philip Haylock, 25 Whitehouse Road, Sawtry, Huntingdon, PE28 5UA
G7 KJX R Tebbutt, 37 Christchurch Drive, Daventry, NN11 4RX
G7 KKW J Marsden, 11 Firethorn Drive, Hyde, SK14 3SN
G7 KLJ Steve Kerr, 2 Shaw Cross, Kennington, Ashford, TN24 9JY
G7 KLN J Abbey, 4 Northway, Curzon Park, Chester, CH4 8BB
G7 KLP C Hartigan, Doonagore, Doolin, Ireland
G7 KLR L Pooley, 51 Lincroft, Cranfield, Bedford, MK43 0HS
G7 KLS A Macaulay, 14 Shipcote Lane, Gateshead, NE8 4JA
G7 KLT T Hassall, 5 Ashworth Street, Bacup, OL13 9LS
G7 KLV G Lovegrove, 64 Vicarage Lane, Great Baddow, Chelmsford, CM2 8HY
G7 KLZ J Fowler, Quinnhaven, Banton Shard, Bridport, DT6 3EB
G7 KMA S Balkham, 70 St. Thomas Road, Hastings, TN34 3LQ
G7 KMD N Hilton, 9 Waterloo Fields, Kingswood, Welshpool, SY21 8LF
G7 KME D Silverton, 49 Brighton Road Holland-on-Sea, Clacton-on-Sea, CO15 5SR
G7 KMF R Lythall, 1 Stotfold Drive, Thurnscoe, Rotherham, S63 0LZ
G7 KMH S Smith, 12 Holgate Close, Malton, YO17 7YP
GM7 KMM S Linksted, 1 Stevenson Avenue, Polmont, Falkirk, FK2 0GU
G7 KMO Peter Butler, 15 Roxby Close, Bessacarr, Doncaster, DN4 7JH
G7 KMP Jonathan Davies, 14 Cullen View, Probus, Truro, TR2 4NY
G7 KMW A Brown, 4 Kimberley Close, Redditch, B98 8RL
G7 KNA Andrew Jenner, 24 The Willows, Nailsea, Bristol, BS48 1JQ
G7 KNK Harry Arrowsmith, 15 Hermitage Close, Frimley, Camberley, GU16 8LP
G7 KNM M Giles, 9 Bower Green, Lords Wood, Chatham, ME5 8TN
G7 KNN B Jones, Rivendell, Heol Llewelyn ROAD, Wrexham, LL11 3PB
G7 KNQ C Martin, 54 Holme Avenue, East Leake, Loughborough, LE12 6QL
G7 KNS Gordon Bubb, Clearways, Hadlow Stair, Tonbridge, TN10 4HD
G7 KNU P Davis, 29 Wiltshire Drive, Trowbridge, BA14 0RX
G7 KOF L Barr, 7 Southwold Gardens, New Silksworth, Sunderland, SR3 1LG
G7 KOI G Russ, 12 Marconi Road, Chelmsford, CM1 1QB
G7 KON William Humphreys, 43 Arundel Street, Bolton, BL1 6RR
G7 KOS S Mccormick, 22 Eric Road, Wallasey, CH44 5RQ
GM7 KPE J Reid, 10 Fernhill Gardens, Windygates, Leven, KY8 5DZ
G7 KPF A Gayne, 119 Lower Lickhill Road, Stourport-on-Severn, DY13 8UQ
G7 KPH M Wood, 2 Ridings Lane, New Mill Road, Huddersfield, HD7 2SQ
G7 KPM J Haywood, 15 Keddington Avenue, Lincoln, LN1 3SU
G7 KPS Christopher Pickles, 1 Sycamore Cottage, West Street, Lincoln, LN5 0JA
G7 KQN Colin Parsons, Little Foxes, Craig Penllyn, COWBRI DGE
G7 KQT S Schrier, 163 West Lane, Hayling Island, PO11 0JW
G7 KRB Stephen Wells, 55 Staverton Road, Daventry, NN11 4EY
G7 KRC KEIGHLEY ARS c/o Kathryn Conlon, 4 Hill Crest Drive, Slack Head, Milnthorpe, LA7 7BB
G7 KRE T Benjamin, 24 Moat Farm Drive, Rugby, CV21 4HG
G7 KRG KEIGHLEY RAY GR c/o T Binns, Crossfarm Cottage, Keighley, BD22 9LE
G7 KRH T Hurley, 22 Honeyhill, Wootton Bassett, Swindon, SN4 7DX
G7 KRI J Tilley, 40 Marlborough Road, Stretford, Manchester, M32 0AN
G7 KRM B Walker, 3 Moorlands Drive, Mayfield, Ashbourne, DE6 2LP
G7 KRO D Willis, 5 St. Andrews Place, Brightlingsea, Colchester, CO7 0RH
GM7 KRQ H Gordon, Hawthorn Cottage, Methlick, Ellon, AB41 7DS
G7 KRS KETTERING & DISTRICT A.R.S c/o Chris Woodward, Flat 3, Burley House, Rockingham Road, Market Harborough, LE16 8XS
G7 KRT Haisan Leong, 38 Woodland Road, Sawston, Cambridge, CB22 3DU
G7 KRY Anthony Ryall, 1 Vine Tree, Rumble Street, Usk, NP15 1QG
G7 KRZ [illegible], 61 Hayfield Road, Chapel-en-le-Frith, High Peak, SK23 0JF
GM7 KSA R Vennard, 4 Braehead, Girdle Toll, Irvine, KA11 1BD
G7 KSE Alexander Hill, 53 Fairladies, St. Bees, CA27 0AR
G7 KSH C Coloman, 16 Greyhound Road, Glemsford, Sudbury, CO10 7SJ
G7 KSP G Hampson, 11 Gladstone Grove, Stockport, SK4 4BX
G7 KSQ Stuart Little, 25 Thrift Wood, Bicknacre, Chelmsford, CM3 4HT
G7 KSS M Watts, 74 Westfield Road, Caversham, Reading, RG4 8HJ
G7 KSV D pickering, 15 Primrose Close, Purley on Thames, Reading, RG8 8DG
G7 KTD W Walker, 6 Romford Street, Burnley, BB12 8AF
G7 KTH Gary Pargeter, 2 Mayfair Drive, Northwich, CW9 8GF
G7 KTL C Lake, 56 Kenilworth Court, Kenilworth Close, New Milton, BH25 6BN
G7 KTP Tim Daniels, Three Yew Trees Newton St. Margarets, Hereford, HR2 0QG
G7 KTQ J Klunder, 58 Windsor Drive, Brinscall, Chorley, PR6 8PX
G7 KTR A Slinn, Santon, Pound Lane, Sevenoaks, TN14 7NA
GM7 KTY P May, 6 Hillpark Way, Edinburgh, EH4 7BJ
G7 KUB R Warrell, Rose Cottage, Brookbottom, High Peak, SK22 3AY
GD7 KUG D Rutherford, 9 College Drive, Ruislip, HA4 8SD
G7 KUM A Yorke, 45 Ling Road, Chesterfield, S40 3HT
GM7 KUN Christine Schofield, Airidh Ghrianach Knock, Carloway, Isle of Lewis, HS2 9AU



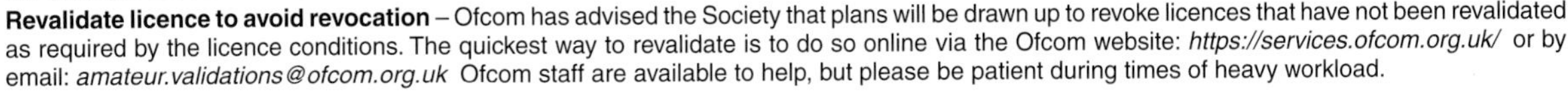
**IMPORTANT NOTE**
**Revalidate licence to avoid revocation** – Ofcom has advised the Society that plans will be drawn up to revoke licences that have not been revalidated as required by the licence conditions. The quickest way to revalidate is to do so online via the Ofcom website: *https://services.ofcom.org.uk/* or by email: *amateur.validations@ofcom.org.uk* Ofcom staff are available to help, but please be patient during times of heavy workload.

G7 KUR Philip Rennison, 30 Millfield Road, Chorley, PR7 1RE
G7 KUU K Bates, Newhaven Cottage, Star Green, Stroud, GL6 6AD
GM7 KVB A Whtye, 3 Glenfield Road, Cowdenbeath, KY4 9EP
GI7 KVR P Mcdonald, 13 Heathfield, Culmore, Londonderry, BT48 8JD
G7 KVT B Moorey, 132 Queensway, Hereford, HR1 1HQ
GM7 KVU George Kilgour, 2/1 6 Thornwood Place, Glasgow, G11 7PP
G7 KVZ J Ashmore, 46 Mease Close, Measham, Swadlincote, DE12 7NA
G7 KWA J Billam, 46 Rugby Road, Rainworth, Mansfield, NG21 0AU
G7 KWD Martin Savin, Flat 3, 30 Thurso Close, Reading, RG30 4YJ
G7 KWF A Richards, 18 Orchard Way, Lower Kingswood, Tadworth, KT20 7AD
G7 KWN A Dance, 8 Eversley Road, Arborfield Cross, Reading, RG2 9PU
G7 KWO J Lewis, 6 Abbots Way, Beckenham, BR3 3RL
G7 KWP G Lewis, 7 Hollam Drive, Dulverton, TA22 9EL
G7 KWQ T Holliday, 131 Skinburness Road, Silloth, Wigton, CA7 4QH
G7 KWS S Riches, 5 Norfolk St., Forest Gate, London, E7 0HN
GD7 KWT J Ruddock, 13a Murray Road, Northwood, HA6 2YP
GM7 KXJ R Donnet, 13 Coranbae Place, Doonfoot, Ayr, KA7 4JB
G7 KXN M Bonser, 24 Meend Garden Terrace, Cinderford, GL14 2EB
G7 KXS P Adam, 50 Lower Edge Road, Rastrick, Brighouse, HD6 3LD
G7 KXT G Belt, Flat 2, 3 King George Avenue, Leeds, LS7 4LH
G7 KXV I Eastham, 51 Chapman Road, Fulwood, Preston, PR2 8NY
G7 KXZ Charles Holdford, 23 Willow Close, Newbury, RG14 7FX
G7 KYD S Walker-Kier, 45 Anstey Road, Peckham, London, SE15 4JX
G7 KYF T Fellows, 38 Bedser Drive, Greenford, UB6 0SE
G7 KYG J Hope, 29 Horner Road, Taunton, TA2 8DZ
G7 KYH J Mann, Hyatts Mead, East End, Banbury, OX15 5LH
G7 KYI M Wilcockson, Conybeare House, Willowbrook, Windsor, SL4 6HL
G7 KYJ Adrian Clark, 10 Garfield Close, Lincoln, LN1 3QP
G7 KYL M Lack, 39 Riverview, CHURCH LANEHAM, Retford, DN22 0FL
GW7 KYT Thomas Hulmes, 12 Blackmill Road, Bryncethin, Bridgend, CF32 9YW
G7 KYW C Mellings, 4 Kiln Lane, Horley, RH6 8JG
G7 KYX G Stones, Ropercroft, Chapel Road, Boston, PE22 9PW
G7 KZG C Cain, Rydal House Audley Road, Newport, TF10 7DT
G7 KZJ M Woodland, 8 Berkeley Crescent, Stourport-on-Severn, DY13 0HJ
GM7 KZL J Mawson, 5 Forth View, Kirknewton, EH27 8AN
G7 KZN H Davies, 47 Vincent Road, Rainhill, Prescot, L35 8PE
G7 KZV C Dodson, 64 Stoneleigh Road, Solihull, B91 1DQ
G7 KZY L Stirrup, 16 Berwyn Grove, St. Helens, WA9 2AR
GM7 LAC P Green, Clochcan School Cottage, Auchnagatt, Ellon, AB41 8UJ
G7 LAF Matthew Kidman, 465 Grove Green Road, London, E11 4AA
G7 LAK Christopher Wilkinson, 9 Cheddar Close, Rainworth, Mansfield, NG21 0HX
G7 LAL I Mazura, 45 Bolingbroke Road, Scunthorpe, DN17 2NQ
G7 LAN D Halsey, 67 Watling Street, Rochester, ME2 3JH
G7 LAS Robert Cridland, 47 Stanhope Road, Swadlincote, DE11 9BQ
GD7 LAV A Gawne, Keristal House, Marine Drive, Douglas, Isle of Man, IM4 1BJ
G7 LAW J Danner, 16 Batemans Acre South, Coventry, CV6 1BE
G7 LAX William Keeys, 9 Broomfield Avenue, Rayleigh, SS6 9EJ
G7 LBD L Lewis, 29 Sefton Avenue, Hove Edge, Brighouse, HD6 2NA
G7 LBH A Champion, 5 Airedale Cliff, Leeds, LS13 1EA
G7 LBL A Batey, 9 Rampton Drift, Longstanton, Cambridge, CB24 3EH
G7 LBM T Howard, 21 Church Lane, Thornhill, Dewsbury, WF12 0JZ
G7 LBO D Wilson, 109 Nightingale Drive, Taverham, Norwich, NR8 6TR
G7 LBP Michael Akiki, 103 Main Street, Tupper Lake, United States, 12986
G7 LCD A Sermons, 18 Crispin Way, Uxbridge, UB8 3WS
G7 LCK J Berry, Roseneath, Walcote Road, Lutterworth, LE17 6EQ
GI7 LCQ Craig Serplus, 14 Claggan Park, Aghadowey, Coleraine, BT51 4BD
G7 LCS A Daniels, Wiscombe, Cleveland Road, Worcester Park, KT4 7JQ
G7 LCV Michael Sims, 4 Arran Close, Stapleford, Nottingham, NG9 8LT
G7 LCW C Simpson, 124 Tattershall Road, Boston, PE21 9LR
G7 LDD R Newton, 114 Kingston Road, Taunton, TA2 7SP
GW7 LDP P Martin, 19 Clos Bevan, Gowerton, Swansea, SA4 3GY
G7 LDR E Woolfenden, 6 Cliff Grange, Bury New Road, Salford, M7 4EZ
GM7 LDU W Adie, 16 Gordon Crescent, Methlick, Ellon, AB41 7DH
G7 LEB F Stevens, 4 Pennine Road, Bedford, MK41 9AS
G7 LED D Miles, 2 Barrington Road, Solihull, B92 8DP
G7 LEL David Hawkins, 93 Buxton Drive, Bexhill-on-Sea, TN39 4AS
G7 LEN WEST LINCS RYNT c/o A Clark, Emergency Planning Department, Fire Brigade Headquarters, Lincoln, LN5 8EL
G7 LET Ian Maughan, 95 York Road, Swindon, SN1 2JR
G7 LEX Stuart Wilkes, The Coach House, Astley Abbotts, Bridgnorth, WV16 4SP
G7 LEY ESSEX PACKET GR c/o G Lloyd, 9 Hornbeam Walk, Witham, CM8 2SZ
G7 LFC Derek Hughes, 86 Colinmander Gardens, Ormskirk, L39 4TF
G7 LFL S Botterill, 7 Plumtree Road, Cotgrave, Nottingham, NG12 3HT
G7 LFM A Cocker, 30 Shaw Road, Rochdale, OL16 4SH
G7 LFQ James White, 56a Clarendon Street, Herne Bay, CT6 8LZ
GM7 LFT A Monk, 36 North Road, Saline, Dunfermline, KY12 9UQ
G7 LFZ William Mumford, Agden Green Farm, The Green, Great Staughton, St. Neots, PE19 5DQ
G7 LGI G BRYCE, 135 Fairbridge Road, Upper Holloway, London, N19 3HF
G7 LGS Nigel Green Green, 2 Whittaker Mews, High Street, Uttoxeter, ST14 5JU
G7 LGY H Abbott, 1 St. Lawrence Close, Heanor, DE75 7AN
G7 LHS S Gray, 26 Hatfield Gardens, Appleton, Warrington, WA4 5QJ
G7 LHT F Wilson, 3a Vernon Road, Kirkby-in-Ashfield, Nottingham, NG17 8EJ
G7 LHV Graham Beaumont, 16 Chelburn View, Littleborough, OL15 9QQ
G7 LIE Brian Lovatt, Daleholme, Masterman Place, Barnard Castle, DL12 0ST
G7 LIH S Warren, 41 Barton Road, Rugby, CV22 7PT
G7 LII Jason Eccles, 30 The Stour, Daventry, NN11 4PR
G7 LIK C Tunbridge, 12 Burnham Road, Latchingdon, Chelmsford, CM3 6EU
G7 LIT G Blaxall, 27 St. Davids Road, Hextable, Swanley, BR8 7RJ
G7 LIW D Kent, 10 Goldgarth, Grimsby, DN32 8QS
G7 LJA P Gibson, 62 Glen Park, Pensilva, Liskeard, PL14 5PW
G7 LJB C Mott-Gotobed, 5 Cotswold Close, Basingstoke, RG22 5BA
GM7 LJE J Freer, 30 Kilmarnock Drive, Cruden Bay, Peterhead, AB42 0NG
GJ7 LJJ Nigel Utting, Oberon, Bagatelle Road, St. Saviour, Jersey, JE2 7TX
G7 LJN Mark Jeffs, 6 St. James House, Fore Street, Teignmouth, TQ14 8HE
G7 LJQ G Roser, 26 Willow Road, Larkfield, Aylesford, ME20 6QZ
G7 LKC J Radtke, 22 Spinney Drive, Banbury, OX16 9TA
G7 LKI D Connor, 101 Millington Road, Birmingham, B36 8BW
G7 LKL S Titterington, 33 Victoria Road, Urmston, Manchester, M41 5BZ
G7 LKR A Maciver, 55 Nordale Park, Rochdale, OL12 7RT
G7 LKV R Spray, 132 Mansfield St., Sherwood, Nottingham, NG5 4BD
G7 LKY D Parkinson, 36 Henley Road, Ipswich, IP1 3SA
G7 LKZ C BOWDEN, 20 Parc Peneglos, Mylor Bridge, Falmouth, TR11 5SL
G7 LLD M Bewley, 21 Duloe Gardens, Pennycross, Plymouth, PL2 3RS
G7 LLY J Wharton, 74 Brompton Park, Brompton on Swale, Richmond, DL10 7JP
G7 LMI J Hollowood, 10 Rossendale Close, Shaw, Oldham, OL2 8JJ
G7 LMR K Lavin, 35 Manor Bend, Galmpton, Brixham, TQ5 0PB
G7 LMT D Pantrey, 10 Columbine Close, East Malling, West Malling, ME19 6ES
G7 LMX L Pearse, Hill Farm, Middleway Road, Shepton Mallet, BA4 6TS
G7 LNB A West, 142 The Street, Kingston, Canterbury, CT4 6JQ
G7 LND Roy Williams, 45 Station Road, Westbury, BA13 3JW
G7 LNG J Tucker, 2 Ivydene Road, Ivybridge, PL21 9BH
G7 LNI S Czarnota, 11 Spring Park, Chapel Road, Ipswich, IP6 9NX
G7 LNJ R Woolridge, 8 Alastair Drive, Yeovil, BA21 3BT
G7 LNK Paul Knox, 4 Nuffield Drive, Banbury, OX16 1BX
G7 LNM D Gilham, 53 The Close, Bradwell, Great Yarmouth, NR31 8DR
G7 LNO Graham Cash, 3 Hallydown Crescent, Eyemouth, TD14 5TB
G7 LNP A Jones, 1 Abbey Way, Rushden, NN10 9HF
G7 LNT P Cundall, 11 Corn Mill, Menston, Ilkley, LS29 6BY
G7 LNU Nathan Sparrow, 46 Thomas Bell Road, Earls Colne, Colchester, CO6 2PF
G7 LNV N Turland, 2 Ludlow Close, Beeston, Nottingham, NG9 3BY
G7 LNY Howard Mascall, 37 Carnival Close, Ilminster, TA19 9DG
G7 LOA L Fisher, 195 Malvern Road, Billingham, TS23 2PJ
G7 LOE J Bhogal, 36 Titford Road, Warley, Oldbury, B69 4QA
G7 LOG T Smallwood, 51 Barlow Road, Barlow, Blaydon-on-Tyne, NE21 6JU
GM7 LOK David Barr, 17 Ballantrae, East Kilbride, Glasgow, G74 4TZ
G7 LOV E Farrar, 23 Grovehill Road, Filey, YO14 9NL
G7 LOY A Powell, 76 Glendale Avenue, Washington, NE37 2JS
G7 LOZ Keith Blackham, 86 Heather Road, Small Heath, Birmingham, B10 9TA
G7 LPB Anthony Sellick, 15 Thorpe Street, Raunds, Wellingborough, NN9 6LS
G7 LPD Peter Wilkinson, 43 Polperro Drive, Freckleton, Preston, PR4 1YD
G7 LPE Stephen Wragg, 26-28 High Street, Market Lavington, Devizes, SN10 4AG
G7 LPF C Hewitt, 9 Alford Fold, Fulwood, Preston, PR2 3UU
G7 LPK R Hilliard, 16 Hood Close, Sleaford, NG34 7WJ
G7 LPM L La Traille, 33 Festival Crescent, New Inn, Pontypool, NP4 0NB
G7 LPN T Snape, 4 Back Street, Abbotsbury, Weymouth, DT3 4JP
G7 LPO A Perry, 63a Brookland Road, Huish Episcopi, Langport, TA10 9TH
G7 LPP F RICE, 42 Donegal Road, Knowle, Bristol, BS4 1PL
G7 LPT A Page, 153 Westfield, Plymouth, PL7 2EQ
G7 LPV Gary Soden, 21 Bracknell Drive, Alvaston, Derby, DE24 0BP
G7 LPW Keith Sharples, 11 West Drove North, Walpole St. Peter, Wisbech, PE14 7HU
G7 LPY N Wootton, 13a Grange Road, Tettenhall, Wolverhampton, WV6 8RQ
G7 LPZ David Williams, 7 Hampton Drive, Great Sankey, Warrington, WA5 1JF
G7 LQD M Baguley, 2 Kensington Way, Northwich, CW9 8GG
G7 LQK R Dunn, 12 Roseberry St., Beamish, Stanley, DH9 0QR
G7 LQN C King, 33 Alexandra Road, Swallownest, Sheffield, S26 4TA
G7 LQO Lawrence Brown, 19 Stephen Drive, Sheffield, S10 5NX
G7 LQP E Livermore, Parkside, High Street, Sudbury, CO10 9DD
G7 LQU Howard Payne, 28 Humber Drive, Bury, BL9 6SJ
G7 LRB P Stevens, 6 The Rocks, Hereford Road, Shrewsbury, SY3 7QU
G7 LSB L Brown, 4 Loraine Gardens, Ashtead, KT21 1PD
G7 LSD P Wainwright, 3 Ashridge Close, Nuneaton, CV11 4XG
G7 LSF J Blain, 91 Deanfield Road, Henley-on-Thames, RG9 1UU
G7 LSG Paul Campbell, 13 Springfield Close, Marden, Hereford, HR1 3EH
GM7 LSI John Stuart, 3 Pringle Road, Elgin, IV30 4HN
G7 LSP Paul Harness, 16 Norfolk Street, Boston, PE21 6PW
G7 LSZ Mark Foreman, 12 Groveland Road, Beckenham, BR3 3QA
G7 LTG P Savage, 60 Colonial Road, Bordesley Green, Birmingham, B9 5NG
G7 LTO Michael Milns, 3 Merlin Court, Batley, WF17 0RG
G7 LTP P Sawyer, 96 Violet Lane, Croydon, CR0 4HG
G7 LTR D Ingham, 19 Recreation Avenue, Ashton-in-Makerfield, Wigan, WN4 8SU
G7 LTT Mark Phillips, 2 Hemwood Road, Windsor, SL4 4YU
G7 LTU G Smith, 36 Sandalwood Road, Loughborough, LE11 3PS
G7 LTW T Metcalfe, 39 Chobham Road, Frimley, Camberley, GU16 8PS
GM7 LTX A Warner, 41 Gaynor Avenue, Loanhead, EH20 9LU
G7 LUB J Broome, Henbant Fach, Penuwch, Tregaron, SY25 6QZ
G7 LUF Gary Whitehouse, 27 Kings End Road, Powick, Worcester, WR2 4RB
G7 LUK P Preston, 45 Saxons Heath, Long Wittenham, Abingdon, OX14 4PU
G7 LUL F Russell, 61a Fleet Street, Plymouth, PL2 2BU
G7 LUN James Keddie, Garrion, Bowland Road, Galashiels, TD1 3ND
G7 LUO N Head, 12 Heston Walk, Redhill, RH1 5JB
G7 LUR J Nolan, 33 Cambridge Road, Langford, Biggleswade, SG18 9PS
G7 LVE D Wright, Flat 1 The Annexe, Uxbridge, UB9 5HJ
G7 LVG John Ashton-Jones, Kiddley Kopse, Mordiford, Hereford, HR1 4LR
G7 LVM J Maule, 12 Edith Cavell Way, Steeple Bumpstead, Haverhill, CB9 7EE
G7 LVN Martin Odam, 89 Balch Road, Wells, BA5 2BX
G7 LVS M Unsworth, 41 Aylesbury Crescent, Hindley Green, Wigan, WN2 4TY
GM7 LWA Steven Leith, 3 County Houses, Roseisle, Elgin, IV30 5YE
G7 LWF J Totten, 28 Newman Road, Devizes, SN10 5LE
G7 LWH E Dalley, 5 Anstey Mill Close, Alton, GU34 2QT
G7 LWU Stuart Porter, 1 Belt Drove, Elm, Wisbech, PE14 0BA
G7 LWY D Northeast, 11 Repton Road, Earley, Reading, RG6 7LJ
G7 LXA C Staff, Chinewood, Pelting Drove, Wells, BA5 3BA
G7 LXB William Roberts, 36 Wray Court, Emerson Valley, Milton Keynes, MK4 2GF
G7 LXH D Hayzen, 79 Swinburne Avenue, Hitchin, SG5 2QZ
G7 LXI J Baines, Pentre Clawdd Cottage, Gobowen, Oswestry, SY10 7AE
G7 LXP D Remnant, 26 Roundway, Watford, WD18 6LB
G7 LXV N Hobbs, 224 Belchers Lane, Bordesley Green, Birmingham, B9 5RY
G7 LXY Justin Hopkins, 7 Montgomery Close, Coventry, CV3 4FS
G7 LYB R Brown, 61 Paddockhurst Road, Gossops Green, Crawley, RH11 8EU
G7 LYH J Briggs, 16 Belmont Place, Colchester, CO1 2HU
G7 LYL Craig Nixon, 52 Gloucester Drive, Basingstoke, RG22 4PH
G7 LYN Steve Laugher, Jasmine Cottage, Healey, Ripon, HG4 4LH
G7 LYS Colin Plummer, Barley House Farm, Biddulph Park, Stoke-on-Trent, ST8 7SW
G7 LZB Antony Howat, 6 Richmond Road, London, N2 8JT
G7 LZM L Mountain, 45 Westway Gardens, Redhill, RH1 2JB
G7 LZY William Eatwell, 45 Admirals Walk, Minster on Sea, Sheerness, ME12 3BB
G7 MAB M Dodson, 64 Stoneleigh Road, Solihull, B91 1DQ
GM7 MAG Paul Budgen, 12 Boggs Holdings, Pencaitland, Tranent, EH34 5BB
G7 MAJ S Hoyle, 79 Reevy Avenue, Bradford, BD6 3RR
GD7 MAN THREE LEGS VHF CONTEST GROUP c/o A Kissack, 30 High View Road, Douglas, Isle of Man, IM2 5BH
G7 MAR John Rivers, 1 Hazelwood Close, Ryde, PO33 2UP
G7 MAT K Hinton, 10 Hillview Road, Basingstoke, RG22 6BQ
G7 MAV A Goodall, 10 Russell Court, Leatherhead, KT22 8AR
GM7 MBB L Millar, 34 Brora Drive, Renfrew, PA4 0XA
G7 MBH Michael Davis, 3 Thornley Close, Ushaw Moor, Durham, DH7 7NN
GI7 MBP W Kane, 21 Mount Coole Gardens, Belfast, BT14 8JY
G7 MBY D Richards, 6 Kingley Close, Wickford, SS12 0EN
G7 MCE David Wilkinson, 56 Cobden Street, Dalton-in-Furness, LA15 8SE
G7 MCK K Singleton, Spring Cottage, Barcombe Lane, Paignton, TQ3 2QS
G7 MCS B Mcshea, 5 Frensham Avenue, Fleet, GU51 3EL
G7 MCT C Taylor, 36 Harewood Road, Shaw, Oldham, OL2 8EA
GI7 MDJ S Clarke, 86 Roddens Crescent, Castlereagh, Belfast, BT5 7JP
GI7 MDK D Robinson, 4 Ballylesson Road, Magheramorne, Larne, BT40 3HL
G7 MDM Stephen Wilkins, 5, BLACKTHORN CLOSE, Gainsborough, DN21 1WB
GI7 MDP S Mcilvenna, 10 Sycamore Court, Drumaness, Ballynahinch, BT24 8QZ
G7 MDT D Limb, 34 Elmwood Avenue, Boston, PE21 7RU
G7 MDV C Prowse, 125 Hill Road, Portchester, Fareham, PO16 8JY
G7 MDY W Collins, 99 Barkers Lane, Bedford, MK41 9TB
G7 MEA R Thomas, 43 Orkney Close, Torquay, TQ2 7DS
G7 MEE A Wood, 14 Anatase Close, Sittingbourne, ME10 5AN
G7 MEG D Cash, 3 Marsh Lane, Wolverhampton, WV10 6RU
G7 MER Christopher Hurst, 28 Hengistbury Road, Barton-on-Sea, BH25 7LU
G7 MES M Stevens, Autumn Cottage, Silver Street, Horncastle, LN9 5NH
G7 MEU David Hughes, 35 Kensington Close, Widnes, WA8 3BA
G7 MEX Mexborough & District ARS c/o James Saiger, 10 Markham Avenue, Armthorpe, Doncaster, DN3 2AZ
G7 MEZ J Arter, 18 Essex Road, Westgate-on-Sea, CT8 8AP
G7 MFA A Sejwacz, 20 Wellington Gardens, Newton-le-Willows, WA12 9LT
G7 MFE Simon Morris, The Grange, Downash Farm, Rosemary Lane, Wadhurst, TN5 7PS
G7 MFH Brian Fifield, Clyro, Lower Coombses, Chard, TA20 2SX
G7 MFO R Parkes, 7 MAin Street, Preston, Hull, HU12 8UB
G7 MFP A Dresser, 7 Torcross Grove, Calcot, Reading, RG31 7AT
G7 MFR Christopher Jenkins-Powell, 43 Cambridge Road, Lee-on-The-Solent, PO13 9DH
G7 MFW D Burdett, 17 Brambledown, Chatham, ME5 0DY
G7 MFX P March, 39 Rochford Garden Way, Rochford, SS4 1QH
G7 MFY A Wakeling, The Willows, Litcham Road, King's Lynn, PE32 2LJ
G7 MFZ M Sherratt, 21 Tweedale Close, Mursley, Milton Keynes, MK17 0SB
G7 MGA R Thorley, 9 Birchendale Close, Tean, Stoke-on-Trent, ST10 4LT
G7 MGC Thomas Gerrard, 41 Auberson Road, Bolton, BL3 3AU
G7 MGG R Shirley, 1 St. Richards Court, Bellingham Crescent, Hove, BN3 7FW
G7 MGM D Barnes, 36 Westbrook Crescent, Cockfosters, Barnet, EN4 9AS
G7 MGQ M Ball, 11 Plantation Road, Thorne, Doncaster, DN8 5EA
G7 MGT P Cox, 17 Hyde Lane, Upper Beeding, Steyning, BN44 3WJ
G7 MGV C Chadburn, 31 Darwin Close, Top Valley, Nottingham, NG5 9LN
G7 MGW Eric Palmer, 102 Park Avenue, Bryn-y-Baal, Mold, CH7 6TP
G7 MGX P Asbury, 67 Orchard Way, Measham, Swadlincote, DE12 7JZ
G7 MGY Stefan Welger, 55 Burford Avenue, Swindon, SN3 1BX
G7 MHB M Burt, 44 Overton Close, Buckley, CH7 2AX
G7 MHD A Thorp, 34 Third Avenue, Hightown, Liversedge, WF15 8JU
G7 MHF R Johnston, Gledrid Cottage, Oaklands Road, Wrexham, LL14 5DW
G7 MHL J Britton, Salters Rest, Salters Mill, Shrewsbury, SY4 5NW
G7 MHO S Fell, 14 Rectory Avenue, Corfe Mullen, Wimborne, BH21 3EZ
G7 MHQ G Taylor, 21 New Road, Kirkheaton, Huddersfield, HD5 0JB
G7 MHV S Stillwell, 130 London Road, Chatteris, PE16 6SF
G7 MID A Haydon, 9 Ash Close, Newport, PO30 5UR
G7 MIE S Hudson, 20 Churchill Road, Gravesend, DA11 7AQ
G7 MIF B Dickenson, 22 Ford Close, Herne Bay, CT6 8AN
G7 MII D Burgin, 7 Bramble Close, Halliford, Shepperton, TW17 8RR
G7 MIM Terry Wheeler, 60 Bredhurst Road, Gillingham, ME8 0PE
G7 MIN A Jones, 17 Maybush Drive, Chidham, Chichester, PO18 8SR
G7 MIP Howard Kneale, 57 Danforth Close, Framlingham, Woodbridge, IP13 9HP
G7 MIS A trott, 8a Wyatt Road, Kempston, Bedford, MK42 7EH
G7 MIT T Good, 11 Moorland Road, Didsbury, Manchester, M20 6BB
G7 MIZ J Locker, Delamere, 8 Concordia Avenue, Wirral, CH49 6JD
G7 MJD Adrian Bruring, 5 Church Lane, Hartford, Huntingdon, PE29 1XP
G7 MJI P Sayers, 23 Roseveare Road, Eastbourne, BN22 8RS
G7 MJJ R Delves, 66 Palmeira Road, Bexleyheath, DA7 4UX
G7 MJP Colin Edwards, 16 Martin Street, Normanton, WF6 1DA
G7 MJS G Davies, 78 Chatsworth Road, Southport, PR8 2QF
G7 MJV Andrew Watts, 35 Coldharbour Lane, Salisbury, SP2 7BY
G7 MJX D Hanson, 64 Laxfield Way, Lowestoft, NR33 7HH
G7 MKB J Humphries, 25 Wrekenton Row, Wrekenton, Gateshead, NE9 7JD
G7 MKF B Stracey, 31 Westfield Road, Margate, CT9 5PA
G7 MKG Peter Bradbury, 40 Titty Ho, Raunds, Wellingborough, NN9 6DF
G7 MKJ Nicholas Austin, 30 Cardinal Avenue, Borehamwood, WD6 1EP
G7 MKP A Brooks, 7 Lindford Drive, Norwich, NR4 6LT
G7 MKQ A Airey, 2 Rossmere, Greenways Estate, Spennymoor, DL16 6TZ
G7 MKV Peter Nicholls, 11 Minsmere Road, Belton, Great Yarmouth, NR31 9NX
G7 MLC G Bunn, 2 Wrench Road, Norwich, NR5 8AS
G7 MLJ P Skinner, 84 Beresford Avenue, Tolworth, Surbiton, KT5 9LW
G7 MLK D Rose, 87 Second Avenue, Sudbury, CO10 1QX
GW7 MLN J Corcoran, 23 Tan yr Allt, Abercrave, Swansea, SA9 1XF
G7 MLO J Large, 5 Raynsford Rise, Stanningfield Road, Bury St. Edmunds, IP30 0TS
G7 MLT Andrew Armstrong-Bednall, 63 Wellington Street, Heanor, DE75 7FW

G7 MLW G Kilbey, 38 Midland Road, Stonehouse, GL10 2DH
G7 MLX G Crisp, Hoppers Farm, Great Kingshill, High Wycombe, HP15 6FY
G7 MMB Stephen Read, 90 Plantation Road, Amersham, HP6 6HL
G7 MML [illegible]
GW7 MMG P Pike, 19 Hillrise Park, Clydach, Swansea, SA6 5DY
GW7 MMH Edmund Cooke, 32 Chapel Road, Three Crosses, Swansea, SA4 3PU
GM7 MMI J Wilson, 32 Silverburn Road, Bridge of Don, Aberdeen, AB22 8RW
G7 MMJ Stephen Pratt, 57 Regency Court, Bradford, BD8 9EX
G7 MMK D Baggaley, 6 Bylands Place, Newcastle, ST5 3PQ
G7 MMW Peter Francis, 14 Fulmar Place, Stoke-on-Trent, ST3 7QF
G7 MND W South, Dufonis, Dorchester Road, Wareham, BH20 6EQ
G7 MNE B Altman, 5 Ridgemount Gardens, Enfield, EN2 8QL
G7 MNG M Whale, 499 Maidstone Road, Wigmore, Gillingham, ME8 0JX
G7 MNK J Lambe, 4 St. Georges Road, Enfield, EN1 4TX
G7 MNL SOUTH DEVON RAYNET GROUP c/o Colin Coker, 46 Clarendon Road, Ipplepen, Newton Abbot, TQ12 5QS
G7 MNO R Nightingale, 58 Nutfield Grove, Filton, Bristol, BS34 7LJ
G7 MNQ P Bland, 19 Sookholme Drive, Warsop, Mansfield, NG20 0DN
G7 MNS Andrew Cartwright, 118 High Road West, Felixstowe, IP11 9AL
G7 MNT B Woods, 64 Yarningale Road, Coventry, CV3 3EQ
G7 MNZ Clinton Gaskin, 2 The Briars, West Kingsdown, Sevenoaks, TN15 6EZ
G7 MOB Philip Thain, 26 Haston Lee Avenue, Blackburn, BB1 9QT
G7 MOD David Rust, 26 Mill Road, Wiggenhall St. Germans, King's Lynn, PE34 3HL
G7 MOH E Middleton, Fairwinds, Southella Road, Yelverton, PL20 6AT
G7 MOK N Rieger-Ridd, 3 Rockland Close, Swaffham, PE37 7SP
G7 MOO K Parker, 11 Ringer Way, Clowne, Chesterfield, S43 4DW
G7 MOW K Starnes, 19 Stoneham Close, South Malling, Lewes, BN7 2ET
G7 MOX W Jones, 62 Mallings Drive, Bearsted, Maidstone, ME14 4HG
G7 MOY B Jenkins, 6 Stuart Drive, Thetford, IP24 3GA
G7 MPF R Ransome, High Winds, High Town Green, Bury St. Edmunds, IP30 0SZ
G7 MPH R Cole, 18 Borrowdale Close, Benfleet, SS7 3HE
G7 MPJ J Tweedy, 15 West Crescent, Gateshead, NE10 8AY
G7 MPV Jeffrey Woods, Rozel, Bigbury Road, Canterbury, CT4 7ND
G7 MPZ C Atkins, 278 Walderslade Road, Chatham, ME5 9AA
G7 MQC C Thomas, 1 George Gent Close, Steeple Bumpstead, Haverhill, CB9 7EW
G7 MQE D Smith, Rhos Newydd, High Street, Wrexham, LL11 5LH
G7 MQF A Kirkham, 49 Macclesfield Road, Leek, ST13 8LD
G7 MQP Simon Richardson, 73 Primrose Copse, Horsham, RH12 5PZ
G7 MQQ H Griffiths, 11 Gensing Road, St. Leonards-on-Sea, TN38 0ER
G7 MQU D Sandever, 57 Hayes Lane, Wimborne, BH21 2JB
G7 MQW Ronald Carroll, 71 Pelham St., Manton, Worksop, S80 2TT
G7 MRF M Farmer, 3 Brackenberry, Cross Heath, Newcastle, ST5 9PS
G7 MRH E Cole, 11 Ainsworth House, Wellington Road, Brighton, BN2 3BG
G7 MRL Norman Williams, 1 Dorset Close, Whitehaven, CA28 8JP
G7 MRO B Bowker, 205 Smallshaw Lane, Ashton-under-Lyne, OL6 8RJ
G7 MRY M Pratt, Pool Cottage, Ringwell Lane, Bath, BA2 7NZ
G7 MRZ R Thompson, 4 Hill Top Road, Birdwell, Barnsley, S70 5QZ
G7 MSC B Knight, 3 Burgess Cottages, Mongeham Road, Deal, CT14 8JW
G7 MSF K Sanderson, 45 Bygrove, New Addington, Croydon, CR0 9DG
G7 MSG Martin Olivant, 43 Jessop Road, Stevenage, SG1 5LQ
G7 MSH Helen Samwells, 2 Dudley Walk, Macclesfield, SK11 8SD
G7 MSK T Mann, 21 Glastonbury Court, Yeovil, BA21 3TW
G7 MSN D Stead, 15 Reeves Close, Porthleven, Helston, TR13 9PB
G7 MSQ G Shelley, 41 Thornley Road, Stoke-on-Trent, ST6 7AL
G7 MSS D Forward, 4b Cowper Road, Deal, CT14 9TW
G7 MST T Bennett, Rose Cottage, High Street, Rotherham, S62 6LN
G7 MTA C Parr, 13 Peartree Avenue, Southampton, SO19 7JN
G7 MTE Mark coote, 22 Tennyson Close, Boston, PE21 8DL
G7 MTF Tristan Foley, Flat 38, Windmill Court, Uxbridge Road, Swindon, SN5 8RT
G7 MTG A Blakeston, 8 Andersen Court, Townville, Castleford, WF10 3HY
G7 MTI Frank Paley, 68 Dennil Road, Leeds, LS15 8SD
G7 MTJ C Chase, Asholt, Ermine Street, Scunthorpe, DN15 0AD
GM7 MTQ Stuart Saunders, 3 Dunkeld Place, Dunkeld Road, Blairgowrie, PH10 6RX
G7 MTT W Johns, 6 Wakefield Avenue, Bournemouth, BH10 6DS
G7 MTV Mark Bourne, 100 Dimsdale View West, Newcastle, ST5 8EL
G7 MTW R Powell, 4 Diana Close, Spencers Wood, Reading, RG7 1HP
G7 MTX P Mullis, 39 Buckingham Road, Lawn, Swindon, SN3 1HZ
G7 MUB R Harcourt, 7 Lightfoot Close, Newark, NG24 2HT
G7 MUD CHRISTCHURCH ARS c/o D Layne, 5 Howe Close, Christchurch, BH23 3JA
G7 MUE S Roper, 1 Holywell Road, Kilnhurst, Mexborough, S64 5UQ
G7 MUH F Horsfall, 7 Broughton Way, Carleton, Poulton-le-Fylde, FY6 7LW
G7 MUN John Smith, 9 Water Meadow Way, Downham Market, PE38 9HA
G7 MUT T Cannon, 5 Barn Close, Upton, Poole, BH16 5RX
G7 MUY Adrian Sadler, 3 Denison Gardens, Chaddesden, Derby, DE21 6RG
G7 MVF M cotton, 2 Redhill View, Castleford, WF10 4QL
G7 MVG H Millington, Arran Clayton Road, Mold, CH7 1SU
G7 MVN B Wallace, 59 St. Judes Road, Englefield Green, Egham, TW20 0BT
G7 MVU Nigel Brown, 6 Hundon Place, Haverhill, CB9 0AP
G7 MVX Garth Trudgill, 61 Lansdowne Road, Coxhoe, Durham, DH6 4DN
G7 MVY R Stockley, 10 Swan Road, Timperley, Altrincham, WA15 6BX
GI7 MWA S Stewart, 3 Killyfaddy Road, Magherafelt, BT45 6EX
G7 MWB W Bone, 217 Bensham Road, Gateshead, NE8 1US
G7 MWC R Moss, 6 Adelaide Gardens, Stonehouse, GL10 2PZ
G7 MWH Philip Cross, Churchill House, Churchill Road, Louth, LN11 7QW
G7 MWI L Hansen, 19 Market St., Appledore, Bideford, EX39 1PW
G7 MWJ A Holloway, 31 Gays Road, Hanham, Bristol, BS15 3JR
GM7 MWL H Murray, 23 Denmore Gardens, Bridge of Don, Aberdeen, AB22 8LJ
G7 MWM A Scott, 22 Planters Grove, Lowestoft, NR33 9QL
G7 MWS P Parrish, 5 Kestrel Lane, Cheadle, Stoke-on-Trent, ST10 1RU
G7 MWU G Haswell, 16 Hither Green, Jarrow, NE32 4LP
G7 MWW J Smith, 19 The Crescent, Mitcheldean, GL17 0SB
GM7 MWX R Raynor, La Pergola, Kilmuir, Inverness, IV1 1XG
G7 MXL T Grange, 7 The Cherries, Canvey Island, SS8 0BB
G7 MXM Christopher Turner, 55 Fordfield Road, Ford Estate, Sunderland, SR4 6XG
G7 MXN L Orchard, 678 Devonshire Road, Blackpool, FY2 0AW
G7 MXQ Brian Gilbraith, 19 Bullcote Green, Royton, Oldham, OL2 6NJ
G7 MXT D Harris, 12 Turner Avenue, Billingshurst, RH14 9PU
GU7 MXZ B Heath, 5 Mansell St., St. Peter Port, Guernsey, GY1 1HP
G7 MYD P Williams, 5 Bright St., Cross Keys, Newport, NP11 7[illegible]
GM7 MYF Colin Dennett, Lyn-Ard, Smollett Street, Alexandria, G83 0DW
G7 MYI Alan Stride, 3 Barnfield Cottages, Edmondsham, Wimborne, BH21 5RD
G7 MYJ Ramon Ball, 5 Miller Fold Avenue, Accrington, BB5 0NT
G7 MYM David Roberts, Chatterbox, 3a, Station Road, Pershore, WR10 1NQ
G7 MYN C George, 22 Elgar Drive, Shefford, SG17 5RZ
G7 MYO Charles McIver, 2 Abbey Meadows, Chertsey, KT16 8RA
G7 MYT Peter Hilton, 40 Megstone Avenue, Whitelea Chase, Cramlington, NE23 6TU
G7 MYY L Fuller, 78c Seal Road, Sevenoaks, TN14 5AT
G7 MZA Roger Loukes, Cortijos Romero 32, La Vinuela, Malaga, Spain, 29712
G7 MZE T Ingle, 68 Wooldale Drive, Filey, YO14 9ER
G7 MZJ I Mitchell, 87 Bluebell Avenue, Penistone, Sheffield, S36 6AF
G7 MZK D Mitchell, 28 Southgate, Penistone, Sheffield, S36 6EA
G7 MZL Neil Baker, The Cottage, North Street, Crowborough, TN6 3LY
G7 MZS Leigh Terry, 8 Carters Close, Slyfield, Guildford, GU1 1FR
G7 MZW A Calvert, 122 Grampian Way, Thorne, Doncaster, DN8 5YW
G7 MZX Robert Barron, 15 Fernhill Close, Poole, BH17 8SQ
G7 MZY I Sharp, 6 Ullswater Drive, Bath, BA1 6NP
GM7 MZZ G Whiting, 21 Leckethill Court, Cumbernauld, Glasgow, G68 9EG
GM7 NAA I Skeoch, 1 Castleton Crescent, Grangemouth, FK3 0BH
G7 NAI J Stock, 31 Grange Road, Wickham Bishops, Witham, CM8 3LT
G7 NAL Harold Wood, 31 Goring Avenue, Gorton, Manchester, M18 8WW
G7 NAO I Langmuir, 2 Nelson Road, Newport, PO30 1QT
G7 NAP D Gee, 28 Rein Road, Morley, Leeds, LS27 0JA
G7 NAR South Gloucestershire Raynet c/o J Davis, 62 Kingscote, Yate, Bristol, BS37 8YE
G7 NBE M Goodwin, 23 Saxon Way, Ashby-de-la-Zouch, LE65 2JR
GD7 NBF A Sadler, 23 Wolsey Road, Moor Park, Northwood, HA6 2HN
G7 NBG L Mcguire, 71 Kingsmead Park, Bedford Road, Rushden, NN10 0NF
G7 NBI Peter Webb, 42 Holland Road, Ampthill, Bedford, MK45 2RS
G7 NBJ David Corfield, 177 Hurst Rise, Matlock, DE4 3EU
G7 NBL C King, 1 Cranfield Place, Somersham, Huntingdon, PE28 3YJ
G7 NBP Christopher Williams, 28 Sundorne Crescent, Shrewsbury, SY1 4JE
G7 NBQ S Ambrose, 3 Three Mile Pond, Sawbridgeworth, CM21 9ED
G7 NBR W Hayward, 15 Whitehouse Road, South Woodham Ferrers, Chelmsford, CM3 5PF
G7 NBU Keith Bassett, Manor Farm, Marsh Green, Exeter, EX5 2EX
G7 NBV F YOUNG, 6 Birchvale Court, Desborough, Kettering, NN14 2UY
G7 NBZ P Millerchip, 6 Washbrook View, Ottery St. Mary, EX11 1EP
G7 NCD Timothy Willis, 15 Cedar Court, Congleton, CW12 3JP
G7 NCE Kerry Derbidge, 1 Batch View, Grange Avenue, Street, BA16 9PE
G7 NCG E Turner, 16 The Rowans, Doddington, March, PE15 0SE
G7 NCP J Merrington, Cartref, Ball Lane, Frodsham, WA6 8HP
G7 NCV Kurt Hobbs, 18 Heneage Road, Grimsby, DN32 9DZ
G7 NCW G Hinton, 25 Linden Close, Prestbury, Cheltenham, GL52 3DX
GU7 NCZ N Turner, Camellia Lodge, L'Aumone, Castel, Guernsey, GY5 7RT
G7 NDB K Marshall, Doveysmead, Chapel Street, Basingstoke, RG25 2BZ
G7 NDC P Hirst, 47a Rowley Lane, Fenay Bridge, Huddersfield, HD8 0JG
G7 NDI D Bunney, 10 Richborough Close, Earley, Reading, RG6 5PW
G7 NDN S Ward, 3 Weysprings, Haslemere, GU27 1DF
G7 NDO Herbert Blackburn, 4 Hawkridge, Furzton, Milton Keynes, MK4 1BQ
G7 NDQ G Bettyes, 44 Springfield Road, Oundle, Peterborough, PE8 4LT
G7 NDS Mark Fry, 14 The Lawns, Collingham, Newark, NG23 7NT
G7 NDT R walker, 46 Lodge Road, Little Houghton, Northampton, NN7 1AE
GI7 NEB Joseph Conlon, 30 Drumglass Way, Dungannon, BT71 4AG
G7 NEC C Watson, Westwood, Laneside, Queensbury, BD13 1NE
G7 NEE Eric Maloney, 56 Westonfields Drive, Longton, Stoke-on-Trent, ST3 5JA
G7 NEG R Smith, 32 Water Lane, Wootton, Northampton, NN4 6HE
G7 NEH Graham Pemberton, 2 Hockenhull Avenue, Tarvin, Chester, CH3 8LP
G7 NEM J Barber, 2 Gresley Way, March, PE15 8QA
G7 NER T Stokes, 33 Talbot Drive, Euxton, Chorley, PR7 6PD
GI7 NET Kieran Nolan, 34 Lisgoole Park, Drumgallan, Enniskillen, BT74 5ND
GI7 NFB M Robinson, 92a Dromore Road, Hillsborough, BT26 6HU
G7 NFF Peter Salmon, Mill House, Monreith, Newton Stewart, DG8 9LJ
G7 NFG Martin Holdsworth, 9 Beaumont Close, Bowburn, Durham, DH6 5QA
G7 NFK J Watmough, Fern Cottage, Fern Road, Buxton, SK17 9NP
G7 NFM E Jones, Maes Y Coed, Vownog Road, Mold, CH7 6ED
G7 NFN R Bailey, 29 Priory Close, Bath, BA2 5AL
G7 NFO M Hall, 30 Kingsley Avenue, Rugby, CV21 4JY
G7 NFR J Thomas, 2 Alexandra Road, Uxbridge, UB8 2PQ
G7 NFT J Parry, Charlbury, Usk Road, Newport, NP18 1LP
G7 NFW E Skoyles, 29 Gordon Avenue, Thorpe St. Andrew, Norwich, NR7 0DW
G7 NFY M Mee, Anncott, Hylas Lane, Rhyl, LL18 5AG
G7 NGB J Alger, Church Hill Cottage Church Hill, Caterham, CR36SA
G7 NGF D Hatcher, 8 Churchfield, Monks Eleigh, Ipswich, IP7 7JH
G7 NGI J Price, 32 Wiltshire Drive, Trowbridge, BA14 0RE
G7 NGN R Williams, 73 Quedgeley Park, Greenhill Drive, Gloucester, GL2 5NZ
G7 NGQ A Bauer, 5 Horse Fayre Fields, Spalding, PE11 3FA
GW7 NGU J Vaughan, Montrose, 5 Trewarren Drive, Haverfordwest, SA62 3TR
G7 NGX Andrew McKenna, 12 Sunnyside Road, Beeston, Nottingham, NG9 4FH
G7 NHB Robert Griffiths, 4 Wolrige Way, Plympton, Plymouth, PL7 2RU
G7 NHC David Scholes, 71 Pelham Street, Ashton under Lyne, OL7 0DU
G7 NHD Thomas Carroll, 32 Marfords Avenue, Wirral, CH63 0JW
G7 NHE J Turnbull, 32 Haydon, Washington, NE38 8PF
G7 NHF K Williams, 1 St. Ives Way, Halewood, Liverpool, L26 7YW
G7 NHL K Mitchell, 2 Ripon Gardens, Buxton, SK17 9PL
G7 NHQ K Cotterill, 1 Molineux Avenue, Broadgreen, Liverpool, L14 3LT
G7 NHR P Dunlop, 4 Birket Avenue, Moreton, Wirral, CH46 1QZ
GM7 NHS Richard Johnson, 3 Hopetoun Green, Bucksburn, Aberdeen, AB21 9QX
GM7 NHU G Devereux, 44 Greenhead Road, Dumbarton, G82 2PN
G7 NHV K Johnson, 43 Glencoe Road, Great Sutton, Ellesmere Port, CH66 4NA
G7 NHW K Mckane, 60 Hazelwood Road, Callington, PL17 7EU
GU7 NHX A Dorrian, Le Petit Jardin, Clos Des Emrais, Castel, Guernsey, GY5 7YB
G7 NHY V Brooker, Flat 6, 46 Foxglove Way, Wallington, SM6 7JU
G7 NHZ A Greatbatch, 46 Java Crescent, Trentham, Stoke-on-Trent, ST4 8RT
G7 NIA H Seldon, 22 Downside Avenue, Plymouth, PL6 5SD
G7 NIB Ian Clark, 1 Hayward Parade, Oakengates, Telford, TF2 6EZ
G7 NID T Thorpe, 12 Berberis Walk, Greenstead Estate, Colchester, CO4 3QA
[illegible] J R[illegible], 1 Fire Station Yard, Rochdale, OL11 1DT
G7 NII D Kalawsky, 23 Brook Lane, Loughborough, LE11 3RA
G7 NIL J Chin, 198 Bermondsey Wall East, London, SE16 4TT
G7 NIN D Townsend, 40 Popes Lane, Sturry, Canterbury, CT2 0JZ
G7 NIR Lester Jones, 53 Ennisdale Drive, Wirral, CH48 9UF
G7 NIU S Tanner, 31 Four Acres, Portland, DT5 2JG
G7 NIW G Durno, Lothlorien, Upper Denbigh Road, St. Asaph, LL17 0BH
G7 NIX C Shurety, O Fran Villa, Camp Road, Norwich, NR8 6LD
G7 NIZ A Bowers, 29 Windflower Place, Northampton, NN3 5HA
G7 NJB G Harris, 58 The Leas, Minster on Sea, Sheerness, ME12 2NL
G7 NJD J Stewart, 22 Garden Road, Kendal, LA9 7ED
G7 NJE C White, 13 Peel Street, Heywood, OL10 4QD
G7 NJG David Godwin, 2 Barncroft Drive, Hempstead, Gillingham, ME7 3TJ
G7 NJI Mark Tribe, The Paddock, Wix Hill, Leatherhead, KT24 6ED
G7 NJM P Martin, 2 Gwarllyn, Tudweiliog, Pwllheli, LL53 8NG
G7 NJP M Neal, 3 Nursery Way, Grimston, King's Lynn, PE32 1DQ
GW7 NJQ S Richardson, Holmleigh, Broughton, Cowbridge, CF71 7QR
GW7 NJT John Jones, 8 Manor Court, Ewenny, Bridgend, CF35 5RH
G7 NJW G Bullen, 24 Meadowside Road, Sutton Coldfield, B74 4SJ
G7 NJX Sean Mullen, 12 Hampton Park, Bristol, BS6 6LH
G7 NJZ J O'Rourke, 39 Rutherglen Road, Corby, NN17 1ER
G7 NKE M Jordan, 17 Gilbert Avenue, Tuxford, Newark, NG22 0JB
G7 NKH D Smith, 7 Salisbury Road, Carshalton, SM5 3HA
G7 NKI Adrian Davin, 58 Bandley Rise, Stevenage, SG2 9NR
G7 NKJ Tony Westbrook, 5 Newlands, Northallerton, DL6 1SJ
G7 NKU C Prout, 1 Westbrook, Lustrells Vale, Brighton, BN2 8EZ
G7 NKV A Taylor, Moonlight Cottage, 4 Alderley Road, Macclesfield, SK11 9AP
G7 NLA C Milburn, Greenleas, Furlongs Lane, Horncastle, LN9 6LD
G7 NLF G Bandara Mieee, 26 Undine Street, London, SW17 8PR
G7 NLJ R Watts, 33 Rockside View, Matlock, DE4 3GP
G7 NLP J Greenacre, 30 Ramsey Grove, Bury, BL8 2RE
G7 NLR NORTH LANCASHIRE RAYNET GROUP c/o David Andrew, 14 Westfield Grove, Morecambe, LA4 4LQ
G7 NLY John James, 14 Fairview Drive, Bayston Hill, Shrewsbury, SY3 0LE
G7 NLZ Richard Bennett, 40 Busticle Lane, Sompting, Lancing, BN15 0DJ
G7 NMB CUMBRIA EMERGENCY PLANNING UNIT c/o K Bennett, 38 Northumberland Street, Workington, CA14 3EY
G7 NME B Caldicott, 1 Naish Road, Burnham-on-Sea, TA8 2LE
G7 NMI Karl Osborne, 42 Barbrook Lane, Tiptree, Colchester, CO5 0EF
GI7 NMK L Breadon, 32 Ashley Crescent, Millisle, Newtownards, BT22 2BG
G7 NMT M Beach, 11 Lawday Link, Farnham, GU9 0BS
G7 NNA A Watkin Ba Hnd, Aarburg, Windsor Close, Oswestry, SY11 2UA
G7 NND J Ranson, 18 Beaufort Avenue, Brooklands, Sale, M33 3WL
G7 NNM J Day, Tynwtra, Bwlch-y-Ffridd, Newtown, SY16 3HX
G7 NNR B Hughes, Dorfstrasse 71, Waldfeucht, Germany, 52525
GM7 NNS A Strachan, Rose Lea, Mormond View, Peterhead, AB42 8HX
G7 NNU John Wood, 19 Arbour Crescent, Macclesfield, SK10 2JB
G7 NNZ Daniel Dukeson, 116 Findon Street, Sheffield, S6 4QP
G7 NOF D Hall, 214 Ashby Road, Loughborough, LE11 3AG
G7 NOI Graham Evans, 20 Bleasdale Court, Longridge, Preston, PR3 3TX
G7 NOQ J Howarth, 61 Poplar Drive, Lamaleach Park, Lamaleach Drive, Preston, PR4 1EG
G7 NOR Michael Bowers, Uprising, Shottendane Road, Margate, CT9 4NE
G7 NOS Andrew Roxburgh, 10 Yewdale Park, Poplar Road, Prenton, CH43 5XD
GI7 NOW R McMaster, 40 Woodlands, Ballycarry, Carrickfergus, BT38 9JD
G7 NPL C Daniel, 24 Canterbury Road, Dewsbury, WF12 7LA
GM7 NPR Graham White, 66 Connor Street, Airdrie, ML6 7DT
G7 NPT Joseph Boyd, 26 Pear Tree Place, Warrington, WA4 1AX
G7 NQJ B Silcocks, 2 Derham Road, Bristol, BS13 7SA
G7 NQR Eugene Purvis, 36 Birchington Avenue, Middlesbrough, TS6 7EZ
G7 NQU J Varnham, 1 Burgin Road, Anstey, Leicester, LE7 7FA
G7 NQX J Bailes, 48 Harlech Close, Eston, Middlesbrough, TS6 9SZ
G7 NQZ D Harrison, 18 The Crescent, Eaglescliffe, Stockton-on-Tees, TS16 0JB
G7 NRG P Atkinson, 8 Thanet Terrace, Appleby-in-Westmorland, CA16 6TU
G7 NRO C Flanagan, 19a High Street, Wolviston, Billingham, TS22 5JY
G7 NRR Philip Morris, Antler Cottage, High Street, Northampton, NN6 9JS
G7 NRS A Saunders, 9 Capitol Close, Bolton, BL1 6LU
G7 NRV R Wheeldon, 10 Mill View Court, School Lane, St. Neots, PE19 8GJ
G7 NSK Peter Blunden, 20 Fiskerton Road, Reepham, Lincoln, LN3 4EB
G7 NSN J Vinters, 106 Halifax Road, Ripponden, Sowerby Bridge, HX6 4AG
G7 NTA T Blunsdon, 3 Railway Terrace, Aberbeeg, Abertillery, NP13 2AD
G7 NTG James Smith, 54 Greenfield Avenue, Kettering, NN15 7LL
G7 NTI Andrew Wood, 23 Cross Ryecroft Street, Ossett, WF5 9EW
G7 NTO D Brown, 5 Ash Close, Watlington, OX49 5LW
G7 NTP P Banks, Ysgol Emrys Ap Iwan, Rhuddlan Road, Abergele, LL22 7HE
G7 NTS M Berry, 25 New Square, South Horrington Village, Wells, BA5 3JS
G7 NTY Martin Sables, Ladymede, East Street, Chulmleigh, EX18 7DD
G7 NUC David Poulet, 58 Fir Street, Sheffield, S6 3TH
G7 NUE A Isted, 22 Tavy Road, Worthing, BN13 3PG
G7 NUG T Brown, 125 Godinton Road, Ashford, TN23 1LN
G7 NUM M Exton, Thorn Cottage, 42 High Street, Bourne, PE10 0SR
G7 NUN S Latham, 4 Shaston Road, Stourpaine, Blandford Forum, DT11 8TA
GM7 NUQ C Main, Riverview, St. James's Place, Inverurie, AB51 3UB
G7 NVB K Fowler, Flat 10, Westwood House Edinburgh Road, Norwich, NR2 3RL
GM7 NVG C Park, Flat 4, Corrow, Cairndow, PA24 8AD
G7 NVI P Lancaster, 2 North Farm Road, Lancing, BN15 9BS
GW7 NVM R Skelton, 45 Hendrefoilan Avenue, Sketty, Swansea, SA2 7NR
G7 NVS David Poulton, 93 Pretoria Road, Ibstock, LE67 6LP
G7 NVZ G Moore, 1 Sibland Way, Thornbury, Bristol, BS35 2EJ
G7 NWR North Wiltshire Raynet Group c/o Andrew Sharman, 3 Deben Crescent, Swindon, SN25 3QB
G7 NXV D Ross, 37 Cartmell Drive, Leeds, LS15 0NQ
GM7 NYB Neil Macfarlane, 3 Kilmore Terrace, Dervaig, Isle of Mull, PA75 6GN
G7 NYD B Collinge, 4 Ash Grove, Preesall, Poulton-le-Fylde, FY6 0EW
G7 NYF Peter Ridley, 11 Thorney Close, Fareham, PO14 3AF

**IMPORTANT NOTE**

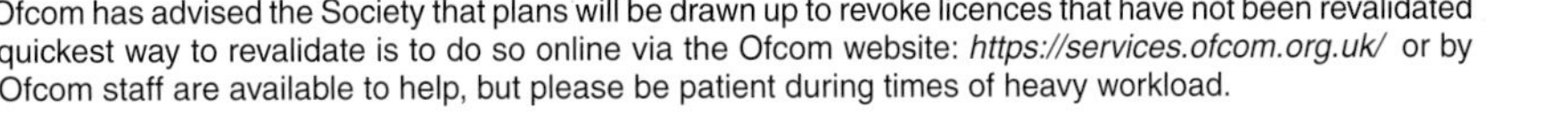
**Revalidate licence to avoid revocation** – Ofcom has advised the Society that plans will be drawn up to revoke licences that have not been revalidated as required by the licence conditions. The quickest way to revalidate is to do so online via the Ofcom website: *https://services.ofcom.org.uk/* or by email: *amateur.validations@ofcom.org.uk* Ofcom staff are available to help, but please be patient during times of heavy workload.

G7 NYP GLOUCESTER REPEATER GROUP c/o Nicholas Negus, Camlan Farm, Camlan, Machynlleth, SY20 9EP
GM7 NZI R Simpson, 2/1 53 Jedworth Avenue, Glasgow, G15 7QE
G7 NZM G Clifton, 21 Park Road, Featherstone, Wolverhampton, WV10 7HS
G7 NZO S Bate, 40 Reeds Avenue, Earley, Reading, RG6 5SR
G7 NZR Anthony Haslam, The Goldings, Hayton, Brampton, CA8 9JA
G7 NZU L Elliott, 4 Oulton Drive, Oulton, Leeds, LS26 8EN
G7 NZV R Easting, 3 Ellistons Yard, Ballingdon Street, Sudbury, CO10 2BU
G7 NZY C Wells, 6 Craister Court, Cambridge, CB4 2SH
G7 NZZ Douglas Poole, 239 Forest Road, Fishponds, Bristol, BS16 3QY
G7 OAA J Davies, 78 Chatsworth Road, Southport, PR8 2QF
GM7 OAF E Capstick, 24 Dalmore Crescent, Helensburgh, G84 8JP
G7 OAH Kenneth Adams, Queena, Bicton, Liskeard, PL14 5RF
G7 OAI John Cannell, 53 Thimble Close, Hurstead, Rochdale, OL12 9QP
G7 OAJ Stephen Walker, 90 Stoneleigh Avenue, Worcester Park, KT4 8XY
G7 OAS A White, 1 Little Orchard, Stanway Road, Broadway, WR12 7NQ
G7 OAV A Holohan, 8 School House Terrace, Kirk Deighton, Wetherby, LS22 4EH
GM7 OAW A Irvine, 41 Craighead Road, Bishopton, PA7 5DT
G7 OAX K Morrison, 29 Abbotsbury Road, Broadstone, BH18 9DB
G7 OBC Roger Hudman, 27 Egerton Road, Streetly, Sutton Coldfield, B74 3PQ
G7 OBD M Peach, 48 Melrose Avenue, Portslade, Brighton, BN41 2LS
G7 OBE David Peach, 48 Melrose Avenue, Portslade, Brighton, BN41 2LS
G7 OBF J Aston, 9 Beaufort Close, Reigate, RH2 9DG
GM7 OBM Duncan Macpherson, 138 Broomhill Crescent, Alexandria, G83 9QL
G7 OBP G Turner, 11 Royds Crescent, Rhodesia, Worksop, S80 3HF
G7 OBR M Biddles, 12 Manor Gardens, Glenfield, Leicester, LE3 8FN
G7 OBS Mike Simkins, 37 St. Andrews Meadow, Harlow, CM18 6BL
G7 OBX N Finbow, 5 Pinners Close, Burnham-on-Crouch, CM0 8QH
G7 OCC A Graham, 19 Talbot Road, Rushden, NN10 9NS
G7 OCH J Doy, 11a Shrubland Avenue, Ipswich, IP1 5EA
G7 OCK Brian Phillipson, 27 Victoria Avenue, Crook, DL15 9DB
G7 OCQ William Horwood, 2a Bellrope Lane, Roydon, Diss, IP22 5RG
GM7 OCU Graham Rule, 105/19 Causewayside, Edinburgh, EH9 1QG
G7 OCX J Turton, 60 Shafton Lane, Leeds, LS11 9RE
G7 OCY D Norton, 52 Letchworth Road, Leicester, LE3 6FG
G7 ODB R Evans, 18 Lilac Close, Keyworth, Nottingham, NG12 5DN
G7 ODG R Beadle, 8 Erica Gardens, Croydon, CR0 8LG
G7 ODM G Wane, 1a Rickyard Close, Polesworth, Tamworth, B78 1DE
G7 ODN Leandro Nicoletti, 6 Laverock Close, Kimberley, Nottingham, NG16 2QX
GW7 ODP M Jones, 11 Brenig Road, Penlan, Swansea, SA5 7BW
G7 ODR D Cartwright, Kenwood, Jaggers Lane, Hope Valley, S32 1AZ
G7 ODT C Wright, Top Farm Bungalow, Ermine Street, Huntingdon, PE28 4EW
G7 ODV Gareth Bagley, Woodcroft, Catton Bottom, Redhill, RH1 3BH
G7 ODZ Bernard Fowler, 4 Langley Street, Derby, DE22 3GL
G7 OEA Philip Foulkes, 60 Hornby Boulevard, Litherland, Liverpool, L21 8HG
G7 OED Richard Stanley, 58 Wells Gardens, Basildon, SS14 3QS
G7 OES William Robinson, 5 North View, Newfield, Chester le Street, DH2 2SD
G7 OET Lorna Hitchen, 40 Methuen Avenue, Fulwood, Preston, PR2 9QX
G7 OEW S Yohn, Ortner House, Abbeystead, Lancaster, LA25 9BD
G7 OEY Geoffrey Hodges, 12 Linwal Avenue, Houghton-on-the-Hill, Leicester, LE7 9HD
G7 OFI Paul Smith, 11 Springwell Close, Maltby, Rotherham, S66 7HG
G7 OFM R Squires, 2 Fishergreen, Ripon, HG4 1NW
G7 OFU N Patterson, 63 Squires Wood, Fulwood, Preston, PR2 9QA
G7 OFV T Wordsworth, 61 Crane Road, Kimberworth, Rotherham, S61 3HN
G7 OGL D Parker, 9 Warwick Gardens, Thrapston, Kettering, NN14 4XB
G7 OGN D Arnold, 18 Pheasant Way, Spring Park, Northampton, NN2 8BJ
G7 OGO K Mann, 89 Wootton Village, Boars Hill, Oxford, OX1 5HW
G7 OGR David Arthurs, 32 Lowfield Avenue, Rotherham, S61 4PD
GM7 OGS D Rushmer, 1a Low St., New Pitsligo, Fraserburgh, AB43 6NQ
G7 OGT Tony McDonald, 3 Widden Close, Sway, Lymington, SO41 6AX
G7 OHD P Martindale, 4 The Crayke, Bridlington, YO16 6YP
G7 OHM R Jarvis, 5 Caldecote Avenue, Cockermouth, CA13 9EQ
G7 OHO John Hislop, 10 Park Wood Close, Broadstairs, CT10 2XN
G7 OHW L Blanchard, 1 Dibden Lane, Alderton, Tewkesbury, GL20 8NT
G7 OIA M Larcombe, 52 Orchard Road, Burgess Hill, RH15 9PL
G7 OIB S Johnson, 85 Bradley Road, Trowbridge, BA14 0QS
GW7 OIK David Todd, Tyrcae, Gwernogle, Carmarthen, SA32 7SA
GM7 OIN John Cowan, 1 Treebank Crescent, Ayr, KA7 3NF
G7 OIR A Grundy, 21 Ribston Close, Shenley, Radlett, WD7 9JW
G7 OIT I Guest, 46 Brasenose Drive, Kidlington, OX5 2EQ
G7 OJA P Mann, Jasmine Cottage, 11 New Mills Road, High Peak, SK22 2JG
GM7 OJJ J Alexander, Newton Of Kinmundy Cottage, Kinmundy, Peterhead, AB42 5AY
G7 OJO R Brown, 34 Fallowfield Road, Solihull, B92 9HH
GW7 OJT Edward Wolfenden, 18 Edison Crescent, Clydach, Swansea, SA6 5JF
G7 OJU F Dixon, 9 Lincoln Road, Fenton, Lincoln, LN1 2EP
G7 OJX K Trigg, 41 Veasey Road, Hartford, Huntingdon, PE29 1TA
G7 OJY V Holyoake, 14 Maudlin Court, De Cham Road, St. Leonards-on-Sea, TN37 6JY
G7 OJZ James Neale, 20 Oakfield Road, Wollescote, Stourbridge, DY9 9DL
G7 OKF M Hawkins, 294 Norton Lane, Earlswood, Solihull, B94 5LP
G7 OKI Wayne Cornish, 21 Centaur Street, Portsmouth, PO2 7HB
G7 OKO F Webb, 50 Hassam Avenue, Newcastle, ST5 9ET
G7 OKR J Campbell, 94 Liscard Road, Wallasey, CH44 8AB
G7 OKT S Smith, 17 Thackers Way, Deeping St. James, Peterborough, PE6 8HP
G7 OKV K Porter, 5 Roberts Drive, Marston Moretaine, Bedford, MK43 0GN
GM7 OKX Geoff Chesworth, Auchinway, Skares, Cumnock, KA18 2RE
G7 OKY D Schofield, 26 The Chase, Coulsdon, CR5 2EG
G7 OLC Paul Broadhead, 45 Priory Close, Dudley, DY1 3ED
G7 OLF W Levick, 50 Wintern Court, Lea Road, Gainsborough, DN21 1NA
G7 OLG Mark Hodge, 271 Marsh Lane, Bootle, L20 5BG
G7 OLH A Barnett, 20 Mortlake Drive, Mitcham, CR4 3RQ
G7 OLU W Wood, The Alley Off Of Gajdoru St, Xaghra, Gozo, Malta, XRA 104
G7 OLW P Newton, 22 Barrow Rise, Weymouth, DT4 9HJ
G7 OMA Michael Heales, 11 Cardinals Walk, Hampton, TW12 2TR
G7 OMF K Gill, 358 Moor End Road, Halifax, HX2 0RH
G7 OMI J Patel, 1 The Glade, Furnace Green, Crawley, RH10 6JS
G7 OMM R Stroud, 22 Marvell Close, Crawley, RH10 3AL
G7 OMN J Eyes, 31 Langdale Road, Crewe, CW2 8RS
G7 OMQ John Gillman, 4 Yeosfield, Riseley, Reading, RG7 1SG
GM7 OMU S Macmillan, 74 Canberra Avenue, Clydebank, G81 4LN
GI7 OMY D O'Buitigh, 11 Rossnareen Avenue, Belfast, BT11 8LP
G7 ONB Christopher Robinson, Jordan, The Green, Stowmarket, IP14 3AB
G7 ONE Ronald Jacobs, Rose Mount, Grove Road, Ventnor, PO38 1TH
G7 ONF Michael Whitley, Apple Tree Cottage, Ratten Row, Driffield, YO25 3TJ
G7 ONI Jonathan Churchill, 30 Brigade Place, Caterham, CR3 5ZU
GM7 ONJ A Martin, The Cairn, Duntrune, by Dundee, DD4 0PP
G7 ONL R Ramsey, 17 Derby Road, Guisborough, TS14 7DP
G7 ONR P Green, 43 Malvern Road, Hull, HU5 5TP
G7 ONV D Das, 4 Farcliff, Sprotbrough, Doncaster, DN5 7RE
G7 OOB Karl Barnes, The Old School House, 32 Church Street, Peterborough, PE6 8DA
G7 OOE John Bone, Rochford Farm, Smithfield Road North Kelsey Moor, Market Rasen, LN7 6HG
G7 OOF E Cottle, 3 Mainstone, Romsey, SO51 8HG
G7 OOH Kevin McAllister, Willow Croft, 109a Kaye Lane, Huddersfield, HD5 8XT
G7 OOI R Phillipson, 22 Bagmere Close, Brereton, Sandbach, CW11 1SG
GI7 OOM D Magowan, 35 Princeton Avenue, Lurgan, Craigavon, BT66 8LW
G7 OOO SCARBOROUGH SEG c/o Roy Clayton, 9 Green Island, Irton, Scarborough, YO12 4RN
G7 OOP A Constantine, Fairways, Birchington Close, Bexhill-on-Sea, TN39 3TF
G7 OOS Jeffrey Smalley, 88 Gledhow Wood Road, Leeds, LS8 4DH
G7 OOT S Godrich, 11 The Ringway, Queniborough, Leicester, LE7 3DN
G7 OOU D Witts, 3 Ormsgill Court, Heelands, Milton Keynes, MK13 7PZ
G7 OOV R Nunn, 49 Lulworth Drive, Roborough, Plymouth, PL6 7DT
G7 OPB Andrew Adams, 44 Berkeley Vale Park, Berkeley, GL13 9TG
G7 OPD P Hundy, 101 Goodway Road, Great Barr, Birmingham, B44 8RS
G7 OPG Robin Palmer, Maypole Dock, Quaker Lane, Southall, UB2 4RG
G7 OPI Drew Pink, 87 Lillybrook Estate, Lyneham, Chippenham, SN15 4AS
G7 OPJ J Buttery, 38 Wigmore Gardens, Worle, Weston-Super-Mare, BS22 9AQ
GM7 OPN W Cairns, 74 Jean Armour Drive, Mauchline, KA5 6DT
G7 OPS Paul Whiting, 77 Melford Way, Felixstowe, IP11 2UH
G7 OPY L Kelly, 8 Solent Hill, Freshwater, PO40 9TG
G7 OQB E Dockray, 2 The Gardens, Farsley, Pudsey, LS28 5HW
GM7 OQE G Murray, 26 Forbes Road, Rosyth, Dunfermline, KY11 2AN
G7 OQG Steve Williams, 7 Wilton Crescent, Macclesfield, SK11 8TH
G7 OQL C Broad, 96 Kingsley Court, Fraddon, St. Columb, TR9 6PD
G7 OQO Alun Thomas, 7 Lon y Wennol, Llanfairpwllgwyngyll, LL61 5JX
G7 OQQ S Ayers, 20 Wytham View, Eynsham, Witney, OX29 4LU
G7 OQT Kim Peacock, 20a Pool View Caravan Park, Buildwas, Telford, TF8 7BS
GW7 ORB D Cullen, 12 Davies Road, Pontardawe, Swansea, SA8 4PH
G7 ORE Essex Raynet c/o Neil Smith, Clare Cottage, White Ash Green, Halstead, CO9 1PD
G7 ORG R Gunner, White House, The Whiteway, Cirencester, GL7 7BA
GM7 ORJ A Ross, 16 Croft Road, Kiltarlity, Beauly, IV4 7HZ
G7 ORK D Love, Woodland View, Lower Street, Shepton Mallet, BA4 6BB
G7 ORN Stephen Tideswell, 1 Ivy Bank Cottage, Brickhill Lane, Burton-on-Trent, DE13 8SW
G7 ORS Jesper Lorenzen, 40 Boundary Road, Ramsgate, CT11 7NW
G7 ORT D Buckley, 22b Anerley Grove, Kingstanding, Birmingham, B44 9QH
G7 ORV S Edmonds, 170 Halton Road, Sutton Coldfield, B73 6NZ
G7 ORW R Garnett-Frizelle, 17 Bridport Avenue, New Moston, Manchester, M40 3WP
GM7 ORX J Lee, 2/5 Heriot Bridge, Edinburgh, EH1 2HR
G7 OSB C Baker, 22 Court Park, Thurlestone, Kingsbridge, TQ7 3LX
G7 OSH WATCOMBE RC c/o H Davies, 33 Sandown Road, Ocean Heights, Paignton, TQ4 7RL
G7 OSJ Ann Ramm, 17 Sharrington Road, Bale, Fakenham, NR21 0QX
G7 OSK D Ramm, 24 Rowan Way, Holt, NR25 6TZ
G7 OSO A Kinnersley, Weathertop, Barthomley Road, Stoke-on-Trent, ST7 8HU
GM7 OSQ D Clark, Benmhor, Baluachrach, Tarbert, PA29 6TF
G7 OSR Graham Parry, 72 France Furlong, Great Linford, Milton Keynes, MK14 5EJ
G7 OST S Timms, 7 Portway Drive, High Wycombe, HP12 4AU
G7 OTE S Barlow, 16 Arundel Avenue, Urmston, Manchester, M41 6NQ
G7 OTQ Andrew Dibbins, 2 Edwards Close, Briggs Lane, Oswestry, SY10 8PS
GM7 OTT I Alexander, Newton Of Kinmundy Cottage, Kinmundy, Peterhead, AB42 5AY
G7 OUZ Brian Donkin, 13 Saddlebow Road, King's Lynn, PE30 5BQ
G7 OVB Geoffrey Hutton, 17 Fonteyn Place, Stanley, DH9 6XE
G7 OVE D Brown, 9 Lancaster Way, East Winch, King's Lynn, PE32 1NY
G7 OVK C Carson, 15 Sudbury Way, Beaconhill Green, Cramlington, NE23 8HG
G7 OVM N Ward, 79 Ulwell Road, Swanage, BH19 1QU
G7 OVS J Dilks, Handley Farm Bungalow, Brant, Beckingham, LN5 0RN
G7 OWB K moorcroft, 96 Mersea Road, Colchester, CO2 7RH
G7 OWP J Oliphant, 16 Sylvias Close, Amble, Morpeth, NE65 0GB
G7 OWQ B Clifton, 3 Kirton Road, Cosham, Portsmouth, PO6 2ES
GM7 OWU Brander Brander, 3 Spartleton Place, Dundee, DD4 0UJ
G7 OWV M Baines, 21 Acre Moss Lane, Kendal, LA9 5QE
G7 OWX D Allen, 130 Seamer Road, Scarborough, YO12 4EY
G7 OXA D Giles, 73 Barsby Drive, Loughborough, LE11 5UJ
G7 OXH R Wilkins, 85 St. Richards Road, Otley, LS21 2AL
G7 OXK J Leach, 2 Andover Close, Feltham, TW14 9XG
G7 OXN B Burdis, Toledillo 19, Malaga, Spain, 29570
G7 OXP R Singleton, 91 Robins Lane, St. Helens, WA9 3NF
G7 OXV R Blott, Chateau Perigord II Bloc E Apt 5, 6 Lacets Saint Leon, Monaco, Monaco, MC98000
G7 OXY Jacqueline Mathew, 139 Dowthorpe Hill, Earls Barton, Northampton, NN6 0PX
G7 OYD P Thompson, Manorside, Whales Lane, Goole, DN14 0SB
G7 OYF B Gilbert, 3 Williams Way, West Row, Bury St. Edmunds, IP28 8QB
G7 OYP S Hollis, 89 Longfield Lane, Cheshunt, Waltham Cross, EN7 6AN
GU7 OYU A Stoaling, Carando, La Petite Mare de Lis Clos, La Rocquette, Castel, Guernsey, GY5 7BN
G7 OYX Brian Dorey, 8 Richmond Road, Swanage, BH19 2PZ
GD7 OZA G Johnston, 94 Abercorn Crescent, Harrow, HA2 0PU
G7 OZE K Baldry, 160 Rover Drive, Castle Bromwich, Birmingham, B36 9LL
G7 OZH David Albury, 40 Mulberry Gardens, Fordingbridge, SP6 1BP
G7 OZI A Morley, 6 Millway, Chudleigh, Newton Abbot, TQ13 0JN
G7 OZJ S Morley, 4 Nebular Court, Leighton Buzzard, LU7 3TT
G7 OZP J Barrett, Tree Tops, Comins Coch, Aberystwyth, SY23 3BL
G7 OZQ T Pluck, 29 Templegate View, Leeds, LS15 0HQ
G7 OZU Mark Knight, 30 Mountbatten Drive, Biggleswade, SG18 0JJ
G7 PAF Robert Scaife, 50 Springbank Road, Gildersome, Leeds, LS27 7DJ
G7 PAG Paul Gould, 10 Heron Park, Lychpit, Basingstoke, RG24 8UJ
G7 PAK Anthony Smith, 153 Seymour Way, Sunbury-on-Thames, TW16 7NL
G7 PAY Carol Wilson, 107 Hamilton Avenue, Uttoxeter, ST14 7FE
GM7 PBB John Gray, 5 North Dell, Isle of Lewis, HS2 0SW
G7 PBC P Sherburn, 70 Briarwood Road, Stoneleigh Park, Epsom, KT17 2NG
G7 PBH H Parrish, 5 Kestrel Lane, Cheadle, Stoke-on-Trent, ST10 1RU
G7 PBK JH Oliver, 27 Rosamund Avenue, Pickering, YO18 7HF
G7 PBO M Sewell, 4 Cherfield, Minehead, TA24 5TD
GW7 PBP V Roberts, 44 Mount Crescent, Morriston, Swansea, SA6 6AP
GI7 PBQ R Young, 8 Glenside Avenue, Drumbo, Lisburn, BT27 5LQ
G7 PBT Richard Spirrell, 32 Churchfield Drive, Castle Cary, BA7 7LA
G7 PBV John Mason, 56 Skegby Road, Sutton-in-Ashfield, NG17 4EZ
G7 PCE K Toop, 10 Hunt Road, Blandford Forum, DT11 7LZ
G7 PCF J Banfield, Highbury, 81 Clophill Road, Bedford, MK45 2AD
G7 PCG Mark Bartlett, 3 Bantocks Road, Great Waldingfield, Sudbury, CO10 0RT
G7 PCT P Treadwell, 22 Meynell Close, Melton Mowbray, LE13 0RA
G7 PCV A Newman, 115 Wolverhampton Road, Cannock, WS11 1AR
G7 PCW Chris Naylor, 127 Samuel Street, London, SE18 5LE
G7 PCX G Bellis, 70 Osborne St., Rhos, Wrexham, LL14 2HT
G7 PDH John Swallow, 26 Balmoral Road, Abbots Langley, WD5 0ST
G7 PDO M Galea, 17 Waterloo Road, Horsham St. Faith, Norwich, NR10 3HS
G7 PDR J Martin, 45 Quail Holme Road, Knott End-on-Sea, Poulton-le-Fylde, FY6 0BT
G7 PDU S Willis, 180 Thisselt Road, Canvey Island, SS8 9BL
G7 PEB J Edwards, 37 The Orchard, Swanley, BR8 7UR
G7 PEC Essex RAYNET c/o G Tiller, 12 Birk Beck, Waveney Drive, Chelmsford, CM1 7PJ
G7 PEE Trevor Griffiths, 1 Alum Close, Trowbridge, BA14 7HD
G7 PEN E Penn, 53 Manfield Avenue, Walsgrave, Coventry, CV2 2QF
G7 PEO Philip Bennett, 14 Harlech Crescent, Prestatyn, LL19 8DG
G7 PER D Boughton, 59 Redland Drive, Kirk Ella, Hull, HU10 7UX
G7 PEU R Chapman, 12 Lynton Road, Chesham, HP5 2BU
G7 PEX K Barker, 72 Keel Drive, Slough, SL1 2XY
G7 PFD A Thompson, Forge House, Newark, Newark, NG22 0PN
G7 PFG James Smith, 23 West Thorpe, Basildon, SS14 1LX
G7 PFI M Green, 167 Bannerdale Road, Sheffield, S11 9FA
GW7 PFK E Gillet, 4 Camrose Court, Caldy Close, Barry, CF62 9DR
G7 PFL Michael Dormer, 5 Kipling Walk, Basingstoke, RG22 6BN
G7 PFT Jon Richardson, Unit F, Tollgate Business Centre, Stafford, ST16 3HS
G7 PFY J Ellis, 11 Moorland Crescent, Guiseley, Leeds, LS20 9EF
G7 PGH M Card, 11 Manifold Road, Eastbourne, BN22 8EH
G7 PGY G Sleeman, 21 Millbank, Kintbury, Hungerford, RG17 9UW
G7 PHB S Beesley, 15 Byron Close, Cheadle, Stoke-on-Trent, ST10 1XB
G7 PHC Mary Porter, 16 The Oval, Scarborough, YO11 3AP
G7 PHD Ian Connor, Flat 2, 9 Cheney Road, Ramsgate, CT12 4BG
G7 PHE Alan Lassman, 69 St. Ladoc Road, Keynsham, Bristol, BS31 2EQ
G7 PHG David Harbron, 48 Sheridan Road, Biddick Hall, South Shields, NE34 9JJ
G7 PHH W Alderman, Outspan, Quethiock, Liskeard, PL14 3SQ
G7 PHI Timothy Larsen, 47 Lizard Lane, Sunderland, SR6 7AL
G7 PHK J Wilkes, 229 Merland Rise, Tadworth, KT20 5JQ
G7 PHL R Marshall, 3 Lawrence Crescent, Sutton-in-Ashfield, NG17 4HX
G7 PHR W Stewart, 1 Laing Close, Bardney, Lincoln, LN3 5XS
G7 PHT K Dennis, 4 Ash Grove, Sheringham, NR26 8PT
G7 PHW S Dobson, 166 Lynfield Drive, Bradford, BD9 6EZ
G7 PHY D Dobson, 166 Lynfield Drive, Bradford, BD9 6EZ
G7 PIB J Challenger, 33 Blossom Close, Langstone, Newport, NP18 2LT
G7 PIG C Wood, 2 Longfield Avenue, Heald Green, Cheadle, SK8 3NH
G7 PIJ M Beeson, 1 Tamar Grove, Cheadle, Stoke-on-Trent, ST10 1QQ
G7 PIK B Bashford, 51 Broadwater Road, Worthing, BN14 8AH
GW7 PIN William Griffiths, 147 High Street, Tonyrefail, Porth, CF39 8PL
G7 PIP R Oswald, 17 Dunclutha Road, Hastings, TN34 2JA
G7 PIR J Briggs, 20 Druids Lane, Maypole, Birmingham, B14 5SN
GI7 PIZ Steve Hewitt, 11 Erindee Close, Donaghadee, BT21 0NS
G7 PJD R Howes, 58 Mayfield Avenue, Orpington, BR6 0AQ
GI7 PJF R Stewart, 1 Portmore Hall, Ballydonaghy Road, Crumlin, BT29 4WT
G7 PJG Dinesh Gohill, Flat 71, Hatton Place, Luton, LU2 0FD
GI7 PJU Christopher Robinson, 19d Divis Tower, Belfast, BT12 4QB
G7 PJW P Dann, 151 West End, March, PE15 8DD
G7 PKD R Brown, 1 Octavian Close, Hatch Warren, Basingstoke, RG22 4TY
G7 PKG B Jenkins, 40 Windmill Drive, Filey, YO14 0FD
G7 PKH P Precious, 99 Sherwood Avenue, St. Albans, AL4 9PW
G7 PKJ D Alway, 79 Landseer Avenue, Bristol, BS7 9YW
G7 PKK Alexander Sharp, 170 Kingshill Road, Swindon, SN1 4LL
G7 PKP John Constance, Flat 2, Littlecote, 41a Blackwater Road, Eastbourne, BN20 7DG
G7 PKQ P Troll, 18 Bowness Road, Millom, LA18 4LS
GM7 PKT R Morrison, Corran Gardens, Corran Gardens, Fort William, PH33 6SJ
G7 PKY E Plant, 22 Bournville Road, London, SE6 4RN
G7 PLE J Goodliffe, 25 Stansgate Avenue, Cambridge, CB2 0QZ
G7 PLP Ian Brown, 9 Larford Walk, Stourport-on-Severn, DY13 0HE
G7 PLS N Partridge, 13 Alderney Way, Immingham, DN40 1RB
G7 PLV A Miller, 10 Limerick Close, Ipswich, IP1 5LR
GW7 PMA J Beach, 11 St Annes, Western Lane, Swansea, SA3 4EW
G7 PMB Richard Cooke, 2 Harvey Court, Warrington, WA2 9SD
G7 PMF V Lennox, 64 Oak Avenue, Blidworth, Mansfield, NG21 0TL
G7 PMG S Rose, 84 Woodhill Park, Pembury, Tunbridge Wells, TN2 4NP
G7 PMI D Williams, 6 Raven Crescent, Billericay, CM12 0JF
G7 PMK W Harrison, 67 Connaught Gardens, Shoeburyness, Southend-on-Sea, SS3 9LR
G7 PMO K Walton, 10 Holme Close, Wellingborough, NN9 5YF
G7 PMQ Paul Slight, 4 Field Close Welton, Lincoln, LN2 3TT

G7 PMU O Lawrence, 95 West Avenue, Clacton-on-Sea, CO15 1HB
G7 PMV C Cimber, 10 Harrowdene Gardens, Teddington, TW11 0DH
G7 PMW [illegible] Sandhill Road, Buckingham, MK18 2[illegible]
G7 PMX M Firth, 5 Courtenays, Seacroft, Leeds, LS14 6JZ
G7 PMY Brian Whittington, Flat 5, Anjou Court, 8 Hereward Road, Eastbourne, BN23 6TQ
G7 PNE D Head, 76 Dryden Crescent, Stevenage, SG2 0JH
G7 PNF M Cleverley, 43 Friesian Gardens, Newcastle, ST5 6BB
G7 PNG Robert Owen, 53 Huntingdon Drive, Castle Donington, Derby, DE74 2SR
G7 PNM P Smith, 41a Thornhill Road, Middlestown, Wakefield, WF4 4RU
G7 PNP P Collins, Summerly, Hollow Road, Saffron Walden, CB11 3SL
GM7 PNX Nicholas Armstrong, 4 Arboretum Road, Edinburgh, EH3 5PD
G7 POA F Greaves, Ratooragh, Schull, Ireland, WEST CORK
G7 POC N Phillips, 9 Symonds Close, Chandler's Ford, Eastleigh, SO53 3TP
G7 POI L Selway, Rua Do Rochio No 7, Poco Redondo, Tomar, Portugal
G7 POL P Duggan, 1 Bracken Way, Blackpool, FY2 0WQ
G7 POS D Ford, 44 Cardinal Square, Beeston, Leeds, LS11 8HR
G7 POT Steven Eastwood, 1 Westlands Mews, Driffield, YO25 5RS
G7 POV A Lickley, 18 Byron Street, Macclesfield, SK11 7PL
G7 POW J Sanderson, 40 Sheldon Close, Bransholme, Hull, HU7 4RU
G7 PPC B Lowe, 19 Wolverhampton Road, Bloxwich, Walsall, WS3 2EZ
G7 PPL A Pardivalla, 2 Mcdowell Way, Narborough, Leicester, LE19 2RA
GM7 PPN H Waugh, 93 Denholm Road, Musselburgh, EH21 6TU
G7 PPS H lodge, 69 Helena Road, Rayleigh, SS6 8LQ
G7 PQB K Hunter, 30 Loxley Road, Lowestoft, NR33 9PG
G7 PQD B Harrison, 145 St. Leonard St., Hendon, Sunderland, SR2 8QB
G7 PQL M Robinson, 1 Selby Close, Baxenden, Accrington, BB5 2TQ
G7 PQM Richard Buchan, 16 Lomond Drive, Kettering, NN15 5DE
G7 PQP D Little, 20 Vicarage Close, Shillington, Hitchin, SG5 3LS
GW7 PQS M Waite, 4 Clos Bryngwyn, Garden Village, Swansea, SA4 4BJ
G7 PQW Neil Wills, 18 Hollis Way, Southwick, Trowbridge, BA14 9PH
G7 PQX P Smith, 4 Stone Lodge Lane, Ipswich, IP2 9PA
G7 PRB Wade Ross, 18 Linseed Avenue, Newark, NG24 2FJ
G7 PRC S Newstead, Rectory Cottage, Church Road, Norwich, NR12 8YL
G7 PRD R Ware, 23 Harmsworth Drive, Stockport, SK4 4RP
G7 PRH D Browne, 92 Keir Hardie Way, Barking, IG11 9NX
G7 PRI Paul Newton, 5 Sandford Road, Winscombe, BS25 1HD
G7 PRK R Zeal, 5 Llanthewy Road, Newport, NP20 4JR
G7 PRO P Julian, 45 Rectory Avenue, Corfe Mullen, Wimborne, BH21 3EZ
G7 PRW M Price, 11 Walnut Crescent, Malvern, WR14 4AX
G7 PSC P Rivers, 39 Ashton Road, Birmingham, B25 8NZ
G7 PSF J Block, 40 Cromwell Road, Cambridge, CB1 3EF
GM7 PSH A Stevens, 69 Polwarth Terrace, Prestonpans, EH32 9PX
G7 PSK Nigel Kingsley-Lewis, Forge Cottage, Rudham Road, Fakenham, NR21 7BY
G7 PSL M Lamb, 52 Crookham Grove, Morpeth, NE61 2XF
G7 PSS P Allnutt, 37 Moss Mead, Chippenham, SN14 0TN
G7 PST T Ellis, 19 Cavendish Avenue, Colchester, CO2 8BP
G7 PSU Alexander Drummond, 10-12 Tottington Road, Turton, Bolton, BL7 0HS
G7 PSV M Bennett, 83 Middlethorpe Road, Cleethorpes, DN35 9PP
G7 PSZ A Laurence, Brookvale, Nooklands, Preston, PR2 8XN
G7 PTA M Masterman, 7 Pond Bank, Blisworth, Northampton, NN7 3EL
G7 PTB R Player, 49 St. Johns Road, Tilney St. Lawrence, King's Lynn, PE34 4QJ
G7 PTC Michael Watson, 3 The Pastures, Coulby Newham, Middlesbrough, TS8 0UJ
G7 PTD J Midwood, 4 Larch Crescent, Holt, NR25 6TU
G7 PTH R Graham, 8 Pecche Place, Chineham, Basingstoke, RG24 8AA
G7 PTM J Taylor, 8 Worsley Avenue, Blackpool, FY4 2DH
G7 PTT Alan Myers, 24 Milburn Street, Crook, DL15 9DY
G7 PTV Michael Howse, 28 Courtiers Drive, Bishops Cleeve, Cheltenham, GL52 8NU
G7 PTX P Castle, 26 Chestnut Walk, Pulborough, RH20 1AW
G7 PTZ R Mold, 134 Kipling Avenue, Brighton, BN2 6UE
G7 PUA John Campbell, 48 Renforth Street, Gateshead, NE11 9BE
GI7 PUG G Clegg, 45 Strandburn Drive, Belfast, BT4 1NA
G7 PUK David Glass, 2 Thorne Square, Sunderland, SR3 4PA
G7 PUL R Hoggard, 1 Whiphill Close, Bessacarr, Doncaster, DN4 6DX
G7 PUN D Russon, 123 Queens Drive, Newton-le-Willows, WA12 0LN
G7 PUP A Hurd, 27 Atlow Close, Chesterfield, S40 4LQ
G7 PUW S frizzell, 85 Gibbon Road, Newhaven, BN9 9ER
G7 PUZ L Martyn, 1 Canewdon Hall Close, Canewdon, Rochford, SS4 3PY
G7 PVE G Eddy, 102 Springfield Close, Andover, SP10 2QT
G7 PVF M Folland, 14 High St., Shoreham, Sevenoaks, TN14 7TD
G7 PVG B Fox, 10 Materman Road, Stockwood, Bristol, BS14 8SS
GU7 PVI M Major, East Liberty, Gibauderie, St. Peter Port, Guernsey, GY1 1XJ
G7 PVL Christopher Watts, 49 Mizzymead Rise, Nailsea, Bristol, BS48 2JN
G7 PVU Graham Rouse, 18 Westfield Avenue, Woking, GU22 9PH
G7 PVZ W Sefton, 87 Lillibrooke Crescent, Maidenhead, SL6 3XL
G7 PWA N Padley, 12a Wey Close, Ash, Aldershot, GU12 6LY
G7 PWI C Thornton, 1 Elizabeth Drive, Tring, HP23 5HL
G7 PWJ James Churchill, 68 Anthony Road, London, SE25 5HB
G7 PWK Michael Bibb, 9 Nelson Court, Old Nelson Street, Lowestoft, NR32 1EH
G7 PWL M Turnbull, 11 Waverley Avenue, Whitley Bay, NE25 8AU
GI7 PWQ Xqi Mccormick, 46 Lany Road, Moira, Craigavon, BT67 0NZ
G7 PWS C Collins, 32 St. Martins Road, New Romney, TN28 8JY
G7 PWU H Tomlinson, 42 Gawsworth Avenue, Crewe, CW2 8PB
G7 PWV P Woolhouse, 21 Coombe Wood Hill, Purley, CR8 1JQ
GM7 PXJ L Michie, 5 Torridon Place, Rosyth, Dunfermline, KY11 2EZ
GM7 PXL Derrick Warner, Torran, Lettorfinlay, Spean Bridge, PH34 4DZ
G7 PXR Peter Gilbert, 4 Ruby Street, Bristol, BS3 3DY
G7 PXS G Mape, 34 Amberwood Drive, Manchester, M23 9NZ
G7 PXX Errol Spires, 8 Regent Terrace, Barrow Road, Barrow-upon-Humber, DN19 7QB
G7 PYB John Bailey, 1 Greenwood, Ascot, SL5 8LL
G7 PYH D Williams, 8 Red Lodge, 1 Hillside Road, London, W5 2JA
G7 PYN A Wentworth, 5 York Avenue, Prestwich, Manchester, M25 0FZ
G7 PYQ N Gell, 1 Lawton Road, Rushden, NN10 0DX
G7 PYR Victor Tuff, 8 Millcroft Court, Blyth, NE24 3JG
G7 PYT G Coleman, 23 Graham Avenue, Patcham, Brighton, BN1 8HA
G7 PYV A Turner, 20 Kipling Gardens, Upper Stratton, Swindon, SN2 7LJ
G7 PYW K Houghton, 42 Pear Tree Avenue, Coppull, Chorley, PR7 4N[illegible]
G7 PZB R Dewsbery, 8 Woodfield Close, Market Harborough, LE16 9DX
G7 PZE P Eastham, 4 [illegible], [illegible] BB0 0DY
G7 PZF W Bailey, 15 Norfolk Road, Congleton, CW12 1NY
GM7 PZH M Drennan, 6 Hillpark Way, Edinburgh, EH4 7BJ
G7 PZL A Morton, 54 Rose Farm Approach, Normanton, WF6 2RZ
G7 PZM Stuart Carter, 6 Bramalea Close, London, N6 4QD
G7 PZQ P Breese, 11 Balham Grove, Birmingham, B44 0NF
G7 PZT John Keen, 30 Fielding Crescent, Blackburn, BB2 4TD
G7 PZU A Haworth, 8 Cornmill Place, Barnoldswick, BB18 5ED
G7 RAB David Evans, 31 Kinsbourne Way, Thornhill, Southampton, SO19 6HB
G7 RAE J Kirkwood, 6 Trinity Court, Rothwell, Kettering, NN14 6YQ
G7 RAF A Corbett, Lan Y Llyn, 10 Rutland Avenue, Waddington, LN5 9FW
G7 RAG J Dennis, The Old Chapel House, Alford, LN13 9PH
GI7 RAH T Tweedie, 17 Schomberg Park, Belfast, BT4 2HH
G7 RAI M Moorhouse, 11 Hazel Grove, Huddersfield, HD2 2JP
G7 RAJ David Eggett, 68 Forest Lane, Kirklevington, Yarm, TS15 9ND
GM7 RAK J Boyd, 102 Provost Milne Grove, South Queensferry, EH30 9PL
G7 RAL LOUGHBOROUGH & DISTRCT ARC c/o Ian Hewitt, 26 Outwoods Drive, Loughborough, LE11 3LT
GI7 RAM J Christie, 3 Victoria Drive, Sydenham, Belfast, BT4 1QT
G7 RAT Reigate Amateur Transmitting Society c/o Peter Tribe, The Paddock, Wix Hill, Leatherhead, KT24 6ED
G7 RAU D edwards, 37 Barton Close, Whippingham, East Cowes, PO32 6LS
G7 RAZ Michael Wager, 115 Queensway, Taunton, TA1 4NL
G7 RBA Mathew Sims, 23 Winding Way, Alwoodley, Leeds, LS17 7RB
G7 RBB A Perkins, 9 St. Martins Close, Canterbury, CT1 1QG
G7 RBC Ivan Rodgers, 89 Braemar Road, Worcester Park, KT4 8SN
G7 RBL Christopher Johnson, 2 Goodwin Avenue, Newcastle, ST5 9EF
G7 RBQ Peter Dodman, 15 Goscote Close, Redditch, B97 6UF
G7 RBR J PAVIA, 703 Wolffs Road, Rangiora Rd6, New Zealand, 7476
G7 RBS A Sercombe, 28 Strumpshaw Road, Brundall, Norwich, NR13 5PA
G7 RBT S Worrall, 41 The Grove, Stourport-on-Severn, DY13 9ND
GM7 RBW Dave Saunders, 172 Colinton Mains Road, Edinburgh, EH13 9DB
G7 RCC R Wendes, 108 Osborne Road, East Cowes, PO32 6RZ
GI7 RCH Gary Walker, 16 Stormount Crescent, Belfast, BT5 4NT
G7 RCK Salim Motala, 28 Fishwick View, Preston, PR1 4YB
G7 RCL R Abbott, 2 Leybourne Drive, Springfield, Chelmsford, CM1 6TX
G7 RCP David Baines, 157 Hall Green Road, West Bromwich, B71 2DY
G7 RCU Andrew Walker, 37 East Road, Brinsford, Wolverhampton, WV10 7NP
G7 RCW J Matthews, The Forge, Norton Heath, Ingatestone, CM4 0LJ
G7 RDA P Brownsett, 10 Great Aldens, Bedford, MK41 8JS
GM7 RDH Roderick Spence, Leyan, Harray, Orkney, KW17 2LQ
G7 RDJ R Middleton, 32 West Busk Lane, Otley, LS21 3LW
G7 RDP B Gawthorpe, 19 Tower Hill, Clitheroe, BB7 1PD
G7 RDQ T Rochford, 41 Lynwood Drive, Blakedown, Kidderminster, DY10 3JZ
G7 RDT DORSET RAYNET c/o Adrian Lambert, 69 Anvil Crescent, Broadstone, BH18 9DZ
GM7 RDY J Mowat, Nether Bigging, Shapinsay, Orkney, KW17 2EB
G7 REC D Allison, 52 Boyn Valley Road, Maidenhead, SL6 4ED
GM7 REF EPPING FOREST RAYNET GROUP c/o Mike Harrington, Mount Pleasant House, North Road, Wick, KW1 4DN
GM7 REG J Robertson, 13 Swanston View, Edinburgh, EH10 7DG
G7 REH P Evans, 45 Chiltern Drive, Charvil, Reading, RG10 9QF
G7 REJ S Hutchinson, 42 Greenham Mill, Mill Lane, Newbury, RG14 5QW
G7 RES Gary O'Neill, 1 Whittingham Place, Avenue Road, Freshwater, PO40 9UR
GM7 REY John MacDonald, 27 Melantee, Fort William, PH33 6PY
G7 RFA A Lord, The Mount, Trefecca, Brecon, LD3 0PW
G7 RFC ESSEX RAYNET c/o Graham Farrell, 95 Washington Road, Maldon, CM9 6JF
G7 RFD P Johnson, Sixpenny Cottage, Farthings Fold, Bourne, PE10 0RN
G7 RFE R Johnson, 8 Merlin Close, Bourne, PE10 0BZ
G7 RFH J Fearns, 23 Homestead Street, Stoke-on-Trent, ST2 0RQ
G7 RFM G Hunt, 7 Kevington Drive, St. Pauls Cray, Orpington, BR5 2NT
G7 RFO Robert Thomson, 3 Harley Street, Todmorden, OL14 5JE
G7 RFS K Abeynayake, 25 Anderson Avenue, Earley, Reading, RG6 1HD
G7 RFT K Whittle, 26 Beachs Drive, Chelmsford, CM1 2NJ
G7 RFX R Oxlade, 3 Thyme Court, Northampton, NN3 8HY
G7 RFZ J Bilmen, 145 The Maples, Harlow, CM19 4RD
G7 RGA Peter Cattanach, 8a Approach Road, London, E2 9LY
G7 RGG S Emmett, 14 Ernle Road, Calne, SN11 9BT
G7 RGI Hazel Yates Jones, 5 Southville Road, Bradford-on-Avon, BA15 1HS
G7 RGJ P Musselwhite, 80 Craven Road, Orpington, BR6 7RT
G7 RGO E Allan, 282 Bilton Road, Rugby, CV22 7EG
G7 RGR Roger Jones, 18 Ash Grove, Burnham-on-Crouch, CM0 8DP
G7 RGU R Oxlade, 3 Thyme Court, Lumbertubs, Northampton, NN3 8HY
G7 RGV Harold Jump, 4 Bankwood, Shevington, Wigan, WN6 8EY
G7 RHD M Clarke, 6 Oldbrook Fold, Timperley, Altrincham, WA15 7PA
G7 RHE Steven Payne, 19 Weavers Lane, Sevenoaks, TN14 5BT
G7 RHF Alan Richards, 3 Marsh Gate, Clee St. Margaret, Craven Arms, SY7 9DU
G7 RHI A Mclocklin, 43 Forbes Avenue, Potters Bar, EN6 5NB
G7 RHM Koon Pang, 30 Barnwood Avenue, Gloucester, GL4 3AH
G7 RHT P Bennett, 47 Bakers Ground, Stoke Gifford, Bristol, BS34 8GD
G7 RHU Andrew Dowker, 11 Bowley Street, London, SW19 1XF
G7 RIA M Cook, 20 Chalton Heights, Chalton, Luton, LU4 9UF
G7 RIB P Nicholls, 11 Ifor Hael Road, Rogerstone, Newport, NP10 9FB
G7 RIE J Glenn, School House, 70 Norwich Road, Norwich, NR12 7EG
G7 RIJ E Devine, 23 Radley Avenue, Wickersley, Rotherham, S66 2HZ
G7 RIO W Care, 29 Wheal Gorland Road, St. Day, Redruth, TR16 5LT
G7 RIU D Kirk, 19 The Meads, Hildersley, Ross-on-Wye, HR9 7NF
GJ7 RIY G Webster, Woolly Mammoth, St. Helier Marina, St. Helier, Jersey, JE2 3ND
GM7 RJG A Forbes, 28 Innes Street, Inverness, IV1 1NS
G7 RJO C Compton, 55 Lulot Gardens, London, N19 5TR
G7 RJW Darren Lamden, 6 Ashbourne Way, Thatcham, RG19 3SH
GW7 RKC Andrew Hall, 29 Ely Street, Tonypandy, CF40 1BY
G7 RKE J Bottomley, Grove House, 2 Woodlane, Falmouth, TR11 4RG
G7 RKJ Charles Hindmarsh, 5 Jackman Drive, Horsforth, Leeds, LS18 4HS
G7 RKO R Kennedy, 6 Oaky Balks, Alnwick, NE66 2QE
G7 RKQ S Rudge, 1 Marl Mews, Marl View Terrace, Conwy, LL31 9BJ
G7 RKT P Jones, 14 Westerleigh Road, Clevedon, BS21 7US
G7 RKU P Dickinson, Haven, The Row, Bury St. Edmunds, IP[illegible] 4BL
G7 RKV D Wilson, 32 Laurel Bank, The Highlands, Whitehaven, CA28 [illegible]
G7 RKW James Hart, 75 Falconers Road, Luton, LU2 9ET
G7 RKX C Hill-Smith, Top Flat, The Warehouse, West Street, Newton Abbot, TQ13 7DU
G7 RLK Stephen Drury, 5 Hawthorn Close, Healing, Grimsby, DN41 7SR
G7 RLO L Van Beers, Cob Cottage, Tram Inn, Hereford, HR2 9AN
G7 RLQ T Winton, 84 Pembroke Road, Clifton, Bristol, BS8 3EG
GW7 RLS CITY & COUNTY OF SWANSEA c/o J Gray, City And County Of Swansea, Emergency Planning Unit, Swansea, SA1 3SN
G7 RLV Christopher Pitchford, 84 New Road, Rubery, Birmingham, B45 9HY
G7 RLX G Coleman, 120 Kidderminster Road South, Hagley, Stourbridge, DY9 0JH
G7 RLZ Gwilym Roberts, 4 Fawnog Wen, Penrhyndeudraeth, GWYN EDD
G7 RMD D Devlin, 5 Kelsall Avenue, Sutton Manor, St. Helens, WA9 4DQ
G7 RMF Matthew Buckley, 8 Highthorne Street, Armley, Leeds, LS12 3LB
GM7 RMF E Walker, 38 Greenbank Gardens, Edinburgh, EH10 5SN
G7 RMG Geoffrey Chapman, Crockers Farm, Stoke Wake, Blandford Forum, DT11 0HF
G7 RMJ Martin Amies, Home Farm, Hulme Walfield, Congleton, CW12 2JJ
G7 RMQ Roland Scarce, 16b Pembroke Road, Framlingham, Woodbridge, IP13 9HA
G7 RMW MID WARKS RAYNET GROUP c/o R Medcalf, 19 All Saints Road, Warwick, CV34 5NL
G7 RMX Neil Taylor, West Mede, Exeter Road, Honiton, EX14 1AX
G7 RMZ EAST CHEHIRE RAYNET GROUP c/o Bruce Williams, 3 Welton Close, Wilmslow, SK9 6HD
G7 RNA OBO NORTH ANGLIA RAYNET c/o Kevin Kent, 5 Jubilee Road, Heacham, King's Lynn, PE31 7AR
G7 RNB Shirley Bieber, Tonkins Quay, Mixtow, Fowey, PL23 1NB
G7 RNC Terrence Heywood-Bell, 4 Aberthaw Close, Newport, NP19 9QA
G7 RNF T Roberts, 5 Parkfield, Osterley Road, Isleworth, TW7 4PF
GM7 RNJ M Dennis, 47 Viewfield Road, Aberdeen, AB15 7XP
G7 RNN NORTH NORFOLK RAYNET c/o Alan Farrow, 18 The Green Trimingham, Norwich, NR11 8ED
G7 RNQ R Young, 12 Elmwood Close, Stokesley, Middlesbrough, TS9 5HX
G7 RNX Alex Linney, 5 Elliscales Avenue, Dalton-in-Furness, LA15 8BW
G7 ROC J Armstrong, 15b Lamberton, Berwick-upon-Tweed, TD15 1XB
G7 ROI J Naylor, 46 Loxley Drive, Mansfield, NG18 4FB
G7 ROM Andrew Boardman, 147 Musgrave Road, Bolton, BL1 4HW
G7 ROP R Sykes, 46 Crescent Road, Netherton, Dudley, DY2 0NW
G7 ROY Roy Clayton, 9 Green Island, Irton, Scarborough, YO12 4RN
G7 RPJ John Barnard, 39 Ecclestone Close, Bradwell, Great Yarmouth, NR31 8RG
G7 RPK L Goffin, The Hollies, Belaugh Green Lane, Norwich, NR12 7AJ
G7 RPP I Gurney, 81 College Road, Isleworth, TW7 5DP
GM7 RPT David Hutchison, 55 Springfield Road, Tarbolton, Mauchline, KA5 5QU
G7 RPW S Pike, 1 Barley Garth, Burton Pidsea, Hull, HU12 9AF
G7 RQD M Folkes, 3 Colindale Road, Ferring, Worthing, BN12 5JF
GW7 RQI D Pearson, 142 Heol Bryngwili, Cross Hands, Llanelli, SA14 6LY
GM7 RQK S Skidmore, 6 Blairlinn View, Cumbernauld, Glasgow, G67 4AD
G7 RQO Bryan Wylie, 54 Cromwell Street, Lincoln, LN2 5LP
G7 RQV N Jenkins, 4 Meadowbrook Avenue, Pontnewydd, Cwmbran, NP44 1BJ
G7 RRC CALDERDALE RAYNET ARC c/o A Baines, 60 Norton Drive, Halifax, HX2 7RB
G7 RRD G Smith, 12 Oakwood Glade, Holbeach, Spalding, PE12 7JS
G7 RRJ Andrew McConnachie, 16 Poplar Avenue, Wyre Piddle, Pershore, WR10 2RJ
GW7 RRM S Whitehouse, 15 Goetre Fach Road, Killay, Swansea, SA2 7SG
G7 RRO G Gardner, 47 Old Road, Stanningley, Pudsey, LS28 6BG
GW7 RRS A Davis, 5 Jubilee Road, Bridgend, CF31 3BA
G7 RRY M Saltmer, 12 Beechings Mews, Whitby, YO21 3DW
G7 RSA Colin Hawkes, 103 Station Road, Roydon, King's Lynn, PE32 1AW
G7 RSE Leonard Clarke, 83 Lancaster Street, Blaina, Abertillery, NP13 3EQ
G7 RSK A Scott, 62 Berry Meade, Ashtead, KT21 1SG
G7 RSM Roger Bloor, 7 Highfield Court, Clayton Road, Newcastle, ST5 3LT
G7 RTA S Harding, 39 Clayton Road, Lidget Green, Bradford, BD7 2LX
GI7 RTB Peter McCrory, 24 Drumcoo Green, Dungannon, BT71 4AJ
G7 RTC G Darby, 69 Churchill Road, Earls Barton, Northampton, NN6 0PQ
G7 RTI K Werner, 85 Brecon Way, Downley, High Wycombe, HP13 5NW
G7 RTJ D Bransby, 7 West Cliff Avenue, Whitby, YO21 3JB
G7 RTL RADIO-TELE LINCOLNSHIRE GROUP c/o M Pell, 7 Churchfleet Lane, Gosberton, Spalding, PE11 4NE
G7 RTN J Burrows, 37 Braydeston Crescent, Brundall, Norwich, NR13 5LD
G7 RTO B Thoaker, 25 Pinewood Drive, Plymouth, PL6 7SP
G7 RTQ M Cowley, 72 Warley Road, Warley, Oldbury, B68 9TB
G7 RTR David Freeman, 59 Verulam Way, Cambridge, CB4 2HJ
G7 RTX Karl Brookes, Kohima, Spout Lane, Stoke-on-Trent, ST2 7LR
G7 RUC D Millen, 42 Gayhurst Drive, Sittingbourne, ME10 1UD
G7 RUH Roger Peggram, Starcroft, Janes Close, Southampton, SO45 1WJ
G7 RUJ D Brain, 26 Exeter Road, Weston-Super-Mare, BS23 4DB
G7 RUN Martin Graves, 20 Stace Way, Worth, Crawley, RH10 7YW
G7 RUQ Lorne Murphy, Flat 3, Evelyn Court, 187 South Coast Road, Peacehaven, BN10 8NS
G7 RUR Sladjan Todorovic, 10 Larchwood Close, Sale, M33 5RP
G7 RUS Richard Parkin, 25 Kent House Lane, Beckenham, BR3 1LE
G7 RUX Jason Gardner, 67 Woodside Road, Tunbridge Wells, TN4 8PY
G7 RUY D Ager, 11 Tilbury Close, St. Pauls Cray, Orpington, BR5 2JR
G7 RVC Peter Sutherland, 9 Lely Close, Bedford, MK41 7LS
G7 RVH Richard Bush, Church View, Overcross Banham, Norwich, NR16 2BY
G7 RVI T Hankins, Cawdor House, Cawdor, Ross-on-Wye, HR9 7DN
GD7 RVP Stephen Rand, 3 Yn Aittin Vooar, Bretney Road, Jurby, Isle of Man, IM7 3EU
GM7 RVR N Moir, 34 Souter Drive, Inverness, IV2 4XJ
G7 RVT T Smith, 9 Crofters Way, Westlands, Droitwich, WR9 9HU
G7 RVW R Crofts, Little Isle, Woodgate Green, Tenbury Wells, WR15 8LX



**IMPORTANT NOTE**

**Revalidate licence to avoid revocation** – Ofcom has advised the Society that plans will be drawn up to revoke licences that have not been revalidated as required by the licence conditions. The quickest way to revalidate is to do so online via the Ofcom website: *https://services.ofcom.org.uk/* or by email: *amateur.validations@ofcom.org.uk* Ofcom staff are available to help, but please be patient during times of heavy workload.

G7 RVY H Branch, 326 Springfield Road, Chelmsford, CM2 6BA
G7 RWC C Halbert, 3 Third Row, Ellington, Morpeth, NE61 5HF
G7 RWF John Buck, 14 Crosstree Walk, Colchester, CO2 8QF
G7 RWN D Taylor, 48 Southcroft Road, Gosport, PO12 3LD
GJ7 RWT A Cutland, Little Gables, La Route Orange, St. Brelade, Jersey, JE3 8GQ
G7 RWW J Kirkham, 100 Prince Charles Avenue, Derby, DE22 4FL
G7 RWY Barry Sankey, 121 Green Lane, Coventry, CV3 6EB
G7 RXB N Larson, 90 Lingfield Ash, Coulby Newham, Middlesbrough, TS8 0SU
G7 RXE Vince Donald, 20 Parkhill Road, Barnby Dun, Doncaster, DN3 1DP
G7 RXI V Ball, 30 Park Drive, Worlingham, Beccles, NR34 7DJ
G7 RXJ J Dyson, 4 Nightingale Cottages, Frantfield, Edenbridge, TN8 5BB
G7 RXK R Thompson, 7 Rufford Close, Sutton-in-Ashfield, NG17 4BX
GM7 RXL D Winton, 273 Hilton Drive, Aberdeen, AB24 4NT
G7 RXO Niels Larsen, 6 Shrewsbury Close, Barwell, Leicester, LE9 8JX
G7 RXW M Lockitt, 19 Roundway Down, Perton, Wolverhampton, WV6 7SX
G7 RXX S Cooper, 30 Pinta Drive, Stourport-on-Severn, DY13 9RY
G7 RXZ D Cooper, 1a Kent Street, Dudley, DY3 1UU
G7 RYA D Tomlin, 154 Court Lane, Erdington, Birmingham, B23 5RG
GM7 RYK G Pollard, 127 Braeside Park, Mid Calder, Livingston, EH53 0TE
G7 RYL D Sheridan, 78 Oaklands Park, Buckfastleigh, TQ11 0BP
G7 RYM R Pugh, 41 East Beach Park, Shoeburyness, Southend-on-Sea, SS3 9SG
G7 RYN D Proctor, 11 Bedford Rise, Winsford, CW7 1NE
G7 RYO K Turner, 34 Amherst Road, Kenilworth, CV8 1AH
GM7 RYT David Weller, 66 Dolphin Road, Currie, EH14 5SA
G7 RYW Frederick Trainer, 23 Woodend Avenue, Hunts Cross, Liverpool, L25 0NY
G7 RZN E Taylor, 8 First Avenue, Prestatyn, LL19 7LP
G7 RZQ Nick Waterman, 1 Wood Lane Close, Sonning Common, Reading, RG4 9SP
G7 RZW Albert Davies, 16 Sutton Road, Bolton, BL3 4QR
G7 SAC Sutton & Cheam RS c/o J Puttock, Sutton & Cheam Rs, 53 Alexandra Avenue, Sutton, SM1 2PA
G7 SAI E Birt, 10 Wilden Lane, Stourport-on-Severn, DY13 9LR
GM7 SAK Alistair Jardine, 17 Louisa Drive, Girvan, KA26 9AH
GM7 SAQ Ron Turner, 26 Skye Crescent, Crieff, PH7 3FB
G7 SAX R Newman, 31 Oval Gardens, Alverstoke, Gosport, PO12 2RA
GI7 SBF John Henderson, 1 Brook Lodge Ballinderry Lower, Lisburn, BT28 2GZ
G7 SBJ E Birtwistle, 29 Church View, Pentre, Deeside, CH5 2DP
G7 SBK David Hunt, 298 Cavendish Road, Carlton, Nottingham, NG4 3QH
G7 SBO R Thomas, 25 Lon Lwyd Isaf, Pentraeth, LL75 8LN
G7 SBP N Hancocks, 9a St. Philip Street, Penzance, TR18 2DN
G7 SBZ M Newton, 24 Chestnut Avenue, York, YO31 1BR
G7 SCE P Farman, 298 Laburnum Grove, Portsmouth, PO2 0EX
GM7 SCJ G Deas, 81 Speirs Road, Bearsden, Glasgow, G61 2LT
G7 SCL J Robinson, 4 Gardner Close, Loughborough, LE11 5YB
G7 SCN Paul Brotherton, 73 Thorneywood Rise, Nottingham, NG3 2PE
G7 SCO D Brooke, 34 Park Road, Burwell, Cambridge, CB25 0ES
G7 SCP Desmond Wain, 51 Foxstone Way, Eckington, Sheffield, S21 4JX
G7 SCR SUFFOLK COASTAL RAYNET c/o R Keen, 13 Mill View Close, Woodbridge, IP12 4HR
G7 SCT G Rutherford, 24 Chestnut Avenue, Hedon, Hull, HU12 8NH
G7 SCU Dogan Ibrahim, 14 Dunvegan Road, London, SE9 1SA
G7 SCV John Straughan, 16 Garner Close, Chapel Park, Newcastle upon Tyne, NE5 1SQ
G7 SCX P O'Rourke, 186 Cottingham Road, Corby, NN17 1SY
G7 SCZ D Kiteley, 13 Chiltern Close, Astley Cross, Stourport-on-Severn, DY13 0NU
G7 SDC D Coe, 105 Raynham Road, Bury St. Edmunds, IP32 6ED
G7 SDD M Smith, 38 Vestry Road, Street, BA16 0HX
GW7 SDE I Jones, 14 Clare Court, Loughor, Swansea, SA4 6UH
G7 SDG R Martin, 8 Short Lane, Bricket Wood, St. Albans, AL2 3SE
G7 SDM G Davies, 11 Ninfield Close, Carlton Colville, Lowestoft, NR33 8SD
GM7 SDP D Ryan, 1 Clashbenny Place, St. Madoes, Perth, PH2 7TS
G7 SDQ M Smith, Sunt Kelda, Weston Town, Shepton Mallet, BA4 6JG
G7 SEG Andrew Harrison, 44 Rosslyn Road, Whitwick, Coalville, LE67 5PT
G7 SEJ M Baskeyfield, 3 Merlewood, Bracknell, RG12 9PA
G7 SEK Richard Newham, 18 Highfields Close, Ashby-de-la-Zouch, LE65 2FN
G7 SEO R Plant, 22 The Woodlands, Wokingham, RG41 4UY
G7 SER SUTTON COLDFIELD & DIST RAYNET c/o J Trickey, 59 Shelley Drive, Sutton Coldfield, B74 4YD
G7 SEU Elizabeth Kershaw, 83 Foxhunter Drive, Oadby, Leicester, LE2 5FH
G7 SEY P Simpson, The Conifers, Woodhouse Lane, Telford, TF4 3BJ
G7 SFA Michael Stevens, Flat 7, 1a Woodstock Road, Croydon, CR0 1JS
G7 SFD Martin King, 4 Keith Avenue, Ramsgate, CT12 6JQ
GM7 SFE R Lawrie, 84 Redlawood Road, Cambuslang, Glasgow, G72 7TP
G7 SFF D Hartshorn, 21 Hucklow Avenue, Chesterfield, S40 2LT
G7 SFI S Merrifield, 2 Larkspur Glade, Telford, TF3 2AQ
G7 SFJ S Pratt, 15 Springwell Close, Cowling, Keighley, BD22 0AP
G7 SFL M James, 49 Church Street, Fontmell Magna, Shaftesbury, SP7 0NY
G7 SFM R Wiltshire, 30 Cearns Road, Oxton, Prenton, CH43 2JP
G7 SFS L Banner, 7 Lowdham Road, Gedling, Nottingham, NG4 4JP
G7 SFY B Purkiss, 99 Westland Road, Yeovil, BA20 2AZ
G7 SGK R Ward, 9 Shelton Avenue, East Ayton, Scarborough, YO13 9HB
G7 SGM R Gifford, 100 Gadebridge Road, Hemel Hempstead, HP1 3EW
G7 SGO Richard Percival, 145 Queen Street, Whitehaven, CA28 7AW
G7 SHI C Conce, 35 Mortimer Drive, Sandbach, CW11 4HS
G7 SHW George Stephens, 46 Newall Drive, Beeston, Nottingham, NG9 6NX
G7 SJD S Fitzpatrick, 21 Corn Close, South Normanton, Alfreton, DE55 2JD
G7 SJK T Masson, Apple Tree Cottage, Neath Gardens, Reading, RG3 4UL
G7 SJP J Pettifer, 7a Catherine Road, Woodbridge, IP12 4JP
G7 SJS P Roberts, 5 Snelston Crescent, Littleover, Derby, DE23 6BL
G7 SJX Brian Shields, 20 Gresley Court, Grantham, NG31 7RH
G7 SKA P Burnett, 4 Lavendon Court, Barton Seagrave, Kettering, NN15 6QH
GM7 SKB D Fortune, 26 Newton Grove, Newton Mearns, Glasgow, G77 5QJ
GW7 SKC OBO WEST GLAMORGAN CC c/o J Gray, City And County Of Swansea, Emergency Planning Unit, Swansea, SA1 3SN
G7 SKF P Morgan, 6 Elmgrove Road East Hardwicke, Gloucester, GL2 4PY
G7 SKH G Murray, Brookside, Thirlby, Thirsk, YO7 2DJ
G7 SKL J Koops, 64 Winchester Avenue, Nuneaton, CV10 0DW
G7 SKR D Tarbatt, 9 Dashwood Close, Warrington, WA4 3JA
G7 SKV D Graham, 11 Hibernia St., Deane, Bolton, BL3 5PQ
G7 SKW B McInnes, 4 Lindrick Road, Hatfield Woodhouse, Doncaster, DN7 6PF
G7 SKX A Wilkinson, 21 Solbys Road, Basingstoke, RG21 7TG
G7 SLJ D Lloyd-Jones, 2 Leyside, Rayne, Braintree, CM77 6DE
G7 SLL G Peach, 120 Craven Road, Newbury, RG14 5NR
GI7 SLN G McAfee, 12 Skerryview, Craigahullier, Portrush, BT56 8NJ
G7 SLP P Hardcastle, 19 Dunkirk Terrace, Halifax, HX1 3RB
GJ7 SLU C Whittaker, Coeur Joyeux, La Rue Des Sapins, St. Peter, Jersey, JE3 7AD
G7 SLV R Walker, 210 London Road, Worcester, WR5 2JT
G7 SLY P Taylor, 46 Ralph Road, Staveley, Chesterfield, S43 3PY
G7 SLZ T Gill, 21 Trevor Smith Place, Taunton, TA1 3RW
G7 SMC G Jameson, 17 Lansbury Avenue, Mastin Moor, Chesterfield, S43 3AG
G7 SME P Helliwell, 1 Beechfield Avenue, Barton, Torquay, TQ2 8HU
G7 SMH G Newton, 8 Lynch Mead, Winscombe, BS25 1AT
G7 SMN Andy Holden, 1 Rose Cottage, Little Bramford Lane, Ipswich, IP1 2PH
G7 SMQ Brian Cottee, 41 Colesbourne Road, Clifton, Nottingham, NG11 8JG
G7 SMT Frederick Claydon, 5 Mill Gardens, Ringmer, Lewes, BN8 5JD
G7 SMV Larry Ashford, 13 Cefn Court, Rogerstone, Newport, NP10 9AH
G7 SMZ R Walker, 24 Colin Street, Alfreton, DE55 7HT
G7 SNB O Newland, 22a Cromwell Road, Basingstoke, RG21 5NR
G7 SNC Ivan Palmer, 182 Salhouse Road, Norwich, NR7 9AD
G7 SNF H Joynes, The Shrubbery, Straits Lane, Newport, NP18 2BY
G7 SNJ Robert Chaytor, 19 Granville Avenue, Hartlepool, TS26 8ND
G7 SNP K Jordan, 7 Park Avenue, Bedlington, NE22 7EH
G7 SNQ S Taylforth, 1 Clough Terrace, Barnoldswick, BB18 5PD
G7 SNR Susan Brodie, Waterloo Cottage, Tanners Green, Norwich, NR9 4QS
G7 SNT Benjamin Jordan, 40 High Street, Coltishall, Norwich, NR12 7HD
G7 SNW John Ward, 68 Moreton Road North, Luton, LU2 9DP
G7 SNX M Pearce, 42 Pine Close, Rudloe, Corsham, SN13 0LB
GI7 SOB Keith Elgin, 50 Ballinteer Road, Macosquin, Coleraine, BT51 4LZ
G7 SOE M Howard, East Dean House, East End Langtoft, Peterborough, PE6 9LP
G7 SOH Charles Brown, 9 Marjorie Street, Rhodesia, Worksop, S80 3HR
G7 SOP S Frank, 36 Melksham Road, Bestwood Park, Nottingham, NG5 5RX
G7 SOV C Howarth, 5 West Mount, Orrell, Wigan, WN5 8LX
G7 SOZ Simon Jude, 9 Winchfield, Great Gransden, Sandy, SG19 3AN
GM7 SPA James Brown, 11 Oak Gardens, Oak Drive, Lenzie, Glasgow, G66 4BF
GM7 SPB Malcolm Garrington, South Orrock Bungalow, Balmedie, Aberdeen, AB23 8XY
G7 SPE R Keep, 14 Foster Road, Kempston, Bedford, MK42 8BU
G7 SPL D Pomfret, 52 Warwick Close, Bury, BL8 1RT
G7 SPM C Jones, Nb Guanche Bradford On Avon Marina, Widbrook Bradford-on-Avon, BA15 1UD
G7 SPN Stephen Townsley, 222 Prince Consort Road, Gateshead, NE8 4DX
G7 SPP Heather Conrad, 22 Low Stobhill, Morpeth, NE61 2SG
G7 SPZ Roland Brown, 19 Comberton Road, Toft, Cambridge, CB23 2RY
G7 SQC P Young, 31 Cygnet Walk, North Bersted, Bognor Regis, PO22 9LY
G7 SQH G Chew, 45 Brackley, Weybridge, KT13 0BL
G7 SQM Nicholas Crawford, 20 Fearnley Crescent, Kempston, Bedford, MK42 8NL
G7 SQW A Woods, 10 Radcliffe Road, Drayton, Norwich, NR8 6XZ
G7 SQY D Colton, 9 Thornemead, Peterborough, PE4 7ZD
G7 SRA SUDBURY AND DISTRICT RADIO AMATEURS c/o Mark Hickford, 3 Ashen Road, Clare, Sudbury, CO10 8LQ
G7 SRB D Shorten, 32 Stoneleigh Drive, Carterton, OX18 1ED
G7 SRC ESSEX RAYNET c/o N Hull, C/O 95 Washington Road, Maldon, CM9 6JF
G7 SRG Sandwell Raynet Group c/o Patrick Skerritt, 39 Bache Street, West Bromwich, B70 7EW
G7 SRH Martin Harper, 31 Lorland Road, Cheadle Heath, Stockport, SK3 0JJ
G7 SRI Maurice Lowe, 2 White Post Bungalows, North Leverton, Retford, DN22 0AS
G7 SRJ S Jones, Smiddy Cottage, Auchencrow, Eyemouth, TD14 5LS
G7 SRK R Carder, 45 Chalklands, Linton, Cambridge, CB21 4JQ
G7 SRL Arthur Gallichan, 4 Wigston Road, Hillmorton, Rugby, CV21 4LT
G7 SRV P Everett, 26 Tennyson Close, Horsham, RH12 5PN
G7 SRZ J Trybulski, 78 Ditchling Road, Brighton, BN1 4SG
G7 SSA M Addicott, Orchardleigh, The Street, Radstock, BA3 4HG
G7 SSB Darren Jones, 429 Redmires Road, Sheffield, S10 4LF
G7 SSD J Edwards, 17 Marlowe Close, Galley Common, Nuneaton, CV10 9QP
G7 SSG Jeffrey Smye, 24 Eastfield Road, Wincanton, BA9 9LT
G7 SSJ David Sutton, 32 Queensway, Euxton, Chorley, PR7 6PW
G7 SSK R Walton, Easingmoor House, Thorncliffe Road, Leek, ST13 7LW
G7 SSN N Cole, 40 Primrose Court, Ty Canol, Cwmbran, NP44 6JJ
G7 SSQ P Cole, 9 Perry Court, Thornhill, Cwmbran, NP44 5UD
G7 SSW J Haywood, 7 Anna Walk, Stoke-on-Trent, ST6 3BX
G7 STC K Gater, 110 Byrds Lane, Uttoxeter, ST14 7NB
G7 STD L Goodridge, 110 Quarrendon Road, Amersham, HP7 9EP
G7 STG Barry Spavins, 8 Berkeley Avenue, Briar Bank Park, Bedford, MK45 3WH
GM7 STI Ian Pearce, 1 Mount Farm Cottage, Cupar, KY15 4NA
G7 STL M Anderson, 94 Tolworth Road, Surbiton, KT6 7SZ
G7 STM M Wyatt, 8 St Mary'S Drive, Sutterton, PE20 2LU
G7 STQ Michael Oura, The Quoins, Gloucester Road, Bath, BA1 8AD
G7 STT J Baker, 20 Homespring House, Pittville Circus Road Roa, Cheltenham, GL52 2QB
G7 SUA D Wiseman, 12 Hamilton Way, Acomb, York, YO24 4LE
G7 SUM G Fewings, 22 Watcombe Road, West Southbourne, Bournemouth, BH6 3LU
G7 SUQ A Jobson, 7 Dunlin Close, Norton, Stockton-on-Tees, TS20 1SJ
G7 SUS R Biss, 1 Fairey Crescent, Gillingham, SP8 4PE
G7 SUT Charles James, (James), Lower Kenneggy Farm, Lower Kenneggy, Rosudgeon, Penzance, TR20 9AR
G7 SUU Robin Wolk, Calle Zaragoza 48, Castalla, Spain, 3420
G7 SUV J patterson, 161 Ringwood Road, Eastbourne, BN22 8UW
G7 SVE A Jackson, 14 West Field Gardens, Sandy, SG19 1HF
G7 SVF Kevin Ingram, 15 Kent Avenue, East Cowes, PO32 6QN
G7 SVM D Bradley, 22 Grosvenor Road, Ettingshall Park, Wolverhampton, WV4 6QY
G7 SVQ Richard Holmes, 18 Dresden Close, Mickleover, Derby, DE3 0RD
G7 SVT D Bultitude, 1 Pembroke Gardens, Northampton, NN5 7ES
G7 SVU Neil Hinchliffe, 19 Grange Road, Blidworth, Mansfield, NG21 0RN
GM7 SWB R Bambrey, 6/2 Admiral Terrace, Edinburgh, EH10 4JH
G7 SWE F Rowbotham, 56 Farnborough Road, Clifton, Nottingham, NG11 8GF
G7 SWH A Howell, 35 Melton Road, Wakefield, WF2 7PR
G7 SWQ Ian Wild, 153 Alexandra Road, Sheffield, S2 3EH
G7 SWR M Prentice, 26 Meir View, Stoke-on-Trent, ST3 6AH
G7 SWV Colin Smith, 11 Woods Close, Haskayne, Ormskirk, L39 7JL
G7 SWW Roy Jones, 92 dale road, normanton, Derby, DE23 6QW
GM7 SWX D Curran, 104 Mcpherson Crescent, Chapelhall, Airdrie, ML6 8XL
G7 SWZ J Halliday, 14 Heath Gardens, Halifax, HX3 0BD
G7 SXB Duane Phillips, 197 Downall Green Road, Ashton-in-Makerfield, Wigan, WN4 0DW
G7 SXG David Dean, 17 Drayton Close, Runcorn, WA7 4TW
GM7 SXI Andrew Williams, 14 St. Phillans Avenue, Ayr, KA7 3BZ
G7 SXJ John Farrow, Flat 6, Hillstone Court, 22-24 Castle Hill Avenue, Folkestone, CT20 2QT
GW7 SXN D Davies, 35 Ty Llwyd Parc Estate, Quakers Yard, Treharris, CF46 5LA
GW7 SXU I Harries, Gwastad, Maenygroes, New Quay, SA45 9RJ
G7 SYC W Jarvill, 66 Gloucester Road, Newbury, RG14 5JN
G7 SYD Sydney Applegate, 180 Logan Street, Nottingham, NG6 9FU
G7 SYE T Laskey, 72 Windermere Avenue, Ramsgate, CT11 0PL
G7 SYJ Martin Hogg, 55 Ardenfield Drive, Wythenshawe, Manchester, M22 5DJ
G7 SYQ Andrew Orchiston, 16 Windsor Close, Collingham, Newark, NG23 7PR
G7 SYS R Baxter, 107 Kendale Road, Bridgwater, TA6 3QE
G7 SYT Chris Denman, 12 Woodland Close, Northampton, NN5 6NH
G7 SYU D Bowers, 88 Stamford Avenue, Springfield, Milton Keynes, MK6 3LQ
G7 SYW R Brown, 18 Gissons, Exminster, Exeter, EX6 8AH
G7 SYY S Howarth, 14 Eaves Lane, Chorley, PR6 0PY
G7 SZA S Mussell, Whitehouse, Priory Road, Boston, PE21 0RD
G7 SZB N Kendal-Ward, 2 King Charles Court, Sunderland, SR5 4PD
G7 SZF N Hartley, 66 Broad Lane, Norris Green, Liverpool, L11 1AN
G7 SZG K Gardner, 27 Lindon Drive Alvaston, Derby, DE24 0LP
G7 SZO R Collinson, 56 Orchard Valley, Hythe, CT21 4EA
G7 SZW Douglas Green, 43 James Street, Selsey, Chichester, PO20 0JG
G7 SZZ Richard Roberts, 9 Birch Close, Woking, GU21 7PR
G7 TAE S Wersby, 24 Idsworth Court, Basingstoke, RG24 9RR
G7 TAF M Hawes, 78 Martyns Way, Bexhill-on-Sea, TN40 2SH
G7 TAJ Steve Duckling, 39 Collington Lane West, Bexhill-on-Sea, TN39 3TD
G7 TAT Jeff Moye, 33 Prince Charles Road, Colchester, CO2 8NS
G7 TAV Steven Houghton, 28 Heron Way, Mayland, Chelmsford, CM3 6TP
G7 TAX F Roullier, 19 Terling Road, Dagenham, RM8 1DS
G7 TBC Peter Stockdale, 77 Fort Hill Road, Sheffield, S9 1BA
G7 TBF N Smith, 47 Kiveton Lane, Todwick, Sheffield, S26 1HJ
G7 TBJ J Kewn, 31 Trescoe Road, Long Rock, Penzance, TR20 8JY
G7 TBM T Goodwin, 41 Wood Road, Prestwich, Manchester, M25 2GP
G7 TBU Samuel Fitzjohn, 10 Samsons Close, Brightlingsea, Colchester, CO7 0RP
G7 TBW T Polain, 22 Hilltop Avenue, Hullbridge, Hockley, SS5 6BN
G7 TBX A Siddle, 5 Neneside, Benwick, March, PE15 0YF
G7 TCB P Hubbersley, 10 Dove Avenue, Penwortham, Preston, PR1 9RP
G7 TCD G Ward, 162 Greenbank Road, Darlington, DL3 6ES
G7 TCH HASTINGS COLLEGE RC c/o D Grandfield, Hastings College, Arts & Technology, St Leonards on Sea, TN38 0HX
G7 TCQ Shawn Preston, 18 Station Road, Great Wyrley, WS6 6LQ
G7 TCW Christopher Haslewood, 66 Hunter Road, Cannock, WS11 0AF
GI7 TDA J McKeever, 19 Corrycroar Road, Pomeroy, Dungannon, BT70 3DY
G7 TDD P Rose, 27 Zealand Road, Canterbury, CT1 3QW
G7 TDN A Baines, 60 Norton Drive, Halifax, HX2 7RB
G7 TDQ D Banister, 41 Tynycoed Road, Great Orme, Llandudno, LL30 2QA
G7 TDR R Smith, 47 Kiveton Lane, Todwick, Sheffield, S26 1HJ
G7 TEA Alan Goddard, 50 Ardmore Walk, Manchester, M22 5QG
GI7 TEB John Mathers, 14 Castlewood Avenue, Coleraine, BT52 1JR
G7 TEG G Fletcher, 171 Obelisk Rise, Northampton, NN2 8TX
G7 TEO Patricia Taylor, 8 First Avenue, Prestatyn, LL19 7LP
G7 TEP Kevin Blain, 17 Hillside Close, Headley Down, Bordon, GU35 8BL
G7 TET I Mowbray, 23 Rhodes Avenue, Bishop's Stortford, CM23 3JN
G7 TEZ G Masters, 85 Petersham Road, Creekmoor, Poole, BH17 7DW
G7 TFA B Wrampling, 18d May Avenue, Canvey Island, SS8 7EE
G7 TFG H Orchel, Gildertofts, Ingleby Greenhow, Middlesborough, TS9 6JF
GI7 TFK Stephen McCormick, 74 Belsize Road, Lisburn, BT27 4BH
G7 TFL S Dodds, 4 Claremont Road, Wisbech, PE13 2JR
GM7 TFN C Paton, 4 Abbeyhill, Dhailling Road, Dunoon, PA23 8FG
G7 TFU B George, 43 Claverton Road West, Saltford, Bristol, BS31 3DU
G7 TFX J Patterson, 11 Elmway, Chester le Street, DH2 2LD
G7 TGB C Bristow, 18 Clarendon Close, Chepstow, NP16 5TL
G7 TGF A Gibson, 100 Top Row, Darton, Barnsley, S75 5JQ
G7 TGG Craig Preston, 34 Forrester Street Precinct, Walsall, WS2 8RE
GI7 TGJ G Heaney, 38 Derryvore Lane, Portadown, Craigavon, BT63 5RS
G7 TGK C Coombe, 123 Farleigh Road, Pershore, WR10 1JY
G7 TGN P Dawson, 1 Eastfield Road, Bridlington, YO16 7DZ
G7 THF Michael Thompson, 4 Saxony Way, Donington, Spalding, PE11 4YA
GI7 THH Terry White, Shallamar, 3a Park Road, Strabane, BT82 8EL
G7 THI F Gillespie, Low Fold, Hoff, Appleby-in-Westmorland, CA16 6TA
G7 THJ Brian Mills, 37 Ashley Road, Hildenborough, Tonbridge, TN11 9ED
G7 THK K Grover, 6 Wren Court, Battle, TN33 0DU
G7 THL Dave Rowlandson, Korevaarstraat 6C, Leiden, Netherlands, 2311JS
GI7 THY R Larimer, 131 Carnalea Road, Seskanore, Omagh, BT78 2PP
G7 THZ J Reid, 12 Marlay Grove, Crownhill, Milton Keynes, MK8 0AT
G7 TIB D Cross, 91 Ilges Lane, Cholsey, Wallingford, OX10 9PA
G7 TIE I Chamberlain, 14 High House Avenue, Wymondham, NR18 0HY
G7 TIK Christopher McQueen, 40 Pen Green Lane, Corby, NN17 1BJ
G7 TIM G Jones, 42 Everard Road, Southport, PR8 6NA
G7 TIN Richard Martin, 2 East View, North Walsham Road, North Walsham, NR28 0PJ
G7 TIR D Thomas, 5 Minster Drive, Urmston, Manchester, M41 5HA
G7 TIV J Askew, 22 Cowslip Grove, Calne, SN11 9QQ
G7 TIW A Morris, 9 Otter Way, Wootton Bassett, Swindon, SN4 7SH
G7 TIX D Price, Sabrina, Pool Road, Newtown, SY16 1DW
G7 TIY D Miller, 139 Town Lane, Bebington, Wirral, CH63 8LB
G7 TJD M Crosfill, Polmennor Farmhouse, Heamoor, Penzance, TR20 8UL

GW7 TJM Martin Roberts, 2 Donnen Street, Port Talbot, SA13 1NE
G7 TJQ Cameron Shaw, [illegible] Southern Way, Stoke on Trent, ST6 1PX
G7 TJY C Ho, Po Box 900, Fanling Post Office, Hong Kong, Hong Kong
G7 TJZ [illegible]
G7 TKB F Coles, Le Bouillo, Estampes, France, [illegible]170
G7 TKG B Mersi, 4 Westdown Road, Bournemouth, BH11 9EQ
G7 TKI R Pettett, 2 Windmill Close, Great Dunmow, Dunmow, CM6 3AX
G7 TKM M Hewitt, 17 Farquhar Road, Maltby, Rotherham, S66 7PD
G7 TKO Michael Smith, 11 Martigny Road, Melksham, SN12 7PG
G7 TKP M Hewitt, 3 Orchard Rise, Bourne Lane, Reading, RG7 5NS
G7 TKT P Ashford, 3 Valley Road, Cheadle, SK8 1HY
G7 TKW M Peppiatt, 31e Llverton St, Kentish Town, London, NW5 2PE
G7 TLC D Benton, Hawthorn Cottage, Penrose, Wadebridge, PL27 7TB
G7 TLD M Clare, 43 Birchfield Close, Oxford, OX4 6DL
G7 TLK K Hemsil, 51 Lynher Drive, Saltash, PL12 4PA
G7 TLL H Hodson, 1 Chevin Avenue, Borrowash, Derby, DE72 3HR
G7 TLR K Marshall, 28 Deerness Grove, Esh Winning, Durham, DH7 9LY
G7 TMC M Conlon, 3 Selside, Brownsover, Rugby, CV21 1PG
G7 TMF T Foster, 98 Station Road, Carlton, Nottingham, NG4 3DA
G7 TMH A Hunt, 63a Toms Lane, Kings Langley, WD4 8NJ
G7 TMM A Kirkham, Flat 6, The Laurels, 14 Marlborough Road, Buxton, SK17 6RD
G7 TMO P Foster, 218 Stoops Lane, Bessacarr, Doncaster, DN4 7JQ
GI7 TMQ J Bell, 72 Coleraine Road, Portrush, BT56 8HN
G7 TMR R Nelson, 15 Poplars Close, Burgess Hill, RH15 9SZ
G7 TMU Victor Swanwick, 43 Hormare Crescent, Storrington, Pulborough, RH20 4QX
G7 TNO D Lunn, 23 Moynton Close, Crossways, Dorchester, DT2 8TX
G7 TNQ Michael Mrzyglod, 8 Beech Road, Shillingford Hill, Wallingford, OX10 8LU
GW7 TNS M Davies, 33 Hazel Mead, Brynmenyn, Bridgend, CF32 9AQ
G7 TNT B Scarsbrook, Salix, 96 Moss Lane, Alderley Edge, SK9 7HW
G7 TNU P Sparke, 18 Gordon Road, Haywards Heath, RH16 1EJ
G7 TNZ Martin Wells, 37 Water Meadows, Worksop, S80 3DF
G7 TOA S Haigh, 2 Locker Avenue, Warrington, WA2 9PS
G7 TOB R Wardell, 1 Enfield Close, Norden, Rochdale, OL11 5RT
G7 TOF I Pardington, 36 Rivermeads Avenue, Twickenham, TW2 5JJ
G7 TOI P Goodayle, 2 Downs Road, Seaford, BN25 4QL
G7 TOO P Crabtree, 106 Sagecroft Road, Thatcham, RG18 3BF
G7 TOU Michael Mussard, 35 Oakfield Gardens, Beckenham, BR3 3AY
G7 TOY A Peet, 95 Recreation Street, Mansfield, NG18 2HP
G7 TOZ J Whytock, 48 Lythe Fell Avenue, Halton, Lancaster, LA2 6NL
G7 TPB John Kilminster, 499 Hagley Road West, Quinton, Birmingham, B32 2AA
G7 TPD T Morton, 28 Turnfields, Ickford, Aylesbury, HP18 9HP
G7 TPG Brian Barber, 114 Scrogg Road, Newcastle upon Tyne, NE6 4HA
G7 TPH R Hand, 70 Flansham Lane, Bognor Regis, PO22 6AH
GI7 TPO Geoffrey Hodgkinson, 675 Crumlin Road, Belfast, BT14 7GD
G7 TPS Barry Seed, The Old Orchard, Main Road Cherhill, Calne, SN11 8UY
G7 TPW A Grigor, 48 Valebridge Drive, Burgess Hill, RH15 0RW
G7 TQA D Legge, 28 Dresser Road, Prestwood, Great Missenden, HP16 0NA
G7 TQC C Banister, York Avenue, East Cowes, PO32 6JT
G7 TQE T Brown, 138 Holmesdale Road, South Norwood, London, SE25 6HY
G7 TQT Ray Denton, 37 Tenby Road, Cheadle Heath, Stockport, SK3 0UN
GU7 TQX E Grisley, Les Clercs, Contree Des Clercs, St Pierre Du Bois, Guernsey, GY7 9DA
G7 TRB Philippe Stevenson, 50 Field Lane, Beeston, Nottingham, NG9 5FJ
G7 TRG Keith Liddle, 36 Vicarage Lane, Grasby, Barnetby, DN38 6AU
G7 TRL Darren Wright, 167 Whitby Avenue, Ingol, Preston, PR2 3GA
G7 TRM Keith White, 20 Agnes Close, Bude, EX23 8SB
G7 TSB E Jones, 26 Wood End, Bluntisham, Huntingdon, PE28 3LE
G7 TSO Kevin Jones, 128 Isabella Road, Queensland, Australia, 4869
G7 TSP C Leman, 92 Queens Crescent, Eastbourne, BN23 6JP
G7 TSQ J Stafford, 89 Mossley Road, Ashton-under-Lyne, OL6 9RH
G7 TTH G Quint, 4 Gibson Grove, Malvern, WR14 1NX
GI7 TTO D Dunlop, 63 Cloyfin Road, Coleraine, BT52 2NY
GM7 TTU Robert Emmott, 81 Coll, Isle of Lewis, HS2 0LR
G7 TTX M Tahla, Penrhiw, Ffestiniog, Blaenau Ffestiniog, LL41 4PN
G7 TTY Andrew Hubbard, 116 Ashland Road West, Sutton-in-Ashfield, NG17 2HS
G7 TUD J Pedley, 4 Tinwald View, Back Road, Dumfries, DG1 1RT
G7 TUG N Mitchell, 49 Kersey Road, Felixstowe, IP11 2UL
G7 TUH P Ferguson, 152 Chestnut Drive, Sale, M33 4HR
G7 TUK J Steel, 10 Green Courts, Winterton-on-Sea, Great Yarmouth, NR29 4AQ
G7 TUM John Moore, Waveney, Abbotts Way, Bush Estate, Norwich, NR12 0TA
G7 TUP Richard Irwin, 7 Hameau Des Peupliers, Rue Du Vert Pre, Lys Lez Lannoy, France, 59390
G7 TUQ B Forhead, 67 Gale Moor Avenue, Gosport, PO12 2SZ
G7 TUS R Munden, 2 Hain Villa, Forest Road, Ruardean, GL17 9XR
G7 TUV J Hewitt, 6 Crawley Walk, Warley, Cradley Heath, B64 5EX
G7 TVL E Roberts, 800 Walsall Road, Great Barr, Birmingham, B42 1EU
G7 TVQ Joseph Gilbert, 1 The Atrium, Higher Warberry Road, Torquay, TQ1 1TJ
G7 TVT L Whiteside, 8 The Orchards, Eaton Bray, Dunstable, LU6 2DD
GI7 TVV Albert McCready, 25 Glendun Park, Bangor, BT20 4UX
G7 TWA Dennis Bullard, 20 Chaney Road, Wivenhoe, Colchester, CO7 0QZ
G7 TWC Michael Ruttenberg, 90 Heath View, London, N2 0QB
G7 TWJ P Edwards, Cleveland, Blackberry Road, Lingfield, RH7 6NQ
GM7 TWM Ian Hipkin, 1 Maclennan Place, Dufftown, Keith, AB55 4EF
G7 TWU F Clarkson, 313 Normanby Road, Middlesbrough, TS6 0BQ
G7 TWW Christos Papaioannou, 2 Temple Lane, Temple, Marlow, SL7 1SA
G7 TXF A Scott, 60 Lowndes Park, Driffield, YO25 5BG
G7 TXR S Loyd, Maple House, Pangbourne Road, Reading, RG8 8LN
G7 TXU Anthony Ling, 3 Hogs Edge, Brighton, BN2 4NQ
G7 TXW Mark Oliver, 14 Harwood Road, Bridgemary, Gosport, PO13 0TT
G7 TXX D Williams, 57 Hillside Avenue, Kidsgrove, Stoke-on-Trent, ST7 4LW
G7 TYB J Hawley, 89 Mansfield Avenue, Denton, Manchester, M34 3NS
G7 TYH Steven Furminger, 24 Royal Road, Teddington, TW11 0SB
G7 TYJ J Pennington, 94 Rutland Avenue, Nuneaton, CV10 8EG
G7 TYO Stephen Birtwhistle, 14 Woodley Street, Bury, BL9 9HZ
G7 TYP B Cook, 40 Preston Avenue, Alfreton, DE55 7JY
G7 TYR OBO WEST KENT RAYNET c/o Denis Collins, 71 Trench Road, Tonbridge, TN10 3HG
G7 TYT M Claxton, 9 Thompson Avenue, Beverley, HU17 0BG
G7 TZB D Vincent, 6 Nathan Gardens, Poole, BH15 4JZ
G7 TZD P Jones, 361 Wellingborough Road, Rushden, NN10 6BA
[illegible]
GW7 TZI M Tonkin, 185 Pentregethin Road, Cwmbwrla, Swansea, SA5 8AU
G7 TZN Stewart Buckingham, 8 Tedder Avenue, Buxton, SK17 9JU
G7 TZO Christopher Turner, 308 North Road, Yate, Bristol, BS37 7LL
G7 TZQ R Darby, 25 Bramley Road, Marsh Lane, Sheffield, S21 5RD
G7 TZU Thomas Stalker, 172 Kirkby Road, Barwell, Leicester, LE9 8FS
G7 TZV Gerard Broughton, 111 Broadway, Manchester, M40 3NL
G7 TZW Roger Wheatley, 288 Bennett Street, Long Eaton, Nottingham, NG10 4JA
G7 TZX David Johnson, 12 Heron Close, Broughton, Chester, CH4 0RL
G7 TZZ Jonathan Eyre, 41 Wood Street, Geddington, Kettering, NN14 1BG
GM7 UAC E Edwards, 12 Highfield Place, Girdle Toll, Irvine, KA11 1BW
G7 UAH Brian Titmarsh, 28 Folly View, Stanstead Abbotts, Ware, SG12 8AX
G7 UAK Stanley Hunter, 30 Adelaide Street, Barrow-in-Furness, LA14 5TX
G7 UAL D Witherall, 221 Poynters Road, Dunstable, LU5 4SH
G7 UAT Jamie Johnson, 10 Croft Avenue, Newcastle, ST5 8EY
G7 UAV Ivan Morris, 60 Moorland Avenue, Lincoln, LN6 7DD
G7 UAY David Pickering, 32 Dorset Drive, Buckshaw Village, Chorley, PR7 7DN
G7 UBB E Knight, 258 Arundel Road West, Peacehaven, BN10 7PP
G7 UBD T Thomas, 166 Bluebell Road, Southampton, SO16 3LP
G7 UBK G Reddecliffe, 5 Stanley Close, Dymchurch, Romney Marsh, TN29 0TY
G7 UBO John Pearson, 22 Ashburnham Close, Norton, Doncaster, DN6 9HJ
G7 UBP Charlotte Howard, 144 Fairfield Road, Heysham, Morecambe, LA3 1LR
G7 UBQ T Bray, 2 Camborne Drive, Fixby, Huddersfield, HD2 2NF
G7 UBX P Pleydell, 6 The Croft, Meriden, Coventry, CV7 7NQ
GI7 UBY C Lunnon, 3 Parkfield, Crumlin, BT29 4SG
G7 UCB P Hudson, 47 Hall Farm Road, Duffield, Belper, DE56 4FJ
G7 UCG J Woodward, 108 Tamworth Road, Sutton Coldfield, B75 6DH
G7 UCL Sally Ann Dixon, 5 Swanmore Road, Havant, PO9 4LG
G7 UCN A Allport, 55 Byrds Lane, Uttoxeter, ST14 7NF
G7 UCO D Reed, 8 Wolverstone Drive, Hollingdean, Brighton, BN1 7FB
G7 UCP D Hornby, 7 Shawfield Grove, Rochdale, OL12 7SU
G7 UCR K Yeo, 48 Great Goodwin Drive, Guildford, GU1 2TY
GI7 UCS Martin Grainger, 22 Castle Oaks, Mountfield, Omagh, BT79 7BN
G7 UCT B Lord, 13 Park Ave, Norden, Timperley, WA14 5AQ
G7 UCZ D Evans, 3 Dalkeith Close, Bransholme, Hull, HU7 5AS
G7 UDE D Clark, 32 Laburnum Grove, Burstead Close, Brighton, BN1 7HX
G7 UDJ C Edwards, 6 Blacksmiths Close, Nether Broughton, Melton Mowbray, LE14 3EW
G7 UDM Derek Bonfield, 49 Linden Grove, Chandler's Ford, Eastleigh, SO53 1LE
G7 UDU John Selwyn, 5 Main Road, Billockby, Great Yarmouth, NR29 3BG
GI7 UDV William Weir, 9 Ripley Terrace, Portadown, Craigavon, BT62 3ED
G7 UDX Christine Harris, 8 Trelawney Rise, Callington, PL17 7PT
G7 UEC D Denyer, 85 Highlands Road, Horsham, RH13 5ND
G7 UEI D Longhurst, Burston, Wood Road, Hindhead, GU26 6PZ
G7 UEJ S Kitchen, 344 Windward Way, Castle Bromwich, Birmingham, B36 0UH
G7 UEK A Jones, 60 Heywood Drive, Starcross, Exeter, EX6 8SD
G7 UEL R Dean, Sandshadow, Stow Road, King's Lynn, PE34 3PF
G7 UET Andrew Levy, 29 Ferndale Avenue, Reading, RG30 3NQ
G7 UEV C Fox, 3 Manor Drive, Wragby, Market Rasen, LN8 5SL
G7 UEX P Cardwell, 2 Hayfield Place, Sheffield, S12 4XH
G7 UFF L Brackstone, 276 Ladyshot, Harlow, CM20 3EY
G7 UFI Barrie Courtenay, 251 Smeeth Road, Marshland St. James, Wisbech, PE14 8ES
GM7 UFN T Graham, 265 Gilmartin Road, Linwood, Paisley, PA3 3SU
GM7 UFO N Bartley, 7 South Quarry Boulevard, Gorebridge, EH23 4GL
G7 UFT R Elliott, 16 Prince Philip Road, Colchester, CO2 8PA
G7 UFV D Riseborough, 2 The Barn, Grigsons Wood, Norwich, NR16 2LW
G7 UFW Nathan Brickwood, 77 Grange Road, Northampton, NN3 2AX
G7 UGA Michael Turner, 14 The Rookery, Barrow upon Soar, Loughborough, LE12 8JZ
G7 UGC Alban Fellows, 343 Wake Green Road, Birmingham, B13 0BH
GI7 UGP David Robertson, 27 Ann Street, Newtownards, BT23 7AD
G7 UGR David Barnett, 124 Gaywood Rd, King Lynn, PE30 2PX
G7 UGW John Smith, 32 Aberdeen Street, Hull, HU9 3JU
G7 UGY Nicholas Thornley, The Vicarage, Broughton-in-Furness, LA20 6HS
G7 UHE G Tiller, 12 Birk Beck, Waveney Drive, Chelmsford, CM1 7PJ
G7 UHG D Tropman, 91 Reindeer Road, Fazeley, Tamworth, B78 3SW
G7 UHL S Yuill, 24 Marigolds, Deeping St. James, Peterborough, PE6 8SN
G7 UHS C Jewell, 43 Rannoch Road, Bristol, BS7 0SA
G7 UHT C Griffiths, 33 Westwood Road, Ryde, PO33 3BJ
G7 UHW Michael McDermott, 4 Tolcairn Court, 28 Lessness Park, Belvedere, DA17 5BT
G7 UHX B Anderson, 22 The Drive, Clacton-on-Sea, CO15 4NN
G7 UHY R Blewitt, Flat 16, Lincoln House, Wolverhampton, WV10 0JB
G7 UID Stuart Clarke, 75 Beaumont St., Netherton, Huddersfield, HD4 7HE
G7 UII C Savage, 20 Croft Crescent, Awsworth, Nottingham, NG16 2QY
G7 UIO N Johnson, 81 Yeo Closse Efford, Plymouth Devon, Plymouth, TA5 2BQ
GI7 UIP Kenneth O'Reilly, 400 Coa Road, Killymittan, Enniskillen, BT94 2FU
GJ7 UIT Christopher Totty, 34 Le Clos Paumelle, Bagatelle Road, St. Saviour, Jersey, JE2 7TW
G7 UIU S Palmer, 54 Hawthorn Road, Exeter, EX2 6EA
G7 UIZ J Hughes, Maes Y Ffynnon, 7 Meadow Gardens, Llandudno, LL30 1UW
G7 UJC G Taylor, 34 Hockley Road, Poynton, Stockport, SK12 1RW
GM7 UJJ J Scott, 1 Carrick Knowe Drive, Edinburgh, EH12 7EB
GM7 UJO S Maxwell, 24 Castle Drive, Airth, Falkirk, FK2 8GD
G7 UJT S Dransfield, Gardener Ground House, West End, Goole, DN14 8RW
G7 UJY M Poole, 184 Woodgates Lane, Swanland, North Ferriby, HU14 3PR
G7 UKA T Collier, 23 The Riggs, Brandon, Durham, DH7 8PQ
G7 UKF M Ellis, 64 Coppice Drive, Dordon, Tamworth, B78 1QZ
G7 UKK Andrew Firth, 59 Station Road, Shepley, Huddersfield, HD8 8DS
G7 UKN K Riley, 27 Limewood Close, Blythe Bridge, Stoke-on-Trent, ST11 9NZ
G7 UKR M Blackburn, 36 Mardale Grove, Barrow-in-Furness, LA13 9QG
G7 ULC C Probert, 25 Elizabethan Way, Rugeley, WS15 2EE
GI7 ULG Sean Murdoch, 4 Seymour Hill Mews, Dunmurry, Belfast, BT17 9PW
G7 ULJ Paul White, 18 Valley Farm Court, Nottingham, NG5 9DQ
G7 ULL P Craig, 6 Marsham Close, Chislehurst, BR7 6JD
G7 ULM Peter Howarth, 4 Ringwood Avenue, London, N2 9NS
G7 ULN J Grundy, 47 Northiam Road, Eastbourne, BN20 8LP
G7 ULS K Hurst, 94 East Park, Harlow, CM17 0SB
[illegible] S75 6JD
G7 UMA Neil Tindall, 87 The Grove, Marton-in-Cleveland, Middlesbrough, TS7 8AN
G7 UMF Derek Griffiths, Home Farm House Cottage, Leebotwood, Church Stretton, SY6 6LX
G7 UMS Kirkley Keepin, 10 Briers Gate, Henllys, Cwmbran, NP44 6EE
G7 UMW Alan Banner, 6 Heol y Wal, Bradley, Wrexham, LL11 4BY
G7 UMY D Rockliffe, 3 Hewell Lane, Barnt Green, Birmingham, B45 8NZ
G7 UNB A Bevington, 54 Pheasant Road, Smethwick, B67 5PD
G7 UNU N Davies, 16 St. Leonards Close, Scole, Diss, IP21 4DW
G7 UNV Evan Jones, Crungoed Farm Llanbister Road, Llandrindod Wells, LD1 5UR
G7 UNW Nicholas Othen, 234a Regents Park Road, London, N3 3HP
G7 UNY John Gabbatiss, 5 Ashtree Road, Watton, Thetford, IP25 6PF
G7 UNZ W Scott, Rose Brae, Lazonby, Penrith, CA10 1AJ
G7 UOD H Golding, 11 Southwold Crescent, Broughton, Milton Keynes, MK10 7RW
G7 UOH S Lupton, Egryn, Ffordd Dewi Sant, Pwllheli, LL53 6EA
G7 UOL R Bennion, 3 Dorrington Close, Ruskington, Sleaford, NG34 9EQ
G7 UOQ N Birt, 60 Church Road Woodley, Reading, RG5 4QB
G7 UOS B Yates, 131 Kingsway North, Leicester, LE3 3BF
G7 UOU Andrew Colville, 34 Great North Road, Welwyn, AL6 0PS
GM7 UPD C Edwards, Hillhead Croft, Chapel of Garioch, Inverurie, AB51 5HE
G7 UPL Sally Northeast, 143 Henderson Road, Southsea, PO4 9JE
G7 UPN C Jackson, 2 Northway, Guildford, GU2 9SB
G7 UPP R Hall, Greenviews, Lower Kingsbury, Sherborne, DT9 5ED
GI7 UPQ Michael Cunningham, 4 Garvaghy Road, Portglenone, Ballymena, BT44 8EF
GI7 UPU F Gillespie, 33 Clonliffe Park, Londonderry, BT48 8NT
G7 UPZ Ian Sansom, 26 Finedon Road, Wellingborough, NN8 4EB
G7 UQA Dean Haigh, 29 Victoria Grove, Wakefield, WF2 8UP
G7 UQG NEWCASTLE DISTRICT SCOUTS RC c/o Roger Bloor, 7 Highfield Court, Clayton Road, Newcastle, ST5 3LT
G7 UQJ M Mee, Cerrig Gwynion, Penisarwaun, Caernarfon, LL55 3PW
GM7 UQM M Horne, 10 Blair Place, Kirkcaldy, KY2 5SQ
G7 UQQ Thomas Wakeling, Flat 17, Kleffens Court, London, SE3 7QX
G7 UQV Mark Willoughby, 30 Kipling Road, Ipswich, IP1 6EW
GI7 UQW B Neill, 81 Orangefield Road, Belfast, BT5 6DD
GI7 URC A Brown, 3 Clara Road, Belfast, BT5 6FN
G7 URJ Jennifer O'Brien, 45 Rossall Promenade, Thornton-Cleveleys, FY5 1LP
G7 URL D Foster, Pentlow, Crowle Bank Road, Scunthorpe, DN17 3HZ
G7 URM T Heartfield, 69 Great Thrift, Petts Wood, Orpington, BR5 1NF
G7 URP David Palmer, Edison House, Bow Street, Attleborough, NR17 1JB
G7 URR Samuel Easter, Flat 11, Saxon Court, Hitchin, SG4 9TB
G7 URS R Bird, 9 Orchard Lane, Wembdon, Bridgwater, TA6 7QY
G7 URT C Langham, 9 Laurence Close, Shurdington, Cheltenham, GL51 4SZ
G7 URW Nigel Tucker, 13a Fordens Lane, Holcombe, Dawlish, EX7 0LD
GI7 USA A Niblock, 2 Inverary Valley, Larne, BT40 3BJ
G7 USB J Ainsworth, 42 Buttfield Road, Hessle, HU13 0AS
G7 USC Gary McKelvie, 63 Oaklands, Guilden Sutton, Chester, CH3 7HE
G7 USG J Sutherland, 4 Cherbury Close, Bracknell, RG12 9HT
G7 USI R Hayselden, 400 Heath End Road, Nuneaton, CV10 7HG
G7 USJ T Cogan, 11 Highgrove Walk, Weston-Super-Mare, BS24 7EF
G7 USM K Reavill, 11 Clarence Road, Beeston, Nottingham, NG9 5HY
G7 USP S Garwood, 42 Fleetdyke Drive, Lowestoft, NR33 9HB
G7 USQ B Siddall, 6 Delside Avenue, Manchester, M40 9LF
G7 USV D Atkins, 14 Ryde Place, Lee-on-The-Solent, PO13 9AU
G7 USX M Woollard, Barnside, Colchester Road, Colchester, CO7 7EG
G7 UTB H Scott-Telford, 9 Squires Close, Rochester, ME2 2TZ
G7 UTC M Bean, Ashmore, Belle Vue Road, Sudbury, CO10 2PP
GM7 UTD David Forrest, 15 Invergarry Avenue, Thornliebank, Glasgow, G46 8UR
G7 UTE B Spencer, 80 Horncastle Road, Boston, PE21 9HY
G7 UTG John Dodds, 84 Borrowdale Avenue, Walkerdene, Newcastle upon Tyne, NE6 4HL
G7 UTH R Banks, 50 Vale Road, Portslade, Brighton, BN41 1GG
G7 UTI G Ducros, 21 Wardlow Gardens, Plymouth, PL6 5PU
G7 UTR George Kelsall, 3 Raven Street, Bingley, BD16 4LB
G7 UTS M James, 7 Greenfield Park, Portishead, Bristol, BS20 6RG
G7 UTT R Pearce, 15 St. Andrews Road, Backwell, Bristol, BS48 3NR
G7 UTY C Lewis, 3 Jacobs Close, Stantonbury, Milton Keynes, MK14 6EJ
G7 UUA P Matthews, 6 West Road, Halstead, CO9 1EH
G7 UUB F Gibbs, 62 Wenvoe Avenue, Bexleyheath, DA7 5BT
G7 UUC Matthew West, 69 Frampton Crescent, Bristol, BS16 4JD
G7 UUD K Matthews, 6 West Road, Halstead, CO9 1EH
G7 UUG N Griffiths, 125 Coleridge Way, Crewe, CW1 5LF
G7 UUH A Hughes, 30 Liddell Drive, Llandudno, LL30 1UH
G7 UUK M Hooper, 12 Meare, Dunster Crescent, Weston-Super-Mare, BS24 9DW
G7 UUL A Riggs, Lower House, Stockland Bristol, Bridgwater, TA5 2PY
G7 UUN Robert Wood, 7 Lilac Grove, Luston, Leominster, HR6 0EF
G7 UUP J Chapman, 8 Oakfield Court, Stanley Common, Ilkeston, DE7 6XB
G7 UUR Nancy Bone, 217 Bensham Road, Gateshead, NE8 1US
G7 UUT A Wilson, 36 Davey Crescent, Great Shelford, Cambridge, CB22 5JF
G7 UUW S Hearn, 28 Noithrop Avenue, Banbury, OX16 2NF
G7 UVB David Anstie, 20 Keyes Road, Norwich, NR1 2JX
G7 UVF Gary Cheetham, 35 South Park Grove, New Malden, KT3 5BZ
G7 UVL Deborah Croot, 58 Dixie Street, Jacksdale, Nottingham, NG16 5JZ
G7 UVN I Cross, 25 Yatesbury Avenue, Blakelaw, Newcastle upon Tyne, NE5 3SZ
G7 UVO Roy Moss, 142 Mold Road, Connah's Quay, CH5 4QP
G7 UVP J Swanwick, Ramblers, Clarks Farm Road, Chelmsford, CM3 4PH
GM7 UVS J Graham, 265 Gilmartin Road, Linwood, Paisley, PA3 3SU
G7 UVV I Perry, Meadow Cottage, Mill Lane, Halstead, CO9 2NW
G7 UVW D Mills, 11 Northfield Road, Dagenham, RM9 5XH
G7 UVY Colin Carr, 10 Bonds Road, Hemblington, Norwich, NR13 4QF
G7 UWB B Wright, 2 Butterfly Gardens, Rushmere St. Andrew, Ipswich, IP4 5TF
G7 UWC C Wright, 16a Worcester Road, Ipswich, IP3 0RR
G7 UWE P Smith, 12a Sandicroft Place, Preesall, Poulton-le-Fylde, FY6 0PB
G7 UWG R Hancox, 13 Regnum Close, Eastbourne, BN22 0XH



**IMPORTANT NOTE**

**Revalidate licence to avoid revocation** – Ofcom has advised the Society that plans will be drawn up to revoke licences that have not been revalidated as required by the licence conditions. The quickest way to revalidate is to do so online via the Ofcom website: *https://services.ofcom.org.uk/* or by email: *amateur.validations@ofcom.org.uk* Ofcom staff are available to help, but please be patient during times of heavy workload.

G7 UWI M Jones, 41 Milton Brow, Weston-Super-Mare, BS22 8DD
G7 UWL D Cottage, 14 Wellington Lane, Farnham, GU9 9BA
G7 UWO G Holland, 15 Rollis Park Road, Oreston, Plymouth, PL9 7LU
G7 UWP P Groves, Flat 4, 147 St. Peters Rise, Bristol, BS13 7ND
G7 UWR C Hillman, Cariad, Yarn Barton, Templecombe, BA8 0JH
G7 UWS D Brunt, 91 Shaftesbury Avenue, Feltham, TW14 9LW
G7 UWV I Brown, 58 Highfields Road, Bilston, WV14 0SF
G7 UWW C Harding, 1 Saddleton Grove, Saddleton Road, Whitstable, CT5 4LY
G7 UWZ Derek Pooley, 25 Wharncliffe Road, Highcliffe, Christchurch, BH23 5DB
G7 UXD R Wade, The Limes, Hunston, Bury St. Edmunds, IP31 3EL
GM7 UXH E Gaunt, 17/1 Crewe Road Gardens, Edinburgh, EH5 2NJ
G7 UXK S Hedges, 25 Rudland Close, Thatcham, RG19 3XW
G7 UXQ J Manwaring, 38 Norfolk Road, Consett, DH8 8DD
G7 UXR M Taylor, 27 Lincoln Road, Newark, NG24 2BU
G7 UXU H Andrews, 24 Belvoir Road, Widnes, WA8 6HR
GW7 UXY Ronald Williams, 16 Tir Dafydd, Pontyates, Llanelli, SA15 5TP
G7 UYB C Major, 17 Jubilee Cottages, Station Road, Bedford, MK43 0PN
G7 UYI B Williamson, 12 Middleton Close, Southampton, SO18 2FP
G7 UYJ J Jardine, 41 Charles Drive, Anstey, Leicester, LE7 7BH
G7 UYT John O'Toole, 4 Lindisfarne Road, Dagenham, RM8 2RA
G7 UYW J Kitchener, 101 Highfield Road, Tring, HP23 4DS
G7 UZA J Rodinson, 21 Graylands Road, Liverpool, L4 9UG
G7 UZG A McWilliam, 43 Hylder Close, Swindon, SN2 2SL
G7 UZI P Pullen, 12 Kimpton Road, Sutton, SM3 9QJ
G7 UZN David Dawson, 12 Thurlow Terrace, Kentish Town, London, NW5 4JB
G7 UZO D Lock, 1 Heaton Avenue, Huddersfield, HD5 0LJ
G7 UZS N Thompson, 30 Dene View, Ashington, NE63 8JF
G7 UZX G Saunders, 63 Holst Avenue, Basildon, SS15 5RH
G7 UZY A Musther, 2 Fakenham Close, Lower Earley, Reading, RG6 4AB
G7 VAB M Richards, Sunnymead, Ardley End, Bishop's Stortford, CM22 7AJ
G7 VAD Michael Beeney, Oakville Farm, Lewes Road, Uckfield, TN22 5JH
G7 VAE James Beeney, 37 Coppice Avenue, Eastbourne, BN20 9PP
G7 VAG G Podmore, 9 Pendlebury Street, Warrington, WA4 1TU
G7 VAH Stephen Rutter, Fairview, Waxham Road, Norwich, NR12 0UX
G7 VAS Martin Kay, 12 The Crescent, Ashton-on-Ribble, Preston, PR2 1JP
G7 VAU Gary Gillies, Fairlawns, Sussex Street, Bedale, DL8 2AN
G7 VAY R Dingle, 87 Eighth Avenue, Bridlington, YO15 2NA
G7 VBD Martin Ennis, 1 Nairn Road, Cramlington, NE23 1RQ
GW7 VBE Diane Harris, 29 Queen Street, Blaengarw, Bridgend, CF32 8AH
G7 VBF J Barwell, Flat 9, Corinth House, 33 Barley Lane, Ilford, IG3 8XE
G7 VBJ D Wager, 162 Harvest Fields Way, Sutton Coldfield, B75 5TJ
G7 VBL J Munday, 20 Highcroft, Wood Road, Hindhead, GU26 6PW
G7 VBN B Richards, 2 Craddock Row, Sandhutton, Thirsk, YO7 4RT
GI7 VBS D Aughey, 239 Bridge Street, Portadown, Craigavon, BT63 5AR
G7 VBU David Firth, 59 Station Road, Shepley, Huddersfield, HD8 8DS
G7 VBY Derek Morrison-Smith, 1 Neptune House, Upper Corris, Machynlleth, SY20 9BQ
G7 VBZ P Bunce, 66 Berry Park, Saltash, PL12 6EN
G7 VCB L Tooze, Flat 1, 91 Harbour Road, Seaton, EX12 2NJ
G7 VCE M Flack, 31 Harebell Close, Cambridge, CB1 9YL
G7 VCF John O'Donnell, 3 Linden Avenue, Altrincham, WA15 8HA
G7 VCG I Firby, 19 St. Georges Drive, Manchester, M40 5HL
G7 VCJ C Hansford, 14 Parsonage Crescent, Castle Cary, BA7 7LT
G7 VCK M Robertson, 67 Oatland Gardens, Leeds, LS7 1SL
G7 VCM C Kemp, 10 Laurel Close, Dartford, DA1 2QL
G7 VCN R Bawley, 52 Pitville Avenue, Liverpool, L18 7JG
G7 VCP P Stubbs, 2 Cynthia Road, Runcorn, WA7 4TX
GI7 VCR S Robertson, 32 Castle Meadows, Carrowdore, Newtownards, BT22 2TZ
G7 VCT G Mould, 25 Kingsley Road, Talke Pits, Stoke-on-Trent, ST7 1RB
GM7 VCV Alan Brown, 14 Laverock Avenue, Greenock, PA15 4NF
G7 VCY D Seymour, 24 Farley Dell, Coleford, Radstock, BA3 5PJ
G7 VCZ Peter Weaver, Stoneway House, Leys Hill, Ross-on-Wye, HR9 5QU
G7 VDA Iain Singer, 197 Rosalind Street, Ashington, NE63 9BB
G7 VDD P McGowan, 11 Pankhurst Gardens, Gateshead, NE10 8EN
G7 VDH J Brook, Windsor Cottage, New Laithe Bank LANE, Holmfirth, HD9 1HL
G7 VDI D Norris, Flat 11, 10 Cromartie Road, London, N19 3SJ
G7 VDJ S Henry, Hertford College, Catte Street, Oxford, OX1 3BW
G7 VDK G Taylor, 4 Brown Crescent, Eighton Banks, Gateshead, NE9 7EX
GM7 VDL William Steele, 1 James Street, Bannockburn, Whins of Milton, Stirling, FK7 0NQ
GM7 VDM J Steele, 35 Devlin Court, Whins of Milton, Stirling, FK7 0NP
G7 VDN H May, 18 Pennant Hills, Bedhampton, Havant, PO9 3JZ
G7 VDQ Donald Butterworth, 6 Fir Grove, Weaverham, Northwich, CW8 3JD
G7 VDS M Hudson, 10 Coppice Close, Madeley, Telford, TF7 4DW
G7 VDT Gordon Baines, 20 Whitehall Rise, Wakefield, WF1 2AL
G7 VDU A Marston, 111 Averil Road, Leicester, LE5 2DE
G7 VDV Neil Keech, 14 Simpson Court, Ashington, NE63 9SD
G7 VDX S Taverner, 8 The Rye Lea, Droitwich, WR9 8SS
G7 VEB D Wilson, 210 Stanks Lane South, Swarcliffe, Leeds, LS14 5PD
G7 VEE A Saunders, 25 Southern Drive, South Woodham Ferrers, Chelmsford, CM3 5NY
G7 VEF Robert Parkin, 17 Roberts Road, Watford, WD18 0AY
G7 VEH J Humphrey, Bryn Ebbw, Beaufort Hill, Ebbw Vale, NP23 5QR
G7 VEI A Stripp, 87 Elthorne Park Road, London, W7 2JH
G7 VEL C Lee, 23 Forest Close, Coed Eva, Cwmbran, NP44 4TE
G7 VEX N Hindle, 19 Barkway Road, Royston, SG8 9EA
G7 VEY John Martin, 3 The Rise, Calne, SN11 0LQ
G7 VFA J Juggins, 5 Charter Close, Helston, TR13 8SR
G7 VFC Owen Dewberry, The Stables, Barrack Street, Manningtree, CO11 2RB
G7 VFE R Wills, 14 Penwood Heights, Penwood, Highclere, Newbury, RG20 9EY
G7 VFJ Stephen MaGee, Lle Da, Cefn Bychan Road, Mold, CH7 5EL
G7 VFL K Sherman, 12 Portland Drive, Stourbridge, DY9 0SD
G7 VFQ Anthony Latham, 273 Adswood Road, Stockport, SK3 8PA
GM7 VFR J Smith, 28 Tollerton Drive, Irvine, KA12 0QE
G7 VFU A Mullord, 296 City Way, Rochester, ME1 2BL
G7 VFV G Somers, 21 Paterson Road, Aylesbury, HP21 8LN
G7 VFX R Watts-Read, 43 Whyteleafe Hill, Whyteleafe, CR3 0AJ
G7 VFY Stephen Walters, Flat 2, 2d Lodge Lane, London, N12 8AF
G7 VGA S Bonney, 22 Gordon Drive, Abingdon, OX14 3SW
G7 VGB David Beynon, 129 Eureka Place, Ebbw Vale, NP23 6LN
G7 VGC Brian Goody, Flat 31, Homeweave House, Robinsbridge Road, Colchester, CO6 1UL
G7 VGE Roger Teague, 18 Aspen Close, Great Blakenham, Ipswich, IP6 0HQ
G7 VGH M Smith, 57 Lynn Road, Ely, CB6 1DD
G7 VGJ Adrian Cole, 58 Stradbroke Drive, Chigwell, IG7 5QZ
G7 VGK William Parrett, 5 Coniston Close, Walton, Liverpool, L9 0NG
G7 VGL D Pemberton, 12 Victor Road, Thatcham, RG19 4LX
G7 VGM S Taylor, 28 Parton Street, Hartlepool, TS24 8NN
G7 VGN A Chamberlain, 7 Mccalmont Way, Newmarket, CB8 8HU
G7 VGO A Hurst, 12 Spilsby Close, Hartlepool, TS25 2RD
GI7 VGR C Dorrian, 47 Albany Drive, Carrickfergus, BT38 8BF
G7 VGT Paul Schranz, 42 South Townside Road, North Frodingham, Driffield, YO25 8LE
G7 VGX H Dolman, 28 The Downs, Middleton, Manchester, M24 1TJ
G7 VGY David Childs, 7 Grange Road, East Cowes, PO32 6EA
G7 VHC Charles Spires, 5 Springhead, Sutton Veny, Warminster, BA12 7AG
G7 VHD Arthur Jones, 57 Dinerth Road, Rhos on Sea, Colwyn Bay, LL28 4YG
G7 VHF EAST ANGLIAN SIX METER GROUP c/o A Durrant, 2 Ramsey Hall Cottages, Wix Road, Harwich, CO12 5LS
G7 VHG S Bryan, 18 Whalley Crescent, Wroughton, Swindon, SN4 9EP
G7 VHJ P Gow, 11 Rodley Square, Lydney, GL15 5AZ
G7 VHN J Hart, 35 Aintree Close, Uxbridge, UB8 3HS
G7 VHO B Hart, 35 Aintree Close, Uxbridge, UB8 3HS
GM7 VHQ I Helie, 25 Ard Road, Renfrew, PA4 9DD
G7 VHU D Beastall, 11 Hopwood Bank, Horsforth, Leeds, LS18 5AW
G7 VHX J Golding, 65 Longworth Avenue, Tilehurst, Reading, RG31 5JU
G7 VHZ E Gilowski, 126 Owlsmoor Road, Owlsmoor, Sandhurst, GU47 0ST
G7 VIA Martin Airs, 67 Croft Road, Wallingford, OX10 0HN
G7 VIB D Airs, Cornerways Cottage, Poffley End, Witney, OX29 9UW
G7 VIE J Cooper, 9 Highfield Crescent, Halesowen, B63 2BD
G7 VIG A Smith, 19 Gibsons Gardens, North Somercotes, Louth, LN11 7QH
G7 VIH Peter Wilson, 117 Naseby Road, Kettering, NN16 0LL
G7 VIK Norman Higgins, 6 Larksfield Avenue, Bournemouth, BH9 3LP
G7 VIL G Mason, Finchale, Fieldhouse Lane, Durham, DH1 4NB
G7 VIO Anthony Delwiche, 13 Bell Meadow, Godstone, RH9 8ED
G7 VIP F Marston, 1 Weaver Road, Leicester, LE5 2RL
G7 VIR A James, 19 Coach Lane, Redruth, TR15 2TP
G7 VIV Alastair Nussey, 9 Brent Street, Brent Knoll, Highbridge, TA9 4DU
GI7 VIW A Harvey, 5 Kilmaine Road, Bangor, BT19 6DT
G7 VIY A Harper, 3 Eskdale Crescent, Blackburn, BB2 5DT
G7 VJA Ken Sharman, 1 The Greenwoods, Hartland, Bideford, EX39 6JA
G7 VJD Diane Skidmore, Weavers, Kingsdale Road, Berkhamsted, HP4 3BS
G7 VJE Christopher Rohrer, Alpenrose, Bedlars Green, Bishop's Stortford, CM22 7TP
G7 VJG P Veitch, 78 Hughes Street, Swindon, SN2 2HG
G7 VJH Terry Scanlon, 11 Caterhouse Road, Framwellgate Moor, Durham, DH1 5HP
G7 VJI G Dawes, 587 Charminster Road, Bournemouth, BH8 9RQ
G7 VJJ T Wood, 39 Baker Road, Bournemouth, BH11 9JD
GW7 VJK Nigel Cole, Tycoch, Llandovery, SA20 0UP
G7 VJM C Margetts, 16 Lahn Drive, Droitwich, WR9 8TQ
G7 VJQ C Radford, 12 Homewood Drive, Kirkby-in-Ashfield, Nottingham, NG17 8QB
G7 VJT Shaun Tibbetts, 113 Highfield Crescent, Halesowen, B63 2AY
G7 VJU Robert Jones, 20 Carnoustie Close, Southport, PR8 2FB
G7 VJY T Hill, 15 Catkin Walk, Rugeley, WS15 2NS
G7 VKA K Wandless, 11 Havanna, Killingworth, Newcastle upon Tyne, NE12 5BL
G7 VKB A Cunnington, 131 Colson Road, Loughton, IG10 3QY
G7 VKG Mark Gibson, 6 Harrison Road, Mansfield, NG18 5RG
G7 VKJ C Buckley, 14 Sunny Drive, Prestwich, Manchester, M25 3JJ
G7 VKK Peter Collings, Clematis, Mill Common, Halesworth, IP19 8RQ
GM7 VKN R Beharie, Isengard, Norseman Village, Firth, KW17 2NY
G7 VKY B Shrimpling, 45 Fairmont Road, Grimsby, DN32 8DZ
G7 VLA M Sandham, 7 Mill Close, Caverswall, Stoke-on-Trent, ST11 9HA
G7 VLB S Kirkbright, 48 Plant Crescent, Stafford, ST17 4EH
GM7 VLC A Haines, 164 North High Street, Musselburgh, EH21 6AR
G7 VLD K Howard, 43 Hazeldell, Watton at Stone, Hertford, SG14 3SN
G7 VLF N Faiz, 48 Cox House, Field Road, London, W6 8HN
G7 VLH Christopher Hill, 14 Blenheim Close, Chandler's Ford, Eastleigh, SO53 4LD
G7 VLJ E Baker, 29 Ashcroft Road, Ipswich, IP1 6AB
G7 VLL J Woodhouse, 5 Dolphin Villas, Hazlerigg, Newcastle upon Tyne, NE13 7NG
G7 VLR LEICESTER RAYNET GROUP c/o Andrew Holmes, 5 Launde Park, Market Harborough, LE16 8BH
GM7 VLZ A Pearce, 105 Gyle Park Gardens, Edinburgh, EH12 8NQ
G7 VME P Schofield, 22 Atherton Court, Meadow Lane, Windsor, SL4 6BN
G7 VML L Alden, 20 Kings Walk, Shoreham-by-Sea, BN43 5LG
G7 VMO Stewart Fawcett, 45 Forresters Close, Norton, Doncaster, DN6 9HX
G7 VMQ T Jones, Ockton House, 24 Station Road, Okehampton, EX20 1EA
GW7 VMT Elaine Wetherall, 38 Argyle Street, Pembroke Dock, SA72 6HL
G7 VNC C Cave, Little Meadow, Brewham Road, Bruton, BA10 0JD
G7 VND G Morris, 17 Bradshaw Road, Inkersall, Chesterfield, S43 3HJ
G7 VNE P Oates, 22 Sunny Bank Walk, Mirfield, WF14 0NH
G7 VNG A Elmes, Pookeezows, 10 Farnham Avenue, Hassocks, BN6 8NS
G7 VNJ Matthew Swain, 29 Huntingdon Gardens, Newbury, RG14 2RT
G7 VNK R Cronshaw, Flat 2, 28 Adelaide Terrace, Blackburn, BB2 6ET
G7 VNL C Lambert, 43 Church Road, Guildford, GU1 4NQ
G7 VNM A Melham, 201 Bridgewood Road, Worcester Park, KT4 8XU
G7 VNN Chris Backhouse, Elm House, The Green Saxlingham Nethergate, Norwich, NR15 1TH
G7 VNO C Brown, 6 Ellesmere Avenue, Derby, DE24 8WD
G7 VNP K Knights, 15a Little Ditton, Woodditton, Newmarket, CB8 9SA
G7 VNQ Kevin Staddon, 1 Aller Grove, Whimple, Exeter, EX5 2TJ
G7 VOA D Hughes, 60 Martingale Place, Downs Barn, Milton Keynes, MK14 7QN
G7 VOH S Holland, 49 Oxland Road, Illogan, Redruth, TR16 4SH
G7 VOI T Nicholas, Talmont, Chester Road, Tarporley, CW6 0SD
G7 VOK A Bracey, 42 Lampton Grove, Bristol, BS13 0QA
G7 VOM J Snelgrove, 22 Plains Avenue, Maidstone, ME15 7AU
G7 VON A Snelgrove, 22 Plains Avenue, Maidstone, ME15 7AU
GW7 VOO Paul Hockey, 98 Meadow Rise, Brynna, Pontyclun, CF72 9TF
G7 VOQ Simon Casey, 5 Willow Road, Leyland, PR26 8NP
G7 VOT A Moseley, 46 Harford Street, Middlesbrough, TS1 4PR
G7 VOX M Ingram, Foxhill, Lower Daggons, Fordingbridge, SP6 3EE
G7 VPA N Muncey, 2 Ladysmith Avenue, Whittlesey, Peterborough, PE7 1XX
G7 VPD R Friend, High Hedges, Church Road, Norwich, NR12 8YL
G7 VPL W Jackson, 4 Beaumaris Avenue, Blackburn, BB2 4TW
G7 VPN A Berry, 13 Collimer Close, Chelmondiston, Ipswich, IP9 1HX
G7 VPQ J Bishop, 27 Southway, Blacon, Chester, CH1 5NW
G7 VPS D Barwood, 5 Kingfisher Drive, Necton, Swaffham, PE37 8NN
GM7 VPT D Leask, Avonmuir, The Loan, Linlithgow, EH49 6LW
G7 VPU P Ansell, White Hatch, Uvedale Road, Oxted, RH8 0EW
G7 VQA N Parker, 47 Rosehill Road, Rhyl, LL18 4TN
GM7 VQB T Roy, 1 Rose Terrace, Leven, KY8 4DF
G7 VQC D Driver, 27 Cricketers Way, Chatteris, PE16 6UR
G7 VQE David Frost, 82 Sheeplands Lane, Sherborne, DT9 4BP
G7 VQI G Burch, 6 The Barracks, Parkend, Lydney, GL15 4HR
G7 VQJ W willmott, 3 Chesterton Road, Cliffe, Rochester, ME3 7QX
G7 VQL M Endean, 17 Dryden Place, Tilbury, RM18 8HQ
G7 VQM Christopher Davis, 94 Greenwood Lane, Wallasey, CH44 1DW
G7 VQO William Good, 53 Harrow Lane, St. Leonards-on-Sea, TN37 7JY
G7 VQR P Mawdsley, 7 Aldebert Terrace, London, SW8 1BH
G7 VQW Peter Jessup, 2 Tile Lodge Cottages, Hoath Road, Canterbury, CT3 4JN
G7 VQX J Hunter, 7 Berry Hill, Nunney, Frome, BA11 4NR
G7 VRJ M Holland, 31 The Avenue, Andover, SP10 3EP
G7 VRK Steven Balding, 13 Church Close, Colby Road, Norwich, NR11 7DY
G7 VRX R Croft, Wallbury Lodge, Dell Lane, Bishop's Stortford, CM22 7SQ
G7 VRY P Bambridge, 8 Temple Lane, Tonwell, Ware, SG12 0HP
GM7 VSB Jason O'Neill, 39 Ardneil Court, Ardrossan, KA22 7NQ
G7 VSE D Briggs, 15 Orkney Close, Manchester, M23 2AT
GW7 VSF W Thomas, 2 Ffordd Trecastell, Llanharry, Pontyclun, CF72 9ND
G7 VSG D Webster, 1 Woodbine Close, Potter Heigham, Great Yarmouth, NR29 5NF
G7 VSJ D Jones, 6 Eastville, Bath, BA1 6QN
G7 VSL R Taylor, 11 Ranscombe Close, Brixham, TQ5 9UR
G7 VSM J Skinner, 85 Main St., Barton Under Needwood, Burton-on-Trent, DE13 8AB
G7 VSN Lee Franklin, 4 Rossington Close, Metheringham, Lincoln, LN4 3DS
GW7 VSO David Lewis, 45 Llewellyn St., Pontygwaith, Ferndale, CF43 3LF
G7 VSP S Wooster, 44 King Johns Road, North Warnborough, Hook, RG29 1EJ
GW7 VST G Davies, 41 Woodlands Road, Barry, CF63 4EF
G7 VSW S Weston, 11 Friars Road, Abbey Hulton, Stoke-on-Trent, ST2 8DQ
G7 VTC Leslie Dodd, Fanharley, London Apprentice, St. Austell, PL26 7AR
G7 VTE G Forster, 33 Deer Valley Road, Holsworthy, EX22 6DA
G7 VTH D Reacher, 33 Cator Crescent, New Addington, Croydon, CR0 0BL
G7 VTJ T Scott, 50 Davison Avenue, Whitley Bay, NE26 1SH
G7 VTL C Davis, 38 Courtney Close, Tewkesbury, GL20 5FB
G7 VTN P McCaulay, 33 Millmoor Way, North Hykeham, Lincoln, LN6 9PJ
G7 VTQ D Parsons, 1 Kent Drive, Congleton, CW12 1SD
G7 VTR Colin Taylor, Tookeys House, Tookeys Drive, Astwood Bank, B96 6BB
G7 VTS P Green, 81 Victoria Road, Farnborough, GU14 7PP
G7 VTT J King, Portland Manor Care Home, Thornhill Road, Newcastle upon Tyne, NE20 9PZ
G7 VTW D Jacques, 33 North Street, Otley, LS21 1AH
G7 VUB R Walker, 30 Ruth Close, Tipton, DY4 0AQ
G7 VUH C Squire, 19 Southfield Road, Burley in Wharfedale, Ilkley, LS29 7PA
G7 VUL J Cook, 32 Ash Bank Road, Stoke-on-Trent, ST2 9DR
G7 VUM D Riches, 118 Drayton Road, Norwich, NR3 2DL
G7 VUP J Milner, Sweetcroft Brentor, Tavistock, PL19 0NJ
G7 VUU K Coe, 5 George St., Enderby, Leicester, LE19 4NQ
G7 VVB P Andrew, 11 Meadow Rise, Giggleswick, Settle, BD24 0EF
G7 VVF D Rossiter, 37 Meadway, Enfield, EN3 6NT
G7 VVK Peter Bradley, 22 Cavalier Close, Romford, RM6 5EJ
G7 VVL N Quest, 21 Neave Crescent, Romford, RM3 8HN
G7 VVO I Anderson, 18 St. Anthonys Drive, Wick, Bristol, BS30 5PW
G7 VVX A Renton, 18 Stoneworks Garth, Crosby Ravensworth, Penrith, CA10 3JE
G7 VWA D Lever, 35 Carteret Road, Luton, LU2 9JZ
G7 VWC R Bleach, Flat 12, Everest House, 7-11 Hogarth Road, Hove, BN3 5RG
G7 VWG R Evans, 113 Highbridge Road, Burnham-on-Sea, TA8 1LW
G7 VWM John Hazell, 7 Higher Road, Woolavington, Bridgwater, TA7 8EA
G7 VWN John Blackwell, 7 Church Road, Darley Dale, Matlock, DE4 2GG
G7 VWO D Lisle, Kent li, Broadmoor Hospital, Crowthorne, RG45 7EG
G7 VWW J Brook, 45 Colonial Court, Senoia, United States, 30276
GI7 VXC A Crozier, 5 Meadowvale, Dromore, BT25 1BF
G7 VXK B Amare, PO BOX 30464, Addis Ababa, Ethiopia
G7 VXQ C Elcombe, 10 Northport Drive, Wareham, BH20 4DR
GM7 VXR P Crankshaw, 3 North Neuk, Troon, KA10 6TT
G7 VXS Glyn Burchell, 23b Luff Meadow, Stowmarket Road, Ipswich, IP6 8DP
G7 VYB R Patel, 30 Buckingham Drive, Luton, LU2 9RA
G7 VYF K Tadesse, PO Box 60229, Addis Ababa, Ethiopia
G7 VYI R Smith, 15 Broxtons Wood, Westbury, Shrewsbury, SY5 9QR
G7 VYN M Jarman, 143 Rotherham Road, Barnsley, S71 2LL
G7 VYQ I Holman, 7 The Silent Woman Park, Tavistock, PL19 9LQ
GM7 VYR I Findlay, 2 Bothwell Road, Uddingston, Glasgow, G71 7ET
G7 VYT J Graver, 15 Cartwright Road, Charlton, Banbury, OX17 3DG
G7 VYW William Moreton, 17 Hadley Road, Bilston, WV14 6RX
G7 VYY S Middleton, 22 Hall Villa Lane, Toll Bar, Doncaster, DN5 0LH
G7 VYZ M Lancastle, 31 Ridgeside, Kirk Merrington, Spennymoor, DL16 7HF
G7 VZD Jamie Payne, 15 Belmont Road, Tiverton, EX16 6AR
G7 VZI Hayden Charles, 6 Bridewell Street, Wymondham, NR18 0AR
G7 VZK S Briggs, 33 Menzies Close, Southampton, SO16 8FX
G7 VZL D Forster, 15 Bracondale, Norwich, NR1 2AL
G7 VZM M blacklock, 39 Birtwistle Avenue, Colne, BB8 9RS
G7 VZQ D Polley, 33 Wye Close, Crawley, RH11 9QZ

| | | |
|---|---|---|
| G7 | VZR | C Gain, 14 Battens Avenue, Overton, Basingstoke, RG25 3NL |
| G7 | [illegible] | Anthony [illegible], 21 Rydal Avenue, [illegible] |
| [illegible] | [illegible] | [illegible] |
| G7 | VZY | M Page-Jones, 2 Chestnut Close, Romsey, [illegible] |
| G7 | WAA | E Donaghy, Mendips, 36 Attwood Road, Salisbury, SP1 3PD |
| G7 | WAB | WORKED ALL BRITAIN AWARDS GROUP c/o Kevin Hale, 58 St. Stephens Road, Saltash, PL12 4BJ |
| G7 | WAC | WYTHALL CONTEST GROUP c/o Lee Volante, Richmond House, Icknield Street, Birmingham, B38 0EP |
| G7 | WAE | Thomas Ward, 127 Lower Lime Road, Oldham, OL8 3NP |
| G7 | WAF | D Keeble, 71 St. Lawrence Avenue, Bolsover, Chesterfield, S44 6HS |
| G7 | WAQ | A Moss, 4 Stock Terrace, Stock Chase, Maldon, CM9 4AB |
| G7 | WAS | S Staines, 6 The Quantocks, Flitwick, Bedford, MK45 1TQ |
| G7 | WAW | David Thompson, 12 Dam Head Road, Barnoldswick, BB18 5NH |
| G7 | WBA | R Grandshaw, Treehaven, South Lane, Salisbury, SP5 2BZ |
| G7 | WBE | D Welch, 38 Little Sammons, Chilthorne Domer, Yeovil, BA22 8RB |
| G7 | WBH | David Small, 17 Claygate Road, Wimblebury, Cannock, WS12 2RN |
| G7 | WBJ | W Naylor, 5 Burman Close, Shirley, Solihull, B90 2DR |
| G7 | WBL | Jim Wheeler, 92 Holford Road, Bridgwater, TA6 7NZ |
| G7 | WBM | Peter Longhurst, Burston, Wood Road, Hindhead, GU26 6PZ |
| G7 | WBO | M Stenning, 56 Hampshire Court, Brighton, BN2 1JZ |
| G7 | WBR | Erwin Davis, 33 Truggers, Handcross, Haywards Heath, RH17 6DQ |
| G7 | WBU | A Hopley, 41 Old Pound Close, Lytchett Matravers, Poole, BH16 6BW |
| G7 | WBW | Adrian Millward, 26 Osprey Close, Scotton, Catterick Garrison, DL9 3RA |
| G7 | WBY | Blake Mulder, 8 Chapel Close, Little Gaddesden, Berkhamsted, HP4 1QG |
| G7 | WBZ | M Malone, 11 Pine Close, Rishton, Blackburn, BB1 4JX |
| G7 | WCB | A Bennett, 12 Barns Close, Walsall, WS9 9BD |
| G7 | WCF | C Rose, The Barn, Killigorrick Farm, Liskeard, PL14 4QP |
| G7 | WCG | D Seabrook, 44 Village Centre, Richmond Letcombe Centre, Letcombe Regis, OX12 9RG |
| G7 | WCN | K Packer, 47 Sheppard Road, Basingstoke, RG21 3JH |
| G7 | WCP | C Sharpe, 30 Mardale Way, Loughborough, LE11 3SS |
| GW7 | WCR | J Pitkin, 29 Dolwerdd Estate, Pen Y Parc, Cardigan, SA43 1RF |
| GI7 | WCS | J Stitt, 199 Gobbins Road, Islandmagee, Larne, BT40 3TX |
| G7 | WDC | M Kiteley, 13 Chiltern Close, Astley Cross, Stourport-on-Severn, DY13 0NU |
| G7 | WDD | B Bird, 4 Berkeley Crescent, Frimley, Camberley, GU16 8YN |
| G7 | WDG | Paul Wyatt, 6 Bridge Road, Coalville, LE67 3PW |
| G7 | WDM | N Feetham, 154 Magdalen Lane, Hedon, Hull, HU12 8LB |
| G7 | WDN | A Goodridge, 14 Fox Lane North, Chertsey, KT16 9HW |
| G7 | WDO | C Barker, 15 Epping Green, Hemel Hempstead, HP2 7JP |
| G7 | WDS | A James, 36 Lemon Hill, Mylor Bridge, Falmouth, TR11 5NA |
| G7 | WEB | D Raxter, 2 Lower Croft, Cropthorne, Pershore, WR10 3NA |
| GM7 | WED | R Feilen, 131 Croftend Avenue, Glasgow, G44 5PF |
| G7 | WEK | C Taylforth, 1 Clough Terrace, Barnoldswick, BB18 5PD |
| G7 | WEM | T Hewitt, 6 Mayfield, Catforth Road, Preston, PR4 0HH |
| G7 | WEN | J Marron, 30 York Road, Nunthorpe, Middlesbrough, TS7 0EZ |
| G7 | WEP | M Williams, 59 Thistledene, Thames Ditton, KT7 0YH |
| G7 | WER | P Fisher, 21 Charlotte Close, Mount Hawke, Truro, TR4 8TS |
| G7 | WEW | A Cossey, 11 Halden Avenue, Norwich, NR6 6UX |
| G7 | WFD | M Dockerty, 16 Valley Way, Stalybridge, SK15 2QZ |
| GW7 | WFI | J Piggott, 32 East View, Bargoed, CF81 8LU |
| G7 | WFK | G Tew, 5 Hill Top Avenue, Tamworth, B79 8QB |
| G7 | WFQ | J Stevens, Springfield Cottage, 57 Brindley Street, Stourport-on-Severn, DY13 8JG |
| GM7 | WFT | G Edwards, 142 Swanston Muir, Edinburgh, EH10 7HY |
| G7 | WFZ | R Stone, 222 Dedworth Road, Windsor, SL4 4JP |
| G7 | WGA | J Potter, 198 Battle Road, St. Leonards-on-Sea, TN37 7AL |
| G7 | WGD | Duncan Price, 15 Heath Green, Dudley, DY1 3TN |
| G7 | WGE | Theresa Forster, 20 Bryant Avenue, Slough, SL2 1LG |
| G7 | WGI | J Gordon, 19 Heywood Gardens, Havant, PO9 4HR |
| G7 | WGK | Dean Rowley, New House, Hereford Road, Shrewsbury, SY3 0EQ |
| G7 | WGL | K Firth, Chimneys, 30 Kingscroft, King's Lynn, PE31 6QN |
| GM7 | WGM | Stuart Andrew, 16 Colthill Road, Milltimber, AB13 0EF |
| G7 | WGO | G Bradshaw, 3 Falmouth Avenue, Haslingden, Rossendale, BB4 6QN |
| G7 | WGP | Anthony Brook, 163 Station Road, Mickleover, Derby, DE3 9FL |
| G7 | WGX | D Sayles, 82 Molineaux Road, Shiregreen, Sheffield, S5 0JY |
| G7 | WGY | D Wilson, 75 Gainsborough Road, Scotter, Gainsborough, DN21 3RU |
| G7 | WGZ | E Morley, 91 Allerton Road, Stoke-on-Trent, ST4 8PQ |
| G7 | WHA | O Morley, 91 Allerton Road, Stoke-on-Trent, ST4 8PQ |
| G7 | WHI | David Page, 23 Wheatlands Drive, Countesthorpe, Leicester, LE8 5RT |
| G7 | WHM | A Howgate, 7 Caledonian Way, Belton, Great Yarmouth, NR31 9PQ |
| G7 | WHP | W Jones, 7 Hampstead Gardens, Hockley, SS5 5HN |
| GM7 | WHQ | Stuart Gray-Thompson, Vagastie, Lairg, IV27 4AD |
| G7 | WHU | M Nock, Mesquida, Lorraine Road, Newhaven, BN9 9QB |
| [illegible] | [illegible] | J [illegible], 19 [illegible] Road, Hove, BN3 5HP |
| G7 | WHZ | T Crane, 15 Belchamps Way, Hawkwell, Hockley, SS5 4NT |
| G7 | WIC | G Probyn, 24 Woollaton Close, Grange Park, Swindon, SN5 6BB |
| G7 | WID | Gary White, 21 Tollfield Road, Boston, PE21 9PN |
| G7 | WIG | Robert Bilsland, 56 Cowleigh Bank, Malvern, WR14 1PH |
| G7 | WIQ | Simon Pack, 245a Beacon Road, Loughborough, LE11 2QZ |
| G7 | WIY | M Downing, 12 Martindale Road, Woking, GU21 3PJ |
| G7 | WJC | Brian Webster, 50 Blackburn Road, Rishton, Blackburn, BB1 4BH |
| G7 | WJE | H Coots, 40 Essex Close, Romford, RM7 8DD |
| G7 | WJJ | Roy Morton, 29 Lanmoor Estate, Lanner, Redruth, TR16 6HN |
| G7 | WJK | J Stephens, 19 Aspen Fold, Oswaldtwistle, Accrington, BB5 4PH |
| G7 | WJP | Andrew Anderson, 232 Annan Road, Dumfries, DG1 3HE |
| G7 | WJV | Robert Stroud, 55 Haymeads Lane, Bishop's Stortford, CM23 5JJ |
| G7 | WJW | G Cripps, 116 Beaver Lane, Ashford, TN23 5NX |
| G7 | WJZ | P Clarke, 21 Long Furlong Road, Sunningwell, Abingdon, OX13 6BL |
| G7 | WKC | T Hasted, Springfield House, Birds End, Bury St. Edmunds, IP29 5HE |
| G7 | WKG | Andrew Roche, Flat 22, Trident Court, Birmingham, B20 2NX |
| G7 | WKH | P Clark, 21 Sandfield Road, Arnold, Nottingham, NG5 6QA |
| G7 | WKP | A Andrew, Thrift, Madles Lane, Ingatestone, CM4 9QA |
| G7 | WKV | B Jewell, 49 Spinney Road, Burton Latimer, Kettering, NN15 5ND |
| G7 | WKW | M Davis, 12 Amesbury Road, Cholderton, Salisbury, SP4 0EP |
| GI7 | WLA | Derek Calvin, 65 Tannaghmore Road, Markethill, Armagh, BT60 1TW |
| G7 | WLC | D Evans, 1 Hill Cottages, Layer Breton Hill, Colchester, CO2 0PR |
| G7 | WLL | Ivan Irlam, 31 Wyatt Road, Dartford, DA1 4SN |
| G7 | WLM | J Tamlyn, Hedge Rise, Sidmouth Road, Exeter, EX2 5QJ |
| G7 | WLO | J Burt, Olivet, Lanton Road, Jedburgh, TD8 6SD |
| G7 | WLV | G Southall, 6 Dudley Wood Avenue, Dudley, DY2 0DG |
| G7 | WLY | R Bagwell, 375 Taylor Street, South Shields, NE33 5AW |
| [illegible] | [illegible] | [illegible] PA7 5DT |
| G7 | WRG | Walsall Raynet Group c/o Steven Glazzard, 109 Highfields Road, Chasetown, Burntwood, WS7 4QS |
| G7 | WSH | Robert Munt, Box 166, Jarfalla, Sweden, SE-177 23 |
| G7 | WWW | B Cole, 6 Parkstone Parade, Hastings, TN34 2PS |
| G7 | XPC | P Chorley, Boone Hill House, Mount Boone Hill, Dartmouth, TQ6 9NZ |
| G7 | ZMS | Mark Larcombe, 65 Western Road, Burgess Hill, RH15 8QW |
| G7 | ZRT | Ronald Thayne, 213 carlton road, boston, Sbl-Aps-00177, PE21 8NG |
| G7 | ZZY | P Pile, Apartment 836, Lagos, Portugal, 8600 |

## G*8

| | | |
|---|---|---|
| G8 | AAC | J Billingham, 14 St. Matthews Court, Sutherland Road, Brighton, BN2 2EX |
| G8 | AAD | B Blight, 43 North Street, Oxon, OX9 3BJ |
| G8 | AAE | D Phillips, 2 Walkers Close, Cholmsford, CM1 6UW |
| G8 | AAF | F Blake, 3 Morfa Gaseg, Llanfrothen, Penrhyndeudraeth, LL48 6BH |
| G8 | AAI | M Bues, 7A Alice Parkins Close, Hadleigh, IP7 6FE |
| G8 | AAL | P Hamblett, 13 Ironside Close, Bewdley, DY12 2HX |
| G8 | AAR | F May, Quatre Vents, Church Road, Sudbury, CO10 0QP |
| G8 | AAT | R Pye, 7 Meadow View, Potterspury, Towcester, NN12 7PH |
| G8 | AAU | N Stanners, 22 Brands Hill Avenue, High Wycombe, HP13 5QA |
| G8 | ABB | Geoffrey Rogers, 10 The Laurels, Bletchley, Milton Keynes, MK1 1BL |
| G8 | ABX | Geraint Catling, 3 The Tene, Baldock, SG7 6DG |
| G8 | ACA | Howard Crockett, 28 Church Lane, Middleton, Tamworth, B78 2AW |
| G8 | ACL | H Cosford, 3 Applewood, Park Gate, Southampton, SO31 7HQ |
| G8 | ACQ | R Whattam, The Aviary No1, Arkwright Rd, Bedford, MK44 1SE |
| G8 | ACT | G Gunn, 18 Barnfield, Hatfield Broad Oak, Bishop's Stortford, CM22 7JR |
| G8 | ADA | J Robinson, 7 Rhyl Street, Liverpool, L8 6QL |
| G8 | ADC | J Haile, 145 Dunstable Road, Caddington, Luton, LU1 4AN |
| G8 | ADD | B Carter, 51 Smirrells Road, Birmingham, B28 0LA |
| G8 | ADH | C Slingsby, Dunelm Lodge, Faceby, TS9 7DA |
| GM8 | ADK | M Ritchie, 11 Cromwell Road, Aberdeen, AB15 4UH |
| G8 | ADQ | Jeffrey Taylor, 21 Launcestone Close, Earley, Reading, RG6 5RY |
| G8 | ADX | Eric Lawley, 3 Barnicott Close, Newton Ferrers, Plymouth, PL8 1BP |
| G8 | ADY | Paul Harrison, 2 The Barns, Bridge End, Bedford, MK43 7LP |
| G8 | ADZ | N Shepherd, 7 High St., Kelvedon, Colchester, CO5 9AG |
| G8 | AEN | Peter Helm, 74 Neston Road, Walshaw, Bury, BL8 3DB |
| G8 | AER | J Tanner, Merlins Mill, Toadsmoor Road, Stroud, GL5 2UG |
| G8 | AEU | J Nightingale, 6 Aubrey Close, Chelmsford, CM1 4EJ |
| G8 | AFA | C Atkins, 2 Eastlands, Yetminster, Sherborne, DT9 6NQ |
| G8 | AFI | P Funnell, 25 Broadyates Road, Yardley, Birmingham, B25 8JF |
| G8 | AFN | P Cleall, 139 Preston Grove, Yeovil, BA20 2DB |
| GI8 | AFS | M Granville, 33 Dunfield Terrace, Londonderry, BT47 2ES |
| G8 | AFU | P Gilby, 191 Send Road, Send, Woking, GU23 7ET |
| G8 | AGB | F Goodwin, 36 Grange Drive, Ryton, NE40 3LF |
| G8 | AGJ | J Evans, 1 Grosvenor Close, Hatch Warren, Basingstoke, RG22 4RQ |
| GM8 | AGM | Martin Collar, Shoemakers Croft, Hatton, Peterhead, AB42 0TB |
| G8 | AGN | B Chambers, 5 The Ridge, Sheffield, S10 4LL |
| GW8 | AHB | P Swinbank, 13 Mundy Place, Cardiff, CF24 4BZ |
| G8 | AHE | Les Arnold, 402 Bournville Gardens, 49 Bristol Road South, Birmingham, B31 2FT |
| G8 | AHK | University of Surrey EARS c/o Laurence Stant, EARS, University of Surrey Student's Union, Guildford, GU2 7XH |
| G8 | AHN | J Barnes, 2 Mappins Road, Catcliffe, Rotherham, S60 5TH |
| G8 | AHR | P Rushworth, 2 Aberdeen Close, Coventry, CV5 7NE |
| G8 | AIE | P Willcocks, 27 Manor Road, Barnet, EN5 2LE |
| G8 | AIM | F Tarver, 14 South View Road, Leamington Spa, CV32 7JD |
| G8 | AIP | Martin Osment, Flat 2, Weavers Court, Shoreham-by-Sea, BN43 5ES |
| GI8 | AIR | William Parkes, 15 Bushfoot Park, Portballintrae, Bushmills, BT57 8YX |
| G8 | AJA | D Hardy, 7 Coed Y Go Cottages, Coed Y Go, Oswestry, SY10 9AU |
| G8 | AJM | C Payne, 27 Cookham Road, Maidenhead, SL6 7EF |
| G8 | AJP | J Eade, White Cottage, Whatlington, Battle, TN33 0NL |
| G8 | AJZ | R Boardall, 9 Oxford Street, Pimhole, Bury, BL9 7EL |
| G8 | AKA | T Wiltshire, Bramblings, Pelican Road, Tadley, RG26 3EL |
| G8 | AKC | C Bell, Croftner, Mary Tavy, Tavistock, PL19 9QD |
| G8 | AKE | J Warrington, 26 Lynton Road, Melton Mowbray, LE13 0NN |
| G8 | AKF | J Ballantyne, Brookeside, Ashwellthorpe Road, Norwich, NR16 1AW |
| G8 | AKL | Gerald Ashcroft, Huntingdonshire Amateur Radio Society, Buckden Village Hall, St. Neots, PE19 5UY |
| G8 | AKM | G Roper, 19 Normay Rise, Newbury, RG14 6RY |
| G8 | AKP | Philippa McQuade, Old Swan, Holt Road, Melton Constable, NR24 2PH |
| G8 | AKQ | Stephen Birkill, St. Anns, Ecclesall Road South, Sheffield, S11 9PX |
| G8 | AKU | B Willson, Hilltop, Cryers Hill Road, High Wycombe, HP15 6LJ |
| G8 | AKX | M Perry, 216 Marlpool Lane, Kidderminster, DY11 5DL |
| G8 | ALD | Michael Lunt, 18 Longhurst Road, Hindley Green, Wigan, WN2 4PL |
| G8 | ALE | Michael Brereton, Gleaston, Ulverston, LA12 0QH |
| G8 | ALQ | A Whitlock, 23 Daly Way, Aylesbury, HP20 1JW |
| G8 | ALR | J Cull, Drybrook Cottage, Amesbury Road, Salisbury, SP4 0ER |
| G8 | ALS | M Stevenson, 15 Wall Hill Road, Allesley, Coventry, CV5 9EN |
| G8 | AMD | H Bate, 88 Darnick Road, Sutton Coldfield, B73 6PG |
| G8 | AMG | M Foster, 9 Norman Way, Irchester, Wellingborough, NN29 7AT |
| G8 | AMJ | D [illegible] |
| G8 | AMK | Leslie Parry, 13 Cannon Hill, Bracknell, RG12 7QA |
| G8 | AMU | C Saveker, 23 Southlands Avenue, Horley, RH6 8BS |
| G8 | ANN | G Townsend, 61 Richmond Park Road, London, SW14 8JU |
| G8 | ANO | D Lawton, Crenchurst, Pinewood Road, High Wycombe, HP12 4DD |
| G8 | ANT | Simon Holland, 14 The Vineries, Eastbourne, BN23 7TP |
| GD8 | ANU | Charles Howard, 5 Ballure Grove, Ramsey, Isle of Man, IM8 1NF |
| GM8 | AOB | James Briscoe, 2 Peebles Place, Fort William, PH33 6UG |
| G8 | AOE | Brian Duffell, 7 Potto Close, Yarm, TS15 9RZ |
| G8 | AOG | M Browne, 143 Thatch Leach Lane, Whitefield, Manchester, M45 6EP |
| G8 | AOI | T Knight, 3 Eaton Close, Rainworth, Mansfield, NG21 0AR |
| G8 | AOJ | George Smith, Forest View Cottage, Gorsty Knoll, Coleford, GL16 7LR |
| G8 | AOK | A Porch, 17 Purcell Close, Brighton Hill, Basingstoke, RG22 4EL |
| G8 | AOO | B Hills, 3 Frithmead Close, Basingstoke, RG21 3JW |
| G8 | AOZ | P Hughes, 247 High Greave, Sheffield, S5 9GS |
| G8 | APB | Christopher Plummer, Barley House Farm, Newtown, Stoke-on-Trent, ST8 7UW |
| [illegible] | [illegible] | [illegible] |
| G8 | APM | G White, 1 Drakes Close, Hythe, Southampton, SO45 6BP |
| G8 | APW | D Taylor, 87 Grasmere Road, Chester le Street, DH2 3EU |
| G8 | APY | J Bond, Folly House, The Reddings, Cheltenham, GL51 6RL |
| G8 | APZ | Sydney Lucas, 84 Woodman Road, Warley, Brentwood, CM14 5AZ |
| G8 | AQA | Paul Nickolls, Holy Mill, Longville, Much Wenlock, TF13 6ED |
| G8 | AQB | M Ballance, 24 Western Road, Wolverton, Milton Keynes, MK12 5BE |
| G8 | AQH | Rodney Hine, 149 Bolton Hall Road, Bradford, BD2 1BQ |
| G8 | AQN | A Hibberd, 20 Barby Lane, Rugby, CV22 5QJ |
| G8 | AQO | Alan Copperwaite, 71 Gladbeck Way, Enfield, EN2 7EL |
| G8 | AQP | S Warner, 14 Andrews Way, Aylesbury, HP19 8WA |
| G8 | ARA | B King, 15 Newstead Road, West Southbourne, Bournemouth, BH6 3HJ |
| GW8 | ARC | Alan Craggs, 15 Pen-y-Groes Avenue, Cardiff, CF14 4SP |
| G8 | ARF | L Thompson, 44 Tillmouth Avenue, Holywell, Whitley Bay, NE25 0NP |
| G8 | ARH | Nigel Blackmore, 35 Weyhill Gardens, Weyhill, Andover, SP11 0QT |
| G8 | ARM | B Pickrell, Perrans, Ludgvan, Penzance, TR20 8AJ |
| G8 | ARR | Peter Edwards, Trevland, Felindre, Knighton, LD7 1YL |
| GW8 | ASA | G Wyatt, 3 Creidiol Road, Mayhill, Swansea, SA1 6TZ |
| G8 | ASC | P Richards, 134 Downhills Park Road, Tottenham, London, N17 6BP |
| G8 | ASD | A Pugh, Willcroft, Mold Road, Wrexham, LL11 4AF |
| G8 | ASG | M Farrell, Hobberley House, Hobberley Lane, Leeds, LS17 8LX |
| G8 | ASJ | G Swan, Morogar, Post Office Lane, Worcester, WR5 3NX |
| G8 | ASP | I Gurton, 28 Bloomfield Road, Harpenden, AL5 4DB |
| G8 | ASW | R Warrender, 102 Turnberry Road, Great Barr, Birmingham, B42 2HT |
| G8 | ASX | A Hoggan, 25 Clingan Road, Bournemouth, BH6 5PY |
| GM8 | AT | W Beattie, Alastrean House, Tarland, Aboyne, AB34 4TA |
| G8 | ATB | S Chettle, The Byre, 2 Park Lane Mews, Hatherton, CW5 7QX |
| G8 | ATC | R Gayton, 20 Barton Close, Exton, Exeter, EX3 0PE |
| G8 | ATD | Andy Barter, 503 Northdown Road, Margate, CT9 3HD |
| G8 | ATE | Robert Turlington, 2 Laithwaite Close, Leicester, LE4 1BX |
| G8 | ATG | M Williamson, 120 Warbreck Hill Road, Blackpool, FY2 0TR |
| G8 | ATK | Michael Hearsey, Halycon, Lawday Link, Farnham, GU9 0BS |
| G8 | ATL | M Lankester, 154 Gorse Lane, Clacton-on-Sea, CO15 4RJ |
| G8 | ATP | K Mintern, 71 Crafts End, Chilton, Didcot, OX11 0SB |
| G8 | ATS | J Reeve, 16 Junction Road, Mildenhall, Bury St. Edmunds, IP28 7BZ |
| G8 | AUJ | G Papworth, 12 Brook Way, Cupernham, Romsey, SO51 7JZ |
| G8 | AUL | P Buck, 41 Marion Street, Brighouse, HD6 2BJ |
| G8 | AUN | R Chiddick, 87 Aylsham Road, Norwich, NR3 2HW |
| G8 | AUU | Christopher Partridge, 6 Blagdon Walk, Teddington, TW11 9LN |
| G8 | AVB | Charles Dickson, 5 Arrow View, Ledbury, HR8 2FR |
| G8 | AVC | Raymond Evans, Mansfield, 1 Horsehead Lane, Chesterfield, S44 6HU |
| G8 | AVK | R Kimberley, 8 Nutwell Road, Weston-Super-Mare, BS22 6EN |
| G8 | AVM | Ian Macdonald, Benvoir, Newton Stewart, DG8 9EE |
| G8 | AVO | J Wainwright, Fairview, The Green, Radstock, BA3 5UY |
| G8 | AVQ | John Florentin, 17 Campden Hill Gardens, London, W8 7AX |
| G8 | AVV | K Bennett, 11 Dunelm Court, South Street, Durham, DH1 4QX |
| G8 | AVZ | Mike Keeping, 8 Calderdale Close, Southgate, Crawley, RH11 8SQ |
| G8 | AWB | R Lawrence, Chapel Cottage, Bray Shop, Callington, PL17 8PZ |
| G8 | AWE | Michael Wellspring, 7 Rue Du 19 Mars 1962, Ruffec, France, 16700 |
| G8 | AWI | Colin Smith, 129 Earls Road, Nuneaton, CV11 5HP |
| G8 | AWM | Frank Evans, Ty Cryr, Chepstow Road, Usk, NP15 1HN |
| G8 | AWN | Barrie Procter, 28 Holme Grove, Burley in Wharfedale, Ilkley, LS29 7QB |
| G8 | AWY | J Ward, 71 Rothschild Avenue, Aston Clinton, Aylesbury, HP22 5LY |
| G8 | AXA | M Wallace, 17 Leamington Avenue, Orpington, BR6 9QA |
| G8 | AXN | C Amery, 9 View Close, Biggin Hill, Westerham, TN16 3XE |
| G8 | AXO | Andrew Nunn, 9 Elmhurst Court, Hamblin Road, Woodbridge, IP12 1HB |
| G8 | AXR | R Moore, 22 Cardan Drive, Ilkley, LS29 8PH |
| G8 | AXV | K Shail, Veeda Glenta, Blackmore Park Road, Malvern, WR13 6NN |
| G8 | AYC | N Walker, 36 Meyrick Drive, Wash Common, Newbury, RG14 6SX |
| G8 | AYJ | J Hanson, 22 Church Way, Falmouth, TR11 4SG |
| G8 | AYM | N Pritchard, 108 Kynaston Avenue, Aylesbury, HP21 9DS |
| G8 | AYV | Julian Lewis, Newnham House, Shurton, Bridgwater, TA5 1QG |
| G8 | AYY | P Gaskin, 58 Elmcroft Road, Yardley, Birmingham, B26 1PL |
| G8 | AZA | J Agar, 291 Overdale, Eastfield, Scarborough, YO11 3RE |
| G8 | AZB | S Smith, 11 Grayshott Laurels, Lindford, Bordon, GU35 0QB |
| G8 | AZM | D Johnson, 195 Staplers Road, Newport, PO30 2DP |
| G8 | AZN | R Barnes, 18 Battle Road, Tewkesbury, GL20 5TZ |
| G8 | AZR | J Dimmock, 93 Barton Road, Harlington, Dunstable, LU5 6LG |
| G8 | AZT | J Jones, 9 Queens Walk, Thornbury, Bristol, BS35 1SR |
| G8 | BAD | D Donati, 53 Smithbarn, Horsham, RH13 6DT |
| G8 | BAG | G Rowley, 7 Hall Farm Close, Castle Donington, Derby, DE74 2NG |
| G8 | BAJ | Philip Southby, 51 Teddington Park, Teddington, TW11 8DE |
| G8 | BAK | P Knight, 4 Dimmock Road, Wootton, Bedford, MK43 9DW |
| G8 | BAQ | Brian Kneller, Mystic Flight, Brackenhill Road, [illegible], Doncaster, DN9 2LR |
| G8 | BAS | D Gardiner, 31 Alexander Drive, Cirencester, GL7 1UG |
| G8 | BAV | J Bosworth, 57 Livingstone Road, Derby, DE23 6PS |
| G8 | BAY | Kenneth Arnold, 22 Silverbirch Court, Friends Avenue, Waltham Cross, EN8 8LZ |
| G8 | BAZ | P Talbot, 19 Bladen Valley, Briantspuddle, Dorchester, DT2 7HP |
| G8 | BBC | ARIEL RADIO GRP c/o Jonathan Kempster, 25 Andersens Wharf, Copenhagen Place, London, E14 7DX |
| G8 | BBK | R Nelson, 10 Wragg Drive, Newmarket, CB8 7SD |
| G8 | BBV | J Goulty, 1 Larksway, Felixstowe, IP11 2PN |
| G8 | BBZ | P Barker, 3 Hudson Fold, Heptonstall, Hebden Bridge, HX7 7PH |
| G8 | BCA | R Chambers, 11 Thetford Road, Mildenhall, Bury St. Edmunds, IP28 7HX |
| G8 | BCF | G Podmore, Crownfield, Kings Lane, Faringdon, SN7 7SS |
| G8 | BCG | Peter Taylor, The Byre Coombe Farm, St. Keyne, Liskeard, PL14 4HS |
| G8 | BCI | Eric Rowlands, Wychanger Cottage, Luccombe, Minehead, TA24 8TA |
| G8 | BCJ | A Unsworth, Meadow View, Clockhouse Lane, Grays, RM16 5UR |
| G8 | BCL | Howard Bottomley, Nerefield, Aylesbury Road, Aylesbury, HP18 0BL |
| G8 | BCO | Cae Boys, 34 Firacre Road, Ash Vale, Aldershot, GU12 5JT |
| G8 | BDF | J Hanney, 16 Parsonage Barn Lane, Ringwood, BH24 1PX |
| G8 | BDM | Jeremy Adams, 1 Powell Close, Creech St. Michael, Taunton, TA3 5TE |
| G8 | BDQ | G Hedley, 260e 100 Sts, Raymond, Alberta, Canada, T0K 2S0 |
| G8 | BDU | Duncan Fisken, Leycroft, Welshmill Road, Frome, BA11 2LA |



**IMPORTANT NOTE**

**Revalidate licence to avoid revocation** – Ofcom has advised the Society that plans will be drawn up to revoke licences that have not been revalidated as required by the licence conditions. The quickest way to revalidate is to do so online via the Ofcom website: *https://services.ofcom.org.uk/* or by email: *amateur.validations@ofcom.org.uk* Ofcom staff are available to help, but please be patient during times of heavy workload.

G8 BDX Alex Scott, 20 Treaty Park, Birgham, Coldstream, TD12 4NG
G8 BDZ Kevin Cowdell, 6 Pearl Street, Bristol, BS3 3EA
G8 BEH Douglas Hill, 5 The Kempsters, Trimley St. Mary, Felixstowe, IP11 0XR
G8 BEK C Dunn, 75 Waddington Avenue, Burnley, BB10 4LA
G8 BEQ K Greenough, 2 Bexley Close, Glossop, SK13 7BG
G8 BFA S Davis, 21 Cordville Close, Chaddesden, Derby, DE21 6WX
G8 BFC P Johnson, 15 Elvaston Lane, Alvaston, Derby, DE24 0PX
G8 BFH J Marriott, 104 Whinbush Road, Hitchin, SG5 1PN
G8 BFK S Ballard, 26 Crafts End, Chilton, Didcot, OX11 0SA
G8 BFL B Jayne, 38 Townfields, Lichfield, WS13 8AA
G8 BFM A Whittaker, 6 Kingsbridge Way, Bramcote, Nottingham, NG9 3LW
G8 BFO Roger Hayter, Glanyrafon, Talywern, Machynlleth, SY20 8NY
G8 BFV David Edwards, 34 Campkin Road, Wells, BA5 2DG
G8 BGI B Hepburn, 52 Hibiscus Grove, Bordon, GU35 0XA
G8 BGL R Gilliatt, 21 Main St., Thorpe On The Hill, Lincoln, LN6 9BG
G8 BGM Michael Lee, 32 Fernham Road, Faringdon, SN7 7LB
G8 BGT A Dermont, 7 Pool Close, Little Comberton, Pershore, WR10 3EL
G8 BGV P Selwood, 43 Keene Way, Galleywood, Chelmsford, CM2 8NT
G8 BHC John Richmond-Hardy, 45 Burnt House Lane, Kirton, Ipswich, IP10 0PZ
G8 BHE Norman Gutteridge, 68 Max Road, Quinton, Birmingham, B32 1LB
G8 BHH Russel Stewart, 424 Wood End Road, Wolverhampton, WV11 1YD
G8 BHK John Vickers, 242b High Road, Trimley St. Martin, Felixstowe, IP11 0RG
GM8 BHR Geoffrey Pearson, 2 Hamilton Terrace, Edinburgh, EH15 1NB
G8 BHX M Berry, 27 Greenway Road, Heald Green, Cheadle, SK8 3NR
G8 BHY A Heath, 7 Coral Close, Coventry, CV5 7AD
G8 BIA F Hopwood, 1 Trem Y Mynydd, Abergele, LL22 9YY
G8 BIG M Stebbings, 15 St. Helena Way, Horsford, Norwich, NR10 3EA
G8 BIH J Akam, 10 Apple Tree Road, Alderholt, Fordingbridge, SP6 3EW
G8 BII Bryan Hunt, 53 The Sands, Milton-under-Wychwood, Chipping Norton, OX7 6ER
G8 BIJ J Batten, 23 Lerowe Road, Wisbech, PE13 3QH
G8 BIR R Harris, 35 Freemantle Road, Eastville, Bristol, BS5 6SY
G8 BIS Paul Lyon, Frogs Hall, Cannon Street, New Romney, TN28 8BJ
G8 BIW R Booth, 16 Darwynn Avenue, Swinton, Mexborough, S64 8DU
G8 BIX A Parcell, Birdies Barn, Minions, Liskeard, PL14 5LE
G8 BJA D Couchy, 8 Chapel St., Wincham, Northwich, CW9 6DA
G8 BJB G King, 62 Heathfield Road, Sholing, Southampton, SO19 1DP
GM8 BJF B Flynn, 15 Riselaw Crescent, Edinburgh, EH10 6HN
GM8 BJJ A Morton, 4 Mountstuart Street, Millport, KA28 0DP
G8 BJO James Barfoot, 21 Richard Crampton Road, Beccles, NR34 9HN
G8 BJQ L Case, 58 Brookdale, Widnes, WA8 4TB
G8 BKD Paul Scotney, 30 Trinity Road, Rothwell, Kettering, NN14 6HY
G8 BKE C Towns, 21 Seafield Close, Barton on Sea, New Milton, BH25 7HR
G8 BKG David Wright, 61 Potton Road, St. Neots, PE19 2NN
G8 BKH George Shepherd, 64 Dawley Road, Arleston, Telford, TF1 2JF
G8 BKL Eric Danks, 18 Lichfield Street, Stourport-on-Severn, DY13 9EU
G8 BKQ Charles Clark, 21a Headland Park Road, Paignton, TQ3 2EN
G8 BLB P Blakeney, 45 Hampden Avenue, Chesham, HP5 2HL
G8 BLD John Draper, 31 Skelton Road, Diss, IP22 4PW
G8 BLK Michael Keightley, 20 Longrood Road, Rugby, CV22 7RG
G8 BLP C Bond, 5 Rushley Close, Sheffield, S17 3EG
G8 BME F Burrow, 51 Stanhope Avenue, Morecambe, LA3 3AJ
G8 BMG D Platt, 14 Dorset Place, Newcastle, ST5 3DG
G8 BMH John Parry, 29 Heath Road, Upton, Chester, CH2 1HT
G8 BMI G Theasby, 115 Bevercotes Road, Sheffield, S5 6HB
G8 BMP M Taylor, 96 Woodhouses Road, Burntwood, WS7 9EJ
G8 BMQ B Cedar, 29 Velsheda Court, Hythe Marina Village, Southampton, SO45 6DW
G8 BMZ P Cowling, 94 Welholme Road, Grimsby, DN32 0NG
G8 BNB R Gibbs, 15 Gosford Hill Court, Bicester Road, Kidlington, OX5 2XP
GI8 BNC J Mccann, 61 Glengawna Road, Glengawna, Omagh, BT79 7WJ
G8 BNE R Kendall, Random Stones, Arkendale Road, Knaresborough, HG5 0QA
G8 BNG A Green, 37 Bramcote Lane, Nottingham, NG8 2NA
GM8 BNH I Gall, Cluaran, Bridge of Don, Aberdeen, AB23 8BD
G8 BNK P Banbury, 16 Gloucester Road, Whitstable, CT5 2DS
G8 BNR R Wells, 279 Hatfield Road, St. Albans, AL4 0DH
G8 BOB Albert Robinson, 29 Thomas Manning Road, Diss, IP22 4HL
G8 BOI M Simpson, 9 Brock House, 2 Batter Street, Plymouth, PL4 0EF
G8 BOJ K Agombar, 54 Julien Road, London, W5 4XA
G8 BOP M Palmer, 109 Longfellow Road, Dudley, DY3 3EF
G8 BOQ Kenneth Phillips, 1140 RIVERBERRY DRIVE, Reno, United States, NV 89509
G8 BPH J Rome, 1 Bridge Cottages, Matching Road, Bishop's Stortford, CM22 7AS
G8 BPN G Wilkerson, Hill House, Newton, Leominster, HR6 0PF
G8 BPQ J Wiseman, 147 Hilton Road, Nottingham, NG3 6AR
G8 BPS Chris Booth, 11 High St., Haxey, Doncaster, DN9 2HX
G8 BPU Harrold Skelhorn, 9 Moss Lane, Bollington, Macclesfield, SK10 5HJ
G8 BPW Anthony Stoker, 35a Church End Lane, Runwell, Wickford, SS11 7JE
G8 BPY P Hollis, 5 Salisbury Road, New Malden, KT3 3HZ
G8 BQF Alan Dixon, 2 Yorkdale Drive, Hambleton, Selby, YO8 9YB
G8 BQH Michael Marsden, Hunters Moon, Buckingham Road, Aylesbury, HP22 4EF
G8 BQK Geoffrey Oatway, 21 Victoria Park, Colwyn Bay, LL29 7AX
G8 BQT I Hudson, Flat 32, Three Crowns House, King's Lynn, PE30 5DT
G8 BQZ P Plunkett, 30 Broadlands Avenue, Shepperton, TW17 9DQ
G8 BRD Christopher Dawson, 33 Rough Common Road, Rough Common, Canterbury, CT2 9DL
G8 BRF Alan Hirst, 11 Yew Tree Lane, Poynton, Stockport, SK12 1PU
G8 BRG Peter Mitchell, 3 Goodwin Court, Farnsfield, Newark, NG22 8LU
G8 BRK D Geldart, 92 Rockingham Street, Barnsley, S71 1JR
G8 BRL B Ward, 10 Upper Moorfield Road, Woodbridge, IP12 4JW
G8 BRU G Gallamore, 30 Orchard Avenue, Partington, Manchester, M31 4DL
G8 BSD J Ceresole, 7 Stokes Bay Home Park, Stokes Bay Road, Gosport, PO12 2QU
G8 BSP A Wicks, 1 Castle Hill Close, Shaftesbury, SP7 8LQ
GM8 BSQ A Shepherd, 2 Westwood Place, Skene, Westhill, AB32 6WS
GM8 BSU Anthony Weller, 18 Froghall Road, Aberdeen, AB24 3JL
G8 BTC B Fenwick, 16 Pine Walk, Uckfield, TN22 1TU
G8 BTD Peter Sladen, 2 Burlea Close, Crewe, CW2 8SZ
G8 BTL H Futcher, Sarum, 12 Thursby Road, Woking, GU21 3NZ
G8 BTU J Dowson, The Granary, St. Peters Road, Leicester, LE8 5WJ
G8 BTV P Marlow, 1 Vineries Close, Leckhampton, Cheltenham, GL53 0NU
G8 BTX Trevor Storeton-West, 8 Cullcott Close, Yoxford, Saxmundham, IP17 3GZ
G8 BTY M Dennis, Thistledown, Yallands Hill, Taunton, TA2 8NA
G8 BUB B Goodall, 10 Westoby Close, Shepshed, Loughborough, LE12 9SS
GD8 BUE Ian Rae, 65 Lezayre Park, Ramsey, Ramsey, Isle of Man, IM8 2PT
G8 BUF Michael Higgins, 59 Clinton Crescent, Ilford, IG6 3AH
G8 BUI Conrad Nowikow, 5 Park Hill, Harlow, CM17 0AE
G8 BUV C Chapman, 6 Pickhurst Green, Hayes, Bromley, BR2 7QT
G8 BUX BUXTON RADIO AMATEURS c/o D CARSON, 21 Harris Road, Harpur Hill, Buxton, SK17 9JS
G8 BUZ John Paine, 1 Elm Close, London, SW20 9HX
G8 BVB P Power, 8 The Fairway, Camberley, GU15 1EF
G8 BVF Jonathan Wearing, 122 Dixon Drive, Chelford, Macclesfield, SK11 9BX
G8 BVL M Porter, Birklands, 16 The Oval, Scarborough, YO11 3AP
G8 BVQ Richard Straker, 26 Constance Crescent, Hayes, Bromley, BR2 7QJ
G8 BVR G Oddy, 2 Manor Farm, Chard, TA20 2EB
G8 BVU Philip Reilly, 21 Russell Crescent, Nottingham, NG8 2BQ
G8 BVY G Spinks, 40 Ferndale Avenue, Walthamstow, London, E17 9EH
G8 BWA M Pollard, 3 Highfield Road, Chertsey, KT16 8BU
G8 BWH R Robinson, 1 John Dixon Lane, Darlington, DL1 1HG
G8 BWP C Jones, 2 Windmill Crescent, Wolverhampton, WV3 8HY
GW8 BWX Alexander Hancock, 38 High Street, Pontycymer, Bridgend, CF32 8HY
G8 BXA Adrian Nicol, 18 Lower End, Swaffham Prior, Cambridge, CB25 0HT
G8 BXC Richard Clark, 41 Avenue Road, Bexleyheath, DA7 4EP
G8 BXD Ray Edgecombe, 48 Birchwood Road, Woolaston, Lydney, GL15 6PE
G8 BXH J Pryke, 52 Oaklands Avenue, Watford, WD19 4LW
G8 BXJ Alan Pullen, 22700 Gault Street, WEST HILLS, Ca, United States, 91307-2306
G8 BXM P Shield, 56 Station Road, Tempsford, Sandy, SG19 2AX
G8 BXO John Stacey, 3 West Park, South Molton, EX36 4HJ
G8 BXQ T Hordley, 9 Newtown, Charlton Marshall, Blandford Forum, DT11 9NN
G8 BYB A Hebden, Reedecraft, Mill Green Road, Spalding, PE11 3PU
G8 BYC Charles Keen, Brighton Road, Radio Relay, Lewes, BN7 3JL
G8 BYI R Burrows, 76 Southfield, Southwick, Trowbridge, BA14 9PW
G8 BZJ A Matheson, 1 St. Edmunds Close, Bromeswell, Woodbridge, IP12 2PL
G8 BZL Graham Lindsay, 71 Woodland Avenue, Hove, BN3 6BJ
G8 BZN D Goadby, Ty Mawr, Bryncroes, Pwllheli, LL53 8EH
GM8 BZP David Joiner, 8 Damask Crescent, Newmachar, Aberdeen, AB21 0NG
G8 BZR Peter Clark, 10 Chez Gueunie, St. Leger Magnazeix, France, 87190
G8 BZT D Allen, 156 Middlecotes, Tile Hill, Coventry, CV4 9AZ
G8 CA AXE VALE ARC c/o P Cross, Balls Farm Cottage, Musbury Road, Axminster, EX13 8TT
G8 CAA C Broomfield, 8 Woodview Crescent, Hildenborough, Tonbridge, TN11 9HD
GD8 CAB J Sawford, 68 Harlyn Drive, Pinner, HA5 2DA
G8 CAH A Parsons, 153 Denman Drive, Ashford, TW15 2AP
G8 CAK P Kenyon, The Elvins, Norton, Presteigne, LD8 2EP
G8 CAM I Foster, 22 Margetts Place, Lower Upnor, Rochester, ME2 4XF
G8 CAU J Borradaile, 25 Inglewood Crescent, Carlisle, CA2 6JJ
G8 CBA Gordon Tipler, Scotts House, Chorley, Bridgnorth, WV16 6PR
G8 CBB C Barnes, 44 Wheatley Drive, North Wootton, King's Lynn, PE30 3QQ
G8 CBE K Quarman, 127 Highfield Lane, Hemel Hempstead, HP2 5JG
G8 CBO K Smith, 6 Hermitage Close, North Mundham, Chichester, PO20 1JZ
G8 CBU R Aldous, 23 Aldhous Close, Luton, LU3 2LZ
G8 CCD John Hodge, 71 Rawcliffe Road, Walton, Liverpool, L9 1AN
G8 CCF Stephen Hall, Knackershole Barn, Dulverton, TA22 9RU
G8 CCJ D Petri, 42 Lucas Road, Snodland, ME6 5PY
G8 CCL Jonathan White, 22 Millfields, Station Road, Burnham-on-Crouch, CM0 8HS
G8 CCN Rex Read, 76 School Road, Downham, Billericay, CM11 1QN
G8 CCO J Hess, 3 Havana Court, Eastbourne, BN23 5UH
G8 CCQ E Peel, Chucks Corner, Deans Lane, Tadworth, KT20 7UD
G8 CCV M O'Donnell, 40 Mercers Drive, Bradville, Milton Keynes, MK13 7AY
G8 CDA Markham Richards, Copperknobs, High Street, Stockbridge, SO20 6HE
G8 CDB P Strudwick, 40 Fifth Avenue, Chelmsford, CM1 4HD
G8 CDC Edward Peter Jones, Tudor House, Stoneleigh Road, Leamington Spa, CV32 6QR
G8 CDD Richard Leman, Crundalls Farmhouse, Gedges Hill, Tonbridge, TN12 7EA
G8 CDG N Broadbent, 2 Market Hill, Clare, Sudbury, CO10 8NN
G8 CDV Terence Jeacock, 9 Parkwood Rise Barnby Dun, Doncaster, DN3 1LY
GM8 CEA Richard Spencer, Pitagown House, Cluny, Newtonmore, PH20 1BS
G8 CEP D Clough, 165 Pilgrims Way, Andover, SP10 5HT
G8 CET W Marsden, 163 Buxton Old Road, Disley, Stockport, SK12 2AY
G8 CEX B Turner, 50 Bosworth Road, Leigh-on-Sea, SS9 5AB
GJ8 CEY A Hearne, 2 Teighmore Park, La Chevre Rue, Grouville, Jersey, JE3 9EF
G8 CEZ R Fuller, 35 Chichester Walk, Merley, Wimborne, BH21 1SL
G8 CFD Robert Rimmer, 6 The Dene, Blackburn, BB2 7QS
GM8 CFS D Slight, Wychwood, Springhill Road, Peebles, EH45 9ER
G8 CGM P Raybould, 115 Curlew Crescent, Bedford, MK41 7HY
G8 CGW J Elliott, 92 Hinckley Road, Barwell, Leicester, LE9 8DN
G8 CHA N Blackburn, 158 Dyas Road, Great Barr, Birmingham, B44 8SW
G8 CHC Brian King, 32 Mayfield, Buckden, St. Neots, PE19 5SZ
G8 CHI Arnold Tidder, 3 Fernway Close, Wimborne, BH21 2ST
G8 CHK Royston King, 28 Jenkinson Road, Towcester, NN12 6AW
G8 CHN G Barber, 666 Bradford Road, Birkenshaw, Bradford, BD11 2EE
G8 CHO Stanley Humm, 235 Felmongers, Harlow, CM20 3DP
G8 CHY K Twort, 39 Mile End Lane, Stockport, SK2 6BN
GM8 CIF D Macdonald, 22 Drummie Road, Devonside, Tillicoultry, FK13 6HT
G8 CIG Philip Tester, Gable Crest, Longburton, Sherborne, DT9 5PD
G8 CIJ F Fyfe, 28 Whitton Close, Greatworth, Banbury, OX17 2EH
G8 CIT W McKillop, 2 Moores Green, Wokingham, RG40 1QG
G8 CIX Martin Maynard, 41 Liverpool Avenue, The Pyramid, Southport, PR8 3NP
G8 CJA M Dowson, The Granary, St. Peters Road, Leicester, LE8 5WJ
G8 CJD C Hutton, 25 Fiddlers Lane, East Bergholt, Colchester, CO7 6SJ
G8 CJG Robert Kirsch, Milntack House, Laurieston, Castle Douglas, DG7 2PW
G8 CJH D Fletcher, 17 Durley Chine Road South, Bournemouth, BH2 5JT
G8 CJL A Dorling, 4 The Pastures, Rushmere St. Andrew, Ipswich, IP4 5UQ
G8 CJM A Croft, 15 Blenheim Avenue, Chatham, ME4 6UU
G8 CJQ Robert Barnes, 3 Ivy Cottages, Church Lane, Knutsford, WA16 7RD
G8 CJT C Coles, 15 Somerdale Avenue, Bath, BA2 2PG
G8 CJW James West Of Stow, Stow Mill, Stow, Galashiels, TD1 2RB
G8 CKB Peter Ebsworth, Olamyra 20, Forland, Steinsland, Norway, 5379
G8 CKJ A Williams, 54 St. Augustine Road, Griffithstown, Pontypool, NP4 5EZ
G8 CKK A Zerafa, 2 Furnwood, St. George, Bristol, BS5 8ST
G8 CKN R Powers, The Dell, Hussell Lane, Alton, GU34 5PF
G8 CKS J Sargent, The Coach House, Speltham Hill, Waterlooville, PO7 4RU
G8 CKV S Dale, 30 Almond Road, Peterborough, PE1 4LT
G8 CLJ I Richmond, 48 Broadstone Road, Harpenden, AL5 1RF
G8 CLK K Woollven, 7 Heatherstone Avenue, Dibden Purlieu, Southampton, SO45 4LR
G8 CLW John Griffin, 185 Eastcote Avenue, West Molesey, KT8 2EX
G8 CLY J Lythgoe, 18 Ranleigh Walk, Harpenden, AL5 1SR
G8 CLZ QRZ AMATEUR RADIO GROUP OF SUSSEX c/o J Eade, White Cottage, Whatlington, Battle, TN33 0NL
G8 CMD A Ashford, 56 Guarlford Road, Malvern, WR14 3QP
G8 CME Mary Crawshaw, 50 Kibble Grove, Brierfield, Nelson, BB9 5EW
G8 CMG R Williams, 18 Woodford Crescent, Plymouth, PL7 4QY
G8 CMK William Blankley, 16 Charles Road, St. Leonards-on-Sea, TN38 0QA
G8 CMO R Grounds, 101 Honeysuckle Way, Witham, CM8 2XQ
G8 CMP Colin Heymans, Chez Heymans, Vernantes, France, 49390
G8 CMU Michael Adcock, Phocle Green, Ross-on-Wye, HR9 7TL
G8 CNF S Biddiscombe, 20 Arlington Close, Malpas, Newport, NP20 6QF
GW8 CNS W Mathias, Grenan Bungalow, Highland Avenue, Bridgend, CF32 9YH
G8 CON Janet Beith, 18 Avenue Road, New Milton, BH25 5JP
G8 COR G Peters, 156 Preston Road, Whittle-le-Woods, Chorley, PR6 7HE
G8 CPA J Vizor, 31 Somerset Road, Swindon, SN2 1NE
G8 CPB Jan Kozminski, Heronsforde, Park View Road, Caterham, CR3 7DL
G8 CPF Michael Edwards, 1 Heron Close, Minehead, TA24 6UL
G8 CPJ I Lever, 23 Anton Road, Andover, SP10 2EN
G8 CPK David Hibbin, 95a Thorpe Acre Road, Loughborough, LE11 4LF
G8 CPN J Hawkins, Westhay Farm, Higher Clovelly, Bideford, EX39 5SH
G8 CPQ V Humphrey, 5 Wistow Road, Luton, LU3 2UR
G8 CPZ A Barth, 83 London Road, Aston Clinton, Aylesbury, HP22 5LD
G8 CQG P Cornell, 22 Ravine Road, Bournemouth, BH5 2DU
G8 CQH Peter Best, 21 Greening Drive, Edgbaston, Birmingham, B15 2XA
G8 CQQ A Paterson, 36 Bracadale Road, Nottingham, NG5 5EE
G8 CQR R Lee, 13 Haley Close, Exmouth, EX8 4PJ
G8 CQV W Hunter, 2 Green Acre, Goosnargh, Preston, PR3 2BQ
G8 CQX John Hawes, Green Trees, 193 Leckhampton Road, Cheltenham, GL53 0AD
G8 CQZ Clifford Powlesland, The Ferns, Broad Street, Gloucester, GL19 3BN
G8 CRB Stephen Blunt, 53 Butt Lane, Milton, Cambridge, CB24 6DG
G8 CRC Clifford Callegari, 16 Rustington Court, St. Johns Road, Eastbourne, BN20 7HS
G8 CRH Ian Troughton, Rhiwbina, Pentre Lane, Cwmbran, NP44 3AP
G8 CRM P Watson, Tall Oak, 6 New Road, Bury St. Edmunds, IP29 5QL
G8 CRV John Christian, 5 Towers Way, Corfe Mullen, Wimborne, BH21 3UA
G8 CRX S Winford, Mayflower, South Hanningfield Road, Chelmsford, CM3 8HJ
G8 CRZ P Hunt, 17 Selfridge Avenue, Southbourne, Bournemouth, BH6 4NB
GM8 CSE H Hogarth, 32 Broomhall Park, Edinburgh, EH12 7PU
G8 CSK S Browning, 12 Sunderland Close, Woodley, Reading, RG5 4XR
G8 CSQ P Benson, Ashbank Bungalow, Bentham, Lancaster, LA2 7HX
G8 CSR John Credland, Lieu-dit Cornier, Prayssas, France, 47360
G8 CSY P Thompson, 26 Carleton Avenue, Blackpool, FY3 7JN
G8 CTB K Chambers, 24 Primrose Close, Flitwick, Bedford, MK45 1PJ
G8 CTD A Tait, Birch Glen, 71 Twemlows Avenue, Whitchurch, SY13 2HD
G8 CTJ M Maxey, 28 Herald Way, Burbage, Hinckley, LE10 2NX
G8 CTR David Upton, Polwin, Budock Water, Falmouth, TR11 5DT
G8 CTX C Havercroft, 28 Anglers Way, Cambridge, CB4 1TZ
G8 CUA R Boittier, 5 The Crescent, Harlow, CM17 0HN
G8 CUB R Ray, Little Mallards, Mallard Way, Brentwood, CM13 2NF
G8 CUG P Cockram, 14 Langshott Close, Woodham, Addlestone, KT15 3SE
G8 CUL M Stevens, 67 New Road, East Hagbourne, Didcot, OX11 9JX
G8 CUN G Rawlings, 109 The Upway, Basildon, SS14 2JD
G8 CUX Denis Stanton, 122 Foxon Lane, Caterham, CR3 5SD
G8 CVF J Dobson, 11a Glenburn Avenue, Eastham, Wirral, CH62 8DJ
GM8 CVN J Struthers, 79 Woodfield Park, Colinton, Edinburgh, EH13 0RA
G8 CVP R Perry, 49 Harwich Road, Little Clacton, Clacton-on-Sea, CO16 9NE
G8 CVQ Andrew Parr, 8 Kingston Avenue, North Cheam, Sutton, SM3 9TZ
G8 CVS J Jenkinson, 16 Leybourne Close, Walderslade, Chatham, ME5 9JN
G8 CVV B Chuter, 27 Tas Combe Way, Eastbourne, BN20 9JA
G8 CWE Terrence Cook, 141 Station Road, Watlington, King's Lynn, PE33 0JG
G8 CWJ J Abbott, 20 Highbury Avenue, Salisbury, SP2 7EX
G8 CWQ G Horsfall, Lancaster New Road, Garstang, Preston, PR3 1AD
G8 CXA David Froggatt, 2 Cobden Avenue, Mexborough, S64 0AD
G8 CXF J Lucas, 48 Sycamore Drive, Ash Vale, Aldershot, GU12 5PR
G8 CXI David Phillips, 13 Bowford Avenue, Bexleyheath, DA7 4ST
G8 CXK Gerald Peck, 45 Bentley Close, Northampton, NN3 5JS
G8 CXT David Coxhill, 19 Long Street Road, Hanslope, Milton Keynes, MK19 7BL
G8 CXV R Brown, 19c Arlington Drive, Mapperley Park, Nottingham, NG3 5EN
G8 CXW P Appleby, 23 Oban Drive, Ashton-in-Makerfield, Wigan, WN4 0SJ
G8 CXZ M Mills, 145 Park St., Haydock, St. Helens, WA11 0BL
G8 CYA N Parker, 10 Lockhart Close, Kenilworth, CV8 1RB
G8 CYE Stephen Cook, 24 Beaufort Court, Beaufort Road, Richmond, TW10 7YG
G8 CYG W Steer, Downside Membury, Axminster, EX13 7AF
G8 CYK William Poel, Hockham Hill, Spring Elms Lane, Chelmsford, CM3 4SD
G8 CYL P Smith, 61 Waverley Drive, Chertsey, KT16 9PF
G8 CYT F White, 12 Burcombe Road, Kinson, Bournemouth, BH10 5JT
G8 CYU P York-Jones, 18 Solway Road, Cheltenham, GL51 0LZ
G8 CYW Stuart Wisher, 17 Kenmore Crescent, Greenside, Ryton, NE40 4QY
G8 CYX D Storey, 43 Harwood Close, Welwyn Garden City, AL8 7ST
G8 CZE Frank Beesley, 9 Northway, Droylsden, Manchester, M43 6EF
G8 CZG David Bell, 196 Whalley Road, Langho, Blackburn, BB6 8AA
G8 CZI D Paterson, 3 Shawcroft Close, Shaw, Oldham, OL2 7DA
G8 CZJ Josie Meredith, 25 Frankel Avenue, Redhouse, Swindon, SN25 2NJ

G8 CZM K Jones, 3 Webb Avenue, Perton, Wolverhampton, WV6 7YH
G8 CZP V Maund, 73 Norwich Road, Barham, Ipswich, IP6 0DH
G8 CZQ Ian [illegible], West Common Lodge, West Common Close, Gerrards Cross, SL9 7QR
GM8 CZU I Davidson, 3 Hillcrest Avenue, Kirkcaldy, KY2 5TU
GM8 DAB Brian Smith, 14 High Shore, Macduff, AB44 1SL
GD8 DAI Alexander Justin, Garth, Park View Road, Pinnor, HA5 3YF
G8 DAM D Goodway, 35 South Avenue, Buxton, SK17 6NQ
G8 DBH C Wallwork, Honeywicke Cottage, Honeywick Lane, Dunstable, LU6 2BJ
G8 DBK Paul Barker, 24 Main Street, South Croxton, Leicester, LE7 3RJ
G8 DBO Kim Smith, Wilson Hall Farm, Slade Lane, Wilson, Derby, DE73 8AG
G8 DBP J Mills, 93 Gays Road, Hanham, Bristol, BS15 3JX
G8 DBU N Greensted, High View Oust Care Home, Poulton Lane, Canterbury, CT3 2NH
G8 DCD J Durrant, 27 Trafford Road, Willerby, Hull, HU10 6AJ
G8 DCJ P McQuail, 3 Post Office Lane, Draycott, Moreton-in-Marsh, GL56 9JZ
G8 DCX Ron Sangster, 10 Addison Road, Banbury, OX16 9DH
G8 DD SOUTH NOTTS ARC c/o David Hill, 86 The Downs, Nottingham, NG11 7EB
G8 DDC DUNSTABLE DWN RD c/o C Asquith, 36 Sunningdale, Luton, LU2 7TE
G8 DDH Mark Lelliott, Well Lane Corner, Lower Froyle, Alton, GU34 4LJ
G8 DDN Philip Bennett, Whitelands, Common Mead Lane, Gillingham, SP8 4RB
G8 DDY P Thompson, Albury, Downside Avenue, Ventnor, PO38 2DE
G8 DEC A Malcolm, 68 Old Birmingham Road, Lickey End, Bromsgrove, B60 1DG
G8 DEJ Thomas Ray, 1 Providence Lane, Leamore, Walsall, WS3 2AQ
G8 DEL D Coppen, 100 Atbara Road, Teddington, TW11 9PD
G8 DEM B Willetts, 11 Albert Road, Warley, Oldbury, B68 0NA
G8 DER R Richardson, Radbourne Cottage Farm, Upper Radbourne, Southam, CV47 1NQ
G8 DET John Bowen, 6 Bishops Court Gardens, Chelmsford, CM2 6AZ
G8 DEX J Hosking, 21 Yeo Valley Way, Wraxall, Bristol, BS48 1PS
G8 DEY D Parr, 58 Ritson St., Toxteth, Liverpool, L8 0UF
GM8 DFC R Cliff, 32 Lochardil Road, Inverness, IV2 4LD
G8 DFI B Oliver, 6 Catherton Road, Cleobury Mortimer, Kidderminster, DY14 8EB
G8 DFU T Lovelock, Edificio Mibemor Apto 12 A, Carretera Espana 28, Santa Ursula, Islas Canarias, Spain, 38390
GM8 DFX J Lincoln, 59 Obsdale Park, Alness, IV17 0TR
GI8 DGB B Moore, 34a Feumore Road, Ballinderry Upper, Lisburn, BT28 2LH
G8 DGC Stephen Hall, 3 Sleepers Delle Gardens, Winchester, SO22 4NU
G8 DGH E Townsend, The Manor House, Leicester, LE8 0AP
G8 DGR R Smallwood, The Island, Hyde End Lane, Reading, RG7 4TH
G8 DGW M Wickham, 43 Bishopstone, Aylesbury, HP17 8SH
G8 DHA D Bishop, Oyston Lodge, Lynstone Road, Bude, EX23 8LR
G8 DHE Geoffery Mather, 72 Cranleigh Road, Worthing, BN14 7QW
G8 DHF S Matthews, 213 Hucclecote Road, Gloucester, GL3 3TZ
G8 DHI G Roberts, 56 Horse Shoes Lane, Sheldon, Birmingham, B26 3HY
G8 DHJ C pickering, 28 George V Avenue, Margate, CT9 5QA
G8 DHQ D Digby, 73 Bedford Street, Crewe, CW2 6JB
G8 DHT John Clifford, Dippers Barn, Pool Quay, Welshpool, SY21 9JY
G8 DHU Michael Baxter, 11b The Leys, Roade, Northampton, NN7 2NR
G8 DHV N Eaton, 3 Thirslet Drive, Heybridge, Maldon, CM9 4YN
GI8 DHW John Hendron, 9 Drumahiskey Road, Bendooragh, Ballymoney, BT53 7QL
G8 DIQ T Hall, 7 Sweetlake Cottage, Nobold, Shrewsbury, SY5 8NH
G8 DIR K Walker, 12 Willow Park, Minsterley, Shrewsbury, SY5 0EH
G8 DIU Brian Cannon, 52 Goodhew Close, Yapton, Arundel, BN18 0JA
G8 DIY P Geeson, 109 Folly Road, Mildenhall, Bury St. Edmunds, IP28 7BT
G8 DJF A Dickson, 7 Sandford Gardens, High Wycombe, HP11 1QT
G8 DJL John Renaut, 4 Brune Way, West Parley, Ferndown, BH22 8QG
G8 DJO Michael Adcock, 37 Ashpole Road, Bocking, Braintree, CM7 5LW
G8 DJT Gary Platts, 1 Blacksmiths Court, Kingham, Chipping Norton, OX7 6GE
G8 DJU J Frisby, 66 Clear Crescent, Melbourn, Royston, SG8 6JD
G8 DJW G Membury, 11 York Terrace, Dorchester, DT1 2DP
GM8 DKB E Taynton, 42 Craigmount Park, Edinburgh, EH12 8EE
G8 DKD Charles Weale, 110 Stoney Lane, Kidderminster, DY10 2LU
GM8 DKG Colin Pegrum, 4 Northampton Drive, Glasgow, G12 0LE
G8 DKI David Lucas, The Old Barn, The Street, Malmesbury, SN16 9DL
G8 DKK B Harber, 45 Brandles Road, Letchworth Garden City, SG6 2JA
G8 DKV Martyn Coldicott, The Old Cottage, Church Lane, Ilkeston, DE7 6DE
GD8 DKW M Solomons, 389b Alexandra Avenue, Harrow, HA2 9EF
G8 DLH A Hall, 19 Crewkerne Road, Chard, TA20 1EZ
G8 DLL M Monro, 6 Yew Tree Road, Hayling Island, PO11 0QE
G8 DLP R Baker, Royal Oak House, Crich, Matlock, DE4 5BH
G8 DLT G Baraclough, 1 Foxgloves, 68 Dorchester Road, Poole, BH16 5NS
G8 DLX M Crampton, 55 Gilbert Avenue, Bilton, Rugby, CV22 7BZ
G8 DLZ P Lea, 7 Cressex Road, High Wycombe, HP12 4PG
G8 DML J Hughes, 12 Plough Garth, Kellington, Goole, DN14 0PD
G8 DMT M Caley, 40 Spenser Way, Jaywick, Clacton-on-Sea, CO15 2QT
G8 DMU Anthony Frazer, Keld House, Harrogate, HG2 0PG
G8 DNH James Webber, 18 Azalea Close, Calne, SN11 0QT
G8 DNL K Smith, 19 Westfield Avenue, South Croydon, CR2 9JY
G8 DNP Peter Donoghue, Hillcrest, The Green, Harlow, CM17 0QD
GW8 DOA G Pollard, 3 Carey Walk, Neath, SA10 7DD
G8 DOB Ian Stuart, 87 Hedgrove Park, Cheltenham, GL51 0QZ
G8 DOF P White, 6 Curzon Court, Curzon Street, Chester, CH4 8PA
G8 DOH A Seeds, 114 Beaufort Street, London, SW3 6BU
G8 DOR Andrew Barrett, 38 Haw Lane, Bledlow Ridge, High Wycombe, HP14 4JJ
G8 DOW B Lee, 19 Lizard Head, Littlehampton, BN17 6RY
G8 DOY Roger Elliott, Flat 27, Queen Mother Court, 151 Sellywood Road, Birmingham, B30 1TH
G8 DPE V Brooks, 19 Malham Avenue, Wigan, WN3 5PR
G8 DPH Tom Booth, 155 Oxford Road, Windsor, SL4 5DX
G8 DPQ D Hendon, 2 Ellis Avenue, Onslow Village, Guildford, GU2 7SR
GM8 DPV J Hunting, 77 Califer Road, Forres, IV36 1JB
G8 DPW D Holden, 63 High Street, Queenborough, ME11 5AG
G8 DQD T Taylor, 15 Kennard Road, Bristol, BS15 8AA
G8 DQE Richard Lees, 6 Library Road, Ferndown, BH22 9JP
G8 DQF L Johnston, 9 Tunbridge Close, Burwell, Cambridge, CB25 0EL
G8 DQK Andrew Symonds, 45 Westfield Road, Dereham, NR19 1JB
G8 DQN N Hunter, 33 Chapel Court, Billericay, CM12 9LX
G8 DQP James Peden, 51a Bewdley Road, Kidderminster, DY11 6RL
G8 DQZ Anthony Lord, 2 Hazelmere, Diss Road, Diss, IP22 1NQ
[illegible]
G8 DRE D Atkinson, 54 Egret Crescent, Colchester, CO4 3FP
G8 DRK Robin Vince, 5 Bay Tree Road, Bath, BA1 6NA
G8 DRQ R Cochrane, 134 Moor Lane South, Ravenfield, Rotherham, S65 4QR
G8 DSG W Jones, Elm Hurst, Station Road, Shrewsbury, SY4 2BB
G8 DSM J Witherspoon, 109 Bromsgrove Road, Redditch, B97 4RL
GW8 DSO Christopher Warwick, 33 Ceri Road, Townhill, Swansea, SA1 6LS
G8 DST G Smith, 23 Whaggs Lane, Whickham, Newcastle upon Tyne, NE16 4PF
G8 DSU Robert Gill, 61 Cross Deep Gardens, Twickenham, TW1 4QZ
G8 DTA A Parsons, 20 Paddocks Lane, Prestbury, Cheltenham, GL50 4NX
G8 DTE Malcolm Pusey, 6 Blagdon Close, Martinstown, Dorchester, DT2 9JT
G8 DTF R Price, 29 Birchfield Drive, Worsley, Manchester, M28 1ND
G8 DTM F Partington, 21 East Road, Wymeswold, Loughborough, LE12 6ST
G8 DTQ Bryan Petifer, 14 Wood Lane, Caterham, CR3 5RT
G8 DTS B Norcliffe, 2 Alexander Drive, Heswall, Wirral, CH61 6XT
G8 DTT W Moore, 20 Richard Moon Street, Crewe, CW1 3AX
G8 DTX Ian Sanderson, 15 Gorse Road, Huddersfield, HD3 4BN
G8 DUF R Bird, 129 Park Road, Formby, Liverpool, L37 6AD
G8 DUI D Cox, 52 Avill Crescent, Taunton, TA1 2PL
G8 DUO I Casewell, 7 Pine Drive, Finchampstead, Wokingham, RG40 3LD
GW8 DUP R Harris, 64 Frederick Place, Llansamlet, Swansea, SA7 9SX
G8 DUT H Orgel, 1 Taunton Grove, Whitefield, Manchester, M45 6TJ
G8 DUV C Zammit, 9 Sandbanks Drive, Basingstoke, RG22 4UL
G8 DUW I Redfern, 8 Lilac Grove, Stourport-on-Severn, DY13 8SR
GW8 DUY Christopher Davies, 14 Twynpandy, Pontrhydyfen, Port Talbot, SA12 9TW
G8 DVF T Jones, 5 Blue Hatch, Frodsham, WA6 7QJ
G8 DVJ G Wilks, 8 Chestnut Grove, East Barnet, Barnet, EN4 8PU
G8 DVN D Smith, 3 Woods Lane, Calverton, Nottingham, NG14 6FF
G8 DVS A Sterry, 9 Finch Avenue, Wakefield, WF2 6SE
G8 DVU R West, 55 Burney Bit, Pamber Heath, Tadley, RG26 3TL
G8 DVW Robin Leadbeater, The Birches, Torpenhow, Wigton, CA7 1JF
G8 DWF Nick Earl, 162 Winchmore Hill Road, London, N21 1QP
G8 DWL THE GRAFTON ARS c/o Brian Bond, 86 Agar Grove, Camden Town, London, NW1 9TL
G8 DWP P Lee, 223 Chelmsford Road, Shenfield, Brentwood, CM15 8SA
G8 DWW C Garcia, Hunts Farm, Chapel Hill, Bristol, BS48 3PR
G8 DWX Graham Haslip, 1 Sea Cottages, 28 Steyne Road, Seaford, BN25 1QF
G8 DX J White, 6 Damy Green, Neston, Corsham, SN13 9TN
G8 DXF C Tarran, Woodlands, School Road, Romsey, SO51 6AR
G8 DXH M Powell, 13 The Spinnaker, South Woodham Ferrers, Chelmsford, CM3 5GL
G8 DXI William O'Connor, 3 Sterndale Close, Desborough, Kettering, NN14 2XL
G8 DXM C Taylor, 45 Greenfield Street, Shrewsbury, SY1 2PY
G8 DXO R Humble, 3 Abbey Gardens, Galhampton, Yeovil, BA22 7AG
G8 DXP A Cheasley, 25 Normanhurst Road, Walton-on-Thames, KT12 3EQ
G8 DXT Christopher Beresford, 2 Moulton Close, Swanwick, Alfreton, DE55 1ES
G8 DXU B Pollard-Wilkins, Systems Integration Electronic, 14 Seabeach Lane, Eastbourne, BN22 7NZ
G8 DXV H King, 11 Priory Mead, Doddinghurst, Brentwood, CM15 0NB
G8 DXZ J Sandys, Tarn Cottage, The Maultway, Camberley, GU15 1PS
G8 DYA C West, 14 Ashleigh Gardens, Wymondham, NR18 0EX
G8 DYG M Marshallsay, 2 Prospect Cottages, Lime Street, Gloucester, GL19 4NX
G8 DYI K Holdway, 18 Pennymore Close, Stoke-on-Trent, ST4 8YQ
GW8 DYR D Garner, 34 Caswell Drive, Caswell, Swansea, SA3 4RJ
G8 DYT John Hotchin, 151 Winchester Road, Grantham, NG31 8RX
G8 DZC P Martin, 58 Hearn Road, Woodley, Reading, RG5 3QG
G8 DZH J Ray, 7 Barnmead, Theydon Bois, Epping, CM16 7ET
G8 DZJ G Booth, 68 Tarragon Drive, Meir Heath, Stoke-on-Trent, ST3 7YE
G8 DZN Bob Bird, 7 Old Kingsdown Close, Broadstairs, CT10 2HG
G8 DZW R Brookes, 29 Ripley Road, Liversedge, WF15 6QE
G8 EAD M Hutchings, 109 Longlands Way, Heatherside, Camberley, GU15 1RU
G8 EAH Ian Carress, 1 Riplingham Road, Skidby, Cottingham, HU16 5TR
G8 EAJ P Cannon, Field Cottage, Mathon Road, Malvern, WR13 6ER
G8 EAN J Cunningham, 62 Kings Hill, Beech, Alton, GU34 4AN
G8 EAX S Herod, 8 Deben Way, Felixstowe, IP11 2NS
G8 EBD G Welch, 18 Alderdale, Wolverhampton, WV3 9JF
G8 EBM Stephen Haseldine, 3 Burland Green Lane, Weston Underwood, Ashbourne, DE6 4PF
G8 EBQ R Martin, 10 Westways, Stoneleigh, Epsom, KT19 0PQ
G8 EBT Robert Lees, Hurlands, Hurlands Lane, Godalming, GU8 4NT
G8 EBX P Starling, 14 Merton Place, Littlebury, Saffron Walden, CB11 4TH
G8 ECG K Montgomery, The Old Village Post Office, High Street, Oxford, OX44 9HP
G8 ECI D Brown, 14 Watts Lane, Louth, LN11 9DG
G8 ECR Peter Jago, 39 Royal Avenue, Flat 2, London, SW3 4QE
G8 ECZ Peter Barker, 14 Elsworth Green, Newcastle upon Tyne, NE5 3YB
G8 EDH Peter Clarke, 111 Cambridge Avenue, Gidea Park, Romford, RM2 6QX
G8 EDN Terrence John Gallagher, 35 Wilhelmina Avenue, Coulsdon, CR5 1NL
G8 EDQ Chris Soundy, 16 Crane Cottages, West Cranmore, Shepton Mallet, BA4 4QN
G8 EDS W Hind, 3 Birds Hill, Letchworth Garden City, SG6 1PH
G8 EDX Carlo Vitiello, 1 North Street, Rothersthorpe, Northampton, NN7 3JD
G8 EEA D Hill, 872 Oldham Road, Rochdale, OL11 2BN
G8 EEK B Bruce, Three Ways, Wisbech Road, Wisbech, PE14 9RF
G8 EEM Christopher Gill, 77 Main Road Hambleton, Selby, YO8 9HW
G8 EEY Anthony Mobbs, 149 The Paddocks, Old Catton, Norwich, NR6 7HR
G8 EFK E Carter, 44 Plattes Close, Shaw, Swindon, SN5 5SA
G8 EFU Clive Bloxidge, 33 Rosemary Hill Road, Sutton Coldfield, B74 4HL
G8 EGE John Denton, 32 Highfields Mead, East Hanningfield, Chelmsford, CM3 8XA
G8 EGG D Hemingway, Conygore Farm, Howell Hill, Yeovil, BA22 7QZ
G8 EGL C Burton, 13 Newells Terrace, Misterton, Doncaster, DN10 4DP
G8 EGM M Booth, 16 Falcon Drive Birdwell, Barnsley, S70 5SN
G8 EGU Michael Smith, 35 Queen Street, Balderton, Newark, NG24 3NS
G8 EHD P Brenton, 40 Furneaux Road, Plymouth, PL2 3ET
G8 EHF J Healen, 12 Primrose Lane, Standish, Wigan, WN6 0NR
G8 EHM Eric Vavasour, 15 Mill Lane, Earl Shilton, Leicester, LE9 7AW
GW8 EHQ John Brown, 106 Marlborough Road, Penylan, Cardiff, CF23 5BY
G8 EHS Anthony Fletcher, 35 Wimborne Avenue, Ipswich, IP3 8QW
G8 EHX Michael Melbourne, 32 Lake Farm Road, Rainworth, Mansfield, NG21 0ED
G8 EIE Roger Forster, 7 Western Way, Alverstoke, Gosport, PO12 2NE
G8 EII Mike Smith, 17 Girton Close, Owlsmoor, Sandhurst, GU47 0UP
G8 EIN Nicoll Shepherd, 166 Chaldon Way, Coulsdon, CR5 1DF
G8 EJC Roger Drew, 9 Sona Merg Close, Heamoor, Penzance, TR18 3QL
G8 EJQ Paul Vaughan, 15 Humber Gardens, Wellingborough, NN8 5WE
GM8 EJS J Gilmour, Barnmill, Mosspark Avenue, Glasgow, G62 8NL
G8 EKD Michael Nilson, 9 Middlemead, Folkestone, CT19 5UB
GM8 EKF F Benson, 53 Warriston Drive, Edinburgh, EH3 5NA
G8 EKG Geoffrey Newstead, 97 Hawthorn Crescent, Burton-on-Trent, DE15 9QN
G8 EKH Ewen Mann, 63 St. James Court, Halifax, HX1 1YP
G8 EKN M Biltcliffe, 19 Kennedy Road, Bicester, OX26 2BE
G8 EKW G Thornton, 4 Fir Tree Close, Exmouth, EX8 4EU
G8 EKZ A Jones, 97a Bakers Ground, Stoke Gifford, Bristol, BS34 8GD
G8 ELG E Joyce, 34 Milton Avenue, Eaton Ford, St. Neots, PE19 7LE
G8 ELH D Fisher, 17 Thrushel Close, Swindon, SN25 3PP
G8 ELP Andy Stockley, Blacksole House, The Boulevard, Herne Bay, CT6 6GZ
G8 ELW Richard Straker, 24 Laton Road, Hastings, TN34 2ES
G8 EMA D Pedley, 1 Mount Pleasant Close, Kingsbridge, TQ7 1NR
G8 EMB W Tickell, 26 Shear Brow, Blackburn, BB1 7EX
G8 EMH D Roebuck, 7 Elm Tree Close, North Anston, Sheffield, S25 4FG
G8 EMU J Wheeler, 4 London Road, Tetbury, GL8 8JL
G8 EMX G Hankins, 92 Sunningdale Road, Birmingham, B11 3QJ
G8 EMY Kent Britain, Blenheim Cottage, Falkenham, Ipswich, IP10 0QU
G8 ENA Edward Fellows, 343 Wake Green Road, Birmingham, B13 0BH
G8 ENB Richard Whitby, 138 Browns Lane, Stanton-on-the-Wolds, Nottingham, NG12 5BN
G8 END I Bodie, Seamark, Penpol Devoran, Truro, TR3 6NW
G8 ENS Jean Morris, 6 Barrowby Gate, Grantham, NG31 7LT
G8 ENW P Baker, Top Of The Hill, Post Office Lane, Cheltenham, GL52 3PS
G8 ENY R Hersey, 7 Tower Close, Brandon, IP27 0LJ
G8 EOH G Simpkins, 26 Shrewsbury Street, Hodnet, Market Drayton, TF9 3NP
G8 EOJ E March, 23 Pebworth Close, Redditch, B98 9JX
G8 EOM D Garrard, 48 Shorefields, Benfleet, SS7 5BQ
G8 EOV Bryan Cross, 34 Harewood Road, Doncaster, DN2 6DF
G8 EOZ Keith Waight, 13 Kilda Road, Highworth, Swindon, SN6 7HS
G8 EPC M Dyke, Cortijo Las Marrojas, Buzon 48 Palancar, 1820 Granada, Spain
G8 EPH C Kilvington, 53 hall st.skegby, Sutton in Ashfield, NG17 3EJ
G8 EPK D Skye, 16 Lulworth Avenue, Poole, BH15 4DQ
G8 EPQ R Prew, 16 Stokenchurch Place, Bradwell Common, Milton Keynes, MK13 8AT
G8 EPR D HICKS, 17 Branches Close, Bewdley, DY12 2HD
G8 EPS G Phelan, 113 Albert Road, Epsom, KT17 4EN
G8 EPZ C Ward, 4 The Hawthorns, Charvil, Reading, RG10 9TS
G8 EQB A Vickers, 3 Wingrove Avenue, Sunderland, SR6 9HJ
G8 EQC D Cliffe, Common Farm, Riley Hill, Lichfield, WS13 8JE
G8 EQD David Wright, 22 West Hill, Rotherham, S61 2HB
G8 EQI John Fellows, 8 The Links, Gwernaffield, Mold, CH7 5DZ
G8 EQO B Tyler, 842 Handsworth Road, North Vancouver, Canada, V7R ZA2
G8 EQY Frederick Butler, 511 Fulbridge Road, Peterborough, PE4 6SB
G8 EQZ Clive Reynolds, 49 Westborough Way, Anlaby Common, Hull, HU4 7SW
G8 ERA Michael Voss, 9 Chapel Close, Garndiffaith, Pontypool, NP4 7QS
G8 ERN R Walker, 12 Foldyard Close, Sutton Coldfield, B76 1QZ
G8 ERQ G Hardwick, 13 Wharfedale Place, Harrogate, HG2 0AY
G8 ERV Keith Blackman, 32 French's Gate, Dunstable, LU6 1BQ
G8 ESK B Kermode, 7 Midgeham Grove, Harden, Bingley, BD16 1DA
G8 ESW Walter Brade, 53 Coventry Gardens, Beltinge, Herne Bay, CT6 6SB
G8 ETD T Rumble, Tanatside, Marsh Road, Holbeach Hurn, Spalding, PE12 8JT
G8 ETI N Foggin, 12 Linnetsdene, Covingham, Swindon, SN3 5AG
G8 ETN S LAST, 72 Humber Road, Chelmsford, CM1 7PG
G8 ETP Michael Furnival, 114 Tilt Road, Cobham, KT11 3HQ
G8 ETR Richard Cooke, 4 New Forest Close, Far Forest, Kidderminster, DY14 9TJ
G8 ETS D Swale, 369 Scalby Road, Scarborough, YO12 6TG
G8 ETU A Metcalf, 10 Manor Bend, Galmpton, Brixham, TQ5 0PB
G8 ETV Peter Richardson, 3 Butlers Close, Amersham, HP6 5PY
G8 EUE M Gasper, The Barn, Back Road, Halesworth, IP19 9DZ
G8 EUF C Hall, The Orchard, Arkholme, Carnforth, LA6 1AX
GM8 EUG N Robertson, 10 Warrenpark Road, Largs, KA30 8EF
GD8 EUH D Pickard, Mont y Mer, St. Georges Crescent, Port Erin, Isle of Man, IM9 6HR
G8 EUV C Fenton-Coopland, 14 Chevril Court, Wickersley, Rotherham, S66 2BN
G8 EUX Peter Saul, 51 Windsor Close, Towcester, NN12 6JB
G8 EVD T Cartwright, 132 Mere Road, Wigston, LE18 3RL
G8 EVI Alan Clark, 3 North Street, Owston Ferry, Doncaster, DN9 1RT
G8 EVR Kenneth Taylor, 39 Bowerfield Crescent, Hazel Grove, Stockport, SK7 6JB
G8 EVY Cambridge & District RC c/o John Bonner, 40 Lyles Road, Cottenham, Cambridge, CB24 8QR
G8 EWC A Rouse, Clinton, Church Road, Colchester, CO7 8HS
G8 EWD M Smith, 47 Salisbury Road, Market Drayton, TF9 1AR
G8 EWF Bernard Gilbert, 1 Wilmington Drive, Sutton-on-Sea, Mablethorpe, LN12 2JU
G8 EWL C Burgess, Jalna, 12 Foley Close, Ashford, TN24 0XA
G8 EWN David Edmonds, Great House Cottage, Ripponden, Sowerby Bridge, HX0 4LQ
G8 EWP R Edney, 63 Annandale Avenue, Bognor Regis, PO21 2ET
G8 EWT Graham Diacon, Raddle Barn, Southleigh Road, Witney, OX29 6UW
G8 EXF Richard Slatter, 1 Angells Meadow, Ashwell, Baldock, SG7 5QS
GD8 EXI S Baker, Ballanarran House, Surby Road, Ballafesson, Port Erin, Isle of Man, IM9 6TE
G8 EXJ B Jones, 38 Wyresdale Road, Lancaster, LA1 3DU
G8 EXK Malcolm Stuart Hatch, 6 Portland Street, Blyth, NE24 1NP
G8 EXN C Briggs, 22 Woodlesford Crescent, Halifax, HX2 0RB
G8 EXQ Tom Connell, 28 Tasman Close, Corringham, Stanford-le-Hope, SS17 7LD
G8 EXS P Atherton, 10 Cheriton Drive, Ravenshead, Nottingham, NG15 9DG



**IMPORTANT NOTE**

**Revalidate licence to avoid revocation** – Ofcom has advised the Society that plans will be drawn up to revoke licences that have not been revalidated as required by the licence conditions. The quickest way to revalidate is to do so online via the Ofcom website: *https://services.ofcom.org.uk/* or by email: *amateur.validations@ofcom.org.uk* Ofcom staff are available to help, but please be patient during times of heavy workload.

GM8 EXU J Steven, Andor, Skitten, Wick, KW1 4RX
G8 EXZ Stephen Warren, 269 Upper Weston Lane, Southampton, SO19 9HY
G8 EYA W Rimmer, 79 Brookhurst Avenue, Wirral, CH63 0LA
G8 EYM Norman Kearey, 73 Wellesley Drive, Crowthorne, RG45 6AL
G8 EYP Andrew Faulkner, 79a West Drive, Highfields Caldecote, Cambridge, CB23 7RY
G8 EYQ J Clee, 34 Knebworth Road, Bexhill-on-Sea, TN39 4JJ
G8 EYY M Hancock, 12 Mellor Road, Hillmorton, Rugby, CV21 4BP
G8 EZB M Whitlock, 85 Antrobus Road, Sutton Coldfield, B73 5EL
G8 EZD A Gifford, Broncroft, Rock Green Bank, Ludlow, SY8 2DT
G8 EZE P Swallow, 1 Auden Crescent, Ledbury, HR8 2UU
G8 EZG Alan Pybus, 10 Plough Close, Rothwell, Kettering, NN14 6YF
G8 EZL T Lambert, 40 Deepdale Road, North Shields, NE30 3AN
G8 EZR K James, 67 Drakes Way, Portishead, Bristol, BS20 6LD
G8 EZT R Elgy, 130 Stebbing House, Queensdale Crescent, London, W11 4TG
G8 EZU K Darbyshire, 24 Neston Road, Walshaw, Bury, BL8 3DB
G8 EZV Graham White, 94 Wingate Road, Luton, LU4 8PY
G8 EZZ R Chambers, 15 Barnfield Close, Braunton, EX33 2HL
G8 FAB Southampton ARC c/o Malcolm Troy, 22 Jackie Wigg Gardens, Totton, Southampton, SO40 9LZ
G8 FAD W Chown, 7840 Sw 136th Avenue, Beaverton, United States, 97008
G8 FAK S Sherratt, 21 Tweedale Close, Mursley, Milton Keynes, MK17 0SB
G8 FAR Roderick Elms, Fernside, Great Burches Road, Benfleet, SS7 3NA
G8 FAS Steven Hotham, 7 Cedar Drive, Everton, Lymington, SO41 0ZB
GD8 FAT B Haines, 20 Westfield Gardens, Harrow, HA3 9EJ
G8 FAW J Cawley, 16 Yorke Road, Dartmouth, TQ6 9HN
G8 FAX E Bye, 16 Daws Heath Road, Rayleigh, SS6 7QH
G8 FBF D Fellows, 10 Benning Way, Wokingham, RG40 1XX
G8 FBK L West-Knights, 4 Paper Buildings, Temple, London, EC4Y 7EX
G8 FBM Michael Bates, 11 The Rise Partridge Green, Horsham, RH13 8JB
G8 FBQ B Corker, 46 Danelaw, Great Lumley, Chester le Street, DH3 4LU
G8 FBW A Williams, 16 Hillside Road, Penn, High Wycombe, HP10 8JJ
G8 FC RAF ARC c/o R Finch, 1 Cherry Tree Cottage, Church Road, High Wycombe, HP10 8LN
G8 FCO G Onions, 3 Tower Rise, Tividale, Oldbury, B69 1NP
G8 FCQ M Lister, 246 Wigston Lane, Aylestone, Leicester, LE2 8DH
G8 FCT Robert Chadwick, Ithaca, Heck Lane, Goole, DN14 0RD
G8 FDE B McManus, 6 Rowley Road, St. Neots, PE19 1UF
G8 FDF J Bastable, 94 Baymead Lane, North Petherton, Bridgwater, TA6 6RN
G8 FDI G Felton, 10 Penbodeistedd, Llanfechell, LL68 0RE
G8 FDJ John Roberts, 2 Lomas Lea, Stannington, Sheffield, S6 6EW
G8 FDR M Bingham, 18 Ladywell Gate, Welton, Brough, HU15 1NL
G8 FDZ D Targett, 10 Thames Mews, Poole, BH15 1JY
G8 FED John Argyle, 23 Edward Road, Kennington, Oxford, OX1 5LH
G8 FEJ M Woudstra, Flat 1, 2 Upper Park Road, St. Leonards-on-Sea, TN37 6SJ
G8 FEK E Gawthorpe, 35 Highfield Way, North Ferriby, HU14 3BG
G8 FET John Guppy, 16 Barnfield Close, Hastings, TN34 1TS
G8 FEZ F Stuart, 70 Peartree Road, Herne Bay, CT6 7EQ
G8 FFA Eric Davis, 24 Redcar Avenue, Hereford, HR4 9TJ
G8 FFC C McManus, 6 Rowley Road, St. Neots, PE19 1UF
G8 FFF Christopher Player, 88 Civray Avenue, Downham Market, PE38 9QP
GM8 FFH D Brown, 14 Barloan Place, Dumbarton, G82 3QW
GM8 FFK G George, 13 Balmoral Terrace, Elgin, IV30 4JH
G8 FFM Barry Jackson, 23 Rylands Heath, Luton, LU2 8TZ
G8 FFN A Blakemore, 20 Derwent Road, Coventry, CV6 2HB
G8 FFU C Burrows, 6 Brook Way Lower Somersham, Ipswich, IP8 4PE
G8 FFW P Rycroft, Shore View House, 100 Pilling Lane, Poulton-le-Fylde, FY6 0HG
GM8 FFX Graham Knight, 6 Findon Road, Findon, Aberdeen, AB12 3RN
G8 FFZ P Ewington, 26 Dickens Road, Rugby, CV22 5RW
G8 FGB S Whitehead, 74 Manchester Road, Haslingden, Rossendale, BB4 5TE
G8 FGQ H Brittan, Meadowhurst Cottage, Woodcock Heath, Uttoxeter, ST14 8QS
G8 FGY P Griffiths, 5 Chestnut Crescent, Carlton Colville, Lowestoft, NR33 8BQ
G8 FGZ C Boon, Corner Cottage, Brackenhill, Nottingham, NG14 7EF
G8 FHC Michael Passam, Birchenbower, Birchendale, Stoke-on-Trent, ST10 4HL
G8 FHI Malwyn Clarke, 2 The Grove Penton Grafton, Andover, SP11 0RS
GM8 FHK John Gallacher, 23 East Avenue, Carluke, ML8 5TS
G8 FIE N McFetridge, 16 Blagrove Lane, Wokingham, RG41 4BA
G8 FIF Des Howlett, 4 Privet Close, Lower Earley, Reading, RG6 4NY
G8 FIG Christopher Cole, 157 Cherry Tree Road, Beaconsfield, HP9 1BD
G8 FJA P Webster, 3 Eden Avenue, Bare, Morecambe, LA4 6QL
G8 FJG R Shoulder, 264 Wennington Road, Rainham, RM13 9UU
G8 FJR D Jowett, 59 Old Road, Thornton, Bradford, BD13 3DQ
G8 FKF C Sargeant, Northview, 20 South Marsh Road, Grimsby, DN41 8AN
G8 FKH D Balharrie, 27 Norfolk Road, Uxbridge, UB8 1BL
G8 FKL Geoffrey Twibell, Greenhill Cottage, Moulsford, Wallingford, OX10 9JD
G8 FKP C Simkins, Finches, Cherry Close, Great Missenden, HP16 0QD
G8 FLL D Roseman, 101 Westbrook, Bromham, Chippenham, SN15 2EE
G8 FLS Iain MacIver, 160 Marsden Road, Burnley, BB10 2QP
G8 FLV A Nicholson, 29 Quaker Lane, Northallerton, DL6 1EE
G8 FMA E Sillars, 34 Sandown Road, Stevenage, SG1 5SF
G8 FMC David Keston, 8 Copse Gate, Winslow, Buckingham, MK18 3HX
G8 FMD C Wells, 5 Hepplewhite Close, Baughurst, Tadley, RG26 5HD
G8 FME A Hilton, 28 Eastern Esplanade, Broadstairs, CT10 1DR
G8 FMI F Steed, 19 Chancery Lane, Debenham, Stowmarket, IP14 6RN
GM8 FMR David Taylor, 14 Fenton Street, Alloa, FK10 2DT
G8 FMT Peter March, Devonholme, Bedford Road, Hitchin, SG5 3RX
G8 FMW R Whitehouse, 92 Willenhall Road, Bilston, WV14 6NP
G8 FMX David Beard, 9 Bowgate, Gosberton, Spalding, PE11 4ND
G8 FMZ Philip McNamara, Sunnybank Cottage, Lower Swell, Cheltenham, GL54 1LG
G8 FNG Paul Robinson, 52 Lea Court, New Road, Crewe, CW3 9DN
G8 FNH Michael Nash, 12 Ruston Park, Rustington, Littlehampton, BN16 2AB
GW8 FNO R Gregory, 5 Bryn Castell, Radyr, Cardiff, CF15 8RA
G8 FNR David Stone, 165 Wellington Hill West, Bristol, BS9 4QW
G8 FOL G Spencer, Tyn Cae, Llanfwrog, Anglesey, LL65 4YL
G8 FOT B Butterworth, 21 Higher Drive, Purley, CR8 2HQ
G8 FOY L Oakes, Flat 2 Brython, 54-56 Lloyd Street, Llandudno, LL30 2YP
G8 FOZ R Evans, 84 The Fairways, Leamington Spa, CV32 6PP
G8 FPA David Hoult, 19a Becksitch Lane, Belper, DE56 1UZ
G8 FPG S Banner, Oedhofstrasse 18, Amstetten, Austria, 3300
G8 FPU R HUTTON, 5 Tollemache Road, Prenton, CH43 8SU
G8 FPW Frank Brown, The Bungalow, Oxcroft Bank, Shepeau Stow, Spalding, PE12 0TY
G8 FQN Robert Schneider, 15 Hope Lane, Upper Hale, Farnham, GU9 0HY
G8 FQS P Simpson, 17 Reynard Close, Horsham, RH12 4GX
G8 FQZ C Stocker, 8 Brook Drive, Astley, Manchester, M29 7HS
G8 FRH P Lyall, 20 Horn Lane, Woodford Green, IG8 9AA
G8 FRI James Lucas, 42 Westerleigh Road, Bath, BA2 5JE
G8 FRS Keith Gurr, 35 Shelley Road, Stratford-upon-Avon, CV37 7JS
G8 FRY N Friday, Annexe, 34 Broadway, Sandown, PO36 9BY
G8 FSJ Rodney Page, 39 Carlton Street, Kettering, NN16 8EB
G8 FSL A Benham, 15 South Lodge Drive, Southgate, London, N14 4XD
G8 FSN Brian Steadman, 8 Machno Place, Denbigh, LL16 3YA
GU8 FSU V Rees, Le Chene Lodge, Le Chene Hill, Forest, Guernsey, GY8 0AJ
G8 FSV Adrian Mason, 7 Queen Bertha Road, Ramsgate, CT11 0ED
G8 FTE Richard Cowley, 18 Mill Road, Willingham, Cambridge, CB24 5UU
G8 FTP Phil Jarrett, 15 Groveside, East Rudham, King's Lynn, PE31 8RL
G8 FTW R Goodchild, 48 Coral Drive, Ipswich, IP1 5HS
G8 FTX Damian Gotch, The Bungalow, West Lane, Shipley, BD17 5DW
G8 FUB L Jones, 52 New Lane, Aughton, Ormskirk, L39 4UD
G8 FUH S Melling, 15 Woodbridge Hill Gardens, Guildford, GU2 8AR
G8 FUI William Raybould, 33 Roberts Green Road, Dudley, DY3 2BB
G8 FUJ Peter French, Oakdene, Forward Green, Stowmarket, IP14 5HJ
G8 FUL J Masterton, 15 Maylins Drive, Sawbridgeworth, CM21 9HG
G8 FUO R britton, 12 Bulkeley Avenue, Windsor, SL4 3LP
G8 FVC Douglas Mclay, 6 Burton Road, Castle Gresley, Swadlincote, DE11 9HD
G8 FVE Kevin Lake, 79 Sherrards Way, Barnet, EN5 2BP
G8 FVI C Reeves, 37 Arnold Gardens, Kinmel Bay, Rhyl, LL18 5NH
G8 FVJ D Still, 133a Feltham Road, Ashford, TW15 1AB
G8 FVM Philip Peake, 34 Blackhalve Lane, Wolverhampton, WV11 1BH
GM8 FVN Graham Adams, Heath Court, Morven Way, Ballater, AB35 5SF
G8 FVT D Bainton, 86 Holywell Avenue, Whitley Bay, NE26 3AD
G8 FWA J Errington, The Woodlands, Station Road, Leicester, LE8 9FP
G8 FWC David Sharp, 85 Chiltern Road, Goole, DN14 6HW
G8 FWD T Mckee, 19 Wall Lane Terrace, Cheddleton, Leek, ST13 7ED
G8 FWE J MAIDMENT, 36 Bannock Road, Whitwell, Ventnor, PO38 2RD
G8 FWF J Bowers, 56 Mendip Vale, Coleford, Radstock, BA3 5PR
G8 FWH J Hill, 21 Somersby Road, Mapperley, Nottingham, NG3 5QB
G8 FWK J Cranfield, 65 Broome Manor Lane, Swindon, SN3 1NB
G8 FXA G Griffiths, 51 Bempton Road, Liverpool, L17 5DB
G8 FXC Martin Bradford, 3 Veysey Close, Hemel Hempstead, HP1 1XQ
G8 FXG Nigel Lay, Orchard Cottage, Parkham, Bideford, EX39 5PL
G8 FXL A Patterson, 139 Lowther Road, Bournemouth, BH8 8NP
G8 FXM D Toombs, 1 Chalgrove, Welwyn Garden City, AL7 2QJ
G8 FXN Raymond Thackeray, 104 Stag Leys, Ashtead, KT21 2TL
G8 FXU D Percival, Trebakken, 11 Lamborne Close, Sandhurst, GU47 8JL
G8 FXV M White, 2 Mill Close, Denmead, Waterlooville, PO7 6PE
G8 FXX R Limb, Charnwood House, Station Road, Henley-on-Thames, RG9 3JS
G8 FYK K Payne, Flat 4, Monton Bridge Court, Eccles, M30 8UW
G8 FYX N Fensch, Glen Cottage, Bowl Road, Ashford, TN27 0HB
G8 FZI M Logsdon, Pilgrims Cottage, Langford Budville, Wellington, TA21 0RH
G8 FZT T Unsworth, Heathview, 15 Fenton Road, Huntingdon, PE28 2SD
G8 FZV David Ryan, Turners Oak, Barrs Lane, Woking, GU21 2JN
G8 FZW J Brown, 16 Greenwood Close, Moulton, Northampton, NN3 7RD
G8 GAJ J Niman, 3 The Meadows, Whitefield, Manchester, M45 7RZ
G8 GAR H Taylor, 21 Windermere Road, Coulsdon, CR5 2JF
G8 GAT M Smith, 241 Sandbanks Road, Poole, BH14 8EY
GM8 GAX P Howson, 1 Howetown, Fishcross, Alloa, FK10 3AW
G8 GBE P Richardson, 50 Amberley Road, Gosport, PO12 4EW
G8 GBM R Head, 29 Kingslea Road, Solihull, B91 1TQ
G8 GBP C Fawdon, 21 Bevan Close, Southampton, SO19 9PE
G8 GBU D Barker, 311 Uttoxeter Road, Mickleover, Derby, DE3 9AH
G8 GBY Hull & District ARS c/o Bryan Aslin-Smith, 161 Sharp St., Newland Avenue, Hull, HU5 2AE
G8 GCK Graham Croome, 10 Axford Close Gedling, Nottingham, NG4 4BB
G8 GCO Nigel Wall, 9 North Close, Ipswich, IP4 2TL
G8 GCS Colin Coker, 46 Clarendon Road, Ipplepen, Newton Abbot, TQ12 5QS
G8 GDC R Laver, 40 Middleton Close, Tysoe, Warwick, CV35 0SS
G8 GDH D Brown, 56 Paddock Road, Staincross, Barnsley, S75 6LE
G8 GDI R Dunn, 48 Stanley Hill, Amersham, HP7 9HL
GM8 GDN Malcolm Brunton, 2 Easter Place, Portlethen, Aberdeen, AB12 4XL
G8 GDZ R Thompson, 23 Fox Hill, Selly Oak, Birmingham, B29 4AG
G8 GEA K Warriner, Windover, 16 The Ridgeway, Eastbourne, BN20 0EU
G8 GEB S Rowsby, 10 Echells Close, Bromsgrove, B61 7EB
G8 GEE R Sherwood, 19 Norton Drive, Warwick, CV34 5FE
G8 GEF S Edwards, Fernlea, Meathop, Grange-over-Sands, LA11 6RB
G8 GET J Shepherd, 6 The Jordans, Coventry, CV5 9JT
G8 GEV John Moore, 17 Kings Grove, Barton, Cambridge, CB23 7AZ
G8 GEZ L Wooller, 4 Old Court Close, Brighton, BN1 8HF
G8 GFA R Marshall, The Village School, Upleatham, Redcar, TS11 8AG
G8 GFB Christopher Jones, 1 Primrose Hill Road, Euxton, Chorley, PR7 6BA
G8 GFF Neil Sanderson, 54 Kelvedon Close, Chelmsford, CM1 4DG
G8 GFS Michael Winiberg, Summerhill, Smallhythe Road, Tenterden, TN30 7NB
G8 GFW Jonathan Douglas, 1030 Shields Road, Walkerville, Newcastle upon Tyne, NE6 4SR
G8 GFY D King, 108 Huddersfield Road, Meltham, Holmfirth, HD9 4AG
G8 GFZ T Cockram, The Bungalow, Dyke Hill, Chard, TA20 2PY
G8 GGI R Geddes, 107 Dukes Avenue, New Malden, KT3 4HR
G8 GGM P Burfoot, 18 Ember Road, Langley, Slough, SL3 8ED
G8 GGO P Carson, 16 Gaynes Park Road, Upminster, RM14 2HJ
G8 GGR Christine Coleman, 18 Chester Street, Coventry, CV1 4DJ
G8 GGS M Clarke, Ruskin, Ashurst Drive, Tadworth, KT20 7LS
G8 GGW N Dudman, Chapel House, Pen y Bryn, Wrexham, LL14 1UA
G8 GHB A Hunt, 21 Plumpton Gardens, Cantley, Doncaster, DN4 6SN
G8 GHH C Gibbs, 10 Waverley Road, Margate, CT9 5QB
G8 GHK W White, 60 Parklands, Rochford, SS4 1SH
G8 GHL S Garland, 53 The Crescent, Horsham, RH12 1NA
G8 GHO Jerry Wood, 17 Yew Tree Park Road, Cheadle Hulme, Cheadle, SK8 7EP
G8 GHQ P Laverock, Telegraph Tower, St Mary's, Isles of Scilly, TR21 0NR
G8 GHR Brian Farey, 5 Ivel View, Sandy, SG19 1AU
G8 GHT J Sansum, 50 Farm Road, Maidenhead, SL6 5JD
GM8 GHV W Sherriffs, Hillcrest, Disblair, Aberdeen, AB21 0RJ
G8 GIF K Turner, 20 Rawdon Way, Faringdon, SN7 7YT
G8 GIG A Patterson, Rose Cottage, Ingatestone Road, Ingatestone, CM4 0RS
G8 GIH K Foster, 52 Bottesford Avenue, Scunthorpe, DN16 3EN
G8 GIK J Hart, 28a Dunton Road, Stewkley, Leighton Buzzard, LU7 0HZ
G8 GIL Michael Dimmock, 14 North Street, Bletchley, Milton Keynes, MK2 2PY
G8 GIN J Walker, 11 Burrett Gardens, Wisbech, PE13 3RP
GM8 GIQ C Wearing, 16 Campbell Drive, Troon, KA10 6XE
G8 GIU J Harman, 13 Linthorpe Court, South Shields, NE34 9BU
G8 GIZ David Ollerhead, 15 Kingsley Road, Chester, CH3 5RR
G8 GJA P Reeves, 77 Cale Way, Wincanton, BA9 9BS
G8 GJC Jonathan Tillin, 37 St. Laurence Gardens, Belper, DE56 1HH
G8 GJG Nigel Giltrow, 7 The Square, Milton-under-Wychwood, Chipping Norton, OX7 6JN
GM8 GJI E Smith, The Steading, Craigmyle, Banchory, AB31 4LS
G8 GJM R Harwood, 9 Cornwall Close, Woose Hill, Wokingham, RG41 3AG
G8 GJO A Heasman, 170 Plum Lane, London, SE18 3HF
G8 GJQ D Grant, 1 Crispe Park Close, Birchington, CT7 9BN
G8 GJU M Bernard, 33 Station Road, Over, Cambridge, CB24 5NJ
G8 GJV T England, 30 Sparrow Way, Burgess Hill, RH15 9UL
G8 GJW C Drouet, Barn Lea, Holcot Road, Northampton, NN6 9BS
G8 GKC C Ridley, 39 Lancelot Road, Welling, DA16 2HX
G8 GKH R Hadley, 36 Folly Lane, Cheltenham, GL50 4BY
G8 GKL C Rauch, 40 Russett Close, King's Lynn, PE30 3HB
G8 GKR Maureen Fellows, 343 Wake Green Road, Birmingham, B13 0BH
G8 GKX David Nicholson, Avenida EspaÃ±A, Edificio Sorrento, Malaga, Spain, 29793
G8 GLB P Brown, 4 King Edgar Close, Ely, CB6 1DP
G8 GLC I Cooper, 77a Benhill Wood Road, Sutton, SM1 3SL
G8 GLD M Bounford, 19 Pear Tree Road, Bignall End, Stoke-on-Trent, ST7 8NH
G8 GLI J Husk, Brandhu, Common Moor, Liskeard, PL14 6EP
G8 GLP B Barnett, 21 Primrose Walk, Maldon, CM9 5JJ
G8 GLS Gregory Wimlett, 3 London Street, Fleetwood, FY7 6JE
G8 GLV A Brown, 23 Vincents Way, Naphill, High Wycombe, HP14 4RA
G8 GLY Alan Higgins, 86b Cranleigh Road, Bournemouth, BH6 5JL
G8 GLZ Andrew Findlay, 7 Market Square, Winslow, Buckingham, MK18 3AB
G8 GMA D Elliott, 56 Lincoln Avenue, Willenhall, WV13 1JQ
G8 GMB S Bradshaw, 82 Arden Way, Market Harborough, LE16 7DD
G8 GML P Melbourne, 2 Jubilee Cottages, Jubilee Lane, Colchester, CO7 7RY
G8 GMU Brian Leathley-Andrew, 4 Robinson Road, Bedworth, CV12 0EL
G8 GNO S Ferdenzi, 4 Ashworth Road, Rossendale, BB4 9JE
G8 GNX J Bartholomew, 33 Manor Way, Woodmansterne, Banstead, SM7 3PN
G8 GNZ Geoffrey Blake, 22 Cannon Leys, Galleywood, Chelmsford, CM2 8PD
GW8 GOC M Black, Mediascene Ltd, Unit A-d, Bowen Industrial Estate, Bargoed, CF81 9AB
G8 GOM Anthony Ireson, 32 The Avenue, Wellingborough, NN8 4ET
G8 GON Alec Jefford, 37 Marions Way, Exmouth, EX8 4LF
G8 GOO Philip Nelson, 15 Hill Street, Gerlan, Bangor, LL57 3TD
G8 GOR A Pearce, 153 Henver Road, Newquay, TR7 3EJ
G8 GOS K Roche, 96 Porter Road, Basingstoke, RG22 4JR
G8 GOT D Parkin, 252 Standbridge Lane, Crigglestone, Wakefield, WF4 3JA
G8 GPF David Clark, 6 Bradley Park Road, Torquay, TQ1 4RD
G8 GPO OFRAC BALDOCK c/o David Thorpe, 70 Willow Way, Ampthill, Bedford, MK45 2SP
GW8 GQE John Moore, Oak House, Falcondale Drive, Lampeter, SA48 7SB
G8 GQF I McEnteggart, 46 Bissley Drive, Maidenhead, SL6 3UZ
G8 GQG J Crow, 58 Cooden Drive, Bexhill-on-Sea, TN39 3AX
G8 GQJ R Clark, 9 Conigre, Chinnor, OX39 4JY
G8 GQS B Summers, 9 Prior Croft Close, Camberley, GU15 1DE
G8 GRB R Day, 20 Linacre Road, Torquay, TQ2 8LF
G8 GRC John Drakeley, Rowan Cottage, Four Crosses Lane, Cannock, WS11 1RU
G8 GRD Lawrence Hetherington, 5 Withey Close West, Bristol, BS9 3SX
GD8 GRE Christopher Wilkinson, 3 Carrick Bay View, Ballagawne Road, Colby, Isle of Man, IM9 4DD
G8 GRL K Edwards, 11 Foxes Road, Ashen, Sudbury, CO10 8JS
G8 GRO Roy Nicholls, 10 Polmeere Road, Penzance, TR18 3PD
G8 GRP E Poole, Ramillies Hall School, Ramillies Avenue, Cheadle, SK8 7AJ
G8 GRQ Alan Plail, 46 Hayling Rise, Worthing, BN13 3AG
G8 GRS Richard Woodward, 158 Highridge Road, Bishopsworth, Bristol, BS13 8HU
G8 GRT R Oakley, 17 Windmill Close, Ellington, Huntingdon, PE28 0AJ
G8 GSL Iain Liston-Brown, 20 Chatterton Avenue, Lichfield, WS13 8EF
G8 GSU Robert Wade, 31 Church Street, Crowthorne, RG45 7PD
G8 GT Francis Clare, Glen View, Newport Road, Caldicot, NP26 3BZ
G8 GTD S Porter, 20 Newbridge Road, Ambergate, Belper, DE56 2GR
G8 GTI Kenneth Barnes, 75 Southmeade, Liverpool, L31 8EG
G8 GTR D Murray, 27 Station Avenue, Walton-on-Thames, KT12 1NF
G8 GTU Peter Stephens, 3 Inett Way, Droitwich, WR9 0DN
G8 GTV Barrie Raby, 10 Bulverton Park, Sidmouth, EX10 9EW
G8 GTZ N Matthews, 12 Petrel Croft, Basingstoke, RG22 5JY
G8 GUA R Wood, 339 Horse Road, Hilperton Marsh, Trowbridge, BA14 7PE
G8 GUH Gerald Ohara, 107 Castlesteads Drive, Carlisle, CA2 7XD
G8 GUJ J Stubbs, 14 The Glen, Langstone, Newport, NP18 2NR
G8 GUN H Parker, 7 The Hollies, Clee Hill, Ludlow, SY8 3NZ
G8 GUS M Board, 48 Skipper Way, Lee-on-The-Solent, PO13 9EY
G8 GUX J Thomson, 2 Wilton Hill, Hawick, TD9 8BA
G8 GVL K E Woods, 7 Ives Close, West Bridgford, Nottingham, NG2 7LU
G8 GVN Edward Shield, 14 Wellwood Street, Amble, Morpeth, NE65 0EL
G8 GVV P Richmond, 57 The Fairway, Daventry, NN11 4NW
G8 GVW P Shillito, Little Orchard, Thorney Road, Martock, TA12 6BG
G8 GVZ L Sullivan, 89 Richmond Crescent, Mossley, Ashton-under-Lyne, OL5 9LQ
G8 GWB N Awcock, 16a Ongar Road, Writtle, Chelmsford, CM1 3NU
G8 GWJ J Vincent, 12 Spelman Road, Norwich, NR2 3NJ
G8 GWK C Cornell, 80 Walcot Avenue, Round Green, Luton, LU2 0PR

G8 GWM Nigel Hay, 20 The Ridgeway, Fetcham, Leatherhead, KT22 9AZ
G8 GWP G Atkinson, Parkland Stables, [illegible], NG11 0NP
G8 GWR [illegible], 104 [illegible]
G8 GWX Roger Howells, Chester Sub Aqua Club, Chester City Baths, [illegible], CH1 1QP
G8 GXF J Ashmore, 3 The Cedars, Stockwell Road, Wolverhampton, WV6 9AZ
G8 GXN K Raynor, 17 Kirkstone Walk, Nuneaton, CV11 6EZ
G8 GXO Peter Rogers, 20 Hall Lane, Sutton, Macclesfield, SK11 0EP
G8 GXS J Hadjioannou, The Vicarage, Wakefield Road, Pontefract, WF9 5BX
G8 GYB V Vesma, Durvale, 5 Jonas Drive, Wadhurst, TN5 6RJ
G8 GYH A Rice, 117 Manners Way, Southend-on-Sea, SS2 6QP
G8 GYI Graham Stanley, 133 Park Lane, Kidderminster, DY11 6TE
G8 GYK G Carter, 19 Wych Elms, Park Street, St. Albans, AL2 2AR
G8 GYL I Bishop, 5 Trent Close, Tolpuddle, Dorchester, DT2 7HA
G8 GYM R Claridge, 124 Pemdevon Road, Croydon, CR0 3QP
G8 GYN Alan Blair, 45 St. Georges Road, North Shields, NE30 3JZ
G8 GYP V Holmes, 104 York Avenue, Hayes, UB3 2TP
G8 GYS P Wright, 1 Lambourne Way, Thruxton, Andover, SP11 8NE
G8 GYV B Simms, 63 Kingfisher Road, Weston-Super-Mare, BS22 8TX
G8 GYX Tom Ellinor, 53 Hillside, Banstead, SM7 1HG
G8 GYY Chris Gregory, 5 Fox Close, Wigginton, Tring, HP23 6ED
G8 GZC G Tew, 73 King Cerdic Close, Chard, TA20 2JB
GI8 GZM John Mawhinney, 12 Shane Park, Lurgan, Craigavon, BT66 7HD
G8 GZN A Hill, 1 Greenways, Highcliffe, Christchurch, BH23 5BA
G8 GZR Robert Langdon, 42 Caldbeck Drive, Woodley, Reading, RG5 4LA
G8 GZV R Duke, 5 Pembroke Close, Billericay, CM12 0PF
G8 GZW A Davis, 8 Roberts Road, Greatstone, New Romney, TN28 8RL
G8 GZX John Eadie, 5 Silver Street, Cublington, Leighton Buzzard, LU7 0LJ
G8 HAM Simon Collins, 2 St. Teresa's Drive, Chippenham, SN15 2BD
G8 HAU R Lambarth, 38 Kirkley Park Road, Lowestoft, NR33 0LG
G8 HAV Peter Fox, 5 Llandovery Close, Winsford, CW7 1NA
GM8 HBB J Edwards, 58 Maxwellton Avenue, East Kilbride, Glasgow, G74 3AF
G8 HBQ P Davies, 24 Upland Grove, Leeds, LS8 2SX
GM8 HBY Crawford Ross, 16 Glebe Crescent, Airdrie, ML6 7DH
G8 HBZ S Stephenson, 6 Livingstone Close, Rothwell, Kettering, NN14 6HT
G8 HCJ Alan Levett, 3 Nottington Court, Nottington, Weymouth, DT3 4BL
G8 HCK A Rutter, The Uplands, Castle Howard Road, Malton, YO17 6NJ
G8 HCL V Menday, Huf House, Horseshoe Ridge, Weybridge, KT13 0NR
G8 HCS H Stratton, 26 Marjorie Road, Chaddesden, Derby, DE21 4HQ
G8 HCW C Morgan, 24 High Mead, Wootton Bassett, Swindon, SN4 8LW
G8 HCZ Peter Iredale, Mayfield, Woodlands Road, Ipswich, IP7 5LJ
GW8 HDH J Dowdall, 56 Goetre Bellaf Road, Dunvant, Swansea, SA2 7RP
G8 HDJ P Muxlow, 17 Station Road, Grasby, Barnetby, DN38 6AP
G8 HDK M Balls, Balmore, Swallow Lane, Wisbech, PE13 5PQ
G8 HDL M Connell, 38 White Close, High Wycombe, HP13 5NG
G8 HDM Ian Arnold, 44 Elwick Avenue, Acklam, Middlesbrough, TS5 8NT
G8 HDP R Jenkins, 10 Ulstan Close, Woldingham, Caterham, CR3 7EH
G8 HDS P MacKimm, 36 Links View, Rochdale, OL11 4DD
G8 HEB T Brady, 8 Cefn Hawys, Red Bank, Welshpool, SY21 7RH
G8 HER A Lambert, 2 Huxley Close, Locks Heath, Southampton, SO31 6RR
G8 HEU Paul Whitehead, 7 Vulcan Road, Freckleton, Preston, PR4 1JN
GW8 HF D Phillips, 34 Graig Terrace, Graig, Pontypridd, CF37 1NH
G8 HFL Lester Caine, 25 Smallbrook Road, Broadway, WR12 7EP
G8 HFW T Hall, 23 Burcott Gardens, Addlestone, KT15 2DE
G8 HGG P Abernethy, Enfield House, Halford, Shipston-on-Stour, CV36 5DA
G8 HGI Martin Warriner, 135 Showfields Road, Tunbridge Wells, TN2 5UN
G8 HGL David Lambert, 4 Tamworth Road, Bedford, MK41 8QY
G8 HGM K Ellis, 11 Ringwood Close, Eastbourne, BN22 8UH
G8 HGN R Harrison, 59 Grange Road, Great Burstead, Billericay, CM11 2RQ
G8 HGP W Plucknett, 17 Skylark Corner, Stevenage, SG2 9NL
GM8 HHC P Dick, Napier House, 8 Colinton Road, Edinburgh, EH10 5DS
G8 HHO M Strange, 60a Manor Road, Dersingham, King's Lynn, PE31 6LH
G8 HHR J Bardell, 239 Meadow Road, Droitwich, WR9 9BZ
G8 HHZ P Woods, 14 Cromwell Road, London, N10 2PD
G8 HI K Burnitt, 15 St. Bedes, East Boldon, NE36 0LE
G8 HIG Lindsay Cole, 151 Carshalton Park Road, Carshalton, SM5 3SF
G8 HIO T Ellis, Hollybush House, Hawley Green, Camberley, GU17 9BP
G8 HIQ Stephen Whitehouse, 2 Lindholme Drive, Rossington, Doncaster, DN11 0UR
G8 HJD Christopher Tubis, Rockleaze Mews, Rockleaze Avenue, Bristol, BS9 1NG
G8 HJF Christopher Williams, Kingsclere House, Fox's Lane, Newbury, RG20 5SL
G8 HJG C Williams, Kingsclere House, Fox's Lane, Newbury, RG20 5SL
G8 HJH M Norton, 179b Kimbolton Road, Bedford, MK41 8DR
G8 HJK Paul Hunt, 40 Leighton Road, Toddington, Dunstable, LU5 6AL
G8 HKF Stephen Down, 1 Dove Close, Honiton, EX14 2GP
G8 HKK Martin North, 69 High St, Thirroul, Australia, NSW 2515
G8 HKN R Meakins, 335 Court Road, Orpington, BR6 9BZ
G8 HKP E Michael Jakins, 2 South View Place, Midsomer Norton, Radstock, BA3 2AX
G8 HKS Michael Booth, 30 Manor Green, Harwell, Didcot, OX11 0DQ
G8 HLE R Marshall, 54 [illegible] Avenue, Maidstone, [illegible]
G8 HLH R Wheeler, 14 Robins Lane, St. Helens, WA9 3NF
G8 HLJ E Edwards, Flat 27, Barncroft, Wirral, CH61 6YH
G8 HLQ Edward Birch, 17 Canalside Cottages, Chester Road, Runcorn, WA7 [illegible]
G8 HMA R Smith, 5 Newton Close, Loughborough, LE11 [illegible]
G8 HMG P Walker, 12 Brownlow Road, Redhill, RH1 6AW
G8 HMJ M Kellett, Wistow Gate, Glen Road, Leicester, LE8 9FH
G8 HMV J Nicholas, 4 Lion Lane, Clee Hill, Ludlow, SY8 3NJ
G8 HMZ P Cheseldine, 6 Lissett Close, Lincoln, LN6 0SY
G8 HNA Sheila Clark, 1 Roman Road, Broadstone, BH18 9DF
G8 HNM R Parker, 1 Whitmore Orchard, Whitmore Lane, Taunton, TA2 6SR
G8 HNS Roger Stanleigh, Shallow Pool Bungalow, Looe, PL13 2ND
G8 HNT T Thompson, 25 Meadow Avenue, Codnor, Ripley, DE5 9QN
G8 HOI Roderick Warner, Barley Hill Farm, Combe St. Nicholas, Chard, TA20 3HJ
G8 HOR Ernie York, Combe Brune, Prayssac, France, 46220
G8 HOS V Mander, Meadowlands, Severn Lane, Welshpool, SY21 7BB
G8 HOU Howard Cox, 21 North Avenue, Hayes, UB3 2JE
G8 HPF Ian McLenaghan, 82 Cheam Road, Epsom, KT17 1QP
G8 HPJ P Beaumont, 1 Byron Road, Mexborough, S64 0DG
G8 HPL W Taylor, Bywell, Chester Road, Wrexham, LL12 0HN
G8 HPN G Staniewicz, Flat 1, Jubilee Farm, Gillingham, SP8 5SJ
G8 HPS A Hancock, 9 Elmside, Willand, Cullompton, EX15 2RN
G8 HPV D Green, 67 Coombe Park Road, Binley, Coventry, CV3 [illegible]
G8 HPW [illegible] 8HT
G8 HPY A Mander, 18 Bridge Avenue, Otley, LS21 2AA
G8 HQM Stephen Bastow, Bryn Goleu, Rhosgadfan, Caernarfon, LL54 7LB
G8 HQO Nigel Johnson, 56 Clarkson Avenue, Wisbech, PE13 2EG
G8 HQP David Kimber, 10 Butler Road, Solihull, B92 7QL
G8 HQW Pauline Kirby, 2 Kneeton Park, Middleton Tyas, Richmond, DL10 6SB
G8 HRA Clive Ryalls, 15 Belmont Way, South Elmsall, Pontefract, WF9 2BT
G8 HRC HAVERING & DISTRICT ARC c/o David Nuttall, 92 Long Road, Lowestoft, NR33 9DH
G8 HRF Kevin Dodman, 10 Newark Road, Lowestoft, NR33 0LY
G8 HRW S Watkin, 9 Longden Close, Haynes, Bedford, MK45 3PJ
G8 HSI John Carey, 7 Church Road, Walton on The Naze, CO14 8DF
G8 HSR Edward Warren, 37 Kingston Drive, Mangotsfield, Bristol, BS16 9BQ
G8 HSS Martin Saxon, 4 The Coppice, Impington, Cambridge, CB24 9PP
G8 HST M Sanders, 19 Brunswick Gardens, Hainault, Ilford, IG6 2QU
GM8 HSY H Reekie, 5 Golf Course Road, Bonnyrigg, EH19 2EU
G8 HTA Keith Parker, 20 River Avenue, Hoddesdon, EN11 0JS
G8 HTB A Barker, Bank Royd Barn, Bank Royd Lane, Halifax, HX4 0EW
G8 HTF D Fletcher, 40 Bentham Drive, Liverpool, L16 5EU
G8 HTM J Taylor, Perry House, 188 Walstead Road, Walsall, WS5 4DN
G8 HTN A Kettlety, 106 Denton Road, Audenshaw, Manchester, M34 5BD
G8 HTO Alan Farrell, 206 London Road, Delapre, Northampton, NN4 8AU
G8 HTZ Stephen Druitt, 25 Holcroft, Orton Malborne, Peterborough, PE2 5SL
GI8 HUD T Huddleston, 29 North Parade, Belfast, BT7 2GF
G8 HUF S Carpenter, Fernlea, Fernhill Lane, Camberley, GU17 9HA
G8 HUG I Coulson, 56 Potterdale Drive, Little Weighton, Cottingham, HU20 3UX
G8 HUH Thomas Rabbitts, Laurel Cottage, Wick Lane, Highbridge, TA9 4BU
G8 HUO D Sharpe, 37 Oulton Avenue, Bramley, Rotherham, S66 2SS
G8 HUR N Mills, 3 Whitfield Close, Wilford, Nottingham, NG11 7AU
G8 HUS A Mead, 12 Wyelands View, Mathern, Chepstow, NP16 6HN
G8 HUT N Onions, Windy Ridge, Dunmow Road, Dunmow, CM6 3PJ
G8 HUV M Rowlands, 3 Littledown View, Great Durnford, Salisbury, SP4 6AU
G8 HUY John Hill, 22b Strait Lane, Hurworth, Darlington, DL2 2AL
G8 HVF C Billson, Knotts End, Bateman Road, Loughborough, LE12 6NN
G8 HVT M Evans, 25 Walnut Close, Nailsea, Bristol, BS48 4YH
G8 HVV Chris Goadby, Heligan, 12 School Road, Newmarket, CB8 9RX
G8 HVX Alan Staniforth, 25 Brown Ct, East Brunswick, United States, 8816
G8 HVZ R Anderson, 23 Callington Road, Saltash, PL12 6DU
G8 HWI John Simons, 7a Walton Way, Stone, ST15 0JF
G8 HWJ Edward Smith, Brickyard House, Wainfleet Road, Skegness, PE24 5AT
GW8 HWL P Jenkins, 20 Dimbath Avenue, Blackmill, Bridgend, CF35 6ED
G8 HWQ David Mappin, 13 Willow Close, Filey, YO14 9NY
GW8 HWS Jonathan Mills, 13 Egerton Street, Cardiff, CF5 1RF
G8 HXD M Ledger, 58 Mount Pleasant Close, Lightwater, GU18 5TR
G8 HXE Keith Haywood, 6 Lydney Road, Urmston, Manchester, M41 8RN
G8 HXR Michael Brooke, 70 Wootton Avenue, Peterborough, PE2 9EG
G8 HXT P Randall, 236 Lower Luton Road, Wheathampstead, St. Albans, AL4 8HN
G8 HXW A Sargent, 22 Duckmill Crescent, Duckmill Lane, Bedford, MK42 0AF
G8 HYI E Whitfield, Erw Mor, Nebo, Amlwch, LL68 9NE
G8 HYK J Brockwell, 10 Tregony Rise, Lichfield, WS14 9SN
G8 HYL H Tuff, 4 Battery Terrace, Mevagissey, St. Austell, PL26 6QS
G8 HYM S Bradley, 247 Filey Road, Scarborough, YO11 3AE
G8 HYP M Peers, 6 Manor Farm Cottages, Warboys Road, Huntingdon, PE28 3DA
GW8 HYT Paul Madden, Ty Gwyn, Rhandirmwyn, Llandovery, SA20 0NT
G8 HYU Denise Pepper, 52 King Style Close, Crick, Northampton, NN6 7ST
G8 HZI Martin Holdsworth, 2 Newman Drive, Branston, Burton-on-Trent, DE14 3DZ
G8 HZJ R Ingamells, Moor View, Small Banks, Moorside Ilkley, LS29 0QQ
G8 HZL D Wildman, 2 Bluecoat Walk, Harmans Water, Bracknell, RG12 9NP
G8 HZN Richard Orchard, Tredinneck Moor, Newmill, Penzance, TR20 8XT
G8 HZQ P Healy, 63 Hazelwood Drive, St. Albans, AL4 0UP
G8 HZS T Storey, 50 Longfield Road, Darlington, DL3 0HX
G8 IAJ J Richardson, 43 Front St., Leadgate, Consett, DH8 7SB
G8 IAK R Thomas, 88 Parkway, London, SW20 9HG
G8 IAM Stephen Lloyd, 4 Cwmdu Court, Cwmdu, Crickhowell, NP8 1RU
G8 IAN M Lees, 175 Overdale Road, Romiley, Stockport, SK6 3EN
G8 IAR P Smith, 35 Garrett Close, Kingsclere, Newbury, RG20 5SD
G8 IBC David Herke, 24 The Lawns, Farnborough, GU14 0RF
G8 IBE R Bailey, 6 Kestrel Close, Horsham, RH12 5WD
G8 IBK Malcolm Murray, Heads Nook Hall, Heads Nook, Brampton, CA8 9AA
G8 IBL H Hallybone, Birch Bassett, 52 Busbridge Lane, Godalming, GU7 1QQ
G8 IBO T Gill, 21 Winn Road, London, SE12 9EX
G8 IBP R May, 10 Lime Close, Wokingham, RG41 4AW
G8 IBR Nyall Davies, 1 Helens Close, The Street, Diss, IP22 1RW
G8 IC Michael Dawson, 60 Ashenhurst Road, Todmorden, OL14 8DD
GM8 ICC Alexander Campbell, 2 Cairndhu Cottage, Cairnbaan, Lochgilphead, PA31 8SQ
GW8 ICD David Davies, 53 Crawshay Street, Ynysybwl, Pontypridd, CF37 3EF
G8 ICT Christopher Hopley, Clayton Cottage, Alltami Road, Mold, CH7 6RW
G8 IDF J Pimlott, 40 Queens Road, Higher St. Budeaux, Plymouth, PL5 [illegible]
G8 IDJ I Judd, 33 Coles Mede, Otterbourne, Winchester, SO21 2EG
G8 IDK R Voisey, 2 Chester Place, Malvern, WR14 1RQ
G8 IDL D Smith, The Old Forge, High Street, Newmarket, CB8 0SE
G8 IEA S Parham, 132 Wrotham Road, Gravesend, DA11 7LB
G8 IEI J McKillop, 2 Moores Green, Wokingham, RG40 1QG
G8 IEL R Tust, 28 Osprey Close, Beechwood, Runcorn, WA7 3JH
GM8 IEM Martin Hall, 199 Clashmore, Lochinver, Lairg, IV27 4JQ
G8 IER Phillip Nice, 5 Walden Close, Doddington, March, PE15 0TW
G8 IEV B Guy, Hawthorn Folly, Cul De Sac, Boston, PE22 8EY
G8 IEW C Davies, Applecroft, St. Johns Road, Gloucester, GL2 7DF
G8 IEZ Christopher Moss, 11 Sheepfold Crescent Barrow, Clitheroe, BB7 9XR
G8 IFF Nigel Gunn, 1865 El Camino Drive, Xenia, United States, OH 45385-1115
G8 IFH K Thomas, 1 Byways, Yateley, GU46 6NE
G8 IFN N Hinderwell, 1 Bower Grove, West Mersea, CO5 8GJ
G8 IFT I Gordon, 40 Grange Crescent, Rubery, Birmingham, B45 9XB
G8 IHA J Gregory, 2 Abbey Dale Close, Kilburn, Belper, DE56 0PY
G8 [illegible] G Styler, 95 Fairoaks Drive, Great Wyrley, Walsall, WS6 6HA
G8 [illegible] David Cochrane, 18 Russell Avenue, Dunchurch, Rugby, CV22 6PX
G8 [illegible] [illegible], 7 Mowbray Road, Northallerton, [illegible]
G8 IIC Rodney Owens, 7 Warrington Road, Ipswich, IP1 3QU
GM8 IID N Paterson, 4 Cambridge Road, Renfrew, PA4 0SL
G8 IIG G Punter, 18 Lodge Road, Sharnbrook, Bedford, MK44 1JP
GM8 IIH W Jarvie, Wester Auchinrivoch, Banton, Glasgow, G65 0QZ
G8 IIK David Hooker, Pennywood, Clarke Road, New Romney, TN28 8PB
GM8 IIO W Robson, 18 Colinton Mains Green, Edinburgh, EH13 9AG
G8 IIS Brian Heaney, 19 Ormonde Drive, Liverpool, L31 7AN
G8 IIZ William Rush, 17 Hagden Lane, Watford, WD18 0HQ
G8 IJC C Phillipson, 24 Wyatt Close, Martin, Lincoln, LN4 3RN
G8 IJE B Laxton, 1 Stoney Lane, Walsall, WS3 3RF
G8 IJG D Adams, 77 Chestnut Crescent, Shinfield, Reading, RG2 9HA
G8 IJI Keith Williamson, 4 Lynwood Drive, Wakefield, WF2 7EF
G8 IJM Howard Wallington, 5 Glebe Road, Royal Wootton Bassett, Swindon, SN4 7DU
G8 IJS R Sayer, Vignouse, Paimpont, France, 35380
G8 IJT M Cawood, 51 Mayflower Drive, Marford, Wrexham, LL12 8LD
G8 IK V Morse, 42 Kingscote Road, Dorridge, Solihull, B93 8RA
G8 IKA D Poll, 66 Southlands Avenue, Orpington, BR6 9NF
G8 IKG K Raper, 26 Lancaster Way, Scalby, Scarborough, YO13 0QH
G8 IKH Robert Rolley, Glas Cwm, Dyffryn Crawnon, Crickhowell, NP8 1NU
G8 IKK J Channon, Tremayne, Chymbloth Way, Helston, TR12 6TB
G8 IKS D Warwick, Orchard Cottage, Colber Lane, Harrogate, HG3 3JR
G8 IKW P Nutt, 40 Parkfield Drive, Middleton, Manchester, M24 4ED
G8 ILB N Allinson, 4 Crooks Barn Lane, Stockton-on-Tees, TS20 1LW
G8 ILD Roger Barrow, 50 Redhill Drive, Bredbury, Stockport, SK6 2HQ
G8 ILG J Law, 29 Brackenwood, Orton Wistow, Peterborough, PE2 6YP
G8 ILJ Stuart Nutt, 23a Hesketh Drive, Southport, PR9 7JX
G8 ILN M Grindrod, 20 Castle Mead, Kings Stanley, Stonehouse, GL10 3LD
G8 ILP T Voller, 179 High Street, Harriseahead, Stoke-on-Trent, ST7 4JU
G8 ILU J Parker, 6 Cedar Drive, Bourne, PE10 9SQ
G8 ILW D Couse, 6 Reading Drive, Sale, M33 5DL
G8 ILZ I Walker, 51 Whitlock Drive, Wimbledon, London, SW19 6SJ
G8 IMB M Stubbs, Crofters, Harry Stoke Road, Bristol, BS34 8QH
G8 IMH M Fereday, 35 Manor House Park, Codsall, Wolverhampton, WV8 1ES
G8 IMI C Kitchener, 4 Cramswell Close, Haverhill, CB9 9QL
G8 IMJ Robert Head, 21 Church Street, Fleetwood, FY7 6JR
G8 IMM R Keeley, 6 Standings Rise, Whitehaven, CA28 6SX
G8 IMS M Stroud, 65 Applegarth Avenue, Guildford, GU2 8LX
G8 IMX D Jones, Borie Du Ritou, Lherm, France, 46150
G8 IMZ Arthur Palfrey, 51 The Village Wigginton, York, YO32 2PR
G8 INA David Harris, 102 Greatmeadow, Northampton, NN3 8DF
G8 INC K Davenport, 10 Woodend Lane, Hyde, SK14 1DT
G8 INL B Miller, 1 The Meadows, Monk Fryston, Leeds, LS25 5PJ
G8 INO A Brown, 25 Birch Lane, Haxby, York, YO32 3RP
G8 INS Paul Williams, 2 Rosamund Road, Crawley, RH10 6QF
G8 INZ T Prentice, 36 Ives Close, Yateley, GU46 7RD
G8 IOA Philip Crockford, 24 High Street, Easton on the Hill, Stamford, PE9 3LN
G8 IOJ D Martin, 54 The Crossway, Portchester, Fareham, PO16 8PB
G8 IOK J Noden, 1 Ashley Court, Providence Hill, Southampton, SO31 8AT
GM8 IOL Richard Thomson, Middlerig Farm, Bathgate, EH482HH
GD8 IOM Island RC c/o Michael Jerrome-Jones, Fairfield, Jurby Road, Ramsey, Isle of Man, IM7 2EB
G8 ION J Hollis, 5 Brierley Close, Dunstable, LU6 3NB
G8 IOS K Evans, 12 Moxhull Drive, Sutton Coldfield, B76 1LZ
G8 IOW P Wright, 70 Hardy Barn, Shipley, Heanor, DE75 7LY
G8 IPA A Powell, 3 Vinings Road, Sandown, PO36 8DU
G8 IPF H Billingham, Tanglewood, Brookside Orchard, Pulborough, RH20 3BD
G8 IPG Alan Shaw, 92 Freemantle Road, Romsey, SO51 0AX
G8 IPK Christopher Knight, The Lodge, 16a Cromptons Lane, Liverpool, L18 3EX
G8 IPN Christopher Foote, 3 Mere Road, Weybridge, KT13 9NU
G8 IPQ Alan Badcock, 7 Heathfield Road, Chandler's Ford, Eastleigh, SO53 5RP
G8 IPT P Hughes, 27 Hemsworth Avenue, Little Sutton, Ellesmere Port, CH66 4SG
G8 IPY Bruno Hewitt, 177 Avery Hill Road, London, SE9 2EX
G8 IQA Fred Hall, 34 Dronfield Road, Eckington, Sheffield, S21 4BR
G8 IQC M White, 5 Marlowe Close, Rogerstone, Newport, NP1 0BT
G8 IQF Christopher Newell, 16a Pembroke Road, Framlingham, Woodbridge, IP13 9HA
G8 IQT T Spicer, 3 Parkers Fields, Quorn, Loughborough, LE12 8EJ
G8 IQX Michael Dixon, 57 Northease Drive, Hove, BN3 8PP
G8 IRC D De Fraine, Block 8 Lot 3, Rosalina Village 1, Upper Libby Road, Davao City, Philippines, 8023
G8 IRL K Brown, 56 Haydock Close, Alton, GU34 2TL
G8 IRM M Emery, 45 Old Pasture Road, Frimley, Camberley, GU16 8RT
G8 IRN A Telford, 9 Fellside, Tower Wood, Windermere, LA23 3PW
G8 IRS John Wiles, 12a Ashling Gardens, Denmead, Waterlooville, PO7 6PR
G8 ISF Graham Sharp, 46 Coronation St., Monk Bretton, Barnsley, S71 2ES
G8 ISI F Breame, 68 Church Road, Bramshott, Liphook, GU30 7SH
G8 ISJ James Witt, 67 [illegible] Avenue, Coventry, CV3 5DS
G8 ISM P Goldsmith, 5 Old School Court, Rock Hill Road, Ashford, TN27 9DW
G8 ITB Richard Perzyna, 29 Lakeside Drive, Bromley, BR2 8QQ
GI8 ITD T Davidson, 26 Lower Parklands, Dungannon, BT71 7JN
GU8 ITE David Eaton, Glenfield, Le Foulon, St. Andrew, Guernsey, GY6 8UF
G8 ITG Peter Levitt, 23 Castello Drive, Birmingham, B36 9TB
G8 ITI J Evans, Rosegarth, Woodbine Road, Blackwood, NP12 1QH
G8 ITJ Michael Admans, 8 Webb Street, Nuneaton, CV10 8JQ
G8 ITU P Wragg, The Whey Inne Cottage, Main Street, Newark, NG22 8EA
G8 ITX Owen Williams, Dalree, Bush Bank, Hereford, HR4 8EN
G8 IUB BIRMINGHAM ARS c/o David Cottam, 14 Barnard Close, Rednal, Birmingham, B45 9SZ
G8 IUC Roger Glover, 8 Woodberry Way, Chingford, London, E4 7DX
G8 IUD Brian Sermons, 17 Well Side, Marks Tey, Colchester, CO6 1XG
G8 IUG P Tewkesbury, 267 York Road, Stevenage, SG1 4HD
G8 IUM M Richardson, 27 Cell Farm Avenue, Old Windsor, Windsor, SL4 2PD
G8 IUN Stephen Tolputt, Walnut Lodge, Annings Lane, Bridport, DT6 4QN

UK Callsigns

**IMPORTANT NOTE**

**Revalidate licence to avoid revocation** – Ofcom has advised the Society that plans will be drawn up to revoke licences that have not been revalidated as required by the licence conditions. The quickest way to revalidate is to do so online via the Ofcom website: *https://services.ofcom.org.uk/* or by email: *amateur.validations@ofcom.org.uk* Ofcom staff are available to help, but please be patient during times of heavy workload.

G8 IUP I Walukiewicz, Louise Cottage, Branksome Avenue, Stockbridge, SO20 6AH
G8 IUQ M Wareing, 20 Middlesex Avenue, Burnley, BB12 6AA
G8 IVB P Samson, 49 Crest View Drive, Petts Wood, Orpington, BR5 1BZ
G8 IVO R Hartland, Three Gables, Crozens Lane, Hereford, HR1 1XY
G8 IWB A Parker, 33 Colerne Drive, Hucclecote, Gloucester, GL3 3SX
G8 IWE R Thomas, 6 Copeland Drive, Poole, BH14 8NW
G8 IWF T Bierney, 5318 N 106 Avenue, Glendale, United States, 85307
G8 IWI J Pearce, 34 Fleetwood Avenue, Westcliff-on-Sea, SS0 9RA
G8 IWJ Gregory Strange, 12 Bronington Avenue, Bromborough, Wirral, CH62 6DT
G8 IWO N Jones, 14 Salcombe Grove, Swindon, SN3 1ER
G8 IWQ A Jacques, 17 Pyrethrum Way, Willingham, Cambridge, CB24 5UX
G8 IWR Thomas Campbell, 18 Cyclops Mews, London, E14 3UA
G8 IWT Richard Shears, 15 Hale Pit Road, Bookham, Leatherhead, KT23 4BS
G8 IWX B Homer, 116 Shorncliffe Road, Folkestone, CT20 2PQ
G8 IXC L Prior, 64 Montfort Road, Walderslade, Chatham, ME5 9HA
G8 IXK Brendan Owen, 21 Marlborough Road, Luton, LU3 1EF
G8 IXL P Baker, Doules Mead, Heath Lane, Farnham, GU10 5PA
G8 IXN Keith Watkins, 23 Mount Ambrose, Redruth, TR15 1NX
G8 IXP R Lister, 8 Carlton Avenue, Wilmslow, SK9 4EP
G8 IXX J Brister, 49 Tiverton Road, Loughborough, LE11 2RU
GM8 IXZ A Legood, 25 Frankfield Place, Dalgety Bay, Dunfermline, KY11 9LR
G8 IYD Dudley Hancock, 4 Elmside, Willand, Cullompton, EX15 2RN
G8 IYE Philip Shore, 1 Whatsill, Hopton Wafers, Kidderminster, DY14 0QB
G8 IYH Allen Bevington, Malthouse, Hoggs Lane, Swindon, SN5 4HQ
G8 IYJ C Buckland, 7 The Maltings, Royal Wootton Bassett, Swindon, SN4 7EZ
G8 IYK Robert Sayers, 3 Riversdale Cottages, The Staithe, Norwich, NR12 9BY
G8 IYN C Marsh, 6 De Burgh Hill, Dover, CT17 0BS
G8 IYS J Simkins, 18 Riding Hill, South Croydon, CR2 9LN
G8 IYZ A Barker, 8 Manor Avenue, Attenborough, Nottingham, NG9 6BP
GI8 IZB BEANEATERS DX GROUP c/o J Crawford-Baker, Georges Nest, 131 Gobbins Road, Larne, BT40 3TX
G8 IZR P Higginson, 18 Park Meadow, Westhoughton, Bolton, BL5 3UZ
G8 IZW P Cain, 22 Ditton Green, Luton, LU2 8RU
G8 IZY S Eldridge, 6 Cobbles Crescent, Crawley, RH10 8HA
G8 JAB A Berriman, Meadowside, Little-in-Sight, St. Ives, TR26 1AX
G8 JAC Andrew Jackson, 59 Leas Road, Warlingham, CR6 9LP
G8 JAD John Townsend, 56 Seymour Road, Northfleet, Gravesend, DA11 7BN
G8 JAI Anthony Livesley, Gates Garth, Barbon, Carnforth, LA6 2LJ
G8 JAN P Biggadike, 49 Willow Road, Downham Market, PE38 9PG
G8 JAW B Heed, 3 Woodcote Green, Downley, High Wycombe, HP13 5UN
G8 JAY A Jay, Jasper, The Reddings, Cheltenham, GL51 6RT
G8 JBC C Jervis, 8 Portobello Close, Willenhall, WV13 3QA
G8 JBD P Godfrey, 3 Lowry Way, Lowestoft, NR32 4LW
G8 JBJ J Berry, 4 Newlands Park Way, Newick, Lewes, BN8 4PG
G8 JBM Spencer Wood, 18 Grange Road, Shanklin, PO37 6NN
G8 JBP G Head, 34 Balds Lane, Stourbridge, DY9 8SG
G8 JBQ R Hughes, Court Church View, South Perrott, Beaminster, DT8 3HU
G8 JBT D Bellingham, 22 Princes Drive, Codsall, Wolverhampton, WV8 2DJ
G8 JBV David Dawe, 7 Princes Road, Romford, RM1 2SR
G8 JCB P Pullinger, 1 Sycamore Cottages, Upper Wield, Alresford, SO24 9RP
G8 JCC C Purchase, 35 Pasture Way, Bridport, DT6 4DW
G8 JCD Michael Northey, Achill Mist House, Kilmeaney, Listowel, Ireland, CO KERRY
GM8 JCF Peter Carnegie, 29 Castle Terrace, Cullen, Buckie, AB56 4SD
G8 JCL J Essex, 40 Lincoln Walk, Heywood, OL10 3JB
G8 JCN W Allen, 143 Cherry Crescent, Rawtenstall, Rossendale, BB4 6DS
G8 JCS Anthony Bunting, Manor House, Market Place, Market Rasen, LN8 6DE
G8 JCV P Hewitt, 28 Amersham Avenue, Langdon Hills, Basildon, SS16 6SJ
GW8 JDB V Grayson, Willow Lodge, Croeslan, Llandysul, SA44 4SJ
G8 JDC Thomas Robinson, 17 Balliol Road, Brackley, NN13 6LY
G8 JDD Robert Kelsall, The Cottage, Denford Road, Stoke-on-Trent, ST9 9QG
G8 JDN Ben Deefholts, 64 Bridge End, London, E17 4ES
G8 JDQ K Few, 35 Whitton Close, Swavesey, Cambridge, CB24 4RT
G8 JEI N Cross, Glan Alaw, Llanddeusant, Holyhead, LL65 4AG
G8 JEM E Cheer, 15 Stibbs Way, Bransgore, Christchurch, BH23 8HG
GM8 JET David Higginson, Glencoe, Hagg Crescent, Johnstone, PA5 8TA
G8 JFC F Wilmott, 2 Manor Close, Misson, Doncaster, DN10 6HE
G8 JFL D Crough, 32 Roundaway Road, Ilford, IG5 0NP
G8 JFT N Hewitt, 36 Princes Terrace, Kemp Town, Brighton, BN2 5JS
G8 JFX T Simmons, Cedar House, Blueberry Close, Northampton, NN6 9XL
GM8 JGB W Fleming, 65 Dundonald Park, Cardenden, Lochgelly, KY5 0DG
G8 JGE Christopher Newbury, 37 Johns Avenue, Hendon, London, NW4 4EN
G8 JGF Paul Walters, 3 Inkerman Street, Selston, Nottingham, NG16 6BQ
G8 JGL N Owen, 59 Fernwood Drive, Leek, ST13 8JA
G8 JGM J Martin, 19c Willow Tree Road, Altrincham, WA14 2EQ
G8 JGU R Hallam, 37 Dingle Avenue, Appley Bridge, Wigan, WN6 9LF
G8 JHA T White, 24 Chapel Street, Tingley, Wakefield, WF3 1RE
G8 JHC I Whitworth, 104 The Dormers, Highworth, Swindon, SN6 7PD
G8 JHE Michael Brogan, 31 Rempstone Road, East Leake, Loughborough, LE12 6PW
G8 JHG John Collins, Hill Crest, The Hill, Millom, LA18 5HB
G8 JHH Martin Baugh, 71 Hatch Lane, Old Basing, Basingstoke, RG24 7EF
G8 JHL J Lovell, 2 Moran Close, Wilmslow, SK9 3UF
G8 JHM I Carney, 39 Blenheim Crescent, Luton, LU3 1HB
G8 JHO P Evans, 5 Hunters Close, Bilston, WV14 7BN
G8 JIE Christopher Riding, 14 The Coppice, Clayton le Moors, Accrington, BB5 5RU
G8 JIP Graeme Miller, 39 Scrivens Mead, Thatcham, RG19 4FQ
G8 JIS T Macey, Whitegates, Histons Hill, Wolverhampton, WV8 2HA
G8 JIT John McKinnon, 142 Hughes Street, Bolton, BL1 3EZ
G8 JIU P Dunham, 19 The Lunds, Kirk Ella, Hull, HU10 7JJ
G8 JJE B Thompson, 21 Birling Place, Corby, NN18 0LZ
G8 JJF John Puddifoot, 49 Richmond Road, Wolverhampton, WV3 9JG
G8 JJK T Barrett, Flat 38, Queens Court, Cheltenham, GL50 2LU
GM8 JJN J Pryde, 7 The Engine Green, Fishcross, Alloa, FK10 3JN
G8 JJP Peter Tabberer, 8 Wynnstay Road, Old Colwyn, Colwyn Bay, LL29 9DS
G8 JJR K McMahon, 27 Marlborough Avenue, Doncaster, DN5 8EH
GW8 JJZ R Merrick-Jenkins, 165 Victoria Road, Port Talbot, SA12 6QJ
G8 JKB C Hemmings, 11 Brookside, Desborough, Kettering, NN14 2UD
G8 JKC J Kendall, 26 Bryn Seiri Road, Conwy, LL32 8NR
G8 JKD Cedric Littman, 70 Orbel Street, London, SW11 3NY
G8 JKV D Leary, 200 Gilbert Road, Cambridge, CB4 3PB
G8 JLA K Turner, 13 Stanhope Street, Saltburn-by-The-Sea, TS12 1AL
G8 JLB B Silver, 280 Britten Road, Brighton Hill, Basingstoke, RG22 4HR
G8 JLD J Garters, Sun Patch, Garfield Road, Hailsham, BN27 2BT
G8 JLM P Higham, 56 Coopers Avenue, Heybridge, Maldon, CM9 4YX
GW8 JLY L Leach, 4 Ollivant Close, Llandaff, Cardiff, CF5 2RJ
G8 JMB J Button, 16 Meadow Rise, Broadstone, BH18 9ED
G8 JMG John Gartland, 175 Talbot Street, Whitwick, Coalville, LE67 5AY
G8 JMK D Butler, 144 Longridge Way, Weston-Super-Mare, BS24 7HS
G8 JMO Brian Justin, 1704 Cottontown, Forest Virginia, United States, 24551
G8 JMP Donald Beech, 8 Copthorne Drive, Lightwater, GU18 5TE
G8 JMS Stuart Miller, 44 Greenway Road, Galmpton, Brixham, TQ5 0LZ
G8 JMU J Potter, 15 Alterton Close, Goldsworth Park, Woking, GU21 3DD
G8 JMY D Hugman, 7 St. Michaels Close, North Waltham, Basingstoke, RG25 2BP
G8 JNI D Bookham, 1 Monks Rise, Fleet, GU51 4HB
G8 JNJ Martin Ehrenfried, 160 Botley Road, Romsey, SO51 5SW
G8 JNO Stephen Munday, 25 Southend Road, Weston-Super-Mare, BS23 4JY
G8 JNR R Hedderley, 17 Linford Close, Handsacre, Rugeley, WS15 4EF
G8 JNZ Karen Crowder, 15 Fleetwood Close, Minster on Sea, Sheerness, ME12 3LN
GI8 JOA D Thompson, 16 Lynden Gate Park, Portadown, Craigavon, BT63 5YJ
G8 JOC E Powell, 49 Normanby Road, Worsley, Manchester, M28 7TS
G8 JOX Joseph Dobson, Home Farm, Mansmore Lane, Kidlington, OX5 2US
G8 JOY T Bowen, 7 Bedford Close, Greenmeadow, Cwmbran, NP44 5HN
G8 JPA J Hunt, Woodstone Farm, High Common ROAD, Diss, IP22 2HS
GI8 JPF T Phillips, 52 Belfast Road, Bangor, BT20 3PU
G8 JPJ D Jones, 184 Harwich Road, Little Clacton, Clacton-on-Sea, CO16 9PU
G8 JPU D Potts, 25 Southlands Road, Congleton, CW12 3JY
G8 JPV M Parnell, 101 Ridgeway, Wellingborough, NN8 4RZ
G8 JPW J Abbott, 11 Red House Road, Bodicote, Banbury, OX15 4BB
G8 JQG J Hough, 77 Pennine Court, Macclesfield, SK10 2RN
G8 JQH P Wright, 33b Slack Lane, Crofton, Wakefield, WF4 1HX
G8 JQS G Greensmith, Japonica, Hawthorne Avenue, Westerham, TN16 3SG
G8 JQV D Marchant, 11 Derehams Lane, Loudwater, High Wycombe, HP10 9RH
G8 JQW Roger Thomas, 24 Trowell Grove, Trowell, Nottingham, NG9 3QH
GI8 JRE J Donnelly, 9 Lomond Heights, Cookstown, BT80 8XW
G8 JRF M Willson, 19 The Willows, Highworth, Swindon, SN6 7PG
G8 JRL C Jones, 21 Hallfield Close, Flint, CH6 5HL
G8 JRN R Stockdale, 53 Brightwalton, Newbury, RG20 7BT
G8 JRW M Austin, The House, Four Seasons Village, Winkleigh, EX19 8DP
G8 JRZ A Mills, 42 Mora Avenue, Chadderton, Oldham, OL9 0EJ
G8 JSC K Austin, 139 Sewall Highway, Coventry, CV2 3NG
G8 JSE Frank Cowlin, Ululantes, 9 Zealand Close, Hinckley, LE10 1TJ
G8 JSF R Williams, 35 Broadhurst Grove, Lychpit, Basingstoke, RG24 8SB
G8 JSL P Smith, 13 Manor Garth, Pakenham, Bury St. Edmunds, IP31 2LB
G8 JSM C Wood, 57 Holly Crescent, Rainford, St. Helens, WA11 8ER
G8 JSN P Bailey, 50 Amis Avenue, New Haw, Addlestone, KT15 3ET
G8 JSR V Hinksman, 1 Shaw Lane, East Woodburn, Hexham, NE48 2SL
G8 JTD OTLEY ARS c/o Richard Leach, 4 Honey Pot Drive, Shipley, BD17 5TJ
G8 JTG E Spanton, 14 Days Lane, Sidcup, DA15 8JN
G8 JTL M Davies, 25 Walker Avenue, Quarry Bank, Brierley Hill, DY5 2LY
G8 JUC J Wheatley, 44 Kingswood Close, Boldon Colliery, NE35 9LG
G8 JUG N Spenceley, 70 Kingston Road, Teddington, TW11 9HY
G8 JUK B Storeton-West, Nazdar, Camps Heath, Lowestoft, NR32 5DW
G8 JUS T Gale, 58 Westwood Road, Newbury, RG14 7TL
G8 JUT Stephen Linton, 41 Long Close, Bristol, BS16 2UF
G8 JUV S York, 10 Beechwood Avenue, Wallasey, CH45 8NX
GM8 JUY R McMillan, 12 Parkthorn View, Dundonald, Kilmarnock, KA2 9EZ
G8 JVA Christopher Arscott, 13 Larwood Place, Oldbrook, Milton Keynes, MK6 2PZ
G8 JVE M Rowe, 97 Old Worthing Road, East Preston, Littlehampton, BN16 1DU
G8 JVI Albert Hicks, 10 Evans Close, Eynsham, Witney, OX29 4QY
G8 JVM R Bown, Park View, Chapel Street, Telford, TF4 3DD
G8 JVS M Fairey, Boston Gates, Whinns Lane, Wetherby, LS23 7AL
G8 JVU A Johnson, Clematis Cottage, Wheatlow Brooks, Stafford, ST18 0EW
G8 JVV Jonathan Burchell, Gooseleys Farm, Harrow Hill, Halstead, CO9 4LX
G8 JVW P Boswell, 10 The Grange, Wombourne, Wolverhampton, WV5 9HX
GM8 JVZ M Nimmo, The Court, 6 Farington Street, Dundee, DD2 1PJ
G8 JWC David Luscombe, 31 Tewkesbury Drive, Prestwich, Manchester, M25 0HR
G8 JWD I Rees, Knowle Cottage, Whittonditch Road, Marlborough, SN8 2PX
G8 JWE J Hickman, 41 Field Road, Ramsey, Huntingdon, PE26 1JP
G8 JWK R Staveley, 52 New Road, Wootton Bassett, Swindon, SN4 7DG
GW8 JWL G Smith, 13 Lapwing Close, Penarth, CF64 5GA
G8 JWP John Griffiths, 1 The Lawn, Rhymney, Tredegar, NP22 5LS
GM8 JWQ K Faloon, Moss-Side Croft, 6 Rothiemay, Huntly, AB54 5NY
G8 JWT Roger Trett, 40 High Street, Stoke Goldington, Newport Pagnell, MK16 8NR
G8 JXG John Dean, 6 Greenleas, Pembury, Tunbridge Wells, TN2 4NS
G8 JXK S Blew, 24 Batts Park, Taunton, TA1 4RE
G8 JXP D McCabe, 78 Oakleigh Road, Stratford-upon-Avon, CV37 0DN
G8 JXS Michael Stephenson, 6 Cedar Road, Tewkesbury, GL20 8PX
G8 JXU Clive West-Bulford, 25 Sunnyside Close, Heacham, King's Lynn, PE31 7DX
G8 JXV T Trew, Stockers Lodge, Bere Farm Lane, Fareham, PO17 6JJ
G8 JYN Basingstoke ARC c/o Paul Cresswell, 108 Hawthorn Way, Basingstoke, RG23 8NH
G8 JYS M Fletcher, 88 Langton Road, Norton, Malton, YO17 9AE
G8 JYV K Dumbill, 30 Caithness Drive, Crosby, Liverpool, L23 0RQ
G8 JYX P Johnson, 42 College Gardens, London, E4 7LG
G8 JZI Peter Smith, 4 Fellstone Vale, Withnell, Chorley, PR6 8UE
G8 JZO Jonathan Gibbs, 6 Southampton Close, Blackwater, Camberley, GU17 0HB
G8 JZT S Osborn, 67 Chessington Avenue, Bexleyheath, DA7 5NP
G8 JZX Christopher Stephenson, Armanby, Main Street, Selby, YO8 8QT
G8 JZZ R Taylor, Higher Priestacott, Belstone, Okehampton, EX20 1QX
G8 KAE Roger Bushell, 102 Winchester Gardens, Northfield, Birmingham, B31 2QB
G8 KAM J Hurnandies, 70 Orchard Rise West, Sidcup, DA15 8SZ
G8 KAP David Patrick, Quarryside, Stockdalewath, Carlisle, CA5 7DP
G8 KAS B Buschl, 27 The Drive, Court Farm Road, Newhaven, BN9 9DJ
G8 KB Philip Johnson, 55 Rodney Hill, Loxley, Sheffield, S6 6SG
G8 KBB D Roberts, 32 Woodbridge Close, Appleton, Warrington, WA4 5RD
G8 KBG A Price, Botterham House, Botterham, Dudley, DY3 4RA
G8 KBH D Ward, 3 Sherbourne Close, Poulton-le-Fylde, FY6 7UB
G8 KCB J Nally, 313 Wyndhurst Road, Stechford, Birmingham, B33 9DL
G8 KCH K Houston, 6 Ashgrove, Llanellen, Abergavenny, NP7 9HP
GW8 KCY Mark Bover, Glynfach Bungalow, Pontyates, Llanelli, SA15 5TF
G8 KDD D Coton, 17 Flambards Close, Meldreth, Royston, SG8 6JX
G8 KDF Martin Sach, Old School, Cambridge Road, St. Neots, PE19 6ST
G8 KDM A Smith, 2a Chesterfield Road, Barlborough, Chesterfield, S43 4TR
G8 KDO P Topham, 5 Kings Road, Cambridge, CB3 9DY
G8 KDU Robert Eager, 45 Fleetwood Avenue, Herne Bay, CT6 8QW
G8 KEA M Sutton, 178 Cole Lane, Borrowash, Derby, DE72 3GN
G8 KED C Mullineaux, 27 Ashfield Avenue, Lancaster, LA1 5EB
G8 KEJ M Johnson, 23 The Crest, Surbiton, KT5 8JZ
G8 KEK P Wilson, 5 Mons Close, Harpenden, AL5 1TD
G8 KEO Peter Dickinson, Halshanger Farm, Ashburton, Newton Abbot, TQ13 7HY
GI8 KEP K Bones, 54 Derryvolgie Park, Lisburn, BT27 4DA
GW8 KEV Kevin Shafto, 67 Boverton Road, Llantwit Major, CF61 1YA
G8 KFD R Gwynn, 36 Woodstock Close, Burbage, Hinckley, LE10 2EG
G8 KFF R Parker, 17 Valley Road, Streetly, Sutton Coldfield, B74 2JE
GI8 KFG P Douglas, 21 Hillhead Road, Ballycarry, Carrickfergus, BT38 9HE
G8 KFJ D Greig, 23 Parsons Walk, Walberton, Arundel, BN18 0PA
G8 KFK P Loten, 15 Hornsea Burton Road, Hornsea, HU18 1TP
G8 KFN R Heron, 46 Bradvue Crescent, Bradville, Milton Keynes, MK13 7AJ
G8 KFS Brian Russell, 56 Kingsmead Avenue, Surbiton, KT6 7PP
G8 KGC NUNSFIELD HSE RD c/o D Barker, 311 Uttoxeter Road, Mickleover, Derby, DE3 9AH
G8 KGE S Bailey, 50 Quantock Close, Warmley, Bristol, BS30 8UT
G8 KGG Andrew Ward, 49 Spielplatz, Lye Lane, St. Albans, AL2 3TD
G8 KGK Gordon Higton, Hillcrest, Cow Brow, Carnforth, LA6 1PJ
G8 KGR R Tidswell, Helloplane, Clubhurn Lane, Spalding, PE11 4BQ
G8 KGS Christopher Suslowicz, 2366 Coventry Road Sheldon, Birmingham, B26 3LS
G8 KGV Paul Jessop, 84 Common Road, Kensworth, Dunstable, LU6 3RG
G8 KHF John Dove, 33 The Haystack, Daventry, NN11 0NZ
G8 KHH C Young, 26 Horsham Avenue, Peacehaven, BN10 8HX
G8 KHI R Partridge, 34 Milestone Close, Stevenage, SG2 9RR
G8 KHU David Fielding, 216 Andover Road, Newbury, RG14 6PY
G8 KHV R Evans, 6 Park End, Lichfield, WS14 9US
G8 KIG Paul Winwood, 2 The Warren, Abingdon, OX14 3XB
G8 KIH J Sargent, 9 Lee Woottens Lane, Basildon, SS16 5HD
G8 KIK D Bland, 17 Knowles Close, Kirklevington, Yarm, TS15 9NL
GM8 KIQ John Harper, 11 Cathburn Holding, Cathburn Road, Wishaw, ML2 9QL
G8 KIW Heather Muller, 118 Park Lane, Northampton, NN5 6PZ
G8 KIZ Samuel Morris, 23 Ellesmere Way, Carlisle, CA2 6LZ
G8 KJI Jonathan Richardson, 4 Torrington Lane, East Barkwith, Market Rasen, LN8 5RY
G8 KJJ L Haywood, 9 Canberra Crescent, West Bridgford, Nottingham, NG2 7FL
G8 KJK Graham Park, 34 Delafield Road, Abergavenny, NP7 7AW
GM8 KJO David Moodie, 1 Lageonan Road, Grandtully, Aberfeldy, PH15 2QY
G8 KJP Peter King, 25 Lamellyn Drive, Truro, TR1 3JR
G8 KJT Roger Burgess, 3 Deeside Avenue, Chichester, PO19 3QF
G8 KKA B Stevens, 2 Hawthorn Crescent, Shepton Mallet, BA4 5XR
G8 KKD David Jones, Garsdon Mill, Garsdon, Malmesbury, SN16 9NR
G8 KKH C Hills, 8 Blackdale, Cheshunt, Waltham Cross, EN7 6DF
G8 KKN D McFarlane, 10 Green Lane, Vicars Cross, Chester, CH3 5LA
G8 KKU J Walker, 21 Garden Hedge, Leighton Buzzard, LU7 1DJ
G8 KLC P Webber, 60 Trowley Hill Road, Flamstead, St. Albans, AL3 8EE
G8 KLE B Jenkins, Bryher, 4 Halt Road, Truro, TR4 9QE
G8 KMK KIRKLESS RAYNET c/o Gerald Edinburgh, 77 Westerley Lane, Shelley, Huddersfield, HD8 8HP
G8 KMM J Bryant, 12 Dale Tree Road, Barrow, Bury St. Edmunds, IP29 5AD
G8 KMP M Pollock, 25 Meadow Lane, Burgess Hill, RH15 9HZ
G8 KMR Mike Davis, 8 Mead Close, Leckhampton, Cheltenham, GL53 7DX
G8 KNC Stephen Wheatley, 18 Orchard Way, Hurstpierpoint, Hassocks, BN6 9UB
G8 KNF Duncan Hawkins, 109 Elphinstone Road Walthamstow, London, E17 5EY
G8 KNJ T Blinco, 9 Powell Close, Forest Hill, Oxford, OX33 1EN
G8 KNN Jonathan Bigwood, 133 Gilbert Road, Cambridge, CB4 3PA
G8 KNS M Jelfs, Adams Acre, Chapel Lane, Wimborne, BH21 3SL
G8 KNU R Jacobs, 1 Coverdale, Northampton, NN2 8UU
G8 KOC Roy Backham, 15 Rushmead Close, South Wootton, Kings Lynn, PE30 3LY
G8 KOD Raymond Adams, Swinneys, Station Road, Carterton, OX18 3PR
G8 KOE Martin Newell, 12 Pooles Close, Nether Stowey, Bridgwater, TA5 1LZ
G8 KOF D M McNaughton, 6 Wilderhaugh Court, Galashiels, TD1 1QL
G8 KOL D Slocombe, 7 Talbot Avenue, Herne Bay, CT6 8AD
G8 KOM D Hanson, 42 Choseley Road, Knowl Hill, Reading, RG10 9YT
G8 KOQ N Morris, 88 Tynesbank, Worsley, Manchester, M28 0SL
G8 KOS Stephen Head, 3 Ripon Gardens, Waterlooville, PO7 8ND
G8 KOV Derek Dunn, Allahoo, Curlew Drive, Kingsbridge, TQ7 2AA
G8 KPD B Fothergill, 53 Meadow Court, Ponteland, Newcastle upon Tyne, NE20 9RA
G8 KPE Edward Howard, 15 Amherst Road, Bexhill-on-Sea, TN40 1QH
G8 KPG G Wright, 58 Lifton Croft, Kingswinford, DY6 8RZ
GM8 KPH Martin Hobson, 17 Well Brae, Pitlochry, PH16 5HH
G8 KPV Graham Hickman, Pine Tree Cottage, Calverton Road, Blidworth, Mansfield, NG21 0NW
G8 KPY D Pratt, 77 Hayfield Road, St. Mary Cray, Orpington, BR5 2DL
G8 KQA R Laslett, Dinnages, Street End Lane, Heathfield, TN21 8SA
G8 KQB Stephen Prior, East Brantwood, Manor Road, Barnstaple, EX32 0JN
G8 KQH O HARVEY, 33 Copthall Way, New Haw, Addlestone, KT15 3TU
G8 KQV S Evans, 4 Holcot Lane, Anchorage Park, Portsmouth, PO3 5TR
G8 KQZ G Dawkins, 8 Chancery Lane, Eye, Peterborough, PE6 7YF

| | | |
|---|---|---|
| G8 | KRB | Keith Barnes, Fallseat Close, Totnes, TQ9 5AN |
| [illegible] | [illegible] | [illegible] Park, Whitefield, Manchester, M45 7NT |
| G8 | KRT | Roger Horne, 51 Welland Court, Higham, [illegible] |
| G8 | KRV | J Cottier, 83 Elizabeth Drive, Tamworth, B79 8DE |
| G8 | KSA | W Hall, 67 Selwyn Drive, Stockton-on-Tees, TS19 8XF |
| G8 | KSC | D Goodwin, 41 Newpool Road, Knypersley, Stoke-on-Trent, ST8 6NT |
| G8 | KSD | Alan Hewett, 1 Mountside, Westfield Lane, Folkestone, CT18 8BY |
| G8 | KSE | Walter Salisbury, 28 Dyke Street, Brymbo, Wrexham, LL11 5AH |
| G8 | KSF | Alan Salisbury, 28 Dyke St., Brymbo, Wrexham, LL11 5AH |
| G8 | KSH | Alison Wilkins, 2 Beechfield Crescent, Banbury, OX16 9AR |
| GM8 | KSJ | David Cowie, 8 Centre Street, Kelty, KY4 0EQ |
| GW8 | KSL | Roland Cleaver, 61 Llewellyn Park Drive Morriston, Swansea, SA6 8PF |
| G8 | KSM | Rick Beament, Midlands Farm, Horndon, Tavistock, PL19 9NQ |
| G8 | KST | T Mayer, 61 Rawley Crescent, New Duston, Northampton, NN5 6PU |
| G8 | KSW | J Wood, 38 Beech Lane, West Hallam, Ilkeston, DE7 6GU |
| G8 | KSX | Andy Thompson, Carloway, Turner Lane, Ilkley, LS29 0LE |
| G8 | KSZ | I Newbold, 40 Heath Close, Stonnall, Walsall, WS9 9HU |
| G8 | KTA | P Thomas, 76 Church Road, Braunston, Daventry, NN11 7HQ |
| G8 | KTC | M Rhys, 2 Sun Lane, Teignmouth, TQ14 8EF |
| G8 | KTE | Colin Price, 4 Greenway Close, Helsby, Frodsham, WA6 0QX |
| G8 | KTG | D Smith, 76 Reigate Road, Brighton, BN1 5AG |
| G8 | KTV | D Adams, 7 Kingston Park, Pennington, Lymington, SO41 8ES |
| G8 | KTX | M Butler, 7 Bassett Road, Coventry, CV6 1LF |
| G8 | KUA | C Bridgland, 10 Eastlands Grove, Stafford, ST17 9BE |
| G8 | KUV | A Simonds, 3 Links Close, Seaford, BN25 4NU |
| G8 | KUZ | J Wiggins, 35 Downing Avenue, Newcastle, ST5 0LB |
| G8 | KVN | Andrew Nelson, 29 Coxford Road, Southampton, SO16 5FG |
| G8 | KVO | C Miller, Broomwood, South Park, Sevenoaks, TN13 1EL |
| G8 | KVU | C Smith, 48 Sherbourne Crescent, Coventry, CV5 8LE |
| G8 | KW | Richard Shears, 15 Hale Pit Road, Bookham, Leatherhead, KT23 4BS |
| G8 | KWD | G Bettley, 1 Dovetrees, Covingham, Swindon, SN3 5AX |
| G8 | KWH | N Liddle, 135 York Road, Acomb, York, YO24 4NP |
| G8 | KWJ | D Barnwell, Bernagh, Duncombe Street, Kingsbridge, TQ7 1LR |
| G8 | KWN | R Bryant, 81 Dukes Drive, Halesworth, IP19 8TJ |
| G8 | KWP | A Darragh, The Gables, Belle Vue Lane, Chester, CH3 7EJ |
| G8 | KWV | J Bailey, 27 West Mead, Ewell, Epsom, KT19 0BJ |
| GM8 | KXF | Gordon Robb, 3 Doonholm Park, Ayr, KA6 6BH |
| G8 | KXO | B Gamble, 79 Humphries House, Lindon Drive, Walsall, WS8 6DL |
| GW8 | KXW | John Watts, 6 Castle View, Haverfordwest, SA61 2JA |
| GI8 | KYI | T Carlisle, 55 North Road, Carrickfergus, BT38 8NA |
| G8 | KYK | C Keens, 3 Kirk Gardens, Totton, Southampton, SO40 9UZ |
| GW8 | KZA | J Wells, 30 St. Andrews Road, Barry, CF62 8BR |
| G8 | KZG | Peter Delaney, 6 East View Close, Wargrave, Reading, RG10 8BJ |
| G8 | KZJ | E Lockyear, 140 Andover Road, Orpington, BR6 8BL |
| G8 | KZN | W Clinton, 5 Moorland Crescent, Castleside, Consett, DH8 9RF |
| G8 | KZO | R Edgeley, 6 Hearne Gardens, Shirrell Heath, Southampton, SO32 2NR |
| G8 | KZY | C Denison, 40 Leysholme Drive, Leeds, LS12 4HQ |
| G8 | LAB | R Harste, 2 Park Drive, Ingatestone, CM4 9DT |
| G8 | LAM | R Lambley, 31 Ridgeway Road, Redhill, RH1 6PQ |
| G8 | LAN | Robert Garner, Flat 4, Beaufort Court, Clevedon, BS21 7PQ |
| G8 | LAU | David Peck, 3 Dearnford Avenue, Wirral, CH62 6DX |
| G8 | LAY | E Hibbett, Trumps Lodge, Broad Street, Ottery St. Mary, EX11 1BY |
| GM8 | LBC | C Dalziel, 2 Alder Avenue, Hamilton, ML3 7LL |
| G8 | LBG | J Cook, Highlands, Littledown, Shaftesbury, SP7 9HD |
| G8 | LBS | C Ranson, 281 Hawthorn Drive, Ipswich, IP2 0QG |
| G8 | LBT | Martin Rigby, 16 Juniper Way, Penrith, CA11 8UF |
| G8 | LCA | J Scott, 123 Cotswold Way, Tilehurst, Reading, RG31 6SR |
| G8 | LCC | C Forster, 48 Woolsington Gardens, Woolsington, Newcastle upon Tyne, NE13 8AR |
| G8 | LCE | M Perrett, Magpie Cottage, Bar Lane, Falmouth, TR11 4BL |
| G8 | LCI | Arthur Goode, 445 Street Lane, Leeds, LS17 6HQ |
| GI8 | LCJ | David Craig, 40 Chilton Road, Carrickfergus, BT38 7JT |
| G8 | LCK | Lee Reynolds, 31 Notre Dame Road, Lille, United States, 4746 |
| G8 | LCL | S Tames, 21 Lind Close, Earley, Reading, RG6 5QX |
| G8 | LCM | K Day, Powys Lodge, 6 Court Road, Worcester, WR8 9LP |
| G8 | LCP | N Jamieson, 1 Langdale Place, Newton Aycliffe, DL5 7DX |
| G8 | LCS | J Monte, 11 Woodfield Avenue, Hyde, SK14 5BB |
| G8 | LCZ | J Sellick, 7 The Boulevard, Lytham St. Annes, FY8 1EH |
| G8 | LDB | K Oldham, 165 Mountsorrel Lane, Rothley, Leicester, LE7 7PU |
| G8 | LDC | J Salthouse, 10 Ramillies Avenue, Cheadle Hulme, Cheadle, SK8 7AL |
| G8 | LDJ | C Douglas, 22 Connaught Road, Sittingbourne, ME10 1EH |
| G8 | LDU | George Noble, 19 Atlas Close, Kings Hill, West Malling, ME19 4PS |
| G8 | LDV | B Harrad, 32 Woodfield Avenue, Northfleet, Gravesend, DA11 7QG |
| G8 | LDW | P Harness, 7 Castlegate, Gipsey Bridge, Boston, PE22 7BS |
| G8 | LDY | Robert TOMPKINS, 16 Garden Close, Watford, WD17 3DP |
| GM8 | LEA | Norman Adam, Bridaig Villa, Gladstone Avenue, Dingwall, IV15 9PG |
| G8 | LEB | R Hill, Rose Lodge, 35 Colne Fields, Huntingdon, PE28 3DL |
| G8 | LED | Northampton RC c/o John Cockrill, 28 Northampton Road, Harpole, Northampton, NN7 4DD |
| G8 | LEG | Keith Hardy, 4 Forest Hill, Maidstone, ME15 6UU |
| G8 | LEM | H Griffith, 9 Devonshire Road, West Kirby, Wirral, CH48 7HR |
| G8 | LES | M Sanders, 39 Telegraph Lane, Four Marks, Alton, GU34 5AX |
| G8 | LF | Edgar Byrne, 40 Wentworth Avenue, Ascot, SL5 8HQ |
| GM8 | LFB | James Rabbitts, 38 Murchison Street, Wick, KW1 5HW |
| GM8 | LFI | M Cartmell, 33 Orrok Park, Edinburgh, EH16 5UW |
| GI8 | LFY | A Penn, 9 Milltown Road, Donaghcloney, Craigavon, BT66 7NE |
| G8 | LGA | R Ward, 1 Horton, Downswood, Maidstone, ME15 8TN |
| G8 | LGC | J Williams, 24 Hilltop Gardens, Denaby Main, Doncaster, DN12 4SB |
| G8 | LGE | P Devine, 3 The Hawthorns, Outwood, Wakefield, WF1 3TL |
| G8 | LGM | Robert Field, 20 Hill Road, Watlington, OX49 5AD |
| G8 | LGP | Keith Harris, 20 Westminster Close, Devizes, SN10 1BF |
| G8 | LGS | P Chitty, 109 Bannings Vale, Saltdean, Brighton, BN2 8DH |
| G8 | LGT | D Blakemore, 20 Northfield Road, Coventry, CV6 2HB |
| G8 | LGU | R Milliken, 15 Lee Grove, Chigwell, IG7 6AD |
| G8 | LGY | R Tyson, 18 Blackthorn Close, Gedling, Nottingham, NG4 4AU |
| G8 | LHD | D Allen, 21 Goldings Close, Haverhill, CB9 0EQ |
| G8 | LHF | P Earl, Holly Cottage, Popes Lane, Colchester, CO6 2DZ |
| G8 | LHI | Martin Levy, Flat 11, Deepdene Court, Kingswood Road, Bromley, BR2 0NW |
| G8 | LHP | A Milne, 49 Cleevemount Road, Cheltenham, GL52 3HF |
| G8 | LHQ | M [illegible], 30 [illegible] Avenue, London, SW4 9HD |
| [illegible] | [illegible] | [illegible], 81 Littlehaven Lane, Horsham, RH12 4JB |
| G8 | LHT | I Harwood, 56 Spring [illegible] |
| G8 | LHW | Philip Cunnington, 7 Torquay Close, Rayleigh, SS6 9PH |
| G8 | LHZ | P Avon, 81 Parsonage Barn Lane, Ringwood, BH24 1PU |
| G8 | LID | Norman Dowler, 1 Cottage Walk, Clacton-on-Sea, CO16 8DG |
| G8 | LIE | Neil Borrell, 12 Nutfield Close, Hemlington, Middlesbrough, TS8 9QQ |
| G8 | LIH | G Storey, 27 Dyche Road, Sheffield, S8 8DQ |
| G8 | LII | J Lee, 225 Avenue Road, Rushden, NN10 0SN |
| G8 | LIK | Steven Hurst, Fareview, Woodhead Road, Holmfirth, HD9 2PX |
| G8 | LIP | B Greenbeck, 10 Campbell Avenue, Bottesford, Scunthorpe, DN16 3SA |
| G8 | LIU | N Clyne, 78 Halford Road, Ickenham, Uxbridge, UB10 8QA |
| G8 | LIX | R Keates, 35 Walsh Grove, Birmingham, B23 5XE |
| G8 | LIY | David Henn, 14 Spring Close, Rode Heath, Stoke-on-Trent, ST7 3TQ |
| GW8 | LJJ | E Edwards, 11 Old Village Road, Barry, CF62 6RA |
| G8 | LJQ | C Asquith, 142b Newbegin, Hornsea, HU18 1PB |
| G8 | LJU | J Spicer, 6 Avenue Road, Worcester, WR2 4ES |
| G8 | LJY | Alan Griffiths, 17 Ferenberge Close, Farmborough, Bath, BA2 0DH |
| G8 | LKA | San Whitehead, 4 Colleton Crescent, Exeter, EX2 4DG |
| G8 | LKB | Ian Rabson, 50 Burwell Meadow, Witney, OX28 5JQ |
| G8 | LKK | Roger Horsford, 2 Old Mill, Mill Lane, Chard, TA20 2ND |
| GM8 | LKL | Alan Hogg, 43 Muir Wood Road, Currie, EH14 5JN |
| G8 | LKP | Joseph Duchscherer, 36 Hamdon Close, Stoke-Sub-Hamdon, TA14 6QN |
| G8 | LKQ | Dermot Falkner, 45 Westwood, Carleton, Skipton, BD23 3DW |
| G8 | LKS | D Burton, 48 West Beeches Road, Crowborough, TN6 2AG |
| G8 | LKW | H Colville, 185 West Heath Road, Northfield, Birmingham, B31 3HD |
| GW8 | LKX | Michael Corrigan, 3 Heathway, Heath, Cardiff, CF14 4JQ |
| G8 | LLD | P Pritchard, 5 Hilldene Close, Flitwick, Bedford, MK45 1AQ |
| G8 | LLJ | M Tutt, 9 Russell Drive, Dunbridge, Romsey, SO51 0RA |
| G8 | LLS | S Perkins, 6 Delamere Road, Malvern, WR14 2BQ |
| G8 | LM | J Jennings, Mill Side, Mill Road, Lutterworth, LE17 5DE |
| G8 | LMC | Robert Lovell, 16 North View, Staple Hill, Bristol, BS16 5RU |
| G8 | LMF | P Rigby, 92 Albany Road, Ansdell, Lytham St. Annes, FY8 4AR |
| G8 | LMI | David Morgan, 23 Banstead Road, Caterham, CR3 5QH |
| G8 | LMW | C Smith, 73 Desford Road, Newbold Verdon, Leicester, LE9 9LG |
| G8 | LMY | D Sweetland, 15 Wasdale Close, Owlsmoor, Sandhurst, GU47 0YQ |
| G8 | LNC | David Golding, 27 Wesermarsch Road, Cowplain, Waterlooville, PO8 8JJ |
| G8 | LNG | D Severn, 20 Somerton Avenue, Wilford, Nottingham, NG11 7FD |
| GM8 | LNH | Roger Pascal, 19 Clach Na Strom, Whiteness, Shetland, ZE2 9LG |
| G8 | LNQ | C Tindill, The Old School, Bellerby, Leyburn, DL8 5QN |
| G8 | LNU | L Tucker, 10 The Meadow, Waterlooville, PO7 6YJ |
| G8 | LOF | Steve Champion, 2 St. Andrews Hill, Waterbeach, Cambridge, CB25 9NA |
| G8 | LOJ | S Dorrington-Ward, Higher Dairy, Stoke Abbott, Beaminster, DT8 3JT |
| GM8 | LON | Robert Bruce, 36 Mallaig Avenue, Dundee, DD2 4TW |
| G8 | LOP | P Coomber, 10 Streeton Way, Earls Barton, Northampton, NN6 0HX |
| G8 | LOU | Philip Mattos, Olive House, Rock Road, Wadebridge, PL27 6NW |
| G8 | LOZ | J Ramsay, Strathmore, 5 Parkhurst Road, Guildford, GU2 8AP |
| G8 | LPA | Neil Hilbery, 16 Albert Road, Ashford, TW15 2LU |
| G8 | LPC | R Cawley, 59 The Horseshoe, Hemel Hempstead, HP3 8QS |
| G8 | LPI | R Bray, 2 Hill Park, Walsall Wood, Walsall, WS9 9RD |
| G8 | LPN | Keith Edwards, 22 Claverton Estate, Stoulton, Worcester, WR7 4RH |
| G8 | LPX | C Morgan, 43 Ferndown Road, Manchester, M23 9AW |
| G8 | LQB | W Morrison, 14 Browns Grove, Kesgrave, Ipswich, IP5 2GP |
| G8 | LQF | J Pettifor, 12 Windmill Road, Atherstone, CV9 1HP |
| GM8 | LQL | W Cowell, High Clachaig, Kilmory, Isle of Arran, KA27 8PG |
| G8 | LQM | Paul Green, Nut House, 2 Warren Barns, Warren Lane, Bedford, MK45 4AS |
| G8 | LQN | George Bryce, 6a Kingfisher Drive, Whitby, YO22 4DY |
| G8 | LQO | C McKenzie, 6 Pasturefield Close, Sale, M33 2LD |
| G8 | LQP | Richard Lines, 5 Dowling Drive, Pershore, WR10 3EF |
| G8 | LQZ | R Banfield, 2 Laleham Close, Eastbourne, BN21 2LQ |
| G8 | LRD | P Hutchings, 59 Braemor Road, Calne, SN11 9DU |
| GW8 | LRO | A Williams, 1 Glyncoch Terrace, Pontypridd, CF37 3BW |
| G8 | LRS | Dave Massey, 28 Rufus Close, Rownhams, Southampton, SO16 8LR |
| G8 | LSA | H Potter, Burwood House, Salisbury Road, Woking, GU22 7UR |
| G8 | LSC | P Wheeler, 3 Oatfield Road, Orpington, BR6 0ER |
| G8 | LSD | A Wyatt, 75 Millbrook Road, Crowborough, TN6 2SB |
| G8 | LSH | Daniel Oakley, 48 Nethercourt Avenue, London, N3 1PT |
| G8 | LSI | R Dungan, 2 Lamorna Close, Orpington, BR6 0TD |
| G8 | LSS | A Tompson, 38 The Crescent, Caddington, Luton, LU1 4JA |
| GI8 | LTB | R McWilliams, 4 Wheatfield Drive, Coleraine, BT51 3RD |
| G8 | LTC | R Hore, 6 Watling Gate, Brockhall Village, Blackburn, BB6 8BN |
| G8 | LTD | S Vaslet, 4 Coniston Crescent, Redmarshall, Stockton-on-Tees, TS21 1HT |
| G8 | LTN | Alan Brown, Casita, The Ridge, Cold Ash, Thatcham, RG18 9HT |
| G8 | LTV | G Snellgrove, 142 Arail St., Six Bells, Abertillery, NP3 2NQ |
| G8 | LTY | A Harman, 107 Kempson Drive, Great Cornard, Sudbury, CO10 0YF |
| G8 | LUL | Roland Myers, 33 Withenfield Road, Manchester, M23 9BT |
| G8 | LUP | A Semark, 11 Fir Tree Close, Thorpe Willoughby, Selby, YO8 9PF |
| GI8 | LUR | Arthur Hewitt, 18 Knockview Avenue, Newtownabbey, BT36 6TZ |
| G8 | LUV | G Fairbrass, 230 Kirkby Road, Barwell, Leicester, LE9 8FS |
| G8 | LVC | P Johnson, 54 Beechwood Close, Chandler's Ford, Eastleigh, SO53 5PB |
| G8 | LVF | A Dicroto, 20 Marder Road, London, W13 9EN |
| G8 | LVL | D Holmes, 80 Roydale Close, Loughborough, LE11 5UW |
| G8 | LVM | Andrew Holmes, 5 Launde Park, Market Harborough, LE16 8BH |
| G8 | LVQ | WHITE ROSE ARS c/o E Hannaby, 34 Woodlea Lane, Meanwood, Leeds, LS6 4SX |
| G8 | LVW | Christopher Snell, 138 Main Road, Great Leighs, Chelmsford, CM3 1NP |
| G8 | LWA | D Tyler, Wayside View, Orsett Road, Stanford-le-Hope, SS17 8PN |
| G8 | LWC | J Stuart, 4 Pine Grove, Havant, PO9 2RW |
| G8 | LWO | Frederick Merritt, 17 Blakes Way, Eaton Socon, St. Neots, PE19 8PU |
| G8 | LWQ | S Wood, Lucerne, Berrycroft, Ely, CB7 5BL |
| G8 | LWS | ARIEL RA GP LWS c/o G Rowlands, C/O Gareth Rowlands, Engineering Pigeon Holes, Acton, London, W3 0RP |
| G8 | LXN | W Askey, 32 Hurst Rise, Matlock, DE4 3EP |
| G8 | LXS | G Pascoe, Newhaye, Broadhempston, Totnes, TQ9 6DB |
| G8 | LXY | S Clarke, 128 Putteridge Road, Luton, LU2 8HQ |
| G8 | LYB | Stephen Tompsett, 9 Ashlawn Road, Rugby, CV22 5ET |
| G8 | LYG | W Leach, 15 Beech Lea, Blunsdon, Swindon, SN26 7DE |
| GM8 | LYO | P Mahood, The Briars, Woodside, Kirriemuir, DD8 4PG |
| GM8 | LYQ | Iain Lindsay, 10/4 Mertoun Place, Edinburgh, EH11 1JZ |
| [illegible] | [illegible] | [illegible], 70 Church Road, Hatfield Peverel, Chelmsford, CM3 2LB |
| G8 | LYW | D [illegible] |
| GD8 | LZE | David Dix, 1 Highfield Crescent, Northwood, HA6 1EZ |
| G8 | LZG | Graham Allen, 21 Dale Road, Welton, Brough, HU15 1PE |
| G8 | LZK | Michael Ball, 46a Daniels Crescent, Long Sutton, Spalding, PE12 9DS |
| G8 | LZO | J Hibbert, 80 High Street, Newchapel, Stoke-on-Trent, ST7 4PT |
| G8 | LZS | P Martin, 35 Martineau Lane, Hurst, Reading, RG10 0SF |
| G8 | LZY | S Brown, Maes Yr Haidd, 8 Glanceulan, Aberystwyth, SY23 3HF |
| GD8 | MAA | G Chaplin, 8 Manor House Drive, Northwood, HA6 2UJ |
| G8 | MAD | Paul Tostevin, 20 Wallace Avenue, Worthing, BN11 5QY |
| G8 | MAF | T Beckham, 2 Sandbanks Place, Ersham Road, Hailsham, BN27 3LJ |
| G8 | MAG | S Blake, 26 Nightingale Drive, Towcester, NN12 6RA |
| G8 | MAR | M Sibley, 10 Ainley Close, Huddersfield, HD3 3RJ |
| G8 | MAV | P Lewis, Westbank, 46 Weyside Road, Guildford, GU1 1HX |
| G8 | MAY | Anne Lake, 9 Grafton Close, King's Lynn, PE30 3EZ |
| G8 | MBE | S Fouracres, Old Oaks, Shillingford, Tiverton, EX16 9AY |
| G8 | MBJ | J Parsons, 34 Mill Hill, Brancaster, King's Lynn, PE31 8AQ |
| G8 | MBK | Pauline Bland, 17 Knowles Close, Kirklevington, Yarm, TS15 9NL |
| G8 | MBM | C Proctor, 15 Chiltern Street, Aylesbury, HP21 8BN |
| G8 | MBQ | R Jones, 46 Wilmington Close, Woodley, Reading, RG5 4LR |
| G8 | MBU | Robert Williams, 14 Coronation Avenue, Northwood, Cowes, PO31 8PN |
| G8 | MBV | I Wood, Tessian Lodge, Lydden Road, Dover, CT15 7HE |
| G8 | MCA | G Bryan, 34 Shelbury Close, Sidcup, DA14 4BE |
| G8 | MCC | C Divall, 22 Knightstone Rise, Bridport, DT6 3DR |
| G8 | MCJ | B Pritchard, 14 Rugby Way, Croxley Green, Rickmansworth, WD3 3PH |
| G8 | MCR | V Eagles, 3 Church Road, Buckhurst Hill, IG9 5RU |
| G8 | MCT | Colin Bate, Apartment 19, Tavinor Place, Tamworth, B78 3HQ |
| G8 | MCW | P Elkins, 615 Blandford Road, Upton, Poole, BH16 5ED |
| G8 | MCY | M Dannatt, 46 Laburnham Road, Biggleswade, SG18 0NX |
| G8 | MDG | D Shaw, 35 Tinshill Lane, Leeds, LS16 6BU |
| G8 | MEA | C Wilson, 27 Cedarwood Drive, St. Albans, AL4 0DN |
| G8 | MEC | D Uttley, 1 Edgeside, Great Harwood, Blackburn, BB6 7JS |
| G8 | MED | P Shirtliff, 2 Birch Avenue, Newton, Preston, PR4 3TX |
| G8 | MEE | K Patman, 5 Lime Grove, Holbeach, Spalding, PE12 7NG |
| G8 | MEH | Leslie Steele, Caprice, Woodville Road, Bude, EX23 9JA |
| G8 | MEI | R Whitby, 24 Macaulay Avenue, Great Shelford, Cambridge, CB22 5AE |
| G8 | MEM | A Lillywhite, 1 Roblin Close, Aylesbury, HP21 9DT |
| G8 | MER | M Busson, 14 Squires Gate, Rogerstone, Newport, NP10 0BP |
| G8 | MEX | I Glenn, 257 Wimpole Road, Barton, Cambridge, CB23 7AE |
| G8 | MFF | R Hedley, 20 Spencer Drive, Tiverton, EX16 4PY |
| G8 | MFH | R Lake, 3 Pembridge Chase, Bovingdon, Hemel Hempstead, HP3 0QR |
| G8 | MFI | Stephen McGuigan, 1 Phoenix Yard, Red Hill, Maidstone, ME18 5LD |
| G8 | MFM | Robert Wood, 36 New England Road, Haywards Heath, RH16 3JS |
| G8 | MFO | T Sorensen, 22 The Cottrells, Angmering, Littlehampton, BN16 4AF |
| GW8 | MFQ | Alan John, 79 Harding Close, Boverton, Llantwit Major, CF61 1GX |
| G8 | MFR | R Irwin, Copperfield, 97 Offerton Lane, Stockport, SK2 5BS |
| G8 | MFU | D Parry, 19 Norton Lane, Great Wyrley, Walsall, WS6 6PE |
| G8 | MFV | R Hickmott, Brisley Cottage, Canterbury Road, Ashford, TN25 4DW |
| GM8 | MFZ | Neil Kennedy, Deveron, North Deeside Road, Pitfodels, Aberdeen, AB15 9PL |
| G8 | MGD | David Marshall, 7 Aesops Orchard, Woodmancote, Cheltenham, GL52 9TZ |
| G8 | MGE | G Young, 30 Degenhardt Streett, South Australia, Australia, 5545 |
| G8 | MGF | John Tait, 2 Bron-y-Coed, Coed-y-Glyn, Wrexham, LL13 7QJ |
| G8 | MGG | W Whiteside, Blenkarn, Leighton Drive, Milnthorpe, LA7 7BE |
| G8 | MGK | J Dosher, 40 Bromfield Road, Redditch, B97 4PN |
| G8 | MGO | John Marshall, 34 Derwent Drive, Swindon, SN2 7NJ |
| G8 | MGP | Anthony Hill, 5 Lilac Walk, Kempston, Bedford, MK42 7PE |
| G8 | MGQ | D Garwood, Appletree House, 13 Market Street, Bradford-on-Avon, BA15 1LL |
| G8 | MGZ | P Haynes, 2 The Chase, Furnace Green, Crawley, RH10 6HW |
| G8 | MHA | Leslie Humphrey, 1 Falkenham Road, Kirton, Ipswich, IP10 0NP |
| G8 | MHD | C Cooper, 16 Paulton Drive, Bishopston, Bristol, BS7 8JJ |
| G8 | MHE | G Cross, 117 Broadway, Eccleston, St. Helens, WA10 5PB |
| G8 | MHI | K Russell, 12 Evans Close, Greenhithe, DA9 9PG |
| G8 | MHN | S Scrase, 5 Clinton Road, Leatherhead, KT22 8NU |
| G8 | MHO | A Fraser, 184 Old Road, Harlow, CM17 0HQ |
| G8 | MHT | Graham Dallaway, Flat 6, Crabtree Court, Buxton Old Road, Stockport, SK12 2RZ |
| GM8 | MHU | Ian Fraser, 12 Auchlea Place, Aberdeen, AB16 6PD |
| G8 | MIA | Andrew Malbon, The Lodge, Blithbury Road, Rugeley, WS15 3HJ |
| G8 | MIC | Martin Williams, Flat 2, High Point, London, N6 4BA |
| G8 | MIF | Francis Golding, 16 Lessness Park, Belvedere, DA17 5BG |
| G8 | MIH | R Green, 33 Bulkington Avenue, Worthing, BN14 7HH |
| G8 | MII | T Ashton, 30 Highfields Road, Chasetown, Burntwood, WS7 4QU |
| G8 | MIN | R Welsh, 14 Drayton Close, High Halstow, Rochester, ME3 8DW |
| G8 | MIT | C Wyatt, 273 Nuthurst Road, Birmingham, B31 4TQ |
| GI8 | MIV | G Hutchinson, 40 Oldstone Hill, Muckamore, Antrim, BT41 4SB |
| G8 | MIW | J West, 21 Gardenia Crescent, Mapperley, Nottingham, NG3 6JA |
| G8 | MJF | K Bottomley, Whispering Winds, 15 Marvell Rise, Harrogate, HG1 3LT |
| G8 | MJH | P Harrison, 154 Cherrydown Avenue, Chingford, London, E4 8DZ |
| GM8 | MJV | Tom Melvin, Blue House, Remote, Pathhead, EH37 5UP |
| G8 | MJX | D Coomber, 1 Brympton Road, Coventry, CV3 1GW |
| G8 | MKC | Milton Keynes ARS c/o David White, 1 Whaddon Road, Shenley Brook End, Milton Keynes, MK5 7AF |
| G8 | MKE | C Rose, 45 Clent Road, Warley, Oldbury, B68 9ES |
| G8 | MKG | G Barraclough, Thorn Tree Farm, Ripponden, Sowerby Bridge, HX6 4LS |
| G8 | MKN | I Wager, 106 Turner Road, Colchester, CO4 5JT |
| G8 | MKO | R Pocock, 3 Brewery Cottages, Netherley Road, Prescot, L35 1QG |
| G8 | MKQ | A Bullock, 35 Parkstone Avenue, Thornton-Cleveleys, FY5 5AE |
| G8 | MKS | Paul Moore, 3a High Street, Mow Cop, Stoke-on-Trent, ST7 3ND |
| G8 | MKT | Robert Maxwell, 24 Jensen, Tamworth, B77 2RH |
| G8 | MKW | J Green, Huntley, Chesham Road, Tring, HP23 6HH |
| G8 | MKX | J Donnithorne, 6 Bulbourne Court, Tring, HP23 4TP |
| G8 | MLA | Philip Richardson, 11 Overstone Road Coldham, Wisbech, PE14 0ND |
| G8 | MLB | Nigel Bourner, 11 Richborough Road, Sandwich, CT13 9JE |

UK Callsigns

**IMPORTANT NOTE**

**Revalidate licence to avoid revocation** – Ofcom has advised the Society that plans will be drawn up to revoke licences that have not been revalidated as required by the licence conditions. The quickest way to revalidate is to do so online via the Ofcom website: *https://services.ofcom.org.uk/* or by email: *amateur.validations@ofcom.org.uk* Ofcom staff are available to help, but please be patient during times of heavy workload.

G8 MLD M Warren, 17 Bolehill Park, Hove Edge, Brighouse, HD6 2RS
G8 MLI Kenneth Huxham, 16 Torridge Road, Plymouth, PL7 2DG
G8 MLK J Owen, The Old Coach House, Callow Hill, Virginia Water, GU25 4LD
G8 MLW D Carr, 39 Fallowfield Road, Walsall, WS5 3DH
G8 MM John Pink, 6 Spencer Walk, Rickmansworth, WD3 4EE
GM8 MMA W Williamson, Leeskol, Camb, Shetland, ZE2 9DA
G8 MMF P Dorrington, 57 Ferring Lane, Ferring, Worthing, BN12 6QS
G8 MMG Donald Bentley, 55 Saddlers Road, Quedgeley, Gloucester, GL2 4SY
G8 MMM G Nicholas, Greenbank, Chester High Road, Neston, CH64 7TR
G8 MMN M Holmes, 8 High Street, Norley, Frodsham, WA6 8JS
G8 MMP M SWAIN, 38 Longdale Lane, Ravenshead, Nottingham, NG15 9AD
GM8 MMW William Dick, 58 Kirkland Road, Glengarnock, Beith, KA14 3AJ
G8 MNC Michael Bilkey, Pebble Flek, The Green, St. Austell, PL25 5TA
GM8 MNG C Raine, Broomhill Edgehead, Pathhead, EH37 5RS
G8 MNL P Carruthers, 16 Wivenhoe Close, Rainham, Gillingham, ME8 7QB
GM8 MNM Richard Hood, Milton of Auchindoir House, Rhynie, Huntly, AB54 4JB
G8 MNO W Stewart, 9 Ashley Road, Marnhull, Sturminster Newton, DT10 1LQ
GM8 MNR David Jenkins, 16 Bentinck Street, Galston, KA4 8HT
G8 MNY John Stockley, 27 Campden Road, South Croydon, CR2 7ER
G8 MOF F Bellamy, 3 Manor Road, Crowle, Scunthorpe, DN17 4ET
G8 MOG David Dale, Blackwood Hall, Felton, Morpeth, NE65 9QW
GM8 MOI C Stirling, 20 Craigford Drive, Bannockburn, Stirling, FK7 8NQ
G8 MOK G McKay, 20 Blandford Road, Eccles, Manchester, M30 8WA
G8 MOL P Marshall, 134 Gladbeck Way, Enfield, EN2 7EN
G8 MOS A Reale, 20 Wickham Close, Alton, GU34 1RR
GI8 MOV F Warwick, 20 Wellington Crescent, Ballymena, BT42 2RZ
G8 MOZ G Elliott, 9 Hove Avenue, St. Julians, Newport, NP19 7QP
G8 MPG George Rigby, 1 Route Danton, Petit Caudos, Mios, France, 33380
G8 MPM W Brock, 15 Picketleaze, Chippenham, SN14 0DN
G8 MQF M Cooper, Woodstock, Snow Hill, Crawley, RH10 3EG
G8 MQK J Lindley, 17 Leyfield Bank, Holmfirth, HD9 1XU
G8 MQT T Smith, 416 Charminster Road, Bournemouth, BH8 9SG
G8 MQX R Eccles, 6 Queens Drive, Barnsley, S75 2QJ
G8 MQY Brian Densham, 47 High Street, Paulerspury, Towcester, NN12 7NA
G8 MRI Roger Davey, 23 Campbell Close, Hunstanton, PE36 5PJ
G8 MRN Melvin Watch, 9 High Drive, Rowner, Gosport, PO13 0QS
GM8 MRW S Mcqueen, Flat 56, Hilltree Court, 96 Fenwick Road, Glasgow, G46 6AA
G8 MSM R Hudson, 12 Hairpin Croft, Peacehaven, BN10 8EQ
GM8 MST G Kelly, 36 Craigleith Drive, Edinburgh, EH4 3JU
G8 MSY J Wilkinson, 11 Wigmore Road, Tadley, RG26 4HH
G8 MTA B Haylett, 5 Riverside Close, Whittlesey, Peterborough, PE7 1DL
G8 MTB Micheal Greenfield, 8 The Spinney, Clayton, Newcastle, ST5 4DA
G8 MTI M Dibsdall, 28 Court Farm Avenue, Ewell, Epsom, KT19 0HF
G8 MTV J Wood, Coach House, Croft on Tees, Darlington, DL2 2SL
G8 MUF J Ames, 16 Vere Gardens, Henley Road, Ipswich, IP1 4NZ
G8 MUV Bernard Clarke, 3 Hingley Street, Cradley Heath, B64 5LA
G8 MUX John Mottram, Church View, New Road, High Peak, SK23 7NH
G8 MVC Richard Westlake, Flat 9, Grosvenor Court, 135-139 The Grove, London, W5 3SL
G8 MVD K Wilks, 72 Grasmere Road, Bradford, BD2 4HX
G8 MVH John Armstrong, 11 Dennis Willcocks Close, Newington, Sittingbourne, ME9 7SE
G8 MVJ C Chambers, Hollybank, Back Street, Driffield, YO25 3TD
G8 MVS N Fuller, 11 Hayes Mead Road, Bromley, BR2 7HR
G8 MVY Edward Phillips, 2 Primrose Cottage, The Street, Reading, RG7 1QY
G8 MWA MEDWAY A.R.T. c/o John Hale, 136 Bush Road, Cuxton, Rochester, ME2 1HB
G8 MWD D Lewing, 7 Routh Court, Feltham, TW14 8SJ
G8 MWE Kevin Knight, 54 Vicarage Lane, Water Orton, Birmingham, B46 1RU
G8 MWN K Harris, Bella Vista, Station Road, Yelverton, PL20 7JS
G8 MWU P Stafford, 5 Westmead Drive, Newbury, RG14 7DJ
G8 MWW W Westlake, West Park, Clawton, Holsworthy, EX22 6QN
G8 MWX A Priestley, 55 Derwent Avenue, Garforth, Leeds, LS25 1HN
G8 MXD G York, 13 Cherwell Close, Thornbury, Bristol, BS35 2DN
G8 MXQ Andrew Taylor, 180 Smeeth Road, Marshland St. James, Wisbech, PE14 8JB
G8 MXR W Pitt, 1 Windy Ridge, James Street, Stourbridge, DY7 6ED
G8 MXT L Mansfield, 25 Carlton Road, Derby, DE23 6HB
G8 MXV Kevin Ayriss, 6 Langstons, Trimley St. Mary, Felixstowe, IP11 0XL
G8 MXW C Down, 100 Lynwood Drive, Merley, Wimborne, BH21 1UQ
G8 MYF Michael Johnson, 42 Marlborough Road, Ryde, PO33 1AB
G8 MYG C Hunt, Rowan Bank, 2 Cranston Rise, Bexhill-on-Sea, TN39 3NJ
G8 MYJ Christopher Drewe, 37 Baker Street, Chelmsford, CM2 0SA
G8 MYK Alan Rowley, Holly Cottage, 368 Highters Heath Lane, Birmingham, B14 4TE
GM8 MYO C Tyler, 26 Sinclair Way, Livingston, EH54 8HW
G8 MYV D Webster, 35 Raymond Road, Maidenhead, SL6 6DF
G8 MZA D Garrett, Brookside Farm, Tonge, Derby, DE73 8BD
G8 MZD P Diggins, 8 Gloucester Gardens, Bagshot, GU19 5NU
G8 MZQ W Katz, The Beacon, Goathland, Whitby, YO22 5AN
G8 MZR R Harris, 15 Quarry Rise, Undy, Caldicot, NP26 3JU
G8 MZW S Adams, 29 Rothbury Grove, Bingham, Nottingham, NG13 8TG
G8 MZY D Cushman, 50 St. Peters St., Syston, Leicester, LE7 1HJ
G8 MZZ P Boam, 36 Copeland Drive, Stone, ST15 8YP
GW8 NAC K Davies, 45 Castle View, Simpson Cross, Haverfordwest, SA62 6EN
G8 NAG M Smith, 18 Manor Lane, Verwood, BH31 6HX
G8 NAI J Lazzari, 3 Terson Way, Weston Coyney, Stoke-on-Trent, ST3 5RQ
GM8 NAL Philip Corbishley, Tweedbank House, Cardrona, Peebles, EH45 9HX
G8 NAM P Buttress, 18 Taffrail Gardens, South Woodham Ferrers, Chelmsford, CM3 5WH
G8 NAP P Beacon, 67 St. Helena Road, Polesworth, Tamworth, B78 1NJ
G8 NBF C Morgan, 84 Treowen Road Newbridge, Newport, NP11 3DP
GW8 NBI Andrew Buxton, 17 Gower Rise, Gowerton, Swansea, SA4 3DZ
G8 NBO L Phillips, 14 Heal Park Crescent, Fremington, Barnstaple, EX31 3AP
GM8 NBV C Davies, 35 Laverock Avenue, Hamilton, ML3 7DD
G8 NCK N Brown, 9 Redhill Close, Tamworth, B79 8EJ
GW8 NCN Desmond Sanford, 35 Summerfield Avenue, Cardiff, CF14 3QA
G8 NCS Michael Green, 21 Hill View Rise, Northwich, CW8 4XA
G8 NCU J Watkins, Llwynteg, Glanwern, Borth, SY24 5LT
G8 NDB Gordon Jarrett, 1 Church Street, Twycross, Atherstone, CV9 3PJ
G8 NDE J Turner, 9 Clifton Avenue, Culcheth, Warrington, WA3 4PD
G8 NDF D Simpson, 10 Buckingham Way, Byram, Knottingley, WF11 9NN
G8 NDK K Lindley, 25 Lindsey Court, Epworth, Doncaster, DN9 1SD
G8 NDN C Keens, Toad Hall, 69 Lillywhite Crescent, Andover, SP10 5NA
G8 NDR Nicholas Burridge, 8 Cedar Close, Ware, SG12 9PG
G8 NDV P Fay, 42 Roberts Road, Salisbury, SP2 9BY
G8 NED Wisbech Amateur Radio & Electronics Club c/o Alan Bridgeland, 17 Oldfield Lane, Wisbech, PE13 2RJ
G8 NEF R Peel, 76 Cypress Grove, Ash Vale, Aldershot, GU12 5QW
G8 NEH C Nunn, 29 Wheatland Close, Winchester, SO22 4QL
G8 NEI K Marsh, 1 Parr Close, Exeter, EX1 2BG
G8 NEL Steve Nightingale, 6 Robinson Way, Burbage, Hinckley, LE10 2EU
G8 NEO David Edwards, 3 Murton Close, Burwell, Cambridge, CB25 0DT
GM8 NET A Fraser, 7 Burnet Rose Court, East Kilbride, Glasgow, G74 4TG
G8 NEY David Millard, Weavern House, Hartham Lane, Chippenham, SN14 7EA
G8 NFD Kelvin Gardiner, 8 Foxlands Drive, Sutton Coldfield, B72 1YZ
GM8 NFG J Aitken, Hamabo, 10 Lynnpark, Kirkwall, KW15 1SL
G8 NFM Frank Turner, 46 Main Street, Kings Newton, Derby, DE73 8BX
G8 NFP Tony Crockett, 57 Upland Road, Sutton, SM2 5HW
G8 NFZ Stephen Sims, 71 Green Street, Eastbourne, BN21 1QZ
G8 NGE Kenneth Ebborn, 18 St. Marys Park, Ottery St. Mary, EX11 1JA
G8 NGF D Stone, Aston Hill Cottage, Aston Hill, Shrewsbury, SY5 9JS
G8 NGJ Patricia Richardson, Brembridge Farm, Shillingford, Tiverton, EX16 9BT
G8 NGM N King, 42 Constance Close, Witham, CM8 1XY
G8 NGZ Vince Edwards, 33 Eyrescroft, Bretton, Peterborough, PE3 8ES
G8 NHD P Mart, 1 Montana Close, Great Sankey, Warrington, WA5 8GB
G8 NHG R Wilkins, Churchways, Speen Lane, Newbury, RG14 1RL
G8 NHM J Graves, Willses, Upper Lane, Newport, PO30 4BA
G8 NHO John Austin, 5 Mercia Road, Baldock, SG7 6RZ
G8 NIE D Sharpe, 5 Drydales, Kirk Ella, Hull, HU10 7JU
G8 NIK M Carena, Armorel, Shire Lane, Rickmansworth, WD3 5NH
G8 NIL D Bales, 30 Railway Road, Wisbech, PE13 2QA
G8 NIU Richard Whiting, 12 Lodge Close, Englefield Green, Egham, TW20 0JF
G8 NJA TORBAY ARS c/o Derrick Webber, 43 Lime Tree Walk, Milber, Newton Abbot, TQ12 4LF
G8 NJI Philip Woodhead, Manor Farm, Newsham Hill Lane, Bridlington, YO15 1HL
G8 NKJ L Reid, 26 Mansion Avenue, Whitefield, Manchester, M45 7SS
G8 NKM A O'Donovan, 2 Mackenzie Road, Beckenham, BR3 4RU
G8 NKN Stephen Gorwits, 29 Howitt Drive, Bradville, Milton Keynes, MK13 7DY
G8 NLF Ian Munro, 30 Willow Close, Bordon, GU35 0TH
G8 NLK M Bennett, 68 Meadow Hill Road, Birmingham, B38 8DA
G8 NLS Stephen O'Brien, Flat 2, 99 Howard Street, North Shields, NE30 1NA
G8 NMH B O'Regan, 10 School Hill, Little Sandhurst, Sandhurst, GU47 8LD
G8 NMK C Eccles, 23 River House, Common Road, Evesham, WR11 4QY
G8 NMM Chris Reid, 138/17, Bang Saray, Pattaya, Thailand, 20230
G8 NMO Doreen Pechey, Jays Lodge, Crays Pond, Reading, RG8 7QG
G8 NMT J Hicks, 14 Oakwood, Flackwell Heath, High Wycombe, HP10 9DW
G8 NNA Barry Crellin, 60 College Fields, Woodhead Drive, Cambridge, CB4 1YZ
GW8 NNF R Galpin, 23 Heol Y Delyn, Lisvane, Cardiff, CF4 5SR
G8 NNP B Gower, 132 Goldsworthy Way, Slough, SL1 6AY
G8 NNS G Stamp, 41 Willoughby Road, Wallasey, CH44 3DZ
G8 NNU T Rowe, 68 Cobourg Road, Montpelier, Bristol, BS6 5HX
G8 NNX M Cohen, 41 South Station Road, Liverpool, L25 3QE
G8 NOB N Bean, 33 Badger Close, Guildford, GU2 9PJ
G8 NOD Michael Stamford, The Old Wheelwrights, East Street, Leominster, HR6 9HB
G8 NOF Ronald Holt, Tile House, Vicarage Hill, Solihull, B94 5EB
G8 NOP P Price, Calwich View, Dove Street, Ashbourne, DE6 2GY
G8 NOS A Swallow, 67a Strines Road, Marple, Stockport, SK6 7DT
GW8 NP Highfields ARC c/o Stephen Williams, 371 Coed Y Gores, Llanedeyrn, Cardiff, CF23 9NR
G8 NPD J Hodnett, 126 Northwood Lane, Newcastle, ST5 4BN
G8 NPH Aidan Arnold, 2 Duck Lane, Haddenham, Ely, CB6 3UE
G8 NPP A Brown, Dunlop Hiflex Powerbend, Pennywell Industrial Estate, Sunderland, SR4 9EN
G8 NPR J Blackshaw, 23 Cherry Orchard, Oakington, Cambridge, CB24 3AY
G8 NPZ P Whiteman, 22 Hartsbourne Road, Earley, Reading, RG6 5PY
G8 NQC P Manser, 61 Galsworthy Drive, Caversham, Reading, RG4 6QB
G8 NQI J Gartside, 12 Starfield Avenue, Hollingworth Lake, Littleborough, OL15 0NG
G8 NQK G English, 25 Powell Gardens, Newhaven, BN9 0PS
G8 NQN Martin Bancroft, 19 Neap House Road, Gunness, Scunthorpe, DN15 8TS
G8 NQO Alex Whyatt, 11 The Perrings, Nailsea, Bristol, BS48 4YD
G8 NQY W Lea, 20 Gloucester Road, Walsall, WS5 3PN
G8 NRC Stephen Deighton, 6 Meadowlands, Bolsover, Chesterfield, S44 6XR
G8 NRF G Wood, 3 Cleveleys Road, Great Sankey, Warrington, WA5 2SR
G8 NRP M Andrew, 80 Hamble Drive, Abingdon, OX14 3TE
G8 NRR R BAMBROOK, 26 Croft Road, Thame, OX9 3JF
G8 NRS N.A.R.S.A. c/o P Smith, Obo Bury Radio Soc, Moses Yth Comm Cntr, Bury, BL9 0BS
G8 NRU D Carr, 78 Kingsleigh Road, Heaton Mersey, Stockport, SK4 3PG
G8 NSD Frank Taylor, 96 Elvaston Road, North Wingfield, Chesterfield, S42 5HH
G8 NSE F Wood, 96 Manchester Road, Astley, Manchester, M29 7EJ
G8 NSK J Barnes, 23 Spenser Road, King's Lynn, PE30 3DP
G8 NSO Stephen Fleetham, 17 Tetbury Hill, Avening, Tetbury, GL8 8LT
G8 NSS P Leach, 5 Capesthorne Close, Werrington, Stoke-on-Trent, ST9 0PF
G8 NST J Leek, 30 Casuarina Road, Bucklands Beach, Auckland, New Zealand, 1706
G8 NSX Ronald Miller, 89 Moorside, Spennymoor, DL16 7DZ
G8 NSZ M Stanway, 72 Sheldons Court, Winchcombe Street, Cheltenham, GL52 2NR
G8 NTD K Johnson, 24 Capers Close, Enderby, Leicester, LE19 4QD
G8 NTG W Howell, 6 Unity Avenue, Sneyd Green, Stoke-on-Trent, ST1 6DE
G8 NTH A Hewat, 41 Summersbury Drive, Shalford, Guildford, GU4 8JG
G8 NTJ K Hand, 75 Hill Street, Hednesford, Cannock, WS12 2DW
G8 NTQ Michael Roper, 6 Ilmington Close, Hatton Park, Warwick, CV35 7TL
GD8 NTR John Williams, 133a Wiltshire Lane, Pinner, HA5 2NB
G8 NTS J Smart, Greystone, High Street, Swindon, SN26 7AR
G8 NTY Christopher Mallows, 9 Chestnut Drive, Shenstone, Lichfield, WS14 0JH
G8 NTZ David Kowalczyk, 5 Priestthorpe Lane, Bingley, BD16 4ED
G8 NVB N Brown, 9a Decoy Drive, Eastbourne, BN22 0AB
G8 NVC Brian Ellis, 7 Highmoor Close, Corfe Mullen, Wimborne, BH21 3PU
GM8 NVE Dave Watters, 28 Bruce Road, Crossgates, Cowdenbeath, KY4 8AZ
GM8 NVG A Wilson, Lochend, Beith, KA15 2LN
G8 NVH S Reynolds, 242 Butchers Lane, Mereworth, Maidstone, ME18 5QH
G8 NVI A Stevens, 67 New Road, East Hagbourne, Didcot, OX11 9JX
G8 NVS Simon Hindle, Oakdene, Shutterton Lane, Dawlish, EX7 0PD
G8 NVT R Hatfield, 1 Slade Close, Ottery St. Mary, EX11 1SY
G8 NVX M Moss, 24 Magna Lane, Dalton, Rotherham, S65 4HH
G8 NVZ G Evans, 4 Holcot Lane, Anchorage Park, Portsmouth, PO3 5TR
G8 NWC Graham Boor, 27 Welbeck Drive, Spalding, PE11 1PD
G8 NWI Jeff Vine, 117 Betterton Road, Rainham, RM13 8ND
G8 NWK G Milner, 3 Briggs Villas, Queensbury, Bradford, BD13 2EP
G8 NWL J Mason, 46 Bradford Street, Chelmsford, CM2 0FJ
G8 NWM V Maxfield, 50 Hanthorpe Road, Morton, Bourne, PE10 0NT
G8 NWS J Caddick, 58 Beachcroft Road, Wall Heath, Kingswinford, DY6 0HX
G8 NWU Michael Wright, 69 Wroxham Drive, Nottingham, NG8 2QR
G8 NWZ M Percy, 73 Ridgeway, Wellingborough, NN8 4RY
G8 NXA T Ehlen, 58b Warriner Gardens, London, SW11 4DU
G8 NXB N Borrett, 15 Holman Road, Epsom, KT19 9PQ
G8 NXD Mike Waterfall, 12a Boskenna Road, Four Lanes, Redruth, TR16 6LS
G8 NXE S Eyles, 2 Salisbury Close, Lichfield, WS13 7SN
G8 NXJ I Livesey, 26 Hilltop Road, Twyford, Reading, RG10 9BN
GW8 NXK G Garner, 31 Clare St., Manselton, Swansea, SA5 9PG
G8 NXQ William Povey, 31 Baddlesmere Road, Whitstable, CT5 2LB
G8 NXS D Stevenson, 86 Kingston Road, Luton, LU2 7SA
G8 NXY W Godwin, 116 Sandicroft Close, Birchwood, Warrington, WA3 7LA
G8 NYB D Reed, 59 Cowley Avenue, Chertsey, KT16 9JJ
G8 NYC J Primmer, 46 Grantham Crescent, Ipswich, IP2 9PD
G8 NYD M Perry, 23 Victors Crescent, Hutton, Brentwood, CM13 2HZ
G8 NYH R Adams, 2 Longwill Avenue, Melton Mowbray, LE13 1UR
G8 NYJ Ian Gibbs, 3 Badger Drive, Lightwater, GU18 5TS
G8 NYK Martin Nicholson, Flat 14, Whyke Court, Chichester, PO19 8TP
G8 NYM Michael Lister, 97 Hightown Road, Liversedge, WF15 8DG
G8 NYR B Rabey, 36 Park Way, St. Austell, PL25 4HR
GM8 NYV Simon Richardson, Rowan Bank, Melvich, Thurso, KW14 7YJ
G8 NYZ N Cooper, 4 Mossfield Crescent, Kidsgrove, Stoke-on-Trent, ST7 4YA
G8 NZB B Durrant, Brymar, 16 Merrymeet, Exeter, EX4 2JP
G8 NZC K Edmunds, 44 Antonia Circuit, Hallett Cove, Australia, SA 5158
G8 NZD C Atkinson, 8 Southwood Road, Dunstable, LU5 4EA
G8 NZK Nicholas O'Hagan, 5 Bankside, Finchampstead, Wokingham, RG40 3QB
GM8 NZL E Hogg, 43 Muir Wood Road, Currie, EH14 5JN
G8 NZN David Roberts, 12 Erw'r Llan, Nannerch, Mold, CH7 5RF
G8 NZO John Crozier, 43 Shepherds Way, Birmingham, B23 5XR
G8 NZR Kay Pullan, 18 Heathfield, Mirfield, WF14 9BJ
G8 OAD Graham Baxter, 4 Deeping Road, Baston, Peterborough, PE6 9NP
GM8 OAH W Easton, 21 Cameron Avenue, Bishopton, PA7 5ES
G8 OBB T Hooker, Inglewood, Woodside Road, Luton, LU1 4DJ
G8 OBK M Bruce-Smith, 28 Belmont Road, Bramhall, Stockport, SK7 1LE
G8 OBP D Payne, 11 Welbeck Close, Blaby, Leicester, LE8 4HF
G8 OBT T Graham, 22 Locker Park, Wirral, CH49 2RZ
G8 OCA J Astle, River View, Brough, Kirkby Stephen, CA17 4BZ
G8 OCE John Hardy, 42 Fir Tree Drive, Wales, Sheffield, S26 5LZ
G8 OCF Richard Harris, 7 Kestrel Road, Flitwick, Bedford, MK45 1RB
G8 OCM E Dubbins, 2 Elizabeth Avenue, Rose Green, Bognor Regis, PO21 3EL
G8 OCO M Hughes, 49 Reedings Road, Barrowby, Grantham, NG32 1AU
GI8 OCR J Mcilveen, 31 Edenaveys Crescent, Armagh, BT60 1NT
G8 OCS D Simpson, 6 St. Martins Close, Stratford-upon-Avon, CV37 9QW
G8 OCT Stephen Terry, 341 Dickard Rd., Seneca, United States, SC 29672
G8 OCV Christopher Smart, Old Queens Head, Ipswich Road, Diss, IP21 4XP
G8 ODK R Varley, 41 Lang Lane, West Kirby, Wirral, CH48 5HQ
GM8 OEG A Swiffin, Glebe House, Kellas, Dundee, DD5 3PD
G8 OEJ E Bray, Rothesay, 6 Empshott Road, Southsea, PO4 8AU
G8 OEK P Brown, Estate Yard House, Beverley, HU17 7PN
G8 OEO John Thompson, 4 The Grove, Ponteland, Newcastle upon Tyne, NE20 9HQ
G8 OEU T Hipwood, 3 Camview, Paulton, Bristol, BS39 7XA
G8 OFA M Cranage, Corris House, West Gomeldon, Salisbury, SP4 6LS
G8 OFI Geoffrey Radivan, 15 Agecroft Road West, Prestwich, Manchester, M25 9RE
G8 OFN R Pashley, 50 Cherry Bank Road, Sheffield, S8 8RD
G8 OFO Rupert Short, Langtree House, Castle Hill, Fordingbridge, SP6 2AX
G8 OFQ Geoffrey Dobson, 9 Fitzpain Road, West Parley, Ferndown, BH22 8RZ
G8 OFR R Coole, Courtyard Cottage, Horse Fair Lane, Swindon, SN6 6BN
G8 OFX A Nelson, 37 Brook Way, Romsey, SO51 7JZ
G8 OFZ Ian McGowan, Meld House, Hawthorn Road, Shrewsbury, SY3 7NB
G8 OGP Sydney Martin, Aldon, The Hayes, Cheddar, BS27 3HS
G8 OGR John Holton, 24 Great Austins, Farnham, GU9 8JQ
G8 OHC G Scholes, 14 Braemar Road, Bulwell, Nottingham, NG6 9HN
G8 OHG Jack Myall, 52 Princethorpe Way, Binley, Coventry, CV3 2HF
G8 OHH John Morgan, 41 Lingen Avenue, Hereford, HR1 1BY
G8 OHM SOUTH BIRMINGHAM RS c/o Norman Gutteridge, 68 Max Road, Quinton, Birmingham, B32 1LB
G8 OHP C Cainsford-Betty, 85 Gatesden Road, Fetcham, Leatherhead, KT22 9QP
G8 OHS Malcolm Emery, 25 Bradgate Drive, Sutton Coldfield, B74 4XG
G8 OID C Vaslet, Little Copse Farm, Heath End, Newbury, RG20 0AT
G8 OIJ Anthony Stark, 5 Ladyhill Road, Newport, NP19 9RY
G8 OIV C Merrell, 40 Fanton Walk, Wickford, SS11 8QT
G8 OIY S Robertson, 249 Ware Road, Hertford, SG13 7EJ
G8 OJK V Willett, 20 The Green, Sharlston Common, Wakefield, WF4 1EF
G8 OJQ Alan Hopkinson, Springfield, Neston Road, Neston, CH64 4AR
G8 OJR W Pearce, 160 Philip Lane, Tottenham, London, N15 4JN
G8 OKB R McCann, Goss House, Clark Street, Stourbridge, DY8 3UF
G8 OKD Malcolm Bailey, 28 St. Pauls Hill Road, Hyde, SK14 2SW
G8 OKE R Brown, 8 Grassmere Way, Waterlooville, PO7 8QD
G8 OKI L Mather, 8 Carnoustie Avenue, Chesterfield, S40 3NN

G8 OKN M [illegible], 30 Warwick Road, Southam, CV47 0HW
G8 OKR Brian Kirkpatrick, 88 [illegible] Park Drive, Newport, NP20 3NH
G8 OKS B Dawson, 9 [illegible], Billingham, TS23 3HG
G8 OKZ D Shillington, 6 Moss Close, Willaston, Neston, CH64 2XQ
GI8 OLH T Lavery, 21 Mussenden Grange, Articlave, Coleraine, BT51 4US
G8 OLK P Smith, 21a Meadow Way, Bracknell, RG42 1UE
G8 OLL D Porter, 9 School Avenue Brownhills, Walsall, WS8 6AG
G8 OLP Michael Matthews, 19 Perrylands, Charlwood, Horley, RH6 0BL
G8 OLY D Curwell, 9 St. Georges Road, Aldershot, GU12 4LD
G8 OMB D Parker, 146 Merlin Avenue, Nuneaton, CV10 9QJ
G8 OMC D Smith, 71 Ashbourne Avenue, Aspull, Wigan, WN2 1HW
G8 OMQ D Bliss, 11 Bubblestone Road, Otford, Sevenoaks, TN14 5PN
G8 OMW F Rowan, 91 St. Nicholas Road Littlemore, Oxford, OX4 4PW
G8 ONH J Sager, Well Cottage, Fen Lane, Woodbridge, IP12 4NZ
G8 ONP J Eastwood, Llys Iwan, Dole, Bow Street, SY24 5AE
G8 ONR M Loader, 20 Edgcumbe Drive, Tavistock, PL19 0ET
G8 ONS K Creighton, 10 Oram Close, Allery Banks, Morpeth, NE61 1XF
G8 ONY B Goodhew, 101 Brier Road, Sittingbourne, ME10 1YL
G8 OO Alan Holdsworth, Millfield, The Green, Dereham, NR20 5LL
G8 OOC J Morecroft, 217a Longhurst Lane, Mellor, Stockport, SK6 5PN
G8 OOF G Ellison, 36 Park Hill, Clapham, London, SW4 9PB
G8 OOQ M Barton, 23 Caledonia Place, Bristol, BS8 4DL
G8 OOS M Reeson, 19 Southlands Avenue, Louth, LN11 8EW
G8 OPA P Barry, 32 Rutland Avenue, Sidcup, DA15 9DZ
G8 OPC D Crawley, 9 Gwynns Walk, Hertford, SG13 8AD
G8 OPE Michael De Rouffignac, 2 Westgate, Old Malton, Malton, YO17 7HE
G8 OPI Jonathan Spooner, 59 Woodlands Park Drive, Dunmow, CM6 1WT
G8 OPO G Bartels, 37 Faircross Avenue, Romford, RM5 3SX
G8 OPP J Birkett, 13 The Strait, Lincoln, LN2 1JD
G8 OPX Tim Willford, 15 Foxglove Close, Broughton Astley, Leicester, LE9 6YU
G8 OPY G winston, 8 Linnet Close, Shoeburyness, Southend-on-Sea, SS3 9YE
G8 OQC J Kent, 106 Victoria Road, Barnet, EN4 9PA
G8 OQG P Jobbins, 35 Keys Avenue, Horfield, Bristol, BS7 0HQ
G8 OQP T Spacagna, 3a Station Road, Romsey, SO51 8DP
G8 OQR Martin Crossman, 31a Eastcote Grove, Southend-on-Sea, SS2 4QA
G8 OQT J Lambert, 125 Tudor Way, Mill End, Rickmansworth, WD3 8HT
G8 OQV William Jackson, Modesgate Tidenham Chase, Chepstow, NP16 7LZ
G8 ORM M Baguley, 42 Kendall Avenue, Shipley, BD18 4DY
G8 ORO Donald Coulter, 15 Woodville Way, Whitehaven, CA28 9LT
G8 ORR C Brown, 179 Bournville Lane, Birmingham, B30 1LY
G8 ORX Keith Rashleigh, 43 Oxshott Way, Cobham, KT11 2RU
G8 OSG N McAlpine, 15 Sparrows Herne, Basildon, SS16 5JH
G8 OSH N Hubbard, 31 Bridlington Crescent, Monkston, Milton Keynes, MK10 9HG
G8 OSJ D Halliwell, 9 Berkeley Avenue, Alsager, Stoke-on-Trent, ST7 2BW
G8 OST Robert Taylor, 8 Vetch Walk, Haverhill, CB9 7YE
G8 OSX K Dawson, Mayfield House, 3 The Green, Tamworth, B78 3HW
G8 OSZ S Ashley, 12 Dene Close, Wellingborough, NN8 5QP
G8 OTA H O'Tani, 8a The Avenue, Keynsham, Bristol, BS31 2BU
G8 OTC C Anderson, 42 Elizabethan Way, Rugeley, WS15 2EE
G8 OTD S Ballard, 28 Hildyard Close, Hardwicke, Gloucester, GL2 4PZ
G8 OTG R Cannon, 111 Brangbourne Road, Bromley, BR1 4LP
G8 OTH C Churchill, 87 Bradley Crescent, Shirehampton, Bristol, BS11 9SR
GM8 OTI John Cooke, 6 Greenbank Terrace, Edinburgh, EH10 6ER
G8 OTS ARIEL RADIO GROUP c/o Tom Ellinor, 53 Hillside, Banstead, SM7 1HG
G8 OTZ D Logan, 33 Foxwood Drive, Kirkham, Preston, PR4 2DS
G8 OUG Laurence Gray, 20 Turner Way, Clevedon, BS21 7YN
G8 OUH I Harfield, White Gates, Crofton Avenue, Lee on The Solent, PO13 9NJ
G8 OUI D Baines, 1 Carole Close, Sutton Leach, St. Helens, WA9 4PW
G8 OUM David Briggs, 43 Monmouth Walk, Markham, Blackwood, NP12 0QR
G8 OUS S Greendale, 15 Bosworth Road, Cambridge, CB1 8RG
G8 OUT B Horrocks, 17 Wood Grove, Whitefield, Manchester, M45 7ST
G8 OUY Dave Smith, 41 Mitcham Road, Camberley, GU15 4AR
G8 OVO Nigel Lihou, 6 St. Laurence Avenue, Warwick, CV34 6AR
G8 OVZ S Gosby, 20 Woodland Mount, Hertford, SG13 7JD
G8 OWA Robert Lewin, 180 Ladybank Road, Mickleover, Derby, DE3 0RR
G8 OWO K Metcalf, 21 Forest Gate, Evesham, WR11 1XZ
G8 OWS J Greenall, 6 Neasham Drive, Darlington, DL1 4LG
G8 OWZ O Cockram, 446 Holdenhurst Road, Bournemouth, BH8 9AE
G8 OXD Paul Brown, 41 School Street, Castleford, WF10 2SB
G8 OXE M Brooks, 3 Wood Side, Wood Street, March, PE15 0SB
G8 OXG N Powell, 42 Sheraton Drive, Kidderminster, DY10 3QR
G8 OXI E Mannix, La Vieille Scierie, Les Allues, France, 73550
G8 OXS Andrew Lambert, Electronic Media Services, Lynchborough Road, Liphook, GU30 7SB
G8 OXU C Madge, 89 Heron Gardens, Rayleigh, SS6 9TU
G8 OXX P Bailey, 236 Sandy Lane, Droylsden, Manchester, M43 7JX
G8 OYB K Armstrong, 2 Blackthorn Grove, Shawbirch, Telford, TF5 0LL
G8 OYF J Popplewell, 6 Roseleigh Avenue, Manchester, M19 2NP
G8 OYL W Shave, 26 Hessle Avenue, Boston, PE21 8DA
G8 OYM Paul Taylor, 41 Cross Road, Southwick, Brighton, BN42 4HG
G8 OYQ M Everitt, 48 Rant Meadow, Hemel Hempstead, HP3 8EQ
G8 OYT B Jones, [illegible], Rhyl, LL18 4JF
G8 OYY J Bishop, No1 Dillhurst Cottage, Plaistow Street, Lingfield, RH7 6EY
G8 OZD A Batty, 23 Sandyshot Walk, Wythenshawe, Manchester, M22 5AQ
G8 OZH John Burrell, 6 Blenheim Croft, Brackley, NN13 7ET
G8 OZP R Platts, 43 Iron Walls Lane, Tutbury, Burton-on-Trent, DE13 9NH
G8 OZQ S Pallett, 6 Lancaster Close, Coalville, LE67 4TG
G8 OZT Neil Morley, Mazongill, Orton, Penrith, CA10 3RZ
G8 OZY P Harrison, 91 Obelisk Rise, Northampton, NN2 8QU
G8 PAB Morton Humphries, 20 Taunton Street, Swindon, SN1 5EE
G8 PAG M Rose, 20 Broad Piece, Soham, Ely, CB7 5EL
GM8 PAH D Schofield, 3 Craiglockhart Grove, Edinburgh, EH14 1ET
G8 PAI D Rout, Two Akers, Wrabness Road, Harwich, CO12 5NE
G8 PAL Peter Hankinson, 37 Victoria Avenue, Whitefield, Manchester, M45 6DP
G8 PAN S Day, 14 The Crescent, Market Harborough, LE16 7JJ
G8 PAT P G McGuinness, 9 Farmdale Road, Carshalton, SM5 3NG
G8 PBH Anthony Kent, 46 Russley Road, Bramcote, Nottingham, NG9 3JE
G8 PBI I Murphy, The Nurseries, Carnon Crease, Truro, TR3 6LJ
GW8 PBM Anthony Royston, 10 Elmgrove Place, Dinas Powys, CF64 4DJ
G8 PMX [illegible] Jones, Olan Goro, Braich Talog, Bangor, LL57 4PD
G8 PBT G [illegible], Dairy Farmhouse, West Winterslow, Salisbury, SP5 1RE
GJ8 PCY P Falle, [illegible] Main Road, Grouville, Jersey, JE3 9EP
G8 PDE W Burin, 7 Sunniside Terrace, Sunderland, SR6 7XE
GI8 PDK David Courtney, 79 Fort Road, Belfast, BT8 8LX
G8 PDM C Commander, 8 Cannon Place, Hampstead, London, NW3 1EJ
G8 PDP R Hinchliffe, 34 Oaklea, Ash Vale, Aldershot, GU12 5HP
G8 PDY S Procter, 8 Pond End Road, Sonning Common, Reading, RG4 9SA
G8 PEA K Wibberley, 5a Marston Road, Croft, Leicester, LE9 3GX
GM8 PEB Paul Fineron, Court Yard, Crauchie, East Linton, EH40 3EB
G8 PEN C Vernon, 48 Long Beach, Hemsby, Great Yarmouth, NR29 4JD
G8 PFL J Turner, 32 Petunia Crescent, Chelmsford, CM1 6YP
G8 PFR Michael Gibson, Eccles Wall Farm, Bromsash, Ross-on-Wye, HR9 7PW
GW8 PFT P Hinson, 7 Awel Tywi, Llangunnor, Carmarthen, SA31 2NL
G8 PFZ R Harrison, Badgers Oak, Redbrook Street, Ashford, TN26 3QU
G8 PGE David Sinclair, 12a Sunnydown Road, Winchester, SO22 4LD
G8 PGF A Price, 11 Gatcombe Gardens, Titchfield, Fareham, PO14 3DR
G8 PGH Kevin James, 44 The Oakfield, Littledean Hill Road, Cinderford, GL14 2DE
G8 PGI P Lord, Beechwood House, Main Street, Lutterworth, LE17 5QA
GI8 PGJ D Campbell, 18 The Counties, Mark Street, Portrush, BT56 8QA
G8 PGO D Carter, 49 Hinckley Road, Sapcote, Leicester, LE9 4LG
G8 PHB P McKenzie, 21 Arnside Walk, Chapel House, Newcastle upon Tyne, NE5 1BT
G8 PHG B Cook, Hinsley Mill House, Hinsley Mill Lane, Market Drayton, TF9 1HP
G8 PHJ M Palmer, 16 Trulock Road, Tottenham, London, N17 0PH
G8 PHM M Kent, Meadow Bank, Rye Lane, Sevenoaks, TN14 5JF
G8 PHQ Christopher Challender, 9 Blick Close, West Winch, King's Lynn, PE33 0UA
G8 PHS E Campbell, 2 Russell Avenue, March, PE15 8EL
G8 PHV R Wetton, 8 St. Moritz Close, Northwick, Worcester, WR3 7ND
G8 PIC C Pomphrett, 47 North Leas Avenue, Scarborough, YO12 6LJ
G8 PIN R Bannister, 14 Amery Close, Worcester, WR5 2HL
G8 PIO O Futter, 25 Amhurst Gardens, Belton, Great Yarmouth, NR31 9PH
G8 PIP P Elwell, 4 Richmond Grove, Wollaston, Stourbridge, DY8 4SF
G8 PIQ P Tagg, 22 Hambledon Road, Waterlooville, PO7 7UB
G8 PIR Lowestoft District and Pye ARC c/o John Elsdon, 15 Union Road, Lowestoft, NR32 2BZ
GM8 PIV E Souter, 3/2 10 James Gray Street, Glasgow, G41 3BS
G8 PIY D Clifton, 10 Scotney Road, Basingstoke, RG21 5SR
G8 PJC John McDonald, 17 Highfield Close, Wokingham, RG40 1DG
G8 PJD P Deffee, 18 Poplar Road, Kensworth, Dunstable, LU6 3RS
G8 PJQ C Cole, 70 Throgmorton Road, Yateley, GU46 6FA
G8 PK B Wilson, 23 The Oaks, Soham, Ely, CB7 5FF
G8 PKB L Rudge, 8 Penrallt Estate, Llanystumdwy, Criccieth, LL52 0SR
G8 PKG I Bosworth, 6 Busbys Close, Stonesfield, Witney, OX8 8EU
G8 PKJ G Rowland, 18 Heights Way, Leeds, LS12 3SN
G8 PKM C Mitchell, 6 Oak Crescent, Ashbourne, DE6 1HR
GW8 PKV Michael James, 28 Bloomfield Gardens, Narberth, SA67 7EZ
G8 PL A M'Garry-Durrant, 2 Ramsey Hall Cottages, Wix Road, Harwich, CO12 5LS
G8 PLI Nikolas Vranic, 30 Mitchell Street, Sheffield, S3 7NL
G8 PLJ J Bailey, 8 Hild Avenue, Cudworth, Barnsley, S72 8RN
G8 PLO R Clubley, Church Hill House, High Street, Braintree, CM7 4BY
GM8 PLR P Paterson, Corse Sands, Kininmonth, Peterhead, AB42 4JU
G8 PMA L Pennell, 182 Northampton Road, Wellingborough, NN8 3PJ
G8 PMJ D Hughes, 18 Bailey Close, Pewsey, SN9 5HU
G8 PMR Law Morris, 17 Kestrel Close, Hornchurch, RM12 5LS
G8 PNE Peter Griffiths, 42 The Hook, New Barnet, Barnet, EN5 1LQ
G8 PNM N Cocking, 114 Heavygate Road, Sheffield, S10 1PF
G8 PNN G Emmerson, 72 The Gables, Widdrington, Morpeth, NE61 5RB
G8 POE J Phillips, 235 Barn Mead, Harlow, CM18 6ST
G8 POG Peter Wood, 23 Shipley Avenue, Newcastle upon Tyne, NE4 9QY
G8 POI Derek Evans, 70 Kingsway, West Wickham, BR4 9JG
G8 POK Graham West, 6 Willerton Close, Chidswell, Dewsbury, WF12 7SQ
G8 POL M Williams, 22 Charlecote Drive, Nottingham, NG8 2SB
G8 POO Simon Robinson, 23 Jameson Drive, Corbridge, NE45 5EX
G8 POP R Mundy, 12 Cantors Way, Minety, Malmesbury, SN16 9QZ
G8 POQ B Salter, 34 Southways, Stubbington, Fareham, PO14 2AQ
G8 POS Alan Axon, 7 Tudor Grove, Groby, Leicester, LE6 0YL
G8 PPA S Ornstein, 19 Colvin Gardens, Barkingside, Ilford, IG6 2LH
G8 PPD A Saunders, Suffolk House, Main Road, Ipswich, IP9 1DX
G8 PPF R Rogers, 24 Treza Road, Porthleven, Helston, TR13 9NB
G8 PPN A Osmond, The Moorings, 10 North Road, Shanklin, PO37 6DB
G8 PPQ Geoffrey Boakes, 16 Terminus Drive, Herne Bay, CT6 6PP
G8 PPR D Bancroft, 31 Thorndene Way, Bradford, BD4 0SW
GD8 PPU Garry Brookes, 44 Magherchirrym, Port Erin, Isle of Man, IM9 6DB
G8 PQA A Gapper, 12 Meadow Mead, Frampton Cotterell, Bristol, BS36 2BQ
G8 PQB G Grantham, 18 Fen End Lane, Spalding, PE12 6AD
G8 PQH F Rowsell, 9 Long Close, Crawley, RH10 7DD
G8 PQJ J Robinson, Rose Cottage, Wavering Lane West, Gillingham, SP8 4NR
G8 PQN Martin Henshaw, 23a Bedford Road, Northill, Biggleswade, SG18 9AH
G8 PQZ C Collier, 6 Copse Close, Tilehurst, Reading, RG31 6HH
G8 PRC PLYMOUTH RADIO COMMUNITY c/o Peter Connor, 20 Longfield, Lutton, Ivybridge, PL21 9SN
G8 PRH A Hartley, 16 Old Thorne Road, Hatfield, Doncaster, DN7 6ER
G8 PRJ S Sanders, 19 Brunswick Gardens, Hainault, Ilford, IG6 2QU
G8 PRK R Holmwood, 3 Stanstead Road, Caterham, CR3 6AD
G8 PRN Robert Morley, 21 Meadow View, Skelmanthorpe, Huddersfield, HD8 9ET
G8 PRP B Youster, 24 Sunningdale Road, Weston-Super-Mare, BS22 6XP
G8 PRU Prudential ARS c/o James Butler, 14 Fairfield Road, Barnard Castle, DL12 8EB
G8 PSC J Benoy, 46 Bickham Road, Plymouth, PL5 1SB
G8 PSF A Ball, 20 Inverness Avenue, Enfield, EN1 3NT
GW8 PSJ Leslie Finch, Hafan Deg, Waungilwen, Llandysul, SA44 5YG
G8 PSO Robert Gould, 20 Southwood Drive, Coombe Dingle, Bristol, BS9 2QU
G8 PSS J Meldrum, 7 Kerryhill Drive, Pity Me, Durham, DH1 5FN
GM8 PSV B Thomson, 51 Main Street, Newmill, Keith, AB55 6UR
G8 PSZ Christopher Wood, Wudum Wic, Farm Close, Market Drayton, TF9 3UH
G8 PTR B [illegible], 10 [illegible] Road, Warley, Oldbury, B68 0NU
G8 PTT A [illegible], 71 [illegible] Way, Northampton, NN2 8PH
G8 PTL M [illegible], Chasetown, Burntwood, WS7 3QL
G8 PTN D Stoney, 7 Sandwell Close, Long Eaton, Nottingham, NG10 3RG
G8 PTS Willam Leddington, 4 Cherry Walk, Monmouth, NP25 5DE
G8 PTW C Wallace, Windy Ridge, Langley Priory, Derby, DE74 2QQ
G8 PTY P Thornton-Evison, Greyfriars, Townsend, Wantage, OX12 0AT
G8 PUB Sydney Lucas, 84 Woodman Road, Warley, Brentwood, CM14 5AZ
G8 PUE John Taylor, The Jays, 5 Watling Close, Bourne, PE10 9XL
G8 PUH B Merrell, 16 Box Close, Broadfield, Crawley, RH11 9QT
G8 PUK S Mann, 1 Blackthorn Avenue, Bramley, Rotherham, S66 2LU
G8 PUN J Keleher, 9 Broadwell Drive, Leigh, WN7 3NE
G8 PUR Terence Rose, 41 Keats Way, Hitchin, SG4 0DP
G8 PUT Chris Townsend, 2 Netherfield Drive, Netherthong, Holmfirth, HD9 3ES
G8 PUY Nicholas Dowsett, 21 St. Marys Road, Burnham-on-Crouch, CM0 8LX
G8 PVG D Hobbs, 46 Gloucester Road, Bridgwater, TA6 6DZ
G8 PVK R Still, The Manor House, 5 Beechwood Avenue, Bournemouth, BH5 1LY
GJ8 PVL Peter Bertram, Roz-Den, La Rue De La Guilleaumerie, St. Saviour, Jersey, JE2 7HQ
G8 PVR J Riggs, 4 Plough Green, Saltash, PL12 4JZ
G8 PWA Dennis Lee, 44 Charnwood Crescent, Newton, Alfreton, DE55 5SH
G8 PWE I Ashford, 53 Gilpin Crescent, Walsall, WS3 4HR
G8 PWK M Forsey, 84 Garner Road, Walthamstow, London, E17 4HH
G8 PWO J Thwaites, 15 Spring Head Road, Kemsing, Sevenoaks, TN15 6QL
G8 PWT L Cook, 16 Florence Road, Maidstone, ME16 8EN
G8 PWU John Crossland, 1 Carter Lane, Flamborough, Bridlington, YO15 1LW
G8 PX OXFORD & DISTRICT ARS c/o E Burrell, 27 Blandford Avenue, Oxford, OX2 8EA
G8 PXA R Lucyk, 106 Peet Street, Derby, DE22 3RG
G8 PXI Michael Parker, 65 Shoreham Drive, Penketh, Warrington, WA5 2HY
G8 PXO Glenis Murray, P9, Atico 3C, Tigaiga 3, Puerto de la Reina, Spain, 38632
G8 PXU B Gascoigne, 108 Blandford Avenue, Castle Bromwich, Birmingham, B36 9JD
G8 PY Craig Bell, 2 The Pastures, Long Bennington, Newark, NG23 5EG
G8 PYD Graham Farrell, 95 Washington Road, Maldon, CM9 6JF
G8 PYE Peter Barrett, 30 Rosslyn Park Road, Plymouth, PL3 4LN
G8 PYU I Walton, Tall Trees, Tredington, Shipston-on-Stour, CV36 4NG
G8 PZD T Wills, 66 Kipling Road, St. Marks, Cheltenham, GL51 7DQ
G8 PZF B Simpson, 14 Priestthorpe Lane, Bingley, BD16 4EE
G8 PZI W Nolan, South Lawn, 77 Reigate Road, Reigate, RH2 0RE
GM8 PZR Nigel Clarke, ROSE CROFT, 1 The Meadows, Dunoon, PA23 7UP
G8 PZS John Coady, 21 Garth Wen, Llanfaes, Beaumaris, LL58 8PT
G8 PZX F Gunn, 8 College Gardens, Hornsea, HU18 1EF
G8 QM V Flowers, Eothen Homes Ltd, 45 Elmfield Road, Newcastle upon Tyne, NE3 4BB
G8 QZ D Sager, 29 Station Road, Mickleover, Derby, DE3 9GH
G8 RAC J Maines, Brick House Farm, Marden, Hereford, HR1 3ET
G8 RAF RAF ARS c/o R Finch, 1 Cherry Tree Cottage, Church Road, High Wycombe, HP10 8LN
G8 RAJ R Shore, 42 King George Avenue, Bournemouth, BH9 1TX
G8 RAK Graham Ogle, Glan Eden, Brynford, Holywell, CH8 8LQ
G8 RAN Kevin Reeman, 4 Alfreda Avenue, Hullbridge, Hockley, SS5 6LT
G8 RAO A Yates, 87 Princess Road, Warley, Oldbury, B68 9PW
GW8 RAS R Neville, 11 Heol Urban, Llandaff, Cardiff, CF5 2QP
G8 RAU Richard Lewis, 66 Derek Gardens, Southend-on-Sea, SS2 6QY
G8 RAV Ken Jackson, 17 Copperfield Close, Kettering, NN16 9EW
G8 RAX Colin Hill, 183 Manchester Road, Swinton, Manchester, M27 4FA
G8 RBI C Allen, 8 Shoulbard, Fleckney, Leicester, LE8 8TX
G8 RBK I Brown, 56 Church Lane, Darley Abbey, Derby, DE22 1EY
GM8 RBR William Egerton, Croft House, Upper Breakish, Isle of Skye, IV42 8PY
G8 RBS P Bickersteth, Tregarth, Fernsplatt, Truro, TR4 8RJ
G8 RBU Trevor Dewey, 93 Calverton Road, Arnold, Nottingham, NG5 8FQ
G8 RBV Derek Deighton, 3 Bluebell Way, Huncoat, Accrington, BB5 6TD
G8 RBW C Ellison, 29 Ashton Road, Clay Cross, Chesterfield, S45 9FA
G8 RBX L Fitzpatrick-Browne, 24 Beechmount Avenue, Hanwell, London, W7 3AG
G8 RBY P Hodson, 43 Thorpe Road, Melton Mowbray, LE13 1SE
G8 RCE Kevin Shergold, 28 Berkeley Crescent, Stourport-on-Severn, DY13 0HJ
G8 RCK Roy Woollard, 68 Trunk Furlong, Aspley Guise, Milton Keynes, MK17 8HX
G8 RCL Graham Whiston, 86 Elmsett Close, Great Sankey, Warrington, WA5 3RX
G8 RCO D Russell, 53 The Campions, Borehamwood, WD6 5QE
G8 RCZ G Fermor, 26 Byron Road, Exeter, EX2 5QN
G8 RDA K Forster, 10 Springfield Oval, Witney, OX28 6EG
G8 RDB R George, Juniper Cottage, Hillesden, Buckingham, MK18 4BX
G8 RDG R Maltby, Meadow Croft, Bishop Lane, Henfield, BN5 9DG
G8 RDJ Jonathan Davies, 11 Bromley Road, Macclesfield, SK10 3LN
G8 RDK L Mayhew, 47 Beeches Avenue, Worthing, BN14 9JE
G8 RDN T Sale, 20 Redwood Drive, Chase Terrace, Burntwood, WS7 2AS
G8 RDP John Webb, 6 Chatsworth Avenue, Fleetwood, FY7 8EG
G8 RDQ Philip Williams, 30 Duchess Drive, Bridgnorth, WV16 4JD
G8 RDT Kevin Williams, Jasmine Cottage, Little Hill Farm, Dodwell, Stratford-upon-Avon, CV37 9ST
G8 REF P Ellis, 15 Alexander Close, Bognor Regis, PO21 4PS
GM8 REG R Bell, Fairview, Main Street, Huntly, AB54 7SY
G8 REO Rod Mitchell, 4 Friendly Fold Road, Halifax, HX3 5QF
G8 REQ F Robinson, 13 Dorset Drive, Wirral, CH61 8SX
G8 RER J Fothergill, 53 Meadow Court, Ponteland, Newcastle upon Tyne, NE20 9RA
G8 RES Malcolm Howard, 105 Tennyson Road, King's Lynn, PE30 5PA
GW8 REV C Marsh, 50 Timothy Rees Close, Cardiff, CF5 2AU
G8 RF F Raby, 20 Lime Tree Road, Codsall, Wolverhampton, WV8 1NT
G8 RFC R Cassell, 1 St. Saviour Close, Colchester, CO4 0PW
G8 RFD Phil Short, 193 Conygre Grove, Bristol, BS34 7HZ
G8 RFE M Wallace, 26 Parsons Drive, Glen Parva, Leicester, LE2 9NS
G8 RFF Keith Richardson, 31 Castlefields Drive, Brighouse, HD6 3XF
G8 RFL D Robinson, 25 Angelica, Amington, Tamworth, B77 3JZ
G8 RFP D Clarke, 29 Haugh Lane, Sheffield, S11 9SB
G8 RFV R Bulmer, 4 Valerian Drive, Stafford, ST16 1FJ



**IMPORTANT NOTE**

**Revalidate licence to avoid revocation** – Ofcom has advised the Society that plans will be drawn up to revoke licences that have not been revalidated as required by the licence conditions. The quickest way to revalidate is to do so online via the Ofcom website: *https://services.ofcom.org.uk/* or by email: *amateur.validations@ofcom.org.uk* Ofcom staff are available to help, but please be patient during times of heavy workload.

G8 RFW Graham Sharpe, 5 Thorpe Avenue, Coal Aston, Dronfield, S18 3BB
G8 RFY Christopher Garner, 22 Old Ashby Road, Loughborough, LE11 4PG
G8 RFZ D Cooke, 27 Goosehill Court, Balby, Doncaster, DN4 8SX
G8 RGN John Harling, 6 Fontwell Road, Little Lever, Bolton, BL3 1TE
GM8 RGO M Robson, Whistlefield Cottage, Loch Eck, Dunoon, PA23 8SG
G8 RGU M Burt, Hartcliff Farm, Okeford Fitzpaine, Blandford Forum, DT11 0EF
G8 RHC J Cranage, Corris House, West Gomeldon, Salisbury, SP4 6LS
G8 RHM K Hoggett, 14 Wyld Court, Allesley, Coventry, CV5 9LQ
G8 RHN M Kirkham, Toll Bar House, 417 Louth Road, Grimsby, DN36 4PX
G8 RHP J Williams, Monte Vista, Llandyrnog, Denbigh, LL16 4HH
G8 RHQ S Ormondroyd, 15 Meadowlands, Blundeston, Lowestoft, NR32 5AS
G8 RHU E Carvill, 61 Midhurst Drive, Ferring, Worthing, BN12 5BQ
G8 RHZ Alastair Robertson, 22 Court Way, Twickenham, TW2 7SN
G8 RIB P Fallon, 17 Blundell Road, Widnes, WA8 8SS
G8 RIC K Murphy, 79 Torkington Road, Hazel Grove, Stockport, SK7 6NR
G8 RIK R Milner, Lyndene, Holyhead Road, Shrewsbury, SY4 1EE
G8 RIM J Boardman, Ivy Dene, Redditch Road, Birmingham, B48 7TL
G8 RIP M Walmsley, 27 Russell Avenue, Preston, PR1 5TP
G8 RIR Simon Newbury, 87 Tower Road, Epping, CM16 5EW
G8 RIS R Merriman, 8 Abbots Meadow, Chittlehampton, Umberleigh, EX37 9QE
G8 RIW B Harvey, 56 Oakwood Drive, Grimsby, DN37 9RN
G8 RJB R Bridgwater, 31 Pembroke Avenue, Worthing, BN11 5QS
G8 RJF K Freer, 54a High Lane East, West Hallam, Ilkeston, DE7 6HW
G8 RJM S Reap, The Staddles, Romsey Road, Stockbridge, SO20 8DB
G8 RJO D Shaw, 31 Windwhistle Circle, Weston-Super-Mare, BS23 3TU
G8 RJQ Michael Corke, 6 Dhow Street, Sun Valley, Cape Town, South Africa, 7975
G8 RJZ M Wills, 9 Allerdale Close, Thirsk, YO7 1FW
G8 RKG D Peck, Flat 1, Shrewsbury Court, 21-23 Manor Road, Worthing, BN11 3RU
G8 RKH Laurence Hunt, 15 Oxford Street, Cowes, PO31 8PT
G8 RKO J Butler, 36 Park Road, Bracknell, RG12 2LU
G8 RKX A Titley, 6 Spring View, Luddendenfoot, Halifax, HX2 6EX
G8 RLD Robert Dowdell, 5614 Front Drive, Holiday, Florida, United States, 34690
GI8 RLE John Ashe, 49 Deans Walk, Richhill, Armagh, BT61 9LD
G8 RLF R Dickerson, 7 Sixpenny Close, Titchfield Common, Fareham, PO14 4SY
GI8 RLG H Emerson, Little Castle Dillo, Co Armagh, BT61 7DF
G8 RLH T Alston, Little Red House, Eaton Bishop, Hereford, HR2 9QT
GW8 RLI Vincent Banfield, New Haven, Main Road, Pontypridd, CF38 1RY
G8 RLN John Barnett, 11 Ridge Street, Stourbridge, DY8 4QF
G8 RLW Gordon Woodward, 6 Lang Road, Huntington, York, YO32 9SD
G8 RMI S Blake, 10 Dimore Close, Hardwicke, Gloucester, GL2 4QQ
G8 RML Michael Juby, Silver Birch, High Road, Diss, IP22 5RU
G8 RMP J Bond, 19 Compton Avenue, Mannamead, Plymouth, PL3 5DA
GM8 RMR E Scott, 81 Rosehaugh Road, Inverness, IV3 8SR
GI8 RNG William Smyth, 11 Alexander Park, Armagh, BT61 7JB
GD8 RNM Ian Lucking, 32 Nolton Place, Edgware, HA8 6DL
G8 RNT Peter Walkling, Flat 36, Highlands House, Southampton, SO19 7GG
G8 RNU Ian Strange, Holly Lane, Tansley, Matlock, DE4 5FF
G8 RNV Neil Jefferies, 8 Cambridge Green, Fareham, PO14 4QX
G8 ROG Alison Johnston, 12 Whitby Court, Caversham, Reading, RG4 6SF
G8 RON R Eyes, 6 Bakers Lane, Southport, PR9 9RN
G8 ROS R Platt, 40 Solent Drive, Darcy Lever, Bolton, BL3 1RN
G8 ROU David Hardy, 15 Riversdale, Ambergate, Belper, DE56 2EU
G8 RPA K Mendum, 23 Eton Avenue, East Barnet, Barnet, EN4 8TU
G8 RPD John Fennell, Broad Park Cottage, Stanbury Copse, Ilfracombe, EX34 8DW
GM8 RPE J Robinson, 18 Craigshannoch Road, Wormit, Newport-on-Tay, DD6 8ND
G8 RPI G Atkinson, 25 Appletrees, Bar Hill, Cambridge, CB23 8SJ
GI8 RPP M Elder, 44 Learmount Road, Claudy, Londonderry, BT47 4AQ
GI8 RPT N Copeland, 34 Glenkyle Park, Newtownabbey, BT36 6SP
G8 RQF J Duffy, 5 Birch Court, Prudhoe, NE42 6PZ
G8 RQH Ian Sherer, 1a Appleyard Drive, Barton-upon-Humber, DN18 5TD
GI8 RQI David Allen, 40 Bramblewood Drive, Banbridge, BT32 4RA
G8 RQN P Needham, 2 Woodridge Close, Bracknell, RG12 9QX
G8 RRC Philip Sharpe, 27 Record Road, Emsworth, PO10 7NS
G8 RRN M Jones, 5 The Pines, Felixstowe, IP11 9SU
GJ8 RRP John Parry, 2 Thornley, Bagatelle Road, Jersey, JE2 7TZ
G8 RRR H Potter, 134 Ifield Road, Crawley, RH11 7BW
G8 RRS M Ellison, 22 Cotebrook Drive, Upton, Chester, CH2 1RD
G8 RSA S Hasko, 105 High Street, Brampton, Huntingdon, PE28 4TQ
GM8 RSC J Chinnock, 3b Dundee Street, Letham, Forfar, DD8 2PQ
G8 RSE J Murphy, 15 Loders Close, Poole, BH17 9BF
G8 RSI G Whitney, 4 Sefton Crescent, Sale, M33 7EN
G8 RSK P Tyrell, 14 Park Farm Road, Horsham, RH12 5EW
G8 RSQ Louise Williamson, 17 Ray Bond Way, Aylsham, Norwich, NR11 6UT
G8 RSV S Staniforth, 4 Moses View, Shireoaks, Worksop, S81 8NH
G8 RSX T Beck, 10 Rookery Close, Hatfield Peverel, Chelmsford, CM3 2DF
G8 RTB R Breeze, 119 Sundorne Road, Shrewsbury, SY1 4RP
GM8 RTI John Grieve, Elhanan, Myrtlefield Lane, Inverness, IV2 5UE
G8 RTK Leighton Man, 4 Back Lane, Yeadon, Leeds, LS19 7SQ
G8 RTN G Smith, 1 Abbey Road, Bedford, MK41 9LG
G8 RUX Richard Gough, 3 Meadowlands, Havant, PO9 2RP
G8 RVO A Parsons, 8 Queen Annes Gardens, Ealing, London, W5 5QD
GJ8 RVT JERSEY ARS c/o M Turner, 4 Le Clos Sara, St. Lawrence, Jersey, JE3 1GT
G8 RVY P Lee, 8 Sandringham Gardens, Ellesmere Port, CH65 9EY
G8 RVZ Ian Martin, Coppilow Barn, Haunton Road, Tamworth, B79 9HP
G8 RW Peter Standley, 9 Capelands, New Ash Green, Longfield, DA3 8LG
G8 RWG Niels Montanana, 91 Coulsdon Road, Coulsdon, CR5 2LD
G8 RWH Ian Jackson, 5 Vivien Close, Chessington, KT9 2DE
G8 RWJ R Adams, Ground Floor Flat, 91 Mount Pleasant Road, Hastings, TN34 3SL
G8 RWM F Box, 11 Cook Avenue, Newport, PO30 2LL
G8 RWN Malcolm McKenzie, Flat 40, Broomfield House, Huddersfield, HD3 4RS
G8 RWU Andrew Capron, 28 Windmill Road, Hemel Hempstead, HP2 4BN
G8 RWZ Kenneth Hodson, 117 High Lane, Brown Edge, Stoke-on-Trent, ST6 8RT
G8 RXB E Hampson, Raynet Group, 21 Marlowe Road, Wallasey, CH44 3DA
G8 RXY G Alcock, 61 Henshall Hall Drive, Congleton, CW12 3TY
G8 RXZ Peter Allgood, 11 Dover Drive, Leegomery, Telford, TF1 6TD
G8 RYE David Cope, 29 Desford Road, Newbold Verdon, Leicester, LE9 9LG
G8 RYJ P Tegg, Glendale, 36 Wrecclesham Hill, Farnham, GU10 4JW
G8 RYK Robert Taylor, 18 Spruce Avenue, Selston, Nottingham, NG16 6DX
G8 RYL I Smith, 12 Windmill Lane, Fulbourn, Cambridge, CB21 5DT
G8 RYO P Sargent, The Old School House, Main Road, Market Rasen, LN8 6JY
G8 RYX Simon Jones, 30 Lindford Chase, Lindford, Bordon, GU35 0TB
G8 RZL T Claydon, 20 Ivy Lane, Royston, SG8 9DQ
G8 RZN David Dunn, Valarms, The Pigeons, Wisbech, PE13 4JU
G8 RZS E Humpston, 2 The Glebe, Hildersley, Ross-on-Wye, HR9 5BL
G8 RZZ Kevin Colman, 10 South Rise, North Walsham, NR28 0EE
G8 SAL SALTASH DIST AR c/o Kevin Hale, 58 St. Stephens Road, Saltash, PL12 4BJ
G8 SAN R Charlton, 31 Meriden Road, Hampton-in-Arden, Solihull, B92 0BS
GM8 SAP D Cooper, 4 Ruskie Avenue, Callander, FK17 8LA
G8 SAR Mark Elliott, 54 Bankhouse Road, Trentham, Stoke-on-Trent, ST4 8EL
G8 SAU Barry Titmarsh, Knole House, 38 Cromer Road, Sheringham, NR26 8RR
G8 SAX P Wilkinson, 60 Whalley Drive, Aughton, Ormskirk, L39 6RF
GM8 SBH H Cromack, Pier View, Kilchattan Bay, Isle of Bute, PA20 9NW
G8 SBJ T Blankley, 16 Charles Road, St. Leonards-on-Sea, TN38 0QA
G8 SBK L Cleak, Danetre, Newport Road, Cwmbran, NP44 3AE
GW8 SBN John Kemp, Poldhu 259 Delffordd, Rhos, Swansea, SA8 3EP
GW8 SBO P Sibert, Glaspant Manor, Capel Iwan, Newcastle Emlyn, SA38 9LS
G8 SBQ David Wooller, 95 Havant Road, Drayton, Portsmouth, PO6 2JE
G8 SBS John Westlake, 41d Shirley Road, Southampton, SO15 3EW
G8 SCG J Downes, 6 Lagonda, Glascote, Tamworth, B77 2RY
G8 SCI Martin Bynorth, 374 Bloxwich Road, Walsall, WS2 7BG
G8 SCY Christopher Rosewall, 12 Treloggan Lane, Newquay, TR7 2JN
G8 SDE Roy Pitts, 84 Prospect Avenue, Pye Nest, Halifax, HX2 7HP
G8 SDN E McIver, 31 Hartshill, Bedford, MK41 9AL
G8 SDS SOUTH DORSET RS c/o W Barton, 4 Hawthorn Flats, Hawthorn Road, Dorchester, DT1 2PE
G8 SDU Robert Clayton, 1 Raymond Road, Norwich, NR6 6PL
G8 SED P Starling, 5 Ash Close, Bacton, Stowmarket, IP14 4NR
G8 SEE R Stone, 10 Rosemullion Gardens, Tolvaddon, Camborne, TR14 0EY
G8 SEK C Watts, 2 Southbrook Cottages, Bayford, Wincanton, BA9 9NL
G8 SEQ John Beech, 124 Belgrave Road, Coventry, CV2 5BH
G8 SEV P Matthews, 10 Norway Close, Corby, NN18 9EG
G8 SEY A Graver, 8 Avenue Road, Bishop's Stortford, CM23 5NU
G8 SFA S Milsom, 30 Beechwood Drive, Prudhoe, NE42 5PN
G8 SFD C Williams, 14 Milton Place, Bideford, EX39 3BN
G8 SFF M Watson, 7 Grange Lane, Willingham by Stow, Gainsborough, DN21 5LB
G8 SFI S Firth, 8 Lyndale Avenue, Osbaldwick, York, YO10 3QB
G8 SFM K Saunders, Chelipaux, Peyrat de Bellac, France, 87300
G8 SFQ T Mcnamara, 12 Scarsdale Road, Great Barr, Birmingham, B42 2JW
G8 SFT D Mansell, 57 Ger y Llan Penrhyncoch, Aberystwyth, SY23 3HQ
G8 SFU David Kirkham, 2 Thames Meadow Drive, Hogsthorpe, Skegness, PE24 5PU
G8 SGB Philip Houseago, 11 Arnstones Close, Colchester, CO4 3AS
G8 SGF P Gilliland, 34 Cavan Drive, St. Albans, AL3 6HP
G8 SGH Paul Marshall, 123 Rochford Garden Way, Rochford, SS4 1QJ
G8 SGI Simon Pascoe, 34 Ravens View, Witham St Hughs, Lincoln, LN6 9JE
G8 SGM A Boyce, 22 Peggotty Close, Chelmsford, CM1 4XU
G8 SGP Guy Wheeler, 14 Sparkmill Terrace, Beverley, HU17 0PA
G8 SGV M Williams, The Hideaway, 39 Terrys Avenue, Victoria, Australia, 3160
G8 SGX S Saltmer, 12 Beechings Mews, Whitby, YO21 3DW
G8 SH John Storey, Hampstead House, Fairfax Road, Birmingham, B31 3QY
G8 SHC P Hammond, 35 West Green, Barrington, Cambridge, CB22 7RZ
G8 SHE Richard Shears, 20 Regency Gardens, Grantham, NG31 9JW
G8 SHF C Scrase, 101 Tower Way, Dunkeswell, Honiton, EX14 4XH
G8 SHR P Goodfellow, 10 St. Agnes Walk, Knowle, Bristol, BS4 2DL
G8 SIE R Stark, Roneragh, Llanrhaeadr, Denbigh, LL16 4NN
G8 SIG Andrew Jeffery, 14 Holly Mount, Shavington, Crewe, CW2 5AZ
G8 SIK Andrew Sturt, 6 Kenley Road, Kingston upon Thames, KT1 3RW
G8 SIM J Green, 7 Russell Road, Runcorn, WA7 4BG
G8 SIT M Shewring, 2 Glan Hafan, Trefechan, Aberystwyth, SY23 1AT
G8 SIU Derek Stillwell, 2b Lesley Owen Way, Shrewsbury, SY1 4RB
G8 SJA P Farrar, 17 Clough Lane, Halifax, HX2 8SG
G8 SJO S Ootam, 9 Harewood Road, Isleworth, TW7 5HB
GI8 SJS R Hoey, 28 Hanwood Heights, Dundonald, Belfast, BT16 1XU
G8 SKA Raymond Holden, 5 Lawrence Grove, Kidderminster, DY11 7DR
G8 SKG Leonard Challis, 30 London Road, Kirton, Boston, PE20 1JA
GI8 SKN David Reid, 2 New Line, Carrickfergus, BT38 9DL
GI8 SKR G Bannister, 65 Osborne Drive, Belfast, BT9 6LJ
G8 SLB P Lockwood, 36 Davington Road, Dagenham, RM8 2LR
G8 SLC M Truman, Cotwood, Ponsanooth, Truro, TR3 7HJ
G8 SLE Graham Leake, Flat 3, Edward May Court, Bournemouth, BH11 8AW
G8 SLM Enid Santer, 3 Barn Park, Liverton, Newton Abbot, TQ12 6HE
G8 SLP J Barry, 18 Hough Green, Chester, CH4 8JG
G8 SLU M Hack, Anmee The Ride, Ifold, Billinghurst, RH14 0TF
G8 SMA C Ward, 25 Blewbury Drive, Tilehurst, Reading, RG31 5HJ
G8 SMH K Hempsall, 69 Wantage Road, Didcot, OX11 0AE
G8 SMR STH MNCHESTER RD c/o John Heath, 19 Anson Road, Swinton, Manchester, M27 5GZ
G8 SMZ Christopher Shaw, 1 Guilford Cottages, East Langdon, Dover, CT15 5JD
GM8 SNB G Allan, 13 Mitchell Drive, Rutherglen, Glasgow, G73 3QP
GM8 SNE P Walter, Allt Beag, Pitconnochie Road, Dunfermline, KY12 8QD
G8 SNF Ian Hewitt, 26 Outwoods Drive, Loughborough, LE11 3LT
G8 SNQ R Knock, 18 The Hawthorns, Eccleston, Chorley, PR7 5QW
G8 SNV M Dey, Windy Lodge, 18 Ripley Road, Hampton, TW12 2JH
G8 SOI D Carter, 35 Upland Road, West Mersea, Colchester, CO5 8DR
GM8 SOK John Sturrock, 4 Ann Street, Edinburgh, EH4 1PJ
G8 SOU R Topping, 47 Celtic Road, Deal, CT14 9EF
G8 SPC Robert Thomas White, 3 Robin Lane, Clevedon, BS21 7EX
G8 SPD M Beevers, Pool House Cottage, Astley, Stourport-on-Severn, DY13 0RH
G8 SPE R Armstrong, 21 Gunnersbury Gardens, London, W3 9AE
G8 SPM P Armitage, 22 Lyncombe Close, Exeter, EX4 5EJ
G8 SPP C Parkinson, 77 Lime Grove, Doddinghurst, Brentwood, CM15 0QX
G8 SPU R Doughty, 47 Red Lion Close, Tividale, Oldbury, B69 1TP
G8 SQA Timothy Povey, 24 Townley Way, Earls Barton, Northampton, NN6 0HR
G8 SQH D Hutchinson, Ryton Villa, Horsecroft Lane, Dymock, GL18 2EJ
G8 SQK L Mcmahon, 25 Belmont Avenue, Warrington, WA4 1LY
G8 SQP R Marquiss, 66 Oakwood Rise, Tunbridge Wells, TN2 3HF
G8 SQY S Cade, 2 Grosvenor Crescent, Louth, LN11 0BD
G8 SQZ S Westlake, 11 Mount Road, Evesham, WR11 3HE
G8 SRC SWINDON ARC c/o Den Forrest, 166 Meadowcroft, Swindon, SN2 7LE
G8 SRN E Day, 10 Carlrayne Lane, Menston, Ilkley, LS29 6HH
G8 SRS STOCKPORT RS c/o Bernard Naylor, 47 Chester Road, Poynton, Stockport, SK12 1HA
G8 SRV Andrew Ashe, 34 College Avenue, Maidenhead, SL6 6AX
G8 SRZ Brian Atkinson, 3 Sandy Close, Whitwell, Worksop, S80 4PY
G8 SSE Kevin Lawrence, 7 Canada Way, Pak House, Worcester, WR2 4DJ
G8 SSL A Marwood, 65 Castleton Avenue, Arnold, Nottingham, NG5 6NH
G8 SSP Christopher Horswell, Hungereckstrasse 60/3, Vienna, Austria, A-1232
G8 SSS Exmoor RC (Inc. North Devon Raynet) c/o John Stacey, 3 West Park, South Molton, EX36 4HJ
G8 SSX D Baker, 99 Repton Road, Wigston, LE18 1GD
G8 SSY E Davies, 5 Cheapside, Horsell, Woking, GU21 4JG
G8 STD ST DUNSTANS ARS c/o E John, Obo St. Dunstans Ars, 52 Broadway Avenue, Wallasey, CH45 6TD
G8 STE D Barber, 12 Hulton Road, Gaywood, King's Lynn, PE30 4QE
G8 STF Thomas Woods, 12 Primrose Court, Egerton Street, Wallasey, CH45 2PE
G8 STI J Maiden, 42 Timberdine Avenue, Worcester, WR5 2BD
G8 STJ David Carter, 4 Sandale Close, Gamston, Nottingham, NG2 6QG
G8 STM Adrian Bilton, 34 Wimberley Way, South Witham, Grantham, NG33 5PU
G8 STR B Beestin, 45 Pinehill Road, Crowthorne, RG45 7JP
G8 STW J Ferguson, Sunway, Snow Hill, Sudbury, CO10 8QE
G8 STY J Holmes, 45 College Avenue, Gillingham, ME7 5HY
G8 SUG Geoffery Peterson, 15 Hindhead Green, Watford, WD19 6TR
G8 SUJ W Shambrook, 49 Beaufort Road, Church Crookham, Fleet, GU52 6AY
G8 SUM Keith Smith, 11 Church Street, Earl Shilton, Leicester, LE9 7DA
G8 SUN S Williams, 11 Cotman Drive, Hinckley, LE10 0GB
G8 SUQ J Corbidge, 11 Berkeley Close, Folkestone, CT19 5NA
G8 SUV B Pont, 56 Ravenhill Road, Bristol, BS3 5BT
G8 SUW N Pont, Maisemoor, 17 Vicarage Lane, Bridgwater, TA7 9LR
GM8 SVB A Duncan, 23 Fife Street, Macduff, AB44 1YA
G8 SVN J Iliffe, Glen Usk, Haisbro Avenue, Newport, NP19 7HY
G8 SVR J Allart, 54 Urban Gardens, Concord, Washington, NE37 3DE
G8 SVT T Ellis, 58 Greenland Drive, Sheffield, S9 5GJ
G8 SVZ F Keeble Buckle, 4 Croft Close, Meeting Green, Newmarket, CB8 8YG
G8 SWC H Moyle, 9 Park Approach, Welling, DA16 2AW
G8 SWK D Taylor, 19 Armley Grange Oval, Leeds, LS12 3QJ
G8 SWL Elisabeth Theodorson, 7 Kingfisher Court, Overstone Lakes, Ecton Lane, Northampton, NN6 0BD
G8 SWM Robert Wright, High Banks, Toot Hill Road, Ongar, CM5 9LJ
G8 SWO Colin Watts, 42 Truscott Avenue, Bournemouth, BH9 1DB
G8 SWW P Copeman, 1 Chestnut Avenue, Welney, Wisbech, PE14 9RG
G8 SXA J Davies, Ballards Piece, Forest Hill, Marlborough, SN8 3HN
G8 SXB D Mullenger, 6 Churchfields, Kingsley, Bordon, GU35 9PJ
G8 SXD Brian Davies, Ballards Piece, Forest Hill, Marlborough, SN8 3HN
G8 SXI G Howells, 53 Abbey Road, Rhos on Sea, Colwyn Bay, LL28 4NR
G8 SXJ F Hutchings, 21 School Lane, St. Ives, Ringwood, BH24 2PF
G8 SXQ A Leigh, 12 Bowmens Lea Aynho, Banbury, OX17 3AG
G8 SXU J Simmons, 167 Bourne Vale, Hayes, Bromley, BR2 7LX
G8 SYA K Parker, 3 Cross Roads, East Stour, Gillingham, SP8 5LW
G8 SYC A Harris, 17 Fir Tree Close, Patchway, Bristol, BS34 5ER
G8 SYD M Thomson, 11 Uranus Road, Hemel Hempstead, HP2 5QF
G8 SYE Nicholas Trotman, 38 Oldbury Road, Nuneaton, CV10 0TD
G8 SYM D Whittle, 16 Garner Drive, Astley, Manchester, M29 7RT
G8 SYS P Evans, 14 Hudson Drive, Burntwood, WS7 0EW
G8 SYV J Morgan, Linden Lea, Fakes Road, Great Yarmouth, NR29 4JL
GW8 SZC P Henry, 1 Afan Valley Road, Neath, SA11 3SS
G8 SZG Clive Just, 2 The Old Rectory, Felmersham, Bedford, MK43 7HN
G8 SZR M Matthews, 11 Church End, Ashdon, Saffron Walden, CB10 2HG
GM8 SZS B McCaffrey, 5 The Old Orchard, Limekilns, Dunfermline, KY11 3HS
G8 SZX D A Towers, 20 Valiant Close, Glenfield, Leicester, LE3 8JH
G8 SZZ S Jackson, 18 Blakesware Gardens, Edmonton, London, N9 9HU
G8 TAE Gordon Wooltorton, 155 El Alamein Way, Bradwell, Great Yarmouth, NR31 8SX
G8 TAQ A Dyce, 26 Forest Road, Winford, Sandown, PO36 0JY
G8 TAU A Fisher, 2 Hillside Mansions, Barnet Hill, Barnet, EN5 5RH
GI8 TAX R McLoughlin, 27 The Manor, Portadown, Craigavon, BT62 3QU
G8 TAY Robert Manning, 9 Irstead Road, Norwich, NR5 8AR
G8 TB B Wynn, 67 Old Lodge Lane, Purley, CR8 4DN
G8 TBB J O'Meara, 117 Little Sutton Lane, Sutton Coldfield, B75 6SN
G8 TBF Robert Jenkins, 11 Westfield Drive, Worksop, S81 0JS
GW8 TBG M Terry, 265 Delffordd, Rhos, Swansea, SA8 3EP
G8 TBL N Mosedale, Flat 4, St. Matthews House, 98 George Street, Croydon, CR0 1PJ
G8 TBU N Doe, Longclose, Langtree, Torrington, EX38 8NR
G8 TBV A Kyle, 6 Mill Hill Drive, Halesworth, IP19 8DB
G8 TBW R Crathorne, 340 Farnborough Road, Castle Vale, Birmingham, B35 7PD
G8 TBX Stephen Pybus, 26 White Laithe Court, Leeds, LS14 2EQ
G8 TBY David Padley, 31 South Drive, Rhyl, LL18 4SU
GM8 TCG John Blackie, Drumcharry, Montrose Road, Auchterarder, PH3 1BZ
G8 TCH Michael Bell, 1 Dow Brae, Town Yetholm, Kelso, TD5 8SA
G8 TCP C Dawe, 21 Portmellon Park, Mevagissey, St. Austell, PL26 6XD
G8 TDP D Cooke, 19 St. Aldwyn Road, Seaham, SR7 0AN
G8 TEB D Clarke, 59 Baden Powell Crescent, Pontefract, WF8 3QD
G8 TEC Geoffrey Cook, Flat 3, Southdown Court, Southdown Road, Winchester, SO21 2BX
G8 TEF A crute, Shangri La, Winsor Estate, Looe, PL13 2JY
G8 TEK K Worley, 33 Lynbrook Close, Netherton, Dudley, DY2 9HE
G8 TEL P Hynes, 3 Holt Park Gardens, Leeds, LS16 7RB
G8 TEO B Jay, 13 Oakhurst Road, West Moors, Ferndown, BH22 0DW
G8 TEQ David Linsdall, 2b Linkswood Road, Burnham, Slough, SL1 8AT
G8 TFB S Haywood, 12 Elm Terrace, Tividale, Oldbury, B69 1UD
G8 TFR Stephen Cottis, 61 Oaken Grove, Maidenhead, SL6 6HN

| | | |
|---|---|---|
| G8 | TFU | Philip Simpson, Flat 5, 582-584 Sheffield Road, Chesterfield, S41 8LX |
| G8 | TFW | L Stamp, 41 Willoughby Road, Wallasey, CH44 3DZ |
| G8 | TFX | N Richards, 38 Parsons Road, Irchester, Wellingborough, NN29 7EA |
| G8 | TGD | [illegible] |
| G8 | TGD | D Troop, 10 Mellowdew Road, Coventry, CV2 5GL |
| G8 | TGH | B Wilmott, 27 Apple Grove, Bognor Regis, PO21 4NB |
| GW8 | TGS | W Williams, 17 Llys yr Onnen, Coity, Bridgend, CF35 6FA |
| G8 | THE | R Hill, 12 Winchelsea Lane, Hastings, TN35 4LG |
| G8 | THH | D Baker, 5 Larkspur Close, Bishop's Stortford, CM23 4LL |
| GW8 | THL | Frederick David Morgan, Glantowy Lodge, Capel Dewi Road, Carmarthen, SA32 8AA |
| GW8 | THM | M Griffin, 3 Pritchard Close, Llandaff, Cardiff, CF5 2QS |
| G8 | THR | P Crossley, Firpark Farm, Fir Park, Market Rasen, LN8 3YL |
| G8 | THZ | A Tipper, 24 Waverley Road Hoylake, Wirral, CH47 3DD |
| G8 | TIA | Derek Trickett, 25 Spring Street, Halesowen, B63 2SY |
| G8 | TIO | P Dear, Flat 3, The Beeches, 13 Wray Park Road, Reigate, RH2 0US |
| G8 | TIU | Andrew Brierley, Church House, Arundel Road, Littlehampton, BN16 4JS |
| G8 | TIX | Gerald May, 19 McLaren Cottages, Abertysswg, Tredegar, NP22 5BH |
| G8 | TJG | F Starkey, 13 Thorncliffe Drive, Darwen, BB3 3QA |
| G8 | TJI | M Oldfield, Willows, Stablebridge Road, Aston Clinnton, Aylesbury, HP22 5ND |
| G8 | TJR | C Colebrook, 14 Yiewsley Drive, Darlington, DL3 9XS |
| G8 | TKD | David Hensby, 28 Moorland Crescent, Whitworth, Rochdale, OL12 8SU |
| G8 | TKQ | J Ackerley, 24 Macaulay Road, Lutterworth, LE17 4XB |
| G8 | TKY | Tom Bootyman, 14 Vale View, Ackworth, Pontefract, WF7 7HQ |
| G8 | TLC | S Parker, 22 Lincoln Drive, Syston, Leicester, LE7 2JW |
| G8 | TLH | R Rogers, 14 Coningsby Drive, Franche, Kidderminster, DY11 5LU |
| G8 | TLL | L Stewart, The Spinney, Holmes Lane, Scunthorpe, DN15 9QY |
| G8 | TLP | R Pearce, Doral, The Grove, Sevenoaks, TN15 6JJ |
| G8 | TLT | J Beveridge, Westwood, Hedgerow, Gerrards Cross, SL9 0HD |
| G8 | TLU | F Laird, 7a Meadow Road, Toddington, Dunstable, LU5 6BB |
| G8 | TMD | Tony Clint, 11 Home Lea, Rothwell, Leeds, LS26 0PP |
| GI8 | TME | J Campbell, 115 Dromore Road, Ballynahinch, BT24 8HU |
| G8 | TMJ | Paul Faulkner, 8 Parkfield Road, Cheadle Hulme, Cheadle, SK8 6EX |
| G8 | TML | J Foster, 14 Braemar Grove, Heywood, OL10 3RR |
| G8 | TMM | E Gilbert, 34 School Lane, Harpole, Northampton, NN7 4DR |
| G8 | TMQ | Paul Stevens, 17 Weaver Close, Brierley Hill, DY5 4QN |
| G8 | TMR | Philip Taylor, 22 Windermere Drive, Rainford, St. Helens, WA11 7LD |
| G8 | TMV | Colin Tuckley, 98 Woodland Road, Sawston, Cambridge, CB22 3DU |
| G8 | TNA | Stephen Thompson, 96 Tregonissey Road, St. Austell, PL25 4DS |
| G8 | TNB | P Thompson, Lyndhurst Cottage, Main Street, Newark, NG23 6ST |
| G8 | TND | C Schiffman, 14 Caspian Way, Purfleet, RM19 1LE |
| G8 | TNE | D Pickford, 80 Hollowood Avenue, Littleover, Derby, DE23 6JD |
| G8 | TNH | P Jeffries, 22 Ingrams Way, Hailsham, BN27 3NP |
| G8 | TNS | Stewart Ward, 2 Nursery Road, Rugeley, WS15 1EZ |
| G8 | TNU | Adrian Lambert, 69 Anvil Crescent, Broadstone, BH18 9DZ |
| G8 | TOI | Roy Hempstead, 21 Lymington Avenue, Clacton-on-Sea, CO15 4PJ |
| G8 | TOP | N Huggins, 4 Kelmscott Close, Goldings, Northampton, NN3 8XN |
| G8 | TOQ | J Jackson, 1 Dolly Garth, Arkengarthdale Road, Richmond, DL11 6QX |
| G8 | TOT | David Lodge, 134 Deyne Road, Huddersfield, HD4 7EP |
| G8 | TOX | K Taylor, Swn Y Don, Llanfaes, Beaumaris, LL58 8RG |
| G8 | TPC | B Taylor, 161 Sidegate Lane, Ipswich, IP4 4JN |
| G8 | TPF | Nicholas Sanvoisin, 11 ter rue Lecoq, St Nom la Breteche, France, 78860 |
| G8 | TPM | Nigel Wellsbury, 15 Woodlands Road, Cookley, Kidderminster, DY10 3TL |
| G8 | TPP | M Strudwick, 65 Neave Crescent, Harold Hill, Romford, RM3 8HN |
| G8 | TQH | Andrew McMullin, Fair View, Rickham East Portlemouth, Near Salcombe, TQ8 8PJ |
| G8 | TQI | P Herod, 4 St. James Road, Little Paxton, St. Neots, PE19 6QW |
| G8 | TQJ | R Markfort, 105 Woodlands Way, Southwater, Horsham, RH13 9TF |
| G8 | TQK | A Mayhew, 51 Upland Road, Sutton, SM2 5HW |
| G8 | TQP | R Healey, 35 Tirlebank Way, Tewkesbury, GL20 8ES |
| G8 | TQV | R Tuckett, 89 Hillbrook Road, Tooting, London, SW17 8SF |
| G8 | TQZ | B Woods, 84 Beauly Way, Rise Park, Romford, RM1 4XR |
| G8 | TRG | TOWER RADIO GROUP c/o R Green, 2 Ragley Walk, Rowley Regis, B65 9NT |
| G8 | TRO | K PROSSER, 12 Sycamore Court, Woodfieldside, Blackwood, NP12 0DA |
| G8 | TRQ | M Walker, 20 Littlewood Lane, Cheslyn Hay, Walsall, WS6 7EJ |
| G8 | TRR | W Pickard, 19 Canham Close, Kimpton, Hitchin, SG4 8SD |
| G8 | TRU | S Lynch, 4 Tanglewood, Welwyn, AL6 0RU |
| G8 | TRY | Gerry Scott, 19 Penkett Road, Wallasey, CH45 7QF |
| G8 | TSC | John Collins, 44 Rosedale Gardens, Thatcham, RG19 3LE |
| G8 | TSG | D Johansen, 45 Marfords Avenue, Bromborough, Wirral, CH63 0JJ |
| GI8 | TSI | I Raine, 48 Gardners Road, Lisburn, BT27 5PD |
| G8 | TSV | Gareth Rowlands, 39 Nelthorpe Street, Lincoln, LN5 7SJ |
| G8 | TSZ | Alan Twyford, 70 Ellesfield Drive, West Parley, Ferndown, BH22 8QW |
| GM8 | TTD | P Palin, Nampara, Roadside, Thurso, KW14 8SR |
| G8 | TTE | E Thomas, Fairfield, St. Marys Road, Oakham, LE15 8SU |
| G8 | TTI | D Kearns, 14 Draycot Cerne, Chippenham, SN15 5LD |
| G8 | TTJ | John Holland, 2 Hadfield Close, Staunton, Gloucester, GL19 3QY |
| G8 | TTP | R Nicholls, 57 Mandalay Court, London Road, Brighton, BN1 8QW |
| G8 | TTU | C Smithson, 4 Calder Avenue, Littleborough, OL15 9JE |
| G8 | TTX | Kenneth Bugg, 28 Well Close, Winscombe, BS25 1HQ |
| G8 | TUH | David George, East Winch Road, Blackborough End, Kings Lynn, PE32 1SF |
| G8 | TUN | C Denton, 34 Brook Lane, Ormskirk, L39 4RE |
| G8 | TUU | R Boyce, 9 Kestrel Close, Bexhill-on-Sea, TN40 1UG |
| G8 | TVC | T Webb, 25 Wheatfield Drive, Ramsey, Huntingdon, PE26 1SH |
| G8 | TVM | K Biggs, 33 Blanford Gardens, West Bridgford, Nottingham, NG2 7UQ |
| G8 | TVU | A COLE, 14 Ellesmere Grove, Stainforth, Doncaster, DN7 5BS |
| GM8 | TVV | N Coote, 130 Castle Gardens, Paisley, PA2 9RD |
| G8 | TVW | D Young, 58 Furzefield Road, Welwyn Garden City, AL7 3RJ |
| GW8 | TVX | Richard Hope, 75 Priors Way, Dunvant, Swansea, SA2 7UH |
| G8 | TVZ | R Midgeley, 2 Digswell Park Cottages, Digswell Park Road, Welwyn Garden City, AL8 7NN |
| G8 | TWA | G Jasper, Clann Farm, Clann Lane, Bodmin, PL30 5HD |
| GI8 | TWB | B Mitchell, 7 Crea Road, Randalstown, Antrim, BT41 3DX |
| G8 | TWR | J Evans, 49 Inverness Avenue, Enfield, EN1 3NU |
| G8 | TWS | M Corbett, 32 Bibury Road, Cheltenham, GL51 6BA |
| G8 | TWT | Brian Fisher, 24 Vessey Road, Worksop, S81 7PG |
| G8 | TWZ | Ivan Goodman, 271 Alcester Road, Hollywood, Birmingham, B47 5HJ |
| G8 | TXA | Ruth Heeley, 4 Cherry Tree Lane, Halesowen, B63 1DU |
| GM8 | TXC | John Hedley, c/o 10 White Street, [illegible] |
| [illegible] | [illegible] | [illegible] |
| G8 | TXK | P Shulver, 6 Swallow Wood, Fareham, PO16 0UP |
| G8 | TXL | John Sillitoe, 42 Marsham Road, Kings Heath, Birmingham, B14 5HD |
| G8 | TXT | B Linkins, Curlew Cottage, Higher Wringworthy Farm, Looe, PL13 1PR |
| G8 | TXW | G Sutcliffe, 41 Rose Avenue, Irlam, Manchester, M44 6AQ |
| G8 | TXX | Jill Taylor, The Jays, 5 Watling Close, Bourne, PE10 9XL |
| G8 | TYD | D Prouse, 137 Queensway, Shiphay, Torquay, TQ2 6BZ |
| G8 | TYF | H Sasse, Flat 8 The Beeches, 43 Queens Road, Leicester, LE2 1WQ |
| G8 | TYH | R Marsh, 58 Statham Avenue, Lymm, WA13 9NL |
| G8 | TYX | J Hodgson, 9 Elm Road, North Moreton, Didcot, OX11 9BB |
| G8 | TYY | T Hopkins, 5 Rochester Close, Bacup, OL13 8RN |
| G8 | TZE | D Pritt, 387 London Road, Clanfield, Waterlooville, PO8 0PJ |
| G8 | TZJ | A Sellers, 2 Dunkenshaw Crescent, Lancaster, LA1 4LQ |
| G8 | TZN | R North, 5 George Road, Guildford, GU1 4NP |
| G8 | TZU | R Collis, 17 Belvedere Close, Guildford, GU2 9NP |
| G8 | TZW | G Dunn, 29 Sundridge Road, Kingstanding, Birmingham, B44 9NY |
| G8 | UAD | R White, 72 Green Lane, Bournemouth, BH10 5LF |
| G8 | UAE | David Scott, Hyde Bungalow, The Hyde, Stourbridge, DY7 6LS |
| G8 | UAF | Gordon Scargill, 98 Southleigh Road, Leeds, LS11 5SG |
| G8 | UAI | D Lockwood, 61 Green Lane, Tickton, Beverley, HU17 9RH |
| GW8 | UAM | Lewis Wright, 19 Lon y Fran, Caerphilly, CF83 2RX |
| GW8 | UAP | T Williamson, Fforch Farm, Treorchy, CF42 6TF |
| G8 | UBD | Andrew Baker, 19 Rockington Way, Crowborough, TN6 2NJ |
| G8 | UBF | G Beadle, 81 Halford Road, Uxbridge, UB10 8QA |
| G8 | UBJ | R Lester, 71 Ronelean Road, Surbiton, KT6 7LL |
| G8 | UBN | Grant Hodgson, East Cottage, Chineham Lane, Basingstoke, RG24 9LR |
| G8 | UBP | Andrew Jacketts, Flat 7, Jenneth Court, 44 Mauldeth Road, Stockport, SK4 3NB |
| G8 | UBU | R Jarvis, 39 Moy Road, Colchester, CO2 8NZ |
| G8 | UBX | R Manning, 2 Reydon Close, Haverhill, CB9 7WG |
| G8 | UCC | I Bradley, 8 Hunt Avenue, Heanor, DE75 7QB |
| G8 | UCK | T Colligan, 53 Datchworth Turn, Hemel Hempstead, HP2 4PB |
| G8 | UCN | A Crookes, 23 Helliwell Lane, Deepcar, Sheffield, S36 2NH |
| G8 | UCP | Martyn Culling, 101 Orchard Drive, Park Street, St. Albans, AL2 2QL |
| G8 | UCR | D Davis, 2 Bowen Road, Rotherham, S65 1LH |
| GI8 | UCS | A Edwards, 126 Merville Garden Village, Newtownabbey, BT37 9TJ |
| G8 | UCV | John Thompson, 37 Plane Tree Road, Wirral, CH63 2NW |
| G8 | UCY | D Walker, Bessbrook, 43 Wimborne Road, Wimborne, BH21 3DS |
| G8 | UCZ | C Wren, 24 Willow Way, Martham, Great Yarmouth, NR29 4SH |
| G8 | UDA | B Watson, 3 Anderton Rise, Millbrook, Torpoint, PL10 1DA |
| G8 | UDG | D Roberts, 25 Metcalfe Road, Cambridge, CB4 2DB |
| G8 | UDI | Paul Newman, 4 Old Barn Court, Ludford, Market Rasen, LN8 6AZ |
| G8 | UDJ | Martin Loach, 82 Honeybottom Lane, Dry Sandford, Abingdon, OX13 6BX |
| G8 | UDS | J Farrant, Partida Barrancs 16, Orba, Alicante, Spain, 3790 |
| G8 | UDV | A Frost, 10 Ramsden Square, Cambridge, CB4 2BJ |
| G8 | UDZ | Andrew Gilbertson, 9 Sun Hill Crescent, Alresford, SO24 9NJ |
| G8 | UEE | S Melvin, 2 Salters Court, Newcastle upon Tyne, NE3 5BH |
| G8 | UEF | P Mobberley, The Willows, Hobro, Kidderminster, DY11 5ST |
| G8 | UEI | Anthony Howells, 27 Swallow Lane, Aylesbury, HP19 7HW |
| G8 | UEK | N Hosker, 4 Alexandra Court, Coronation Close, Chester, CH2 3PZ |
| G8 | UEY | J Rice, 15 The Muntings, Stevenage, SG2 9DW |
| G8 | UEZ | C Southall, 40 Tathall End, Hanslope, Milton Keynes, MK19 7NF |
| G8 | UFF | A Patis, Flat 13, Fleur de Lis 41 High Street, Christchurch, BH23 1AS |
| G8 | UFO | Eric Charlton, 2 Bullock Road, Washingley, Peterborough, PE7 3SH |
| G8 | UFX | I Fowler, 1 Mayfields, Shefford, SG17 5AU |
| G8 | UGK | P Warburton, 4384 Henneberyy Road, Manilus, United States, 13104 |
| G8 | UGL | James Wakenell, 15 Cuckoo Oak Green, Madeley, Telford, TF7 4HT |
| GM8 | UGO | W Kay, 22 Linton Terrace, Perth, PH1 1LE |
| G8 | UGS | P Marks, 106 Darlton Drive, Arnold, Nottingham, NG5 7LW |
| G8 | UHJ | Patrick Couch, 6 Quantock Gardens, Ramsgate, CT12 6SW |
| G8 | UHK | A Rolls, 9 Mareschal Road, Guildford, GU2 4JF |
| G8 | UHM | W Rosser, Fanshawgate House, Fanshaw, Dronfield, S18 7WA |
| G8 | UHO | D Reay, 78 Wyresdale Road, Lancaster, LA1 3DY |
| G8 | UHT | P Shaw, Poole Bank Cottage, Poole, Nantwich, CW5 6AL |
| G8 | UHV | G Watson, 22 School Road, Laughton, Sheffield, S25 1YP |
| G8 | UHW | C Mobbs, 5 Garth Avenue, Leeds, LS17 5BH |
| G8 | UID | Raymond Nock, 83 Coles Lane, West Bromwich, B71 2QW |
| G8 | UIG | M Chedzoy, Helena, Picts Hill, Langport, TA10 9EZ |
| G8 | UIL | Jim Gale, Barn End, Highampton, Beaworthy, EX21 5LT |
| G8 | UIO | David Salter, 9 Old Milverton Road, Leamington Spa, CV32 6BA |
| GI8 | UIU | P Moore, 13 Ballygallum Road, Downpatrick, BT30 7DA |
| G8 | UIV | Paul Morton-Thurtle, 23 Wife of Bath Hill, Canterbury, CT2 8PQ |
| G8 | UIW | S Threlfall-Rogers, 43 Nanpantan Road, Loughborough, LE11 3ST |
| G8 | UJF | D Headland, Hazelwood, Haywards Lane, Cheltenham, GL52 6RF |
| G8 | UJO | B Leveton, Orman House, 17a Grove Avenue, Norwich, NR5 0JD |
| G8 | UJQ | Roger Halford, 3 Praed Place, Lelant, St. Ives, TR26 3DX |
| G8 | UJS | Alan McDermott-Roe, 16 Heathlands Avenue, W Parley, Ferndown, BH22 8RP |
| G8 | UJV | M Johnson, 9 South Road, Brampton, Huntingdon, PE28 4PX |
| G8 | UKH | L Luck, 1313 Rodney Lane, Winchester, Canada, K0C 2K0 |
| G8 | UKI | Timothy Gawn, 15 Barradon Close, Torquay, TQ2 8QE |
| G8 | UKO | P Coupe, 10a West End Avenue, Brundall, Norwich, NR10 5RF |
| G8 | UKV | M Vincent, 9 Sleapford, Long Lane, Telford, TF6 6HQ |
| G8 | UKY | R Mills, 3 Hallam Moor, Liden, Swindon, SN3 6LS |
| G8 | UKZ | Andrew Walker, Glandenys, Cross Inn, Llanon, SY23 5NA |
| G8 | ULH | James Wilson, 2 Reston Court, Cleethorpes, DN35 0JQ |
| G8 | ULJ | G Sowter, 21 Seawell Road, Bude, EX23 8PD |
| G8 | ULL | M Williams, 9 Fir Tree Drive, West Winch, King's Lynn, PE33 0PR |
| G8 | ULM | David Petty, 16 Audley End, Saffron Walden, CB11 4JB |
| G8 | ULQ | Claire Copsey, Flat 17, Cherrywood Court, Solihull, B92 8QS |
| G8 | UMA | J Spencer, 78 Copse Avenue, Farnham, GU9 9EA |
| G8 | UMB | Ronald Kennedy, Little Thatch, Canterbury Road, Dover, CT15 7HJ |
| G8 | UML | M Spencer, 79 Salisbury Close, Alton, GU34 2TP |
| GM8 | UMN | J Norrie, 13 Pentland Crescent, Dundee, DD2 2BU |
| G8 | UMO | M Walker, Winterfell, Fen Road, Boston, PE22 8HA |
| GD8 | UMY | P Mahoney, 2 Elm Lodge, Elm Avenue, Ruislip, HA4 8PH |
| G8 | UNO | R Clarke, 58 Orpin Road, Merstham, Redhill, RH1 3EY |
| G8 | UNP | Roger Burningham, 7 Gilder Way, Fishtoft, Boston, PE21 0QS |
| G8 | UOJ | A Abrahams, 69 Culverhouse Road, Luton, LU3 1PY |
| [illegible] | [illegible] | [illegible] TR16 [illegible] |
| GD0 | UOZ | M Frootone, 12 St. Martins Approach, Ruislip, HA4 7QD |
| G8 | UPD | R Payne, 1 Mill Hill, Horning, Norwich, NR12 8LQ |
| G8 | UPF | K Hutchinson, 8 Innage Crescent, Bridgnorth, WV16 4HU |
| G8 | UPI | David McAlpin, Birchwood, Belzies, Lockerbie, DG11 1SA |
| G8 | UPJ | Peter Nunn, 1 Talwrn Court, Coedpoeth, Wrexham, LL11 3NN |
| G8 | UPK | D Akester, 19 Bracken Park, Bingley, BD16 3LG |
| G8 | UPO | D Steele, 43 Lancastria Mews, Boyndon Road, Maidenhead, SL6 4SA |
| GW8 | UQC | D Bolton, Mill Farm, Manorowen, Fishguard, SA65 9PT |
| G8 | UQR | T Riley, 3 Hoefield Crescent, Nottingham, NG6 8AY |
| G8 | UQV | F Hall, 478 Darwen Road, Bromley Cross, Bolton, BL7 9DX |
| G8 | UQY | Graham Waywell, 14 Causeway, Great Harwood, Blackburn, BB6 7HU |
| G8 | URB | Fergus McGilp, 10 Walters Close, Cold Ash, Thatcham, RG18 9PU |
| G8 | URG | P Ridgeon, 10 Warwick Close, Amberstone, Hailsham, BN27 1NS |
| G8 | URI | G Cross, 11 Highfield App/Ch, Billericay, CM11 2PD |
| G8 | URU | R Drew, Clattering Ford, Roadhead, Carlisle, CA6 6NT |
| G8 | URZ | N Bird, 15 Bramley Close, Powick, Worcester, WR2 4SR |
| G8 | USA | M Dawes, 20 Alden Street, Danvers, United States, 1923 |
| G8 | UST | M Freeman, Sunnyside, Long Lane, Mansfield, NG20 8AZ |
| G8 | UTH | Graham Sunderland, 28 Tillotson Avenue, Sowerby Bridge, HX6 1BX |
| G8 | UTK | B Davies, Rhosyr, Llanfair Pg, LL61 5JB |
| G8 | UTQ | W Vander Byl, 45 Scotby Road, Scotby, Carlisle, CA4 8BD |
| G8 | UTW | H Mirams, 58 Ing Head Terrace, Shelf, Halifax, HX3 7LB |
| G8 | UTY | J Wardle, Spring Cottage, Chapel Road, Hayle, TR27 6BA |
| G8 | UUC | G York, 10 Beechwood Avenue, Wallasey, CH45 8NX |
| G8 | UUG | A Lenton, 45 Holland Gardens, Fleet, GU51 3NF |
| GI8 | UUN | G Curtis-Smith, 19a Glenavy Road, Lisburn, BT28 3UT |
| G8 | UUR | Philip Dicken, 72 Shobnall Street, Burton-on-Trent, DE14 2HJ |
| G8 | UUS | P Owen, 2 Plantation Road, Wollaton, Nottingham, NG8 2ER |
| G8 | UUV | M Copsey, 1 Cooper Terrace, Dereham Road, Dereham, NR19 2BJ |
| G8 | UUW | C Fyfe, 20 Larch Grove, Galashiels, TD1 2LB |
| G8 | UVF | T Cassell, 3 Rose Hill, Waterlooville, PO8 9QU |
| G8 | UVG | C Parker-Larkin, 42 Brett Green, Layham, Ipswich, IP7 5LX |
| G8 | UVN | Barrie Rigby, 76 Woodland Road, Rode Heath, Stoke-on-Trent, ST7 3TL |
| G8 | UVU | M Soble, Whitethorn Farm, Carey, Hereford, HR2 6NG |
| G8 | UVY | John Haste, 11 Corporation Road, Chelmsford, CM1 2AR |
| G8 | UVZ | B Hart, 63 Newcastle Road, Congleton, CW12 4HL |
| G8 | UWD | R Hornby, 63 Shearwater Drive, Bicester, OX26 6YR |
| G8 | UWE | M Jefford, 37 Marions Way, Exmouth, EX8 4LF |
| G8 | UWG | A Kay, 19 Chesnut Grove, Higher Tranmere, Birkenhead, CH42 0LB |
| G8 | UWI | A Downing, 21 Firfield Road, Thundersley, Benfleet, SS7 3UU |
| G8 | UWL | D Cooper, 18 Stockfield Road, Stoke-on-Trent, ST3 7AP |
| G8 | UWM | Martin Crossley, 3 Derby Street, Stockport, SK3 9HF |
| G8 | UWS | J Stopford, Bracken, The Street, Dover, CT15 7BH |
| G8 | UXB | B Payne, 45 Kellaway Avenue, Westbury Park, Bristol, BS6 7XS |
| G8 | UXD | A Gill, 4 Cornubia Close, Hayle, TR27 4RL |
| G8 | UXL | Peter Nicholson, 3 Eastleigh, Skelmersdale, WN8 6AX |
| G8 | UXW | John Benton, Emiviz, The Ridge, Salisbury, SP5 2LQ |
| G8 | UXX | K Brazington, 38 Tamworth Road, Amington, Tamworth, B77 3BT |
| G8 | UXY | K Bone, 69 Tone Hill, Tonedale, Wellington, TA21 0AY |
| G8 | UYB | K Reece, Winsford Grange Nursing Home, Station Road by Pass, Winsford, CW7 3NG |
| G8 | UYF | R Ritchie, Flat 2, Sundon Park Parade, Luton, LU3 3BH |
| G8 | UYK | F Rowntree, 5 Cornel House, Osborne Road, Windsor, SL4 3SQ |
| G8 | UYL | R Rumbelow, The Spinney, The Chase, Leatherhead, KT22 0HR |
| G8 | UYM | Jennifer Sanderson, 5 Babbacombe Drive, Ferryhill, DL17 8DA |
| G8 | UYR | B Smith, 3 Harwin Close, Wolverhampton, WV6 9LF |
| G8 | UYW | T Carrig, 12 Longmoor Drive, Liphook, GU30 7XA |
| G8 | UYY | M Daish, 27 Westbourne Road, Portsmouth, PO2 7LB |
| G8 | UZM | R Jefferson, 23 Valerian Avenue, Heddon-on-the-Wall, Newcastle upon Tyne, NE15 0EA |
| G8 | UZQ | Stephen Haywood, 19 Crich Way, Newhall, Swadlincote, DE11 0UU |
| G8 | UZV | J Dowie, 19 Brooklands Drive, Wolverley, Kidderminster, DY11 5EB |
| G8 | UZW | J Durrant, 114 Rosebank Avenue, Hornchurch, RM12 5QS |
| G8 | UZY | Damian Fisher, 8 Beech Road, Stibb Cross, Torrington, EX38 8HZ |
| G8 | UZZ | J Fogg, 17 Whitelands Meadow, Wirral, CH49 2RJ |
| G8 | VAD | J Goodings, 133 Lache Lane, Chester, CH4 7LU |
| G8 | VAE | A Griffiths, 11 The Dreys, Sewards End, Saffron Walden, CB10 2LL |
| G8 | VAF | K Chittenden, Ponders, Hall Road, Colchester, CO6 3DX |
| GM8 | VAM | George Brazier, 117a East Clyde Street, Helensburgh, G84 7PL |
| G8 | VAN | Maurice Goodwin, 15 Meadow Close, Repton, Derby, DE65 6GT |
| G8 | VAR | B Harper, 51 Cross Lane, Scarborough, YO12 6DQ |
| G8 | VAT | Graham Denton, Highfield Lodge, High Eggborough, Nr Goole, DN14 0PX |
| G8 | VBA | R Webb, 78 Station Road, Rolleston-on-Dove, Burton-on-Trent, DE13 9AR |
| G8 | VBC | R Timms, 20 Driftside, Blackfordby, Swadlincote, DE11 8BD |
| G8 | VBE | Clive Thomas, Apartment 231, Bournville Gardens Village, Birmingham, B31 2FS |
| G8 | VBI | David Spruson, 1 The Old Orchard, Whitehall, South Petherton, TA13 5AQ |
| G8 | VBK | R Penver, 56 Cottesmore Avenue, Ilford, IG5 0TG |
| G8 | VBW | Micheal Corbett, 2 Templar Close, Stenson Fields, Derby, DE24 3EL |
| G8 | VBX | D Coulthart, 23 Larchfield Road, Dumfries, DG1 4HU |
| GW8 | VCA | D Dyer, Ty Newydd, 24d Fforest Hill, Neath, SA10 8HD |
| G8 | VCH | J Grevatt, 17 Foxdale Drive, Angmering, Littlehampton, BN16 4HF |
| G8 | VCI | C Gwynne, 5 Stanstead Avenue, Nottingham, NG5 5BL |
| G8 | VCJ | Christopher Gunn, 57 Treffry Road, Truro, TR1 1WL |
| G8 | VCL | R German, 29 Glenthorne Gardens, Sutton, SM3 9NL |
| G8 | VCN | Michael Newport, 32 Stavordale Road, Weymouth, DT4 0AB |
| G8 | VCO | Brian Nicholls, 35 Lynn Road, Downham Market, PE38 9NJ |
| G8 | VCQ | W Norman, 27 Newport Road, Barnstaple, EX32 9BG |
| G8 | VCU | S Morgan, 5 Parklands, Ufford, Woodbridge, IP13 6ES |
| G8 | VDJ | Robert Pauley, Lamplight, Casterton Lane, Tinwell, Stamford, PE9 3UQ |
| G8 | VDP | K Roberts, 35a Rockley Avenue, Birdwell, Barnsley, S70 5QY |

UK Callsigns

**IMPORTANT NOTE**

**Revalidate licence to avoid revocation** – Ofcom has advised the Society that plans will be drawn up to revoke licences that have not been revalidated as required by the licence conditions. The quickest way to revalidate is to do so online via the Ofcom website: *https://services.ofcom.org.uk/* or by email: *amateur.validations@ofcom.org.uk* Ofcom staff are available to help, but please be patient during times of heavy workload.

G8 VDQ Christopher Parnell, 213a Northfield Avenue, London, W13 9QU
G8 VEE G Blore, Ty Newydd, Cymau, Wrexham, LL11 5EU
G8 VEM A Challis, 12 Moorland Crescent, Boultham Moor, Lincoln, LN6 7NL
G8 VEN H Chapman, 24 Croft Road, Cosby, Leicester, LE9 1SE
G8 VEQ A Stone, 47 Oakford Villas, North Molton, South Molton, EX36 3HJ
G8 VER VERULAM ARC c/o Robert Heath, 26 Lancaster Avenue, Hadley Wood, Barnet, EN4 0EX
G8 VEZ T Wagg, 15 Barncroft Way, Havant, PO9 3AA
GW8 VFF A Wilkins, 11 Redhouse Road, Ely, Cardiff, CF5 4FG
G8 VFI D Franklin, 49 Hope Road, Benfleet, SS7 5JQ
G8 VFL E Bury, 26 Northfield Avenue, Hanham, Bristol, BS15 3RB
G8 VFM James Callaghan, 271 Belvoir Road, Coalville, LE67 3PL
GW8 VFQ R Elliott, 19 Pencoed, Dunvant, Swansea, SA2 7PQ
G8 VG A Windle, Flat 11, Parham House, 15 King George's Drive, Liphook, GU30 7GB
GW8 VGB Robert Morgan, 4 Underhill Lane, Horton, Swansea, SA3 1LB
G8 VGI Albert Lilly, 47 Horton Street, Frome, BA11 3DP
G8 VGQ P Andrews, 1 Waite Meads Close, Purton, Swindon, SN5 4ET
G8 VGU T Burgess, 14 Shrubcote, Tenterden, TN30 7BA
G8 VGY E Davis, 850 Lantana Rd, Lantana, United States, 33462
G8 VHB Michael Fitzgibbons, 8 Lundhill Close, Wombwell, Barnsley, S73 0RW
G8 VHF BEWLAY BROS ARC c/o J Goodier, 20 Poleacre Lane, Woodley, Stockport, SK6 1PG
G8 VHG I Gower, 10 Homethorpe, Hull, HU6 9EU
G8 VHI Reg Woolley, 103 Mancetter Road, Nuneaton, CV10 0HP
G8 VHK Michael Stanford, Fircroft, Mutton Hall Lane, Heathfield, TN21 8NR
G8 VHL S Price, 35 Western Road, Goole, DN14 6QW
G8 VHN G Prentice, The Willow, Barretts Lane, Ipswich, IP6 8RZ
G8 VHO Shaun Pratt, 4 Bussex Square, Westonzoyland, Bridgwater, TA7 0HD
G8 VHX Gordon Shering, 66 Oliver Street, Ampthill, Bedford, MK45 2QL
G8 VIB J Sim, 22 Dene View, Ashington, NE63 8JT
G8 VIC Michael Rump, 24 Stoneleigh Avenue, Brighton, BN1 8NP
G8 VIV Kevin Bilke, 8 Ibworth Lane, Fleet, GU51 1AU
G8 VJG K Halls, Little Rema, Cray Road, Swanley, BR8 8LP
G8 VJO Colin Green, 12 Spenser Grove, Great Harwood, Blackburn, BB6 7JU
G8 VJP Lawrence French, 14 Manor Farm Court, Thrybergh, Rotherham, S65 4NZ
G8 VJR D Fowles, 52 Lucy Lane South, Stanway, Colchester, CO3 0HY
G8 VJU Kevin Earl, 210 Churchill Avenue, Chatham, ME5 0JS
G8 VJW G Davey, 147 Deeds Grove, High Wycombe, HP12 3PA
G8 VJY A Crafer, 155 Upham Road, Swindon, SN3 1DR
GI8 VKA R Coulter, 34 Toberdowney Valley, Ballynure, Ballyclare, BT39 9TS
G8 VKI Roger Tetchner, 4 Glebe Close, Weymouth, DT4 9RL
GM8 VKN M Tarr, Westholme, 1 Methven Drive, Dunfermline, KY12 0AH
G8 VKO G Tandy, 13 St. Marys Avenue, Bramley, Tadley, RG26 5UU
G8 VKQ Clive Sparke, 47 Jobes, Balcombe, Haywards Heath, RH17 6AF
G8 VKS John Williams, Glenview, Penrhos, Usk, NP15 2LF
GM8 VL GMDX Group c/o R Ferguson, 19 Leighton Avenue, Dunblane, FK15 0EB
G8 VLL A Kett, 476 Earlham Road, Norwich, NR4 7HP
G8 VLP M Clark, 6 Shalcross Drive, Cheshunt, Waltham Cross, EN8 8UX
G8 VLR R Law, 19 Central Drive, Bramhall, Stockport, SK7 3JU
G8 VLS David Leeder, 60 Montagu Avenue, Newcastle upon Tyne, NE3 4JN
G8 VLY R Macbeth, 9 Woodside, Stroud, GL5 1PL
G8 VLZ R Manning, 1 Homer Park, Plymstock, Plymouth, PL9 9NN
G8 VMF M Clayton, 54 Banks Road, Golcar, Huddersfield, HD7 4RE
G8 VML L Gibson, 57a Heritage Park, Hatch Warren, Basingstoke, RG22 4XT
G8 VMP K Webster, 5 Ridgmont Road, St. Albans, AL1 3AG
G8 VMQ A Parker, 78 Whitbarrow Road, Lymm, WA13 9BA
G8 VMY D Whitfield, Framingham, Manor Road, Hayling Island, PO11 0QR
G8 VMZ M Walpole, 9 The Paddocks, Brandon, IP27 0DX
G8 VNF Bill Benson, 31 Helston Close, Brookvale, Runcorn, WA7 6AA
G8 VNL J Darlington, 111 Maas Road, Northfield, Birmingham, B31 2PP
G8 VNN A Dowsett, 70 Warren Drive, Broughton, Chester, CH4 0PT
G8 VNO G Edmonds, 29a Manor Park, Woolsery, Bideford, EX39 5RH
G8 VNP R Elden, 124 Larchcroft Road, Ipswich, IP1 6PQ
G8 VNX J Dixon, 19 Pheasant Drive, Wincham, Northwich, CW9 6PX
G8 VOB Kimberley Fisher, 50 Queen Street, Henley-on-Thames, RG9 1AP
G8 VOC Vincent Prank, Daisy Cottage, Long Green, Woodbridge, IP13 7JD
G8 VOH Philip Renshaw, Hayes Pond Cottage, Hayes Plat, Rye, TN31 6HQ
G8 VOI Robert Reeves, 4 Elmwood Avenue, Waterlooville, PO7 7LG
G8 VOQ R Rogalewski, 47 Partridge Crescent, Dewsbury, WF12 0HT
G8 VOY John McParlane, 47 Aragon Road, Kingston upon Thames, KT2 5QB
G8 VPD I Morley, 2 Livingstone Walk, Park Wood, Maidstone, ME15 9JB
G8 VPE J Noy, 14 Poplar Drive, Filby, Great Yarmouth, NR29 3HU
G8 VPG S O'Sullivan, 15 Witney Close, Saltford, Bristol, BS31 3DX
G8 VPH B Aveling, 6 Brambling, Wilnecote, Tamworth, B77 5PQ
G8 VPO James Chalmers, 10 Hornbeam Close, Wokingham, RG41 4UR
G8 VPR Brian Gallear, 5 Oak Avenue, Cannock, WS12 4QA
G8 VPX A Davies, 14 Primrose Grove, Keighley, BD21 4NP
G8 VQA S Foulser, 9 Oak Coppice Close, Eastleigh, SO50 8PH
G8 VQE B Haylett, 160 Hookfield, Harlow, CM18 6QN
G8 VQH M Russ, 71 Farriers Close, Martlesham Heath, Ipswich, IP5 3SN
G8 VQJ Michael Otterson, 161 Tollgate Lane, Bury St. Edmunds, IP32 6DF
G8 VQK M Nightingale, 31 Cradge Bank, Spalding, PE11 3AB
G8 VQN Richard Martin, 12 Rectory Road, Duxford, Cambridge, CB22 4RZ
G8 VQQ J Locke, 2 Norton Close, Daventry, NN11 4GW
G8 VQS R Kugler, 96 Sanforth St., Whittington Moor, Chesterfield, S41 8RU
G8 VQX Alan Hyde, 57b Tan Lane, Caister-on-Sea, Great Yarmouth, NR30 5DT
G8 VR Kerry Rochester, 22 Langford Road, Cockfosters, Barnet, EN4 9DS
G8 VRN Derek Sutton, 14 Brocklesby Close, Gainsborough, DN21 1TT
GW8 VRS David Fone, 29 South Rise, Cardiff, CF14 0RF
G8 VRV G Dyer, Chez Nous, 11 Fore Street, St. Austell, PL26 7NN
G8 VRW Michael Davis, 4 Joel Close, Earley, Reading, RG6 5SN
G8 VSF J Williams, Staithe Marsh House, The Staithe, Norwich, NR12 9DA
G8 VSH P Taylor, 62 Westfield Avenue, Ashchurch, Tewkesbury, GL20 8QP
G8 VSI M Sutcliffe, 2 Adam Croft, Cullingworth, Bradford, BD13 5JF
G8 VSN G Tennant, 85 Coronation Drive, South Normanton, Alfreton, DE55 2HS
G8 VSR J Rowley, 10 Friars Close, Cheadle, Stoke-on-Trent, ST10 1AT
G8 VSV David Petty, 7 Luscombe Close, Ipplepen, Newton Abbot, TQ12 5QJ
G8 VSX R Hammond, 43 Witham Road, Woodhall Spa, LN10 6RG
GI8 VTK A Boston, 14 Galloway Point, Donaghadee, BT21 0ES
G8 VTV Andrew Hart, 78 Shepherds Way, Rickmansworth, WD3 7NR
G8 VTX I King, Mill Ghyll, Low Lane, Kendal, LA8 8AT
G8 VTY Nigel Knight, 36a Abbey Street, Rugby, CV21 3LH
G8 VU D Blair, 121 Longstomps Avenue, Chelmsford, CM2 9BZ
G8 VUG I Wilkinson, 6 Cwm Teg, Old Colwyn, Colwyn Bay, LL29 8ZA
G8 VUK Alun Palmer, 32 Woodstock Road, Carshalton, SM5 3DZ
G8 VUM P Reade, 30 Hayleigh House, Silcox Road, Bristol, BS13 0JG
G8 VUN R Roberts, 5480 Laburnum Ave., British Columbia, Canada, V8A 4MB
G8 VUS A Branton, 20 Sling Lane, Malvern, WR14 2TU
G8 VUU R Clifford, 1 Darell Croft, Sutton Coldfield, B76 1HU
GW8 VUV A Gravell, 49 Rehoboth Road, Five Roads, Llanelli, SA15 5DJ
G8 VVB C Heath, 3 Trebellan Drive, Hemel Hempstead, HP2 5EL
G8 VVC C Haver, 31 Edenham Road, Hanthorpe, Bourne, PE10 0RB
G8 VVG S Hackett, 11 Fairfield Avenue, Upminster, RM14 3AZ
G8 VVM David Merrick, 259 Wigmore Road, Gillingham, ME8 0LZ
G8 VVP P North, 84a Park Road, Great Sankey, Warrington, WA5 3ET
G8 VVR Chris Key, 23 Oxford Road, Kesgrave, Ipswich, IP5 1EL
G8 VVX R Williams, The Basement, 9 Charlton Street, Llandudno, LL30 2AA
G8 VVY R Shelley, 1 Cornfield Drive, Bishops Cleeve, Cheltenham, GL52 7YR
G8 VVZ A Stephens, 12 Sutherland Walk, Aylesbury, HP21 7NS
G8 VWH S Hall, 31 Somerton Gardens, Earley, Reading, RG6 5XG
G8 VWU J Marriott, 9 Albany Walk, Peterborough, PE2 9JN
G8 VWV Andrew Ball, White Cottage, Barracks Lane, Reading, RG7 1BB
G8 VXB D Young, 66 Porchester Road, Kingston upon Thames, KT1 3PS
G8 VXR C Hunt, 41 Maylands Way, Harold Wood, Romford, RM3 0BQ
G8 VXU D Llewelyn, 56 Marlpit Lane, Seaton, EX12 2HN
G8 VXY P Nicol, 38 Mitten Avenue, Rubery, Birmingham, B45 0JB
G8 VYK Selex Galileo Sports & Leisure Club c/o Michael Purser, 17 Firecrest Road, Chelmsford, CM2 9SN
G8 VYO A Swain, 6 Abbots Grove, Belper, DE56 1BX
G8 VYP C Syms, 24 Warmdene Road, Patcham, Brighton, BN1 8NL
G8 VYQ G Todd, 100 Avebury Drive, Washington, NE38 7DB
G8 VYT D Tombs, 24 Ferndale Road, Northville, Bristol, BS7 0RP
GM8 VYZ A Raine, 10 Castle View, Airth, Falkirk, FK2 8GE
G8 VZB A Poole, 6 Rutland Avenue, Willsbridge, Bristol, BS30 6EZ
G8 VZD C Ramsey, 10 Lindley Road, London, E10 6QT
G8 VZI M Warren, 39 St. Marks Road, Weston-Super-Mare, BS22 7PF
G8 VZJ M Webb, 24 College Avenue, Grays, RM17 5UW
G8 VZR Nigel Giddings, Colliford Lake Park, St. Neot, Liskeard, PL14 6PZ
G8 VZS I Chapman, 188 Goodhart Way, West Wickham, BR4 0HA
G8 VZT David Hall, 4 Steventon Road, Wellington, Telford, TF1 2AS
G8 VZY Barry Levie, 51 Budges Road, Wokingham, RG40 1PL
G8 VZZ P Marks, Flat 3, 47 The Thoroughfare, Woodbridge, IP12 1AH
G8 WAJ N Treanor, 23 Norton Avenue, Penketh, Warrington, WA5 2RB
G8 WAL Paul Taylor, 14 Cedern Avenue, Elborough, Weston-Super-Mare, BS24 8PA
G8 WAM G Weeks, Forge Cottage, The Bury, Hook, RG29 1ND
G8 WAP Richard Warren, 22 Tyndale, North Wootton, King's Lynn, PE30 3XD
G8 WAV C Jacobs, 133 Fordham Road, Isleham, Ely, CB7 5QX
G8 WAW Ian Howard, 85 Mollington Avenue, Liverpool, L11 3BQ
G8 WBG S Netherton, 33 Bethel Road, St. Austell, PL25 3HB
G8 WBK A Maufe, 28 Dale View, Ilkley, LS29 9BP
G8 WBL T Mole, 20 Horns Park, Bishopsteignton, Teignmouth, TQ14 9RP
G8 WBN D Neale, 24 Addison Road, Reading, RG1 8EN
G8 WBO S Holley, 37 Bouverie Avenue, Salisbury, SP2 8DU
G8 WBP P Humphreys, 910 High Lane, Stoke-on-Trent, ST6 6HE
G8 WBT A Farnborough, 9 Mitchelmore Road, Yeovil, BA21 4BA
G8 WBU A Greenall, 44 Hardy Road, Blackheath, London, SE3 7NN
G8 WBY Jonathan Osborne, Little Martins, Langham, Colchester, CO4 5PY
GI8 WBZ A Smith, 12 Sandringham Heights, Carrickfergus, BT38 9EG
G8 WCA K Winter, Derwen, Hillside, Monmouth, NP25 4LY
G8 WCH R Shepherd, 299 West Wycombe Road, High Wycombe, HP12 4AA
G8 WCQ V W McClure, 43 Roman Way, Seaton, EX12 2NT
G8 WCT A Grindrod, 54 Priestley Drive, Pudsey, LS28 9NQ
G8 WCX A Essex, 32 Crossfield Drive, Skellow, Doncaster, DN6 8RJ
G8 WDC WIRRAL&DIST ARC c/o Gerry Scott, 19 Penkett Road, Wallasey, CH45 7QF
G8 WDX R Lamkin, 3 Homestead Close, Upton, Aylesbury, HP17 8XQ
G8 WEM M O'Neill, Coolrake, Moone, Co Kildare, Ireland
GW8 WEY T Jones, 80 Taff Embankment, Cardiff, CF11 7BG
G8 WFP C Kershaw, 50 Wellgarth, Halifax, HX1 2BJ
G8 WFS J Lawson-reay, The Nook, Conway Road, Llandudno, LL30 1PY
G8 WGD P Randall-Cook, 3 Wellmeadow, Staunton, Coleford, GL16 8PQ
G8 WGE Ian Robinson, 26 Wick Road, Teddington, TW11 9DW
G8 WGN J Marks, Stam 69, Huizen, Netherlands, 1275 CG
G8 WGP Gilwell Park Scout RC c/o Stuart Barber, Homedale, St. Monicas Road, Tadworth, KT20 6ET
G8 WGQ D Onione, 19 Chapman Close, Kempston, Bedford, MK42 8RU
GM8 WGU A Irving, 23 Woodlea Park, Sauchie, Alloa, FK10 3BG
G8 WHB H Couchman, Pond Cottage, Woodside Green, Maidstone, ME17 2EU
G8 WHD P Whittington, 7 Bowden Rise, Seaford, BN25 2HZ
GI8 WHP S Craig, 8 Andrew Avenue, Larne, BT40 1EB
G8 WHR S Wood, 90 Plymyard Avenue, Bromborough, Wirral, CH62 6BR
G8 WIM WIMBLEDON&DIS RD c/o G Cripps, 115 Bushey Road, Raynes Park, London, SW20 0JN
G8 WIR J Vousden, 44 Castle Road, Tankerton, Whitstable, CT5 2DY
GI8 WIU S Douthart, 75 Market Street, Ballycastle, BT54 6DS
G8 WJB J Geer, 31 The Beeches, Salisbury, SP1 2JH
GM8 WJK James Nicolson, Clickhimin, Serrigar, Orkney, KW17 2RL
GI8 WJN Albert Humphreys, 20 Ballyreagh Road, Tempo, Enniskillen, BT94 3EH
G8 WJY M Garton, 13 Damaskfield, Worcester, WR4 0HY
G8 WKA R Reich, Stream Cottage, Laurel Grove, Farnham, GU10 4UA
G8 WKE John Bloxham, 15 Windmill Road, Breachwood Green, Hitchin, SG4 8PG
G8 WKH I Jones, 2 Castle Keep Mews, Newcastle, ST5 2SD
G8 WKK M Daniels, 6 Middlemead, Stratton-on-the-Fosse, Radstock, BA3 4QH
G8 WKL DOWNSIDE SCL AR c/o M Daniels, 6 Middlemead, Stratton-on-the-Fosse, Radstock, BA3 4QH
G8 WKT John Larkins, 38 Hurst Road, East Molesey, KT8 9AF
G8 WKX R Denton, 18 Sealand Court, Esplanade, Rochester, ME1 1QH
G8 WKZ Kenneth Spragg, 88 Low Lane, Middlesbrough, TS5 8EB
G8 WLB S Austen, Shiralee, The Plain Road, Ashford, TN25 6RA
G8 WLD William Parrott, 11 St. Georges, Chester, CH1 3HG
G8 WLL S Lown, 50 Fall Birch Road, Lostock, Bolton, BL6 4LG
G8 WLV R Barber, 10 St. Leonards Close, Upper Minety, Malmesbury, SN16 9QB
G8 WLY J List, 41 Westbury Crescent, Dover, CT17 9QQ
G8 WMC E Holman, Weavers Cottage, The Shoe, Chippenham, SN14 8SA
G8 WMF Anthony Dawe, Highcroft, Upper House Lane, Guildford, GU5 0SX
G8 WMG James Bassnett, 105 Edgemoor Drive, Crosby, Liverpool, L23 9UF
G8 WMK G Bessant, 4 Sleigh Road, Sturry, Canterbury, CT2 0HR
G8 WMW A Rowell, 25 Headcorn Gardens, Cliftonville, Margate, CT9 3ES
G8 WNB K Phillips, Lluest Y Coed, 39 Llwyn Ynn, Talybont, LL43 2AG
G8 WNK John Davies, 6 Grendale Avenue, Stockport, SK1 4BL
G8 WNQ R Harrison, Margaty, Pencoys, Redruth, TR16 6LR
G8 WOX Anthony Hartland, 11 Linden Avenue, Kidderminster, DY10 3AA
G8 WOZ D Hopkins, 14 Abraham Drive, Silver End, Witham, CM8 3SP
G8 WPA Barry Lloyd, 238 Brecknock Road, London, N19 5BQ
G8 WPF A Middleton, 13 Ragleth Road, Church Stretton, SY6 7BN
G8 WPL D Hughes, 12 Spencer St., Reddish, Stockport, SK5 6UH
G8 WPO S Pateman, Kingswood House, 32 Balls Chase, Halstead, CO9 1NY
G8 WPU Ian Rivett, 30 Millside Close, Kingsley, Northampton, NN2 7TR
G8 WPV A Reason, 71 Cavendish Road, Hazel Grove, Stockport, SK7 6HU
G8 WQ Weymouth and District Short Wave Club c/o Geoffrey Watts, 3 Maple Grove Knightsdale Road, Weymouth, DT4 0FE
G8 WQC J Maxworthy, 3 Hoylake Close, Slough, SL1 5UR
G8 WQE A Vaughan, 12 Kingsley Road, Frodsham, WA6 6SG
G8 WQT T Rickard, 137 Hugin Avenue, Broadstairs, CT10 3HN
G8 WQW J Shergold, 35 Orchard Grove, New Milton, BH25 6NZ
G8 WQZ D Mead, 9 Abraham Drive, Silver End, Witham, CM8 3SP
G8 WRB David Kirkby, Stokes Hall Lodge, Burnham Road, Chelmsford, CM3 6DT
GW8 WRC T Harston, Ogilvie House, St. Ishmaels, Haverfordwest, SA62 3TD
G8 WRG WARKS RAYNET GR c/o David Salter, 9 Old Milverton Road, Leamington Spa, CV32 6BA
G8 WRI W Lawrence, 15 Rissington Road, Tuffley, Gloucester, GL4 0HP
G8 WRL R Williamson, 35 Villiers Avenue, Twickenham, TW2 6BL
G8 WRV R Bygrave, 35 East St., St. Neots, Huntingdon, PE19 1JU
G8 WRY G Brock, 54 Lord Haddon Road, Ilkeston, DE7 8AW
G8 WSB Michael Beardsley, 121 Wood Road, Lower Gornal, Dudley, DY3 2LR
G8 WSC R Burg, 20 Rowan Way, Witham, CM8 2LJ
G8 WSF Francis Price, 26 Teviot Gardens, Pensnett, Brierley Hill, DY5 4QL
G8 WSH M French, 32 St. Michaels Road, Long Stratton, Norwich, NR15 2PH
G8 WSM Weston Super Mare RS c/o David Dyer, 26 Locking Road, Weston Super Mare, BS23 3DF
G8 WSP Peter Arup, Alma House, Broadway Road, Windlesham, GU20 6BU
G8 WSQ A Beeston, 8 Meadow Close, Repton, Derby, DE65 6GT
G8 WSR WIRRAL SCHOOLS RC c/o S Wood, 90 Plymyard Avenue, Bromborough, Wirral, CH62 6BR
G8 WSS M Blair, 12 Medoc Close, Pitsea, Basildon, SS13 1NR
G8 WSU James Hoggarth, Cotherstone, Rockingham Paddocks, Kettering, NN16 9JR
G8 WSV M Cartwright, 9 Montgomery Close, Kettering, NN15 5BY
G8 WSW Roger Carter, 46 Arterial Road, Leigh-on-Sea, SS9 4DA
G8 WSX CHICHESTER ARC c/o G Goodyer, Flat, 54 Wyndham Road, Petworth, GU28 0EQ
G8 WSY P Bloor, 216 Waterloo Street, Burton-on-Trent, DE14 2NB
G8 WSZ J Foster, 1 Thorn Court, Four Marks, Alton, GU34 5BY
G8 WTB D Crowe, 10 Bryn Teg, Arddleen, Llanymynech, SY22 6PZ
G8 WTM R Britt, 2 Lindisfarne Court, Maldon, CM9 6UQ
G8 WTN J Capon, 24 Furness Close, Chadwell St. Mary, Grays, RM16 4JB
G8 WTZ D Holland, 42 Front Street, Sunniside, Bishop Auckland, DL13 4LW
G8 WUF D Legg, 2 Birkbeck Road, Wimbledon, London, SW19 8NZ
G8 WUG Roy Spence, 8 Stoneleigh, Sawbridgeworth, CM21 0BT
G8 WUM Harold Matthews, 24 Clos Y Berllan, Rhuddlan, Rhyl, LL18 2UL
G8 WUO K Baker, 57 Hedingham Road, Hornchurch, RM11 3QH
G8 WUR Stephen Browning, 360 Aureole Walk, Newmarket, CB8 7AZ
G8 WUS P Besley, Parklyn House, Foxbury Lane, Emsworth, PO10 8RN
G8 WUU J Cooper, 156 Church Road, Benfleet, SS7 4EN
G8 WUY Ronald Dowthwaite, 13 Lynton Close, Knutsford, WA16 8BH
G8 WVB S Ayer, 335 Ings Road, Kingston Upon Hull, Hull, HU7 4UY
G8 WVH John Bull, 12 Eastfield Crescent, Laughton, Sheffield, S25 1YT
G8 WVO P Dawson, 35 Crofton Road, Ipswich, IP4 4QP
G8 WVZ C Edwards, 9 Bradworth Close, Osgodby, Scarborough, YO11 3PZ
G8 WW L Carter, 65 Parry Road, Wyken, Coventry, CV2 3LW
G8 WWC G Ludlow, 48 Clifford Avenue, Walton Cardiff, Tewkesbury, GL20 7RW
G8 WWD Gordon Hunter, 151 Norwich Drive, Wirral, CH49 4GD
G8 WWF P O'Ryan, 12 Minton Close, Congleton, CW12 3TD
G8 WWI P Leverington, 28 Burymead, Stevenage, SG1 4AY
G8 WWJ J Kirton, 13 Saltersford Road, Grantham, NG31 7HH
G8 WWM A Morgan, 316 Middle Road, Southampton, SO19 8NT
G8 WWO J Jackson, 12 Lower Laith Avenue, Todmorden, OL14 5RU
G8 WWW Mark Harrington, 17 Church Road, Penponds, Camborne, TR14 0QE
GM8 WWY W Kemp, 35 Quarry Drive, Kirkintilloch, Glasgow, G66 3RY
GW8 WXP R Hadland, 41 Colby Road, Burry Port, SA16 0RH
G8 WXU Glyn Evans, 24 Beaufort Road, Billericay, CM12 9JL
G8 WXV Alan Faulkner, 8 Wayside, Trull, Taunton, TA3 7HS
G8 WYB M Jefferson, 16 Edge Dell, Stoney Haggs, Scarborough, YO12 4LL
G8 WYI Paul Herring, 52 Mellowship Road, Eastern Green, Coventry, CV5 7BY
G8 WYR LEEDS & DIS ARS c/o M Howes, Yarnbury Rufc, Brownberrie Lane, Leeds, LS18 5HB
G8 WYW C Burn, Ynysfallen, Church Street, Wrexham, LL14 2RL
G8 WZJ Alan Collier, 44 Cockington Close, Leigham, Plymouth, PL6 8RQ
G8 WZK Denis Collins, 71 Trench Road, Tonbridge, TN10 3HG
G8 WZO P Evans, 63 Broadfield Road, London, SE6 1NQ
G8 WZR D Gale, 8 Buccaneer Way, Duffryn, Newport, NP10 8ER
G8 WZW Ken Aspden, Langriggs, Goose House Lane, Darwen, BB3 0EH
G8 XAA BRISTOL RAYNET c/o A Williams, 38 Seneca Street St. George, Bristol, BS5 8DX

| | | |
|---|---|---|
| G8 | XAJ | Terence Sherman, 44 Cleveland Avenue, Weymouth, DT3 6AG |
| G8 | XAN | Roger Woods, [illegible] Place, Long Eaton, Nottingham, NG10 1LS |
| G8 | XAO | Graham Woodman, [illegible] |
| G8 | XAS | G Evans, Wynona, Esplanade, Penmaenmawr, LL34 6LY |
| G8 | XAX | Keith Tully, 225 Main Road, Harwich, CO12 3PL |
| G8 | XBY | Philip Allwood, 24 Summerhill Road, Lyme Regis, DT7 3DT |
| G8 | XCE | David Baker, 48 Elmwood Street, Burnley, BB11 4BP |
| G8 | XCJ | I Coton, 77 Lockesfield Place, London, E14 3AJ |
| G8 | XCL | I Davis, 28 Sycamore Close, Lydd, Romney Marsh, TN29 9LE |
| G8 | XCR | C Girling, 20 Fore Street, Praze, Camborne, TR14 0JX |
| G8 | XCW | Rob Thomson, Shire Jee Neevas, Cold Ash Hill, Thatcham, RG18 9PH |
| G8 | XCY | A Worsfold, 5 Turner Close, Langney, Eastbourne, BN23 7PF |
| G8 | XDD | David Lucas, 43 Larcombe Road, Petersfield, GU32 3LS |
| G8 | XDL | R Medcalf, 19 All Saints Road, Warwick, CV34 5NL |
| G8 | XDM | Peter Mutter, 129 Demesne Road, Wallington, SM6 8EW |
| G8 | XDR | C Johnstone, 24 Elibank Road, Eltham, London, SE9 1QH |
| G8 | XDU | G Howe, 49 Hadham Road, Bishop's Stortford, CM23 2QU |
| G8 | XDV | S Huyton, 33 Hide Gardens Rustington, Littlehampton, BN16 3NP |
| G8 | XEC | G Murray, 3 Domoney Close, Thatcham, RG19 4DY |
| G8 | XEF | M McIver, 31 Hartshill, Bedford, MK41 9AL |
| G8 | XEI | V Nolan, 127 Martins Lane, Blakehall, Skelmersdale, WN8 9BQ |
| G8 | XEN | H Hughes, 101 Mousehold Avenue, Norwich, NR3 4RX |
| G8 | XER | J Smith, 32 Station Crescent, Lidlington, Bedford, MK43 0SD |
| G8 | XET | C Street, Russets, Isle Brewers, Taunton, TA3 6QN |
| G8 | XEU | R Stephens, 21 St. James Avenue, Lancing, BN15 0NN |
| G8 | XEZ | M Ward, 11 Rogate Gardens, Portchester, Fareham, PO16 8DS |
| G8 | XFK | Ron Young, 34 Wharfedale Drive, Bridlington, YO16 6FB |
| G8 | XFY | Ian Downie, 17 Clyfton Crescent, Immingham, DN40 2AZ |
| G8 | XGB | K Dickson, 29 Sunnyfields Drive, Minster on Sea, Sheerness, ME12 3DH |
| G8 | XGG | S Gwilliam, 40 Falcon Close, Droitwich, WR9 7HF |
| G8 | XGK | P Manford, Smithy Hay, Hay Lane, Rugeley, WS15 4QG |
| G8 | XGO | P Mckellow, 155 Pittmans Field, Harlow, CM20 3LE |
| G8 | XGS | John Hindmarsh, Roseworth Cottage, Roseworth Cottage West, Hexham Road, Newcastle upon Tyne, NE15 9EB |
| G8 | XGT | M Saul, 49 Lukins Drive, Dunmow, CM6 1XQ |
| G8 | XGW | N Shearing, 51 Mill Lane, Huthwaite, Sutton-in-Ashfield, NG17 2SJ |
| G8 | XHD | P Riebold, 58 Woodcrest Road, Purley, CR8 4JB |
| G8 | XHK | K Prior, 9 Tangmere Road, Crawley, RH11 0JJ |
| G8 | XHN | R Harman, 17 Coldharbour Lane, Bushey, WD23 4NP |
| G8 | XHU | G Arrowsmith, 2 Orchard Drive, Bishops Hull, Taunton, TA1 5ES |
| G8 | XIJ | John Bryant, 5 Lismore Avenue, Ladybridge, Bolton, BL3 4NR |
| G8 | XIM | I Churchill, 12 Wyedale Avenue, Coombe Dingle, Bristol, BS9 2QQ |
| G8 | XIN | M Chapman, 4 Amberley Court, Sidcup, DA14 6JT |
| G8 | XIR | K Church, 11 Cambria Crescent, Riverview Park, Gravesend, DA12 4NJ |
| G8 | XIY | A Tee, 136 Burstellars, St. Ives, PE27 3TJ |
| G8 | XIZ | H Tillotson, 30 St. Laurence Road, Northfield, Birmingham, B31 2AX |
| G8 | XJB | Brian Simmons, 88 Wellcome Avenue, Dartford, DA1 5JW |
| GW8 | XJC | Richard Smith, 6 Lavender Court, Brackla, Bridgend, CF31 2ND |
| G8 | XJE | J Williams, 32 Fair Street, Broadstairs, CT10 2JL |
| G8 | XJK | D Hamilton, 4 Lilley Close, Bury St. Edmunds, IP33 2HZ |
| G8 | XJL | Martin Halford, 35 The Limes, Stony Stratford, Milton Keynes, MK11 1ET |
| G8 | XJN | W Hefferman, 74 Balmoral Drive, Borehamwood, WD6 2RB |
| G8 | XJO | S Hedicker, 1 Hares Close Cottages, Selborne Road, Liss, GU33 6HG |
| G8 | XJT | J Clarkson, 57 Chaigley Road, Longridge, Preston, PR3 3TQ |
| G8 | XKD | Robert Dance, 402 Wimborne Road East, Ferndown, BH22 9NB |
| G8 | XKH | W Flood, 3 March Meadow, Wavendon Gate, Milton Keynes, MK7 7TB |
| G8 | XKI | Gary Fowler, 12 Laughton Road, Hexthorpe, Doncaster, DN4 0BT |
| G8 | XKT | D Last, 77 Brunswick Road, Ipswich, IP4 4BS |
| G8 | XKW | James Ness, Fenway, Dalbeattie Road, Dumfries, DG2 7PL |
| G8 | XLA | Lesley Mayes, Stone House Goathland, Whitby, YO22 5AN |
| G8 | XLB | J Martin, Thatched Cottage, Thaxted Road, Saffron Walden, CB11 3BJ |
| G8 | XLE | W Metcalf, 30 Rosemary Road, Waterbeach, Cambridge, CB25 9NB |
| G8 | XLG | C Proctor, 69 Goodrington Road, Paignton, TQ4 7HZ |
| G8 | XLH | Alan Ralph, 15 Portchester Close, Stanground, Peterborough, PE2 8UP |
| G8 | XLI | James Rigby, 93 Birch Grove, Ashton-in-Makerfield, Wigan, WN4 0QX |
| G8 | XLL | R Stubbs, 35 Laburnum Drive, Rhyl, LL18 4JH |
| G8 | XLZ | K Riley, 122 Dryden Road, Gateshead, NE9 5TX |
| G8 | XMH | D Higgins, 80 Hill Morton Road, Sutton Coldfield, B74 4SG |
| G8 | XML | J Hopper, 21 Knowles Avenue, Crowthorne, RG45 6DU |
| G8 | XMO | H Houghton, 21 John Gwynn House, Newport Street, Worcester, WR1 3NY |
| G8 | XMS | L Sellar, 'Kiawah', Ringfield Drive, Hereford, HR1 4PR |
| G8 | XMU | E Jones, 3 Byland Close, Boston Spa, Wetherby, LS23 6PU |
| GW8 | XMW | D Jones, 7 Llys y Godian, Trimsaran, Kidwelly, SA17 4BQ |
| G8 | XMZ | R Linton, 11 Keats Lane, Wincham, Northwich, CW9 6PP |
| G8 | XNA | J Lane, 12 Penarwyn Woods, St. Blazey Gate, Par, PL24 2DG |
| G8 | XNB | Richard Lelliott, Smugglers Cottage, Oreham Common, Henfield, BN5 9SB |
| G8 | XNC | R Lacey, 12 Melville Avenue, Frimley, Camberley, GU16 8NA |
| G8 | XND | D Lucas, 6 Holborns Site, Main Road, Spalding, PE12 9PF |
| G8 | XNH | H Pearce, 28 Windsor Grove, Bodmin, PL31 2BP |
| G8 | XNL | John Rigby, 43a Corser Street, Stourbridge, DY8 2DE |
| G8 | XNN | H Vadgama, 20 Hollies Walk, Wootton, Bedford, MK43 9LB |
| G8 | XNO | P Lambert, 92 Winterslow Drive, Leigh Park, Havant, PO9 5DZ |
| G8 | XOB | P Ashcroft, Fendley Corner, Common Lane, Harpenden, AL5 5DW |
| G8 | XOC | D Bird, 119 Brandon Road, Watton, Thetford, IP25 6LL |
| G8 | XOE | B Baker, Linden Lea, Fivehead, Taunton, TA3 6PU |
| G8 | XOM | Paul Cook, Orchard Cottage, New Road, Elmswell, IP30 9BS |
| G8 | XOR | David Sparrow, 23 Tranmere Grove, Ipswich, IP1 6DU |
| G8 | XOU | I Spinks, 28 Chantry Court, Necton, Swaffham, PE37 8HA |
| G8 | XOV | L Sedgwick, 28 Fairhaven Road, Redhill, RH1 2LA |
| G8 | XOX | R Sneath, 16 Wavish Park, Torpoint, PL11 2HJ |
| G8 | XPB | K Chadwick, 5 Mason Close, Great Sutton, Ellesmere Port, CH66 2GU |
| G8 | XPD | M Dawkins, 24 Beaufort Drive, Barton Seagrave, Kettering, NN15 6SF |
| G8 | XPQ | Brian Whitehead, 17a Home Close, Histon, Cambridge, CB24 9JL |
| G8 | XPZ | S Lovell, 98b Baker Road, Newthorpe, Nottingham, NG16 2DP |
| G8 | XQA | P Lineham, 10 Streetsbrook Road, Shirley, Solihull, B90 3PL |
| G8 | XQD | Tom Miller, 35 Caudle Avenue, Lakenheath, Brandon, IP27 9AU |
| G8 | XQH | E Massey, 21 Arlington Drive, Macclesfield, SK11 8QL |
| G8 | XQI | Philip Nightingale, 15 North Drive, Thornton-Cleveleys, FY5 3AQ |
| G8 | XQL | J Alcock, Shirley Cottage, Welland Road, Worcester, WR8 0SJ |
| G8 | XQN | A [illegible], 37 Alledge Drive, Woodford, Kettering, NN14 4JQ |
| G8 | XQS | Martin Chapple, 34 Deepdale Way, Darlington, DL1 2TA |
| G8 | XQT | C Dodds, 6 Cascadia Close, High Wycombe, HP11 1JW |
| G8 | XQZ | Geoff Farmer, 39 Plough Rise, Upminster, RM14 1XR |
| G8 | XRG | R Margetts, Mowbray, Arbor Road, Leicester, LE9 3GE |
| G8 | XRL | R Mills, 131 High Road East, Felixstowe, IP11 9PS |
| G8 | XRP | R Pryor, 27 Hollickwood Avenue, London, N12 0LS |
| G8 | XRS | G Nuttall, 120 Cleevelands Avenue, Cheltenham, GL50 4PX |
| G8 | XRW | D Owen, 18 Bushey Close, Capel St. Mary, Ipswich, IP9 2HW |
| G8 | XSA | William Ash, 53 Waxland Road, Halesowen, B63 3DN |
| GI8 | XSB | F Aughey, 239 Bridge Street, Portadown, Craigavon, BT63 5AR |
| G8 | XSD | James Atkinson, 8 Grove End, Luton, LU1 5PF |
| G8 | XSF | M Ainley, 152 Bourne View Road, Huddersfield, HD4 7JS |
| G8 | XST | William Butchers, 12 Church Road, St. Marychurch, Torquay, TQ1 4QY |
| G8 | XSU | M Bond, 58 Street Pauls Street, Clitheroe, BB7 2LS |
| GI8 | XSY | K Steenson, 100 Morgans Hill Road, Cookstown, BT80 8BW |
| G8 | XTD | R Cavendish, 66 Coachmans Drive, Liverpool, L12 0HX |
| G8 | XTE | Peter Connor, 20 Longfield, Lutton, Ivybridge, PL21 9SN |
| G8 | XTJ | J Fitzgerald, 21 Honor Road, Prestwood, Great Missenden, HP16 0NJ |
| G8 | XTO | Royston Evans, 48 St. Marys Rise, Writhlington, Radstock, BA3 3PD |
| G8 | XTR | P Emmans, 16 Foresters Close, Rags Lane, Waltham Cross, EN7 6TF |
| G8 | XTU | Michael Fowler, 28 St. Hildas Road, Doncaster, DN4 5EE |
| G8 | XTW | P Seaford, 14 Nevis Close, Leighton Buzzard, LU7 2XD |
| G8 | XUB | Nicholas Reddish, 15 Drakes Close, Redditch, B97 5NG |
| G8 | XUE | L Radcliffe, 25 Oakleigh Drive, Codsall, Wolverhampton, WV8 1JP |
| G8 | XUH | J Pearson, 14 Gorse Close, Brampton Bierlow, Rotherham, S63 6HW |
| GM8 | XUK | Darren King, 59 South Knowe, Crossgates, Cowdenbeath, KY4 8AW |
| G8 | XUL | David James, 19 Estuary Drive, Felixstowe, IP11 9TL |
| G8 | XUM | P Jeavons, Manora, Penisarwaun, Caernarfon, LL55 3PW |
| G8 | XUN | M Hickman, 24 Calverley Road, Kings Norton, Birmingham, B38 8PW |
| G8 | XUU | E White, 97 Fillongley Road, Meriden, Coventry, CV7 7LW |
| G8 | XUW | D Shields, 54 Wildmoor Lane, Catshill, Bromsgrove, B61 0PA |
| G8 | XVJ | Erik Gedvilas, 33 Parkdale Road, Paddington, Warrington, WA1 3EN |
| G8 | XVO | C Hetherington, 23 Falkland Court, Braintree, CM7 9LL |
| G8 | XWH | C Langham, 27 Fyfield Avenue, Swindon, SN2 5ED |
| G8 | XWR | Malcolm Izzard, 17 Greenfields Avenue, Alton, GU34 2ED |
| G8 | XXA | J Harrison, 10 Gaia Lane, Lichfield, WS13 7LW |
| G8 | XXC | Peter Prince, 21 Ash Close, Appley Bridge, Wigan, WN6 9HU |
| G8 | XXG | S Richardson, 52 Nailsea Park, Nailsea, Bristol, BS48 1BB |
| G8 | XXI | J Akines, 105 Sutcliffe Avenue, Grimsby, DN33 1EZ |
| G8 | XXJ | J Allchin, 40 Vale Road, Seaford, BN25 3EZ |
| G8 | XXM | Chris Beecher, 25 Holborn Crescent, Tattenhoe, Milton Keynes, MK4 3EQ |
| G8 | XXU | M Caulton, 115 Delves Green Road, Walsall, WS5 4NH |
| G8 | XXV | G Clarke, 28 Little Potters, Bushey, WD23 4QT |
| G8 | XXZ | P Grace, 6 Davis Grove, Yardley, Birmingham, B25 8LQ |
| G8 | XYA | N Southorn, 20 Bratton Avenue, Devizes, SN10 5BA |
| G8 | XYJ | Matthew Porter, 8 Stanton Drive, Ludlow, SY8 2PH |
| G8 | XYQ | D Stanford, Laurel House, Top Road, Woodbridge, IP13 6JF |
| G8 | XYR | Roy Tiller, Wayside, Ockley Lane, Hassocks, BN6 8NU |
| G8 | XYS | R Travett, 39 Amwell Road, Cambridge, CB4 2UH |
| G8 | XYU | G Powell, 203 Queenborough Road, Minster on Sea, Sheerness, ME12 3EL |
| G8 | XZB | J Payne, 25 Ringwood Road, Bath, BA2 3JL |
| G8 | XZC | A Pinder, 2 Eleanor Road, Woodlands, Harrogate, HG2 7AJ |
| G8 | XZQ | M Fowler, 1 Mayfields, Shefford, SG17 5AU |
| G8 | XZX | J Tyler, 16 Stratton Road, Bude, EX23 8AE |
| G8 | YAE | C Wenn, 11 Bysouth Close, Ilford, IG5 0XN |
| GM8 | YAQ | R Wroblewski, 1 Normandy Place, Rosyth, Dunfermline, KY11 2HJ |
| G8 | YAS | A Miller, 113 West Front Road, Bognor Regis, PO21 4TB |
| G8 | YAT | I Naylor, 8 Churchill Close, Uttoxeter, ST14 8BB |
| G8 | YAU | R Newton, Cascades, Top Road, Brigg, DN20 0NN |
| G8 | YAZ | G Oates, 21 Churchill Mansions, Cooper Street, Runcorn, WA7 1DH |
| G8 | YBH | A Bristow, 2 Nursery Cottages, Staplehurst Road, Tonbridge, TN12 9BS |
| G8 | YBO | Robert Colebrook, 21 Hillclose Avenue, Darlington, DL3 8BH |
| G8 | YBR | Ivan Davidson, 1 Mooracre Lane, Bolsover, Chesterfield, S44 6ER |
| G8 | YBT | Norman Dilley, 26 Linhey Close, Kingsbridge, TQ7 1LL |
| GI8 | YBU | M Dunne, 26 Duncreggan Road, Londonderry, BT48 0AD |
| G8 | YBY | Enid Santer, 3 Barn Park, Liverton, Newton Abbot, TQ12 6HE |
| G8 | YBZ | M Hampson, 7 Merryfield Close, Bransgore, Christchurch, BH23 8BS |
| G8 | YCI | A Lewis, 8 Arundel Road, Hartford, Huntingdon, PE29 1YW |
| G8 | YCK | K Tomlinson, 27 Brackens Lane, Alvaston, Derby, DE24 0AQ |
| G8 | YCL | S Turner-Smith, 26 Ash Church Road, Ash, Aldershot, GU12 6LX |
| G8 | YCP | J Sergeant, 5 Jedburgh Close, North Shields, NE29 9NU |
| G8 | YCQ | N Storey, 15 Tower Avenue, Upton, Pontefract, WF9 1ED |
| G8 | YDB | Rosalie Merry, Havenwood, Oak Farm Lane, Sevenoaks, TN15 7JU |
| G8 | YDD | John Jebb, 99 [illegible], Nunthorpe, Middlesbrough, TS7 0QL |
| G8 | YDE | Stuart Inns, 11 Hodds Wood Road, Chesham, HP5 1SQ |
| G8 | YDI | D Belcher, 20 Gibson Drive, Upper Benefield, Peterborough, PE8 5AW |
| G8 | YDJ | Mark Alexander, 101 Richmond Street, Stoke-on-Trent, ST4 7DZ |
| G8 | YDN | Dylan Catlough, 18 Tyn y Celyn, Glan Conwy, Colwyn Bay, LL28 5NN |
| GM8 | YEC | P Eunson, Sandwick Cottage, Bridge End, Shetland, ZE2 9LD |
| G8 | YEF | A Eaton, 16 Wood Road, Godalming, GU7 3NN |
| G8 | YEJ | J Glover, 5 Meadow Rise, Wymondham, Melton Mowbray, LE14 2AP |
| G8 | YEN | M Stevens, Staging Post, Abbotskerswell, Newton Abbot, TQ12 5NX |
| G8 | YEO | Yeovil ARC c/o Richard Spirrell, 32 Churchfield Drive, Castle Cary, BA7 7LA |
| G8 | YEP | Brian Moyer, 6 Barrington Road, Sutton, SM3 9PP |
| G8 | YEQ | N Littleboy, 22 Sylvaner Court, Vyne Road, Basingstoke, RG21 5NZ |
| G8 | YFA | Anthony Regnart, 3 Preston Avenue, North Shields, NE30 2BW |
| G8 | YFH | David Oliver, 30 Lipscombe Rise, Alton, GU34 2HP |
| G8 | YFK | James Mason, 80 Swallow Drive, Milford on Sea, Lymington, SO41 0XG |
| G8 | YFP | Joseph Wells, 21 Main Street, Ewerby, Sleaford, NG34 9PH |
| GI8 | YGG | P Foley, 5 Woodland Drive, Cookstown, BT80 8PL |
| GM8 | YGI | P Sime, 29 Huntingtower Road, Baillieston, Glasgow, G69 7BH |
| G8 | YGK | W Standing, 72 Ivydore Avenue, Durrington, Worthing, BN13 3JD |
| G8 | YGM | D Southward, 3a Carnoustie Close, West Derby, Liverpool, L12 9NE |
| G8 | YGO | G Tarr, 40 The Garth, Coniston, LA21 8EQ |
| G8 | YGT | B Senior, 1 Bedale Close, Coalville, LE67 3BE |
| G8 | YHF | Stephen Kenyon, 8 Dunedin Gardens, Ferndown, BH22 0EQ |
| G8 | YIG | C Fawcett, 24 Quarry Rise, [illegible] |
| GM8 | YIK | Andrew Robson, Flat 12, 37 Hesperus Broadway, [illegible] |
| G8 | YIN | S Wood, 246 Rush Green Road, Romford, RM7 0LA |
| GI8 | YJD | R Perver, 6 Gransha Road, Bangor, BT20 4TG |
| GI8 | YJF | D Roxburgh, 5 Forestbrook Park, Rostrevor, Newry, BT34 3DX |
| GW8 | YJN | A Price, 45 Baring Gould Way, Haverfordwest, SA61 2SB |
| G8 | YJQ | Philip Holt, Flat 13, Norbiton Hall, Kingston upon Thames, KT2 6RA |
| G8 | YJS | Gerald Hammond, 21 Cawston Road, Reepham, Norwich, NR10 4LU |
| G8 | YJT | Colin Jarvis, 516 Kingsbury Road, Erdington, Birmingham, B24 9NF |
| GI8 | YJV | P Lloyd, 18 Demesne Road, Holywood, BT18 9NB |
| G8 | YJZ | P Rayson, 26 Leys Road, Pattishall, Towcester, NN12 8JZ |
| G8 | YKE | C Andrew, 17 St. James Close, Kettering, NN15 5HB |
| G8 | YKG | M Armour, 22 Langcliffe Close, Culcheth, Warrington, WA3 4LR |
| G8 | YKM | Alex Browne, 140 Tongham Road, Aldershot, GU12 4AT |
| G8 | YKO | S Bardsley, 73 Highlands, Royton, Oldham, OL2 5HL |
| G8 | YKS | Derek Barton, Salisbury Hill Barn, St. Brides Netherwent, Caldicot, NP26 3AT |
| GM8 | YKT | E Brumby, 141 Morriston Road, Elgin, IV30 4NB |
| G8 | YKV | A Cragg, 28 Damian Way, Hassocks, BN6 8BJ |
| G8 | YKY | D canham, 82 Rugby Road, Binley Woods, Coventry, CV3 2AX |
| G8 | YLA | R Cato, Orrell House, Winterpit Lane, Horsham, RH13 6LZ |
| GW8 | YLK | Benjamin Evans, Mynyddmelin, Pontfaen, Fishguard, SA65 9SL |
| G8 | YLM | Michael Farnworth, 16 Lees Court, Ribble Avenue, Darwen, BB3 0HW |
| G8 | YLR | Robert Foss, 4 Sandy Close, Wimborne, BH21 2NG |
| G8 | YLS | D Fox, 4 Lacey Grove, Annesley, Nottingham, NG15 0EG |
| G8 | YMD | James Carins, 26 Roman Way, St. Margarets-at-Cliffe, Dover, CT15 6AH |
| G8 | YMM | Paul Stevenson, 6 Dighton Gate, Stoke Gifford, Bristol, BS34 8XA |
| G8 | YMN | M Shorter, 10 Lodgefield Road, Chestfield, Whitstable, CT5 3RF |
| G8 | YMR | Allen Snow, 20 Blenheim Drive, Bredon, Tewkesbury, GL20 7NQ |
| G8 | YMS | P Swarbrook, 14 The Willows, Leek, ST13 8XF |
| G8 | YMT | Dennis Smith, 7 Peterdale Road, Brimington, Chesterfield, S43 1JA |
| G8 | YMU | L Shaw, 108 Brookvale Road, Solihull, B92 7JA |
| G8 | YMW | A Sneath, 21 Garrick Close, Lincoln, LN5 8TG |
| G8 | YMZ | J Trent, The Hollies Bourne Road, West Bergholt, Colchester, CO6 3EP |
| G8 | YNC | P Tuck, 30 Brownlow Road, New Southgate, London, N11 2DE |
| G8 | YNE | S Horner, 15 Newhouse Road, Huddersfield, HD2 1ED |
| G8 | YNF | G Holman, 62 The Ridge, Kennington, Ashford, TN24 9EU |
| G8 | YNG | A Hall, 33 Deanwood Road, Dover, CT17 0NT |
| G8 | YNH | M Hall, 20 Cubitt House, Black Bull Road, Folkestone, CT19 5SH |
| G8 | YNI | J Hancock, 78 Bridle Lane, Streetly, Sutton Coldfield, B74 3HF |
| G8 | YNK | M Higton, 12 Chestnut Avenue, Mickleover, Derby, DE3 9FT |
| G8 | YNP | J Hill, Coach House Cottage, 15 Pike Lane, Rugeley, WS15 4AF |
| G8 | YOC | M Witchard, 110 Bradley Road, Huddersfield, HD2 1QY |
| G8 | YOE | C Victory, Pennros, Treworgans, Cuberts, TR8 5HH |
| G8 | YOG | J Woodard, 213 Leicester Road, Ibstock, LE67 6HP |
| G8 | YOK | J Ward, 3 Sherbourne Close, Poulton-le-Fylde, FY6 7UB |
| G8 | YOX | A Munday, 77 Postland Road, Crowland, Peterborough, PE6 0JB |
| G8 | YOY | M Maxwell, 962 Bury Road, Bolton, BL2 6NX |
| G8 | YPH | T McKnight, 31 Cavendish Road, Eccles, Manchester, M30 9EE |
| G8 | YPK | Vincent Maddex, 140a Kents Hill Road, Benfleet, SS7 5PH |
| G8 | YPL | P Martin, 23 Molyneux Road, Maghull, Liverpool, L31 3DX |
| G8 | YPN | P Lutman, 47 Conan Drive, Richmond, DL10 4PQ |
| G8 | YPQ | Malc Waring, Woodside Cottage, Mansfield Road, Ollerton, Newark, NG22 9DX |
| G8 | YPR | R Williams, 54 Woodlands Avenue, Talgarth, Brecon, LD3 0AT |
| G8 | YPV | G Williams, 54 Greenacre, Wembdon, Bridgwater, TA6 7RF |
| G8 | YPY | D Wilson, 35 Darbishire Road, Fleetwood, FY7 6QA |
| G8 | YQA | D Arnold, 10 Shaw Place, Leek, ST13 6ES |
| G8 | YQC | Michael Beetlestone, 19 Tenbury Road, Birmingham, B14 6AD |
| G8 | YQH | V Carter, 69 Angela Crescent, Horsford, Norwich, NR10 3HE |
| G8 | YQN | Peter Gebbie, 76 Muston Road, Filey, YO14 0AN |
| G8 | YQO | D Henderson, Reverie Pennys Lane, Margaretting, Ingatestone, CM4 0HA |
| G8 | YQU | G Lenihan, 28 Paddock Crescent, Sheffield, S2 2AR |
| GM8 | YRE | J Firth, 6 Upper Burnside Drive, Thurso, KW14 7XB |
| G8 | YRF | R Foxley, 20b Alder Copse, Horsham, RH12 1LD |
| G8 | YRL | B Trim, Endon Cottage, 63b Rose Street, Wokingham, RG40 1XS |
| GM8 | YRT | W Stewart, 20 Corrie Place, Scone, Perth, PH2 6QE |
| G8 | YRW | R Williams, 29 Woodfield Road, Bude, EX23 8JB |
| GM8 | YRX | E Saxon, 73 Upper Burnside Drive, Thurso, KW14 7XB |
| G8 | YRY | Charles Rourke, 116 Brands Hill Avenue, High Wycombe, HP13 5PX |
| G8 | YSA | Paul Powers, 5 Bracken Close, Hugglescote, Coalville, LE67 2GP |
| G8 | YSH | Leslie Jannetta, 1 Lake Road, Hadston, Morpeth, NE65 9TF |
| G8 | YSJ | William Bannerman, [illegible] The Cornfield, Langham, Buttercups, Holt, NR25 7DQ |
| G8 | YTF | Gerald McGowan, 281 Ashgate Road, Chesterfield, S40 4DR |
| GI8 | YTH | S Moore, 7 Cyprus Avenue, Belfast, BT5 5NT |
| GW8 | YTO | A Ham, 40 Celtic Way, Rhoose, Barry, CF62 3FT |
| G8 | YTP | Stephen Holgate, 91 Valley Road, Stockport, SK4 2DB |
| G8 | YTR | S Higgs, 5 Lawnswood Close, Cowplain, Waterlooville, PO8 0RU |
| G8 | YTU | F Adams, 27 Challenger Close, Malvern, WR14 2NN |
| G8 | YTX | K Bagshaw, 36 St. Peters Road, Buxton, SK17 7DX |
| GM8 | YUI | George McClintock, 13 St. Andrews Drive, Gourock, PA19 1HY |
| G8 | YUJ | John Milburn, Orme View, Anglesey, LL73 8PE |
| G8 | YUK | A White, 10 Stott Drive, Urmston, Manchester, M41 6WA |
| GM8 | YUM | George Walker, 24 George Street, Cellardyke, Anstruther, KY10 3AU |
| G8 | YUO | Margaret Taylor, 4 Yew Tree Court, Botley Road, Southampton, SO31 1EA |
| G8 | YUP | Bernie Stevens, 77 Dean Lane, Hazel Grove, Stockport, SK7 6EJ |
| G8 | YUR | M Robelou, 12 Cooks Drove, Earith, Huntingdon, PE28 3QG |
| G8 | YVC | M Smith, 31 Burringham Road, Scunthorpe, DN17 2BD |
| G8 | YVM | Neil Matthes, 24 Albany Road, Fleet, GU51 3LY |
| G8 | YVP | M Nicholson, 33 Painshawfield Road, Stocksfield, NE43 7PX |
| G8 | YVQ | C Harper, Chusan, Farley Court, Church Road, Reading, RG7 1TT |
| G8 | YVS | Roy Hillan, 128A Bridge Street, Deeping St. James, Peterborough, PE6 8EH |
| G8 | YVW | C Stacey, 157 Ormond Road, Sheffield, S8 8FT |
| GI8 | YWE | M Anderson, 17 Leydene Court, Lisburn, BT28 3LL |

**IMPORTANT NOTE**

**Revalidate licence to avoid revocation** – Ofcom has advised the Society that plans will be drawn up to revoke licences that have not been revalidated as required by the licence conditions. The quickest way to revalidate is to do so online via the Ofcom website: *https://services.ofcom.org.uk/* or by email: *amateur.validations@ofcom.org.uk* Ofcom staff are available to help, but please be patient during times of heavy workload.

G8 YWJ Alan Frost, 76 Tregrea Estate, Beacon, Camborne, TR14 7SU
G8 YWK William Gleave, 6 Sidlaw Avenue, Chester le Street, DH2 3DD
G8 YWL G Pitt, 17 Penfound Gardens, Bude, EX23 8FF
G8 YXI D Shemeld, 13 Arran Road, Sheffield, S10 1WQ
G8 YXJ R Skells, 31 Perry Road, Leverington, Wisbech, PE13 5AE
G8 YXQ D Chatterton, 27 Victoria Road, Folkestone, CT19 5AT
G8 YXR E Ferris, Karravas, Osborne Road, Deal, CT14 8BT
G8 YXZ R Dominy, 8 Meadow Road, Claygate, Esher, KT10 0RZ
G8 YYA H Duesbury, 4 Harbour View Close, Poole, BH14 0PF
G8 YYC G Miller, 93 Shepherds Grove Park, Stanton, Bury St. Edmunds, IP31 2BN
GW8 YYF K Jones, 3 Penffordd, Pentyrch, Cardiff, CF15 9TJ
G8 YYL Guinevere Johnson, Kilmurry House, Kilmurry Fermoy, Co Cork, Ireland
GI8 YYM Ian Ferris, 48 Abbey Gardens, Belfast, BT5 7HL
G8 YYW Michael Freeman, 2 Poolthorne Farm Cottage, Cadney, Brigg, DN20 9HU
G8 YYX A Layton, 7 Higher Saxifield, Harle Syke, Burnley, BB10 2HB
G8 YZA K Sawday, 15 Moorland View, Buckfastleigh, TQ11 0AF
G8 YZC R Smith, 86 Manor Road, Borrowash, Derby, DE72 3LN
G8 YZF M Bishop, 6 Tiverton Close, Kingswinford, DY6 8PD
G8 YZL P Thackeray, Little Oaks, Slough Lane, Wimborne, BH21 7JL
G8 YZY D Spencer, 28 Watery Lane, Minehead, TA24 5NZ
G8 ZAD R Mantle, 37 Willis Road, Stockport, SK3 8HQ
G8 ZAJ Christopher French, 26 Wood Street, Ash Vale, Aldershot, GU12 5JG
GM8 ZAK Hugh GEMMELL, 53 Southesk Avenue, Bishopbriggs, Glasgow, G64 3AD
G8 ZAT John Haslip, 18 Downsview Drive, Wivelsfield Green, Haywards Heath, RH17 7RW
G8 ZAU D Hoodless, 21 Meadow Close, Eastwood, Nottingham, NG16 3DQ
G8 ZAX R Rees, 69 Pewley Way, Guildford, GU1 3PZ
G8 ZBC C Lucas, 8 Hawker Close, Broughton, Chester, CH4 0SQ
G8 ZBJ W Sheldon, 15 Hawthorn Place, Walsall, WS2 0HZ
G8 ZBN T Nye, 18 Kingsway, Chandler's Ford, Eastleigh, SO53 2FE
G8 ZCJ J Skidmore, 55 Elmsleigh Road, Heald Green, Cheadle, SK8 3UD
G8 ZCK Christopher Wilson, 19 Chace Avenue, Potters Bar, EN6 5LX
GM8 ZCS Andrew Westerman, 2 Whim Square, West Linton, EH46 7BD
G8 ZCV G Byars, 31 Roman Reach, Caerleon, Newport, NP18 3SQ
GI8 ZDB Robert Logue, 46 Brunswick Road, Londonderry, BT47 5SZ
G8 ZDS Philip Hocking, 10 South Terrace, Camborne, TR14 8ST
G8 ZDT P Langford, 1 Kingaby Gardens, Rainham, RM13 7PH
G8 ZEE A Hudson, 1 Laburnum Court, Cheltenham, GL51 0XE
G8 ZEI E Whitham, 44 Tyddyn Isaf, Menai Bridge, LL59 5DA
GM8 ZEJ J Borland, 4 Shanter Place, Kilmarnock, KA3 7JB
G8 ZEK P Jacobi, Highbury, Furzehill, Wimborne, BH21 4HD
GM8 ZEQ M Smith, Haremuir Bungalow, Benholm, Montrose, DD10 0HX
G8 ZES P Street, 50 Dickson House, Ridgway Road, Stoke-on-Trent, ST1 3BA
G8 ZEV C Hartt, 9 Laura Grove, Paignton, TQ3 2LR
G8 ZEW A Joy, 15 Wymersley Close, Great Houghton, Northampton, NN4 7PT
G8 ZEX Stephen Lauritson, 42 Woodstock Road, Kingswood, Bristol, BS15 9UE
G8 ZFD C Askin, 54 York Road, Hull, HU6 9RA
G8 ZFI P Bryant, 21 Devonshire Close, Stevenage, SG2 8RY
G8 ZFL A Butcher, 4 Maple Close, Oldland Common, Bristol, BS30 9PX
G8 ZFQ Melvyn Kanelis, 57 Ringwood Avenue, Redhill, RH1 2DY
G8 ZFS P Wiley, 9 Simpson Avenue, Hunmanby, Filey, YO14 0LB
G8 ZFT R Thompson, 329 Prestbury Road, Prestbury, Cheltenham, GL52 3DF
G8 ZFU G Taylor, 8 Ullathorne Road, Streatham, London, SW16 1SN
GM8 ZFW John Morris, 1 Wealthyton Cottages, Keig, Alford, AB33 8BH
G8 ZFX Paul Blake, 35 Kings Court 71-76 Wright Street, Hull, HU2 8JR
GI8 ZFZ David Alexander, 33 Greenan Road, Newry, BT34 2PJ
GM8 ZGC C Dowers, 38 Ascot Avenue, Glasgow, G12 0AX
G8 ZGF R Mackrell, 17 Townfield Avenue, Worsthorne, Burnley, BB10 3JG
G8 ZGK Alfred Mockford, 58 Wendover Heights, Old Tring Road, Aylesbury, HP22 6PH
G8 ZGM Stephen Berks, 14 Austen Way, Hastings, TN35 4JH
G8 ZGQ A Longuet, 10 Severnmead, Grovehill, Hemel Hempstead, HP2 6DX
G8 ZGS J Holden, 128 Greenways, Norwich, NR4 6HA
G8 ZGY R Bareham, 49 Wharf Road, Crowle, Scunthorpe, DN17 4HU
G8 ZHA R Morrall, 32 Broadstone Avenue, Walsall, WS3 1EW
G8 ZHN P Gibbons, 13 Canon Park, Berkeley, GL13 9DF
G8 ZHR N Lawes, 87 Glebelands, Crayford, Dartford, DA1 4RY
G8 ZHS P Lester, 1c Eastwood Road, London, E18 1BN
GI8 ZHW J McDonnell, 6 Sandhurst Park, Bangor, BT20 5NU
G8 ZIA A Bowman, Evergreen, Durham Road, Stockton-on-Tees, TS21 3LT
G8 ZIC C Harrison, 2 Bridgemere Close, Radcliffe, Manchester, M26 4FS
G8 ZID M Sisley, 18 Willowsmere Drive, Lichfield, WS14 9XF
G8 ZIH J Eady, Pytchley Lodge, Pytchley, Kettering, NN14 1EE
G8 ZIK E Serwa, 102 Cornwall Road, Wolverhampton, WV6 8UZ
GW8 ZIL I Bell, 102 Ewenny Road, Bridgend, CF31 3LN
G8 ZIP K Lake, 22 Chapmans Close, Stirchley, Telford, TF3 1ED
G8 ZIW G Ludar-Smith, 2 Springmead, Queenborough Lane, Braintree, CM77 7PX
G8 ZIY P Eyre, 27 Holborn View, Codnor, Ripley, DE5 9RB
G8 ZJE D Webb, 51 Garden Road, Walton on The Naze, CO14 8RR
G8 ZJH B McCourt, 3 Elm Drive, Greasby, Wirral, CH49 3NP
G8 ZJK Robin Cole, Flat 6, Barton Court, 19 Southwood Road, Hayling Island, PO11 9PS
G8 ZJO S Tomschey, 21 Momus Boulevard, Coventry, CV2 5LL
GM8 ZJS J Thomson, 23 Douglas Road, Longniddry, EH32 0LQ
G8 ZK ZK CONTEST GROUP c/o C Archer, 118 Cator Lane, Beeston, Nottingham, NG9 4BB
GM8 ZKF D Robson, 6 Ladywood Estate, Milngavie, Glasgow, G62 8BE
G8 ZKG R Roberts, 93 Newtown Road, Malvern, WR14 1PD
GM8 ZKN Ian Diment, 22 Academy Place, Bathgate, EH48 1AS
GM8 ZKU Simon Hawley, Hill Of Ardiffery, Hatton, Peterhead, AB42 7TB
G8 ZLF T Gilleard, 3 Paul Crescent, Humberston, Grimsby, DN36 4DF
G8 ZLL I Thomas, 30 Alcot Close, Crowthorne, RG45 7NE
G8 ZLN Peter Thompson, 81 Ashmead Road, Banbury, OX16 1AA
G8 ZLT M Chambers, 3 Manod Road, Blaenau Ffestiniog, LL41 4DD
G8 ZLU M Wright, 17 Colwyn Crescent, Stockport, SK5 7LL
G8 ZMC A McCalden, 127 Kings Road, Godalming, GU7 3EU
G8 ZME M O'Toole, Daffodil Cottage, Dunsmore, Aylesbury, HP22 6QH
GM8 ZMF Martyn Osborn, 3 Lovers Lane, South Queensferry, EH30 9UP
G8 ZMG S Watson, 61 Glenview Road, Shipley, BD18 4AR
G8 ZML B Ewart, 36 Sycamore Rise, Holmfirth, HD9 7TJ
G8 ZMM Roger Bunney, 35 Grayling Mead, Romsey, SO51 7RU
G8 ZMQ P Burnley, 45 Ashwell Road, Heaton, Bradford, BD9 4AX
G8 ZNB A Harris, 55 Frenchgate, Richmond, DL10 7AE
G8 ZNK G Barnes, 18 Wellesley Avenue, Goring-by-Sea, Worthing, BN12 4PN
G8 ZNL M Speight, 12 Kinmel Close, Redcar, TS10 2RY
G8 ZOE Stephen Trott, 6 Mounton Drive, Chepstow, NP16 5EH
G8 ZOJ Gordon Barrett, The Old Chapel, 5 Tappers Lane, Bridgwater, TA6 6SJ
G8 ZOO John Molinghen, 16 Dumpers Lane, Chew Magna, Bristol, BS40 8SS
G8 ZOV R Nicholson, 24 Barnmead, Haywards Heath, RH16 1UZ
GM8 ZOW P Oram, 24 John Smith Place, Kelty, KY4 0NL
G8 ZOY G Page, 1a Montagu Gardens, Wallington, SM6 8EP
G8 ZPD Paul Davies, 46 Spring Street, Colley Gate, Halesowen, B63 2SZ
G8 ZPE P Cooper, The Bungalow, Clopton, Kettering, NN14 3DZ
G8 ZPH D BUCKNELL, 46 Heath Row, Bishop's Stortford, CM23 5DE
G8 ZPO R Blackwell, 46 Wyatts Drive, Thorpe Bay, Southend-on-Sea, SS1 3DG
G8 ZPW A Martin, 23 Portfield Road, Christchurch, BH23 2AF
G8 ZQA Peter Stonebridge, 207 Henley Road, Ipswich, IP1 6RL
G8 ZQB J Smith, 7 Mill Hill Close, Whetstone, Leicester, LE8 6NF
G8 ZQG S Wood, 8a Glendale Avenue, Glenfield, Leicester, LE3 8GF
G8 ZQJ Derek Young, 9 Larchfield House, Highbury Estate, London, N5 2DE
G8 ZQM K Pascoe, 21 Cotswold Avenue, Sticker, St. Austell, PL26 7ER
GM8 ZQY S Frey, 2 Balgeddie Gardens, Glenrothes, KY6 3QR
G8 ZRD Ivan Gilzean, 35 Pieces Terrace, Waterbeach, Cambridge, CB25 9NE
G8 ZRE David Hewitt, 31 Broadmead, Vicars Cross, Chester, CH3 5PT
G8 ZRG B Hawes, 129 Wycombe Lane, Wooburn Green, High Wycombe, HP10 0HJ
G8 ZRM R Myers, 9 Romney Road, Rottingdean, Brighton, BN2 7GG
G8 ZRN G John, 29 Park Road, Northville, Bristol, BS7 0RH
G8 ZRQ R Knight, 9 Crispin Road, Strood, Rochester, ME2 3TW
G8 ZRU D Moger, 47 Powys Grove, Banbury, OX16 0UG
G8 ZRV G Sargant, 9 Orchard Way, Reigate, RH2 8DS
G8 ZSD I Worthington, 7 Bowness Close, Gamston, Nottingham, NG2 6PE
G8 ZSK Alan Allcock, 30 Clyde Grove, Crewe, CW2 8NA
G8 ZSM L Barlow, 4 Bucknell Place, Thornton-Cleveleys, FY5 3HZ
G8 ZSP A Blanchard, 41 Deane Drive, Galmington, Taunton, TA1 5PQ
G8 ZSZ I Dickinson, 16 Heathfield Grove, Beeston, Nottingham, NG9 5EB
G8 ZTB S Fenn, 21 Waarem Avenue, Canvey Island, SS8 9DS
G8 ZTD J Francis, 9 Holland Close, Bognor Regis, PO21 5TW
G8 ZTF J Hargraves, 321 Northway, Maghull, Liverpool, L31 0BW
G8 ZTG J Harman, 20 Sunview Avenue, Peacehaven, BN10 8PJ
G8 ZTM N Ledeux, 14 Jubilee Close, Cam, Dursley, GL11 5JQ
G8 ZTN Paul Lock, Monks Rest, The Street, Bridport, DT6 6PE
G8 ZTR J Macdonald, 74 Bradford Road, Boston, PE21 8BJ
G8 ZTT MID CHESHIRE AR c/o Peter Fox, 5 Llandovery Close, Winsford, CW7 1NA
GM8 ZTV F Millar, 13 Edzell Park, Kirkcaldy, KY2 6YB
G8 ZUF K Rogers, 36 Goodacre Road, Ullesthorpe, Lutterworth, LE17 5DL
G8 ZUI G Shaw, 8 Nightingale Place, Buckingham, MK18 1UF
G8 ZUL Richard Yates, 16 Arnold Grove, Shirley, Solihull, B90 3JR
G8 ZUU M Smith, 2 Newbury Close, Mapperley, Nottingham, NG3 5QW
G8 ZUZ D Unwin, 1 Bentinck Close, Nuncargate, Nottingham, NG17 9ET
G8 ZVI L Hart, 28a Dunton Road, Stewkley, Leighton Buzzard, LU7 0HZ
G8 ZVK Brian Ackroyd, 91 Bulford, Wellington, TA21 8DH
G8 ZVM M Atkinson, Menamber Farm, Trenear, Helston, TR13 0HE
G8 ZVS R Bird, 80 Clearmount Road, Weymouth, DT4 9LE
G8 ZVX Anthony Breeds, 26 Heighton Road, Newhaven, BN9 0JU
G8 ZVZ I Collins, Knapp Cottage, Pixley, Ledbury, HR8 2QB
G8 ZWA P Collins, 40 Shacklegate Lane, Teddington, TW11 8SH
G8 ZWC L Curtis, 34 Gaisford Road, Worthing, BN14 7HW
G8 ZWF Robert Cowling, 20 Claremont Hill, Shrewsbury, SY1 1RD
G8 ZWN Michael Davies, Sunningdale, Sulhamstead Hill, Reading, RG7 4DE
G8 ZWU K Graham, 670 Stafford Road, Ford Houses, Wolverhampton, WV10 6NW
G8 ZXL R Pretty, 77a High Street, Ewell, Epsom, KT17 1RX
G8 ZXQ James Mc Dermott, Milking Green Gate, Eliock, Sanquhar, DG4 6LD
G8 ZXT J Marshall, 58 Sandbed Court, Leeds, LS15 8JJ
G8 ZXU P McGuinness, 83 Beaconsfield, Telford, TF3 1NH
G8 ZXY W Mason, 365 Heath Road South, Birmingham, B31 2BJ
G8 ZXZ David Holmes, 17 Spring Hall Close, Halifax, HX3 7NE
G8 ZYC ZYCOMM ELECT LT c/o Ian Sneap, 51 Nottingham Road, Ripley, DE5 3AS
G8 ZYH E Hitch, 35 Hawthorndene Road, Hayes, Bromley, BR2 7DY
G8 ZYI Norman Hitch, 1b Greenlands, Platt, Sevenoaks, TN15 8LL
G8 ZYM I Hammond, 1 Old Rectory Close, Barham, Ipswich, IP6 0PY
G8 ZYR Phil Hodgkinson, 25 Polisken Way, St. Erme, Truro, TR4 9RB
G8 ZYT S Higlett, 28 Oak Crescent, Potton, Sandy, SG19 2PY
G8 ZZB Dean Kellett, April Cottage, 10 Yorkdale Drive, Selby, YO8 9YB
G8 ZZG T Lock, 40 Chertsey Road, Ashford Common, Ashford, TW15 1SQ
G8 ZZK D Lee, 14 Woodview Close, West Kingsdown, Sevenoaks, TN15 6HP
G8 ZZL P Lake, 125 Woodward Road, Dagenham, RM9 4ST
G8 ZZR Peter Vince, 19 Links Road, Ashtead, KT21 2HB
G8 ZZS Dominic Vaughan, Orchard Farm House, Framsden, Stowmarket, IP14 6HD
G8 ZZT J Tonks, Flat, 3 Greystone Passage, Dudley, DY1 1SL
G8 ZZV Alan Tye, 3 Parkwood Court, Forest Park, Nottingham, NG6 9FB
G8 ZZW Ian Shepherd, 12 Grains Road, Delph, Oldham, OL3 5DS
G8 ZZY A Smart, 101 Bardon Road, Coalville, LE67 4BF

## M*0

M0 AAA Reading and District ARC c/o Vincent Robinson, 4 Hilltop Road, Caversham, Reading, RG4 7HR
M0 AAC Pat Bergin, 15 Monks Way, Harmondsworth, West Drayton, UB7 0LE
M0 AAD Mark Stockton, 37 Ney Street, Ashton-under-Lyne, OL7 9NL
M0 AAF D Hodgson, 1b Court Farm Avenue, Epsom, KT19 0HD
M0 AAK Martin Pearson, 56 Parkwood Green, Parkwood, Gillingham, ME8 9PP
M0 AAM R Armstrong, 71 Bradshaw View, Queensbury, Bradford, BD13 2FF
M0 AAN W Glover, 21 West End Way, Lancing, BN15 8RL
M0 AAP Ian Parker, 60 Claremont Crescent, Newbury, RG14 2FE
M0 AAR John Kemp, 394 Great Thornton Street, Hull, HU3 2LT
M0 AAS John Whittaker, 10 Pownall Court, Wilmslow, SK9 5QE
MD0 AAV Simon Bates, 6 Foxdell, Northwood, HA6 2BU
MI0 AAW S Blakley, 123 Mount Merrion Avenue, Belfast, BT6 0FN
MI0 AAZ John Anderson, 1 Claragh Hill Drive, Kilrea, Coleraine, BT51 5YR
M0 ABA Thomas Hackett, 157 Caulfield Road, Shoeburyness, Southend-on-Sea, SS3 9LU
MM0 ABB Chris Kane, 46 Hillmoss, Kilmaurs, Kilmarnock, KA3 2RS
MI0 ABD J McCarrison, 11 Boretree Island Park, Newtownards, BT23 7BW
M0 ABF K Molyneux, 220 Woodlands Holiday Homes Pk, Dowles Road, Bewdley, DY12 3AE
M0 ABG Andrew Powell, 2 Ormsby Close, Hopton, Great Yarmouth, NR31 9TY
M0 ABI Michael Lennon, 4 The Lees, Faringdon, SN7 7BB
MM0 ABJ C Ewart, 13 Princes Street, Innerleithen, EH44 6JT
M0 ABK M Gray, 19 Marsh View, Newton, Preston, PR4 3SX
MI0 ABN N Crawford, 10 White Mountain Road, Lisburn, BT28 3QY
M0 ABO Jose Valle Espin, 203 Broadway, Horsforth, Leeds, LS18 4HL
M0 ABP J Barker, Karma, 6 Acredykes, Bridlington, YO15 1LY
M0 ABQ W Couse, 68/29 Moo 3 Rattanapron Village, Tambon Khungkong, Chiang Mai, Thailand, 50230
M0 ABT S Little, 46 Marine Drive, Seaford, BN25 2RU
M0 ABU IOTA Chasers International c/o Christopher Colclough, 52 Alexandra Street, Nuneaton, CV11 5RL
MW0 ABV paul plummer, hill road, Neath Abbey, sa108nd
M0 ABW W Johnson, 74 High Meadows, Romiley, Stockport, SK6 4QE
M0 ABY Adrian Soane, 24 Nurseries Road, Wheathampstead, St. Albans, AL4 8TP
M0 ABZ Anthony Allbright, Greenacre, Carne Road Newlyn, Penzance, TR18 5QA
M0 ACA E Morley, 91 Allerton Road, Stoke-on-Trent, ST4 8PQ
M0 ACB E McDonald, 32 Butterwick Road, Messingham, Scunthorpe, DN17 3PB
M0 ACC Arthur Dixon, 17 Coppice Court, Weymouth, DT3 5SA
M0 ACI Stanley Stacey, Trehill, Trekenner, Launceston, PL15 9PH
M0 ACK Mike Jackson, 121 Kiln Lane, Eccleston, St. Helens, WA10 4RH
M0 ACL Liz Jones, 47 Pine Crescent, Chandler's Ford, Eastleigh, SO53 1LN
M0 ACM Den Forrest, 166 Meadowcroft, Swindon, SN2 7LE
M0 ACN John Green, 9 Armorial Road, Coventry, CV3 6GH
MM0 ACR L Skinner, Manse Hall, Drumoak, Banchory, AB31 5HA
MM0 ACT R Skinner, Manse Hall, Drumoak, Banchory, AB31 5HA
M0 ACU M Eddyvean, 41 Liddell Road, Cowley, Oxford, OX4 3QU
M0 ACV T Bevan, 6 Buttermere Grove, West Auckland, Bishop Auckland, DL14 9LG
M0 ACW OVER THE HILL DX GROUP c/o Reginald Williams, Dyffryn Coed, Union Road, Coleford, GL16 7QB
M0 ADB N Pringle, 21 Petersmiths Drive, New Ollerton, Newark, NG22 9RZ
M0 ADG D Morris, 86 Richardson Street, Carlisle, CA2 6AG
M0 ADJ THE SQUAREBASHERS EXPEDITION GROUP c/o Walter Davidson, 30 Tirlebank Way, Tewkesbury, GL20 8ES
M0 ADN H Epps, 10 Eastbury Court, Smiths Wharf, Wantage, OX12 9GS
M0 ADR Graham Galbraith, 24 Airedale, Hadrian Lodge West, Wallsend, NE28 8TL
M0 ADW R Latham, 47 Oldfield Park, Westbury, BA13 3LQ
M0 ADY A Grundy, 21 Ribston Close, Shenley, Radlett, WD7 9JW
M0 AEC S Roper, 15 St. Gerards Road, Solihull, B91 1TZ
M0 AEJ V Trend, 64 Shutlock Lane, Moseley, Birmingham, B13 8NZ
M0 AEK J Sloan, 141 Bridgemere Road, Eastbourne, BN22 8TY
MW0 AEL S Townsend, 42 Burns Crescent, Bridgend, CF31 4PY
M0 AEN M Austen, 11 Corn Avill Close, Abingdon, OX14 2ND
M0 AEP Graham Dawes, 11 Ferriby Road, Barton-upon-Humber, DN18 5LE
M0 AEQ M Bardell, 47 Calverleigh Crescent, Furzton, Milton Keynes, MK4 1HY
M0 AET Kenneth Jones, Ferny Hoolet, Pale Lane, Basingstoke, RG27 8SW
M0 AEU Frank Heritage, 50 Laurel Close, North Warnborough, Hook, RG29 1BH
M0 AEV E Jones, 18 Madryn Terrace, Llanbedrog, Pwllheli, LL53 7PF
MI0 AEX Jeffrey Smith, 54a Blackstaff Road, Kircubbin, Newtownards, BT22 1AF
M0 AEZ M Herpe, 11 Manor Way, Sutton-in-Craven, Keighley, BD20 7PN
M0 AFC T Boon, 27 Meadowside Avenue, Clayton le Moors, Accrington, BB5 5XF
M0 AFD S Edwards, 59 St. Andrews Road, Colwyn Bay, LL29 6DL
M0 AFF Frank Hallsworth, 15 Stokesay Drive, Hazel Grove, Stockport, SK7 5PW
M0 AFJ Tim Hague, 10 Oxford Street, Wolverton, Milton Keynes, MK12 5HP
M0 AFQ B Eagleton, 3 Coldridge Close, Pendeford, Wolverhampton, WV8 1XZ
M0 AFR Philip Walker, 1 Vicarage Lane, Fordington, Dorchester, DT1 1LH
M0 AFS P Whiteley, 53 Sharp Lane, Almondbury, Huddersfield, HD4 6SS
MI0 AFT J Stewart, 16 Maritime Drive, Carrickfergus, BT38 8GQ
M0 AFV R Rippin, Copperfields, Bourton on the Hill, Moreton-in-Marsh, GL56 9AE
M0 AFW C Parkinson, 4 Campion Drive, Killamarsh, Sheffield, S21 1TG
M0 AFX D Waters, Station House, Station Road, Manningtree, CO11 2LH
M0 AFY R Ford, 70 Jubilee Road, Darnall, Sheffield, S9 5EH
M0 AFZ P Nairne, 137 Barden Road, Tonbridge, TN9 1UX
M0 AGA K Gunstone, 67 Woodside, Skegby, NG17 3EB
MW0 AGE J Chinnock, 22 Mill Road, Pyle, Bridgend, CF33 6AP
M0 AGJ Alan Bowker, 120 Broomhouse Lane, Doncaster, DN4 9DB
M0 AGL J Monks, 2 Low Hutton Park, Huttons Ambo, York, YO60 7HH
M0 AGO H Wray, 22 Askew Dale, Guisborough, TS14 8JG
M0 AGP Michael Weber, 50 Wandle Road, London, SW17 7DW
M0 AGR M Bray, 2 Camborne Drive, Fixby, Huddersfield, HD2 2NF
M0 AGS E Smeaton, 27 Sandringham Avenue, Burton-on-Trent, DE15 9BJ
M0 AGT R Markham, 51 Park View, Crewkerne, TA18 8HT
M0 AGU J Shorthouse, 84 Mount Pleasant, Ackworth, Pontefract, WF7 7HU
M0 AGV T Mcguigan, 14 Ash Grove, Heald Green, Cheadle, SK8 3JA
M0 AGW W Mason, 104 Chester Road, Poynton, Stockport, SK12 1HG
M0 AGY Mark Griffin, 15 Victoria Road, St. Austell, PL25 4QF
MM0 AHC M Collins, Redwoods, Barcaldine, Oban, PA37 1SG
M0 AHF G May, 14 Tennyson Avenue, Dukinfield, SK16 5DP
MI0 AHH C Doris, 9 Gortalowry Park, Cookstown, BT80 8JH
MI0 AHI J Doris, 92 Coolnafranky Park, Cookstown, BT80 8PW
M0 AHJ C John, 5 Highfield Gardens, Aldershot, GU11 3DB
M0 AHS M Nicholas, 4 Chesterfield Mews, Chesterfield Road, Ashford, TW15 3PF

| | | |
|---|---|---|
| M0 | AHT | Wilfred Burt, 3 Edward St, Hetton Le Hole, Houghton le Spring, DH5 9EL |
| M0 | AHV | H Banks, 104 Viking Road, Bridlington, YO16 6TB |
| M0 | AHY | G Parsons, Gull Cottage, Briar Close, Hastings, TN35 4DP |
| M0 | AHZ | Robert Brown, 17 Ridgeway, North Seaton, Ashington, NE63 9TJ |
| M0 | AIB | Stephen Budd, 19 Queen Street, Worthing, BN14 7BL |
| M0 | AIC | S Deary, 43 Old Road, Tintwistle, Glossop, SK13 1LH |
| M0 | AID | Kelvin Marsh, Highgrove, Creech Heathfield, Taunton, TA3 5EW |
| MW0 | AIE | H Duncombe, 1 Pennar Court, Pembroke Dock, SA72 6NW |
| MI0 | AIH | David Martin, 34 Lower Kildress Road, Cookstown, BT80 9RN |
| M0 | AIJ | Charles Blake, 30 Pine Tree Walk, Poole, BH17 7EH |
| MM0 | AIK | SCOTTISH DX CONTEST CLUB c/o Brian Devlin, Borrodale, Main Street, Stirling, FK8 3PW |
| M0 | AIR | Stephen Meynell, 1 Hawks Mead, Liss, GU33 7SN |
| M0 | AIS | Arnold Benns, 7 Brooklands Road, Burnley, BB11 3PR |
| M0 | AIT | R Holt, 41 Garden Avenue, Ilkeston, DE7 4DF |
| M0 | AIX | L Hutchinson, 20 Steeple Court, Ireland Street, Bingley, BD16 2QD |
| M0 | AIY | R Carter, 16 Holts Lane, Clayton, Bradford, BD14 6BL |
| MW0 | AIZ | R Ramm, Mor Wolir, Sarnau, Llandysul, SA44 6QY |
| M0 | AJB | THE NORTH WEST 320 DX CLUB c/o A Birch, 6 Crescent Road, Wallasey, CH44 0BQ |
| M0 | AJC | M McInally, Flat 7, 32-33 Edgar Road, Margate, CT9 2EJ |
| M0 | AJD | M Saxton, Cartref, Church Lane, Horncastle, LN9 6NN |
| MW0 | AJH | J Donnell, 42 Wentworth Crescent, Mayals, Swansea, SA3 5HT |
| M0 | AJI | Stephen Nursey, 1 Lyddington Road, Gretton, Corby, NN17 3DA |
| M0 | AJJ | Paul Olson, 23 Dennett Close, Liverpool, L31 5PD |
| MM0 | AJQ | J Stone, 1 Seafield Crescent, Bilston, Roslin, EH25 9TD |
| M0 | AJT | C Towle, 116 Stainton Drive, Grimsby, DN33 1JB |
| M0 | AJX | Gordon Jones, 7 Hardwick View, Skegby, Sutton-in-Ashfield, NG17 3BW |
| M0 | AKA | G Powell, 203 Queenborough Road, Minster on Sea, Sheerness, ME12 3EL |
| M0 | AKD | G Dublon, 25 Carr Lane, Sandal, Wakefield, WF2 6HJ |
| M0 | AKE | R Johnson, 24 Balmoral Avenue, Stanford-le-Hope, SS17 7BD |
| M0 | AKF | M Temblett, 42 Westward Road, Bristol, BS13 8DB |
| M0 | AKI | E Woollen, 6 Back Lane, Kington Magna, Gillingham, SP8 5EL |
| M0 | AKJ | Robert Hunt, 8 Spicer Close, Cullompton, EX15 1QD |
| M0 | AKK | S Elden, 124 Larchcroft Road, Ipswich, IP1 6PQ |
| MM0 | AKM | J Hood, 88/2 Craighouse Gardens, Edinburgh, EH10 5LW |
| M0 | AKQ | R gawan, 39 The Filberts, Fulwood, Preston, PR2 3YS |
| M0 | AKR | K Daniels, 17 Berry Park Road, Plymouth, PL9 9AG |
| M0 | AKS | R Lusty, 483 Bacup Road, Rossendale, BB4 7JA |
| MM0 | AKX | ATC SCOTLAND & N.IRELAND REGION ARC c/o J Ramsay, 150 City Road, Dundee, DD2 2PW |
| M0 | AKY | T Money, 119 Twyford Way, Canford Heath, Poole, BH17 8SR |
| M0 | AKZ | R Taylor, 46 Crescent Road, Netherton, Dudley, DY2 0NW |
| M0 | ALB | Norman Hixson, Flat 35, Milward Court, Reading, RG2 7BG |
| M0 | ALC | A Barth, 83 London Road, Aston Clinton, Aylesbury, HP22 5LD |
| M0 | ALD | John Britten, 10 Broadgate Avenue, Horsforth, Leeds, LS18 5DT |
| M0 | ALE | P Johnson, 91 Highlands Road, Andover, SP10 2PZ |
| M0 | ALF | R Faithfull, 5 Hadleigh Road, Portsmouth, PO6 3RD |
| MW0 | ALG | D Burge, Ucheldir, Maenygroes, New Quay, SA45 9TH |
| M0 | ALH | Stephen Case, 5 Haldon Grove, Birmingham, B31 4LN |
| M0 | ALK | R Cook, 3 Mill Close, Hartford, Huntingdon, PE29 1YL |
| MM0 | ALM | David Wood, West Raedykes, Rickarton, Stonehaven, AB39 3SY |
| M0 | ALN | Andrew Lane, 14 Hertford Place, Newport, NP19 7SN |
| M0 | ALO | Donald Hooper, 21 High Street, Great Linford, Milton Keynes, MK14 5AX |
| MD0 | ALQ | G Denby, 14 Talman Grove, Stanmore, HA7 4UQ |
| M0 | ALR | T Knight, 117 Ennerdale Road, Cleator Moor, CA25 5LR |
| MI0 | ALS | Edward Stanford, 33 Glenview Gardens, Belfast, BT5 7LY |
| M0 | ALT | Ian Halliwell, 61 Cliffe Road, Shepley, Huddersfield, HD8 8AG |
| M0 | ALX | Mark Feasey, 4 Abbeydale, Carlton Colville, Lowestoft, NR33 8WJ |
| MM0 | ALY | A Brown, 4 Averon Park, Blackburn, Aberdeen, AB21 0LH |
| M0 | ALZ | Terry Thompson, 23 Oaklands, Paulton, Bristol, BS39 7RP |
| M0 | AMB | B Metcalfe, 5 Oakdale Avenue, Bradford, BD6 1RP |
| M0 | AME | D Draper, 36 Highfield Gardens, Combe Martin, Ilfracombe, EX34 0HQ |
| M0 | AMF | Ronald Jefferies, 38 Towbury Close, Redditch, B98 7YZ |
| MW0 | AMI | R Hall, 33 Heol Y Garreg Las, Llandeilo, SA19 6EB |
| MW0 | AMJ | Lyn Carter, 13 Maes Dolau, Idole, Carmarthen, SA32 8DQ |
| M0 | AMM | G Smith, East Lodge, Woodlands Drive, Bradford, BD10 0NX |
| MW0 | AMN | G Thomas, Stonehall Mill Farm, Wolfscastle, Haverfordwest, SA62 5NT |
| M0 | AMP | A Davies, 27 Foxley Grove, Bicton Heath, Shrewsbury, SY3 5DF |
| MW0 | AMQ | G Thomas, Stonehall Mill Farm, Wolfscastle, Haverfordwest, SA62 5NT |
| M0 | AMS | M Burke, 53 Valley Drive, Great Sutton, Ellesmere Port, CH66 3QB |
| MM0 | AMV | RW Moodie, 26 Seton Court, Port Seton, Prestonpans, EH32 0TU |
| MM0 | AMW | D Gillies, 10 Killeonan, Campbeltown, PA28 6PL |
| M0 | AMX | J Howell, Orchard House, Blennerhasset, Wigton, CA7 3QX |
| MM0 | AMY | J Gillespie, Flat 15, 30 Kennington Road, Glasgow, G12 0NY |
| M0 | AMZ | J Williams, 61 Longfield Road, South Woodham Ferrers, Chelmsford, CM3 5JJ |
| M0 | ANB | R Palmer, 115 Francis Avenue, Ilford, IG1 1TT |
| M0 | ANC | Royston Jones, 31 Main Street, Awsworth, Nottingham, NG16 2HH |
| M0 | ANH | J Waller, 56 Daventry Road, Dunchurch, Rugby, CV22 6NS |
| M0 | ANK | S Cotterill, 320 Hamstead Road, Great Barr, Birmingham, B43 5EH |
| M0 | ANL | A King, 79 Spring Hill Road, Accrington, BB5 0EX |
| M0 | ANN | G Wardale, 25 The Crescent, Huyton, Liverpool, L36 6ER |
| M0 | ANO | Robert Spencer, 19 Trafalgar Road, Cirencester, GL7 2EJ |
| M0 | ANP | Nigel Crooks, 3 Grove Court, Settle, BD24 9QR |
| M0 | ANQ | James Chadwick, 4 Toronto Street, Bolton, BL2 6PF |
| M0 | ANS | A Rawlings, Open University, Walton Hall, Milton Keynes, MK7 6AA |
| M0 | ANU | G Coolledge, 49a Enfield Avenue, New Waltham, Grimsby, DN36 4RB |
| M0 | ANV | R Davies, 4 Maes Derlwyn, Llanberis, Caernarfon, LL55 4TW |
| MW0 | ANX | J Jensen, Pistyll Canol Farm, Llandeilo Road, Ammanford, SA18 2LQ |
| M0 | AOA | D Young, 6 Gus Walker Drive, Pocklington, York, YO42 2WA |
| M0 | AOB | J Allen, 149 Penistone Road, Waterloo, Huddersfield, HD5 8RP |
| M0 | AOD | D Kay-Newman, Kay-Spray, Pottery Road, Ilminster, TA19 9QN |
| MM0 | AOF | David Henry, 25 Claremont Street, Aberdeen, AB10 6QQ |
| M0 | AOG | G Dyson, 32 Farleigh Fields, Orton Wistow, Peterborough, PE2 6YB |
| M0 | AOH | John Barber, 9 Chiswick Street, Carlisle, CA1 1HQ |
| M0 | AOI | J Russell, 46 Eastleigh Drive, Tingley, Wakefield, WF3 1PF |
| M0 | AOJ | Alan Elliott, 26 WATERY LANE, Minehead, TA24 5NZ |
| M0 | AOK | S Millar, 4 Broomfield, Benfleet, SS7 2ST |
| MM0 | AOL | R Bloomfield, Tipi Ska, South Lethans, Dunfermline, KY12 9TE |
| M0 | AOM | M Goodrich, Urb Les Basetes B3, Adsubia, Alicante, Spain, 3786 |
| MM0 | AOQ | Colin Greig, 5 Mitchell Place, Stuartfield, Peterhead, AB42 5WE |
| M0 | AOT | Miles Stanley, 16 Fenton, Keswick, CA12 4AZ |
| MM0 | AOY | D Stephen, 16 The Square, Portlethen, Aberdeen, AB12 4QA |
| M0 | AOZ | Mark Boothman, 24 Anvil Way, Kennett, Newmarket, CB8 8GY |
| M0 | APC | Michael Brown, 6 Rose Court, Garforth, Leeds, LS25 1NS |
| M0 | APD | J Udall, 4 Council House, Church Lane, Swadlincote, DE12 8DL |
| MM0 | APF | INVERCLYDE CONTEST GROUP c/o Jim Fisher, High Birches, Culbokie, Dingwall, IV7 8JS |
| M0 | APH | A Gilbert, 79a Station Road, Brimington, Chesterfield, S43 1LJ |
| M0 | APK | D Allen, 162 Wood Lane, Newhall, Swadlincote, DE11 0LY |
| M0 | APL | B Tucker, 2 Hundall Court, Grasscroft Close, Chesterfield, S40 4HN |
| M0 | APN | Andrew Nelson, 29 Coxford Road, Southampton, SO16 5FG |
| M0 | APW | E Komp, 118 Marine Crescent, Goring-by-Sea, Worthing, BN12 4HR |
| M0 | APY | R Arey, 4 Iveson Lawn, Leeds, LS16 6NA |
| M0 | APZ | F Piper, 6 Russell Street, Little Hulton, Manchester, M38 0LW |
| M0 | AQA | G Shaw, 6 Bromstone Road, Broadstairs, CT10 2HA |
| M0 | AQE | Ernest Entwistle, 43 Brock Road, Chorley, PR6 0DB |
| M0 | AQF | T Davies, 20 The Coppice, Impington, Cambridge, CB24 9PP |
| M0 | AQH | E Blackburn, 2 Stockwell Drive, Knaresborough, HG5 0LW |
| MJ0 | AQJ | N Jones, 1 Cornucopia Court, Le Mont Pinel, St. Helier, Jersey, JE2 4RS |
| M0 | AQK | Kevin Hesketh, 13 Elm Road, St. Helens, WA10 3NE |
| M0 | AQO | G Willson, 40 Grace Gardens, Bishop's Stortford, CM23 3EX |
| M0 | AQP | A Bellamy, 8 Dorothy Road, Kettering, NN16 0PH |
| M0 | AQQ | K Evans, 5 Garswood Avenue, Rainford, St. Helens, WA11 8JW |
| MW0 | AQT | E Lucocq, 96 Carisbrooke Way, Cardiff, CF23 9HX |
| M0 | AQW | PROCESSED AUDIO GROUP c/o M Storkey, 9 Waterman Court, Acomb, York, YO24 3FB |
| MI0 | AQX | J May, 8 Oak Vale Avenue, Newry, BT34 2BQ |
| M0 | AQZ | Leslie Griffiths, Tros Y Garreg, Plas Road, Holyhead, LL65 2LU |
| M0 | ARA | Jeffrey Layton, 6 Granby Road, Cheadle Hulme, Cheadle, SK8 6LS |
| M0 | ARC | East Yorkshire Contest c/o Victor Lindgren, 143 Hull Road, Anlaby, Hull, HU10 6ST |
| MW0 | ARD | A Davies, 19 Maes-y-Dderwen, Dinas Cross, Newport, SA42 0XF |
| M0 | ARH | Neil Ravilious, 17 Halls Green, Weston, Hitchin, SG4 7DR |
| M0 | ARK | B Shepherd, 17 Huntock Place, Brighouse, HD6 2NW |
| M0 | ARL | Huw Davies, 1 Bryn Siriol, Coedpoeth, Wrexham, LL11 3PZ |
| M0 | ARM | Leslie Hill, 1 Thatched Cottage, Camp Road, London, SW19 4UR |
| M0 | ARO | J Parsons, 36 Gainsborough, Milborne Port, Sherborne, DT9 5BD |
| M0 | ARQ | John Churchill, 59 Highfield, Letchworth Garden City, SG6 3PY |
| M0 | ART | B Addis, 22 Percy Street, Cramlington, NE23 6RG |
| M0 | ARV | M Thomas, 12 School Road, Rhosllanerchrugog, Wrexham, LL14 1BB |
| M0 | ARX | M Richardson, 39 Wilson Avenue, Deal, CT14 9NL |
| M0 | ARY | Mary O'Rourke, Brookside Farm, Walpole, Halesworth, IP19 9BH |
| M0 | ARZ | S Hurst, 25 Florence Road, Northampton, NN1 4NA |
| MM0 | ASB | Robert Barbour, 40 Mannerston Holdings, Linlithgow, EH49 7ND |
| M0 | ASC | Alan Clayton, 7 Salisbury Avenue, Broadstairs, CT10 2DT |
| M0 | ASD | Arthur Gallichan, 4 Wigston Road, Hillmorton, Rugby, CV21 4LT |
| M0 | ASE | C Humphreys, 40 Baffins Road, Copnor, Portsmouth, PO3 6BG |
| M0 | ASF | U Nehmzow, 26 Woodlands, Colchester, CO4 3JA |
| M0 | ASG | C Nehmzow, 26 Woodlands, Colchester, CO4 3JA |
| M0 | ASI | Neil Johns, 85 South Hill, Hooe, Plymouth, PL9 9PT |
| M0 | ASJ | Simon Griggs, 2 Titchwell Drive, Ipswich, IP3 9GB |
| M0 | ASL | J Phillips, 57 Ffordd Llanerch, Penycae, Wrexham, LL14 2ND |
| M0 | ASN | John Marron, 190 Cotswold Crescent, Billingham, TS23 2QH |
| M0 | ASO | P Bluthner, 31 Westway, Garforth, Leeds, LS25 1DA |
| M0 | ASR | Domingo Campanario, 3 Foxearth Hall, Leek Road, Stoke-on-Trent, ST9 0DG |
| M0 | AST | Alan Stanley, 22 Brookhouse Road, Walsall, WS5 3AD |
| M0 | ASU | Karl Hamer, Flat 7, Red Court, 66 Upper Park Road, Salford, M7 4JA |
| MI0 | ASV | G Best, 1 Bensons Road, Lisburn, BT28 3QX |
| M0 | ASY | Sheldon Werner, 4225 Place Sainte-Helene, Laval, Canada, H7W 1P3 |
| M0 | ATA | A Rundle, 9 Windsor Terrace, East Herrington, Sunderland, SR3 3SF |
| M0 | ATB | R Hutton, 57 Sandy Lane, Upton, Poole, BH16 5EJ |
| M0 | ATC | AIR TRAINING CORPS c/o Christoper Hoare, 16 Shrivenham Road, Highworth, Swindon, SN6 7BZ |
| M0 | ATD | Arne Holzapfel, Flat 1-4, 4 Tanner Street, London, SE1 3LD |
| MW0 | ATG | H Thomas, 15 Coronation Terrace, Pontypridd, CF37 4DP |
| M0 | ATI | Geoff Roberts, 3 Bryn Nebo, Bwlchgwyn, Wrexham, LL11 5YB |
| MW0 | ATK | S Brewer, 16 Oxwich Close, Cefn Hengoed, Hengoed, CF82 7JB |
| M0 | ATL | P Nash, 110 Cranborne Road, Potters Bar, EN6 3AJ |
| M0 | ATQ | James Torry, 41 Nevill Road, Rottingdean, Brighton, BN2 7HH |
| MW0 | ATR | D Williams, 17 Brynawelon, Llanelli, SA14 8PU |
| M0 | ATS | John Ammundsen, 62 Linden Avenue, Broadstairs, CT10 1HR |
| M0 | ATT | Emlyn Cooke, Anelog, Rhewl Fawr Road, Holywell, CH8 9HJ |
| M0 | ATV | Anthony Reilly, 19 The Ridgway, Romiley, Stockport, SK6 3EE |
| M0 | ATX | Elisa Williams, Dyffryn Coed, Union Road, Coleford, GL16 7QB |
| M0 | ATY | Christopher Kirkland, 9 Holland Way, Newport Pagnell, MK16 0LL |
| M0 | ATZ | Colin Hardy, Flat 1, Stoneway Court, 29 Pensby Road, Wirral, CH60 7RA |
| M0 | AUA | D Hazel, 7 Arley Close, Upton, Chester, CH2 1NW |
| M0 | AUF | Neal Handforth, 10 High Street, Bromborough, Wirral, CH62 7HA |
| M0 | AUG | G Ashton, 53 Dorset Avenue, Shaw, Oldham, OL2 7EG |
| M0 | AUH | J Everson, 35 Brynwern, Pontypool, NP4 6HH |
| M0 | AUK | John Sporton, 199 Glaisdale Drive West, Nottingham, NG8 4GY |
| MM0 | AUP | P Laird, Kanlee, Carness Road, Kirkwall, KW15 1UE |
| M0 | AUR | Alan Taylor, 18 Chestnut Road, Glemsford, Sudbury, CO10 7PS |
| M0 | AUS | A Dowie, 12 Malvern Drive, Gonerby Hill Foot, Grantham, NG31 8GA |
| M0 | AUW | R Hull, 1 Northfield Cottage, Withington Road, Cheltenham, GL54 4LL |
| M0 | AUY | Laurence Jeffries, 73 Poole Lane, Bournemouth, BH11 9DY |
| M0 | AVA | D Salsbury, 1 Somerset Avenue, Tyldesley, Manchester, M29 8LQ |
| M0 | AVF | Antonio De Araujo, 13 Fifth Avenue Shaws Trailer Park, Knaresborough Road, Harrogate, HG2 7NJ |
| M0 | AVH | David Eaton-Watts, 129 Blake Road, West Bridgford, Nottingham, NG2 5LA |
| MI0 | AVI | NEWRY HIGH SCHOOL RC c/o G Millar, 1 Mullybrannon Road, Dungannon, BT71 7ER |
| M0 | AVK | D Swift, 8 Grove Lane, Buxton, SK17 9HG |
| M0 | AVL | David Meakin, 47 Sheridan Street, Walsall, WS2 9QX |
| M0 | AVN | A Oatey, Robin Hill, Blackpost Lane, Totnes, TQ9 5RF |
| M0 | AVP | A Baughan, Camino Tigalate No. 31-33, Villa de Mazo, St. Cruz de Tenerife, Spain, 38730 |
| M0 | AVQ | J Worthington, 23 Sefton Avenue, Congleton, CW12 3DB |
| M0 | AVS | V Saundercock, 14 Rashleigh Avenue, Plymouth, PL7 4DA |
| M0 | AVU | M Scott, 36 Glebe Crescent, Newcastle upon Tyne, NE12 7JR |
| M0 | AVW | C Spence, 32 Woodford Walk, Thornaby, Stockton-on-Tees, TS17 0LT |
| M0 | AVY | A Johnson, 131 Rylands Road, Kennington, Ashford, TN24 9LU |
| M0 | AVZ | D Clutterbuck, 2 Spring Valley Drive, Leeds, LS13 4RN |
| M0 | AWB | A Boom, Oakthorpe House, 8a Peterborough Road, Peterborough, PE6 0BA |
| M0 | AWD | M Mansfield, Piso 4 (IZQ), Avda Jaime I - 14, Altea (Alicante), Spain, 3590 |
| M0 | AWE | Anthony Ellis, 50 Taylors Crescent, Cranleigh, GU6 7EN |
| M0 | AWH | P Bush, 144 Stoke Lane, Westbury-on-Trym, Bristol, BS9 3RN |
| M0 | AWI | J Ross, 42 Stanwell Drive, Westward Ho, Bideford, EX39 1HE |
| MM0 | AWJ | Kenneth Gray, 18 Greenmantle Place, Glenrothes, KY6 3QQ |
| MI0 | AWL | A Smith, 12 Sandringham Heights, Carrickfergus, BT38 9EG |
| M0 | AWN | C Gladman, 24 Priory Road, Chessington, KT9 1EF |
| MW0 | AWO | S Jones, 6 Heol Will Hopkin, Llangynwyd, Maesteg, CF34 9ST |
| M0 | AWP | Paul Oliver, 21 Charlotte Close, Mount Hawke, Truro, TR4 8TS |
| MM0 | AWU | G Moffat, 16/1 Laichpark Loan, Edinburgh, EH14 1UH |
| M0 | AWX | G Schoof, 5 Canal Row, Haigh, Wigan, WN2 1NA |
| M0 | AWY | D Ashdown, Cartwheels, 4 Honeysuckle Close, Hailsham, BN27 3TP |
| MW0 | AXA | W Townsend, 133 Hazeldene Avenue, Brackla, Bridgend, CF31 2JR |
| M0 | AXC | D Russell, 8 Norburton, Burton Bradstock, Bridport, DT6 4QL |
| M0 | AXE | David Marshall, 34 Brentwood Road, Sheffield, S11 9BU |
| M0 | AXG | K Wheeler, 26 Melverton Avenue, Wolverhampton, WV10 9HN |
| M0 | AXJ | Alan Clay, 22 Park Street, Wallasey, CH44 1AT |
| M0 | AXL | John Cook, 36 Kotuku Street, Coffs Harbour, Australia, NSW2450 |
| M0 | AXN | John Davis, 60 West Bar Street, Banbury, OX16 9RZ |
| M0 | AXO | Graham Morris, 7 Rowley View, Bilston, WV14 8DE |
| MM0 | AXR | Tudor Rees, 23 Doune Road, Dunblane, FK15 9AT |
| M0 | AXV | M Amplett, 44a Darby Road, Coalbrookdale, Telford, TF8 7EW |
| M0 | AXW | J Davies, 243a Bradford Road, Winsley, Bradford-on-Avon, BA15 2HL |
| M0 | AXX | E Moody, The Apiaries, Rufford Lane, Newark, NG22 9DG |
| M0 | AXZ | P Morgan, 20 Bishops Way, Buckden, St. Neots, PE19 5TZ |
| M0 | AYA | S Sellman, Ireland Farm, Banbury Road, Warwick, CV35 0HH |
| M0 | AYB | Travis Davies, 95 High Brigham, Brigham, Cockermouth, CA13 0TJ |
| M0 | AYC | Jonathan Soakell, 162 Manor Road, New Milton, BH25 5ED |
| MM0 | AYE | J Welsh, 7 South Cathkin Cottage, Rutherglen, Glasgow, G73 5RG |
| M0 | AYF | D Kostryca, 9 Cherry Tree Road, Gainsborough, DN21 1RG |
| M0 | AYG | Martin Wood, Weavers, Kingsdale Road, Berkhamsted, HP4 3BS |
| M0 | AYI | G Waring, 7 Tynedale Terrace, Stanley, DH9 7TZ |
| M0 | AYO | Howard Parker, 21 Mayfield Street, Hull, HU3 1NS |
| M0 | AYS | Charles Pocock, 4 Broadfields, Harpenden, AL5 2HJ |
| M0 | AYU | I Gibson, 7 Peverells Wood Close, Chandler's Ford, Eastleigh, SO53 2FY |
| M0 | AYV | Andrew Eyles, 25 Chatham Road, Winchester, SO22 4EE |
| M0 | AYX | Anthony King, 28 King Street, Cirencester, GL7 1JT |
| M0 | AYY | R Corfield, 35 Taplings Road, Winchester, SO22 6HE |
| M0 | AZB | R Goddard, 6 Upper Ley Dell, Chapeltown, Sheffield, S35 1AL |
| M0 | AZC | P Niel, 19 Fountains Close, Whitby, YO21 1JS |
| M0 | AZE | M Surplice, 43a Cremorne Road, Sutton Coldfield, B75 5AQ |
| M0 | AZG | J Kisiel, Wayside Cottage, South Stoke Road, Reading, RG8 0PL |
| M0 | AZJ | G Gould, 32 Archer Road, Kenilworth, CV8 1DJ |
| M0 | AZK | David Hill, 35 Bridle Lane, Sutton Coldfield, B74 3QE |
| M0 | AZN | R Cullis, 20 Larch Close, New Inn, Pontypool, NP4 0RT |
| M0 | AZP | Paul LeMasonry, 7 Eastwood Road, Sittingbourne, ME10 2LZ |
| M0 | AZR | S Gale, 39 Thorley Park Road, Bishop's Stortford, CM23 3NG |
| M0 | AZS | R Buckle, 25 Portsmouth Close, Rochester, ME2 2QY |
| M0 | AZT | M Thomas, 36 Seaview Avenue, Peacehaven, BN10 8SA |
| M0 | AZV | N Devine, 46 Tytton Lane West, Wyberton, Boston, PE21 7HL |
| M0 | AZW | K Coe, 5 George St., Enderby, Leicester, LE19 4NQ |
| M0 | AZY | F Willis, 99 Kenilworth Court, Coventry, CV3 6JB |
| M0 | AZZ | Anthony Bond, 6 Meadway, Knebworth, SG3 6DN |
| M0 | BAA | BLACKSHEEP CONTEST + DX GROUP c/o Stephen Purser, 80 John Bold Avenue Stoney Stanton, Leicester, LE9 4DN |
| MM0 | BAC | C Mackay, 27 Barleyknowe Terrace, Gorebridge, EH23 4EQ |
| M0 | BAE | Arthur Radford, 25 Priory Avenue, Kirkby-in-Ashfield, Nottingham, NG17 9BU |
| MM0 | BAG | G Craig, 9 Green Drive, Inverness, IV2 4EX |
| M0 | BAH | Andy Tyler, 15 Chanctonbury Close, Washington, Pulborough, RH20 4AR |
| M0 | BAI | P Byrne, 9 Glenluce Road, Liverpool, L19 9BX |
| M0 | BAJ | D Nelson, 46 The Cunnery, Kirk Langley, Ashbourne, DE6 4LP |
| M0 | BAK | K Williams, Cranesbie, 6 Dore Road, Sheffield, S17 3NB |
| M0 | BAL | F Johnson, 7 Pharos Court, Pharos Street, Fleetwood, FY7 6BG |
| M0 | BAM | Ashley Tobin, 17 Brockhampton, Cheltenham, GL54 5XH |
| M0 | BAO | A Edwards, 53 Priory Glade, Yeovil, BA21 3SQ |
| M0 | BAP | William Stewart, Hillfield Bungalow, Grunt Lane, Stroud, GL6 6PH |
| M0 | BAR | Brian Bartley, 5 Cookes Wood, Broompark, Durham, DH7 7RL |
| MI0 | BAT | S Gilmore, 8 Fortfield, Dromore, BT25 1DD |
| M0 | BAU | G Hoyle, 60 Randle Meadow, Great Sutton, Ellesmere Port, CH66 2BC |
| M0 | BAV | Leslie Evans, 16 Kynaston Drive, Wem, Shrewsbury, SY4 5DE |
| M0 | BAW | D Rose, 31 Mount Crescent, Warley, Brentwood, CM14 5DB |
| M0 | BAY | G Bilson, Fieldgate, 55 Littlemoor Lane, Alfreton, DE55 5TY |
| M0 | BAZ | J Waterfield, 287 Turves Green, Birmingham, B31 4BS |
| M0 | BBE | J Hayward, 76 Lincoln Road, Skegness, PE25 2EE |
| MI0 | BBF | D Doherty, 175 Bridge Road, Glarryford, Ballymena, BT44 9QA |
| M0 | BBH | M Redman, 19 Richmond Road, Rugby, CV21 3AB |
| M0 | BBK | J Meakin, White House Farm, Osmotherley, Northallerton, DL6 3QA |
| MW0 | BBL | A Hadden, 164 Derwen Fawr Road, Sketty, Swansea, SA2 8DP |
| M0 | BBM | B Meredith, 27 Hyde Place, Llanhilleth, Abertillery, NP13 2RT |
| M0 | BBO | S Woodford, 31 Seaborough View, Crewkerne, TA18 8JB |
| M0 | BBQ | K Taylor, 241 Daventry Road, Cheylesmore, Coventry, CV3 5HH |
| M0 | BBR | Reginald Rogers, 3 Ross House, 43 Ashley Road, Salisbury, SP2 7DD |
| M0 | BBT | T Pirrie, Walnut Thatch, Tysoe Road, Warwick, CV35 0UE |
| MW0 | BBU | Stephen Lloyd, 41 Coombs Drive, Milford Haven, SA73 2NU |
| M0 | BBV | R Lyford, 4 Wrentham Estate, Old Tiverton Road, Exeter, EX4 6ND |

**IMPORTANT NOTE**

**Revalidate licence to avoid revocation** – Ofcom has advised the Society that plans will be drawn up to revoke licences that have not been revalidated as required by the licence conditions. The quickest way to revalidate is to do so online via the Ofcom website: *https://services.ofcom.org.uk/* or by email: *amateur.validations@ofcom.org.uk* Ofcom staff are available to help, but please be patient during times of heavy workload.

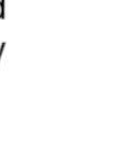

M0 BBW S Gleadall, 59 Old Chapel Road, Warley, Smethwick, B67 6HU
M0 BCC R Clapp, 11 Kensington Gardens, Ilkeston, DE7 5NZ
M0 BCE W Johnstone, 67 Station Lane, Birkenshaw, Bradford, BD11 2JE
M0 BCF R Cranwell, 7 Central Drive, Elston, Newark, NG23 5NT
M0 BCG Ian Williams, Alma Cottage, South Marston, Swindon, SN3 4SN
M0 BCH C Chadburn, 31 Darwin Close, Top Valley, Nottingham, NG5 9LN
M0 BCI Nicholas Armstrong, 112 Chandos Street, Netherfield, Nottingham, NG4 2LW
M0 BCJ Glenn Lewis, 42 Ladywood Road, Ilkeston, DE7 4NE
M0 BCK Kevin Bell, 71 Wheatfield Road, Stanway, Colchester, CO3 0YA
M0 BCL P Williams, 37 Winyards View, Crewkerne, TA18 8JA
M0 BCN David James, 54 Woolacombe Lodge Road, Birmingham, B29 6PX
M0 BCQ Craven Radio Amateur Group c/o Francis Peel, Nuttercote Cottage, Thornton in Craven, Skipton, BD23 3TT
MM0 BCR L Haynes, 29 Invercauld Road, Aberdeen, AB16 5RP
M0 BCT Martin Danfer, The Nook, Mill Common, Woodbridge, IP12 2ED
M0 BCV Stuart Graham, 4 Oakland Avenue, Ellenborough, Maryport, CA15 7BU
M0 BCW P Mason, 15 Granton Avenue, Clifton, Nottingham, NG11 9AL
M0 BCZ R Burton, 1 Avenue Court, Mount Avenue, London, W5 1PY
MM0 BDA R August, Smiddyhill House, Stracathro, Brechin, DD9 7QE
M0 BDB Roland Taylor, 86-88 Hillside Crescent, Leigh-on-Sea, SS9 1HQ
M0 BDD Derrick Webster, 210 Walesby Lane, New Ollerton, Newark, NG22 9UU
M0 BDE Bruce Thorburn, 3 Victoria Road, Bexhill-on-Sea, TN39 3PD
M0 BDF James Reid, Rosebury, Soldridge Road, Alton, GU34 5JF
M0 BDH P Fisher, 21 Charlotte Close, Mount Hawke, Truro, TR4 8TS
M0 BDJ R Hawkins, 10 Mackenzie Close, Swindon, SN3 6JR
M0 BDL D Ferris, 167 Lonsdale Avenue, Doncaster, DN2 6HF
M0 BDQ Konstantine Kisselev, 37 Stanley Avenue, Barking, IG11 0LD
M0 BDS G Butler, 53 Farm Road, Beeston, Nottingham, NG9 5DA
M0 BDU D goodwin, 15 Tennyson Road, Bentley, Doncaster, DN5 0EG
M0 BDW Paul Hayes, 1 Stile Plantation, Royston, SG8 9HP
MI0 BDX Alex Patterson, 33 Marlborough Park, Carryduff, Belfast, BT8 8NL
MI0 BDZ M Chancellor, 55 Brae Hill Park, Belfast, BT14 8FP
M0 BEA Colin Lewis, 60 Caeau Gleision, Rhiwlas, Bangor, LL57 4UA
M0 BEC R Millerchip, 16 Kennedy Crescent, Gosport, PO12 2NN
MM0 BED James Macdonald, 22 New Parliament Place, Campbeltown, PA28 6GY
M0 BEE WHITEHAVEN ARC (T.S.BEE) c/o Norman Williams, 1 Dorset Close, Whitehaven, CA28 8JP
M0 BEH Peter Mutter, 129 Demesne Road, Wallington, SM6 8EW
M0 BEJ G Moody, 25 Norbiton Common Road, Kingston upon Thames, KT1 3QB
M0 BEK C Dunn, 75 Waddington Avenue, Burnley, BB10 4LA
MW0 BEL A Owen, Rafael Fawr House, The Fraich, Fishguard, SA65 9QJ
M0 BEM M Taperell, 16 Parkhall Croft, Birmingham, B34 7BU
M0 BEO K Anderson, 53 Priory Grove, Hull, HU4 6LU
M0 BEQ H Walsh, 38 Potter Hill, Greasbrough, Rotherham, S61 4PA
M0 BER David Jones, 3 Tai Clwch, Rhosmeirch, Llangefni, LL77 7SJ
MI0 BES J May, 8 Oak Vale Avenue, Newry, BT34 2BQ
M0 BET V Hughes, Manley, 1 Garden Drive, Llandudno, LL30 3LL
M0 BEV R Knapp, 85 Eastern Avenue, Liskeard, PL14 3TD
M0 BEX J Hrycan, 40 Marina Drive, Marple, Stockport, SK6 6JL
MW0 BEY M Beynon, Sunrise, Kilgetty Lane, Narberth, SA67 8JL
M0 BFA Derek Wilson, 30 Little Avenue, Swindon, SN2 1NL
M0 BFB K Francks, 63 Parc Godrevy, Pentire, Newquay, TR7 1TY
M0 BFT Steven Smith, 225 South Drove, Lutton Marsh, Spalding, PE12 9NT
M0 BFV L Papazoglou, 37a Eaton Road, Wirral, CH48 3HE
M0 BGE T Parker, 24 Burrows Close, Lawford, Manningtree, CO11 2HE
MM0 BGO R Herd, 4 Smithy Lane, Balmullo, St. Andrews, KY16 0FG
M0 BGR Henry Howard, 10 Lawnside, London, SE3 9HL
M0 BGS Geoffrey Steedman, 5 Allerton Grange Gardens, Leeds, LS17 6LL
M0 BGT G Dyson-Bawley, Grahil, 33 Ridgetor Road, Liverpool, L25 6DG
M0 BGU J Moran, 27 Mellor Grove, Bolton, BL1 6DA
MM0 BGW Andrew Munro, 2 Woodlands View, Inshes Wood, Inverness, IV2 5AQ
M0 BHA C Baker, 1 Astley Green, Darleyhall, Luton, LU2 8TS
M0 BHE Malcolm Sadler, Hill View, Horton, Ilminster, TA19 9QU
M0 BHG B Giles, 171 St. Stephens Road, Saltash, PL12 4NJ
M0 BHH D Newing, 13 Maxwell Road, Broadstone, BH18 9JG
M0 BHJ Peter Worlledge, 181 Roselands Drive, Paignton, TQ4 7RN
M0 BHK G Robertson, 22 Carlton Villas, Hatt, Saltash, PL12 6PS
M0 BHM Adrian Whitehouse, 19 Cleeve Road, Marlcliff, Alcester, B50 4NX
M0 BHN Paul Jarvis, 26 Nally Drive, Woodcross, Bilston, WV14 9UT
M0 BHO Stephen Coe, 113 Highfield Road, Yeovil, BA21 4RJ
M0 BHP Graeme King, 12 Bracken Close, Blackburn, BB2 5AH
M0 BHQ F Lugg, 4 Newbury Close, Walsall, WS6 6DF
M0 BHR Geoff King, 8 Oak Lane, Burghill, Hereford, HR4 7QP
M0 BHV J Cocks, 12 Birch Pond Road, Plymouth, PL9 7PG
M0 BHW G Jeckells, Hogals End, Mill Street, Thetford, IP25 7QN
MM0 BHX T Costford, Hillmont, Covenanter Road, Shotts, ML7 5PA
M0 BIC J Brocklebank, Springfield House, Sixhills Lane, Market Rasen, LN8 6AN
M0 BIH R Deakin, 40 Brussels Road, Stockport, SK3 9QG
M0 BII S Keightley, 11 Sandringham Avenue, Wisbech, PE13 3ED
M0 BIJ Christopher Harden, 24 Fuchsia Gardens, Southampton, SO16 6TY
M0 BIK B Bourne, 19 The Crescent, Beeston, Sandy, SG19 1PQ
M0 BIN Charles Rodgers, 37 Heatherset Gardens, Norbury, London, SW16 3LS
MM0 BIR J Brady, 19 Mill Gardens Powmill, Dollar, FK14 7LQ
M0 BIT P smith, Karinya, Rectory Road Haddiscoe, Norwich, NR14 6PG
MM0 BIX E Cameron, 5 King Street, Ferryden, Montrose, DD10 9RR
M0 BIZ Maurice Thomas, 39 Treworder Road, Truro, TR1 2JZ
M0 BJD Brain Duffy, 8 Mirfield Close, Halewood, Liverpool, L26 9XP
M0 BJE A Cockram, 70 Arlington Drive, Marston, Oxford, OX3 0SJ
M0 BJJ Shoji Miyake, Hakata Radio, P.O.Box 232, Hakata, Japan, 812-8799
M0 BJK G Scothorn, School House, Kirk Balk, Barnsley, S74 9HU
M0 BJL Shaun Jarvis, Kellow, Old Lyndhurst Road, Southampton, SO40 2NL
MD0 BJM Michael Rodgers, 1 Kings Court, Ramsey, Ramsey, Isle of Man, IM8 1LJ
M0 BJN Frank Humphris, 169 Bloxham Road, Banbury, OX16 9JU
M0 BJO D Greenway, 37 Primrose Hill Park Homes, Primrose Hill, Somerton, TA11 7AP
M0 BJP R Pearce, Kolner, 86 Thrupp Lane, Stroud, GL5 2DG
M0 BJR Martin Brown, 3 The Vines, Kelsale, Saxmundham, IP17 2PU
M0 BJS P Hall, 25 Ham Green, Pill, Bristol, BS20 0EY
M0 BJT Kevin Davison, 2 Sitwell Close, Spondon, Derby, DE21 7GT
MJ0 BJU A Mourant, Little Mead, Claremont Road, St. Saviour, Jersey, JE2 7RT
M0 BJX Robert Glover, 89 Cambridge Road, Linthorpe, Middlesbrough, TS5 5LD
M0 BKA John Slough, 2554 Hamilton Rd., Lebanon Oh, United States, 45036-8849
M0 BKD P McCormack, 3 Greenway Close, Torquay, TQ2 8EF
M0 BKF C Brodrick, 6 Carlton Street, Hartlepool, TS26 9ES
M0 BKG G Rundle, 15 Sandown Road, Paignton, TQ4 7RL
M0 BKK J Rowe, The Old Rectory, Wickenby, Lincoln, LN3 5AB
M0 BKL S Passmore, 35 David Road, Paignton, TQ3 2QF
M0 BKN Susan Sherwin, 1 Nursery Close, Wroughton, Swindon, SN4 9DR
M0 BKS K Sim, 3 Thorngate Close, Penwortham, Preston, PR1 0XN
M0 BKV D Kamm, Delabole Head, Week St. Mary, Holsworthy, EX22 6UU
M0 BKX E WILLOX, 3 Pound Gate, Hassocks, BN6 9LU
M0 BLD S Tsuzuki, Flat 4, Lancing House, Watford, WD24 4RL
M0 BLF Dominic Smith, 67 Lambs Lane, Cottenham, Cambridge, CB24 8TB
M0 BLH Scott Laddiman, 31 Gordon Godfrey Way, Horsford, Norwich, NR10 3SG
M0 BLI J Crangle, 43 Scarfell Close, Peterlee, SR8 5PF
M0 BLM Arthur Evans, Maescolwyn, Old Hall, Llanidloes, SY18 6PS
M0 BLN K Hopps, 100 Etherley Lane, Bishop Auckland, DL14 6TU
M0 BLO R Jackson, 4 Hornbrook Gardens, Plymouth, PL6 6LS
M0 BLR Adrian Cresswell, 31 New Street, Doddington, March, PE15 0SP
M0 BLS G Hickford, Sanclare, 94 Manor Road, Fleetwood, FY7 7HY
M0 BLT M Waldron, 32 Windmill Street, Upper Gornal, Dudley, DY3 2DQ
M0 BLU William Jepson, 3 Marchog, Holyhead, LL65 2HD
M0 BLV George Peacock, 7 Penscose, Witney, OX28 2EG
M0 BLY S Young, 126 Stevens Road, Dagenham, RM8 2QL
M0 BLZ Ann Blackburn, 6 Victoria Street, Cullingworth, Bradford, BD13 5AE
MM0 BMA W Irwin, 15 Inchcolm Place, East Kilbride, Glasgow, G74 1DR
M0 BMB B Bentham, 89 Westborough Way, Hull, HU4 7SW
M0 BMD J Green, 788 The Ridge, St. Leonards-on-Sea, TN37 7PS
MI0 BME P Maile, 3 Cairnmore Avenue, Lisburn, BT28 2DW
M0 BMF D Anger, 17 Dell Road, Andover, SP10 3JT
MM0 BMG Norman Stewart, 160 Carrick Knowe Drive, Edinburgh, EH12 7EW
M0 BMJ Iain Singer, 197 Rosalind Street, Ashington, NE63 9BB
MI0 BML Patrica Doris, 92 Coolnafranky Park, Cookstown, BT80 8PW
MI0 BMM B Moore, 8 Orken Lane, Aghalee, Craigavon, BT67 0ED
M0 BMN Paul Webb, 40 Links Road, Penn, Wolverhampton, WV4 5RF
M0 BMR P Pirrazzo, 30 Coronation Road, Middlewich, CW10 0DL
M0 BMT D Bunting, 6 Mill Gardens, Worksop, S80 3QG
M0 BMU J Moritz, Carillon, 6 Bell Lane, Hatfield, AL9 7AY
M0 BMW K Wrack, 18 Carrs Road, Cheadle, SK8 2EE
M0 BMX M Fitchett, Barkenroy, Ludgvan, Penzance, TR20 8AJ
M0 BMY L Bilson, Fieldgate, 55 Littlemoor Lane, Alfreton, DE55 5TY
M0 BMZ A Martin, 11 The Mount, Worcester Park, KT4 8UD
M0 BNB Thomas Rogers, The Willows, 48 Hillock Lane, Wrexham, LL12 8YL
M0 BNC Norton Clark, 28 Thickthorn Close, Kenilworth, CV8 2AF
M0 BNF I Copping, 54 Hartley Road, Kirkby-in-Ashfield, Nottingham, NG17 8DP
M0 BNO G Ryder, 11 Claremont Gardens, Farsley, Pudsey, LS28 5BF
M0 BNP J Taylor, 3 Inhams Close, Murrow, Wisbech, PE13 4HS
M0 BNR N Rodley, 268 Grovehill Road, Beverley, HU17 0HP
M0 BNS BRITISH NATURIST ARS c/o C Beesley-Reynolds, Kaos Roams, Palmerston Close, Leicester, LE8 0JJ
M0 BNZ D Brooks, The Elms, Trewoon Road, Helston, TR12 7DS
M0 BOB Robert Adlington, 33 Columbine Way, Romford, RM3 0XN
M0 BOC G Tomlinson, 7 Cronkeyshaw Road, Rochdale, OL12 0QR
M0 BOH Sharon Saiger, 10 Markham Avenue, Armthorpe, Doncaster, DN3 2AZ
M0 BOI B Johnson, 10 Saffron Road, Tickhill, Doncaster, DN11 9PW
MI0 BOK P Connolly, 94 North Parade, Belfast, BT7 2GJ
M0 BOL R Rose-Round, 16 Forshaw Avenue, Blackpool, FY3 7PW
M0 BOM R Wilson, 22 Leadhills Way, Bransholme, Hull, HU7 4ZA
M0 BOQ Robert Slater, 79a Ainsworth Road, Radcliffe, Manchester, M26 4FA
MI0 BOU J Orr, 17 Argyll View, Larne, BT40 2JR
M0 BOX Simone Wilson, 21 Plumian Way, Balsham, Cambridge, CB21 4EG
MI0 BPB A Mulholland, 83 Tullyrain Road, Donaghcloney, Craigavon, BT66 7PP
M0 BPC Kenneth Hunt, 4 Oak Avenue, Willington, Crook, DL15 0BJ
MM0 BPF R Armstrong, 98 Burnbank Road, Ayr, KA7 3QJ
M0 BPM D Barclay, 13 Avondale Terrace, Chester le Street, DH3 3ED
M0 BPN N Taplin, 149 Frindsbury Road, Strood, Rochester, ME2 4JD
M0 BPO J McKinney, La Congerie, Miallet, France, 24450
M0 BPP K Lindsay, 34 Park Crescent, Newtown St. Boswells, Melrose, TD6 0QS
M0 BPQ S Bunting, 17 Sunnydene Avenue, Highams Park, London, E4 9RE
M0 BPS D Lawrence, 42 Upper Packington Road, Ashby-de-la-Zouch, LE65 1UL
M0 BPT Robert Walker, P.O.Box 6743, Tipton, DY4 4AU
M0 BPU T Lyne, The Ark, 10 Blackfields Avenue, Bexhill-on-Sea, TN39 4JL
MM0 BPV A Finlayson, 10b Flesherin, Isle of Lewis, HS2 0HE
M0 BPW Adam Shelswell, Waterway, Dock Lane, Woodbridge, IP12 1PE
M0 BPX K Scott, Kirklands, Craigend Road, Galashiels, TD1 2RJ
M0 BPY Paul Henson, 1 Eaton Close, Rainworth, Mansfield, NG21 0AR
M0 BQB T Lee, 91 Old Vicarage Park, Narborough, King's Lynn, PE32 1TG
M0 BQC D Bradley, 1 Traddles Court, Chelmsford, CM1 4XZ
M0 BQD Elizabeth Lee, 16 Phoenix Chase, North Shields, NE29 8SS
M0 BQE C Margetts, 16 Lahn Drive, Droitwich, WR9 8TQ
M0 BQF M Preece, 51 Drancy Avenue, Willenhall, WV12 5RD
M0 BQH K Goodacre, 22 New Ferry Road, Wirral, CH62 1BJ
MM0 BQI Jim Martin, 3 Lismore Avenue, Edinburgh, EH8 7DW
MM0 BQJ T Cassidy, 14 Hillshaw Green, Bourtreehill South, Irvine, KA11 1EQ
MM0 BQL A Cromack, 10 Manse Road, Ardersier, Inverness, IV2 7SR
M0 BQO U Rose, 45 Ringstead Crescent, Weymouth, DT3 6PT
M0 BQT S Smith, 55 Market Street, Ilkeston, DE7 5RB
M0 BQZ J Romanis, 23 Old Farm Lane, Stubbington, Fareham, PO14 2BZ
M0 BRA BRACKWELL RADIO AMATEUR CONTEST GROUP c/o G Leonard, 65 Qualitas, Bracknell, RG12 7QG
M0 BRB Brian Jewell, 44 Clovelly Road, Bideford, EX39 3DF
M0 BRE Peter Mann, 10 Hawkins Way, Wootton, Abingdon, OX13 6LB
MM0 BRG R Harman, 2 Dornoch Court, Kilwinning, KA13 6QN
M0 BRH T Linham, 44 Vestry Road, Street, BA16 0HX
M0 BRI P Walker, 20 Arbury Banks, Chipping Warden, Banbury, OX17 1LU
M0 BRL Timothy Wells, 38 Stow Park Avenue, Newport, NP20 4FN
M0 BRM E Turner, 16 The Rowans, Doddington, March, PE15 0SE
MW0 BRO M Lewis, 3 St. Patricks Hill, Llanreath, Pembroke Dock, SA72 6XQ
M0 BRP Mark Wastie, 5 Penscose, Witney, OX28 2EG
M0 BRT J Johnson, 16 The Grove, Abingdon, OX14 2DQ
M0 BRU W Griffin, 48 Wardle Way, Kidderminster, DY11 5UJ
M0 BSB W Carr, 37 Keats Road, Greenmount, Bury, BL8 4EP
M0 BSC Peter Bonsey, Wood View, Lower Kelly, Calstock, PL18 9RY
M0 BSD Brian Daly, 10 Parfitts Close, Farnham, GU9 7DH
M0 BSF N Rigazzi-Tarling, 3 The Planes, Bridge Road, Chertsey, KT16 8LE
M0 BSH W Kerslake, 42 Silverdale Road, Newcastle, ST5 2TB
M0 BSI L Preston, Hillside, Trewennack, Helston, TR13 0PQ
M0 BSJ Philip Poore, 42 Kibblewhite Crescent, Twyford, Reading, RG10 9AX
M0 BSL M Pell, 7 Churchfleet Lane, Gosberton, Spalding, PE11 4NE
MM0 BSM Stuart McQuillian, 68 Strathmore Drive, Cornton, Stirling, FK9 5BE
M0 BSP E Doyle, 33 Bodenham Road, Northfield, Birmingham, B31 5DP
M0 BSQ J Doyle, 33 Bodenham Road, Northfield, Birmingham, B31 5DP
MI0 BSU D Stanley, 59 Gransha Road, Kircubbin, Newtownards, BT22 1AJ
M0 BSV Anthony Ilett, 34 Westbrook Park Road, Woodston, Peterborough, PE2 9JG
M0 BSW Paul Collins, 6 Haywardsfield, Peterborough, PE3 6FB
MM0 BSX Graham Scattergood, 14 Market Street, Forfar, DD8 3EY
M0 BSZ A Wanford, 4 Willows Close, Tydd St. Mary, Wisbech, PE13 5QR
MM0 BTD J Ganson, 1 Beechwood Terrace West, St. Fort, Newport-on-Tay, DD6 8JH
M0 BTG Gillian Barrett, 114 William Street, Long Eaton, Nottingham, NG10 4GD
M0 BTI E Thomas, Craig-Y-Don, Conwy, LL27 0JJ
MI0 BTK A Cobb, 62 Katesbridge Road, Dromara, Dromore, BT25 2PN
M0 BTL A Withers, 5 Tintern Road, Skelton-in-Cleveland, Saltburn-by-The-Sea, TS12 2YN
MI0 BTM G Duffy, Carrageen Cottage, Enniskillen, BT74 6ET
M0 BTN John Escreet, Colfield, Carlton Lane, Hull, HU11 4RA
M0 BTO S Collins, 69 Walkers Heath Road, Kings Norton, Birmingham, B38 0AL
M0 BTP E Edmunds, 5 Nelsons Quay, St. Helens, Ryde, PO33 1TA
M0 BTR J Springett, 31 Mountbatten Court, Andover Road, Winchester, SO22 6BA
M0 BTU G Rowlands, Dalar Wen, Rhosmeirch, Llangefni, LL77 7SJ
M0 BTX R Elliott, 8 Bridge Place, Amersham, HP6 6JF
M0 BTY P Fincham, 36 Bradley Road, Nuffield, Henley-on-Thames, RG9 5SG
M0 BTZ Roscoe Harrison, 55 Stapleford Close, Romsey, SO51 7HU
M0 BUA R Miller, 1 Mews Cottages, Penview Crescent, Helston, TR13 8RX
M0 BUE J Page, Highcroft Farmhouse, Gay Street, Pulborough, RH20 2HJ
M0 BUF E Smith, Lamorna, Broadmead Road, Woking, GU23 7AD
M0 BUG N Jamieson, 1 Langdale Place, Newton Aycliffe, DL5 7DX
MM0 BUH Nigel Smith, Nether Dallachy, Boyndie, Banff, AB45 2JT
M0 BUR N Armer, 17 Keswick Road, Lancaster, LA1 3HJ
M0 BUT T Gale, 33 Watson Close, Upavon, Pewsey, SN9 6AF
M0 BUV A Zerafa, 2 Furnwood, St. George, Bristol, BS5 8ST
M0 BUY John Swann, 1 Sunnindale Drive, Tollerton, Nottingham, NG12 4ES
M0 BVD Christopher Turner, North Holme, Drain Lane, York, YO43 4DQ
M0 BVF G Kapranos, 89 Spohr Terrace, South Shields, NE33 3LQ
MI0 BVG T Wedlock, 13 Drumawhey Road, Newtownards, BT23 8RS
M0 BVI P Rogers, Flat 4 Holmdale, 2 Osborne Road, Poole, BH14 8SD
M0 BVM Dudley Hancock, 4 Elmside, Willand, Cullompton, EX15 2RN
M0 BVN A Tavender, 38 Hesley Grove, Chapeltown, Sheffield, S35 1TX
M0 BVO J Cull, Drybrook Cottage, Amesbury Road, Salisbury, SP4 0ER
M0 BVQ G Stokes, 9 The Haven, Harwich, CO12 4LA
M0 BVT John Colles, The Orangery, Ufford Place, Woodbridge, IP13 6DP
M0 BVU Stephen Noble, 30 Flude Road, Coventry, CV7 9AQ
M0 BVV F Hibberd, 58 Stoke Green, Coventry, CV3 1AN
M0 BVW B Cox, 13 The Graylands, Coventry, CV3 6EW
M0 BVX D Poulton, 14 George Street, Gun Hill, Coventry, CV7 8HL
M0 BVY N Marshall, 8 Ellesboro Road, Harborne, Birmingham, B17 8PT
M0 BVZ P Poulton, 14 George Street, Gun Hill, Coventry, CV7 8HL
M0 BWB J Ridout, 136 Church Hill Road, Cheam, Sutton, SM3 8NA
M0 BWC John Barton, 27 Francis Way, Salisbury, SP2 8EF
M0 BWF D Surman, 27 Stanley Road, Hinckley, LE10 0HP
M0 BWH S Jones, 34 Bury Green, Little Downham, Ely, CB6 2UH
M0 BWI William Wilson, 8 Nora Street, South Shields, NE34 0RA
M0 BWL S Vinnicombe, 8a Cross Road, Cholsey, Wallingford, OX10 9PE
MW0 BWM P Lane, 51 Maesgwyn, Cwmdare, Aberdare, CF44 8TH
M0 BWN H Seldon, 22 Downside Avenue, Plymouth, PL6 5SD
M0 BWO B Watts, 24 Carisbrooke Way, Redcar, TS10 2LJ
M0 BWP R Bebbington, 64 Stafford Road, Toll Bar, St. Helens, WA10 3JH
M0 BWQ H Robinson, 5 Coppice Close, Haxby, York, YO32 3RR
M0 BWS David Pacheco III, 35 Baker Close, Caversfield, Bicester, OX27 8FQ
M0 BWU Andy Hall, 16 Brushfield Avenue, Sileby, Loughborough, LE12 7NX
M0 BWV R Pugsley, 156 Thistledown Road, Clifton, Nottingham, NG11 9ED
M0 BWW K Allen, 55 Brandish Crescent, Clifton, Nottingham, NG11 9JZ
M0 BWY David Hill, 86 The Downs, Nottingham, NG11 7EB
M0 BXA Lisbeth Jensen, 17 Middleton Close, Tysoe, Warwick, CV35 0SS
M0 BXB E Fuller, 36 North Road, Hull, HU4 6LJ
M0 BXC S Coulston, 15 West View, Clitheroe, BB7 1DG
M0 BXD C Robinson, Flat 1, St. Michaels Court, Worcester, WR2 5QR
M0 BXF P Pickup, 13 Siddows Avenue, Clitheroe, BB7 2NX
M0 BXG Jonathan King, 19 Fawnbrake Avenue, London, SE24 0BE
M0 BXJ Anthony Chalk, 42 Erskine Road, Colwyn Bay, LL29 8EU
M0 BXM David Grimshaw, 83 Ripon Street, Blackburn, BB1 1TW
M0 BXQ V Barnes, 16 Aldingbourne Park, Hook Lane, Chichester, PO20 3YR
M0 BXU P White, 61 North Street, Pewsey, SN9 5ES
M0 BYB A Lee, 91 Old Vicarage Park, Narborough, King's Lynn, PE32 1TG
M0 BYI D Moorey, 35 Broadway Manor, The Broadway, Hull, HU9 3PN
M0 BYJ R Stoddart, 4 Belmont Road, Rednal, Birmingham, B45 9LW
M0 BYL British Young Ladies Club c/o Jenni Jones, 69 Pound Street, Warminster, BA12 8NW
M0 BYM James Robson, Flat 6, 57 Lewisham Park, London, SE13 6QP
MI0 BYR David Christie, 5 Moneydig Park, Garvagh, Coleraine, BT51 5JP
MW0 BYS William Reed, 2 St. Marys Park, Jordanston, Milford Haven, SA73 1HR
M0 BYT Gwynfor Roscoe, 45 Bro Infryn, Glasinfryn, Bangor, LL57 4UR
M0 BYU C Dodshon, 62 Moor Road, Melsonby, Richmond, DL10 5PE

M0 BYV J Goldsbrough, 63 Aske Road, Redcar, TS10 2DP
M0 BYY J Elliott, 21 [illegible] Road, Hampton Hill, Hampton, TW12 1DW
M0 BYZ [illegible]
M0 BZA R Stoddart, 163 Flatts Lane, Middlesbrough, TS6 0PP
M0 BZB S Thirlaway, 10a Sea View, Blackhall Colliery, Hartlepool, TS27 4AX
M0 BZC A Phillips, 39 Stonechat Road, Billericay, CM11 2NZ
M0 BZE B Allen, 1 St. Marys Close, Mursley, Milton Keynes, MK17 0HP
M0 BZH M Wilkinson, 124 Doncaster Road, Darfield, Barnsley, S73 9JA
M0 BZI F Western, 12 St. Oswalds Crescent, Brereton, Sandbach, CW11 1RW
M0 BZK D Mapeley, 6 Green Lane, Wolverton, Milton Keynes, MK12 5HB
M0 BZN L Evans, 184 West Street, Dunstable, LU6 1NX
M0 BZO G Major, 17 Jubilee Cottages, Station Road, Bedford, MK43 0PN
M0 BZQ Julian Doxey, 34 Lime Tree Crescent, New Rossington, Doncaster, DN11 0BT
M0 BZR M Taylor, 38 Manor Road, Slyne, Lancaster, LA2 6LB
M0 BZS R Cornthwaite, 18 Slaidburn Drive, Accrington, BB5 0JJ
M0 BZU N BURKILL, 4 Giles Street, Cleethorpes, DN35 8EA
M0 BZV P Wade, 11 Hillside Crescent, Puriton, Bridgwater, TA7 8AP
M0 BZX S Turner, 75 Keir Hardie Avenue, Stanley, DH9 6JU
M0 BZY M Haywood, 4 Wentworth Gate, Birmingham, B17 9EB
M0 BZZ Robert Bunker, 12 Bedford Avenue, Birkenhead, CH42 4QX
M0 CAA Mike Bennett, 48 Orchard Grove, Fareham, PO16 9DX
MW0 CAB I Jones, 4 Crowhill, Haverfordwest, SA61 2HL
MI0 CAC H Mcgoldrick, 2 Carsdale, Mullanahoe Road, Dungannon, BT71 5GA
M0 CAD Brian Lovatt, Daleholme, Masterman Place, Barnard Castle, DL12 0ST
MM0 CAE James Gauson, 112a High Street, New Pitsligo, Fraserburgh, AB43 6NN
M0 CAG Ian Solly, 4 Goodwin Road, Ramsgate, CT11 0LP
M0 CAJ D Tysoe, 21 Burnt Close, Luton, LU3 3SU
M0 CAM GRANTA CONTEST GROUP c/o M Marsden, 38 Lambert Cross, Saffron Walden, CB10 2DP
M0 CAN John Wallis, 8 Kennedy House Hainworth Lane, Keighley, BD21 5BD
M0 CAR S Robertson, 5 Sear Hills Close, Balsall Common, Coventry, CV7 7QL
M0 CAS Mark Cassidy, 6 Mosley Street, Blackburn, BB2 3ST
M0 CAV J Pearson, 82 Devoke Avenue, Worsley, Manchester, M28 7EN
M0 CAX D deakin, Restholme Cottage, Mosham Road, Doncaster, DN9 3BA
M0 CAZ A Dahalay, The Manor, 80 Beach Road, Weston Super Mare, BS22 9UU
M0 CBA G Gunn, The Old Rectory, Coreley, Ludlow, SY8 3AW
M0 CBD W Eldridge, Minafon, Llangeitho, Tregaron, SY25 6TT
M0 CBF N Lock, 57 Western Way, Basingstoke, RG22 6DF
M0 CBG Allan Godney, 38 Botley Drive, Havant, PO9 4QY
M0 CBI J Isaacs, 9 Rowan Close, Shaftesbury, SP7 8RG
M0 CBK H Purves, 2 Bourtree Close, Wallsend, NE28 9AA
MM0 CBL G Black, 9 Mcculloch Road, Girvan, KA26 0EF
M0 CBM R Newman, 11 Pine View Close, Woodfalls, Salisbury, SP5 2LR
M0 CBN Roy Liversidge, 17 Millbank Close, High Green, Sheffield, S35 4NS
M0 CBP Beverley Muizelaar, BIRTLEY CB SERVICES, PHOENIX COMMUNICATIONS, 33 PENSHAW VIEW, PORTOBELLO ROAD, Birtley, DH3 2JL
M0 CBQ W Salt, 89 Woodhall Drive, Waltham, Grimsby, DN37 0UX
M0 CBT L Betts, 33 Four Wells Drive, Sheffield, S12 4JB
MI0 CBX R Graham, 21 Meadowvale Crescent, Bangor, BT19 1HQ
M0 CCA Stephen Ramsden, 168 Aberdeen Avenue, Plymouth, PL5 3UW
MM0 CCC John MacLean, 15 Muirpark Terrace, Tranent, EH33 2AS
M0 CCD J Newsome, 241 Skellow Road, Skellow, Doncaster, DN6 8JL
M0 CCF FirePower Museum c/o Michael Buckley, Springfield, 12 Ranmore Avenue, Croydon, CR0 5QA
MI0 CCG S Hutton, 2 Darkfort Crescent, Portballintrae, Bushmills, BT57 8WH
MW0 CCK K Dancer, 63 Romilly Crescent, Cardiff, CF11 9NQ
MW0 CCL K Dancer, 118 Fairwater Grove West, Cardiff, CF5 2JR
M0 CCN J Jones, Clayton View, Francis Road, Wrexham, LL11 6EH
M0 CCQ Paul Burgess, 27 Watergate Street, Ellesmere, SY12 0EX
M0 CCS A Patterson, 3 Trem Y Foryd, Kinmel Bay, Rhyl, LL18 5JE
M0 CCU J Buckley, 14 Eastfield, Foxholes, Driffield, YO25 3QW
M0 CCV L Parsons, Gull Cottage, Briar Close, Hastings, TN35 4DP
M0 CCW Martin Thornton, Bramble Lodge, Youngers Lane, Skegness, PE24 5JQ
M0 CCX D Hughes, 121 Felixstowe Road, Ipswich, IP3 8EA
M0 CCZ A Cubitt, 132 Cauldwell Hall Road, Ipswich, IP4 5BP
M0 CDB Trevor Faris, 7 Mount Close, Fetcham, Leatherhead, KT22 9EF
M0 CDC Peter Russell, 57 Norburn Park, Witton Gilbert, Durham, DH7 6SG
M0 CDF Maurice Brooks, 9 The Green, Blaby, Leicester, LE8 4FQ
M0 CDG Chris Gozzard, Craig Dulas, Rhydyfoel Road, Abergele, LL22 8EG
M0 CDJ S Spevack, Pips Hill, Garden Close, Leatherhead, KT22 8LR
MM0 CDK D Dickson, C/O, 115 Hatton Gardens, Glasgow, G52 3PU
M0 CDL John Griffin, 35 Cottage Street, Kingswinford, DY6 7QE
M0 CDN N Bullough, 29 Redfern Road, Stone, ST15 0LE
MW0 CDO P Tomlinson, 17 Heol Yr Orsedd, Margam, Port Talbot, SA13 2HL
M0 CDQ John Grant, 47 Coneyford Road, Shard End, Birmingham, B34 7AY
M0 CDS S Sanders, 82 Hazelwood Road, Callington, PL17 7EU
M0 CDU E Linley, 40 Belvoir Road, Cleethorpes, DN35 0SE
MM0 CDW Andrew Bryce, 23 Primrose Avenue, Inverkip, Greenock, PA16 0DS
M0 CDX CHILTERN DX CLUB-THE UK DX FOUNDATION c/o E Cheadle, Lower Withers Barns, Middleton on the Hill, Leominster, HR6 0HY
M0 CDY J Elsworth, 15 Elm Avenue, Christchurch, BH23 2HJ
M0 CDZ K Mountford, 22 Hollington Drive, Oxford, Stoke-on-Trent, ST6 6TZ
M0 CEB M Bridge, 100 Pennine Road, Bacup, OL13 9PH
M0 CEC T Ditchfield, 13 Alexandra Road, Waterloo, Liverpool, L22 1RJ
M0 CEG Andrew Shipp, 12 Rainbow Court, Paston Ridings, Peterborough, PE4 7UP
M0 CEM D Wells, 8 Corrigan Close, Bletchley, Milton Keynes, MK3 6BP
M0 CEO Brian Lewin, 68 Brackley Square, Woodford Green, IG8 7LS
M0 CEQ N Taylor, 18 Chestnut Road, Glemsford, Sudbury, CO10 7PS
M0 CER Toshihiko Oka, 65 Curzon Street, London, W1J 8PE
M0 CES Donald Hadden, 52 Brant Road, Lincoln, LN5 8SH
M0 CEU R Pinborough, 57 Ousebank Way, Stony Stratford, Milton Keynes, MK11 1LA
M0 CEW J lehane, 174 Stamfordham Drive, Liverpool, L19 6PZ
M0 CEX P Newell, Thwaites Bank, Spring Avenue, Keighley, BD21 4TD
MM0 CEZ Peter Moran, 3 Dunottar Ave, Coatbridge, ML54LL
M0 CFB A Bennett, 24 Alder Avenue, Fenham, Newcastle upon Tyne, NE4 9TB
M0 CFD Nicholas Hould, 50 Laurel Avenue, Forest Town, Mansfield, NG19 0DW
MM0 CFE [illegible]
M0 CFF [illegible]
M0 CFH Alan Hewett, 1 Mountside, Westfield Lane, Folkestone, CT18 8BY
MD0 CFM Andrew Hosking, 30 Edrick Road, Edgware, HA8 9JD
MW0 CFQ P Brennan, 1 Gerddi Mair, St. Clears, Carmarthen, SA33 4ET
M0 CFR R Gould, 22 Vereker Drive, Sunbury-on-Thames, TW16 6HF
M0 CFT K Pye, 5 Teme Avenue, Wellington, Telford, TF1 3HU
M0 CFX C Moss, Dickenson Barn Farm, Old Meadows Road, Bacup, OL13 8PX
M0 CFZ B Farrington, 15 Derwent Avenue, Morecambe, LA4 5PT
M0 CGA A Desoer, 53 Highfield Road South, Chorley, PR7 1RH
M0 CGB A Wiseman, 28 Inchfield, Worsthorne, Burnley, BB10 3PS
M0 CGE G Corneloues, Klosterle, Harwich Road, Harwich, CO12 5AD
M0 CGF I Schofield, 9 Ashdene Road, Heaton Mersey, Stockport, SK4 3AD
M0 CGO R Pratt, 11 Park Road, Ryde, PO33 2BG
MW0 CGP T Peters, 37 Lon Coed Bran, Cockett, Swansea, SA2 0YD
M0 CGR J Clarey, 2 Sebastopol Cottages, Redmere, Ely, CB7 4SS
M0 CGS S Hancock, Monrad, Back Street, Gainsborough, DN21 3DL
M0 CGT Geoffrey Thompson, 160 Rempstone Road, Wimborne, BH21 1SX
MI0 CGV Thomas Keery, 51 School Road, Ballyroney, Banbridge, BT32 5JF
M0 CGW C Selwyn-Smith, Miranda, East Molesey, KT8 9AN
MM0 CGZ Carl Cregan, 4 Fowlers Court, Prestonpans, EH32 9AT
M0 CHD James Neale, 20 Oakfield Road, Wollescote, Stourbridge, DY9 9DL
MW0 CHI Malcolm Broxtom, 4 Owen Street, Pembroke, SA71 4EP
M0 CHJ S Birbeck, 9 Shelley Drive, Accrington, BB5 2QS
M0 CHL D Fitzpatrick, 21 Stambridge Road, Clacton-on-Sea, CO15 3JR
MU0 CHN Jeremy Gardner, The Ferns, Rue De La Girouette, St. Saviour, Guernsey, GY7 9NN
M0 CHO G Burton, 41 Etchingham Road, Langney, Eastbourne, BN23 7DS
M0 CHR D Whitelock-Wainwright, 21 Whelan Gardens, St. Helens, WA9 5TD
M0 CHS M Stanley, Harby, Brightlingsea Road, Colchester, CO7 8JH
M0 CHU T Hodby, 1 Hawksworth Close, Rotherham, S65 3JX
MM0 CHV W Adamson, 47 Clarinda Gardens, Dalkeith, EH22 2LW
MW0 CIA N Lemon, 75 Tai Llwyd Road, Neath, SA10 7DY
MI0 CIB Peter Bell, 1 Knockbracken Drive, Coleraine, BT52 1WN
M0 CIC Roy Partington, 47 Sandmoor Road, New Marske, Redcar, TS11 8DJ
M0 CIE Michael Coles, 1 Church Street, Taunton, TA1 3JE
M0 CIF D Rosewarn, 16 Charles Crescent, Taunton, TA1 2XN
M0 CIH John Jones, 14 Heol Tywysog, Pentre Halkyn, Holywell, CH8 8HA
M0 CIK David Cox, Seyehethen, Barr Road, Dumfries, DG4 6JZ
M0 CIO R Wyatt, 8 Millbrook Road, Bushey, WD23 2BU
M0 CIP L Thompson, 9 Elmwood Drive, Ponteland, Newcastle upon Tyne, NE20 9QQ
M0 CIR Clive Ryalls, 15 Belmont Way, South Elmsall, Pontefract, WF9 2BT
MW0 CIS A Bray, Coedcelyn, Talley, Llandeilo, SA19 7YR
MW0 CIT V Bray, Coedcelyn, Talley, Llandeilo, SA19 7YR
M0 CIW Howard Delafield, 205 South Avenue, Abingdon, OX14 1QU
M0 CIY I Seabrook, 44 Village Centre, Richmond Letcombe Centre, Letcombe Regis, OX12 9RG
MW0 CJB D Newton-Goverd, 2 Blaen Y Morfa, Morfa, Llanelli, SA15 2BG
M0 CJC Geoffrey Fowle, 12 Lytham Road, Broadstone, BH18 8JS
M0 CJD C Rhodes, Wayside, Hardstoft, Chesterfield, S45 8AH
M0 CJE N Murphy, 9 Eliot Gardens, Newquay, TR7 2QE
MM0 CJF James Smith, 41 Dickie Drive, Peterhead, AB42 1HB
M0 CJG D Jesinger, C/O B Jesinger, 29 Breakspeare Close, Watford, WD24 6DA
MM0 CJH Norman Fowler, 59 Milnefield Avenue, Elgin, IV30 6EJ
M0 CJI R Turner, 53 Queens Drive, Sandbach, CW11 1BN
M0 CJJ R Brown, 1 Octavian Close, Hatch Warren, Basingstoke, RG22 4TY
M0 CJK S coulthard, 90 Rochester Crescent, Crewe, CW1 5YQ
M0 CJM N Toombes, 46 The Vale, Oakley, Basingstoke, RG23 7LD
M0 CJN J Gilmore, 30 Fairhaven Road, Caversfield, Bicester, OX27 8TU
M0 CJO A Kay, Pear Tree Cottage, Hale House Lane, Farnham, GU10 2JG
M0 CJR N Porter, 27 Severn Road, Aveley, South Ockendon, RM15 4NR
M0 CJS M Elliott, 32 Kingsbridge Road, Nuneaton, CV10 0BY
MM0 CJT Alexander Mctaggart, 147/7 Lower Granton Road, Edinburgh, EH5 1EX
M0 CJY Ian James, 56a Ridgeway, Rotherham, S65 3NN
M0 CJZ G Dobson, 12 Ash Grove, Auckley, Doncaster, DN9 3LN
M0 CKA Peter Webb, 42 Holland Road, Ampthill, Bedford, MK45 2RS
M0 CKB K Prior, 74 Walden Way, Frinton-on-Sea, CO13 0BQ
M0 CKC Peter Wakelam, 6 Frankland Road, Durham, DH1 5HZ
M0 CKE James Balls, 7 Rowan Close, Holbeach, Spalding, PE12 7BT
MM0 CKF J Lewis, 9 Cessnock Road, Troon, KA10 6NJ
M0 CKG D Kilburn, Shepherds Cottage, Burradon, Morpeth, NE65 7HF
M0 CKI S Edwards, 5 Chalk Lane, Sutton Bridge, Spalding, PE12 9YF
MM0 CKK A McLuckie, 35 Churchill Avenue, Kilwinning, KA13 7JN
M0 CKL Gordon Sell, 135 Northfields, Norwich, NR4 7ET
M0 CKM K Moro, 18 Douglas Close, Ford, Arundel, BN18 0TG
M0 CKO S Westall, 4 South View, Great Harwood, Blackburn, BB6 7NL
M0 CKP David Warner, 6 Desborough Road, Rushton, Kettering, NN14 1RG
M0 CKS R Cockings, 4 Freeman Gardens, High Green, Sheffield, S35 4NT
M0 CKU W Melsenbach, 5 Byland Court, Whitby, YO21 1JJ
M0 CKV R Wiltshire, 30 Cearns Road, Oxton, Prenton, CH43 2JP
M0 CKX Frederick Graseley, 11 Wilberforce Road, South Anston, Sheffield, S25 5EG
M0 CLB M Youlden, Coed Llai, Lon St. Ffraid, Holyhead, LL65 2YR
M0 CLD I Arnold, 4 Larkspur Close, Tanfield Lea, Stanley, DH9 9UH
M0 CLE A Pascoe, 53 Priory, Bovey Tracey, Newton Abbot, TQ13 9HP
M0 CLG Graeme Cundry, 101 Stoneleigh Avenue, Worcester Park, KT4 0YA
M0 CLH T Martin, 40 Palmers, Wantage, OX12 7HB
M0 CLI E Donaghy, Mendips, 36 Attwood Road, Salisbury, SP1 3PR
M0 CLJ John Townsend, 28 West Street, Darfield, Barnsley, S73 9NF
M0 CLK P Fitzpatrick, 16 Lichens Crescent, Oldham, OL8 2NS
M0 CLL J Nixon, 25 Grove Court, Alsager, Stoke-on-Trent, ST7 2DS
M0 CLM M Bromley, Chartley, Norton Lea, Warwick, CV35 8JX
M0 CLN Paul Tose, 5 California Terrace, Whitby, YO22 4EE
M0 CLO K Farthing, 86 Coldnailhurst Avenue, Braintree, CM7 5PY
MI0 CLP M Hunter, 11a Maydown Road, Drumsollen, Armagh, BT61 8BU
M0 CLR W Huddleston, 17 Moorside Road, Brookhouse, Lancaster, LA2 9PJ
MW0 CLT T Jones, Pen-Y-Bryn, Pen-Y-Lan, Swansea, SA4 3LJ
MW0 CLU Leslie Crompton, 8 Morgans Terrace, Pontrhydyfen, Port Talbot, SA12 9TD
M0 CMA [illegible]
M0 CMC M Veary, 4a Trident Road, Watford, WD25 7AN
M0 CME David Viney, 5 Waters Edge, Bognor Regis, PO21 4AW
M0 CMF T Beaumont, Po Box 109, Camberley, GU15 4ZF
M0 CMH Martin Hemmings, 117 Kingston Hill Avenue, Romford, RM6 5QP
M0 CMI D Lane, 230 Raeburn Avenue, Eastham, Wirral, CH62 8BB
M0 CMK L Taylor, 56 Fenland Village, Osborne Road, Wisbech, PE13 3JR
M0 CMN Barry Pearce, 12 Bean Avenue, Bracebridge, Worksop, S80 2EW
MM0 CMO E Skea, Craigard, Craigton, Inverness, IV1 3YG
M0 CMP J Miller, 3 Surbiton Road, Camberley, GU15 4BW
M0 CMQ E Brendish, Flat 1, 19 Blake Hall Road, London, E11 2QQ
M0 CMS J Lewis, 3 Jacobs Close, Stantonbury, Milton Keynes, MK14 6EJ
M0 CMT J Donald, 67 Cradley, Widnes, WA8 7PL
M0 CMW John Hudson, 3 Vogan Avenue, Crosby, Liverpool, L23 0SG
M0 CMZ I Okanoue, 142-1 Yoshida, Okoh-Cho, Kochi, Japan, 783-0045
MW0 CNA M Evans, 322 Heol Gwyrosydd, Penlan, Swansea, SA5 7BR
MW0 CNB B Polgreen, 344 Heol Gwyrosydd, Penlan, Swansea, SA5 7BP
MW0 CNC Terence Symons, Brynchwyth Farm, Fairyland Road, Neath, SA11 3QE
MW0 CND Martin Davies, 10 Torrington Road, Gendros, Swansea, SA5 8DU
M0 CNE P Black, 43 Malvern Avenue, Rugby, CV22 5JN
MM0 CNF J Clark, 16 Bonnyton Avenue, Drongan, Ayr, KA6 7DG
M0 CNG David Pearce, 2 Mell Avenue, Hoyland, Barnsley, S74 9HF
M0 CNH B Cockerill, 4 Foxglove Close, Rugby, CV23 0TS
MI0 CNI W Hamilton-Sturdy, 319 Old Glenarm Road, Larne, BT40 1TU
M0 CNL P Glover, 29 Ferndale Close, Clacton-on-Sea, CO15 4TP
M0 CNM Andrew Williams, 18 Sturdee Close, Thetford, IP24 2LF
M0 CNN S McNally, 14 Cornelius Drive, Pensby, Wirral, CH61 9PR
M0 CNP David Edwards, 3 Murton Close, Burwell, Cambridge, CB25 0DT
M0 CNU Thomas Norman, 113 Dracaena Avenue, Falmouth, TR11 2ER
MM0 CNV W Aitken, 63 Newlands Road, Grangemouth, FK3 8NT
M0 CNW H Martin, 23 St. Marks Road, Gorefield, Wisbech, PE13 4QQ
M0 CNX P Frampton, 118 Ramnoth Road, Wisbech, PE13 2JD
M0 CNY S Hayes, 36 Mayfield Road, Chaddesden, Derby, DE21 6FW
M0 CNZ P Enrico, Trengrouse House, Polhorman Lane, Helston, TR12 7JD
MW0 COB Royston Cobb, 107 High Street, Neyland, Milford Haven, SA73 1TR
M0 COC R Dean, 10 Livingstone Road, Ellesmere Port, CH65 2BE
MW0 COD C Queeley, 63 Tonna Road, Caerau, Maesteg, CF34 0RU
MW0 COE R Thomas, 11 Heol-y-Parc, North Cornelly, Bridgend, CF33 4LT
MW0 COF Linda Thomas, 11 Heol-y-Parc, North Cornelly, Bridgend, CF33 4LT
M0 COI N Burgess, 12 Glenfield Road, Grimsby, DN37 9EE
M0 COJ D Clenshaw, 4 Spring Meadow, Glemsford, Sudbury, CO10 7PN
M0 COM David Fryer, 16 Elston Place, Aldershot, GU124HY
M0 CON J Bell, 5 Burntwood Close, London, SW18 3JU
M0 COO M Goulbourne, 9 Hawksworth Close, Liverpool, L37 7EX
M0 COP Peter Wesley, Stalden, Ludlow Road, Church Stretton, SY6 6RB
M0 COQ Chris Cane, 69 Stoughton Drive North, Leicester, LE5 5UD
M0 COT John Pears, 8 The Hawthorns, Ellesmere, SY12 9ER
M0 COV Victor Fairhurst, 44 Harold Road, Coventry, CV2 5LG
MW0 COZ J Gearey, 46 Priory Street, Carmarthen, SA31 1NN
M0 CPB D tucker, 137 Seaford Road, London, W13 9HS
M0 CPC C Cook, Hilbry Cottage, Douglas Road, Crowborough, TN6 3QT
M0 CPD P Jennings, 5 Pantydwr, Nantybwch, Tredegar, NP22 3RZ
M0 CPE Herman Broyles, 16 Clifton Road, Shefford, SG17 5AE
M0 CPF Kevin Payne, 20 Laburnum Road, Exeter, EX2 6EG
M0 CPK A Childs, 5 Barnes Wallis Drive, Leegomery, Telford, TF1 6XT
M0 CPL W Harrison, 2 Mount House Road, Formby, Liverpool, L37 3LB
M0 CPN M Watkins, Highwinds, Bryn Pydew, Llandudno Junction, LL31 9QF
MM0 CPS COCKENZIE & PORT SETON ARC c/o Robert Glasgow, 7 Castle Terrace, Port Seton, Prestonpans, EH32 0EE
M0 CPT P Smith, 3 Glenfield Square, Farnworth, Bolton, BL4 7TG
M0 CPU N Bartlett, 12 Woodpecker Close, Verwood, BH31 6JY
M0 CPW Paul Thompson, 73 Erlstoke Close, Plymouth, PL6 5QN
M0 CQD Ivan Booley, 4 Furness Avenue, Littleborough, OL15 9HU
M0 CQF Clive Warhurst, 26 Cherry Tree Close, Ilkeston, DE7 4HQ
M0 CQH I Hirst, 30 Lincoln Road, Skellingthorpe, Lincoln, LN6 5UU
M0 CQN N Whitton, 157 Adeyfield Road, Hemel Hempstead, HP2 5JZ
M0 CQQ D Newman, 89 Sea Place, Goring-by-Sea, Worthing, BN12 4BH
M0 CQR D CONDE, 13 Broxton Road, Wrexham, LL13 9BA
MM0 CQT S Forsyth, West Park, Innes Road, Fochabers, IV32 7NL
M0 CQV J Hubbard, 99 Tuckers Road, Loughborough, LE11 2PH
M0 CQW J Drummond, 60 Park Lane, Exeter, EX4 9HP
M0 CRA Michael Clarke, Gresford, 58 Mold Road, Deeside, CH5 4QN
M0 CRD R Mayne, 17 School Street, Cottingley, Bingley, BD16 1QP
M0 CRG CALDERDALE RAYNER ARG c/o A Baines, 60 Norton Drive, Halifax, HX2 7RB
M0 CRH A Roberts, 23 Church St., Great Harwood, Blackburn, BB6 7NF
MW0 CRI D Marston, Rosario, High Street, Cardigan, SA43 3EF
M0 CRJ R Jones, 8 Downing Avenue, Newcastle, ST5 0JY
M0 CRM Norman Williams, 1 Dorset Close, Whitehaven, CA28 8JP
M0 CRN S Corbett, 80 Helmsdale Lane, Great Sankey, Warrington, WA5 1SY
M0 CRO G Mansell, 87 Bifield Road, Stockwood, Bristol, BS14 8TT
M0 CRP A Harradine, 4 Hesketh Drive, Lostock Gralam, Northwich, CW9 7QJ
MI0 CRQ Kevin McAuley, Layde View, 19 Rathlin Avenue, Ballycastle, BT54 6DQ
MI0 CRR Paul Quinn, 11 Blackpark Road, Ballyvoy, Ballycastle, BT54 6QZ
M0 CRU James Bradley, 46 Brunswick Park Road, Wednesbury, WS10 9HH
M0 CRW John Roebuck, 2 Royston Close, Walton, Chesterfield, S42 7NE
M0 CRY R Scott, 198 Slade Green Road, Erith, DA8 2JG
M0 CRZ S Crabtree, 107 Rochdale Road, Shaw, Oldham, OL2 7JT
M0 CSB M Tait, 2 Wilsford Close, Walsall, WS4 1QP
M0 CSC C Osborn, Lowfield, Station Road, Usk, NP15 2EP
M0 CSD Thomas Quinn, 11 Meadowfield Stokesley, Middlesbrough, TS9 5EL
M0 CSE John Burdett, 27 Hayfield Road, Woolston, Warrington, WA1 4PE
M0 CSF A Prandoczky, 18 Kestrel Court, Birtley, Chester le Street, DH3 2PT
M0 CSK R Roberts, 24 Ffordd Pentre, Johnstown, Wrexham, LL14 1PP
M0 CSO C McKenzie, 15 Belmont Drive, Saltney Ferry, Chester, CH4 0AL
M0 CSP E Watson, 49 Beechfield Road, Bolton, BL1 6HZ
M0 CSQ R Bates, 51 Boyton Road, Ipswich, IP3 9PD

**IMPORTANT NOTE**

**Revalidate licence to avoid revocation** – Ofcom has advised the Society that plans will be drawn up to revoke licences that have not been revalidated as required by the licence conditions. The quickest way to revalidate is to do so online via the Ofcom website: *https://services.ofcom.org.uk/* or by email: *amateur.validations@ofcom.org.uk* Ofcom staff are available to help, but please be patient during times of heavy workload.

M0 CSR J Gardiner, 18 Granville Terrace, Guiseley, Leeds, LS20 9DY
M0 CST A Holbrook, 6 Birch Tree Way, Maidstone, ME15 7RR
M0 CSU M Deacon, 26 Brecon Chase, Minster on Sea, Sheerness, ME12 2HX
M0 CSV Robert Tavener, 22 St. Michaels Drive, Roxwell, Chelmsford, CM1 4NU
M0 CSZ S Clarey, 36 Birchin Bank, Elsecar, Barnsley, S74 8DP
M0 CTC P Ridgeon, 10 Warwick Close, Amberstone, Hailsham, BN27 1NS
M0 CTF D Kaye, 20 Tryfan Close, Redbridge, Ilford, IG4 5JX
M0 CTI B Johnson, 268 Badsley Moor Lane, Rotherham, S65 2QP
M0 CTJ G Tepper, 5 Herbert Street, Mexborough, S64 0JZ
M0 CTK M Stocking, 125 Bassnage Road, Halesowen, B63 4HD
M0 CTL Stephen Ball, 9 Willeton Street, Stoke-on-Trent, ST2 9JA
M0 CTM Ray Pearson, 21 West Parade, Spalding, PE11 1HD
M0 CTN D Haughton, 31 Holmes Carr Road, New Rossington, Doncaster, DN11 0QF
M0 CTP G Hyde, 21 Hawling Road, Market Weighton, York, YO43 3JR
M0 CTQ C Greaves, 2 Attlee Avenue, New Rossington, Doncaster, DN11 0QX
M0 CTR Andrew Smith, 25 Hill Corner Road, Chippenham, SN15 1DW
MM0 CTT D Stewart, 45 Kilwinning Road, Irvine, KA12 8RZ
MM0 CTU C Stewart, 45 Kilwinning Road, Irvine, KA12 8RZ
MW0 CTX B Jones, 87 Heol Llanelli, Pontyates, Llanelli, SA15 5UB
MW0 CUA D Herbert, 50 Clos Cilsaig, Dafen, Llanelli, SA14 8QU
M0 CUD S Marsh, 6 Mayland Drive, Streetly, Sutton Coldfield, B74 2DG
M0 CUF S Hall, 28 Pugneys Road, Wakefield, WF2 7JT
MM0 CUG Gary Grant, 11 Auchriny Circle, Bucksburn, Aberdeen, AB21 9JJ
M0 CUH D Coisson, 21 Medalls Path, Stevenage, SG2 9DX
M0 CUI K Lucas, 7 Wales Road, Kiveton Park, Sheffield, S26 6RA
M0 CUK Mark Slade, 7 Glebe Field, Chaddleworth, Newbury, RG20 7EZ
M0 CUL M Stevens, 67 New Road, East Hagbourne, Didcot, OX11 9JX
MI0 CUN Paul Alexander, 59a Lismurn Park, Ahoghill, Ballymena, BT42 1JW
M0 CUP M Phillips, Roseneath, 4c Valley Road, Kenley, CR8 5DG
M0 CUQ G Cooke, 37 Hertford Close, Woolston, Warrington, WA1 4EZ
M0 CUS G mack, 1085 Evesham Road, Astwood Bank, Redditch, B96 6EB
M0 CUT S Cruise, 18 Morris Court Close, Bapchild, Sittingbourne, ME9 9PL
M0 CUU Colin Wardell, 8 Lower End, Bricklehampton, Pershore, WR10 3HL
M0 CUY S Odell, 41 Pevensey Park Road, Westham, Pevensey, BN24 5HW
M0 CVA Victor Dewey, 100d Bromley High Street, London, E3 3EG
M0 CVB J Sherbourne, 4 Chelston Terrace, Chelston, Wellington, TA21 9HT
M0 CVC C Blackburne, 1 Watson Road, Leeds, LS14 6AE
M0 CVG B Watmough, 28 Aspin Oval, Knaresborough, HG5 8EL
MM0 CVH I Cowie, 23 Shetland Walk, Aberdeen, AB16 6WD
M0 CVJ C Dale, The Common, Alsager, Stoke-on-Trent, ST7 2TQ
M0 CVK Robert Henshall, 19 Townson Road, Ashmore Park, Wolverhampton, WV11 2PP
M0 CVO Nigel Booth, 2 East Street, Grantham, NG31 6QW
M0 CVP Bruce Sutherland, 9 Park Drive South, Hoole, Chester, CH2 3JT
M0 CVR P Gurney, Perhams Green, Plymtree, Cullompton, EX15 2LW
M0 CVS Frederick Sadler, 12 Yokecliffe Drive, Wirksworth, Matlock, DE4 4EX
M0 CVT R Evans, The Brae, Coed-Cae-Ddu Road, Blackwood, NP12 2DA
M0 CVU Alan Forrest, 2 Otterburn Grove, Blyth, NE24 4QP
M0 CVW P Wright, 145 Park Avenue, Bryn-y-Baal, Mold, CH7 6TR
M0 CVZ Donovan Haynes, 113 The Glade, Croydon, CR0 7QP
MM0 CWB J Benson, 15 Hawkhill Place, Stevenston, KA20 4HN
M0 CWC Robin Lawrence, Ukutaba, Burton Road, Wareham, BH20 6EY
MW0 CWF Stephen Beer, 4 Churchfields, Barry, CF63 1FP
MM0 CWI C Mcgowan, 1 Hatton Lodge, Hatton Farm Road, Peterhead, AB42 0LN
MM0 CWJ James Cameron, 407 Smerclate, Isle of South Uist, HS8 5TU
M0 CWN J Thorne, Willow Lodge, The Street, Bury St. Edmunds, IP29 5AP
MW0 CWS B Chapman, Gwyddfan, Mountain Road, Kidwelly, SA17 4EY
M0 CWT Michael Kozakowski, 18 Town Barn Road, Crawley, RH11 7EB
M0 CWX A Morris, 133 Shuttlewood Road, Bolsover, Chesterfield, S44 6NX
M0 CWY G Andrews, 18 Vyne Road, Sherborne St. John, Basingstoke, RG24 9HX
M0 CWZ Kevin Summers, 30 David Street, Kirkby-in-Ashfield, Nottingham, NG17 7JW
MM0 CXA Andrew Burns, 52 Threewells Drive, Forfar, DD8 1EP
MM0 CXB K Gillen, Flat 1/R, 12 Fergus Drive, Glasgow, G20 6AG
MI0 CXE G Rea, 50 Culrevog Road, Dungannon, BT71 7PY
MW0 CXH P Evans, 35 Trinity Road, Llanelli, SA15 2AB
M0 CXL D Findlay, 78 South View Road, Bradford, BD4 6PJ
M0 CXO A Preece, 1 Springfield Close, Thirsk, YO7 1FH
M0 CXQ C Zdziech, 200 Kensington Street, Rochdale, OL11 1QS
MW0 CXW I Gray, Maesyderi, Pontycleifion, Cardigan, SA43 1DR
M0 CXY E Threadingham, 2 Cullum Close, Chichester, PO19 6GG
MM0 CXZ V Madden, 38 Hunters Avenue, Dumbarton, G82 2RZ
M0 CYB READING UNIVERSITY ARC c/o Thomas Cannon, 35 Loddon Bridge Road, Woodley, Reading, RG5 4AP
M0 CYD Simon Bourne, 1 Humewood Grove, Stockton-on-Tees, TS20 1JU
M0 CYE J Taylor, 46 Lever House Lane, Leyland, PR25 4XL
M0 CYF S Austen-Jones, 8 Kent Road, Fleet, GU51 3AH
M0 CYG P Davies, Devildchies, West Bay Road, Bridport, DT6 4EH
M0 CYJ Denis Nicole, 53 Cobbett Road, Southampton, SO18 1HJ
M0 CYM E Tewsley, 20 Crookhorn Lane, Waterlooville, PO7 5QF
MM0 CYR P Palin, Nampara, Roadside, Thurso, KW14 8SR
M0 CYT Stephen Warrillow, PO Box 1455, Camberwell East, Victoria, Australia, 3126
M0 CYU D Holdroyd, 2 Vicarage Lane, Naburn, York, YO19 4RS
M0 CYX G Yoxall, 33 Glasgow Road, Southsea, PO4 8HR
M0 CZA Alan Moss, Lime Kiln Basin, Whitebridge Estate, Stone, ST15 8LQ
M0 CZB John Webber, 5 Leda Mews, Achilles Close, Hemel Hempstead, HP2 5WR
M0 CZC M Wade, 42 New Road, Burnham-on-Crouch, CM0 8EH
M0 CZE M Garton, 13 Damaskfield, Worcester, WR4 0HY
MI0 CZF William Belshaw, 9 Ashmount Gardens, Lisburn, BT27 5BZ
MM0 CZH David Mitchell, 3 Lade Crescent, Bucksburn, Aberdeen, AB21 9HJ
MM0 CZK Charles Rogers, 18 Primrose Lane, Rosyth, Dunfermline, KY11 2SL
MM0 CZM Archibald Stewart, G/1 290 Dumbarton Road, Old Kilpatrick, Glasgow, G60 5LJ
M0 CZN W Martin, 17 Dickens Road, Maidstone, ME14 2QW
M0 CZP T Ostley, 30 Ashley Way Brighstone, Newport, PO30 4HH
M0 CZQ J Chaldecott, 20 Haynes Avenue, Poole, BH15 2ED
M0 CZR B Richtering, 20 Shaftesbury Avenue, Hornsea, HU18 1LX
M0 CZT D Ward, 42 Fern Grove, Cherry Willingham, Lincoln, LN3 4BG
M0 CZU Don Gregson, 15 Alice Street, Oswaldtwistle, Accrington, BB5 3BL
M0 CZX F Boele, 301 Cell Barnes Lane, St. Albans, AL1 5QB
MM0 DAA Stuart Mackie, Kierycraigs Lodge, Blairadam, Kelty, KY4 0JF
M0 DAB D Bowles, 23 Broughton Way, Rickmansworth, WD3 8GW
M0 DAC D Tanner, 55 Arundel Drive, Bramcote, Nottingham, NG9 3FN
M0 DAD D Roper, 84 Tynedale Drive, Cowpen, Blyth, NE24 4DS
M0 DAE M Haladij, 50 Liberty Drive, Duston, Northampton, NN5 6TU
M0 DAG Dean Godden, 94 The Common, South Normanton, Alfreton, DE55 2EP
M0 DAH S Brown, 4 Foundry Lane, Manchester, M4 5LB
M0 DAL A Pounder, 6 Barbondale Grove, Knaresborough, HG5 0DX
M0 DAN D Black, 8 Cornwood Close, Finchley, London, N2 0HP
MW0 DAR P Drew, 14 Longfield Court, Hirwaun, Aberdare, CF44 9NG
M0 DAS A Cockburn, Grangewood, Blenheim Road, New Romney, TN28 8RD
MM0 DAT David Thain, 20 Spey Street, Fochabers, IV32 7EH
M0 DAW D Wilcox, Medlar Cottage, Faringdon Road, Swindon, SN6 8AJ
M0 DAX C Evans, Y Rhosfa, High Street, Llanfyllin, SY22 5AF
M0 DAY B Aizlewood, Stoner Rise, Stoner Hill, Petersfield, GU32 1AG
M0 DAZ D Drake, 38 Frinton Road, Nottingham, NG8 6GQ
M0 DBA Malcolm Brown, 2 Beacon Road, Marazion, TR17 0HF
M0 DBB J Hayhurst, 3 Bryn Trewan, Caergeiliog, Holyhead, LL65 3LS
MM0 DBC D Brown, 10 Culmore Place, Falkirk, FK1 2RP
M0 DBD L Dean, 24 Spennithorne Avenue, Leeds, LS16 6JA
MM0 DBF W Callanan, 3 Eden Place, Aberdeen, AB25 2YF
M0 DBG Geoffrey Hodges, 12 Linwal Avenue, Houghton-on-the-Hill, Leicester, LE7 9HD
M0 DBH Stephen Palik, 10 Clare Road, Northborough, Peterborough, PE6 9DN
M0 DBI A Wylie, 15 Worcester Gardens, Greenford, UB6 0BH
M0 DBJ O Peters, 3 Churchill Close, Sutton, Ely, CB6 2QF
MI0 DBK A Quinn, 11 Derrynaught Road, Collone, Armagh, BT60 1LZ
M0 DBM Clive Bloxidge, 33 Rosemary Hill Road, Sutton Coldfield, B74 4HL
M0 DBO N Head, 12 Heston Walk, Redhill, RH1 5JB
MM0 DBR Alan Butler, 68 Laws Road, Aberdeen, AB12 5LJ
M0 DBT C Gregson, 11 Coupe Green, Hoghton, Preston, PR5 0JR
MW0 DBV Brian Davies, 2a Berwick Road, Bynea, Llanelli, SA14 9SS
M0 DBX Richard Batten, 118 Marryat Road, New Milton, BH25 5JF
M0 DBY A Burton, Normanby View, Otby Lane, Market Rasen, LN8 3UT
M0 DCB David Brown, 7 Limber Hill, Cheltenham, GL50 4RJ
MM0 DCC R Rutherford, 3 Stevenson Street, Oban, PA34 5NA
M0 DCD Andrew Ripley, 11 Gallows Hill Drive, Ripon, HG4 1UP
M0 DCG Thomas Spence, 38 Burtonwood Road, Great Sankey, Warrington, WA5 3AJ
M0 DCH Derek Hyde, 13 Rainwall Court, Sutterton, Boston, PE20 2EG
M0 DCM David Maydew, 128 Thorne Road, Willenhall, WV13 1AW
M0 DCO B Tutty, 14 Nursery Walk, Canterbury, CT2 7TF
M0 DCP Yusuke Ochiai, 1-8-7 Tamagawa Den-en-chofu, SETAGAYA-KU, Tokyo, Japan, 1580085
MW0 DCQ W Ashton, 44 Bryn Awel, Bettws, Bridgend, CF32 8SA
M0 DCS David Taylor, Garth Farm, Hull Road, Selby, YO8 6NH
MW0 DCT Paul Grace, 48 Linden Avenue, West Cross, Swansea, SA3 5LA
M0 DCU Simon Quantrill, 7 Clare Road, Kessingland, Lowestoft, NR33 7PS
M0 DCV Peter Howell, 18 High St., Foxton, Cambridge, CB2 6SP
M0 DCW D Eggleton, 79 Hazel Close, Twickenham, TW2 7NP
M0 DCY G Tarr, 40 The Garth, Coniston, LA21 8EQ
M0 DCZ J Farrer, 3 Pitt Garth, Haggs Lane, Grange-over-Sands, LA11 6PH
M0 DDA B Waterloo, 55 Solent Road, Hill Head, Fareham, PO14 3LB
M0 DDB C Parr, 13 Peartree Avenue, Southampton, SO19 7JN
M0 DDC Alan Copperwaite, 71 Gladbeck Way, Enfield, EN2 7EL
M0 DDE Andrew Bullock, 2 Pen y Geulan, Revel, Welshpool, SY21 8AH
M0 DDI P Cassidy, 8 Sheldrake Road, Newark, NG24 2JX
M0 DDK Robert Milne, 8 Connaught Walk, Rayleigh, SS6 8UY
M0 DDT Colin Potter, 12 Beech Road, Headington, Oxford, OX3 7RR
M0 DDU L Shepherd, 334 Copnor Road, Portsmouth, PO3 5EL
M0 DDV M Watts, 51 Chester Road, Sidcup, DA15 8RX
MI0 DDW S Cooper, 31 Kilbroney Valley, Rostrevor, Newry, BT34 3SR
M0 DDY J Todd, 108 Clee Road, Grimsby, DN32 8NX
M0 DEA Nestor Jacovides, 2 Dionysou Street, Nicosia, Cyprus, 2123
M0 DEB 759 SQN(BECCLES)ATC c/o E Lugmayer, 17 Borough End, Beccles, NR34 9YW
MM0 DEC I Street, 5 Calder Road, Bellsquarry, Livingston, EH54 9AA
M0 DEF Gary Thacker, 4 Duffy Place, Rugby, CV21 4EF
M0 DEI AEROVENTURE ARS c/o Roy Liversidge, 17 Millbank Close, High Green, Sheffield, S35 4NS
M0 DEJ A Neumann, 5 Denfield Avenue, Halifax, HX3 5NL
M0 DEK D Dyde, 10 Essex Close, Worcester, WR2 5RW
M0 DEL Andrew Lindley, 2 Mickle Hill Farm Cottage, Mickle Hill Road, Blackhall Colliery, Hartlepool, TS27 4DF
M0 DEN Dennis Miller, 9 Brenden Avenue, Somercotes, Alfreton, DE55 4JD
M0 DEO C Wheldon, 39 Felton Avenue, South Shields, NE34 6RY
M0 DEP Russell Speed, 48 Stony Lane, Burton, Christchurch, BH23 7LE
M0 DEQ Michael Clarke, 21 Sycamore Road, Greenstead Estate, Colchester, CO4 3NF
M0 DER R Hannigan, 4 Westlands Avenue, Tetney, Grimsby, DN36 5LP
M0 DES Richard Beck, 7 Swansley Lane, Lower Cambourne, Cambridge, CB23 6ER
M0 DEV M Twells, Camels, Annscroft, Shrewsbury, SY5 8AN
M0 DEW D Walters, 17 Heol Islwyn, Llanrhystud, SY23 5BW
M0 DEX John Constable, 438 Old Road, Clacton-on-Sea, CO15 3SB
M0 DEY Helen Watt, 40 Long Wood Road, Bristol, BS16 1FD
M0 DFA David Crake, Kentolop, Holyhead Road, Shrewsbury, SY4 1EE
M0 DFD Stephen Sparkes, Flat 1, 14 Hall Road, Wilmslow, SK9 5BN
M0 DFF Mark Bywater, 16 Grove Road, Cromer, NR27 0BY
M0 DFH C Miles, 26 Meadowside, Grindleton, Clitheroe, BB7 4RR
M0 DFL C Houghton, 20 St. Peters Way, Thurston, Bury St. Edmunds, IP31 3RZ
MW0 DFN D Thomas, 48 Gilbert Road, Llanelli, SA15 3RA
MI0 DFO Raymond Kennedy, 83 Craigstown Road, Randalstown, Antrim, BT41 2PN
M0 DFQ Paul Dickman, 1 Old Hall Close, Henley, Ipswich, IP6 0RJ
M0 DFW David Wright, 61 Potton Road, St. Neots, PE19 2NN
M0 DFX D Riley, 9 Century Avenue, Mansfield, NG18 5EE
M0 DFY M Watts, 14 Beverley Close, Lowestoft, NR33 8QQ
MM0 DFZ A Fraser, 29 Seafield Gardens, Aberdeen, AB15 7YB
M0 DGA Andrew Bevins, 12 Wheatstone Road, Formby, Liverpool, L37 6BF
M0 DGB D Balharrie, 27 Norfolk Road, Uxbridge, UB8 1BL
MM0 DGI Steven Spence, Halley, Deerness, Orkney, KW17 2QL
M0 DGJ G Hannam, 26 Cornflower Way, Melksham, SN12 7SW
M0 DGK Mark Luby, 19 Robin Lane Bentham, Lancaster, LA2 7AB
M0 DGQ Barry Zarucki, 26 Heathfield Road, Kings Heath, Birmingham, B14 7DB
MM0 DGR SCOTTISH-RUSSIAN ARS c/o Jurij Phunkner, 7 Plenshin Court, Glasgow, G53 6QW
M0 DGT P Kelly, 3 Orchard Lea Close, Woking, GU22 8QW
M0 DGU C Thompson, 80 Aston Road, Willerby, Hull, HU10 6SG
MI0 DGX N McCully, 12 Cargygray Road, Hillsborough, BT26 6BL
M0 DHE J Coxon, Links View Farm, Fairy Lane, Sale, M33 2JT
MW0 DHF Philip King, 11 Lord Street, Penarth, CF64 1DD
M0 DHI B Phillips, 79 Leen Valley Drive, Shirebrook, Mansfield, NG20 8BJ
M0 DHM E Dean, 26 Silverdale, Maidstone, ME16 9JG
M0 DHN G Kirkpatrick, 23 Hornhatch, Chilworth, Guildford, GU4 8AY
M0 DHO D Honey, Bluebell Cottage, Crondall Road, Fleet, GU51 5SU
M0 DHP R Benitez, 4 Raphael Drive, Thames Ditton, KT7 0BL
MM0 DHQ A Clark, 20 Church Street, Kilwinning, KA13 6BE
M0 DHU Mario Tachibana, 81334076949, Tokyo, Japan, 1070062
M0 DHX Howard Wood, 402 Chemin De Peyrebelle, Valbonne, France, 6560
MM0 DHY A Hart, Tigh na Coille, Daviot, Inverness, IV2 5EP
M0 DID SEAREG c/o J Williams, 24 Hilltop Gardens, Denaby Main, Doncaster, DN12 4SB
M0 DIG J O'Mahoney, 36 Scotton Gardens, Catterick Garrison, DL9 4HX
M0 DIJ J Allen, 2 Chichester Walk, Chichester Road, Ramsgate, CT12 6NX
M0 DIL S Lowe, 31 Court Farm Road, Bristol, BS14 0EH
M0 DIN Andrew Bottrill, Woodside, 1 Gilling Close, Richmond, DL10 5AY
M0 DIQ R Mullen, 18 Chandlers Ridge, Nunthorpe, Middlesbrough, TS7 0JL
MM0 DIS I Elder, 24 Birniehill Avenue, Bathgate, EH48 2RR
M0 DIT Joshua Tildesley, 16 Calver Grove, Keighley, BD21 2RX
M0 DIW D Walls, 15 Brant Road, Lincoln, LN5 8RL
M0 DJA Ambrose Garner, The Chestnuts, Surfleet, Spalding, PE11 4BA
M0 DJB A White, 368 Hall Lane, Whitwick, Coalville, LE67 5PF
M0 DJD D Gould, 9 Holly Lane, Barwell, Leicester, LE9 8BT
M0 DJF David Fradley, 70 Springfield Road, Rugeley, WS15 2NH
M0 DJH Dennis Hale, 51 Lynhurst Avenue, Sticklepath, Barnstaple, EX31 2HY
M0 DJI J Ford, 1 Cherry Road, Enfield, EN3 5SE
M0 DJQ B Cushing, 55 Friday Street, Eastbourne, BN23 8AX
M0 DJT D Turner, 3 Manor Park Road, London, N2 0SN
M0 DJW D Westland, 8 Faris Barn Drive, Woodham, Addlestone, KT15 3DZ
M0 DKD D Bains, 2 Arundel Road, Brighton, BN2 5TD
M0 DKJ I Rose, 26 Beckford Road, Cowes, PO31 7SG
M0 DKL E Whittle, 15 Weavers Court, Scorton, Preston, PR3 1NQ
M0 DKN D Turney, 2 Beult Meadow, Cage Lane, Ashford, TN27 8PZ
M0 DKP J Jones, 3 St. Judes Walk, Cheltenham, GL53 7RU
M0 DKS D Kees, 23 Walsingham Way, Billericay, CM12 0YE
M0 DKT D Knott, 80 Melville Court, Chatham, ME4 4XJ
M0 DKU Peter Giles, Fenimora Cottage, Paines Hill, Bicester, OX25 4SQ
M0 DKV D Bowers, Oak House, Church Stile Lane, Exeter, EX5 1HP
M0 DKX J Horsfield, 91 Harlington Road, Mexborough, S64 0DT
M0 DLB S Orange, 20 Borrowdale Avenue, Fleetwood, FY7 7LF
MD0 DLC P Bartlett, 43 Chamberlain Way, Pinner, HA5 2AU
M0 DLE R Tingay, 1 Ullswater Road, Sompting, Lancing, BN15 9UF
M0 DLG J Conway, 27 Victoria Road, Gorleston, Great Yarmouth, NR31 6EF
MM0 DLH A Dunsmore, 21 East Croft, Ratho, Newbridge, EH28 8PD
M0 DLI CHATHAM HOUSE RC c/o John Hislop, 10 Park Wood Close, Broadstairs, CT10 2XN
M0 DLL D Gray, 68 Sixth Cross Road, Twickenham, TW2 5PD
M0 DLM D Mann, Hunters Lodge, Grange Road, Bedford, MK44 3NT
M0 DLP D Birch, 4 Avon Bank Cottages, Avon Bank, Pershore, WR10 3JP
M0 DLR Kenneth Jones, 41 Denison Street, Beeston, Nottingham, NG9 1AY
M0 DLT Barrie Holland, 2 Blythorpe, Hull, HU6 9HG
M0 DLX D Cleal, 3 Mulberry Court, Guildford, GU4 7EQ
M0 DLY C Kerrison, 18 Parks Road, Dunscroft, Doncaster, DN7 4AH
M0 DLZ P Parry, 31 Swinburne Way, Daybrook, Nottingham, NG5 6BX
M0 DMA D Marshall, 75 North Hill Road, Sheffield, S5 8DT
M0 DMB C Simpson, 6 Dalton Close, Driffield, YO25 6YE
M0 DMD Michael Coe, 2 Burnslack Road, Ribbleton, Preston, PR2 6EX
M0 DME Andrew Pell, 5 Gallery Close, Northampton, NN3 5NT
M0 DMF C Price, Our House, The Green, Belper, DE56 2FW
M0 DMI J Enderby, 42 Claremont Avenue, Chorley, PR7 2HL
M0 DMJ Colin Peters, 9 Evelyn Close, Twickenham, TW2 7BL
MM0 DMK D McKenzie, 67 Alexander Drive, Dedridge, Livingston, EH54 6DF
M0 DMR D Rolf, 40 Gunton Drive, Lowestoft, NR32 4QB
M0 DMS Daniel Schofield, 9 Manor View, Shafton, Barnsley, S72 8NQ
MI0 DMT John Hyndman, 4 Larchfield Gardens, Kilrea, Coleraine, BT51 5SB
MM0 DMU J Stewart, 76 Caroline Terrace, Edinburgh, EH12 8QU
MD0 DMV P Baillie-Searle, 2 Marguerite Place, Foxdale, Douglas, Isle of Man, IM4 3HE
M0 DMX B Donnachie, 9 Sir Henry Brackenbury Road, Ashford, TN23 3FJ
M0 DMY C Hicken, 76 Peaksfield Avenue, Grimsby, DN32 9QG
M0 DMZ M Hughes, 88 Church Meadow Road, Rossington, Doncaster, DN11 0YD
MI0 DNB E Hyndman, 4 Larchfield Gardens, Kilrea, Coleraine, BT51 5SB
M0 DND D Brown, 97 Hewett Road, Portsmouth, PO2 0QS
M0 DNF C Owen, 58 Bedwellty Road, Cefn Fforest, Blackwood, NP12 3HB
MM0 DNH B Shippey, 15/6 Goldenacre Terrace, Edinburgh, EH3 5QP
M0 DNJ D Cook, Flat 6, 7 Heath Court, Felixstowe, IP11 0YQ
M0 DNK Robert Law, 12 Gwel y Llan, Llandegfan, Menai Bridge, LL59 5YH
MI0 DNM William Wilson, 68 Ballygowan Park, Banbridge, BT32 3AW
M0 DNN A Wharton, 184 Surbiton Road, Stockton-on-Tees, TS19 7SH
M0 DNO Charles Anderson, Flora, 61 Merrilees Crescent, Clacton-on-Sea, CO15 5XY
M0 DNP D Pettitt, 98 Wheble Drive, Woodley, Reading, RG5 3DU
M0 DNR Robert Dickson, 102 Blakemere Crescent, Portsmouth, PO6 3SH

M0 DNU A Bateman, 3 Church Croft, Bramshall, Uttoxeter, ST14 5DE
M0 DNV A Truman, 92 Spring Meadow, Sutton Hill, Telford, TF7 4AQ
M0 DNW S Dyson, 18 Coniston Road, Chorley, PR7 2JA
[illegible]
M0 DNY Philip Crump, 99a Mayfield Road, Southampton, SO17 3SY
M0 DNZ D Pauley, Charente, Westerfield Road, Ipswich, IP6 9AJ
M0 DOA David Brace, 10 Sawles Road, St. Austell, PL25 4UD
M0 DOB Stuart Dobbs, Firbeck House Farm Cottage, Steetley, Worksop, S80 3EB
M0 DOC J Jones, 16 Laurel Avenue, Darwen, BB3 3AG
M0 DOD T Bodily, Flat 5, Denbeigh House, Rushden, NN10 0AT
M0 DOH R Golsby, 19 Glascote Close, Shirley, Solihull, B90 2TA
M0 DOK A Abrams, The Cottage, Victoria Road, Rushden, NN10 0AS
M0 DOL Chris Darlington, 24 West Ridge, Northampton, NN2 7RA
M0 DOM S Cassidy, 71 Kensington Avenue, Penwortham, Preston, PR1 0EE
M0 DON LEICESTER AMATEUR RADIO SHOW c/o John Theodorson, 7 Kingfisher Court, Overstone Lakes, Ecton Lane, Northampton, NN6 0BD
M0 DOP Allen Hoskins, 38 Tasmania Close, Basingstoke, RG24 9PQ
M0 DOR E Roberts, 6 Trem Y Moelwyn, Tanygrisiau, Blaenau Ffestiniog, LL41 3SS
M0 DOS A Yearp, 29 Humber Road, Ferndown, BH22 8XN
MM0 DOT D MacNaughton, 24 Kepplehills Drive, Bucksburn, Aberdeen, AB21 9PQ
M0 DOW Andrew Holden, 21 East View Meadowfield, Durham, DH7 8RY
M0 DOY R Kendrick, Forest View, 126 Ameysford Road, Ferndown, BH22 9QE
MW0 DOZ Heads of the Valleys ARC c/o Aeronwen Sneddon, 3 Marigold Close, Gurnos, Merthyr Tydfil, CF47 9DA
M0 DPF S Thompson, 49 Audley Place, Sutton, SM2 6RW
MD0 DPG P Taylor, 14 Royal Park, Ramsey, Isle of Man, IM8 3UF
M0 DPH Paul Tullock, 17 Owthorne Walk, Bridlington, YO16 7GB
M0 DPJ B Crawshaw, 112 Waller Road, Sheffield, S6 5DQ
M0 DPQ Robert Meadley, 2 Lower Hollacombe Cottages, Torquay Road, Paignton, TQ3 2DP
M0 DPS C Seager, 77 Stonewood, Bean, Dartford, DA2 8BZ
M0 DPV A Clark, 195 Ivyhouse Road, Dagenham, RM9 5RS
M0 DPW D Wakefield, Rosebank, 109 Spring Road, Stoke-on-Trent, ST3 7JA
M0 DPY K Wilks, 23 Ryemoor Road, Haxby, York, YO32 2GX
M0 DQB J Brown, Ladythorn, Cleeve Hill, Cheltenham, GL52 3QB
M0 DQH G Waugh, 5008 Spartanburg Cove, Austin, Texas, United States, 78730
M0 DQK H Beckett, 5 Chatsworth, Benfleet, SS7 3BB
M0 DQL David Porteus, 2 Olive Grove, Blackpool, FY3 9AS
M0 DQN Roger Richards, Broadsands House, Dartmouth Road, Churston, Brixham, TQ5 0JZ
M0 DQO C Bloy, 7 Eagle Close, Fareham, PO16 8QX
MM0 DQP J Mckay, 28 Lumsden Crescent, St. Andrews, KY16 9NQ
M0 DQZ Anthony Lord, 2 Hazelmere, Diss Road, Diss, IP22 1NQ
MM0 DRA Alistair Cattanach, 8 Auchencairn Place, Monifieth, Dundee, DD5 4TS
M0 DRB F Dawson, Silverthorne, Lower Sea Lane, Bridport, DT6 6LR
M0 DRE J Barnett, 5 Manknell Road, Chesterfield, S41 8LZ
M0 DRF Dermot Falkner, 45 Westwood, Carleton, Skipton, BD23 3DW
M0 DRG Darryl Green, 39 Heritage Park, Hatch Warren, Basingstoke, RG22 4XT
M0 DRI Alan Lyons, 3 Croes Bleddyn Cottages, Itton, Chepstow, NP16 6BN
M0 DRK Derek Carman, 30 Parsonage Close, Burwell, Cambridge, CB25 0ER
M0 DRL D Lane, The Coltmoor, Peterchurch, Hereford, HR2 0SW
M0 DRM ROSE & CROWN RC c/o James Mahoney, 61 Wood Street, Barnsley, S70 1NA
M0 DRN A Small, 28 Capel Street, Capel-le-Ferne, Folkestone, CT18 7LZ
M0 DRO R Bell, Sandgate House, Gough Road, Folkestone, CT20 3BE
M0 DRQ M Munn, 54 Longbeech Park, Canterbury Road, Ashford, TN27 0HA
M0 DRS S Sampathkumar, 52 Crowstone Road, Westcliff-on-Sea, SS0 8BD
MM0 DRT D Taylor, Hillcrest, Woodside Place, Banchory, AB31 5XW
MW0 DRU D Underwood, 891 Heol Y Ffynon, Penrhys, Ferndale, CF43 3RN
M0 DSB D Brunton, 29 Norfolk Road, Wangford, Beccles, NR34 8RE
M0 DSC D Westwood, 60 Selwyn Drive, Stockton-on-Tees, TS19 8XF
M0 DSF Douglas Fenna, 84 High Park Road, Ryde, PO33 1BX
M0 DSI D Shaw, 35 Tinshill Lane, Leeds, LS16 6BU
MM0 DSM Ellie McNeill, 3 Sunnybrae Terrace Maddiston, Falkirk, FK2 0LP
M0 DSN D Tomlinson, 3 Holgate Road, Nottingham, NG2 2EB
M0 DSO R De Ieso, 9 Quilter Meadow, Old Farm Park, Milton Keynes, MK7 8QD
M0 DSR N Passam, 177 Uttoxeter Road, Blythe Bridge, Stoke-on-Trent, ST11 9HQ
M0 DSS David Smith, Church House Farm, Church Terrace, Alnwick, NE66 2YD
MW0 DSV Roland Price, 13 West Street, Pembroke, SA71 4ET
M0 DSW John Wright, Castlemead, Pewsey Road, Marlborough, SN8 1NQ
M0 DSX Dave Mir, 2 Chadhurst Cottages, Coldharbour Lane, Dorking, RH4 3JH
M0 DSY Dylan Bailey, 28c Cliff Road, Dovercourt, Harwich, CO12 3PP
M0 DSZ D Church, Elm Cottage, Pant, Oswestry, SY10 9RB
M0 DTA R Cheverall, 1 Clarkwood Cottages, Twitty Fee, Chelmsford, CM3 4PG
M0 DTB Terence Belton, Flat 6, 28 First Avenue, Hove, BN3 2FF
M0 DTD R Edwards, 59 Allt-yr-yn Close, Newport, NP20 5EE
M0 DTH R Howes, 8 Birbeck Road, Caldicot, NP26 4DX
M0 DTI D Donohoe, 38 Coast Drive, Lydd on Sea, Romney Marsh, TN29 9NL
M0 DTJ John Holloway, 23 Yew Tree Close, 23 Yew Tree Close, Hatfield Peverel, CM3 2SG
M0 DTK Dudley Cox, 45 Victoria Road, Walderslade, Chatham, ME5 9HB
MM0 DTL ABERDEENSHIRE CONTEST GROUP c/o Ian Ross, Idlewild, Kintore, Inverurie, AB51 0XA
MD0 DTQ DANUM SCHOOL RC c/o Daniel Wood, The Hawthorns, Droghadfayle Road, Port Erin, Isle of Man, IM9 6EL
M0 DTS Robert Swinbank, Oxhill Farm, Hilton, Yarm, TS15 9LD
MM0 DTW R Clark, 52 Loons Road, Dundee, DD3 6AQ
MM0 DUN Martin Higgins, 11 Strathyre Place, BROUGHTY FERRY, Dundee, DD5 3WN
M0 DUP S Cox, 63 Netherfield Avenue, Eastbourne, BN23 7BT
M0 DUQ Alan Bridgeland, 17 Oldfield Lane, Wisbech, PE13 2RJ
MM0 DUR M McKay, Tryggo, Sarclet, Wick, KW1 5TU
M0 DUT Jeremy Barley, 2 Little Hobbyvines, Duckend, Stebbing, CM6 3BP
M0 DUU David Allen, 20 Evenley Road, Northampton, NN2 8JR
M0 DUV D Tootill, 17 Newington Way, Craven Arms, SY7 9PS
M0 DUY D Bates, Bold Gate Lodge, Praze, Camborne, TR14 0NQ
MM0 DVB A Paton, 17 Union Terrace, Keith, AB55 5EQ
M0 DVD Paul Barnes, 20 Benbow Drive, South Woodham Ferrers, Chelmsford, CM3 5FP
M0 DVF I Dummer, 29 Chisholm Close, Southampton, SO16 8GU
M0 DVG William Bosworth, 87 Tregorrick Road, Exhall, Coventry, CV7 9FH
MI0 DVH M McLaughlin, 22 Duncreggan Road, Londonderry, BT48 0AD
[illegible]
MW0 DVM David Morris, 14 Church Terrace, Porth, CF63 6EY
M0 DVQ A Yates, 12 Rowsley Avenue, Derby, DE23 6JY
M0 DVR P King, 46 Ranelagh Road, Redhill, RH1 6BJ
M0 DVT J Adlington, 23 Newstead Road, Abbey Hulton, Stoke-on-Trent, ST2 8HU
M0 DVW Gordon Mallinson, 6 Deerplay Court, Bacup, OL13 8GE
MM0 DVZ John Craig, Birchwood Cottage, Birchwood, Dornoch, IV25 3PW
M0 DWB M Tidman, 3 Stannet Way, Wallington, SM6 8BE
M0 DWC D Wraight, 2 Silver Mead, Congresbury, Bristol, BS49 5EX
MI0 DWD D Hamilton, 7 Bolea Park, Limavady, BT49 0SH
MI0 DWE David Eames, Idlewild, Farnamullan, Enniskillen, BT94 5EA
MM0 DWF L Boehme, 26 Sandylands Road, Cupar, KY15 5JS
M0 DWG R Gooch, 14 Cotterill Road, Surbiton, KT6 7UN
M0 DWK Christopher Jewell, 4 Springfield, Bentham, Lancaster, LA2 7BA
M0 DWM D Cartlidge, 2 Walton Road, Walsall, WS9 8HN
M0 DWP D Peters, 5 Riverside, Buntingford, SG9 9HJ
M0 DWQ Roger Spensley, 30 Kilsyth Close, Fearnhead, Warrington, WA2 0SQ
M0 DWR Derek Dukes, 8 The Village, Jacobstowe, Okehampton, EX20 3RF
M0 DWS D Summerwill, 52 Lanmoor Estate, Lanner, Redruth, TR16 6HN
M0 DWT David Todd, 20 Wasdale Close, Plymouth, PL6 8TL
M0 DWU A Fleming, 3 Surrey Avenue, Wirral, CH49 6NL
M0 DWW Marc Giffin, 17 Wolfe Road, Maidstone, ME16 8NX
M0 DWX D McIntosh, 23 Charles Road, Solihull, B91 1TS
M0 DWZ R Webster, 74 Bescar Brow Lane, Scarisbrick, Ormskirk, L40 9QG
MM0 DXC C Stevenson, 8 Carlaverock Grove, Tranent, EH33 2EB
MM0 DXD J Wilson, 20 Ballumbie Gardens, Dundee, DD4 0NR
M0 DXF P Binswanger, 115 Hawthorn Bank, Spalding, PE11 1JQ
MM0 DXH James Hume, 8/11 Leslie Place, Edinburgh, EH4 1NH
M0 DXJ C Humphris, Glebe House, School Lane, Spilsby, PE23 4AU
MM0 DXK RENFREW ARS c/o R Camley, 16 Ferness Oval, Balornock, Glasgow, G21 3SQ
M0 DXM Guenter Pesch, Nikolaus-Jansen-St. 10, Simmerath, Germany, D52152
M0 DXN T Green, 9 Craiglands Park, Ilkley, LS29 8SX
M0 DXP D Poole, 17 London Street, Chertsey, KT16 8AP
M0 DXQ Peter Dougherty, 79 Beverly Road, West Caldwell, New Jersey, United States, 07006-6532
M0 DXR Mark Haynes, 196 The Downs, Harlow, CM20 3RH
M0 DXS David Sheppard, 75 St. Nicholas Road, Littlestone, New Romney, TN28 8QA
M0 DXT William Tinnion, 3 Brayton Road, Aspatria, Wigton, CA7 3DJ
M0 DXV M Whitehead, 29 Coulsons Road, Bristol, BS14 0NN
MW0 DXX S Pettipher, Min Yr Afon, Llandysul, SA44 5AT
M0 DXZ J Arnell, Jersey Farm, Little London, Bishop's Stortford, CM23 1BD
M0 DYA O Haselden, 15 Broadmeadow Close, Totton, Southampton, SO40 8WB
M0 DYB S Kemp, 5 Oakdale Avenue, Frodsham, WA6 6PY
M0 DYG P Swan, 44 Gillingham Road, Gillingham, ME7 4RR
M0 DYH P Lockley, 2 Valley View, Bowling Green, Falmouth, TR11 5AP
M0 DYO M Blair, 20 Hazel Drive, Burn Bridge, Harrogate, HG3 1NY
M0 DYQ P Davis, 9 Chartwell Road, Kirkby-in-Ashfield, Nottingham, NG17 7HB
M0 DYR S Durham, 24 Morgan Close, Yaxley, Peterborough, PE7 3GE
MW0 DYS A Davies, 17 Queens Road, Merthyr Tydfil, CF47 0NB
M0 DYU W Walker, 9 Malthouse Lane, Dorchester-on-Thames, Wallingford, OX10 7LF
M0 DYV R Hitchins, 64 Stratford Road, Salisbury, SP1 3JN
M0 DYW D Donnelly, 11 Anton Street, London, E8 2AD
MM0 DYX Dave Francis, 2 Morlich Crescent, Dalgety Bay, Dunfermline, KY11 9UW
MW0 DYZ R Jeffery, 66 Redhouse Road, Cardiff, CF5 4FH
M0 DZA M Brinnen, 82 Victoria Road, Mablethorpe, LN12 2AJ
M0 DZB Keith Johnson, 16 Lamberts Close, Weasenham, King's Lynn, PE32 2TE
M0 DZC R Barton, 57 Croxteth Drive, Rainford, St. Helens, WA11 8LA
M0 DZD M Smith, 16 Thornton Drive, Brierley Hill, DY5 2BS
M0 DZG B Banks, 1 Worthington Road, Dunstable, LU6 1PN
M0 DZH P Holloway, 87 Buckle Place, Houndstone, Yeovil, BA22 8SG
M0 DZL Michael Fitzpatrick, 3 Orchard Close, Yealmpton, Plymouth, PL8 2JQ
M0 DZM Kieran Enright, Flat 11/A, Cold Springs Farm, Buxton, SK17 6ST
M0 DZO A Hambidge, 25 Lock Drive, Stechford, Birmingham, B33 8AB
M0 DZT R Clark, 10 Clarendon Road, Bournemouth, BH4 8AL
M0 DZV J Maynard, 5 Farm Close, Crowthorne, RG45 6SE
M0 DZW I Massey, 129 Church Street, Milnthorpe, LA7 7DZ
M0 DZX Anthony Nolan, 41 Taylor Street, Rochdale, OL12 0HX
M0 EAB Slawomir Pochojka, 38 West Cotton Close, Northampton, NN4 8BY
M0 EAD John Hart, 1 Cuckoo Nest, Harden, Bingley, BD16 1BD
M0 EAE K Kotarba, 14 North Park, Bristol, BS15 1UW
M0 EAF Richard Astbury, 12 Southall Road, Ashmore Park, Wolverhampton, WV11 2PZ
MM0 EAI D Jamieson, Drumrao, Barbour Road, Helensburgh, G84 0JN
M0 EAK I Cappleman, 8 The Woodlands, Lilleshall, Newport, TF10 9LN
M0 EAL G Gauld, 16 Swallow Close, Thornton-Cleveleys, FY5 2JN
M0 EAM G Bourne, 72 Cornish Way, Royton, Oldham, OL2 6JY
MW0 EAN R Westcott, 8 Pen Y Bigyn, Llanelli, SA15 1PB
M0 EAO Colin McDonnell, Sunholme, Witham Bank West, Boston, PE21 8PU
M0 EAQ T Gale, Flat 15 Browning Apartments, 140 Hamlets Way, London, E3 4GS
MM0 EAR Dalriada ARC c/o Kieran Carroll, 32a Meadowburn Place, Campbeltown, PA28 6ST
M0 EAS P Humphrey, Sirmione, Cookshall Lane, High Wycombe, HP12 4AL
MW0 EAT A Phillips, 85 Gorseinon Road, Penllergaer, Swansea, SA4 9AB
M0 EAU D Harrison, 72 North Farm Road, Lancing, BN15 9BU
MM0 EAX D Thomson, Boat House, Finstown, Isle of Orkney, KW17 2EH
M0 EAY Christopher Knight, Shamrock, Stow Lane, Wisbech, PE13 2JU
M0 EAZ R Bennett, Hawthorn Cottage, Mill Lane, Norwich, NR12 8HP
M0 EBD R Chew, 1 Exeter Close, Aintree, Liverpool, L10 8LU
M0 EBG Peter Good, 80 Meredith Road, Stevenage, SG1 5QS
M0 EBI N Billingham, 19 Tumulus Road, Saltdean, Brighton, BN2 8FR
M0 EBJ S Rope, 51 Medeswell Close, Brundall, Norwich, NR13 5QG
M0 EBN S Bywater, Birch Wood, Norwich Road, Cromer, NR27 0HG
M0 EBO I Bryant, 17 Kent Road, Southampton, SO17 2LJ
M0 EBP Georgina Joyce, 8 Christ Church Street, Preston, PR1 8PJ
M0 EBQ A Medhurst, 44 Battle Road, Hailsham, BN27 1DS
M0 EBR Mark Shuttleworth, 19 Edgerton Drive, Tadcaster, LS24 0QW
M0 EBT J Larden, 26 Cragson Walk, Dawley, Telford, TF4 3GA
[illegible]
M0 EBV H Hepworth, The Leas, Main Street, York, YO42 1RX
M0 EBX M Tween, 79 West Close, Fernhurst, Haslemere, GU27 3JS
M0 ECC Ernest Cassidy, 3 The Elms, Great Chesterford, Saffron Walden, CB10 1QD
M0 ECF D Edwards, 25 Bryn Coed, Gwersyllt, Wrexham, LL11 4UE
M0 ECK HMS CAVALLER RC c/o Brian Lucas, 8 Gilbert Close, Hempstead, Gillingham, ME7 3QQ
M0 ECL E Lerpiniere, The Windmill, Millwrights, Colchester, CO5 0LQ
M0 ECM Christopher Martin, 14 Freeston Terrace, St. Georges, Telford, TF2 9HD
M0 ECP K Fujita, C/O TANITA INTERNATIONAL, 301 Wing-on Plaza 62 Mody Road TST-East, Kowloon, Hong Kong
M0 ECQ Ian Nutt, Green Acres, Chapple Road, Newton Abbot, TQ13 9JY
M0 ECR East Cheshire Radio Group c/o Geoffrey Hannan, 20 Arlington Drive, Stockport, SK2 7EB
M0 ECS S Seabrook, 29 Gadby Road, Sittingbourne, ME10 1TJ
M0 ECW East Cheam Wireless Society c/o Timothy Watts, 26 Woodger Close, Guildford, GU4 7XR
M0 ECX Neil Rothwell Hughes, Cefn Glass, Clyro, Hereford, HR3 5JT
MW0 ECY Heinz Fingerhut-Holland, Osnok, 1 South Cliff Street, Tenby, SA70 7EB
M0 ECZ E Williams, 22 Sherwood Avenue, Melksham, SN12 7HL
M0 EDA S Hemmings, Leylands, Leigh Road, Frome, BA11 3LR
M0 EDE Alison Holmes, 127 Lower Oxford Street, Castleford, WF10 4AG
MI0 EDF P Hawthorne, 77 Pollock Drive, Lurgan, Craigavon, BT66 8JP
MU0 EDN B Gray, 21 Auderville, Alderney, Guernsey, GY9 3XE
M0 EDO S Williams, 6 Grasmere Road, Dewsbury, WF12 7PU
M0 EDP J Barr, 41 Salisbury Road, London, E12 6AA
M0 EDQ N Heyne, Croxdale, Chiddingly Road, Heathfield, TN21 0JH
M0 EDR S Pritchard, 18 Cumberland Avenue, Basingstoke, RG22 4BG
M0 EDS G Edwards, Ogwen Terrace, High Street, Bangor, LL57 3AY
M0 EDU Mike Wade, Watts Palace Cottage, Chitcombe Road, Rye, TN31 6EX
M0 EDX Anton Koval, Hen DY Newydd, Sarnau, Llanymynech, SY22 6QL
M0 EEB Michael Brady, 3 Bransdale, Worksop, S81 0XY
M0 EEG Chris Pomfrett, 17 Manifold Close, Sandbach, CW11 1XP
M0 EEH P Halpin, 50 Celtic Road, Deal, CT14 9EF
M0 EEK David Edgar, 31 Albany Villas, Hove, BN3 2RT
M0 EEL S Connelly, 79 Pettycot Crescent, Gosport, PO13 0SJ
M0 EEP J Nethercott, 30 Goldcrest Road, Chipping Sodbury, Bristol, BS37 6XF
M0 EFE D Houldridge, 1 Merlin Close, Longhill, Hull, HU8 9UY
MM0 EFI F wenseth, 2 Sunnybank Cottage, Logie Coldstone, Aboyne, AB34 5PQ
MM0 EFJ M Donnachie, Roselea, Cluny Road, Dingwall, IV15 9NJ
MI0 EFM E Mulligan, 27 Hillside Park, Belfast, BT9 5EL
MU0 EFR Denzil Robert, Nos Treis Liberation Drive, 7 Route Des Clos Landais, Guernsey, Guernsey, GY7 9PH
MM0 EFW E Currie, 59a South Street, Fochabers, IV32 7EF
M0 EGA Richard Price, 2 Wordsworth Avenue, Easington Lane, Houghton le Spring, DH5 0NR
M0 EGC East of Greenwich Radio Amateur Club c/o Thomas Jackson, 86 Lascelles Avenue, Withernsea, HU19 2EB
MD0 EGL Conor Dunne, 48 Drury Road, Harrow, HA1 4BW
M0 EGN C Lewis, 3 Sovereign Way, Calcot, Reading, RG31 4US
M0 EGV I Robinson, 5 The Meadows, Bempton, Bridlington, YO15 1LU
M0 EHA G Beam, 35a Moor Lane, York, YO24 2QX
M0 EHF Essex RAYNET c/o G Tiller, 12 Birk Beck, Waveney Drive, Chelmsford, CM1 7PJ
M0 EHL M Longbottom, 32 Anns Hill Road, Gosport, PO12 3JY
M0 EHS P Shaw, 16 Sutherland Road, Cradley Heath, B64 6EA
M0 EIW John Walsh, 35 Carisbrooke Road, Bushbury, Wolverhampton, WV10 8AB
M0 EJB D Banks, 9 Woodbank, Egremont, CA22 2HL
M0 EJF C Briggs, 21 Peak View Road, Chesterfield, S40 4NW
M0 EJG J Freeman, High Meadow, Martens Lane, Colchester, CO6 5AG
M0 EJL Peter Kendall, 3 Hurstwood Close, Lincoln, LN2 4TX
M0 EJW Martin Bishop, 22 Herringston Road, Dorchester, DT1 2BS
M0 EKB G Patrick, Athena, 121 Ringmer Road, Worthing, BN13 1DX
M0 ELA A McKenzie, Little Wishmore, Whitbourne, Worcester, WR6 5SR
M0 ELC L Chesters, 8 Conway Grove, Blacon, Chester, CH1 5RU
MM0 ELF William McCue, 188 Redburn, Alexandria, G83 9BU
M0 ELO Craig Bootz, 10 Davenport Avenue, Nantwich, CW5 5QJ
MM0 ELP C Maxwell, 29 Ambleside Rise, Hamilton, ML3 7HJ
MM0 EMC E McPherson, 12 Lambourn, Wolfhill, Perth, PH2 6TQ
M0 EMD E Deeley, 26 Eversley Crescent, Isleworth, TW7 4LS
M0 EME P Tomlinson, 217 Old Hall Road, Tapton, Chesterfield, S40 1HQ
M0 EMM David Martin, 14 Freeston Terrace, St. Georges, Telford, TF2 9HD
M0 EMR Colin Wright, 8 Kendal Road, Sheffield, S6 4QG
M0 EMW E Wheeler, 3 Brazo Road, Porthleven, Helston, TR13 9LR
MI0 ENR R McFadden, 36 Trinity Drive, Ballymoney, BT53 6EQ
M0 EOT B Podmore, 78 Ridge Road, Stoke-on-Trent, ST6 5LP
M0 EOU John Hinds, 69 Carshalton Grove, Wolverhampton, WV2 2QZ
MM0 EPC EUROPEAN PSK CLUB c/o Jurij Phunkner, 7 Plenshin Court, Glasgow, G53 6QW
M0 EPH Edward Rippon, 10 Camelot Avenue, Nottingham, NG5 1DW
M0 EPX Liam Stone, 6 Lyntons, Pulborough, RH20 1AZ
M0 EQD David Wright, 22 West Hill, Rotherham, S61 2HB
MM0 EQE Alan Thompson, 22 Lochend Road, Carnoustie, DD7 7QF
MW0 EQL James Sneddon, 3 Marigold Close, Gurnos, Merthyr Tydfil, CF47 9DA
M0 EQM Raymond Agacy, 23 Highgate Lane, Bolton-upon-Dearne, Rotherham, S63 8HR
M0 EQY M Campbell, 377 Bushbury Lane, Wolverhampton, WV10 8JZ
M0 ERG EAGLE RADIO GROUP c/o Terry Stow, 38 The Strand, Mablethorpe, LN12 1BQ
M0 ERJ E Jones, 37 Sluice Road, Denver, Downham Market, PE38 0DY
MM0 ERK B Murray, Sherwood Cottage, Farnell, Brechin, DD9 6UH
M0 ERN Ernest Coleby, 13 Farm Close, Sunniside, Newcastle upon Tyne, NE16 5PP

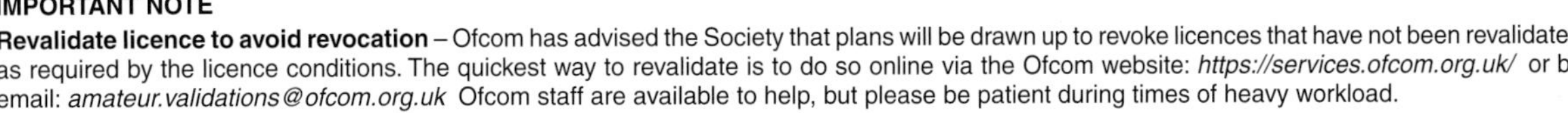

**IMPORTANT NOTE**

**Revalidate licence to avoid revocation** – Ofcom has advised the Society that plans will be drawn up to revoke licences that have not been revalidated as required by the licence conditions. The quickest way to revalidate is to do so online via the Ofcom website: *https://services.ofcom.org.uk/* or by email: *amateur.validations@ofcom.org.uk* Ofcom staff are available to help, but please be patient during times of heavy workload.

UK Callsigns

| | | |
|---|---|---|
| M0 | ERS | J Hauton, 15 Bourne Close, Lincoln, LN6 7DR |
| M0 | ERY | M Samborskyy, St Johns College, St Johns Street, Cambridge, CB2 1TP |
| M0 | ESB | David Baldwin, 46 Muirfield Road, Watford, WD19 6LN |
| M0 | ESP | ILERA c/o Lenio Marobin, Flat 60, Tudor Court, London, N1 4NU |
| M0 | ESR | ALDERLEY EXPLORER SCOUT A R U c/o P Phillips, 2 Millstream Close, Goostrey, Crewe, CW4 8JG |
| M0 | ESU | M Bown, 47 Ullswater Crescent, Weymouth, DT3 5HF |
| M0 | ESW | Etienne Swanepoel, Bridge Cottage, Tinhay, Lifton, PL16 0AH |
| M0 | ESZ | M Owens, 66 Woodlands Road, Bishop Auckland, DL14 7LZ |
| M0 | ETA | Gavin Andrews, 158 Latchmere Road, Kingston upon Thames, KT2 5TU |
| M0 | ETE | Ronald Home, Beech Cottage, The Green, Huntingdon, PE28 9NA |
| M0 | ETP | J Bolton, 2 Patterdale Street, Hetton-le-Hole, Houghton le Spring, DH5 0BH |
| M0 | ETQ | David Bolton, 2 Patterdale Street, Hetton-le-Hole, Houghton le Spring, DH5 0BH |
| M0 | ETS | C lyon, 3 Doodstone Avenue, Lostock Hall, Preston, PR5 5TY |
| M0 | ETY | S Hindle, 35 Heyhead Street, Brierfield, Nelson, BB9 5BN |
| M0 | EUI | G Plant, 11 Shinwell Grove, Stoke-on-Trent, ST3 7UG |
| M0 | EUK | Graeme Stoker, 1 Mitford Way, Dinnington, Newcastle upon Tyne, NE13 7LW |
| M0 | EUS | A Jones, 3 Warren Houses, Tile Lodge Road, Ashford, TN27 0BX |
| M0 | EUY | A Swan, 47 Warren Close, Whitehill, Bordon, GU35 9EX |
| M0 | EVE | P Beier, 20 Markham Avenue, Armthorpe, Doncaster, DN3 2AZ |
| M0 | EVG | GIRLGUIDING UK. ELLEN HOLME DIV c/o Dennis Martin, 70 Moorlands Drive, Stainburn, Workington, CA14 4UJ |
| M0 | EVI | A Butler, 88 Manor Road, New Milton, BH25 5EJ |
| M0 | EVK | R Kidd, 4 Oakfields Close, Norwich, NR4 6XH |
| M0 | EWG | B Read, 111 Fitzpain Road, West Parley, Ferndown, BH22 8SF |
| M0 | EWW | R Moreton, 25 Holyoake Place, Rugeley, WS15 2NP |
| M0 | EXM | Brian Wheeler, 2 Rose Street, Houghton le Spring, DH4 5BB |
| MW0 | EYE | Cherian Varghese, 15 Crestacre Close, Newton, Swansea, SA3 4UR |
| M0 | EYT | Paul Marsh, 10 Pardys Hill, Corfe Mullen, Wimborne, BH21 3HW |
| M0 | EZO | Raymond Griffin, 25 Dowell Street, Honiton, EX14 1LT |
| M0 | EZP | David Brewerton, Apartment 16, Miller Court, Axminster Drive, Brighouse, HD6 4FP |
| M0 | FAK | R Chick, 15 Bonfire Close, Chard, TA20 2EG |
| MU0 | FAL | C Fallaize, Lorbert, Pleinheaume Road, Vale, Guernsey, GY6 8NR |
| M0 | FAT | Andrew Moffatt, 10 Chaddesdon Walk, Denaby Main, Doncaster, DN12 4EL |
| M0 | FAZ | D Fower, 31 Hillswood Avenue, Leek, ST13 8EQ |
| M0 | FBB | S Smale, 230 Wareham Road, Corfe Mullen, Wimborne, BH21 3LW |
| M0 | FBM | Alan Lock, 2 Knutscroft Lane, Thurloxton, Taunton, TA2 8RL |
| MU0 | FBO | Richard Stockwell, Fleurs Des Champs, La Colline Des Bas Courtils, Saint Saviours, Guernsey, GY7 9YQ |
| M0 | FCA | W Mannerfelt, 16 Suffolk Road, London, SW13 9NB |
| M0 | FCB | Frederick Brunt, 74 Bardley Crescent, Tarbock Green, Prescot, L35 1RJ |
| M0 | FCD | Mike Christieson, September Cottage, Rushlake Green, Heathfield, TN21 9PP |
| M0 | FCG | Michael Hardy, 4 Kirk Balk, Hoyland, Barnsley, S74 9HU |
| M0 | FCI | D Houghton, 23 Westminster Crescent, Doncaster, DN2 6JH |
| MM0 | FCM | Colin Sheridan, Townhead Cottage, Newbigging, Carnwath, Lanark, ML11 8NB |
| M0 | FCP | F Parsons, 17 Hannah More Close, Wrington, Bristol, BS40 5QG |
| M0 | FCR | Albert Crespo, 266 Trinity Road, London, SW18 3RQ |
| M0 | FCT | Paul Ryder, 6 Lodge Drive, Moulton, Northwich, CW9 8RQ |
| M0 | FCW | M Ballard, 41 Middlefield Avenue, Halesowen, B62 9QJ |
| M0 | FCY | D Thornton, 63 Houghtonside, Houghton le Spring, DH4 4BW |
| MW0 | FDG | Janusz Myszka, 55 Cefn Glas Road, Bridgend, CF31 4PJ |
| M0 | FDX | G Marsden, 71 Sedgley Avenue, Rochdale, OL16 4TY |
| M0 | FEU | Matthew Pullan, Hauptstrasse 26/3, St Radegund Bei Graz, Austria, 8061 |
| M0 | FEY | E Fey, 18 Mead Close, East Huntspill, Highbridge, TA9 3NF |
| MM0 | FFC | Ian Douglas, 15 Henderson Crescent, Broxburn, EH52 6HA |
| M0 | FFS | B Tuffill, Rainbows End, 1 Loggans Close, Hayle, TR27 5BD |
| M0 | FFX | T Wootten, Trinity Hall, Cambridge, CB2 1TJ |
| M0 | FGA | David McCarty, PO Box :4910, The Woodlands, United States, TX 77387 |
| M0 | FGB | F Buck, 89 Marsh Street, Barrow in Furness, LA14 2AD |
| M0 | FGC | Timothy Seed, Al Hudd, Hudd Trading & Services, Ruwi, Oman, PC 112 |
| M0 | FHM | Donald Sunderland, 1 Allfield Cottages, Condover, Shrewsbury, SY5 7AP |
| M0 | FIL | Philip Graham, 29 Lancaster Street, Colne, BB8 9AZ |
| M0 | FIS | P Fisher, 12 St. Anns Avenue, Grimsby, DN34 4PW |
| MD0 | FIX | N Wallace, 85 Erin Vale, Lezayre Park, Ramsey, Isle of Man, IM8 3PU |
| M0 | FJM | J Hudson, 55 William Street, Churwell, Leeds, LS27 7RD |
| M0 | FJS | F Stevenson, 33 Highfield Close, Amersham, HP6 6HG |
| M0 | FLC | I Pollard, Ilderton Glebe Cottage, Ilderton, Alnwick, NE66 4YD |
| M0 | FMT | Peter March, Devonholme, Bedford Road, Hitchin, SG5 3RX |
| M0 | FMY | J Mallichan, 17 Napier Road, Gillingham, ME7 4HB |
| M0 | FOG | Nigel Brereton, 10 Coverdale Close, Stoke-on-Trent, ST3 7RZ |
| M0 | FOR | George Mcinnes, 32 Bis Rue D'ezy, 32 Bis Rue D'ezy, Ivry la Bataille, France, 27540 |
| M0 | FOX | P Leicester, 30 Knighton Street, North Wingfield, Chesterfield, S42 5JA |
| M0 | FPA | R Etchells, 6 Woodbank Court, Canterbury Road, Manchester, M41 7DY |
| M0 | FPQ | C Groves, Wyandell Hailsham Road, Heathfield, TN21 8AS |
| M0 | FRA | Terence Fray, 20 St. Catherines Road, Blackwell, Bromsgrove, B60 1BN |
| M0 | FRC | FRANKLIN RADIO GROUP c/o Robert Topliss, 12 Dorothy Avenue, Skegness, PE25 2BP |
| M0 | FRG | Andrew Howard, 4 Woodgarth Avenue, Manchester, M40 1QE |
| M0 | FRH | Ian Fraser, Flat 2, 77 Bayford Road, Littlehampton, BN17 5HN |
| M0 | FRS | Cae Boys, 34 Firacre Road, Ash Vale, Aldershot, GU12 5JT |
| MW0 | FRY | R Fry, Old Police Station, Parkmill, Swansea, SA3 2EQ |
| M0 | FSH | N Harris, 23 Winchester Avenue, Chatham, ME5 9AR |
| M0 | FSK | F Kennedy, 3 Brookfield View, Bolton le Sands, Carnforth, LA5 8DJ |
| M0 | FSN | P Wells, 7 Kings Meadow, Overton, Basingstoke, RG25 3HP |
| M0 | FTL | R Metcalfe, 33 Midland Terrace, Hellifield, Skipton, BD23 4HJ |
| M0 | FTR | FIVE TOWNS ARC c/o J Adlington, 23 Newstead Road, Abbey Hulton, Stoke-on-Trent, ST2 8HU |
| M0 | FUN | N Fisher, 101 Avocet Way, Bridlington, YO15 3NT |
| M0 | FVD | Meeko Kittika, 51 Overlea Drive, Burnage, Manchester, M19 1QY |
| M0 | FVV | A Hanner, 8 Countryside Farm Park, Church Lane, Steyning, BN44 3HF |
| MM0 | FWG | Roger Gaisford, 9 Rattray Street, Boness, EH51 9PE |
| M0 | FWM | F Mifflin, Windsor House, Harras Road, Whitehaven, CA28 6SG |
| M0 | FWO | J Thomson, 51 Birch Avenue, Cuerden Residential Park, Leyland, PR25 5PD |
| M0 | FXB | A Macrides, 16 Newtons Road, Kewstoke, Weston-Super-Mare, BS22 9LG |
| M0 | FXX | R Limb, Charnwood House, Station Road, Henley-on-Thames, RG9 3JS |
| M0 | FYA | A Young, 39 Thornton Drive, Hoghton, Preston, PR5 0LX |
| M0 | FZR | Robin Wickenden, Selwood Cottage, Moor Lane, Wincanton, BA9 9EJ |
| M0 | FZU | C Walcott, 28 Balfour Road, London, W13 9TN |
| MM0 | FZV | G Bourhill, 30c Salters Road, Wallyford, Musselburgh, EH21 8AA |
| M0 | FZW | Norman Crampton, 7 Barneveld Avenue, Canvey Island, SS8 8NZ |
| M0 | FZX | S norman, 27 Ashburton Road, Ickburgh, Thetford, IP26 5JA |
| M0 | GAC | G A'Court, Three Oaks, Greenhill Lane, Winscombe, BS25 5PE |
| M0 | GAD | A Grove, 66 Hazel Way, Crawley Down, Crawley, RH10 4EU |
| M0 | GAE | Graham Errington, 22 Willoughby Drive, Whitley Bay, NE26 3DY |
| M0 | GAG | Richard Burrell, 38 Standhill Crescent, Barnsley, S71 1SU |
| M0 | GAH | Adrian Cunningham, 23 Heathgate Close, Birstall, Leicester, LE4 3GW |
| MM0 | GAI | Thomas Spencer, 3 Tramore Crescent, Prestwick, KA9 1LT |
| M0 | GAN | P Street, 50 Dickson House, Ridgway Road, Stoke-on-Trent, ST1 3BA |
| M0 | GAQ | Kevin Ingram, 15 Kent Avenue, East Cowes, PO32 6QN |
| M0 | GAV | Andrew Burton, 51 Wilcox Road, Sheffield, S6 1BQ |
| M0 | GAX | Craig Ponder, 12 Wood Street, Doddington, March, PE15 0SA |
| M0 | GBA | Graham Allison, 16 Copse Road, Plymouth, PL7 1PZ |
| M0 | GBB | BRIGG AND DISTRICT ARC c/o David Ogg, 36 Cliff Road, Winteringham, Scunthorpe, DN15 9NQ |
| M0 | GBC | W Jefferies, 26 Norcutt Road, Twickenham, TW2 6SR |
| M0 | GBF | Bryan Findler, 1 Gordon Avenue, Stoke-on-Trent, ST6 2LY |
| M0 | GBH | C Johnston-Stuart, 35 Robbins Close, Bradley Stoke, Bristol, BS32 8AS |
| M0 | GBK | S Nash, 6 Berry Road, Meltham, Holmfirth, HD9 5PL |
| M0 | GBO | J Zemlicka, 37 Ascot Gardens, Southall, UB1 2SA |
| MI0 | GBU | Causeway RC c/o Neil Bolt, 32 Bush Gardens, Bushmills, BT57 8AE |
| MW0 | GBW | Bernie Collins, 58 Brockhill Way, Penarth, CF64 5QD |
| M0 | GBZ | Euan McPherson, 138 Shephall View, Stevenage, SG1 1RR |
| M0 | GCA | Thomas Sheridan, 4 Stane Close, Bishop's Stortford, CM23 2HU |
| M0 | GCB | Thomas Kim, 9 Temeraire Heights, Folkestone, CT20 3TL |
| M0 | GCC | G Cook, 61 Mortomley Lane, High Green, Sheffield, S35 3HS |
| MW0 | GCD | G Day, 20 St. Johns Drive, Pencoed, Bridgend, CF35 5NF |
| MM0 | GCF | John Brown, 78 Egilsay Street, Glasgow, G22 7RG |
| M0 | GCH | Glenn Holmes, 11 Claudeen Close, Southampton, SO18 2HQ |
| MD0 | GCI | Maurice Rowley, Flat 23, Minstrel Court, 170 High Street, Harrow, HA3 7AX |
| M0 | GCR | G Rumsey, 13 Greenhills Road, Northampton, NN2 8EL |
| MW0 | GCS | L Powell, 2 Gelliderw, Pontardawe, Swansea, SA8 4NB |
| M0 | GCT | Stuart tweddle, 3 Bron Ffinan, Pentraeth, LL75 8UT |
| M0 | GCU | James Jordan, The Cottage, Papcastle, Cockermouth, CA13 0LA |
| MI0 | GCV | T Conlon, 7 Waringfield Gardens, Moira, Craigavon, BT67 0FQ |
| M0 | GCX | V Narinian, 15 Headley Gardens, Great Shelford, Cambridge, CB22 5JZ |
| MM0 | GDG | A Curlis, 94 Kirkhill Road, Aberdeen, AB11 8FX |
| M0 | GDH | George Hogg, 7 Elbra Farm Close, Ellenborough, Maryport, CA15 7RG |
| MM0 | GDI | Brian Massie, Flat 4/r, 74 Commercial Street, Dundee, DD1 2AP |
| M0 | GDJ | Brian Holt, 36 Tenter Hill Lane, Sheepridge, Huddersfield, HD2 1EJ |
| MM0 | GDL | D Lindsay, 114 Strathblane Road, Milngavie, Glasgow, G62 8HD |
| MW0 | GDM | P De Mengel, Fern Cottage, 1 The Gail, Haverfordwest, SA62 4HJ |
| M0 | GDP | Raymond Parkinson, 18 Lime Tree Gardens, Lowdham, Nottingham, NG14 7DJ |
| M0 | GDT | Dean HOLLAND, 13 Linley Drive, Boston, PE21 7EJ |
| M0 | GDU | Richard North, 24 Gadesden Road, Epsom, KT19 9LB |
| M0 | GDV | D HARBRON, 6 West View, Penshaw, Houghton le Spring, DH4 7HP |
| M0 | GDX | D Hayes, 43 Linden Avenue, Sheffield, S8 0GA |
| M0 | GEB | G Beesley, Stone Barn, Black Dog, Crediton, EX17 4QX |
| M0 | GEC | G Clennell, 69 Seventh Row, Ashington, NE63 8HX |
| M0 | GED | Fred Holt, 8 Pleasant View, Coppull, Chorley, PR7 4PH |
| M0 | GEF | G Freeman, 11 Westward Road, Malvern, WR14 1JX |
| MW0 | GEI | S Walmsley, 29 Shelley Court, Machen, Caerphilly, CF83 8TT |
| M0 | GEK | Leslie Gange, 15 Ham Close, Worthing, BN11 2QE |
| M0 | GEL | Simon Attwood, 60 Underwood Avenue, Ash, Aldershot, GU12 6PL |
| M0 | GEN | G Barusevicus, 6 Middlebrook Crescent, Bradford, BD8 0EN |
| MM0 | GEO | George Muir, 50 Davidson Way, Livingston, EH54 8HQ |
| M0 | GEP | Anthony Holdup, Tunnel Farm, Tunnel Rd, Imbil (Po 155), Australia, 4570 |
| M0 | GES | G Stollard, Apt 6 Falaise, 14 West Overcliff Drive, Bournemouth, BH4 8AA |
| M0 | GEU | Allan Nicholson, 14 Rossinyol, Los Arcos, Alicante, Spain, 3530 |
| M0 | GEX | C Farley, 1 Wesley Cottages, Mutley, Plymouth, PL3 4RB |
| M0 | GEY | M Spinks, 26 Church Hill, Royston, Barnsley, S71 4NH |
| MM0 | GFA | C ROY HILL CLUB c/o Patrick McBride, 1 Hillside, Croy, Glasgow, G65 9HJ |
| M0 | GFD | Julian Smith, 131 Steadman House, Bow Common Lane, London, E3 4HT |
| MI0 | GFE | ANTRIM & DISTRICT ARS c/o Robert Robinson, 31 Brantwood Gardens, Antrim, BT41 1HP |
| M0 | GFF | Brian Courtney, 44 Uxbridge Road, Rickmansworth, WD3 7AR |
| M0 | GFJ | David Russell, Halfway House, Holbrook Road, Ipswich, IP9 1BP |
| MD0 | GFK | Adam Ochot, 10 Farman Terrace, Hinkler Road, Harrow, HA3 9BD |
| M0 | GFM | G Dawson, Bramwell, Winchester Road, Southampton, SO32 2LG |
| M0 | GFN | J Bell, 50 Colchester Terrace, Sunderland, SR4 7RY |
| M0 | GFO | R Mcdermott, 2 Monument Close, Wellington, TA21 9AL |
| MM0 | GFP | Andrew Stewart, 21 Mansfield Avenue, Newtongrange, Dalkeith, EH22 4SJ |
| MM0 | GFR | George Forster, 4 Kirk Brae, Morvern, Oban, PA80 5XW |
| M0 | GFX | Peter Hull, 1 Sawpits Close, Stogumber, Taunton, TA4 3TX |
| M0 | GFZ | Ivan Vano, 3 Westbourne Road, Feltham, TW13 4LX |
| MI0 | GGB | Samuel Quigg, 100 Whispering Pines, Limavady, BT49 0UF |
| MM0 | GGD | G Duncan, 5 Jarvis Place, Carnoustie, DD7 7BR |
| MM0 | GGG | D Banks, 60 Leander Crescent, Bellshill, ML4 1JB |
| M0 | GGH | Juraj Gubric, 279 Nottingham Road, Eastwood, Nottingham, NG16 2AP |
| M0 | GGK | David Lawson, 30 Meadowcroft, St. Helens, WA9 3XQ |
| M0 | GGL | Christopher Lester, 21 Mortimer Way, Witham, CM8 1SZ |
| M0 | GGM | G Markey, Trebrown Farm, Horningtops, Liskeard, PL14 3PU |
| M0 | GGO | C loughran, 8 Douglas Road, Dover, CT17 0BD |
| M0 | GGP | Angel Of The North ARC c/o Stephen Townsley, 222 Prince Consort Road, Gateshead, NE8 4DX |
| M0 | GGQ | Andrew Shaw, 43 Borough Avenue, Radcliffe, Manchester, M26 2QG |
| M0 | GGT | Bryan Ashton, 138 Christchurch Road, Norwich, NR2 3PG |
| M0 | GGU | Graham Arthur Medlicott, Lower Medlicott Farm, Wentnor, Bishops Castle, SY9 5EL |
| M0 | GGW | Gordon Milsom, Flat 8, Sovereign Court, High Wycombe, HP13 6XL |
| M0 | GGX | J Patient, 4 Bucklebury Heath, South Woodham Ferrers, Chelmsford, CM3 5ZU |
| M0 | GGZ | Anthony Chaplin, 33 The Crofts, Little Wakering, Southend-on-Sea, SS3 0JS |
| M0 | GHA | Michael Dudley, 4 Peppermint Grove, Skegness, PE25 3LJ |
| M0 | GHC | Marcin Wojcik, 43 Connaught Road, London, W13 0TF |
| M0 | GHE | L Nordgren, 41 Forest Road, London, E7 0DN |
| MI0 | GHI | A Murphy, 13 Torrens Park, Lislagan Upper, Ballymoney, BT53 7DE |
| M0 | GHK | Thomas Lee, 54 Shielfield Terrace, Tweedmouth, Berwick-upon-Tweed, TD15 2EE |
| MM0 | GHM | Graham Cochrane, 33 Portland Road, Galston, KA4 8EA |
| MM0 | GHN | Norman Inglis, 35 Portland Park, Hamilton, ML3 7JY |
| M0 | GHO | G Hopkins, 27 The Templars, Worthing, BN14 9JT |
| M0 | GHR | Ian Millar, 3a South Street, Wiveliscombe, Taunton, TA4 2LZ |
| MM0 | GHT | Kenneth Brown, 21 Strain Crescent, Airdrie, ML6 9ND |
| M0 | GHV | S Young, 27a Norton Road, London, E10 7LQ |
| M0 | GHW | Frederick Mole, Five Oaks, Sponden Lane, Cranbrook, TN18 5NR |
| M0 | GHX | Chris Painter, 45 Meadow Lane, Beeston, Nottingham, NG9 5AE |
| M0 | GHY | Peter Hollas, 46 Askham Fields Lane, Askham Bryan, York, YO23 3PS |
| M0 | GHZ | David Millard, Weavern House, Hartham Lane, Chippenham, SN14 7EA |
| M0 | GIA | Sean Amesbury, 13 Haddon Close, Macclesfield, SK11 7YG |
| M0 | GIB | David Gibbons, 4 Ivychurch Mews, Runcorn, WA7 5AR |
| M0 | GID | G Dunne, Three Ways, Northbourne Road, Deal, CT14 0HJ |
| M0 | GIE | P Ellis, 40 Grasmere Road Royton, Oldham, OL2 6SR |
| M0 | GIF | Robin Manser, Flat 38, St. Johns Court, Portsmouth, PO2 8NA |
| M0 | GIG | D Wharlley, 15 Crampton Court, Grosvenor Road, Broadstairs, CT10 2XU |
| MI0 | GIJ | James Thompson, 119 Rathkyle, Antrim, BT41 1LN |
| M0 | GIL | Gillian Wildman, 55 Hill Street, Bradley, Bilston, WV14 8SB |
| M0 | GIM | Grazyna Mitchener, Cabins, Wenham Road, Ipswich, IP8 3EY |
| MW0 | GIN | S Peel, 28 Dan yr Allt, Llanelli, SA14 8AT |
| M0 | GIP | W James, 3 Midfield Close, Gillow Heath, Stoke-on-Trent, ST8 6RD |
| M0 | GIQ | G Griffiths, 16 Back Lane, Winteringham, Scunthorpe, DN15 9NW |
| M0 | GIU | P Tier, 16a Burcombe Road, Bournemouth, BH10 5JT |
| M0 | GIW | David Ryan, 21 Brooke Street, Thorne, Doncaster, DN8 4AX |
| M0 | GIY | Paul Swansbury, 119a Trelowarren Street, Camborne, TR14 8AW |
| M0 | GIZ | C Melia, 3 Ramshead Grove, Leeds, LS14 1PL |
| M0 | GJA | Kort Nyquist, 25 Marsh View, Newton, Preston, PR4 3SX |
| MM0 | GJC | G Costa, 54 High Street, Dollar, FK14 7BA |
| M0 | GJD | James Farrant, Orchard Cottage, Claycastle, Crewkerne, TA18 7PB |
| M0 | GJH | Andrew Vine, Hilden, Woodland Avenue, Cranleigh, GU6 7HZ |
| M0 | GJJ | Gareth Johnson, 199 Lynwood, Folkestone, CT19 5TA |
| M0 | GJK | G Knight, 20 Crossway, Welwyn Garden City, AL8 7EE |
| M0 | GJL | Robert Brodie, 7 Island Street, Salcombe, TQ8 8DP |
| MI0 | GJN | M Edwards, 21 Old Grange Avenue, Carrickfergus, BT38 7UE |
| M0 | GJS | Gary Suter, 11 Summerdown Close, Durrington, Worthing, BN13 3QG |
| M0 | GJU | James Freeman, 12 Norfolk Terrace, Cambridge, CB1 2NG |
| M0 | GJV | E Jones, 43 Wesley Road, Wimborne, BH21 2QB |
| M0 | GJX | A Jordan, 21 Madison Avenue, Exeter, EX1 3AH |
| M0 | GKA | Scott Len, 13 Griffiths Gardens, Caversfield, Bicester, OX27 8FL |
| MM0 | GKB | Kenneth Mackintosh, Allt Dubh, Scatwell, Strathconon, Muir of Ord, IV6 7QG |
| M0 | GKD | Alan Mockford, 29 Kingston Close, Blandford Forum, DT11 7UQ |
| M0 | GKG | Robin Ley, 23 Heronbridge Close, Westlea, Swindon, SN5 7DR |
| M0 | GKJ | Robert Frencham, 1 Eggerslack Cottages, Windermere Road, Grange-over-Sands, LA11 6EX |
| M0 | GKK | SURREY SPACE CENTRE c/o David Fishlock, 93 Shackstead Lane, Godalming, GU7 1RL |
| MI0 | GKL | BUSHVALLEY ARC c/o Samuel Quigg, 100 Whispering Pines, Limavady, BT49 0UF |
| MM0 | GKN | John Hogg, 31 Woodlea Court, Crosshouse, Kilmarnock, KA2 0ES |
| M0 | GKO | Gordon Thorpe, 81 knoll drive, Coventry, CV3 5PJ |
| M0 | GKP | Peter Haydn Smith, 3 King Henrys Road, Lewes, BN7 1BT |
| M0 | GKR | A Steel, 78 Water Meadows, Worksop, S80 3DB |
| MM0 | GKT | David Bushby, Coach House, Dalginross, Crieff, PH6 2HB |
| MM0 | GKU | Thomas McCall, 119 Claremont, Alloa, FK10 2ER |
| M0 | GKV | Michael Williams, 30 Elm Drive, Risca, Newport, NP11 6HJ |
| M0 | GKW | M Sullivan, 9 Roundacre, Halstead, CO9 1XE |
| M0 | GLF | Joseph Stanford, 1941 Ute Creek Drive, LONGMONT, Co, United States, 80504 |
| MI0 | GLG | Tyrone Currie, 26 High Street, Portaferry, Newtownards, BT22 1QT |
| M0 | GLI | Mark Shasby, 19 Crawshaw Grange, Crawshawbooth, Rossendale, BB4 8LY |
| M0 | GLJ | Adrian Craig, 133 Green Lane, Ilkeston, DE7 5PP |
| MD0 | GLK | Andrew Dorman, 1 Sprucewood Rise, Foxdale, Douglas, Isle of Man, IM4 3JP |
| M0 | GLL | Colin Smith, 2 Blankney Close, Fareham, PO14 3RX |
| M0 | GLP | G Parker, 420 Meadow Lane, Nottingham, NG2 3GD |
| M0 | GLQ | Christopher Senior, 29 Ilex Way, Goring-by-Sea, Worthing, BN12 4UY |
| M0 | GLS | S Day, 2 Cae Job, Piercefield Lane, Aberystwyth, SY23 1RJ |
| M0 | GLT | Marinus Rosenbrand, 12 Greville Road, Cambridge, CB1 3QL |
| M0 | GLU | Antal Vincz, 3 Bowman Court London Road, Crawley, RH10 8XG |
| M0 | GLV | Marcin Jusko, 13 Ellerby Grove, Hull, HU9 3PR |
| MM0 | GLX | Brian Burt, 182 Old Inverkip Road, Greenock, PA16 9JG |
| M0 | GMA | Graham May, 95 Moorfield Avenue, Denton, Manchester, M34 7TX |
| M0 | GMC | G Collier, 6 Copse Close, Tilehurst, Reading, RG31 6RH |
| M0 | GMD | Duncan Gray, 68 Endeavour Way, Hythe Marina Village, Southampton, SO45 6LA |
| M0 | GME | Gary Ellis, 46 The Uplands, Scarborough, YO12 5HX |
| M0 | GMG | Roger Bell, 92 Dean Drive, Wilmslow, SK9 2EY |
| M0 | GMH | Owen Williams, 39 Camden Road, Maes-Y-Coed, Brecon, LD3 7RT |
| M0 | GMI | Christopher Woodbridge, 53 Baffins Road, Portsmouth, PO3 6BE |

M0 GMK Colin Dawson, 9 Mulberry Close, Poringland, Norwich, NR14 7WF
M0 GMN William Owen, 8 Sandhurst Avenue, Lytham St. Annes, FY8 2DA
M0 GMO P Cheshire, 29 Madison Avenue, Exeter, EX1 3AH
M0 GMQ P Hall, 13 Sheard Avenue, Ashton-under-Lyne, OL6 8DS
M0 GMS S Smith, 5 Melhuish Close, Witheridge, Tiverton, EX16 8AZ
M0 GMT Daniel Clapp, 35 St. Elmo Road, Worthing, BN14 7EJ
M0 GMU P Sweatman, 14 Clover Court, Jasmine Grove, Waterlooville, PO7 8BP
M0 GMW C Watts, 10 Kemble Gardens, Bristol, BS11 9RY
M0 GMZ Huw Hughes, Llecyn y Llan, Llanerchymedd, Llannerch-Y-Medd, LL71 8EH
M0 GNA A Shaw, 21 Laburnum Road, Prenton, CH43 5RP
M0 GNB Marcin Dreszer, 47 Lydgate Court, Nuneaton, CV11 5RR
M0 GNC Adrian Ellison, 24 The Grove, Brentwood, CM14 5NS
M0 GNF Cyril Hughes, 26 Tan y Bryn, Valley, Holyhead, LL65 3ES
MM0 GNH Kenneth Foreman, 16 Beveridge Place, Kinross, KY13 8QY
M0 GNJ David Coventry, 1 Seacrest Avenue, Fleetwood, FY7 6FG
M0 GNK Robert Jennings, 8709 Crcongrass Way, Parker, United States, CO 80134
M0 GNL Colin Price, Byways, Taylors Lane, Chichester, PO18 8QQ
M0 GNM H Donnelly, 6 Famet Walk, Purley, CR8 2DY
M0 GNO Edward Whiten, 17 Scott Close, Ashby-de-la-Zouch, LE65 1HT
M0 GNP Douglas Salter, 142 Brays Road, Birmingham, B26 2PP
MM0 GNS Charles Stewart, 9 Rousay Wynd, Kilmarnock, KA3 2GP
M0 GNU Dennis White, 3 West Street, South Normanton, Alfreton, DE55 2AJ
MM0 GNX Annette Messner, 6 Elistoun Drive, Tillicoultry, FK13 6NT
M0 GNY Michal Zlobinski, 16 Birkdale Avenue Atherton, Manchester, M46 9PY
M0 GOA J Goacher, 41 Clay Hill, Two Mile Ash, Milton Keynes, MK8 8AY
M0 GOB Brian Holland, 11 Silverlands Park, Buxton, SK17 6QX
M0 GOC Tony Ward, 1 Darrismere Villas, Edinburgh Street, Hull, HU3 5AS
M0 GOF J McCulloch, 2 Riverbank Wynd, Gatehouse of Fleet, Castle Douglas, DG7 2EA
MM0 GOG Duncan Baillie, 126 Main St., Fauldhouse, Bathgate, EH47 9BW
M0 GOH Peter Preston, 49 Cowpasture Lane, Sutton-in-Ashfield, NG17 5AR
M0 GOI K Hornby, 9 Woodland View, Silkstone Common, Barnsley, S75 4SA
M0 GOK David Richards, 73 Greenfields Avenue, Alton, GU34 2EW
M0 GOL T Goldsmith, 37 Cowdray Road, Sunderland, SR5 3PG
M0 GOM Jason Roissetter, 48 St. Julians Avenue, Newport, NP19 7JU
MM0 GON G Craig, 1 Butt Avenue, Helensburgh, G84 9DA
M0 GOO John Brook, The Clock Tower, Rectory Lane, Chichester, PO20 9DT
M0 GOP G Oliver, 17 Jack Stephens Estate, Penzance, TR18 2QE
M0 GOQ Jose Barbieri, 20 Gilbard Court, Chineham, Basingstoke, RG24 8RG
M0 GOT Simon Martin, 3 Houndsmill, Horsington, Templecombe, BA8 0ED
MW0 GOV Colin Davis, Denant Mill, Dreenhill, Haverfordwest, SA62 3TS
M0 GOW C Gowing, Barbosa, Remembrance Road, Newbury, RG14 6BA
M0 GOX Janet Davies, Penralt, Abercaseg Road, Bangor, LL57 3SP
M0 GOY Roman Klima, 54a Pickering Road, Hull, HU4 6TL
MI0 GOZ V Maksimavicius, 19 Ballyronan Road, Magherafelt, BT45 6BS
MI0 GPB G Bunting, 8 Moor Park Avenue, Belfast, BT10 0QE
M0 GPC Stuart Withnall, 2 Lansdown Close, Cheltenham, GL51 6QP
M0 GPD Jerry Fuller, Bramble Cottage, Leggatt Hill, Petworth, GU28 9DP
M0 GPE Timothy Loker, 24 St. Albans Hill, Hemel Hempstead, HP3 9NG
MI0 GPF Grey Point For Military Radio Group c/o Samuel Baird, 11 Laral Park, Newtownabbey, BT37 0LH
M0 GPG Graham Dyson, 6 Twynersh Avenue, Chertsey, KT16 9DE
M0 GPH Terence Hall, 18 Common Lane, New Haw, Addlestone, KT15 3LH
M0 GPJ Danny Waite, 1 Naseby Court, Bradville, Milton Keynes, MK13 7EP
M0 GPK William Gasser, 76 Empress Road, Derby, DE23 6TE
MM0 GPL Christopher Jones, Croy Lodge, Shandon, Helensburgh, G84 8NN
M0 GPN Briney Taylor, 47 Sandy Drive, Victoria Point, Australia, 4165
M0 GPO G Otter, 3 Glen Park Avenue, Glenfield, Leicester, LE3 8GH
M0 GPP Brian Doyle, 3 Bryn Road, Flint, CH6 5HU
M0 GPQ Adam Wieckowski, 110 Scrubs Lane, London, NW10 6QY
M0 GPU Alan Norrie, 45 Eastern Way, Ponteland, Newcastle upon Tyne, NE20 9RD
M0 GPV William Tommasini, 49 Taverner Close, Poole, BH15 1UP
M0 GPW Peter Andrew, 13 Luke Road, Aldershot, GU11 3BW
M0 GPX Boguslaw Wagiel, 116 Hurworth Avenue, Slough, SL3 7FQ
M0 GPY Howard Schmidt, 88 Candlemas Lane, Beaconsfield, HP9 1AE
MM0 GPZ Gordon Paterson, 20 Craigmuir Road, Blantyre, Glasgow, G72 9UA
M0 GQB Martin Cox, Haugh Shaw Hall, Haugh Shaw Road, Halifax, HX1 3LE
M0 GQD Jonny Parrett, 6 Shelley Road, East Grinstead, RH19 1TA
M0 GQE Gary Moss, 10 Thistlegreen Road, Dudley, DY2 9JT
MM0 GQF Zbigniew Biorka, 21 Whitehills Lane South, Cove, Aberdeen, AB12 3SU
MI0 GQG R Crozier, 33 Cullentragh Road, Poyntzpass, Newry, BT35 6SD
MI0 GQI Melvyn Crozier, 33 Cullentragh Road, Poyntzpass, Newry, BT35 6SD
M0 GQJ David Downer, 19 Watergate Road, Newport, PO30 1XN
M0 GQM Tamsin Kidwell, 2 Batts Farmyard, Wilton, Marlborough, SN8 3SS
M0 GQP Benjamin Sims, 4 New Cottages, Cranwich Road, Thetford, IP26 5EQ
M0 GQR Antony Matheson, 21 Warren Hill Road, Woodbridge, IP12 4DU
M0 GQS Daniel Roguszczak, 4 Home Farm Close, Reading, RG2 7TD
M0 GQU Marcin Burzynski, 2 Somerset Avenue, Luton, LU2 0PJ
M0 GQV Philip Langabeer, 1 Newfield Crescent, Middlesbrough, TS5 8RE
M0 GQW Edward Tart, Sunnybank Farm, Wattlesborough Heath, Shrewsbury, SY5 9EG
M0 GRA G Hickford, 56 Alexander Close, Abingdon, OX14 1XB
M0 GRB N harris, 45 Sleigh Road, Sturry, Canterbury, CT2 0HT
M0 GRE W Greenall, 356 Warrington Road, Abram, Wigan, WN2 5XA
M0 GRF D Blyth, 45 Clarence Road, Bilston, WV14 6NZ
MI0 GRG Michael McGrory, 110 Kilmore Road, Kilmore, Armagh, BT61 8NR
M0 GRH G Hart, 55 Runswick Drive, Nottingham, NG8 1JE
M0 GRI R Ingham, 84 Marian Court, Gateshead, NE8 2JB
M0 GRJ G Jones, 12 Field Close, Flint, CH6 5RQ
MI0 GRN A Cartin, 64 Ashgrove Park, Magherafelt, BT45 6DN
M0 GRO Walter Chance, 29 Myrtle Avenue, Kings Heath, Birmingham, B14 5DU
M0 GRP Graham Priestley, 53 Millfield Gardens, Crowland, Peterborough, PE6 0HA
M0 GRR Sam Turner, 12 Park Street, Morecambe, LA4 6BN
M0 GRT Niculita Rotari, 81 School Road, Dagenham, RM10 9QD
M0 GRU Andrew Webb, 17 Dickins Way, Horsham, RH13 6BQ
M0 GRV Gary Parks, 26 South Avenue, Elstow, Bedford, MK42 9YS
M0 GRW Christopher [illegible], 110 Barkham Ride, Finchampstead, Wokingham, RG40 4EN
M0 GRX Aldridge & Barr Beacon ARC c/o Edward Roberts, 117 Walstead Road, Walsall, WS5 4LU
M0 GRY G Collis, 16 Hill Grove, Barrow Hill, Chesterfield, S43 2NW
MW0 GRZ Grzegorz Woloszun, 44 Cowbridge Road, Bridgend, CF31 3DA
M0 GSC Michael Bracci, 12 Bowling Green Close, Bognor Regis, PO21 4HB
M0 GSI C Nelmes, 119 Exeter Road, Dawlish, EX7 0AN
M0 GSK Michael Silver, 52 Park Crescent, Elstree, Borehamwood, WD6 3PU
M0 GSL Gerrard Walker, 6 Tenbury Drive, Shrewsbury, SY2 5YB
M0 GSN P Newton, 61 Ashbourne Crescent, Taunton, TA1 2RA
M0 GSO Robert Harris, 1 Hollinhey Close, Bootle, L30 7RN
M0 GSP S Palmer, 58 Highlands Way, Whiteparish, Salisbury, SP5 2SZ
MM0 GSQ Arthur Young, 4/4 Prestonfield Terrace, Edinburgh, EH16 5EE
M0 GSR S Poyser, 80 Derby Road, Long Eaton, NG10 4LB
MM0 GSS G Smith, 40 Pirleyhill Drive, Shieldhill, Falkirk, FK1 2EA
M0 GSV Andrew Nesbitt, 9 Manor Road, Richmond, TW9 1YD
MM0 GSW I Wishart, 7 Cairngorm Crescent, Kirkcaldy, KY2 5RF
M0 GSX P Stocking, 6 Royal Oak Road, Rowley Regis, B65 8NX
MU0 GSY L Roithmeir, La Rance, Kimberley Estate, Sandy Hook, St. Sampson, Guernsey, GY2 4EW
M0 GSZ Graham Starling, 4 Three Corner Drive, Norwich, NR6 7HA
MM0 GTB Neil Hirst, 25 Conifer Road, Mayfield, Dalkeith, EH22 5BY
M0 GTE P Allen, 6 Helston Close, Wigston, LE18 2JH
MM0 GTG G Stoddart, 1 Barrs Brae, Kilmacolm, PA13 4DE
M0 GTH Richard Killen, 3 Great Charles Close, St. Stephen, St. Austell, PL26 7PW
MI0 GTI A Jamison, 11 Richmond Gardens, Newtownabbey, BT36 5LA
M0 GTJ Richard Henderson, 14 Oxford Avenue, St. Albans, AL1 5NS
M0 GTL Keith Cossey, 34 Pinewood Road, Hordle, Lymington, SO41 0GP
MI0 GTM Joseph Sills, 145 Ballycolman Estate, Strabane, BT82 9AJ
M0 GTN Derek Embrey, 21 Rockfield Glade, Parc Seymour, Caldicot, NP26 3JF
M0 GTO J Bateman, 26 Thackeray Road, East Ham, London, E6 3BW
M0 GTP G Perry, 36 Poplar Avenue, Wolverhampton, WV11 1DL
M0 GTQ Neil Bennett, 16 Dickens Road, Worksop, S81 0DP
M0 GTR Peter Henderson, Riverside, Ravens Bank, Holbeach, Spalding, PE12 8RW
M0 GTS Dennis Dearman, 12 Woodgate Road, Moulton Chapel, Spalding, PE12 0XF
M0 GTT Ross Wilcox, 19 Lower Chapel Lane, Frampton Cotterell, Bristol, BS36 2RL
MM0 GTU Adrian Cumming, Woodhead, Linlithgow, EH49 7RJ
MM0 GTX B Thomson, 21 Sound of Kintyre, Machrihanish, Campbeltown, PA28 6NZ
MW0 GTY G Jones, 182 Pontardulais Road, Tycroes, Ammanford, SA18 3RD
M0 GUC Mark Elkington, Warner Leys, Iron Hill Farm, Daventry, NN11 6YJ
M0 GUD G Gash, 61 Beaconsfield Road, Rotherham, S60 3HB
MM0 GUE James McMorland, 382 Maryhill Road, Glasgow, G20 7YQ
M0 GUF Geoffrey Jones, 57 Oxford Road, Banbury, OX16 9AJ
M0 GUG Eric Govan, 9 Willowbank, Sandwich, CT13 9QA
M0 GUH Edwin Cobb, 8 Manor Avenue, Poole, BH12 4LD
M0 GUJ David Tarrant, 17 Orchard Close, Corfe Mullen, Wimborne, BH21 3TW
M0 GUK George Skea, 15 The Shires, Gilwern, Abergavenny, NP7 0EX
M0 GUL Charles Repton, 47 Beechwood Cl;ose, Amersham, HP6 6QU
M0 GUM Dave Adshead, 16 Moat Way, Swavesey, Cambridge, CB24 4TR
M0 GUN John Symons, 120 St. Edmunds Walk, Wootton Bridge, Ryde, PO33 4JE
M0 GUO Peter Fry, 4 Stretham Road, Wicken, Ely, CB7 5XH
MI0 GUQ Castles and Stately Homes on Air - NI c/o Roberta Wadey, 92 Shore Road, Kircubbin, Newtownards, BT22 2RP
M0 GUR James Harris, Flat 3, Herstmonceux Place, Church Road, Herstmonceux, Hailsham, BN27 1RL
M0 GUU George Moore, 40 Main Street, South Rauceby, Sleaford, NG34 8QG
MW0 GUV Andy Hubbard, Pant-y-Meillion, Velindre, Penboyr, Llandysul, SA44 5JA
MM0 GUW Murray McCabe, 15 Laggan Road, Glasgow, G43 2SY
MM0 GUX Michael Potts, 6 Strathearn Grove, Kirkintilloch, Glasgow, G66 2PL
M0 GUZ Dominic Webb, 9 Dunsfold Close, Crawley, RH11 8EY
MI0 GVC Castlerock ARS c/o Kathryn Mullan, 19 Parklea, Portstewart, BT55 7HA
M0 GVE Mark Kentell, C/O Marian Swetman, 24 Kendal Court, Congleton, CW12 4JN
M0 GVI David Capon, 144 Stow Road, Magdalen, King's Lynn, PE34 3BD
M0 GVK Derek Lyon-McKeil, 1372 Turnstone Way, Ca, United States, 94087-3736
M0 GVL Paul Smith, 24 Bede Crescent, Benington, Boston, PE22 0DZ
M0 GVN Patrick Keane, Pembroke Lodge, Byes Lane, Reading, RG7 2QB
M0 GVP LV21 Lightship Museum c/o Colin Turner, 182 Station Road, Rainham, Gillingham, ME8 7PR
M0 GVQ Andrew Sibley, 27 Sherwood Road, Tetbury, GL8 8BU
M0 GVT Chris Lee, Spey Cottage, Doctors Commons Road, Berkhamsted, HP4 3DW
M0 GVW M Wibberley, 5 The Row, Broadwell, Rugby, CV23 8HF
M0 GVX Allan Farrar, [illegible] Wensley Street, Thurnscoe, Rotherham, S63 0PX
M0 GVY Michael Hall, 20 The Spinney, Finchampstead, Wokingham, RG40 4UN
M0 GVZ Conor Turton, 32 Northfield Crescent, Driffield, YO25 5ES
M0 GWA G Rodmell, 2 Meadow Way, Walkington, Beverley, HU17 8SD
M0 GWB Michael Baker, 39 Exham Close, Warwick, CV34 5UL
M0 GWC G Chaloner, 9 Fairthorne Rise, Old Basing, Basingstoke, RG24 7EH
M0 GWD Michael Ponsford, 83 Grant Road, Portsmouth, PO6 1DU
M0 GWE Kevin Graffham, 15 Hayes Road, Clacton-on-Sea, CO15 1[illegible]
M0 GWF Arthur Randles, 62 Brookside Avenue, Poynton, Stockport, SK12 1PW
M0 GWG Mirfield ATC 868 Squadron c/o James Thornton, 29 Farrar Avenue, Mirfield, WF14 9ED
M0 GWH David Iveson, 11 Newport Road, North Cave, Brough, HU15 2NU
M0 GWK N Green, 11 Wythburn Way, Rugby, CV21 1PZ
MW0 GWL John Gwilliam, 39 Wyndham Street, Glynfach, Porth, CF39 9HT
M0 GWM Roger Poole, 57 Loxley Avenue, Shirley, Solihull, B90 2QF
MM0 GWO Hamish Storie, 33 Harbour Street, Plockton, IV52 8TN
M0 GWQ John Baker, 4 Pine Close, Landford, Salisbury, SP5 2AW
M0 GWR John Akinin, 70 Valley Road, West Bridgford, Nottingham, NG2 6HQ
MW0 GWT G Thomas, 20 Ael Y Bryn, Caerau, Maesteg, CF34 0YG
M0 GWV Allan Warner, 35 Lon y Berllan, Abergele, LL22 7JF
M0 GWY Iwan Williams, 19 Stryd y Brython, Ruthin, LL15 1JA
M0 GWZ Lewis Hicks, Front Basement, 11a Ventnor Villas, Brighton, BN3 3DD
M0 GXB George Bichard, 9 Kelburne Close, Winnersh, Wokingham, RG41 5JG
M0 GXC Gennaro Zaza, 42 Borras Road, Wrexham, LL12 7EP
M0 GXE Timothy Banks, 18 Leicester Road, Newport, NP19 7ER
M0 GXH J Hayward, 55 Hill Crest, Hoyland, Barnsley, S74 0BU
M0 GXK Jose Rodriguez Comillan, Flat 51, Waxham, London, NW3 2JJ
M0 GXM Michael Roe, 68 Argyle Street, Cambridge, CB1 3LR
M0 GXN Stephen Woodmore, 66 Imperial Way, Chislehurst, BR7 6JR
M0 GXO Ian Sheppard, 1 Frederick Street, Grassmoor, Chesterfield, S42 5AR
MM0 GXQ Gary Milne, 139 Rannoch Drive, Cumbernauld, Glasgow, G67 4ES
MM0 GXU G Sutherland, 7 Abbotsgrange Road, Grangemouth, FK3 9JD
M0 GXV Colin Berry, 60 Copthorne Road, Leatherhead, KT22 7EE
M0 GXW Robert Lee, 24 Wheatfields, Seaton Delaval, Whitley Bay, NE25 0PZ
MM0 GXY Royston Mannifield, 2 Plewlands Avenue, Edinburgh, EH10 5JY
M0 GYA Richard Moody, 372 Walsall Road, Perry Barr, Birmingham, B42 2LX
M0 GYB M Peterson, 4 Allen Drive, Mansfield, NG18 3HZ
M0 GYC Dan Fletcher, Flat 2, Redmires Court, Salford, M5 4US
MM0 GYD Alexander Young, 21 Corrour Road, Glasgow, G43 2DY
M0 GYF 610 Sqn City of Chester Air Cadets ARC c/o Mark Buxton, 610 SQN Air Traning Corps, Cadet Training Centre, Chester, CH1 4AN
MM0 GYG Andrew Fletcher, 164 Mayfield Road, Edinburgh, EH9 3AR
M0 GYH Michael Pearce, 38 Salisbury Road, Beaconsfield Upper, Victoria, Australia, 3808
M0 GYI David Leigh, 39 Hill Chase, Chatham, ME5 9HE
M0 GYK Andrew Roberts, 14 Cowper Close, Newport Pagnell, MK16 8PG
M0 GYL Michael Redman, 91 St. Andrews Road, Malvern, WR14 3PU
M0 GYM P McEwen, 26 Walton Avenue, North Shields, NE29 9BS
M0 GYN Kieron Hulme, Sutherland Road, Longsdon, Stoke-on-Trent, ST9 9QD
M0 GYO Darren Parker, 53 Brisbane Way, Cannock, WS12 2GR
M0 GYP Susan Gillard, 1 Chevening Close, Stoke Gifford, Bristol, BS34 8NJ
M0 GYR East Yorkshire Emergency Communications Group c/o Andrew Russell, 3 St. Nicholas Close, North Newbald, York, YO43 4TT
M0 GYS David Garner, Flat 4, Joseph Nye Court, Portsmouth, PO1 3RD
M0 GYU Lachizar Karchev, 6 Croombs Road, London, E16 3RY
MW0 GYV Peter Oseland, 6 Oaklands Close, Bridgend, CF31 4SJ
MM0 GYX Ian Watson, 10 Christie Place, Elgin, IV30 4HX
M0 GYY Gary Lewis, 93 Eastcliff, Portishead, Bristol, BS20 7AD
MM0 GZA Stephen Hargreaves, 4 Oxenfoord Avenue, Pathhead, EH37 5QD
M0 GZB Anthony Armitage, 6 Rosebery Avenue Hythe, Southampton, SO45 3HJ
M0 GZC Ian Coulson, 2 Marl Hurst, Edenbridge, TN8 6LN
M0 GZD Sherwood ARC c/o Edward Rippon, 19 Camelot Avenue, Nottingham, NG5 1DW
M0 GZE Piotr Slup, 1 The Meadow, Copthorne, Crawley, RH10 3RG
M0 GZF Martin Lamport, Bartwood, Dancing Green, Ross-on-Wye, HR9 5TE
M0 GZH Nigel Smith, 2 Norton Villas, Vicarage Road, Maidstone, ME18 6DX
M0 GZI Medway RS c/o Michael Sharp, Pentober, Firmingers Road, Orpington, BR6 7QG
M0 GZK Chun Sum Yung, Flat 8, Bridge Court, London, E10 7JS
M0 GZL Alan Burleton, 27 Doncaster Road, Bristol, BS10 5PN
M0 GZM John Oldman, 94 Mornington Road, London, E4 7DT
M0 GZN Fort Purbrook ARC c/o Geoff Wiggins, Cherry Trees Thorney Road, Emsworth, PO10 8BN
M0 GZS Graeme Hendry, 55 Central Avenue, Southport, PR8 3EQ
M0 GZU Regional Access Group c/o Adam Willis, Ledwyche Farm, Bleathwood, Ludlow, SY8 4LF
M0 GZW Thomas Lorn, 152 Brougham Court, Peterlee, SR8 1PZ
M0 GZX Allan Stromstedt, 3 Whisperdale Manor, Weydale Avenue, Scarborough, YO12 6AN
M0 GZZ Duncan Taylor, 1 Mayfield Farm Cottages, Reston, Eyemouth, TD14 5LG
MW0 HAB Mark Mainwaring, 36 Oak Street, Gilfach Goch, Porth, CF39 8UG
M0 HAF Robert Hambly, 144 Station Road, Irchester, Wellingborough, NN29 7EW
M0 HAG Graham Hill-Adams, 6 Broadleaze Way, Winscombe, BS25 1JX
M0 HAH Martin Summers, 21 Quantock Avenue, Caversham, Reading, RG4 6PY
M0 HAL P Musselwhite, 80 Craven Road, Orpington, BR6 7RT
M0 HAM C Bays, 116 Rochester Road, Durham, DH1 5PN
M0 HAN Peter Maennel, 4 Central Buildngs, Market Place, York, YO61 3AB
M0 HAO Maximo Martin De La Fuente, 8 Waterside Gardens, Reading, RG1 6QE
M0 HAP Alan Phillips, 3 Pen y Llys, Rhyl, LL18 4EH
MM0 HAR Harry Stuart, 31 Robertson Road, Lhanbryde, Elgin, IV30 8PE
M0 HAS Brian Hayes, 22 Urban Way, Biggleswade, SG18 0HT
MW0 HAT Richard Hatfield, 35 Victoria Road, Penarth, CF64 3HY
M0 HAU John Goodale, 82 Farnborough Road, Farnborough, GU14 6TH
MM0 HAY Scott Hay, 20 Woodside Way, Glenrothes, KY7 5DF
M0 HAZ Anthony Freeman, 34 Marmion Road, Coningsby, Lincoln, LN4 4RG
M0 HBC Brian Broad, 22 Minchin Acres, Hedge End, Southampton, SO30 2BJ
M0 HBE Marc Robins, 17 Old Turnpike, Fareham, PO16 7HB
M0 HBH Ewan Mathieson, 30 Lynfield Road, Frome, BA11 4JB
M0 HBI Stuart Payne, 3 Abbey Fields, Telford, TF3 2AH
M0 HBJ Stephen Blaikie, 22 Juno Close, Goring-by-Sea, Worthing, BN12 4UB
M0 HBL Peter Richmond, 7 Softley Drive, Norwich, NR4 7SE
M0 HBM Barry Denyer-Green, Dunsley South, Park Road, Forest Row, RH18 5BX
M0 HBN John Bell, 255 Willington Street, Maidstone, ME15 8EP
M0 HBO K Such, 38 Hornby Grove, Hull, HU9 4PG
M0 HBT David Nelson, 110 Chandag Road, Keynsham, Bristol, BS31 1QF
M0 HBU Ian Duffie, Trebeighan Farm, Saltash, PL12 5AE
M0 HBV David Ingrey, 1 Pondero Road, Fordham, Colchester, CO6 3LX
M0 HBW Barry Adby, 26 Love Lane, Wallington, OX10 6RA
M0 HBX Jonathan Pelham, 20 Merchants Court, Bedford, MK42 0AT
M0 HBY Peter Watkins, 135 Lodge Road, Writtle, Chelmsford, CM1 3JB
M0 HCA Frederick Price, 538 Abergele Road, Old Colwyn, Colwyn Bay, LL29 9LD
M0 HCC Cristian Dumitrescu, The Rectory, Halkyn, Holywell, CH8 8BU
M0 HCE Marcel De Jong, Grachtstraat 64, Oirsbeek, Netherlands, 6938 HP
M0 HCI G Burton, 26 Church View, Egremont, CA22 2DT
MI0 HCK Conor Robinson, 71 Eglantine Road, Lisburn, BT27 5RQ
M0 HCM Lukasz Michalowski, 11 St. Giles Park, Catterick Garrison, DL9 4XA
M0 HCN Dan Mills, 261 West Wycombe Road, High Wycombe, HP12 3AS
MM0 HCO Stuart McKenzie, 0/2 69 Glenkirk Drive, Glasgow, G15 6AU
M0 HCP John Hunt, 14 Nevill Close, Hanslope, Milton Keynes, MK19 7NY

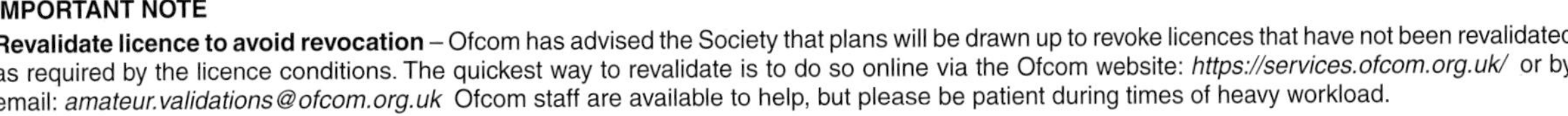

**IMPORTANT NOTE**

**Revalidate licence to avoid revocation** – Ofcom has advised the Society that plans will be drawn up to revoke licences that have not been revalidated as required by the licence conditions. The quickest way to revalidate is to do so online via the Ofcom website: *https://services.ofcom.org.uk/* or by email: *amateur.validations@ofcom.org.uk* Ofcom staff are available to help, but please be patient during times of heavy workload.

M0 HCT Martin Fitzjohn, 96 Nightingale Gardens, Nailsea, Bristol, BS48 2BN
M0 HCV Christopher Wallace, 16 Morley Square, Bristol, BS7 9DW
M0 HCW John Morgan, 3 Maes Yr Hebog, Penrhyn Bay, Llandudno, LL30 3EY
M0 HCY Blackwater Radio Contest Group c/o Alan Copperwaite, 71 Gladbeck Way, Enfield, EN2 7EL
M0 HCZ Colin Lycett, 2 Royce Avenue, Hucknall, Nottingham, NG15 6FU
MM0 HDA Robert Fairfull, 26 Inchconnachan Avenue, Balloch, Alexandria, G83 8JN
M0 HDC Norfolk County Raynet c/o Stuart Lucas, 31 Lilian Close, Norwich, NR6 6RZ
M0 HDE A Morris, 71 Lurdin Lane, Standish, Wigan, WN6 0AQ
M0 HDG HALLAM DX GROUP c/o Nicholas Totterdell, Moscar Cross House, Hollow Meadows, Sheffield, S6 6GL
M0 HDJ David Hall, 8 Colston Close, Bristol, BS16 4PQ
M0 HDK Edward Erbes, 488 Birkfield Drive, Ipswich, IP2 9JE
M0 HDN Bernd Richter, C/O Dr Steffen Grant, Wolfson College, Oxford, OX2 6UD
M0 HDP P Bolton, 2 Alexander Court, Chute Lane, St. Austell, PL26 6NU
M0 HDQ Gerard Van Breemen, 58 Horseshoe Lane, Bromley Cross, Bolton, BL7 9RR
M0 HDR Raymond Scholey, Barleycroft, Lower Road, Ipswich, IP6 9AR
M0 HDS HINCKLEY DISTRICT SCOUTS c/o M Smith, Hinckley District Scout Hq, St Marys Road, Hinckley, LE10 1EQ
M0 HDT Derek Simpson, 50 Castle Hill, Berkhamsted, HP4 1HF
M0 HDU John Legrain, 22 Cromwell Drive, Didcot, OX11 9RB
M0 HDV David Cowling, 11 Shakespeare Avenue, Scunthorpe, DN17 1SA
MM0 HDW James Duncan, 36 Bank Row, Wick, KW1 5EY
M0 HDX James Dexter, 96 Wilsthorpe Road, Chaddesden, Derby, DE21 4QS
MD0 HEB H Blackburn, 32 Close Rushen, Castletown, Isle of Man, IM9 1NN
M0 HEF Kevin Allen, 55 Fall Road, Heanor, DE75 7PQ
M0 HEJ George Hatt, 4H Colman House, Earlham Road, Norwich, NR4 7TJ
M0 HEM John O'Toole, 4 Lindisfarne Road, Dagenham, RM8 2RA
M0 HEP Giacomo Zorzi, 3b Ambleside Avenue, Telscombe Cliffs, Peacehaven, BN10 7LS
M0 HET Stephen Cordner, 29 Buxton Road, Aylsham, Norwich, NR11 6JD
M0 HEW Tony Johnson, 81 Welbeck Street, Whitwell, Worksop, S80 4TN
M0 HEX John Ash, 47 Stein Road, Emsworth, PO10 8LB
M0 HEY M Hickford, 56 Alexander Close, Abingdon, OX14 1XB
M0 HFA Alex Birkett, 67a Branston Road, Burton-on-Trent, DE14 3BY
M0 HFB Pawel Szewczyk, 514 Whaddon Way, Bletchley, Milton Keynes, MK3 7LD
M0 HFC Humber Fortress DX ARC c/o John Cunliffe, 142 Hall Road, Hull, HU6 8SB
M0 HFE Barnsley And District ARC c/o Jan Sobanski, 10 Robert Avenue, Barnsley, S71 5RB
M0 HFF E Bray, 28 Henshall Avenue, Latchford, Warrington, WA4 1PY
M0 HFH John Rowden, 18 Dyrham Close, Thornbury, Bristol, BS35 1SX
M0 HFI Clan Maclean ARS c/o James McLean, 24 Durham Drive, Oswaldtwistle, Accrington, BB5 3AT
M0 HFO Mark Jessop, Department Of Electronic Engineering, Claverton Down, BA2 7AY
M0 HFQ Cameron Lai, Storeys Way, Cambridge, CB3 0DG
M0 HFR G Hall, 93 Berkeley Avenue, Bexleyheath, DA7 4TZ
MM0 HFU Edward Horn, 3 McKay Place, Newton Mearns, Glasgow, G77 6UZ
M0 HFW de Havilland Heritage Radio Group c/o James Newton, 28 Dunstan Road, London, NW11 8AA
M0 HFX A Walker, 17 Carr House Road, Halifax, HX3 7QY
M0 HFY Barry Eames, 22 Ashgrove Close, Hardwicke, Gloucester, GL2 4RT
M0 HFZ Bryan Cox, 7 Wolsey Avenue, London, E6 6HG
M0 HGA Dean Corless, 10 The Tannery, Shipston-on-Stour, CV36 4EH
M0 HGD David Molloy, 187 Babylon Lane, Heath Charnock, Chorley, PR6 9ET
M0 HGG Christopher Regan, 1 Fairways, Birkenhead, CH42 8JZ
MW0 HGK Wai Ming Tse, Marino Room, Fulton Houose, Swansea, SA2 8PP
MW0 HGM Andrew Pritchard, 12 Llys le Breos, Mayals, Swansea, SA3 5DL
MM0 HGN David Higgins, 1 Peacehill Farm Cottages, Wormit, Newport-on-Tay, DD6 8PJ
M0 HGS Michael Shepherd, Oak View, Church Road, Halstead, CO9 4PR
M0 HGV R Dodds, West Villa, The Green, Wallsend, NE28 7PG
M0 HGY James Read, 31 Merebrook Road, Macclesfield, SK11 8RH
M0 HHA Michael Meehan, 14 Grosvenor Road, Walton, Liverpool, L4 5RB
M0 HHB Graham Willard, 4 Varrier Jones Place, Papworth Everard, Cambridge, CB23 3XP
M0 HHC Kenneth Jackson, 4 Milfoil Close, Marton-in-Cleveland, Middlesbrough, TS7 8SE
M0 HHD Peter Rogers, 16 Begonia Close, Basingstoke, RG22 5RA
M0 HHF Clive Greenwood, 1 Bentinck Close, Boughton, Newark, NG22 9HP
M0 HHG Gary Aldridge, Greenridge, Fore Street, Teignmouth, TQ14 9QR
MD0 HHH Henry Dorman, 1 Sprucewood Rise, Foxdale, Douglas, Isle of Man, IM4 3JP
M0 HHI John Hughes, Milestone House, Easole Street, Dover, CT15 4HE
M0 HHM William Roberts, 113 Somerset Avenue, Luton, LU2 0PL
M0 HHP Marcin Kasprzyk, 420d Streatham High Road, London, SW16 3SN
M0 HHR Michael Lee, 34 Astley Road, Liverpool, L36 8DA
M0 HHT HAMPSHIRE HILLTOPPERS ARC c/o Andrew Digby, 4 Paddock Close, South Wonston, Winchester, SO21 3EQ
MI0 HHU Richard Benko, 23 Six Mile Water Mill Drive, Antrim, BT41 4FG
MI0 HHV Bryan Craney, 8a Drumhoy Drive, Carrickfergus, BT38 8NN
M0 HHW William White, 15 St. Walstans Road, Taverham, Norwich, NR8 6NF
M0 HHX Andrew Currie, 20 Portal Road, Eastleigh, SO50 6AY
M0 HIA Alan Bulman, 21 Stannington Road, North Shields, NE29 7JY
M0 HIC HOUNSLOW A.R. INSTRUCTION CENTRE c/o M De Silva, 31 Rosemary Avenue, Hounslow, TW4 7JQ
M0 HID Savino Leo, 37 The Coldra, Newport, NP18 2LS
M0 HIE Edward Harman, 53 Anthony Road, Borehamwood, WD6 4NB
M0 HIG Brian Hultquist, 37 New Road, Tiptree, Colchester, CO5 0HN
M0 HIH Kevin Manos, 102 Goodwood Avenue, Sale, M33 4QL
M0 HIJ Norfolk & Suffolk 4x4 Response c/o James Whiteside, The Old Antique Shop, Bank Street Pulham Market, Diss, IP21 4TG
M0 HIL D Hill, 11 Paddock Lane, Metheringham, Lincoln, LN4 3YG
M0 HIM Peter Newsome, 49 Danby Close, Washington, NE38 9JB
M0 HIN Hinckley Sea Cadets c/o Vincent Hopkins, 109 Smith Street, Coventry, CV6 5EH
M0 HIO David Bednarski, 52 Seabridge Lane, Newcastle, ST5 3EY
M0 HIP Hippings Methodist Primary School Amateur Radio Cl c/o James McLean, 24 Durham Drive, Oswaldtwistle, Accrington, BB5 3AT
M0 HIQ Derek Cotton, 1 Fieldfare Close, Penwortham, Preston, PR1 9NG
M0 HIW Philip Jones, 10 Moulton Road, Tivetshall St. Margaret, Norwich, NR15 2AJ
M0 HIX Alan Holmes, 2 Park Farm Cottages, Park Lane, Chichester, PO20 3TL
M0 HIY Ashley Thomas, 1 Millers Close, Ruardean Hill, Drybrook, GL17 9AU
M0 HIZ William Easdown, 38 North Street, Barming, Maidstone, ME16 9HF
M0 HJA Patrick Chong, 320 Glenalmond Avenue, Cambridge, CB2 8DT
M0 HJB Mark Stillman, 58 Highfield Road, Bognor Regis, PO22 8PH
MM0 HJC Clydebank Joint Cadet Centre c/o Joseph Connelly, 9 Glenhead Crescent, Hardgate, Clydebank, G81 6LW
M0 HJD D HARBRON, 6 West View, Penshaw, Houghton le Spring, DH4 7HP
M0 HJE Paul Frost, 11 Church Road, Swainsthorpe, Norwich, NR14 8PH
M0 HJF Howard Felstead, Rosmede, Windmill Drive, Littlehampton, BN16 3HW
M0 HJG Norman Williams, 27 Meadway, Rogiet, Caldicot, NP26 3SA
M0 HJI Richard Havart, 2 Holly Farm Road, Reedham, Norwich, NR13 3TH
M0 HJJ Andrea Wierdis, 39 Milton Road, London, SW19 8SF
M0 HJL Richard Taylor, 27 The Holt, Hailsham, BN27 3ND
M0 HJN Wlodzimierz Tomczyk, 3L, 76 Berry Street, New York, United States, 11249
M0 HJO J Brooks, Treven House, Treven, Tintagel, PL34 0DT
M0 HJQ Peter Garrett, 21 Wychbury Road, Wolverhampton, WV3 8DN
MD0 HJR David Vale, 21 Chelston Road, Ruislip, HA4 9SA
M0 HJW Ian Ftaiha, 44b The Broadway, London, NW7 3LH
M0 HJY Robert Green, 44 Aldwyn Place, Larchwood Drive, Egham, TW20 0RZ
MW0 HKA Max Day, 11 Troedrhiw-Trwyn, Pontypridd, CF37 2SE
M0 HKB Kieron Brunning, 70 Cobbold Road, Felixstowe, IP11 7QR
M0 HKC Kevin Cullum, 13 Baker Road, Shotley Gate, Ipswich, IP9 1RT
M0 HKE Andrew Mullin, 111 Arps Road, Codsall, Wolverhampton, WV8 1SG
M0 HKG M Clarke, 14 Tower Court, Haverhill, CB9 9DD
M0 HKH Antonio Fronters, Flat 54, Central Quay North, Bristol, BS1 4AU
M0 HKI Leslie Todman, 17 Hall Road, St. Dennis, St. Austell, PL26 8BE
M0 HKJ Warrington Sea Cadets ARC c/o Lee Layland, 3 Thirlmere Road, Golborne, Warrington, WA3 3HH
M0 HKK Alan Doe, 26 Beachfield Road, Bembridge, PO35 5TN
M0 HKL Gabriele Alberti, Kreuzerweg 33, München, Germany, 81825
M0 HKP Dale Potts, 30a Tower Hill, Gomshall, Guildford, GU5 9LS
M0 HKS Michael Booth, 30 Manor Green, Harwell, Didcot, OX11 0DQ
M0 HKT Mexborough ARS c/o Allan Farrar, 8 Wensley Street, Thurnscoe, Rotherham, S63 0PX
MM0 HKU Edmund Duncan, 3 George Street, Banff, AB45 1HS
M0 HKV Paul Bull, 87 Braemor Road, Calne, SN11 9DU
M0 HKW Adrian Brand, 6 Walnut Close, Milton, Cambridge, CB24 6ET
M0 HLA A Smith, 186 Longmead Drive, Nottingham, NG5 6DJ
M0 HLB David Slater, 13 Longford Close, Rainham, Gillingham, ME8 8EW
M0 HLC Colin Taylor, 1 Jasmine Gardens, Warrington, WA5 1GU
M0 HLD D Hicks, 36 Middlesex Road, Maidstone, ME15 7PL
M0 HLF Amanda Higton, 4 Paddocks Close, Pinxton, Nottingham, NG16 6JR
M0 HLI Darren Hyde, 136 Station Road, Woodmancote, Cheltenham, GL52 9HN
MM0 HLK 1777 Squadron Air Training Cadets ARS c/o Barry Spink, 9 St. Andrews Crescent, Dumbarton, G82 3ER
M0 HLM Klaus Schmidt, Church, Corner, Mareham-le-Fen, PE22 7RA
MM0 HLN Helen Mason, 20 David's Crescent, Kilwinning, KA13 6JJ
M0 HLO Graham Belson, 19 Hamilton Road, Grantham, NG31 9QG
M0 HLP George Bunting, 31 Hardwick Avenue, Allestree, Derby, DE22 2LN
MM0 HLQ Graeme Gilmour, 100 Main St., Milngavie, Glasgow, G62 6JN
M0 HLR Anthony Brown, 3 Alston Road, New Hartley, Whitley Bay, NE25 0ST
M0 HLS Christopher Murphy, 17 Shepherd Street, Littleover, Derby, DE23 6GA
MM0 HLU Ioannis Konstas, 25/4 Milton Street, Edinburgh, EH8 8HA
M0 HLV Christopher Hicks, Withymore Cottage, Day House Lane, Wotton-under-Edge, GL12 7QY
M0 HLW Wallace Maxwell, 25 Laurel Drive, Buckley, CH7 2QP
M0 HLX Douglas Bailey, 2b Queens Road, Enfield, EN1 1NE
M0 HLY Andrew Nisbet, 65a Hamilton Road, Felixstowe, IP11 7BE
M0 HLZ Michael Meachen, 20 Wilkinson Road, Rackheath, Norwich, NR13 6SG
M0 HMB Richard Stratford, 32a Priory Avenue, High Wycombe, HP13 6SW
MI0 HMC H McErlean, 24 Mullaghboy Heights, Magherafelt, BT45 5NU
M0 HME Richard Campbell, 17 Elgar Road, Southampton, SO19 0JG
M0 HMF Michael Smith II, The White House, Old Avenue, West Byfleet, KT14 6AE
M0 HMI Giora Tamir, 3rd Floor, Future House, Egham, TW20 9AH
M0 HMJ Dainius Sadauskas, 4 Priory Road, Tiverton, EX16 6TQ
M0 HMO Heather Lomond, Holy Mill, Longville, Much Wenlock, TF13 6ED
M0 HMR C Harmer, Spring Corner, Rockness Hill, Stroud, GL6 0PJ
M0 HMS Eugene Purvis, 36 Birchington Avenue, Middlesbrough, TS6 7EZ
M0 HMU Fleetwood Radio Enthusiasts Group c/o John Earnshaw, 128 Shakespeare Road, Fleetwood, FY7 7HJ
MW0 HMV C Josey, 157 Waterloo Road, Penygroes, Llanelli, SA14 7PU
M0 HMX Richard Sykes, 11 Lodwells Orchard, North Curry, Taunton, TA3 6DX
MI0 HMY A Hamill, 15 Maythorn Avenue, Coleraine, BT52 2EU
M0 HMZ Pavel Iljin, 63 Phillimore Road, Birmingham, B8 1PS
M0 HNA Southern Microwave Group c/o D Austen, Tudorlands, Silchester Road, Tadley, RG26 5DG
M0 HNC Antonio Ribeiro, 38a Galpins Road, Thornton Heath, CR7 6EB
M0 HND Technical Experimenters Group c/o Brian Smith, 73 Devon Street, Hull, HU4 6PL
M0 HNE Roger Ashley, 15 Wimbourne Drive, Gillingham, ME8 9EN
M0 HNF Paul Dickson, 49 Signal Road, Grantham, NG31 9BL
M0 HNG A Douglas, Gobbins Cottage, Sandy Lane, Ormskirk, L40 5TU
M0 HNH Arron Reason, 1 Iles Cottages, St. Marys, Stroud, GL6 8NX
M0 HNI Richard Weaver, 15 Sharps Field, Headcorn, Ashford, TN27 9UF
M0 HNJ Phillip Evans, 144 Grenville Street, Stockport, SK3 9ET
M0 HNK Richard Tofts, Elmcroft, Redhill Road, Ross-on-Wye, HR9 5AU
M0 HNL R Campbell, 2 Hesketh Bank, York, YO10 5HH
MW0 HNM Louise Williams, 96 Shelone Road, Neath, SA11 2PU
M0 HNN Thomas Walsh, 2 Ashfield Mews, Ashington, NE63 9GJ
M0 HNO Hideaki Nishio, 28 New Lane, Havant, PO9 2NQ
MI0 HNQ Hilltop ARC Co.Down c/o Andrew McGarvey, 66a Scaddy road, Downpatrick, BT30 9BS
M0 HNT Alan Angus, 51 Osprey Drive, Blyth, NE24 3QS
M0 HNX Richard Raeburn, 145 Paddock Road, Basingstoke, RG22 6QQ
M0 HOB Garell Brotherhood, 17 Baldwin Close, Forest Town, Mansfield, NG19 0LR
MD0 HOF Besim Ajeti, 88 Bushfield Crescent, Edgware, HA8 8XJ
M0 HOH Stephen Dalton, Flat 31, Matheson Lang Gardens, London, SE1 7AN
M0 HOI Stephen Musgrave, Orchard Cottage, Stalmine, Poulton le Fylde, FY6 0LZ
M0 HOJ Francisco Costa, 9 Moyne Close, Cambridge, CB4 2TA
M0 HOK Lee Carberry, 23 Greens Beck Road, Stockton-on-Tees, TS18 5AR
MM0 HOL C King, 19 Gleneagles Way, Deans, Livingston, EH54 8EW
M0 HOM M Hotchin, 122 Buckingham Avenue, Scunthorpe, DN15 8NS
M0 HOP A Hopson, 1 Hall Lane, Leicester, LE2 8SF
M0 HOQ James Johnson, 226 Preston New Road, Southport, PR9 8NY
M0 HOT Halam Rose, 10 St. Vincents Close, Girton, Cambridge, CB3 0PE
M0 HOU Hooman Atifeh, 9 Stagshaw Close, East Hunsbury, Northampton, NN4 0WE
M0 HOV Angel Hristov, Flat A, 71 Beckenham Lane, Bromley, BR2 0DN
M0 HOY Stephen Curtis, 354 St. Helens Road, Leigh, WN7 3PQ
MI0 HOZ Michael Conaghan, 94 Curlyhill Road, Strabane, BT82 8LS
M0 HPB Darren Bisbey, 17 Benson Close, Lichfield, WS13 6DA
MI0 HPE Paul Dorris, 29 Eia Street, Belfast, BT14 6BT
M0 HPF Glitsun Cheeran, 201 Eastcombe Avenue, London, SE7 7LH
M0 HPG Cumbria Raynet Group c/o Paul Woodburn, 21 The Row, Silverdale, Carnforth, LA5 0UG
MW0 HPH Ystrad Mynach College c/o Philip Jones, 23 Pinecroft Avenue, Aberdare, CF44 0HY
M0 HPJ James Whiteside, The Old Antique Shop, Bank Street Pulham Market, Diss, IP21 4TG
M0 HPL R Masshedar, 6 Hutton Avenue, Hartlepool, TS26 9PN
M0 HPP Gerard Fleming, 1 Balmoral Drive, Methley, Leeds, LS26 9LE
M0 HPR Hugh Richardson, 7 Regent Road, Leyland, PR25 2LJ
M0 HPS H Powell, 6 Sowbury Park, Chieveley, Newbury, RG20 8TZ
M0 HPT David Prior, 10 Birley Close, Appley Bridge, Wigan, WN6 9JL
M0 HPU David Rudling, Rose Cottage, Ludwells Lane, Southampton, SO32 2NP
M0 HPV D Green, 67 Coombe Park Road, Binley, Coventry, CV3 2NW
M0 HPW Michael Phillips, 59 Bradeley Road, Haslington, Crewe, CW1 5PX
MI0 HPX Anthony Sweeney, 117 Lisnablagh Road, Coleraine, BT52 2HD
M0 HPZ Leslie Edmonds, 3 Waterlow Road, London, N19 5NJ
M0 HQA Peter Massolt, 26 Redgate Heights, Hunstanton, PE36 5EA
M0 HQB Krzysztof Kulpinski, 44 Redlake Drive, Taunton, TA1 2RS
M0 HQC Maureen Copse, 3 The Limes, Market Overton, Oakham, LE15 7PX
MM0 HQD David Searle, 32 Linwood Terrace, Hamilton, ML3 9AF
M0 HQE Anthony Clark, 106 Bruce Avenue, Hornchurch, RM12 4HZ
M0 HQG South Normanton and District ARC c/o John Mason, 56 Skegby Road, Sutton-in-Ashfield, NG17 4EZ
M0 HQH Stephen Pettit, 88 Abbotsbury, Great Hollands, Bracknell, RG12 8QX
MM0 HQI Laurie Richings, 2 St. Margarets Place, Edinburgh, EH9 1AY
M0 HQJ Henry Quigg, Station Cottage, Ripon Road, Thirsk, YO7 4PS
M0 HQL Stephen Mitchell, 42 Fairfield Avenue, Felixstowe, IP11 9JJ
M0 HQM Robert Givens, 13 Wakehurst Drive, Crawley, RH10 6DL
M0 HQO Peter Freeman, 57 Ruffa Lane, Pickering, YO18 7HN
M0 HQP Neil Marley, Penstemons, Chapel Lane Pen Selwood, Wincanton, BA9 8LY
M0 HQQ Paul Taylor, 54 Church Road, Stanley, Liverpool, L13 2BA
M0 HQR Praful Naik, 82 Misbourne Road, Uxbridge, UB10 0HW
M0 HQU Cesar Lombao, 6 Privet Close, Lower Earley, Reading, RG6 4NY
M0 HQZ James Widdowson, 26 Woodville Gardens West, Boston, PE21 8BW
M0 HRA Henry Alderson, 66 Houghtonside ESTATE, Houghton le Spring, DH4 4BW
M0 HRC Wayne Nicholas, 16 Withymoor Road, Netherton, Dudley, DY2 9LA
M0 HRD C Hughes, 88 Derlwyn Street, Phillipstown, New Tredegar, NP24 6BA
MI0 HRG Hill Top Radio Group c/o Bronwin Vaughan, 29 Crew Road, Victoria Bridge, Strabane, BT82 9LS
M0 HRH C Morley, 191 Purbrook Way, Havant, PO9 3RS
MM0 HRL Ian Gourlay, 76 Largo Road, St. Andrews, KY16 8NJ
M0 HRM Colin Greenwood, 44 Fountain Street, Heckmondwike, WF16 9HS
MI0 HRO Charles Stockdale, 3 Hightown Drive, Newtownabbey, BT36 7TG
M0 HRP Robin Huelin, 15 Hill Chase, Walderslade, Chatham, ME5 9HE
M0 HRT Robert Bryan, 1 White Cottage, Old Warwick Road, Solihull, B94 6LN
MI0 HRV Tommy Darrah, 42 Pinewood Avenue, Carrickfergus, BT38 8EW
M0 HRW Alan Parker, 9 Milecastle Court, Newcastle upon Tyne, NE5 2PA
M0 HRY Steve Wheeler, 98 Charterhouse Road, Orpington, BR6 9EW
M0 HRZ David Irving, Noctorum House, Noctorum Road, Prenton, CH43 9UQ
MM0 HSA Hugh Steele, 44 Waverley Crescent, Livingston, EH54 8JN
MM0 HSB William Forrester, 149 Whyterose Terrace, Methil, Leven, KY8 3AR
M0 HSG Peter Scrimshaw, 38 Fourdrinier Way, Hemel Hempstead, HP3 9RP
M0 HSH James Brooks, 8 Moorhaven Close, Torquay, TQ1 4AA
M0 HSI CLWUD PORTABLE OPERATING GROUP c/o Melfyn Allington, 3 Wynnes Parc Cottages, Brookhouse, Denbigh, LL16 4YB
M0 HSJ Howard Jones, 116 Dark Lane, Bedworth, CV12 0JH
M0 HSQ Ioannis Tsimperidis, 5 Somersham, Welwyn Garden City, AL7 2PZ
MM0 HSR Brannock High Radio Group c/o Peter Bainbridge, 46 Rigghouse View, Bathgate, EH47 0SE
M0 HSS Alan Perrow, 16 Bannister Walk, Cowling, Keighley, BD22 0NU
M0 HSU Stewart Challis, 73 Rivenhall Way, Hoo, Rochester, ME3 9GF
MM0 HSV Ken Baird, 24 Main Street, Sorn, KA5 6HU
M0 HSW H Scott Whittle, 92 The Grove, London, W5 5LG
M0 HSX M Josi, 10 Robert Close, Billericay, CM12 9DS
M0 HSZ John Merritt, 41 Great Grove, Bushey, WD23 3BQ
M0 HTA Ian Cooke, 11 Farriers Gate, Chatteris, PE16 6AY
M0 HTB Henryk Banasiak, 11 Westfield Road, Backwell, Bristol, BS48 3NE
M0 HTE John Taylor, 90 Village Road, Gosport, PO12 2LG
M0 HTF Colin Rose, 132 Golf Green Road, Jaywick, Clacton-on-Sea, CO15 2RW
M0 HTG Goronwy Edwards, 17 Glan y Mor Road, Penrhyn Bay, Llandudno, LL30 3NL
M0 HTI S Storey, 11 Enderby Road, Sunderland, SR4 6BA
M0 HTJ Hamtests.co.uk c/o Paul Gibson, 7 Greenfields Road, Horley, RH6 8HW
M0 HTK Hans Kassier, 26 Higher Port View, Saltash, PL12 4BX

M0 HTL [illegible]
M0 HTO Brecon and Radnor ARS c/o David Bowen, 25 Maendu Terrace, Brecon LD3 9HH
M0 HTQ Kaoru Numata, 28 Guildhouse Street, London, SW1V 1JJ
M0 HTR Ashton in Makerfield ARC c/o Peter Williams, 35 Cansfield Grove Ashton-in-Makerfield, Wigan, WN4 9SE
M0 HTS Craig Mellor, 104 Rocky Lane, Eccles, Manchester, M30 9LY
M0 HTU John Stokoe, 15 Robin Close, Market Deeping, Peterborough, PE6 8PQ
M0 HTV Martin Gregson, 10 Eden Avenue, Consett, DH8 6EZ
M0 HTX David J Anderson, 53 Collywell Bay Road, Seaton Sluice, Whitley Bay, NE26 4RG
M0 HTY Mark Tointon, 13 Ridgeway, Broadstone, BH18 8DY
M0 HUA Aaron Brown, 7 Coombewood Drive, Romford, RM6 6AA
M0 HUD Stephen Pantony, 40 Park Avenue, Redhill, RH1 5DP
MM0 HUF Stewart Harvey, 63 Darley Road, Cumbernauld, Glasgow, G68 0JR
M0 HUG S Fyre, St. Michael Mead, The Common, Norwich, NR12 8BA
M0 HUH Ping Liang Tan, St Edmund's College, Cambridge, CB3 0BN
M0 HUI Ian Magness, Orchard House, Ravenswood Drive, Camberley, GU15 2BU
M0 HUM 633 (West Swindon) Squadron ATC c/o Robin Ley, 23 Heronbridge Close, Westlea, Swindon, SN5 7DR
M0 HUN John Hunt, Flat 2, Winton Court, 25 Goldsel Road, Swanley, BR8 8DU
MM0 HUQ Magnus Flaws, West Voe, Sumburgh, Shetland, ZE3 9JN
M0 HUS Hugh Steers, 39 Upwood Road, London, SE12 8AE
MW0 HUU Martin Pope, 4 Croft Villas, Narberth, SA67 7DY
M0 HUV Eduardo Valdez, 65 Broken Cross, Charminster, Dorchester, DT2 9QB
M0 HUW Gabriele Gentile, Via B, Vecchia, Preganziol, Italy, 31022
MW0 HUY K Saltmarsh, 15 Colbourne Road, Beddau, Pontypridd, CF38 2LN
M0 HUZ Clifford Warwick, 104 Church Road, Formby, Liverpool, L37 3NH
MM0 HVA John Hawkins, 1f2 13 Watson Crescent, Edinburgh, EH11 1HB
MW0 HVB Howard Bancroft, Stop and Call, Goodwick, SA64 0EX
M0 HVC Robin Barnard, 3 Heaths Close, Enfield, EN1 3UP
M0 HVD David King, 25 Church Road, Worthing, BN13 1ET
M0 HVE Lawrence Sargent, 13 Park View, Thornton, Liverpool, L23 4TD
MM0 HVF Zoe Bak, 62/6 North Gyle Loan, Edinburgh, EH12 8LD
M0 HVI Mieczyslaw Kurczab, 159 Huddersfield Road, Halifax, HX3 0AH
M0 HVK David Ackrill, Flat 60, Chamberlaine Court, Banbury, OX16 2PA
MW0 HVL Emrys England, 2 Luton Street, Blaenllechau, Ferndale, CF43 4PB
M0 HVM Joerg Vollbrecht, Reinsdorf, Steingasse 3, Nebra (Unstrut), Germany, 6642
M0 HVN David Connolly, 2 Layton Close, Birchwood, Warrington, WA3 6PT
M0 HVO I Bailey, 8 Willow Drive, Ringwood, BH24 3BE
M0 HVP 1466 Holmfirth c/o Neil Tindall, Royds Mount, Linthwaite, Huddersfield, HD7 5QX
M0 HVQ D Holland, 7 Hayward Close, Walkington, Beverley, HU17 8YB
M0 HVR John Brawn, 9 Westbury Road, Westbury-on-Trym, Bristol, BS9 3AY
M0 HVS Neil Bethell, 4 Magazine Road, Wirral, CH62 3LH
MM0 HVU David Smith, 25 High Academy Street, Armadale, Bathgate, EH48 3HG
M0 HVV Mark Hickman, 13 Millfields Avenue, Rugby, CV21 4HJ
MM0 HVW David Plummer, 39 St. Nicholas Drive, Banchory, AB31 5YG
M0 HWC Hadley Wood Contest Group c/o Michael Ruttenberg, 90 Heath View, London, N2 0QB
M0 HWD Daniel Levy, Flat 36, Claydon House, London, NW4 1LS
MI0 HWG Peter Moore, 32 Kinnegar Rocks, Donaghadee, BT21 0EZ
M0 HWH Kevin Quigley, 12 Silver Lane, Billingshurst, RH14 9RJ
M0 HWI Andrzej Trzepietowski, 78 Dennis Road, Coventry, CV2 3HR
M0 HWJ Andrzej Boldireff Strzeminski, 47 Waters Edge, Canterbury, CT1 1WX
M0 HWL Robert Riches, Flat 21, Hawthornden, Otley, LS21 3LE
M0 HWM S Baddeley, 50 Western Esplanade, Herne Bay, CT6 8JA
M0 HWN E Wilcockson, Conybeare House, Willowbrook, Windsor, SL4 6HL
M0 HWO George Phillips, 73 Gotham Road, Wirral, CH63 9NG
M0 HWP Jeremy Tarrant, 70 Sunnymead, Midsomer Norton, Radstock, BA3 2SD
M0 HWQ Peter Browne, 151 North Road, St. Andrews, Bristol, BS6 5AH
M0 HWS Griffith Hewis, 10 Albert Road, New Malden, KT3 6BS
M0 HWT George Mutch, 94 Abbotswood Road, Brockworth, Gloucester, GL3 4PF
MW0 HWU RAYNET Pembrokeshire c/o Ian Baker, 28 Kensington Road, Neyland, Milford Haven, SA73 1TL
M0 HWV Paul Heiney, 6 Arthur Street, Oxford, OX2 0AS
M0 HWW Bojan Vemic, Flat 3, Oldbury House, London, W2 5HA
M0 HWY Henry Kennedy, 11 Green Road, High Wycombe, HP13 5BD
M0 HXA Adrian Crosland, 18 Duncan Crescent, Bovington, Wareham, BH20 6NN
MI0 HXB Thomas Browne, 7 Hawthorn Park, Greysteel, Londonderry, BT47 3YE
M0 HXC Ferdinand Kroon, 7 Old Drum Mews, Chapel Street, Petersfield, GU32 3DP
M0 HXE D Hill, 109 Hitchin Close, Romford, RM3 7EQ
M0 HXF Richard Thorpe, 6 Millthorpe, Sleaford, NG34 0LD
M0 HXG Andrew Carden, Hazelgrove, South Allington, Kingsbridge, TQ7 2NB
M0 HXH John [illegible], [illegible] Road, New Ollerton, Newark, NG22 0UL
M0 HXI Cyril Baker, 29 Green Lane, Bristol, BS11 9JD
M0 HXK Henry Carruthers, 31 Baden Street, Hartlepool, TS26 9BJ
M0 HXM Daniel Estevez, Oceano Atlantico, 38, Tres Cantos, Spain, 20760
M0 HXN John Orme, 42 Dovecote, Newport Pagnell, MK16 8BD
M0 HXO John Neal, 48 Mansfield Road, South Normanton, Alfreton, DE55 2ER
M0 HXS Edgar Haonor, 110 Great Stone Road, Manchester, M16 0HD
M0 HXV Andrew Yeomans, 65 Grove Road, Tring, HP23 5PB
MW0 HXX David Machon, 22 Albert Street, Caerau, Maesteg, CF34 0UF
M0 HYA Jonathan Starbuck, 8 Plas Panteidal, Aberdyfi, LL35 0RF
M0 HYC T Rutt, Granthorpe, Hull Road, Hull, HU11 5RN
M0 HYD F Hyde, 10 Devonshire Drive, Barnsley, S75 1EE
M0 HYE Thomas Byers, 1 Hazelwood Avenue, Sunderland, SR5 5AH
M0 HYG Harry Hope, 51 Margravine Gardens, London, W6 8RN
M0 HYH Colin Glass, The Old Homestead, Havikil Lane, Knaresborough, HG5 9HN
M0 HYJ Alan Rand, 17 Fairways Drive, Harrogate, HG2 7ES
M0 HYK Kym Dutfield-Cooke, Tan yr Efail, Segurinside, Llandudno Junction, LL31 9QE
M0 HYL Alan Robnett, 38b Woodmere Avenue, Watford, WD24 7LN
MM0 HYM William Jackson, 3 Annick Road, Dreghorn, Irvine, KA11 4EY
M0 HYN Barnaby Davies, 12 Scalebor Gardens, Burley-in-Wharfedale, LS29 7BX
MW0 HYP D Thomas, 67 Crynallt Road, Neath, SA11 3RN
MI0 HYQ Artur Zakrzewski, 11 Millbrook Gardens, Kilrea, Coleraine, BT51 5RZ
M0 HYX Magnetic Fields Control group c/o Paul Merchant, 16 Melrose Drive, Peterborough, PE2 [illegible]
M0 HZA Gary Charlesworth, 51 Reservoir Road, Surfleet, Spalding, PE11 4DH
M0 HZB Zheng Yao, 56 The Spinney North Cray, Sidcup, DA14 5NF
M0 HZC Vasilije Perovic, Trinity College, Cambridge, CB2 1TQ
MI0 HZD Karol Mikicki, 429 Beersbridge Road, Belfast, BT5 5DU
M0 HZE Philip Preston, 12 Backney View, Greytree, Ross-on-Wye, HR9 7JP
M0 HZF Gheorghe Craioveanu, 56 Springhead Parkway, Northfleet, Gravesend, DA11 8BF
M0 HZH Razvan Fatu, 46 Garfield Street, Watford, WD24 5HB
MM0 HZI Tim Johnston, The Old Schoolhouse, Luggate Burn, Haddington, EH41 4QA
MM0 HZJ Stephen Thomas, Sunnybrae, New Aberdour, Fraserburgh, AB43 6NA
M0 HZK David Pearson, 37 Elmridge, Leigh, WN7 1HN
MM0 HZL Hazel McKay, 44a Torbane Drive, East Whitburn, Bathgate, EH47 0JQ
M0 HZM D Greenland, 1 Hilltop, Tuesley Lane, Godalming, GU7 1SB
M0 HZN Karel Hagemans, 4 Redwood Drive, Aylesbury, HP21 7TN
MM0 HZO Neil Clark, 30 Davidson Place, St. Cyrus, Montrose, DD10 0BS
M0 HZP David Morrow, 73 Manor Road, Fleetwood, FY7 7LJ
M0 HZR Nigel Barker, 17 Pippin Walk, Hardwick, Cambridge, CB23 7QD
M0 HZT Jenni Jones, 69 Pound Street, Warminster, BA12 8NW
M0 HZU David Eate, 69 Dunyeats Road, Broadstone, BH18 8AE
M0 HZV Miroslav Mirchev, 82 Collingwood Road, Uxbridge, UB8 3EL
M0 HZW Walter Dawkins, 2 Nativity Close, Sittingbourne, ME10 1ET
M0 HZX Mark Stephens, 123 Church Street, Westhoughton, Bolton, BL5 3SF
M0 HZY Jeffrey Strandberg, Apartment 201, Satin House, 15 Piazza Walk, London, E1 8PW
M0 IAA Ian Astley, 1 Howard Crescent, Durkar, Wakefield, WF4 3AJ
M0 IAD Ian MacDonald, Broomhill, Mill Lane, Worthing, BN13 3DH
M0 IAE ANGLO-EUROPEAN SCHOOL RC c/o Michael Adcock, 37 Ashpole Road, Bocking, Braintree, CM7 5LW
M0 IAF Ian Fletcher, 19 Church Street, St. Day, Redruth, TR16 5JY
M0 IAG Sandor Donath, 12a Comerford Road, London, SE4 2AX
M0 IAH Ian Pryke, 9 Charles Avenue, Grundisburgh, Woodbridge, IP13 6TH
M0 IAJ Iain Jones, 8a Orchard Close, Longford, Gloucester, GL2 9BB
M0 IAK Ibrahim Karbhari, Flat B, 226 Westbourne Park Road, London, W11 1EP
MM0 IAL Iain Lindsay, 265 Stirling Street, Denny, FK6 6QJ
M0 IAM C Collins, 31 Warren Road, Godalming, GU7 3SH
M0 IAS George Reywer, 1 Tiverton Close, Houghton le Spring, DH4 4XR
M0 IAT Ian Chick, 55 Wills Avenue, Paignton, TQ3 2RG
M0 IAW M Walker, 123 Poplar Avenue, Bentley, Walsall, WS2 0EW
M0 IAX Mark Bumstead, Windmill Girl III, Riverside Boatyard, Southampton, SO31 1AA
M0 IAZ Richard Dykes, 1 Streeters Close, Godalming, GU7 1YY
M0 IBD William Thiele, 50B, The Highway, London, E1W 2BG
MM0 IBE Paul Woods, 92 Preston Crescent, Prestonpans, EH32 9RD
M0 IBH Keith Davies, 25 Kinmel Avenue, Abergele, LL22 7LR
MW0 IBI TAFF VALE ARC c/o Ashley Burns, 34 Lakeside Gardens, Merthyr Tydfil, CF48 1EN
MM0 IBJ 3 Towns Technology Group c/o Marcus Hazel-McGown, 27 Ashdale Avenue, Saltcoats, KA21 6AA
M0 IBK Hans Schröder, Hamptstrabe 8, Dieblich, Germany, 56332
MM0 IBL C Niven, 30 Motray Crescent, Guardbridge, St. Andrews, KY16 0XD
M0 IBN William Parish, 46 Bannard Road, Maidenhead, SL6 4NR
MM0 IBO Jorge Moreno, 1-19 Albion Street, Glasgow, G1 1LH
M0 IBQ Michael Savage, 3 Marlborough Close, Cheltenham, GL53 7RY
M0 IBR Brian Clayton, 26 Wood Walk, Mexborough, S64 9SG
M0 IBT David Jones, 19 Ffordd Hebog, Y Felinheli, LL56 4QZ
M0 IBW Stuart Harrison, 8 St. Michaels Close, Buckland Dinham, Frome, BA11 2QD
MM0 IBX Allan Beacham, 16 Queen Victoria Park Inchmarlo, Banchory, AB31 4AL
M0 IBY UTC Sheffield ARC c/o Mark Rigby, 75 Manchester Road, Deepcar, Sheffield, S36 2QX
MW0 IBZ Ian Baker, 28 Kensington Road, Neyland, Milford Haven, SA73 1TL
M0 ICA P Rushby, 16 Foxhill Lane, Selby, YO8 9AR
M0 ICB I Buchner, 52 Hillview, Coldstream, TD12 4ED
MW0 ICE C Evans, 25 Beech Drive, Hengoed, CF82 7JP
M0 ICG Gary Tagg, Tinkers Cottage, Nevendon Road, Wickford, SS12 0QB
M0 ICI Jacobus Le Roux, 20 Varsity Drive, Twickenham, TW1 1AG
M0 ICJ Lukasz Zywicki, 18 Springbank, Brigg, DN20 8PW
M0 ICK Michael Heywood, 16 Edinburgh Drive, Hindley Green, Wigan, WN2 4HL
M0 ICL JOZEF SALEK, 19 Eskmont Ridge, London, SE19 3PZ
M0 ICO Paul Jones, 76 Pengwern, Llangollen, LL20 8AS
M0 ICP Ian Pass, 69 Cotswold Road, Bath, BA2 2DL
M0 ICS Carl Schofield, 1187 Manchester Road, Castleton, Rochdale, OL11 2XZ
M0 ICT M Gascoyne, 31 Dale View, Hemsworth, Pontefract, WF9 4TA
M0 IDJ David Jones, Drove Farm, Sheepdrove, Hungerford, RG17 7UN
M0 IDC J Clark, 27 The Gabriels, Newbury, RG14 6PZ
M0 IDG Ian Garrard, 33 Uplands Road, Hockley, SS5 4DL
M0 IDI P Ackew, 6 Claremont Avenue, Newcastle upon Tyne, NE15 7LB
M0 IDK Ian King, 7 Greenacres Avenue, Blythe Bridge, Stoke-on-Trent, ST11 9HU
M0 IDL Geoffrey Stockley, Flat 1, The Pentagon, 94 Stanley Green Road, Poole, BH15 3AG
M0 IDM Alexander Ferriroli, 142 Hillbury Road, Warlingham, CR6 9TD
M0 IDR Ian Reeve, 36 Stone Pippin Orchard, Badsey, Evesham, WR11 7AA
MW0 IDT Ian Booth, 4 Church Meadow, Boverton, Llantwit Major, CF61 2AT
M0 IDX R Dallimore, 8 Parc Gwellyn, Kinmel Bay, Rhyl, LL18 5HN
M0 IED Andy Wedge, 30 Primrose Way, Locks Heath, Southampton, SO31 6WX
MM0 IEJ J MacDonald, 24 St. Pauls Drive, Armadale, Bathgate, EH48 2LT
MM0 IEL Ivor Lee, Roehill, Crossroads, Keith, AB55 6LQ
M0 IEO M Sanderson, 2 East Crescent, Canvey Island, SS8 9HL
M0 IET C Blount, 55 Silverthorne Drive, Caversham, Reading, RG4 7NR
M0 IFH Ian Harley, 1 Portland Crescent, Meden Vale, Mansfield, NG20 9PJ
M0 IFT Robert David Hodson, 99 Alcester Road, Hollywood, Birmingham, B47 5NR
M0 IGB Ian Bennett, 44 Haig Avenue, Whitley Bay, NE25 8JG
M0 IGG Stephen Wright, 23 Kitchener Street, Walney, Barrow-in-Furness, LA14 3QW
M0 IGR B Jackson, 32 Seymour Street, Peterlee, SR8 4EN
M0 IGY J Hardman, 45 Doncaster Avenue, Manchester, M20 1DH
MM0 IIL [illegible], Drente Cottage, Inverugie, Peterhead, AB42 3DN
M0 IIL W Humphries, 47 [illegible] Crescent, [illegible], Bristol, [illegible]
M0 IHM Ian Millman, 70 Springdale Avenue, Broadstone, BH18 8EX
M0 IIE John Bennet, 53 Haven Road, Barton-upon-Humber, DN18 5BS
MI0 IIG Ian Gibb, 1 Shankill Road, Garvary, Enniskillen, BT94 3DB
M0 IIM C Gordon, 90 Sunholme Drive, Wallsend, NE28 9YW
M0 IJZ Russell Meech, 26 Priory Street, Tonbridge, TN9 2AN
M0 IKB A Young, 15 Shelton Avenue, East Ayton, Scarborough, YO13 9HB
M0 IKD Michael Draper, 160 Chanctonbury Road, Burgess Hill, RH15 9HA
M0 IKE David Bilson, 31 Middleton Drive, Inkersall, Chesterfield, S43 3HS
M0 IKM T Palmer, 29 Field End, Maresfield, Uckfield, TN22 2DJ
M0 IKT David Capstick, 3 Andrew Close, Dibden Purlieu, Southampton, SO45 4LS
M0 IKW Gary Cannon, 30 Main Street, Flixton, Scarborough, YO11 3UB
M0 ILM Michael Miller, Barn Cottage, Wingfield Hall, Manor Road, Alfreton, DE55 7NH
M0 ILN David Bee, 25 Blatcher Close, Minster on Sea, Sheerness, ME12 3PG
M0 ILT P Evans, 12 Cottage Corner, Ilton, Ilminster, TA19 9ER
M0 IMD Ian Douglas, 13 Castlereagh Street, New Silksworth, Sunderland, SR3 1HJ
M0 IME Martin Scobie, Suncourt, Meadfoot Sea Road, Torquay, TQ1 2LQ
M0 IMJ Ivor Morgan Jones, 43 Crewes Avenue, Warlingham, CR6 9NZ
M0 IML Barry Vile, 24 Hudson Close, Dover, CT16 2SG
M0 IMM Shaun Imms, 26 Greenwood Avenue, Rowley Regis, B65 9NJ
M0 IMP I Pollard, 24 Terminus Road, Littlehampton, BN17 5BX
M0 IMS Mark Sims, 5 Sandy Leaze, Bradford-on-Avon, BA15 1LX
M0 IMT Ian Turner, 1 Elmwood Rise, Dudley, DY3 3QJ
M0 IMW I Walker, 24 Hawthorn Road, Norwich, NR5 0LP
M0 INB Ian Barraclough, Maru, 25 Blaithroyd Lane, Halifax, HX3 9PS
M0 IND Peter Ind, 30 Thompson Road, Stroud, GL5 1SY
M0 INF Andrew Fugard, Flat 4, 1 Hazelmere Road, London, NW6 6PY
M0 INI Mark Smith, Church Farm, Market Drayton, TF9 4DN
M0 INP I Popgueorguiev, 239 Westborough Road, Westcliff-on-Sea, SS0 9PR
MM0 INS Cephas Ralph, 37 Seaview Terrace, Edinburgh, EH15 2HE
M0 INY Ian Davis, Top Pub Brown Edge, Hill Top, Stoke-on-Trent, ST6 8TX
M0 IOA Avalon ARC c/o Martin Wheeler, 114 Boundary Way, Glastonbury, BA6 9PH
MM0 IOB Angus Macleod, 2 Eoligarry, Isle of Barra, HS9 5YD
M0 IOC Ian O'Connor, 7 Grove Court, Shotton Colliery, Durham, DH6 2QD
M0 IOI Stuart Leask, 1 Collington Street, Beeston, Nottingham, NG9 1FJ
M0 IOK David Proctor, 4 The Green, Sproatley, Hull, HU11 4XR
MM0 IOL Angus Morrison, 6a Upper Barvas, Isle of Lewis, HS2 0QX
MD0 IOM M Perry, 18, Station Park, Colby, Isle of Man, IM9 4NH
M0 IOT Chris Norris, 53 Station Road, Castlethorpe, Milton Keynes, MK19 7HF
MI0 IOU Tom Herbison, 22 Dernaveagh Road, Ballymena, BT43 6SX
M0 IOW B Cant, 15 Mountbatten Drive, Newport, PO30 5SG
MM0 IPD Leslaw Flis, 22 Crown Street, Inverness, IV2 3AX
M0 IPR Ian Ridings, 25 Mond Road, Irlam, Manchester, M44 6QA
M0 IPS Andrew Hollings, 39 Rendham Road, Saxmundham, IP17 1EA
M0 IPX BRIMHAM CONTEST GROUP c/o N Clarke, Brimham Lodge Farm, Brimham Rocks Road, Harrogate, HG3 3HE
M0 IQX Alan Emmerson, 8 Weston Close, Cannock, WS11 7YX
MM0 IRC Charles Fraser, Rockside, Locheport, Isle of North Uist, HS6 5EU
M0 IRD Ian Day, 137 Tuffley Lane, Tuffley, Gloucester, GL4 0NZ
M0 IRI Roger Trelease, 23 Torridon Close, Woking, GU21 3DB
MM0 IRJ I Johnstone, 14 Carledubs Crescent, Uphall, Broxburn, EH52 6TH
M0 IRK Phil Holmes, 1 Leonards Place, Bingley, BD16 1AD
M0 IRN SUTHRIGE CONTEST GROUP c/o J Warburton, 31 Greenwood Road, Thames Ditton, KT7 0DU
M0 IRP Ian Pipe, 8 Glebe Drive, Stottesdon, Kidderminster, DY14 8UF
M0 IRS David Spinks, 15 Brunlees Drive, Telford, TF3 2NH
M0 IRT Ivan Thomas, 47 Salisbury Avenue, Coventry, CV3 5DA
MI0 IRZ Davy Gregg, 9 Willowfield, Tandragee, Craigavon, BT62 2EJ
M0 ISF Christopher Astbury, Old Police House, Llanegryn, Tywyn, LL36 9SL
M0 ISL Andreas Paulick, Wormbacher Weg 27, Berlin, Germany, 12207
M0 ISN CHORLEY & DISTRICT A.R.S c/o Ernest Entwistle, 43 Brock Road, Chorley, PR6 0DB
M0 ISQ R Lilley, 3 Coultshead Avenue, Billinge, Wigan, WN5 7HS
M0 IST N Lasseter, 7 Fordstone Avenue, Preesall, Poulton-le-Fylde, FY6 0EB
M0 ISW Iain Singlehurst-Ward, 39 Nadder Close, Tisbury, Salisbury, SP3 6JL
M0 ITA Roberto Ritossa, 9, Rue Veronese, Paris, France, 75013
M0 ITI Michael Bruce, 28 Pheasants Way, Rickmansworth, WD3 7ES
M0 ITV E Taylor, 14 Sycamore Grove, Doncaster, DN4 6NX
M0 ITX N White, 33 Beach Road, Lee-over-Sands, Clacton-on-Sea, CO16 8EX
M0 ITY J [illegible], 100 Dunmow Road, Bishop's Stortford, CM23 5HN
M0 IUK David Grayson, 79 Errington Avenue, Sheffield, S2 2EA
M0 IUM Graeme Clark, 65 Chyvelah Vale, Gloweth, Truro, TR1 3YJ
MW0 IUN Ieuan Jones, 21 Albert Street, Maesteg, CF34 0UF
M0 IVE Ivelin Valkov, Flat 10, Warlingham House, London, SE16 3DQ
M0 IVO W Gissing, 2 Yeo Moor, Clevedon, BS21 6UQ
M0 IVW D Barrett, 183 Wilson Avenue, Brighton, BN2 5PD
M0 IWA D Brooks, 61 Carisbrooke High Street, Newport, PO30 1NR
M0 IWZ A Hanna, 35 Orchard Drive, Mayland, Chelmsford, CM3 6EP
M0 IZS David Sexton, 21 Rosedale Crescent, Earley, Reading, RG6 1AS
M0 JAD Peter Holland, 30 Knighton Park Road, London, SE26 5HJ
M0 JAE J Allen, 20 Spa Hill, Kirton Lindsey, Gainsborough, DN21 4BA
M0 JAF John King, 22 Latchmere Gardens, Leeds, LS16 5DN
M0 JAG A Pegg, 18 Blythe Way, Shanklin, PO37 7NJ
M0 JAI Peeyush Gaur, 34 Queensberry Avenue, Copford, Colchester, CO6 1YN
M0 JAJ J Stedman, 60 Sandown Road, Ipswich, IP1 6RE
M0 JAK J Swain, 84 Sunnymead Drive, Waterlooville, PO7 6BX
M0 JAM John Mortimer, 4 Nethercliffe Crescent, Guiseley, Leeds, LS20 9HN
MW0 JAN J Day, 20 St. Johns Drive, Pencoed, Bridgend, CF35 5NF
M0 JAO Jonathan Sansom, 4 Vicarage Road, Eastbourne, BN20 8AU
M0 JAP David James, Bramble Cottage, Tray Lane, Atherington, Umberleigh, EX37 9HY
M0 JAQ John Malia, 47 Clent Way, Longbenton, Newcastle upon Tyne, NE12 8QG

UK Callsigns

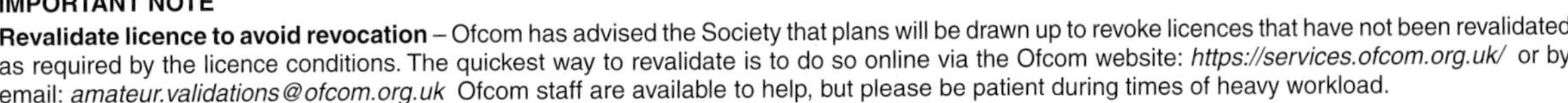

**IMPORTANT NOTE**

**Revalidate licence to avoid revocation** – Ofcom has advised the Society that plans will be drawn up to revoke licences that have not been revalidated as required by the licence conditions. The quickest way to revalidate is to do so online via the Ofcom website: *https://services.ofcom.org.uk/* or by email: *amateur.validations@ofcom.org.uk* Ofcom staff are available to help, but please be patient during times of heavy workload.

MI0 JAR J Rice, 42 The Crescent, Ballymoney, BT53 6ES
MI0 JAT Joey McGoldrick, 23 Lettercarn Road, Clare, Castlederg, BT81 7QY
M0 JAV John Rogers, 9 Cherry Tree Avenue, Shireoaks, Worksop, S81 8PH
MW0 JAW Helen Stevens, 59 Arfryn Avenue, Llanelli, SA15 3RW
M0 JAX John Edwards, 45 Bramshaw Gardens, Bournemouth, BH8 0BT
MI0 JAY William Graham, 19 Margaret Square, Ballymoney, BT53 6BZ
M0 JAZ Jim Sadler, 10 Spindle Warren, Havant, PO9 2PU
M0 JBC J Crank, 38 Harley Avenue, Harwood, Bolton, BL2 4NU
M0 JBD J Day, 124 Radstock Road, Southampton, SO19 2HU
M0 JBF John Cobb, 32 Dellmont Road, Houghton Regis, Dunstable, LU5 5HU
MI0 JBK John Mackenzie, 30 Dalriada Gardens, Ballycastle, BT54 6DZ
MM0 JBS J Summers, 1 Main Road, Fairlie, Largs, KA29 0DP
MI0 JBT James Traynor, 8 Roeville Terrace, Limavady, BT49 0BH
M0 JBW J McLaughlin, 34 Cambridge Road, Birstall, Batley, WF17 9JF
M0 JBZ Jonathan Chalmers, 19 Brettenham Crescent, Ipswich, IP4 2UB
M0 JCC Ian Jefferson, 125 Telscombe Way, Luton, LU2 8QP
M0 JCD J Dalgliesh, 61 Clonners Field, Stapeley, Nantwich, CW5 7GU
M0 JCE Julian Crewe, 22 Myrtle Tree Crescent, Weston-Super-Mare, BS22 9UL
M0 JCH Jeremy Paul, Mimosa Lodge, 59 Baring Road, Cowes, PO31 8DW
M0 JCK Eric Beechill, Belleroyd Farm Blackshaw Head, Hebden Bridge, HX7 7JP
M0 JCL J Plant, 67 Kenley Road, London, SW19 3JJ
M0 JCM John Murray, 2 The Cuttings, Hampstead Norreys, Thatcham, RG18 0RR
M0 JCQ James Stevens, 23 Edlyn Close, Berkhamsted, HP4 3PQ
M0 JCR Jude Reynolds, 64 Albury Road, Merstham, Redhill, RH1 3LL
M0 JCS John Stevenson, 18 Drakehouse Lane, Sheffield, S20 1FW
M0 JCT J Townsend, 47 Main St., Wolston, Coventry, CV8 3HH
M0 JCZ Marcin Ciechan, 109 Rose Vale, Liverpool, L5 3PD
M0 JDA J Dale, Corydon, Church Street, Sevenoaks, TN14 7SW
M0 JDB J Pollard, 72 Windy Arbour, Kenilworth, CV8 2BB
M0 JDD J Dedier, 10 Lowry Close, Haverhill, CB9 7GH
M0 JDE David Foxall, 1 Doe Hey Grove, Farnworth, Bolton, BL4 7HS
M0 JDL John Dowdeswell, 18 Lechlade Gardens, Fareham, PO15 6HF
M0 JDP J Page, 5 Riddimore Avenue, Hereford, HR2 7LJ
M0 JDR J Reed, 99 Norsey Road, Billericay, CM11 1BU
M0 JDS J Sweatman, 14 Clover Court, Jasmine Grove, Waterlooville, PO7 8BP
M0 JDW J Williams, 1 Nant Terrace, Pentraeth, LL75 8YE
M0 JEA Martin Augustus, 84 Wright Way, Stapleton, Bristol, BS16 1WH
M0 JEC Norman Bland, 3 Kennet Road, Newbury, RG14 5JA
M0 JEK Andre Skarzynski, 93 Richmond Walk, St. Albans, AL4 9BB
M0 JEM J Mayo, 134 Bromsgrove Road, Redditch, B97 4SP
M0 JEP J Price, Oxted Place East, Broadham Green Road, Oxted, RH8 9PF
MJ0 JER R Taylor, 21 Samares Avenue, La Grande Route De St. Clement, St. Clement, Jersey, JE2 6NY
MM0 JET 49F SON AIR CADET RC c/o Brian Burt, 182 Old Inverkip Road, Greenock, PA16 9JG
M0 JEZ Jeremy Powell, 46 Woodmancote, Yate, Bristol, BS37 4LL
M0 JFB Jon Button, 1 Amber Close, Rainworth, Mansfield, NG21 0FU
M0 JFD J Dixon, 23 Dee Way, Winsford, CW7 3JB
M0 JFE John Earnshaw, 128 Shakespeare Road, Fleetwood, FY7 7HJ
M0 JFM John Marsh, 14 Eyam Road, Hazel Grove, Stockport, SK7 6HP
M0 JFP James Preece, 17 Cherry Tree Avenue, Staines, TW18 1JB
M0 JFW John Wheeler, 428 Bromsgrove Road, Hunnington, Halesowen, B62 0JL
M0 JGB J Greenway-Brown, 207 Lowe Avenue, Wednesbury, WS10 8NS
MW0 JGE N Lewis, 1 Clyne Drive, Blackpill, Swansea, SA3 5BU
M0 JGH Jonathan Hunt, 15 Greenway, London, SW20 9BQ
M0 JGM James Marlett, 6 Delamere Avenue, Sutton Manor, St. Helens, WA9 4AP
MM0 JGP John Pirie, Mill House, Watermill Cottage, Fraserburgh, AB43 7ED
M0 JGS Joseph Seaton, 52 Shrubbery Street, Kidderminster, DY10 2QY
M0 JHB John Butcher, 3 Basket Gardens, London, SE9 6QP
M0 JHC James Clarke, 29 Sisial y Mor, Rhosneigr, LL64 5XB
M0 JHD Julia Hardy, LAMBDA HOUSE, SEANOR LANE, Chesterfield, S45 8DH
M0 JHF James Foster, 23 High Street, Cumnor, Oxford, OX2 9PE
M0 JHG John Ginever, 66 London Road, Maidstone, ME16 8QU
MM0 JHL Jonathan Hutchinson, Hawthorn Cottage, Addiewell, West Calder, EH55 8NL
M0 JHM James Rymer, 23 Chetwode Road, Tadworth, KT20 5PS
M0 JHP Hector Alava Moreira, 19b Lessness Park, Belvedere, DA17 5BG
M0 JHW James Wheeldon, 11 Stathern Walk, Grantham, NG31 7XG
M0 JIB Jamie Bickers, 3 The Old Brickyard, West Haddon, Northampton, NN6 7GP
M0 JIL G Heyes, 5 Ashgarth Way, Harrogate, HG2 9LD
M0 JIU John Uren, 4 Killivose Road, Camborne, TR14 7RN
M0 JJA Robert James, 51 The Lampreys, Gloucester, GL4 6QU
M0 JJB John Barton, 93 Cardigan Road, Bridlington, YO15 3JU
M0 JJC Alexander Coghlan, Charterhouse, Orchard Road, Salisbury, SP5 2JA
M0 JJD John Dignan, 754 Leigh Road, Leigh, WN7 1TF
M0 JJE M Chisholm, 15 Summerfield Avenue, Waltham, Grimsby, DN37 0NQ
M0 JJH G Cavie, Dawn, Maypole Road, Colchester, CO5 0EN
M0 JJK James King, 18 Ross Road, Wallington, SM6 8QB
M0 JJM Darren Oliver, 20 Five Oaks Close, Malvern, WR14 2SW
M0 JJN James Nicholls, 4 Sadler Close, Colchester, CO2 7LU
M0 JJR J Reilly, 22 Charlecote Gardens, Sydenham, Leamington Spa, CV31 1GE
MM0 JJV J Vennard, 4 Braehead, Girdle Toll, Irvine, KA11 1BD
M0 JKB Stuart Lucas, 31 Lilian Close, Norwich, NR6 6RZ
M0 JKF John Ferrol, 29 Westlands, Haltwhistle, NE49 9BS
M0 JKG Justin Gaskin, Badgers Barn, Canterbury Road, Folkestone, CT18 8DF
M0 JKN Jennifer Wilson, Flat 5, Blake House, London, SE1 7DX
M0 JKP M Howden, 11 Marsh Lane Gardens, Kellington, Goole, DN14 0PG
M0 JKQ C Poulson, 9 Scattergate Green, Appleby-in-Westmorland, CA16 6SP
MI0 JLC S Mulligan, 65 Glebe Walk, Lisburn, BT28 1PZ
M0 JLE Julie Cafe, Flat 6, Smiths Court, 73 East Borough, Wimborne, BH21 1PJ
M0 JLM J Mitchell, 5 Orchard Close, Truro, TR1 3PA
MW0 JLN N Howells, 21 Coed Bach, Pencoed, Bridgend, CF35 6TF
M0 JLP Jeremy Powell, 23 Park Road, Norton, Malton, YO17 9DZ
M0 JLR Joseph Redhead, 28 Sandfields, Frodsham, WA6 6PT
M0 JLT L Taylor, 33 Priestley Avenue, Darton, Barnsley, S75 5LG
M0 JLW J Welford, 26 Templewood, Welwyn Garden City, AL8 7HX
M0 JLY Andrew Page, 207 Brooklyn Road, Cheltenham, GL51 8DZ
MM0 JMB John Brown, 11 Fairway Avenue, Elgin, IV30 6XF
M0 JMC J Mccutcheon, 15 Maytrees, Hitchin, SG4 9LT
M0 JME Jamie Blundell, 21 Walmsley Street, Fleetwood, FY7 6LJ
MM0 JMI Jamie Davies, 5, Orlit Houses, North Berwick, EH39 5JE
M0 JMJ John Jukes, 22 Hazelmere Road, Creswell, Worksop, S80 4HS
MM0 JMK J McKechnie, 8 Waulker Avenue, Stirling, FK8 1SA
MI0 JML James McCaw, 62 High Street, Ballymena, BT43 6DT
M0 JMN R Andrews, Mount View, Park Lane, Worcester, WR2 6PQ
M0 JMP J Poole, 18 Grosvenor Avenue, Kidderminster, DY10 1SS
M0 JMS J Karlstad, Flat B, 28 Market Place, North Walsham, NR28 9BS
M0 JMV D Vanstone, Tymperley Farm, Great Henny, Sudbury, CO10 7LX
M0 JMY James McMullan, 17 Banbury Close, Accrington, BB5 4BZ
M0 JNP John Perry, Flat 7 Lancaster House, Belle Vue Rd, Paignton, TQ4 6HD
MD0 JNS John Shatford, 31 Pinner Park Avenue, Harrow, HA2 6LG
M0 JNX J Hall, 1 Nash Close, Earley, Reading, RG6 5SL
M0 JOB J O'Brien, 76 Berkeleys Mead, Bradley Stoke, Bristol, BS32 8AU
M0 JOD J Preece, 51 Drancy Avenue, Willenhall, WV12 5RD
M0 JOH John Godfrey, 17 Lichfield Road, Sneinton, Nottingham, NG2 4GF
MM0 JOK John Burgoyne, 5 Shankston Crescent, Cumnock, KA18 1HA
M0 JOL Marcus O'Leary, 4 Park Farm Close, Martinstown, Dorchester, DT2 9TW
MM0 JOM John Mann, 10 Brimmond Walk, Westhill, AB32 6XH
M0 JOO Anthony Williams, 12 St. Wilfrids Crescent, Brayton, Selby, YO8 9EU
M0 JOR John Orr, 13 Haldane Close, Brierley, Barnsley, S72 9LL
M0 JOY Joyce Bilson, 31 Middleton Drive, Inkersall, Chesterfield, S43 3HS
M0 JPA John Wake, 60 Cloverville Approach, Odsal, Bradford, BD6 1ET
M0 JPB J Bull, 91 Lime Road, Wednesbury, WS10 9NF
MI0 JPC James Cosgrove, 91 Church Street, Newtownards, BT23 4AN
MI0 JPD John Doyle, 25 Parkmore Road, Magherafelt, BT45 6PF
M0 JPG John Gleeson, 124 rushes mead, Harlow, CM18 6QE
MI0 JPL J Jones, 107 Belfast Road, Whitehead, Carrickfergus, BT38 9SU
M0 JPM Jan Meijer, Birchwood East End, Gooderstone, King's Lynn, PE33 9DB
M0 JPN T Nakagawa, 7 Milton Street, Barrowford, Nelson, BB9 6HE
MI0 JPO J Alexander, 24 Alexandra Avenue, Ballymoney, BT53 6EX
M0 JPP Sean Clark, 20 Herbert Street, Loughborough, LE11 1NX
M0 JPS J Styles, 42 Brook Street, Woodbridge, IP12 1BE
M0 JPT Johnny Tan, 22, Jalan Geikie, Miri, Malaysia, 98000
M0 JPW Julian Woolvin, 62 Whitewood Park, Liverpool, L9 7LG
M0 JQK T Firth, 126 Tombridge Crescent, Kinsley, Pontefract, WF9 5HE
M0 JQW Jieqiong Wang, Flat 31, 74 Arlington Avenue, London, N1 7AY
M0 JRA A Jessop, 4 Katherine St., Thurcroft, Rotherham, S66 9LG
M0 JRE John Eaton, 38 Litchford Road, New Milton, BH25 5BQ
MM0 JRF John fyfe, 53a Ware Road, Glasgow, G34 9AR
M0 JRJ John Jenkins, 31 Pendrell Street, London, SE18 2PH
M0 JRL James Lynn, 2 The Fairways, Redhill, RH1 6LP
M0 JRQ Christopher Pearson, 4 Brentwood Close, Thorpe Audlin, Pontefract, WF8 3ES
MM0 JRR John Rayne, 8 Bankton Grove, Livingston, EH54 9DW
M0 JRW J Wilson, 1 Locarno Avenue, Runwell, Wickford, SS11 7HX
M0 JRX Oliver Bross, 8 Queens Drive, Buckley, CH7 2LJ
M0 JRZ John Robb, 37 Wroxham Road, Woodley, Reading, RG5 3AX
M0 JSA Alan Jones, 5 Meadowlands, Kirton, Ipswich, IP10 0PP
M0 JSD J Delaney, 33 Deepdale Close, Ibstock, LE67 6LW
M0 JSE Malcolm Egan, 32 Hawksworth Avenue, Guiseley, Leeds, LS20 8EJ
MM0 JSG John Galloway, 144 Strathkinnes Road, Kirkcaldy, KY2 5PZ
M0 JSH Josiah Yan, St Edmund's College, Mount Pleasant, Cambridge, CB3 0BN
MI0 JSJ Jane Smith, 54a Blackstaff Road, Kircubbin, Newtownards, BT22 1AF
M0 JSN Jonathan Sowman, 53 Newton Wood Road, Ashtead, KT21 1NN
M0 JSP J Swiffen, Dragonelle, Marina Drive, Shireoaks, S81 8NQ
M0 JSR John Street, 22 Roman Acre, Wick, Littlehampton, BN17 7HN
M0 JSW Jeremy Woodland, 14 Kelham Green, Nottingham, NG3 2LP
M0 JSX Jonathan Sawyer, 9 Waller Court, Caversham, Reading, RG4 6DB
M0 JSZ F Jackson, 5 Chalmers Avenue, Haversham, Milton Keynes, MK19 7AG
M0 JTB J Brown, 55 Barrington Road, Rubery, Birmingham, B45 9EU
MI0 JTE J Elliott, 183 Kilraughts Road, Ballymoney, BT53 8NL
M0 JTH J Thomas, 77 Hawthorn Avenue, Lowestoft, NR33 9BB
M0 JTJ J Talbot-Jones, 21 Downsview Drive, Wivelsfield Green, Haywards Heath, RH17 7RN
M0 JTN Martin Chivers, 4 Hunters Lodge, Fareham, PO15 5NF
M0 JTQ Joshua Cook, 37 Leigham Court Drive, Leigh-on-Sea, SS9 1PT
M0 JUK J Allison, 72 Chantry Croft, Kinsley, Pontefract, WF9 5JL
MM0 JUL R Mitchell, Broadhills, Isle of Coll, PA78 6TB
M0 JVC Kevin Francis, 203 Colchester Road, Lawford, Manningtree, CO11 2BU
M0 JVG Joachim Gesiau, Huelchratherstr 37, Koeln, Germany, D-50670
M0 JVM David Turton, 8 Lightwoods Road, Warley, Smethwick, B67 5AY
MM0 JVQ Satoshi Imamura, 1/1 11 Comiston Gardens, Edinburgh, EH10 5QH
M0 JVT John Turner, 4 Viewlands, Upper Luton Road, Chatham, ME5 7BE
M0 JVV John Waddy, 70 Linden Avenue, Prestbury, Cheltenham, GL52 3DS
M0 JVW John Wild, 11 Whitefield Avenue, Newton-le-Willows, WA12 8BY
M0 JWA J Arrow, 29 Billington Gardens, Hedge End, Southampton, SO30 2AX
M0 JWE J williamson, 64 Sandy Lane, Irlam, Manchester, M44 6WJ
MM0 JWH James Hosea, 61 John Street, Helensburgh, G84 9JZ
M0 JWJ John Jordan, 31 Rotherham Road, Dinnington, Sheffield, S25 3RG
M0 JWL M Lee, Up To Date House, Shore Road, Boston, PE22 0NA
M0 JWM Johnny McCown, 3 Villiers Way, Weeting, Brandon, IP27 0GA
M0 JWP John Pritchard, 1 Tan y Coed, Maesgeirchen, Bangor, LL57 1LU
M0 JWR John Richardson, 6 Clarence Road, Scorton, Richmond, DL10 6EE
M0 JXE D Ennion, 347 Parkgate Road, Chester, CH1 4BE
M0 JXG James Guess, Flat 257, Helen Gladstone House, London, SE1 0QB
MM0 JXI J Innes, 33 Monktonhall Place, Musselburgh, EH21 6RR
M0 JXM Dennis Easterling, 14 Brunswick Close, Biggleswade, SG18 0DA
M0 JXS Colin Ember, 35 Mattock Lane, Ealing, London, W5 5BH
MW0 JYC M Coleman, 96 Shelone Road, Briton Ferry, Neath, SA11 2PU
M0 JYM J Sangster, 139 Higher Road, Liverpool, L26 1UN
MW0 JZE A David, 45 Amanwy, Llanelli, SA14 9AH
M0 JZG Arthur Loukes, 14 Batchwood View, St. Albans, AL3 5TD
M0 JZK John Howlett, 29 Little London, Heytesbury, Warminster, BA12 0ES
MW0 JZM Steven Kedward, 9 Lawrence Avenue, Aberdare, CF44 9EW
M0 JZT Martin Hunt, 189 Gibbins Road, Birmingham, B29 6NH
MI0 JZZ C McLelland, 7 Alderbrook Gardens, Coleraine, BT51 3PZ
M0 KAA R Cook, 28 Ring Road, Lancing, BN15 0QE
M0 KAB Kenneth Bull, 111 Hinksford Mobile Home Park, Kingswinford, DY6 0BB
M0 KAC 1127 (KENDAL) SQN AIR CADETS c/o Roy Walker, 35 Romany Close, Letchworth Garden City, SG6 4LA
M0 KAD K Allen, 48 Beaumont Rise, Worksop, S80 1YG
M0 KAE K Ely, 6 Baxter Square, Town End Farm Estate, Sunderland, SR5 4ND
MI0 KAG Elizabeth Rantin, 8a Buchanans Road, Newry, BT35 6NS
M0 KAI Karl Boothman, 5 Millwood Road, Doncaster, DN4 9DA
M0 KAJ David Ramsell, 36 West Street, Burton-on-Trent, DE15 0BW
MW0 KAK David Knight, Ty-Onnen, Station Road, Haverfordwest, SA62 5RZ
MI0 KAM Kathryn Mullan, 19 Parklea, Portstewart, BT55 7HA
M0 KAN Keith Nicholson, 11 Lancaster Way, Skellingthorpe, Lincoln, LN6 5UF
M0 KAO Kevin Jeffery, 9 Gordon Road, Tunbridge Wells, TN4 9BL
M0 KAP Paul Smith, 3 Watts Road, Colchester, CO2 9DZ
M0 KAR K Lott, 6 Centurion Close, College Town, Sandhurst, GU47 0HH
M0 KAU Kevin Sharpe, 18 Dudhill Road, Rowley Regis, B65 8HT
M0 KAW Kevin Wells, 2 Holmefield, Farndon, Newark, NG24 3TZ
MW0 KAY J Harvey, 17 Heol-y-Plwyf, Ynysybwl, Pontypridd, CF37 3HU
M0 KBA Ashish Bhakoo, 4 Bryden Cottages, High Street, Uxbridge, UB8 2NY
M0 KBB Keith Bright, 20 Radley Road, Bristol, BS16 3TL
M0 KBC K Aird, 11 Minter Avenue, Densole, Folkestone, CT18 7DS
M0 KBD Paul Smiths, 12 Lambton Road, Stockton-on-Tees, TS19 0ER
M0 KBH Nigel Kimber, 72 Churchfield Road, Scunthorpe, DN16 3DW
M0 KBP Bisher Al-rawi, Flat 17, Harrow Lodge, London, NW8 8HR
M0 KBT D Bisset, Jordieland Cottage, Kirkcudbright, DG6 4XT
M0 KBW Michal Cerveny, Apartment 214, Piccadilly Heights Wain Avenue, Chesterfield, S41 0GF
M0 KCA John Cater, 5 Shady Grove, Hilton, Derby, DE65 5FX
M0 KCC Kevin Cartwright, 53 Sedgley Road, Dudley, DY1 4NE
MM0 KCD Ken Davies, 1 Myreton Way, Falkirk, FK1 5NZ
M0 KCE Thomas Peterson, 59 Fleet Street, Holbeach, Slane Lodge, Spalding, PE12 7AU
M0 KCF George McCaffery, 7 Cliffe Court, Sunderland, SR6 9NT
M0 KCO Kevin Cornmell, 19 Forest Road, Chandler's Ford, Eastleigh, SO53 1NA
M0 KCP Robert Hutton, 22a Victoria Road, Maldon, CM9 5HF
M0 KCR KENT COUNTY RAYNET c/o Dennis Spalding, 171 Minster Road, Minster on Sea, Sheerness, ME12 3LH
MM0 KCS Michael Brunsdon, 25 Buckstone Lea, Edinburgh, EH10 6XE
M0 KCW Chris Wade, 31 Melton Green, Wath-upon-Dearne, Rotherham, S63 6AA
M0 KCZ Raymond Robinson, 22 Riddings Court, Timperley, Altrincham, WA15 6BG
M0 KDA Shaun Bolton, 201 Lime Tree Avenue, Crewe, CW1 4HZ
M0 KDE A Lee, 14 Bernice Avenue, Chadderton, Oldham, OL9 8QJ
M0 KDH D Hughes, 75 Suncote Avenue, Dunstable, LU6 1BN
MW0 KDL K Lewis, 21 Wheatley Place, Merthyr Tydfil, CF47 0TA
M0 KDM Paul Sephton, 11 Moss Avenue, Leigh, WN7 2HH
M0 KDR K Roberts, Burnwithian Cottage, Burnwithian, Redruth, TR16 5LG
M0 KDT J Hobbs, 2 Eccles Road, Wittering, Peterborough, PE8 6AU
M0 KDU Michael Shingler, 4 Church Lane, Checkley, Stoke-on-Trent, ST10 4NJ
M0 KDV D Craven, 69 Markham Avenue, Rawdon, Leeds, LS19 6NE
M0 KDX Paul Martin, 82 Lesbourne Road, Reigate, RH2 7JX
MM0 KDY D Latto, 8 Aspen Avenue, Glenrothes, KY7 5TA
M0 KEB Kevin Legg, Bennetts, High Street, Clacton-on-Sea, CO16 0EG
M0 KED Andrew Keddie, 6 Vulcan Crescent, North Hykeham, Lincoln, LN6 9SB
M0 KEE John Charlton, Hillside House, Ham Lane, Bristol, BS41 8JA
M0 KEF Peter Munson, 8 Longley Lane, Spondon, Derby, DE21 7AT
M0 KEG Ian Bain, 45 Larpool Crescent, Whitby, YO22 4JD
M0 KEJ Kinga Borszlak, 37 Bank Lane, Little Hulton, Manchester, M38 9UH
M0 KEL Kelvan Gale, 4 Field Court, Sea Road, Littlehampton, BN16 1JS
M0 KEP Tim Keep, 119 Radley Road, Abingdon, OX14 3RX
M0 KEQ Kevin Mogford, 49 Cefn Road, Rogerstone, Newport, NP10 9AQ
M0 KER R Kerswill, 38 Longbridge Road, Bramley, Tadley, RG26 5AN
M0 KEV K Dawson, 57 Fair View, Blackwood, NP12 3NR
M0 KEY B Wagstaff, 6 Willingham Road, Fillingham, Gainsborough, DN21 5BN
M0 KFB K Bell, 11 Mill Lane, Hogsthorpe, Skegness, PE24 5NF
M0 KFO Alastair Smith, 58 Wellesbourne Road, Barford, Warwick, CV35 8DS
M0 KFU David Sweeney, 3 Robin Hood Close, Woking, GU21 8SS
M0 KFW K Whittaker, 32 Ashleigh Mount Road, Exeter, EX4 1SW
MM0 KFX Adam Hutchison, 24 Tanna Drive, Glenrothes, KY7 6FX
M0 KGA Stefan Pollak, 96 Lansbury Avenue, Feltham, TW14 0JR
M0 KGK Duncan Munro, 48 Laburnum Way, Witham, CM8 2NY
M0 KGM K Graham, 22 Repton Avenue, Oldham, OL8 4JB
M0 KGP P Kelly, Arosfa, Westminster Road, Wrexham, LL11 6DN
MM0 KGS George Sinclair, 33 Keptie Road, Arbroath, DD11 3EF
M0 KGV Clive Moulding, 28 Queens Avenue, Highworth, Swindon, SN6 7BA
M0 KGY K Yearsley, Garth Lea, Lon St. Ffraid, Holyhead, LL65 2YH
M0 KHA Colin Wale, 23 Castleton Avenue, Bournemouth, BH10 7HW
M0 KHO David Plurighi, 21 Buscot Drive, Abingdon, OX14 2BJ
M0 KHS K Shuttleworth, 27 Queens Drive, Fulwood, Preston, PR2 9YJ
M0 KHW Kenneth Wright, 12 Bushmead Road, Luton, LU2 7EU
M0 KHZ K Wheatley, 1 Braithwaite Court, Egremont, CA22 2DN
M0 KIB C McKinney, 168 Salisbury Avenue, Barking, IG11 9XU
M0 KID Neil Fairbairn, 15 Hewitt Road, Dover, CT16 1TH
M0 KIG K Gamble, 67 Queen Street, Burntwood, WS7 4QQ
MW0 KIJ Nicholas Sugg, 18 Dolgynog, Penderyn, Aberdare, CF44 9JT
M0 KIL Keith Lockstone, Oceana, The Parade, Pevensey, BN24 6LX
M0 KIM K Rawlings, Prebbles Hill Cottage, Pluckley, Ashford, TN27 0PE
M0 KIN Alexander Clarke, 57 Welland Avenue, Grimsby, DN34 5JP
M0 KIR Mike Kirkman, 8 Ashington Drive, Arnold, Nottingham, NG5 8GH
M0 KJC K Cole, 4 Marsham Road Hazel Grove, Stockport, SK7 5JB
MM0 KJG K Glacken, 14 Hailes Avenue, Edinburgh, EH13 0NA
M0 KJK Karol Jan Kolesnik, 15 Steer Road, Swanage, BH19 2RU
MM0 KJM K Martin, 4 Hunter Crescent, Troon, KA10 7AH
M0 KJT Keith Todman, 12 Winscombe, Bracknell, RG12 8UD
M0 KKA Crispin Wheatley-Hince, Jasmine Cottage, Main Street, Newbury, RG20 7EH
M0 KKB Stephen King, Cupertino House, Giles Lane, Canterbury, CT2 7NA
M0 KLA S Collins, 3 Sedgemoor Road, Camps Bay, South Africa
M0 KLB Ernesto Gomez Lozano, Flat D, 212 Kennington Road, Oxford, OX1 5PG
M0 KLH K Hawes, 837 Garratt Lane, London, SW17 0PG

M0 KLJ J Athersmith, 7 Byron Street, Ulverston, LA12 [illegible]

M0 KLK William Foster, 55 Drake Avenue Minster on Sea, Sheerness, ME12 [illegible]

M0 KLL K Lloyd, 1 Fordenbridge Square, Sunderland, SR4 0BA

M0 KLM B Whiteley, 2a Beechfield Close, Thorpe Willoughby, Selby, YO8 9QJ

M0 KLN Kevin Nevins, 4 Ubbanford, Norham, Berwick-upon-Tweed, TD15 2LA

MM0 KLR Kilmarnock and Loudoun ARC c/o A Clark, 20 Church Street, Kilwinning, KA13 6BE

M0 KLT G Clark, 28 Manor Road, Woolton, Liverpool, L25 8QG

M0 KLW Andrew Jones, 2 Erw Terrace, Bethel, Caernarfon, LL55 1YT

M0 KMB A Bailey, 58 Billy Buns Lane, Wombourne, Wolverhampton, WV5 9BP

M0 KMI K Mills, 6 West Coombe, Bristol, BS9 2BA

MI0 KMJ A McGuinness, 42 Downshire Road, Carrickfergus, BT38 7LD

M0 KMR Medway Raynet c/o Raymond Sohst, 2 Shaftesbury Drive, Maidstone, ME16 0JS

M0 KMS K Smith, 62 Waterloo Road, Talywain, Pontypool, NP4 7HJ

M0 KMT Mark Sadler, 12 John Corbett Drive, Amblecote, Stourbridge, DY8 4BW

M0 KMW M Ballard, 41 Middlefield Avenue, Halesowen, B62 9QJ

M0 KNB M Honan, 49 Dorset Street, Nottingham, NG8 1PU

MM0 KNE Thomas Kane, 40b Brisbane Street, Greenock, PA16 8NP

M0 KNH Keith Holman, 39 Trellech Court, Yeovil, BA21 3TE

MM0 KNN C Kennedy, 16 New Garrabost, Isle of Lewis, HS2 0PL

M0 KNX Maurizio Mattiello, Via Luigi Gaudio, 21, Faedis, Udine, Italy, 33040

M0 KOG Colin Haw, 72 Mayflower Road, Boston, PE21 0EZ

M0 KOH Ken Hough, 15 Moorside Road, Endmoor, Kendal, LA8 0EN

M0 KOI P Burrows, 53 Falmouth Place, Murdishaw, Runcorn, WA7 6JF

M0 KOM Leonard Wilson, 6 Marrick Road, Middlesbrough, TS3 7RX

M0 KOO Kristian Milone, 20 Farmers Way, Copmanthorpe, York, YO23 3XU

M0 KOT Nicholas Baulf, 1 Lower Chart Cottages, Brasted Chart, Westerham, TN16 1LS

MM0 KOZ R Thomson, 30 Sealstrand, Dalgety Bay, Dunfermline, KY11 9NG

MI0 KPA S Frazer, 2 Cavanballaghy Road, Killylea, Armagh, BT60 4NZ

M0 KPB K Blanshard, 30 Torquay Crescent, Symonds Green, Stevenage, SG1 2RS

M0 KPC Paul Gagliardi, 7 Saxon Way, Jarrow, NE32 3QA

M0 KPK Duncan Smith, 333A, Forton Road, Gosport, PO12 3HF

M0 KPO Steven Warren, 1 Morley Close, Stapenhill, Burton-on-Trent, DE15 9EW

M0 KPT Kelvin Towler, 3 Elm Grove, Barnham, Bognor Regis, PO22 0HF

M0 KPW Christopher Leviston, 13 Pryors Walk, Askam-in-Furness, LA16 7JG

MI0 KQU Steve Homer, 10 Bann Drive, Londonderry, BT47 2HW

M0 KRA Stanley King, 136 Cane Creek Road, Walhalla, United States, SC29691-3937

MM0 KRC Kieran Carroll, 32e Meadowburn Place, Campbeltown, PA28 6ST

M0 KRD Dirk Niggemann, 35 Holm Court, Twycross Road, Godalming, GU7 2QT

M0 KRL K Fell, 5 Henry Road, Wath-upon-Dearne, Rotherham, S63 7NF

M0 KRM Medway Raynet c/o Raymond Sohst, 2 Shaftesbury Drive, Maidstone, ME16 0JS

M0 KRP Rod Bullen, 2 Redlands Cottages, East Coker, Yeovil, BA22 9HF

M0 KRR Alastair Kerr, 23 Manor Park, Duloe, Liskeard, PL14 4PT

MW0 KRS C Young, 34 Penlan Crescent, Uplands, Swansea, SA2 0RL

M0 KRU Peter Crewe, The White House, 81 Seas End Road, Spalding, PE12 6LD

M0 KRW I Bricknell, 171 Springthorpe Road, Birmingham, B24 0SN

M0 KRX Kevin Rosema, Apartment 801, 25 Goswell Road, London, EC1M 7AJ

M0 KSA Roy Love, 48 Langland Drive, Dudley, DY3 3TH

M0 KSC STAR CENTRE (KEIGHLEY COLLEGE) c/o Philip Cole, 48 Emily Street, Keighley, BD21 3HY

M0 KSG Matthew Wood, 26 Parkfield Crescent, Kimpton, Hitchin, SG4 8EQ

M0 KSO The Kings School RS c/o Dave Lee, 188 Manstone Avenue, Sidmouth, EX10 9TJ

M0 KSR K McInnes, 4 Lindrick Road, Hatfield Woodhouse, Doncaster, DN7 6PF

MM0 KSS Kevin Scott, 139 Farns Heugh Circle, Cove Bay, Aberdeen, AB12 3RW

MW0 KST Steven Davies, 5 Maldwyn Street, Cardiff, CF11 9JR

MM0 KTE K Taylor, 77 Queen Margaret Fauld, Dunfermline, KY12 0RL

MM0 KTL Kit Lane, 23 Mayfield Avenue, Tillicoultry, FK13 6HB

M0 KTR Anatole Rowbottom, 8 Hedge Drive, Colchester, CO2 9DT

M0 KTT Christopher Jacobs, Flat 33, The Lodge, Waterlooville, PO7 8BX

M0 KTV Bipin Chauhan, 45 Burnham Drive, Whetstone, Leicester, LE8 6HY

M0 KUL Karl Richards, 44 Curly Bridge Close, Farnborough, GU14 9AU

M0 KUP Adrian Anderson, 89a Malmesbury Park Road, Bournemouth, BH8 8PS

M0 KUR S Campion, Flat 6, Carousel Steps, 10 Hawtree Close, Southend-on-Sea, SS1 2TZ

M0 KUY Giuseppe Michilin, 46 Via N.Sauro, Preganziol, Italy, 31022

M0 KVA A Dokic, 28 Tudor Gardens, Shoeburyness, Southend-on-Sea, SS3 9JG

M0 KVF Konrad Emery-Ford, 15 Anson Close, Grantham, NG31 7EN

M0 KVK Kevin Oim, 10 St Julians Wells, Kirk Ella, Hull, HU10 7AF

M0 KVM Kevin Mills, 106 Goodwin Crescent, Swinton, Mexborough, S64 8QD

M0 KVN K Finn, 132 Lansdowne Grove, Wigston, LE18 4LY

M0 KVR Andrew Burfield, 4 Eastern Crescent, Chelmsford, CM1 4JQ

M0 KWA Headcorn Aerodrome c/o Patrick Blunt, 17 Ottens Drive, Staplehurst, Tonbridge, TN12 0LR

M0 KWD K Bradd, Flat 2, Montrose Court, London, NW9 5BS

M0 KWK David Hall, 1 Pendreth Place, Cleethorpes, DN35 7UR

M0 KWM David Wright, 203 Winn Street, Lincoln, LN2 5EY

M0 KWN Roger Skinner, 20 Kenyon Road, Portsmouth, PO2 0JZ

M0 KWP Jeremie Simon, 14 LE PEYNEYUC [illegible], Daignac, France, 33420

M0 KWR Kent Royce, 11 Church Lane, Stibbington, Peterborough, PE8 6LP

M0 KWS Shirley Kendrick, 29 Waterside, Silsden, Keighley, BD20 0LQ

M0 KWV M Evans, 16 Colville Grove, Sale, M33 4FW

M0 KWW Keith Willson, Ludpit Cottage, Ludpit Lane, Etchingham, TN19 7DB

M0 KWY David Wells, 27 Victoria Avenue, Camberley, GU15 3HT

M0 KXD D Rivron, Spring Cottage, Bellerby, Leyburn, DL8 5QN

M0 KXK R Britt, Thoroughfare House, South Burlingham Road, Norwich, NR13 4FA

M0 KXQ Phillip Hardacre, 13 St. Johns Street, Bridlington, YO16 7NL

M0 KYI Keith Armstrong, 29 Thorntree Avenue, Crofton, Wakefield, WF4 1NU

M0 KYL A Kyle, Greengates, Ainsworth Street, Ulverston, LA12 7EU

M0 KYR Kyriakos Orfanidis, Flat 36, Cumberland Court, London, W1H 7DP

MM0 KZA Antonio Marques Gomes, 34 Campbell Close, Hamilton, ML3 6BF

M0 KZB Eric Arkinstall, 79 Sundorne Road, Shrewsbury, SY1 4RU

M0 KZC Martin Clack, 42a Provost Street, Fordingbridge, SP6 1AY

M0 KZH David Taylor, 49 Boggart Hill Gardens, Seacroft, Leeds, LS14 1LJ

MM0 KZJ A Thomson, 5 [illegible] Grove, Dunfermline, KY11 8DU

M0 KZM M Osborne, Flat 5, 7 St Catherines Road, Littlehampton, BN17 6HS

M0 KZP Neil Simmonds, 3 Noneley Hall Barns, Noneley, Shrewsbury, SY4 5SL

M0 LAA S Jefferson, 8 Panturner Road, Stoke-on-Trent, ST2 0SX

M0 LAB M Labourn, 6 Healey Drive, Ossett, WF5 8NA

M0 LAE Lee Clark, 30 Warwick Square, London, SW1V 2AD

M0 LAF Neil Smith, 40 Fairdale Drive, Newthorpe, Nottingham, NG16 2FG

M0 LAG D Whelan, 431 Leeds Road, Huddersfield, HD2 1XT

M0 LAH Shaun Chng, Wolfson College, Cambridge, CB3 9BB

M0 LAI Peter Tolcher, 15 Langstone Close, Torquay, TQ1 3TX

M0 LAL Chris Mole, 6 Clements Road, Chorleywood, Rickmansworth, WD3 5JT

M0 LAO Andrew Powell, 31 Highmead, Pontllanfraith, Blackwood, NP12 2PF

M0 LAS Lorraine Milford, 82b Oakley Lane, Oakley, Basingstoke, RG23 7JX

M0 LAT Andrew Laity, 9 Haverhill Road, Stapleford, Cambridge, CB22 5BX

M0 LAW M Martin, 2821 Bissonnet St, Houston, United States, 77005-4014

M0 LAY Austin Brooks, Halleluya Cottage, Wash Lane, Halesworth, IP19 0RB

M0 LAZ Anthony Burton, 54 West Ashton Road, West Ashton, Trowbridge, BA14 6FG

MI0 LBA CHURCH ISLAND AMATEUR RADIO GROUP c/o T Flanagan, 18 Hunters Park, Bellaghy, Magherafelt, BT45 8JE

M0 LBB B Bush, 2 Penrhiw Cottages, Brynithel, Abertillery, NP13 2AU

M0 LBD Damien Lagan, 24 Milvain Close, Gateshead, NE8 3RS

MM0 LBF Robert Bertram, Noroc, 46 Main Street, Pathhead, EH37 5QB

M0 LBJ B Jenson, 10 Tintern Close, Portsmouth, PO6 4LS

M0 LBK L Karthauser, 17 Manor Close, Abbotts Ann, Andover, SP11 7BJ

M0 LBL Marine Radio Museum Society c/o William Cross, 31 Joshua Close, Liverpool, L5 0TD

M0 LBM P Matthews, 15 Tennyson Way, Melton Mowbray, LE13 1LJ

M0 LBR Frederic Labrosse, 72 Ger y Llan Penrhyncoch, Aberystwyth, SY23 3HQ

MI0 LBS L Stewart, 22 Edenmore Court, Newtownabbey, BT37 0ZA

MM0 LBX John Cattigan, Lunan Home Farm Cottage, Lunan Bay, Arbroath, DD11 5ST

M0 LBY Laurence Lay, 17 Herbert Road, Hornchurch, RM11 3LD

M0 LCA Ewen Taylor, 32 Knoll Drive, Warwick, CV34 5YQ

M0 LCC Lymington Community Association RC c/o Keith Cromar, 17 Ipley Way, Hythe, Southampton, SO45 3LG

MW0 LCH L Holder, 133 Maple Drive, Brackla, Bridgend, CF31 2PR

M0 LCK M Heathcote, Ingledown, Trelogan, Holywell, CH8 9BZ

M0 LCM Lawrence Micallef, 132 Kings Hedges Road, Cambridge, CB4 2PB

M0 LCR Vincent Lynch, 16 Okehampton Crescent, Sale, M33 5HR

M0 LCW LIDS CW & Data Club c/o Alexander Hill, 53 Fairladies, St. Bees, CA27 0AR

M0 LCY M Taylor, 23 Hestham Crescent, Morecambe, LA4 4QF

M0 LDC Lawrence Spriggs, 19 Mackenzie Square, Stevenage, SG2 9TT

M0 LDG Lundy DX Group c/o John Edmunds, Caroline Cottage, New Passage Road, Bristol, BS35 4LZ

M0 LDH L Hawkins, 121 Selmeston Road, Eastbourne, BN21 2TL

M0 LDI Darryl Burden, 16 Milnthorpe Lane, Wakefield, WF2 7DE

M0 LDJ Lee Jessup, Boat Farm, Mill Lane Govilon, Abergavenny, NP7 9SD

M0 LDQ David Arnold, The Chase, Rectory Road, Penzance, TR19 6BB

M0 LDR Lyndon Reynolds, 49 Westborough Way, Anlaby Common, Hull, HU4 7SW

M0 LDV David Curtis, 7 Neale Close, Aylsham, Norwich, NR11 6DJ

M0 LDX Stephen Cape, 92 Davison Avenue, Whitley Bay, NE26 3SY

M0 LDY Jonathan Davies, 5 Beauchamp Road, Kenilworth, CV8 1GH

M0 LDZ Laurence Cook, 43 Midge Hall Drive, Rochdale, OL11 4AX

MW0 LEA Paul Price, Ravenscroft, Blaenporth, Cardigan, SA43 2AS

M0 LEB Abdul Sadka, 39 Hilliers Avenue, Uxbridge, UB8 3JQ

M0 LED L Dixon, 23 Gipsy Lane, Buckfastleigh, TQ11 0DL

M0 LEE L Jones, 200 Upper Eastern Green Lane, Coventry, CV5 7DP

M0 LEF Massimiliano Verardi, 21 Snowdon Street, Y Felinheli, LL56 4HQ

M0 LEH Martin Reynolds, 24 Burton Close, Corringham, Stanford-le-Hope, SS17 7SB

M0 LEK LEEK AND DISTRICT ARC c/o S Jefferson, 8 Panturner Road, Stoke-on-Trent, ST2 0SX

MJ0 LEL Leslie Langlois, Farleyer, La Rue Des Platons, Jersey, JE3 5AA

MM0 LEN L Cochrane, 2 Muir Terrace, Paisley, PA3 4LT

M0 LEO L Boberschmidt, 3928 Denfeld Court, Maryland, United States

M0 LEP Richard Hewett, 118 Lovibonds Avenue, Orpington, BR6 8EN

MM0 LER Martin Dickeson, 41 Hossack Drive, Elgin, IV30 6JY

M0 LET Robert Collis, 20 Little Meadows, Haxby, York, YO32 3YY

MW0 LEW Lewis Thomas, 2 Goytre Crescent, Goytre, Port Talbot, SA13 2YD

M0 LEX Roydan Styles, Padcroft, Weir, OL13 8QL

M0 LEY C Kirby, 1 Church Farm Lane, South Marston, Swindon, SN3 4SR

M0 LEZ L Robinson, 15 Seldown Lane, Poole, BH15 1UA

M0 LFC Friskney + East Lincolnshire Communications Club c/o Brendan Derbin-Cykoc, 1 Lentons Lane, Friskney, Boston, PE22 8RR

M0 LFS L Spacek, 33 Wellesley Road, Colchester, CO3 3HE

M0 LGA O Parker, 37 Springbank Crescent, Gildersome, Leeds, LS27 7DN

M0 LGB G Benson, 2 Guisborough Road, Nunthorpe, Middlesbrough, TS7 0LB

M0 LGC Letchworth Garden City ARC c/o Mark Russell, 107 Cambridge Road, Hitchin, SG4 0JH

M0 LGE Richard Samphire, Courtlands, Newport Road, Caldicot, NP26 3BZ

M0 LGL Lee Leyland, 3 Thirlmere Road, Golborne, Warrington, WA3 3[illegible]

M0 LGN Ben Fitzgerald-O'Connor, 24 Routh Street, London, E6 5[illegible]

M0 LGP L Porter, 52 Mary Towneley Fold, Burnley, BB10 4LZ

MM0 LGR David Boden, 42, Kirkwynd, Maybole, KA197AE

MM0 LGS Jaroslaw Gorczynski, 54 Lindsay Gardens, Bathgate, EH48 1DU

MM0 LGT Malcolm McKay, 75 Cardross Road, Dumbarton, G82 4JL

M0 LHA Miles Burton, 139 Avenue Road, Erith, DA8 3BA

M0 LHB L Bramley, 140 Nevill Road, Hove, BN3 7QB

M0 LHK L King, Orchard Meadow, Coombe Cross, Newton Abbot, TQ13 9EP

M0 LHR P Lonsdale, 77 Burtons Road, Hampton Hill, Hampton, TW12 1DE

M0 LHS Robert Silcox, 103 Oakdale Road, Downend, Bristol, BS16 6EG

M0 LIE Edward St Quinton, Mill Cottage, The Thorofare, Woodbridge, IP13 8BB

M0 LIJ Dawn Smout, Sunrays, Warbage Lane, Bromsgrove, B61 9BH

M0 LIO Licolnshire Scout RC c/o Alan Hull, 1 Occupation Lane, New Bolingbroke, Boston, PE22 7LW

M0 LIS Elizabeth Crossley, 3 Delibes Road, Basingstoke, RG22 4LZ

M0 LIT Clive Martin, 20 Hall Green Road, West Bromwich, B71 3LA

MM0 LJA I Auchterlonie, 2 Lawhead Road East, St. Andrews, KY16 9ND

M0 LJC Leonard Carpenter, Shadlow Marina, London [illegible], DE72 [illegible]

M0 LJD Laura Goldsmith, 7 Fengate Drove, Weeting, Brandon, IP27 0PW

M0 LJK Laura Marriott, 94 Lyndhurst Road, Worthing, BN11 2DW

M0 LJL Lee Lewis, Flat 4, 41 Marine Parade, Lowestoft, NR33 0QN

MI0 LJM S Hutton, 2 Darkfort Crescent, Portballintrae, Bushmills, BT57 8WH

M0 LJT Albert Tranter, 122 Summerhill Road, Bristol, BS5 8JU

M0 LKD Lutz Kahlbau, 7 Hamilton Court, De la Warr Road, Lymington, SO41 0PR

M0 LKE Leslie Kett, 52 Northgate, Hornsea, HU18 1EU

M0 LKL Leslie Brigham, 42 Cayley Close, Clifton, York, YO30 5PT

M0 LKS Robert Boruch, 4 Hafton Road, Salford, M7 3TF

M0 LKT Lee Taylor, Apartment 10, The Church Apartments 47a Seamer Road, Scarborough, YO12 4EF

M0 LKY Anne Lee, 27 Victoria Avenue, Camberley, GU15 3HT

MI0 LLG Stephen Horner, 10 Meadow Court, Bushmills, BT57 8SD

M0 LLK Christopher Tanner, Pen y Gogarth, Llaneilian, Amlwch, LL68 9NH

M0 LLO Christopher Hill, 9 Oliver Road, Newport, NP19 0HU

M0 LLS I Powell, 9 Cardinal Crescent, Bromsgrove, B61 7PR

M0 LLW A Bostock, 26 Ingham Road, Bawtry, Doncaster, DN10 6NW

M0 LLY Richard Martin, Llan Owen, Rhulen, Builth Wells, LD2 3UY

M0 LMB Bruce Savage, 33 Sky End Lane, Hordle, Lymington, SO41 0HG

MM0 LMC Logan Chan, 5 Lansdowne Drive, Cumbernauld, Glasgow, G68 0JB

M0 LMH Lee Hudson, 68 Eleanor Road, Harrogate, HG2 7AJ

M0 LMI Lubomir Mikolka, Flat, 1 Scotney Court, Romney Marsh, TN29 9JP

M0 LMN David Robinson, Height End Farm, Kirk Hill Road, Rossendale, BB4 8TZ

M0 LMO M Moody, 16 Tyersal Court, Bradford, BD4 8EN

M0 LMR Dorothy Stanley, 58 Wells Gardens, Basildon, SS14 3QS

M0 LMS D Partridge, 44 Trumpet Terrace, Cleator, CA23 3DY

MW0 LMW D Jenkins, 36 BRYNAWEL ROAD, Gorseinon, SA4 4UX

M0 LNE Dave Bruce, Nunfield House, Bull Lane, Sittingbourne, ME9 7SL

M0 LNX L Worton, 52 Buttermere Road, Stourport-on-Severn, DY13 8NX

M0 LOB Michael Garry, 34 Conway Road, Paignton, TQ4 5LH

M0 LOG W Stuart, 9 Charminster Close, Great Sankey, Warrington, WA5 1JY

M0 LOU D Cave, 26 Longsight Road, Mapplewell, Barnsley, S75 6HB

M0 LOW Derek Barnes, 11 Yewside, Gosport, PO13 0ZD

MM0 LOZ David Leech, The Croft House, 9, Ruilick, Beauly, IV4 7AB

M0 LPA Neil Hilbery, 16 Albert Road, Ashford, TW15 2LU

M0 LPF B Hall, 30 Worcester Road, Dudley, DY2 9LN

M0 LPG L Parsons, Ty Crwn, Rhosgadfan, Caernarfon, LL54 7HU

M0 LPK Michal Napieralski, Flat 2, Reeves House, Crawley, RH10 7SW

M0 LPL Garry Roberts, 25 Chalfont Way, Liverpool, L28 3QB

M0 LPT Panagiotis Perreas, 180a St. Ann's Road, London, N15 5RP

M0 LPW Louis Walker, 55 Silverlands Road, St. Leonards-on-Sea, TN37 7DF

M0 LQR Thomas Longmore, 3 Dairy Farm Cottages, Northlands Road, Gainsborough, DN21 5DN

M0 LRB Ralf Baechle, Dr.-Schuhwerk-Strasse 32A, St. Blasien, Germany, 79837

MI0 LRC Stephen Davison, 60 Coronation Place, Craigavon, BT66 7AN

M0 LRD Ron Taylor, 89 St. Johns Road, Pelsall, Walsall, WS3 4EZ

M0 LRG Leicester Radio Group c/o Frederick Barkhouse, 312 Humberstone Lane, Leicester, LE4 9JP

M0 LRO Mark Street, Flat 6, Derwent Court, Solihull, B92 7BU

M0 LRS L Smith, 20 Street Loyes Street, Bedford, MK40 1ZL

M0 LSA Lisa Mossop, 4 Brookdale Way, Waverton, Chester, CH3 7NT

M0 LSE LONDON AND SOUTH EAST REGION AIR CADETS c/o D Sharp, 8 Beechfield, Hoddesdon, EN11 9QH

M0 LSI Nigel Highfield, 298 Mersea Road, Colchester, CO2 8QY

MM0 LSM Andrew Halcrow, Da Cro, Branchiclate, Burra Isle, ZE2 9LA

M0 LSN Derek Harding, 9, Gilbert Street, Blenheim, New Zealand, 7201

M0 LSS Luke Storry, The Chimes, Madeira Drive, Bude, EX23 0AJ

M0 LSV R Derham, Netherwood, Copse Lane, Hook, RG29 1SX

M0 LSX Alan Dale, 37 Bussey Road, Norwich, NR6 6JF

M0 LSY Yunfei Li Song, The Colony, Chesterton Lane, Cambridge, CB4 3AA

M0 LTA Andrew Tokely, 17 Sycamore Avenue, Horsham, RH12 4TP

M0 LTD C Brink, 138 Brookside, Burbage, Hinckley, LE10 2TN

M0 LTK James Forsyth, 2 Littleworth, Oxford, OX33 1TR

M0 LTN Alan Rademaker, 26 Elm Park Close, Houghton Regis, Dunstable, LU5 5PN

M0 LTO W Jones, 13 The Coppice, Enfield, EN2 7BY

M0 LTP Zalewski Lukasz, 19 Harlington Road, Uxbridge, UB8 3HX

M0 LTS Laurence Stant, EARS, University of Surrey Student's Union, Guildford, GU2 7XH

M0 LTT Mark Lovatt, 3 Withington Close, Atherton, Manchester, M46 0EZ

M0 LTW Lionel Farrant, 21 Lime Tree Walk, Newton Abbot, TQ12 4LF

M0 LUD Graham Stanley, 11 Marlborough Street, Ossett, WF5 8JW

M0 LUK Christopher Moreton, 20 Millbrook Court, Little Mill, Pontypool, NP4 0HT

MM0 LUP A Young, Broomloan, Staffin, Portree, IV51 9JX

M0 LUS Colin Campbell, 5 Ryebank, Holmfirth, HD9 1EU

M0 LUT Andrew Lutley, Springfield, Rookery Hill, Ashtead, KT21 1HY

M0 LUV Eric Curling, 919 Oxford Road, Tilehurst, Reading, RG30 6TP

M0 LUY Lucy Isaac Sneath, 21 Garrick Close, Lincoln, LN5 8TG

M0 LVR Oliver De Peyer, Flat 5, Molasses House, London, SW11 3TN

M0 LVW Louie Van Wezel, 1 Waveney Road, Felixstowe, IP11 2NT

M0 LWM Barry Blackstone, 2 Sneating Hall Cottages, Sneating Hall Lane, Frinton-on-Sea, CO13 0EW

MM0 LWS Mark Strachan, 62 Charleston Drive, Dundee, DD2 [illegible]

M0 LWT Carl Jenkins, 25 Longmeadow Grove, Birmingham, B31 4SU

M0 LXA C Staff, Chinewood, Pelting Drove, Wells, BA5 3BA

M0 LYD Robert Beck, Moorings, Pleasance Road Central, Romney Marsh, TN29 9NP

M0 LYI Matthew Lynn, 5 Woodlands Court, Crook, DL15 8QN

M0 LYN G Molyneaux, 206 Chamber Road, Oldham, OL8 4DJ

M0 LYQ Edmundas Gudziunas, 66 Packwood Close, Bentley Heath, Solihull, B93 8AW

M0 LZM Peter Radford, 43 Bells Lane, Nottingham, NG8 6EX

M0 LZZ Christopher Stubbs, 50 Laburnum Close, Rogerstone, Newport, NP10 9JQ

M0 MAC James McGowan, 72B Adelphi Crescent, Hornchurch, RM12 4JZ



**IMPORTANT NOTE**

**Revalidate licence to avoid revocation** – Ofcom has advised the Society that plans will be drawn up to revoke licences that have not been revalidated as required by the licence conditions. The quickest way to revalidate is to do so online via the Ofcom website: *https://services.ofcom.org.uk/* or by email: *amateur.validations@ofcom.org.uk* Ofcom staff are available to help, but please be patient during times of heavy workload.

M0 MAF Michael Milne, Flambards, Manor Road, Dunmow, CM6 2JR
M0 MAH M Arnett, 5 Ffordd Cerrig Mawr, Caergeiliog, Holyhead, LL65 3LU
M0 MAI M Mahoney, Hurdle Cottage, Mannington, Wimborne, BH21 7JZ
M0 MAJ M Jones, 20 Chelsea Drive, Sutton Coldfield, B74 4UG
M0 MAL M Elliott, 4 Maple Close, Keelby, Grimsby, DN41 8EL
MD0 MAN Robert Cunningham, 3 Kellets Cottage, Lhergy Cripperty, Union Mills, Isle of Man, IM4 4NF
M0 MAO Maurice Boland, 24 Hallam Close, Moulton, Northampton, NN3 7LB
MI0 MAP James Phillips, 22 Mill Brae, Dromara, Dromore, BT25 2QJ
M0 MAQ Eric MacGurk, 10 Elmore Road, Lee on Solent, PO139DU
MD0 MAR A El Khalidi, 66 Trevor Crescent, Ruislip, HA4 6ND
M0 MAT M Jeffery, 31b High Street, Staple Hill, Bristol, BS16 5HB
MW0 MAU Mark Uphill, 1 Brynview Avenue, Ystrad Mynach, Hengoed, CF82 7DB
M0 MAW Norman Cheesewright, 5 Duberly Close, Perry, Huntingdon, PE28 0BP
M0 MAX M Skinley, 69 George Street, Wellington, TA21 8HZ
M0 MAY Max Stokes, 15 Parc Terrace, Newlyn, Penzance, TR18 5AS
M0 MAZ Mario Stevenson, 127 Walton Road, Chesterfield, S40 3BX
M0 MBA Zoltan Derzsi, 217 Bensham Road, Bensham, Gateshead, NE8 1US
M0 MBB Mark Bowell, 39 Gledwood Drive, Hayes, UB4 0AQ
MM0 MBC John Curtis, 11 Haston Crescent, Perth, PH2 7XD
M0 MBD David De La Haye, 4 Nicola Mews, Ilford, IG6 2QE
M0 MBE Steven Holt, 14 Fir Street, Cadishead, Manchester, M44 5AU
M0 MBG Michael Cooper, 9 Conway Close, Crewe, CW1 3XN
MM0 MBH Mark Holbrook, Lintbrae Cottage, Stewarton, Kilmarnock, KA3 5JT
M0 MBI A Fox, 53 Shakespeare Way, Taverham, Norwich, NR8 6SL
M0 MBM Michael Bridgehouse, 43 Age Croft, Oldham, OL8 2HG
M0 MBO M Bay, 94 Russell Road, Toddington, Dunstable, LU5 6QF
M0 MBR Malcolm Mutkin, 13 The Grove, Radlett, WD7 7NF
M0 MBS M Hyman, 6 Belvedere Court, St. Anns Road, Manchester, M25 9LB
M0 MBT Mark McKenna, 1 Kennet Avenue, Jarrow, NE32 4DB
M0 MBV Martin Hennessey, 57 Northern Road, Aylesbury, HP199QT
M0 MBZ Michael Bray, 26 South Park Close, Redruth, TR15 3AR
M0 MCA Andrew Howden, 7 West Vale, Filey, YO14 9AY
MI0 MCB J McBride, 21 Mosside Gardens, Mosside, Ballymoney, BT53 8QQ
MI0 MCC C McClelland, 2 Stuart Park, Ballymoney, BT53 7BE
M0 MCE D McEwan, 29 St. Andrews Avenue, Weymouth, DT3 5JS
M0 MCG MOORS CONTEST GROUP c/o G Matthews, 11 Beaver Way, Woodley, Reading, RG5 4UD
M0 MCH Melvin Chapman, 3 Whitton Close, Doncaster, DN4 7RB
M0 MCI Paul Illidge, 55 East Park Road, Spofforth, Harrogate, HG3 1BH
M0 MCL Kevin Winton, 130 George V Avenue, Worthing, BN11 5RX
M0 MCN Bryan Wilson, 119 Fountains Close, Washington, NE38 7TQ
M0 MCO Mark Tinsell-Stanton, 38 Comberton Road, Kidderminster, DY10 3DT
M0 MCP Robert Van-der-Wijst, 6 Willow Street, Romford, RM7 7LJ
M0 MCT Matthew Bradbury, 30 Hazel Road, Maltby, Rotherham, S66 8BD
M0 MCV R Treacher, 93 Elibank Road, London, SE9 1QJ
M0 MCW K Phillips, 31b Waterloo Close, Blackburn, BB2 4RQ
M0 MCY Michael Rolph, 7 Broom Crescent, Ipswich, IP3 0EE
M0 MDC Michael Clements, 21 Mallard Place, Twickenham, TW1 4SW
M0 MDE David Edmondson, 21 Hawthorne Close, Heathfield, TN21 8HP
M0 MDG John Ivory, 25 Dabbs Hill Lane, Northolt, UB5 4AH
MM0 MDH Marc Herridge, The Hollies, Petticoat Lane, Orkney, KW17 2RP
M0 MDJ Malcolm Johns, 151 Somerset Street, Abertillery, NP13 1DR
M0 MDO Douglas McAuslan, Casa Arco Iris, Via Variante Nascente, Santa Barbara de Nex, Portugal, 8005-491
M0 MDP Philip Murphy, 41 Tower Street, Sunderland, SR2 8NF
M0 MDT Marc Griffiths, Mandalay, Bromfield Street, Wrexham, LL14 1NF
M0 MEA M Attlesey, 1 The Landway, Borough Green, Sevenoaks, TN15 8RG
M0 MEB Eamonn Bias, 10 Riverdale Road, Shrewsbury, SY2 5TA
M0 MED D Creighton, 8 Stockton Road West, Hawthorn, Seaham, SR7 8RS
M0 MEG A Stephenson, 1 Northrop Close, Sunnybrow, Crook, DL15 0NS
M0 MEH M Horton, 12 Kelburn Close, Chandler's Ford, Eastleigh, SO53 2PU
M0 MEI Maurice Eilec, 32 Bond Close, Tadley, RG26 4EW
M0 MEL M Kirk, 128 Perry Hill Road, Oldbury, B68 0BJ
M0 MEN Mike Norris, 35 Sudbrooke Road, London, SW12 8TQ
M0 MEO Mexborough & District ARS c/o Sharon Saiger, 10 Markham Avenue, Armthorpe, Doncaster, DN3 2AZ
MI0 MEV David Lynas, 7 Regency Avenue, Dollingstown, Craigavon, BT66 7TY
M0 MEW T Garcia-Quismondo, 11 Half Moon Lane, Worthing, BN13 2EN
MW0 MEX G Johnson, 47 Heol Fawr, Penyrheol, Caerphilly, CF83 2JU
M0 MEY Terence Hart, 17a Meyrick Park Crescent, Bournemouth, BH3 7AG
M0 MFA Frank Alfrey, 16 Walls Road, Bembridge, PO35 5RA
M0 MFB Martin Brown, 35 Oak Drive, Colwyn Bay, LL29 7YP
M0 MFC Michael Chester, 84 Edinburgh Drive, Spalding, PE11 2RT
MI0 MFI Edward Taylor, 17 Rutherglen Street, Belfast, BT13 3LR
M0 MFL Karl Bantock, 22 Deepdale Drive, Consett, DH8 7EH
M0 MFP Christopher Reed, 126 North Road, Withernsea, HU19 2AY
M0 MFT Michael Tweedie, 14 St. Cuthbert Drive, Romanby, Northallerton, DL7 8JF
M0 MGA Martin Smyth, 111 Forest Road Whitehill, Bordon, GU35 9BA
MM0 MGB A Britton, 15 Glenbrook, Balerno, EH14 7JE
M0 MGF Jason Gower, 10 Dann Court, Hedon, Hull, HU12 8GT
M0 MGI Matthew Isbell, 20 Woodland Crescent, Wolverhampton, WV3 8AS
MI0 MGJ M James, 6 Portaferry Road, Newtownards, BT23 8NN
M0 MGK Gerard Marley, 55 Hardie Drive, West Boldon, East Boldon, NE36 0JJ
M0 MGS M Smith, 313 Stourbridge Road, Dudley, DY1 2EF
M0 MHC Crewe Museum & Heritage Centre ARC c/o J Dalgliesh, 61 Clonners Field, Stapeley, Nantwich, CW5 7GU
M0 MHY James Mahoney, 61 Wood Street, Barnsley, S70 1NA
MM0 MHZ BACKPACKERS RADIO ACTIVITY GROUP c/o Paul Thompson, 31 St. Marys Drive, Perth, PH2 7BY
M0 MID Peter Staite, Chestnut Farm, Eastville, Boston, PE22 8LX
M0 MIE Matthew Ireland, Pen y Gadlas, Ffordd Bryniau, Prestatyn, LL19 8RD
M0 MIG M Ortega Navarro, 10 Junction Close, Burgess Hill, RH15 0NZ
M0 MIK Mike Jameson, 3 OLYMPIAN WAY, Matlock, DE4 2GX
M0 MIM Michael Pearce, 1 Briars Wood, Horley, RH6 9UE
M0 MIQ Muhammad Iqbal, 6 Hobart Road, High Wycombe, HP13 6UD
M0 MIR Miroslaw Lesniowski, 3 Woodland Avenue, Worksop, S80 2RB
M0 MIT B Mitchell, 34 St. Marys Avenue, Gosport, PO12 2HX
MW0 MJB Mark Lee, 1 Maes y Frenni, Crymych, SA41 3QJ
M0 MJD M Davis, The Innings, Cricketts Lane, Chippenham, SN15 3EG
M0 MJF Michael Firth, 209 High Street, Wickham Market, Woodbridge, IP13 0RQ
M0 MJG M Garrett, 489 Dorchester Road, Weymouth, DT3 5BP
M0 MJH Mark Hickford, 3 Ashen Road, Clare, Sudbury, CO10 8LQ
M0 MJK Martin Keyte, 3 Lower High St., Mow Cop, Stoke-on-Trent, ST7 3PB
M0 MJS M Sykes, Hope Cottage, 8 Brookside Road, Wimborne, BH21 2BL
M0 MJT Neil Tyerman, 44 Hawkstone Close, Guisborough, TS14 7PE
M0 MJW Michael Whitfield, 10 Bede Haven Close, Bude, EX23 8QF
MM0 MJY Martin Yarrow, Lomond Villa, Downies Village, Aberdeen, AB12 4QX
M0 MKE KING EDWARD VII SCHOOL c/o P Treadwell, 22 Meynell Close, Melton Mowbray, LE13 0RA
M0 MKG Mark Gray, 15 The Circle, Cwmbran, NP44 7JP
M0 MKH Michael Hadfield, 22 Mansfield Road, Clowne, Chesterfield, S43 4DH
M0 MKO Martin O'Connor, 28 Cardigan Road, Southport, PR8 4SF
M0 MKR Milton Keynes Raynet c/o John Breen, 68 Honeysuckle Way, Bedford, MK41 0TF
M0 MKV Andrew Crawford, 4 Trimpley Drive, Kidderminster, DY11 5LB
MM0 MLB M Burgess, 11 Cromar Drive, Dunfermline, KY11 8GE
MM0 MLD William Lawson, 60 Inglis Avenue, Port Seton, Prestonpans, EH32 0AQ
M0 MLE John Statham, Oakwoods, School Lane, Reading, RG8 8LT
M0 MLG M Goff, 27 Harley Road, Oxford, OX2 0HS
M0 MLH F Peters, 60a Clarendon Road, London, E17 9AZ
M0 MLJ Raymond Tattersall, 56 Larch Road, New Ollerton, Newark, NG22 9SX
M0 MLK Marie Kipling, 12 Jolly Brows, Bolton, BL2 4LZ
M0 MLM Mike Millen, Flat 9, Sussex Court, Bognor Regis, PO21 2PY
MM0 MLO M Olesen, 19 Holmston Crescent, Ayr, KA7 3JJ
M0 MLT M Titcombe, 1a Langdale Avenue, Harpenden, AL5 5QU
M0 MLV Terry Jones, Flat 92, Berkeley Court, London, NW1 5ND
M0 MLW M Wren, Church Farm, Pointon Fen, Sleaford, NG34 0LF
M0 MLY John Malley, 18 Park View, Seaton Delaval, Whitley Bay, NE25 0AL
M0 MLZ Martin Mills, 17 Hornby Street, Plymouth, PL2 1JD
M0 MMC K Sansom, 23 Victoria Crescent, Poole, BH12 2JQ
MM0 MMG M Gourlay, 14 Holmes Holdings, Broxburn, EH52 5NS
M0 MMJ Manmeet Majhail, 3 Poynders Hill, Hemel Hempstead, HP2 4PQ
M0 MMO James Summerhill, 43 Rangers Walk, Bristol, BS15 3PW
M0 MMR Sean Quinn, 17 Cleveland Road, Southampton, SO18 2AP
M0 MMS Mark Mattingley-Scott, Bergheimer Strasse 28, Heidelberg, Germany, 69115
M0 MMT M Johnson, Chy-an-Gwelva, Foundry, Truro, TR3 7BU
M0 MMX Dennis Watt, 4 Spring Gardens Terrace, Padiham, Burnley, BB12 8JB
M0 MNG Edmund Spicer, 3 Golden Avenue Close, East Preston, Littlehampton, BN16 1QS
M0 MNO David Edge, 122 Aldbanks, Dunstable, LU6 1AJ
MM0 MNS Matthew Stuart, 3 Arran View, Largs, KA30 9ER
M0 MNU Peter Richardson, 14 Portland Street, Worksop, S80 1RZ
M0 MNV Piotr Bienko, 94 Foster Street, Lincoln, LN5 7QF
M0 MNX ARTIE MOORE ARS c/o K Dawson, 57 Fair View, Blackwood, NP12 3NR
MM0 MOB M Overthrow, 63 Primrose Avenue, Larkhall, ML9 1JX
MM0 MOC MUSEUM OF COMMUNICATION ARC c/o A Dailey, 82 Don Drive, Livingston, EH54 5LP
MI0 MOD Thomas Thompson, 8 Knockburn Avenue, Lisburn, BT28 2QF
M0 MOI Stephen Pettitt, 11 Derling Drive, Raunds, Wellingborough, NN9 6LF
M0 MOL G Mollard, 1 Barnard Street, Barrow-in-Furness, LA13 9TD
M0 MON Derlwyn Williams, 10 Bronllys, Gaerwen, LL60 6JN
M0 MOR Anthony Turner, 29 Welling Road, Orsett, Grays, RM16 3DW
M0 MOS Matthew Beckett, 59 Broadacre, Caton, Lancaster, LA2 9NH
MM0 MOT Alastair Smith, 56 Ayr Road, Douglas, Lanark, ML11 0QA
MM0 MPA Douglas Panton, 64 Dochart Crescent, Polmont, Falkirk, FK2 0RE
M0 MPB Mark Budd, 28 Ladymeadow Court, Middleton, Milton Keynes, MK10 9HZ
M0 MPF Michael Finn, 23 Spa Lane, Hinckley, LE10 1JA
M0 MPI Michael Ibbett, 40 Hunt Hill Close, Stevenage, SG1 6DS
M0 MPM Michael Meerman, 24 Horseshoe Crescent, Burghfield Common, Reading, RG7 3XW
M0 MPS B Hopkins, 28 Dean Lodge Grange Road, Southbourne, Bournemouth, BH6 3ND
M0 MPT M Travis, Cherrydayle, 6 Lingwood Close, Southampton, SO16 7GJ
M0 MPY Stuart Gray, 2 Gloucester Road, Pilgrims Hatch, Brentwood, CM15 9ND
M0 MRC Clifford Robinson, 9 Chatsworth Avenue, Culcheth, Warrington, WA3 4LD
MI0 MRG MARCONI RADIO GROUP c/o Paul Quinn, 11 Blackpark Road, Ballyvoy, Ballycastle, BT54 6QZ
M0 MRI Andrew Titmus, 5 Kithurst Crescent, Goring-by-Sea, Worthing, BN12 6AJ
M0 MRJ M Jebbett, 16 Hastings Meadow Close, Kirby Muxloe, Leicester, LE9 2DR
M0 MRK Mark Newton, Hall Farm Bungalow, Holbeck, Worksop, S80 3NF
M0 MRL David Weight, 12 Durrants Path, Chesham, HP5 2LH
M0 MRM A Moe, 43 Huntlyburn Terrace, Melrose, TD6 9BH
M0 MRN 247 (ASHTON) SQUADRON ATC c/o Ian Kilkenny, 23 Hazelhurst Road, Stalybridge, SK15 1HD
MM0 MRO C Munro, 11 Craigleith Hill Green, Edinburgh, EH4 2ND
M0 MRP Matthew Phillips, 71 Stour View Gardens, Corfe Mullen, Wimborne, BH21 3TL
M0 MRQ Philip Lawrence, The Hollies, The Street, Bungay, NR35 2LZ
M0 MRS Marches ARS c/o Malcolm Bobby, Hafan, Church Street, Penycae, LL14 2RL
M0 MRT K Turner, 3 Park Close, Pinxton, Nottingham, NG16 6QQ
MI0 MRV Marvi Kashkoush, 41 Dunamallaght Road, Ballycastle, BT54 6PF
M0 MRW Michael Williams, 19 Fourfields Way, Arley, Coventry, CV7 8PX
M0 MRY John Mullins, 61 St. Johns Road, Slough, SL2 5EZ
M0 MSA Mid Somerset ARC c/o Terry Thompson, 23 Oaklands, Paulton, Bristol, BS39 7RP
MI0 MSB Barry Campbell, 3b Crewe Road, Ballinderry Upper, Lisburn, BT28 2PL
M0 MSC M Capper, 12 Otterbury Close, Bury, BL8 2TY
M0 MSE Mark Edmonds, 60 Shenstone Road, Maypole, Birmingham, B14 4TJ
M0 MSF T Reed, Seafield, Charing Hill, Ashford, TN27 0NG
M0 MSG Malcolm Gibbons, 117 Ettingshall Road, Bilston, WV14 9XF
MM0 MSH Benjamin McCosh, 10 Woodilee, Broughton, Biggar, ML12 6GB
MI0 MSM D Dellett, 5 Larchfield Gardens, Kilrea, Coleraine, BT51 5SB
MI0 MSO Philip Hosey, 13 Glenelly Gardens, Omagh, BT79 7XG
M0 MSS M Simpson, 19 Owens Quay, Bingley, BD16 4DX
M0 MSX Michael Smith, 6 Neeps Terrace, Middle Drove, Wisbech, PE14 8JT
M0 MSZ Martin Strange, 101 Southbroom Road, Devizes, SN10 1LY
M0 MTA Mark Atfield, 42 Pauls Croft, Cricklade, Swindon, SN6 6AJ
M0 MTC Wirral & District ARC WADARC c/o Geoffrey Brown, 13 Francis Avenue, Moreton, Wirral, CH46 6DH
M0 MTD T Davidson, 2 Ridgeway, Rotherham, S65 3PQ
M0 MTF Marcin Tomasz Falkowski, 23 Casson Street, Crewe, CW1 3EG
M0 MTI Matti Juvonen, 7 Alphin Brook, Didcot, OX11 7FG
M0 MTJ Mike Smith, 6 Peverill Road, Perton, Wolverhampton, WV6 7PH
M0 MTN C MARTIN, 14 Campbell Road, Eastleigh, SO50 5AD
MM0 MTO Raymond Foulds, 83 Croftfoot Road, Glasgow, G44 5JU
M0 MTR M Roynon, 16 Greenwood Avenue, Pontnewydd, Cwmbran, NP44 5JE
M0 MTS Christopher Small, Riddings Barn, Hope Bagot, Ludlow, SY8 3AE
M0 MTW John Bailey, 22 Wilford Drive, Ely, CB6 1TL
M0 MTX A Price, 67 Mansfield Road, Glapwell, Chesterfield, S44 5QA
M0 MUC Mitchell Wolfson, 4 Crabmill Lane, Easingwold, York, YO61 3DE
MM0 MUL A Jackson, Union Farm, Craigrothie, Cupar, KY15 5PJ
MW0 MUM Aeronwen Sneddon, 3 Marigold Close, Gurnos, Merthyr Tydfil, CF47 9DA
MM0 MUN Edward Munro, 55 Abergeldie Road, Aberdeen, AB10 6ED
MM0 MUR Gordon Murray, The Barn House, Springfield Farm, Carluke, ML8 4QZ
M0 MUZ M Hickman, 40 Tredington Grove, Caldecotte, Milton Keynes, MK7 8LR
M0 MVB Steven Norman, 38 The Croft, Christchurch, Wisbech, PE14 9PU
M0 MVK M BANCROFT, 1 John Street, Knutton, Newcastle, ST5 6DT
M0 MVL Mark Lloyd, 17 Williams Mead, Bartestree, Hereford, HR1 4BT
M0 MVM Anthony Holt, Tyshoni, New Street, Llandrindod Wells, LD1 6BU
M0 MVO Ryszard Zakrzewski, 31 Kingston Crescent, Chelmsford, CM2 6DN
MI0 MVP Alexander Simpson, 10 Woodview Park, Tandragee, Craigavon, BT62 2DD
M0 MVS Maritime Volunteer Service RC c/o Leslie Miller, 28 Arthur Road, Margate, CT9 2EN
MW0 MWA Arron Nisbet, 8 Pen Dinas, Tonypandy, CF40 1JD
M0 MWK M Kirby, 1 Dugdale Avenue, Bidford-on-Avon, Alcester, B50 4QE
MW0 MWL D Mead, 35 Holly Street, Rhydyfelin, Pontypridd, CF37 5DA
M0 MWR Martin Redstall, 56 Westmorland Road, Felixstowe, IP11 9TJ
M0 MWS Martin Smith, Ashcroft, Black Horse Lane, Winterbourne Earls, Salisbury, SP4 6HW
M0 MWT R Wilkes, 157 Saltwells Road, Dudley, DY2 0BN
MM0 MWW ORKNEY ARC c/o C Penna, North Windbreck, Deerness, Orkney, KW17 2QL
M0 MXC Mark Craven, 78 Connaught Road, Brookwood, Woking, GU24 0HF
M0 MXO Chertsey RC c/o James Preece, 17 Cherry Tree Avenue, Staines, TW18 1JB
MW0 MXT C Fisher, 22 Troed Y Bryn, Upper Tumble, Llanelli, SA14 6BP
M0 MXX M Day, 33 Ryndle Walk, Scarborough, YO12 6JT
M0 MYA David Passey, Blue House Cottage, Blue House Lane Albrighton, Wolverhampton, WV7 3AA
M0 MYB H Ibbitson, Tor View, Whitstone, Holsworthy, EX22 6TB
M0 MYC Ralph Browne, 2 Martham Close, London, SE28 8NF
M0 MYE D Myers, 1 Dalton Cottages, Shildon, DL4 2LH
M0 MYJ Andrew Frost, 12 Nightingale Gardens, Nailsea, Bristol, BS48 2BH
M0 MYK Michael Knowles, 86 West Shore Road, Walney, Barrow-in-Furness, LA14 3UD
MM0 MYL Charles Williamson, 31 Medrox Gardens, Cumbernauld, Glasgow, G67 4AJ
M0 MYN C George, 22 Elgar Drive, Shefford, SG17 5RZ
M0 MZX A Watts, 12 Duchy Close, Dorchester, DT1 2EL
M0 NAA Gordon Porter, Higher Bramble, Trusham, Newton Abbot, TQ13 0NW
M0 NAB Nicholas Hockenhull, 6 Bryn Gannock, Deganwy, Conwy, LL31 9UG
M0 NAC 90(SPEKE) SQUADRON ATC AMATEUR RADIO CLB c/o N Walker, 79 Arklow Drive, Hale Village, Liverpool, L24 5RR
M0 NAE Barrie Smith, Maple Lodge, Burtoft Lane South, Boston, PE20 2PF
M0 NAF Nigel Auckland, Glenfield, The Avenue, Southampton, SO32 1BP
M0 NAG Nicholas Parry, Worlingham Court, Marsh Lane, Beccles, NR34 7PE
M0 NAI Richard Neufeld, 19 Douai Grove, Hampton, TW12 2SR
M0 NAK Christopher Marshall, 51 Hedgerow Close, Redditch, B98 7QF
M0 NAL Philip Shaw, 25 Headcorn Road, Platts Heath, Maidstone, ME17 2NH
M0 NAM Neil Matthes, 24 Albany Road, Fleet, GU51 3LY
M0 NAO N Tateishi, 4-24-14-402 TAIHEI, SUMIDA-KU, Tokyo, Japan, 1300012
M0 NAP A Newell, 17 Southlands Grove, Thornton, Bradford, BD13 3BG
M0 NAQ Robert Smith, 15 Hollybush Road, North Walsham, NR28 9XT
M0 NAR NORTHWEST ARC c/o R Seddon, 255 Westleigh Lane, Leigh, WN7 5PN
M0 NAS Neil Smith, Clare Cottage, White Ash Green, Halstead, CO9 1PD
M0 NAU Natalie Coventry, 1 Seacrest Avenue, Fleetwood, FY7 6FG
M0 NAW Neil Carey, 16 Cannamanning Road, Penwithick, St. Austell, PL26 8UX
M0 NAX Gerald Mosner, Zum Roehrbrunnen 16, Dreieich, Germany, 63303
M0 NAY Christopher Pegrum, 3 Bretland Road, Tunbridge Wells, TN4 8PS
M0 NAZ Andrew Davies, 4 Capella Path, Hailsham, BN27 2JY
M0 NBA Benjamin Chalmers, 19 Brettenham Crescent, Ipswich, IP4 2UB
M0 NBC North bristol AmateurRC c/o Paul Stevenson, 6 Dighton Gate, Stoke Gifford, Bristol, BS34 8XA
M0 NBJ Neil Jones, 26b Wellington Road, Wallasey, CH45 2NG
M0 NBK Gareth Hicks, 8 Mill Lane, Stockton-on-Tees, TS20 1LG
M0 NBL Fergus Noble, 1045, 45th Street Apartment A, California, United States, 94608
M0 NCA Norfolk Coast ARS c/o S Appleyard, Plumtree House, Mill Lane, Cromer, NR27 9PH
M0 NCC Northampton DX c/o Gary Bansil, 32 Nethermead Court, Northampton, NN3 8NE
M0 NCE Neil Irvine, 100 Cavendish Road, Sunbury-on-Thames, TW16 7PL
M0 NCG Mark Dumpleton, 23 Watermans Yard, Norwich, NR2 4SD
M0 NCK Nick Jewitt, 45 North Road, Kirkburton, Huddersfield, HD8 0QH
M0 NCN Michael Gillingham, Frecheville Rectory, Brackenfield Grove, Sheffield, S12 4XS
M0 NCZ Neikolas Czernuszka, 12 Durham Drive, Ashton-under-Lyne, OL6 8BP
M0 NDA Nuneaton & District ARC c/o D Parker, 146 Merlin Avenue, Nuneaton, CV10 9QJ
M0 NDC Nigel Caulfield, Smallbrook Road, Whitchurch, SY13 1BS
M0 NDE Nigel Evans, 17 Nightingale Close, Burnham-on-Sea, TA8 2QJ
M0 NDJ Denis Noe, 21 Gale Crescent, Banstead, SM7 2HZ
MI0 NDK Nigel Jameson, 15a Ednagee Road, Castlederg, BT81 7QF
M0 NDL S Emary, Mallards, Fishers Lane, Highbridge, TA9 4LZ

MM0 NDM N MacLucas, Lochnell Lodge, Benderloch, Oban, PA37 1QS
M0 NDO A Hall, 22 Park Grove, Swillington, Leeds, LS26 8UN
M0 NDN [illegible], 11 Bluebell Gardens, Grappenhall, Warrington, WA4 3LE
M0 NDT David Farrant, 13 Bramham Down, Guisborough, TS14 7HY
M0 NDU John Knight, 30 Ash Meadow, Lea, Preston, PR2 1RX
MM0 NDX Colin McGowan, 21 Franchi Drive, Stenhousemuir, Larbert, FK5 4DX
M0 NDY R Potter, 8 Hansard Way, Kirton, Boston, PE20 1QN
M0 NDZ Tinko Daskalov, 4 Arden Walk, Rugeley, WS15 4ER
M0 NEC N Eccles, 55 New Street, Lymington, SO41 9BP
M0 NED Peter Kelly, 20 Fareham Close, Walton-le-Dale, Preston, PR5 4JX
M0 NEG Kevin Metcalfe, 33 Corsican Drive, Hednesford, Cannock, WS12 4SS
M0 NEH Neil Hoare, 5 Kelsey Head, Port Solent, Portsmouth, PO6 4TA
M0 NEO Neil Thomson, 2 Doon Terrace, Dumfries, DG2 9EE
M0 NER Alan Fraser, 18 Donside Close, Boldon Colliery, NE35 9BS
M0 NEU D Newgas, 32 Merton Lane, Highgate, London, N6 6NB
M0 NEV L Creek, 382 Ripon Road, Stevenage, SG1 4NQ
MM0 NEW B Newcombe, 9 Calder Road, Bellsquarry, Livingston, EH54 9AA
M0 NEX Leigh Jepson, 16 Golborne Street, Newton-le-Willows, WA12 9TH
M0 NFB Neill Bisiker, 31 Lansdowne Avenue, Waterlooville, PO7 5BL
M0 NFD Northern Fells Contest Group c/o Clive Davies, 28 Neville Road, Darlington, DL3 8HY
M0 NFI Neil Mooney, 60 Rhyddings Street, Oswaldtwistle, Accrington, BB5 3EY
M0 NFR New Forest ARS c/o Richard Ferguson, 31 Barton Court Road, New Milton, BH25 6NW
M0 NFY Neville Young, 139 Northumberland Street, Norwich, NR2 4EH
M0 NGB N O'Brien-Bird, 18 Milsted Close, Sunderland, SR3 2RF
M0 NGC Craig Austin, 26 De Montfort Road, Reading, RG1 8DL
M0 NGI Peter Strachan-Buckley, 9 Short Street, Aldershot, GU11 1HA
M0 NGL Nigel Nash, Roann, Bedmond Road, Hemel Hempstead, HP3 8SH
M0 NGS NORTHAMPTONSHIRE GRAMMAR SCHOOL ARC c/o Robert Tickle, 5 Bramley Court, Harrold, Bedford, MK43 7BG
M0 NGY Jon Fautley, 71 Pullman Lane, Godalming, GU7 1YB
M0 NHK Newbury and District Hackspace c/o Norman Bland, 3 Kennet Road, Newbury, RG14 5JA
MM0 NHM Neil Morris, 23 sedgebank, Sedgebank, Livingston, EH54 6HE
M0 NIB Nicholas Bown, 18a Warley Hill, Warley, Brentwood, CM14 5HA
M0 NIC N bellamy, 9 Maybank Road, Yate, Bristol, BS37 4BS
MI0 NID N.I Dxer's Group c/o Simon Barnes, 191 Marlacoo Road, Portadown, Craigavon, BT62 3TD
M0 NIE Boguslaw Niewiadomski, 41 The Crescent, Keresley End, Coventry, CV7 8LB
M0 NIF G Calder, 41 Wood End Way, Chandler's Ford, Eastleigh, SO53 4LN
M0 NIG N Howe, 45 Kettering Road, Islip, Kettering, NN14 3JT
M0 NIL Robert Blackwell, Vikings Hall, Baylham, Ipswich, IP6 8JS
M0 NIW Norman Wiseman, 23 Kingsway, Langley Park, Durham, DH7 9TB
M0 NJE N Eustice, 22 Lower Wear Road, Exeter, EX2 7BQ
M0 NJJ Neil Pipkin, 46 Charles Avenue, Albrighton, Wolverhampton, WV7 3LF
M0 NJM P Martin, 2 Gwarllyn, Tudweiliog, Pwllheli, LL53 8NG
M0 NJP N Pettefar, 44 Duck Lane, Laverstock, Salisbury, SP1 1PU
M0 NJS Nigel Sheridan, Cemetery Lodge, Lochmaben, Lockerbie, DG11 1RL
M0 NJW Nigel Wears, 25 Topcliffe Mews, Morley, Leeds, LS27 8UL
M0 NJX Matthew Nassau, 1A Burford Road, Bromley, BR1 2EY
M0 NKA Krassian Atanassov, 250 Dyas Avenue, Birmingham, B42 1HG
M0 NKE Neil Yorke, 21 Braemar Way, Nuneaton, CV10 7LF
M0 NKR Andrew Goldsmith, 7 Fengate Drove, Weeting, Brandon, IP27 0PW
M0 NKS B Maggs, 44 Coldharbour Road, Hungerford, RG17 0AZ
M0 NKY J Atkinson, 17 Agricola Gardens, Hadrian Park, Wallsend, NE28 9RX
M0 NLP Colin Bowman, 26 Albany Hill, Tunbridge Wells, TN2 3RX
M0 NLR Alan Clark, 3 North Street, Owston Ferry, Doncaster, DN9 1RT
M0 NLW Newton-Le-Willows Raynet c/o Peter Williams, 2 Sycamore Avenue, Newton-le-Willows, WA12 8LT
MI0 NLY Seamus Carlin, 9 Mullandra Park, Kilcoo, Newry, BT34 5LS
M0 NMC N Mcintyre, 27 Chapel Close St Ann's Chapel, Gunnislake, PL18 9JB
M0 NMD N Davison, 1 Retford Close, Derby, DE21 4DX
M0 NMH Nigel Hilton, 20 Darbyshire Close, Deeping St. James, Peterborough, PE6 8SF
M0 NMI David Blake, Pound Farm, Swan Lane, Leigh, Swindon, SN6 6RD
M0 NMO MONITORING MONTHLY c/o Kevin Nice, 19 Southill Road, Poole, BH12 3AW
M0 NNB Alan Hopper, 7 Holmesdale Villas, Swallow Lane, Dorking, RH5 4EY
M0 NNH Gary Bansil, 32 Nethermead Court, Northampton, NN3 8NE
M0 NNL N Lutte, Long Durford, Durford Wood, Petersfield, GU31 5AW
M0 NOA A De Broise, 67 Astley Lane, Swillington, Leeds, LS26 8UE
M0 NOC Paul Bolton, 1 Acorn Rise, Hollesley, Woodbridge, IP12 3JT
M0 NOE M Hodgson, 10a Myrtle Grove, Enfield, EN2 0DZ
M0 NOI Carl Dirkin, 16 Marystow Close, Allesley, Coventry, CV5 0EA
M0 NOK Barry Handley, 68 Northfield Avenue, Rothwell, Leeds, LS26 0SW
MI0 NOR Norman McKee, 54 Castlemore Park, Belfast, BT6 9RP
M0 NOS A Hayward, 4 Alton Court Cottages, Penyard Lane, Ross-on-Wye, HR9 5NR
M0 NOV Edward Lane, 50 Oakhurst Close, Belper, DE56 2TR
M0 NOW N Walker, 79 Arklow Drive, Hale Village, Liverpool, L24 5RR
M0 NOZ J Norrington, 32 Fulfen Way, Saffron Walden, CB11 4DW
M0 NPA Nicholas Aleksander, 3 Elm Walk, London, NW3 7UP
M0 NPD Nicholas Du Pre, Vine Cottage, Willington Street, Maidstone, ME15 8ED
M0 NPL Neil Livingstone, 2 Mickleton, Wilnecote, Tamworth, B77 4QY
M0 NPQ Nerijus Ubonis, 8 Burleigh Close, Great Yarmouth, NR30 2RU
M0 NPT Abdelghani Mesbah, 121 Hood Street, Nottingham, NG5 4AQ
M0 NQB N Berrie, 12 Packenham Road, Basingstoke, RG21 8XT
M0 NQU E Wagner, 3 Sarre Road, London, NW2 3SN
MM0 NQY Peter Davis, 2 Virkie Cottages, Virkie, Shetland, ZE3 9JS
M0 NRC NEWTON LE WILLOWS ARC c/o Keith Horsfield, 59 Queens Drive, Newton-le-Willows, WA12 0LY
M0 NRG N Grigsby, 67 Abshot Road, Fareham, PO14 4NB
M0 NRH Nicholas Hickson, 27 Cressing Road, Witham, CM8 2NP
M0 NRJ Nick Johnson, Belair, Western Road, Crediton, EX17 3NB
M0 NRP Anne Andrew, 80 Hamble Drive, Abingdon, OX14 3TE
M0 NRS Norman Stoker, 11 Hewley Crescent, Throckley, Newcastle upon Tyne, NE15 9AT
M0 NRW Matt Reeve, 3 Princess Anne Terrace, Loddon, Norwich, NR14 6LL
M0 NRY L Brackstone, 276 Ladyshot, Harlow, CM20 3EY
MW0 NCC NEATH & DISTRICT [illegible] CADET UNIT c/o J [illegible], [illegible] Woodland Road, Neath, [illegible]
M0 NSI Brian Taylor, 15 Gledhall Street, Stalybridge, SK15 1LE
M0 NSP Gediminas Jurgaitis, 70b Ingleby Road, Ilford, IG1 4RY
M0 NSR NORFOLK SCOUT RADIO c/o C Rolph, The Hollies, Back Lane, Eastgate, Norwich, NR10 4HL
M0 NTA T Brookes, 94 Newton Road, Lowton, Warrington, WA3 1DG
M0 NTC Gerard Bull, 9 Kilburn Place, Dudley, DY2 8HP
M0 NTG 93CONTEST GROUP c/o J Spurgeon, Whitgift House, Whitgift, Goole, DN14 8HL
M0 NTH D Higgs, 4 Rowsley Road, Stretford, Manchester, M32 9QA
M0 NTI Thomas Ward, 127 Lower Lime Road, Oldham, OL8 3NP
M0 NTK John Carrington, 15 Astley Court, Newcastle upon Tyne, NE12 6YR
M0 NTN Norman Norris, 6 Tell Grove, London, SE22 8RH
M0 NTT John Naylor, 15 Cawder Road, Skipton, BD23 2QE
M0 NTY Christopher Shane, 21 Avon Walk, Leighton Buzzard, LU7 3DE
M0 NUC BREDE STEAM AHS c/o Steve Stewart, 4 Westminster Crescent, Hastings, TN34 2AW
M0 NUG N Lewis, 81 Long Lane, Upton, Chester, CH2 1JG
M0 NUK NESTLE UK EMPLOYEE RC c/o Michael Brown, 6 Rose Court, Garforth, Leeds, LS25 1NS
M0 NUX Julian Horn, 8 Princess Close, Watton, Thetford, IP25 6XA
M0 NUZ A Charlton, 26 Saundergate Lane, Wyberton, Boston, PE21 7BZ
M0 NVJ C Drury, 129 Greenhill Road, Mossley Hill, Liverpool, L18 7HQ
M0 NVQ Robert Lynch, 2 Launceston Close, Oldham, OL8 2XE
M0 NVS Phillip Rees, 3 Nash Green, Hemel Hempstead, HP3 8AA
M0 NVY William Oliver, Pwllmeyric, Chepstow, NP16 6LE
MI0 NWA Joseph Baker, 324 Clonmeen, Drumgor, Craigavon, BT65 4AT
M0 NWC North West ARC c/o James McInnes-Boylan, 54 Fernbeck Close, Farnworth, Bolton, BL4 8BR
M0 NWI Peter Martin, 5 Shropshire Drive, Wilpshire, Blackburn, BB1 9NF
M0 NWK Adrian Leggett, 5 Syerston Way, Newark, NG24 2SU
MI0 NWO Dorothy Adams, 65 Rose Park, Limavady, BT49 0BF
M0 NWT James Turner, 2 South Drive, Padiham, Burnley, BB12 8SH
M0 NWW Nigel Warner, 12 Bay Road, Harwich, CO12 3JZ
M0 NWY Simon Newhouse, 28 Hillmorton Lane, Lilbourne, Rugby, CV23 0SS
M0 NXP M Whitaker, 5 Horns Drove, Rownhams, Southampton, SO16 8AH
MI0 NYC J Murdie, 9 Henderson Park, Bangor, BT19 1NS
M0 NYP Stuart Vzor, 40 Henlow Road, Birmingham, B14 5DS
M0 NYW Jason Woodman, 139 Central Avenue, Hayes, UB3 2BT
M0 NYX J Hyde, The Grove, 7 Mill Lane, Kidderminster, DY10 3ND
M0 NZA V Vesma, Durvale, 5 Jonas Drive, Wadhurst, TN5 6RJ
M0 NZL Donald Foster, Flat 3, Edenthorpe Lodge, 7 St. Johns Road, Eastbourne, BN20 7JA
M0 NZR D Bond, 4 Alfred Road, Haydock, St. Helens, WA11 0QD
M0 OAA John Chatterton, 6 Bayliss Road, Wargrave, Reading, RG10 8DR
M0 OAB Brian Hodson, 176 Carters Mead, Harlow, CM17 9EU
M0 OAC David Ion, 78 Blackmore Street, Derby, DE23 8AX
M0 OAE Quentin Wright, 9 Browning Avenue, Warwick, CV34 6JQ
M0 OAL Alastair Weller, 43 Bramley Road, Worthing, BN14 9DS
M0 OAR P Wallace, 7 Trinity View, Ketley Bank, Telford, TF2 0DX
M0 OAT Graeme Walker, 2 Cliffe Bank Cottages, Piercebridge, DL2 3SX
MI0 OBC David Best, 13 Cranley Green, Bangor, BT19 7FE
MI0 OBE J Watt, 23 Riverview Park, Ballymoney, BT53 7QS
M0 OBL Mark Orbell, 21 Reedings Road, Barrowby, Grantham, NG32 1AU
MI0 OBR Andrew Savage, 469 Old Belfast Road, Bangor, BT19 1RQ
M0 OBU John Haddleton, 6 Pembridge Road, Stoke-on-Trent, ST3 3BX
M0 OBW D Wilson, 12 New Street, Elworth, Sandbach, CW11 3JF
M0 OBY David Clavey, 32 Apollo Close, Dunstable, LU5 4AQ
M0 OBZ James McInnes-Boylan, 54 Fernbeck Close, Farnworth, Bolton, BL4 8BR
M0 OCC Oxo Contest Club c/o Charles Wilmott, 60 Church Hill, Hoyston, Barnsley, S71 4NG
MI0 OCG Orchard County DX Club c/o Alexander Simpson, 10 Woodview Park, Tandragee, Craigavon, BT62 2DD
M0 OCK Alan Mock, 115 Swanfield Drive, Chichester, PO19 6TD
M0 OCL Leanne Hendry, 109 Grove Avenue, New Costessey, Norwich, NR5 0HZ
M0 OCT CHESTERFIELD REPEATER GROUP c/o Stephen Brown, 24 Braemar Residential Park, Kirkby Green, Lincoln, LN4 3PD
M0 ODD Ian Rotheram, 60 Whitewood Park, Liverpool, L9 7LG
M0 ODE S Hawkins, Forest Edge, Deer Park, Blandford Forum, DT11 0AY
MM0 ODI R Kelly, 11 Kelvin Drive, Chryston, Glasgow, G69 0LZ
MM0 ODL Fred Gordon, Croft of Torrancroy, Strathdon, AB36 8US
M0 ODM David Merridale, THE GRANARY, FALLEDGE LANE, Upper Denby, HD8 8TT
M0 ODS D Eccles, 64 Brookwood Drive, Stoke-on-Trent, ST3 6HY
M0 ODX Glyn Evans, 24 Beaufort Road, Billericay, CM12 9JL
M0 OFR Marco Gianni, 121 Springfield Park Avenue, Chelmsford, CM2 6EW
M0 OFI Duncan Gunn, 40 The Pastures, Oadby, Leicester, LE2 4QD
M0 OFM Peter Joyce, 92 Essex Road Halling, Rochester, ME2 1AX
M0 OGI Mark Shopland, 120 Whitewood Park, Liverpool, L9 7LG
M0 OGS Martin Jordan, 10 Wilmot Green, Great Warley, Brentwood, CM13 3DD
M0 OGX Kazuhiko Fujita, 3-21 Denenchofu Honcho, Ota-ku, Tokyo, Japan, 1450072
M0 OGY David Ogg, 36 Cliff Road, Winteringham, Scunthorpe, DN15 9NQ
M0 OHI G Chaffey, 63 Underwood Road, Eastleigh, SO50 6FX
M0 OIC Bryan Downes, 6 Greenland Crescent, Beeston, Nottingham, NG9 5LB
MI0 OIM Martin Edwards, 58 Rosemount Park, Newtownabbey, BT37 0NL
M0 OJC Christopher Curry, 41 Bargate, Richmond, DL10 4QY
M0 OJG John Canning, The Mount, Birmingham Road, Alcester, B49 5EG
M0 OJO Nicholas Hudson, Woodpecker Cottage Red Lane, Aldermaston, Reading, RG7 4PA
M0 OJX Tadao Yamamoto, 1141-5, KOZUKUE, KOUHOKU, Kanagawa, Japan, 222-0036
M0 OKB B Bailur, 27 Russell Road, Felixstowe, IP11 2BG
MM0 OKG Jonathan Bowes, 1 Greendyke Cottage, Falkirk, FK2 8PP
M0 OKK Gary Cooper, Holmfield, Chelmorton, Buxton, SK17 9SG
M0 OKS Souradip Mookerjee, 6 Dipper Drive, Altrincham, WA14 5YF
M0 OKT Cathryn Law, 23 Yeldersley Close, Chesterfield, S40 4LG
M0 OLD Steven Old, Firtrees, Main Street, Scarborough, YO11 3UD
M0 OLE Owain Thomas, Garth Celyn, St. Davids Road, Aberystwyth, SY23 1EU
M0 OLG Malcolm Hirst, 69 Pulton Road, St. Neots, PE19 2NN
M0 OLO Christopher Hennie, 28 Foxwell Drive, Hucclecote, Gloucester, [illegible]
M0 OMA Matthew Hocking, 52 Ashbourne Drive, Newcastle, ST5 6HL
M0 OMC HOLSWORTHY ARC c/o Donald Roomes, View Field, Milton Damerel, Holsworthy, EX22 7NY
M0 OMD D Haigh, 80 Saddlers Road, Quedgeley, Gloucester, GL2 4SY
MM0 OMG Gordon Robinson, 3 Ivy Lane, Dysart, Kirkcaldy, KY1 2XD
M0 OMI J Jones, 1 Knebworth Road, Bexhill-on-Sea, TN39 4JH
MM0 OML T Cockayne, 2b Bogleshole Road, Cambuslang, Glasgow, G72 7PR
MM0 OMS M Scullion, 24 Langmuir Road, Kirkintilloch, Glasgow, G66 2QE
M0 OMT Stuart Thompson, 30 Southport Parade, Hebburn, NE31 2AQ
M0 OMV J Barton, 37 Lytton Road, Sheffield, S5 8AX
M0 OND James Isherwood, 11 Manor Crescent, Chesterfield, S40 1HU
M0 ONI Gary Swift, 43 Storth Lane, Kiveton Park, Sheffield, S26 5QS
M0 ONQ A Richardson, The Chalet, Lincoln Road, Lincoln, LN4 2EX
M0 ONS C Stone, 26 Chesham Road North, Weston-Super-Mare, BS22 8AD
M0 ONY P Bown, 19 Victory Villas, Hatherop Road, Fairford, GL7 4JU
M0 ONZ Andrew Cross, 12 Appleby Drive, Langdon Hills, Basildon, SS16 6NU
M0 OOD J Newman, Reeds, The Street, Cranbrook, TN17 4DB
M0 OOO Andrew White, 1 The Red House, Old Gallamore Lane, Market Rasen, LN8 3US
M0 OOT James Logan, Whetstead, Grange Road, Gillingham, ME7 2UN
M0 OPG Owain Griffiths, 272 Worcester Road, Malvern, WR14 1BD
M0 OPK Peter Kirby, 102 Waterloo Road, Crowthorne, RG45 7NW
MI0 OPM David Kirkwood, 1 Rural Cottages, Front Road, Lisburn, BT27 5LF
M0 OPS H Willott, 14 Warwick Close, Chepstow, NP16 5BU
M0 OPY Douglas Pingel, 7 Duffryn Close, Bassaleg, Newport, NP10 8PD
M0 ORC Jonathan King, 60 Harwood Avenue, Bromley, BR1 3DU
M0 ORE G Moore, 2 Spinacre, Barton on Sea, New Milton, BH25 7DF
M0 ORF A Froom, Lindfield, 84 Fambridge Road, Maldon, CM9 6AF
M0 ORI Daniel Dart, Ticklebelly Cottage Lower Charlton Trading Estate, Shepton Mallet, BA4 5QE
MM0 ORK 59 DEGREES NORTH ARC c/o Edmund Holt, Ashwell, Cannigall, Kirkwall, KW15 1SX
M0 ORN S Pafrey, 75 New Queen Street, Bristol, BS15 1DE
M0 ORR C Hale, 24 Wolverhampton Road, Kidderminster, DY10 2UT
M0 ORS David Holman, 38 Polyear Close, Polgooth, St. Austell, PL26 7BH
M0 ORY Benjamin Shephard, 74 Harcourt Street, Kirkby-in-Ashfield, Nottingham, NG17 8DD
M0 OSB Graham Webster, 15 Bridge Road, Chichester, PO19 7NW
M0 OSE Robert Allan, 1 Grosvenor Road, Borehamwood, WD6 1BT
M0 OSH Wojciech Rogalski, 9 Medwin Grove, Birmingham, B23 5DY
M0 OSM Cecil Penfold, 14 Romney Road, Tetbury, GL8 8JU
M0 OSX A Logan, 23 Cherry Tree Rise, Walkern, Stevenage, SG2 7JL
M0 OSY Steve Houssart, Flat 3, Virginia Court, London, SE16 6PU
M0 OTA Gordon Hutchinson, 128 Crescent Drive North, Brighton, BN2 6SF
M0 OTE D Barlow, 7 Peter Street, Eccles, Manchester, M30 0JF
M0 OTF 124 Hereford City Squadron ATC c/o Ian Hince, Colcombe Coach House, Hampton Bishop, Hereford, HR1 4JS
M0 OTH Alex Hodgson, Bryngwyn Bach Rhuallt, St. Asaph, LL17 0TH
M0 OTL Gary Humphrey, 324 Snarlton Lane, Melksham, SN12 7QW
M0 OTO Martin Pesendorfer, 13 Blake Road, London, N11 2AD
M0 OTS 126 (CITY OF DERBY) SQN AIR TRAINING CRP c/o R Bateman, 81 Stanton Street, Derby, DE23 6NF
M0 OTT C Darby, Brookfield, Forest Green, Dorking, RH5 5SG
MW0 OUC J Bidwell, 26 Lone Road, Clydach, Swansea, SA6 5HR
M0 OVB Robert Gowers, 43 Tungstone Way, Market Harborough, LE16 9GA
MM0 OVD Derek Adamson, 5 Central Quadrant, Ardrossan, KA22 7DY
M0 OVI Ovidiu Popa, 14 Mulberry Close, Horsham, RH12 2NH
MM0 OVK M McAllister, 36 Girvan Crescent, Newmilns, KA16 9HZ
MM0 OVV Alan Bernard, 200 Carden Avenue, Cardenden, Lochgelly, KY5 0EN
MM0 OWL C Barclay, 3 Kildrummy Drive, Gartcosh, Glasgow, G69 8LE
M0 OWO A Roworth, 17 Davidson Avenue, Congleton, CW12 2EQ
M0 OWS Paul Stocks, 12 Bredbury Drive, Farnworth, Bolton, BL4 7QD
M0 OXD Cristian Romocea, 21 Hurst Lane, Cumnor, Oxford, OX2 9PR
M0 OXO Charles Wilmott, 60 Church Hill, Royston, Barnsley, S71 4NG
M0 OXR S Wyatt, 55 Ridgefield Road, Oxford, OX4 3BX
MM0 OXX Alex Berry, 41 Bruce Drive, Stenhousemuir, Larbert, FK5 4DD
M0 OXZ P Rodley, 27 Tollgate Close, Northampton, NN2 6RP
M0 OYZ Geoffrey Brierley, 35 Ochrewell Avenue, Deighton, Huddersfield, HD2 1LL
M0 OZD Robert Pounder, 65 Stubsmead, Swindon, SN3 3TB
M0 OZH M Flynn, 20 Manwood Avenue, Canterbury, CT2 7AH
M0 OZI Colin Osborne, Gwinwydden, Tremont Road, Llandrindod Wells, LD1 5BH
M0 OZJ D Alchelor, Rose Cottage, Chilsworthy, Holsworthy, EX22 7BQ
MM0 OZY D Leiper, 168 Carmuirs Avenue, Camelon, Falkirk, FK1 4PA
M0 PAA Paul Thompson, 3 Floyers Field, West Stafford, Dorchester, DT2 8FJ
M0 PAC P De Camps, 22 Osier Road, Spalding, PE11 1UU
M0 PAF Graham Bally, 85 Cubcar Lane, Elsecar, Barnsley, S74 8RW
M0 PAG Trevor Pagden, 199 Woad Farm Road, Boston, PE21 0EN
M0 PAI Adrian Dodd, 68 Windlehurst Road, High Lane, Stockport, SK6 8AE
M0 PAJ Peter Alley, 58 Osprey Close, Watford, WD25 9AH
M0 PAL B Bock, 21 Winston Grove, Retford, DN22 6SQ
M0 PAM Armando Martins, 0 Thornhurst, Churchill Avenue, Herne Bay, CT6 6SQ
M0 PAO Peter McFadden, Maple Cottage, Great Gap, Leighton Buzzard, LU7 9DZ
M0 PAQ Philip Meerman, 24 Horseshoe Crescent, Burghfield Common, Reading, RG7 3XW
M0 PAR A Holland, 18 Mason Close, Malvern, WR14 2NF
M0 PAV S Richards, 18 Lowfields, Staxton, Scarborough, YO12 4SR
M0 PAW K Pawley, 12 Barchington Avenue, Torquay, TQ2 8LB
M0 PAX Graham Levine, 65 Clitheroe Road, Romford, RM5 2SL
M0 PAY John Houghton, 7 West View, Skeffling, Hull, HU12 0US
MM0 PAZ S McKinnon, 8 Rowanlea Avenue, Paisley, PA2 0RP
M0 PBD C Smith, 10 The Row, Weeting, Brandon, IP27 0QG
M0 PBN P Biggin, Galadean, Farriers Way, Newport, PO30 3JP
M0 PBO M Waistell, 23 Halton Court, Sheffield, S12 4ND
M0 PBR P Rawlinson, 15 Elmbourne Drive, Belvedere, DA17 6JE

**IMPORTANT NOTE**

**Revalidate licence to avoid revocation** – Ofcom has advised the Society that plans will be drawn up to revoke licences that have not been revalidated as required by the licence conditions. The quickest way to revalidate is to do so online via the Ofcom website: *https://services.ofcom.org.uk/* or by email: *amateur.validations@ofcom.org.uk* Ofcom staff are available to help, but please be patient during times of heavy workload.

M0 PBT Paul Burgess, 61 Grosvenor Avenue, Torquay, TQ2 7JX
M0 PBX Pamela Batson, 71 North Parade, Falmouth, TR11 2TE
M0 PBZ P Bond, 16 Little Avenue, Swindon, SN2 1NL
M0 PCA P Asbury, 67 Orchard Way, Measham, Swadlincote, DE12 7JZ
M0 PCB Iain Kelly, 261 Bodiam Avenue, Tuffley, Gloucester, GL4 0XW
M0 PCH Colin Morgan, 28 Tewther Road, Bristol, BS13 0NL
MI0 PCJ Peter Hume, 2 Seabourne Parade, Belfast, BT15 3NP
M0 PCK Paul Clay, Sylvesterweg 35, Viktring, Austria, 9073
M0 PCR Peter Rudd, 27 Loxwood Avenue, Worthing, BN14 7QY
M0 PCS Philip Sefton, 27 Donovan Avenue, London, N10 2JU
MW0 PCT Stephen Gau, Disgwylfa, The Downs, Cardiff, CF5 6SB
M0 PCX Panagiotis Chronopoulos, Flat 11, Eton Hall, London, NW3 2DW
M0 PCZ Paul Colyer, 23 Florida Road, Torquay, TQ1 1JY
M0 PDA P Stallibrass, 12 Sheerwater Close, Bury St. Edmunds, IP32 7HR
M0 PDB B Beed, 72 Looseleigh Lane, Plymouth, PL6 5HH
M0 PDC P Collins, 59 Portman Road, Scunthorpe, DN15 8PE
MM0 PDD Samuel Burnside, Woodend Farm, Buchlyvie, Stirling, FK8 3PD
M0 PDE Christine Clark, 2 Orchard Close, Elmstead, Colchester, CO7 7AS
M0 PDF A Clark, 2 Orchard Close, Elmstead, Colchester, CO7 7AS
M0 PDG John Nicholls, 2 Karen Rise, Arnold, Nottingham, NG5 8GE
M0 PDH P Hardwick, 2 Cliffe Cottages, Sandy Lane, Liss, GU33 7JE
M0 PDL S Symonds, 301 North Fairlee Farm, Fairlee Road, Newport, PO30 2JU
M0 PDP Stephen Martin, 77 Chatford Drive, Shrewsbury, SY3 9PH
M0 PDQ Christopher Almey, 152 Queensgate, Bridlington, YO16 6RW
MW0 PDR Philip Randall, 24 Ffordd-y-Goedwig, Pyle, Bridgend, CF33 6HY
M0 PDU Leslie Fuller, Rosemar Lodge Westford, Wellington, TA21 0DX
M0 PDV Paul Devlin, Brynteg, Fron Bache, Llangollen, LL20 7BP
M0 PDW P Whiteley, Grantham House, Grantham Road, Halifax, HX3 6PL
M0 PDX Niall Pagdin, 74 Thelwall New Road, Thelwall, Warrington, WA4 2HY
M0 PDY Paul Dyer, 19 Church Road, Evesham, WR11 2NE
M0 PDZ Peter Harper, 7 Duncan Gardens, Bath, BA1 4NQ
M0 PEA Glenn Pearson, 41 Myrica Grove, Hoole, Chester, CH2 3EW
M0 PEB Philip Burke, 38 Bosworth Square, Rochdale, OL11 3QG
M0 PEG P Grainger, 36 Orchard Road, Wigton, CA7 9JL
MW0 PEH Brian Sellers, 86 St. John Street, Ogmore Vale, Bridgend, CF32 7BB
M0 PEM Clement Rawlin, 5 Japonica Hill, Immingham, DN40 1LT
M0 PER Alan Perkins, 3 Intake Close, Willaston, Neston, CH64 2XG
M0 PES R Blacker, 1 Ashwindham Court, Woking, GU21 8AW
M0 PET P Gough, 32 Roebuck Road, Walsall, WS3 1AL
M0 PEW Paul Woolley, 84 Bowthorpe Road, Norwich, NR2 3TP
MM0 PFH Pentland Firth Radio Hams c/o Donald Morrison, 4 West Murkle, Murkle, Thurso, KW14 8YT
M0 PFO Paul Noble, 14 Park Street, Swallownest, Sheffield, S26 4UP
M0 PFW Paul Borer, 88 Beechings Way, Gillingham, ME8 6LX
M0 PFX Paul Fuller, 6 Annalee Road, South Ockendon, RM15 5DJ
M0 PGB P Barnard, Hawthorns, Old Church Road, Chelmsford, CM3 8BG
M0 PGC P Corley, 90 Hill Road, Benfleet, SS7 1AL
M0 PGD P Dann, 42 Brandon Road, Birmingham, B28 8DX
M0 PGH Gary Hart, 11 Sadlers Ride, West Molesey, KT8 1SU
M0 PGI Geoffrey Hartless, 32 Long Acre, Mablethorpe, LN12 1JF
M0 PGL Ahmad Md Ali, 11 Kings Avenue, Manchester, M8 5AS
M0 PGM P Meadows, 6a College Road, Maidenhead, SL6 6BE
M0 PGS P Smith, The Dingle, 27 Habberley Road, Bewdley, DY12 1JH
M0 PGW Peter Whiffing, 12 Murton Street, Newcastle upon Tyne, NE13 9AF
M0 PGX Paul Graham, 19 Pontop View, Rowlands Gill, NE39 2JP
MM0 PHD Philip Dutton, 31 Summerfield Place, Edinburgh, EH6 8BA
M0 PHE Richard Phillips, 39 Thornhill Road, Littleover, Derby, DE23 6FZ
M0 PHL Philip Stephens, 100a Foxglove Road, Eastbourne, BN23 8BX
M0 PHM P Matthews, 16 Roman Drive, Bodmin, PL31 1EL
M0 PHO Paul Honey, 3 Peterswood, Harlow, CM18 7RJ
M0 PHP C Rodway, 39 Megstone, Pimlico Court, Gateshead, NE9 5HG
M0 PHX Phoenix Radio Group c/o Alan Clayton, 6 Albert Road, Bunny, Nottingham, NG11 6QE
M0 PIA Cornelis Kolderman, 6 Flanders Close, Kemsley, Sittingbourne, ME10 2PX
M0 PIB P Badley, 37 Martins Lane, Dorchester-on-Thames, Wallingford, OX10 7JE
MW0 PIC Robert Miles, 63 Phillip Street, Caegarw, Mountain Ash, CF45 4BG
MM0 PID Victoria Hamilton, 10/3 Fox Street, Edinburgh, EH6 7HN
M0 PIE Robert Cockroft, 8 Lumb Lane, Huddersfield, HD4 6SZ
M0 PIK Bernard Pike, 19 Cardigan Gardens, Reading, RG1 5QP
M0 PIP C Sidey, 10 Gerrans Close, St. Austell, PL25 3DN
M0 PIT Phil Hayes, 4 London Road, Roade, Northampton, NN7 2NL
M0 PIX R Bibby, 40 Morval Crescent, Runcorn, WA7 2QS
M0 PJA Paul Archer, 31 Stoney Bank Drive, Kiveton Park, Sheffield, S26 6SJ
M0 PJC P Crabtree, 106 Sagecroft Road, Thatcham, RG18 3BF
M0 PJD Peter Davies, 53 Lammas Road, Cheddington, Leighton Buzzard, LU7 0RY
M0 PJF P Franklin, 1 Aberdeen Court, Newcastle upon Tyne, NE3 2XU
M0 PJG Peter Galer, 62 Court Mount Park, Birchington, Birchington, CT7 0BU
MW0 PJJ Philip Jones, 23 Pinecroft Avenue, Aberdare, CF44 0HY
M0 PJK Peter Knappett, Hope Cottage, The Green, Clacton-on-Sea, CO16 0BU
MI0 PJL P Letters, 24 Old Grange Avenue, Carrickfergus, BT38 7UE
M0 PJM Peter Mcmillan, 28 Front Street, Tudhoe Colliery, Spennymoor, DL16 6TG
MM0 PJQ P Quinn, 24 Highfield Avenue, Paisley, PA2 8LG
MW0 PJR P Rees, 8 Pencae Terrace, Llanelli, SA15 1NZ
MI0 PJS P Smiley, 100 Lislaban Road, Cloughmills, Ballymena, BT44 9HZ
M0 PJT P Tomlinson, 11 Haynes Close, Clifton, Nottingham, NG11 8JN
M0 PJX D Dickson, 6 St. Johns View, Old Hutton, Kendal, LA8 0NG
M0 PJY P Yarwood, Kia Mena, Downderry, Torpoint, PL11 3JA
M0 PKE Patrick Walsh, 181 Hermes Close, Hull, HU9 4DR
M0 PKH Peter Halloway, 82 Northwall Road, Deal, CT14 6PP
M0 PKL Coryan Wilson-Shah, 42 Glenthorne Road, London, N11 3HJ
M0 PKV Peter Slade, End Cottage, Monyash Road, Bakewell, DE45 1FG
M0 PKW Paul Watson, 10 Whitelands Crescent, Baildon, Shipley, BD17 6NN
MI0 PLC Raymond Thomson, 1 Litchfield Park, Coleraine, BT51 3TN
M0 PLG Peter Gyngell, 54 Association Walk, Rochester, ME1 2XD
M0 PLH P Hamnett, 13 Breakwater Court West, Berry Head Road, Brixham, TQ5 9AG
M0 PLN John Kendrick, 29 Waterside, Silsden, Keighley, BD20 0LQ
M0 PLO Pawel Lesiecki, 165 New Road, Stoke Gifford, Bristol, BS34 8TG
M0 PLP Peter Hallson, 4 Cranbrook Drive, Esher, KT10 8DL
M0 PLR David White, 10 Meaux Road, Wawne, Hull, HU7 5XD
M0 PLS Jacek Walczak, 18 Heathfield, Chippenham, SN15 1BQ
M0 PLT Gary Myers, 6 Ullswater Close, Biggleswade, SG18 8LX
M0 PLV Paul Le Vallois, 14 London Row, Arlesey, SG15 6RX
M0 PLX Jacek Telecki, 5 Stonegate, Cowbit, Spalding, PE12 6AH
M0 PLY Ronald Austen, Holly Tree Cottage, Station Road, Immingham, DN40 3AX
MJ0 PMA Paul Ahier, Les Trois Carres, La Rue D'Aval, Jersey, JE3 6ER
M0 PMC P Curnow, 13a Warden Close, Maidstone, ME16 0JL
M0 PMH Paul Holmquest, 6 Rhyme Hall Mews, Fawley, Southampton, SO45 1FX
M0 PMJ Peter Mullen, 14 Anderson Road, Hemswell Cliff, Gainsborough, DN21 5XP
M0 PML R Laight, 380 Tile Hill Lane, Coventry, CV4 9DJ
M0 PMM P Levetsky, 9 Moat Walk, Pound Hill, Crawley, RH10 7ED
MD0 PMN P Best, 5 The Willows, Ballasalla, Isle of Man, IM9 2EW
M0 PMR Shaun Macauley, 1 Moricambe Crescent, Anthorn, Wigton, CA7 5AS
M0 PMV P Mansfield, 27 Popplechurch Drive, Swindon, SN3 5DE
MM0 PMW M Mclauchlan, 8 Craigie St., Ballingry, Lochgelly, KY5 8NS
M0 PNA Paul Fulbrook, 167 Droitwich Road, Fernhill Heath, Worcester, WR3 7TZ
M0 PNB Panagiotis Bozikis, 336 Higham Hill Road, London, E17 5RG
M0 PNC Marc Bloore, 6a Lovatt Close, Stretton, Burton-on-Trent, DE13 0HZ
M0 PNN Paul Bowen, 12 Powell Place, Newport, TF10 7BS
M0 PNZ Robert Maddock, 48 Collygree Parc, Goldsithney, Penzance, TR20 9LY
M0 POA Andrea Polesel, 33 The Maltings, Leighton Buzzard, LU7 4BS
MW0 POB Jamie Lewis, 2 Tymaen Crescent Cwmavon, Port Talbot, SA12 9EA
MM0 POD A Conlon, Kilrae, Barrpath, Glasgow, G65 0EX
M0 POG Stafford Portable Operating Group c/o Peter Wilkes, 8 Cloverdale, Stafford, ST17 4QJ
M0 POI Points of Historical Interest c/o Maurice Boland, 24 Hallam Close, Moulton, Northampton, NN3 7LB
M0 POP M Carey, 52 Malvern Road, Bournemouth, BH9 3AJ
M0 POQ Richard Finch, 19b Kiln Road, Newbury, RG14 2LS
MI0 PPA Frank Kearney, 45 Sperrin Park, Omagh, BT78 5BA
M0 PPG Gregory Beacher, 22 Trowbridge Gardens, Luton, LU2 7JY
M0 PPM Frederick Miers, Pentre Isaf, Bryneglwys, Corwen, LL21 9NA
M0 PPO V Frostick, Clawdd Llwyd, Ceunant, Caernarfon, LL55 4RR
M0 PPP Graham Batty, 85 Cobcar Lane, Elsecar, Barnsley, S74 8BW
M0 PPR Pawel Rozenek, 224 Bowers Avenue, Norwich, NR3 2PR
M0 PPS I Underwood, 12 Forge Lane, Gillingham, ME7 1UG
MI0 PPW Jonathan MacFarlane, 1 Main Street, Uttony, Magheraveely, Enniskillen, BT92 6NB
M0 PPZ P Zanek, Mulberry Hill, Violet Lane, Tadley, RG26 5JX
M0 PQI Thomas Pearsall, 16 Langdale Road, Leyland, PR25 3AR
MI0 PQR GreenIsland Electronics ARS c/o Brian McKeen, 27 Old Grange Drive, Carrickfergus, BT38 7HG
M0 PRA David Runyard, Tilecroft, Shortheath Crest, Farnham, GU9 8SA
MM0 PRB Paul Bacon, 12 The Greens, Maddiston, Falkirk, FK2 0FN
MW0 PRC Philip Randall, 24 Ffordd-y-Goedwig, Pyle, Bridgend, CF33 6HY
M0 PRD P Denham, 48 Marl Court, Thornhill, Cwmbran, NP44 5TY
M0 PRF Jeffrey Petch-Harrison, 13 Church Lane, Shepton Mallet, BA4 5LE
M0 PRI David Price, 11 Cefn Melindwr, Capel Bangor, Aberystwyth, SY23 3LS
MI0 PRM Edith Simpson, 10 Woodview Park, Tandragee, Craigavon, BT62 2DD
M0 PRN Paul Norman, Mellon, Hincaster, Milnthorpe, LA7 7ND
M0 PRO J White, 6 Damy Green, Neston, Corsham, SN13 9TN
MW0 PRP Peter Pugh, 27 Bank Street, Tonypandy, CF40 1PJ
M0 PRT B Dey, 15 Bradenham Road, Grange Park, Swindon, SN5 6EB
M0 PRV Darren Parvin, 11 Stanhope Way, Sevenoaks, TN13 2DZ
MM0 PSA P Smith, 13 Newmills Grove, Balerno, EH14 5SY
M0 PSB Joe Bell, 8 Firsleigh Park, Roche, St. Austell, PL26 8JN
M0 PSC Philip Croxford, 1 Meteor Close, Bicester, OX26 4YA
M0 PSD Peter Davies, 2 Lynfords Drive, Runwell, Wickford, SS11 7PP
M0 PSE P Glandfield, Flat 5, 7 Cargate Avenue, Aldershot, GU11 3EP
MW0 PSG 1ST PENCOED SCOUT GROUP c/o G Day, 20 St. Johns Drive, Pencoed, Bridgend, CF35 5NF
M0 PSH P Blythe, 4 Stonehaven Road, Aylesbury, HP19 9JQ
M0 PSI Ali Al-Azzawi, 33 Clare Mead, Rowledge, Farnham, GU10 4BJ
M0 PSK Christopher Gibson, 1 Ryelands Orchard, Leominster, HR6 8QQ
MM0 PSM Sam Milne, 5 Moriston Court, Grangemouth, FK3 0JJ
M0 PSR Robert Tickle, 5 Bramley Court, Harrold, Bedford, MK43 7BG
M0 PSS P Shaw, 10 Godbold Road, London, E15 3AL
M0 PST P Stevens, 19 Elmfield Place, Newton Aycliffe, DL5 7BD
M0 PSW John Godfrey, 4 Cherry Close, Houghton Conquest, Bedford, MK45 3LQ
M0 PSY D Shuttleworth, 27 Union St., Egerton, Bolton, BL7 9SP
M0 PSZ Susan MacDonald, Woodside Cottage, Horton Way, Verwood, BH31 6JJ
MM0 PTE Peter Bainbridge, 46 Rigghouse View, Bathgate, EH47 0SE
M0 PTG Paul Threakall, 83 Gregory Avenue, Birmingham, B29 5DG
M0 PTO Matt Reeve, 3 Princess Anne Terrace, Loddon, Norwich, NR14 6LL
M0 PTR P Clifford, 69a Higher Blandford Road, Broadstone, BH18 9AE
M0 PTS Phillip Boultwood, 32 Makepiece Road, Bracknell, RG42 2HJ
M0 PTT R Winthrop, 50 Ullswater Road, Carlisle, CA2 5RG
M0 PUB A Rowe, 12 The Knapps, Semington, Trowbridge, BA14 6JG
M0 PUC J Woods, Haycocks Farm, Haycocks Lane, Colchester, CO5 8SS
M0 PUD A Eyre, St. Michael Mead, The Common, Norwich, NR12 8BA
M0 PUN R Punter, 35 Devonshire Rise, Tiverton, EX16 4QR
M0 PUT Keith Puttock, 12 Beechfields, School Lane, Petworth, GU28 9DH
M0 PVA Michael Dixon, 55 Henthorn Road, Clitheroe, BB7 2LD
M0 PVI P Handley, 97 Applegarth Avenue, Guildford, GU2 8LX
M0 PVN Paul Nicholls, 23 Bishops Gate, Birmingham, B31 4AJ
M0 PVP Brian Mendham, 252 Gregson Lane, Hoghton, Preston, PR5 0LA
MW0 PVW Phillip Witts, 82 Park View, Llanharan, Pontyclun, CF72 9SB
M0 PWB Peter Booth, 12 Heathgate, Wickham Bishops, Witham, CM8 3NZ
M0 PWC Paul Clark, 60 Somerville Avenue, Newcastle, ST5 0LH
M0 PWD Paul Woodburn, 21 The Row, Silverdale, Carnforth, LA5 0UG
M0 PWF Wendy Feather, Long House Farm, Ellers Road, Keighley, BD20 7BH
MD0 PWI Colin Ingles, 1 Hillberry View, Onchan, Isle of Man, IM3 3GB
M0 PWL M Mynn, 15 Shearling Drive, Lower Cambourne, Cambridge, CB23 6BZ
MM0 PWM P Mackie, 8 Letham Avenue, Pumpherston, Livingston, EH53 0NG
M0 PWS P Snelson, 103 Queens Road, Vicars Cross, Chester, CH3 5HF
M0 PWT David Garnett, Hill View, Snailbeach, Shrewsbury, SY5 0NS
M0 PWY Lyndon Jones, Ty'r Ysgol, Holland Street, Ebbw Vale, NP23 6HT
M0 PXD Paul Donaghy, 67 Brockenhurst Way, Bicknacre, Chelmsford, CM3 4XN
M0 PXI Lucy Evans, 1 Leave Acre, New Buildings, Crediton, EX17 4PL
M0 PXM Paul Matthew, 24 Jubilee Close, Pamber Heath, Tadley, RG26 3HP
M0 PXP John Maudsley, Knight Stainforth Hall, Little Stainforth, Settle, BD24 0DP
M0 PXS Philip Simpson, 7 Hawthorn Close, Wootton, Ulceby, DN39 6RB
M0 PXY Andy Schofield, 24 Oldbrook, Bretton, Peterborough, PE3 8SH
M0 PXZ Chris Fox, 45 Park Road, Wivenhoe, Colchester, CO7 9LS
M0 PYA W Allen, 109 Barston Road, Oldbury, B68 0PU
M0 PYE N Atkins, The Old Rectory, Church Lane, Chippenham, SN14 6DE
M0 PYG Geoffrey Fielding, Chapel Court, Chapel Lane, Malvern, WR13 5HX
MI0 PYN Stefan Pynappels, 38 Dora Avenue, Newry, BT34 1JW
MM0 PYS ELDERSLIE ARS c/o K Gillen, Flat 1/R, 12 Fergus Drive, Glasgow, G20 6AG
M0 PYT P Kimberlee, 24 Jacey Road, Shirley, Solihull, B90 3LJ
M0 PZC Chris Hulbert, Dyers Farm, Edge, Malpas, SY147DN
M0 PZD Arnoldas Jakstas, Flat 9, Kendal Court, 112 Godstone Road, Kenley, CR8 5GE
M0 PZR Philip Hanman, 7 Tremenheere Road, Penzance, TR18 2AH
M0 RAB C Finnegan, 249 Winchester Road, Basingstoke, RG22 6EP
M0 RAC R Cochrane, 7 Lawn Terrace, Blackheath, London, SE3 9LJ
M0 RAD AVON VALLEY ARA c/o Peter Badham, 75 Newtown Road, Worcester, WR5 1HH
MM0 RAG J Hutson, 13 Greenan Road, Ayr, KA7 4ET
MM0 RAI Theo Vanderydt, 33 Portland Road, Galston, KA4 8EA
M0 RAL Alan Hopkinson, Springfield, Neston Road, Neston, CH64 4AR
MM0 RAM D Stevenson, 51 Shannon Drive, Falkirk, FK1 5HU
M0 RAN M Moran, 27 Burnet Close, Padgate, Warrington, WA2 0UH
M0 RAP R Peech, 17 Chestnut Avenue, Crossgates, Leeds, LS15 8ED
M0 RAR Nicholas Booth, Greenfield, Westmancote, Tewkesbury, GL20 7EP
M0 RAT N McWilliam, 13 Rawlins Street, Liverpool, L7 0JE
M0 RAU R Johnson, 3 Lindens Close, Thorney Toll, Wisbech, PE13 4AR
M0 RAW R Wyeth, 112 Main Road, Crockenhill, Swanley, BR8 8JL
M0 RAX Martin Pike, 5 Rowan Drive, Heybridge, Maldon, CM9 4BW
M0 RAZ Richard Bowman, 3 Hodders Way, Cargreen, Saltash, PL12 6NY
MW0 RBA R Arnould, 13 Laurel Place, Sketty, Swansea, SA2 8JL
M0 RBB R Brier, 48 Burton Rise, Kirkby-in-Ashfield, Nottingham, NG17 9BR
M0 RBC R Colman, 197 Coppins Road, Clacton-on-Sea, CO15 3LA
M0 RBD Mikael Czerski, 3 Ladymead Close, Whaddon, Milton Keynes, MK17 0LL
M0 RBE R Smith, 4 London Road, Lindal, Ulverston, LA12 0LL
M0 RBF Richard Ferguson, 31 Barton Court Road, New Milton, BH25 6NW
M0 RBG Richard Blandford, 60 Benomley Road, Almondbury, Huddersfield, HD5 8LS
M0 RBH R Hookham, 32 Atherley Court, Southampton, SO15 7NG
M0 RBI Clive Bennett, The Old Cottage, Waterside Road, Southminster, CM0 7QT
M0 RBJ Nigel Roberts, 40 Armour Road, Tilehurst, Reading, RG31 6HN
M0 RBK Roger Bleaney, 40 Broadstone Road, Harpenden, AL5 1RF
M0 RBL Robert Lovesey, 33 Ty Isaf Park Avenue, Risca, Newport, NP11 6NB
M0 RBM R Medland, 5 Bay Tree Cottages, Hospital Road, Bude, EX23 9BP
MM0 RBN Robert Corkey, 11 Golf View Cardenden, Lochgelly, KY5 0NW
M0 RBQ Richard Simpson, 22 Kenworthy Road, Stocksbridge, Sheffield, S36 1BZ
MM0 RBR Robert Hutton, 2 Watson Place, Dunfermline, KY12 0DR
M0 RBT R Hunt, 7 Knotley Hall Cottages, Chiddingstone Causeway, Tonbridge, TN11 8JH
M0 RBU David Perrin, 54 High Street, West Wratting, Cambridge, CB21 5LU
M0 RBV Richard Briant, Talarvor, Llanon, SY23 5HG
M0 RBX Rodney Buckland, 34 Beechwood Drive, Meopham, Gravesend, DA13 0TX
M0 RBY Robert Hall, Concorde Cottage, Ellingstring, Ripon, HG4 4PW
M0 RCC R Chadwick, 4 Gleneagles Drive, Haydock, St. Helens, WA11 0YS
M0 RCD E Capstick, 130 Ship Lane, Farnborough, GU14 8BJ
M0 RCE G Morse, 34 Headford Road, Bristol, BS4 1QE
MI0 RCF Lough Erne ARC c/o Herbie Graham, Cavancarragh, Lisbellaw, Co Fermanagh, BT94 5GL
M0 RCH CHRISTLETON HIGH SCHOOL A.R.C c/o S Smith, 102 Gresford Road, Llay, Wrexham, LL12 0NW
M0 RCI Robert Chappell, 11 Highfields Road, Darton, Barnsley, S75 5ER
M0 RCK Robert Wells, 27 Victoria Avenue, Camberley, GU15 3HT
M0 RCL A Jackson, 4 Orchard Close, Wilberfoss, York, YO41 5RW
M0 RCM R Moxham, 8 Dunroyal Close, Helperby, York, YO61 2NH
M0 RCN Richard Hart, 4 Glade Mews, Guildford, GU1 2FB
M0 RCP R Peterson, 9 Moseley Wood View, Leeds, LS16 7ES
M0 RCR R Room, 197 Newbridge Road, Bath, BA1 3HH
M0 RCT R Tomkinson, 24 Beech Drive, Wistaston Green, Crewe, CW2 8RE
M0 RCU David Littlewood, 50 Industry Road, Sheffield, S9 5FQ
M0 RCV SOUTH EAST HAMPSHIRE RAYNET c/o Paul Raxworthy, 32 St. Marys Avenue, Alverstoke, Gosport, PO12 2HX
M0 RCW MAESGEIRCHEN ARC c/o E Barnes, 2 Trem Y Garnedd, Bangor, LL57 1NA
M0 RCX Robert Rawson, 68 Broom Nook, Leeds, LS10 3LR
M0 RCY Darren Osborne, 12 Sandringham Close, Brackley, NN13 6JQ
M0 RCZ Richard Shipman, 1 Lledfair Place, Heol Pentrerhedyn, Machynlleth, SY20 8DL
M0 RDA A Rivers, 34 Brookfield, Mawdesley, Ormskirk, L40 2QJ
M0 RDB Robert De Savigny-Bower, 55 Fenwick Close, Woking, GU21 3BZ
M0 RDC R Cooke, 1 Fiona Walk, Fazakerley, Liverpool, L10 4YW
MM0 RDD R Duncan, 12 Douglas Loan, Kirkwall, KW15 1FU
MW0 RDF A Williams, 18 Lamb Lane, Killay, Swansea, SA2 7ES
M0 RDI Adrian Riddick, 30 Britannia Road, Banbury, OX16 5DW
M0 RDR R Rawlinson, 17 Walmer Place, Winsford, CW7 1HA
M0 RDS R Staniland, The Cottage, Martin Moor, Lincoln, LN4 3BQ
MM0 RDT Robert Tripney, 7 Sunnyside St., Camelon, Falkirk, FK1 4BJ
M0 RDV R Vincent, 141 Timberleys, Littlehampton, BN17 6QD
M0 RDW Richard Wyatt, 297 Weston Road, Stoke-on-Trent, ST3 6HA
M0 RDX John Scott, 443 Ford Green Road, Stoke-on-Trent, ST6 8LX
M0 RDY Roy Hawkins, 3 Fairways Drive, Harrogate, HG2 7ES

M0 RDZ Richard Dell, 18 Greenacres, Fulwood, Preston, PR2 7DA
M0 REB D Hook, 28 Rifford Road, Exeter, EX2 5JT
M0 REC H Clare, 26 Hall Park, Swanland, North Ferriby, HU14 3NI
M0 RED G Sharp, 18 Whitbred Road, Salisbury, SP2 9PE
M0 REG WORTHING RADIO EVENTS GROUP c/o Nigel Thrower, 8 Upton Gardens, Worthing, BN13 1DA
M0 REH R Harlow, Swyn-Y-Mor, Penrallt Road, Holyhead, LL65 2UG
M0 REJ R Jones, 39 Dalton Lane, Barrow-in-Furness, LA14 4LE
M0 REK Ruth Kelly, 42 Hinton Wood Avenue, Christchurch, BH23 5AH
M0 REM Matthew King, 126 Blythsford Road, Hall Green, Birmingham, B28 0UT
M0 REV James Drake, 37 Weston Lane, Southampton, SO19 9GN
M0 REX Rex Duffy, 45 Chatham Road, Winchester, SO22 4EE
M0 REZ Steven Brown, 4 Dorado Gardens, Orpington, BR6 7TD
MM0 RFA Robert Aird, 42 Kelvin Walk, Largs, KA30 8SJ
M0 RFH R Hill, Jarrah, 26 Yealmpstone Drive, Plymouth, PL7 1HG
M0 RFK D Gardner, 122 All Saints Avenue, Maidenhead, SL6 6LT
M0 RFM R Mannock, Craybourne, Higher Brill, Falmouth, TR11 5QG
M0 RFU Jonathan Byrne, 316 Turncroft Lane, Stockport, SK1 4BP
M0 RFW Richard White, 2 Uplands Cottages, Rattle Road, Pevensey, BN24 5DT
M0 RFY Derek Murray, 53 Waverley Avenue, Sutton, SM1 3JX
M0 RGB Gary Baker, Ling Cottage, Crag Foot, Carnforth, LA5 9SA
M0 RGC G Henshall, 43 Cumberworth Road, Skelmanthorpe, Huddersfield, HD8 9AB
M0 RGD Ronald Dale, 17 Spencer Gardens, Brackley, NN13 6AQ
M0 RGE Rodney Edwards, 46 Lavers Oak, Martock, TA12 6HG
M0 RGF Anthony Mobbs, 149 The Paddocks, Old Catton, Norwich, NR6 7HR
M0 RGI I Reichenfeld, 7 Hazelbank Close, Liphook, GU30 7BY
M0 RGL Gloucestershire County Raynet c/o Andrew Webb, 47 Granville Street, Gloucester, GL1 5HL
MW0 RGM Richard Meal, 8 Rhodfa Llwyn-Eithin, Llanelli, SA15 4HN
M0 RGN Peter Williams, 35 Cansfield Grove Ashton-in-Makerfield, Wigan, WN4 9SE
M0 RGO D Robertson, 53 Moor Lane, Weston-Super-Mare, BS22 6RA
MJ0 RGR Roger Bisson, Apartment 184, Block 4 Spectrum, Gloucester Street, Jersey, JE2 3DE
M0 RGS LRGS ARC c/o D Saul, 78 Ingleton Drive, Lancaster, LA1 4QZ
MI0 RGX Richard Gilmore, 86 Lylehill Road, Templepatrick, Ballyclare, BT39 0HL
M0 RGY Anthony Mobbs, 149 The Paddocks, Old Catton, Norwich, NR6 7HR
M0 RHB Roderick Burton, 23 Freston, Paston, Peterborough, PE4 7EN
M0 RHD Robert Burton, 6 Troed Y Garn, Llangybi, Pwllheli, LL53 6DQ
M0 RHE R Head, 24 Beaufort Road, Church Crookham, Fleet, GU52 6AZ
M0 RHG Daniel Bines, 39 School Close, Bretton, Peterborough, PE3 9FS
MM0 RHH R Henry, 164 Auchenbothie Road, Port Glasgow, PA14 6JE
M0 RHI R Hunt, 5 Rue de Wiltz, L-2734, Luxembourg-Bonnevoie, Luxembourg
MM0 RHL Edinburgh Hacklab c/o Timothy Hawes, 1 Summerhall, Edinburgh Hacklab, Edinburgh, EH9 1PL
M0 RHO Rhodri Morgan, 14 Ash Road, Ashurst, Southampton, SO40 7AT
M0 RHQ John Middleton, 8 Cullen Close, Newark, NG24 1DF
M0 RHR Ashley Thomas, 1 Millers Close, Ruardean Hill, Drybrook, GL17 9AU
M0 RHS Richard Hawkins, Forest Edge, Milton Abbas, DT11 0AY
M0 RHT R Titcombe, 82 Liverpool Road, Buckley, CH7 3NB
M0 RHW William Westlake, 2 Chegwin Court, Newquay, TR7 2DE
M0 RIA T Lea, 1 Roseland Close, Keyworth, Nottingham, NG12 5LQ
MI0 RIB William Nicholl, 58 Dunnalong Road, Bready, Strabane, BT82 0DW
M0 RIC D Moore, 379 Main Road, Harwich, CO12 4DW
M0 RIG Andrew Rigg, 12 Bydales Drive, Marske-by-the-Sea, Redcar, TS11 7HJ
M0 RIK F Woodhams, Greenways, Mill Lane, Fareham, PO15 5DU
M0 RIS M Baines, 21 Acre Moss Lane, Kendal, LA9 5QE
M0 RIU David Simmons, 8 Lower Grange, Huddersfield, HD2 1RU
M0 RIV Riviera ARC c/o Alan Wyatt, 78 Marldon Road, Torquay, TQ2 7EH
M0 RJB R Blaney, 16 Pages Close, Wymondham, NR18 0TU
MD0 RJE Ray Edwards, 23 Queens Walk, Ruislip, HA4 0LX
M0 RJG Roger Green, 35 Harrier Mill, Henlow, SG16 6BQ
M0 RJH John Reynolds, 38 Spring Lane, Hockley Heath, Solihull, B94 6QY
MM0 RJJ Rolfe James, The Garret, Alyth, Blairgowrie, PH11 8HQ
M0 RJK R Kavanagh, Chatslea, 57 Mill Lane, Littlehampton, BN16 3JP
M0 RJM Roger Millington, Quaintways, The Avenue, Tarporley, CW6 0BA
MI0 RJN R Neill, 15 Hawthorn Place, Coleraine, BT52 2ES
M0 RJO R Odell, 41 Pevensey Park Road, Westham, Pevensey, BN24 5HW
MM0 RJR Robert Renshaw, Smithy House, Scotscalder, Halkirk, KW12 6XJ
M0 RJS R Somerville Roberts, Bank House, 7 Mill Lane, Stoke-on-Trent, ST7 3LD
M0 RJT R Gilbert, 61 Coltstead, New Ash Green, Longfield, DA3 8LN
MI0 RJW R Wylie, 17 Massey Park, Belfast, BT4 2JX
M0 RJX Robert Harrison, 18-20 Hall Lane, Kirkburton, Huddersfield, HD8 0QW
MI0 RJY James Young, 7 Dunmore Close, Cookstown, BT80 8AS
M0 RJZ N Noda, 14 Widmer Court, Vicarage Farm Road, Hounslow, TW3 4NL
M0 RKA M Hawker, 47 Lynher Drive, Saltash, PL12 4PA
M0 RKB Mark Brady, 24 Gregory Avenue, Colwyn Bay, LL29 7ND
M0 RKD R Hark, 5 Victoria Park, Bagillt, CH6 6JS
M0 RKF Christopher Whitelaw, 18 Marine Drive, Bishopstone, Seaford, BN25 2RT
M0 RKH R Hayes, 59 Elm Road, Folksworth, Peterborough, PE7 3SX
MD0 RKI R Kijak, 13 Falcon Cliff Court, Douglas, Douglas, Isle of Man, IM2 4AH
MM0 RKN Carrie Welsh, 28 Peacock Wynd, Motherwell, ML1 4ZL
M0 RKR Graham Hirst, 15 Hengist Close, Horsham, RH12 1SB
MM0 RKT R Towers, 34 South Park Road, Hamilton, ML3 6PN
M0 RKW R Watson, 5 Angrove Gardens, Sunderland, SR4 7TB
M0 RKX Mark Hemming, 11 Blackberry Way, Evesham, WR11 2AH
M0 RKY Richard Brown, 62 Charlecote Park, Telford, TF3 5HD
MD0 RLA R Allcote, 77 Ballanorris Crescent, Ballabeg, Castletown, Isle of Man, IM9 4ER
M0 RLC Roger Coley, Dorcasia, Church Lane, Ipswich, IP6 9BE
MW0 RLD Barbara Shelley, Sunray, Pendine, Carmarthen, SA33 4PD
M0 RLI D Thomas, 130 Norwich Avenue, Southend-on-Sea, SS2 4DH
MW0 RLJ Robert Johns, Llanferran, St Nicholas, Goodwick, SA64 0LL
M0 RLM Rui Lima Matos, Flat 9, 55 Shepherds Hill, London, N6 5QP
MM0 RLN David Nixon, 17 Semple Place, Linwood, Paisley, PA3 3RT
M0 RLO Denis Holland, 87 Grovelands Avenue, Hitchin, SG4 0RA
M0 RLP R Le Piez, 279 Oakley Road, Southampton, SO16 4NR
M0 RLW R Williams, 16 Irving Road, Norwich, NR4 6RA
MI0 RMD Raymond Nelson, 65, 65 Dernawilt Road, Rosslea, BT92 7FN
M0 RMF G Coupe, 112 Greenwood Drive, Kirkby-in-Ashfield, Nottingham, NG17 0GH
M0 RMG Geoffrey Chapman, Crockers Farm, Stoke Wake, Blandford Forum, DT11 0HF
M0 RMH R Halliwell, 38 Larch Grove, Kendal, LA9 6AU
M0 RMI Richard Miller, 23 Clarendon Road, Sevenoaks, TN13 1EU
M0 RMJ Richard Jeffs, 45 Forest Road, Bingham, Nottingham, NG13 8RL
MI0 RMK Ann Mackenzie, 30 Dalriada Gardens, Ballycastle, BT54 6DZ
M0 RML RADIO MILLENIUM LODGE c/o P Greenhalgh, 13 Primrose Avenue Urmston, Manchester, M41 0TY
M0 RMN Marion Watmough, 6 Blair Park, Knaresborough, HG5 0TH
M0 RMO Clive Larner, 98 Allandale, Hemel Hempstead, HP2 5AT
M0 RMP R Perkin, 26 Hall Avenue, Leek, ST13 6BU
M0 RMS Denley Isaac, Bryn Ddu Farm, Babell, Holywell, CH8 8PR
M0 RMT Jacek Gorlinski, 63 Crosby Avenue, Scunthorpe, DN15 8PA
M0 RMW Roger Williams, 4 Larkfield Close, Farnham, GU9 7DA
M0 RMY Thomas Rowlands, 7 Northfield Crescent, Beeston, Nottingham, NG9 5GR
M0 RMZ Richard Mansfield, 8 Haysoms Drive, Greenham, Thatcham, RG19 8EY
M0 RNC M Brigham, 21 Overdale Close, York, YO24 2RT
M0 RND Adam Greig, 3 Fir Grange Avenue, Weybridge, KT13 9AR
M0 RNI David Harris, 6 Baker Lea, Monkland, Leominster, HR6 9DB
M0 RNP Paul Rainer, 3 St. Martins Road, Folkestone, CT20 3LA
M0 RNR Brian Pickup, 3 Mews Court, Houghton le Spring, DH5 8GB
M0 RNU Melville Nutt, 110 Birkinstyle Lane, Shirland, Alfreton, DE55 6BT
M0 RNW Ronald Wellsted, 127 Goldthorn Hill, Wolverhampton, WV2 4PS
M0 ROC Mark Russell, 107 Cambridge Road, Hitchin, SG4 0JH
M0 ROJ R Reeves, Goldford House, Goldford Lane, Malpas, SY14 8LL
M0 ROK Martin Harrison, 91 Rye Road, Hastings, TN35 5DH
M0 ROM Romano Pasika, 192 Longfield Lane, Cheshunt, Waltham Cross, EN7 6AQ
M0 RON Andrew Eustace, 3 Linworth Road, Bishops Cleeve, Cheltenham, GL52 8PF
M0 ROO Robert Smith, South Cottage, Radley Green Road, Chelmsford, CM1 4NW
MM0 ROR Iain Learmonth, 85 Lord Hay's Grove Old Aberdeen, Aberdeen, AB24 1WT
MM0 ROV M Gerrard, 10 Whinhill Gardens, Aberdeen, AB11 7WD
M0 ROW TS Gambia / Thorne Sea Cadets c/o Phil Ormsby, Wood Cottage, Little Heck, Goole, DN14 0BU
M0 ROY Roy Henson, 2 Byron Street, Shirebrook, Mansfield, NG20 8PJ
M0 RPB R Bennett, 28 Neyland Path, Fairwater, Cwmbran, NP44 4PX
M0 RPD I Handley, Rosedale, Chapman Street, Market Rasen, LN8 3DS
M0 RPE Ronald Evans, Brackenwood, 30 Northop Country Park, Mold, CH7 6WD
M0 RPF Robert Fullagar, 6 Locke Way, Stafford, ST16 3RE
M0 RPI David Akerman, The Brick Barn, Coppice Farm, Ross-on-Wye, HR9 7QW
M0 RPJ Rene Jepsen, 20 The Mount, Aspley Guise, Milton Keynes, MK17 8EA
MW0 RPK R King, 87 Matthysens Way, St. Mellons, Cardiff, CF3 0PL
M0 RPO R Powell, 26 Fenwick Avenue, South Shields, NE34 9AJ
M0 RPR Martyn Roper, 13 St. Cuthbert Street, Worksop, S80 2HN
MI0 RPT Richard Tomalin, 22 Drumfad Road, Millisle, Newtownards, BT22 2JQ
M0 RQK B Dare, 1 St. Johns Villas, Sivell Place, Exeter, EX2 5ES
M0 RQN Joseph Foster, 23 High Street, Cumnor, Oxford, OX2 9PE
M0 RRC RUSTYRADIOS A.R.C.G. c/o S Williams, 32 Waterdell Lane, St. Ippolyts, Hitchin, SG4 7QZ
M0 RRD Risca and District ARS c/o Clive Jenkins, 10 Marsh Court, Abergavenny, NP7 5HQ
MI0 RRE Robert Rantin, 8a Buchanans Road, Newry, BT35 6NS
M0 RRG RICHMOND RAYNET GROUP c/o John Kirby, 2 Kneeton Park, Middleton Tyas, Richmond, DL10 6SB
MM0 RRM Rachael Murray, The Barn House, Springfield Farm, Carluke, ML8 4QZ
M0 RRN Derek Barker, 12 The Weavers, Denstone, Uttoxeter, ST14 5DP
M0 RRR Reginald Reeves, 15 Higher Albert Street, Chesterfield, S41 7QE
M0 RRX Robin Ridge, Roskellan House, Maenlay, Helston, TR12 7QR
M0 RRY Gordon Sherry, 22 York Street, Oswestry, SY11 1LX
M0 RSA D Silburn, 34 Northfields, Strensall, York, YO32 5XW
M0 RSC CHESTERFIELD & DISTRICT SCOUTS ARC c/o Keith Greatorex, 54 Lilac Grove, Glapwell, Chesterfield, S44 5NG
M0 RSD Keith Winwood, 146 Chapel Street, Pensnett, Brierley Hill, DY5 4EQ
M0 RSE RS of Great Britain c/o Graham Coomber, 2 Bracken Grove, Catshill, Bromsgrove, B61 0PB
M0 RSF C Darlow, 418 Broad Lane, Bramley, Leeds, LS13 3DF
M0 RSG Edward Flint, The Bell House, Kingston Deverill, Warminster, BA12 7HE
M0 RSH R Hansford, 17 Dolver Close, Corby, NN18 8NB
MM0 RSI Robert Inglis, 13 Princes Street, California, Falkirk, FK1 2BX
M0 RSJ Keith Dunstan, 53 Church View Road, Camborne, TR14 8RQ
M0 RSM R Mahon, 7 Grassendale Avenue, North Prospect, Plymouth, PL2 2JL
MI0 RSN R Robinson, 30 Trasnagh Drive, Newtownards, BT23 4PD
MI0 RSO S Mcclean, 28A Ashfield Court, Donaghadee, BT20 0BF
M0 RSP Richard Padon, 21 The Rookery, Balsham, Cambridge, CB21 4EU
M0 RST C Taylor, 48 Northdown Park Road, Cliftonville, Margate, CT9 3PT
M0 RSV Graham Jones, 31 Liverpool Road, Buckley, CH7 3LH
M0 RSW R Weatherup, Sidney Sussex College, Cambridge, CB2 3HU
M0 RSY A Davies, Penthouse Caravan, Shutt Green LANE, Stafford, ST19 9LX
M0 RTC R Cunningham, 47 Westfield Avenue, Skelmanthorpe, Huddersfield, HD8 9AH
MM0 RTD D Robertson, 17 Keswick Drive, Hamilton, ML3 7HN
M0 RTE A Green, 40 Claines Road, Northfield, Birmingham, B31 2EE
M0 RTL Andrew Garthwaite, 278 Carlton Road, Barnsley, S71 2BA
M0 RTM Christian Bell, 4 Main Street, Newbold, Rugby, CV21 1HW
M0 RTO Rougham Tower Museum c/o R Coleman, 16 Mouse Lane Rougham, Bury St. Edmunds, IP30 9JB
M0 RTP Rael Paster, 8 Rachaels Lake View, Warfield, Bracknell, RG42 3XU
M0 RTQ Kieron Jones, 1 Mowbray Street, Epworth, Doncaster, DN9 1HR
MM0 RTT Robert Turpie, 11 Ashkirk Place, Dundee, DD4 0TN
M0 RTV R Coombs, 55 Highfield Road, Hemsworth, Pontefract, WF9 4EA
M0 RTW Catherine Colless, 128 Ditton Lane, Fen Ditton, Cambridge, CB5 8SS
MI0 RTY M Strawbridge, 9 Wheatfield Crescent, Coleraine, BT51 3RA
MI0 RUC Neil Bolt, 32 Bush Gardens, Bushmills, BT57 8AE
MW0 RUH David Thomas, 23 Merthyr Dyfan Road, Barry, CF62 9TG
M0 RUK C Lote, 8 Warren Place, Walsall, WS8 6BY
M0 RUM Martine Symmonds, 24 Woodville Grove, Stockport, SK5 7HU
M0 RUZ Russell Brierley, 34 Hamlet Road, Alvaston, Derby, DE24 0DU
M0 RVC Robin Gripp, 23 Edmond Locard Court, Chepstow, NP16 6FA
MI0 RVH Thomas Nelson, 1, 1 Annashanco, Rosslea, BT92 7PT
M0 RVI Ravi Miranda, Flat 56, Amelia House 11 Boulevard Drive, London, NW9 5JP
M0 RVJ John Goodman, St. Francis House Highlands, Banbury, OX16 1FA
M0 RVT Davy Rajanayagam, 87 Riffel Road, London, NW2 4PG
M0 RWA Richard Anderson, 56a Cheriton Avenue, Adwick-le-Street, Doncaster, DN6 7BT
M0 RWB Robert Broadbridge, 8 Moreton Road, Bournemouth, BH9 3PR
M0 RWD D Eastwood, 13 Riverwood Drive, Halifax, HX3 0TH
M0 RWG Richard Grout, 5 Branton Close, Great Ouseburn, York, YO26 9SF
M0 RWH R Hornby, 61 Fulwood Heights, Fulwood, Preston, PR2 9AW
MM0 RWJ Robert Welsh, 28 Peacock Wynd, Motherwell, ML1 4ZL
M0 RWK West Kent Raynet c/o Darren Parvin, 11 Stanhope Way, Sevenoaks, TN13 2DZ
M0 RWL Robert Lane, 9 Hartoft Road, Hull, HU5 4JZ
M0 RWM Robert Mayfield, 75 Cartwright Street, Loughborough, LE11 1JW
M0 RWN R Nock, 83 Coles Lane, West Bromwich, B71 2QW
M0 RWR Riverway ARS c/o Ernest Reynolds, 4 Underwood Close, Stafford, ST16 1TB
M0 RWS R Stokes, 44 Broxhead Road, Havant, PO9 5LA
M0 RWW R Wells, 44 Woodlea, Leybourne, West Malling, ME19 5QY
M0 RXB Roy Badami, 373 Camden Road, London, N7 0SH
M0 RXD Raymond Dutton, Burn Naze, Old Mill Road, Penmaenmawr, LL34 6TE
M0 RXM Andrzej Matynka, 28 Balmoral Close, Chippenham, SN14 0UT
M0 RXV Roger Mansell, 1412 Warwick Road, Knowle, Solihull, B93 9LG
M0 RXX Stuart Southern, 37 Conway Road, Calcot, Reading, RG31 4XP
M0 RXZ B Lupton, 124 Wolsey Crescent, New Addington, Croydon, CR0 0PF
M0 RYA J Kay, Uplands Farm, Dallington, Heathfield, TN21 9NG
M0 RYB Peter Lock, The Firs, The Butts, Norwich, NR16 2EQ
M0 RYK Michael Granatt, 16 Culverden Avenue, Tunbridge Wells, TN4 9RF
MI0 RYM Ryan Murphy, 40 Stoneypath, Londonderry, BT47 2AF
M0 RYS Ryan Sayre, 8 Lorne Road, Richmond, TW10 6DS
M0 RZC George Bodley, 34 Claremont Road, Newbridge, Newport, NP11 5DL
MJ0 RZD Robert Luscombe, Flat, 1 Rouge Bouillon, St. Helier, Jersey, JE2 3ZA
M0 RZE Ian Laidler, 5 South St., West Rainton, Houghton le Spring, DH4 6PA
M0 RZX Ben Forrest, 32 Idonia Road, Perton, Wolverhampton, WV6 7NQ
M0 RZY Ewen Moore, 23 Woodland Road, Rode Heath, Stoke-on-Trent, ST7 3TJ
M0 SAA Barry Matthews, 30 Oaklands Drive, Brandon, IP27 0NR
M0 SAB A Brackstone, 3 Petunia Close, Basingstoke, RG22 5NX
M0 SAC Michael Couchman, 20 Belmont Road, Gillingham, ME7 5JB
M0 SAD D Platt, 50 Poplars Road, Stalybridge, SK15 3EN
MM0 SAH S Henderson, 13 Dunnottar Place, Kirkcaldy, KY2 5YX
MI0 SAI Simon Barnes, 191 Marlacoo Road, Portadown, Craigavon, BT62 3TD
MM0 SAJ Stephen Smith, 10 Munro Street, Stenhousemuir, Larbert, FK5 4QF
MM0 SAK Alistair Jardine, 17 Louisa Drive, Girvan, KA26 9AH
M0 SAL Douglas Salter, 142 Brays Road, Birmingham, B26 2PP
MI0 SAM S Christie, 17 Kilburn Street, Belfast, BT12 6JS
M0 SAO Altaf Dossa, 24 Warwick Drive, Cheshunt, Waltham Cross, EN8 0BW
MI0 SAP Stephen Murray, 117 Knockview Drive, Tandragee, Craigavon, BT62 2BL
M0 SAQ D Astley, 34 Church Terrace, Glossop, SK13 7RL
M0 SAR Stuart Roy, 28 Kingston Rise, New Haw, Addlestone, KT15 3EY
M0 SAT D Remnant, 26 Roundway, Watford, WD18 6LB
M0 SAV Anthony Smithies, 35 Dialstone Lane, Stockport, SK2 6AA
MM0 SAX G Sproul, 132 Muirdrum Avenue, Glasgow, G52 3AP
M0 SAY D Sayles, 82 Molineaux Road, Shiregreen, Sheffield, S5 0JY
M0 SAZ Michael Parker, Ridgeways, Mill Common, Halesworth, IP19 8RQ
M0 SBA Stephen Walker, 33 Parkside, Somercotes, Alfreton, DE55 4LA
M0 SBB Anthony Southwell, 56 Lambrook Road, Taunton, TA1 2AF
M0 SBC Kenneth Smith, 7 Rosebery Avenue, Morecambe, LA4 5RU
M0 SBD Michael Denut, 17 Quillet Road, Newlyn, Penzance, TR18 5QR
M0 SBF S Larkins, 4 Water Lane, Greenham, Thatcham, RG19 8SS
M0 SBH Subash Nandalan, 22 The Common, Parbold, Wigan, WN8 7DA
MW0 SBJ David John, 29 Eleanor Street, Tonypandy, CF40 1DW
M0 SBK Shane Johnson, 2 North Square, Edlington, Doncaster, DN12 1ED
M0 SBL Patrick Trembath, 48 Treveneth Crescent, Newlyn, Penzance, TR18 5NG
MM0 SBO Stephanie Boyd, 1 St. Marks Lane, Edinburgh, EH15 2PX
M0 SBR Hector Hamilton, Flat B, 9 Cambridge Drive, London, SE12 8AG
M0 SBT Simon Burtsal, 69a Pewley Way, Guildford, GU1 3PZ
MW0 SBX Marc Price, 9 Grandison Street, Swansea, SA1 2HQ
M0 SBY Stephen Bassett, 3 Lower Merryfield, Anchor Road, Radstock, BA3 5PG
M0 SBZ David Smith, 105 Princes Street, Dunstable, LU6 0AG
M0 SCA Simon Light, 16 Cabot Road, Yeovil, BA21 5FQ
M0 SCB T Bacon, Norreum, Church Road, Reading, RG7 1TJ
M0 SCG SANDS AMATEUR RADIO CONTEST GROUP c/o Brian Walson, 7 Branksome Drive, Morecambe, LA4 5UJ
M0 SCO Simon Court, Eastgate Cottage, Perrys Lane, Norwich, NR10 4HJ
M0 SCP Dennis Purbrick, 88 Nabbs Lane, Hucknall, Nottingham, NG15 6NS
M0 SCR Cornwall Raynet Group c/o Keith Harris, 8 Trelawney Rise, Callington, PL17 7PT
M0 SCS Simon Smith, 110 Deaconsfield Road, Hemel Hempstead, HP3 0JA
M0 SCT G Rutherford, 24 Chestnut Avenue, Hedon, Hull, HU12 8NH
M0 SCU Stewart Culshaw, 37 Netherby Road, Wigan, WN6 7PU
M0 SCW S Warren, 1 Morley Close, Stapenhill, Burton-on-Trent, DE15 9EW
M0 SCX Stevan Wing, Flat 32, Wilkinson Drop, Benfleet, SS7 2BG
M0 SCY Sandringham School ARC c/o Alan Gray, 5 Meadow Close, St. Albans, AL4 9TG
M0 SDA Erik Gedvilas, 33 Parkdale Road, Paddington, Warrington, WA1 3EN
M0 SDB Daniel Bower, 89 Halifax Road, Sheffield, S6 1LA
M0 SDC Sheffield DX Net c/o Colin Wilson, 82 Lennox Road, Sheffield, S6 4FN
MW0 SDD Swansea & District ARC c/o John William Bidwell, 26 Lone Road, Clydach, Swansea, SA6 5HR
M0 SDE SUGAR DELTA A.R.C c/o S Preston, The Chapel, Robson Street, Shildon, DL4 1EB
M0 SDG M Torrington, 4 Aylesby Gardens, Grimsby, DN33 1SB



**IMPORTANT NOTE**

**Revalidate licence to avoid revocation** – Ofcom has advised the Society that plans will be drawn up to revoke licences that have not been revalidated as required by the licence conditions. The quickest way to revalidate is to do so online via the Ofcom website: *https://services.ofcom.org.uk/* or by email: *amateur.validations@ofcom.org.uk* Ofcom staff are available to help, but please be patient during times of heavy workload.

M0 SDJ Danny Wild, 10 The Green, Lydd, Romney Marsh, TN29 9ES
M0 SDM Stewart Mason, 8 Barrowby Gate, Grantham, NG31 7LT
M0 SDP S Plows, Ivy House Farm, Main Street, Nuneaton, CV13 6BZ
MI0 SDR Derek Reid, 13 Gelvin Grange, Londonderry, BT47 2LD
M0 SDS S Stocker, 2 Peveril Avenue, Borrowash, Derby, DE72 3JJ
M0 SDT S Theaker, 10 Grange Fields Mount, Leeds, LS10 4QN
M0 SDU Liviu Soldan, 35 Lingfield Gate, Leeds, LS17 6DB
M0 SDW S Willoughby, Bella Cottage, 111 Radwinter Road, Saffron Walden, CB11 3HY
M0 SDY Paul Cattermole, Blaxhall Hall Crossing, Little Glemham, Woodbridge, IP13 0BP
M0 SEA A Newns, 3 Fox's Yard, Harbour Village, Penryn, TR10 8GF
M0 SEB Sebastian Banach, Apartment 124, Advent House, 2 Isaac Way, Manchester, M4 7EB
M0 SEC Leslie Hayward, Cefn Gribyn, Carmel, Llanerchymedd, LL71 7BU
M0 SED D Cockburn, 4 Tranmere Avenue, Heysham, Morecambe, LA3 2BB
MM0 SEK J McPhillips, 86 Glenburn Avenue, Motherwell, ML1 5EF
M0 SEL Steven Elliott, 50 West End Road, Mortimer Common, Reading, RG7 3TH
M0 SEM M Skinner, 5 Sycamore Avenue, Upminster, RM14 2HR
M0 SEO Ritsu Seo, Flat 130, Oslo Court, London, NW8 7EP
M0 SER Carl Lewis, 9 Chatsworth Gardens, Sydenham, Leamington Spa, CV31 1WA
M0 SET Paul Harvey, 22 Meredale Road, Liverpool, L18 5EX
M0 SEV Paul Holmes, 53 Bishops Hull Road, Bishops Hull, Taunton, TA1 5EP
MD0 SEW John Sewell, 8 Anglesmede Crescent, Pinner, HA5 5SP
MM0 SEY Nigel Rogers, 108 Beechwood Road, Cumbernauld, Glasgow, G67 2NP
M0 SEZ S Ezard, 59 Station Farm, Croesyceiliog, Cwmbran, NP44 2JW
M0 SFA Stephen Astbury, 131 Denton Avenue, Grantham, NG31 7JG
M0 SFD Fadel Derry, 13 Fraucup Close, Ford, Aylesbury, HP17 8XU
M0 SFI Filip Sidzhimov, 51 Barnwood Road, Guildford, GU2 8JD
MM0 SFM S Forrest-Mcneill, 34 Maitland Hog Lane, Kirkliston, EH29 9DX
M0 SFR Alex Shafarenko, 42 Church Street, Baldock, SG7 5AF
M0 SFT Dave Swift, 15 Gloucester Walk, Westbury, BA13 3XF
M0 SGA A Suttle, Delvillewood House, 81 Albert Street, Shildon, DL4 2DN
MW0 SGD Simon Doherty, 104 Cromwell Road, Milford Haven, SA73 2EN
M0 SGE S Gearey, 32 Bridgeside, Deal, CT14 9SS
M0 SGF Sydney Francis, 17 Garden Close, Rough Common, Canterbury, CT2 9BP
M0 SGH Stephen Hall, Orchard End, Asenby, Thirsk, YO7 3QR
M0 SGJ Stephen James, Wardens House, Kirkstone Close, Doncaster, DN5 9QZ
M0 SGK S Knott, 24 John Street, Leek, ST13 8BL
MM0 SGQ Stephen Gill, 5 Ramornie Place, Kingskettle, Cupar, KY15 7PT
MW0 SGR James Jenkins, 30 Mayflower Avenue, Llanishen, Cardiff, CF14 5HQ
M0 SGS Stuart Priestley, 49 Victoria Crescent, Pudsey, LS28 7SS
M0 SGV Stuart Vanstone, 2 Walker Crescent, Weymouth, DT4 9AU
M0 SGW S Whalley, 1 Cambridge Road, Gatley, Cheadle, SK8 4AE
MW0 SGX John William Bidwell, 26 Lone Road, Clydach, Swansea, SA6 5HR
M0 SGZ Jonathan Bennette Alincastre, 90 York Crescent, Durham, DH1 5PT
M0 SHA SURBITON HERITAGE AMATEUR RADIO c/o T Fell, 24 Ardmay Gardens, Surbiton, KT6 4SW
M0 SHD Steve Hyde-Dryden, 90 Broadoaks Grange, Carlisle, CA1 2TA
M0 SHF N Newman, 1 Hadham Park Cottages, Cradle End, Ware, SG11 2EH
M0 SHI M Joshi, 14 Doyle Close, Erith, DA8 3QT
M0 SHK Kenneth Holloway, 6 Britons Lane Close Beeston Regis, Sheringham, NR26 8SH
M0 SHM Stephen Marriott, 4 Stone Cross Gardens, Catterall, Preston, PR3 1YQ
M0 SHN Abdullah Al-Shakarchi, 17 Fairfax Place, London, NW6 4EJ
M0 SHP S Shepherd, 24 Brayton Road, Whitehaven, CA28 6EF
M0 SHQ Stephen Hedgecock, 37 Tennyson Road, Maldon, CM9 6BE
M0 SHR ST. HELENS RAYNET GROUP c/o Paul Gaskell, 131 Greenfield Road, Dentons Green, St Helens, WA10 6SH
M0 SHV Amrit Sidhu-Brar, White Gates, Main Road, Northampton, NN7 3NA
M0 SHY S Wildman, 55 Hill Street, Bradley, Bilston, WV14 8SB
M0 SII Salford University ARS c/o Vincent Lynch, 16 Okehampton Crescent, Sale, M33 5HR
MM0 SIL John Connelly, 60 Frankfield Street, Glasgow, G33 1BU
M0 SIN T Brundrett, 45 Talbot Crescent, Whitchurch, SY13 1PH
MW0 SIP Anthony Ferguson, Mount, Salem, Llandeilo, SA19 7HD
M0 SIR Andrew Blount, 267 Merritts Brook Lane, Northfield, Birmingham, B31 1UJ
MJ0 SIT Stephen Whitfield, Ceylon Cottage, Journeaux Street, St Helier, Jersey, JE2 3XQ
M0 SIY Simon Shaul, Shepherds Cottage, Middle Street, Gainsborough, DN21 5BU
M0 SJD S Davies, 1 Holly Drive, Stafford, ST17 0NH
M0 SJG Stephen Goodwin, 9 Downsview, Warminster, BA12 9DU
MM0 SJH S Harvey, West Waterhall, Dounby, Orkney, KW17 2JE
M0 SJJ S Jones, 39 Dalton Lane, Barrow-in-Furness, LA14 4LE
M0 SJK S Kearley, 36 Priory Road, Wirral, CH48 7EU
M0 SJL Sarah Low, 11 Bitterley Close, Ludlow, SY8 1XP
M0 SJR S Roberts, 7 Alberta Grove, Prescot, L34 1PX
M0 SJV S Viney, 5 Hawthorne Grove, Dudley, DY3 2QQ
M0 SJW S Whitehead, 55 Crombie Road, Sidcup, DA15 8AT
M0 SJY Steven Yearley, 5 Gilda Terrace, Rayne Road, Braintree, CM77 6RE
M0 SKA D Bennett, 2 Broadway, Blackburn, BB1 8QZ
M0 SKC S Clay, Akers Lodge, 6 Penn Way, Rickmansworth, WD3 5HQ
MW0 SKD E Edwards, 6 Kerslake Terrace, Tonypandy, CF40 1EQ
M0 SKG STROOD KENT CONTEST GROUP c/o B Howard, 15 Cambridge Road, Strood, Rochester, ME2 3HW
M0 SKI Jordan Skittrall, 14 Tamarin Gardens, Cambridge, CB1 9GH
M0 SKM Stephen Marshall, 96 Bidwell Hill, Houghton Regis, Dunstable, LU5 5EP
M0 SKO Adam Skolik, 98 Rushdene, London, SE2 9RU
M0 SKV Mark Sherrey, 14 The Grove, Hallatrow, Bristol, BS39 6ES
M0 SKY L Sparks, 9 Hawk Place, Moresby Parks, Whitehaven, CA28 8YG
MM0 SLB STROMNESS ACADEMY ARC c/o Derek Smith, Yeldavale, Harray, Orkney, KW17 2LE
M0 SLC Karol Molnar, 201 Fold Croft, Harlow, CM20 1SW
MI0 SLE Dariusz Tarnowski, 11 Millbrook Gardens, Kilrea, Coleraine, BT51 5RZ
M0 SLF S Farnell, 16 Lily Way, Lowestoft, NR33 8NN
M0 SLH James Hewitt, 1 Highfield, Gloucester Road, Chepstow, NP16 7DF
M0 SLP Stephen Richardson, 89 Mead End, Biggleswade, SG18 8JR
M0 SLR South Lancashire ARC c/o Jason Bridson, 10 Clegg Street, Astley, Manchester, M29 7DB
M0 SMA Brian Beckett, 38a Whinney Banks Road, Middlesbrough, TS5 4HG
MM0 SMB B McSherry, 3 Taylor Road, Whitburn, Bathgate, EH47 0NL
M0 SMC S McGregor, 16 Dibbins Green, Wirral, CH63 0QF
MM0 SMD James Nicol, 18 Tininver Street, Dufftown, Keith, AB55 4AZ
M0 SME George Bystryakov, 20 Elmhurst Gardens, Leeds, LS17 8BG
M0 SMG Alan Booth, 16 Coronation Street, Wessington, Alfreton, DE55 6DX
M0 SMH Syed Hassan, 69 Waltham Close, West Bridgford, Nottingham, NG2 6LD
M0 SMJ Michael Seaward, 7 St. Olafs Road, Stratton, Bude, EX23 9AF
M0 SMN Darren Richardson, 25 Comptons Lane, Horsham, RH13 5NL
M0 SMP S Peel, 21 Fairfield Avenue, Ormesby, Middlesbrough, TS7 9BB
M0 SMS Wolfgang Pinkhardt, 43 Cambrian Way, Calcot, Reading, RG31 7DD
M0 SMT S Tasker, 217 Humberston Fitties, Humberston, Grimsby, DN36 4HE
MI0 SMV Stephen Mcveigh, 28 Waringfield Avenue, Moira, Craigavon, BT67 0FA
M0 SMW Shigetaka Watanabe, 48 Adlington Road, Wilmslow, SK9 2BJ
MI0 SMY Sam Dallas, 101 Coagh Road, Stewartstown, Dungannon, BT71 5JL
M0 SMZ Octavian Carp, Rothera Research Station, Stanley, Antarctica, FIQQ 1ZZ
M0 SNB SECRET NUCLEAR BUNKER CONTEST GROUP c/o George Smart, Old Queens Head, Ipswich Road, Diss, IP21 4XP
M0 SND James Popple, 10 Kingsmead Park, Waterbeach, Cambridge, CB25 9PF
MI0 SNG Stephen Gilmour, 14g Malcolm Road, Lurgan, Craigavon, BT66 8DF
MM0 SNK John Dow, 52 Muirfield Way, Deans, Livingston, EH54 8EN
M0 SNT Alan Blake, Northfield Cottage, Droxford Road, Fareham, PO17 5AZ
M0 SNW Simon Wheeldon, 32 Beech Grove Terrace, Garforth, Leeds, LS25 1EG
M0 SNX Thanawit Lertruengpanya, Flat 1, Mallow Court, London, SE13 7PR
M0 SOA G McCourty, The Orchard, Eaton, Tarporley, CW6 9AJ
M0 SOC Second Class Operators Club (UK) c/o Ryan Pike, 63 Bishopstone, Aylesbury, HP17 8SH
M0 SOE Bruce Macmillan, 27 Fiveways Rise, Deal, CT14 9QN
M0 SOL Solway DX Group c/o C Wolf, 35a Moorhouse Road, Carlisle, CA2 7LU
M0 SOT Andrew Cowan, 217 South Park Road, Wimbledon, London, SW19 8RY
M0 SOU John Lovelock, Sea Spray, The Lizard, Helston, TR12 7NU
M0 SOX G Galliver, 29 Archery Fields, Odiham, Hook, RG29 1AE
M0 SPA Staffordshire ARC c/o Neville Briggs, 20 Broad Lane, Pelsall, Walsall, WS4 1AP
M0 SPB R evans, 21 Quilter Close, Bilston, WV14 9AX
M0 SPC Colin Smith, 175 Church Road, Three Legged Cross, Wimborne, BH21 6RG
M0 SPD S Davies, 10 Knutsford Green, Wirral, CH46 8TT
M0 SPH S Hodkinson, 17 Thorn Well, Westhoughton, Bolton, BL5 2PJ
M0 SPJ Paul Snook, 7 Sandhurst Avenue, Kwazulu Natal, South Africa, 3610
M0 SPK Peter Susa, 3 Ainsdale Drive, Whitworth, Rochdale, OL12 8QB
MM0 SPL Scott Ling, Leadburnlea, Leadburn, West Linton, EH46 7BE
M0 SPM R Jones, 51 Tennyson Drive, Ormskirk, L39 3PJ
M0 SPN Steven Netting, 39 Poulton Street, Swindon, SN2 1BH
M0 SPS Andrew Hutley, 90 Main Road, Crick, Northampton, NN6 7TX
M0 SPX Spixworth Scout Radio Group c/o Paul Burgess, 26 William Peck Road, Spixworth, Norwich, NR10 3QB
M0 SQC Polish ARC c/o Boguslaw Niewiadomski, 41 The Crescent, Keresley End, Coventry, CV7 8LB
M0 SRA SIMPSON ARS c/o Ian Ridings, 25 Mond Road, Irlam, Manchester, M44 6QA
M0 SRB S Britten, 10 Second Avenue, Wolverhampton, WV10 9PP
M0 SRC SEND RC c/o M Constantine, 19 Elmcroft, Fairview Avenue, Woking, GU22 7NX
M0 SRJ R Shenton, 2 The Croft, Stramshall, Uttoxeter, ST14 5AG
MI0 SRM S Mccormick, 8 New Close, Portavogie, Newtownards, BT22 1DZ
M0 SRN P Holland, 2 Blythorpe, Hull, HU6 9HG
M0 SRO D Coupe, 6 Berry Avenue, Kirkby-in-Ashfield, Nottingham, NG17 8GE
MD0 SRP Paul Skidmore, 36 Princes Drive, Harrow, HA1 1XH
MI0 SRR D Poots, 18 Upper Quilly Road, Dromore, BT25 1NP
M0 SRS Sidney Smith, 3 Apple Close, Offord D'Arcy, St. Neots, PE19 5SE
MM0 SRX Strathclyde 4x4 Response c/o Thomas Kane, 40b Brisbane Street, Greenock, PA16 8NP
M0 SSB I Rowlands, 22 Maes William Williams VC, Amlwch, LL68 9DS
M0 SSD George Birkby, 8 Kestrel Drive, Dalton-in-Furness, LA15 8QA
M0 SSE J Mossman, 12 Cheviot Crescent, Hadston, Morpeth, NE65 9SP
M0 SSF Peter Midwood, 4 Larch Crescent, Holt, NR25 6TU
MM0 SSG Craig Haldane, 72A Coatbridge Road, Glenmavis, Airdrie, ml6 0nj
M0 SSH S Herman, Barbary House, California Lane, Bushey, WD23 1EX
M0 SSJ Paul Dekkers, 21 Nodens Way, Lydney, GL15 5NP
M0 SSK K Baker, 64 Pendle Drive, Basildon, SS14 3LZ
M0 SSM Stuart McMurtrie, 5 Hill Road, Carshalton, SM5 3RA
M0 SSN Brian Woods, 28 Delph Drive, Burscough, Ormskirk, L40 5BE
M0 SSO Martin Slater, 8, Oldham, OL1 4QB
M0 SSP Waterlooville ARC c/o Richard Shillabeer, 29 Newlease Road, Waterlooville, PO7 7BX
M0 SSR Steve Stewart, 4 Westminster Crescent, Hastings, TN34 2AW
M0 SST SOUTH STAFFORDSHIRE AR TUTORS GRP c/o Richard Finch, 12 Simcox Street, Hednesford, Cannock, WS12 1BG
M0 SSV Stephen Vickers, 35 Lanchester Road, Birmingham, B38 9AG
M0 SSW SILCOATES SCHOOL AR & ELEC.CLUB c/o Nigel Wears, 25 Topcliffe Mews, Morley, Leeds, LS27 8UL
M0 SSX Sussex 4x4 Response c/o David Green, St. Annes, Poundfield Road, Crowborough, TN6 2BG
M0 SSY R Moss, 52 Queen St., Audley, Stoke-on-Trent, ST7 8HB
M0 STA R Stafford, 1 Riverside Cottages, Swinstead Road, Grantham, NG33 4PZ
M0 STF Stuart Binns, 174 Enfield Chase, Guisborough, TS14 7LQ
M0 STI David Smith, Heath Farm, Heath Road, Bury St. Edmunds, IP30 9RL
M0 STL Andrew Palmer, 127 Gloucester Road, Brighton, BN1 4AF
M0 STN Stephen Neale, 48 Five Acres Fold, Northampton, NN4 8TQ
M0 STO Spencer Tomlinson, 8 Levett Road, Stanford-le-Hope, SS17 0BB
M0 STS Gordon Sowden, The Grange Lodge, Rodley Lane, Pudsey, LS28 5QH
M0 STT Scott Gordon, 15 Turnstone Road, Chatham, ME5 8NE
MM0 STU Stewart Macpherson, 0/1 78 Banff Road, Greenock, PA16 0EL
M0 STV Stephen Ridgeon, 7 Southlands, Haxby, York, YO32 2PB
M0 SUF Simon Batley, 2 Boulge Road, Hasketon, Woodbridge, IP13 6LA
M0 SUG David Eastlake, 148 Pursey Drive, Bradley Stoke, Bristol, BS32 8DP
M0 SUN Charles Tate, 11a Nether Lea, Cranage, Crewe, CW4 8HX
M0 SUR St George's Academy ARC (SGA-ARC) c/o Paul Dickson, 49 Signal Road, Grantham, NG31 9BL
MM0 SUS I Macdonald, The Cottage, High Craigton, Glasgow, G62 7HA
M0 SUU Wendy Malcolm-Brown, Flat 11, Chiltern Court, Harpenden, AL5 5LY
M0 SUZ Susan Coombes, 33 Clarence Park Road, Bournemouth, BH7 6LF
M0 SVA Tim Papadopoulos, 77 Cottrell Road, Bristol, BS5 6TN
M0 SVB Stephen Bell, 1 Cherwell Road, Aylesbury, HP21 8TW
MM0 SVE S Shaw, 2 Highfield, Dalry, KA24 4HP
M0 SVR Steven Ring, 35 Sturmer Close, Yate, Bristol, BS37 5UR
M0 SVV Simon Wade, 42 Beauclerk Green, Winchfield, Hook, RG27 8BF
MW0 SWB A Jones, 69 Hendre Gwilym, Tonypandy, CF40 1HF
M0 SWC David Brough, 38 Tynedale Avenue, Crewe, CW2 7NY
M0 SWD B Marshland, 2 Tunstall Hill Close, Sunderland, SR2 9DU
M0 SWE M Sweeney, 3 Orchard Cottages, Asenby, Thirsk, YO7 3QW
M0 SWF Sean Fry, 3 Mariners View, Gillingham, ME7 2RW
M0 SWH Stephen Heard, 29 Grange Farm Road, Yatton, Bristol, BS49 4RB
M0 SWL Brian Bosson, 1 Broomsgrove, Pewsey, SN9 5LE
M0 SWO Anthony Sword, 100 Eaton Road, Norwich, NR4 6PS
MW0 SWR George Waters, 23 Fenwick Drive, Brackla, Bridgend, CF31 2LD
M0 SWT Murray Colpman, Kirk House, Goodworth Clatford, Andover, SP11 7RN
M0 SWZ Ian Swindells, 69 Danby Close, Newton Moor, Hyde, SK14 4AF
M0 SXA Essex Ham c/o Pete Sipple, 52 Fillebrook Avenue, Leigh-on-Sea, SS9 3NT
M0 SXH Steven Hunter, 9 Gelt Burn, Didcot, OX11 7TZ
M0 SXM Stephen Morris, 23 De Courtenai Close, Bournemouth, BH11 9PG
M0 SYG Antony Sygerycz, 75 O'Brien Road, Cheltenham, GL51 0UP
M0 SYJ Piotr Krzeminski, Flat 46, Polden House, Bristol, BS3 4LG
M0 SYM Simon Ludlam, 34 Sussex Place, London, W2 2TH
M0 SYR South Yorkshire Repeater Group c/o Chris Turnbull, 16 Crown Avenue, Cudworth, Barnsley, S72 8SE
M0 SYS Simon Strange, 94 Digby Avenue, Nottingham, NG3 6DY
M0 SYY Colin Gibson, 3 Conway Drive, Billinge, Wigan, WN5 7LH
M0 SZD Stephen Denman, 12 Dyke Vale Road, Sheffield, S12 4ER
M0 SZQ S Jones, Marvin House, Ryhill Pits Lane, Wakefield, WF4 2DU
M0 TAA Edward Slevin, Woodcock Hall, Cobbs Brow Lane, Wigan, WN8 7NB
M0 TAB Anthony Brotherhood, 5 Longcliffe Road, Shepshed, Loughborough, LE12 9LW
M0 TAD B Catchpoole, 8 Buckland Avenue, Basingstoke, RG22 6JL
MW0 TAF Ernest Brookes, 40 Llancayo Street, Bargod, Bargoed, CF81 8TG
M0 TAJ Terry Kemp, 30 Tawny Sedge, King's Lynn, PE30 3PW
M0 TAK Timothy Cooper, Flat 6, Smiths Court, 73 East Borough, Wimborne, BH21 1PJ
M0 TAL Catherine Travis, 4 Kingsdale, Worksop, S81 0XJ
M0 TAM Merv Cox, 61 Barton Close, East Cowes, PO32 6LS
M0 TAN T Nichols, 12 Ivy Grove, Shipley, BD18 4JZ
M0 TAO Oliver Bock, Okenstr. 34, Jena, Germany, D-07745
M0 TAP William Cooper, 20 Staple Close, Waterlooville, PO7 6AH
M0 TAQ Frank Clements, 40 Ellison Fold Terrace, Darwen, BB3 3EB
MD0 TAT Frederick Felix, Flat 41, King Edward Court, Wembley, HA9 7DQ
M0 TAV Vincent Hopkins, 109 Smith Street, Coventry, CV6 5EH
M0 TAW T Woodhouse, The Old Granary, 12 Limekiln Lane, Newport, TF10 9EZ
M0 TAX Edward Underhill, 61 Goldthorne Avenue, Sheldon, Birmingham, B26 3LA
M0 TAY A H Ayres, Brynhyfryd, Phocle Green, Ross on Wye, HR9 7TW
M0 TAZ David Cutts, 38 Berkeley Drive, Hornchurch, RM11 3PY
M0 TBA A Baker, 6 Bayliss Avenue, Wolverhampton, WV4 6NW
M0 TBB Carl Morris, 17 Percy Road, Wrexham, LL13 7EA
MI0 TBD Anthony Kelly, 16 Union Street Mews, Coleraine, BT52 1EN
MI0 TBE Edward Hill, 24 Whitehouse Park, Newtownabbey, BT37 9SQ
M0 TBG Team Thunderbox c/o Clive Moulding, 28 Queens Avenue, Highworth, Swindon, SN6 7BA
MM0 TBH James Kelly, 41 Glenshee Street, Glasgow, G31 4RT
MW0 TBI S Smith, 36 Jones Street, Tonypandy, CF40 2BY
M0 TBJ Terry Buck, 6 Lynn Road, Terrington St. Clement, King's Lynn, PE34 4JX
M0 TBK E Cree, 24 Old Lincoln Road, Caythorpe, Grantham, NG32 3EJ
MI0 TBN S Donnelly, 14 Derryloste Road, Derrytrasna, Craigavon, BT66 6PS
M0 TBQ Donald Nicholls, 62 Queen Elizabeth Way, Telford, TF3 2JW
M0 TBR R Thorogood, 4 Deerhurst Close, Calcot, Reading, RG31 7RX
M0 TBS Toby Tiesdell-Smith, 4 Godwin Close, West Ewell, Epsom, KT19 9LD
MI0 TBV Trevor McKee, 4 Earlford Heights, Newtownabbey, BT36 5WZ
M0 TBW Richard East, 6 Ashley Road, Worcester, WR5 3AY
MM0 TBY S Turnbull, 15 Woodruff Gait, Dunfermline, KY12 0NL
M0 TCB Daniel Howarth, 32 Cotswold Drive, Rothwell, Leeds, LS26 0QZ
M0 TCC Tuck Choy, 39 Netherton Road, Manchester, M14 7FN
M0 TCD Adrian Allen, Milverton, Mill Road, Pulborough, RH20 2PZ
M0 TCE C Eaglen, 46 Sark Close, Hounslow, TW5 0PZ
M0 TCF Linden Allen, 481 Topsham Road, Exeter, EX2 7AQ
MW0 TCJ T Jones, 2 Glyndefaid Cottage, Ynysymond Road, Swansea, SA7 9JA
M0 TCL David Mort, 7 Sheldon Avenue, Congleton, CW12 3LD
M0 TCM THORPE CAMP MUSEUM RADIO GROUP c/o Anthony Nightingale, 42 Spilsby Road, Horncastle, LN9 6AW
M0 TCN Colin Lyne, 4 Bridge Close, Catterick Garrison, DL9 4PG
MM0 TCP Keith Brown, 36 Donald Wynd, Largs, KA30 8TH
MM0 TCQ Thomas Campbell, 10 Barra Gardens, Old Kilpatrick, Glasgow, G60 5HR
M0 TCR T Rozier, 124 Deansfield Road, Wolverhampton, WV1 2LD
M0 TCT Terry Collins, 11 Joseph Gardens, Silver End, Witham, CM8 3SN
MD0 TCX Piotr Sniezek, 204 Quadrant Court, Empire Way, Wembley, HA9 0EY
MM0 TDB Derek Gartshore, 85 Springhill Street, Douglas, Lanark, ML11 0NZ
M0 TDC Robert Stevenson, 97 Queen Street, Crewe, CW1 4AL
M0 TDD Mashuai Xian, 39 Belson Road, London, SE18 5PU
M0 TDE Triode Amateur Radio Group c/o Nigel Knapton, 4 Crabmill Lane, Easingwold, York, YO61 3DE
M0 TDF William Welch, Kenilworth, School Lane, Oswestry, SY11 3LD
M0 TDG T Grant, 89 Greenside, Borehamwood, WD6 4JD
M0 TDK Anthony Tyrwhitt-Drake, Holly Cottage, Church Lane, Beccles, NR34 0AU
M0 TDM R Hydes, 60 Handsworth Grange Road, Sheffield, S13 9HH

M0 TDP Anthony Pickett, 4 Trembel Road, Mullion, Helston, TR12 7DY
MW0 TDQ Jerzy Grzywaczewski, 11 Oxford Court, Ogmore Vale, Bridgend, CF32 7EL
M0 TDW Bill Woodcotte, 2 Little Mead, Shalbourne, Marlborough, SN8 3QB
M0 TDX Trevor Clapp, Windrush, One Pin Lane, Slough, SL2 3QW
M0 TEA Alan Goddard, 50 Ardmore Walk, Manchester, M22 5QG
M0 TEB Martyn Bell, 36 Schneider Road, Barrow-in-Furness, LA14 5DW
M0 TEF A Smith, 101 Chaucer Drive, Lincoln, LN2 4LT
M0 TEG David Horner, 21 Ainsworth Road, Little Lever, Bolton, BL3 1RG
M0 TEI Alexander Wright, Hills Road, Cambridge, CB2 8PH
M0 TEK Edward Moore, 44 Bridge Street, Oxford, OX2 0BB
M0 TEN Ernest Williams, 15 Tenth Street, Peterlee, SR8 4NE
M0 TER B Ashcroft, 16 Edge Lane, Crosby, Liverpool, L23 9XE
M0 TES C Brown, Town End House, Ulverston Road, Ulverston, LA12 0PZ
M0 TET Alan Ford, 5 Prenede, Roches, France, 23270
M0 TEX Ralph Rushlow, 94 Tennyson Street, Guiseley, Leeds, LS20 9LW
M0 TEY Stephen Cole, 109 Maidstone Road, Rochester, ME1 1RN
M0 TEZ T Mullaney, 8 Westerham Close, Macclesfield, SK10 3BG
M0 TFC Thanet Radio and Electronics Club c/o Patrick Kirkden, 22 Leas Green, Broadstairs, CT10 2PL
M0 TFH Edward Hull, Flat 17, Parade Court, Portsmouth, PO2 9RB
MI0 TFK Robin Vage, 80 Chinauley Park, Banbridge, BT32 4JL
M0 TFN T Nolan, 2 Shore Road, Cowes, PO31 8LB
M0 TFO Robert Styles, 52 Vernham Grove, Bath, BA2 2TB
M0 TFS T Smith, Keepers Lodge Cottage, Norton, Runcorn, WA7 1QZ
MM0 TFU Iain Macalister, 33 King Street, Crosshill, Maybole, KA19 7RE
M0 TFX Tom Fisk, 2 Hall Farm Cottage, Caston Road, Attleborough, NR17 1BW
M0 TFY David Butler, Church Cottage, Church Road, Badminton, GL9 1HT
MM0 TGB Tam Brown, 11 Approach Row, East Wemyss, Kirkcaldy, KY1 4LB
M0 TGC James David Hay, 15a Somer Fields, Lyme Regis, DT7 3EZ
M0 TGF Morris Leach, 64 Grove Street, Wantage, OX12 7BG
MM0 TGG George Jamieson, 6 Maryville Park, Aberdeen, AB15 6DU
M0 TGM D Pask, Apartment 403, 1314 Tower Road, Halifax, Canada, NS B3H 4S7
M0 TGN Daniel Trudgian, 18 Hart Close, Wootton Bassett, Swindon, SN4 7FN
MI0 TGO Brian Burns, 24 Lisburn Road, Moira, Craigavon, BT67 0JR
M0 TGS ALTRINCHAM GRAMMAR SCHOOL FOR BOYS c/o Garry Binns, 22 Carlyn Avenue, Sale, M33 2EA
M0 TGT Simon Faulkner, Mount Pleasant, Elkstones, Buxton, SK17 0LU
M0 TGV Giles Cooke, 12 Marcus Road, Dartford, DA1 3JX
M0 TGW Mark Rigby, 75 Manchester Road, Deepcar, Sheffield, S36 2QX
M0 TGX T Green, 35 Park Road, Allington, Grantham, NG32 2EB
M0 TGY Timothy Guy, 16 Cogdeane Road, Poole, BH17 9AS
M0 THA T Hurren, 257 Norwich Road, Wroxham, Norwich, NR12 8SL
M0 THB Tom Barratt, 17 Main Road, Collyweston, Stamford, PE9 3PF
MM0 THE Archie Lang, 202 Devonside Road, Carmichael, Biggar, ML12 6PQ
M0 THJ Anthony Howell-Jones, Savenay House, Poltimore, Exeter, EX4 0AP
M0 THM Tim McConnell, 51 Langney Road, Eastbourne, BN21 3QD
M0 THN Richard Blane, Redfield, Buckingham Road, Buckingham, MK18 3LZ
M0 THO Alessandro Boato, Via A Diaz 20, Marcon, Venezia, Italy, 30020
M0 THT Laser ATC RAC (South) c/o Thomas Toon, 9 Boundstone Lane, Sompting, Lancing, BN15 9QL
M0 THY Hanying Tang, Flat 31, 74 Arlington Avenue, London, N1 7AY
MM0 TIA Samuel Martin, 104 The Braes, Tullibody, Alloa, FK10 2TT
MM0 TIE Robert Adamson, 6 Camdean Crescent, Rosyth, Dunfermline, KY11 2TJ
M0 TIF J Housego, 16 Ligo Avenue, Stoke Mandeville, Aylesbury, HP22 5TX
M0 TIL Coalhouse Fort c/o John Parker, 76 Elm Road, Grays, RM17 6LD
M0 TIN David Le Grove, Apartment 3, Beechwood, Ilkley, LS29 8AH
MI0 TIP W Thompson, 25 Darby Road, Carrickfergus, BT38 7XU
MM0 TIR Rosemary Fearsaor-Hughes, 12 Donald Street, Dunfermline, KY12 0BY
M0 TIU A Beaumont, 11 Essex Road, West Thurrock, Grays, RM20 3JA
M0 TIW A Thornton, Little Gables, Rosemary Lane, Ryde, PO33 2UX
M0 TIX R Cowles, Bonnie Rock, 76 Fordham Road, Ely, CB7 5AL
M0 TJB Terry Barnes, Flat 38, Mill Court, Harlow, CM20 2JG
M0 TJC James Hill, 15 Law Close, Littleport, Ely, CB6 1TS
MW0 TJD T Davies, 58a Ynyswen Road, Treorchy, CF42 6ED
M0 TJL Tracey Leavold, 129 Aylsham Road, Norwich, NR3 2AD
MI0 TJM T Mulholland, 215 Finaghy Road North, Belfast, BT11 9ED
MM0 TJR T Thorne, Top Flat, 26 Mary Elmslie Court, Aberdeen, AB24 5BE
M0 TJS T Scott, Ael y Bryn, St. Harmon, Rhayader, LD6 5LG
MM0 TJT THE JAGGY THISTLES c/o William Findlay, 46 Rowallan Drive, Kilmarnock, KA3 1TU
M0 TJU Evan Duffield, 92 Crosby Street, Stockport, SK2 6SP
M0 TJV C Vernon, 29 Alice St., Deane, Bolton, BL3 5PJ
M0 TJW T Beardwood, Flat 9, Alma House, Ripon, HG4 1NG
M0 TKA T Kay, 64 Cowcliffe Hill Road, Huddersfield, HD2 2PE
M0 TKD Keith Raistrick, 2 Greenacres Grove, Shelf, Halifax, HX3 7RN
MM0 TKE Tim Korby, 1 St Marks Lane, Edinburgh, EH15 2RY
M0 TKS T Kyriacou, 54 Sutton Avenue, Silverdale, Newcastle, ST5 6TB
M0 TKT Robert Bradshaw, 272 Councillor Lane, Cheadle Hulme, Cheadle, SK8 5PN
M0 TKX Amar Sood, Parima, Sewardstone Road, London, E4 7RA
M0 TLC R Ainsworth, 181 Carlton Road, Boston, PE21 8NG
MI0 TLF T Flanagan, 18 Hunters Park, Bellaghy, Magherafelt, BT45 8JE
M0 TLM Ian Williams, 36 Telford Road, Tamworth, B79 8EY
M0 TLN Sergei Moissejev, 50 Filey Road, Reading, RG1 3QQ
M0 TLO Raymond Hunter, 3 Sandyway, Croyde, Braunton, EX33 1PP
M0 TLR Gary Taylor, 39 Stafford Road, St Helens, WA10 3JH
M0 TLX David Burdsall, 37 Fulmar Walk, Whitburn, Sunderland, SR6 7BW
M0 TLY M Casey, 7 Cobham Avenue, Manchester, M40 5QW
M0 TMA Toby Moncaster, 4 Crossways Gardens, Cambridge, CB2 9JT
M0 TMB Robin Cook, 6 Aster Road, Ipswich, IP2 0NQ
M0 TMC E Newby, 22 Acton Road, Liverpool, L32 0TT
M0 TMF Anthony Fullwood, 16 Hollands Place, Walsall, WS3 3AU
MM0 TMG Kevin Cussick, 15a Finlow Terrace, Dundee, DD4 9ND
M0 TMH Tom Mitchell, 9 Rhiw Grange, Colwyn Bay, LL29 7TT
MW0 TMI Dean Willis, 51 Fforchaman Road, Cwmaman, Aberdare, CF44 6NG
M0 TMJ Thomas James, Penrallt, Mountain, Holyhead, LL65 1YR
M0 TMM Michael Elliott, 60a Forest Street, Shepshed, Loughborough, LE12 9DA
M0 TMN Thoa Nguyen, 9 Green Street, Cambridge, CB2 3JU
M0 TMO Keith Chadwick, 17 Nettlebed Nursery, New Road, Shaftesbury, SP7 8QS
M0 TMP Terry McElwee, Little Borough, Borough Farm Road, Godalming, GU8 5JZ
M0 TMS Tomasz Schwabe, 112 Clarkson Court, Hatfield, AL10 9GW
M0 TMT Robert Aldridge, 6 Sheppards Court, [illegible] Lane North, Greenford, UB6 7QJ
MI0 TMW T Wylie, 17 Whinsmoor Park, Broughshane, Ballymena, BT42 4JG
M0 TMX Declan McGlone, 32 Shipley Mill Close, Kingsnorth, Ashford, TN23 3NR
MM0 TMZ Anthony Miles, 9 Buchanan Drive, Lenzie, Glasgow, G66 5HS
M0 TNB Marcin Jakubowski, 75 Ashcombe Road, London, SW19 8JP
M0 TNC Ashley Burton, 12 Munden Grove, Watford, WD24 7EE
M0 TNE Terry Newman, 10 Dereham Road, Garvestone, Norwich, NR9 4AD
M0 TNG Stuart Adaway, 20 Foundry Street, Barnsley, S70 1PL
M0 TNL Vladimir Behal, 21 Bromley Road, London, E17 4PR
M0 TNT A Roberts, Chy Kerenza, Parc Morrep, Penzance, TR20 9TE
M0 TNV Mark Clough, 8 Skeldyke Road, Kirton, Boston, PE20 1LR
M0 TNX Kevin Haworth, 11 Petersfield Close, Bootle, L30 1SG
MM0 TOB Toby Burnett, 16 Iona Drive, Oban, PA34 5AR
M0 TOF John Morgan, Glas y Dorlan, Pontrhydfendigaid, Ystrad Meurig, SY25 6EJ
M0 TOG David White, Woodpeckers, Top Green, Romsey, SO51 0JP
M0 TOL Tolmers Scout Campsite c/o A Rixon, 17 Brimmers Way, Aylesbury, HP19 7HR
M0 TOP Anthony Topsfield, Wild Willow Cottage, Hancock Lane, Truro, TR2 5DD
M0 TOR John Dearden, 7 Wadworth Street, Denaby Main, Doncaster, DN12 4EN
M0 TPA Anthony Patrick, The Woodlands, Nantwich Road, Chester, CH3 9JH
M0 TPC Central Raynet Telpac Group c/o Peter Fox, 5 Llandovery Close, Winsford, CW7 1NA
MM0 TPD James Watson, 64 Anstruther Street, Law, Carluke, ML8 5JG
M0 TPG Anthony Gravell, 21 Wickridge Close, Stroud, GL5 1ST
M0 TPH George Emsden, Flat 47, Cedar Court, London, N10 1EG
M0 TPJ Terry Mallaband, 29 Ferndale Road, Burgess Hill, RH15 0HB
M0 TPW T Winyard, 48 Windsor Drive, Yate, Bristol, BS37 5DY
MM0 TQH Richard Hay, Roddach Cottage East, Cummingston, Elgin, IV30 5XY
M0 TQV R Tuckett, 89 Hillbrook Road, Tooting, London, SW17 8SF
M0 TRB Trenchard Bowden, Carmel, Swallowcliffe, Salisbury, SP3 5PW
MI0 TRC P Clarke, 26 Derryhale Lane, Portadown, Craigavon, BT62 4HL
M0 TRE A Kicman, 15 Leaden Close, Leaden Roding, Dunmow, CM6 1SD
M0 TRF Trevor Knox, 3a Queens Way, Ringwood, BH24 1QB
M0 TRK Raymond Tarling, Moonrakers, Ashley, Corsham, SN13 8AN
M0 TRN Thomas Horsten, Kastelsvej 4, 2.Tv, Copenhagen E, Denmark, 2100
M0 TRO A Roberts, 54 Greenfields Avenue, Alton, GU34 2EE
M0 TRP A Pursglove, 78 Alfreton Road, Westhouses, Alfreton, DE55 5AJ
MM0 TRS Sascha Troscheit, 20 James Street, St. Andrews, KY16 8YA
M0 TRV T Hammett, 9 Coral Close, Aughton, Sheffield, S26 3RB
M0 TRW T Wormald, 12 Church Lane, Owermoigne, Dorchester, DT2 8HS
M0 TRY Robert Barnes, 275 Oregon Way, Chaddesden, Derby, DE21 6UR
M0 TSA Ian Macfarlane, 70 Ashby Drive, Rushden, NN10 9HH
MM0 TSB James Morris, 42b Church Street, Borve, Isle of Lewis, HS2 0RT
M0 TSD Steve Smith, 103 Comberford Road, Tamworth, Tamworth, B79 8PE
M0 TSM M Bull, Sunrise, Ram Lane, Norwich, NR15 2DG
M0 TSN M Lee, 46 Little Lane, Huthwaite, Sutton-in-Ashfield, NG17 2RA
M0 TSW Timothy Walker, 11 Banburies Close, Bletchley, Milton Keynes, MK3 6JP
M0 TTB Andrew Bright, 86 Fourth Avenue, Watford, WD25 9QQ
M0 TTE Simon Fairbourn, 17 Perry's Lane, Wroughton, Swindon, SN4 9AX
M0 TTF D D'Mellow, 164 The Gore, Basildon, SS14 2DA
M0 TTG TALL TREES CONTEST GROUP c/o Brian Gale, Tall Trees Farm, Noah's Ark Lane, Great Warford, WA16 7AX
M0 TTH Thomas Haley, 3 Orchard View, Cropredy, Banbury, OX17 1NR
M0 TTI S White, Upton Farm, Upper Strode, Bristol, BS40 8BG
M0 TTK Mark Buxton, 610 SQN Air Traning Corps, Cadet Training Centre, Chester, CH1 4AN
M0 TTL Andrew Dickinson, 5 Brentwood Villas, Perry Street, Hull, HU3 6AL
M0 TTO G Grant, 15 Watson Close, Rugeley, WS15 2PE
MW0 TTR Aberdare ARS c/o Barry Werrell, 26 Glynhafod Street, Cwmaman, Aberdare, CF44 6LD
M0 TTT R Morgan, 153 Beanfield Avenue, Coventry, CV3 6NY
M0 TTU M Evans, The Brae, Coed-Cae-Ddu Road, Blackwood, NP12 2DA
M0 TTX G Watkins, 21 Comberton Avenue, Kidderminster, DY10 3EG
M0 TTY Daryl Spence, 30 Chestnut Drive, Shirebrook, Mansfield, NG20 8NH
MI0 TUB David Given, 15 Middle Road, Lisburn, BT27 6UU
M0 TUK P Ray, 136 Haselbury Road, London, N18 1QD
M0 TUN Geo BERGERET, 20 rue Labrouste, Paris, France, 75015
M0 TUR Dogan Biyikli, Basement, 300 Portobello Road, London, W10 5TA
M0 TUT S Prescott, 210 Inver Road, Blackpool, FY2 0LW
M0 TUV Andrew Dingwall, 48 Village Farm Caravan Site, Bilton Lane, Harrogate, HG1 4DL
M0 TUW Richard Harris, 7 Fosse Lane, Shepton Mallet, BA4 4PS
M0 TUX B Sutton, 25 Mead Road, Folkestone, CT19 5QY
M0 TVA Christopher Beresford, 13 Chaseside Avenue, Twyford, Reading, RG10 0BT
M0 TVC TRENT VALE ARC c/o Paul ryder, 4 edgeway, Nottingham, NG00LY
M0 TVG Michael Shurley, 43 Charles Close, Wroxham, Norwich, NR12 8TU
M0 TVL C Mackay, 665a Edenfield Road, Rochdale, OL11 5XE
M0 TVR Trevor Parker, 100 Horsebridge Hill, Newport, PO30 5TL
M0 TVT Keith Wilson, 26 Mill Field, Sutton, Ely, CB6 2QB
M0 TVU Paul Swingewood, 6 Goodall Grove, Great Barr, Birmingham, B43 7PQ
M0 TVV Mark Hillman, Flat 5, 32 South Terrace, Littlehampton, BN17 5NU
M0 TVX Raymond Taylor, 10 Barnwell Lane, Cromford, Matlock, DE4 3QY
M0 TWC Travelling Wave Contest Group c/o Keith Haywood, 6 Lydney Road, Urmston, Manchester, M41 8RN
M0 TWG P Hallewell, 32 Shaldon Grove, Aston, Sheffield, S26 2DH
M0 TWJ William Twemlow, Flat 6, 27 Marmion Road, Liverpool, L17 8TT
MM0 TWK Christopher Hall, 3 Academy Street, Tain, IV19 1ED
M0 TWL Terence Larman, 861 London Road, Westcliff-on-Sea, SS0 9SZ
M0 TWM Jonathan Nethercott, 6 Glade Mews, Guildford, GU1 2FB
M0 TWO P Dunn, 13 Stanton Avenue, Newsham Farm Estate, Blyth, NE24 4PL
M0 TWR T Robinson, 35 Stoneham Lane, Swaythling, Southampton, SO16 2NU
M0 TWS Trevor Wood, 44 Wincobank Lane, Sheffield, S4 8AA
M0 TWW Timothy Larman, 861b London Road, Westcliff on Sea, SS0 9SZ
MM0 TWX Piero Calvi-Parisetti, 1 Aytoun Road, Glasgow, G41 5RL
M0 TXD G Rowberry, 32 Tiree Avenue, Worcester, WR5 3UA
M0 TXK Maurice Fletcher, 7 Richard Street, Bacup, OL13 8QJ
M0 TXL Philip Dunnicliffe, 19 Woodland Road, Chelmsford, CM1 2AT
MI0 TXM Andrew McGarvey, 66a Scaddy road, Downpatrick, BT30 9DS
MM0 TXO A Reid, Johnston Farm, Leslie, Insch, AB52 6PD
M0 TXP Patrick Cassells, 5 Saxon Way, Liverpool, L33 4DW
M0 TXR Paul McDonough, 91 Lever Street, Little Lever, Bolton, BL3 1BA
M0 TXS Mandy Townsend, 25 BARTON ROAD, Bedford, MK42 0NA
M0 TXX Greg Acton, 39 Craig Road, Macclesfield, SK11 7YH
M0 TYG D Moore, Camara, 379 Main Road, Harwich, CO12 4DW
M0 TYK J Shephard, 19 Duffryn, Telford, TF3 2BU
M0 TYN Gary Cockburn, 20 Hexham Avenue, Hebburn, NE31 2HN
MM0 TYR C Taylor, 47 Seaview, Knock, Isle of Lewis, HS2 0PD
M0 TYW W Hibberd, 169 Highbury Grove, Cosham, Portsmouth, PO6 2RL
M0 TZO Paul Gibson, 7 Greenfields Road, Horley, RH6 8HW
M0 TZT S Emmett, Middle Farm, East Side, Evesham, WR11 8QW
M0 TZY Steven Crabb, 1 Council Houses, Hall Lane, Norwich, NR12 7BB
M0 TZZ Philip Moore, 24 Plough Road, Dormansland, Lingfield, RH7 6PS
M0 UAA David Bowen, 25 Maendu Terrace, Brecon, LD3 9HH
M0 UAC David Carter, 9 Grange Close, Ipplepen, Newton Abbot, TQ12 5RX
M0 UAS Dom Williams, 2 Tyning Road, Peasedown St. John, Bath, BA2 8HT
M0 UAT Ian Marsh, 56b Oliver Crescent, Farningham, Dartford, DA4 0BE
M0 UAV Tyler Ward, James Barn, Horham, Eye, IP21 5ER
M0 UCD John Turner, 17 Beechwood Road, Dronfield, S18 1PW
M0 UCH Colin Howard, 1 Beale Road, Cheltenham, GL51 0JN
M0 UCK Adrian Manning, 14 Baxter Gardens, Kidderminster, DY10 2HD
M0 UDA Andrew Cattell, 2 St. James Close, Ruscombe, Reading, RG10 9LJ
MM0 UDI R Duncan, South Backieley, Turriff, AB53 4GS
M0 UEZ Ronald Scholefield, 4 Minnie Street, Haworth, Keighley, BD22 8PR
M0 UFA Mark Atherton, 39 Fairmead Road, Wirral, CH46 8TU
M0 UFC Mark Bryant, 284 Brantingham Road, Chorlton cum Hardy, Manchester, M21 0QU
M0 UGD Derrick Underwood, 24 Wheatcroft Road, Rawmarsh, Rotherham, S62 5ED
M0 UGH Andrew Stevenson, Klaustaler Strasse 1, Berlin, Germany, 13187
M0 UGL James Wakenell, 15 Cuckoo Oak Green, Madeley, Telford, TF7 4HT
M0 UGR Clive Luckett, 257 Folkestone Road, Dover, CT17 9LL
MM0 UIG M Mackinnon, 17 Valtos, Miavaig, Isle of Lewis, HS2 9HR
M0 UJD Kevin Colman, 10 South Rise, North Walsham, NR28 0EE
M0 UKA Richard Wood, 7 Wishart Green, Old Farm Park, Milton Keynes, MK7 8QB
M0 UKC UK YOUNG CONTESTERS GROUP c/o S Pearson, 8 The Pastures, Edlesborough, Dunstable, LU6 2HL
M0 UKI UK Islands Group c/o Charles Wilmott, 60 Church Hill, Royston, Barnsley, S71 4NG
M0 UKM Michael Busch, Dammstrabe 4, Neuwied, Germany, 56564
M0 UKO Andrew Shaw, 20 Hillcrest Close, Thrapston, Kettering, NN14 4TB
M0 UKS J Banham, Timandra, Mill Road, Norwich, NR15 2ST
MM0 UKW C Houston, 1 Macredie Place, Perceton, Irvine, KA11 2BF
M0 ULC W Biernacki, 2a Tewkesbury Terrace, London, N11 2LT
M0 ULD M Elford, 10 Meadowlands, Lymington, SO41 9LB
MJ0 ULE Steve Huelin, Stebezel, La Petite Rue De La Pointe, St Peter, Jersey, JE3 7YZ
MI0 ULK Stephen Morrow, 769 Farranseer Park, Macosquin, Coleraine, BT51 4NB
M0 ULR Jan Gromadzki, 13 Merrill Heights, Maidenhall Approach, Ipswich, IP2 8GA
MM0 UMH Leslie Mitchell Hynd, Smithy House Bruichladdich, Isle of Islay, PA49 7UN
MI0 UNA U Murray, 80 Canterbury Park, Londonderry, BT47 6DU
M0 UND Kelvin Harding, Flat 2, Close House, Blandford Forum, DT11 7HA
M0 UNI Geoff Rigby, Gas House Farm, Shavington Park, Market Drayton, TF9 3SY
M0 UNJ Artur Perek, 28 Sefton Avenue, Plymouth, PL4 7HB
M0 UNN Sarunas Jukna, 85 St. Davids Crescent, Aspull, Wigan, WN2 1SZ
MW0 UNU Horia Ilie, Flat 1 30 Alexandra Road, Swansea, SA1 5DQ
M0 UOE University of Essex ARS (EARS). c/o U Nehmzow, University of Essex, Department of Biological Sciences, Colchester, CO4 3SQ
M0 UOG THE UNIVERSITY OF GREENWICH c/o P Smith, 1 Lambourne Place, London, SE3 7BH
M0 UOK Barry Eddy, 58 Meadow Way, Plymouth, PL7 4JB
M0 UOO Richard Bone, 16 Gray Close, Warsash, Southampton, SO31 9TB
M0 UPA Jan Van Der Elsen, SULA Lightship, Llanthony Road, Gloucester, GL2 5HH
M0 UPH Aled Williams, 8 Old Tanymanod Terrace, Blaenau Ffestiniog, LL41 4BU
M0 UPU Anthony Stirk, 5 Hall Stone Court, Shelf, Halifax, HX3 7NY
M0 URF Andrew Vincent, Station Road, Andoversford, Cheltenham, GL54 4HP
M0 URI S Spencer, 19 Coseley Avenue, Hartshorne, Swadlincote, DE11 7EZ
M0 URL Peter Gavin, 11 Campbell Close, Yateley, GU46 6GZ
MM0 URN Ian Quinnell, Acarsaid, Kinlochbervie, Lairg, IV27 4RP
M0 URX Tim Beaumont, PO Box 17, Kenilworth, CV8 1SF
M0 USD Richard Davies, 24 Evesham Avenue, Whitley Bay, NE26 1QR
M0 USK Curtis Burke, 523 Monnow Way, Bettws, Newport, NP20 7DW
MI0 UST Shaun Dockery, 58a Blenheim Drive, Newtownards, BT23 4RB
M0 USV Denis Soames, 40 Woodland Drive, North Anston, Sheffield, S25 4EP
M0 USY P Shields, 34 Dryden Close, Grantham, NG31 9QS
M0 UTA Alex Emmerson, 61 Culver Road, Stockport, SK3 8PG
M0 UTD Mark Jones, 110 Becconsall Drive, Crewe, CW1 4RP
M0 UTG John Dodds, 84 Borrowdale Avenue, Walkerdene, Newcastle upon Tyne, NE6 4HL
M0 UTH G Guinan, 5a Temple Lane, Silver End, Witham, CM8 3QY
M0 UTT Bernard Bull, Swan Cottage, Swan Road, Welshpool, SY21 0RH
M0 UTX John Swift, 2 Pear Tree Avenue, Long Drax, Selby, YO8 8NQ
MI0 UTY David Cartin, 6 Grange Avenue, Magherafelt, BT45 5RP
M0 UUU A Sheard, 15 Bent Lanes, Urmston, Manchester, M41 8PB
M0 UWD Geoffrey Deacon, 32 Gloucester Road, Exwick, Exeter, EX4 2EF
M0 UWS Ian Lindsay, 17 Middleforth Green, Penwortham, Preston, PR1 9TB
M0 UXB D Coomber, 14 Francis Green Lane, Penkridge, Stafford, ST19 5HF
M0 UXO Michael Gritton, 53 Brinkburn Grove, Banbury, OX16 3WX
M0 UXS Carol Dutton, 7 Ellery Grove, Lymington, SO41 9DX

**IMPORTANT NOTE**

**Revalidate licence to avoid revocation** – Ofcom has advised the Society that plans will be drawn up to revoke licences that have not been revalidated as required by the licence conditions. The quickest way to revalidate is to do so online via the Ofcom website: *https://services.ofcom.org.uk/* or by email: *amateur.validations@ofcom.org.uk* Ofcom staff are available to help, but please be patient during times of heavy workload.

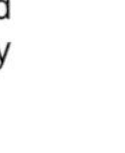

MI0 UYD Peter Page, 259 Bridge Street, Portadown, Craigavon, BT63 5AR
M0 UYR R Brooker, 18 Honeybourne Way, Petts Wood, Orpington, BR5 1EZ
MW0 UZO Daniel White, 222 St. Fagans Road, Cardiff, CF5 3EW
M0 VAA Gerald McGowan, 281 Ashgate Road, Chesterfield, S40 4DB
MI0 VAC Victor Crothers, 5 Thornleigh Park, Ballymoney, BT53 7BX
M0 VAD Denis Cook, 44 Statfold Lane, Fradley, Lichfield, WS13 8NY
M0 VAG A Grant, 26 Fountains Avenue, Boston Spa, Wetherby, LS23 6PX
M0 VAH Edward Whitehouse, 16 rue Gaston de Caillavet, Paris, France, 75015
M0 VAI Dimitris Vainas, 51 Magister Road, Bowerhill, Melksham, SN12 6FD
M0 VAM Martyn Medcalf, 47 Paddock Drive, Chelmsford, CM1 6UX
M0 VAP Manuel Alcaino Pizani, Flat 45, Brian Redhead Court, 123 Jackson Crescent, Manchester, M15 5RR
M0 VAR Belvoir Vale AR c/o Brian Hiley, 9 Pinfold Lane, Harby, Melton Mowbray, LE14 4BU
M0 VAS V Papanikolaou, 104 West Drive Gardens, Soham, Ely, CB7 5EX
M0 VAT A Rodgers, 123 Mill Lane, Northfield, Birmingham, B31 2RP
M0 VAU M Vaughan, c/o Vaughan Industries Ltd, Unit 3 Sydney House, Truro, TR4 8HH
M0 VAW V Werrett, 3 Hardingham Drive, Sheringham, NR26 8YE
M0 VBD Robin Darby, 4 Whately Mews, Whately Road, Lymington, SO41 0XS
M0 VBR Jess Baughan, Chestnut Farm, Eastville, Boston, PE22 8LX
M0 VBT Martynas Kveksas, 29 Saxon Way, Reigate, RH2 9DH
M0 VBW Brian Whall, 3 Farrow Close, Great Moulton, Norwich, NR15 2HR
M0 VBY D Potter, 30 Mersham Gardens, Goring-by-Sea, Worthing, BN12 4TQ
M0 VCA J Davis, 29 Willow Tree Rise, Bournemouth, BH11 8EE
M0 VCC Haydn Morris, Tafarn Pennionyn Groeslon, Caernarfon, LL54 7DE
M0 VCE Nick Baker, 56 Chalklands, Bourne End, SL8 5TJ
M0 VCP Simon Pryke, Pately, School Lane, Woodbridge, IP13 6DX
M0 VCR P Woodhouse, 8 Greenhill Road, Halesowen, B62 8EZ
M0 VCS V Stocker, 25 Davies Drive, Uttoxeter, ST14 7EQ
M0 VDM Jeffrey Savage, Rufford, Barnes Lane, Lymington, SO41 0RR
M0 VDX GROUP TWO c/o J White, 6 Damy Green, Neston, Corsham, SN13 9TN
M0 VEC Robert Trevan, 35 Oaktree Drive, Hook, RG27 9RA
M0 VED Anthony Elliott, 32 Tredegar Walk, Hartlepool, TS26 0TP
M0 VES M Thompson, 2 Old Coastguard Cottages, Holmpton, Withernsea, HU19 2QU
M0 VET M Williams, Jurys, Fore Street, South Molton, EX36 3HL
M0 VEY P Sidwell, 7 Spring Field Close Sigglesthorne, Hull, HU11 5QP
M0 VFC Robert Chipperfield, 13 Harlestones Road, Cottenham, Cambridge, CB24 8TR
M0 VFG Patrick Hawkins, Broadaford Farm, Bittaford, Ivybridge, PL21 0LD
M0 VFR Steve Tomlinson, 7 Springwell Close, Crewe, CW2 6TX
MI0 VFW Mid Ulster ARC c/o James Lappin, 46 Grange Road, Kilmore, Armagh, BT61 8NX
M0 VGA David Silkstone, 169 Otley Road, Harrogate, HG2 0DA
M0 VGC Richard West, 557 East Bank Road, Sheffield, S2 2AG
M0 VGG J Tricklebank, 1 Hewell Road, Barnt Green, Birmingham, B45 8NG
MW0 VGH R Ashworth, The Vicarage, Crymych, SA41 3RN
M0 VGV Gautham Venugopalan, 3 Southwater Close, London, E14 7TE
M0 VHC Thomas Oliver, 17 East Lea, Newbiggin-by-The-Sea, NE64 6BQ
M0 VHG Vincent Greatwood, 11 The Green, Long Preston, Skipton, BD23 4PQ
M0 VIG A Smith, 19 Gibsons Gardens, North Somercotes, Louth, LN11 7QH
M0 VIN C Vincent, 64 Park End Road, Romford, RM1 4AU
M0 VIR Daniel Smith, 48 Shirley Gardens, Tunbridge Wells, TN4 8TH
M0 VIT Jeremy Franks, 14 The Hamlet, Slades Hill, Templecombe, BA8 0HJ
M0 VJX Bradley Walker, 255 Packington Avenue, Birmingham, B34 7RU
M0 VKC N Williams, 17 Sunnyside, Malpas, SY14 7AA
M0 VKG Andrew Smith, 32 Cotswold Drive, Rothwell, Leeds, LS26 0QZ
M0 VKJ Andreas Yiangou, 153 Hoppers Road, London, N21 3LP
M0 VKK Richard Cresswell, Meadow View, Hulver Road, Beccles, NR34 7UW
M0 VKR Lee Bullen, 2 Rowley Cottages, Hermitage Road, Upton, Langport, TA10 9NP
M0 VKS D Vickers, 178 Bakewell Road, Matlock, DE4 3BA
M0 VKX R Routledge, Silvermoor Cottage, Denwick, Alnwick, NE66 3RG
M0 VKY Simon Billingham, 6 Swinford Leys Wombourne, Wolverhampton, WV5 8HT
M0 VLA A Howsen, Oakland Villa, Seaton Road, Maryport, CA15 8ST
M0 VLC Frederick Harwood, 1, South Highall Cottage, Woodhall Spa, LN10 6UR
M0 VLF James Sales, 10 Wolsey Drive, Walton-on-Thames, KT12 3AY
M0 VLI William Toher, The Chapel, Station Road, Darlington, DL2 1JG
M0 VLL Victor Leppard, 39 Queensland Drive, Colchester, CO2 8UD
M0 VLN Michael Jones, 29 Highbridge Road, Burnham-on-Sea, TA8 1LL
M0 VLP G QRP Club c/o Peter Barville, Felucca, Pinesfield Lane, West Malling, ME19 5EN
M0 VLT Alan MacDonald, Woodside Cottage, Horton Way, Verwood, BH31 6JJ
M0 VMC D Burkin, 26 Rampton Road, Cottenham, Cambridge, CB24 8UL
MD0 VMD Peter Birchall, 7 Richmond Close, Douglas, Isle of Man, IM2 6HR
M0 VMH V Hocking, 80 Barton Tors, Bideford, EX39 4HA
M0 VMV Rod Vale, 611 College Road, Birmingham, B44 0AY
M0 VMW VINTAGE & MILITARY ARS c/o Stuart McKinnon, 145 Enville Road, Kinver, Stourbridge, DY7 6BN
M0 VNG Max White, 7 Overthwart Crescent, Worcester, WR4 0JW
M0 VNK David Bishop, 1 Charnwood Drive, Barton Seagrave, Kettering, NN15 6TU
M0 VNO Darryl Harwood, 36 Seaview Drive, Great Wakering, Southend-on-Sea, SS3 0BE
M0 VNR Nick Ramsey, Dalestones, Lansdown Road, Bath, BA1 5TB
M0 VOG VINTAGE OPERATING GROUP c/o Michael Buckley, Springfield, 12 Ranmore Avenue, Croydon, CR0 5QA
M0 VOK R remnant, 172 Burnham Road, Highbridge, TA9 3EH
M0 VOL Colin Brayshaw, 11 Fire Station Yard, Castle Road, Scarborough, YO11 1TL
M0 VOM Noel Curran, 8 Daneswood Close, Whitworth, Rochdale, OL12 8UX
M0 VOS Simon Devos, Applecross Cottage, Main Road, Newark, NG23 7HR
M0 VOZ Michael Crockford, Centre Cottage Kelk, Driffield, YO25 8HL
M0 VPC James Elstone, 54 Oakfield, Woking, GU21 3QS
M0 VPE I Stirzaker, 16 Belvoir Avenue, Emerson Valley, Milton Keynes, MK4 2AB
MM0 VPF Hedley Phillips, Maplebank, Leithen Road, Innerleithen, EH44 6NJ
M0 VPG Robert Killington, 5 Ladymead Close, Maidenbower, Crawley, RH10 7JH
M0 VPK Mark Smith, 18 Hawthorn Road, Old Leake, Boston, PE22 9NY
M0 VPL Matthew Wells, 23 Eastmead, Bognor Regis, PO21 4QT
MM0 VPR Paul Rice, 36 Namur Road, Penicuik, EH26 0LL
M0 VQJ RAF Holmpton ARA c/o John Swift, 2 Pear Tree Avenue, Long Drax, Selby, YO8 8NQ
M0 VQP Arkadiusz Majoch, 66 Boughton Green Road, Northampton, NN2 7SP
M0 VRG Vintage Radio Group c/o Alan Clayton, 6 Albert Road, Bunny, Nottingham, NG11 6QE
MW0 VRQ Steven Trahearn, 148 Gladstone Road, Barry, CF62 8ND
M0 VRS John Strange, Culloden, Ulting Road, Chelmsford, CM3 2LU
M0 VRT L Hummerstone, 70 Salisbury Road, Plymouth, PL4 8TA
M0 VRW Paul Wilson, 45 Newquay Close, Hartlepool, TS26 0XG
M0 VSD Laurie Kirkcaldy, 62 West Garth Road, Exeter, EX4 5AN
M0 VSE Philip Taylor, 104 Winstanley Drive, Leicester, LE3 1PA
MM0 VSG VITAL SPARKS GROUP c/o Anthony Cushley, 12 Achaphubil, Fort William, PH33 7AL
M0 VSP Neville Briggs, 20 Broad Lane, Pelsall, Walsall, WS4 1AP
M0 VSQ Vulture Squadron Contest Group c/o Iain Kelly, 261 Bodiam Avenue, Tuffley, Gloucester, GL4 0XW
M0 VSR Tse Wai Ming, 313 Devizes Road, Salisbury, SP2 9LU
MM0 VSU Leslie Bradley, Amon Sul, Kiltarlity, Beauly, IV4 7HT
M0 VSW Stuart Whall, 17 Vicarage Road, Deopham, Wymondham, NR18 9DR
M0 VTA Duncan McNicholl, 186 Coldhams Lane, Cambridge, CB1 3HH
M0 VTG David Howlett, 21 Chandlers, Orton Brimbles, Peterborough, PE2 5YW
M0 VTJ T Scott, 50 Davison Avenue, Whitley Bay, NE26 1SH
M0 VTK John Martin, 62 Llwyn Ynn, Talybont, LL43 2AL
M0 VTR Martyn Newell, 55 Station Road, Brimington, Chesterfield, S43 1JU
M0 VTS Peter Wilkes, 8 Cloverdale, Stafford, ST17 4QJ
MM0 VTV Robin Farrer, 23 Upper Craigour, Edinburgh, EH17 7SE
M0 VUE Christopher Suddell, Lynhurst, Littleworth Lane, Horsham, RH13 8JX
MM0 VUV Robert Fraser, 72 Ferguson Drive, Denny, FK6 5AG
M0 VVA Andrew Amos, 19 Poets Gate, Cheshunt, Waltham Cross, EN7 6SB
M0 VVC Matthew Walker, 3 Finch Close, Tadley, RG26 3YJ
M0 VVG Elkstones ARS c/o Raymond King, 8 Rydal Court, Congleton, CW12 4JL
M0 VVM T Aldred, 31 Cock Road, Bristol, BS15 9SH
MW0 VVO Stuart Barry, 7 Redhill Park, Haverfordwest, SA61 2HA
M0 VVQ Nigel Ham, 59 Thorpe Gardens, Alton, GU34 2BQ
M0 VVR Chun Yin Chak, 1507, Blk G, 5Butterfly Valley Rd, Kowloon, Hong Kong
M0 VVT Malcolm McGregor, 141 Herne Road Ramsey St. Marys, Ramsey, Huntingdon, PE26 2SY
M0 VVV J Worthington, The Old Hundred, Farm Lane, Farnham, GU10 5QE
M0 VVZ Phil Haywood, 5 Mayfield Drive, Kenilworth, CV8 2SW
MW0 VWC W Wiggans, Bronysgawen, Llanboidy, Whitland, SA34 0EX
M0 VWK M Poole, 15 Roberts Place, Dorchester, DT1 2JJ
MM0 VWR D Green, 35 Douglas Avenue, Brightons, Falkirk, FK2 0HB
M0 VWW Jaroslaw Bielen, Flat 8, Clara Grant House, London, E14 8PH
M0 VXX Tristan Quiney, 20 Britannia Gardens, Stourport-on-Severn, DY13 9NZ
M0 VYW Antony Willsher, 1 Tolputt Court, Gladstone Road, Folkestone, CT19 5NE
M0 VZR C Gain, 14 Battens Avenue, Overton, Basingstoke, RG25 3NL
M0 VZS David Clewer, 45 Ashfield Road, Andover, SP10 3PE
M0 VZT R Clay, 75 Trinity View, Ketley Bank, Telford, TF2 0DY
M0 WAB W Baxter, 19 Westbury Road, Nottingham, NG5 1EP
M0 WAD A Waddington, 8 Redbrook Close, Bromborough, Wirral, CH62 6EA
M0 WAE Lon Severe, 5655 Guincho CT, California, United States, 5655
M0 WAF Paul Marchant, 16 Melrose Drive, Peterborough, PE2 9DN
M0 WAG Oliver Prin, 19 The Colliers, Heybridge Basin, Maldon, CM9 4SE
M0 WAH W Horsewell, 15 Highcroft Lane, Waterlooville, PO8 9NX
M0 WAI Carrie Lam, 58 Sparrow Hill, Loughborough, LE11 1BU
M0 WAJ A Hagland, 11 Coppice View, Heathfield, TN21 8YS
MM0 WAK W Laurie, 306 Lanark Road West, Currie, EH14 5RR
M0 WAM D Beet, 1 Shottesford Avenue, Blandford Forum, DT11 7XU
M0 WAO Biton Walstra, Flat 2, 147 Brighton Road, Redhill, RH1 6PS
MM0 WAP Frederick Pudsey, 21/2 Bathfield, Edinburgh, EH6 4DU
M0 WAQ Mark Welland, 76 Lovel Road, Chalfont St. Peter, Gerrards Cross, SL9 9NX
M0 WAR D Warwick, 36 Annetts Hall Borough Green, Sevenoaks, TN15 8DZ
M0 WAS O Staines, 6 The Quantocks, Flitwick, Bedford, MK45 1TQ
M0 WAU John Lynch, Beechway, Raddel Lane, Warrington, WA4 4EE
M0 WAV Alan Snelson, 6 Rayleigh Close, Braintree, CM7 9TX
MM0 WAX Brian Hendry, 9 Glen Aray View, Inveraray, PA32 8TW
M0 WAY W Thomas, 5 Thornley Road, Wolverhampton, WV11 2HR
M0 WAZ Warren Payne, 2a Victoria Road, Leighton Buzzard, LU7 2NT
M0 WBB W Brown, 126 Alexandra Road, Ashington, NE63 9LU
M0 WBC J Phillips, 56 Rosemary Avenue, Hounslow, TW4 7JG
M0 WBD D Blake, The Bramleys, Gaysfield Road, Boston, PE21 0SF
M0 WBF Wayne Millington, 93 Feiashill Road, Trysull, Wolverhampton, WV5 7HT
M0 WBG Neil Challis, 48 Brunsfield Close, Wirral, CH46 6HE
M0 WBJ Benjamin Webb, 4 Edale Avenue, Audenshaw, Manchester, M34 5TU
M0 WBK Wayne Knapp, 32 Turner Close, Shoeburyness, Southend-on-Sea, SS3 9TL
M0 WBR Robert Walker, 1a Winifred Way, Caister-on-Sea, Great Yarmouth, NR30 5AB
M0 WBS William Bennison, 21 Ashdene Close, Chadderton, Oldham, OL1 2QG
M0 WBY John Willby, 10 Sunbury Road, Birmingham, B31 4LJ
M0 WCA Matthew Bostock, 86 Beauvale Drive, Ilkeston, DE7 8SJ
M0 WCB Wessex Contest Group ARS c/o Daniel Trudgian, 18 Hart Close, Wootton Bassett, Swindon, SN4 7FN
MM0 WCG WOODPECKER CONTEST GROUP c/o Ron Fraser, Hopefield Cottage, Gladsmuir, Tranent, EH33 2AL
M0 WCK Christos Kakoutas, Trinity College, Trinity Street, Cambridge, CB2 1TQ
M0 WCL Clayton Lonie Jr, 41 De la Hay Avenue, Plymouth, PL3 4HS
M0 WCM W Maddox, 28a Redcar Avenue, Ingol, Preston, PR2 3YY
M0 WCR M McSherry, 5 Briery Croft, Stainburn, Workington, CA14 1XJ
M0 WCS Derek Sewell, 19 St. Leonards Way, Ashley Heath, Ringwood, BH24 2HS
MM0 WCT Trevor Woods, Marsden, Lochard Road, Stirling, FK8 3SZ
M0 WDC WEST DEVON CLUB c/o Zoltan Ritter, 64 Thames Gardens, Plymouth, PL3 6HE
M0 WDG David Wressell, 30 Monarch Close, Chatham, ME5 7PD
M0 WDJ David Watson, 56 Lambton Avenue, Delves Lane Industrial Estate, Consett, DH8 7JE
M0 WDL D Lee, Meadowside, Kingstone, Ilminster, TA19 0NT
M0 WDP W Phillips, 55 Kilton Crescent, Worksop, S81 0AX
M0 WDU Duncan Walsh, 8 Prestwold Way, Aylesbury, HP19 8GZ
M0 WDZ Simon Horne, 29 Shaftesbury Street, Fordingbridge, SP6 1JF
M0 WEB B Munro-Smith, 8 Billings Way, Cheltenham, GL50 2RD
M0 WEC P Wagstaff, 49 The Paddock, Earlsheaton, Dewsbury, WF12 8BY
MW0 WEE A Brown, Oakridge, 6 Bro Hafan, Llandysul, SA44 6NQ
MM0 WEI Edward Ireland, The Steading, Blairmains, Shotts, ML7 5TJ
M0 WEL David Wells, 34 Bramble Way, Wymondham, NR18 0UN
M0 WEN C Owen, Garden Cottage, Holbeck Woodhouse, Worksop, S80 3NQ
M0 WET T Clarke, 80 Bendall Road, Birmingham, B44 0SN
M0 WEV John Wedge, 18 Reapers Walk, Pendeford, Wolverhampton, WV8 1TS
M0 WFA A Walker, 14 Maritime Avenue, Hartlepool, TS24 0XF
MW0 WFB Laurie Bowman, Chanrick, Penderyn Road, Aberdare, CF44 9RU
M0 WFF Devon RC c/o David Atkinson, 596 Wolseley Road, St. Budeaux, Plymouth, PL5 1UX
M0 WFK Peter Ashton, 14 Poppy Close, Boston, PE21 7TJ
M0 WFM Mark Deeley, Unit 8, West Cannock Way, Cannock Chase Enterprise Centre, Tachosoft UK Limited, Cannock, WS12 0QW
M0 WFN W Newton, 7 Moss Close, Bridgwater, TA6 4NA
M0 WFO Steven Harris, 1 Eastbank Drive, Worcester, WR3 7BH
M0 WFR Frank Waller, 249 Summer Lane, Wombwell, Barnsley, S73 8QB
M0 WFX Christian Bolton, 201 Lime Tree Avenue, Crewe, CW1 4HZ
M0 WGA Ray mahorney, Walnut Cottage, Church Lane, Wallingford, OX10 0SD
M0 WGB Gerald Beale, 34 Teville Road, Worthing, BN11 1UG
M0 WGC Christopher Watkins, 25 Citadilla Close, Gatherley Road, Richmond, DL10 7JE
M0 WGI S sugihara, Southfield, Park Lane, Wokingham, RG40 4PY
MI0 WGL William Leonard, 57 Mullanavehy, Enniskillen, BT92 2EW
MI0 WGM Graeme McCusker, 41 The Granary, Waringstown, Craigavon, BT66 7TG
M0 WGO Ian Paterson, 11 Ocho Rios Mews, Eastbourne, BN23 5UB
M0 WGS Wings Museum c/o Barrie Bloomfield, 2 Walstead Manor Cottages, Scaynes Hill Road, Haywards Heath, RH16 2QG
MI0 WGW Ernest Kyle, 2, wattstown, Coleraine, BT521SP
M0 WHA W Anderton, 15 Queens Crescent, Lockerbie, DG11 2BA
M0 WHB William Bray, 46 Alexandra Road, Lostock, Bolton, BL6 4BB
M0 WHC W Clayton, 10 Springbank Road, Liverpool, L4 2QR
MI0 WHG Windy Hill Contest Group c/o S Frazer, 2 Cavanballaghy Road, Killylea, Armagh, BT60 4NZ
M0 WHO Michael Sims, 133 Canterbury Road, Hawkinge, Folkestone, CT18 7BS
M0 WHP Ryszard Hoppe, 18 Poplars Close, Luton, LU2 8AE
M0 WHQ Norfolk County Raynet c/o Anthony Mobbs, 149 The Paddocks, Old Catton, Norwich, NR6 7HR
M0 WHR Dale Williams, 76 Quince, Amington, Tamworth, B77 4EU
M0 WHY A Bell, Montem, Jail Lane, Westerham, TN16 3AU
M0 WIA William Armes, 11 Rutland Road, Broadheath, Altrincham, WA14 4HW
M0 WIK Kerry Morris, 44 Leamington Road, Weymouth, DT4 0EZ
MW0 WIL William Howe, 78 Coychurch Road, Pencoed, Bridgend, CF35 5NA
M0 WIN Owen Prosser, 2 Caroline Close, Ventonleague, Hayle, TR27 4EX
M0 WIS Darryl Powis, 18 Merlin Court, Huddersfield, HD4 7SP
M0 WIT Darren Whitley, 10 Kenmore Drive, Cleckheaton, BD19 3EJ
M0 WIZ Ian Moore, Sun House, 33 Church Lane, Trowbridge, BA14 0TE
MI0 WJC William Campbell, 9 Rochester Court, Coleraine, BT52 2JL
MI0 WJM W Murray, 80 Canterbury Park, Londonderry, BT47 6DU
M0 WJT William Taylor, 99 St. Marys Close, Littlehampton, BN17 5QQ
M0 WJW W Wellington, 57 Hillcrest, Whitley Bay, NE25 9AF
MM0 WKJ W Jenkins, 3a Manse Grove, Stoneyburn, Bathgate, EH47 8EW
M0 WKO Peter Holton, 66 Mill Road, Gillingham, ME7 1JB
M0 WKR N Clarke, Brimham Lodge Farm, Brimham Rocks Road, Harrogate, HG3 3HE
M0 WKT Geoffrey Williams, 18 Elmsleigh Road, Farnborough, GU14 0ET
M0 WLA WEST COAST ROLLERS (SCIENCE AND ENGINEERING CLUB) c/o Rob Wynne, South Graceholme, High Lorton, Cockermouth, CA13 9UQ
M0 WLD Benjamin Wild, 1 Sunnymount, Midsomer Norton, Radstock, BA3 2AS
M0 WLF I Prater, 470 Bishport Avenue, Bristol, BS13 0HS
M0 WLH William Lionheart, Marsham, Start Lane, High Peak, SK23 7BP
M0 WLK R Readman, 1 Millside Close, Kilham, Driffield, YO25 4SF
MM0 WLL W Fleming, 65 Dundonald Park, Cardenden, Lochgelly, KY5 0DG
M0 WLS Warren Le Serve, 120 Cheam Road, Sutton, SM1 2EB
MU0 WLV Adam Prosser, Woodlands, La Vassalerie, St. Andrew, Guernsey, GY6 8XL
M0 WLY Ahmed Omar, 84 Beaumont Hill, Darlington, DL1 3ND
M0 WMB Mark Chanter, 7 Woodford Crescent, Plymouth, PL7 4QY
M0 WML Richard Davison, 2 Marlow Terrace, Mold, CH7 1HH
M0 WMO Douglas Tordoff, 49 Dale Edge, Eastfield, Scarborough, YO11 3EP
M0 WMR W Ross, 62 Derwent Drive, Tewkesbury, GL20 8BB
M0 WMT Mark Lawrence, 62 Church Road, Swindon Village, Cheltenham, GL51 9RG
M0 WMX Dave Colver, 85 Whitemoor Lane, Belper, DE56 0HD
M0 WNF Neil Fellingham, 23 Brooklands, Colchester, CO1 2WA
M0 WNI Robert Karpinski, 55 Cambridge Avenue, New Malden, KT3 4LD
MW0 WNL R Bartrum, 1a Bakers Way, Bryncethin, Bridgend, CF32 9RJ
M0 WNT Albert Taylor, 90 Coppice Avenue, Eastbourne, BN20 9QJ
M0 WNV Terry Higginson, 109 London Road, Biggleswade, SG18 8EE
M0 WNW N White, 2 Appleby Cottages, Whithorn, Newton Stewart, DG8 8DQ
MM0 WOA Graham Woan, 6 Sandpiper Road, Lochwinnoch, PA12 4NB
M0 WOB D Bowden, 58 Southville, Yeovil, BA21 4JF
M0 WOD G Norgrove, 161 New Road, Bromsgrove, B60 2LH
M0 WOJ Alexander Landless, 2 Aspen Way, Banstead, SM7 1LE
M0 WOS Warwick Barnes, Cushendall, Lyngate Road, North Walsham, NR28 0DH
M0 WOW D Dunne, 1 Burton Gardens, Brierfield, Nelson, BB9 5DR
M0 WPA Scott Robbins, 4 Central Buildngs, Market Place, York, YO61 3AB
M0 WPL Piotr Loda, St. Albans Court, Sandwich Road Nonington, Dover, CT15 4HH
M0 WPN W Nichols, Newcourt Farmhouse, Silverton, Exeter, EX5 4HT
M0 WPR William Rees, 41 Chine Walk, West Parley, Ferndown, BH22 8PS
M0 WPS Wayne Phillips, 36 Beeches Road Great Barr, Birmingham, B42 2HF
M0 WPT Jose Rosa, Flat 7, Wick Hall, Abingdon, OX14 3NF

M0 WPX Data - Dxers c/o Kenneth Holloway, 6 Britons Lane Close Beeston Regis, Sheringham, NR26 8SH
M0 WQK Drew Blackie, 10 Queens Avenue, Ilfracombe, EX34 9LQ
M0 WQN V Keeley, Hawthorne, Cowbit Drove, Pinchbeck, Spalding, PE11 3TG
M0 WRA R Wray, 10 Winstanley Road, Sale, M33 2AR
M0 WRC WORKINGTON DISTRICT ARC c/o S Topping, 7 Beckstone Close, Harrington, Workington, CA14 5QR
M0 WRD David Whitehouse, 6 Larch Close, Heathfield, TN21 8YW
M0 WRI Peter Hodge, 141 Linden Place, Newton Aycliffe, DL5 7BQ
MM0 WRL Louis Urlings, Tongue, Lairg, IV27 4XD
MM0 WRO Mateusz Lorenowicz, 5d Burnbank Terrace, Breadalbane Street, Oban, PA34 5PB
MW0 WRP W Powell, 40 Heol Ty Newydd, Cilgerran, Cardigan, SA43 2RT
MW0 WRQ TonyPandy Scout Group c/o Brian Jones, 10 Hughes Street, Penygraig, Nr Tonypandy, CF40 1LX
M0 WRS William Smith, 41 Bush Street, Wednesbury, WS10 8LE
MM0 WRX K McCormick, 4 Birch Way, Renfrew, PA4 8FB
MW0 WRY John Richardson, 15 Calland Street, Plasmarl, Swansea, SA6 8LE
M0 WSA Stacy Williams, Flat 35, Winterton House, London, E1 2QR
M0 WSB A Craddock, 58 Vicarage Road, Mickleover, Derby, DE3 0ED
M0 WSC Tak Kwok, 313 Devizes Road, Salisbury, SP2 9LU
MW0 WSD Neils Orchard, The Burrows, Spring Gardens, Whitland, SA34 0HL
M0 WSE Alice Champion, 2 St. Andrews Hill, Waterbeach, Cambridge, CB25 9NA
MM0 WSK John Muchowski, 71 The Braes, Tullibody, Alloa, FK10 2TT
M0 WSN Ron Swinburne, 32 Hollywell Road, Birmingham, B26 3BX
M0 WSR B Harrison, 43a Rumbridge Street, Totton, Southampton, SO40 9DR
MM0 WST D Caiden, 9 Forthview Terrace, Edinburgh, EH4 2AE
M0 WSW Stuart Whittaker, 25 Cleveleys Road, Blackburn, BB2 3JS
M0 WTC Joe Paradas, 13 Meadow Road, Hemel Hempstead, HP3 8AH
M0 WTG Duncan Cooper, Little Heath, Bradfield Common, North Walsham, NR28 0QR
M0 WTH Alan Hawrylyshen, Rosecroft, Round Grove, Croydon, CR0 7PP
M0 WTV Winter Hill Television Society c/o Darren Storer, 527 Chesterfield Road, Sheffield, S8 0RW
M0 WTW William Walker, Keepers Cottage, 247 Forest Road, Tunbridge Wells, TN2 5HT
M0 WTX Stuart Jackson, 64 Main Road, Moulton, Northwich, CW9 8PB
M0 WTY Robert Clare, Kimberley, Boston Road, Boston, PE20 3AP
M0 WUL William Stewart, 5 St. Catherines Close, Uttoxeter, ST14 8EF
M0 WUS S Burns, 22 Pendle Close, Peterlee, SR8 2JS
MW0 WVR Western Valleys Raynet c/o Brian Jones, 10 Hughes Street, Penygraig, Nr Tonypandy, CF40 1LX
MI0 WWB William Bradley, 14 Ardmore Grange, Ballygowan, Newtownards, BT23 5TZ
M0 WWD James Godfrey, 6 Moor Lane, Croyde, Braunton, EX33 1NN
MI0 WWF Steven Nash, 45 Parkfield Road, Ahoghill, Ballymena, BT42 1LY
M0 WWH Nicholas Robertson, Craigenveoch Farm, Glenluce, Newton Stewart, DG8 0LD
MM0 WWM Roy Jowett, Fearnoch, Ardentallen, Oban, PA34 4SF
MW0 WWR WEST WALES RADIO GROUP c/o I Gray, Maesyderi, Pontycleifion, Cardigan, SA43 1DR
M0 WWV Anthony Norden, 10 School Lane, Watton at Stone, Hertford, SG14 3SF
MM0 WXD Duncan Fisher, 1 Inverleith Row, Edinburgh, EH3 5LP
MM0 WXE Anthony Barclay, 21 Netherlea, Scone, Perth, PH2 6QA
M0 WXF Andrew England, 12 Bronte Court, Swinburne Road, Wellingborough, NN8 3BF
M0 WXO Makoto Shibata, 18-41 Moegino Aobaku, Yokohama, Japan, 156-0045
MM0 WXT Alan Salkeld, 6/3 Oxgangs Gardens, Edinburgh, EH13 9BE
M0 WXU P Elsey, 62b Coleraine Road, London, SE3 7PE
M0 WXY Stephen Baldwin, 143 Oxford Road, Swindon, SN3 4JA
M0 WYB John Scully, 10 Eckweek Road, Peasedown St. John, Bath, BA2 8EQ
M0 WYC The RC c/o D Jones, St. Marys Centre, Main Street, Pontefract, WF9 1AF
M0 WYE H Burnham, 13 The Close, Wye, Ashford, TN25 5BD
M0 WYH Darren Lester, 171 Glenavon Road, Birmingham, B14 5BT
M0 WYM Charles Ivermee, 42 Daniell Street, Truro, TR1 2DN
M0 WYN Dulyn Davies, 2 Hendre Ddu, Manod, Blaenau Ffestiniog, LL41 4BH
M0 WYR WYRE AMATEUR RADIO GROUP c/o Kevin Haworth, 11 Petersfield Close, Bootle, L30 1SG
M0 WYT Trevor Webster, 1 Fen Close, Newton, Alfreton, DE55 5TD
M0 WYZ Keith Winwood, 146 Chapel Street, Pensnett, Brierley Hill, DY5 4EQ
M0 WZM Martin Kitt, 18 Brickmakers Road, Colden Common, Winchester, SO21 1TT
M0 WZT Mark Fulbrook, 2 Cob Place, Westbury, BA13 3GS
MW0 WZX C Davies, 28 James Street, Pontarddulais, Swansea, SA4 8HZ
M0 WZZ W RAMSAY, 1 Northburn Road, Eyemouth, TD14 5AU
M0 XAB Alisdair Lark, 20 Lawfield, Coldingham, Eyemouth, TD14 5PB
M0 XAO Gary Dean, 62 Baptist Close, Abbeymead, Gloucester, GL4 5GD
M0 XAI Richard Eyre, 123 Baden Powell Road, Chesterfield, S40 2RL
M0 XAJ Andrew Calvert, 11 Pine Tree Walk, Poole, BH17 7EH
M0 XAK Alan Kent, 4 Sellerdale Drive, Wyke, Bradford, BD12 9DA
M0 XAL Alan Briscoe, 69 Sharpe Street, Tamworth, B77 3HZ
M0 XAM A Morgan, 10 Keysworth Drive, Wareham, BH20 7BD
M0 XAT Malcolm Harwood, 36 Coronation Avenue, Seaton, Workington, CA14 1DW
MM0 XAU Hans Stoeteknuel, C/O Marriott PARKVIEW, DUNROSSNESS, Shetland, ZE2 9JQ
M0 XAW Richard Watson, 8 Bourne Close, Warminster, BA12 9PT
MI0 XAX Eoghan Murray, 33 Orpen Avenue, Belfast, BT10 0BS
MM0 XBD B Donnelly, 19 Douglas Drive, Dunfermline, KY12 9YG
M0 XBI Alexei Romanov, 10 Gloucester Walk, Westbury, BA13 3XG
M0 XBM C Atkinson, 7 Hamilton Road, Grantham, NG31 9QG
M0 XBN Brian Johnson, 6 Trevor Road, Swinton, Manchester, M27 0YH
M0 XBR Andrew Brade, Sand Gap, Bursea Lane, York, YO43 4DF
M0 XBS Bruce Saunders, 88 Bramwoods Road, Chelmsford, CM2 7LT
M0 XBW Bradley Woollett, 24 Earlsworth Road, Willesborough, Ashford, TN24 0DN
M0 XBY Bromley & District ARS c/o Richard Perzyna, 29 Lakeside Drive, Bromley, BR2 8QQ
M0 XCH C Harding, 27 Eston Avenue, Malvern, WR14 2SR
M0 XCJ Colin Jackson, 84 Ogley Road, Walsall, WS8 6BB
M0 XCO Matthew MacDonell, 54 Cinque Foil, Peacehaven, BN10 8DZ
MM0 XCP James Reid, 69 Limekiln Wynd, Mossblown, KA6 5BE
M0 XCR C Raphael, 90 Main Street, South Rauceby, Sleaford, NG34 8QQ
M0 XCT David Aldred, 14 The Meadows, Radcliffe, Manchester, M26 4NS
M0 XCX Peter Houghton, 151c London Road, Calne, SN11 0AQ
M0 XDA Dean Sullivan, 15 Market Lane, Witham, CM8 1GF
M0 XDC Derek Copsey, Fairview, Mill Lane, Brentwood, CM15 0PP
M0 XDF David Ferrington, The Redwoods, 20 Innings Lane, Bracknell, RG42 3TR
M0 XDJ K Gribben, 44 Fern Close, Birchwood, Warrington, WA3 7NU
M0 XDK Corrie Griffiths, 12 Bank View, Northampton, NN4 0RS
M0 XDL Raymond Coles, 10 Littlemoor Road, Weymouth, DT3 6AA
MW0 XDN A Newsome, 20 Agamemnon House, Nelson Quay, Milford Haven, SA73 3AY
M0 XDS D Sharp, 8 Beechfield, Hoddesdon, EN11 9QH
MW0 XDT R Snape, Ashdale, Broadmoor, Kilgetty, SA68 0RN
M0 XDV Alessandro Dalla-Volta, 88 Claygate Lane, Esher, KT10 0BJ
M0 XDX P Dumpleton, 20 Cambridge Road North, Mablethorpe, LN12 1QR
M0 XDY Rami Abousaid, 136 Sixth Cross Road, Twickenham, TW2 5PE
MM0 XEA J Dunlop, 5 Loudon Road, Glasgow, G33 6NJ
M0 XED C Butcher, 60 Barton Road, Canterbury, CT1 1YH
M0 XEE Martin Toher, The Chapel, Station Road, Darlington, DL2 1JG
M0 XEK Edwin Kemp, 4 Foundry Flats, Foundry Square, Hayle, TR27 4AE
M0 XER Leo Bodnar, 47 Alchester Court, Towcester, NN12 6RL
M0 XEY C Blain, 8 Fern Way, Weaverham, Northwich, CW8 3EZ
M0 XFU Christopher Jenkins, Flat 5, The Lawns, Usk, NP15 1BA
M0 XFX J Hawkes, 183 Borden Lane, Sittingbourne, ME10 1DA
M0 XGB K Dickson, 29 Sunnyfields Drive, Minster on Sea, Sheerness, ME12 3DH
M0 XGD Graeme Davies, 10 Leaway, Prudhoe, NE42 6QE
M0 XGG Mike Bayliff, Glenbracken, Coxpark, Gunnislake, PL18 9AZ
M0 XGK James Haig, 1 Vallibus Close, Lowestoft, NR32 3DS
M0 XGR William Smith, 29 Peasemore Road, Sunderland, SR4 0HN
M0 XGT Trevor Thomas, 55 Bath Street, Southampton, SO14 6GR
M0 XGW Gareth Whall, 10 Hillcrest Court, Ipswich Road, Diss, IP21 4YJ
M0 XID Gary Hurst, 13 Flamstead End Road, Cheshunt, Waltham Cross, EN8 0HL
M0 XIG John Wakefield, Oakhurst, Lower Common Road, Romsey, SO51 6BT
M0 XIK Joshua Lambert Hurley, 19 Hill Close, West Bridgford, Nottingham, NG2 6GQ
M0 XJM James Meek, 30 Cleveland Square, London, W2 6DD
M0 XJP Martin Juhe, 75 Pondcroft Road, Knebworth, SG3 6DE
M0 XKD Lindsay Booth, 8 Rowthorne Close, Northampton, NN5 4WB
M0 XKL Roger Jones, Flat 2, Tan y Geraint, 33 Princess Street, Llangollen, LL20 8RD
M0 XKO Paul Goodridge, 22 Horefield, Porton, Salisbury, SP4 0LE
M0 XKW Keith Williams, 35 Lord Street, Coventry, CV5 8DA
M0 XKX Kent County Raynet c/o Michael Granatt, 16 Culverden Avenue, Tunbridge Wells, TN4 9RF
M0 XLB Stefan Borrell, Rose Cottage, Colchester Main Road, Colchester, CO7 8DD
MI0 XLK Samuel Baird, 11 Laral Park, Newtownabbey, BT37 0LH
M0 XLT Kevin Jackson, 7 River Place, Gargrave, Skipton, BD23 3RY
M0 XLX H Knight, 10 Welford Road, Barton, Alcester, B50 4NP
M0 XLY Mark Oxley, 49 Dalton Crescent, Shildon, DL4 2LE
M0 XMC M Coad, House 2, VLA, Woodham Lane, Addlestone, KT15 3NB
M0 XMD Michael Davidson, 19 Mason Street, Workington, CA14 3EH
MW0 XMG P Provis, Dingle Gardens, Croesbychan, Aberdare, CF44 0EJ
M0 XMH Mark Hopewell, 4 Cotes Crescent, Bicton Heath, Shrewsbury, SY3 5AS
MW0 XMI R Lacey, 7 Oak Tree Drive, Cefn Hengoed, Hengoed, CF82 8FN
M0 XMK Michael Rose, 115 New Street, Brightlingsea, Colchester, CO7 0DJ
M0 XML EX-MILITARY LAND ROVER ASSOC. c/o John Butcher, Mount Pleasant, Trampers Lane, Fareham, PO17 6DG
M0 XMS Malcolm Smith, 21 Buckden Close, Woodley, Reading, RG5 4HB
M0 XOC M Cox, 55 Malvern Crescent, Ince, Wigan, WN3 4QA
M0 XOL Trevor Brownen, 43 Great Rea Road, Brixham, TQ5 9SW
M0 XOM Rob Smith, 21 Canal Road Crossflatts, Bingley, BD16 2SR
M0 XON Keith Handscombe, 8 Fletcher Road, Ipswich, IP3 0LF
M0 XOR Michael Hauser, 27 Abbey Street Ickleton, Saffron Walden, CB10 1SS
M0 XOS Oliver Snowdon, Churchill College, Cambridge, CB3 0DS
M0 XOT John Messenger, 34 Goylands Close, Llandrindod Wells, LD1 5RB
M0 XOU I Spinks, 28 Chantry Court, Necton, Swaffham, PE37 8HA
M0 XOX James Hill, 45 Venus Street, Congresbury, Bristol, BS49 5HA
M0 XPA Peter Hekman, 28 Beechcroft Avenue, Crewe, CW2 6SQ
M0 XPB Peter Bannon, 73 London Road, Worcester, WR5 2DU
M0 XPD Paul Darlington, 8 Uplands Road, Urmston, Manchester, M41 6PU
MM0 XPG David Hulin, 18 Bartonholm Gardens, Irvine, KA12 8TD
M0 XPJ Julian Parfitt, 5 Sheridan Road, Frimley, Camberley, GU16 7DU
M0 XPL Christopher Lawrence, CROFT HOUSE, STATION ROAD, Lancaster, LA2 8ER
M0 XPM Patrick Mullen, 12 Poplar Grove, Conisbrough, Doncaster, DN12 2JG
M0 XPS Peter Standley, 9 Capelands, New Ash Green, Longfield, DA3 8LG
MM0 XPT Charles Watkinson, The Hillock Farmhouse, Lumphanan, Banchory, AB31 4QL
MM0 XPZ Stephen Groves, 1/1 99 Delville Street, Greenock, PA15 4SX
MD0 XRA Jeffrey Batsman, 141 Bury Street, Ruislip, HA4 7TQ
M0 XRC Axholme RC c/o John Fennell, Bajamar House, Belton Road, Doncaster, DN9 1JL
MM0 XRI Robert Irvine, 9 Pearce Grove, Edinburgh, EH12 8SP
M0 XRM Dennis Bingham, 33 Sheffield Road, Creswell, Worksop, S80 4HN
M0 XRU T Nutbeem, 6 Morris Rise, Blaenavon, Pontypool, NP4 9PA
M0 XRZ Michael Nicholls, Grahams Onsett Farm, Newcastleton, TD9 0TT
M0 XSD Colin Catlin, 27 Main Street, Frizington, CA26 3SA
M0 XSG Craig Braisby, 4 Langmans Way, Woking, GU21 3QY
M0 XSM Colin Couston, Bridge Cottage, Stanlake Lane, Reading, RG10 0BL
M0 XSR Stuart Roberts, 4 Portland Place, Liverpool, L5 3PJ
M0 XTD Ciaran Morgan, 5 Montgomery Avenue, Hampton-on-the-Hill, Warwick, CV35 8QP
MM0 XTW Andrew Wallis, Pine Cottage, South Smallburn, Peterhead, AB42 5BL
M0 XTX Alberto Cristofoletti, Via Marzars 107, Gemona Del Friuli, Italy, 33014
MW0 XTZ M Digby, 40 Waterloo Road, Ammanford, SA18 3SF
M0 XUB Boruo Xu, 138 King's College, Cambridge, CB2 1ST
M0 XUH Glyn Thomas, Lowside Barn, Rothersyke, Egremont, CA22 2UD
M0 XUU Ravi Gopan, 81 Hilmanton, Lower Earley, Reading, RG6 4HN
M0 XVF Jeremy Smith, 8 Mayfields, Spennymoor, DL16 0RN
M0 XVI Matthew Collins, Chilterns, Little Frieth, Henley-on-Thames, RG9 6NR
M0 XVL Gerhard Elsigan, Traunuferstr 143 A, Haid, Austria, A-4053
M0 XVX A Smith, 10 Borrowdale Road, Malvern, WR14 2DS
M0 XWD J Brown, 8 Chatsworth Street, Sutton-in-Ashfield, NG17 4GG
M0 XWS M Swann, 56 Mansel Drive, Old Catton, Norwich, NR6 7NB
M0 XXJ Jonathan Creaser, 8 Millwood Road, Hounslow, TW3 2HH
M0 XXK Kam Mitchell, 1 Denstroude Cottages, Denstroude Lane, Canterbury, CT2 9JX
M0 XXL Douglas Tinn, 6 Billiemains Farm Cottage, Duns, TD11 3LG
M0 XXO DAVID CRACKNELL, 120 WOODHILL, London, SE18 5JL
MM0 XXP Alan Pitkethley, 99 Margaretvale Drive, Larkhall, ML9 1EH
MM0 XXW M Whyte, 147/2 Lower Granton Road, Edinburgh, EH5 1EX
M0 XXX H Taylor, Sunnyside Well, Chaingate Lane, Bristol, BS37 9XN
M0 XYD Essex AR DX Group c/o Paul Fuller, 6 Annalee Road, South Ockendon, RM15 5DJ
M0 XYL M Turner, 2 Oakleigh Road, Droitwich, WR9 0RP
M0 XYX Anthony Loyd, Maple House, Pangbourne Road, Reading, RG8 8LN
M0 XYZ Adrian Clark, 10 Garfield Close, Lincoln, LN1 3QP
M0 XZG Geoffrey Welch, Amazonas, Sandy Lane, Liverpool, L38 3RP
M0 XZX Simon Lowe, 14 Windmill Rise, York, YO26 4TX
MM0 YAB Christopher Phillips, 8 The Square, Newtongrange, Dalkeith, EH22 4QD
M0 YAC Kenneth Smith, 3 Pendoylan Walk, Cwmbran, NP44 7JX
M0 YAD CWMBRAN ARS c/o Kenneth Smith, 3 Pendoylan Walk, Cwmbran, NP44 7JX
MW0 YAE Gary Thatcher-Sharp, 20 Dilys Street, Blaencwm, Treorchy, CF42 5DT
MW0 YAG A Graham, 2 Heol Undeb, Beddau, Pontypridd, CF38 2LB
M0 YAH William Coburn, 42 Hinton Wood Avenue, Christchurch, BH23 5AH
M0 YAL W Dowkes, Woodlea, Gillamoor Road, York, YO62 6EL
MI0 YAM D Foley, 14 Chestnut Hall Court, Maghaberry, BT67 0GJ
M0 YAV William Jones, 8 Oakbrook Close, Ewyas Harold, Hereford, HR2 0NX
M0 YAY David Young, 75 Broadlands Road, Southampton, SO17 3AP
M0 YBC David Croft, 33 Roughaw Road, Skipton, BD23 2PY
MW0 YBZ Paul Smith, 29 Heol Cwarrel Clark, Caerphilly, CF83 2NE
M0 YCB Christopher Button, 37 Smith Square, Harworth, Doncaster, DN11 8HW
M0 YCG Yorkshire Dales Contest Group c/o Dermot Falkner, 45 Westwood, Carleton, Skipton, BD23 3DW
MM0 YCJ Colwyn Jones, 11b Ettrick Road, Edinburgh, EH10 5BJ
MI0 YCK Declan Mayock, 10 Gortcille, Cladymore Road, Armagh, BT60 2FF
M0 YCQ S Burgess, 10b Scotland Street, Ellesmere, SY12 0EG
M0 YCS Christian Steuwe, Moor Barn Farm, Madingley Road, Cambridge, CB23 7PG
M0 YDB David Breed, 8 Tudor Street, New Rossington, Doncaster, DN11 0JG
M0 YDC Dennis Conway, 5 St. Cuthmans Road, Steyning, BN44 3RH
M0 YDF D Fowler, 5 Highfield Cottages, Everingham, York, YO42 4JG
M0 YDH David Holman, 20 Green Drive, Wolverhampton, WV10 6DW
M0 YDJ Darran Jackson, 3 Laburnum Road, Cadishead, Manchester, M44 5AS
M0 YDK A Morrell, 19 Nairn Road, Stamford, PE9 2YR
M0 YDW Dereck White, 9 Wyatts Lane, Tavistock, PL19 0EU
M0 YDX B Dallimore, 4 Llys Dyffryn, St. Asaph, LL17 0SX
M0 YEE Alan Chapman, 1 Fortunes Way, Bedhampton, Havant, PO9 3LX
M0 YEP Dominic Stinson, 1 The Croft, Earls Colne, Colchester, CO6 2NH
MM0 YEQ Gordon Pearce, 1 Inchbelle Farm Cottage, Kirkintilloch, Glasgow, G66 1RS
M0 YES Peter Shaw, 32 Hardwick Road East, Worksop, S80 2NT
MM0 YET Gordon Burnett, 1b Craig Road, Troon, KA10 6DA
M0 YFT Anthony Murdoch, 2 Birtwistle Terrace, Langho, Blackburn, BB6 8BT
M0 YGB Andrew Birch, 3 Partridge Way, High Wycombe, HP13 5JX
M0 YGG Andrew Mansfield, 3 The Coppice, Thrapston, Kettering, NN14 4QA
M0 YGJ Gareth Owen, 14 Bideford Road, Newport, NP20 3BJ
M0 YGM Ivan Ivanov, 12 Cole Court, Coventry, CV6 1PY
M0 YGW James Newton, 28 Dunstan Road, London, NW11 8AA
M0 YHA YHA Amateur Radio Group c/o Alan Clayton, 6 Albert Road, Bunny, Nottingham, NG11 6QE
M0 YIG Gary Coleman, 19 Grunmore Drive, Stretton, Burton-on-Trent, DE13 0GZ
M0 YIJ Geoffrey Chesters, 50 Primrose Chase, Goostrey, Crewe, CW4 8LJ
M0 YIM M Ellwood, 27 Bath Meadow, Halesowen, B63 2XH
M0 YJO John Hatton, 49 Buxton Street, Morecambe, LA4 5SR
M0 YJT Colin Jarvis, 516 Kingsbury Road, Erdington, Birmingham, B24 9NF
M0 YJW Jason Williams, 10 Masefield Avenue Eaton Ford, St. Neots, PE19 7LS
M0 YKB Dilawar Yakub, 42 Swift Close, Blackburn, BB1 6LF
M0 YKR Yorkshire Resistors c/o George Bystryakov, 20 Elmhurst Gardens, Leeds, LS17 8BG
M0 YKO Simon Davison, 6 Denby Drive, Baildon, Shipley, BD17 7DQ
M0 YLA R Cato, Orrell House, Winterpit Lane, Horsham, RH13 6LZ
M0 YLG Glenys Roddis, 61 Everton Road, Potton, Sandy, SG19 2PD
M0 YLS S Smith, 102 Gresford Road, Llay, Wrexham, LL12 0NW
MI0 YLT Summer McCormick, 46 Lany Road, Moira, Craigavon, BT67 0NZ
M0 YLY Stuart Myland, 2 Willhays Close, Kingsteignton, Newton Abbot, TQ12 3YT
M0 YMA Andrew Banks, 2 Holt Close, Farnborough, GU14 8DG
MI0 YMF Martin Foley, 44 Gallows Street, Dromore, BT25 1BD
MM0 YMG Malcolm Gibson, 18 Pentland View, Edinburgh, EH10 6PS
M0 YMJ P Copplin, 3 Firtree Close, Rough Common, Canterbury, CT2 9DD
M0 YMM N Trangmar, 8 Maxstoke Close, Meriden, Coventry, CV7 7NB
M0 YNK Yanick Watkins, 1 St. Saviour Close, Colchester, CO4 0PW
M0 YOJ James Boone, Amberley, Pinewood Road, High Wycombe, HP12 4DA
M0 YOL David Parker, 5 Beam Avenue, Dagenham, RM10 9BS
M0 YOM James Thresher, 328 Gospel Lane, Birmingham, B27 7AJ
M0 YOT J Partington, 56 Rutherford Drive, Bolton, BL5 1DL
M0 YPJ Paul Kirby, 30 New Street, Eccleston, Chorley, PR7 5TW
M0 YPW Paul Woodfin, Laurel Cottage, Barrow Street, Much Wenlock, TF13 6EN
M0 YRF YORKSHIRE RADIO FRIENDS c/o Richard Potter, Hatfield Wood House, Village Hall, Doncaster, DN7 6BP
M0 YRG Adrian Tring, 12 Ainsdale Close, Orpington, BR6 8DJ
M0 YRM Martin Radulov, 60 St. Marks Avenue, Northfleet, Gravesend, DA11 9LW
MM0 YSK Sohan Ram, 28 Craigievar Gardens, Kirkcaldy, KY2 5SD



**IMPORTANT NOTE**

**Revalidate licence to avoid revocation** – Ofcom has advised the Society that plans will be drawn up to revoke licences that have not been revalidated as required by the licence conditions. The quickest way to revalidate is to do so online via the Ofcom website: *https://services.ofcom.org.uk/* or by email: *amateur.validations@ofcom.org.uk* Ofcom staff are available to help, but please be patient during times of heavy workload.

M0 YSR Richard Moys, 12a Palmerston Avenue, Fareham, PO16 7DP
M0 YUG G Bates, 230 Brook Street, Erith, DA8 1DZ
M0 YUX Sergio Fabris, Via Marchesan 43, Treviso, Italy, 31100
M0 YVG Derek Cordes, 22 Holywell Avenue, Newcastle upon Tyne, NE6 3RY
M0 YVK E Howells, 72 Holywell Crescent, Abergavenny, NP7 5LG
M0 YVT Christopher Williams, 1 South View, Freeholdland Road, Pontypool, NP4 8LL
M0 YVX Barbara Tomlinson, 7 Springwell Close, Crewe, CW2 6TX
M0 YYA Geoffrey Bridge, 49 Wilton Gardens, Radcliffe, Manchester, M26 2UP
M0 YYV Melvyn Lomax, 7 Planetree Road, West Derby, Liverpool, L12 6RE
M0 YYY YATE CONTEST GROUP c/o H Taylor, Sunnyside Well, Chaingate Lane, Bristol, BS37 9XN
M0 YZA C Wilson, 28 Warren Crescent, Marsh Lane, Sheffield, S21 5RW
M0 YZF Carl Preece, 14 Dock Street, Widnes, WA8 0QX
M0 YZV Lewis Sadler, 2 Birkdale Avenue, Dinnington, Sheffield, S25 2SX
M0 ZAA John Wellard, 19 South Motto, Kingsnorth, Ashford, TN23 3NJ
M0 ZAB Adrian Bubb, 2 Hill Top House, Hutton Conyers, Ripon, HG4 5DX
M0 ZAE Hermfried Ehm, 17 Stuart Road, Kempston, Bedford, MK42 8HS
M0 ZAF Robert Barter, 17 West Gate, Plumpton Green, Lewes, BN7 3BQ
M0 ZAI Matthew Grice, 48 St. Ives Road, Coventry, CV2 5FZ
M0 ZAK J Steel, 6 Central Avenue, Shepshed, Loughborough, LE12 9HP
MM0 ZAL B Keiller, Da Cro, Branchiclate, Burra Isle, ZE2 9LA
M0 ZAM Gary Campbell, 10 Welbeck Road, Rochdale, OL16 4XP
MW0 ZAP James Davies, Rose Villa, Creigiau, Cardiff, CF15 9NN
M0 ZAQ Charles Rayment, Brambles, Alltami Road, Mold, CH7 6RT
M0 ZAR Sydney Smith, PO Box 446, Clanwilliam, Clanwilliam, South Africa, 8135
M0 ZAV Ricky Amos, 6 Eccles Road, Wittering, Peterborough, PE8 6AU
MM0 ZAW Alan Woodford, Nordkette, Levenwick, Shetland, ZE2 9GY
M0 ZAY Haydn Jones, Flat 18, Wentworth House, London, N1 2FP
MM0 ZBD David Brown, 181/1 (Gf) Gorgie Road, Edinburgh, EH11 1TT
MM0 ZBH Paul McLaren, 1 Morayvale, Aberdour, Burntisland, KY3 0XE
M0 ZBT S Green, 6 Garth Villas, Rimswell, Withernsea, HU19 2DB
M0 ZBZ Mike Carvell, 10 Burns Close, Stevenage, SG2 0JN
M0 ZCE M Douglas, 486 Malpas Road, Newport, NP20 6NB
MM0 ZCG SHETLAND CONTEST GROUP c/o Hans Stoeteknuel, C/O Marriott PARKVIEW, DUNROSSNESS, Shetland, ZE2 9JG
M0 ZCJ Charles Jonas, 1 St. Johns Road, Stansted, CM24 8JP
M0 ZCM Andrew Adams, 45 Four Oaks Road, Tedburn St. Mary, Exeter, EX6 6AP
M0 ZCO Conor O Broin, Dewhurst, Orchard End, Weybridge, KT13 9LS
M0 ZCP Chris Parker, 40 Holman Way, Ivybridge, PL21 9TE
MM0 ZCT C Thompson, Lochview West, 2 St. Ninians Avenue, Linlithgow, EH49 7BP
M0 ZCW P Smith, 101 Brunel Avenue, Newthorpe, Nottingham, NG16 3RE
M0 ZDB David Brownsea, 47 Southill Road, Bournemouth, BH9 1SH
M0 ZDC Dunstan Cooke, Apartment 9, 27 Sheldon Square, London, W2 6DW
M0 ZDD Alan Henderson, 5 Snipe House Cottages, Alnwick, NE66 2JD
M0 ZDE Denis Kirkden, 57 Crow Hill Road, Margate, CT9 5PF
M0 ZDG David Griffin, 4 Willowside Way, Royston, SG8 5ET
M0 ZDH D Hardwick, 30 Halfcot Avenue, Stourbridge, DY9 0YB
MD0 ZDJ Stephen Scott, 13 Silver Close, Harrow, HA3 6JT
M0 ZDO Andrew Hipkiss, 2 Brooklands, Walsall, WS5 4DJ
M0 ZDU Alan Hawkes, Coachmans Emsworth Road, Lymington, SO41 9BL
M0 ZEB David Featherby, 14 Station Road, Sutton, Ely, CB6 2RL
M0 ZED P Phillips, 2 Millstream Close, Goostrey, Crewe, CW4 8JG
M0 ZEE Juan Rufes, Flat 1 & 3-8, 12 Smyrna Road, London, NW6 4LY
M0 ZEH Stephen Hendy, Flat 2, 33 Kingston Road, Leatherhead, KT22 7SL
M0 ZEL Selwyn James, 35 Prospect Road, Dronfield, S18 2EA
M0 ZEM R Donaldson, 10 Berry Avenue, Trimdon Grange, Trimdon Station, TS29 6EE
M0 ZEN A Bolton, Valley Lodge, Upper Redbrook, Monmouth, NP25 4LU
M0 ZEQ Francis Trigg, Pendle, 2 Langley Common Road, Wokingham, RG40 4TS
MM0 ZET Eshaness RC c/o Hans Hassel, Sumra, Eshaness, Shetland, ZE2 9RS
M0 ZEY Neil Horton, 51 Walsingham Gardens, Epsom, KT19 0LS
M0 ZFF David Almond, 55 Forde Park, Yeovil, BA21 3QP
MM0 ZFG Steven Street, 13 Cobbler View, Arrochar, G83 7AD
M0 ZGB Jonathan Hobbs, 82 Perry's Lane, Wroughton, Swindon, SN4 9AP
M0 ZGT Eleri Ayre, 1 Spring Gardens, Broadmayne, Dorchester, DT2 8PP
M0 ZID Sidney Frampton, 20 Winslow Close, Boldon Colliery, NE35 9LR
MM0 ZIF Marcus Hazel-McGown, 27 Ashdale Avenue, Saltcoats, KA21 6AA
M0 ZIG John Stoppard, 14 Brookside Bar, Chesterfield, S40 3PJ
M0 ZIM Mark Raynor, 68 Cambridge Street, South Elmsall, Pontefract, WF9 2AR
M0 ZIP Cross Border Contest Group c/o Kenneth Pritchard, 9 Golf Close, Pyrford, Woking, GU22 8PE
M0 ZJB Mark Collier, 8 Masefield Mews, Dereham, NR19 2SY
M0 ZJO Jonathan Rawlinson, Westfield Farm, Risden Lane, Cranbrook, TN18 5DU
M0 ZJQ Adam Rawlinson, Westfield Farm, Risden Lane, Hawkhurst, Sandhurst, Cranbrook, TN18 5DU
M0 ZJV Sybrand De Vries, 46 Chaulden House Gardens, Hemel Hempstead, HP1 2BP
M0 ZKA F Hruszka, 20 Winchester Avenue, Leicester, LE3 1AU
M0 ZKK Matthew Bayman, 4 South Road, Beccles, NR34 9NN
M0 ZLE Mark Holmes, 6 Wells Court, Saxilby, Lincoln, LN1 2GY
M0 ZLF Thomas Lovell, 42 Sonja Crest, Immingham, DN40 2EQ
M0 ZLH Adam Stabler, 11 Lincolns Avenue, Gedney Hill, Spalding, PE12 0PQ
M0 ZLI David Challis, 3 West Leaze Place, Bradley Stoke, Bristol, BS32 8AF
M0 ZLK C Forber, 32 Larch Avenue, Newton-le-Willows, WA12 8JF
M0 ZLP Richard Edmondson, Kamway, Stanhoe Road, King's Lynn, PE31 8NJ
M0 ZMB Paul Smart, 142 Finch Road, Chipping Sodbury, Bristol, BS37 6JB
M0 ZMM R Hodgkinson, 39 Oxford Road, Carlton-in-Lindrick, Worksop, S81 9BD
M0 ZMO Louis Whitfield, 72 Macaulay Road, Luton, LU4 0LP
M0 ZMS Matthew Strickland, Ancoats, Piercy End, York, YO62 6DQ
M0 ZMT M Thompson, 133 Redford Avenue, Horsham, RH12 2HH
M0 ZMX M Hardingham, Prospect House, High Street, Rochester, ME3 0BS
M0 ZNP Chantelle Gray, 2 Gloucester Road, Pilgrims Hatch, Brentwood, CM15 9ND
M0 ZNZ Guy Richardson, Berwick Cottage, Bailes Lane, Guildford, GU3 2AX
M0 ZOE Zofia Dunne, 1 Burton Gardens, Brierfield, Nelson, BB9 5DR
MM0 ZOG Peter Riddle, Carngeal, Pitlochry, PH16 5JL
M0 ZOM Colin Langdon, 6 Glebe Close, Stockton, Southam, CV47 8LG
M0 ZOO WORCESTER RADIO AMATEURS ASSOCIATION c/o Peter Badham, 75 Newtown Road, Worcester, WR5 1HH
M0 ZOR Bruce Trayhurn, 15 Wight Drive, Caister-on-Sea, Great Yarmouth, NR30 5UN
M0 ZOV J Renmans, 17 Cartmel Crescent, Chadderton, Oldham, OL9 8DA
M0 ZPA Paul Davies, 67a James Street, Stoke-on-Trent, ST4 5HR
M0 ZPD Paul Davies, 46 Spring Street, Colley Gate, Halesowen, B63 2SZ
M0 ZPG P Morris, 10 Haslam Avenue, Sutton, SM3 9ND
M0 ZPK Patrick Kirkden, 22 Leas Green, Broadstairs, CT10 2PL
M0 ZPL Dariusz Janowicz, 20 Salisbury Road, St. Leonards-on-Sea, TN37 6RX
M0 ZPM Pamela McGillewie, 64 Caradoc View, Hanwood, Shrewsbury, SY5 8ND
M0 ZPU Robert Compton, 18 Drove Road, Gamlingay, SG193NY
M0 ZPZ Charles Parry, 27 Tynedale Close, Stockport, SK5 7NA
MM0 ZRC Robert Chroston, Stonganess, Cullivoe, Shetland, ZE2 9DD
M0 ZRD Darren Hind, 19 Ellington Road, Arnold, Nottingham, NG5 8SJ
M0 ZRF Robert Fidler, 44 Windermere Avenue, Ramsgate, CT11 0PF
M0 ZRG Gene Reynolds, 43 Orchard Drive, Watford, WD17 3DX
M0 ZRR Timothy Cooper, 9 Websters Close, Shepshed, Loughborough, LE12 9AT
M0 ZRS Richard Styles, 4 Coningsby Close, Gainsborough, DN21 1SS
M0 ZRX Karl Bianchini, 10 St. Leonards Road, Headington, Oxford, OX3 8AA
MI0 ZSC John Sinclair, 29 Tern Crescent, Carrickfergus, BT38 7RU
M0 ZSJ John Gibson, 22 Woodburn Drive, Chapeltown, Sheffield, S35 1YS
M0 ZSM Stephen Sissens, 20 Fallow Drive, Eaton Socon, St. Neots, PE19 8QL
M0 ZSS Mark Chamberlain, 79 Riddy Lane, Luton, LU3 2AJ
M0 ZTD Tim Digman, 74 Baddlesmere Road, Whitstable, CT5 2LA
M0 ZTE Steve Broom, 128 Springhill Road, Wolverhampton, WV11 3AQ
M0 ZTG A Hill, 5 Park Road, Thurnscoe, Rotherham, S63 0TG
M0 ZUB Daniel Zubrzycki, 16 Oldfield Avenue, Hull, HU6 7UN
M0 ZUI David Bill, 5 Kennington Road, Wolverhampton, WV10 9RJ
M0 ZUS Terry Lewis, 34 Erw Goch, Ruthin, LL15 1RR
M0 ZVB P Carpenter, 11 Lakeside, Beckenham, BR3 6LX
M0 ZVF Robert Baxter, 30 Croft Gate, Bolton, BL2 3JJ
MW0 ZVR Barry Bateman, Galltygog Farm, Llwydcoed, Aberdare, CF44 0DJ
MU0 ZVV J Bligh, The Bounty, Salines Lane, St. Sampson, Guernsey, GY2 4FL
M0 ZVX Michael Smith, 5 Cresswell Road, Ellington, Morpeth, NE61 5HR
MW0 ZWR G Spicer, 6 Cromwell Road, Neath, SA10 8DR
M0 ZWT Ian Lonsdale, 23 Hunts Field, Clayton-le-Woods, WLC Scout Council, Chorley, PR6 7TT
M0 ZWW William Warwicker, 13 Elm Tree Avenue, Tile Hill, Coventry, CV4 9EU
M0 ZXG Gareth Carless, Silver Cottage, Silver Street, South Petherton, TA13 5BY
MM0 ZXI John Stewart, 6 Sutherland Way, East Kilbride, Glasgow, G74 3DL
M0 ZXJ Jon Wildsmith, 7 Doctors Hill, Stourbridge, DY9 0YE
M0 ZXQ Ian Talbot, 41 Elmwood Close, Cannock, WS11 6LX
M0 ZXW David Bambrough, 7 Barnwell View, Herrington Burn, Houghton le Spring, DH4 7FB
M0 ZXY Daniel Pugh, 8 Clos Deiniol, Llanbadarn Fawr, Aberystwyth, SY23 3TX
M0 ZYD Denis Moger, 23 Elmsleigh Road, Paignton, TQ4 5AN
M0 ZYF WESSEX DX GROUP c/o Terence Langdon, 58 Upper Marsh Road, Warminster, BA12 9PN
M0 ZYT John McColl, 84a Perth Street, Hull, HU5 3NZ
M0 ZZA Anthony De Maillet, Brock Cottage, The Park, Banbury, OX15 5JB
M0 ZZE Stephen Sims, 8 Hurdles Way, Duxford, Cambridge, CB22 4PA
M0 ZZI A Downing, Claybrook Cottage, Mill Lane, Newport, PO30 2LA
MM0 ZZO James McGinty, 1/1 119 Neilston Road, Paisley, PA2 6ER
M0 ZZT S Webber, 59 Mincinglake Road, Exeter, EX4 7DY

## M*1

M1 AAC Lindsay Healy, 31 Roach Road, Sheffield, S11 8UA
MW1 AAH D Creber, 8 George Manning Way, Gowerton, Swansea, SA4 3HB
M1 AAS A Jackson, 29 Bramble Avenue, Birkenhead, CH41 0AX
MM1 ABA Ian Hopley, 53 Redmoss Road, Aberdeen, AB12 3JJ
M1 ABC B Johnson, 20 Valleyside, Hemel Hempstead, HP1 2LN
M1 ABF K Perkin, 25 Rownall View, Leek, ST13 8JN
M1 ABG Hamish Mallin, Riverside House, Rope Walk, Southampton, SO31 4HD
M1 ABM M Chapman, Woodcroft, Windmill Green, Pevensey, BN24 5DY
M1 ABT Robert Macleod, 24 Heol Powis, Gungrog Hill, Welshpool, SY21 7TP
M1 ABU S Norman, 36 Saddlers Park, Eynsford, Dartford, DA4 0HA
M1 ABV Barend Breet, 23 Mitchell Street, Eccles, Manchester, M30 8AJ
M1 ABX S Painting, Claytons, Inkpen, Newbury, RG17 9QE
M1 ABY Tony Highams, 16 Lumley Road, Sutton, SM3 8NN
M1 ACA G Barnett, 63 Sandcroft, Sutton Hill, Telford, TF7 4AB
M1 ACB Stephen Thomas, 2 Myrtle Cottages, Sandy Lane, Saxmundham, IP17 1HR
M1 ACC J Chambers, 9 Farnborough Road, Swindon, SN3 2DR
M1 ACF ACF/CCF RC c/o Michael Buckley, Springfield, 12 Ranmore Avenue, Croydon, CR0 5QA
M1 ACJ Stephen Shearing, 42 Meadow Park, Wesham, Preston, PR4 3DN
M1 ACK R Mackay, Conifers, The Street, Guildford, GU4 7TJ
M1 ACL Keith Green, 42 Dartmouth Street, Stoke-on-Trent, ST6 1HB
M1 ACN M Goom, 47 Sandringham Court, Slough, SL1 6JU
M1 ACO R Hankin, The Brewhouse, High Swinside Farm, Cockermouth, CA13 9UA
M1 ACQ S Thomas, 111 Jersey Avenue, Bristol, BS4 4QX
M1 ACT C LEEDS, 23 Fairfax Road, Norwich, NR4 7EZ
M1 ADK G Pratt, 2 Houghton Road, Newbottle, Houghton le Spring, DH4 4EF
M1 ADN T Whiting, 28 Legarde Avenue, Hull, HU4 6AP
M1 ADP Kenneth Eastwood, 19 Sandby Drive, Sheffield, S14 1DF
M1 ADT R Vickerstaff, 16 Sewell Wontner Close, Kesgrave, Ipswich, IP5 2GB
M1 ADV I Owen, 11 Thatchers Walk, Stowmarket, IP14 2DR
M1 ADX Jonathan Towns, 72 Longfields Road, Norwich, NR7 0NA
M1 ADZ N Davis, 71 Brettenham Road, London, E17 5AZ
M1 AEA M Waldron, 32 Windmill Street, Upper Gornal, Dudley, DY3 2DQ
M1 AEB Gordon Haughie, 100 Henton Road, Edwinstowe, Mansfield, NG21 9LE
M1 AED M Cattell, 12 Fairway Road, Warley, Oldbury, B68 8BE
M1 AEG Andrew Green, Michigan, North Road, Whitemoor, St. Austell, PL26 7XN
M1 AEH D Cossey, 11 Halden Avenue, Norwich, NR6 6UX
M1 AEI David Bowyer, East Foldhay, Zeal Monachorum, Crediton, EX17 6DH
M1 AEJ Ashley Benjamin, Fieldview, Field Common Lane, Walton-on-Thames, KT12 3QH
M1 AEK David Mulliner, 23 Nostell Way, Bridlington, YO16 6FY
MM1 AEL C Haswell, 6 Lochlann Road, Culloden, Inverness, IV2 7HB
M1 AEO R East, 27 Caddywell Meadow, Torrington, EX38 7NZ
M1 AEP M Griffin, 76 Waylands, Swanley, BR8 8TN
M1 AEQ Franklin Allenby, 9 Church View, Holme-on-Spalding-Moor, York, YO43 4BG
M1 AEV Antony Aiello, Llamedos, Walnut Road, Wisbech, PE14 7NP
M1 AEX R Pyman, Broumfield, North Street, Maidstone, ME16 9HF
M1 AEZ Robert Barrett, Kennoway, Hay Lane, Slough, SL3 6HJ
M1 AFF R Dyer, 4 Downleaze, Durrington, Salisbury, SP4 8AB
M1 AFP P Jefford, 61 Willow Way, Flitwick, Bedford, MK45 1LN
M1 AFQ Alan Brooks, 86 Violet Road, Norwich, NR3 4TS
M1 AFU S Derwin, 5 Hawthorne Grove, Yarm, TS15 9EZ
M1 AFV S Wells, 1 Neath Gardens, Leeds, LS9 6RG
MW1 AFW C Jones, Arosfa House, 7 Wood Street, Bargoed, CF81 8NW
M1 AFX David Hill, 3 Morcar Road, Stamford Bridge, York, YO41 1PR
M1 AFZ C Grizzell, 16 Lower Park, Minehead, TA24 8AX
M1 AGA Reginald Taylor, 5 Thirlmere Drive, Bury, BL9 9QE
M1 AGE D Thorley, 4 Bateman Close, Leominster, HR6 9NW
M1 AGH D Hirst, 2 The Birches, Marlborough Road, Swindon, SN3 1PT
M1 AGK Richard Large, 5 Jasmine Close, Abbeydale, Gloucester, GL4 5FJ
M1 AGP John Davies, 10 Gorselands, Hollesley, Woodbridge, IP12 3QL
M1 AGR M Skinner, 44 Westminster Crescent, Sheffield, S10 4EX
M1 AGW Stephen Whitehouse, 47 Mulberry Road, Bloxwich, Walsall, WS3 2NG
M1 AGY J Cuddy, 4 Thames Gardens, Plymouth, PL3 6HD
M1 AHA S Lefevre, 9 Old Barn Crescent, Hambledon, Waterlooville, PO7 4SW
M1 AHF M Byatt, 15 Lower Farm Road, Plympton, Plymouth, PL7 1JJ
M1 AHJ P Clarke, 36 Eldred Drive, Orpington, BR5 4PF
MM1 AHL D McArthur, 12 Laburnum Grove, Lenzie, Kirkintilloch, Glasgow, G66 4DF
M1 AHN Geoff Boyce, 10 Quarry Close, Ross-on-Wye, HR9 7DR
M1 AHR K Peters, 82 Blackmoor Road, Moortown, Leeds, LS17 5JP
M1 AHT L Russell, 106 Stambridge Road, Rochford, SS4 1DP
M1 AHU Godfrey Armstrong, 61 Victoria Drive, Llandudno Junction, LL31 9PF
M1 AHY M Rhodes, 1 Chetwode, Overthorpe, Banbury, OX17 2AB
MI1 AIB Paul Lewis, 15 Foyle Drive, Ballykelly, Limavady, BT49 9PG
M1 AIK T Bardgett, 49 Street James Street, South Petherton, TA13 5BN
M1 AIM A Moore, Silver Trees, Woodlands Lane, Pulborough, RH20 3HG
M1 AIN D White, 19a Gravenhurst Road, Campton, Shefford, SG17 5NY
M1 AIS Jean White, Pathways, Down Barton Road, Birchington, CT7 0PY
M1 AIX W Greenall, 356 Warrington Road, Abram, Wigan, WN2 5XA
M1 AIY M Bastin, 14 Golvers Hill Road, Kingsteignton, Newton Abbot, TQ12 3BP
M1 AJA L Mason, 42 Linden Way, Thorpe Willoughby, Selby, YO8 9ND
M1 AJG M Ramskill, 7 Hobart Road, Dewsbury, WF12 7LS
M1 AJM S Hoskins, 21 Wicken House, London Road, Maidstone, ME16 8QP
M1 AJQ W Clarke, 10 Athol Close, Sinfin, Derby, DE24 9LZ
M1 AJT P Amos, 16 Eastry Road, Erith, DA8 1NN
M1 AJU S Bradford, 28 Downs Road, Walmer, Deal, CT14 7SY
M1 AKF P Wilson, 9 The Brooklands, Wrea Green, Preston, PR4 2NQ
M1 AKH Derrick Mulvana, 17 Wildlake Orton Malborne, Peterborough, PE2 5PG
M1 AKL Vincent Parsons, 20 Old Top Road, Hastings, TN35 5DJ
M1 AKN R Day, 21 Chepstow Road, Bury St. Edmunds, IP33 2ES
M1 AKT David Thomas, 82 Fir Tree, Thurgoland, Sheffield, S35 7BS
M1 AKV R Kowalski, 2 Newcross Park, Kingsteignton, Newton Abbot, TQ12 3TJ
M1 ALA David Mawson, 84 Walnut Avenue, Weaverham, Northwich, CW8 3DX
M1 ALE A Ratcliff, 71 Spring Gardens, Leek, ST13 8DD
M1 ALF R Coston, 272 Warley Road, Blackpool, FY2 0UG
M1 ALH K Whinney, 6 Eve Balfour Way, Haughley, Stowmarket, IP14 3NW
M1 ALM R Hodgkins, 38 Byron Road, Gillingham, ME7 5QH
M1 ALO Paul Rogers, Flat 11, Green Court, Lewes, BN7 1HY
M1 ALR C Moore, Colchester House, Farrington Road, Bristol, BS39 7LW
M1 ALT Michael Oram, 43 Peverell Avenue West, Poundbury, Dorchester, DT1 3SU
M1 ALU T Summers, The Mistress, East Bank, Doncaster, DN7 5JF
M1 ALX David Philip, 15 Eastfield Way, St. Austell, PL25 4HS
M1 AMA Anthony Day, 9 Arundel Road, Tewkesbury, GL20 8AS
M1 AMB D Snow, 7 Aynsley Close, Cheadle, Stoke-on-Trent, ST10 1DP
M1 AMI D Chambers, 94 Hawthorn Avenue, Colchester, CO4 3JR
M1 AMJ D Bonnett, 254 Norwich Road, Wisbech, PE13 3UT
M1 AMP H Withers, 23 Fernie Road, Guisborough, TS14 7LZ
M1 AMW C Whitehead, 18 Victoria Quay, Ashton-on-Ribble, Preston, PR2 2YW
M1 AMZ K Birch, 16 Brentwood Avenue, Thornton-Cleveleys, FY5 3QR
M1 ANC C McLean, 18 Chatfield Road, Gosport, PO13 0TN
MW1 AND Melvyn Griffiths, 32 Hill Street, Aberdare, CF44 6YG
M1 ANK Roland Taylor, 86-88 Hillside Crescent, Leigh-on-Sea, SS9 1HQ
M1 ANL D Clapp, Wenlock Edge, Park Hill, Shepton Mallet, BA4 4AZ
M1 ANN Alan Webb, 207 Evesham Road, Redditch, B97 5EN
M1 ANO C Worlledge, 181 Roselands Drive, Paignton, TQ4 7RN
MM1 ANP John Tobias, Gowanpark House, Gowanpark, Cumnock, KA18 2NZ
M1 ANQ D Hirst, 66 Turncroft Lane, Stockport, SK1 4AB
M1 ANR R Hall, 12 West Lane, Edwinstowe, Mansfield, NG21 9QT
M1 ANT D Simcock, 51 Broadway, Stockport, SK2 5SF
M1 AOB R Pentney, 11 Beech Park, Holsworthy Beacon, Holsworthy, EX22 7NB
M1 AOD V Bolger, Little Annaside, Bootle, Millom, LA19 5XL
M1 AOF A Wainwright, 23 Old School House, Shotley Gate, Ipswich, IP9 1QP
M1 AOG S Moriarty, 31 Guernsey Way, Banbury, OX16 1UE
M1 AOL J Pepper, 16 Chartwood, Loggerheads, Market Drayton, TF9 4RJ
M1 AOR S Darrigan, 76 Mersey Road, Birkenhead, CH42 1LW
M1 AOU Russell Pinchen, 9 Orwell Close, Swindon, SN25 3LZ
M1 AOX Jonathan O'Neill, 4 Heathlands Road, Little Sutton, Ellesmere Port, CH66 5PB
M1 APB S Sanders, 52 Hazelwood Road, Callington, PL17 7EU
M1 APC James Hulme, 28 Chapel Close, Gunnislake, PL18 9JB
M1 APF J Thompson, 8 Roman Drive, Leeds, LS8 2DR
M1 APH P Hildebrand, 82 Reed Drive, Redhill, RH1 6TB
M1 APL A Speakman, 12 Allerton Avenue, Leeds, LS17 6RF
M1 APQ Martin hodson, 93 Deer Park Road, Fazeley, Tamworth, B78 3SZ
M1 APS Colin Stuart, 5 Deloraine Court, Hawick, TD9 7QE

M1 APT I Waterhouse, Little Bracken, 9 Willingdon Drove, Eastbourne, BN23 8AL
M1 APX M Simpson, 19 Owens Quay, Bingley, BD16 4DX
M1 AQI M Sanderson, 14 Hazelwood Avenue, York, YO10 0PD
M1 [illegible]
M1 AQP C Chapman, 9 Edinburgh Avenue, Sawston, Cambridge, CB22 3DW
M1 AQX A Pell, 1 Oak Grove, Daventry, NN11 0XG
M1 AQY G Cottam, 26 Ayton Court, Bedlington, NE22 6NS
M1 ARF D Riley, 9 Century Avenue, Mansfield, NG18 5EE
M1 ARH William Mountford, 3 Spurstow Close, Prenton, CH43 2NQ
M1 ARI Michael Axford, Flat 32, Brampton Tower, Southampton, SO16 7FB
M1 ARL D Edwards, 28 Solingen Estate, Blyth, NE24 3EP
M1 ARM D Rees, Y Coed, Tan Lan Hill, Holywell, CH8 9JB
M1 ARS Harry Cawley, 11 Cleveland Way, Winsford, CW7 1QL
M1 ART F Melhuish, Alverdean, Mile End Road, Coleford, GL16 7QD
M1 ARU R Stanley, 113 Upper Brents, Faversham, ME13 7DL
M1 ARX S Russell, 13 Burdett Road, Crowborough, TN6 2EN
MI1 ASN Philip McMahon, 26 Ballycraigy Road, Newtownabbey, BT36 5ST
M1 ASR G Jefferies, 6 Main Street, Skipsea, Driffield, YO25 8SJ
M1 ASS T Summerfield, 48 Sandells Avenue, Ashford, TW15 1AL
M1 ASV A Evans, 14a New Road, Tiptree, Colchester, CO5 0HJ
M1 ATA A Betts, 96 Maidstone Road, Paddock Wood, Tonbridge, TN12 6DX
M1 ATB G Gale, 2 Manston Crescent, Crossgates, Leeds, LS15 8QZ
M1 ATC AIR TRAINING CORPS c/o R Courtney, 5 Bute Close, Highworth, Swindon, SN6 7HN
M1 ATI R Proctor, 11 Bedford Rise, Winsford, CW7 1NE
M1 ATJ P Whitby, 90 Manor Road, Martlesham Heath, Ipswich, IP5 3SY
M1 ATP A Plitsch, 64 Oxford Road, Lowestoft, NR32 1TP
MM1 ATR L Robinson, 17 Burn Brae Avenue, Westhill, Inverness, IV2 5RG
M1 ATU M Bloss, 20 Barry Walk, Brighton, BN2 0HP
MM1 ATY D Shirley, 17 Carlaverock Terrace, Tranent, EH33 2PL
MM1 AUF C McClintock, 13 St. Andrews Drive, Gourock, PA19 1HY
MM1 AUG M McClintock, 30 Findhorn Road, Inverkip, Greenock, PA16 0HX
M1 AUH A Easton, 1 Wood Street, Warrington, WA1 3AY
MI1 AUI V Hughes, 7 Craiglands Manor, Newtownabbey, BT36 5FG
M1 AUK B Pittaway, 66 Montrose Avenue, Leamington Spa, CV32 7DY
M1 AUN J Yarnall, 85 Wombourne Park, Wombourne, Wolverhampton, WV5 0LX
M1 AUO D Eady, The Rectory, Rectory Lane, Cheltenham, GL51 9RD
M1 AUP Alan Bailey, 47 Whiteridge Road, Kidsgrove, Stoke-on-Trent, ST7 4TH
M1 AUR M Garlick, 59 Foundry Avenue, Leeds, LS9 6BY
MW1 AUV P Davies, 4 Caradog Place, Townhill, Swansea, SA1 6NH
M1 AUW Tim Roberts, 109 Gordon Avenue, Norwich, NR7 0DS
M1 AUY G West, 15 Hildens Drive, Tilehurst, Reading, RG31 5HW
M1 AUZ G Wright, School House Farm, The Gravel, Mere Brow, Preston, PR4 6JX
M1 AVB Jason Pidgeon, 5 Brook Terrace, Church Street, Seaton, EX12 4AG
MI1 AVH T Allingham, 17 Coagh Road, Cookstown, BT80 8RL
M1 AVM S Cooper, 93 Langton Road, Norton, Malton, YO17 9AE
M1 AVU Geoffrey Purrier, Archways, Forge Hill, Lydbrook, GL17 9QS
M1 AVV Simon Linney, 3 Severn Road, Walney, Barrow-in-Furness, LA14 3TS
M1 AVW R Hedges, Flat 6, Devington Court, Falmouth, TR11 4PD
M1 AWC M Worrall, 15 Whitegate Drive, Bolton, BL1 8SF
MI1 AWM Darren Reid, 179 Melmount Road, Sion Mills, Strabane, BT82 9LA
M1 AWN Jolyon Sanders, 135 Windmill Avenue, Kettering, NN15 7DZ
M1 AWS A Jones, 35 St. Marys Close, Aspull, Wigan, WN2 1RL
M1 AWT P McCarthy, 43 Bodnant Road, Llandudno, LL30 1LT
MM1 AWV Robert Lynch, 21 Carnoustie Avenue, Gourock, PA19 1HF
M1 AWX S Yendell, 35 Chester Road, Newquay, TR7 2RH
M1 AXD P Gartell, 1 Springfields, Richards Castle, Ludlow, SY8 4EP
M1 AXE Garry Ecclestone, 6 Laurel Drive, Rugby, CV22 7TL
M1 AXG D Tucker, 2 New Cottages, Bill Hill, Wokingham, RG40 5QU
M1 AXM Dominic Russell, 15 Morton Way, Boxfield Road, Axminster, EX13 5LE
M1 AXP B Cornall, Fernholm, Taylors Lane, Preston, PR3 6AB
M1 AXX R Brotherton, Richanchor, Mill Lane, York, YO23 2UL
M1 AYA P Booth, 61 Coalpit Lane, Rugeley, WS15 1EW
M1 AYC A Booth, 35 Gillamore Drive, Whitwick, Coalville, LE67 5PA
MI1 AYL Brendan Byrne, 6 Holymount Road, Gilford, Craigavon, BT63 6AT
M1 AYN J Hewlett, 28 Coombs Road, Coleford, GL16 8AY
M1 AYR K Miller, 15a Holly Close, Cherry Willingham, Lincoln, LN3 4BH
M1 AYU Richard Freeman, 6 Sutton Road, Leverington, Wisbech, PE13 5DW
M1 AZA M Gould, 36 Wistaria Road, Wisbech, PE13 3RH
M1 AZB J Titterton, Longfield Farm, Ifield Road, Horley, RH6 0DR
M1 AZF Shaun Arbuckle, 8 Hungerford Road, Calne, SN11 9BG
M1 AZG A Ruston, 42 The Straits, Dudley, DY3 3BH
M1 AZI S Dunlop, 13 Queen St., Nantyglo, Brynmawr, NP3 4LZ
M1 AZJ A Bottrell, 36 Tremodrett Road, Roche, St. Austell, PL26 8JA
M1 AZM P Jefferson, 20 Buckstone Grove, Leeds, LS17 5HW
M1 AZO K Austen, 12b Downs Road, Folkestone, CT19 5PW
M1 AZQ Alan Beale, Flat 2, Brabstone House, Medway Drive, Greenford, UB6 8LN
M1 AZR GWENT RAYNET GROUP c/o R Snelling, 91 Oakfield Road, Newport, NP20 4LP
M1 AZV Robert Aston, 2 Brockwood Crescent, Keyworth, Nottingham, NG12 5HQ
M1 AZY J Drummond, 60 Park Lane, Exeter, EX4 9HP
M1 BAA Mark Crow, 100 Bois Moor Road, Chesham, HP5 1SS
M1 BAC J Booth, 35 Gillamore Drive, Whitwick, Coalville, LE67 5PA
M1 BAD David Redding, 11 Camley Gardens, Maidenhead, SL6 5JW
M1 BAI A Saunders, 128 Foxcroft Drive, Wimborne, BH21 2LA
MW1 BAJ James Alexander, 34 Trefelin Street, Port Talbot, SA13 1DQ
M1 BAN Tamara Baldwin, 17 Warn Crescent, Oakham, LE15 6LZ
M1 BAR BAR-PACKERS CONTEST GROUP c/o N Roscoe, 35 Kenilworth Road, Cheadle Heath, Stockport, SK3 0QL
M1 BAS C Bastin, 42 Peterborough Road, Exeter, EX4 2EG
M1 BAV G Noble, 96 Foxroyd Lane Estate, Dewsbury, WF12 0BD
M1 BBB P Marshall, 1 Prospect Cottages, St. Anns Chapel, Gunnislake, PL18 9HH
M1 BBH C Tan, 62a Jalan Sepah Puteri 5/6, Kota Damansara, Selangor, Malaysia
M1 BBR Marcus Deglos, 405 Armidale Place, Bristol, BS6 5BQ
M1 BBS Daniel Hawkes, 23 Gloucester Avenue, Margate, CT9 3NN
M1 BBU G Price, 118 Broadstone Road, Heaton Chapel, Stockport, SK4 5HS
M1 BCB D Ball, 14 Ayot Path, Borehamwood, WD6 5BJ
M1 BCM Jeffrey Worthing, 27 Mayfield Close, Shrewsbury, SY1 4BF
M1 BCR Andrea Richards, 3 Beeston Close, Watford, WD19 6LF
M1 BCU Andrew Howe, 6 Crabapple Close, Wymondham, NR18 0XT
M1 BCY T Keeler, 72 Grotton Road, Selsey, Chichester, PO20 0[illegible]
M1 BCZ Roger Craggs, 1 Hylton Court, Bowmonts Road, Tadley, RG26 3SH
M1 BDD E Reynolds, Cloonagh, Ballinagore, Ireland
M1 BDH Tim Woods, 7 Holywell Gutter Lane, Hereford, HR1 1XA
M1 BDJ G Hamlin, Down Farm Bungalow, Stockbridge, SO20 8EA
M1 BDL A Statham, 24 Fulton Close, High Wycombe, HP13 5SP
M1 BDO W Lodeweegs, 68 Totterdown Lane, Weston-Super-Mare, BS24 9NJ
M1 BDR ESSEX RAYNET (BRAINTREE DISTRICT) c/o Graham Farrell, 95 Washington Road, Maldon, CM9 6JF
M1 BDS P Colwell, 56 Hamelin Road, Gillingham, ME7 3EX
MW1 BDV W Davis, Cartref, Blaenannerch, Cardigan, SA43 1SN
M1 BDY Valerie Beard, 13 Mayesford Road, Romford, RM6 4NU
M1 BEC ETON COLLEGE ARC c/o M Wilcockson, Conybeare House, Willowbrook, Windsor, SL4 6HL
M1 BED Bedford and District ARC c/o Robert Leask, 80 Mill Road, Sharnbrook, Bedford, MK44 1NP
M1 BEO D Tatlow, Mulberry House, Bettys Grave, Cirencester, GL7 5ST
M1 BEP A Amos, 118 Mount Hill Road, Bristol, BS15 8QR
M1 BEQ A Strange, 8 Tregarn Close, Langstone, Newport, NP18 2JL
M1 BEW C Sampson, Stable Cottage, Lower House Farm, Welshpool, SY21 8LA
M1 BEX G Olsen, 36 Bluebell Way, South Shields, NE34 0BZ
MM1 BFE James Murray, 27 Wellpark Road, Banknock, Bonnybridge, FK4 1TP
M1 BFF Andrew Buckley, 68 Abingdon Way NE, Calgary Ab, Canada, T2A6R8
M1 BFG H Tribe, The Paddock, Wix Hill, Leatherhead, KT24 6ED
M1 BFI Zane Billington, 63 Westfield Road, Dunstable, LU6 1DN
M1 BFO P Aplin, 30 Cheviot Drive, Charvil, Reading, RG10 9QD
M1 BFR Antony Day, 21 Coronation Road, Prestbury, Cheltenham, GL52 3DA
M1 BFV F Richardson, 3 Carl Moult House, Regent Street, Swadlincote, DE11 9PH
M1 BFX John Belcher, 101 Colne Drive, Romford, RM3 9LA
M1 BFY A Davey, 34 Monkswood, Littleport, Ely, CB6 1JD
M1 BGF M Shearman, 4 Mcdonough Close, Fitton Hill, Oldham, OL8 2PD
M1 BGK J Lewis, 12 Eastleigh Road, Staple Hill, Bristol, BS16 4SQ
M1 BGL George Langford, 22 Kensington Way, Oakengates, Telford, TF2 6NA
M1 BGS M Elvers, 3 St. Michaels Road, Maidstone, ME16 8BS
M1 BGT Robert Williams, 19 Venice Close, Chellaston, Derby, DE73 5BX
M1 BGY I Townson, 53 Brompton Road, Bradford, BD4 7JD
M1 BHC M Lee, 11 Sturrocks, Vange, Basildon, SS16 4PQ
M1 BHE B Vickers, 17 Linden Close, Dewsbury, WF12 8PL
M1 BHN J Chambers, 16 Wood Way, Huntington, York, YO32 9QG
M1 BHO Richard Hopkins, 15 Station Drive, Dalbeattie, DG5 4FA
M1 BHP D Hartley, 10 Tamworth Grove, Clifton, Nottingham, NG11 8JA
M1 BHW P Frier, 58 Parklands, Rochford, SS4 1SH
M1 BHZ T Shepherd, 40 Pheasant Way, Cirencester, GL7 1BL
M1 BIB P Brooke, 34 Park Road, Burwell, Cambridge, CB25 0ES
M1 BIG Tom Read, 57 Ollard Avenue, Wisbech, PE13 3HF
M1 BIK Christopher Blackmur, 9 Cameron Close, Chatham, ME5 0DD
M1 BIL C Pugh, 16 Park Gate, Somerhill Road, Hove, BN3 1RL
M1 BIX Gregory Perrins, 1 Cornhill Gardens, Leek, ST13 5PZ
M1 BIY Jonathan Laker, 33 Rowan Road, Havant, PO9 2UX
M1 BJB Sisir Mandal, 9 Bron y Dre, Wrexham, LL13 7RW
M1 BJC P Marshall, 75 Drewstead Road, London, SW16 1AA
M1 BJE Stephen Robinson, 140 The Street, Kirtling, Newmarket, CB8 9PD
MM1 BJG Osvaldo Ferula, 4 Castledyke Road, Carstairs, Lanark, ML11 8SU
MM1 BJP Allan Mcdermid, 38 Steading Drive, Alexandria, G83 9EB
M1 BJS G Ausher, 94 New Road, Ditton, Aylesford, ME20 6AE
MM1 BJT D Smith, 1076 Aikenhead Road, Glasgow, G44 4TJ
MM1 BJZ Paul Fraser, 8 Devon Walk, Cumbernauld, Glasgow, G68 9NT
M1 BKE P Hewlett, 28 Coombs Road, Coleford, GL16 8AY
M1 BKF W Hill, 492 Earlham Road, Norwich, NR4 7HP
M1 BKI James Carins, 26 Roman Way, St. Margarets-at-Cliffe, Dover, CT15 6AH
M1 BKL Paul Coddington, 2 Canal View, Chemistry, Whitchurch, SY13 1BZ
M1 BKQ Terence Beadman, Cottage Farm, Shackerstone, Nuneaton, CV13 6NL
M1 BKS M Kevern, Wheal Bal, Trewellard, Penzance, TR19 7SP
M1 BKW S Plant, 17 New Road, Driffield, YO25 5DJ
MW1 BLE Colin Beech, 9 Wesley Court, Pembroke Dock, SA72 6NE
M1 BLJ S Brion, 165 Kings Head Hill, London, E4 7JG
M1 BLO P Hoggard, 41 Malpas Close, Bransholme, Hull, HU7 4HH
M1 BLW E Banks, 165 Burstall Hill, Bridlington, YO16 7NH
M1 BLX Paul Buxton, Bower House, Thornham Road, Eye, IP23 8HP
MI1 BLZ D Kyle, Sea Breezes, Rathlin Island, Ballycastle, BT54 6RT
M1 BMC K Termie, 14 Hollins Lane, Marple Bridge, Stockport, SK6 5BB
M1 BMD W Curtis, 5 Cambridge Road, Kesgrave, Ipswich, IP5 1EN
MM1 BMK Ian Mitchell, Carradale Yondertonhill, Hatton, Peterhead, AB42 0RE
M1 BMQ John Waters, 71 Sixth Avenue, Blyth, NE24 2SU
M1 BMR Chris Robinson, 76 Ingrams Way, Hailsham, BN27 3NX
M1 BMU Elizabeth Woodward, 6 Lang Road, Huntington, York, YO32 9SD
M1 BMV J Fox, Stonecroft, Horley, Banbury, OX15 6BJ
M1 BMW J Burrill, 3 Town Farm Close, Pinchbeck, Spalding, PE11 3SG
M1 BNG Richard Smith, 45 The Avenue, Mortimer Common, Reading, RG7 3QU
M1 BNH Patrick Walton, 106 Aberford Road, Oulton, Leeds, LS26 8SN
M1 BNI M Clarke, 1 Burbury Close, Bedworth, CV12 8DU
M1 BNK A Wood, 1 Minch Road, Hartlepool, TS25 3QY
MI1 BNO B Bonnar, 49 Cullycapple Road, Aghadowey, Coleraine, BT51 4AR
M1 BNR HOLSWORTHY COMMUNITY COLLEGE c/o G Forster, 33 Deer Valley Road, Holsworthy, EX22 6DA
MW1 BNY B Fitzpatrick, 1 Pistyll Newydd Mynyddygarreg, Kidwelly, SA17 4NW
M1 BOA Gary Heard, The Saddlers, The Street, Norwich, NR11 7PD
M1 BOB R Allen, 43 Vowell Close, Bristol, BS13 9HS
M1 BOD P Hanfrey, 49 Allotment Road, Niton, Ventnor, PO38 2DZ
MI1 BOE A Prenter, 5 Knockview Gardens, Newtownabbey, BT36 6UA
M1 BOL D Harding, Po Box 11755 Apo, Grand Cayman, Cayman Islands
M1 BOP M Riley, 63 Clifford Road, Ipswich, IP4 1PJ
M1 BOZ A Thompson, 26 Balmoral Avenue, Clitheroe, BB7 2QH
M1 BPD Alexander Collins, Old Orchards, Romsey Road, Stockbridge, SO20 6PR
M1 BPK John Bloor, Hillview House, Whitegates, Bromyard, HR7 4ES
M1 BPN A Burchell, 25 Cherbury Close, London, SE28 8PG
M1 BPS A Wellman, Lyndale, Northleach, Cheltenham, GL54 3JJ
M1 BPU Brian Lake, 1 Eunice Grove, Chesham, HP5 1RL
M1 BPW P Williams, 20 Downs View Road, Westbury, BA13 3AQ
M1 BPY D Dixey, 102 Blackberry [illegible], Chichester, [illegible]
M1 BQC S Sparks, 36 Tormynton Road, Worle, Weston-Super-Mare, BS22 9HT
M1 BQD K Sparks, 29 Pennycress, Weston-Super-Mare, BS22 8QH
M1 BQE C Sparks, 36 Tormynton Road, Worle, Weston-Super-Mare, BS22 9HT
M1 BQF John Stewart, 45 The Cross, Wivenhoe, Colchester, CO7 9QH
M1 BQM G Preedy, 12a Grange Court, Prescot Road, Stourbridge, DY9 7LA
M1 BQO Endaf Buckley, 3 Cae Eithin, Minffordd, Penrhyndeudraeth, LL48 6EF
M1 BQS F Gibson, 125 Chelveston Drive, Corby, NN17 2QJ
M1 BQT F Lloyd, 2 Larkfield, Cholsey, Wallingford, OX10 9QT
M1 BQU Simon Daniels, 9 Beechwoods, Burgess Hill, RH15 0DE
M1 BQW V edwards, Elder Cottage, Skeyton Common, Norwich, NR10 5BB
M1 BQY Trowbridge & District ARC c/o Ian Carter, 12 Bobbin Lane, Westwood, Bradford-on-Avon, BA15 2DL
M1 BQZ W Rowley, 6 Sea King Crescent, Colchester, CO4 9RJ
MI1 BRA BELFAST ROYAL ACADEMY AMATEUR c/o N Moore, 164 Ardenlee Avenue, Belfast, BT6 0AE
MI1 BRS R Dickey, 8 Coachmans Way, Hillsborough, BT26 6HQ
M1 BRU D Pope, 32 Barn Crescent, Newbury, RG14 6HD
M1 BRX D Seddon, 22 Newton Road, Lowton, Warrington, WA3 1EB
M1 BRY John Fisher, 18 The Smooting, Tealby, Market Rasen, LN8 3XZ
M1 BRZ D Lee, 53 Portmellon Park, Mevagissey, St. Austell, PL26 6XD
M1 BSB W Charman, 23 The Vineries, Wimborne, BH21 2PU
M1 BSE Jeffrey Wharton, Flat 12, Brent Court, Leicester, LE3 2XQ
M1 BSF Steven Ogden, 2 Eagle Close, Heysham, LA3 2LY
M1 BSI C Bowen, 45 Morris Road, Nottingham, NG8 6NE
M1 BSM Gunther Meyer, 447-449 Manchester Road, Stockport, SK4 5DJ
M1 BSN J Riley, Peacehaven, Great Street, Stoke-Sub-Hamdon, TA14 6SH
M1 BSO B Ford, 7 Courtwick Road, Wick, Littlehampton, BN17 7NE
M1 BSP Mark Ford, 27 The Street, Rustington, Littlehampton, BN16 3PA
M1 BSU Tony Osborne, 134 Merridale Road, Wolverhampton, WV3 9RJ
M1 BSV M Black, 37 Castle Street, Nelson, BB9 0TW
M1 BSX Alexander May, 18 Pennant Hills, Bedhampton, Havant, PO9 3JZ
M1 BSY W Ginger, Wenick House, 152 Hawks Road, Hailsham, BN27 1NA
M1 BTA G Haines, 12 Tyn Rhos Estate, Gaerwen, LL60 6HL
M1 BTD D Wilkinson, 89 The Northern Road, Liverpool, L23 2RD
M1 BTI David Piggin, 26 Moss Lane, Timperley, Altrincham, WA15 6SZ
MW1 BTM C Jones, 17 Grove House Court, Pontygwaith, Ferndale, CF43 3LJ
M1 BTO N Martin, 28 Churchmead Close, Lavant, Chichester, PO18 0AY
M1 BTR J Charles, Ash Tree, Priory Lane, Louth, LN11 8SP
M1 BTU P Mason, 23 Glade Close, Little Billing, Northampton, NN3 9SN
M1 BUC A Benson, 12 Longfellow Road, Caister-on-Sea, Great Yarmouth, NR30 5RH
M1 BUG M Dugdale, 57 Macauley Avenue, Blackpool, FY4 4YF
M1 BUJ I Lewis, 15 Margaret Close, Thurmaston, Leicester, LE4 8GL
M1 BUL D Atkins, 79 Roe Lane, Southport, PR9 7HR
MW1 BUN David Luke, 56 Maerdy Park, Pencoed, Bridgend, CF35 5HX
M1 BUP N Foyen, Sherborne Valley Kennels, Sherborne, DT9 4SZ
M1 BUQ I Houghton, 39 Fiskerton Way, Oakwood, Derby, DE21 2HY
M1 BUU Colin Evans, 10 Albion Street, Cross Roads, Keighley, BD22 9EB
M1 BUX Steven Leaker, 166 Beckett Road, Doncaster, DN2 4BB
M1 BVI K Worrall, 66 Elm St., Hollingwood, Chesterfield, S43 2LH
M1 BVP M Taylor, 56 Newgate Lane, Mansfield, NG18 2LQ
M1 BVT Michael Beardsley, 121 Wood Road, Lower Gornal, Dudley, DY3 2LR
M1 BVX S Simpson, 20 Staveley Grove, Keighley, BD22 7DH
M1 BWH F Laycock, 8 Melling Way, Liverpool, L32 1TP
M1 BWJ J Bottle, 15b Elizabeth House, Alexandra Street, Maidstone, ME14 2BX
M1 BWN S Jarrett, 17 Wolmers Hey, Great Waltham, Chelmsford, CM3 1DA
M1 BWR E Oakley, Brooklands Lodge, Park View Close, Ventnor, PO38 3EQ
M1 BWS A Kerr, 14 Glamorgan Close, St. Helens, WA10 3XT
M1 BWZ P Quick, 70 Trent Avenue, Maghull, Liverpool, L31 9DE
M1 BXC A Blakeney, 7 Gayton Road, Eastcote, Towcester, NN12 8NG
M1 BXD Mark Cross, Hans Napp, Broadhempston, Totnes, TQ9 6BD
M1 BXF Gavin Nesbitt, 19 Ditton Green, Woodditton, Newmarket, CB8 9SQ
M1 BXJ M Ellis, 62 Peterborough Road, Crowland, Peterborough, PE6 0BA
M1 BXM M Forster, 10 Weaver Valley Road, Winsford, CW7 3JU
M1 BXO D Ellis, 67 Ambersham Crescent, East Preston, Littlehampton, BN16 1AJ
M1 BXQ J Squire, 57 The Avenue, Chinnor, OX39 4PE
M1 BXU D Napper, 47 Mallard Walk, Sidcup, DA14 6SG
MW1 BXX Malcolm Broxtom, 4 Owen Street, Pembroke, SA71 4EP
M1 BYG A Chatel, 17 Star Holme Court, Star Street, Ware, SG12 7EA
M1 BYH Andrew Moss, 3 Haddon Close, Macclesfield, SK11 7YG
M1 BYI P Stockley, 41 Fairway Court, Cleethorpes, DN35 0NN
M1 BYQ Robert Josephs, 113 Patrick Street, Grimsby, DN32 0PQ
M1 BYT H Bloomfield, 49 Oak Crescent, Garforth, Leeds, LS25 1PW
M1 BZF S Gore, 10 Cambridge Street, Guiseley, Leeds, LS20 9AU
M1 BZG L Haybould, 7 Tenbury House, Highfield Lane, Halesowen, B63 4RN
M1 BZI F Lee, 26 Lache Hall Crescent, Chester, CH4 7NF
M1 BZJ P Buer, 71 Belvedere Road, Ashton-in-Makerfield, Wigan, WN4 8RX
M1 BZK David Riches, 15 Ashton Way, Saltash, PL12 6JE
M1 BZR D Wright, 18 Allensway, Stanford-le-Hope, SS17 7HE
M1 BZZ P Lonsdale, 14 Dunne Close, Wirral, CH63 9YJ
MM1 CAC Graeme Mathers, 46 Castle Street, Fraserburgh, AB43 9DH
M1 CAE R Naylor, 93 Woodland Road, Halton, Leeds, LS15 7DN
M1 CAH Christopher Bennet, 32 Angelica Avenue, Stotfold, Hitchin, SG5 4HH
M1 CAK P Prior, 17 Layton Avenue, Malvern, WR14 2ND
M1 CAO R Cameron, 23 Ravenscroft, Hook, RG27 9NP
M1 CAR R Griffiths, 22 Quarry Road, Hereford, HR1 1SS
M1 CAX K Biggs, 22 Wallingford Close, Bracknell, RG12 9JE
M1 CAY M Harris, Weathercock Cottage, East Mersea Road, Colchester, CO5 8SL
M1 CBC Patrick Wainwright, 5 Pulcroft Road, Hessle, HU13 0ND
M1 CBH J Tomlins, 3 Turnstone Crescent, Mansfield, NG18 3SP
M1 CBI R Bird, 37 Beachwood Avenue, Kingswinford, DY6 0HL
M1 CBO Roger Appleby, 3 St. Judes Way, Burton-on-Trent, DE13 0LR
M1 CBT C Taylor, 48 Northdown Park Road, Cliftonville, Margate, CT9 3PT
M1 CBU R Stansfield, Sundene, 157 Hollin Lane, Wakefield, WF4 3EG

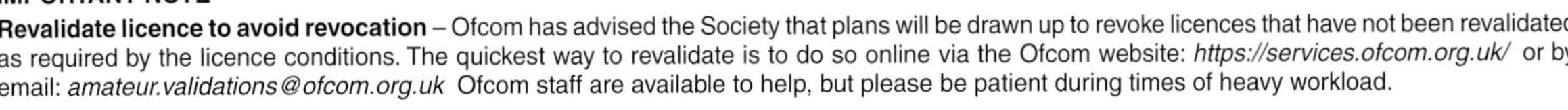
**IMPORTANT NOTE**

**Revalidate licence to avoid revocation** – Ofcom has advised the Society that plans will be drawn up to revoke licences that have not been revalidated as required by the licence conditions. The quickest way to revalidate is to do so online via the Ofcom website: *https://services.ofcom.org.uk/* or by email: *amateur.validations@ofcom.org.uk* Ofcom staff are available to help, but please be patient during times of heavy workload.

M1 CBV G Leeder, 89 Chesterton Avenue, Harpenden, AL5 5ST
M1 CBY D Howse, 24 Sandown Road, Bishops Cleeve, Cheltenham, GL52 8BY
M1 CBZ A Howse, 24 Sandown Road, Bishops Cleeve, Cheltenham, GL52 8BY
M1 CCA B Thomas, Hazel Mount, Lockhams Road, Southampton, SO32 2BD
M1 CCF Michael Buckley, Springfield, 12 Ranmore Avenue, Croydon, CR0 5QA
M1 CCG George Smales, 6 Chestercourt Cottages, Camblesforth, Selby, YO8 8HZ
M1 CCL Roger Chantler, 68 Chilton Lane, Ramsgate, CT11 0LQ
M1 CCN P Terry, 13 Lamsey Lane, Heacham, King's Lynn, PE31 7LA
M1 CCQ A Bantoft, 110 St. Peters Road, Wiggenhall St. Peter, King's Lynn, PE34 3HF
MM1 CCR A Annan, Easter Cottage, Blairlogie, Stirling, FK9 5PX
MI1 CCT Bronwin Vaughan, 29 Crew Road, Victoria Bridge, Strabane, BT82 9LS
MI1 CCU Ian Morrow, 90 Bracky Road, Sixmilecross, Omagh, BT79 9PH
M1 CCX P Tully, 9 Beechcroft, Rothbury, Morpeth, NE65 7RA
M1 CCY R Coxon, 7 Elworthy Road, Longhoughton, Alnwick, NE66 3LS
M1 CDJ F Waite, 91 Priors Hill, Wroughton, Swindon, SN4 0RL
M1 CDL David Hall, 4 Burns Close, Peterborough, PE1 3JJ
M1 CDP A Mcdade, 20 Westbury Walk, Corby, NN18 0AE
M1 CDQ R Damm, 18 Mayfair Crescent, Waltham, Grimsby, DN37 0EE
M1 CDT Charles Behan, St Chads Close, Hornjnglow, Burton-on-Trent, DE13 0ND
M1 CDV A McEwen, 23 Cobholm Road, Great Yarmouth, NR31 0BU
M1 CDX K Leach, 6 Tewkesbury Avenue, Blackpool, FY4 2NF
M1 CEA Terry Gladman, 43 Queens Road, New Malden, KT3 6BY
M1 CEC C Thompson, 27 Sycamore Road, Chorley, PR6 0JD
M1 CEM Brian Harper, 36 Percy Street, Oswaldtwistle, Accrington, BB5 4LY
M1 CEW Michael Trueblood, 44 Wallgate Road, Liverpool, L25 1PR
M1 CEY J Page, 23 Meadowsweet Close, Thatcham, RG18 4DS
M1 CFA K thorley, Helston, Mountain, Holyhead, LL65 1YR
MM1 CFC David Goodfellow, 4 West Grange Street, Monifieth, Dundee, DD5 4LD
MW1 CFE Anthony Evans, Flat 5, Nanthir Lodge, Nanthir Road, Bridgend, CF32 8BL
M1 CFG S Appleby, 2 Stella Farm, Narborough, King's Lynn, PE32 1HY
M1 CFW Richard Powell, 151 Bury Hill Close, Anna Valley, Andover, SP11 7LL
M1 CFZ A Moore, Wheatcroft House, Wheatcroft, Matlock, DE4 5GU
M1 CGB M Jones, 69 Brompton Drive, Brierley Hill, DY5 3NZ
M1 CGF Leslie Griffiths, 91 Worrall Road, wadsley, Sheffield, S8 4BA
M1 CGI A Fishwick, Causeway House Farm, Coppice Lane, Chorley, PR6 9DA
M1 CGJ S Fishwick, Causeway House Farm, Coppice Lane, Chorley, PR6 9DA
M1 CGM Kevin Foster, 48 The Street, Newbourne, IP12 4NY
M1 CGO N Hewgill, 40 Lime Tree Place, Stowmarket, IP14 1BT
M1 CGQ Nicholas Mulryan, Flat 31, Chatsworth Lodge, Buxton, SK17 6XX
M1 CGR P Bowles, 25 North Down, Staplehurst, Tonbridge, TN12 0PG
M1 CHF B Crossley, 19 Westwick Close, Walsall, WS9 9EA
M1 CHM M Bailey, 10 Argyll Avenue, Doncaster, DN2 6LG
MM1 CHQ David Wildridge, 1 Glamis Gardens, Dalgety Bay, Dunfermline, KY11 9TD
M1 CHS J Arundale, 29 Deepdale Avenue, Scarborough, YO11 2UQ
M1 CHU W Llewellyn, 105 Sandford Avenue, Church Stretton, SY6 7AB
M1 CIE J Morgan, 15 Town Head, Dearham, Maryport, CA15 7JW
M1 CIG S Spurr, 12 Rushmoor Close, Rickmansworth, WD3 1NA
M1 CIJ J Turk, 25 Berkeley Road, Newbury, RG14 5JE
M1 CIM B Kidane, Po Box 10130, Addis Ababa, Ethiopia
MM1 CIR Peter Merckel, 1 Mortimer Court, Dalgety Bay, Dunfermline, KY11 9UQ
M1 CIS P Jameson, 1 White Acres Road, Mytchett, Camberley, GU16 6EY
M1 CJB L Holyer, 45 Crabble Hill, Dover, CT17 0RX
M1 CJE Andrew Eastland, 4 Bergamot Close, Manton, Marlborough, SN8 4HT
M1 CJF R Barrett, Upland, Tidings Hill, Halstead, CO9 1BJ
M1 CJM R Wallis, 26 Heather Bank, Osbaldwick, York, YO10 3QH
M1 CJN S Ogiela, 107 Osbaldwick Lane, York, YO10 3AY
M1 CJT Anthony Brotherhood, 5 Longcliffe Road, Shepshed, Loughborough, LE12 9LW
M1 CJX Tony Cotterell, 52 The Crofts, Hatch Warren, Basingstoke, RG22 4RF
M1 CJZ A Roberts, 3 Jaynes Close, Banbury, OX16 9ES
M1 CKA P Geier, 24 Deirdre Avenue, Wickford, SS12 0AX
M1 CKJ G Reeds, 26 Holme Leaze, Steeple Ashton, Trowbridge, BA14 6EH
M1 CKK S Lowe, 2 Bryn Eglwys, Llanfachreth, Dolgellau, LL40 2EF
M1 CKO S Chapman, 9 Edinburgh Avenue, Sawston, Cambridge, CB22 3DW
M1 CKQ B Roth, 10 Bernard Road, Brighton, BN2 3EQ
M1 CKU malcolm brooks, 13 Weatherside, Blaydon-on-Tyne, NE21 5QL
MM1 CKW A Johnson, 20 Falkirk Road Glen Village, Falkirk, FK1 2AG
M1 CKZ M O'Dwyer, Westlands, 19 High Street, Southminster, CM0 7AY
M1 CLI M Poulter, 26 West Crescent, Duckmanton, Chesterfield, S44 5HE
M1 CLO G Capon, 24 Beech Drive, Brackley, NN13 6JH
MM1 CLR R Vause, 100 Carmuirs Avenue, Camelon, Falkirk, FK1 4PB
M1 CLW E Brown, 76 West Park Drive, Swallownest, Sheffield, S26 4UY
M1 CLX A Robinson, 11 The Crescent, Whalley, Clitheroe, BB7 9JW
M1 CLZ J Taylor, 307 Birmingham Road, Lickey End, Bromsgrove, B61 0ER
M1 CML K Grout, 36 Churchill Road, Exmouth, EX8 4DN
M1 CMM Jeffrey Timmis, Upper House, Abdon, Craven Arms, SY7 9HX
M1 CMN Matt Curtis, 20 Alder Road, Folkstone, CT19 5BZ
M1 CMR Bryan Jarvis, 26 Longhouse Road, Halifax, HX2 8RE
MM1 CMU J Freeland, 12 Mccathie Drive, Newtongrange, Dalkeith, EH22 4BW
M1 CMW Roger Cottington, 3 Dickens Drive, East Malling, West Malling, ME19 6SJ
MJ1 CNB Neil Fryer, 25 Walter Benest Court, La Route des Quennavais, St. Brelade, Jersey, JE3 8NS
M1 CND Tim Wooldridge, 12 Redwood Avenue, Leyland, PR25 1RN
M1 CNE Matthew Howard, 18 Hydehurst Close, Crowborough, TN6 1EN
M1 CNG Philip Keeler, 13 Wynnstay, Oak Hall Park, Burgess Hill, RH15 0TD
M1 CNH J Gibb, 25 Ferndown Gardens, Cobham, KT11 2BH
M1 CNI G Lambley, Jasmic, Main Road, Spilsby, PE23 4BE
M1 CNJ F Manley, 37 Goodrington Close, Banbury, OX16 0DB
M1 CNK P Wilton, 217 Chamberlayne Road, Eastleigh, SO50 5HZ
M1 CNL Rodrick Tew, 66 St. Nicholas Estate, Baddesley Ensor, Atherstone, CV9 2EZ
M1 CNN D Hayward, 1 Elidyr Road, Newbridge, Newport, NP11 3EE
M1 CNP Clifford Robinson, 9 Chatsworth Avenue, Culcheth, Warrington, WA3 4LD
M1 CNS E Mathias, 17 St. Johns Terrace, Lewes, BN7 2DL
M1 CNX Stephen Russell, 11 Lowgate Avenue, Bicker, Boston, PE20 3DF
M1 CNY K Wilson, 12 New Street, Elworth, Sandbach, CW11 3JF
MW1 COB Kenneth Sands, 5 Ynysgau Street, Ystrad, Pentre, CF41 7UE
M1 COE Robert Alford, 24035 Gray Road, Box 103, Dennis, United States, 67341
M1 COJ Anthony Jones, 22 Wendover Avenue, Towyn, Abergele, LL22 9LP
M1 COL Colchester Radio Amateurs Club c/o Herbert Yeldham, 19 Wade Reach, Walton on The Naze, CO14 8RG
MJ1 COO Dennis Gallichan, Apartment 9, Millbrook Crescent, La Route de St. Aubin, Jersey, JE3 1LY
M1 COQ M Panton, Lavers, Preston Road, Sudbury, CO10 9QD
MM1 COS S Laepong, 29d Hill Street, Montrose, DD10 8AZ
M1 COV Sydney Rollinson, College Farm Cottage, Humber Lane, Hull, HU12 0UX
MW1 COY Stephen Beer, 4 Churchfields, Barry, CF63 1FP
M1 CPB Keith Ellison, 33 Priory Grove, Sunderland, SR4 7SU
M1 CPC F Moy, 86 Manningford Road, Birmingham, B14 5LX
M1 CPD F Marrai, 19 Hind Close, Chigwell, IG7 4EA
M1 CPL R Ramsay, Fairview, Briar Close, Hastings, TN35 4DP
MM1 CPP I Mcleary, 23 Dalhousie Road, Dalkeith, EH22 3AT
M1 CQC S Eglinton, 2 Victoria Road, Saltash, PL12 4DL
M1 CQF M Barnes, 58 Prince Street, Dalton in Furness, LA15 8EU
M1 CQI A Oxlade-Gotobed, 22 St. Peters Road, Basingstoke, RG22 6TD
M1 CQK Neil Ore, Willowdene, Rode Lane, Norwich, NR16 1NW
M1 CQL A Ore, Willowdene, Rode Lane, Norwich, NR16 1NW
M1 CQM B Suyat, 24 Lynmouth Road, London, E17 8AF
M1 CQN Barry Adkins, 4 Orion Close, Ward End, Birmingham, B8 2AU
M1 CQP K Blackwell-Chambers, 37 St. Johns Road, Oakley, Basingstoke, RG23 7JP
M1 CQR R Field, 3 Waveney Drive, Belton, Great Yarmouth, NR31 9JU
M1 CQS Graham Browne, 30 Dereham Road, Easton, Norwich, NR9 5EJ
M1 CQT A Wheeler, 60 Bredhurst Road, Gillingham, ME8 0PE
M1 CQU N Anderson, 19 Berrylands, Liss, GU33 7DB
M1 CQX M Jones, 44 Purley Road, Sunderland, SR3 1QS
M1 CRA WACRAL (World Association of Christian Radio Amate c/o Peter Jackson, 24 Woodfield Park, Walton, Wakefield, WF2 6PL
M1 CRE Stephen Woolley, 139 Prince of Wales Avenue, Flint, CH6 5JU
M1 CRF Neill Ovenden, 1 Bridge Cottages, Sea Lake Road, Lowestoft, NR32 3LQ
M1 CRL J Edwards, 44 Hunter Road, Norwich, NR3 3PY
M1 CRO Colechester Contest Group c/o J Lemay, Carlton House, White Hart Lane, Colchester, CO6 3DB
M1 CRP P Booth, 39 New Close, Eyam, Hope Valley, S32 5QX
M1 CRQ J Nuttall, 114 Plumstead Road, Norwich, NR1 4JX
M1 CRZ A Tudge, 7 Moreton Avenue, Whitefield, Manchester, M45 8GG
MI1 CSA J Higgins, 43 Temple Road, Garvagh, Coleraine, BT51 5BJ
M1 CSC S Swancutt, 100 Oundle Road, Birmingham, B44 8EN
M1 CSE Chris Childs, 43 Eastdale Road, Burgess Hill, RH15 0NJ
M1 CSG G Millsott, 11 Kingsthorn Road, Poundbury, Dorchester, DT1 3RR
M1 CSI D Avery, 38 Junction Road, Burgess Hill, RH15 0JN
M1 CSL Dean Cooke, 125 Glenhills Boulevard, Leicester, LE2 8UH
M1 CSU I Bliss, 3 Ford Road, Ashford, TW15 2RF
M1 CSZ S Eggleton, 5 Ladywood Grange, Lady Margaret Road, Ascot, SL5 9QH
M1 CTB C Dale, 3 Ivatt Close, Bawtry, Doncaster, DN10 6QF
M1 CTG Mark Hopkins, The Black Swan, Burn Bridge Road, Harrogate, HG3 1PB
M1 CTJ T Joyes, 58 Ellan Hay Road, Bradley Stoke, Bristol, BS32 0HB
M1 CTK D Hunt, 4 Warmdene Road, Brighton, BN1 8NL
M1 CTM A McMullen, 70 Sylvan Avenue, Timperley, Altrincham, WA15 6AB
M1 CTO L Chung, 104 Penland Road, Haywards Heath, RH16 1PH
MI1 CTQ Joseph Murphy, 19 Fernagh Road, Omagh, BT79 0HX
M1 CUC James McGowan, 72B Adelphi Crescent, Hornchurch, RM12 4JZ
MI1 CUS J Woods, 18 Mullaghdrin Road, Dromara, Dromore, BT25 2AF
M1 CUX Richard Matthews, 18 Hawkins Close, Daventry, NN11 4JQ
M1 CUY Jonathan Wood, 3 Harold Collins Place, Colchester, CO1 2GQ
M1 CVB P Bull, 8 Mayfield Lane, Martlesham Heath, Ipswich, IP5 3TZ
M1 CVF Catherine Block, 10 Beatrice Road, Capel-le-Ferne, Folkestone, CT18 7LL
M1 CVG D Wood, 17 St. Peters Close, Henley, Ipswich, IP6 0RH
M1 CVH Suzanne Blewitt, 9 Durlston Close, Amington, Tamworth, B77 3QG
M1 CVK Kelvin Bennett, 34 Shrubbery Close, Barnstaple, EX32 9DG
M1 CVL Michael Crossley, Lower Park Road, Manchester, M14 5RB
M1 CVM D Giles, 9 Ty Newydd Court, Pontnewydd, Cwmbran, NP44 1LJ
M1 CVT A Mallett, 8 Shaws Close, Prestwood, Great Missenden, HP16 0SL
M1 CVU Krunoslav Smolkovic, 26 Keeling Way, Attleborough, NR17 1YF
M1 CVX S Taylor, 17 York Close, Clayton le Moors, Accrington, BB5 5RB
M1 CWA Julian Gough, 52 Kingston Road, Bristol, BS3 1DP
M1 CWB I Bryant, 17 Kent Road, Southampton, SO17 2LJ
M1 CWD David Taberer, 4 Hillfields Road, Brierley Hill, DY5 2NG
M1 CWG Terry Foreman, 13 Hill Rise, Dartford, DA2 7HX
M1 CWO J Collinson, 50 Willoughby Park, Alnwick, NE66 1ET
M1 CWV A Dykes, 149 Mayfield Road, Chaddesden, Derby, DE21 6FZ
M1 CWW A Thomson, 43 Vale Crescent, Southport, PR8 3SZ
M1 CWY Oliver White, 35 Drage Street, Derby, DE1 3RW
M1 CXA Jason Marcus, 115 Kimberley Road, Solihull, B92 8QA
M1 CXI J Wiltshire, Sunny Bank, Alkham Valley Road, Folkestone, CT18 7EH
M1 CXK A Cordier, 49 Laburnum Avenue, Dartford, DA1 2QN
M1 CXN H Neal, 141 Manor Road, Erith, DA8 2AQ
MM1 CXO J donnelly, 21 Mcdonald Drive, Irvine, KA12 0QS
M1 CXP R Gill, 45 Biggin Lane, Ramsey, Huntingdon, PE26 1NB
M1 CXV Graham Allison, 16 Copse Road, Plymouth, PL7 1PZ
M1 CXW D Hines, 31 Clegge Street, Warrington, WA2 7AT
M1 CXX J Cooksey, 93 New Barns Avenue, Manchester, M21 7DB
M1 CXY R Borrow, 32 Cherrywood Drive, Northfleet, Gravesend, DA11 8PL
MJ1 CYD Claus-Dieter Paland, 19 Maison St. Louis, St. Saviour, Jersey, JE2 7LX
M1 CYJ G Jenkinson, 4 Brundish House, Braithwell Road, Rotherham, S66 8JT
M1 CYK E Doran, 57 Guildford Road, Colchester, CO1 2RZ
M1 CYL K Langhamer, 26 Maple Drive, Burgess Hill, RH15 8AW
M1 CYM Maurice Husband, 31 Crescent Road, Colwall, Malvern, WR13 6QW
M1 CYN WBP Hamilton, 22 Rayford Close, Dartford, DA1 3AJ
M1 CYP B Millard, 11a Fourways, Tetney, Grimsby, DN36 5NF
M1 CYR K Scott, 362 Cannock Road, Heath Hayes, Cannock, WS12 3HA
M1 CYT S Davey, 16 Tilecroft, Welwyn Garden City, AL8 7QY
M1 CYX D Bryan, 14 Fairfield Way, Totland Bay, PO39 0EF
M1 CZA K Churchill, 76 Preston Drive, Bexleyheath, DA7 4UE
M1 CZF M Lewis, The Manor House, The Green, Banbury, OX17 1BU
M1 CZH A Price, 26 Churchward Close, Stourbridge, DY8 4HX
M1 CZI Iain Johnson, 25 Florence Street, Swindon, SN2 1BA
M1 CZL S Bond, Powell Cottage, 1 Powell Close, Leamington Spa, CV33 9PX
M1 CZM J Stocks, 10 Hollycroft Road, Emneth, Wisbech, PE14 8AY
M1 CZO D Charles, 29 Acacia Gardens, Upminster, RM14 1HT
M1 CZX Trevor Ruane, Ventnor, High Lane, Haslemere, GU27 1AZ
M1 CZY D Mageehan, 37 Gosbecks Road, Colchester, CO2 9JR
M1 CZZ Darren Gallier, 86 Pine Tree Road, Oldham, OL8 3LQ
M1 DAB M McPhail, 126 Welbeck St., Creswell, Worksop, S80 4AN
M1 DAH James Saiger, 10 Markham Avenue, Armthorpe, Doncaster, DN3 2AZ
MM1 DAK I Mcdonald, 5 Well Street, Rosehearty, Fraserburgh, AB43 7NW
M1 DAM Anthony Cartwright, 7 Pen Parc, Malltraeth, Bodorgan, LL62 5BG
M1 DAN D Black, 8 Cornwood Close, Finchley, London, N2 0HP
M1 DAP Michael Purcell, 14 Adelaide Road, Blacon, Chester, CH1 5SY
M1 DAS D Nicolson, Woodbridge House, Wembworthy, Chulmleigh, EX18 7SN
M1 DAU C Cater, 40 Frances Avenue, Wrexham, LL12 8BN
MI1 DAW R bamber, 15 Ladybrook Parade, Belfast, BT119ER
M1 DBA P Kinley, 18 Larchwood Road, Wrexham, LL12 7SG
M1 DBC R Carroll, 27 Sheraton Grange, Stourbridge, DY8 2BE
M1 DBF Glyn Jones, 6 Birks Holt Drive, Maltby, Rotherham, S66 7JZ
M1 DBK M Lawrance, 18 The Green Road, Sawston, Cambridge, CB22 3LP
M1 DBL J Makkinje, 4 Sedbergh Drive, Kendal, LA9 6BJ
M1 DBM B Barrett, Kite Hill Camping Park, Firestone Copse Road, Wootton Bridge, PO33 4LQ
M1 DBW Paul Roberts, 7 Boscombe Road, Swindon, SN25 3EZ
M1 DCE Peter Rollinson, 4 Turmarr Villas, Easington, Hull, HU12 0TJ
M1 DCF G Hopkins, 132 Laurel Road, Bassaleg, Newport, NP10 8PT
M1 DCH Peter Wakefield, Flat 3, 21 Priests Road, Swanage, BH19 2RG
MW1 DCI P Bevan, 61 Dinas St., Plasmarl, Swansea, SA6 8LQ
M1 DCK W Curry-Peace, 14 Springfield Road, Stoke-on-Trent, ST4 6RU
M1 DCV M McBride, 127 Leicester Causeway, Coventry, CV1 4HL
M1 DCX T Dore, 18 Evenlode Gardens, Moreton-in-Marsh, GL56 0JF
M1 DCY A Dore, 18 Evenlode Gardens, Moreton-in-Marsh, GL56 0JF
M1 DDB Timothy Smith, 69 Sunningdale, Grantham, NG31 9PF
M1 DDF O Spevack, 16 Ranmore Road, Dorking, RH4 1HD
M1 DDI K Skidmore, 239 Alfreton Road, Blackwell, Alfreton, DE55 5JN
M1 DDR D Carter, 30 Swift Way, Sandal, Wakefield, WF2 6SR
M1 DDW James Reed, 9 Mercer Drive, Harrietsham, Maidstone, ME17 1AY
M1 DDY T Reed, Seafield, Charing Hill, Ashford, TN27 0NG
MM1 DEA G Leadbetter, 8 Tomtain Brae, Cumbernauld, Glasgow, G68 9ER
MM1 DEE G Hall, 84 Queen Street, Kirkintilloch, Glasgow, G66 1JW
M1 DEG R Smith, 41 Middle Deal Road, Deal, CT14 9RG
M1 DEJ M Hibbert, 5 Cliff View Road, Cliffsend, Ramsgate, CT12 5ED
M1 DEY K Armstrong, 8 Caxton Garth, Threshfield, Skipton, BD23 5EZ
MI1 DEZ Trevor Reid, 15 Gillistown Road, Ahoghill, Ballymena, BT42 2RJ
M1 DFB A Dunster, 113 Canterbury Road, Folkestone, CT19 5NR
M1 DFC S Bell, 14 Charlotte Avenue, Wickford, SS12 0DX
M1 DFK Craig Ansell, 51 East Road, Brinsford, Wolverhampton, WV10 7NP
M1 DFM Karl Davies, 58 Popes Lane, Sturry, Canterbury, CT2 0LA
M1 DFO A Bruce, 4 Drayton Manor, 507 Parrswood Road, Manchester, M20 5GJ
MW1 DFQ B Howard, 64 Lawrenny St., Neyland, Milford Haven, SA73 1TB
M1 DFW K Prakash, 14 Masham Road, Harrogate, HG2 8QF
M1 DGE D Cockayne, 32 Shaw Close, Garforth, Leeds, LS25 2HA
M1 DGK J KIRKHAM, Flat 31, 123 St. Anns Road, London, W11 4BT
M1 DGL R Walsh, 10 Standen Road Bungalows, Clitheroe, BB7 1LA
M1 DGP C Anderson, 15 John Gunn Close, Chard, TA20 1DG
M1 DGQ Peter Talbot, 5 Stones Walk, Burghfield Common, Reading, RG7 3JA
M1 DGS David Snell, 154 Oaks Cross, Stevenage, SG2 8NA
M1 DGW Michael Wharton, 9 Orchard View, Linton Colliery, Morpeth, NE61 5SP
M1 DGX J Griffiths, 10 Cote Road, Telford, TF5 0NQ
M1 DGY Herbert Shemming, 6 Smiths Place, Kesgrave, Ipswich, IP5 2YR
M1 DHA Alan Davis, 19 Grange Street, Barnoldswick, BB18 5LB
M1 DHC Anthony Killing, 102 Coquet Grove, Newcastle upon Tyne, NE15 9LH
M1 DHG M Hilton, 40 Megstone Avenue, Whitelea Chase, Cramlington, NE23 6TU
M1 DHI John Edwards, Willows, Sunray Avenue, Whitstable, CT5 4EQ
M1 DHJ Ian maltas, 20 Suddaby Close, Hull, HU9 3RG
M1 DHM R Fraser, 12 Birchen Road, Halewood, Liverpool, L26 9TL
M1 DHO R Bloxam, 39 Claremont Drive, Ravenstone, Coalville, LE67 2ND
M1 DHT D Shackleton, 29 Windmill Green, Ditchingham, Bungay, NR35 2QP
MM1 DHU Patrick McBride, 1 Hillside, Croy, Glasgow, G65 9HJ
M1 DHV Graham Turner, 35 Horncastle Road, Wragby, Market Rasen, LN8 5RB
M1 DHW J Simlat, 7 Coventry Close, Wroughton, Swindon, SN4 9BB
M1 DHY D Sandell, 29 Manor Road, Herne Bay, CT6 6RF
M1 DIB D Beck, 25c Lickless Gardens, Horsforth, Leeds, LS18 5QU
M1 DIE John Halsall, 83 Poole Road, Leeds, LS15 7HD
M1 DIL Barrie Jones, 39 Rosewood Avenue, Burnham-on-Sea, TA8 1HE
M1 DIM K Ingram, 19 Charsley Close, Amersham, HP6 6QQ
M1 DIN D BLYTHE, 51 Lea Side, Halton-Lea-Gate, Brampton, CA8 7LA
M1 DIR R Duncan, 41 Sambourne Road, Warminster, BA12 8LL
M1 DJA Ambrose Garner, The Chestnuts, Surfleet, Spalding, PE11 4BA
M1 DJB C Meakin, 56 Coronation Walk, Gedling, Nottingham, NG4 4AQ
M1 DJC G Griffiths, 8 Grays Lane, Paulerspury, Towcester, NN12 7NW
M1 DJG K Miller, 52 Stanway Road, Shirley, Solihull, B90 3JE
M1 DJI Andrew Waddington, 5 Glenview Avenue, Bradford, BD9 5PA
MM1 DJJ Gregory Waddington, Wester Lathallan, Leven, KY8 5QP
M1 DJN R Francis, 50 Parsonage Chase, Minster on Sea, Sheerness, ME12 3JX
M1 DJO B Kynaston, 76 Thorncliffe Avenue, Dukinfield, SK16 4UD
M1 DJP A Saltmarsh, 55 Wentworth Grove, Winsford, CW7 2LJ
M1 DJS Ian Turk, 11 Medway Crescent, North Hykeham, Lincoln, LN6 8UB
MI1 DJW James Campbell, 2 Lakeview, Crumlin, BT29 4YA
M1 DJX A Baker, 35 Greenover Road, Brixham, TQ5 9NA
M1 DKA A Lewis, 111a Cheltenham Road, Longlevens, Gloucester, GL2 0JG
M1 DKF R Saward, Vigeland, 61a Old Main Road, Boston, PE20 2BU
M1 DKK Nathan Hall, 98 Newbiggin Road, Ashington, NE63 0TH
M1 DKL D Liddard, 81 Tattersall Gardens, Leigh-on-Sea, SS9 2QS
M1 DKM A Williams, 25 Tre Rhosyr, Newborough, Llanfairpwllgwyngyll, LL61 6TG
M1 DKP Alan Maylin, 221 Branksome Avenue, Stanford-le-Hope, SS17 8DD
M1 DKV R Illman, 35 Courtenay Park, South Brent, TQ10 9BT
M1 DKY H Sanders, Copper Coins, Deans Drove, Poole, BH16 6EQ
M1 DKZ Darren Forster, 3 West View, Middleton, Ludlow, SY8 3ED

M1 DLE M Gifford, 22 St. Agnes Way, Kesgrave, Ipswich, IP5 1JZ
M1 DLG D Smith, 19 Victoria Grove, Newbury, RG14 7RA
M1 DLM D Russell, 6 Deansgate Lane North, Liverpool, L37 7ER
M1 DLR R Wood, Bank View, Bilham Road, Huddersfield, HD8 9PA
M1 DLX CASTLE HOUSE SCHOOL ARC c/o J Griffiths, 10 Cote Road, Telford, TF5 0NQ
M1 DMB S Atton, A1 Manor Park, Happisburgh, NR12 0PW
MM1 DME R Fuggle, 14 Quebec Drive, East Kilbride, Glasgow, G75 8SA
M1 DMH J Hardman, 2 Well Orchard, Bamber Bridge, Preston, PR5 8HJ
M1 DMN R Gilbert, 8 Church Road, West Kingsdown, Sevenoaks, TN15 6LL
M1 DMR B Pilcher, 283 London Road, Portsmouth, PO2 9HE
M1 DMT J Hull, 68 Meadow Avenue, West Bromwich, B71 3EE
MM1 DMU D McCann, 69 Davies Drive, Lomond Industrial Estate, Alexandria, G83 0UF
M1 DMX M Grainger, 9 Fox Hollow, East Goscote, Leicester, LE7 3WZ
M1 DNA D Bruce, 15 St. Richards Road, Deal, CT14 9JR
M1 DNC L Hall, 15 Fullwood Avenue, Newhaven, BN9 9SP
M1 DNE S Moppett, 59 piccadilly, Tamworth, B78 2ER
M1 DNG R Steward, Long Meadow, Seven Acres Lane, Southwold, IP18 6UL
M1 DNJ David Houbart, 10 Lancelot Close, Rochester, ME2 2YT
M1 DNQ Jon Golding, Flat 6, Daver Court, London, SW3 3TS
MD1 DNT D Hughes, 13 Julian Road, Douglas, Isle of Man, IM2 6HW
M1 DNY D Reid, 11 Caer Delyn, Bodffordd, Llangefni, LL77 7EJ
M1 DNZ D Herridge, 93 Freshbrook Road, Lancing, BN15 8DE
M1 DOA S Jefferson, 8 Panturner Road, Stoke-on-Trent, ST2 0SX
MI1 DOG S McAuley, Layde View, 19 Rathlin Avenue, Ballycastle, BT54 6DQ
M1 DOO V Lee, 11 Mead Lane, Cwmbran, NP44 1NW
M1 DOR D Stuart, 58 Woodplace Lane, Coulsdon, CR5 1NF
M1 DOS Carl Smith, 13 West Winds Road, Winterton, Scunthorpe, DN15 9RU
M1 DOT S Myall, 71 Barnes Avenue, Fearnhead, Warrington, WA2 0BL
M1 DOU Martin Jones, 3 St. Catherines Close, Llanfaes, Beaumaris, LL58 8LH
M1 DOZ David Sampson, Spirits Hall, Mountains Road, Maldon, CM9 8BY
MM1 DPC M Mclauchlan, 8 Craigie St., Ballingry, Lochgelly, KY5 8NS
M1 DPE Leonard Stockwell, 167 Hathaway Road, Grays, RM17 5LW
MM1 DPH Jim Crichton, 1 Glenmuir Road, Ayr, KA8 9RD
M1 DPI E Thompson, 9 Elmwood Drive, Ponteland, Newcastle upon Tyne, NE20 9QQ
M1 DPJ Anthony Bonner, Flat 15, Enderleigh House, Havant, PO9 1LQ
MI1 DPL J Stewart, 45 Mull Road, Antrim, BT41 2TR
M1 DPO J Gould, 14 Homestead Road, Orpington, BR6 6HW
M1 DPQ Peter Orr, 74 Amalfi Tower, Lakeside Village, Sunderland, SR3 3AL
M1 DPU P Cain, 108 Spencer Road, Norwich, NR6 6DG
M1 DPW Peter Whiffing, 12 Murton Street, Newcastle upon Tyne, NE13 9AF
M1 DPX Dennis Collins, 1 New Street Close, Stradbroke, Eye, IP21 5JH
M1 DPY J Bowes, 40 Nursery Road, Angmering, Littlehampton, BN16 4FH
MI1 DQB C Gardner, 50 Kirkliston Park, Belfast, BT5 6ED
M1 DQE I Fletcher, Priory House, 56 Fairfield Road, Saxmundham, IP17 1BA
M1 DQG Ronald Kennedy-Bright, Sandiacre, Orchard Lane, Hanwood, Shrewsbury, SY5 8LE
M1 DQH Alec Duffield-Dyche, 1a The Hawthorns, Brockton, Shrewsbury, SY5 9JY
M1 DQI Mark Jones, 8 Sunfield Gardens, Bayston Hill, Shrewsbury, SY3 0LA
M1 DQQ S Stanley, 35 Statham Close, Lymm, WA13 9NN
M1 DQU A Bedford, 44 Kirtling Place, Haverhill, CB9 0AU
MM1 DQV Andrew Gibbs, Cathlawhill Farm, Torphichen, Bathgate, EH48 4NW
MM1 DQW Glen Steven, 36 Springhill Terrace, Springside, Irvine, KA11 3AL
M1 DQX Paul Ormerod, 14 Fanny Moor Crescent, Huddersfield, HD4 6PL
M1 DRB Kate Glover, 14 Crawley Crescent, Eastbourne, BN22 9RN
M1 DRK Andrew Thomson, 9 Patrington Garth, Bransholme, Hull, HU7 4NZ
M1 DRL D Luff, 12 Swan Lane, Sellindge, Ashford, TN25 6EP
M1 DRM M Bird, Driftwood, 37 Beachwood Avenue, Kingswinford, DY6 0HL
MI1 DRP P McDaid, 66 Laurel Drive, Strabane, BT82 9PN
M1 DRZ J Stevens, Springfield Cottages, 57 Brindley Street, Stourport-on-Severn, DY13 8JG
MM1 DSD G Duncan, 5 Jarvis Place, Carnoustie, DD7 7BR
M1 DSE Paul Gibson, 46 Seacrest Avenue, North Shields, NE30 3DP
M1 DSQ N Taylor, 36 Bodmin Avenue, Slough, SL2 1SL
M1 DSU James Henderson, 12 Chathill Terrace, Newcastle upon Tyne, NE6 3BB
M1 DSV A Newell, Thwaites Bank, Spring Avenue, Keighley, BD21 4TD
MM1 DSX J Spiers, 29b Carlyle Gardens, Haddington, EH41 3LS
M1 DSZ S Turner, 27 Huntspill Road, Highbridge, TA9 3DQ
M1 DTC N Chapman, 43 Meadow View Road, Exmouth, EX8 4ET
M1 DTG M Whitchurch, 94 Hundred Acres Lane, Amersham, HP7 9BN
MM1 DTN W Gray, 1 Regent Court, Regent Street, Keith, AB55 5ED
M1 DTO T Jones, 40 Chester Road South, Kidderminster, DY10 1XJ
M1 DTS E Cawte, Woodpeckers, Rectory Gardens, Church Stretton, SY6 6DP
MW1 DTT Simon Walters, 9 Salisbury Road, Abercynon, Mountain Ash, CF45 4NU
M1 DTU J Weddell, 10 High Street, Eyemouth, TD14 5EU
M1 DUA Darren Riley, Flat 2, Crown Crest Court, Sevenoaks, TN14 5AS
M1 DUB Donald Stalley, 42 Gadby Road, Sittingbourne, ME10 1TJ
M1 DUC John Parker, 70 Elm Road, Grays, RM17 6LD
M1 DUD R Burrows-Ellis, Pateley, School Lane, Woodbridge, IP13 6DX
MW1 DUJ D Jones, 13 Pontardulais Road, Cross Hands, Llanelli, SA14 6NT
M1 DUO Robert Easthope, 18 Desmond Avenue, Cherry Hinton, Cambridge, CB1 9JS
MM1 DVC I Hendry, 47 Fraser Place, Keith, AB55 5EB
M1 DVJ Christopher Wood, Grammar School Bungalow, Greenway Road, Brixham, TQ5 0LW
M1 DVO R Waters, Romosco, Mill Lane, Manningtree, CO11 2QP
M1 DVV P Wise, 13 Waltham Road, Newton Abbot, TQ12 1LH
M1 DWQ I Lowcock, Sunflower Cottage, Loddiswell, Kingsbridge, TQ7 4QJ
M1 DWT D Hamilton, 120 Hall Road, Hull, HU6 8SB
MM1 DWU G McVittie, 46 Mote Hill Road, Girvan, KA26 0EB
M1 DWW Wade Bennett, 23 Halstead Road, Earls Colne, Colchester, CO6 2NG
M1 DXB B Smith, 39b Palace Avenue, Paignton, TQ3 3EQ
M1 DXG R Williamson, Beverley, Swineshead Road, Boston, PE20 1SG
M1 DXL Clive Churchward, Church Ward Logistics Ltd, 75 Atlantic Business Centre, Altrincham, WA14 5NQ
M1 DXN Howard Temperley, 40 Wycombe Close, Urmston, Manchester, M41 7ND
M1 DXO N Onions, 34 Redwing Court, Southsea, PO4 8PB
M1 DXQ M Rhead, 11 Shelley Road, Stoke-on-Trent, ST2 8JN
MM1 DXU M Richards, Minerva, Stronsay, Orkney, KW17 2AS
MD1 DXW William Griffiths, 7 Cooyrt Shellagh, Ballasalla, Isle of Man, IM9 2EU
M1 DYC James Guilford, 2 Lacey Close, Ilkeston, DE7 9LJ
M1 DYD F Frost, Flat 34, Kingsley Court, 21 Pincott Road, Bexleyheath, DA6 7LA
M1 DYE D Ejugue, Po Box 62449, Addis Ababa, Ethiopia
M1 DYF A Teffera, PO Box 819, Addis Abba, Ethiopia
M1 DYG N Teklehaimanot, PO Box 21866, Addis Abba, Ethiopia
M1 DYH M Belete, PO Box 181922, Addis Ababa, Ethiopia
M1 DYI E Melaku, 77 Chaucer Drive, Lincoln, LN2 4LT
M1 DYJ A Williams, Alwent Farm, Staindrop, Darlington, DL2 3NS
M1 DYK H Davison, 15 High St., Rippingale, Bourne, PE10 0SR
M1 DYL D Davison, 15 High St., Rippingale, Bourne, PE10 0SR
M1 DYO A Cruise, Badgers Holt, Park Grove, Chalfont St. Giles, HP8 4BG
M1 DYP K Suddes, 10 Tilecroft, Welwyn Garden City, AL8 7QY
M1 DYS Robert Broadbridge, 8 Moreton Road, Bournemouth, BH9 3PR
M1 DYU G Rogers, 55 Upholland Road, Billinge, Wigan, WN5 7JA
M1 DYW O Barnes, 14 Caroline Close, Wivenhoe, Colchester, CO7 9SD
M1 DZM J Nichols, 51 Ashbourne Road, Barnsley, S71 3QD
M1 DZP C Hastwell, 6 Barn Close, Worthing, BN13 2BE
M1 DZR R Daglish, Beck Lea, Pasture Road, Frizington, CA26 3XN
M1 DZT Kenneth Burnell, 27 Manners Gardens, Seaton Delaval, Whitley Bay, NE25 0DW
MM1 DZW R Heath, 73 King Street, Inverbervie, Montrose, DD10 0RB
MW1 EAA Gordon Tucker, 18 Plymouth Road, Penarth, CF64 3DH
M1 EAB Andrew Thornton, 11 St. Nicholas Close, Richmond, DL10 7SP
M1 EAI A Beale, 6 Meadow View, Belper, DE56 1UT
M1 EAJ Y Bessell-Baldwin, 10 Tudor Close, Barton-le-Clay, Bedford, MK45 4NE
M1 EAK Colin Day, 35 Rochford Road, St. Osyth, Clacton-on-Sea, CO16 8PH
M1 EAN B Authers, 91 Hay Green Lane, Bournville, Birmingham, B30 1RF
M1 EAW K Gallacher, 63 Holst Avenue, Basildon, SS15 5RH
M1 EAZ John Barker, 53 Derby Street, Colne, BB8 9AA
M1 EBC J Best, Longview, Central Road, Maryport, CA15 7ER
M1 EBD D Best, Longview, Central Road, Maryport, CA15 7ER
M1 EBH Larry Emmerson, Topsie, Eastsands, Marlborough, SN8 3AN
M1 EBI Philip Bird, 10a Shackleton Road, Bloxwich, Walsall, WS3 3BZ
M1 EBK M Rowley, 20 Long Leasow, Selly Oak, Birmingham, B29 4LT
M1 EBL C Venables, Deepdene, Rickford, Guildford, GU3 3PQ
M1 EBN Rose Bunce, 4 Newlands, Gainsborough, DN21 1QZ
M1 EBS Paul Beckwith, 4 Hunters Yard, Riseley, Bedford, MK44 1EN
M1 EBU Warren Mitchell, 8 Woodland Crescent, Burgess Hill, RH15 0LJ
M1 EBW I Merrill, 26 Catkin Drive, Giltbrook, Nottingham, NG16 2UB
M1 EBY Peter Clark, 6 Haynes House, Booker Place, High Wycombe, HP12 4QD
M1 ECB K Cronin, 9 Marriott Close, Beeston, Nottingham, NG9 4JB
M1 ECC D Wright, 74 Witchards, Basildon, SS16 5BN
M1 ECD NORTH-NORTHANTS RAYNET GROUP c/o Michael Wright, 25 St. Matthews Road, Kettering, NN15 5HE
M1 ECH Stephen Kirby, 2 Kneeton Park, Middleton Tyas, Richmond, DL10 6SB
M1 ECI A Funnell, 15 Hendham Road, London, SW17 7DH
M1 ECM M White, 100 Burnham Road, Coventry, CV3 4BQ
M1 ECQ C Hamilton, 101 Gipsy Lane, Swindon, SN2 8DL
M1 ECT Michael Procter, 141 Ruddington Lane, Nottingham, NG11 7BY
M1 ECV David Bould, 38 Curlew Grove, Bridlington, YO15 3NX
M1 ECW N Mcmahon, 23 St. James Close, Bramley, Tadley, RG26 5XH
M1 ECY S Williams, 32 Waterdell Lane, St. Ippolyts, Hitchin, SG4 7QZ
M1 EDA S Fabian, 3 Manor Cottages, Horley, Banbury, OX15 6BJ
M1 EDF Geoffrey Powell, Sycamore Cottage, Church Lane, Tamworth, B79 0LD
M1 EDL A Wolverson, 28 Thorness Close, Alvaston, Derby, DE24 0UY
M1 EDO J Hurst, 13 Peregrine Road, Hainault, Ilford, IG6 3SR
M1 EDW P Tinkler, 27 Cavendish Drive, Carlton, Nottingham, NG4 3DX
MM1 EDY J Goldstraw, 26 Craigmill Gardens, Carnoustie, DD7 6HT
M1 EEN G Butterfield, Pasadena, St. Helens Road, Norwich, NR12 0LU
M1 EEP J Nethercott, 30 Goldcrest Road, Chipping Sodbury, Bristol, BS37 6XF
M1 EEQ A Waite, 221 Hasler Road, Poole, BH17 9AH
M1 EER G Ball, 11 Jersey Close, Congleton, CW12 3TW
M1 EEW Alan Beckwith, 19 Westmorland Avenue, Dukinfield, SK16 5JA
M1 EEY N Beckley, 76 Keir Hardie Way, Barking, IG11 9NY
M1 EEZ A Kypriadis, 119 Whitfield Villas, South Shields, NE33 5NH
M1 EFP John Carter, 5 Hastings Avenue, Seaford, BN25 3LB
M1 EFT Paul Swanton, 54 South Avenue, Warrington, WA2 8BQ
M1 EGC G Hancock, 1 Lypiatt Mead, Corsham, SN13 9JL
M1 EGD S Sykes, 12 Banksville, Holmfirth, HD9 1XP
M1 EGG Pauline Bird, 37 Beachwood Avenue, Kingswinford, DY6 0HL
M1 EGL Peter Ritchley, 34 Chesildene Avenue, Throop, Bournemouth, BH8 0DS
M1 EGM Barnaby Ritchley, 25 Branwell Close, Christchurch, BH23 2NP
M1 EGN J Eyres, 13 Newburn Crescent, Swindon, SN1 5ES
M1 EGP R Argent, 122 Church Road, Hadleigh, Benfleet, SS7 2UA
MM1 EGS T May, 34 Dee Place, East Kilbride, Glasgow, G75 8RZ
M1 EGV C Mills, 6 Levisham Gardens, Bewsey, Warrington, WA5 0GD
M1 EGW D Green, 5 St. Benedicts Close, Cranwell Village, Sleaford, NG34 8DB
M1 EGX M Beddard, 86 Walmley Road, Sutton Coldfield, B76 2QH
M1 EGZ W Gravestock, 23a Murrell Road, Ash, Aldershot, GU12 6ST
M1 EHB J Nixon, Coates Cottage, Main Road, Hull, HU12 9AX
M1 EHD Peter Leadill, 7 Keldale, Haxby, York, YO32 3GG
M1 EHF G Gore-Thorne, 19 Poplar Way, Ringwood, BH24 1UY
M1 EHI Ian Croasdale, 16 Buttermere Avenue, Chorley, PR7 2JG
M1 EHJ R Roychoudhuri, 62a Parkway, Eastbourne, BN20 9DY
MM1 EHO S McNeil, 35 Sutors Avenue, Nairn, IV12 5AZ
M1 EHV Christopher Aram, 1 Snuggs Lane, East Hanney, Wantage, OX12 0HU
M1 EHW Michael Palmer, Chedburgh Llandevaud, Newport, NP18 2AE
M1 EHZ J Brown, 3 Malton Close, Blyth, NE24 5AS
M1 EIE S Stephenson, 31 Sherbrooke Avenue, Hull, HU5 4AG
MI1 EIH G Brennan, 15 Kinnegar Rocks, Donaghadee, BT21 0EZ
M1 EIJ Stuart Sweetlove, 36 Park Avenue, Corsham, SN13 0JT
M1 EIO A Rixon, 17 Brimmers Way, Aylesbury, HP19 7HR
M1 EIR E Fishbourne, 8 Somers Walk, Tupsley, Hereford, HR1 1QX
M1 EIU Colin Martin, Flat 3, Dodds House, Vicarage Lane, Tarporley, CW6 9BP
M1 EIW George Kinney, 1 Eden Park, Brixham, TQ5 9LS
M1 EIZ Lisa Rutherford, 197 Rosalind Street, Ashington, NE63 9BB
M1 EJD Danny Pickering, 2 Priory Green, Highworth, Swindon, SN6 7NU
M1 EJE D Clarke, 85 Bolton Street, Brixham, TQ5 9DJ
M1 EJG John Clarke, 21 Mill Lane, Blakedown, Kidderminster, DY10 3ND
M1 EJI D Hunt, 2a Cliff Road, Radcliffe on Trent, Nottingham, NG12 2GA
M1 EJJ M Laurie, 2 The Steps, Phocle Green, Ross-on-Wye, HR9 7TW
M1 EJO P Matthews, The Stables, Alkham Road, Dover, CT16 3EE
M1 EJQ Melanie Cross, 7 Hallside Road, Enfield, EN1 4AD
M1 EJS S Birchall, 11 Rosebery Road, Felixstowe, IP11 7JR
M1 EJX Martin Heley, 22 St. Lawrence Road, Dunscroft, Doncaster, DN7 4AS
M1 EKA B Parker, 38 Cross St., Thurcroft, Rotherham, S66 9NJ
M1 EKD S Pounder, 4 Otley Mount, East Morton, Keighley, BD20 5TD
M1 EKH David Bowker, 54 Edward Street, Middleton, Manchester, M24 6BN
M1 EKK Keith George, 2 Larchmont, Clayton, Bradford, BD14 6AB
M1 EKL A Sanderson, 186 Kentmere Avenue, Leeds, LS14 1BN
M1 EKM Graham Harden, 13 Greenfield Road, Coleford, GL16 8BY
M1 EKU P Lancaster, 16 Wiltshire Close, Bury, BL9 9EY
M1 ELB Craig Mitchell, A 3, Pakkalanrinne 14, Vantaa, Finland, 1510
MM1 ELE D Marshall, 10 Spencer Crescent, Carnoustie, DD7 6DQ
M1 ELI D Welch, 41 Mersey Way, Bletchley, Milton Keynes, MK3 7PS
M1 ELM A Dent, 22 Moorside Road, Bournemouth, BH11 8DF
M1 ELN Michael Pike, 1 Sevelm, Up Hatherley, Cheltenham, GL51 3RZ
M1 ELQ P Houghton, 19 Gilthwaites Lane, Denby Dale, Huddersfield, HD8 8SG
M1 ELR C Davies, 94 Alnwick Drive, Moreton, Wirral, CH46 6ET
M1 ELS J Matias, 23 The Approach, London, W3 7PA
M1 ELW H Watson, 5 Arbroath, Ouston, Chester le Street, DH2 1QY
M1 EMB Stuart Buckley, 1 Chouler Gardens, Stevenage, SG1 4TB
M1 EMC K Heselton, 3 Winterslow Road, Penhill, Swindon, SN2 5JJ
M1 EMG D Woodcroft, 23 Wilkin Walk, Cottenham, Cambridge, CB24 8TS
M1 EMO P Warden, 12a Landscape View, Saffron Walden, CB11 4AU
M1 EMP N Douglas, 2 Huntingdon Close, Fareham, PO14 4JP
M1 EMR J Dixon, 5 Laburnum Avenue, Moorends, Doncaster, DN8 4SF
M1 EMU P Gilmore, 2 Ridgeway, Billericay, CM12 9NT
M1 EMX H Harvey, 153 Stradbroke Grove, Clayhall, Ilford, IG5 0DL
M1 ENA John Long, 1 Tangway, Chineham, Basingstoke, RG24 8SU
M1 ENE S Marcot, 5 The Crescent, West Wickham, BR4 0HB
M1 ENJ C Berry, 29 Marlborough Crescent, Long Hanborough, Witney, OX29 8JP
M1 ENK J Berry, 29 Marlborough Crescent, Long Hanborough, Witney, OX29 8JP
M1 ENQ R Townsend, 56 Seymour Road, Northfleet, Gravesend, DA11 7BN
M1 ENX S Caine, Magnolia House, The Larches, East Grinstead, RH19 3QL
M1 ENZ Richard Rouse, 7 Bangalay Place, Leonay, Australia, 2750
M1 EOK Peter Davies, 9 Cramer Court, Rhyl, LL18 2BX
M1 EOO Harold Matthews, 24 Clos Y Berllan, Rhuddlan, Rhyl, LL18 2UL
M1 EOP M Jurkiewicz, 21 Porlock Avenue, Stafford, ST17 0HS
M1 EOR M Edwards, 2 The Twyn, Fleur de Lis, Blackwood, NP12 3UL
M1 EOU Thomas Nadin, 4 Firtree Rise, Chapeltown, Sheffield, S35 1QG
M1 EOV P Spurgeon, 15 Ketts Close, Wymondham, NR18 0NB
M1 EOZ Clynton Cartwright, 8 Hudson Road, Blackpool, FY1 6LY
MJ1 EPG B Allchin, Pont Marquet Cottage, La Rue Des Mans, St. Brelade, Jersey, JE3 8BL
M1 EPI D Williams, Rhilin, Rhydwyn, Holyhead, LL65 4EA
M1 EPK Keith Hichisson, 2 Tithe Farm Close, Langford, Biggleswade, SG18 9NE
M1 EPN R Shaddick, 5 Shrewsbury Bow, Weston-Super-Mare, BS24 7SB
M1 EPR Alex Wilson, 22 Ormesby Road, RAF Coltishall, Norwich, NR10 5JY
M1 EPU SOUTH DEVON RAYNET c/o Geraldine Coker, 46 Clarendon Road, Ipplepen, Newton Abbot, TQ12 5QS
M1 EPX Michael Clark, 8 Willow Close, Clevedon, BS21 6HR
M1 EQA N Trewin, 70 Trelowen Drive, Penryn, TR10 9WS
M1 EQB G Trouse, 1 Amanda Close, Bexhill-on-Sea, TN40 2TB
M1 EQD P Burton, 99 Western Avenue, Blacon, Chester, CH1 5QX
MM1 EQE Alan Thompson, 22 Lochend Road, Carnoustie, DD7 7QF
MI1 EQI Chris Blake, 9 Mullaghbrack Road, Hamiltonsbawn, Armagh, BT60 1JU
M1 EQN Darren Whittle, 27 Convent Close, Aughton, Ormskirk, L39 4XP
M1 EQO Clive Hayward, 12 Trouvere Park, Hemel Hempstead, HP1 3HY
M1 EQV M Edwards, 225 Monkmoor Road, Monkmoor, Shrewsbury, SY2 5SW
M1 EQW S Harrison, 20 Wisewood Avenue, Wisewood, Sheffield, S6 4WG
M1 ERA S Trimble, Pentreath, Cury Cross Lanes, Helston, TR12 7BJ
M1 ERD Adrian Trimble, Pentreath, Cury Cross Lanes, Helston, TR12 7BJ
M1 ERF F Tatlow, 45 Pasture Road, Stapleford, Nottingham, NG9 8HR
M1 ERH S Bird, 12 Commercial Road, Shepton Mallet, BA4 5DH
M1 ERJ P Chandler, 94 Shrubland Street, Leamington Spa, CV31 3BD
MI1 ERL Peter Cranston, 135 Saintfield Road, Lisburn, BT27 6YW
M1 ERN J Baugh, 172 Pontefract Road, Cudworth, Barnsley, S72 8BE
M1 ERO David Eastope, 9 St. Davids House, Willow Way, Redditch, B97 6PG
M1 ERP S Carruthers, 13 Belah Road, Carlisle, CA3 9RE
M1 ERU S Carrington, 137 Richmond Park Road, Bournemouth, BH8 8UA
M1 ERV S Chambers, 1 Northleaze, Corsham, SN13 0QW
M1 ERY K Sylvester, 8 Beacon Park Close, Skegness, PE25 1HQ
M1 ESD J Smith, 68 Golden Cross Lane, Catshill, Bromsgrove, B61 0LG
M1 ESH William Inch, 17 Grantley Close Copford, Colchester, CO6 1YP
M1 ESI D Crane, 132 Windermere Drive, Warndon, Worcester, WR4 9JD
M1 ESM L Wallace, 20 Radworthy, Furzton, Milton Keynes, MK4 1JH
M1 ESV R Scotland, 11 Edwards Court, Slough, SL1 2HY
M1 ESW S D'Sylva, 56 Fenby Gardens, Scarborough, YO12 5LB
M1 ETC M Ribton, 80 Trafalgar Street, Gillingham, ME7 4RN
M1 ETM A Hayward, 67 Pinecroft, Carlisle, CA3 0DD
M1 ETN Daniel Allen, 9 Paddock Road, Buntingford, SG9 9EX
M1 ETS Ernest Coleby, 13 Farm Close, Sunniside, Newcastle upon Tyne, NE16 5PP
M1 ETT Thomas Hindson, 73d Leigh Road, Wimborne, BH21 2AA
M1 ETW Ade Talabi, 1 Crealock Grove, Woodford Green, IG8 9QZ
M1 ETX Andrew Toomer, 40 Newland Avenue, Driffield, YO25 6TX
M1 EUE J Underwood, 27 Woodville Road, London, E17 7ER
M1 EUF John Cunningham, 16 Welbeck Road, Doncaster, DN4 5EY
M1 EUL B Fielding, 16 The Horseshoe, York, YO24 1LX
M1 EUM Peter Thorne, 23 Doxey Fields, Stafford, ST16 1HJ
M1 EUN John Fletcher, 66 Deightonby Street, Thurnscoe, Rotherham, S63 0JA
M1 EUR P Allen, 32 Milward Road, Loscoe, Heanor, DE75 7JX
M1 EUX Christopher Bartlett, 25 Westfields, Buckingham, MK18 1DZ
MI1 EVD T Carlisle, 12 Drumawhey Gardens, Bangor, BT19 1SR

**IMPORTANT NOTE**

**Revalidate licence to avoid revocation** – Ofcom has advised the Society that plans will be drawn up to revoke licences that have not been revalidated as required by the licence conditions. The quickest way to revalidate is to do so online via the Ofcom website: *https://services.ofcom.org.uk/* or by email: *amateur.validations@ofcom.org.uk* Ofcom staff are available to help, but please be patient during times of heavy workload.

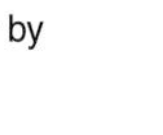

M1 EVF Stephen Larden, Flat 42, Worcester House, Halesowen, B63 4TJ
M1 EVH Kenneth Wright, 60 Ashley Road, Walsall, WS3 2QF
MM1 EVJ J Conway, 26 Kerse Avenue, Dalry, KA24 4DJ
M1 EVN Joseph Haughey, 31 Woodfield Drive, West Mersea, Colchester, CO5 8PX
M1 EVP BATH AND NORTH EAST SOMERSET RAYNET c/o F Smedley, 13 Justice Avenue, Saltford, Bristol, BS31 3DR
M1 EVZ Stewart Punch, 23 Blythe Way, Maldon, CM9 6UE
MM1 EWA Michael Macleod, 1/3 43 Garnethill Street, Glasgow, G3 6QD
M1 EWD T Beevers, 4 Cloud Avenue, Stapleford, Nottingham, NG9 8BN
M1 EWF Roy Read, 111 Fitzpain Road, West Parley, Ferndown, BH22 8SF
M1 EWJ Edwin Williams, 82 Maes Llwyn, Amlwch, LL68 9BG
M1 EWM A O'Hea, 12 De Vitre Place, Grove, Wantage, OX12 0DA
M1 EWP B Lester, 37 Cormorant Drive, St. Austell, PL25 3BB
M1 EWT Chris Berry, Berriscot, 7 Gloweth Villas, Truro, TR1 3LU
M1 EWV Kevin Trench, 10 Victoria Road, Morley, Leeds, LS27 9DS
M1 EXJ M Wohlgemuth, 39 Great Mead, Waterlooville, PO7 6HH
M1 EXL C Constable, 9 Ridgeway Close, Heathfield, TN21 8NS
M1 EXO S Andrews, 64 Bradgate Road, Markfield, LE67 9SN
M1 EXQ Michael Peters, Yare House, Thuxton, Norwich, NR9 4QJ
M1 EXS Graham Burton, 2 Derwent Street, Darwen, BB3 1EF
M1 EXW M Gardner, 21 Tiptree, Castlehaven Road, London, NW1 8TL
M1 EYA Richard Neale-Gardner, 72 Queensway, Barwell, Leicester, LE9 8AP
M1 EYG R Barrow, Fern House, Ripponden Old Lane, Sowerby Bridge, HX6 4PA
M1 EYH Frank Bailey, 28 Coopers Field, St. Martins, Oswestry, SY11 3BU
MM1 EYI Nigel Kingon-Rouse, 148 Oldwood Place, Livingston, EH54 6UX
M1 EYL A Banks, 12 Taylor Court, Weston-Super-Mare, BS22 7LU
M1 EYO Alan Poxon, 34 Conduit St., Tintwistle, Glossop, SK13 1LR
M1 EYP Thomas Read, 31 Merebrook Road, Macclesfield, SK11 8RH
M1 EYQ B Davis, 104 Lever House Lane, Leyland, PR25 4XP
M1 EYS E Shears, 22 Richborough Drive, Charlton, Andover, SP10 4EZ
M1 EYT A Kok, 14 Throop Road, Templecombe, BA8 0HR
M1 EYU D Kok, 14 Castle View, Fron Goch, Caernarvon, LL55 4LE
MM1 EYZ D Bruce, 32 Jesmond Ave North, Bridge Of Don, Aberdeen, AB22 8WL
M1 EZB Chris Coonick, 36 Magdalen Way, Weston-Super-Mare, BS22 7PG
M1 EZC G Sawyer, 101 Southern Drive, Loughton, IG10 3BY
M1 EZD G Wesson, 12 Lea Close, Alcester, B49 6AP
M1 EZE S Davies, 35 Queensland Crescent, Chelmsford, CM1 2DZ
M1 EZG M Lane, 21 Winterbourne Road, Poole, BH15 2ES
M1 EZH L Copley, Elmford, Mount Pleasant South, Whitby, YO22 4RQ
M1 EZJ Mark Skelton, 16 Southfield, Sutton Hill, Telford, TF7 4HP
M1 EZK K Sheehan, 43 Central Avenue, Beverley, HU17 8LL
M1 EZL John Anderson, 16 Hanham Road, Corfe Mullen, Wimborne, BH21 3PZ
M1 EZP M White, 1 Nursery Close, Wroughton, Swindon, SN4 9DR
M1 EZR David Birkenshaw, 33 Fosse Close Braunstone, Leicester, LE3 2RY
M1 EZT Matthew Marston, Fair View Farm, Carnkie, Helston, TR13 0DZ
M1 EZX D Bridle, 8 Cowleaze Road, Broadmayne, Dorchester, DT2 8EW
MI1 EZZ R Bradley, 45 Alexandra Park Avenue, Belfast, BT15 3ER
M1 FAA T Atkinson, 10 Invicta Close, Chislehurst, BR7 6SJ
M1 FAF T Jones, 354 Bridgeman Street, Bolton, BL3 6SJ
M1 FAI Stevan Taylor, 20 Eastview Road, Wargrave, Reading, RG10 8BH
M1 FAJ D Green, 7 Greenside Court, Mickleover, Derby, DE3 0RG
MI1 FAR A Bryce, 123 Newtownards Road, Comber, Newtownards, BT23 5LD
MM1 FAS R Krawczyk, 7 Anderson Crescent, Bishopmill, Elgin, IV30 4HJ
M1 FAT Clifford Jayne, 65a Park Crescent, Abergavenny, NP7 5TL
M1 FAX D Jones, 87 Forster Street, Warrington, WA2 7AX
M1 FAY Kenneth Rowsell, Elmtree Cottage, Chilworthy, Chard, TA20 3BH
M1 FBF A Wilson, 2 Briar Close, Newhall, Swadlincote, DE11 0RX
M1 FBI Stephen Plews, 70 Baulkham Hills, Penshaw, Houghton le Spring, DH4 7RZ
M1 FBL Jeremy Rowe, Hunters Brook, Fine Lane, Newport, PO30 3JY
M1 FBN I Dalton, 10 Durkar Fields, Durkar, Wakefield, WF4 3BY
M1 FBS David Faul, PH 1113, 1200 The Esplanade North, Ontario, Canada, L1V6V3
M1 FBW J Machalski, 28 Longdales Road, Lincoln, LN2 2JU
MI1 FCB Thomas Kilgore. MBE, 8 Slieve Shannagh Park, Newcastle, BT33 0HW
M1 FCC C Chuter, 35 Longford Road, Bognor Regis, PO21 1AB
M1 FCE D Sampson, 32 The Gannets, Stubbington, Fareham, PO14 3SY
M1 FCF D Patrick, 72 Bracken Crescent, Eastleigh, SO50 8ND
M1 FCG R Boyns, 65 Alma Road, Plymouth, PL3 4HE
M1 FCH T Johnson, Orchard House, Tollerton, York, YO61 1PS
MI1 FCQ Patrick McCauley, 5 Brookmount Rise, Omagh, BT78 5AL
M1 FCV Clive Roberts, 57 Chandos Road, Lightpill, Stroud, GL5 3QT
M1 FCW M Yeomans, 60 Tickton Grove, Hull, HU6 8NJ
M1 FCX A Rudnicki, 1 Willoughby Court, London Colney, St. Albans, AL2 1HL
M1 FCZ E Snowdon, 22 Twizziegill View, Easington, Saltburn-by-The-Sea, TS13 4NX
MM1 FDF Steven Taylor, Sunny Brae, Kinellar, Aberdeen, AB21 0TY
M1 FDH Frank Howker, 65 Boxley Drive, West Bridgford, Nottingham, NG2 7GN
M1 FDK G Charnock, Oldhouse, Newton St Margarets, Hereford, HR2 0QR
M1 FDN David Morgan, Castle Cottage, Newbridge, Newport, NP11 3NT
M1 FDO C Heading, 27 Broadlands Avenue, Eastleigh, SO50 4PP
M1 FEK David Stewart, 1 Twelve Acres, Welwyn Garden City, AL7 4TG
M1 FEM F Priborsky, Fohrenstrasse 49, Tuttlingen, Germany, 78532
MM1 FEO Peter Gaskin, Tibertich, Kilmartin, Lochgilphead, PA31 8RQ
M1 FEQ K Watson, Bolehall Manor Club, Amington Road, Tamworth, B77 3LH
M1 FER G Watson, 85 Thomas Street, Tamworth, B77 3PP
M1 FES L Watson, 85 Thomas Street, Tamworth, B77 3PP
M1 FET Edward Dodd, 2 Chichester Crescent, Chadderton, Oldham, OL9 0RW
MW1 FEU Matthew Williams, 68 Hengoed Road, Penpedairheol, Hengoed, CF82 8BR
M1 FEW A Franklin, 11 Harden Hills, Shaw, Oldham, OL2 8NE
M1 FEX Catherine Wells, 37 Water Meadows, Worksop, S80 3DF
M1 FEY Ian Trail, 10 Hillary Drive, Crowthorne, RG45 6QE
M1 FEZ Simon West, 11 Moot Way, Woodhurst, Huntingdon, PE28 3BJ
M1 FFA D Byfield, 23 New Cross, Longburton, Sherborne, DT9 6EJ
M1 FFC P Francis, 2 Holly Close, Bridgwater, TA6 4XP
M1 FFE D Evans, 5 Brunel Road, Fairwater, Cwmbran, NP44 4QT
M1 FFF David Leech, 9 Little Gransden Lane, Great Gransden, Sandy, SG19 3BA
M1 FFG G Siviter, Flat 106, Lancaster House, Rowley Regis, B65 0QE
M1 FFM M Dawson, 140a Healey Road, Scunthorpe, DN16 1HT
M1 FFN J Willingham, 49 Creek Road, Hayling Island, PO11 9RA
M1 FFO Ernest Sanderson, Flat, Old Post Office, High Street, Yeovil, BA22 7NQ
M1 FFP B White, 7 Park Street, Castle Cary, BA7 7EH
MD1 FFR Kamal Singam, 44 Forty Lane, Wembley, HA9 9HA
M1 FFS James Bates, Flat 7, Colyer House, London, SE2 0AJ
M1 FFV J Cobb, 39 Bank House Road, Sheffield, S6 3TL
M1 FFX C Burgess, 13 Glyne Drive, Pebsham, Bexhill-on-Sea, TN40 2PW
M1 FFY R Davies, 9 Ramsay Road, Headington, Oxford, OX3 8AX
MW1 FGB R Tremelling, 45 Bryntawe Road, Ynystawe, Swansea, SA6 5AD
M1 FGH Mike James, Oak Tree House, St. Matthews Terrace, Leyburn, DL8 5EL
M1 FGM T Beswick, 37 Grovewood Road, Misterton, Doncaster, DN10 4EF
M1 FGO A Kerr-Munslow, 28 Swallow Court, St. Neots, PE19 1NP
MW1 FGV J Rowe, 41 Station Road, Ammanford, SA18 2DB
M1 FHA Thomas Earp, 14 Drakes Avenue, Devizes, SN10 5AZ
M1 FHB R Earp, The Croft, Westbury, BA13 4NY
M1 FHC Clive Earp, 84 High Street, Littleton Panell, Devizes, SN10 4EU
MI1 FHE K Lunney, 20 Inniskeen Close, Enniskillen, BT74 6HD
M1 FHJ J Bolton, 11 Forest Drive, Lytham St. Annes, FY8 4PF
MM1 FHL J Crockett, 70 Inchview Terrace, Edinburgh, EH7 6TH
MM1 FHO Leonard Norman, 16 Cotton Street, Balfron, Glasgow, G63 0PF
M1 FHP Edward Parr, 36 Ridley Drive, Great Sankey, Warrington, WA5 1HP
M1 FHQ I Anderson, 32 Wentworth Drive, Lancaster, LA1 3RJ
MM1 FHR G Hunt, 1 Love Street, Kilwinning, KA13 7LQ
MM1 FHS Neil Sampson, 47 Muirend Road, Perth, PH1 1JD
M1 FHT Antony Di Domenico, 11 Mowbreck Court, Wesham, Preston, PR4 3AG
M1 FHX D Jordan, 38 Weston Lane, Otley, LS21 2DB
MM1 FHZ David Bickle, Lon Mhor, 10 Grean, Isle of Barra, HS9 5XU
M1 FIB Paul Heathcote, Flat 6, Balmoral House, 12 Balmoral Road, Westcliff-on-Sea, SS0 7AZ
M1 FIE Alan Phelan, Calle Misericordia 14, Cadiz, Spain, 11330
M1 FIG Steven Hunt, 1 Trefusis Cottages, Flushing, Falmouth, TR11 5TE
M1 FII C Parker, 3 Lyon Close, Abingdon, OX14 1PT
M1 FIL P Smith, 12 Philip Lane, Werrington, Stoke-on-Trent, ST9 0ER
M1 FIP F O'Sullivan, 10 Hampton Close, London, NW6 5LR
M1 FIR J Coles, 45 Common Lane, Titchfield, Fareham, PO14 4BX
MI1 FIS D Stewart, 16 Weavers Lodge, Donaghcloney, Craigavon, BT66 7LE
M1 FJA R Clifford-Smith, 133 Ringwood Drive, North Baddesley, Southampton, SO52 9HF
M1 FJB Samuel Hunt, Maxxwell House, Hill Lane Business Park, Markfield, LE67 9PY
M1 FJC J Soltysik, 24 Cottage Close, Hednesford, Cannock, WS12 1BS
M1 FJD James Coleman, 52 Brook Street, Colchester, CO1 2UT
M1 FJF P Griffiths, 60 Greengate Street, Barrow-in-Furness, LA14 1EZ
M1 FJG Andrew Roberts, 15 North River Road, Great Yarmouth, NR30 1JY
M1 FJH F Bate, 42 Portefields Road, Worcester, WR4 9RF
M1 FJJ Andrew Parkinson, 21 Lambton Street, Bolton, BL3 3LG
MW1 FJK Kevin Hughes, 33 Brynglas, Penygroes, Llanelli, SA14 7PY
M1 FJL R Collins, Perch Hill Cottage, Perch Hill, Wells, BA5 1JA
MM1 FJM R Moore, Ard Na Marca, North Shurton, Gulberwick, ZE2 9TX
M1 FJP P Blackman, 73 St. Marks Road, Chester, CH4 8DE
M1 FJQ D Tinker, 116 Longley Avenue West, Sheffield, S5 8WF
M1 FLY Richard Eyre, Old Cottage, Portskewett, Caldicot, NP26 5TU
M1 FMC A Bailey, 2 Kingswood Road, Shrewsbury, SY3 8UX
M1 FMJ Shaun Smith, 44 St. Johns Road, Warminster, BA12 9LY
M1 FNE George Scott, 19 Witton Gardens, Jarrow, NE32 5YJ
M1 FRB Frederick Barnes, 4 Pound Close, Ducklington, Witney, OX29 7TH
M1 FRH F Haddock, Flat 1, 305a London Road South, Lowestoft, NR33 0DX
MI1 FRM F Montgomery, 4 Thornbrook, Lisburn, BT27 5LW
M1 FUR COULSDON AMATEUR TRANSMITTING SOCIETY c/o Andrew Briers, 33 Deans Walk, Coulsdon, CR5 1HR
M1 FWD S Davies, 17 Priory Gardens, Pilton, Barnstaple, EX31 1PT
M1 FZL Peter Haskins, 62 Peartree Road, Broomfield, Herne Bay, CT6 7EE
MM1 FZR S Gillies, 49 Meadowside Road, Queenzieburn, Glasgow, G65 9EJ
M1 GAP A Prince, 29 St. Stephens Road, West Bromwich, B71 4LR
MM1 GAR A Gardner, West Winds, Dennistoun Road, Port Glasgow, PA14 6XH
MM1 GBS D James, 9 Dunbar Lane, Duffus, Elgin, IV30 5QN
M1 GCS G Steedman, 61 Granville Street, Barnsley, S75 2TQ
M1 GDB G Bell, 4 Fairley Way, Cheshunt, Waltham Cross, EN7 6LG
M1 GDE G Edgar, 61 Winchester Avenue, Lancaster, LA1 4HX
M1 GDH D Hayward, 16 Heathway, Dagenham, RM10 9PP
M1 GEO George Smart, Old Queens Head, Ipswich Road, Diss, IP21 4XP
M1 GFE F Ettridge, 17 Head Street, Goldhanger, Maldon, CM9 8AY
M1 GGG Geoffrey Ma, 26 Church Lane, Chalgrove, Oxford, OX44 7TA
M1 GHT David Buckerfield, 62 Springfield Crescent, Somercotes, Alfreton, DE55 4LH
M1 GIZ S Bridger, 80 Springhill Crescent, Madeley, Telford, TF7 4DP
MW1 GLD G Davies, 77 Tydraw Street, Port Talbot, SA13 1BR
M1 GMO M Hastry, 56 Kilsyth Close, Fearnhead, Warrington, WA2 0SQ
M1 GOH R Horry, 1 Nidds Lane, Kirton, Boston, PE20 1LZ
M1 GPC G Carpenter, 67 Angela Crescent, Horsford, Norwich, NR10 3HE
M1 GPE G Emmerson, 29 Dulsie Road, Talbot Woods, Bournemouth, BH3 7DY
M1 GRA Graham Stephens, 2 Limousin Way, Bridgwater, TA6 6GR
M1 GSM S Watson, 6 Mount Pleasant, Stanley, Crook, DL15 9SF
M1 GTI D Burgin, 15 Birch Grove, Chippenham, SN15 1DD
M1 GUR Peter Gurney, Flat 1215, De Montfort House, Leicester, LE1 5XS
M1 GUS J Batchelor, 5 Gladden Fields, South Woodham Ferrers, Chelmsford, CM3 7AH
M1 GWA G Warburton, 50 Clarendon Road, Sheffield, S10 3TR
M1 GXL Terry Higginson, 109 London Road, Biggleswade, SG18 8EE
M1 HFM M Poole, 18 Lockway, Drayton, Abingdon, OX14 4LG
M1 HFX R Ayers, Flat 2, 35 Commercial Road, Weymouth, DT4 7DY
M1 HGV Martin Rule, 23 Rue Du Puits Doux, Sacy, St Christophe A Berry, France, 2290
M1 HHL B Miller, Oakhurst, 1 Southern Oaks Barton on Sea, New Milton, BH25 7JT
M1 HJE S Elliott, 7 Manor Close Harston, Cambridge, CB22 7QF
M1 HLG Hilary Glover, 9 Willingdon Drove, Eastbourne, BN23 8AL
M1 HLL S Batchelor, 2 Belmont Avenue, Atherton, Manchester, M46 9RR
M1 HMP P Grech, 108 Hind Grove, London, E14 6HU
MM1 HMV Brian Shearer, Latheron, 113 Auchamore Road, Dunoon, PA23 7JJ
MM1 HMZ B Allison, 5 Mayfield Drive, Howwood, Johnstone, PA9 1BJ
M1 HOP A Hopson, 1 Hall Lane, Leicester, LE2 8SF
M1 HQX William Hammond, 28 Fengate Mobile Home Park, Peterborough, PE1 5XD
M1 HVJ Alan Jefferiss, 27 Sherbourne Drive, Maidenhead, SL6 3EP
MM1 HWB P Oldham, 2 Old Bar Road, Nairn, IV12 5BX
M1 HZR C Best, 25 Park Way, St. Austell, PL25 4HR
M1 HZZ Andrew Barbour, 36 Roseacre Drive, Elswick, Preston, PR4 3UQ
M1 IAN I Tennent, Flat 2, 97 Rydens Road, Walton-on-Thames, KT12 3AW
MM1 ICE A Somerville, 8 Craiglockhart Park, Edinburgh, EH14 1HE
M1 IDE C Edis, 21 Stocks Road, Kimberley, Nottingham, NG16 2QF
M1 IFT A Bartle, 10 Holme Dene, Haxey, Doncaster, DN9 2JX
M1 IHM Dylan Jones, 32 Darliston, Telford, TF3 2DP
M1 IKE Mike Collins, 3 Beacon View, Grayrigg, Kendal, LA8 9BT
M1 IOS J Goody, 9 Garrison Lane, St. Mary's, Isles of Scilly, TR21 0JD
M1 IOW P Legg, 20 Arthur Moody Drive, Newport, PO30 5JR
M1 IRB Ian Bush, 17 Queens Place, Shoreham-by-Sea, BN43 5AA
M1 IRM Philip Rowley, 6 Duesbury Green, Longton, Stoke-on-Trent, ST3 2RZ
MM1 JAA A Asbury, 18 Ballaig Avenue, Bearsden, Glasgow, G61 4HA
MM1 JAC J Campbell, 119 Campbell Avenue, Dumbarton, G82 3PB
M1 JAK Alan Hanson-Brown, 35 York Avenue, Bedworth, CV12 9EL
M1 JAN J Carfoot, 11 Parc Sychnant, Conwy, LL32 8SB
MM1 JAS J Shankland, 2 Strathdoon Place, Ayr, KA7 4PB
M1 JCB Timothy Wightman, Laithbutts Farm, Cowan Bridge, Carnforth, LA6 2JL
M1 JCL J Plant, 67 Kenley Road, London, SW19 3JJ
M1 JCS Christopher Starr, 64 Green Lane, Lambley, Nottingham, NG4 4QE
M1 JDW J Mitchell, Sheet Hill Farmhouse, Winfield Lane, Sevenoaks, TN15 0LZ
M1 JEC J Cook, 35 Holdbrook Way, Romford, RM3 0JD
M1 JES J Gilbert, Thr Oaktree, Ellenbrook Lane, Hatfield, AL10 9NT
M1 JHG John Green, 33 Edenvale Crescent, Lancaster, LA1 2NW
M1 JHL Surjit Jouhal, 35 Cherrywood Gardens, Nottingham, NG3 6LR
M1 JIM J O'Hea, 12 De Vitre Place, Grove, Wantage, OX12 0DA
M1 JJN J Nicholson, 6 Mill Gardens, West End, Southampton, SO18 3AG
M1 JJS P Springate, 10 Pipers Close, Burnham, Slough, SL1 8AW
M1 JKB J Brown, 10 Lomond Avenue, Sinfin, Derby, DE24 3HH
M1 JLM B Murfitt, 21 Priors Drive, Norwich, NR6 7LJ
M1 JMB J Bevan, 1 Condor Close, Weston-Super-Mare, BS22 8SE
MI1 JOE W Murray, 80 Canterbury Park, Londonderry, BT47 6DU
M1 JON J Godding, 58 Dukeswood Road, Longtown, Carlisle, CA6 5UJ
M1 JPS John Patterson, 39 Coquet Drive, Ellington, Morpeth, NE61 5LN
M1 JSS J Smout, Sunrays, Warbage Lane, Bromsgrove, B61 9BH
M1 JTA John Tyers, Sheldawn, Trent Lane, Newark, NG23 7HL
MM1 JWF James Frame, 24 Douglas Crescent, Erskine, PA8 6BJ
M1 JWM J Machin, 42 Woodstock Road, Loxley, Sheffield, S6 6TG
M1 JWR J Rutherford, Nook On Lyne, Longtown, Carlisle, CA6 5TS
M1 JWS J Smith, 134 Blaker Court, Fairlawn, London, SE7 7EU
M1 KAZ A Forrest, 261 East End Road, London, N2 8AY
M1 KCB K Crank, 319 Manchester Road, Clifton, Manchester, M27 6PT
M1 KDH K Harvey, 29 The Hobbins, Bridgnorth, WV15 5HH
M1 KDJ K Jowett, 4 Crosslanes, Purton Stoke, Swindon, SN5 4JN
M1 KDP M Heron, Heron House, Park Road, Gwynedd, LL42 1PL
M1 KEJ Richard Jeffs, 45 Forest Road, Bingham, Nottingham, NG13 8RL
M1 KES M Oconnor, 13 Ashburnham Road, Southend-on-Sea, SS1 1QB
M1 KEV K Mahoney, 11 Leyland Walk, Bristol, BS13 8PY
M1 KEY Michael Hardy, 4 Kirk Balk, Hoyland, Barnsley, S74 9HU
M1 KGL K Large, 6 Sylvden Drive, Wisbech, PE13 3UD
M1 KIP L Kipling, 16 Northampton Close, Bracknell, RG12 9EF
M1 KMC Andrew Coathup, 54 Rydal Road, Kendal, LA9 6LB
M1 KOS K Tsioumparakis, 10 Lavender Close, Leatherhead, KT22 8LZ
M1 KPW K Whitmarsh, 7 Foxs Furlong, Chineham, Basingstoke, RG24 8WN
M1 KSB Keith Best, 42 Falmer Avenue, Goring-by-Sea, Worthing, BN12 4TD
M1 KTA Robert Baines, 34 Bury Road, Stapleford, Cambridge, CB22 5BP
M1 KTY Katie Mallows, 57 Top Road, Kingsley, Frodsham, WA6 8DA
M1 KVN K Finn, 132 Lansdowne Grove, Wigston, LE18 4LY
M1 KWH K Hargreaves, Langton Lodge, Fordcombe Road, Tunbridge Wells, TN3 0RB
M1 LAN A Worsley, 10 Millfield View, Worksop, S80 3QB
M1 LAP L Pollard, 45 Nanny Marr Road, Darfield, Barnsley, S73 9AB
MM1 LBA A Bulloch, 4 Cartleburn Gardens, Kilwinning, KA13 7ND
M1 LCL Alfons Kvilums, 78 Wagon Lane, Solihull, B92 7PN
M1 LEO A Bennett, 16 Manor Avenue, Crewe, CW2 8BD
M1 LES Leslie Rodger, 19 South Walk, West Wickham, BR4 9JA
M1 LIP A Lippett, 2 Ralph Court, Stafford, ST17 9FR
MM1 LJB C Newman, Upper Flat, 3 Lindsay Gardens, Alexandria, G83 0US
M1 LLL M Greatorex, Cwm Pennant, Moel View Road, Prestatyn, LL19 9SU
M1 LMJ L Jones, 47 Pine Crescent, Chandler's Ford, Eastleigh, SO53 1LN
M1 LMO Neil Waring, 3 Sampson St., Eastoft, Scunthorpe, DN17 4PQ
M1 LOL A King, 8 Rydal Court, Congleton, CW12 4JL
M1 LOU Robert Cave, 26 Longsight Road, Mapplewell, Barnsley, S75 6HB
M1 LRX David Horwood, 42 Southlands Drive, Timsbury, Bath, BA2 0HB
M1 LSD Lee Dawes, 52 Ridley Road, Carlisle, CA2 4LD
M1 LSG A Jenkins, 25 Maes Hyfryd, Flint, CH6 5LN
M1 LTS Liam Stone, 6 Lyntons, Pulborough, RH20 1AZ
M1 LXM Alex May, 7 Stanton Close, Blandford Forum, DT11 7RT
M1 LYE F Lye, 5 New Road, Hextable, Swanley, BR8 7LS
M1 LYN L Asbury, 67 Orchard Way, Measham, Swadlincote, DE12 7JZ
M1 MAB I Patrick, 17 Stamford Way, Fair Oak, Eastleigh, SO50 7JJ
M1 MAD M Cottrell, 9 Woodland Terrace, Kingswood, Bristol, BS15 9PU
M1 MAJ Martyn Johnson, 2a St. Margarets Road, Girton, Cambridge, CB3 0LT
M1 MAL Malcolm Cadman, Flat 17, Harmon House, London, SE8 3AS
M1 MBZ M Stephens, 45 Ham Farm Lane, Emersons Green, Bristol, BS16 7BW
M1 MCL Martyn Lawson, 9 Headingley Mews, Wakefield, WF1 3AB
M1 MCW Stuart Lace, 19 Methuen Street, Walney, Barrow-in-Furness, LA14 3PS
M1 MDE D Elwood, High Farm Cottage 7, Newport Road, Market Drayton, TF9 2TH
M1 MDP Michael Palmer, 57 Bemersley Road, Stoke-on-Trent, ST6 8JF
MW1 MFY D Lee, 13 Yr Efail, Treoes, Bridgend, CF35 5EG

MM1 MHD P Overton, Cluanie, Cairnballoch, Alford, AB33 8HQ
M1 MHZ C Offer, Chapel Yard Cottage, Quadring Eaudyke, Spalding, PE11 4QB
M1 MIC M Hodgson, 17 Athlone Terrace, Armley, Leeds, LS12 1UA
M1 MIJ W Waddington, 117 Dominion Ct, Walney, Barrow in Furness, LA14 3DP
M1 MKL M Livesey, 33 Carrington Close, Birchwood, Warrington, WA3 7QA
M1 MLM A Lomas, 32 Crestway Road, Baddeley Green, Stoke-on-Trent, ST2 7LD
M1 MNR MID NORTON RAYNET GROUP c/o L Knighton, 5 Quidenham Road, East Harling, Norwich, NR16 2JD
M1 MOB M Dimambro, 26 Fetcham Court, Bank Top, Newcastle upon Tyne, NE3 2UL
M1 MOD A Straw, 348 London Road, Charlton Kings, Cheltenham, GL52 6YT
M1 MOG M Taylor, 136 Lenthall Avenue, Grays, RM17 5AB
MM1 MOY John Dye, Allt na Slanaichd, Moy, Inverness, IV13 7YE
M1 MPA M Allgar, 13 Deacon Avenue, Kempston, Bedford, MK42 7DU
M1 MPB Mark Burfield, 78 Glebe Hey Road, Wirral, CH49 8HQ
M1 MPK M Kassai, 6 Cranhill Close, Littleover, Derby, DE23 3XU
M1 MPW Mark Wooldridge, 26 Little Meadow, Bar Hill, Cambridge, CB23 8TD
M1 MRB Martin Butler, 210 Green Wrythe Lane, Carshalton, SM5 2SP
M1 MRS Robert Shepperley, Flat F, London, N3 1QL
M1 MSF Michael Forster, 4 West Street, Top Flat, Horncastle, LN9 5JF
M1 MST C Walker, Fairfield House, Brimfield, Ludlow, SY8 4ND
M1 MTV A Mason, 76 Burlington Way, Mickleover, Derby, DE3 9BD
M1 MUM P Worledge, 181 Roselands Drive, Paignton, TQ4 7RN
M1 MUS M Suleyman, 26 Old Park Road South, Enfield, EN2 7DB
M1 MVX Anthony Webb, 5 Highfield Avenue, Bristol, BS15 3RA
M1 NAD D Barratt, Chapel Cottage, Briantspuddle, Dorchester, DT2 7HX
M1 NAS Nigel Swann, 11 Erdyngton Road, Leicester, LE3 1JF
M1 NCC N Cordell, 392 Laceby Road, Grimsby, DN34 5LX
M1 NEW Karl Mason, School House, Redwing Drive, Weston-Super-Mare, BS22 8XJ
M1 NHR NORTH HERTFORDSHIRE RAYNET ASSOC. c/o K Edwards, 289 Monks Walk, Buntingford, SG9 9DZ
M1 NIS B Ashcroft, 16 Edge Lane, Crosby, Liverpool, L23 9XE
M1 NIZ Mark Austin, 11a Beverley Road, Ipswich, IP4 4BU
M1 NMG N Gunnell, 8 Upper Dingle, Madeley, Telford, TF7 5RX
M1 NNN John Earnshaw, Dunelm, Ayton Road, Scarborough, YO12 4RQ
M1 NPH Nigel Holland, 40 Marlborough Road, Castle Bromwich, Birmingham, B36 0EH
M1 NSC NATIONAL SPACE CENTRE AMATEUR RADIO SOC. c/o Geoffrey Griffiths, 14 Mansion House Gardens, Melton Mowbray, LE13 1LE
M1 NTV N Tartt, 47 Leatham Park Road, Featherstone, Pontefract, WF7 5DP
M1 NTY D Basson, Penrhyn, Stonehouse Road, Sevenoaks, TN14 7HW
M1 NUS R Cartwright, 14 Bromsgrove Avenue, Eccles, Manchester, M30 8WB
M1 NXX J Lynch, 14 The Pastures, Cayton, Scarborough, YO11 3UU
M1 OBR David O'Brien, 14 Lower Bettesworth Road, Ryde, PO33 3EL
M1 OCN C Wilson, M/V Wolf, Chandlers Quay, Maldon, CM9 4LF
M1 ONE C Chambers, 75a Main Street, Sedbergh, LA10 5AB
MI1 OPM David Kirkwood, 1 Rural Cottages, Front Road, Lisburn, BT27 5LF
M1 OXR Andrew Garthwaite, 278 Carlton Road, Barnsley, S71 2BA
M1 PAB P Bush, 1a Sherwood Close, Kennington, Ashford, TN24 9PT
M1 PAC Philip Cole, 48 Emily Street, Keighley, BD21 3HY
M1 PAF Paul Fletcher, 7 Gatesyde Place, Eskdale, Holmrook, CA19 1UD
M1 PAH D Hardy, 1 Upper Steeping, Desborough, Kettering, NN14 2SQ
M1 PAM P Rodger, 19 South Walk, West Wickham, BR4 9JA
M1 PAS W James, 3 Midfield Close, Gillow Heath, Stoke-on-Trent, ST8 6RD
M1 PFS P Symonds, 29 Gladstone Road, London, SW19 1QU
M1 PGH P Howell, 16 Everard Road, Bedford, MK41 9LD
M1 PGT Peter Tomlin, 10 The Martins, Thatcham, RG19 4FD
M1 PJB Peter Backx, 43 Tindale Avenue, Cramlington, NE23 2BP
M1 PJH Peter Hill, Flat 12, Danecourt 14 St. Peters Road, Poole, BH14 0PA
M1 PKB P Booth, 19 Gunville Crescent, Bournemouth, BH9 3PZ
M1 PKW P Wilton, Downsview Cottage, Wappingthorn Farm Lane, Steyning, BN44 3AG
M1 PLC Mathew Woods, 76 Kings Road, Evesham, WR11 3BS
M1 PMR A Marshall, 53 Birch Close, Corfe Mullen, Wimborne, BH21 3TB
M1 PRC Peterborough & District ARC c/o Ronald Smith, 29 George Street, Peterborough, PE2 9PD
M1 PRO Jason Woodman, 139 Central Avenue, Hayes, UB3 2BT
M1 PTE P Merrick, 12 Wilkinson Road, Wednesbury, WS10 8SH
M1 PTR P Ridley, 6 Elm Close, Poynton, Stockport, SK12 1QH
M1 PTT Chris Morrison, 28 Grindle Close, Thatcham, RG18 3PD
M1 PUW P Rogers, 39 Irby Road, Bristol, BS3 2LZ
M1 PVC Paul Craven, Hithe, Colemans Hatch, Hartfield, TN7 4EN
M1 PVF P Flavell, 26 Hulles Way, North Baddesley, Southampton, SO52 9NS
M1 PWT P Telco, 7 Brockswood Lane, Welwyn Garden City, AL8 7BA
M1 PXB J Hopkins, Millinder House, Westerdale, Whitby, YO21 2DE
M1 PYE J Pye, 19 Nellan Crescent, Stoke-on-Trent, ST6 1PS
M1 RAD Daniel Clapp, 35 St. Elmo Road, Worthing, BN14 7EJ
MM1 RAH R Hemesley, 32 Meadowbank Road, Kirknewton, EH27 8BS
M1 RAL Richard Leach, 4 Honey Pot Drive, Shipley, BD17 5TJ
MI1 RAY R Maguire, 139 Carrowshee Park, Drumhaw, Enniskillen, BT92 0FS
MI1 RDR R Ross, 13 Eureka Drive, Belfast, BT12 5NR
M1 RDX A Casson, 111 Risedale Road, Barrow-in-Furness, LA13 0QY
M1 REC Gary Doughty, 1a The Crescent, Ketton, Stamford, PE9 3SY
M1 REJ R Jacklin, 63 Ventnor Rise, Heathfield, Nottingham, NG5 1NW
M1 REK Raymond King, 8 Rydal Court, Congleton, CW12 4JL
M1 RES F Garrett, 11 Third Avenue, Flint, CH6 5LT
M1 RGW R Warren, 23 Bramshaw Close, Winchester, SO22 6LT
M1 RIC R Booth, 108 Newlands Gardens, Workington, CA14 3PE
M1 RIG David Gleadell, 4 Smith House Avenue, Brighouse, HD6 2LE
MM1 RIK R Irvine, 83 Glenacre Road, Cumbernauld, Glasgow, G67 2NT
M1 RJG Robert Gooderham, 3 School Meadow, Barnby, Beccles, NR34 7QL
M1 RJJ J Fowler, 48 Grangefields Road, Jacob's Well, Guildford, GU4 7NP
M1 RJL Robin Lapwood, Rose Cottage, Challow Road, Wantage, OX12 9DN
M1 RJS R Speak, 87 Sea View Avenue, West Mersea, Colchester, CO5 8BX
M1 RKB R Browning, 575 Rayleigh Road, Leigh-on-Sea, SS9 5HR
M1 RKY M Harriott, 20 Old Road, Tean, Stoke-on-Trent, ST10 4EG
MM1 RMS R Scott, Morven, Isle of Lewis, HS2 0QX
M1 RMW Richard Warry, 51 South Street, Crewkerne, TA18 8DB
M1 ROD R Jones, 7 Warner Avenue, North Cheam, Sutton, SM3 9RH
M1 ROE Alex Roebuck, Holford Farm, Chester Road, Knutsford, WA16 0TZ
MD1 RPC R Woolley, Moaralyn, King Williams Road, Castletown, Isle of Man, IM9 1DL
M1 RSB R Bain, Oak Tree Cottage, Long Barn Road, Sevenoaks, TN14 6NH
M1 RSJ R Bayliss, 10 Quarry Road, Dudley, DY2 0EF
M1 RST R Dixon, 22 Hobbs Court, Hardings Lane, Eastleigh, SO50 8AG
M1 RTT R Thong, 10 Lowry Close, Haverhill, CB9 7GH
M1 RWB R Blears, The Manor House, Warrington Road, Chester, CH2 4EA
M1 SAB S Armatage, Hayleazes, Lincoln Hill, Hexham, NE46 4BE
M1 SAC S Clark, 22 Church Street, Alwalton, Peterborough, PE7 3UU
M1 SAM S Evans, 44 Edwards Drive, Plymouth, PL7 2SU
M1 SAN Roger Sanders, Magnolia Cottage, Lanreath, Looe, PL13 2NX
MW1 SAS A Jones, 4 Crowhill, Haverfordwest, SA61 2HL
M1 SAZ S Haynes, 4 Monkend Terrace, Croft on Tees, Darlington, DL2 2SQ
M1 SCS Simon Smith, 113 Deaconsfield Road, Hemel Hempstead, HP3 9JA
M1 SCW S Whiteman, 19 St. Peters St., Syston, Leicester, LE7 1HL
M1 SEM Shane Rear, 18 Brook Lane Cottages, Sellindge, Ashford, TN25 6HG
M1 SFS SUMMERFIELDS ARC c/o Robin Lapwood, Rose Cottage, Challow Road, Wantage, OX12 9DN
M1 SGA S Ash, 1 Twyford Mill, Pig Lane, Bishops Stortford, CM22 7PA
M1 SHA S Creber, 4 Joyce Close, Swindon, SN25 4GX
M1 SHE J Little, 7 Deerfern Close, Great Linford, Milton Keynes, MK14 5BZ
M1 SIM Christine Simcock, 51 Broadway, Stockport, SK2 5SF
M1 SIN T Brundrett, 45 Talbot Crescent, Whitchurch, SY13 1PH
M1 SJA A Stockwell, 28 Scholars Walk, Chalfont St. Peter, Gerrards Cross, SL9 0EJ
M1 SJE Sarah Elliott, 2 Pytchley Close, Leicester, LE4 2PZ
M1 SJH S Harrison, 64 Douglas Road, Leigh, WN7 5HG
M1 SKA D Spencer, 4 Douglas Road, Copnor, Portsmouth, PO3 6AU
M1 SKI A Grabianski, 29 Lismore Road, Highworth, Swindon, SN6 7HU
M1 SKY Anthony Johnston, 17 Andover Street, Barrow-in-Furness, LA14 2SZ
M1 SLH Kieran Taylor, 23 The Chestnuts, Abingdon, OX14 3YN
M1 SMF S Flanagan, 33 Ullswater Road, Chorley, PR7 2JB
M1 SNM M Wilson, 3 Brookhurst Close, Chelmsford, CM2 6DX
M1 SPW N Walker, 79 Arklow Drive, Hale Village, Liverpool, L24 5RR
M1 SPY Stephen Pybus, Grewgrass Farm, Grewgrass Lane, Redcar, TS11 8EB
M1 SRC SURREY RAYNET c/o Timothy Dabbs, 4 Caverleigh, Cadogan Road, Surbiton, KT6 4DH
M1 SRH S Howlett, 20 Long Perry, Capel St. Mary, Ipswich, IP9 2XD
M1 SRP S Pickin, 138 Boreham Field, Warminster, BA12 9EF
M1 SSB Steve Bygrave, 19 Kent Road, Lowestoft, NR32 2HW
M1 STI B Hall, 5 Perche Court, Midhurst, GU29 9TE
M1 SUE S Gunn, 5 School Villas, Broxted, Dunmow, CM6 2BS
M1 SUM D Sumner, 30e Malvern Avenue, Ellesmere Port, CH65 5AD
M1 SWB Steve Bainbridge, 6 Sandyville Grove, Liverpool, L4 8UL
M1 SWL INTERNATIONAL SHORTWAVE LEAGUE c/o Arthur Kinson, 6 Uplands Park, Broad Oak, Heathfield, TN21 8SJ
M1 SWR S Rigby, 29 Grovefields, Leegomery, Telford, TF1 6YL
M1 SWS S Southworth, 157 Birkwood Avenue, Cudworth, Barnsley, S72 8JB
MM1 SYD S Mccance, 34 Calside, Paisley, PA2 6DB
M1 SYG Antony Sygerycz, 75 O'Brien Road, Cheltenham, GL51 0UP
M1 TAD T Denby, 22 Okehampton Crescent, Welling, DA16 1DE
M1 TAF M Williams, 1 Gomer Court, Abergele, LL22 7UU
M1 TAP Alan Prince, 1 Woodside, Inglesbatch, Bath, BA2 9DZ
M1 TAT Jonpaul Pymm, Larkfield, Goxhill Road, Barrow-upon-Humber, DN19 7EE
M1 TAZ Colin Curtis, 69 Lerwick Croft, Bicester, OX26 4XX
M1 TCI T Cheeseman, 1 Queens Avenue, Birchington, CT7 9QN
M1 TCP Peter Braidwood, 77 Pheasant Way, Cirencester, GL7 1BJ
M1 TCQ James Cheese, Unit 2, Cinnabar House, Morecambe, LA4 5BW
M1 TCR T Rozier, 124 Deansfield Road, Wolverhampton, WV1 2LD
M1 TDD Trevor Denham, 41 Brimfield Road, Purfleet, RM19 1RQ
M1 TEM T Murphy, 13 Avon Road, Melton Mowbray, LE13 0EJ
M1 TES James Crawford, 23 Meadow Road, Bungay, NR35 1LE
M1 TET C Causby, 5 Blenheim Drive, Higher Folds, Leigh, WN7 2YR
M1 TLK Robert Sanderson, 7 Faraday Drive, Milton Keynes, MK57DE
M1 TMF Anthony Fullwood, 16 Hollands Place, Walsall, WS3 3AU
M1 TOD R Todd, 26 Dene Road, Guildford, GU1 4DD
M1 TOM T Boardman, 9 Elm Grove, Farnworth, Bolton, BL4 0AY
M1 TRC Chris Marren, 7 Mill Lane, East Ardsley, Wakefield, WF3 2BL
M1 TSC TRINITY SCHOOL RC c/o R Evans, 7 Westland Drive, Hayes, Bromley, BR2 7HE
M1 TSU I James, 21 Evelyn Way, Irchester, Wellingborough, NN29 7AP
M1 TUG J WILSON, 36a Havelock Road, Maidenhead, SL6 5BJ
M1 TVR M Bryan, 16 Walesmoor Avenue, Kiveton Park, Sheffield, S26 5RG
M1 TXT P Dimambro, 26 Fetcham Court, Bank Top, Newcastle upon Tyne, NE3 2UL
M1 TZR P Hadfield, 68 Foxholes Road, Hyde, SK14 5AW
M1 UKC B Welland, 171 Hillcrest Road, Newhaven, BN9 9EZ
M1 ULD Geoff Auld, Victoria Villa, Sheepwash, Choppington, NE62 5NG
MI1 UNA U Murray, 80 Canterbury Park, Londonderry, BT47 6DU
M1 VCD John Roberts, 13 Maes-y-Coed, Gwersyllt, Wrexham, LL11 4PF
M1 VGH A Hutchinson, 118 York Road, Haxby, York, YO32 3EG
M1 VHF B Burden, 18 Challenor Close, Finchampstead, Wokingham, RG40 4UJ
M1 VHT Keith Morrison, 31 Simonside Crescent, Hadston, Morpeth, NE65 9YA
M1 VIP Andrew Evans, Apartment 28, Stocks Court, 2 Harriet Street, Manchester, M28 3JW
M1 VLS B Wilson, 131 Denmark Road, Beccles, NR34 9DW
MI1 VOX D McCullough, 21 Ophir Gardens, Belfast, BT15 5EP
M1 VPL A Westland, 8 Faris Barn Drive, Woodham, Addlestone, KT15 3DZ
M1 VPN D Coombes, The Old School, Combe Raleigh, Honiton, EX14 4UL
M1 VRC R Broadley, The Smithy, Elstronwick, Hull, HU12 9BP
M1 VSR C Worthington, 63 Torkington Road, Hazel Grove, Stockport, SK7 4RL
MM1 VTB C Budas, 20 Oak Avenue, Bearsden, Glasgow, G61 3HD
M1 WAW W Waller, 4 New Road, Sheerness, ME12 1BW
M1 WAZ S Eggleton, 12 Cedar Grove, Trowbridge, BA14 0HS
M1 WDK S Papworth, Spring Cottage, Gringley Road, Doncaster, DN10 4HT
M1 WDX F Buck, 89 Marsh Street, Barrow in Furness, LA14 2AD
M1 WEH M Harrold, 29 Barnford Crescent, Warley, Oldbury, B68 8PP
M1 WEJ W Jones, 53 Bro Enddwyn, Dyffryn Ardudwy, LL44 2BG
M1 WHO S Calver, 82 Kristiansand Way, Letchworth Garden City, SG6 1UE
M1 WIN Christopher Higgins, 52 Pittsfield, Cricklade, Swindon, SN6 6AW
MM1 WKH Alan Cartledge, Ard Shonas, Rogart, IV28 3XE
M1 WRX J Mallows, 57 Top Road, Kingsley, Frodsham, WA6 8DA
M1 WTL W Leverett, 514 Arleston Lane, Stenson Fields, Derby, DE24 3AG
M1 WVS E Brown, Rose Cottage, Grindlow, Buxton, SK17 8RJ
M1 WWW Andrew Varley, Bank Farm, Matlock Road, Spitewinter, Chesterfield, S45 0LL
MW1 WYN Wyn Britten-Jones, 101 Mill View Estate, Maesteg, CF34 0DE
M1 XCG Christopher Gaskell, 109 Soughers Lane, Ashton-in-Makerfield, Wigan, WN4 0JT
MM1 XJS Kenneth Brown, 21 Strain Crescent, Airdrie, ML6 9ND
M1 XRC T Crellin, 2 Senlac Green, Uckfield, TN22 1NN
M1 XTN Stuart Morris, Undercliffe Villas, 4 Carnarvon Road, Reading, RG1 5SD
M1 XXT C Willetts, The Retreat, Wood Lane, Colchester, CO3 9TR
M1 XZG R Mckenzie, 19 Goldfinch Drive, Cottenham, Cambridge, CB24 8XY
MM1 YAM Clive Allanson, Five Acres, Lairg, IV27 4DG
M1 YOW D Bland, 16 Tennyson Avenue, Grays, RM17 5RG
M1 ZAR J Wilkinson, 26 Hazelwood Grove, South Croydon, CR2 9DU
M1 ZEM James Ogden, 14 Bishops Close, Little Downham, Ely, CB6 2TQ
M1 ZXG Nigel Moffat, 22 Churchill Way, Acklington, Morpeth, NE65 9DB
M1 ZXZ A Clarke, 31 Northern Rise, Great Sutton, Ellesmere Port, CH66 4QY
M1 ZZA C Thomas, 55 High Street, Aylburton, Lydney, GL15 6BZ
M1 ZZY T Quinn, 43 Stirtingale Road, Bath, BA2 2NG

## M*3

MD3 AAI M Rish, 3 Lime St., Port St. Mary, Isle of Man, IM9 5ED
M3 AAQ Lyndon Handley, 4 Manor Close, Draycott, Stoke-on-Trent, ST11 9AZ
M3 AAS P Rixon, 52 New Road, Hatfield Peverel, Chelmsford, CM3 2JA
M3 AAY Neil Sanderson, 54 Kelvedon Close, Chelmsford, CM1 4DG
M3 ABQ J Sharp, 3 Inkerman Road, Eton Wick, Windsor, SL4 6LE
M3 ABY J Turnbull, 32 Haydon, Washington, NE38 8PF
M3 ACA G Reeds, 26 Holme Leaze, Steeple Ashton, Trowbridge, BA14 6EH
M3 ACF Michael Buckley, Springfield, 12 Ranmore Avenue, Croydon, CR0 5QA
M3 ACU W Ashley, 23 Lavenham Close, Clacton-on-Sea, CO16 8BZ
M3 ACY J Bailey, 13 Newark Road, Mexborough, S64 9EZ
M3 ADB Andrew Bottrill, Woodside, 1 Gilling Close, Richmond, DL10 5AY
M3 ADJ B Waterloo, 55 Solent Road, Hill Head, Fareham, PO14 3LB
M3 ADL P Jefford, 61 Willow Way, Flitwick, Bedford, MK45 1LN
M3 ADT R Vickerstaff, 16 Sewell Wontner Close, Kesgrave, Ipswich, IP5 2GB
M3 ADX Glyn Evans, 24 Beaufort Road, Billericay, CM12 9JL
M3 AEA P Ewington, 26 Dickens Road, Rugby, CV22 5RW
M3 AEE A Buckman, 116 Ashling Park Road, Waterlooville, PO7 6EG
M3 AEJ Alan Pearce, 8 Carworgie Way, St. Columb Road, St. Columb, TR9 6PT
M3 AEZ S Hall, 122 Norwich Road, New Costessey, Norwich, NR5 0EH
M3 AFF A Barnes, 3 Sparks Villas, Black Torrington, Beaworthy, EX21 5PX
M3 AFR K Barry, Flat 1, 24 Vale Street, Denbigh, LL16 3BE
M3 AFS I Dwyer, 39 Berry Road, Newquay, TR7 1AS
MW3 AFX L Cook, 8 Swn Yr Afon Flats, Cefn Coed, Merthyr Tydfil, CF48 2SA
M3 AGA P Bore, 29 Edgerton Road, Lowestoft, NR33 9BG
M3 AGB A Bennett, 16 Manor Avenue, Crewe, CW2 8BD
M3 AGE David Doroba, Flat 3, 305a London Road South, Lowestoft, NR33 0DX
M3 AGH M Arnold, 27 Poplar Avenue, Bentley, Walsall, WS2 0NT
M3 AGI M Taylor, 29 Ferndale Avenue, Reading, RG30 3NQ
MI3 AGR P McDaid, 66 Laurel Drive, Strabane, BT82 9PN
M3 AHJ R Selwood, 33 Chandlers, Sherborne, DT9 3RT
M3 AHL D Furness, 9 Ouzel Drive, Bradford, BD6 3YN
M3 AHO M Coulson, 64 Craddock Street, Spennymoor, DL16 7TA
M3 AHQ James Maloney, 196 Finchale Road, Hebburn, NE31 2BW
M3 AHR T Hamilton, 16 Weardale Street, Spennymoor, DL16 6ER
M3 AHS Ann Stevenson, 97 Queen Street, Crewe, CW1 4AL
M3 AHU Gerard Taylor, 63 Millbrook Towers, Stockport, SK1 3NL
M3 AHZ Alma Hardman, 47 Oatlands Road, Manchester, M22 1AH
M3 AIE Christopher Gibson, 11 Parkside Avenue, Queensbury, Bradford, BD13 2HQ
M3 AIG Keith Rendell, 18 Bonfire Close, Chard, TA20 2EG
M3 AIL G Langdon, 43 Daniel Street, Ryde, PO33 2BH
MI3 AIN R Martin, 23 Scaddy Road, Downpatrick, BT30 9BW
M3 AIR Philip Robinson, 13 Carrfield Avenue, Liverpool, L23 9SS
M3 AIS Steven Martin, 25 Wellswood Road, Ellesmere Port, CH66 1JX
M3 AIZ J Smith, 44 Knapp Way, Malvern, WR14 1SG
M3 AJA J Crowhurst, 5 Hampshire Road, Canterbury, CT1 1SJ
M3 AJC A Charbit, 65 Bourne Street, London, SW1W 8JW
M3 AJD I Humberstone, 20 Kingswood Road, Colchester, CO4 5JX
MI3 AJK D Poots, 18 Upper Quilly Road, Dromore, BT25 1NP
M3 AJN S Panczol, 11 Chauncy Road, Manchester, M40 0GG
M3 AJS J Hunt, 11 Vicarage Lane, Poynton, Stockport, SK12 1BG
M3 AJU R Smith, 41 Middle Deal Road, Deal, CT14 9RG
M3 AJV Anthony Brown, 23 Yew Tree Drive, Chesterfield, S40 3NB
M3 AJW A Walmough, Apartment 31, Wyatville House, Buxton, SK17 6WJ
M3 AKE R Hoey, 225 King Avenue, Bootle, L20 0DY
M3 AKH Andrew Hanson, Pilgrim Cottage, South Road, Truro, TR3 7AD
M3 AKQ R Vickerstaff, 16 Sewell Wontner Close, Kesgrave, Ipswich, IP5 2GB
M3 AKR A Beers, 5 Wayside Estate, Christchurch, Wisbech, PE14 9NY
M3 AKW K Pentney, 21 Windmill Court, North Street, Tunbridge Wells, TN2 4SU
M3 ALB A Borda, 4 Bow Arrow Lane, Dartford, DA1 1YY
MM3 ALG J Freeland, 12 Mccathie Drive, Newtongrange, Dalkeith, EH22 4BW
M3 ALX B HARRISON, 15 Helmington Terrace, Hunwick, Crook, DL15 0LQ
M3 ALZ S Matley, 67 Alexandra Road, Chandlers Ford, SO53 2BP
M3 AMA A Ayling, 58 Lower Derby Road, Portsmouth, PO2 8EX
M3 AMB Sean Hegarty, 4 Whitland Avenue, Bristol, BS13 9QQ
M3 ANE R Odle, 24 Longfellow Road, Gillingham, ME7 5QG
M3 ANH G Gore-Thorne, 19 Poplar Way, Ringwood, BH24 1UY
M3 ANW A Dennis, Menlo Park, Salisbury Road, Marlborough, SN8 3RP
M3 ANX Russell Meech, 26 Priory Street, Tonbridge, TN9 2AN
M3 AOC D Fawcett, 6 Wand Hill, Boosbeck, Saltburn-by-The-Sea, TS12 3AW
M3 AOM R Adkins, 91 Fernbank Road, Birmingham, B8 3LL



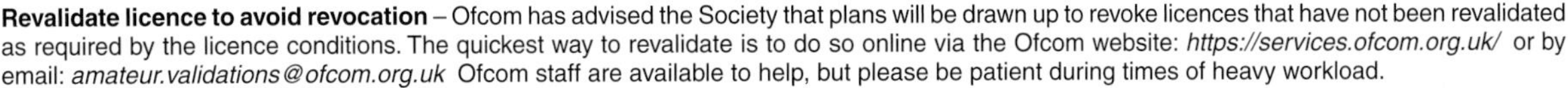

**IMPORTANT NOTE**

**Revalidate licence to avoid revocation** – Ofcom has advised the Society that plans will be drawn up to revoke licences that have not been revalidated as required by the licence conditions. The quickest way to revalidate is to do so online via the Ofcom website: *https://services.ofcom.org.uk/* or by email: *amateur.validations@ofcom.org.uk* Ofcom staff are available to help, but please be patient during times of heavy workload.

M3 AOP B Smith, 45 Branson Avenue, Stoke-on-Trent, ST3 5LA
M3 AOQ Ian Sephton, 131 Smeaton Road, Upton, Pontefract, WF9 1LG
M3 APA A Sims, 127 Cooks Spinney, Harlow, CM20 3BW
M3 APE David Baines, 157 Hall Green Road, West Bromwich, B71 2DY
M3 APO James Watson, 11 Burdock Walk, Morecambe, LA3 3QJ
M3 APQ D Hawken, The Old House, Hophurst Place, Hophurst Lane, Crawley, RH10 4LN
M3 AQF J Paul, 38 Avon Road, Upminster, RM14 1QU
M3 AQG P Harrison, 55 Hudson Close, Worcester, WR2 4DP
MW3 AQI W Morse, 32 Marysfield Close, Marshfield, Cardiff, CF3 2TY
M3 AQJ L Jesson, 17 Omaha Drive, Hinckley, LE10 0WU
M3 AQK Peter O'Shea, 37 Barclay Court, Ilkeston, DE7 9HJ
MM3 AQM W Fullerton, 28 Kerr Avenue, Saltcoats, KA21 5PS
M3 AQN G MacAuley, 1 Moricambe Crescent, Anthorn, Wigton, CA7 5AS
M3 AQP J Burns, 34 Fleswick Avenue, Whitehaven, CA28 9PB
MM3 AQW C MacLean, 16 Glamis Avenue, Elderslie, Johnstone, PA5 9NR
M3 ARB Andrew Bronze, 215 Ivyhouse Road, Dagenham, RM9 5RS
M3 ARM D Rees, Y Coed, Tan Lan Hill, Holywell, CH8 9JB
M3 ARS A Simpson, 36 Little Sammons, Chilthorne Domer, Yeovil, BA22 8RB
M3 ARU Alan Smith, 130 Watkin Street, Warrington, WA2 7DN
M3 ASC D Goldsbrough, 45 Tithe Barn Road, Stockton-on-Tees, TS19 8SZ
MW3 ASG John James, 67 Eyre Street, Cardiff, CF24 2JT
MI3 ASH Amber Nicholl, 58 Dunnalong Road, Bready, Strabane, BT82 0DW
M3 ASI D Dutton, 24 Pear Tree Avenue, Newhall, Swadlincote, DE11 0NB
M3 ASN D Hartless, 2 Brendon, Wilnecote, Tamworth, B77 4JW
M3 ASZ B Gilligan, 4 Orion Close, Ward End, Birmingham, B8 2AU
M3 ATB T Brierley, 6 Bridle Avenue, Wallasey, CH44 7BJ
M3 ATC Ian Kilkenny, 23 Hazelhurst Road, Stalybridge, SK15 1HD
MM3 ATI B Rodger, 95a Main Street, Coaltown, Glenrothes, KY7 6HX
MI3 ATT D Cooke, 7 Killyclooney Road, Dunamanagh, Strabane, BT82 0LZ
M3 AUB D Albison, 6 Rossendale Way, Shaw, Oldham, OL2 7TX
M3 AUC David Edney, 49 Burns Road, Loughborough, LE11 4ND
M3 AUF R Cassell, 10 Palmer Close, Branston, Burton-on-Trent, DE14 3DY
M3 AUK M Blake, 125 Ludlow Road, Portsmouth, PO6 4AF
M3 AUL K Rickard, 7 Thorngate Close, Penwortham, Preston, PR1 0XN
M3 AUN L Rowley, 10 Derby Place, Newcastle, ST5 3DX
M3 AUP S Burns, 22 Pendle Close, Peterlee, SR8 2JS
MM3 AUX J Milne, 24 Lorne Street, Edinburgh, EH6 8QP
M3 AVF J Hancox, 3 Hillfoot Road, Liverpool, L25 7UJ
MI3 AVJ D Johnson, 6 Sand Pits, Fenaghy Road, Ballymena, BT42 1JL
M3 AVU R Taylor, 48 Clark Avenue, Pontnewydd, Cwmbran, NP44 1RZ
M3 AVZ Robert Henshall, 14 Greenway, Congleton, CW12 4PS
MM3 AWC John Harrington, 1 Kynoch Terrace, Keith, AB55 5FX
MM3 AWD Scott Mcleman, 2 Coplandhill Road, Peterhead, AB42 1GS
M3 AWI David Davies, Penrallt, Abercaseg Road, Gerlan, Bangor, LL57 3SP
M3 AWN Barrie Procter, 28 Holme Grove, Burley in Wharfedale, Ilkley, LS29 7QB
M3 AWQ Matthew Stables, 58 Cedar Street, Derby, DE22 1GE
M3 AWS A Smith, Woodlands, Old School Lane, Biggleswade, SG18 9JL
M3 AXB S Balkham, 70 St. Thomas Road, Hastings, TN34 3LQ
M3 AXT S Plant, Columbell, 43 Fairfield Road, Bromsgrove, B61 9JW
M3 AXW D Evans, 7 Bryn Piod, Llanfachreth, Dolgellau, LL40 2EE
M3 AXX D Monnington, 77 Stanley Gardens, Paignton, TQ3 3NX
M3 AXZ K Wood, 17 Coster View, Great Bedwyn, Marlborough, SN8 3NS
M3 AYC T Symonds, 68 Manor Crescent, Pan, Newport, PO30 2BH
M3 AYJ A Parkman, 28 Shamblers Road, Cowes, PO31 7HF
M3 AYL D Stone, Bridor, 12 Robertson Avenue, Sleaford, NG34 8NJ
M3 AYP A Allen, 43 Marriott Road, Dudley, DY2 0JY
M3 AYQ D Payne, 31 Cockering Road, Canterbury, CT1 3UP
M3 AYS Gregory McGann, 2 Comlongon Mains Cottages, Clarencefield, Dumfries, DG1 4NA
M3 AYT W Wilson, 35 Darbishire Road, Fleetwood, FY7 6QA
M3 AZE P Harris, 18 The Broadway, Wombourne, Wolverhampton, WV5 0HY
M3 AZF D Hastings, 43 Delmar Avenue, Leverstock Green, Hemel Hempstead, HP2 4LZ
M3 AZH P Beresford, 58 Plumptre Way, Eastwood, Nottingham, NG16 3LR
M3 AZP R Butland, 4 Park Close, Sonning Common, Reading, RG4 9RY
M3 AZR David Jones, 31 Summerhill Drive, Liverpool, L31 3DN
M3 AZS R Mallender, 252 Kilton Road, Worksop, S80 2EB
M3 BAA M Eggleton, 78 Toronto Avenue, Blackpool, FY2 0PD
M3 BAH B Holmes, 11 Deerness Road, Bishop Auckland, DL14 6UB
M3 BAL A McCarthy, Lockinuar House, Treswell Road, Retford, DN22 0HU
M3 BAN J Sergeant, 15 Tennyson Place, Walton-le-Dale, Preston, PR5 4TT
M3 BAO L Walker, Little Chapple, Skilgate, Taunton, TA4 2DP
M3 BAS B Squance, 4 Glenholt Road, Plymouth, PL6 7JA
M3 BAT P Batty, 134 Plymouth Road, Scunthorpe, DN17 1TS
M3 BAW B Welthy, 8 Du Cane Place, Witham, CM8 2UQ
M3 BBA R Banfield, 2 Laleham Close, Eastbourne, BN21 2LQ
M3 BBB M Goodwin, 23 Saxon Way, Ashby-de-la-Zouch, LE65 2JR
M3 BBC J Holme, 17 Oxlea Grove, Westhoughton, Bolton, BL5 2AF
M3 BBF John Mackett, 49 Tennyson Road, Cowes, PO31 7PY
M3 BBL B Blackham, 5 Reedham Drive, Bramley, Rotherham, S66 2SW
MW3 BBQ G Richards, 3 Pen Y Mynydd, Bettws, Bridgend, CF32 8SE
M3 BBS R Pilgrim, 36 Wessex Gardens, Twyford, Reading, RG10 0AY
M3 BBY R Taylor, 93 Blue Dolphin Park, Reculver Lane, Herne Bay, CT6 6SS
MM3 BCA A MacInnes, 377 South Boisdale, Isle of South Uist, HS8 5TE
MM3 BCC Robert Sutherland, Tigh - Na - Coille, Mill Road, Nairn, IV12 5EW
M3 BCH B Heirene, 9 Ryecroft Crescent, Barnet, EN5 3BP
M3 BCM Andrew Birkhead, 9 Parkfield Terrace, Branscombe, Seaton, EX12 3DD
M3 BCQ M Shepherd, 47 Ripley Grove, Barnsley, S75 2RX
MI3 BCR Trevor Washbourne, 15 Apsley Street, Belfast, BT7 1BL
MD3 BCS B Smith, 98 Orange Hill Road, Burnt Oak, Edgware, HA8 0TW
M3 BCW S Peake, 3 Marigold Walk, Bermuda Park, Nuneaton, CV10 7SW
M3 BDA Andrew Yates, Kingsomborne, Broadway, Totland Bay, PO39 0BL
M3 BDC C Ciotti, 6 Bascott Road, Bournemouth, BH11 8RH
M3 BDH Tim Woods, 7 Holywell Gutter Lane, Hereford, HR1 1XA
M3 BDQ Julian Harvey, Flat 15, Gloucester House, Felixstowe, IP11 9DE
M3 BEE P Sykes, 2 Thornton Villas, Barrow Road, Barrow-upon-Humber, DN19 7QG
MI3 BEG Amber Nicholl, 58 Dunnalong Road, Bready, Strabane, BT82 0DW
M3 BEK R Cowlishaw, 23 Aldrich Drive, Willen, Milton Keynes, MK15 9HP
M3 BER S Berry, 4 Newlands Park Way, Newick, Lewes, BN8 4PG
M3 BET A Waters, 12 Anvil Court, Whittonstall, Consett, DH8 9JU
M3 BFB M Bellamy, 23 Hazelwood, Benfleet, SS7 4NW
M3 BFG R Hogwood, 22 Queen Elizabeth Drive, Easington Lane, Houghton le Spring, DH5 0NW
M3 BFJ E Harvey, 62 Archibald Road, Romford, RM3 0RH
M3 BFU L Buttriss, 5 Church Close, Upper Sheringham, Sheringham, NR26 8UB
M3 BFX M Jones, 138 Brompton Farm Road, Rochester, ME2 3RE
M3 BFY D Bowler, 43 Stirtingale Road, Bath, BA2 2NG
M3 BGE R Allen, 77 Highfield Road, Stroud, GL5 1ES
M3 BGN G Wilson, 28 Cross Stile, Ashford, TN23 5EH
MW3 BGP Thomas Davies, Brynteifi, Alltyblacca, Llanybydder, SA40 9ST
M3 BGT J Ahmed, 59 Ramsgate, Lofthouse, Wakefield, WF3 3PX
MM3 BHD R Freeland, 12 Mccathie Drive, Newtongrange, Dalkeith, EH22 4BW
MM3 BHG E Hay, 11 Lovat Road, Glenrothes, KY7 4RU
M3 BHI D Wilkinson, 139 Church Road, Jackfield, Telford, TF8 7ND
M3 BHK S Prinnett, 29 Ford Park Road, Plymouth, PL4 6RD
M3 BHP Dean Rugen, 19 Jacksons Close, Haskayne, Ormskirk, L39 7LD
M3 BIB A McGoff, 55 Knights End Road, March, PE15 9QA
M3 BIC Tony Humphries, 10 Cropthorne Avenue, Leicester, LE5 4QL
MI3 BIE D Hamilton, 7 Bolea Park, Limavady, BT49 0SH
M3 BIK Duncan Newell, 54 Galsworthy Road, Stoke-on-Trent, ST3 5UB
M3 BIO A Ness, 15 Scotforth Road, Lancaster, LA1 4TS
M3 BIR D Cupit, Hazelnut Cottage, Weston Road, Newark, NG22 0HB
M3 BIZ M Turner, 2 Higher Farm Cottages, Swallowcliffe, Salisbury, SP3 5PE
M3 BJB A Berry, 4 Newlands Park Way, Newick, Lewes, BN8 4PG
M3 BJE R Thorpe, 20 Bitham Lane, Stretton, Burton-on-Trent, DE13 0HA
M3 BJH B Hill, 26 Providence Close, Leamore, Walsall, WS3 2AL
M3 BJJ Brian Jenkinson, 7 Chestnut Avenue, Thorngumbald, Hull, HU12 9LD
M3 BJL William Lewis, 3 St. Martins Road, Folkestone, CT20 3LA
M3 BJR B Reynolds, 8 Stanwick Road, Higham Ferrers, Rushden, NN10 8LE
M3 BJV B Adkins, 4 Orion Close, Ward End, Birmingham, B8 2AU
M3 BJW B Moseley, 232 West Bromwich Road, Walsall, WS1 3HL
M3 BJZ D Koch, 83 Springfield Park, Maidenhead, SL6 2YU
MI3 BKA J Calvert, 29 Recreation Road, Larne, BT40 1EW
M3 BKB Antony Smith, Harrowstones, Harrowbeer Lane, Yelverton, PL20 6EA
M3 BKI M Davenport, 44 Vicarage Road, Hastings, TN34 3LY
M3 BKJ T Davenport, 44 Vicarage Road, Hastings, TN34 3LY
M3 BKU P Bridger, 2 Marline Avenue, St. Leonards-on-Sea, TN38 9HP
M3 BKV K Blanch, 2c Brunswick Terrace, North Street, Sandown, PO36 8BG
M3 BLF D Saunders, 42 Peascroft, Long Crendon, Aylesbury, HP18 9AU
M3 BLG L Grainger, 81 Windmill Rise, Tadcaster, LS24 9HR
MI3 BLN W Coates, 76 Chippendale Avenue, Bangor, BT20 4PY
M3 BLO M Herdman, 23 Marshall Avenue, Huncoat, Accrington, BB5 6NB
M3 BLR D Griffiths, 48 Conygar View, Dunster, Minehead, TA24 6PW
M3 BLX M Mina, 15 Manor Drive, Manchester, M21 7QG
MW3 BMF B Francis, 105 Cilmaengwyn Road, Pontardawe, Swansea, SA8 4QN
M3 BMH F Patrovits, 30 St. Ronans Drive, Seaton Sluice, Whitley Bay, NE26 4JQ
M3 BMI D Richmond, 37 Coopers Rise, Godalming, GU7 2NH
M3 BMQ J Nicholls, 55 Moat Avenue, Coventry, CV3 6BT
M3 BMU R Scott, 157 Fairview Road, Stevenage, SG1 2NE
M3 BMV G Brown, 3 Willow Lane, Goostrey, Crewe, CW4 8PP
M3 BMW B Ashcroft, 16 Edge Lane, Crosby, Liverpool, L23 9XE
M3 BNU M Hall, 20 Cubitt House, Black Bull Road, Folkestone, CT19 5SH
M3 BOB Frederica Kennedy, 19 High Street, East Hoathly, Lewes, BN8 6DR
MW3 BOC F Hart, 60 Heol Bryncwils, Sarn, Bridgend, CF32 9UE
MM3 BOJ D Boden, 42 Kirkwynd, Maybole, KA19 7AE
M3 BOO Owen Williams, 39 Camden Road, Maes-Y-Coed, Brecon, LD3 7RT
MW3 BOP K Stimson, 10 Heol Twyn Du, Merthyr Tydfil, CF48 1LU
M3 BOR R Parfitt, 7 Water Lane Close, Barnstaple, EX32 9JX
M3 BOU M Mullis, 18 Springfield Grove, Southam, CV47 0ES
M3 BOV D Waters, 17 Lower Herne Road, Herne Bay, CT6 7NA
M3 BPC L Bailey, 59 Church Road, Newick, Lewes, BN8 4JY
M3 BPF Louise Gaspar, 18 North Hill Close, Burton Bradstock, Bridport, DT6 4RY
M3 BPG P Robertson, 64 Castle Street, Frome, BA11 3DY
M3 BPL A Browell, 67 Stadium Avenue, Blackpool, FY4 3QA
M3 BPN Martyn Newell, 55 Station Road, Brimington, Chesterfield, S43 1JU
MM3 BPR S McLaughlin, 21 Shirrel Road, Motherwell, ML1 4RD
MI3 BPS G Brennan, 15 Kinnegar Rocks, Donaghadee, BT21 0EZ
M3 BPY P Hollis, 5 Salisbury Road, New Malden, KT3 3HZ
M3 BPZ N Heyne, Croxdale, Chiddingly Road, Heathfield, TN21 0JH
MM3 BQK M Cleland, 85 Carfin Street, Motherwell, ML1 4JL
M3 BQM Paul Alborough, Flat 8, Morey Court, Ryde, PO33 1HA
M3 BQN R Coleman, 77 Millstrood Road, Whitstable, CT5 1QF
M3 BQT J Rabbitt, 66 Parkfield Avenue, Delapre, Northampton, NN4 8QB
MI3 BRJ S Molloy, 6 Glenloch Park, Coleraine, BT52 1TY
M3 BRL P Long, 84 Manor Crescent, Newport, PO30 2BH
M3 BRQ T Leaver, 132 Old London Road, Hastings, TN35 5LZ
MM3 BRR Richard Hall, 13 Cleat, Castlebay, Isle of Barra, HS9 5XX
M3 BRT B Ingham, 19 Recreation Avenue, Ashton-in-Makerfield, Wigan, WN4 8SU
M3 BRU M Brunsdon, 7 Oldberg Gardens, Brighton Hill, Basingstoke, RG22 4NP
M3 BRV Daniel Prout, 2 Pine Crest Way, Bream, Lydney, GL15 6HG
M3 BRW P Buttery, 103 Elsie Street, Goole, DN14 6DY
MI3 BRX Henry Mairs, 9 Eureka Drive, Belfast, BT12 5NR
MM3 BSC B Clark, 6 The Links, Cumbernauld, Glasgow, G68 0EP
M3 BSF G Mate, 200 Chesterholm, Carlisle, CA2 7XY
M3 BSH B Hunt, 15 High St. North, West Mersea, Colchester, CO5 8JU
M3 BSI A Storer, 40 Danetre Drive, Daventry, NN11 4GY
MW3 BSJ B Jones, 4 Old Tanymanod, Blaenau Ffestiniog, LL41 4BU
M3 BSM L Cookman, The Flat Above Cobblers Corner, Ewhurst Road, Cranleigh, GU6 7AA
MI3 BSN J Murphy, 19 Kilburn Park, Armagh, BT61 9HA
M3 BTG G Simpson, 17 Temple Crescent, Leeds, LS11 8BG
M3 BTI John Roddam, Birches Farm, Button Street, Preston, PR3 2LH
M3 BTJ J Meredith, 42 Hollins Crescent, Talke, Stoke-on-Trent, ST7 1JY
M3 BTN Michael Luxton, 3 The Paddocks, Newgate Street, Brecon, LD3 8DJ
M3 BTZ P Bull, 8 Mayfield Lane, Martlesham Heath, Ipswich, IP5 3TZ
M3 BUA J Bull, 8 Mayfield Lane, Martlesham Heath, Ipswich, IP5 3TZ
M3 BUH A Brownley, 5 Skipton Road, Sheffield, S4 7DD
MI3 BUT K Martin, 19 North Street, Ballycastle, BT54 6BW
M3 BUU J Olive, 8 Mead Road, Chipping Sodbury, Bristol, BS37 6DQ
MM3 BUZ B Cameron, 6/4 Parkgrove Green, Edinburgh, EH4 7RQ
M3 BVA P Booth, 19 Gunville Crescent, Bournemouth, BH9 3PZ
M3 BVK E Shirley, Ham House, Ham Lane, Shepton Mallet, BA4 5JW
M3 BVL Gordon Sowden, The Grange Lodge, Rodley Lane, Pudsey, LS28 5QH
M3 BVM C Timm, 14 Little Copse Chase, Chineham, Basingstoke, RG24 8GL
M3 BVP R Hirst, 105 Haregate Road, Leek, ST13 6PX
M3 BVQ Adam Waller, 30 Avocet Crescent, College Town, Sandhurst, GU47 0XN
M3 BVX J Pickard, 39 Eafield Avenue, Milnrow, Rochdale, OL16 3UN
M3 BVY A Spinks, 10 Foxley Close, West Earlham, Norwich, NR5 8DQ
M3 BWF P Cook, 4 Church View, Highworth, Swindon, SN6 7ER
M3 BWT B Williams, 6 Aulton Terrace, Thornhill, DG3 5AN
MM3 BWV A Smith, 4 The Terrace, Lhanbryde, Elgin, IV30 8NY
M3 BWZ K Scully, 2 St. Michaels & All Angels Church, Canada Road, Deal, CT14 7BL
M3 BXC William Bray, 46 Alexandra Road, Lostock, Bolton, BL6 4BB
M3 BXE D Morris, 17 Mallory Way, Daventry, NN11 0UN
M3 BXG H Chick, 15 Bonfire Close, Chard, TA20 2EG
M3 BXH C Wheeler, 17 Orchard Way, Timberscombe, Minehead, TA24 7UL
M3 BXN B Holland, 23 White Avenue, Langold, Worksop, S81 9PT
MI3 BXQ B Hannigan, 4 Silverhill Road, Strabane, BT82 0AE
M3 BXS R Wake, 55 Bearsdown Road, Eggbuckland, Plymouth, PL6 5TR
M3 BXX J Smith, 8 Albert Gardens, Halifax, HX2 0HT
M3 BXY K Walker, 77 Blackwood Grove, Halifax, HX1 4QG
M3 BXZ D Burgess, 67 Fair Close, Beccles, NR34 9QT
M3 BYA David Passey, Blue House Cottage, Blue House Lane Albrighton, Wolverhampton, WV7 3AA
M3 BYF J Milne, 9 Roman Road, Colchester, CO1 1UR
MI3 BYJ D Bell, 1 Knockbracken Drive, Coleraine, BT52 1WN
M3 BYL Belynda Lewis, 20 Annes Walk, Caterham, CR3 5EL
MI3 BYQ J Martin, 19 North Street, Ballycastle, BT54 6BW
M3 BYS T Scott, 157 Fairview Road, Stevenage, SG1 2NE
M3 BYX Norman Clayton, 280a Loughborough Road, Leicester, LE4 5LH
MD3 BZA A Radcliffe, Cronk-Vue, Ballayockey, Andreas, Isle of Man, IM7 3HP
M3 BZC S Galloway, 1 Mount Pleasant, Leeds, LS10 3TB
M3 BZQ R Bostock, 5 Hethersett Way, New Rossington, Doncaster, DN11 0RZ
M3 CAA T Groves, 113 Rylands Road, Kennington, Ashford, TN24 9LR
MI3 CAB Philip Gibson, 109 Magheramenagh Drive, Portrush, BT56 8SY
M3 CAD David Coubrough, 7 Porchester Road, Billericay, CM12 0UG
M3 CAE J Chantler, 1 Glencoe Road, Margate, CT9 2SL
M3 CAM R Ingate, 158 Springfield Road, Chelmsford, CM2 6LG
M3 CAP P Barusevicus, 6 Middlebrook Crescent, Bradford, BD8 0EN
M3 CAQ Carlo Hosegood, 4 The Orchard, Sixpenny Handley, Salisbury, SP5 5QL
M3 CAW Simon Court, Eastgate Cottage, Perrys Lane, Norwich, NR10 4HJ
M3 CAZ C Mitchell-Watson, 144 Shakespeare Crescent, Dronfield, S18 1ND
M3 CBC Paul Eden, 22 Greenside, Stoke Prior, Bromsgrove, B60 4EB
M3 CBH B Kenny, 37 Coningswath Road, Carlton, Nottingham, NG4 3SF
MI3 CBJ N Mckittrick, The Coach House, 74 Lyle Road, Bangor, BT20 5LT
MI3 CBL S Wylie, 38 Elmfield Park, Donaghadee, BT21 0AX
M3 CBN K Hughes, High Lane Cottage, Congleton Road, Macclesfield, SK11 9RR
MM3 CBO J McCash, 11 Tintagel Gardens, Chryston, Glasgow, G69 0PH
MW3 CBS Robert Hulme, 33 Clos Rhandir, Loughor, Swansea, SA4 6UF
M3 CBV B Robertshaw, 14 Lawrence Avenue, Mansfield Woodhouse, Mansfield, NG19 8DJ
M3 CBW Alan Edwards, 6 St. Pauls Road, Nuneaton, CV10 8HL
M3 CBX K Roberts, 8 Bronant, Lixwm, Holywell, CH8 8NG
M3 CBY David Smith, 58 Marriott Road, Leicester, LE2 6NT
MI3 CCA Gary Wright, 38 Fern Grove, Bangor, BT19 1FG
M3 CCB A Gunn, 6 St. Pauls Road, Nuneaton, CV10 8HL
MW3 CCE C Evans, 25 Beech Drive, Hengoed, CF82 7JP
M3 CCF Thomas Toon, 9 Boundstone Lane, Sompting, Lancing, BN15 9QL
M3 CCJ S Grainger, 62 Meadowleaze, Longlevens, Gloucester, GL2 0PS
MI3 CCN R Reilly, 220 Ardmore Road, Londonderry, BT47 3TE
M3 CCQ David Burnett, 16 Church Lane, Reepham, Lincoln, LN3 4DQ
M3 CCS Tony Sutton, View House, 11 Arnhill Road, Corby, NN17 3DN
MI3 CCT T Vaughan, 20 Killen Park, Killen, Castlederg, BT81 7TJ
M3 CCY Jonathan White, 22 Millfields, Station Road, Burnham-on-Crouch, CM0 8HS
MI3 CDA C Adjey, 1 Foyle Park, Portstewart, BT55 7DL
M3 CDE Stephen Martins, 46 Ruskin Road, Mansfield, NG19 7LX
M3 CDI Carl Pearson, Greystones Farm Cottage, Richmond, DL11 7AJ
M3 CDL J Roberts, 1 Seymour Drive, Rhuddlan, Rhyl, LL18 5PP
MJ3 CDP Claus-Dieter Paland, 19 Maison St. Louis, St. Saviour, Jersey, JE2 7LX
M3 CDV A Willoughby, 25 Maple Close, Louth, LN11 0DW
M3 CDY Jack Bennett, 39 West View, Parbold, Wigan, WN8 7NT
M3 CEB B Cooke, 2 Harvey Place, Andover, SP10 2BU
MI3 CEM C Murray, 80 Canterbury Park, Londonderry, BT47 6DU
M3 CEN C Nowell, Crofters Cottage, Back Lane, York, YO23 3SH
M3 CER M Clews, 16 Chestnut Street, Worcester, WR1 1PA
M3 CET M Tibbits, 8 Holly Road, Northampton, NN1 4QR
M3 CEV R Murphy, 20 Chatsworth Road, Rhyl, LL18 2JJ
M3 CEZ Barry Stratford, 5 The Sycamores, Peacehaven, BN10 8AB
M3 CFI Carmel Isherwood, 32 Franklin Close, Old Hall, Warrington, WA5 8QL
M3 CFJ Steven Taylor, 12 Minehead Avenue, Burnley, BB10 2NP
M3 CFM K Sawyers, 27 Dukeswood Road, Longtown, Carlisle, CA6 5UJ
M3 CFP Charles Penfold, 149 Shuttlewood Road, Bolsover, Chesterfield, S44 6NX
M3 CFU P Flook, 17 Valentine Close, Bristol, BS14 9ND
MI3 CGA R Doherty, 48 Drumard Park, Londonderry, BT48 0RL
MM3 CGC J Stanway, 38 Temple, Gorebridge, EH23 4SQ
M3 CGH Clive Horne, 56 Pilkington Road Braunstone, Leicester, LE3 1RA
M3 CGI D Hunter, 39 Nicholas Avenue, Rudheath, Northwich, CW9 7LD
M3 CGJ J Nenova, 18 Longtown Road, Romford, RM3 7QL
M3 CGM S Allen, 9 Tiled House Lane, Brierley Hill, DY5 4LG
M3 CGO C Oliver, 25 Mary Peters Drive, Greenford, UB6 0SS
M3 CGP B Townsend, 21 Royds Drive, New Mill, Holmfirth, HD9 1LH
M3 CGS C Shirley, Ham Houes, Ham Lane, Shepton Mallet, BA4 5JW
MI3 CGT C Brown, 33 Broadlands Drive, Carrickfergus, BT38 7DJ
MI3 CGU A Brown, 33 Broadlands Drive, Carrickfergus, BT38 7DJ

MI3 CGZ S Buchanan, 162 Victoria Road, Bready, Strabane, BT82 0DZ
M3 CHU John Waterhouse, 12 Orchard Way, Marcham, Abingdon, OX13 6PP
MW3 CHZ M Bowen, 7 Ael y Bryn, Beddau, Pontypridd, CF38 2AL
M3 CIE C Barker, 14 Hall Road, Wilmslow, SK9 5BN
M3 CIG J Smith, 9 Trafalgar Road, Newport, PO30 1QD
MW3 CII J Binny, 27 Porthamal Road, Oakdene, Cardiff, CF14 6AQ
M3 CIJ C Martin, 92 Thrupp Lane, Thrupp, Stroud, GL5 2DG
M3 CIO Thomas Brookes, 370 Broxtowe Lane, Nottingham, NG8 5ND
M3 CIP C Powis, 28 Kington Gardens, Birmingham, B37 5HS
M3 CIS C Smith, 18 Poynings Place, Old Portsmouth, Portsmouth, PO1 2PB
MI3 CIV M Semple, 58 Green Drive, Larne, BT40 2ER
MI3 CIW David Ritchie, 58 Green Drive, Larne, BT40 2ER
MI3 CIZ C Mccord, 23 Blackthorn Green, Larne, BT40 2JE
M3 CJA C Alexander, 25 Diamedes Avenue, Stanwell, Staines-upon-Thames, TW19 7JE
M3 CJE Richard Wilberforce, 106 Marlborough Road, Slough, SL3 7JY
M3 CJH C Houghton, 47 Aldersleigh Crescent, Hoghton, Preston, PR5 0BB
M3 CJI D Sharp, 9 Portland Street, Wakefield, WF1 5HE
MM3 CJP C Paton, 4 Abbeyhill, Dhallling Road, Dunoon, PA23 8FG
M3 CJX Christopher Cross, 7 Anne Close, Norwich, NR7 0PH
M3 CKD Peter Loomes, 107 Main Street, Sedgeberrow, Evesham, WR11 7UE
MI3 CKF D Hamilton, 1 Meadow Bank, Ballysally Road, Coleraine, BT52 2QA
M3 CKH Cheryll Hammett, 63 Treffry Road, Truro, TR1 1WL
M3 CKM S Plumb, 94 Wellington Street, New Whittington, Chesterfield, S43 2BG
M3 CKO C Plumb, 94 Wellington Street, New Whittington, Chesterfield, S43 2BG
MM3 CKP S Aron, 2/3, 14 Woodend Road, Glasgow, G73 4DX
M3 CKU G Moorhouse, 29 Dee Road, Walsall, WS3 1NW
MM3 CLA C Henderson, 22 Bowmont Place, East Kilbride, Glasgow, G75 8YG
M3 CLO K Williams, 36 Castle Street, Tiverton, EX16 6RG
M3 CLP C Palmer, 21 Ibbett Close, Kempston, Bedford, MK43 9BT
M3 CLT Charles Turner, 1b Amberbanks Grove, Blackpool, FY1 6DW
MJ3 CMB Chris Boudier, 253 Le Marais, St. Clement, Jersey, JE2 6GH
M3 CMG Christian Griffiths, Cambrian View, Holywell Road, St. Asaph, LL17 0TD
M3 CMI H Dickinson, 1 Larch Close, Kirkheaton, Huddersfield, HD5 0NJ
M3 CMK A Saville, 4 Shannon Court, Downs Barn, Milton Keynes, MK14 7PP
M3 CMM B Perryman, 55 Alderfield, Penwortham, Preston, PR1 9HD
M3 CMP Carl Pidd, 36 Hunster Close, Doncaster, DN4 6RE
M3 CMW C Willimot, 5 Green Lane, Upton, Huntingdon, PE28 5YE
M3 CMX J roberts, 52 School Lane, Toft, Cambridge, CB23 2RE
M3 CNC Andrew Wood, 38 Hartfield Close, Hasland, Chesterfield, S41 0NU
M3 CND Steven Bradley, 6 Downing Street, South Normanton, Alfreton, DE55 2HE
MW3 CNL D Emanuel, 98 Moorland Road, Neath, SA11 1JL
MM3 CNS Graham Kennedy, 1 Martins Buildings, High Street, Perth, PH2 7QP
M3 COE Robert Alford, 24035 Gray Road, Box 103, Dennis, United States, 67341
MI3 CON A Mclernon, 520 Carneety Terrace, Castlerock, Coleraine, BT51 4SZ
M3 CPH Christopher Harrap, 68 Stafford Road, Weston-Super-Mare, BS23 3BS
MD3 CPK C Kelly, 16 Viking Road, Douglas, Isle of Man, IM2 6PB
M3 CPN Daniel Dunford, 6 Island Close, Rotherham, S60 3JZ
M3 CPX Nigel Thorne, Barford Stream, Churt Road, Farnham, GU10 2QU
M3 CPY Lee Goodby, 21 Kiniths Way, Hurst Green, Halesowen, B62 9HJ
MI3 CQB T Elliott, 183 Kilraughts Road, Ballymoney, BT53 8NL
MI3 CQO R Brennan, 15 Kinnegar Rocks, Donaghadee, BT21 0EZ
M3 CQP J Gilbert, Sweden End, Ambleside, LA22 9EX
MI3 CQR R Hendy, 23 Captains Road, Forkhill, Newry, BT35 9RR
M3 CQT A Loasby, 89 Jubilee Crescent, Wellingborough, NN8 2PQ
M3 CQW D Wooding, 3 Chichele Court, North Street, Rushden, NN10 6BU
MI3 CQX M Finnegan, 18 Springfarm Heights, Newry, BT35 8XA
M3 CRD Clive Matthews, The Lawns, Ridsale Street, Darlington, DL1 4EG
M3 CRL C Finnis, 44 Disraeli Road, Christchurch, BH23 3NB
M3 CRV D Priestner, 4 Oak Street, Northwich, CW9 5LJ
M3 CSF C Finnis, 44 Disraeli Road, Christchurch, BH23 3NB
M3 CSH C Hastwell, 6 Barn Close, Worthing, BN13 2BE
M3 CSI C Massimo, 2 Beattie Close, Bookham, Leatherhead, KT23 3JF
M3 CSK C Newton, 7 Moss Close, Bridgwater, TA6 4NA
M3 CSM M Lawson, 233 Southwell Road West, Mansfield, NG18 4HF
M3 CSN Janet Prichard, 1 Polton Dale, Swindon, SN3 5BN
M3 CSR D Hunter, 9 Sleigh Road, Sturry, Canterbury, CT2 0HR
MI3 CSS Robert Ennis, 91 Main Street, Carrowdore, Co Down, BT27 2HW
MI3 CST D Browne, 26 Brooklands Gardens, Dundonald, Belfast, BT16 2PQ
M3 CSV P Southern, 32 Mayville Avenue, Scarborough, YO12 7NP
MM3 CSX A Thomson, 5 Gib Grove, Dunfermline, KY11 8DH
M3 CSZ David Croft, 33 Roughaw Road, Skipton, BD23 2PY
M3 CTB M Burnett, 16 Church Lane, Reepham, Lincoln, LN3 4DQ
M3 CTN M McEwen, 31 Holmes Carr Road, New Rossington, Doncaster, DN11 0QF
M3 CTO Alfons Kvilums, 78 Wagon Lane, Solihull, B92 7PN
M3 CTT Michael Holroyd, 9 Coniston Green, Aylesbury, HP20 2AJ
M3 CUH S Twynam, 129 Poplar Drive, Herne Bay, CT6 7QA
M3 CUK M Wilkins, 2 Evans Road, Eynsham, Witney, OX29 4QD
M3 CUU D Taylor, 8 Curlew Close, Winsford, CW7 1SW
MM3 CVB C Budas, 20 Oak Avenue, Bearsden, Glasgow, G61 3HD
M3 CVD D Dewberry, 66 Churchill Avenue, Brigg, DN20 8DY
M3 CVH C Holley, 28 White Horse, Uffington, Faringdon, SN7 7SE
M3 CVL Christopher Egan, 20 Woodhatch Road, Brookvale, Runcorn, WA7 6BJ
M3 CVM Dave Healey, 28 Witley Drive, Sale, M33 5NQ
M3 CVO Andrew Beal, 29 Bennetts Road North, Keresley End, Coventry, CV7 8JX
M3 CVS C Vincent-Squibb, 47 Hoskyn Close, Rugby, CV21 4LA
M3 CVW Robin Twose, 10 Galingale Close, Bicester, OX26 3FD
M3 CWA T Watkins, 6 Linnet Close, Waterlooville, PO8 9UY
M3 CWC John Harrison, 12 Hillside Crescent, Skipton, BD23 2LE
M3 CWH C Harlow, 12 Penhurst Court, Grove Road, Worthing, BN14 9DG
MM3 CWO John Mills, 65 Strathkinnes Road, Kirkcaldy, KY2 5PX
M3 CWV C Harding, 9 Westbourne Road, Middlesbrough, TS5 5BN
M3 CWZ A Robinson, 30 Cope Street, Walsall, WS3 2AT
MW3 CXA Mark Jennings, 30 Gelligaer Road, Cefn Hengoed, Hengoed, CF82 7HL
MI3 CXD W Duffy, 4 Deramore Drive, Strathfoyle, Londonderry, BT47 6XL
MI3 CXM Paul Boyd, 2 Fernwood Park, Antrim, BT41 1QF
M3 CXX A Boyle, 15 Slatter, Satchell Mead, London, NW9 5UQ
M3 CYJ G Jenkinson, 4 Brundish House, Braithwell Road, Rotherham, S66 8JT
M3 CYM Debbie Taylor, 7 Trellewelyn Close, Rhyl, LL18 4NF
M3 CYQ D Hughes-Burton, 6 Troed Y Garn, Llangybi, Pwllheli, LL53 6DQ
M3 CYS Kyle Limbert, 7 Acacia Avenue, Liverpool, L36 5TL
M3 CYU E Hughes Burton, 6 Troed Y Garn, Llangybi, Pwllheli, LL53 6DQ
M3 CYW Norman Duke, 15 Plover Gardens, Barrow-in-Furness, LA14 3AY
M3 CZB Matthew Hillary, 29 Gatcombe Avenue, Portsmouth, PO3 5HG
M3 CZE Mark Gray, 14 Florence Crescent, Gedling, Nottingham, NG4 2QJ
M3 CZJ G Whitear, 19 Camelia Close, Littlehampton, BN17 6UT
M3 CZL Ronald Morley, 6 Ford Drive, Yarnfield, Stone, ST15 0RP
M3 CZM J Stocks, 10 Hollycroft Road, Emneth, Wisbech, PE14 8AY
M3 CZW C Walton, Old Drive, Sulby, Northampton, NN6 6EZ
M3 CZX Paul Godfrey, Bramley End, Old Lane, Chester, CH3 6QX
M3 CZY D Mageehan, 37 Gosbecks Road, Colchester, CO2 9JR
M3 DAB D Bambrook, 18 Vervain Close, Bicester, OX26 3SR
M3 DAE D Edge, 20 Parkers Court, Hallwood Park, Runcorn, WA7 2FP
M3 DAF Dale Fearn, 15 Broome Acre, Broadmeadows, Alfreton, DE55 3AW
M3 DAM A Petts, 19 Sandwell Avenue, Darlaston, Wednesbury, WS10 7RH
M3 DAO David Jones, Frondeg, Penrhyndeudraeth, LL48 6BU
M3 DAV R Halsall, 50 Leominster Drive, Manchester, M22 5DH
M3 DAY David Young, 126 Milne Road, Hull, HU9 4UL
M3 DBA K Miller, 8 Cooks Close, Ashburton, Newton Abbot, TQ13 7AN
MI3 DBB D Brown, 17 Parkmore Drive, Strathfoyle, Londonderry, BT47 6XA
M3 DBF J McConnell, Ty Cerrig, Tremeirchion, St. Asaph, LL17 0UP
M3 DBG B Miller, The Piglet, Saddleback Barn, Staverton, TQ9 6AN
M3 DBS D Baines, 21 Vera Road, Norwich, NR6 5HU
M3 DBU D Burrows, Pen Parc, 9 Clough Hall Road, Stoke-on-Trent, ST7 1AR
M3 DBX D Billinge, 29 Stanley Avenue, Wallasey, CH45 8JN
M3 DBY Debra Pratt, 4 Bussex Square, Westonzoyland, Bridgwater, TA7 0HD
M3 DBZ S Issatt, 7 Birch Road, Doncaster, DN4 6PD
M3 DCJ John Whiffin, 335 High Street, Eastleigh, SO50 5NE
M3 DCL M Phillips, Orchards, Brains Green, Blakeney, GL15 4AJ
MI3 DCM D Christie, 1 Marino Park, Ballymoney, BT53 7BB
MM3 DCN Angus McLellan, 57 Hunter Street, Kirn, Dunoon, PA23 8JR
M3 DCP David Palmer, Edison House, Bow Street, Attleborough, NR17 1JB
M3 DCS R Cowles, Bonnie Rock, 76 Fordham Road, Ely, CB7 5AL
MI3 DDK E Kashkoush, 41 Dunamallaght Road, Ballycastle, BT54 6PF
MM3 DDQ Paul Lucas, 69a Broomhill Crescent, Beechwood, Alexandria, G83 9QT
M3 DDY M Rote, 71b Headland Crescent, Exeter, EX1 3NP
M3 DDZ E Perez-Mendez, 56 Gaunt Close, Sheffield, S14 1GD
M3 DEA D Adams, 9 Brancaster Avenue, Charlton, Andover, SP10 4EN
M3 DEB D Stanley, Harby, Brightlingsea Road, Colchester, CO7 8JH
MM3 DEC Ellie McNeill, 3 Sunnybrae Terrace Maddiston, Falkirk, FK2 0LP
M3 DEE Derek Lewis, 10 Addington Road, Bolton, BL3 4QZ
M3 DEI Dion Jones, 21 Ralph Street, Borth-y-Gest, Porthmadog, LL49 9UA
M3 DEJ D Balls, 7 Rowan Close, Holbeach, Spalding, PE12 7BT
M3 DEL A Siddle-Ward, 40 Wynn Avenue North, Old Colwyn, Colwyn Bay, LL29 9RH
MW3 DEM M Dancer, 281 Fishguard Road, Llanishen, Cardiff, CF14 5PW
M3 DFB A Dunster, 113 Canterbury Road, Folkestone, CT19 5NR
M3 DFC D Chambers, 94 Hawthorn Avenue, Colchester, CO4 3JR
MM3 DFG A Graham, 72 India Street, Montrose, DD10 8PW
M3 DFL D Little, 3 Swallow Dale, Thringstone, Coalville, LE67 8LY
M3 DFM Karl Davies, 58 Popes Lane, Sturry, Canterbury, CT2 0LA
M3 DFP D Wicks, Friars Piece, Parham, Woodbridge, IP13 9LY
MI3 DFR P Clarke, 26 Derryhale Lane, Portadown, Craigavon, BT62 4HL
M3 DFS R Mordaunt, 330 Harborough Avenue, Sheffield, S2 1UU
M3 DFU J Mole, 11 Branfill Road, Upminster, RM14 2YX
M3 DFV G Jelley, 28 Blanches Road, Partridge Green, Horsham, RH13 8HZ
M3 DFW Darren Ferrow, 1 Temple Avenue, Blyth, NE24 5ET
MM3 DFZ A Zimnowlocki, 44 Dunbar Place, Kirkcaldy, KY2 5SE
M3 DGD C Densham, 27 Lloyds Crescent, Exeter, EX1 3JQ
M3 DGJ D Gardiner, 28 Winfield, Newent, GL18 1QB
M3 DGN A Lister, 15 Elmwood Drive, Breadsall, Derby, DE21 4GB
M3 DGP P Davison, 30 The Meadows, Sedgefield, Stockton-on-Tees, TS21 2DH
M3 DGR D Riley, 7 High St., Bolsover, Chesterfield, S44 6HF
M3 DHA S Webber, 124a Exeter Road, Kingsteignton, Newton Abbot, TQ12 3LY
MM3 DHE R Murray, 11 Castleview, Dundonald, Kilmarnock, KA2 9HZ
MM3 DHG J Blades, 58 Hunter Road, Crosshouse, Kilmarnock, KA2 0LD
MM3 DHN P Traill, 38 Burnside Road, Gorebridge, EH23 4EU
MI3 DHR David Richards, 70 Cherryhill Avenue, Dundonald, Belfast, BT16 1JD
M3 DHS D Sherwin, 5 North Road, Buxton, SK17 7EA
M3 DHV S Mccron, 72 Yeoman Way, Trowbridge, BA14 0QP
M3 DHW J Mccron, 72 Yeoman Way, Trowbridge, BA14 0QP
M3 DIB K Dibben, 82 Lenthay Road, Sherborne, DT9 6AF
M3 DIG Keith Dignall, 11 Mottershead Road, Widnes, WA8 7LD
MW3 DIL D Jenkins, 10 Ty Fry Close, Brynmenyn, Bridgend, CF32 8YB
M3 DIM H Brace, 56a Patching Hall Lane, Chelmsford, CM1 4DA
M3 DIS Rachel Hensman, 24 Belchmire Lane, Gosberton, Spalding, PE11 4HG
M3 DIT C Jacklin, 69 Prince William Drive, Butterwick, Boston, PE22 0JG
M3 DIU John Neenan, 11 Shaftesbury Square, West Bromwich, B71 1DX
M3 DIW D Brooks, 61 Carisbrooke High Street, Newport, PO30 1NR
M3 DIY Colin Ashworth, 79 Stonehouse Road, Rugeley, WS15 2LJ
MM3 DIZ D McArthur, 12 Laburnum Grove, Lenzie, Kirkintilloch, Glasgow, G66 4DF
M3 DJG K Miller, 52 Stanway Road, Shirley, Solihull, B90 3JE
MI3 DJM Joe McBride, 22 Birchwood, Omagh, BT79 7RA
MW3 DJV T Lewis, 22 Munro Place, Barry, CF62 8BU
M3 DJW D Wilson, 158 Higher Lane, Rainford, St. Helens, WA11 8BH
MW3 DJZ K Lewis, 22 Munro Place, Barry, CF62 8BU
MM3 DKA J Blair, 47 Chapelhill Mount, Ardrossan, KA22 7LU
M3 DKC L Johnson, 222 Norwich Road, Norwich, NR5 0EZ
M3 DKG David Morris, 4 Carnarvon Road, Reading, RG1 5SD
M3 DKK Adam Milne, 73 Warren Way, Barnham, Bognor Regis, PO22 0LR
MI3 DKN H Gillespie, 30 Groarty Road, Rosemount, Londonderry, BT48 0JX
M3 DKO Leonard Best, 97 Pine Tree Avenue, Canterbury, CT2 7TA
M3 DKT D Knott, 80 Melville Court, Chatham, ME4 4XJ
M3 DKZ G Tetley, 21 Lowther Crescent, Leyland, PR26 6QA
M3 DLB David Beach, 2 Millward Close, Telford, TF2 8AR
M3 DLC David Cole, Amber Lights, Market Lane, Wisbech, PE14 7LT
M3 DLE D Edwards, 5 Chalk Lane, Sutton Bridge, Spalding, PE12 9YF
M3 DLJ D McDougall, 15 Caldew Drive, Dalston, Carlisle, CA5 7NS
MI3 DLO M Harte, 17 Main St., Carrickmore, Omagh, BT79 9AY
M3 DLP C Thulborn, 37 Lewisham Road, Liverpool, L11 1EE
M3 DLT J Hanlon, 271 Windrows, Skelmersdale, WN8 8NP
M3 DLU M Salt, 84 Hayfield, Stevenage, SG2 7JR
M3 DLZ E Harvey, 125 North Road, Clowne, Chesterfield, S43 4PQ
M3 DMG D Glazebrook, 12 Kestrel Road, Haverhill, CB9 0PH
M3 DMJ D Jodrell, 2 Charlesworth Street, Crewe, CW1 4DE
MI3 DMM D McAuley, Layde View, 19 Rathlin Avenue, Ballycastle, BT54 6DQ
M3 DMW Roger Weir, 130 Alexander Square, Eastleigh, SO50 4BX
M3 DMY Allen Hoskins, 38 Tasmania Close, Basingstoke, RG24 9PQ
MM3 DMZ A Perks, The Lodge, Cemetery Drive, Dumbarton, G82 5HD
M3 DNB N Brown, 241 Bury Road, Tottington, Bury, BL8 3DY
M3 DNC B Brown, 241 Bury Road, Tottington, Bury, BL8 3DY
MI3 DNN W Crozier, 3 Carnhill Avenue, Newtownabbey, BT36 6LE
M3 DNX J Cox, 5 Golden Avenue Close, East Preston, Littlehampton, BN16 1QS
M3 DOA K Hodder, 12 Garden Crescent, Barnham, Bognor Regis, PO22 0AR
MI3 DOD K Campbell, 11 Caldwell Drive, Portrush, BT56 8ST
M3 DOM D Pritchard, 8 The Paddock, Dawlish, EX7 0EJ
M3 DOO J Dooley, 29 The Drive, Alsagers Bank, Stoke-on-Trent, ST7 8BB
MM3 DOP J Shields, 4 Rhindmuir Drive, Baillieston, Glasgow, G69 6ND
M3 DOR G Membury, 11 York Terrace, Dorchester, DT1 2DP
M3 DOX A Bajjon, 35a Blackford Road, Shirley, Solihull, B90 4BU
M3 DPB M Oram, Lancroft, West End Road, Boston, PE21 7NQ
M3 DPF D Fitzgerald, 60 The Links, Trevethin, Pontypool, NP4 8DQ
M3 DPG David Ganner, 58 Kilngate, Lostock Hall, Preston, PR5 5UW
M3 DPI J Spencer, 19 Meadowside Avenue, Bolton, BL2 2SS
M3 DPP D Pitchfork, 57 Calvary Crescent, Bentilee, Stoke-on-Trent, ST2 0AQ
M3 DPQ R Palmer, 34 Ditmas Avenue, Kempston, Bedford, MK42 7DP
M3 DPS D Sager, 29 Station Road, Mickleover, Derby, DE3 9GH
M3 DPV D Vaughan, 40 Meadow Terrace, Phillipstown, New Tredegar, NP24 6BW
M3 DPX Andrew Holden, 21 East View Meadowfield, Durham, DH7 8RY
M3 DPY W Ellis, 16 Furlong Drive, Tean, Stoke-on-Trent, ST10 4LD
MW3 DQB Robert Cotterell, 49 Graham Court, Caerphilly, CF83 1RF
M3 DQJ A Barker, 21 Raysmith Close, Southwell, NG25 0BG
M3 DQQ David Horsley, 1 Mead Close, Swanley, BR8 8DQ
MM3 DQV B McRae, 29 Woodneuk Road, Gartcosh, Glasgow, G69 8AG
MM3 DQX K McRae, 29 Woodneuk Road, Gartcosh, Glasgow, G69 8AG
M3 DRA D Abbott, 29 Malvern Road, Peterborough, PE4 7TT
M3 DRH D Horner, 38 Lyndhurst Drive, Bicknacre, Chelmsford, CM3 4XL
M3 DRM D Mostyn, 3 Woodlands Avenue, Cheadle Hulme, Cheadle, SK8 5DD
M3 DSA D Bartlett, 9 Macdonald Avenue, Hornchurch, RM11 2NF
M3 DSE D Essery, Hilgay, First Avenue, Watford, WD25 9PS
M3 DSI A Clarke, 2a Albany Road, Sittingbourne, ME10 1EB
MI3 DSM D McCord, 24 Craigstown Road, Moorfields, Ballymena, BT42 3DF
M3 DSU S Jenner, 4 Christie Close, Chatham, ME5 7NG
M3 DSW D Wadey, 15 Canberra Place, Tangmere, Chichester, PO20 2WB
M3 DTC R Charge, Llifon, Waunfawr, Caernarfon, LL55 4YY
M3 DTD George Asher, Hillside, 23 Mill Road, Saxmundham, IP17 1DP
M3 DTH R Hart, 11 Ivy Hall Road, Sheffield, S5 0GX
MW3 DTO Dale Owens, 16 Blanche Street, Dowlais, Merthyr Tydfil, CF48 3PE
M3 DUL R Clark, 10 Clarendon Road, Bournemouth, BH4 8AL
M3 DUO D Flaherty, 24 Ansdell Drive, Eccleston, St. Helens, WA10 5DW
M3 DUR M Durrant, 21 Sycamore Crescent, Macclesfield, SK11 8LL
MD3 DUZ S Kelly, 44 Westhill Avenue, Castletown, Isle of Man, IM9 1HY
M3 DVA S Maskrey, The Hayloft, Stamford Lane, Chester, CH3 7QD
M3 DVB J Moreton, 71 Bevendean Avenue, Saltdean, Brighton, BN2 8PF
MM3 DVD P Gazinski, 14 Corstorphine Road, Edinburgh, EH12 6HN
M3 DVG D Green, 89 Upper Ratton Drive, Eastbourne, BN20 9DJ
M3 DVH David Swaby, 34 Lambourne Road, Barking, IG11 9PS
M3 DVM J Diston, 15 Colletts Gardens, Broadway, WR12 7AX
M3 DVN D Noel, 58 Easenhall Lane, Redditch, B98 0BJ
M3 DVP A Whyman, 8 Staplers Close, Great Totham, Maldon, CM9 8UN
M3 DVQ A Cree, 24 Old Lincoln Road, Caythorpe, Grantham, NG32 3EJ
M3 DVU J Bennell, 146 Hales Crescent, Warley, Smethwick, B67 6QX
M3 DWA B Hollis, 212 Rye Lane, Halifax, HX2 0QP
M3 DWD D Wraight, 8 Embleton Road, Bristol, BS10 6DT
MI3 DWQ Desmond McGlone, 10 O'Neill Terrace, Dromore, Omagh, BT78 3AW
M3 DWV D Simmons, 39 Crosier Court, Upchurch, Sittingbourne, ME9 7AS
MW3 DWZ B Chapman, Gwyddfan, Mountain Road, Kidwelly, SA17 4EY
M3 DXD A Clark, Brookside, Milford, Bakewell, DE45 1DX
M3 DXI T Townson, 4 Crawford Street, Bradford, BD4 7JJ
M3 DXL Leo Sucharyna Thomas, Dame School House, 103 High Street, Milton Keynes, MK11 1AT
M3 DXN Brian Worthington, 9 Greenway, Penwortham, Preston, PR1 0TD
M3 DXR Rose Bunce, 4 Newlands, Gainsborough, DN21 1QZ
MD3 DXW William Griffiths, 7 Cooyrt Shellagh, Ballasalla, Isle of Man, IM9 2EU
M3 DYF P Beier, 20 Markham Avenue, Armthorpe, Doncaster, DN3 2AZ
M3 DYL W Malin, 729 Wellingborough Road, Northampton, NN3 3JE
M3 DYR P Burns, 395 Hastilar Road South, Sheffield, S13 8EH
MM3 DYT P McKenzie, 76 East Bankton Place, Livingston, EH54 9BZ
M3 DYU L Horn, 9 Musson Close, Irthlingborough, Wellingborough, NN9 5XW
M3 DYY D Ellison, 48 Keenan Drive, Bootle, L20 0AL
M3 DZC M Blakely, 2957 Wilderness Blvd, Florida, United States, 34219
MI3 DZD W Brown, 16 Wallace Park, Rasharkin, Ballymena, BT44 8QH
MM3 DZG A McAlpine, 7 Lothian Road, Ayr, KA7 3BU
M3 DZK H Cartwright, 142 Ferness Road, Hinckley, LE10 0SE
M3 DZN D Newman, 11 Pine View Close, Woodfalls, Salisbury, SP5 2LR
M3 DZQ Raymond Sharpe, 38 Burlington Close, London, W9 3LZ
M3 DZT A Dominy, 58c Church St, Harwich, CO12 3DS
MM3 DZW C Graham, 4 Dodridge Cottages, Pathhead, EH37 5UJ
M3 EAE H Carter, 23 Brookdale Avenue, Marple, Stockport, SK6 7HP
M3 EAI Alan Whitburn, 14 Westgil Pen Ffordd, Blackwood, NP12 3QS
MI3 EAQ S Flanagan, 18 Hunters Park, Bellaghy, Magherafelt, BT45 8JE
M3 EAT T Bowskill, 522 New St., Hilcote, Alfreton, DE55 5HU
M3 EBA Allan Barnett, 53a Walkford Road, Walkford, Christchurch, BH23 5QD
M3 EBF E Fury, 1 Wigley Drive, Wigley, Ludlow, SY8 3DR
M3 EBG H Crossley, 196 Middleton Road, Heywood, OL10 2LH
M3 EBK M Hall, 22 Leam Road, Lighthorne Heath, Leamington Spa, CV33 9TE

**IMPORTANT NOTE**

**Revalidate licence to avoid revocation** – Ofcom has advised the Society that plans will be drawn up to revoke licences that have not been revalidated as required by the licence conditions. The quickest way to revalidate is to do so online via the Ofcom website: *https://services.ofcom.org.uk/* or by email: *amateur.validations@ofcom.org.uk* Ofcom staff are available to help, but please be patient during times of heavy workload.

M3 EBO J Hann, 2 Leighton Green, Westbury, BA13 3PN
M3 EBU Edward Burrows, 9 Clough Hall Road, Kidsgrove, Stoke-on-Trent, ST7 1AR
M3 EBZ Anne Cattermole, Blaxhall Hall Crossing, Little Glemham, Woodbridge, IP13 0BP
M3 ECD D Snowdon, Holly Cottage, Romsey Road, Romsey, SO51 0HG
M3 ECF R Maltby, 15 Haglane Copse, Pennington, Lymington, SO41 8DT
M3 ECJ P Stringfellow, 5 Cowslip Way, Romsey, SO51 7RR
MM3 ECO W Davidson, 44 Abercromby Crescent, Helensburgh, G84 9DX
M3 ECQ J Ranger, 9 Mitchells Close, Romsey, SO51 8DY
M3 ECS Richard Parkhouse, 3 Thornfield Close, Seaton, EX12 2SS
M3 ECU A Caws, 61 Kinver Close, Romsey, SO51 7JW
M3 ECV David Bould, 38 Curlew Grove, Bridlington, YO15 3NX
M3 ECW R Parsons, 145 Middlemarch Road, Coventry, CV6 3GJ
M3 ECZ M Douglas, 486 Malpas Road, Newport, NP20 6NB
M3 EDC Edward Cook, 13 High Street, Cawston, Norwich, NR10 4AE
M3 EDI C Edis, 21 Stocks Road, Kimberley, Nottingham, NG16 2QF
M3 EDS S Elliott, 25 Staunton Heights, 27 Dunsbury Way, Havant, PO9 5AR
M3 EDU M Islam, 158 Somerville Road, Chadwell Heath, Romford, RM6 5AT
MM3 EDW Charles Martin, 82 Biggart Road, Prestwick, KA9 2EQ
M3 EEJ J Maguire, 15 Brown Lane, Heald Green, Cheadle, SK8 3RR
MD3 EEW E Wood, The Hawthorns, Droghadfayle Road, Port Erin, Isle of Man, IM9 6EL
MU3 EFB K Le Boutillier, Tiverton, Bailiffs Cross Road, St. Andrew, Guernsey, GY6 8RT
M3 EFC P France, 29 Seymour Road, Broadgreen, Liverpool, L14 3LH
M3 EFL M Arnold, 334 Stourbridge Road, Halesowen, B63 3QR
M3 EFQ D Hillman, 132 Vicarage Road, Oldbury, B68 8HY
M3 EFV S Leathes, Harrogate Ladies' College, Clarence Drive, Harrogate, HG1 2QG
M3 EFW C Broadbent, 9 Orchard Road, Bromley, BR1 2PR
M3 EFX R Lockyear, 114 Wishaw Close, Redditch, B98 7RF
M3 EGB E Bateman, 32 Park Avenue, Bodelwyddan, Rhyl, LL18 5TB
M3 EGC E Colley, 14 Hawthorne Close, Tyldesley, Manchester, M29 8PH
MI3 EGD A Crawford, Cullena, 10 Gulf Road, Londonderry, BT47 3TW
M3 EGF R Mahoney, 1 Warner Avenue, Barnsley, S75 2EQ
MI3 EGJ C Hazlett, 13 Faughanview Park, Claudy, Londonderry, BT47 4HQ
M3 EGM T Brain, 47 Bankwood Crescent, New Rossington, Doncaster, DN11 0PU
M3 EGU R Knight, 35 Bayswater Road, Headington, Oxford, OX3 9PB
M3 EGV Melanie Johnson, 5 Blackbird Close, Thurston, Bury St. Edmunds, IP31 3PF
M3 EGY J Johnson, 5 Blackbird Close, Thurston, Bury St. Edmunds, IP31 3PF
MW3 EGZ C Cutcliffe, 12 Heol Fargoed, Bargoed, CF81 8PP
M3 EHA Tom Williams, 110 Brindley Avenue, Grange Estate, Winsford, CW7 2EG
M3 EHF D Austen, Tudorlands, Silchester Road, Tadley, RG26 5DG
M3 EHH C Mason, Riding Lea, Middleton Road, Barnard Castle, DL12 0AQ
M3 EHJ Simon Cooper-Hutley, 5 Greenacres Park, Adbolton Lane, West Bridgford, Nottingham, NG2 5AX
M3 EHK Martin Ellaway, 20 Nuttingtons, Leckhampstead, Newbury, RG20 8QL
MM3 EHM R Allan, 30 Woodside Way, Glenrothes, KY7 5DF
M3 EHP D Moses, 121 Badger Avenue, Crewe, CW1 3JN
M3 EHY R Cheesley, 1 Lechlade Road, Inglesham, Swindon, SN6 7RB
M3 EIA R Southworth, 37 Pound Close, Lyneham, Chippenham, SN15 4PJ
M3 EIJ C Bailey, 13 Newark Road, Mexborough, S64 9EZ
MM3 EJB John Burgoyne, 5 Shankston Crescent, Cumnock, KA18 1HA
M3 EJL E Lawrence, 4 Malvern Road, Gillingham, ME7 4BA
M3 EJR E Trueman, 2 Nursery Close, Saxilby, Lincoln, LN1 2JD
MM3 EJV Stephen Waldron, 9 Fullarton Avenue, Dundonald, Kilmarnock, KA2 9DX
M3 EJX J Bennett, 21 Scott Avenue, Sutton Manor, St. Helens, WA9 4AN
M3 EKA Erika Denman, 12 Woodland Close, Northampton, NN5 6NH
M3 EKC Andrew Taylor, 106 Raeburn Avenue, Surbiton, KT5 9EA
MM3 EKL R Harrigan, 7 Almond Crescent, Paisley, PA2 0NG
M3 EKP J Virdee, 29 Larkfield Crescent, Houghton le Spring, DH4 4PE
M3 EKR N harris, 45 Sleigh Road, Sturry, Canterbury, CT2 0HT
M3 EKU P Lancaster, 16 Wiltshire Close, Bury, BL9 9EY
M3 EKY D Osbourne, 34 Lambourne Road, Barking, IG11 9PS
M3 EKZ D Smedley, 27 Stirling Avenue, Loughborough, LE11 4LJ
M3 ELD Emma Dalton, 120 Goodway Road, Birmingham, B44 8RG
M3 ELN H Van Schie, 135 Mellish Court, Bletchley, Milton Keynes, MK3 6PE
M3 ELP A Ogburn, 88 Castle Rise, Runcorn, WA7 5XW
M3 ELS M Skinner, 5 Sycamore Avenue, Upminster, RM14 2HR
M3 ELT Alistair Kerr, 161 Dalswinton Avenue, Dumfries, DG2 9NU
M3 ELV L Tremble, 7 Allerton Grove, Birkenhead, CH42 5LR
M3 EMA E Holmes, 11 Deerness Road, Bishop Auckland, DL14 6UB
M3 EMN K Alexander, 25 Diamedes Avenue, Stanwell, Staines, TW19 7JE
M3 EMO E O'Neal, 22 Hill Lane, Birmingham, B43 6NA
M3 EMS S Calver, 82 Kristiansand Way, Letchworth Garden City, SG6 1UE
M3 EMU R Searle, Hollies, 27 Cuckmere Rise, Heathfield, TN21 8PG
M3 EMX P Hardwick, 2 Cliffe Cottages, Sandy Lane, Liss, GU33 7JE
M3 ENE Raymond Evans, 53 Faygate Road, Eastbourne, BN22 9RR
M3 ENF Tammie Evans, 53 Faygate Road, Eastbourne, BN22 9RR
M3 ENJ C Berry, 29 Marlborough Crescent, Long Hanborough, Witney, OX29 8JP
M3 ENO E Cross, 17 Nicholson Court, Tideswell, Buxton, SK17 8PX
MM3 ENP W Wilson, Laurieston Farm, Hollybush, Ayr, KA6 6HB
M3 ENS R Nelson, 10 Westmorland Avenue, Willington Quay, Wallsend, NE28 6SN
M3 ENY N Wootton, 54 York Road, Harlescott, Shrewsbury, SY1 3RA
MI3 EOD N Crawford, 10 White Mountain Road, Lisburn, BT28 3QY
MI3 EOH Brian McCalmont, 19 Drumsesk Place, Warrenpoint, Newry, BT34 3NL
M3 EOL B Jenson, 10 Tintern Close, Portsmouth, PO6 4LS
M3 EOQ D Horton, Glen View, New Road, Bude, EX23 9LE
M3 EOT G Gidman, 8 Minerva Close, Knypersley, Stoke-on-Trent, ST8 6SZ
M3 EOX J Allen, 2 Chichester Walk, Chichester Road, Ramsgate, CT12 6NX
M3 EOY S McLaughlin, 7 Marine Terrace, Criccieth, LL52 0EF
M3 EOZ A Hammond, 52 Esther Avenue, Wakefield, WF2 8BX
M3 EPC Nigel G Ball Ball, 12 Dixons Farm Mews, Clifton, Preston, PR4 0PA
MW3 EPJ S McDonald, 56 Scotchwell View, Haverfordwest, SA61 2RE
MW3 EPK G Llewellyn, Hazeldene, Abercrave, Swansea, SA9 1SP
M3 EPQ D Caldwell, 44 Maxwell Road, Littlehampton, BN17 7BW
M3 EPR Alex Wilson, 22 Ormesby Road, RAF Coltishall, Norwich, NR10 5JY
MW3 EQE O Richards, 57 Maesgwyn, Aberdare, CF44 8TL
M3 EQL S Suresh, 2 Amberley Walk, Kingsmead, Milton Keynes, MK4 4AX
M3 EQP T Thompson, 7 West Bank, Dorking, RH4 3BZ
M3 EQQ J Laney, 18 Dyrham Close, Thornbury, Bristol, BS35 1SX
MI3 EQS T McDonnell, 52 Moira Road, Glenavy, Crumlin, BT29 4JL
M3 EQW M Savage, 23 Queen Mary Road, Salisbury, SP2 9LD
M3 EQY Stephen Heard, 42 Hallowell Down, South Woodham Ferrers, Chelmsford, CM3 5FS
MM3 ERD E Davidson, 44 Abercromby Crescent, Helensburgh, G84 9DX
MM3 ERP Peter Smith, 1 Hillside Cottages, Tillymorgan, Insch, AB52 6UN
M3 ERR Derek Barnes, 11 Yewside, Gosport, PO13 0ZD
MW3 ESE Sian Reed, 2 St. Marys Park, Jordanston, Milford Haven, SA73 1HR
MW3 ESF Stuart Reed, 2 St. Marys Park, Jordanston, Milford Haven, SA73 1HR
M3 ESG Andrew Pickersgill, 6 Berry Croft, Abingdon, OX14 1JL
M3 ESH E Hoy, 39 Blackbird Road, Caldicot, NP26 5RE
M3 ESK K Crane, 15 Leighton Road, Ipswich, IP3 0LJ
M3 ESN J Carragher, High Gorses, Henley Down, Battle, TN33 9BP
M3 ESQ M Peters, 9 Evelyn Close, Twickenham, TW2 7BL
M3 ESS M Stirling, 3 Rother Croft, New Tupton, Chesterfield, S42 6BE
M3 ETB F Llewellyn, 47 St. Teilos Road, Abergavenny, NP7 6HB
M3 ETH J Goodyear, 30 Ashburton Road, Alresford, SO24 9HH
M3 ETI D Mcspadden, 37 Halliday Crescent, Southsea, PO4 9JU
M3 ETQ N Greene, 308 Cedar Road, Nuneaton, CV10 9DY
M3 EUE J Underwood, 27 Woodville Road, London, E17 7ER
M3 EUF W Atherton, 64 Dam Lane, Rixton, Warrington, WA3 6LB
M3 EUM Norman Miller, 1 Alanbrooke Road, Colchester, CO2 8EG
M3 EUP C Cutler, 18 Berkeley Road, Peterborough, PE3 9PA
M3 EUR F Watt, 5 Brambling Road, Horsham, RH13 6AX
M3 EUU Dennis Broad, 34 Arderne Avenue, Crewe, CW2 8NS
M3 EUW A Lomas, Scanderlands Farm, Gloves Lane, Alfreton, DE55 5JJ
M3 EUY S Swan, 47 Warren Close, Whitehill, Bordon, GU35 9EX
M3 EVB E Munn, 32a Brunswick Street, Wakefield, WF1 4PW
M3 EVC Marc Lorimer, 1a Lingford Street, Hucknall, Nottingham, NG15 7SJ
M3 EVE D Watson, 14 Gawber Road, Barnsley, S75 2AF
M3 EVF D Redmayne, 2 Park Road West, Curzon Park, Chester, CH4 8BG
M3 EVI Alan Bowron, 11 Lealholme Grove, Fairfield, Stockton-on-Tees, TS19 7AP
M3 EVJ B Green, 12 The Ridgeway, Coal Aston, Dronfield, S18 3BY
M3 EVM S Burnand, 53 Sidley Road, Eastbourne, BN22 7JL
M3 EVN D Evans, 35 Caroline Road, Llandudno, LL30 2TY
M3 EVP Alvin Munton, 56 Jacklin Drive, Leicester, LE4 7SU
M3 EVR P Smith, 77 Holymoor Road, Holymoorside, Chesterfield, S42 7EA
M3 EVV M Harrison, 43 Erskine Road, South Shields, NE33 2TH
MD3 EVY J Phillips, 1 Cronk Elfin, Ramsey, Isle of Man, IM8 2EX
MM3 EWI Jim Woods, 12 Westbank Terrace, MacMerry, Tranent, EH33 1QE
M3 EWN E Nevard, Millinder House, Westerdale, Whitby, YO21 2DE
M3 EWQ E Quinn, 20 Greenfield Road, Rotherham, S65 3NX
M3 EWR E Roberts, 10 Ael Y Bryn, Waunfawr, Caernarfon, LL55 4AZ
M3 EWU T Bruty, 6 Moorhayes, Moorside, Sturminster Newton, DT10 1HL
M3 EWV J Kooner, 44 Headingley Road, Birmingham, B21 9QD
M3 EWW S Cooper, 53 Queensway, Warton, Preston, PR4 1XU
M3 EWY N Kellow, 17 Queensway, Warton, Preston, PR4 1XT
M3 EWZ R Dobson, 1 Auster Crescent, Freckleton, Preston, PR4 1JL
M3 EXJ Leslie Taylor, 17 Lacy Street, Hemsworth, Pontefract, WF9 4NW
M3 EXK J Wilkes, 47 Greenwood Park, Hednesford, Cannock, WS12 4DQ
MM3 EXW G Rotherham, 12 Industry Lane, Edinburgh, EH6 4EZ
M3 EXY Roger Shuttleworth, 5 Eastwood Drive, Marple, Stockport, SK6 7PW
MI3 EYB P McKeown, 7 Knockoneill Road, Maghera, BT46 5NX
M3 EYH A Carter, 37 Seathorne, Withernsea, HU19 2BB
M3 EYK J Dodsworth, 12 Fowlmere Road, Birmingham, B42 2EA
MM3 EYM C Somerville, 39 Edgehead Village, Pathhead, EH37 5RL
MM3 EYN R Somerville, 39 Edgehead Village, Pathhead, EH37 5RL
M3 EYP James Read, 31 Merebrook Road, Macclesfield, SK11 8RH
M3 EYR C Greene, 308 Cedar Road, Nuneaton, CV10 9DY
M3 EYS C Lewis, 41 Hazelholt Drive, Havant, PO9 3DL
M3 EYW S Thornton, 2 Sceptre Grove, New Rossington, Doncaster, DN11 0RW
M3 EYX W Thornton, 2 Sceptre Grove, New Rossington, Doncaster, DN11 0RW
M3 EYY E Driver, 99 Queens Road, North Weald, Epping, CM16 6JQ
M3 EYZ Christopher Board, Pinmoor, Moretonhampstead, Newton Abbot, TQ13 8QA
M3 EZB G Leake, 154 Wareham Road, Lytchett Matravers, Poole, BH16 6DT
MI3 EZF Patrick Rice, 11 Kirkwood Park, Saintfield, Ballynahinch, BT24 7DP
M3 EZH L Copley, Elmford, Mount Pleasant South, Whitby, YO22 4RQ
M3 EZJ Manuel Pires, 7 Felstead Close, Earley, Reading, RG6 5TP
MI3 EZK B Flanagan, 50 Towncastle Road, Strabane, BT82 0AJ
M3 EZY M James, 82 Hill Crescent, Sutton-in-Ashfield, NG17 4JA
M3 FAA Declan McGlone, 32 Shipley Mill Close, Kingsnorth, Ashford, TN23 3NR
M3 FAC C Perkins, Havasu, Treragin, Callington, PL17 8BL
M3 FAE J Modha, 95 Stanway Road, Shirley, Solihull, B90 3JF
M3 FAK M O'Brien, 4 Teal Close, Hawkinge, Folkestone, CT18 7TG
M3 FAL Franklyn Van Den Langenberg, Flat 26, Yew Tree Court, Shifnal, TF11 9BF
M3 FAY F Eavis, 61 Hitchmead Road, Biggleswade, SG18 0NL
M3 FBG B Jennings, 6 The Bungalow, St. Johns Road, Ventnor, PO38 3EL
M3 FBJ T McCann, 21 Ladyseat, Longtown, Carlisle, CA6 5XX
M3 FBN David Whitehead, 89 Cowpes Close, Sutton-in-Ashfield, NG17 2BU
M3 FBR Michael Morris, 12 Redver Gardens, Newport, PO30 5JJ
MI3 FBW P Coulter, 6 Skelton Close, Carrickfergus, BT38 8GP
MI3 FBX Campbell Gardner, 10 Abbington Manor, Bangor, BT19 1ZQ
MI3 FCA S Churchill, 20 Killen Park, Killen, Castlederg, BT81 7TJ
MM3 FCG William McCue, 188 Redburn, Alexandria, G83 9BU
MI3 FCK James Morgan, 6 Gannet Way, Carrickfergus, BT38 7RT
M3 FCN P Norman, 25 Hillswood Avenue, Leek, ST13 8EQ
M3 FCO B Dawson, 29 Hillswood Avenue, Leek, ST13 8EQ
M3 FCR L Moreland, 25 St. Georges Avenue, Bridlington, YO15 2ED
M3 FCS R Horne, 1 Ireland Road, Ipswich, IP3 0EJ
M3 FDB J Johnson, 5 Oakey Ley, Bradfield St. George, Bury St. Edmunds, IP30 0AU
M3 FDM M Goss, 80 Merryhills Drive, Enfield, EN2 7PD
M3 FDO W Corbett, 32 Fairview Avenue, Risca, Newport, NP11 6HU
M3 FDQ A Proctor, 3 The Courtyard, Tattingstone Park, Ipswich, IP9 2NF
M3 FDV A Goodwin, 36 Cambridge Street, Bridlington, YO16 4JZ
M3 FEA R Morley, 191 Purbrook Way, Havant, PO9 3RS
M3 FEC Frank Curry, 22 Caernarvon Close, Towcester, NN12 6UP
M3 FED Fiona Dunn, 71 Redfield Road, Midsomer Norton, Radstock, BA3 2JH
M3 FEG J Restall, 1 Johndory, Dosthill, Tamworth, B77 1NY
M3 FEL Eleanor Fellows, 95 Arnold Road, Eastleigh, SO50 5AS
MI3 FEO R Robinson, 92 Groomsport Road, Bangor, BT20 5NT
M3 FES F Shirley, Ham House, Ham Lane, Shepton Mallet, BA4 5JW
MM3 FET Alexander Galbraith, 22 Jeffrey Street, Kilmarnock, KA1 4EB
MI3 FEX D Rantin, 8 Buchanans Road, Newry, BT35 6NS
M3 FEY J Brinnen, 134 Victoria Road, Mablethorpe, LN12 2AJ
M3 FFA T Johnson, 43 Cherry Orchard Avenue, Halesowen, B63 3RZ
M3 FFE Winifred Johnson, 43 Cherry Orchard Avenue, Halesowen, B63 3RZ
M3 FFI D Ross, 27 The Meadows, Skegness, PE25 2JA
M3 FFK D Lythall, 71 Bennett Street, Kimberworth, Rotherham, S61 2JZ
MW3 FFL B Kendrick, 77 Heolddu Crescent, Bargoed, CF81 8US
M3 FFO P Hoe, 12 Ashbridge Rise, Chandler's Ford, Eastleigh, SO53 1SA
M3 FFU D Deakin, 75 Dairyground Road, Bramhall, Stockport, SK7 2QW
M3 FFV P lloyd, 71 Grove Road, Stourbridge, DY9 9AE
M3 FGG Stephen James, Wardens House, Kirkstone Close, Doncaster, DN5 9QZ
MM3 FGH N MacAulay, 68 Lorn Road, Dunbeg, Oban, PA37 1QQ
MM3 FGI Colin Gillespie, 18 Roslin Crescent, Rothesay, Isle of Bute, PA20 9HT
MI3 FGK Nigel Craig, 29 Oughtagh Road, Killaloo, Londonderry, BT47 3TR
MM3 FGL Archibald MacDonald, Manderley, Benvoullin Road, Oban, PA34 5EF
M3 FGO J Taylor, 8 Orchard Grove, Dudley, DY3 2UU
M3 FGQ I Prior, 81 Ladymeade, Ilminster, TA19 0EA
M3 FGR D Rootes, 1 Shelfinch, Toothill, Swindon, SN5 8AR
M3 FGU Steven Fellows, 8 Cardale Street, Rowley Regis, B65 0LY
MW3 FGV J Rowe, 41 Station Road, Ammanford, SA18 2DB
M3 FGX J Wells, 54 Queens Road, Everton, Liverpool, L6 2NG
M3 FHI R Norwood, 5 Galadriel Spring, South Woodham Ferrers, Chelmsford, CM3 7BD
M3 FHK Martin Tompkins, 4 Prospect View, Rawtenstall, Rossendale, BB4 8JG
MI3 FHM C Patton, 13 Oldpark Avenue, Ballymena, BT42 1AX
M3 FHO Graham Flack, 20 The Pastures, Hardwick, Cambridge, CB23 7XA
M3 FHP D Haines, 29 Parks Road, Mitcheldean, GL17 0DQ
M3 FHQ C Price, 10 St. James Park, Lower Milkwall, Coleford, GL16 7LG
M3 FHV Brendan Cahill, 56 Dene Road, Headington, Oxford, OX3 7EE
MI3 FHZ E Mccrystal, 33 Richmond Park, Omagh, BT79 7SJ
M3 FIB Simon Watling, 1 Chediston Green, Chediston, Halesworth, IP19 0BB
M3 FIH G Street, Flat 9, Weavers Cottages, Congleton, CW12 1AG
M3 FIK S Bowkett, 4 May Street, Newport, NP19 0EG
M3 FIM K Meredith, 3 Abbots Road, Abbey Hulton, Stoke-on-Trent, ST2 8DU
M3 FIP J Shingler, 19 Cherry Tree Avenue, Runcorn, WA7 5JJ
M3 FIW J Watson, 32 Franklin Close, Old Hall, Warrington, WA5 8QL
M3 FIX A Thompson, 51 Kempe Way, Weston-Super-Mare, BS24 7DZ
M3 FIY E Watson, 32 Franklin Close, Old Hall, Warrington, WA5 8QL
M3 FIZ A Finn, 105 Lynmouth Close, Biddulph, Stoke-on-Trent, ST8 6LS
MM3 FJA G Cull, 2 Pitairlie Farm Cottages, Pitgaveny, Elgin, IV30 5PQ
M3 FJB Joseph Bell, 2 Rake Lane, Milford, Godalming, GU8 5AB
M3 FJC B Hinchliffe, 272 South Street, Rotherham, S61 2NP
M3 FJD A Lythall, 71 The Crescent, Bolton-upon-Dearne, Rotherham, S63 8HQ
M3 FJE Maurice Jones, 6d Terrace Road, Walton-on-Thames, KT12 2SU
M3 FJN A Siebert, Po Box 127, Nantwich, CW5 8AQ
M3 FJP John Park, 18 Ladgate Grange, Middlesbrough, TS3 7SL
M3 FJQ Mohammed Rafique, 21 Syddall Avenue, Heald Green, Cheadle, SK8 3AA
M3 FJR L Siebert, Po Box 127, Nantwich, CW5 8AQ
MW3 FJW F Finch, Porthgwyh, Lamb Road, Aberdare, CF44 9JU
MM3 FJX J O'Connor, 23 Osborne Terrace, Cockenzie, Prestonpans, EH32 0BY
M3 FKA Jonathan Davies, 5 Beauchamp Road, Kenilworth, CV8 1GH
MI3 FKI G Kane, 83d Killymeal Road, Dungannon, BT71 6LG
M3 FKK Jared Waddington, 2 Heron Court, Daventry, NN11 0XT
M3 FKL M Rose, 128 Boultham Park Road, Lincoln, LN6 7TG
M3 FKM G Rose, 128 Boultham Park Road, Lincoln, LN6 7TG
M3 FKN M Mellish, 302 Belvedere Road, Burton-on-Trent, DE13 0RD
MM3 FKO C Lorimer, 70a Morningside Drive, Edinburgh, EH10 5NU
M3 FKS Brian Hoare, 2 St. Peters Close, South Newington, Banbury, OX15 4JL
M3 FKV R Johnson, 30 Thorpe Downs Road, Church Gresley, Swadlincote, DE11 9FB
M3 FKW S Papworth, 103 Station Road, Quainton, Aylesbury, HP22 4BX
M3 FLA G Backhouse, De10 Isaf, Bryneglwys, Corwen, LL21 9NP
M3 FLB A Bean, 25 Riverfield Grove, Bolehall, Tamworth, B77 3NB
M3 FLC F Hanmore, 7 Tarbert Walk, London, E1 0EE
M3 FLE D Wallstone, 128 Maltby Road, Mansfield, NG18 3BL
M3 FLI C BAINBRIDGE, The Brindles, Primrose Hill, Deeside, CH5 4QA
M3 FLJ G Jackson, 5 Woodside Close, Siddington, Macclesfield, SK11 9LQ
M3 FLK G Badham, 13 Maesglas Close, Newport, NP20 3BD
M3 FLL I Hamilton, 36 North Parade, Hoylake, Wirral, CH47 3AJ
M3 FLP Simon Dec, 101 Cranford Road, Northampton, NN2 7QY
M3 FLU A Yates, 4 High Street, Abergele, LL22 7AR
M3 FLV Joshua Showell, 14a Station Approach, Hayes, Bromley, BR2 7EH
M3 FLZ M McCormick, Sarnia, 73 Abelia, Tamworth, B77 4EZ
MM3 FMB Steven Markey, 232 Main Street, Renton, Dumbarton, G82 4QA
M3 FME D Bennett, 29 Margraten Avenue, Canvey Island, SS8 7JD
M3 FMI N Ashley, 24 Wingfield Road, Trowbridge, BA14 9ED
M3 FMK A Jones, 36 Sutherland Drive, Newcastle, ST5 3NZ
MD3 FMN T Hardwick, 3 Poplar Terrace, Douglas, Isle of Man, IM2 4AR
M3 FMP D Gilhooly, 50 Hillborough Crescent Houghton Regis, Dunstable, LU5 5NX
M3 FMQ A Ritchie, 50 Hillborough Crescent Houghton Regis, Dunstable, LU5 5NX
M3 FMV C Hatter, 14 Morland Avenue, Bromborough, Wirral, CH62 6BE
MM3 FMY L Dickenson, 9 Naver Road, Thurso, KW14 7QA
M3 FNA A Brooks, 93 Durham Road, Stockton-on-Tees, TS19 0DE
M3 FNC F Chance, 128 Chapel St., Pensnett, Brierley Hill, DY5 4EQ
M3 FNH W Stopforth, 52 Cypress Road, Southport, PR8 6HF
M3 FNM P Hewitt, 166 Sheringham Avenue, London, E12 5PQ

M3 FNO J Scott-Brown, 2 Haddon Close, Fareham, PO14 1PH
M3 FNR S Pitchford, 419 Chell Heath Road, Stoke-on-Trent, ST6 6PB
M3 FNT Sarah Williams, 11 Hilda Street, Leigh, WN7 5DG
M3 FNY J Eagle, 1b Kingsley Avenue, Daventry, NN11 4AN
M3 FOD [illegible] Newby, 22 Acton Road, Liverpool, L32 0TT
MM3 FOE S Espie, 70 Everard Rise, Livingston, EH54 6JD
MI3 FOJ D Kane, 22 Rowan Road, Ballymoney, BT53 7AQ
M3 FOK J Old, 33 Rookhill Road, Pontefract, WF8 2BY
MI3 FOL C Sloan, 5 Laurel Hill Road, Coleraine, BT51 3AY
M3 FOQ John Woods, 3 Ingle Avenue, Morley, Leeds, LS27 9NP
M3 FOR Michael Baldwin, 52 Salisbury Road, Chatham, ME4 5NN
M3 FOS Ian Woods, 3 Ingle Avenue, Morley, Leeds, LS27 9NP
M3 FOV D Read, L'Eglise, Durley Street, Southampton, SO32 2AA
M3 FPA R Etchells, 6 Woodbank Court, Canterbury Road, Manchester, M41 7DY
MI3 FPB I Buchanan, 162 Victoria Road, Bready, Strabane, BT82 0DZ
MI3 FPE J Rice, 42 The Crescent, Ballymoney, BT53 6ES
MW3 FPF J Briers, 117 Heath Mead, Cardiff, CF14 3PL
M3 FPG A Jennings, 6 The Bungalow, St. Johns Road, Ventnor, PO38 3EL
M3 FPH P Harris, Flat 33, Buckingham Court Shrubbs Drive, Bognor Regis, PO22 7SF
MM3 FPI Christopher Jones, Croy Lodge, Shandon, Helensburgh, G84 8NN
M3 FPM Layla Noel, 58 Easenhall Lane, Redditch, B98 0BJ
MI3 FPN C McIntyre, 18 Glanroy Crescent, Newtownabbey, BT37 9JZ
M3 FPS J Reid, 5 Hamlet Road, Fleetwood, FY7 7HW
M3 FPT Peter Turner, 92 Lancashire Street, Leicester, LE4 7AE
M3 FPU S Cash, 6 The Mariners, Valetta Way, Rochester, ME1 1FB
M3 FPZ D Turner, 11 Weetwood Road, Congresbury, Bristol, BS49 5BN
M3 FQA S Wadsworth, 47 Kilnhurst Road, Todmorden, OL14 6AX
M3 FQG John Heagren, 84 Avon Drive, Alderbury, Salisbury, SP5 3TH
MM3 FQI G Robinson, 12 Hannahston Avenue, Drongan, Ayr, KA6 7AU
M3 FQM J Dunning, 16 Shaggs Meadow, Lyndhurst, SO43 7BN
M3 FQN S Dunning, 16 Shaggs Meadow, Lyndhurst, SO43 7BN
M3 FQT C Inwood, 7 The Poplars, George Street, Mablethorpe, LN12 2BP
M3 FQX J Wadeson, 75 Bedford Drive, Sutton Coldfield, B75 6AX
M3 FRB D Munday, 29 Coombe Park, Wroxall, Ventnor, PO38 3PH
M3 FRD F Mcloughlin, 128 Windrows, Church Farm, Skelmersdale, WN8 8NW
M3 FRE J French, Eypes Mouth Country Hotel, Eypes, Bridport, DT6 6AL
M3 FRJ R Fearnley, 8 Keepside Close, Ludlow, SY8 1BQ
M3 FRQ S Smith, 7 Rosebery Avenue, Morecambe, LA4 5RU
M3 FRS B Page, 21 Catherington Way, Havant, PO9 2BS
M3 FRT G Lowe, 12 Willow Court, Calow, Chesterfield, S44 5AP
M3 FRU G Ison, 2 Hayes Road, Nuneaton, CV10 0NH
M3 FRX M Breffit, 10 Garrard Place, Ixworth, Bury St. Edmunds, IP31 2EP
M3 FSB S Babic, 17 Ashwood Drive, Broadstone, BH18 8LN
M3 FSC F Creese, 69 Locksley Drive, Ferndown, BH22 8JX
M3 FSD Douglas Babic, 17 Ashwood Drive, Broadstone, BH18 8LN
M3 FSE D Edge, Lymn Bank Cottage, Lymn Bank, Skegness, PE24 4PJ
M3 FSQ David Gornall, 40 Welbrow Drive, Longridge, Preston, PR3 3TB
MI3 FSR J Brown, 4 Stratford Gardens, Bangor, BT19 6ZH
M3 FSS Anthony Goodchild, 62 Chestnut Drive, Sale, M33 4HL
M3 FSU G Lewis, 57 Oakwood Road, Sutton Coldfield, B73 5EH
M3 FSV P Murray, 2 Thurlow Gardens, Bishop Auckland, DL14 7GH
MI3 FSW Fiona White, 28 Lord Warden's Parade, Bangor, BT19 1YU
MI3 FSX T Mulholland, 215 Finaghy Road North, Belfast, BT11 9ED
M3 FSY K Doorbar, 23 Oaktree Road, Rugeley, WS15 1AD
M3 FTA M Everall, 17 Golden Park Avenue, Torquay, TQ2 8LR
MW3 FTB S Robson, 16 Dunraven Road, Sketty, Swansea, SA2 9LG
MW3 FTC D Robson, 16 Dunraven Road, Sketty, Swansea, SA2 9LG
M3 FTE S Manley, 25 Acland Park, Feniton, Honiton, EX14 3WA
M3 FTI F Gibbs, 62 Wenvoe Avenue, Bexleyheath, DA7 5BT
M3 FTJ J Lightly, 8 Smithville Close, St. Briavels, Lydney, GL15 6TN
M3 FTK C Gale, 51 Heron Way, Horsham, RH13 6DW
M3 FTP J McKenna, 33 Low Islwyw, Prestatyn, LL19 8HQ
M3 FTU P Panayiotou, 7 Aireville Rise, Bradford, BD9 4ES
M3 FTV Angela Dunham, 28 Kingfisher Close, Chatteris, PE16 6TP
M3 FTW Christopher David Hughes, 1 Guernsey Avenue, Buckshaw Village, Chorley, PR7 7AG
MW3 FTY R Edwards, 86 Priors Way, Dunvant, Swansea, SA2 7UJ
M3 FTZ A De Vries, 13 St. Valerie Road, Worthing, BN11 3LL
M3 FUB Nicholas Phillips, First Floor Flat, 116 Lodge Road, Croydon, CR0 2PF
M3 FUD Stephen Merison, 9 Beechfield Crescent, Banbury, OX16 9AR
MM3 FUG W Gillespie, 33 Lochnell Road, Dunbeg, Oban, PA37 1QJ
M3 FUH Chris Pomfrett, 17 Manifold Close, Sandbach, CW11 1XP
M3 FUQ S Day, The Lodge, Attleborough Fish Farm Norwich Road, Attleborough, NR17 2LA
M3 FUR P Henderson, 214 Marsh Street, Barrow-in-Furness, LA14 1BQ
M3 FUV P Spowart, Ruggs Hall, Clatterway Hill, Matlock, DE4 2AH
M3 FVA Steven Grimbleby, 96 Waldeck Street, Reading, RG1 2RE
M3 FVC J Huntington, 87 Dinerth Road, Rhos on Sea, Colwyn Bay, LL28 4YH
M3 FVE D Millard, 114 Ainsdale Drive, Werrington, Peterborough, PE4 6RP
MW3 FVH M Hallett, 24 Brynhyfrydst, Clydach Vale, Tonypandy, CF40 2DZ
M3 FVJ C Brown, 44 Stanley Avenue, Inkersall, Chesterfield, S43 3SY
MI3 FVW D Shaw, 9 The Ten Cottages, Newtownards Road, Donaghadee, BT21 0PU
M3 FVX Paul Sarratt, Unit 20f, Brooke Business Park, Lowestoft, NR33 9LZ
M3 FWA P Matthew-Brown, 57 The Limes Avenue, London, N11 1RD
M3 FWJ C Green, 160 Ashbrook Road, London, N19 3DJ
M3 FWO K Beckett, 95 Warrens Hall Road, Dudley, DY2 8DH
M3 FWR D MARSH, 1 Caunts Crescent, Sutton-in-Ashfield, NG17 2FH
M3 FWS J Steele, 70 The Crescent, Andover, SP10 3BU
M3 FWT J Cooper, 1 Dearing Close, Lyndhurst, SO43 7JP
M3 FWU Raymond Wilson, 7 Cornwall Close, Kirton Lindsey, Gainsborough, DN21 4DF
MI3 FXE J Higginson, 47 Ballycorr Road, Ballyclare, BT39 9DD
M3 FXM D Toombs, 1 Chalgrove, Welwyn Garden City, AL7 2QJ
M3 FXQ H Gregory, 178 Over Lane, Belper, DE56 0HL
M3 FXU Freda Siviter, Flat 76, Lancaster House, Rowley Regis, B65 0QE
M3 FXX V Hocking, 80 Barton Tors, Bideford, EX39 4HA
MW3 FYA C Alloway, 9 Millands Park, Llanmaes, Llanwitmajor, CF61 3XR
MM3 FYF J Fyfe, 5 Beaufort Avenue, Newlands, Glasgow, G43 2YL
M3 FYM A Reynolds, 44a Mill Lane, Codnor, Ripley, DE5 9QG
MM3 FYN D Innes, 39 Mormond Place, Strichen, Fraserburgh, AB43 6SY
M0 FYD [illegible], 30 [illegible] Avenue, Newport, NP19 9TA
M0 FYV [illegible], 4 [illegible], Middlesbrough, TS0 6LW
M3 FYZ A Shollam, 1 Trafalgar House, Nelson Drive, Cannock, WS12 2GH
M3 FZB P Paduch, 291 Rochfords Gardens, Slough, SL2 5XH
M3 FZC J Charter, 36 Northumberland Avenue, London, E12 5HD
M3 FZE R Humphreys, 19 Monks Green, Fetcham, Leatherhead, KT22 9TL
MM3 FZI Russell McDonald, 12 Queen Street, Tayport, DD6 9NE
M3 FZJ Lewis Larkins, 34 Guycroft, Otley, LS21 3DS
M3 FZM Clive Ramsdale, 87 Mill Lane, Kirk Ella, Hull, HU10 7JN
M3 FZO B Shepherd, 19 Washfield Lane, Treeton, Rotherham, S60 5PU
M3 FZS A Green, 10 Howard Close, Teignmouth, TQ14 9NW
M3 FZV M Reed, Channel Pool, Armathwaite, Carlisle, CA4 9QY
MD3 GAB J Espey, 9a Hilltop View, Douglas, Isle of Man, IM2 2LA
M3 GAE J Law, 5 Sudbury Close, Chesterfield, S40 4RS
M3 GAF G Allen, 39 Hallam Road, Newton Heath, Manchester, M40 2SY
M3 GAG Paul Gagliardi, 7 Saxon Way, Jarrow, NE32 3QA
M3 GAP G Porter, 65 Bartlett St., Wavertree, Liverpool, L15 0HN
M3 GAV C Tomlinson, 9 Wells Close, Astley, Manchester, M29 7WF
M3 GBA G Barlow, Ingleneuk, Hammersley Hayes Road, Stoke-on-Trent, ST10 2DW
M3 GBB Gail Bartley, 19 SOUTH AVENUE, SHADFORTH, Durham, DH61LB
M3 GBC M Casey, 7 Cobham Avenue, Manchester, M40 5QW
M3 GBD B Dallimore, 4 Llys Dyffryn, St. Asaph, LL17 0SX
MJ3 GBJ S Boudier, 253 Le Marais, St. Clement, Jersey, JE2 6GH
MM3 GBL C Galbraith, 77 Netherwood Park, Deans, Livingston, EH54 8RW
M3 GCD E Elsworth-Wilson, 31 Douglas Avenue, Brixham, TQ5 9EL
M3 GCH Shaun Shreeves, 6 Bowshaw Avenue, Batemoor, Sheffield, S8 8EZ
M3 GCJ G Johnson, 30 Trinidad Close, Basingstoke, RG24 9PY
M3 GCM G Masters, 85 Petersham Road, Creekmoor, Poole, BH17 7DW
M3 GCN G Newton, 18 Parks Road, Dunscroft, Doncaster, DN7 4AH
M3 GCP G Papworth, 70 Edward Road, West Bridgford, Nottingham, NG2 5GB
M3 GCR G Watt, 5 Brambling Road, Horsham, RH13 6AX
M3 GCS G Sadler, 43 Laurel Grove, Stafford, ST17 9EF
M3 GCT P Wylie, 40 Sheepwash Avenue, Choppington, NE62 5NN
MM3 GDC Graham Cochrane, 33 Portland Road, Galston, KA4 8EA
M3 GDE G Edgar, 61 Winchester Avenue, Lancaster, LA1 4HX
M3 GDI T Whittam, 27 Dimples Lane, Garstang, Preston, PR3 1RD
M3 GDK Philip Weaver, 1 Madeley Street, Newcastle, ST5 6LS
MW3 GDL G Jones, 31 Parcy Mynach, Pontyberem, Llanelli, SA15 5EN
M3 GDQ N Yates, 9 Osborn Close, Ipplepen, Newton Abbot, TQ12 5XB
M3 GDV D Bearne, 59 Foxhole Road, Foxhole Estate, Paignton, TQ3 3TD
M3 GDX J Ormerod, 14 Fanny Moor Lane, Hall Bower, Huddersfield, HD4 6PJ
M3 GDY Garry Dealey, 69 Upper Belmont Road, Chesham, HP5 2DD
MI3 GEI G Convery, 12 Linen Grove, Belfast, BT14 8PP
M3 GEK C Poolman, 5 Slessor Street, Waddington, Lincoln, LN5 9NE
M3 GEN B Jones, 39 Dalton Lane, Barrow-in-Furness, LA14 4LE
MI3 GER G Reilly, 220 Ardmore Road, Londonderry, BT47 3TE
MM3 GEW R Brooks, 6 Amochrie Drive, Paisley, PA2 0BE
M3 GEX E Rogers, Maes Gwersyll, Garthmyl, Montgomery, SY15 6RS
M3 GEZ D Fowles, 2 The Red House, Gallamore Lane, Market Rasen, LN8 3UB
MI3 GFA G Freeman, 817 Windyhall Park, Coleraine, BT52 1TU
M3 GFE T Dunn, Rakers Rest, 31 Orleigh Avenue, Newton Abbot, TQ12 2TP
M3 GFH M Dunn, 6 Hamilton Drive, Newton Abbot, TQ12 2TL
M3 GFO Nigel Fox, 15 Hawthorne Grove, Bentley, Doncaster, DN5 0PQ
M3 GFW P Seaman, 18 Earlsford Road, Mellis, Eye, IP23 8DY
M3 GFZ M Hewson, 27 Grange Crescent, Lincoln, LN6 8BT
M3 GGA K Mccarthy, 260 Whalley Drive, Bletchley, Milton Keynes, MK3 6PJ
M3 GGE Graham Townsend, 19 Landor Crescent, Rugeley, WS15 1LP
M3 GGN H Staples, 79 High St., Scotter, Gainsborough, DN21 3TL
M3 GGV T White, 25 Vicarage Close, Shillington, Hitchin, SG5 3LS
M3 GHA G Halls, 17 Ellesborough Grove, Two Mile Ash, Milton Keynes, MK8 8NP
M3 GHD D Bell, 27 Kings Coombe Drive, Kingsteignton, Newton Abbot, TQ12 3YU
M3 GHE M Barnes, 49 Harrowden Road, Bedford, MK42 0RS
M3 GHF G Creed, Great House Farm, Croesypant, Pontypool, NP4 0JD
M3 GHG P Walton, 15 Arbour Close, Northwich, CW9 7BF
M3 GHH G Hazlewood, 102 Throne Road, Rowley Regis, B65 9JX
M3 GHI John Haslam, 25 Lulworth Road, Eccles, Manchester, M30 8WP
M3 GHL G Law, 14 Sandpit Lane, Hilton, Bridgnorth, WV15 5PH
M3 GHO G Cliffe, 5 Laurel Cottages, Ongar Hill Road, King's Lynn, PE34 4JB
M3 GHR R Gill, 84 Leypark Road, Exeter, EX1 3NT
M3 GHS Stephen Stanhope, 61 Heathfield Street, Manchester, M40 1LF
MI3 GHW G McLernan, 20 Drumcor Hill, Enniskillen, BT74 6BQ
MI3 GHY Ian Gibb, 1 Shankill Road, Garvary, Enniskillen, BT94 3DB
M3 GID G Dunne, Three Ways, Northbourne Road, Deal, CT14 0HJ
M3 GIE R Harper, 19 Tennyson Avenue, King's Lynn, PE30 2QG
M3 GIF E Roberts, 8 Skamacre Crescent, Lowestoft, NR32 2QG
M3 GIH E Peck, 11 Blake Road, Stapleford, Nottingham, NG9 7HN
M3 GIK M Haughey, 10 Sharp Street, Hull, HU5 2AB
MM3 GIR Kevin Gibson, 136 Henrietta Street, Girvan, KA26 0AE
M3 GIX A Scrutton, 35 Gainsborough Road, Warrington, WA4 6DA
M3 GIY J Eaton, 10 Motcombe Farm Road, Heald Green, Cheadle, SK8 3RW
MI3 GJG Gerald McGill, 4 Grainan Park, Londonderry, BT48 7UA
MI3 GJI P Bingham, 28 Carnew Road, Katesbridge, Banbridge, BT32 5PC
M3 GJN P Marriott, 38 Westfields, Tilney St. Lawrence, King's Lynn, PE34 4QS
M3 GJW G Watson, 2 Bow St., Mansfield Woodhouse, Mansfield, NG19 9PJ
M3 GKB D Khan, 20 Cae Penrallt, Trearddur Bay, Holyhead, LL65 2WA
M3 GKE B luetchford, 89 Lime Grove, Gayton, King's Lynn, PE32 1QU
M3 GKG A Boag, 53 Castlewood Road, London, N16 6DJ
M3 GKH G Buxton, 11 The Green, Northfield, Birmingham, B31 5HT
M3 GKI N Axon, 30 Rating Row, Beaumaris, LL58 8AF
M3 GKJ S Willis, 1 Whiffins Orchard, Coopersale Common, Epping, CM16 7HT
M3 GKK M Ahmed, 75 Drove Road, Swindon, SN1 3AE
M3 GKX James Boot, 110 Wallace Road, Bilston, WV14 8AU
M3 GKY Anne Reed, 32 Hollis Garden, Cheltenham, GL51 6JQ
M3 GLA G Astbury, 12 Southall Road, Ashmore Park, Wolverhampton, WV11 2PZ
M3 GLC G Colclough, Little Hallands, Norton, Seaford, BN25 2UN
MM3 GLH G Bruce, 60 Kingsmills, Elgin, IV30 4BU
M3 GLM G Parkins, 73 Orwell View Road, Shotley, Ipswich, IP9 1NW
M3 GLT G Talbot, 28 Chevalier Grove, Crownhill, Milton Keynes, MK8 0EJ
M3 GMB C Bradshaw, 02 Arden Way, Market Harborough, LE16 7DD
M3 GMG G McGeough, 57 Stonehouse Park, Thursby, Carlisle, CA5 6NS
MI3 GMI Melvyn Crozier, 33 Cullentragh Road, Poyntzpass, Newry, BT35 6SD
M3 GML G Linfield, 82 Claremont Road, Swanley, BR8 7QT
MM3 GMP J Williams, Baptist Manse, Balemartine, Isle of Tiree, PA77 6UA
M3 GMY A Pickering, 16 Chestnut Grove, Accrington, BB5 0ND
M3 GNB J Carfoot, 11 Parc Sychnant, Conwy, LL32 8SB
M3 GNM Stuart McLoughlin, 40 Rowlandson Gardens, Bristol, BS7 9UH
M3 GNN A Glover, 103a Latimer Street, Liverpool, L5 2RF
M3 GNY Alan Hunt, Chestnut Cottage, Hine Town Lane, Blandford Forum, DT11 0SN
MM3 GOE Tearlach MacDonald, Main Road Farm, Balephuil, Isle of Tiree, PA77 6UE
MM3 GOI S Adam, 231/1 Gogarloch Syke, Edinburgh, EH12 9JF
MM3 GOT E Griffiths, Achnamara, Heanish, Isle of Tiree, PA77 6UL
M3 GOV Andrew Ward, 81 Northbrooks, Harlow, CM19 4DB
MM3 GOX Callum Williams, Ormer, Kirkapol, Isle of Tiree, PA77 6TW
MM3 GOY E Williams, Ormer Cottage, Kirkapol, Scarinish, PA77 6TW
M3 GOZ N Gostling, 49 Roundhouse Road, Dudley, DY3 2AX
MM3 GPB A Williams, Ormer Cottage, Kirkapol, Scarinish, PA77 6TW
MW3 GPG S Griffiths, 8 Heol Cynwyd, Llangynwyd, Maesteg, CF34 9TB
M3 GPJ G Jackson, 40 Ellis Avenue, Old Colwyn, Colwyn Bay, LL29 9LB
MM3 GPL Gavin Lawrie, 3 Anderson Court, Dornoch, IV25 3RT
M3 GPM Peter Mansfield, 106 Field Lane, Burton-on-Trent, DE13 0NN
M3 GPN Russell Orton, 18 Clarel Street, Penistone, Sheffield, S36 6AU
M3 GPP A Ault, 89 Southbourne Coast Road, Bournemouth, BH6 4DX
M3 GPR G Richards, 1 Maple Close, Seaton, EX12 2TP
M3 GPX Stephen Russell, 11 Lowgate Avenue, Bicker, Boston, PE20 3DF
M3 GQB James Gyton, 10 Longcroft, Southdown Road, Shoreham-by-Sea, BN43 5AY
M3 GQD J Reynolds, 3 Ardleigh, Basildon, SS16 5RA
MW3 GQE J Doyle, 18 The Paddocks, Tonna, Neath, SA11 3FD
M3 GQI K Cramp, 20 Combeland Road, Minehead, TA24 6BT
M3 GQL Peter Ridgers, 231 The Greenway, Epsom, KT18 7JE
M3 GQM T Sheppard, 1 Waveney Walk, Crawley, RH10 6RL
M3 GQP A Hall, 21 Eardulph Avenue, Chester le Street, DH3 3PR
MM3 GQR Stella McIver, 9 Balvicar Road, Oban, PA34 4RP
M3 GQS Christopher Williams, Pen Y Cae, Bodeiliog Road, Denbigh, LL16 5PA
MM3 GQT Gordon Mc Gregor, 34 Cairn Road, Cumnock, KA18 1HN
M3 GQW Edward Tart, Sunnybank Farm, Wattlesborough Heath, Shrewsbury, SY5 9EG
MM3 GQY H Dineley, Banks, Burray, Orkney, KW17 2ST
MW3 GRC G Coombes, 25 Afan Valley Road, Cimla, Neath, SA11 3SS
M3 GRF A Grace, 2 St. Peters Crescent, Bicester, OX26 4XA
M3 GRI T Griffiths, 56 The Avenue, Totland Bay, PO39 0DN
M3 GRY Graham Crane, 35 Betjeman Avenue, Wootton Bassett, Swindon, SN4 8JY
M3 GSI C Nelmes, 119 Exeter Road, Dawlish, EX7 0AN
MM3 GSL G Shaw, 1 Fir Park, Sorn, Mauchline, KA5 6HY
M3 GSM P Taylor, 17 Ladstone Towers, Sowerby Bridge, HX6 2QW
M3 GSQ W Howarth, 12 Church Terrace, Outwell, Wisbech, PE14 8RQ
M3 GSR William Chave, 91 Newman Road, Exeter, EX4 1PQ
MI3 GSW G Heggan, 18 Glen View, Moira, Craigavon, BT67 0AP
M3 GTA D Langmead, 38 Milton Grove, London, N11 1AX
M3 GTB G Bland, 20 Brereton Close, Castlefields, Runcorn, WA7 2LR
MM3 GTF Frank Davidson, 27 Gordon Way, Livingston, EH54 8JG
M3 GTG R Chisholm, 162 Ardington Road, Northampton, NN1 5LT
M3 GTH E Jones, 43 Wesley Road, Wimborne, BH21 2QB
M3 GTK J Walker, 34 Vian Road, Waterlooville, PO7 5TW
MW3 GTM G Mainwaring, 3 Elias Street, Neath, SA11 1PP
MI3 GTO George Shaw, 49 Cloughey Road, Portaferry, Newtownards, BT22 1NQ
M3 GTQ G Thompson, 28 St. Georges Road, Atherstone, CV9 3BP
M3 GTT G Hines, 126 Linacre Lane, Bootle, L20 6ES
M3 GTV Andrew Burfield, 4 Eastern Crescent, Chelmsford, CM1 4JQ
M3 GUF R Glossop, 21 Elizabeth Avenue, Tattershall Bridge, Lincoln, LN4 4JJ
M3 GUG K Bell, 12a Mill Lane, Carlton, Goole, DN14 9NG
M3 GUH S Tolhurst, Gwernynydd Fach, Nantmel, Llandrindod Wells, LD1 6EW
M3 GUJ A Brimble-Brice, 15 Egremont Road, Exmouth, EX8 1RX
M3 GUM K Waterhouse, 74 Clifford Road, West Bromwich, B70 8JY
M3 GUO S Shaw, 4 Perry Hill, Chelmsford, CM1 7RD
M3 GUQ M Reynolds, 15 Michelham Close, Eastbourne, BN23 8JD
M3 GUU S Whiting, 76 Norwich Drive, Bracebridge Heath, Lincoln, LN4 2TF
M3 GVC Philip Wayer, 4 Chatburn Avenue, Waterlooville, PO8 8UB
MM3 GVE Christopher Brown, 9 Newton Crescent, Rosyth, Dunfermline, KY11 2QW
M3 GVF Adam Roberts, 9 Llys Hendre, Rhuddlan, Rhyl, LL18 5YF
M3 GVJ L Ballinger, 9 Somerville Court, Cirencester, GL7 1TG
M3 GVN J [illegible], 32 Essex Road, London, E12 6RE
M3 GVT G Finney, 78 Lockley Street, Stoke-on-Trent, ST1 6PQ
MW3 GVU J Brennan, 1 Gerddi Mair, St. Clears, Carmarthen, SA33 4ET
M3 GWC Stephen Clarkson, Carisbrooke, Poolhouse Road, Wolverhampton, WV5 8AZ
M3 GWH G Haines, 12 Tyn Rhos Estate, Gaerwen, LL60 6HL
MI3 GWQ T Cosgrove, 301 Russell Court, Claremont Street, Belfast, BT9 6JX
M3 GWW G Wheelhouse, 86 Severn Street, Hull, HU8 8TQ
M3 GWZ Philip French, 4 Acacia Avenue, Newport, NP19 9AT
M3 GXB G Beaver, 23 West Drive Gardens, Soham, Ely, CB7 5EF
M3 GXG M Sartorius, 7 Pinehurst, London Road, Egham, TW20 0HQ
M3 GXI B Saunders, 4 Mendip Road, Worthing, BN13 2LP
M3 GXX Ian Bardell, 32 Bridle Road, Watton, Thetford, IP25 6NA
M3 GYA Helen Denmead, 47 Holland Road, Clevedon, BS21 7YJ
M3 GYB M Peterson, 4 Allen Drive, Mansfield, NG18 3HZ
M3 GYH R Blake, Taita, Linnards Lane, Northwich, CW9 6ED
M3 GYI Lionel Horton, 36 Merevale Crescent, Morden, SM4 6HL
MM3 GYU T Bloomfield, Midyard House, Carnwath, Lanark, ML11 8LH
M3 GZD D Blackmore, 67 Morval Crescent, Runcorn, WA7 2QS
M3 GZE S Lee, 154 Grangeway, Runcorn, WA7 5JA



**IMPORTANT NOTE**

**Revalidate licence to avoid revocation** – Ofcom has advised the Society that plans will be drawn up to revoke licences that have not been revalidated as required by the licence conditions. The quickest way to revalidate is to do so online via the Ofcom website: *https://services.ofcom.org.uk/* or by email: *amateur.validations@ofcom.org.uk* Ofcom staff are available to help, but please be patient during times of heavy workload.

MM3 GZG Kyle Cunningham, 11 Glendoune Street, Girvan, KA26 0AA
M3 GZI A Farmar, Hawkes Place, Horslett Hill, Holsworthy, EX22 6RS
M3 GZJ M Stroweger, 88 Castle Rise, Runcorn, WA7 5XW
M3 GZP I Plain, 18 Arundel Road, Bath, BA1 6EF
M3 GZQ M Parris, 19b Milfoil Drive, Eastbourne, BN23 8BR
M3 GZT A Cain, 55 Lytham Green, Muxton, Telford, TF2 8SQ
M3 GZU R Lang, 89 Dodthorpe, Hull, HU6 9HA
M3 GZW R Shadbolt, 58 Westfield Road, Manea, March, PE15 0LN
M3 HAC J Hilton, 32 Dowry St., Fitton Hill, Oldham, OL8 2LP
M3 HAD H Rhymes, 12 Reedling Drive, Southsea, PO4 8UF
M3 HAE C M Davies, Afallon, 3 Penygraig, Aberystwyth, SY23 2JA
MM3 HAF Matthew Hoskin, 16 Rossie Woods, Rossie, Montrose, DD10 9TS
M3 HAI G Young, 47 Birdhill Road, Woodhouse Eaves, Loughborough, LE12 8RP
M3 HAJ I Griffiths, 147 Greenlawns, St. Marks Road, Tipton, DY4 0SU
M3 HAK R Siebert, Po Box 127, Nantwich, CW5 8AQ
M3 HAL Stephen Hall, Orchard End, Asenby, Thirsk, YO7 3QR
M3 HAM Garry Roberts, 25 Chalfont Way, Liverpool, L28 3QB
MW3 HAQ C Olding, 10 Ty Nant, Caerphilly, CF83 2RA
MW3 HAS H Mustafa, 17 Furness Close, Ely, Cardiff, CF5 4PG
M3 HAT Harriet Kennedy, 19 High Street, East Hoathly, Lewes, BN8 6DR
M3 HAU W Kent, Long Spring Cottage Gracious Lane, Sevenoaks, TN13 1TJ
M3 HAW S Avery-Hawkins, 41 Daniels Welch, Coffee Hall, Milton Keynes, MK6 5DA
M3 HAZ H Flower, 17 Scott Grove, Morecambe, LA4 4LN
M3 HBB G Smith, 4 Sweden Park, Ambleside, LA22 9EY
M3 HBC M Lewis, 96 Roundhouse Close, Nantyglo, Ebbw Vale, NP23 4QY
MW3 HBF D Williams, 54 Howell Street, Pontypridd, CF37 4NR
M3 HBG A Hoe, 12 Penshurst Way, Eastleigh, SO50 4RJ
M3 HBM J Baxter, 10 Speedwell Close, Bedworth, CV12 0NS
M3 HBP S Bethell, 35 Fulford Road, Bristol, BS13 9RL
M3 HBS Jonathan Bodie, 4 Trewartha Vean, Merther Lane, Truro, TR2 4AG
MD3 HBT Thomas Richley, 30 Chicheley Road, Harrow, HA3 6QL
M3 HBX David Glenn, 84 Cambridge Street, Normanton, WF6 1ER
M3 HCA E Foster, 12 Dunham Grove, Leigh, WN7 3DS
M3 HCB H Benton, Emiviz, The Ridge, Salisbury, SP5 2LQ
M3 HCE Anthony Macnauton, 27a Lincoln Road, Poole, BH12 2HT
M3 HCG C Hawkins, 118 Aldebury Road, Maidenhead, SL6 7HE
M3 HCL C Lott, 6 Centurion Close, College Town, Sandhurst, GU47 0HH
M3 HCP D Hounslow, 3 Hengrave Green, Ivington, Leominster, HR6 0JL
M3 HCW Paul Webb, 7 Chapel Road, Prestatyn, LL19 7TH
M3 HDL R Guess, 69 Rowan Drive, Kirkby-in-Ashfield, Nottingham, NG17 8FP
M3 HDV B Hampson, 38 Parley Road, Bournemouth, BH9 3BB
M3 HEC Andrew Spencer, 45 Long Close, Chippenham, SN15 3JZ
M3 HEE A Totterdell, 35 Meadow Bank Avenue, Sheffield, S7 1PB
M3 HEI S Collett, 81 Wycombe Road, Prestwood, Great Missenden, HP16 0HW
M3 HEJ E Haycock, 55 Ashbourne Road, Rocester, Uttoxeter, ST14 5LF
M3 HEO Alexander Fagan, 77 Watling Street West, Towcester, NN12 6AG
M3 HER F Nation, 1 Claydon Path, Aylesbury, HP21 9EF
M3 HET H Thomas, 15 Ashgrove Way, Bridgwater, TA6 4UB
M3 HEV Andrew Sturgess, Hawks Barn, Long Lane, Shaftesbury, SP7 0BJ
M3 HFA E Gainford, 10 The Spinney, Ashford, TN23 3LF
M3 HFH Steven OVERALL, Flat 74, Douglas Buildings, London, SE1 1EL
M3 HFO H Foster, 11 Rosedale Gardens, Rhyl, LL18 4TY
M3 HFT E Rogers, Maes Gwersyll, Garthmyl, Montgomery, SY15 6RS
M3 HFU Gareth Johnson, 199 Lynwood, Folkestone, CT19 5TA
M3 HFX B Mccann, 21 Ladyseat, Longtown, Carlisle, CA6 5XX
M3 HGA T Winwood, 2 The Warren, Abingdon, OX14 3XB
M3 HGE A Hitchens, 16 Harrisons Place, Northwich, CW8 1HX
M3 HGH Kenneth Stewart, 17 Delamere Street, Bury, BL9 6NE
M3 HGL B Peck, 60 Richmond Road, Ipswich, IP1 4DP
M3 HGM N Peters, 57 High Street, Collingtree, Northampton, NN4 0NE
M3 HGO R Ward, Beech Cottage, Saron Road, Goytre, NP4 0BN
M3 HGP P Bland, 5 Pembroke Way, Winsford, CW7 1QZ
M3 HGR Hugh Roberts, Hen Ddol, Northfield Road, Barmouth, LL42 1PT
M3 HGT D Leverton, 21 Laburnum Grove, Killamarsh, Sheffield, S21 1GR
M3 HGW Martin Bancroft, Old Stables, Jeffrey Lane, Doncaster, DN9 1LT
M3 HGX D Clark, 12 Wilson Crescent, Lostock Gralam, Northwich, CW9 7QH
M3 HGZ J Sejwacz, Flat 9, Mayrick Court, Newton le Willows, WA12 9GB
M3 HHB M Clarke, 14 Tower Court, Haverhill, CB9 9DD
M3 HHC S Crossley, 29 Rycroft Avenue, Bingley, BD16 1PU
M3 HHN A Highfield, 29 Blewitt Street, Brierley Hill, DY5 4AW
M3 HHQ D Sejwalz, 4 Ash Avenue, Newton-le-Willows, WA12 8HJ
M3 HHX M Arnott, 2 Hambleton Close, Elsecar, Barnsley, S74 8DS
M3 HIE Brian Edwards, 4 Hythe Crescent, Seaford, BN25 3TU
M3 HIG T Higgins, 15 Ellen Street, Warrington, WA5 0LY
M3 HIM Felim Doyle, 1 Claydon Path, Aylesbury, HP21 9EF
M3 HIN Alan Watkinson, 34 Marble House, Felspar Close, London, SE18 1LN
M3 HIO S Jarvis, 11 The Green, Northfield, Birmingham, B31 5HT
M3 HIP J Dixon, 6 Howland Close, Eastbourne, BN23 5AJ
M3 HIT A Bryan, 16 Walesmoor Avenue, Kiveton Park, Sheffield, S26 5RG
M3 HIX P Owen, 24 Sirhowy Court, Green Meadow, Tredegar, NP22 4PL
M3 HJB H Beier, 20 Markham Avenue, Armthorpe, Doncaster, DN3 2AZ
MM3 HJC C Hazle, 6 Dalneigh Road, Inverness, IV3 5AH
M3 HJD L Dixon, 55 Henthorn Road, Clitheroe, BB7 2LD
M3 HJE Harriet Evans, Littlefield House, Bolney Road, Haywards Heath, RH17 5AW
M3 HJF S Gilchrist, Kening, Ashleigh Crescent, Barnstaple, EX32 8LA
M3 HJG A Howe, 18 Co-operation Street, Crawshawbooth, Rossendale, BB4 8AG
M3 HJJ S Norris, 15 East View, Choppington, NE62 5UF
M3 HJN D Newton, 84 Ameysford Road, Ferndown, BH22 9QB
M3 HJU Peter Webster, Tredavros Farm, Bodmin, PL30 5BE
M3 HJV Carl Lishman, 6 Clarence Road, Accrington, BB5 0NA
M3 HJW R Smith, 61 Waverley Drive, Chertsey, KT16 9PF
MM3 HKE L Higgins, 11 Strathyre Place, BROUGHTY FERRY, Dundee, DD5 3WN
MM3 HKG Darryl Menzies, 12 Dean Avenue, Dundee, DD4 7LH
M3 HKH S Crighton, 12 Cwm Road, Waunlwyd, Ebbw Vale, NP23 6TR
M3 HKM Kevin Monaghan, The Bulstone Hotel, Branscombe, Seaton, EX12 3BL
M3 HKT R Woodley, 1 Melton Drive, Didcot, OX11 7JP
M3 HKV L Flawn, Autumns, Jacks Lane, Bishop's Stortford, CM22 6NT
M3 HLA J Meeks, 64 Belford Street, Burnley, BB12 0DF
M3 HLD R Metcalfe, 33 Midland Terrace, Hellifield, Skipton, BD23 4HJ
MM3 HLG S Paul, 10 Beechwood Gardens, Westhill, AB32 6YE
M3 HLN Hayley Noel, 58 Easenhall Lane, Redditch, B98 0BJ
M3 HLP P Hallas, 37 Oakfield Road, Bromborough, Wirral, CH62 7BA
M3 HLV J Ferguson, 41 Brunswick Street, Burnley, BB11 3NX
M3 HLX Brian Buskin, 6 Elgin Close, Bedlington, NE22 5HJ
M3 HLZ Raymond Davies, Parclands, Raglan, Usk, NP15 2BX
MJ3 HMA M Haddon, Balik Pulau, Bradford Ave, La Route Des Genets, St. Brelade, Jersey, JE3 8DP
M3 HMC H Chambers, 354 Townsend Avenue, Norris Green, Liverpool, L11 5AJ
M3 HME Edward Cotton, 98 Severn Street, Hull, HU8 8TQ
M3 HMK H Knighton, 5 Quidenham Road, East Harling, Norwich, NR16 2JD
M3 HML B Just, 2 The Old Rectory, The High Road, Bedford, MK43 7HN
MM3 HMM D Macmillan, 2 Fladda Road, Oban, PA34 4HZ
M3 HMT H Tate, 52 Marlborough Road, London, N22 8NN
M3 HND Joseph Robinson, 6 Hubert Road, Winchester, SO23 9RG
M3 HNE M Ellis, 58 Egghill Lane, Northfield, Birmingham, B31 5NT
M3 HNK Terry Lee, 68 Wharton Drive, Springfield, Chelmsford, CM1 6BF
M3 HNL J Stone, 27 The Heathlands, Warminster, BA12 8BU
M3 HNM N Evans, 38 Cockster Road, Longton, Stoke-on-Trent, ST3 2EG
MW3 HNP J Nelson, 31 Y Drim, Ponthenry, Llanelli, SA15 5NY
M3 HNQ I Rowland, 45 Birks Road, Mansfield, NG19 6JU
M3 HNU Glen Sandell, 20 Kirkby View, Sheffield, S12 2NB
M3 HNV P Breckell, 45 Gordon Avenue, Mansfield, NG18 3AZ
M3 HOD Adam Hodson, 22 Walmley Ash Road, Sutton Coldfield, B76 1HY
M3 HOE A Hoe, 12 Ashbridge Rise, Chandler's Ford, Eastleigh, SO53 1SA
M3 HOM J Homsey, 105 Lynwood, Folkestone, CT19 5DD
M3 HOU E Edwards, 5 Brindley Road, Silsden, Keighley, BD20 0LD
M3 HOV Brian Brown, 3 Swaledale, Worksop, S81 0UY
M3 HOY M Hoy, 39 Blackbird Road, Caldicot, NP26 5RE
M3 HPF C Jamieson, 19 Melton Road, Whissendine, Oakham, LE15 7EU
M3 HPM D Woodward, 140 Ewe Lamb Lane, Bramcote, Nottingham, NG9 3JW
M3 HPN W Hodgson, 11 Tudor Court, Hitchin, SG5 2BE
M3 HPO Andrew Riches, 84 Elgar Drive, Shefford, SG17 5RA
M3 HPT P Taylor, Fenway Farm, Ten Mile Bank, Downham Market, PE38 0EU
M3 HPY Katherine Tokley, 9 Peel Road, Springfield, Chelmsford, CM2 6AQ
M3 HPZ A Bidwell, 134 Milton Road, Weston-Super-Mare, BS23 2US
M3 HQB C Smith, 71 Connaught Road, Luton, LU4 8ER
MM3 HQC Jason Henry, 7 Wenlock Road, Paisley, PA2 6UJ
M3 HQD Rob Wilkes, 33 Pembroke Road, Bulwark, Chepstow, NP16 5AF
MM3 HQL G Fuller, 20 Drumellan Road, Ayr, KA7 4XA
M3 HQN Michael Haworth, 26 Willowhey, Marshside, Southport, PR9 9TW
M3 HQP John Parrott, 2 Boyd Close, Wirral, CH46 1RX
M3 HQQ Howard Dixon, 45 Penkhull Terrace, Stoke-on-Trent, ST4 5DH
M3 HQS M williamson, 5 Fernbank Close, Crewe, CW1 6ES
M3 HQU M Copeman, 1 Chestnut Avenue, Welney, Wisbech, PE14 9RG
M3 HQV E Carter, 34 Wrexham Road, Brynteg, Wrexham, LL11 6HR
M3 HQW A Hillbeck, 28 Darent Avenue, Walney, Barrow-in-Furness, LA14 3NU
M3 HRC P Edwards, 5 Brindley Road, Silsden, Keighley, BD20 0LD
M3 HRE S Esp, 34 Wrexham Road, Brynteg, Wrexham, LL11 6HR
M3 HRM D Morgan, 171 Town Road, London, N9 0HJ
M3 HRN C Fower, 31 Hillswood Avenue, Leek, ST13 8EQ
M3 HRT S Brashill, 42 Bannister Street, Withernsea, HU19 2DT
M3 HRV David Porter, 39 Panama Road, Burton-on-Trent, DE13 0SQ
M3 HRY P Odle, 24 Longfellow Road, Gillingham, ME7 5QG
M3 HSC J Taggart, 250 Thomas Drive, Liverpool, L14 3LF
M3 HSE C Hoyle, 43 Helme Drive, Kendal, LA9 7JB
MM3 HSG Mark Douglas, 195 Dumbuck Road, Dumbarton, G82 3NU
M3 HSH K White, 30 Nuneaton Road, Bedworth, CV12 8AL
M3 HSI Alistair McGann, 8 Hertford Close, Whitley Bay, NE25 9XH
M3 HSJ D Teasdale, 43 Easington Road, Stockton-on-Tees, TS19 8ES
M3 HSM Christopher Hayes, 7 Harries Court, Waltham Abbey, EN9 3NS
M3 HSR P Hilton, 14 Masefield Road, Thatcham, RG18 3AF
M3 HSS J Little, 41 Sevenoaks Road, Portsmouth, PO6 3JP
M3 HSV C Srinivasan, 2 Hall Drive, Burley in Wharfedale, Ilkley, LS29 7LL
MI3 HSW Hazel White, 28 Lord Warden's Parade, Bangor, BT19 1YU
MW3 HSZ P Bishop, 76 Heol Homfray, Cardiff, CF5 5SB
M3 HTA S Fuller, 29 Beckley Road, Wakefield, WF2 9QB
M3 HTE E Townley, Beetham, Water Hill Lane, Halifax, HX2 7SG
M3 HTF D O'Flanagan, 16 Corbett Road, London, E11 2LD
M3 HTG A Murphy, 22 Shenley Fields Drive, Birmingham, B31 1XH
M3 HTO Paul Hardy, 21 West Avenue, Boston Spa, Wetherby, LS23 6EJ
M3 HTR S Adlam, 50 High St., Westtown, Dewsbury, WF13 2QF
MM3 HTY D Cunningham, 53 Hillhouse Avenue, Bathgate, EH48 4BB
M3 HUB A Hubbard, 26 Lydgate Drive, Wingerworth, Chesterfield, S42 6TF
M3 HUG R Hughes, 7 Willow Place, Darlington, DL1 5LX
M3 HUS Nigel Payne, 19 Sid Park Road, Sidmouth, EX10 9BW
M3 HUW H Weatherhead, 39 Meadow Park, Dawlish, EX7 9BU
M3 HUX N Gibson, 19 Nene Side Close, Badby, Daventry, NN11 3AD
M3 HUY David Dolan, 29 Byland Way, Monk Bretton, Barnsley, S71 2JY
M3 HVA P Robinson, Wall Lane Bank Cottage, Leek Road, Leek, ST13 7HH
M3 HVE R Dolman, 3 Cloonmore Avenue, Orpington, BR6 9LE
M3 HVH M Bell, 10 Margaret Close, Darfield, Barnsley, S73 9QE
M3 HVL K Phizacklea, 23 High Duddon Close, Askam-in-Furness, LA16 7EW
M3 HVN P Harris, 17 Seymour Avenue, Great Yarmouth, NR30 4BB
M3 HVO G Omar, 140 Twickenham Road, Isleworth, TW7 7DJ
M3 HVP T Moss, 3 Haddon Close, Macclesfield, SK11 7YG
M3 HVS Andrew Hunt, 14 Offranville Close, Leicester, LE4 8NR
M3 HVU Kevin Jessop, 61 Fountayne Street, Goole, DN14 5HQ
M3 HVV K Ozwell, 109 Abbey Road, Grimsby, DN32 0HN
M3 HVW J Naylor, 12 Princess Avenue, Wesham, Preston, PR4 3BA
M3 HVX D Proctor, 58 Hornby Drive, Newton, Preston, PR4 3SU
M3 HVY C Hacker, 49 Lamaleach Drive, Freckleton, Preston, PR4 1AJ
MJ3 HWC H Carrel, 5 Belmont Road, St. Helier, Jersey, JE2 4SA
M3 HWH Graham Jackson, Pathways, Down Barton Road, Birchington, CT7 0PY
M3 HWN D Wardman, 2 Silver St., Scruton, Northallerton, DL7 0QR
M3 HWP Daniel Potts, 19 Clay Street, Workington, CA14 2XZ
M3 HWS J Cleverley, 4a Godfrey Street, Netherfield, Nottingham, NG4 2JG
M3 HWV A Hicks, 30 Manna Drive, Elton, Chester, CH2 4RP
M3 HWW H Wilson, 2 Railway Close, Burwell, Cambridge, CB25 0DW
M3 HWX D Beer, 46 The Mailyns, Gillingham, ME8 0DZ
M3 HWY Geoffrey Elsworthy, 40 Moorfield Way, Wilberfoss, York, YO41 5PL
M3 HXB B Wheat, 23 Stead Street, Eckington, Sheffield, S21 4FY
M3 HXF Joe Merchant, 186 Manor Hall Road, Southwick, Brighton, BN42 4NH
M3 HXG D Haestier, 18 Midhurst Rise, Brighton, BN1 8LP
M3 HXH J Clarke, 144 St. Johns Avenue, Kidderminster, DY11 6AU
M3 HXM W Cole, The Spinney, Holmes Chapel Road, Congleton, CW12 4SN
M3 HXO M Shaw, 10 Beechwood Avenue, Shevington, Wigan, WN6 8EH
M3 HXQ T Johnson, 27 Fonthill Road, Bristol, BS10 5SR
M3 HXS Matthew Campbell, 71 SAGES LEA, Woodbury Salterton, EX5 1RA
M3 HXT M Bower, 21 Raglans, Exeter, EX2 8XN
M3 HXW Kostas Bouris, 3 Suffolk Court, Vicarage Road, Maidenhead, SL6 7DT
M3 HXZ N Mcintyre, 27 Chapel Close St Ann's Chapel, Gunnislake, PL18 9JB
M3 HYD M Douglas, 13 Castlereagh Street, New Silksworth, Sunderland, SR3 1HJ
M3 HYE D Bruce, 6 Princes Way, King's Lynn, PE30 2QL
M3 HYF A Sargent, 15 Wilton Road, Balsall Common, Coventry, CV7 7QW
MM3 HYG P McArthur, 22 Bridgeway Terrace, Kirkintilloch, Glasgow, G66 3HJ
M3 HYI A Reynolds, Fairview, Coombe Way, Teignmouth, TQ14 9QA
M3 HYO E Hearne, 6 Hillview Road, Basingstoke, RG22 6BQ
M3 HYQ W Oakley, 1 Southern Avenue, Henlow, SG16 6EY
M3 HYV K Jones, Court House Farm, Holmes Chapel Road, Crewe, CW4 8AS
M3 HZA D Pryor, 10 Thornton Crescent, Church Langton, Market Harborough, LE16 7TA
MW3 HZB K Clark, 56 Morris Avenue, Llanishen, Cardiff, CF14 5JW
M3 HZC C Chen, Conville And Cains College, Trinity Street, Cambridge, CB2 1TA
M3 HZD D Evans, 7 Bowerwood Road, Fordingbridge, SP6 1BJ
M3 HZE D Mannion, 17 Balmoral Road, Haslingden, Rossendale, BB4 4EA
M3 HZH D Hubbard, 99 Tuckers Road, Loughborough, LE11 2PH
M3 HZK R Gooch, 14 Cotterill Road, Surbiton, KT6 7UN
M3 HZM M Holden, 26 Valebridge Drive, Burgess Hill, RH15 0RW
M3 HZN R Rippin, 28 Ridgeway West, Market Harborough, LE16 7LG
M3 HZO P Sarll, 81 Austendyke Road, Weston Hills, Spalding, PE12 6BX
M3 HZP B Holden, 26 Valebridge Drive, Burgess Hill, RH15 0RW
M3 HZW J Trevarrow, 17 Harland Road, Elloughton, Brough, HU15 1JT
M3 IAA Denis Rose, 21 Whites cresecent, Market Haborough, LE16 8GJ
M3 IAC I Crabb, 9a Lonsdale Road, Southend-on-Sea, SS2 4LZ
M3 IAE S Wilkinson, 144 West End Road, Morecambe, LA4 4EF
M3 IAF I Firby, 19 St. Georges Drive, Manchester, M40 5HL
MM3 IAG I Gerrard, 10 Station Road, Ardersier, Inverness, IV2 7ST
MI3 IAI T Scott, 25 Lisavon Drive, Belfast, BT4 1LJ
M3 IAO A Hirst, 34 Woodhall Avenue, Bradford, BD3 7BU
M3 IAP Graham Evans, 57 Lock Crescent, Kidlington, OX5 1HF
M3 IAQ Philip King, Philinda, Carmen Street, Saffron Walden, CB10 1NR
M3 IBE M Hagan, 186 Saltwell Road, Gateshead, NE8 4XH
M3 IBJ D Dickinson, 8 East Grange Garth, Leeds, LS10 3EJ
MM3 IBM C Mckillop, 7 Auchneagh Farm Lane, Greenock, PA16 7BJ
M3 IBS F Carter, 26 Union Road, Shirley, Solihull, B90 3DQ
M3 IBT J Ferrol, 29 Westlands, Haltwhistle, NE49 9BS
M3 IBY J Leaman, 40 Higher Budleigh Meadow, Newton Abbot, TQ12 1UL
M3 IBZ D Langridge, 12 Battles Lane, Kesgrave, Ipswich, IP5 2XF
M3 ICA Angelika Schmidt, Church Corner, Fieldside, Boston, PE22 7RA
MM3 ICD W Ton, 24 Craigmount Hill, Edinburgh, EH4 8DL
M3 ICF Colin Sole, 55 Hearth Street, Market Harborough, LE16 9AQ
M3 ICH Matthew Green, 4 Tudor Court, Grimethorpe, Barnsley, S72 7NA
M3 ICN A Groat, Rose Cottage, Ivy Dene Lane, East Grinstead, RH19 3TN
M3 ICO Lianne Widdowson, 11 Belmont Drive, Staveley, Chesterfield, S43 3PQ
M3 IDA D Cothey, Summerley House, Skircoat Moor Road, Halifax, HX3 0HA
M3 IDB N Dallen, 77 Hazon Way, Epsom, KT19 8HG
M3 IDC F Harris, 51 Hillmans Road, Newton Abbot, TQ12 1AA
M3 IDD D Barker, 21 Boundary Crescent, Lower Gornal, Dudley, DY3 2HJ
M3 IDF J Kelly, 1 Bramble Close, New Ollerton, Newark, NG22 9TN
M3 IDH K Reynolds, 3 Lilac Close, Chelmsford, CM2 9NY
M3 IDJ Michael Schonborn, 116 Hough Lane, Wombwell, Barnsley, S73 0EF
M3 IDK D Norman, 22 Stirling Street, Hull, HU3 6SL
M3 IDO A Thompson, 3 Rufford Road, Long Eaton, Nottingham, NG10 3FP
M3 IDQ A Stevenson, 2 Diddington Close, Bletchley, Milton Keynes, MK2 3EB
MM3 IDR Kelly Tait, 33 Bankton Avenue, Livingston, EH54 9LD
M3 IDW J Valle Espin, 203 Broadway, Horsforth, Leeds, LS18 4HL
M3 IDY J Taylor, Roseneath, 4c Valley Road, Kenley, CR8 5DG
M3 IEA P Jays, 138 Lower Wear Road, Exeter, EX2 7BD
MM3 IEC E Cohen, 234 Allison Street, Glasgow, G42 8RT
M3 IEF Susan Painting, 15 Surrey Walk, Walsall, WS9 8JG
M3 IEG M Luxton, 6 Tumbling Field Lane, Tiverton, EX16 4LN
M3 IEL R Tust, 28 Osprey Close, Beechwood, Runcorn, WA7 3JH
M3 IEM T Rogers, 45 Church Road, Westoning, Bedford, MK45 5LP
M3 IEP N Freeman, 27 Montpelier Drive, Caversham, Reading, RG4 6QA
M3 IEQ J Wheeler, 41 Winnards Park, Sarisbury Green, Southampton, SO31 7BX
M3 IET C Blount, 55 Silverthorne Drive, Caversham, Reading, RG4 7NR
M3 IEU T Bougourd, 1 Poplar Close, Newton Abbot, TQ12 4PG
M3 IEV M Blount, 55 Silverthorne Drive, Caversham, Reading, RG4 7NR
M3 IEW Paul Mellish, 302 Belvedere Road, Burton-on-Trent, DE13 0RD
M3 IFA Aaron Harrison, 16 Ingshead Avenue, Rawmarsh, Rotherham, S62 5BH
M3 IFB D Symonds, 2 Montgomery Cottages, Stonham Road, Stowmarket, IP14 5LS
M3 IFE M Ferguson, 80 Chester Road, Holmes Chapel, Crewe, CW4 7DR
M3 IFF Peter Webster, 15 Napier Street, Workington, CA14 2PT
M3 IFG Frank Gatenby, 6 Telford Close, Audenshaw, Manchester, M34 5FB
MI3 IFI D Sloan, 15 Deramore Drive, Portadown, Craigavon, BT62 3HH
M3 IFJ B Royce, 82 Ridge Lane, Watford, WD17 4TA
M3 IFK K Holloway, 26 Aldgate Drive, Brierley Hill, DY5 3NT
MI3 IFO E McClements, 5 Eastbank, Strathfoyle, Londonderry, BT47 6UW
MW3 IFZ N Bruines, 24 Trenel, Burry Port, SA16 0UT
M3 IGA L Igali, 22 Mile End Road, Norwich, NR4 7QY
M3 IGN I Nicholls, 8 Northcroft Road, Corsham, SN13 0LS
MI3 IGO A Blythe, 159 Victoria Road, Bready, Strabane, BT82 0DZ

M3 IGZ Malcolm Matthias, 18 Brynmally Park, Pentre Broughton, Wrexham, LL11 6BP
M3 IHA Shaun Duffy, 38 Well Lane, Newton, Chester, CH2 2HL
M3 IHB G Jones, Wern Lodge, Gobowen, Oswestry, SY10 7JY
M3 IHC Gary Ward, 4 Walnut Drive, Whitchurch, SY13 1UD
MW3 IHD I Davies, 221 Trowbridge Green, Rumney, Cardiff, CF3 1RF
M3 IHN R Hargeaves, 58 Horsewell Lane, Wigston, LE18 2HQ
M3 IHO Keith Brown, 143 Princes Road, Ellesmere Port, CH65 8EP
M3 IHQ T Wall, 50 Higham Gobion Road, Barton-le-Clay, Bedford, MK45 4LT
M3 IHR D Tilley, 15 Dowhills Park, Liverpool, L23 8SS
M3 IHS C Ivers, 11 Twelve Acre Crescent, Farnborough, GU14 9PW
M3 IHU J Smith, Wilson Hall Farm, Slade Lane, Derby, DE73 1AG
M3 IHV H Donnelly, 6 Famet Walk, Purley, CR8 2DY
M3 IHX Daniel Eacott-Palfrey, 165 High Street, Blaina, Abertillery, NP13 3AW
MI3 IHY Samuel Quigg, 100 Whispering Pines, Limavady, BT49 0UF
MI3 IHZ K McDonald, 37 Ardgarvan Cottages, Limavady, BT49 0NF
M3 IIA Ronald Dale, 17 Spencer Gardens, Brackley, NN13 6AQ
MM3 IIG Mark Pentler, 19/4 Wardlaw Street, Edinburgh, EH11 1TN
MI3 IIH T Mcconnell, 41 Moyra Road, Doagh, Ballyclare, BT39 0SQ
M3 IIJ R Parry, 5 Accar Y Forwyn, Denbigh, LL16 3PW
MI3 IIL C McConnell, 41 Moyra Road, Doagh, Ballyclare, BT39 0SQ
M3 IIN B Stokes, 19 Hall Park, Barrow-on-Trent, Derby, DE73 7HD
M3 IIP Kenneth Hunt, 13 Beaumaris Court, Spondon, Derby, DE21 7RG
MM3 IIT G Saunders, Tower Guest House, 32 James Street, Stornoway, HS1 2QN
M3 IIV John Hurkett, 9 Fair Field Park, Five Lanes, Launceston, PL15 7RQ
M3 IIW G Whittle, 22 Warwick Street, Leigh, WN7 2NH
M3 IJD Joseph Doyle, 10 Greenall Court, Prescot, L34 1NH
M3 IJE I Ewen, 26 Court Road, Eastbourne, BN22 9EZ
M3 IJF E Livesey, Reevsmoor, Hollington, Ashbourne, DE6 3AG
M3 IJH Ian Harrop, 35 Langdale Crescent, Dalton-in-Furness, LA15 8NR
MM3 IJI M Carmichael, 39 Longsdale Crescent, Oban, PA34 5JR
M3 IJO P Chaney, 246 Agar Road, Illogan Highway, Redruth, TR15 3NJ
M3 IJS I Sapstead, 18 Rib Close, Standon, Ware, SG11 1QS
M3 IJT C King, 1 Victoria Court, Hadley, Telford, TF1 5FL
M3 IJV James Millichip, 33 Lincoln Road, Stevenage, SG1 4PJ
M3 IJZ C Young, 15 Shelton Avenue, East Ayton, Scarborough, YO13 9HB
MW3 IKC J Kenchington, 36 Lando Road, Pembrey, Burry Port, SA16 0UR
M3 IKD J Hockedy, 22 Victoria Road, Frome, BA11 1RR
M3 IKE Michael Murray, PO Box 55, Calle San Jaime, Benijofar, Spain, 3178
M3 IKI J Batson, 19 Seaview Road, Canvey Island, SS8 7PB
M3 IKJ Antony Knitter, 15 Thompson Drive, Hatfield, Doncaster, DN7 6JX
M3 IKM T Palmer, 29 Field End, Maresfield, Uckfield, TN22 2DJ
M3 IKN Christopher Staite, 85 Pierce Avenue, Solihull, B92 7JY
M3 IKR D Powis, Fircroft, Pound Lane, Woodbridge, IP13 0LN
MM3 IKS G Frew, 20 Achlonan, Taynuilt, PA35 1JJ
M3 IKV E Smith, Grange Farm, Main Street, Newark, NG23 5PX
M3 ILA Kenneth Burnell, 27 Manners Gardens, Seaton Delaval, Whitley Bay, NE25 0DW
M3 ILB N Silveston, 115 Noreen Avenue, Minster on Sea, Sheerness, ME12 2EJ
M3 ILG Brian Evans, 12 The Mead, Thaxted, Dunmow, CM6 2PU
M3 ILJ E Copper, 238 Canterbury Road, Kennington, Ashford, TN24 9QL
M3 ILM Nicholas Hickson, 27 Cressing Road, Witham, CM8 2NP
M3 ILR I Roper, 1 Holywell Road, Kilnhurst, Mexborough, S64 5UQ
M3 ILV Peter Hinchliffe, 21 Prospect Hill, Haslingden, Rossendale, BB4 5EF
M3 ILY Michael Andrews, 27 Bramble Avenue, Norwich, NR6 6LN
M3 ILZ B McAndrew, 8 Springhill Walk, Morpeth, NE61 2JT
M3 IMB I Berry, 4 Newlands Park Way, Newick, Lewes, BN8 4PG
MM3 IMC Iain McCuaig, 20 Kirk Street, Dunoon, PA23 7DP
M3 IME Graham Kerr, 6 Penn Kernow, Launceston, PL15 9TN
M3 IMJ P Coppin, 3 Firtree Close, Rough Common, Canterbury, CT2 9DB
MM3 IMK I MacKinnon, 6 Glencruitten Rise, Oban, PA34 4RX
M3 IMM I Margetts, 16 Lahn Drive, Droitwich, WR9 8TQ
MI3 IMO Thomas McNaughter, 36 Elms Park, Coleraine, BT52 2QE
M3 IMP Ian Phillpott, 14 Buttercup Close, Paddock Wood, Tonbridge, TN12 6BG
M3 IMR A Ashworth, 22 Crow Lane, Ramsbottom, Bury, BL0 9BR
M3 INC A Nicholls, 6 Curtin Drive, Moxley, Wednesbury, WS10 8RJ
M3 IND V Lowe, 35 Elm Place, Armthorpe, Doncaster, DN3 2DE
M3 INH E Hayes, 11 Ashleigh Wood, Monaleen, Ireland
M3 INJ A Hayes, 11 Ashleigh Wood, Monaleen, Ireland
M3 INL I Lockyer, 11 Lorina Road, Ramsgate, CT12 6DD
M3 INO G Patterson, 28 Highcliffe, Spittal, Berwick-upon-Tweed, TD15 2JH
M3 INQ Z Ardern, 9 Chaucer Avenue, Mablethorpe, LN12 1DA
MI3 INS D Quigg, 9 Springhill Terrace, Limavady, BT49 9BS
MM3 INY Lewis Affleck, 1 Fank Brae, Mallaig, PH41 4RQ
M3 IOC R Mitchell, 2 Corbar Road, Stockport, SK2 6EP
MM3 IOF M Mitchell, Raithburn Farm, Glasgow Road, Kilmarnock, KA3 6ES
MI3 IOH W Bradley, 16 Mullaghanagh Road, Dungannon, BT71 7AY
MJ3 IOJ Nicole Taylor, 21 Samares Avenue, La Grande Route do St. Clement South, St. Clement, Jersey, JE2 6NY
M3 IOK R Peel, 34 Pagdin Drive, Styrrup, Doncaster, DN11 8LU
MM3 IOM J Grundey, 8 Fraser Avenue, Blairgowrie, PH10 6QJ
M3 IOQ B Reynard, 90 Barnsley Road, Darton, Barnsley, S75 5NS
M3 IOT I Bewley, 21 Duloe Gardens, Pennycross, Plymouth, PL2 3RS
M3 IOX S Bridgoc, 140 Highbridge Road, Burnham-on-Sea, TA8 1LW
M3 IPD K Barron, 80 Primrose Crescent, Norwich, NR7 0SF
MW3 IPK T Vincent, 88 Lake Street, Ferndale, CF43 4HE
M3 IPM S Jackson, 1 The Avenue, Burton-upon-Stather, Scunthorpe, DN15 9EX
M3 IPQ Alan Badcock, 7 Heathfield Road, Chandler's Ford, Eastleigh, SO53 5RP
M3 IPT William Bull, 117 Walton Road, Wednesbury, WS10 0EU
M3 IPY L Earnshaw, 63 Manor Road, Fleetwood, FY7 7LJ
M3 IPZ P Wyles, Casa De La Rosa, Torbay Road, Torquay, TQ2 6RG
MM3 IQA G Robbins, 33 Moffat Court, Glenrothes, KY6 1JR
MM3 IQD M Gourlay, 14 Holmes Holdings, Broxburn, EH52 5NS
MW3 IQE K Lowther, Cynon Villa, Main Road, Mountain Ash, CF45 4BX
M3 IQF D Green, 12 Nostell Road, Ashton-in-Makerfield, Wigan, WN4 9XD
M3 IQG S Parris, 3 Manor Close, Ringmer, Lewes, BN8 5PA
M3 IQJ C Evans, Bridge Farm, Shrawley, Worcester, WR6 6TQ
M3 IQN M Davis, 63 Glebe Road, Barrington, Cambridge, CB22 7RP
M3 IQP L Bailey, 18 Dudley Place, St. Helens, WA9 1BL
M3 IQQ A Harris, 32 King Edward Road, Gillingham, ME7 2RE
M3 IQS Michael Cox, 7 Wilson Close, Daventry, NN11 9WH
MM3 IQU W Curry, 35 Tarvit Terrace, Springfield, Cupar, KY15 5SE
MW3 IQY S Beer, 49 Central Street, Pwllypant, Caerphilly, CF83 2NJ
M0 IRF [illegible], 10 Gloucester Close, Little Hulton, Manchester, [illegible]
M0 IRH [illegible]
M3 IRJ G Rogers, 51 Abingdon Road, Urmston, Manchester, M41 0GW
M3 IRM J Richardson, 3 Aylesbury Avenue, Urmston, Manchester, M41 0SB
M3 IRP Philip Cannam, 1 Field Close, Hinckley, LE10 1TH
M3 IRQ S Cannam, 82 Barwell Lane, Hinckley, LE10 1SS
M3 IRR Andrew Gregory, 140 Alder Street, Newton-le-Willows, WA12 8HP
M3 IRS D Mountford, 189 Lloyd Street, Stockport, SK4 1NH
M3 IRU P Dawson, 88 Urmson Road, Wallasey, CH45 7LQ
MI3 IRV E Mercer, 8 Woodside Gardens, Portadown, Craigavon, BT62 1EW
M3 IRX A Ryan, 60 Stanstead Road, Halstead, CO9 1YB
MI3 IRY W Cooney, 30 Clanbrassil Park, Portadown, Craigavon, BT63 5XT
MM3 ISA I McDermid, 52 Main Street, Pathhead, EH37 5QB
MI3 ISC P England, 30 Kernan Grove, Portadown, Craigavon, BT63 5RX
M3 ISG A Evans, 58 Lime Tree Avenue, Crewe, CW1 4HL
M3 ISH Sarah Brooks, 7 Mayfield Road, Northwich, CW9 7AS
M3 ISI S Russon, 165 Billington Avenue, Newton-le-Willows, WA12 0AU
M3 ISJ M Turner, 65 Neville Street, Newton-le-Willows, WA12 9DB
M3 ISN A Dickinson, 77 Ullswater Avenue, Warrington, WA2 0NQ
M3 ISO I Butler, 23 Owen Way, Basingstoke, RG24 9GH
M3 ISQ R Lilley, 3 Coultshead Avenue, Billinge, Wigan, WN5 7HS
MI3 ISX S Butler, 25 Chippendale Avenue, Bangor, BT20 4PX
M3 ISY M Edwards, Rouse Farm, Normans Lane, Warrington, WA4 4PY
MM3 ITA J Veal, 6 Morrison Avenue, Tranent, EH33 2AR
M3 ITH Peter Cowin, 16 West Lane, Shap, Penrith, CA10 3LT
M3 ITI S Roberts, 7 Alberta Grove, Prescot, L34 1PX
M3 ITK D Willson, Flat 26, King Charles Place, Shoreham-by-Sea, BN43 5JH
M3 ITL I Buckton, 67 Tennyson Avenue, Middlesbrough, TS6 7ND
M3 ITM S garthwaite, 278 Carlton Road, Barnsley, S71 2BA
M3 ITT David Jones, 77 Brinkburn Grove, Banbury, OX16 3WX
M3 ITU M Stephenson, 15 Springwood Road, Hoyland, Barnsley, S74 0AZ
M3 ITZ S Heywood, 16 Edinburgh Drive, Hindley Green, Wigan, WN2 4HL
M3 IUC M Sturt, 2 Golden Villas, Heathfield Road, Freshwater, PO40 9LQ
M3 IUH A Morris, 22 Dixon Avenue, Newton-le-Willows, WA12 0NE
M3 IUK A Mason, 16 Newstead View, Fitzwilliam, Pontefract, WF9 5DP
MW3 IUS I Canterbury, Brynllethryd Bungalow, Senghenydd, Caerphilly, CF83 4HJ
M3 IUV P Watson, 32 Shrewsbury Way, Saltney, Chester, CH4 8BY
M3 IUX R Higham, 17 Walkmill Gardens, Wellington, Seascale, CA20 1EF
M3 IUZ B Homer, 7 King Street, Quarry Bank, Brierley Hill, DY5 2DH
M3 IVA K Yates, 103 Raleigh Crescent, Stevenage, SG2 0EB
M3 IVD A Taplin, 2 Old London Road, Rawreth, Wickford, SS11 8TZ
M3 IVI W Ivison, 11 Durham St., Fence Houses, Houghton le Spring, DH4 6LA
M3 IVN G Newman, 32 Pilgrim Street, Sheffield, S3 9GX
M3 IVO W Gissing, 2 Yeo Moor, Clevedon, BS21 6UQ
M3 IVV G Merrington, Cartref, Ball Lane, Frodsham, WA6 8HP
M3 IVX R Balm, 250 Coppice Road, Arnold, Nottingham, NG5 7HF
M3 IVY Sheena Dixon, 5 Swanmore Road, Havant, PO9 4LG
MW3 IWC M Phillips, 45 Lewis St., Aberbargoed, Bargoed, CF81 9DZ
M3 IWG J Pusey, 29 Arthur Moody Drive, Newport, PO30 5JR
M3 IWJ B Larman, Cornhill, Mount Bovers Lane, Hawkwell, SS5 4JE
M3 IWK D Judge, 12 Heelas Road, Wokingham, RG41 2TL
M3 IWN S Reynolds, 2 Lawson Court, Boldon Colliery, NE35 9NH
M3 IWO W Hall, 16 Barrington Close, Chelmsford, CM2 7AX
M3 IWR J Chapman, South View, Mill End, Buntingford, SG9 0SU
M3 IWT M Champness, 10 Isaac Square, Great Baddow, Chelmsford, CM2 7PP
M3 IWX Samantha Evans, 7 Gloster Ropewalk, Dover, CT17 9ES
M3 IWZ A Hanna, 35 Orchard Drive, Mayland, Chelmsford, CM3 6EP
M3 IXC D Watson, 10 Gimson Close, Tuffley, Gloucester, GL4 0YQ
M3 IXD D Watson, 45 Kennel Lane, Brockworth, Gloucester, GL3 4NP
M3 IXE H Hector, 71 Edinburgh Drive, North Anston, Sheffield, S25 4HB
M3 IXF M Lucas, 38 Hazel Avenue, Braunton, EX33 2EZ
M3 IXH R Maas, 15 Pine Court, Attleborough, NR17 2HU
M3 IXJ Andrew Littleford, 1 Cynlas, Kinmel Bay, Rhyl, LL18 5LP
M3 IXK Anthony Pickett, 4 Trembel Road, Mullion, Helston, TR12 7DY
M3 IXM Paul Bell, 7 GREENWOOD WAY, Norwich, NR7 9HW
M3 IXO D Lowe, 5 Daisy Street, Bury, BL8 2QG
M3 IXT Sarah Widdowson, 11 Belmont Drive, Staveley, Chesterfield, S43 3PQ
M3 IXU D Skinner, 77 Rolleston Avenue, Petts Wood, Orpington, BR5 1AL
M3 IXY A Guest, 1 Green Meadows, Cannock, WS12 3YA
MW3 IXZ H Alban, 15 Catherine Close, Abercanaid, Merthyr Tydfil, CF48 1YY
M3 IYE Philip Ellwood, 13 Wensley Drive, Manchester, M20 3DD
M3 IYG A Dixon, Flat 4, Farmer House, London, SE16 4BY
MI3 IYH Albert Wilson, 108a Salia Avenue, Carrickfergus, BT38 8NE
M3 IYO Kenneth Peabody, 23 Grange Mount, West Kirby, Wirral, CH48 6ET
MI3 IYP S Kelly, 1 Ardnamoyle Park, Londonderry, BT48 8HN
M3 IYX P Chapman, 4 The Street, Sutton, Pulborough, RH20 1PS
M3 IYY J Side, Railway Crossing Cottage, Ash Road, Sandwich, CT13 9JB
M3 IZB Paul Lewis, 37 Speedwell Close, Melksham, SN12 7TE
M3 IZD George Curtis, 11 Bloomery Way, Maresfield, Uckfield, TN22 2DP
M3 IZH J Winson, 11 Windsor Crescent, Ilkeston, DE7 4HD
M3 IZI D Groom, 10 Sunnymead Road, Burntwood, WS7 2LL
M3 IZJ J Winson, 12 Hydal Ave, Long Eaton, NG10 4EB
M3 IZM K Winson, 41 Windsor Crescent, Ilkeston, DE7 4HD
M3 IZN A Lodge, 45 Laneside Avenue, Sutton Coldfield, B74 2BU
MM3 IZO Edwin Stuart, 92 Linefield Road, Carnoustie, DD7 6DT
M3 IZP B Johnson, 199 Lynwood, Folkestone, CT19 5TA
M3 IZQ K Coad, 17 Dilly Lane, Barton on Sea, New Milton, BH25 7DQ
M3 IZV C Sadler, 44 Caraway Road, Fulbourn, Cambridge, CB21 5DU
M3 IZW J Reynolds, Fairview, Coombe Way, Teignmouth, TQ14 9QA
M3 JAB J Butler, 219 Ridge Avenue, Burnley, BB10 3JF
M3 JAC J Sefton, 87 Lillibrooke Crescent, Maidenhead, SL6 3XL
M3 JAL A Lloyd, 10 Makepeace Close, Vicars Cross, Chester, CH3 5LU
M3 JAP John Phillips, 39 Bryn Glas, Rhosllanerchrugog, Wrexham, LL14 2EA
M3 JAZ J Hudspeth, 108 Fir Tree Lane, Burtonwood, Warrington, WA5 4NE
M3 JBB John Benson, 51 Hollowood Avenue, Littleover, Derby, DE23 6JD
M3 JBE J Brocklesby, 34 Sinnington End, Highwoods, Colchester, CO4 9RE
M3 JBF N Morphew, 5 Canterbury Close, Canterbury Road, Folkestone, CT19 5EL
M3 JBH J Glass, 18 Norman Way, Colchester, [illegible]
M0 JBM M Butcher, 67 Keepers Coombe, Bracknell, RG12 0TW
MJ3 JBQ J Daniells, Le Belon, 2 Clos Vallios, St. Lawrence, Jersey, JE3 1GP
M3 JBW J Wolohan, 24 Granby Close, Corby, NN18 0AG
M3 JBZ A Bell, 28 Haven Baulk Avenue, Littleover, Derby, DE23 4BJ
M3 JCA Colin Ashman, 40 St. Matthews Road, Kettering, NN15 5HE
MI3 JCB M Crawford-Baker, George's Nest, 131 Gobbins Road, Larne, BT40 3TX
M3 JCE T Higgins, 7 Nobles Close, Coates, Peterborough, PE7 2BT
M3 JCH J Hill, Anglecroft, Somerford Booths, Congleton, CW12 2JU
M3 JCQ J Bond, Oakley, 19 Poplar Road, Tenterden, TN30 7NT
M3 JCS John Sanderson, 54 Kelvedon Close, Chelmsford, CM1 4DG
M3 JCT Lee Jenner, Flat 1, 28a Park Road, Tunbridge Wells, TN4 0NX
M3 JCU J Turton, 2 Elkstone Road, Chesterfield, S40 4UT
M3 JCY John Connolly, 2 Waring Avenue, St. Helens, WA9 2QG
MW3 JDA Julie Ruck, 286 Barry Road, Barry, CF62 8HF
M3 JDF J Feather, 21 Cedar Avenue, Wickersley, Rotherham, S66 2NT
M3 JDG J Godding, 58 Dukeswood Road, Longtown, Carlisle, CA6 5UJ
M3 JDJ J Handley, 4 Manor Close, Draycott, Stoke-on-Trent, ST11 9AZ
M3 JDN J Dobson, 17 Cadley Causeway, Fulwood, Preston, PR2 3RU
MI3 JDQ J Quigg, 9 Springhill Terrace, Limavady, BT49 9BS
M3 JDS J Smith, Clare Cottage, White Ash Green, Halstead, CO9 1PD
M3 JDX J Porteious, 62b Church Close, Stilton, Peterborough, PE7 3RG
M3 JEE J Edmondson, 11 Cedar Terrace, Fencehouses, Houghton le Spring, DH4 5ND
M3 JEH J Hutt, 48 Hill Crest, Swillington, Leeds, LS26 8DL
MW3 JEK C Gilbert, 11 Parc Y Deri, Neath, SA10 6BQ
M3 JEM J Carvill, THE LODGE, OLDBURY ROAD, Worcester, WR2 6AA
M3 JEP M Clarke, 138 Colne Road, Halstead, CO9 2HJ
M3 JER James Higgins, 6 Larksfield Avenue, Bournemouth, BH9 3LP
M3 JFA P Cowan, 230 High Street, Felixstowe, IP11 9DS
M3 JFB J Beezer, 23 Milburn Street, Sunderland, SR4 6AU
M3 JFF J Neighbour, 14 The Drive, Wallsend, NE28 8DQ
M3 JFP J Payne, Jomayne, Farm Lane, Evesham, WR11 8TL
M3 JFS J Skinner, 12 Hanbury House, Cardy Close, Redditch, B97 6LP
MM3 JFW A Wilson, 20 Ballumbie Gardens, Dundee, DD4 0NR
M3 JGH John Hewitt, 166 Ormskirk Road, Rainford, St. Helens, WA11 8SW
M3 JGI Jeffrey Ives, 87 Sheepwalk, Paston, Peterborough, PE4 7BJ
M3 JGJ A Skinner, Chevington, Carlton Road, Godstone, RH9 8LD
M3 JGN Richard Wild, 90 Broadway East, Redcar, TS10 5DP
M3 JGQ A Greenland, 19 The Ridgeway, Potton, Sandy, SG19 2PS
MM3 JGR J Gracie, 34 Ayr Road, Dalmellington, Ayr, KA6 7SJ
MD3 JGS A Foxon, 39 Droghadfayle Road, Port Erin, Isle of Man, IM9 6EN
MM3 JGT John Thomson, 26 South Dean Road, Kilmarnock, KA3 7RB
M3 JGU M Connell, 17 The Crescent, Stockport, SK3 8SL
M3 JGW W Holroyd, 8 Carr Dene Court, Preston Street, Preston, PR4 2XA
M3 JGX A Hadfield, 50 Eastbourne Road, Southport, PR8 4DT
M3 JHC J Clark, 27 The Gabriels, Newbury, RG14 6PZ
M3 JHJ John HICKEY, 11 Greenfield Avenue, Hodthorpe, Worksop, S80 4XT
M3 JHL J Locke, 2 Fairnley Road, Nottingham, NG8 4AH
M3 JHR J Richardson, 44 Cross Tree Road, Wicken, Milton Keynes, MK19 6BT
MM3 JHS James Hume, 8/11 Leslie Place, Edinburgh, EH4 1NH
M3 JHT J Tarver, 14 South View Road, Leamington Spa, CV32 7JD
M3 JHV C Smith, 11 Chesterton Road, Thatcham, RG18 3UH
M3 JHW John Warren, 1 Acre Close, Rochester, ME1 2RE
M3 JIA J Allanson, Tresco, Hampton, Swindon, SN6 7RL
M3 JIC P Baker, 25 Regency Court, Winsford, CW7 1FE
M3 JID J Douglas, 26 Walker Drive, Bootle, L20 6NN
M3 JIE P Stanhope, 53 Shackleton Court, Croydon Drive, Manchester, M40 2NP
M3 JIH Daniel Marsland, 154 Moss Lane, Litherland, Liverpool, L21 7NN
M3 JII Amy Riley-Marsland, 154 Moss Lane, Litherland, Liverpool, L21 7NN
M3 JIJ K Powell, 11 Terrig Street, Shotton, Deeside, CH5 1XU
M3 JIK S Sanders, 3 Edmunds Square, Mickleover, Derby, DE3 0DU
M3 JIL Gillian Hinsley, 38 Swindon Lane, Prestbury, Cheltenham, GL50 4NY
MM3 JIN J Nicol, 18 Tininver Street, Dufftown, Keith, AB55 4AZ
M3 JIR S Buckley, 31 Rose Avenue, Irlam, Manchester, M44 6AQ
M3 JIT Wayne Carty, 49 Princess Gardens, Blackburn, BB2 5EJ
M3 JIU G Rigby, 106 Broadway Crescent, Binstead, Ryde, PO33 3QS
M3 JIV R French, 19 Melstone Avenue, Stoke-on-Trent, ST6 6EX
M3 JIW J Biggin, Galadean, Farriers Way, Newport, PO30 3JP
M3 JJB John Barton, 93 Cardigan Road, Bridlington, YO15 3JU
MM3 JJC A Curlis, 94 Kirkhill Road, Aberdeen, AB11 8FX
M3 JJH A Williams, 78 Hales Crescent, Smethwick, B67 6QS
M3 JJM J Martin, 1 Collins Lane, West Harting, Petersfield, GU31 5NZ
M3 JJN J Nicholson, 6 Mill Gardens, West End, Southampton, SO18 3AG
M3 JJS J Stanton, Waters & Stanton Plc, 22 Main Road, Hockley, SS5 4QS
M3 JJT Daniel Thompson, 34 Drake Avenue, Worcester, WR2 5HH
M3 JJU M Slee, 88 West Avenue, Lightcliffe, Halifax, HX3 8TJ
M0 JKA D Moffatt, 5 Florence Place, Decoy Road, Newton Abbot, TQ12 1DX
M3 JKB J Arkinstall, 4 Severn Way, Four Crosses, Llanymynech, SY22 6NQ
M3 JKE Nigel Knapton, 4 Crabmill Lane, Easingwold, York, YO61 3DE
M3 JKG T Shaughnessy, 220 Ladybank Road, Mickleover, Derby, DE3 0RS
M3 JKI Daryl Evans, 69 Westbourne, Honeybourne, Evesham, WR11 7PT
M0 JKJ S Asling, 18 Cecilia Grove, St. Peters, Broadstairs, CT10 3DE
M3 JKM K McLaughlin, 34 Cambridge Road, Birstall, Batley, WF17 9JF
M3 JKP S Asling, 18 Cecilia Grove, St. Peters, Broadstairs, CT10 3DE
M3 JKT John Phillips, The Manor, Blackwoods, York, YO61 3ER
MM3 JKX J Kirkpatrick, Brims School House, Longhope, Stromness, KW16 3NZ
M3 JKZ J Hawkins, 294 Norton Lane, Earlswood, Solihull, B94 5LP
M3 JLA J Astbury, 12 Southall Road, Ashmore Park, Wolverhampton, WV11 2PZ
M3 JLB J Bevan, 18 Martin Road, Diss, IP22 4HR
M3 JLD J Denny, 9 Hawthorn Way, Macclesfield, SK10 2DA
M3 JLE J Isard-Brown, 5 Grove Crescent, Croxley Green, Rickmansworth, WD3 3JT
M3 JLF K Bailey, 37 Cherry Tree Drive, Filey, YO14 9UZ
M3 JLH Jason HOOD, 17 Grays Close, Motcombe, Shaftesbury, SP7 9QB
M3 JLI J Nicholas, Greenbank, Chester High Road, Neston, CH64 7TR

**IMPORTANT NOTE**

**Revalidate licence to avoid revocation** – Ofcom has advised the Society that plans will be drawn up to revoke licences that have not been revalidated as required by the licence conditions. The quickest way to revalidate is to do so online via the Ofcom website: *https://services.ofcom.org.uk/* or by email: *amateur.validations@ofcom.org.uk* Ofcom staff are available to help, but please be patient during times of heavy workload.

M3 JLK J Knowles, 72 Uplands Avenue, Connah's Quay, Deeside, CH5 4LG
M3 JLR J Ramsay, Lane Cottage, Weymore Cottages, Bucknell, SY7 0EP
MM3 JLS S Bence, 14 Stein Terrace, Ferniegair, Hamilton, ML3 7FR
M3 JLV S Matthews, 13 Princes Close, Sidcup, DA14 4RH
M3 JLW D Collins, 29 Brook Drive, Verwood, BH31 6DH
M3 JLX S Crouch, 152 Thornhill Road, Brighouse, HD6 3AH
MI3 JMC James McCaw, 62 High Street, Ballymena, BT43 6DT
M3 JMI Jason Ioannou, Flat 3, Alexandra Court, Coventry, CV3 1FF
M3 JMJ J Jenkinson, 7 Chestnut Avenue, Thorngumbald, Hull, HU12 9LD
M3 JMK J Keegan, The Cottage, 11 Condor Grove, Lytham St. Annes, FY8 2HE
M3 JMQ D Baines, 3 Dunkirk Avenue, Houghton le Spring, DH5 8HN
M3 JMU M Crowley, 133 Jessop Road, Stevenage, SG1 5LH
M3 JMW J Moxley-Wyles, 7 Gidley Way, Horspath, Oxford, OX33 1RQ
M3 JMX J Moore, 51a High Mount St., Hednesford, Cannock, WS12 4BL
M3 JMY J Martin, 70 Moorlands Drive, Stainburn, Workington, CA14 4UJ
M3 JNB J Kearney, 12 Forshaw Lane, Burtonwood, Warrington, WA5 4ES
M3 JND J Dukes, 79 Jubilee Avenue, Boston, PE21 9LE
M3 JNJ A Campbell, 3 North Shawbost, Isle of Lewis, MS2 9BD
M3 JNQ J Clark, 1 Brooklime Road, Liverpool, L11 2YH
M3 JNR Kenneth Challoner, 23 Chapel Lane, Queensbury, Bradford, BD13 2QA
M3 JNT J Foulds, 7 Bridge Road, Little Sutton, Spalding, PE12 9EG
M3 JNU Stuart Robinson, 29 Grange Lane, Mountsorrel, Loughborough, LE12 7HY
M3 JNX M January, Blaenau Ucha Farm, Treudoyn, Mold, CH7 4NS
M3 JNY J Hutton, Cassiobury, The Street, Diss, IP22 2PS
M3 JOA N Lambert, 3 Nightingale Walk, Stockton-on-Tees, TS20 1SZ
M3 JOC R Denim, 15 Saxon Rise, Collingbourne Ducis, Marlborough, SN8 3HQ
M3 JOF J Miles, 11 Enborne Gate, Newbury, RG14 6AZ
M3 JOJ J Smith, 5 Manifold Gardens, Plymouth, PL3 6HL
M3 JOM C loughran, 8 Douglas Road, Dover, CT17 0BD
M3 JOS D Bown, 34 Kings Gardens, Bedworth, CV12 8JG
M3 JOW J Fletcher, 32 Chapel Lane, Barwick in Elmet, Leeds, LS15 4EJ
MW3 JPF J Freelove, 12 Honeyborough Road, Neyland, Milford Haven, SA73 1RE
M3 JPG G Bowen, 93 Pelham Road, Bexleyheath, DA7 4LY
M3 JPI J Pickering, Batemill, Batemill Lane, Macclesfield, SK11 9BW
M3 JPM G Meyer, 37 Marina Village, Preston Brook, Runcorn, WA7 3BH
M3 JPP H Tonge, 38 Colemeadow Road, Billesley Common, Birmingham, B13 0JL
MM3 JPS Jake McKenzie, Flat D, 15 Glenbervie Road, Aberdeen, AB11 9JD
M3 JPU J Patient, 4 Bucklebury Heath, South Woodham Ferrers, Chelmsford, CM3 5ZU
MW3 JQC Mark Breakwell, 9 Llys y Dderwen, New Quay, SA45 9SY
MI3 JQD B Young, 1 Oakleigh Grove, Castlederg, BT81 7WD
M3 JQG D Johnson, 11 Horseshoe Avenue, Dove Holes, Buxton, SK17 8DP
M3 JQJ L Berry, 6 Warren Park Close, Brighouse, HD6 2RU
M3 JQK E Williams, Criafol, Upper Llandwrog, Caernarfon, LL54 7PU
M3 JQM Lee Ross, 2 Bedford Street, Blackburn, BB2 4EU
M3 JQN Vincenzo Greco, 5 Council House, Halton Fen, Spilsby, PE23 5BE
M3 JQS J Hellowell, Upper Hole Head Farm, Ash Hall Lane, Soyland, HX6 4NU
M3 JQT S Watson, 32 Bradley View, Holywell Green, Halifax, HX4 9DN
M3 JQV P Handy, 30 Kingfisher Drive, Cheltenham, GL51 0WN
M3 JQW D Roscoe, 28a Princess Street, Chorley, PR7 3AP
M3 JQX T Leaworthy, 7 Maesderwen Rise, Stafford Road, Pontypool, NP4 5SS
M3 JQY J Johnson, 30 Thorpe Downs Road, Church Gresley, Swadlincote, DE11 9FB
M3 JRA A Jessop, 4 Katherine St., Thurcroft, Rotherham, S66 9LG
M3 JRF J Francis, 11 Middle Stream Close, Bridgwater, TA6 6LF
M3 JRI J Rowe, 45 Durham Road, Wilpshire, Blackburn, BB1 9NH
MI3 JRJ J Johnston, The Farm House, Tully, Enniskillen, BT92 7AR
MM3 JRK G Smith, 40 Pirleyhill Drive, Shieldhill, Falkirk, FK1 2EA
M3 JRM J Marter, 4 Meadow Way, Seaford, BN25 4QT
M3 JRN J Newman, 25 Milebush Road, Southsea, PO4 8NF
M3 JRQ Laura Pearson, 4 Brentwood Close, Thorpe Audlin, Pontefract, WF8 3ES
M3 JRR J Read, 26 Chaucer Road, Walsall, WS3 1DF
MM3 JSB J Bence, 5 Braeside Gardens, Hamilton, ML3 7PN
M3 JSF J Flores-Watson, 10 Bramwell Gardens, Coventry, CV6 6NB
MI3 JSH Samuel Hutchinson, 21 Lord Warden's Grange, Bangor, BT19 1YN
M3 JSK J Killian, 7 Dankworth Road, Basingstoke, RG22 4LJ
M3 JSM A McLaughlin, 34 Cambridge Road, Birstall, Batley, WF17 9JF
M3 JSO J O'Shea, 56 Crummock Gardens, London, NW9 0DJ
M3 JSQ M Woodruff, 14 Primatt Crescent, Shenley Church End, Milton Keynes, MK5 6AS
M3 JST J Taylor, 5 Pyman Close, Martham, Great Yarmouth, NR29 4UR
MI3 JTB J Black, 17 Fairymount Terrace, Taylors Avenue, Carrickfergus, BT38 7HN
M3 JTI Stephen Smith, 72 Throstle Lane, Leeds, LS10 4EY
MW3 JTJ J Jones, Bronydd, Blaenffos, Boncath, SA37 0HZ
MI3 JTM J Monteith, 58 Bells Hill, Limavady, BT49 0DQ
M3 JTO J bosworth, 10 Aston Street, Leeds, LS13 2BJ
MD3 JTT J Talbot, 43 Harcroft Meadow New Castletown Road, Douglas, Isle of Man, IM2 1JT
M3 JTU A Bailey, 58 Billy Buns Lane, Wombourne, Wolverhampton, WV5 9BP
M3 JTZ Robert Smith, 17 Julian Road, Spixworth, Norwich, NR10 3QA
M3 JUC K Marsh, 21 Edward Road, Eynesbury, St. Neots, PE19 2QF
M3 JUF J Schofield, 6 Robin Royd Avenue, Mirfield, WF14 0LF
M3 JUL J Townsend, 56 Seymour Road, Northfleet, Gravesend, DA11 7BN
M3 JUM R Jones, 10 Erw Wen Road, Colwyn Bay, LL29 7SD
M3 JUO S Jacklin, 3 Houston Road, Rugby, CV21 1BS
M3 JUW Adrian Davis, 34 Novers Park Drive, Bristol, BS4 1RG
M3 JUY R Thomas, 52 Victoria Road, Saltney, Chester, CH4 8SS
M3 JUZ R Shams-Nia, 1090 Eastern Avenue, Ilford, IG2 7SF
MW3 JVH Elaine Chell, Mesen Fach, Llanybydder, SA40 9TY
MI3 JVJ P Hannigan, 4 Silverhill Road, Strabane, BT82 0AE
M3 JVK James Kirkham, 35 Central Avenue, Woodlands, Doncaster, DN6 7NW
M3 JVP J Papworth, Flat 1, 70 Edward Road, Nottingham, NG2 5GB
M3 JVR Brian Gibbs, Flat 6, Castleton Court, Southsea, PO5 3AU
MI3 JVV K Hannigan, 4 Silverhill Road, Strabane, BT82 0AE
M3 JVW John Wheway, 20 Radnor Street, Derby, DE21 6DZ
MI3 JVX E McGowan, 83 Strabane Old Road, Londonderry, BT47 2QB
M3 JWJ D Shields, 42 Studland Park, Westbury, BA13 3HL
M3 JWM J Watts, 10 Lacy Road, Ludlow, SY8 2NS
M3 JWN J Newell, 7 Talbot, Tamworth, B77 2RS
M3 JWQ Wayne Johnson, 10 Archdale Road, Nottingham, NG5 6EB
M3 JWV L Percival, Blue Cedars, Gresford, Wrexham, LL12 8RN
M3 JWW J Wainwright, 8 Common Lane, Cutthorpe, Chesterfield, S42 7AN
M3 JWZ S Sewell, The Old Vicarage, Church Bank, Crewe, CW4 8PG
M3 JXE D Ennion, 347 Parkgate Road, Chester, CH1 4BE
MI3 JXG C Birney, 40 Forthill Park, Irvinestown, Enniskillen, BT94 1FJ
M3 JXI H Southall, 12 Prescot Close, Mickleover, Derby, DE3 0TB
M3 JXN Paul Jackson, Langsmead Barn, Eastbourne Road, Lingfield, RH7 6JX
MI3 JXO David Burke, 7 Edinburgh Villas, Omagh, BT79 0DW
M3 JXV Carolyne Trew, Ringstone Lodge, 66 Oakwood Road, Horley, RH6 7BX
M3 JXX J Driver, 99 Queens Road, North Weald, Epping, CM16 6JQ
M3 JXY A Gowans, 38 Beech Way, Twickenham, TW2 5JT
M3 JYA D Kemp, 7 Hillhurst Grove, Birmingham, B36 9TS
M3 JYE B Edwards, 16 Whitland Close, Rednal, Birmingham, B45 8SJ
M3 JYG D Batty, 168 Rotherhithe New Road, London, SE16 2AP
M3 JYH Gary Swain, 3 Flaxfield Drive, Crewkerne, TA18 8DF
M3 JYO D Perks, 6 Old School Gardens, Yatton Keynell, Chippenham, SN14 7BB
M3 JYP Christopher Lester, 21 Mortimer Way, Witham, CM8 1SZ
M3 JYW X Chen, Harrogate Ladies' College, Clarence Drive, Harrogate, HG1 2QG
M3 JYZ C Williamson, 53a High St., Whitwell, Hitchin, SG4 8AJ
M3 JZA Keith Armstrong, 29 Thorntree Avenue, Crofton, Wakefield, WF4 1NU
M3 JZD K Hart, 70 Hatfield Crescent, Stoke-on-Trent, ST3 3JQ
M3 JZE S Peregrine, 18 Gisborne Close, Mickleover, Derby, DE3 9LU
M3 JZF D Wall, 96 Albert Street, Wigan, WN5 9EF
M3 JZI Martin Rolls, 49 St. Bedes, 14 Conduit Road, Bedford, MK40 1FD
M3 JZK M Stinton, 57 Wildfields Road, Clenchwarton, King's Lynn, PE34 4DE
M3 JZL David Atkins, 20 Nappsbury Road, Luton, LU4 9AL
M3 JZM Brian Hayes, 22 Urban Way, Biggleswade, SG18 0HT
M3 JZN M Johnson, 25 Rowan Drive, Kirkby-in-Ashfield, Nottingham, NG17 8FU
M3 JZO S Bacon, 34 Fishers St., Kirkby In Ashfield, Nottingham, NG17 9AH
M3 JZP Anthony Bacon, 34 Fishers St., Kirkby In Ashfield, Nottingham, NG17 9AH
M3 JZT Wayne Soffe, 96 Urban Road, Doncaster, DN4 0EP
M3 JZV J Jenkins, 13 Birch Hill, Newport, NP20 6JD
M3 JZX M Cooper, 69 Leicester Road, Kibworth Harcourt, Leicester, LE8 0NP
M3 KAC David Carr, 19 Kingsmead Walk, Speedwell, Bristol, BS5 7RL
M3 KAE B McFarlane, 11 Hill Street, Barnsley, S71 5AL
M3 KAK K Harden, 59 Violet Avenue, Edlington, Doncaster, DN12 1NW
M3 KAL Keith Lawton, Meadowbank, Sutton St. Nicholas, Hereford, HR1 3BJ
M3 KAN K Hudson, 20 Cranmer Grove, Mansfield, NG19 7JR
M3 KAQ P Cotton, 33 Rowley Street, Ashton under Lyne, OL6 8DT
M3 KAU H Leaver, 1 Litcham Close, Litcham, King's Lynn, PE32 2QX
M3 KAX D Larkin, 19 Elizabeth Court, Hemsworth, Pontefract, WF9 4TQ
M3 KAY K Limbert, 7 Acacia Avenue, Liverpool, L36 5TL
M3 KBB K Barrow, Fenway Farm, Ten Mile Bank, Downham Market, PE38 0EU
M3 KBE J Kelly, 1 Bramble Close, New Ollerton, Newark, NG22 9TN
M3 KBF C Harley, 1 Portland Crescent, Meden Vale, Mansfield, NG20 9PJ
M3 KBG J Crowther, 16 Linden Avenue, Tuxford, Newark, NG22 0JR
M3 KBL Kevin Brendan Lee, 10 Queens Way, Pontefract, WF8 2LX
MU3 KBP K Pratt, Avalon Le Clos Des Sablon, Sandy Lane, St. Sampson, Guernsey, GY2 4RN
MJ3 KBQ C Daniells, Le Belon, 2 Clos Vallios, St. Lawrence, Jersey, JE3 1GP
M3 KBY Martin Kirby, 76 Burton Road, Overseal, Swadlincote, DE12 6JJ
M3 KBZ R Griffiths, 7 Macnaghten Road, Tankersley, Barnsley, S75 3DD
M3 KCA C Jameson, Flat 9, Britannia Court, Poole, BH12 3HF
M3 KCC C Evans, 158 Delamore Street, Liverpool, L4 3SX
M3 KCG R Jones, 79 Turpins Rise, Stevenage, SG2 8QZ
MW3 KCL M Brennan, 37 Marguerites Way, Cardiff, CF5 4QW
M3 KCO Kevin Cornmell, 19 Forest Road, Chandler's Ford, Eastleigh, SO53 1NA
M3 KCP K Preen, 12 Sandpit Lane, Hilton, Bridgnorth, WV15 5PH
M3 KCQ T Stansfield, 40 Rushmore House, Rubery, Birmingham, B45 9RU
MD3 KCT J Kennaugh, White Gables, 25 Kissack Road, Castletown, Isle of Man, IM9 1NW
M3 KCU M Johnson, 25 Rowan Drive, Kirkby-in-Ashfield, Nottingham, NG17 8FU
M3 KDK S Allington, 137 Marshall Lane, Northwich, CW8 1LA
M3 KDL C Lote, 8 Warren Place, Walsall, WS8 6BY
M3 KDM D Langley, 2 Holly Bush Cottages, Holmesdale Road, Sevenoaks, TN13 3XN
MM3 KDN Andrew Traynor, 56 Craigard Road, Dundee, DD2 4PT
M3 KDO R Smith, 3 Vernon Road, Southport, PR9 7EZ
MI3 KDR K Dickson, 66 Lisnabreeny Road, Belfast, BT6 9SR
M3 KDV D Craven, 69 Markham Avenue, Rawdon, Leeds, LS19 6NE
M3 KDY C Lindsay, 152 Dinmore Avenue, Blackpool, FY3 7QS
M3 KEC S Conlon, 6 Cardigan St., Ashton On Ribble, Preston, PR2 2AS
M3 KEF K Forster, Meadow View, Cracow Moss, Crewe, CW3 9BS
M3 KEJ K Jefferson, 125 Telscombe Way, Luton, LU2 8QP
M3 KEL K Watwood, 57 Cliveden Road, Stoke-on-Trent, ST2 8LP
M3 KER M Springett, 31 Mountbatten Court, Andover Road, Winchester, SO22 6BA
M3 KEV Kevin Graffham, 15 Hayes Road, Clacton-on-Sea, CO15 1TX
M3 KEW J Taylor, 90 Aldam Road, Doncaster, DN4 9EL
M3 KEY Kelly Calderbank, 6 Heathfield, Heath Charnock, Chorley, PR6 9LA
M3 KEZ D Brough, 57 Francis Road, Ashford, TN23 7UP
M3 KFE S Evans, 54 Stafford Crescent, Newcastle, ST5 3EA
M3 KFH R Drake, 42 Sunningdale Close, Doncaster, DN4 6UR
MI3 KFI T Warmington, 57 Carbet Road, Portadown, Craigavon, BT63 5RJ
M3 KFK R Head, 24 Beaufort Road, Church Crookham, Fleet, GU52 6AZ
M3 KFL Peter Hooper, 4 Castlemead Close, Saltash, PL12 4LF
M3 KFO W Cromack, 45 Southroyd Park, Pudsey, LS28 8AX
M3 KFP J Rouse, Nettleden, Galane Close, Northampton, NN4 9YR
M3 KFQ N Camp, 1 Higher Tresillian Cottages, Tresillian, Newquay, TR8 4PL
M3 KFR F Coles, 8 Moore Close, Church Crookham, Fleet, GU52 6JD
M3 KFT John Ferrol, 29 Westlands, Haltwhistle, NE49 9BS
M3 KGE Stephen Angove, 3 Bar View Lane, Hayle, TR27 4AJ
M3 KGG K Gordon, 308 Claremont Road, Swanley, BR8 7QZ
M3 KGJ K Weston, 114 Morland Road, Ipswich, IP3 0LZ
M3 KGK Gordon Higton, Hillcrest, Cow Brow, Carnforth, LA6 1PJ
M3 KGO Richard Dunn, 15 Catkins Close, Catshill, Bromsgrove, B61 0TT
M3 KGP P Kelly, Arosfa, Westminster Road, Wrexham, LL11 6DN
M3 KGQ J Kelly, Arosfa, Westminster Road, Wrexham, LL11 6DN
M3 KGV Clive Moulding, 28 Queens Avenue, Highworth, Swindon, SN6 7BA
M3 KHA Anthony Hughes, 8 Bullens Green Lane, Colney Heath, St. Albans, AL4 0QS
M3 KHC Alan Vick, Flat 8, Kingshill Court, Newport, NP20 4DT
M3 KHE Kevin Bradley, 9 Spruce Grove, Kirkby-in-Ashfield, Nottingham, NG17 7QB
MW3 KHH A Peake, 24 Rhigos Gardens, Cardiff, CF24 4LS
M3 KHI C Webb, 5 Pound Lane, Preston Bissett, Buckingham, MK18 4LX
M3 KHJ C Ashman, 56 Farriers Close, Swindon, SN1 2QT
M3 KHK S Rosser, 25 Clos Tir Ypwll, Pantside, Newport, NP11 5GE
M3 KHM L Lebaldi, 11 Artle Place, Lancaster, LA1 2QP
M3 KHT H Kwan, Harrogate Ladies' College, Clarence Drive, Harrogate, HG1 2QG
MD3 KHW K Stocker, 50 Mount Drive, Harrow, HA2 7RP
MW3 KHY J Robinson, 41 Ashbrook, Brackla, Bridgend, CF31 2AT
MI3 KIL P Hill, Flat 12 Kilcreggan Homes, Elizabeth Avenue, Carrickfergus, BT38 7EP
M3 KIN D Kinsey, 161 Heath Road South, Weston, Runcorn, WA7 4RP
M3 KIO E Smith, The Cabin, Nothe Parade, Weymouth, DT4 8TX
M3 KIQ M Lebaldi, 11 Artle Place, Lancaster, LA1 2QP
M3 KIR K Romang, 27 Elley Green, Neston, Corsham, SN13 9TX
M3 KIT Kate Cattermole, Blaxhall Hall Crossing, Little Glemham, Woodbridge, IP13 0BP
M3 KIU Steven Moore, 10 Strathcona Avenue, Hull, HU5 4AD
M3 KIZ P Lewis, 16 Valley Road, St. Albans, AL3 6LR
M3 KJB K Brooks, 24 Morris Drive, Weaverham, Northwich, CW8 3LP
M3 KJC K Cole, 4 Marsham Road Hazel Grove, Stockport, SK7 5JB
M3 KJD Kevin Davies, 20a Hart Road, Wolverhampton, WV11 3QJ
M3 KJE William Taylor, 99 St. Marys Close, Littlehampton, BN17 5QQ
MM3 KJG K Glacken, 14 Hailes Avenue, Edinburgh, EH13 0NA
M3 KJK J Mason, 22 Eskdale Avenue, Halifax, HX3 7NH
M3 KJM K Marsh, 11 Apollo Road, Stourbridge, DY9 8YG
M3 KJS Keith Sealey, 8 Esplanade, Burnham-on-Sea, TA8 1BE
M3 KJV D Baker, 65 Madison Street, Tunstall, ST6 5HS
M3 KJY C Atkins, 87 Wentworth Road, Doncaster, DN2 4DA
M3 KKA N Holdridge, 15 Ballam Avenue, Doncaster, DN5 9DY
M3 KKB S Silvers, 39 Hickinwood Crescent, Clowne, Chesterfield, S43 4AQ
M3 KKF R Taylor, 38 Edleston Road, Crewe, CW2 7HD
M3 KKG Kenneth Gledhill, 19 Palmers Terrace, Treknow, Tintagel, PL34 0EH
M3 KKI Iris Todorovic, 26 Orwell Close, Bury, BL8 1UU
M3 KKN Gary Allen, 60 Danefield Road, Northwich, CW9 5PX
M3 KKO M O'Neill, 15 School Road, Hockley Heath, Solihull, B94 6QH
MI3 KKP A Temple, 10 Clagan Cottages, Claudy, Londonderry, BT47 4DA
M3 KKQ George Thomas, 15 Buckley Avenue, Byley, Middlewich, CW10 9NW
M3 KKS Bruce John Roaf, 8 Weare Close, Portland, DT5 1JP
M3 KKX A Jones, 27 Fishpond Lane, Holbeach, Spalding, PE12 7DQ
M3 KKZ R Kirby, 44 Wilby Avenue, Little Lever, Bolton, BL3 1QE
M3 KLB Mark Crombie, 12 Sir James Reckitt Haven, Hull, HU8 8QR
M3 KLF Mark Muldowney, 37 Norham Avenue, Southampton, SO16 6PS
MM3 KLO S Dobie, 26 Kaims Gardens, Livingston Village, Livingston, EH54 7DY
M3 KLS K Symonds, 68 Manor Crescent, Pan, Newport, PO30 2BH
M3 KLT M Rogers Jones, Glanva, 20 Birchwood, Leyland, PR26 7QJ
M3 KLU N Andrews, Temple View House, Shopland Road, Rochford, SS4 1LH
M3 KLY K Lingham, 102 Chancery Lane, St. Helens, WA9 1SQ
MI3 KMB Kara Barr, 64 Owenreagh Drive, Strabane, BT82 9DT
M3 KMH Keith Haywood, 6 Lydney Road, Urmston, Manchester, M41 8RN
M3 KML K Iball, Highcroft, 18 Tan y Coed, Mold, CH7 6TU
M3 KMN K Macnauton, 27a Lincoln Road, Poole, BH12 2HT
M3 KMO K Owen, 10 Pitcher Lane, Leek, ST13 5DB
M3 KMS K Stanley, 3 Hale Way, Colchester, CO4 5BD
M3 KMT Terrence Ramsden, 37 Hyde Abbey Road, Winchester, SO23 7DA
MW3 KMU G Cattle, 39 Park View, Abercynon, Mountain Ash, CF45 4TP
M3 KMW K Ward, 127 Lower Lime Road, Oldham, OL8 3NP
MM3 KMX S Mclachlan, 531 Blair Avenue, Glenrothes, KY7 4RF
MW3 KNE A McTaggart, Brick Hall, Hundleton, Pembroke, SA71 5QX
M3 KNF R Curno, 19 Beckwith Road, Yarm, TS15 9TG
M3 KNK Harry Arrowsmith, 15 Hermitage Close, Frimley, Camberley, GU16 8LP
MW3 KNR R Seal, 5 Millfield, Lisvane, Cardiff, CF14 0RW
M3 KNT Kent Royce, 11 Church Lane, Stibbington, Peterborough, PE8 6LP
M3 KNV A Lebaldi, 11 Artle Place, Lancaster, LA1 2QP
MM3 KNY Kenneth Brown, 21 Strain Crescent, Airdrie, ML6 9ND
M3 KOA J Beecroft, 9 Link Road, Alton, GU34 2PE
M3 KOF R Johnson, 30 Avenue Road, Coalville, LE67 3PB
M3 KOJ K Lowe, 6 Markland Crescent, Clowne, Chesterfield, S43 4NG
M3 KOL C Cresswell, 38 Hindley Crescent, Barnton, Northwich, CW8 4LL
M3 KOR H Ross, 9 First Avenue, Edwinstowe, Mansfield, NG21 9NZ
M3 KOU A ROSE, 133 Petersmith Drive, New Ollerton, Newark, NG22 9SG
M3 KPB K Bromley, 40 Winfrith Road, Fearnhead, Warrington, WA2 0QE
M3 KPF J Simmons, 246 Ruskin Road, Crewe, CW2 7JY
M3 KPG K Stretton, 6 Highfields, Hilltop Drive, Rye, TN31 7HT
M3 KPL J Kelly, 1 Bramble Close, New Ollerton, Newark, NG22 9TN
M3 KPO Steven Warren, 1 Morley Close, Stapenhill, Burton-on-Trent, DE15 9EW
M3 KPQ Anthony Jermyn, 3 Tudor Walk, Carlton Colville, Lowestoft, NR33 8NE
M3 KPU A Parkes, 59 Wellington Gardens, Battle, TN33 0HD
M3 KPZ T Wood, 33 Somerdale Avenue, Bristol, BS4 2XN
M3 KQB P Broughton, 23 Ivy Place, Tantobie, Stanley, DH9 9PT
M3 KQC S Chandler, 4 Gladstone House, Horton Crescent, Epsom, KT19 8BW
M3 KQD B Minks, 132 North Road, Clowne, Chesterfield, S43 4PF
M3 KQF P Woodcock, 27, Knighton, Stafford, ST20 0QH
MM3 KQI R Houston, 19 Deansloch Place, Aberdeen, AB16 5SB
M3 KQP A Eastwell, 11 Middlebere Drive, Wareham, BH20 4SD
M3 KQR B Hall, 126 Eton Road, Burton-on-Trent, DE14 2SN
M3 KQS Nicholas Ralph, 24 Back Street, Laxton, Goole, DN14 7TP
M3 KQT D Berry, Flat5, 106 Braybrooke Road, Hastings, TN34 1TG
M3 KQW Howard Malpas, 148 Queen Street, Crewe, CW1 4AU
M3 KQY R BRISLEY, 15 Elm Fields, Old Romney, Romney Marsh, TN29 9SN
M3 KRB L Kirby, 76 Burton Road, Overseal, Swadlincote, DE12 6JJ

M3 KRD K Dukes, 127 Carlton Road, Boston, PE21 8LL
M3 KRE R Jacobs, 35 Edgar Road, Canterbury, CT1 1NR
MI3 KRL K McCrystal, 85 Tamlaght Road, Omagh, BT78 5BB
M3 KRM D Whitlock, Railway Crossing Cottage, Ash Road, Sandwich, CT13 0JD
M3 KRN C Richards, 2 Castle Lodge Crescent, Caldicot, NP26 4JL
M3 KRO Geoff Ticehurst, 118 Old Roman Bank, Terrington St. Clement, King's Lynn, PE34 4JP
M3 KRP K Taylor, 3 The Drive, Lichfield, WS14 9QT
M3 KRQ N Gadalla, 26 South Parade, Boston, PE21 7PN
M3 KRR D Corbett, 12 The Crescent, Middlesbrough, TS5 6SQ
M3 KRS C Cheverall, 1 Clarkwood Cottages, Twitty Fee, Chelmsford, CM3 4PG
M3 KRX D Shires, 1 West Close, High Coniscliffe, Darlington, DL2 2LN
M3 KRY Roger Dyson, 4 Royston Lane, Royston, Barnsley, S71 4NL
M3 KRZ Bertram Renowden, Hilrowenick, Polwithen Drive, St. Ives, TR26 2SP
M3 KSE N Lane, 13 Traston Road, Newport, NP19 4RQ
M3 KSG K Gordon, 308 Claremont Road, Swanley, BR8 7QZ
M3 KSH Alison Wilkins, 2 Beechfield Crescent, Banbury, OX16 9AR
M3 KSI Mark Giudice, 31 Woodfield Cross, Tredegar, NP22 4JG
M3 KSK M Sheppard, 107 Queen St., Swinton, Mexborough, S64 8NF
MD3 KSN F Kelly, 5 Maynrys, Castletown, Isle of Man, IM9 1HP
M3 KSP S Eldridge, 20 Edendale Road, Melton Mowbray, LE13 0EW
M3 KSS A Eades, Violet Bank, 18 Hillside Road, Leigh-on-Sea, SS9 2DT
MM3 KSV P McCluskey, 119 Tower Drive, Gourock, PA19 1SG
M3 KTA Robert Baines, 34 Bury Road, Stapleford, Cambridge, CB22 5BP
M3 KTD Katie Davidson, 5 Hanover Parc, Indian Queens, St. Columb, TR9 6ER
M3 KTH K Howard, 5 St Nicholas Street, Dereham, NR19 2BS
M3 KTT B Gardner, 40 Wynall Lane South, Stourbridge, DY9 9AH
M3 KTV B Bean, 46 Grand Drive, Herne Bay, CT6 8JS
M3 KUB C Barber, 22 Robinson Court, Chilwell, Nottingham, NG9 6RF
M3 KUE B Hall, 65 Cavendish Road, Worksop, S80 2ST
M3 KUG Mark Amos, 233b Abington Avenue, Northampton, NN1 4PU
M3 KUH A Grannon, The Chestnuts, Church Lane, Hull, HU11 4PR
M3 KUJ Danielle Russell, 72 Langholm Drive, Cannock, WS12 2EZ
M3 KUK J Jones, 15 Kinnaird Road, Sheffield, S5 0NN
M3 KUM C Sellors, 5 Neale Close, Leiston, IP16 4HJ
M3 KUN Joseph Wilson, 20 Elgitha Drive, Thurcroft, Rotherham, S66 9PD
M3 KUO D Holden, 2 Beacon Road, Bickershaw, Wigan, WN2 4AF
M3 KUQ T Bush, 19 Spring Vale, Waterlooville, PO8 9DA
M3 KUS P Connelly, 3 Finch Close, Weston-Super-Mare, BS22 8XS
MM3 KUU G White, 119 Waggon Road, Brightons, Falkirk, FK2 0EJ
M3 KUV P Baines, 4 Oxford Crescent, Hetton-le-Hole, Houghton le Spring, DH5 9LJ
M3 KUY S Church, The Willows, Warboys Road, Huntingdon, PE28 3AH
M3 KUZ K McKeown, 27 Lusty Glaze Road, Newquay, TR7 3AE
M3 KVC A Stacey, 311 Hyde End Road, Spencers Wood, Reading, RG7 1DD
M3 KVD D Speed, 137 Church Road North, Skegness, PE25 2QQ
M3 KVG D Hannon, 20 High Street, Whittlebury, Towcester, NN12 8XJ
M3 KVH K Harrison, 55 Hudson Close, Worcester, WR2 4DP
M3 KVI D Swann, 37 Burgh Road, Skegness, PE25 2RA
M3 KVJ K King, 16 Clare Way, Bexleyheath, DA7 5JU
M3 KVK J McKeown, 27 Lusty Glaze Road, Newquay, TR7 3AE
M3 KVL K Martin, 19 Comrie Crescent, Burnley, BB11 5HX
MM3 KVN K Clark, 38 Dunsinane Drive, Perth, PH1 2DU
M3 KVR K Robertson, 189 Harrowby St., Farnworth, Bolton, BL4 7DF
M3 KVU E Smith, 18 Poynings Place, Old Portsmouth, Portsmouth, PO1 2PB
MM3 KVV W morrison, 7 Knowehead Crescent, Kirriemuir, DD8 5ab
M3 KVW M Keelan, 16 North Drive, Harwell, Didcot, OX11 0PE
MM3 KVY M McConnell, 6 Langlaw Road, Mayfield, Dalkeith, EH22 5AX
M3 KWF Samuel Mainzer, Lillypool House, Waldersea, Wisbech, PE14 0NR
M3 KWL Kevin Langdon, Flat 9, Aspects Park Gate, Nuneaton, CV11 6DY
M3 KWR Miroslav Sutty, 14 Sedgwick Street, Cambridge, CB1 3AJ
M3 KWS S Dunn, 64 Stucley Road, Bideford, EX39 3EQ
M3 KWZ D Leese, 41 Woolston Avenue, Congleton, CW12 3DZ
M3 KXB R Walsh, 117 Westbourne, Telford, TF7 5QN
M3 KXD C Collins, 2 Kew Crescent, Sheffield, S12 3LP
M3 KXE P Beresford, 23 High Lowe Avenue, Congleton, CW12 2EP
M3 KXF J Gregory, 9 Longfields Crescent, Hoyland, Barnsley, S74 9HZ
M3 KXG Raymond Robinson, 18 O'Connell Road, Liverpool, L3 6JF
M3 KXI C Day, 4 Marlborough Way, Market Harborough, LE16 7LW
M3 KXS E Booth, 18 Maple Road, Kiveton Park, Sheffield, S26 5PH
M3 KXV D Hollinrake, 4 Sandwood Avenue, Broughton, Chester, CH4 0RJ
M3 KXY David Preston, Home View, Paradise Lane, Reading, RG7 6NU
M3 KXZ P Millis, 26 Chalkland Rise, Brighton, BN2 6RH
M3 KYD P Hummerstone, 70 Salisbury Road, Plymouth, PL4 8TA
M3 KYG S Humphreys, 52 Llys Owain, Bangor, LL57 1SH
M3 KYH J List, 41 Westbury Crescent, Dover, CT17 9QQ
M3 KYK J Hall, 1 Nash Close, Earley, Reading, RG6 5SL
MM3 KYO K Rafferty, 13 Robin Crescent, Buckhaven, Leven, KY8 1EZ
M3 KYQ K Rennison, 49 Syston Avenue, St. Helens, WA11 9JJ
M3 KYV Saranne Littlewood, Townside Lodge, Townside, Immingham, DN40 3PS
M3 KYZ D Whitelock, 22 Anne Crescent, Waterlooville, PO7 7NA
M3 KZB C Cascarino, 87 Esther Grove, Wakefield, WF2 8EX
M3 KZC Martin Clack, 42a Provost Street, Fordingbridge, SP6 1AY
MM3 KZD S Corstorphine, 33 Springfield, West Barns, Dunbar, EH42 1UF
M3 KZI Arron Pateman, 37 Hemans Road, Daventry, NN11 9AL
M3 KZJ P Lamb, 13 Pool End, St. Helens, WA9 3RE
M3 KZP K Page, 62 Farndon Avenue, Sutton Manor, St. Helens, WA9 4DN
M3 KZR A Bailey, 9 Park View, Abram, Wigan, WN2 5QR
M3 KZS P Allen, 4 The Links, Northam, Bideford, EX39 1LS
M3 KZT D Baugh, 72 Langdale Close, Plymouth, PL6 8SP
M3 KZV Paul Camplin, 16 Green Street, Hoyland, Barnsley, S74 9RF
M3 KZW S Grainger, 15 Carr House Lane, Wirral, CH46 6EN
M3 LAG Lee Gething, 106 Westgate, Elland, HX5 0BB
M3 LAJ L Jarman, 53 Enderby Crescent, Gainsborough, DN21 1XQ
M3 LAP Alexander Clarke, 57 Welland Avenue, Grimsby, DN34 5JP
M3 LAQ Terry Mynors, 6 Walcott Avenue, Christchurch, BH23 2NG
M3 LBC R Kenton, 65 Warren Drive, Broughton, Chester, CH4 0PU
M3 LBD T Donegan, 19 Teesgate, Thornaby, Stockton-on-Tees, TS17 9AN
M3 LBG David Davis, 36 Arbour Street, Southport, PR8 6SQ
M3 LBJ I Blundell, 43 Ponsonby Place, London, SW1P 4PS
M3 LBK L Karthauser, 17 Manor Close, Abbotts Ann, Andover, SP11 7BJ
M3 LBM N Waller, 16 Rother Croft, New Tupton, Chesterfield, S42 6BE
M3 LBN G Lowe, 8 Markland Crescent, Clowne, Chesterfield, S43 4NG
M3 LBP David Clarke, 10 Siward Road, Bromley, BR2 9JZ
M3 LBQ Marilyn Bradshaw, 342 Manchester Road, Blackrod, Bolton, BL6 5BG
M3 LBR S Cross, 31 Parkfields, Abram, Wigan, WN2 5XR
M3 LBT R Carter, 43 Sheldon Avenue, Standish, Wigan, WN6 0LW
M3 LBX D Roberts, 118 Ffordd Ddyfrdwy, Mostyn, Holywell, CH8 9PQ
M3 LBY P Wallstone, 3 Wilson Street, Mansfield, NG19 7JW
M3 LBZ Jonathan Fletcher, Paradise Barn, Bounds Lane, Chard, TA20 2TJ
M3 LCE R Armstrong, 26 Lancaster Road, Carnforth, LA5 9LD
M3 LCF D Lovell, 109 Aylesbury Crescent, Plymouth, PL5 4HX
M3 LCI Michael Davidson, 19 Mason Street, Workington, CA14 3EH
M3 LCL C Lewis, 19 Elgar Close, Great Sutton, Ellesmere Port, CH65 7AZ
M3 LCP L Pentney, 4 Caley Road, Tunbridge Wells, TN2 3BL
M3 LCS M Croxford Simmons, 37 Queens Road, Askern, Doncaster, DN6 0LU
M3 LCU P Loose, Tae Ping, Main Road, King's Lynn, PE31 8BP
M3 LCW P Bradley, 4 Paddocks Close, Pinxton, Nottingham, NG16 6JR
M3 LCZ Russell Humpage, 10 Whalley Road, Lancaster, LA1 2HA
M3 LDC Lucy Cattermole, Blaxhall Hall Crossing, Little Glemham, Woodbridge, IP13 0BP
M3 LDD D Darling, 10a South St., Portslade, Brighton, BN41 2LE
M3 LDE L Davis, Romano House, Gorefield Road, Wisbech, PE13 5AS
M3 LDF S Brown, 21 Woborrow Road, Heysham, Morecambe, LA3 2PW
M3 LDH L Holmes, 61 Maryland Lane, Wirral, CH46 7TS
M3 LDI J Bircumshaw, 39 Woodthorpe Lane, Sandal, Wakefield, WF2 6JG
M3 LDJ J Harvey, 125 North Road, Clowne, Chesterfield, S43 4PQ
M3 LDL J Parker, 1 Schoose Caravan Park, Workington, CA14 4JA
M3 LDM T Craddock, 12 Wold Road, Burton Latimer, Kettering, NN15 5PN
MI3 LDO P Logan, 18 Castle Lane, Lisnaskea, Enniskillen, BT92 0FW
M3 LDQ L Paddon, 21 Oak Park Drive, Havant, PO9 2XE
MM3 LDR Allan Sloan, 36 Paterson Avenue, Irvine, KA12 9JJ
M3 LDS R Walker, 2 Evie Place, Kings Road, Dorchester, DT1 1NJ
M3 LDT Robert Taylor, 40 Ashwood Drive, Chelmsley Wood, Birmingham, B37 6TN
M3 LDX K Siviter, 27 Queensway Close, Mark, Highbridge, TA9 4PH
MW3 LDY Nigel Cole, Tycoch, Llandovery, SA20 0UP
M3 LEB S Bell, 221 Horninglow Road, Sheffield, S5 6SG
M3 LEF J Peace, 237a Mapperley Plains, Nottingham, Ng3 5RG
MD3 LEG H Leslie, 2 Close Lhergy, Union Mills, Douglas, Isle of Man, IM4 4LU
M3 LEK E Little, 7 Deerfern Close, Great Linford, Milton Keynes, MK14 5BZ
M3 LEL L Sargeant, 99 Pot Kiln Road, Great Cornard, Sudbury, CO10 0DX
M3 LEN L Brackstone, 276 Ladyshot, Harlow, CM20 3EY
MD3 LEP Andrew Le Prevost, 58 Meadow Crescent, Douglas, Douglas, Isle of Man, IM2 1QX
MW3 LEW L Jenkins, 10 Ty Fry Close, Brynmenyn, Bridgend, CF32 8YB
M3 LEX A Green, Croft House, Welbeck Road, Chesterfield, S44 6DH
M3 LEY S Oakley, 9 Goldsmith Road, Walsall, WS3 1DL
M3 LFC David Hughes, 86 Colinmander Gardens, Ormskirk, L39 4TF
MI3 LFE Anthony Boylan, 36 Callan Bridge Park, Armagh, BT60 4BU
M3 LFG L Gill, 2 Loxton Court, Mickleover, Derby, DE3 0PH
M3 LFH S Gregory, 1 St Martin Street, Atherton, Manchester, M29 9DN
MM3 LFI D Pomphrey, Flat 2/9, 109 Bell Street, Glasgow, G4 0TQ
M3 LFL Keith Morgan, Gwel-Yr-Afon, Penrhyncoch Road, Aberystwyth, SY23 3EA
M3 LFO Stella Yeldham, 19 Wade Reach, Walton on The Naze, CO14 8RG
M3 LFP R Brown, 9 Bayleaf Crescent, Oakwood, Derby, DE21 2UG
M3 LFQ W Mcgill, 49 Anthony Close, Colchester, CO4 0LD
M3 LFU A Heyes, 528 Manchester Road, Paddington, Warrington, WA1 3TZ
M3 LFV D Mear, 10 Peters Court, Hatton, Derby, DE65 5JG
M3 LFZ Melvyn Haseldine, 4 Triangle Building, Wolverton Park Road, Milton Keynes, MK12 5FJ
M3 LGF D Sporton, 40 Eastwood Park Drive, Hasland, Chesterfield, S41 0BD
M3 LGH Stuart Hallam, 18a Market Street, Hoylake, Wirral, CH47 2AE
M3 LGI I Scott, 13 Fairmount Road, Bexhill-on-Sea, TN40 2HN
M3 LGJ Peter Bishop, 37 Church Street, Bradenham, Thetford, IP25 7QL
MI3 LGL L Logue, 21 Moyagh Road, Cullion, Londonderry, BT47 2SL
M3 LGM M Steeples, 17 Windsor Avenue, Thurlstone, Sheffield, S36 9RX
MW3 LGS S Lewis, 11 Treseder Way, Cardiff, CF5 5NW
MM3 LGU Ross Pennykid, 50 Queen Street, Edinburgh, EH2 3NS
M3 LGX G Cash, 28 Bramblewood Close, Prenton, CH43 9YT
M3 LGY M Cash, 28 Bramblewood Close, Prenton, CH43 9YT
MW3 LHA L Hailstone, 1 Hornbeam Close, Cimla, Neath, SA11 3XA
M3 LHE J Gammer, 12 West Rise, Tonbridge, TN9 2PG
M3 LHF D Dunne, 1 Burton Gardens, Brierfield, Nelson, BB9 5DR
M3 LHG G Gammer, 12 West Rise, Tonbridge, TN9 2PG
M3 LHI Zofia Dunne, 1 Burton Gardens, Brierfield, Nelson, BB9 5DR
M3 LHM L Marshall, Thistledome, First Avenue, Watford, WD25 9PS
M3 LHQ K Brice, 10a Nelson Drive, Exmouth, EX8 2PU
M3 LHU Chris jenkins, 36 Greenwich Road, Hailsham, BN27 2PE
M3 LHW A England, Beech House, Vicarage Gardens, Bradford, BD11 2EF
M3 LHX Anthony Woodsford, 5 Eliot Close, Wickford, SS12 9ED
M3 LHZ S Yates, 14 Rushden Road, Sandon, Buntingford, SG9 0QR
M3 LIB J Bradburn, 4 Lathkil Grove, Buxton, SK17 7PH
M3 LIN Linda Chesters, 8 Conway Grove, Blacon, Chester, CH1 5RU
M3 LIU I Cartmell, 10 Derwent Avenue, Burnley, BB10 1HZ
M3 LIV Z Bayliss, 16 Oakmere Close, Sandbach, CW11 1WN
M3 LIW J Brown, 8 Chatsworth Street, Sutton-in-Ashfield, NG17 4GG
M3 LIX M Knell, 8 Wimborne Gardens, Kirby Cross, Frinton-on-Sea, CO13 0TH
M3 LIY S Mason, 38 Wheaton Vale, Birmingham, B20 1AJ
M3 LJA Linda Abel, 121 Angela Road, Horsford, Norwich, NR10 3HF
MD3 LJB Laura Bazley, 24 Ballahane Close, Port Erin, Port Erin, Isle of Man, IM9 6EG
M3 LJF L Gudgeon, Shillingsworth Cottage, Leckhampstead Road, Milton Keynes, MK19 6BY
M3 LJI G Billington, 47 Smithy Leisure Park, Cabus Nook Lane, Preston, PR3 1AA
M3 LJJ J Lightfoot, Apple Tree Cottage, Flowers Hill, Reading, RG8 7BD
M3 LJK A Smith, 46 Mulberry Close, Goldthorpe, Rotherham, S63 9LB
MI3 LJQ William Phair, 98 Cedar Grove, Holywood, BT18 9QB
MD3 LJS J Keig, 60 Garth Avenue, Surby, Port Erin, Isle of Man, IM9 6QZ
M3 LJX G Jones, 7 St. Ives Road, Weston-Super-Mare, BS23 3XX
M3 LJZ L Jennings, 29 Mountbatten Drive, Newport, PO30 5SJ
M3 LKD J Dixon, 20 Dee Way, Winsford, CW7 0JD
M3 LKE E Smith, 30 Teignmouth Road, Torquay, TQ1 4EA
M3 LKJ Philip Manning, 1 Waverley Gardens, Ash Vale, Aldershot, GU12 5JP
M3 LKM E Spurr, 20 Mannington Way, West Moors, Ferndown, BH22 0JE
M3 LKO Kevin Cope, 64 Queen St., Pensnett, Brierley Hill, DY5 4HA
MM3 LKR F wenseth, 2 Sunnybank Cottage, Logie Coldstone, Aboyne, AB34 5PQ
M3 LKU A Head, 34 Balds Lane, Stourbridge, DY9 8SG
MM3 LKV C Duncan, 131 Croftend Avenue, Glasgow, G44 5PF
M3 LKY S Smith, 10 Parkway South, Doncaster, DN2 4JS
M3 LLB Lorraine D'Aubray-Butler, 16 Francis Road, Frodsham, WA6 7JR
M3 LLC L Steele, 14 Rowley View, West Bromwich, B70 8QR
M3 LLK A Soper, 16 Queen Elizabeth Drive, Crediton, EX17 2EJ
M3 LLM Janet Proudman, 61 Iffley Road, Oxford, OX4 1EB
M3 LLN Andrew Sibley, 27 Sherwood Road, Tetbury, GL8 8BU
M3 LLQ J Beards, 175 Blackhalve Lane, Wolverhampton, WV11 1AH
M3 LLT C Moseley, 15 Holden Crescent, Walsall, WS3 1PY
MM3 LLU B Gaudie, Sunnyside, Harray, Orkney, KW17 2JS
MW3 LLV H Leonard, 11 Newton Road, Grangetown, Cardiff, CF11 8AJ
M3 LLX J Wright, 2 Regent Road, Church, Accrington, BB5 4AR
M3 LLZ Peter Davies, 53 Lammas Road, Cheddington, Leighton Buzzard, LU7 0RY
M3 LMA L Adkins, 4 Orion Close, Ward End, Birmingham, B8 2AU
M3 LMB Philip Breslin, 8a Mountfield, Tennyson Road, Yarmouth, PO41 0PS
M3 LMC L McLaughlin, 34 Cambridge Road, Birstall, Batley, WF17 9JF
M3 LMD B Dake, 100 Lodge Road, West Bromwich, B70 8PL
M3 LME G Beardmore, 9 Ashmore Drive, Gnosall, Stafford, ST20 0RP
M3 LMH L Holme, 11 Oxlea Grove, Westhoughton, Bolton, BL5 2AF
M3 LML M Carney, 2 Lilac Meadows, Lawley Village, Telford, TF4 2NX
M3 LMQ R Hines, 159 Langstone Drive, Exmouth, EX8 4JE
MI3 LMR R Spence, 299 Moyarget Road, Mosside, Ballymoney, BT53 8DL
M3 LMU E Meek, 7 High Tree Rise, Oakdale, Blackwood, NP12 0DP
M3 LMV Adrian Harris, Flat 11, Tyn-y-Coed, Cwmbran, NP44 4PQ
M3 LMX L Pearse, Hill Farm, Middleway Road, Shepton Mallet, BA4 6TS
M3 LMZ Michael Williams, 30 Elm Drive, Risca, Newport, NP11 6HJ
MI3 LNC H Davis, 29 Fir Park, Broughshane, Ballymena, BT42 4DH
MM3 LNF I Mcgurk, 40 Beechwood Drive, Alexandria, G83 9NP
M3 LNH Rhosyn Celyn, 20 Aylesbury Street, Wolverton, Milton Keynes, MK12 5HZ
M3 LNJ R Woolridge, 8 Alastair Drive, Yeovil, BA21 3BT
M3 LNM S Lawton, 4 Astland Gardens, Tarleton, Preston, PR4 6SX
M3 LNN J Tomlinson, 1 Thirlmere Avenue, Burnley, BB10 1HU
M3 LNQ C Woodruff, 14 Primatt Crescent, Shenley Church End, Milton Keynes, MK5 6AS
M3 LNR L Naylor, 23 Lilla Close, Whitby, YO21 3LY
MM3 LNT Matthew Paterson, 20 Loch Street, Rosehearty, Fraserburgh, AB43 7JT
M3 LNU Marian Durban, 62 Westfield Way, Charlton, Wantage, OX12 7EP
M3 LNY L Goff, 27 Harley Road, Oxford, OX2 0HS
M3 LOA R Loader, Sunnyside, Main Street, Leyburn, DL8 4LU
M3 LOE D Gosling, 19 Alcester Close, Plymouth, PL2 1EA
MM3 LOF Hazel Paterson, 20 Loch Street, Rosehearty, Fraserburgh, AB43 7JT
M3 LOI L Pring, 42 The Links, Trevethin, Pontypool, NP4 8DQ
M3 LOT D Filby, 14 Jeffcut Road, Chelmsford, CM2 6XN
M3 LOX Daren Loxley, 33 Longwood Road, Tingley, Wakefield, WF3 1UG
M3 LOY Briney Taylor, 47 Sandy Drive, Victoria Point, Australia, 4165
M3 LPE J Wright, 26 Walmsley Close, Church, Accrington, BB5 4HL
M3 LPF S Hemmings, Leylands, Leigh Road, Frome, BA11 3LR
M3 LPI R Brierley, 26 Jacobsen Avenue, Hyde, SK14 4DW
M3 LPJ L Renmans, 70 Burman Road, Liverpool, L19 6PW
M3 LPK G Brierley, 26 Jacobsen Avenue, Hyde, SK14 4DW
M3 LPN B Kersey, 61 Crown Road, Portslade, Brighton, BN41 1SJ
M3 LPQ R Spooner, 45 Shaftesbury Avenue, Southport, PR8 4NH
M3 LPR J main, 15 Byron Road, Lydiate, Liverpool, L31 0DB
M3 LPT L Towler, 8 Stowehill Road, Peterborough, PE4 7PY
M3 LPU Philip Higgins, 9 Claremont Grove, Exmouth, EX8 2JW
MD3 LPW L Wernham, Rogane Cottage, Church Lane, Santon, Isle of Man, IM4 1EZ
M3 LQA P Ellis, 40 Grasmere Road Royton, Oldham, OL2 6SR
M3 LQB A Foulds, 4 Kropacz Court, South Street, Doncaster, DN6 7JL
M3 LQC J Bhart, 47 Fitzroy Avenue, Broadstairs, CT10 3LS
M3 LQD S Moakes, 46 Parsonage Street, Stockport, SK4 1HZ
M3 LQE M Clarke, 48 Delves Wood Road, Huddersfield, HD4 7AS
M3 LQF W Fox, The Old Mill Cottage, Hoby Road, Leicester, LE7 4TJ
M3 LQI Steven Briggs, 24 Mulberry Close, Taunton, TA1 2LT
M3 LQJ Paul Langford, 39 Hodnell Drive, Southam, CV47 1GQ
MM3 LQK William Caithness, 36 Wards Drive, Muir of Ord, IV6 7PX
M3 LQL P Hickling, 49 Roundhill Close, Syston, Leicester, LE7 1PP
MI3 LQN H McErlean, 24 Mullaghboy Heights, Magherafelt, BT45 5NU
M3 LQO Rodney Claydon, 3 Birch Trees, Ambleside Road, Windermere, LA23 1EU
M3 LQP D Brown, 21 Woborrow Road, Heysham, Morecambe, LA3 2PW
M3 LQV P Newton, 61 Ashbourne Crescent, Taunton, TA1 2RA
M3 LQW Rodney Edwards, 46 Lavers Oak, Martock, TA12 6HG
M3 LQX M Hall, 2 Hallcroft Road, Haxey, Doncaster, DN9 2HP
M3 LQY Robert Vigors, 12 Sandfield Park, Lichfield Road, Brownhills, WS8 6LN
M3 LRF P Callaghan, 41 Higher Ash Road, Talke, Stoke-on-Trent, ST7 1JN
M3 LRI D Hill, 11 Paddock Lane, Metheringham, Lincoln, LN4 3YG
M3 LRJ A Smith, 101 Chaucer Drive, Lincoln, LN2 4LT
M3 LRK M Davey, 67 Rotherham Baulk, Carlton-in-Lindrick, Worksop, S81 9LE
M3 LRN A Page, 148 Waleton Acres, Carew Road, Wallington, SM6 8PY
M3 LRP R Plater, Garsides, Keeling Street, Louth, LN11 7QU
MI3 LRR W Turtle, 35 Buckna Road, Broughshane, Ballymena, BT42 4NJ
M3 LRU J Fitzpatrick, 29 Delmar Road, Knutsford, WA16 8BG
M3 LRW L Wilson, The Rectory, Church Close, Thetford, IP25 7LX
M3 LRX David Horwood, 42 Southlands Drive, Timsbury, Bath, BA2 0HB
M3 LRZ Lyndon Reynolds, 49 Westborough Way, Anlaby Common, Hull, HU4 7SW
M3 LSE M Newton, 43 Hayfield Road, Minehead, TA24 6AD
M3 LSF L Caslin, 30d Holmewood, Holme, Peterborough, PE7 3PG

**IMPORTANT NOTE**

**Revalidate licence to avoid revocation** – Ofcom has advised the Society that plans will be drawn up to revoke licences that have not been revalidated as required by the licence conditions. The quickest way to revalidate is to do so online via the Ofcom website: *https://services.ofcom.org.uk/* or by email: *amateur.validations@ofcom.org.uk* Ofcom staff are available to help, but please be patient during times of heavy workload.

UK Callsigns

**M3 LSK** T Newton, 43 Hayfield Road, Minehead, TA24 6AD
**MW3 LSL** K Summers, 11 Bro Hawen, Rhydlewis, Llandysul, SA44 5RF
**M3 LSO** Scott Arnott, 40 deanhead road, Eyemouth, TD14 5SA
**M3 LSS** L Stephenson, 15 Springwood Road, Hoyland, Barnsley, S74 0AZ
**M3 LSU** J Evans, 7 Westland Drive, Hayes, Bromley, BR2 7HE
**M3 LSX** Alan Dale, 37 Bussey Road, Norwich, NR6 6JF
**M3 LTA** L Talbot, 26 Chevalier Grove, Crownhill, Milton Keynes, MK8 0EJ
**M3 LTG** L Gear, 26 Woodlands Way, Denaby Main, Doncaster, DN12 4LR
**M3 LTH** M Preston, 53 Links Road, Birmingham, B14 4TW
**M3 LTP** P Whittall, 165a High Street, Brierley Hill, DY5 3BU
**M3 LTR** D Ingham, 19 Recreation Avenue, Ashton-in-Makerfield, Wigan, WN4 8SU
**M3 LTT** A Old, 3 Middle Down Close, Plymouth, PL9 9TX
**M3 LTU** Mark Brady, 24 Gregory Avenue, Colwyn Bay, LL29 7ND
**M3 LTV** A Walker, 76 Greenway, Birmingham, B20 1EQ
**M3 LTW** J Bennett, 2 Victoria St., Pensnett, Brierley Hill, DY5 4LB
**M3 LUA** John Stevens, 6 The Mount, Litlington, Royston, SG8 0QG
**M3 LUD** P Ludders, 280 Hopewell Road, Bilton Grange, Hull, HU9 4HH
**M3 LUK** L Mandeville, 6 Oxford Road, Benson, Wallingford, OX10 6LX
**M3 LUO** David Shoubridge, 19 Manor Road, East Grinstead, RH19 1LP
**M3 LUP** S Ingram, 39 Doyle Road, Bolton, BL3 4SA
**M3 LUU** A Gillett, 441 Radipole Lane, Weymouth, DT4 0QF
**M3 LUW** G Stocks, 62 Ridge Park Avenue, Plymouth, PL4 6QA
**M3 LUZ** A Bent, 14 Pleasant Road, Eccles, Manchester, M30 0FS
**M3 LVA** L Adlington, 21 Newstead Road, Stoke-on-Trent, ST2 8HU
**M3 LVF** R Hawkins, Nook Cottage, Common-y-Coed, Caldicot, NP26 3AX
**M3 LVK** C Street, Russetts, Isle Brewers, Taunton, TA3 6QN
**M3 LVL** Carwyn Edwards, 3 Millfield Court, Millfield Lane, York, YO10 3AW
**M3 LVM** Kevin West, 36 Watlington Road, Cowley, Oxford, OX4 6SS
**M3 LVP** J Gilbert, 148 Purcell Road, Coventry, CV6 7LB
**M3 LVR** E Mills, 76 Main St., Burton Joyce, Nottingham, NG14 5EH
**MM3 LVT** J Dock, 75 Ferguslie Park Avenue, Paisley, PA3 1BE
**M3 LVX** P Hampton, 4 Moorland View, Plymstock, Plymouth, PL9 8NW
**M3 LVY** A Kendrick, 13 Queens Drive, Middlewich, CW10 0DG
**MI3 LVZ** Martin McCloy, 4 Audleys Park, Newtownards, BT23 8UA
**M3 LWG** Jeff Arrowsmith, 45 Hilderic Crescent, Dudley, DY1 2EU
**MM3 LWJ** Robert Bertram, Noroc, 46 Main Street, Pathhead, EH37 5QB
**MM3 LWO** G Mallolm, 32 Holly Crescent, Dunfermline, KY11 8BT
**M3 LWP** J Clowes, 52 Pennine Drive, St. Helens, WA9 2BU
**MD3 LWQ** M Corlett, 18 Queens Drive, Peel, Isle of Man, IM5 1BQ
**MM3 LWT** S Conway, 26 Rennie Street, Kilmarnock, KA1 3AR
**MI3 LWU** A Geary, 10 Jubilee Park, Armagh, BT60 1JA
**M3 LWV** M Powell, 37 Newnham Close, Mildenhall, Bury St. Edmunds, IP28 7PD
**M3 LWX** W Alder, 21 Manor Gardens, London, SW20 9AB
**MM3 LWZ** C Coore, 14 Craigs Drive, Edinburgh, EH12 8UW
**M3 LXA** S Jones, 10 Litchborough Grove, Whiston, Prescot, L35 7NE
**M3 LXB** J Wynne, 43 Lansdown Road, Broughton, Chester, CH4 0NZ
**MI3 LXE** I Stevenson, 55 Churchtown Road, Downpatrick, BT30 7AZ
**M3 LXF** S Mills, 27 Boscow Crescent, St. Helens, WA9 3SX
**M3 LXH** S Wright, Flat 2, 19 Cearns Road, Prenton, CH43 2JL
**MI3 LXJ** B Crozier, 33 Cullentragh Road, Poyntzpass, Newry, BT35 6SD
**M3 LXK** C Wynne, 43 Lansdown Road, Broughton, Chester, CH4 0NZ
**MI3 LXN** D Bryans, 1 Meadowvale Avenue, Bangor, BT19 1HG
**M3 LXP** P Enfield, 1 Horton Park, Blyth, NE24 4JD
**M3 LXR** L Dexter, 27 Underwood Avenue, Worsbrough, Barnsley, S70 4AU
**M3 LXS** M Woolley, 4 Robert Street, Warrington, WA5 1TQ
**M3 LXU** T Moscrop, 64 Gresham Road, Norwich, NR3 2NG
**M3 LXV** John Thompson, 114 Corisande Road, Selly Oak, Birmingham, B29 6RP
**MI3 LXW** M Lewis, 7 Liester Park, Ballyrobert, Ballyclare, BT39 9RZ
**MI3 LXZ** A McDowell, 10 Lord Wardens Vale, Bangor, BT19 1GH
**M3 LYA** Carole Cooper, The Haven, Ipswich Road, Norwich, NR15 2TA
**M3 LYC** L Bentley, 1 Cotswold Road, Lupset, Wakefield, WF2 8EL
**M3 LYG** R Dunkley, 25 St. Andrews Crescent, Wellingborough, NN8 2ES
**MM3 LYH** Colin Rodger, 23 Harrysmuir Road, Pumpherston, Livingston, EH53 0NT
**M3 LYP** I Hallatt, 11 Cheshire St., Audlem, Crewe, CW3 0AH
**M3 LYQ** Robert Lovesey, 33 Ty Isaf Park Avenue, Risca, Newport, NP11 6NB
**M3 LYR** T Martin, 46 Hayes Crescent, Frodsham, WA6 7PG
**MM3 LYS** E Smith, 27 Elm Lane, Foresters Lodge, Glenrothes, KY7 5TD
**M3 LYU** C Arner, 6 Welshampton Close, Great Sutton, Ellesmere Port, CH66 2WL
**M3 LYV** C Walsh, 133 Belfield Road, Accrington, BB5 2JD
**M3 LYX** D Shaw, 37 Smirthwaite View, Normanton, WF6 1AW
**M3 LYZ** Fiona Martin, 1 Marsh Street, Strood, Rochester, ME2 4BB
**MI3 LZA** T Conway, 203 Garrymore, Moyraverty, Craigavon, BT65 5JF
**MW3 LZC** T Rule, 35 Pill Street, Penarth, CF64 2JS
**MM3 LZD** J Dinning, South Brae, Aiket Road, Kilmarnock, KA3 4BP
**MI3 LZF** J Steele, 46 Circular Road, Newtownards, BT23 4BN
**M3 LZK** J Delves, 34 Tatton Road, Crewe, CW2 8QA
**M3 LZL** F Shields, 4 Occupation Lane, Earlsheaton, Dewsbury, WF12 8PY
**M3 LZR** N Finlay, 50 Melchett Crescent, Rudheath, Northwich, CW9 7EP
**M3 LZT** David Young, 20 Summerhouse, Tickenham, Clevedon, BS21 6SN
**MM3 LZU** C Tait, 39 Baleshrae Crescent, Kilmarnock, KA3 2GN
**M3 MAA** M Ahmed, 59 Ramsgate, Lofthouse, Wakefield, WF3 3PX
**MD3 MAN** A Espey, 9a Hilltop View, Douglas, Isle of Man, IM2 2LA
**M3 MAR** Margaret Jeffery, 14 Holly Mount, Shavington, Crewe, CW2 5AZ
**M3 MBC** M Bridgeland, 17 Oldfield Lane, Wisbech, PE13 2RJ
**M3 MBF** A Fryer, 9 The Oval, Guildford, GU2 7TS
**M3 MBG** Michael Ford, 119 Neerings, Coed Eva, Cwmbran, NP44 6UL
**M3 MBH** David Hemmings, 1 Sunray Grove, Hucknall, Nottingham, NG15 6RF
**M3 MBI** S Stuart, 46 Breach Road, Heanor, DE75 7NJ
**MI3 MBM** Martin Buchanan, 49 Glengiven Avenue, Limavady, BT49 0RW
**MJ3 MBQ** M Daniells, Le Belon, 2 Clos Vallios, St. Lawrence, Jersey, JE3 1GP
**M3 MBR** M Roberts, 13 St. Michaels Court, Stevenage, SG1 5TB
**M3 MBV** Huawei Su, 135 Devana Road, Leicester, LE2 1PN
**M3 MBZ** M Stephens, 45 Ham Farm Lane, Emersons Green, Bristol, BS16 7BW
**M3 MCA** A Matthews, 70 Branksome Hall Drive, Darlington, DL3 9SR
**MD3 MCB** Brian Perrin, 18 Bellevue Park, Peel, Isle of Man, IM5 1UF
**M3 MCF** Malcolm Frame, 125 Park Road, Stanley, DH9 7QE
**M3 MCG** M Gaston, Ellena, Lochans Mill Avenue, Stranraer, DG9 9BZ
**M3 MCU** J McCallum, 48 Heber Street, Bristol, BS5 9JT
**MM3 MDB** Michael Brunsdon, 25 Buckstone Lea, Edinburgh, EH10 6XE
**M3 MDI** J Gleeson, 64 Swift Road, Oldham, OL1 4QU
**M3 MDK** K Sawyers, 27 Dukeswood Road, Longtown, Carlisle, CA6 5UJ
**M3 MDN** M Dobson, 17 Cadley Causeway, Fulwood, Preston, PR2 3RU
**M3 MDS** A Flemming, 20 Chatham Hill, Chatham, ME5 7AA
**MI3 MDV** A Jess, 25 Gransha Road, Dundonald, Belfast, BT16 2HB
**M3 MDW** M Webster, 2 Brook Close, Nottingham, NG6 8NL
**M3 MDY** M de Young, 97 Kingfisher Road, Larkfield, Aylesford, ME20 6RE
**M3 MEB** Michael Collins, 73 Westholme Road, Bidford-on-Avon, Alcester, B50 4AN
**M3 MEE** Sandra Parker, 100 Horsebridge Hill, Newport, PO30 5TL
**M3 MEF** M Fry, 46 Butt Parks, Crediton, EX17 3HE
**M3 MEG** D Cash, 3 Marsh Lane, Wolverhampton, WV10 6RU
**MM3 MEH** M Mumford, 13 Galloway Drive, Culloden, Inverness, IV2 7ND
**M3 MEI** G Childs, 144 Sturdee Avenue, Gillingham, ME7 2HL
**M3 MEO** D Blake, 3 Carrside, Eastfield, Scarborough, YO11 3DE
**M3 MEP** M Porteious, 17 Church Walk, Yaxley, Peterborough, PE7 3YD
**M3 MER** R Fox, 3 Cherry Blossom Close, Harlow, CM17 0EX
**M3 MES** Nicholas Messenger, Sunnydell, 55 Lower Road, Uxbridge, UB9 5ED
**M3 MEU** David Cowman, 23 Kirk Flatt, Great Urswick, Ulverston, LA12 0TB
**M3 MEW** M Wright, 9 Dinmore Avenue, Blackpool, FY3 7RR
**M3 MEY** G Alker, Bryn Y Mor, Lon Ganol, Menai Bridge, LL59 5YA
**MI3 MFD** F Doherty, 52 Madison Avenue, Eglinton, Londonderry, BT47 3PW
**M3 MFE** Philip Penfold, 2 The Leas, Essenden Road, St. Leonards-on-Sea, TN38 0PU
**M3 MFF** M Frohnsdorff, 75 Alexander Drive, Faversham, ME13 7TA
**M3 MFG** C Goulty, Seletar, 22 Western Avenue, Felixstowe, IP11 9TS
**MM3 MFN** A Harkess, 7 Gardiner Road, Prestonpans, EH32 9HF
**M3 MFR** M Mcminn, 74 Findlater Court Lochside, Dumfries, DG2 0NB
**M3 MFS** S Spencer, 55 Witton Lane, West Bromwich, B71 2AA
**M3 MFT** Alan Hill, 1 Rochester Close, Mountsorrel, Loughborough, LE12 7UH
**M3 MFU** D Mutlow, Dunvegan, Wood Lane, Nuneaton, CV13 0AU
**M3 MFX** C Richardson, 16 Church Road, Boreham, Chelmsford, CM3 3EF
**M3 MFZ** J Gosling, 14 Turnor Close, Colsterworth, Grantham, NG33 5JH
**M3 MGD** D Riggs, 37 Moot Gardens, Downton, Salisbury, SP5 3LG
**M3 MGI** T Rogers, 18 Field Road, Bridlington, YO16 4AU
**M3 MGJ** M Minshull, 12 Dunnett Close, Attleborough, NR17 2NG
**MM3 MGK** S Brown, 21 Whiteford Avenue, Dumbarton, G82 3JU
**M3 MGL** M Talbot, 26 Chevalier Grove, Crownhill, Milton Keynes, MK8 0EJ
**M3 MGN** M Naylor, 96 New Meadows, Rawmarsh, Rotherham, S62 7FE
**M3 MGO** Mark Butler, Wood Green, Astley, Stourport-on-Severn, DY13 0RU
**M3 MGP** Mark Champion, 155 Walton Road, Walton on The Naze, CO14 8NF
**M3 MGQ** J Browne, 29 Longbridge Close, Tring, HP23 5HG
**M3 MGU** R Gregory, Town End, Kirkby Road, Askam-in-Furness, LA16 7EY
**M3 MGZ** M Curtis, 10 Woodstock Gardens, Blackpool, FY4 1JP
**M3 MHD** M Downes, 38 Queensway, Warton, Preston, PR4 1XU
**MW3 MHG** G Rees, 47 Loftus Street, Cardiff, CF5 1HL
**M3 MHL** A Parrish, 5 Kestrel Lane, Cheadle, Stoke-on-Trent, ST10 1RU
**M3 MHN** M Hurren, 257 Norwich Road, Wroxham, Norwich, NR12 8SL
**M3 MHP** E Skinner, 11 Finch Crescent, Leighton Buzzard, LU7 2PE
**MM3 MHQ** A Mccurdy, 5 Kestrel Place, Greenock, PA16 7BL
**M3 MHR** M Reavell, 85 Mccarthy Close, Birchwood, Warrington, WA3 6RS
**MM3 MHS** M Shearer, 113 Auchamore Road, Dunoon, PA23 7JJ
**M3 MHV** M Vaughan, 12 Kingsley Road, Frodsham, WA6 6SG
**M3 MHZ** D Lawrence, 23 Parkmead Road, Wyke Regis, Weymouth, DT4 9AL
**M3 MIB** M BANCROFT, 1 John Street, Knutton, Newcastle, ST5 6DT
**M3 MID** Kenneth Middleton, 1 Campbell Court, Lochmaben, Lockerbie, DG11 1NF
**MI3 MIE** Y Wilson, 59 Crew Road, Upperlands, Maghera, BT46 5TU
**M3 MIF** J Tranter, 64 Geneva Drive, Newcastle, ST5 2QH
**M3 MIG** Paul Cattermole, Blaxhall Hall Crossing, Little Glemham, Woodbridge, IP13 0BP
**M3 MIH** Andrew Nesbitt, 9 Manor Road, Richmond, TW9 1YD
**M3 MII** G Elsworth, 367 West Dyke Road, Redcar, TS10 4PS
**M3 MIJ** J Dean, 12 Abbeydale Road South, Sheffield, S7 2QN
**M3 MIN** A Jones, 17 Maybush Drive, Chidham, Chichester, PO18 8SR
**M3 MIO** Alan Cox, 1 Low House Cottages, Coniston, LA21 8ER
**M3 MIP** R Parrish, 5 Kestrel Lane, Cheadle, Stoke-on-Trent, ST10 1RU
**M3 MIQ** M Fradley, 43 Grange Drive, Penketh, Warrington, WA5 2JN
**M3 MIR** S Carruthers, 13 Belah Road, Carlisle, CA3 9RE
**M3 MIS** J Greatrix, West Cottage, Main Road, Boston, PE20 3PZ
**M3 MIU** F Lie, Harrogate Ladies' College, Clarence Drive, Harrogate, HG1 2QG
**M3 MIV** T Skinner, Flat 4, 25-27 Bridge Street, Leighton Buzzard, LU7 1AH
**M3 MIX** Michael Cole, 9 Troopers Drive, Romford, RM3 9DE
**M3 MJD** M Dennison, 27 Chapel St., Cawston, Norwich, NR10 4BG
**M3 MJE** Martin Edwards, 3 George Street, Bourne, PE10 9HE
**M3 MJH** Mark Hickford, 3 Ashen Road, Clare, Sudbury, CO10 8LQ
**MI3 MJI** R Wylie, 69 Rubane Road, Kircubbin, Newtownards, BT22 1AU
**M3 MJJ** Jane Miller, Flat 1 Block 2, St. Phillips Place, Eastbourne, BN22 8LW
**M3 MJL** M Lee, Up To Date House, Shore Road, Bourne, PE22 0NA
**M3 MJM** M Marter, 4 Meadow Way, Seaford, BN25 4QT
**M3 MJN** Michael Noon, 97 Cherrycroft, Skelmersdale, WN8 9EF
**M3 MJV** Michael Verrechia, 7 Willow Lane, Great Cambourne, Cambridge, CB23 6AB
**M3 MJY** M Kirby, 2 Morton Crescent, Bradwell, Great Yarmouth, NR31 8NT
**M3 MKB** M Baxter, 5 Farnborough Street, Farnborough, GU14 8AG
**M3 MKD** M Davies, 57 Ladybrook Lane, Mansfield, NG18 5JF
**M3 MKH** Michael Hall, 10 Darwin Walk, Northampton, NN5 6LR
**M3 MKJ** M Allen, 8 Green Close, South Wonston, Winchester, SO21 3EE
**M3 MKK** M Kilkenny, 23 Hazelhurst Road, Stalybridge, SK15 1HD
**MW3 MKN** Mark Joseph, 32 Charles Street, Trealaw, Tonypandy, CF40 2UN
**M3 MKO** J Duffield, 4 Church Hill, Easingwold, York, YO61 3JS
**M3 MKV** J Beech, 124 Belgrave Road, Coventry, CV2 5BH
**M3 MKY** J Harsley, 42 Bowbridge Road, Newark, NG24 4BZ
**M3 MKZ** J Swift, 52 North Street, Burwell, Cambridge, CB25 0BB
**M3 MLA** Philip Richardson, 11 Overstone Road Coldham, Wisbech, PE14 0ND
**MD3 MLB** M Bazley, 9B Cronk Y Berry View, Douglas, Isle of Man, IM2 6HH
**M3 MLF** M Firth, 126 Tombridge Crescent, Kinsley, Pontefract, WF9 5HE
**M3 MLG** M Goff, 27 Harley Road, Oxford, OX2 0HS
**M3 MLI** M Litt-Wilson, 14 Wastwater Rise, Seascale, CA20 1LB
**M3 MLK** S Ellison, 23 Murphy Grove, St. Helens, WA9 1QY
**MM3 MMB** M Baird, Creag Saval, Lairg, IV27 4ED
**MI3 MMC** M McClure, 12 St. Patricks Park, Ballymoney, BT53 6JG
**M3 MMG** M Glover, 96 Byron Street, Macclesfield, SK11 7QA
**MM3 MMI** F Millar, 13 Edzell Park, Kirkcaldy, KY2 6YB
**M3 MMJ** M Jones, 47 Maes Derw, Llandudno Junction, LL31 9AN
**M3 MML** Max Lemer, 1 Holmbush Court, Brent Street, London, NW4 2NS
**M3 MMN** G Larrigan, 9 Sandpiper Gardens, Chippenham, SN14 6YH
**MM3 MMO** Duncan Frost, Flat 6, 88 Albion Street, Glasgow, G1 1NY
**M3 MMP** P Evans, 4 Havelock Court, Havelock Street, Aylesbury, HP20 2NU
**M3 MMZ** Matthew Morse, 5 Northload Terrace, Glastonbury, BA6 9JW
**M3 MND** Amanda Harrop, 35 Langdale Crescent, Dalton-in-Furness, LA15 8NR
**MU3 MNG** R Bougourd, 3 Bartholemew, Victoria Road, St. Peter Port, Guernsey, GY1 1JB
**M3 MNQ** Barney Gould, 26 Trembel Road, Helston, TR12 7DY
**M3 MNR** J Redrup, 58 Shaftesbury Road, Bournemouth, BH8 8ST
**M3 MNT** T Williamson, 286 Glynswood, Chard, TA20 1BX
**M3 MNU** Peter Richardson, 14 Portland Street, Worksop, S80 1RZ
**M3 MNV** N Hill, 53 Broadmead, Pontllanfraith, Blackwood, NP12 2NJ
**M3 MNY** K King, Chad Lane Farm, Chad Lane, St. Albans, AL3 8HW
**M3 MOB** J Howarth, 5 Sydenham Building, Bath, BA2 3BS
**M3 MOC** F Mocatta, 19 St. Petersburgh Place, London, W2 4LA
**M3 MOF** Janet Jones, The Studio, Ferney Hoolet, Hook, RG27 8SW
**M3 MOH** G Worrall, 94 Scotia Road, Stoke-on-Trent, ST6 4ET
**MW3 MOJ** Kierion Pitt, 21 Maes-yr-Onen, Nelson, Treharris, CF46 6LF
**M3 MON** Roy Monk, 3 Little Orchard Way, Shalford, Guildford, GU4 8JY
**M3 MOP** A Barden, 38 Silver Close, Tonbridge, TN9 2UY
**M3 MOQ** S Hutchinson, 28 Willow Road, New Balderton, Newark, NG24 3DA
**MI3 MOT** A Thompson, 23 Causeway End Park, Lisburn, BT28 2HX
**M3 MOV** Mark Greenhow, 39 Boston Avenue, Runcorn, WA7 5XE
**M3 MPC** M COLES, 29 Sydney Road, Exeter, EX2 9AH
**MM3 MPK** M Kilday, 129 Lenzie Avenue, Deans, Livingston, EH54 8NS
**MI3 MPL** H Currie, 58 Duneden Park, Belfast, BT14 7NF
**M3 MPM** Mark Murrell, 11 Canberra Street, Hull, HU3 2HT
**M3 MPT** P Denehy, 17 Coverdale, Hull, HU7 4AL
**M3 MQA** L Woolley, 4 Robert Street, Warrington, WA5 1TQ
**M3 MQB** J Hing, 6 Peartree Walk, Billericay, CM12 0PY
**M3 MQC** J Taylor, 6 Hawks Close, Walsall, WS6 7LE
**M3 MQH** Nigel Hilton, 20 Darbyshire Close, Deeping St. James, Peterborough, PE6 8SF
**M3 MQI** A Paxton, Havencroft, Dalbury Lees, Ashbourne, DE6 5BE
**M3 MQJ** G Jackson, Long Pools Farm, Marsh Lane, Market Drayton, TF9 2TG
**M3 MQM** K Rooney, Red Cap Farm, Green Fairfield, Buxton, SK17 7JF
**M3 MQP** M Richards, 57 Coronation Close, Broadstairs, CT10 3DL
**M3 MQR** David Rogers, 9 Prospect Place, Stafford, ST17 4HZ
**M3 MQX** R Rowe, 16 Orchard Road, Plymouth, PL2 2QY
**M3 MQY** G Robinson, Crowstone Mews, Syke House Lane, Greetland, HX4 8PA
**M3 MRA** N Rogers, 4 Lawson Court, Millfield Avenue, Market Harborough, LE16 8XR
**M3 MRC** M Clayton, 12 The Broadway, Abergele, LL22 7DF
**MI3 MRF** T McCullough, 23 Edenvale Park, Antrim, BT41 1AY
**MI3 MRG** A Colligan, 8 Mourneview Crescent, Lisburn, BT28 3HD
**M3 MRJ** J Johnson, 3 Rumbold Road, Hoddesdon, EN11 0LP
**M3 MRK** M Knowles, 72 Uplands Avenue, Connah's Quay, Deeside, CH5 4LG
**MW3 MRL** M Williams, 5a Derllwyn Close, Tondu, Bridgend, CF32 9DH
**M3 MRM** James Milner, Stone Cottage, Wistanstow, Craven Arms, SY7 8DG
**M3 MRN** J Orange, 20 Borrowdale Avenue, Fleetwood, FY7 7LF
**M3 MRO** M Ross, 143 Rose Lane, Romford, RM6 5NR
**M3 MRQ** N Thain, 24 Wilmington Road, Hastings, TN34 2BT
**M3 MRS** Lorna Henley, 5 Gosselin Street, Whitstable, CT5 4LA
**M3 MRU** Duncan Edwards, 10 Queens Avenue, Ilfracombe, EX34 9LN
**MM3 MRX** G Robertson, 24 Tippet Knowes Court, Winchburgh, Broxburn, EH52 6UW
**M3 MRZ** T Hall, 8 Vicarage Close, Billesdon, Leicester, LE7 9AN
**M3 MSB** S Bridge, 59 Alder Hey Road, St. Helens, WA10 4DN
**M3 MSC** L Nicklin, 59 Laurel Road, Armthorpe, Doncaster, DN3 2ES
**M3 MSH** M Hunt, 57 Colsterdale, Worksop, S81 0XH
**M3 MSJ** M Stocker, 1 Rectory Close, Carlton, Bedford, MK43 7JT
**M3 MSL** M Witter, 55 William Street, Churwell, Leeds, LS27 7RD
**M3 MSN** M Austin, 37 Oxenden Road, Cheriton, Folkestone, CT20 3NJ
**M3 MSP** M Skinner, 4 Florence Road, Canvey Island, SS8 7EH
**M3 MSQ** E Bartlett, 86 Usk Road, Tilehurst, Reading, RG30 4HU
**M3 MST** Matthew Wilson, 234 Aylsham Drive Ickenham, Uxbridge, UB10 8UF
**M3 MSU** Lee Wardle, 41 Otter Way, Barnstaple, EX32 8PS
**M3 MSX** M Jenkins, 1 Green End Road, Sawtry, Huntingdon, PE28 5UX
**M3 MSY** M Spicer, 13 Strawberry Path, Oxford, OX4 6RA
**M3 MSZ** M Swift, 8 Grove Lane, Buxton, SK17 9HG
**M3 MTB** Mark Buxton, 610 SQN Air Traning Corps, Cadet Training Centre, Chester, CH1 4AN
**M3 MTC** Catherine Mathewson, 33 Thornton Road, Bootle, L20 5AN
**M3 MTE** Mike Lambert, 94 Cleveland Road, Worthing, BN13 2HE
**M3 MTL** M Price, 9 Herbarth Close, Liverpool, L9 1JZ
**MM3 MTM** M McLeary, 146 Captains Road, Edinburgh, EH17 8DX
**M3 MTP** M Powell, 2a Park Avenue, Uttoxeter, ST14 7AX
**MM3 MTQ** Stuart Saunders, 3 Dunkeld Place, Dunkeld Road, Blairgowrie, PH10 6RX
**M3 MTR** M Bryan, 16 Walesmoor Avenue, Kiveton Park, Sheffield, S26 5RG
**M3 MUA** James Anderson, 121 Barton Road, Stretford, Manchester, M32 9AF
**M3 MUB** S Sheath, 13 Sandpipers, Watermead Road, Portsmouth, PO6 1LB
**M3 MUF** N Wilson, 24d Southlands Road, Weymouth, DT4 9LQ
**M3 MUI** Michael Bromfield, The Cottage, Huttoft, LN13 9RF
**M3 MUO** P Riddle, 8 Lower Dingle, Oldham, OL1 4PB
**M3 MUP** Norman Speight, Flat 14, Cranbrook, London, NW1 0LJ
**M3 MUQ** W Mitchell, Flat 12, 1 Benwell Road, London, N7 7AY
**M3 MUU** S Langton, 51 Fairfield Avenue, Datchet, Slough, SL3 9NF
**M3 MUX** D Baseden, 27 Bayfield, Painters Forstal, Faversham, ME13 0EF
**M3 MVI** M Pick, 290 Horsley Road, Washington, NE38 8HS
**M3 MVJ** R Jefferiss, 26 Welby Close, Maidenhead, SL6 3PY
**M3 MVK** Mark Egerton, 1 Belgrave Road, Northwich, CW9 8DB

M3 MVM J Goulding, 79 Dalston Drive, Manchester, M20 5LQ
M3 MVN C Harding, 27 Eston Avenue, Malvern, WR14 2SR
M3 MVO D Campbell, Beeches, Hammersley Lane, High Wycombe, HP10 8H[illegible]
M3 MVR A Hobbs, 2 The Mead, Beaconsfield, HP9 1AW
MW3 MVT R Williams, Royston, 18 Grove Street, Maesteg, CF34 0HY
M3 MVV J Mullarkey, 41 Foyle Avenue, Chaddesden, Derby, DE21 6TZ
MM3 MVY M Gerrard, 10 Whinhill Gardens, Aberdeen, AB11 7WD
MI3 MWA B Allen, 48 Kevlin Gardens, Omagh, BT78 1QS
M3 MWE Geraint Rowlands, 8 Cleveland Avenue, Tywyn, LL36 9EG
M3 MWG Michael Gosling, 1 Zion Street, Plymouth, PL1 2HX
M3 MWM M Martin, 80 Waveney Road, Hull, HU8 9LY
M3 MWO Angela Rowlands, 8 Cleveland Avenue, Tywyn, LL36 9EG
M3 MWQ Ben Hall, 64 Synehurst Crescent, Badsey, Evesham, WR11 7XX
M3 MWT Robert Waters, 7 The Orchards, Beadlam, York, YO62 7SH
M3 MWV M Williams, 6 Richmond Terrace, Barrow-in-Furness, LA14 5LH
M3 MXA M Anderson, 5 Saffron Court, Wakefield, WF2 0FQ
M3 MXC J Baker, 43 Clyde St., Risca, Newport, NP11 6BP
M3 MXF R Mason, 27 Meadway, Malvern, WR14 1SB
M3 MXG S Turner, 28 Fox Lea, Kesgrave, Ipswich, IP5 2YU
M3 MXH M Price, 43 Heckington Drive, Nottingham, NG8 1LF
M3 MXI S Delaney, 26 Forest Close, Waterlooville, PO8 8JE
M3 MXJ Paul Smith, 1 Grappenhall Hall School House, Church Lane, Warrington, WA4 3ES
M3 MXM R Hardy, 35 Chilton Road, Ipswich, IP3 8PD
MM3 MXN S Reid, 14 St. Marys, Monymusk, Inverurie, AB51 7HH
M3 MXO Rowan Kemp, 4 Scoones Close, Bapchild, Sittingbourne, ME9 9SW
M3 MXP A MacGregor, 53 Napier Place, Orton Wistow, Peterborough, PE2 6XN
M3 MXV K Moulder, 51a Aston Cantlow Road, Wilmcote, Stratford-upon-Avon, CV37 9XN
M3 MXW D Platt, 50 Poplars Road, Stalybridge, SK15 3EN
M3 MXX T Chapman, 16 Andover Place, Cannock, WS11 6EH
M3 MXZ M Rowe, 15 Atlantic Close, Treknow, Tintagel, PL34 0EL
M3 MYE D Sykes, 2 The Street, Claxton, Norwich, NR14 7AS
M3 MYG K Toner, 17 Crooked End Place, Ruardean, GL17 9YN
M3 MYI C Toner, 17 Crooked End Place, Ruardean, GL17 9YN
M3 MYK M Lees, 28 Pleasant Avenue, Bolsover, Chesterfield, S44 6LL
M3 MYM D Cox, 3 Besley Court, Lethbridge Road, Wells, BA5 2FE
M3 MYQ R Jenkins, Glascoed, Garthmyl, Montgomery, SY15 6RT
M3 MYT N Groat, Rose Cottage, Ivy Dene Lane, East Grinstead, RH19 3TN
M3 MYW M EDMOND, 12 Yeoman Close, Worksop, S80 2RR
M3 MYZ M Jones, 65 Montgomery Avenue, Bournemouth, BH11 8BN
M3 MZA R Nicholson, 24 Barnmead, Haywards Heath, RH16 1UZ
M3 MZC A Nicholson, 24 Barnmead, Haywards Heath, RH16 1UZ
M3 MZG D Butterfield, 57 Holmes Road, Retford, DN22 6QU
M3 MZN S Hamilton, 25 Keefe Close, Chatham, ME5 9AG
M3 MZO I Graham, 17 Royal Avenue, Stranraer, DG9 8ET
M3 MZP T Burcombe, 49 Huntingdon Close, Mitcham, CR4 1XJ
M3 MZR E Moon, Moon Marine, Rock Channel, Rye, TN31 7HJ
M3 MZT L Jones, 13 Bracewell Close, Sutton, St. Helens, WA9 3SH
M3 MZW Martin Weller, 27 Rochester Avenue, Woodley, Reading, RG5 4NA
MM3 MZX A Bowers, 4a Pine Street, Greenock, PA15 4HW
M3 NAE Nathan Edwards, 37 Rifle Green, Blaenavon, Pontypool, NP4 9QN
M3 NAF Nigel Foster, 18 Austen Ave, Sawley, Nottingham, NG103GG
M3 NAH Nicholas Higham-Hook, 31 Ringwood, Bracknell, RG12 8YG
M3 NAL C Corbishley, 15 High St., Hardingstone, Northampton, NN4 7BT
M3 NAO D Bradshaw, 65 Lichfield Court, Sheen Road, Richmond, TW9 1AX
M3 NAQ S Marles, 4 Maes Y Llan, Conwy, LL32 8NB
M3 NAR B Johnson, 15 Oak Avenue, Willington, Crook, DL15 0BJ
M3 NAT Nathaniel Poate, 15 Phillips Close, Chippenham, SN14 0TH
M3 NAW N White, 10 Elm Crescent, Alderley Edge, SK9 7PQ
M3 NBB N Beith, 18 Avenue Road, New Milton, BH25 5JP
M3 NBD N Draper, 107 Arkwrights, Harlow, CM20 3LY
M3 NBG L Mcguire, 71 Kingsmead Park, Bedford Road, Rushden, NN10 0NF
M3 NBH Toby Dunne, 5 Lambsickle Lane, Weston, Runcorn, WA7 4QZ
M3 NBI K Fell, 5 Henry Road, Wath-upon-Dearne, Rotherham, S63 7NF
M3 NBK A Rooney, 44 Heritage Drive, Gillingham, ME7 3EH
M3 NBL Norman Bland, 3 Kennet Road, Newbury, RG14 5JA
MW3 NBN E Mcmurray, 30 St. Martins Crescent, Llanishen, Cardiff, CF14 5QA
M3 NBQ S Cross, 138 Crow Lane West, Newton-le-Willows, WA12 9YL
M3 NBU J Shufflebotham, 316 Stockport Road, Hyde, SK14 5RU
M3 NBX V Pomfrett, 17 Manifold Close, Sandbach, CW11 1XP
M3 NBZ N Birnie, 61 Pipers Croft, Dunstable, LU6 3JZ
M3 NCB David Lawson, 30 Meadowcroft, St. Helens, WA9 3XQ
MI3 NCC P Haughey, 10 Captains Road, Forkhill, Newry, BT35 9RR
M3 NCD N Welsh, 7 Dunlin Close, Beechwood, Runcorn, WA7 3JR
M3 NCE M Palmer, 22 Nightingale Drive, Poulton-le-Fylde, FY6 7UQ
M3 NCG Mark Dumpleton, 23 Watermans Yard, Norwich, NR2 4SD
M3 NCH P Blackie, 30 Queens Avenue, Ilfracombe, EX34 9LS
M3 NCL N Lees, 31 Cosford Drive, Dudley, DY2 9JN
MM3 NCM N Cunningham, 11 Glendoune Street, Girvan, KA26 0AA
M3 NCN Michael Gillingham, Frecheville Rectory, Brackenfield Grove, Sheffield, S12 4XS
M3 NCO M Collingswood, School House, Norton Canes High School, Cannock, WS11 3SP
M3 NCP L Steer, 51 Kings Chase, East Molesey, KT8 0DG
M3 NCQ M Williams, 136 Courtfield Road, Quedgeley, Gloucester, GL2 4UF
MW3 NCS N Sedgebeer, 16 metcafe street, caerau, Maesteg, CF34 0TB
M3 NCT N croft, 22 King Edward Crescent, Leeds, LS18 4BE
M3 NDC N Crowley, 133 Jessop Road, Stevenage, SG1 5LH
M3 NDF Dawood Fard, 187 Fleetwood Road South, Thornton-Cleveleys, FY5 5NS
M3 NDJ C Belham, 1 Kenmare Bank, Northwich, CW9 8BN
M3 NDO John Hoskins, 18 Bryn Yr Onnen, Southsea, Wrexham, LL11 6RG
M3 NDR Nigel Nash, Roann, Bedmond Road, Hemel Hempstead, HP3 8SH
M3 NDU R Williams, Bryn Mawr, Gwalchmai, Holyhead, LL65 4PY
M3 NDZ Andrew Humphriss, 44 Bishops Close, Stratford-upon-Avon, CV37 9ED
M3 NEA J Rice, 15 The Muntings, Stevenage, SG2 9DW
M3 NEC N Chisholm, 162 Ardington Road, Northampton, NN1 5LT
M3 NEE Michael Jay, Web-Stile Farm, Coley Hill, Bristol, BS39 5ED
M3 NEG Pat McGarry, 10 Douglas Avenue, Soothill, Batley, WF17 6HG
M3 NEI N McLoughlin, 19 Byron Road, Newport, NP20 3HJ
M3 NEL Neil Snape, 6 Heathfield, Heath Charnock, Chorley, PR6 9LA
MI3 NEN N Nicholl, 54 Derryhill Road, Artigarvan, Strabane, BT82 0HN
M3 NEP Neil Payne, 7 Lane House, Eastfield Close, Worcester, WR3 7TT
M3 NER A Sumner, 10 Dorset Road, Lytham St. Annes, FY8 2ED
M3 NFA J McVay, 38 Robert Hall Street, Leicester, LE4 5RB
M3 NFB Simon Billingham, 6 Swinford Leys Wombourne, Wolverhampton, WV5 8HT
M3 NFE I Jenkin, 50 Maswell Park Road, Hounslow, TW3 2DW
M3 NFF B Armstong, 3 Walton Cres, Llandudno Junction, LL31 9RR
M3 NFG N Gonzalez, 46 Whitton View, Rothbury, Morpeth, NE65 7QN
M3 NFH E Ashley, The Chapel, Ashford, Ludlow, SY8 4BX
M3 NFJ A Williams, 327 Locking Road, Weston-Super-Mare, BS23 3LY
M3 NFK M Watmough, 39 Ripon Gardens, Buxton, SK17 9PL
M3 NFL Neil Leddington, 20 Bewell Head, Bromsgrove, B61 8HY
M3 NFQ S Grainger, 132 Honiton Way, Hartlepool, TS25 2PY
M3 NFU A Scarlett, 87 Coronation Avenue, Shildon, DL4 2AZ
M3 NFW M Millward, 35 Sandown Close, Blackwater, Camberley, GU17 0EN
M3 NFZ D Matthews, 42 College Road, Oswestry, SY11 2SG
M3 NGC Garry Champion, 20 Greenfields Edenside, Kirby Cross, Frinton-on-Sea, CO13 0SW
M3 NGE N Taggart, 61 Well Lane, Curbridge, Witney, OX29 7PB
M3 NGF R Dickerson, 68 Chestnut Avenue, Spixworth, Norwich, NR10 3QQ
M3 NGG Neil G Clare, 123 Cunningham Road, Tamerton Foliot, Plymouth, PL5 4PU
MM3 NGJ A Bernard, 200 Carden Avenue, Cardenden, Lochgelly, KY5 0EN
M3 NGK N Kaye, 33a Aggborough Crescent, Kidderminster, DY10 1LQ
M3 NGM J Gregory, 17 Meadowgarth, Belford, NE70 7PA
M3 NGN B Cook, 218 Ffordd Pennant, Mostyn, Holywell, CH8 9NZ
M3 NGO M Hirst, 21 Manor Farm Court, Thrybergh, Rotherham, S65 4NZ
M3 NGP A Cook, 218 Ffordd Pennant, Mostyn, Holywell, CH8 9NZ
M3 NGU R Bunting, 9 Hammond Way, Attleborough, NR17 2RQ
MM3 NGV M Baird, 28 Loch Road, Bridge of Weir, PA11 3NB
M3 NGY A Gromen-Hayes, 95 Maypole Road, Ashurst Wood, East Grinstead, RH19 3RB
M3 NGZ D Boot, 10 Madehurst Rise, Sheffield, S2 3BJ
M3 NHA Brian Ratcliff, 27 Furlong Road, Manchester, M22 1UD
MW3 NHC I Meek, 30 St. Peters Road, Penarth, CF64 3PP
M3 NHD N Harley, 1 Portland Crescent, Meden Vale, Mansfield, NG20 9PJ
M3 NHE J Kelly, 12 Park Road, Milford on Sea, Lymington, SO41 0QU
M3 NHI S Whitehead, 55 Crombie Road, Sidcup, DA15 8AT
M3 NHN R Williams, 92 Bowleaze, Greenmeadow, Cwmbran, NP44 4LF
M3 NHP Norman Powell, 4 Holme Court Avenue, Biggleswade, SG18 8PF
M3 NHS Nicholas Smith, 248a South Street, Romford, RM1 2AD
M3 NHU L Gale, 9 Ely Close, Worthing, BN13 1BH
M3 NHV R Gale, 9 Ely Close, Worthing, BN13 1BH
M3 NHW A Hill, 39 Lambs Row, Lychpit, Basingstoke, RG24 8SL
M3 NHZ N Hubbard, 7 Creake Road, Syderstone, King's Lynn, PE31 8SF
M3 NIA H Samuels, 11 Bennions Road, Wrexham, LL13 7AW
M3 NIC Nick Rowland, 93 Lockington Crescent, Stowmarket, IP14 1DA
MI3 NIE Chriss Morton, 29 Lackaboy View, Enniskillen, BT74 4DY
M3 NIF F Radford, 3 Pierpoint Terrace, Brighton Road, Hassocks, BN6 9TR
M3 NII C Sims, 7 Ainthorpe Lane, Ainthorpe, Whitby, YO21 2JN
M3 NIT Mitchell Paris, 13 Butfield, Lavenham, Sudbury, CO10 9SD
M3 NIZ G Myers, Janians, 30b Gladstone Road, Ashtead, KT21 2NS
M3 NJA N Atrill, 22 Lester Close, Plymouth, PL3 6PX
M3 NJC N Cook, Chinthurst, Springfield Road, Woolacombe, EX34 7BX
M3 NJD N Darby, 60 Pine St., Grange Villa, Chester le Street, DH2 3LX
M3 NJJ D Wharlley, 15 Crampton Court, Grosvenor Road, Broadstairs, CT10 2XU
M3 NJK R McAllister, 57 Wigan Road, Standish, Wigan, WN6 0BE
M3 NJM N Marsh, 16 Daytona Quay, Eastbourne, BN23 5BN
M3 NJO Nicola O'Hara, 41 Exeter Street, Blackburn, BB2 4AU
M3 NJQ C Johnston, 15 Queens Road, Haydock, St. Helens, WA11 0RH
MI3 NJU P Mcmahon, 9 Sycamore Court, Drumaness, Ballynahinch, BT24 8QZ
MM3 NJV P Vernon, 54 Brothock Way, Arbroath, DD11 4BH
M3 NJY N Young, 5 Winslow Road, Boston, PE21 0EJ
M3 NKB J Banks, 11 Church Street, Banwell, BS29 6EA
M3 NKC D Ansell, 30 Curzon Avenue, Horsham, RH12 2LB
M3 NKF S Harrison, 13 Gillsike House, Thornbury Road, Wakefield, WF2 8BN
MW3 NKG R Rooker, 37 High Close, Nelson, Treharris, CF46 6HJ
M3 NKL D Toyne, 19 Poachers Rest, Welton, Lincoln, LN2 3TR
M3 NKN John Hickman, Ardoch, Harlestone Road, Northampton, NN6 8AW
M3 NKO N March, 25 Emlyn Road, London, W12 9TF
M3 NKP J Keeble, 2 Astley Cooper Place, Brooke, Norwich, NR15 1JB
M3 NKU C Prout, 1 Westbrook, Lustrells Vale, Brighton, BN2 8EZ
M3 NKW J Dale, 37 Bussey Road, Norwich, NR6 6JF
M3 NKX D Doggett, 82 Hurst Road, Kennington, Ashford, TN24 9RS
M3 NKZ M Ashton, 138 Christchurch Road, Norwich, NR2 3PG
M3 NLA N King, 232 Prince Consort Road, Gateshead, NE8 4DX
M3 NLE J Grosvenor, 10 Neves Close, Lingwood, Norwich, NR13 4AW
MM3 NLH S Anderson, Newbigging Toll House, Drumsturdy Road, Dundee, DD5 3RF
M3 NLI D Bale, 22 Highgrove Court, Rushden, NN10 0DH
M3 NLJ N Jeffery, 7 Corfe Way, Winsford, CW7 1LU
M3 NLK N Lake, 64 Womersley Road, Norwich, NR1 4QB
M3 NLM Paul Mayhin, 16 Appleby Road, London, E16 1LQ
M3 NLN Andrew Paulizky, 17a Angles Road, London, SW16 2UU
M3 NLP N Pearson, 34 Downside Road, Sutton, SM2 5HP
M3 NLQ M Thompson, 4 Oat Hill Road, Towcester, NN12 6EZ
M3 NLW L Boull, 80 Ascot Road, Baswich, Stafford, ST17 0AQ
M3 NLX R Mcdermott, 2 Monument Close, Wellington, TA21 9AL
M3 NMB N Beech, 94 Victoria Road, Runcorn, WA7 5ST
MI3 NMG N McGonigle, 27 Rousky Road, Dunamanagh, Strabane, BT82 0SF
MM3 NMI Daniel Small, 30 Caledonia Crescent, Ardrossan, KA22 8LW
M3 NMJ Spencer Martin, 12 Essex Road, Weymouth, DT4 0BA
M3 NMK A Kemplay, Les Chaumes, Amailloux, France, 79350
M3 NMM J Jones, 55 Layton Road, Gosport, PO13 0JG
M3 NMP M Pedley, 60 Ack Lane East, Bramhall, Stockport, SK7 2BY
M3 NMR C Purkiss, Flat 6, 220 Greenheys Lane West, Manchester, M15 5AF
M3 NMU A Greenhough, 14 The Dale, Wirksworth, Matlock, DE4 4EJ
M3 NMV L Yates, 108 Hawthorn Avenue, Lowestoft, NR33 9BB
M3 NMX [illegible], 7 Normandy Way, Fordingbridge, SP6 1NW
M3 NNA Ann Lydford, 93 Hardwick Avenue, Chepstow, NP16 5ED
M3 NNG R Simms, 3 The Byeway, London, SW14 7NL
M3 NNH Kyla Bansil, 65 Hervey Street, Northampton, NN1 3QL
M3 NNI A Mason, 4 Quay Mill Walk, Great Yarmouth, NR30 1JG
M3 NNJ Jayne Moore, 2 Newsons Meadow, Lowestoft, NR32 2NW
M3 NNM P Morling, 7 Hobill Close, Leicester Forest East, Leicester, LE3 3PS
MM3 NNO C Lewis, 9 Cessnock Road, Troon, KA10 6NJ
M3 NNQ J Blamey, 46 First Avenue, Canvey Island, SS8 9LP
M3 NNV Jeremy Paul, Mimosa Lodge, 59 Baring Road, Cowes, PO31 8DW
M3 NNY J Brough, 10 Linnet Close, Huntington, Cannock, WS12 4TP
M3 NNZ B James, 19 Dukes Crescent, Sandbach, CW11 1BL
M3 NOD N Lightfoot, 4 Prospect Close, Hatfield Peverel, Chelmsford, CM3 2JE
M3 NOE Denis Noe, 21 Gale Crescent, Banstead, SM7 2HZ
M3 NOF D Donnelly, 72 Bagots Oak, Stafford, ST17 9SB
MI3 NOH N O'Hagan, 55 Meadowside, Antrim, BT41 4HD
M3 NOJ J Reynolds, 4 Perriclose, Chelmsford, CM1 6UJ
M3 NOM R Smallman, 128 Bevan Lee Road, Cannock, WS11 4PT
M3 NON N O'Sullivan, The Hollies, The Street, Bungay, NR35 2LZ
M3 NOR Andrew Norman, 3 Chorlton Villas, Malpas, SY14 7JJ
M3 NOS D Leggett, Fools Watering, London Road, Beccles, NR34 8AQ
M3 NOW S Wilson, 55 Kent Road, Reading, RG30 2EJ
M3 NOY Stephen James, 124 Alcock Avenue, Mansfield, NG18 2NF
M3 NPA A Nicholson, 7 Lingfoot Crescent, Sheffield, S8 8DA
M3 NPC N Collins, Hill Farm, Broadheath, Tenbury Wells, WR15 8QN
M3 NPE J Nicholson, 7 Lingfoot Crescent, Sheffield, S8 8DA
MM3 NPG M Maltman, 30 Haldane Place, Dundee, DD3 0JR
M3 NPH Aidan Arnold, 2 Duck Lane, Haddenham, Ely, CB6 3UE
M3 NPI M Harrell, 34 Nelson Drive, Cannock, WS12 2GF
M3 NPK N Kerner, Headingley Cottage, Ryehurst Lane, Bracknell, RG42 5QZ
M3 NPO D Shuttleworth, 27 Union St., Egerton, Bolton, BL7 9SP
MI3 NPR N Robinson, 21 Bracken Brae, Dungannon, BT71 4DW
M3 NPS Nigel Sharpe, 23 Cheney Road, Faversham, ME13 8DG
M3 NPW P Webster, 39 Farndale Terrace, Leeds, LS14 5BQ
M3 NPX S Taylor-Mccormick, North View Boarding Kennels, Skitham Lane, Preston, PR3 6BD
M3 NPZ A Paterson, 35 Darlington Road, Richmond, DL10 7BG
M3 NQA R Simmonds, The Mill House, Great Ponton, Grantham, NG33 5DX
MW3 NQE S Walmsley, 29 Shelley Court, Machen, Caerphilly, CF83 8TT
MW3 NQH S Pope, 11 Haman Place, Gelligaer, Hengoed, CF82 8EG
M3 NQI P Denham, 48 Marl Court, Thornhill, Cwmbran, NP44 5TY
M3 NQK J Argent, 7 Lloyds Hill, Buckley, CH7 3ER
M3 NQL Lee Kelsey, 111-113 George Street, Mablethorpe, LN12 2BS
M3 NQN K ROBERTS, 40 Portland Drive, Skegness, PE25 1HF
M3 NQO Alex Whyatt, 11 The Perrings, Nailsea, Bristol, BS48 4YD
M3 NQS N Turner, 1 Mannings Close, Saffron Walden, CB11 4BD
MM3 NQT M Simon, 100 Findhorn Place, Edinburgh, EH9 2NZ
M3 NQU E Wagner, 3 Sarre Road, London, NW2 3SN
M3 NQY G Eklund, 26-28 Zulu Road, Nottingham, NG7 7DR
MI3 NRB Neil Bolt, 32 Bush Gardens, Bushmills, BT57 8AE
M3 NRI A Miles, 5 Pershore Road, Basingstoke, RG24 9BE
M3 NRJ Nigel Johnson, 27 Redford Crescent, Bristol, BS13 8SA
M3 NRK Nigel Kind, 18 Cunningham Road, Bentley, Walsall, WS2 0AY
M3 NRQ Andrew Nicholson, 7 Lingfoot Crescent, Sheffield, S8 8DA
M3 NRV D Giering, 1 Church Street, Chulmleigh, EX18 7BU
M3 NRW E Cromwell, 92 Hatch Road, Pilgrims Hatch, Brentwood, CM15 9QA
MM3 NRX Jason Williams, 2/L 17 St. Clement Place, Dundee, DD3 9NZ
M3 NSB A Oatey, Robin Hill, Blackpost Lane, Totnes, TQ9 5RF
MI3 NSF R Foley, 6 Lislane Drive, Saintfield, Ballynahinch, BT24 7HU
M3 NSG N Garry, 4 Fairstead, Skelmersdale, WN8 6RD
M3 NSH S Shelley, 8 Harewood Close, Eastleigh, SO50 4NZ
M3 NSJ K Ingham, 57 Cleaver Street, Burnley, BB10 3BS
M3 NSM G Coyle, 19 Mounsey Road, Bamber Bridge, Preston, PR5 6LS
M3 NSO Nik Walch, 52 Marsh House Road, Sheffield, S11 9SP
M3 NSQ Stephen Beedham, 27 Malpas Close Bransholme, Hull, HU7 4HH
MI3 NSR Graeme McCullough, 32 Thistlemount Park, Lisburn, BT28 2UN
M3 NSS M Price, 25 School Crescent, Lydney, GL15 5TA
M3 NST N Stirling, 3 Rother Croft, New Tupton, Chesterfield, S42 6BE
M3 NSX I Thompson, 33 Longsight Road, Mapplewell, Barnsley, S75 6HD
M3 NSZ R Stanway, 72 Sheldons Court, Winchcombe Street, Cheltenham, GL52 2NR
MW3 NTE Heinz Fingerhut-Holland, Osnok, 1 South Cliff Street, Tenby, SA70 7EB
MU3 NTH N Thomas, 6 Tunstall Terrace, Gibauderie, St. Peter Port, Guernsey, GY1 1XJ
M3 NTI M Gough, 57 Ravenglass Road, Westlea, Swindon, SN5 7BN
M3 NTJ N Thompson, 33 Longsight Road, Mapplewell, Barnsley, S75 6HD
M3 NTQ Michael Gough, 58 Church Street, Brierley, Barnsley, S72 9JG
M3 NTR Martin Pratchett, Adastra Cottage, Letcombe Regis, Wantage, OX12 9J[illegible]
M3 NTW C Northwood, Apartment 50, 2 Munday Street, Manchester, M4 7DD
MM3 NTX G Askew, 49 Kittlegairy Road, Peebles, EH45 9LX
M3 NTZ S Woodward, 19 Beech Court, Spondon, Derby, DE21 7TP
M3 NUB N Vichitcheep, Oriel College, Oriel Square, Oxford, OX1 4EW
M3 NUC W Groves, 3 Tetbury Close, Newport, NP20 5HX
M3 NUE A Lunn, 57 Greets Green Road, West Bromwich, B70 9ES
M3 NUH M Gladders, 2 Albion Mansions, Saltburn-by-The-Sea, TS12 1JP
M3 NUI S Scotson, 86 Derrydown Road, Birmingham, B42 1RT
M3 NUL Stuart Heaton, 2 Hunmanby Road, Reighton, Filey, YO14 9RT
M3 NUM Anne Ogle, 22 Warwick Street, Daventry, NN11 4AL
M3 NUO K Holdt, 18 Garrard Road, Banstead, SM7 2ER
MW3 NUP M Lewis, 4 Coldwell Terrace, Pembroke, SA71 4QL
M3 NUQ Michael Dickenson, 6 The Pavilions, Blandford Forum, DT11 7GF
M3 NUU Derek Copsey, Fairview, Mill Lane, Brentwood, CM15 0PP
MI3 NUW Aubrey Kincaid, 428 Cushendall Road, Ballymena, BT43 6QE
MW3 NUX D James, 10 Hafan Deg, Pencoed, Bridgend, CF35 6YG
M3 NVA A Caine, 116b Hill Street, Hednesford, Cannock, WS12 2DR
M3 NVC Philip Brown, 13 Rydal Road, Weston-Super-Mare, BS23 3RT
MM3 NVD Duncan Baillie, 126 Main St., Fauldhouse, Bathgate, EH47 9BW

**IMPORTANT NOTE**

**Revalidate licence to avoid revocation** – Ofcom has advised the Society that plans will be drawn up to revoke licences that have not been revalidated as required by the licence conditions. The quickest way to revalidate is to do so online via the Ofcom website: *https://services.ofcom.org.uk/* or by email: *amateur.validations@ofcom.org.uk* Ofcom staff are available to help, but please be patient during times of heavy workload.

M3 NVE Kevin Whiteley, 14 Milton Street, Goole, DN14 6EL
M3 NVF N Fletcher, 2 Handforth Road, Crewe, CW2 8PL
M3 NVG John Webster, 72 Grosvenor Street, Derby, DE24 8AT
M3 NVH J Boull, 80 Ascot Road, Baswich, Stafford, ST17 0AQ
M3 NVK Laurence Bolton, 59 Picquets Way, Banstead, SM7 1AB
M3 NVL J Jones, 5 Cranleigh Road, Liverpool, L25 2RP
M3 NVO L Clift, 8 Kendal Road, Gloucester, GL2 0NB
M3 NVP M Weir, 153 Tyndale Crescent, Birmingham, B43 7HX
M3 NVQ R Wright, 12 Bryn Teg, Arddleen, Llanymynech, SY22 6PZ
M3 NVR P Gutteridge, 75a Collingwood Drive, Birmingham, B43 7JW
M3 NVS J Caddick, 135 Broadway, Dunscroft, Doncaster, DN7 4HB
M3 NVV C Elliot, 106 Occupation Road, Corby, NN17 1EG
MI3 NVX Martin McCay, 2 Riverview, Spamount, Castlederg, BT81 7NA
M3 NWD A Broll, 17 Broadway, Farcet, Peterborough, PE7 3AY
MM3 NWF K Whyte, 27 Queens Road, Inverbervie, Montrose, DD10 0RY
M3 NWH F Gear, 251 Abington Avenue, Northampton, NN3 2BU
M3 NWK S Lofthouse, 32 Westbrook Park Road, Peterborough, PE2 9JG
MI3 NWO Dorothy Adams, 65 Rose Park, Limavady, BT49 0BF
M3 NWQ F Stone, Rrt, Gosw, 2 Rivergate, Bristol, BS1 6EH
MI3 NWU G McKeever, 45 Blackthorn Court, Coleraine, BT52 2EX
M3 NWY T Bown, 16 Sandringham Court, Queen Elizabeth Road, Nuneaton, CV10 9AR
M3 NWZ Anthony Smith, 17 High Street, Stranraer, DG9 7LL
M3 NXA A Davenport, 10 Woodend Lane, Hyde, SK14 1DT
M3 NXC Peter waring, 5 Berry Avenue, Eckington, Sheffield, S21 4AR
MW3 NXD A Downing, 3a Pant Hirgoed, Pencoed, Bridgend, CF35 6YD
M3 NXE S Burrows, 78a Coronation Road, Earl Shilton, Leicester, LE9 7HJ
M3 NXF N Forbes, 55 The Henrys, Thatcham, RG18 4LS
M3 NXH I Rimell, 51 Woodlands Avenue, Woodley, Reading, RG5 3HF
M3 NXJ I Livesey, 26 Hilltop Road, Twyford, Reading, RG10 9BN
M3 NXK S Dudley, 28 Walnut Lane, Wednesbury, WS10 0BH
M3 NXO B Bradford, 3 Beverly Close, Thornton-Cleveleys, FY5 5DR
M3 NXQ V Stokes, 52 Brantley Avenue, Wolverhampton, WV3 9AR
MM3 NXY R Hay, 12 Mitchell Brae, Balmedie, Aberdeen, AB23 8PW
M3 NXZ S Robinson, 1 Woodgate Road, Manchester, M16 8LX
M3 NYA M Hoggan, 19 The Drive, Uckfield, TN22 1BY
MI3 NYB N Throne, 12 Mason Road, Magheramason, Londonderry, BT47 2RY
M3 NYF A Probst, 37 Devonshire Street, Skipton, BD23 2ET
M3 NYG N Carson, Gov Office For The South West, 2 Rivergate, Bristol, BS1 6EH
M3 NYI A Parradine, 76 Parsonage Road, Rainham, RM13 9LF
M3 NYM Gary Whitehurst, 28 Vicarage Fields, Warwick, CV34 5NJ
M3 NYQ M Hives, 36 Partridge Way, Chadderton, Oldham, OL9 0NT
M3 NYX Jennifer Mock, 29 Tavistock Place, Paignton, TQ4 7NZ
MM3 NYY G Cleary, 2 Merlinford Avenue, Renfrew, PA4 8XS
M3 NZG A Wheeler, 14 Sparkmill Terrace, Beverley, HU17 0PA
M3 NZK J Bayliss, 39 Elms Avenue, Littleover, Derby, DE23 6FB
M3 NZN C Jones, 20 Freeman Road, Wednesbury, WS10 0HQ
M3 NZR David Oddie, 5 The Bridleway, Forest Town, Mansfield, NG19 0QJ
M3 NZV L Williams, 2 The Tannery, Dol-y-Bont, Borth, SY24 5LX
M3 NZW A Woods, 58 Lawley Bank Court, Telford, TF4 2PP
MM3 NZX Michael McDonald, 106 Stamperland Gardens, Clarkston, Glasgow, G76 8NR
M3 NZZ I Powe, 69 Park Place, Risca, Newport, NP11 6BN
M3 OAB A Barker, 43 Ploughmans Drive, Shepshed, Loughborough, LE12 9SG
M3 OAC G Dray, 17 Lime Tree Avenue, Malvern Wells, Malvern, WR14 4XE
M3 OAG W Scott, 8 Woodacre, Whalley Range, Manchester, M16 8QQ
M3 OAJ S Clay, Akers Lodge, 6 Penn Way, Rickmansworth, WD3 5HQ
MD3 OAK D Smith, 98 Orange Hill Road, Burnt Oak, Edgware, HA8 0TW
M3 OAL Michelle Edwards, 30 Morrison Road, Tipton, DY4 7PU
M3 OAM M Parkin, 51 Far Lane, Rotherham, S65 2HQ
M3 OAQ S Tingay, 9 Cottage Homes, Wakefield Road, Huddersfield, HD5 9XT
M3 OAS I Miller, 64 Queens Road, Vicars Cross, Chester, CH3 5HD
M3 OAT J Judson, 559 Colne Road, Burnley, BB10 2LG
MM3 OAW A Irvine, 41 Craighead Road, Bishopton, PA7 5DT
M3 OAX F Smith, 32 Amesbury Drive, London, E4 7PZ
M3 OAZ R Wertheim, 127 Higher Lane, Whitefield, Manchester, M45 7WH
M3 OBB Owen Boar, 19 Blyford Road, Lowestoft, NR32 4PZ
M3 OBD A Dow, 5 Verney Crescent, Liverpool, L19 4UR
M3 OBL L Kenton, 24 Penygraig Road, Brymbo, Wrexham, LL11 5AD
M3 OBM G Scarr, 15 Church Park, Overton, Morecambe, LA3 3RA
M3 OBN L Stott, 70 Elizabeth Street, Ashton under Lyne, OL6 8SX
M3 OBO Robert Goody, 113 Kenneth Road, Basildon, SS13 2BH
M3 OBQ G Stephens, 8 New Molinnis, Bugle, St. Austell, PL26 8QL
M3 OBS Mike Simkins, 37 St. Andrews Meadow, Harlow, CM18 6BL
M3 OBU C Camsey, 1 Park Close, Milford on Sea, Lymington, SO41 0QT
M3 OBX Duncan Sanderson, 65 Holm Flatt Street, Parkgate, Rotherham, S62 6HJ
M3 OBZ D Thomas, 51 Sandringham Avenue, Vicars Cross, Chester, CH3 5JF
M3 OCA P Bainbridge, 24 Peckers Hill Road, Sutton, St. Helens, WA9 3LW
M3 OCJ S Rafter, 30 Monmouth Grove, St. Helens, WA9 1QB
M3 OCL H Lister, 68 Spring Avenue, Gildersome, Leeds, LS27 7BT
M3 OCP M Cave, 26 Longsight Road, Mapplewell, Barnsley, S75 6HB
M3 OCQ R Hall, 5 Lea Hill Road, Birmingham, B20 2AS
M3 OCR O Crump, 3 Vosper Road, Southampton, SO19 9SS
M3 OCS A Owen, 5 Croft Close, Rowton, Chester, CH3 7QQ
MM3 OCY D McClelland, 5 Cambusmoon Terrace, Gartocharn, Alexandria, G83 8RU
M3 ODC P Hatter, 14 Morland Avenue, Bromborough, Wirral, CH62 6BE
M3 ODH D Hughes, 75 Suncote Avenue, Dunstable, LU6 1BN
M3 ODK L Bedford, 29 Kent Road, Brookenby, Market Rasen, LN8 6EW
M3 ODL K bedford, 29 Kent Road Brookenby, Binbrook, Market Rasen, LN8 6EW
M3 ODN M Cave, 26 Longsight Road, Mapplewell, Barnsley, S75 6HB
M3 ODO Elwyn White, 3 Davy Drive, Maltby, Rotherham, S66 7EN
MM3 ODV A Cairns, 57 Miller Street, Dumbarton, G82 2JA
M3 OEB D Norris, 24 Northway, Fulwood, Preston, PR2 9TP
MW3 OEC L Williams, 35 Heol Y Sheet, North Cornelly, Bridgend, CF33 4EU
MD3 OED Rhett Britton, 3 Meadowfield, Port Erin, Isle of Man, IM9 6PH
M3 OEE B Hood, 52 Kent Street, Preston, PR1 1RY
M3 OEF S Tattum, 20 Drew Close, Poole, BH12 5ET
M3 OEG John Cook, 42 Pampas Close, Colchester, CO4 9ST
M3 OEH Anthony McLean, 47 Tarn Drive, Bury, BL9 9QB
MW3 OEJ M Kay, 3 Protheroe Avenue, Pen-y-Fai, Bridgend, CF31 4LU
M3 OEK B Horton, 44 Chamberlain Street, Street Helens, WA10 4NL
M3 OEN J Edwards, 36 Westerley Lane, Shelley, Huddersfield, HD8 8HP
M3 OEO M Hammersley, 47 Shaftesbury Avenue, Timperley, Altrincham, WA15 7NP
MI3 OEQ A Jamison, 11 Richmond Gardens, Newtownabbey, BT36 5LA
M3 OER Simon Warner, 28 Jameson Bridge Street, Market Rasen, LN8 3EW
M3 OEV Martin Cuff, 14 The Mount, Ringwood, BH24 1XX
M3 OFA N Shaw, Greenacres Poultry, Three Lowes, Stoke-on-Trent, ST10 3BW
M3 OFB J Woodruff, 62 Burton St., Rishton, Blackburn, BB1 4PD
M3 OFC S Dunne, 1 Burton Gardens, Brierfield, Nelson, BB9 5DR
M3 OFD D Dunne, 1 Burton Gardens, Brierfield, Nelson, BB9 5DR
MM3 OFE J Jackson, 25 Lomond Crescent, Alexandria, G83 0RJ
M3 OFH J Wright, 12 Bryn Teg, Arddleen, Llanymynech, SY22 6PZ
M3 OFJ M Jones, 20 Chelsea Drive, Sutton Coldfield, B74 4UG
M3 OFN Christian Bamford, 47 Trent Road, Shaw, Oldham, OL2 7YQ
M3 OFS E Marsh, 16 Laurel Close, North Warnborough, Hook, RG29 1BH
M3 OFU T Smith, 151 Halfords Lane, West Bromwich, B71 4LQ
M3 OFV M Kempson, 67 Esther Avenue, Wakefield, WF2 8BY
MI3 OFX T Conlon, 7 Waringfield Gardens, Moira, Craigavon, BT67 0FQ
M3 OGC L Li, Harrogate Ladies' College, Clarence Drive, Harrogate, HG1 2QG
M3 OGD Richard Whiteside, Hill Crest, Farley Hill, Matlock, DE4 5LT
M3 OGL T Woodhouse, The Old Granary, 12 Limekiln Lane, Newport, TF10 9EZ
M3 OGM Brian Debenham, 80 Stewart Road, Chelmsford, CM2 9BD
MM3 OGS R Keay, 26 Cherrywood Drive, Beith, KA15 2DZ
M3 OGU Tracey Mussell, Whitehouse, Priory Road, Boston, PE21 0RD
M3 OGV R Bicknell-Thompson, 4 Linden Court, Greenfrith Drive, Tonbridge, TN10 3LW
M3 OHC M Holgate, 10 Brecon Crescent, Ashton-under-Lyne, OL6 8UA
MM3 OHD Darren Hague, 13 North Dell, Ness, Isle of Lewis, HS2 0SW
MI3 OHE E Paterson, 1 Sycamore Grove, Belfast, BT4 2RB
MI3 OHF M Graham, 21 Meadowvale Crescent, Bangor, BT19 1HQ
MI3 OHG N wylie, 26 Lisnoe Walk, Lisburn, BT28 1QD
M3 OHI G Chaffey, 63 Underwood Road, Eastleigh, SO50 6FX
M3 OHJ B Stanfield, 24 Rowan Close, Kingsbury, Tamworth, B78 2JR
M3 OHL B Widdowson, 28 Highfield Lane, Chesterfield, S41 8AU
M3 OHN Peter Martin, Flat 9, Paddock Court, Graham Avenue, Brighton, BN41 2WU
M3 OHO M Murray, 2 Meadway, Penwortham, Preston, PR1 0JL
MI3 OHP Jim Leetch, 30 Murob Park, Ballymena, BT43 6JG
M3 OHQ W Kong, Harrogate Ladies' College, Clarence Drive, Harrogate, HG1 2QG
M3 OHR L Gray, 29 Longview Road, Liverpool, L36 1TA
M3 OHX D Taylor, 58 Shenstone Road, Great Barr, Birmingham, B43 5LN
M3 OHY H Chan, Harrogate Ladies' College, Clarence Drive, Harrogate, HG1 2QG
M3 OHZ H Fleming, 29 Model Village, Creswell, Worksop, S80 4BN
M3 OIA S Gearey, 32 Bridgeside, Deal, CT14 9SS
MI3 OIB Darren Couser, 48c Cloyne Crescent, Newtownabbey, BT37 0HH
M3 OIC L Crane, 32 Clarke Avenue, Newark, NG24 4NY
M3 OII T Hoggan, 9 Nursery Field, Buxted, Uckfield, TN22 4NG
M3 OIK William Jaggard, 2 Aled Drive, Rhos on Sea, Colwyn Bay, LL28 4UU
M3 OIL Owen Hutley, 1 John Ray Street, Braintree, CM7 9DZ
M3 OIN H List, 41 Westbury Crescent, Dover, CT17 9QQ
MD3 OIS M Wallace, 61 Vernon Road, Ramsey, Isle of Man, IM8 2EG
M3 OIV Andrew Webb, 47 Granville Street, Gloucester, GL1 5HL
MM3 OIX N White, 29 Forgie Crescent, Maddiston, Falkirk, FK2 0LY
M3 OIY P Yuen, Harrogate Ladies' College, Clarence Drive, Harrogate, HG1 2QG
MM3 OIZ J McMonigle, 19 Monach Gardens, Dreghorn, Irvine, KA11 4EB
MM3 OJE Tommy McFarlane, 23 West Edith Street, Darvel, KA17 0EE
M3 OJJ A CHANCE, 24 Doddsfield Road, Slough, SL2 2AD
M3 OJK N Schall, 39 Emery Avenue, Chorltonville, Manchester, M21 7LE
M3 OJN Peter Moule, 30 Hillview Road, Chelmsford, CM1 7RX
M3 OJP David Moulding, 5 Chalk Lane, Sutton Bridge, Spalding, PE12 9YF
MM3 OJR John Rae, Bowhouse Farm Cottage, Auchtermuchty, Cupar, KY14 7ES
M3 OJS Peter Rowlands, 314 Stourbridge Road, Catshill, Bromsgrove, B61 9LH
M3 OJU P Watling, 1 Chediston Green, Chediston, Halesworth, IP19 0BB
MM3 OJV C McGougan, 6 Calder Place, Kilmarnock, KA1 3QL
M3 OJW S Earl, 9a Florida Street, Daws Hill Lane, High Wycombe, HP11 1QA
M3 OJX L Parkman, 9 Malim Way, Gonerby Hill Foot, Grantham, NG31 8QF
M3 OJY Barbara Mason, 4108 Hingston Avenue, Montreal, Canada, H4A 2J7
M3 OKC P Barnes, 75 Cedar Drive, Sutton at Hone, Dartford, DA4 9EW
M3 OKE A Ewence, 9 Mount Pleasant, Bradford-on-Avon, BA15 1SJ
MD3 OKG K Glaister, 42 Barrule Drive, Onchan, Douglas, Isle of Man, IM3 4NR
MD3 OKH B Glaister, 42 Barrule Drive, Onchan, Douglas, Isle of Man, IM3 4NR
M3 OKP R Bancroft, Newraine, Hallin Croft, Penrith, CA10 1EW
M3 OKY R Eglington, 33 Bradley Lane, Bilston, WV14 8EW
M3 OKZ A Cammish, 6 West Vale, Filey, YO14 9AY
M3 OLD Sam Old, 33 Rookhill Road, Pontefract, WF8 2BY
M3 OLE S Jervis, 45 Wyndham Road, Stoke-on-Trent, ST3 3LX
M3 OLF I Scott, Croft Cottage, Cumwhinton, Carlisle, CA4 8ER
M3 OLI O Palmer, 21 Ibbett Close, Kempston, Bedford, MK43 9BT
MI3 OLM C Doole, 110 Moyagall Road, Knockloughrim, Magherafelt, BT45 8PJ
M3 OLN E Hall, 5 The Paddocks, Thursby, Carlisle, CA5 6PB
M3 OLQ P Oliver, 12 Walkmill Crescent, Carlisle, CA1 2WF
MW3 OLT L Thomas, 15 Blaenwern, Newcastle Emlyn, SA38 9BE
M3 OLU A Shepherd, 39 Minehead Road, Dudley, DY1 2NZ
M3 OLW A mills, 11 Brighton Avenue, Southend-on-Sea, SS1 2QN
MW3 OLX W Hance, 27 Maesybont, Glanamman, Ammanford, SA18 2AY
M3 OLZ Lee Bullen, 2 Rowley Cottages, Hermitage Road, Upton, Langport, TA10 9NP
M3 OME A Sadanandam, 23 Clare Avenue, Hoole, Chester, CH2 3HT
M3 OMF O Fury, 1 Wigley Drive, Wigley, Ludlow, SY8 3DR
MM3 OMI Brian Bowman, 15 Aboyne Road, Aberdeen, AB10 7BS
MM3 OML T Cockayne, 2b Bogleshole Road, Cambuslang, Glasgow, G72 7PR
M3 OMT T Baggley, 16 Seaton Road, Seaton, Workington, CA14 1DT
M3 OMU Jason Fletcher, 217 Homefield Road, Sileby, Loughborough, LE12 7TG
M3 OMX K Boulton, 440a Crownhill Road, Plymouth, PL5 2QS
M3 OMZ D Jarvis, 7 Leonard Road, Greatstone, New Romney, TN28 8UJ
M3 ONB I Woollen, 33 The Oaks, Taunton, TA1 2QX
M3 OND Ian Turner, 1 Elmwood Rise, Dudley, DY3 3QJ
M3 ONE C Chambers, 75a Main Street, Sedbergh, LA10 5AB
MW3 ONG John Brydges, 9 Twynygarreg, Treharris, CF46 5RL
M3 ONH S Fooks, 11 Breck Road, Darlington, DL3 8NH
MM3 ONI O Stein, Library House, Stafford Street, Tain, IV19 1AZ
M3 ONK R Bray, 49 Montacute Way, Wimborne, BH21 1TZ
M3 ONM P Connor, 244 Gregory Avenue, Birmingham, B29 5DR
MD3 ONP J Taylor, Ballafayle Cottage, Ballafayle, Ramsey, Isle of Man, IM7 1ED
M3 ONV John Bonar, 40 Quarry Close, Minehead, TA24 6EE
MM3 ONX L Paget, 40 Davaar Drive, Kilmarnock, KA3 2JG
M3 OOA T Durant, 39 Snydale Road, Normanton, WF6 1NY
M3 OOC B Cooper, 71 High Street, Birstall, Batley, WF17 9RG
M3 OOE G Kelly, Elthor, 30 Station Road, Aylesbury, HP22 5UL
M3 OOH Simon Howroyd, Aeronautical & Automative, Stewart Miller Building, Loughborough University, Loughborough, LE11 3TU
M3 OOL T Peterson, 9 Moseley Wood View, Leeds, LS16 7ES
M3 OOP J Dietsch, 21 Lake View Avenue, Chesterfield, S40 3DR
M3 OOQ A Shaw, 2 Montrose Avenue, Montrose Street, Hull, HU8 7RY
MM3 OOT J ferrans, 77 Knockinlaw Road, Kilmarnock, KA3 2AS
M3 OOU Adrian Hoskins, 38 Tasmania Close, Basingstoke, RG24 9PQ
M3 OOX F Spencer, 29 Kliffen Place, Halifax, HX3 0AL
M3 OOY C Spencer, 29 Kliffen Place, Halifax, HX3 0AL
M3 OPA N Van-Den-Langenberg, 2 Grove Crescent, Bridgnorth, WV15 5BS
M3 OPB O Blackburn, 128 High St., Crigglestone, Wakefield, WF4 3EF
M3 OPC J Spencer, 29 Kliffen Place, Halifax, HX3 0AL
M3 OPD K Spencer, 29 Kliffen Place, Halifax, HX3 0AL
M3 OPG Paul Reed, 32 London Road, Warmley, Bristol, BS30 5JH
M3 OPM J Newby, 22 Acton Road, Liverpool, L32 0TT
M3 OPN J Shemwell, 4 Darvel Close, Bolton, BL2 6UD
M3 OPS A Hindle, 41 Seedfield, Staveley, Kendal, LA8 9NJ
M3 OPT David Wicks, 148 Long Lane, Staines, TW19 7AJ
M3 OPU K Morris, 80 Bridge Street, Chatteris, PE16 6RN
M3 OPV K Ashcroft, 9 Aldermere Crescent, Urmston, Manchester, M41 8UE
M3 OPW H Foot, 1 South View, Piddletrenthide, Dorchester, DT2 7QS
M3 OPX A Robinson, 75 De La Pole Avenue, Hull, HU3 6RD
M3 OQD H Nehmzow, 26 Woodlands, Colchester, CO4 3JA
M3 OQG A Chruscinski, 39 Sherwood Rise, Mansfield Woodhouse, Mansfield, NG19 7NP
M3 OQH Dennis Cooper, 52 Meadow Lane, Birkenhead, CH42 3YE
M3 OQI Graham Woodward, 5 Barnard Road, Chelmsford, CM2 8RR
M3 OQJ B Male, 44 Lakefields, West Coker, Yeovil, BA22 9BT
M3 OQK Gary Williamson, 4b Havelock Place, Bridlington, YO16 4JN
M3 OQL S Davin, 39 Nags Head Hill, Bristol, BS5 8LN
M3 OQQ T Garvey, 21 Oak Park Drive, Havant, PO9 2XE
MM3 OQR Paul Taylor, 23 Linksfield Gardens, Aberdeen, AB24 5PF
M3 OQS Brian Daley, 129a Kingsway South, Warrington, WA4 1RW
MM3 OQV J Campbell, 6 Dunard Court, Carluke, ML8 5RX
M3 OQZ P Hall, 13 Sheard Avenue, Ashton-under-Lyne, OL6 8DS
M3 ORB A Goold, 6 The Elms, Kempston, Bedford, MK42 7JN
M3 ORE Gerald Beale, 34 Teville Road, Worthing, BN11 1UG
M3 ORL A Price, 67 Mansfield Road, Glapwell, Chesterfield, S44 5QA
M3 ORN O Newton, 84 Ameysford Road, Ferndown, BH22 9QB
M3 ORP J Marlow, West Bulthy, Bulthy, Welshpool, SY21 8ER
M3 ORQ Matthew Winch, 2 Cranleigh Gardens, Cowes, PO31 8AS
M3 ORT Bruce Williams, 3 Welton Close, Wilmslow, SK9 6HD
M3 ORU C Baker, 10 Kirton Close, Coventry, CV6 2PG
M3 ORV R Mccolm, 7 Southwell Street, Portland, DT5 2DP
MW3 ORY N Lewis, 1 Clyne Drive, Blackpill, Swansea, SA3 5BU
M3 ORZ George Marshall, 12 Arthur Avenue, Caister-on-Sea, Great Yarmouth, NR30 5PQ
M3 OSA J Ross, 142 Bridle Road, Croydon, CR0 8HJ
M3 OSC E Walker, 2 The Green, Blencogo, Wigton, CA7 0DF
M3 OSF J Cobbold, 2 The Green, Blencogo, Wigton, CA7 0DF
MW3 OSI Robert Johnson, 25 Lon Tyrhaul, Llansamlet, Swansea, SA7 9SF
MM3 OSK Andrew Twort, 17 Balallan, Isle of Lewis, HS2 9PN
M3 OSP P Moss, 26 Woodlands Avenue, Farnham, GU9 9EY
M3 OSQ Robbie Phillips, 188 Charston, Greenmeadow, Cwmbran, NP44 4LD
M3 OSS C Dell, 18 Greenacres, Fulwood, Preston, PR2 7DA
M3 OSU T Pollard, 35 Weatherall St. North, Salford, M7 4TH
M3 OSW K Sheldon, 35 Weatherall St. North, Salford, M7 4TH
M3 OSY W Shiu, Harrogate Ladies' College, Clarence Drive, Harrogate, HG1 2QG
M3 OTB T Jones, 10 Putnams Drive, Aston Clinton, Aylesbury, HP22 5HH
M3 OTE D Barlow, 7 Peter Street, Eccles, Manchester, M30 0JF
M3 OTG A Hill, 5 Park Road, Thurnscoe, Rotherham, S63 0TG
M3 OTI Daniel Scotcher, 17 St. Dominics Square, Luton, LU4 0UN
M3 OTM O Morris, 1 Crawford Avenue, Peterlee, SR8 5EG
M3 OTP R Taifford, 28 Paddock Wood, Prudhoe, NE42 5BJ
M3 OTQ T Chapman, 17 Trevor Road, Swinton, Manchester, M27 0YH
M3 OTR Scott Taylor, 49 Chestnut Avenue, West Drayton, UB7 8BU
M3 OTS G Rees, 32 Glencroft Close, Burton-on-Trent, DE14 3GJ
M3 OTU R Woolgar, 2 Solent Way, Milford on Sea, Lymington, SO41 0TE
M3 OTZ M Wright, 8 St. Wilfrids Road, Oundle, Peterborough, PE8 4NX
MW3 OUC J Bidwell, 26 Lone Road, Clydach, Swansea, SA6 5HR
M3 OUF R Crewe, 12 Dimple Gardens, Ossett, WF5 8LJ
M3 OUG C Rickwood, 7 Bromley Mount, Wakefield, WF1 5LB
M3 OUH Paul Loxton, 32 Parkhill Crescent, Wakefield, WF1 4EZ
M3 OUI P Marsh, 16 Laurel Close, North Warnborough, Hook, RG29 1BH
M3 OUL George Martin, 12 Poolside, Phase 1, St. Joseph, Trinidad and Tobago
MI3 OUN S Fulton, 120 Dunnalong Road, Bready, Strabane, BT82 0DP
M3 OUQ W Speak, 20 Pear Tree Drive, Wincham, Northwich, CW9 6EZ
M3 OUS E McFalls, 119 Park Avenue, Shelley, Huddersfield, HD8 8JZ
M3 OUU Robert Hobbs, 11 Woodview, Hoath, Canterbury, CT3 4LD
M3 OUV S Clarke, 49 Torr View Avenue, Plymouth, PL3 4QN
M3 OVA Tony Durant, 2 Tamar Way, North Hykeham, Lincoln, LN6 8TZ
M3 OVC Paul Goodburn, 2 Sunray Cottages, Holt Street, Dover, CT15 4HZ
M3 OVE M Love, 9 Firswood Drive, Swinton, Manchester, M27 5QY

M3 **OVF** M Clay-Burley, 90 Station Road, Hednesford, Cannock, WS12 4DL
M3 **OVG** Thomas Speak, 20 Pear Tree Drive, Wincham, Northwich, CW9 6EZ
M3 **OVM** Kevin Nelson, 40 Staunton Road, Newark, NG24 4EX
M3 **OVO** A Howe, Southern Point, Grange View, Houghton le Spring, DH4 [illegible]
M3 **OVT** J Jones, 18 Ffordd [illegible], Caernarfon, LL55 2EP
MM3 **OVV** S Monaghan, 13 Ballyhennan Crescent, Tarbet, Arrochar, G83 7DB
M3 **OVX** J Alexander, 14 Barlow Road, Stretford, Manchester, M32 0RG
M3 **OVZ** D Smith, 21 Station Drive, Wisbech St. Mary, Wisbech, PE13 4RX
M3 **OWF** A Holter, 1 Glynde Court, Westfield Close, Polegate, BN26 6EE
M3 **OWN** O Dixon, 28 Manchester Road, Audenshaw, Manchester, M34 5GB
M3 **OWO** Matt Middleditch, 8 Royal Close, Yeovil, BA21 4NX
M3 **OWQ** P Mcspirit, 26 Horridge Avenue, Newton-le-Willows, WA12 0AS
M3 **OWU** David Adlam, 31 Coxons Close, Huntingdon, PE29 1TS
M3 **OWZ** S Waller, 17 Vere Road, Peterborough, PE1 3DZ
M3 **OXB** B Rodriguez, Sprouston House, Newtown St. Boswells, Melrose, TD6 0RY
M3 **OXD** Cristian Romocea, 21 Hurst Lane, Cumnor, Oxford, OX2 9PR
M3 **OXN** P JODRELL, 2 Greggs Avenue, Chapel-en-le-Frith, High Peak, SK23 9TU
MM3 **OXQ** S McKinnon, 8 Rowanlea Avenue, Paisley, PA2 0RP
M3 **OXV** J Evans, 4 Birchwood Close, Rhostyllen, Wrexham, LL14 4DD
M3 **OXY** O Bazar, 1 Claremont Road, London, NW2 1BP
MM3 **OYB** S Morgan, 23 Duncan Road, Glenrothes, KY7 4HS
M3 **OYC** S Poyser, 80 Derby Road, Long Eaton, NG10 4LB
M3 **OYE** C Windsor, 44 Paragon Place, Norwich, NR2 4BL
M3 **OYJ** C Rose, 32 Hobart Place, Thornton-Cleveleys, FY5 3DQ
MM3 **OYL** Andrew Shearman, 4 Millbrae Crescent, Clydebank, G81 1EH
M3 **OYN** R Watson, 60 Beresford Avenue, Surbiton, KT5 9LJ
MI3 **OYP** C Bronnan, 1 Ballyscullion Lane, Bellaghy, Magherafelt, BT45 8NQ
M3 **OYQ** Noel Loughran, 22 Edulf Road, Borehamwood, WD6 5AD
M3 **OYR** Anthony Kirby, 36 Baron Street, Darwen, BB3 1NP
M3 **OYS** A Parkes, 52/53 Tonning Street, Lowestoft, NR32 2AN
M3 **OYU** N Carr, 6 Baldwin Avenue, Eastbourne, BN21 1UJ
M3 **OYW** J Waugh, 9 Kedar Bank, Mouswald, Dumfries, DG1 4LU
M3 **OYZ** Michael Ward, 50 St. Wilfrids Road, Doncaster, DN4 6AD
M3 **OZB** J Robinson, 34 High St., Dragonby, Scunthorpe, DN15 0BE
M3 **OZC** H Derbyshire, 12 Trinity Homes, St. Clare Road, Deal, CT14 7PX
M3 **OZD** Robert Pounder, 65 Stubsmead, Swindon, SN3 3TB
M3 **OZE** J Baldry, 160 Rover Drive, Castle Bromwich, Birmingham, B36 9LL
M3 **OZH** M Flynn, 20 Manwood Avenue, Canterbury, CT2 7AH
M3 **OZI** S Bird, 9 Almery Drive, Carlisle, CA2 4EX
MI3 **OZK** Joseph Sills, 145 Ballycolman Estate, Strabane, BT82 9AJ
M3 **OZN** P Davies, 92 Thirlmere Road, Hinckley, LE10 0PF
MI3 **OZT** R Hepburn, 34 Pinewood Crescent, Claudy, Londonderry, BT47 4AD
MM3 **OZU** G Craig, 1 Butt Avenue, Helensburgh, G84 9DA
MM3 **OZW** W Pauley, 43 Pringle Avenue, Tarves, Ellon, AB41 7NZ
M3 **OZY** O Morris, 44 Leamington Road, Weymouth, DT4 0EZ
M3 **PAE** C Hume, Sundhopeburn, Yarrow, Selkirk, TD7 5NF
M3 **PAI** M Norton, Springfield, Back Lane, Kingston, Sturminster Newton, DT10 2DT
M3 **PAP** Joan Parrott, 2 Boyd Close, Wirral, CH46 1RX
M3 **PAU** P Laing, 5 Talisman Close, Barrow-in-Furness, LA14 2UT
M3 **PAX** Nathan Haigh, 10 Moor Park Gardens, Dewsbury, WF12 7AS
M3 **PBA** P Alce, 1/2 Arawa street, Christchurch, New Zealand, 8013
M3 **PBB** R Bannon, 18 Clavell Road, Liverpool, L19 4TR
M3 **PBE** R Rudd, 11 Woodlands Way, Lepton, Huddersfield, HD8 0JA
M3 **PBK** B Kellner, 95 Shakespeare Road, Ipswich, IP1 6ET
M3 **PBP** D Parsons, Barn Owl Cottage, Stoke St. Mary, Taunton, TA3 5BY
M3 **PBQ** M Hopkins, 15 Station Drive, Dalbeattie, DG5 4FA
M3 **PBR** Trevor Cumming, 2 Ash Grove, Perth Street, Hull, HU5 3PF
M3 **PBU** Wayne Wilkinson, 35, Fitzgerald Court, Haughton Green, M34 7LB
MW3 **PBV** Spencer Williams, 11 Beyron Clos, Pontypridd, CF37 5HW
M3 **PBW** Paul Roberts, 7 Boscombe Road, Swindon, SN25 3EZ
M3 **PCC** P Crossley, Firpark Farm, Fir Park, Market Rasen, LN8 3YL
MI3 **PCF** P Ford, 25 Carnhill, Londonderry, BT48 8BA
M3 **PCP** Paul Papper, 50 Lincoln Road, Stevenage, SG1 4PL
M3 **PCQ** J Campbell, 22 Horsewhim Drive, Kelly Bray, Callington, PL17 8GL
M3 **PCW** A Maxwell, Tysties, Tile Barn, Newbury, RG20 9UY
M3 **PCX** G Bellis, 70 Osborne St., Rhos, Wrexham, LL14 2HT
MM3 **PDC** P Cooper, Ambleside, Lauriston, Montrose, DD10 0DJ
M3 **PDD** P Bennett, 94 Queensway, Taunton, TA1 5QT
M3 **PDE** Paul Eckersley, 18 Mulberry Court, Guildford, GU4 7EQ
M3 **PDG** E Aitken, 20 Plover Drive, Bury, BL9 6JH
M3 **PDH** D Hutton, 57 Pheasant Wood Drive, Thornton-Cleveleys, FY5 2AW
M3 **PDK** James Fuller, Rosemar Lodge, Westford, Wellington, TA21 0DX
MI3 **PDL** Patrick Burns, 25 Orchard Road, Strabane, BT82 9QS
MM3 **PDM** Peter McKay, 7 Buchanness Drive, Boddam, Peterhead, AB42 3AT
MI3 **PDN** R Neill, 84 Carnreagh, Craigavon, BT64 3AN
M3 **PDP** J Clarkson, 56 Edward Bailey Close, Binley, Coventry, CV3 2LZ
M3 **PDU** Leslie Fuller, Rosemar Lodge Westford, Wellington, TA21 0DX
M3 **PDY** Paul Dyer, 19 Church Road, Evesham, WR11 2NE
MW3 **PEH** Brian Sellers, 66 St. John Street, Ogmore Vale, Bridgend, CF32 7DD
M3 **PEQ** M vincent, 8 Waldemar Park, Norwich, NR6 6TD
MD3 **PER** S Perry, 1 Cronk Grianagh Estate, Strang, Douglas, Isle of Man, IM4 4QP
MM3 **PEV** R Stevenson, 17 Springbank Gardens, Lawthorn, Irvine, KA11 2BY
MM3 **PEY** D Oates, 14 Craighlaw Avenue, Eaglesham, Glasgow, G76 0EU
MM3 **PFA** R Maddock, 21 Marywell Brae, Kirriemuir, DD8 4BJ
M3 **PFE** Richard Laverick, 55 Bondicar Terrace, Blyth, NE24 2JW
M3 **PFF** A Fisher, 17 Spicers Way, Totton, Southampton, SO40 9AX
M3 **PFK** P Fisk, 6 Piccadilly Square, Burnley, BB11 4QG
M3 **PFL** K Gallagher, Flat 1, 59 Trinity Road, Bridlington, YO15 2HF
M3 **PFM** Paul Morris, Canary Cottage, Eye Road, Eye, IP23 7JX
M3 **PFN** John Fisk, The Cottage In The Croft, The Croft, Norwich, NR8 5DT
M3 **PFU** S Silver, 2 Brandon Close, Grange Park, Swindon, SN5 6AA
M3 **PFY** P Murthwaite, 34 Cambridge Street, Bridlington, YO16 4JZ
M3 **PGB** DT Brierley, 639 Borough Road, Birkenhead, CH42 9QA
M3 **PGD** P Danvers, Heath Farm, Aylsham Road, North Walsham, NR28 0JP
M3 **PGH** P Howell, 16 Everard Road, Bedford, MK41 9LD
M3 **PGI** G Pollard, 24 Terminus Road, Littlehampton, BN17 5BX
M3 **PGK** Allan Farrar, 8 Wensley Street, Thurnscoe, Rotherham, S63 0PX
M3 **PGL** P Lockwood, 6 Pounteys Close, Middleton St. George, Darlington, DL2 1LF
MW3 **PGN** A Gazi, 51 Cyncoed Road, Cardiff, CF23 5SB
M3 **PGV** P Goodchild, 577 Parrs Wood Road, East Didsbury, Manchester, M20 [illegible]
M3 **PGS** Peter Stevenson, 6 Dighton Gate, Stoke Gifford, Bristol, BS34 8XA
M3 **PGU** G Farrar, 174 Houghton Road, Thurnscoe, Rotherham, S63 0SA
M3 **PGY** C Graham, 19 Pontop View, Rowlands Gill, NE39 2JP
MM3 **PHC** T Given, 26 Campbell Court, Cumnock, KA18 1NP
M3 **PHF** J Austin, 66 Homewood Avenue, Sittingbourne, ME10 1XJ
M3 **PHG** P Greenway, 26 Coleridge Gardens, Burnham-on-Sea, TA8 2QA
M3 **PHJ** P Malone, Wilmer, Richards Lane, Llogan, TR16 4DQ
M3 **PHO** J Wilson, 448 Hythe Road, Willesborough, Ashford, TN24 0JH
M3 **PHP** P Goodhall, 2 Manor Place, Oxford, OX1 3UN
M3 **PHQ** Nicholas benes, 22 Park Lane, High Ercall, Telford, TF6 6BA
M3 **PHR** P Norman, 22 Stirling Street, Hull, HU3 6SL
M3 **PHS** P Saben, Tredinneck Moor, Newmill, Penzance, TR20 8XT
M3 **PHX** C Duffill, 181 Foden Road, Great Barr, Birmingham, B42 2EH
M3 **PHZ** Alan Billings, 46 Thorley Drive, Cheadle, Stoke-on-Trent, ST10 1SA
M3 **PIA** G Mears, 59 Hastoe Park, Aylesbury, HP20 2AB
M3 **PIH** D Huckle, 1 Glebe Road, Biggleswade, SG18 0PE
M3 **PIK** Michael Gould, 57 Fowler Close, Leicester, LE4 0SF
M3 **PIL** R Harvey, 50 Warren Walk, Ferndown, BH22 9LY
M3 **PIO** T Nakagawa, 7 Milton Street, Barrowford, Nelson, BB9 6HE
M3 **PIQ** Thomas Bourne, 100 Dimsdale View West, Newcastle, ST5 8EL
M3 **PIW** J Witchell, 43 Elm Way, Shepton Mallet, BA4 5JX
M3 **PIY** D Dewsbury, 62 Yew Tree Drive, Leicester, LE3 6PL
M3 **PJG** P Goodayle, 2 Downs Road, Seaford, BN25 4QL
M3 **PJI** Paul Jones, 63 Regent Street, Rotherham, S61 1HW
M3 **PJJ** P Johnson, 7 Harrington Court, Hertford Heath, Hertford, SG13 7QT
MI3 **PJM** P McCausland, 31 Oakleigh Fold, North Street, Craigavon, BT67 9BS
M3 **PJN** P Northover, 66 Howard Drive, Letchworth Garden City, SG6 2DQ
M3 **PJS** P Seabrook, 29 Gadby Road, Sittingbourne, ME10 1TJ
MW3 **PJU** M Sawford, 62 Heol Briwnant, Cardiff, CF14 6QH
M3 **PJV** Peter Vipond, The Old Forge, Nentsbury, Alston, CA9 3LH
MW3 **PKC** L Hill, 25 Heol Cae Derwen, Bargoed, CF81 8QB
M3 **PKE** R North, 11 Tintagel Close, Keynsham, Bristol, BS31 2NL
M3 **PKH** Connie Bell, 12a Mill Lane, Carlton, Goole, DN14 9NG
M3 **PKL** Barbara North, 11 Tintagel Close, Keynsham, Bristol, BS31 2NL
M3 **PKM** P Bliss, 6 Jubilee Gardens, Biggleswade, SG18 0JW
M3 **PKQ** J Blackburn, 64 Marsh Lane, Birmingham, B23 6PJ
M3 **PKU** Keven Kenton, 24 Penygraig Road, Brymbo, Wrexham, LL11 5AD
M3 **PKZ** M Woolley, 84 Bowthorpe Road, Norwich, NR2 3TP
M3 **PLB** P Beier, 20 Markham Avenue, Armthorpe, Doncaster, DN3 2AZ
MM3 **PLC** Adrian Greig, 22 Dothan Road, Kirkcaldy, KY2 6GZ
M3 **PLI** Paul Lister, 73 Seabrook Court, Seabrook, Hythe, CT21 5RY
M3 **PLN** P Lewin, The Hawthorns, Hawthorne Drive, Stafford, ST19 9NQ
M3 **PLP** P Price, 4 Priory Avenue, North Ferriby, HU14 3AE
M3 **PLU** Tammy Colman, Edison House, Bow Street, Great Ellingham, NR17 1JB
M3 **PLV** C McCollum, 25 Byron Road, Locking, Weston-Super-Mare, BS24 8AG
M3 **PMI** J Segrove, 87 Henry Road, West Bridgford, Nottingham, NG2 7ND
M3 **PMK** P Kidd, 78 Studfield Road, Sheffield, S6 4SU
M3 **PML** P Lines, 56 Old Hall Close, Amblecote, Stourbridge, DY8 4JQ
M3 **PMN** P Nicholls, 30 Nailbourne Court, Palm Tree Way, Lyminge, CT18 8LX
M3 **PMO** P Brindle, 5 Showfield Close, Sherburn in Elmet, Leeds, LS25 6LW
MI3 **PMR** R Catney, 32 Cairndore Avenue, Newtownards, BT23 8RF
M3 **PMU** D Jones, 34 Pen y Bryn, Rassau, Ebbw Vale, NP23 5AJ
MI3 **PMW** M Pollock, 5 St. Marys Terrace, Stream Street, Newry, BT34 1HL
M3 **PMX** A Marlow, 66 Woodborough Road, Winscombe, BS25 1BA
M3 **PMY** Trevor Wright, 11 Ash Close, Daventry, NN11 0XH
M3 **PNA** Neil Phillpott, Eescroft, Stombers Lane, Folkestone, CT18 7AP
M3 **PNB** Ben Crosswell, 5 Harty Ferry View, Whitstable, CT5 4TE
M3 **PNF** G Taylor, 31 Ashfurlong Crescent, Sutton Coldfield, B75 6EN
M3 **PNH** N Hoyle, 34 The Drive, Halifax, HX3 8NJ
M3 **PNI** E Bishop, 40 Auburn Grove, Blackpool, FY1 5NJ
M3 **PNO** Carl Turner, 28 Fox Lea, Kesgrave, Ipswich, IP5 2YU
M3 **PNR** R Williams, Plaen Cottage, Bodfari, Denbigh, LL16 4BS
M3 **PNV** J Nelmes, 118 Silcoates Lane, Wrenthorpe, Wakefield, WF2 0PE
M3 **PNY** P Robinson, 19 St. Wilfrids Crescent, Brayton, Selby, YO8 9EU
M3 **PNZ** Robert Maddock, 48 Collygree Parc, Goldsithney, Penzance, TR20 9LY
MI3 **POB** P O'Brien, 71 Whitepark Road, Ballycastle, BT54 6LP
M3 **POH** D Dean, 119 Queens Drive, Newton-le-Willows, WA12 0LN
MM3 **POI** T Penna, North Windbreck, Deerness, Orkney, KW17 2QL
M3 **POP** J Morris, Lurdin Lodge, 71 Lurdin Lane, Wigan, WN6 0AQ
M3 **POQ** Richard Finch, 19b Kiln Road, Newbury, RG14 2LS
M3 **POV** N Trangmar, 8 Maxstoke Close, Meriden, Coventry, CV7 7NB
M3 **POW** M Davies, 5 Twyford Avenue, Great Wakering, Southend-on-Sea, SS3 0EZ
MM3 **PPA** Sam Parsons, Linksview, Barrock, Thurso, KW14 8SY
MI3 **PPD** R Reilly, 220 Ardmore Road, Londonderry, BT47 3TE
M3 **PPG** John Godfrey, 4 Cherry Close, Houghton Conquest, Bedford, MK45 3LQ
MI3 **PPI** P Pearson, 1 Rock Cottages, Springwell Road, Bangor, BT19 6LZ
M3 **PPK** N Green, 11 Wythburn Way, Rugby, CV21 1PZ
M3 **PPO** A Wade, 40 Throxenby Lane, Scarborough, YO12 5HW
M3 **PPQ** P Sharrock, 31 Burland Grove, Winsford, CW7 2JE
M3 **PPR** P Saving, Room 1, Abbeyfield, The Globe Field, Sevenoaks, TN13 3DR
M3 **PPU** Mathew Whitten, 75 Regent Street, Whitstable, CT5 1JQ
M3 **PPY** J Evans, 21 Quilter Close, Bilston, WV14 9AX
M3 **PPZ** R Bullen, 67 Abberley Road, Liverpool, L25 9QY
M3 **PQB** Donald Fagg, 62 Hawkins Road, Folkestone, CT19 4JA
M3 **PQE** Adrian Paffey, 1 St. Vincent Road, Newport, NP19 0AN
M3 **PQF** E Paffey, 1 St. Vincent Road, Newport, NP19 0AN
M3 **PQG** J Goldfinch, 138 Palmerston Road, Chatham, ME4 5SJ
M3 **PQI** M Bhatia, Swaynes House, Room 2 Flat 2, Wirenhoe Park, CO4 3SQ
M3 **PQJ** J Pratt, 4 Bussex Square, Westonzoyland, Bridgwater, TA7 0HD
M3 **PQL** A Mellor, 112 Allerton Road, Stoke-on-Trent, ST4 8PL
MI3 **PQM** P Millar, 37 Thorncroft, Ahoghill, Ballymena, BT42 1RX
M3 **PQN** A Phillips, 15 Hertford Close, Woolston, Warrington, WA1 4EZ
M3 **PQQ** Robbie FERN, 3 Park Road, Featherstone, Wolverhampton, WV10 7HS
M3 **PQS** Eric Curling, 919 Oxford Road, Tilehurst, Reading, RG30 6TP
M3 **PQT** S Broadbent, 86 Inverness Road, Dukinfield, SK16 5AB
M3 **PQU** Aaron Milton-Eldridge, 2 Partridge Close, Didcot, OX11 6AB
M3 **PQV** M Nolan, [illegible], Potterne, Devizes, SN10 5NJ
M3 **PRA** P Ramsay, 18 Sherwood Avenue, Abingdon, OX14 3NL
M3 **PRN** S Martin, 14 Mount Road, Thatcham, RG18 4LA
M3 **PRS** G Manchester, 251 Osmaston Park Road, Allenton, Derby, DE24 8DA
M3 **PRU** P Broadbere, 65 Bramwell Avenue, Prenton, CH43 0RQ
M3 **PRY** S Mariott, 4 Stone Cross Gardens, Catterall, Preston, PR3 1YQ
M3 **PRZ** P Radmall, Appleford, Bowcombe Road, Kingsbridge, TQ7 2DJ
M3 **PSB** S Birch, 6 Crescent Road, Wallasey, CH44 0BQ
M3 **PSC** P Cattel, 21 School Hill, Chickerell, Weymouth, DT3 4BA
M3 **PSD** Paul Sheppard, 107 Queen Street, Swinton, Mexborough, S64 8NF
M3 **PSE** Paul Elliott, 11 Forgefields, Herne Bay, CT6 7TB
M3 **PSF** Robert Baroch, 29 Salters Lane, Redditch, B97 6JY
M3 **PSI** K Morton, 47 Trinity Court, Halstead, CO9 1PY
MM3 **PSL** P Leech, The Croft House, 9, Ruilick, Beauly, IV4 7AB
M3 **PSO** W Weaver, Challacombe House, Perrinpit Road, Bristol, BS36 2AT
M3 **PSR** Pauline Roberts, 43 Ashbourne Crescent, Sale, M33 3LQ
M3 **PSS** Philip Swanepoel, Bridge Cottage, Tinhay, Lifton, PL16 0AH
M3 **PSU** Jo Davidson, 5 Hanover Parc, Indian Queens, St. Columb, TR9 6ER
M3 **PSZ** A Laurence, Brookvale, Nooklands, Preston, PR2 8XN
M3 **PTA** Paul Chambers, 257 Kings Acre Road, Hereford, HR4 0SR
M3 **PTB** T Bishop, 4 Walnut Grove, Worlington, Bury St. Edmunds, IP28 8SF
M3 **PTG** T Green, Huntley, Chesham Road, Tring, HP23 6HH
M3 **PTI** B Parton, 51 Marston Grove, Stoke-on-Trent, ST1 6EF
M3 **PTQ** W Hughes, 9 North Brook Close, Greetham, Oakham, LE15 7SD
M3 **PTR** P Dryden, 27 Delaval Crescent, Blyth, NE24 4AZ
MM3 **PTS** P Stronach, 76 Lilac Grove, Inverness, IV3 5RE
M3 **PTV** B Fitzakerley, 38 Hazel Grove, Armthorpe, Doncaster, DN3 3HG
M3 **PTX** Christine Hunter, 14 Southdown Way, Storrington, Pulborough, RH20 3NS
M3 **PUB** Pauline Swynford, 6 The Rise, Cold Ash, Thatcham, RG18 9PD
M3 **PUE** A Hannon, 8 Circular Road West, Liverpool, L11 1AZ
MI3 **PUH** J Dunlop, 118 Ardenlee Avenue, Belfast, BT6 0AD
M3 **PUI** T Chappelow, 12 Topcliffe Court, Morley, Leeds, LS27 8UG
M3 **PUL** P Stead, 36 Reeds Avenue East, Wirral, CH46 1RQ
M3 **PUN** John Rideout, 4 Treetops, Northampton, NN3 8XA
M3 **PUQ** J Hunt, 104 Hamilton Avenue, Sutton, SM3 9RL
M3 **PUT** Chris Townsend, 2 Netherfield Drive, Netherthong, Holmfirth, HD9 3ES
M3 **PUU** B Hill, 97 Maesglas Grove, Newport, NP20 3DN
M3 **PUZ** Susan Turford, 1 Portland Crescent, Bolsover, Chesterfield, S44 6EG
M3 **PVB** T Ireland, 114 Alder Lane, Warrington, WA2 8AW
M3 **PVC** M Cook, 9 Drenewydd, Park Hall, Oswestry, SY11 4AH
M3 **PVI** P Handley, 97 Applegarth Avenue, Guildford, GU2 8LX
M3 **PVP** A Botley, Flat 1/B, 46 Trull Road, Taunton, TA1 4QH
M3 **PVQ** E Rhodes, The Old Forge, Stoke Gabriel, Totnes, TQ9 6RL
M3 **PVU** J Watson, 20 St. Marys Gardens, Hilperton Marsh, Trowbridge, BA14 7PG
M3 **PVV** T Crisp, 6 Tumlins, All Cannings, Devizes, SN10 3PQ
M3 **PVX** Darren Jones, 23 Earnshaw Street, Hollingworth, Hyde, SK14 8PE
M3 **PWE** P Ward, 69 Woodlands Avenue, Tadcaster, LS24 9HP
M3 **PWK** S Platts, 59 Sea View Road, Drayton, Portsmouth, PO6 1EW
M3 **PWL** P Lane, 70 West End Road, Epworth, Doncaster, DN9 1LB
M3 **PWM** P Mitchell, 13 Ashorne Close, Matchborough, Redditch, B98 0EY
M3 **PWO** D Robertson, 53 Moor Lane, Weston-Super-Mare, BS22 6RA
M3 **PWS** P Sykes, 2 Thornton Villas, Barrow Road, Barrow-upon-Humber, DN19 7QG
M3 **PWW** Paul Wright, 16 Hainault Avenue, Giffard Park, Milton Keynes, MK14 5PA
M3 **PWZ** B PEARSON, 28 Hanmer Way, Staplehurst, Tonbridge, TN12 0PA
M3 **PXE** K Peel, 123 Cunningham Road, Tamerton Foliot, Plymouth, PL5 4PU
M3 **PXF** T Gabriel, 57 West Down Road, Delabole, PL33 9DT
MM3 **PXG** S Simpson, 9 Finavon Place, Dundee, DD4 9DZ
M3 **PXK** R Ellery, No 14 Roughtor View, Planet Park, Delabole, PL33 9BX
M3 **PXL** P Houghton, 37 Cedar Avenue, Cottingham, HU16 4AL
MM3 **PXO** E Mccook, 6 Elms Place, Stevenston, KA20 4EF
M3 **PXP** M Williams, 9 Clarence Place, Stonehouse, Plymouth, PL1 3JN
M3 **PXQ** Nicky Kendall, 19 Clowance Lane, Mount Wise, Plymouth, PL1 4HU
M3 **PXT** P Mutavdzic, 1 Hawthorne Drive, Kingwood, Henley-on-Thames, RG9 5WE
M3 **PXU** V Parton, 51 Marston Grove, Stoke-on-Trent, ST1 6EF
M3 **PXW** Barry Smith, 25 Lancing Road, Ellesmere Port, CH65 5BB
M3 **PXY** Camilla Fox, 45 Park Road, Wivenhoe, Colchester, CO7 9LS
M3 **PXZ** Chris Fox, 45 Park Road, Wivenhoe, Colchester, CO7 9LS
MD3 **PYB** D Furlong, 6a Glebe Avenue, Ruislip, HA4 6QZ
M3 **PYD** Michael Smith, 6 Neeps Terrace, Middle Drove, Wisbech, PE14 8JT
M3 **PYG** Antony M Webb, 104 Birds Nest Avenue, Leicester, LE3 9ND
M3 **PYI** W McBain, Willow Cottage, Gedney Broadgate, Spalding, PE12 0DE
M3 **PYJ** S Randall, 23 Onslow Road, Plymouth, PL2 3QG
M3 **PYO** David Horner, 21 Ainsworth Road, Little Lever, Bolton, BL3 1RG
M3 **PYR** P Rushby, 16 Foxhill Lane, Selby, YO8 9AR
M3 **PYS** Edward Ransom, 10 Gillercomb, Redcar, TS10 4SG
M3 **PYT** Claire Harris, 23 Washbourne Close, Plymouth, PL1 4ST
M3 **PYW** B Simmonds, 55 Pepys Road, St. Neots, PE19 2EN
MM3 **PYX** D Oakden, 9 Forthview Terrace, Edinburgh, EH4 2AE
M3 **PYY** C Thomas, 22 Sea Road, Abergele, LL22 7BU
M3 **PZC** M Marston, The Retreat, The Catch, Holywell, CH8 8DU
M3 **PZF** J Bealey, 17 Chelston Road, Newton Abbot, TQ12 2NN
MM3 **PZJ** Scott Ling, Leadburnlea, Leadburn, West Linton, EH46 7BE
M3 **PZL** R McKenzie, 4 Simpkin Street, Abram, Wigan, WN2 5QD
M3 **PZN** L Mckenzie, 4 Simpkin Street, Abram, Wigan, WN2 5QD
M3 **PZO** Sean Connor, 8 Bro Arfon, Upper Llandwrog, Caernarfon, LL54 7BH
MI3 **PZV** C Stewart, 99 Hillmount Road, Cullybackey, Ballymena, BT42 1NZ
M3 **PZX** P Seabrook, 29 Gadby Road, Sittingbourne, ME10 1TJ
M3 **RAA** Amanda Gordon, 5 Parc Hendy, Mold, CH7 1TH
M3 **RAE** John Webb, 6 Chatsworth Avenue, Fleetwood, FY7 8EG
M3 **RAK** Roy King, 79 Holmside Avenue, Minster on Sea, Sheerness, ME12 3EZ
M3 **RAU** Thomas Rowlands, 3, POOL, Llanfairfechan, LL330TN
M3 **RBF** D Gemmell, 36 Church Street, Dumfries, DG2 7AS
M3 **RBI** R Gilbert, 61 Coltstead, New Ash Green, Longfield, DA3 8LN
MM3 **RBJ** B Johnston, 71 Upper Mastrick Way, Aberdeen, AB16 5QG

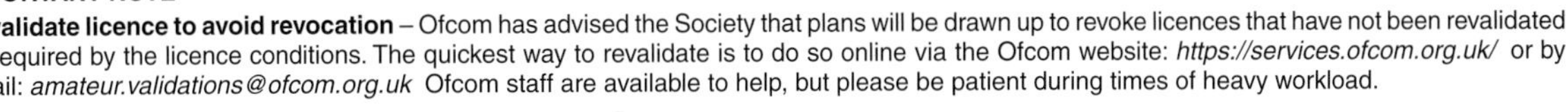

**IMPORTANT NOTE**

**Revalidate licence to avoid revocation** – Ofcom has advised the Society that plans will be drawn up to revoke licences that have not been revalidated as required by the licence conditions. The quickest way to revalidate is to do so online via the Ofcom website: *https://services.ofcom.org.uk/* or by email: *amateur.validations@ofcom.org.uk* Ofcom staff are available to help, but please be patient during times of heavy workload.

MI3 RBM R Abraham, 9 Milfort Gardens, Waringstown, Craigavon, BT66 7PD
M3 RBP R Peacock, 27 Greenside, Kendal, LA9 5DU
M3 RBQ Colin Boarer, 37 The Martlets, Rustington, Littlehampton, BN16 2UB
M3 RBT R Kerr, The Dower House, Church Square, Derby, DE73 8JH
M3 RBU B Upton, 1 Sunningdale Close, Eastleigh, SO50 8PU
M3 RBX Rebecca Swynford, 6 The Rise, Cold Ash, Thatcham, RG18 9PD
M3 RCC N Prescott, 3 View Fields, Station Road, Doncaster, DN9 3AE
M3 RCD R Cooke, 22 Shepperton Close, Great Billing, Northampton, NN3 9NT
M3 RCE Robert Edwards, 15 Burghley Street, Bourne, PE10 9NS
M3 RCI K Steele, Flat 22, Bradgate Court, Staunton Avenue, Derby, DE23 1PR
M3 RCQ M Snowden, Amber Lights, Market Lane, Wisbech, PE14 7LT
MM3 RCR Jessica McMartin, 19 Bruce Street, Bannockburn, Stirling, FK7 8UF
M3 RCS R Swietlik, 3 Tarvin Close, Sutton Manor, St. Helens, WA9 4DL
M3 RCT Mark Russell, 107 Cambridge Road, Hitchin, SG4 0JH
M3 RCV R Treacher, 93 Elibank Road, London, SE9 1QJ
M3 RCW R Wiggins, 68 Beaconsfield Road, Burton-on-Trent, DE13 0NT
MM3 RCX Kieran Carroll, 32e Meadowburn Place, Campbeltown, PA28 6ST
MM3 RCZ A Conlon, Kilrae, Barrpath, Glasgow, G65 0EX
M3 RDA R Astbury, 12 Southall Road, Ashmore Park, Wolverhampton, WV11 2PZ
M3 RDH R Hastings, 43 Delmar Avenue, Leverstock Green, Hemel Hempstead, HP2 4LZ
MM3 RDP David Moore, 47 Lockhart Street, Germiston, Glasgow, G21 2AP
M3 RDS R Smith, 445 Flixton Road, Urmston, Manchester, M41 6JL
M3 RDV R Dewes, 31 Woodlea Avenue, Lutterworth, LE17 4TU
M3 RDW Rebecca Wells, 37 Water Meadows, Worksop, S80 3DF
M3 RDY Ricky Young, 4 Hammond Court, Mablethorpe, LN12 2EL
MI3 REA W Rea, 3 Carwood Way, Newtownabbey, BT36 5JT
M3 REJ R Jones, 13 Tir Estyn, Deganwy, Conwy, LL31 9PY
M3 REL R Lewis, 53 Harewood Crescent, Louth, LN11 0JD
M3 REM A Kernick, 40 Leyster Street, Morecambe, LA4 5NF
M3 REP A Wilkinson, 6 Humbledon View, Sunderland, SR2 7RX
M3 REQ Andrew Williamson, 25 Manor Road, Rugby, CV21 2SZ
M3 RET M Parkes, 12 Penderel Street, Walsall, WS3 3DX
M3 REX Stephanie Thompson, Rutland, Quaker Lane, Wirral, CH60 6RD
M3 REZ A Adkins, 91 Fernbank Road, Birmingham, B8 3LL
M3 RFF Peter Richardson, 31 Castlefields Drive, Brighouse, HD6 3XF
M3 RFG R Gray, Upper Bisterne Farmhouse, Bisterne, Ringwood, BH24 3BP
M3 RFH Roger Henderson, 9 Green Mead, South Woodham Ferrers, Chelmsford, CM3 5NL
M3 RFI K Hilton, 199a Sale Lane, Tyldesley, Manchester, M29 8PG
M3 RFK Robert Eardley, Bridge Cottage, Martin, Fordingbridge, SP6 3LD
M3 RFO J McCue, 40 Bradbury Road, Stockton-on-Tees, TS20 1LE
M3 RFQ D Brown, 38 Tamworth Road, Sutton Coldfield, B75 6DG
M3 RFR Stephen Fallows, 23 Howard Street, Burnley, BB11 4BJ
M3 RFW R Ford, 30 Cartmel Close, Worcester, WR4 9NT
MW3 RFX R Ashworth, The Vicarage, Crymych, SA41 3RN
M3 RGC R Cummings, Juan Rodriguez El Cusques, No. 25 (plot 20a), Alicante, Spain
M3 RGD R Hogben, The Steppes, Presteigne Road, Knighton, LD7 1HY
M3 RGE M Tate, 52 Marlborough Road, London, N22 8NN
M3 RGG Jennifer Brown, 339 Manor Road, Brimington, Chesterfield, S43 1NU
MM3 RGH R Heath, 73 King Street, Inverbervie, Montrose, DD10 0RB
M3 RGJ R Jamieson, 3 Waterpark Road, Prenton Park, Birkenhead, CH42 9NZ
M3 RGK K Harley, 5 Saltrens Cottages, Monkleigh, Bideford, EX39 5JP
M3 RGN P Frampton, 118 Ramnoth Road, Wisbech, PE13 2JD
M3 RGP R Prangnell, 124 St. Marys Road, Cowes, PO31 7SR
M3 RGU Aaron Oxlade, 27 Spenfield Court, Northampton, NN3 8LZ
MM3 RGZ John Cairney, 5 James Street, Bannockburn, Stirling, FK7 0NQ
MM3 RHA William Hawthorn, 8 Drummond Place, Stirling, FK8 2JE
M3 RHB James Palmer, 2 Dagonet Road, Bromley, BR1 5LR
M3 RHG R Greatrix, 24 Berwick Drive, Cannock, WS11 1NS
MM3 RHH R Henry, 164 Auchenbothie Road, Port Glasgow, PA14 6JE
M3 RHI R Chalk, 42 Erskine Road, Colwyn Bay, LL29 8EU
M3 RHJ Mark Dennis, 10 Welland Court, Burton Latimer, Kettering, NN15 5ST
M3 RHK Surjit Bharrich, 8 Ferrers Ave, Tutbury, DE13 9JR
M3 RHL R Looker, 165 Mollison Drive, Wallington, SM6 9GX
M3 RHO R Frylinck, 46 Buckingham Road, Richmond, TW10 7EQ
M3 RHP Roy Montague, 71 Middlethorpe Road, Cleethorpes, DN35 9PP
M3 RHR Gary Kensett, 12 Rustics Close, Calvert, Buckingham, MK18 2FG
MM3 RHT G Fyfe, 7 Coralmount Gardens, Kirkintilloch, Glasgow, G66 3JW
M3 RIA Robert Wilkinson, 18 Green Road, Kendal, LA9 4QR
M3 RIE M Hatton, Elisha Cottage, St. Peters Walk, Hull, HU7 5FB
MI3 RIF James Smyth, 37 Ardfreelin, Newry, BT34 1JG
MI3 RIL L Scott, 4 Killycor Avenue, Claudy, Londonderry, BT47 4BX
M3 RIP M Pearce, 104 Sea Lane, Goring-by-Sea, Worthing, BN12 4PU
M3 RIU Mark Manser, 17 Emperor Way, Kingsnorth, Ashford, TN23 3QY
MI3 RIV Santhoshkumar Datchanamourty, 22 Craigmore Road, Bessbrook, Newry, BT35 6LF
MM3 RIX Rick Guthrie, 27 Meadowbank Road, Kirknewton, EH27 8BH
M3 RJB R Bird, 78 Arden Road, Hockley, Tamworth, B77 5JE
M3 RJF R Fitzgerald, 18 Humber Crescent, St. Helens, WA9 4HD
M3 RJH R Hicks, 31 Arundel Road, Great Yarmouth, NR30 4LD
M3 RJI Paul Cummings, 24 Spindle Road, Malvern, WR14 2WB
M3 RJK R Kelso, 55d Lewisham Hill, London, SE13 7PL
M3 RJO D Shaw, 31 Windwhistle Circle, Weston-Super-Mare, BS23 3TU
M3 RJP R Page, 156 High Road, Newton, Wisbech, PE13 5ET
M3 RKE K Simmons, 26 Red Hill Close, Studley, B80 7BZ
MM3 RKF Royston Mannifield, 2 Plewlands Avenue, Edinburgh, EH10 5JY
M3 RKJ James McColl, 6 Grenville Close, Bodmin, PL31 2FB
M3 RKK Edward Whiten, 17 Scott Close, Ashby-de-la-Zouch, LE65 1HT
M3 RKN R Neville, 4 Danson Gardens, Blackpool, FY2 0XH
M3 RKR R Rudd, 43 Greenlands Road, East Cowes, PO32 6HT
M3 RKV A Gallop, 19 Springfield Gardens, Deanshanger, Milton Keynes, MK19 6HX
M3 RKZ Peter Lewin, 12a Station Street, Chatteris, PE16 6NB
MI3 RLA A Holmes, 5 Cambrai Cottages, Belfast, BT13 3PS
M3 RLB A Marlborough, Maximillian Cottage, Manswood Common, Wimborne, BH21 5BH
MM3 RLG Matthew Geldart, 13b Greystone Place, Newtonhill, Stonehaven, AB39 3UL
M3 RLH Michael Thompson, 7 Kilsby Drive, Swindon, SN3 4EQ
M3 RLM M Rose, 71 Old Street, Ludlow, SY8 1NS
M3 RLO C Rodway, 11 Cleveland Avenue, Bishop Auckland, DL14 6AR
M3 RLS R Simms, 21 Hatch Lane, Old Basing, Basingstoke, RG24 7EA
M3 RLT Bernard Tarpey, 54 lowforce, wilnecote, Tamworth, B77 4LU
M3 RLX G Cox, 6 Bullfinch Close, Poole, BH17 7UP
M3 RMD M Rout, 2 Woods End Cottages, Kirby Bedon, Norwich, NR14 7EB
M3 RMG Geoffrey Chapman, Crockers Farm, Stoke Wake, Blandford Forum, DT11 0HF
M3 RMH R Hunt, 5 Rue de Wiltz, L-2734, Luxembourg-Bonnevoie, Luxembourg
M3 RMI J Salmon, 25 Helston Road, Chelmsford, CM1 6JF
M3 RMQ J Weston, 29 Langdale Road, Orrell, Wigan, WN5 0EB
M3 RMS Robert Stevenson, 97 Queen Street, Crewe, CW1 4AL
M3 RMU Andrew Teed, 21 Sheen Close, Salisbury, SP2 9PJ
M3 RMV Robin Ley, 23 Heronbridge Close, Westlea, Swindon, SN5 7DR
M3 RMX R Moore, 47 Darwin Road, Walsall, WS2 7EN
M3 RMZ P Randall, 289 Wilson Avenue, Rochester, ME1 2SS
M3 RNG Philip Davies, 1 Wroxhall Cottage, Oldwich Lane East, Kenilworth, CV8 1NR
M3 RNK Vanessa Penprase, 62 California Gardens, Plymouth, PL3 6SZ
M3 RNL R Van Den Bogaerde, Pilgrims End, Bullinghope, Hereford, HR2 8EB
M3 RNM M James, 7 Pixey Place, Oxford, OX2 8BB
MI3 RNN Rebecca Nicholl, 58 Dunnalong Road, Bready, Strabane, BT82 0DW
M3 RNO Ronald Baker, 12 Byland Road, Skelton-in-Cleveland, Saltburn-by-The-Sea, TS12 2NJ
M3 RNS Paul Rainey, 27 School Road, Silver End, Witham, CM8 3RZ
M3 RNU R Hargate, 79 Boundary Road, Beeston, Nottingham, NG9 2QZ
M3 RNW Robert Whitehead, 1 Easton Town Cottage Easton Town, Hornblotton, Shepton Mallet, BA4 6SG
M3 RNX Alfred Cleal, 38 Hazelwood Avenue, Bolton, BL2 3NR
M3 RNY R Kirk, 5 Sidcup Court, Southgate Way, Chesterfield, S43 2NR
M3 ROF D Johnson, 4 Armadale Close, Arnold, Nottingham, NG5 8RG
M3 ROI W Chorlton, 25 Ash Grove, Orrell, Wigan, WN5 8NG
M3 ROQ R Bird, 12 Windsor Road, Loughborough, LE11 4LL
M3 ROU Peter Curnow, 22 greengate close, wardle, Rochdale, OL129PX
MM3 ROV D Brown, Courtyard Cottage, Letters Farm, Argyll, PA27 8BX
M3 ROW F Webley, 2 Octavian Drive, Bancroft, Milton Keynes, MK13 0PN
M3 ROX I Davies, 2 Dinas Terrace, Aberystwyth, SY23 1BT
M3 RPA O Akanyeti, Sq/H8/4/A University Quays, Lightship Way, Colchester, CO2 8GY
M3 RPD I Handley, Rosedale, Chapman Street, Market Rasen, LN8 3DS
M3 RPE RP Evenden, 20 Sussex Road, Tonbridge, TN9 2TR
M3 RPF Robert Fullagar, 6 Locke Way, Stafford, ST16 3RE
M3 RPH R Hales, 8 Barton Close, Kingsbridge, TQ7 1JU
M3 RPK H Friberg, 19 Holmcroft, Newbiggin-by-The-Sea, NE64 6DQ
M3 RPQ J Faulkner, 3 Britannia Quay, 37 River Road, Littlehampton, BN17 5DB
M3 RPR R Roebuck, 50 Henson Avenue, Blackpool, FY4 3LY
M3 RPS P Cooper, 10 Headon Gardens, Exeter, EX2 6LE
M3 RPX A Davies, 24 Ash Lane, Mancot, Deeside, CH5 2BR
M3 RPZ Max Stokes, 15 Parc Terrace, Newlyn, Penzance, TR18 5AS
M3 RQB Raymond Wood, 1 Kildare Garth, Kirkbymoorside, York, YO62 6LN
MM3 RQC James Livingstone, 17 Livingstone Drive, Bo'ness, EH51 0BQ
M3 RQG Harry Smith, Ryefield, Windyknowe Road, Galashiels, TD1 1RG
M3 RQJ Monica Powell, 2 Walton Avenue, Twyford, Banbury, OX17 3LB
M3 RQO Daniel Rouse, 46 Frensham Drive, Bradford, BD7 4AS
MM3 RQP Frederick Pudsey, 21/2 Bathfield, Edinburgh, EH6 4DU
M3 RQQ Robin Baldwin, Flat H, 13 Clement Attlee Way, King's Lynn, PE30 4EJ
M3 RQR Les Robinson, 19 Adur Avenue, Shoreham-by-Sea, BN43 5NN
M3 RQW Laurence Lay, 17 Herbert Road, Hornchurch, RM11 3LD
M3 RQY James Colderwood, 34 Desborough Way, Norwich, NR7 0RR
MI3 RRE Robert Rantin, 8a Buchanans Road, Newry, BT35 6NS
M3 RRJ J Roughley, 42 Thistledown Close, Wigan, WN6 7PA
M3 RRN David Dunstan, 2 Trevarren Avenue, Four Lanes, Redruth, TR16 6NH
M3 RRU Geoffrey Clements, 9 Esgair y Gog, Bronllys, Brecon, LD3 0HY
M3 RRV Roy Taylor, 2 Chadwick Road, Moorends, Doncaster, DN8 4NG
MW3 RRW Gordon Tucker, 18 Plymouth Road, Penarth, CF64 3DH
M3 RRZ Steven Garrett, 44 Wardle Crescent, Leek, ST13 5PW
M3 RSH R Hodgkinson, 39 Oxford Road, Carlton-in-Lindrick, Worksop, S81 9BD
M3 RSN Carole Keeley, 3a St. Marks Road, Huyton, Liverpool, L36 0XA
MI3 RST J Donaldson, 12 Drumcrow Road, Glenanne, Armagh, BT60 2JQ
M3 RSX Ray Shippey, 43 Westbury Street, Bradford, BD4 8PB
M3 RTE R Turner, 2 Gate House Cottages, Hunton Road, Tonbridge, TN12 9SG
MM3 RTH William Fitzsimons, 34 Caledonian Road, Stevenston, KA20 3LG
M3 RTI R Brew, 45 Stephenson Road, Braintree, CM7 1DL
M3 RTP Sandra Crawford, 9 Cottage Homes, Wakefield Road, Huddersfield, HD5 9XT
M3 RTR S Davis, 104 Cairo Avenue, Peacehaven, BN10 7LA
M3 RTU Matthew Kidner, 4 Tonypistyll Road, Newbridge, Newport, NP11 4HJ
MW3 RUH David Thomas, 23 Merthyr Dyfan Road, Barry, CF62 9TG
M3 RUI R Wang, 86 Sunnyside Road, Beeston, Nottingham, NG9 4FG
M3 RUK Kevin Moody, 114 Acomb Road, York, YO24 4EY
M3 RUL C Rule, 109 Carshalton Park Road, Carshalton, SM5 3SJ
M3 RUO Sean Mcguinness, 64 Newshaw Lane, Hadfield, Glossop, SK13 2AT
M3 RUR Jill Stimpson, 2 Church Avenue, Kings Sutton, Banbury, OX17 3RJ
MI3 RUV Declan McCloskey, 1 Dernaflaw Cottages, Dernaflaw Road, Londonderry, BT47 4PP
M3 RUW Dave Jaynes, 81 Bude Crescent, Stevenage, SG1 2QL
MM3 RUZ G Ruzgar, 22 Ochil Terrace, Dunfermline, KY11 4BW
M3 RVE Gordon Thorpe, 81 knoll drive, Coventry, CV3 5PJ
M3 RVJ Dave Purser, 11 Barnards Close, Malvern, WR14 3NJ
M3 RVK Graeme Myall, 418 Chester Road, Warrington, WA4 6ES
M3 RVM David Morley, 5 Pelham Close, Westham, Pevensey, BN24 5NL
M3 RVN S Jones, 15 Corn Hill, Porthmadog, LL49 9AT
MW3 RVP Robin Latsbury, 57 Westbourne Road, Whitchurch, Cardiff, CF14 2BR
M3 RVQ Robert Lester, 17 Clarence Road, Capel-le-Ferne, Folkestone, CT18 7LW
M3 RVS Raymond Sohst, 2 Shaftesbury Drive, Maidstone, ME16 0JS
M3 RVX M Brandon, 9 Holly Drive, Winsford, CW7 1DZ
M3 RWC R Cornwall, 9 Bishop Close, Dunholme, Lincoln, LN2 3US
M3 RWD Rodney Davidson, 3 Eastridge Drive, Bishopsworth, Bristol, BS13 8HQ
M3 RWI John Marshall, 18 Dunnett Road, Folkestone, CT19 4BX
M3 RWK Jonathan Lecaille, Tezlan, Colton Road, Norwich, NR9 5BB
M3 RWN R Nock, 83 Coles Lane, West Bromwich, B71 2QW
M3 RWR D Griffiths, 1 Ballard Crescent, Dudley, DY2 9EZ
M3 RWV R Chown, 7 Foden Walk, Wilmslow, SK9 2HQ
M3 RWZ Richard Zieba, 14 Sisial Y Mor, Rhosneigr, LL64 5XB
M3 RXD Robert Dryburgh, 21 Glebe Close, Stow on the Wold, Cheltenham, GL54 1DJ
MI3 RXF Ian Smyth, 42 Mullintill Road, Claudy, Londonderry, BT47 4JN
M3 RXG Robert Jolly, 102 Swanstree Avenue, Sittingbourne, ME10 4LF
M3 RXH Roger Rimmer, 8 Greensward Close, Standish, Wigan, WN6 0RY
MW3 RXK Sidney Merrifield, 37 South View Drive, Rumney, Cardiff, CF3 3LX
MM3 RXM Robert May, 12 Clochbar Gardens, Milngavie, Glasgow, G62 7JP
M3 RXO Luke Milburn, 55 Hyde Heath Court, Crawley, RH10 3UQ
M3 RXP R Whittle, 20 Marlbrook Lane, Marlbrook, Bromsgrove, B60 1HN
M3 RXQ Michael Milne, Flambards, Manor Road, Dunmow, CM6 2JR
M3 RXT Jason Bridson, 10 Clegg Street, Astley, Manchester, M29 7DB
MI3 RXU I Ophert, 5 Cloghboy Road, Bready, Strabane, BT82 0DN
M3 RXW R Webb, 17 St. Marys Close, Chudleigh, Newton Abbot, TQ13 0PL
M3 RYA R Petts, 19 Sandwell Avenue, Darlaston, Wednesbury, WS10 7RH
MI3 RYD Stephen Davison, 60 cornation place, Craigavon, BT66 7AN
M3 RYG Ryan Hughes, 117 Liverpool Road, Irlam, Manchester, M44 6EH
M3 RYI S Ashcroft, 9 Aldermere Crescent, Urmston, Manchester, M41 8UE
MI3 RYJ James Hazlett, 25 Gorteen Crescent, Limavady, BT49 9EW
M3 RYN R Fowler, Ryland, Back Lane, Doncaster, DN9 3AJ
M3 RYO Mark Shasby, 19 Crawshaw Grange, Crawshawbooth, Rossendale, BB4 8LY
M3 RYR Robert Farrar, 41 Newtown Avenue, Cudworth, Barnsley, S72 8DY
M3 RYT Jason Abbott, 22 Brent Close, Witham, CM8 1TJ
M3 RYY Richard Crowther, 6 Kaliton, Church Street, Callington, PL17 7GB
M3 RYZ Alan Clunnie, 19 Griffin Road, Warwick, CV34 6QX
M3 RZB R Brittain, 159 Caledonia Road, Wolverhampton, WV2 1JA
MJ3 RZD Robert Luscombe, Flat, 1 Rouge Bouillon, St. Helier, Jersey, JE2 3ZA
M3 RZE Craig Russell, 255 Leeds Road, Shipley, BD18 1EH
M3 RZF Simon Harris, Cross House, Mill Lane, Preston, PR3 2JX
M3 RZG John Plant, The Cottage, Back Springfield Road, Lytham St. Annes, FY8 1TN
M3 RZI Owen Rabbitt, 20 Lysander Drive, Padgate, Warrington, WA2 0GL
M3 RZJ Bruce Trayhurn, 15 Wight Drive, Caister-on-Sea, Great Yarmouth, NR30 5UN
M3 RZL A Linden, 12 Godstone House, Pardoner Street, London, SE1 4DT
M3 RZM Graham Kingstone, 17 Ullswater Drive, Leighton Buzzard, LU7 2QR
M3 RZN Ivor Seaman, 6 Aylsham Road, Buxton, Norwich, NR10 5EX
M3 RZO Neal Bardell, 1 Walshs Manor, Stantonbury, Milton Keynes, MK14 6BU
M3 RZP Rebecca Powell, 53 St. Marys Road, Adderbury, Banbury, OX17 3HA
MI3 RZT James Thompson, 119 Rathkyle, Antrim, BT41 1LN
M3 RZU Unwana Ekpe, Cathedral Court, University Campus, Guildford, GU2 7JH
M3 RZV Roger Millington, Quaintways, The Avenue, Tarporley, CW6 0BA
M3 RZW Robert Lancaster, 10 Railway Terrace, Tirphil, New Tredegar, NP24 6EY
M3 RZY Sarah Trotter, 62 Regent Street, Whitstable, CT5 1JQ
M3 SAA S Atkinson, 10 Pond Lane, New Tupton, Chesterfield, S42 6BG
M3 SAB S Hughes, 9 Melverton Avenue, Wolverhampton, WV10 9HN
M3 SAI R Blore, Ty Nwydd, Cymau, Wrexham, LL11 5EU
MM3 SAK A McNeil, 21 Dumbreck Terrace, Queenzieburn, Glasgow, G65 9EA
M3 SAO A Osmond, 36 Knowles Road, Leicester, LE3 6LU
M3 SAR Sarah Abraham, 12 Graham Road, Halesowen, B62 8LJ
M3 SAY S Yapp, Dickers Farm, Beechy Road, Uckfield, TN22 5JG
M3 SAZ Sarah Greatorex, 54 Lilac Grove, Glapwell, Chesterfield, S44 5NG
M3 SBA Adam Savory, 33 Bretch Hill, Banbury, OX16 0LE
M3 SBB B Stoneley, 44 Ilthorpe, Hull, HU6 9ER
M3 SBE S Edwards, 5 Gorse Hill Road, Brickfields, Worcester, WR4 9TU
M3 SBJ S Inman, 9 Colbert Avenue, Ilkley, LS29 8LU
M3 SBP S Palin, Rose Tree Cottage, 17 Rowland Lane, Thornton-Cleveleys, FY5 2QX
M3 SBQ Kenneth Walsh, Interval, Liverpool Marina, Liverpool, L3 4BP
M3 SBS D Green, 144 Dilloways Lane, Willenhall, WV13 3HJ
M3 SBT Brian Lockley, 35 High Street, Blackpool, FY1 2BN
M3 SBY Brett Young, 25 rombalds drive, Skipton, BD23 2SP
M3 SCA S Ahmed, 59 Ramsgate, Lofthouse, Wakefield, WF3 3PX
M3 SCF H Fish, New House Peaton, Peaton, Craven Arms, SY7 9DW
M3 SCJ C Short, 8 Whitley Willows, Lepton, Huddersfield, HD8 0GD
MM3 SCO G Macleod, 12a Loyal Terrace, Tongue, Lairg, IV27 4XQ
M3 SCQ David MacGregor, Willows Halt, Will Row, Mablethorpe, LN12 1PJ
M3 SCX S Williamson, 19 Alcester Close, Plymouth, PL2 1EA
M3 SDB Simon Bennett, 17 Knox Close, Norwich, NR1 4LN
M3 SDJ S Hackwood, 7 Marshall Avenue, Brown Edge, Stoke-on-Trent, ST6 8SD
M3 SDK J Donald, 11 Row Brow Park, Dearham, Maryport, CA15 7JU
M3 SDN N Hurst, 74 Holden Road, Salterbeck, Workington, CA14 5LZ
MM3 SDP Isaac Lipkowitz, Chuccaby, Longhope, Stromness, KW16 3PQ
M3 SDQ Matthieu Behrooz-kafshdooz, 20 Byron Road, London, W5 3LL
M3 SDV Jonathan Pelham, 20 Merchants Court, Bedford, MK42 0AT
M3 SEE S England, 4 Ouse Close, Chandler's Ford, Eastleigh, SO53 4RW
M3 SEJ J Shepherd, 9 Wrea Head Close, Scalby, Scarborough, YO13 0RX
MI3 SEK Raymond Thomson, 1 Litchfield Park, Coleraine, BT51 3TN
MI3 SEO Stephen Murray, 117 Knockview Drive, Tandragee, Craigavon, BT62 2BL
MM3 SES S Smart, 6 Alton Bank, Nairn, IV12 5PJ
M3 SET S Taylor, 43 Toronnen, Bangor, LL57 4TG
MI3 SEV M Severn, 99 Crawfordsburn Road, Bangor, BT19 1BJ
M3 SEY Michael Howes, 1 The Meadows, Herne Bay, CT6 7XB
M3 SEZ S Ezard, 59 Station Farm, Croesyceiliog, Cwmbran, NP44 2JW
M3 SFC Arthur Woodward, 19 Hazel Grove, Winchester, SO22 4PQ
M3 SFJ F Smith, 9 Bramwell Street, Street Helens, WA9 2DP
M3 SFK R Birkitt, 34 Santon Downham, Brandon, Ipswich, IP27 0TG
MI3 SFL Stephen Leitch, 212 Belfast Road, Muckamore, Antrim, BT41 2EY
M3 SFN Albert Passey, 3 The Yard, Bayton, Kidderminster, DY14 9LH
MW3 SFP Simon Parry, Aukland Terrace, Crymych, SA41 3QG
M3 SFZ Stephen Free, Mill Farm, Hargham Road, Attleborough, NR17 1DT

M3 SGE C Sargent, Bradley, Holcombe Village, Dawlish, EX7 0JT
M3 SGF S Blount, 55 Silverthorne Drive, Caversham, Reading, RG4 7NR
M3 SGG Natalie Evans, 17 [illegible] Close, Burnham on Sea, TA8 2QJ
[illegible] WF15 7HW
M3 SGJ John Scott, The Parsonage, 102A, Nutley Lane, Reigate, RH2 9HA
MM3 SGQ Stephen Gill, 5 Ramornie Place, Kingskettle, Cupar, KY15 7PT
M3 SGS Stephen Salmon, 35 Westgate Road, Lytham St. Annes, FY8 2SG
M3 SGV Rob Greaves, 7 Eller Brook Close, Heath Charnock, Chorley, PR6 9NQ
MW3 SGX John William Bidwell, 26 Lone Road, Clydach, Swansea, SA6 5HR
M3 SGZ John Bentham, 18 Cauldon Avenue, Swanage, BH19 1PQ
M3 SHB S Brown, 6 Good Avenue, Trimdon Grange, Trimdon Station, TS29 6EF
M3 SHI Agnus Shillabeer, 29 Newlease Road, Waterlooville, PO7 7BX
M3 SHJ S Hughes, 4 Cobden Court, Birkenhead, CH42 3YH
M3 SHK Robert Silversides, 7 Earles Lane, Kelsall, Tarporley, CW6 0QR
M3 SHN S Neale, 28 Needham Drive, Sutton St. James, Spalding, PE12 0EG
M3 SHQ K Browne, 24 Oaktree Avenue, Cuerden Residential Park, Leyland, PR25 5PJ
MM3 SHT David McClure, 10 Greystone Close, Strathaven, ML10 6FW
M3 SHW K Shaw, 2 Montrose Avenue, Montrose Street, Hull, HU8 7RY
M3 SHX A Davies, 13 The Close, Stalybridge, SK15 1HU
M3 SHZ P Bennett, 1 Queens Road, Carterton, OX18 3YB
M3 SII Kaidei Borthwick, 15 Thomas Close, Ixworth, Bury St. Edmunds, IP31 2UQ
MI3 SIL S Linton, 68 Old Frosses Road, Cloughmills, Ballymena, BT44 9NA
M3 SIM S Lord, 34 Alsop Street, Leek, ST13 5NZ
M3 SIS L Simmons, 2 Blakemere Way, Sandbach, CW11 1XU
M3 SIY Simon Shaul, Shepherds Cottage, Middle Street, Gainsborough, DN21 5BU
M3 SIZ Joan Easdown, 38 North Street, Barming, Maidstone, ME16 9HF
M3 SJD Susan Darby, 4 Whately Mews, Whately Road, Lymington, SO41 0XS
M3 SJH S Hewitt, 4 Carrow Road, Dagenham, RM9 4TJ
M3 SJK S Kerrison, 18 Parks Road, Dunscroft, Doncaster, DN7 4AH
M3 SJL J Lowe, 46 Runshaw Avenue, Appley Bridge, Wigan, WN6 9JN
M3 SJM S Whitaker, 34 Alder Grove, Poulton-le-Fylde, FY6 8EH
M3 SJQ J Cleaver, 27 Lawton Crescent, Biddulph, Stoke-on-Trent, ST8 6EH
M3 SJV P McCarthy, 38 Lyndhurst Drive, Leyton, London, E10 6JD
M3 SJW Stephen Wills, 8 Frobisher Road, Yeovil, BA21 5FP
M3 SJX Brian Shields, 20 Gresley Court, Grantham, NG31 7RH
M3 SJY Kenneth Young, 14 Beechwood Avenue, Chatham, ME5 7HH
M3 SKB S Brown, Rushbrook, Holly Grange Road, Lowestoft, NR33 7RR
M3 SKC Deanne Bryan, 3 George Street, Bourne, PE10 9HE
M3 SKD S Kidd, 27 Hillswood Avenue, Leek, ST13 8EQ
M3 SKN Philip Probst, 37 Devonshire Street, Skipton, BD23 2ET
M3 SKQ Peter Henderson, Riverside, Ravens Bank, Holbeach, Spalding, PE12 8RW
M3 SKT Simone Taylor-Toms, 34 Larkspur Drive, Chandler's Ford, Eastleigh, SO53 4HU
M3 SKU Valerie Fitzpatrick, 20 Bewell Head, Bromsgrove, B61 8HY
M3 SKV J Hawkes, 183 Borden Lane, Sittingbourne, ME10 1DA
MW3 SKW M Barber, 1 Gwernant, Cwmllynfell, Swansea, SA9 2FT
M3 SKY S Keevil, Gamekeepers Cottage, Snarehill, Thetford, IP24 2QA
M3 SKZ Jack Parfitt, 5 Sheridan Road, Frimley, Camberley, GU16 7DU
M3 SLB S Berry, 4 Newlands Park Way, Newick, Lewes, BN8 4PG
M3 SLD J Bradley, Tongue Of Bombie, Kirkudbright, DG6 4QD
M3 SLF R Cave, 26 Longsight Road, Mapplewell, Barnsley, S75 6HB
M3 SLI J Backhouse, De10 Isaf, Bryneglwys, Corwen, LL21 9NP
MW3 SLL S Lawton, Bryngarw Lodge, Brynmenyn, Bridgend, CF32 8UU
M3 SLO D Dash, 36 Rockvilla Close, Varteg, Pontypool, NP4 7QF
M3 SLQ Nick Jones, 10 Leamington Close, Cannock, WS11 1PW
MI3 SLT S Whitten, 2 Springwell Manor, Castlederg, BT81 7DR
M3 SLZ T Gill, 21 Trevor Smith Place, Taunton, TA1 3RW
M3 SMD Anthony Davies, 7 Windermere Grange, Edlington, Doncaster, DN12 1NQ
M3 SMI John Smith, 9 Water Meadow Way, Downham Market, PE38 9HA
M3 SMK S Mackimm, 16 Stanneybrook Close, Rochdale, OL16 2YH
M3 SML S Lowe, 59 Knight Avenue, Gillingham, ME7 1UE
M3 SMM S Mole, 17a Marlborough, Seaham, SR7 7SA
M3 SMN S Kent, 4 Arden Close, Chesterfield, S40 4NE
M3 SMR Robert Shepperley, Flat F, London, N3 1QL
M3 SMY S Harkness, 114 Morland Road, Ipswich, IP3 0LZ
M3 SMZ S Rdwards, 59 Laburnum Road, Tipton, DY4 9QS
MM3 SNB George McGeouch, 49 Auckland Street, Glasgow, G22 5NY
M3 SNF Ian Hewitt, 26 Outwoods Drive, Loughborough, LE11 3LT
MW3 SNH Brian Jones, Browerdd, Llangybi, Lampeter, SA48 8NH
MW3 SNJ S Jones, 14 Lower Cross Road, Llanelli, SA15 1NQ
M3 SNL R James, 50 Andrew Allan Road, Rockwell Green, Wellington, TA21 9DY
M3 SNN Nigel Chapman, 8 Pennine Drive, Edith Weston, Oakham, LE15 8HY
M3 SNO R Snowden, 5 Eastfield Road, Wisbech, PE13 3EJ
M3 SNQ Lee Thornton, 11 Polruan Road, Truro, TR1 1QR
MW3 SNW S Williams, 56 Heol Llansantffraid, Sarn, Bridgend, CF32 9NH
M3 SNX Garry Rigden, Corner House, Ashford Road, Ashford, TN27 0EE
M3 SNY Roy Beardshall, 41 Hill Crest, Hoyland, Barnsley, S74 0BU
M3 SNT S [illegible], 91 Winchester Avenue, Lancaster, LA1 4HX
M3 SOF S Vaux, 171 Foxon Lane, Caterham, CR3 5SH
M3 SOG R Stearn, 18 Kings Avenue, Chippenham, SN14 0UJ
M3 SOQ Peter Swann, 2 Little Walton, Eastry, Sandwich, CT13 0DW
M3 SOT S Gregory, 11 Ribblesdale Avenue, Congleton, CW12 2BS
M3 SOV Peter Fernie, 39 North Parade, Falmouth, TR11 2TE
M3 SOY Jack Rolph, The Hollies, Back Lane, Norwich, NR10 4HL
M3 SPA Ian Beresford, 16a Holbeck Hill, Scarborough, YO11 2XD
M3 SPG S garthwaite, 278 Carlton Road, Barnsley, S71 2BA
M3 SPJ Shirley Hinds, 69 Carshalton Grove, Wolverhampton, WV2 2QZ
M3 SPL A Ladell, 25 Harwood Avenue, Thetford, IP24 2LY
M3 SPP R Penrose, 41 Milton Road, Eastbourne, BN21 1SH
M3 SPQ Sharon Schonborn, 116 Hough Lane, Wombwell, Barnsley, S73 0EF
M3 SPR S McLaughlin, 34 Cambridge Road, Birstall, Batley, WF17 9JF
M3 SPU Paul Saunders, 62 Parkfield Avenue, Eastbourne, BN22 9SF
M3 SPY R Gardner, 6 Meade Road, Liverpool, L13 9AA
MW3 SQA Colin Davis, Denant Mill, Dreenhill, Haverfordwest, SA62 3TS
M3 SQE David McDonald, 3 Lindley Street, Mansfield, NG18 1QE
M3 SQG Mark Breslin, 15 Acorn Gardens, East Cowes, PO32 6TD
M3 SQH A Lawrence, 75 Church Street, Ilkeston, DE7 8QP
M3 SQI N Thake, Easton House, Water Street, Shaftesbury, SP7 0HS
MM3 SQJ James Morris, 10 Middlemas Road, Dunbar, EH42 1GJ
MM3 SQM D Mchardy, 486 Kilmarnock Road, Glasgow, G43 2BW
M3 SQO Philip Burke, 38 Bosworth Square, Rochdale, OL11 3QG
M3 SQP Stuart Scotching, 26 Newton Way, Leighton Buzzard, LU7 4YU
M3 SQQ S Oxenham, 10 Arnside Close, Plymouth, PL6 8UU
M3 SQS Russell Garland, 113 The Drive, Feltham, TW14 0AH
M3 SQT Christopher Eyre, 23 Nelson Street, Congleton, CW12 4BS
M3 SQU Marijan Van Den Bergh, The Parsonage, Masefield Drive, Tamworth, B79 8JB
M3 SQV Stacey Sandford, 11 Browning Close, Tamworth, B79 8NB
M3 SQX Innocent Okorji, 107 Hillside Avenue, Borehamwood, WD6 1HH
M3 SQZ Nigel Swift, 59 Milton Avenue, Malton, YO17 7LB
MM3 SRF Robin Farrer, 23 Upper Craigour, Edinburgh, EH17 7SE
MI3 SRG E Coates, 148 Springwell Road, Groomsport, Bangor, BT19 6LX
M3 SRH S Hubball, 24 Newstead Road, Stoke-on-Trent, ST2 8HX
M3 SRI S Issatt, 69 St. Lawrence Avenue, Snaith, Goole, DN14 9JH
MM3 SRK A Ross, 16 Croft Road, Kiltarlity, Beauly, IV4 7HZ
MI3 SRL Samuel Hea, 70 Raloo Road, Larne, BT40 3DU
M3 SRQ John Stoppard, 14 Brookside Bar, Chesterfield, S40 3PJ
M3 SRT Stuart Thompson, 30 Southport Parade, Hebburn, NE31 2AQ
M3 SRV J Matthews, 23 Elmhurst, Bridgnorth, WV15 5DJ
M3 SRY P Seymour, 34 Northcliffe Rd, Grantham, NG31 8DP
M3 SSG A Butler, 12 South Bank Cottages, South Stoke, Reading, RG8 0HX
M3 SSI S Bangalore, 2 Amberley Walk, Kingsmead, Milton Keynes, MK4 4AX
M3 SSL M Belcher, 52 Kynaston Road, Didcot, OX11 8HD
M3 SSO Brian Hawes, 3 Orchard Close, Cassington, Witney, OX29 4BU
M3 SSU E Little, 41 Sevenoaks Road, Portsmouth, PO6 3JP
M3 STJ S Jordan, 3 Keir Road, Wednesbury, WS10 0HL
M3 STQ Stuart Fox, 34 Lynwood Avenue, Felixstowe, IP11 9HS
M3 STR N Soltysik, 24 Cottage Close, Hednesford, Cannock, WS12 1BS
MI3 STW R Bradley, 45 Alexandra Park Avenue, Belfast, BT15 3ER
MI3 STY S Nicholl, 89 Glenshane Road, Londonderry, BT47 3SF
M3 SUF Nina Smith, 7 Hawthorne Avenue, Connah's Quay, Deeside, CH5 4TF
M3 SUI S Allen, 33 Rookhill Road, Pontefract, WF8 2BY
M3 SUJ Jordan Cook, 40 Preston Avenue, Alfreton, DE55 7JY
M3 SUK A Holland, 40 Sunnyside Road, Poole, BH12 2LQ
MM3 SUS S Holt, Ashwell, Cannigall, Kirkwall, KW15 1SX
M3 SUT C Sutherland, 17 Walton Garth, Drighlington, Bradford, BD11 1HW
MM3 SUV Arthur McCaig, 46 Patterson Drive, Law, Carluke, ML8 5LT
M3 SUW Peter Hopkins, 40, Grange Close, Condover, Shrewsbury, SY5 7AT
M3 SUY Tobias Van Den Bergh, 19 Perrycrofts Crescent, Tamworth, B79 8UA
M3 SVB Scott Black, 7 Harwood Close, Gosport, PO13 0TY
M3 SVC S Cox, 19 Exbury Way, Andover, SP10 3UH
M3 SVD M Hewitt, Redwood House, Adbury Holt, Newbury, RG20 9BW
M3 SVF L Leung, Harrogate Ladies' College, Clarence Drive, Harrogate, HG1 2QG
M3 SVH Kevin Henderson, 42 Chartwell Avenue, Wingerworth, Chesterfield, S42 6SP
M3 SVJ Robert Gee, Flat 1D, Quarmby Road, Huddersfield, HD3 4HQ
MI3 SVM S Murray, 80 Canterbury Park, Londonderry, BT47 6DU
M3 SVN S Adkins, 4 Orion Close, Ward End, Birmingham, B8 2AU
M3 SVO Lee Birdsall, 8 North Cote, Ossett, WF5 9RE
M3 SVP Anton Chapman, 24 Eaton Grange Drive, Long Eaton, Nottingham, NG10 3QE
M3 SVT S Taylor, 17 Mendip Drive, Bolton, BL2 6LQ
M3 SVZ Geoffrey Brierley, 35 Ochrewell Avenue, Deighton, Huddersfield, HD2 1LL
MM3 SWA S Anderson, 33 Dryden Avenue, Loanhead, EH20 9JT
MI3 SWD S McAuley, Layde View, 19 Rathlin Avenue, Ballycastle, BT54 6DQ
M3 SWF R Jenkinson, Esperance, West End Road, Doncaster, DN9 1LB
MM3 SWG Stephen Groves, 1/1 99 Belville Street, Greenock, PA15 4SX
M3 SWJ Gareth Davies, 66 Allt-yr-yn View, Newport, NP20 5GG
M3 SWK Simon Walker, 64 Belmont Road, Rugby, CV22 5NY
MW3 SWO Stephen Owen, 500 Cowbridge Road West, Cardiff, CF5 5DA
M3 SWQ Alan Marks, grosvenor hotel, 51 grosvenor road, Scarborough, YO112LZ
M3 SWS S Lowe, 31 Court Farm Road, Bristol, BS14 0EH
MM3 SWU Stephen Jenkins, 66 Spruce Avenue, Johnstone, PA5 9RG
M3 SWV Jacqueline Moppett, 59 piccadilly, Tamworth, B78 2ER
MM3 SWW Duncan Elliot, thisleycrook, Torphins, AB214NR
M3 SXF S Forbes, 55 The Henrys, Thatcham, RG18 4LS
MI3 SXI I McKeown, 19 Castlehill, Comber, Newtownards, BT23 5XA
M3 SXJ Patrick Duckles, 8 Railway Cottages, Skillings Lane, Brough, HU15 1EN
M3 SXK Chris Little, 22 Edinburgh Road, Broseley, TF12 5PE
MI3 SXM Geoffrey Hutton, 13 Meadowbank, Sepatrick, Banbridge, BT32 4PZ
M3 SXP Stephen Porring, 26 Celandine Grove, Thatcham, RG18 4FF
MI3 SXQ Chris Cunningham, 1 Ballykeel Court, Ballymartin, Newry, BT34 4XW
MI3 SXR A Robb, 10 Rosepark East, Belfast, BT5 7RL
MM3 SXT Stanley Thorogood, 38 Forres Drive, Glenrothes, KY6 2JU
M3 SXU Julie Jackson, 90 Horne Street, Bury, BL9 9HS
M3 SXV Shaun Simms, 31 Chestnut Road, Cawood, Selby, YO8 3TB
M3 SXZ David George, 9 winscombe court, Frome, BA11 2DZ
MM3 SYB D Nicholson, 4 Upper Barvas, Isle of Lewis, HS2 0QX
M3 SYC Anthony Chaplin, 33 The Crofts, Little Wakering, Southend-on-Sea, SS3 0JS
MI3 SYF David Bates, 31 Drumard Park, Lisburn, BT28 2HU
M3 SYH David Horton, 2 Brampton Way, Bulkington, Bedworth, CV12 9PR
MI3 SYI Brendan McDonald, 20 Aughan Park, Poyntzpass, Newry, BT35 6TW
M3 SYL S Stratford, 23 The Fairway, Banbury, OX16 0RR
M3 SYN Simon Lanaway, 1 Clovers Cottages, Faygate Lane, Horsham, RH12 4SH
MM3 SYO Stephen Angus, 20 Norlands, Errol, Perth, PH2 7QU
MM3 SYQ Alistair Clark, 17 Glentilt Terrace, Perth, PH2 0AE
MM3 SYU Carol Higgins, 11 Strathyre Place, BROUGHTY FERRY, Dundee, DD5 3WN
M3 SYV Thomas Symons, Southgate, The Commons, Mullion, TR12 7HZ
M3 SYW Craig Voke, 16 Exton Road, Chichester, PO19 8BP
M3 SYY B Sweeney, 14 Eaves Lane, Chorley, PR6 0PY
M3 SYZ S Symonds, 68 Manor Crescent, Pan, Newport, PO30 2BH
M3 SZC S Crabtree, 107 Rochdale Road, Shaw, Oldham, OL2 7JT
M3 SZD Mark Musgrave, Hillside Cottage, Hiraddug Road, Rhyl, LL18 6HS
M3 SZF Glenn Williams, Ty Newydd, Rhyd, Penrhyndeudraeth, LL48 6ST
MJ3 SZI Michael Brown, 200 La Marais, St. Clement, Jersey, JE2 6[illegible]
M3 SZK Matthew Gridley, 40 Canton Drive, Bridgwater, TA6 [illegible]
M3 SZM Sean Macdonald, 157 Delapre Drive, Banbury, OX16 3WS
M3 SZO Roy Savery, 75 Bramley Road, Tewkesbury, GL20 8AQ
M3 SZQ Stephen Snelson, 212 Dickson Road, Blackpool, FY1 2JS
M3 SZS David Forster, 23 Field Street, Padiham, Burnley, BB12 7AU
M3 SZT James Smith, 32 Youlgreave Drive, Sheffield, S12 4SE
M3 SZY S Holt, 108 Blandford Avenue, Castle Bromwich, Birmingham, B36 9JD
M3 TAE T Eadon, Chapel Cottage, Newcastle Road South, Sandbach, CW11 1RS
MW3 TAF C Williams, 96 Bryn Road, Swansea, SA2 0AT
M3 TAG A Aldred, 78 The Drive, Horley, RH6 7NH
M3 TAN S Greenfield, 4 Charlesworth Square, Gomersal, Cleckheaton, BD19 4NX
MM3 TAV A McConchie, 15 Slains Crescent, Cruden Bay, Peterhead, AB42 0PZ
M3 TAW T Whittam, 27 Dimples Lane, Garstang, Preston, PR3 1RD
M3 TBF Thomas Ferguson, Willowmead, Church End, Bedford, MK44 2RP
M3 TBG Adrian Archer, 23 St. Ives Road, Somersham, Huntingdon, PE28 3ER
M3 TBH T Hobbs, 2 The Lynch, West Stour, Gillingham, SP8 5RN
M3 TBK E Cree, 24 Old Lincoln Road, Caythorpe, Grantham, NG32 3EJ
MI3 TBL T Littler, 15 Belmont Grove, Lisburn, BT28 3YB
M3 TBP C Parker, Red Leas, 22 Bent Lane, Colne, BB8 7AA
M3 TBQ Ivor Hill, 74 Clarence Road, Torpoint, PL11 2LT
M3 TBU T Burnham, Creedy Barn, Kennerleigh, Crediton, EX17 4RU
M3 TBV David Parker, 5 Beam Avenue, Dagenham, RM10 9BS
M3 TBW Jacqueline Humphrey, Flat 1, Kingswood House, 10 Lewes Road, Eastbourne, BN21 2BX
MM3 TBY S Turnbull, 15 Woodruff Gait, Dunfermline, KY12 0NL
M3 TBZ D Heathcote, 154 High St., Harriseahead, Stoke-on-Trent, ST7 4JX
M3 TCD Raymond Taylor, 10 Barnwell Lane, Cromford, Matlock, DE4 3QY
M3 TCG T Graham, First Floor Flat, 43 Belgrave Crescent, Bath, BA1 5JU
M3 TCR P Ryall, Windsor Lodge, Pantile Hill, Southminster, CM0 7BA
M3 TCT S Farrar, 20 Cleveland Grove, Lupset, Wakefield, WF2 8LD
M3 TCU Glenn Read, Flat 3, Parkmead Court, Ryde, PO33 2HD
M3 TCX Thomas Carroll, 14 Glenpark Drive, Southport, PR9 9FA
M3 TCY T Earnshaw, 63 Manor Road, Fleetwood, FY7 7LJ
M3 TDB T Berry, Roseneath, Walcote Road, Lutterworth, LE17 6EQ
M3 TDH T Hewitt, 6 Mayfield, Catforth Road, Preston, PR4 0HH
M3 TDM T Reddington, 174 Home Farm Road, Wirral, CH49 7LH
M3 TDP T Packham, Bradstow Lodge, 19 Crow Hill, Broadstairs, CT10 1HN
M3 TDT Anthony Dockerill, 8 Bennett Road, Swanton Morley, Dereham, NR20 4LY
M3 TEE Darron West, 42 Scholars Green, Wigton, CA7 9QW
M3 TEG Tom Mason, 31 Manor Park Road, Hailsham, BN27 3AT
M3 TEI Robert McKnight, Gortadrohid, Reengaroga, Co. Cork, Ireland
M3 TEL T Pink, 11 Harmony Meadow, Roche, St. Austell, PL26 8EJ
MI3 TEM Stephen Kirkwood, 1 Rural Cottages, Front Road, Lisburn, BT27 5LF
M3 TEN T Newman, Sometimes (The Workshop), South Pew, Dorchester, DT2 9HZ
M3 TEP Terry Payne, 2 Greenleas, Waltham Abbey, EN9 1SZ
MM3 TEQ Aimee Leiper, 6 inchyra place, Grangemouth, FK39EQ
M3 TET Rita Mills, 61 Thetford Road, Great Barr, Birmingham, B42 2JA
M3 TEV Stephen Ball, 13 Yew Tree Road, Hayling Island, PO11 0QE
M3 TEY John Hartshorne, 8 Ashbee Street, Bolton, BL1 6NT
M3 TFA Anh Minh Tran, Flat 4, Room 9, Rayleigh Tower, Colchester, CO4 3SQ
M3 TFB D Parkinson, 4 Meadow View, Sherburn in Elmet, Leeds, LS25 6BY
M3 TFE Jonathan King, 60 Harwood Avenue, Bromley, BR1 3DU
MI3 TFF Francis Finlay, 62 Slieveboy Road, Claudy, Claudy, BT47 4AS
M3 TFG Benjamin Jones, 75 Mamble Road, Stourbridge, DY8 3SY
M3 TFI Deborah Woods, 3 Brook Street, Port Sunlight, Wirral, CH62 5DB
M3 TFK Peter Hemsley, 140 Greenhill Lane, Riddings, Alfreton, DE55 4EX
M3 TFM S Lytollis, 8 St. Martins Court, Brampton, CA8 1PL
M3 TFO Robert Styles, 52 Vernham Grove, Bath, BA2 2TB
M3 TFP T Plummer, 33 East Street, Sudbury, CO10 2TU
M3 TFS Richard Shoubridge, 2 Copestake Drive, Burgess Hill, RH15 0LD
M3 TFW T Harlow, 10 Fraser Road, Poole, BH12 5AY
M3 TFX Tom Fisk, 2 Hall Farm Cottage, Caston Road, Attleborough, NR17 1BW
M3 TFZ Edmund Miller, 8 Arthur Avenue, Caister-on-Sea, Great Yarmouth, NR30 5PQ
M3 TGA Georgia Arnold, 2 Duck Lane, Haddenham, Ely, CB6 3UE
M3 TGC G Cooper, 10 Granary Court Magdalene Lawn, Barnstaple, EX32 7FA
M3 TGD James Park, 3 Flaxfield Drive, Crewkerne, TA18 8DF
M3 TGE Roger Lindon, 134 Station Road, Sutton Coldfield, B73 5LD
M3 TGJ James Alexander, 335 Canterbury Road, Birchington, CT7 9TY
M3 TGK P Wale, Munstead Oaks, Hascombe Road, Godalming, GU8 4AB
M3 TGL Gary Thorne, 72 Devonshire Road, London, E16 3NJ
M3 TGO G Hope, 3 Bean Close, Great Chart, Ashford, TN23 3BG
M3 TGP T Porter, 208 Clapgate Lane, Ipswich, IP3 0RG
M3 TGS A Sayers, 4 Roughley Avenue, Warrington, WA5 1BL
M3 TGT H Doman, 13 Cumbria Close, Maidenhead, SL6 3DD
M3 TGW T Willis, 32 Sandover, Northampton, NN4 0TG
M3 TGZ David Lines, 68 Rugby Place, Brighton, BN2 5JA
M3 THE M Peck, 60 Riverside Drive, Tern Hill, Market Drayton, TF9 3QU
MW3 THI Michael Price, 18 Rhiw Tremaen, Brackla, Bridgend, CF31 2JA
M3 THJ Brian Mills, 37 Ashley Road, Hildenborough, Tonbridge, TN11 9ED
M3 THN P Cubley, 58 John Street, Newhall, Swadlincote, DE11 0CR
MD3 THQ John Peerless, 503 Honeypot Lane, Stanmore, HA7 1JH
M3 THY T Lupton, 81 Home Farm Lane, Bury St. Edmunds, IP33 2QL
M3 TIC Mark Tinsell-Stanton, 38 Comberton Road, Kidderminster, DY10 3DT
M3 TID B Maddox, 72 Church Road, Hartshill, Nuneaton, CV10 0LY
M3 TIE L Morrell-Cross, Delta Lodge, 14 Rushton Crescent, Bournemouth, BH3 7AF
M3 TIF C Robertson, 7 Richmond Street, Bury, BL9 9BS
M3 TII Bethany Aylward, 53 Overdown Rise, Portslade, Brighton, BN41 2YF
M3 TIJ James Aylward, 53 Overdown Rise, Portslade, Brighton, BN41 2YF
M3 TIK Matthew Richardson, 1 Cedar Drive, Lowestoft, NR33 9HA
M3 TIL J Tillson, 23 The Fitches, Knodishall, Saxmundham, IP17 1UX
M3 TIQ David Perkins, 56 Cliff Street, Rishton, Blackburn, BB1 4EE
M3 TIY Adam Green, 65 Rosamond Road, Bedford, MK40 3UG

**IMPORTANT NOTE**

**Revalidate licence to avoid revocation** – Ofcom has advised the Society that plans will be drawn up to revoke licences that have not been revalidated as required by the licence conditions. The quickest way to revalidate is to do so online via the Ofcom website: *https://services.ofcom.org.uk/* or by email: *amateur.validations@ofcom.org.uk* Ofcom staff are available to help, but please be patient during times of heavy workload.

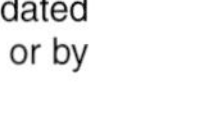

M3 TIZ D Marsh, 16 Laurel Close, North Warnborough, Hook, RG29 1BH
MW3 TJG T Gwyther, 15 Denbigh Court, Caerphilly, CF83 2UN
M3 TJI T Adams, 11 St. Georges Crescent, Gravesend, DA12 4AR
M3 TJJ John Jones, 19 Southbank Street, Leek, ST13 5LS
MI3 TJK John Mackenzie, 30 Dalriada Gardens, Ballycastle, BT54 6DZ
M3 TJL T Lake, 85 Clarkson Road, Norwich, NR5 8ED
MI3 TJM T Moore, 43 Woodburn Park, Londonderry, BT47 5PS
M3 TJO Terry Jones, 21 Lynwood Gardens, Croydon, CR0 4QH
M3 TJQ Andrew Newton, 114 Kingston Road, Taunton, TA2 7SP
MI3 TJR T Ruddell, 30 Ballynacor Meadows, PORTADOWN, Craigavon, BT63 5UU
M3 TJT Toby Ticehurst, 37 New Road, Ridgewood, Uckfield, TN22 5TG
M3 TJU Evan Duffield, 92 Crosby Street, Stockport, SK2 6SP
MI3 TJV Gary Harkin, 22 Bracken Vale, Omagh, BT78 5RS
M3 TKE Russell McKie, 16 silver street, creetown, Newton Stewart, DG8 7HU
MW3 TKI Ian Hoyle-Jackson, 21 Kimberley Close, Sketty, Swansea, SA2 9DZ
MI3 TKK Patrick Wylie, 6 Collinvale House, Green Road, Ballyclare, BT39 9PJ
M3 TKN Mark Roche, 1 Lancaster Close, London, NW9 5RE
M3 TKO Keith Smart, 33 East Street, Littlehampton, BN17 6AU
M3 TKP Cyril Mokes, 22 Oxclose Lane, Arnold, Nottingham, NG5 6GA
M3 TKQ Jonathan Macey, 62 Uttoxeter Road, Hill Ridware, Rugeley, WS15 3QU
M3 TKT M Turowski, 57 Millwood Road, Orpington, BR5 3LQ
M3 TKU Paul Loveden, 57 St. Marys Road, Rawmarsh, Rotherham, S62 5BD
M3 TKV T King, 24 Royston Avenue, Basildon, SS15 4EW
M3 TKW Keith Rowley, 10 Mount Close, Wombourne, Wolverhampton, WV5 9ER
M3 TLB D Weller, 6 Aldervale, Fermor Road, Crowborough, TN6 3BY
M3 TLD David Tattersall, 17 Badger Close, Durkar, Wakefield, WF4 3QD
MD3 TLG A Dann, 18 Salcombe Way, Ruislip, HA4 6BA
MM3 TLH T Holt, Ashwell, Cannigall, Kirkwall, KW15 1SX
M3 TLJ Barry Hawkins, 4 Hastings Drive, Barwell, Leicester, LE9 8AE
M3 TLK Robert Sanderson, 7 Faraday Drive, Milton Keynes, MK57DE
M3 TLL Louise Nilon, 5 Denby Drive, Baildon, Shipley, BD17 7PQ
M3 TLM T Lockett, 14 Tildsley Crescent, Weston, Runcorn, WA7 4RN
M3 TLN C Dean, 119 Queens Drive, Newton-le-Willows, WA12 0LN
M3 TLO Douglas Livings, 30, Grenfell Avenue, Holland-on-Sea, Clacton-on-Sea, CO15 5XH
M3 TLP A Buckley, 53 Derlwyn St., Phillipstown, New Tredegar, NP24 6AZ
MM3 TLQ David Field, 7 Admiralty Street, Portknockie, Buckie, AB56 4NB
M3 TLT A McGregor, 41 Breedon Close, Corby, NN18 9PG
M3 TLU Mark Bailey, 71 Somerfield Road, Walsall, WS3 2EG
MI3 TLV M Nicholl, 34 Berryhill Road, Artigarvan, Strabane, BT82 0HN
M3 TLW Rik Crook, 80 Kings Road, Biggin Hill, Westerham, TN16 3XY
M3 TLX David Burdsall, 37 Fulmar Walk, Whitburn, Sunderland, SR6 7BW
M3 TLY Ian Bain, 45 Larpool Crescent, Whitby, YO22 4JD
M3 TLZ David Pointon, 1 Cross Cottages, Alsager Road, Audley, Stoke-on-Trent, ST7 8JQ
M3 TME M Trick, 24 King Street, Tiverton, EX16 5JE
M3 TMG G Thorpe, Jasmine, Sutton Road, Mablethorpe, LN12 2PT
M3 TMI David Redfern, 39 St. Marks Close, Cromford, Matlock, DE4 3QD
M3 TMM T McKain, 21 Oakhurst Grove, East Dulwich, London, SE22 9AH
MI3 TMN A McNulty, 2 Devenish Crescent, Devenish, Enniskillen, BT74 4RB
M3 TMQ M Mutton, 74 Alexandra Road, Sheerness, ME12 2AT
M3 TMR R Thomas, 35 Under Ffrydd Wood, Knighton, LD7 1EF
M3 TMX Jordan Harrop, 35 Langdale Crescent, Dalton-in-Furness, LA15 8NR
M3 TMY Thomas Mulraney, 3 Salvia Close, Churchdown, Gloucester, GL3 1LL
M3 TMZ Andrew Laister, 2 Warlow Crest, Greenfield, Oldham, OL3 7HD
M3 TNB Mark Anthony, The Bungalow, Magpies Cottage, Redruth, TR16 5JL
M3 TND N Tam, Room 102, 30 Evelyn Gardens, London, SW7 3BG
M3 TNE Cynthia Pickford, 80 Hollowood Avenue, Littleover, Derby, DE23 6JD
MM3 TNF John ODonnell, 33 Broomward Drive, Johnstone, PA5 8HR
MM3 TNG D Stewart, 45 Kilwinning Road, Irvine, KA12 8RZ
M3 TNH Kennth Stockley, 357 Clements Road, Ramsgate, CT12 6UG
M3 TNJ J Armstrong-Taylor, Driftwood, Station Road, Yelverton, PL20 7JS
M3 TNK N Fong, Harrogate Ladies' College, Clarence Drive, Harrogate, HG1 2QG
M3 TNL Peter Dawson, 400 Ropery Road, Gainsborough, DN21 2TH
M3 TNM Thoa Nguyen, 9 Green Street, Cambridge, CB2 3JU
M3 TNN Thomas Ellis, 84 Revelstoke Road, London, SW18 5PB
M3 TNO Joshua Dean, 25 Chantry Avenue, Bexhill-on-Sea, TN40 2EA
M3 TNV Mark Clough, 8 Skeldyke Road, Kirton, Boston, PE20 1LR
M3 TNW Peter Stanford, 4 Barkway Road, Royston, SG8 9EA
M3 TNY T Limbert, 7 Acacia Avenue, Liverpool, L36 5TL
M3 TOB Harold Matthews, 24 Clos Y Berllan, Rhuddlan, Rhyl, LL18 2UL
M3 TOE A Ward, 39 Linley Close, Bridgwater, TA6 4HL
M3 TOF D Holyoake, 281 Causeway, Green Road, Oldbury, B68 8LT
M3 TOI Martin Francis, 72 Bro Ednyfed, Llangefni, LL77 7WD
M3 TOJ Helen Clough, 8 Skeldyke Road, Kirton, Boston, PE20 1LR
M3 TOR C Jones, PO Box 293, Ford, Plymouth, PL2 1WT
M3 TOT E Brown, Rose Cottage, Grindlow, Buxton, SK17 8RJ
MM3 TOV E Wilson, The Old Schoolhouse, Fife, KY15 4NB
M3 TOY Adam Ashworth, 79 Stonehouse Road, Rugeley, WS15 2LL
M3 TPD T Dooley, 32 Coult Avenue, North Hykeham, Lincoln, LN6 9RG
MM3 TPF Nigel Mann, Ramsburn Cottage, Knock, Huntly, AB54 7LQ
M3 TPG T Greenall, Hall Lane Farm, Hall Lane, Warrington, WA4 4AF
M3 TPH Tim Hazel, 84 Rodwell Avenue, Weymouth, DT4 8SQ
M3 TPI Anthony Paxton, 20f Green End, Granborough, Buckingham, MK18 3NT
MW3 TPJ Trevor Price, 5 Rhodfa'r Pant, Pant, Merthyr Tydfil, CF48 2DG
M3 TPN T Norrington, 32 Fulfen Way, Saffron Walden, CB11 4DW
MI3 TPR R Thompson, 14 Gelvin Grange, Londonderry, BT47 2LD
M3 TPU Raymond Morgan's, 7 Tennyson Close, Braintree, CM7 1AB
M3 TPW Tim Wooldridge, 12 Redwood Avenue, Leyland, PR25 1RN
M3 TPY Trevor Purcell, 18 Millberg Road, Seaford, BN25 3ST
M3 TPZ D Windus, 6 Blacklands Court, 40 St. Helens Park Road, Hastings, TN34 2DN
M3 TQA Andrew Madge, Flat, 17 Newcomen Road, Dartmouth, TQ6 9BN
M3 TQB Derek Holmes, 4 Council House, Nidds Lane, Boston, PE20 1LZ
M3 TQD James Lear, 6 South View Green, Bentley, Ipswich, IP9 2DR
M3 TQF C Bailey, 37 Cherry Tree Drive, Filey, YO14 9UZ
M3 TQG Graham Joy, Fair Oak, Higher Furzeham Road, Brixham, TQ5 8QP
MM3 TQH Richard Hay, Roddach Cottage East, Cummingston, Elgin, IV30 5XY
MM3 TQI Duncan George, D C George, 91 Regent Street, Keith, AB55 5ED
M3 TQJ Terry Kemp, 30 Tawny Sedge, King's Lynn, PE30 3PW
M3 TQN Nicholas Davies, 156 Britannia Avenue, Dartmouth, TQ6 9LQ
M3 TQP Robin Messingham, 2 The Lodge, Sotherington Lane, Liss, GU33 6DA
M3 TQT Francis Goodall, 1 Parkfield Grove, Leeds, LS11 7LS
M3 TQU Joseph Bingham, 31 Wyre Close, Paignton, TQ4 7RU
M3 TQX Aden Basterfield, Red 3, Purn Holiday Park, Bridgwater Road, Weston-Super-Mare, BS24 0AN
M3 TQY Geoffrey Rogers, 26 Chaucer Close, Waterlooville, PO7 6AQ
M3 TRC P Lee, 15 Talkin Drive, Middleton, Manchester, M24 5LS
M3 TRJ T Jones, 5 Broomfields Road, Appleton, Warrington, WA4 3AE
M3 TRO S Phillips, 37 Wensley Road, Barnsley, S71 1SB
M3 TRP S Walker, 26 The Warren, Hardingstone, Northampton, NN4 6EW
MI3 TRR R Elliott, 183 Kilraughts Road, Ballymoney, BT53 8NL
MI3 TRY R Scott, 46 Monnaboy Road, Eglinton, Londonderry, BT47 3HP
MM3 TRZ T Reilly, 21 North Street, Motherwell, ML1 1LQ
M3 TSA T Hickson, 27 Cressing Road, Witham, CM8 2NP
M3 TSE E Hughes, 20 Elmsdale Avenue, Coventry, CV6 6ES
M3 TSF Paul Shayler, 38 Maryside, Slough, SL3 7ET
M3 TSG A Ryder, 4 Edgeway, Strelley, Nottingham, NG8 6LY
M3 TSI Paul Hewson, 30 Princess Road, Kirton, Boston, PE20 1JW
M3 TSJ S Trott, 6 Mounton Drive, Chepstow, NP16 5EH
M3 TSN M Lee, 46 Little Lane, Huthwaite, Sutton-in-Ashfield, NG17 2RA
M3 TSO T Owens, 74 Tees Crescent, Stanley, DH9 6JD
M3 TSV K Reason, 28 St. Marys Grove, Swindon, SN2 1RQ
M3 TTA E Davies, 58 Popes Lane, Sturry, Canterbury, CT2 0LA
M3 TTH Thomas Haley, 3 Orchard View, Cropredy, Banbury, OX17 1NR
M3 TTK Stewart Ridley, 123 Lanercost Drive, Newcastle upon Tyne, NE5 2DL
M3 TTS D Brook, 140 Dearne Hall Road, Barugh Green, Barnsley, S75 1LX
MW3 TUB D Butler, 34 The Beeches, Llandysul, SA44 4HS
M3 TUC T Gerrard, 16 Haig Road, Carlisle, CA1 3AS
M3 TUD C Copeman, 1 Chestnut Avenue, Welney, Wisbech, PE14 9RG
M3 TUF Julian Caswell, 3 Pavilion Court, Roydon, Diss, IP22 5SP
M3 TUH Brian Tucker, 12 Alpha Place, Appledore, Bideford, EX39 1QY
M3 TUJ Lee Taylor, Apartment 10, The Church Apartments 47a Seamer Road, Scarborough, YO12 4EF
M3 TUL Adrian Dodd, 68 Windlehurst Road, High Lane, Stockport, SK6 8AE
M3 TUO Andrew Browning, 11 Heather Close, Sirhowy, NP22 4PW
M3 TUQ Ronald Spiers, 7 Laurel Drive, Bognor Regis, PO21 3ND
MM3 TUR P Turner, 99 Maitland Hog Lane, Kirkliston, EH29 9DU
MI3 TUS W Donnell, 71 Niblock Oaks, Antrim, BT41 2DP
M3 TUU Gordon Milsom, Flat 8, Sovereign Court, High Wycombe, HP13 6XL
M3 TUW Sam Tucker, 11 Maple Drive, Killamarsh, Sheffield, S21 1GA
MI3 TUZ Alice Gault, 7 Gardenmore Place, Larne, BT40 1SE
M3 TVC R Evans, 7 Westland Drive, Hayes, Bromley, BR2 7HE
M3 TVD D Steele, 22 Grindle Close, Thatcham, RG18 3PD
M3 TVJ Joshua Evans, 16 Longfield Place, Poulton-le-Fylde, FY6 7DB
M3 TVK John Hodgson, 5 Clifton Place, Freckleton, Preston, PR4 1RQ
M3 TVN T Bunce, 31 Kensington Avenue, Middlesbrough, TS6 0QQ
MM3 TVQ Craig Smith, 37 Glebe Road, Mosstodloch, Fochabers, IV32 7JH
M3 TVV David Render, 4 Station Terrace, Allerton Bywater, Castleford, WF10 2BS
M3 TVZ David Darby, 8 Mulberry Close, Blackfield, Southampton, SO45 1FH
MM3 TWA I Whiteford, 54 Bilby Terrace, Irvine, KA12 9DT
M3 TWB S Bourdon, 35 Main Road, Woolverstone, Ipswich, IP9 1BA
MM3 TWG T Galbraith, 77 Netherwood Park, Deans, Livingston, EH54 8RW
M3 TWK W Tam, Harrogate Ladies' College, Clarence Drive, Harrogate, HG1 2QG
MI3 TWM C Smallwoods, 73 Knightsbridge, Londonderry, BT47 6FE
M3 TWP D Jenner, 116 Trench Road, Tonbridge, TN10 3HQ
MW3 TWQ Philip Jones, 36 Hopkin Street, Treherbert, Treorchy, CF42 5HL
M3 TWS T Stanford, 27 Mill Gardens, Elmswell, Bury St. Edmunds, IP30 9DQ
M3 TWV John Benbow, 20 Clifton Close, Thornton-Cleveleys, FY5 4NG
MM3 TWW E Wallace, 57 Henderson Park, Windygates, Leven, KY8 5DL
M3 TWY Mark Sewell, Flat 2-4, 6 Augusta Road, Ramsgate, CT11 8JP
M3 TXA Trevor Ruddick, Hazel Gill, Croglin, Carlisle, CA4 9RR
M3 TXF Hugh Northcote, 58 Warren Avenue, Wakefield, WF2 7JN
M3 TXG David Mardlin, 13 Churchill Crescent, Sonning Common, Reading, RG4 9RU
M3 TXH Andrew Hensman, 20 St. Marys Road, Braintree, CM7 3JR
MI3 TXI Robert Carlin, 10 Top Of The Hill, Londonderry, BT47 2HA
M3 TXJ Richard Williams, 198 Leverington Common, Leverington, Wisbech, PE13 5BP
M3 TXL T Graham, Woodtown, Sampford Spiney, Yelverton, PL20 6LJ
M3 TXM Mark Bristow, 9 Chadwick Drive, Harold Wood, Romford, RM3 0ZA
M3 TXP Alan Barker, 37 Newbarns Road, Barrow-in-Furness, LA13 9SF
M3 TXQ Phillip Holman, 13 Morsefield Lane, Redditch, B98 0EH
M3 TXR Elaine Smith, 13 Eagle Avenue, Waterlooville, PO8 9UB
M3 TXS H Huckle, 43 The Baulk, Biggleswade, SG18 0PX
MI3 TXT Michael Kashkoush, 41 Dunamallaght Road, Ballycastle, BT54 6PF
M3 TXU John Fesel, 3 Brook Street, Port Sunlight, Wirral, CH62 5DB
M3 TXV Ross Fuller, 183 Nottingham Road, Alfreton, DE55 7FL
MM3 TYA M Anthoney, 10 Cedar Road, Kilmarnock, KA1 2HP
M3 TYC Gary Lewis, 93 Eastcliff, Portishead, Bristol, BS20 7AD
M3 TYG Maria Kyriacou, 54 Sutton Avenue, Silverdale, Newcastle, ST5 6TB
M3 TYI Ian Morris, 6 St. Nicholas Court, Gloucester, GL1 2QZ
M3 TYL M Tyler, 40 Bullards Lane, Woodbridge, IP12 4HE
M3 TYM T Martin, 14 Campbell Road, Eastleigh, SO50 5AD
M3 TYO Gary Aldridge, Greenridge, Fore Street, Teignmouth, TQ14 9QR
M3 TYQ David Brown, 9 Chisholm Close, Standish, Wigan, WN6 0QP
M3 TYS K Sanchez-Garci, 74 Gorthorpe, Hull, HU6 9EZ
M3 TYU Matthew Rawlings, 3 Greenlands, Woolton Hill, Newbury, RG20 9TB
M3 TYV David Atkinson, 133 Lingmoor Rise, Kendal, LA9 7PL
M3 TYW David Gray, 7 Montague Street, Cleethorpes, DN35 7AP
M3 TYX Daniel McGrath, 48 Willersley Avenue, Orpington, BR6 9RS
M3 TYZ Darren Prince, 29 St. Stephens Road, West Bromwich, B71 4LR
M3 TZB D Vincent, 6 Nathan Gardens, Poole, BH15 4JZ
M3 TZE Roydon TSE, 1 Oaklands, Gallows Lane, Westham, BN24 5AW
M3 TZF Frederick Wells, 12 Portelet Place, Hedge End, Southampton, SO30 0LZ
M3 TZI John Cowell, Mount Rivers, Bootle, Millom, LA19 5XN
M3 TZN Mark Robins, 13 Sarum Way, Hungerford, RG17 0LJ
M3 TZO Paul Gibson, 7 Greenfields Road, Horley, RH6 8HW
MM3 TZP Stuart Wright, 22 Beechwood, Linlithgow, EH49 6SF
M3 TZQ Gerald Wright, 2 Hillcrest Drive, Castleford, WF103QN
M3 TZS Qobolwakhe Mdlongwa, 27 Faifax Avenue, Bierley, Bradford, BD4 6JY
M3 TZX Dave Wright, 130 Keedonwood Road, Bromley, BR1 4QN
MW3 UAA Helen Lee, 3 Summerhill Park, Simpson Cross, Haverfordwest, SA62 6EU
M3 UAE Terence Baines, 10 Croydon Avenue, Leigh, WN7 1TP
M3 UAG David Garland, 8 Ladywell Gate, Welton, Brough, HU15 1NL
M3 UAJ Adam Neale, 5 Millside, Wombourne, Wolverhampton, WV5 8JJ
M3 UAK M Brown, Welbeck, 15 Stanton Drive, Morpeth, NE61 6YW
M3 UAM C Byrne, 31 Graham Drive, Castleford, WF10 3EY
M3 UAO Michael Lewis, 6 Remembrance Road, Newbury, RG14 6BA
M3 UAP Glyn Davies, 45 Greensway, Abertysswg, Tredegar, NP22 5AR
M3 UAQ Ashley Quinton, 102 Downside Close, Ipswich, IP2 9YQ
M3 UAR Andrew Riches, 20 Western Gardens, Crowborough, TN6 3EB
M3 UAV Gareth Preece, 22 Crofters Green, Bradford, BD10 8RZ
M3 UAW Thomas Arrow, Crystalwood, Stonemans Hill, Newton Abbot, TQ12 5PZ
M3 UAX David Foyston, 10 Ash Grove, Wickersley, Rotherham, S66 2LJ
M3 UAY Matthew Scarr, 15 Church Park, Overton, Morecambe, LA3 3RA
M3 UBB G Goddard, 13 Inkerman Road, Darfield, Barnsley, S73 9NB
MM3 UBD Dennis Branson, Derelochy, Kingsteps, Nairn, IV12 5LF
M3 UBE Andrew Jones, 49 Newton Road, Dalton-in-Furness, LA15 8NQ
M3 UBF John Bonney, 37 Watery Lane, Brackley, NN13 7NJ
M3 UBG Glen Chambers, 26 Parkin Close, Cropwell Bishop, Nottingham, NG12 3DG
M3 UBH C Davis, 1 Ashland Court, North Street, Crewkerne, TA18 7AP
M3 UBK Barry Oakley, 8a Crooked Mile, Waltham Abbey, EN9 1PS
M3 UBL David Hutchinson, 91, Pentland Avenue, Billingham, TS23 2RF
M3 UBQ John Robinson, Edenmoor, Market St, Whitworth, OL12 8RU
MM3 UBR James Branson, East Bank, South Road, Fochabers, IV32 7LU
M3 UBS David Foord, 28 Ferndale, Teversham, Cambridge, CB1 9AL
M3 UBT Colin Bowes, 11 burghwallis Lane, Sutton, Doncaster, DN69JU
M3 UBU Katherine Kettle, 19 St. Trinians Drive, Richmond, DL10 7SS
MW3 UBY M Davies, 70 Heol Bryncwils, Sarn, Bridgend, CF32 9UE
M3 UCA Francis Bano, 14 Norman Trollor Court, Cromer, NR27 9RR
M3 UCC Katrina Wilton, 71 Aston Clinton Road, Weston Turville, Aylesbury, HP22 5AB
M3 UCF Terry Austin, 51 Ashburnham Road, Ramsgate, CT11 0BH
M3 UCH David Francis, 1905 London Road, Leigh-on-Sea, SS9 2SY
MM3 UCI Margaret McCallum, 15 Quarry Road, Law, Carluke, ML8 5HB
M3 UCJ Christopher Hewett, 20 Cornwallis Avenue, Herne Bay, CT6 6UQ
M3 UCL Xiao Liu, Harrogate Ladies' College, Clarence Drive, Harrogate, HG1 2QG
M3 UCO Daniel Lawson, 2 The Blossoms, Fulwood, Preston, PR2 9RF
MI3 UCS M Edwards, 21 Old Grange Avenue, Carrickfergus, BT38 7UE
M3 UCU Stephen Finch, 25 Perth Avenue, Ince, Wigan, WN2 2HJ
M3 UCV Thomas Kilroy, 55 Summerfield Crescent, Brimington, Chesterfield, S43 1HB
M3 UCY Lucy Long, 25 St. Matthias Road, Deepcar, Sheffield, S36 2SG
M3 UCZ Beverly Smallbone, 46 Lesters Road, Cookham, Maidenhead, SL6 9LS
M3 UDA Gareth Lloyd, 2 Bryn y Coed, Holywell, CH8 7AU
MM3 UDB D Brown, 18 Louisa Drive, Girvan, KA26 9AH
M3 UDC Daniel Moran, 47 Radcliffe Park Road, Salford, M6 7WP
M3 UDD Matthew Fisher, 12 Abbey Mews, Pontefract, WF8 1TD
M3 UDF Desmond Fagan, 58 Main Street, Linton, Swadlincote, DE12 6PZ
M3 UDG David Gaut, 14 Bedale Close, Belmont, Durham, DH1 2BB
MM3 UDI Peter Pirie, Willowbank, Kirkton of Tough, Alford, AB33 8ER
M3 UDJ Mark Pharoah, 116 Chatsworth Street, Barrow-in-Furness, LA14 5TP
M3 UDK James Cunningham, 56 Askam Avenue, Pontefract, WF8 2PN
MM3 UDL B Nielsen, House Of Shannon, Fortrose, IV10 8RA
M3 UDN Paul Thompson, 3 Floyers Field, West Stafford, Dorchester, DT2 8FJ
MW3 UDO Raymond Oliver, 6 Clevedon Avenue, Sully, Penarth, CF64 5SX
MM3 UDQ John Smith, 10 High Street, Portknockie, Buckie, AB56 4LD
M3 UDS Wen SUN, 10 Scholfield Way, Eastbourne, BN23 6HQ
M3 UDU Samuel Hadfield, 56 Risborough Road, Bedford, MK41 9QW
MM3 UDV David Herd, 4 West Fairbrae Drive, Edinburgh, EH11 3SY
M3 UDW Megan Evans, 9 st teilos close, EBBW VALE, Blaenau Gwent, NP23 6NE
M3 UDZ Darren Mellor, 28 Winster Road, Staveley, Chesterfield, S43 3NJ
MM3 UEA E Ewing, Arisaig, Priestland, Darvel, KA17 0LP
M3 UEC Richard Griffiths, 8 Mount Street, King's Lynn, PE30 5NH
M3 UED G Henstridge, 23 Queen Mary Road, Salisbury, SP2 9LD
M3 UEE Jamie Price, 3 Perlethorpe close, Gedling, Nottingham, NG4 4GF
M3 UEF Paul Marchant, 16 Melrose Drive, Peterborough, PE2 9DN
M3 UEG Ailsa Evans, 9 st teilos close, EBBW VALE, Blaenau Gwent, NP23 6NE
M3 UEJ Chris Liversidge, Flat 3-8 Cromwell Terrace, Scarborough, YO11 2DT
M3 UEK N Hosker, 4 Alexandra Court, Coronation Close, Chester, CH2 3PZ
M3 UEL Susan Long, 25 St. Matthias Road, Deepcar, Sheffield, S36 2SG
M3 UEN Aaron Sammut, The Quaker Cottage, Wainfleet Bank, Skegness, PE24 4JP
M3 UEP Andrew Hudders, 22 Third Avenue, Wolverhampton, WV10 9PQ
M3 UEQ Gary Lambert, 17 Starcross Road, Weston-Super-Mare, BS22 6NY
M3 UER Peter Stone, 32 Worcestershire Lea, Warfield, Bracknell, RG42 3TQ
MM3 UET M Henderson, 22 Bowmont Place, East Kilbride, Glasgow, G75 8YG
MW3 UEU Keith Rowney, 14a Fagwr Road, Craig-Cefn-Parc, Swansea, SA6 5TB
M3 UEW Avril Allanson, 5 Kingsley Avenue, Crofton, Wakefield, WF4 1RN
M3 UEX David Carr, 36 Rookwood Mount, Leeds, LS9 0LL
M3 UEY Anthony Holder, 12a Evenlode Road, Gloucester, GL4 0JT
M3 UEZ Wayne Sargeant, 44a Nelson Street, Buckingham, MK18 1DA
M3 UFA Stephen Leggett, 246 Delaware Road, Shoeburyness, Southend-on-Sea, SS3 9NT
MI3 UFD Robert Stirrup, 211 Leckagh Drive, Magherafelt, BT45 6ND
M3 UFE Julia Chau, Harrogate Ladies' College, Clarence Drive, Harrogate, HG1 2QG
M3 UFF A Mullen, 81 Worcester Street, Stourbridge, DY8 1AX
M3 UFG Duncan Gunn, 40 The Pastures, Oadby, Leicester, LE2 4QD
MW3 UFH Christopher Livingstone-Lawn, 72 Ty Mawr Avenue, Rumney, Cardiff, CF3 3AG

M3 UFJ Stephen Jarvis, 10 Wood Lane, Wolverhampton, WV10 8HJ
M3 UFK Nigel Nash, 36 North Street, Walsall, WS2 8AT
M3 UFL William Oakey, 11 Boaford Grove, London, SW20 9LB
M3 UFS Adrian Brand, 6 Walnut Close, Milton, Cambridge, CB24 6LT
M3 UFT Derek Redman, 5 Dee Road, Lancaster, LA1 2QX
M3 UFU Kenneth Hillbeck, 28 Darent Avenue, Walney, Barrow-in-Furness, LA14 3NU
M3 UFW Nicholas McLean, 21 Matlock Avenue, Wigston, LE18 4NA
M3 UFX S Blackwell, 6 Buckingham Drive, Heanor, DE75 7TY
M3 UFY Jeffrey Poland, 14 malleson road, Liverpool, L139DF
M3 UFZ Anthony Saddington, 7 Baker Court, Thrapston, Kettering, NN14 4XA
M3 UGA Thomas Barnard, 8 Argyle Road, Poulton-le-Fylde, FY6 7EW
M3 UGB G Boast, 118 Barnsley Road, Moorends, Doncaster, DN8 4QR
M3 UGD Derrick Underwood, 33 Meadow Avenue, Rawmarsh, Rotherham, S62 7EE
M3 UGF Michael Sims, 133 Canterbury Road, Hawkinge, Folkestone, CT18 7BS
M3 UGH Michael Cole, 25 Freemans Road, Minster, Ramsgate, CT12 4EL
MI3 UGI Mervyn Downey, 18 Castlevue Park, Moira, Craigavon, BT67 0LN
M3 UGJ Gillian Douch, High House, Ulverston Road, Ulverston, LA12 0PZ
M3 UGK Michael Rimmer, 15 Brade Street, Southport, PR9 8LS
MM3 UGL Nigel Rogers, 108 Beechwood Road, Cumbernauld, Glasgow, G67 2NP
M3 UGO Darren Hill, 19 Farren Road, Birmingham, B31 5HH
M3 UGQ Peter Woodyard, 65 Raglan Street, Lowestoft, NR32 2JS
M3 UGR Clive Luckett, 257 Folkestone Road, Dover, CT17 9LL
M3 UGT Gary Taylor, 39 Stafford Road, St Helens, WA10 3JH
M3 UGV Stephen Jawor, 5 Cotteswold Rise, Stroud, GL5 1HD
M3 UGW Grace Wilcock, 42 Erskine Road, Colwyn Bay, LL29 8EU
M3 UGX William Owen, 8 Sandhurst Avenue, Lytham St. Annes, FY8 2DA
MD3 UGY M Wernham, Fair Isle, Lhoobs Road, Douglas, Isle of Man, IM4 3JB
M3 UGZ Robert Dearden, 218 South Street, Highfields, Doncaster, DN6 7JQ
M3 UHB Anthony Hartley, 47 Windyways, Little Sutton, Ellesmere Port, CH66 1JG
M3 UHC John Bramley, 17 oakholme rise, Worksop, S81 7LJ
M3 UHG Robert Smith, 20a Waverley, Skelmersdale, WN8 8BD
M3 UHH Hannah Hopkins, 3 Colegrave Road, Bloxham, Banbury, OX15 4NT
MI3 UHI Ernest Kyle, 2, wattstown, Coleraine, BT521SP
M3 UHJ Philip Hopkinson, 28 Stockdove Way, Thornton-Cleveleys, FY5 2AR
MM3 UHK William Spiers, 19/2 150 Charles Street, Glasgow, G21 2QF
MI3 UHL William Mark, 82 Glenleslie Road, Clough, Ballymena, BT44 9RH
M3 UHN Paul Sherratt, 39 vimy road, Leighton Buzzard, LU7 1FQ
M3 UHO Bethan Cayford, 13 Harford Square, Newtown, Ebbw Vale, NP23 5FA
M3 UHP C Rodway, 39 Megstone, Pimlico Court, Gateshead, NE9 5HG
M3 UHQ Lawrie Richardson, 4 The Elms, Plymouth, PL3 4BR
M3 UHS Ian Stephens, 2 Boniface Walk, Burnham-on-Sea, TA8 1RE
M3 UHV Daniel Lisi, 60 Middlecroft Road, Staveley, Chesterfield, S43 3XH
M3 UHW Marie Troth, 21 Willow Road, Bromsgrove, B61 8PN
M3 UHX Adrian Anderson, 89a Malmesbury Park Road, Bournemouth, BH8 8PS
M3 UHY Russell Bond, 21 Coleridge Close, Bletchley, Milton Keynes, MK3 5AF
MI3 UIA Paddy Dallas, 12 Glendun Crescent, Coleraine, BT52 1UJ
M3 UIB Bernard Wilde, Seaways, Cliff Rise, Fowey, PL23 1QQ
M3 UIC David Simpson, 140 Church Road, Redfield, Bristol, BS5 9HN
M3 UIF Martin Wheeler, Homeleigh, Station Road, Rochester, ME3 7RN
M3 UII Stephen ODonoghue, 15 Chandlers Close, New Waltham, Grimsby, DN36 4WH
M3 UIJ Mark Rogers, 22 Robson Drive, Hoo, Rochester, ME3 9EA
M3 UIK David McCrae, 37 burnside, Wigton, CA7 9RE
M3 UIL T Brice, 10a Nelson Drive, Exmouth, EX8 2PU
MI3 UIM Patrick McKeever, 7 St Canices Park, Eglinton, Londonderry, BT47 3AQ
MW3 UIN Liam Collins, 283 Graig Road, Godrergraig, Swansea, SA9 2NZ
M3 UIP J Dunkin, 6 Kingsley Grove, Grimsby, DN33 1NL
MW3 UIQ Angela Henderson, 45 brynamman road, lower brynamman, Ammanford, SA18 1TR
M3 UIS Andrew Marshall, 13 The Markhams, New Ollerton, Newark, NG22 9QX
M3 UIU Ubong Ukommi, 30 Oregano, Room 3, Hazel Farm, Guildford, GU2 9TY
MI3 UIV Victor Madden, 2, rathbeg, Limavady, BT49 0AT
MI3 UIW J Smyth, 39 Whitehill Park, Limavady, BT49 0QF
MM3 UIX Calum Dyer, 55 Duthie Road, Gourock, PA19 1XS
MW3 UIY Paul Henderson, 45 brynamman road, lower brynamman, Ammanford, SA18 1TR
M3 UJD Kevin Colman, 10 South Rise, North Walsham, NR28 0EE
M3 UJE Michael Corbett, 23 Heathfield Road, Fleetwood, FY7 7LY
M3 UJF Jacob Finney, 40 Tempest Avenue, Darfield, Barnsley, S73 9BJ
M3 UJH J Harbron, 48 Sheridan Road, Biddick Hall, South Shields, NE34 9JJ
M3 UJJ Scott Leazell, 15 Dawn Crescent, Upper Beeding, Steyning, BN44 3WH
M3 UJL J Bailey, 38 Barlow Drive, Sheffield, S6 5HQ
M3 UJM Ian Blackmore, 3 Backetts Oasts, Edenbridge, TN8 7AU
M3 UJN Christopher Hall, 10 First Street, Pont Bungalows, Consett, DH8 6JG
M3 UJO Jonathan Phillips, Flat L 15, International House, Guildford Court, Guildford, GU2 7JL
M3 UJP James Fletcher, 2 Sunflower Meadow, Irlam, M44 6TD
M3 UJQ Allan Hackman, The Herdsman Cottage, Brighouse Bay, Kirkcudbright, DG6 4TT
M3 UJR Janet Roberts, 93 Earlshall Road, Eltham, London, SE9 1PP
M3 UJS Joshua Suter, 11 Summerdown Close, Durrington, Worthing, BN13 3QG
M3 UJT I Dransfield, Gardener Ground House, West End, Goole, DN14 8RW
M3 UJV George Roth, 121 St. Annes Road, Wolverhampton, WV10 6SL
M3 UJX Louise Parish, 0 Courtwick Road, Wick, Littlehampton, BN17 7NE
M3 UJY Charles Meakin, 102 Ryknield Road, Kilburn, Belper, DE56 0PF
M3 UJZ James Cooke, Iolanthe, Chidham Lane, Chichester, PO18 8TH
M3 UKB C Price, 9 Arlington Avenue, Aston, Sheffield, S26 2AA
M3 UKD John Parfrey, 47 Ford Lane, Rainham, RM13 7AS
M3 UKF Paul Seaton, Sunnymead, Newland, Barnstaple, EX32 0ND
MM3 UKG David Gilmour, 35 Hailes Gardens, Edinburgh, EH13 0JH
M3 UKH Gareth Nelson, 812 Hessle Road, Hull, HU4 6RD
M3 UKJ John Boyd, 46 Tunbridge Road, Sunderland, SR3 4BG
M3 UKK Michael Davison, Maes y Neuadd Hotel, Talsarnau, LL47 6YA
M3 UKN N Hewitt, Redwood House, Adbury Holt, Newbury, RG20 9BW
M3 UKO Neil Batchelor, 1 Withies Close, Withington, Hereford, HR1 3PS
M3 UKP Geoffery Stratford, 23 The Fairway, Banbury, OX16 0RR
M3 UKR Anthony Rogers, 260 Griffiths Drive, Wolverhampton, WV11 2JS
M3 UKU Marwan Qassim, Winchester Road, Kings Somborne, Stockbridge, SO20 6NY
M3 UKV T Vincent, 9 Sleapford, Long Lane, Telford, TF6 6HQ
MI3 UKW Martin McErlean, 38a Culbane Road, Portglenone, Ballymena, BT44 8NZ
M3 UKX J Cornish, 2 Micklehill Drive, Shirley, Solihull, B90 2PU
M3 UKY Debbie Gordon, 32 Claremont Road, Stockport, SK2 7AR
M3 ULB S Owen, 21 Market Place, Hingham, Norwich, NR9 4AF
M3 ULE L Lawrence, Rookery Rise, 8 Woodside, Brede, TN31 6DS
M3 ULI Catherine Dunn, 13 Springfield Road, Leyland, PR25 1AR
M3 ULL M Bull, 5 Roach Place, Rochdale, OL16 2DD
M3 ULM Ian Cottom, 8 Bridgewater Rise, Brackley, NN13 6DA
M3 ULN Richard Fricker, 2 Buttermere Drive, Allestree, Derby, DE22 2SN
M3 ULO Sharon Clarke, 2 Dawn Crescent, Upper Beeding, Steyning, BN44 3WH
M3 ULQ Jan Gromadzki, 13 Merrill Heights, Maidenhall Approach, Ipswich, IP2 8GA
M3 ULS John Firth, 36 Howley Grange Road, Halesowen, B62 0HW
M3 ULT S Dickinson, 6 Mill Hill, Wombwell, Barnsley, S73 8SJ
M3 ULU Adrian Eizzard, 41 Sycamore Drive, Waddington, Lincoln, LN5 9DR
M3 ULW Leigh Weston, 114 Morland Road, Ipswich, IP3 0LZ
M3 ULX Gordon Quilter, 83 Jameson Street, Wolverhampton, WV6 0NT
M3 ULZ Andrew Nokes, 3 Eastview, Ditcheat, Shepton Mallet, BA4 6PN
M3 UMA Michael Abel, 121 Angela Road, Horsford, Norwich, NR10 3HF
M3 UMB Michael Bryans, The Lodge, The Warren Croydon Road, Bromley, BR2 7AL
MI3 UMC Jim Mcauley, 29 Ilse Court, Larne, BT40 3NT
M3 UMD Benjamin Walton, 40 Princess Street, Mapplewell, Barnsley, S75 6ET
M3 UMJ Mitchell Hockin, 7 Gourders Lane, Kingskerswell, Newton Abbot, TQ12 5DZ
M3 UML Michael Layton, 16 Gwelmor, Camborne, TR14 7BP
M3 UMM Margaret Case, 6 BOLDVENTURE, CLOSE, St Austell, PL25 3DY
MD3 UMN D Kneale, 4 Glashen Terrace, Ballasalla, Isle of Man, IM9 2ET
M3 UMR Sarah Rutt, Granthorpe, Hull Road, Hull, HU11 5RN
M3 UMV Paul Shaves, 33 Derwent Drive, Bletchley, Milton Keynes, MK3 7BG
M3 UMW Michael Wilkins, 11 Brockwell Lane, Kelvedon, Colchester, CO5 9BB
M3 UMX Russell Bates, 61 Park View, Crowmarsh Gifford, Wallingford, OX10 8BN
MM3 UMY Caroline Brogan, 13 Mitchell Avenue, Cambuslang, Glasgow, G72 7SQ
MM3 UMZ John Paul Bain, 13 Mitchell Avenue, Cambuslang, Glasgow, G72 7SQ
M3 UNB Keith Puttock, 12 Beechfields, School Lane, Petworth, GU28 9DH
M3 UNF Hugh Coram, Flat 3 19 Courtland Road, Paignton, TQ3 2AB
M3 UNG Neil Graham, 8 Owlcotes Terrace, Pudsey, LS28 7PF
M3 UNH Andrew Hitchcott, 121 Oakhurst Road, Acocks Green, Birmingham, B27 7PB
M3 UNI Neil Bradshaw, 2 Lynton Avenue Anlaby Park Road South, Hull, HU4 7DA
M3 UNK M Wright, 2 Regent Road, Church, Accrington, BB5 4AR
M3 UNL Paula Webb, 104 Birds Nest Avenue, Leicester, LE3 9ND
M3 UNP Nandan Patel, 16 Camellia Court, 18 Copers Cope Road, Beckenham, BR3 1NB
MM3 UNQ Nicki McIntyre, 27 Broadstone Avenue, Port Glasgow, PA14 5AT
M3 UNR Andrew Worlledge, 181 Roselands Drive, Paignton, TQ4 7RN
M3 UNT Peter Morris, 689 Tonge Moor Road, Bolton, BL2 3BW
M3 UNY Luke Bain, 45 Larpool Crescent, Whitby, YO22 4JD
MW3 UNZ Stephen valentine, 12 Cwrt y Glyn, Carmel, Llanelli, SA14 7SA
M3 UOB Christopher Bradley, 6 Copeland Row, Evenwood, Bishop Auckland, DL14 9PY
M3 UOC Jamal Mohammed, 2 Moffat Avenue, Ipswich, IP4 3JH
M3 UOD John Cooke, 2 church street close, thurnscoe, Rotherham, S63 0QT
MM3 UOE Robert Wilson, 12 Queen Road, Irvine, KA12 0XA
M3 UOF Kevin Mills, 106 Goodwin Crescent, Swinton, Mexborough, S64 8QR
M3 UOG Tony Cater, 14, britannia road marshgreen, Wigan, WN5OEN
M3 UOJ Kevin Barry, 25 Delabole Road, Merstham, Redhill, RH1 3PB
M3 UOK A Abraham, Flat 2, 41 Francis Road, Birmingham, B33 8SL
M3 UOL L Haworth, 139 Manchester Road, Accrington, BB5 2NY
M3 UOM Matthew Catterall, 2 Cedar Road, Bishop Auckland, DL14 6ET
M3 UON Sarder Kamal, 40 Catterick Way, Borehamwood, WD6 4QT
M3 UOO Richard Bone, 16 Gray Close, Warsash, Southampton, SO31 9TB
MM3 UOR Richard Paterson, 1/r 162, glasgow street, Ardrossan, KA228HA
MM3 UOS Davy Young, 81 leaven place, Irvine, KA12 9PA
M3 UOX Adam Tate, 24 Brentingby Close, Melton Mowbray, LE13 1ES
M3 UOY Greg Fripp, 35 Kiln Close, Bovey Tracey, Newton Abbot, TQ13 9YL
M3 UOZ A Knight, 39 Thurnview Road, Evington, Leicester, LE5 6HL
M3 UPB Leo Banahan, 18 Lynn Road, Ely, CB6 1DA
M3 UPF Mark Elkington, Warner Leys, Iron Hill Farm, Daventry, NN11 6YJ
M3 UPJ Paul Harley, 16 Clover Drive, Rushden, NN10 0TZ
M3 UPK John Addy, 12 Wortley Avenue, Swinton, Mexborough, S64 8PT
M3 UPL Brenda Banks, 44 Manor Road, Swinton, Mexborough, S64 8PY
M3 UPN Peter Needham, 9 Westwood, Broughton, Brigg, DN20 0AU
M3 UPO Dean Clarke, 2 Dawn Crescent, Upper Beeding, Steyning, BN44 3WH
M3 UPP Lori-Ann Wilkes, 24 Ilminster, Dunster Crescent, Weston-Super-Mare, BS24 9EB
M3 UPQ M Davis, 3 Pollards Court, Rochford, SS4 1GH
M3 UPT S Thomas, 103 Liverpool Road, Upton, Chester, CH2 1BB
M3 UPX Mark Davies, 11 High Street, Malltraeth, Bodorgan, LL62 5AS
MM3 UPY Scott Dunbar, 34 Windyknowe Crescent, Bathgate, EH48 2BU
M3 UQA Philip Allison, Flat 11, Forest Court, 5-11 Salisbury Road, Fordingbridge, SP6 1EG
M3 UQB Christopher Kerridge, 2 Allerdale Close, Thirsk, YO7 1FW
M3 UQD Eduardo Subbutella Ricciardi, 97 Harp Island Close, London, NW10 0DQ
M3 UQE Martin Burstow, 40 Wyndham Road, Petworth, GU28 0EQ
M3 UQF Gail Hillbeck, 28 Darent Avenue, Walney, Barrow-in-Furness, LA14 3NU
M3 UQG Harry Roxbrough, 17 Stanwell Close, Sheffield, S9 1PZ
M3 UQH Jonny Parrett, 6 Shelley Road, East Grinstead, RH19 1TA
M3 UQI Geoffrey Jones, 57 Oxford Road, Banbury, OX16 9AJ
M3 UQJ Graham Dyson, 6 Twynersh Avenue, Chertsey, KT16 9DE
M3 UQL David Smith, 131 canons walk, Thetford, IP24 3PT
M3 UQN Duncan Taylor, 1 Mayfield Farm Cottages, Reston, Eyemouth, TD14 5LG
M3 UQO Graham Smith, 92 Brighton Road, Banstead, SM7 1BU
M3 UQP Kenneth Eccleston, 53 Oxford Drive, Woodley, Stockport, SK6 1JE
MM3 UQT C Page, 37 Pine Crescent, Hamilton, ML3 8TZ
M3 UQY R Cox, 17 Hyde Lane, Upper Beeding, Steyning, BN44 3WJ
M3 UQZ Darren Hartley, Flat 4, The Old Mill, Station Road, Ellesmere Port, CH66 1NY
M3 URA Maura Barber, 4 Oatlands Road, Tadworth, KT20 6BS
M3 URD Richard Dell, 18 Greenacres, Fulwood, Preston, PR2 7DA
M3 URE Robert Eddy, 15 Western Place, Penryn, TR10 8HQ
M3 URF Richard Freeman, 9 Bramley Road, Wisbech, PE13 3PA
MW3 URG R Grayson, Willow Lodge, Croeslan, Llandysul, SA44 4SJ
M3 URH L Harman, 7 Loughborough Road, Walton on the Wolds, Loughborough, LE12 8HT
M3 URO Daniel Fullick, 69 Shakespeare Road, St. Dials, Cwmbran, NP44 4LW
M3 URQ Stewart Urquhart, 1 Hatchery Cottage, Station Road, Duns, TD11 3HS
M3 URS Rebecca Singh, Stainburn House, Barrowby Lane, Harrogate, HG3 1HY
M3 URT Stephen Mather, 104 Appley Lane North, Appley Bridge, Wigan, WN6 9DS
M3 URV R Earnshaw, 53 Blue Waters Drive, Paignton, TQ4 6JF
M3 URW Roy Woodford, SOUTH CALVADNACK, CARNMENELLIS, Redruth, TR166PN
M3 URX David Craig, Pear Tree Cottage, Cripps Corner Road, Robertsbridge, TN32 5QS
M3 URZ Carl Evans, 1 Rialto Road, Mitcham, CR4 2LT
M3 USB M Percevic, 12 Ash Road, Crewe, CW1 4DU
M3 USC Stephen Coleman, 32 Southwell Road, Wisbech, PE13 3LQ
M3 USF Andrew Willis, 17 Ladypit Terrace, Whitehaven, CA28 6AQ
M3 USH S Hall, 14 Nicholson Place, East Hanningfield, Chelmsford, CM3 8UT
M3 USJ Sarah Hall, 5 Ropery Lane, Barton-upon-Humber, DN18 5TW
M3 USK Curtis Burke, 523 Monnow Way, Bettws, Newport, NP20 7DW
MM3 USN J Challis, Bay Villa, Strachur, Cairndow, PA27 8DE
MW3 USP Steven Barwell, 50 Mill Close, Caerphilly, CF83 2LL
M3 USQ Lawrence King, 1 Kingsbury Drive, Old Windsor, Windsor, SL4 2NQ
MW3 USS D Provis, Dingle Gardens, Croesbychan, Aberdare, CF44 0EJ
M3 UST Graham Evans, 2 Tower Farm Cottages, Featherbed Lane, Hemel Hempstead, HP3 0BT
MM3 USV Alan Dunn, 9 Glen Roy Drive, Neilston, Glasgow, G78 3QJ
M3 USW Pamela Jenkins, 48 The Pantiles, Bexleyheath, DA7 5HG
MW3 USX Stephen Tozer, 110 Glanffornwg, Wild Mill, Bridgend, CF31 1RL
M3 USZ Robert Lane, 9 Hartoft Road, Hull, HU5 4JZ
MM3 UTH P Dower, 1670 Maryhill Road, Glasgow, G20 0HJ
M3 UTJ Stuart Crichton, 27 Rosewood Ave, Stockport, SK4 2DQ
M3 UTK Colin Couston, Bridge Cottage, Stanlake Lane, Reading, RG10 0BL
M3 UTM Alan Williams, 1 Nimmings Close, Birmingham, B31 4TA
M3 UTN Kenneth Bailey, The Firs, South Chard, Chard, TA20 2RX
M3 UTP Trish Pentz, 30 Lindrick Way, Harrogate, HG3 2SU
M3 UTQ Daniel Bell, 18 Julius Hill, Warfield, Bracknell, RG42 3UN
MM3 UTU Margaret Magee, 30 Burnfield Drive, Mansewood, Glasgow, G43 1BW
M3 UTV M Compagno, 18 Bromford Crescent, Birmingham, B24 9RJ
MI3 UTY Mark Regan, 80 Killowen Drive, Magherafelt, BT45 6DS
M3 UTZ Sulaiman Abdullah, 24 The Grove, Walsall, WS5 4BX
M3 UUE Helen Roberts, 7 Hungerford Road, Stourbridge, DY8 3AB
M3 UUF Robert Dicker, 38 Inkerman Road, Southampton, SO19 9DA
M3 UUG Sue Lenton, 45 Holland Gardens, Fleet, GU51 3NF
M3 UUL Gary Rowe, 10 Alexander Avenue, Selston, Nottingham, NG16 6FW
M3 UUN Thomas Harding, 7 Chiltern Avenue, Poulton-le-Fylde, FY6 7DY
M3 UUO Daryl Lynch, 12 Shipley Close, Blackpool, FY3 7UJ
M3 UUS Simon Mallinson, 63 Celandine Avenue, Locks Heath, Southampton, SO31 6WZ
MM3 UUT Lewis Thomson-Best, 5 Gib Grove, Dunfermline, KY11 8DH
M3 UUW John Douch, High House, Ulverston Road, Ulverston, LA12 0PZ
M3 UUY Derlwyn Williams, 10 Bronllys, Gaerwen, LL60 6JN
M3 UUZ Robert Chandler, 26 Chalky Bank, Gravesend, DA11 7NY
M3 UVA Jacqueline Grainger, 32 Ellenfoot Drive, Maryport, CA15 7DB
M3 UVC Ramon Milton, 66 Hoo Marina Park, Vicarage Lane, Rochester, ME3 9TG
M3 UVD David Tregear, 3 Mossbourne Road, Poulton le Fylde, Blackpool, FY6 7DU
MM3 UVF William McBlain, 9 Reid's Avenue, Stevenston, KA20 4BB
M3 UVJ J Binfield, 55 Gladstone Road, Broadstairs, CT10 2HY
M3 UVK William Penny, 159 COXFORD ROAD, MAYBUSH, Southampton, SO16 5JX
M3 UVM Robert Murphy, 23 Lowndes Close, Stockport, SK2 6DW
M3 UVO Andrew Ruocco, 83 Epsom Road, Morden, SM4 5PR
M3 UVQ Andrew Barton, 11 Grove Avenue, Beeston, Nottingham, NG9 4ED
M3 UVS Sarah Thorne, 72 Devonshire Road, London, E16 3NJ
M3 UVT Michael Garry, 34 Conway Road, Paignton, TQ4 5LH
M3 UVV Andrew Mackay, 16 Vicarage Close, Chard, TA20 2HH
M3 UVX D Housden, 12 Regent House, Cheltenham Gardens, Southampton, SO30 2UD
M3 UVY Anthony Marrison, 271 Reeth Place, Newton Aycliffe, DL5 7NA
M3 UWB Martin Tromans, 10 Crofters View, Little Wenlock, Telford, TF6 5AU
M3 UWE Douglas Scott, 7 Teal Close, Brookside, Telford, TF3 1NY
M3 UWF Steven Roberts, 92 Upper Horsebridge, Hailsham, BN27 1NY
M3 UWI K Jones, 41 Milton Brow, Weston-Super-Mare, BS22 8DD
M3 UWJ R Wilmot, 41 Milton Brow, Weston-Super-Mare, BS22 8DD
M3 UWM Terry Kitto, 4 Pennard, St. Breock, Wadebridge, PL27 7LL
M3 UWR Sheena Dawson, 51 St. Edwards Road, Gosport, PO12 1PW
M3 UWT Andrew Chapman, 10 Derwent Road, Seaton Sluice, Whitley Bay, NE26 4JH
M3 UWU Mark Redfern, 31 Malvern Crescent, Dudley, DY2 0RZ
M3 UWV Colin King, 5 Manor Road, Tamworth, B77 0PE
M3 UWW Rosario Massimino, 115 Trelowarren Street, Camborne, TR14 8AW
MU3 UWX Malcolm Barker, Rosee Terres, Les Effards Road, St Sampson, Guernsey, GY2 4YW
M3 UWY Timothy Baddeley, 44 Lowry Close, Willenhall, WV13 3BD
M3 UWZ John Evans, 2 Collfryn Cottages, Bethesda Bach, Caernarfon, LL54 5SF
M3 UXC A Spaxman, 12 Stanhope Gardens, Barnsley, S75 2QB
M3 UXE Emily Bruce, 26 Queens Road, Wilbarston, Market Harborough, LE16 8QJ
M3 UXF Michael Bridgehouse, 43 Age Croft, Oldham, OL8 2HG
M3 UXG Michael Leech, 11 Westlake Close, Torpoint, PL11 2BZ
M3 UXH Kieran Sowter, 55 Ward Street, New Tupton, Chesterfield, S42 6XR
M3 UXI Lewis Allanson, 5 Kingsley Avenue, Crofton, Wakefield, WF4 1RN

**IMPORTANT NOTE**

**Revalidate licence to avoid revocation** – Ofcom has advised the Society that plans will be drawn up to revoke licences that have not been revalidated as required by the licence conditions. The quickest way to revalidate is to do so online via the Ofcom website: *https://services.ofcom.org.uk/* or by email: *amateur.validations@ofcom.org.uk* Ofcom staff are available to help, but please be patient during times of heavy workload.

M3 UXK Ken Jones, 14 Second Avenue, Garston, Watford, WD25 9PX
M3 UXL Andrew Bond, 21 Coleridge Close, Bletchley, Milton Keynes, MK3 5AF
M3 UXM Matthew Bruce, 26 Queens Road, Wilbarston, Market Harborough, LE16 8QJ
M3 UXN Paul Overton, 39 Bridle Road, Madeley, Telford, TF7 5HB
MM3 UXO Liam Aitken, 92b Belville Street, Greenock, PA15 4TA
M3 UXR Marc Griffiths, 11 Frogwell Park, Chippenham, SN14 0RB
M3 UXS Carol Dutton, 7 Ellery Grove, Lymington, SO41 9DX
M3 UXU D Almond, 2 Farm Veiw, New Tupton, Chesterfield, S42 6BD
M3 UXX Michele Selvey, 52 Coppice Close, Cheslyn Hay, Bella Casa, Walsall, WS6 7EZ
M3 UYC David Griffin, 4 Willowside Way, Royston, SG8 5ET
M3 UYF Martin Heritage, High View, Common Lane, Corley, Coventry, CV7 8AQ
M3 UYG Andrew Goldsmith, 7 Fengate Drove, Weeting, Brandon, IP27 0PW
M3 UYH Gillian Broadbent, 7 James Close, Llanon, SY23 5HP
MW3 UYJ Joeseph Davies, 19 Falcon Place, Blaenymaes, Swansea, SA5 5NX
M3 UYK Yan Ki Chiu, Harrogate Ladies' College, Clarence Drive, Harrogate, HG1 2QG
M3 UYL George Richards, 1 Brisbane Road, Weymouth, DT3 6RB
M3 UYO Roy Dewis, 6 St Nicolas Close, Pevensey, BN245LB
M3 UYQ Edward Hull, Flat 17, Parade Court, Portsmouth, PO2 9RB
M3 UYU Barry Elderbrant, 20 Loxley Road, Southport, PR8 6NR
M3 UYV Julian Rudd, 5 St Andrews Close, Blofield, Norwich, NR13 4JX
M3 UYW Penny Underhill, 35 Windermere Road, Reading, RG2 7HU
MW3 UYX Ashley Harvey, 15 Pen y Lan, Penclawdd, Swansea, SA4 3LL
M3 UYY Marrianne Dale, 37 Bussey Road, Norwich, NR6 6JF
M3 UZA Peter Greenhalgh, 33 Shepherds Lane, Chester, CH2 2DH
M3 UZB Simon Bailey, 23 Maple Avenue, Tolladine, Worcester, WR4 9RD
M3 UZE Benjamin Monksummers, 29 Cloverfields, Peacemarsh, Gillingham, SP8 4UP
MW3 UZH James Alfei, 9 Brookside, Gowerton, Swansea, SA4 3AY
M3 UZK Adrian Lancefield, 19 Tawny Sedge, King's Lynn, PE30 3PW
M3 UZL Susan Clarke, Brimham Lodge Fm, Harrogate, HG3 3HE
M3 UZN Roland Truelove, 104 Malines Avenue, Peacehaven, BN10 7RL
MW3 UZO Daniel White, 222 St. Fagans Road, Cardiff, CF5 3EW
MW3 UZP Jason Young, 10 Heol Fion, Gorseinon, Gorseinon, SA4 4PN
M3 UZV John Porter, 29 Dainton Grove, Birmingham, B32 3EJ
M3 UZW j Boag, 60 Harebell, Amington, Tamworth, B77 4NA
M3 UZZ Gerrard Hamilton, 5 Eascott Common, Eastcott, Devizes, SN10 4PL
M3 VAE C Johns, 6 Cranham Close, Bristol, BS15 4QB
M3 VAF Keith Brown, 41 Church Street, Swinton, Mexborough, S64 8EF
M3 VAG B Gittings, 29 Highdown Way, Swindon, SN25 4YD
MM3 VAH Victoria Hamilton, 10/3 Fox Street, Edinburgh, EH6 7HN
M3 VAM Martyn Medcalf, 47 Paddock Drive, Chelmsford, CM1 6UX
M3 VAQ James Bowley, 2 Cottage, Middle Battenhall Farm, Worcester, WR5 2JL
M3 VAR A Dokic, 28 Tudor Gardens, Shoeburyness, Southend-on-Sea, SS3 9JG
M3 VAS V Papanikolaou, 104 West Drive Gardens, Soham, Ely, CB7 5EX
M3 VAT D Fielding, 2 Christchurch Road, Bradford-on-Avon, BA15 1TB
MW3 VAY Alan Davies, 19 Williams Place, Merthyr Tydfil, CF47 9YH
M3 VBD Robin Darby, 4 Whately Mews, Whately Road, Lymington, SO41 0XS
MM3 VBF Scott Campbell, 78 Liddel Road, Cumbernauld, Glasgow, G67 1JE
M3 VBG Bridget Garry, 34 Conway Road, Paignton, TQ4 5LH
M3 VBH A Townsend, 42 Grove Avenue, Yeovil, BA20 2BD
M3 VBI John Broadhurst, Flat 2, Pennant Court, Rowley Regis, B65 8DW
M3 VBL Jeff Williams, 3 Corner Field, Kingsnorth, Ashford, TN23 3NH
M3 VBM V Maynard, 34 Heath Farm Park, Barford St. Martin, Salisbury, SP3 4BQ
M3 VBN Keith Sloan, Woodland Halt, Old Station Road, Winchester, SO21 1BA
M3 VBP George Bramham, 1 Watson Avenue, Dewsbury, WF12 8PZ
M3 VBT V Be-Dard, 53 Cottingley Crescent, Leeds, LS11 0HZ
M3 VBV Gillian Foster, 18 Austen Avenue, Long Eaton, Nottingham, NG10 3GG
M3 VBY D Potter, 30 Mersham Gardens, Goring-by-Sea, Worthing, BN12 4TQ
M3 VBZ Anthony Bolton, 26, St Margarets Avenue, Sutton, SM3 9TT
M3 VCB Christopher Burbridge, 9 Victoria Road, Stirchley, Birmingham, B30 2LS
MI3 VCI Gary Lyttle, 37 Cloyfin Park, Coleraine, BT52 2BL
M3 VCK Michael Brown, 9 Warsop Road, Barnsley, S71 3NR
M3 VCM Charles Mallory, 11 Baymead Meadow, North Petherton, Bridgwater, TA6 6QW
M3 VCO A Knowles, 260 Haunchwood Road, Nuneaton, CV10 8DL
M3 VCP S Pryke, 85 Ascot Drive, Ipswich, IP3 9BY
M3 VCQ Sharon Wilson, 82 Lennox Road, Sheffield, S6 4FN
M3 VCV L smith, 11 Appleby Avenue, Timperley, Altrincham, WA15 7HY
M3 VCW C West, 1 Willetts Mews, Hoddesdon, EN11 9DX
M3 VCY Graham Shakespeare, 6 Waterworks Cottages, Clough Road, Hull, HU6 7QB
M3 VCZ Mark Rigby, 75 Manchester Road, Deepcar, Sheffield, S36 2QX
M3 VDA Andrew Crowther, 11 Goodman Court, Central Drive, Chesterfield, S44 5BA
M3 VDE Richard Ferguson, 31 Barton Court Road, New Milton, BH25 6NW
M3 VDF Gary Bertola, 17 Caraway Drive, Branston, Burton-on-Trent, DE14 3FQ
M3 VDH David Hind, 116 Gowthorpe, Selby, YO8 4HA
M3 VDL M Gosling, 15 Ridgeway Close, Studley, B80 7PL
M3 VDN John Owen, 90 Granville Drive, Kingswinford, DY6 8LW
M3 VDO David Oliver, 11 Crooked Creek Road, Rendlesham, Woodbridge, IP12 2GL
M3 VDP D Powell, 36 Beechwood Drive, Stone, ST15 0EH
M3 VDT D Cattermole, Blaxhall Hall Crossing, Little Glemham, Woodbridge, IP13 0BP
M3 VDU Ben Marston, 111 Averil Road, Leicester, LE5 2DE
M3 VDV Gary Haggas, 30 Bracknell Road, Thornaby, Stockton-on-Tees, TS17 9AU
M3 VDY Robert Drake, 14 Park Road, Hunstanton, PE36 5BP
M3 VDZ George Yuill, 14 Gardyn Croft, Taverham, Norwich, NR8 6UZ
MM3 VEE J Cochrane, 72 Clarkwell Road, Hamilton, ML3 9RQ
MM3 VEG J Wilson, Bank Cottage, 186 Kilsyth Road Banknock, Bonnybridge, FK4 1HX
MW3 VEH David Thomas, 57 Brynhyfryd Street, Treorchy, CF42 6DT
M3 VEJ R Buckwell, 75 Brookside Avenue, Polegate, BN26 6DQ
MW3 VEL L Wilson, 48 The Woodlands, Brackla, Bridgend, CF31 2JG
M3 VEM C Vernon, 80 Shirley Drive, Worthing, BN14 9BB
M3 VEN P Hoy, 39 Blackbird Road, Caldicot, NP26 5RE
MI3 VEQ Brian McConnell, 108 Moss Road, Lambeg, Lisburn, BT27 4NU
M3 VEW Ashley Allanson, 5 Kingsley Avenue, Crofton, Wakefield, WF4 1RN
M3 VEX K Fox, 39 Felton Avenue, South Shields, NE34 6RY
M3 VEY Anthony Selvey, 52 Coppice Close, Cheslyn Hay, Bella Casa, Walsall, WS6 7EZ
MW3 VFB Brian Pugh, Plas Newydd, 12 Fair Meadow Close, Milford Haven, SA73 3TF
M3 VFC Michael Derringer, 19 Skylark Road, Trumpington, Cambridge, CB2 9AQ
M3 VFE Patrick Goddard, 62, Woodlands Drive, Thetford, IP24 1JJ
MI3 VFF James Dunlop, 34 Keel Park, Moneyrea, Newtownards, BT23 6DE
MI3 VFJ Noel Orr, 22 Shrewsbury Drive, Bangor, BT20 3JF
MM3 VFK Jon Scally, 28 Seymour Avenue, Kilwinning, KA13 7PQ
M3 VFL Anthony Senior, 38 haslemere, way, Crewe, CW1 4JZ
M3 VFM V Millard, 20 Droveway Gardens, St. Margarets Bay, Dover, CT15 6BS
MW3 VFN Emlyn Thomas, 29 Maes y Wern, Carway, Kidwelly, SA17 4HF
M3 VFP Nigel Roberts, 40 Armour Road, Tilehurst, Reading, RG31 6HN
M3 VFS Martin Toher, The Chapel, Station Road, Darlington, DL2 1JG
M3 VFU M Whitehead, 19 Wrose Brow Road, Shipley, BD18 2NT
MI3 VFZ Tyrone Currie, 26 High Street, Portaferry, Newtownards, BT22 1QT
M3 VGF Adrian Lewington, 6 Brookhill Road, Darton, Barnsley, S75 5EL
M3 VGH Gavin Hunter, 5 Charlton Grove, Silsden, Keighley, BD20 0QG
MW3 VGJ Alexander Lloyd, 4 Gladstone Terrace, Miskin, Mountain Ash, CF45 3BS
M3 VGK B Lunn, 204a Main Street, Horsley Woodhouse, Ilkeston, DE7 6AX
M3 VGP Peter Temple, 136 Roborough Close, Bransholme, Hull, HU7 4RP
M3 VGT Alexander Andover, Bishoper Farmhouse, Brokenborough, Malmesbury, SN16 9SR
M3 VGX Michael Price, 52 Newmarket Street, Norwich, NR2 2DW
M3 VGZ Dave Adshead, 16 Moat Way, Swavesey, Cambridge, CB24 4TR
M3 VHA Richard Parker, 29 Hill Lea Gardens, Cheddar, BS27 3JH
M3 VHB Stephen Allen, 27 Cottons Meadow, Kingstone, Hereford, HR2 9EW
M3 VHC W Ho, Harrogate Ladies' College, Clarence Drive, Harrogate, HG1 2QG
M3 VHE Juha Heinonen, Riittiontie 155, Vampula, Finland, 32610
M3 VHH Kenneth Foster, 10 Bleaswood Road Oxenholme, Kendal, LA9 7EY
M3 VHI Robert Silcox, 103 Oakdale Road, Downend, Bristol, BS16 6EG
MM3 VHM Vikki Moran, 31 Hermitage Crescent, Coatbridge, ML5 4NE
M3 VHO Fai Ling Vania Ho, Harrogate Ladies' College, Clarence Drive, Harrogate, HG1 2QG
M3 VHQ Joe Winson, 35 Newington Avenue, Southend-on-Sea, SS2 4RD
M3 VHU John Redfearn, 3 Taylor Hill Road, Huddersfield, HD4 6HN
M3 VHV Luke Kelly, 9 Ham Lane, Farrington Gurney, Bristol, BS39 6TW
MI3 VHW Tom Boyd, 40 Walnut Park, Larne, BT40 2WF
M3 VHZ Stuart Southern, 37 Conway Road, Calcot, Reading, RG31 4XP
M3 VIA Nigel Purkiss, 357 Fair Oak Road, Eastleigh, SO50 8AA
M3 VIB Maciej Michalak, 100 Nursery Lane, Northampton, NN2 7TJ
M3 VIG Danielle Potter, 1 Wentworth Road, Rugby, CV22 6BG
M3 VII Andrew Price, 49 Pigeon Bridge Way, Aston, Sheffield, S26 2GX
M3 VIJ Don Williams, 18 Lower Greave Road, Meltham, Holmfirth, HD9 4DY
M3 VIO Charlotte Skinner, Beeston Marina Ltd, 1a the Quay, Beeston Marina, Riverside Road, Nottingham, NG9 1NA
M3 VIR Soon-Young Kim, 9 Magness Road, Deal, CT14 9JF
MM3 VIS J Dowson, 19 Tweed Crescent, Wishaw, ML2 8QR
M3 VIU John Bettles, 2 Ellfield Close, Bristol, BS13 8EF
MM3 VIV V Smith, 10 Quilco, Dounby, Orkney, KW17 2HW
M3 VIW C Wall, 26 Wallace Lane, Whelley, Wigan, WN1 3XT
M3 VJC J Cooper, 5 Crosswalla Fields, Helston, TR13 8XH
M3 VJE John Edmunds, 22 Horsewhim Drive, Kelly Bray, Callington, PL17 8GL
M3 VJI Declan McEvoy, 33 Heathcote Drive, Hasland, Chesterfield, S41 0BB
M3 VJJ V Bowkett, 9 Gwealmayowe Park, Helston, TR13 0PE
M3 VJL Vanessa Lea, 30 Cardiff Road, Pwllheli, LL53 5NU
M3 VJM James Ball, 26 Verona Court, Yeo Vale Road, Barnstaple, EX32 7EN
MW3 VJN Tristan Thomas, Emlyn House Cawdor Terrace, Newcastle Emlyn, SA38 9AS
M3 VJO Jonathan Sawyer, 9 Waller Court, Caversham, Reading, RG4 6DB
M3 VJS V Sansom, 70 Valley Road, West Bridgford, Nottingham, NG2 6HQ
M3 VJW Janice Whittington, Watertown Farm, Landcross, Bideford, EX39 5JA
M3 VJX Philip Buckley, 9 Carton Close, Rochester, ME1 2QF
MW3 VKA Anthony Vincent, 88 Lake Street, Ferndale, CF43 4HE
M3 VKB V Britton, Badgers Hollow, Witt Road, Salisbury, SP5 1PL
M3 VKF Robert Mottershead, 10 St. Mary Close, Blackpool, FY3 7UB
M3 VKJ K Jones, Railway Crossing Cottage, Ash Road, Sandwich, CT13 9JB
M3 VKM R Morgan, 14 Woodland Road, Pontllanfraith, Blackwood, NP12 2LS
M3 VKN Ian Astley, 1 Howard Crescent, Durkar, Wakefield, WF4 3AJ
MM3 VKO S MacDonald, 366 Millcroft Road, Cumbernauld, Glasgow, G67 2QW
MM3 VKP Cameron Allan, Ardroy, Kinclaven Road, Perth, PH1 4EY
M3 VKS D Vickers, 178 Bakewell Road, Matlock, DE4 3BA
M3 VKT Evangelos Kottis, Guildford Court Reception, University Campus, Guildford, GU2 7JL
M3 VLB J Goodman, 4 Maloren Way, West Moors, Ferndown, BH22 0BQ
M3 VLG Charity Periam, 5 Elliott Walk, Preston, PR1 7TP
M3 VLH P Harlow, 92 Eton Road, Burton-on-Trent, DE14 2SW
M3 VLI William Toher, The Chapel, Station Road, Darlington, DL2 1JG
M3 VLJ Lyndon Jones, Ty'r Ysgol, Holland Street, Ebbw Vale, NP23 6HT
M3 VLL Lisa Burbidge, 33 Burcote Fields, Towcester, NN12 6TH
M3 VLN V Grimmer, 48 Bingham Avenue, Sutton-in-Ashfield, NG17 3AR
M3 VLO P Todd, 93 Derwent Drive, Tibshelf, Alfreton, DE55 5LT
M3 VLT Paul Honey, 3 Peterswood, Harlow, CM18 7RJ
M3 VMA Matt Collins, 8 Pictor Grove, Buxton, SK17 7TQ
MD3 VMD Peter Birchall, 7 Richmond Close, Douglas, Isle of Man, IM2 6HR
M3 VME Clint Frost, 6 Link Way, Arborfield Cross, Reading, RG2 9PD
MD3 VMN Voirrey Matthewman, Monte Rosa, 7 Ballaughton Close, Isle of Man, IM2 1JE
M3 VMQ John Farrer, 37 Priory Grove, Ditton, Aylesford, ME20 6BB
M3 VMU John Easterbrook, 2 Warden Road, Eastchurch, Sheerness, ME12 4EJ
M3 VMV Rod Vale, 611 College Road, Birmingham, B44 0AY
MW3 VMY Terrance Peters, 74 Maes y Capel, Pembrey, Burry Port, SA16 0EG
M3 VNG Julia Hardy, LAMBDA HOUSE, SEANOR LANE, Chesterfield, S45 8DH
M3 VNH Natalie Halford, 20 Albany Avenue, Manchester, M11 1HQ
M3 VNI James Preece, 25 Broadmead, Catford, London, SE6 3TG
M3 VNK David Bishop, 1 Charnwood Drive, Barton Seagrave, Kettering, NN15 6TU
M3 VNL C Lockyear, 26 Wentworth Gardens, Exeter, EX4 1NH
M3 VNM Alan Crawford, 39 Fownhope Close, Redditch, B98 0LA
M3 VNN Victor Nikolaidis, 35-46 Ernst Chain Road, Manor Park, Guildford, GU2 7YW
M3 VNO Darryl Harwood, 36 Seaview Drive, Great Wakering, Southend-on-Sea, SS3 0BE
M3 VNP Graham Simcock, 11 Bannatyne Close, Manchester, M40 3TD
M3 VNQ Terence McBride, 53 Blackdown Grove, St. Helens, WA9 2BD
MW3 VNR C Mukans, 2 Ffordd Cottages, Johnstown, Carmarthen, SA33 5BL
M3 VNS S Sampathkumar, 52 Crowstone Road, Westcliff-on-Sea, SS0 8BD
MM3 VNT Michael Robertson, 1a Church Street, Lochgelly, KY5 9JS
MM3 VNU Robert Robb, 8 Morven View, Tarland, Aboyne, AB34 4UH
MW3 VNV Mathew Williams, 39 Heol Cennen, Ffairfach, Llandeilo, SA19 6UL
MM3 VNW Allan Sim, 44 Hillmoss, Kilmaurs, Kilmarnock, KA3 2RS
M3 VNX Lorraine Wolfe, 90 Alderney Road, Erith, DA8 2JD
MW3 VNZ Raymond blackmore, 96 Vachell Road, Cardiff, CF5 4HJ
M3 VOA A Sartorius, 7 Pinehurst, London Road, Egham, TW20 0HQ
M3 VOB P Sheargold, 7 Mendip Close, Rough Hills, Wolverhampton, WV2 2HF
M3 VOI Robert Reeves, 4 Elmwood Avenue, Waterlooville, PO7 7LG
M3 VOJ Adam Kreissl, 382 Oldfield Road, Altrincham, WA14 4QT
M3 VOL J Garner, 2 Coniston Grove, Haresfinch, St. Helens, WA11 9NH
M3 VON A Ellerington, 4 Wathcote Close, Richmond, DL10 7DX
M3 VOR I Mitchell, 4 Walhouse Drive, Penkridge, Stafford, ST19 5SP
M3 VOU Michael Carr, 25 Malvern Avenue, Fareham, PO14 1QF
M3 VOW N Pettefar, 44 Duck Lane, Laverstock, Salisbury, SP1 1PU
M3 VOY P Williams, 45 Blackgate Lane, Tarleton, Preston, PR4 6US
M3 VOZ J Morris, 4 Pleasant Terrace, Lincoln, LN5 8DA
M3 VPA Andrew Penfold, 179 byron rd, thornhill, Southampton, SO19 6FB
M3 VPB Theophilus Horsoo, 5 Kelmarsh Court, Great Holm, Milton Keynes, MK8 9EN
M3 VPD Justin Bridges, 65 Abbots Gate, Bury St. Edmunds, IP33 2GB
M3 VPH Paul Hanson, 34 20th Avenue, Hull, HU6 9JH
M3 VPJ Peter Allen, 21 Chase Vale, Burntwood, WS7 3GD
MM3 VPK C Hebenton, 43 East Avenue, Uddingston, Glasgow, G71 6LG
M3 VPM Paul McDonough, 91 Lever Street, Little Lever, Bolton, BL3 1BA
M3 VPN Michael Thornton, 50 newtown avenue, cudworth, Barnsley, S72 8DY
MI3 VPO D Smith, 164 Ballygowan Road, Hillsborough, BT26 6EG
M3 VPP Ludovic Reeves, 38 Eagle View, Aston, Sheffield, S26 2GL
M3 VPQ Stephen Woods, 30 Kenilworth Drive, Earby, Barnoldswick, BB18 6NA
M3 VPT Graham Taylor, 57 Edinburgh Avenue, Walsall, WS2 0JD
M3 VPX Philip Dodds, 22 Wheatear Lane, Ingleby Barwick, Stockton-on-Tees, TS17 0TB
M3 VPY Patrick McDonough, 91 Lever Street, Little Lever, Bolton, BL3 1BA
M3 VQD Kenneth Morgan, 43 Kenilworth Drive, Earby, Barnoldswick, BB18 6NA
M3 VQF Andrew White, 88 Southville, Yeovil, BA21 4JF
M3 VQG S Walcot, 1 South Avenue, Kidlington, OX5 1DE
MI3 VQH John Kane, 5 Woodlawn Court, Carrickfergus, BT38 8DP
M3 VQI Jasmin Watts, 6 Elm Court, Newhaven, BN9 9NR
M3 VQJ Graham Ellis, 3 Cae Bach, Talybont, Bangor, LL57 3YJ
M3 VQL T Morton, 43 Hob Moor Drive, York, YO24 4JU
M3 VQN Brian Roberts, 102 Brougham Road, Marsden, Huddersfield, HD7 6BJ
M3 VQP Arkadiusz Majoch, 66 Boughton Green Road, Northampton, NN2 7SP
M3 VQQ Charles Unwin, Mansells Farm, Mansells Lane, Hitchin, SG4 8TJ
M3 VQS Brian Jamieson, 14 Ridgeway, Ashington, NE63 9TJ
M3 VQV David Page, 19 LAMORNA DRIVE, Callington, PL17 7QH
M3 VQZ Nicholas Long, 25 Blendworth Lane, Southampton, SO18 5GY
M3 VRA Vera Tomlinson, 180 Kendal Drive, Castleford, WF10 3QZ
M3 VRB John Banks, 73 Buckthorn Avenue, Stevenage, SG1 1TN
M3 VRD R Elias, Roganann, Dol Y Bont, Borth, SY24 5LX
MM3 VRI M McCallum, 15 Quarry Road, Law, Carluke, ML8 5HB
M3 VRL Sakthivel Sethuraman, 9 Bramcote close, Aylesbury, HP20 1QE
M3 VRM M Veal, 2 Bernards Close, Christchurch, BH23 2EH
M3 VRN Roger Noake, 44 Loxton Square, Bristol, BS14 9SF
M3 VRP Susan Broadhurst, Flat 8 Pennant Court, Ross Heights, Rowley Regis, B65 8DW
M3 VRU Tom Baxendale, 3 Greylands Close, Sale, M33 6GS
M3 VRV Marvin Hemstock, 6 Hucknall Crescent, Gedling, Nottingham, NG4 4HZ
M3 VRX Neil Thomson, 2 Doon Terrace, Dumfries, DG2 9EE
M3 VRY John Docherty, 208 Thornton Close, Newton Aycliffe, DL5 7NP
M3 VSB Keigley-Anne Dunne, 1 Burton Gardens, Brierfield, Nelson, BB9 5DR
MM3 VSC S Clark, 75 Treeswoodhead Road, Kilmarnock, KA1 4PB
M3 VSF Belinda Nicholls, 5 Golden Miller Close, Newmarket, CB8 7RT
M3 VSG David Riley-Kydd, 26 Talwrn Road, Wrexham, LL11 3PG
M3 VSH A Freedman, Rivermeade, Irwell Vale, Bury, BL0 0QA
M3 VSL V Cronin, 4 Carnarvon Road, Reading, RG1 5SD
M3 VSO Robert Langmuir, 24 Briar Road, Bexley, DA5 2HN
M3 VSQ Michael Wilde, 18 Ledston Luck Cottages, Kippax, Leeds, LS25 7BX
M3 VST F McDermott, 6 Bruce Street, Swindon, SN2 2EL
MM3 VSU Susan Rodwell, Bourtree, Kennethmont, Huntly, AB54 4NN
M3 VSW Stuart Whall, 17 Vicarage Road, Deopham, Wymondham, NR18 9DR
MM3 VSX Leslie Forbes, Woodside, The Muirs, Huntly, AB54 4GD
M3 VSZ Luke Schofield, 23 The Mount, Wrenthorpe, Wakefield, WF2 0NZ
M3 VTA Carla Duffield, 32 Mount Close, Honiton, EX14 1QZ
MM3 VTB Alexander Hamilton, 10/3 Fox Street, Edinburgh, EH6 7HN
M3 VTG David Howlett, 21 Chandlers, Orton Brimbles, Peterborough, PE2 5YW
M3 VTH Alan Ball, 10 Stifford Road, Aveley, South Ockendon, RM15 4AA
MI3 VTJ Terence Dorrian, 29 Sperrin Road, Limavady, BT49 0AS
M3 VTK John Martin, 62 Llwyn Ynn, Talybont, LL43 2AL
M3 VTL Peter Wilson, 2 Mary Rose Close, Cheslyn Hay, Walsall, WS6 7BE
M3 VTN Natasha Mayall, 10 South Rise, North Walsham, NR28 0EE
M3 VTO S Tanner, Flat 2, The Granary, Totnes, TQ9 5GN
M3 VTP Michael Tuffs, izzyinn 87, foxhills road, Scunthorpe, DN158LL
M3 VTQ Eric MacGurk, 10 Elmore Road, Lee on Solent, PO139DU
M3 VTS Peter Wilkes, 8 Cloverdale, Stafford, ST17 4QJ
M3 VTV Andrew Bedford, 1 Carder Crescent, Bilston, WV14 0JT

M3 VTX Paula Hind, 116 Gowthorpe, Selby, YO8 4HA
M3 VUA J Hunt, 101 Kinoulton Court, Grantham, NG31 7XR
M3 VUC Nathan Lay, 33 School Road, Billericay, CM12 9LU
M3 VUH Stephen Colmon, 22 Shoaswater Way, Stowmarket, IP14 5UG
M3 VUJ Royston Williams, 34 Maendu Terrace, Brecon, LD3 9HH
M3 VUK Brian Lewis, 68 Irwin Avenue, Rednal, Birmingham, B45 8QU
M3 VUN George Diggins, 7 Minterne Road, Bournemouth, BH9 3EH
M3 VUO Graham Twigg, 4 Crossway, Widnes, WA8 8SQ
M3 VUP Andrew Lowe, 65 North Road, Clowne, Chesterfield, S43 4PG
MD3 VUQ Graham Grimshaw, 1 Hardy Close, Pinner, HA5 1NL
M3 VUS Vin Shen Ban, Christs College, Cambridge, CB2 3BU
M3 VUV Paul Larner, 1a West Street, Horncastle, LN9 5JE
M3 VUX Timothy Loker, 24 St. Albans Hill, Hemel Hempstead, HP3 9NG
M3 VUY Sandy Heard, 2 Merrifield Road, Bideford, EX39 4BX
M3 VUZ Vaughan Ball, 24 Carr Lane, Warsop, Mansfield, NG20 0BN
M3 VVA Kieran Jones, 4 Hawthorne Road, Castle Bromwich, Birmingham, B36 0HH
M3 VVB Luke Cunningham, 96 Kingsleigh Drive, Castle Bromwich, Birmingham, B36 9DY
M3 VVH Gordon Spencer, 7 Squadron Close, Castle Vale, Birmingham, B35 7PF
MM3 VVI John Mason, 27 Niddrie Marischal Gardens, Edinburgh, EH16 4LX
M3 VVJ W Jones, Bryn Golau, Mynytho, Pwllheli, LL53 7RL
MW3 VVO Stuart Barry, 7 Redhill Park, Haverfordwest, SA61 2HA
M3 VVQ Neil McDougall, 15 Answell Avenue, Manchester, M8 4GG
MM3 VVS Iain Lindsay, 265 Stirling Street, Denny, FK6 6QJ
MW3 VVW S Smith, 36 Jones Street, Tonypandy, CF40 2BY
M3 VVY M Soper, 16 Queen Elizabeth Drive, Crediton, EX17 2EJ
MM3 VVZ Frederick Coombes, 44 lochfield rd, Paisley, PA2 7RL
M3 VWD Gary Clarke, 11 Blackfordby Lane, Moira, Swadlincote, DE12 6EX
MW3 VWE E Jones, Afon Lodge Caravan Park, Parciau Bach, Carmarthen, SA33 4LG
M3 VWF Trevor Fentiman, 64 St. Nicholas Road, Faversham, ME13 7PD
M3 VWG P Brown, 102 Lang Avenue, Lundwood, Barnsley, S71 5PT
MI3 VWH Richard Hunter, 5 Castle Rise, Tandragee, Craigavon, BT62 2NE
M3 VWJ Paul Kavanagh, 83 Imperial Avenue, Southampton, SO15 8PT
M3 VWK M Poole, 15 Roberts Place, Dorchester, DT1 2JJ
M3 VWM Mark Williams, 37 Pilton Vale, Newport, NP20 6LG
M3 VWO David Best, 8 Carno Street, Rhymney, Tredegar, NP22 5EA
M3 VWP Carl Johnson, 42 Reap Lane, Portland, DT5 2JX
M3 VWR A Bartlett, 62 Kewstoke Road, Bath, BA2 5PU
M3 VWW Stuart Hegarty, 10 New Street, Ash, Canterbury, CT3 2BH
M3 VWY Nick Jewitt, 45 North Road, Kirkburton, Huddersfield, HD8 0QH
M3 VXB Barry Walker, 18 Seals Green, Kings Norton, Birmingham, B38 9UW
M3 VXC Dominic Stinson, 1 The Croft, Earls Colne, Colchester, CO6 2NH
M3 VXG Glynis Kavanagh, 83 Imperial Avenue, Southampton, SO15 8PT
M3 VXH Samuel Liles, 62 Southwood Drive, Surbiton, Surbiton, KT5 9PH
MI3 VXI Stephen Henry, 105 Ramsey Park, Macosquin, Coleraine, BT51 4NG
M3 VXK James Diplock, 8 Lodge Road, Messing, Colchester, CO5 9TU
MM3 VXL Julian May, 12 Clochbar Gardens, Milngavie, Glasgow, G62 7JP
M3 VXM John Stephenson, 54 Elizabeth Road, Haydock, St. Helens, WA11 0PP
M3 VXN M Parkinson, 4 Meadow View, Sherburn in Elmet, Leeds, LS25 6BY
M3 VXO Oliver Carpenter-Beale, 6 Betherinden Cottages, Bodiam Road, Cranbrook, TN18 5LW
MM3 VXP Patrick McKay, 35a Charlotte Street, Helensburgh, G84 7SE
MI3 VXQ Ashley Stone, 16 Sealands Parade, Belfast, BT15 3NT
M3 VXX Tristan Quiney, 20 Britannia Gardens, Stourport-on-Severn, DY13 9NZ
M3 VXY Stephen Hill, 1 Meadow Crescent, Wesham, Preston, PR4 3BB
MM3 VYA Andrew Rodgers, 13 Mill Street, Caldercruix, Airdrie, ML6 7QB
M3 VYB Peter Anstis, 112 West Street, Hartland, Bideford, EX39 6BQ
M3 VYD Kevin Lowcock, 43 Larch Street, Nelson, BB9 9RH
M3 VYE H Ewer, 89 Bridle Close, Enfield, EN3 6EB
M3 VYF Michael Ward, 93 Sandsfield Lane, Gainsborough, DN21 1BQ
M3 VYK Les Shallcross, 6 Wimbrick Close, Wirral, CH46 9RY
M3 VYM Lorne Clark, 16 Kibblewhite Crescent, Twyford, Reading, RG10 9AX
M3 VYN Vinnie Roberts, 17 Houldsworth Crescent, Coventry, CV6 4HL
MM3 VYR I Findlay, 2 Bothwell Road, Uddingston, Glasgow, G71 7ET
M3 VYS Michael Clifford, 100 Cromwell Road, Hounslow, TW3 3QJ
M3 VYT Steven Ferguson, 12 Summerfields, Dalston, Carlisle, CA5 7NW
MM3 VYU William Young, 26 Needle Green, Carluke, ML8 4AF
M3 VYV Jeffrey Arblaster, 22 Wood Lane, Carlton, Barnsley, S71 3JJ
MM3 VYY Timothy Mason, 11 St Serf Road, Glenrothes, KY74EA
M3 VZC Christopher McNulty, 91 Barn Hey Crescent, Wirral, CH47 9RW
M3 VZH Scott Prichard, 4 Morecambe Road, Scale Hall, Lancaster, LA1 5JA
MM3 VZI Daniel O'Kane, 0/1 51 Girvan Street, Glasgow, G33 2DP
M3 VZL Stuart Haycock, 51 South Crescent, Southend-on-Sea, SS2 6TB
M3 VZN Robin Hales, Greenroofs, 4 Beach Road, St. Osyth, CO16 8ET
M3 VZP Salvatore Arpino, 24 Ashfield Lane, Milnrow, Rochdale, OL16 4EW
M3 VZQ David Riddick, 36 Shadygrove Road, Carlisle, CA2 7LD
M3 VZS David Clewer, 45 Ashfield Road, Andover, SP10 3PE
M3 VZU Adrian Murphy, 16 Spencer Avenue, Peterborough, PE2 8QH
M3 VZV Stephen Thompson, 64 Church Road, Fordham, Colchester, CO6 3NJ
M3 VZZ Charles Halls, 2 Cock Fen Road, Lakesend, Wisbech, PE14 9QE
M3 WAC W Clarke, 41 Upton Road, Atherton, Manchester, M46 9RQ
M3 WAF W Morgan, Little Sandyhurst House, 186 Sandyhurst Lane, Ashford, TN25 4NX
M3 WAL S Patchett, Ty Ucha Farm, Nantyr, Llangollen, LL20 7DD
M3 WAP Alexander Cosic, 35 Betteridge Drive, Sutton Coldfield, B76 1FN
M3 WAV Andrew Swain, 1 George Street, Brimington, Chesterfield, S43 1HG
M3 WAY P Jones, 22 Blair Road, Trowbridge, BA14 9JZ
M3 WBA Paul Allin, 25 Castleton Road, Hope, Hope Valley, S33 6SB
MD3 WBC J Wernham, Fair Isle, Lhoobs Road, Douglas, Isle of Man, IM4 3JB
M3 WBI William Whitcher, 17 Watermead, Stratton St. Margaret, Swindon, SN3 4WE
M3 WBJ B Williams, 2 Pokas Cottages, Chelveston, Wellingborough, NN9 6AL
M3 WBK Wayne Knapp, 32 Turner Close, Shoeburyness, Southend-on-Sea, SS3 9TL
MI3 WBL B Lockhart, 5 Lisnalee Park, Mountnorris, Armagh, BT60 2UP
M3 WBM Mark Leake, 20 Witchampton Road, Broadstone, BH18 8HZ
M3 WBN Richard Cranston, Hyrton House, Middle Street, Lincoln, LN1 2RG
M3 WBQ Wendie Argyle, 62 Yew Tree Drive, Leicester, LE3 6PL
M3 WBR H Lowthian, West Brownrigg, Penrith, CA11 9PF
M3 WBS S Bridgeman, 32 Brooklands, Brinkworth, Chippenham, SN15 5BA
M3 WBT Brian Weston, 10 Clement Drive, Peterborough, PE2 0RQ
MI3 WBU Shane Hanlin, 8a Buchanans Road, Newry, BT35 0NJ
MM3 WBV Stephan Verth, 23 the quilts, leith, Edinburgh, EH65RY
MD3 WCA William Alexander, 53 Woodlands Drive, Stanmore, HA7 3PB
M3 WCE J Thomas-Jones, 10 Greenwood Avenue, Gwersyllt, Wrexham, LL11 4EB
M3 WCI A Hand, 150 Curtin Drive, Moxley, Wednesbury, WS10 8RN
M3 WCM Wayne Cornish, 21 Centaur Street, Portsmouth, PO2 7HB
M3 WCO T Purcell, 28 Millberg Road, Seaford, BN25 3ST
M3 WCQ J Winter, Flat 23, Knightlow Lodge Knightlow Avenue, Coventry, CV3 3HH
M3 WCR Clive Wilson, 234 Aylsham Drive Ickenham, Uxbridge, UB10 8UF
M3 WCS Evan Dobson, 12 Mawnog Fach, Bala, LL23 7YY
M3 WCX Carol Whall, 52 Spitfire Road, Upper Cambourne, Cambridge, CB23 6FN
M3 WCY C Wong, Harrogate Ladies' College, Clarence Drive, Harrogate, HG1 2QG
M3 WCZ Matthew Beckett, 59 Broadacre, Caton, Lancaster, LA2 9NH
M3 WDB Carl Jones, 34 Cadle Road, Wolverhampton, WV10 9SJ
M3 WDC W Carless, 39 Harrison Road, Cannock, WS11 0AQ
M3 WDH W Henderson, 14 Highfield Road, Newcastle upon Tyne, NE5 5HS
MI3 WDI D Wiggins, 12 Vauxhall Park, Belfast, BT9 5GZ
M3 WDK P Shaw, 45 Wood End Road, Wolverhampton, WV11 1NW
M3 WDN E Temple, 32 Lower Barresdale, Alnwick, NE66 1DW
MI3 WDO Dermot Dallas, 28 Kemp Park, Ballycastle, BT54 6LE
MM3 WDT W Thom, 1 Bennan, Mossdale, Castle Douglas, DG7 2NG
M3 WDU Alan Stanmore, 198 heathfield road, Southport, PR8 3HE
M3 WDV Dennis Golding, Windrush Cottage, 84-85 Bradenstoke, Chippenham, SN15 4EL
M3 WDY Wendy Duffield, 26 Mount Close, Honiton, EX14 1QZ
M3 WDZ W Davies, 17 Oakdale Avenue, Harrogate, HG1 2JN
M3 WEA T Bradley, 1 Park Close, North Weald, Epping, CM16 6BP
M3 WEF Paul Brown, 8 Hobbes Close, Malmesbury, SN16 0DA
MM3 WEI Edward Ireland, The Steading, Blairmains, Shotts, ML7 5TJ
M3 WEJ James Varley, 54 Richmond Park Road, Kingston upon Thames, KT2 6AH
M3 WEQ Adrian Schuler, 6 Tatham Court, Taunton, TA1 5QZ
MI3 WES William Spence, 8 Kilmahamogue Road, Moyarget, Ballycastle, BT54 6JH
MM3 WEV George Weir, 95 White Street, Whitburn, Bathgate, EH47 0BH
M3 WEZ L Wayman, Oak Tree Lodge, Redbridge Road, Dorchester, DT2 8BG
M3 WFB D Read, 9 Meadow Road, Albrighton, Wolverhampton, WV7 3DZ
M3 WFC Joe Paradas, 13 Meadow Road, Hemel Hempstead, HP3 8AH
M3 WFE Alison Kitney, 3 Wordsworth Close, Torquay, TQ2 6EA
MW3 WFF Richard Jones, 36 Heol-y-Cae, Cefn Coed, Merthyr Tydfil, CF48 2RT
M3 WFG W Griffiths, 68 Altcar Lane, Formby, Liverpool, L37 6AY
MW3 WFH W Harries, 18 Bro Teify, Alltyblacca, Llanybydder, SA40 9SR
MD3 WFJ Jeanie Hill, 54 Wybourn Drive, Onchan, Isle of Man, IM3 4AT
M3 WFK Peter Ashton, 14 Poppy Close, Boston, PE21 7TJ
M3 WFL Michael Rimmer, 7 Brookdale, Southport, PR8 3UA
M3 WFO Alex Jamieson-Colville, 77 Salters Way, Dunstable, LU6 1UG
MI3 WFT C Mooney, 12 Curragh Walk, Londonderry, BT48 8HX
MM3 WFU David Green, 1 The Square, Tomintoul, Ballindalloch, AB37 9ET
M3 WFY Christopher Nelson, 14 Windy Harbour Road, Southport, PR8 3DU
M3 WGB K Moore, Flat 8, Lindis Court, Boston, PE21 8SX
M3 WGI S sugihara, Southfield, Park Lane, Wokingham, RG40 4PY
M3 WGK Leslie Mobley, 2 Boxhedge Road West, Banbury, OX16 0BS
M3 WGM M Brown, 27 Greenfield Close, Dunstable, LU6 1TS
M3 WGO Ian Paterson, 11 Ocho Rios Mews, Eastbourne, BN23 5UB
M3 WGV Victor Wright, Beech Cottage, Baron Wood, Carlisle, CA4 9TP
MM3 WGW Grant Wallace, 29 Dunlop Street, Stewarton, Kilmarnock, KA3 5AT
M3 WGY D Wilson, 75 Gainsborough Road, Scotter, Gainsborough, DN21 3RU
M3 WGZ Graham Wells, 22 Mill Road, Deal, CT14 9AA
M3 WHA A Ballinger, 9 Somerville Court, Cirencester, GL7 1TG
M3 WHB B Balchin, 301 New Hall Lane, Preston, PR1 5XE
M3 WHG T Tunstell, 23 Swallow Crescent, Innsworth, Gloucester, GL3 1BL
M3 WHH John Cook, 20 Huntingdon Close, Totton, Southampton, SO40 3NX
M3 WHL D Dawson, 11 Aukland Grove, St. Helens, WA9 5LR
M3 WHN William Northcote, 58 Warren Avenue, Wakefield, WF2 7JN
M3 WHQ George Woods, 8 Wareham Road, Lytchett Matravers, Poole, BH16 6DP
M3 WHR Dale Williams, 76 Quince, Amington, Tamworth, B77 4EU
MM3 WHS D McLean, 72 Bowfield Crescent, Glasgow, G52 4HJ
M3 WHV Stephen Everson, 41 Westminster Lane, Newport, PO30 5ZF
M3 WHX Helen Smith, 6 Tynemouth Place, North Shields, NE30 4BJ
M3 WHY G Cahill, 81 Albemarle Road, Willesborough, Ashford, TN24 0HJ
M3 WIA William Armes, 11 Rutland Road, Broadheath, Altrincham, WA14 4HW
M3 WIC R Ashwick, 98 Woodbury Avenue, East Grinstead, RH19 3UX
M3 WID J Free, Flat 6, 60 Wyncroft Road, Widnes, WA8 8QE
M3 WIJ H Parkinson, 61 Cinnamon Lane, Fearnhead, Warrington, WA2 0AG
M3 WIT John Withers, 16 Tamworth Close, Etherley Dene, Bishop Auckland, DL14 0RN
M3 WIV Leah Elston, 11 Woodland Walk, Blaina, Abertillery, NP13 3JB
M3 WIX Malcolm Wilson, 12 Gorsey Lane, Great Wyrley, Walsall, W30 0JA
M3 WJA A Whitelam, 107 Wellholme Road, Grimsby, DN32 0NQ
MM3 WJD D Wishart, Curcum, Swannay, Orkney, KW17 2NS
M3 WJK J Knowles, 10 Grove Hill, Hessle, HU13 0RT
M3 WJN John Gorman, 1 Patterdale Road, Ashton-in-Makerfield, Wigan, WN4 0EF
MI3 WJQ William Jordan, 24 Spelga Place, Newtownards, BT23 4ND
MW3 WJP P watson, Eirianfa, Cwmduad, Carmarthen, SA33 6XJ
M3 WJU Julian Mowlam, 46 walpole street, Weymouth, DT4 7HQ
M3 WJV Christopher Cherry, 12 Scarisbrick New Road, Southport, PR8 6PY
M3 WJY Arthanari Margaswamy, 59 Grants Yard, Station Road, Burton-on-Trent, DE14 1BW
MM3 WJZ Iain Cogle, 67 Dryburgh Avenue, Rutherglen, Glasgow, G73 3EU
M3 WKC Alasdair Campbell, Gate House, The Bog, Shrewsbury, SY5 0NG
MM3 WKF Dave Goodwin, Little Dens, Stuartfield, Peterhead, AB42 5DG
M3 WKK Richard Horgan, 74 Inglewhite Road, Longridge, Preston, PR3 2NA
M3 WKL Carl Winch, 2 Cranleigh Gardens, Cowes, PO31 8AS
M3 WKM B Howard, 5 St Nicholas Street, Dereham, NR19 2BS
M3 WKV Kevin Dale, 26 Warwick Place, Langdon Hills, Basildon, SS16 6DU
M3 WKZ Ben Drury, 6 Ellen Grove, Harrogate, HG1 4RH
MM3 WLA Martin Dickeson, 41 Hossack Drive, Elgin, IV30 6JY
MW3 WLB Lewis Pearce, 31 high street, abertridwr, Caerphilly, CF83 4DD
M3 WLD W Douglas, 1 Sleetbeck Road, Roadhead, Carlisle, CA6 6PA
M3 WLH Danny Flynn, [illegible]
M3 WLL W Burnett, [illegible]
M3 WLO M Anderson, 38 Shellard Road, Filton, Bristol, BS34 7LU
MW3 WLS D Elias, 31 Banc Y Gors, Upper Tumble, Llanelli, SA14 6BR
M3 WLU John Hepburn, 32 Green Croft, Ashington, NE63 8EF
M3 WLV Darryl Burden, 16 Milnthorpe Lane, Wakefield, WF2 7DE
MI3 WLW Lynette Norton, 19 Gosford Road, Collone, Armagh, BT60 1LQ
M3 WLX Michael Gooch, Flat 16, Townfield Court, 32 Horsham Road, Dorking, RH4 2JE
M3 WLY R Readman, 1 Millside Close, Kilham, Driffield, YO25 4SF
M3 WMC W Carr, 18 Whiteway Close, Bristol, BS5 7QZ
M3 WMH M Hughes, 79 Ffordd Pentre, Mold, CH7 1UY
M3 WMI N Mitchell, 37 Brookside, Glan Y Mor, Fairbourne, LL38 2BX
MI3 WMK M McKeen, 27 Old Grange, Carrickfergus, BT38 7HQ
M3 WMO R Laughlin, 7 Catherine Hunt Way, Colchester, CO2 9HN
M3 WMP W Phillips, Annfield, Penrhyndeudraeth, LL48 6LS
MM3 WMQ Neil Davidson, 25 Hopetoun Court, Bucksburn, Aberdeen, AB21 9QS
M3 WMS Wilfred Stone, 79 Woodlands Road, Allestree, Derby, DE22 2HH
M3 WMU S Breese, 11 Balham Grove, Birmingham, B44 0NF
M3 WMV David Messenger, 18 Glebelands, Harlow, CM20 2PA
M3 WNC Stephen Willmott, Emborough Grove, Radstock, BA34SF
M3 WNF Neil Fellingham, 23 Brooklands, Colchester, CO1 2WA
MM3 WNH Andrew Maitland, 6 Thorn Avenue, Coylton, Ayr, KA6 6NL
M3 WNI Robert Karpinski, 55 Cambridge Avenue, New Malden, KT3 4LD
M3 WNM Nik Edwards, 15 Penderry Rise, Catford, London, SE6 1EZ
MM3 WNP Steven Murray, 25 Braefoot, Girdle Toll, Irvine, KA11 1BY
M3 WNT Tom Corker, North Side, Wingerworth Hall Estate, Chesterfield, S42 6PL
M3 WNV James Giffard, 5 Hazelwood Road, Oxted, RH8 0JA
M3 WNX Sergei Moissejev, 50 Filey Road, Reading, RG1 3QQ
M3 WNZ Jim Strawbridge, 36 St. Dunstans Road, Salcombe, TQ8 8AN
M3 WOC C Bellis, Cliffe Bungalow, Barnsley Road, Barnsley, S72 9JX
M3 WOD David Wood, 27 St. Mildreds Avenue, Ramsgate, CT11 0HT
M3 WOI Richard Burlong, 20 Jockey Mead, Horsham, RH12 1LF
M3 WOK B Burden, 18 Challenor Close, Finchampstead, Wokingham, RG40 4UJ
M3 WOL Pamela Bickley, Smithy Cottage, Old Post Office Road, Bury St. Edmunds, IP29 5RD
M3 WOQ Katie Meyer, 42 Sandcross Lane, Reigate, RH2 8EL
M3 WOS Warwick Barnes, Cushendall, Lyngate Road, North Walsham, NR28 0DH
M3 WOW C Constable, 9 Ridgeway Close, Heathfield, TN21 8NS
M3 WOX Julie Ward, 64 Laxey Road, Blackburn, BB2 3LQ
M3 WOY Clifford Dunstan, 67 Knights Way, Mount Ambrose, Redruth, TR15 1PA
M3 WPC R Brett, 3 Rectory Close, Chingford, London, E4 8BG
M3 WPH S Harrison, 2 Hendre, Newtown, Ebbw Vale, NP23 5FE
M3 WPI Oliver Prin, 19 The Colliers, Heybridge Basin, Maldon, CM9 4SE
M3 WPJ Peter Corbin, 26a Padnell Avenue, Waterlooville, PO8 8DY
M3 WPK Harold Burch, 46 School Lane, Horton Kirby, Dartford, DA4 9DQ
M3 WPM Marta Almeida, 20 Gresley Court, Grantham, NG31 7RH
M3 WPN Robert Blackett, 10 Acton Gardens, Wrexham, LL12 8DD
M3 WPO Paul Woolley, 84 Bowthorpe Road, Norwich, NR2 3TP
M3 WPP Desmond Rayner, 42 Chapelgate, Sutton St. James, Spalding, PE12 0EE
M3 WPS W Snowden, 5 Eastfield Road, Wisbech, PE13 3EJ
M3 WPU Michael McHugh, 51 Rutland Street, Hyde, SK14 4SY
M3 WPV Mark Hardy, 66 Exeter Road, Doncaster, DN2 4LF
M3 WPW William Whyatt, 11 The Perrings, Nailsea, Bristol, BS48 4YD
M3 WQA Mark Herbert, 31 Mayfield Avenue, New Haw, Addlestone, KT15 3AQ
MI3 WQC Sean Dillon, 2 Otter Park, Strathfoyle, Londonderry, BT47 6YU
MW3 WQE Peter Pritchard, 1a Pant Hirgoed, Pencoed, Bridgend, CF35 6YD
M3 WQF David Robinson, 49 Meldon Drive, Bradley, Bilston, WV14 8BQ
M3 WQG Derrick Pearson, 49 Longford Road, Twickenham, TW2 6EB
M3 WQI Kerry Squires, 10 Markham Avenue, Armthorpe, Doncaster, DN3 2AZ
M3 WQL Heather Short, 71 Lilac Crescent, Burnopfield, Newcastle upon Tyne, NE16 6QF
M3 WQN Jason Gordon, 21 Barras Avenue, Annitsford, Cramlington, NE23 7QX
MM3 WQO Edward Hughes, 12 Cults Drive, Tomintoul, Ballindalloch, AB37 9HW
MI3 WQT A McBride, 2 Glenbrook Cottage, Lugan, Craigavon, BT66 8QT
M3 WQV Erica Morgan, Holly Cottage, Old Racecourse, Oswestry, SY10 7PQ
M3 WQX James Neal, 75 Park Lane, Castle Donington, Derby, DE74 2JG
M3 WRA J Turner, 35 Horncastle Road, Wragby, Market Rasen, LN8 5RB
M3 WRB R Wheatley, 46 Victory Road, Steeple Claydon, Buckingham, MK18 2NY
MW3 WRH Waine Hucker, 14 Greenway Court, Barry, CF63 2FE
M3 WRJ R Waghorne, 5 Freelands Drive, Church Crookham, Fleet, GU52 0TE
M3 WRK Sheharyar Sarwar, REDFERN 57A, UNI OF WARWICK, Coventry, CV47AL
M3 WRM Ralph Dadge, 14 North Roskear Village, Camborne, TR14 0AS
M3 WRN Jonathan Hills, 67 Thornham Road, New Milton, BH25 5AE
M3 WRO O Woods, 8 Fairway Close, Croydon, CR0 7SH
M3 WRQ Clive Smith, 66 Minard Road, London, SE6 1NR
M3 WRS C Webber, 60 Mincinglake Road, Exeter, EX4 7DY
M3 WRZ Leslie Alyson Coyne, 75 Newtown Road, Worcester, WR5 1HH
M3 WSC Owen Owen, 8 Masshyfrmd, Garndolbenmarn, LL51 9SX
M3 WSE John Cullen, Flat 14, Maybury Mews, 121 Maybury Road, Woking, GU21 5JQ
M3 WSH S Holmes, 11 Holford Rise, Bremilham Road, Malmesbury, SN16 0FA
M3 WSI Sam Warren, 9 Warning Tongue Lane, Doncaster, DN4 6TB
M3 WSJ J Woodroof, 37 Danefield Road, Northampton, NN3 2LT
M3 WSN J Spillett, Mockbeggar Cottage, Mockbeggar, Ringwood, BH24 3NQ
M3 WSO R Pitman, 10 Somerville Way, Bridgwater, TA6 5SA
M3 WSQ Louise Simpson, 462 Leeds Road, Wakefield, WF1 2DU
M3 WSR B Harrison, 43a Rumbridge Street, Totton, Southampton, SO40 9DR
M3 WSS L Shand, 52 Ten Acre Way, Rainham, Gillingham, ME8 8TL
M3 WSU Bruce Stewart-Whyte, 29 Pilgrims Way, Dover, CT16 2DF
M3 WSV T Kyriacou, 54 Sutton Avenue, Silverdale, Newcastle, ST5 6TB
M3 WSW S Ngai, Harrogate Ladies' College, Clarence Drive, Harrogate, HG1 2QG
M3 WTA D Whitton, Sea View, Baycliff, Ulverston, LA12 9RL
M3 WTB B Walden, 59 Brook View Drive, Keyworth, Nottingham, NG12 5RA
M3 WTC William Caine, 116b Hill Street, Hednesford, Cannock, WS12 2DR



**IMPORTANT NOTE**

**Revalidate licence to avoid revocation** – Ofcom has advised the Society that plans will be drawn up to revoke licences that have not been revalidated as required by the licence conditions. The quickest way to revalidate is to do so online via the Ofcom website: *https://services.ofcom.org.uk/* or by email: *amateur.validations@ofcom.org.uk* Ofcom staff are available to help, but please be patient during times of heavy workload.

M3 WTD R BURROW, 162 Broadway, Horsforth, Leeds, LS18 4HQ
M3 WTG Duncan Cooper, Little Heath, Bradfield Common, North Walsham, NR28 0QR
M3 WTL W Leverett, 514 Arleston Lane, Stenson Fields, Derby, DE24 3AG
M3 WTN Bernard Watkin, 48 Peel Park Crescent, Little Hulton, Manchester, M38 0BU
M3 WTO S Liu, Harrogate Ladies' College, Clarence Drive, Harrogate, HG1 2QG
M3 WTP E Lowe, 21 Sherwood Avenue, Creswell, Worksop, S80 4DL
M3 WTR W Randall, 3 Penygraig, Aberystwyth, SY23 2JA
MI3 WTT A McDonnell, 52 Moira Road, Glenavy, Crumlin, BT29 4JL
M3 WTU Jeffrey Townsend, 124 Rough Common Road, Rough Common, Canterbury, CT2 9BU
M3 WTY Robert Clare, Kimberley, Boston Road, Boston, PE20 3AP
M3 WUA Andrew Smith, 20 South Terrace, Northampton, NN1 5JY
M3 WUB Christopher Garner, 30 Pendula Road, Wisbech, PE13 3RR
M3 WUE David Tidswell, 1 Cherrytree Grove, Spalding, PE11 2NA
M3 WUG Joe Stainton, 24 Clifton Road, Huddersfield, HD1 4LL
M3 WUH John Middleton, 16 Kyme Road, Boston, PE21 8NQ
MM3 WUI Paul Bingham, 129 Livingstone Terrace, Irvine, KA12 9ER
M3 WUJ Joshua Garner, 30 Pendula Road, Wisbech, PE13 3RR
M3 WUK Wayne Norwood, Flat 5, 20 Upperton Gardens, Eastbourne, BN21 2AH
M3 WUM R Miles, Haseley Lodge, Birmingham Road, Warwick, CV35 7HF
M3 WUN Ian McMahon, 8 Thackeray Close, Liverpool, L8 8NE
M3 WUO Ian Taylor, Flat 65, Kemsley, London, SE13 6QW
MM3 WUP William Steele, 1 James Street, Bannockburn, Whins of Milton, Stirling, FK7 0NQ
M3 WUQ Iakovos Petropouleas, 16 Amfissis Street, Holargos, Athens, Greece, 155 62
M3 WUS Martin Blenkinsop, 23 Pilmoor Drive, Richmond, DL10 5BJ
M3 WUV Frederick Harwood, 1, South Highall Cottage, Woodhall Spa, LN10 6UR
M3 WUW Matthew Noakes, 26 Box Lane, Pontefract, WF8 2JW
M3 WUX K Bailey, 58 Billy Buns Lane, Wombourne, Wolverhampton, WV5 9BP
M3 WVB Peter George, 4 Mandelbrote Drive, Littlemore, Oxford, OX4 4XG
M3 WVC Sean Howard, 17 Webdell Court, Norwich, NR1 2NB
M3 WVD M Bradshaw, 118 Queens Road, Vicars Cross, Chester, CH3 5HE
M3 WVF G Bradshaw, 118 Queens Road, Vicars Cross, Chester, CH3 5HE
M3 WVG Keith Pain, 200 Manor Road, Mitcham, CR4 1JF
M3 WVI Jack Hanley, 5 Timline Green, Bracknell, RG12 2QP
M3 WVJ Russell Brown, 8 Eliot Walk, Kidderminster, DY10 3XP
MI3 WVL Jamie Mooney, 12 Curragh Walk, Londonderry, BT48 8HX
MM3 WVN Sharon Clark, 5b Ladykirk Road, Prestwick, KA9 1JW
M3 WVO D Willey, 17 Bridge Place, Saxilby, Lincoln, LN1 2QA
MM3 WVP Rae Hutton, 1 Grianairigh, Northton, Isle of Harris, HS3 3JA
MM3 WVQ Rachel Robinson, 12 Hannahston Avenue, Drongan, Ayr, KA6 7AU
M3 WVT A Bradshaw, 118 Queens Road, Vicars Cross, Chester, CH3 5HE
M3 WVX Levi Gaynor, 225 Watson Court, Stadium Way, Watford, WD18 0FA
M3 WVY Erika Pinvisase, 225 Watson Court, Stadium Way, Watford, WD18 0FA
M3 WWD L Hornby, 4 Shakespeare Road, Prestwich, Manchester, M25 9GW
M3 WWH Rob Fraser, 7 Hawthorne Avenue, Fleetwood, FY7 7PY
MI3 WWJ Edith Simpson, 10 Woodview Park, Tandragee, Craigavon, BT62 2DD
MM3 WWM Roy Jowett, Fearnoch, Ardentallen, Oban, PA34 4SF
M3 WWN W Northover, 13 Dagenham Avenue, Dagenham, RM9 6LD
MW3 WWO H Golaszewski, 16 Wingate Drive, Llanishen, Cardiff, CF14 5LR
MM3 WWP C Wood, 5 Bridgend Gardens, Windygates, Leven, KY8 5BP
MM3 WWQ Andrew Patrick, 30 Flatt Road, Largs, KA30 9EA
M3 WWR W Witham, 4 King George Road, Colchester, CO2 7PE
M3 WWU Graham Armitage, Windmill Cottage, Greens Gardens, Nottingham, NG2 4QD
MM3 WWV Roderick Kennedy, 45 Rodney Road, Gourock, PA19 1XG
M3 WWY George Miller, Silvermine, Cooks Lane, Axminster, EX13 5SQ
M3 WWZ R Alexander, 14 Ashfield Terrace, Appley Bridge, Wigan, WN6 9AG
M3 WXB B Williamson, 114 Radburn Road, New Rossington, Doncaster, DN11 0SH
M3 WXD R Fenn, 43 Grantley Close, Ashford, TN23 7UE
M3 WXG Graham Lewis, Millbrook, Church Street, Market Drayton, TF9 2TF
M3 WXH Colin Renouf, 27 Ashburton Road, Croydon, CR0 6AP
M3 WXI Christopher Bossons, 31 Hanbridge Avenue, Newcastle, ST5 8HH
MW3 WXN B Chandler, 100 Shakespeare Avenue, Penarth, CF64 2RX
M3 WXP Dean Brookes, 13 Princess Street, Woodlands, Doncaster, DN6 7LX
MD3 WXS Clare Ashworth, Ravenscourt Lodge, Peel Road, Douglas, Isle of Man, IM1 5EQ
M3 WXU David Grundy, 25 Albert Street, Bignall End, Stoke-on-Trent, ST7 8QB
M3 WXW C Glitsun, 152 St. Awdrys Road, Barking, IG11 7QE
M3 WXX J CHILD, 12 Beachill Road, Havercroft, Wakefield, WF4 2EJ
M3 WXY R Kerswill, 38 Longbridge Road, Bramley, Tadley, RG26 5AN
M3 WYA K Willoughby, 11 Hardistry Drive, Pontefract, WF8 4BU
M3 WYF R Wright, 4 Wynne Close, Broadstone, BH18 9HQ
M3 WYG Paul Engledow, 62 Purland Road, Norwich, NR7 9DZ
MM3 WYI Neil Stewart, 220 Grieve Road, Greenock, PA16 7AL
M3 WYJ Barbara Gale, Barn End, Highampton, Beaworthy, EX21 5LT
M3 WYL S Wylde, The Spinney, Mill Lane, Louth, LN11 7HU
MM3 WYM Matthew Stewart, 42 Ailsa Road, Gourock, PA19 1DY
M3 WYQ Peter Holton, 66 Mill Road, Gillingham, ME7 1JB
M3 WYR Christopher Smith, 199a Richardshaw Lane, Stanningley, Pudsey, LS28 6AA
M3 WYT R White, 2 Chambers Manor Cottages, Epping Upland, Epping, CM16 6PJ
M3 WYV Ewan Scott, 31 South Croft, Upper Denby, Huddersfield, HD8 8UA
M3 WYZ C Pearman, Wicken Cottage, Mill Hill, Edenbridge, TN8 5DB
M3 WZF David Walker, 8 Wescoe Avenue, Great Houghton, Barnsley, S72 0DW
M3 WZG Wayne Power, 23 Drawbridge Close, Maidstone, ME15 7PD
MM3 WZH Samuel Armstrong, 85 Blantyre Court, Erskine, PA8 6BP
M3 WZJ Richard Davies, 43 Woodfield Road, Holt, NR25 6TX
MM3 WZL J Scott, 5 Barnwood Gate, Galston, KA4 8NA
M3 WZN John Mitchell, 27 Tanager Close, Norwich, NR3 3QD
M3 WZP Mark Williams, 22 Molland Lea, Ash, Canterbury, CT3 2JF
M3 WZR Paul Fry, 76 Mount Pleasant Road, New Malden, KT3 3LB
M3 WZS Susan Thorne, 2 Ellfield Close, Bristol, BS13 8EF
M3 WZT Mark Fulbrook, 2 Cob Place, Westbury, BA13 3GS
M3 WZV Stephen Norman, 44 Martival, Leicester, LE5 0PH
M3 WZY J Dunn, 9 Wakefield Road, Stoke-on-Trent, ST4 5PU
M3 WZZ S Bobby, 56 Ffordd Offa, Rhosllanerchrugog, Wrexham, LL14 2EY
M3 XAC Alan Curry, 30 Hillside Road, Norton, Stockton-on-Tees, TS20 1JG
MM3 XAF Andrew Ferries, Cairnbeathie, Lumphanan, AB31 4QA
M3 XAG C Tame, 28 Tyrrells Way, Sutton Courtenay, Abingdon, OX14 4DF
M3 XAH A Holmes, 614 City Road, Manor, Sheffield, S2 1GH
M3 XAI Harry Parfitt, 5 Sheridan Road, Frimley, Camberley, GU16 7DU
M3 XAJ A Morgan, 46 Greensway, Abertysswg, Tredegar, NP22 5AR
M3 XAK Matthew Gaunt, 12 Glastonbury Abbey, Bedford, MK41 0TX
M3 XAM A Morgan, 18 Keysworth Drive, Wareham, BH20 7BD
M3 XAN A Brooks, 52 Houldsworth Drive, Chesterfield, S41 0BS
M3 XAO Alex Hughes, 80c Royle Green Road, Manchester, M22 4WB
M3 XAR M Roberts, 15 Pineside Avenue, Cannock Wood, Rugeley, WS15 4RG
M3 XAU Elliott Landon, 24 Larchwood Close, Sale, M33 5RP
M3 XAV Richard Hartle, 7 Boggard Lane, Charlesworth, Glossop, SK13 5HL
M3 XAW John Scott, 47 Bowness Street, Sunderland, SR5 4LA
M3 XAY A Yorkston, 26 Hamilton Road, London, NW10 1PA
M3 XBC B Chamberlain, 10 Scott Road, Bishop's Stortford, CM23 3QH
M3 XBE Shane Best, 38 Greensway, Abertysswg, Tredegar, NP22 5AR
M3 XBF Martin Fisher, 25 Tennyson Road, Diss, IP22 4PY
M3 XBH B Harrison, 24 Alderton Road, Nottingham, NG5 6DX
M3 XBL Lawrence Rabone, 6 Cranwell Grove, Kesgrave, Ipswich, IP5 2YN
M3 XBN Ovidiu Popa, 25 Wells Park Road, London, SE26 6JQ
M3 XBO John Kaby, 3 Kexby Mill Close, North Hykeham, Lincoln, LN6 9TB
M3 XBS Owen Rogers, 22 Robson Drive, Hoo, Rochester, ME3 9EA
M3 XBT B Totterdell, 35 Meadow Bank Avenue, Sheffield, S7 1PB
M3 XBZ Martin Stead, 38 Park Road, Bracknell, RG12 2LU
M3 XCA Anthony Clarke, 14 Tower Court, Haverhill, CB9 9DD
M3 XCB John Gardiner, 146 Durley Drive, Prenton, CH43 3BB
M3 XCF Roger Cockayne, 20 The Shrubbery, Rugeley, WS15 1JJ
M3 XCH C Howard, 44 Corhampton Road, Bournemouth, BH6 5PB
M3 XCJ Joel Fergusson, 11 Castle Lane, Solihull, B92 8DB
M3 XCN Chris Norton, 34 The Grove, Little Aston, Sutton Coldfield, B74 3UD
M3 XCP Chris Parker, 40 Holman Way, Ivybridge, PL21 9TE
M3 XCS Bill Porter, 5 Ribble Road, Liverpool, L25 5PN
M3 XCU John Callis, 51 Pipistrelle Way, Oadby, Leicester, LE2 4QA
M3 XCV Mark Abberley, 10 Cranesbill Close, Featherstone, Wolverhampton, WV10 7TY
M3 XCX David Michael, 4 Sunningdale Drive, Lincoln, LN6 7UD
MW3 XDB D Barnett, 49 Parcyrhun, Ammanford, SA18 3HD
M3 XDD David Baseden Butt, 24 Lowry Way, Stowmarket, IP14 1UF
M3 XDH D Harris, 23 Shearer Road, Portsmouth, PO1 5LL
M3 XDI Jonathan Peain, 29 Wild Flower Way, Ditchingham, Bungay, NR35 2SF
M3 XDM D Matthews, 57 Rhea Hall Estate, Highley, Bridgnorth, WV16 6LD
MM3 XDP D Paterson, 42 Third Avenue, Alexandria, G83 9BJ
M3 XDQ Adam Green, 37 Fisher close, worsley mesnes, Wigan, WN3 5UT
M3 XDV J Hall, 9 Stone Court, South Hiendley, Barnsley, S72 9DL
MM3 XDW D Woods, 39 Northfield, Tranent, EH33 1HU
M3 XDZ Patrick Neal, 14 Hilltop Close, Desborough, Kettering, NN14 2LQ
M3 XEA Kevin Bindley, 56 Iona Close, Beaumont Leys, Leicester, LE4 0QY
MI3 XEB R Throne, 12 Mason Road, Magheramason, Londonderry, BT47 2RY
M3 XEF Max Goodwin, Bramble Cottage, Well Hill Lane, Orpington, BR6 7QJ
M3 XEG Bradley Pearce, 4 Mary Chapman Close, Norwich, NR7 0UD
M3 XEI Wolfgang Walther, 139 East Street, Epsom, KT17 1EJ
M3 XEJ Emma-Jane Ellison, 5 Darwin Terrace, Darwin Street, Shrewsbury, SY3 8QQ
M3 XEL M Morris, 8 Millfield, Lambourn, Hungerford, RG17 8YQ
M3 XEN Jon Fautley, 71 Pullman Lane, Godalming, GU7 1YB
M3 XEO Barbara Hay, Riverside Road, Great Yarmouth, NR31 6PZ
M3 XEQ Paul Whalan, The Old Post Office, South Street, Faversham, ME13 9NR
M3 XEX Neil Farrow, 30 Highdown, Southwick, Brighton, BN42 4QS
MI3 XEY Cieva Cartin, 64 Ashgrove Park, Magherafelt, BT45 6DN
M3 XFA Fitzroy Alexis, 44 Osborne Road, Enfield, EN3 7RW
M3 XFB Jane Emery, 63 Warren Road, Orpington, BR6 6JF
M3 XFC Matthew Calvert, 8 Brixham Drive, Wigston, LE18 1BH
M3 XFD N Field, 9 Shepherds Fold Drive, Winsford, CW7 2UE
M3 XFG Thomas Large, 5 Raynsford Rise, Stanningfield Road, Bury St. Edmunds, IP30 0TS
M3 XFH Mike Ashton, Lodge Farm Bungalow, Wattisham Road, Ipswich, IP7 7LU
M3 XFI Martin Radford, 3 Cockshott Drive, Armley, Leeds, LS12 2RL
MM3 XFM Barry Burrows, 27 Bughtknowes Drive, Bathgate, EH48 4DP
M3 XFN Karl Cross, 31 Parkfields, Abram, Wigan, WN2 5XR
MM3 XFP Catherine-Anne Lee-Marr, 65 Barry Road, Carnoustie, DD7 7QQ
M3 XFS J Sewell, 56 Victoria Court, Luddesdown Road, Swindon, SN5 8HL
M3 XFT B Dixon, 21 Pankhurst Road, Hoo, Rochester, ME3 9DF
MM3 XFX Douglas Sandilands, Cuil moss cottage, Ardgour, Fort William, PH33 7AB
M3 XFZ Andy Coop, 47 Amy Street, Rochdale, OL12 7NJ
M3 XGA G White, 89 Kings Drive, Thingwall, Wirral, CH61 9QA
M3 XGB I Bennett, 4 Frances Close, Wivenhoe, Colchester, CO7 9RP
M3 XGC K Emlay, 195 Barrington Street, Manchester, M11 4FB
M3 XGD Jackie Challinor, 69 Brimrod Lane, Rochdale, OL11 4QF
M3 XGI Edward Brook, 30 Pitchstone Court, Farnley, Leeds, LS12 5SZ
M3 XGL Geoffrey Clarke, 28 Mayfield Way, Mendlesham, Stowmarket, IP14 5SH
MM3 XGP G Kelly, 203 Meldrum Court, Glenrothes, KY7 6UP
MI3 XGR J Doherty, 62 Coolessan Walk, Limavady, BT49 9EN
MM3 XGS G Suttie, 9b Pentland Crescent, Dundee, DD2 2BU
M3 XGU Carl Frizzell, 85 Gibbon Road, Newhaven, BN9 9ER
M3 XGV Michael Wills, 23 Moat Avenue, Coventry, CV3 6BT
M3 XGW Gareth Whall, 10 Hillcrest Court, Ipswich Road, Diss, IP21 4YJ
M3 XGY Stephen Kay, 476 North Drive, Thornton-Cleveleys, FY5 2HX
M3 XGZ John Murray, 2 The Cuttings, Hampstead Norreys, Thatcham, RG18 0RR
M3 XHB John Kelly, 66 Denison Road, Feltham, TW13 4QG
M3 XHC Joshua Akinin, 70 Valley Road, West Bridgford, Nottingham, NG2 6HQ
M3 XHH S Kiley, 178 Kingfisher Drive, Woodley, Reading, RG5 3LQ
M3 XHK Paul Goodall, 61 Turf Hill Road, Rochdale, OL16 4XG
M3 XHL John Edwards, 19 Bryntirion, Henllan, Denbigh, LL16 5YL
M3 XHM Patricia Aitken, 25 Clunbury Road, Northfield, Birmingham, B31 3SY
M3 XHN Shaun Hutchinson, 17 Monsom Lane, Repton, Derby, DE65 6FX
M3 XHQ David Jewitt, 26 Sands Lane, Barmston, Driffield, YO25 8PG
M3 XHT Mark Shipham, 1 The Farmhouse, Farmhouse Lane, Hemel Hempstead, HP2 7AR
M3 XHU Clive Hall, 28 Tidebrook Place, Stoke-on-Trent, ST6 6XF
M3 XHV Stuart Elliott, 21 Somerville, Didcot, OX11 8UD
M3 XHW H Wright, 168 Spinney Hill Road, Northampton, NN3 6DN
M3 XHY Bernadette Smith, 7 Kestrel Avenue, Bransholme, Hull, HU7 4ST
M3 XHZ Mark Tickner, 111 Fennel Crescent, Crawley, RH11 9DT
MM3 XIA Iain Anderson, Cantyhaugh, Ogscastle, Lanark, ML11 8NE
M3 XID John Roberts, 51 Bradfield Road, Broxtowe, Nottingham, NG8 6GP
M3 XIE David Robinson, 19 Meadow Lane, Newcastle, ST5 9AJ
M3 XIF Jonathan Williams, 41 Overton Lane, Hammerwich, Burntwood, WS7 0LQ
M3 XIG Sharon Abberley, 10 Cranesbill Close, Featherstone, Wolverhampton, WV10 7TY
M3 XIH Ashleigh Williams, 41 Overton Lane, Hammerwich, Burntwood, WS7 0LQ
MW3 XIJ A Pritchard, 20 St. Malo Road, Cardiff, CF14 4HN
M3 XIK Joshua Lambert Hurley, 19 Hill Close, West Bridgford, Nottingham, NG2 6GQ
M3 XIL Stephen Hoy, 114 Sheppey Beach Villas, Manor Way, Sheerness, ME12 4QY
M3 XIM David Hackling, 20 Millers Lane, Norwich, NR3 3LU
M3 XIO Mario Stevenson, 127 Walton Road, Chesterfield, S40 3BX
M3 XIP Andrew Pomfrey-Jones, 46 Hampton Road, Erdington, Birmingham, B23 7JJ
MI3 XIU Michael Sinton, 34 West Link, Holywood, BT18 9NX
M3 XIV Richard Treherne, 58 Cherry Orchard, Tewkesbury, GL20 8PJ
MM3 XIW Thomas Hunter, 1a Glen Avenue, Largs, KA30 8RQ
M3 XIY Benjamin Kerry, 1, churchway, Diss, IP22 1RN
MM3 XIZ Ian Hepworth, Bronte Cottage, Inverugie, Peterhead, AB42 3DN
MM3 XJA James Arthur, West Lodge, Murdoustoun, North & South Road, Motherwell, ML1 5LB
M3 XJE D Holdsworth, 3 Briardale Road, Bradford, BD9 6PU
M3 XJF Justin Ferrington, The Redwoods, 20 Innings Lane, Bracknell, RG42 3TR
M3 XJG James Gaskin, Badgers Barn, Canterbury Road, Folkestone, CT18 8DF
M3 XJH Albert Hylton, 3 Jubilee Cottages, Tring Road, Dunstable, LU6 2JU
M3 XJK Alan Weaver, 77 East Acres, Widdrington, Morpeth, NE61 5NT
M3 XJL John Landless, 2 Aspen Way, Banstead, SM7 1LE
M3 XJO J Boids, 2 Crown Street, Hoyland, Barnsley, S74 9HS
M3 XJP James Plows, 187 Whitebeam Road, Birmingham, B37 7PA
M3 XJQ Robert Barter, 17 West Gate, Plumpton Green, Lewes, BN7 3BQ
MM3 XJW Jeanette Wood, 5 Damhead Steading, Kinloss, Kinloss, IV36 3UA
M3 XJX Andrew Reay, 12 Victoria Avenue, South Hylton, Sunderland, SR4 0QZ
M3 XJZ Joseph Gabriel, 60 Goodwin Road, Ramsgate, CT11 0JJ
M3 XKD Lindsay Booth, 8 Rowthorne Close, Northampton, NN5 4WB
M3 XKF N Tideswell, 19 Wish Court, Ingram Crescent West, Hove, BN3 5NY
MM3 XKH K Hail, 70 Nobleston Estate, Alexandria, G83 9DB
M3 XKI Lawrie Taute, 4 Mendelssohn Grove, Browns Wood, Milton Keynes, MK7 8DH
M3 XKJ James Lewis, Millbrook, Church Street, Market Drayton, TF9 2TF
M3 XKL Roger Jones, Flat 2, Tan y Geraint, 33 Princess Street, Llangollen, LL20 8RD
M3 XKM Kevin Mountford, 7 Flaxman Close, Barlaston, Stoke-on-Trent, ST12 9BD
M3 XKN Laura Griffiths, 90 Keats Road, Wolverhampton, WV10 8NB
M3 XKO Paul Goodridge, 22 Horefield, Porton, Salisbury, SP4 0LE
M3 XKP Ian Pass, 69 Cotswold Road, Bath, BA2 2DL
M3 XKY Gail Leverton, 24 Saxton Avenue, Bradford, BD6 3SW
M3 XLB P Bailey, 44 Shelley Road, Wellingborough, NN8 3DB
M3 XLC Graham Auld, 6 Pheabens Field, Bramley, Tadley, RG26 5BX
MD3 XLJ L Justin, Garth, Park View Road, Pinner, HA5 3YF
M3 XLK Robyn Crerar, 60 Gloucester Drive, London, N4 2LN
M3 XLM L Matthewman, 2 St. Margaret Road, Ludlow, SY8 1XN
MM3 XLO John Nattress, 44 Broadlands, Carnoustie, DD7 6JY
MM3 XLQ Stefan Rennie, 27 Whiting Road, Wemyss Bay, PA18 6EB
M3 XLR J Percival, Blue Cedars, Gresford, Wrexham, LL12 8RN
M3 XLS Leslie Turner, 16 Woodland Place, Scarborough, YO12 6EP
M3 XLW K Weston, 44 Shelley Road, Wellingborough, NN8 3DB
M3 XLY Roger Suffling, 44 Belgrave Manor, Woking, GU22 7TW
M3 XMA Maaruf Ali, 12 Hazeleigh Gardens, Woodford Green, IG8 8DX
M3 XMB M Brittain, 159 Caledonia Road, Wolverhampton, WV2 1JA
M3 XME M Elmer, 10 Mona Terrace, Llanfairfechan, LL33 0RE
MW3 XMG P Provis, Dingle Gardens, Croesybchan, Aberdare, CF44 0EJ
M3 XMH Mark Higham, 30 Broome Road, Southport, PR8 4EQ
M3 XMJ Sean Spicer, 34 Hillcrest Avenue, Halesowen, B63 2PR
M3 XMK Michael Rose, 115 New Street, Brightlingsea, Colchester, CO7 0DJ
M3 XMO Alan Reed, 94 Moor Lane, Loughborough, LE11 1BA
M3 XMP Gamma Prasad, 15 Newby Gardens Oadby, Leicester, LE2 4UG
M3 XMQ David White, 10 Meaux Road, Wawne, Hull, HU7 5XD
M3 XMS F Gavins, 26 Upton Avenue, Cheadle Hulme, Cheadle, SK8 7HX
MM3 XMT Alan Stevenson, Starwood Croft, Craigellachie, Aberlour, AB38 9SQ
M3 XMU Gary Hunter, 58 Repps Road, Martham, Great Yarmouth, NR29 4QT
M3 XMY Darryl Worthington, 19 Shoobridge Street, Leek, ST13 5LA
M3 XMZ Robert Honeybourne, Flat 40, Napier Court West, Southend-on-Sea, SS1 1NH
M3 XNA Hayley Ball, Manor House, Tolgus Hill, Redruth, TR15 1AX
M3 XNB N Newby, 22 Acton Road, Liverpool, L32 0TT
M3 XNC George Reywer, 1 Tiverton Close, Houghton le Spring, DH4 4XR
M3 XNE A Warr, 2 Fairfield Road, Bournheath, Bromsgrove, B61 9JN
M3 XNK Neil Price, 22 Hanover Road, Warley, Rowley Regis, B65 9DZ
M3 XNM Ian Stevenson, 79 Lunedale Road, Darlington, DL3 9AT
M3 XNN David Elliott, 54 Grisedale Gardens, Gateshead, NE9 6NP
M3 XNO Jason Manning, 9 Belmont Road, Taunton, TA1 5NS
MM3 XNP Nick Page, 2 Spey Drive, Fochabers, IV32 7QS
M3 XNR Kirk Batt, 14 Milne Park West, New Addington, Croydon, CR0 0DN
M3 XNT Paul Johannessen, 72 Duncombe Road South, Garston, Liverpool, L19 1QJ
M3 XNU Anthony Hollis, 9 St. Georges Road, Donnington, Telford, TF2 7NP
M3 XNV David Dunn, 69 Broadwaters Drive, Kidderminster, DY10 2RY
MW3 XNW N Williams, 1 Picton Terrace, Pontlottyn, Bargoed, CF81 9PT

| | | |
|---|---|---|
| M3 | XNX | N Stubbs, 5 Newland Street, Wakefield, WF1 5AH |
| M3 | XNZ | S Edgar, 61 Winchester Avenue, Lancaster, LA1 4HX |
| M3 | XOA | William Dunstan, 57 Orchard Vale, Flushing, Falmouth, TR11 5TT |
| M3 | XOD | Ian Donnelly, 17 Jessop Close, Horncastle, LN9 [illegible] |
| M3 | XOE | David Coe, 199 Newark Road, North Hykeham, Lincoln, LN6 8QS |
| M3 | XOH | Michael Finn, 23 Spa Lane, Hinckley, LE10 1JA |
| MI3 | XOI | G Gorman, 4 Springwell Park, Groomsport, Bangor, BT19 6LF |
| M3 | XOJ | Malcolm Benson, 11 Hield Grove, Aston by Budworth, Northwich, CW9 6LN |
| MM3 | XOK | Barry Hughes, 49 Marmion Drive, Kirkintilloch, Glasgow, G66 2BH |
| M3 | XOQ | Dean Vale, 10 Elsworth Road, Birmingham, B31 3BT |
| M3 | XOR | M Nelmes, 119 Exeter Road, Dawlish, EX7 0AN |
| M3 | XOT | E Blake, Ty Capel, Tynygraig, Ystrad Meurig, SY25 6AE |
| M3 | XOU | Jason Howell, 56 Prouds Lane, Bilston, WV14 6PU |
| M3 | XOV | Trevor Harris, 94 Marigold Crescent, Dudley, DY1 3NX |
| M3 | XOW | Rob Woodworth, 2 Harrington Court, Meltham, Holmfirth, HD9 4ED |
| M3 | XOY | Ka Yan Chu, Harrogate Ladies' College, Clarence Drive, Harrogate, HG1 2QG |
| M3 | XOZ | Wing Hang Veronica Yung, 17 York Road, Harrogate, HG1 2QL |
| M3 | XPF | Paul Collins, 14 Roundel Way, Marden, Tonbridge, TN12 9TW |
| M3 | XPH | Paul Hennessey, 11 Monmouth Drive, Eaglescliffe, Stockton-on-Tees, TS16 9HU |
| M3 | XPI | Alan Ackroyd, Prospect House, Causeway, Weymouth, DT4 9RX |
| M3 | XPJ | Julian Parfitt, 5 Sheridan Road, Frimley, Camberley, GU16 7DU |
| M3 | XPK | Colin Park, 197 Occupation Road, Albert Village, Swadlincote, DE11 8HD |
| M3 | XPL | Peter Rabone, 6 Cranwell Grove, Kesgrave, Ipswich, IP5 2YN |
| M3 | XPM | Paul Murray, 45 Commercial Street, Risca, Newport, NP11 6AW |
| M3 | XPN | John Shannon, 16 Croft Drive, Tickhill, Doncaster, DN11 9UL |
| M3 | XPP | Paul Jarvis, 24 St. Peters Gardens, Leeds, LS13 3EH |
| M3 | XPR | Paul Ryan, 9 Fleet Close, Wokingham, RG41 3UE |
| M3 | XPS | P Scarratt, 339 Utting Avenue East, Norris Green, Liverpool, L11 1DF |
| M3 | XPU | Robert Wellburn, 86 Granville Street, Grimsby, DN32 9NU |
| M3 | XPW | Jonathan Oglesby, 22 Elm Drive, Finningley, Doncaster, DN9 3EG |
| M3 | XPY | Dimitrios Chatzikos, 53 Benbow Court, Shenley Church End, Milton Keynes, MK5 6JE |
| M3 | XQB | Moira Harbron, 39 Raleigh Road, Sunderland, SR5 5RD |
| M3 | XQE | Victor Wallace, 10 Maes Llydan, Benllech, Tyn-Y-Gongl, LL74 8RD |
| M3 | XQG | Vincent Littlewood, 31 Herriot Drive, Chesterfield, S40 2UR |
| M3 | XQH | John Wilson, 46 Redwood Drive, Maltby, Rotherham, S66 8DL |
| M3 | XQJ | Paul Jones, 76 Pengwern, Llangollen, LL20 8AS |
| M3 | XQK | Dennis Goodfellow, 60 Pickering Green, Gateshead, NE9 7DX |
| M3 | XQL | Micheal Watson, 5 Birchwood Avenue, Whickham, Newcastle upon Tyne, NE16 5QS |
| M3 | XQM | Alistair Macrae, White Lodge, Verwood Road, Three Legged Cross, Wimborne, BH21 6RR |
| M3 | XQO | Sandra Lewis, 40 Bridle Road, Burton Latimer, Kettering, NN15 5QP |
| M3 | XQQ | D Burman, 6 Goodyers Avenue, Radlett, WD7 8BA |
| M3 | XQT | Judith Thompson, 13 Wentworth Avenue, Luton, LU4 9EN |
| M3 | XQV | Robert Dutton, 473 Manchester Road, Lostock Gralam, Northwich, CW9 7QD |
| M3 | XQW | Edward Slevin, Woodcock Hall, Cobbs Brow Lane, Wigan, WN8 7NB |
| M3 | XQX | Bruce Laker, Rose Cottage, Haughley Green, Stowmarket, IP14 3RQ |
| M3 | XQY | Josephine Dutton, 473 Manchester Road, Lostock Gralam, Northwich, CW9 7QD |
| M3 | XQZ | James Freeman, 12 Norfolk Terrace, Cambridge, CB1 2NG |
| M3 | XRD | D rogers, 44 County Street, Oldham, OL8 3RN |
| M3 | XRG | Gavin Duffy, 34 Twentyfifth Avenue, Blyth, NE24 2QW |
| M3 | XRH | Sandra Lawford, 26 Venetian Crescent, Darfield, Barnsley, S73 9PL |
| M3 | XRI | John Jones, Greystones, Rhewl, Oswestry, SY10 7AS |
| M3 | XRK | David Richards, 73 Greenfields Avenue, Alton, GU34 2EW |
| M3 | XRO | R Doughty, 1 Woodland Road, Wakefield, WF2 9DR |
| M3 | XRP | Rosemary Potter, 1 Wentworth Road, Rugby, CV22 6BG |
| M3 | XRQ | Ben Benson Jnr, 12 South Drive, Rudheath, Northwich, CW9 7JQ |
| M3 | XRR | Ian Moule, 88 Redstone Lane, Stourport-on-Severn, DY13 0JG |
| MI3 | XRT | L Murray, 80 Canterbury Park, Londonderry, BT47 6DU |
| M3 | XRV | Robert Peacock, 21 Breamish Drive, Washington, NE38 9HS |
| M3 | XRW | Ronald Wainwright, 69 George A Green Road, Wakefield, WF2 8HA |
| M3 | XRY | G Jones, 31 Cranage Close, Halton Lodge, Runcorn, WA7 5YN |
| M3 | XSA | K Sproates, 36 Manor Park, Writhlington, Radstock, BA3 3NB |
| M3 | XSD | P Davies, 15 Kingsley Road, Chester, CH3 5RR |
| MM3 | XSF | S Turnbull, 15 Woodruff Gait, Dunfermline, KY12 0NL |
| MD3 | XSG | D Levey, Heriots Wood, The Common, Stanmore, HA7 3HT |
| M3 | XSI | S I'Anson, 2 Osborne Close, North Walsham, NR28 0SX |
| M3 | XSJ | Jon Wildsmith, 7 Doctors Hill, Stourbridge, DY9 0YE |
| M3 | XSK | Jordan Skittrall, 14 Tamarin Gardens, Cambridge, CB1 9GH |
| M3 | XSN | Andrew Thompson, 7 Lammermoor Road, Liverpool, L18 4QP |
| M3 | XSP | S Purkiss, 19 The Hurstings, Maidstone, ME15 6YN |
| M3 | XSR | T Came, 15 Brookland Road, Langport, TA10 9TA |
| M3 | XST | K Wood, 52 Trench Road, Tonbridge, TN10 3HB |
| M3 | XSU | James Martin, 3 Pipitsmead House, Alder Court, Fleet, GU51 5AH |
| M3 | XSV | Mark Ridpath, The Grange, Main Street, Hull, HU12 0JF |
| M3 | XSY | Christine Throup, Willow House, O'Keys Lane, Worcester, WR3 8RL |
| M3 | XSZ | Maiza Bekara, 9 southwood Court Pine Grove, Weybridge, KT13 9AT |
| M3 | XTA | David Smith, 28a Bagshot Green, Bagshot, GU19 5JR |
| M3 | XTE | Jonathan Ransom, 4 Vicarage Road, Eastbourne, BN20 0AU |
| M3 | XTF | William Welch, Kenilworth, School Lane, Oswestry, SY11 3LD |
| M3 | XTG | Terence Greenaway, 11 Gribben Close, Tregonissey, St Austell, PL254EA |
| M3 | XTK | P Mortiboy, 72 Uplands, Stevenage, SG2 7DW |
| M3 | XTL | Matthew Porter, 8 Stanton Drive, Ludlow, SY8 2PH |
| M3 | XTM | J Maguire, 14 Botha Road, St. Eval, Wadebridge, PL27 7TS |
| M3 | XTP | Kevin Hemmings, 11 Collenswood Road, Stevenage, SG2 9ER |
| M3 | XTR | P Britton, 71 Upper Forster Street, Walsall, WS4 2AB |
| M3 | XTT | Nigel Pearson, 116 The Stour, Daventry, NN11 4PT |
| M3 | XTV | Terence Benson, 83 Glovers Road, Birmingham, B10 0LE |
| MD3 | XUA | Carola James, 75 Silverburn Crescent, Ballasalla, Isle of Man, IM9 2DY |
| MI3 | XUC | James McCollum, 26 Corkey Road, Loughgiel, Ballymena, BT44 9JJ |
| M3 | XUE | Sidney Leadbetter, 11 Cogos Park, Mylor Bridge, Falmouth, TR11 5SF |
| M3 | XUF | Fiona Leung, Harrogate Ladies' College, Clarence Drive, Harrogate, HG1 2QG |
| M3 | XUG | Robert Barnes, 275 Oregon Way, Chaddesden, Derby, DE21 6UR |
| M3 | XUH | Glyn Thomas, Lowside Barn, Rothersyke, Egremont, CA22 2UD |
| MM3 | XUI | G Taylor, 15 Ronaldsvoe, Kirkwall, KW15 1XF |
| M3 | XUJ | Paul Hodgkinson, 66 Meadow Lane, Newhall, Swadlincote, DE11 0UW |
| M3 | XUO | Jonathan Steven, 1 Tree Terrace, Tree Road, Brampton, CA8 1TY |
| M3 | XUR | Philip Andrews, 15 Park Lane, Bath, BA1 2XH |
| MI3 | XUS | Simon Barnes, 191 Marlacoo Road, Portadown, Craigavon, BT62 3TD |
| M3 | XUT | R Udall, 139 Leicester Road, Measham, Swadlincote, DE12 7JG |
| M3 | XUU | Ravi Gopan, 84 Hilmanton, Lower Earley, Reading, RG6 4HN |
| M3 | XUV | Elizabeth Paddison, 3 Westacre Gardens, Ormesby, Great Yarmouth, NR29 3SP |
| M3 | XUW | Graham Marsh, Apartment 76, 874 Wilmslow Road, Manchester, M20 5AB |
| MM3 | XUX | John Munro, 5 Wallace Gait, Perth, PH1 2NS |
| MM3 | XUY | Grant Nicholson, 4 John Street, Oban, PA34 5NS |
| MW3 | XVB | David Jones, 26 Ffos y Cerridden, Nelson, Treharris, CF46 6HQ |
| M3 | XVC | Vernon Couchman, 4 Fairfield Road, St. Leonards-on-Sea, TN37 7UA |
| MM3 | XVD | Irene Woods, 12 Westbank Terrace, MacMerry, Tranent, EH33 1QE |
| M3 | XVF | Jeremy Smith, 8 Mayfields, Spennymoor, DL16 6RN |
| M3 | XVJ | Matthew Emmott, 1 Swallow Close, Kendal, LA9 7SN |
| M3 | XVK | Leslie Ward, 20 North Street, Maryport, CA15 6HR |
| M3 | XVQ | John Mitchell, 27 Watts Close, Southampton, SO16 9WA |
| MW3 | XVR | Alan Jones, 11 Bigyn Road, Llanelli, SA15 1NT |
| M3 | XVU | B Lovius, 10 Templemore Avenue, Liverpool, L18 8AH |
| M3 | XVW | Victoria Walker, Kirby Welch & Co, West View, Longlands Lane, Wetherby, LS22 4BB |
| M3 | XVZ | Phil Read, 53 Hill Top Road, Oldbury, B68 9DU |
| M3 | XWB | John McColl, 84a Perth Street, Hull, HU5 3NZ |
| M3 | XWC | John Conway, 18 Headland Close, Welford on Avon, Stratford-upon-Avon, CV37 8EU |
| M3 | XWE | Timothy Banks, 18 Leicester Road, Newport, NP19 7ER |
| M3 | XWF | John Coogan, 1 Langsett Rise, Sheffield, S6 2TY |
| M3 | XWK | Liam Thompson, 34 Broadway, Gateshead, NE9 5PY |
| M3 | XWM | Tony Sayers, 12 Hutton Terrace, Willington, Crook, DL15 0DS |
| M3 | XWN | Christopher Cartwright, 8 Hawes Grove, Bradford, BD5 9AN |
| M3 | XWP | Andrew Wright, 149 Burton Road, Overseal, Swadlincote, DE12 6JL |
| M3 | XWR | Jake Halsall, 8 Woodcock Street, Wakefield, WF1 5LG |
| MM3 | XWS | W Anderson, Hillview, North Street, Johnstone, PA6 7HJ |
| M3 | XWV | Jose Barbieri, 20 Gilbard Court, Chineham, Basingstoke, RG24 8RG |
| MI3 | XWW | David Bradley, 33 Lilac Avenue, Limavady, BT49 0HS |
| M3 | XWX | Scott Kerslake, 4 Guipavas Road, Callington, PL17 7PL |
| M3 | XWY | Euan Jennings, 17 Manor Way, Worcester Park, KT4 7PH |
| M3 | XXA | Julian Smith, 131 Steadman House, Bow Common Lane, London, E3 4HT |
| M3 | XXB | S Bunce, 15 Downs View Road, Bembridge, PO35 5QS |
| M3 | XXE | Nicholas Dwyer, 82 Staunton Road, Kingston upon Thames, KT2 5TL |
| M3 | XXG | Graham Collins, 110 Hawthorn Crescent, Burton-on-Trent, DE15 9QW |
| MM3 | XXI | J Redmond, 14 Bankfaulds Avenue, Kilbirnie, KA25 6AB |
| M3 | XXK | Kam Mitchell, 1 Denstroude Cottages, Denstroude Lane, Canterbury, CT2 9JX |
| M3 | XXL | Steve Pearce, 20 Barcote Walk, Plymouth, PL6 5QE |
| M3 | XXM | Michael Jennings, Springfield Farm, The Causeway, King's Lynn, PE34 3PP |
| M3 | XXO | Steven Rogan, 21 Montrose Place, Selkirk, TD7 5BH |
| MM3 | XXP | Alan Pitkethley, 99 Margaretvale Drive, Larkhall, ML9 1EH |
| M3 | XXS | Stephen Mellor, 11 Bolton Meadow, Leyland, PR26 7AJ |
| M3 | XXU | Ashley Friswell, 142 Aldermans Green Road, Coventry, CV2 1PP |
| M3 | XXY | K Dobson, 152 Foryd Road, Kinmel Bay, Rhyl, LL18 5LS |
| M3 | XYA | Romano Pasika, 192 Longfield Lane, Cheshunt, Waltham Cross, EN7 6AQ |
| MI3 | XYB | J Throne, 12 Mason Road, Magheramason, Londonderry, BT47 2RY |
| M3 | XYC | Michael Cowan, Oak Haven, Smugglers Lane, Chichester, PO18 8QW |
| M3 | XYH | Craig Bell, 60 East Vines, Sunderland, SR1 2DP |
| M3 | XYI | William Cooper, 20 Staple Close, Waterlooville, PO7 6AH |
| M3 | XYJ | Simon Wright, Emergency Planning Unit, NYCC County Hall, Northallerton, DL7 8AD |
| M3 | XYK | Alan Raby, 209 Duke of York Avenue, Wakefield, WF2 7DH |
| M3 | XYM | Mark Leonard, 5 Nettleton Garth, Burstwick, Hull, HU12 9DY |
| M3 | XYN | Lyn Marsh, 14 Herrick Road, Barnby Dun, Doncaster, DN3 1AW |
| M3 | XYO | Jennie Douglas, 14 Mountfields Walk, South Kirkby, Pontefract, WF9 3SJ |
| M3 | XYP | G Thompson, 24 Fairmead Way, Sunderland, SR4 0NA |
| M3 | XYT | Matthew Carter, 17a Goodramgate, York, YO1 7LW |
| M3 | XYU | Charles Coverley, Flat 1, Bridge House, 9 Kingsbridge Lane, Newton Abbot, TQ13 7DX |
| M3 | XYW | Annmarie Rosser, 25 Clos Tir-y-Pwll Newbridge, Newport, NP11 5GE |
| M3 | XYX | W Slater, 47 Broom Road, Lakenheath, Brandon, IP27 9EZ |
| M3 | XYZ | R Hill, 12 Winchelsea Lane, Hastings, TN35 4LG |
| M3 | XZB | Samuel Gardner, 92 Gladstone Street, Abertillery, NP13 1NE |
| M3 | XZD | John Kelly, 2 Tamar Close, Higham, Barnsley, S75 1PS |
| M3 | XZE | Peter Webb, Picture This, The Studio, 152 Bearton Road, Hitchin, SG5 1UA |
| M3 | XZF | X Fang, Harrogate Ladies' College, Clarence Drive, Harrogate, HG1 2QG |
| M3 | XZG | David Mestel, 41 Glisson Road, Cambridge, CB1 2HA |
| M3 | XZH | Paul Snook, 7 Sandhurst Avenue, Kwazulu Natal, South Africa, 3610 |
| M3 | XZJ | Leon Aldred, 1 Eaton Grange Cottages, Eaton, Grantham, NG32 1FL |
| M3 | XZK | Alan Smith, 20 Linden Road, Coxheath, Maidstone, ME17 4QS |
| M3 | XZN | Martin Robinson, 10 Bramley Gardens, Poulton-le-Fylde, FY6 7RD |
| MW3 | XZP | Chris Maggs, 15 Stuart Street, Treorchy, CF42 6SN |
| M3 | XZR | Davis Davis, 9 Park Gate Mews, Upper Norwich Road, Bournemouth, BH2 5RA |
| M3 | XZS | Sharon Greaves, 9 Park Gate Mews, Upper Norwich Road, Bournemouth, BH2 5RA |
| M3 | XZT | Kenneth Lewinton-Smith, 4 Old School Road, Barnstaple, EX32 9DP |
| M3 | XZV | Walden Jones, 2 Derwen Close, Connah's Quay, Deeside, CH5 4AU |
| M3 | XZY | Michael Reilly, Flat 59, The Keep, Stafford, ST17 9TW |
| M3 | YAA | J Wellard, 19 South Motto, Kingsnorth, Ashford, TN23 3NJ |
| M3 | YAD | Adrian Rossant, 18 North Hill Close, Burton Bradstock, Bridport, DT6 4RY |
| MW3 | YAE | Gary Thatcher-Sharp, 20 Dilys Street, Blaencwm, Treorchy, CF42 5DT |
| M3 | YAI | H Litten, 55 Downton View, Ludlow, SY8 1JE |
| M3 | YAJ | A Jay, Jasper, The Reddings, Cheltenham, GL51 6RT |
| M3 | YAL | B Loughran, 26 Squirrels Field, Mile End, Colchester, CO4 5YA |
| M3 | YAO | Eric Bicknell, 12 Victory Road, Southampton, SO15 [illegible] |
| M3 | YAP | A Sheppard, 1 Waveney Walk, Crawley, RH10 6RL |
| M3 | YAS | Gemma Cummings, 18 Castleton Boulevard, Skegness, PE25 2TX |
| M3 | YAV | Peter Meredith, Bottom Flat, 74 Earl Street, Grimsby, DN31 2PW |
| M3 | YAW | Stephen Johnson, 43 Terry Gardens, Kesgrave, Ipswich, IP5 2EP |
| M3 | YAX | Leslie Mason, 9 Trenethick Avenue, Helston, TR13 8LU |
| MW3 | YBB | P Ryalls, 3 Bryn Terrace, Blaenclydach, Tonypandy, CF40 2RY |
| MM3 | YBD | W Doull, 9 Mcdowall Avenue, Ardrossan, KA22 7AJ |
| M3 | YBF | P Teszner, 21 Sprinkwood Grove, Stoke-on-Trent, ST3 6EQ |
| MM3 | YBG | Christiana Gerrard, 10 Whinhill Gardens, Aberdeen, AB11 7WD |
| MI3 | YBI | Thomas Quin, 165 Marlacoo Road, Portadown, Craigavon, BT62 3TD |
| M3 | YBJ | Robert Marsh, 56b Oliver Crescent, Farningham, Dartford, DA4 0BE |
| M3 | YBK | T Jones, 25 Pritchard Terrace, Phillipstown, New Tredegar, NP24 6BS |
| M3 | YBL | Brenda Lace, 19 Methuen Street, Walney, Barrow-in-Furness, LA14 3PS |
| M3 | YBN | Kevin Stokes, 21 Victoria Grove, Bideford, EX39 2DN |
| MM3 | YBQ | Keith Verrall, 7 Roshven View, Arisaig, PH39 4NX |
| M3 | YBR | A Knight, 3 Hawthorn, Appledore, Ashford, TN26 2AH |
| M3 | YBT | Michael Fearon, 70 George Street, Heywood, OL10 4PW |
| M3 | YBU | Mark Shields, 43 Thompson Avenue, Richmond, TW9 4JP |
| M3 | YBW | Bea Warner, 15 Grosvenor Gardens, Shifnal, TF11 8EB |
| MW3 | YBX | Daniel Smethurst, Glengarth, Portfield Gate, Haverfordwest, SA62 3LS |
| M3 | YCB | Charles Boston, 53 Bullock Road, Terrington St. Clement, King's Lynn, PE34 4PR |
| M3 | YCD | Daniel Clarke, 57 Glanville Place, Kesgrave, Ipswich, IP5 1NQ |
| MM3 | YCG | Callum Graham, 5 Ashkirk Road, Strathaven, ML10 6JT |
| MM3 | YCI | Sadie Burt, 182 Old Inverkip Road, Greenock, PA16 9JG |
| M3 | YCJ | Violet Powell, 35 Bramber Close, Banbury, OX16 0XF |
| M3 | YCK | Cho Kwan Sergius Fung, Harrogate Ladies' College, Clarence Drive, Harrogate, HG1 2QG |
| MW3 | YCL | Christopher Steer, 1 Park Way, Park, Merthyr Tydfil, CF47 8RH |
| M3 | YCM | Martin Hemmings, 166 Kimbolton Crescent, Stevenage, SG2 8RW |
| M3 | YCN | William Fowler, 20 The Court, Anderby Creek, Skegness, PE24 5YQ |
| M3 | YCO | D Cope, 14 Burland Road, Newcastle, ST5 7ST |
| M3 | YCR | Charles Rayment, Brambles, Alltami Road, Mold, CH7 6RT |
| M3 | YCS | Charles Sutton, 56 Neatherd Road, Dereham, NR20 4AY |
| M3 | YCT | Peter Howells, 20 Warwick Street, Stourport on Severn, DY13 8JB |
| M3 | YCU | Ailing Wang, Harrogate Ladies' College, Clarence Drive, Harrogate, HG1 2QG |
| M3 | YCV | Christine Watts, 35 Coldharbour Lane, Salisbury, SP2 7BY |
| M3 | YCZ | Carla Landless, 2 Aspen Way, Banstead, SM7 1LE |
| M3 | YDA | A Collins, Robin Post House, Robin Post Lane, Hailsham, BN27 3RA |
| M3 | YDB | D Bush, 19 Spring Vale, Waterlooville, PO8 9DA |
| MI3 | YDF | D Foley, 14 Chestnut Hall Court, Maghaberry, BT67 0GJ |
| M3 | YDH | D Hume, Sundhopeburn, Yarrow, Selkirk, TD7 5NF |
| M3 | YDI | Melvyn Siddle, 8 Coleridge Close, Oulton, Leeds, LS26 8ET |
| M3 | YDJ | D Wilkinson, 6 Brambledown Road, South Croydon, CR2 0BL |
| MM3 | YDK | Dawne Kilgour, 56 Green Street, Rothes, Aberlour, AB38 7BD |
| M3 | YDL | S Thornton, 29 Farrar Avenue, Mirfield, WF14 9ED |
| M3 | YDM | T Yardley, 19 Elms Close, Shareshill, Wolverhampton, WV10 7JT |
| M3 | YDS | Shane Morgan, 20 Cwrt y Babell, Cwmfelinfach, Newport, NP11 7NR |
| M3 | YDT | E Gittins, 40 Melyd Avenue, Prestatyn, LL19 8RN |
| M3 | YDV | Dean Brame, 7 Roche Garden, Exeter, EX2 6LS |
| M3 | YDW | Dereck White, 9 Wyatts Lane, Tavistock, PL19 0EU |
| M3 | YDY | D Murfitt, 10 Benefield Road, Moulton, Newmarket, CB8 8SW |
| M3 | YEA | L Pollard, 11 Alfriston Road, Worthing, BN14 7QU |
| MM3 | YEC | J Dupont, 11 Golf View Cardenden, Lochgelly, KY5 0NW |
| M3 | YEE | H Ngi, Harrogate Ladies' College, Clarence Drive, Harrogate, HG1 2QG |
| MW3 | YEG | J Thorne, 11 Dowland Road, Penarth, CF64 3QX |
| M3 | YEJ | Jason Marsh, 31 Clay Street, Soham, Ely, CB7 5HJ |
| M3 | YEK | Andrew Taylor, 16 Bellmans Road Whittlesey, Peterborough, PE7 1TY |
| M3 | YEM | Ruth Browne, 30 Cromwell Road, Southowram, Halifax, HX3 9SE |
| MM3 | YEQ | Gordon Pearce, 1 Inchbelle Farm Cottage, Kirkintilloch, Glasgow, G66 1RS |
| M3 | YET | A Reilly-Cooper, 40 Clough Lane, Northwich, CW8 1JR |
| M3 | YEU | James Mobbs, 6 School View, Banbury, OX16 4SD |
| M3 | YEZ | J Dearden, 218 South Street, Highfields, Doncaster, DN6 7JQ |
| M3 | YFG | Guy Walton, 11 Redwood Close, Hoyland, Barnsley, S74 0EJ |
| M3 | YFH | Andrew Sherman, 31 Peartree Avenue, Kingsbury, Tamworth, B78 2LG |
| M3 | YFI | Terence Hall, 18 Common Lane, New Haw, Addlestone, KT15 3LH |
| M3 | YFJ | Alan King, 6 Dunsfold Close, Crawley, RH11 8EY |
| M3 | YFL | James Solomon, 17 Chadwick Terrace, Macclesfield, SK10 2DQ |
| M3 | YFM | Adam Reader, 12 Valleyside, Swindon, SN1 4NB |
| M3 | YFN | Michael Green, Fire Beacon Cottage, East Hill, Sidmouth, EX10 0LR |
| MM3 | YFR | David Cockburn, 88 Knockmarloch Drive, Kilmarnock, KA1 4QN |
| MM3 | YFT | Peter Finnie, 20 St. Margarets Road, Ardrossan, KA22 7ER |
| M3 | YFW | Richard Stevens, Durham House, Cavendish Road, Sudbury, CO10 8PJ |
| M3 | YFX | John Edward Loveridge, 96 High Road, Islington, King's Lynn, PE34 3BN |
| M3 | YFY | Daniel Rolfe, 40 Hillrise Avenue, Sompting, Lancing, BN15 0LU |
| M3 | YFZ | Joseph Blower, 4 Lamorna Close, Luton, LU3 2TH |
| M3 | YGB | Andrew Birch, 3 Partridge Way, High Wycombe, HP13 5JX |
| M3 | YGC | Graham Cowley, 2 Manor Close, Farcet, Peterborough, PE7 3AA |
| M3 | YGD | Gary Hyland Davis, 31 Melody Close, Warden, Sheerness, ME12 4PU |
| M3 | YGF | Gillian Ferguson, 31 Barton Court Road, New Milton, BH25 6NW |
| MM3 | YGI | Ronald Stratton, 18 Dunnock Park, Perth, PH1 5FN |
| MD3 | YGK | Yogarajah Gopikrishna, 29 Alandale Drive, Pinner, HA5 3UP |
| M3 | YGL | Donna Gribben, 44 Fern Close, Birchwood, Warrington, WA3 7NU |
| M3 | YGO | Chris Chew, 10 Bruce Drive, South Croydon, CR2 8SL |
| M3 | YGQ | Martin Smith, 3 Fleetwood Close, March, PE15 9NN |
| M3 | YGR | Julius Katz, 8 Astor Drive, Birmingham, B13 9QR |
| M3 | YGS | Safwan Dingmar, 10 Kertland Street, Savile Town, Dewsbury, WF12 9PU |
| M3 | YGT | Georgios Giannakopoulos, 3 Vauxhall Quay, Plymouth, PL4 0EZ |
| M3 | YGU | Brett Chamberlain, 2 Stocks Loke, Cawston, Norwich, NR10 4BS |
| M3 | YGV | James Harris, Flat 3, Herstmonceux Place, Church Road, Herstmonceux, Hailsham, BN27 1RL |
| M3 | YGY | Corey Lee-Koo, Flat 5, 211 Sussex Gardens, London, W2 2RJ |
| M3 | YGZ | Ian McCourt, 181 Fircroft Road, Ipswich, IP1 6PS |



**IMPORTANT NOTE**

**Revalidate licence to avoid revocation** – Ofcom has advised the Society that plans will be drawn up to revoke licences that have not been revalidated as required by the licence conditions. The quickest way to revalidate is to do so online via the Ofcom website: *https://services.ofcom.org.uk/* or by email: *amateur.validations@ofcom.org.uk* Ofcom staff are available to help, but please be patient during times of heavy workload.

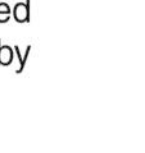

MM3 YHA Donald Morrison, 4 West Murkle, Murkle, Thurso, KW14 8YT
M3 YHC D Jones, Bryn Hyfryd, Pandy Tudur, Abergele, LL22 8UL
M3 YHD Keith Marshall, 38 Staunton Road, Newark, NG24 4EX
M3 YHF Georgina Hand, Hollinhurst Farm, Park Lane, Stoke-on-Trent, ST9 9JB
M3 YHG John Stringer, 31 Pipit Lane, Birchwood, Warrington, WA3 6NY
M3 YHH Bernard Hand, Hollinhurst Farm, Park Lane, Stoke-on-Trent, ST9 9JB
M3 YHJ Azad Hussain, 789 Scarborough Street, Dewsbury, WF12 9AY
M3 YHL Hon Ying Janet Lee, Harrogate Ladies' College, Clarence Drive, Harrogate, HG1 2QG
M3 YHM Wing Tung Mok, 17 York Road, Harrogate, HG1 2QL
M3 YHN Matthew Roberts, 7 Maxwell Place, Stoke-on-Trent, ST4 6RE
M3 YHP Heather Pentz, 30 Lindrick Way, Harrogate, HG3 2SU
M3 YHQ R Mason, 9 Farmfields Rise, Woore, Crewe, CW3 9SZ
M3 YHR Andrew Rolland, Flat 29, Renfrew Court, Eastbourne, BN22 7SZ
MM3 YHS Hugh Steele, 44 Waverley Crescent, Livingston, EH54 8JN
M3 YHT Geoffrey Winterbottom, 35 Abingdon View, Worksop, S81 7RT
M3 YHU Chris Cleverley, 4a Godfrey Street, Netherfield, Nottingham, NG4 2JG
M3 YHV M Chidgey, 46 Station Road, Shirehampton, Bristol, BS11 9TX
MW3 YHW John Loughlin, 453 Heol-y-Waun, Penrhys, Ferndale, CF43 3NW
M3 YHY Fergus Noble, 1045, 45th Street Apartment A, California, United States, 94608
M3 YHZ Craig Amos, 33 Douglas Road, Newcastle, ST5 9BP
M3 YIC L Crabtree, 23 Ava Crescent, Richmond Hill, Ontario, Canada, L4B 2X1
M3 YIE Anthony Fuller, Flat 5 Maple House, 3 Fairfield Road, Havant, PO9 1AG
M3 YIF Nick Spooner, 14 Glebe Road, Ongar, CM5 9HW
MM3 YIG Craig Doolan, 56 Forfar Road, Greenock, PA16 0YL
MM3 YIH Calum Rodgers, 3 Merrylee Avenue, Port Glasgow, PA14 5UT
M3 YII Annamarie Boag, 60 Harebell, Amington, Tamworth, B77 4NA
M3 YIL Alan Bloor, Corner House, Cross in Hand, Heathfield, TN21 0SR
MM3 YIQ Ross Doolan, 56 Forfar Road, Greenock, PA16 0YL
M3 YIT Tching-Yee Yip, 6 Fulwith Grove, Harrogate, HG2 8HN
M3 YIV Andrew Little, 60a Murray Road, Horndean, Waterlooville, PO8 9JL
M3 YIX Mike Norris, 35 Sudbrooke Road, London, SW12 8TQ
M3 YIY Ross Egan, 9 Wilbye Grange, Wellingborough, NN8 3PS
M3 YJA James Appleby, 79 Glenwoods, Newport Pagnell, MK16 0NG
M3 YJB J Birch, 6 Crescent Road, Wallasey, CH44 0BQ
M3 YJD John Dowdeswell, 18 Lechlade Gardens, Fareham, PO15 6HF
MI3 YJE J Elliott, 30 Moyle Road, Ballycastle, BT54 6AN
M3 YJF David Moran, 23 Abbotsfield Crescent, Tavistock, PL19 8EY
M3 YJG Raymond Lawton, 41 Almond Avenue, Armthorpe, Doncaster, DN3 2HE
M3 YJH Julia Hume, Sundhope Farm, Selkirk, TD7 5NF
M3 YJJ Leon Gleed, 9 Medlock Close, Bettws, Newport, NP20 7EJ
M3 YJL John Cairns, 17 Alfred Avenue, Worsley, Manchester, M28 2TX
M3 YJM J Patrick-Gleed, 6 Julius Way, Lydney, GL15 5QS
M3 YJN N Allen, 5 Limecroft View, Wingerworth, Chesterfield, S42 6NR
M3 YJP G Patrick-Gleed, 6 Julius Way, Lydney, GL15 5QS
M3 YJQ Andrew Graham, 11 Bettws Close, Bettws, Newport, NP20 7YA
M3 YJT H Taylor, 21 Charlecote Drive, Chandler's Ford, Eastleigh, SO53 1SF
M3 YJU Paul Boreham, 67 Brent Lane, Dartford, DA1 1QT
M3 YJW Jason Williams, 10 Masefield Avenue Eaton Ford, St. Neots, PE19 7LS
M3 YJY Leslie Bourne, 9a Partridge Croft, Lichfield, WS13 6SD
M3 YKA A Francis, 33 Libeneth Road, Newport, NP19 9AP
M3 YKC K Comben, 9 West Lane, North Baddesley, Southampton, SO52 9GB
M3 YKF Ka Man Fung, Harrogate Ladies' College, Clarence Drive, Harrogate, HG1 2QG
M3 YKH A Harding, Sunnydene, Wellmead, Axminster, EX13 7SQ
M3 YKI Ahned Sulieman, 22 Warren Court, 80 Charlton Church Lane, London, SE7 7AD
MW3 YKL C Warburton, 71 Richards Terrace, Cardiff, CF24 1RW
M3 YKN David Bright, 103b Langer Road, Felixstowe, IP11 2EA
M3 YKO Tsz Ki Joffee Chan, Harrogate Ladies' College, Clarence Drive, Harrogate, HG1 2QG
MM3 YKR Ronald Murray, 18 Braids Road, Kirkcaldy, KY2 6JE
M3 YKS Brenda Shackleton, 54a Blueleighs Park Homes, Ipswich, IP6 0ND
M3 YKT Linwood Jones, 16 Oxland Road, Illogan, Redruth, TR16 4SH
M3 YKZ David Warren, 36 Milner Road, Heswall, Wirral, CH60 5RZ
M3 YLB R Beck, 73 Crowborough Road, Southend-on-Sea, SS2 6LW
M3 YLJ Gill Whitehead, 29 Coulsons Road, Bristol, BS14 0NN
M3 YLK John Swift, 56 Leymoor Road, Huddersfield, HD3 4SW
M3 YLL John Swift, 56 Leymoor Road, Huddersfield, HD3 4SW
M3 YLM Yan Ling Lam, Harrogate Ladies' College, Clarence Drive, Harrogate, HG1 2QG
M3 YLN C Mills, 118 Saxon Gardens, Shoeburyness, Southend-on-Sea, SS3 9PX
M3 YLO A Powell, 31 Highmead, Pontllanfraith, Blackwood, NP12 2PF
MM3 YLP Colin Edwards, 18 Gelshfield, Halkirk, KW12 6UZ
M3 YLQ John Painter, 58 Glenfield Road, Plymouth, PL6 7LN
M3 YLR David Trevelyan, 35 St. Kingsmark Avenue, Chepstow, NP16 5LY
M3 YLT Lee Timmins, 83 Loxdale Sidings, Bilston, WV14 0TN
M3 YLU R Noon, Forest Hill Cottage, Rushall Lane, Wimborne, BH21 3RT
M3 YLV Kyran Bell, 2 Hill Street, Risca, Newport, NP11 6QH
MD3 YLX David Cain, 7 Cronk y Berry Mews, Douglas, Isle of Man, IM2 6HQ
M3 YLZ Maxim Brewster, Blackthorn Farm, Common Road, Diss, IP21 4PH
M3 YMC M Comben, 9 West Lane, North Baddesley, Southampton, SO52 9GB
M3 YMD M Dudley, 418 Sandon Road, Stoke-on-Trent, ST3 7LH
MI3 YMF Martin Foley, 44 Gallows Street, Dromore, BT25 1BD
M3 YMG Michael Crawley, 16 The Meadows, Herne Bay, CT6 7XF
M3 YMH Michael Hurst, 20 Albany Avenue, Manchester, M11 1HQ
M3 YMI John Ellis, Tegfan, Rhostryfan, Caernarfon, LL54 7NF
MM3 YMM Margaret Holmes, 1 Lauren Way, Paisley, PA2 9JW
MM3 YMN Jonathan Henderson, 7 Rowanhill Close, Port Seton, Prestonpans, EH32 0SY
MM3 YMQ Neil Hirst, 25 Conifer Road, Mayfield, Dalkeith, EH22 5BY
M3 YMS Mark Statham, 17 Nicholas Meadow, Higher Metherell, Callington, PL17 8DE
MM3 YMU Rona Morrison, 4 West Murkle, Murkle, Thurso, KW14 8YT
M3 YMX James Lewis, 4 Moor Park, Clevedon, BS21 6EH
M3 YMY Matthew Ireland, Pen y Gadlas, Ffordd Bryniau, Prestatyn, LL19 8RD
MW3 YNA Michael Seagrave, The Firs, Llannon, Llanelli, SA14 6AP
M3 YNB Nicholas Buttery, 22 Mallard Road, Rowlands Castle, PO9 6HN
M3 YNC Charles Mackintosh, 18 Park Avenue, Castleford, WF10 4JT
M3 YND Bryan Anderson, 41 Lower Meadow, Harlow, CM18 7RE
M3 YNE J Godfrey, 36 Greenwich Road, Hailsham, BN27 2PE
M3 YNH Christopher Arundel, 54 Broadmead, Castleford, WF10 4SE
M3 YNI Robert Pike, 66 Prowses, Hemyock, Cullompton, EX15 3QG
M3 YNJ Alex Richards, 5 Bloomfield Drive, Bracknell, RG12 2JW
M3 YNK Gerald Lane, 32 Caellepa, Bangor, LL57 1HF
M3 YNM Ryan Johnson, 50 Barnaby Rudge, Chelmsford, CM1 4YG
M3 YNN Stephen Linton, 89 Ragpath Lane, Stockton-on-Tees, TS19 9JS
MM3 YNP Craig Haldane, 72A Coatbridge Road, Glenmavis, Airdrie, ml6 0nj
M3 YNS Mike Gibson, 58 Byron Street, Barrow-in-Furness, LA14 5RL
M3 YNX Daniel Brewer, 15 Morella Road, London, SW12 8UQ
M3 YNY Neil Oldrid, 125c Denby Dale Road, Wakefield, WF2 8EB
MM3 YOC Raymond Munro, 20 County Cottages, Piperhill, Nairn, IV12 5SE
M3 YOE A Yorke, 33 Avon Crescent, Stratford-upon-Avon, CV37 7EX
M3 YOG Kathryn Cartledge, Oysterber Farm, Burton Road, Lancaster, LA2 7ET
M3 YOH Lianne Mason, 432 Lichfield Road, Sutton Coldfield, B74 4BL
MM3 YOL D Davidson, 44 Abercromby Crescent, Helensburgh, G84 9DX
M3 YOM James Thresher, 328 Gospel Lane, Birmingham, B27 7AJ
M3 YOO Adam Williams, 54 Longbridge, Willesborough, Ashford, TN24 0TA
M3 YOP Timothy Court, Eastgate Cottage, Perrys Lane, Norwich, NR10 4HJ
M3 YOQ Michael Harris, Rodmarton House, Broad Town, Swindon, SN4 7RG
M3 YOT J O'Malley, 140 Allerburn Lea, Alnwick, NE66 2QP
M3 YOU M Young, 19 Tallents Close, Sutton at Hone, Dartford, DA4 9HS
M3 YOW George Eycott, 1 Ham Road, Wanborough, Swindon, SN4 0DF
M3 YOX William Hopkins, 30 Charles Darwin Road, Plymouth, PL1 4GU
M3 YOZ Stephen Hughes, 117 Liverpool Road, Irlam, Manchester, M44 6EH
M3 YPA Stephen Watts, 29 Brook Drive, Corsham, SN13 9AU
M3 YPB Christopher Bond, Tryfan, Vicarage Lane, Neston, CH64 5TJ
M3 YPD Jacob Scholz, 272 W. Academy St., CLAYTON, New Jersey, United States, 8312
M3 YPG W Gratton, Park House, Brimham Rocks Road, Harrogate, HG3 3HE
MM3 YPH Elizabeth Bertram, 20 Kyles View, Largs, KA30 9ET
M3 YPI B Deakin, 8 Patey Street, Manchester, M12 5RP
M3 YPJ Paul Kirby, 30 New Street, Eccleston, Chorley, PR7 5TW
MM3 YPN Stephen Hargreaves, 4 Oxenfoord Avenue, Pathhead, EH37 5QD
M3 YPP Charles Snow, 19 Salters Road, Haylands, Ryde, PO33 3HU
M3 YPR Pamala Ruocco, 8 Birchenall Street, Manchester, M40 9ND
M3 YPS L Brady, 9 Wordsworth Close, Wootton Bassett, Swindon, SN4 8HJ
M3 YPU Taylor Galloway, 63 Molloy Road, Shadoxhurst, Ashford, TN26 1HR
M3 YPW Ronald Coleman, 5 Meeting Lane, Burton Latimer, Kettering, NN15 5LS
M3 YPX David Parkhouse, 5 Long Yard, Briston, Melton Constable, NR24 2LB
M3 YQC Peter Rogers, 16 Begonia Close, Basingstoke, RG22 5RA
M3 YQG Alexander Bandtock, 28 Campion Road, Westoning, Bedford, MK45 5LB
M3 YQH Gordon Smith, 6 Grange Crescent, Childer Thornton, Ellesmere Port, CH66 5NB
M3 YQL John Edwards, 2 Maes Merddyn, Gaerwen, LL60 6DG
M3 YQM D Morbey, 44 Browning Road, Plymouth, PL2 3AP
MM3 YQO Glen Moir, 38 Forest Park, Stonehaven, AB39 2GF
MM3 YQP Christopher Pate, 75 Castings Avenue, Falkirk, FK2 7BJ
M3 YQQ N Lutte, Long Durford, Durford Wood, Petersfield, GU31 5AW
M3 YQR Charles Hay, Sea Cadets, Riverside Road, Great Yarmouth, NR31 6PX
M3 YQT John Best, 24 Suggitts Lane, Cleethorpes, DN35 7JJ
M3 YQU John Jones, 55 London Road, Holyhead, LL65 2NS
MM3 YQX Kathryn McBride, 1 Cowal Place, Gourock, PA19 1EJ
M3 YRB D Grantham, 7 Goodwin Close, Sandiacre, Nottingham, NG10 5FF
M3 YRC S Evans, 2 Firbeck Crescent, Langold, Worksop, S81 9SB
M3 YRH Rina Horner, 21 Ainsworth Road, Little Lever, Bolton, BL3 1RG
M3 YRJ Thomas Lamont, 30 Shackleton Close, Old Hall, Warrington, WA5 9QE
M3 YRM Richard Moles, 14 Dorsett Road, Stourport-on-Severn, DY13 8EL
M3 YRO Paul Harvey, 22 Meredale Road, Liverpool, L18 5EX
M3 YRR Connor Reid, 128 Main Street, Hensingham, Whitehaven, CA28 8PX
M3 YRS T Godfrey, 12 Beacon House, Chulsa Road, London, SE26 6BP
M3 YRV Samuel Honywood, 169 Primrose Lane, Croydon, CR0 8YQ
M3 YRW John Woods, 21 Appleyard Crescent, Norwich, NR3 2QN
M3 YRX Marion McKone, 12 Hawkshead Road, Knott End-on-Sea, Poulton-le-Fylde, FY6 0QE
M3 YRZ Kevin Lovell, wimbledon court, 3, Miiddlesbrough, TS5 5JP
M3 YSA James Hallam, Woodlands, Walls Hill Road, Torquay, TQ1 3LZ
M3 YSC Steven Croucher, 17 Sundridge Road, Woking, GU22 9AU
M3 YSD S Dudley, 365 Sandon Road, Stoke-on-Trent, ST3 7LJ
M3 YSF Stephen Forrest, 66 Amberwood Drive, Manchester, M23 9BW
MM3 YSJ Jenna Sinclair, 21 Oxford Avenue, Gourock, PA19 1XU
M3 YSL C Brooks, 61 Boxfield Green, Stevenage, SG2 7DR
M3 YSM S Marshall, 43 Glenkerry House, 98 Burcham Street, London, E14 0SL
M3 YSN Martin Beardsley, 2 Wingrove Avenue, Newcastle upon Tyne, NE4 9AL
M3 YSQ Jeremy Powell, 46 Woodmancote, Yate, Bristol, BS37 4LL
M3 YSS Samuel Stewart, 8 Craig Street, Peterborough, PE1 2EJ
M3 YSU G Cummings, 18 Castleton Boulevard, Skegness, PE25 2TX
M3 YSV D Benwell, 1452a London Road, Leigh-on-Sea, SS9 2UW
M3 YSW Glenn Thrower, 8 Upton Gardens, Worthing, BN13 1DA
M3 YSY Aleksandrs Zabalujevs, 30 Miles Close, London, SE28 0NJ
M3 YSZ Christina Gao, 17 York Road, Harrogate, HG1 2QL
M3 YTA J Marland, 8 Dulverton Gardens, Edinburgh Road, Bolton, BL3 1TR
MM3 YTB Michael Martin, Flat A, 11 Craigpark Street, Clydebank, G81 5BS
M3 YTE A Duffield, 4 Crabmill Lane, Easingwold, York, YO61 3DE
M3 YTF Gary Garman, 11 Rye Close, Norwich, NR3 2LF
M3 YTG Andrew Clark, 330 Stafford Road, Caterham, CR3 6NJ
MI3 YTH Alan Shilliday, 26 Iskymeadow Road, Armagh, BT60 3JS
MM3 YTI Graham White, 75 Burnside Terrace, Polbeth, West Calder, EH55 8SU
M3 YTL T Lee, Harrogate Ladies' College, Clarence Drive, Harrogate, HG1 2QG
M3 YTP Tony Philpott, 32 Windermere, Faversham, ME13 8JQ
M3 YTQ John Hughes, 84 Rodwell Avenue, Weymouth, DT4 8SQ
M3 YTV D Telford, 37 Swillington Lane, Swillington, Leeds, LS26 8QF
M3 YTZ J Chaplin, 101 St. Cuthberts Drive, Gateshead, NE10 9AB
M3 YUA Michael Boyd, 4 Crowton Cottages, Winsford Road, Winsford, CW7 4DP
M3 YUB David George, 13 Cheltenham Way, Mablethorpe, LN12 2AX
M3 YUC Adam Waudby, 7 Forest Grove, York, YO31 1BL
M3 YUD Stuart Nutt, 23a Hesketh Drive, Southport, PR9 7JX
M3 YUH Pui Ying Lai, Harrogate Ladies' College, Clarence Drive, Harrogate, HG1 2QG
M3 YUK John Burman, 6 Goodyers Avenue, Radlett, WD7 8BA
M3 YUN Raymond Spalding, 7 Kingfisher Close, Scawby Brook, Brigg, DN20 9FN
M3 YUP Bernie Stevens, 77 Dean Lane, Hazel Grove, Stockport, SK7 6EJ
MD3 YUQ David Kelly, 41 High Street, Port St Mary, Isle of Man, IM9 5DN
M3 YUR Charlotte Potter, 4 Tomlinson Street, Stoke-on-Trent, ST6 4NW
MM3 YUS Paul McCann, 3 Exmouth Place, Gourock, PA19 1JE
MI3 YUT David Elliott, 15 Derrychara Park, Enniskillen, BT74 6JP
MM3 YUU Mark Morrison, 8 Garallan, Kilwinning, KA13 6LU
M3 YUV Lee Hudson, 68 Eleanor Road, Harrogate, HG2 7AJ
MM3 YUW Duncan Williamson, 31 Medrox Gardens, Cumbernauld, Glasgow, G67 4AJ
MM3 YUX Charles Williamson, 31 Medrox Gardens, Cumbernauld, Glasgow, G67 4AJ
MM3 YUY David Bendoris, 35 St. Michael's Wynd, Kilwinning, KA13 6WH
MI3 YVB Bryan Craney, 8a Drumhoy Drive, Carrickfergus, BT38 8NN
M3 YVD Thomas Munro, 71 Zig Zag Road, Liverpool, L12 9EQ
M3 YVE Yvette Neary, 3 Wordsworth Close, Torquay, TQ2 6EA
M3 YVF Michael Lowe, 22 Ryelands Close, Market Harborough, LE16 7XE
M3 YVG Matthew Lowe, 22 Ryelands Close, Market Harborough, LE16 7XE
M3 YVJ I Priest, 11 Dunlin Close, Kingswinford, DY6 8XP
M3 YVK David Bowen, 25 Maendu Terrace, Brecon, LD3 9HH
M3 YVN Ronan Wall, 30 Church Street, Skerries, Co. Dublin, Ireland
M3 YVR Linda Palir, 116 Carville Crescent, Brentford, TW8 9RD
M3 YVT Jonathan Storey, 3 Woodside Road, Poole, BH14 9JH
MM3 YVU Ricardo Corrieri, 226 Telford Road, East Kilbride, Glasgow, G75 0DL
M3 YVW Luke Brisco, 1 Bescot Way, Thornton-Cleveleys, FY5 3QA
M3 YVY Jasper Hudd, South Crofty Cottage, North Pool Road, Redruth, TR15 3JQ
M3 YVZ Jarrett Smith, 98 Dorset Road, Coventry, CV1 4EB
M3 YWC Anthony Johnston, 44 Cradoc Road, Brecon, LD3 9LH
M3 YWD William Disney, 98, Widney Lane, Solihull, B91 3LL
M3 YWE John Watkins, 6 Glebelands, Biddenden, Ashford, TN27 8EA
M3 YWF Simon Griffiths, 22 Manor Rise, Arleston, Telford, TF1 2ND
M3 YWG Warren Gradwell, 9 Nottingham Drive, Bolton, BL1 3RH
M3 YWH Kenneth Skerry, 18 Park Avenue, Cheadle, Stoke-on-Trent, ST10 1LZ
M3 YWI Francesca Ingram, 19 Charsley Close, Amersham, HP6 6QQ
M3 YWJ Emma Mcmahon, 120 New Ferry Road, Wirral, CH62 1DY
M3 YWM Paul Higgins, 59 Hillwood Road, Halesowen, B62 8NQ
M3 YWO Kevin Lovell, 2 Beckingham Hall Cottages, Tolleshunt Major, Maldon, CM9 8EH
M3 YWP Douglas Spooner, 30 Clover Road, Norwich, NR7 8TF
M3 YWR John Towner, 4 The Copse, Scarborough, YO12 5HG
MI3 YWT V Crichton, 10 Bann Drive, Londonderry, BT47 2HW
M3 YWU John Morris, 15 New Wanstead, London, E11 2SH
MM3 YWZ C Dinning, South Brae, Aiket Road, Kilmarnock, KA3 4BP
M3 YXB Richard Barrett, 18 Bullstake Close, Oxford, OX2 0HN
M3 YXC David Jones, 18 Maxwell Drive, Hazlemere, High Wycombe, HP15 7BX
M3 YXD Darren Wilson, 24 Hallamshire Mews, Wakefield, WF2 8YB
M3 YXE Joshua Peters, 11 Clockhouse Lane, Ashford, TW15 2EP
M3 YXF J Barber, 13 Dock Road, Sharpness, Berkeley, GL13 9UA
M3 YXH Gordon Sweet, 12 Old Harrow Road, St. Leonards-on-Sea, TN37 7EG
M3 YXJ Andrew McGreish, 36 Eastmoor Road, Oxborough, King's Lynn, PE33 9PX
M3 YXK Graham McGreish, 36 Eastmoor Road, Oxborough, King's Lynn, PE33 9PX
M3 YXL A Needham, 49 Macclesfield Road, Buxton, SK17 9AG
M3 YXM Janet Pick, 178 Alcester Road South, Kings Heath, Birmingham, B14 6DE
MM3 YXN Dennis Cowie, 69 Broomfield Park, Portlethen, Aberdeen, AB12 4XT
M3 YXP John White, 15 Norham Drive, Newcastle upon Tyne, NE5 5PR
M3 YXQ Steve Pye, 23 Dene Way, Donnington, Newbury, RG14 2JL
M3 YXR Xiaolu Ren, Lincoln House, Clarence Drive, Harrogate, HG1 2QD
M3 YXS Lynn Reddall, 68 Broadhurst Green, Hednesford, Cannock, WS12 4LF
M3 YXT Nathan Wall, 6 Ashton Lane, Braithwell, Rotherham, S66 7AJ
M3 YXU Peter Dossett, 20 Vineyard Close, Southampton, SO19 7DD
M3 YXV Christopher Reynolds, 9 Skeyton Road, North Walsham, NR28 0BS
M3 YXW Douglas Cox, 9 Northbrook Copse, Bracknell, RG12 0UA
MI3 YXX Lawrence Bradley, 4 Rathbeg Drive, Limavady, BT49 0BB
M3 YYE Adam Hinckley, 114 Lawn Lane, Hemel Hempstead, HP3 9HS
M3 YYG Scott Senior, 4 Flowers Meadow, Liverton, Newton Abbot, TQ12 6UP
M3 YYJ Jamie Whitford-Robson, 13 Perryman Close, Plymouth, PL7 4BP
M3 YYK Keith Yardley, 4 park road, hillton, Wolverhampton, WV107HS
M3 YYM Ian McPherson, 86 Fletemoor Road, St. Budeaux, Plymouth, PL5 1UH
M3 YYO Miles Scott-Martin, 7 East Quay Road, Poole, BH15 1RD
M3 YYQ Robert Smith, 13a Elwy Road, Rhos on Sea, Colwyn Bay, LL28 4SB
M3 YYR Nick Brown, 58 Molesworth Road, Plympton, Plymouth, PL7 4NU
M3 YYS Simon Hurrell, 4 Woodland Drive, Plympton, Plymouth, PL7 1SN
MI3 YYT David Bradley, 33 Lilac Avenue, Limavady, BT49 0HS
M3 YYU Ruth Woolley, 84 Bowthorpe Road, Norwich, NR2 3TP
M3 YYV Sow Chiew, 11 Svenskaby, Orton Wistow, Peterborough, PE2 6YZ
M3 YYW John Elsmore, 8 Clos Aberconway, Prestatyn, LL19 9HU
M3 YZA Scott Russell, 11 Morville Road, Dudley, DY2 9HR
M3 YZC Daniel Matheson, 21 Warren Hill Road, Woodbridge, IP12 4DU
M3 YZE Tracy Knight, 25 Allcot Road, Portsmouth, PO3 5DE
M3 YZH Jack Kelly, Martins, Fairwarp, Uckfield, TN22 3BE
M3 YZI Benjamin Sims, 4 New Cottages, Cranwich Road, Thetford, IP26 5EQ
M3 YZJ John King, 2 Perys Court, Cracknore Hard Lane, Southampton, SO40 4UT
M3 YZK David Lisi, 56 Gipsy Lane, Old Whittington, Chesterfield, S41 9JB
M3 YZN John Murray, 6 Sheridan Court, St. Edmunds Road, Dartford, DA1 5NF
M3 YZO Mark Lewis, Hillside, Caldy, Craven Arms, SY7 8QR
M3 YZP Melanie Tydeman, 6 Colin Close, Corfe Mullen, Wimborne, BH21 3QG
M3 YZQ Jeremy Thompson, 32 Church Street, Warnham, Horsham, RH12 3QR
M3 YZU Ian Dennison, 18 Tredington Grove, Caldecotte, Milton Keynes, MK7 8LR
M3 YZV Michael Hughes, 19 Pendine Crescent, North Hykeham, Lincoln, LN6 8UW
M3 ZAA M Walker, 4 Lemonroyd Marina, Fleet Lane, Leeds, LS26 9AJ

M3 ZAE Richard Giles, 75 Bradstocks Way, Sutton Courtenay, Abingdon, OX14 4DA
M3 ZAI Jack Elderfield, 2 Westwood Close, Amersham, HP6 6RP
M3 ZAL J Million, 6 Pacefield Square, Thornley, Durham, DH6 3DB
M3 ZAM Nampasa Tweed, 42 Uphill Road, Worthing, BN11 [illegible]
M3 ZAN S Tokley, 9 Peel Road, Springfield, Chelmsford, CM2 6AQ
MW3 ZAQ Z David, 45 Amanwy, Llanelli, SA14 9AH
M3 ZAR Roger Whitehouse, 29 Greswolde Road, Solihull, B91 1DY
M3 ZAV Aaron Cooper, 23 Ash Street, Manchester, M9 5XY
M3 ZAW Christopher Beresford, 13 Chaseside Avenue, Twyford, Reading, RG10 9BT
M3 ZAY Terry Hooper, 71 Collins Parc, Stithians, Truro, TR3 7RB
M3 ZAZ Z Ripley, 11 Gallows Hill Drive, Ripon, HG4 1UP
M3 ZBF John Evans, 112 Seagrave Crescent, Sheffield, S12 2JP
M3 ZBG B Gibbons, 35 Lavant Road, Stone Cross, Pevensey, BN24 5EZ
M3 ZBH B Hudson, 34 Eastwood Road, Bexhill-on-Sea, TN39 3PS
M3 ZBI Alastair Yarnold, 33 Abbotts Park, Cornwood, Ivybridge, PL21 9PP
M3 ZBJ Z Jennings, 29 Mountbatten Drive, Newport, PO30 5SJ
M3 ZBL Robert Leese, 22 Southlands Road, Congleton, CW12 3JY
M3 ZBQ Ian Marsh, 56b Oliver Crescent, Farningham, Dartford, DA4 0BE
M3 ZBR Ryan Offord, 16 Clive Avenue, Ipswich, IP1 4LU
M3 ZBS Alan Knight, 37 Crispe House, 72 Dovehouse Mead, Barking, IG11 7EB
M3 ZBV A Higham, 12 Lakenheath Drive, Sharples, Bolton, BL1 7RJ
M3 ZBW Robert Weaver, 116 Carville Crescent, Brentford, TW8 9RD
M3 ZBX John Gleeson, 124 rushes mead, Harlow, CM18 6QE
M3 ZBZ Mike Carvell, 10 Burns Close, Stevenage, SG2 0JN
M3 ZCA Alex Bevan, 5 Streetly End, West Wickham, Cambridge, CB21 4RS
M3 ZCB Caroline Blackmun, 2a St. Margarets Road, Girton, Cambridge, CB3 0LT
M3 ZCE Roger Sutton, 56 Neatherd Road, Dereham, NR20 4AY
M3 ZCG C Gregory, 81 Fiskerton Way, Oakwood, Derby, DE21 2HY
M3 ZCI Paul Probert, 54 New Hall Road, Ruabon, Wrexham, LL14 6AT
M3 ZCJ Jacey Ching Chi Ma, Harrogate Ladies' College, Clarence Drive, Harrogate, HG1 2QG
M3 ZCM Darren Griffin, 16a Kent Road, Fleet, GU51 3AH
M3 ZCN Michael Barnes, 51 Lower Way, Great Brickhill, Milton Keynes, MK17 9AG
M3 ZCO Carol Owen, 97 Maesglas Grove, Newport, NP20 3DN
M3 ZCR C Zarucki, 26 Heathfield Road, Kings Heath, Birmingham, B14 7DB
MM3 ZCS Stephen Kirkbride, 18 North Roundall, Limekilns, Dunfermline, KY11 3JY
M3 ZCU David Cook, 19 Almond Avenue, Risca, Newport, NP11 6PF
M3 ZCW Carl Lewis, 9 Chatsworth Gardens, Sydenham, Leamington Spa, CV31 1WA
M3 ZCX David Stringer, 18 Townfield Close, Ravenglass, CA18 1SL
MI3 ZCY Chaoyang Wang, 32 Broadlands, Carrickfergus, BT38 7BL
M3 ZDF D Fleetwood, 9 Reynolds Close, Swindon, Dudley, DY3 4NQ
MM3 ZDG Z Graham, 15 Stone Crescent, Mayfield, Dalkeith, EH22 5DT
MM3 ZDI Roger Cook, 5 Monkstadt, Linicro, Portree, IV51 9YN
M3 ZDK David Osmand, Flat 5, Long Barn Rosevidney, Penzance, TR20 9BX
MW3 ZDQ Anthony Davies, 25 Llanfair Road, Tonypandy, CF40 1TA
M3 ZDS D Scott, 33 Manor Crescent, Honiton, EX14 2DF
M3 ZDT Ian Arrow, Crystalwood, Stonemans Hill, Newton Abbot, TQ12 5PZ
M3 ZDV D Vasey, 22 Rickleton Village Centre, Washington, NE38 9ET
M3 ZDW Daniel Whyatt, 11 The Perrings, Nailsea, Bristol, BS48 4YD
M3 ZED Alan Henderson, 5 Snipe House Cottages, Alnwick, NE66 2JD
M3 ZEH Stephen Hendy, Flat 2, 33 Kingston Road, Leatherhead, KT22 7SL
M3 ZEI Duncan McTaggart, 59 Gainsborough Road, Richmond, TW9 2DZ
M3 ZEJ Rosemary Fagg, 62 Hawkins Road, Folkestone, CT19 4JA
M3 ZER D Nazer, 20 College Road, Ringwood, BH24 1NX
MM3 ZET H Dally, 3 Gremmasgaet, Lerwick, Shetland, ZE1 0NE
M3 ZEV Barrie Dexter, 13 Guildford Avenue, Sheffield, S2 2PJ
M3 ZEW Ellis Melman, 177 Grantham Road, London, E12 5NB
M3 ZEY Michael Blagg, 17 Flint Avenue, Forest Town, Mansfield, NG19 0DS
M3 ZFB James Creed, 11 Athelstan Road, Faversham, ME13 8QL
M3 ZFH Stephen Recht, 1 Ireton Close, Chalgrove, Oxford, OX44 7RZ
M3 ZFI Charles Godfrey, 97 Whalley Drive, Bletchley, Milton Keynes, MK3 6HX
M3 ZFJ Jonathan Tomlinson, 12 Brook Drive, Whitefield, Manchester, M45 8FR
M3 ZFK William Taylor, Garth Wood, Fishers Brae, Eyemouth, TD14 5NJ
M3 ZFL Cameron Rycott, 7 Crescent Grove, London, SW4 7AF
M3 ZFN Darren Hoare, 47 High Street, Chalgrove, Oxford, OX44 7SJ
M3 ZFO Daniel Davies, Winter House, Beccles Road, Norwich, NR14 6RE
M3 ZFS Sam Fisher, 4 Beaufont Gardens, Bawtry, Doncaster, DN10 6RT
M3 ZFV Peter Whiteley, 1 Newton Close, Fareham, PO14 3LF
M3 ZFW Douglas Tinn, 6 Billiemains Farm Cottage, Duns, TD11 3LG
M3 ZFY R Fye, 201 North Wing The Residence, Kershaw Drive, Lancaster, LA1 3SY
M3 ZGA Gary Campbell, 10 Welbeck Road, Rochdale, OL16 4XP
M3 ZGC V Jolliffe, 54 Glendale Avenue, Wash Common, Newbury, RG14 6RU
M3 ZGD G Donaldson, Moddershall House, Moddershall, Stone, ST15 8TG
M3 ZGE Christopher Walton, 6 Hilltop Road, Bearpark, Durham, DH7 7DP
M3 ZGF Perry Daniel, 57 Jephson Road, Sutton-in-Ashfield, NG17 5EH
M3 ZGG T Turner, 88 Bevan Close, Huntingdon, PE29 1TJ
M3 ZGH G Houlton, 7 Bartletts Hillside Close, Chalfont St. Peter, Gerrards Cross, SL9 0HH
M3 ZGI Michael Jurczyszyn, 115 The Twitchell, Sutton-in-Ashfield, NG17 5AX
MM3 ZGK Gordon Munn, 28 Broomfield Park, Portlethen, Aberdeen, AB12 4XT
M3 ZGM G Machin, 7 Lansdown, Yate, Bristol, BS37 4LS
M3 ZGO Casper Ikeda-Chew, 10 Bruce Drive, South Croydon, CR2 8SL
M3 ZGR G Renshaw, 46 Forge Close, Caerleon, Newport, NP18 3PW
M3 ZGS Graham Somerville, 22 Woolven Close, Burgess Hill, RH15 9RR
M3 ZGT Jamie Throup, Willow House, O'Keys Lane, Worcester, WR3 8RL
M3 ZGU Leslie Retford, 111 Lander Close, Old Hall, Warrington, WA5 9PL
M3 ZGX P Beltrami, 15 Woodroffe Square, Calne, SN11 8PW
M3 ZGY Gerald Lane, 23 Parkesway, Saltash, PL12 4AL
M3 ZHC Xzanthy Beltrami, 20 Brunel Way, Calne, SN11 9FN
MD3 ZHD William Callister, 13 Fairfield Avenue, Onchan, Isle of Man, IM3 4BG
MU3 ZHF Henry Fletcher, Le Villocq House, Le Villocq, Castel, Guernsey, GY5 7SA
M3 ZHG Chris Barnes, 23 South Street, Crewe, CW2 6HN
M3 ZHI James Brookes, 85 St. Johns Road, Rotherham, S65 1LT
M3 ZHM Lesley Martin, Little Acre, Swan Lane, Edenbridge, TN8 6AJ
M3 ZHP Dawid Pastwik, 451 Chorley Old Road, Bolton, BL1 6AH
M3 ZHQ John Martin, Little Acre, Swan Lane, Edenbridge, TN8 6AJ
M3 ZHU Michael Hilliar-Mills, 1-3 Queens Road, Criccieth, LL52 0EG
M3 ZHV Tracy Gorbutt, 26 Whitethorn Avenue, Withernsea, HU19 2LN
M3 ZHW Thomas Hulme, 157 Birkinstyle Lane, Stonebroom, Alfreton, DE55 6LD
M3 ZHX Ken Baker, 6 Manor Park, Duloe, Liskeard, PL14 4PT
M3 ZHY J Mason, 9 Farmfields Rise, Woore, Crewe, CW3 9SZ
M3 ZHZ Laurence Kay, 27 Millbrook Drive, Shawbury, Shrewsbury, SY4 4PQ
M3 ZIA Zia Ul Haq, 7 Alden Walk, Stockport, SK4 5NW
M3 ZID Richard Farrington, 25 Monks Walk, Bridge Street, Evesham, WR11 4SL
M3 ZIE Hermfried Ehm, 17 Stuart Road, Kempston, Bedford, MK42 8HS
M3 ZIF James Whittick, 91 Godfrey Way, Dunmow, CM6 2SQ
M3 ZIH Clifford Warwick, 104 Church Road, Formby, Liverpool, L37 3NH
M3 ZII Stephen Savastano, 571 Lonsdale Road, Stevenage, SG1 5EA
M3 ZIL E Greatorex, 22 Marlborough Way, Uttoxeter, ST14 7HL
M3 ZIM Teresa Fuller, 49 Scotby Avenue, Chatham, ME5 8ER
M3 ZIN P Bruce, Pilgrims, Broadmead, Lymington, SO41 6DH
M3 ZIO Marie Denon, 3 Duke Street, Clowne, Chesterfield, S43 4RZ
M3 ZIR Darryl Powis, 18 Merlin Court, Huddersfield, HD4 7SP
M3 ZIV Iain Vickers, 3 Nesbit Road, St. Marys Bay, Romney Marsh, TN29 0SF
M3 ZIX Steven Norris, 95 Waterloo Road, Ashton-on-Ribble, Preston, PR2 1BH
M3 ZIZ Kathryn Druce, 20 Manwood Avenue, Canterbury, CT2 7AH
M3 ZJB Mark Collier, 8 Masefield Mews, Dereham, NR19 2SY
M3 ZJD Francis Stevens, 191 Parkinson Drive, Chelmsford, CM1 3GW
M3 ZJE Martin Lake, 48 Sedgemoor Road, Bath, BA2 5PL
M3 ZJF Peter Hayward, 14 Micklewright Avenue, Crewe, CW1 4DF
M3 ZJG Peter Gibbs, 9 Walton Heath, Darlington, DL1 3HZ
M3 ZJH J Herant, 75 Victoria St., Chesterton, Newcastle, ST5 7EP
M3 ZJJ Gloria Chiu, Harrogate Ladies' College, Clarence Drive, Harrogate, HG1 2QG
M3 ZJK Simon Knight, 1 Hayne Bungalows, Bolham, Tiverton, EX16 7RE
M3 ZJL J Logan, 13 Hornel Road, Kirkcudbright, DG6 4LH
M3 ZJM James Leavesley, 37 Western Road, Stourbridge, DY8 3XU
MI3 ZJN Gerry Connolly, 54 Granemore Park, Keady, Armagh, BT60 2GP
M3 ZJO Jonathan Rawlinson, Westfield Farm, Risden Lane, Cranbrook, TN18 5DU
M3 ZJQ Adam Rawlinson, Westfield Farm, Risden Lane, Hawkhurst, Sandhurst, Cranbrook, TN18 5DU
M3 ZJS G Stokes, 24 Dayslondon Road, Waterlooville, PO7 5NN
MI3 ZJT Jonathan Tripathy, 3 Lough Derg Park, Carryduff, Belfast, BT8 8PH
M3 ZJV Griffith Hewis, 10 Albert Road, New Malden, KT3 6BS
MM3 ZJY George Boyter, 36 Main Street, Springfield, Cupar, KY15 5SQ
M3 ZKA K Allgar, 13 Deacon Avenue, Kempston, Bedford, MK42 7DU
M3 ZKD Kenneth Dungey, 247 Park Road, Sittingbourne, ME10 1ER
M3 ZKE K Pemberton, 38 Milford Drive, Bournemouth, BH11 9HJ
M3 ZKF Kate Franklin, 1 Aberdeen Court, Newcastle upon Tyne, NE3 2XU
M3 ZKI I King, 7 Shardlow Close, Haverhill, CB9 7RF
M3 ZKL Ian Jutting, 68 The Ridgeway, Tonbridge, TN10 4NN
M3 ZKO Lauren Bodie, Harrogate Ladies' College, Clarence Drive, Harrogate, HG1 2QG
MM3 ZKQ Gerald Wareham, 12 The Knowe, Leven Road, Leven, KY8 5JH
M3 ZKT Marie Sargeant, Bryn Hyfryd, Parc Bel, Caernarfon, LL54 6LF
M3 ZKU David Boardman, 57 Sunningdale Road, Huddersfield, HD4 5DX
MW3 ZKW K Watt, 28 Llysgwyn, Morriston, Swansea, SA6 6BJ
MW3 ZKX Albert Griffiths, 6 South View, Taffs Well, Cardiff, CF15 7SE
M3 ZKY Jonathan Jones, 3 Rhedyw Road, Llanllyfni, Caernarfon, LL54 6SN
M3 ZLA Glyn Burrows, Aubrietia, Malpas, SY14 8AY
M3 ZLE Liam Ellerington, 120 Stagsden Road, Bromham, Bedford, MK43 8QJ
M3 ZLI Paul Dee, 3 Cherry Orchard, Upton-upon-Severn, Worcester, WR8 0LR
M3 ZLJ Joan Priestman, 198 Felmongers, Harlow, CM20 3DW
M3 ZLL J Giovinazzo, 3 Eleanor Avenue, Epsom, KT19 9HD
M3 ZLO Mark Dring, 1 Hannaford Close, St Columb Rd, TR9 6FH
M3 ZLP Zoe Gribben, 44 Fern Close, Birchwood, Warrington, WA3 7NU
M3 ZLR L Rodrigues, The Tek Doctor, Pump Square, Boston, PE21 6QW
M3 ZLS Lee Sammut, 16 Queen Marys Road, New Rossington, Doncaster, DN11 0TS
M3 ZLV Monika Ferenc, 2a Rosedene Avenue, London, SW16 2LT
M3 ZLW D Wakefield, 36 Finch Road, Earley, Reading, RG6 7JU
MW3 ZLX Robert Miles, 63 Phillip Street, Caegarw, Mountain Ash, CF45 4BG
M3 ZLY Tony Farr, 125 Rochford Way, Walton on The Naze, CO14 8SP
M3 ZLZ L Wilmott, 60 Church Hill, Royston, Barnsley, S71 4NG
M3 ZMB M Bateup, 18 The Quadrant, Hassocks, BN6 8BP
M3 ZME Martin Davison, 1 Chancel Way, Barnsley, S71 2HS
MI3 ZMJ M Johnston, 52 Lansdowne Road, Newtownards, BT23 4NT
M3 ZMM David Mainwaring, 1 Buckingham Close, Didcot, OX11 8TX
M3 ZMN M Nash, 11 Frederick Street, Warrington, WA4 1HX
M3 ZMO Andrew Maguire, 132 Wigan Road, Ormskirk, L39 2BA
MI3 ZMP M Paterson, 121 Ballybunden Road, Killinchy, Newtownards, BT23 6RZ
M3 ZMQ Michael Tew, Willowell, Spring Valley Lane, Colchester, CO7 7SD
M3 ZMR Iain Nicholson, 2 Broom Close, Leyland, PR25 5RQ
M3 ZMS Matthew Strickland, Ancoats, Piercy End, York, YO62 6DQ
M3 ZMT M Thompson, 100 Redford Avenue, Horsham, RH12 2HH
M3 ZMU John Taylor, 60 Rockfield Way, Undy, Caldicot, NP26 3FD
M3 ZMV Peter Hayward, 14 Micklewright Avenue, Crewe, CW1 4DF
M3 ZMX Ian Botham, 12 Lairgill, Bentham, Lancaster, LA2 7JZ
M3 ZMZ M Bryant, 26 Coronation Road, Melksham, SN12 7PF
M3 ZNC David Crosland, Simmonds Green, Varley Road, Huddersfield, HD7 5TY
M3 ZNF Geoffrey Panton, Dale View, Thorpe Fendykes, Skegness, PE24 4QN
M3 ZNG Neil Galbraith, 213 Queens Road, Portsmouth, PO2 7LX
M3 ZNJ Adrian Harvey, 28 Langdown Road, Hythe, Southampton, SO45 6EW
M3 ZNL Justin Ho, 231 Rush Green Road, Romford, RM7 0JP
M3 ZNM Alan Sivyer, 63 Sugden Road, Worthing, BN11 2JG
MM3 ZNN Paul Holmes, Maraval, Doune Road, Dunblane, FK15 9AT
M3 ZNO Tian Cao, Harrogate Ladies' College, Clarence Drive, Harrogate, HG1 2QG
M3 ZNP Ronald Young, 26 Silent Woman Park, Coldharbour, Wareham, BH20 7PE
MM3 ZNQ William Beaton, 4 Moorfield Gardens, Springfield, Cupar, KY15 5SH
M3 ZNR David Belton, 4 Sandown Road, Toton, Nottingham, NG9 6GN
M3 ZNT Irene Govan, 9 Willowbank, Sandwich, CT13 9QA
M3 ZNV Neal Edwards, 36 Joseph Luckman Road, Bedworth, CV12 8BQ
M3 ZNX Terence Hayward, 11 Radnor Close, Bodmin, PL31 2BZ
M3 ZNY Peter Robinson, Flat 2, 46 Cliff Road, Sheringham, NR26 8BJ
M3 ZNZ Leo Burke, 24 Pinecliffe Avenue, Bournemouth, BH6 3PZ
M3 ZOG J Hyde, The Grove, 7 Mill Lane, Kidderminster, DY10 3ND
M3 ZOH Neil Terry, 6 Edinburgh Cottages, West Newton, King's Lynn, PE31 [illegible]
M3 ZOI Jack Ogden, 295 Church Road, St. Annes, Lytham St. Annes, FY8 3NP
M3 ZON C Holdt, 18 Garrard Road, Banstead, SM7 2ER
M3 ZOO M Lucas, 20 Collin Road, Kendal, LA9 5HN
M3 ZOP C Poulson, 9 Scattergate Green, Appleby-in-Westmorland, CA16 6SP
M3 ZOR Zoe Rabbitt, 66 Parkfield Avenue, Delapre, Northampton, NN4 8QB
M3 ZOU E Nicholson, 24 Barnmead, Haywards Heath, RH16 1UZ
M3 ZPB P Burgess, Tally Ho Cottage, High Street, Swindon, SN4 0AE
M3 ZPE Phil Evans, 22 Northumberland Road, Wigston, LE18 4WL
M3 ZPJ Peter Smyth, 91beechcroft ave, darcy lever, Bolton, BL26HB
M3 ZPM Paul May, 95 Moorfield Avenue, Denton, Manchester, M34 7TX
M3 ZPO Luke Gravel, 16 North Street, Crowle, Scunthorpe, DN17 4NB
M3 ZPR P Rogers, 11 Beech Crescent, Mexborough, S64 9EH
M3 ZPT Peter Tatham, 54 High Street, Cleckheaton, BD19 3PX
M3 ZPW Barry Matthews, 30 Oaklands Drive, Brandon, IP27 0NR
M3 ZPY Leslie Dobson, 8 Coronation Street, Darfield, Barnsley, S73 9HA
M3 ZPZ Paul Shurmer, 1 The Glebe, East Harling, Norwich, NR16 2SZ
M3 ZQA Peter Stonebridge, 207 Henley Road, Ipswich, IP1 6RL
M3 ZQB Matthew Plummer, 27 Bowness Court, Congleton, CW12 4JR
M3 ZQC Marc Hackett, 26 Wantage Road, Didcot, OX11 0BP
M3 ZQF Matthew Ashworth, 123 Forest Road, Liss, GU33 7BP
M3 ZQG Daniel Bates, 80 St. Leonards View, Polesworth, Tamworth, B78 1JY
M3 ZQJ Benjamin Yarwood, 12 Charminster Close, Waterlooville, PO7 7RP
M3 ZQM K Pascoe, 21 Cotswold Avenue, Sticker, St. Austell, PL26 7ER
M3 ZQN Anthony Wheeler, 8 Elsworth Grove, Birmingham, B25 8EJ
MM3 ZQP Joseph Docherty, 23 Turret Drive, Polmont, Falkirk, FK2 0QW
M3 ZQV Lee Marshall, 62 Bacons Lane, Chesterfield, S40 2TN
MM3 ZQW Colin Bryson, 29 Roull Road, Edinburgh, EH12 7JW
MM3 ZQX Alex Falconer, 61 Mountcastle Drive North, Edinburgh, EH8 7SP
M3 ZRA Richard Alford, 1 School Lane, Winmarleigh, Preston, PR3 0JY
M3 ZRB Robin Brown, 194 Wymersley Road, Hull, HU5 5LN
MM3 ZRF Robert Fletcher, Balnacraig Cottage, Mid Balnacraig, Alness, IV17 0XL
M3 ZRG Roy Gladman, 18 Willingdon, Ashford, TN23 5YF
M3 ZRH R Harrison, 16 Curlew Rise, Morley, Leeds, LS27 8US
M3 ZRK Douglas Rowlands, 1 Forge Lane, Bassaleg, Newport, NP10 8NF
M3 ZRM Richard McGregor, 84 Churchill Way, Burton Latimer, Kettering, NN15 5RS
M3 ZRO B Stoner, Montrose, Wesley Road, Whitby, YO22 4RW
M3 ZRP Elvis Stooke, 11 Westgate, Grantham, NG31 6LT
M3 ZRQ Glenn Crane, 22 Brewery Street, Burgh le Marsh, Skegness, PE24 5LG
M3 ZRR Colin Taylor, 1 Jasmine Gardens, Warrington, WA5 1GU
M3 ZRS R Stokes, 44 Broxhead Road, Havant, PO9 5LA
M3 ZRV Richard Allwood, 1, 46, Guildford, GU2 7JN
M3 ZRW David Williams, 15 Charwood Road, Wokingham, RG40 1RY
M3 ZRX C Waterworth, 4 Mossdale Road, Ashton-in-Makerfield, Wigan, WN4 0EQ
M3 ZRY Liam Acton, 39 Craig Road, Macclesfield, SK11 7YH
MM3 ZRZ David Hibberd, 21u Riverside Drive, Aberdeen, AB11 7DG
M3 ZSA Simon Arthur, 17 Bromeswell Road, Ipswich, IP4 3AS
M3 ZSB Steven Bannister, 162 Dobcroft Road, Sheffield, S11 9LH
M3 ZSC Sharon Colman, 197 Coppins Road, Clacton-on-Sea, CO15 3LA
M3 ZSD E Worthington, 7 Bowness Close, Gamston, Nottingham, NG2 6PE
MM3 ZSF S Faccenda, 21 Gairdoch Drive, Carronshore, Falkirk, FK2 8AQ
M3 ZSH Shaun Hampson, 12 Flying Fields Drive, Macclesfield, SK11 7GE
M3 ZSI Simon Airs, 6 The Willows, Culham, Abingdon, OX14 4NN
M3 ZSJ John Gibson, 22 Woodburn Drive, Chapeltown, Sheffield, S35 1YS
MM3 ZSK Christopher Brash, 4 Union Street, Lossiemouth, IV31 6BA
MW3 ZSM Peter Leyshon, 34 South Street, Porth, CF39 0EG
MW3 ZSO Rhys Madson, 44 Manor Court, Church Village, Pontypridd, CF38 1DW
M3 ZST C Butters, 4 The Dovecote, Pitsford, Northampton, NN6 9SB
M3 ZSU Peter Westwell, Roden House, Dobsons Bridge, Whitchurch, SY13 2QL
M3 ZSV Timothy Bendelow, 3 St. Giles Close, Thirsk, YO7 3BU
M3 ZSW Shannon White, 1 The Red House, Old Gallamore Lane, Market Rasen, LN8 3US
M3 ZSY Denis Bendelow, 4 St. Michaels View, Kirklington, Bedale, DL8 2NH
MW3 ZTB Thomas Beach, 97 Van Road, Caerphilly, CF83 1LA
M3 ZTD George Berry, 5 Oakholme Rise, Worksop, S81 7LJ
MW3 ZTH Joss Evans, 8 Hengoed Crescent, Cefn Hengoed, Hengoed, CF82 7HF
M3 ZTK T King, 215 Hartland Road, Reading, RG2 8DN
M3 ZTL Robert Hincks, 79 Forest Street, Shepshed, Loughborough, LE12 9BZ
M3 ZTN Alexander Romanov, 132 Latchmere Drive, Leeds, LS16 5DY
M3 ZTP J Wilson, 14 Elms Drive, Morecambe, LA4 6DQ
M3 ZTR T Reed, Channel Pool, Armathwaite, Carlisle, CA4 9QY
MM3 ZTS Christopher Owens, 77 Hill Street, Alloa, FK10 2LW
M3 ZTT Andrew Jeffery, 14 Holly Mount, Shavington, Crewe, CW2 5AZ
M3 ZTU Darren Stewart, 79 Eastfield Road, Driffield, YO25 5EZ
M3 ZTW A Mackenzie, 31 Beverley Rise, Billericay, CM11 2HU
M3 ZTX James Thornhill, 47 Hopton Lane, Mirfield, WF14 8JP
M3 ZUA Paul Coleman, 10 Carr Road, Fleetwood, FY7 6QJ
M3 ZUB David Burrows, 19 Fleming Avenue, Bottesford, Nottingham, NG13 0ED
M3 ZUC Alan O'Keeffe, 1 Elmfield Road, Liverpool, L9 3BL
M3 ZUD David Marshall, 117 Ogley Hay Road, Chase Terrace, Burntwood, WS7 2HU
M3 ZUF David Harris, 15 Mill Road, Pontllanfraith, Blackwood, NP12 2GE
M3 ZUG Danny Duncan, 12 Carrington Avenue, Poppleton Road, York, YO26 4SH
M3 ZUJ Alan Trudgett, 103 Shandon Road, Worthing, BN14 9EA
M3 ZUL David Stinson, 234 Pelsall Lane, Rushall, Walsall, WS4 1NG
MM3 ZUP Callum Webster, 18 St. Fort Road, Wormit, Newport-on-Tay, DD6 8LA
M3 ZUS Susanna Skinner, Chevington, Carlton Road, Godstone, RH9 8LD
M3 ZUT Pierre Streatfield, 66 Ockendon Road, London, N1 3NW
M3 ZUU Robert Duncan, 12 carrington avenue, poppleton road, York, YO26 4SH
M3 ZUW Julian Caithness, 6 Station Road, Catworth, Huntingdon, PE28 0PE
M3 ZUY Bernard Pike, 19 Cardigan Gardens, Reading, RG1 5QP
M3 ZUZ Aidan Finn, 29 Argyle Road, Weymouth, DT4 7LX
M3 ZVA Sarah Toher, The Chapel, Station Road, Darlington, DL2 1JG
M3 ZVB Graham Spencer, Erw Uchaf, Gorad, Holyhead, LL65 3BT



**IMPORTANT NOTE**

**Revalidate licence to avoid revocation** – Ofcom has advised the Society that plans will be drawn up to revoke licences that have not been revalidated as required by the licence conditions. The quickest way to revalidate is to do so online via the Ofcom website: *https://services.ofcom.org.uk/* or by email: *amateur.validations@ofcom.org.uk* Ofcom staff are available to help, but please be patient during times of heavy workload.

M3 ZVD David Jones, 7 Bryn Rhedyw, Llanllyfni, Caernarfon, LL54 6SS
M3 ZVF Benjamin Lane, 26 Deben Valley Drive, Kesgrave, Ipswich, IP5 2FB
M3 ZVH Clifford Nicholls, 26 Maes Geraint, Pentraeth, LL75 8UR
M3 ZVI Frank Lees, 5 St. Winifred Road, Rainhill, Prescot, L35 8PY
M3 ZVK V King, 215 Hartland Road, Reading, RG2 8DN
M3 ZVS Keith Brown, 142 Moor Lane, Woodford, Stockport, SK7 1PJ
M3 ZVT Martin Joynson-Ellis, Roadside Cottage, Craiglemine, Newton Stewart, DG8 8NE
M3 ZVU M Flanagan, 804 New Hey Road, Huddersfield, HD3 3YW
M3 ZVW John Purgal-Woods, Invicta Cottage, Carbrooke Road, Thetford, IP25 6SD
M3 ZVX Jonathan Barker, 26 Ardley Road, Fewcott, Bicester, OX27 7PA
MM3 ZVY Kirsty Bourhill, 30c Salters Road, Wallyford, Musselburgh, EH21 8AA
M3 ZWB Nigel Watts, 404 March Road, Turves, Peterborough, PE7 2DW
M3 ZWD Zoe Dixon, 18 Norfolk Close, Plymouth, PL3 6DB
M3 ZWF Eric Watson, 4 Moorland Avenue, Blackburn, BB2 5EQ
M3 ZWG Ian Garratt, 2 Hayclose Crescent, Kendal, LA9 7NT
M3 ZWH Robin Simmons, 10 Whitehall Road, Kingswinford, DY6 9DY
M3 ZWI Christopher Riley, 15 Keddington Crescent, Louth, LN11 0AP
M3 ZWJ Graham Hill-Adams, 6 Broadleaze Way, Winscombe, BS25 1JX
M3 ZWK Colin Powell, 37 Newnham Close, Mildenhall, Bury St. Edmunds, IP28 7PD
M3 ZWL Ricky Amos, 6 Eccles Road, Wittering, Peterborough, PE8 6AU
M3 ZWM Ian Coulson, 2 Marl Hurst, Edenbridge, TN8 6LN
M3 ZWN Nigel Woodstock, 8 Fernheath Close, Bournemouth, BH11 8SL
MW3 ZWO Stephen Gibbon, 1 Graig Terrace, Senghenydd, Caerphilly, CF83 4HN
M3 ZWP Karl Bainbridge, 29 Bluebell Grove, Calne, SN11 9QH
M3 ZWQ Kenneth Toohey, 197 Broad Oak Road, St. Helens, WA9 2AQ
MW3 ZWR G Spicer, 6 Cromwell Road, Neath, SA10 8DR
MW3 ZWS Stephen Todd, 4 uplands road, pontardawe, Swansea, SA8 4AH
M3 ZWW William Warwicker, 13 Elm Tree Avenue, Tile Hill, Coventry, CV4 9EU
M3 ZWX Jack Turley, 19 Ibstock Drive, Stourbridge, DY8 1NW
M3 ZXA Neil Gleaden, 25 Ridgway Avenue, Darfield, Barnsley, S73 9DU
M3 ZXB Benjamin Kerridge, 2 Allerdale Close, Thirsk, YO7 1FW
M3 ZXC S Makins, 10 Lower Mill Street, Ludlow, SY8 1BH
M3 ZXE Nicki Egginton, Emergency Planning Unit, Shropshire County Council, Shrewsbury, SY2 6ND
M3 ZXG Gareth Carless, Silver Cottage, Silver Street, South Petherton, TA13 5BY
M3 ZXH George Cross, 45 Foundry Street, Horncastle, LN9 6AG
MM3 ZXL Steve Fradley, 30 Polmont Park, Polmont, Falkirk, FK2 0XT
M3 ZXN Anthony Walton, 65 Broadway East, Rotherham, S65 2XA
M3 ZXQ Ian Talbot, 41 Elmwood Close, Cannock, WS11 6LX
M3 ZXX Kelvin Willets, 138 Hayward Avenue, Donnington, Telford, TF2 8DD
M3 ZYE Peter Mason, 20 Coronation Road, Six Bells, Abertillery, NP13 2PJ
M3 ZYF Harry Armstrong, 83 Hillary Grove, Carlisle, CA1 3JQ
M3 ZYG D Zygadllo, 56 Scarisbrick Crescent, Liverpool, L11 7DW
M3 ZYI Ian Ball, 17 Homecroft Road, Goldthorpe, Rotherham, S63 9DX
M3 ZYK Mark Edge, 19 Burton Avenue, Rushall, WS4 1NH
M3 ZYM David Lavender, 39 Albany Crescent, Bilston, WV14 0HT
M3 ZYN Robert Townsend, 23 The Lanes, Cheltenham, GL53 0PU
M3 ZYO Brian Hall, 6 Marshall Close, Parkgate, Rotherham, S62 6DB
M3 ZYQ Meuryn Daymond, 6 Vale View Place, Bath, BA1 6QW
M3 ZYR Alan Caulfield, 1 Carleton Close, Amesbury, Salisbury, SP4 7TU
MM3 ZYS Seonag Robertson, 20 Knockard Place, Pitlochry, PH16 5JF
M3 ZYT Nicholas Bond, 21 Coleridge Close, Bletchley, Milton Keynes, MK3 5AF
MI3 ZYU Anthony Kelly, 16 Union Street Mews, Coleraine, BT52 1EN
M3 ZYV Alex Brown, 7 Whilton Crescent, West Hallam, Ilkeston, DE7 6PE
M3 ZYW B Pemberton, 38 Pond Green Way, St. Helens, WA9 3SD
M3 ZYX T Parker, 2 Kipling Close, Grantham, NG31 9ND
M3 ZYY P Day, 46 Beatrice Avenue, Saltash, PL12 4NG
M3 ZYZ Charles Wilmott, 60 Church Hill, Royston, Barnsley, S71 4NG
M3 ZZA Mike Daniels, 10 Downland Avenue, Peacehaven, BN10 8TH
M3 ZZD Daniel Smith, 7 Kestrel Avenue, Bransholme, Hull, HU7 4ST
M3 ZZE Elizabeth Olver, 41 Mount Tamar Close, Plymouth, PL5 2AL
M3 ZZF Damon Webb, 7 Bittern Road Iwade, Sittingbourne, ME9 8FR
M3 ZZH Hannah Metson, Higher Churchtown Barn, North Hill, Launceston, PL15 7PQ
M3 ZZI Debbie Brooker-Evans, 379 Crownhill Road, Plymouth, PL5 2LN
M3 ZZJ Lynda Haley, 17 Oak Drive, Crownhill, Plymouth, PL6 5TZ
M3 ZZL Paul Stretton, 38 queens way melbourne, Derby, DE73 8FG
M3 ZZN Aron Hobson, 8 Sycamore Road, Colchester, CO4 3NF
M3 ZZQ J Snape, 2 Orchard Close, Fort Avenue, Preston, PR3 3YS
M3 ZZS Jeff Skinner, 36 Milton Road, Waterloo, Liverpool, L22 4RF
MW3 ZZU Elgan Jones, 39 Ger-y-Llan, Velindre, Llandysul, SA44 5YB
M3 ZZV Graham Jones, 39 Thurlow Way, Barrow-in-Furness, LA14 5XP
M3 ZZW G Smith, 7 Kestrel Avenue, Bransholme, Hull, HU7 4ST
M3 ZZX Ian Curtress, 7 Gwinnett Court, Shurdington, Cheltenham, GL51 4GQ

## M*5

M5 ABC David Last, Hillview, New Road, Bridport, DT6 4NY
M5 ABH David Drew, 14 Greensfields, Skegby, Nottingham, NG17 3DN
M5 ABJ J Bodle, 48 Bolsover Road, Hove, BN3 5HP
M5 ABN Peter Herbert, Flat 1, 7 Cockington Lane, Paignton, TQ3 1EE
M5 ABR P Lovelock, 82 Chaworth Road, West Bridgford, Nottingham, NG2 7AD
M5 ABT R Clark, 36 Southfields, Stanley, DH9 7PH
M5 ACD G Steabler, 1 Westhill Road, Grimsby, DN34 4SG
M5 ACF E McLusky, 11 Ripon Road, Killinghall, Harrogate, HG3 2DG
M5 ACJ R Payne, 58 Sheepcote Lane, Amington, Tamworth, B77 3JW
M5 ACR Andrew Jakusz-Gostomski, 15 Goodliffe Gardens, Tilehurst, Reading, RG31 6FZ
M5 ACS M Arnfield, Cleabarrow, Plumley Moor Road, Knutsford, WA16 0TU
M5 ACT E Bluer, 8 Cedar Close, Waterlooville, PO7 7LN
M5 ACX M Anderson, 1 Thames Close, Ferndown, BH22 8XA
M5 ADA Christopher Goodhand, 22 Somin Court, Doncaster, DN4 8TN
M5 ADD A Dimmock, Gwyndy, Llandegfan, Menai Bridge, LL59 5PW
M5 ADE A Deane, Flat 1-6, 76 Church Street, Tewkesbury, GL20 5RX
M5 ADF D Hook, 15 Wordsworth Avenue, Sutton-in-Ashfield, NG17 2GG
M5 ADI Darren Williams, 4 Dunkirk Rise, College Bank Way, Rochdale, OL12 6UH
M5 ADL Adrian Lambert, 69 Anvil Crescent, Broadstone, BH18 9DZ
M5 ADM K Marriott, 99 Stapleford Lane, Toton, Nottingham, NG9 6FZ
M5 ADQ B Woolnough, 57 Cranborne Road, Potters Bar, EN6 3AB
M5 ADW P Booth, 68 Trem Eryri, Llanfairpwllgwyngyll, LL61 5JF
M5 AEC M Drinkwater, Green Quarter, Ward Green Old Newton, Stowmarket, IP14 4EZ
M5 AEE A Steadman, 4 Vineyard Way, Buckden, St. Neots, PE19 5SR
M5 AEF R Burrows-Ellis, Pateley, School Lane, Woodbridge, IP13 6DX
M5 AEH Colin Shackleton, 54a Blueleighs Park Homes, Ipswich, IP6 0ND
M5 AEI Ellis Howe, 22 Freston, Paston, Peterborough, PE4 7EN
M5 AEO Jonathan Kempster, 25 Andersens Wharf, Copenhagen Place, London, E14 7DX
MM5 AES John Robertson, 138 East Main St., Armadale, Bathgate, EH48 2PB
M5 AFE W Higgs, 955 Oldham Road, Rochdale, OL16 4SE
M5 AFG David Hall, 4 Steventon Road, Wellington, Telford, TF1 2AS
M5 AFH J Denmead, 47 Holland Road, Clevedon, BS21 7YJ
MI5 AFL I McCrum, 12 Bishops Court Road, Downpatrick, BT30 7NU
MI5 AFM William Weir, 9 Ripley Terrace, Portadown, Craigavon, BT62 3ED
M5 AFV James Jones, 10 Huntington Close, Winyates, Redditch, B98 0NF
M5 AFX Gordon Rhodes, Hill Farm, Upton Hill, Gloucester, GL4 8DA
M5 AFY Sue Hall, 58 Lower Meadow Court, Northampton, NN3 8AX
M5 AGB Anthony Koeller, 116 Parham Road, Gosport, PO12 4UE
M5 AGG Christopher Ellis, Broken Ridge, Fir Tree Close, Ringwood, BH24 2QW
M5 AGI J Addison, 20 St. Davids Drive, Callands, Warrington, WA5 9SB
MM5 AGM C Campbell, 18 Parkview Avenue, Falkirk, FK1 5JX
M5 AGR Gilchrist MacAdam, Acacia Lodge, 10 The Green, Royston, SG8 7AD
M5 AGS John Lightfoot, Flat 18, The Cloister, Wokingham, RG40 1AW
M5 AGV D Adkins, 196 High Road, North Weald, Epping, CM16 6EF
M5 AGW William Green, 2 Irkdale Avenue, Enfield, EN1 4BD
M5 AGY A Dermont, 7 Pool Close, Little Comberton, Pershore, WR10 3EL
M5 AGZ M Gill, The Cottage, Barrowell Green, London, N21 3AU
M5 AHF M Latimer-Sufit, Flat 615, Jacqueline House, 52 Fitzroy Road, London, NW1 8UB
MI5 AHG James Campbell, 7 Desert Road, Mayobridge, Newry, BT34 2JB
MM5 AHM G Welch, 43 Lady Nairne Road, Dunfermline, KY12 9YD
MM5 AHO G Crowley, 3 Park View, Westfield, Bathgate, EH48 3PP
M5 AIB H Whiteoak, 18 Gregory Springs Mount, Mirfield, WF14 8LG
MM5 AII L McCay, 4 South Mound, Houston, Johnstone, PA6 7DX
M5 AIO J Dixon, 8 East View, St. Ippolyts, Hitchin, SG4 7PD
M5 AIQ Alan Bain, 5 Norgrove Park, Gerrards Cross, SL9 8QT
MM5 AIR GLASGOW WEST OF SCOTLAND AIR CADETS c/o Keith Ross, 9 Bennan Place, East Kilbride, Glasgow, G75 9NR
M5 AJB J Button, 1 Ross Cottages, Southey Green, Halstead, CO9 3RN
MI5 AJH E Holmes, 7 Bamford Park, Dundrod, Crumlin, BT29 4JW
M5 AJK Thomas Dawson, 54 Graeme Road, Enfield, EN1 3UT
M5 AJO M Woodhouse, 5 Redhill Close, Bristol, BS16 2AH
M5 AJP D Hone, 9 Marshall Close, Farnborough, GU14 8RY
MM5 AJW D McKay, Tryggo, Sarclet, Wick, KW1 5TU
M5 AJZ S Coles, 54 Brasslands Drive, Portslade, Brighton, BN41 2PN
M5 AKT M Ellis, 4 Magna Crescent, Great Hale, Sleaford, NG34 9JX
M5 AKW L Howell, 18 High St., Foxton, Cambridge, CB2 6SP
M5 AKY Don Vosper, 5 Franklyn Terrace, Farrington Gurney, Bristol, BS39 6UD
M5 AKZ Ian McClelland, Parkburn, Dumfries, DG1 1RB
M5 ALA A Shone, 50 Whitefield Avenue, Norden, Rochdale, OL11 5YG
M5 ALC R Hatcher, 61 Holland Road, Oxted, RH8 9AU
M5 ALG Andrew Levy, 29 Ferndale Avenue, Reading, RG30 3NQ
MI5 ALJ STRABANE ARS c/o C Hannigan, 4 Silverhill Road, Strabane, BT82 0AE
MI5 ALO E Clementson, 84 Portaferry Road, Newtownards, BT23 8SN
M5 ALS D Munro, 16 Gullimans Way, Leamington Spa, CV31 1LA
M5 ALU A Horsfield, 45 Burchnall Close, Deeping St James, PE6 8QJ
MM5 ALX D Ross, 37 Shillinghill, Alness, IV17 0SZ
MM5 AMM Paul Adams, 19 Fernbank, Stirling, FK9 5AD
M5 AMN A Waddington, 41 Willow Way, Farnham, GU9 0NU
MI5 AMO J Houston, 9 Lismore Drive, Dundonald, Belfast, BT16 1SL
M5 AMV H Cardwell, 3 Old Talbot, Llanwnog, Caersws, SY17 5JG
M5 AOI Neville Pollard, 5 The Stackfield, Wirral, CH48 9XS
MM5 AON Robert Henry, Woodyard House, Woodyard Road, Dumbarton, G82 4BG
M5 ARC WISBECH AR & ELECTRONICS CLUB c/o James Balls, 7 Rowan Close, Holbeach, Spalding, PE12 7BT
M5 ASK W Booker, 3 Hollybank Avenue, Sheffield, S12 2BL
M5 ASR J Richardson, 9 Hilton Avenue, Aylesbury, HP20 2EX
M5 ATR Tracey Ralph, 15 Portchester Close, Stanground, Peterborough, PE2 8UP
M5 AXA Ian Bassett, 47 Queensdown Gardens, Brislington, Bristol, BS4 3JD
M5 BAD C Leese, 15 Ewden Way, Barnsley, S75 2JW
M5 BAZ B Carter, 25 Chester Close, Chafford Hundred, Grays, RM16 6ET
M5 BFL S Shenstone, Jubilee Cottage, Marlow Road, High Wycombe, HP14 3JP
M5 BGR D Wrigley, 32 Avon Road, Chadderton, Oldham, OL9 0PH
M5 BIL William Cooper, 16 Beaumont Hill, Dunmow, CM6 2AP
M5 BJC B Crighton, 2 The Lake House, Savage Cat Farm, Gillingham, SP8 5QR
M5 BMW Roy Read, 111 Fitzpain Road, West Parley, Ferndown, BH22 8SF
M5 BOP M Riley, 63 Clifford Road, Ipswich, IP4 1PJ
M5 BRY Bryan Ray, 18 Manor Close, Wyton, Huntingdon, PE28 2AG
M5 BTB P Brown, 42 Foxglove Close, Burgess Hill, RH15 8UY
M5 BUF R Hanney, 74 Avon Road, Bournemouth, BH8 8SF
M5 BXB S Burrows, 33 Pettys Close, Cheshunt, Waltham Cross, EN8 0EW
M5 CAB William Daly, 85 Lordens Road, Huyton, Liverpool, L14 9PA
M5 CAD Glyn Cadwaladr, Madog Yacht Club, Pen Y Cei, Porthmadog, LL49 9AT
M5 CBR A Hutton, 29 Manor Road, Ashford, TW15 2SL
M5 CBS M Beale, 8 Blakeney Avenue, Swindon, SN3 3NW
MM5 CFA C Allan, 35 Hamewith Court, Alford, AB33 8QW
MI5 CFM Lee Baine, 5 Bute Park, Dundonald, Belfast, BT16 2NU
M5 CHH C Hollins, 56 Lovell Road, Cambridge, CB4 2QR
M5 CJH C Hindle, 15 Kirkstone Court, Kirk Merrington, Spennymoor, DL16 7XJ
M5 CKN S Swinden, 4 Uwch Y Maes, Dolgellau, LL40 1GA
M5 CMO Brian Armstrong, 35 Northfields Crescent, Settle, BD24 9JP
M5 COL H Craven, 4 Amanda Drive, Louth, LN11 0AZ
M5 CYM Ian Taylor, 7 Trellewelyn Close, Rhyl, LL18 4NF
M5 DAD P Stevenson, Nant Fach Cerrigydrudion, Corwen, LL21 0SB
M5 DAP D Parker, 12 Sedge Close, Ivybridge, PL21 0WD
MI5 DAW R bamber, 15 Ladybrook Parade, Belfast, BT119ER
M5 DHB David Brattle, 51 George Street, Bedford, MK40 3RY
M5 DIK R King, 10 Bucks Avenue, Watford, WD19 4AS
M5 DJC M Cressey, 33 Parklands Drive, Harlaxton, Grantham, NG32 1HX
M5 DJO E Owen, Pant-Y-Fedwen, 39 Glanrafon Estate, Caernarfon, LL55 2UW
M5 DLA R Felds, 93 Bancroft Lane, Mansfield, NG18 5LL
M5 DND N Read, 29 Welsh Street, Bishops Castle, SY9 5BS
M5 DNK D Kennedy, Holmcroft, Lewis Road, Chichester, PO20 0RG
MM5 DOG Kenny MacDonald, 5 The Stances, Kilmichael Glassary, Lochgilphead, PA31 8QA
M5 DRW Jonathan Noel, 58 Easenhall Lane, Redditch, B98 0BJ
M5 DUO P Richards, 114 Northleach Close, Redditch, B98 8RD
MM5 DWW David Wishart, Curcum, Swannay, Orkney, KW17 2NS
M5 DZH J Lynch, 21 Worthington Avenue, Hopwood, Heywood, OL10 2LN
M5 EAY Geoffery Hobson, 30 Leigh Road, Westbury, BA13 3QL
M5 ECX T Watts, 6 Keynsham Walk, Swindon, SN3 2AL
MI5 EEM William Hesketh, 49 Mount Michael Park, Belfast, BT8 6JX
M5 EHG R Bateman, 81 Stanton Street, Derby, DE23 6NF
M5 ENM Roger Gurowich, 1 St. Cuthberts Villas, Haybridge, Wells, BA5 1AH
M5 ERN Ernest Coleby, 13 Farm Close, Sunniside, Newcastle upon Tyne, NE16 5PP
M5 EXY Mario Brashill, 42 Bannister Street, Withernsea, HU19 2DT
M5 FAB Steven Sawyers, 72 Langrigg Road, Carlisle, CA2 6DH
M5 FOX D Ross, 37 Cartmell Drive, Leeds, LS15 0NQ
M5 FUN J Constable, 9 Ridgeway Close, Heathfield, TN21 8NS
MM5 FWD S Robertson, Grove Cottage, 30 Commerce Street, Insch, AB52 6HX
M5 GAC Geoffrey Pendrick, 23 Hazel Drive, Spondon, Derby, DE21 7DS
M5 GHT George Thompson, 15 Caithness Road, Hylton Castle Estate, Sunderland, SR5 3RE
M5 GJO Grayhame Orlebar, 21 Field Lane, Willersey, Broadway, WR12 7QB
M5 GUS R Guscott, 19 Springfield Way, Threemilestone, Truro, TR3 6BJ
M5 GUY Guy Austin, 23 Ngunguru Heights Rise, Ngunguru, Whangarei, New Zealand, 173
M5 GVY N Drumm, 8 Harpur Place, Thornhill, Egremont, CA22 2SG
M5 HDF MIDLAND CONTEST GROUP c/o M Waldron, 32 Windmill Street, Upper Gornal, Dudley, DY3 2DQ
M5 HFJ J Glover, 22 Hampden Road, Birkenhead, CH42 5LH
MI5 HIL B Hill, 5 Heathers Close, Magheralin, Craigavon, BT67 0RN
MI5 HNA C Archibald, 37 Jellicoe Drive, Belfast, BT15 3LA
MW5 HOC Darren Warburton, 71 Richards Terrace, Cardiff, CF24 1RW
M5 HOT M Palmer, 21 Ibbett Close, Kempston, Bedford, MK43 9BT
M5 IAN Catherine-Alison Hunt, 105 Cropston Road, Anstey, Leicester, LE7 7BQ
M5 IEP G Tomkins, The Close, Broomfield, Bradford, BD14 6PJ
M5 IGE D Russell, 90 Halleys Way, Houghton Regis, Dunstable, LU5 5HZ
M5 IMI C Wilson, The Rectory, Church Close, Thetford, IP25 7LX
MM5 ISS Graeme Milne, 19 Fairview Crescent, Danestone, Aberdeen, AB22 8ZB
M5 ITE P Hayler, 27 Birch Way, Heathfield, TN21 8BB
M5 JAO J Owen, 10 Pitcher Lane, Leek, ST13 5DB
M5 JON John Edmunds, Caroline Cottage, New Passage Road, Bristol, BS35 4LZ
M5 JWR John Richardson, 6 Clarence Road, Scorton, Richmond, DL10 6EE
M5 JWS J Summers, 3 Thatchers Close, Burgess Hill, RH15 0QU
MI5 JYK Peter Lowrie, 15 Elderburn, Newtownabbey, BT36 5NF
M5 KEN J Sharples, 21 Alexandra Pavilions, Stanleyfield Close, Preston, PR1 1QW
M5 KHH C Young, 26 Horsham Avenue, Peacehaven, BN10 8HX
M5 KJM K Murphy, 79 Torkington Road, Hazel Grove, Stockport, SK7 6NR
M5 KVK G Howell, 19 Constable Avenue, Eaton Ford, St. Neots, PE19 7RH
M5 KZI Michael Hillary, 45 Frances Road, Purbrook, Waterlooville, PO7 5HH
M5 LAR S Pounder, 4 Otley Mount, East Morton, Keighley, BD20 5TD
M5 LLT W Mocroft, 42 Sheraton Grange, Stourbridge, DY8 2BE
M5 LMG L Griffiths, 8 Golfa Close, Middletown, Welshpool, SY21 8EZ
M5 LMY D Sweetland, 15 Wasdale Close, Owlsmoor, Sandhurst, GU47 0YQ
M5 LRO Christopher Pickett, 49 Scotby Avenue, Chatham, ME5 8ER
MI5 LYN E Lynn, 60 Lurgan Tarry, Lurgan, Craigavon, BT67 9HN
M5 MDH M Hampton, 58 Cranbury Road, Eastleigh, SO50 5HA
M5 MDX STOCKPORT RS c/o Bernard Naylor, 47 Chester Road, Poynton, Stockport, SK12 1HA
MI5 MTC Michael Clarke, 19 Ardlougher Road, Irvinestown, Enniskillen, BT94 1RN
M5 MUF Michael Johnson, 1 Ferndale Drive, Ratby, Leicester, LE6 0LH
MW5 MWR M Randall, 15 Erw Wen, Pencoed, Bridgend, CF35 6YF
M5 NEV N Bridle, 95 Kings Stone Avenue, Steyning, BN44 3FJ
M5 OOO H Clayton, 3 Waterden Court, Queensdale Place, London, W11 4SQ
M5 PIP Phillip Nicolson, 56 Serpentine Road, Harborne, Birmingham, B17 9RE
M5 PLY I Peters, 7 Rougemont Close, Plymouth, PL3 6QY
M5 POO Simon Robinson, 23 Jameson Drive, Corbridge, NE45 5EX
MM5 PSL Peter Leybourne, 13 Sandblister Place, Virkie, Shetland, ZE3 9JX
M5 PSW Peter Walking, Flat 36, Highlands House, Southampton, SO19 7GG
M5 PWR Paul Collins, 50 Seacroft Esplanade, Skegness, PE25 3BE
M5 PYE E Boyd, 7 Fritton Court, Haverhill, CB9 8LX
M5 RAG R Mullen, 4 Bay View Grove, Barrow-in-Furness, LA13 0EQ
M5 REG R Barber, 35 Lower Park Crescent, Bishop's Stortford, CM23 3PU
M5 REV R Moll, Penn Cottage, Green End, Buckingham, MK18 3NT
M5 RFD Charles Wardale, 18 Wolsey Way, Lincoln, LN2 4QH
M5 RHG Richard Gower, 21 Saltings Crescent, West Mersea, Colchester, CO5 8GG
M5 RIC Richard Brokenshaw, 64 Southview Road, Weymouth, DT4 0JE
MI5 RJS R Scott, 4 Killycor Avenue, Claudy, Londonderry, BT47 4BX
M5 RMF R Fisher, 65 Kylemore Avenue, Mossley Hill, Liverpool, L18 4PZ
M5 ROB Robert Johnson, 4 Ton Lane, Lowdham, Nottingham, NG14 7AR
M5 RPT Robert Tickle, 5 Bramley Court, Harrold, Bedford, MK43 7BG
M5 RST D Warren, 10 Meadow Bank, Meadow Lane, Alfreton, DE55 2BR
M5 SAV C Bown, 37 Firthwood Road, Coal Aston, Dronfield, S18 3BW
M5 SJS J Covel, 25 Epsom Mews, Bury New Road, Salford, M7 2BZ
M5 SLC Stewart Collis, 14 Mills Drive, Wellington, TA21 9ED
M5 SRE P Scott, 15 Victoria Drive, Blackwell, Alfreton, DE55 5JL
M5 SSB David Rogers, 74 Station Road, Marlow, SL7 1NX
M5 SUE S Coombs, 10 Horseshoe Walk, Widcombe, Bath, BA2 6DE
M5 TAM J Bore, 14 Westwood Avenue, Lowestoft, NR33 9RH
MI5 TCC Trevor Campbell, 27 Silverbrook Park, Newbuildings, Londonderry, BT47 2RD

M5 TLA Martin Robinson, 19 St. Wilfrids Crescent, Brayton, Selby, YO8 9EU
M5 TLE Terence Evans, Gwauntrebeddau, Tregynon, Newtown, SY16 3ER
M5 TMC T Green, 1 Forest Road, Blidworth, Mansfield, NG21 0SJ
M5 TNT S Purdy, Boxwood House, Plumpton, Penrith, CA11 9PA
M5 TTT J Chisholm, 162 Ardington Road, Northampton, NN1 5LT
MM5 TUW Glen Collie, Newton Cottage, Newton Avenue Elderslie, Johnstone, PA5 9BE
M5 TWO C Van Zuilen, Stiermarkenweg 12, Alkmaar, Netherlands, 1827 EK
M5 TXJ David Shaw, Cotehouse, Bleatarn, Appleby-in-Westmorland, CA16 6PX
M5 WAH William Hetherington, 32 Almond Way, Lutterworth, LE17 4XJ
M5 WGD W Dalzell, 9 Pyms Lane, Crewe, CW1 3PJ
M5 WIZ R Wiseman, 12a Breadcroft Lane, Harpenden, AL5 4TE
MI5 WJB W Blanchflower, 7 Casaeldona Park, Belfast, BT6 9RB
M5 WJF Wayne Faulkner, 49 Oakfield Road, Shrewsbury, SY3 8AD
M5 WSS WELLINGTON SCHOOL RS c/o Pete Norman, 3 The Gables, Waterloo Road, Wellington, TA21 8JB
M5 XYZ C Edgar, 61 Winchester Avenue, Lancaster, LA1 4HX
M5 YEX R Hooper, 41 Lady Meers Road, Cherry Willingham, Lincoln, LN3 4BW
MM5 YLO N Marriott, Parkview, Dunrossness, Shetland, ZE2 9JG
M5 ZAP A Morgan, 153 Beanfield Avenue, Coventry, CV3 6NY
M5 ZZZ Stephen Burke, 17 The Crescent, Wragby, Market Rasen, LN8 5RF

## M*6

MM6 AAA Allan Martin, Cathcart, Farr, Inverness, IV2 6XJ
M6 AAC Roy Johns, 27 Stoneby Drive, Wallasey, CH45 0LG
M6 AAD Alan Mullinder, 15 Withington Close, Oakengates, Telford, TF2 6JR
M6 AAE Andrew Ellsom, 23 High Street, Isle of Grain, Rochester, ME3 0BJ
M6 AAG Ashley Taylor, 4 Oxford Street, Carnforth, LA5 9LG
M6 AAJ John O'Brien, 16 Arnold Road, Darlington, DL1 1JG
M6 AAK Alan Comber, 7 Quantock Close, Rushmere St. Andrew, Ipswich, IP5 1AS
M6 AAL Manuel Castro, 11 Monfa Avenue, Stockport, SK2 7BH
MM6 AAM Kieran McClung, 21 Mochrum Avenue, Maybole, KA19 8AX
M6 AAO Jamal Holmes, 15 Nash Close, Farnborough, GU14 0HL
M6 AAP Kevin Metcalfe, 33 Corsican Drive, Hednesford, Cannock, WS12 4SS
MM6 AAR Robert Donaghy, 66 Newton Road, Dundee, DD3 0LT
M6 AAU Craig Todd, 36 Bainton Grove, Clifton, Nottingham, NG11 8LG
M6 AAV Michael Cushing, 12 Coltsfoot Drive, Broadheath, Altrincham, WA14 5JY
M6 AAX Alex Garman, 50 Heys Road, Prestwich, Manchester, M25 1JY
M6 ABB Arran Bocutt, 15 Peel Avenue, Frimley, Camberley, GU16 8YT
M6 ABE Anthony Egan, 4 Rutter Avenue, Warrington, WA5 0HP
M6 ABG Joshua Larcombe, 3 Archer Terrace, Plymouth, PL1 5HD
M6 ABN Paul Fulbrook, 167 Droitwich Road, Fernhill Heath, Worcester, WR3 7TZ
MM6 ABO Kirsty Balfour, 55b Cockels Loan, Renfrew, PA4 0NE
M6 ABQ Ryan Fiddy, 6 Goose Lane, Sutton, Norwich, NR12 9SE
M6 ABR Aidan Cost, 55 Whitcliffe Grange, Richmond, DL10 4ET
M6 ABS Andre Skarzynski, 93 Richmond Walk, St. Albans, AL4 9BB
MD6 ABT Ernest Johnston, 57 Bishop Ken Road, Harrow, HA3 7HU
M6 ABV M Magnall, 18 Osprey Avenue, Westhoughton, Bolton, BL5 2SL
M6 ABX Stuart Townsend, 17 White Hart Street, East Harling, Norwich, NR16 2NE
M6 ABZ Abigail Jones, 41 Milton Brow, Weston-Super-Mare, BS22 8DD
M6 ACA William Shelley, 91 Canterbury House, Stratfield Road, Borehamwood, WD6 1NT
M6 ACB Patrick Flood, 115 Court Farm Road, Newhaven, BN9 9DY
M6 ACC Alistair Cheng, Pastures Green, Firwood Road, Virginia Water, GU25 4NG
M6 ACD Thomas Barker, 93 Hughenden Road, St. Albans, AL4 9QN
M6 ACE Deiniol Murphy, 23 Lowndes Close, Stockport, SK2 6DW
M6 ACF Adrian Fulton, 117 Rokeby Park, Hull, HU4 7QE
MM6 ACI David Simpson, Bridgefoot of Ironside Cottage, New Deer, Turriff, AB53 6UP
M6 ACJ Adam Petrie, 17 Brecon Close, Ashington, NE63 0HT
MM6 ACM Andrew Crawford, Hillview, Kintore, Inverurie, AB51 0XX
M6 ACP Gemma Dale, 26 Kensington Close, Houghton Regis, Dunstable, LU5 5TJ
M6 ACQ Max Ward, 22 Old Road, Leighton Buzzard, LU7 2RE
M6 ACV Robert Moore, 96 Queen Street, Castle Douglas, DG7 1EG
MM6 ACW Alan Woodford, Nordkette, Levenwick, Shetland, ZE2 9GY
M6 ACX Adrian Clark, 386 Wold Road, Hull, HU5 5QG
M6 ACY Albert Connelly, 40 Queen Street, Castle Douglas, DG7 1HS
M6 ACZ Sandra Holmes, 26 Station Road, Llanrwst, LL26 0EP
M6 ADB Andrew Banks, 2 Holt Close, Farnborough, GU14 8DG
M6 ADD Andrew Dade, 15 Barton Road, Berrow, Burnham-on-Sea, TA8 2LT
M6 ADE Adrian Smith, 93 Sheriffs Highway, Gateshead, NE9 6QN
M6 ADF Edward Pierce, 26 Station Road, Aberconwy Aerials, Llanrwst, LL26 0EP
M6 ADG Alan Gladman, 19 Colchester Road, Wymering, Portsmouth, PO6 3RH
M6 ADI Warren Knight, 1 Caravan Site, Drakes Drive, St. Albans, AL1 5AE
M6 ADJ Harvey Murray, 39 Warneford Way, Leighton Buzzard, LU7 4JG
M6 ADM Adam McCallum, 15 Leabank, Newcastle upon Tyne, NE15 7LN
MJ6 ADQ Joe Crowder, 90 Hue Court, Hue Street, Jersey, JE2 3RX
MW6 ADS Andrew Scott, 11 Clive Road, St. Athan, Barry, CF62 4JD
M6 ADT Andrew Carman, 26 Coronation Close, Happisburgh, Norwich, NR12 0RL
M6 ADU David Jones, 1 Brig y Nant, Llangefni, LL77 7QD
M6 ADX David Richards, Flat 40, Leander Court, Teignmouth, TQ14 8AQ
M6 ADY Adrian Slim, Troublesome Reach, Playford Road, Woodbridge, IP13 6ND
MW6 ADZ Adrian Lewis, 3 Aster View, Port Talbot, SA12 7ED
M6 AEB Kevin Hay, Irvine House Lodge, Canonbie, DG14 0XF
M6 AEC Bradley Myler-Cook, 11a York Street, Boston, PE21 6JN
M6 AEE Stephen Hedgecock, 37 Tennyson Road, Maldon, CM9 6BE
M6 AEG Charles Holmes, 8 Byron Way, Caister-on-Sea, Great Yarmouth, NR30 5RW
M6 AEJ Thomas Roberts, 62 Ullswater Avenue, Jarrow, NE32 4FY
M6 AEK John Clark-McIntyre, 184 Quebec Road, Blackburn, BB2 7DP
M6 AEL John Hofman, Brookside, High Street, Stockbridge, SO20 6EY
MW6 AEM David Barraclough, 6 Bryn Terrace, Llangynwyd, Maesteg, CF34 0EA
M6 AEN Daniel Field, 5 Clos Crugiau, Rhydyfelin, Aberystwyth, SY23 4RN
M6 AEQ Byrom Livesey, 14 Sycamore Avenue, Tyldesley, Manchester, M29 8WQ
M6 AER Christopher Fielden, 44 Hylion Road, Leicester, LE2 6JE
M6 AES Anthony Southwell, 56 Lambrook Road, Taunton, TA1 2AF
M6 AET Graham Hunt, 35 Outram Street, Sutton-in-Ashfield, NG17 4BA
M6 AEU Amy Seymour, Home Farm, Lodge Lane, Northampton, NN6 7PQ
M6 AEW Alasdair Cockett, 10 San Marcos Drive, Chafford Hundred, Grays, RM16 6LT
MW6 AEX Andrew Bayliss, 2 Hillside Terrace, Tonypandy, CF40 2HJ
M6 AEY James Smith, 2 Quibble Court, 22b Church Street, Ripley, DE5 3BU
M6 AEZ Robert Gore, 4 Westaway Park, Yatton, Bristol, BS49 4JU
M6 AFA Andrew Lambert, 1 Glebelands, Lympstone, Exmouth, EX8 5JD
M6 AFB Alex Brown, 45 Saffron Park, Kingsbridge, TQ7 1RW
M6 AFC Lee Hillier, 10 Buttermere Close, Folkestone, CT19 5JH
M6 AFF John Hone, 12 Marlborough Close, Exmouth, EX8 4NA
M6 AFG Clive Stacey, 9 Nile Road, Southampton, SO17 1PF
M6 AFH Matthew Leslie Hillam, Common Farm, Swinefleet, Goole, DN14 8DW
MI6 AFI Glen Todd, DP1, Shanaghy, Enniskillen, BT92 0EQ
M6 AFK Adam Studdart, 11 Degas Close, Connah's Quay, Deeside, CH5 4WQ
M6 AFL Adrian Lawrence, Lillooah, Watling Street, Hinckley, LE10 3ED
MI6 AFM Jarlath Rice, Shanaghy, Lisnaskea, Enniskillen, BT92 0EQ
M6 AFN Johnhenry Hitchens, 6 Oak Road, Alderholt, Fordingbridge, SP63bl
MM6 AFQ James Watt, 10 Marshall Gardens, Luncarty, Perth, PH1 3YX
M6 AFS Paul Carter, 18 Park Road, Allington, Grantham, NG32 2EB
M6 AFU Timothy Turner, 23 Pankhurst Drive, Bracknell, RG12 9PS
M6 AFW Stephen Coombes, Pantiles, South Crescent, Skegness, PE24 5RQ
M6 AFX Simon O'Donnell, 26 Park Avenue, Skegness, PE25 2TF
M6 AFY William Haddock, 5 Bradley Close, Middlewich, CW10 0PF
M6 AGA Gary Wright, 21 Larkfield Road, Redditch, B98 7PL
M6 AGD Jason Salter, 20 Burrow Road, Chigwell, IG7 4HQ
MD6 AGF John Kaighin, 10 Kerrocruin, Kirk Michael, Isle of Man, IM6 1AF
M6 AGG Adam Greig, 3 Fir Grange Avenue, Weybridge, KT13 9AR
M6 AGH Alan Holt, 36 The Maltings, Malmesbury, SN16 0RN
M6 AGI Rodney Coombes, Rose Cottage, Choice Hill Road, Chipping Norton, OX7 5PZ
M6 AGK Robert Hambly, 144 Station Road, Irchester, Wellingborough, NN29 7EW
MM6 AGL R Perkins, 6 Pettens Close, Balmedie, Aberdeen, AB23 8WZ
M6 AGO Barry Denyer-Green, Dunsley South, Park Road, Forest Row, RH18 5BX
M6 AGQ Kenneth Jones, 40 Sandrock Hill Road, Wrecclesham, Farnham, GU10 4RJ
M6 AGR Arthur Rayner, 12 Newhaven Drive, Lincoln, LN5 9UF
MW6 AGS David Evans, 1 Heol Glyndwr, Fishguard, SA65 9LN
M6 AGT Ronald MacDonald, 20 The Peacocks, Warwick, CV34 6BS
MI6 AGV Albert Hamilton, 9 Slievenamaddy Avenue, Newcastle, BT33 0DT
M6 AGY David Pearson, 73 Stackwood Avenue, Barrow-in-Furness, LA13 9HJ
M6 AGZ Jason Gore, 88 Rowan Drive, Kirkby-in-Ashfield, Nottingham, NG17 8FR
M6 AHA Philip Whitmore, 3 Crown Bank, Talke, Stoke-on-Trent, ST7 1PT
MM6 AHB Andrew Hearty, 27 Newton Drive, Newmains, Wishaw, ML2 9DB
M6 AHD Altaf Dossa, 24 Warwick Drive, Cheshunt, Waltham Cross, EN8 0BW
M6 AHE Emily Yohn, Ortner House Farm, Abbeystead, Lancaster, LA2 9BD
M6 AHF William Starkey, Flat 13, Ramsden Court, Barrow-in-Furness, LA14 2HH
M6 AHH Antony Hall, 14 Stanelow Crescent, Standon, Ware, SG11 1QF
MM6 AHJ William Noon, 0/1 445 Royston Road, Glasgow, G21 2DE
MM6 AHK Robert Renshaw, Smithy House, Scotscalder, Halkirk, KW12 6XJ
M6 AHL Rowan Butterfield, 35 Bede Crescent, Benington, Boston, PE22 0DZ
MM6 AHN Ailie MacDougall, 8 Mount Stuart Drive, Wemyss Bay, PA18 6DX
MI6 AHO Adrian Ismay, 21 Hillsborough Drive, Belfast, BT6 9DS
M6 AHP Amos Meekins, 12 Myrtle Road, Kettering, NN16 9TW
MW6 AHQ David Williams, 23 Parc y Ffynnon, Ferryside, SA17 5TQ
MI6 AHR Tomasz Grzybek, 75 Kilburn Street, Belfast, BT12 6JT
M6 AHS Anthony Strong, 55 Coley View, Halifax, HX3 7EB
MW6 AHT Oliver Davis, 18 Ty Gwyn Drive, Brackla, Bridgend, CF31 2QF
M6 AHU Kevin Lambert, 38 Whittleford Road, Nuneaton, CV10 9HU
MW6 AHV Dan Burton, 11 Cwrt-Ucha Terrace, Port Talbot, SA13 1LD
M6 AHW Anthony Whitehead, 82 High Park Avenue, Stourbridge, DY8 3NA
MM6 AHX Michael Mason, 16 Avalon Gardens, Linlithgow Bridge, Linlithgow, EH49 7QE
MM6 AHY Joshua Liddell, 49 Inchbrae Road, Glasgow, G52 3HA
M6 AHZ Phil Dawes, 49 Altofts Lodge Drive, Altofts, Normanton, WF6 2LB
M6 AIA Andrew Barton, 51 Fieldhead Gardens, Dewsbury, WF12 7SN
M6 AIB Colin Durrant, 12 Witney Close, Ipswich, IP3 9QF
MW6 AIC Rheinallt Morgan, Spinning Wheel, Derwydd Road, Ammanford, SA18 2LX
M6 AIE Christopher Ring, 29 Shelley Close, Newport Pagnell, MK16 8JB
M6 AIF Kieron Ball, 29 Heather Grove, Wigan, WN5 9PJ
M6 AIH Andrew Bailey, 49 Grange Crescent, Gosport, PO12 3DS
M6 AII Bruce Ansell, 7 Bramley Green Road, Bramley, Tadley, RG26 5UE
M6 AIJ Raymond Tarling, Moonrakers, Ashley, Corsham, SN13 8AN
MM6 AIK Douglas Seivwright, 35 Invercauld Gardens, Aberdeen, AB16 5RR
M6 AIL Marguerita Barba, 5 Vetch Way, Andover, SP11 6RR
M6 AIN Colin Campbell, 5 Ryebank, Holmfirth, HD9 1EU
M6 AIO Andy Hadfield, 12 Manor Road, Caister-on-Sea, Great Yarmouth, NR30 5HG
M6 AIP Keith McFadden, 40 Brookfield Road, Northampton, NN2 7LP
M6 AIQ Len Fletcher, 110 West Street, North Creake, Fakenham, NR21 9LH
M6 AIR Michael Ward, Fair View, Queen Street, Winkleigh, EX19 8JB
M6 AIT Jack Barker-Gunn, 5 De Montfort Road, Lewes, BN7 1SP
M6 AIU David Crashley, 59 Brookdale Road, Nuneaton, CV10 0DL
M6 AIV Anthony Barrett, 4 Wood Cottages, Cummings Cross, Newton Abbot, TQ12 6HJ
M6 AIW Amanda Nutt, 9 Hereford Road, Southport, PR9 7DX
M6 AIZ Mary Burbeck, 5 Wouldham Terrace, Saxville Road, Orpington, BR5 3AT
MI6 AJA John Bingham, 27 Carrickdale Gardens, Portadown, Craigavon, BT62 3BN
M6 AJB Andrew Brown, Ponsharden Cottage, High Offley Road, Stafford, ST20 0LG
M6 AJC Andrew Cosham, 81 Howard Road, Sompting, Lancing, BN15 0LP
M6 AJF Aaron Fysh, 74 Kingsway, King's Lynn, PE30 2EL
M6 AJH Amy Higham, 30 Broome Road, Southport, PR8 4EQ
MM6 AJI Michael McGrorty, 17 Fernbank, Stirling, FK9 5AD
M6 AJJ Michael Baker, 8 Higher Polsham Road, Paignton, TQ3 2SY
M6 AJL Andrew Lashbrook, 1 Fortescue Road, Exeter, EX2 8LA
M6 AJM Adam Mitchell, 18 Holly Leys, Stevenage, SG2 8JA
MI6 AJN Andrew Ruddell, 16 Beechfield Manor, Aghalee, Craigavon, BT67 0GB
MI6 AJO Ivan Hoey, 58 Tullynamullan Road, Shankbridge, Ballymena, BT42 2LR
M6 AJP Andrew Pilkington, 26 Ryelands Close, Market Harborough, LE16 7XE
M6 AJX Rita Newstead, 97 Hawthorn Crescent, Burton-on-Trent, DE15 9QN
M6 AKA Glyn Adgie, Flat 1, 6 Grayfield Avenue, Birmingham, B13 9AD
M6 AKE Ian Brayley, 9 Church View Close, Southampton, SO19 8SJ
M6 AKF Paul Edwards, 8 Aughton Way, Broughton, Chester, CH4 0QL
M6 AKG Helen Turner, 39 Court Crescent, Kingswinford, DY6 9RJ
M6 AKH Stephen Ward, 11 Blenheim Drive, Hawkinge, Folkestone, CT18 7FA
M6 AKI Barrie Smith, Maple Lodge, Burtoft Lane South, Boston, PE20 2PF
M6 AKJ Leon Lee, 8 William Avenue, Margate, CT9 3XT
M6 AKK Andrew Forsythe, 24 The Welkin, Lindfield, Haywards Heath, RH16 2PH
M6 AKO Simon Clarke, 27 Coronation Road, Callington, PL17 7BX
M6 AKP Angela Plastow, Bradcar Cottage, Bradcar Road, Attleborough, NR17 1EQ
MM6 AKQ Sean Burns, 3/R 170 Lochee Road, Dundee, DD2 2NH
M6 AKT Alan Millin, 79 Court View, Stonehouse, GL10 3PJ
M6 AKV Andrew Vivian, 15 Kew Klavji, Rhind Street, Bodmin, PL31 2FE
M6 AKW Ian Duffie, Trebeighan Farm, Saltash, PL12 5AE
M6 AKY Roger Bleaney, 40 Broadstone Road, Harpenden, AL5 1RF
MM6 AKZ Derek Henderson, 10 Rye Crescent, Glasgow, G21 3JS
M6 ALB Anthony Burns, 76, 76, Morprth, NE65 0TF
MW6 ALC Adam Cotter, 3 George Street, Aberdare, CF44 6RY
M6 ALE Dave Trollope, 80 Azalea Drive, Trowbridge, BA14 9GG
M6 ALG Michael Yates, 8 Johnsons Street, Ludham, Great Yarmouth, NR29 5NZ
M6 ALH Abbiegail Higham, 30 Broome Road, Southport, PR8 4EQ
M6 ALJ Albert Jones, 23 Cranley Road, Hersham, Walton-on-Thames, KT12 5BT
M6 ALK Alan Kent, 4 Sellerdale Drive, Wyke, Bradford, BD12 9DA
MI6 ALL Alister Armstrong, 45 Rathmena Drive, Ballyclare, BT39 9HZ
M6 ALM Thomas Hamilton, Flat 14, Kingsmead Court, Dunstable, LU6 1NQ
M6 ALO Annie N Nugorski, 49 Buxton Street, Morecambe, LA4 5SR
M6 ALP Robert McLeod, 75 Davis Street, Stanley, Falkland Islands, FIQQ 1ZZ
M6 ALQ Stephen Hassall, 21 Bridgnorth Grove, Newcastle, ST5 7QP
M6 ALT David Ingrey, 1 Ponders Road, Fordham, Colchester, CO6 3LX
M6 ALU Jack Harrington, 69 Lexden Road, Colchester, CO3 3QE
M6 ALW John Anderson, 15 Dickson Road, London, SE9 6RA
M6 ALX Alex Jones, 5 Meadowlands, Kirton, Ipswich, IP10 0PP
M6 ALY Anthony Hall Hall, 254 Walton Road, Walton on The Naze, CO14 8LT
MM6 ALZ Fred Gordon, Croft of Torrancroy, Strathdon, AB36 8US
MM6 AMA Anne Marie Campbell, 1b Craig Road, Troon, KA10 6DA
M6 AMC Adrian Cottrell, 1 Wilberforce Road, Doncaster, DN2 4RW
M6 AMD Angela Dingle, 29 Castle View, Witton le Wear, Bishop Auckland, DL14 0DH
M6 AMG Joshua Smith, 2 Sunfields Close, Polesworth, Tamworth, B78 1LW
M6 AMI Terence Gladman, 39 Fairview Avenue, Rainham, RM13 9RL
M6 AMJ Bernard Jones, 9 St. James Close, Hanslope, Milton Keynes, MK19 7LF
M6 AML Andrew Lennon, 12 Lockgate East, Windmill Hill, Runcorn, WA7 6LB
M6 AMN Clifford Johnson, 58 Cheviot Road, London, SE27 0LG
M6 AMO Allan Fleming, 39 Urswick Green, Barrow-in-Furness, LA13 0BH
M6 AMR Adrian Riddick, 30 Britannia Road, Banbury, OX16 5DW
M6 AMT Andrew Taylor, 11 Fillingfir Drive, Leeds, LS16 5EG
M6 AMV Geoffrey Lyon, 1 Eckersley Street, Wigan, WN1 3PP
M6 AMW Aaron Webb, 3 Blackmore Close, Thame, OX9 3ZH
M6 AMX Timothy Brown, 6 Wolstanholme Close, Congleton, CW12 3RX
MM6 ANB Paul Rice, 36 Namur Road, Penicuik, EH26 0LL
M6 AND Andrew Faulkner, 45 Broad Leys Road, Barnwood, Gloucester, GL4 3YW
M6 ANE Don Mansfield, 1 Vale View, Alexandra Road, Heathfield, TN21 8EF
M6 ANI Philip Hanman, 7 Tremenheere Road, Penzance, TR18 2AH
M6 ANM Ellen Seymour, Home Farm, Lodge Lane, Northampton, NN6 7PQ
M6 ANN Eric Mann, 6 London Road, St. Georges, Telford, TF2 9LQ
M6 ANO Julie Seymour, Home Farm, Lodge Lane, Northampton, NN6 7PQ
M6 ANP Andrew Coulthard, 47 Keston Crescent, Stockport, SK5 8NQ
M6 ANT Albert Taylor, 90 Coppice Avenue, Eastbourne, BN20 9QJ
M6 ANV George Parkinson, 11 Curtis Road, Poole, BH12 3AQ
M6 ANX Peter Tillotson, 9 Holker Street, Barrow-in-Furness, LA14 5RQ
M6 ANY mitchell Tuffill, 1 Madden Close, Nottingham, NG5 5US
M6 AOB Ashley Back, 34 Willow Road, Redhill, RH1 6LW
M6 AOE Stewart Parfitt, 14 Heol y Twyn, Rhymney, Tredegar, NP22 5DW
M6 AOF Stephen Salisbury, 6 Ryan Close, Leyland, PR25 2XW
M6 AOG Michael Whelan, 134 FIELDS FARM RD, Hyde, SK14 3QW
MM6 AOH Stuart McKenzie, 0/2 69 Glenkirk Drive, Glasgow, G15 6AU
M6 AOJ John Burnham, 43 Oatlands Walk, Birmingham, B14 5QD
M6 AOK Callum Macleod, 2 Welford Road, Chapel Brampton, Northampton, NN6 8AF
M6 AOL Peter Butler, 38 Oak Hill, Hollesley, Woodbridge, IP12 3JY
MM6 AON Adam Watson, 10 Christie Place, Elgin, IV30 4HX
M6 AOP Oliver Ernster, 19 Rose Gardens, Farnborough, GU14 0RW
M6 AOQ Mark Keilty, 17 Cliff Road, Wallasey, CH44 3DJ
MI6 AOR Aaron O'Reilly, 15 Killowen Point, Rostrevor, Newry, BT34 3AN
M6 AOS Andrew Ashton, 10 Lindsey House, Church, Accrington, BB5 4AG
M6 AOT Fulu Fang, 82 Mill Hill Road, Norwich, NR2 3DS
M6 AOV Richard Simpson, 22 Kenworthy Road, Stocksbridge, Sheffield, S36 1BZ
M6 AOW Adam Edwards, 2 Keystone Gardens, Steventon New Road, Ludlow, SY8 1LE
MI6 AOX Tommy Darrah, 42 Pinewood Avenue, Carrickfergus, BT38 8EW
MW6 AOY James Regan, 30 Tynybedw Terrace, Treorchy, CF42 6RL
MI6 AOZ Darryl Mudd, 14 Bloomfield Road, Belfast, BT5 5LT
M6 APA Angel Armstrong, 30 Tennyson Avenue, Hull, HU5 3TW
M6 APB Alan Bradshaw, 130 Low Lane, Morecambe, LA4 6PS
M6 APC Adrian Cook, 37 Railway Road, Stretford, Manchester, M32 0RY
M6 APG Neville Briggs, 20 Broad Lane, Pelsall, Walsall, WS4 1AP
M6 API Peter Bishop, 18 Holmwood Avenue, South Croydon, CR2 9HY
M6 APM A Munford, 16 Broadhurst Way, Brierfield, Nelson, BB9 5HG
MM6 APO Arjunsingh Bais, 7 Marine Terrace, Aberdeen, AB11 7SF
M6 APR Alexander Ralph, 3 The Leys, St. Albans, AL4 9HD
M6 APS Thomas Bell, 13a Church Road, Cinderford, GL142ED
M6 APU Derek Bolton, 88 Goldsborough, Wilnecote, Tamworth, B77 4DF
M6 APW Anthony Woodhouse, 4 Grafton Close, St. Albans, AL4 0EX
M6 AQA Albert Wright, 69 Thomas Street, Tamworth, B77 3PR
M6 AQD Paul Hateley, 44 Painters Croft, Coseley, Bilston, WV14 8AP
M6 AQE Malamkunnu Mohammed Shafi, 142 Whitethorn Street, London, E3 4DB



**IMPORTANT NOTE**

**Revalidate licence to avoid revocation** – Ofcom has advised the Society that plans will be drawn up to revoke licences that have not been revalidated as required by the licence conditions. The quickest way to revalidate is to do so online via the Ofcom website: *https://services.ofcom.org.uk/* or by email: *amateur.validations@ofcom.org.uk* Ofcom staff are available to help, but please be patient during times of heavy workload.

RSGB

M6 AQG Paul Frost, 86 Grantham Road, Sleaford, NG34 7NW
M6 AQI Christopher Croasdale, 76 Kingsway, Euxton, Chorley, PR7 6PP
MW6 AQJ Adam Hanley, 11 Heol Pearetree, Rhoose, Barry, CF62 3LB
M6 AQK Thomas Pearsall, 16 Langdale Road, Leyland, PR25 3AR
M6 AQL Simon Melton, 2 The Orchard, Bishopthorpe, York, YO23 2RX
MM6 AQM Craig McIntyre, 2/1, 6 Castle Street, Glasgow, G81 4HH
M6 AQN Edward Aksamit, 14 Popplewell Gardens, Gateshead, NE9 6TU
M6 AQO Gordon Foster, 22 Bradley Cottages, Consett, DH8 6JZ
M6 AQQ Andrew Northall, 40 Rowan Grove, Wirral, CH63 2NH
M6 AQR Kevin McCarthy, 34 Shawley Way, Epsom, KT18 5PB
M6 AQT Leigh Matthews, 20 Harbridge Road Broughton, Chester, CH4 0FT
M6 AQU Richard Davison, 2 Marlow Terrace, Mold, CH7 1HH
M6 AQV Valko Yotov, 25 Cross Gates Close, Bracknell, RG12 9TY
M6 AQW Terence Newman, 39 Claremont Road, West Byfleet, KT14 6DY
M6 AQY Lynn Gibbons, 18 Langdale Road, Ribbleton, Preston, PR2 6AN
M6 ARB Andrew Boucher, 46 Langton Hill, Horncastle, LN9 5AH
M6 ARC Steve Froggatt, 140 Greenlea Court, Huddersfield, HD5 8QB
M6 ARD William Peel, 15 Brockhurst Close, Horsham, RH12 1UY
M6 ARF Michael Dailey, 58 Waincliffe Mount, Leeds, LS11 8AH
M6 ARH Antony Hargreaves, 27 Meadow Head Close, Blackburn, BB2 4TY
M6 ARI Brian Ansdell, 8 Oakland Drive, Dudley, DY3 2SH
M6 ARJ Harley Clough, 8 Skeldyke Road, Kirton, Boston, PE20 1LR
M6 ARK Mark Harris, 29 Queen Street, Halesowen, B63 3TZ
M6 ARL Simon Skirving, 1 Hallington Close, Bolton, BL3 6YH
M6 ARM Andrew Martyn, 6 North Side, New Tupton, Chesterfield, S42 6BW
MM6 ARN Arran O'Neill, 39 Ardneil Court, Ardrossan, KA22 7NQ
M6 ARP Alan Holmes, 2 Park Farm Cottages, Park Lane, Chichester, PO20 3TL
M6 ARQ John Whitehead, 20 Seamill Park Crescent, Worthing, BN11 2PN
M6 ARR Iain Clark, 21 James Street, Epping, CM16 6RR
M6 ARS Brian Holland, 11 Silverlands Park, Buxton, SK17 6QX
MM6 ART Arthur Young, 4/4 Prestonfield Terrace, Edinburgh, EH16 5EE
M6 ARU James Cherry, 12 Scarisbrick New Road, Southport, PR8 6PY
M6 ARV Lawrence Williams, Broomstreet Farm, Porlock, Minehead, TA24 8JR
M6 ARW A Winkley, 77 Lechlade Road, Birmingham, B43 5ND
MW6 ARX Kurt Bevan, 76 Tennyson Close, Pontypridd, CF37 5ER
M6 ARY Harold Watts, 696 Knowsley Lane, Knowsley, Prescot, L34 9EH
M6 ASC Andrew Davies, 68 Wood Street, Castleford, WF10 1LN
M6 ASD Amardeep Dhillon, 13 Weston Road, Guildford, GU2 8AU
M6 ASE Arron Everett, 16 Robyns Road, Beeston Regis, Sheringham, NR26 8YJ
M6 ASF Alan Foote, Flat One, Kimber's Close Kennet Road, Newbury, RG14 5JF
M6 ASH David Amos, 86 Royal Military Avenue, Folkestone, CT20 3EJ
M6 ASI Teresa Williams, Broomstreet Farm, Porlock, Minehead, TA24 8JR
M6 ASJ Raymond Bawden, 43 Enys Road, Camborne, TR14 8TW
M6 ASK Alexander Keatley, 156 Earlswood Way, Colchester, CO2 9NE
M6 ASM Alan McLachlan, 4 Stratton Close, Bexleyheath, DA7 4AJ
M6 ASO Timothy Roper, 48 Rowthorne Lane, Glapwell, Chesterfield, S44 5QD
M6 ASP Timothy Chapman, 1 East Dean Road, Lockerley, Romsey, SO51 0JL
M6 ASQ Allen Hefford, 9 Chamomile Gardens, Farnborough, GU14 9XY
M6 AST Benjamin Herrick, Honing road, Dilham, NR28 9PL
M6 ASV C Redmond, 6 Apsley Road, Southsea, PO4 8RH
M6 ASW Anthony Williams, Brockwell House, The Street, Diss, IP22 1BX
M6 ASZ garry wall, Flat 2, Coniston House Holyoake Road, Worsley, Manchester, M28 3DH
M6 ATA Stephen Waldock, 102 Fraser Road, London, N9 0BY
M6 ATB Daniel Zubrzycki, 16 Oldfield Avenue, Hull, HU6 7UN
MW6 ATC Alexandra Hare, 243 Heritage Park, St. Mellons, Cardiff, CF3 0DU
M6 ATD Ernesto Gomez Lozano, Flat D, 212 Kennington Road, Oxford, OX1 5PG
M6 ATF Neill Bisiker, 31 Lansdowne Avenue, Waterlooville, PO7 5BL
M6 ATG Alan Jones, 3 Manor Way, Kinmel Bay, Rhyl, LL18 5BP
M6 ATH Adrian Hicks, The Granary, Vann Lake Road, Dorking, RH5 5JB
M6 ATI Attila Paricsi, Flat 2, Felton House Farm, 20 Upper Town Lane, Bristol, BS40 9YF
M6 ATJ Martin Brasher, 48 Eldertree Road, Thorpe Hesley, Rotherham, S61 2TQ
M6 ATK Robert Atkins, 2 Sandpiper Crescent, Malvern, WR14 1UY
M6 ATL George Hatt, 4H Colman House, Earlham Road, Norwich, NR4 7TJ
M6 ATM Andrew Malin, 50 Leicester Road, Sharnford, Hinckley, LE10 3PR
M6 ATP Derek Gibson, 25 Middleham Close, Ouston, Chester le Street, DH2 1TA
M6 ATQ Ian Cooper, 6 Back Lane East, Great Bromley, Colchester, CO7 7UB
MM6 ATR Allan Robertson, 6c Fergusson Road, Cumbernauld, Glasgow, G67 1LR
M6 ATS John Watts, 70 Castleway North, Leasowe, Wirral, CH46 1RW
MM6 ATU Thomas Mcbride, 27 Cassillis Road, Maybole, KA19 7HF
M6 ATW Anthony Wilkinson, 17 Stewkins, Audnam, Stourbridge, DY8 4YW
M6 ATX Gareth Williams, 19 Holden Walk, Wigan, WN5 9JQ
M6 ATY Fred Payne, 45 Foxhill, Shaw, Oldham, OL2 7NQ
M6 ATZ Damien Hargreaves, 42 Amesbury Avenue, London, SW2 3AA
M6 AUA Philip Cooke, 44 Brooklands Park, Craven Arms, SY7 9RL
M6 AUB Sean Drake-Brockman, 13 St. Johns Place, Bury St. Edmunds, IP33 1SW
M6 AUC Sara Hall, 3 Cedar Grove, Prestwich, Manchester, M25 3DY
MJ6 AUD Paul Ahier, Les Trois Carres, La Rue D'Aval, Jersey, JE3 6ER
M6 AUE Gary Carter, 19 Brathay Crescent, Barrow-in-Furness, LA14 2BG
M6 AUF Emlyn Washbrook, 47 Westgate, Leominster, HR6 8SA
MM6 AUG Brian Moerman, 19/3 Pirniefield Bank, Edinburgh, EH6 7QQ
M6 AUH Harry Beaumont, 1 Ashley Walk, Orleton, Ludlow, SY8 4HD
MM6 AUI David Cunningham, 21 Constable Acre, Cupar, KY15 4AE
MM6 AUJ Frankie Linn, 14 Elphinstone Road, Tranent, EH33 2HR
M6 AUK Timothy Taylor, Flat 9, Arundel Keep, 14 Arundel Road, Eastbourne, BN21 2EW
M6 AUL Howard Hylton, 214 School Road, Hall Green, Birmingham, B28 8PF
M6 AUM William Saint, Flat 74, Ferrier Point, London, E16 1QW
M6 AUO Jeremy McMahon, 39 Hobmoor Croft, Birmingham, B25 8TJ
M6 AUP Thomas Mundell, 98 Westley Road, Bury St. Edmunds, IP33 3SD
M6 AUQ Peter Thearle, 12 Grange Walk, Bury St. Edmunds, IP33 2QB
MI6 AUU Marttin Rushbrooke, 22 Dublin Road, Omagh, BT78 1ES
MW6 AUX David Jones, 17 Miners Row, Aberdare, CF44 0TP
M6 AUZ Susan Michaelis, Fieldgate, Coltstaple Lane, Horsham, RH13 9BB
M6 AVA Chris Hughes, 10 Langford Road, Stockport, SK4 5BR
M6 AVD Malcolm Darwen, 31 Jacks Key Drive, Darwen, BB3 2LG
MM6 AVE John Rankin, 17 Dippin Place, Saltcoats, KA21 6AB
M6 AVF Derek Bailey, 2 Southey Crescent, Maltby, Rotherham, S66 7LY
M6 AVG Lindsay Scott, 28 Cavendish Place, New Silksworth, Sunderland, SR3 1JW
M6 AVH Christopher Lee, 67 Hill Park Road, Fareham, PO15 6HT
M6 AVJ Aline Johnson, 3 Lindens Close, Thorney Toll, Wisbech, PE13 4AR
M6 AVK Robin Gripp, 23 Edmond Locard Court, Chepstow, NP16 6FA
M6 AVL Paul Smart, 142 Finch Road, Chipping Sodbury, Bristol, BS37 6JB
M6 AVM Adrian Hunter, 9 Gelt Burn, Didcot, OX11 7TZ
M6 AVN John Brownbill, 18 Milestone Road, Stratford-upon-Avon, CV37 7HH
M6 AVO Anthony Craven, 45 Benhams Drive, Horley, RH6 8QT
M6 AVQ Collis Brown, 77 Grange Road, Ramsgate, CT11 9LP
M6 AVR Alasdair Mackay, 11 Greatheed Road, Leamington Spa, CV32 6ES
M6 AVS Robin Paxman, 11 Gibsons Gardens, North Somercotes, Louth, LN11 7QH
M6 AVT Rory Whalley, 188 Astley Street, Astley, Manchester, M29 7AX
M6 AVU Antony Pawlak, 8 Healey Close, Crewe, CW1 4RS
M6 AVV James Noon, 108 Cardinal Avenue, Morden, SM4 4SX
M6 AVW Anthony Wilson, 11 Headland Way, Alton, Stoke-on-Trent, ST10 4AN
M6 AVY Ian Clark, 22 Rosemary Avenue, Grimsby, DN34 4NJ
M6 AVZ Keith Nicholson, 11 Lancaster Way, Skellingthorpe, Lincoln, LN6 5UF
M6 AWA Allen Warnes, 13 Warren Avenue, Fakenham, NR21 8NP
M6 AWB Anthony Baker, 2 Stileway, Meare, Glastonbury, BA6 9SH
M6 AWC Andrew Coombes, 3 Marshall Close, Purley on Thames, Reading, RG8 8DQ
M6 AWE Dominic Barrett, 208 Doncaster Road, Rotherham, S65 2UE
M6 AWG Mark Feltham, Flat 5 Rebecca Court, 9, Beckenham, BR3 1NN
M6 AWI Ben Sturman, 26 Howes Avenue, Thurston, Bury St. Edmunds, IP31 3PY
M6 AWJ Michael Harland, Challenger Quay, Falmouth, TR11 3YL
M6 AWL Michael Shepherd, Oak View, Church Road, Halstead, CO9 4PR
M6 AWN Christopher Smith, 44 Brooksfield, Bildeston, Ipswich, IP7 7EJ
M6 AWO Michael Nelson, 30 Unicorn Place, Bury St. Edmunds, IP33 1YP
M6 AWP Lamek Amunyela, 1 Artillery Street, Colchester, CO1 2JJ
M6 AWQ Philip Vickers, 37 Faraday Road, Ipswich, IP4 1PU
M6 AWR Adam Wright, 52 Thorpe Way, Cambridge, CB5 8UB
M6 AWS Adam Stabler, 11 Lincolns Avenue, Gedney Hill, Spalding, PE12 0PQ
M6 AWU John Harris, 4 Burgh Old Road, Skegness, PE25 2LN
MW6 AWV Arron Williams, 88 Heol Homfray, Cardiff, CF5 5SB
M6 AWY Jonathan Van-Boques-Tal, 9 Stubbins Lane, Gazeley, Newmarket, CB8 8RL
M6 AWZ Andrew Wilson, 4 Oxford Street, Doe Lea, Chesterfield, S44 5PH
M6 AXB Andrew Blamire, 21 The Laurels, Banstead, SM7 2HG
M6 AXC Daniel Howarth, 32 Cotswold Drive, Rothwell, Leeds, LS26 0QZ
M6 AXD Stephen Denman, 12 Dyke Vale Road, Sheffield, S12 4ER
M6 AXF Michael Atherton, 25 Forest Road, Sutton Manor, St. Helens, WA9 4AY
M6 AXG Thomas Winter, 8 Thorpe Street, Hartlepool, TS24 0DX
M6 AXH Adam Hanson, 14 Braithwaite Avenue, Keighley, BD22 6EU
M6 AXI Alan Chapman, 1 Fortunes Way, Bedhampton, Havant, PO9 3LX
M6 AXJ Stephanie McCluskey, 29 Hotspur Avenue, Bedlington, NE22 5TD
M6 AXL Adam Lowery, 21 Westlea Avenue, Riddlesden, Keighley, BD20 5EJ
M6 AXM Manish Rai, Coldham Hall, Stanningfield, Bury St. Edmunds, IP29 4SD
M6 AXP Jenny Christoforou, 55 Wood Street, Taunton, TA1 1UW
M6 AXR Rupert Williamson, 2 Hobbs Road, Shepton Mallet, BA4 4LS
M6 AXS Nick Sutherland, 34 Little Heath Road, Chobham, Woking, GU24 8RL
MM6 AXT Stuart Glen, 5/3 Renfrew Chambers, 136 Renfield Street, Glasgow, G2 3AU
M6 AXU Guy Briggs, 54 Behind Berry, Somerton, TA11 6JY
M6 AXW Paul Hearnshaw, Flat 10, 83 Swallows Meadow, Solihull, B90 4PH
M6 AXX Adam Marsden, 38 Sandhill Road, Rawmarsh, Rotherham, S62 5NT
MW6 AYA Amy Young, 14 Ramsons Way, Cardiff, CF5 4QY
M6 AYC Thomas Jagger, 50 North Street, Lower Hopton, Mirfield, WF14 8PN
M6 AYE Amy Yap, Harrogate Ladies' College, Clarence Drive, Harrogate, HG1 2QG
M6 AYG Brian Leckey, 76 Cardigan Lane, Leeds, LS4 2LN
M6 AYH Colin Wilson, 87 Levensgarth Avenue, Fulwood, Preston, PR2 9FP
M6 AYI Reg Unsworth, 22 Meadow House Park, Badcocks Lane, Tarporley, CW6 9RT
M6 AYJ Michael West, Flat 1, 32 High Street, Dawlish, EX7 9HP
M6 AYK Denis Moger, 23 Elmsleigh Road, Paignton, TQ4 5AX
M6 AYL Adam Sharam, 30 Heywood Avenue, Maidenhead, SL6 3JA
M6 AYM Patrick Armitage, 250 Abbeydale Road South, Totley Rise, Sheffield, S17 3LL
M6 AYN Alex Taylor, 65 Teign Bank Road, Hinckley, LE10 0ED
M6 AYP Richard Styles, 4 Coningsby Close, Gainsborough, DN21 1SS
M6 AYQ Justin Fudge, 25 Virginia Orchard, Ruishton, Taunton, TA3 5LP
MM6 AYR James McMorland, 382 Maryhill Road, Glasgow, G20 7YQ
M6 AYS Walter Abbott, 12 Yew Tree Gardens, Birchington, CT7 9AJ
M6 AYU Charlie Green, 14 St. Andrews Road, Bletchley, Milton Keynes, MK3 5DR
M6 AYV Andrew Lidster, 51 Plantation Drive, North Ferriby, HU14 3BD
M6 AYW Kieran Reeves, 5 Westby Crescent, Whiston, Rotherham, S60 4EA
M6 AYY Christopher Leviston, 13 Pryors Walk, Askam-in-Furness, LA16 7JG
M6 AYZ Mark Tarrant, Wayside Cottage, Gabber Lane, Plymouth, PL9 0AW
M6 AZA Andrew Adams, 45 Four Oaks Road, Tedburn St. Mary, Exeter, EX6 6AP
M6 AZD Andrew Lines, 26 McIntyre Walk, Bury St. Edmunds, IP32 6PF
M6 AZE Axel Seedig, 8 Barton Court, Cambridge Road West, Farnborough, GU14 6QA
M6 AZF Sharon Case, 136 Old Basin, Bridgwater, TA6 6LJ
M6 AZG Paul Culverwell, 29 Elland Road, Brierfield, Nelson, BB9 5RX
M6 AZH Edward Cross, 12b Oakridge, Three Rivers Country Park, Clitheroe, BB7 3JW
M6 AZK Stephen Tyler, 11 Windmill Cottages, Dilmore Lane, Worcester, WR3 7RX
M6 AZM Penelope Comley-Ross, 48 High Street, Topsham, Exeter, EX3 0DY
M6 AZN George Villiers, 88 Redwald Road, Rendlesham, Woodbridge, IP12 2TE
MW6 AZP Alexander Dighton, 84 Trefelin, Aberdare, CF44 8LF
M6 AZQ MARK OSBAND, 22, SAMIAN CRESCENT, Folkestone, CT194JW
M6 AZR Karl Braisher, 7 Ormond Road, Thame, OX9 3XN
M6 AZS Peter McFadden, Maple Cottage, Great Gap, Leighton Buzzard, LU7 9DZ
M6 AZT Stewart Mason, 8 Barrowby Gate, Grantham, NG31 7LT
M6 AZU Alan Bulman, 21 Stannington Road, North Shields, NE29 7JY
M6 AZV Derek Simpson, 50 Castle Hill, Berkhamsted, HP4 1HF
M6 AZX Neville Robinson, G19, Grange Country Park, Straight Road, Colchester, CO7 6UX
M6 AZY Wayne Millington, 93 Feiashill Road, Trysull, Wolverhampton, WV5 7HT
M6 BAA Bob Gutteridge, 121 Station Road South, Walpole St. Andrew, Wisbech, PE14 7LZ
M6 BAC Ben Warner, 22-23 St. Georges Terrace, Herne Bay, CT6 8RH
M6 BAD Benjamin Barnes-Martin, 145 Farm Road, Barnsley, S70 3DW
M6 BAG Darren Roberts, 27 Nairn Street, Jarrow, NE32 4HX
M6 BAH Stephen Bassett, 5 The Terrace, The Green, Stratford-upon-Avon, CV37 0JD
MI6 BAI Brian Baird, 12 Manse Park, Newtownards, BT23 4TN
MI6 BAJ Andrew Bell, 4 Mount Pleasant View, Newtownabbey, BT37 0ZY
M6 BAK Christopher Baker, 60 Belvedere Road, Danbury, Chelmsford, CM3 4RB
M6 BAM Benjamin Butler, 42 Station Road, Stanbridge, Leighton Buzzard, LU7 9JF
M6 BAN B Banner, 99 Swingate, Kimberley, Nottingham, NG16 2PU
M6 BAQ Brian Williamson, 23 Tower Hamlets Street, Dover, CT17 0DY
M6 BAR Daniel Barry, Coasters, Station Road, Colchester, CO7 8LH
M6 BAS Beth-Ann Sweet, 14 Bryn Celyn, Colwyn Bay, LL29 6DH
MW6 BAU Keith Burgess, 18 Fairmeadows, Maesteg, CF34 9JL
M6 BAV David Woodbine, 29 Compass Tower, Munnings Road, Norwich, NR7 9TW
M6 BAW John Middleton, 17 Woods Loke West, Lowestoft, NR32 3DN
M6 BAX Rodney Baxter, Rose Dene, Hornsby, Brampton, CA8 9HF
M6 BAY Kym Abela, 32 King Edward Road, Gillingham, ME7 2RE
M6 BAZ Barry Pike, 36 Larkfield Avenue, Sittingbourne, ME10 2DP
M6 BBA Jonathan Bennette Alincastre, 90 York Crescent, Durham, DH1 5PT
M6 BBB william kenway, 40 Grove Avenue, Gosport, PO12 1JX
M6 BBC Andrew Bright, 86 Fourth Avenue, Watford, WD25 9QQ
M6 BBD Paul Symonds, 10 Cowper Court, Willunga, Australia, 5172
M6 BBE Elizabeth Puttock, 12 Beechfields, School Lane, Petworth, GU28 9DH
M6 BBF George Symonds, 10 Cowper Ct, Willunga, Australia, 5172
M6 BBG Peter Holland, 30 Knighton Park Road, London, SE26 5RJ
M6 BBH Myles Ramsey, 21 Goldsmith Road, Eastleigh, SO50 5EN
MM6 BBK Paul Robertson, 4 Mona Terrace, Elgin Street, Kirkcaldy, KY2 5HS
M6 BBL Nigel Stanley, 253 Brownley Road, Manchester, M22 9UX
M6 BBM Martin Harrison, 91 Rye Road, Hastings, TN35 5DH
M6 BBO Alan Applegate, 13 Deacons Close, KINGS STANLEY, Stonehouse, GL10 3JA
M6 BBP Andrew Page, 207 Brooklyn Road, Cheltenham, GL51 8DZ
M6 BBQ Alan Smith, 311 Albion Street, Southwick, Brighton, BN42 4AT
M6 BBR Charlie John, 27 Berberis Walk, West Drayton, UB7 7TZ
M6 BBS Peter Hyde, Flat 4, 1 Longhorn Avenue, Gloucester, GL1 2AR
M6 BBT Mark Boswell, 5 Woods Avenue, Marsden, Huddersfield, HD7 6JX
M6 BBU Reginald Boardman, 12 St. Margarets Road, Alderton, Tewkesbury, GL20 8NN
M6 BBX Alexander Wright, Hills Road, Cambridge, CB2 8PH
M6 BBZ Paul Lettington, 21 Bideford Close, Woodley, Reading, RG5 3SE
M6 BCB Ray Balmforth, 33 Lees Hall Road, Dewsbury, WF12 0RH
M6 BCC Andrew Boots, 36a Church Street, Charlton Kings, Cheltenham, GL53 8AR
MM6 BCF Colin Flynn, 6c White Street, Ayr, KA8 9BW
M6 BCG Andrew Pickles, 87a Laburnum Road, Waterlooville, PO7 7EW
M6 BCH Bipin Chauhan, 45 Burnham Drive, Whetstone, Leicester, LE8 6HY
M6 BCJ Jack Bullock, 49 Gallimore Close, Stoke-on-Trent, ST6 4DZ
M6 BCK Thomas Humphries, The Nook, Yeldham Road, Halstead, CO9 3QJ
M6 BCN Peter Kingston, 2 Deepdale, Great Easton, Market Harborough, LE16 8SS
M6 BCQ Paul Chester Chester, 33 Salehurst Road, London, SE4 1AS
M6 BCU Beren Miles, 35 Plantation Drive, Walkford, Christchurch, BH23 5SG
M6 BCV Michael Cooper, 9 Conway Close, Crewe, CW1 3XN
M6 BCW Graham Fearnhead, 27 Lukins Drive, Dunmow, CM6 1XQ
M6 BCY John Edwards, 45 Bramshaw Gardens, Bournemouth, BH8 0BT
M6 BCZ Kevin Percy, 55 Buxton Avenue, Heanor, DE75 7UN
MM6 BDA Bruce Adams, 18 Bellfield Road, North Kessock, Inverness, IV1 3XU
M6 BDD Jeffrey O'Brian, 83 Bramdean Crescent, London, SE12 0UJ
M6 BDG Ben Greenberg, 94 Rivermead Court, Ranelagh Gardens, London, SW6 3SA
M6 BDI Alan Bucknell, 12 Cliveden Grove, Hereford, HR4 0NE
MJ6 BDJ Leslie Langlois, Farleyer, La Rue Des Platons, Jersey, JE3 5AA
M6 BDL Benjamin Little, 25 Thrift Wood, Bicknacre, Chelmsford, CM3 4HT
M6 BDM David Mead, 32 Sherborne Road, Farnborough, GU14 6JT
M6 BDO Antony Butcher, 31 Wittonwood Road, Frinton-on-Sea, CO13 9JZ
M6 BDQ David Nelson, 110 Chandag Road, Keynsham, Bristol, BS31 1QF
M6 BDR Bryan Roberts, 10 Morningside Way, Liverpool, L11 1BD
MW6 BDS Dagmar Bancroft, Stop and Call, Goodwick, SA64 0EX
M6 BDV Barry Vile, 24 Hudson Close, Dover, CT16 2SG
M6 BDX Theresa Scott, 39 Neil Avenue, Holt, NR25 6TG
M6 BEB Simon Smedley, Spring Cottage, Frys Well, Radstock, BA3 4HA
M6 BEC Brian Beckett, 21 Horseshoes Lane, Langley, Maidstone, ME17 1SR
M6 BED Josh Bratchley, Marrowbone House, Calstock Road, Gunnislake, PL18 9BU
M6 BEE Bridget Azzaro, 5 Rye Hill Close, Bere Regis, Wareham, BH20 7LU
MI6 BEF William Tosh, 38 Ballycastle Road, Coleraine, BT52 2DY
MW6 BEG Philip Sherwood, High Croft Jeffreyston, Kilgetty, SA68 0RG
M6 BEH Barry Humphrey, 45 Rose Avenue, Hazlemere, High Wycombe, HP15 7PH
M6 BEI Iskander Whitlock, 109 Sorrell Drive, Newport Pagnell, MK16 8TZ
M6 BEJ James Preston, 25 Hamlet Road, Haverhill, CB9 8EH
M6 BEK Rebecca Clare, Kimberley, Boston Road, Boston, PE20 3AP
M6 BEL Kendra Whitelaw, 18 Marine Drive, Seaford, BN25 2RT
M6 BEM Michael Hooper, Oakview, Oxford Road, Newbury, RG20 8RU
M6 BEN Ben Robinson, 13 Dene View, Ellington, Morpeth, NE61 5HQ
M6 BEQ Craig Bradley, 22 Park Street, Skipton, BD23 1NS
M6 BEU Robert Bartha, 6 Chappell Close, Aylesbury, HP19 9QA
M6 BEX Rebecca Whitehead, 29 Coulsons Road, Bristol, BS14 0NN
M6 BEZ Peter Bailey, 16 Manchester Road, Holland-on-Sea, Clacton-on-Sea, CO15 5PL
M6 BFB Vincent Walsh, 11 Coronation Drive Crosby, Liverpool, L23 3BN
M6 BFC Benjamin Spaxman, 12 Stanhope Gardens, Barnsley, S75 2QB
M6 BFF Keith Rivett, 89 Maidstone Road, Felixstowe, IP11 9EE

M6 BFG Colin Bowman, 26 Albany Hill, Tunbridge Wells, TN2 3RX
MM6 BFH Barry Haynes, 18 Drummond Street, Greenock, PA16 9DN
M6 BFI Abbie Jobbling, 3 Orchard Close, Cranfield, Bedford, MK43 0HX
MI6 BFJ Joseph Cromer, 9 Colinbrook Park, Dunmurry, Belfast, BT17 0N?
M6 BFK Alan Eaton, 18a Summer Street, Leighton Buzzard, LU7 1HT
M6 BFL Keith Goldsworthy, Flat 2, Jordan House, Biggleswade, SG18 8FS
M6 BFM Philip McLaren, 10 Haulfryn, Ruthin, LL15 1HB
M6 BFO Christopher Murphy, 17 Shepherd Street, Littleover, Derby, DE23 6GA
M6 BFP Ronald Roberts, 25 Carlisle Avenue, Bootle, L30 1PX
M6 BFQ Trevor Dodd, 11 Park Meadow, Princes Risborough, HP27 0EB
MM6 BFR Peter Riddle, Carngeal, Pitlochry, PH16 5JL
M6 BFS Nick Grudgings, North Lodge, Templewood Lane, Slough, SL2 3HW
M6 BFU Andrew Elwin, 30 Kingsland Road, Aylesbury, HP21 9SL
M6 BFV Oliver Gledhill, 20 Curtis Way, Kesgrave, Ipswich, IP5 2FX
M6 BFX Steven Cave, Weymess Farm, Park Lane, Banbury, OX17 2RX
M6 BFZ Neil Edwards, 11 Sandringham Road, Eccleston, Chorley, PR7 5SN
M6 BGA Kevin Baker, 27 St. Matthews Close, Cherry Willingham, Lincoln, LN3 4LS
M6 BGB Amandeep Singh Ajmani, Apartment 605, 7 Anchor Street, Ipswich, IP3 0BW
M6 BGC J Moody, 16 Lingcrest, Gateshead, NE9 6SN
MI6 BGD Bertie Gilliland, 28 Baird Avenue, Donaghcloney, Craigavon, BT66 7LP
M6 BGE Simon Saunders, 5 Park Court, Woking, GU22 7NW
M6 BGF Philip Martin, 56 Devonshire Gardens, Bursledon, Southampton, SO31 8HE
M6 BGG Marcelo Chiesa, 56 Waddon Road, Croydon, CR0 4JD
M6 BGH Brian Higgins, 2 Bishops Yard, High Street, Huntingdon, PE28 3JB
M6 BGI Tracey-Ann Biss, 64 Tower Hill, Williton, Taunton, TA4 4JR
M6 BGJ Adam Jones, 62 Butterthwaite Cresent, Sheffield, S5 0DY
M6 BGK Alan Yates, 19 Lauriston Park, Cheltenham, GL50 2QL
M6 BGL Kevin Oliver, 28 King Richards Hill, Earl Shilton, Leicester, LE9 7EY
M6 BGM Paul James, 44 Narbonne Avenue, Eccles, Manchester, M30 9DL
MW6 BGO Donald Lacaman, 6 Falcon Road, Haverfordwest, SA61 2UE
M6 BGP Benjamin Peacock, 17 Herril Ings, Tickhill, Doncaster, DN11 9UE
M6 BGR Clive Weal, 12 Alpine Crescent, The Elms, Lincoln, LN1 2EX
M6 BGS Andrew Somerville, 106 Bush Hill, Northampton, NN3 2PG
M6 BGT James Summerhill, 43 Rangers Walk, Bristol, BS15 3PW
MM6 BGV Allan Timmins, 16 Queens Crescent, Garelochhead, Helensburgh, G84 0DW
M6 BGW Mary Jane Walters, 110 Slade Road, Portishead, Bristol, BS20 6BB
M6 BGZ Terence Chapman, 74 Kidderminster Road, Bewdley, DY12 1BY
M6 BHA Terry Harris, 12 Maple Close, Stourport-on-Severn, DY13 8TA
M6 BHB Roger Smith, Five Elms, Lullington Road, Tamworth, B79 9JA
M6 BHC Allan Tipler, 27 Clumber Street, Hucknall, Nottingham, NG15 7PJ
MW6 BHD David Rees, 49 Fair View, Hirwaun, Aberdare, CF44 9SA
M6 BHE Simon Verity, 29 Patterdale Avenue, Fleetwood, FY7 8NW
M6 BHF Mark Atherton, 39 Fairmead Road, Wirral, CH46 8TU
M6 BHH Geoffrey Hartless, 32 Long Acre, Mablethorpe, LN12 1JF
M6 BHI Ronald Smyth, 7 Hallam Crescent East, Leicester, LE3 1DD
M6 BHJ Benjamin Hall, 1 Lytham Road, Leicester, LE2 1YD
M6 BHK Cameron Lai, Storeys Way, Cambridge, CB3 0DG
M6 BHM Marek Biadon, 57 Fern Hill Road, Oxford, OX4 2JW
M6 BHN Jamie Hunter, 35 Inglefield, Hartlepool, TS25 1RN
MW6 BHO Aled Edwards, Tir Brwyn, Rhydargaeau, Carmarthen, SA33 6BL
M6 BHP Robert Earland, 11 Roseberry Avenue, Great Ayton, Middlesbrough, TS9 6EN
M6 BHQ Lee Copeland, 78 Penderyn Crescent, Ingleby Barwick, Stockton-on-Tees, TS17 5HQ
MM6 BHS Jamie Read, Knockenny Farm, Glamis, Forfar, DD8 1UE
M6 BHT T Scales, 77 Upper Eastern Green Lane, Coventry, CV5 7DA
M6 BHU Paul Chamberlain, 61 Balloch Road, London, SE6 1SP
M6 BHY Michael Worley, 12 Hall Farm Close, Melton, Woodbridge, IP12 1RL
M6 BHZ Patricia Palmer, 2 Dagonet Road, Bromley, BR1 5LR
M6 BIA Robert Kijewski, 14 East Street, Heanor, DE75 7NE
M6 BIB Neil Foster, Four Beeches, Gribthorpe, Goole, DN14 7NT
M6 BID Michael Knights, 3 East View Cottages, Church Road, Woodbridge, IP13 0AT
M6 BIE Julian Lucas, 64 Fitzroy Road, Whitstable, CT5 2LE
M6 BIF Donald Sobey, Flat 2 73 Park Road, Blackpool, FY1 4JQ
M6 BIG david robson, 24 sperrin close, Hull, Hu94af
M6 BIH Gregory Fordyce, 2 Church Street, East End, Earlston, TD4 6HS
MI6 BII Paul Floyd, 9 Killybrack Mews, Omagh, BT79 7FB
M6 BIJ Ivan Jarvis, 6 Mullett Road, Wednesfield, Wolverhampton, WV11 1DD
M6 BIM Harold Salter, La Cachette, LES MALPIERRES, Charroux, France, 86250
MM6 BIO Neil Mackenzie, 59 Plasterfield, Stornoway, HS1 2UR
MM6 BIP Adam Stanley, 40 Coll, Isle of Lewis, HS2 0LP
M6 BIR Bartosz Jakubowski, 120 Chandos Street, Coventry, CV2 4HT
M6 BIS Darren Bisbey, 17 Benson Close, Lichfield, WS13 6DA
M6 BIU Chris Andrews, 34, 34 Russell St, Kettering, N16 0EL
M6 BIV Oscar Hall, 2 Beverley Lodge, Paradise Road, Richmond, TW9 1LL
M6 BIX James Coward, 31 Oxford Road, Hyde, SK14 5QZ
MM6 BIY John Corrigan, 27 Stonecraig Road, Wishaw, ML2 8BZ
M6 BJA Brian Ardrey, 13 Grebe Avenue, St. Helens, WA10 3QL
M6 BJB Kerry Richards, 5 Ramsay Close, Birchwood, Warrington, WA3 6PS
M6 BJC Malcolm Holding, 5 Bowness Avenue, Blackpool, FY4 4TF
M6 BJE Ben Emmerson, 1 Tivydale Drive, Darton, Barnsley, S75 5PG
M6 BJF Benjamin Froggatt, 11 Goldsmith Road, Walsall, WS3 1DL
MI6 BJG Richard Gilmore, 86 Lylehill Road, Templepatrick, Ballyclare, BT39 0HL
M6 BJI Ian Taylor, 111 Kings Road, Lancing, BN15 8EQ
MM6 BJJ Mark Batchelor, Flat 11, 21 Albany Terrace, Dundee, DD3 6HR
M6 BJK Keith Ashton, 23 Wood Close, Wells, BA5 2GA
M6 BJL Michael Blackmore, Timbers, Wolvershill Road, Banwell, BS29 6DG
M6 BJM Matthew Bunting, 31 Hardwick Avenue, Allestree, Derby, DE22 2LN
M6 BJN BRIAN POTTER, 55 LINDSWORTH ROAD, KING'S NORTON, Birmingham, B30 3RP
M6 BJO Kevin Hawkins, 2 Couford Grove, Huddersfield, HD2 1TH
M6 BJP Patricia Blackmore, Timbers, Wolvershill Road, Banwell, BS29 6DG
M6 BJQ Ryan Stokes, 3 Parham Walk, Grange Park, Swindon, SN5 6EQ
M6 BJT Barry Claydon, 45 Riverside, Horley, RH6 7LN
M6 BJW Brian White, 53 Dacre Road, Brampton, CA8 1BN
M6 BJX Charles Thorpe-Morgan, 31 Dinglederry, Olney, MK46 5ES
M6 BJY Matthew Walker, 20 Fernhurst Road, Mirfield, WF14 9LJ
MI6 BJZ Paul Donnelly, 43 Ashfield Gardens Fintona, Omagh, BT78 2DD
MI6 BKA Macauley Hynes, 81 Furze Close, Peatmoor, Swindon, SN5 5DU
MM6 BKB Paul Blauper, 1 Lindertis Cottages, Kirriemuir, DD8 5NT
MI6 BKD John Tierney, 48 Ashfield Gardens Fintona, Omagh, BT78 2DD
M6 BKE Colin Opie, 354 Beaumont Road, Plymouth, PL4 9EN
M6 BKF Cathryn Lane, 47 Glenarm Road, London, E5 0LY
M6 BKH Douglas Storer, 13 The Square, Lower Burraton, Saltash, PL12 4SH
M6 BKI Steven Shaw, 2 Daleside, Todmorden, OL14 7NE
M6 BKJ Pennie Crow, 31 Lakeside, Overstone Park, Northampton, NN6 0QS
M6 BKK Ben Tompkins, 63 Chapnall Road, Wisbech, PE13 3TU
M6 BKL Kenneth Jackson, 4 Milfoil Close, Marton-in-Cleveland, Middlesbrough, TS7 8SE
M6 BKM Darren Turnbull, 63 Brecklands, Mundford, Thetford, IP26 5EG
M6 BKN Stephen Walker, 73 Sunnybank Road, Halifax, HX2 8RL
M6 BKO John Legrain, 22 Cromwell Drive, Didcot, OX11 9RB
MM6 BKP Robert Adamson, 6 Camdean Crescent, Rosyth, Dunfermline, KY11 2TJ
MM6 BKQ Samuel Burnside, Woodend Farm, Buchlyvie, Stirling, FK8 3PD
M6 BKS Neil Adam, Tan Ffordd, Mynydd Llandygai, Bangor, LL57 4LX
M6 BKT Sam Keeble, 38 Sandford Rise, Sandy, SG19 1ED
M6 BKU Bryan Cox, 7 Wolsey Avenue, London, E6 6HG
M6 BKV Michael Archer, 4 The Bungalows, Mill Lane, Grays, RM20 4YD
M6 BLA Jennifer Forder, 157 Kennington Road, Kennington, Oxford, OX1 5PE
M6 BLB Julian Sell, 30 Plantation Close, Saffron Walden, CB11 4DS
M6 BLC Byron Cripps, 215 Bournemouth Road, Poole, BH14 9HU
M6 BLE Stewart Robertson, 43 Kindar Drive, New Abbey, Dumfries, DG2 8DA
M6 BLF Michael Featherstone, 62 Poles Hill, Chesham, HP5 2QR
M6 BLG Thomas Clayton, 4 Kingscote Close, Nine Elms, Swindon, SN5 5UP
M6 BLH Richard Pringle, 14 Marjorie Street, Cramlington, NE23 6XQ
M6 BLI Steven Algar, 17 Riseway Close, Norwich, NR1 4NJ
M6 BLK Paul Blake, 70 Front Street, South Hetton, Durham, DH6 2RG
M6 BLM Kenneth Gallop, Airedale House, Airedale Drive, Castleford, WF10 2QA
M6 BLN Martin Halliday, 1 Everard Close, Bury St. Edmunds, IP32 6RU
M6 BLO Louis Monshall, 7 Sweden Close, Harwich, CO12 4JU
M6 BLP Bernard Pearn, 230 Lloyds Avenue, Kessingland, Lowestoft, NR33 7TU
M6 BLQ Nigel Fahey, 5 Hillside, Felmingham, North Walsham, NR28 0LE
M6 BLS Tabraze Malik, Flat 19, Edinburgh House, Crawley, RH11 9BZ
M6 BLT William Thomson, 72 Hurstwood Avenue, Bexleyheath, DA7 6SG
M6 BLV John Moore, 53 The Boulevard, Great Sutton, Ellesmere Port, CH65 7DX
MI6 BLW Eammon MacGra, 12c Glenabbey Drive, Londonderry, BT48 8SU
M6 BLX Robert Bowles, 7 Vineside, Gosport, PO13 0ZU
MM6 BLY William Robb, 18 Buckie, Erskine, PA8 6EE
M6 BLZ James Blezard, 10 North Row, Barrow-in-Furness, LA13 0HE
M6 BMB John Bell, 6 Highfields, Fetcham, Leatherhead, KT22 9XA
MI6 BMC William McCormick, 6 Church Street, Rosslea, Enniskillen, BT92 7DD
M6 BME Paul Toal, 35 Black Lane, Whiston, Stoke-on-Trent, ST10 2JQ
M6 BMF Anthony Ledger, 23 Wentworth Park, Freshbrook, Swindon, SN5 8QX
M6 BMI Peter Davies, 2 Lynfords Drive, Runwell, Wickford, SS11 7PP
M6 BMJ Ben Adamson, 21 The Hatches, Frimley Green, Camberley, GU16 6HG
M6 BMK Paul Hardy, 50 Harridge Road, Leigh-on-Sea, SS9 4HA
MW6 BMM William Murphy, 148 Caergynydd Road, Waunarlwydd, Swansea, SA5 4RE
M6 BMN Derek Hagan, 8 Charles Close, Westcliff-on-Sea, SS0 0EU
M6 BMO Billy Morris, 131 Littlehampton Road, Worthing, BN13 1QX
MM6 BMP Alexander Moerman, 11 Cupar Road, Kettlebridge, Cupar, KY15 7QD
MM6 BMQ John Bannerman, 77/2 Park Avenue, Edinburgh, EH15 1JP
M6 BMT Brendan Thistlethwaite, 29 Lamorna Drive, Callington, PL17 7QH
M6 BMV Richard Gocher, 20 Mulberry Crescent, South Shields, NE34 8DD
M6 BMW Brian Martin, 11 Alpha Street, Toll Bar, Doncaster, DN5 0RA
M6 BMY Syed Maqbool, 69 Waltham Close, West Bridgford, Nottingham, NG2 6LD
M6 BMZ John Kay, 36 Winnington Road, Marple, Stockport, SK6 6PT
M6 BNA Anthony Piper, 3 Bakers Court, Bakers Court Lane, Lynton, EX35 6EW
M6 BNB James Duffy, 18 Burnmoor Road, Bolton, BL2 5NH
M6 BNC Brian Chambers, 13 Cherry Tree Crescent, Walton, Wakefield, WF2 6LQ
M6 BNF Sarah Morgan, Holly Cottage, Old Racecourse, Oswestry, SY10 7PQ
M6 BNG Richard Watson, 8 Bourne Close, Warminster, BA12 9PT
M6 BNH Graham Thomas, 4 Oak Tree Close, Buckley, CH7 3JU
M6 BNJ Gary Hartley, 1 Manor View, Shafton, Barnsley, S72 8NQ
M6 BNK Michael Wilsher, 70 Norris Road, Blacon, Chester, CH1 5DZ
M6 BNL David Beck, 94 Shaldon Crescent, Plymouth, PL5 3RB
M6 BNM James Sanders, 76 Fullerton Road, Plymouth, PL2 3AX
MM6 BNN Owen Mckenzie, 179 Gordons Mills Road, Aberdeen, AB24 2XS
MM6 BNO Karen Mckenzie, 179 Gordons Mills Road, Aberdeen, AB24 2XS
M6 BNP Daniel Brookes, 177 Charnwood Close, Rubery, Birmingham, B45 0JY
M6 BNQ Christopher Marshall, 51 Hedgerow Close, Redditch, B98 7QF
MM6 BNS Tim Donaldson, Brawview Cottage, Maybole, KA19 8EN
M6 BNT Lawrence Waller, 1 Anne Arundel Court, Heathhall, Dumfries, DG1 3SL
M6 BNU Stephen Ditchburn, 10 St. Hilda Avenue, Waterlooville, PO8 0JF
M6 BNV Itziar Balboa, 7 Edenhall Close, Tilehurst, Reading, RG31 6RH
M6 BNW Anthony New, 41 White Lodge Close, Wirral, CH62 0DN
M6 BNX Wayne Evans, Ninewar Farm, Duns, TD11 3PP
M6 BNY J Beeney, 17 Norton Avenue, Herne Bay, CT6 7TA
M6 BOA Timothy Cocks, 9 Mountfield Way, Westgate-on-Sea, CT8 8HR
M6 BOB Robert Manning, 15 Gurston Rise, Northampton, NN3 5HY
MW6 BOC Gerald Williams, 36 Park Street, Taibach, Port Talbot, SA13 1TD
MM6 BOD David Lightbody, Glenorchy, Brownrigg Road, Falkirk, FK1 3BA
M6 BOE Neil Marley, Penstemons, Chapel Lane Pen Selwood, Wincanton, BA9 8LY
M6 BOF Bruce Savage, 80 Sky End Lane, Hordle, Lymington, SO41 0HG
M6 BOI Kyle Mecca, 38 Abbots Road, Faversham, ME13 8DE
M6 BOJ Jeffrey Savage, Rufford, Barnes Lane, Lymington, SO41 0RR
M6 BOK Vic Casambros, 5 Roman Way, Folkestone, CT19 4JS
M6 BOP D Holland, 7 Hayward Close, Walkington, Beverley, HU17 8YB
M6 BOQ Alan Hill, 2 Liddon Road, Chalgrove, Oxford, OX44 7YH
MM6 BOQ Leigh-Ann Mitchell, CARRADALE, YONDERTONHILL HATTON, Peterhead, AB42 0RE
M6 BOR Gyles Wren, Ashleigh, Yew Tree Hill, Matlock, DE4 5AR
MM6 BOS Adrian Manson, Clochcan Schoolhouse, Auchnagatt, Ellon, AB41 8UJ
M6 BOT Tadas Gedvygas, 9 The Mill, Kirton, Boston, PE20 1LB
MW6 BOW Trevor Bowen, 7a Heol Maes y Cerrig, Loughor, Swansea, SA4 6SW
M6 BOX Christopher Pegrum, 3 Bretland Road, Tunbridge Wells, TN4 8PS
M6 BOY Colin Butler, 210 Green Wrythe Lane, Carshalton, SM5 2SP
M6 BOZ Elizabeth Bosley, 42 Snowdon Street, Porthmadog, LL49 9DF
M6 BPA Anna Parnell-Brookes, 14 Greens Close, Hullavington, Chippenham, SN14 6EG
M6 BPB Keith Furlong, 3 Oak Meadow, South Molton, EX36 4EY
M6 BPC Gary Colson, 3 Dartford Road, Dartford, DA1 3EE
M6 BPD David Taylor, 143 Sandhurst Road, London, SE6 1UR
M6 BPE Paul Skinner, 27 Westcots Drive, Winkleigh, EX19 8JW
M6 BPG Benjamin Patrick-Gleed, 6 Julius Way, Lydney, GL15 5QS
M6 BPH Brian Parsons, 29 Hillside Crescent, Buckley, CH7 2JS
M6 BPI William Colquhoun, 68 The Fairway, Bromley, BR1 2JY
M6 BPJ Bruce Saunders, 88 Bramwoods Road, Chelmsford, CM2 7LT
M6 BPK James Johnson, 4 Wallace Close, Hullbridge, Hockley, SS5 6NE
M6 BPL Mark Lefton, 35 Hawkstone Avenue, Whitefield, Manchester, M45 7PR
M6 BPM Ian Wightman, 13 Gillsland, Eyemouth, TD14 5JF
M6 BPO Richard Extrance, 1 Morngate Caravan Park, Bridport Road, Dorchester, DT2 9DS
M6 BPQ John Chatterton, 6 Bayliss Road, Wargrave, Reading, RG10 8DR
M6 BPR Keith Butt, 15 Hamble Park, Fleet End Road, Southampton, SO31 9JU
M6 BPS Paul Dent, 33 Cavalier Close, Dibden, Southampton, SO45 5TU
M6 BPU Clive Shepherd, 1 Holley Park, Okehampton, EX20 1PL
M6 BPV John McRobie, 6 Southill Gardens, Bournemouth, BH9 1SJ
M6 BPW Bonnie Pao Nan Wang, Harrogate Ladies' College, Clarence Drive, Harrogate, HG1 2QG
M6 BPX Jason Revell, 37 Tennyson Street, Goole, DN14 6EB
MM6 BPY Lijo Joseph, 3/4 60 Wilson Street, Glasgow, G1 1HD
MW6 BQA Richard Staples, 16 Wellington Street, Aberdare, CF44 8EW
M6 BQB Richard Babb, 15 Sylvan Lane, Hamble, Southampton, SO31 4QG
M6 BQC Mervyn Powis, 11 Temeside Estate, Ludlow, SY8 1JR
M6 BQD Harvey Brewster, 1 Pinewood Drive, Camblesforth, Selby, YO8 8JU
M6 BQE Mark Simpson, 32 Underhill Lane, Wolverhampton, WV10 8NS
M6 BQF George Evans, 13 Lydgate Road, Sale, M33 3LW
MM6 BQG Barry Wetton, Tigh Air Achnoc, West Helmsdale, Helmsdale, KW8 6HH
MM6 BQH Kenneth Monaghan, 10 Sauchiewood Cottages, Mintlaw, Peterhead, AB42 5LR
M6 BQJ Ross Thompson, 16 West Leys Court, Moulton, Northampton, NN3 7UB
M6 BQK Chris Holmes, 10 Southampton Street, Farnborough, GU14 6AX
M6 BQL John Griffiths, Llain Bach, Beach Road, Porthmadog, LL49 9YA
M6 BQM Kenneth Williamson, 1 Rhug Gardens, Corwen, LL21 0EH
M6 BQN Thomas Gilmour, 83 Billington Road, Leighton Buzzard, LU7 4TG
M6 BQO William Martin, 54 Merritt Road, Greatstone, New Romney, TN28 8SZ
M6 BQP Jase Griffiths, 7 Tynedale Road, Blackpool, FY3 7UE
M6 BQQ Steven Murray, 79 Nightingale Road, Liverpool, L12 0QN
M6 BQR Thomas Walton, 72 Burleigh Road, Frimley, Camberley, GU16 7EB
M6 BQS Richard Brabazon, 15 Albany Road, St. Leonards-on-Sea, TN38 0LP
M6 BQV Adam Brackpool, 2 Stocks Fields, Stocks Hill, Steyning, BN44 3DU
M6 BQW Kevin Jeffery, 9 Gordon Road, Tunbridge Wells, TN4 9BL
M6 BQZ Muhammad Iqbal, 6 Hobart Road, High Wycombe, HP13 6UD
M6 BRB Leslie Bethell, 30 Finger Road, Dawley, Telford, TF4 3LB
MW6 BRF Simon Phillips, 58 Taff Embankment, Cardiff, CF11 7BG
M6 BRH Simon Plummer, 2 Langdale Avenue, Outwood, Wakefield, WF1 3TX
M6 BRI Brian Millard, 19 Snowdrop Close, Abbeymead, Gloucester, GL4 4DZ
M6 BRJ Robert Green, 44 Aldwyn Place, Larchwood Drive, Egham, TW20 0RZ
M6 BRK Anthony Freshwater, 90a New Road, Minster on Sea, Sheerness, ME12 3PT
M6 BRL Mark Burrell, 16 Atholl, Ouston, Chester le Street, DH2 1RS
M6 BRN Mark Brundrit, 144 Reginald Road, Southsea, PO4 9HP
M6 BRO Jason Smith, 38 The Vineries, Burgess Hill, RH15 0NF
M6 BRP Stephen Fiske, 12 Meliden Crescent, Bolton, BL1 6AJ
M6 BRQ Adam Clements, 49 Canberra Court, West Avenue, Huntingdon, PE26 1EY
M6 BRR Stephane Ray, 28 Stenbury View, Wroxall, Ventnor, PO38 3DB
M6 BRS Denis Glover, 11 Collbrook Avenue, Odsal, Bradford, BD6 1HL
M6 BRT Albert Williams, 101 Horsebridge Hill, Newport, PO30 5TL
M6 BRU Andrew Dean, 14 Harvest Close, Worsbrough, Barnsley, S70 5AY
MM6 BRV David Drysder, 37 Farburn Drive, Stonehaven, AB39 2BZ
M6 BRY Bryan Allen, 13 Woodgrove Road, Rotherham, S65 3RW
M6 BRZ Bernard Ager, 6 South Dibberford Farm, Beaminster, DT8 3HD
M6 BSA Christopher Jacobs, Flat 33, The Lodge, Waterlooville, PO7 8BX
M6 BSB Sam Hamill, 3 Pear Tree Croft, Brede, Rye, TN31 6EJ
M6 BSC Ben Sewell, 12 Haylands Square, South Shields, NE34 0JB
M6 BSG Beverley Garton, 13 Damaskfield Road, Lyppard, Kettleby, WR4 0HY
M6 BSI Adam Butler, 50 Scafell Way, West Bromwich, B71 1DQ
MM6 BSK Ian Bannerman, 39 Muirkirk Drive, Glasgow, G13 1BZ
M6 BSL Tayfun Bilsel, 16 Hide Close, Sawston, Cambridge, CB22 3UR
M6 BSO Bradley Scott, 6 Congleton Edge Road, Congleton, CW12 3JJ
M6 BSP Barry Smith, 28 Newhill Road, Wath-upon-Dearne, Rotherham, S63 6JY
M6 BSQ Paul Bull, 87 Braemor Road, Calne, SN11 0DU
MW6 BSR Ben Sturgess, 22 Heol Pant y Deri, Cardiff, CF5 5PL
M6 BSS James Chambers, 10 Derwent Street, Astley, Manchester, M29 7AT
M6 BST Thomas Hughes, 1 Sunnybank Road, Astley, Manchester, M29 7BJ
M6 BSU Anthony Buckley, Cliff House, Sudbrook, Caldicot, NP26 5TB
M6 BSV Christopher Hayes, 7 Hadstock Close, Sandiacre, Nottingham, NG10 5LQ
M6 BSW Barry Stone, 27 Mountbatten Close, Stretton, Burton-on-Trent, DE13 0FD
M6 BSX John Cummins, Flat 30, St. Giles, Moor Hall Lane, Chelmsford, CM3 8AR
M6 BSY Daniel Carter, 36 Sanderling Drive, Leigh, WN7 1HU
M6 BSZ Chris Batchelor, 8 Howards Court, Stevenage, SG1 3DF
M6 BTA Leslie Brown, 28 Farley Way, Stockport, SK5 6JD
M6 BTB Robert Dalley, Birchley, Pine Avenue, Camberley, GU15 2LY
MW6 BTC Vincent Kennedy, Gerynant, Felin Ban Farm Estate, Cardigan, SA43 1PG
M6 BTE Bradley Ellerton, 4 Darfield Avenue, Owlthorpe, Sheffield, S20 6SU
M6 BTF Pradthana Likitplug, Harrogate Ladies' College, Clarence Drive, Harrogate, HG1 2QG
M6 BTG Malcolm Mutkin, 13 The Grove, Radlett, WD7 7NF
M6 BTH Dwayne D'Souza, 3b Friend Street, London, EC1V 7NS
M6 BTK Joseph Barton, 18 Jeffrey Avenue, Longridge, Preston, PR3 3TH
M6 BTM Benedict Gillett, 65 Kingsway, Wallasey, CH45 4PN

**IMPORTANT NOTE**

**Revalidate licence to avoid revocation** – Ofcom has advised the Society that plans will be drawn up to revoke licences that have not been revalidated as required by the licence conditions. The quickest way to revalidate is to do so online via the Ofcom website: *https://services.ofcom.org.uk/* or by email: *amateur.validations@ofcom.org.uk* Ofcom staff are available to help, but please be patient during times of heavy workload.

M6 BTN Robert Nicholson, 57 Barnbridge, Tamworth, B77 1DF
M6 BTP Paul Wilmot, 12 Brierholme Close, Hatfield, Doncaster, DN7 6EH
M6 BTQ Christopher Manning, 12 Whitehill Close, Camberley, GU15 4JR
MM6 BTR John Rayne, 8 Bankton Grove, Livingston, EH54 9DW
M6 BTS Brendan Seward, 21 Chapel Close, Gunnislake, PL18 9JB
M6 BTT Philip Moye, 13 Post Mill Gardens, Grundisburgh, Woodbridge, IP13 6UP
M6 BTY Patrick Matthews, The Old Chapel, High Street, Huntingdon, PE28 0PF
M6 BUA Harry Bennett, 59 Scott Road, Bishop's Stortford, CM23 3QN
M6 BUB Rebecca Hughes, 86 Colinmander Gardens, Ormskirk, L39 4TF
M6 BUC Jack Chambers, 2 Farm Cottages, Magdalen Laver, Ongar, CM5 0ES
M6 BUD Jeremy Woodland, 14 Kelham Green, Nottingham, NG3 2LP
M6 BUE Jake Shepherd, 12 Glebelands, Harlow, CM20 2PA
M6 BUF John Hurst, 1 Castle Cottage, Aberedw, Builth Wells, LD2 3UL
M6 BUG James Millar, The Kennels, Lanton, Jedburgh, TD8 6SU
MW6 BUH Ivan Baylis, 248 Trebanog Road, Porth, CF39 9EL
M6 BUI Derek Sewell, 19 St. Leonards Way, Ashley Heath, Ringwood, BH24 2HS
M6 BUJ Garry Sheppard, 32 Bramble Drive, Hailsham, BN27 3EG
M6 BUL Joshua Ball, Conifers, Main Road, Spilsby, PE23 4BY
M6 BUR Tom Skinner, 13 Sawbrook, Fleckney, Leicester, LE8 8TR
M6 BUS george robinson, 91 Tilstock Crescent, Shrewsbury, SY2 6HH
M6 BUU D Barrow, 7 Church Lea, Burton Lazars, Melton Mowbray, LE14 2UB
M6 BUW Stephen Bishton, 18 Galloway Road, Poole, BH15 4JX
M6 BUX Paul Kerr, 35 Coppice Gardens, Stone, ST15 8BL
M6 BUY Charlie Marlow, 59 Purford Green, Harlow, CM18 6HN
M6 BVC Christopher Norwood, 29 Wrockwardine Road, Wellington, Telford, TF1 3DA
M6 BVD Ronald Owen, 4 Aldersleigh Drive, Stafford, ST17 4RY
M6 BVE Albert Hamilton, 56 Wyvern, Telford, TF7 5QH
MW6 BVG John Campbell, 1b Bush Road, Mountain Ash, CF45 3BY
M6 BVI William White, 15 St. Walstans Road, Taverham, Norwich, NR8 6NF
M6 BVK Gareth Kennedy-Brown, 4 Seymour Gardens, Brockley, London, SE4 2DN
M6 BVM Simon Baynton, 50 Briton Way, Wymondham, NR18 0TT
MI6 BVN Peter McCullagh, 6 Striff Lane, Omagh, BT79 0WA
M6 BVP Brendon Hull, 3 Cavalry Crescent, Eastbourne, BN20 8NT
M6 BVQ Ryan Unsworth, 20 Arran Drive, Rhyl, LL18 2NS
MW6 BVR Jack Cowles, 76 Pendine Close, Barry, CF62 9DE
M6 BVS Benjamin Watkins, 41 Kingshill Avenue, St. Albans, AL4 9QH
M6 BVU Matthew Ryall, 2 Greenacres, Trinant, Newport, NP11 3AE
M6 BVV John Morgan, Cedars, Springhill, Abingdon, OX13 5HL
M6 BVW David Edwards, 29 Larch Road, Maltby, Rotherham, S66 8AZ
M6 BVX Andrew Booth, 32 Acacia Avenue, Maltby, Rotherham, S66 8DS
M6 BVZ Robert Ede, 14 Elm Close, Kidsgrove, Stoke-on-Trent, ST7 4HR
M6 BWB William Beston, 79 Priestlands, Romsey, SO51 8FJ
M6 BWC Wayne Tunstall, 89 Lever Street, Little Lever, Bolton, BL3 1BA
M6 BWE Gerald Brown, 51 Arncliffe Drive, Knottingley, WF11 8RH
M6 BWF Robert Leverington, 130 Osborn Road, Barton-le-Clay, Bedford, MK45 4NY
M6 BWG Allan Madden, 51 Valley View, Cwmtillery, Abertillery, NP13 1JE
M6 BWH Vijayalatha Venugopalan, 16 Willowfield Drive, Stoke-on-Trent, ST4 8FR
M6 BWJ Liam Wale, 4 Essex Gardens, Market Harborough, LE16 9JS
M6 BWK Steven Casey, Flat 4, Belmont Court, Plymouth, PL3 4DN
M6 BWL Zacharia Burningham, 255 Welland Park Road, Market Harborough, LE16 9DP
M6 BWN Nicholas Bown, 18a Warley Hill, Warley, Brentwood, CM14 5HA
M6 BWO Martin Walters, 65 Bannawell Street, Tavistock, PL19 0DP
M6 BWP Patrick Walsh, 181 Hermes Close, Hull, HU9 4DR
M6 BWQ Andrew Carden, Hazelgrove, South Allington, Kingsbridge, TQ7 2NB
MW6 BWR Peter Smith, 19 Grandison Street, Neath, SA11 2PG
M6 BWV Amanda Forber, 32 Larch Avenue, Newton-le-Willows, WA12 8JF
M6 BWW Shane Laurence, 23 West Street, Bridlington, YO15 3DX
M6 BWZ STEPHEN THIRLWALL, 2 Crossfield Avenue Blythe Bridge, Stoke-on-Trent, ST11 9PL
M6 BXA Jack Lambert, Birchwood Norwich Road, Cromer, NR27 0HG
M6 BXB Mick Stopper, 29 Daisy Dale, Boston, PE21 6DS
M6 BXD Paul Hargreaves, 9 Croston Road, Lostock Hall, Preston, PR5 5LA
M6 BXF Damian Hook, The Delvine, Amisfield, Dumfries, DG1 3LH
MM6 BXH Jamie Curran, 355d Charleston Drive, Dundee, DD2 4HP
M6 BXI Richard Briant, Talarvor, Llanon, SY23 5HG
M6 BXJ Mark Stillman, 58 Highfield Road, Bognor Regis, PO22 8PH
M6 BXK John Street, 22 Roman Acre, Wick, Littlehampton, BN17 7HN
M6 BXL James Lugsden, 21 Overhill Way, Beckenham, BR3 6SN
M6 BXM Hylton Phillips, Flat 10, Fitch Court, 59-63 Effra Road, London, SW2 1DD
M6 BXN Claus Barthel, 176 Lumb Lane, Droylsden, Manchester, M43 7LJ
M6 BXO Kevin Cartwright, 53 Sedgley Road, Dudley, DY1 4NE
M6 BXP Brett Pearson, 9 Dunbar Close, Kidderminster, DY10 3XS
MM6 BXQ Adam Main, 1 Sunnyside Drive, Portlethen, Aberdeen, AB12 4LZ
MM6 BXR Darrell Reid, 111 Oswald Road, Ayr, KA8 8NX
M6 BXS Alan Parker, 9 Milecastle Court, Newcastle upon Tyne, NE5 2PA
MM6 BXT Brian Templeton, 43 Elm Park, Ardrossan, KA22 7BZ
M6 BXU Boruo Xu, 138 King's College, Cambridge, CB2 1ST
M6 BXV Christopher Astbury, Old Police House, Llanegryn, Tywyn, LL36 9SL
M6 BXW Kenneth Morris, 95 Murrayfield Drive, Wirral, CH46 3RR
M6 BXX Jonathan Baker, 29 Ravensmoor Close, North Hykeham, Lincoln, LN6 9AZ
M6 BXY Phillip Arnold, 20 Upper Seagry, Chippenham, SN15 5EX
M6 BXZ Jacek Walczak, 18 Heathfield, Chippenham, SN15 1BQ
M6 BYA Kevin Stowe, 7 Glebe Way, Corsham, SN13 9UL
M6 BYE Thomas Byers, 1 Hazelwood Avenue, Sunderland, SR5 5AH
M6 BYF Dean Rogers, 5 Semple Gardens, Chatham, ME4 6QD
M6 BYH Keith Broscomb, Flat 34, White Willows, 70 Dyche Road, Sheffield, S8 8DS
MM6 BYJ Marcin Stypka, 28/9 Halmyre Street, Edinburgh, EH6 8QD
M6 BYL William Charlton, 2 Mallory Street, Earl Shilton, Leicester, LE9 7PH
M6 BYM Thomas Flynn, 15 Dale Garth, Scarborough, YO12 5NB
M6 BYN Richard Jones, 62 Nathaniel Road, Long Eaton, Nottingham, NG10 1GB
M6 BYO Stephen Downe, 9 Danesway, Exeter, EX4 9ES
M6 BYP Paul Hoyle, Flat 3, 84 Beverley Road, Hull, HU3 1YD
M6 BYQ Mohammed Abid, 16 Cliff Gardens, Scunthorpe, DN15 7PJ
M6 BYR Jonathan Byrne, 316 Turncroft Lane, Stockport, SK1 4BP
M6 BYS Adrian Marriott, 81 Victoria Road, Pinxton, Nottingham, NG16 6NH
M6 BYT Geraint Pierce, Ley Farm, Green Lane, Halton, Wrexham, LL14 5BG
M6 BYU Graham Webster, 15 Bridge Road, Chichester, PO19 7NW
M6 BYV Robert Potts, 4 Maes Glan, Rhosllanerchrugog, Wrexham, LL14 2DT
M6 BYX Neil Williams, 60 Denbigh Close, Wrexham, LL12 7TW
M6 BYY Simon Bradley, 9 Crofton Road, Southsea, PO4 8NX
M6 BYZ Gordon Flinn, 38 Fir Grove, Whitehill, Bordon, GU35 9ED
M6 BZA David Rudling, Rose Cottage, Ludwells Lane, Southampton, SO32 2NP
M6 BZB Yiyi Zhan, 65a Mason Street, Edge Hill, Liverpool, L7 3EN
MI6 BZC Richard Benko, 23 Six Mile Water Mill Drive, Antrim, BT41 4FG
M6 BZD P Elsey, 62b Coleraine Road, London, SE3 7PE
M6 BZE Berkley Evelyn, 19 Laudsdale Road, Rotherham, S65 3LG
M6 BZF Keith Cunningham, 18 Dovenby Fold, Ince, Wigan, WN2 2PS
M6 BZG Robert Paddock, 1 Harris Road, Bexleyheath, DA7 4QD
M6 BZH Ronald Packman, 42 Shad Thames, London, SE1 2YD
MI6 BZI Charles McCormick, Flat 4, Legacorry House, Main Street, Armagh, BT61 9RW
M6 BZK Richard Boyes, 63 Larch Road, New Ollerton, Newark, NG22 9SX
M6 BZM Kathleen Slater, 63 29th Avenue, Hull, HU6 8DG
M6 BZN Graham Slater, 63 29th Avenue, Hull, HU6 8DG
MW6 BZP Matthew Clifford, 27 Primrose Street, Tonypandy, CF40 1BW
MM6 BZQ Alan Catterall, Asta House, Scalloway, Shetland, ZE1 0UQ
M6 BZT Andrew Smith, 32 Cotswold Drive, Rothwell, Leeds, LS26 0QZ
M6 BZU Edward St Quinton, Mill Cottage, The Thorofare, Woodbridge, IP13 8BB
M6 BZV Samuel Barker, Strait Hey Farm, Stock Hey Lane, Todmorden, OL14 6HB
M6 BZW Frederick Phillips, 34 Park Drive, Maldon, CM9 5JQ
MW6 BZX James Palmer, Clettwr Hall, Pontshaen, Llandysul, SA44 4TU
M6 BZY Inga Kazlauskaite, 4 Spencer Way, Redhill, RH1 5LY
M6 CAA Christopher Sparrey, 166 Abberley Avenue, Stourport-on-Severn, DY13 0LT
MI6 CAD Andrew Cahalan, 131 The Meadows, Randalstown, Antrim, BT41 2JD
M6 CAE Christopher Catley, 40 Rockvilla Close, Varteg, Pontypool, NP4 7QF
M6 CAF Christopher Furlong, 22 Swaisland Road, Dartford, DA1 3DE
M6 CAG Micheal Sinclaire, 38 Ford Court, Winsford, CW7 1NJ
M6 CAH Brian Heath, 108 Cow Lane, Bramcote, Nottingham, NG9 3BB
M6 CAI Craig Ingamells, 2 St. Mary's Drive, Sutterton, Boston, PE20 2LU
M6 CAJ Paul Lewickyj, 37 Maple Street, Lincoln, LN5 8QS
MI6 CAK Colm Doyle, 13 Tobin Park, Cookstown, BT80 0JL
MW6 CAN Colin Davies, 70 Farm Drive, Port Talbot, SA12 6TF
M6 CAP Carl Preece, 14 Dock Street, Widnes, WA8 0QX
M6 CAQ Christopher Allen, 43 Leyside, Christchurch, BH23 3RE
M6 CAR Carrie-Ann Taylor, 212 Plantation Hill, Worksop, S81 0HD
M6 CAS shaun hill, 1 Beresford Road, Walsall, WS3 1JX
M6 CAT David Welford, 24 Hawthorn Crescent Quarrington Hill, Durham, DH6 4QW
MI6 CAV Clara McCammick, 23 Atkinson Avenue, Portadown, Craigavon, BT62 3HY
M6 CAW Clive Waldron, 2 The Bourne, Eastleach, Cirencester, GL7 3NN
MI6 CAY Mark Mullaney, 21 Aghagay Meadows, Aghagay, Enniskillen, BT92 8AE
M6 CBA Thomas Ingleby, 27 Burley Wood Lane, Leeds, LS4 2SU
M6 CBC Carl Kent, 19 Coppice Rise, Harrogate, HG1 2DP
M6 CBD Charlotte Arblaster, 22 Wood Lane, Carlton, Barnsley, S71 3JJ
MM6 CBE Colin Ellison, 1 Newton Road, St. Fergus, Peterhead, AB42 3DD
MI6 CBG Ciaran Gorman, 5 Linden Gardens, Bangor, BT19 6EB
MM6 CBH Colin Smith, 3 Distillery Cottages, Aberfeldy, PH15 2EB
M6 CBH Colin Hunt, 105 Worlds End Lane, Weston Turville, Aylesbury, HP22 5RX
MM6 CBI Claire Addison, 8d Thomson Street, Johnstone, PA5 8RZ
M6 CBJ Chris Jerome, Kennel Cottage, Hobland Road, Great Yarmouth, NR31 9AR
MW6 CBL Shawn Morgan, 31 Church Street, Briton Ferry, Neath, SA11 2JG
M6 CBM Liam Clark, 18 Cutters Close, Narborough, Leicester, LE19 2FY
M6 CBN Camille Nunnen, 16 Meden Avenue, Warsop, Mansfield, NG20 0PS
M6 CBO Alan Carter, 18 Maynard Terrace, Clutton, Bristol, BS39 5PL
M6 CBP Susan Bennett, 154 Humberstone Road, Grimsby, DN32 8HR
M6 CBQ Andrew Cranston, 19 Chatsworth Crescent, Ipswich, IP2 9BS
M6 CBU William Porter, 92 Turner Road, Tonbridge, TN10 4AJ
M6 CBY Barry Eames, 22 Ashgrove Close, Hardwicke, Gloucester, GL2 4RT
M6 CCA Marc Landon, 29 Portland Road, Hucknall, Nottingham, NG15 7SL
M6 CCB Geoff Winchester, 35 Hammond Court, Preston, PR1 7LL
M6 CCE Christopher Etchells, 7 Woodlands Drive, Sandford, Wareham, BH20 7QA
M6 CCF Christopher Finbow, Medlars Cottage, The Street, Woodbridge, IP13 7JP
MW6 CCG Steven Williams, 1 Lawrence Terrace, Llanelli, SA15 1SW
M6 CCH Alan Fisher, 63 Soloman Drive, Bideford, EX39 5XY
M6 CCI Finn Roberts, 17 Ansisters Road, Ferring, Worthing, BN12 5JG
M6 CCJ Jacob Davidge, 45 Ferring Street, Ferring, Worthing, BN12 5JW
M6 CCM David James, 123 Bruce Road, Woodley, Reading, RG5 3DY
M6 CCN Martin Tucker, 182 Salisbury Road, Amesbury, Salisbury, SP4 7HW
MM6 CCO William Hannah, 71 Herriot Avenue, Kilbirnie, KA25 7JB
M6 CCR Ashish Bhakoo, 4 Bryden Cottages, High Street, Uxbridge, UB8 2NY
MM6 CCS Charles Stewart, 7 Hawkhill Place, Stevenston, KA20 4HN
MI6 CCU Nathan Morrow, 90 Bracky Road, Sixmilecross, Omagh, BT79 9PH
M6 CCV Paul Marshall, 21 Lymn Avenue, Gedling, Nottingham, NG4 4EA
MM6 CCW Julie Ransome, Evanton, Novar, Dingwall, IV16 9XH
M6 CCX Michael Perkins, 2 Buckingham Orchard, Chudleigh Knighton, Newton Abbot, TQ13 0EW
MM6 CCY Alexander Douglas, 124 Lunderston Drive, Glasgow, G53 6BS
M6 CCZ Mark Berry, 522 Nacton Road, Ipswich, IP3 9QD
M6 CDB Andrew Welch, 18 Monk Close, Tipton, DY4 7TP
M6 CDC Cameron Chalmers, 2 Canterbury Road, Bracebridge Heath, Lincoln, LN4 2TD
M6 CDD Marc Wollaston, 20 Hall Street, Church Gresley, Swadlincote, DE11 9QU
M6 CDE Gary Leedham Hawkes, 103 Wood Lane, Hednesford, Cannock, WS12 1BW
M6 CDG Delin Chang, Montefiore House, Wessex Lane, Southampton, SO18 2NU
M6 CDH Carl Hailstone, 2 Thornfield Avenue, Thornton-Cleveleys, FY5 5BH
M6 CDK John Rowe, 22 Treaty Road, Glenfield, Leicester, LE3 8LU
M6 CDL Raymond Winstanley, 173 Wimborne Road, Poole, BH15 2EF
M6 CDN Dale Watkiss, 54 Rolston Close, Plymouth, PL6 6TN
M6 CDO Alan Copse, 3 The Limes, Market Overton, Oakham, LE15 7PX
M6 CDQ Alan Camp, 3 Acre Close, Witnesham, Ipswich, IP6 9EU
M6 CDR Michael Mason, 7 Langland Close, Malvern, WR14 2UY
M6 CDT John Schleswick, Flat 46, Fitzroy House, 55-59 Great Pulteney Street, Bath, BA2 4DW
M6 CDU Edmund Mortimer, 6 Lanes End, Gastard, Corsham, SN13 9QS
MW6 CDV Kenneth House, Liddington, Dehewydd Lane, Pontypridd, CF38 2EN
M6 CDW Chris Wade, 31 Melton Green, Wath-upon-Dearne, Rotherham, S63 6AA
M6 CDY Lee Johnson, Flat 4, 1 Howe Street, Salford, M7 2EU
MW6 CDZ Garry Saltmarsh, 15 Colbourne Road, Beddau, Pontypridd, CF38 2LN
M6 CEA Martin Anderson, 27 Laing Road, Colchester, CO4 3UT
M6 CEB Matthew Bamber, 10 Sedgeley Mews, Freckleton, Preston, PR4 1PT
M6 CEE Tim Massey, 2 Cannon Heath Farm Cottages, Cannon Heath, Basingstoke, RG25 3EJ
M6 CEF Colin Davies, 1 Meadow Close, Bagworth, Coalville, LE67 1BR
M6 CEG John Martin, 2 Pant Heulog, Dyffryn Ardudwy, LL44 2BU
M6 CEH Charlotte Halloway, 82 Northwall Road, Deal, CT14 6PP
M6 CEI Keith Knapp, 46 Robin Hood Road, St. Johns, Woking, GU21 8SY
MI6 CEJ Connor Jeffrey, 197 Finvola Park, Dungiven, Londonderry, BT47 4ST
MM6 CEL Roy Wood, 42 Moorhouse Avenue, Paisley, PA2 9NY
MM6 CEM Catherine MacDonald, 2 Porterfield Road, Inverness, IV2 3HW
M6 CEN R Bird, 431 Ombersley Road, Worcester, WR3 7DQ
M6 CEO Anna Humphries, 3 Suffolk Drive, Worcester, WR3 8QT
M6 CEP Charlotte Pritchard, 8 Hoon Avenue, Newcastle, ST5 9NY
M6 CEQ Dennis Brown, 2 Kibworth Grove, Stoke-on-Trent, ST1 5QP
M6 CES Colin Canning, 17 Blackadder Crescent, Greenlaw, Duns, TD10 6XN
M6 CET Michael Knowles, 86 West Shore Road, Walney, Barrow-in-Furness, LA14 3UD
M6 CEU Andrew Burkitt, 53 Westwood Drive, Bourne, PE10 9PY
MM6 CEW Carole Watson, 20 Norlands, Errol, Perth, PH2 7QU
MW6 CEX Mark Brittle, 11 Haig Place, Gendros, Swansea, SA5 8BT
M6 CEY Melvin Green, 26 Drake Crescent, Kidderminster, DY11 6EE
M6 CEZ ivan hill, 3 beresford rd, Walsall, WS3 1JX
MM6 CFA Stewart Mackenzie, 20 Meadowhouse Road, Edinburgh, EH12 7HP
M6 CFD Jeffrey Seaton, 3 Tunstall Street, Middlesbrough, TS3 6PE
M6 CFE Duane Cowdrey, 264 Stamford Road, Brierley Hill, DY5 2QF
MM6 CFH Charles Fraser-Hopewell, 2/1 70 Albert Road, Glasgow, G42 8DW
M6 CFI Kate Davis, 26 Mendip Drive, Nuneaton, CV10 8PT
M6 CFL Michael Shortall, 16 Darnaway Close, Birchwood, Warrington, WA3 6TR
M6 CFN Nathan Layland, 3 Thirlmere Road, Golborne, Warrington, WA3 3HH
M6 CFO Martin Summers, 21 Quantock Avenue, Caversham, Reading, RG4 6PY
M6 CFQ Robert Smith, 15 Hollybush Road, North Walsham, NR28 9XT
M6 CFR John Mather, 33a Forest Road, Southport, PR8 6JD
MM6 CFS Fiona Sturrock, 16 Carlyle Crescent, Buckhaven, Leven, KY8 1DW
M6 CFT Chris Mole, 6 Clements Road, Chorleywood, Rickmansworth, WD3 5JT
M6 CFU Stephen Watson, 8 Church Close, Overstrand, Cromer, NR27 0NY
M6 CFV Andrew Stride, 2 Bailey Close, Pewsey, SN9 5HU
M6 CFW A Comerford, 21 New Cross Road, Headington, Oxford, OX3 8LP
M6 CFX Michael Lawrence, 61 Jack Warren Green, Cambridge, CB5 8US
M6 CFY Allan Collison, Bramble Lodge, Walkford Lane, New Milton, BH25 5NL
M6 CFZ Jason Gilpin, 17 Roundstone Crescent, East Preston, Littlehampton, BN16 1DG
M6 CGB Clifford Bradley, 3 Trecarne Gardens, Delabole, PL33 9DP
M6 CGC Charles Cheadle, 55 Ellison Street, Sheffield, S3 7JH
M6 CGF Stephen Gateson, Flat 6, Kenley House, Croydon, CR0 6AQ
MW6 CGH Morgan Jones, 102 Thomas Street, Tonypandy, CF40 2AH
M6 CGJ Richard Wheeler, 12 Drake Avenue, Didcot, OX11 0AD
M6 CGK Darren Richardson, 25 Comptons Lane, Horsham, RH13 5NL
M6 CGL Gary Livesey, 48 Kingsway, Leyland, PR25 1BL
M6 CGM Charlie Martin, 63 Oversetts Road, Newhall, Swadlincote, DE11 0SL
MI6 CGQ Charles Glenn, 40 Deanfield, Londonderry, BT47 6HY
M6 CGS Colin Smith, 17 Sunningdale Avenue, Sale, M33 2PJ
M6 CGV Connor Briant, Talarvor, Llanon, SY23 5HG
M6 CGW Christopher Watkins, 25 Citadilla Close, Gatherley Road, Richmond, DL10 7JE
M6 CGX Richard Coy, 35 Windmill Way, Kirton Lindsey, Gainsborough, DN21 4FE
M6 CGY Adam Lycett, 2 Royce Avenue, Hucknall, Nottingham, NG15 6FU
M6 CGZ Alex Dalgliesh, 62 Newbury Street, Wantage, OX12 8DF
M6 CHC Michael Corrigan, 33 Westbourne Road, Knott End-on-Sea, Poulton-le-Fylde, FY6 0BS
M6 CHD Roger Kergozou, Lilac Cottage, The Street, Cheddar, BS27 3TH
M6 CHF Catreena Ferguson, Royd Moor, Royd Moor Lane, Pontefract, WF9 1AZ
M6 CHG Lionel Hartley, 206 Latimer Road, Eastbourne, BN22 7JF
M6 CHH John Wishart, 19 Chepstow Close, Chippenham, SN14 0XP
M6 CHI Malcolm Boon, 45 Carlton Road, Wickford, SS11 7ND
M6 CHJ James Chaplin-Madden, 96 Salisbury Road, Great Yarmouth, NR30 4LS
M6 CHL David Cobbold, 65 St. Olaves Road, Bury St. Edmunds, IP32 6RR
MM6 CHM Robert Russell, 1/2 37 Dunagoil Road, Glasgow, G45 9UR
MM6 CHN Charles Denny, Darnaway, Castleton Place, Ballater, AB35 5ZQ
M6 CHO Simon Wheeldon, 32 Beech Grove Terrace, Garforth, Leeds, LS25 1EG
M6 CHP Mandy Sycamore, 86 Grove Road, Tiptree, Colchester, CO5 0JG
M6 CHQ Andrew Bowlzer, 3 Moreia Terrace, Harlech, LL46 2YW
M6 CHS Michael Drury, 19 Cuffley Avenue, Watford, WD25 9RB
M6 CHT Adam Trowse, 110 New Road, Hethersett, Norwich, NR9 3HQ
M6 CHU Robert Last, 30 Abbot Road, Bury St. Edmunds, IP33 3UB
MM6 CHV Robert Gilchrist, 5 Glenburn, Leven, KY8 5BD
M6 CHW Christopher Wright, 14 Orchard Close, Poughill, Bude, EX23 9ES
M6 CHX Michael Harrison, 2a Arundel Road, Camberley, GU15 1DL
MM6 CHY Colin Hay, 30 Kincardine Place, East Kilbride, Glasgow, G74 3DN
M6 CHZ Henry Eager, 45 Fleetwood Avenue, Herne Bay, CT6 8QW
MM6 CIA Charles Anderson, 11 Willowpark Place, Aberdeen, AB16 6XY
M6 CIE William Canavan, 9 The Ridings, Deanshanger, Milton Keynes, MK19 6JD
M6 CIF Ian Constable, 19 Rugby Road, Dunchurch, Rugby, CV22 6PG
M6 CIG Edmund Black, 34 White Bank Road, Oldham, OL8 3JH
MM6 CIJ Craig Thomson, 54 Alderman Road, Glasgow, G13 3YE
M6 CIK Paul Turner, 43 Nelson Way, Mundesley, Norwich, NR11 8JD
M6 CIM Philip Stuart, 5 Welbeck Gardens, Woodthorpe, Nottingham, NG5 4NX
M6 CIP Alexander Rowson-Brown, The Fox, Station Road, Baldock, SG7 5RN
M6 CIQ Martin Walls, 38 Poplar Drive, Royston, SG8 7ER

M6 CIS Benjamin Rush, Springside, Uploders, Bridport, DT6 4NU
M6 CIU Garell Brotherhood, 17 Baldwin Close, Forest Town, Mansfield, NG19 0LR
M6 CIW Craig Wolte, 13 Hempbridge Road, Selby, [illegible]
MD6 CIX Andrew Skinner, 17 Hawkins Close, Harrow, HA1 4DJ
MW6 CIY Michael Bennett, 24 Bryn Street, Merthyr Tydfil, CF47 0TG
M6 CJA Colin Apps, 16 Lismore Park, 5 Waterloo Road, Southport, PR8 2FY
MM6 CJC James Scanlan, Aitnoch Farmhouse, Grantown-on-Spey, PH26 3PX
M6 CJD Colin Dawson, 9 Mulberry Close, Poringland, Norwich, NR14 7WF
M6 CJE David Chidgey, 42 Half Acre, Williton, Taunton, TA4 4NZ
M6 CJF Carl Foster, 8 William Iliffe Street, Hinckley, LE10 0LY
M6 CJG Christopher Groves, 3 Hudson Davies Close, Pilley, Lymington, SO41 5PA
M6 CJH Christopher Hill, 9 Oliver Road, Newport, NP19 0HU
M6 CJI Joseph Power, 43 Valley Road, Melton Mowbray, LE13 0DU
M6 CJJ Patrick Spry, 8 Maun Green, Newark, NG24 2HA
M6 CJK Roger Tuffin, 133 Shirley Drive, Hove, BN3 6UJ
M6 CJM Christopher Martin, 2 Whitethorn Cottages, Dark Lane, Cheltenham, GL51 9RW
M6 CJN Carl Jenkins, 15 Were Close, Warminster, BA12 8TB
M6 CJP Christopher Petrie, 14 Rotherfield Avenue, Eastbourne, BN23 8JQ
M6 CJQ Brian Hiley, 9 Pinfold Lane, Harby, Melton Mowbray, LE14 4BU
M6 CJR Christopher Rundle, 1 Trezaise Close, Roche, St. Austell, PL26 8HW
MW6 CJS Calum Sweeney, 67 Blandon Way, Cardiff, CF14 1EH
M6 CJT Chris Wright, 16 Baseley Way Longford, Coventry, CV6 6QA
M6 CJU Johnathan Brown, 8 Herald Drive, Chichester, PO19 8DE
M6 CJV Stephen Clarke, 20 Woodlands Way, Southwater, Horsham, RH13 9HZ
M6 CJW Chris Moss, 19 Tozer Close Wallisdown, Bournemouth, BH11 8RB
M6 CJX Ian Ftaiha, 44b The Broadway, London, NW7 3LH
MM6 CJY Craig Yohn, G-8 King's Hall, College Bounds, Aberdeen, AB24 3TS
MM6 CJZ John Wright, 43 Spey Court, Stirling, FK7 7QZ
MM6 CKC Matthew Thomson, 30 Pladda Road, Saltcoats, KA21 6AQ
M6 CKD Robert Basford, 4 Renoir Close, St. Ives, PE27 3HF
M6 CKE Liam Hardy, 221 Rookery Lane, Lincoln, LN6 7PJ
M6 CKF Chris Farmer, 3 Laxton Way, Banbury, OX17 1GJ
M6 CKH Steven Tobin, 57 Barnetby Road, Hessle, HU13 9HE
M6 CKI Roy Baines, 27 Lichfield Road, Bloxwich, Walsall, WS3 3LT
M6 CKJ Edward Taylor, Lanthorn close, Broxbourne, EN107NR
M6 CKL Paul Ridley, 218 Lichfield Road, Rushall, Walsall, WS4 1SA
M6 CKM John Jones, Isfryn Bungalow, Glan-y-Nant, Llanidloes, SY18 6PQ
M6 CKO Elliot Redmond, 28 Common Lane, Polesworth, Tamworth, B78 1LS
M6 CKP Clarence Prior, 38 Windmill Road, Wombwell, Barnsley, S73 8PP
M6 CKQ Andrew Colman, 5 Burn Heads Road, Hebburn, NE31 2TB
M6 CKR Michael Sadler, 14 Woodlands Avenue, Water Orton, Birmingham, B46 1SA
M6 CKT Craig Jenkings, 8 Tolworth Hall Road, Birmingham, B24 9NE
M6 CKU Praful Naik, 82 Misbourne Road, Uxbridge, UB10 0HW
M6 CKV Paul Pain, 12 Maple Drive, Bamber Bridge, Preston, PR5 6RA
M6 CKY Nicolai Gamulea Schwartz, 17 Harbour Way, Hull, HU9 1PL
M6 CKZ Darren Osborne, 12 Sandringham Close, Brackley, NN13 6JQ
M6 CLA Susan Clark, 43 Age Croft, Oldham, OL8 2HG
M6 CLB Elaine Williams, 35 Cansfield Grove, Ashton-in-Makerfield, Wigan, WN4 9SE
MM6 CLC Charlotte Collins, Redwoods, Barcaldine, Oban, PA37 1SG
M6 CLF Caroline Studdart, 33 Linden Avenue, Connah's Quay, Deeside, CH5 4SN
M6 CLG John Hurlbutt, 55 Prospect Avenue, Seaton Delaval, Whitley Bay, NE25 0EL
M6 CLI C Herlingshaw, 48 Keats Road, Normanby, Middlesbrough, TS6 0RW
M6 CLJ Ceri Jones, 19 Crud y Castell, Denbigh, LL16 4PQ
M6 CLK Michael Clark, 34 Magdalene Road, Owlsmoor, Sandhurst, GU47 0UT
M6 CLL Andrew McCall, 40 Nethway Avenue, Blackpool, FY3 8JU
MM6 CLM Chris Houston, 34 Briarhill Road, Prestwick, KA9 1HY
M6 CLO Paul Clough, 101 Cathedral View, Houghton le Spring, DH4 4HN
MW6 CLP R Piper, 16 Elm Rise, Bryncethin, Bridgend, CF32 9SX
M6 CLQ Morgan Eccleston, 15 Wood Road North, Manchester, M16 9GQ
M6 CLV Craig Vernau, 62 Princethorpe Road, Ipswich, IP3 8NX
M6 CLW Aidan Sockett, 37 Windsor Road, Thorpe Hesley, Rotherham, S61 2QS
M6 CLX Phillip Cooke, 3 St. Stephens Court, Congleton, CW12 1QW
M6 CLY Mark Poole, 37 Gordon Road, Stoke-on-Trent, ST6 5PZ
M6 CLZ Cameron Sinclair, 43 Newton Way, Leighton Buzzard, LU7 4SU
M6 CMB Lee Gorecki, 6 Robinhood Lane, Winnersh, Wokingham, RG41 5LX
M6 CMF Bartholomew Hill, 41 Heath Road, Leighton Buzzard, LU7 3AB
M6 CMG Charles Goodhand, 37 Westwick Gardens, Lincoln, LN6 7RQ
M6 CMI David Lyons, 2 Goswick Farm Cottages, Berwick-upon-Tweed, TD15 2RW
M6 CMK Christopher Kennedy, 30 Tatton Close, Cheadle, SK8 2LZ
MM6 CMM Calum McKinlay, 19 Ash Grove, Blackburn, Bathgate, EH47 7QJ
M6 CMO Ian Steggles, 4 Hawley Vale, Hawley Road, Dartford, DA2 7RL
MI6 CMQ R Donnan, 71 Victoria Avenue, Newtownards, BT23 7ED
M6 CMR Carl Millar, 60 Rodney Way, Ilkeston, DE7 8PW
M6 CMT Catherine Travis, 4 Kingsdale, Worksop, S81 0XJ
MI6 CMU Trevor Crawford, 21 Ardranny Drive, Newtownabbey, BT36 6BD
M6 CMW Connor Walsh, 8 Kemp Road, Leicester, LE3 9PS
MM6 CMY Campbell Matheson, 322 Millfield Hill, Erskine, PA8 6JN
M6 CMZ Mark Jones, 80 Gloucester Road, Coleford, GL16 8BN
M6 CNA Malcolm Bradley, 11 Pike Road, Coleford, GL16 8DE
MM6 CNC Catherine Comrie, 11 Glendoune Street, Girvan, KA26 0AA
M6 CND David Ward, 43 Spencer Street, Accrington, BB5 6SY
M6 CNG Gareth Southall, 28 Manor Road, Woodford Halse, Daventry, NN11 3QP
M6 CNI Richard Johnson, 24 Fairfields, Upper Denby, Huddersfield, HD8 8UB
M6 CNK Darren Stockton, 78b Alderfield, Penwortham, Preston, PR1 9HA
M6 CNL Harry Hope, 51 Margravine Gardens, London, W6 8RN
M6 CNN Dave Bedford, 28 Farne Avenue, South Shields, NE34 7HG
MM6 CNO Stephen Leighton, 75 Dovecot Road, Tullibody, Alloa, FK10 2QU
M6 CNP Mark Pratt, 1 Ashvale Gardens, Romford, RM5 3QA
M6 CNQ Paul Phillips, 49 Lower Northam Road, Hedge End, Southampton, SO30 4HE
M6 CNR John Taylor, 90 Village Road, Gosport, PO12 2LG
M6 CNS Maurice Phillips, 66 Minden Way, Winchester, SO22 4DU
MM6 CNV Michael Jamieson, 5 Straid Bheag, Barremman, Helensburgh, G84 0QX
M6 CNX Jonathan Mooney, 107 Tedder Road, South Croydon, CR2 8AR
M6 CNY Richard East, 6 Ashley Road, Worcester, WR5 3AY
M6 CNZ James Haddow, 28 Church Marks Lane, East Hoathly, Lewes, BN8 6EQ
M6 [illegible] Robert Williams, 11 Trinity Buildings, [illegible]
MW6 COD David Codd, Gwachal-tagy Farm, Haverfordwest, SA62 6HF
M6 COE Christopher Bussell, 3 Downshire Terrace, Street Lane, Haywards Heath, RH17 6UL
MM6 COF Graham Steele, 1/2 50 Motehill Road, Paisley, PA3 4ST
M6 COH Sebastian Parris-Hughes, 23 Hobney Rise, Westham, Pevensey, BN24 5NN
M6 COI Martin Bernasinski, 1 Elizabeth Place, Clyde Road, London, N15 4LA
M6 COL Colin Spicer, 3 School View, Tunstall, Sittingbourne, ME9 8DX
MI6 COM Seamus Carroll, Drumguiff Lane, Enniskillen, BT92 7HP
M6 CON Conor Morgan, 33 Congella Road, Torquay, TQ1 1JU
M6 COP Laurence Elphick, 30 The Meadows, Heskin, Chorley, PR7 5NR
M6 COU Paul Coulsey, 25 Weavers Close, Horsham St. Faith, Norwich, NR10 3HY
M6 COV Alison Hughes, 86 Colinmander Gardens, Ormskirk, L39 4TF
M6 COX Ashley Cox, 20 Dunns Dale, Maltby, Rotherham, S66 7NR
M6 COY Craig Burden, 24b Coombe Road, London, W4 2HR
M6 CPA C Godding, 11 Oakleaze Road, Thornbury, Bristol, BS35 2LL
M6 CPC Carlos Cook, 95 Station Road, Eccles, Manchester, M30 0PZ
MM6 CPE Christopher English, Easter Backlands, Roseisle, Elgin, IV30 5YD
M6 CPH Christopher Holden, 49 Whalley Road, Lancaster, LA1 2HE
M6 CPJ David Martin, 3 Hillside Avenue, Rowley Regis, B65 0EZ
MM6 CPK Euan Dungavel, 9c Anderson Crescent, Ayr, KA7 3RL
M6 CPM Christopher Marsh, 16 Drake Head Lane, Conisbrough, Doncaster, DN12 2AA
M6 CPN Chris Norris, 53 Station Road, Castlethorpe, Milton Keynes, MK19 7HF
M6 CPO Simon Parker, 10 Wheelwrights Close, Sixpenny Handley, Salisbury, SP5 5SA
M6 CPP Nick Barnard, 10 Whites Lane Kessingland, Lowestoft, NR33 7TF
M6 CPQ Thomas Walsh, 2 Ashfield Mews, Ashington, NE63 9GJ
M6 CPR Gavin Doxey, 2 Nettlecroft, Barnsley, S71 5SD
M6 CPS Courtney Stratford, 15 Ferndale Road, Banbury, OX16 0RZ
M6 CPT Darren Baker, 21 Dolman Road, Gosport, PO12 1RB
MU6 CPV Philippe Martin, 1 Hazeldene, Grandes Maisons Road, Guernsey, Guernsey, GY2 4JS
M6 CPW Peter Winkley, 38 St. Georges Terrace, Plymouth, PL2 1HS
M6 CPX R Gleave, 52 Cranborne Avenue, Warrington, WA4 6DE
M6 CPZ Merlin Spiers, 99 Chapel Street, Tiverton, EX16 6BU
M6 CQB Ian McKean, 142b High Street, Cranfield, Bedford, MK43 0EL
M6 CQD Alex Fitton, 72 Newnham Court, Ipswich, IP2 9UE
M6 CQE David Bishop, 62 Brindley Crescent, Hednesford, Cannock, WS12 4DS
M6 CQF John Crill, 92 Waterside, Exeter, EX2 8GZ
M6 CQG Mark Chanter, 7 Woodford Crescent, Plymouth, PL7 4QY
M6 CQH Maurice Fletcher, 7 Richard Street, Bacup, OL13 8QJ
M6 CQJ Albert Tranter, 122 Summerhill Road, Bristol, BS5 8JU
M6 CQK Samuel Collins, 7 Fir Grove, Macclesfield, SK11 7SF
MW6 CQL Chris Summerfield, 11 Woodland Park, Penderyn, Aberdare, CF44 9TX
M6 CQO Colin Osborne, Gwinwydden, Tremont Road, Llandrindod Wells, LD1 5BH
M6 CQR Paul Dunn, 9 Parkside Gardens, Widdrington, Morpeth, NE61 5RP
MI6 CQS Stephen Allen, 21 Derrymore Meadows, Bessbrook, Newry, BT35 7GA
MM6 CQU Andrew Prentice, 24 Victoria Road, Grangemouth, FK3 9JN
M6 CQV Colin Norman, 15 Maple Close, Sedbergh, LA10 5JE
M6 CQX Gareth Davies, 28 Danum Close, Hailsham, BN27 1UX
M6 CRA Carl Ancill, 8 Ipley Way, Hythe, Southampton, SO45 3LJ
M6 CRE Cyril Etches, Round Bays, Grasmere Avenue, Skegness, PE24 5TZ
M6 CRF Charles Faulkner, Mount Pleasant, Elkstones, Buxton, SK17 0LU
M6 CRG Carl Gloess, 5 Elmers Lane, Kesgrave, Ipswich, IP5 2GW
M6 CRH Conor Hogan, 82 Abbott Road, Didcot, OX11 8HY
M6 CRJ Callum Jackson, 30 Coronation Avenue, Mile Oak, Tamworth, B78 3NW
M6 CRO Christopher Ordish, 37 Mill Lane, Earl Shilton, Leicester, LE9 7AY
M6 CRP Charles Pulford, 41 Morris Street, Sheringham, NR26 8JY
MM6 CRQ Carrie Welsh, 28 Peacock Wynd, Motherwell, ML1 4ZL
M6 CRR Robert Flevill, 62 Grosvenor Way, Horwich, Bolton, BL6 6DJ
M6 CRT Colin Thomas, 7 Hillside Road, Nether Heyford, Northampton, NN7 3JU
MW6 CRU Craig Uphill, 167 Nantgarw Road, Caerphilly, CF83 1AN
MM6 CRW Charles Watkinson, The Hillock Farmhouse, Lumphanan, Banchory, AB31 4QL
M6 CRX Andrew Haynes, 16 Hills Crescent, Colchester, CO3 4NU
M6 CRZ Carl Roberts, 25 Queens Lea, Willenhall, WV12 4JA
M6 CSA Stephen Clark, 1 Hanover Cottages, Lippen Lane, Southampton, SO32 3LE
M6 CSC Ralph Metcalfe, Little Shernden, Shernden Lane, Edenbridge, TN8 5PS
M6 CSE Charlotte Cartwright, 8 Charles Close, Westcliff-on-Sea, SS0 0EU
M6 CSF Richard Powell, 19 Normandie Close, Ludlow, SY8 1UJ
M6 CSG Colin Greenwood, 44 Fountain Street, Heckmondwike, WF16 9HS
M6 CSI Christopher David Andrew Morris, 38 Beltony Drive, Crewe, CW1 4TX
M6 CSJ Bryan Appleby, 64 Lundy Close, Southend-on-Sea, SS2 6HB
M6 CSK David Polley, 6 Coneygear Road, Hartford, Huntingdon, PE29 1QJ
M6 CSL Christopher Langmaid, Flat 4, Woodlawn High Street, Partridge Green, Horsham, RH13 8HR
M6 CSM Cameron Moppett, 13 Piccadilly, Tamworth, B78 2ER
M6 CSN Christine Snelling Nash, 22 Church Lane, Kimpton, Hitchin, [illegible]
M6 CSP Chris Pritchard, 6 St. Catherines Road, Crawley, RH10 3TA
M6 CSR Carl Robinson, 5 Minsmere Rise, Middleton, Saxmundham, IP17 3PA
M6 CSS Karl Latham, 20 Kenyon Avenue, Wrexham, LL11 2ST
M6 CSU Chingiz Sharif, 38 King George Road, Ware, SG12 7DT
M6 CSW Jennifer Cheng, Harrogate Ladies' College, Clarence Drive, Harrogate, HG1 2QG
M6 CSX Andrew Askam, 8 The Pastures, Weston-on-Trent, Derby, DE72 2DQ
M6 CSZ Andrew Chaplin, 10 St. Leonards Road, Malinslee, Telford, TF4 2EB
M6 CTA Paul Kerton, 21 Appledore, Bracknell, RG12 8QY
M6 CTB John Pritchard, 1 Tan y Coed, Maesgeirchen, Bangor, LL57 1LU
MW6 CTD Tony Cooper, Dorset Cottage, Alltyblacca, Llanybydder, SA40 9SU
MW6 CTE Mark Sunderland, 5 Grosvenor Street, Cardiff, CF5 1NH
M6 CTG Margaret Adam, Tan Ffordd, Mynydd Llandygai, Bangor, LL57 4LX
MM6 CTH Shane Fenton, 19 Rowan Road, Girvan, KA26 0BY
M6 CTI John Roberts, 4 Oaks Road, Staines-upon-Thames, TW19 7LG
M6 CTJ Craig Johnson, 2 Dawber Street, Worksop, S81 7DS
MM6 CTL Christine Livingstone, 391 Dyke Road, Glasgow, G13 4QE
M6 CTN [illegible]
M6 CTO Dallan McGleenan, Greenfields, Ellonby, Penrith, CA11 9SJ
MM6 CTQ Allan Montgomery, 13 Rathad A'Mhaoir, Plasterfield, Stornoway, HS1 2UP
MW6 CTS Derek Hooper, Nant y Dryslwyn Cottage, Ty Mawr, Llanybydder, SA40 9RD
M6 CTT Clifford Rogers, 65 Darwin Close, Taunton, TA2 6TR
MD6 CTU Piotr Sniezek, 204 Quadrant Court, Empire Way, Wembley, HA9 0EY
M6 CTV Cedric Leek, 36 Cheltenham Way, Mablethorpe, LN12 2AX
M6 CTW Charles Winnan, 133 Deepcut Bridge Road, Deepcut, Camberley, GU16 6SD
M6 CTX Thomas Rowlands, 7 Northfield Crescent, Beeston, Nottingham, NG9 5GR
M6 CTY Peter Keane, 18 Main Road, Cannington, Bridgwater, TA5 2JN
M6 CTZ Jack Richards, 5 Dumfries Place, Weston-Super-Mare, BS23 4LQ
MW6 CUA Gerwyn Young, 104 Ystrad Road, Pentre, CF41 7PW
M6 CUC R Keast, 7 The Finches, Newport, PO30 5GU
M6 CUD Sarah Kneale, 59 Mayflower Avenue, Saxmundham, IP17 1BU
M6 CUE Neil Connor, 28 Church Street, Hungerford, RG17 0JE
M6 CUJ James Parsons, 5 Shawfields, 41 Cranley Road, Guildford, GU1 2JE
M6 CUL Daniel Richards, 58 Holm Lane, Oxton, Prenton, CH43 2HS
M6 CUN Justin Colledge-Wiggins, Raffles, Southcombe, Chipping Norton, OX7 5QH
M6 CUP Thomas Cavanagh, 186 Cole Valley Road, Birmingham, B28 0DQ
M6 CUQ Gavan Fantom, 6 Middle Street, Farcet, Peterborough, PE7 3AX
M6 CUR James Troughton, Rhiwbina, Pentre Lane, Cwmbran, NP44 3AP
M6 CUS Stephen Williams, 16 Pinewood Avenue, New Haw, Addlestone, KT15 3AA
M6 CUU Kane Boaler, 12 Belmont, Slough, SL2 1SU
M6 CUV A Douglas, Gobbins Cottage, Sandy Lane, Ormskirk, L40 5TU
M6 CUW Adam Allan, 13 Albert Street, Cowes, PO31 7ND
M6 CUX Allan Stott, 2 Douglas Drive, Maghull, Liverpool, L31 9DG
MI6 CUZ Gavin Angus, 59 Millisle Road, Donaghadee, BT21 0HZ
M6 CVB Jonathon Adler, 1 Searles Meadow, Dry Drayton, Cambridge, CB23 8BW
MW6 CVC Charlotte Smith, 29 Heol Cwarrel Clark, Caerphilly, CF83 2NE
M6 CVE Paul Herron, 102 Garden City Villas, Ashington, NE63 0EU
M6 CVF Leo Bishop, Franklin House, Canford School, Wimborne, BH21 3AF
M6 CVH Christopher Hodgetts, 16 Myrtle Drive, Rogerstone, Newport, NP10 9EA
M6 CVJ John Smith, 9 Kingsmead Mews, Coventry, CV3 3NA
M6 CVK Christopher Saxton, 7 Stoney Way, Tetney, Grimsby, DN36 5PG
M6 CVN Martin Beecroft, Hafod Y Wennol, Llanddoged, Llanrwst, LL26 0TY
MI6 CVO Roy Lambe, 39 Tildarg Avenue, Belfast, BT11 9LU
M6 CVP Lee Stamper, 22 Douglas Road, Workington, CA14 2QY
M6 CVQ Ian Nicholls, 34 West Close, Bath, BA2 1PY
M6 CVU Zoltan Szikszai, 135 Devon Road, Newark, NG24 4JL
M6 CVV Jr Zoltan Szikszai, 135 Devon Road, Newark, NG24 4JL
MI6 CVW Andrew Logan, 15 Park Lane, Saintfield, Ballynahinch, BT24 7PR
M6 CWA Gordon Moon, 9 Blackstock Court, Bootle, L30 0PN
MM6 CWB Colin Beedie, 21 Marywell Village, Arbroath, DD11 5RH
MI6 CWC Clifford Campbell, 21 Elms Park, Coleraine, BT52 2QF
M6 CWD Brian Ledson, 16 Caton Close, Southport, PR9 9XF
M6 CWF Anthony Mills, 6 Kildare Drive, Peterborough, PE3 9TS
M6 CWG Daniel Levy, Flat 36, Claydon House, London, NW4 1LS
M6 CWI Samuel Chapman, 5 Bracton Drive, Nottingham, NG3 2LN
M6 CWJ Ben Harrison, 104 Jeavons Lane, Great Cambourne, Cambridge, CB23 5FN
M6 CWK Charles Austin, 1a Arundel Road, Peacehaven, BN10 8TE
MW6 CWL K Saltmarsh, 15 Colbourne Road, Beddau, Pontypridd, CF38 2LN
MM6 CWN Colin Cowan, 23 Park Road, Hamilton, ML3 6PD
M6 CWO Jay Cox, 87 Richmond Road, Leighton Buzzard, LU7 4RF
M6 CWP Charlie Hicks, 1 Elers Road, London, W13 9QA
M6 CWQ Attila Toth, 14 Runcorn Road, Sunderland, SR5 5ET
M6 CWR Colin Ralphson, 20 Monsal Grove, Buxton, SK17 7TF
M6 CWU Jodie Henningway, 64 South Cliff, Bexhill-on-Sea, TN39 3EE
M6 CWV Natasha Symes, 23 Elm Road, Portslade, Brighton, BN41 1SA
M6 CWW Charlie Warhurst, 30 Nether Royd View, Silkstone Common, Barnsley, S75 4QQ
M6 CWX Tim Digman, 74 Baddlesmere Road, Whitstable, CT5 2LA
M6 CWZ Liam Starrett, 50 Danes Road, Bicester, OX26 2LP
MM6 CXA Stewart Harvey, 63 Darley Road, Cumbernauld, Glasgow, G68 0JR
M6 CXB Christopher Burnham, 2 Barry House, Elmleigh Road, Bristol, BS16 9AG
M6 CXC Barry Burr, Flat 87, Horatia House, Southsea, PO5 4AL
MM6 CXD Alistair Donald, 10 Fraser Road, Burghead, Elgin, IV30 5YN
M6 CXF Alan Angus, 51 Osprey Drive, Blyth, NE24 3QS
M6 CXH Richard Pilkington, 64 School Lane, Higher Bebington, Wirral, CH63 2LW
M6 CXI Gary Megson, 147 Duke of York Avenue, Wakefield, WF2 7DA
MM6 CXJ Robin Inglis, Roadside, Skirza, Wick, KW1 4XX
M6 CXM Joseph Van De Vondel, 50 Holmsley Lane, South Kirkby, Pontefract, WF9 3JF
M6 CXN Frederick Strickland, Hinds Cottage, Beverley Road, Driffield, YO25 9PF
M6 CXO Matthew Hickey, [illegible] Hollow, [illegible] Drive, Prestbury, SK10 4BG
M6 CXP Ben Fitzgerald-O'Connor, 24 Routh Street, London, E6 5XX
M6 CXS Peter Sanderson, 2 East Crescent, Canvey Island, SS8 0HL
MI6 CXU Stephen Carter, 1 Carmavy Road, Nutts Corner, Crumlin, BT29 4TF
M6 CXV Alex Todd, 16 Worcester Close, Bracebridge Heath, Lincoln, LN4 2TY
M6 CXW Adam Little, 19 Lancaster Drive, Padiham, Burnley, BB12 7DP
M6 CXZ Martyn Hack, 25 Horse Field View, Melton Mowbray, LE13 0TF
M6 CYA Charlotte Rowe, 1 Avondale Street, Stoke-on-Trent, ST6 4NN
M6 CYB Kevin Williams, 18 Bye Road, Lidlington, Bedford, MK43 0RU
M6 CYC Colin Carter, 142 Hall Street, Briston, Melton Constable, NR24 2LQ
M6 CYD Angela McInnes, 87 Sovereigns Quay, Bedford, MK40 1TF
M6 CYE Adrian Forrester, 16 East Street, Hebburn, NE31 1HL
M6 CYF Robert Silcock, 18 Saxon Road, Southampton, SO15 1JJ
M6 CYG Craig Ashworth, 12 North Terrace, Tebay, Penrith, CA10 3XH
M6 CYL Yik Lam Janice Chong, Harrogate Ladies' College, Clarence Drive,

**IMPORTANT NOTE**

**Revalidate licence to avoid revocation** – Ofcom has advised the Society that plans will be drawn up to revoke licences that have not been revalidated as required by the licence conditions. The quickest way to revalidate is to do so online via the Ofcom website: *https://services.ofcom.org.uk/* or by email: *amateur.validations@ofcom.org.uk* Ofcom staff are available to help, but please be patient during times of heavy workload.

Harrogate, HG1 2QG
**M6 CYM** Glynn Lewis, Upper Cwm Farm, Llantilio Crossenny, Abergavenny, NP7 8TG
**M6 CYN** George Couzens, 8 Haven Close, East Cowes, PO32 6GL
**M6 CYO** James Mason, 77 Albutts Road, Walsall, WS8 7ND
**M6 CYP** Simon Mouradian, 29 Colin Park Road, London, NW9 6HT
**MM6 CYQ** Geoffrey Lewin, Larch Cottage, Lein Road, Fochabers, IV32 7NW
**M6 CYS** Adrian Owen, 8 Rose Court, Rose Street, Wigan, WN1 3DQ
**M6 CYT** D Elias, 12 Dagmar Terrace, London, N1 2BN
**M6 CYU** Kieron Evans, Maesyronnen, Sarnau, Llanymynech, SY22 6QL
**M6 CYV** Kenneth Nicholson, 68 Brunswick Street, Leigh, WN7 2PL
**M6 CYW** Joshua Wood, 17 Harrington Avenue, Lincoln, LN6 7UP
**M6 CYX** Jack Shanahan, 23 Pentre Gwyn, Trewern, Welshpool, SY21 8DY
**M6 CYY** Andrew Brown, 114 Tavistock Road, Birmingham, B27 7LA
**M6 CZA** Michael Singfield, 8 Barnes Crescent, Sutton-in-Ashfield, NG17 5BL
**M6 CZB** Nithin Shajan, 19 Sturgess Avenue, London, NW4 3TR
**M6 CZC** Becky Crowhurst, 33 Clarendon House, Clarendon Road, Hove, BN3 3WW
**M6 CZD** Thomas Ayland, 225 Stroud Road, Gloucester, GL1 5JU
**M6 CZF** S Amos, 172 Radcliffe Road, West Bridgford, Nottingham, NG2 5HF
**MM6 CZH** Sohan Ram, 28 Craigievar Gardens, Kirkcaldy, KY2 5SD
**M6 CZI** John Prodger, Brindles, Warren Hill Lane, Aldeburgh, IP15 5QB
**M6 CZJ** Jonathan Chalmers, 19 Brettenham Crescent, Ipswich, IP4 2UB
**M6 CZL** Timothy Watson, Montrose Farm, Bury Road, Diss, IP22 2PY
**M6 CZR** Roy Houghton, Kirklea, Sunk Island Road, Hull, HU12 0DS
**M6 CZS** Paul Hampton, Caretakers Flat, T.A. Centre, Newport, NP20 5XE
**M6 CZU** Lloyd Taylor, 140 Cotswold Way, Risca, Newport, NP11 6RG
**M6 CZV** Craig Parry, 91 Grove Road, Risca, Newport, NP11 6GL
**M6 CZW** Terry Grayson, 20 Harrison Street, Tow Law, Bishop Auckland, DL13 4EE
**M6 DAC** David Clough, 8 Skeldyke Road, Kirton, Boston, PE20 1LR
**M6 DAD** John Armstrong, Sherrylea, Coventry Road, Coventry, CV7 8BY
**MM6 DAF** Darryl Hannah, 7 Katrine Place, Irvine, KA12 9LU
**M6 DAF** Darren Flunder, 17 Hampden Crescent, Kettering, NN16 0LA
**M6 DAG** David Walker, Flat 5, Seward Court, 380-396 Lymington Road, Christchurch, BH23 5HD
**MW6 DAI** David Evans, 29 Mount Pleasant, Bedlinog, Treharris, CF46 6SD
**M6 DAK** Colin Wood, 18 Tufa Close, Chatham, ME5 9LU
**M6 DAL** Derek Cannon, HOOK 2 SISTERS, MAUTBY SITE MAUTBY, Great Yarmouth, NR29 3JB
**M6 DAM** John Malia, 47 Clent Way, Longbenton, Newcastle upon Tyne, NE12 8QG
**M6 DAN** Daniel Trudgian, 18 Hart Close, Wootton Bassett, Swindon, SN4 7FN
**M6 DAP** David Pasika, 192 Longfield Lane, Cheshunt, Waltham Cross, EN7 6AQ
**MM6 DAQ** David Rooney, 19 Aurs Drive, Barrhead, Glasgow, G78 2LR
**M6 DAR** Darren May, 12 Marl Crescent, Llandudno Junction, LL31 9HS
**M6 DAS** Derinda Starling, Long Field Barn, Clarkes Lane, Beccles, NR34 8HR
**MM6 DAT** Ian Mclachlan, Railway Cottage, Nether Falla, Peebles, EH45 8QZ
**M6 DAU** Dane Skates, 137 Potton Road, Biggleswade, SG18 0ED
**M6 DAW** David Harris, 6 Baker Lea, Monkland, Leominster, HR6 9DB
**M6 DAX** Dax Blackhorse-Hull, 1 Occupation Lane, New Bolingbroke, Boston, PE22 7LW
**MI6 DAY** Andrew Galbraith, 62 Millbrook Gardens, Castlederg, BT81 7DF
**M6 DAZ** Darren Coleman, 9 Hogan Close, Newport, PO30 5UF
**M6 DBA** Dominic Austin, 50 Lowestoft Road, Gorleston, Great Yarmouth, NR31 6LZ
**M6 DBB** Stephen Elias, 20 Attlee Way, Cefn Golau, Tredegar, NP22 3TA
**M6 DBD** Malcolm Williams, 31 Waundeg, Nantybwch, Tredegar, NP22 3SN
**M6 DBE** David Bateson, 19 Rothesay Road, Heysham, Morecambe, LA3 2UR
**M6 DBF** Philip Allen, Flat 32, Whitworth Court, 9 Whitworth Road, Southampton, SO18 1JR
**M6 DBG** David Bentley, 36 Byron Road, Mexborough, S64 0DG
**M6 DBH** Darren Coote, 4 Hunters Oak, Watton, Thetford, IP25 6HL
**M6 DBI** Rodney Buckland, 34 Beechwood Drive, Meopham, Gravesend, DA13 0TX
**MM6 DBJ** Adam Brown, 17 Glamis Drive, Dundee, DD2 1QN
**M6 DBK** David Wells, 27 Victoria Avenue, Camberley, GU15 3HT
**M6 DBL** David Smith, 186 Weekes Drive, Slough, SL1 2YR
**MM6 DBN** William Brannan, 16 Cairngorm Gardens, Cumbernauld, Glasgow, G68 9JD
**M6 DBP** David Robinson, Gorge View, Birch Hill, Cheddar, BS27 3JN
**M6 DBQ** Daniel Baldwin, 47 Charlcote Crescent, Crewe, CW2 6UH
**M6 DBS** Sean Ward, 22 St. Margarets Close, Horstead, Norwich, NR12 7ER
**MM6 DBT** John Ross, 159 Grahams Road, Falkirk, FK2 7BQ
**MW6 DBU** David Humphreys, 99 Park Road, Treorchy, CF42 6LB
**M6 DBV** Adrian Higgins, The Brambles, Tewsley Close, Barnstaple, EX31 2JT
**M6 DBW** David Woodridge, 8 Goold Close, Corston, Bath, BA2 9AF
**M6 DBX** Michael Hall, 29 The Spinney, Finchampstead, Wokingham, RG40 4UN
**M6 DCA** Devon Cotz, 159 Ecclesfield Road, Sheffield, S5 0DH
**M6 DCB** Phil Jones, 102 Manor House Lane, Preston, PR1 6HP
**M6 DCD** Graham Clarke, 2 Beare Green Cottages, Horsham Road, Dorking, RH5 4PE
**MI6 DCE** eamon macgra, 12c Glenabbey Drive, Londonderry, BT48 8SU
**MI6 DCH** David Hickey, Flat 2, 123 Woodvale Road, Belfast, BT13 1BP
**M6 DCI** David Morgan, Castle Cottage, Newbridge, Newport, NP11 3NT
**M6 DCJ** Colin Davies, 83 Freeston Avenue, St. Georges, Telford, TF2 9EN
**M6 DCK** Derrick Renshaw, 25 Ashley Road, Worksop, S81 7JS
**M6 DCL** David Lane, 46 Berkeley Vale Park, Berkeley, GL13 9TQ
**MM6 DCM** Derek Mifsud, 25 Priory Road, Linlithgow, EH49 6BP
**M6 DCN** Daniel Davies, 49 Heol y Wal, Bradley, Wrexham, LL11 4BY
**M6 DCO** Dean Close, 22 Station Road, Dodworth, Barnsley, S75 3JE
**M6 DCP** Alfred Pearson, 29 Broadoak Road, Langford, Bristol, BS40 5HD
**M6 DCQ** Dawn Manning, 153 Pavilion Road, Worthing, BN14 7EG
**M6 DCS** David Shephard, 17 Grimsby Road, Laceby, Grimsby, DN37 7DF
**MM6 DCT** David Tourish, 1 Langside Drive, Kilbarchan, Johnstone, PA10 2EL
**M6 DCU** Daniel Matthews, 81 Kipling Avenue, Goring-by-Sea, Worthing, BN12 6LH
**M6 DCW** David Wetton, 1 Monksway, Birmingham, B38 9LW
**M6 DCY** Alastair Kerr, 23 Manor Park, Duloe, Liskeard, PL14 4PT
**M6 DCZ** David Mooney, 107 Tedder Road, South Croydon, CR2 8AR
**M6 DDB** David Beavis, 17 Kingsbere Crescent, Dorchester, DT1 2DY
**M6 DDC** Derek Chebsey, 21 Shortlands Lane, Walsall, WS3 4AG
**M6 DDD** Vilnis Vesma, Pound House, Market Square, Newent, GL18 1PS
**M6 DDE** Michael Smith, 5 Cresswell Road, Ellington, Morpeth, NE61 5HR
**M6 DDF** David Poulton, 115 Highview, Vigo, Gravesend, DA13 0TQ
**M6 DDH** Daniel Muirhead, 13 Berry Street, Skelmersdale, WN8 8QZ
**M6 DDI** David Barnett, 49 Loundes Road, Unstone, Dronfield, S18 4DE
**M6 DDL** Peter Smith, 156 Esther Grove, Wakefield, WF2 8ET
**M6 DDM** Derek Maycroft, 100 Benwick Road, Doddington, March, PE15 0UH
**MM6 DDN** Daniel Hornal, 95 Traprain Crescent, Bathgate, EH48 2BD
**M6 DDO** David Daniels, 128 Woodcock Road, Norwich, NR3 3TD
**M6 DDP** Denise Stallibrass, 12 Sheerwater Close, Bury St. Edmunds, IP32 7HR
**M6 DDQ** David James, 13 Lincoln Road, Fenton, Lincoln, LN1 2EP
**MW6 DDR** Dennis Reeves, 27 Beaufort Road, Pembroke, SA71 4PX
**M6 DDT** David Cross, 4 Burns Avenue, Gloucester, GL2 5BJ
**M6 DDU** Daniel Hind, 116 Gowthorpe, Selby, YO8 4HA
**M6 DDV** David Borrett, 84 Kingsway, Mapplewell, Barnsley, S75 6EX
**M6 DDW** Dominic Webb, 9 Dunsfold Close, Crawley, RH11 8EY
**MM6 DDX** Mark Dougan, 41d Balmerino Road, Dundee, DD4 8RP
**M6 DDY** James Betteridge, 57 Wood Road, Chaddesden, Derby, DE21 4LY
**M6 DEC** Declan Hunter, 9 Gelt Burn, Didcot, OX11 7TZ
**MI6 DED** Alistair McCann, 6 Bowens Meadow, Lurgan, BT66 7UT
**M6 DEE** Delphine Daniels, 16 Bainbridge Road, Warsop, Mansfield, NG20 0ND
**M6 DEF** Anthony Clarke, 9 Birbeck Way, Frettenham, Norwich, NR12 7LG
**M6 DEG** Derek Edge, 18 Sandringham Avenue, Whitehaven, CA28 6XL
**M6 DEI** Daniel Beresford, 40 Fern Crescent, Congleton, CW12 3HQ
**M6 DEJ** David Thomas, Stud Farm Bungalow, Stud Farm Drive, Tamworth, B78 3HS
**M6 DEK** Derek Knighton, Holme Lea, Church Lane, Chesterfield, S44 5AL
**M6 DEL** Derek Millard, 112 Avenue Road, Sandown, PO36 8DZ
**M6 DEM** Leanne Demirkaya, 11 Beech Park, Holsworthy Beacon, Holsworthy, EX22 7NB
**M6 DEO** Derek Sproston, 22 Oakland Avenue, Haslington, Crewe, CW1 5PB
**M6 DER** Ian Dermondy, 4 West Bank Close, Keighley, BD22 6HQ
**M6 DES** Denis Kirkden, 57 Crow Hill Road, Margate, CT9 5PF
**M6 DEV** Mark Raynor, 68 Cambridge Street, South Elmsall, Pontefract, WF9 2AR
**MJ6 DEY** James Bryant, 5 Louiseberg Court Queen's Road, St. Helier, Jersey, JE2 3GQ
**M6 DEZ** James Butters, 7 Brick Street, Derby, DE1 1DU
**M6 DFC** Darryl Gee, 20 Davies Avenue, Brymbo, Wrexham, LL11 5AS
**M6 DFD** David Pennison, 69 Caneland Court, Waltham Abbey, EN9 3DS
**M6 DFE** Jonathan Wood, 3 Lion Lane, Haslemere, GU27 1JF
**M6 DFG** Daniel Gardner, 106 Willclare Road, Birmingham, B26 2NY
**M6 DFI** Eileen Garry, 34 Conway Road, Paignton, TQ4 5LH
**M6 DFJ** Terence Saunders, 40 Southdown Avenue, Brixham, TQ5 0AN
**M6 DFK** John Bailey, 22 Wilford Drive, Ely, CB6 1TL
**M6 DFL** David Loveys, 36 Harrowby Street, Stafford, ST16 3TY
**M6 DFM** Douglas McAuslan, Casa Arco Iris, Via Variante Nascente, Santa Barbara de Nex, Portugal, 8005-491
**M6 DFP** Darren Parker, 53 Brisbane Way, Cannock, WS12 2GR
**M6 DFQ** Ajay Singh, 19 Severn Crescent, Slough, SL3 8UU
**M6 DFS** Dennis Slade, 22 Oaklands Road, Mangotsfield, Bristol, BS16 9EY
**M6 DFT** Ian Lonsdale, 23 Hunts Field, Clayton-le-Woods, WLC Scout Council, Chorley, PR6 7TT
**M6 DFU** Kenneth Pownall, 17 Horsebridge Road, Blackpool, FY3 7BQ
**M6 DFW** Darren Whitley, 10 Kenmore Drive, Cleckheaton, BD19 3EJ
**MW6 DFX** Manuel Jorge Lima Barbosa, 65 Precelly Place, Milford Haven, SA73 2BW
**M6 DFY** Robert Medway, 54 Peasland Road, Torquay, TQ2 8PA
**M6 DFZ** Daniel Antcliffe, 3 Wiltshire Mews, Cottam, Preston, PR4 0NP
**M6 DGA** Dave Ashford, 56 Finch Close, Shepton Mallet, BA4 5GL
**MM6 DGC** David Baillie, 77 Main Street, Fauldhouse, Bathgate, EH47 9AZ
**M6 DGD** Douglas Bailey, 2b Queens Road, Enfield, EN1 1NE
**M6 DGG** Mark Harrison, 43 Second Avenue, Woodlands, Doncaster, DN6 7QQ
**M6 DGI** Peter Kirby, 102 Waterloo Road, Crowthorne, RG45 7NW
**M6 DGJ** Donald Cole, 4 Rosedale Road, Margate, CT9 2TD
**M6 DGM** Denis Mouland, 2 Chafy Cottages, Holnest, Sherborne, DT9 6HX
**M6 DGN** Iain Coleman, 24 Westwood Avenue, Plymouth, PL6 7HS
**M6 DGO** Anthony Thacker, 47 Hamilton Street, Walsall, WS3 3EN
**M6 DGP** Paul Dransfield, 7 Washburn Close, Filey, YO14 0DL
**M6 DGQ** Alexander Evans, 27 Oaklands Park Drive, Rhiwderin, Newport, NP10 8RB
**M6 DGR** Aleks Deaves, 46-48 Charlemont Drive, Manea, March, PE15 0GA
**M6 DGS** David Sole, 10 Glyn Place, East Melbury, Shaftesbury, SP7 0DP
**M6 DGU** Richard Jones, Mole Corner, Red Shute Hill, Thatcham, RG18 9QW
**MW6 DGW** Dave Whitcombe, 4 Fairview Road, Llangyfelach, Swansea, SA5 7JJ
**M6 DGY** Douglas Coyle, 3 Claddyburn Terrace, Cairnryan, Stranraer, DG9 8RD
**M6 DGZ** Danyel Baillie, 69 Main Street, Kirkcowan, Newton Stewart, DG8 0HQ
**M6 DHB** Alan Curd, 44 Elbridge Avenue, Bognor Regis, PO21 5AD
**MM6 DHF** Don Forsyth, Solano, Stirling Road, Dumbarton, G82 2PF
**M6 DHG** Darren Lappage, 37 Matlock Drive, Cannock, WS11 6EN
**MM6 DHI** Robert Buchan, 5 Fairview Terrace, Danestone, Aberdeen, AB22 8ZH
**M6 DHK** David Drake, 16 Orchard Street, Stafford, ST17 4AN
**MM6 DHS** Kevin Paterson, 28 Eglinton Street, Saltcoats, KA21 5DG
**M6 DHT** Adam Herd, 7 Water Lane, Greenham, Thatcham, RG19 8SH
**M6 DHU** David Maton, 41 Bemerton Gardens, Kirby Cross, Frinton-on-Sea, CO13 0LQ
**M6 DHV** David Vincent, 38 Methuen Street, Walney, Barrow-in-Furness, LA14 3PR
**M6 DHW** David Collins, 50 Woodville, Barnstaple, EX31 2HL
**MM6 DHZ** Iain Smith, 32 Kaimes Avenue, Kirknewton, EH27 8AU
**M6 DIB** Derrick Bloxsome, 74 Dunclair Park, Plymouth, PL3 6DE
**M6 DIF** S Goldsbrough, 2 Stutte Close, Louth, LN11 8YN
**M6 DIG** D Gozzard, Craig Dulas, Rhydyfoel Road, Abergele, LL22 8EG
**M6 DIH** Andrew Arnold, 1 Fairford Gardens, Wordsley, Stourbridge, DY8 5RF
**M6 DII** Ricky Wilson, 54 Curling Lane, Badgers Dene, Grays, RM17 5JB
**M6 DIJ** Duane Yates, 16 Sunnyfield Road, Prestwich, Manchester, M25 2RD
**M6 DIL** Dilawar Yakub, 42 Swift Close, Blackburn, BB1 6LF
**M6 DIM** Dimitris Vainas, 51 Magister Road, Bowerhill, Melksham, SN12 6FD
**M6 DIO** Jonathan Cant, 42 Copandale Road, Beverley, HU17 7BW
**M6 DIQ** Andrew Collins, 14 Double Corner, Mendlesham Road, Cotton, Stowmarket, IP14 4RF
**M6 DIT** Nicholas Bauff, 1 Lower Chart Cottages, Brasted Chart, Westerham, TN16 1LS
**M6 DIU** Paul Sewell, 17 Chatham Close, Coventry, CV3 1LY
**M6 DIV** John Inwood, Flat 359, Hagley Road Retirement Village, 330 Hagley Road, Birmingham, B17 8BP
**M6 DIZ** Dianne Balsdon, 25 Hanover Road, Plymouth, PL3 6BY
**M6 DJA** David Aspital, 47 St. Augustines Park, Ramsgate, CT11 0DF
**MM6 DJC** Declan Caveney, Westerhill, Clashnamuiach, Tain, IV20 1XP
**M6 DJG** Zoran Zmajkovic, 67 Oakington Avenue, Little Chalfont, Amersham, HP6 6SX
**M6 DJI** Darren Oliver, 20 Five Oaks Close, Malvern, WR14 2SW
**M6 DJJ** Ben Dodson, 3 Bradley Road, Patchway, Bristol, BS34 5LF
**M6 DJK** Leslie Oxberry, 72 Crowhall Towers, Crowhall Lane, Gateshead, NE10 0NG
**M6 DJL** Daniel Reader, 190 Barnby Dun Road, Doncaster, DN2 4RF
**M6 DJN** Davie Neville, 2 Oak View, Luddendenfoot, Halifax, HX2 6HQ
**M6 DJP** Darren Parvin, 11 Stanhope Way, Sevenoaks, TN13 2DZ
**M6 DJQ** Marcus Hodges, Simba Cottage, Lower Road, Bishop's Stortford, CM22 7RA
**M6 DJS** David Smith, 173 Leicester Road, Shepshed, Loughborough, LE12 9DG
**MW6 DJT** David Terrell, 82 Baglan Street, Treherbert, Treorchy, CF42 5AR
**M6 DJW** Dennis Wheeler, 58 Frankley Beeches Road, Northfield, Birmingham, B31 5AE
**M6 DJX** David Jones, 102 Bryce Road, Brierley Hill, DY5 4ND
**M6 DJY** Marc Jones, 3 John Street, Loftus, Saltburn-by-The-Sea, TS13 4JD
**M6 DKC** David Cowan, 30 Crucian Way, Liverpool, L12 0AW
**M6 DKF** Brian Hodgson, 28 Grove Drive, Woodhall Spa, LN10 6RT
**MM6 DKI** Sean Young, 6 Ramsey Cottages, Bonnyrigg, EH19 3JG
**M6 DKK** Peter Weston, 9 Clarendon Road, Smethwick, B67 6DA
**M6 DKL** Paul Shaw, 21 Urmson Street, Oldham, OL8 2AN
**M6 DKM** W Molloy, 32 Millers Barn Road, Jaywick, Clacton-on-Sea, CO15 2QB
**MM6 DKN** Neil Ford, 11 Kenmore Way, Coatbridge, ML5 4FN
**M6 DKO** Jack Greenwood, 18 Rookery Lane, Lincoln, LN6 7PY
**M6 DKR** Darren Reeves, Flat 2, Challonsleigh, Blandford Forum, DT11 7HB
**M6 DKS** Graham Hutton, 8 Popples Drive, Halifax, HX2 9SQ
**M6 DKT** David Carter, 85 Dingle Street, Oldbury, B69 2DZ
**M6 DKU** Brian Burgess, 26 Bakers Way, Morton, Bourne, PE10 0XW
**M6 DKW** Craig Nicholl, 36 Eylewood Road, London, SE27 9NA
**M6 DKY** Oscar Silva, 1 Grangewood Terrace, London, SE25 6TA
**M6 DLA** David Aldred, 19 Birch Avenue, Bacton, Stowmarket, IP14 4NT
**M6 DLB** Phillip Booth, 7 Handley Crescent, East Rainton, Houghton le Spring, DH5 9QX
**M6 DLC** Doreen Smith, 106 Middle Street, Blackhall Colliery, Hartlepool, TS27 4EB
**M6 DLD** Hessam Rasooli Nia, 14 Larch Close, London, N11 3NN
**M6 DLH** David De La Haye, 4 Nicola Mews, Ilford, IG6 2QE
**M6 DLI** Jamie Walters, 138 Lowe Avenue, Wednesbury, WS10 8NU
**M6 DLJ** Daniel Caldicott, 564 Fulbridge Road, Peterborough, PE4 6SA
**M6 DLL** Daniel Hallsworth, 29a Stephenson Court, Station Road, Stockport, SK5 6LE
**M6 DLM** Derek Le Mare, The Sycamore, Church Bank, Barnard Castle, DL12 0AH
**M6 DLO** Raymond Coles, 10 Littlemoor Road, Weymouth, DT3 6AA
**M6 DLP** Michael Palmer, New Haven, Stoneraise, Carlisle, CA5 7AX
**M6 DLQ** Peter Leng, The Barn, Gildersleets, Settle, BD24 0AH
**M6 DLU** Simon Bland, 61a Terminus Terrace, Southampton, SO14 3FE
**M6 DLV** Callum Tompkins, 19 Windermere Road, Gloucester, GL2 0NH
**M6 DLW** Daniel Endean, 11 Forrester Drive, Brackley, NN13 6NE
**M6 DLX** Barry Tufnell, 1 Moorlands Court Wath-upon-Dearne, Rotherham, S63 6DD
**M6 DLY** Ahmed Omar, 84 Beaumont Hill, Darlington, DL1 3ND
**M6 DMA** David Stockton, 19 Chadwick Road, Middlewich, CW10 0EA
**M6 DMB** Darren Booth, 75 Meynell Road, Sheffield, S5 8GL
**M6 DMC** Derek McCrae, 4 Castramont Road, Gatehouse of Fleet, Castle Douglas, DG7 2JE
**M6 DMD** Dylan Mathieson-Dodd, 1 Dag Lane, North Kilworth, Lutterworth, LE17 6HD
**M6 DME** Duane Elmy, 2 Mill Road Drive, Purdis Farm, Ipswich, IP3 8UT
**M6 DMF** Joshua Wilkins, 58 High Road, Wormley, Broxbourne, EN10 6JN
**M6 DMG** David Gillingham, 3 Rosier Close, Thatcham, RG19 4FN
**M6 DMN** Joseph Robbins, 5 South Close, Greatworth, Banbury, OX17 2DZ
**M6 DMO** Liam Wilburn, 23 Sutton Road, Kirk Sandall, Doncaster, DN3 1NY
**M6 DMP** David Allen, 9 The Crescent, Cootham, Pulborough, RH20 4JU
**M6 DMQ** David McQuirk, 200 Farthing Grove, Netherfield, Milton Keynes, MK6 4HW
**M6 DMW** Denise Carey, 78 Bentley Road, Bramley, Rotherham, S66 1UH
**M6 DMY** Alison Hoskins, 38 Tasmania Close, Basingstoke, RG24 9PQ
**M6 DNB** Daniel Eltham, 47 Glenville Close, Royal Wootton Bassett, Swindon, SN4 7EU
**M6 DNC** Dave Swift, 15 Gloucester Walk, Westbury, BA13 3XF
**M6 DNF** David Featherby, 14 Station Road, Sutton, Ely, CB6 2RL
**M6 DNJ** Paige Hendry, 19 Parsons Close, Portsmouth, PO3 5LN
**M6 DNL** Paul Swainson, 11 Conway Drive, Banbury, OX16 0QW
**M6 DNM** David Mitchell, 4 Millstream Close, Goostrey, Crewe, CW4 8JG
**M6 DNN** Dennis Baker, 117 Locks Road, Locks Heath, Southampton, SO31 6LJ
**M6 DNO** Cydney Sheath, 53 Manor Road, Trowbridge, BA14 9HS
**M6 DNP** David Morgan, Ty Bettws, Kilgwrrwg, Chepstow, NP16 6PN
**M6 DNQ** Brian Southern, 25 Chilgrove Avenue, Blackrod, Bolton, BL6 5TR
**M6 DNR** Lee Childs, 354 Linnet Drive, Chelmsford, CM2 8AL
**M6 DNS** Michael Carter, 61 Catherine Avenue, Swallownest, Sheffield, S26 4RQ
**M6 DNV** Giles Cater, 7 Seymour Street, Chelmsford, CM2 0RX
**M6 DNX** Daniel Garnham, Shardleas, Cock Road, Halstead, CO9 2SH
**M6 DNZ** David Hart, 163 Wakering Road, Shoeburyness, Southend-on-Sea, SS3 9TN
**M6 DOA** David Morris-Jones, 121 Plas Dinas, Blacon, Chester, CH1 5SW
**M6 DOB** Alan Cowan, 111 Oaks Drive, St. Leonards, Ringwood, BH24 2QS
**MM6 DOC** Anthony Davis, 13 High Road, Auchtermuchty, Cupar, KY14 7BE
**MI6 DOD** Alan Henry, 10 Drumbreda Crescent, Armagh, BT61 7PE
**M6 DOF** Adam Barker, 49 Rockingham Avenue, York, YO31 0TD
**M6 DOG** Harley Godzisz, 1 Wyvern Place Green Lane, Addlestone, KT15 2UD
**M6 DOJ** Alexander Deery, 25 Ribblesdale Place, Preston, PR1 3NA

M6 DOM Dominic Chrumka, 94 Clock House Road, Beckenham, BR3 4JT
M6 DON Donald Ashcroft, Don Ashcroft, Iken House, Woodbridge, IP12 2GA
M6 DOW James Hogg, 31 Roseberry Grove, York, YO30 4SU
M6 [illegible] [illegible], Warrington, WA3 3JJ
M6 [illegible] [illegible]
M6 DPA Craig Cooke, 42 Biddle Road, Leicester, LE3 9HG
M6 DPF Peter Fletcher, 106 Stubley Lane, Dronfield, S18 1PH
M6 DPG David Glover, 21 Monastery Road, Paignton, TQ3 3BU
M6 DPH David Hancock, 2 Trevine Meadows, Indian Queens, St. Columb, TR9 6NB
M6 DPI Steven Hammond, 89 Beeston Road, Sheringham, NR26 8EJ
M6 DPJ Paul Jones, 50 Clay Lane, Doncaster, DN2 4RJ
M6 DPL Graham Payne, Jay Close, Eastbourne, Bn23 7RW
M6 DPN Roy Fripp, 41 Sweyns Lease, East Boldre, Brockenhurst, SO42 7WQ
M6 DPO Peter Beck, Castle Clanyard Farm, Drummore, Stranraer, DG9 9HF
M6 DPP Daniel Parkinson, 24 New Road, Chatteris, PE16 6BW
M6 DPQ Marylyn McMath, 12 Townhead Crescent, Dalry, Castle Douglas, DG7 3UR
M6 DPR David Runyard, Tilecroft, Shortheath Crest, Farnham, GU9 8SA
M6 DPU Cesar Lombao, 6 Privet Close, Lower Earley, Reading, RG6 4NY
M6 DPV Donald Peacock, Wyncum, Bridge Road, Castle Douglas, DG7 1TN
M6 DPW David Wilde, 9 Redstone Park, Redhill, RH1 4AS
M6 DPX Andrew Potts, 28 Thistlebarrow Road, Salisbury, SP1 3RT
M6 DPY Darren Hansford, 62 Bays Road, Pennington, Lymington, SO41 8HN
M6 DPZ David Stansfield, 19 Cotman Fields, Norwich, NR1 4EN
M6 DQC Robert Blackwell, Vikings Hall, Baylham, Ipswich, IP6 8JS
M6 DQD Jonathan Earye, 28 Halls Drift, Kesgrave, Ipswich, IP5 2DE
M6 DQF Jordan Pace, 25 Heaton Close, London, E4 6UF
MM6 DQG Ayden Stewart, 94 Dick Crescent, Burntisland, KY3 0BT
M6 DQH Ben Hartland, 2 Brookland Close, Pevensey Bay, Pevensey, BN24 6RT
M6 DQK Andrew Walker, 4 Pretymen Crescent, New Waltham, Grimsby, DN36 4NS
M6 DQN James Davis, 23 Blueberry Gardens, Andover, SP10 3XD
M6 DQO David Stocker, 2 Paull Road, Bodmin, PL31 1QJ
M6 DQP Anne Lewis, Four Winds Cottage, Main Street, Brough, HU15 1RJ
M6 DQQ Dai Jones, 6 Frondeg, Tredegar, NP22 3NT
M6 DQR Gary Youll, 4 Shaftsbury Court, Barnstaple Road, Scunthorpe, DN17 1YB
M6 DQS Thomas Meaker, 1 Chemin Des Brugues, Roquevidal, France, 81470
M6 DQT Wesley Martindale, 57 Limefield Street, Accrington, BB5 2AF
M6 DQV Paul Patterson, 3 Barnes Close, Southampton, SO18 5FE
M6 DQW Alfred Hose, 9 Cothey Way, Ryde, PO33 1QY
MM6 DQY John Dow, 52 Muirfield Way, Deans, Livingston, EH54 8EN
M6 DQZ Selwyn James, 35 Prospect Road, Dronfield, S18 2EA
M6 DRG Graeme Lythgoe, 137 Dore Avenue, Fareham, PO16 8DU
M6 DRH Donald Hughes, Warrenwood, Granville Rise, Totland Bay, PO39 0DX
M6 DRI Terry McKinley, 7 May Close, Godshill, Ventnor, PO38 3HB
MM6 DRJ Mark Cormack, 2 Coghill Street, Wick, KW1 4PN
M6 DRK David Kurn, 10 Dymoke Road, Mablethorpe, LN12 2BF
MW6 DRM David Machon, 22 Albert Street, Caerau, Maesteg, CF34 0UF
M6 DRO David Drought, 14 McMinnis Avenue, St. Helens, WA9 2PL
M6 DRP Daniel Rogers, 24 Bramblewood Way, Halesworth, IP19 8JT
M6 DRR David Roderick, 88 Broadway, Wakefield, WF2 8LY
M6 DRT Daniel Chamberlain, The Bungalow, Bure Valley Lane, Norwich, NR11 6UA
MW6 DRV Daniel Jones, 59 Llewelyn Street, Aberdare, CF44 8LA
M6 DRW David Williams, 2 Cublington Cottages, Madley, Hereford, HR2 9NX
M6 DRZ Alison Potts, 103 Etherstone Street, Leigh, WN7 4HY
MM6 DSC Kerr Henderson, 66 Rashierigg Place, Longridge, Bathgate, EH47 8AT
MM6 DSD Dennis McMillan, 45 Eton Avenue, Dunoon, PA23 8DG
M6 DSE Stephen Clarke, 23 Sinclair Court, Scarborough, YO12 7SD
M6 DSH Daniel Sheehan, 2 Pheasant Close, Thurston, Bury St. Edmunds, IP31 3TR
M6 DSJ Dominique Johnson, 5 Blackbird Close, Thurston, Bury St. Edmunds, IP31 3PF
M6 DSO David Osborne, 1 Bramble Close, Wilnecote, Tamworth, B77 5GG
MW6 DSP David Pitman, 60 Tudor Estate, Maesteg, CF34 0SW
M6 DSQ Matthew Bayman, 4 South Road, Beccles, NR34 9NN
M6 DSR Damien Rhodes, 66 Lindale Gardens, Blackpool, FY4 3PQ
M6 DSS Walter Stewart, 43 Newlands Drive, Halesowen, B62 9DX
M6 DST David Stubbs, 39 Torcross Way, Redcar, TS10 2RU
M6 DSU John Wynne, 8 Coed Artro, Llanbedr, LL45 2LA
M6 DSV Shaun Townsend, 13 Cornwall Drive, Bury, BL9 9ET
M6 DSW David Weight, 12 Durrants Path, Chesham, HP5 2LH
M6 DSY Sarah Waters, 28c Cliff Road, Dovercourt, Harwich, CO12 3PP
M6 DTA Douglas Easden, 20 Brunel Way, Calne, SN11 9FN
M6 DTC Don Cane, 98 Lancaster Road, Northolt, UB5 4TL
M6 DTD Kenneth Worton, 39 Staite Drive, Cookley, Kidderminster, DY10 3UA
MI6 DTE David Best, 13 Cranley Green, Bangor, BT19 7FE
M6 DTF Martyn Bell, 36 Schneider Road, Barrow-in-Furness, LA14 5DW
M6 DTG David Griffiths, 12 Lanceside Road, Grange over Sands, LA11 7DX
M6 DTH David Hodgson, 13 Merlin Way, Bicester, OX26 6YG
M6 DTJ Daniel Brown, 1 North Dalton Cottages, Dalton-le-Dale, Seaham, SR7 8PY
M6 DTL Brian Jarvis, [illegible] Aldermans Green Road, Coventry, CV2 1NN
M6 DTM Daniel McCarrigle, 40 The Glade, Waterlooville, PO7 7PE
M6 DTN David Thurman-Newell, 76 Prince Charles Avenue, Minster on Sea, Sheerness, ME12 3PP
M6 DTO Guy Howe, 3 Halton Close, Lincoln, LN6 0YZ
M6 DTP Matthew Topham, Scardale, Thorn Bank, Hebden Bridge, HX7 5HS
M6 DTR Steve Rhenius, Baythorne Cottage, Baythorne End, Halstead, CO9 4AB
M6 DTS Diane Mason, 15 Lake Street, Dudley, DY3 2AU
M6 DTT Derrick Timbrell, 15a Firgrove Crescent, Yate, Bristol, BS37 7AH
M6 DTU Michael Woodruff, 21 Cross Green Close, Formby, Liverpool, L37 4BP
MM6 DTV Ian Campbell, 1 Carbostbeg, Carbost, Isle of Skye, IV47 8SH
M6 DTX Peter Rose, 10 Milton Street, Worthing, BN11 3NE
M6 DTY Colin Bell, 28 New Forest Motel, 230 Hurn Road, Ringwood, BH24 2BT
M6 DUB Tim Price, 40 East Street, Kidderminster, DY10 1SE
M6 DUD Mark Pugh, 31 Cambridge Street, Reading, RG1 7PA
M6 DUE David Duell, Manor House, 144 Stonhouse Street, London, SW4 6BE
M6 DUF Lee Duffy, 46 Church Road, Worcester, WR3 8NU
M6 DUH Benjamin Brown, 40 Stansfield Road, London, SW9 9RZ
M6 DUI Graeme Walton, 7 Burlington Street, Ulverston, LA12 7JA
M6 DUJ Alan Arnold, 83 Grange Crescent, Lincoln, LN6 8BY
M6 DUK Sarah House, 12 Glebe Close, Abbotsbury, Weymouth, DT3 4LD
[illegible]
M6 DUM John Rowe, 9 Corfield Close, Finchampstead, Wokingham, RG40 4PA
M6 DUN Alan Rigler, 10 The Ball, Dunster, Minehead, TA24 6SD
M6 DUO Anita Richards, 114 Northleach Close, Redditch, B98 8RD
MI6 DUP Robert McAuley, 37 Ladyhill Road, Antrim, BT41 2RF
MI6 DUR Andrew Savage, 469 Old Belfast Road, Bangor, BT19 1RQ
M6 DUT Terry Collins, 56 Grasvenor Avenue, Barnet, EN5 2DB
MM6 DUV Douglas Loughren, 16 Merlewood Road, Inverness, IV2 4NL
M6 DUW Terence Brooks, 200 Kingsway, College Estate, Hereford, HR1 1HE
M6 DUX Samuel Jones, 3 Brockton, Lydbury North, SY7 8BA
M6 DUY Brian Lewis, 10 Healey Avenue, Knypersley, Stoke-on-Trent, ST8 6SQ
M6 DUZ Barnaby Davies, 12 Scalebor Gardens, Burley-in-Wharfedale, LS29 7BX
M6 DVA James Bligh-Wall, 3 George Street, Elworth, Sandbach, CW11 3BL
M6 DVE David Thomson, 24, jubilee creasent, Clowne, S43 4NB
M6 DVF Keith George, Wylye, Auberrow, Hereford, HR4 8AN
M6 DVH Desmond Harrison, 30 Horseshoe Drive, Cannock, WS12 0FH
MI6 DVM David Simpson, 16 Scarvagh Locks Scarva, Craigavon, BT63 6NB
MI6 DVN Mark Devlin, 17 Moninna Park, Newry, BT35 8PP
MM6 DVO David Plummer, 39 St. Nicholas Drive, Banchory, AB31 5YG
M6 DVP David Price, 11 Cefn Melindwr, Capel Bangor, Aberystwyth, SY23 3LS
MW6 DVQ Albert Jones, 31 Russell Terrace, Carmarthen, SA31 1SZ
MM6 DVR Richard Dover, Brooklet, Dodside Road, Glasgow, G77 6PZ
M6 DVS Ethan Johnson, 3 Conifer Way, Dunmow, CM6 1WU
M6 DVT Daniel Truscott, 37 Langley, Chulmleigh, EX18 7BQ
M6 DVU Michael Hughes, 183 Station Road, Hednesford, Cannock, WS12 4DP
M6 DVV S Storey, 11 Enderby Road, Sunderland, SR4 6BA
M6 DVW Stewart Challis, 73 Rivenhall Way, Hoo, Rochester, ME3 9GF
M6 DVX Nick Malyon, 19 Swallow Cliffe, Shoeburyness, Southend-on-Sea, SS3 8BL
M6 DVY James Sawyer, 18 The Mead, Dunmow, CM6 2PD
M6 DVZ John Haywood, 8 Cedar Close, Market Rasen, LN8 3BE
M6 DWA Andrew Dickson, The Rowans, Pwllmeyric, Chepstow, NP16 6LA
MM6 DWC Kyle Mclachlan, 13 Dick Terrace, Penicuik, EH26 8BW
M6 DWE David Webb, 52 Simpkin Close, Eaton Socon, St. Neots, PE19 8PD
M6 DWF David Waring, 12 Mary Street, Farnhill, Keighley, BD20 9AU
M6 DWG Darren Gaskell, 10 Freshford, St. Helens, WA9 3WT
M6 DWI Colin Hopper, 31 Mary Road, Deal, CT14 9HW
M6 DWJ David Johnston, 5 Moorfield Crescent, Hemsworth, Pontefract, WF9 4EQ
M6 DWM Dylan Mitchard, 49 Gladstone Road, Broadstairs, CT10 2HY
M6 DWO Aaron Martin, 97 The Maltings, Dunmow, CM6 1BY
MM6 DWP Donald Park, 54 Coblecrook Gardens, Alva, FK12 5BL
M6 DWS David Saunders, 17 Sandy Lane Prestwich, Manchester, M25 9RU
M6 DWT David Potts, 103 Etherstone Street, Leigh, WN7 4HY
M6 DWV Mark Tointon, 13 Ridgeway, Broadstone, BH18 8DY
M6 DWW David Ward, 26 Binsted Road, Sheffield, S5 8LL
M6 DWY R Friar, 16 Upper Field Close, Redditch, B98 9LE
M6 DWZ Dylan Whitaker, 67 Oakington Avenue, Amersham, HP6 6SX
M6 DXA Alice Kendrick, 29 Waterside, Silsden, Keighley, BD20 0LQ
M6 DXC Sara-Jayne Crawford, 63 Birch Grove Crescent, Brighton, BN1 8DP
M6 DXG Daniel Gaffney, 9 Tudor Road, Newton Abbot, TQ12 1HT
M6 DXH Iain Craig, Lowry Hill, Irthington, Carlisle, CA6 4PE
M6 DXI John Mee, 42 Potters Mead, Wick, Littlehampton, BN17 7HY
MM6 DXJ Steven Lewington, 2 Coulliehare Cottages, Udny, Ellon, AB41 7PH
M6 DXK Colin Rose, 132 Golf Green Road, Jaywick, Clacton-on-Sea, CO15 2RW
M6 DXL Marshall Groves, 15 Plains Lane, Littleport, Ely, CB6 1RJ
M6 DXN David Curtis, 7 Neale Close, Aylsham, Norwich, NR11 6DJ
M6 DXQ David Baker, 26 Lighthouse Close, Happisburgh, Norwich, NR12 0QE
M6 DXS Daniel Martin, 25 Broadmead Road, Blaby, LE8 4AB
M6 DXV Gary Frayne, 98 De Lacy Court, New Ollerton, Newark, NG22 9RW
M6 DXW Dave Sawyer, 60 Greenway Lane, Chippenham, SN15 1AE
M6 DXZ Peter Jones, 18 Meadowbrook Road, Wirral, CH46 0RS
M6 DYA Derek Greaves, 2 Willow Walk, Keynsham, Bristol, BS31 2TR
M6 DYB David Brough, 38 Tynedale Avenue, Crewe, CW2 7NY
M6 DYC Philip Richards, 2 The Mayflowers, Norwich, NR11 6FZ
M6 DYD Mark Jones, 1 Elizabeth Court, Elizabeth Avenue, Norwich, NR7 0GY
M6 DYF John Jones, 40 Maes Mona, Amlwch, LL68 9AT
M6 DYG Stephen Aspinall, 3 Grasswood Road, Wirral, CH49 7NT
M6 DYH David Houston, The Knowe, Hardgate, Castle Douglas, DG7 3LD
M6 DYI Leslie Edmonds, 3 Waterlow Road, London, N19 5NJ
MW6 DYL Dylan Witts, 82 Park View, Llanharan, Pontyclun, CF72 9SB
MD6 DYM Joshua Dunbar, 3 Glenview Terrace, Port Erin, Isle of Man, IM9 6HA
M6 DYN Chris Hall, 14 Marsh Road, Cowes, PO31 8JQ
M6 DYO Dylan Osborne, 3 Low Farm Road, Tunstall, Norwich, NR13 3PU
M6 DYP Rachel Laver, 20 Hall Street, Church Gresley, Swadlincote, DE11 9QU
M6 DYR Antony Hateley, 75 Allanville, Camperdown, Newcastle upon Tyne, NE12 9XT
M6 DYU Heidi Coghlan, Charterhouse, Orchard Road, Salisbury, SP5 2JA
M6 DYV Richard Dunn, 29 Hawe Lane, Sturry, Canterbury, CT2 0LL
M6 DYW Stephen Jones, 35 Sturminster Road, Bristol, BS14 8BQ
M6 DYX Mark Carter, 15 Lashbrooks Road, Uckfield, TN22 2AY
M6 DYY Nigel Christopher, 161 Manor Road, Verwood, BH31 6DX
M6 DZA Darren Gilbert, 34 Sullivan Way, Elstree, Borehamwood, WD6 3DH
M6 DZB Michael Bennetts, 2 Chywoone Terrace, Newlyn, Penzance, TR18 5NR
MM6 DZC Tom Whyte, Upper Hillockhead, Glass, Huntly, AB54 4XS
M6 DZF Alastair Tiling, 9 Coombe Street, Coventry, CV3 1GG
M6 DZH Hans Kassier, 26 Higher Port View, Saltash, PL12 4BX
M6 DZN Andrea Bertoneri, Flat 1, 1 Albert Road, Nottingham, NG9 2GU
M6 DZP Mike Marsh, 25 Southdown Road, Seaford, BN25 4PD
M6 DZQ Edward Zieba, 12b Chingford Avenue, London, E4 6RP
MW6 DZR David Malins, 25 Earl Street, Cardiff, CF11 7DQ
M6 DZS Zoltan Derzsi, 217 Bensham Road, Bensham, Gateshead, NE8 1US
M6 DZT John Emery, Mulberry Cottage, Quarry Lane, Chard, TA20 3PH
M6 DZU Andrew Stopford, 1 Campden Road, Cheltenham, GL51 6AA
M6 DZV David Abbott, 3 Brewhouse Lane, Soham, Ely, CB7 5JD
M6 DZX Arthur Norton, Flat 9, Brownhill Court, Southampton, SO16 9LB
M6 DZY Mark Parker, 35 Prescelly Close, Nuneaton, CV10 8QA
M6 DZZ Darren Hunt, 6 Lewisham Terrace, Newtown, Berkeley, GL13 9NP
M6 EAA Eva Johnston, 67 Eversfield Road, Horsham, RH13 5JS
MI6 EAC [illegible]
[illegible]
M6 EAE James Collier, 133 Woodstock Road, Moston, Manchester, M40 0DG
MI6 EAF Gordon Monteith, 66 Allen Park, Dunamanagh, Strabane, BT82 0PD
M6 EAH Melanie Bell, 2 Hawthorne Road, Blyth, NE24 3DT
MI6 EAI Edward Taylor, 17 Rutherglen Street, Belfast, BT13 3LR
M6 EAJ John Lavery, Shottendane Road, Birchington, CT7 0HD
M6 EAL Eric Westwood, 17 Ennerdale Drive, Congleton, CW12 4FR
M6 EAN Ian Parsons, 44 Hungerford Crescent, Bristol, BS4 5HQ
M6 EAO James Johnson, 1, GOLYGFAR EGLWYS RUABON, Wrexham, LL14 6TD
M6 EAP Robert Taylor, 18a Barnhall Road, Tolleshunt Knights, Maldon, CM9 8HA
MI6 EAS John Henderson, 1 Brook Lodge Ballinderry Lower, Lisburn, BT28 2GZ
M6 EAT Ronald Carr, Rambler House Hill, Berkeley, GL13 9EB
M6 EAW Edward Whitehouse, 16 rue Gaston de Caillavet, Paris, France, 75015
M6 EAX Fay Davies, 64 Main Road, Moulton, Northwich, CW9 8PB
MI6 EAZ Albert Ioner, 10 Heatherlea Avenue, Portstewart, BT55 7HF
M6 EBA Alex Thompson, 30 Birchwood Avenue, Lincoln, LN6 0JB
M6 EBB Eamonn Bias, 10 Riverdale Road, Shrewsbury, SY2 5TA
M6 EBC Mark Petchey, 74 Avondale, Ellesmere Port, CH65 6RW
MM6 EBD Alasdair Connell, 39 Glebe Crescent, Maybole, KA19 7HZ
M6 EBE Eamonn Fogarty, Noke Farm, Hogscross Lane, Coulsdon, CR5 3SJ
M6 EBG Jenny Thomas, 128 Nuns Way, Cambridge, CB4 2NS
M6 EBI Cyrus Bakes, Container City Building, 48 Trinity Buoy Wharf, London, E14 0FN
MM6 EBJ Kenneth McCuish, 6 Lighthouse Buildings, Birch Drive, Isle of Islay, PA43 7HZ
M6 EBL Robert Clarke, 37b Northdown Park Road, Margate, CT9 2NH
M6 EBN Peter Singleton, 7 Queen Street, Glossop, SK13 8EL
M6 EBO Brian Clayton, 26 Wood Walk, Mexborough, S64 9SG
M6 EBP Dale Morton, 3 Pritchett Road, Birmingham, B31 3NL
M6 EBQ Dorothy Stanley, 58 Wells Gardens, Basildon, SS14 3QS
M6 EBR Andrew Buckland, 21 Malton Close, Monkston, Milton Keynes, MK10 9HR
MI6 EBS Edwin McKnight, 14 Marlacoo Beg Road, Portadown, Craigavon, BT62 3TF
M6 EBU Lloyd Wells, 54 Giffords Cross Avenue, Corringham, Stanford-le-Hope, SS17 7NH
M6 EBW Christopher Rowles, 15 Rockhampton Walk, Colchester, CO2 8UJ
M6 EBX Sam Lucas, Sunnyside, Church Road, Woodbridge, IP13 7NU
M6 EBY Christine Barrett, 5 Oakapple Drive, Dereham, NR19 2SR
MM6 EBZ Ewan Ball, Rottenrow Farm, Ochiltree, Cumnock, KA18 2RJ
M6 ECB Emily Brady, 24 Gregory Avenue, Colwyn Bay, LL29 7ND
MI6 ECC Gareth Haslem, 124 Castle Rise, Tandragee, Craigavon, BT62 2NF
M6 ECD Craig Dennis, 1 West Villa, Crathorne, Yarm, TS15 0BA
M6 ECH Edward Hearn, 41 Romney Avenue, Newcastle, ST5 7JR
M6 ECJ Aaron Meszaros, 9 Dunster Gardens, Cheltenham, GL51 0QT
M6 ECM Gareth Stapylton, 5 Coxcomb Walk, Crawley, RH11 8BA
MW6 ECR Kenneth Gulliford, 5 The Square Abertridwr, Caerphilly, CF83 4DH
M6 ECT Eric Singleton, 1 Ermin Park, Brockworth, Gloucester, GL3 4BD
MI6 ECV Christina Rafferty-Floyd, 9 Killybrack Mews, Omagh, BT79 7FB
M6 ECW Cliff Wilson, 31 Violet Road, South Woodford, London, E18 1DG
M6 ECX Jacques Steventon, 16 Tarrant Rushton, Blandford Forum, DT11 8SD
M6 ECZ James Hart, Flat 4, 17 Trinity Gardens, Folkestone, CT20 2RP
M6 EDA Angharad Bache, 62 Whittingham Road, Halesowen, B63 3TP
M6 EDB James Beck, 32 Oakfield Close, Potters Bar, EN6 2BE
M6 EDD Eddie Young, 210 high street, Gt Wakering, SS3 0LS
M6 EDH Edward Hitchins, 50-52 Moorland Road, Weston-Super-Mare, BS23 4HR
M6 EDI Jason Cherry, 12 Scarisbrick New Road, Southport, PR8 6PY
M6 EDJ Daniel Jeffery, 4 Sandhurst Drive, Beeston, Nottingham, NG9 6NH
MI6 EDK Kenneth Pearson, 17, KNOCKNAMOE ROAD, Omagh, BT79 7LB
M6 EDL Michelle Leigh, 310 Carlton Road, Barnsley, S71 2BQ
M6 EDM Philip Short, 15 Longcroft Court, Birchwood Crescent, Chesterfield, S40 2HT
M6 EDN Andrew Strange, 38 Manor Road, Martlesham Heath, Ipswich, IP5 3SY
M6 EDO Derek Rasbarry, 27 Royle Close, Romford, RM2 5PS
MI6 EDP Anthony Connolly, 68 Willowbank Gardens, Belfast, BT15 5AJ
MW6 EDQ James Blaxland, 28 Limewood Close, St. Mellons, Cardiff, CF3 0BU
MW6 EDR Ernest Rees, 23 Northlands Park, Bishopston, Swansea, SA3 3JW
M6 EDS Edwin Kaye, 119 St. Bernards Avenue, Louth, LN11 8AS
M6 EDV Paul Tranter, Swn y Gwynt, Trefeglwys, Caersws, SY17 5PU
M6 EDW Jake Tranter, Swn y Gwynt, Trefeglwys, Caersws, SY17 5PU
M6 EDY Edward Harman, 53 Anthony Road, Borehamwood, WD6 4NB
M6 EDZ Edson Twagirayezu, 33 Morton Street, Stoke-on-Trent, ST6 3PN
M6 EEA Markus Rabl, 14 Bramble Close, Durrington on Sea, BN13 3HZ
M6 EEB Joseph Cameron, 29 Webster Road, Stanford-le-Hope, SS17 0BE
MI6 EEC Seamus Carlin, 9 Mullandra Park, Kilcoo, Newry, BT34 5LS
M6 EED Craig Wright, 74 Charlemont Crescent, West Bromwich, B71 3DA
M6 EEF Asad Al Busaidi, 18 Sirdar Road, Southampton, SO17 3SJ
M6 EEG Andrew Gow-Barber, Plum Tree Villa, Butchers Lane, Ormskirk, L39 6SY
M6 EEH Christine Branch, 62 Turnpike Road, Connor Downs, Hayle, TR27 5DT
MW6 EEJ Joshua Nicholas, 41 Heritage Drive, Cardiff, CF5 5QD
M6 EEL Lee Layland, 3 Thirlmere Road, Golborne, Warrington, WA3 3HH
M6 EEM David Roberts, 3 Heather Avenue, Melksham, SN12 6FX
M6 EEN Peter Farren, 89 Fosterd Road, Newbold, Rugby, CV21 1DE
M6 EEO Stuart Davey, 38 Wordsworth Street, Barrow-in-Furness, LA14 5SE
M6 EEP Matthew Bartlett, 7 Thoresby, Tamworth, B79 7SQ
MM6 EEQ Graham Brown, 21 Russell Road, Lanark, ML11 7HL
MM6 EER Euan Grant, 6 Lindsay Place, Wick, KW1 4PF
M6 EET Paul Sladen, 25 Linden Grove, Beeston, Nottingham, NG9 2AD
M6 EEU Alexander Bullard, 15 Rowan Drive, Lutterworth, LE17 4SP
M6 EEW Alan George, 15 Ely Road, Croydon, CR0 2LW
MM6 EEX Leslie Grant, 6 Lindsay Place, Wick, KW1 4PF
M6 EEZ Pete Murphy, 44 Samuel Road, Sheffield, S2 3UF
M6 EFC David Shannon, 68 Thornhill Road, Claydon, Ipswich, IP6 0DZ
MI6 EFD Elizabeth Hudson, 10 Kiloanin Crescent, Banbridge, BT32 4NU
M6 EFE Thomas Barnden, 27 Milton Road, Wokingham, RG40 1DE
M6 EFF Emma Phipps, 12 Salisbury Close, Wokingham, RG41 4AJ

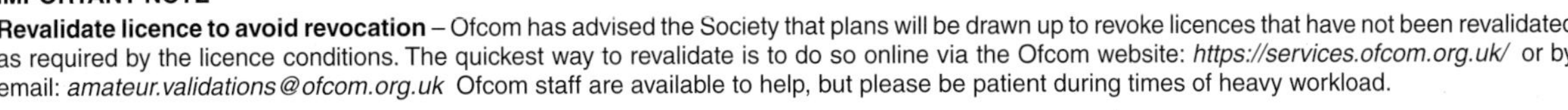

**IMPORTANT NOTE**

**Revalidate licence to avoid revocation** – Ofcom has advised the Society that plans will be drawn up to revoke licences that have not been revalidated as required by the licence conditions. The quickest way to revalidate is to do so online via the Ofcom website: *https://services.ofcom.org.uk/* or by email: *amateur.validations@ofcom.org.uk* Ofcom staff are available to help, but please be patient during times of heavy workload.

UK Callsigns

M6 EFG Jonathan Jones-Robinson, 32 Verity Close, London, W11 4HE
MM6 EFH Peta Donachie, 53 Seaforth Avenue, Wick, KW1 5NE
M6 EFJ Alexander Davies, 20 Hope Street, Halesowen, B62 8LU
MW6 EFK Allan Williams, 62 Wern Road, Llanelli, SA15 1SR
M6 EFL Aikaterini Miariti, 19 Camelot Avenue, Nottingham, NG5 1DW
M6 EFM Eva Falomir Montanes, 31 Albany Road, Birmingham, B17 9JX
M6 EFP Peter Williamson, 22 Earls Road, Shavington, Crewe, CW2 5EZ
M6 EFR Denis Richman, 48 Whinfield Avenue, Fleetwood, FY7 7NE
M6 EFU Simbarashe Nyakabau, 4 Vale View, Charvil, Reading, RG10 9SJ
M6 EFV John Dyer, 32 Brynystwyth, Penparcau, Aberystwyth, SY23 1SS
M6 EFW Christopher Steele, 40 Landor Road, Whitnash, Leamington Spa, CV31 2JX
M6 EFY Richard Bell, 2 Hawthorne Road, Blyth, NE24 3DT
MD6 EGA Varun Santhosh, 119 Vaughan Road, Harrow, HA1 4EF
M6 EGB Ian Kendrick, 10 Bankhouse Drive, Congleton, CW12 2BH
MM6 EGC James Flannigan, 21 Kirkbean Avenue, Rutherglen, Glasgow, G73 4EA
M6 EGD Patrick Jones, 2 Fourth Avenue, Gwersyllt, Wrexham, LL11 4EE
M6 EGE Jeremy Powell, 23 Park Road, Norton, Malton, YO17 9DZ
M6 EGF Harvey Reeves, 15 Mill Rise, Kidsgrove, Stoke-on-Trent, ST7 4UR
M6 EGG Matthew Jackson, 1 Kindred Barns, Ludlow, SY8 4LF
MW6 EGH Elliot Barker, 7 Upper High Street, Bedlinog, CF46 6RY
M6 EGJ Anish Chudasama, 19 Walton Court, Bolton, BL3 6QP
M6 EGK Christopher Suddell, Lynhurst, Littleworth Lane, Horsham, RH13 8JX
M6 EGL Karl Hendricks, Stranton, Darsham Road Westleton, Saxmundham, IP17 3AH
M6 EGM Leo Karaalp, 45 Baird Grove, Kesgrave, Ipswich, IP5 2DQ
M6 EGN Antony Blackburn, 81 Belvedere Road, Ipswich, IP4 4AD
M6 EGO Peter Bailey, 103 Jarden, Letchworth Garden City, SG6 2NZ
M6 EGP John Churchill, West Winds, Brandheath Lane, New End, Redditch, B96 6NG
M6 EGS Michael Sparrow, 21 Langwell Crescent, Ashington, NE63 8AB
M6 EGT Jonathan Drake, 38 Fawcett Road, Stevenage, SG2 0EJ
M6 EGU Akash Sharma, 152 Ladybarn Lane, Manchester, M14 6RW
M6 EGV Chris Cousins, 43 Avon Close, Little Dawley, Telford, TF4 3HP
M6 EGW Michael Delaney, 24 Lonsdale Road, Manchester, M19 3FL
M6 EGX Michael Cross, 11 Polyplatt Lane, Scampton, Lincoln, LN1 2TL
M6 EGZ Matthew Rozier, 5 Pond Piece, Brandeston, Woodbridge, IP13 7AW
M6 EHA Malcolm Philpott, Garmisch, Hazel Road, Aldershot, GU12 6HP
M6 EHB Neil Livingstone, 2 Mickleton, Wilnecote, Tamworth, B77 4QY
M6 EHC Emily Harris, 147 Longdown Road, Congleton, CW12 4QH
M6 EHD Edward Delasalle, 31 West Hill Road, Hoddesdon, EN11 9DL
M6 EHE Leslie Hayward, The Chimes, Farley Way, Hastings, TN35 4AS
M6 EHF Simon Bateson, 2 Green Crescent, Coxhoe, Durham, DH6 4BE
M6 EHH Nigel Hunt, 21 Reams Close, Fishtoft, Boston, PE21 0LL
M6 EHI David Humm, 15 Sherborne Road, Farnborough, GU14 6JS
M6 EHK Sander Buruma, 40, 40, Groningen, Netherlands, 9746PL
MM6 EHL Stuart Nicol, 9 Cowley Street, Methil, Leven, KY8 3QG
M6 EHM Jack Jackson, 49 Leafield Rise, Two Mile Ash, Milton Keynes, MK8 8BX
M6 EHO Gregory Gibbs, 11 Fieldway Avenue, Leeds, LS13 1ED
M6 EHP P Garraway, The Poplars, Crowell Road, Chinnor, OX39 4HP
M6 EHQ Paul Simmen, 7 Thorpe Road, Thornton, Bradford, BD13 3AT
M6 EHR Paul Marshall, 3 Woodlands Croft, Kippax, Leeds, LS25 7RN
M6 EHU David Hodgson, 11 Harmony Place, Mountain, Bradford, BD13 1LD
M6 EHV William Bradley, 4 forest view avenue, London, E10 6DX
M6 EHW Paul Fellows, Flat 8, Gilbert Wilkinson House, Pontefract, WF8 1QA
M6 EHX Bernard Bull, Swan Cottage, Swan Road, Welshpool, SY21 0RH
M6 EHY Kevin Brundle, 17 The Paddocks, Hailsham, BN27 3AQ
M6 EIA John Rutter, 19 Shaftesbury Avenue, Great Harwood, Blackburn, BB6 7ST
M6 EIB Alan Timms, 37 Tan y Llan, Tregynon, Newtown, SY16 3HA
M6 EIC Stuart Hammond, 1 Elmer Close, Bognor Regis, PO22 6JU
M6 EIE Richard Rose, 30 Brudenell Close, Cawston, Rugby, CV22 7GN
M6 EIF Keith Goss, 57 Nursery Road, Leicester, LE5 2HQ
M6 EIG Richard Webb, Norbury, Terrills Lane, Tenbury Wells, WR15 8DD
M6 EIH Kevin Harris, 78 Dovedale Road, Thurmaston, Leicester, LE4 8NB
M6 EII John Rodgers, 5 Richil House, 7 Ayston Road, Oakham, LE15 9RL
M6 EIJ Jack Iason, 23 Baydon Grove, Calne, SN11 9AT
M6 EIK Goronwy Edwards, 17 Glan y Mor Road, Penrhyn Bay, Llandudno, LL30 3NL
M6 EIL Simon Wilkes, 24 Shelley Grove, Droylsden, Manchester, M43 7YG
M6 EIM John Darmont, 114 Bowers Avenue, Norwich, NR3 2PS
M6 EIO Michael Topple, 41 Whitehall Close, Colchester, CO2 8AJ
M6 EIP Christopher Newton, 5 St. Andrews Crescent, Harrogate, HG2 7RT
M6 EIR Alan Mulcahy, 85 Grifon Road, Chafford Hundred, Grays, RM16 6NP
M6 EIS Isabella Mahoney, 74 Radegund Road, Cambridge, CB1 3RS
M6 EIU Guy Farnbank, 2 Market Hill, Foulsham, Dereham, NR20 5RU
M6 EIW David Blake, Pound Farm, Swan Lane, Leigh, Swindon, SN6 6RD
M6 EIX Gerard Edgar, 98 Barnton Road, Dumfries, DG1 4HN
M6 EIY Jamie Lee, 30 Wentworth, Yate, Bristol, BS37 4DJ
M6 EIZ George Denman, 123 Links Avenue, Norwich, NR6 5PQ
M6 EJA Andy Ashton, 46 Kingsland, Harlow, CM18 6XL
M6 EJB Patrick State, 53 Long Lane, Shirebrook, Mansfield, NG20 8AZ
M6 EJC Ethan Crosby, 16 Tweed Avenue, Ellington, Morpeth, NE61 5ES
M6 EJF Robert Glynn, 106 Fairway Avenue, West Drayton, UB7 7AP
M6 EJG Andre Zennadi, 54 Llys Gwyrdd, Henllys, Cwmbran, NP44 7LS
M6 EJI Robert Colson, 1 Wildon Cottages, Twyn Allwys Road, Abergavenny, NP7 9RS
M6 EJJ Matthew Corr, Flat K, Windsor Court, 14 Winn Road, Southampton, SO17 1EN
MI6 EJK Joshua Milligan, 12 Rose Park, Limavady, BT49 0BF
M6 EJL Mark Leggett, 22 St. Marys Drive, Diss, IP22 4PT
MM6 EJO Bryan Holderness, 13 Glencairn Street, Camelon, Falkirk, FK1 4LY
M6 EJP Jean Siddle, 7 Farebrother Street, Grimsby, DN32 0NH
M6 EJR Ashley Graham, 34 Balmoral Road, Stockport, SK4 4EB
M6 EJS Edward Scott, 10 Spratton Road, Brixworth, Northampton, NN6 9DS
MI6 EJT Christopher Rafferty, 9 Killybrack Mews, Omagh, BT79 7FB
M6 EJV Arthur Bowman, The Glen, Vicarage Road, Bude, EX23 8LN
M6 EJW Edward Williamson, 70 Douglas Drive, Stevenage, SG1 5PH
M6 EJX Robert Dickerson, 61 Highfield Terrace, Queensbury, Bradford, BD13 2BE
M6 EJY Marcus Boddy, 4 Witton Close, Reedham, Norwich, NR13 3HJ
M6 EKA Eric Armstrong, 30 Tennyson Avenue, Hull, HU5 3TW
M6 EKB Evangeline Beech, 124 Evering Avenue, Poole, BH12 4JH
M6 EKC Sergey Klymenko, 32 Oxford Road, London, NW6 5SL
M6 EKD Michael D'Arcy, 1 Moorfield Avenue, Denton, Manchester, M34 7TF
M6 EKE James Welch, 130 Sandy Lane, Upton, Poole, BH16 5LY
M6 EKF Graeme Haime, 26 Pipit Close, Weymouth, DT3 5RT
M6 EKI Graham Lloyd, 1 Holmside Terrace, Stanley, DH9 6ET
M6 EKK Andrew Powell, 2 Ormsby Close, Hopton, Great Yarmouth, NR31 9TY
M6 EKL Simon Ross, 3 Highlands, Lakenheath, Brandon, IP27 9EU
M6 EKM Malcolm Collyer, 26 Beaumont Drive, Northampton, NN3 8PS
M6 EKO Stuart Johnson, Willow End Cottage, Willow Corner, Thetford, IP25 6SS
M6 EKP Stephen Milner, Pavilion House, School Lane, Ormskirk, L40 3TG
M6 EKQ Matthew Isbell, 20 Woodland Crescent, Wolverhampton, WV3 8AS
M6 EKS Jessica Goodson, 56 Chestnut Avenue, Spixworth, Norwich, NR10 3QQ
M6 EKT Adam Price, 141 Attlee Way, Cefn Golau, Tredegar, NP22 3TE
M6 EKV Andrew Rimmer, 9 Brook Meadow, Wirral, CH61 4YS
M6 EKW David Rimmer, 41 Ashburton Road, Wallasey, CH44 5XB
M6 EKX Colin Evenden, 36 Castle View Road, Fareham, PO16 9LA
M6 EKY Adrian Irwin, 145 Coppermill Lane, London, E17 7HD
M6 ELB William Astill, 14 Barlow Road, Exton, Oakham, LE15 6BL
M6 ELC Jean-Paul Parkes, 24 Kenilworth Road, Lichfield, WS14 9DP
M6 ELD Andrew Titmus, 5 Kithurst Crescent, Goring-by-Sea, Worthing, BN12 6AJ
M6 ELE Eleanor Nichols, 20 Holly Blue Road, Wymondham, NR18 0XJ
M6 ELF Katherine Windass, 58 Nicholas Gardens, High Wycombe, HP13 6JG
M6 ELH Neil Henderson, 14 Herbert Street, Carlisle, CA1 2QE
M6 ELI Robert Wells, 27 Victoria Avenue, Camberley, GU15 3HT
M6 ELJ Stephen Girdwood, 8 Blaydon Walk, Wellingborough, NN8 5YU
M6 ELL Kenneth Ellison, 3 Rutherglen Square, Sunderland, SR5 5LH
M6 ELN David Shuttleworth, 11 Bladen Close, Countesthorpe, Leicester, LE8 5SB
M6 ELR Emma Reeve, 12 Sime Street, Worksop, S80 1TD
M6 ELS James Fay, 16 Foxhill, Whissendine, Oakham, LE15 7HP
MM6 ELU Greg Ross, 5 Main Street, Alford, AB33 8QA
M6 ELW Paul Thornley, 94 Lindel Road, Fleetwood, FY7 7LX
M6 ELX Eiddwen Davies, 26 Heol Sant Gattwg, Llanspyddid, Brecon, LD3 8PD
M6 ELY Elysia Cockett, 2a Priory Avenue, Petts Wood, Orpington, BR5 1JF
M6 ELZ Eleanor Luckett, 180 Clarendon Place, Dover, CT17 9QF
M6 EMA Emma Bansil, 65 Hervey Street, Northampton, NN1 3QL
M6 EMC James Landless, 2 Aspen Way, Banstead, SM7 1LE
M6 EME Ellen Dunstan, 57 Orchard Vale, Flushing, Falmouth, TR11 5TT
M6 EMF Paul Cotton, 29 Peake Avenue, Nuneaton, CV11 6DW
M6 EMG Duncan Gray, 68 Endeavour Way, Hythe Marina Village, Southampton, SO45 6LA
M6 EMH Neil Asling, 9 First Avenue, Halifax, HX3 0DL
MI6 EMI Karol Mikicki, 429 Beersbridge Road, Belfast, BT5 5DU
M6 EMJ Edward Jubb, 59 Buckingham Road, Conisbrough, Doncaster, DN12 3DG
M6 EMK Michael Klimaszewski, 1 Woodwards Cottages, New Brighton, Wrexham, LL11 3ED
M6 EML Robin Copus, 44 Lyndhurst Road, Tilehurst, Reading, RG30 6UE
M6 EMM Emma Barton, 86 Forge Lane, Kingswinford, DY6 0LG
M6 EMO Joseph Carey, 68 Queen Elizabeth Way, Woking, GU22 9AJ
M6 EMP Peter Holman, 20 Green Drive, Wolverhampton, WV10 6DW
M6 EMQ Alan Johnson, Avonworth, Grange Road, Bedford, MK43 7HJ
M6 EMS Christopher Shaw, 52 Margaret Avenue, Halesowen, B63 4BX
M6 EMT Kevin Titmarsh, 3 Meadow Cottages, Banningham, Norwich, NR11 7ED
M6 EMV Eulah Varley, Flat 8, Brookside Court, Liverpool, L23 0TT
M6 EMW Steven Hunter, 9 Gelt Burn, Didcot, OX11 7TZ
M6 EMX Benjamin Di-Giulio, 4 Highlands, Lakenheath, Brandon, IP27 9EU
M6 EMY Emily Porter, 16 The Oval, Scarborough, YO11 3AP
M6 EMZ Emma Stubbs, 39 Torcross Way, Redcar, TS10 2RU
M6 ENB Tony Gurney, 24 Langley Way, Kettering, NN15 6HL
M6 ENC Frederick Pauling, Kingswood Farm House, Dalehouse Lane, Kenilworth, CV8 2JZ
M6 END Dhirendra Kataria, 87 Oakhill Road, Horsham, RH13 5LH
M6 ENE Vincent Roets, 7 Thorne Close, Harworth, Doncaster, DN11 8SN
M6 ENF Elishia Briggs, 20 Broad Lane, Pelsall, Walsall, WS4 1AP
M6 ENH Frank Baker, 275 Bye Pass Road, Beeston, Nottingham, NG9 5HS
M6 ENI Charlie Passey, 1 Forest House, Baxworthy, Bideford, EX39 5SF
M6 ENJ Oliver Karaalp, 45 Baird Grove, Kesgrave, Ipswich, IP5 2DQ
M6 ENK Alexander Seelig, Flat 67, Regents Riverside, Reading, RG1 8QS
M6 ENM Thomas Smythe, 19 Lime Close, Witham, CM8 2PA
M6 ENN Sean Burton, 20 Flowerdown Avenue, Cranwell, Sleaford, NG34 8HZ
M6 ENP Kevin Matthews, St. Helens Cottage, Flimby, Maryport, CA15 8RX
M6 ENQ Michael McKenna, 25 Heaton Place, Norton Road, Colwyn Bay, LL28 4TL
MI6 ENR Mark Carruthers, 8 Ruskey Road, Cookstown, BT80 0AA
M6 ENS David Spencer, 38 Town House Road, Nelson, BB9 9LL
M6 ENU Carole Gordon, 9 Park Road, Camberley, GU15 2SP
M6 ENV Donald Lock, 22 The Towers, Southgate, Stevenage, SG1 1HE
M6 ENW Rui Ferreira, 22 Vereker Road, London, W14 9JS
M6 ENX Michael Reid, 3 Beeston Hall Mews, Brook Lane, Tarporley, CW6 9TZ
MW6 ENY Kirsty Miles, 25 Park Street, Penrhiwceiber, Mountain Ash, CF45 3YW
M6 ENZ Steven Margrave, 30 Newtown Road, Bedworth, CV12 8QU
M6 EOA Callum Shaw, 26 Grant Road, Spixworth, Norwich, NR10 3NN
M6 EOB Bernard Bland, 12 Park Lane, Pickmere, Knutsford, WA16 0JX
M6 EOC Paul Flanagan, 71 Fellway, Pelton Fell, Chester le Street, DH2 2BY
MM6 EOE Adrian Mather, 38 Shandon Crescent, Balloch, Alexandria, G83 8EX
M6 EOK Paul Wilson, 45 Newquay Close, Hartlepool, TS26 0XG
M6 EOM Dieudonne Kawayida, Flat 7, Centre Point, London, SE1 5NU
M6 EON Ian Patrick, Beech Cottage, Church Road, Boston, PE22 8RD
MW6 EOP Lewis Rowlands, 14 Claerwen, Gelligaer, Hengoed, CF82 8EW
M6 EOQ Richard Edmonds, Flat 3, Elvington Lodge, 40 Reigate Hill, Reigate, RH2 9NG
M6 EOR Don Kershaw, 2 Croftlands, Neddy Hill, Carnforth, LA6 1JE
M6 EOS Anthony Beresford, 22 Tennyson Road, Rotherham, S65 2LR
M6 EOT Luke Seaton, 4 Highfield Close, Empingham, Oakham, LE15 8QB
MM6 EOU Alec Gorman, 36 Harestanes Road, Armadale, Bathgate, EH48 3LA
M6 EOW Emma Taylor, 11 Carlton Street, Featherstone, Pontefract, WF7 6AA
M6 EOX Ian Smith, White House Farm, Sandy Lane, Market Rasen, LN8 3YF
M6 EOY Ioan Jones, 90 Preston, Cirencester, GL7 5PR
MM6 EOZ Garry Headridge, 79 Sheriffs Park, Linlithgow, West Lothian, EH49 7SR
M6 EPA Timothy Bannister, 6 Tanners Road, North Baddesley, Southampton, SO52 9FD
M6 EPD Declan Oshea, 37 Barclay Court, Ilkeston, DE7 9HJ
MW6 EPE Paul Plummer, 26 Hill Road, Neath Abbey, Neath, SA10 7NR
M6 EPF Andrew Cook, 84 Clent View Road, Birmingham, B32 4LW
M6 EPH Dale Hollis, 192 Rodbourne Road, Swindon, SN2 2AF
M6 EPJ Philip Jensen, 36 Douglas Street, Derby, DE23 8LH
M6 EPL Nigel Wachs, 59 Broad Oak Way, Cheltenham, GL51 3LL
M6 EPN David Cocks, 30 Great Arler Road, Leicester, LE2 6FF
M6 EPO Ashley Smith, 6 Chi Rio Close, Shepton Mallet, BA4 4TR
M6 EPQ David Tomlinson, 74 Bradshaw Avenue, Riddings, Alfreton, DE55 4AA
M6 EPS Stephen Cook, 270 Heneage Road, Grimsby, DN32 9NP
M6 EPT Jose Rosa, Flat 7, Wick Hall, Abingdon, OX14 3NF
MM6 EPV Alastair Fraser, GARDEN COTTAGE, MOY, Inverness, IV13 7YQ
M6 EPW Elizabeth Waller, 27 Main Street, Frizington, CA26 3SA
MW6 EPX Douglas Bentley, 9 Pen-y-Lan Place, Cardiff, CF23 5HE
M6 EPZ Nigel Highfield, 298 Mersea Road, Colchester, CO2 8QY
MI6 EQA James McGoldrick, 45 Stewarts Road, Dromara, Dromore, BT25 2AN
M6 EQB Michael Prince, Crantock, Fourth Avenue, Ross-on-Wye, HR9 7HR
MI6 EQC Estelle Hamill, 24 Beechvalley, Dungannon, BT71 7BN
MI6 EQD Anthony Clearn, 182a Clonmore Road, Dungannon, BT71 6HX
M6 EQE Keith Missenden, 47 Roseacre Drive, Elswick, Preston, PR4 3UQ
M6 EQF Peta McGuinness, The Coach House, Cwmdauddwr, Rhayader, LD6 5HA
M6 EQG Andrew Nicholson, 147 Watling Avenue, Seaham, SR7 8JG
M6 EQJ Andrew Sherwin, 10 Closes Side Lane, East Bridgford, Nottingham, NG13 8NA
M6 EQK Jeremy Tarrant, 70 Sunnymead, Midsomer Norton, Radstock, BA3 2SD
M6 EQL Robert Paul, Mayo, Raleigh Park, Barnstaple, EX31 4JD
MM6 EQM David Frost, Lanfair, Kennethmont, Huntly, AB54 4NN
M6 EQN David Akerman, The Brick Barn, Coppice Farm, Ross-on-Wye, HR9 7QW
M6 EQO Amy Walton, 11 Parkfield Road, Northwich, CW9 7AR
M6 EQQ Mark Stiles, 49 Cattedown Road, Plymouth, PL4 0PL
M6 EQR Danny Leonard, 19 Nancy Street, Manchester, M15 4FZ
M6 EQS Jahanzeb Khan, 123 Highfield Road, Hall Green, Birmingham, B28 0HR
M6 EQT Derek Taylor, Flat 207, The Metropole, Folkestone, CT20 2LU
M6 EQV Adam Gould, 57 Northfield Road, Onehouse, Stowmarket, IP14 3HE
MM6 EQW Calum Hutton, 1 Hillend Cottage, Roberton, Biggar, ML12 6RR
MM6 EQY Kevin Archibald, 31 Westwood Park, Deans, Livingston, EH54 8QP
M6 EQZ Samuel Carlisle, 103 Crossway, Plymouth, PL7 4HZ
M6 ERA Carl Jenkins, 25 Longmeadow Grove, Birmingham, B31 4SU
M6 ERC Eric Thornes, 11 Ringtale Place, Baldock, SG7 6RX
M6 ERD Raymond Overy, 62 Dykelands Road, Sunderland, SR6 8ER
M6 ERF Emily Felton, 29 Pavitt Meadow, Galleywood, Chelmsford, CM2 8RQ
M6 ERH Ellis Halford, 69 Thirlmere Avenue, Astley, Manchester, M29 7PZ
M6 ERI Maciej Konstantynowicz, 12 Long Wall, Haddenham, Aylesbury, HP17 8DL
M6 ERN Raymond Springall, 27 Westbourne Park, Scarborough, YO12 4AS
MM6 ERO Zoe Bak, 62/6 North Gyle Loan, Edinburgh, EH12 8LD
M6 ERP simon jones, Queens Court, Flat 20, Cottage Lane, Burntwood, WS7 4XY
M6 ERQ David Lockyer, 19b Drury Lane, Buckley, CH7 3DU
M6 ERR Nigel Wood, 10 Perriclose, Chelmsford, CM1 6UJ
M6 ERS Thomas Glenday, 52 Hollow Road, Bury St. Edmunds, IP32 7AZ
M6 ERU Malcolm Cook, 194 Exeter Road, Kingsteignton, Newton Abbot, TQ12 3NJ
M6 ERW Ian Westby, 5 Rosklyn Road, Chorley, PR6 0NJ
M6 ERZ Esther Gibson, 8 Llanthewy Close, Croesyceiliog, Cwmbran, NP44 2PF
MW6 ESA Steven Beavis, 65 Old Road, Baglan, Port Talbot, SA12 8TU
M6 ESH Michael Young, 4 Priestburn Close, Esh Winning, Durham, DH7 9NF
M6 ESJ Andrew Fugard, Flat 4, 1 Hazelmere Road, London, NW6 6PY
MM6 ESL Gordon Robinson, 3 Ivy Lane, Dysart, Kirkcaldy, KY1 2XD
M6 ESO Peter Melling, 23 Appletree Close, Cottenham, Cambridge, CB24 8UJ
M6 ESP Eric Phiri, 26 The Drove, Andover, SP10 3DL
M6 ESR Antony Hunt, 4 Weedon Road, Swindon, SN3 4EE
M6 ESS Mark Chamberlain, 79 Riddy Lane, Luton, LU3 2AJ
M6 EST Robert Aldridge, 6 Sheppards Court, Horsenden Lane North, Greenford, UB6 7QJ
M6 ESV Philip Taylor, 104 Winstanley Drive, Leicester, LE3 1PA
MW6 ESW Ewan Sweeney, 67 Blandon Way, Cardiff, CF14 1EH
M6 ESY Stephen Prior, 17 Queens Walk, London, NW9 8ES
M6 ESZ Edward Palo, 13 Welwyn Close, St. Helens, WA9 5HL
M6 ETA David Wall, 227 Wayfield Road, Chatham, ME5 0HJ
M6 ETC Phil Hayes, 4 London Road, Roade, Northampton, NN7 2NL
M6 ETD Richard Sindall, 16 Chantrell Road, Wirral, CH48 9XP
MI6 ETE William Curry, 7 Ballyversal Road, Coleraine, BT52 2ND
M6 ETG Phillip Loades, 126 Wealcroft, Gateshead, NE10 8QS
M6 ETH Edward Heath, 63 Meadway, Dunstable, LU6 3JT
M6 ETJ Toby Widdowson, 18 Newland Road, Banbury, OX16 5HQ
M6 ETL David Arthur, 10 Lyndhurst Court, Lyndhurst Road, Hove, BN3 6FZ
M6 ETN Robert Bradshaw, 272 Councillor Lane, Cheadle Hulme, Cheadle, SK8 5PN
M6 ETP Emlyn Price, 47 Albany Road, Reading, RG30 2UL
M6 ETR Stephen Guest, 19 Ellesmere Avenue, Ashton-under-Lyne, OL6 8UT
M6 ETS Nicholas Luckett, Flat 3, 25 Upton Park, Slough, SL1 2DA
M6 ETU Peeyush Gaur, 34 Queensberry Avenue, Copford, Colchester, CO6 1YN
M6 ETW Richard Atkins, 19 Vale View, Abergavenny, NP7 6BE
MW6 ETY Dean Pesticcio, Ty Ffynnon Farm, St. Mellons Road, Cardiff, CF3 2TX
M6 ETZ Malcolm Harvey-Ross, Flat 4, Harley Court, Church Road, Southampton, SO31 9GD
MM6 EUA Charles Lockerbie, 11 Bedford Place, Aberdeen, AB24 3PA
M6 EUB Oliver Wilson, 99 Farnham Road, Durham, DH1 5LN
M6 EUC Terrence Coghlan, 13 Michaels Road, Rhyl, LL18 4SH
M6 EUE Nigel Austerfield, 22b Princes Avenue, Withernsea, HU19 2JA
M6 EUH David Levett, 11 Love Lane, London, SE25 4NG
M6 EUI Damien Nolan, Flat 7, Fonthill Court, London, SE23 3SJ
M6 EUK Stephen Taylor, 9 Crud yr Awel, Prestatyn, LL19 8YQ
M6 EUL Nick Ramsey, Dalestones, Lansdown Road, Bath, BA1 5TB
MW6 EUO Alan Jones, 75 Hollybush Road, Cardiff, CF23 6SZ
M6 EUP James Wilson, 125 Langroyd Road, Colne, BB8 9ED

**M6 EUU** Robert Scholey, 2 Newfield Crescent, Wath-upon-Dearne, Rotherham, S63 6JN
**M6 EUV** Andrey Shitov, Flat 4, 31 St. Leonards Road, Exeter, EX2 4LR
**M6 EUW** [illegible]
**M6 EUY** [illegible]
**M6 EUZ** Richard Smith, 2 Queen Street, Boston, PE21 8XB
**M6 EVA** Michael Knowles, 12 Dalestorth Avenue, Mansfield, NG19 6NT
**M6 EVB** Joseph Barnes, Glebe Farm, Billington, Stafford, ST18 9DQ
**M6 EVC** James Chapelle, 7 Elizabeth Way, Stowmarket, IP14 5AX
**M6 EVD** Clive Harper, 4 Bentley Avenue Jaywick, Clacton-on-Sea, CO15 2JW
**M6 EVE** Bryony Howard Evered, 2-3 Lower Downside, Downside, Shepton Mallet, BA4 4JP
**M6 EVF** Mark Cadman, 7 Horsham Avenue, Stourbridge, DY8 5LU
**M6 EVH** Harold Woodfin, 8 Bank Hall Close, Bury, BL8 2UL
**M6 EVI** Christopher Charles, 83 Dibdale Road, Dudley, DY1 2RX
**M6 EVK** Paul Hadley, Narrow Boat Oregon The Moreings, Kinver, DY7 6LG
**M6 EVL** Adrian Hoile, 4 Lindale Mount, Wakefield, WF2 0BH
**M6 EVM** John Thompson, Rose Cottage, Mickleton, Barnard Castle, DL12 0JD
**MM6 EVO** James Rickerby, 12/10 Hermand Street, Edinburgh, EH11 1LR
**M6 EVR** Stephen Rodgers, Flat 2, Rossiter Wood Court, Bristol, BS11 0RW
**MM6 EVS** Stephen Chisholm, 111 Philips Wynd, Hamilton, ML3 8PH
**M6 EVU** Matthew Marrs, 43 Ely Close, Toothill, Swindon, SN5 8DB
**M6 EVW** Eric Woodward, 309 Hartfields Manor, Hartfields, Hartlepool, TS26 0NW
**M6 EVX** Charles Haynes, 25 Barnards Hill Lane, Seaton, EX12 2EQ
**M6 EVY** Russell Dalton, 1 Ballard Estate, Four Lanes, Redruth, TR16 6QL
**M6 EVZ** David Lynch, 7 Dollant Avenue, Canvey Island, SS8 9EJ
**M6 EWC** Philip Jenkins, 137 Hawkhurst Road, Brighton, BN1 9EB
**M6 EWD** Graham Dobson, 4 Durley Gardens, Orpington, BR6 9LL
**M6 EWH** Michael Lovering, 16 Portland Avenue, Sittingbourne, ME10 3QY
**M6 EWI** Richard Farhall, 2 Banks Cottages, Mountfield, Robertsbridge, TN32 5JZ
**M6 EWJ** Richard Scott, 9 Burrows Close, Lawford, Manningtree, CO11 2HE
**M6 EWK** Simon Dempster, 18 Salisbury Street, Swindon, SN1 2AN
**M6 EWL** Carl Donovan, 222 Long Riding, Basildon, SS14 1RS
**M6 EWN** David Carmichael, 22 California Close Great Sankey, Warrington, WA5 8WU
**M6 EWO** Paul Chadwick, 112 Sandy Lane, Warrington, WA2 9JA
**M6 EWQ** Albert Williams, 5 Burleys Road, Crawley, RH10 7DB
**MM6 EWR** Allan Robson, 100 Dawson Avenue, East Kilbride, Glasgow, G75 8LH
**M6 EWT** Anton Grounds, 27 Longridge, Colchester, CO4 3FD
**M6 EWU** Lee Schofield, 21 Wyche Close, Rudheath, Northwich, CW9 7TY
**M6 EWV** Shaun Wagstaff, 20 Hawfinch Road, Cheadle, Stoke-on-Trent, ST10 1RX
**M6 EWW** Edward Whitewood, 2 Finn Farm Cottage, Finn Farm Road, Ashford, TN23 3EX
**MM6 EWX** Angel Jr Talplacido, 15 Park Avenue, Thurso, KW14 8JP
**MM6 EWY** Alan Kinnersley, 5a Regent Terrace, Dunshalt, Cupar, KY14 7HB
**M6 EWZ** Emad Wali Zangana, 78 Hivings Hill, Chesham, HP5 2PG
**M6 EXA** Stuart Morris, 14 Englefield Close, Crewe, CW1 3YN
**M6 EXB** Terence Thompson, 14 Queen Street, Northwich, CW9 5JL
**M6 EXC** Andrew Hawksworth, 17 St. Clements Court, Weston, Crewe, CW2 5NS
**M6 EXD** Mark Chapman, 7 Dragons Lane, Shipley, Horsham, RH13 8GD
**M6 EXE** Christopher Rose, 47 Ramsons Avenue, Conniburrow, Milton Keynes, MK14 7BB
**M6 EXF** Stephen Brown, 53 Prestwich Avenue, Worcester, WR5 1QF
**M6 EXG** Alan Holt, 27 Ramsey Road, Middlestown, Wakefield, WF4 4QF
**M6 EXH** Daniel Phelps, 728 Sewall Highway, Coventry, CV6 7JJ
**M6 EXI** Jon Edgson, 59 Gilmour Crescent, Worcester, WR3 7PJ
**M6 EXJ** Stephen Baldwin, 143 Oxford Road, Swindon, SN3 4JA
**M6 EXL** Louise Mead, 5 Hill Top, Ebbw Vale, NP23 6PJ
**M6 EXN** Alfred Gaskin, 44 Vale Road, Sutton, SM1 1QH
**M6 EXO** Thomas Crocker, 32 Godmanston Close, Poole, BH17 8BU
**MI6 EXP** Adrian Boyd, 27 The Meadows, Dungannon, BT71 6PW
**M6 EXQ** David Harrison, 104 Gracemere Crescent, Birmingham, B28 0TZ
**M6 EXR** Mitch Jackson, 35 Marshall Road, Willenhall, WV13 3PB
**M6 EXS** Jeffrey Harrison, 69 The Cliff, Wallasey, CH45 2NN
**M6 EXT** Richard Murray, 41 East Street, Rochdale, OL16 2EG
**MI6 EXU** Talula Reid Bamford, 2 New Line, Carrickfergus, BT38 9DL
**MW6 EXV** Elizabeth McMorrow, 164 Maes Glas, Caerphilly, CF83 1JW
**M6 EXW** Tempie Williams, 5 Burleys Road, Crawley, RH10 7DB
**M6 EXX** Bruce Jackson, 2a Scrooby Street, Rotherham, S61 4PL
**M6 EXY** Sharon Lake, 85 Clarkson Road, Norwich, NR5 8ED
**M6 EXZ** Gary Cockerell, 21 Coningsby, Bracknell, RG12 7BE
**M6 EYA** Martin Rasell, 21 Avenue Sucy, Camberley, GU15 3EB
**M6 EYB** Shane Crouch, 24 Church Road, Stanfree, Chesterfield, S44 6AQ
**M6 EYD** Steve Houssart, Flat 3, Virginia Court, London, SE16 6PU
**M6 EYF** Andrew Screen, Greenglade, Frith End, Bordon, GU35 0RA
**M6 EYG** Paul Cooper, 9 Handby Street, Hasland, Chesterfield, S41 0AT
**M6 EYH** Stephen Brown, 3 Grazebrook Croft, Birmingham, B32 3NL
**M6 EYI** Bryan Ravenhill-Lloyd, 7 Tegarn, Pencaenewydd, Pwllheli, LL53 6RA
**M6 EYJ** Richard Miller, 23 Clarendon Road, Sevenoaks, TN13 1EU
**M6 EYK** Jamie Peden, 07 Gwanhill Lane, Pontefract, WF8 8QN
**M6 EYL** Daniel Humphrey, 11 Colborne Close, Poole, BH15 1UR
**M6 EYM** David Smith, 3 Charles Close, Abergavenny, NP7 6AP
**M6 EYO** Richard Sansom, 72 Wannock Lane, Eastbourne, BN20 9SQ
**M6 EYP** Simon Strange, 94 Digby Avenue, Nottingham, NG3 6DY
**M6 EYR** Kevin Sim, 49 St. Julians Wells, Kirk Ella, Hull, HU10 7AF
**M6 EYS** Paul Smith, 29 Ford Hayes Lane, Stoke-on-Trent, ST2 0HB
**M6 EYT** Gareth Thomas, 8 Bankside, Headington, Oxford, OX3 8LT
**M6 EYU** Eifion Parry, 22 Park Road, Tanyfron, Wrexham, LL11 5SH
**M6 EYV** Jonathan Allen, Croston, Old Hall Road, Ulverston, LA12 7DL
**M6 EYW** Gordon Littlechild, 85 Long Road, Canvey Island, SS8 0JB
**M6 EYX** Christopher Taylor, 23 Heol Derw, Brynmawr, Ebbw Vale, NP23 4TT
**M6 EYY** Dione Wilkinson, 20 Manchester Road, Barnoldswick, BB18 5PR
**M6 EYZ** David Mullard, 46 Green Lane, Clanfield, Waterlooville, PO8 0JX
**M6 EZA** Graham Iredale, Ship Cottage, Main Street, Maryport, CA15 7DX
**M6 EZC** Mark Bentley, 5 Stokewell Road, Wath-upon-Dearne, Rotherham, S63 6EL
**M6 EZE** Michael Wood, 22 Prickett Road, Bridlington, YO16 4AT
**M6 EZF** John Holland, 1 Ravenwood, Swadlincote, DE11 9AQ
**M6 EZG** Richard Zerafa, 2 Furnwood, Bristol, BS5 8ST
**M6 EZK** John Power, 12 Campbell Gordon Way, London, NW2 6RS
**MD6 EZM** a Metselaar, 5a Canons Corner, Edgware, HA8 8AE
**M6 EZN** Andrew Barrett-Sprot, 1 Malting End, Wickhambrook, Newmarket, CB8 [illegible]
**MM6 EZO** [illegible]
**M6 EZP** Jonathan Crabb, 23 Chaffinch Crescent, Billericay, CM11 2YX
**M6 EZR** Peter Harper, 7 Duncan Gardens, Bath, BA1 4NQ
**M6 EZS** James Marks, Chantry End, Oak Hill, Epsom, KT18 7BU
**M6 EZT** Alexander Wareham-Kirk, Orchard Cottage, Church Street, Haverhill, CB9 7SG
**M6 EZV** Keith Gosney, 76 Westgate Lane, Lofthouse, Wakefield, WF3 3NS
**M6 EZX** Dennis Buchan, 40 Bradstone Road, Winterbourne, Bristol, BS36 1HQ
**M6 EZY** Craig Cowley, Rushbrook, Leafield Road, Chipping Norton, OX7 6EA
**M6 EZZ** Carl Garratt, 1 Billinge Crescent, St. Helens, WA11 9DQ
**M6 FAB** Lynette Rolinson, 534 Haslucks Green Road, Shirley, Solihull, B90 1DS
**M6 FAC** Brian Whelan, 147 Lawsons Road, Thornton-Cleveleys, FY5 4PL
**M6 FAE** Stephen Moorcroft, 57 Town Green Lane, Aughton, Ormskirk, L39 6SE
**M6 FAF** Jaqueline Taylor, 19 Juniper Walk, Kempston, Bedford, MK42 7SX
**M6 FAH** Simon Wilson, 4 Fenn Street, Tamworth, B77 2LP
**M6 FAJ** Tom Loveland, 32 Whateley Lane, Whateley, Tamworth, B78 2ET
**M6 FAK** Adam Ladd, 42 Plovers Way, Bury St. Edmunds, IP33 2NJ
**MW6 FAM** Alan Williams, 211a New Road, Skewen, Neath, SA10 6ET
**MW6 FAN** Adam Burgess, 18 Fairmeadows, Maesteg, CF34 9JL
**M6 FAS** Frederick Southgate, 52 Jeffrey Lane, Belton, Doncaster, DN9 1LT
**MI6 FAU** William Montgomery, 56 Hazelbank Road, Drumahoe, Londonderry, BT47 3NY
**M6 FAW** Stephen Fawley, 122 Neerings, Coed Eva, Cwmbran, NP44 6UL
**M6 FAX** Raymond Stewart, 20 Siddal Street, Halifax, HX3 9BH
**M6 FAY** Michael Furnivall, 10 Wilwick Lane, Macclesfield, SK11 8RS
**MW6 FBA** Morgan Young, 104 Ystrad Road, Pentre, CF41 7PW
**M6 FBB** Michael Carr, 51 Langton Road, Holton-le-Clay, Grimsby, DN36 5BH
**M6 FBD** David Smith, Heath Farm, Heath Road, Bury St. Edmunds, IP30 9RL
**M6 FBE** Michael Money, 2 Lodge Farm Cottage, Black Horse Road, Norwich, NR10 5DJ
**M6 FBF** Alex Robson, 6 Wood View, Maltby, Rotherham, S66 7PA
**M6 FBH** F Hatfull, 16b Church Street, Easton on the Hill, Stamford, PE9 3LL
**M6 FBJ** Ben Jordan, 605 Monnow Way, Bettws, Newport, NP20 7DJ
**M6 FBK** Alan Blake, Northfield Cottage, Droxford Road, Fareham, PO17 5AZ
**M6 FBM** Robert Proudman, 7a Fithern Close, Dudley, DY3 1YA
**M6 FBO** John Borthwick, 62a Rea Valley Drive, Birmingham, B31 3XE
**M6 FBP** Benjamin Preskey, 115 Selwyn Street, Hillstown, Chesterfield, S44 6LS
**M6 FBQ** John Hewitt, 13 Sullivan Grove, South Kirkby, Pontefract, WF9 3RJ
**M6 FBR** Christopher Chew, 9 Raven Park, Haslingden, Rossendale, BB4 4HN
**M6 FBT** Isabell Long, Flat 6, Kingfisher Court, Woking, GU21 6DQ
**MM6 FBU** Anthony Miller, 100 Kerrylamont Avenue, Glasgow, G42 0DW
**M6 FBV** Sean Cooney, 56 Manor House Lane, Preston, PR1 6HN
**M6 FBW** Sean Richards, 28 Lincoln Way, Thetford, IP24 1DG
**M6 FBY** Stuart Chau, Foley House, Heath Lane, Stourbridge, DY8 1QX
**M6 FBZ** Andrew Raven, 14 Paddock Close, Belton, Great Yarmouth, NR31 9NT
**MM6 FCA** Robert McDonald, 135 Pappert, Alexandria, G83 9LG
**M6 FCC** Fred Cornes, 18 Barke Street, Highley, Bridgnorth, WV16 6LQ
**M6 FCD** Nathan Fisher, 23 Matlock Road, Ferndown, BH22 8QT
**M6 FCE** Colin Smith, 50 Oakwood Avenue, Wakefield, WF2 9JS
**M6 FCF** Benjamin Taylor, 12 Clovelly Road, Stockport, SK2 5AZ
**M6 FCG** Gregory Miles, 14 Woodlands Road, Great Shelford, Cambridge, CB22 5LW
**M6 FCH** Adam Finch, 6 Clover Way, Thetford, IP24 1LQ
**M6 FCI** Alwin Renny, 216 Malvern Road, Bournemouth, BH9 3BX
**M6 FCJ** Darren Vincelli, 20 Worthington Close, Hyde, SK14 3QX
**M6 FCM** Christopher Boyle, 8 Westlees Close, North Holmwood, Dorking, RH5 4TN
**M6 FCN** Alwin John, Flat 10, Melbourne Court, 46 Seabourne Road, Bournemouth, BH5 2HT
**M6 FCQ** Dominic Aissa, 28 Sirdar Road, London, N22 6RG
**M6 FCR** Bernard Scannell, 60 Burnside Road, Dagenham, RM8 1XD
**M6 FCS** Freddie Spickernell, Stockstreet Farm, Mile Elm, Calne, SN11 0NE
**M6 FCT** Michael White, 41 Pendragon Park, Glastonbury, BA6 9PG
**M6 FCV** gary allen, 38 Edenside Kirby Cross, Frinton-on-Sea, CO13 0TQ
**M6 FCW** Lewis Heaney, 2 The Spinney, Eastleigh, SO50 8PF
**M6 FCX** Lewis Palmer, 39 Baird Grove, Kesgrave, Ipswich, IP5 2DQ
**M6 FCZ** Chris Cooper, 25 Waterside Close, Loughborough, LE11 1LP
**M6 FDD** Brian Evans, 2 Hastings Road, Eccles, Manchester, M30 8JR
**M6 FDF** David Sankey, 6 Paulls Close, Martock, TA12 6DE
**M6 FDG** David Chapman, 27 Cuff Crescent, London, SE9 5RF
**M6 FDH** Kevin Wells, 45 Laburnum Avenue, Yaxley, Peterborough, PE7 3YQ
**M6 FDI** William Dawson, 7 Field Close, Warboys, Huntingdon, PE28 2UT
**M6 FDK** Antoinette Patmore, 5 Milton Close, Ramsey, Huntingdon, PE26 1LU
**M6 FDL** Phillip Dixon, 34 Lodge Road, Wolverhampton, WV10 6TH
**M6 FDN** Michael Martin, 7 Carlton Drive, Priorslee, Telford, TF2 9SH
**M6 FDR** Alan Raine, 91 Lulworth Avenue, Jarrow, NE32 3SB
**M6 FDS** Steven Jackson, 5 Duchess Park Close, Shaw, Oldham, OL2 7YN
**M6 FDT** Aidan Ascroft, Caretakers Cottage, School Road, Worksop, S81 9PX
**M6 FDU** Kenneth Gilpin, 39 Medina View, East Cowes, PO32 6LG
**M6 FDW** Thyagarajagopalan KRISHNAMURTHY, 64 Ingleby Road, Ilford, IG1 4RY
**M6 FDX** Duncan McMorrin, 24 Katherine Close, Addlestone, KT15 1NX
**M6 FDY** Colin Sparks, 160 Wetmore Road, Burton-on-Trent, DE14 1QS
**MD6 FEA** Florence Agboma, Flat 6, Mayfair Court, Edgware, HA8 7UH
**MI6 FEB** Fergal Beattie, 19 Derrin Road, Enniskillen, BT74 6AZ
**M6 FEC** Paul Mills, 212 Carsic Road, Sutton-in-Ashfield, NG17 2BS
**M6 FED** Howard Clark, 8 Snowberry Avenue, Belper, DE56 1RE
**M6 FEE** Fiona Mather, 98 Heathgate, Norwich, NR3 1PN
**M6 FEF** Matthew Lewis, 6 Criccieth Close, Buckley, CH7 3QF
**M6 FEH** Robert Harris, 11 Maes Morgan, Llanrhaeadr ym Mochnant, Oswestry, SY10 0LH
**M6 FEI** William Darvill, 35 Allard Close, Northampton, NN3 5LZ
**M6 FEJ** Andrew Croston, 3 Fore Street, Chulmleigh, EX18 7BR
**M6 FEK** Rolf Fekete, 22 St. Andrews Road, Ellesmere Port, CH65 5DG
**M6 FEL** Martin Bryant, 19 Brooklands Road, Havant, PO9 3NS
**M6 FEM** Samantha Blackham, 5 Rogate Road, Worthing, BN13 2DT
**M6 FEN** Stephen Morris, 108 Templeside, Temple Ewell, Dover, CT16 3BA
**M6 FEO** Raymond Hunter, 3 Sandyway, Croyde, Braunton, EX33 1PP
**M6 FEP** Ross Phillips, 162 Glebelands, Pulborough, RH20 2JL
**M6 FEQ** Christopher Ford, 61 Gould Road, Barnstaple, EX32 8ET
**M6 FER** [illegible]
**M6 FEW** Steven Church, Sapphire Ridge, Mill Lane, Brackley, NN13 5JS
**MM6 FEX** Thomas McCallum, 3 Hillington Gardens, Glasgow, G52 2TP
**M6 FEZ** Thomas Ryder, 9a London Road, Slough, SL3 7RL
**M6 FFB** Michael Hall, 67 Darlinghurst Grove, Leigh-on-Sea, SS9 3LF
**M6 FFD** Andrew Ramsbottom, 1 Barn Gill Close, Blackburn, BB2 3HU
**M6 FFE** Edward Durkin, 30 Douglas Road West, Stafford, ST16 3NX
**M6 FFF** Andrew Beardsley, 10 Moreton Close, Church Crookham, Fleet, GU52 8NS
**M6 FFI** Christopher Lowe, 45 Winster Road, Chaddesden, Derby, DE21 4JY
**M6 FFJ** Jacob Preston, 35 First Avenue, Kidsgrove, Stoke-on-Trent, ST7 1DN
**M6 FFK** Michael Ross, 11 Queens Place, Otley, LS21 3HY
**M6 FFM** Stephen Saville, Little Cranebrook, St. Michaels Road, Verwood, BH31 6JA
**M6 FFN** Glen Wilkins, 7 Byron Road, Newport, NP20 3HJ
**M6 FFO** Binoy Issac, 9b Poplar Grove, Stockport, SK2 7JD
**MD6 FFP** Tyler Skerton, 60 Oakington Avenue, Harrow, HA2 7JJ
**MM6 FFQ** Dylan Harvie, 58 Esk Drive, Livingston, EH54 5LE
**M6 FFR** Richard Moys, 12a Palmerston Avenue, Fareham, PO16 7DP
**M6 FFS** Ryan Young, 48 Sussex Street, Cleethorpes, DN35 7NP
**M6 FFT** Samantha Merridale, THE GRANARY, FALLEDGE LANE, Upper Denby, HD8 8YH
**M6 FFU** Lynette Smith, Flat 3, Granville Court, Ramsgate, CT11 8DD
**M6 FFV** Nicholas Calderley, 48 Woodlands, Horbury, Wakefield, WF4 5HH
**M6 FFW** Frederick Harvey, 86 McCrae Street, Swan Hill, Victoria, Australia, 3585
**M6 FFY** Mark Clarke, 4 Mill Lane, Brant Broughton, Lincoln, LN5 0RP
**M6 FFZ** Robert Radley, 131 North Marine Road, Scarborough, YO12 7HU
**M6 FGA** Peter Meanwell, 20 Crow Park Avenue, Sutton-on-Trent, Newark, NG23 6QG
**M6 FGC** Daryl Newsome, 1 Freemans Wharf, Plymouth, PL1 3RN
**M6 FGE** Martin Swan, 35 Colston Close, Plymouth, PL6 6AY
**M6 FGG** Kevin Legg, Bennetts, High Street, Clacton-on-Sea, CO16 0EG
**M6 FGH** Nicholas Speller, 3 Homestall Close, Oxford, OX2 9SW
**M6 FGI** Kevin Phillips, 6 Bellingham Crescent, Plymouth, PL7 2QP
**MW6 FGJ** Michael Roberts, 38 Gwilliam Court, Monkton, Pembroke, SA71 4JL
**M6 FGK** Joby Akira, 61 Dinsdale Gardens, Rustington, Littlehampton, BN16 3NT
**M6 FGM** Brent Wilcock, 32 Mallard Road, Scotton, Catterick Garrison, DL9 3NP
**MW6 FGN** Neil Thomas, 8 Western Terrace, Blaengwynfi, Port Talbot, SA13 3YE
**MW6 FGQ** Pamela Passmore, 127 High Street, Neyland, Milford Haven, SA73 1TR
**MM6 FGR** Andrew Morrison, 4 West Murkle, Murkle, Thurso, KW14 8YT
**MD6 FGW** David Crouch, 87 Nibthwaite Road, Harrow, HA1 1TD
**M6 FGY** Barrie Bestwick, 185 Ashbourne Road, Turnditch, Belper, DE56 2LH
**MD6 FGZ** Thomas Crawley, 41 Lynmouth Drive, Ruislip, HA4 9BY
**M6 FHB** David Bell, 27 Parkfields Avenue, London, NW9 7PG
**M6 FHD** Daniel Hartropp, 185 Leighton Road, London, NW5 2RD
**M6 FHE** John Hawbrook, 7 Birkdale, Norwich, NR4 6AF
**M6 FHF** Alexander Bailie, 2 Loch Lane, Watton, Thetford, IP25 6HE
**MI6 FHG** Philip Donaghy, 24 Chasewood Close, Portadown, Craigavon, BT63 5TY
**M6 FHH** Freddie Stisted, 19 Danehurst Street, London, SW6 6SA
**M6 FHI** Catherine Colless, 128 Ditton Lane, Fen Ditton, Cambridge, CB5 8SS
**MI6 FHJ** Nat Cully, 42 Omerbane Road, Cloughmills, Ballymena, BT44 9PE
**MW6 FHK** Nick Williams, 20 Railway Terrace, Pontyberem, Llanelli, SA15 5HN
**M6 FHM** Bridget Hodgkinson, 26 Daywell Rise, Rugeley, WS15 2RE
**M6 FHN** Ilija Popovic, 13 Walden Avenue, Arborfield, Reading, RG2 9HR
**M6 FHO** Kevin Hunt, Pound Farm, Gallows Hill, Diss, IP22 1RZ
**M6 FHP** Paul Driver, 68 Ripon Road, Dewsbury, WF12 7LG
**M6 FHQ** Jonathan Mount, Forge Cottage, The Street, Goodnestone, Canterbury, CT3 1PQ
**M6 FHR** George Daymond, 30 Elizabeth Drive, Newcastle upon Tyne, NE12 9QP
**M6 FHS** Robert Goldup, 57 Partridge Way, Old Sarum, Salisbury, SP4 6PX
**M6 FHV** Harry Peberdy, 53 Far Lane, Normanton on Soar, Loughborough, LE12 5HA
**M6 FHX** Stephen Powell, 29 Coppice End Road, Derby, DE22 2TA
**M6 FHY** David Pearson, 37 Elmridge, Leigh, WN7 1HN
**M6 FIA** Xiting Sinfia Zhang, Harrogate Ladies' College, Clarence Drive, Harrogate, HG1 2QG
**M6 FIB** Martin Sutton, 5 All Saints Court, Didcot, OX11 7NG
**M6 FID** Philippa Hack, Catte Street, Oxford, OX1 3BW
**MM6 FIF** Mark Cook, 7 Donald Gardens, Dundee, DD2 2RZ
**M6 FIG** Michael Fisher, 96 Reepham Road, Norwich, NR6 5PD
**M6 FII** John Thurman, Warbanks Farm, Cockfield, Bury St. Edmunds, IP30 0JP
**M6 FIJ** Donard De-Cogan, 52 Gurney Road, New Costessey, Norwich, NR5 0HL
**MI6 FIK** Damien Wilson, 81 Parknasilla Way, Aghagallon, Craigavon, BT67 0AU
**M6 FIL** Philip Dunnicliffe, 19 Woodland Road, Chelmsford, CM1 2AT
**MW6 FIN** Finley Toomey-Langford, 8 Heol Dinas Isaf, Williamstown, Tonypandy, CF40 1NG
**M6 FIO** Diana Smith, 62 Fulwoods Drive, Leadenhall, Milton Keynes, MK6 5LB
**M6 FIR** Michael Firth, 209 High Street, Wickham Market, Woodbridge, IP13 0RQ
**M6 FIT** Michael Bryan, 13 Elmwood Avenue, Sunderland, SR5 5AW
**MI6 FIU** Peter Heid, 1 Nettlehill Mews, Lisburn, BT28 3HN
**M6 FIV** Alan Foley, 23 Church Lane, Wymington, Rushden, NN10 9LW
**M6 FIW** Robert Nicholson, 4 Morris Court, Aylesbury, HP21 9QT
**MW6 FIX** Robert Hemming, 21 New Road, Jersey Marine, Neath, SA10 6JR
**MW6 FIY** Simon Binnion, 55 Dythel Park, Pen-y-Mynydd, Llanelli, SA15 4RR
**M6 FIZ** Richard Faulkner, 31 Faulkland View, Peasedown St. John, Bath, BA2 8TG
**M6 FJA** Alexander Ferriroli, 142 Hillbury Road, Warlingham, CR6 9TD
**M6 FJC** Calum Featherstone, 3 Pogmoor Road, Barnsley, S75 2EW
**M6 FJD** Karl Dobson, 1 Howarth Road, Ashton-on-Ribble, Preston, PR2 2HH
**M6 FJE** Christopher Hinds, 15 Logan Way, Hemyock, Cullompton, EX15 3RD
**M6 FJF** John Slater, 1 Highfield Court, Swinton, Mexborough, S64 8RF
**M6 FJI** John Tonkyn, Flat 6, Loyd Court, St. Albans, AL4 0AZ
**M6 FJJ** Fabienne Johnson, 12 Hollins Road, Harrogate, HG1 2JF
**M6 FJL** John Robb, 37 Wroxham Road, Woodley, Reading, RG5 3AX
**M6 FJM** David Marden, 1 Stroma Gardens, Hailsham, BN27 3AZ

**IMPORTANT NOTE**

**Revalidate licence to avoid revocation** – Ofcom has advised the Society that plans will be drawn up to revoke licences that have not been revalidated as required by the licence conditions. The quickest way to revalidate is to do so online via the Ofcom website: *https://services.ofcom.org.uk/* or by email: *amateur.validations@ofcom.org.uk* Ofcom staff are available to help, but please be patient during times of heavy workload.

M6 FJN Michael Singer, 1 Bentley Road, Slough, SL1 5BB
M6 FJO Ian Waddingham, 102 Nethershire Lane, Sheffield, S5 0QE
M6 FJQ Amy-Marie Lawson, Hafryn, Dyserth Road, Rhyl, LL18 5RB
MM6 FJS Owen Sharp, 18 Urquhart Court, Kirkcaldy, KY2 5TX
M6 FJT Tabitha Cook, 19 Cae Bach Aur Estate, Bodffordd, Llangefni, LL77 7JS
M6 FJU Polly Appleton, The Old Rectory, Station Road, Grimsby, DN36 5SQ
M6 FJV William Little, Burnside, Main Street, Lochans, Stranraer, DG9 9AW
M6 FJW Simon Michael, 191 Sutton House, Scunthorpe, DN15 6SN
M6 FJX Caian Williams, 18 Caer Delyn, Llannerch-Y-Medd, LL71 8EJ
M6 FJZ Sarah Turner, 8 Market Place, Hingham, Norwich, NR9 4AF
M6 FKA Lee Hall, 51 Bridgeacre Gardens, Coventry, CV3 2NQ
M6 FKB Steven Broll, 23 St. Medans, Monreith, Newton Stewart, DG8 9LL
MW6 FKC G Stevenson, 3 Llwyn Rhosyn, Cardiff, CF14 6NS
MM6 FKD Emma Rattray, Blashieburn Cottage, Blairingone, Dollar, FK14 7NT
M6 FKE Karen Collings, 97 Jackmans Place, Letchworth Garden City, SG6 1RF
M6 FKF Antonino Cucchiara, 172 Gonville Crescent, Stevenage, SG2 9LZ
M6 FKG Justin Shears, 161 Park Road, Keynsham, Bristol, BS31 1AS
M6 FKH James Fensom, Foxdown House, Blandford Camp, Blandford Forum, DT11 8BP
M6 FKI William Tennison, 85 Primrose Field, Harlow, CM18 6QT
M6 FKJ Francis Johnson, 13 Honeymead Lane, Sturminster Newton, DT10 1EW
M6 FKN Arthur Blackwell, 27 Prince Charles Crescent, Farnborough, GU14 8DJ
M6 FKO Gareth Carver, 4 Andrews Road, Farnborough, GU14 9RY
M6 FKP Craig Robinson, 3 Folly Hall Road, Bradford, BD6 1UL
M6 FKQ Anthony Smith, 116 Pilling Lane, Preesall, Poulton-le-Fylde, FY6 0HG
M6 FKR Guy Howard, 8 Paddock Road, Woodford, Kettering, NN14 4FL
M6 FKS Thorolf Nash, 32 Collington Rise, Bexhill-on-Sea, TN39 3RS
M6 FKU Lauren Harris-Pugh, 13 Myddelton Park, London, N20 0HT
M6 FKV Mark Sherrey, 14 The Grove, Hallatrow, Bristol, BS39 6ES
M6 FKW Alan Metcalf, 71 Harper Road, Coventry, CV1 2AL
M6 FKY Mark Bumstead, Windmill Girl III, Riverside Boatyard, Southampton, SO31 1AA
M6 FKZ Carl Spencer, 18 Coatsby Road, Kimberley, Nottingham, NG16 2TH
M6 FLA Frederic Labrosse, 72 Ger y Llan Penrhyncoch, Aberystwyth, SY23 3HQ
M6 FLB Fiona Buss, 44 Courtenay Road, Maidstone, ME15 6UL
M6 FLC Colin Horridge, 6 Back Sreet, East Stockwith, Gainsborough, DN21 3DL
M6 FLE Paul Threakall, 83 Gregory Avenue, Birmingham, B29 5DG
M6 FLF Craig Wilson, 12 Desmond Avenue, Hornsea, HU18 1AF
MM6 FLG Mark Bradshaw, 32 Greycraigs, Cairneyhill, Dunfermline, KY12 8XL
M6 FLH Daniel Adkin, 31 Fieldway Broad Oak, Rye, TN31 6DL
M6 FLJ Sarah Bumstead, Riverside Boatyard, Blundell Lane, Southampton, SO31 1AA
M6 FLK John Kelly, 74 Hatfield Crescent, Bedford, MK41 9RB
M6 FLL Diane Dobson, 1 Howarth Road, Ashton-on-Ribble, Preston, PR2 2HH
M6 FLN Julian Horn, 8 Princess Close, Watton, Thetford, IP25 6XA
M6 FLQ Kelvin Hobbs, 75 Coronation Walk, Gedling, Nottingham, NG4 4AS
M6 FLR Jason Bicknell, 37 Nicholsfield, Loxwood, Billingshurst, RH14 0SR
M6 FLS D Greenland, 1 Hilltop, Tuesley Lane, Godalming, GU7 1SB
M6 FLU John Salter, 103 Cootes Avenue, Horsham, RH12 2AF
M6 FLW Simon Miller, 32 Edward Street, Swinton, Mexborough, S64 8NL
M6 FLX Thomas Clarke, 20 Cliff Road, Felixstowe, IP11 9PJ
M6 FLY Jake Hall, 43 Norwood Drive, Brierley, Barnsley, S72 9EG
M6 FLZ R Sutton, 80 Fishbourne Lane, Ryde, PO33 4EU
M6 FMC Michael Straughan, 71 Silcoates Lane, Wrenthorpe, Wakefield, WF2 0PA
M6 FME Freddie Moulsdale, 63 Windermere Avenue, St. Helens, WA11 7AG
M6 FMF Christopher Bassi, Cherries, Hadham Road, Ware, SG11 1LH
M6 FMG Philip Booker, 17 Colton Copse, Chandler's Ford, Eastleigh, SO53 4HQ
M6 FMI Oliver Earl, 11 Danbury Road, Shirley, Solihull, B90 2BU
M6 FMJ Kenneth Browne, 4 Darcey Drive, Brighton, BN1 8LF
M6 FML Fiona Lowndes, 46 Danes Road, Manchester, M14 5JS
M6 FMN Giselle Morris, 6 Eastcourt Road, Worthing, BN14 7DB
M6 FMO Owen Cook, 38 Redbourn Way, Scunthorpe, DN16 1NE
M6 FMP Felix Morson-Pate, 3 Tudor Lawns, Carr Gate, Wakefield, WF2 0UU
M6 FMS Frank Mace, 40 Quarry House Gardens, East Rainton, Houghton le Spring, DH5 9RD
M6 FMT Adam Webber, 73 Greenwood Road, Yeovil, BA21 3LF
M6 FMU John D'Aubray-Butler, 16 Francis Road, Frodsham, WA6 7JR
MW6 FMV Martin Plowman, Ffynnonwen, Plwmp, Llandysul, SA44 6EY
M6 FMW Christopher Jones, 32 Vale Street, Denbigh, LL16 3BE
M6 FMY Nigel Mountford, 78b Meeting House Lane, London, SE15 2TX
M6 FMZ Karl Bianchini, 10 St. Leonards Road, Headington, Oxford, OX3 8AA
M6 FNA Edwin Flikkema, 7 St. James Mews, Great Darkgate Street, Aberystwyth, SY23 1DW
M6 FNB Nicholas Berry, The Mount, Deerfold, Bucknell, SY7 0EF
M6 FNE Robin Merrifield, 6 Miersfield, High Wycombe, HP11 1TX
M6 FNF Michael Byrd, 17a Castle Gates, Shrewsbury, SY1 2AB
M6 FNG Matthew Grice, 48 St. Ives Road, Coventry, CV2 5FZ
M6 FNK Frank Williams, 5 Tyn Rhos Estate, Gaerwen, LL60 6HL
M6 FNL Glen Cunningham, 21 Dunmuir Road, Castle Douglas, DG7 1LQ
M6 FNM Robert Bound, 24 Buttington Road, Shrewsbury, SY2 5TS
MI6 FNO Cecil Johnston, 32 Maghereagh Road, Randalstown, Antrim, BT41 4NS
MM6 FNQ Keoni Jones, 6 Traill Street, Castletown, Thurso, KW14 8UG
M6 FNR Frank Riches, 4 Priory Close, Chelmsford, CM1 2SY
MW6 FNT Phillip Taylor, 11 Watford Close, Caerphilly, CF83 1NQ
M6 FNU Kathleen Cole, 30 Wood Road, Rotherham, S61 3RQ
M6 FNV Dale Robins, 12 Kestrel Way, Duffryn, Newport, NP10 8WF
M6 FNX Nicholas Farrington-Smith, 3 Milton Road, Wokingham, RG40 1DE
M6 FNY David Chatterton, 3 Hunt Close, South Wonston, Winchester, SO21 3HY
MI6 FNZ Robert McKay, 31 Squires Hill Crescent, Belfast, BT14 8RE
M6 FOD Christopher Dempster, 5 Wydean Close, Lydney, GL15 5BS
M6 FOE Leslie Downs, 18 Burnopfield Gardens, Newcastle upon Tyne, NE15 7DN
M6 FOJ Dave Sutherland, 78 Holmden Avenue, Wigston, LE18 2EF
M6 FOK Christopher Fletcher, 1 New Row, Barnetby, DN38 6DX
M6 FOL William Bartle, 6 The Cottages, Eccles Road, High Peak, SK23 0EZ
M6 FOR Ian Forester, 35 Thackeray Street, Sinfin, Derby, DE24 9GY
M6 FOT Stuart Moore, 24 Cattell Drive, Sutton Coldfield, B75 7LQ
M6 FOW Christopher Davis, Flat 3 Mansell Court, Haywards Heath, RH16 3LG
M6 FOX S Spence, 8 Teasdale Street, Consett, DH8 6AF
M6 FOY Frank Foy, 4 The Square, East Rounton, Northallerton, DL6 2LB
M6 FOZ Alan Foster, 47 Westmorland Drive, Desborough, Kettering, NN14 2XB
M6 FPC John Lord, 5 Langworthy Avenue Little Hulton, Manchester, M38 9GQ
M6 FPD Linda Coveney, 72 Bonney Road, Leicester, LE3 9NH
M6 FPE Robert Coveney, 72 Bonney Road, Leicester, LE3 9NH
M6 FPF Frank Clifton, Battery Road, Paull, Hull, HU12 8FP
M6 FPG David Worth, 33 Lime Grove Close, Leicester, LE4 0UG
MM6 FPI Andrew McClements, 25 Dundonald Crescent, Auchengate, Irvine, KA11 5AX
M6 FPJ Dave Lavell, 51 Kingfisher Close, Newport, PO30 5XS
MI6 FPK Frank Kearney, 45 Sperrin Park, Omagh, BT78 5BA
MW6 FPL Andrew Muldoon, 29 Eider Close, St. Mellons, Cardiff, CF3 0DF
M6 FPN Levente Varga, 53 Kempton Avenue, Northolt, UB5 4HF
M6 FPP Janusz Misztela, 15 Severus House, 155 Varcoe Gardens, Hayes, UB3 2FJ
M6 FPQ Eric Turner, 2 Shepherds Close, Fen Ditton, Cambridge, CB5 8XJ
M6 FPU Edward Duell, 28 Watergall, Bretton, Peterborough, PE3 8NA
M6 FPV Tristan Van Den Bosch, 8 Stopford Garth, Wakefield, WF2 6RT
M6 FPW Steven Bale, 78 Highfield Road, Rushden, NN10 9QJ
MM6 FPX Thomas McConnell, 163 Strathaven Road, Stonehouse, Larkhall, ML9 3JN
MM6 FPY Iain Blackstock, 149d Carlisle Road, Crawford, Biggar, ML12 6TP
M6 FQA Eifion Parry, Blaen Gwenin, Llanrhystud, SY23 5BZ
M6 FQC Megan Roberts, 41 Mcneill Avenue, Crewe, CW1 3NW
MW6 FQD Martin Roberts, 20 Upper Robinson Street, Llanelli, SA15 1TR
M6 FQE Kenneth De La Hunty, Bangers Whistle, Yarmouth Road, Newport, PO30 4LZ
MD6 FQF Adam Goldstein, 33 Hughenden Avenue, Harrow, HA3 8HA
M6 FQH Craig Goldsmith, 15 Stephenson Road, Cowes, PO31 7PP
M6 FQJ Barry Boxall, 10 Honor Avenue, Wolverhampton, WV4 5HH
M6 FQK Dale Briggs Briggs, 5 Links Avenue, Cromer, NR27 0EQ
M6 FQL George Milner, 17 Haragon Drive, Amesbury, Salisbury, SP4 7FS
M6 FQM Anthony James, 27a Lilliana Way, Bridgwater, TA5 2GG
M6 FQN Robert Vickers, 3 Grenville Avenue, Goring-by-Sea, Worthing, BN12 6JE
M6 FQO Andrew Lipson, Field House, The Haven, Cambridge, CB21 5BG
M6 FQP Anthony Phillips, 43 Jutland Road, Hartlepool, TS25 1LP
M6 FQR Aaron Coote, 148 Clarendon Street, Dover, CT17 9RB
M6 FQU Raymond Parker, 53 Tunstall Road, Canterbury, CT2 7BX
M6 FQW Luke Jones, 206 Lowe Avenue, Wednesbury, WS10 8NS
M6 FQX Michael Rowland, 30 Tunnmeade, Harlow, CM20 3HL
MI6 FQY Barry Maguire, 39 Carland Road, Dungannon, BT71 4AA
M6 FQZ Alan Williams, Corner Cottage, South Brent, TQ10 9JF
M6 FRC Frank Clement, 40 Ellison Fold Terrace, Darwen, BB3 3EB
M6 FRD Finley Redden, 31 Lawfield, Coldingham, Eyemouth, TD14 5PB
M6 FRF Robert Abel, 31 Derwent Road, Birmingham, B30 2UY
M6 FRG Jon Marshall, 1 Farriers Reach, Bishops Cleeve, Cheltenham, GL52 7UZ
M6 FRI John Hewart, 14 Kestrel Close, Marple, Stockport, SK6 7JS
M6 FRJ Dariusz Janowicz, 20 Salisbury Road, St. Leonards-on-Sea, TN37 6RX
M6 FRK Frank Waller, 249 Summer Lane, Wombwell, Barnsley, S73 8QB
M6 FRO Neil Froggatt, 13 Stroudes Close, Worcester Park, KT4 7RB
M6 FRP Nigel Williams, The Bothy, Norton Bavant, Warminster, BA12 7BB
MI6 FRQ Joseph Coleman, 418 Antrim Road, Flat 3, Belfast, BT15 5GA
M6 FRR John Hingley-Hickson, The White House, School Lane, Grantham, NG32 2ES
M6 FRS Timothy Warner, 6 Cotswold court, Skelmersdale road, Clacton-on-Sea, CO15 6EN
M6 FRT Jane Parsons, 4 Saunders Close, Uckfield, TN22 2BX
M6 FRV Francis Raven-Vause, 624 Hillbutts, Wimborne, BH21 4DS
M6 FRW Andrew Middleton, 56 Caradoc Road, Prestatyn, LL19 7PF
M6 FRY Christopher Fryer, 14 Perks Road, Wolverhampton, WV11 2ND
M6 FRZ Jordan Potter, 21 Ulverston Crescent, Lytham St. Annes, FY8 3RZ
M6 FSA Simon Taylor, 18 Cycle Street, York, YO10 3LJ
MM6 FSB David Kelly, 21 Dhailling Road, Dunoon, PA23 8EA
M6 FSC Richard Cooper, 86 Grove Road, Tiptree, Colchester, CO5 0JG
M6 FSD Michael Tayler-Grint, 94 Dairymans Walk, Guildford, GU4 7FF
M6 FSE Lisa Beaney, Penns Cottage, Horsham Road, Steyning, BN44 3LJ
M6 FSG Christopher Moore, 25 Stonybeck Close, Westlea, Swindon, SN5 7AQ
M6 FSI Dorian Woolger, 8 Old Cottages, Horsham Road, Worthing, BN14 0TQ
M6 FSJ Stephen Burgess, 101 Brendon Way, Nuneaton, CV10 8NW
M6 FSM Huw Morgan, 18 Chrystel Close, Tipton St. John, Sidmouth, EX10 0AY
M6 FSN Anton Lillis, 4 Sinclair Close, Gillingham, ME8 9JQ
M6 FSO Justin Forbes, Weald Barkfold Farm, Plaistow, Billingshurst, RH14 0PJ
MM6 FSP Stephen Paterson, 14-16 New Street, Findochty, Buckie, AB56 4PS
MM6 FSQ Carl Mackenzie, 89c Needless Road, Perth, PH2 0LD
MM6 FSR Sandy Riley, 67 Wallacebrae Wynd, Danestone, Aberdeen, AB22 8YD
M6 FST William Morris, Fairview, Trefonen, Oswestry, SY10 9DP
M6 FSU Michael Cain, 59 Bailey Road, Leigh-on-Sea, SS9 3PJ
MM6 FSV Paul Hadley, 27 William Street East Wemyss, Kirkcaldy, KY1 4PG
M6 FSW Polly Bowen, Arbor Tree Bungalow, Mill Street, Craven Arms, SY7 8EN
M6 FSZ Daniel Walker, 24 Redwood, Esh Winning, Durham, DH7 9AG
MM6 FTA Lukasz Pinkowski, 73 Willow Grove, Livingston, EH54 5NA
MW6 FTC Huw Lloyd, Apartment 126, Woodlands, Hayes Road, Penarth, CF64 5QE
M6 FTF Michael Goodman, 20b Orchard Estate, Little Downham, Ely, CB6 2TU
MM6 FTG Fraser Gorman, 11 Maxwell Drive, Baillieston, Glasgow, G69 6JB
M6 FTI Trevor Wetherill, Upper House Cottage, Holme Marsh, Kington, HR5 3JS
M6 FTK Matthew Botham, 17 Laburnum Drive, Oswestry, SY11 2QW
M6 FTN Dan Nathan, 138 Anchor Lane, Hemel Hempstead, HP1 1NS
M6 FTO Gary Lawlor, 14 Woodkirk Close, Seghill, Cramlington, NE23 7TZ
M6 FTP Richard Denham, 179 Pandon Court, Shield Street, Newcastle upon Tyne, NE2 1XY
M6 FTQ Anthony Street, 110 Magdalen Street, Colchester, CO1 2LF
M6 FTT Matthew Knowles, 11 Thorneycroft Avenue, Birkenhead, CH41 8HJ
M6 FTU Rebecca Pownall, 17 Horsebridge Road, Blackpool, FY3 7BQ
M6 FTV Steven Barber, 82 Prunus Road, Crewe, CW1 4HB
M6 FTW Milo Bowman, 26 Albany Hill, Tunbridge Wells, TN2 3RX
M6 FTX Victor Reeve, 29 Benfield Way, Portslade, Brighton, BN41 2DN
M6 FTY Neil Baker, 13 Elm Way, Melbourn, Royston, SG8 6UH
M6 FUA Carolyn Fear, 13 Frome Road, Chipping Sodbury, Bristol, BS37 6LD
M6 FUD Adam Bentley, 40 Wyndale Road, Leicester, LE2 3WR
M6 FUE Alfred Davies, 15 The Crescent, Woodside Park, Poulton-le-Fylde, FY6 0QW
M6 FUF Neil Irvine, 100 Cavendish Road, Sunbury-on-Thames, TW16 7PL
M6 FUH Jacob Reich, Flat 1-3, 16 North Pole Road, London, W10 6QL
M6 FUJ Helen Braithwaite, 20 Richard Moon Street, Crewe, CW1 3AX
M6 FUL Marc Fuller, Rose Cottage, Wickmoor, Bridgwater, TA7 0JR
M6 FUM Ian Kent, 111 Sinclair Avenue, Banbury, OX16 1BQ
M6 FUN Philip Hawkes, 22 Longfield Road, Bristol, BS7 9AG
M6 FUO Bernice Harrop, 7 Haythorne Way Swinton, Mexborough, S64 8SQ
M6 FUQ Andrew Howard, 24 Ladybower Lane, Poulton-le-Fylde, FY6 7FY
M6 FUR Dawn Esdale, The Bell Inn, Central Lydbrook, Lydbrook, GL17 9SB
M6 FUT Ronald Harris, 63 The Drive, High Barnet, Barnet, EN5 4JG
M6 FUU Samuel Bloomfield, 20 Farmers Way, Copmanthorpe, York, YO23 3XU
M6 FUW Chris Wilkinson, 13 Kimpton Close, Lee-on-The-Solent, PO13 8JY
M6 FVA Stephen Matthews, Aberbran Fawr, Aberbran, Brecon, LD3 9NG
M6 FVC Gerald Gee, 17 Portherras Villas, Pendeen, Penzance, TR19 7TJ
MM6 FVD Andrew McDonald, 8 North Rayne Cottages, Meikle Wartle, Inverurie, AB51 5BY
M6 FVE Andy Day, 47 Rembrandt Way, Bury St. Edmunds, IP33 2LT
M6 FVH James McMullen, 25 Dene Road, North Shields, NE30 2JW
M6 FVJ Phillip Randerson, 7 Roman Crescent, Swindon, SN1 4HH
MM6 FVK R Rothon, 112 Ravenswood Rise, Livingston, EH54 6PG
M6 FVL Edmund Smethurst, 4 Lower New Row, Worsley, Manchester, M28 1BE
M6 FVM Joby Poriyath, 18 Howard Close, Cambridge, CB5 8QU
M6 FVN Andrew Storey, 1 Mill Hill Road, Bingham, Nottingham, NG13 8YR
M6 FVO Michael Oates, 3 Browning Drive, Rotherham, S65 2NT
M6 FVP Benjamin Straker, 37 Beech Grove Avenue, Garforth, Leeds, LS25 1EF
M6 FVR Jason Rawley, 33 Dorset Way, Billericay, CM12 0UD
M6 FVS Rupert Sage, 9 North Hall Farm Bungalows, Barley Road, Royston, SG8 7PZ
M6 FVT Kevin Young, 29 High Street, Coedpoeth, Wrexham, LL11 3RY
M6 FVV Mark Roberts, 463, Brighton Road, Lancing, BN15 8LF
M6 FVW Conrad Fox, Millstone Cottage, Prior Wath Road, Scarborough, YO13 0AZ
M6 FVZ Emil Preda, 59 Cambridge Street, Stockport, SK2 6PY
M6 FWB Clive Andrews, 16 Highfield Gardens, Aldershot, GU11 3DE
M6 FWC Helen Wilcox, 3 Hutton Street Hutton Wandesley, York, YO26 7ND
MI6 FWD James Tipping, 16 The Oaks, Portadown, Craigavon, BT62 4HX
M6 FWE Michael Scambell, 8 South Bank Road, East Cowes, PO32 6JE
M6 FWG Simon Barclay, 64 Deepdene Avenue, Dorking, RH5 4AE
M6 FWL Craig Pugsley, 86 West Town Lane, Bristol, BS4 5DZ
M6 FWM Maddy Tonkin, 8 Artis Avenue, Wroughton, Swindon, SN4 9BP
M6 FWO Jorge Capovila, 30 Church Street, Barrow-in-Furness, LA14 2JG
M6 FWP Lee Churchill, 1 Fowlswick Cottages, Allington, Chippenham, SN14 6LU
M6 FWQ Michael Duque, 2 Spekehill, London, SE9 3BN
M6 FWS Millie Smith, 151 Rowlands Road, Worthing, BN11 3LE
M6 FWT Matthew Goslin, 44 Teasdale Road, Walney, Barrow-in-Furness, LA14 3SF
M6 FWV Roger Sage, 29 Rosewarne Park, Connor Downs, Hayle, TR27 5LJ
M6 FWW Dale Longson, 1 Beckside, Plumpton, Penrith, CA11 9PD
M6 FWX David Jones, Drove Farm, Sheepdrove, Hungerford, RG17 7UN
M6 FWY David Smith, 7 Oakley Grove, Wolverhampton, WV4 4LN
M6 FWZ Federico Da Dalt, 13 Johnstone Street, Bath, BA2 4DH
M6 FXA Susan Hodge, 2 Parc Cemlyn, Prestatyn, LL19 9NX
MU6 FXB Robert Batiste, Asile de Paix, Clos Des Sablons, Sandy Lane, St. Sampson, Guernsey, GY2 4RN
M6 FXD Daniel Sawyer, 2 St. Johns Court, Palmerston Mews, Bournemouth, BH1 4JH
M6 FXE Richard Hall, 72 Guy Road, Kenilworth, CV8 1FX
M6 FXG Thomas Chapman, 37 Pheasant Way, Cirencester, GL7 1BJ
M6 FXI David Thomas, 51 Barrhill Avenue, Brighton, BN1 8UE
M6 FXJ Darryl Parkes, 4 Round Saw Croft, Rubery, Birmingham, B45 9TT
M6 FXL Thomas Goodenough, 12 Jerounds, Harlow, CM19 4HE
MM6 FXM Tim Johnston, The Old Schoolhouse, Luggate Burn, Haddington, EH41 4QA
M6 FXO Kevin Burton, Knox Stones View, Fellbeck, Harrogate, HG3 5ET
MM6 FXQ Julian Anderson, The Grange, Leslie Road, Kinross, KY13 9JE
M6 FXU Richard Ball, 77 Old Brumby Street, Scunthorpe, DN16 2AJ
MM6 FXW Robert Bisset, 5 Woodlands Court, 44 Barnton Park Avenue, Edinburgh, EH4 6EY
M6 FXX Carl Freckelton, 447 Newark Road, North Hykeham, Lincoln, LN6 9SP
M6 FXY Lorraine Gregory, 11 Roundhill Road, Castleford, WF10 5AF
MM6 FXZ William Goodfellow, 1 Yester Place, Haddington, EH41 3BE
M6 FYB Peter Amond, Treetops, Gedding, Bury St. Edmunds, IP30 0QD
M6 FYD Leslie Clark, 43 Grange View, Harworth, Doncaster, DN11 8QP
M6 FYE Paul Escott, 84 Salisbury Avenue, Bootle, L30 1PZ
MM6 FYF Tom Burnett, 45 The Murrays Brae, Edinburgh, EH17 8UF
M6 FYG Tim Weston, 4 The Pightle, Peasemore, Newbury, RG20 7JS
M6 FYH Colin Cosgrove, 5 Kingswell Avenue, Wakefield, WF1 3DY
M6 FYJ Sam Lo, Upper Maisonette, 41 Park Street, Bath, BA1 2TD
M6 FYK Eddie Dutson, 19 MAYTHORN DRIVE, Cheltenham, GL51 0QH
M6 FYL Marie Kipling, 12 Jolly Brows, Bolton, BL2 4LZ
M6 FYM Andrew Armstrong, View Firth, Brewery Brow, Whitehaven, CA28 6PE
M6 FYN Ross Gunson, 3 Kenilworth Drive, Halifax, HX3 8XP
M6 FYQ Peter Moore, 22 Audit Hall Road, Empingham, Oakham, LE15 8PH
MM6 FYR David Fraser, Applecross, Easterton, Peterhead, AB42 0TQ
M6 FYT Paul Jones, 186 Crowmere Road, Shrewsbury, SY2 5LA
M6 FYV Vanessa Stanley, 82 Sycamore Grove, Bracebridge Heath, Lincoln, LN4 2RD
MM6 FYW Amy Booth, 29 Golf Terrace, Insch, AB52 6JY
MW6 FYY Christopher Marchant, 8 Morningside Walk, Barry, CF62 9TE
M6 FZA Robin Carr, 14 Southwell Close, Kirkby-in-Ashfield, Nottingham, NG17 8GP
M6 FZB Peter Cox, 25 Coronation Crescent, Margate, CT9 5PN
M6 FZC Ian Skeggs, 24 Kendall Road, Beckenham, BR3 4PZ
MM6 FZD Steven Boyd, 163 Upper Craigour, Edinburgh, EH17 7SQ
M6 FZE Anthony Korben, 219 Leyland Road, Penwortham, Preston, PR1 9SY
M6 FZF Robert Waterson, 43 Highland Road, Twerton, Bath, BA2 1DY
M6 FZG John Williamson, 7 Dale Close, Wrecclesham, Farnham, GU10 4PQ
M6 FZJ James Foster, 23 High Street, Cumnor, Oxford, OX2 9PE

M6 FZK Christopher Bowler, 42a Honor Road, Prestwood, Great Missenden, HP16 0NL
M6 FZM Ian Jones, 8 Hengrove Avenue, Bristol, BS14 9TB
M6 FZN Cy Williams, 21 [illegible] Close, South Ockendon, RM15 [illegible]
M6 FZQ Christopher Tacon, 83 Upper Shaftesbury Avenue, Southampton, SO17 3RU
M6 FZR Christopher Bowskill, 5 Trent Walk, Daventry, NN11 4QF
M6 FZS Matthew Holmes, 18 Danube House, Darwin Close, York, YO31 9PE
M6 FZT Robert Clow, 25 Scott Street, Newcastleton, TD9 0QQ
M6 FZV Neal Walker, Wayfield Farm, Calstock Road, Gunnislake, PL18 9BY
M6 FZW Teresa Crampton, 7 Barneveld Avenue, Canvey Island, SS8 8NZ
M6 FZX Timothy Cash, 7 Park Lane, Wolverhampton, WV10 9QE
M6 FZY Jane Langdon, 6 Glebe Close, Stockton, Southam, CV47 8LG
MW6 GAA George Green, 110 Garden Suburbs, Trimsaran, Kidwelly, SA17 4AE
M6 GAB Graham Barnes, 24 Gainsborough Court, Andover, SP10 3SS
M6 GAC George Moore, 40 Main Street, South Rauceby, Sleaford, NG34 8QG
M6 GAD Gary Briggs, 5 Links Avenue, Cromer, NR27 0EQ
M6 GAE Philip Shaw, 25 Headcorn Road, Platts Heath, Maidstone, ME17 2NH
M6 GAF Gareth Furlong, 22 Swaisland Road, Dartford, DA1 3DE
M6 GAG Anthony Court, 3 Forsythia Close, Hythe, Southampton, SO45 3DJ
M6 GAH George Hardill, 107 Leicester Road, Whitwick, Coalville, LE67 5GN
M6 GAI Geoffrey Robinson, 16 Stanley Road South, Rainham, RM13 8AA
M6 GAK Gary Cockburn, 20 Hexham Avenue, Hebburn, NE31 2HN
M6 GAL Tara Kimberlee, 24 Jacey Road, Shirley, Solihull, B90 3LJ
M6 GAM Allan Mawson, 14 Pontop View, Consett, DH8 7JB
M6 GAN Grahame Rudley, 37 Cherry Way, Shepperton, TW17 8QQ
M6 GAO Geoffrey Webster, 24 Martin Avenue, Barrow upon Soar, Loughborough, LE12 8LG
MI6 GAQ Gerard O'Reilly, 20 Lower Clonard Street, Belfast, BT12 4NH
M6 GAR Gary Veale, 8 Duchy Cottages, Stoke Climsland, Callington, PL17 8PA
M6 GAS Brent Hammond, 35 Stratford Avenue, Newcastle, ST5 0JS
M6 GAT Gordon Buckley, 2 Brothertoft Road, Boston, PE21 8HD
MI6 GAU Gary Ferguson, 22 Cloneen Drive, Ballymoney, BT53 6PT
M6 GAV Javed Ali, 5 Darent Close, Bettws, Newport, NP20 7SQ
M6 GAW Gary Watson, 70 Garden Hey Road, Moreton, Wirral, CH46 5NE
M6 GAX Gareth Austin, 24 Newcomen Road, Tunbridge Wells, TN4 9PA
M6 GBA Andrew Jones, 2 Erw Terrace, Bethel, Caernarfon, LL55 1YT
M6 GBB David Nock, 1 Nelson Close, Bradenham, Thetford, IP25 7RB
MI6 GBC Gareth Boyes, 50 Halfpenny Gate Road, Moira, Craigavon, BT67 0HW
M6 GBD Gregory Davage, 18 Wensum Way, Fakenham, NR21 8NZ
M6 GBE Henriks Vecenans, 155 Upper Dale Road, Derby, DE23 8BP
M6 GBF Gerald Bouchier, Flat 24, Hine House, St. Albans, AL4 0EY
MU6 GBG Gavin Lanoe, 1 Pinetrees Estate, Route de L'Islet, Guernsey, Guernsey, GY2 4EX
M6 GBH Ian Appleby, 96 Cranbrook Road, Poole, BH12 3BT
M6 GBI Carl Gibson, 17 Clyde Court, Grantham, NG31 7RB
M6 GBJ George Walker, 25 Bear Close, Woodstock, OX20 1JT
M6 GBK Ian Fraser, 3 Plover House, Mears Beck Close, Morecambe, LA3 1FL
M6 GBL Lucas Ulvenmoe, 6 Havelock Street, Blackpool, FY1 4BN
M6 GBM Neil Cunningham, 115 Trafalgar Road, Washington, NE37 3DJ
M6 GBN Garry Barton, 9 Tees Crescent, Stanley, DH9 6HX
M6 GBO Bruce Goodman, 6 Victoria Court, Penzance, TR18 2EX
MM6 GBP Gerald Holford, 5 Deerpark Cottages, Evanton, Dingwall, IV16 9XH
M6 GBQ Andrew Self, 10 Cecil Road, Hertford, SG13 8HR
M6 GBR Peter Chamberlain, 22 Stanedge Grove, Wigan, WN3 5PL
MM6 GBS Graham Somerville, 4 Kirkhill Way, Penicuik, EH26 8HH
M6 GBU David Carter, 9 Grange Close, Ipplepen, Newton Abbot, TQ12 5RX
M6 GBV Andrew Finn, 202 Northgate Road, Stockport, SK3 9NJ
M6 GBW George Wright, Flat 27, Maudland House, Preston, PR1 2YJ
MM6 GBX George McFarlane, 59 Millburn Road, Bathgate, EH48 2AF
M6 GBY Robin Hughes, 96 Retallick Meadows, St. Austell, PL25 3BZ
M6 GBZ Daniel Smith, 48 Shirley Gardens, Tunbridge Wells, TN4 8TH
M6 GCB Graham Beynon-Fisher, 21 Scopsley Green, Whitley, Dewsbury, WF12 0NF
M6 GCC Grant Clements, 4 Hobart Square, Norwich, NR1 3JB
M6 GCD George Lees, 16 Kingfisher Close, Congleton, CW12 3FF
M6 GCE Jason Berry, 4 Millfield Road, Chorley, PR7 1RE
MU6 GCI James Littlewood, Wayland, LES MARTIN, L'islet, Guernsey, GY2 4XW
M6 GCJ Gregory Cooke, 25 Avill, Hockley, Tamworth, B77 5QE
M6 GCK Gary Cornish, 78 Kerry Avenue, Ipswich, IP1 5LD
M6 GCM Nick Baker, 56 Chalklands, Bourne End, SL8 5TJ
M6 GCN Geoff Leedham, 15 Vale Park, Rhyl, LL18 2EN
M6 GCO Scott Gilbert, 1 Woodlark Drive, Cottenham, Cambridge, CB24 8XT
M6 GCP Mark Poulter, 32 Woburn Street, Hull, HU3 5LW
M6 GCQ Geraint Jones, 6 Tre Ambrose, Holyhead, LL65 1LR
M6 GCS Graham Smith, Glencairn, Gaston Lane, Malmesbury, SN16 0LY
M6 GCT Graeme Taylor, 24 Otter Close, Bletchley, Milton Keynes, MK3 7QP
M6 GCV Grant Adey, 24 Burycroft, Welwyn Garden City, AL8 7AW
M6 GCW Graham Wager, Dovecote, Turbary, Doncaster, DN9 1DY
M6 GCX Mark Parnham, 9 The Close, Addington, West Malling, ME19 5BL
M6 GCY Ben Thomson, 50 Thomson Street, Stockport, SK3 9DR
M6 GDA John Kemp, 9 Chequers Orchard, Stone Street, Canterbury, CT4 5PN
M6 GDB Geoffrey Buckley, 28 St. Leonards Court, House Lane, St. Albans, AL4 9UY
M6 GDC David Cowling, 11 Shakespeare Avenue, Scunthorpe, DN17 1SA
MI6 GDD Graeme Drummond, 44 Camphill Park, Ballymena, BT42 2DJ
M6 GDI Jamie Williams, 41 Overton Lane, Hammerwich, Burntwood, WS7 0LQ
M6 GDJ Graham Johnson, 31 Hillcrest Avenue, Grays, RM20 3DA
M6 GDL Craig Foulkes, 5 Kennedy Close, Chaddesden, Derby, DE21 6LW
MI6 GDN Gemma Nelson, 65 Dernawilt Road, Annagolgan, Enniskillen, BT92 7FN
M6 GDO Colin Stallard, 72 Jutland Road, Hartlepool, TS25 1LW
M6 GDP Andrew Parker, 18 Lincoln Close, Keynsham, Bristol, BS31 2LJ
M6 GDQ Graham Whatmough, Stud Bungalow, Wakefield Lodge Estate, Towcester, NN12 7QX
M6 GDS George Sole, 16 Beech Crescent, Hythe, Southampton, SO45 3QG
M6 GDT Grahame Thomas, 14 Lower Meadow, Cheshunt, Waltham Cross, EN8 0QU
M6 GDV Graham Butterworth, 11 St. Davids Road, Robin Hood, Wakefield, WF3 3TG
MM6 GDY Donna Collins, 6, BLACKBERRY FARM, Throntonloch, EH42 1QT
M6 GDZ Gary Dean, 62 Baptist Close, Abbeymead, Gloucester, GL4 5GD
M6 GEA George Abraham, 90 Louise Road, Northampton, NN1 3RR
M6 GEE Harry Hee, 21 Kenmore Close, Whitley, Gateshead, NE10 [illegible]
M6 GEF Geoffrey Bridge, 49 Wilton Gardens, Radcliffe, Manchester, M26 2UP
M6 GEG Andrew Gleave, 21 Ferndale Close, Stokenchurch, High Wycombe, HP14 3NT
M6 GEJ Glynis Johnson, 4 Wallace Close, Hullbridge, Hockley, SS5 6NE
M6 GEK Stephen Sissens, 20 Fallow Drive, Eaton Socon, St. Neots, PE19 8QL
M6 GEP Stephen Ingledew, 34 Sunningbrook Road, Tiverton, EX16 6EB
M6 GEQ Martin Emmett, 37 Newnham Close, Mildenhall, Bury St. Edmunds, IP28 7PD
M6 GEU Gordon Walker, 43 Wordsworth Crescent, Kidderminster, DY10 3EY
M6 GEV Ralph Heslop, 7 Fieldfare Close, Clanfield, Waterlooville, PO8 0NQ
M6 GEY Andrew Callaghan, 46 Highfields, Great Yeldham, Halstead, CO9 4QQ
M6 GFA Christopher Claypole, 1 Richmond Crescent, Leominster, HR6 8RX
M6 GFE Gary Teale, 97a Wokingham Road, Reading, RG6 1LH
M6 GFF Geoff Gibbs, 2 Salesfrith Cottages, Bicknacre Road, Chelmsford, CM3 8AP
MM6 GFG Denis Speirs, 45 Elmbank Crescent, Arbroath, DD11 4EZ
M6 GFH Gijsbert Molendijk, 47 Lodge Road, Scunthorpe, DN15 7EN
M6 GFM Geoffrey Cummins, 10 Baugh Gardens, Bristol, BS16 6PN
MI6 GFO Geoffrey Craig, 103 Moyle Parade, Larne, BT40 1ET
M6 GFP Gareth Stephens, 2, Lon Y Wylan, LL22 9YL
MW6 GFQ Jeffrey Cooper, 24 Radyr Court Close, Cardiff, CF5 2QG
MU6 GFR Denzil Robert, Nos Treis Liberation Drive, 7 Route Des Clos Landais, Guernsey, Guernsey, GY7 9PH
M6 GFS Glyn Williams, 186 Columbia Road, Bournemouth, BH10 4DT
M6 GFV Paul Dorman, 2 Moises Hall Road, Wombourne, Wolverhampton, WV5 0LF
M6 GFW Graham Watson, The Dell, Nova Scotia Road, Great Yarmouth, NR29 3QD
M6 GFY Geoffrey Waring, 18 The Mead, Beaconsfield, HP9 1AW
M6 GFZ Gerald Linnett, 62 Melrose Avenue, Burtonwood, Warrington, WA5 4NW
M6 GGD Glen Durham, 47 Jervis Road, Hull, HU9 4BT
MM6 GGE George Semple, 28 Newton Brae, Cambuslang, Glasgow, G72 7UW
MI6 GGF Andrew Dowling, 74 Ashmount Gardens, Lisburn, BT27 5DA
M6 GGG Daniel Rosenschein, 101 Christchurch Road, London, SW14 7AT
M6 GGH George Heard, Mill View, Mill Road, Braintree, CM7 4QG
M6 GGI Graham Ridley, 12 Garforth Avenue, Steeton, Keighley, BD20 6SP
M6 GGJ Graham Jacks, 2 Corve View, Ludlow, SY8 2QD
M6 GGL Sean Cooke, Flat 25, Attree Court, Brighton, BN2 0FZ
M6 GGM Andrew Mansfield, 3 The Coppice, Thrapston, Kettering, NN14 4QA
MI6 GGN Gerald Scullion, 50 Tober Road, Pharis, Ballymoney, BT53 8NY
M6 GGO Leslie Jones, 44 Althorpe Drive, Loughborough, LE11 4QU
M6 GGQ Gary Wilson, 28 Tannsfeld Road, London, SE26 5DF
M6 GGT David Turford, 51 Moorfield Avenue, Bolsover, Chesterfield, S44 6EJ
M6 GGU Callum Haines, 10 Inhams Court, Whittlesey, Peterborough, PE7 1TS
M6 GGV Glyn Guest, 90 Pearson Crescent, Wombwell, Barnsley, S73 8SG
M6 GGW George Warnock, 31 Greycote, Shortstown, Bedford, MK42 0XD
M6 GGX Christosfari Ogidih, 89 West Road, Birmingham, B43 5PG
M6 GGZ Thomas Holman, 20 Green Drive, Wolverhampton, WV10 6DW
MI6 GHA Elizabeth Forde, 35 Torr Gardens, Larne, BT40 2JH
M6 GHB Graham Brooks, 14 Chalton Crescent, Havant, PO9 4PT
M6 GHC Luis Arenas Martinez, 22 Brickfield Road, Southampton, SO17 3AE
M6 GHD Alan Burleton, 27 Doncaster Road, Bristol, BS10 5PN
MM6 GHF Glyn Farrer, 23 Upper Craigour, Edinburgh, EH17 7SE
MM6 GHH David Livingstone, The Bungalow, Stewarton, Campbeltown, PA28 6PG
M6 GHI Adam Haslam, 25 Lulworth Road Eccles, Manchester, M30 8WP
M6 GHJ John O'Donnell, 24 Foxhill, Watford, WD24 6SY
M6 GHM Graham Bell, 34 Manor Road, Eastham, Wirral, CH62 8BN
M6 GHN John Oldham, 5 Amersham Rise, Nottingham, NG8 5QG
M6 GHP Glenn Pearson, 41 Myrica Grove, Hoole, Chester, CH2 3EW
M6 GHQ Brandon Gamble, 67 Queen Street, Burntwood, WS7 4QQ
M6 GHR John Greenfield, 12 Firbeck Road, Nottingham, NG8 2FB
M6 GHS Mark Roper, 11 East Close, Beverley, HU17 7JN
M6 GHT Gary Hart, 11 Sadlers Ride, West Molesey, KT8 1SU
M6 GHV John Harris, 108 Gresley Wood Road, Church Gresley, Swadlincote, DE11 9QN
M6 GHW George Harrison-Webb, Picture This, The Studio, 152 Bearton Road, Hitchin, SG5 1UA
MM6 GHX Rolfe James, The Garret, Alyth, Blairgowrie, PH11 8HQ
M6 GHY Carl Beamish, 15 Pen yr Hwylfa, Harlech, LL46 2UW
M6 GIA David Owen, Tanrallt, Blaenpennal, Aberystwyth, SY23 4TP
M6 GIB Ronald Gibbs, 7 Thornhill, Eastfield, Scarborough, YO11 3LY
M6 GIC Anibal Oliveira, 13 A Lakefield Road, London, N22 6RR
MM6 GID Wayne Elder, 3 Mossgiel Place, Dundee, DD4 8AP
MI6 GIF Graham Clarke, 12 Church Green, Dromore, BT25 1LL
M6 GIH Gordon Hutchinson, 128 Crescent Drive North, Brighton, BN2 6SF
M6 GII Colin Daines, 19 Kingfisher Road, Mountsorrel, Loughborough, LE12 7FQ
M6 GIL Gillian Coleman, 5 Meeting Lane, Burton Latimer, Kettering, NN15 5LS
M6 GIM James Isherwood, 11 Manor Crescent, Chesterfield, S40 1HU
MI6 GIN Bob Emerson, 67 Castlemore Avenue, Belfast, BT6 9RH
M6 GIP Paul Duffield, 19 Chorley Lane, Charnock Richard, Chorley, PR7 3Q[illegible]
M6 GIQ Daniel Lynch, Wessex House, Drake Avenue, Staines-upon-Thames, TW18 2AP
M6 GIS Julian Goodman, 70 Bradford Road, Eccles, Manchester, M30 9FT
MW6 GIU A Sprott, 61 Brynymor, Three Crosses, Swansea, SA4 3PE
MW6 GIV Alistair Vowles, 232 Pentregethin Road, Gendros, Swansea, SA5 8AW
MM6 GIW Cameron Ferguson, 18 The Braes, Lochgelly, KY5 9QH
M6 GIX Alex Hodgson, Bryngwyn Bach Rhuallt, St. Asaph, LL17 0TH
M6 GIY Brian Smith, 9 Epperstone Court, West Bridgford, Nottingham, NG2 7QR
M6 GIZ Mark Staley, 164 Rotherham Road, Maltby, Rotherham, S66 8NA
M6 GJA Anthony Gallagher, 174 Queensway, West Wickham, BR4 9DZ
M6 GJB Gareth Baigent, 21 Burton Crescent, Leeds, LS6 4DN
M6 GJC Garreth Cregg, 17 Rake Way, Aylesbury, HP21 9AL
M6 GJD Derek Toller, Field Cottage, Ash Lane, Derby, DE65 6HT
M6 GJE Gary Groves, 5 Beech Road, Ashurst, Southampton, SO40 7AY
M6 GJG Sean Kelly, 4 Berrells Court, Olney, MK46 4AR
M6 GJH Gary Henderson, 4 Castell Road, Loughton, IG10 2LT
M6 GJI Thomas Kelly, 50 Ivanhoe Road, Herne Bay, CT6 6EQ
M6 GJL Jess Paine, 28 Laurel Way, Bottesford, Nottingham, NG13 0FP
M6 GJM Graham Morris, 7 Hawley View, Bilston, WV14 8[illegible]
M6 GJN Gareth Noble, 6 Sturrocks, Basildon, SS16 4PQ
M6 GJO Gareth Owens, 45 Molesworth Terrace, Millbrook, Torpoint, PL10 1DH
M6 GJP Graham Priest, Lilac Cottage, School Lane, Tamworth, B79 9JJ
M6 GJQ Emma Wilkinson, 40 Charmouth Road, St. Albans, AL1 4SN
M6 GJS Gary Shaw, 6 Vickers Close, Woodley, Reading, RG5 4PA
M6 GJT G Tyler, Crofton, Stoney Ley, Worcester, WR6 5NG
M6 GJV Ashley Wesselby, 16 Hudson Way, Grantham, NG31 7BX
M6 GJX Andrew Hunt, 76 Andrew Street, Bury, BL9 7HB
M6 GJY Thomas Garrett, 12 Poulders Gardens, Sandwich, CT13 0BE
MI6 GKB Gareth Black, 41a Meeting House Lane, Lisburn, BT27 5BY
M6 GKC Goodwell Kapfunde, Gunnels Wood Road, Stevenage, SG1 2AS
MM6 GKE Liam Scott, 5 Links Road, Saltcoats, KA21 6BE
M6 GKF Gary Morris-Roe, 77 Bridlebank Way, Weymouth, DT3 5RP
M6 GKG Gemma Gordon, 40 Grange Crescent, Rubery, Birmingham, B45 9XB
M6 GKH Richard Curant, 18 Ramley Road, Lymington, SO41 8GQ
M6 GKI Jack McGowan, 2 Turnstone Drive, Liverpool, L26 7WR
M6 GKK Anthony Maclean, 10 Elizabeth Close, West Hallam, Ilkeston, DE7 6LW
M6 GKM Julian Harrison, 28 Cherry Tree Avenue, Belper, DE56 1FR
M6 GKN Glyn Davies, 7 Bayleys Close, Empingham, Oakham, LE15 8PJ
M6 GKQ Robert Buchan-Terrey, Godre'r Coed, Aberhosan, Machynlleth, SY20 8RA
M6 GKS George Skea, 15 The Shires, Gilwern, Abergavenny, NP7 0EX
M6 GKT Matthew Gunn, 7 New Street, Talybont, SY24 5HD
MW6 GKU Paul Daniel, 71 Coed Isaf Road, Pontypridd, CF37 1EN
M6 GKV David Crook, 72 Rushleigh Avenue, Cheshunt, Waltham Cross, EN8 8PS
M6 GKW Graham Wade, 43 Green Park, Cambridge, CB4 1SX
M6 GKX Elwyn York, Flat 23, Paul Stacey House, Coventry, CV1 5GU
MI6 GLC Geoff Crabbe, 39 Arran Avenue, Ballymena, BT42 4AP
M6 GLD Robert Broughton, Swimbridge, Barnstaple, EX32 0PY
MM6 GLI Graham Irvine, 11 Hazel Road, Cumbernauld, Glasgow, G67 3BN
M6 GLJ Shaun Solomon, 71 Ashby Road, Moira, Swadlincote, DE12 6DN
MW6 GLK Stephen Bateman, 26 Kenneth Treasure Court, Bethania Row, Cardiff, CF3 5UD
M6 GLM Stephen Weeks, Edithmead, Highbridge, TA9 4HE
M6 GLN Glyn Brittleton, 1 Littler Lane, Winsford, CW7 2NE
MM6 GLO Gloria Burns, 25 Dunskey Road, Kilmarnock, KA3 6FJ
M6 GLP Gary Parker, 24 Burrows Close, Lawford, Manningtree, CO11 2HE
M6 GLS Dean Hendricks, 60 Heath View, Leiston, IP16 4JP
M6 GLU Max Morgan-Lucas, 41 Ash Drive, Eye, IP23 7DA
M6 GLV Christopher Tanner, Pen y Gogarth, Llaneilian, Amlwch, LL68 9NH
M6 GLW Glynne Williams, 7 The Grove, Patchway, Bristol, BS34 6PE
M6 GLX Christopher Cowen, Rosita, White Street Green, Sudbury, CO10 5JN
M6 GLZ Gerald Pardoe, 3 Bar Meadow, Shobdon, Leominster, HR6 9BZ
M6 GMA Glenn Marsden, 38 Sandhill Road, Rawmarsh, Rotherham, S62 5NT
M6 GMC Terence Smith, 109 Ferriston, Banbury, OX16 1XA
M6 GMD Mark Davison, 27 Ford Street, Consett, DH8 7AE
M6 GMF Glyn Fildes, 10 Windmill Gardens, St. Helens, WA9 1EN
MW6 GMJ Colette Jones, 28 Ivor Street, Maesteg, CF34 9AH
MI6 GMK Gerard McKinley, 26 Newry Street, Warrenpoint, Newry, BT34 3JZ
M6 GMM Michael Dickinson, 16 Shearwater Avenue, Newcastle upon Tyne, NE12 8PH
M6 GMO Maurice Boland, 24 Hallam Close, Moulton, Northampton, NN3 7LB
M6 GMP Charles Crank, 12 Boundary Lane North, Cuddington, Northwich, CW8 2PL
M6 GMQ Brian Healey, 14 Orchard Close, Ferring, Worthing, BN12 6QP
M6 GMR Graham Mather, Shawdene House, Donnington, Newbury, RG14 3AJ
M6 GMS Grahame Moss, 125 Lavender Avenue, Mitcham, CR4 3RS
M6 GMT Wendy Pattison, 30 Main Street, Flixton, Scarborough, YO11 3UB
M6 GMU Liam Lewis, 33 Boscobel Road, Buntingsdale, Market Drayton, TF9 2HG
M6 GMV Mark Shepherd, 38 Pryors Lane, Bognor Regis, PO21 4LH
M6 GMY Laurence Allison, 25 Saddington Road, Fleckney, Leicester, LE8 8AX
MW6 GNA Angela Jones, 23 Pinecroft Avenue, Aberdare, CF44 0HY
M6 GNC Brian Chandler, 1 Rambridge Farm Cottages, Weyhill, Andover, SP11 0QF
M6 GND Andrew Holden, 4 Gilberts Drive, East Dean, Eastbourne, BN20 0DJ
MI6 GNF Robert Simpson, 3 Portadown Road, Tandragee, Craigavon, BT62 2BB
M6 GNG Christopher Wilsher, 70 Norris Road, Blacon, Chester, CH1 5DZ
M6 GNH George Harris, 100 Bennett Lane, Batley, WF17 6DB
M6 GNJ Georgia Hance, 23 Catalina Avenue, Chafford Hundred, Grays, RM16 6RE
M6 GNM Kieran Markham, Hampers Cottage, Hampers Lane, Pulborough, RH20 3HZ
M6 GNN G Norris, Trade Winds, Preston Road, Preston, PR4 0TT
M6 GNO Andrew Vanderahe, 28 Percival Close, Norwich, NR4 7EA
MI6 GNP Michael Parke, 39 Crannog Park, Strathfoyle, Londonderry, BT47 6NF
M6 GNR Paul Wollen, 41 Bernadette Close, Exeter, EX4 0DU
M6 GNS Stephanie Hance, 23 Catalina Avenue, Chafford Hundred, Grays, RM16 0RE
M6 GNU Patrycjusz Myszka, 55 Broadbent Avenue, Ashton-under-Lyne, OL6 8RL
M6 GNV Andrew Stoll, 6 Dovestone Gardens, Littleover, Derby, DE23 4EJ
M6 GNW Adrian Land, 27 Peaks Lane, New Waltham, Grimsby, DN36 4LG
M6 GNX James Elliot, 82 Hollinside Road, Sunderland, SR4 8BG
M6 GNY Aiffah Ali, 42 Blease Close, Staverton, Trowbridge, BA14 8WD
MI6 GOA Jack Proctor, 16 Lisnavaragh Road, Scarva, Craigavon, BT63 6NX
M6 GOB Ian Pashley, 3 Princess Street, Brimington, Chesterfield, S43 1HP
M6 GOC Jacob Adams, 10 Leckford Road, Oxford, OX2 6HY
M6 GOE Paul Everton, 110 Sandstone Road, Sheffield, S9 1AG
M6 GOF Mark Stitt, 165 Millbrook Close, Skelmersdale, WN8 8QS
M6 GOG Mark Henman, 22 Elizabeth Avenue, Tattershall Bridge, Lincoln, LN4 4JJ
M6 GOI Eoin Walsh, 12 Ock Meadow, Stanford in the Vale, Faringdon, SN7 8LN
M6 GOL Michael Goldthorpe, 84 Park Lane, Allerton Bywater, Castleford, WF10 2AP
MM6 GON Jamie Loughren, 16 Merlewood Road, Inverness, IV2 4NL
M6 GOO Jonathan Goolden, Northdene, Church Lane, Louth, LN11 0QD
M6 GOQ Jonathan Walsh, Flat 4, 14 Chantry Road, Bristol, BS8 2QD

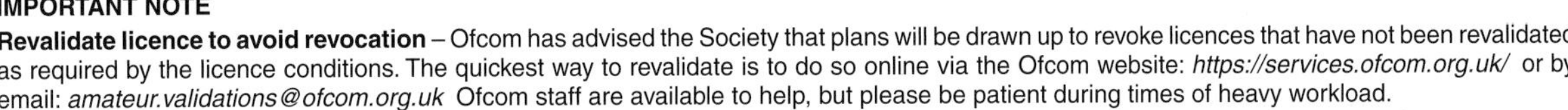

**IMPORTANT NOTE**

**Revalidate licence to avoid revocation** – Ofcom has advised the Society that plans will be drawn up to revoke licences that have not been revalidated as required by the licence conditions. The quickest way to revalidate is to do so online via the Ofcom website: *https://services.ofcom.org.uk/* or by email: *amateur.validations@ofcom.org.uk* Ofcom staff are available to help, but please be patient during times of heavy workload.

UK Callsigns

**MM6 GOR** Gordon Campbell, 98 Netherton Road, East Kilbride, Glasgow, G75 9LB
**M6 GOS** Stuart Forshaw, 8 Stavesacre, Leigh, WN7 3LD
**M6 GOU** Richard Harlow, 28 Dovecliff Crescent, Stretton, Burton-on-Trent, DE13 0JH
**M6 GOW** Norman Gowans, 1 Dunmuir Road, Castle Douglas, DG7 1LG
**M6 GOX** Darren Chadwick, 19 Regent Crescent, Failsworth, Manchester, M35 0LR
**MW6 GOY** Christopher Rees, 5 Dare Villas, Aberdare, CF44 8AH
**M6 GOZ** Chris Gozzard, Craig Dulas, Rhydyfoel Road, Abergele, LL22 8EG
**M6 GPA** Mark Phillips, 1 The Vale, Oakham, LE15 6JQ
**M6 GPB** Gregory Beacher, 22 Trowbridge Gardens, Luton, LU2 7JY
**M6 GPC** Gary Coleman, 19 Grunmore Drive, Stretton, Burton-on-Trent, DE13 0GZ
**MM6 GPD** Lauren Macdonald, 193 Den Walk, Buckhaven, Leven, KY8 1DJ
**M6 GPE** Gillian Davison, Lyncroft House, Swan Lane, Edenbridge, TN8 6AJ
**M6 GPF** Gerard Fleming, 1 Balmoral Drive, Methley, Leeds, LS26 9LE
**M6 GPG** G Bates, 230 Brook Street, Erith, DA8 1DZ
**M6 GPI** Tiberiu Patatu, 103 Compair Crescent, Ipswich, IP2 0EJ
**M6 GPJ** Michael Stevens, 4 Wellwood Close, Horsham, RH13 6AL
**M6 GPK** Glenn Kendall, 6 Badger Wood, Todmorden, OL14 6BB
**M6 GPM** Gary Mayell, Flat 17, Eagle House Goldsmiths, Grays, RM17 6PX
**M6 GPM** George McCaffery, 7 Cliffe Court, Sunderland, SR6 9NT
**M6 GPN** Gareth Morris, 5 Arundel Road, Bath, BA1 6EF
**M6 GPO** Peter Jones, 2 Herons Court, Doncaster, DN2 4GD
**M6 GPP** Graham Powell, 7 Donstan Road, Highbridge, TA9 3LA
**MI6 GPQ** David Boyd, 11 Abbey Gardens, Belfast, BT5 7HL
**M6 GPR** Peter Riley, Sunrise, Field Lane, Swadlincote, DE11 7BT
**M6 GPS** Gary Stevens, 17 Manston Close, Ernesettle, Plymouth, PL5 2SN
**M6 GPT** Richard Hampson, 12 Oakhays, South Molton, EX36 4DB
**M6 GPU** David McLean, 38 New Burlington Road, Bridlington, YO15 3HS
**MI6 GPV** Gabriel Cassidy, 90 Ardmeen Green, Downpatrick, BT30 6JL
**MI6 GPZ** Brian Cousins, 58 Bannview Heights, Banbridge, BT32 4NA
**M6 GQA** Leslie Warren, 85 Greenwood Avenue, Blackpool, FY1 6PR
**M6 GQB** Samuel McCormack, 3 Pickover Gate, Butts Lane, Todmorden, OL14 8RJ
**M6 GQC** Benjamin Angus, Network Rail Advanced Apprenticeship, Faraday Building, Hms Sultan Gosport, PO12 3BY
**MW6 GQD** Brian Lee, 46 Knowling Mead, Tenby, SA70 8EB
**M6 GQF** Dina Thomas, 18 Howard Close, Cambridge, CB5 8QU
**M6 GQG** Eric Thresher, 18 Sandy Lane, Preesall, Poulton-le-Fylde, FY6 0EH
**M6 GQH** John Fitzpatrick, 56 Littlehaven Lane, Horsham, RH12 4JB
**MI6 GQI** Jens Steuerwald, 50 Myrtlefield Park, Belfast, BT9 6NF
**M6 GQJ** Mark Horn, 105 Wards Hill Road, Minster on Sea, Sheerness, ME12 2LH
**M6 GQK** Ethan Beckett, 4 Princes Avenue, Ramsgate, CT12 6DW
**MM6 GQP** Peter Hunter, 3 Promenade, Leven, KY8 4HZ
**M6 GQR** John Townsend, 47 Cosgrove Avenue, Sutton-in-Ashfield, NG17 3JY
**M6 GQS** Steven Klee, 4 Abbots Road, Pershore, WR10 1LL
**M6 GQT** Ian Alderman, 107 Manton Drive, Luton, LU2 7DL
**M6 GQU** Alex Scott, 37 Pikestone Close, Hayes, UB4 9QT
**M6 GQV** Julie (Joolz) Durkin, Selsdon House, 23 Jameson Road, Bexhill-on-Sea, TN40 1EG
**M6 GQW** Harry Beavis, 8 Hookfield, Harlow, CM18 6QG
**M6 GQX** Melvyn Martin, 19 Elmsleigh Road, Farnborough, GU14 0ET
**M6 GQY** Lakota Brearley, Ash Tree Lodge, Snaith Road, Goole, DN14 0AT
**M6 GQZ** Peter Collier, Flat 9, Henry House, London, SW8 2TF
**MW6 GRB** Gareth Ralls, 2 Yew Close, Merthyr Tydfil, CF47 9SD
**M6 GRC** Geoffrey Clarke, 6 Coverdale Road, Scunthorpe, DN16 2RP
**M6 GRD** Carl Schofield, 1187 Manchester Road, Castleton, Rochdale, OL11 2XZ
**MI6 GRF** Conor Robinson, 71 Eglantine Road, Lisburn, BT27 5RQ
**M6 GRH** Graham Hooper, 16 Trentham Drive, Orpington, BR5 2EP
**M6 GRJ** Richard Davison, Lyncroft House, Swan Lane, Edenbridge, TN8 6AJ
**M6 GRK** George Kenyon, 2 Langdale Terrace, Stalybridge, SK15 1EX
**M6 GRN** Adam Young, 48 Sussex Street, Cleethorpes, DN35 7NP
**M6 GRO** Phillip Clark, 446 Ringwood Road, Ferndown, BH22 9AY
**M6 GRP** Graham Priestley, 53 Millfield Gardens, Crowland, Peterborough, PE6 0HA
**M6 GRQ** Daniel McClurg, 48 Braybrooke Drive, Furzton, Milton Keynes, MK4 1AF
**MD6 GRR** Graydon Rodwell, 87 Park Avenue, Ruislip, HA4 7UL
**M6 GRT** Jason Grant, 80 Fishbourne Road West, Chichester, PO19 3JL
**M6 GRU** James Stewart, 43 Balcombe Gardens, Horley, RH6 9BY
**M6 GRV** Clement Rawlin, 5 Japonica Hill, Immingham, DN40 1LT
**M6 GRX** Spencer Tomlinson, 8 Levett Road, Stanford-le-Hope, SS17 0BB
**M6 GRZ** Grzegorz Kober, 1 Sawston, King's Lynn, PE30 4XT
**M6 GSC** Garry Casey, 10 Windermere Road, Dukinfield, SK16 4SJ
**M6 GSD** Christopher Riley, 6 Inworth Walk, Colchester, CO2 8LP
**M6 GSF** Graham Ferns, Oxlea House, Meerbrook, Leek, ST13 8SL
**MI6 GSG** Gary Gregg, 30 Claremont Avenue, Moira, Craigavon, BT67 0SS
**M6 GSH** Paul Pritchard, Flat 148, Nine Acre Court, Salford, M5 3HU
**M6 GSJ** Graham Sawyer, 432 Rowood Drive, Solihull, B92 9LQ
**M6 GSK** Michael Silver, 52 Park Crescent, Elstree, Borehamwood, WD6 3PU
**MW6 GSL** Anastassia Cassar, 48 High Street Abertridwr, Caerphilly, CF83 4FD
**M6 GSO** Graeme Wren, 14 Overdale, Triangle, Sowerby Bridge, HX6 3HZ
**M6 GSP** Paul Lumb, 6 Toronto Street, Wallasey, CH44 6PR
**M6 GSQ** Joshua Creese, Matrice, Church Road, Billericay, CM11 1RR
**M6 GSR** Gary Rogers, 24 Bevan Place, Swanley, BR8 8BH
**M6 GSS** Gary Smith, Ty Clyd, Llanfihangel-Nant-Bran, Brecon, LD3 9NA
**M6 GST** Graham Starling, 4 Three Corner Drive, Norwich, NR6 7HA
**M6 GSW** John Perry, Flat 7 Lancaster House, Belle Vue Rd, Paignton, TQ4 6HD
**M6 GSX** Derek Tinkler, 19 Askew Dale, Guisborough, TS14 8JG
**M6 GSY** Nicholas Blampied, Beech Cottage, Beech Way, Lydney, GL15 6NB
**M6 GSZ** Douglas Gray, 1 The Crossway, Fareham, PO16 8PE
**M6 GTB** Glyn Bolan, 82 Calve Croft Road, Manchester, M22 5FU
**M6 GTC** Jason Tanner, 7 Larch Crescent Eastwood, Nottingham, NG16 3RB
**M6 GTD** David Garratt, 215 Coalpool Lane, Walsall, WS3 1RF
**M6 GTE** Graham Tomkins, 22 Arundel Drive, Orpington, BR6 9JG
**M6 GTH** George Hofman, Brookside, High Street, Stockbridge, SO20 6EY
**M6 GTK** Gary Tagg, Tinkers Cottage, Nevendon Road, Wickford, SS12 0QB
**M6 GTO** Benjamin Thomas, 26 Corfe Crescent, Torquay, TQ2 7QX
**M6 GTQ** Benjamin Partridge, 14 Whitewells Road, Bath, BA1 6NZ
**M6 GTQ** Tracey Haswell, 65 Eastlea Crescent, Seaham, SR7 8EE
**M6 GTR** Michael Bailey, 17 Sparrowhawk Way, Hartford, Huntingdon, PE29 1XE
**M6 GTT** John Owen, 8 Highridge Crescent, Bristol, BS13 8HN
**M6 GTU** Rick Burton, 1 Broad Street, Long Eaton, Nottingham, NG10 1JH
**M6 GTV** Glen Burns, 17 Heather Close, Canvey Island, SS8 0GX
**MI6 GTY** George Mamijs, 33 Glenmore Walk, Lisburn, BT27 4RY
**M6 GUA** John Hammond, 8 Rowntree Way, Saffron Walden, CB11 4DG
**M6 GUB** Nicholas Emberson, Lion House, Audley End, Saffron Walden, CB11 4JB
**M6 GUC** Russell Millen, 21 Sunnymead, Tyler Hill, Canterbury, CT2 9NW
**MU6 GUE** S Tostevin, Hillside, Le Francais, Vale, Guernsey, GY3 5NL
**M6 GUF** Alistair Cook, 35 Park Road, Cheveley, Newmarket, CB8 9DF
**M6 GUH** Darren Hughes, 32 Achille Road, Grimsby, DN34 5RB
**M6 GUJ** Stephen Noller, 3 Thor Road, Norwich, NR7 0JS
**M6 GUM** Andrew Bell, 36 Schneider Road, Barrow-in-Furness, LA14 5DW
**M6 GUO** Marcus Wilson, 116 Aylestone Hill, Hereford, HR1 1JJ
**M6 GUS** Gustav Frisholm, 99 Greenfell Mansions, Glaisher Street, London, SE8 3EX
**M6 GUT** Gerhard Taljaard, Flat 140, 105 London Street, Reading, RG1 4QD
**M6 GUU** Emma Shaw, Upper Floor Flat, 21 Norfolk Road, Littlehampton, BN17 5PW
**M6 GUV** David Connolly, 2 Layton Close, Birchwood, Warrington, WA3 6PT
**M6 GUW** Dean Kemp, 9 Luscombe Way, Rackheath, Norwich, NR13 6SS
**M6 GUX** David Wittering, 156 Langland Road, Netherfield, Milton Keynes, MK6 4HX
**M6 GUY** Guy Westbrook, The Haven, St. Johns Road, Norwich, NR12 9BE
**M6 GUZ** Andrew Burridge, 6 St. Paul Street, Plymouth, PL1 3RZ
**M6 GVA** Peter Kelly, 388 Aylsham Road, Norwich, NR3 2RL
**M6 GVC** Graham Clayton, The Forge, High Street, Moreton-in-Marsh, GL56 0LL
**M6 GVD** Myles McSweeney, Barn House, Stone Quarry Road, Haywards Heath, RH17 7LP
**MD6 GVF** Giovanni Vailati Facchini, 8 Orchard Crescent, Edgware, HA8 9PW
**M6 GVI** John Day, 120 Goring Road, Colchester, CO4 0DB
**M6 GVJ** David Gough, 29 Belvedere Road, Biggin Hill, Westerham, TN16 3HX
**M6 GVL** Brian Alston, 9 Central Avenue, Church Stretton, SY6 6EE
**M6 GVM** Andrew Spurling, 62 Swains Meadow, Church Stretton, SY6 6HT
**M6 GVN** Gavin Tilling, 1 Bellington Cottages, Worcester Road, Kidderminster, DY10 4NE
**M6 GVP** Geoffrey Haynes, 25 Ladbroke Road, Bishops Itchington, Southam, CV47 2RA
**M6 GVR** Gareth Spalding-Reffold, 30 Almond Avenue, Risca, Newport, NP11 6PF
**M6 GVS** Gary Greaves, 183 Wordsworth Avenue, Sheffield, S5 8NE
**M6 GVT** Lucien Edwards, 10 Marsh View, Beccles, NR34 9RT
**M6 GVU** Euan Hammond, 14 Wannock Gardens, Polegate, BN26 5PA
**M6 GVW** Grenville Weston, 131 Ringwood Road, Eastbourne, BN22 8TQ
**M6 GVY** Ruan Kendall, 24 Scotland Road, Cambridge, CB4 1QG
**M6 GWB** George Bunting, 31 Hardwick Avenue, Allestree, Derby, DE22 2LN
**M6 GWC** David Roberts, 22 Bolster Moor Road, Golcar, Huddersfield, HD7 4JU
**M6 GWE** George Wells, 1 Seagrove Way, Seaford, BN25 3QY
**M6 GWF** Gary Waterfall, 18 Sandbed Lane, Belper, DE56 0SH
**M6 GWG** Gavin Weston, 94 Redhill Road, Northfield, Birmingham, B31 3LA
**M6 GWI** Nicholas Shears, 12 Westlees Close, North Holmwood, Dorking, RH5 4TN
**M6 GWK** Keith Jones, Gorswen, Brynrefail, Caernarfon, LL55 3NT
**M6 GWL** Simon Gwillym, The Willows, Llantrisant, Usk, NP15 1LG
**M6 GWM** Michael Martin, 1 Y Gorlan, Bryn Street, Newtown, SY16 2HN
**M6 GWN** Gareth Wynne, 19 Monks Orchard, Nantwich, CW5 5TX
**M6 GWO** Wayne Jones, 20 Beaumaris Way, Grove Park, Blackwood, NP12 1DE
**M6 GWQ** Michael Priest, 35 Albert Road, Chaddesden, Derby, DE21 6SJ
**M6 GWS** George Salter, 9 Spring Gardens, Malvern Link, Malvern, WR14 1AP
**M6 GWT** George Toomer, 7 Rosewood Drive, Barnby Dun, Doncaster, DN3 1BJ
**M6 GWU** Stephen Barrett, 43 The Ridgeway, Meols, Wirral, CH47 9RZ
**M6 GWW** Nicholas Robertson, Craigenveoch Farm, Glenluce, Newton Stewart, DG8 0LD
**M6 GWZ** Kevin Webb, 52 Princes Avenue, Walsall, WS1 2DH
**M6 GXC** Trinity Hales, Flat 6, St. Georges Court, Cambridge, CB1 7UP
**M6 GXD** John Puddy, 26 Lakewood Road, Bristol, BS10 5HH
**MU6 GXE** Adam Prosser, Woodlands, La Vassalerie, St. Andrew, Guernsey, GY6 8XL
**M6 GXF** Guy Fernando, 1 Rosemary Avenue, West Molesey, KT8 1QF
**M6 GXG** Ben Hulstone, 7 McDonna Street, Bolton, BL1 3LP
**M6 GXH** Brian Gates, 3 Highfield, Taunton, TA1 5JE
**M6 GXI** David Burt, 2 Cae Masarn, Pentre Halkyn, Holywell, CH8 8JY
**M6 GXJ** Daniel Ashton, Flat 3, 12 Christ Church Road, Folkestone, CT20 2SL
**M6 GXK** Peter Blagden, 24 Ashgrove Avenue, Gloucester, GL4 4NE
**M6 GXL** Alan Jones, 1a Invicta Road, Folkestone, CT19 6EY
**M6 GXP** Henry Butcher, 12 Bath Road, Willesborough, Ashford, TN24 0BJ
**M6 GXR** Steven Thresher, 4 Huntersway, Culmstock, Cullompton, EX15 3HJ
**M6 GXS** James Allen, 29 Wood Cottage Lane, Folkestone, CT19 4QG
**M6 GXU** Simon Gordon, 8 Maesteg, Cymau, Wrexham, LL11 5EP
**M6 GXV** Mark Butler, Longuenesse, Dodwell Lane, Southampton, SO31 1AD
**M6 GXW** Alec Ashby, 3 The Caravan, Heather Bank, Maryport, CA15 6PB
**M6 GXZ** Maurice George, 26 Yew Tree Road, Ormskirk, L39 1NU
**M6 GYA** Alan Bairstow, 12 Danesfield Avenue, Waltham, Grimsby, DN37 0QE
**MW6 GYB** Laurence Brown, 13 Station Road Loughor, Swansea, SA4 6TR
**M6 GYC** Grant Codrai, 6 Lavender Lane, Rowledge, Farnham, GU10 4AX
**M6 GYD** James Gardiner, 31 Rodway Road, Tilehurst, Reading, RG30 6EH
**M6 GYE** Derek Voak, 63 Green Lane, Crawley, RH10 8JX
**M6 GYF** Stuart Robottom-Scott, 73 St. Bernards Road, Solihull, B92 7DF
**M6 GYG** George Christison, 9 Victoria Avenue, Market Harborough, LE16 7BQ
**M6 GYH** Martin Thompson, 305 Highters Heath Lane, Birmingham, B14 4NX
**M6 GYI** Andrew Hofstedt, 25 Keith Connor Close, London, SW8 3DD
**M6 GYK** Stephen Woodfield, 1 Kingsley Court, Church Road, Birmingham, B25 8XS
**MM6 GYL** Maria Glasper, 1 Lindertis Cottages, Kirriemuir, DD8 5NT
**M6 GYM** James Martin, 20 Hall Green Road, West Bromwich, B71 3LA
**M6 GYP** Alan Dobie, 32 Meadow View, Castle Douglas, DG7 1HF
**M6 GYQ** Keith Holloway, 59 Darrell Way, Abingdon, OX14 1HG
**MM6 GYR** Gregory Brass, 53 Cardowan Drive, Cumbernauld, Glasgow, G68 9PA
**M6 GYS** Ian Cook, 93 Cathedral View, Houghton le Spring, DH4 4HN
**M6 GYT** Ashley Bottomley, 14 Queens Terrace, Dukinfield, SK16 4NU
**M6 GYU** David Perry, Manor Garth, Wesley Road, Whitby, YO22 4RW
**M6 GYV** Daniel Harris, Gatehouse 19, Skitfield Road, Dereham, NR20 5QN
**M6 GYY** Tim Keep, 119 Radley Road, Abingdon, OX14 3RX
**M6 GYZ** David Murray, 39 Eliotts Drive, Yeovil, BA21 3NN
**M6 GZA** Gerald Watson, 20 Windermere Drive, West Auckland, Bishop Auckland, DL14 9LF
**M6 GZB** Ryan Appleby, Flat 3, 198 Comberton Road, Kidderminster, DY10 1UE
**M6 GZC** Cassandra Ezard, 59 Station Farm, Croesyceiliog, Cwmbran, NP44 2JW
**MI6 GZD** Ronald Bishop, 2 Alexander Park, Carrickfergus, BT38 7LL
**M6 GZE** Rees Adams, Aston View, Brownshill, Stroud, GL6 8AG
**MI6 GZF** Vincent Kinney, 49 Lanntara, Ballymena, BT42 3BE
**M6 GZG** John Causer, 2 Kidd Croft, Tipton, DY4 0AF
**M6 GZI** William Jones, 37 Sedgefield Close, Wirral, CH46 9RW
**M6 GZJ** Nicolas Ngan, 6 Wynton Grove, Walton-on-Thames, KT12 1LW
**M6 GZK** Nicholas Heywood, 38 Thurne Rise, Martham, Great Yarmouth, NR29 4PU
**M6 GZL** Callum Snowden, 11 Marion Drive, Shipley, BD18 2EY
**M6 GZN** Michael Armstrong, 5 Aireside, Cononley, Keighley, BD20 8LT
**M6 GZO** Robert Bruce, 19 Spindle Beams, Rochford, SS4 1EH
**M6 GZR** Gary Foster, 248 Harbour Lane, Milnrow, Rochdale, OL16 4EL
**MM6 GZS** George Sinclair, 33 Keptie Road, Arbroath, DD11 3EF
**M6 GZT** Tony Marshall, 63a Newport Road, Ventnor, PO38 1BD
**M6 GZU** Christopher Waters, 45 Elmdale Road, Bedminster, Bristol, BS3 3JF
**M6 GZW** Christopher Hughes, 41 Rotherham Road, Dinnington, Sheffield, S25 3RG
**M6 GZX** Lesley Kurdi, 8 Gwel Afon, Penparcau, Aberystwyth, SY23 3PL
**M6 GZY** Paul Garrett, April Cottage, Castle Avenue, Blandford Forum, DT11 0RY
**M6 GZZ** Gareth Williams, Flat 137, Rosser, Aberystwyth, SY23 3LH
**M6 HAC** Paul Selby, 24 Juniper Close, Guildford, GU1 1PA
**MI6 HAD** Alan McMillen, 55 Northwood Road, Belfast, BT15 3QS
**M6 HAE** Andrew Watts, 175 Ber Street, Norwich, NR1 3HB
**MI6 HAF** Adam McKinley, 36 Grangewood Drive, Londonderry, BT47 5WN
**M6 HAG** Stephen Haigh, 17 Glebe Street, Swadlincote, DE11 9BW
**MM6 HAH** Hugh Halley, 1 Grant Crescent, Renton, Dumbarton, G82 4NH
**M6 HAK** Robert Singer, 19 Rosalind Avenue, Bebington, Wirral, CH63 5JR
**M6 HAM** Darren Trotter, 48 Swindon Road, Sunderland, SR3 4EE
**M6 HAR** Kris Harbour, 43 Falcon Drive, Stanwell, Staines-upon-Thames, TW19 7EU
**M6 HAS** Heather Searle, 2 Tukes Avenue, Gosport, PO13 0SE
**M6 HAT** Richard Hatton, 1 Bowman Mews, Southfields, London, SW18 5TN
**M6 HAU** Alexander Coghlan, Charterhouse, Orchard Road, Salisbury, SP5 2JA
**MM6 HAV** Herve Venries, 24 Lady Place, Livingston, EH54 6TB
**M6 HAY** Hayden Harding, 1 Saddleton Grove, Saddleton Road, Whitstable, CT5 4LY
**M6 HBB** Harry Buckley, 10 Lower Hey Lane, Mossley, Ashton-under-Lyne, OL5 9DE
**M6 HBC** Jonathan Moore, 118 Heneage Road, Grimsby, DN32 9JQ
**M6 HBD** Keith Mills, 72 Sycamore Road, Ecclesfield, Sheffield, S35 9YW
**M6 HBE** Horace Broadhurst, 4 Corfield Crescent, Telford, TF2 6HD
**MM6 HBF** John Hawkins, 1f2 13 Watson Crescent, Edinburgh, EH11 1HB
**M6 HBG** Hagorly Hutasuhut, Hawkridge, Warden Road, London, NW5 4SA
**M6 HBH** Brandon Wilson, 38 Cotleigh Drive, Sheffield, S12 4HU
**MI6 HBI** Beth Huddleston, 4 Knightsbridge Court, Bangor, BT19 6SD
**M6 HBK** A Curry, 58 Greenfields, St. Martins, Oswestry, SY11 3AH
**MW6 HBM** John Hodder, 39-40 Richmond Terrace, Carmarthen, SA31 1HG
**M6 HBN** Robert Tongs, 9 Woodland Drive, Winterslow, Salisbury, SP5 1SZ
**M6 HBP** Heather Pascall, 60 Weyland Road, Witnesham, Ipswich, IP6 9ET
**M6 HBQ** Ivor Newton, 16 Cross Close, Newquay, TR7 3LB
**MM6 HBR** Heather Ling, Leadburnlea, Leadburn, West Linton, EH46 7BE
**M6 HBS** Jonathan Hobbs, 82 Perry's Lane, Wroughton, Swindon, SN4 9AP
**M6 HBT** Sarah Hopkins, 3 Butts Road, Wolverhampton, WV4 5QD
**M6 HBU** John McDonald, 222 Bristol Avenue, Farington, Leyland, PR25 4QZ
**M6 HBV** James Haynes, 16 Mountsfield, Frome, BA11 5AR
**MM6 HBY** Joseph Church, 2 Gorse Loan, Perth, PH1 2SE
**M6 HCA** Christopher Edwards, 47 Victory Street, Plymouth, PL2 2BY
**M6 HCB** David McHugh, 38 Quarryhill Road, Wath upon Dearne, S63 7TD
**M6 HCD** Daniel Humphries, 100 Sunnyside Avenue, Stoke-on-Trent, ST6 6EB
**M6 HCE** Shaun Ellis, 9 Deanwood Close, Whiston, Prescot, L35 3UX
**M6 HCG** Christopher Loud, 24 Harrington Avenue, Lowestoft, NR32 4JU
**M6 HCI** Leslie Call, 9 Hyperion Avenue, Polegate, BN26 5HT
**M6 HCJ** Steven Shone, 22 Fenwick Road, Great Sutton, Ellesmere Port, CH66 4UF
**MM6 HCK** Christopher Northcott, 14/3 Marytree House, 12 Craigour Green, Edinburgh, EH17 7RP
**M6 HCO** Michael Moore, 46 Scholes Park Road, Scarborough, YO12 6QY
**MI6 HCP** Scott Mawhinney, 1 Bramble Lane, Dungannon, BT71 6FF
**M6 HCR** Harry Rogers, 127 Estella Road, Portsmouth, PO2 7SN
**M6 HCS** Robert Hicks, 1 Shenstone Road, Maypole, Birmingham, B14 4TH
**M6 HCT** John Poulter, 71H, HIGHSTREET, Canvey Island, SS8 7RD
**M6 HCU** Paul Lockwood, 80 Falmouth Road, Leicester, LE5 4WH
**M6 HCV** David Poulter, 1 Deacon Drive, Laindon, Basildon, SS15 5FY
**M6 HCW** Cyril Haynes, 4 Thorn Close, Rugby, CV21 1JN
**M6 HCY** Richard Porcher, 9 Blenheim Close, Oswestry, SY11 2UN
**M6 HCZ** Charles Young, 21a Union Crescent, Margate, CT9 1NS
**M6 HDA** David Harley, 39 Tweedale Crescent, Madeley, Telford, TF7 4EA
**MW6 HDB** Howard Bancroft, Stop and Call, Goodwick, SA64 0EX
**M6 HDD** Matthew Hurst, 48 Radcot Close, Woodley, Reading, RG5 3BG
**M6 HDE** David Edmondson, 21 Hawthorne Close, Heathfield, TN21 8HP
**M6 HDG** Eryk Majoch, 66 Boughton Green Road, Northampton, NN2 7SP
**MI6 HDH** Glenn Plunkett, 18 Carnhill Place, Carrickfergus, BT38 7RL
**M6 HDI** Simon Ruddy, 27 Grove Park Walk, Harrogate, HG1 4BP
**M6 HDK** Hanna Karpuk, Harrogate Ladies' College, Clarence Drive, Harrogate, HG1 2QG
**M6 HDM** Donna Money, 2 Lodge Farm Cottage, Black Horse Road, Norwich, NR10 5DJ
**M6 HDO** Ian Johnson, 35 Church Parade, Canvey Island, SS8 9RQ
**MW6 HDP** Michelle Waldman, 892 Llangyfelach Road, Treboeth, Swansea, SA5 9AU
**M6 HDS** David Stapleton, 179 Woodcock Road, Norwich, NR3 3TQ
**MW6 HDT** Dean Jenkins, 82 Hilltop, Llanelli, SA14 8DB
**M6 HDU** Sandra Chipperfield, 3 Clayton Avenue, Upminster, RM14 2EZ
**MW6 HDV** Johnathon Hodson, 6 Heol Pentwyn, Tonyrefail, Porth, CF39 8DF

M6 HDW Darren Wilson, Flat 20, Limerick House, Woking, GU22 7JF
M6 HDY David Hardy, 61 Westbourne Road, Handsworth, Birmingham, B21 8AU
MM6 HDZ Hans De Zeeuw, Abriachan, Monaltrie Avenue, Ballater, AB35 5RX
M6 HFA Terrence Heath, 10 Deacons Park, [illegible], LD3 9BB
M6 HFB Steven Drakeley, 1 Council Houses, Churchtown, Wadebridge, PL27 7QA
MW6 HED Malcolm Hedley, North Vatson Farm, Devonshire Drive, Saundersfoot, SA69 9EE
M6 HEE Stuart Marr, 49 Gallows Hill, Ripon, HG4 1RG
M6 HEF David Robinson, Height End Farm, Kirk Hill Road, Rossendale, BB4 8TZ
M6 HEG Phil Sellick, 34 Atherton Street, London, SW11 2JE
MW6 HEI Richard Williams, 88 Parc Pendre, Kidwelly, SA17 4TE
M6 HEJ Niklas Lunden, Flat 15 Western House, 8 Woodfield Place, London, W9 2BJ
M6 HEK Ernest Wright, 351 Market Street, Droylsden, Manchester, M43 7EA
M6 HEN Brian Henshaw, 4 Cumberland Close, Darwen, BB3 2TR
M6 HEO Lance Davis-Edmonds, Bladnoch Cottage, Bladnoch, Newton Stewart, DG8 9AB
M6 HEP Hayley Purdy, 13 St. Cuthbert Street, Worksop, S80 2HN
MM6 HEQ Robert Wilson, 3 St. Peters Park, Stromness, KW16 3EH
M6 HET Ivan Szabo, 95 Westgate, Grantham, NG31 6LE
M6 HEW Wayne Fearby, 165 Brandsfarm Way, Telford, TF3 2JJ
M6 HEX John Ash, 47 Stein Road, Emsworth, PO10 8LB
MW6 HEY Corinne Hey, 84 Trefelin, Aberdare, CF44 8LF
MM6 HEZ William Demczur, 25 Maitland Court, Helensburgh, G84 7EE
M6 HFA Jennifer Wilson, Flat 5, Blake House, London, SE1 7DX
MM6 HFC Thomas Crowther, 74 Commerce Street, Lossiemouth, IV31 6QQ
M6 HFF Michael Collins, The Haven, Kettleby Lane, Brigg, DN20 8SW
M6 HFG Harold Furniss, 3 Byron Avenue, Chapeltown, Sheffield, S35 1SQ
M6 HFI Gordon Mitchell, 49 Nash Road, Romford, RM6 5JP
M6 HFJ Adam Cheung, 197 St. Lukes Avenue, Ramsgate, CT11 7HS
M6 HFK Christopher Boal, 92 West Street, Millbrook, PL10 1AF
M6 HFM Harry Mcevoy, 18 Brookfield Gardens, Wirral, CH48 4EL
M6 HFN David Kelly, 6 Sandham Walk, Bolton, BL3 6RA
M6 HFO Graham Allen, 14 The Parsonage, Sixpenny Handley, Salisbury, SP5 5QJ
M6 HFP Justin Darley, 159 Main Road, Hawkwell, Hockley, SS5 4EL
M6 HFQ Alex White, 82 Shakespeare Street, Sinfin, Derby, DE24 9HE
M6 HFR Raymond Heffer, 36 Raven Avenue, Tibshelf, Alfreton, DE55 5NR
M6 HFT David Merridale, THE GRANARY, FALLEDGE LANE, Upper Denby, HD8 8YH
M6 HFU James Russell, 1 West Street, Bishops Lydeard, Taunton, TA4 3AU
M6 HFV Matthew Johnson, Charnwood, The Close, Ringwood, BH24 2PE
M6 HFX Tony Crawshaw, 208 Ovenden Road, Halifax, HX3 5QG
MW6 HFY Nathaniel James, 8 Garth Lwyd, Caerphilly, CF83 3QB
MM6 HFZ Peter McNally, 0/3 6 Carillon Road, Glasgow, G51 1QL
M6 HGA Leslie Brookhouse, 16 Clockmill Road, Walsall, WS3 4AH
M6 HGD Lydia Brookhouse, 16 Clockmill Road, Walsall, WS3 4AH
M6 HGE James Webber, 14 Raleigh Street, Scarborough, YO12 7JZ
M6 HGF Adrian Highfield, 38 Brunswick Gardens, Garforth, Leeds, LS25 1HF
MD6 HGG Stephen Entwisle, 30 Arden Mhor, Pinner, HA5 2HR
M6 HGH Hector Hamilton, Flat B, 9 Cambridge Drive, London, SE12 8AG
MI6 HGI Andrew Glasgow, 17b Loy Street, Cookstown, BT80 8PZ
M6 HGJ Michael Hadley, 75 Glendower Avenue, Coventry, CV5 8BD
M6 HGM Barry Holgate, 9 Nursery Grove, Bridlington, YO16 4QS
M6 HGN Jonathan Allen, 43 Borrowdale Road, Stockport, SK2 6DX
M6 HGO Piotr Niewiadomski, 79a Dartmouth Road, London, SE23 3HT
M6 HGR Steven Yardley, 22 Wedgwood Road, Clifton, Manchester, M27 8RT
MI6 HGS Michael Doogan, 54 Birchdale Manor, Lurgan, Craigavon, BT66 7SY
M6 HGU Virginia Keith, Valentine Cottage, Frankton Road, Rugby, CV23 9QT
MI6 HGV Eneas Rainey, 22 Cherry Gardens, Ballymoney, BT53 7AS
M6 HGW Gary Whitton, 40 Louville Avenue, Withernsea, HU19 2PB
M6 HGX Euan Keith, Valentine Cottage, Frankton Road, Rugby, CV23 9QT
M6 HGY Anthony Flintoff, 44 Newsham Way, Northallerton, DL7 8HT
MW6 HGZ Rhodri Taylor, 4 Castell Morgraig, Caerphilly, CF83 3JH
M6 HHA George Coldham, 27 Welsby Road, Leyland, PR25 1JA
MM6 HHB Duncan Burgess, Quendale Farm, Quendale, Shetland, ZE2 9JD
M6 HHC Kenneth Townsend, Flat 3, Drake House, Bexhill-on-Sea, TN39 3TS
M6 HHD Keith Davies, 25 Kinmel Avenue, Abergele, LL22 7LR
M6 HHE Matthew Seabrook, 1 The Covers, Morpeth, NE61 2RU
M6 HHF John Sim, 11 Haven Close, Istead Rise, Gravesend, DA13 9JR
M6 HHH Jake Hoare, Flat 3, 6 High Street, Watlington, OX49 5PR
MM6 HHJ David Jappy, 21 Primrose Avenue, Grangemouth, FK3 8YG
M6 HHM Hannah McCarthy, 80 Vaughan Williams Way, Warley, Brentwood, CM14 5WT
M6 HHO Dawid Przybylski, 108 Windrows, Skelmersdale, WN8 8NW
M6 HHQ David Harvey, Bennettshayes Barn, Awliscombe, Honiton, EX14 3PY
MM6 HIA Aileen Crichton, 11 Baillie Court , Sauchie, Alloa, FK10 3FG
M6 HIB Denise Bedworth, 96 Balmoral Road, Stourbridge, DY8 5JB
MM6 HIG Harry Glennie, 97 Smithfield Crescent, Blairgowrie, PH10 6UE
M6 HIJ Mark Willis, 7 Belvawney Close, Chelmsford, CM1 4YR
M6 HIK Andrew Gretton, 67 Hawthorn Crescent Arnold, Nottingham, NG5 8RF
M6 HIL Hilary Penfold, 15 Carmans Close, Loose, Maidstone, ME15 0DR
M6 HIM Paul Green, 15 Dickenson Road, Chesterfield, S41 0RX
M6 HIP Stephen Admans, 77 Beaumont Road, Birmingham, B30 2FR
M6 HIT I Petrie, 88 Vicarage Road, Henley-on-Thames, RG9 1JT
M6 HIU jean leatherd, 115 Rothesay Road, Blackburn, BB1 2ER
MM6 HIZ Daniel Aitken, 64 Brunton Street Cathcart, Glasgow, G44 3NQ
M6 HJA Kevin Rouse, 42 Berkley Close, Highwoods, Colchester, CO4 9RR
M6 HJB Hannah Barton, 86 Forge Lane, Kingswinford, DY6 0LG
M6 HJC Holly Christie, Flat B, 9 Britannia Road, Westcliff-on-Sea, SS0 8BS
M6 HJD Joshua Hawley, 11 Upper Green Way, Tingley, Wakefield, WF3 1TA
M6 HJE Matthew Jessop, 3 Albert Avenue Mayfield Street, Hull, HU3 1NY
M6 HJF Stephen Rattley, 2 Burnt Cottages, Beanacre, Melksham, SN12 7PT
M6 HJG Anthony Lawrence, 6 Beaver Court, Ashford, TN23 5QR
M6 HJH desmond money, Flat 1, Lewin Court 24b Plumstead High Street, London, SE18 1SL
M6 HJI Neil Crudgington, Appledore Blackness Lane, Keston, BR2 6HL
M6 HJK Alison Scott, 19 Estuary Drive, Felixstowe, IP11 9TL
M6 HJM Harry McNeill, 44 Anglesey Road, Wirral, CH48 5EG
M6 HJQ Gabriel Cairns Thomas, 121 London Road, Bagshot, GU19 5DH
M6 HJR David Riman, 22 Princess Road, Hinckley, LE10 1EB
M6 HJT Harry Hughes, 27 The Holt, Hailsham, BN27 3ND
M6 HJU Christian Cairns Thomas, 121 London Road, Bagshot, GU19 5DH
M6 HJV Jason Moore, 5 The Grange 259 Hillbury Road, Warlingham, CR6 9TL
M6 HJX [illegible], 00 Water Lane, London, [illegible]
M6 HJZ [illegible]
M6 HKA John Daniels, 27 Hammerwater Drive, Warsop, Mansfield, NG20 0DJ
M6 HKB Dave Lock, 20 Jasmine Close Trimley St. Martin, Felixstowe, IP11 0UY
M6 HKC Andrew Chambers, 19 Marina Road, Durrington, Salisbury, SP4 8DB
M6 HKI MICHAEL DAVIES, 2 Ellins Terrace, Normanton, WF6 1BL
M6 HKJ David Killingley, 17 Colbert Drive, Leicester, LE3 2JB
M6 HKK Darren Wetherilt, 3 Egdon Crescent, Cheltenham, GL51 6GF
MW6 HKL Luster Chang, 14 Greenfield Gardens, Pentrebach, Merthyr Tydfil, CF48 4BQ
M6 HKN Dale Potts, 30a Tower Hill, Gomshall, Guildford, GU5 9LS
M6 HKO Steven Marden, 65 Hedley Way, Hailsham, BN27 3FZ
M6 HKQ Antony Lamont, 89 Newlands Whitfield, Dover, CT16 3ND
M6 HKS David Collier, 7 Compass Close, Ashford, TW15 1UT
M6 HKT Ben Scott, 3 Chaplin Close, Basildon, SS15 4EJ
M6 HKU David Holdbrook, 9 Johns Terrace Colchester Road, Romford Essx, RM3 0AW
M6 HKW Helen Knowles, 80 Holborn Avenue, Coventry, CV6 4FZ
M6 HKX Humayun Khayer, 24 Clyde Road, Stoke-on-Trent, ST6 3DJ
M6 HKY Anthony Hickey, 144 Gisburn Road, Barnoldswick, BB18 5LQ
M6 HKZ James Poole, Ramillies Hall School, Ramillies Avenue Cheadle Hulme, Cheadle, SK8 7AJ
M6 HLA Alan Harvey, 20 Fellowes Place, Plymouth, PL1 5NB
MI6 HLC Samantha Savage, 469 Old Belfast Road, Bangor, BT19 1RQ
M6 HLE Paul Bray, 24 Eldon Terrace, Bristol, BS3 4NZ
M6 HLF John Taylor, The Old Mission Hall Orton Avenue, Peterborough, PE2 9HL
MW6 HLG Helene Griffiths, 5 Heol-y-Sarn, Llantrisant, Pontyclun, CF72 8DA
M6 HLL Mathew Beharrell, 110 Scotforth Road, Lancaster, LA1 4SQ
M6 HLP Ann-Marie Wilson, 39 Rochford Garden Way, Rochford, SS4 1QH
MW6 HLQ lee stevens, 75 Long Mains Monkton, Pembroke, SA71 4HX
M6 HLR Lauren Richardson, 35 Vidgeon Avenue, Hoo, Rochester, ME3 9DE
M6 HLS Peter Hollis, 89 Longfield Lane, Cheshunt, Waltham Cross, EN7 6AN
M6 HLT Alexander Tomkins, 28 Newborough Close, Austrey, Atherstone, CV9 3EX
MW6 HLU Alan Garner, 15 Midland Place Llansamlet, Swansea, SA7 9QU
M6 HLW christian robinson, 69 Sanger Avenue, Chessington, KT9 1BY
MI6 HLY Robert Lawrence, 8 Wynford Park, Lisburn, BT27 5HJ
MM6 HLZ Hayley Ross, 16 Myreton Drive, Bannockburn, Stirling, FK7 8PX
MM6 HMB Hugh Brown, 2 Kincaid Way, Milton of Campsie, Glasgow, G66 8DT
MM6 HMC Hugh Campbell, 10 Stewart Avenue, Linlithgow, EH49 6DQ
M6 HMD Emma Gibbs, 35 St. Michaels, Houghton le Spring, DH4 5NR
M6 HME wayne mayall, 17 Norrington Grove, Birmingham, B31 5NY
M6 HMG Mark Hughes, 58 Grange Lane North, Scunthorpe, DN16 1RW
M6 HMK Helen Melhuish, 22 Mayflower Close, Glossop, SK13 8UD
M6 HML Kian Lees, 24 Marks Road, Wokingham, RG41 1NN
M6 HMM philip bromley, Pinewood Lodge, Blandford Road Coombe Bissett, Salisbury, SP5 4LH
M6 HMP Michael Hughes, 58 Grange Lane North, Scunthorpe, DN16 1RW
M6 HMQ James webb, 49 Perth Avenue, Leicester, LE3 6QQ
M6 HMR Hope Pittard, 21 The Oaklands, Church Eaton, Stafford, ST20 0BA
M6 HMS Matthew Taylor, 21 Ravenstone, Wilnecote, Tamworth, B77 4JZ
M6 HMU Matteo Gosi, 49 Elms Drive, Oxford, OX3 0NW
M6 HMV Oliver Wood, 26 Parkfield Crescent, Kimpton, Hitchin, SG4 8EQ
M6 HMW Jason Walker, 5 Preston Avenue, Alfreton, DE55 7JX
M6 HMZ Ross Dalziel, 24 Horringford Road, Liverpool, L19 3QX
M6 HNA Mike Rosewell, 54 Alder Drive, Chelmsford, CM2 9EZ
M6 HNB Harley Baird, 35 St. Peters Road, Wolvercote, Oxford, OX2 8AX
M6 HND Alan Haylor, 33 Crimp Hill Road, Old Windsor, SL4 2QY
M6 HNF Colin Seymour, 12 Silver Street Riccall, York, YO19 6PB
M6 HNG neil kerry, 47 Harpley Dams Hillington, King's Lynn, PE31 6DP
M6 HNH Sebastian Kozlowski, 28 Osney Crescent, Paignton, TQ4 5EY
M6 HNI Michael Lawson, 131 Windermere Avenue, Ilkeston, DE7 4EZ
M6 HNJ Ewan Potter, 3 Thomson Court Chadwick Close, Crawley, RH11 9LH
M6 HNN darren gibson, 38 Newellhill, Tenby, Sa708en
M6 HNO Philip Byrne, 18 St. Aidans Square, Bingley, BD16 2BN
MM6 HNQ Alexander Macintyre, Bunillidh Sinclair Street, Halkirk, KW12 6XT
M6 HNS Heather Stanley, 253 Brownley Road, Manchester, M22 9UX
M6 HNT Matthew Hunt, 13 Pine Halt, Station Road, Cheltenham, GL54 4JX
M6 HNV paul sargeant, 6 Meldon Way, Blaydon-on-Tyne, NE21 6HJ
M6 HNW Thomas Fletcher, 15a Les maisonette, Penryhn road, Colwyn Bay, LL29 8LG
M6 HNX Adam Lorne, 8 campbell close, Grantham, Ng31 8aw
M6 HNZ Michael Bruce, 28 Pheasants Way, Rickmansworth, WD3 7ES
M6 HOF Steven Baines, Slowe Cottage, Riverside Lane, Newnham, GL14 1JE
M6 HOG Gary Snape, 3 Jasper Close, Barlaston, Stoke-on-Trent, ST12 9BL
MW6 HOH Peter Gostelow, Coedmor Llangrannog, Llandysul, SA44 6AG
M6 HOI Matthew Simkins, 37 St. Andrews Meadow, Harlow, CM18 6BL
M6 HOK Ken Hough, 15 Moorside Road, Endmoor, Kendal, LA8 0EN
M6 HOM Seyed Hossein Mirjalili Mohanna, Apartment 18b, White Croft Works, 69 Furnace Hill, Sheffield, S3 7AH
MM6 HOO Garry Freeburn, 31 Courthill, Rosneath, Helensburgh, G84 0RN
M6 HOP Richard Hope, 32 Winstanley Place, Rugeley, WS15 2QB
M6 HOQ david hampson, 29 Holywell Road Kilnhurst, Mexborough, S64 5UQ
M6 HOS Brendon Mulholland, 6 Burnside, Longhoughton, Alnwick, NE66 3JQ
M6 HOT Daniel Hubbard, 14 Parkfield Crescent, Kimpton, Hitchin, SG4 8EQ
M6 HOU Raymond Houlton, L11 Woodlands Caravan Park, The Marshes Lane, Preston, PR4 6JS
M6 HOV Nikki James Fox, 3 Dog Rose Drive, Bourne, PE10 0FG
M6 HOY Jamie Wilson, 5 Queens Road, Hoylake, Wirral, CH47 2AG
M6 HPC Heather Cooper, 3 Waunddu, Pontnewynydd, Pontypool, NP4 6QZ
M6 HPD Robert David Hodson, 99 Alcester Road, Hollywood, Birmingham, B47 5NR
M6 HPF raymond stringer, 9 Pershore Close, Walsall, WS3 2UQ
M6 HPG Harry Bell, 7 Rosecomb Way, Haxby, York, YO32 3ET
M6 HPH Harry Heathfield, 17 Exchange Street South Elmsall, Pontefract, WF9 2RD
M6 HPK Rhys Hopkins, 132 Laurel Road Bassaleg, Newport, NP10 8PT
M6 HPL Carl Gorse, 34 Ellison Street, Hartlepool, TS26 9AN
MW6 HPN Adam Hampson, 48 Heol Cwm Ifor, Caerphilly, CF83 2EU
M6 HPR Ian Hooper, 25 Honey Lane, Buntingford, SG9 9BQ
M6 HPS Hayden Partridge, 19 Dickens Drive, Melton Mowbray, LE13 1HZ
M6 HPT [illegible], Cambridge, [illegible]
[illegible] HPU [illegible], OV10 0DQ
M6 HPX William Rainbow, 69 Abbotsweld, Harlow, CM18 6TG
M6 HPY Paul Clark, 43a Eastbury Avenue, Rochford, SS4 1SE
M6 HQB joe bristow, 55 Haldon Close, Bristol, BS3 5LR
MM6 HQC Gordon Fleming, 28 Fenwick Drive, Hamilton, ML3 7YG
M6 HQD John Hobbs, 1 Hotground Cottage, Branfield, Hertford, SG14 2QG
MM6 HQE Alasdair McCormick, Flat 2 16 Marine Drive, Edinburgh, EH5 1FD
MM6 HQI William McBain, 9/12 Tower Place, Edinburgh, EH6 7BZ
MM6 HQK B McCabe, 173 Marmion Road Cumbernauld, Glasgow, G67 4AW
M6 HQL marcus dimmick, 16 Bushell Way Kirby Cross, Frinton-on-Sea, CO13 0TW
M6 HQM Russell Shulver, 35 Osborne Villas, Hove, BN3 2RA
M6 HQN Wayne Hamlet, 18 Bridle Lane, Alfreton, DE55 1LG
M6 HQO Robert Olive, Lorien, The Ridge, Thatcham, RG18 9HZ
M6 HQR Peter Nolan, 14 Woodlands Road, Stafford, ST16 1QR
M6 HQW Koji Kimura, 51 Nelson Road, Newport, PO30 1RE
M6 HQX Clive Poole, 1 Ripon Gardens, Ilford, IG1 3SL
M6 HQZ Holly Bamford, 14 Calf Close, Haxby, York, YO32 3NS
M6 HRC Peter Humphreys, 30 The Chestnuts, Hinstock, Market Drayton, TF9 2SX
M6 HRD Vincent Greatwood, 11 The Green, Long Preston, Skipton, BD23 4PQ
M6 HRE Shaun Bligh-Wall, 81 Warmingham Road, Leighton, Crewe, CW1 4PS
MM6 HRF Martin Reynolds, 22 Fergus Place, Dyce, Aberdeen, AB21 7DD
M6 HRG Stuart Probert, 11 Wilwick Lane, Macclesfield, SK11 8RS
M6 HRI Hari Hughes, Cefn Glass, Clyro, Hereford, HR3 5JT
M6 HRJ Robin Huelin, 15 Hill Chase, Walderslade, Chatham, ME5 9HE
M6 HRK Gareth Hughes, 19 Gernant Braichmelyn, Bethesda, Bangor, LL57 3RE
M6 HRL Howard Russell, 4 Dearnsdale Close, Stafford, ST16 1SD
M6 HRM Henry McBrien, Hamilton House, Hayes Lane, Wokingham, RG41 4TA
M6 HRN Benjamin Hurren, 29 Chalk Lane, Ixworth, Bury St. Edmunds, IP31 2JQ
M6 HRO John Mills, 24 Charles Street, Ryhill, Wakefield, WF4 2BU
M6 HRQ shane duncan, 5 Spring Vale Bilton, Hull, HU11 4DN
M6 HRR H Roberts, 12 Rheidol Terrace, Aberystwyth, SY23 1JU
M6 HRS Hayden Richardson, 103 Marys Mead, Hazlemere, High Wycombe, HP15 7DT
M6 HRT Jakub Krol, 40 Hampton Gardens, Southend-on-Sea, SS2 6RW
M6 HRV Harvey Harkishin, 2 Kingfisher Close, Bournemouth, BH6 5BB
M6 HRW Hannah Woolley, 84 Bowthorpe Road, Norwich, NR2 3TP
M6 HRX Raymond May, 6 Gordon Court Well Street, Loose, Maidstone, ME15 0QF
MM6 HRZ William McEwan, 240 Turriff Brae, Glenrothes, KY7 6UT
M6 HSA Paulo Sousa, 11 Broom Crescent, Ipswich, IP3 0EE
M6 HSB Paul Henry, 22 Huddleston Close, Wirral, CH49 8JP
MM6 HSC Hugh Campbell, 8b Hawthorn Place, Uphall, Broxburn, EH52 5BX
M6 HSE Henry Evans, 26 Peartree Court, Welwyn Garden City, AL7 3XN
M6 HSF Farzad Hayati, Apartment 19, 29 Longleat Avenue, Birmingham, B15 2DF
M6 HSG Katharine Tabor, 11 Lordsbridge Court, Mervyn Road, Shepperton, TW17 9HE
M6 HSH Michael Williams, 21 Elmbrook Close, Basildon, SS14 2FH
M6 HSI Jan Ostapiuk, 49 Rectory Place, Gateshead, NE8 1XN
MM6 HSK Hollie King, 19 Gleneagles Way, Deans, Livingston, EH54 8EW
MI6 HSL Neil Davis, 19 Toberhewny Hall, Lurgan, Craigavon, BT66 8JZ
M6 HSP Neville Hawkins, Deganwy Hardwick Road, King's Lynn, PE30 5BB
M6 HSQ Roy Terry, 31 Barnwood Road, Birmingham, B32 2LY
M6 HSR Hugo Stewart-Roberts, Cinderfield House, Cornwells Bank, Lewes, BN8 4RH
MM6 HSS Sarah Skerratt, 3/2 18 Mardale Crescent, Edinburgh, EH10 5AG
M6 HST Michael Ghost, S/R Mess HMS Collimgwood, Newgate Lane, Fareham, PO14 1AS
M6 HSX Heider Sati, 27 Hanover Road, London, SW19 1EB
M6 HSY George Kokinis, 40 Cruise Road, Sheffield, S11 7EF
M6 HTB Stewart Crane, 5 Buchanan Road, Wigan, WN5 9SB
M6 HTC Alan Porter, 35 St. Andrews Crescent, Hindley, Wigan, WN2 3EQ
M6 HTF Steven Griffiths, 111 Windmill Road, Hemel Hempstead, HP2 4BP
M6 HTG David Quinney, 8 Crabwood Road, Southampton, SO16 9EZ
M6 HTI Jason Trimmer, The Lodge, Horsemoor Lane Winchmore Hill, Amersham, HP7 0PL
M6 HTJ Timothy Hare, 2 Primula Drive, Norwich, NR4 7LZ
M6 HTK John Ormston, 15 Thackeray End, Aylesbury, HP19 8JE
M6 HTM Joel Clyne, Ravenswood, Green Lane, Wisbech, PE14 7BJ
M6 HTN Alice Horton, 51 Walsingham Gardens, Epsom, KT19 0LS
M6 HTQ philip day, 1 Pine Close, Lutterworth, LE17 4UT
MM6 HTS Stephen Spencer, 54 Maclennan Crescent, Inverness, IV3 8DN
MW6 HTT Holly Thomas, 2 Ffordd Donaldson, Copper Quarter, Swansea, SA1 7FJ
M6 HTU tony leatherbarrow, 17 Egerton, Skelmersdale, WN8 6AA
M6 HTW Christopher Brooks, 318 Fleetwood Road North, Thornton-Cleveleys, FY5 1LD
M6 HTX Henryk Banasiak, 11 Westfield Road, Backwell, Bristol, BS48 3NE
M6 HTZ Lee Shearson, 23 Thumpers, Hemel Hempstead, HP2 5SL
M6 HUC Colin Lycett, 2 Royce Avenue, Hucknall, Nottingham, NG15 6FU
MM6 HUE James Hughes, 2 Forest Place, Townhill, Dunfermline, KY12 0EP
M6 HUF Stephen Pennell, 450 Cog Lane, Burnley, BB11 5HR
M6 HUH Wing Lam, Harrogate Ladies' College, Clarence Drive, Harrogate, HG1 2QG
M6 HUI Anthony Windle, 10 Longshaw Street, Blackburn, BB2 4HS
M6 HUK Alan Thomas, 34 Buscot Drive, Abingdon, OX14 2BL
M6 HUL Sean Lyon, 10 Sycamore Close, Preston, Hull, HU12 8TZ
M6 HUM Daniel Wood, School Farm, Brock Road, Preston, PR3 0XD
M6 HUN Julie Collier, 133 Woodstock Road, Moston, Manchester, M40 0DG
M6 HUP Muhammad Arif, 171 Henley Road, Bedford, MK40 4FZ
M6 HUQ Rachel Harwood, 27 Barnards Hill Lane, Seaton, EX12 2EQ
M6 HUR Jonathan Jefferies, Millfield Cottage, 1 Bolnhurst Road, Bedford, MK44 2LF
M6 HUS Muhammad Ali Hussain, 10 Mercia Crescent, Stoke-on-Trent, ST6 3JB
MM6 HUT Adam Hutchison, 24 Tanna Drive, Glenrothes, KY7 6FX
MD6 HUV Ian Barnes, 35 Copley Road, Stanmore, HA7 4PF

**IMPORTANT NOTE**

**Revalidate licence to avoid revocation** – Ofcom has advised the Society that plans will be drawn up to revoke licences that have not been revalidated as required by the licence conditions. The quickest way to revalidate is to do so online via the Ofcom website: *https://services.ofcom.org.uk/* or by email: *amateur.validations@ofcom.org.uk* Ofcom staff are available to help, but please be patient during times of heavy workload.

M6 HUW James McLean, 24 Durham Drive, Oswaldtwistle, Accrington, BB5 3AT
M6 HUX Malcolm Lisle, 16 Collegiate Crescent, Sheffield, S10 2BA
MW6 HUY david johnson, 12 Bro Dedwydd Dunvant, Swansea, SA2 7PR
M6 HUZ Luke Hughes, 32 Calder Road, Blackpool, FY2 9TX
MD6 HVA Jack Wills, 1 Aberdeen Road, Harrow, HA3 7NF
M6 HVD Lyndon Evans, Manor Grove 24 Leopold Grove, Blackpool, FY1 4LD
M6 HVE Martin Simonsohn, 5 Pitt Close, Blandford St. Mary, Blandford Forum, DT11 9PS
M6 HVF Stuart Pearson, 3 Berkeley Road, Shirley, Solihull, B90 2HS
M6 HVH Michael Rose, 149 Claremont Road, Blackpool, FY1 2QJ
M6 HVI Anthony Corbett, 122 Harrowby Road, Stoke-on-Trent, ST3 7AN
M6 HVK KEVIN ORCHARD, 47 Trezaise Road Roche, St. Austell, PL26 8HD
M6 HVL Hong Ly, 18 Mullway, Letchworth Garden City, SG6 4BH
M6 HVM Paul Ashton, 32 Sycamore Road, New Ollerton, Newark, NG22 9PS
M6 HVN Gareth Edwards, 7 Maple Crescent, Leigh, WN7 5QX
M6 HVO alan jarvis, 10 West Park, Wadebridge, PL27 6AN
M6 HVP Craig Harwood, 27 Barnards Hill Lane, Seaton, EX12 2EQ
M6 HVR malcolm murray, 31 Feeny Street, Sutton Manor, St. Helens, WA9 4BJ
M6 HVS Brian Smithers, 16 Potters Grove, New Malden, KT3 5DE
M6 HVU Thomas Raymond, 12 Mill Race Wolsingham, Bishop Auckland, DL13 3BW
M6 HVX Simon Bourne, 1 Humewood Grove, Stockton-on-Tees, TS20 1JU
M6 HVY Emma Money, 4 Cromes Place, Badersfield, Norwich, NR10 5JT
M6 HWC Harry Cheesman, 49 Front Street, Chirton, North Shields, NE29 7QN
M6 HWD julian redgrave, 24 Burnham Close Trimley St. Mary, Felixstowe, IP11 0XG
M6 HWE Andrew Jepson, Edmonton road, Mansfield, NG21 9ah
M6 HWG Cori Haws, 5 Mallow Close, Locks Heath, Southampton, SO31 6XF
M6 HWH Massimo Milioto, 7 Bennett Green, Colchester, CO4 5ZR
M6 HWJ Tony Armstrong, 12 Mayhouse Road, Burgess Hill, RH15 9RF
M6 HWL Jonathan Dyson, FY8 3TL, Lytham St. Annes, FY8 3TL
M6 HWM Douglas Hatchman, 14 Rudyard Close, Brighton, BN2 6UA
M6 HWN John Laws, 47 Hampshire Place, Peterlee, SR8 2HE
M6 HWO michael wells, 14, werrington grove, Peterborough, pe46nt
M6 HWQ neil williams, 245 Central Drive, Bilston, WV14 8JE
M6 HWT George Scholey, LN9 6JH, Horncastle, LN9 6JH
MI6 HWV Gordon Osprey, 24 The Hollies, Carrickfergus, BT38 8HA
M6 HWW Alan Graham, Shore Gate House, Bowness-on-Solway, Wigton, CA7 5BH
M6 HWX Gary McRitchie, 21 St Marys field, Colchester, CO3 3BP
M6 HXA Paul Barden, 17 Chapel Fields, Charterhouse Road, Godalming, GU7 2BS
M6 HXB Darryl Jones, 91 Laburnum Road, High Wycombe, HP12 3LP
M6 HXC Herschel Chawdhry, Trinity College, Cambridge, CB2 1TQ
M6 HXD Mark Shelley, 21 Ripley Close New Addington, Croydon, CR0 0RP
M6 HXE adrian moss, Winstons, Mayfield Lane Durgates, Wadhurst, TN5 6DG
M6 HXF Jeremy Phillips, 27 New Road Blackwater, Camberley, GU17 9AY
M6 HXG PAUL STOKES, 26 Ashford Road, Hastings, TN34 2HA
M6 HXI Ronald Bowen, 4 Crossley Gardens, Halifax, HX1 5PU
M6 HXK Richard Calvert, 2 Coneyburrow Road, Tunbridge Wells, TN2 3NA
M6 HXN Robert Hammond, 28 Birch Way, Hastings, TN34 2JZ
M6 HXO Geoffrey John Dyson, 4 Davenport Avenue, Blackpool, FY2 9EP
M6 HXU Adam Loader, Rowan Tree House, Crowfield, Brackley, NN13 5TW
MM6 HXY Gaynor Towell, 265 Stirling Street, Denny, FK6 6QJ
M6 HYJ Holly Jeram, 6 Lavender Lane, Rowledge, Farnham, GU10 4AX
M6 HYM A Mears, Flat 248, 5 Charter House, Portsmouth, PO1 2SN
MW6 HYS Christopher Lowes, 3 Castle Close Creigiau, Cardiff, CF15 9NJ
M6 HYT Peter Carr, 9 Hollydene Villas, Hythe, Southampton, SO45 4HU
M6 HYW Ho Wong, Harrogate Ladies' College, Clarence Drive, Harrogate, HG1 2QG
M6 HYX sarah short, 16 Melrose Drive, Fletton, Peterborough, pe2 9dn
M6 HZD Harry Donovan, 2 All Saints, Weeting, IP27 0QH
M6 HZL Hazel Smith, 12 Lockgate East, Windmill Hill, Runcorn, WA7 6LB
MM6 HZO Arturs Artamonovs, 4/1 Clerk Street, Edinburgh, EH8 9HX
M6 HZZ Stuart Conway, 21 Milcote Avenue, Hove, BN3 7EJ
MM6 IAB John Joyce, 7 Leitch Street, Greenock, PA15 2HJ
M6 IAC Zach Cole, 31 High Street, Kimpton, Hitchin, SG4 8RA
M6 IAF Sanjay Shambhu, 34 Gascoigns Way, Patchway, Bristol, BS34 5BY
MW6 IAG Joshua Richards, 210a Pandy Road, Bedwas, Caerphilly, CF83 8EP
MM6 IAI Iain Brown, 28 Garden Road, Cults, Aberdeen, AB15 9RE
M6 IAJ Iain Jones, 8a Orchard Close, Longford, Gloucester, GL2 9BB
M6 IAL Isabel Lambert, 69 Anvil Crescent, Broadstone, BH18 9DZ
M6 IAN Ian Shires, 19 Prince Charles Avenue, Sittingbourne, ME10 4NA
M6 IAO Ian Phillips, 324 The Meadway Tilehurst, Reading, RG30 4PD
M6 IAQ Alison Instone, 63 Larch Road, New Ollerton, Newark, NG22 9SX
MM6 IAR Andrew Rees, Mains of Atherb, Maud, Peterhead, AB42 4RD
M6 IAS Ian Scott, 21 Field Avenue, Shepshed, Loughborough, LE12 9SH
M6 IAT David James, 53 Whittingham Road, Ilfracombe, EX34 9LL
M6 IAV Ian Avery, 4 Southampton Drive, Liverpool, L19 2HE
M6 IAW Ian Ferguson, 4 Queens Court, Castle Douglas, DG7 1HR
M6 IAX Ian Lawton, 38 Battershall Close, Plymouth, PL9 9UU
MM6 IAY Alistair MacLennan, 7 Treaslane, Portree, IV51 9NX
M6 IAZ Darrin Goldthorpe, 30 Morrissey Close, St. Helens, WA10 4JW
MM6 IBB Iain Bainbridge, 2 Courthill Road Cottage, Arbroath, DD11 4UX
M6 IBC Ian Barber, 8 Newlands Close, Lowestoft, NR33 7EY
MW6 IBD Ivor Daniel, 35 New Road, Upper Brynamman, Ammanford, SA18 1AF
M6 IBF Ian Graham, 49 Eagle Close, Leighton Buzzard, LU7 4AT
M6 IBG Ivan Bruno-Gaston, 83 Althorne Gardens, London, E18 2DB
M6 IBH Cliff Tate, 87 Overdale Road, Middlesbrough, TS3 7NQ
M6 IBI Alan Douglas, 3 Beech Avenue, Bilsborrow, Preston, PR3 0RH
M6 IBL Geoffrey Eibl-Kaye, 1 Main Road, Littleton, Winchester, SO22 6PS
M6 IBO Aedan Lawrence, 28 Broad Oak Lane, Bexhill-on-Sea, TN39 4HE
MI6 IBR Barry Rocks, 4 Millview, Randalstown, Antrim, BT41 3BA
M6 ICA Colin Irons, 11 Elm Grove, Moira, Swadlincote, DE12 6HH
M6 ICB Ian Bushnell, 4 Upper Tail, Watford, WD19 5DF
MI6 ICD Ian Cairns, 18 Molyneaux Avenue, Larne, BT40 2TU
M6 ICH Michael Pullen, 2 Wyatts Lane, Little Cornard, Sudbury, CO10 0NT
M6 ICK Michael Ginty, 34 High Street, Branston, Lincoln, LN4 1NB
MW6 ICM Allan Moody, Perthiteg, Cwmhiraeth, Llandysul, SA44 5XJ
M6 ICO John Stevenson, 18 Drakehouse Lane, Sheffield, S20 1FW
M6 ICP Ian Pears, 10 Pixley Dell, Consett, DH8 7DB
M6 ICQ Darren Banks, 41 East Road, Rotherham, S65 2UX
M6 ICR Sean Barlow, Apartment 34, Jet Centro, Sheffield, S2 4AH
MW6 ICU Alexander Jones, 8 Arles Road, Cardiff, CF5 5AP
M6 IDB Melanie Parker, 1 Ham Road, Wanborough, Swindon, SN4 0DF
M6 IDC Ian Cosham, 54 Hawkins Crescent, Shoreham-by-Sea, BN43 6TP
M6 IDD Paul Wilcox, 2 Merryhill Terrace, Belmont, Hereford, HR2 9RT
M6 IDE Andrew Currie, 20 Portal Road, Eastleigh, SO50 6AY
M6 IDF Ian Firth, 124 Viking Road, Bridlington, YO16 6TB
M6 IDG Ian Garrard, 33 Uplands Road, Hockley, SS5 4DL
MI6 IDJ Kieran McLaverty, 123a Castle Road, Antrim, BT41 4ND
M6 IDK Ian King, 7 Greenacres Avenue, Blythe Bridge, Stoke-on-Trent, ST11 9HU
M6 IDM Emmanuel Ogbua, 11 Coltness Crescent, London, SE2 0UY
M6 IDN Ian Norfolk, Arwelfa, High Street, Uckfield, TN22 3LP
M6 IDO Yuan Wang, 25 Cunningham Avenue, Hatfield, AL10 9LR
M6 IDP Ian Nelson, 30a Walnut Road, Torquay, TQ2 6HS
M6 IDR Ian Reeve, 36 Stone Pippin Orchard, Badsey, Evesham, WR11 7AA
M6 IDX Jonathan Ledger, 18 Claremont Street, Rotherham, S61 2LT
M6 IEA Edward Aspden, 9 Cledford Crescent, Middlewich, CW10 0EZ
M6 IEM John Paine, 6 Penwood Court, Allenby Road, Maidenhead, SL6 5BW
M6 IEO Leo Metcalfe, 40 St. Anns Court, Hartlepool, TS24 7HY
M6 IER Alasdair Unwin, 152 Epsom Road, Guildford, GU1 2RP
M6 IEW Ian Williams, 23 Symons Close, Blackwater, Truro, TR4 8ER
M6 IFF Joanna Sharrad, 52 Springwood Drive, Ashford, TN23 3LQ
M6 IFH Iain Harrison, 8 Jeffrey Avenue, Longridge, Preston, PR3 3TH
M6 IFI Ian Iremonger, 2 Harbord Road, Cromer, NR27 0BP
M6 IFO David Harris, 35 Itchenor Road, Hayling Island, PO11 9SN
M6 IFT Gary Saunders, 140 Highbridge Road, Burnham-on-Sea, TA8 1LW
M6 IFW Ivan Warman, 8 Burley Road, Bishop's Stortford, CM23 3LR
MW6 IGC Ian Curnock, Penlan Fron, Cynwyl Elfed, Carmarthen, SA33 6UD
M6 IGH Ian Garforth, 63 Upper Perry Hill, Bristol, BS3 1NJ
M6 IGJ Ian Gareth Jackson, 22 Greenwood Avenue, Congleton, CW12 3HH
M6 IGK Isabel Wideman, Silcoates Lane, Wrenthorpe, Wakefield, WF2 0PD
M6 IGM Gary Benford, 34 Victoria Gardens, Colchester, CO4 9YD
M6 IGS Nigel Hawkins, 17 Meddins Lane, Kinver, Stourbridge, DY7 6BZ
M6 IGW Andrew Wilson, 38 Cotleigh Drive, Sheffield, S12 4HU
M6 IGZ Aaron Hogg, 17 Kedleston Close, Stretton, Burton-on-Trent, DE13 0FN
M6 IHC Ian Clement, 24 Millais, Horsham, RH13 6BS
M6 IHH Ian Hutchinson, Bridgend, Mill Lane, North Hykeham, Lincoln, LN6 9PA
MM6 IHQ Tim Rogers, Marypark Farm, Marypark, Ballindalloch, AB37 9BG
M6 IIG Robert Blair, Springfield, Pewsey Road, Pewsey, SN9 6EN
M6 IIL Liam Furr, 158 Eastern Avenue, Southend-on-Sea, SS2 4AZ
MM6 IIO Ruari Treble, 2 Baidland Meadow, Dalry, KA24 5HP
M6 IJB Ian Boddy, 5 Boverton Avenue, Brockworth, Gloucester, GL3 4ER
M6 IJD Nicholas Dimonaco, 41 Brongwinau, Comins Coch, Aberystwyth, SY23 3BQ
M6 IJH Ian Holdford, 46 Hildreth Road, Prestwood, Great Missenden, HP16 0LY
M6 IJJ Ian Johns, Flat 14, Baker Street House, James Street, Pontypool, NP4 9EH
M6 IJM Ian Morgan, 30 Farm Road, Hutton, Weston-Super-Mare, BS24 9RH
MM6 IJP Ian Purkis, Lochaber Croft, Tullynessle, Alford, AB33 8QQ
M6 IJQ Robert Bedford, 29 Kent Road, Brookenby, Market Rasen, LN8 6EW
M6 IJW Paul Williams, 52 Rhoslan, Tredegar, NP22 4PF
M6 IKA Isaac Abraham, 12 Graham Road, Halesowen, B62 8LJ
MM6 IKB Katherine Breimann, 5 Leaside, Mossbank, Shetland, ZE2 9TF
M6 IKD Michael Draper, 160 Chanctonbury Road, Burgess Hill, RH15 9HA
M6 IKE Michael Smith, 9 Coniston Street, Salford, M6 6BG
M6 IKI Michael Anostalgia, 136 Avenue Road Extension, Leicester, LE2 3EH
M6 IKM Michael Moffat, 19 Croftfield Road, Seaton, Workington, CA14 1QW
MD6 IKR David Cain, Flat, Ballavagher, Main Road, Union Mills, Isle of Man, IM4 4AR
M6 IKY Michael Rawson, 221 Carter Street, Fordham, Ely, CB7 5JU
M6 ILB Richard Brocklehurst, 12 Harriers Close, Christchurch, BH23 4SL
M6 ILC Chun Hong Teddy Ng, Battersea Court University Campus, Guildford, GU2 7JQ
MI6 ILF Ian Forsythe, 45 Kensington Park, Portadown, Craigavon, BT63 5PQ
M6 ILH Ian Hobbs, 115 Adams Way, Croydon, CR0 6XR
M6 ILM David Sharpe, 31 Malyons, Basildon, SS13 1PJ
M6 ILN David Bee, 25 Blatcher Close, Minster on Sea, Sheerness, ME12 3PG
M6 ILO Milo Noblet, 1 Lingdale Road, Wirral, CH48 5DG
M6 ILP Ian Patterson, 63 Orchard Road, South Ockendon, RM15 6HP
M6 ILR Valerie Seabright, 208 Park Way, Rubery, Birmingham, B45 9WA
M6 ILS John Anthony, 21 Belgrave Street, Denton, Manchester, M34 3WP
M6 ILY Daniel Stuart, 6 Ross Crescent, Watford, WD25 0DB
M6 IMA Ian Maley, 6 Meadow View Close, Newport, TF10 7NN
MM6 IMB Michael Breimann, 5 Leaside, Mossbank, Shetland, ZE2 9TF
MM6 IMF Ian Fairbairn, 2 The Steadings, Slackend, Buckie, AB56 5BS
M6 IMH Ian Hickinbottom, Clover Cottage, Snead, Montgomery, SY15 6EB
M6 IMI Adam Rayner, 78 Maximus Road, North Hykeham, Lincoln, LN6 8JU
M6 IMM Charles Milne, 24 Berton Close, Blunsdon, Swindon, SN26 7BE
MM6 IMP Caroline McConnachie, Flat C, 5 Whitefaulds Crescent, Maybole, KA19 8AY
M6 IMR Iain Rich, 39 Wren Close, Heathfield, TN21 8HG
M6 IMS Mark Sims, 5 Sandy Leaze, Bradford-on-Avon, BA15 1LX
MW6 IMT Ian Thornton, 18 Margaret Terrace, Blaengwynfi, Port Talbot, SA13 3UU
M6 IMW I Walker, 24 Hawthorn Road, Norwich, NR5 0LP
M6 IMZ Ian Ross, 48 Henry Drive, Leigh-on-Sea, SS9 3QF
M6 INA Geoffrey Allen, 13 Strathmore Avenue, Hull, HU6 7HJ
MM6 INC Andrew McMath, 57 Hillhouse Avenue, Bathgate, EH48 4BB
M6 IND Peter Ind, 30 Thompson Road, Stroud, GL5 1SY
M6 ING Damanjit Singh, 28 Chadview Court, Chadwell Heath Lane, Romford, RM6 4BF
M6 INI Mark Smith, Church Farm, Market Drayton, TF9 4DN
M6 INM Michael Davies, 91 Ameysford Road, Ferndown, BH22 9QD
MM6 INN Andi MacInnes, 1 Brent Place, Glenrothes, KY7 6TA
MM6 INS Cephas Ralph, 37 Seaview Terrace, Edinburgh, EH15 2HE
M6 INT Matt Pomfret, 5 Malvern Crescent, Ince, Wigan, WN3 4QA
M6 INV Michael Basford, 4 Renoir Close, St. Ives, PE27 3HF
M6 INW David Inwood, 78 Lower Thrift Street, Northampton, NN1 5HP
M6 INX Janet Porter, 14 Longwestgate, Scarborough, YO11 1QB
M6 IOA Ioana Dumitrescu, The Rectory, Halkyn, Holywell, CH8 8BU
M6 IOG Fred Cooper, Needhams Farm House, Spittal Hill Road, Boston, PE22 0PA
M6 IOI Leslie Emanuel, 20 Wychwood Drive, Redditch, B97 5NW
M6 IOL Mark Scott, 19 Saltburn Road, Sunderland, SR3 4DJ
MD6 IOM Peter Morgan, Thal'loo Glass, Nassau Road, The Dog Mills, Mwyljyn Moddey, Isle of Man, IM7 4AQ
M6 ION Laurence Rimington, 3 Amicombe, Wilnecote, Tamworth, B77 4JJ
M6 IOS Geoffrey Baker, 56 Chalklands, Bourne End, SL8 5TJ
M6 IOW Frank Alfrey, 16 Walls Road, Bembridge, PO35 5RA
M6 IPA Robert Howitt, 7 Badgers Close, Chelmsford, CM2 8QB
MI6 IPB David Neill, 8 Castle Meadows, Carrowdore, Newtownards, BT22 2TZ
M6 IPH James Dunn, 39 Bramble Lane, Wye, TN25 5AB
M6 IPJ Peter Allen, 45 Under Knoll, Peasedown St. John, Bath, BA2 8TY
M6 IPL Ian Pilton, Rainbow Barn, Pool Foot Farm, Ulverston, LA12 8AA
MM6 IPP Ian Patterson, 8 Jane Street, Dunoon, PA23 7HX
M6 IQO Peter Taylor, 32 Heliers Road, Liverpool, L13 4DH
M6 IRC Ian Crowson, 19 Burgoyne Road, Southsea, PO5 2JJ
MI6 IRE Paul McAleer, 24 Wansbeck Street, Belfast, BT9 5FQ
M6 IRJ M Impey, 20 Chalton Road, Luton, LU4 9ER
M6 IRK Ooan Kirton, Woodland, Moretonhampstead, Newton Abbot, TQ13 8SD
M6 IRL David Lovejoy, 9 West View Close, Middlezoy, Bridgwater, TA7 0NP
M6 IRM Miriam Raine, 91 Lulworth Avenue, Jarrow, NE32 3SB
M6 IRP Ian Pipe, 8 Glebe Drive, Stottesdon, Kidderminster, DY14 8UF
M6 IRU Sean Airey, 22 Primrose Street, Lancaster, LA1 3BN
M6 IRW Ieuan Wilkes, 7 Oriel Close, Dudley, DY1 2UW
M6 ISA Isaiah Stone, 169 Booth Road, Wednesbury, WS10 0EW
M6 ISB Stephen Brown, 18 Goring Avenue, Manchester, M18 8WW
M6 ISD Samuel Dodd, 21 Sunningdale, Grantham, NG31 9PF
M6 ISG Terry Holland, 68 Church Street, Billericay, CM11 2TS
M6 ISH Ruth Kinder, 21 Oakdene, Chobham, Woking, GU24 8PS
M6 ISJ Sarah Jones, 5 Meadowlands, Kirton, Ipswich, IP10 0PP
M6 ISR Ian Robinson, 1 Low Bank Cottages, Bywell, Stocksfield, NE43 7AF
M6 ISZ Iain Townsend, 256 Fishponds Road, Eastville, Bristol, BS5 6PY
M6 ITD Christopher Knowles, 10 The Drive, Southwick, Brighton, BN42 4RR
M6 ITI Christopher Johnson, Suite 204, 33 Queen Street, Wolverhampton, WV13AP
M6 ITL James White, 4 Kings Road, New Milton, BH25 5AY
M6 ITM Ian MacDonald, 2 Garland Place, Hexham, NE46 3QG
M6 ITN Robert Goodall, 76 Beaconfield Road, Plymouth, PL2 3LF
M6 ITQ Jonathan Swales, 90 Earlswood Road, Dorridge, Solihull, B93 8RN
M6 ITU Paul James, 25 Brookfield Road, Churchdown, Gloucester, GL3 2PQ
M6 ITV Ivor Roberts, 15 Broadcroft, Hemel Hempstead, HP2 5YX
M6 ITW Kenneth Young, 51 Haven Road, Barton-upon-Humber, DN18 5BS
M6 ITX Martin Reynolds, 24 Burton Close, Corringham, Stanford-le-Hope, SS17 7SB
M6 ITY Aditya Praveen, 28 Long Deacon Road, London, E4 6EG
MD6 IUH Kai Payne, 2 Dreeym Balley Cubbon, Ballacubbon, Isle of Man, IM9 4PR
M6 IUK David Grayson, 79 Errington Avenue, Sheffield, S2 2EA
MW6 IUN Ieuan Jones, 21 Albert Street, Maesteg, CF34 0UF
M6 IVE Celso Cavalcante Pinheiro Filho, Flat 6, 2a Trumans Road, London, N16 8BD
M6 IVI John Knights, 18 Kenilworth Gardens, Blackpool, FY4 1JJ
M6 IVN Ivan Clarke, 86 Charlton Road, Andover, SP10 3JY
MD6 IVO Ivelin Yovchev, 11 Beverley Drive, Edgware, HA8 5NQ
MM6 IVP Simon Morrison, 11 Culriach, Bogmoor, Fochabers, IV32 7PX
M6 IVR Ivor Goodman, 26 Vallansgate, Stevenage, SG2 8PY
M6 IVS Nathan Bookham, 116 Clare Gardens, Petersfield, GU31 4EU
M6 IWA Ivan Wright, 1 Greaves Avenue, Old Dalby, Melton Mowbray, LE14 3QE
M6 IWB Ian Bunting, 6 Forster Close, Aylsham, Norwich, NR11 6BD
M6 IWF Ian Francis, 34 Furlong Road, Bourne End, SL8 5AA
MW6 IWM Ian Miles, 40 Seymour Street, Mountain Ash, CF45 4BL
M6 IWP Ian Perkins, 28 Newstead, Tamworth, B79 7UU
M6 IXI Richard Forss, Lower Conghurst Oast, Conghurst Lane, Cranbrook, TN18 4RW
M6 IZA Iain Brunt, 62 Greenwood Drive, Watford, WD25 0HX
M6 IZE Andrew Hampson, 12 Oakhays, South Molton, EX36 4DB
MD6 IZI Isabel Dorman, 1 Sprucewood Rise, Foxdale, Douglas, Isle of Man, IM4 3JP
M6 IZP Arnoldas Jakstas, Flat 9, Kendal Court, 112 Godstone Road, Kenley, CR8 5GE
M6 IZW Isabel Whiteley, 2 The Meade, Manchester, M21 8FA
M6 IZY Elizabeth Coffey, 4 Castlehall, Tamworth, B77 2EG
M6 JAB Jayne Burrows, 78a Coronation Road, Earl Shilton, Leicester, LE9 7HJ
MI6 JAD Steven Mullan, 135 Ballyavelin Road, Limavady, BT49 0QB
MM6 JAE James Dalgety, 5 East Court, Edinburgh, EH16 4ED
M6 JAF Douglas Oliphant, 9 West Park Road, South Shields, NE33 4LB
M6 JAG Richard Price, 92 Lincoln Road, Ingham, Lincoln, LN1 2XF
M6 JAJ Jonathan Smith, 46 Mulberry Close, Goldthorpe, Rotherham, S63 9LB
M6 JAK James King, 18 Ross Road, Wallington, SM6 8QB
M6 JAL Michael Leggett, 7 Barley Way, Thetford, IP24 1LG
M6 JAM Jade Hunter, 164 Grange Road, Newark, NG24 4PP
MM6 JAN Janis McClure, Bridgefoot Croft, Udny, Ellon, AB41 6RT
M6 JAQ Joshua Snell, 32 Meadow Halt, Ogwell, Newton Abbot, TQ12 6FA
M6 JAR John Revell, 63 Mountbatten Road, Bungay, NR35 1PP
M6 JAS Jason Wells, 18 Roewood Road, Holbury, Southampton, SO45 2JH
M6 JAU John Goldsmith, 20 Trinity Mews, Bury St. Edmunds, IP33 3AT
M6 JAV Michael Sharman, 33 Bungalow Estate, Lady Lane, Coventry, CV6 6BD
MM6 JAW John White, 63 Allershaw Tower, Wishaw, ML2 0LP
M6 JAZ James Cleeter, 49 Hunters Field, Stanford in the Vale, Faringdon, SN7 8LZ
M6 JBA John Ashley, Rowborough Farm Cottages, Brading, Sandown, PO36 0BA
M6 JBB John Berry, 31 New Hall Way, Flockton, Wakefield, WF4 4AX
M6 JBE Jason Nicholson, 32 Home Orchard, Yate, Bristol, BS37 5XQ
M6 JBG Jonathan Bethell, 47 Montgomery Road, Ipswich, IP2 8QB
MI6 JBH Jim Monaghan, 6 Barranderry Heights, Enniskillen, BT74 6JW
M6 JBI Jay Innes, Trelawny, Marine Drive, Bude, EX23 0AH
M6 JBL Blazej Jedryka, 31 Backhold Lane, Halifax, HX3 9DR
M6 JBM Jonathan Russell, 75 Swallow Dale, Thringstone, Coalville, LE67 8LY

MM6 JBN Joseph Breen, 26 Maxwelton Road, Glasgow, G33 1LR
M6 JBO Jonathan Davies, 10 Leaway, Prudhoe, NE42 6QE
MW6 JBQ Graham Smith, 49 Coed Cae, Caerphilly, CF83 1BU
M6 JBR Jake Whitton, 11 Dursley Road, Bristol, BS11 9XB
M6 JBT Joseph Hawkins, 24 Green Lane, Stourbridge, DY9 7EW
M6 JBU John Burnett, 218 High Street, Clapham, Bedford, MK41 6BS
M6 JBV Michael Redmore, 8 Hambledon Rise, Northampton, NN4 8TT
M6 JBX Jon Blower, 8 The Laundry, Seifton, Ludlow, SY8 2DH
M6 JBZ Jill Given, 29 Littlecote Gardens, Appleton, Warrington, WA4 5DL
M6 JCA James Aubury, 27 Gravel Walk, Tewkesbury, GL20 5NH
M6 JCB Christina Daniels, 19 Gerrard Street, Rochdale, OL11 2EB
M6 JCC Jodie Clare, Kimberley, Boston Road, Boston, PE20 3AP
M6 JCD Jeremy Pearson, 29 Broadoak Road, Langford, Bristol, BS40 5HD
M6 JCE Julian Crewe, 22 Myrtle Tree Crescent, Weston-Super-Mare, BS22 9UL
M6 JCF Josiah Faulkner, Mount Pleasant, Elkstones, Buxton, SK17 0LU
M6 JCG James Anderson, The Firs, Crapstone, Yelverton, PL20 7PJ
MW6 JCH John Hudson, 27 Brynelli, Dafen, Llanelli, SA14 8PW
M6 JCJ Joshua Ward, 6 Fairfield, Telegraph Hill, Redruth, TR16 5AH
MM6 JCL John Littlefair, 30 Maple Crescent, Cambuslang, Glasgow, G72 7NN
M6 JCN Jason Barma, 28 Briarfield Road, Timperley, Altrincham, WA15 7DB
M6 JCP Jason Preece, 42 Henderson Road, Widnes, WA8 7LR
M6 JCQ James Stevens, 23 Edlyn Close, Berkhamsted, HP4 3PQ
M6 JCR Jude Reynolds, 64 Albury Road, Merstham, Redhill, RH1 3LL
M6 JCS Jason Shettler, 504 Leeds Road, Huddersfield, HD2 1YW
M6 JCT John Anderson, 139 Cromwell Road, Rushden, NN10 0EG
MM6 JCU James Coubrough, 41 Bridge Court, Alexandria, G83 0BZ
MW6 JCV John Baldwin, 94 King Street, Abertridwr, Caerphilly, CF83 4BG
M6 JCW Jack Wright, 19 Halstead Close, Woodley, Reading, RG5 4LD
M6 JCX James Coxon, Flat 6, Neptune House, London, SE16 7AU
MM6 JCZ James Mackenzie, 40 Deanswood Park, Deans, Livingston, EH54 8NX
M6 JDA J Dale, Corydon, Church Street, Sevenoaks, TN14 7SW
M6 JDC James Collier, Stradishall Manor, The Street, Newmarket, CB8 8YW
M6 JDD Denis Riley, 14 Mond Road, Widnes, WA8 7NB
M6 JDF Jack Firth, 7 Manor Avenue, Derby, DE23 6EB
M6 JDL James Harding, 181 South Coast Road, Peacehaven, BN10 8NS
M6 JDM John Briggs, 47 Greenland Avenue, Derby, DE22 4AQ
MM6 JDN Ian Colborn, Gardeners Cottage, Craighouse, Isle of Jura, PA60 7XG
M6 JDQ Josephine-Louise Hole, 17 Cromford Street, Sheffield, S2 4BP
M6 JDU John Clarke, 160 Hall Lane Estate, Willington, Crook, DL15 0PP
MW6 JDY Judith Evans, 311 Delffordd, Rhos, Swansea, SA8 3ER
MM6 JEA James Addison, Lambhill Bungalow, St. Katherines, Inverurie, AB51 8TS
M6 JED Paul Carberry, 203 Fitzwilliam Road, Rotherham, S65 1NB
M6 JEG Derek Stephens, 15 South Lodge, Fareham, PO15 5NQ
M6 JEH Jackie Harris, 45 Sleigh Road, Sturry, Canterbury, CT2 0HT
M6 JEJ John Brennan, 12 Berne Road, Thornton Heath, CR7 7BG
M6 JEK James Durey, 7 Staplers Close, Great Totham, Maldon, CM9 8UN
M6 JEL James Christensen, 29 Caerphilly Close, Rhiwderin, Newport, NP10 8RF
M6 JEM Julie Shaw, 2a Priory Avenue, Petts Wood, Orpington, BR5 1JF
M6 JEO Jacob Saunders, 123 Medway Road, Ferndown, BH22 8UR
M6 JEP David Jepson, 104 Norris Street, Warrington, WA2 7RW
M6 JEQ Jack McDermott, 2 Denton Terrace, Castleford, WF10 4LN
M6 JER Jerry Harley, 170 Windsor Road, Hull, HU5 4HH
M6 JET Tracey Edwards, 5 Sienna Mews, Plumstead Road, Norwich, NR1 4LR
M6 JEV Jade Gourley, 88 Redstone Lane, Stourport-on-Severn, DY13 0JG
M6 JEY James Johnson, 226 Preston New Road, Southport, PR9 8NY
M6 JEZ Jez Mitchell, 11 Brookside Drive, Oadby, Leicester, LE2 4PB
M6 JFA John Worsley, 3 Sheephouse Road, Hemel Hempstead, HP3 9LW
M6 JFB James Houston Copland, Janefield, Colvend, Dalbeattie, DG5 4QN
M6 JFC James Carroll, 103 Brays Lane, Coventry, CV2 4DS
M6 JFE David Foley, 1 Hill Rise Close, Harrogate, HG2 0DQ
M6 JFF James Carpenter, 20 Barton Grove, Kedington, Haverhill, CB9 7PT
M6 JFH John Feltham, 12 Penrith Way, Eastbourne, BN23 8NS
MW6 JFI John Power, 27 Seaview Crescent, Goodwick, SA64 0AZ
M6 JFJ John Spratley, 10 The Willows, Jarrow, NE32 4QN
M6 JFK John Knight, 30 Ash Meadow Lea, Preston, PR2 1RX
MW6 JFL John Lewis, 9 Llwyn Bedw, Cefn Pennar, Mountain Ash, CF45 4DZ
M6 JFN Steven Butler, 41 Angus Close, Chessington, KT9 2BN
MI6 JFO William Forde, 35 Torr Gardens, Larne, BT40 2JH
M6 JFP Jonathan Poole, 82 Merchants Way, Canterbury, CT2 8PN
M6 JFR Adam Bass, 7 Woodlands, Horbury, Wakefield, WF4 5HH
M6 JFS Richard Jones, 32 Remington Drive, Sheffield, S5 9AH
MM6 JFU Jim Frew, 57 Ford Avenue, Dreghorn, Irvine, KA11 4BN
M6 JFV Johnathan Hardingham, 11 All Saints Close, Weybourne, Holt, NR25 7HH
M6 JFW John Leonard, 51 Molyneux Drive, Bodicote, Banbury, OX15 4AX
M6 JFY Christopher Holmes, Old Vicarage Farmhouse, Course Lane, Wigan, WN8 7LA
MW6 JGC Jeffrey Sollis, 15 Llanwonno Road, Mountain Ash, CF45 3NB
M6 JGF John Glover, 12 Willow Street, London, E4 7EG
M6 JGG Jordan Goodwin, 6 Worrall Street, Congleton, CW12 1DT
M6 JGH Raymond Hart, Jays, South Street, Gillingham, SP8 5ET
M6 JGI John Griffiths, 257 Bolton Road, Ashton-in-Makerfield, Wigan, WN4 8TG
M6 JGJ John Watson, 11 Globe Close, Upton Pyne, Exeter, EX5 5JB
MI6 JGK Gregory Gardiner, 60 Limestone Meadows, Moira, Craigavon, BT67 0UT
M6 JGM Jordan Gould-Martin, 203 Maple Crescent, Leigh, WN7 5SW
M6 JGN Joshua Elsey, 6 Lewis Avenue, Walthamstow, London, E17 5BL
M6 JGP Jonathan Paradi, 168 Castle Road, Northolt, UB5 4SG
M6 JGQ John Chapman, Boundary Farm, Garlic Street, Diss, IP21 4RL
M6 JGR Janice White, 62 Dalebrook Road, Burton-on-Trent, DE15 0AD
M6 JGS Joseph Seaton, 52 Shrubbery Street, Kidderminster, DY10 2QY
M6 JGT John Jeffryes, 15 Shirley Road, St. Albans, AL1 5ES
M6 JGU Julie Borrett, 84 Kingsway, Mapplewell, Barnsley, S75 6EX
M6 JGV John Chown, 40 Kemps Green Road, Balsall Common, Coventry, CV7 7QF
M6 JGY Jose Diez, 174 Humber Avenue, Coventry, CV1 2AR
M6 JGZ James Gilbert, 101 Eastbrook Road, Lincoln, LN6 7EW
M6 JHB John Burdett, 5 Winston Drive, Wainscott, Rochester, ME2 4LJ
M6 JHC Matthew Clark, 10 Priory Close, Sporle, King's Lynn, PE32 2DU
M6 JHD David Capstick, 3 Andrew Close, Dibden Purlieu, Southampton, SO45 4LS
MI6 JHE Jenyth Evans, 404 Foreglen Road, Dungiven, Londonderry, BT47 4PN
M6 JHF Philip Attwater, 42 Danescourt Crescent, Sutton, SM1 3EA
M6 JHG James Guess, Flat 257, Helen Gladstone House, London, SE1 0QB
MM6 JHH John Hutchinson, 100 Glen Avenue, Largs, KA30 8QQ
M6 JHI Andrew Randall, 90 Derby Road, Golborne, Warrington, WA3 [illegible]
MW6 JHJ Joshua Cook, 40 Cemaes Crescent, Rumney, Cardiff, CF3 1TA
M6 JHK Jacqueline Hele Kergozou de la Boessiere, Lilac Cottage, The Street, Cheddar, BS27 3TH
M6 JHL John Lumm, 25 Knowsley Way, Hildenborough, Tonbridge, TN11 9LG
M6 JHN John Curwen, Corner Cottage, Tailors Green, Stowmarket, IP14 4LL
M6 JHO Douglas Carey, Flat 7, Pelman House, Epsom, KT19 8HH
M6 JHP John Palmer, 14 Linnet Close, Luton, LU4 0XJ
M6 JHQ James Rimmer, 33 New Cut Lane, Southport, PR8 3DW
M6 JHR Julie Ross, Foundry Cottage, Crowders Lane, Battle, TN33 9LP
M6 JHS James Harris, 36 Northmoor Way, Wareham, BH20 4SJ
M6 JHT James Hill, 45 Venus Street, Congresbury, Bristol, BS49 5HA
M6 JHU Brett Jones, 64 Church Lane, Barwell, Leicester, LE9 8DG
M6 JHV John Hancox Hancox, 10 Dunham Close, Newton-on-Trent, Lincoln, LN1 2LH
M6 JHZ John Hazeltine, 21 Hassock Way, Wimblington, March, PE15 0PJ
M6 JIC Robert Price, 17 Treleven Road, Bude, EX23 8SA
M6 JID James Darwin, 14 Croftwood Grove, Whiston, Prescot, L35 3UT
MI6 JIE Joseph Evans, 404 Foreglen Road, Dungiven, Londonderry, BT47 4PN
M6 JIF Jozef Miazek, 2 Oak Grove, Armthorpe, Doncaster, DN3 2DJ
M6 JIG Colin Ember, 35 Mattock Lane, Ealing, London, W5 5BH
M6 JIH James Horsley, 33 Amalfi Tower, Sunderland, SR3 3AN
M6 JII John Donovan, 18 Ellesmere Street, Eccles, Manchester, M30 0JN
M6 JIK Bruce Ferry, 43 North Leigh Tanfield Lea, Stanley, DH9 9PA
M6 JIL Jillian Ullersperger, 60 Reeds Avenue, Earley, Reading, RG6 5SR
M6 JIN Jinesh Ramachandran, 1 Keplerlaan, ESTEC, TEC-SWS, Noordwijk, Netherlands, 2201AZ
M6 JIQ James Ritson, 38 Hawkshead Road, Burtonwood, Warrington, WA5 4PW
M6 JIR John Barrett, 9 Hook Road, Goole, DN14 5JB
M6 JIS Malcolm Furby, 14 Larch Avenue, Wickersley, Rotherham, S66 2PQ
M6 JIT John Topping, 10 St. Pauls Road, Blackpool, FY1 2NY
M6 JIW Jason Barker, Pearl Bungalow, Killerby Cliff, Scarborough, YO11 3NR
M6 JIX Jago Packer, 20 Shipman Road, Market Weighton, York, YO43 3RB
M6 JIY David Baker, 22 Cleveland Road, Plymouth, PL4 9DF
M6 JIZ James Watson, 82 Glendale Avenue, Washington, NE37 2JS
MM6 JJB John Barclay, 24 Wellcroft Road, Hamilton, ML3 9SG
M6 JJD Jonathan Smyth, 5 Lime Close, Lakenheath, Brandon, IP27 9AJ
M6 JJE Scott Kelly, 7 Cedar Grove, Greetland, Halifax, HX4 8HT
M6 JJF Jake Fradley, 9 Hagley Park Gardens, Rugeley, WS15 2GY
M6 JJG Jeremy Greenhalgh, 7 Swynford Close, Kempsford, Fairford, GL7 4HN
M6 JJH John Holbrook, 24 Birks Holt Drive, Maltby, Rotherham, S66 7JZ
M6 JJI James Roberts, 27 Pike Purse Lane, Richmond, DL10 4PS
M6 JJK Clifford Webb, 16 Bexfield Road, Foulsham, Dereham, NR20 5SB
M6 JJL John Jackson, 23 Hawkley Drive, Tadley, RG26 3YH
M6 JJM Jordan Marriott, 55 Ellis Avenue, Stevenage, SG1 3SL
M6 JJN James Nicholls, 4 Sadler Close, Colchester, CO2 7LU
MW6 JJO James Owens, 16 Blanche Street, Dowlais, Merthyr Tydfil, CF48 3PE
MM6 JJQ James McCrae, 50 Cuilmuir View, Croy, Glasgow, G65 9HQ
M6 JJS James Stewart, 29 Cornwall Close, Leamore, Walsall, WS3 2AR
M6 JJV Steven Virgo, 41 Lynch Road, Berkeley, GL13 9TE
M6 JJW John Wallace, 323 High Street, Dalbeattie, DG5 4DX
M6 JJX Jacqueline Adams, 18 Vanguard Court, Sleaford, NG34 7WL
MI6 JJZ Jonathan Clark, Apartment 16, Pipers Field, 16b Comber Road, Belfast, BT16 2AB
M6 JKA Andrew Kirby, 1 Mount Pleasant, Halstead, CO9 1DX
M6 JKB Stuart Lucas, 31 Lilian Close, Norwich, NR6 6RZ
M6 JKD John Davies, 135 Silvercourt Gardens, Brownhills, Walsall, WS8 6EZ
M6 JKH Jason Horry, 20 Churchgate, Sutterton, Boston, PE20 2NS
M6 JKM Jack England, 88 Kempton Avenue, Hereford, HR4 9TY
M6 JKQ Ryan Poulson, 9 Scattergate Green, Appleby-in-Westmorland, CA16 6SP
M6 JKR Joe Krinks, 29 Swaledale Avenue, Congleton, CW12 2BY
M6 JKS James Shaw, 25 High Street, Gorleston, Great Yarmouth, NR31 6RT
M6 JKT Cameron Baines, Donative Farm, Warton, Tamworth, B79 0JR
M6 JKW John Gelder, 36 Westcombe Court, Wyke, Bradford, BD12 8PT
M6 JKX Jon Kitto, 306 Lewisham Road, London, SE13 7PA
M6 JKY Kevin Young, 48 Sussex Street, Cleethorpes, DN35 7NP
M6 JLA Julia Tayler, 22 Wheatley Road, Leicester, LE4 2HN
M6 JLB Jamie Blundell, 21 Walmsley Street, Fleetwood, FY7 6LJ
M6 JLD Jamie Drinkell, 64 Pioneer Avenue, Burton Latimer, Kettering, NN15 5LH
M6 JLE Julie Wiskow, 15 Ferndale Close, Sandbach, CW11 4HZ
M6 JLH Jordan Hughes, 27 Mitchell Street, Stoke-on-Trent, ST6 4EX
M6 JLL Jacqueline Caswell, 10 Beech Close, Scole, Diss, IP21 4EH
M6 JLM James Mould, 45 Kent Road, Houghton Regis, Dunstable, LU5 5NZ
M6 JLR Jim Ridley, 73 The Markhams, New Ollerton, Newark, NG22 9QY
M6 JLT James Trunks, 37 Carlton Street, Haworth, Keighley, BD22 8JY
M6 JLW James Hart, 420 Butts Road, Southampton, SO19 1DD
MM6 JLX John Leith, 13 Chesterhall Avenue, MacMerry, Tranent, EH33 1QJ
M6 JLY Jordan Lyall, 11 Bowmont, Ellington, Morpeth, NE61 5LT
M6 JLZ John [illegible], 16 Lambert Avenue, Shepshed, Loughborough, LE12 9QH
M6 JMB Marc Bloore, 6a Lovatt Close, Stretton, Burton-on-Trent, DE13 0HZ
MI6 JMC John McCloskey, 19 Kinnyglass Road, Coleraine, BT51 3SN
MI6 JMD Joe McDonald, 70 Allen Park, Dunamanagh, Strabane, BT82 0PD
M6 JMF Jack Fearn, 16 Sandringham Road, Retford, DN22 7QW
M6 JMG Jamie Mobbs, 6 School View, Banbury, OX16 4SD
M6 JMH James Hewitt, 1 Highfield, Gloucester Road, Chepstow, NP16 7DF
MM6 JMI John Wilson, 1 West Long Cottages, Livingston, EH54 7AB
M6 JMJ Malcolm Grimsley, 80 Holborn Avenue, Coventry, CV6 4FZ
M6 JMM Jamie Munton, Flat 42, Fore Street, Callington, PL17 7AQ
M6 JMN Jack Mason, 9 Little Warton Road, Warton, Tamworth, B79 0HR
M6 JMO James Monahan, 48 Church Road, Earley, Reading, RG6 1HS
M6 JMP James Pearse, 23 Buckingham Drive, Knutsford, WA16 8LH
M6 JMR Julian Mark Randall, 18 Thirty First Avenue, Kingston Up on Hull, HU6 8DB
M6 JMX Jonathon Matthews, 50a South Farm Road, Worthing, BN14 7AE
MM6 JNB Janice Beedie, 21 Marywell Village, Arbroath, DD11 5RH
M6 JNC John Clay, Willow Brook, Bairstow Lane, Sowerby Bridge, HX6 2SY
M6 JND James Philps, 130 Turkey Road, Bexhill-on-Sea, TN39 5HH
M6 JNE Jane Nobes, 22 Mansfield Road, Edwinstowe, Mansfield, NG21 9NJ
M6 JNH [illegible]
M6 JNJ [illegible] 5LR
M6 JNM Joan Nofrerias Mondejar, 31 Emmendingen Avenue, Newark, NG24 2FX
MM6 JNN James Needham, 154 Ravenswood Rise, Livingston, EH54 6PQ
MW6 JNP Jason Powell, 31 Laburnum Close, Merthyr Tydfil, CF47 9SN
M6 JNQ Jonathan Nicholas, 35 The Copse, Bridgwater, TA6 4DW
M6 JNR John Reynolds, 38 Spring Lane, Hockley Heath, Solihull, B94 6QY
MD6 JNS John Shatford, 31 Pinner Park Avenue, Harrow, HA2 6LG
M6 JNW Joshua Walker, 232 Bideford Green, Leighton Buzzard, LU7 2TS
M6 JNX Belinda Sanderson, 2 East Crescent, Canvey Island, SS8 9HL
M6 JOD Joanne O'Driscoll, 48 St. Ives Road, Coventry, CV2 5FZ
M6 JOE Joe Bell, 8 Firsleigh Park, Roche, St. Austell, PL26 8JN
M6 JOG Stuart Mayor, 12 Yealand Avenue, Heysham, Morecambe, LA3 2LT
MM6 JOH Joshua Hutton, 2 Watson Place, Dunfermline, KY12 0DR
MM6 JOK John Stewart, 1 Barns Park, Dalgety Bay, Dunfermline, KY11 9XX
M6 JON John Oakley, 59 Bewsey Street, Warrington, WA2 7JQ
M6 JOO Julian Fairhall, 122 Thornhill Rise, Portslade, Brighton, BN41 2YL
M6 JOR Jeff Oliver, 16 Eastdale Road, Burgess Hill, RH15 0NH
MI6 JOS Joshua Millar, 3 Lismurn Park, Ahoghill, Ballymena, BT42 1JN
M6 JOU Joe Best, 7 Lawns Court, Carr Gate, Wakefield, WF2 0UT
M6 JOW Joe Whitmore, 11 Fir Tree Drive, West Winch, King's Lynn, PE33 0PR
MM6 JOX Joanne Greer, 2/2 49 Strathcona Drive, Glasgow, G13 1JH
MI6 JOY Joy Ruddell, 16 Beechfield Manor, Aghalee, Craigavon, BT67 0GB
M6 JPC P Crossley, 4 Bartons Garth, Selby, YO8 9RR
M6 JPD Denise Bache, 62 Whittingham Road, Halesowen, B63 3TP
M6 JPF Jeremy Franks, 14 The Hamlet, Slades Hill, Templecombe, BA8 0HJ
M6 JPH Jacqueline Higginson, 187 Birmingham road, Ansley village, Nuneaton, CV10 9PQ
M6 JPI Joshua Pinder, 11 Andrews Close, Louth, LN11 0BP
M6 JPK Julian Knight, Flat 1, 45 High Street, Lowestoft, NR32 1HZ
M6 JPL John Lynch, Beechway, Raddel Lane, Warrington, WA4 4EE
MM6 JPM Jordan Moffat, 6 Park Grove, Belhelvie, Aberdeen, AB23 8YG
M6 JPO John Owen, 21 Marlborough Road, Luton, LU3 1EF
M6 JPR John Raine, Braemar, Sandy Lane, Crawley, RH10 4HS
M6 JPS Jessica Potts, 103 Etherstone Street, Leigh, WN7 4HY
M6 JPT John Thompson, 7 Hawthorne Terrace, Crosland Moor, Huddersfield, HD4 5RP
M6 JPW Paul Whitehead, 29 Cleveland, Bradville, Milton Keynes, MK13 7AZ
M6 JPX Jamie Powell, Temple Cottage, Monkey Island Lane, Maidenhead, SL6 2ED
M6 JQE Joseph Drea, 25 Loxley Gardens, Burnley, BB12 6PW
M6 JQW Jieqiong Wang, Flat 31, 74 Arlington Avenue, London, N1 7AY
M6 JRB Jennifer Brown, 13 Waterlakes, Edenbridge, TN8 5BX
M6 JRC John Clay, Riverside Mews, Nichols Yard, Sowerby Bridge, HX6 2EE
M6 JRE James Redfern, Flat 11, Poplar Court, Poplar Street, Manchester, M34 5EJ
M6 JRH Jake Harrison, 20 Somerton Avenue, Wilford, Nottingham, NG11 7FD
M6 JRJ Jonathan Rhodes, 13 Blake Hall Drive, Mirfield, WF14 9NL
M6 JRL Joseph Peacock, 8 Alphingate Close, Stalybridge, SK15 3RL
M6 JRO John Oldman, 94 Mornington Road, London, E4 7DT
M6 JRS Jim Sadler, 10 Spindle Warren, Havant, PO9 2PU
M6 JRT Neville Wing, 39 Whittington Road, Hutton, Brentwood, CM13 1JX
M6 JRW Jonathan Wallis, 20 Green Leys, West Bridgford, Nottingham, NG2 7RX
M6 JSH Joshua Horner, 36 Chadsfield Road, Rugeley, WS15 2QP
M6 JSJ Matthew Burrows, 1 Hedgemere, Taverham, Norwich, NR8 6GG
M6 JSM Joanne McClure, 25 Perth Avenue, Ince, Wigan, WN2 2HJ
MM6 JSN Jason Drummond, 6 Nith Street, Glasgow, G33 2AF
M6 JSP John Parris, Starsmead Farmhouse, Haresfield, Stonehouse, GL10 3EG
MI6 JSQ John Quinn, 5 Woodhill Heights, Lurgan, Craigavon, BT66 7DJ
M6 JSR Ian Smith, 2 George Street, Somercotes, Alfreton, DE55 4JT
M6 JSS Jason Stones, EADS Astrium Ltd, Anchorage Road, Portsmouth, PO3 5PU
M6 JST John Tidmarsh, 16 Birch Road, Wellington, TA21 8EP
M6 JSV Jamie Slade, 199 Durham Road, Stevenage, SG1 4JP
M6 JSW Jessica Welsh, 16 Colton Crescent, Dover, CT16 2EP
M6 JSX Cameron Burridge, 43 Rackenford Road, Tiverton, EX16 5AF
M6 JTA Jacob Allen, 15 Wessington Drive, Hereford, HR1 1AH
M6 JTC Peter Askey, The Maltings, Brewery Yard, Kettering, NN14 3BT
M6 JTD David Connor, 145 Welsby Road, Leyland, PR25 1JH
M6 JTE John Elgar, 37 Ness Road, Lydd, Romney Marsh, TN29 9EL
MM6 JTG Josephine McDowall, 50 Dailly Road, Maybole, KA19 7AU
M6 JTH Jamie Thomas, 14 Lower Meadow, Cheshunt, Waltham Cross, EN8 0QU
M6 JTI James Smith, 8 Westbourne Terrace, Thirsk, YO7 1QD
M6 JTL Martin Simons, 123 Main Street, Little Harrowden, Wellingborough, NN9 5BA
M6 JTM Jack Morgan, Leahurst, Holyhead Road, Wrexham, LL14 5NA
M6 JTN Martin Chivers, 4 Hunters Lodge, Fareham, PO15 5NF
M6 JTO John O'Reilly, 22 Abbey Place, Crewe, CW1 4JR
M6 JTP James Payne, 27 Newton Hall Gardens, Rochford, SS4 3EP
M6 JTQ Joshua Cook, 37 Leigham Court Drive, Leigh-on-Sea, SS9 1PT
MW6 JTR Jonathan Roscoe, Ty Newydd, Ludchurch, Narberth, SA67 8JH
M6 JTU Jordan Bridge, 4 Knights Way, Camberley, GU15 1EQ
M6 JTW James Waring, 12 Mary Street, Farnhill, Keighley, BD20 9AU
M6 JUB Andrew Casson, 7 Leebrook Court, Owlthorpe, Sheffield, S20 6QJ
M6 JUC Bridget Marsh, 21 Edward Road, Eynesbury, St. Neots, PE19 2QF
MM6 JUE Ryan Cormack, 2 Coghill Street, Wick, KW1 4PN
M6 JUG Steven Jones, 25 Kents Lane, Crewe, CW1 4PX
M6 JUK James Rhodes, 73 Keats Way, Hitchin, SG4 0DP
M6 JUX Adrian Brown, 17 Quail Ridge, Ford, Shrewsbury, SY5 9LF
M6 JVB John Cobb, 32 Dellmont Road, Houghton Regis, Dunstable, LU5 5HU
MI6 JVC Jordan Heyburn, 20 Victoria Street, Armagh, BT61 9DT
M6 JVK Jamie Kinsey, 1 Keystone Close, Goring-by-Sea, Worthing, BN12 6GA
M6 JVV John Waddy, 70 Linden Avenue, Prestbury, Cheltenham, GL52 3DS
M6 JVW Mark Fogerty, 25 Noel Street, Gainsborough, DN21 2RY
M6 JWC John Cater, 5 Shady Grove, Hilton, Derby, DE65 5FX



**IMPORTANT NOTE**

**Revalidate licence to avoid revocation** – Ofcom has advised the Society that plans will be drawn up to revoke licences that have not been revalidated as required by the licence conditions. The quickest way to revalidate is to do so online via the Ofcom website: *https://services.ofcom.org.uk/* or by email: *amateur.validations@ofcom.org.uk* Ofcom staff are available to help, but please be patient during times of heavy workload.

M6 JWD James Wilson, 35 Lawson Avenue, Jarrow, NE32 5UF
M6 JWE Julian Loveday, 27 New Road, Chatteris, PE16 6BJ
M6 JWF Anthony Marsh, 140 Church Road, Redfield, Bristol, BS5 9HN
M6 JWG John Whitworth, 31 Shirley Close, Chesterfield, S40 4RJ
M6 JWK James Knight, 13 Church Walk, Harrold, Bedford, MK43 7DG
M6 JWL Darren Jenkins, 15 Homefield Close, Winscombe, BS25 1JE
M6 JWO James Whalley, 28 Hatton Lane, Stretton, Warrington, WA4 4NG
M6 JWP John Wade, 44 Newton Park Homes, Newton St. Faith, Norwich, NR10 3LP
M6 JWQ John Oldfield, 2 Bailie Cross Cottages, Poole Road, Wimborne, BH21 4AE
M6 JWT John Thompson, 19 Taverner Road, Boston, PE21 8NL
M6 JWW James Whiteside, The Old Antique Shop, Bank Street Pulham Market, Diss, IP21 4TG
M6 JWX Joshua Wade, 105 Western Avenue, Woodley, Reading, RG5 3BL
M6 JWY John Willby, 10 Sunbury Road, Birmingham, B31 4LJ
M6 JXD Jason Darragh, 20 Templar Place, Hampton, TW12 2NE
M6 JXF Douglas Wells, 96 Tennyson Avenue, Rugby, CV22 6JF
MM6 JXH John Hinchcliffe, 30 Craigmount Bank, Edinburgh, EH4 8HH
M6 JXN Jon Culshaw, 13 Ravens Close, Knaphill, Woking, GU21 2LD
M6 JXX Jonathan Creaser, 8 Millwood Road, Hounslow, TW3 2HH
M6 JXY John Yendole, 15 Borgie Place, Weston-Super-Mare, BS22 9HG
M6 JYB Jason Brown, 37 Calshot Avenue, Chafford Hundred, Grays, RM16 6NS
M6 JYI Mark Piatkowski, 152 Anchorway Road, Coventry, CV3 6JG
M6 JYL Yui Yee Jessica Lai, Harrogate Ladies' College, Clarence Drive, Harrogate, HG1 2QG
M6 JYO Sajjan Jyothi Sivaraman Valsala, 41 Vine Street, Romford, RM7 7LH
M6 JYS John Pengilly, 8 Willard Close, Eastbourne, BN22 8SX
M6 JYZ John Sanders, 2 Manor Court Manor Grove, Mangotsfield, Bristol, BS16 9LF
M6 JZC Jonathan Clark, 26 Heron Way, Sandbach, CW11 3AU
M6 JZK John Howlett, 29 Little London, Heytesbury, Warminster, BA12 0ES
M6 JZU Paul Holmes, 82 Moore Avenue, Norwich, NR6 7LG
M6 JZY Jeffrey Smale, 30 Gillian Close, Aldershot, GU12 4HU
MW6 JZZ Jazz Trahearn-O'Brien, 148 Gladstone Road, Barry, CF62 8ND
M6 KAA Kevin Parry, 80 Cripps Avenue, Cefn Golau, Tredegar, NP22 3PB
MW6 KAB Kevin Boulter, 16 Danygraig, Pontlottyn, Bargoed, CF81 9RS
MW6 KAC Mark Woodington, 44 Glas y Gors, Aberdare, CF44 0BQ
M6 KAE John Withall, Humphreys Cottage, Fords Green, Uckfield, TN22 3LJ
M6 KAG Keith Goodyer, 5a Station Road, Bow Brickhill, Milton Keynes, MK17 9JU
M6 KAH Kenneth Holloway, 6 Britons Lane Close Beeston Regis, Sheringham, NR26 8SH
M6 KAM Kerry-Ann McGill, 45 Fir Terraces, Esh Winning, DH7 9JQ
M6 KAN Kevin Sharpe, 18 Dudhill Road, Rowley Regis, B65 8HT
M6 KAO Keith Gibbs, 40a Oakwood Road, Hollywood, Birmingham, B47 5DX
M6 KAQ Kevin Sidaway, 19 Larkwhistle Walk, Havant, PO9 4JA
M6 KAR Karol Slotwinski, 51 Moorland Gate, Heathfield, Newton Abbot, TQ12 6TX
M6 KAS John Gilhespy, 106 Durham Drive, Jarrow, NE32 4QY
M6 KAT Catherine Gibson, 16a Hillside Road, Wool, Wareham, BH20 6DY
MM6 KAU Martin Krawczyk, 19 Wishart Archway, Dundee, DD1 2JA
M6 KAV Keith Booth, 28 Farndale Gardens, Lingdale, Saltburn-by-The-Sea, TS12 3EW
M6 KAX James Killman, 19 Moorland Avenue, Walkeringham, Doncaster, DN10 4LG
M6 KAZ Karen Hamilton, 102a Wilbury Road, Letchworth Garden City, SG6 4JQ
M6 KBA Darren Reeve, 5 Antelope Avenue, Grays, RM16 6QT
MI6 KBB Jonathan Elliott, 26 North Road, Newtownards, BT23 7AN
M6 KBC Stuart Williams, 88 Bloomfield Avenue, Bath, BA2 3AE
M6 KBD Kevin Cook, 7 Grenville Terrace, Bideford, EX39 4BE
M6 KBE Karl Bantock, 22 Deepdale Drive, Consett, DH8 7EH
M6 KBF Barry Wiggins, 103 Lowlands Avenue, Sutton Coldfield, B74 3RF
M6 KBG Keith Glaysher, 66 Talbot Road, Farnham, GU9 8RR
M6 KBH Keith Burt, 27 Rockland Villas, Doncaster Road, Rotherham, S65 4AG
M6 KBI Martine Symmonds, 24 Woodville Grove, Stockport, SK5 7HU
M6 KBJ Kieron Jewell, Flat B, 27 Church Street, Liskeard, PL14 3AQ
MW6 KBK Brian Hopkins, 28 Ivor Street, Maesteg, CF34 9AH
M6 KBO Richard Waterhouse, 10 Falconers Drive, Battle, TN33 0DT
M6 KBS Keith Burness, 4 Fenwick Street, Boldon Colliery, NE35 9HU
M6 KBT Robert Mundy, 39 Wenlock Road, Liverpool, L4 2UU
M6 KBV Kevin Clarke, 41 St. Hilda Street, Bridlington, YO15 3EE
MD6 KBW Kevin Whittle, 113 Ballaquark, Douglas, Isle of Man, IM2 2EU
M6 KBX Joseph Greenwood, 38 Baskerville Road, Sonning Common, Reading, RG4 9LS
M6 KBY Andrew Davies, 96 Broad Lane, Kirkby, Liverpool, L32 6QQ
M6 KCA Kevin Millward, 133 Birchfield Way, Telford, TF3 5HN
MM6 KCB Steven Gallagher, 29 Roslyn Drive Bargeddie, Baillieston, Glasgow, G69 7QZ
M6 KCC Kenneth Cole, 16 Minster Road, Westgate-on-Sea, CT8 8BP
M6 KCD Keith Crosby, 28 Exning Road, London, E16 4NA
M6 KCE Kevin Bradford, 3 Haven Close, Sutton-in-Ashfield, NG17 2DG
M6 KCF Lesley Bee, 25 Blatcher Close, Minster on Sea, Sheerness, ME12 3PG
M6 KCG Karen Allen, 20 Rookery Road, Innsworth, Gloucester, GL3 1AT
M6 KCI David Orme, 47 Poplar Avenue, Oldham, OL8 3TZ
MW6 KCJ Kelvin Uprichard, 111 Fernhill, Mountain Ash, CF45 3EF
M6 KCK Luke Wood, 2 The Bungalows, North Green, Woodbridge, IP13 9NP
M6 KCL John Jopson, 12 Charing Close, Ringwood, BH24 1FA
MM6 KCM Kevin Mair, 26 Hawthorn Drive, Wishaw, ML2 8JS
M6 KCP Andrzej Ostatek, 38 Coronation Way, Keighley, BD22 6HF
MW6 KCQ Michael Cook, 40 Cemaes Crescent, Rumney, Cardiff, CF3 1TA
M6 KCR Keith Riddick, Davah, Port Road, Castle Douglas, DG7 3JW
M6 KCS Arron Stooke, 128 Stamford Street, Grantham, NG31 7BP
M6 KCT Karl Coton, 162 Meadow Way, Jaywick, Clacton-on-Sea, CO15 2SF
M6 KCU Christopher Pavey, 143 Queen Elizabeth Way, Colchester, CO2 8LT
MM6 KCV Cameron Vines, 49 Annandale Gardens, Glenrothes, KY6 1JD
M6 KCY Roddy Fisher, 14 Fraser Avenue, Caversham, Reading, RG4 6RT
M6 KCZ Raymond Robinson, 22 Riddings Court, Timperley, Altrincham, WA15 6BG
MW6 KDA David Henderson, 1 Court Place, Tonypandy, CF40 2RE
M6 KDB Kaylem Beech, 8 Gateford Drive, Worksop, S81 7HL
M6 KDC Dave Stewart, 67 Little Moss Hey, Liverpool, L28 5RJ
M6 KDD Kevin Daniels, 64 Skelton Road, Norwich, NR7 9UH
M6 KDF John Nash, 124 Tanhouse Avenue, Birmingham, B43 5AG
M6 KDJ Keith Hall, 11 Hinton Villas, Hinton Charterhouse, Bath, BA2 7SS
M6 KDK Kristian Khan, 3 Marshall Close, Frimley, Camberley, GU16 9NY
M6 KDM Karina Middlehurst, 7 Statham Drive, Lymm, WA13 9NW
M6 KDO Kerwin Porter, 58 Panama Drive, Atherstone, CV9 3HJ
M6 KDP Robert Hutton, 22a Victoria Road, Maldon, CM9 5HF
M6 KDQ Alastair Bowles, Berrys Cottage, Southend Road, Reading, RG7 6HA
M6 KDR Christopher Pearcey, 17 Peppercorn Close, Christchurch, BH23 3BL
M6 KDV Mark Pulling, 21 Heathlands Avenue, West Parley, Ferndown, BH22 8RW
M6 KDW Kevin Dillow, 101 Martins Lane, Hardingstone, Northampton, NN4 6DJ
M6 KDZ Ian Whiteley, 29 Harvey Avenue, Wirral, CH49 1RT
M6 KEA Adrian Wheeler, 10 Handsacre Crescent, Rugeley, WS15 4DQ
M6 KEB Michael Byrne, 15 Norton Avenue, Canvey Island, SS8 8LG
M6 KEC Daniel Preston, 25 Wymark View, Grimsby, DN33 1RF
MW6 KED Steven Kedward, 9 Lawrence Avenue, Aberdare, CF44 9EW
M6 KEE Micheal Keeler, 15 Grove Park, Kingswinford, DY6 9AD
M6 KEF Christopher Thompson, 5 Oak Avenue, Charlton Kings, Cheltenham, GL52 6JG
M6 KEG Paul Selwood, 5 Sorrel Close, Padgate, Warrington, WA2 0UF
M6 KEH Kerry Dalby, 52 Narborough Road South, Leicester, LE3 2FN
MW6 KEL Kelly Gemmell, 93 North Road, Ferndale, CF43 4RG
M6 KEM Kevin Morris, 46 Raikes Avenue, Bradford, BD4 0QU
M6 KEP Kevin Applegarth, 28 South View Gardens, Pontefract, WF8 2HW
M6 KEQ Kevin Mogford, 49 Cefn Road, Rogerstone, Newport, NP10 9AQ
MM6 KER Kerr Hamilton, 25 Abbotsford Street, Falkirk, FK2 7NH
M6 KES Martin Shore, 24 Gale Street, Rochdale, OL12 0SQ
M6 KET John Daws, 1157 Evesham Road, Astwood Bank, Redditch, B96 6DY
MD6 KEU Yue Law, 59 Lake View, Edgware, HA8 7SA
M6 KEV Kevin Taylor, 23 Arminers Close, Gosport, PO12 2HB
M6 KFA Kelly Atkins, 278a West End Lane, London, NW6 1LJ
M6 KFB Francis Bejoy Kuttikkate, 34 Shetland Crescent, Rochford, SS4 3FJ
M6 KFC David Aldred, 14 The Meadows, Radcliffe, Manchester, M26 4NS
M6 KFE Keith Farnington, 52 Glebe Park, Duns, TD11 3EE
M6 KFF Steven Coupe, 47 Burn Street, Sutton-in-Ashfield, NG17 4LL
M6 KFG James Stokes, 30 Anglers Reach, Grove Road, Surbiton, KT6 4EX
MD6 KFH Kim Gascoyne, 22 Broogh Wyllin, Kirk Michael, Isle of Man, IM6 1HU
MM6 KFJ Aidan Keogh, 251 Main Street, Plains, Airdrie, ML6 7JH
M6 KFK Krzysztof Kozlowski, WOKING HOMES / FLAT 2, ORIENTAL ROAD, Woking, GU22 7BE
M6 KFP Keith Prentice, 24 Sulgrave Close, Liverpool, L16 6AD
M6 KFS Kate Fraser-Smith, 28 Stafford Road, Southampton, SO15 5EA
M6 KFW Kevin Wells, 5 Cook Avenue, Newport, PO30 2LL
M6 KGA Kieran Allgar, 13 Deacon Avenue, Kempston, Bedford, MK42 7DU
M6 KGB William Oliver, Pwllmeyric, Chepstow, NP16 6LE
M6 KGC Keith Cossey, 34 Pinewood Road, Hordle, Lymington, SO41 0GP
M6 KGD Mark Tabberer, 29 Chase Vale, Chasetown, Burntwood, WS7 3GD
M6 KGE Kevin Gibbins, 11a Wood End, Banbury, OX16 9ST
M6 KGF Glyn Flew, 58 Streamside, Mangotsfield, Bristol, BS16 9EA
M6 KGK Kane Goddard, 49 Chalet Hill, Bordon, GU35 0EF
M6 KGL M Horwell, 3 Vanity Close, Oulton, Stone, ST15 8TZ
MM6 KGM Kenneth Mitchell, 72 Laing Gardens, Broxburn, EH52 6XT
M6 KGN Jeff Badham, Cornwillgur House Ross Road, Longhope, GL17 0LP
M6 KGR Kevin Carter, 50 Elliman Avenue Bottom flat, Slough, SL2 5BG
M6 KGS Kenneth Stow, 39 Biverfield Road, Prudhoe, NE42 5ER
M6 KGV Gwyn Carter, 54 Wood Street, Taunton, TA1 1UW
MM6 KHA Karl Adrian, 1c Oliphant Court, Paisley, PA2 0DP
M6 KHB Michael Heaton-Bentley, 65 Brookfield Road, Thornton-Cleveleys, FY5 4DR
MM6 KHM Kenny Macrae, 2 Netherhill Avenue, Glasgow, G44 3XG
M6 KHS Kallum Shaw, 3 The Grange, Woolley Grange, Barnsley, S75 5QP
M6 KIA Axel Taylor, 130a Hazelwood Avenue, Eastbourne, BN22 0UX
M6 KIE Kieran Alejo-Blanco, 23 Southern Road, Thame, OX9 2EE
M6 KIF Mike Chalkley, 36 Cowper Road, Bournemouth, BH9 2UJ
M6 KIG Kenneth Greenfield, 49 Railway Street, Northfleet, Gravesend, DA11 9DU
M6 KIK William Dover, Silverdale, Fox Lane, Basingstoke, RG23 7BB
M6 KIL Benjanin Fryer, 16 Elston Place, Aldershot, GU12 4HY
M6 KIN Martin Makin, 34 Carlton Gardens, Farnworth, Bolton, BL4 7TH
M6 KIP Kenneth Davies, 29 Corser Street, Stourbridge, DY8 2DE
M6 KIR Kira-Lee Halloway, 41 Trenoweth Estate, North Country, Redruth, TR16 4AQ
M6 KIT Chris Ellis, 1 Rugby Road, Stockton-on-Tees, TS18 4AZ
MM6 KIW Craig Ross, 4 Weir Place, Greenock, PA15 2JD
M6 KIX Christine Hill, 3 Beechmount Rise, Stafford, ST17 4QR
M6 KJB Kieran Black, 1 Woodcock Close, Haxby, York, YO32 3NQ
M6 KJD Kurt Davison, 1 Chancel Way, Barnsley, S71 2HS
M6 KJI Matthew Bidwell, 3 Walsingham Place, London, SW4 9RR
M6 KJK Karol Jan Kolesnik, 15 Steer Road, Swanage, BH19 2RU
M6 KJM Kellieann Mitchell, 5 Finch Crescent, Linslade, Leighton Buzzard, LU7 2PE
M6 KJP Kenneth Parker, Flat 3, Gleadless Court, Sheffield, S2 3AE
M6 KJQ Keenan Freer, 1 Masefield Flats, Masefield Road, Rotherham, S63 6NQ
M6 KJR James Cogman, 6 Stockholm Way, Dereham, NR19 1XF
MI6 KJW Kenneth Watt, 2A Drumalane Road, Newry, BT35 8AP
M6 KJY Kevin Younger, 29 Eagle Walk, Bury St. Edmunds, IP32 6RJ
M6 KKA Robert Pallister, 15 High Row, Washington, NE37 2LZ
M6 KKD Kevin Edmands, 61 Somers Road, Keresley End, Coventry, CV7 8LE
M6 KKF Frank Fitton, 6 Leaford Close, Denton, Manchester, M34 3QH
M6 KKG Kouame Komenan, 113 Wightman Road, Finsbury Park, London, N4 1RJ
M6 KKI Rodney Young, 10 Hareholme Lane, Rossendale, BB4 7JZ
M6 KKJ Nick Dyer, 79 Saltmarsh Drive, Bristol, BS11 0NL
M6 KKL Tsz Ying Kitty Cheng, Harrogate Ladies' College, Clarence Drive, Harrogate, HG1 2QG
M6 KKM Darren Linnett, 12 Somersal Close, Shelton Lock, Derby, DE24 9QT
M6 KKO Terence Stack, 31a Chester Road South, Kidderminster, DY10 1XJ
M6 KKY Mark Douglas, 26 Clumber Drive, Northampton, NN3 3NX
M6 KLA Darren Bourne, 4 Market Street, Cheltenham, GL50 3NH
M6 KLC Kevin Chappuis, Priory Cottage, Priory Lane, Bridport, DT6 3RW
M6 KLD Thomas Lovell, 42 Sonja Crest, Immingham, DN40 2EQ
M6 KLF Kevin Francis, 203 Colchester Road, Lawford, Manningtree, CO11 2BU
M6 KLH Karen Hadfield, Flat 8, Maple House, Havant, PO9 1AG
M6 KLK Jia Yu Li, Harrogate Ladies' College, Clarence Drive, Harrogate, HG1 2QG
MW6 KLL Kieran Linahan, 19 Heol Bryn Hebog, Merthyr Tydfil, CF48 1HH
M6 KLM Kerry Mason, 18 Goose Lane, Sutton, Norwich, NR12 9SE
M6 KLN Christian Dzundza, 67 Sidegate Lane, Ipswich, IP4 4HY
M6 KLR Stephen St George, 37 Highfields, Bromsgrove, B61 7DA
MM6 KLT Campbell Sharp, 16 Wood Lane, Monifieth, Dundee, DD5 4HS
M6 KLW Lee Woods, 193 Wimberley Street, Blackburn, BB1 8HU
M6 KLX Andrew Swain, 43 Stretton Road, Morton, Alfreton, DE55 6GW
MM6 KLZ Nial Stewart, 35 Newbattle Gardens, Dalkeith, EH22 3DR
M6 KMA Karl Machen, 55 Aketon Road, Castleford, WF10 5DN
M6 KMC Scott Day, 4 Belfield Road, Etwall, Derby, DE65 6JP
M6 KMD Keith Deans, 31 Northcroft, Sandy, SG19 1JJ
M6 KMF Neil Powell, 21 Hitchen, Merriott, TA16 5QX
M6 KMG Paul Matthew, 24 Jubilee Close, Pamber Heath, Tadley, RG26 3HP
M6 KMI Neville Dobson, 2 Hills Road, Breaston, Derby, DE72 3DF
M6 KMJ Mark Beasley, 292 North Road, Yate, Bristol, BS37 7LL
M6 KMK Andrew Platt, 43 The Butts, Frome, BA11 4AB
M6 KML Keith Lavelle, 8 Burrington Drive, Leigh, WN7 5EU
M6 KMN Kathleen Nilan, 15 Broomhall Road, Pendlebury, Manchester, M27 8XP
M6 KMR Kevin Riding, 18 Cormorant Close, Bransholme, Hull, HU7 4SP
M6 KMS M Harrison, 70 Hope Avenue, Goldthorpe, Rotherham, S63 9EA
M6 KMW Kevin White, 12 Lakefields, West Coker, Yeovil, BA22 9BT
M6 KMZ Kirk Martinez, Forest View, Forest Road, Salisbury, SP5 2BP
M6 KNB Mukundan Narayanankutty, Flat 10, Memorial Heights Monarch Way, Ilford, IG2 7HR
M6 KNC Jacob Richardson, Berwick Cottage, Bailes Lane, Guildford, GU3 2AX
M6 KND David Simmons, 23 Fairey Street, Cofton Hackett, Birmingham, B45 8GU
M6 KNG Alan King, 23 Tower Crescent, Lincoln, LN2 5QF
M6 KNL Glenn Knowles, 29 Delamere Crescent, Cramlington, NE23 3FY
M6 KNM Mark Hill, 109 Kitchener Street, St. Helens, WA10 4LU
MM6 KNO Thomas Knox, 23 Hill Street, Alness, IV17 0QL
M6 KNS Kevin Fletcher, 54 Shipton Road, Scunthorpe, DN16 3HQ
M6 KNU Richard Weightman, 17 Shaw Drive, Knutsford, WA16 8JG
M6 KNW Keeva Woolsey, 1 Park Farm Cottage, Westhorpe Road, Stowmarket, IP14 4SP
M6 KNY Roger Dunnaker, 12 Dagger Lane, West Bromwich, B71 4BA
M6 KOA Leslie Lemmon, 4 Honington Close, Wickford, SS11 8XB
MI6 KOB Kevin Boyle, 764 Springfield Road, Belfast, BT12 7JD
M6 KOH David Elcock, 27 Harefield Road, Southampton, SO17 3TG
M6 KOI Carl Young, 13 Bryn Mawr Road, Holywell, CH8 7AP
M6 KOM David Slater, 13 Longford Close, Rainham, Gillingham, ME8 8EW
M6 KON Simon Rouse, 7 Cranbrook Road, Thurnby, Leicester, LE7 9UA
MM6 KOS Jakub Kosarzecki, 96 Colinton Mains Drive, Edinburgh, EH13 9BL
M6 KOZ Radoslaw Koziolek, 30 Lammas Beanhill, Milton Keynes, MK6 4LA
MW6 KPA Horia Ilie, Flat 1 30 Alexandra Road, Swansea, SA1 5DQ
M6 KPC Kallum Cooper, 12 Waverley Crescent, Brighton, BN1 7BG
M6 KPD Kevin Dance, 20 Harmers Hay Road, Hailsham, BN27 1SU
M6 KPF Keith Foster, Prince of Wales House, Short Bridge Street, Llanidloes, SY18 6AD
M6 KPG Kevin Gallery, 89 Mavis Drive, Coppull, Chorley, PR7 5AE
M6 KPI Jason Barnes, Landhill Farm, Halwill, EX21 5TX
M6 KPK Keith Killingback, Flat 4 Shorland House, 6 Elm Grove Road, Dawlish, EX7 0BZ
M6 KPL Ashley Lloyd, 194 Chartwell, Weymouth, DT4 9SP
MW6 KPN Adrian Thomas, 10 Chapel Street, Gorseinon, Swansea, SA4 4DT
M6 KPO Alastair Smith, 58 Wellesbourne Road, Barford, Warwick, CV35 8DS
M6 KPT Kirsty Thrower, 19 Blyford Road, Lowestoft, NR32 4PZ
M6 KPX Dawn Cooper, Needhams Farm House, Spittal Hill Road, Boston, PE22 0PA
M6 KQJ John Clarke, 49 Brunel Close, Hartlepool, TS24 0UF
M6 KQL Mark Castle, 10 West Place, Gobowen, Oswestry, SY11 3NR
M6 KQW Even Almas, 10, Rindal, Norway, 6657
M6 KRD Dirk Niggemann, 305 Holm Court, Twycross Road, Godalming, GU7 2QT
M6 KRF Keith Furlong, 22 Swaisland Road, Dartford, DA1 3DE
M6 KRH Kevin Hall, 130 Aylestone Lane, Wigston, LE18 1BA
M6 KRI Kenneth Irvin, 21 Fremantle Crescent, Middlesbrough, TS4 3HR
M6 KRK James Birch, 3 Partridge Way, High Wycombe, HP13 5JX
M6 KRL Paul Dudding, 14 St. Levan Close, Marazion, TR17 0BP
M6 KRM Kalvin McLeod, 7 Priory Place, Sporle, King's Lynn, PE32 2DT
MI6 KRP Kyle Pritchard, 18 Ashbrooke, Donaghadee, BT21 0EY
M6 KRR Gary Newton, 20 East Avenue, Syston, Leicester, LE7 2EH
M6 KRV Kevin Bott, 294 Walthall Street, Crewe, CW2 7LE
M6 KRW Keith Wiles, 24 Cromwell Way, Witham, CM8 2ES
M6 KRX Kevin Rosema, Apartment 801, 25 Goswell Road, London, EC1M 7AJ
M6 KRZ Kevin Brazier, 3 Chaffes Terrace, Chaffes Lane, Sittingbourne, ME9 7BQ
M6 KSA Charles Sibley, 57 Palatine Road, Thornton-Cleveleys, FY5 1EY
M6 KSC David Clark, 34 Magdalene Road, Owlsmoor, Sandhurst, GU47 0UT
M6 KSD Karen Darwin, 38 Springbank Road, Gildersome, Leeds, LS27 7DJ
M6 KSG Katrina Stevens, 61a Main Road, Hoo, Rochester, ME3 9AA
M6 KSH Krystyna Haywood, 126 Derby Street, Sheffield, S2 3NF
M6 KSI Aaron Thompson, 10 Belgrave Close, Hersham, Walton-on-Thames, KT12 5PH
MM6 KSJ John Blick, 1 Kennel Cottage, Mount Stuart, Isle of Bute, PA20 9LP
MW6 KSL Steven Davies, 5 Maldwyn Street, Cardiff, CF11 9JR
M6 KSP Simon Pegg, 24 Fleetwood Close, Minster on Sea, Sheerness, ME12 3LN
M6 KSS Shawn Evennett, The Homestead, Pound Green Lane, Thetford, IP25 7LS
M6 KSV Svetlana Karpukhina, 13 Rushmon Court, Barker Road, Chertsey, KT16 9HP
M6 KSW Maria Hancock, 6a Lovatt Close, Stretton, Burton-on-Trent, DE13 0HZ
M6 KTA Kenneth Tilly, 14 McNally Place, Durham, DH1 1JE
M6 KTC Isaac McEwen, 4 The Pantyles, Nightingale Lane, Sevenoaks, TN14 6BX

M6 KTH Karl Thompson, 13 Southfield Drive Sutton Courtenay, Abingdon, OX14 4AY
M6 KTI K[illegible], 6 New Street, Castleford, WF10 2DN
M6 KTJ Kevin Jones, 7 Castlemans Cottages, Hinton St Mary, Sturminster Newton, DT10 1NA
M6 KTK Ninesh Edwards, 42 Cambrai Avenue, Chichester, PO19 7UY
M6 KTM Paul Jones, 16 Severn View, Garndiffaith, Pontypool, NP4 7SN
M6 KTN Kevin Turner, 3 East Street, Sutton-in-Ashfield, NG17 4GQ
M6 KTO Kevin Oxford, 5 Hazel Close, Rendlesham, Woodbridge, IP12 2UR
M6 KTP Anthony Purnell, 46 The Rookery, Deepcar, Sheffield, S36 2NA
M6 KTS Ifor Williams, 5 Fron Goch, Llanberis, Caernarfon, LL55 4LE
M6 KTT Kevin Thistlethwaite, 140 Kingstown Road, Carlisle, CA3 0AY
M6 KTW Kristofer Wise, 4 Cherrydown, Rayleigh, SS6 9ND
M6 KTX Keith Taylor, 26 Elmbridge Road, Birmingham, B44 8AB
M6 KTY Katrina Heron, 26 Lochancroft Lane, Wigtown, Newton Stewart, DG8 9JA
M6 KTZ Karen Tasker, 16 Chopin Road, Basingstoke, RG22 4JN
M6 KUD Lee Ansell, 114 Bowleaze, Greenmeadow, Cwmbran, NP44 4LG
M6 KUH Sarah Terry, 201a Urmston Lane, Stretford, Manchester, M32 9EF
M6 KUT Adam Roberts, 29 Manor Lane, Stourbridge, DY8 3ER
M6 KVA Peter Barnes, 13 Lavender Close, Thornbury, Bristol, BS35 1UL
M6 KVB Kevin Bushell, 4 Birch Grove, Harrogate, HG1 4HR
M6 KVD John Miller, 5 Gilver Lane, Hanley Castle, Worcester, WR8 0AT
M6 KVF Konrad Emery-Ford, 15 Anson Close, Grantham, NG31 7EN
M6 KVG Dean Miles, 133 Marston Lane, Nuneaton, CV11 4RE
MM6 KVI George Hepburn, 33 Saxon Road, Glasgow, G13 2YQ
M6 KVJ David Boyes, 1 Fluxton Cottages, Fluxton, Ottery St. Mary, EX11 1RL
M6 KVK Gary Kirk, 15 Underwood Avenue, Ash, Aldershot, GU12 6PP
M6 KVL Graham Dougherty, 22 Mayplace Avenue, Dartford, DA1 4PZ
M6 KVM David Palmer, 19 Herbert Road, Bath, BA2 3PP
M6 KVN Kevin Sewell, 12 Haylands Square, South Shields, NE34 0JB
M6 KVO Kenneth Owen, 27c Waen Fawr Estate, Holyhead, LL65 1LT
M6 KVS Martynas Kveksas, 29 Saxon Way, Reigate, RH2 9DH
M6 KVT Robert Simpson, 5 Rogate Road, Worthing, BN13 2DT
M6 KVV Graham Easton, Cowpen Road, Blyth, NE24 5TS
M6 KVX Roy Timmons, 36 Worrall Road High Green, Sheffield, S35 3LP
M6 KWA Kenneth Mackenzie, Graceland, Crosshill, Duns, TD11 3UF
M6 KWC Kevin Chapman, 227 Raglan Street, Lowestoft, NR32 2LA
M6 KWG Winton Wightman, 36 Holyoake Avenue, Woking, GU21 4PW
M6 KWH Antony Watts, 21 Ladbroke Hall, Ladbroke, CV47 2DF
M6 KWI Kevin Irwin, 11 The Crofts, Silloth, Wigton, CA7 4EU
M6 KWK David Barnett, 72a Clough Hall Road, Kidsgrove, Stoke-on-Trent, ST7 1AW
M6 KWL Ciaran Henniker, 1 Brook House Drive, Fairfield, Buxton, SK17 7HW
M6 KWM David Wright, 203 Winn Street, Lincoln, LN2 5EY
M6 KWN Roger Skinner, 29 Kenyon Road, Portsmouth, PO2 0JZ
M6 KWP Karl Pritchard, 124 Milburn Road, Ashington, NE63 0PQ
M6 KWT Joseph Caulfield, 2 Thornley Road, Tow Law, Bishop Auckland, DL13 4ED
M6 KWW Barry Covill, Walnut Tree Cottage, Holt Street, Dover, CT15 4HX
M6 KWX Karl Woodard, 11 Bury Road, Stanton, Bury St. Edmunds, IP31 2BZ
MI6 KXM Keith Mitchell, 80 Markethill Road, Collone, Armagh, BT60 1LE
M6 KXQ Brian Wogden, 3 Gordonstoun Place, Blackburn, BB2 2PT
M6 KXS Robert Rigden, 36a Atherston, Bristol, BS30 8YB
M6 KXX Bruce Cook, 11 Arbrook Lane, Esher, KT10 9EG
M6 KYB Kenneth Greenshields, 3 Lovers Walk, Wells, BA5 2QL
M6 KYC Michael Cook, 12 Davison Street, Lingdale, Saltburn-by-The-Sea, TS12 3DX
M6 KYE Aaron Letchford, 64 Medway Road, Sheerness, ME12 1DR
M6 KYK James Slim, Troublesome Reach, Playford Road, Woodbridge, IP13 6ND
M6 KYN Kynan O'Brien, 2 Alfred Street, Newport, NP19 7FJ
M6 KYO Asia Noriega, 363 PRACHA UTHIT ROAD, Don Mueang, Thailand, BANGKOK 10210
M6 KYS Efthimios Panayiotou, 7 Oldfield Drive, Cheshunt, Waltham Cross, EN8 0JL
M6 KZE Paul Orwin, 108 Cordwell Avenue, Chesterfield, S41 8BN
M6 KZM Kevin Maddy, 56 Coachwell Close, Telford, TF3 2JB
M6 KZR D Ellens, 150 Lumley Avenue, South Shields, NE34 7DJ
MI6 KZS Kieran Sullivan, 149 Largy Road Ahoghill, Ballymena, BT42 2RG
M6 KZU Martin Macrae, 91 Chosen Way, Hucclecote, Gloucester, GL3 3BX
M6 LAA Louis Atkins, Mouse Hall, Low Row, Richmond, DL11 6PY
M6 LAC Lee Ainger, 41 Gilbert Road, Camberley, GU16 7RD
MM6 LAD John Cattigan, Lunan Home Farm Cottage, Lunan Bay, Arbroath, DD11 5ST
M6 LAE Lauren Thompson, 43 Manor Road, Horsham St. Faith, Norwich, NR10 3LF
M6 LAF Stephen Dooley, 28 Hilbre Drive, Ellesmere Port, CH65 9JQ
M6 LAG Lyn Burgess, 40 Sheridan Terrace, Hove, BN3 5AF
MM6 LAH Lesley Hall, 70 Nobleston Estate, Alexandria, G83 9DD
M6 LAI Lisa Lewczenko, 10 Saxon Court, Swaffham, PE37 7TP
MM6 LAK Scott Ramsay, 6 Cross Road, Peebles, EH45 8DH
M6 LAL Lara Cowley, 3 Park Villas, Keswick, CA12 5LQ
M6 LAM Malcolm Quemby, 19 Oak Close, Coalville, LE67 4JU
MM6 LAO William Welsh, 25 Currievale Drive, Currie, EH14 5RN
M6 LAP Shane Potts, 22 Valebridge Road, Burgess Hill, RH15 0QY
M6 LAQ Liam Jhon Briggs Briggs, 5 Links Avenue, Cromer, NR27 0EQ
M6 LAS Lorraine Scambell, 8 South Bank Road, East Cowes, PO32 6JE
M6 LAV Lily Hand, 168 Barcroft Street, Cleethorpes, DN35 7DX
M6 LAY Sidney Lay, 7 Hunt Street, Swindon, SN1 3HW
M6 LAZ Graham Sinclair, 23 Cummings Square, Wingate, TS28 5JF
M6 LBC Lindsey Nicholas, 9 Hermitage Gardens, Cotton End, Bedford, MK45 3AY
M6 LBE Steven Blackburn, 20 Seascale Close, Blackburn, BB2 3TP
M6 LBI Ian Hyde, 3 Hibbert Avenue, Denton, Manchester, M34 3NZ
M6 LBK Ian MacDonald, Broomhill, Mill Lane, Worthing, BN13 3DH
M6 LBL Liam Bell, 56 Boyd Road, Wallsend, NE28 7SQ
M6 LBM Luke Mason, 86 The Street, Rockland St. Mary, Norwich, NR14 7AH
M6 LBN Peter Tolcher, 15 Langstone Close, Torquay, TQ1 3TX
M6 LBQ Bing Qiong Li, Harrogate Ladies' College, Clarence Drive, Harrogate, HG1 2QG
M6 LBR Samuel Hayward, 20 Abbey Square, Walsall, WS3 2RJ
MM6 LBS William Thomson, 20 Greenfield Road, Glasgow, G32 0LP
M6 LBT Linda Thomas, 66 Sturdee Avenue, Great Yarmouth, NR30 4HL
M6 LBU Liam Baldwin, 24 Rockrose Way, Portsmouth, PO6 4EZ
M6 LBV Elizabeth Price, Little Acre, Eardisley, Hereford, HR[illegible] 6LX
M6 LBW Elizabeth White, 16 Illingworth Way, Foxton, Cambridge, CB22 6RY
M6 LBX Andrew Walker, 27 Fielding Avenue, Poynton, Stockport, SK12 1YX
M6 LBY Alan Smith, 21 Collard Avenue, Newcastle, ST5 9LH
M6 LCC Peter Harris, 16 Laxton Gardens, Baldock, SG7 6DA
M6 LCF Matthew MacDonell, 54 Cinque Foil, Peacehaven, BN10 8DZ
M6 LCG Eleri Ayre, 1 Spring Gardens, Broadmayne, Dorchester, DT2 8PP
M6 LCH Peter Cairns, 16 East Avenue, Heald Green, Cheadle, SK8 3DL
M6 LCI Liam O'Brien, 172 Cheriton High Street, Folkestone, CT19 4HN
M6 LCK Lindsey Kerr, 13 Tramside Way, Carlisle, CA1 2FH
MM6 LCL Lucy Clark, 9 Aileymill Gardens, Greenock, PA16 0QF
M6 LCP Lisa Bourn, 4 Fell Wilson Street, Warsop, Mansfield, NG20 0PT
M6 LCQ Lisa Man, 115 Northdown Park Road, Margate, CT9 3PX
MI6 LCR Lesley Robinson, 32 Corrycroar Road, Pomeroy, Dungannon, BT70 3DY
M6 LCT Chui Lai, Harrogate Ladies' College, Clarence Drive, Harrogate, HG1 2QG
M6 LCU Christian Keszei, 2 Blackmore Hill Farm Cottages, Calvert Road, Buckingham, MK18 2HA
M6 LCV Adam Grant, 47 Coneyford Road, Birmingham, B34 7AY
M6 LCW Luke Weeks, 11 Brock Close, Deepcut, Camberley, GU16 6GA
MW6 LCX Jason Hawkins, 104 Ty Fry, Aberdare, CF44 7PP
M6 LCY Lucy Heron, 301 Marton Road, Middlesbrough, TS4 2HG
M6 LDB Liam Hoddinott, 30 Deans Mead, Bristol, BS11 0QX
M6 LDD Edwin Daniels, 2 Garstons Close, Fareham, PO14 4EN
M6 LDE David Simmons, 8 Lower Grange, Huddersfield, HD2 1RU
M6 LDF Lee Ferguson, 67 Knowlton Road, Poole, BH17 9EE
M6 LDG Raymond Hazel, 3 Westlands Drive, Hedon, Hull, HU12 8BY
M6 LDH Oi Yee Christy Lai, Harrogate Ladies' College, Clarence Drive, Harrogate, HG1 2QG
MI6 LDI Paul Dorrian, 12 Gortnamona Place, Belfast, BT11 8PP
M6 LDJ Leigh Jepson, 16 Golborne Street, Newton-le-Willows, WA12 9TH
M6 LDK Lee Kelley, 54 Grasmere Street, Liverpool, L5 6RJ
M6 LDL John Hatton, 49 Buxton Street, Morecambe, LA4 5SR
M6 LDM Louis Martin, 78 Llwyn Ynn, Talybont, LL43 2AG
M6 LDQ Liam Dobinson, 20 Newholme Crescent, Evenwood, Bishop Auckland, DL14 9RY
M6 LDR Lee Roworth, 27 Bury Road, Dagenham, RM10 7XR
MW6 LDS Tudor Jones, 13 Bond Street, Aberdare, CF44 7HA
M6 LDU Lance Dumbleton, 9 Wareham Road, Rubery, Birmingham, B45 0JS
M6 LDW Terence Evans, 70 Tremarle Home Park, North Roskear, Camborne, TR14 0AR
M6 LDY Alison Collins, 15 North River Road, Great Yarmouth, NR30 1JY
M6 LDZ Trevor Clapp, Windrush, One Pin Lane, Slough, SL2 3QY
M6 LEA Jenna Bridgehouse, 43 Age Croft, Oldham, OL8 2HG
M6 LEC Lindsey Collinson, 26 Westway Avenue, Hull, HU6 9SA
M6 LEE Lee Davies, 94 Worsley Road, Farnworth, Bolton, BL4 9LX
M6 LEF Lucy Faulkner, Mount Pleasant, Elkstones, Buxton, SK17 0LU
M6 LEG Kevin Foulger, 89 Blaney Crescent, London, E6 6BB
M6 LEH Louise Hargreaves, 32 Bank Road, Carrbrook, Stalybridge, SK15 3JX
M6 LEJ Colin Calvert, 1 Moorsholme Avenue, Manchester, M40 9BW
M6 LEQ Leslie Pinkney, 18 Bridlington Road, Driffield, YO25 5HZ
MD6 LET David Holohan, 22 Ballacannell Estate, Laxey, Laxey, Isle of Man, IM4 7HH
M6 LEU Lee Chadwick, 78 Blakemore, Telford, TF3 1PT
MM6 LEW Mark Strachan, 62 Charleston Drive, Dundee, DD2 2EZ
M6 LEX Kevin Rowland, 9 Churchlands, North Bradley, Trowbridge, BA14 0TD
M6 LEY Christopher Harris, 21 Findlay Street, Leigh, WN7 4EQ
M6 LFB Leslie Bell, 42 Ocean Road, Walney, Barrow-in-Furness, LA14 3DX
M6 LFC Daniel Smith, 5 Verbena Close, Beechwood, Runcorn, WA7 3JA
M6 LFD Linda Drew, 7 Bronte Court, Tamworth, B79 8DN
M6 LFE Brian Lewin, 68 Brackley Square, Woodford Green, IG8 7LS
M6 LFG Jonathan Beavan, 21 Llannerch Road West, Rhos on Sea, Colwyn Bay, LL28 4AU
M6 LFJ Melvin Stevens, The Glen, Sunnyside Avenue, Sheerness, ME12 2RA
MI6 LFK Leonard Corr, 72 Balnamore Road, Ballymoney, BT53 7PT
M6 LFL Louise Dawn Theobold, 25 Aysgarth Road, Leicester, LE4 0ST
M6 LFM Alan Campbell, 35 Goodwood Road, Gosport, PO12 4HN
MM6 LFN Charles Bolton, 30 Brackenhill Drive, Hamilton, ML3 8AY
M6 LFO Paul Noble, 30 Whitewater Rise, Dibden Purlieu, Southampton, SO45 4BY
M6 LFQ K J Blatch, 61 Linden Way, Haddenham, Ely, CB6 3UG
M6 LFR Lee Davison, 58 Priestley Court, South Shields, NE34 9NQ
MM6 LFS Anthony Miles, 9 Buchanan Drive, Lenzie, Glasgow, G66 5HS
M6 LFT Marco Cianni, 121 Springfield Park Avenue, Chelmsford, CM2 6EW
MI6 LFU Clive McCartney, 62 Lakeview, Crumlin, BT29 4YA
M6 LFW Morgan O'Donovan, 58 Ocean View, Cirencester, GL76PA
M6 LFX John Lamb, 9a Matlock Road, Canvey Island, SS8 0EW
M6 LFY John Wells, 27 Victoria Avenue, Camberley, GU15 3HT
M6 LFZ Edward Fish, 16 Cartmel Place, Ashton-on-Ribble, Preston, PR2 1TY
M6 LGA Richard Camphire, Courtlands, Newport Road, Caldicot, NP26 3BZ
M6 LGB Lewis Brazier, 165 Avon Road, Worcester, WR4 9AH
M6 LGD Lancelot Graham Peter Thomas, Fair View, Close Hill, Redruth, TR15 1EP
M6 LGF Dennis Riches, 48 Turner Road, Ipswich, IP3 0LX
M6 LGH John Abraham, 18 Ferneley Crescent, Melton Mowbray, LE13 1HZ
M6 LGI Lee Grover, 12 Winchdells, Hemel Hempstead, HP3 8HZ
M6 LGJ Lucy Woolridge, 40 Cambridge Road, Eastbourne, BN22 7BT
M6 LGL Marius Rusu, 9 Nash Close, Corby, NN18 0QU
M6 LGM Leonard Mann, 14 Orchehill Avenue, Gerrards Cross, SL9 8PX
M6 LGQ Michael Lee, 4 Cluny Court, Wavendon Gate, Milton Keynes, MK7 7TT
MM6 LGS Charles Sloan, 7 Clova Street, Thornliebank, Glasgow, G46 8NA
M6 LGT James Leggett, 15 Catalpa Way, Gorleston, Great Yarmouth, NR31 8LD
M6 LGV Leslie Vickers, 24 Hearnes Meadow, Seer Green, Beaconsfield, HP9 2YJ
MI6 LGX Charles Sheppard, 4 Fairview Drive, Whitehead, Carrickfergus, BT38 9NT
M6 LHA Andrew Crawford, 4 Trimpley Drive, Kidderminster, DY11 5LB
M6 LHB Stephen Hillman, 8 Brigantine Grove, Duffryn, Newport, NP10 8ET
M6 LHD Kathryn Newbould, 47 Old Barber, Harrogate, HG1 3DF
M6 LHE Lee Heppenstall, 15 Gibraltar Road, Hemswell Cliff, Gainsborough, DN21 5XJ
M6 LHF Michael Pickering, 30 Hotspur Avenue, Whitley Bay, NE25 8RP
M6 LHG Lloyd Hagan, 7 Betjeman Mews, [illegible]
M6 LHI Bradley Cummings, 6 Carshalton Way, Lower Earley, Reading, RG6 4EP
M6 LHJ Lawerance Smith, 17 Grove Street, Kirton Lindsey, DN21 4BY
M6 LHM Lee Mason, 2 Iris Close, Widnes, WA8 4GA
M6 LHN Lee Hulse, 202a Shooters Hill Road, London, SE3 8RP
M6 LHO Anthony Rowan, 14 Craven Lea, Liverpool, L12 0NF
M6 LHP Lesley Pass, 14a Elm Avenue, Hucknall, Nottingham, NG15 6GE
M6 LHR Lewis Rich, 39 Wren Close, Heathfield, TN21 8HG
M6 LHS Haotian Ren, 50 Highwoods Drive, Marlow, SL7 3PY
M6 LHV Gary McCarthy, 14 Cosedge Crescent, Croydon, CR0 4DN
MW6 LHW Ho Wa Lau, 40 Newton Road, Mumbles, Swansea, SA3 4BQ
M6 LIB Elizabeth Bayliss, 19 Rugby Road, Dunchurch, Rugby, CV22 6PG
M6 LID Jamie Eades, 128 Russells Hall Road, Dudley, DY1 2NN
M6 LIE Lorna Gregory, 14 Anderson Road, Hemswell Cliff, Gainsborough, DN21 5XP
M6 LIH Nicholas Hindl, 67 Yarmouth Road, Ellingham, Bungay, NR35 2PH
M6 LIK Thomas Russell, 38 Speedwell Close, Witham, CM8 2XL
MM6 LIL Lilaine Clark, 17 Lewis Rise, Broomlands, Irvine, KA11 1HH
M6 LIN Lynn Briggs, 20 Broad Lane, Pelsall, Walsall, WS4 1AP
M6 LIP Steven Shakespeare, 132 Chapel Street, Pensnett, Brierley Hill, DY5 4EQ
M6 LIS Melissa Ramsbottom, 1 Barn Gill Close, Blackburn, BB2 3HU
MI6 LIT Cameron Brush, 56 Larchwood, Banbridge, BT32 3UT
M6 LIV Jamie Hay, 13 Windsor Road, Workington, CA14 5BQ
MW6 LIW David Evans, 23 Wellington Street, Aberdare, CF44 8EW
M6 LIZ Elizabeth Martin, 62 Llwyn Ynn, Talybont, LL43 2AL
M6 LJA Martin Duxbury, 32 Radford Street, Darwen, BB3 2PB
M6 LJB Aaron Billingham, 6 Kemble Close, Lincoln, LN6 0NR
M6 LJD Lee Denham, 8 Kings Field, Bursledon, Southampton, SO31 8EN
MM6 LJE Louise Treble, 2 Baidland Meadow, Dalry, KA24 5HP
M6 LJG Laura Goldsmith, 7 Fengate Drove, Weeting, Brandon, IP27 0PW
M6 LJJ Lee Jones, Brimham Lodge Farm, Harrogate, HG3 3HE
M6 LJK Louis Kirkpatrick, The Cleave, Nine Oaks Estate, Yelverton, PL20 6ND
M6 LJM Laura Marriott, 94 Lyndhurst Road, Worthing, BN11 2DW
MI6 LJO Oisin Conaghan, 94 Curlyhill Road, Strabane, BT82 8LS
M6 LJP Lee Passam, Glanceiro, Llandre, Bow Street, SY24 5BS
M6 LJR Clayton Lonie Jr, 41 De la Hay Avenue, Plymouth, PL3 4HS
M6 LJS Lauren Smith, 177 Waterloo Road, Stoke-on-Trent, ST6 2ER
M6 LJT Marlon Rushton, 15 Oakdene Avenue, Accrington, BB5 6HP
M6 LKA Max Wilkinson, 3 Balsams Close, Hertford, SG13 8BN
M6 LKE Luke Huddart, 1 Rydal Court, Penrith, CA11 8PN
M6 LKF Leslie Chapman, Unit 1, The Crafty Goat, Mersea Road, Langenhoe, Colchester, CO5 7LG
M6 LKI Keith Lockstone, Oceana, The Parade, Pevensey, BN24 6LX
M6 LKL Georg Urban, 33 High Meadow, Hathern, Loughborough, LE12 5HW
M6 LKM Luke McDonnell, 108 Long Lane, Garston, Liverpool, L19 6PQ
M6 LKS Lawrence Spriggs, 19 Mackenzie Square, Stevenage, SG2 9TT
M6 LKT Leslie Kett, 52 Northgate, Hornsea, HU18 1EU
M6 LKY Anne Lee, 27 Victoria Avenue, Camberley, GU15 3HT
M6 LLB Liam Brigham, 42 Cayley Close, Clifton, York, YO30 5PT
M6 LLC Louis Clark, 1 Brooklime Road, Liverpool, L11 2YH
M6 LLD Anthony Clark, 1 Brooklime Road, Liverpool, L11 2YH
M6 LLE Ryan Clark, 1 Brooklime Road, Liverpool, L11 2YH
MI6 LLG Stephen Horner, 10 Meadow Court, Bushmills, BT57 8SD
M6 LLH Anthony Hoyte, 43 Orchard Drive, Mayland, Chelmsford, CM3 6EP
MI6 LLI Roy Hetherington, 112 Screeby Road, Fivemiletown, BT75 0LG
M6 LLL Leia Foulkes, 43 Mill Hayes Road, Stoke-on-Trent, ST6 4JB
M6 LLM Lee Meredith, 131 Trimdon Avenue, Middlesbrough, TS5 8RY
M6 LLO Stuart Hamilton, 16 Faringdon Avenue, Blackpool, FY4 3QQ
MI6 LLS Brian Kelly, 153 Ardanlee, Ballynagard, Londonderry, BT48 8RT
M6 LLU Jake Mottram, 196 Queens Drive, Nantwich, CW5 5LA
M6 LLW Hui Liu, Harrogate Ladies' College, Clarence Drive, Harrogate, HG1 2QG
MI6 LLZ Karl Dorman, 25 Blackthorn Road, Newtownabbey, BT37 0GH
M6 LMB Louise Bate, 87 Dunsheath, Telford, TF3 2BY
M6 LMC Lewis Chatt, Rosemary Cottage, Causeway End Road, Dunmow, CM6 3LU
M6 LMG John Porter, 3 The Walks, Main Road, Woodbridge, IP12 3DZ
MM6 LMH Leslie Mitchell Hynd, Smithy House Bruichladdich, Isle of Islay, PA49 7UN
M6 LMI Lubomir Mikolka, Flat, 1 Scotney Court, Romney Marsh, TN29 9JP
M6 LMJ Luc Mathlin, 29 Wagtail Drive, Stowmarket, IP14 5GH
M6 LMT Gary Wilson, 29 Mill Lane, London, NW6 1NT
M6 LMW Alex Thomson, Shire Jee Neevas, Cold Ash Hill, Thatcham, RG18 9PH
M6 LNC David Lancaster, Linkhill View, Frith Common, Eardiston, Tenbury Wells, WR15 8JX
M6 LND Laurence Dawson, 5 Harbour View, Roker, Sunderland, SR6 0NL
M6 LNE Conner Bruce, Nunfield House, Bull Lane, Sittingbourne, ME9 7SL
M6 LNK William Youd, Keepers Cottage, Woodmanton, Exeter, EX5 1HG
M6 LN[illegible] David Williams, Apartment 64, 7 Tiltman Place, London, N7 7EL
M6 LNX Stephen Wareham, 8 Simon Road, Longlevens, Gloucester, GL2 0ER
M6 LOB Andrew Brash, 44 Broadway East, Chester, CH2 2DP
M6 LOC Leon Barker, 88 Cecilia Road, London, E8 2ET
M6 LOD Lucy Broadhurst, 4 Lilburne Drive, Newport, NP19 0ET
M6 LOE Nicola Chaplin, 5 Maxwell Street, Bury, BL9 7QA
M6 LOF Simon McIlwaine, 70 Autumn Drive, Sutton, SM2 5BA
M6 LOG Ian Van Der Linde, 77 Port Vale, Hertford, SG14 3AF
M6 LOK Aidan Clayton, 7 Church Terrace, Reading, RG1 6AS
M6 LOL Ryan Cooper, 11a Ambleside, Gamston, Nottingham, NG2 6NA
MM6 LON Lorna Nicoll, 15 Redford Walk, Edinburgh, EH13 0AF
M6 LOS John Hawthorn, 1 Tudor Close, Leigh-on-Sea, SS9 5AR
MI6 LOT Leo Treanor, 1 Granemore Park, Keady, Armagh, BT60 2GP
M6 LOW James Lowenthal, 133 Marshalswick Lane, St. Albans, AL1 4UX
M6 LOZ Lawrence Shaw, 21 Dyke Street, Stoke-on-Trent, ST1 2DF
M6 LPB Nigel Petit-Brown, 6 Cope Avenue, Nantwich, CW5 5JE
M6 LPD Darren Lester, 171 Glenavon Road, Birmingham, B14 5BT
M6 LPF Arihant Kuba, Flat 291-295, Jellicoe Court, Southampton, SO16 3UJ
M6 LPI Neil Stone, 43 Common View, Stedham, Midhurst, GU29 0NX
M6 LPK Leon Kiddell, 1 Sparham Hill, Sparham, Norwich, NR9 5QT

**IMPORTANT NOTE**

**Revalidate licence to avoid revocation** – Ofcom has advised the Society that plans will be drawn up to revoke licences that have not been revalidated as required by the licence conditions. The quickest way to revalidate is to do so online via the Ofcom website: *https://services.ofcom.org.uk/* or by email: *amateur.validations@ofcom.org.uk* Ofcom staff are available to help, but please be patient during times of heavy workload.

M6 LPN Jacob Leach, 9 Crown Point Drive, Ossett, WF5 8RQ

M6 LPO Robert Cowperthwaite, 30 Glover Place, Bootle, L20 4QR

M6 LPP peter petersen, 15 Kent Gardens, Birchington, CT7 9RS

M6 LPS Daniel Clement, 24 Millais, Horsham, RH13 6BS

MM6 LPT Julie McKinnon, 8 Rowanlea Avenue, Paisley, PA2 0RP

M6 LPW Louis Walker, 55 Silverlands Road, St. Leonards-on-Sea, TN37 7DF

M6 LQO Judith Thompson, 32 Coult Avenue, North Hykeham, Lincoln, LN6 9RG

M6 LRA Lee Akred, 25 Kitchener Street, Walney, Barrow-in-Furness, LA14 3QW

M6 LRB Daniel Williamson, 4 King Edward Road, Northampton, NN1 5LU

M6 LRF Robert Deller, Four Winds Farm, Buckworth Road, Huntingdon, PE28 4JX

M6 LRG Michael Parker, Ridgeways, Mill Common, Halesworth, IP19 8RQ

M6 LRH Liam Hancock, 106 Hoyle Street, Warrington, WA5 0LW

M6 LRK Alisdair Lark, 20 Lawfield, Coldingham, Eyemouth, TD14 5PB

M6 LRL Simon Vane, 17 Knights Walk, Abridge, Romford, RM4 1DR

MI6 LRN Kevin Bell, 3 Alexandra Crescent, Larne, BT40 1NE

M6 LRO Owain Thomas, Garth Celyn, St. Davids Road, Aberystwyth, SY23 1EU

M6 LRU Robert Walker, 125 Devereux Road, West Bromwich, B70 6RQ

M6 LRV Heather Moore, 52 Limefield Street, Accrington, BB5 2AF

M6 LRW Jonathan Welch, 49 Walshs Manor, Stantonbury, Milton Keynes, MK14 6BU

MM6 LRX Grant Morrison, 11 Goodman Place, Maddiston, FK2 0NB

M6 LSA Lewis Allcock, 26 Castleton Grove, Inkersall, Chesterfield, S43 3HU

M6 LSB Lance Catterall, 14 Dunham Drive, Whittle-le-Woods, Chorley, PR6 7DN

M6 LSE Craig Stoten, 12 Boyd Avenue, Dereham, NR19 1LU

M6 LSG Louis Spong, 2 Strathmore Drive, Charvil, Reading, RG10 9QT

M6 LSH Lyndon Shaw, 47 Beechfields, Eccleston, Chorley, PR7 5RF

M6 LSJ Lionel Sawkins, 20 Nye Close, Cheddar, BS27 3PB

M6 LSK Carl Hare, Flat 6, Elswyn House, 64 Hatherley Road, Sidcup, DA14 4AW

M6 LSP Leah Phillips, 5 Barnes Green, Wirral, CH63 9LU

M6 LST Dorothy Lui, Harrogate Ladies' College, Clarence Drive, Harrogate, HG1 2QG

M6 LSV Sergejs Ludziss, 82 Trinity Avenue, Mildenhall, Bury St. Edmunds, IP28 7LS

MW6 LSW Derrick Johns, 23 Holly Road, Llanharry, Pontyclun, CF72 9JB

MI6 LSY Louis Stock, 15 Mahon Drive, Portadown, Craigavon, BT62 3JB

M6 LTB Liam Burke, teviot, malthouse lane, Peasmarsh, TN31 6TA

M6 LTC Tracy-Anne Craig, Cemetery Lodge, Lochmaben, Lockerbie, DG11 1RL

M6 LTD Paul Asher, 124 Bath Street, Market Harborough, LE16 9JL

M6 LTL Liam Layland, 3 Thirlmere Road, Golborne, Warrington, WA3 3HH

M6 LTM Lauren Simons, 123 Main Street, Little Harrowden, Wellingborough, NN9 5BA

M6 LTO Leslie Trend, 140 Ardleigh, Basildon, SS16 5RW

MW6 LTR Christopher Rowe, 21 Graig Terrace, Abercwmboi, Aberdare, CF44 6AH

M6 LTS Laurence Stant, 3 Uffa Fox Place, Cowes, PO31 7NX

M6 LTU Mantas Brazinskas, 25 Elswick Road, London, SE13 7SP

M6 LUA Tania Goddard, 217 Speedwell Road, Bristol, BS5 7SP

M6 LUC Dion Boden, 249 Nottingham Road, Ilkeston, DE7 5AT

M6 LUD Keith Willson, Ludpit Cottage, Ludpit Lane, Etchingham, TN19 7DB

M6 LUG Peter Schoenmaker, 24 Greenheys Drive, London, E18 2HB

M6 LUI Gary Conboy, 9 Hart Street, Droylsden, Manchester, M43 7AN

M6 LUK Luke Johnson, 7 Southover Way, Hunston, Chichester, PO20 1NY

M6 LUM Lukasz Medza, 16 Huntroyde Avenue, Bolton, BL2 2ET

M6 LUT Andrew Lutley, Springfield, Rookery Hill, Ashtead, KT21 1HY

M6 LUZ Gordon Luscombe, 28 St. Giles Gate, Doncaster, DN5 8PQ

M6 LVA Rachel Scullion, 41 Myrica Grove, Hoole, Chester, CH2 3EW

M6 LVC Iain Collins, 19 Peel Street, Kidderminster, DY11 6UG

M6 LVE Cheryl Johnson, 25 Pelham Street, Worksop, S80 2TW

M6 LVJ Ian Humphries, 13 Malvern Close, Banbury, OX16 9EL

M6 LVK Kevin Jones, 7 Fazan Court, Wadhurst, TN5 6BT

MM6 LVV Richard Holtom, Old Post Office, Church Road, Laurencekirk, AB30 1YS

M6 LVW Laura Walker, 17 Carr House Road, Halifax, HX3 7QY

M6 LVX Tarek Von Bergmann, 110 High Street, Blunsdon, Swindon, SN26 7AB

M6 LWA Lewis Alderson, 62Rusland Park, 62, Kendal, LA9 6AJ

MM6 LWB Leslie Bradley, Amon Sul, Kiltarlity, Beauly, IV4 7HT

M6 LWF Leslie Fish, Iddon Cottage, Bronygarth, Oswestry, SY10 7NF

M6 LWJ Louis Webb, Fern Bank, Wood Lea, Rotherham, S66 8NN

M6 LWM Colin Smithen, 10 High Street, Temple Ewell, Dover, CT16 3DU

M6 LWP Alex Lawler, 6 Woodpecker Lane, Cringleford, Norwich, NR4 7LS

M6 LWR Stephen Ross, 51 Claypiece Road, Bristol, BS13 9DR

M6 LWS William Sawyer, 20 Park Terrace, Willington, Crook, DL15 0QL

M6 LWT Martin Hennessey, 57 Northern Road, Aylesbury, HP199QT

M6 LXC Lee Clark, 30 Warwick Square, London, SW1V 2AD

M6 LXE Alex Elena, 19 Ashburton Gardens, Bournemouth, BH10 4HP

M6 LXH Sarah Li, Harrogate Ladies' College, Clarence Drive, Harrogate, HG1 2QG

M6 LXM Andrew Brighton, 67 Wilks Farm Drive, Sprowston, Norwich, NR7 8RG

M6 LXP Michael Ling, Flat 14, Rowan Court, London, SW20 0BA

M6 LXR Lee Rhodes, 14 Iris Crescent, Bexleyheath, DA7 5QD

MD6 LXW Chris Stickley, 152 Fore Street, Pinner, HA5 2NE

M6 LXX Alexander Erlank, 28 Ashenden Road, Guildford, GU2 7XE

M6 LXY Mark Oxley, 49 Dalton Crescent, Shildon, DL4 2LE

M6 LYA Aleksejs Polakovs, 76 Sandringham Crescent, Leeds, LS17 8DF

M6 LYD Ronald Lyddall, 102 Chapel Road, Brightlingsea, Colchester, CO7 0HE

M6 LYN Linda Groves, 3 Hudson Davies Close, Pilley, Lymington, SO41 5PA

M6 LYO Oliver Lyon, Splinters, Nelson Park Road, Dover, CT15 6HL

M6 LYP Marius Jonusas, 70 Methuen Street, Southampton, SO14 6FR

M6 LYS Alicia Booth, 27 Shenstone Road, Rotherham, S65 2JR

M6 LYY Andrew Allgood, 39 Eastwood, Chatteris, PE16 6XW

M6 LZM Martin Radulov, 60 St. Marks Avenue, Northfleet, Gravesend, DA11 9LW

M6 LZP John Lovelock, Sea Spray, The Lizard, Helston, TR12 7NU

M6 LZT Carl Plant, 6 Leadbeater Avenue, Stoke on Trent, ST4 5HE

M6 LZX Brian Siddle, 7 Farebrother Street, Grimsby, DN32 0NH

M6 LZY Christopher Hillcox, 2 New Hall Drive, Sutton Coldfield, B75 7UU

M6 MAA Maurice Meadowcroft, 210 Dickinson Close, Blackburn, BB2 2LT

M6 MAD Devon Binnall, 21 Appletree Road, Featherstone, Pontefract, WF7 5EA

M6 MAG Clive Lavery, 2 Barley Mow Cottages, Malting Lane, Woodbridge, IP13 6TE

M6 MAH Mark Hyett, 1 Darell Close, Quedgeley, Gloucester, GL2 4YR

M6 MAJ Michael Kealey, 24 Ben Nevis Road, Birkenhead, CH42 6QY

M6 MAK Paul McGrath, 24 Broadoak Drive, Lanchester, Durham, DH7 0QA

M6 MAL Malcolm Wallace, 4 Windmill Court, Edmund Street, Kettering, NN16 0HU

M6 MAM Michael McDougall, 122 Lee Lane, Horwich, Bolton, BL6 7AF

M6 MAO Iain Connors, 3 Wheatfield Way, Chelmsford, CM1 2QZ

M6 MAP Mark Peters, 25 Windsor Court, Falmouth, TR11 3DZ

M6 MAS Samantha Shailes, 9 Ingham Street, Padiham, Burnley, BB12 8DR

M6 MAT Matthew Pye, 2 Kingsley Street, Nelson, BB9 8SA

M6 MAW Max Ansell-Wood, Sanju, Old Lane, Nethertown, Bradford, BD11 1LU

M6 MAX M Trivett, 36 Edward Street, Hartshorne, Swadlincote, DE11 7HG

M6 MAY Michael Buist, 23 St. Chads Drive, Gravesend, DA12 4EL

M6 MBB Matthew Bennett-Blacklock, Theatre View Apartments, 19 Short Street, London, SE1 8LJ

M6 MBC Stephen Cook, 114 Caerphilly Road, Bassaleg, Newport, NP10 8LJ

M6 MBE Barry Adby, 26 Love Lane, Watlington, OX49 5RA

M6 MBF Mark Bailey, 34 Jephson Drive, Birmingham, B26 2HW

M6 MBG Christopher Collins, The Coppice, Old Coach Road, Sheffield, S6 6HX

M6 MBH Mark Creedy, 25 Ryton Close, Redditch, B98 0EW

M6 MBI Martin Collis, 35 Fishergate, Norwich, NR3 1SE

M6 MBK Jason Seaman, 6 Gibbets, Hale Road, Thetford, IP25 7QX

M6 MBO Michael Siddall, 10 Foston Drive, Chesterfield, S40 4SJ

M6 MBP Neil Challis, 48 Brunsfield Close, Wirral, CH46 6HE

M6 MBQ Matthew Ball, 22 Wheatley Drive, Mirfield, WF14 8NW

M6 MBR Michael Burr, 49 Knightsbridge Way, Hemel Hempstead, HP2 5ES

M6 MBS Michael Smith, 78 New Croft, Weedon, Northampton, NN7 4RL

M6 MBU Michael Burnett, 218 High Street, Clapham, Bedford, MK41 6BS

M6 MBX Malcolm Bell, 68 Hereford Drive, Bootle, L30 1PR

M6 MBY Matthew Cox, 120 Helmsdale, Bracknell, RG12 0TB

MW6 MBZ Matthew Argyle, 17 Heol Cae-Rhys, Cardiff, CF14 6AN

MM6 MCA James McArdle, 1 Queen Street, Hamilton, ML3 9JR

M6 MCB Mark Cooper, 6 The Crescent, Cookley, Kidderminster, DY10 3RY

M6 MCC Susan Allaker, 61 West Street, Winterton, Scunthorpe, DN15 9QG

M6 MCE Graeme McEwen, 37 Malvern Way, Twyford, Reading, RG10 9PY

MI6 MCF Jonathan MacFarlane, 1 Main Street, Uttony, Magheraveely, Enniskillen, BT92 6NB

M6 MCH Michael Hill, 10 The Moorings, Littlehampton, BN17 6RG

M6 MCJ Michael Coiley, 25 Spring Garden Street, Queensbury, Bradford, BD13 2AE

MI6 MCK Stephen McKay, 27 Rathbeg Crescent, Limavady, BT49 0AT

M6 MCL Michael Chaffey, 46 Bartlett Way, Poole, BH12 4FD

M6 MCM Stuart McMurtrie, 5 Hill Road, Carshalton, SM5 3RA

M6 MCO Mark Denham, 2 Shorts Corner, Frithville, Boston, PE22 7EA

M6 MCP Marie-Claire Pennington, Brede Court, Brede, Rye, TN31 6EJ

M6 MCR Kara Wills, 24 Bitten Court, Northampton, NN3 8HH

M6 MCS Michael Statham, Broad Oak Bungalow, Manston, Sturminster Newton, DT10 1EZ

MM6 MCT John Leitch, 25 Lime Street, Grangemouth, FK3 8LZ

M6 MCU Michael Barker, 18 Nickleby Road, Waterlooville, PO8 0RH

M6 MCW Michael Wilson, 11a St. Julians Road, London, NW6 7LA

MM6 MCX Barclay Bannister, 12 Dalnottar Terrace, Old Kilpatrick, Glasgow, G60 5DE

M6 MCY Michael Attree, 52 The Ridgeway, St. Albans, AL4 9PS

M6 MCZ Alan Davis, Old Malt Kiln House, Barden, Leyburn, DL8 5JS

M6 MDB Tom Brown, 19 Clover Crescent, Oldham, OL8 2EZ

M6 MDC Denis Smith, 193 Brooke Road, Oakham, LE15 6HQ

M6 MDG David Green, 89 Standhill Crescent, Barnsley, S71 1SS

M6 MDH Mark Hubbard, 14 Parkfield Crescent, Kimpton, Hitchin, SG4 8EQ

M6 MDJ Darren Jefferson, 74 Cloisters Avenue, Barrow-in-Furness, LA13 0BB

M6 MDL Matthew Luttrell, 5 Swallow Drive, Bury, BL9 6JQ

M6 MDM Steven Clarke, 27 Netherhouse Moor, Church Crookham, Fleet, GU51 5TZ

M6 MDN Michael Norman, 28 Cumberland Close, Twickenham, TW1 1RS

M6 MDR Michael Riley, 16 Dudley Avenue, Leicester, LE5 2EE

M6 MDT Matt Taylor, 47 Whisperwood Drive, Balby, Doncaster, DN4 8SB

M6 MDU Mark Duchar, 4 Miller Gardens, Pelton Fell, Chester le Street, DH2 2NX

M6 MDX Mark Abraham, 12 Graham Road, Halesowen, B62 8LJ

M6 MDZ Matthew Smith, 31 Atlantic Crescent, Sheffield, S8 7FW

M6 MEA Mark Atfield, 42 Pauls Croft, Cricklade, Swindon, SN6 6AJ

M6 MEB Paul Shires, 30 Philip Garth, Wakefield, WF1 2LS

M6 MEC Michael Carroll, 11 Old Hall Court, Old Hall Street, Malpas, SY14 8NE

MD6 MED Richard Muswell, 7 Stoneyfields Gardens, Edgware, HA8 9SP

M6 MEJ Michael Bray, 26 South Park Close, Redruth, TR15 3AR

M6 MEK Andrew Walters, 28 St. Giles Close, Retford, DN22 7XA

M6 MEL Mark Lewis, 73 Addenbrooke Street, Wednesbury, WS108HJ

M6 MEN Mark Lucas, Whitegates, Chatteris Road, Huntingdon, PE28 2UQ

M6 MEO Barry Clements, 61 Marcus Avenue, Southend-on-Sea, SS1 3LE

M6 MEP Martin Lawton, 20 Wharfedale Walk, Stoke-on-Trent, ST3 2RS

M6 MEQ Andrew Riley, 35 Ross Avenue, Wirral, CH46 2SA

M6 MES James Reeve, 5 Antelope Avenue, Grays, RM16 6QT

M6 MEU Agnes Sharif, 10 The Boundary, Seaford, BN25 1DG

M6 MEV Michael Sanderson, 20 East View, Castleford, WF10 1PZ

MM6 MFA Magnus Henry, Selkie Stanes, Scatness, Shetland, ZE3 9JW

M6 MFB James Murray, 17 Bro Dawel, Bodedern, Holyhead, LL65 3TB

M6 MFC Christopher Parkes, 3 Greenham Close, Middlesbrough, TS3 9NT

M6 MFD Graham Mansfield, 2 School Street, Syston, Leicester, LE7 1HN

M6 MFF Mike File, Flat 1, 4 Priory Courtyard, Ramsgate, CT11 9PW

M6 MFG Marcos Gainza, Stanhope, High Street, Saxmundham, IP17 3EP

M6 MFJ Michael Coleman, 3 Tummon Road, Sheffield, S2 5FD

M6 MFK Amanda Powell, 37 Newnham Close, Mildenhall, Bury St. Edmunds, IP28 7PD

M6 MFL Matthew Ross, 52 Hermitage Street, Rishton, Blackburn, BB1 4NL

M6 MFM Mamatha Maheshwarappa, R43 Room 2 International House, University Of Surrey, Guildford, GU2 7JL

M6 MFN Aaron Evans, Maesyronnen, Sarnau, Llanymynech, SY22 6QL

M6 MFO Michael Foster, 21 The Bourtons, Newton Road, Totnes, TQ9 6LS

MI6 MFR Adam Morrow, 769 Farranseer Park, Macosquin, Coleraine, BT51 4NB

M6 MFS Sean Finlayson, 41 Low Catton Road, Stamford Bridge, York, YO41 1DZ

M6 MFU Keith Ledson, 202 Brodick Drive, Bolton, BL2 6UE

M6 MFV Darren Baker, 39 Taylor Road, Wallington, SM6 0AZ

M6 MFZ Martin Fitzgerald, Flat 35, Winterton House, London, E1 2QR

M6 MGA Nigel Valvona, 63 Vale Road, Ash Vale, Aldershot, GU12 5HR

M6 MGC Mark Carwardine, Buttington Lodge, Sedbury, Chepstow, NP16 7EX

M6 MGD Lynda Addison, 45 Fir Terraces, Esh Winning, DH7 9JQ

M6 MGF Keith Holman, 39 Trellech Court, Yeovil, BA21 3TE

M6 MGG Roger Wenlock, 8 Dinchope Drive, Telford, TF3 2ES

M6 MGH Mark Hopewell, 4 Cotes Crescent, Bicton Heath, Shrewsbury, SY3 5AS

M6 MGJ Barry Hardy, 10 Spring Farm Road, Burton-on-Trent, DE15 9BN

M6 MGM Mark Margetts, Central House, Llanfechain, SY22 6UJ

M6 MGN Norman Cook, 210 Cemetery Road, Wath-upon-Dearne, Rotherham, S63 6HZ

M6 MGO Mark Wenlock, 3 Kennels Cottages, Hall Lane, Stone, ST15 0RD

MD6 MGP Alan Breen, 1 Snugborough Close, Union Mills, Isle of Man, IM4 4NZ

M6 MGQ Paul Loader, 201 Paddock Road, Basingstoke, RG22 6QG

M6 MGR Malcolm Reeks, 33 Madresfield Village, Madresfield, Malvern, WR13 5AA

M6 MGT John Dickenson, 2 Kirkleys Avenue North, Spondon, Derby, DE21 7FX

M6 MGU Marcus Golding, 11 Southwold Crescent, Broughton, Milton Keynes, MK10 7BW

M6 MGV Michael Pike, 21 Watersmeet Close, Guildford, GU4 7NQ

M6 MGW Mark Walker, 50 College Grove Road, Wakefield, WF1 3RL

M6 MGX Michael Gillard, 66 West End Road, Bradninch, Exeter, EX5 4QP

M6 MGY Mark Gray, 15 The Circle, Cwmbran, NP44 7JP

MW6 MGZ Mark Bannister, 45 Queens Drive, Llantwit Fardre, Pontypridd, CF38 2NT

MD6 MHA Mohini Hersom, 26 Bourne Court, Station Approach, Ruislip, HA4 6SW

M6 MHD Matthew Jodrell, 2 Charlesworth Street, Crewe, CW1 4DE

M6 MHE Todd Harvey, 12 Woodkirk Avenue, Tingley, Wakefield, WF3 1JL

MI6 MHI Aiden Menzies, 21 Woodview Park, Tandragee, Craigavon, BT62 2DD

M6 MHJ Max Jackson, 64 Main Road, Moulton, Northwich, CW9 8PB

MI6 MHK Chloe Moonie, 21 Woodview Park, Tandragee, Craigavon, BT62 2DD

M6 MHL Martyn Lacey, 82 Bowerings Road, Bridgwater, TA6 6HF

M6 MHM Maymun Hashim, 1 Cheylesmore Drive, Frimley, Camberley, GU16 9BL

MM6 MHN John Mulhern, 10 Fisher Court, Knockentiber, Kilmarnock, KA2 0DS

M6 MHO Michael Hossell, 80 Murray Road, Sheffield, S11 7GG

M6 MHQ Michael Rea, 15 Wensleydale Close, Royton, Oldham, OL2 5TQ

M6 MHU Michael Humphries, 5 Coppice Mead, Stotfold, Hitchin, SG5 4JX

M6 MHV Michael Clarke, 54 Stafford Grove, Shenley Church End, Milton Keynes, MK5 6AZ

M6 MHW Mark Hall, 20 Diamond Drive, Oakwood, Derby, DE21 2JP

M6 MHY Martin Williams, 37 Clarendon Road, Weston-Super-Mare, BS23 3EE

M6 MIA Mark Andrews, 286 Huddersfield Road, Mirfield, WF14 9PY

MI6 MIB Phillip Cobain, 53 Oakland Avenue, Belfast, BT4 3BW

M6 MIC Michael Taylor, 24 Crowley Lane, Oldham, OL4 2PN

M6 MID Ilan Shiradski, 69 Masefield Avenue, Borehamwood, WD6 2HG

M6 MIE M Johnson, Chy-an-Gwelva, Foundry, Truro, TR3 7BU

M6 MIF Michelle Vaughan, 22 Arundel Close, Tuffley, Gloucester, GL4 0TW

M6 MIG Michael Greenwood, 8 Coltsfoot Court, Killinghall, Harrogate, HG3 2WW

MI6 MIH Jake Mercer, 32 Templemore Avenue, Belfast, BT5 4FT

M6 MII Matthew Bell, 2 Fox Close, Dunton, Biggleswade, SG18 8RF

MD6 MIK Michael Harris, 8 York Avenue, Stanmore, HA7 2HS

M6 MIL Jamie Milbourne, 102 Bells Marsh Road, Gorleston, Great Yarmouth, NR31 6PR

M6 MIN Dana Mitchell, Flat 2, Weavers Court, 51 Unwin Street, Sheffield, S36 6EH

M6 MIO Frankie Miocinovic, 22 Malvern Grove, Northampton, NN5 6AY

M6 MIP Michael Payne, 14 Linnell Road, Rugby, CV21 4AN

MW6 MIQ Vincent McKendley, 9 Mary Street, Aberdare, CF44 7NF

M6 MIR Henna Mir, 13-15 Wain Street, Stoke-on-Trent, ST6 4ES

MM6 MIS Jonathan Marsh, 8 Hazelton Way, Broughty Ferry, Dundee, DD5 3BT

M6 MIT Mitchell Tarling, 16 Cross Walk, Bristol, BS14 0RX

M6 MIU John Marsh, 14 Eyam Road, Hazel Grove, Stockport, SK7 6HP

M6 MIV Brian Davies, 60 Queensway, Blackburn, BB2 4QT

M6 MIY Paul Billingham, 393 Landseer Road, Ipswich, IP3 9LT

M6 MIZ Mitosz Kwiatkowski-Zelazny, 56 York Road, Hove, BN3 1DL

M6 MJA Matthew Austwick, 6 Worlaby Road, Grimsby, DN33 3JY

MM6 MJC Marcus Clifford, Bridgeton Castle, St. Cyrus, Montrose, DD10 0DN

M6 MJD Paul Smith, 67 Gipsy Lane, Old Whittington, Chesterfield, S41 9JD

M6 MJF Michael Fysh, 3 Jeffrey Close, Kings Lynn, PE30 2HX

MM6 MJG Martin Gilbert, Avonlea, Dounby, Orkney, KW17 2JA

M6 MJI Mark Greensmith, 14 Fountain Road, Draycott-in-the-Clay, Ashbourne, DE6 5HP

M6 MJL Michael Lawrance, 17 Wren Crescent, Scartho Top, Grimsby, DN33 3RA

M6 MJM Michael McCormack, Flat 13, 29 Stoneygate Road, Leicester, LE2 2AE

M6 MJN Matthew Neale, 41 Langford Road, Weston-Super-Mare, BS23 3PQ

M6 MJO Mark O'Loughlin, 2 Hen Ysgol, Forge Road, Crickhowell, NP8 1LU

M6 MJP Stan Parker, 36 Eton Close, Lincoln, LN6 0YF

MM6 MJR M Robertson, Tigh Jenny, Strath, Gairloch, IV21 2BX

M6 MJS Stephen Shields, 9 Berrington Drive, Newcastle upon Tyne, NE5 4BG

M6 MJV David Morley, The Old Mill, Mill Lane, Loughborough, LE12 7UX

MM6 MJY Martin Yarrow, Lomond Villa, Downies Village, Aberdeen, AB12 4QX

M6 MJZ Michael Shepley, 16 Heulwen Close, Hope, Wrexham, LL12 9PR

M6 MKB Michael Buchanan, 36 Church Lane, Manby, Louth, LN11 8HL

M6 MKD Dorian Bell-Stephens, 33 Alcock Crest, Warminster, BA12 8NE

M6 MKE Michael Gregory, 65 Nursery Crescent, North Anston, Sheffield, S25 4BR

M6 MKF Christopher McNaughton, 20 Victoria Avenue, Stockton-on-Tees, TS20 2QB

M6 MKH Andy Freeth, 33 Argus Close, Sutton Coldfield, B76 2TG

M6 MKJ Krzysztof Juszczak, 68 College Road, Sandy, SG19 1RH

M6 MKK Matthew Kendall, 53 Ellerker Rise, Willerby, Hull, HU10 6EU

M6 MKM James Mckie, 59 Leaholme Terrace, Blackhall Colliery, Hartlepool, TS27 4AB

M6 MKN Nigel Driscoll, 42 Adelaide Square, Shoreham-by-Sea, BN43 6LN

M6 MKO Mark Kent, 7 Lockyers Drive, Ferndown, BH22 8AJ

M6 MKU Dennis McDonald, 29 Highfield Crescent, Rayleigh, SS6 8JP

M6 MKV Michael Vardy, 60 Hucklow Avenue, North Wingfield, Chesterfield, S42 5PU

M6 MKW Mitchell Wharton, Sea Cadets, Riverside Road, Great Yarmouth, NR31 6PX

M6 MKX Stephen McGuckian, 71 Heathfield Drive, Tyldesley, Manchester, M29 8PJ

M6 MKY Matthew King, Flat 6, Derwent Court, Solihull, B92 7BU
M6 MLA Mark Lovatt, 3 Withington Close, Atherton, Manchester, M46 0EZ
M6 MLE Derek Pilkington, 107 Saltings Road, Snodland, ME6 5HP
M6 MLF Jeanette Kiely, 35 Chestnut Avenue, Todmorden, OL14 5PH
M6 MLG Peter Arnold, 25 Alliston Drive, Woodville, Swadlincote, DE11 8FS
M6 MLH Michael Hoyland, 3 Telford Street, Barrow-in-Furness, LA14 2ER
M6 MLI Roderick Parker, 58 Bryncastell, Bow Street, SY24 5DF
M6 MLK Mary-Jane Lake, 64 Womersley Road, Norwich, NR1 4QB
M6 MLL Danny Neumunn, 92 Miner Street, Walsall, WS2 8QL
M6 MLM Michael Milano, 35 Orion Road, Rochester, ME1 2UL
M6 MLN Andrew McDermid, 49 Jubilee Street, Irthlingborough, Wellingborough, NN9 5RL
M6 MLO Michael Broyd, 93 Normandy Way, Plymouth, PL5 1NN
M6 MLP Mark Le-Petit, 35 Ellis Avenue, Stevenage, SG1 3SL
M6 MLQ Michael Byard, 1 Fieldside, Long Wittenham, Abingdon, OX14 4QB
M6 MLR Lee Rolt, 11 Noble Hop Way, Halifax, HX2 0SN
MM6 MLT Paul Connon, 4 Highfield Court, Stonehaven, AB39 2PL
M6 MLU Paul Rath, 60 Elstree Road, Bushey Heath, WD23 4GL
M6 MLV David Malcolm Malcolm, 66 Bracken Bank Grove, Keighley, BD22 7AU
M6 MLX Matthew Pacitti-Lamb, 41 Cowell Grove, Highfield, Rowlands Gill, NE39 2JQ
M6 MLY Malcolm Livesey, 24 St. Marys Road, Bamber Bridge, Preston, PR5 6TD
M6 MMB Michael Parkes, 2 Woodhouse Mount, Normanton, WF6 1BN
M6 MMC Malcolm McIntyre, 18 Norlands Crescent, Chislehurst, BR7 5RN
M6 MMF James Hobson, 5 Maes Briallen, Llandudno, LL30 1JJ
MM6 MMG Donald Anderson, Dail Darach, Monydrain Road, Lochgilphead, PA31 8LG
M6 MMH Michael Houghton, 18 Leopold Way, Blackburn, BB2 3UE
M6 MMI Ryan Hewson, Tad-Cu, Almond Avenue, Llandrindod Wells, LD1 6DH
MM6 MML Jane Lucas, 7 Rysland Avenue, Newton Mearns, Glasgow, G77 6EA
M6 MMM Michael Hunter, 126 Turner Street, Stoke-on-Trent, ST1 2NE
M6 MMN Michael Newbury, 2 Rowan Close, Clacton-on-Sea, CO15 2DB
M6 MMP Molly Chu, Harrogate Ladies' College, Clarence Drive, Harrogate, HG1 2QG
M6 MMQ Manmeet Majhail, 3 Poynders Hill, Hemel Hempstead, HP2 4PQ
M6 MMR Stephanie Wellsted, 127 Goldthorn Hill, Wolverhampton, WV2 4PS
M6 MMS Martin Strange, 101 Southbroom Road, Devizes, SN10 1LY
MI6 MMT Michael Torley, 4 Yew Tree Park, Newry, BT34 2QP
MW6 MMU Gareth Edwards, 54 Old Street, Tonypandy, CF40 2AF
M6 MMX Andrew Iggulden, 78 Wrensfield Road, Stockton-on-Tees, TS19 0BD
M6 MMY Michael Barber, 3 Baxter Road, Sunderland, SR5 4LH
M6 MNC Ashleigh Cockburn, 20 Hexham Avenue, Hebburn, NE31 2HN
M6 MND David Rogers, 21 Belmont Park, Pensilva, Liskeard, PL14 5QT
MM6 MNE Colin MacNee, 7 Church Street, Chapelton, Strathaven, ML10 6SD
M6 MNG Neal Giuliano, 13 Walton Drive, Derby, DE23 1GN
M6 MNH Martin Hunt, 37 Shortlands Avenue, Ongar, CM5 0BL
M6 MNI Mark Walton, 38 Wingate Road, Grimsby, DN37 9DU
M6 MNK Peter Roberts, 17 Cannon Hill, Prenton, CH43 4XR
MI6 MNL James Johnston, 19 Killowen Grange, Lisburn, BT28 3HQ
M6 MNO Adrian Hood, 109 Trotters Field, Braintree, CM7 3NW
M6 MNP Michael Killoran, 59 Mount Pleasant Street, Pudsey, LS28 7AY
M6 MNQ Robert Boan, 6 Philip Avenue, Newton Stewart, DG8 6HF
M6 MNS Timothy Hodson, 25 The Rise, Amersham, HP7 9AG
M6 MNT Stephen Hamer, Flat 5, 19 Frimley Road, Camberley, GU15 3EN
M6 MNU Jack Williams, 1 Lower Meadow Drive, Congleton, CW12 4UX
M6 MNV Frederick Brunt, 74 Bardley Crescent, Tarbock Green, Prescot, L35 1RJ
MM6 MNW David Ryan, 3 Barra Place, Stevenston, KA20 3BF
M6 MNX Mark Norfolk, 185 Heath Road, Leighton Buzzard, LU7 3AD
M6 MNZ Michael Bartlett, 34 Yarrow Drive, Birmingham, B38 9QR
M6 MOB Peter Hawes, 6 Robert Street, Sunderland, SR4 6EY
M6 MOC Thomas Forss, Lower Conghurst Oast, Conghurst Lane, Cranbrook, TN18 4RW
M6 MOD Michael O'Driscoll, 17 Petherton Gardens, Bristol, BS14 9BT
M6 MOF Caine Moffitt, 5 Foxton Terrace, Horstead Avenue, Brigg, DN20 8QR
M6 MOG Nicola Saville, 4 Shannon Court, Downs Barn, Milton Keynes, MK14 7PP
M6 MOH Martin Hinds, 10 Lustrells Close, Saltdean, Brighton, BN2 8AS
MI6 MOI Sharon Lewis, 15 Foyle Drive, Ballykelly, Limavady, BT49 9PG
M6 MOJ Anthony Mckie, 5 Greenway, Northenden, Manchester, M22 4LW
M6 MOK Mark Lowin, 1a Burnside Avenue, Blackpool, FY4 4AF
M6 MOM Jane Clare, Kimberley, Boston Road, Boston, PE20 3AP
M6 MON Phillip Montgomery, 14 Saxon Crescent, Horsham, RH12 2HU
M6 MOP David Pomeroy, 73 Pinewood Gardens, North Cove, Beccles, NR34 7PG
M6 MOQ Keith Hamilton, 22 Brinkburn Road, Stockton-on-Tees, TS20 2DF
M6 MOS Malcolm Steele, 10 Green Lane, Houghton, Carlisle, CA3 0NT
M6 MOU David Dormer, 69 Favell Drive, Furzton, Milton Keynes, MK4 1AX
M6 MOV Christopher Moverly, 45 Quinnell Drive, Hailsham, bn27 1qn
M6 MOW Joe Horry, 5 Donington Road, Bicker, Boston, PE20 3EF
M6 MOX Michael Cox, 3 Cromwell Road, Hertford, SG13 7DP
MM6 MOY Peter McDonald, 9 Partan Skelly Way, Cove Bay, Aberdeen, AB12 3PH
M6 MOZ Maurice Meadowcroft, 8 Lamlash Road, Blackburn, BB1 2AS
M6 MPB Michael Bridger, 11, Beecham Close, Newcastle upon Tyne, NE15 6LG
M6 MPD Richard Yarrow, 27 Staplers Road, Newport, PO30 2DB
M6 MPE Mark Evans, 48 Paddock Lane, Aldridge, Walsall, WS9 0BP
M6 MPF Jim Dunn, 3 Hobbs Way, Rustington, Littlehampton, BN16 2QU
MI6 MPH John Martin, 22 Lisbane Road, Saintfield, Ballynahinch, BT24 7BS
M6 MPK Julian Garwood, 4 Ryedale, Carlton Colville, Lowestoft, NR33 8TB
M6 MPL Graeme Clark, 65 Chyvelah Vale, Gloweth, Truro, TR1 3YJ
M6 MPM Michael Mccall, 1 Jaunty Road, Sheffield, S12 3DT
M6 MPO Martin Woolger, 25 Rookwood Park, Horsham, RH12 1UB
M6 MPP Marcin Michalowski, 39 Towan Avenue, Fishermead, Milton Keynes, MK6 2DS
M6 MPQ Paul Harris, 123 St. Georges Court, Tredegar, NP22 3DD
M6 MPR Paul Rodgers, 5 Church Glebe, Sheffield, S6 1XA
MD6 MPS Stephen O'Riordan, 46 Grange Road, London, HA20LW
M6 MPT Michael Thompson, 35 Princes Avenue, Desborough, Kettering, NN14 2RQ
M6 MPV Mark Varley, 50 Gorse Valley Road, Hasland, Chesterfield, S41 0JP
M6 MPW Michael Whotton, 5 Orchard Street, Ibstock, LE67 6LL
M6 MPX Maxwell Phillips, The Well House, Eastbury, Hungerford, RG17 7JL
M6 MPY Richard Greenwood, The Oast, Hazel Street Farm, Spelmonden Road, Tonbridge, TN12 8EF
M6 MPZ Michelle Edmonds, 20 Tomline Road, Ipswich, IP3 8BZ
M6 MQA Matthew Flute, Granville, Canal End, [illegible], WA3 6AP
M6 MQB Matthew Bailey, 1 [illegible] Drive, Glastonbury, BA6 9LA
M6 MQC Martin Le Moine, 115 Rothesay Road, Blackburn, BB1 2ER
M6 MQD Maria Oliver, 14 Ash Road, Ashurst, Southampton, SO40 7AT
M6 MQE Malcolm Ayres, 32 Kinterbury Close, Hartlepool, TS25 1GQ
MI6 MQF Declan Mulligan, 10 Seaview, Ardglass, Downpatrick, BT30 7SQ
M6 MQH Aleksandar Jovanovic, 33 Seward Road, London, W7 2JS
M6 MQJ Michael Trathen, 2 Kempton Close, Benfleet, SS7 3SG
M6 MQK David Ferguson, 94a Moss Lane, Litherland, Liverpool, L21 7RF
M6 MQL Michael Boyle, 64 Spencerfield Crescent, Middlesbrough, TS3 9HD
M6 MQM Michael Tozer, 4 The Grange Dousland, Yelverton, Pl206nn
M6 MQN Josh Milner, 30 Rowena Drive, Thurcroft, Rotherham, S66 9HT
M6 MQP Peter Marlow, 59 Kinross Crescent, Beechdale, Nottingham, NG8 3FT
M6 MQS Merlin Skinner, 8 Addison Close, Caterham, CR3 5LX
M6 MQT Tomasz Mloduchowski, Flat 4, Gwynne House, London, E1 2AG
MI6 MQX Max Elliott, 17 Milebush Road, Dromore, BT25 1RT
M6 MRC Michael Bridges, 65 Abbots Gate, Bury St. Edmunds, IP33 2GB
M6 MRD Paul Sephton, 11 Moss Avenue, Leigh, WN7 2HH
M6 MRG Harry Martin, 27 Gordon Road, Fleetwood, FY7 6UE
M6 MRH Mark Hayward, 15 Easton End, Basildon, SS15 6QB
MI6 MRI Frank Rafferty, 37 Hollybrook Crescent, Newtownabbey, BT36 4ZW
MI6 MRJ John Martin, 23 Winters Gardens, Omagh, BT79 0DZ
M6 MRK Mark McKenna, 1 Kennet Avenue, Jarrow, NE32 4DB
M6 MRL Brian Collins, 15 Bonds Meadow, Lowestoft, NR32 3QL
M6 MRM Matthew Barnfather, 2 Brazenhill Lane, Haughton, Stafford, ST18 9HS
M6 MRN Philip Fisher, 24 Gatacre Street, Walney, Barrow-in-Furness, LA14 3PY
M6 MRP Philip Boxx, 14 Kingsgate Terrace, Hexham, NE46 3EP
M6 MRR Martin Rutter-DaCosta, 144 Bellingdon Road, Chesham, HP5 2HF
M6 MRS Caroline Davies, 68 Wood Street, Castleford, WF10 1LN
M6 MRT Andrew Blamires, 2 Foldings Grove, Scholes, Cleckheaton, BD19 6DQ
M6 MRU Adrian Owen, 3 Chesham Grove, Goole, DN14 6RR
M6 MRV Robert Martin, 21 Lonsdale Crescent, Dartford, DA2 6LQ
M6 MRW Michael Willison, 6 The Paddock, Lady Street, Dulverton, TA22 9BY
M6 MRX Robert Brough, 36 Salstar Close, Aston, Birmingham, B6 4PP
M6 MRY Ben Murray, 23 Tillotson Close, Crawley, RH10 7WQ
MM6 MSA Mark Saddler, 38 Baberton Mains Wynd, Edinburgh, EH14 3EE
M6 MSC Martin Colman, 4 Northmead Drive, North Walsham, NR28 0AU
MW6 MSE Martin Roblin, 6 Gethin Street, Briton Ferry, Neath, SA11 2LU
M6 MSF Mark Farnham, 94 Rochester Way, Crowborough, TN6 2DU
MI6 MSG George Graham, Flat C, 86 Sunningdale Gardens, Belfast, BT14 6SL
M6 MSH Maxim Hatfull, 16b Church Street, Easton on the Hill, Stamford, PE9 3LL
M6 MSI Mark Scott, 14 Masefield Avenue, Swalwell, Newcastle upon Tyne, NE16 3EZ
M6 MSJ Carraeanne Gibson, 17 Clyde Court, Grantham, NG31 7RB
M6 MSM Michael Smith, 24 Fifth Avenue, Portsmouth, PO6 3PE
M6 MSN Paul Evans, 1 Solent Apartments, 16-17 South Parade, Southsea, PO5 2AZ
M6 MSO Stuart Marsh, 37 Springford Gardens, Southampton, SO16 5SW
M6 MSP Stephen Palmer, Tall Trees, Christian Street, Maryport, CA15 6HT
MI6 MSR Mary Ruddy, 204 Alliance Avenue, Belfast, BT14 7NX
M6 MST Martyn Streeter, Fairway, West Chiltington Road, Pulborough, RH20 2EE
M6 MSU S Hodder, 32 Stubbs Close, Wellingborough, NN8 4UQ
MI6 MSV Matthew Steele, 134 Knock Road, Dervock, Ballymoney, BT53 8AB
M6 MSW Mark Wyatt, 18 Eastcote Lane, Hampton-in-Arden, Solihull, B92 0AS
M6 MSX Mark Street, Flat 6, Derwent Court, Solihull, B92 7BU
M6 MSY Martin Saysell, Gordano Valley Riding Centre, Moor Lane, Bristol, BS20 7RF
M6 MTA Muhammad Tauseef Ansari, 37 Lizmans Court, Silkdale Close, Oxford, OX4 2HF
M6 MTC David Godfrey, 10 Settle Road, Romford, RM3 9XR
M6 MTD Peter Williams, 4 Red Gables, Shap, Penrith, CA10 3NL
M6 MTE Eifion Thomas, 13 Cwrt Dolafon, Dolafon Road, Newtown, SY16 2HU
MW6 MTG Mark Blomfield, 99 Mountain Road, Upper Brynamman, Ammanford, SA18 1AN
M6 MTH Mathew Horton, 8 Liptraps Lane, Tunbridge Wells, TN2 3BS
M6 MTI Matthew Ilsley, 38 Coleridge Road, Ottery St. Mary, EX11 1TD
M6 MTJ Michael Jones, 18 Cleveleys Avenue, Heald Green, Cheadle, SK8 3RH
M6 MTK Alan Morris, 4 Saunders Close, Uckfield, TN22 2BX
M6 MTL Dave Baldwin, 19 Bramble Grove, Wigan, WN5 9PR
M6 MTM Marcus Tyler-Moore, 8 Chesworth Gardens, Horsham, RH13 5AR
M6 MTN Michael Banks, 37 Havelock Road, Southsea, PO5 1RU
MI6 MTO Tom Mehaffey, 33 Lenaderg Road, Banbridge, BT32 4PT
M6 MTR Paul March, 46 Christchurch Road, Tilbury, RM18 8XP
M6 MTS Matthew Smith, 5 Newland Avenue, Stafford, ST16 1NL
M6 MTT Paul Crosweller, Flat 3, 18 Pelham Road, Seaford, BN25 1ES
M6 MTU David Atkins, 32 Braybrook, Orton Goldhay, Peterborough, PE2 5SH
M6 MTV Paul Bailey, 4 Roving Bridge Rise, Clifton, Manchester, M27 8AL
M6 MTW Matthew Ellis, Timbers, Fernhill Park, Woking, GU22 0DL
M6 MTZ Reinhard Lenicker, 10 Wellington Grove, Bradford, BD2 3AL
M6 MUB Margaret Masters, 49 St. Johns Avenue, Bridlington, YO16 4ND
MI6 MUC John Morrison, 70 Ravenswood, Banbridge, BT32 3RD
M6 MUD Corrina Brock, 3 Morley Street, Norwich, NR3 1ND
M6 MUF Wayne Burridge, 20 Archer Close, Kingston upon Thames, KT2 5NE
M6 MUJ Ian Fores, 27 Southfield Lane, Whitwell, Worksop, S80 4NS
M6 MUK Martin King, Gate House, Lower Bentham, Lancaster, LA2 7DD
M6 MUP Carol Meredith, 6 North End, Shortstown, Bedford, MK42 0XB
MM6 MUR Gordon Murray, The Barn House, Springfield Farm, Carluke, ML8 4QZ
M6 MUS Adrian Sutton, 3 Grotes Buildings, London, SE3 0QG
M6 MUT Miles Northwood, 34 Whitehead Drive, Wellesbourne, Warwick, CV35 9PW
M6 MUZ Murray Colpman, Kirk House, Goodworth Clatford, Andover, SP11 7RN
M6 MVA Maria Johnson, 143 Swan Lane, Wickford, SS11 7DG
M6 MVB Mark Bradley, 13 Elizabeth Avenue, Bilston, WV14 8EA
M6 MVD Joseph Dilworth, 808 Liverpool Road, Southport, PR8 3QF
M6 MVF Michael Findon, 35 Birchdale Road, Birmingham, B23 7DG
MM6 MVI Martin Vaci, Tigh-na-Sith, Connel, Oban, PA37 1PJ
M6 MVK Jonathan Hart, 57 Brentleigh Way, Stoke-on-Trent, ST1 3GX
MM6 MVM Vincent McGowan, 112 Oronsay Avenue, Port Glasgow, PA14 6EF
M6 MVN Stephen Forshaw, 22a Barley Hall Street, Heywood, OL10 4DH
MM6 MVQ Iain Learmonth, 86 Lord Hay's Grove, Old Aberdeen, Aberdeen, AB24 1WT
M6 MVT Martin Bell, 7 Shiregreen Lane, Sheffield, S5 6AA
M6 MVV Margot McArthur, 25 Lingfield Road, Edenbridge, TN8 5DS
M6 MVW Matthew Ward, 28 Branksome Avenue, Hockley, SS5 5PF
M6 MWA Shelley Hyland-Davis, 34 Melody Close, Warden, Sheerness, ME12 4PU
M6 MWB Mark Bryant, 284 Brantingham Road, Chorlton cum Hardy, Manchester, M21 0QU
M6 MWC Michael Curwen, 40 Grange Street, Morecambe, LA4 6BW
M6 MWD Martin White, 27 Winstone Close, Redditch, B98 8JS
MM6 MWF Michael Flynn, 15 Riselaw Crescent, Edinburgh, EH10 6HN
MW6 MWG John Davies, 1 South View, Pontycymer, CF32 8LE
MW6 MWN William Noble, 2 Harriet Town, Troedyrhiw, Merthyr Tydfil, CF48 4HJ
M6 MWP Michael Poole, 22 Padstow Gardens, Leeds, LS10 4NQ
MW6 MWS Aeronwen Sneddon, 3 Marigold Close, Gurnos, Merthyr Tydfil, CF47 9DA
M6 MWT Michael Tolmie, 18 Park Court, Langer Road, Felixstowe, IP11 2BZ
M6 MWW Mike Watts, 47 Westbury Crescent, Weston-Super-Mare, BS23 4RF
M6 MXA Alistair O'Reilly, 3b Summerleys, Edlesborough, Dunstable, LU6 2HR
M6 MXB Martin Boddy, 26 Tulip Tree Road, Bridgwater, TA6 4XD
M6 MXC Mark Craven, 78 Connaught Road, Brookwood, Woking, GU24 0HF
M6 MXD Morgan Dalziel, 4 Meadow Close, St. Albans, AL4 9TG
M6 MXH Mark Head, 123 High Street, Dunsville, Doncaster, DN7 4BT
M6 MXM Michael Meehan, 14 Grosvenor Road, Walton, Liverpool, L4 5RB
M6 MXO Victoria Adedeji, 3 Royal Troon Mews, Wakefield, WF1 4JL
M6 MXR Mark Russell, Cowmans Cottage Spring Lane, Flintham, Newark, NG23 5LB
M6 MXX Christopher Morrow, 23 Samuel Street, Doncaster, DN4 9AF
MI6 MXZ Michael Masterson, 2 Pinley Drive, Banbridge, BT32 3TZ
M6 MYB Malcolm Bannister, 12 Fineburn Caravan Park, Frosterley, Bishop Auckland, DL13 2SY
M6 MYC Michael Beddall, 11 Sinodun Road, Wallingford, OX10 8AD
M6 MYD Andrew Middleton, 2 Moor View, Godshill, Ventnor, PO38 3HW
M6 MYF Lawrence Thompson, 33 Dalton Crescent, Shildon, DL4 2LE
M6 MYH Stephen Elliott, 79 Somerton Road, Bolton, BL2 6LN
MM6 MYK Mike Cheetham, Plowvent, Muir of Fowlis, Alford, AB33 8NX
M6 MYL Jeffrey Swann, 5 Lanark Close, Hazel Grove, Stockport, SK7 4RU
M6 MYM Mark Beniston, Min-y-Mor, Treleigh, Redruth, TR16 4AY
M6 MYN Ronald Eaton, 31 Pinfold Lane Ruskington, Sleaford, NG34 9EU
M6 MYR Mark Moss, 6 Orchard Close, Watford, WD17 3DU
M6 MYS Arthur Smart, 7 Hinton Grove, Hyde, SK14 5ST
M6 MYT Brian Hilton, 17 Bellwood, Westhoughton, Bolton, BL5 2RT
MI6 MYW Martin McWilliams, 84 Syerla Road, Dungannon, BT71 7ET
M6 MZB Martin Bauer, Flat 21, 5 Queensland Road, London, N7 7FE
M6 MZI Jonathan Williams, 17 Kingshead Close, Castlefields, Runcorn, WA7 2JF
M6 MZJ Michael Juniper, 2 Cranbourne Drive, Hoddesdon, EN11 0QH
M6 MZL Graham Johnson, 22 Beechwood Close, Blythe Bridge, Stoke-on-Trent, ST11 9RH
M6 MZN Jensen Forshaw, 22a Barley Hall Street, Heywood, OL10 4DH
M6 MZU Terence Baldwin, Rose Cottage, High Street, Pontypool, NP4 6HE
M6 MZY Kirsty Phillips, 12 Copland Avenue, Minster on Sea, Sheerness, ME12 3PJ
M6 MZZ Michael Driscoll, 59 Havendale, Hedge End, Southampton, SO30 0FD
MM6 NAA Norman Mcdonald, 8 Newton Place, Perth, PH1 2QJ
MM6 NAB Donald R Bell, 82 Campbell Avenue, Stevenston, KA20 4BP
M6 NAC Nick Carter, 25 Breachfield, Burghclere, Newbury, RG20 9HY
MM6 NAD Norman Anderson, The Cedars, Church Street, Keith, AB55 4AR
MW6 NAG Nicholas Berrall, 41 Nantgarw Road, Caerphilly, CF83 3FB
MM6 NAI Alfred Anderson, 18 Selkirk Street, Wishaw, ML2 8RA
M6 NAJ Nigel Auckland, Glenfield, The Avenue, Southampton, SO32 1BP
M6 NAK Nicola Leech, 59 Lakeside Court, Brierley Hill, DY5 3RQ
M6 NAL Nicholas Parry, Worlingham Court, Marsh Lane, Beccles, NR34 7PE
M6 NAM Mark Cody Cody, 139 Vicarage Road, Watford, WD18 0HA
M6 NAN Leah-Nani Alconcel, Top Lock Cottage, Stoke Pound Lane, Bromsgrove, B60 4LH
M6 NAO Neil Griffiths, 67 Warstones Drive, Wolverhampton, WV4 4PF
M6 NAQ A Cowley, 7 Harwood Road, Gosport, PO13 0TU
M6 NAS Simon Nash, Ashmead, Hemel Hempstead, HP3 0BU
M6 NAT Nathan Jones, 5 Montgomery Crescent, Quarry Bank, Brierley Hill, DY5 2HB
M6 NAU Nicholas Alders, 14 Forest Rise, Crowborough, TN6 2ES
M6 NAV Joshua Glicklich, 86 Ainsdale Road, Bolton, BL3 3ER
M6 NAW Neil White, 5 Badgers Walk, Burgess Hill, RH15 0AE
MW6 NAX Nicholas Jones, 7 Dyffryn, Burry Port, SA16 0TE
MW6 NAZ Dennis Tippett, 30 Berw Road, Tonypandy, CF40 2HD
M6 NBG James McCosh, The Mill House, Moorlands Road, Merriott, TA16 5NF
M6 NBH Aaron Barrett, 2 Friars Close, Clacton-on-Sea, CO15 4EU
MM6 NBI Roderick Mackay, 12 Robertson Square, Wick, KW1 5NF
M6 NBL Annabelle Mackendrick, Cecily Cottage, Brockhill, Wareham, BH20 7NH
M6 NBN Robert Anderson, 26 Bowness Avenue, Warrington, WA2 9NQ
M6 NBO Barry Vickers, 52 Edward Street, Grimsby, DN32 9HJ
M6 NBP Norman Williams, 114 Essex Place, Montague Street, Brighton, BN2 1LL
M6 NBQ Pablo Fabrega, Flat 3, Post Office Court, Whitchurch, SY13 1QT
M6 NBS Nigel Barker, 17 Pippin Walk, Hardwick, Cambridge, CB23 7QD
M6 NBU Nickolas Bruetsch, The Firs, Penglais Road, Aberystwyth, SY23 2EU
M6 NBV Diane Whitelock, 2 Shippards Road, Brighstone, Newport, PO30 4BG
M6 NBW Neil Warden, 1 Forge House, The Street, Woodbridge, IP13 7RT
M6 NBX Norman Cohen, 8 Henry Gepp Close, Adderbury, Banbury, OX17 3FE
M6 NBY Neill Thompson, 10 Belgrave Close, Hersham, Walton-on-Thames, KT12 5PH
M6 NCA Nick Curry, 58 Greenfields, St. Martins, Oswestry, SY11 3AH
M6 NCB Norman Bettridge, 37 Princess Avenue, Warsop, Mansfield, NG20 0PY
MI6 NCC Noreen Corbett, 10 Main Street, Rosslea, Enniskillen, BT92 7PP
M6 NCD Daniel Senior, 16 Cherry Tree Close, Billingshurst, RH14 9NG
M6 NCE Claire-Louise McLennan, 17 Pluto Road, Eastleigh, SO50 5GD
M6 NCF Nigel Froude, 6 Park Road West, Chester, CH4 8BG
MI6 NCG Noel Griffin, 327 Clonmeen, Drumgor, Craigavon, BT65 4AT

**IMPORTANT NOTE**

**Revalidate licence to avoid revocation** – Ofcom has advised the Society that plans will be drawn up to revoke licences that have not been revalidated as required by the licence conditions. The quickest way to revalidate is to do so online via the Ofcom website: *https://services.ofcom.org.uk/* or by email: *amateur.validations@ofcom.org.uk* Ofcom staff are available to help, but please be patient during times of heavy workload.

UK Callsigns

M6 NCI Bernard Dowley, 120 Capel Street, Capel-le-Ferne, Folkestone, CT18 7HB
M6 NCK Nick Taylor, 212 Plantation Hill, Worksop, S81 0HD
M6 NCL Christopher Braddock, 22 Anncroft Road, Buxton, SK17 6UA
M6 NCM Neal McVeagh, 252 Braddon Road, Loughborough, LE11 5YX
M6 NCO Kevin Tonge, 98 Trescott Road, Northfield, Birmingham, B31 5QB
MM6 NCP Nicola Pollard, 30 Abbeyhill Crescent, Edinburgh, EH8 8DZ
M6 NCR George Paton, 29 Hazelmere Road, Stevenage, SG2 8RX
M6 NCS N Sunley, 1 East Lea View, Cayton, Scarborough, YO11 3TN
M6 NCU Colin Johnson, 22 Carleton Close, Great Yeldham, Halstead, CO9 4QJ
M6 NCX Nicholas Coady, 17 Cherry Lane Gardens, Ipswich, IP4 4QQ
M6 NCY Jeanpierre Mooneapillay, 354 Upper Elmers End Road, Beckenham, BR3 3HG
M6 NDB Neil Brown, 9 Devonshire Avenue, Wigston, LE18 4LP
M6 NDC Nick Charlotte, 26 Nettleton Avenue, Mirfield, WF14 9AN
M6 NDE Nicholas Evans, 46 Furzehill Road, Plymouth, PL4 7LA
M6 NDF Daniel Arnold, 91 Matchams Lane, Hurn, Christchurch, BH23 6AW
M6 NDG Nigel Graven, 33 Sheldrake Road, Broadheath, Altrincham, WA14 5LJ
M6 NDI Arosha Kaluarachchi, 103 Bentinck Road, Newcastle upon Tyne, NE4 6UX
M6 NDK Christopher Lucas, 15 Higher Moor, Ruan Minor, Helston, TR12 7JJ
M6 NDM Neil Mason, 30 Mayfield Crescent, Rowley Regis, B65 8HU
M6 NDN Keith Hutchens, 7 Lapwing Close, Thurston, Bury St. Edmunds, IP31 3PW
M6 NDO Andrew Ashmore, 42 Holme Road, Chesterfield, S41 7JF
M6 NDP Neil Plunkett, 11 Stoneleigh Gardens, Grappenhall, Warrington, WA4 3LE
M6 NDR Nicholas Reeve, 4 Ash Grove, Swindon, SN2 1RX
M6 NDT Rees Thatcher, 83 Westfield Drive, North Greetwell, Lincoln, LN2 4RE
M6 NDY Mandy Twitchen, 72 Finedon Road, Burton Latimer, Kettering, NN15 5QB
M6 NEA Naomi Asher, 17 Ashby Road, Cleethorpes, DN35 9PF
M6 NEC Paul Bean, 14 St. Andrews Lane, Necton, Swaffham, PE37 8HY
MM6 NED Edward Brophy, 44 Annieshill View, Plains, Airdrie, ML6 7NT
M6 NEE Julie Todd, 20 Hexham Avenue, Hebburn, NE31 2HN
M6 NEF Natasha Chapman, 13 Clayton Grove, Bracknell, RG12 2PT
M6 NEG Andy Brown, 26 Castle Close, Leconfield, Beverley, HU17 7NX
M6 NEH Neil Hoare, 5 Kelsey Head, Port Solent, Portsmouth, PO6 4TA
M6 NEI Neil Yorke, 21 Braemar Way, Nuneaton, CV10 7LF
M6 NEL Neil Price, 68 Powke Lane, Rowley Regis, B65 0AG
M6 NEM Emily Chance, 33 Larkfield Avenue, Kirkby-in-Ashfield, Nottingham, NG17 9FE
M6 NEO Chrisrtian Radford, 67 Preston Avenue, Alfreton, DE55 7JX
M6 NES Nessa Preval, 63 Dudley Avenue, Leicester, LE5 2EF
M6 NET Diane Bridges, 65 Abbots Gate, Bury St. Edmunds, IP33 2GB
M6 NEV Neville Chambers, 16a Hillside Road, Wool, Wareham, BH20 6DY
M6 NEW William Chesworth, 28 Chapel Close, Gunnislake, PL18 9JB
M6 NEY Michael Bullions, 25 Kirby Road, Dartford, DA2 6HE
M6 NEZ Neil Jones, 46 Devon Street, Barrow-in-Furness, LA13 9PX
M6 NFC Alan Cockburn, 52 Devon Road, Hebburn, NE31 2DW
M6 NFE Grahame Singleton, 129 Thursby Road, Burnley, BB10 3EG
MW6 NFG Nicola Terrell, 82 Baglan Street, Treherbert, Treorchy, CF42 5AR
M6 NFH Neil Holloman, The Cloisters, Llanvihangel Crucorney, Abergavenny, NP7 8DH
M6 NFI Neil Mooney, 60 Rhyddings Street, Oswaldtwistle, Accrington, BB5 3EY
M6 NFL Richard Wraith, 103 Parkway, New Addington, Croydon, CR0 0JA
M6 NFN Colin Davies, Foelallt, North Road, Aberystwyth, SY23 2EL
M6 NFO Louis Finlayson, 6 Popes Court, Whelford, Fairford, GL7 4DZ
M6 NFR Susan Sanderson, 2 East Crescent, Canvey Island, SS8 9HL
M6 NFW Paul Setter, 199 Southbourne Grove, Westcliff-on-Sea, SS0 0AN
M6 NFX Nicholas Furneaux, Hill View, Highridge Road, Bristol, BS41 8JU
M6 NGA Steven Howell, 95 Victoria Road, Bradmore, Wolverhampton, WV3 7HA
M6 NGD Natalie Goode, 46 Robert Road, Tipton, DY4 9BJ
MW6 NGE Gavin Dixon, 9 The Glen, Bryncethin, Bridgend, CF32 9LX
M6 NGF Stephen Dale, 76 Houldsworth Drive, Stoke-on-Trent, ST6 6TJ
M6 NGI Peter Strachan-Buckley, 9 Short Street, Aldershot, GU11 1HA
M6 NGK Gary Wheeler, 39 Woodbine Close, Newport, PO30 1AE
MI6 NGM Andrew McKay, 17 Thorn Hill Road, Banbridge, BT32 3TL
M6 NGO Charlie Stevens, 12 Praetorian Court, Vesta Avenue, St. Albans, AL1 2PP
M6 NGR Nicky Griffiths, 85 Foljambe Road, Chesterfield, S40 1NJ
M6 NGU Rebecca Gwillym, 9 Short Street, Aldershot, GU11 1HA
M6 NGW Nigel Lang, 1 Peartree Court, Old Orchards, Lymington, SO41 3TF
M6 NHA Sara Ratcliff, 27 Furlong Road, Manchester, M22 1UD
MW6 NHC Alexander Taylor, 74 Fidlas Avenue, Cardiff, CF14 0NZ
M6 NHD David Baker, 34 Farnham Road, Durham, DH1 5LA
M6 NHJ Nicholas Jones, 18 Cleveleys Avenue, Heald Green, Cheadle, SK8 3RH
M6 NHK Nicholas Bates, 40 First Street, Bradley Bungalows, Consett, DH8 6JT
MM6 NHM Neil Morris, 23 sedgebank, Livingston, EH54 6HE
M6 NHN Nick Collins, 65 Greenleaf Gardens, Polegate, BN26 6PF
M6 NHP Norman Pettitt, 2a The Oval, Bulford Road, Tidworth, SP9 7SB
M6 NHS Neal Dodge, Crossways, Culford, Bury St. Edmunds, IP28 6DT
M6 NHT Mark Sheppard, 10 Booth Crescent, Mansfield, NG19 7LG
M6 NHX Neil Hurlock, 9 Little Meadow, Exmouth, EX8 4LU
M6 NHZ Neil Hutton, Fordfields, Deadmoor Lane, Newbury, RG20 9DY
MM6 NIA Niamh Hague, 11 Auchriny Circle, Bucksburn, Aberdeen, AB21 9JJ
M6 NIB Nigel Bennett, 44 Glenmoor Road, Buxton, SK17 7DD
M6 NIC Nicholas Bowker, 16 Farncombe Close, Wivelsfield Green, Haywards Heath, RH17 7RA
MI6 NID Peter Moore, 32 Kinnegar Rocks, Donaghadee, BT21 0EZ
M6 NIE Phillip Martin, 26 Kingfisher Close, Chatteris, PE16 6TP
M6 NIK Nicola Armstrong, 1 Sea View Terrace, Churchtown, Helston, TR12 7BZ
M6 NIL Joseph Stacey, Springfields, Laurels Farm, Laurels Road, Great Yarmouth, NR29 5BX
M6 NIN James Dewhirst, Flat 12, Lewis Court, Tamworth, B79 8BE
M6 NIQ Nicholas Stokes, 618a Thorne Road, Netheravon, Salisbury, SP4 9QG
M6 NIS Fawad Nisar, 19a Cromwell Road, Basingstoke, RG21 5NR
M6 NIT Thomas Dixon, Lawn Cottage, Wyver Lane, Belper, DE56 2EF
M6 NIV Joseph Snowden, 10 Woodcroft, Wakefield, WF2 7LS
M6 NIX Anthony Wilkinson, Central House, Main Road, Hull, HU11 4DJ
M6 NJB Nicholas Bennett, 35 West Shepton, Shepton Mallet, BA4 5UD
M6 NJD Nicola Dixon, 39 Urswick Green, Barrow-in-Furness, LA13 0BH
M6 NJE Nigel Spencer, 47 Tyne Road, Oakham, LE15 6SJ
M6 NJG Nik Grey, 1 Norwich Road, Little Plumstead, Norwich, NR13 5JQ
M6 NJH Natasha Hall, 24c Oakleigh Court, Church Hill Road, Barnet, EN4 8UX
M6 NJK Daniel Hawes, Flat 7, Merivale, Hastings, TN35 4PA
M6 NJM Nicola Morris, Fairview, Trefonen, Oswestry, SY10 9DP
M6 NJO Natalie Owen, 32 Westfield Road, Dudley, DY2 8LE
M6 NJP Neil Pipkin, 46 Charles Avenue, Albrighton, Wolverhampton, WV7 3LF
M6 NJS Nicholas Sandy, 5 High Ercal Avenue, Brierley Hill, DY5 3QH
M6 NJT Nigel Jones, 63 Bernwell Road, London, E4 6HX
M6 NJX Matthew Nassau, 1A Burford Road, Bromley, BR1 2EY
M6 NJZ Carol Dodds, 33 Westgate, Oldbury, B69 1BA
M6 NKB Noel Booth, 25 Tetbury Road, Manchester, M22 1GW
M6 NKC Neil Carey, 16 Cannamanning Road, Penwithick, St. Austell, PL26 8UX
M6 NKH Nick Hammond, 453 Smorrall Lane, Bedworth, CV12 0LD
M6 NKM Nigel Morse, 33 Tower Close, Bassingbourn, Royston, SG8 5JX
M6 NKP Nicholas Palin, 21 Ford Lane, Crewe, CW1 3EQ
M6 NKY Nicola Brown, 12 Forest Close, Newport, PO30 5SF
M6 NLA Robert Williams, Bardsville, Porthdafarch Road, Holyhead, LL65 2LL
M6 NLB Nick Burnet, 27 Mackenzie Way, Tiverton, EX16 4AW
M6 NLF Darren Hydes, 31 Ridgehill Avenue, Sheffield, S12 2GL
MW6 NLG Nikki Gladding, 19 Laleston Close, Nottage, Porthcawl, CF36 3HW
M6 NLO Philip Burt, 56 Winslade Road, Sidmouth, EX10 9EX
MM6 NLP Martin Lawson, 23 Kirkfield View, Livingston Village, Livingston, EH54 7BP
M6 NLR David Miller, 127 Thorpe Road, Norwich, NR1 1TR
M6 NLW Stephen Jones, 30 Crown Fields Close, Newton-le-Willows, WA12 0JW
M6 NMD E Martin, 61 Uffington Avenue, Lincoln, LN6 0AG
M6 NME Emma Nudd, 14 Birkbeck Way, Norwich, NR7 0XZ
M6 NMG Michael Glenn, 31 South Drive, Rhyl, LL18 4SU
M6 NMN Norman Norsworthy, Pippins, Trusham, Newton Abbot, TQ13 0NW
M6 NMR Kevin Wright, 6 Windsor Park, Dereham, NR19 2SU
M6 NMS James Blackwell, 2 Camellia Road, Minster on Sea, Sheerness, ME12 3FD
M6 NMT Mariam Afolabi, Harrogate Ladies' College, Clarence Drive, Harrogate, HG1 2QG
M6 NMW Clifford Marcus, 43 Townsend Square, Oxford, OX4 4BB
M6 NNA Les Wilkinson, 20 Coniston Road, Chorley, PR7 2JA
M6 NNB Alan Hopper, 7 Holmesdale Villas, Swallow Lane, Dorking, RH5 4EY
M6 NNC Nigel Hanson, 8 Oak Street, Skegby, Sutton-in-Ashfield, NG17 3FF
M6 NND Nathan Davies, 29 Burns Road, Congleton, CW12 3EE
M6 NNE David Hanwell, 28 Chipperfield Road, Norwich, NR7 9RR
M6 NNJ Paul Johnson, 25 Pelham Street, Worksop, S80 2TW
M6 NNK Paul Banks, 110 Cherwell Drive, Walsall, WS8 7LL
M6 NNU Darren Johnson, 25 Pelham Street, Worksop, S80 2TW
M6 NNX Daniel Austin, 1002 Marsden House, Marsden Road, Bolton, BL1 2JX
M6 NNY Daniel Richardson, 14 Mill Street, Penrith, CA11 9AQ
M6 NOA James Laszlo, 80 Cornwall Gardens, London, SW7 4AZ
M6 NOC Carl Singfield, 35 Leamington Drive, Sutton-in-Ashfield, NG17 5BA
M6 NOD Adrian Smith, Flat 19, Crown Terrace, 10 High Street, Leamington Spa, CV31 3AN
MI6 NOE Gary McCann, 2 Mahon Close, Portadown, Craigavon, BT62 3JF
M6 NOH Jasmine Fletcher, 74 Devonshire Road, Maltby, Rotherham, S66 7DQ
M6 NOJ Clive Denman, 16 Upper Oak Street, Windermere, LA23 2LB
M6 NOK Nathan Kibble, 5 Dunbar Drive, Thame, OX9 3YD
M6 NOL Mark Roys, Flat 13, Gregory House, Lister Avenue, Rotherham, S62 7JA
M6 NOM Neil O'Mahony, 23 Main Road, Broomfield, Chelmsford, CM1 7BU
M6 NON Alison Rosser, 25 Clos Tir-y-Pwll, Newbridge, Newport, NP11 5GE
MM6 NOR Norman Turnbull, 16 Barntongate Terrace, Edinburgh, EH4 8BA
MI6 NOS Joseph Baker, 324 Clonmeen, Drumgor, Craigavon, BT65 4AT
M6 NOT Stuart Leask, 1 Collington Street, Beeston, Nottingham, NG9 1FJ
M6 NPD Norman Dagger, 23 Vassall Road, Bristol, BS16 2LH
M6 NPF Nigel Fairhurst, 7 Heatherlands, Sunbury-on-Thames, TW16 7QU
M6 NPL Andrew Lyman, 12 Chicheley Street, Newport Pagnell, MK16 9AR
M6 NPN Richard Sandwell, 11 Halse Manor, Halse, Taunton, TA4 3AE
MW6 NPW Nicholas Pitt, 33 Scotchwell View, Haverfordwest, SA61 2RD
M6 NPX Neil Paxman, 128 Coggeshall Road, Braintree, CM7 9ES
M6 NQR Ben Somerville Roberts, 21 Regency Way, Ponteland, Newcastle upon Tyne, NE20 9AU
M6 NQT Christopher Gamble, 4 Parracombe Way, Northampton, NN3 3ND
M6 NRA Richard Nagy, 40 Oakhampton Road, London, NW7 1NH
M6 NRB Matthew Beckett, 4 Sandcross Close, Orrell, Wigan, WN5 7AH
M6 NRC Cliff Willson, 17 Knightons Way, Brixworth, Northampton, NN6 9UE
MM6 NRE Ross Hudson, 23 Ross Way, Livingston, EH54 8LA
M6 NRF Conner Halloway, 41, 41 Basset Road, Redruth, TR16 4AQ
M6 NRG Nick Genge, 21 Castle Mead, Washford, Watchet, TA23 0PZ
M6 NRH Nicolas Holmes, 32 Spinney Close, Kidderminster, DY11 6DQ
M6 NRJ Nick Johnson, Belair, Western Road, Crediton, EX17 3NB
M6 NRM Nathaniel Miles, 2 Newland Mill, Witney, OX28 3HH
M6 NRO Nicholas Rostant, 19 Pentre Jane Morgan, Penglais, Aberystwyth, SY23 3TE
MM6 NRQ Anthony Yates, 25 Keith-Hall Road, Inverurie, AB51 3UA
M6 NRS John Swain, 35 Heygate Close, Baildon, Shipley, BD17 6RT
MW6 NRT Neville Tanner, 3 Maes y Tyra, Resolven, Neath, SA11 4NN
M6 NRW Nicholas Waters, 9 Shirley Road, Droitwich, WR9 8NR
M6 NRX Sarah Cook, Deganwy Hardwick Road, King's Lynn, PE30 5BB
M6 NSC Charlotte Kemp, Forest Edge, Deer Park, Blandford Forum, DT11 0AY
M6 NSD Christopher Coppins, 65a Station Road, Herne Bay, CT6 5QQ
M6 NSE Henry Kennedy, 11 Green Road, High Wycombe, HP13 5BD
M6 NSJ Neil Inglis, 74 Runswick Avenue, Whitby, YO21 3UE
M6 NSK Tracey Pennell, 99 Westheath Avenue, Sunderland, SR2 9LQ
MM6 NSM Catherine Morris, 23 Sedgebank, Livingston, EH54 6HE
M6 NSO Nigel Osborne, 12 Spiller Road, Chickerell, Weymouth, DT3 4AX
M6 NSR Edward Parrish, 89 Delamere Drive, Macclesfield, SK10 2PS
MD6 NSS Nick Smith, 4 Cooil Farrane, Douglas, Douglas, Isle of Man, IM2 1NX
M6 NST Keith Theobald, 21 Stirling Close, Rochester, ME1 3AJ
M6 NSW Simon Wilson, 37 New Road, Minster on Sea, Sheerness, ME12 3PU
M6 NSX Dean Sullivan, 15 Market Lane, Witham, CM8 1GF
M6 NTA Alan Briscoe, 69 Sharpe Street, Tamworth, B77 3HZ
M6 NTB Nicholas Turrell, 7 Fern Gardens, Belton, Great Yarmouth, NR31 9QY
M6 NTJ Nicholas Jones, 1 Olaf Close, Andover, SP10 5NJ
M6 NTL Bryan Porter, 74 Whalley Road, Heywood, OL10 3JG
M6 NTM Nigel McNiece, 23 Hempdyke Road, Scunthorpe, DN15 8LA
M6 NTN Nathan Hazlehurst, 4 Titchfield Close, Wolverhampton, WV10 8UN
MI6 NTP Nathan Prentice, 26 Claranagh Road, Claranagh, Enniskillen, BT94 3FJ
M6 NTR Nigel Reeve, 124 Greenhills Road, Eastwood, Nottingham, NG16 3FR
M6 NTT Simon Black, 55 North Road, Hertford, SG14 1NE
M6 NTW Nathan Davies, 99 Sandpiper Way, Duffryn, Newport, NP10 8WY
M6 NTY Darren Balderson, 19 Valley Road, Wellingborough, NN8 2PH
M6 NTZ David Atkinson, 6 Wembley Road, Moorends, Doncaster, DN8 4PR
M6 NUF David Metcalfe, 25 St. Abbs Walk, Hartlepool, TS24 7NW
M6 NUG Duncan Cheadle, Millbrook, Aldersey Lane, Chester, CH3 9EH
M6 NUL Thomas Moye, 33 Prince Charles Road, Colchester, CO2 8NS
MI6 NUM William Millar, 55 Parklands, Antrim, BT41 4NH
MM6 NUP John Mack, 5 Glen Affric Court, Dumbarton, G82 2BN
M6 NUR David Harrod, 14 The Fairways, Danesmoor, Chesterfield, S45 9BG
M6 NUW Lisa Newton, 45 Commercial Street, Risca, Newport, NP11 6AW
M6 NVA Martin Phillips, 2 Millstream Close, Goostrey, Crewe, CW4 8JG
M6 NVB Neal Brown, 6 Beech Avenue, New Mills, High Peak, SK22 4HU
M6 NVG Revanth Adiga, 41 St. Pauls Road, Staines-upon-Thames, TW18 3HQ
MI6 NVM Noeleen McCann, 3 Portadown Road, Tandragee, Craigavon, BT62 2BB
M6 NVR Nicholas Stewart Ross Ross, 38 Jamaica Road, Malvern, WR14 1TU
M6 NVT Nicholas Tennant, 22 The Lizard, Wymondham, NR18 9BH
M6 NWA Nicholas Wong, Montefiore House, Wessex Lane, Southampton, SO18 2NU
M6 NWB John Benbow, 44 Copthorne Park, Shrewsbury, SY3 8TJ
M6 NWC Nigel Clark, Denemead, Cromwell Road, Waltham Cross, EN7 6AS
MM6 NWH Neil Holgate, Flat 25/F, 151 Wyndford Road, Glasgow, G20 8EB
M6 NWI Andrew Howard, 42 Staveley Avenue, Stalybridge, SK15 1BU
M6 NWM Nikolas Miles, 58 High Street, Pentwynmawr, NP114HN
M6 NWO David Etherington, 20 Sandringham Drive, Leeds, LS17 8DA
M6 NWP Colin Jeary, 2 Baker Close, North Walsham, NR28 9JE
M6 NWT James Turner, 2 South Drive, Padiham, Burnley, BB12 8SH
M6 NWY Simon Newhouse, 28 Hillmorton Lane, Lilbourne, Rugby, CV23 0SS
M6 NXA Andrew Milner Smith, 31 Rose Way, Cirencester, GL7 1PS
M6 NXT Laura Stevens, 55 Silverlands Road, St. Leonards-on-Sea, TN37 7DF
M6 NXY Morris Leach, 64 Grove Street, Wantage, OX12 7BG
M6 NXZ Nicola Dexter, 49 Kennedy Road, Horsham, RH13 5DB
M6 NYA Natalia Fill, 3 St. Albans Crescent, London, N22 5NB
MW6 NYE Aneurin Minton, 22 Heol Serth, Caerphilly, CF83 2AN
M6 NYF Dave Lamble, 4 Laburnum Road, Chorley, PR6 7BG
M6 NYL R Davis, 9 Mossdale Road, Liverpool, L33 1UQ
M6 NYM Samuel Braunstein, Crossways, Roundhay Park Lane, Leeds, LS17 8AR
M6 NYY David Paterson, 48 Sissons Road, Leeds, LS10 4JT
M6 NYZ Niall McGroarty, 8 Bakers Lane, Weldon, Corby, NN17 3LR
M6 NZL Peter Payne, 3 Queens Court, Woking, GU22 7NE
M6 NZN Christopher Naylor, 118 Victoria Road East, Thornton-Cleveleys, FY5 3SU
M6 OAB Estelle Colbert, 176 Carters Mead, Harlow, CM17 9EU
M6 OAD Timothy Waters, 21 Daniels Crescent, Long Sutton, Spalding, PE12 9DS
M6 OAE Michael Snow, 6 Evelyn Close, Doncaster, DN2 6PA
M6 OAF Ian Parbery, 38 Moor End, Maidenhead, SL6 2YJ
MM6 OAG Michael Scott, 71 Craigie Road, Perth, PH2 0BL
MM6 OAI Sajimon Chacko, 210 Hillington Road South, Glasgow, G52 2BB
M6 OAJ Adrian Johnson, 37 Coldharbour Road, Hungerford, RG17 0AZ
M6 OAL Shaun Chng, Wolfson College, Cambridge, CB3 9BB
M6 OAN Ashley Neale, 4 Elizabeth Road, Rothwell, Kettering, NN14 6AJ
M6 OAO John Wilkinson, 40 Station Road, Kenilworth, CV8 1JD
M6 OAP Michael Deary, 7 Newbold Avenue, Sunderland, SR5 1LG
M6 OAQ Darran Byng, 1 Ambell Close, Rowley Regis, B65 8PB
M6 OAT James Bullock, Kingswood, Dedham Road, Colchester, CO7 7QB
M6 OAU Lee Boylan, 30 Pembrey Way, Liverpool, L25 9SN
M6 OAV Joshua Mennell, 16 Trafalgar Street West, Scarborough, YO12 7AU
M6 OAW Alistair Willis, 10 Cheviot Close, Hereford, HR4 0TF
M6 OAX Andre Edmonds, 20 Tomline Road, Ipswich, IP3 8BZ
MI6 OAZ Norman Armstrong, 1 Diamond Cottages, Ardmore Road, Crumlin, BT29 4QU
M6 OBB Bruce Beard, 1 Friars Walk, Newcastle, ST5 2HA
M6 OBC Christopher Bridges, 53 St. Margarets London Road, Guildford, GU1 1TL
M6 OBH Malcolm Lachs, 3 King Henrys Walk, Epping, CM16 6FH
M6 OBJ James Bookham, 116 Clare Gardens, Petersfield, GU31 4EU
M6 OBK David Cull, 6 Compass Way, Bromsgrove, B60 3GP
M6 OBO Blake Bentham, 7 Maypole Crescent, Abram, Wigan, WN2 5YL
M6 OBR Brendan O'Brien, 48 Wright Crescent, Bridlington, YO16 4RG
M6 OBS Stephen Martin, 77 Chatford Drive, Shrewsbury, SY3 9PH
M6 OBY Peter Cairns, Thorverton, Exeter, EX5 5NB
M6 OBZ Philip Dimes, 5 Meadowbrook, Oxted, RH8 9LT
M6 OCB Martin Wilsher, Flat K, Wellington Court, Bedford, MK40 2HY
MW6 OCC Barry Werrell, 26 Glynhafod Street, Cwmaman, Aberdare, CF44 6LD
M6 OCD James Gilbody, Winter Meadows, Puxton, Weston-Super-Mare, BS24 6TH
M6 OCH Clive Howe, 21 Gotham Road, Birmingham, B26 1LB
M6 OCJ Owen Jones, Preswylfa, High Street Bryngwran, Holyhead, LL65 3PP
M6 OCK Alan Mock, 115 Swanfield Drive, Chichester, PO19 6TD
M6 OCO David Jones, 28 Fairway Rise, Chard, TA20 1NT
MM6 OCP Harry Adkins, 524 Boydstone Road, Thornliebank, Glasgow, G46 8HW
M6 OCR Daniel Meakin, 27 Spencer Road, Long Buckby, Northampton, NN6 7YP
M6 OCS Oskar Smith, 19 Tedder Road, Bournemouth, BH11 8BT
MW6 OCT Anne Davies, 107 Heol Llanelli, Pontyates, Llanelli, SA15 5UH
M6 OCU Edward Joynson, 15 Home Farm Lane, Bury St. Edmunds, IP33 2QJ
M6 OCV Nigel Powis, 24 Rosemullion Close, Exhall, Coventry, CV7 9NQ
MD6 OCZ Benjamin Salt, 1 Chantry Close, Harrow, HA3 9QZ
M6 ODA Brian Naylor, 9 Withington Drive, Astley, Tyldesley, Manchester, M29 7NW
M6 ODC Glenn Wellstead, 14a Hardy Road, West Moors, Ferndown, BH22 0EX
M6 ODD Jacob Saunders, 9 Capitol Close, Bolton, BL1 6LU
M6 ODE Kathryn Hadley, 5 The Mead, Clutton, Bristol, BS39 5RF
M6 ODF Laura Halloway, 82 Northwall Road, Deal, CT14 6PP

MI6 ODG Danny Gallagher, 33 Mulnafye Road, Omagh, BT79 0PG
M6 ODL mark cavanagh, 97 Denecroft Crescent, Uxbridge, UB10 9HZ
M6 ODM John Barrett, 5 Oakapple Drive, Dereham, NR19 2SR
MI6 ODN Thomas Pashler, 67 [illegible], Ballymoney, BT43 [illegible]
M6 ODP Oliver De Peyer, Flat 5, Molasses House, London, SW11 3TN
M6 ODS Ashley Royds, 3a Fairfield Avenue, Rossendale, BB4 9TG
M6 ODY Peter Lane, 21 Rycroft Avenue, St. Neots, PE19 1DT
MM6 OEC John Maclean, 42b Coll, Isle of Lewis, HS2 0LR
M6 OEM Martin Faulkner, 6 Stanley Avenue, Queenborough, ME11 5DT
M6 OEN Therese Cummings, 45 Sutton Way, Shrewsbury, SY2 6EE
M6 OEP Elwyn Powell, 28 Frederick Avenue, Hereford, HR1 1HL
M6 OFF Colin Gibson, 3 Conway Drive, Billinge, Wigan, WN5 7LH
M6 OFM Jane Joyce, 92 Essex Road, Halling, Rochester, ME2 1AX
MI6 OFN Ciaran May, 20 Harryville Street, Drumgoon, Enniskillen, BT94 4QX
M6 OGK Valerie Terry, 49 St. Julians Wells, Kirk Ella, Hull, HU10 7AF
M6 OHL Martin O'Connor, 28 Cardigan Road, Southport, PR8 4SF
M6 OHN John Jones, 3 Orchard Way, Oxted, RH8 9DJ
M6 OIC Gladius Chinnappa, 34 Barker Road, Chertsey, KT16 9HX
MM6 OIR Patricia Riddiough, 1 Cedar Road, Ayr, KA7 3PE
M6 OJB Oliver Beck, 11 Clarefield Drive, Maidenhead, SL6 5DW
M6 OJC Andrew Bruton, 29 Helyers Green, Wick, Littlehampton, BN17 7HB
M6 OJD Kevin Winton, 130 George V Avenue, Worthing, BN11 5RX
M6 OJI Kaye Schofield, 18 Berrow Walk, Bristol, BS3 5ES
MI6 OJK Jonathan Kavanagh, Flat 1, 161 Andersonstown Road, Belfast, BT11 9EA
M6 OJM John Marriott, 66 Latimer Road, Cropston, Leicester, LE7 7GN
M6 OJO Errol Mehmet, 8 Hailsham Road, London, SW17 9EN
M6 OJT Oliver Trehearne, Claywood House, East Mascalls Lane, Haywards Heath, RH16 2QJ
M6 OKC Andrew Armson, 2 Windmill Gardens, St. Helens, WA9 1EN
M6 OKH Owain Hopkins, Apartment 17, White Croft Works, Sheffield, S3 7AH
M6 OKH Kevin Humble, 2 Woodford Close, Sunderland, SR5 5SA
M6 OKI Oliver Kelland, 16 Esher Place Avenue, Esher, KT10 8PY
MW6 OKJ Julia Orchard, The Burrows, Spring Gardens, Whitland, SA34 0HL
M6 OKK Gary Cooper, Holmfield, Chelmorton, Buxton, SK17 9SG
M6 OKM Thomas Banham, 16 Mallard Court, Oakham, LE15 6RQ
M6 OKO Spencer Neale, Flat 11, The London Court, Ivybridge, PL21 0AS
M6 OKR Marc Cole, 4 Park Manor, Britton Street, Gillingham, ME7 5EX
MI6 OKS Ann-Marie Hanna, 21 Elms Park, Coleraine, BT52 2QF
M6 OKY Christopher Bietz, 5 Consort House, Brewery Lane, Wymondham, NR18 0BD
M6 OLD Graham Waters, 7 Roeselare Close, Torpoint, PL11 2LP
M6 OLE Oliver Ward, 33 Seventh Avenue, Oldham, OL8 3RY
M6 OLF Martin Mills, 17 Hornby Street, Plymouth, PL2 1JD
M6 OLI James Barnett, 16 Beryl Avenue, Blackburn, BB1 9RR
M6 OLL Oliver Booth, Oak Cottage, Knockin, Oswestry, SY10 8HQ
M6 OLP Keith Baker, 26 Shepperton Road, Petts Wood, Orpington, BR5 1DN
M6 OLS John Myers, 43 Boggart Hill Crescent, Leeds, LS14 1LF
M6 OLT Adam Hicks, 15 West Road, Ruskington, Sleaford, NG34 9AL
M6 OLW Oscar Wood, 2 The Bungalows, North Green, Woodbridge, IP13 9NP
M6 OLY Darrin Goodman Goodman, 44 Roberts Road, Madeley, Telford, TF7 5JJ
MI6 OMA Aidan O'Brien, 15 Oldcastle Road, Newtownstewart, Omagh, BT78 4HX
MD6 OMG Nandesh Patel, 78 Wesley Close, South Harrow, Harrow, HA2 0QE
M6 OMH Michael Hawkridge, 27 Northdale Road, Bradford, BD9 4HG
MM6 OMJ Micahel Reid, 16 Hayfield Road, Kirkcaldy, KY2 5DG
M6 OML Mike Lewis, 1 Kingsmead, Stretton, Burton-on-Trent, DE13 0FQ
M6 OMR Benjamin Withers, 29 Yeoman Way, Trowbridge, BA14 0QL
M6 OMS Mark Stradling, 25 Maple Court, Acacia Grove, New Malden, KT3 3BX
MM6 OMT Charles Ingram, 22 The Square, Mintlaw, Peterhead, AB42 5EH
M6 OMX Max Amos, 19 Poets Gate Cheshunt, Waltham Cross, EN7 6SB
M6 OMZ Samuel Smith, 28 Newhill Road, Wath-upon-Dearne, Rotherham, S63 6JY
M6 OND Carol Howard, 6 Robinson Way, Burbage, Hinckley, LE10 2EU
M6 ONL Peter Destoop, 73, oeselgemstraat, Wakken, Belgium, 8720
M6 ONO Richard Cobern, 21 Hopgarden Close, Lamberhurst, Tunbridge Wells, TN3 8DY
M6 ONS Jeffrey Wood, 3 Onslow Mews, Cranleigh, GU6 8FD
MW6 ONZ David Cheeseman, 61 Ffordd Y Millenium, Barry, CF61 5BD
M6 OOB Lisa Atkinson, 55 Warkworth Crescent, Seaham, SR7 8JT
M6 OOC Sean Kiely, 35 Chestnut Avenue, Todmorden, OL14 5PH
M6 OOD Matthew Holbrook-Bull, 66 Wayman Road, Corfe Mullen, Wimborne, BH21 3PN
M6 OOJ Jamie Edwards, 12 Oak Avenue, Norwich, NR7 0PD
MM6 OOK Andrew Currie, 2/2 44 Robertson Street, Greenock, PA16 8QB
M6 OOL Stefan Latimer, 40 Petersham Road, Long Eaton, Nottingham, NG10 4DD
M6 OOM Andrew Shepherd, 33 Meadow Way, EDGWORTH (TURTON), Bolton, BL7 0DE
M6 OOO Pierre France, Flat 39, Netley House, Birmingham, B32 2BT
MM6 OOP Austen Brown, 11 Bowfield Road, West Kilbride, KA23 9LB
M6 OOW Wendy Waterson, 43 Highland Road, Bath, BA2 1DY
M6 OOX Mark Emans, 212 Hilldene Avenue, Romford, RM3 8DB
M6 OOZ John Maclean, 1 Sherwood Close, Exeter, EX2 5DX
M6 OPA Georgios Mourelatos, 13 Stirling Road, London, N22 5BL
M6 OPC Owen Campbell, 3 Hillside Close, Helsby, Frodsham, WA6 9LB
MI6 OPD Jonathan Reilly, 220 Ardmore Road, Londonderry, BT47 3TE
M6 OPJ Jose Viswambaran, 103 Inglehurst Gardens, Ilford, IG4 5HA
M6 OPL Reginald Topley, 85 Stuart Road, Aylsham, Norwich, NR11 6HW
M6 OPM Orla Murphy, 7 Charlotte Street, Leamington Spa, CV31 3EB
M6 OPO Ottilia Pochat, Hame, Bealswood Road, Gunnislake, PL18 9DA
M6 OPS Brian Hillson, Flat 2, Heatherton Park House, Heatherton Park, Taunton, TA4 1EU
MI6 OPT Andrew Bailie, 130 John Street, Newtownards, BT23 4NA
M6 ORB Paul Thomas, 17 Vine Court, St. Pauls Road, Cheltenham, GL50 4LL
M6 ORC Andrew Cross, 12 Appleby Drive, Langdon Hills, Basildon, SS16 6NU
M6 ORE Ewen Moore, 23 Woodland Road, Rode Heath, Stoke-on-Trent, ST7 3TJ
MJ6 ORG Roy Spencer, 1 Excelsior Villas, Le Mont Les Vaux, Jersey, JE3 8LS
MW6 ORH Owen Hopkin, The Forge, Rock Road, Barry, CF62 4PG
M6 ORI Daniel Dart, Ticklebelly Cottage Lower Charlton Trading Estate, Shepton Mallet, BA4 5QE
MM6 ORK Marc Herridge, The Hollies, Petticoat Lane, Orkney, KW17 2RP
M6 ORM Ross Turner, 73 The Yard, Braintree, CM7 3TY
M6 ORO Olive Gascoigne, 19 Shaftesbury Close, Nailsea, Bristol, BS48 2QH
M6 ORT Gregory Wallington, [illegible], London, [illegible]
M6 ORW Owain Wilson, [illegible]
M6 OSF Julian Clover, 174 Sturton Street, Cambridge, CB1 2QF
M6 OSI Michael Kent, 5 Jacklins Close, Hilperton Marsh, Trowbridge, BA14 7UY
M6 OSL Robert Kemish, 32 Glenbrook Walk, Fareham, PO14 3AH
M6 OSM Edward Pavelin, 35 St. Marys Park, Ottery St. Mary, EX11 1JA
M6 OSO James Coombe, 21 Lesnewth, Par, PL24 2DE
M6 OST Robert Frost, 40 Greenwich Close, Downham Market, PE38 9TZ
M6 OSU Owen Summerfield, 2 Walnut Close, Broughton Astley, Leicester, LE9 6PY
M6 OSW James Halford, 20 Protheroe Field, Old Farm Park, Milton Keynes, MK7 8QS
M6 OSX Mark Rawson, 47 Beech Avenue, Kearsley, Bolton, BL4 8SB
M6 OSY David Coppenhall, 55 Vicarage Lane, Elworth, Sandbach, CW11 3BU
M6 OTB David Tate, 14 Wordsworth Avenue, Hartlepool, TS25 5NG
M6 OTH Mark Riches, Fransham Road Farm, Beeston, King's Lynn, PE32 2LZ
M6 OTK Jayne Gilmour, Llyswen, Crick, Caldicot, NP26 5UW
M6 OTL Andrew Rawlins, 7 Kiln Hill, Slaithwaite, Huddersfield, HD7 5JS
MI6 OTP Tommy Pearson, 20 Lammy Walk, Omagh, BT78 5JE
M6 OTR Timothy Rood, 5 Sunny View, Rise, Hull, HU11 5BW
M6 OTT Mark Locke, 25 Tre Rhosyr, Newborough, Llanfairpwllgwyngyll, LL61 6TG
MI6 OTW Paul Fallon, 18 Church View, Killough, Downpatrick, BT30 7RJ
M6 OUD Andrew Cattell, 2 St. James Close, Ruscombe, Reading, RG10 9LJ
M6 OUI Bruce Gimbert, 99-99a Aylestone Road, Leicester, LE2 7LN
M6 OUS Madlen McKean, 3b Summerleys, Edlesborough, Dunstable, LU6 2HR
M6 OVB Robert Gowers, 43 Tungstone Way, Market Harborough, LE16 9GA
M6 OVE Garry Whale, 9 Noster Street, Leeds, LS11 8QJ
M6 OVI Ovidiu Rominger, 27 Paddock Close, Sixpenny Handley, Salisbury, SP5 5NZ
M6 OVR Michael Barry, 58 Bury Road, Radcliffe, Manchester, M26 2UU
M6 OWA Jack Wright, 10 Whalley Road, Heskin, Chorley, PR7 5NY
M6 OWC Stuart Iles, 12 St. Peters Road, Burntwood, WS7 0DJ
MI6 OWD Andrew Kirkpatrick, 33d Ballyferris Walk, Bangor, BT19 1QL
MM6 OWL Brian Ewart, 94 Kirkness Street, Airdrie, ML6 6ET
M6 OWM John Thorpe, 57 Bratch Lane, Wombourne, Wolverhampton, WV5 8DL
M6 OWT Rebecca Harwood, 24 Firle Crescent, Lewes, BN7 1QG
M6 OXB Derek Whittaker, Flat 1, Clement Court, Manchester, M11 1EQ
M6 OXF Daniel East, 43 Conduit Hill Rise, Thame, OX9 2EJ
MM6 OXX Alex Berry, 41 Bruce Drive, Stenhousemuir, Larbert, FK5 4DD
M6 OXY Andrew Oxborrow, 24 Ickworth Crescent, Rushmere St. Andrew, Ipswich, IP4 5PQ
M6 OYA Kenneth Peakman, 32 Mob Lane, Walsall, WS4 1BB
MI6 OYB Paul Magee, 11 Elm Corner, Dunmurry, Belfast, BT17 9PZ
M6 OYD Michael Boyd, 1 Harris Close, Brackley, NN13 6NS
M6 OYF Grant Bailey, 28 Station Road, Polesworth, Tamworth, B78 1BQ
M6 OYZ Stephen Bushnall, 58 Ash Cresent, Durham, SR7 7UF
M6 OZB David Gardner, 1 Neath Court, Thornhill, Cwmbran, NP44 5UH
MW6 OZF Christopher Lyle, 41 Mount Pleasant Avenue, Llanrumney, Cardiff, CF3 5SY
M6 OZI Scott Carpenter, 52 Mewstone Avenue, Wembury, Plymouth, PL9 0JZ
M6 OZM Mark Osborne, 6 Walnut Tree Way, Worthing, BN13 3QQ
MW6 OZO Sheridan Hayward, 22 Dewsland Street, Milford Haven, SA73 2AU
M6 OZT Christopher Moore, 2 Chapel Rise, Cross Hill, Filey, YO14 0JA
M6 OZY Kevin Johnson, 32 Redmire Close, Bransholme, Hull, HU7 5AQ
M6 OZZ Michael Crockford, Centre Cottage Kelk, Driffield, YO25 8HL
M6 PAA Peter Sadler, 52 Kent Avenue, Weston-Super-Mare, BS24 7FH
MW6 PAC Paul Smith, 3 Islington Road, Bridgend, CF31 4QY
M6 PAD Paul Darlington, 8 Uplands Road, Urmston, Manchester, M41 6PU
M6 PAE Peter Lobo-Kazinczi, Flat 1, 52 Park Road, Hull, HU5 2TA
M6 PAF Patrick Bigsby, 14 Rutland Avenue, Sidcup, DA15 9DZ
M6 PAG Warren Johnson, 15 Sanders Road, Hemel Hempstead, HP3 9UB
M6 PAI Peter Blackburne, 16 College View, Connah's Quay, Deeside, CH5 4BY
M6 PAJ Pete Moore, 112 Westbury Lane, Bristol, BS9 2PU
M6 PAM Graham Williams, 99 Maes Llwyn, Amlwch, LL68 9BG
M6 PAN Peter Naylor, 38 Piggott Grove, Stoke-on-Trent, ST2 9BZ
M6 PAO Peter Alley, 58 Osprey Close, Watford, WD25 9AR
M6 PAS Paul Simcox, 67 Mervyn Road, Bilston, WV14 8DB
MI6 PAT Patrick Coogan, Flat B, 44 Ramoan Gardens, Belfast, BT11 8LL
MM6 PAU Malcolm Scott, 28b Highfield Place, Birkhill, Dundee, DD2 5PZ
M6 PAV Peter Villette, 62 Hallwood Road, Kettering, NN16 9RF
M6 PAW Philip Wallis, 28 Munro Street, Stoke-on-Trent, ST4 5HA
M6 PAX Shane Pashley, 3 Princess Street, Brimington, Chesterfield, S43 1HP
M6 PAY Denis Potter, The Lines, Commonside, Boston, PE22 9PR
M6 PBE Paul Carter, 18 Maynard Terrace, Clutton, Bristol, BS39 5PL
MW6 PBF Peter Martin, 6 Herrick Place, Machen, Caerphilly, CF83 8TA
MI6 PBI Linda Thompson, 28 Kirkdale, Newtownabbey, BT36 5BX
M6 PBK Peter Browne, Ham Cottage, Hammingden Lane, Haywards Heath, RH17 6SR
M6 PBL Paul Parkin, Hawksworth House, Main Street, Frolesworth, Lutterworth, LE17 5EG
M6 PBM Peter Mills, 10 Laurel Estate, Cowes, PO31 7HW
M6 PBO W Dickson, The Rowans, Pwllmeyric, Chepstow, NP16 6LA
M6 PBP Scott Bradley, 90 Occaiso House, 90 Play House Square, Harlow, ME20 1AP
M6 PBR Michael Massie, 3 East Mount, North Ferriby, HU14 3BX
M6 PBT Paul Burgess, 61 Grosvenor Avenue, Torquay, TQ2 7JX
M6 PBU Peter Haynes, 105 Mettesford, Matlock, DE4 3EB
M6 PBV Jamie Brown, 26 Lynnes Close, Blidworth, Mansfield, NG21 0TU
MI6 PBW Paul White, 46 Pine Cross, Dunmurry, Belfast, BT17 9QY
M6 PBX Daniel Wainwright, Flat 1, Earlsland, 7 Kendrick Road, Reading, RG1 5DU
M6 PBY Ian Maltby, 22 Fern Avenue, Doncaster, DN5 9QX
MI6 PBZ Philip Bell, 3 Alexandra Crescent, Larne, BT40 1NE
M6 PCB Carol Vincent, 81 Trethannas Gardens, Praze, Camborne, TR14 0LL
M6 PCC Paul Cumbers, Church Cottage, Church Lane, Diss, IP21 4AG
M6 PCD Philip Cheek, 55 Turpin Green Lane, Leyland, PR25 3HA
M6 PCE Clare Fryer, 16 Elston Place, Aldershot, GU12 4HY
M6 PCF Paul Faulkner, 32 Manvers Road, Beighton, Sheffield, S20 1AY
MI6 PCJ Phillip James, 14 Brookmount Crescent, Omagh, BT78 5HG
M6 PCL Phillip [illegible], Weston-Super-Mare, [illegible]
M6 PCM Peter Mason, 1 Keepers Coombe, Bracknell, RG12 0TN
M6 PCN Philip Newth, 20 Barrow Close, Redditch, B98 0NL
M6 PCP Patrick Couch, 28 Polgover Way, St. Blazey, Par, PL24 2DL
M6 PCQ Peter Morris, 14 Marina Road, Darlington, DL3 0AL
M6 PCR Paul Raine, 29 Beech Gardens, Crawley Down, Crawley, RH10 4JB
MW6 PCT Stephen Gau, Disgwylfa, The Downs, Cardiff, CF5 6SB
M6 PCU Thomas Prince, 6 Hyperion Avenue, South Shields, NE34 9AE
M6 PCV Barry Eastman, 3 Ribble Court, Southampton, SO16 9JQ
M6 PCW Paul Cullen, 77 Rising Brook Stafford ST17 9DH, Stafford, ST17 9DH
M6 PCX Philip Coombes, 2 Bissoe Cottages, Bissoe, Truro, TR4 8SU
M6 PCZ Paul Colyer, 23 Florida Road, Torquay, TQ1 1JY
MM6 PDA Simon Haynes, 86 Barrowfield Street, Coatbridge, ML5 4BJ
M6 PDB Piers de Basto, 27 Cloudesley Square, London, N1 0HN
M6 PDC Paul Chell, 174d South Road, Stourbridge, DY8 3RN
M6 PDD Andrew Coleman, Talavera, Bromley Green Road, Ashford, TN26 2EF
M6 PDE Paul Devlin, Brynteg, Fron Bache, Llangollen, LL20 7BP
MW6 PDG Paul Gough, 21 Clos Tyclyd, Cardiff, CF14 2HP
M6 PDI Mike Lidster, 83 Stroma Gardens, Hailsham, BN27 3AZ
M6 PDJ Peter Davies, 15 Carlyle Road, Wolverhampton, WV10 8SL
M6 PDK Peter Coote, 251 Alfreton Road, Pye Bridge, Alfreton, DE55 4PB
M6 PDL Michael James, 42 Doone Way, Ilfracombe, EX34 8HS
M6 PDM Paul Moffitt, Wrawby Farm, Star Carr Lane, Brigg, DN20 8SG
M6 PDN David Ebbs, 28 Betterton Court, Chapmangate, York, YO42 2ET
M6 PDO Andrew Dingwall, 48 Village Farm Caravan Site, Bilton Lane, Harrogate, HG1 4DL
M6 PDP Duncan James, Coombe House, Coombe, Presteigne, LD8 2HL
M6 PDQ Martin Summers, 22a Footners Lane, Burton, Christchurch, BH23 7NT
M6 PDR Peter Roberts, 1 Winston Close, Eastleigh, SO50 4NS
M6 PDS Peter Stean, 2b Deanery Road, London, E15 4LP
M6 PDT Caroline Devlin, 32 Kestrel Drive, Dalton-in-Furness, LA15 8QA
M6 PDU Llewellyn Davies, Tow Path House, Riverside Road, Staines-upon-Thames, TW18 2LE
M6 PDV Fiona Farrer, 16 High Street, Eagle, Lincoln, LN6 9DH
M6 PDW Peter Willis, 21 Alton Close, Swindon, SN2 5HF
MM6 PDX William Davison, 16 Millfore Court, Bourtreehill North, Irvine, KA11 1LT
MI6 PDY Peter Davis, 35 Culross Drive, Dundonald, Belfast, BT16 2SQ
M6 PDZ Peter Bromfield, 32 Mount Pleasant, Halesworth, IP19 8JF
MM6 PEA Andrew Pemberton, 111 Henderson Street, Bridge of Allan, Stirling, FK9 4HH
M6 PEB Peter Beckwith, 1 Whincroft Drive, Ferndown, BH22 9LH
M6 PEC Barry Clayton, 5 Greycourt Close, Halifax, HX1 3LR
MW6 PED James Pedley, Apartment 31, St. Thomas Lofts, Kilvey Terrace, Swansea, SA1 8BG
M6 PEF Paul Phipps, Meakers Cottage, Long Load, Langport, TA10 9JX
M6 PEG Paul Boast, 104 Whittier Road, Nottingham, NG2 4AS
M6 PEI Joseph Brown, The Chapel, Heath Green, Leighton Buzzard, LU7 0AB
MI6 PEJ James Beckett, 20 Old Forge, Banbridge, BT32 4AH
MW6 PEL Paul Day, 15-16 Troedrhiw-Trwyn, Pontypridd, CF37 2SE
M6 PEM Paul Metters, 11 Horton Avenue, South Shields, NE34 8NL
M6 PEN Penny Kiley, 178 Kingfisher Drive, Woodley, Reading, RG5 3LQ
M6 PEO Peter Owen, 17 Jameston, Bracknell, RG12 7WZ
M6 PEP Stephen Hill, 35 Longs Way, Wokingham, RG40 1QW
M6 PEQ Paul Scarlett, 75 Ilfracombe Road, Southend-on-Sea, SS2 4PA
M6 PER Paul Rimmington, 28 Skipton Road Swallownest, Sheffield, S26 4NQ
M6 PEU Graham Wale, 39 Rainham Way, Frinton-on-Sea, CO13 9NR
M6 PEW Paul Woodburn, 21 The Row, Silverdale, Carnforth, LA5 0UG
M6 PEX Callum Rumball, 4 Hasted Close, Bury St. Edmunds, IP33 2UA
MW6 PEY Anthony Fey, 28 Bryn Rhedyn, Caerphilly, CF83 3BT
M6 PEZ Craig Perrin, 66 Central Avenue, Farnworth, Bolton, BL4 0AU
M6 PFA Daniel Rees, 21 Glenthorne Road, Hereford, HR4 9RW
M6 PFB Philip Payne, 5 Hurstwood Close, Bexhill-on-Sea, TN40 2TA
MI6 PFD Pearse Dallas, 12 Glendun Crescent, Coleraine, BT52 1UJ
M6 PFG Peter Roberts, 26 Park Road, Wallasey, CH44 9EB
M6 PFH Paul Holmes, 18 Raleigh Avenue, Whiston, Prescot, L35 3PL
MI6 PFI Peter Sweeney, 14 The Glade, Newtownabbey, BT36 5NW
M6 PFK Paul Lewis, 9 The Hill, Glapwell, Chesterfield, S44 5LX
M6 PFL Paul Flatt, 8 Lingfield Crescent, Queensbury, Bradford, BD13 2SA
M6 PFM Tamas Sarosi, The Grange, Warwick Road, London, W5 3XH
M6 PFO Paul Noble, 14 Park Street, Swallownest, Sheffield, S26 4UP
M6 PFP Brian Robinson, 11 Wimbledon Drive, Stockport, SK3 9RZ
M6 PFR Paul Reeves, 18 Ploughmans Lea, East Goscote, Leicester, LE7 3ZR
MM6 PFT David Barclay, 45 Milton View, Gatehead, Kilmarnock, KA2 0AY
M6 PFU Paul Ogle, 29 Patrick Street, Grimsby, DN32 0JQ
M6 PFV Florin Popa, Flat 403, The Galley, 3 Basin Approach, London, E16 2QW
M6 PFW Peter Woodmass, 104 Calf Close Lane, Jarrow, NE32 4DU
M6 PFX Paul Fuller, 6 Annalee Road, South Ockendon, RM15 5DJ
M6 PFY Paul Knox, 1 Newby Close, Halesworth, IP19 8TU
M6 PFZ Paul Catterall, 117 Beech Hill Lane, Wigan, WN6 8PJ
M6 PGA Philip Burrows, 31 Spencer Road, Lutterworth, LE17 4PG
M6 PGB Phillip Boultwood, 32 Makepiece Road, Bracknell, RG42 2HJ
M6 PGC Graham Shepherd, 48 Lasgarn View, Varteg, Pontypool, NP4 7RZ
MW0 PGD Peter David, 5 The Bungalows, Oakfield Terrace, Bridgend, CF32 7SP
M6 PGF Philp Bagely, 28 Tenacre Lane, Dudley, DY3 1XQ
M6 PGH Paul Hill, 14 Drovers Way, Woodlands, Ivybridge, PL21 9XA
MI6 PGI Peter Brown, 2 Legaterriff Road, Ballinderry Upper, Lisburn, BT28 2EY
M6 PGL Philip Lewis, 154 Meadow Head, Sheffield, S8 7UF
M6 PGN Dominique Martin, 8 Garden City, Langport, TA10 9ST
M6 PGP Paul Pearce, 41 Tennyson Avenue, Boldon Colliery, NE35 9EP
M6 PGQ Philip Challans, Flat 4, Sandringham Court, 2 Chandos Square, Broadstairs, CT10 1QN
MM6 PGT Mary Paget, 40 Davaar Drive, Kilmarnock, KA3 2JG
M6 PGU Philip Hall, 11 Middleton Court, Mansfield, NG18 3RN
M6 PGW Peter Gonczarow, 25 Ribchester Avenue, Burnley, BB10 4PD
M6 PGX Bruce Williams, 8 Lindberg Way, Woodley, Reading, RG5 4XE



**IMPORTANT NOTE**

**Revalidate licence to avoid revocation** – Ofcom has advised the Society that plans will be drawn up to revoke licences that have not been revalidated as required by the licence conditions. The quickest way to revalidate is to do so online via the Ofcom website: *https://services.ofcom.org.uk/* or by email: *amateur.validations@ofcom.org.uk* Ofcom staff are available to help, but please be patient during times of heavy workload.

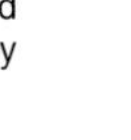

M6 PGY Paul Chapman, 2 Tighe Close, Wantage, OX12 9GD
M6 PGZ Paul Barkley, 32 Tudor Road, Doncaster, DN2 6EN
M6 PHA Paul Biggs, 4 Pendle Close, Southampton, SO16 4QT
M6 PHC Paul Conduit, 16 Rectory Avenue, High Wycombe, HP13 6HW
M6 PHE Phoebe Swinyard, 72 London Road, Wokingham, RG40 1YE
M6 PHF Steven Shaw, 36 Church Street, Moor Row, CA24 3JQ
MM6 PHG Paul Groundwater, Vollbekk, 14 Burnside, Kirkwall, KW15 1TF
MI6 PHH Peter Kinney, 23 Glenariff Crescent, Ballymena, BT43 6ET
M6 PHJ David Collier, 133 Woodstock Road, Moston, Manchester, M40 0DG
M6 PHK Paul Bentley, Fenwick Field, Simonburn, Hexham, NE48 3EL
M6 PHL Philip Hughes, 111 Wisbech Road, Littleport, Ely, CB6 1JJ
M6 PHM Philip Meerman, 24 Horseshoe Crescent, Burghfield Common, Reading, RG7 3XW
MM6 PHO C Hunter, 1 North Gate Lodge, Erines, Tarbert, PA29 6YL
M6 PHP Jamie Dando, 70 Lydgate Road, Southampton, SO19 6NG
MI6 PHQ Paul Wilkin, 55 Clare Heights, Ballyclare, BT39 9SB
M6 PHS Philip Swannick, 11 Derwent Road, Scunthorpe, DN16 2PA
M6 PHT Phillip Hart, 11 Sadlers Ride, West Molesey, KT8 1SU
M6 PHU John Steel, 4 Station Terrace, Boroughbridge, York, YO51 9BU
M6 PHV Phillip Lowe, 4 St. Michaels Cottages, Church Street, Bedale, DL8 2PZ
M6 PHW Paul Walsh, 55 Fore Street, St. Marychurch, Torquay, TQ1 4PU
M6 PHX Paul Hunter, 160 Pembroke Road, Northampton, NN5 7ER
M6 PHY Jack Scoble, 53 Russell Gardens, Poole, BH16 5BF
M6 PHZ Joseph Smith, 51 Myrtle Avenue, Peterborough, PE1 4LR
M6 PIA Michael Moriarty, 30 Longmead, Abingdon, OX14 1JQ
MM6 PIB Maxime Thiebaut, 1/1 Block A, 2 Barrack Street, Hamilton, ML3 0HZ
M6 PIC David Berry, 60 Copthorne Road, Leatherhead, KT22 7EE
M6 PID Philip Hodgson, 19 Raymonds Drive, Benfleet, SS7 3PL
M6 PIH Paul Humphreys, 14 Woodside, Oswestry, SY11 1EP
M6 PII Sarah Whitwell, 12 Lambton Road, Stockton-on-Tees, TS19 0ER
M6 PIK David Pike, 46 Haymans Close, Cullompton, EX15 1EH
M6 PIP Paul Pritchard, 12 Easton Crescent, Billingshurst, RH14 9TU
MI6 PIR Dean Hamilton, 9 Parterre Crescent, Dundrum, Newcastle, BT33 0WJ
M6 PIT Raymond Peacock, 24 Vicarage Estate, Wingate, TS28 5BP
M6 PJA Paul Archer, 31 Stoney Bank Drive, Kiveton Park, Sheffield, S26 6SJ
M6 PJB Peter Bacon, 124 Aughton Road, Swallownest, Sheffield, S26 4TH
M6 PJD Paul Doyle, 27 Brockenhurst Avenue, Havant, PO9 4NS
M6 PJE Peter Elmore, 8 Gray Street, Elsecar, Barnsley, S74 8JR
M6 PJH Paul Higginson, 187 Birmingham road, Ansley village, Nuneaton, CV10 9PQ
MW6 PJJ Philip Jones, 23 Pinecroft Avenue, Aberdare, CF44 0HY
M6 PJM Paul Miller, 46 Great Brooms Road, Tunbridge Wells, TN4 9DH
M6 PJN Peter Noakes, 8 Ellis Avenue, Old Colwyn, Colwyn Bay, LL29 9LB
MW6 PJP Philip Parsons, 31 Clos Tan-y-Fron, Bridgend, CF31 2BY
MM6 PJR Paul Russell, 21 St. Andrews Drive, Law, Carluke, ML8 5GB
M6 PJX Stuart Rodgers, 22 Crabtree Lane, Sutton-on-Sea, Mablethorpe, LN12 2RT
MM6 PKC Iain Macnab, 4/4 27 St. Andrews Crescent, Glasgow, G41 5SD
M6 PKD Colin Barker, 52 Hazelmoor, Hebburn, NE31 1DH
M6 PKL Darren Cadet, 2 Paddockside, Middleton, Ludlow, SY8 3EB
M6 PKN Philip Winterbottom, Fourways Birchcourt, Swinton, S64 8DQ
MM6 PKO George Rothera, Greystones, Watten, Wick, KW1 5UG
M6 PKR Pravin Kumar Karuppannan Rajan, 1 Jacquard Close, Coventry, CV3 5NG
M6 PKT Patrick Hall, 112 Osmaston Park Road, Derby, DE24 8EX
M6 PKU Patrick Kirkden, 22 Leas Green, Broadstairs, CT10 2PL
M6 PLB David Smith, 13a, Barnes Road, Skelmersdale, WN8 8HN
MW6 PLC Philip Carroll, 22 Llanbad, Brynna, Pontyclun, CF72 9QQ
M6 PLE Peter Jenkins, 4 Boulton Avenue, West Kirby, Wirral, CH48 5HZ
M6 PLH Peter Simmonds, 25 Dimlington Bungalows, Easington, Hull, HU12 0TH
M6 PLI Philippa Hartley, 99 Cyprus Street, Stretford, Manchester, M32 8BE
M6 PLK Artur Jedryka, 71 West Royd Drive, Shipley, BD18 1HL
M6 PLO Pedro Thomas, 77 Hawthorn Avenue, Lowestoft, NR33 9BB
MW6 PLP Peter Price, 80 Maesglas, Pontyates, Llanelli, SA15 5SH
M6 PLR Preben Rasmussen, 7 Portal Drive North, Upper Heyford, Bicester, OX25 5TH
M6 PLS Alastair Hornby, 54 Amberley Road, Horsham, RH12 4LN
M6 PLV Paul Le Vallois, 14 London Row, Arlesey, SG15 6RX
M6 PLZ Louisa Perry, 23 Victors Crescent, Hutton, Brentwood, CM13 2HZ
M6 PMA Paul Ambrose, 61 Greenleas Road, Wallasey, CH45 8LR
M6 PMB Peter McMullan-Bell, 57 Melford Avenue, Barking, IG11 9HS
M6 PMD Paul Davies, 67a James Street, Stoke-on-Trent, ST4 5HR
M6 PMF Peter Fenney, 44 Birch Avenue, Oldham, OL8 3TX
M6 PMG Paula Gorman, 1 Hill Top, Stourbridge, DY9 9BZ
M6 PMH Michael Jones, 29 Highbridge Road, Burnham-on-Sea, TA8 1LL
M6 PMJ Peter Mullen, 14 Anderson Road, Hemswell Cliff, Gainsborough, DN21 5XP
M6 PMK Thomas Karpasitis, 4 East Crescent, Enfield, EN1 1BS
MI6 PML Paul McLarnon, 8c Greenview Way, Antrim, BT41 4EG
MI6 PMM Paul Maguire, 139 Carrowshee Park, Drumhaw, Enniskillen, BT92 0FS
MW6 PMN Paul Norman, 56 Granogwen Road, Mayhill, Swansea, SA1 6UW
M6 PMP Paul Parker, 3 Little Charlton, Basildon, SS13 2EJ
M6 PMR Ben Rubery, 142 Chapel Street, Pensnett, Brierley Hill, DY5 4EQ
M6 PMT Peter Troth, Beuna Vista, Hawford Wood, Droitwich, WR9 0EZ
M6 PMZ P Mansfield, 27 Popplechurch Drive, Swindon, SN3 5DE
M6 PNC Paul Griffiths, Cherry House Farm, School Lane, Woodbridge, IP12 2PX
MM6 PND Charlotte Lucking, 39 Overdale Street, Glasgow, G42 9PZ
M6 PNG David Golik, 42 Jonathan Road, Stoke-on-Trent, ST4 8LP
M6 PNJ Paul Jackson, 1 Osbern Road, Preston, Paignton, TQ3 1HN
M6 PNK Charlotte Taylor, 212 Plantation Hill, Worksop, S81 0HD
M6 PNL Paul Neil, 1 High Graham Street, Sacriston, Durham, DH7 6LZ
M6 PNP Andrew Pepler, Flat 2, 5 Fore Street, Westbury, BA13 3AU
M6 PNR Pawel Nowak, 38 Brook Road, Craven Arms, SY7 9RF
M6 PNT Alan Huby, 38 Ferryhill Road, Irlam, Manchester, M44 6DD
MW6 PNW Paul White, 47 Long Acre, North Cornelly, Bridgend, CF33 4BE
M6 PNX Craig Peace, 49 Rossefield Way, Leeds, LS13 3RS
M6 PNY Penny Sayles, 11 Malton Close, Monkston, Milton Keynes, MK10 9HR
MW6 PNZ Claire Stewart, 40 The Pines, Cilfrew, Neath, SA10 8AL
M6 POA Jonathan Lambert, 205 Reading Road, Wokingham, RG41 1LJ
M6 POB Brian Campbell, 9 Granams Croft, Bootle, L30 0PH
M6 POC Zachariah Schwingen, 1 Elizabeth Lockhart Way, Braintree, CM7 9RH
M6 POD Daniel Pochat, Hame, Bealswood Road, Gunnislake, PL18 9DA
M6 POE Saskia Barber, 82 Prunus Road, Crewe, CW1 4HB
MI6 POF Angela Stewart, 35 West Wind Terrace, Hillsborough, BT26 6BS
M6 POG Pollyxeni Gupta, 36 Slimmons Drive, St. Albans, AL4 9AP
MI6 POH Peter O'Hare, 15a Lisdoo Road, Clady, Strabane, BT82 9RQ
M6 POI Lorna Smart, 139 Northumberland Street, Norwich, NR2 4EH
M6 POM Paul Cowper, 12 Galway Road, Leicester, LE4 2PJ
M6 POP Paul Barrow, Beach View, North Promenade, Withernsea, HU19 2DS
M6 POQ Hannah Thomas, 5 Silver Drive, Frimley, Camberley, GU16 9QN
MM6 POV Brian Barclay, 90 Craiglea Crescent, Dalmellington, Ayr, KA6 7UA
M6 POW Liam Patton, 50 Fourgates Road, Dorchester, DT1 2NL
M6 POZ Paul Osborne, 3 Low Farm Road, Tunstall, Norwich, NR13 3PU
M6 PPD James Lawson, 138 Fosse Road, Newport, NP19 4TB
M6 PPH Paul Holmquest, 6 Rhyme Hall Mews, Fawley, Southampton, SO45 1FX
M6 PPJ Paul Pearson, 22 Norris Street, Darwen, BB3 3DR
M6 PPK Paul Kay, 30 Broadway, Grange Park, St. Helens, WA10 3RX
M6 PPL Graham Smith, 559 New Ashby Road, Loughborough, LE11 4EX
M6 PPP Roger Parker, 4 St. Matthews Close, Cherry Willingham, Lincoln, LN3 4LS
M6 PPS John Eyre, 57c Valley Crescent, Wrenthorpe, Wakefield, WF2 0JB
M6 PPT Gareth Owen, 38 Trentham Drive, Bridlington, YO16 6ES
M6 PPV John Smith, 48 London Road, Wymondham, NR18 9BP
M6 PPY Jason Dales, 6 Woodfield Drive, Sawtry, Huntingdon, PE28 5TZ
M6 PPZ Peter Lloyd, 8 Maydor Avenue, Saltney Ferry, Chester, CH4 0AH
M6 PQD Paul Coxon, 14 Barwick Street, Peterlee, SR8 3SA
M6 PQF Pedro Ferreira, Flat 48, Lambert Court, Bushey, WD23 2HF
M6 PQR Neil Richardson, 33 Estcourt Terrace, Leeds, LS6 3EX
M6 PRA Andrew McEwen, 4 The Pantyles, Nightingale Lane, Sevenoaks, TN14 6BX
M6 PRB P Bowes, 72 Keswick Road, Worksop, S81 7PS
M6 PRC Lee Piercy, 35 Claremont Road, Grimsby, DN32 8NU
M6 PRD Pradeep Dheerendra, Flat 3, 379 Caledonian Road, London, N7 9DQ
M6 PRE Phillip Evans, 144 Grenville Street, Stockport, SK3 9ET
M6 PRF Joanna Rowland-Stuart, 86 Wiltshire House, Lavender Street, Brighton, BN2 1LE
MM6 PRH Paul Higgins, Tigh Larich, Trinafour, Pitlochry, PH18 5UG
M6 PRM Peter Munson, 8 Longley Lane, Spondon, Derby, DE21 7AT
M6 PRN Robert Raine, 91 Lulworth Avenue, Jarrow, NE32 3SB
M6 PRO Alan Forrest, 1 Errington Bungalows, Sacriston, Durham, DH7 6NE
M6 PRP Prakash Punjabi, 62 Cleveland Road, London, W13 8AJ
MD6 PRS Paul Skidmore, 36 Princes Drive, Harrow, HA1 1XH
M6 PRV Stephen Bullock, 35 Hillside Road, Haslingden, Rossendale, BB4 5NW
M6 PRW Philip White, 63 Garden Drive, Brampton, Barnsley, S73 0TN
MM6 PRZ Pierre Misson, 2 Sauchie Street, Stirling, FK7 0QW
M6 PSC Philip Croxford, 1 Meteor Close, Bicester, OX26 4YA
M6 PSG Phil Griffith, Long Field Barn, Clarkes Lane, Beccles, NR34 8HR
M6 PSJ Dom Williams, 2 Tyning Road, Peasedown St. John, Bath, BA2 8HT
M6 PSM Paul Mansfield, 56 Sunningdale, Waltham, Grimsby, DN37 0UG
M6 PSN Gareth James, 28 Redcar Road, Romford, RM3 9PT
M6 PSO Pal Szabo, 93 Norham Avenue, Southampton, SO16 6QB
M6 PSR Darren Rowe, 18 Burman Road, Wath-upon-Dearne, Rotherham, S63 7ND
M6 PST P Tugwell, 71 Dunkellin Way, South Ockendon, RM15 5ES
M6 PSV Malcolm Beard, 4a St. Aidans Grove, Liverpool, L36 8JE
MM6 PSX Alistair Ross, 29 East Banks, Wick, KW1 5NL
M6 PSY M O'Halloran, 7 Waver Close, Corby, NN18 8LL
M6 PSZ Susan MacDonald, Woodside Cottage, Horton Way, Verwood, BH31 6JJ
MM6 PTE Peter Davis, 2 Virkie Cottages, Virkie, Shetland, ZE3 9JS
M6 PTF Matthew Haddleton, 42 Grove Road, Pontefract, WF8 2AB
M6 PTH Paul Thomasson, 13 Ringway, Neston, CH64 3RS
M6 PTM Paul Millington, 125 Telford Way, High Wycombe, HP13 5SZ
M6 PTO Ronald Morgan, 17 Forbes Close, Glenfield, Leicester, LE3 8LF
M6 PTP Matthew Taylor, 84 Elgar Crescent, Brierley Hill, DY5 4JJ
M6 PTR Peter Owen, 4 Doonhill Wood, Newton Stewart, DG8 6NU
M6 PTV Charlotte Brough, 40 Denby Grange, Harlow, CM17 9PZ
M6 PTX Peter Hutchins, 25 The Paddock, Maidenhead, SL6 6SD
M6 PTY Pete Ryan, 3 St. Quentin Close, Derby, DE22 3JT
M6 PTZ Peter Twaites, 40 Lower Lane Chinley, High Peak, SK23 6BD
M6 PUD Joanna Crudgington, 29 Wild Flower Way, Ditchingham, Bungay, NR35 2SF
M6 PUF Richard Bath, 10 Tan Rhiw, Penrhyn Bay, Llandudno, LL30 3RB
M6 PUG Andrew Ward, 29 Mainwaring Road, Wallasey, CH44 9DN
M6 PUL David Pullen, 5 Weldon Close, Shotton Colliery, Durham, DH6 2YJ
M6 PUN Jason Schofield, 21a Millgate, Thirsk, YO7 1AA
M6 PUP Sarah Nichols, 20 Holly Blue Road, Wymondham, NR18 0XJ
M6 PUQ Rebecca Vincent, 81 Trethannas Gardens, Praze, Camborne, TR14 0LL
M6 PUS Alexander Russell, 26 Diamond Ridge, Camberley, GU15 4LD
M6 PUT Kirsty Duggan, 97 Maesglas Grove, Newport, NP20 3DN
MI6 PUX Noel Jenkinson, 35 Scarvagh Heights, Scarva, Craigavon, BT63 6LY
M6 PUY Elvin Barrett, 54 The Parade, Greatstone, New Romney, TN28 8SU
M6 PVC Christopher Constantine, 257 Kings Road, Ashton-under-Lyne, OL6 9EG
M6 PVL Peter Lansdown, 60 Craydon Grove, Bristol, BS14 8EY
M6 PVM Paul Massey, 11 Shakestone Close, Writtle, Chelmsford, CM1 3HS
M6 PVN Paul Nicholls, 23 Bishops Gate, Birmingham, B31 4AJ
M6 PVR Paul Richards, 46a Kimbolton Road, Bedford, MK40 2NX
M6 PVV Bernard Woods, 193 Wimberley Street, Blackburn, BB1 8HU
M6 PVZ Daniel Till, 157 Malkin Drive, Church Langley, Harlow, CM17 9HL
M6 PWB Paul Barrett, 8 Durban Road, Portsmouth, PO1 5RR
M6 PWC Paul Castle, 3 Wye Road, Brockworth, Gloucester, GL3 4PP
M6 PWD Paul Deeprose, 2 Denehurst Gardens, Hastings, TN35 4PB
M6 PWE Gary Corkett, 28 Underwood, Hawkinge, Folkestone, CT18 7NT
M6 PWG Philip Green, 3 Beech Court, Long Stratton, Norwich, NR15 2WY
M6 PWH Peter Whitworth, 34 Rye Crescent, Danesmoor, Chesterfield, S45 9HH
M6 PWK Peter Martin, 5 Shropshire Drive, Wilpshire, Blackburn, BB1 9NF
M6 PWL M Mynn, 15 Shearling Drive, Lower Cambourne, Cambridge, CB23 6BZ
M6 PWM James Pattinson, 28 Dunley Close, Swindon, SN25 2BL
MW6 PWO Peter Oseland, 6 Oaklands Close, Bridgend, CF31 4SJ
MI6 PWR Paul Warriner, 25 St. Marys Road, Omagh, BT79 7JX
MW6 PWS Philip Spong, 209 Neath Road, Briton Ferry, Neath, SA11 2BJ
M6 PWT John Girard, 49 Beech Crescent, Hythe, Southampton, SO45 3QF
M6 PXD Paul Donaghy, 67 Brockenhurst Way, Bicknacre, Chelmsford, CM3 4XN
M6 PXF Peter Forbes, 55 The Henrys, Thatcham, RG18 4LS
M6 PXG Paul Guest, 88 Windsor Drive, Wigginton, York, YO32 2YE
M6 PXK Christopher Smith, 83 Sea View Street, Cleethorpes, DN35 8HY
M6 PXL Chris Hembrow, 5 Wetherby Road, Stoke-on-Trent, ST4 8AZ
M6 PXP John Maudsley, Knight Stainforth Hall, Little Stainforth, Settle, BD24 0DP
M6 PXS Philip Simpson, 7 Hawthorn Close, Wootton, Ulceby, DN39 6RB
M6 PXT Lee O'Connor, Newton, Maldon Road, Witham, CM8 1HP
M6 PXW Peter Wright, 4a Alma Street, Melbourne, Derby, DE73 8GA
M6 PXY Andy Schofield, 24 Oldbrook, Bretton, Peterborough, PE3 8SH
M6 PYD Duncan Robinson, 27 Abbeylea Drive, Westhoughton, Bolton, BL5 3ZD
M6 PYE David Pye, 12 Buchanan Drive, Hindley Green, Wigan, WN2 4HJ
M6 PYF Alex Whitmore, 12 Rutland Close, Congleton, CW12 1LT
M6 PYG Geoffrey Fielding, Chapel Court, Chapel Lane, Malvern, WR13 5HX
M6 PYH Gareth Round, 3 Pen yr Hwylfa, Harlech, LL46 2UW
MI6 PYN Stefan Pynappels, 38 Dora Avenue, Newry, BT34 1JW
M6 PYO Philip Pritchard, 3 Melrose Drive, Wolverhampton, WV6 7XH
M6 PYP Richard Hall, 8 Adam Street, Abertillery, NP13 1EX
M6 PYR Pauline Robinson, 15 Cornelius Drive, Wirral, CH61 9PY
MM6 PYX David Ewart, 2 School Quadrant, Airdrie, ML6 6SP
M6 PZA Richard Hawes, 40 Nightingale Way, Thetford, IP24 2YN
M6 PZB Paul Brown, 64 St. Johns Road, Swinton, Mexborough, S64 8QW
M6 PZF Philippa Fairbourn, 17 Perry's Lane, Wroughton, Swindon, SN4 9AX
M6 PZM Adil Aslam, 101 The Oval, Guildford, GU2 7TP
M6 PZP Sharon Owen, 8 Old Tanymanod Terrace, Blaenau Ffestiniog, LL41 4BU
M6 PZW Mark Heenan, 15 Woodacre Green, Bardsey, Leeds, LS17 9AB
M6 PZY Ciaren Puzey, 17 Iverdale Close, Iver, SL0 9RJ
M6 PZZ Linden Evans, 1a South View, Fryston, Castleford, WF10 2QF
MI6 RAC Reece McDaid, 29 Linen Green, Sion Mills, Strabane, BT82 9TL
MI6 RAD Robert Todd, 58 Kilrea Road, Portglenone, Ballymena, BT44 8JB
M6 RAH Raymond Harriman, 9 Millers Close, Rushden, NN10 9RP
M6 RAL Cheryl Jewell, 3 Marsh Gate, Clee St. Margaret, Craven Arms, SY7 9DU
MM6 RAM Robert Law, 4a Burnside Court, Dundee, DD2 3AF
MU6 RAN Richard Phibbs, Aqeb, St Martins, Guernsey, GY4 6AD
M6 RAO Ronald Rider, 25 Kimber Close, Lancing, BN15 8QD
M6 RAP Richard Payne, 110 Keeble Way, Braintree, CM7 3JY
MD6 RAQ Robert Shooter, 14 Cooyrt Shellagh, Ballasalla, Ballasalla, Isle of Man, IM9 2EU
M6 RAR Raymond Calleja, 3 Stoutsfield Close, Yarnton, Kidlington, OX5 1NX
MI6 RAS Gaibriel O'Neill, 46 Ashgrove Road, Newtownabbey, BT36 6LJ
MI6 RAV Raymond Nelson, 65, 65 Dernawilt Road, Rosslea, BT92 7FN
MW6 RAW Ashley Cotter, 3 George Street, Aberdare, CF44 6RY
M6 RAZ Ronald Booker, 6 Kipling Road, Dursley, GL11 4QB
M6 RBA Robert Bristow, Flat 4, 1 Laton Road, Hastings, TN34 2ET
M6 RBB Brett Plackett, 36 Dartmouth Crescent, Brinnington, Stockport, SK5 8BG
MM6 RBE James Howison, 54 Whiteside, Bathgate, EH48 2RG
M6 RBF Robert Fry, 20 St. Andrews Road, Ellesmere Port, CH65 5DG
M6 RBH Richard Haynes, 9 Nant y Felin, Abermule, Montgomery, SY15 6NQ
MJ6 RBI Robin Rumboll, Windsor House, La Grande Route de St. Laurent, Jersey, JE3 1NL
M6 RBK Bill Kalogerakis, Inglewood, Madingley Road, Cambridge, CB23 7PH
M6 RBN Richard Allen, 91 Sea Lane, Goring-by-Sea, Worthing, BN12 4PR
M6 RBO Richard Axtell, 74 Elmshott Lane, Slough, SL1 5QZ
MI6 RBP Robert Lyttle, 14 ROSEHEAD DRIVE, County Antrim, BT14 7BF
M6 RBQ Robert Loseby, 22 Chelmsford Drive, Doncaster, DN2 4JN
MM6 RBT Robert Turpie, 11 Ashkirk Place, Dundee, DD4 0TN
M6 RBU Roger Byard, Wynbourne, Wynolls Hill Lane, Coleford, GL16 8BP
M6 RBV Richard Ford, 6 Briar Court, Wickersley, Rotherham, S66 1AF
M6 RBY Ruby Johnston-Stuart, 35 Robbins Close, Bradley Stoke, Bristol, BS32 8AS
MW6 RBZ Richmond Bishop, 96 Heol Homfray, Cardiff, CF5 5SB
M6 RCA Richard Broadwith, Keepers Cottage, Rookwith, Ripon, HG4 4AY
M6 RCB Richard Brignall, 53 Grange Crescent, St. Michaels, Tenterden, TN30 6DY
M6 RCD Roy Coates, 123 Mansfield Crescent, Armthorpe, Doncaster, DN3 2AR
M6 RCE Clifford Mandville, Garth Hill College, Bull Lane, Bracknell, RG42 2AD
M6 RCF Ronald Goodier, 56 Chariot Street, Manchester, M11 1DP
M6 RCG Paul Ballington, 7 Links Close, Sinfin, Derby, DE24 9PF
M6 RCI Lee Rodgers, 27 Main Street Normanby-by-Spital, Market Rasen, LN8 2HE
MM6 RCK Thomas McGurk, 11 Palmer Road, Currie, EH14 5QH
M6 RCL Robert Lynch, 2 Launceston Close, Oldham, OL8 2XE
M6 RCN Sydney Francis, 17 Garden Close, Rough Common, Canterbury, CT2 9BP
MM6 RCO Ross Nicoll, 15 Redford Walk, Edinburgh, EH13 0AF
M6 RCP Carwyn Edwards, 5 Old School Cottages, Lords Hill, Coleford, GL16 8BD
MI6 RCR Rachel Reid, 13 Gelvin Grange, Londonderry, BT47 2LD
M6 RCS Ross Codrai, 27 Howard Avenue, West Wittering, Chichester, PO20 8EX
MI6 RCV Harry Ashe, 74 Markville, Portadown, Craigavon, BT63 5SZ
M6 RCW Richard Weaver, 15 Sharps Field, Headcorn, Ashford, TN27 9UF
M6 RCY Colin Radley, 7 Garforth Crescent, Droylsden, Manchester, M43 7SW
MM6 RCZ Robert Taylor, Station Cottage, Main Street, Bathgate, EH48 3BU
M6 RDA Alexander Hill, 7 Knebworth Court, Congleton, CW12 3SW
M6 RDC Robbie Cole, 14 Inner Loop Road, Beachley, Chepstow, NP16 7HF
M6 RDD Barry Barwell, 48 Locke King Road, Weybridge, KT13 0TB
M6 RDF Ian Sharman, 63 Duston Road, Northampton, NN5 5AR
M6 RDI Eriks Vecenans, 155 Upper Dale Road, Derby, DE23 8BP
M6 RDK Russell Starr, 43 Newington Avenue, Southend-on-Sea, SS2 4RD
MW6 RDL Richard Green, Hillside, Meinciau Road, Kidwelly, SA17 4RA
MM6 RDM Raymond Matthews, 18 Tweed Street, Grangemouth, FK3 8HJ
M6 RDN Ronald Vickers, 57 Cecil Road, Selly Park, Birmingham, B29 7QQ
M6 RDP Adam Toynton, Key West Flat 2, 7 Sea Lawn Terrace, Dawlish, EX7 0AD
M6 RDQ Richard Ellison, 66 Coronation Road, Wingate, TS28 5JW
MM6 RDR Richard Rowe, 5 Lever Road, Helensburgh, G84 9DP
M6 RDS Rupert Sutton, Yew Tree Farm, Paddol Green, Shrewsbury, SY4 5QZ
MM6 RDT Robin Tourish, 8 Linnpark Gardens, Johnstone, PA5 8LH
MM6 RDU Robert Drummond, 11 Firwood Drive, Bo'ness, EH51 0NX
M6 RDV Raymond Marco De Vries, Corner Cottage, Hillcrest Close, Sturminster Newton, DT10 2DL

M6 RDW Iris Hartless, 32 Long Acre, Mablethorpe, LN12 1JF
M6 RDX Alan Griffiths, 61 Hawarden Road, Penyffordd, Chester, CH4 0JD
M6 RDZ Robert Sidwell, 27 Gressingham Drive, Lancaster, LA1 4RF
M6 REB Robert Beardsley, 10 Moreton Close, Church Crookham, Fleet, GU52 [illegible]
M6 RED Ian Sanderson, 9 Haigh Street, Cleethorpes, DN35 [illegible]
M6 REE Reece Garvey, 2 Link Lane, Oldham, OL8 3AD
M6 REF Christopher Newsam, 15 Devonshire Avenue North, New Whittington, Chesterfield, S43 2DF
M6 REH Roger Ashley, 15 Wimbourne Drive, Gillingham, ME8 9EN
M6 REI Carl Reid, 28 Albion Road, London, N16 9PH
M6 REJ Robert James, 31 Tehidy Gardens, Camborne, TR14 0ET
M6 REK Ruth Kelly, 42 Hinton Wood Avenue, Christchurch, BH23 5AH
M6 REL Rupert Allen, 44 Overing Avenue, Great Waldingfield, Sudbury, CO10 0RJ
MI6 REM Ruth Richey, 10 Crest Road, Enniskillen, BT74 6JJ
M6 REO Damien Curry, 2 Norham Avenue South, South Shields, NE34 7LP
M6 REQ Russell Saunders, 128 Foxcroft Drive, Wimborne, BH21 2LA
M6 RER Robin Ridge, Roskellan House, Maenlay, Helston, TR12 7QR
M6 RES Richard Strong, 2 Dean Avenue, Thornbury, Bristol, BS35 1JJ
MM6 REV Trevor Paterson, Free Church Manse, Church Street, Golspie, KW10 6TT
M6 REW Rachel Wells, 37 Water Meadows, Worksop, S80 3DF
M6 REX Razvan Moldoveanu, 17 Lynchford Road, Farnborough, GU14 6AR
M6 REY Thomas Surrey, 11 Chirton Avenue, North Shields, NE29 0JS
M6 RFA Robert Allen, Milverton, Mill Road, Pulborough, RH20 2PZ
M6 RFB Richard Woodrow, Smallfield, Pebble Road, Pevensey, BN24 6NH
M6 RFC R Crockford, 17 Tadcroft Walk, Calcot, Reading, RG31 7JR
MI6 RFD Andrew Hudson, 31 Knollwood, Seapatrick, Banbridge, BT32 4PE
M6 RFE Andrew Braeman, 59 Freshbrook Road, Lancing, BN15 8DE
M6 RFF Rodney Hancock, 125 Fairham Road, Stretton, Burton-on-Trent, DE13 0BT
M6 RFG Rhodri Morgan, 14 Ash Road, Ashurst, Southampton, SO40 7AT
M6 RFI Gregory Sullivan, 9 Hine Close, Gillingham, SP8 4GN
M6 RFJ David Jacobs, 7 Coppice Close, Ravenstone, Coalville, LE67 2NS
M6 RFK Robert Brookes, 5 Radstock Road, Stretford, Manchester, M32 0AJ
M6 RFL Richard Long, 45 Dean Street, Low Fell, Gateshead, NE9 5XL
M6 RFM Rowan Corney, Lavender Cottage, Worlds End, Waterlooville, PO7 4QU
M6 RFN Steven Hinton, 33 Park Street, Kidderminster, DY11 6TP
M6 RFO Alan Trohear, 3 Frampton Crescent, Bristol, BS16 4JA
M6 RFP Matthew Bruce, 27 Blaenant, Emmer Green, Reading, RG4 8PH
M6 RFQ Robert Newton, 38 Bedford Road, Denton, Northampton, NN7 1DR
M6 RFR Ross Wheddon, 10 Penharget Close, Pensilva, Liskeard, PL14 5SA
M6 RFS Richard Shipman, 1 Lledfair Place, Heol Pentrerhedyn, Machynlleth, SY20 8DL
M6 RFV Richard Manser, 39 Long Meadow, Markyate, St. Albans, AL3 8JN
M6 RFW Peter Wilson, 103 Thors Oak, Stanford-le-Hope, SS17 7BZ
M6 RFZ Roy Finch, Garth Cottage, North Cowton, Northallerton, DL7 0HL
MW6 RGA Roy Anderson, 156 Cockett Road, Cockett, Swansea, SA2 0FQ
M6 RGC David Clavey, 32 Apollo Close, Dunstable, LU5 4AQ
M6 RGD Robin Yates, 8 Ferrybridge Road, Knottingley, WF11 8JF
M6 RGE Wayne Ridge, 91 Ridgeway, Rotherham, S65 3NL
M6 RGF Russ Gott, 21 Broughton Road, Crewe, CW1 4NW
M6 RGH Robert Hall, Concorde Cottage, Ellingstring, Ripon, HG4 4PW
M6 RGI Raymond Griffiths, 2 Old Market Close Acle, Norwich, NR13 3EY
MW6 RGK Tim Bourner, 23 St. Michaels Road, Pembroke, SA71 5JQ
MI6 RGM Ryan Murphy, 40 Stoneypath, Londonderry, BT47 2AF
M6 RGO Rachel Hunter, 9 Gelt Burn, Didcot, OX11 7TZ
M6 RGP Robert Pearson, 38 Bevis Walk, Bury St. Edmunds, IP33 2NS
M6 RGQ Raymond Hobbs, 2 Jenkyns Close, Botley, Southampton, SO30 2UQ
M6 RGR Robert Reeves, 20 Warren Close, Irchester, Wellingborough, NN29 7HF
M6 RGS Robert Sneddon, 33 Egerton Road, South Shields, NE34 0QH
MM6 RGT Robert Thomson, Inchully, Myreriggs Road, Blairgowrie, PH13 9HS
M6 RGU Richard Emery, 115 Humberstone Road, Grimsby, DN32 8DR
MW6 RGW Roger Woodland, 2 Waun Las, Neath, SA10 7RW
M6 RGX William Morrison-Bates, 95 Beaulieu Close, Toothill, Swindon, SN5 8AJ
M6 RGY Paul Jay, 113 Stanks Lane North, Leeds, LS14 5AS
M6 RGZ Robert Hughes, 3 Wood Avens Close, Northampton, NN4 9TX
M6 RHC Richard Cook, 62 High Hazel Road, Moorends, Doncaster, DN8 4QN
M6 RHD R Hughes-Burton, 6 Troed Y Garn, Llangybi, Pwllheli, LL53 6DQ
MM6 RHI Alex Hunsley, 42 Waverley Place, Edinburgh, EH7 5SA
M6 RHK Roger Kirby, 50 Harcourt Avenue, Harwich, CO12 4NT
MI6 RHL Andrew Calvin, 65 Tannaghmore Road, Markethill, Armagh, BT60 1TW
MD6 RHN Daniel Heaton, Highcrest, Perwick Road, Port St. Mary, Isle of Man, IM9 5LP
M6 RHO Rhory Hendry, 3 Claddyburn Terrace, Cairnryan, Stranraer, DG9 8RD
MM6 RHQ Robert Hepburn, 33 Saxon Road, Glasgow, G13 2YQ
M6 RHR Nigel Oldham, Arhosfa, Carmel, Llannerch-Y-Medd, LL71 7DH
M6 RHT Ralph Harkness, 4 Cassalands, Dumfries, DG2 7NS
M6 RHU Mike Ricketts, 3 Lauder Close, Northolt, UB5 5JQ
M6 RHX Andrew Richardson, 18 Warkworth Road, Durham, DH1 5PR
M6 RHY Rhys Cooper, 3 Waunddu, Pontnewynydd, Pontypool, NP4 6QZ
M6 RIA Josh Aldersley, 6 Cavendish Close, Kingswinford, DY6 9PR
M6 RIE Emma Soames, 40 Woodland Drive, North Anston, Sheffield, S25 4EP
M6 RIH R Hayes, Leaf House, Rede Road, Bury St. Edmunds, IP29 [illegible]
M6 RIJ Sam Blades, 18 Hall Cliffe Road, Horbury, Wakefield, WF4 6BX
M6 RIK Richard Langford, 27 Castlegate House Huddersfield Road, Elland, HX5 0RN
M6 RIL Richard Studeny, 36 Maples Street, Nottingham, NG7 6AD
M6 RIN Andrew Barker, 3a Hillcrest Close, Castleford, WF10 3QS
M6 RIO John Roberts, Worlds Wonder, Warehorne, Ashford, TN26 2LU
M6 RIQ Rory Gallagher, 56 Chesterton Square, London, W8 6PJ
MI6 RIR Edward Stevenson, 25 Woodlands Manor, Portadown, Craigavon, BT62 4JP
M6 RIS Richard Sowden, 19 Cramfit Crescent, Dinnington, Sheffield, S25 2XT
M6 RIT Rita Williams, Wold View, Carlton Road, Louth, LN11 8UF
M6 RIU Roger May, 18 The Glebe, Camborne, TR14 7EW
M6 RIV Charmaine Jenkin, 64 Rivers Road, Yeovil, BA21 5RJ
M6 RIY Anita Holland, Four Winds, Nebo, Llanon, SY23 5LF
M6 RIZ Roger Harris, 19 Nightingale Walk, Stevenage, SG2 0QE
M6 RJB Ronald Beardmore, 37 Fernhill Road, Solihull, B92 7RU
MD6 RJE Ray Edwards, 23 Queens Walk, Ruislip, HA4 0LX
M6 RJF Gareth Ropinski, 38 The Leys, Little Eaton, Derby, DE21 5AR
M6 RJH Robert Harrison, [illegible] Hall Lane, [illegible], Huddersfield, HD8 [illegible]
[illegible]
M6 RJK Richard Killnor, Oak Cottage, Priors Byne Farm, Lock Lane, Horsham, RH13 8EF
M6 RJL Jamie Rose, 9 Denewood, New Barnet, Barnet, EN5 1LX
M6 RJM Russell Meehan, 27 Edward Street, Macclesfield, SK11 8JD
M6 RJN Norman Robertson, 3 Wallach Brae, Dalbeattie, DG5 4GY
M6 RJO Richard Hart, 4 Glade Mews, Guildford, GU1 2FB
M6 RJP Roger Parsons, Aintree, Hurn Bridge Road, Lincoln, LN4 4XT
M6 RJQ Rumaisah Munir Munir, 29 Hamilton Drive, Guildford, GU2 9PL
M6 RJR Simon Slater, 4 Churchill Avenue, Middleton, Matlock, DE4 4NG
M6 RJS John Statham, Oakwoods, School Lane, Reading, RG8 8LT
M6 RJW Roderick Wakelam, 1 Cleeve Park Mews, Cleeve Park, Minehead, TA24 6JH
MW6 RJX M Burns, 40 Matthysens Way, St. Mellons, Cardiff, CF3 0PS
MW6 RJY Jon Reason, 158 Caerau Lane, Cardiff, CF5 5JS
M6 RJZ Roger Duthie, 14 Kettles Close, Oakington, Cambridge, CB24 3XA
M6 RKA Mark Halliburton, 7 Penrose Avenue West, Liverpool, L14 6UT
M6 RKB Keith Raymond Bache Bache, 49 Cheviot Way, Halesowen, B63 1HD
M6 RKC Mark Callow, 4 The Firs, Canvey Island, SS8 9TW
M6 RKD Mark Davey, Romea, Long Street, Attleborough, NR17 1LW
M6 RKE Ryan Elger, 18 The Green, Ockley, Dorking, RH5 5TR
MW6 RKF William Lewis, 53 Leyshon Road, Gwaun Cae Gurwen, Ammanford, SA18 1EN
M6 RKI Ricki Hogan, 10 Forest Way, High Wycombe, HP13 7JF
M6 RKK Robert William Dick, 15 Havenwood, Arundel, BN18 0AH
M6 RKL Lee Hall, 180 Moor Road, Chorley, PR7 2NT
M6 RKM Roger Mason, 22 Coronation Close, Happisburgh, Norwich, NR12 0RL
M6 RKO Roy Oczerklewicz, 4 Grisedale Place, Chorley, PR7 2JW
M6 RKP Richard Page, 11 George Street, Cleethorpes, DN35 8PX
M6 RKQ Neil Monaghan, 7 Hoyle Road, Wirral, CH47 3AG
M6 RKR Graham Hirst, 15 Hengist Close, Horsham, RH12 1SB
M6 RKS Mark Swanston, 309 Main Road, Wharncliffe Side, Sheffield, S35 0DQ
MM6 RKT Jeffery Browne, 33 Pilgrims Hill, Linlithgow, EH49 7LN
M6 RKU Adrian Rickwood, 10 Wardrop Road, Catterick Garrison, DL9 3BW
M6 RKW Mark Ware, 20 Brentwood Avenue, Thornton-Cleveleys, FY5 3QR
M6 RKX Mark Hemming, 11 Blackberry Way, Evesham, WR11 2AH
M6 RLA Matthew Lewis, 10 Belle Vue Road, Cwmbran, NP44 3LE
M6 RLB Richard Brown, 62 Charlecote Park, Telford, TF3 5HD
M6 RLC Robert Lee, 28 Champion Way, Oxford, OX4 4NS
MM6 RLD Darren Beeton, Struan, Shiskine, Isle of Arran, KA27 8EW
M6 RLE R Lewis, 33 Sunnycroft, Portskewett, Caldicot, NP26 5RX
M6 RLG Jack Berrisford, 8 Streatham Road, Derby, DE22 4AY
M6 RLK Richard King, 33 Phoenix Court, Wakefield, WF2 9NZ
MM6 RLL James Young, 101 Fleming Way, Hamilton, ML3 9QH
M6 RLN David Carroll, 48 Foster Road, Trumpington, Cambridge, CB2 9JR
M6 RLP Ryan Palmer, New Haven, Stoneraise, Carlisle, CA5 7AX
M6 RLQ Ross Skingley, Tregwin, The Commons, Helston, TR12 7HZ
M6 RLR Roger Hanson, 3 Lower Collier Fold, Cawthorne, Barnsley, S75 4HT
M6 RLT Richard Trim, 23 Coleman Road, Bournemouth, BH11 8EQ
MI6 RLU Francis Burgess, 65 Sunnylands Avenue, Carrickfergus, BT38 8JT
M6 RLV Richard Oliver, 78 High Street Oakington, Cambridge, CB24 3AG
M6 RLW Richard Wood, 7 Wishart Green, Old Farm Park, Milton Keynes, MK7 8QB
M6 RLX Ralph Lang, 31 Ewan Close, Barrow-in-Furness, LA13 9HU
M6 RLZ Kenneth Lewis, 15 Gilbert Scott Way, Kidderminster, DY10 2EZ
M6 RMA Roman Armstrong, 11 High Ditch Road, Fen Ditton, Cambridge, CB5 8TE
M6 RMD Rebecca Dockray, 54 Kelsick Park, Seaton, Workington, CA14 1PY
MI6 RME Robin McDonald, 99 Long Commons, Coleraine, BT52 1LJ
M6 RMG Richard Grout, 5 Branton Close, Great Ouseburn, York, YO26 9SF
M6 RMI Richard Le Feuvre, 14 Elm Drive, Brockworth, Gloucester, GL3 4DH
M6 RMK Roderic Keith-Hill, 32 Thornhill Way, Plymouth, PL3 5NP
M6 RML John Hodson, 77 Waltham Road, Woodford Green, IG8 8DW
M6 RMN Timothy Ahern, 39 Essex Road, Romford, RM7 8BE
M6 RMO Clive Larner, 98 Allandale, Hemel Hempstead, HP2 5AT
M6 RMR Richard Russell, 1 Horeb Cottages, Rhiw Road, Colwyn Bay, LL29 7TL
M6 RMS Robert Somerville, 28 Yeathouse Road, Frizington, CA26 3QJ
M6 RMT Roberta Titmarsh, 38 Cromer Road, Sheringham, NR26 8RR
M6 RMU Robert Mitchell, 8 Prestwood Close, Benfleet, SS7 3LD
M6 RMV Julian Vine, 2 Chapel Meadow Cottages, Chapel Road, Deal, CT14 0JF
M6 RMW Matthew Willmott, 29 Avon Road, Kidderminster, DY11 7PE
M6 RNA Peter Bannon, 73 London Road, Worcester, WR5 2DU
M6 RNC Richard Blow, 2 Stonefall Place, Harrogate, HG2 7QL
M6 RND Bethany Starling, 4 Three Corner Drive, Norwich, NR6 7HA
M6 RNE Stewart Wilson, 2 Bydales Drive, Marske-by-the-Sea, Redcar, TS11 7HJ
M6 RNF Douglas Meekins, 267 Cell Barnes Lane, St. Albans, AL1 5PZ
M6 RNI Robert Hillier, 10 Buttermere Close, Folkestone, [illegible]
M6 RNM Andrew Atkinson, 2E Bagridge Road, Wolverhampton, WV3 8HR
M6 RNN Nigel Renny, 14 Laburnum Road, Dudley, DY1 4EP
M6 RNO Robert Nookes, 31 Pilot Road, Hastings, TN34 2AP
M6 RNQ Raymond Stevens, 7 Dunbreck Grove, Sunderland, SR4 7LL
M6 RNS John Grint, 10 Paddock Gardens, Attleborough, NR17 2EW
M6 RNU Melville Nutt, 110 Birkinstyle Lane, Shirland, Alfreton, DE55 6BT
M6 RNV Andrew Green, Flat 20, Millers Wharf, Stalybridge, SK15 2EA
M6 RNX Jeremy Faulks, 11 Fishguard Spur, Slough, SL1 1TS
M6 RNZ Richard Baldwin, 98 Rosemary Avenue, Braintree, CM7 2TA
M6 ROA Rowan Parrott, 37 Wren Place, Gillingham, SP8 4WE
M6 ROB Rob Smith, 21 Canal Road Crossflatts, Bingley, BD16 2SR
M6 ROC Christopher Martin, kiln close, main road, Lincoln, LN4 4QH
M6 ROE Peter Roe, 36 Rutland Crescent, Harworth, Doncaster, DN11 8HZ
M6 ROF Adam Martin, 81 Langstone Road, Dudley, DY1 2NL
MM6 ROH Robert Hutton, 2 Watson Place, Dunfermline, KY12 0DR
M6 ROI Peter King, 18 Paddock Close, Wantage, OX12 7EQ
M6 ROJ Rory Gudgeon, 6 The Choakles, Wootton, Northampton, NN4 6AP
M6 ROL Richard Rollinson, Beauty Bank Farm, Six Ashes, Bridgnorth, WV15 6ER
M6 ROM Claudiu Romocea, 14 Foxgrove Path, Watford, WD19 6YL
M6 ROP Ronald Pochat, Hame, Bealswood Road, Gunnislake, PL18 9DA
MM6 ROT Stuart McIntosh, 1 Ivybank Road, Port Glasgow, PA14 5LH
M6 ROU Simon-Pierre Courree, 4-6 Alhambra Road, Southsea, PO4 0RL
M6 ROV Robert Van-der-Wiel, 6 Willow Street, Romford, RM7 [illegible]
[illegible]
M6 ROX Samantha Baker, Sunville, Village Road, Mold, CH7 6HT
M6 ROY Roy Trent, 62 Dean Street, Radcliffe, Manchester, M26 3TZ
M6 RPC Robert Cobb, 57 ADAMS DRIVE, Willesborough, Ashford, TN24 0FX
M6 RPD Davy Rajanayagam, 87 Riffel Road, London, NW2 4PG
M6 RPE Andrew Wallman, 30 Elmsway, Bramhall, Stockport, SK7 2AE
M6 RPH Robert Hunt, 7 Tinsley Close, Luton, LU1 5QD
M6 RPJ Taylor Lovato, 52 Ridgeway Road, Tipton, DY4 0TU
M6 RPL Richard Lee, 13 Earle Street, Barrow-in-Furness, LA14 2PZ
M6 RPM Daniel Williams, 11 Berkeley Gardens, London, N21 2BE
MM6 RPN Paul Toner, 1 Milton Mains Road, Clydebank, G81 3NF
M6 RPO W Donnelly, 4 Mayfield Road, Bentham, Lancaster, LA2 7LP
M6 RPP John Anderson, 19a Buttermarket, Thame, OX9 3EP
M6 RPQ Richard Back, 4 St. Johns Close Buckwell, Wellington, TA21 8TF
M6 RPR Martyn Roper, 13 St. Cuthbert Street, Worksop, S80 2HN
MW6 RPS Chris Hurley, 14 Magazine Street, Maesteg, CF34 0TG
M6 RPU Andrew Smith, 20 Bishops Avenue, Worcester, WR3 8XA
M6 RPV Rob Connolly, 29 Gayer Street, Coventry, CV6 7EU
M6 RPW Craig Allan-McWilliams, 2 Sandford Crescent, Crewe, CW2 5GJ
M6 RPX Richard Price, 1 Blakiston Close, Ashington, Pulborough, RH20 3GL
M6 RPY Lawrence Percival-Alwyn, 3 Harraton Cottages, Ducks Lane, Newmarket, CB8 7HQ
M6 RPZ Simon Clarke, 46 Accommodation Road, Horncastle, LN9 5AP
M6 RQB Richard Brindley, 56 Thornton Place, Horley, RH6 8RZ
M6 RQB Robert Gower, 30 Westerham Road, Sittingbourne, ME10 1XF
M6 RQD Richard Dean, 1 Grange Street, Barnoldswick, BB18 5LB
M6 RQE Robert Fakes, 5 Brydges Road, Ludgershall, Andover, SP11 9SJ
M6 RQH Robert Hawkes, The Old Oak Bungalow, Crossway Green, Stourport-on-Severn, DY13 9SJ
M6 RQM John Morris, 1 Minafon, Newtown, SY16 1RH
M6 RQN Joseph Foster, 23 High Street, Cumnor, Oxford, OX2 9PE
M6 RRA Raymond Gibbons, 18 Langdale Road, Ribbleton, Preston, PR2 6AN
M6 RRC Karl Robinson, 103 Recreation Street, Mansfield, NG18 2HP
M6 RRD Richard Davies, 76 Berry Park, Saltash, PL12 6EN
M6 RRG Robert Grainger, Castell-Coryn, Ebbw Vale, NP23 5BB
M6 RRH Stephen Walters, 87 Fairbourne Close, Bransholme, Hull, HU7 5DH
M6 RRJ Robin Jobber, 79 Falcon Way, Ashford, TN23 5UR
M6 RRK Ram Rao, 3 Weller Mews, Enfield, EN2 8FG
M6 RRL Kathleen Woodhams, 83 Langdale Place, Newton Aycliffe, DL5 7DY
M6 RRR Robin Capon, 49 Littlemoor Road, Illingworth, Halifax, HX2 9EF
M6 RRU Christopher McGonigall, 62 Hare Lane, Crawley, RH11 7PU
M6 RRV Ronald Oxley, 17 Hardhurst Road, Alvaston, Derby, DE24 0LF
M6 RRW Keith Hazlewood, 22 Lon Cwm, Llandrindod Wells, LD1 6BE
M6 RSB Robert Blackman, Pilgrim Cottage, The Green, North Walsham, NR28 9SR
M6 RSD Rhys Davies, 19 Prince Charles Avenue, Sittingbourne, ME10 4NA
MI6 RSH Michael Rush, 25 O'Donoghue Park, Bessbrook, Newry, BT35 7AA
M6 RSI Russell Best, 14 Charles Street, Bugle, St. Austell, PL26 8PS
M6 RSJ Stanislaw Rebisz, 29 Rosecroft Drive, Nottingham, NG5 6EH
M6 RSL Barry Leeson, 23 Newports, Crockenhill, Swanley, BR8 8LE
M6 RSO Roger Oliver Oliver, 49 Albany Way, Skegness, PE25 2NB
MM6 RSP Russell Phillips, 32 Parkview, Ayr, KA7 4QF
M6 RSR Sam Richtering, 22 Darwin Drive, Driffield, YO25 5PF
M6 RSS Ronald Brooks, 52 Harebell Drive, Portslade, Brighton, BN41 2UZ
M6 RST Michael O'Donoghue, 24 Whitelock Road, Abingdon, OX14 1NZ
MD6 RSV Neil Hazell, 65 Willaston Crescent, Douglas, Douglas, Isle of Man, IM2 6LL
M6 RSY Ryan Sayre, 8 Lorne Road, Richmond, TW10 6DS
M6 RTA Stephen Himsworth, 17 Forber Road, Middlesbrough, TS4 3HJ
M6 RTC Robert Churchill, 39 Tarring Road, Worthing, BN11 4EP
M6 RTD Rod Wilson, 18 Baldwin Road, Bewdley, DY12 2BP
M6 RTE Stephen Warde, 36 Cornwallis Road, London, E17 6NN
M6 RTG Ian Sharpe, 4 Low Dowfold, Crook, DL15 9AE
M6 RTI Rachel Todd, 42 K D Tower, Cotterells, Hemel Hempstead, HP1 1AS
M6 RTM Richard Townsend, 28 Steele Street, Hoyland, Barnsley, S74 0PS
M6 RTN Martyn Fletcher, 8 Brigham Hill Mansion, Brigham, Cockermouth, CA13 0TL
M6 RTP Rael Paster, 8 Rachaels Lake View, Warfield, Bracknell, RG42 3XU
M6 RTQ Rinaldo Tempo, 35 Warminster Road, Bath, BA2 6XG
M6 RTR Alan Gill, 21 Glenburn Avenue, Wirral, CH62 8DJ
M6 RTU Arthur Loukes, 14 Batchwood View, St. Albans, AL3 5TD
M6 RTX Benjamin Purvis, 23 Deans Gardens, St. Albans, AL4 9LS
M6 RTZ Andrew Lewis, 43 High Street, Colney Heath, St. Albans, AL4 0NS
M6 RUB Ruth Beck, 13 Chaseside Avenue, Twyford, Reading, RG10 9BT
MI6 RUC William Rafferty, 65 Cable Road, Whitehead, Carrickfergus, BT38 9SJ
M6 RUE Rupert King-Evans, 70 Blenheim Park, Aldershot, GU11 2HT
M6 RUG Dave Thorpe, 45 Lord Street, Crewe, CW2 7DH
M6 RUH Ruth Plumley, 9 Haynes Road, Kettering, NN16 0NG
M6 RUK Mark Shockness, 38 Park Grange Court, Sheffield, S2 3SY
M6 RUM Leslie Rumbelow, 71 Anchorage Lane, Doncaster, DN5 8EB
M6 RUN Matthew Saunders, 68 Heywood Road, Prestwich, Manchester, M25 1FN
M6 RUP Alan Tomlinson, 10 Beech Street, Hollingwood, Chesterfield, S43 2HN
M6 RUS Russell Trickey, 19 Eastfields, Folkestone, CT19 5RU
M6 RUT Ruth Williams, 11 Berkeley Gardens, London, N21 2BE
M6 RUZ Russell Brierley, 39 Hatfield Road, Alvaston, Derby, DE24 0BU
M6 RVA Vincent Davies, 24 Holden Road, Brighton-le-Sands, Liverpool, L22 6QE
M6 RVD Roger Davies, 4 Haven Villas, Ferry Road, Exeter, EX3 0JW
M6 RVE Lillian Garrod, 121 Totteridge Lane, High Wycombe, HP13 7PH
M6 RVH Ruth Hughes, 19 Pendine Crescent, North Hykeham, Lincoln, LN6 8UW
M6 RVI Ravi Miranda, Flat 56, Amelia House 11 Boulevard Drive, London, NW9 5JP
M6 RVJ Jacobo Roa Vicens, 80 Queensway, London, W2 3RL
M6 RVM Richard Moore, 22 Chase Close, Nuneaton, CV11 6AJ
M6 RVN Ian Smith, 2 Frederick Street, Woodville, Swadlincote, DE11 8BX
M6 RVR Rajesh Varasani, 34 Freemans Rd, Minster, Ramsgate, CT12 4EL



**IMPORTANT NOTE**

**Revalidate licence to avoid revocation** – Ofcom has advised the Society that plans will be drawn up to revoke licences that have not been revalidated as required by the licence conditions. The quickest way to revalidate is to do so online via the Ofcom website: *https://services.ofcom.org.uk/* or by email: *amateur.validations@ofcom.org.uk* Ofcom staff are available to help, but please be patient during times of heavy workload.

**M6 RVY** Rob Harvey, 15 Abbey Park Way Weston, Crewe, CW2 5NR
**MM6 RWA** Robert Walmsley, 95 Race Road, Bathgate, EH48 2AU
**M6 RWB** Wayne Stobbs, 19 Stonecliffe Bank, Leeds, LS12 5BL
**M6 RWE** Dale Woodhouse, 18 Kirk Close, Ripley, DE5 3RY
**MM6 RWI** Ryszard Wilinski, Ground Flat, Mingulay, Oban, PA34 5ED
**M6 RWJ** Ann Nelson, 30a Walnut Road, Torquay, TQ2 6HS
**M6 RWP** Ronald Whalley, 65 Stanley Street, Nelson, BB9 7ET
**M6 RWS** Keith Saville, Flat 4, Birchitt Court, Sheffield, S17 4QX
**M6 RWT** Roger Trueman, 83 Morton Street, Middleton, Manchester, M24 6AX
**M6 RWV** Adam Durrant, 22 Supple Close, Norwich, NR1 4PP
**M6 RWX** Christopher Anderson, 191 Waveney Road, Hull, HU8 9NA
**MM6 RWZ** Robert Walker, 10 Westpark Gate, Saline, Dunfermline, KY12 9US
**M6 RXA** Robin Arnold, 20 South View, Frampton Cotterell, Bristol, BS36 2HT
**MI6 RXC** Taylor Masterson, 2 Pinley Drive, Banbridge, BT32 3TZ
**M6 RXE** Rebecca Ebdon, 6 Bankside, Headington, Oxford, OX3 8LT
**MM6 RXJ** Robert Johnstone, 20 Barnflat Court, Rutherglen, Glasgow, G73 1JX
**M6 RXL** Harry Leach, 34 Forest Close, Crawley Down, Crawley, RH10 4LU
**M6 RXN** Julian Cartwright, 37a Station Road, Whitwell, Worksop, S80 4UF
**M6 RXR** Roy Rich, 9 Spring Close, Verwood, BH31 6LB
**M6 RXS** Rachel Sanders-Hewett, 26 Verulam Road, Hitchin, SG5 1QE
**M6 RXT** Robert Thomas, 82 Hillcrest Road, Bromley, BR1 4SD
**MW6 RXZ** Justin Davies, 8 Coronation Road, Upper Brynamman, Ammanford, SA18 1BB
**M6 RYA** Ryan Tarr, 16 Stoneleigh Court, Frimley, Camberley, GU16 8XH
**MI6 RYC** Ryan Carmichael, 44 Union Street, Ballymoney, BT53 6HT
**MW6 RYD** Ryan Davies, 63 Maes y Gwernen Road, Cwmrhydyceirw, Swansea, SA6 6LL
**M6 RYK** Geoffrey Mackay, 5 Furze Street, Carlisle, CA1 2DL
**M6 RYL** Rita Harris, 183a Painswick Road, Gloucester, GL4 4AG
**M6 RYN** Ryan Clough, 8 Skeldyke Road, Kirton, Boston, PE20 1LR
**M6 RZA** Matthew Box, 273 Broad Lane, Birmingham, B14 5AF
**M6 RZB** Ken Slade, 7 Cottage Farm Mews, The Street, King's Lynn, PE33 9JQ
**M6 RZL** Robert Higgins, 44 Maeshyfryd Road, Holyhead, LL65 2AL
**M6 RZO** Dawn Bardell, 32 Bridle Road, Watton, Thetford, IP25 6NA
**M6 RZP** Richard Williams, 10 Bramley Close, Twickenham, TW2 7EU
**M6 RZW** Russel Laye, 24 Blackbird Way, Frome, BA11 2UR
**M6 RZX** Ben Forrest, 32 Idonia Road, Perton, Wolverhampton, WV6 7NQ
**M6 RZZ** Richard Reynolds, 22a Beech Avenue, Shepton Mallet, BA4 5XW
**M6 SAA** Sian Llewellyn, 53 Cripps Avenue, Cefn Golau, Tredegar, NP22 3PF
**M6 SAC** Morris Clark, 99 Cheylesmore Drive, Frimley, Camberley, GU16 9BW
**M6 SAD** Shane Abraham, 12 Graham Road, Halesowen, B62 8LJ
**M6 SAE** Stuart Everett, 12 Broadlands, Netherfield, Milton Keynes, MK6 4HL
**M6 SAF** Sally Anne Fisher, 25 Tennyson Road, Diss, IP22 4PY
**M6 SAH** Sarah Painter, 47 Longmead Drive, Nottingham, NG5 6DP
**M6 SAI** Timothy Beighton, 24 Wensley Close, Sheffield, S4 8HL
**M6 SAK** Shirley Kendrick, 29 Waterside, Silsden, Keighley, BD20 0LQ
**M6 SAL** Stuart Lester, 71 Ronelean Road, Surbiton, KT6 7LL
**M6 SAM** Mark Sampson, 45 Carron Street, Stoke-on-Trent, ST4 3DT
**M6 SAN** Samuel Weightman, Rose Cottage, Ellingstring, Ripon, HG4 4PW
**M6 SAQ** Shaun Wills, 8 Amherst Road, Newcastle upon Tyne, NE3 2QQ
**M6 SAS** Anthony Speight, 112 Fern Avenue, Staveley, Chesterfield, S43 3RA
**M6 SAU** Colin Saunders, 68 Heywood Road, Prestwich, Manchester, M25 1FN
**M6 SAV** Stephen Morrell, 20 Brocklebank Close, Bassingham, Lincoln, LN5 9LJ
**MW6 SAW** Philip Campigli, 43 Waterloo Road, Penylan, Cardiff, CF23 9BJ
**M6 SAX** Joshua Elsmore, 8 Clos Aberconway, Prestatyn, LL19 9HU
**M6 SAY** Steven York, 1 The Cottage, Dogdyke Bank, Lincoln, LN4 4JQ
**M6 SBB** Stephen Jeffery, 79 Greenbank Road, Watford, WD17 4FJ
**M6 SBD** Stephen Radford, Littlehampton Marina, Ferry Road, Littlehampton, BN17 5DS
**M6 SBE** Stephen Bennett, 154 Humberstone Road, Grimsby, DN32 8HR
**M6 SBF** Samantha Raine, 91 Lulworth Avenue, Jarrow, NE32 3SB
**M6 SBH** Tracey Thomas, 269 Church Road, St. Annes, Lytham St. Annes, FY8 3NP
**M6 SBI** Stephen Bassett, 3 Lower Merryfield, Anchor Road, Radstock, BA3 5PG
**M6 SBJ** Stewart Crombie, 8 Hazel close, Haverhill, CB9 9LY
**M6 SBK** Roger Colman-Whaley, 37 Suters Drive, Taverham, Norwich, NR8 6UU
**M6 SBL** Stephanie Blaney, 213 Albert Road, Poole, BH12 2EZ
**M6 SBM** Steven Brightmore, 36 Laburnum Road, Langold, Worksop, S81 9RR
**M6 SBN** Stephen Hanson, 4 Flodden Crescent, Branxton, Cornhill-on-Tweed, TD12 4SP
**M6 SBO** Barry McGlynn, 22 Bracken Bank Way, Keighley, BD22 7AB
**M6 SBR** Samantha Flowers, 56 Pilkington Road Braunstone, Leicester, LE3 1RA
**M6 SBS** Simon Stiddard, 21 Studland Park, Westbury, BA13 3HQ
**MW6 SBT** Alan Jones, 10 Dan y Bryn, Caerau, Maesteg, CF34 0UW
**M6 SBU** Stuart De Chastelain, 31 Campion Court, Northampton, NN3 9BW
**M6 SBV** David Hindle, 18 Haig Street, Selby, YO8 4BY
**M6 SBW** J Stephenson, 4 Carlow Drive, West Sleekburn, Choppington, NE62 5UT
**M6 SBY** Sam Hollis, 24 Bosley View, Congleton, CW12 3TU
**M6 SBZ** Darren Sibley, 25 Avery Close, Leighton Buzzard, LU7 4UP
**M6 SCA** Stuart Frampton, 4 Sussex Gardens, Fleet, GU51 2TL
**M6 SCB** Stephen Burnage, 124 Mayfields, Spennymoor, DL16 6TT
**MI6 SCC** Shannon Cole, 16 Otterbank Road, Strathfoyle, Londonderry, BT47 6YB
**MW6 SCD** Stephen Davies, 107 Heol Llanelli, Pontyates, Llanelli, SA15 5UH
**M6 SCF** Simon Faulkner, Mount Pleasant, Elkstones, Buxton, SK17 0LU
**MM6 SCG** Steve Greenland, 0/1 22 Linden Street, Anniesland, Glasgow, G13 1DQ
**M6 SCH** Stephen Hall, 50 Charnock Wood Road, Sheffield, S12 3HN
**M6 SCI** Martin Pope, 94 Hitchin Street, Biggleswade, SG18 8BL
**M6 SCM** Shaun Cumberland, 11 Cleveleys Road, Blackburn, BB2 3JS
**M6 SCN** Roger Rothery, 15 Broomspath Road, Stowupland, Stowmarket, IP14 4DB
**M6 SCP** Stephen Pettit, 88 Abbotsbury, Great Hollands, Bracknell, RG12 8QX
**M6 SCQ** Aaron Chamberlain, 35 Wexham Close, Luton, LU3 3TU
**M6 SCU** Daniel Willetts, 3 Stephens Close, Lewes Road, Lewes, BN8 5ET
**M6 SCV** Steve Chandler, 7 Clinton Road, Redruth, TR15 2LL
**M6 SCW** Stuart Whittaker, 25 Cleveleys Road, Blackburn, BB2 3JS
**M6 SCX** Steven Carpenter, Field View, Old Lyndhurst Road, Southampton, SO40 2NL
**M6 SDA** Ondrej Suda, 52 Coltman Street, Hull, HU3 2SG
**M6 SDE** Darius Ezard, 59 Station Farm, Croesyceiliog, Cwmbran, NP44 2JW
**M6 SDF** Adam Hulok, 3 Dickson Court, Sittingbourne, ME10 3LG
**M6 SDG** Steve Gibbs, 61a Main Road, Hoo, Rochester, ME3 9AA
**M6 SDH** Shane Hopkins, 100 Crawford Avenue, Tyldesley, Manchester, M29 8LS
**M6 SDI** Steven Legg, Woodview, Clay Lane, Clacton-on-Sea, CO16 8HH
**M6 SDL** J Layden, 51 Markham Road, Langold, Worksop, S81 9SH
**M6 SDN** A McConkey, 7 Darwin Road, Stevenage, SG2 0DE
**M6 SDP** Shane Porter, 18 The Crescent, Bircotes, Doncaster, DN11 8DT
**M6 SDQ** Simon Dean, 39 Low Grange View, Leeds, LS10 3DT
**M6 SDS** Sandra Brown, 62 Massingham Park, Taunton, TA2 7TG
**M6 SDT** Daniel Grace, 107 Bush Avenue, Little Stoke, Bristol, BS34 8NG
**M6 SDU** Stuart Seddon, 3 Kinsley Close, Ince, Wigan, WN3 4PQ
**MW6 SDV** Solomon Price, 27 Gilfach Road, Penygraig, Tonypandy, CF40 1EN
**M6 SDW** Richard Wilson, 84 Sir Thomas Whites Road, Coventry, CV5 8DR
**M6 SDX** Steven Page, Flat 25 The Riverfront , Eastern Esplanade, Canvey Island, SS8 7DN
**M6 SDZ** Spyridon Dimopoulos, 1 Oakhurst, Langley Burrell, Chippenham, SN15 4LG
**M6 SEA** Simon Aspinall, 33 Covertside Road, Scarisbrick, Southport, PR8 5HB
**M6 SEB** Sebastian Plowman, 7 Birkdale Close, Cudworth, Barnsley, S72 8EW
**M6 SEC** Timothy Summers, 5 Fairclough Place, Adlington, Chorley, PR7 4AN
**M6 SEE** Andrew Norton, 20 Cloverbank, Kings Worthy, Winchester, SO23 7TP
**MW6 SEF** Marc Williams, 2 St. Andrews Road, Wenvoe, Cardiff, CF5 6AF
**M6 SEG** Selwyn Todd, 24 Pentland Avenue, Redcar, TS10 4HD
**MM6 SEI** David Crane, Otterburn, Dervaig, Isle of Mull, PA75 6QL
**M6 SEJ** Sanderly Jeronimo, Apartment 11, Hydro House, Chertsey, KT16 8JQ
**M6 SEK** Andre Collins, 5 Grove Close, Basingstoke, RG21 3AS
**M6 SEN** Amanda Snelling, 6 Aylsham Road, Buxton, Norwich, NR10 5EX
**M6 SEO** Andrew Gardner, 1a Connaught Place, Weston-Super-Mare, BS23 2QA
**MW6 SEP** Darren Jones, 46 East Avenue, Caerphilly, CF83 2SR
**M6 SER** Stephen Richards, 5 Bloomfield Drive, Bracknell, RG12 2JW
**M6 SES** Susan Stoney, 2 Fishers Mead, Dulverton, TA22 9EN
**M6 SEU** Matthew Fairbairn, 36 Avebury Place, Cramlington, NE23 2UR
**M6 SEV** Steven Greaves, 409 Beaumont Leys Lane, Leicester, LE4 2BH
**M6 SEY** Stefan Borrell, Rose Cottage, Colchester Main Road, Colchester, CO7 8DD
**MI6 SEZ** Peter Wilson, 2 Tweskard Lodge, Belfast, BT4 2RH
**M6 SFB** Paul Latham, 20 Kenyon Avenue, Wrexham, LL11 2ST
**M6 SFC** John Ellery, 7 Midanbury Crescent, Southampton, SO18 4FN
**M6 SFC** Paul Craig, 4 Poolside, Burston, Stafford, ST18 0DR
**MM6 SFF** Steven Ferguson, 17 Brown Street, Shotts, ML7 5HW
**MI6 SFH** Shauna Hand, 12 Church Street, Rosslea, Enniskillen, BT92 7DD
**M6 SFJ** Adam Morton, 6 Chessar Ave, Chessar Ave Blakelaw, Newcastle, NE5 3RE
**MI6 SFK** Sean McElmurray, 43 Main Street, Sixmilecross, Omagh, BT79 9NH
**MD6 SFL** David Webber, 11 Waghorn Road, Harrow, HA3 9ET
**M6 SFM** Steven South, Ivy Cottage, Finkle Street Lane, Sheffield, S35 7DH
**M6 SFP** Stuart Evans, 84 Gainsborough Road, Cefn Golau, Tredegar, NP22 3TH
**M6 SFQ** Barry Cross, 22 Park Avenue, Washingborough, Lincoln, LN4 1DB
**M6 SFR** Allan Marshall, 6 Heron Way, Minnigaff, Newton Stewart, DG8 6PZ
**M6 SFS** Simon Spooner, 51 Lewisham Court, Morley, Leeds, LS27 8QB
**M6 SFT** Simon Foster, 16 Birchcroft Road, Ipswich, IP1 6PA
**M6 SFV** Sam Fry, Piccadilly Farm, Aggs Hill, Cheltenham, GL54 4ET
**M6 SFW** Joshua Walker, 6 Wellington Terrace, Islip, Kettering, NN14 3LJ
**M6 SGA** Simon Gregory, 9 Croftlands Road, Wythenshawe, M22 9YE
**M6 SGC** Stephen Latter, 1539 Great Cambridge Road, Enfield, EN1 4SY
**M6 SGD** Scott Gibbs, 35 St. Michaels, Houghton le Spring, DH4 5NR
**MM6 SGE** Nikolay Pulev, 8/6 Prestonfield Terrace, Edinburgh, EH16 5EE
**MM6 SGF** Scott Gray, 9 Caledonian Crescent, Prestonpans, EH32 9GF
**M6 SGG** Stephen Gillett, 33 School Lane, Northwold, Thetford, IP26 5LL
**M6 SGH** Steven Holt, 14 Fir Street, Cadishead, Manchester, M44 5AU
**M6 SGJ** Susan Powell, 2 Ormsby Close, Hopton, Great Yarmouth, NR31 9TY
**M6 SGL** Susan Gillard, 1 Chevening Close, Stoke Gifford, Bristol, BS34 8NJ
**M6 SGM** Stephen Roberts, 17 East View, Marsh, Huddersfield, HD1 4NU
**M6 SGN** Barry Frith, 159 Milton Road, Grimsby, DN33 1DN
**MM6 SGO** Steven Green, 4 Mid Avenue, Port Glasgow, PA14 6PL
**M6 SGP** Stuart Phipps, 53 Pioneer Avenue, Burton Latimer, Kettering, NN15 5LJ
**MM6 SGQ** Stephen Gilruth, 20 The Aspens, Carberry Crescent, Dundee, DD4 0XJ
**M6 SGS** Graham Street, 105 Jeals Lane, Sandown, PO36 9NS
**M6 SGT** Alan Williams, 11 Broadsmith Avenue, East Cowes, PO32 6QW
**M6 SGV** Manuel Alcaino Pizani, Flat 45, Brian Redhead Court, 123 Jackson Crescent, Manchester, M15 5RR
**M6 SGY** Shane Johnson, 2 North Square, Edlington, Doncaster, DN12 1ED
**MM6 SGZ** Billy Fitzsimmons, 50 Misk Knowes, Stevenston, KA20 3PQ
**M6 SHA** Matthew Shaw, 65 Vicarage Road, Amblecote, Stourbridge, DY8 4JE
**MM6 SHB** Xstuart Bradshaw, 32 Greycraigs, Cairneyhill, Dunfermline, KY12 8XL
**M6 SHC** Sharon Corden, 59 Brindles Field, Tonbridge, TN9 2YR
**M6 SHD** Geoffrey Fisk, 22 Church Street, Wangford, Beccles, NR34 8RN
**M6 SHH** Roseanna Devos, 72 Swinderby Road, Collingham, Newark, NG23 7PB
**M6 SHI** Simon Shillabeer, The Holding, Redbrook Maelor, Whitchurch, SY13 3AD
**M6 SHJ** Stephanie Jones, 56 Ffordd Offa, Rhosllanerchrugog, Wrexham, LL14 2EY
**M6 SHK** Abdullah Al-Shakarchi, 17 Fairfax Place, London, NW6 4EJ
**MD6 SHL** Sajjad Lalji, 76 Abbotts Drive, Wembley, HA0 3SG
**MM6 SHM** Ian McEwan, 4 Sim Street, Stewarton, Kilmarnock, KA3 3BP
**M6 SHP** Andrew Sharp, 30 Burbidge Close, Calcot, Reading, RG31 7ZU
**M6 SHQ** John Butcher, 77 Oulton Road, Lowestoft, NR32 4QW
**M6 SHU** Susan Maton, 41 Bemerton Gardens, Kirby Cross, Frinton-on-Sea, CO13 0LQ
**M6 SHV** Siobhan Roberts, 10 Morningside Way, Liverpool, L11 1BD
**M6 SHW** Simon Wraith, 47 Alma Road, Weymouth, DT4 0AJ
**M6 SHZ** Sharon Dainton, 1 The Woodlands, Stroud, GL5 1QE
**M6 SID** Sidney Jones, 25 The Crescent, Tredegar, NP22 3HN
**M6 SIE** Ben Brown, 114 Woodhorn Road, Ashington, NE63 9EN
**M6 SIF** Simon Frost, 5a Chestnut Grove, Purley on Thames, Reading, RG8 8BU
**M6 SIH** Sheila Bull, Flat 2, 18 Brandon Way, Birchington, CT7 9XE
**M6 SII** Simon Gray, 12 Bell Street, Henley-on-Thames, RG9 2BG
**M6 SIJ** Jacqueline Simkins, 37 St. Andrews Meadow, Harlow, CM18 6BL
**M6 SIK** Wendy Henson, 24 Grimshaw Close, North Road, London, N6 4BH
**MM6 SIM** Simon Beeson, Muir Cottage, The Muirs, Huntly, AB54 4GD
**M6 SIP** Amy Davies, 32 Kinross Road, Wallasey, CH45 8LH
**M6 SIR** Barry Greaves, 2 Peel Street, Padiham, Burnley, BB12 8RP
**MI6 SIS** Gary McCaughey, 6 Killeaton Crescent, Dunmurry, Belfast, BT17 9HD
**M6 SIU** Susan Murdoch-McKay, 19 Beachy Road, Crawley, RH11 9HN
**MM6 SIV** Mairi Sives, 4 Fir Grove, Livingston, EH54 5JP
**M6 SIW** Sian Cartwright-Proctor, 448 Tuttle Hill, Nuneaton, CV10 0HR
**M6 SIX** Brian North, 54 Parklands, Mablethorpe, LN12 1BY
**M6 SIY** Susan Lilley, 34 Rye Crescent, Danesmoor, Chesterfield, S45 9HH
**M6 SIZ** William Easdown, 38 North Street, Barming, Maidstone, ME16 9HF
**M6 SJA** Stephanie Bridges, 65 Abbots Gate, Bury St. Edmunds, IP33 2GB
**M6 SJB** Sean Buckley, 64 Wolseley Road, Rugeley, WS15 2ES
**M6 SJC** Steve Charles, 29 Woolford Close, Winchester, SO22 4DN
**MI6 SJD** Sam Dallas, 101 Coagh Road, Stewartstown, Dungannon, BT71 5JL
**M6 SJE** Jemima Hill, 7 Knebworth Court, Congleton, CW12 3SW
**M6 SJF** Stephen Fox, 77 The Grove, London, N13 5JS
**M6 SJH** Sophie Hayden, 5 Blackbird Close, Thurston, Bury St. Edmunds, IP31 3PF
**M6 SJJ** Sean Jackson, 54 Sefton Avenue, Poulton-le-Fylde, FY6 8BL
**MM6 SJK** Dean Krauskopf, Cocklehaa, Urafirth, Shetland, ZE2 9RH
**M6 SJN** Derek Lawson, 56 River Bank East, Stakeford, Choppington, NE62 5XA
**M6 SJO** Samuel Boniface, 19 Toronto Drive, Smallfield, Horley, RH6 9RB
**M6 SJQ** Charlie Gyngell, 54 Association Walk, Rochester, ME1 2XD
**M6 SJR** Stephen Ray, 18 Crescent Way, Cholsey, Wallingford, OX10 9NE
**M6 SJU** Justin Searle, 18 The Avenue, Bloxham, Banbury, OX15 4QU
**MI6 SJV** David Parkinson, 16 Beechwood Gardens, Moira, Craigavon, BT67 0LB
**M6 SJW** Edmund Watson, 4 Glenluce Drive, Preston, PR1 5TB
**MD6 SJX** Simon Bates, 6 Foxdell, Northwood, HA6 2BU
**M6 SJZ** Steven Lampard, 63 Broadmayne Road, Poole, BH12 4EH
**M6 SKD** Mark Powell, 98 Provan Court, Ipswich, IP3 8GG
**M6 SKF** Simon Froggett, 1 The Paddocks, Pilsley, Chesterfield, S45 8ET
**M6 SKH** Scott Hemmings, 26 Austin Drive, Banbury, OX16 1DJ
**M6 SKL** Sarah Maqbool, 69 Waltham Close, West Bridgford, Nottingham, NG2 6LD
**M6 SKN** Tony Watkins, 72a St. Clements Road, Keynsham, Bristol, BS31 1BA
**M6 SKP** Sean Pryer, 16 Wayside Avenue, Worthing, BN13 3JU
**M6 SKQ** Maxwell Berrisford, 5 Branwell Drive, Haworth, Keighley, BD22 8HG
**M6 SKR** Simon Keith Rogers Rogers, 30 Coed Celynen Drive, Abercarn, Newport, NP11 5AU
**M6 SKT** Timothy Robinson, 39 Cemetery Road, Laceby, Grimsby, DN37 7ER
**M6 SKU** Martin Baker, 14 Thornberry Drive, Dudley, DY1 2PL
**M6 SKV** David Jones, 19 Ffordd Hebog, Y Felinheli, LL56 4QZ
**M6 SKW** Sean Kneeshaw, 40 The Broadwalk, Otley, LS21 2RL
**M6 SKX** Simon Key, 29 Ellesmere Crescent, Brackley, NN13 6BP
**M6 SKY** Richard Coles, Bay Cottage, St. Catherines Road, Ventnor, PO38 2NE
**M6 SKZ** Terence Barham, 88 Rundells, Harlow, CM18 7HD
**M6 SLA** Steve Harvey, 40 Thales Drive, Arnold, Nottingham, NG5 7NF
**M6 SLB** Stuart Berry, 9 Magnolia Street, Winnington, Northwich, CW8 4EH
**M6 SLC** David Morphew, 8 Blinco Lane, George Green, Slough, SL3 6RQ
**M6 SLD** Simon Light, 8 Bennetts Road, Horsham, RH13 5LA
**M6 SLE** Stuart Lee, 360 Ringley Road, Stoneclough, Manchester, M26 1EP
**M6 SLG** Simon Gash, 22 Wood View, Rugeley, WS15 1AT
**M6 SLI** Samuel Lisi, 60 Middlecroft Road, Staveley, Chesterfield, S43 3XH
**M6 SLJ** Lewis Jones, 137 Breck Road, Poulton-le-Fylde, FY6 7HJ
**M6 SLK** Susan Lake, 64 Womersley Road, Norwich, NR1 4QB
**M6 SLL** Lucy Butler, 42 Station Road, Stanbridge, Leighton Buzzard, LU7 9JF
**M6 SLN** Andrew Nisbet, 65a Hamilton Road, Felixstowe, IP11 7BE
**M6 SLO** Stephen Collins, 23 Burrowlee Road, Sheffield, S6 2AT
**M6 SLR** Samantha Perry, 44 Ashcombe Road, Dorking, RH4 1NA
**M6 SLT** Sharon Townsend, 28 Steele Street, Hoyland, Barnsley, S74 0PS
**M6 SLU** Kayleigh Cockburn, 20 Hexham Avenue, Hebburn, NE31 2HN
**MM6 SLV** Kyle Slaven, Drymuir, Drymuir, Peterhead, AB42 5RR
**M6 SLZ** Simon Eloie, 26 Halsbrook Road, London, SE3 8QY
**M6 SMA** Steven Hindmarsh, 55 Warkworth Crescent, Seaham, SR7 8JT
**M6 SMB** Stewart Bide, 9 Greenway, Watchet, TA23 0BP
**M6 SMD** Sioned Davies, 2 Llyfni Terrace, Pontllyfni, Caernarfon, LL54 5ER
**M6 SME** Stephen Elliott, 74 Preston Avenue, Alfreton, DE55 7JX
**M6 SMF** Clive Smith, 21 Mill House Drive, Cheltenham, GL50 4RG
**M6 SMG** Sam Hampshire, Rose Cottage, Heath Road, Norwich, NR12 0SU
**M6 SMK** Susan Kendrick, 103a Latimer Street, Liverpool, L5 2RF
**MM6 SML** James MacLeod, 388 Garrynamonie, Lochboisdale, Isle of South Uist, HS8 5TX
**MI6 SMM** Sean Maguire, 139 Carrowshee Park, Drumhaw, Enniskillen, BT92 0FS
**MM6 SMN** Stewart Nicoll, 15 Redford Walk, Edinburgh, EH13 0AF
**M6 SMQ** Benjamin McGowan, 200 Thomas Drive, Liverpool, L14 3LE
**M6 SMR** Gail Tudor, 31 Church Road, Little Sandhurst, Sandhurst, GU47 8HY
**M6 SMX** Mark Feast, 10 Brackendale Road, Swanwick, Alfreton, DE55 1DJ
**MM6 SMY** Simon Young, 103 Feorlin Way, Garelochhead, Helensburgh, G84 0EB
**M6 SNC** James Sinclair, 19 Ridge Way, Edenbridge, TN8 6AU
**M6 SNE** Sneha Solanki, 2 Churchill Mews, Newcastle upon Tyne, NE6 1BH
**M6 SNF** Stephen Wall, Flat 1, 41 Alexandra Road, Cleethorpes, DN35 8LE
**M6 SNG** Helena Pike, 14 Milton Avenue, Barnet, EN5 2EX
**M6 SNH** Stuart Hawkes, 20 Hilltop, Loughton, IG10 1PX
**M6 SNJ** Samuel Briggs, 3 Chapel Road, Southrepps, Norwich, NR11 8UW
**M6 SNN** Sebastian Ballard, 77 Lanchester Road, Birmingham, B38 9AG
**M6 SNO** Xenia Christofi, 19 Kingsland Avenue, Northampton, NN2 7PP
**M6 SNP** Almont Alkhateb, 51 Mendip Crescent, Bedford, MK41 9ER
**M6 SNQ** Samuel Harrison, 38 Alma Road, Bournemouth, BH9 1AN
**MM6 SNR** Steven Russell, 3 Rankin Road, Wishaw, ML2 8PG
**M6 SNS** Shaun Button, 6 Farfield, Retford, DN22 7TL
**M6 SNU** Tana Lewis, 33 Boscobel Road, Buntingsdale, Market Drayton, TF9 2HG
**MD6 SNV** Archibald Elliott, Round Table House Ronague, Castletown, Isle of Man, IM9 4HJ
**M6 SNW** Daniel Snowden, 10 Woodcroft, Wakefield, WF2 7LS
**M6 SNX** Thanawit Lertruengpanya, Flat 1, Mallow Court, London, SE13 7PR
**M6 SNY** A Dobie, 7 Urr Terrace, Castle Douglas, DG7 1BL
**M6 SNZ** Sanaz Roshanmanesh, 123 The Vale, Edgbaston, Birmingham, B15 2RU
**M6 SOC** Andrew Wallace, 17 Dennis Road, Liskeard, PL14 3NS
**M6 SOE** Paul Strickland, 7 School Lane, Offley, Hitchin, SG5 3AZ
**M6 SOF** Derek Parry, 45 Taff Court, Thornhill, Cwmbran, NP44 5UU

M6 SOG Sara Jackson, 50 Leicester Road, Sharnford, Hinckley, LE10 3PR
M6 SON Greg Stewart, 35 Castle Crescent, Dewsbury, WF12 0EQ
M6 SOO Rachel Landragin, 101 Linden Gardens, Enfield, EN1 4DY
MM6 SOR Ross Ewing, Kildonan House, Caerlaverock Farm, Crieff, PH5 2DD
M6 SOT Craig Scrivens, 28 Bank Hall Road, Stoke-on-Trent, ST6 7DL
M6 SOU Aidarus Nur, 92 Bramble Avenue, Conniburrow, Milton Keynes, MK14 7AP
M6 SOV Michael McDonald, 55 Bournemouth Avenue, Middlesbrough, TS3 0NN
M6 SOZ Gregory Knowles, 31 Gibb Lane, Catshill, Bromsgrove, B61 0JP
M6 SPA Stuart Etheridge, 50 Pond Road, Horsford, Norwich, NR10 3SW
M6 SPC Johnathan Lyon, 53 Hepworth Close, Andover, SP10 3TD
M6 SPG Sarah Coxon, 31 Marden Road, Staplehurst, Tonbridge, TN12 0NE
M6 SPH Stuart Harrison, 26 Lambert Road, Lancaster, LA1 2NA
M6 SPJ Steven Johnston, 67 Eversfield Road, Horsham, RH13 5JS
M6 SPK Stephen Kay, 32 Glossop Street, Derby, DE24 8DU
M6 SPL Stephen Lycett, 58 Hazel Grove, Hucknall, Nottingham, NG15 6ED
M6 SPM Stephen Morris, 23 De Courtenai Close, Bournemouth, BH11 9PG
M6 SPN Stephen Nelson, 13 Llwyn Briscoe, Holyhead, LL65 1HT
M6 SPP Sam Fawcett, Hollins Farm, Marske, Richmond, DL11 7NH
M6 SPS Sioni Summers, 450 Baddow Road, Chelmsford, CM2 9RD
M6 SPT James Gardner, Silverdale, Vicarage Lane, Ormskirk, L40 6HQ
M6 SPU Geoff Mason, 21 Albert Street, Cowes, PO31 7ND
MI6 SPY Jason Woods, 39 Shetland Street, Antrim, BT41 2TG
M6 SPZ Samuel Pendlebury, 6 Normanby Close, Bewsey, Warrington, WA5 0GJ
M6 SQB Slade Stevens, 40 Heath Road, Exeter, EX2 5JX
M6 SQC Stephen Burton, 2 West Batter Law Farm Cottages, Hawthorn, Seaham, SR7 8RZ
M6 SQD Philip Riding, 160 Capel Road, London, E7 0JT
M6 SQE Stanley Tudor, 201 Cruddas Park, Westmorland Road, Newcastle upon Tyne, NE4 7RG
M6 SQF Sean Pearce, 15 Hillfield Court Road, Gloucester, GL1 3QS
M6 SQI Hazel Blythe, 47 Brook Road, Rubery, Birmingham, B45 9UH
M6 SQJ Slawek Kubecki, 139-141 Lapwing Lane, Manchester, M20 6US
M6 SQK Charles Southey, 31 Great North Road, Welwyn Garden City, AL8 7TJ
M6 SQL Ashley Burton, 12 Munden Grove, Watford, WD24 7EE
MI6 SQN Alexander Reid, 40 Hillfoot Street, Belfast, BT4 1PR
M6 SQO Stanley MacMurray, 21 Dymoke Green, St. Albans, AL4 9LX
M6 SQU Brian Turner, 35 Gosforth Lane, Watford, WD19 7AY
M6 SRA Sarah Antill, Lodge Farm, Mansfield Road, Worksop, S80 3DL
M6 SRB Stephen Robinson, 2 Old Hall Crescent, Bentley, Doncaster, DN5 0DW
M6 SRC Susan Cash, The Warren, Kent Hatch Road, Edenbridge, TN8 6SX
M6 SRD Reuben Strong, 82 Oakmount Road, Chandler's Ford, Eastleigh, SO53 2LL
M6 SRE Stephen Bradley, 4 Forest View Avenue, London, E10 6DX
M6 SRF John Cranston, 7 Cowen Gardens, Gateshead, NE9 7TY
M6 SRG Samantha Gibson, 66 Kinoulton Court, Grantham, NG31 7XP
M6 SRI Jason Bibby, 12 James Street, Burton-on-Trent, DE14 3SB
MM6 SRL Stephen Leeman, Beech Cottage, Murrial, Insch, AB52 6NU
M6 SRN Robert Dunn, 15a Connaught Drive, Chapel St. Leonards, Skegness, PE24 5YS
M6 SRO Christopher King, 8a Barton Road, Bedford, MK42 0NA
M6 SRS Stuart Tait, 29 Hotspur Avenue, Bedlington, NE22 5TD
M6 SRV Stephen Vickers, 35 Lanchester Road, Birmingham, B38 9AG
M6 SRZ Phillip Willetts, 223 Eaves Lane, Chorley, PR6 0AG
M6 SSA Stephen Aldersley, 245 St. Johns Road, Chesterfield, S41 8PE
M6 SSB Martin Saywell, 8 Stanley Street North, Bristol, BS3 3LU
M6 SSC Speed Azzaro, 5 Rye Hill Close, Bere Regis, Wareham, BH20 7LU
M6 SSD Susan Shephard, 17 Grimsby Road, Laceby, Grimsby, DN37 7DF
MJ6 SSF Sarah Foot, 4 Aubin Place, Aubin Lane, Jersey, JE2 7PP
M6 SSM Samantha Millard, 20 Spanners Close, Chale Green, Ventnor, PO38 2HY
M6 SSN Brian Woods, 28 Delph Drive, Burscough, Ormskirk, L40 5BE
M6 SSO Sam Jones-Martin, 146 Winchester Road, Four Marks, Alton, GU34 5HZ
M6 SSP Shane Pashley, 3 Princess Street, Brimington, Chesterfield, S43 1HP
M6 SSR Stan Rymel, 3 Southview, Smalldale, Hope Valley, S33 9JQ
M6 SSS Simon Rowe, 9 Corfield Close, Finchampstead, Wokingham, RG40 4PA
M6 SST Stephen Slapper, 1 Standards Keep, Standards Road, Bridgwater, TA7 0EZ
M6 SSW Benjamin Hood, 168 Shay Lane, Walton, Wakefield, WF2 6NP
M6 SSX John Prout, 110 Dorothy Avenue North, Peacehaven, BN10 8DP
M6 SSY Nina Hancock, 3 Market Street, Shipdham, Thetford, IP25 7LY
M6 STA Shaun Ancill, 45 Tristan Close, Calshot, Southampton, SO45 1BN
M6 STC Samuel Cousins, West Mill, Wareham Common, Wareham, BH20 6AA
M6 STF Stephen Briant, 13 Southall Avenue, Brighton, BN2 4BA
M6 STH Alec Huckle, 36 Weldbank Close, Beeston, Nottingham, NG9 5FU
M6 STI Benjamin Richards, 58 Holm Lane, Oxton, Prenton, CH43 2HS
M6 STJ Stacey Sheppard, 256 Brandwood Road, Birmingham, B14 6LD
MU6 STK Catherine Stockwell, Fleurs Des Champs, La Colline Des Bas Courtils, St. Saviours, Guernsey, GY7 9YQ
MM6 STM Steven McMillan, 56 Mount Pleasant, Armadale, Bathgate, EH48 3HB
MI6 STN Steven Nash, 45 Parkfield Road, Ahoghill, Ballymena, BT42 1LY
M6 STP Sam Phythian, 24 Water Grove Road, Dukinfield, SK16 5QS
M6 STR Michael Stroud, 1 Sefton Court, Welwyn Garden City, AL8 6WW
M6 STU Stuart Vzor, 40 Henlow Road, Birmingham, B14 5DS
M6 STV Gordon Wadsworth, 21 Appletree Road, Featherstone, Pontefract, WF7 5EA
M6 STY Tyler Fletcher, 9 Lawn Avenue, Kimpton, Hitchin, SG4 8QD
MM6 SUB Stephen Boyd, Gowanbank Chalet, Garelochhead, Helensburgh, G84 0AF
M6 SUC Joe Murphy, Wessex House, Drake Avenue, Staines-upon-Thames, TW18 2AP
M6 SUE Susan Hadley, 60 chapel street, pensnett, Brierley Hill, DY5 4EF
M6 SUF Stuart Fotheringham, 8 Ivanhoe Court, Ulrica Drive, Thurcroft, S66 9QP
M6 SUH Susan Halewood, 12 Silver Street Riccall, York, YO19 6PB
M6 SUI Shaun Brooker, 4 Fleet End Close, Havant, PO9 5ED
M6 SUJ S Jones, Marvin House, Ryhill Pits Lane, Wakefield, WF4 2DU
M6 SUK Ian Ridsdale, 15 Carlton Road, Hough-on-the-Hill, Grantham, NG32 2BG
M6 SUL Richard Sullivan, 14 Colleton Drive, Twyford, Reading, RG10 0AU
M6 SUM Pete Smith, 14 Highfield Crescent, Kettering, NN15 6JS
M6 SUN Rocio Beaumont, 61 Mitcham Road, Camberley, GU15 4AR
M6 SUP Daniel Markey, 7 Knightsway, Wakefield, WF2 7EG
MM6 SUR Russell Fair, 6 Fairways, Stewarton, Kilmarnock, KA3 5DA
M6 SUU Susan Coombes, 33 Clarence Park Road, Bournemouth, BH7 6LF
M6 SUV David Arnold, The Chase, Rectory Road, Penzance, TR19 6BB
M6 SUX Stephen Harris, 17 Swindale, Wilnecote, Tamworth, B77 4LD
M6 SUZ Suzanne Trigg, 2 Langley Common Road, Barkham, Wokingham, RG40 4TS
M6 SVF Laura Garnett, 32 George Fox Way, Norwich, NR5 8BJ
M6 SVG Joshua Savage, 44 Hastings Road, Maidstone, ME15 7SP
M6 SVJ Stuart Carpenter, Flat 17, Henley Court, Ipswich, IP1 3SD
M6 SVM Philip Edwards, Delfryn, Capel Dewi, Aberystwyth, SY23 3HU
M6 SVN Sharon Jackson, 18 Kentish Gardens, Tunbridge Wells, TN2 5XU
M6 SVT Stephen Tayler, 22 Wheatley Road, Leicester, LE4 2HN
M6 SVV Steven Berrow, 40 Priorsgate Oakdale, Blackwood, NP12 0EL
M6 SVY Darren Southernwood, 24 Silver Gardens, Belton, Great Yarmouth, NR31 9PD
M6 SVZ James Pauline, 54 Laurel Road, Bassaleg, Newport, NP10 8NY
M6 SWA Simon Worger, 6 Glendale Terrace, Mornington Road, Whitehill Bordon, GU35 9AJ
MI6 SWB Stanley Beatty, 132 Joanmount Gardens, Belfast, BT14 6NZ
MM6 SWC Scott Caldwell, 45 Pappert, Alexandria, G83 9LE
M6 SWD Charlie Rich, Red Oak House, Summer Lane, Woodbridge, IP12 2QA
M6 SWE Morgan Nilsson, 11 Holly Hedge Road, Frimley, Camberley, GU16 8ST
M6 SWG George Wicks, 4 Bedford Street, Barnstaple, EX32 8JR
M6 SWH Stephen Hughes, 104 Thornley Road, Stoke-on-Trent, ST6 7BA
M6 SWI Andrew Waller, 64 Heaton Road, Billingham, TS23 3GP
M6 SWL Paul Dekkers, 21 Nodens Way, Lydney, GL15 5NP
MW6 SWN Stephen Butler, 33 Heol Penlan, Neath, SA10 7LB
M6 SWO Steven Woods, 92 Rubens Avenue, South Shields, NE34 8JT
M6 SWP Stephen Plume, The Mallards, Green Lane, Hitchin, SG4 0BU
M6 SWS Ian Franklin, 23 Ingle Drive, Ashby-de-la-Zouch, LE65 2LW
MM6 SWT Zoe McKinnon, 8 Rowanlea Avenue, Paisley, PA2 0RP
M6 SXA Nicholas Carter, Bernette, New Works Lane, Te, TN22 1TT
M6 SXB Maureen Comber, 9 Blackwell Road, East Grinstead, RH19 3HP
M6 SXC Michael Revell, 13 Mount Pleasant, Framlingham, Woodbridge, IP13 9HQ
M6 SXD Jake Brookes, 177 Charnwood Close, Rubery, Birmingham, B45 0JY
M6 SXI Samantha Robinson, 47 Platt Hill Avenue, Bolton, BL3 4JU
M6 SXM Michael Cook, 30 St. Peters Crescent, Selsey, Chichester, PO20 0NA
M6 SXN Stuart Neale, 43 Crompton Road, Pleasley, Mansfield, NG19 7RG
M6 SXO Adam Cooke, Iolanthe, Chidham Lane, Chichester, PO18 8TH
M6 SXP Stephen Price, 43 Charter Road, Weston-Super-Mare, BS22 8LN
M6 SXT Matthew Dengate, 15 Barn Close, Pease Pottage, Crawley, RH11 9AN
MM6 SXY Leanne Ritchie, Braiklaw, Blackhills, Peterhead, AB42 3LA
M6 SYG Stella Dibben, 2 Taynton Covert, Birmingham, B30 3QR
M6 SYH Brian Hooper, 55 Gildas Avenue, Birmingham, B38 9HS
M6 SYK Paul Sykes, 16 Hill Fold, South Elmsall, Pontefract, WF9 2BZ
M6 SYW Sonny Ward, 24 Deloney Road, Norwich, NR7 9DQ
M6 SYX Andrew Davies, 4 Capella Path, Hailsham, BN27 2JY
MM6 SZA Gail Inglis, Roadside, Skirza, Wick, KW1 4XX
M6 SZP William Davies, 61a Lightridge Road, Huddersfield, HD2 2HF
M6 SZZ Malcolm Smith, 21 Buckden Close, Woodley, Reading, RG5 4HB
M6 TAA Tamer Akay, 87 Swithland Avenue, Leicester, LE4 5BQ
M6 TAD Gillian Tew, 66 St. Nicholas Estate, Baddesley Ensor, Atherstone, CV9 2EZ
M6 TAF Andrew Cornelius, 16 Crown House, North Street, Bristol, BS48 4SX
M6 TAG David Cutter, 92 Hillcrest, Bar Hill, Cambridge, CB23 8TQ
M6 TAH Tom Hutchinson, 134 Wingate Square, London, SW4 0AN
M6 TAJ Darren Jarvice, 15 Meden Avenue, Warsop, Mansfield, NG20 0PS
M6 TAK Christopher Smith, 7 Betws Avenue, Kinmel Bay, Rhyl, LL18 5BN
M6 TAL Theresa Lewis, 154 Meadow Head, Sheffield, S8 7UF
M6 TAO Paul Washbrook, 1 Berry Hill Cottages, Berry Hill, Seaton, EX12 3BD
MM6 TAT Tatiana McArthur, 160 Lamond Drive, St. Andrews, KY16 8JP
M6 TAU Andrew West, 33 Mundays Row, Waterlooville, PO8 0HF
MI6 TAW Thomas Wilmot, 4 Esdale Park, Bushmills, BT57 8RB
M6 TAY A H Ayres, Brynhyfryd, Phocle Green, Ross on Wye, HR9 7TW
M6 TAZ Scott Turner, 41 Fox Street, Scunthorpe, DN15 7LE
M6 TBC Allan Doyle, 54 Bro Syr Ifor, Tregarth, Bangor, LL57 4AS
M6 TBD Aaron Smith, 1 Fields Park Road, Newport, NP20 5BA
M6 TBG Anthony Gravell, 21 Wickridge Close, Stroud, GL5 1ST
M6 TBH Portia Bowman, 26 Albany Hill, Tunbridge Wells, TN2 3RX
M6 TBJ Louise Perry, 56 Lambrook Road, Taunton, TA1 2AF
M6 TBK Anthony Green, Flat 19, 19-25 Marine Parade East, Clacton-on-Sea, CO15 1UX
M6 TBL Thomas Lander, 20 Greenfield Square, Morda, Oswestry, SY10 9NZ
M6 TBM Tom Behan, 48 Montrose Avenue, Datchet, Slough, SL3 9NJ
M6 TBN Steven Nicholls, 101 Manchester Road, Worsley, Manchester, M28 3NT
M6 TBO Matthew Bradbury, 30 Hazel Road, Maltby, Rotherham, S66 8BD
M6 TBQ Alex Trick, 2 Newell Close, Aylesbury, HP21 7FE
M6 TBR Cameron Herd, 182 Hungerhill Road, Nottingham, NG3 3LL
M6 TBS Dominic Fletcher, 2 Hillside Close, Heddington, Calne, SN11 0PZ
M6 TBT Alan Talbot, 30 Irwell Road, Walney, Barrow in Furness, LA14 3UZ
M6 TBU Malcolm Johns, 151 Somerset Street, Abertillery, NP13 1DR
M6 TBV Roger Day, 152 Swievelands Road Biggin Hill, Westerham, TN16 3QX
M6 TBX Trevor Burkinshaw, 16 Farnley Avenue, Sheffield, S6 1JP
M6 TCB Peter Deluce, 114 Townsfield Road, Westhoughton, Bolton, BL5 2NT
M6 TCD Thomas Denny, 20 School Lane, Surbiton, KT6 7QH
M6 TCE Robert Hannant, 24 Tower Hill Park Costessey, Norwich, NR8 5AT
M6 TCF Ian Moore, 128 Dalestorth Street, Sutton-in-Ashfield, NG17 4FY
M6 TCG Anthony Gilborts, 22 Granby Road, Buxton, SK17 7TW
M6 TCH Trevor Hewlett, Tuckers Cottage, Alfold Road, Cranleigh, GU6 8NB
M6 TCI Steve Friend, 1 Orford Road, Tunstall, Woodbridge, IP12 2JH
M6 TCK Paul Pritchard, 11 Beacon Avenue, Dunstable, LU6 2AD
M6 TCL Walter Hand, 168 Barcroft Street, Cleethorpes, DN35 7DX
M6 TCM Remington Fowler, 579 Wheatley Lane Road, Fence, Burnley, BB12 9EE
M6 TCN Dudley Woodhams, 83 Langdale Place, Newton Aycliffe, DL5 7DY
MI6 TCO James Allen, 3 Malwood Close, Belfast, BT9 6QX
MM6 TCS Thomas Moffat, 11 Mansfield Road, Prestwick, KA9 2DL
M6 TCX Trevor Atherton, 16 Steeple View, Ashton-on-Ribble, Preston, PR2 2PX
M6 TCY Alan Todd, 27 Woodleigh Crescent, Ackworth, Pontefract, WF7 7JG
M6 TCZ Thomas Clarke, 168 Hykeham Road, Lincoln, LN6 8AP
M6 TDA Michael Aubrey, 4 Broadleaf Close, Oakwood, Derby, DE21 2DH
M6 TDB Lee Brookes, Tyn Llidiart, Llanfairpwllgwyngyll, LL61 6EQ
MW6 TDC Dawn Cowling, Dorset Cottage, Alltyblacca, Llanybydder, SA40 9SU
M6 TDD Timothy Douglas, 72 Oakdale Road, Poole, BH15 3LG
MW6 TDE Peter Phillips, Sedgemoor, Station Road, Kilgetty, SA68 0XS
M6 TDF Trevor Peck, 7 Byron Road, Mablethorpe, LN12 1JD
M6 TDI Adam Jones, 67 Fakenham Road, Beetley, Dereham, NR20 4ET
M6 TDJ John Stringer, 21 Ladywalk, Maple Cross, Rickmansworth, WD3 9YZ
M6 TDK Ian Radford, 7 Eastmount Avenue, Hull, HU8 9EW
MW6 TDL Tyrone Lawrence, 14 Railway Terrace, Caerau, Maesteg, CF34 0UE
M6 TDM Steven Brown, 4 Dorado Gardens, Orpington, BR6 7TD
M6 TDO Steven Hogg, 57 The Grange, Burton-on-Trent, DE14 2EX
M6 TDP Peter Thorley, 57 Riverside Drive, Hambleton, Poulton-le-Fylde, FY6 9EH
M6 TDR Tracey Robertson, 6 Sandringham Court, Bircotes, Doncaster, DN11 8QU
M6 TDS Thomas Stewart, 91 Chequers Field, Welwyn Garden City, AL7 4TX
MD6 TDU Kevin Dodds, 16 Second Avenue, Onchan, Isle of Man, IM3 4LE
M6 TDV Thomas McNamara, 19 Abbey Mews, Pontefract, WF8 1TD
M6 TDY Edward Ashford, 56 Finch Close, Shepton Mallet, BA4 5GL
M6 TDZ Amy Thomas, 5 Thornley Road, Wolverhampton, WV11 2HR
M6 TEE Phillip Hardacre, 13 St. Johns Street, Bridlington, YO16 7NL
M6 TEF Raymond Ashman, Cartref, Salisbury Road, Salisbury, SP4 9QZ
M6 TEG Timothy Guy, 16 Cogdeane Road, Poole, BH17 9AS
M6 TEJ Tim Jones, Formula Cars, Wellington, TA21 9HW
M6 TEK Amar Sood, Parima, Sewardstone Road, London, E4 7RA
M6 TEL David Jones, 2 Trewen, Llandinam, SY17 5BU
M6 TEM John Turner, 34 Vaughan Road, Stotfold, Hitchin, SG5 4EH
M6 TEO Matthew Prentice, 2 Wickenden Road, Sevenoaks, TN13 3PJ
M6 TEP Tracy Jones, Flat 1, Studley Manor, 270 Frome Road, Trowbridge, BA14 0DT
MM6 TEQ Krzysztof Ruchomski, 53a Barnton Avenue, Edinburgh, EH4 6JJ
M6 TER Benjamin Lye, 55 South Avenue, Sherborne, DT9 6AR
MW6 TES Anthony Bailey, 6 Trenos Gardens, Bryncae, Pontyclun, CF72 9SZ
M6 TEV Trevor Howson, 82 Bank End Avenue, Worsbrough, Barnsley, S70 4QN
MM6 TEW Tracey Warr, 407 Smerclate, Isle of South Uist, HS8 5TU
M6 TEY Anthony Brown, 3 Alston Road, New Hartley, Whitley Bay, NE25 0ST
M6 TEZ Terence Archer, 241 Beaver Lane, Ashford, TN23 5PA
M6 TFB Tom Blanchard, 17 Heol Tyn-y-Fron, Penparcau, Aberystwyth, SY23 3RP
M6 TFD David Lyes, 2 Thelnetham Road, Blo Norton, Diss, IP22 2JQ
M6 TFE Tom Fletcher, 3 Moorend Glade, Charlton Kings, Cheltenham, GL53 9AT
M6 TFF Peter Ratcliffe, 61 Queens Avenue, Ilfracombe, EX34 9LS
MI6 TFG Trevor McKee, 4 Earlford Heights, Newtownabbey, BT36 5WZ
M6 TFJ Timothy Johnsen, 5 Willow Lane, Billinghay, Lincoln, LN4 4FN
M6 TFK Tim Kightly, Ferry Hill Farm, London Road, Chatteris, PE16 6SG
MW6 TFL Tony Fletcher, 16 Cefneithin Road, Gorslas, Llanelli, SA14 7HT
MI6 TFN Ryan Taylor, 18 Rose Park Tandragee, Craigavon, BT62 2LZ
M6 TFO Andrew Theobold, 25 Aysgarth Road, Leicester, LE4 0ST
M6 TFP Thomas Chambers, 20 Leysholme Terrace, Leeds, LS12 4HL
M6 TFQ Daisy Thomson, 11 Uranus Road, Hemel Hempstead, HP2 5QF
M6 TFT Hugo Salter, 1 Rock Farm Cottages, Gibbs Hill, Maidstone, ME18 5HT
M6 TFV David Forshaw, 14 Hope Carr Road, Leigh, WN7 3ET
M6 TFW Francis Worrall, 297 Tamworth Road, Amington, Tamworth, B77 3DG
MM6 TFY Thomas Yates, 12 Fulmar Court, Newtonhill, Stonehaven, AB39 3QG
M6 TFZ Emma Walton, 11 Parkfield Road, Northwich, CW9 7AR
MM6 TGB Tam Bell, 22 Queens Road, Elderslie, Johnstone, PA5 9LJ
M6 TGC Brian Grainger, 42 Madeira Avenue, Leigh-on-Sea, SS9 3EB
MM6 TGD Ian Currie, 4 Greendyke Cottage, Falkirk, FK2 8PP
M6 TGF Michael Malone, 5 Brook Park Avenue, Prestatyn, LL19 7HH
M6 TGG Thomas Sutton, Yew Tree Farm, Paddol Green, Shrewsbury, SY4 5QZ
M6 TGH Anthony Hardy, 3 Cornwall Road, Leicester, LE4 0BB
M6 TGI Thomas Hunt, Mad Bess Cottage, Breakspear Road North, Uxbridge, UB9 6LZ
M6 TGJ Terry Woolvin, 34 Baker Street, Tipton, DY4 8JX
M6 TGK Richard Willis, 12 Robartes Road St. Dennis, St. Austell, PL26 8DS
M6 TGL Christopher Dolphin, 24 Saughall Road, Wirral, CH46 6DS
M6 TGM Shane Thorpe, 35 Hunters Grove, Swindon, SN2 1HE
MM6 TGN Anthony Barclay, 21 Netherlea, Scone, Perth, PH2 6QA
M6 TGP Ross Chambers, 3 Westby Way, Poulton-le-Fylde, FY6 8AD
M6 TGQ Toby Webb, 5 Holst Avenue, Manchester, M8 0LS
M6 TGR Helen McPhillips, 160 Pasley Street, Plymouth, PL2 1DT
M6 TGS Alexander Stubbs, 20 Mossfields, Crewe, CW1 4TD
M6 TGU Matthew Bostock, 86 Beauvale Drive, Ilkeston, DE7 8SJ
M6 TGV Ben Mountford, 189 Lloyd Street, Stockport, SK4 1NH
M6 TGY Tony Hopkins, Flat, Horrabridge Stores, Commercial Road, Yelverton, PL20 7QB
M6 TGZ Shaun Richardson, Flat 10, Auburn Mansions, Poole, BH12 1BW
M6 THA K Thacker, 2 New Cottages, Selham, Petworth, GU28 0PJ
M6 THB Tom Barratt, 17 Main Road, Collyweston, Stamford, PE9 3PF
M6 THC Michael Hellon, 75 Brassey Street, Birkenhead, CH41 8BZ
M6 THI Keith Barker, 10 Corn Mill Close, Rochdale, OL12 9UW
M6 THJ Timothy Hunt, 108 Gainsborough Close, Salisbury, SP2 9LU
M6 THO Andrew Thornett, Ty Bedwen, Birch Grove, Lichfield, WS13 6EP
MJ6 THP Thomas Pallot, Biltmore, La Grande Route de St. Laurent, Jersey, JE3 1NH
M6 THQ Christian Taylor, 9 Cesson Close, Chipping Sodbury, Bristol, BS37 6NJ
M6 THS Taras Shevchenko, 25 Kirby Road, Dartford, DA2 6HE
MM6 THU Alan Gray, 28 Le Roux Drive, Oldmeldrum, Inverurie, AB51 0PJ
M6 THV Terence Hunt, 14 Wenlock Road, South Shields, NE34 9BA
MW6 THW Terry Woodley, 2 Parc Onen, Neath, SA10 6AA
M6 THY Hanying Tang, Flat 31, 74 Arlington Avenue, London, N1 7AY
M6 TIA Colin Lyne, 4 Bridge Close, Catterick Garrison, DL9 4PG
M6 TIC Rohan Wainwright, 38 Stanway Close, Worcester, WR4 9XL
M6 TID Alan Gilham, 38 Trojan Way, Waterlooville, PO7 8AL
M6 TIF Christopher Townsend, 4 Chatsworth Close, Aston, Sheffield, S26 2GA
M6 TIG Gary Spiers, 68 Thamesmead, Walton-on-Thames, KT12 2SJ
M6 TIH Trevor Mackenzie, 2 Newcastle Street, Carlisle, CA2 5UH
M6 TII Thomas Ince, 18 Holly Walk, Keynsham, Bristol, BS31 2TU
MI6 TIJ Christopher Thompson, 86 Huguenot Drive, Lisburn, BT27 4YD
M6 TIL Paul Athersmith, 5 Aqueduct Lane, Stirchley, Telford, TF3 1BW
M6 TIO Timothy Allsopp, 413 Boulton Lane, Derby, DE24 9DL

**IMPORTANT NOTE**

**Revalidate licence to avoid revocation** – Ofcom has advised the Society that plans will be drawn up to revoke licences that have not been revalidated as required by the licence conditions. The quickest way to revalidate is to do so online via the Ofcom website: *https://services.ofcom.org.uk/* or by email: *amateur.validations@ofcom.org.uk* Ofcom staff are available to help, but please be patient during times of heavy workload.

**MW6 TIQ** Richard Baker, 20 Hawthorne Terrace, Aberdare, CF44 7HE

**MM6 TIR** Stuart Taylor, 17 Broomhill Close, Kingseat, Aberdeen, AB21 0AH

**M6 TIU** David Firth, 7 Martinet Drive, Lee-on-The-Solent, PO13 8GP

**M6 TIW** Thomas Heath, 6 The Sycamores, Scawthorpe, Doncaster, DN5 7UH

**M6 TIY** Tim Cooper, 11 Warwick Road, Totton, Southampton, SO40 3QP

**M6 TJA** Thomas Atkin, 15 Coxway, Clevedon, BS21 5AQ

**M6 TJB** Tom Bexon, 51 Hookstone Drive, Harrogate, HG2 8PR

**M6 TJD** Tracy Davis, 11 Hinton Villas, Hinton Charterhouse, Bath, BA2 7SS

**MW6 TJE** Gary Hodges, 79 Trefelin, Aberdare, CF44 8LF

**MI6 TJF** Trevor Ferguson, 3 Wheatfield Park, Ballybogy, Ballymoney, BT53 6NT

**M6 TJG** Daniel Watts, 16 Park View Gardens, Bassaleg, Newport, NP10 8JZ

**MW6 TJH** Ryan Orchard, The Burrows, Spring Gardens, Whitland, SA34 0HL

**M6 TJI** Thomas Dyer, 180 Seaton Lane, Hartlepool, TS25 1HF

**M6 TJJ** Tony Hempsall, 16 Central Avenue, Warrington, WA2 8AJ

**M6 TJL** Trevor Lake, 64 Lytchett Drive, Broadstone, BH18 9LB

**M6 TJN** Christine Austen, Fenbank House, Roman Bank, Holbeach Clough, Spalding, PE12 8DH

**M6 TJO** Richard Austen, Fenbank House, Roman Bank, Holbeach Clough, Spalding, PE12 8DH

**M6 TJQ** Daniel Roberts, 43 Minions Close, Atherstone, CV9 2BD

**MW6 TJS** Terence Skerritt, 94 Turberville Street, Maesteg, CF34 0LU

**M6 TJX** Toni Dalby, 52 Narborough Road South, Leicester, LE3 2FN

**M6 TJY** Francis Grimley, 37 Reginald Road, Bexhill-on-Sea, TN39 3PH

**M6 TJZ** Tony Johnson, 81 Welbeck Street, Whitwell, Worksop, S80 4TN

**M6 TKA** Karl Thompson, 184 Tickhill Road, Doncaster, DN4 8QS

**M6 TKC** Thomas Corcoran, 191 Queensway, Rochdale, OL11 2NA

**M6 TKE** Nicola Crabb, 1 Council Houses, Hall Lane, Norwich, NR12 7BB

**M6 TKG** Thomas Bishop, Flat 1, 75 Devonshire Street South, Manchester, M13 9DA

**M6 TKI** Julie Lewis, 93 Eastcliff, Portishead, Bristol, BS20 7AD

**M6 TKK** Thomas Kosteletos, 10 Church Lane, Southwick, Brighton, BN42 4GD

**M6 TKP** Anthony Pickavance, 19 Merton Bank Road, St. Helens, WA9 1DY

**M6 TKR** Timothy Berisford, 18 Cambridge Avenue, Winsford, CW7 2LL

**M6 TKU** Sharon-Ann Heys, 15 Heathfield Road, Fleetwood, FY7 7LY

**M6 TKW** Tony Ward, 1 Darrismere Villas, Edinburgh Street, Hull, HU3 5AS

**M6 TKX** Tom Keable, 5 Redhills Way, Hetton-le-Hole, Houghton le Spring, DH5 0ES

**M6 TLB** Robert Lawrence, 20 Coronation Drive, Forest Town, Mansfield, NG19 0AJ

**M6 TLC** Tara Christopher, 27 Brinkhill Crescent, Nottingham, NG11 8GN

**M6 TLF** Tomas Ford, Clwt Joli, Llanfrothen, Penrhyndeudraeth, LL48 6DU

**MI6 TLG** Robert Greer, 11 Mullaghcarton Road, Lisburn, BT28 2TE

**M6 TLK** Adrian Kyte, 42 Radford Street, Alvaston, Derby, DE24 8NS

**M6 TLL** Andy Lear, 38 The Roundway, Claygate, Esher, KT10 0DW

**M6 TLM** Ian Williams, 36 Telford Road, Tamworth, B79 8EY

**MW6 TLN** Nerys Weightman, 39 Brynamman Road, Lower Brynamman, Ammanford, SA18 1TR

**M6 TLP** Thomas Porter, 30 Woodbridge Close, Appleton, Warrington, WA4 5RD

**M6 TLR** Tomos Rogers, 9 Maryland Road, Risca, Newport, NP11 6BB

**M6 TLS** Justin Olver, 43 Heron Gardens, Portishead, Bristol, BS20 7DH

**M6 TLT** Benjamin Cooper, 80 Holland Road, Maidstone, ME14 1UT

**M6 TLX** Matthew Egan, 4 Rutter Avenue, Warrington, WA5 0HP

**M6 TLY** Andrew Thompson, 25 Ardingly Road, Cuckfield, Haywards Heath, RH17 5HD

**M6 TMA** Robert Smith, 3 Plane Tree Close, Marple, Stockport, SK6 7RJ

**M6 TMC** Stuart Oram, 9 Springbrook, Eynesbury, St. Neots, PE19 2DT

**M6 TMD** Paul Kirby, 36 Durham Road, London, E12 5AX

**M6 TMF** Darren Fletcher, 97 Wallace Crescent, Carshalton, SM5 3SU

**M6 TMG** Mark Galloway, 2 Edendale Terrace, Horden, Peterlee, SR8 4RD

**M6 TMH** Tony Hoyle, 60 Greenbank Crescent, Marple, Stockport, SK6 7PB

**M6 TMJ** Michael Jones, 11 Lower Glen Park, Pensilva, Liskeard, PL14 5PP

**M6 TMK** Timothy Walker, 11 Banburies Close, Bletchley, Milton Keynes, MK3 6JP

**M6 TML** Thomas Longmore, 3 Dairy Farm Cottages, Northlands Road, Gainsborough, DN21 5DN

**M6 TMM** Terry Marsland, 34 Wakefield Road, Stalybridge, SK15 1AJ

**M6 TMO** Thomas Williams, Moor Farm, Moor Lane, Lincoln, LN3 4EG

**M6 TMP** Trevor James, 87a Beaconfield Road, Epping, CM16 5AT

**MI6 TMR** Michael Graham, 32 North Street, Ballinderry Upper, Lisburn, BT28 2ER

**MM6 TMS** Steven Scott, 32 Dixon Terrace, Whitburn, Bathgate, EH47 0LH

**M6 TMY** Mark Thackery, 1 Bockings Grove, Clacton-on-Sea, CO16 8DL

**MI6 TMZ** James Allen, 192 Joanmount Gardens, Belfast, BT14 6PA

**M6 TNA** Anthony Golden, 3 Ampleforth Road, Middlesbrough, TS3 7PU

**M6 TNB** Panagiotis Bozikis, 336 Higham Hill Road, London, E17 5RG

**M6 TNC** James Winstone, 31 Setterfield Way, Rugeley, WS15 1BJ

**MI6 TND** Robert Dobson, 2 Mourne View, Crossgar, Downpatrick, BT30 9HW

**M6 TNH** Richard MacDougall-Smith, 160 Thingwall Park, Bristol, BS16 2BU

**MI6 TNI** Thomas Nelson, 1, 1 Annashanco, Rosslea, BT92 7PT

**MM6 TNO** Diamantino De Freitas, 14 York Street, Clydebank, G81 2PH

**M6 TNR** Thomas Robinson, 39 Cemetery Road, Laceby, Grimsby, DN37 7ER

**M6 TNU** Julie Lambert, 69 Anvil Crescent, Broadstone, BH18 9DZ

**M6 TNY** Anthony Deeming, 16 Marshall Road, Exhall, Coventry, CV7 9BX

**MI6 TNZ** Alan McCullough, 45 Eglantine Park, Hillsborough, BT26 6HL

**MI6 TOA** Thomas McPolin, 40 Olympia Drive, Belfast, BT12 6NH

**MI6 TOC** Tessa Carvill, 18 Hospital Road, Newry, BT35 8PW

**M6 TOD** Thomas Sandham, 96 South Road, Morecambe, LA4 6JS

**M6 TOE** Anthony Smyth, 2 North Edge, Leigh, WN7 1HW

**M6 TOF** John Morgan, Glas y Dorlan, Pontrhydfendigaid, Ystrad Meurig, SY25 6EJ

**M6 TOG** Peter Joyce, 92 Essex Road Halling, Rochester, ME2 1AX

**M6 TOK** Steven Halliday, 8 Newby Farm Road, Scarborough, YO12 6UN

**M6 TOL** Christopher Higgins, 27 Chetwynd Road, Birmingham, B8 2LB

**M6 TOM** Thomas Oliver, 17 East Lea, Newbiggin-by-The-Sea, NE64 6BQ

**M6 TON** Tony Ralph, 54 Clacton Road, Portsmouth, PO6 3QY

**M6 TOR** Darrell Marjoram, 81 Wellington Road, Eccles, Manchester, M30 9GW

**M6 TOT** Steven Dix, 8 Beaumont Road, Longlevens, Gloucester, GL2 0EJ

**M6 TOV** Martin Murdoch, 19 Beachy Road, Crawley, RH11 9HN

**M6 TOW** Andrew Smith, 5 Lode Hill Caravan Site, Downton, Salisbury, SP5 3QH

**M6 TOY** Antony Mepham, 6 Claremont Road, Bexhill-on-Sea, TN39 5BX

**M6 TOZ** David Torrance, 30 St. Norbert Drive, Ilkeston, DE7 4EH

**M6 TPA** Anthony Austin, 4 Cornwall Avenue, Oldbury, B68 0SW

**MI6 TPC** Thomas Crozier, 201d Linn Road, Larne, BT40 2AJ

**M6 TPD** Trevor Davies, 7 Crescent Road, Warley, Brentwood, CM14 5JR

**M6 TPE** Allan Smith, Grasmere, Malthouse Lane, Burgess Hill, RH15 9XA

**M6 TPG** Thomas Gollins, 18 John Street, Stafford, ST16 3PJ

**M6 TPI** Tiffany Kilfeather, Flat 1, 57 Chalk Hill, Watford, WD19 4DA

**MW6 TPM** Timothy McKeown, 13 George Street, Treherbert, Treorchy, CF42 5AH

**M6 TPO** STEVE Bunworth, 21 Slattsfield Close, Selsey, Chichester, PO20 0EB

**M6 TPR** Elizabeth Butler, Tanglewood, Elms Lane, Wolverhampton, WV10 7JS

**M6 TPT** Thomas Taylor, 74 Sandbeck Avenue, Skegness, PE25 3JS

**M6 TPV** Philip Watson, 5 Welsford Close, Wells, BA5 2JE

**M6 TPX** Anthony Patrick, The Woodlands, Nantwich Road, Chester, CH3 9JH

**M6 TPY** Timothy Pollett, 10 Bridport Road, Poole, BH12 4BS

**M6 TQF** Kim Taylor, 44 Main Street, Willoughby, Rugby, CV23 8BH

**M6 TQW** Su-Fay Moss, 6 Orchard Close, Watford, WD17 3DU

**M6 TRC** Nicholas Kent, Flat 1, Manor House, Redruth, TR15 1AX

**M6 TRD** Mark Read, 133 Somersall Street, Mansfield, NG19 6EL

**M6 TRE** Rob Evered, 2-3 Lower Downside, Downside, Shepton Mallet, BA4 4JP

**MI6 TRF** daragh foy, 125 Drumbeg Tullygally, Craigavon, BT65 5AE

**M6 TRH** Donald Constable, 7 Mill Field, Sutton, Ely, CB6 2QB

**M6 TRI** David Freeman, 37 Bedale Road, Nottingham, NG5 3GL

**MW6 TRK** Trevor Keane, 19 Williton Road, Llanrumney, Cardiff, CF3 5QE

**MW6 TRL** Paul Terrell, 82 Baglan Street, Treherbert, Treorchy, CF42 5AR

**MM6 TRO** Sascha Troscheit, 20 James Street, St. Andrews, KY16 8YA

**M6 TRP** Troy Green, 33 Livingstone Road, Broadstairs, CT10 2UF

**M6 TRQ** Ian Troughton, Rhiwbina, Pentre Lane, Cwmbran, NP44 3AP

**M6 TRS** Toby Steel, Barrowfield House, Much Hadham, SG10 6BD

**M6 TRT** Jon Fry, 104 Freemantle Road, Rugby, CV22 7HY

**M6 TRU** Graham Truman, 9 Riddell Avenue, Langold, Worksop, S81 9SS

**M6 TRV** Trevor Meakin, 33 The Markhams, New Ollerton, Newark, NG22 9QX

**M6 TRW** Tom Wheatley, 20 Rugby Road Rainworth, Mansfield, NG21 0AT

**MM6 TRX** Alexander McNeill, 13 Spinkhill, Laurieston, Falkirk, FK2 9JR

**M6 TSA** Tony Stamp, 41 Glamorgan Close, Mitcham, CR4 1XG

**MM6 TSC** Thomas Couper, 10 Sclandersburn Road, Denny, FK6 5LP

**M6 TSG** Timothy Goh, Wolfson College, Cambridge, CB3 9BB

**M6 TSI** Steve Gilham, 7 Weeton Road, Wesham, Preston, PR4 3BQ

**M6 TSM** Thomas Keay, 12 Quarryfields, Seahouses, NE68 7TB

**MM6 TSN** Rorie Thomson, Barthol Hill Cottage, Barthol Chapel, Inverurie, AB51 8TB

**M6 TSP** Christopher Power, 23 Manor Road, Killamarsh, Sheffield, S21 1BU

**M6 TSR** Thomas Scopes, The Garden House, West Common Road, Keston, BR2 6AJ

**M6 TST** Trevor Tate, 339 West Dyke Road, Redcar, TS10 4PS

**MD6 TSW** David Williamson, 45 Bluebell Close, Peel, Isle of Man, IM5 1GH

**M6 TSZ** Matthew Wickens, St. James Vicarage, The Parade, Dudley, DY1 3JA

**M6 TTB** Trevor Bate, 87 Dunsheath, Telford, TF3 2BY

**M6 TTE** Thomas Battelle, 23 King Edward Avenue, Blackpool, FY2 9TA

**M6 TTF** Max Lanham, 3 Park Cottages, New Common, Bishop's Stortford, CM22 7RT

**M6 TTH** Thomas Horsten, Kastelsvej 4, 2.Tv, Copenhagen E, Denmark, 2100

**M6 TTI** Matti Juvonen, 7 Alphin Brook, Didcot, OX11 7FG

**M6 TTK** Robert Rushton, 77 Boroughbridge Road, Knaresborough, HG5 0ND

**M6 TTL** Matthew Walker, 3 Finch Close, Tadley, RG26 3YJ

**M6 TTM** Jack Aldwinckle-Day, 45 Felicia Way, Grays, RM16 4JF

**M6 TTN** Matt Newham, 3 Laurel Drive, Brockworth, GL3 4GF

**M6 TTO** Francis Armstrong, 38 Dovecote Drive, Haydock, St. Helens, WA11 0SD

**M6 TTP** Mark McGowan, 48 Alderley Road, Thelwall, Warrington, WA4 2JA

**M6 TTR** Santina Chuck, 82 Redmayne Drive, Chelmsford, CM2 9AG

**MW6 TTS** Martyn Johns, 39 Tyla Coch, Llanharry, Pontyclun, CF72 9LT

**M6 TTT** Aidan Allen, 65 Hanbury Road, Dorridge, Solihull, B93 8DN

**M6 TTV** Darren Hargrove, 24 Slade Road, Wolverhampton, WV10 6QS

**M6 TTW** Matthew Berlyn, 13 Hopping Jacks Lane, Danbury, Chelmsford, CM3 4PN

**M6 TTY** Tosi Ying Catherine Tong, Lincoln House, Clarence Drive, Harrogate, HG1 2QD

**M6 TUD** Christopher Barker, 18 Nickleby Road, Waterlooville, PO8 0RH

**MM6 TUG** A Cairns, Tigh Bruadair, Achmore, Strome Ferry, IV53 8UX

**M6 TUH** Richard Taylor, Penhawger Park, Liskeard, PL14 3LW

**M6 TUK** C Yunnie, Lighthouse, Park Lane, Totnes, TQ9 7BD

**MI6 TUM** Shane Tumilty, 18 Cluain-Air, Newry, BT34 1PW

**M6 TUN** Andrew Tunney, 79 Scott Street, Burnley, BB12 6NJ

**M6 TUR** David Slee, Turnpike, Brearton, Harrogate, HG3 3BX

**M6 TUT** Adrian Mars, 14 Woodyard Close, London, NW5 4BX

**MI6 TUV** Charles Bailie, 26 Moatview Park, Dundonald, Belfast, BT16 2BE

**MI6 TUX** Eoghan Murray, 33 Orpen Avenue, Belfast, BT10 0BS

**M6 TVB** Stephen Brett, 10 Approach Road, Broadstairs, CT10 1QT

**M6 TVC** Lance Lovell, 28 Park Lane, Blunham, Bedford, MK44 3NJ

**M6 TVE** Steve Rowland, 6 Peach Hall, Tonbridge, TN10 3HD

**M6 TVH** Edward Grabham, 2 The Old School, Horton, Ilminster, TA19 9QS

**M6 TVI** Liam Dillane, 48 Ripple Road, Birmingham, B30 2RB

**M6 TVM** Terence Melia, 6 Burnley Close, Blackburn, BB1 3HL

**M6 TVS** Neil Simmonds, 3 Noneley Hall Barns, Noneley, Shrewsbury, SY4 5SL

**M6 TVV** Scott Baggott, 2 Ellison Street, West Bromwich, B70 7ES

**M6 TVW** Trevor Hawkesford, 24 Greenaway, Morchard Bishop, Crediton, EX17 6PA

**M6 TVX** David Mills, 91 Harp Road, London, W7 1JQ

**M6 TVY** Michael Reaney, Odessa Marine, Little London, Newport, PO30 5BS

**M6 TWA** Trevor Alvey, 2 Sunny View, Back Road, Saxmundham, IP17 3NY

**M6 TWD** Jake Tweed, 24 Windsor Road, Polesworth, Tamworth, B78 1DA

**M6 TWE** Stephen Tweedie, 3 Edgar Road, Westruther, Gordon, TD3 6ND

**M6 TWH** Jennifer Hughes, Midfield Farm, Midfield Caravan Site, Aberystwyth, SY23 4DX

**M6 TWI** Jessica Truman, 9 Riddell Avenue, Langold, Worksop, S81 9SS

**M6 TWL** Thomas Willis, 143 Clarence House, Leeds, LS10 1LH

**M6 TWQ** John Attwood, 13 John Winter Court, Euston Road, Great Yarmouth, NR30 1DU

**M6 TWW** Thomas Ward, 20 Ollerton Road, Edwinstowe, Mansfield, NG21 9QG

**MM6 TWZ** Thomas Dodd, 2 White Wisp Gardens, Dollar, FK14 7BH

**M6 TXA** Adrian Fairclough, Forders House, Forders Lane, Bedworth, CV12 9SG

**M6 TXH** Jason Marriott-Levett, 37 Christleton Drive, Ellesmere Port, CH66 3NN

**M6 TXI** Brian Cave, 2 Beaufort Close, Newcastle upon Tyne, NE5 3XL

**M6 TXJ** John Hopkins, 53 Sprules Road, London, SE4 2NL

**MM6 TXK** Tim Kerby, 1 St. Marks Lane, Edinburgh, EH15 2PX

**M6 TXM** Anthony Miller, Ashtree Farm, Rugby Road, Rugby, CV23 9PN

**MM6 TXN** Sarah Moore, 35 Niddrie House Park, Edinburgh, EH16 4UH

**M6 TXP** Patrick Cassells, 5 Saxon Way, Liverpool, L33 4DW

**M6 TXR** Richard Withers, 50 Coneygear Road, Hartford, Huntingdon, PE29 1QL

**MI6 TXS** Richard Bates, 61 Starbog Road, Kilwaughter, Larne, BT40 2TL

**M6 TXT** Michael Bookham, 116 Clare Gardens, Petersfield, GU31 4EU

**M6 TXU** Alan Gillard, 58 Queens Road, Thame, OX9 3NQ

**M6 TXX** Andrew Gibbins, Flat 120, Greenhill, London, NW3 5TY

**M6 TXZ** Robert Bolton, 1 Beehive Terrace, Trefechan, Aberystwyth, SY23 1BW

**M6 TYB** Rhys Thornett, Ty Bedwen, Birch Grove, Lichfield, WS13 6EP

**M6 TYC** Tyrone Corcoran, 50 Grange Road, Bracebridge Heath, Lincoln, LN4 2PW

**M6 TYE** Stuart Connolly, 82 Cheswood Drive, Minworth, Sutton Coldfield, B76 1YE

**M6 TYG** Carl Morris, 17 Percy Road, Wrexham, LL13 7EA

**M6 TYK** Joshua Goddard, 217 Speedwell Road, Bristol, BS5 7SP

**M6 TYL** Tyler Ward, James Barn, Horham, Eye, IP21 5ER

**MM6 TYM** Timothy Hawes, 139/8 Great Junction Street, Edinburgh, EH6 5JB

**M6 TYN** Michael Tynan, 57 Alpine Drive, Wardle, Rochdale, OL12 9NY

**MI6 TYR** Vincent Brogan, 7 Richmond Park, Omagh, BT79 7SJ

**M6 TYS** Steve Tyrrell, 14 Town Orchard, Southoe, St. Neots, PE19 5YJ

**M6 TYT** Dean Redhead, 2 Richardson Place, Colney Heath, St. Albans, AL4 0NW

**M6 TZA** Anthony Zaulincy Adams, 131 Briar Gate, Long Eaton, Nottingham, NG10 4DH

**M6 TZD** Nicholas Cripps, 66 Forest Road, London, E8 3BT

**M6 TZE** Sarah Johnston, 62 Charlecote Park, Telford, TF3 5HD

**M6 TZO** James Morton, 39 Cuttys Lane, Stevenage, SG1 1UP

**MI6 TZP** Christopher Gault, 7 Gardenmore Place, Larne, BT40 1SE

**M6 TZR** Terry Crouch, 2 Park Farm Close, Horsham, RH12 5EU

**M6 TZU** Stephen Moore, 32 Broadgreen Close, Leyland, PR25 2XA

**MM6 TZX** Scott McCorkell, 28 Leven Road, Hamilton, ML3 7WS

**M6 TZY** Steven Crabb, 1 Council Houses, Hall Lane, Norwich, NR12 7BB

**M6 TZZ** Philip Moore, 24 Plough Road, Dormansland, Lingfield, RH7 6PS

**MI6 UAB** A Pritchard, 16 Ballymaconnell Road South, Bangor, BT19 6DQ

**M6 UAD** Martin Kimber, 2 Church Road, Brandon, IP27 0EN

**M6 UAE** Paul Potter, 21 Ulverston Crescent, Lytham St. Annes, FY8 3RZ

**M6 UAF** James Glynn, 106 Fairway Avenue, West Drayton, UB7 7AP

**M6 UAJ** Aaron Ibbotson, 323 Brincliffe Edge Road, Sheffield, S11 9DE

**M6 UAP** Shunichi Ando, 84 New Hey Road, Cheadle, SK8 2AQ

**M6 UAR** David Randles, 111 East Pines Drive, Thornton-Cleveleys, FY5 3RY

**M6 UAS** Anthony Cudworth, 30 Compass Tower, Munnings Road, Norwich, NR7 9TW

**M6 UAX** Alan MacDonald, Woodside Cottage, Horton Way, Verwood, BH31 6JJ

**M6 UBC** BENJAMIN CLIFFORD, 5 Spirit Quay, London, E1W 2UT

**MI6 UBE** Linda Calderwood, 43 Rathview Park, Mullybritt, Enniskillen, BT94 5EW

**M6 UBH** Russell Tolman, 10 Woodcote Way, Abingdon, OX14 5NE

**M6 UBI** Christine Cotton, 49 Cornwall Road, Portsmouth, PO1 5AR

**M6 UBN** Benjamin Shephard, 74 Harcourt Street, Kirkby-in-Ashfield, Nottingham, NG17 8DD

**M6 UBR** mark moore, 3 St. Pauls Road Birkenshaw, Bradford, BD11 2JY

**M6 UBS** Nicholas Evetts, 35 Wood End Road, Kempston, Bedford, MK43 9BB

**MI6 UBT** Thomas McWilliams, 221 Kings Road, Belfast, BT5 7EH

**M6 UCP** Dennis Ali, 20 Millwall Close, Manchester, M18 8LL

**M6 UCS** Matthew Coote, 4 Hunters Oak, Watton, Thetford, IP25 6HL

**M6 UCY** Lucy Byrne, 15 Norton Avenue, Canvey Island, SS8 8LG

**M6 UDA** Lotte Symonds, 10 Cowper Court, Willunga, Australia, 5172

**M6 UDC** Daniel Cole, 39 Hillside Road, Southminster, CM0 7AL

**M6 UDM** Darren Meehan, 47 Clinton Road, Shirley, Solihull, B90 4RN

**MI6 UDR** Robert Dunwoody, 16 Dernalea Road, Milford, Armagh, BT60 4DZ

**M6 UDS** Gary Chapman, 253 St. Pauls Road, Preston, PR1 6NS

**M6 UDX** Roger Berwick, 10 Hall Lane, Wacton, Norwich, NR15 2UH

**M6 UDY** Dawn Smout, Sunrays, Warbage Lane, Bromsgrove, B61 9BH

**M6 UEA** Martin Robinson, 58 Haworth Road, Cross Roads, Keighley, BD22 9DL

**M6 UEB** James Masters, 31 Lower Beeches Road, Birmingham, B31 5JB

**M6 UED** Masayuki Tanji, Apartment 63, Westside One, Birmingham, B1 1LS

**M6 UEH** Stephen Smith, 39 Wimborne Road, Southend-on-Sea, SS2 5JG

**MM6 UEN** Ian MacDonald, 20 Newbigging Terrace, Auchtertool, Kirkcaldy, KY2 5XL

**M6 UES** Richard Hawes, 40 Nightingale Way, Thetford, IP24 2YN

**M6 UFC** Alex Hunt, 46 Devereaux Crescent, Ebley, Stroud, GL5 4PU

**M6 UFF** Toby George, 19 Princes Crescent, Margate, CT9 1LY

**M6 UFO** Caroline Sims, 133 Canterbury Road, Hawkinge, Folkestone, CT18 7BS

**M6 UFR** Stuart Sinderbury, 30 North Lonsdale Street, Stretford, Manchester, M32 0PG

**M6 UGA** George Amos, Goffs Oak, Waltham Cross, EN7 6SB

**M6 UGM** Graham Mountain, 34 Albert Road, Warlingham, CR6 9EP

**M6 UGN** Stuart Nesling, 64 Ruskin Avenue, Lincoln, LN2 4BT

**M6 UGX** Malcolm Street, 7 Salisbury Street, Sowerby Bridge, HX6 1EE

**M6 UHF** Jayne Smith, 24 Monk Road, Wallasey, CH44 1AJ

**M6 UHN** Brian Marks, 167 Linnet Drive, Chelmsford, CM2 8AH

**M6 UHT** Paul Lunn, 179 Coventry Road, Nuneaton, CV10 7BA

**M6 UHU** James Willetts, 102 Welch Road, Cheltenham, GL51 0EG

**MI6 UIM** Fidelma Hand, 12 Church Street, Rosslea, Enniskillen, BT92 7DD

**M6 UIR** Ross Muir, 4 Blandford Avenue, Worsley, Manchester, M28 2JE

**M6 UJM** Robert Banks, 3 Parkhayes, Woodbury Salterton, Exeter, EX5 1QS

**M6 UKA** Gary Curtis, 49 Dorset Street, Blandford Forum, DT11 7RF

**M6 UKB** Jonathan Lightfoot, 5 Market Hill, Rothwell, Kettering, NN14 6EP

**M6 UKC** Paul Williams, 53 Lilac Avenue, Cannock, WS11 0AR

**M6 UKG** Jean-Paul English, 1 Niton Cottage Pound Lane, Meonstoke, Southampton, SO32 3NP

**M6 UKH** Andrew Shaw, 20 Hillcrest Close, Thrapston, Kettering, NN14 4TB

**M6 UKI** Uros Trnjakov, 63 Welford Gardens, Abingdon, OX14 2BH

**M6 UKJ** James Pithers, 77 Victoria Road, Laindon, Basildon, SS15 6RA

**M6 UKM** Martin Reeves, 41 Hogarth Road, Whitwick, Coalville, LE67 5GF

**M6 UKO** Leslie Salt, 4 Thames Road, Walsall, WS3 1PJ

**M6 UKT** Geoffrey Eason, Whitegates, Parsonage Road, Bishop's Stortford, CM22 6QX

**M6 UKX** George Radulescu, 41 Sherard Road, London, SE9 6EX

**M6 ULA** Jan Van Der Elsen, SULA Lightship, Llanthony Road, Gloucester, GL2 5HH

MM6 ULL Kenneth Mackenzie, Alderwood, Braes, Ullapool, IV26 2TB
MW6 ULX Cathy Griffiths, 20 Lon Heddwch, Clydach, Swansea, SA6 5RE
M6 ULY Bryan Page, 12 Hitchens Close, Hemel Hempstead, HP1 2PP
M6 ULZ Wayne Morrell, 80 Fernwood Rise, Westdene, Brighton, BN1 5EP
M6 [illegible] Paul [illegible], 60 [illegible], Heywood, OL10 [illegible]
M6 UMM Fergus Colquhoun, 2 Heslop Road, London, SW12 8EG
M6 UMR Umar Munir, 100 Ranelagh Road, Southall, UB1 1DG
M6 UNA Jonathan Taberner, 32 Bell Lane, Sutton Manor, St. Helens, WA9 4BD
MI6 UNC Colin Harper, 48 Downpatrick Street Rathfriland, Newry, BT34 5DQ
M6 UNI Mark Knapton, 33 The Queens Drive, Mill End, Rickmansworth, WD3 8LN
M6 UNS Paul Unsworth, 83 Newbold Avenue, Sunderland, SR5 1LL
M6 UNY Lee Betts, 12a Maesgwyn, Pontnewydd, Cwmbran, NP44 1BQ
M6 UPE Robert Saddler, Flat 17, Priestley Court, High Wycombe, HP13 7WZ
M6 UPH Aled Williams, 8 Old Tanymanod Terrace, Blaenau Ffestiniog, LL41 4BU
M6 UPS Andrew Morehen, 20 Castleton Grove, Inkersall, Chesterfield, S43 3HU
M6 UPU Anthony Stirk, 5 Hall Stone Court, Shelf, Halifax, HX3 7NY
M6 URA Dangis Kveksas, 36, Purley, CR8 3AQ
M6 URC Glenn Pritchard, Coedcwm, Glasbury, Hereford, HR3 5NX
M6 URG Susan Kitchen, 16 Crown Avenue, Cudworth, Barnsley, S72 8SE
M6 URM Alan Stansfield, 59 Prankerds Road, Milborne Port, Sherborne, DT9 5BX
MM6 URP Samantha Gray-Jones, Flat C, 7 Nelson Street, Aberdeen, AB24 5EP
M6 URR Michael Riddick, Davah, Port Road, Castle Douglas, DG7 3JW
M6 URS Sorin Banda, 11 Harwood Court, Kent Road, Grays, RM17 6DJ
M6 URX Matej Urban, 59 Sterling Gardens, London, SE14 6DU
M6 USA Mary Dexter, 70 Rutland Street, Derby, DE23 8PR
MI6 USC William Bradley, 14 Ardmore Grange, Ballygowan, Newtownards, BT23 5TZ
M6 USD Juan Rufes, Flat 1 & 3-8, 12 Smyrna Road, London, NW6 4LY
M6 USE David Hardy, 186 Sneyd Hill, Stoke-on-Trent, ST6 1RA
M6 USG Barry Featherstone, 16 Neroche View, Hatch Beauchamp, Taunton, TA3 6AD
M6 USK Christopher Moreton, 20 Millbrook Court, Little Mill, Pontypool, NP4 0HT
M6 USM Joseph Wright, 32 Carlton Road, Long Eaton, Nottingham, NG10 3LF
M6 USO ILYA USOV, 17 Marsh Farm Lane, Swindon, SN1 2GL
M6 USP Ashley Brighton, 38 Greenfield Crescent, Nailsea, Bristol, BS48 1HR
M6 USV Denis Soames, 40 Woodland Drive, North Anston, Sheffield, S25 4EP
M6 UTB Edward Harrington, 53 Kingscroft Road, Banstead, SM7 3NA
M6 UTC Antony Driscoll, 82 Station Road, Langford, Biggleswade, SG18 9PQ
M6 UTI Jonathan Blay, 31 Woodcote House, Queen Street, Hitchin, SG4 9TL
M6 UTP Conner Walker-Riley, 1 Farmcote Court, Hemlington, Middlesbrough, TS8 9LJ
M6 UTT Caroline Dyer, 74 Coldstone Road, Lingfield, RH7 6BT
MI6 UTV Tom Maclaine, 172 Moylagh Road, Seskanore, Omagh, BT78 2PN
M6 UTX Peter Begg, 20 Wilsford Avenue, Uttoxeter, ST14 8XG
M6 UUA Declyn Sealey, 6 Mizzen Road, Hull, HU6 7AG
M6 UUE nick busley, Horseshoe Lodge South Drove, Spalding, PE11 3BD
M6 UUU Adam Castell, 47 Markyate Road, Slip End, Luton, LU1 4BU
M6 UVB Benjamin Axcell, 16 Broomfield Road, Herne Bay, CT6 7AY
M6 UVD dennis smith, 17 Stonepound Road, Hassocks, BN6 8PP
M6 UVF Edward Goodwin, 55 Twickenham Road, Sunderland, SR3 4JN
MI6 UVS Quinton Church, 51 Houston Park, Broughshane, Ballymena, BT42 4LB
M6 UWK John Hadley, 75 Glendower Avenue, Coventry, CV5 8BD
M6 UWS Marie-Ann Sharman, 3 Deben Crescent, Swindon, SN25 3QB
M6 UXB Carl Hurley, 12 Grocott Road, Wednesbury, WS10 8RQ
M6 UZH Umer Hussain, Cardwell Barn, Hostingley Lane, Dewsbury, WF12 0QH
M6 UZK Alan Sweet, 3 Beechwood Grove, Blackpool, FY2 0DZ
M6 UZZ David Willis, 7 Shenstone Drive, Rotherham, S65 2JU
MJ6 VAA Victoria Atherton, 4 Clos de la Mer, La Route de Noirmont, Jersey, JE3 8AL
MM6 VAB Victor Julius, 100 Hogarth Drive, Cupar, KY15 5YU
M6 VAG Keith Lane, 243 Edwin Road, Gillingham, ME8 0JL
M6 VAH Vance Downes, 55 Ashfield Road, Bromborough, Wirral, CH62 7EE
MI6 VAI Les Hambleton, 9 Springvale Road, Ballywalter, Newtownards, BT22 2PE
M6 VAJ Afonwen Coomber, 2 Bracken Grove, Catshill, Bromsgrove, B61 0PB
M6 VAK Derrick Evans, 1 Copeland Street, Hyde, SK14 4TD
M6 VAL Albert Lander, 37 Berry Drive, Paignton, TQ3 3QW
M6 VAM Elizabeth Driscoll, 42 Adelaide Square, Shoreham-by-Sea, BN43 6LN
M6 VAN Evan Tsang, Harrogate Ladies' College, Clarence Drive, Harrogate, HG1 2QG
M6 VAP Andrew Mason, 13 Welton Gardens, Lincoln, LN2 2AY
M6 VAR Tasos Varoudis, 33 Arlington Road, London, NW1 7ES
M6 VAU Simon Fairbourn, 17 Perry's Lane, Wroughton, Swindon, SN4 9AX
M6 VAV Andrew Vasarhelyi, 1 Eldon Close, Langley Park, Durham, DH7 9FR
M6 VAW Darren Vassie, Teddards, Filching, Polegate, BN26 5QA
M6 VAY Alex Vassie, Teddards, Filching, Polegate, BN26 5QA
M6 VAZ James Bevan, 18 Martin Road, Diss, IP22 4HR
MI6 VBB Darren Twaddle, 19 Lisanduff Park, Portballintrae, Bushmills, BT57 8RY
M6 VBD Trevor Hillier, 16 Priory Walk, Leicester Forest East, Leicester, LE3 3PP
MW6 VBE Joseph Jones, 7 St. Mary Street, Trelewis, Treharris, CF46 6AL
M6 VBF Peter [illegible], Chestnut Farm, Eastville, Boston, PE22 8LX
M6 VBJ Jason Vanbesien, Flat 8, Suffolk House, Chester, CH1 3BZ
M6 VBP Sandra Vyner, 20 Ryde Lands, Cranleigh, GU6 7DD
M6 VBR Jess Baughan, Chestnut Farm, Eastville, Boston, PE22 8LX
M6 VBS Alexander Cairns, 202 The Ridgeway, St. Albans, AL4 9XJ
M6 VBX Peter Otterwell, 50 Hythe Road, Staines-upon-Thames, TW18 3EE
M6 VBZ David Scott, 13 Grey Close, Bredbury, Stockport, SK6 1HA
M6 VCA Vishal Chady, 12 Brent Place, Barnet, EN5 2DP
M6 VCB Daniel Bonney, 14 Jersey Close, Congleton, CW12 3TW
M6 VCC Oliver Blacklock, 15 Kings Crescent, Lymington, SO41 9GT
M6 VCK Vicki Corcoran, 50 Grange Road, Bracebridge Heath, Lincoln, LN4 2PW
M6 VCM Victoria Millson, Harrogate Ladies' College, Clarence Drive, Harrogate, HG1 2QG
M6 VCN David Williams, 6 Homestone Gardens, Leicester, LE5 2LJ
M6 VCP Colin Pinder, 70 Highfield Road, Beverley, HU17 9QR
M6 VCS Vanessa Newell, 15 The Grove, Luton, LU1 5PE
M6 VCU K Beswick, 29 Hops Lane, Halifax, HX3 5FB
M6 VDC Neil Fairbairn, 15 Hewitt Road, Dover, CT16 1TH
M6 VDE David Evans, 12 Caerfallwch, Rhosesmor, Mold, CH7 6PN
M6 VDH Victor Harrington, 25 Victoria Street, Norwich, NR1 3QX
MM6 VDP David Pounder, 13 Stornoway Drive, Kilmarnock, KA3 2GJ
M6 VDR C Johnson, The Hollies, Belaugh Green Lane, Norwich, NR12 7AJ
M6 VDX Peter Martin, 108 Headlands Grove, Swindon, SN2 7HP
M6 VFT Matthew Broadhurst, 23 Hucklow Avenue, North Wingfield, Chesterfield, [illegible]
M6 VED Rodney Peter Dawson Dawson, 6 Harleys Field, Abbeymead, Gloucester, GL4 4RN
M6 VEG Jack Waldron, 55 Sheringham Road, Poole, BH12 1NS
M6 VEL David Rogers, 6 Guildford Street, Plymouth, PL4 8DS
M6 VEN Andrew Fisher, 15 Washdyke Lane, Osgodby, Market Rasen, LN8 3PB
M6 VET Alexander Williamson, 4 Garden End, Melbourn, Royston, SG8 6HD
M6 VFA Andrew Colcombe, 217 Church Drive, Quedgeley, Gloucester, GL2 4US
M6 VFC Maureen Chipperfield, 5 Lullingstone Close, Hempstead, Gillingham, ME7 3TS
MM6 VFL Benjamin Sonnet, 12 Winton Circus, Saltcoats, KA21 5DA
M6 VFN Ben Bewick, 35 Waveney Drive, Hoveton, Norwich, NR12 8DP
M6 VFR Tim Ward, 25 Chislehurst Road, Carlton Colville, Lowestoft, NR33 8BY
M6 VGA Joshua Duffain, 53 Maes y Ffynnon, Brecon, LD3 9PL
M6 VGE Stuart Coates, 27 Primula Way, Chelmsford, CM1 6QT
M6 VGR Ian Brodie, 38 Chatsworth Place, Stoke-on-Trent, ST3 7DP
MM6 VGS Raymond Adams, 1/R 61 Capelrig Street, Thornliebank, Glasgow, G46 8LP
M6 VGU Kevin Haythornwhite, 148 Nairne Street, Burnley, BB11 4NP
M6 VGV Gautham Venugopalan, 3 Southwater Close, London, E14 7TE
M6 VHA Vhaire Gudgeon, 6 The Choakles, Wootton, Northampton, NN4 6AP
M6 VHM Liam Mooney, Calde Cottage, Mannings Lane, Chester, CH2 2PB
M6 VHZ James Telfer, 50 Agraria Road, Guildford, GU2 4LF
M6 VIA Laurie Kirkcaldy, 62 West Garth Road, Exeter, EX4 5AN
M6 VIE Steven CROSTON, 12 Sefton Street, London, SW15 1LZ
M6 VIG Christopher Harvey, 1 New Pit Cottages, Bridge Place Road, Bath, BA2 0PE
M6 VIO Simon Morgan, 17 Redwood Close, Witham, CM8 2PL
M6 VIT Antonio Vitiello, 8 Pegasus Road, Leighton Buzzard, LU7 3NJ
M6 VIV Viv Lee, 56 Gordon Road, Fishersgate, Brighton, BN41 1PT
M6 VIX Wendy Reynolds, 4 Underwood Close, Stafford, ST16 1TB
M6 VJG Vincent Gallagher, 4 Kingsdown Way, Bromley, BR2 7PT
MI6 VJR Raymond Russell, 70 Allen Park, Dunamanagh, Strabane, BT82 0PD
M6 VJX Bradley Walker, 255 Packington Avenue, Birmingham, B34 7RU
M6 VKA Steven Rush, 88 Mountview, Borden, Sittingbourne, ME9 8JZ
M6 VKE Daeun Lee, 26 Church Lane, Chalgrove, Oxford, OX44 7TA
M6 VKG Mandy Gibson, 6 Harrison Road, Mansfield, NG18 5RG
M6 VKH Steven Hardy, 55 Westwood Road, East Peckham, Tonbridge, TN12 5DB
M6 VKK Richard Cresswell, Meadow View, Hulver Road, Beccles, NR34 7UW
M6 VKN Raymond Reilly, 29 Burwell Close, Plymouth, PL6 8QD
M6 VKW Victoria Williams, Moor Farm, Moor Lane, Lincoln, LN3 4EG
M6 VKY Victoria Fleming, 1 Balmoral Drive, Methley, Leeds, LS26 9LE
M6 VKZ George Delaforce, 29 Littlebridge Meadow Bridgerule, Holsworthy, EX22 7DU
M6 VLB Kelvin Earwicker, 21 Fitzpain Road, West Parley, Ferndown, BH22 8RZ
M6 VLD Vlad Mereuta, 62 Sutherland Avenue, London, W9 2QU
MM6 VLG William Guy, 121 Stewarton Drive, Cambuslang, Glasgow, G72 8DH
MW6 VLR Rennie Blunden, Philbeach House, Dale, Haverfordwest, SA62 3QU
M6 VLS Steven Morris, 21 Heritage Way Badgers Green, Llanymynech, SY226LL
M6 VMA Chantel Skupski, 57 Three Nooks, Bamber Bridge, Preston, PR5 8EN
M6 VMC Terry McElwee, Little Borough, Borough Farm Road, Godalming, GU8 5JZ
M6 VMG Vincent McGowan, 11 Lime Close, Nuthall, Nottingham, NG16 1FD
M6 VMJ Richard Atkinson, 34 Longley Ings, Oxspring, Sheffield, S36 8ZS
M6 VMP Ian Rotheram, 60 Whitewood Park, Liverpool, L9 7LG
M6 VMR Martin Roberts, 13 Stanley Road, Portslade, Brighton, BN41 1SW
M6 VNL Matthew Harris, 5 Lynmore Close, Northampton, NN4 9QU
MW6 VNP Susan Beer, 67 Killan Road, Dunvant, Swansea, SA2 7TH
M6 VNT Vincent Sheppard, 56 Hansart Way, Enfield, EN2 8ND
MW6 VOC Aubrey Parsons, 21 Rectory Drive, St. Athan, Barry, CF62 4PD
MI6 VOF Patrick McFadden, 35 West Wind Terrace, Hillsborough, BT26 6BS
M6 VOG Bart Rutkowski, 18 Squirrel Close, Coventry, CV2 1FP
M6 VOL Sarah Hill, 86 Gilbert Road, Chichester, PO19 3NL
M6 VOR Matt Darley, 5 Princes Place, Princes Risborough, HP27 0EJ
M6 VOS Simon Devos, Applecross Cottage, Main Road, Newark, NG23 7HR
MW6 VOW Bethan Moore, Lower Bulford Farm, Bulford Road, Haverfordwest, SA62 3ET
M6 VOX Fergus Riche, 1 Lovelstaithe, Norwich, NR1 1LW
M6 VOY Martin Harrison, 2 Broad Farm Cottages, North Street, Hailsham, BN27 4DS
MI6 VOZ Jacek Kacprzyk, 51a Knocknagin Road, Desertmartin, Magherafelt, BT45 5LQ
MM6 VPF Hedley Phillips, Maplebank, Leithen Road, Innerleithen, EH44 6NJ
M6 VPS Mateusz Partyka, Ridgebourne Cottage, Ridgebourne Road, Kington, HR5 3EG
M6 VPW V Williams, 11 Priory Green, Highworth, Swindon, SN6 7NU
M6 VPX Steven Chapman, Wigrams Turn Marina, Shuckburgh Road, Southam, CV47 8NL
M6 VPZ Tommy Godfrey, 43b Broadbury Road, Bristol, BS4 1JT
M6 VQC Charles Powell, 52 Primley Road, Sidmouth, EX10 9LF
M6 VQV Tim Page, 19 LAMORNA DRIVE, Callington, PL17 7QH
M6 VRC Andrew Newbould, 20 Gorsemoor Road, Heath Hayes, Cannock, WS12 3TG
M6 VRD Vicky Bowen, 4 Crossley Gardens, Halifax, HX1 5PU
M6 VRE Emmanouil Vrentzos, Flat 264, The Circle, London, SE1 2JW
MM6 VRH Caitlin Harvey, Hillside, Sarclet, Wick, KW1 5TU
M6 VRO Robin Sheen, 29 Aldermoor Avenue, Southampton, SO16 5GJ
M6 VRR Kemlo Bird, 1 South View, Hill Top Road, Chesterfield, S45 0DA
M6 VRT Andrew Calvert, 11 Pine Tree Walk, Poole, BH17 7EH
M6 VSA Alex Chubb, 70 Goldcrest Road, Chipping Sodbury, Bristol, BS37 6XQ
M6 VSB Bryn-Rhys Rhodes, 81 Witherston Way, Eltham, London, SE9 3JL
M6 VSC Vadim Schultz, 223 Long Down Avenue, Bristol, BS16 1GE
M6 VST Jordan Baker, 70 Vane Close, Norwich, NR7 0US
M6 VTA John Roberts, Hafan Llewelyn, Llanystumdwy, Criccieth, LL52 0SS
M6 VTB Luke Rudge, 10 St. Johns Close, Aldingbourne, Chichester, PO20 3TH
M6 VTC Simon Dillon, 33a Main Road, Cleeve, Bristol, BS49 4NS
M6 VTG David Coles, 36 York Hill, Loughton, IG10 1HT
MM6 VTS Steven Davidson, 6 Sutherland Walk, Mintlaw, Peterhead, AB42 5GT
M6 VTT David Bedford, 55 Sycamore Avenue, Wickersley, Rotherham, S66 2NS
[illegible]
MM6 VUV Robert Fraser, 72 Ferguson Drive, Denny, FK6 5AG
M6 VVA Andrew Amos, 19 Poets Gate, Cheshunt, Waltham Cross, EN7 6SB
M6 VVB Colin Harper, 21 Holly Close Farnham Common, Slough, SL2 3QT
M6 VVE Richard Cook, 62 High Hazel Road, Moorends, Doncaster, DN8 4QN
M6 VVN Vivien Walker, 291 Park Road, Blackpool, FY1 6RR
M6 VVT Marcus Kenny, 5 Fallowfield, Hazlemere, High Wycombe, HP15 7RP
M6 VWA William Vinnicombe, 22 Victoria Park, Cambridge, CB4 3EL
M6 VWD John Bartley, 29 Cheltenham Road, Blackburn, BB2 6HR
M6 VWE christopher Brown, 123 Godinton Road, Ashford, TN23 1LN
M6 VWN Sojan Mathew, 64 Magnolia Way, Costessey, Norwich, NR8 5EH
M6 VWP Edward Byrne, 859 Rochdale Road, Middleton, Manchester, M24 2RA
MM6 VWT William Scott, 6f Drumcross Road, Bathgate, EH48 4HG
M6 VWW Andrew Wilson, 28 Langham Road, Bristol, BS4 2LJ
MM6 VXB Majed Al Saeed, 9 Appin Place, Edinburgh, EH14 1NJ
M6 VXI Darren Holden, 24 Penny Gate Close, Hindley, Wigan, WN2 3DP
MM6 VXR Audrey Latto, 8 Aspen Avenue, Glenrothes, KY7 5TA
M6 VXT Thomas Searle, 18 Witton Lane, Little Plumstead, Norwich, NR13 5DL
MD6 VYN Martin Roberts, 11 Oakleigh Road, Pinner, HA5 4HB
MM6 VZA Jennifer Lauxman McCorkell, 28 Leven Road, Hamilton, ML3 7WS
M6 VZF Edmund Clarke, 1 Farm Drive, Croydon, CR0 8HX
M6 WAA Stephen Myciunka, 67 Templeton Drive, Fearnhead, Warrington, WA2 0WR
MI6 WAB William McDonald, 14 Edenmore Park, Limavady, BT49 0RG
M6 WAD Paul Waddington, 26 Highlands Avenue, Barrow-in-Furness, LA13 0AU
MI6 WAF William Hawkes, 12 Meadow Court, Newtownards, BT23 8YE
MI6 WAG Margaret Barr, 2 Willowvale Close, Islandmagee, Larne, BT40 3SD
MM6 WAI William Inglis, 13 Princes Street, California, Falkirk, FK1 2BX
M6 WAJ Jack Watts, 65 Church Road, Hayling Island, PO11 0NR
MI6 WAM Wayne McCormick, Dernawilt Road, Roslea, BT92 7GG
MI6 WAN William Nelson, 17 Killygullan Drive, Killygullan, Lisnaskea, Enniskillen, BT92 0HJ
M6 WAP William Patterson, 12 Broadleigh Way, Crewe, CW2 6TT
M6 WAQ James Carn, 12 Woodcock Hill Estate Harefiled Road, Rickmansworth, WD3 1PQ
M6 WAS Darryl Simpson, 20 Orwell Road, Harwich, CO12 3LD
M6 WAT Paul Watson, 10 Whitelands Crescent, Baildon, Shipley, BD17 6NN
M6 WAV Alan Snelson, 6 Rayleigh Close, Braintree, CM7 9TX
MM6 WAW Michael Beaton, 76 Fergus Avenue, Howden, Livingston, EH54 6BG
M6 WAX Paul Simmonds, 10 Castlewood Mobile Home Park, Hinckley Road, Leicester, LE9 4JZ
M6 WAY Benjamin Waymark, 13 Beech Ave, Nottingham, NG7 7LJ
M6 WBA Jerome Edridge, 36 Conyngham Lane, Bridge, Canterbury, CT4 5JX
M6 WBB James Whitehead, 12 Polkerris Road, Carharrack, Redruth, TR16 5RJ
M6 WBE David Buckley, 66 Tharp Road, Wallington, SM6 8LE
M6 WBF Ian Westwood, 1 Cook Avenue, Maltby, Rotherham, S66 8QZ
M6 WBG Ben Hall, 139 Somerfield Road, Walsall, WS3 2EN
M6 WBH David Thompson, 17 Sandpiper Close, Blyth, NE24 3QN
MM6 WBJ William Jackson, 3 Annick Road, Dreghorn, Irvine, KA11 4EY
M6 WBK David Beck, 13 Chipstead, Chalfont St. Peter, Gerrards Cross, SL9 9JZ
M6 WBM Adam Buckley, 25 Queensway, Pilsley, Chesterfield, S45 8EJ
MI6 WBN Martin Parke, 222 Lecky Road, Londonderry, BT48 6NS
M6 WBO William Beacroft, 3 Rose Farm Rise, Normanton, WF6 2PL
MM6 WBP William Ferguson, 58o Fraser Walk, Kilmarnock, KA3 7NT
M6 WBR James Webber, 16 Marythorne Road, Bere Alston, Yelverton, PL20 7BZ
M6 WBS William Bennison, 21 Ashdene Close, Chadderton, Oldham, OL1 2QG
MI6 WBT William Turkington, 8a Drummullan Road, Moneymore, Magherafelt, BT45 7XS
MM6 WBU James Mclelland, 24 Hyslop Street, Airdrie, ML6 0ES
M6 WBV Wayne Reeves, 33 Pond Bank, Blisworth, Northampton, NN7 3EL
M6 WBX Dawn Mills, 79 Eastbourne Avenue, Gosport, PO12 4NX
M6 WBZ Keith Hunt, Flat 32, Greenford House, West Bromwich, B70 6DX
M6 WCA Brian Neal, 6 Canterbury Street, Chaddesden, Derby, DE21 4LG
M6 WCB Alan Williams, 74 Broadfield Road, Bristol, BS4 2UW
M6 WCC Paul Homer, 20 Sunningdale Road, Middlesbrough, TS4 3HU
M6 WCE Wayne Steingold, 36 Hailsham Road, Romford, RM3 7SP
M6 WCF Peter Wade, 85 Penymynydd Road, Penyffordd, Chester, CH4 0LF
M6 WCG Colin Jones, 26 Leighlands, Crawley, RH10 3DW
MW6 WCI Wayne Johns, 5 Heol Pantgwyn, Llanharry, Pontyclun, CF72 9HU
MM6 WCK Daniel Ross, 29 East Banks, Wick, KW1 5NL
M6 WCL Oliver Fallon, 26 Central Avenue, Corfe Mullen, Wimborne, BH21 3JD
M6 WCN Danny Jones, 85 Black Butts Lane, Walney, Barrow-in-Furness, LA14 3JL
MW6 WCQ Eon Edwards, 1 Brynhyfryd, Sarn, Bridgend, CF32 9UR
M6 WCQ John Daly, 77 Broomfield Road, Swanscombe, DA10 0LU
M6 WCR Charlotte Bavister, 10 Pheasant Grove, Wixams, Bedford, MK42 6AH
M6 WCT Mark Cooper, 3 [illegible], [illegible], Pontypool, NP4 0QZ
M6 WCU Daniel Crowe, 2 Severnside Cottages, Canal Road, Newtown, SY16 2JN
M6 WCV Wayne Marshall, 24 Union Road, Thorne, Doncaster, DN8 5EL
M6 WCX Wayne Dix, 21 Pine Vale Crescent, Bournemouth, BH10 6BG
M6 WCY Alan Whyman, 2 Wilson Close, Thelwall, Warrington, WA4 2ET
M6 WCZ David Howden, 46 Chestnut Avenue, Armthorpe, Doncaster, DN3 2EP
MI6 WDB William Bradshaw, 206 Manor Street, Belfast, BT14 6ED
M6 WDC Michael Moseley, 86 Sissons Terrace, Leeds, LS10 4LH
MI6 WDD David Milligan, 30 Belgrano Ahoghill, Ballymena, BT42 2QQ
M6 WDF Darren Leyland, 20 Newton Heath, Middlewich, CW10 9HL
M6 WDG Denis Goodwin, 33 York Road, Dunscroft, Doncaster, DN7 4LZ
M6 WDH Wayne Hunt, 4 Four Winds Road, Dudley, DY2 8BY
M6 WDI Keith Lynch, Medindie, Woodside, Ryton, NE40 4SY
M6 WDL William Leake, 61 Jubilee Road, Sutton-in-Ashfield, NG17 2DD
MI6 WDM William McCormick, 1 Mullagreenan Court, Rosslea, Enniskillen, BT92 7PR
M6 WDN Anderson Brito da Silva, 21 Grosvenor Gardens, Newcastle upon Tyne, NE2 1HQ
M6 WDO Trevor Smith, Chy Crowshensy, Clifton Road, Redruth, TR15 3UD
M6 WDQ William Carey, 6 Gainsborough Road, Bexhill-on-Sea, TN40 2UL

UK Callsigns

**IMPORTANT NOTE**

**Revalidate licence to avoid revocation** – Ofcom has advised the Society that plans will be drawn up to revoke licences that have not been revalidated as required by the licence conditions. The quickest way to revalidate is to do so online via the Ofcom website: *https://services.ofcom.org.uk/* or by email: *amateur.validations@ofcom.org.uk* Ofcom staff are available to help, but please be patient during times of heavy workload.

M6 WDR Alan Wallace, 46 Heathfield Road, Grantham, NG31 7NH
M6 WDS Wendy Malcolm-Brown, Flat 11, Chiltern Court, Harpenden, AL5 5LY
M6 WDV Walter Cain, Flat 1, Holme Lodge, Godalming, GU7 3AQ
M6 WDW Joe Summers, Little Trembroath, Stithians, Truro, TR3 7DT
MW6 WDX Paul Dalton, 89 Hillcrest, Brynna, Pontyclun, CF72 9SL
M6 WDY Colin Johnson, 45 Gordon Road, Chelmsford, CM2 9LN
MM6 WEB Jack Webster, 125 Bloomfield Court, Aberdeen, AB10 6DT
M6 WEE Abigail Morris, Fairview, Trefonen, Oswestry, SY10 9DP
M6 WEJ Claire Buswell, 11 Succombs Place, Southview Road, Warlingham, CR6 9JQ
M6 WEK Keith Alabaster, 16 Butlers Road, Horsham, RH13 6AJ
M6 WEL David Wells, 34 Bramble Way, Wymondham, NR18 0UN
M6 WEN Wendy Jefferson, 125 Telscombe Way, Luton, LU2 8QP
M6 WEO Steven Kiel, 32 Weavers Avenue, Frizington, CA26 3AT
M6 WEP Ethan Raine, Stable Cottage, St. Martins, Richmond, DL10 4SJ
MM6 WER James Weir, 36 Main Rd, Ferguslie, Paisley, PA1 2QT
M6 WET David Bilton, 9 Ashby Street, Allenton, Derby, DE24 8JR
M6 WEU Chiara Massimiani, Flat 2 37 York Road, Guildford, GU1 4DN
M6 WEV Anthony Weatherall, The Old Telephone Exchange, The Street, Canterbury, CT3 1ED
M6 WEW Luke Wnekowski, 22 Higgs Lane, Bagshot, GU19 5DP
MI6 WEZ Wesley Todd, 2c Knockwood Crescent, Belfast, BT5 6GE
M6 WFA Wendy Durrant, 19 Rydal Rd, Gosport, PO12 4ES
MW6 WFB Laurie Bowman, Chanrick, Penderyn Road, Aberdare, CF44 9RU
M6 WFC Paul Morgan, 25 Junction Street, Dudley, DY2 8XT
M6 WFE Tom McKenna, 67 Raynsford Road, Great Whelnetham, Bury St. Edmunds, IP30 0TN
MW6 WFF Wayne Webb, 70 Beech Court, Bargoed, CF81 8NS
M6 WFG Stephen Woodruff, 172 Windsor Street, Wolverton, Milton Keynes, MK12 5DR
M6 WFH Martin Sharp, Woodgate Farm, Livery Road, Salisbury, SP5 1RJ
M6 WFI Ian Warnecke, 12 Caxton Road, Margate, CT9 5NP
MW6 WFJ Thomas Penman, 21 Maelog Road, Cardiff, CF14 1HP
MD6 WFK Richard Kissack, 6 Falcon Cliff Court, Douglas, Isle of Man, IM2 4AQ
M6 WFL Christopher Coles, 2 Fern Square, Chickerell, Weymouth, DT3 4NZ
M6 WFM David Silkstone, 169 Otley Road, Harrogate, HG2 0DA
M6 WFR Robert Pickwoad, 2 The Tannery, Barrowden, Oakham, LE15 8EA
M6 WFY Kadriye Payne, Eastern Esplanade, Broadstairs, CT10 1DR
M6 WGB Gary Birch, 23 Hanson Street, Great Harwood, Blackburn, BB6 7LP
M6 WGC Terry Buck, 6 Lynn Road, Terrington St. Clement, King's Lynn, PE34 4JX
M6 WGD Rodney Ward, 35 Mill Street, Drummore, Stranraer, DG9 9PS
M6 WGJ William Joyce, 2 Palmers Cottage, Main Street, Oakham, LE15 8DH
MI6 WGL William Leonard, 57 Mullanavehy, Enniskillen, BT92 2EW
MI6 WGM Graeme McCusker, 41 The Granary, Waringstown, Craigavon, BT66 7TG
MM6 WGT Mervyn Gilchrist, 17 Colinslee Crescent, Paisley, PA2 6SD
M6 WHA Adrian Whadcoat, 38 Edwin Panks Road, Hadleigh, Ipswich, IP7 5JL
M6 WHF David Cracknell, 120 Woodhill, London, SE18 5JL
MD6 WHG William Hogg, Medhamstead, Lhergydhoo, Isle of Man, IM5 2AE
M6 WHN David Donnelly, 23 St. Andrews Avenue, Washington, NE37 1AG
M6 WHO Brian Weller, 86 High Street, Blackpool, FY1 2DW
M6 WHS David Whitehouse, 6 Larch Close, Heathfield, TN21 8YW
M6 WHT Kerry Cullen, 29 Colman Road, London, E16 3JY
M6 WHU Antony Hodgson, 515 Ashingdon Road, Rochford, SS4 3HE
MM6 WHW William Woods, 1 Nursery Lane, Kilmacolm, PA13 4HP
M6 WHZ Max Underwood, 39 Westbury Crescent, Oxford, OX4 3SA
M6 WIB Rob Wilson, 171 Cooden Drive, Bexhill-on-Sea, TN39 3AQ
M6 WIC Bradley Jackson, 46 Princess Road, Market Weighton, York, YO43 3BR
M6 WID Glynn Holland, 6 Moorfield Road, Widnes, WA8 3JE
M6 WIF Maria Landragin, 101 Linden Gardens, Enfield, EN1 4DY
M6 WIK Wiktor Trofimiuk, 120a The Fairway, Northolt, UB5 4SW
M6 WIL William Outram, 7 Bakewell Road, Baslow, Bakewell, DE45 1RE
M6 WIM Anson Forrester, 14 Calder Close, Lytham St. Annes, FY8 3NH
M6 WIQ William Wright, 5 Willow Brook Halsall, Ormskirk, L39 8TL
M6 WIR Derek Daniels, 60 Hawthorn Way, Northway, Tewkesbury, GL20 8TQ
M6 WIS Dorian Wiskow, 15 Ferndale Close, Sandbach, CW11 4HZ
M6 WIT Bryan Witt, 20 Foxglove Close, Newton Aycliffe, DL5 4PF
M6 WIX Jon Unwin, 59 Hempstalls Lane, Newcastle, ST5 0SN
M6 WIZ Steven Keen, 13 Ivy Road, Kettering, NN16 9TG
M6 WJA Andrew Woodhouse, 52 Princess Street, Barnsley, S70 1PJ
M6 WJB Wayne Buss, 83 Gorse Avenue, Chatham, ME5 0UP
M6 WJC Walter Chance, 29 Myrtle Avenue, Kings Heath, Birmingham, B14 5DU
M6 WJF William Foster, 55 Drake Avenue Minster on Sea, Sheerness, ME12 3SA
M6 WJJ Kristina Jones, 98 Common View, Stedham, Midhurst, GU29 0NU
MI6 WJK Jim Kelso, 32 Old Park Manor, Ballymena, BT42 1RW
M6 WJL Gordon Hayers, 87 Bradleigh Avenue, Grays, RM17 5RH
M6 WJM John Wheeler, 79 Keary Road, Swanscombe, DA10 0BX
M6 WJN William Naughton, 14 Carisbrooke, Frimley, Camberley, GU16 8XR
M6 WJP Wilfred Paterson, 1 Burnside Terrace, Stranraer, DG9 8HH
MM6 WJS Stuart Wilson, 2 Kinnear Court, Guardbridge, St. Andrews, KY16 0UE
MI6 WJW William Wilson, 9 Benbradagh Avenue, Limavady, BT49 0AP
M6 WKB Keith Brett, 40 Uppingham Road, Houghton-on-the-Hill, Leicester, LE7 9HH
MM6 WKC Mark Cormack, 2 Coghill Street, Wick, KW1 4PN
M6 WKD Roger Hammond, 3 Hunt Road, Earls Colne, Colchester, CO6 2NX
MI6 WKE Ross Wilkinson, 11 Fairview Park, Dromore, BT25 1PN
M6 WKF Andrew Rowland-Stuart, 86 Wiltshire House, Lavender Street, Brighton, BN2 1LE
M6 WKG Andrew Cole, 104 Newport Road, Cowes, PO31 7PS
M6 WKH james KENYON, 95 Openshaw Drive, Blackburn, BB1 8RB
M6 WKI M Hawker, 47 Lynher Drive, Saltash, PL12 4PA
M6 WKL Peter Houghton, 151c London Road, Calne, SN11 0AQ
MI6 WKN William Gamble, 11 Anderson Park, Limavady, BT49 0RH
M6 WKR Graeme Walker, 2 Cliffe Bank Cottages, Piercebridge, DL2 3SX
M6 WKS Authur Wilkins, 58 High Road, Wormley, Broxbourne, EN10 6JN
M6 WKT Geoffrey Williams, 18 Elmsleigh Road, Farnborough, GU14 0ET
M6 WKW Kevin Davies, 23 Egmanton Road, Meden Vale, Mansfield, NG20 9QN
M6 WKY Chris Wilkinson, 40 Lumley Drive, Consett, DH8 7DT
M6 WKZ Anthony Norden, 10 School Lane, Watton at Stone, Hertford, SG14 3SF
M6 WLA Stephen Fearnhead, 1b Hampden Grove, Eccles, Manchester, M30 0QU
MW6 WLB Kelvin Clarke, 13 Caerphilly Close, Dinas Powys, CF64 4PZ
M6 WLC William Cogdon, 15 Strafford Avenue, Worsbrough, Barnsley, S70 6SU
M6 WLD William Daley, 27 Rosebery Street, Manchester, M14 4UR
M6 WLE Emma Williams, 24 Astbury Street, Congleton, CW12 4EQ
M6 WLF Karl Smith, 3 Hawley Vale, Hawley Road, Dartford, DA2 7RL
M6 WLR Alan Coats, 57 Mill Hill, Boulton Moor, Derby, DE24 5AF
M6 WLS Warren Le Serve, 120 Cheam Road, Sutton, SM1 2EB
M6 WLY Liam Lane-Wells, Gwaelod, Pool Quay, Welshpool, SY21 9LH
M6 WMA Margaret Wall, 227 Wayfield Road, Chatham, ME5 0HJ
M6 WME William Eden, Well Summer Orchard, Farlow, Kidderminster, DY14 0RG
M6 WMH Wendy Clarke, 1 Alvey Terrace, Nottingham, NG7 3DF
M6 WMJ Malcolm Watts, 48 Tasburgh Street, Grimsby, DN32 9LB
MM6 WMK Martin Walker, 2/2 98 Pettigrew Street, Glasgow, G32 7XR
MM6 WMM William Moodie, 1/R, 13 Barend Street, Isle of Cumbrae, KA28 0BL
MI6 WMN William Mcmullen, 69 Lissize Avenue Rathfriland, Newry, BT34 5DE
M6 WMP Michael Weaver, 16 Avocet Drive, Kidderminster, DY10 4JT
M6 WMV John Hancock, Westover House, Yeovil Marsh, Yeovil, BA21 3QU
M6 WMW Michael Suddaby, Bryn Eiddion, Rhydymain, Dolgellau, LL40 2AS
M6 WMZ Alice Law, 2 The Bank, Somersham, Huntingdon, PE28 3DJ
M6 WNB Jeffery Hocking, 26 Musket Road, Heathfield, Newton Abbot, TQ12 6SB
M6 WNC Carl Lindley, 187 Alexandra Road, Sheffield, S2 3EH
M6 WND Elaine Gosal-Tooby, East Wing, Durkar House, Durkar Lane, Wakefield, WF4 3AS
M6 WNE Wayne Gough, 42 Merevale Avenue, Hinckley, LE10 0PY
M6 WNM Wayne McCoo, 8 Newlands Road, Parson Drove, Wisbech, PE13 4LB
M6 WNN Wayne Newey, 3 Forrester Close, Flanderwell, Rotherham, S66 2NL
M6 WNR Nick Rapson, 15 School Close, Bampton, Tiverton, EX16 9NN
M6 WNW Richard Gowler, Merlins Lodge, Church Road, Norwich, NR12 0JP
M6 WOB James Blow, 23 Leaders Way, Lutterworth, LE17 4YS
MM6 WOC William O'Rourke, 6 Busby Place, Kilwinning, KA13 7BA
M6 WOD William Davies, Foelallt, North Road, Aberystwyth, SY23 2EL
M6 WOE Graham Needle, 195 Kingsley Avenue, Kettering, NN16 9ET
MI6 WOF Andrew Pulman, 69 Islandarragh Road, Cape Castle, Ballycastle, BT54 6HS
M6 WOK Michael Smith, 7 Cherry Tree Close, Everton, Lymington, SO41 0ZG
M6 WOL Matthew Walters, 39 Portland Place, Coseley, Bilston, WV14 9TB
M6 WOM Michael Wilkes, 1 Hawburn Close, Bristol, BS4 2PB
M6 WOO Julian Wooldridge, 7 Heather Gardens, Belton, Great Yarmouth, NR31 9PP
M6 WOT David Bowers-Edgley, 8 The Slopes, Lower Henley Road, Reading, RG4 5LE
M6 WOV Virginia Lyall, 29 King Alfreds Road, Sedbury, Chepstow, NP16 7AQ
M6 WOW David Holman, 20 Green Drive, Wolverhampton, WV10 6DW
M6 WPI Tim Hobson-Smith, 15 Henconner Lane, Chapel Allerton, Leeds, LS7 3NX
M6 WPN Shaun Comper, 105 Westway, Copthorne, Crawley, RH10 3QS
MI6 WPP Jonathan Quigley, 99 Boghill Road, Templepatrick, Ballyclare, BT39 0HS
M6 WPS Wayne Phillips, 36 Beeches Road Great Barr, Birmingham, B42 2HF
MI6 WPT Wendy Turkington, 230 Whitechurch Road, Ballywalter, Newtownards, BT22 2LB
MI6 WPW William Wilson, 9 Benbradagh Avenue, Limavady, BT49 0AP
M6 WRB Wayne Bellamy, Flat 57, Eros House, London, SE6 2EG
MW6 WRC William Lockyer, 10 Bryn Moreia Llwydcoed, Aberdare, CF44 0TT
M6 WRD Wayne Davis, 5 Curzon Street, Hull, HU3 6PH
M6 WRE William Rendell, 71 Hardwick Bank Road, Northway, Tewkesbury, GL20 8RP
M6 WRG Mark Richardson, 80 Byron Road, West Bridgford, Nottingham, NG2 6DX
M6 WRH Wayne Hartley, 30 Coltman Avenue, Beverley, HU17 0EY
M6 WRJ William Jones, Lakeside, Roman Drive, Norwich, NR13 5LU
MI6 WRM William Campbell, 9 Rochester Court, Coleraine, BT52 2JL
M6 WRN Martin Wren-Hilton, 28 Arwenack Avenue, Falmouth, TR11 3JW
M6 WRO Jon Wrobel, Flat 27, Tenney House, Grays, RM17 6SG
M6 WRS William Sankey, 14 Carter Grove, Hereford, HR1 1NT
MI6 WRT Reginald Jameson, 64 Blackhill Road, Dromore, Omagh, BT78 3HL
M6 WRW Derek Hampton, 17 Hamilton Road, Worcester, WR5 1AG
M6 WRX Ian Scholey, 27 Newstead Avenue, Fitzwilliam, Pontefract, WF9 5DT
M6 WRY Lawrence Curtis, 39 Mount Stewart Street, Seaham, SR7 7NG
M6 WSB Arthur Mcallister-Bowditch, 24 Morse Close, Chippenham, SN15 3FY
M6 WSC Deborah Waterhouse, 10 Falconers Drive, Battle, TN33 0DT
MM6 WSG Geoffrey Greenwood, 92 Porterfield, Comrie, Dunfermline, KY12 9XG
M6 WSH Scott Halsey, 15 Rectory Drive, Yatton, Bristol, BS49 4HF
M6 WSK Dorothy Clayton, 171 Warning Tongue Lane, Doncaster, DN4 6TU
M6 WSM William Gee, 2 Milton Walk, Worksop, S81 0DH
M6 WSR Ryan Emery, 15 Shaw Crescent, Hutton, Brentwood, CM13 1JD
M6 WST Richard West, 557 East Bank Road, Sheffield, S2 2AG
M6 WSU Susan Blackburn, 36 Mardale Grove, Barrow-in-Furness, LA13 9QG
M6 WSW Denise Willingham, 42 Childers Court, Ipswich, IP3 0DU
M6 WTD Teresa Dallas, 74 Main Street, Asfordby, Melton Mowbray, LE14 3SA
M6 WTE Andrew White, 56 Raleigh Close, Churchdown, Gloucester, GL3 1NT
M6 WTF Samuel Searles-Bryant, 14 Canuden Road, Chelmsford, CM1 2SX
M6 WTG William Jones, 8 Oakbrook Close, Ewyas Harold, Hereford, HR2 0NX
M6 WTL Angela Page, 12 Hitchens Close, Hemel Hempstead, HP1 2PP
M6 WTM David Parsloe, 57 Shanklin Close, Chatham, ME5 7QL
M6 WTP Wayne Pascoe, 34 Jasper Road, London, E16 3TR
M6 WTR Steven Whittaker, 7 Green Lane, Horstead, Norwich, NR12 7EL
M6 WTT Wayne Barnard, 34 Forsyth Drive, Braintree, CM7 1AR
M6 WTW Thomas White, 20 Rosemary Drive, Banbury, OX16 1EZ
M6 WTX Stuart Jackson, 64 Main Road, Moulton, Northwich, CW9 8PB
MI6 WTZ Tammy Saunderson, 32 Seacourt Road, Larne, BT40 1TE
M6 WUB Aaron Best, 140 Heath Road, Ipswich, IP4 5SR
M6 WUG Carey Humphries, 44 Linksway, Folkestone, CT19 5LS
M6 WUH Clive Hopkins, 24 Battle Road, Tewkesbury, GL20 5TZ
MW6 WUK Gareth Reason, 454 Cowbridge Road West, Cardiff, CF5 5BZ
MM6 WUL William Fulton, 15 Staffa Avenue, Port Glasgow, PA14 6DT
M6 WUN Hingwan Cheung, Harrogate Ladies' College, Clarence Drive, Harrogate, HG1 2QG
M6 WVB Vicky Brett, 40 Uppingham Road, Houghton-on-the-Hill, Leicester, LE7 9HH
M6 WVC Thomas Rowlands, 39 Maes Merddyn, Gaerwen, LL60 6DG
M6 WVE Mervyn Huggett, 12 West View Cottages, Lewes Road, Haywards Heath, RH16 2LJ
M6 WVH Paul Wood, 26 Wamil Way, Mildenhall, Bury St. Edmunds, IP28 7JU
M6 WWB Wasily Buczkowski, 54 Woodfield Heights, Wolverhampton, WV6 8PT
M6 WWD Reginald Howard, 13 Top Common, East Runton, Cromer, NR27 9PW
M6 WWE Stephen George, 84 Kenilworth Crescent, Enfield, EN1 3RG
M6 WWF Martin Johnson, 6 Swainson Road, Leicester, LE4 9DQ
M6 WWJ John Harrison, 20 Kelcliffe Avenue, Guiseley, Leeds, LS20 9EW
M6 WWK Martin Wilks, Flat 16, Overton Court, Cheltenham, GL50 3BW
M6 WWM Michael Barnard, 47 Springfields, Dunmow, CM6 1BP
M6 WWR Ian Coleman, 15 St. Andrews Close, Holme Hale, Thetford, IP25 7EH
M6 WWT Allan Walls, 7 Waveney Grove, York, YO30 6EQ
M6 WWZ Andrew Harris, 4 Dol Elian, Old Colwyn, Colwyn Bay, LL29 8YZ
M6 WXD Benjamin Wild, 1 Sunnymount, Midsomer Norton, Radstock, BA3 2AS
M6 WXN Stephen Harcourt, 71 Ingleby Road, Long Eaton, Nottingham, NG10 3DG
M6 WXP Mike Howes, 8 Oat Hill Drive, Northampton, NN3 5AL
MM6 WXS Allan McCall, 1 Finlayson Drive, Airdrie, ML6 8LU
M6 WXY Gary Hogan, 68 Ardleigh, Basildon, SS16 5RB
M6 WYD Gordon Holt, 6 Highfields, Holmfirth, HD9 2PZ
M6 WYG David Griffiths, 5 New Grange Terrace, Pelton Fell, Chester le Street, DH2 2PB
M6 WYN David Griffith, The Old Stables, Briantspuddle Dairy, Dorchester, DT2 7HT
M6 WYR Paul Attwood, 17 Robins Corner, Evesham, WR11 4RJ
M6 WYT Trevor Webster, 1 Fen Close, Newton, Alfreton, DE55 5TD
M6 WYW Michael Whatling, 34 Cannon Park Road, Coventry, CV4 7AY
M6 WYX Neil Soane, 32 Pangdene Close, Burgess Hill, RH15 9UT
M6 WYY Liam O'Neill, Flat 6, Freshwater Court, Lee-on-The-Solent, PO13 9BB
M6 WYZ Aydin Tanseli, 157 Warwick Road, Rayleigh, SS6 8SG
M6 WZB Sarah Burdis, 10 Johnston Avenue, Hebburn, NE31 2LJ
M6 WZK Penny-Louise Goodhand, 2a Moor Road, Sutton, Norwich, NR12 9QN
MU6 WZY S Kirkpatrick, Ste Helene Manor, St Andrew, Guernsey, GY6 8XN
M6 XAA Joey Daniels, 33 Park View, Crewkerne, TA18 8HS
M6 XAB Annabel Clay, 4 Craiglands Park, Ilkley, LS29 8SX
M6 XAC Anthony OConnell, 31 North Avenue, Tredegar, NP22 3HF
M6 XAD Jeffrey Hyde, 1 Constant Row, Risca, Newport, NP11 6QA
MW6 XAE Jonathan Smith, Llanstinan Fach, Letterston, Haverfordwest, SA62 5XD
M6 XAF Terence Chapman, 12 Greenways, Chilcompton, Radstock, BA3 4HT
M6 XAG Steven Broderick, 179 Malpas Road, Newport, NP20 5PP
M6 XAI Aimee Rumsby, 31 Howell Road, Drayton, Norwich, NR8 6BU
M6 XAK Susan Ferguson, 80 London Road, Retford, DN22 7DX
M6 XAL Alexander Darlington, 92 Lancaster Road, St. Albans, AL1 4ES
MI6 XAM Maxine Green, 7 Liester Park, Ballyclare, BT39 9RZ
M6 XAO Andrew Gray, 18 St. Botolphs Green, Leominster, HR6 8ER
M6 XAQ Miles Dennis, 40 Windsor Road, Linton, Swadlincote, DE12 6PL
M6 XAS Andrew Smith, 91 Norcliffe Road, Blackpool, FY2 9EN
M6 XAT Trevor Whelan, 243 Duke Street, Barrow-in-Furness, LA14 1XU
M6 XAV Brian Smith, 24 Oakwood Park, Leeds, LS8 2PJ
M6 XAW Andrew Wardle, 2 Deer Park Place, Sheffield, S6 5ND
M6 XAX Kevin Poulton, 21 East View, London, E4 9JA
M6 XAY Aaron Laxton, 7 St. Christophers Green, Broadstairs, CT10 2SS
MI6 XBA Thomas Agnew, 28 Knockdhu Park, Larne, BT40 2EJ
M6 XBB Justin Clarke, 8 Waveney Heights, Brockdish, Diss, IP21 4LD
MM6 XBD Stephanie Boyd, 1 St. Marks Lane, Edinburgh, EH15 2PX
M6 XBE Thomas Dean, 9 Petrel Close, Bridgwater, TA6 4ET
M6 XBJ Dennis McCarthy, 7 Shenley Close, Wirral, CH63 7QU
MI6 XBL Dale Mooney, 12 Curragh Walk, Londonderry, BT48 8HX
M6 XBM Brian Johnson, 6 Trevor Road, Swinton, Manchester, M27 0YH
M6 XBO Sam Wilson, 7 Dalglish Drive, Blackburn, BB2 4FU
M6 XBP Brendon Pettit, 35 Lakeside Rise, Blundeston, Lowestoft, NR32 5BE
M6 XBQ Michael Doherty, 10 Maple Close, Salford, M6 7AR
M6 XBR James Breward, 40 Malvern Road, Birmingham, B27 6EH
M6 XBS Barry Scott, 8 Chelsea Close, Tilehurst, Reading, RG30 6EP
M6 XBV Craig Somerville, Close House, Gretton, Cheltenham, GL54 5EP
M6 XBW Bradley Woollett, 24 Earlsworth Road, Willesborough, Ashford, TN24 0DN
M6 XBX Matthew Wheeler, 6 Severn Road, Melksham, SN12 8BQ
M6 XCA Miroslav Jurisic, Flat 7, 27 Slade Way, Mitcham, CR4 2GA
M6 XCC Stefan Derner, 14 Elmhurst Drive, Huthwaite, Sutton-in-Ashfield, NG17 2NP
M6 XCF Colin Fish, New House Farm, Peaton, Craven Arms, SY7 9DW
M6 XCO Ronald Kempton, Choices Folly, Marsh Road, Spalding, PE12 9PJ
M6 XCP Christopher Poole, 15 Devon Close, Macclesfield, SK10 3HB
M6 XCZ Mark Hartley, 24 Burnham Avenue, Bognor Regis, PO21 2JU
MW6 XDA Michael Herbert, 6 Yew Street, Taffs Well, Cardiff, CF15 7PT
M6 XDB David Buttle, 29 Upthorpe Drive, Wantage, OX12 7DF
M6 XDG Geoffrey Welch, Amazonas, Sandy Lane, Liverpool, L38 3RP
MD6 XDJ Stephen Scott, 13 Silver Close, Harrow, HA3 6JT
M6 XDO David Witt, 20 Foxglove Close, Newton Aycliffe, DL5 4PF
M6 XDR Chris Darby, 2 Lindsey Court Alfred Street, Lincoln, LN5 7PZ
M6 XDW Dennis Tams, 2 Jacksons Lane, Hazel Grove, Stockport, SK7 6EL
M6 XDX Sonia James, 124 Alcock Avenue, Mansfield, NG18 2NF
M6 XDZ Graham Parsons, 2 The Close, East Grinstead, RH19 1DQ
M6 XEE Stephanie Morton, 77 Hepworth Road, Stanton, Bury St. Edmunds, IP31 2UA
M6 XEL Lawrence Cook, Luckfield, Ellis Road Boxted, Colchester, CO4 5RN
MI6 XEM Alexandra McCusker, 41 The Granary, Waringstown, Craigavon, BT66 7TG
M6 XEP Daniel Heath, 3 Elm Road, Congleton, CW12 4PR
M6 XER Simon Wood, 3 Lion Lane, Haslemere, GU27 1JF
MI6 XEX Alan Rowan-Jenkins, 143 Glenkeen Avenue, Greenisland, Carrickfergus, BT38 8ST
M6 XFM Craig King, Nyth Dedwydd, Llywernog, Aberystwyth, SY23 3AB
M6 XFU Christopher Jenkins, Flat 5, The Lawns, Usk, NP15 1BA
M6 XGB Geoffrey Bogg, The Shires, Main Street, Pickering, YO18 7PG
MW6 XGD Gareth Jukes, 52 Beach Road, Cardiff, CF33 6AS
M6 XGF Sacha Wellborn, 27 Lidgett Park Court, Leeds, LS8 1ED
MM6 XGJ George Jamieson, 6 Maryville Park, Aberdeen, AB15 6DU
MI6 XGN Graham Houston, 51 Rockfield Heights, Connor, Ballymena, BT42 3LH
M6 XGS Gary Stanley, 95 Old Vicarage, Westhoughton, Bolton, BL5 2EG
MM6 XGT Gordon Taylor, 21 Glenesk Avenue, Montrose, DD10 9AQ

M6 XHF Heather Beer, 12 Cross Street, Northam, Bideford, EX39 1BS
M6 XIE Stephanie Reveley, 32 Brynystwyth, Penparcau, Aberystwyth, SY23 1SS
M6 XIP Paul Armstrong, 10 Shirdley Avenue, Liverpool, L32 7QG
M6 XIT David Bond, 33 Cotswold Grove, St. Helens, WA9 2JD
M6 XJC Jeffrey Connell, 81 Holwick Close, Consett, DH8 7UJ
M6 XJM Jason McFarlane, 95 Langley Grove Sandridge, St. Albans, AL4 9DZ
M6 XJP Christopher Dennis, Hillsdene, Plex Lane, Ormskirk, L39 7JY
M6 XJR Joseph Redhead, 28 Sandfields, Frodsham, WA6 6PT
M6 XJS David Washington, 13 Cheddar Waye, Hayes, UB4 0DZ
M6 XJW James Wraith, 6 Ashfield Road, Chippenham, SN15 1QQ
MI6 XKE William Duncan, 74 Bunderg Road, Douglas Bridge, Strabane, BT82 8QQ
M6 XKG Karl Abel, 7 Foldgate View, Ludlow, SY8 1NB
M6 XKN Kirk Northrop, 134 Carlyle Road, London, W5 4BJ
M6 XKT Keith Todman, 12 Winscombe, Bracknell, RG12 8UD
M6 XLG Lewis Graham, Yews Avenue, Barnsley, S70 4BW
M6 XLP Lee Pawson, 32 Cross Street, Upton, Pontefract, WF9 1EU
M6 XLR Nicholas Sayle, 6 Glen View, Wigmore, Leominster, HR6 9UU
M6 XLS Denis Bailey, 18 Hilltop Lane, Chaldon, Caterham, CR3 5BG
M6 XLX John Gascoigne, 64 Prestwold Way, Aylesbury, HP19 8GZ
M6 XMA Dave Jasper, 38 Ingleway Avenue, Blackpool, FY3 8JJ
M6 XMB Benjamin Wale, 23 Castleton Avenue, Bournemouth, BH10 7HW
M6 XMC Martin Callis, 1 Webb Close, Letchworth Garden City, SG6 2TY
MI6 XMG Andrew McGarvey, 66a Scaddy road, Downpatrick, BT30 9BS
M6 XMH Mark Hoult, 43 Hutcliffe Wood Road, Sheffield, S8 0EY
M6 XMN Micheal O'Brien, Flat 9, Neilson Court, Manchester, M23 1LE
M6 XMS Philip Kentish, 17 Rose Drive, Walsall, WS8 7EB
M6 XMX David Lyon, 1 Garnsgate Road, Long Sutton, Spalding, PE12 9BT
M6 XNL Neil Lamerton, 51 High Street, Knaphill, Woking, GU21 2PX
M6 XNO John Mirfield, 14 King Edward Avenue, Horsforth, Leeds, LS18 4BD
M6 XNU Stephen Grace-Bolton, 26 Fairway Road, Blackpool, FY4 4AZ
MI6 XOD Hazel McDowell, 51 Rockfield Heights, Connor, Ballymena, BT42 3LH
M6 XON Trevor Brownen, 43 Great Rea Road, Brixham, TQ5 9SW
M6 XOR Michael Hauser, 27 Abbey Street Ickleton, Saffron Walden, CB10 1SS
MI6 XOS Jamye McGoldrick, 45 Stewarts Road, Dromara, Dromore, BT25 2AN
MI6 XOX Sara McCartan, 13 Lecale Park, Downpatrick, BT30 6ST
M6 XPB Marcus Partner, 22 Moordale Avenue, Bracknell, RG42 1RT
M6 XPC Peter Cox, 75 Rosecroft Gardens, Swadlincote, DE11 9AF
M6 XPN Daniel Rutter, 61 Daubney Street, Cleethorpes, DN35 7BB
M6 XPT Tom Parfitt, 5 Sheridan Road, Frimley, Camberley, GU16 7DU
M6 XQX Simon Lowe, 14 Windmill Rise, York, YO26 4TX
M6 XRA Marc Corbett, 143 Attlee Road, Nantyglo, Ebbw Vale, NP23 4UZ
MI6 XRC Delwyn McLaughlin, 134 Castleroe Road, Coleraine, BT51 3RW
M6 XRD Ross Daniell, Wits End, Barley Mow Lane, Woking, GU21 2HY
M6 XRF Richard Ford, 13 Green Street, Hereford, HR1 2QG
MM6 XRI Robert Irvine, 9 Pearce Grove, Edinburgh, EH12 8SP
MW6 XRO Alan Howard, 71 Tudor Gardens, Merlins Bridge, Haverfordwest, SA61 1LB
MM6 XRS Christopher Rankin, 2 Thomson Green, Livingston, EH54 8TA
M6 XRT Richard Tofts, Elmcroft, Redhill Road, Ross-on-Wye, HR9 5AU
M6 XRX Richard Tooley, 18 Toronto Road, Petworth, GU28 0QX
MM6 XSB Sarah Bellis, 3 Ivy Lane, Dysart, Kirkcaldy, KY1 2XD
M6 XSD Colin Catlin, 27 Main Street, Frizington, CA26 3SA
M6 XSF Steven Dockray, 54 Kelsick Park, Seaton, Workington, CA14 1PY
M6 XSI Neil Jones, 30 Cardiff Road, Pwllheli, LL53 5NU
M6 XSR Simon Mansfield, 32 Great Oaks Park Rogerstone, Newport, NP10 9AT
M6 XSS Anita Corless, 4 Mayfield Road, Bentham, Lancaster, LA2 7LP
M6 XST Sam Harper, 61 Elmhurst, Tadley, RG26 3LF
M6 XSZ Ben Staszewski, 10 Priory Road, Stanford-le-Hope, SS17 7EW
M6 XTA Tracey Windle, 48 Sheridan Road, South Shields, NE34 9JJ
M6 XTB Tim Beale, 28 Redhouse Way, Swindon, SN25 2AZ
M6 XTD Stuart Brookes, 71 Friends Road, Norwich, NR5 8HW
M6 XTF Sreekar Ganti, 59 Fitzroy Road, Blackpool, FY2 0RJ
M6 XTK Connor Walsh, 2 Farnogue Terrace, Wexford, Ireland, Y35 C6X8
M6 XTL David Baker, 32 Richardson Crescent, Hethersett, Norwich, NR9 3HS
M6 XTM Tristan Mercer, Ralph Allen House, Railway Place, Bath, BA1 1SR
M6 XTP Adrian Coleman, 4 St. Mellion Close, Leicester, LE4 1EJ
M6 XTR Mark Moggeridge, 19 Sherwood Close, Fetcham, Leatherhead, KT22 9QT
M6 XTT Edward Field, 4 Redhouse Drive, Sonning Common, Reading, RG4 9NT
M6 XTY David Stott, 19 Canberra Way, Rochdale, OL11 2EL
M6 XVC Natalie Booth, 30 Kingfisher Drive, Cheltenham, GL51 0WN
M6 XVJ Cate Wallwork, 8 Daneswood Close, Whitworth, Rochdale, OL12 8UX
M6 XVS Scott Cooksey, 22 Fitzgerald Place, Brierley Hill, DY5 2SZ
M6 XVT Christopher Williams, 1 South View, Freeholdland Road, Pontypool, NP4 8LL
M6 XVX Michael Lawrence, 16 Timson Close, Market Harborough, LE16 7UU
M6 XWB Paul Forrest, 6 Scarisbrick Place, Liverpool, L11 7DJ
M6 XWD Neil Bishop, 15 Katrine Road, Stourport-on-Severn, DY13 8QB
M6 XWQ Paul Matthewson, 90 Harvey Road, Rugeley, WS15 4UF
M6 XXB Brian Bentham, 7 Maypole Crescent, Abram, Wigan, WN2 5YL
MW6 XXC Dean Clark, 137 Llanedeyrn Road, Penylan, Cardiff, CF23 9DW
M6 XXD Paul Jones, 18 Roundway, Shrewsbury, SY3 7TG
M6 XXI Bernard Littlechild, 85 Long Road, Canvey Island, SS8 0JB
MM6 XXV James Kerr, 2 Burngrange Cottages, West Calder, EH55 8EW
M6 XXX Glyn Pike, 36 Harborough Close, Sheffield, S2 1RH
M6 XYH David Hatch, 18 Victory Park Road, Addlestone, KT15 2AX
M6 XYL Clare Harris, PO Box 114, Norwich, NR7 9XU
M6 XYY Pei Liu, International Hall, University of London, London, WC1N 1AS
M6 XZK Enoch Arthur, 7 Magpie Close, Coulsdon, CR5 1AT
M6 XZY Adrian Mclean, 141 Crawford Avenue, Tyldesley, Manchester, M29 8LS
M6 YAD Gerald Boam, 36 Merlin Way, Woodville, Swadlincote, DE11 7QU
MW6 YAE Denise Bannister, 45 Queens Drive, Llantwit Fardre, Pontypridd, CF38 2NT
M6 YAF Alan Garn, 5 Bassett Street, Walsall, WS2 9PZ
MW6 YAG A Graham, 2 Heol Undeb, Beddau, Pontypridd, CF38 2LB
M6 YAH Charlie Bowden, 12 Butts Green, Stoke-on-Trent, ST2 8EH
M6 YAJ Roy Harrison, Flat 1, 268 Central Drive, Blackpool, FY1 5JB
MI6 YAM John Wilkinson, 11 Fairview Park, Dromore, BT25 1PN
M6 YAN Yanick Watkins, 1 St. Saviour Close, Colchester, CO4 0PW
MM6 YAP Christopher Phillips, 8 The Square, Newtongrange, Dalkeith, EH22 4QD
M6 YAR Ray Parker, 38 Elizabeth Avenue, Tattershall Bridge, Lincoln, LN4 4JJ
M6 YAS Paul Savage, 63 Ashby Road, Braunston, Daventry, NN11 7HE
M6 YAT Linda Jones, 8 Oakbrook Close, Ewyas Harold, Hereford, HR2 0NX
M6 YAV Andrew Vaile, 76 Stanbury Road, London, SE15 2DB
M6 YAW Alan Nicholls, 2 Park End, Forsbrook, Stoke-on-Trent, ST11 9DR
M6 YAY Chris Jack, 18 Swan Close, Martlesham Heath, Ipswich, IP5 3SD
M6 YBA Mark Adams, 18 Vanguard Court, Sleaford, NG34 7WL
M6 YBB Lee West, 153 Chartist House, Mount Street, Hyde, SK14 1RP
M6 YBD Richard Cook, 8 Kingsdown Way New Marske, Redcar, TS11 8JJ
MD6 YBE Helen Jones, Newhaven, Mill Road, Ballasalla, Isle of Man, IM9 2EG
M6 YBT Matthew Harrison, 2 Rosemount Court, Holly Bank Road, York, YO24 4EG
M6 YBV Beverley Young, 11 Gainsborough Avenue, Washington, NE38 7EF
MW6 YBZ Paul Smith, 29 Heol Cwarrel Clark, Caerphilly, CF83 2NE
M6 YCA Ashley Young, 53 Arnold Grove, Newcastle, ST5 8LD
MM6 YCB James Williamson, Clunie Cottage, Tullibardine Road, Auchterarder, PH3 1LX
MM6 YCJ Colwyn Jones, 11b Ettrick Road, Edinburgh, EH10 5BJ
M6 YCR William Jones, 50 Bridge Place, Croydon, CR0 2BB
M6 YCT David Ashton-Hilton, 14 Weetwood Road, Congresbury, Bristol, BS49 5BN
M6 YDA Karl Abel, 7 Foldgate View, Ludlow, SY8 1NB
M6 YDB Daniel Bower, 89 Halifax Road, Sheffield, S6 1LA
M6 YDC Darren Cook, 8 Challis Avenue, Chaddesden, Derby, DE21 6LG
M6 YDD Richard Howson, 37 Ghyllroyd Drive, Birkenshaw, Bradford, BD11 2ET
M6 YDG David Lyall, Royal Hospital Road, London, SW3 4SR
M6 YDN Paul Norman, 11 Windmill Road, Irthlingborough, Wellingborough, NN9 5RJ
MW6 YDP Phillip Warburton, 71 Richards Terrace, Cardiff, CF24 1RW
M6 YDV David Powell, 29 Kelston View, Bath, BA2 1NW
M6 YDW Robert Dean, 15 Gorge Road, Dudley, DY3 1LF
M6 YEB Michael Clayton, 2 West Street, Wroxall, Ventnor, PO38 3BU
M6 YEG James Fenton, 4 Forest Hills, Newport, PO30 5NG
M6 YEH Anne Bate, 16 East Avenue, Heald Green, Cheadle, SK8 3DL
M6 YEL Joe Callaghan, 46 Highfields, Great Yeldham, Halstead, CO9 4QQ
M6 YEO Alexandra Yeo, Bridge House, Tresillian, Truro, TR2 4AU
M6 YEQ Georgina Browning, 6 The Drive, Rickmansworth, WD3 4EB
M6 YES Harry Crabb, 12a Connaught Avenue, Frinton-on-Sea, CO13 9PW
M6 YEY Thomas Windass, 58 Nicholas Gardens, High Wycombe, HP13 6JG
MM6 YEZ Andrew Ewing, Kildonan House, Caerlaverock Farm, Crieff, PH5 2BD
M6 YFT Martin Hubbard, Millstone, Highfield Road, Truro, TR4 8DZ
M6 YFX Iain Cox, 6 Blackman Close, Kennington, Oxford, OX1 5NU
M6 YGD David Young, 75 Broadlands Road, Southampton, SO17 3AP
M6 YGL georgina hansell, 196 The Downs, Harlow, CM20 3RH
MM6 YGT Gary Christie, 10 Calcots Crescent, Elgin, IV30 6GL
MM6 YHO David Chisholm, 111 Phillips Wynd, Hamilton, ML3 8PH
M6 YHQ Paul May, 1 Ballowall Terrace St. Just, Penzance, TR19 7BG
M6 YHR Rhys Boulton, 41 St. Andrews Drive, Libanus Fields, Blackwood, NP12 2ET
M6 YHS David Stevens, 67 New Road, East Hagbourne, Didcot, OX11 9JX
M6 YHV Conor Strange, 14 North Bush Furlong, Didcot, OX11 9DY
M6 YIN Diane Horton, 27 Fishers Lane, Wirral, CH61 9NT
M6 YJB Julie Buck, 55 Chestnut Avenue, Kirkby-in-Ashfield, Nottingham, NG17 8BA
M6 YJD Duncan Sanders, 4 Hodgsons Court, Scotch Street, Carlisle, CA3 8PL
M6 YJK Joseph Williams, 56 Hillingdon Rise, Sevenoaks, TN13 3SB
M6 YKI Kevin Griffiths, 11 Hatfield Meadow, Knighton, LD7 1RY
M6 YKS Barbara Roberts, 6 Trem Y Moelwyn, Tanygrisiau, Blaenau Ffestiniog, LL41 3SS
MM6 YLA William Lawson, 60 Inglis Avenue, Port Seton, Prestonpans, EH32 0AQ
M6 YLC Catherine Rawlinson, Westfield Farm, Risden Lane, Cranbrook, TN18 5DU
M6 YLD Deanna Braithwaite, 300 Arnold Estate Druid Street, London, SE1 2XN
M6 YLE Anne Kelly, 62 Old Warren, Taverham, Norwich, NR8 6GA
MI6 YLG Grace McCormick, 46 Lany Road, Moira, Craigavon, BT67 0NZ
M6 YLH Hazel Earnshaw, 16 Estill Close, Cayton, Scarborough, YO11 3TA
M6 YLI Emily Fisher, 1 Eaton Place, Kingswinford, DY6 8JU
M6 YLJ Robert Whitehead, 29 Coulsons Road, Bristol, BS14 0NN
MM6 YLO Gail Davis, 2 Virkie Cottages, Virkie, Shetland, ZE3 9JS
M6 YLP Rupert Campbell-Black, 10 Wren Close, Towcester, NN12 6RD
M6 YLR David Walker, 290 Shannon Road, Hull, HU8 9RY
M6 YLS Sarah Woolley, Babbington House, Church Road, Crowborough, TN6 3LG
MI6 YLT Summer McCormick, 46 Lany Road, Moira, Craigavon, BT67 0NZ
M6 YMA Anne Andrew, 80 Hamble Drive, Abingdon, OX14 3TE
M6 YME Mark Smith, 2 Tullig, Cahirciveen, County Kerry, Ireland
M6 YMN Michael Lee, 4 Addison Place, Bilston, WV14 7BD
M6 YMR Kevin Humphreys, 16 Thames Close, Ferndown, BH22 8XA
M6 YMY Katya Nolhamo Wright, The Gatehouse, Frogmore Lane, Aylesbury, HP10 9DZ
M6 YNA Bethany Wright, 5 Ridge Road, Kingswinford, DY6 9RB
MM6 YND Amanda Savage, 43 Oak Wynd, Cambuslang, Glasgow, G72 7GS
MM6 YNG Colin Young, 0/2 27 Cardwell Road, Gourock, PA19 1UW
M6 YNK Sam Higton, 2 Sandford Brook, Hilton, Derby, DE65 5HQ
M6 YNL Owen Price, 1 King Street, Odiham, Hook, RG29 1NN
M6 YNN Archie Green, 10 Coopers Hill, Eastbourne, BN20 9HX
M6 YNT Andre Ashby, 5 New Street, Osbournby, Sleaford, NG34 0DL
M6 YNY Keith Edwards, 352 Plessey Road, Blyth, NE24 3HD
M6 YOB Christopher Burns, 29 Foxglove Fold, Castleford, WF10 5UJ
M6 YOG Samantha Lee, 61 Roberts Road, Barton Stacey, Winchester, SO21 3HU
M6 YON Adrian Parkhouse, 3 St. Margarets Avenue, Ashford, TW15 1DR
MD6 YOO Michael Akena, 6 Concord Terrace, Coles Crescent, Harrow, HA2 0HJ
M6 YOS Lee Shackleton, 1 Mayfield Road, Wooburn Green, High Wycombe, HP10 0HG
M6 YOT Charles Watson, 51 Lodge Way, Weymouth, DT4 9UU
M6 YOU Benjamin Roots, 212 Hunt Road, Tonbridge, TN10 4BJ
M6 YOX Carl Yoxall, 136 Chilton Street, Bridgwater, TA6 3HZ
MM6 YOY James Moir, 41 Brisbane Terrace, East Kilbride, Glasgow, G75 8DL
M6 YPA Adrian Cattell, 40 Collyweston Road, Northampton, NN3 5ET
M6 YPD Derrick Hawkes, 121 Berrymoor Road, Wellingborough, NN8 2HR
M6 YPE Joshua Pye, 17 Langden Brook Mews, Morecambe, LA3 3SN
M6 YPG Gary Evans, 1 Hill Top Road, Little Harrowden, Wellingborough, NN9 [illegible]
MM6 YPN Gregory Lailvaux, 4 Oxenfoord Avenue, Pathhead, EH37 5QD
M6 YPS Paul Smith, 16 Cornwood Way, Haxby, York, YO32 2GU
M6 YPW Paul Woodfin, Laurel Cottage, Barrow Street, Much Wenlock, TF13 6EN
M6 YPX Andrew Gillard, 4 Horton Avenue, Thame, OX9 3NJ
MI6 YPY Gregory Kelly, 50 Tullyrain Road, Beagh, Enniskillen, BT94 2AS
M6 YPZ Martin Meyer, 72 Trevelyan, Bracknell, RG12 8YD
MM6 YQP Andrew Sharples, 49 Ramsay Road, Banchory, AB31 5TS
M6 YRL Ryan Yerrell, 88 Woodcock Road, Norwich, NR3 3TD
MM6 YRO Daniel Kennedy, 7 Burns Terrace, Cowie, Stirling, FK7 7BS
M6 YRS John Stallard, 6 Richmond Crescent, Leominster, HR6 8RX
M6 YRW Rose Weber, 11 Harewood Road, Calstock, PL18 9QN
M6 YSB Paul Bunting, 29 Marion Avenue, Wakefield, WF2 0BJ
M6 YSD Stuart Daniels, 48 Gason Hill Road, Tidworth, SP9 7JX
M6 YSM Marc Young, 39 Hollins Lane, Sheffield, S6 5GQ
MM6 YST Alexander Thomson, 15 Waverley Road, Nairn, IV12 4RH
MI6 YSW Donna Lappin, 46 Grange Road, Kilmore, Armagh, BT61 8NX
M6 YTC Thomas Campbell, 3 Hillside Close, Helsby, Frodsham, WA6 9LB
M6 YTI Hayley Cropp, 26 Coventry Street, Brighton, BN1 5PQ
M6 YTS John Stacey, 130 Stocks Lane, East Wittering, Chichester, PO20 8NT
M6 YTU Aharon Coward, 27 Rothley Chase, Haywards Heath, RH16 3PE
MM6 YUJ Craig Duncan, 36 Commore Drive, Glasgow, G13 3TT
M6 YUK Derek Barker, 12 The Weavers, Denstone, Uttoxeter, ST14 5DP
M6 YUL Gyula Szabo, 6 Stanway Road, Gloucester, GL4 4RE
M6 YUM Thomas Benson, 3 Tey Road, Coggeshall, Colchester, CO6 1SY
MM6 YUP Brian Rankin, 5 Glen Affric Court, Dumbarton, G82 2BN
M6 YVN Andrew Lowe, 17 Burnside, Telford, TF3 1SS
M6 YVR Steven Preece, 14 Bettespol Meadows, Redbourn, St. Albans, AL3 7EW
M6 YWA Martyn Smith, 51 Town Lane, Shepton Mallet, BA4 5LX
M6 YWG Roy Holland, 431 Beverley Road, Hull, HU5 1LX
M6 YWO Marc Bruyneel, 14 Riversmead, St. Neots, PE19 1HA
M6 YWX Diane Hartnell, 3 Fairmead, Sidmouth, EX10 9SU
MM6 YYB Iain Nicolson, 1 Gullane Place, Dundee, DD2 3BF
M6 YYD Andrew Edge, 40 Canary Road, Stoke-on-Trent, ST2 0SS
M6 YYF Gerald Finney, 121 School Lane, Caverswall, Stoke-on-Trent, ST11 9EN
M6 YYG Daren Harwood, 5 Willows Close, Washington, NE38 7BB
M6 YYK Steven Hoyle, 20 Brandsby Grove, Huntington, York, YO31 9HL
M6 YYL Rhiannon Ashton-Cox, 1 Church Road, Laindon, Basildon, SS15 4EH
M6 YYM Mikey Thomas, 13 Christchurch Gardens, Reading, RG2 7AH
M6 YYT Gary Finney, 121 School Lane, Caverswall, Stoke-on-Trent, ST11 9EN
M6 YYU John Underwood, 15 Fawsley Road, Northampton, NN4 8NR
M6 YYY Storm Christofi, 19 Kingsland Avenue, Northampton, NN2 7PP
M6 YZF Graham Jones, 31 Liverpool Road, Buckley, CH7 3LH
M6 YZH Benny Yu, 210 Ramsgate Road, Broadstairs, CT10 2EW
M6 YZM Graham Brown, 134 Skipper Way, Lee-on-The-Solent, PO13 8HD
M6 YZX Stephanie Forber, 32 Larch Avenue, Newton-le-Willows, WA12 8JF
M6 ZAB Andrew Fletcher, 74 Devonshire Road, Maltby, Rotherham, S66 7DQ
M6 ZAC Mark Holmes, 6 Wells Court, Saxilby, Lincoln, LN1 2GY
M6 ZAD Adil Ghafoor, 20 Maureen Avenue, Manchester, M8 5AR
M6 ZAF Adrian Barter, 17 West Gate, Plumpton Green, Lewes, BN7 3BQ
M6 ZAH T Hayden, The Ferns, Yarmouth Road, North Walsham, NR28 9LX
MW6 ZAN Peter Smith, 3 Digby Street, Barry, CF63 4NP
M6 ZAQ Zachary Gale, 73 Bowler Street, Levenshulme, Manchester, M19 2UA
M6 ZAS Noor Saudah Zulkfli, Harrogate Ladies' College, Clarence Drive, Harrogate, HG1 2QG
M6 ZAU Chris Sommers, 11 Westlands Grove, Fareham, PO16 9AA
M6 ZAV Benajmin Appleby, 8 Pastures Avenue, Littleover, Derby, DE23 4BE
M6 ZAW David Packham, 5 The Avenue, Yate, Bristol, BS37 4PN
M6 ZAX Matthew Southgate, 107 Englands Lane, Loughton, IG10 2QL
M6 ZAY Alasdair Carter, 19 Burn Lane, Newton Aycliffe, DL5 4HX
MM6 ZAZ Graeme Fyvie, Bridgefoot of Gaval, Mintlaw, Peterhead, AB42 4HA
M6 ZBA John Merritt, 41 Great Grove, Bushey, WD23 3BQ
M6 ZBB Mark Palmer, 116 Claverham Road, Yatton, Bristol, BS49 4LE
M6 ZBD Isaac Painter, 47 Longmead Drive, Nottingham, NG5 6DP
M6 ZBE Sara Berrow, 40 Priorsgate Oakdale, Blackwood, NP12 0EL
MM6 ZBG Christopher Halcrow, Hellia, Cunningsburgh, Shetland, ZE2 9HG
M6 ZBL Zachary Raybould, 31 Meadow Road, Quinton, Birmingham, B32 1AY
M6 ZBM Colin Abbott, 38 Foxcover, Linton Colliery, Morpeth, NE61 5SR
M6 ZBP Nikolaos Tsakonas, 4 Hudson Apartments, Chadwell Lane, London, N8 7RW
M6 ZBQ Alexander Kerr, 9 Martindale Way, Sawston, Cambridge, CB22 3BT
MW6 ZBR Bryan Morgan, 65 Tennyson Close, Pontypridd, CF37 5EP
M6 ZBS KL Florence, 30 Lancaster Gardens, Ealing, London, W13 9JY
M6 ZBT Barbara Tomlinson, 7 Springwell Close, Crewe, CW2 6TX
M6 ZBW Brian Waddingham, 11 Chandlers Court, Tidworth, SP9 7FN
M6 ZCA Carol Archer, 31 Stoney Bank Drive, Kiveton Park, Sheffield, S26 6SJ
M6 ZCB Christopher Burdett, 44 Emmett Carr Lane, Renishaw, Sheffield, S21 3UL
MM6 ZCD Colin Docherty, 25 George Grieve Way, Tranent, EH33 2QT
M6 ZCL Christopher Lynch, 14 Warwick Street, Rochdale, OL12 9SW
M6 ZCM Charles May May, 18 Ravenscroft, Salisbury, SP2 8DL
M6 ZCP Chandith Palawinna, 41 Lealands Drive, Uckfield, TN22 1DW
M6 ZCR Colin Rhodes, 93 Southwark Close, Stevenage, SG1 4PH
M6 ZCT Christopher Tsoi, Fairview, Llanbadarn Fawr, Aberystwyth, SY23 3QU
M6 ZDA Darren Harris, 27 Ashley Road, Poole, BH14 9BS
M6 ZDB David Brownsea, 47 Southill Road, Bournemouth, BH9 1SH
M6 ZDC Dunstan Cooke, Apartment 9, 27 Sheldon Square, London, W2 6DW
M6 ZDF Dean Forbes, 1 Raynham Road, London, W6 0HY
MM6 ZDG David Gillies, 43 Oak Wynd, Cambuslang, Glasgow, G72 7GS
M6 ZDL Andrew Abson, 117 Lysander Road, Rubery, Birmingham, B45 0EN
MM6 ZDW Denise Woodford, Nordkette, Evnabrek, Shetland, ZE2 9GY
MM6 ZDY David Young, Fada Fuireach, Erbusaig, Kyle, IV40 8BB
M6 ZDZ Michael Clark, 25 Westonfields, Albury, Guildford, GU5 9AR
M6 ZEA Christopher Scott, 40 The Close, Skipton, BD23 2BZ
M6 ZEB Adam Hulme, 6 Asmall Close, Ormskirk, L39 3PX
MD6 ZEE Christopher Glaister, Balleigh Villa, Jurby Road, Ramsey, Isle of Man, IM8 3NZ



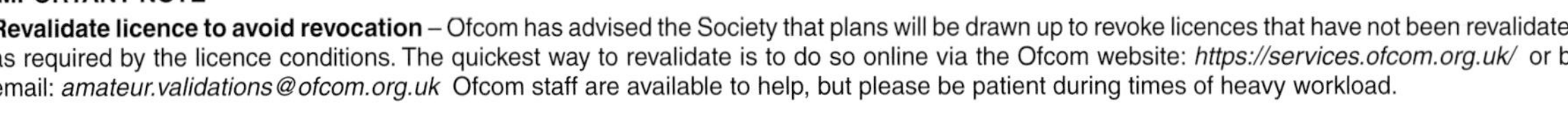

**IMPORTANT NOTE**

**Revalidate licence to avoid revocation** – Ofcom has advised the Society that plans will be drawn up to revoke licences that have not been revalidated as required by the licence conditions. The quickest way to revalidate is to do so online via the Ofcom website: *https://services.ofcom.org.uk/* or by email: *amateur.validations@ofcom.org.uk* Ofcom staff are available to help, but please be patient during times of heavy workload.

**M6 ZEF** John Ferry, 21 Hope Avenue, Goldthorpe, Rotherham, S63 9DT
**M6 ZEJ** Marc Jeffrey, 17 Beweys Park, Lower Burraton, Saltash, PL12 4SW
**M6 ZEK** Evan Short, 2 Richmond Street, Kings Sutton, Banbury, OX17 3RS
**M6 ZEL** Stephen King, 8 Primrose Walk, Pillmere, Saltash, PL12 6XP
**M6 ZEN** Jonathan Charters-Reid, Woodlands Farm, Flaxton, York, YO60 7RJ
**M6 ZEP** Simon Redford, 78 Boyds Walk, Dukinfield, SK16 4AU
**M6 ZES** Gero Dimitrov, Flat 6, Blenholme Court, Hampton, TW12 2BL
**M6 ZET** Adrian Balmer, 11 China Street, Darlington, DL3 0EJ
**M6 ZEW** Wesley Woodhead, 1 Hayle Road, Oldham, OL1 4NP
**M6 ZFE** James McInnes-Boylan, 54 Fernbeck Close, Farnworth, Bolton, BL4 8BR
**MM6 ZFG** Steven Street, 13 Cobbler View, Arrochar, G83 7AD
**M6 ZFX** Stephen Pantony, 40 Park Avenue, Redhill, RH1 5DP
**M6 ZGR** Cristian Panaitescu, 131 Stafford Road, Croydon, CR0 4NN
**MM6 ZGS** Gary Sneddon, 26 Linkwood Road, Airdrie, ML6 6GP
**M6 ZGY** Terrence Talbot-humphries, 9 Conway Avenue, West Bromwich, B71 2PB
**M6 ZGZ** Nicky Glazzard, Snitton Gate Cottage, Snitton, Ludlow, SY8 3JX
**M6 ZIA** Zainab Sati, 27 Hanover Road, London, SW19 1EB
**M6 ZIB** Sherbanu Barker-Mawjee, 23 Raleigh Road, London, N8 0JB
**M6 ZIP** Adam Bradley, Flat 3, 56 Chase Green Avenue, Enfield, EN2 8EN
**M6 ZIY** J Medland, 34 Allderidge Avenue, Hull, HU5 4EQ
**M6 ZIZ** A Morgan, 2 Parc Cambria, Old Colwyn, Colwyn Bay, LL29 9AJ
**M6 ZJB** Julie Brownsea, 47 Southill Road, Bournemouth, BH9 1SH
**M6 ZJH** John Hughes, Herons Way, Munslow, Craven Arms, SY7 9ET
**M6 ZJP** Joanne Povall, Elsich Barn Farm, Seifton, Craven Arms, SY7 9LF
**M6 ZJT** Thomas Heyes, The Coach House, Mossley Hall, Congleton, CW12 3LZ
**M6 ZJW** Zak Wilcoxen, 5 Douglas Lane, Wraysbury, Staines-upon-Thames, TW19 5NF
**M6 ZKA** David Owings, 11 Thingwall Road East, Thingwall, Wirral, CH61 3UY
**MD6 ZKK** Alex Kissack, 6 Falcon Cliff Court, Douglas, Douglas, Isle of Man, IM2 4AQ
**M6 ZLA** Zena Lucas, 31 Lilian Close, Norwich, NR6 6RZ
**M6 ZLC** Sarah Rice, 30 Oveton Way, Bookham, Leatherhead, KT23 4ND
**M6 ZLD** Patrick Frost, 26 Hollies Court, Britannia Road, Banbury, OX16 5DR
**M6 ZLL** Tom Wiggins, 158 Prince Charles Avenue, Derby, DE22 4LQ
**M6 ZLM** Mark Swindells, 35 Rivington Street, St. Helens, WA10 4BL
**M6 ZLN** Zoe Newell, 15 The Grove, Luton, LU1 5PE
**M6 ZLP** Zachary Pepper, 2 Greenways, Walton on the Hill, Tadworth, KT20 7QE
**M6 ZLR** Adam Zeller, Flat 1, 57 Chalk Hill, Watford, WD19 4DA
**M6 ZMC** Michael Clayton, 21 Green Leys, Maidenhead, SL6 7EZ
**M6 ZMI** Thomas Kelly, 2 Weaver House, Chester Road, Runcorn, WA7 3EG
**M6 ZMR** Mathew Richards, 33 Daleside, Dewsbury, WF12 0PJ
**M6 ZNF** Nicholas Farrington, The Old Hall, Main Street, Pershore, WR10 3HS
**M6 ZNZ** Guy Richardson, Berwick Cottage, Bailes Lane, Guildford, GU3 2AX
**M6 ZOC** Miller Cozens, 17 Ash Grove, Norwich, NR3 4BE
**MW6 ZOD** Callum Williams, 1 Lawrence Terrace, Llanelli, SA15 1SW
**M6 ZOG** Stuart Littlechild, 539 Tyburn Road, Birmingham, B24 9RX
**M6 ZOL** Rebecca Bowen, 25 Maendu Terrace, Brecon, LD3 9HH
**M6 ZOM** Scott Corbishley, 86 Boundary Lane, Congleton, CW12 3JA
**M6 ZOO** Nigel Weale, Flat 85, Redbridge Tower, Southampton, SO16 9AW
**M6 ZPB** Pauline Bannon, Rose House, 73 London Road, Worcester, WR5 2DU
**M6 ZPE** Mark Hills, 19 Wilkes Road, Broadstairs, CT10 2HL
**M6 ZPH** Christopher Ward, 1 Granary Close, Codford, Warminster, BA12 0PR
**M6 ZPL** Richard Manning, 21 Whitethorn Way, Oxford, OX4 6ER
**M6 ZPM** Christopher Williams, 39 Meldrum Court, Temple Herdewyke, Southam, CV47 2UF
**M6 ZPS** Zoltan Svanda, 76 Rochester Drive, Westcliff-on-Sea, SS0 0NL
**M6 ZPT** Anthony Pattison-Turner, 47 Jefferson Way, Coventry, CV4 9AN
**M6 ZPY** Richard Pyner, 1 Avon Court, 63 Shakespeare Road, Bedford, MK40 2DS
**M6 ZPZ** Ping Zhou Zhang, Harrogate Ladies' College, Clarence Drive, Harrogate, HG1 2QG
**M6 ZRG** Gene Reynolds, 43 Orchard Drive, Watford, WD17 3DX
**M6 ZRJ** Stella Rogers, 28 Damian Way, Hassocks, BN6 8BJ
**M6 ZRL** Garri Lockett, 15 Deepdale Street, Hetton-le-Hole, Houghton le Spring, DH5 0DQ
**M6 ZRO** Matthew Brough, 40 Denby Grange, Harlow, CM17 9PZ
**M6 ZRT** Timothy Cooper, 9 Websters Close, Shepshed, Loughborough, LE12 9AT
**MM6 ZRX** Kevin Scollay, Nirvana, Orphir, Orkney, KW17 2RB
**M6 ZRZ** Darren Gill, 25 Winston Close, Spencers Wood, Reading, RG7 1DW
**M6 ZSA** Sian Evans, Firswood, Trafford Road, Great Missenden, HP16 0BT
**M6 ZSB** Samuel Bache, 62 Whittingham Road, Halesowen, B63 3TP
**M6 ZSD** Stephen Adams, 31 Northway, Dudley, DY3 3PH
**M6 ZSE** Paul Holmes, 53 Bishops Hull Road, Bishops Hull, Taunton, TA1 5EP
**M6 ZSH** Mohit Gupta, Flat 26, Bassett Court, Southampton, SO16 7DR
**M6 ZSK** Seth Kneller, 366a Kingsland Road, London, E8 4DA
**M6 ZST** Steve Harris, 61 Monks Park Avenue, Bristol, BS7 0UA
**M6 ZSV** Shaun Trevor, 8 Aldin Grange Terrace, Bearpark, Durham, DH7 7AN
**M6 ZTA** Henry Glendinning, 9 Spinners Avenue, Wakefield, WF1 3QD
**M6 ZTB** Richard Janezko, 10a Lens Road, Allestree, Derby, DE22 2NB
**M6 ZTD** Allan Maiden, 79 Green End Road, Manchester, M19 1LE
**MI6 ZTM** Michael Meagher, 42 Mourne View Park, Newry, BT35 6BZ
**M6 ZTO** Kieran Balls, 7 Rowan Close, Holbeach, Spalding, PE12 7BT
**M6 ZTP** Edward Pritchard, Lilliput, Doctors Commons Road, Berkhamsted, HP4 3DR
**M6 ZTS** Graham Harris, Sunnyside Lodge, Mongeham Road, Deal, CT14 8JW
**M6 ZUB** Robert Landragin, 101 Linden Gardens, Enfield, EN1 4DY
**M6 ZUF** Peter Norman, Bungalow Farm, West Haddon, Northampton, NN6 7BH
**M6 ZUT** Mehmet Beyoglu, 177 Nags Head Road, Enfield, EN3 7AD
**MM6 ZUY** John Woods, 33a Dalrymple Street, Girvan, KA26 9EU
**M6 ZVD** Mark Ratcliffe, 50 Pine Road, Barrow-in-Furness, LA14 5EL
**M6 ZVL** Vincent Lynch, 16 Okehampton Crescent, Sale, M33 5HR
**M6 ZWB** Marilyn Wilson, 10 Willow Lane, Langar, Nottingham, NG13 9HL
**M6 ZWP** Jean Scarcliffe, 17 Fillingfir Drive, Leeds, LS16 5EG
**MM6 ZWT** Alasdair Gow, 0/1 319 Glasgow Harbour Terraces, Glasgow, G11 6BL
**M6 ZXB** Penelope Holloway, Flat 26, Swan House, Romford, RM7 8AZ
**M6 ZXC** Thomas Hunter, 5 Regal Court, Dewsbury, WF12 7DE
**M6 ZXH** Peter Baker, 29 Farm Road, Brierley Hill, DY5 2XH
**M6 ZXI** James Clark, 54 Coleshill Place, Bradwell Common, Milton Keynes, MK13 8DP
**M6 ZXL** Christopher Cox, 47 Thurlstone Road, Penistone, Sheffield, S36 9EF
**M6 ZXO** Richard Zlotnicki, Gwynfryn, Goginan, Aberystwyth, SY23 3PD
**M6 ZXQ** Alex Cockle, 32 Old Coach Road, Playing Place, Truro, TR3 6ES
**M6 ZXR** Darrell Thomas, 10 Priory Drove, Great Cressingham, Thetford, IP25 6NJ
**MI6 ZXT** Craig McKinley, 36 Grangewood Drive, Londonderry, BT47 5WN
**M6 ZXX** Oliver Spurway, 89 Woolpitch Wood, Chepstow, NP16 6DR
**M6 ZXY** Daniel Pugh, 8 Clos Deiniol, Llanbadarn Fawr, Aberystwyth, SY23 3TX
**M6 ZXZ** James Richards, 41 Mount Pleasure, Camborne, TR14 7RR
**M6 ZYE** Gary Winnett, 24 Underleys, Beer, Seaton, EX12 3LT
**M6 ZYG** Colin Richardson, 12 Ingsgarth, Pickering, YO18 8DA
**M6 ZYH** Zarah Hill, 101 Abbeydale Road, Sheffield, S7 1FE
**M6 ZYK** Edward Matwiejczyk, 3 Tippett Avenue, Swindon, SN25 2GQ
**M6 ZYQ** Jack Moorhouse, 33 Goylands Close, Llandrindod Wells, LD1 5RB
**M6 ZYX** William Coburn, 42 Hinton Wood Avenue, Christchurch, BH23 5AH
**MM6 ZYZ** Raymond McGlynn, 35 Barrie Quadrant, Clydebank, G81 3EH
**M6 ZZA** John Isles, 128 Ditton Lane, Fen Ditton, Cambridge, CB5 8SS
**M6 ZZB** Simon Bird, Milestone Cottage, Station Road Kimberley, Wymondham, NR18 9HQ
**M6 ZZC** Simon Alexander, 13 Padgate, Thorpe End, Norwich, NR13 5DG
**M6 ZZD** John Nixon, East View, Cloudside, Congleton, CW12 3QG
**M6 ZZE** Christopher Hogan, 26 Mowbray Avenue, St. Helens, WA11 9JD
**MM6 ZZG** George Runcie, Smithy Cottage, Brotherton, Montrose, DD10 0HW
**M6 ZZK** John Gooch, 38 Wards Crescent, Bodicote, Banbury, OX15 4DY
**M6 ZZQ** Steve Collins, 5 Fernleigh Gardens, Stafford, ST16 1HA
**M6 ZZS** Alexander Barrett, 7 Burhill Way, St. Leonards-on-Sea, TN38 0XP
**M6 ZZV** Anrdew Siddall, 12 Russell Gardens, Sipson, West Drayton, UB7 0LS
**M6 ZZY** Robert Soar, 31 Sidlesham Close, Hayling Island, PO11 9ST

# United Kingdom

## 'Details withheld'

Withheld

The callsigns in this list are those of active licences for which the owner has requested details not be given.

### 2*0

2I0 AAD
2W0 AAE
2E0 AAG
2E0 AAH
2E0 AAL
2W0 AAM
2E0 AAP
2E0 AAR
2E0 AAS
2E0 ABW
2E0 ACO
2E0 ACS
2E0 ADD
2E0 ADE
2E0 ADK
2J0 ADM
2E0 ADO
2E0 AER
2E0 AFA
2E0 AFO
2E0 AGT
2I0 AHO
2E0 AHX
2E0 AHY
2E0 AIH
2M0 AIV
2E0 AJS
2E0 AKG
2E0 ALC
2I0 ALE
2E0 ALI
2E0 ALK
2E0 ALM
2W0 APR
2E0 AQR
2U0 ARE
2E0 ARI
2E0 ARO
2E0 ARW
2E0 ASJ
2E0 AUG
2E0 AUH
2E0 AWA
2E0 AWF
2E0 AWL
2E0 AWN
2E0 AWO
2E0 AWP
2M0 AXX
2E0 AYA
2E0 AYD
2E0 AYR
2M0 AZC
2E0 AZN
2E0 AZQ
2E0 AZT
2E0 AZV
2M0 BAE
2E0 BAS
2E0 BAW
2E0 BBB
2E0 BBC
2E0 BBD
2E0 BBP
2E0 BBV
2E0 BCA
2E0 BCI
2E0 BCN
2E0 BCP
2E0 BCU
2M0 BCV
2E0 BCY
2M0 BDA
2E0 BDC
2E0 BDF
2E0 BDG
[illegible] BDI
2W0 BDX
2E0 BDY
2E0 BDZ
2E0 BEJ
2E0 BEK
2E0 BEN
2E0 BEQ
2E0 BES
2E0 BEU
2E0 BEX
2E0 BEY
2I0 BEZ
2I0 BFH
2M0 BFI
2E0 BFL
2E0 BFP
2E0 BFR
2E0 BFU
2M0 BFX
2E0 BFY
2E0 BGA
2E0 BGB
2E0 BGF
2E0 BGG
2E0 BGR
2E0 BHE
2E0 BHI
2I0 BHK
2M0 BHL
2E0 BHM
2M0 BHO
2M0 BHR
2E0 BHV
2E0 BHX
2E0 BIF
2E0 BIJ
2E0 BIV
2E0 BIX
2I0 BIZ
2E0 BJC
2W0 BJG
2E0 BJJ
2E0 BJN
2E0 BJO
2E0 BKC
2E0 BKF
2E0 BKR
2E0 BKS
2E0 BKW
2E0 BLB
2M0 BLE
2W0 BLH
2E0 BLO
2E0 BLP
2E0 BLU
2E0 BLV
2E0 BMA
2I0 BMC
2E0 BMV
2I0 BMX
2E0 BMZ
2E0 BNL
2E0 BNM
2E0 BNN
2M0 BNP
2E0 BNQ
2E0 BNR
2E0 BNS
2E0 BNU
2E0 BOA
2E0 BOE
2M0 BOO
2E0 BOQ
2E0 BOU
2I0 BOW
2E0 BPA
2E0 BPB
2E0 BPC
2E0 BPD
2E0 BPQ
2E0 BPZ
2E0 BQS
2E0 BRA
2E0 BRG
2E0 BRO
2E0 BRV
2E0 BSD
2E0 BSL
2E0 BSV
2E0 BSZ
2E0 BTG
2E0 BTH
2E0 BTK
2E0 BTL
2E0 BTY
2E0 BUB
2E0 BUG
2E0 BUL
2E0 BUR
2E0 BUS
2E0 BUW
2E0 BVA
2E0 BVC
2E0 BVM
2E0 BVP
2E0 BVX
2E0 BVY
2M0 BWB
2E0 BWC
2E0 BWE
2E0 BWF
2E0 BWG
2E0 BWR
2E0 BWS
2E0 BWW
2E0 BWZ
2E0 BXH
2E0 BXL
2M0 BXP
2E0 BXR
2W0 BYE
2E0 BYP
2M0 BYS
2E0 BYV
2E0 BYX
2E0 BYZ
2M0 BZD
2E0 BZK
2E0 BZR
2E0 BZY
2E0 CAB
2E0 CAC
2E0 CAH
2E0 CAI
2W0 CAM
2E0 CAM
2E0 CAN
2E0 CAY
2E0 CAZ
2E0 CBD
2W0 CBK
2W0 CBM
2E0 CBP
2E0 CBR
2E0 CBS
2E0 CBW
2E0 CBY
2E0 CCD
2E0 CCM
2E0 CCN
2E0 CCP
2E0 CCS
2E0 CCV
2E0 CCX
2E0 CDB
2E0 CDC
2M0 CDD
2E0 CDG
2E0 CDH
2E0 CDN
2E0 CDP
2E0 CDX
2E0 CEC
2E0 CEL
2E0 CET
2E0 CEW
2E0 CEZ
2M0 CFF
2E0 CFN
2E0 CFR
2E0 CFS
2E0 CFU
2I0 CGA
2E0 CGC
2E0 CGF
2E0 CGY
2E0 CHB
2E0 CHD
2I0 CHF
2E0 CHG
2E0 CHI
2E0 CHJ
2E0 CHM
2E0 CHO
2E0 CHP
2E0 CHS
2E0 CIC
2E0 CIL
2E0 CIP
2E0 CIW
2E0 CIY
2E0 CIZ
2E0 CJC
2E0 CJH
2E0 CJN
2E0 CJR
2E0 CJT
2E0 CJU
2E0 CJY
2E0 CKA
2E0 CKD
2E0 CKF
2E0 CKG
2E0 CKK
2M0 CKU
2I0 CKZ
2E0 CLA
2E0 CLB
2E0 CLC
2E0 CLG
2E0 CLK
2I0 CLM
2E0 CLO
2E0 CLQ
2E0 CLR
2E0 CLV
2J0 CMB
2E0 CMG
2E0 CMI
2E0 CML
2E0 CMM
2E0 CMS
2E0 CMU
2E0 CMV
2E0 CMW
2E0 CMX
2E0 CNF
2E0 CNJ
2E0 CNK
2E0 CNM
2E0 COO
2E0 COP
2E0 COR
2I0 COS
2W0 COU
2E0 CPA
2E0 CPI
2E0 CPJ
2E0 CPM
2E0 CPP
2E0 CPU
2E0 CPW
2E0 CQA
2E0 CQE
2E0 CQP
2E0 CQW
2M0 CQY
2E0 CRE
2E0 CRF
2E0 CRG
2E0 CRJ
2E0 CRK
2E0 CRL
2E0 CRP
2E0 CRT
2E0 CRZ
2E0 CSA
2E0 CSC
2E0 CSI
2E0 CSM
2E0 CSR
2W0 CSS
2E0 CSW
2E0 CTG
2E0 CTH
2E0 CTP
2E0 CTS
2E0 CTV
2W0 CTY
2E0 CUF
2E0 CUI
2I0 CUN
2E0 CUP
2E0 CUQ
2E0 CUR
2E0 CUZ
2I0 CVA
2E0 CVI
2E0 CVT
2E0 CWD
2E0 CWG
2E0 CWN
2E0 CWP
2E0 CWT
2M0 CWU
2E0 CWX
2E0 CWY
2E0 CXC
2E0 CXR
2E0 CXS
2E0 CXY
2E0 CXZ
2E0 CYA
2E0 CYB
2E0 CYD
2E0 CYH
2E0 CYJ
2E0 CYK
2E0 CYN
2I0 CYQ
2E0 CZF
2E0 CZH
2E0 CZO
2E0 CZQ
2E0 CZV
2E0 CZX
2E0 CZY
2E0 DAD
2W0 DAF
2M0 DAG
2E0 DAK
2E0 DAM
2E0 DAS
2E0 DAZ
2E0 DBC
2E0 DBD
2E0 DBE
2E0 DBJ
2E0 DBR
2E0 DBT
2E0 DBU
2E0 DBV
2E0 DBX
2E0 DCB
2E0 DCC
2E0 DCE
2E0 DCG
2W0 DCO
2E0 DCQ
2E0 DCU
2M0 DDS
2E0 DDT
2E0 DDV
2E0 DDY
2E0 DEB
2E0 DED
2E0 DEL
2E0 DEM
2E0 DET
2E0 DFK
2E0 DFR
2E0 DFU
2E0 DFW
2W0 DFY
2E0 DGE
2E0 DGF
2E0 DGO
2E0 DGP
2E0 DGX
2E0 DGY
2E0 DGZ
2E0 DHH
2E0 DHL
2E0 DHN
2E0 DHU
2E0 DIA
2E0 DIR
2E0 DIS
2E0 DIW
2E0 DIY
2E0 DIZ
2E0 DJD
2M0 DJE
2E0 DJG
2E0 DJN
2E0 DJO
2E0 DKC
2E0 DKH
2E0 DKK
2E0 DKL
2E0 DKN
2E0 DLI
2E0 DLM
2E0 DLQ
2E0 DLS
2E0 DLT
2E0 DLX
2E0 DLY
2I0 DMF
2E0 DMH
2E0 DML
2M0 DMO
2M0 DMT
2E0 DNA
2E0 DNK
2E0 DNN
2E0 DNQ
2E0 DNT
2E0 DNY
2W0 DNZ
2E0 DOA
2E0 DOB
2E0 DOK
2E0 DOO
2M0 DOS
2E0 DOT
2M0 DOU
2E0 DPE
2E0 DPG
2E0 DPJ
2E0 DPK
2E0 DPP
2E0 DPQ
2E0 DPT
2E0 DPV
2E0 DPW
2E0 DQA
2E0 DQE
2E0 DQF
2E0 DQH
2E0 DQI
2E0 DQK
2E0 DQM
2E0 DQR
2E0 DQV
2E0 DQW
2E0 DQZ
2E0 DRD
2E0 DRH
2E0 DRL
2E0 DRP
2E0 DRR
2E0 DRS
2E0 DRU
2E0 DRX
2E0 DSA
2E0 DSD
2E0 DSE
2E0 DSF
2E0 DSM
2E0 DSR
2E0 DTA
2E0 DTI
2E0 DTK
2E0 DTU
2E0 DUC
2W0 DUG
2E0 DUK
2E0 DUT
2E0 DUX
2E0 DUY
2E0 DVA
2E0 DVB
2E0 DVE
2E0 DVJ
2M0 DVL
2E0 DVQ
2E0 DVR
2E0 DVS
2E0 DVT
2E0 DVZ
2E0 DWH
2E0 DWI
2E0 DWN
2W0 DWO
2E0 DWQ
2E0 DWX
2E0 DWY
2E0 DXB
2E0 DXD
2E0 DXE
2E0 DXG
2W0 DXH
2E0 DXI
2E0 DXM
2E0 DXN
2E0 DXP
2E0 DXQ
2E0 DXR
2E0 DXS
2E0 DXT
2E0 DXU
2E0 DXX
2E0 DXY
2W0 DYC
2E0 DYE
2I0 DYF
2E0 DYH
2E0 DYI
2E0 DYL
2E0 DYO
2E0 DYR
2E0 DYS
2E0 DYT
2E0 DYU
2E0 DYW
2E0 DZB
2E0 DZD
2E0 DZG
2E0 DZH
2E0 DZI
2E0 DZL
2E0 DZM
2E0 DZP
2E0 DZU
2M0 DZW
2M0 DZX
2E0 EAB
2E0 EAE
2E0 EAG
2E0 EAH
2E0 EAJ
2E0 EAK
2E0 EAP
2E0 EAQ
2E0 EAX
2E0 EBC
2E0 EBE
2E0 EBM
2E0 EBT
2E0 EBY
2E0 ECA
2E0 ECB
2E0 ECE
2E0 ECL
2E0 ECN
2E0 ECR
2M0 ECS
2E0 EDB
2E0 EDJ
2E0 EDK
2I0 EDT
2M0 EDU
2W0 EDW
2E0 EDZ
2E0 EEA
2E0 EEE
2E0 EEY
2E0 EFM
2E0 EFT
2E0 EGF
2E0 EGG
2E0 EHF
2E0 EHG
2E0 EHZ
2E0 EII
2E0 EIO
2E0 EIS
2E0 EIV
2M0 EJB
2E0 EJF
2E0 EJH
2E0 EJS
2E0 EJX
2E0 EKF
2E0 EKG
2M0 EKL
2E0 ELF
2E0 ELG
2E0 ELL
2E0 ELM
2E0 ELS
2E0 ELW
2M0 EMA
2E0 EMC
2E0 EMJ
2E0 EMO
2E0 EMR
2E0 EMT
2E0 EMU
2E0 EMW
2I0 ENR
2E0 ENT
2E0 EOB
2E0 EOE
2E0 EOG
2E0 EOJ
2E0 EOM
2E0 EOO
2E0 EOW
2E0 EPO
2E0 EQX
2E0 ERA
2E0 ERC
2E0 ERI
2E0 ESD
2E0 ESM
2E0 ESR
2E0 EST
2E0 ETM
2E0 ETO
2M0 ETR
2M0 EUA
2E0 EVD
2E0 EVG
2E0 EVJ
2E0 EVO
2M0 EVS
2E0 EWN
2E0 EWW
2E0 EYQ
2E0 EYT
2E0 EYZ
2E0 EZB
2E0 EZD
2E0 EZE
2E0 EZN
2E0 EZQ
2E0 EZW
2M0 FAD
2W0 FAF
2E0 FAI
2E0 FAL
2E0 FAO
2E0 FAW
2E0 FAX
2E0 FAZ
2W0 FBB
2E0 FBF
2E0 FBG
2E0 FBK
2M0 FCA
2E0 FCC
2W0 FCI
2E0 FCK
2E0 FCL
2E0 FCN
2M0 FCR
2E0 FCU
2E0 FCY
2E0 FDL
2E0 FDM
2E0 FDR
2E0 FDX
2E0 FDY
2E0 FED
2E0 FEI
2E0 FEK
2M0 FES
2E0 FET
2E0 FEW
2E0 FFA
2E0 FFE
2E0 FFG
2E0 FFQ
2E0 FFX
2E0 FGC
2E0 FGI
2E0 FGN
2I0 FHL
2E0 FHS
2E0 FHT
2E0 FHU
2E0 FIG
2E0 FIN
2E0 FIQ
2E0 FJR
2E0 FLC
2E0 FLE
2E0 FLP
2E0 FLT
2E0 FLX
2E0 FLY
2E0 FMC
2W0 FMD
2E0 FNW
2E0 FNX
2M0 FOB
2I0 FOJ
2E0 FON
2E0 FOS
2E0 FOY
2E0 FPE
2E0 FPV
2E0 FQD
2E0 FQH
2E0 FQV
2E0 FRE
2E0 FRN
2E0 FRT
2W0 FRW
2E0 FTE
2E0 FTJ
2E0 FTN
2E0 FTR
2E0 FXI
2E0 FXR
2E0 FXT
2E0 FYB
2E0 FYI
2E0 FZP
2E0 GAB
2E0 GAD
2E0 GAE
2E0 GAI
2E0 GAM
2E0 GAT
2E0 GAW
2E0 GBC
2E0 GBL
2E0 GBQ
2E0 GBR
2W0 GBX
2E0 GBY
2E0 GBZ
2E0 GCA
2E0 GCK
2E0 GCR
2J0 GCU
2E0 GCV
2E0 GCZ
2E0 GDC
2E0 GDD
2D0 GDI
2E0 GDK
2E0 GDP
2E0 GDQ
2M0 GDR
2E0 GDS
2E0 GDU
2E0 GDX
2E0 GED
2E0 GFI
2E0 GFL
2E0 GGF
2E0 GGL
2E0 GGR
2E0 GHL
2E0 GHM
2E0 GHO
2E0 GHQ
2E0 GHS
2E0 GHT
2W0 GHU
2E0 GHW
2W0 GIB
2E0 GIJ
2W0 GIN
2E0 GIO
2I0 GIQ
2E0 GIR
2E0 GJC
2E0 GJD
2M0 GJH
2E0 GJL
2W0 GJM
2E0 GKY
2E0 GKZ
2E0 GLB
2E0 GLE
2E0 GLF
2W0 GLH
2E0 GLK
2E0 GLN
2E0 GLO
2E0 GLX
2W0 GME
2E0 GMH
2E0 GMK
2E0 GMT
2E0 GMV
2E0 GMX
2M0 GND
2I0 GNR
2E0 GNT
2E0 GOA
2E0 GOB
2E0 GOT
2M0 GOV
2E0 GPO
2E0 GPW
2E0 GQB
2E0 GQQ
2E0 GRC
2E0 GRD
2E0 GRG
2E0 GRT
2E0 GRU
2E0 GSD
2W0 GSM
2E0 GSU
2E0 GSX
2E0 GTF
2E0 GTG
2E0 GTI
2E0 GTK
2E0 GTP
2E0 GTV
2W0 GTW
2E0 GUD
2E0 GUX
2E0 GVE
2E0 GVT
2E0 GVX
2E0 GWL
2E0 GWN
2E0 GYO
2E0 GZR
2E0 HAD
2E0 HAE
2E0 HAI
2E0 HAO
2E0 HAR
2E0 HAU
2E0 HAX
2E0 HBA
2E0 HBD
2M0 HBF
2E0 HBG
2E0 HCB
2M0 HCD
2E0 HCH
2E0 HDK
2E0 HEJ
2E0 HEL
2D0 HEN
2E0 HER
2E0 HEW
2I0 HEZ
2M0 HFB
2E0 HFM
2E0 HGB
2E0 HGC
2E0 HGJ
2E0 HGL
2E0 HGV
2M0 HHH
2E0 HII
2E0 HIM
2E0 HIN
2E0 HIX
2E0 HJA
2E0 HKD
2E0 HKS
2E0 HLB
2E0 HMB
2W0 HMF
2E0 HMI
2M0 HMV
2E0 HMW
2E0 HMX
2E0 HMY
2E0 HMZ
2E0 HNA
2E0 HNO
2E0 HNP
2M0 HNY
2E0 HOC
2E0 HOF
2E0 HOP
2E0 HOZ
2E0 HPG
2E0 HPO
2E0 HPT
2E0 HPX
2E0 HPY
2E0 HQH
2E0 HRA
2E0 HRF
2E0 HRO
2E0 HSN
2E0 HSS
2E0 HTA
2E0 HTH
2E0 HTL
2E0 HTT
2E0 HUE
2E0 HUV
2E0 HVX
2E0 HWF
2E0 HWX
2E0 HXD
2E0 HYA
2E0 HYL
2E0 HYR
2E0 HYT
2E0 IAA
2E0 IAB
2E0 IAE
2E0 IAI
2E0 IAP
2E0 IAR
2E0 IAW
2E0 IBC
2E0 IBE
2E0 IBF
2M0 IBO
2M0 IBX
2E0 IBZ
2I0 ICA
2E0 ICG
2E0 ICL
2E0 ICM
2W0 ICO
2E0 ICQ
2E0 IDD
2E0 IDI
2E0 IDP
2E0 IDV
2E0 IEG
2E0 IFH
2E0 IFJ
2E0 IFM
2E0 IGJ
2E0 IIC
2M0 IIE
2E0 IJA
2E0 IJM
2E0 IJS
2W0 IJY
2E0 IKD
2E0 IKF
2E0 IKK
2E0 IKR
2E0 IKY
2I0 ILB
2E0 ILH
2E0 ILT
2E0 ILY
2E0 IMI
2E0 IMK
2E0 IML
2E0 INF
2E0 IOC
2M0 IOT
2E0 IPE
2E0 IPP
2E0 IPR
2E0 IPT
2E0 IPZ
2E0 IRP
2E0 IRQ
2M0 IRV
2E0 ISL
2E0 ISW
2E0 ITT
2E0 IUB
2E0 IUZ
2M0 IVG
2W0 IWA
2E0 IXD
2M0 IZC
2E0 IZL
2E0 JAD
2E0 JAH
2M0 JAS
2E0 JAU
2E0 JAV
2E0 JBB
2E0 JBH
2E0 JBN
2E0 JBO
2E0 JBP
2E0 JBR
2E0 JBU
2E0 JBV
2E0 JCJ
2E0 JCL
2E0 JDN
2M0 JDT
2E0 JDZ
2E0 JEB
2E0 JEF
2E0 JEL
2E0 JER
2E0 JEU
2E0 JEX
2E0 JFA
2E0 JFG
2E0 JFM
2E0 JFQ
2E0 JFS
2E0 JFT
2E0 JFZ
2E0 JGC
2E0 JHB
2E0 JHL
2E0 JHM
2E0 JHR
2E0 JII
2E0 JJD
2E0 JJM
2E0 JJO
2E0 JJX
2E0 JKC
2E0 JKE
2E0 JKJ
2E0 JKK
2M0 JKO
2E0 JKZ
2E0 JLA
2E0 JLC
2E0 JLG
2E0 JLH
2W0 JLL
2E0 JLY
2E0 JMA
2I0 JMG
2E0 JMM
2M0 JMN
2E0 JMP
2E0 JMS
2I0 JMT
2E0 JMZ
2E0 JNI
2E0 JNK
2W0 JNL
2E0 JNQ
2E0 JNS
2E0 JNT
2E0 JOB
2E0 JOJ
2E0 JPG
2E0 JPI
2E0 JPJ
2E0 JPK
2E0 JPN
2E0 JQQ
2E0 JQY
2E0 JQZ
2E0 JRC
2E0 JRF
2E0 JRG
2E0 JRH
2E0 JRM
2E0 JRX
2E0 JSA
2E0 JSC
2E0 JSL
2E0 JSP
2E0 JSW
2E0 JTC
2E0 JTD
2E0 JTF
2E0 JTL
2E0 JTP
2E0 JTR
2E0 JUB
2E0 JUE
2E0 JUG
2E0 JUK
2E0 JUZ
2M0 JVZ
2E0 JWA
2E0 JWF
2E0 JWI
2I0 JWK
2E0 JWZ
2M0 JXC
2E0 JXE
2E0 JXL
2E0 JXP
2E0 JXR
2E0 JXV
2E0 JYK
2E0 JYL
2E0 JYY
2E0 JZF
2E0 JZX
2W0 JZZ
2E0 KAA
2E0 KAD
2E0 KAE
2E0 KAH
2E0 KAI
2E0 KAJ
2E0 KAM
2E0 KAO
2E0 KAQ
2E0 KAV
2W0 KAW
2E0 KAZ
2E0 KBH
2E0 KBI
2E0 KBT
2E0 KCC
2E0 KCE
2E0 KCK
2E0 KCM
2E0 KDD
2E0 KDK
2E0 KDL
2E0 KDM
2E0 KDN
2M0 KDP
2E0 KDW
2E0 KDX
2I0 KDY
2E0 KEB
2E0 KEE
2E0 KEF
2E0 KEH
2E0 KEL
2E0 KEM
2M0 KET
2E0 KFC
2E0 KFU
2E0 KGG
2E0 KGS
2E0 KGX
2E0 KHS
2E0 KHZ
2E0 KIB
2E0 KID
2E0 KIF
2W0 KIP
2E0 KIR
2E0 KJB
2E0 KJG
2M0 KJM
2E0 KJP
2E0 KJS
2E0 KJV
2E0 KJW
2E0 KKA
2E0 KKB
2E0 KKP
2E0 KKS
2E0 KKW
2E0 KLO
2E0 KLT
2W0 KMC
2E0 KME
2E0 KMK
2E0 KMM
2M0 KMO
2I0 KMR
2W0 KMV
2E0 KMW
2E0 KMX
2E0 KNG
2E0 KNJ
2E0 KNO
2M0 KNY
2E0 KOL
2E0 KON
2E0 KOO
2E0 KOW
2E0 KOY
2E0 KPG
2E0 KRJ
2E0 KRL
2E0 KRW
2W0 KSA
2E0 KSB
2M0 KTB
2E0 KTC
2E0 KTE
2E0 KTF
2E0 KTT
2D0 KTY
2M0 KTZ
2E0 KUB
2E0 KUX
2E0 KVL
2E0 KVT
2E0 KWH
2E0 KWK
2E0 KWN
2E0 KWR
2E0 KXO
2E0 KXQ
2E0 KYA
2I0 KYE
2E0 KYO
2E0 KYR
2E0 LAF
2E0 LAG
2E0 LAJ
2E0 LAM
2E0 LAN
2E0 LAP
2M0 LAT
2E0 LAU
2E0 LBC
2E0 LBD
2E0 LBF
2E0 LBO
2E0 LBT
2E0 LBX
2E0 LCO
2I0 LCT
2E0 LCZ
2E0 LDA
2E0 LDB
2E0 LDK
2M0 LDO
2M0 LDT
2E0 LEC
2M0 LEJ
2E0 LER
2E0 LES
2E0 LFJ
2E0 LFO
2E0 LFV
2E0 LGA
2I0 LGD
2E0 LGM
2E0 LHB
2E0 LHP
2E0 LIB
2M0 LIK
2E0 LIS
2W0 LJA
2E0 LJM
2E0 LJU
2E0 LJZ
2E0 LKI
2E0 LKL
2E0 LKP
2E0 LKQ
2E0 LKV
2E0 LKY
2E0 LLR
2E0 LLS
2E0 LMA
2E0 LMB
2E0 LMC
2E0 LMF
2E0 LMX
2M0 LNA
2E0 LNY
2E0 LOK
2E0 LOM
2E0 LOT
2E0 LOV
2E0 LPA
2E0 LPB
2W0 LPF
2E0 LPI
2E0 LPK
2E0 LPT
2E0 LRM
2M0 LRT
2E0 LRY
2E0 LSD
2M0 LSM
2E0 LSN
2E0 LSP
2E0 LSQ
2E0 LSY
2E0 LTE
2E0 LTN
2E0 LTY
2E0 LUC
2E0 LUN
2E0 LUV
2E0 LUX
2E0 LVO
2E0 LWC
2E0 LWG
2E0 LXH
2W0 LXL
2E0 LYL
2E0 LYO
2E0 LYT
2W0 LZZ
2I0 MAG
2E0 MAI
2E0 MAK
2E0 MAM
2I0 MAR
2E0 MAT
2E0 MAU
2E0 MAW
2E0 MBC
2E0 MBF
2E0 MBH
2E0 MBL
2E0 MBP
2E0 MBU
2E0 MBY
2E0 MCE
2E0 MCO
2E0 MCU
2E0 MCX
2E0 MCZ
2E0 MDD
2E0 MDI
2E0 MDL
2E0 MDM
2E0 MDQ
2E0 MDW
2E0 MDY
2E0 MEB
2M0 MEC
2E0 MEE
2E0 MEF
2E0 MEJ
2E0 MEM
2E0 MEN
2E0 MEP
2E0 MER
2E0 MEW
2E0 MFE
2E0 MFH
2E0 MFL
2E0 MFO
2E0 MFP
2E0 MFQ
2E0 MFW
2W0 MGB
2E0 MGD
2E0 MGF
2E0 MGG
2E0 MGH
2I0 MGK
2W0 MGN
2E0 MGS
2E0 MGZ
2M0 MHA
2E0 MHB
2E0 MHS
2E0 MHU
2E0 MHX
2E0 MHZ
2M0 MIA
2E0 MIE
2E0 MII
2E0 MIN
2E0 MIR
2M0 MJB
2E0 MJF
2E0 MJN
2E0 MJR
2E0 MJU
2E0 MJV
2E0 MKD
2E0 MKL
2E0 MKM
2E0 MKQ
2E0 MKR
2E0 MKS
2E0 MKY
2E0 MLN
2E0 MLO
2E0 MLP
2W0 MLW
2E0 MLZ
2E0 MMI
2E0 MMK
2E0 MMR
2M0 MMS
2E0 MMY
2E0 MNE
2E0 MNI
2E0 MNK
2E0 MNM
2E0 MNO
2E0 MNR
2E0 MOA
2E0 MOC
2W0 MOD
2I0 MOE
2E0 MOG
2E0 MOH
2E0 MOI
2W0 MON
2E0 MOP
2E0 MOS
2E0 MPH
2M0 MPK
2E0 MPL
2E0 MPM
2E0 MPU
2E0 MQK
2E0 MRB
2E0 MRE
2E0 MRH
2E0 MRK
2E0 MRL
2I0 MRN
2E0 MRS
2E0 MRV
2E0 MRW
2E0 MRX
2E0 MSD
2D0 MSF
2E0 MSG
2E0 MSJ
2M0 MSK
2E0 MST
2I0 MSV
2E0 MSX
2E0 MTF
2E0 MTI
2M0 MTJ
2E0 MTU
2E0 MTV
2E0 MTW
2E0 MUF
2E0 MUG
2I0 MUL
2E0 MVA
2E0 MVF
2E0 MVG
2E0 MVK
2E0 MVM
2E0 MVR
2E0 MWI
2E0 MWM
2E0 MWR
2E0 MXI
2E0 MXJ
2E0 MXS
2E0 MXV
2E0 MXX
2E0 MXZ
2E0 MYC
2E0 MYJ
2E0 MYN
2E0 MYO
2E0 MYP
2E0 MYR
2E0 MZA
2E0 MZH
2E0 MZI
2E0 MZJ
2M0 MZO
2M0 MZW
2E0 NAC
2E0 NAH
2E0 NAK
2E0 NAL
2E0 NAO
2E0 NAV
2W0 NAW
2E0 NBA
2E0 NBT
2E0 NCH
2E0 NCL
2E0 NCP
2E0 NCW
2E0 NDB
2E0 NDM
2E0 NDS
2E0 NDX
2M0 NEA
2E0 NEP
2E0 NET
2E0 NEW
2E0 NFI
2E0 NFR
2E0 NFY
2E0 NGI
2E0 NGQ
2E0 NGT
2E0 NGU
2E0 NGY
2E0 NHT
2E0 NHX
2E0 NIC
2E0 NIH
2E0 NIK
2E0 NIM
2E0 NIN
2E0 NIV
2E0 NJA
2E0 NJH
2E0 NJP
2E0 NJR
2E0 NJT
2E0 NJZ
2E0 NKD
2E0 NKE
2W0 NKV
2E0 NKY
2M0 NLC
2E0 NLD
2E0 NLF
2E0 NLG
2E0 NMH
2E0 NMP
2W0 NNA
2W0 NNN
2E0 NNS
2E0 NOS
2E0 NPD
2E0 NPH
2E0 NPK
2E0 NRC
2E0 NRD
2E0 NRM
2M0 NRO
2E0 NRP
2E0 NRZ
2E0 NSD
2E0 NSM
2E0 NST
2E0 NSU
2E0 NTE
2E0 NUE
2I0 NUM
2E0 NVR
2E0 NWG
2E0 NWM
2E0 NXR
2E0 NXS
2E0 NZL
2E0 OAD
2E0 OAE
2E0 OAF
2E0 OAL
2M0 OAN
2E0 OAT
2E0 OAV
2E0 OBA
2E0 OBJ
2E0 OCA
2E0 OCC
2E0 OCD
2E0 OCE
2E0 OCK
2E0 OCR
2E0 ODY
2E0 OFC
2I0 OFN
2E0 OFR
2E0 OFX
2E0 OGC
2E0 OGP
2D0 OHM
2E0 OID
2W0 OJA
2E0 OKE
2E0 OLB
2E0 OLD
2E0 OLI
2E0 OLL
2E0 OLX
2E0 OMM
2E0 OMZ

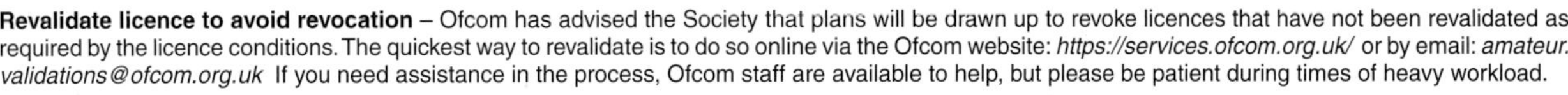

**IMPORTANT NOTE**

**Revalidate licence to avoid revocation** – Ofcom has advised the Society that plans will be drawn up to revoke licences that have not been revalidated as required by the licence conditions. The quickest way to revalidate is to do so online via the Ofcom website: *https://services.ofcom.org.uk/* or by email: *amateur.validations@ofcom.org.uk* If you need assistance in the process, Ofcom staff are available to help, but please be patient during times of heavy workload.

Withheld

2E0 OOB
2E0 OOD
2W0 OOF
2E0 OOK
2E0 OOP
2E0 OOZ
2E0 OPA
2E0 OPR
2E0 ORB
2E0 ORN
2E0 ORO
2E0 ORU
2E0 OSL
2E0 OSM
2W0 OTK
2E0 OTL
2E0 OTO
2E0 OUC
2E0 OUE
2E0 OUN
2E0 OUT
2E0 OUZ
2E0 OVA
2E0 OVE
2E0 OWX
2W0 OXI
2E0 OZV
2E0 OZZ
2E0 PAI
2E0 PAM
2E0 PAQ
2E0 PAR
2E0 PAS
2E0 PAW
2E0 PAZ
2E0 PBA
2E0 PBD
2E0 PBG
2I0 PBI
2E0 PBQ
2E0 PCB
2E0 PCJ
2E0 PCK
2E0 PCM
2E0 PCS
2E0 PDA
2E0 PDC
2E0 PDD
2E0 PDE
2E0 PDF
2E0 PDI
2E0 PDS
2E0 PDW
2M0 PDY
2E0 PEA
2E0 PEB
2E0 PEK
2E0 PEO
2I0 PER
2E0 PET
2E0 PFC
2E0 PFM
2E0 PGA
2E0 PGW
2E0 PGY
2E0 PHD
2E0 PHH
2E0 PHI
2E0 PHJ
2E0 PHW
2E0 PIC
2E0 PIE
2E0 PIV
2M0 PJA
2E0 PJB
2W0 PJP
2E0 PJW
2E0 PKE
2E0 PKG
2E0 PKP
2E0 PKT
2W0 PLD
2E0 PLF
2E0 PLG
2E0 PLM
2E0 PLW
2E0 PMH
2E0 PMT
2W0 PNT
2E0 POA
2E0 POP
2E0 POS
2E0 POW
2I0 POY
2E0 PPC
2E0 PPG
2E0 PPJ
2E0 PPN
2E0 PPP
2E0 PPS
2E0 PRB
2E0 PRF
2E0 PRH
2E0 PRI
2E0 PRJ
2E0 PRP
2E0 PRT
2E0 PRW
2E0 PRX
2M0 PSB
2E0 PSI
2E0 PSJ
2E0 PSL
2E0 PST
2E0 PSU
2E0 PSV

2E0 PSY
2E0 PTM
2E0 PTP
2W0 PTT
2E0 PTU
2E0 PTW
2I0 PTX
2E0 PUP
2E0 PUX
2E0 PVC
2E0 PVO
2E0 PVX
2M0 PWA
2E0 PWE
2E0 PWN
2E0 PWW
2E0 PXC
2E0 PYE
2E0 PYG
2E0 PYU
2E0 PZA
2E0 PZP
2E0 PZW
2E0 RAC
2M0 RAE
2E0 RAN
2E0 RAO
2E0 RAR
2E0 RAT
2E0 RAV
2E0 RAW
2E0 RAX
2E0 RAY
2E0 RAZ
2E0 RBB
2E0 RBD
2E0 RBF
2E0 RBJ
2E0 RBL
2E0 RBM
2E0 RBS
2E0 RCH
2E0 RCK
2E0 RCQ
2E0 RCS
2E0 RCY
2E0 RDB
2E0 RDM
2E0 RDO
2E0 RDY
2E0 REA
2E0 RED
2E0 REE
2E0 REF
2E0 REI
2E0 RET
2E0 REV
2E0 REZ
2E0 RFA
2E0 RFB
2E0 RFV
2E0 RFZ
2E0 RGE
2I0 RGG
2E0 RGH
2E0 RGN
2E0 RGP
2E0 RHD
2E0 RHG
2E0 RHH
2E0 RHR
2E0 RHV
2W0 RHX
2E0 RHZ
2E0 RIB
2E0 RIC
2E0 RIG
2E0 RIP
2E0 RIX
2E0 RJB
2E0 RJC
2E0 RJI
2E0 RJN
2E0 RJS
2E0 RJT
2E0 RKA
2E0 RKD
2E0 RKL
2E0 RKT
2E0 RKY
2W0 RLD
2E0 RLI
2E0 RLL
2E0 RLM
2I0 RMA
2E0 RMG
2E0 RNL
2M0 RNY
2E0 ROL
2M0 ROW
2E0 RPG
2E0 RPI
2E0 RPS
2E0 RQF
2W0 RRG
2E0 RRR
2E0 RRV
2E0 RSF
2E0 RSJ
2E0 RSO
2E0 RSQ
2E0 RSS
2E0 RST
2E0 RSW
2E0 RSX
2E0 RTB

2M0 RTO
2E0 RTR
2E0 RTS
2M0 RTT
2E0 RTV
2E0 RTX
2W0 RTZ
2E0 RUC
2E0 RUE
2W0 RUH
2W0 RUK
2E0 RUM
2E0 RUN
2W0 RVL
2E0 RVO
2M0 RVR
2E0 RVY
2W0 RWD
2E0 RXB
2M0 RXF
2M0 RXM
2E0 RXS
2E0 RYE
2E0 RYO
2E0 RYW
2E0 RYZ
2E0 RZG
2E0 RZK
2E0 RZR
2E0 RZZ
2E0 SAB
2E0 SAE
2E0 SAJ
2E0 SAM
2D0 SAO
2E0 SAQ
2E0 SAV
2E0 SBA
2E0 SBE
2E0 SBF
2E0 SBG
2I0 SBI
2E0 SBQ
2E0 SBT
2E0 SBU
2E0 SBV
2E0 SBW
2E0 SBY
2W0 SCT
2I0 SCY
2E0 SCZ
2E0 SDB
2E0 SDF
2E0 SDH
2E0 SDI
2E0 SDL
2E0 SDN
2E0 SDX
2E0 SEI
2E0 SEM
2M0 SEN
2E0 SEQ
2E0 SFM
2E0 SFO
2E0 SFU
2E0 SGC
2E0 SGE
2E0 SGF
2E0 SGP
2E0 SGR
2E0 SGT
2M0 SGU
2E0 SGX
2E0 SGZ
2E0 SHD
2E0 SHE
2E0 SHL
2E0 SHT
2E0 SIE
2E0 SIG
2M0 SIN
2E0 SIR
2E0 SJC
2E0 SJL
2E0 SJW
2E0 SJY
2M0 SKB
2I0 SKC
2E0 SKF
2M0 SKJ
2E0 SKS
2E0 SKT
2E0 SKW
2E0 SKX
2E0 SLB
2E0 SLC
2E0 SLF
2E0 SMC
2E0 SME
2E0 SMF
2E0 SMH
2E0 SMP
2E0 SMU
2E0 SMW
2E0 SNH
2E0 SNI
2M0 SNO
2E0 SNQ
2E0 SNR
2E0 SOA
2E0 SOL
2E0 SON
2E0 SPC
2E0 SPI
2E0 SPK
2E0 SPW

2M0 SPX
2E0 SQM
2W0 SRE
2E0 SRK
2E0 SRL
2M0 SRM
2E0 SRS
2E0 SRW
2E0 SSH
2M0 SSS
2I0 SSW
2E0 STC
2E0 STE
2E0 STF
2M0 STG
2E0 STJ
2M0 STM
2W0 STO
2E0 STR
2E0 STU
2E0 STW
2E0 STY
2E0 SUJ
2E0 SUL
2W0 SUM
2E0 SUN
2E0 SUR
2E0 SUV
2E0 SVD
2E0 SVR
2E0 SVS
2E0 SVY
2E0 SWG
2M0 SWL
2E0 SWW
2E0 SWX
2E0 SXB
2E0 SXD
2I0 SXG
2E0 SXL
2E0 SXN
2E0 SXO
2E0 SXR
2M0 SXT
2E0 SYK
2E0 SYR
2W0 SZE
2M0 SZN
2E0 SZS
2E0 SZY
2E0 TAB
2E0 TAD
2E0 TBB
2E0 TBC
2E0 TBG
2E0 TBK
2E0 TBM
2E0 TBN
2E0 TBO
2W0 TBP
2E0 TBU
2M0 TBY
2E0 TCE
2W0 TCH
2E0 TCS
2E0 TCZ
2E0 TDA
2E0 TDB
2E0 TDG
2W0 TDK
2E0 TDM
2W0 TDQ
2E0 TDW
2E0 TDX
2E0 TDY
2E0 TDZ
2E0 TEA
2I0 TEB
2E0 TEC
2E0 TEL
2E0 TEM
2M0 TEQ
2E0 TER
2E0 TET
2E0 TEV
2E0 TEY
2E0 TFC
2E0 TFL
2E0 TFR
2W0 TFU
2E0 TFZ
2E0 TGE
2I0 TGH
2E0 TGW
2E0 THB
2E0 THC
2E0 THI
2E0 THL
2E0 THM
2E0 THO
2E0 THT
2E0 THW
2E0 THY
2W0 TID
2E0 TIE
2E0 TIM
2E0 TIX
2E0 TJA
2E0 TJB
2E0 TJH
2E0 TJW
2W0 TKA
2E0 TKC
2E0 TKI
2M0 TKN
2E0 TKO

2E0 TKR
2E0 TKW
2E0 TKZ
2E0 TLK
2E0 TMG
2E0 TMI
2W0 TMJ
2E0 TMP
2I0 TMR
2M0 TMZ
2W0 TNA
2E0 TNG
2M0 TNK
2M0 TNM
2E0 TNR
2E0 TNW
2W0 TNX
2E0 TNY
2E0 TOD
2E0 TOE
2E0 TOH
2E0 TPJ
2E0 TPS
2E0 TPW
2E0 TPX
2E0 TPZ
2E0 TQR
2E0 TRB
2E0 TRC
2E0 TRG
2E0 TRN
2E0 TRQ
2M0 TRT
2E0 TRV
2E0 TSB
2E0 TSF
2E0 TSG
2E0 TSI
2E0 TSL
2M0 TST
2E0 TSW
2E0 TTK
2M0 TTO
2E0 TTR
2E0 TTV
2E0 TTX
2M0 TTZ
2E0 TUP
2E0 TVC
2E0 TVE
2E0 TVI
2E0 TVP
2I0 TWA
2E0 TWC
2D0 TWM
2M0 TXF
2M0 TXN
2E0 TXT
2W0 TYD
2E0 TYF
2E0 TYK
2E0 TYM
2E0 TYN
2E0 TYO
2E0 TYP
2E0 TYU
2M0 TZA
2E0 TZT
2E0 TZX
2E0 UAF
2E0 UAG
2E0 UBD
2E0 UCK
2E0 UCW
2E0 UDL
2E0 UDX
2E0 UEM
2E0 UEX
2E0 UGD
2E0 UGX
2E0 UKH
2E0 UKL
2E0 UKO
2E0 UKP
2E0 UKQ
2E0 UKR
2E0 UKV
2E0 ULA
2E0 ULF
2I0 ULL
2E0 UMB
2E0 UMF
2E0 UMI
2E0 UND
2E0 UNE
2E0 UNX
2E0 UOO
2M0 UOS
2M0 UPG
2E0 URL
2E0 URM
2M0 URN
2M0 USE
2E0 USR
2E0 UTM
2E0 UTO
2E0 UUC
2E0 UUU
2E0 UVV
2E0 UXA
2E0 UYP
2E0 VAB
2E0 VAD
2E0 VAE
2E0 VAL
2E0 VAP

2E0 VAR
2I0 VAX
2E0 VAZ
2E0 VBC
2E0 VBD
2E0 VBE
2E0 VCF
2E0 VCH
2E0 VCO
2E0 VCT
2E0 VDD
2E0 VDS
2E0 VDX
2E0 VEC
2M0 VEG
2E0 VEN
2I0 VEP
2E0 VER
2I0 VEV
2I0 VGW
2E0 VGX
2E0 VIC
2E0 VII
2I0 VIM
2E0 VIP
2E0 VIR
2E0 VKH
2E0 VKV
2E0 VLS
2E0 VLU
2E0 VLV
2E0 VMF
2E0 VMM
2I0 VMR
2E0 VMS
2E0 VMZ
2E0 VNB
2E0 VOL
2E0 VOM
2E0 VOS
2E0 VOV
2E0 VOW
2E0 VOX
2E0 VPC
2E0 VPR
2E0 VPX
2M0 VPZ
2E0 VRG
2E0 VRH
2E0 VRM
2E0 VRN
2E0 VRS
2E0 VSV
2E0 VSX
2E0 VTG
2E0 VTL
2E0 VTM
2E0 VTX
2E0 VVQ
2E0 VVV
2E0 VWB
2E0 VWI
2I0 VXN
2E0 VYV
2E0 VZJ
2E0 VZK
2E0 WAB
2M0 WAC
2E0 WAM
2E0 WAN
2E0 WAR
2E0 WAW
2E0 WAZ
2E0 WBA
2E0 WBC
2E0 WBI
2E0 WBK
2E0 WBM
2E0 WBN
2E0 WBP
2E0 WBY
2E0 WCA
2E0 WCD
2E0 WCH
2E0 WCK
2E0 WCT
2E0 WCW
2E0 WDF
2E0 WDJ
2E0 WDX
2E0 WDZ
2E0 WEN
2E0 WEP
2E0 WFX
2E0 WGA
2E0 WGH
2E0 WGZ
2E0 WHG
2E0 WHR
2E0 WHY
2E0 WIA
2E0 WIB
2E0 WIM
2E0 WIN
2E0 WIW
2E0 WKA
2E0 WKD
2E0 WKI
2E0 WLC
2E0 WLF
2E0 WLJ
2E0 WLL
2E0 WLR
2E0 WLT
2M0 WMR

2E0 WMS
2E0 WNF
2E0 WNJ
2M0 WOA
2W0 WOD
2E0 WOE
2E0 WOM
2E0 WON
2E0 WOO
2E0 WOT
2E0 WPA
2E0 WRA
2E0 WRC
2E0 WRP
2E0 WRZ
2W0 WSA
2E0 WSB
2M0 WSD
2E0 WSF
2W0 WSL
2E0 WTB
2E0 WTC
2E0 WTF
2W0 WTX
2E0 WUT
2E0 WVB
2E0 WVZ
2M0 WWW
2E0 WXV
2E0 WYB
2I0 WYP
2E0 WZD
2E0 WZO
2E0 XAB
2E0 XAC
2E0 XAD
2M0 XAF
2W0 XAJ
2E0 XAK
2E0 XAL
2E0 XAQ
2E0 XAS
2M0 XAT
2E0 XBF
2E0 XBH
2E0 XBI
2E0 XBJ
2E0 XBS
2E0 XDB
2E0 XDJ
2E0 XDL
2E0 XDW
2E0 XFS
2E0 XFU
2E0 XGP
2W0 XHD
2E0 XHF
2E0 XHP
2M0 XIE
2E0 XIH
2E0 XIM
2E0 XIX
2E0 XJC
2E0 XJS
2E0 XKN
2E0 XLD
2E0 XLL
2E0 XMJ
2E0 XML
2E0 XNA
2E0 XNR
2E0 XOB
2E0 XOM
2E0 XOO
2E0 XOS
2E0 XOX
2E0 XPA
2E0 XPB
2I0 XPL
2W0 XPN
2E0 XPR
2E0 XQU
2E0 XRA
2E0 XRC
2E0 XRE
2E0 XRF
2M0 XRI
2E0 XRK
2E0 XRT
2E0 XRY
2E0 XSR
2E0 XTX
2E0 XUB
2E0 XWT
2E0 XWZ
2E0 XXL
2E0 XYL
2E0 XZL
2E0 XZQ
2E0 YAA
2E0 YAH
2E0 YAK
2E0 YAN
2M0 YAT
2E0 YAU
2E0 YAZ
2E0 YBG
2E0 YBO
2E0 YBT
2E0 YBV
2M0 YCB
2W0 YCT
2E0 YCU
2E0 YDL
2E0 YDW

2E0 YFO
2E0 YFT
2E0 YGG
2E0 YHB
2E0 YJM
2E0 YKF
2W0 YLO
2E0 YNE
2E0 YNG
2E0 YNV
2E0 YOB
2M0 YOC
2E0 YOR
2E0 YOT
2E0 YOU
2E0 YPM
2I0 YPT
2E0 YPZ
2E0 YQF
2I0 YQN
2E0 YRD
2E0 YRK
2M0 YRN
2E0 YRO
2E0 YSH
2M0 YTS
2E0 YUB
2E0 YUV
2E0 YVN
2E0 YVO
2E0 YVS
2E0 YVX
2E0 YZE
2E0 YZF
2E0 YZV
2E0 ZAD
2E0 ZAK
2W0 ZAN
2E0 ZAR
2E0 ZAT
2E0 ZAV
2E0 ZAW
2E0 ZAY
2E0 ZAZ
2M0 ZBA
2E0 ZBG
2E0 ZBU
2E0 ZCL
2E0 ZCR
2E0 ZDF
2I0 ZDK
2E0 ZDY
2W0 ZEC
2W0 ZEM
2E0 ZEO
2E0 ZEY
2E0 ZFM
2E0 ZFS
2E0 ZFU
2E0 ZGB
2E0 ZGN
2E0 ZIG
2E0 ZIL
2M0 ZIM
2E0 ZIX
2W0 ZKX
2E0 ZLG
2E0 ZLK
2E0 ZLR
2E0 ZLT
2E0 ZMX
2E0 ZNH
2M0 ZOG
2E0 ZON
2W0 ZOO
2M0 ZPS
2E0 ZQR
2E0 ZRA
2E0 ZRN
2E0 ZRO
2E0 ZSP
2E0 ZTE
2E0 ZTP
2E0 ZTX
2E0 ZUA
2E0 ZWH
2E0 ZWZ
2E0 ZXA
2E0 ZXP
2E0 ZXW
2E0 ZYF
2E0 ZYQ
2E0 ZZI
2E0 ZZM
2E0 ZZR
2E0 ZZW
2E0 ZZX

## 2*1

2E1 AAH
2E1 ACX
2E1 ADB
2E1 AFO
2E1 AJK
2E1 AJL
2E1 AKC
2E1 AKO
2E1 AKU
2E1 ALF
2E1 ALQ
2E1 ANP
2E1 AOQ
2E1 AOR
2E1 APT
2E1 AQG
2E1 AQO

2E1 ATB
2E1 AWM
2E1 AWW
2E1 AWX
2I1 AXG
2E1 AXN
2E1 AXX
2E1 AYY
2E1 AZD
2E1 AZE
2E1 AZL
2E1 BBN
2E1 BCW
2E1 BDF
2E1 BDL
2E1 BDM
2E1 BDU
2W1 BDZ
2E1 BEB
2E1 BEI
2E1 BFO
2E1 BHS
2E1 BJF
2E1 BKA
2E1 BKU
2E1 BKW
2E1 BNA
2E1 BNB
2E1 BNF
2E1 BNP
2E1 BOF
2E1 BOS
2D1 BPJ
2E1 BRE
2E1 BRM
2E1 BTD
2E1 BTH
2E1 BTS
2E1 BTX
2E1 BVM
2E1 BVX
2E1 BWZ
2E1 BZU
2E1 CAD
2E1 CBC
2E1 CBL
2E1 CCB
2E1 CCM
2E1 CDJ
2E1 CDL
2E1 CEL
2E1 CES
2E1 CFA
2E1 CFU
2E1 CGT
2E1 CIC
2E1 CIH
2E1 CJK
2M1 CKE
2E1 CMH
2E1 CMP
2E1 CNT
2E1 COD
2E1 CSN
2E1 CTB
2E1 CTV
2E1 CTZ
2E1 CUK
2E1 CUZ
2E1 CVG
2E1 CXB
2E1 CYH
2E1 CYR
2E1 CYT
2E1 CZI
2E1 CZY
2E1 DAA
2E1 DAG
2E1 DAH
2E1 DCF
2M1 DCH
2E1 DCJ
2E1 DCU
2W1 DDD
2E1 DDN
2E1 DDU
2E1 DEU
2E1 DEV
2E1 DEW
2E1 DFI
2E1 DFR
2E1 DHH
2E1 DHL
2E1 DHN
2E1 DIK
2M1 DIQ
2E1 DJM
2E1 DJR
2E1 DJZ
2E1 DKI
2E1 DMM
2E1 DMW
2E1 DNH
2E1 DNZ
2E1 DOX
2E1 DPJ
2E1 DPU
2E1 DQR
2E1 DQU
2E1 DSG
2E1 DSM
2E1 DSV
2E1 DTI
2E1 DTT
2E1 DUE
2E1 DUN

2E1 DUU
2E1 DVE
2E1 DWP
2E1 DWY
2E1 DWZ
2E1 DXY
2E1 DYR
2E1 DZF
2E1 DZR
2E1 DZY
2E1 EAR
2E1 EBK
2E1 EBO
2E1 EBS
2E1 ECQ
2M1 ECY
2W1 EDF
2E1 EER
2E1 EEU
2E1 EFH
2E1 EFZ
2E1 EGQ
2E1 EIA
2E1 EIL
2E1 EJN
2E1 EKU
2M1 ELH
2E1 ELM
2E1 ELR
2E1 ELY
2E1 EMM
2E1 EMS
2E1 EMT
2E1 ENF
2E1 EQM
2E1 ESU
2E1 ESX
2E1 ESY
2E1 ETP
2E1 EUB
2E1 EUG
2M1 EUT
2E1 EUW
2E1 EVB
2E1 EWF
2E1 EXC
2E1 EYO
2E1 EZF
2E1 FAV
2E1 FCN
2E1 FCV
2E1 FDZ
2E1 FEK
2E1 FFC
2W1 FGK
2E1 FGY
2E1 FHJ
2E1 FIY
2E1 FJU
2E1 FKA
2E1 FKE
2E1 FOJ
2E1 FOP
2E1 FPD
2E1 FPR
2E1 FQP
2E1 FQU
2E1 FTP
2E1 FTU
2E1 FUE
2E1 FUF
2E1 FUN
2E1 FUO
2E1 FUZ
2W1 FVC
2E1 FVG
2E1 FWL
2E1 FXO
2E1 FYA
2E1 FYE
2E1 FZF
2E1 FZL
2E1 GAJ
2E1 GCT
2E1 GCX
2E1 GDC
2E1 GEF
2E1 GEH
2E1 GEN
2E1 GEO
2E1 GEV
2E1 GEW
2E1 GFD
2E1 GFH
2E1 GFL
2E1 GFY
2E1 GGW
2E1 GGX
2E1 GHW
2E1 GIA
2E1 GIB
2E1 GIP
2E1 GIR
2E1 GJN
2E1 GJW
2E1 GKA
2E1 GLA
2E1 GLO
2W1 GMH
2E1 GMS
2E1 GNW
2E1 GNZ
2E1 GOG
2E1 GON
2E1 GOO
2E1 GOQ

2E1 GOS
2E1 GPF
2E1 GPT
2E1 GQA
2E1 GQK
2E1 GQL
2E1 GQM
2E1 GQX
2E1 GQZ
2E1 GRD
2E1 GRE
2E1 GRY
2E1 GSJ
2E1 GSN
2E1 GSV
2E1 GSY
2E1 GTQ
2E1 GTU
2E1 GTZ
2E1 GUA
2E1 GUB
2E1 GUT
2E1 GUU
2E1 GVY
2E1 GWB
2E1 GWI
2E1 GXD
2E1 GXN
2E1 GXP
2E1 GXT
2E1 GYJ
2E1 GZB
2E1 GZE
2E1 GZU
2E1 GZW
2E1 HAA
2E1 HAB
2E1 HAK
2E1 HAT
2E1 HAV
2E1 HBI
2E1 HBV
2M1 HCJ
2E1 HCL
2E1 HDL
2E1 HDW
2E1 HEC
2E1 HEH
2E1 HEL
2E1 HEN
2E1 HIB
2W1 HIE
2E1 HIG
2E1 HIT
2E1 HJK
2E1 HJP
2E1 HJQ
2E1 HJT
2E1 HJY
2E1 HKO
2E1 HKQ
2E1 HKR
2E1 HKT
2E1 HLB
2E1 HLR
2E1 HMH
2E1 HMW
2E1 HNK
2E1 HOJ
2E1 HOL
2E1 HOM
2W1 HOW
2E1 HPE
2E1 HPN
2E1 HPX
2E1 HQB
2E1 HQM
2W1 HQS
2E1 HRA
2E1 HRT
2E1 HRZ
2E1 HSF
2E1 HSK
2E1 HSQ
2E1 HSU
2M1 HSW
2E1 HTG
2E1 HTW
2E1 HTX
2E1 HVG
2E1 HVN
2E1 HWD
2E1 HXG
2E1 HYA
2E1 HYG
2E1 HYK
2E1 HYP
2E1 HYU
2E1 HYV
2E1 HZA
2E1 HZE
2E1 HZH
2E1 HZW
2E1 IAK
2E1 IAV
2M1 IBG
2W1 IBQ
2E1 IBW
2E1 IBY
2E1 ICA
2E1 ICH
2E1 ICK
2E1 ICY
2E1 ICZ
2E1 IDA
2E1 IDD

2E1 IDJ
2E1 IDK
2E1 IDL
2E1 IDN
2E1 IDP
2E1 IDQ
2E1 IDT
2E1 IDU
2E1 IEC
2E1 IED
2E1 IEH
2E1 IEU
2E1 IEW
2E1 IEY
2E1 IFC
2E1 IFF
2E1 IFG
2E1 IFH
2E1 IFK
2E1 IFO
2E1 IFP
2E1 IFS
2E1 IGD
2E1 IGE
2E1 IGH
2E1 IGW
2E1 IHC
2E1 IHH
2E1 IHI
2E1 IHL
2E1 IHP
2E1 IHZ
2E1 IIH
2E1 IIK
2E1 IIN
2E1 IIQ
2E1 IIS
2E1 IIY
2E1 IIZ
2E1 IJB
2E1 IKC
2E1 IKD
2E1 IPA
2E1 IPS
2E1 JAJ
2E1 JAN
2E1 JAY
2E1 JBA
2E1 JDT
2E1 JFJ
2E1 JJE
2E1 JME
2E1 JMK
2E1 JTL
2E1 KGB
2E1 KLM
2E1 LNJ
2E1 LOK
2E1 LOM
2E1 LOU
2E1 LTW
2E1 MDD
2E1 MEX
2E1 MHH
2E1 MJG
2E1 MMH
2E1 MNP
2E1 MOG
2E1 MPC
2E1 MPH
2E1 MTA
2E1 NAI
2E1 NBC
2E1 NBY
2E1 NOJ
2E1 NOP
2E1 NOW
2E1 OKR
2E1 OKT
2E1 PAK
2E1 PAN
2E1 PAT
2E1 PEF
2E1 PJT
2E1 PRK
2E1 RCL
2E1 RCM
2E1 RDX
2E1 REK
2E1 RIC
2E1 RNY
2E1 ROY
2E1 RWH
2E1 SIM
2E1 SKJ
2E1 SLE
2E1 SPS
2E1 SRM
2W1 SWW
2E1 TAF
2E1 TEC
2E1 TFJ
2E1 THX
2E1 TMP
2E1 VAL
2E1 VHS
2E1 VIN
2E1 VIX
2E1 WOW
2E1 YOT
2E1 ZFA

## G*0

G0 AAR
G0 AAV
G0 ABY

G0 ABZ
G0 ACM
G0 ACZ
G0 ADR
G0 AEI
G0 AFB
G0 AFD
G0 AGH
G0 AGM
G0 AGQ
G0 AHN
G0 AJC
G0 AJQ
G0 AKA
G0 AKQ
GU0 ALD
G0 ALV
GM0 AMB
G0 ANC
G0 APE
G0 APM
G0 APQ
G0 APW
G0 AQL
G0 AQP
G0 ARL
G0 ARW
G0 ASC
G0 ASE
G0 ATR
G0 ATV
G0 AUQ
G0 AVO
GM0 AVR
G0 AWD
G0 AWL
G0 AWV
G0 AXN
G0 AYE
G0 AYN
G0 AYU
G0 AYZ
G0 AZS
G0 AZZ
G0 BBH
G0 BBS
GM0 BBU
G0 BBZ
G0 BCK
GI0 BDX
G0 BDY
G0 BEA
G0 BEO
G0 BEQ
G0 BGC
G0 BGD
G0 BGN
G0 BGQ
G0 BGW
G0 BHD
G0 BHX
G0 BIS
G0 BJF
G0 BJO
G0 BJV
G0 BJW
G0 BJX
G0 BKO
G0 BLD
G0 BLJ
G0 BLN
G0 BMA
G0 BMX
G0 BNC
G0 BND
G0 BNI
G0 BOB
G0 BPB
G0 BPI
G0 BQN
G0 BRP
G0 BRV
G0 BRY
G0 BSG
G0 BSS
G0 BSU
G0 BTD
G0 BTO
G0 BUY
G0 BVH
G0 BVX
G0 BVZ
G0 BWS
GM0 BWT
G0 BXT
G0 BYI
G0 BZI
G0 BZK
G0 BZY
G0 CAA
G0 CAC
G0 CAF
G0 CAR
G0 CAU
G0 CAW
G0 CBE
G0 CBH
G0 CCI
G0 CCK
G0 CCP
GW0 CCR
G0 CCW
G0 CCX
G0 CDD
G0 CED
G0 CET

G0 CFL
G0 CFR
G0 CFS
G0 CFV
G0 CFY
G0 CGL
G0 CHK
G0 CHT
G0 CIB
GW0 CID
G0 CIF
G0 CIU
G0 CJM
G0 CKG
G0 CKP
G0 CKT
G0 CKW
G0 CLB
G0 CLK
GM0 CLN
G0 CLP
GM0 CME
G0 CMF
G0 CNE
G0 CNH
G0 CNR
G0 CNS
GM0 COD
G0 CPK
G0 CPW
G0 CRH
G0 CRI
G0 CRQ
G0 CTO
G0 CTX
G0 CUD
G0 CUK
G0 CUQ
G0 CUT
G0 CVL
G0 CVT
G0 CVZ
G0 CWM
G0 CWT
GI0 CWV
G0 CXM
G0 CXP
GM0 CXQ
G0 CXT
G0 CYA
G0 CYP
G0 CYT
G0 CZA
G0 CZB
G0 CZE
G0 CZP
G0 CZT
G0 DBH
G0 DCA
G0 DCB
G0 DCD
G0 DCK
G0 DDB
G0 DDD
G0 DDX
GW0 DFK
G0 DFP
G0 DGV
G0 DGX
G0 DHN
G0 DIN
G0 DIY
G0 DJE
G0 DJH
G0 DJR
G0 DKL
G0 DLD
G0 DLO
G0 DLV
G0 DLY
G0 DMI
G0 DML
G0 DMO
G0 DNC
G0 DNE
G0 DNJ
G0 DNM
G0 DOI
G0 DOT
GM0 DPH
G0 DPZ
GW0 DQA
G0 DQL
G0 DQX
GM0 DTU
G0 DTX
G0 DUE
G0 DUU
G0 DVF
G0 DVR
G0 DVV
G0 DVZ
G0 DWT
G0 DXA
G0 DXC
G0 DXD
GW0 DXQ
G0 DXS
G0 DYB
G0 DYI
G0 DYZ
G0 EAC
G0 EAD
G0 EAK
G0 EAZ
G0 EBB

G0 EBR
G0 EBU
GI0 ECE
GW0 ECO
G0 EED
G0 EFX
G0 EGX
G0 EGY
G0 EHF
G0 EHG
G0 EHH
G0 EHU
G0 EHY
GW0 EIK
GI0 EIO
G0 EIU
G0 EIX
G0 EJG
G0 EJW
G0 EKJ
GM0 EKL
G0 EKP
G0 EKQ
G0 EKR
GM0 EKS
G0 ELE
G0 ELQ
G0 ELS
G0 ELT
GW0 EME
G0 EMG
G0 ENC
G0 ENE
GM0 ENG
G0 ENH
G0 ENR
G0 ENS
G0 EOA
G0 EOB
G0 EOC
G0 EOQ
G0 EOU
G0 EPC
GM0 EPG
G0 EPH
G0 EPX
G0 EPZ
GM0 EQA
G0 EQG
G0 EQP
GM0 EQZ
GM0 ERM
G0 ESB
G0 ESC
G0 ESE
G0 ESF
G0 ESV
G0 ETB
GM0 ETC
G0 ETJ
GW0 ETN
GD0 ETO
G0 ETW
G0 ETY
G0 EUB
G0 EUF
G0 EUH
G0 EUI
G0 EUT
G0 EWM
G0 EWN
G0 EWS
G0 EXF
G0 EXH
G0 EXJ
G0 EXX
G0 EXZ
G0 EYJ
G0 EZA
G0 EZN
GM0 EZP
GI0 FAF
G0 FAI
G0 FAJ
GM0 FAL
G0 FAQ
G0 FAY
GD0 FBI
G0 FBR
G0 FBY
G0 FCD
GM0 FCP
G0 FDM
G0 FDN
G0 FEA
G0 FEF
G0 FEN
G0 FEX
G0 FFH
G0 FFR
G0 FHM
G0 FHR
G0 FHV
GI0 FII
G0 FIK
G0 FIL
G0 FIM
G0 FIR
G0 FIZ
G0 FJI
G0 FJT
GW0 FJV
G0 FJY
G0 FKH
G0 FKL
GI0 FKM

G0 FLC
G0 FLN
GM0 FLO
G0 FMM
G0 FMY
G0 FMZ
G0 FNK
GM0 FNL
G0 FNY
G0 FOA
G0 FOD
G0 FOJ
G0 FOQ
G0 FOZ
G0 FPD
G0 FQB
G0 FQT
G0 FRE
G0 FRJ
G0 FRK
GM0 FRQ
GM0 FRT
GW0 FSA
G0 FSO
G0 FTD
G0 FTM
G0 FUQ
G0 FVG
G0 FVV
G0 FWK
G0 FWL
G0 FWO
GM0 FWV
G0 FXU
G0 FXW
G0 FXZ
G0 FYG
G0 FYN
G0 FYS
G0 FYV
G0 GAC
G0 GAD
GM0 GAV
G0 GBD
G0 GBO
G0 GBX
G0 GCB
G0 GCD
G0 GCH
G0 GCL
G0 GCP
GW0 GCV
G0 GCW
G0 GCX
G0 GCY
G0 GDA
G0 GDO
GI0 GDP
G0 GDU
G0 GDW
G0 GDY
G0 GDZ
G0 GEM
G0 GEY
G0 GFX
G0 GGF
G0 GGU
G0 GGX
G0 GHI
G0 GHK
GW0 GHQ
G0 GIC
G0 GIG
GW0 GIP
GW0 GIQ
G0 GIR
G0 GIS
G0 GIV
G0 GIW
G0 GIY
G0 GJK
G0 GJP
GM0 GJU
G0 GJZ
G0 GKW
G0 GLM
G0 GLT
G0 GMF
GI0 GMG
G0 GMM
G0 GMR
G0 GMU
G0 GNL
G0 GNM
G0 GNR
G0 GNS
G0 GNX
GM0 GOA
G0 GOT
G0 GPD
G0 GQF
G0 GQM
G0 GQZ
G0 GRA
G0 GRK
G0 GRN
G0 GRP
G0 GSD
G0 GTK
G0 GTM
G0 GTO
G0 GTR
G0 GTS
GM0 GTU
G0 GUH
G0 GUK

G0 GUP
G0 GUS
G0 GVC
G0 GVI
G0 GVM
GW0 GVQ
G0 GVY
G0 GWK
G0 GXL
G0 GYD
G0 GYZ
G0 GZP
G0 GZS
G0 HAF
G0 HAH
G0 HCS
G0 HDA
G0 HDN
G0 HEG
G0 HEH
G0 HFH
G0 HFW
G0 HGK
G0 HGX
G0 HHS
G0 HIA
G0 HIO
G0 HIQ
G0 HIS
G0 HIX
G0 HJE
G0 HJP
G0 HJQ
GM0 HKA
G0 HKP
GM0 HKS
G0 HLD
G0 HLM
G0 HLN
GI0 HLO
GW0 HLR
G0 HLT
G0 HLX
G0 HMW
G0 HMY
G0 HMZ
G0 HNA
G0 HNF
G0 HOK
GW0 HOL
G0 HON
G0 HOO
G0 HOR
G0 HPO
G0 HPZ
G0 HQB
G0 HQC
GW0 HQD
G0 HQE
G0 HQJ
G0 HQW
G0 HRA
G0 HSO
G0 HSS
G0 HSZ
G0 HUP
GW0 HUS
GM0 HUU
G0 HUZ
G0 HVL
G0 HVM
G0 HVU
G0 HVV
G0 HVW
G0 HWJ
G0 HWN
G0 HWW
G0 HXO
G0 HXT
G0 HXW
G0 HXX
G0 HXY
G0 HXZ
G0 HYF
G0 HYI
GM0 HYY
G0 HZP
G0 HZR
G0 HZU
G0 HZV
GW0 HZW
G0 HZX
G0 IAM
G0 IAN
G0 IAT
G0 IBB
G0 IBD
G0 IBP
GM0 ICF
G0 ICI
G0 ICX
G0 ICZ
G0 IDA
G0 IFC
G0 IFI
GM0 IFM
G0 IFP
G0 IFZ
G0 IHB
GM0 IIO
GW0 IIR
G0 IIS
G0 IJJ
G0 IJM
G0 IJO
G0 IKA

G0 IKG
G0 IKH
G0 IKS
G0 IKT
G0 IKV
G0 ILF
G0 ILG
G0 ILS
G0 IMN
G0 IMT
G0 INP
G0 IOX
G0 IPJ
G0 IPS
G0 IPU
G0 IPY
G0 IQX
GM0 IRG
G0 IRN
G0 IRO
G0 IRV
G0 IRW
G0 ISD
G0 ISV
G0 ISW
G0 ISX
G0 ITA
G0 ITB
G0 ITH
G0 IUQ
G0 IVA
G0 IVM
G0 IWA
G0 IWM
G0 IWT
G0 IXB
G0 IXE
G0 IXI
G0 IYF
G0 IYZ
G0 IZA
G0 IZM
G0 IZU
G0 IZX
G0 JAY
G0 JBD
G0 JBK
G0 JBW
G0 JCM
G0 JDF
G0 JDI
G0 JDJ
G0 JDP
G0 JDU
G0 JED
G0 JEI
G0 JFN
G0 JGP
GW0 JGZ
GW0 JHA
GM0 JHF
G0 JHV
G0 JHY
G0 JIP
G0 JIZ
G0 JJL
G0 JKB
G0 JKN
G0 JKQ
G0 JKV
G0 JKX
G0 JLA
G0 JLO
G0 JLW
G0 JLY
G0 JMC
G0 JMP
G0 JMT
G0 JMV
G0 JMX
G0 JNH
G0 JNL
G0 JNN
G0 JNP
G0 JNS
G0 JOH
G0 JOO
G0 JOW
G0 JOY
G0 JPA
G0 JPN
GI0 JPO
G0 JPS
G0 JPX
GM0 JQG
G0 JQJ
G0 JQL
G0 JQM
G0 JQN
G0 JQO
G0 jrk
G0 JRP
G0 JSB
G0 JSH
GW0 JSS
G0 JSV
G0 JTB
G0 JUU
G0 JUX
G0 JVA
G0 JVY
G0 JWB
G0 JWQ
G0 JXF
G0 JXM
GW0 JXW

G0 JYB
G0 JYM
G0 JYO
G0 JYP
G0 JYT
G0 JZG
G0 JZP
G0 KAC
G0 KAD
GM0 KAE
G0 KAF
G0 KAG
GM0 KAI
GM0 KAJ
G0 KAL
G0 KAO
G0 KAR
G0 KAV
G0 KAW
G0 KBE
G0 KCQ
G0 KDJ
G0 KDZ
G0 KEH
G0 KEN
GM0 KET
G0 KFA
G0 KFX
G0 KGH
G0 KGK
G0 KGO
G0 KGV
G0 KHB
G0 KHC
G0 KHD
G0 KHG
G0 KHO
G0 KHV
G0 KHW
G0 KIJ
G0 KIS
G0 KIW
G0 KIZ
G0 KJV
G0 KJY
GJ0 KKB
G0 KKI
G0 KLH
G0 KLI
G0 KLM
G0 KLR
G0 KLX
G0 KME
G0 KMX
G0 KNK
G0 KOA
G0 KOM
G0 KOR
GM0 KPK
GW0 KPM
G0 KPR
G0 KQL
GW0 KQZ
G0 KRA
G0 KSB
G0 KSE
G0 KSP
G0 KST
G0 KTM
G0 KUM
G0 KUN
G0 KUO
G0 KVH
G0 KVL
G0 KVT
G0 KVY
G0 KWT
G0 KWV
G0 KXA
GM0 KXE
G0 KXM
G0 KYC
G0 KZC
G0 KZZ
G0 LAH
G0 LAI
G0 LAT
G0 LBW
G0 LCL
G0 LCQ
G0 LCW
G0 LDD
G0 LDM
GI0 LDW
GM0 LEG
G0 LEQ
G0 LET
GM0 LFK
G0 LFU
G0 LFW
G0 LFX
G0 LGD
GW0 LGN
G0 LGS
G0 LGT
G0 LGU
G0 LHF
GM0 LHK
G0 LHT
G0 LIF
G0 LIJ
G0 LIL
G0 LJN
G0 LJX
G0 LKM
G0 LKQ

G0 LKZ
G0 LLW
G0 LMF
G0 LMI
G0 LMM
G0 LMT
G0 LNH
G0 LNR
G0 LNY
G0 LOG
G0 LOQ
G0 LOS
G0 LOV
G0 LPO
G0 LPS
G0 LPW
GD0 LQL
GM0 LQY
GI0 LRB
G0 LRD
G0 LRX
G0 LRY
G0 LSA
GI0 LSB
G0 LSH
G0 LTG
G0 LTN
G0 LTU
G0 LTW
G0 LUJ
G0 LUS
G0 LUZ
G0 LVB
G0 LVQ
G0 LVV
G0 LVZ
G0 LWH
G0 LWK
G0 LWS
G0 LWX
G0 LWY
G0 LWZ
G0 LXA
G0 LXJ
G0 LXS
G0 LXY
G0 LXZ
G0 LYL
G0 LYV
G0 LZB
G0 LZO
G0 MAB
G0 MAO
G0 MBE
G0 MBI
G0 MBK
G0 MBT
GD0 MCU
G0 MDA
G0 MDG
G0 MDI
G0 MDL
G0 MDZ
G0 MEG
G0 MEJ
G0 MEK
G0 MEL
G0 MEP
G0 MES
G0 MEU
G0 MGA
GM0 MGE
G0 MGF
G0 MGK
G0 MGZ
G0 MIN
G0 MIQ
G0 MIY
G0 MJI
G0 MJM
G0 MJN
G0 MJQ
G0 MJS
G0 MJW
G0 MKB
G0 MKI
G0 MKJ
GM0 MKM
G0 MKT
G0 MLD
G0 MLG
G0 MLS
G0 MLY
GM0 MMN
G0 MMR
G0 MNE
G0 MNM
G0 MOD
G0 MOK
GM0 MOP
G0 MPD
G0 MPG
G0 MPX
GW0 MQA
G0 MQF
G0 MQG
G0 MQP
G0 MQS
G0 MRH
G0 MRV
G0 MSL
G0 MSP
G0 MTR
G0 MTX
G0 MTZ
GM0 MUO

G0 MUT
G0 MUU
GM0 MUV
G0 MUW
G0 MUY
GM0 MWR
G0 MXF
G0 MXN
G0 MYW
G0 MYY
G0 MZA
GM0 MZB
G0 MZT
G0 MZV
G0 MZX
G0 NAB
G0 NBF
GM0 NBO
G0 NCF
G0 NCK
G0 NCM
GW0 NCN
G0 NCR
G0 NDN
G0 NDQ
G0 NDT
G0 NEH
G0 NEW
G0 NEY
G0 NFQ
G0 NGL
G0 NGT
GW0 NGU
G0 NGV
G0 NGX
G0 NHF
G0 NHN
G0 NHY
G0 NIH
G0 NII
G0 NIM
G0 NIS
G0 NIV
G0 NJF
G0 NJM
G0 NJY
G0 NKE
G0 NKR
G0 NKV
G0 NLE
G0 NLR
G0 NLS
G0 NME
G0 NMK
G0 NML
G0 NMQ
G0 NMT
G0 NND
GW0 NNY
G0 NOK
G0 NOS
G0 NOT
G0 NOY
G0 NPH
G0 NQZ
G0 NRL
G0 NSX
G0 NSY
G0 NTE
G0 NTK
GW0 NTO
G0 NTP
G0 NTQ
GW0 NTX
G0 NVF
GW0 NVN
G0 NVP
G0 NVW
G0 NWB
G0 NWX
G0 NXR
G0 NYG
G0 NZO
G0 NZQ
G0 NZW
G0 OAH
G0 OAI
G0 OAK
G0 OAN
G0 OAQ
G0 OBS
GM0 OBX
GM0 OCD
G0 OCI
G0 ODF
G0 ODV
G0 OEF
G0 OEP
G0 OFC
G0 OFJ
G0 OGO
GW0 OGT
G0 OGU
G0 OHB
G0 OHE
G0 OHI
G0 OIL
G0 OJA
GI0 OJD
G0 OJH
G0 OJI
G0 OJZ
G0 OKB
GI0 OKM
G0 OKR
G0 OKS

G0 OKW
G0 OLB
G0 OLH
G0 OLL
G0 OLY
G0 OMG
G0 OMK
G0 OMP
G0 OMS
G0 ONC
G0 ONK
G0 OOG
G0 OOH
G0 OOT
G0 OOW
G0 OOY
G0 OPD
G0 OPH
G0 OPR
G0 OPW
G0 OQB
G0 OQC
G0 OQD
G0 OQJ
G0 OQM
G0 OQT
G0 OQU
G0 OQW
G0 ORA
G0 ORI
G0 OSY
G0 OTA
G0 OTO
G0 OUF
G0 OUI
G0 OUQ
G0 OVI
G0 OVU
GI0 OXG
G0 OXI
G0 OXJ
G0 OXO
G0 OXQ
G0 OYB
G0 OZK
G0 OZU
G0 OZW
G0 OZX
G0 OZY
G0 OZZ
G0 PAT
G0 PDL
G0 PDR
G0 PDS
G0 PDU
G0 PFC
G0 PFD
G0 PGE
G0 PHO
G0 PHU
G0 PIF
G0 PIG
G0 PIH
G0 PIM
G0 PIP
G0 PIX
G0 PIZ
GM0 PJD
G0 PJK
G0 PJX
G0 PKH
G0 PKS
G0 PKU
G0 PLM
GW0 PLN
G0 PLV
G0 PMA
G0 PMH
G0 PNY
G0 PNZ
G0 POH
G0 POI
GM0 PPB
G0 PPF
G0 PPN
G0 PPO
G0 PPP
G0 PPU
G0 PQE
G0 PQK
G0 PQQ
G0 PQS
G0 PQT
G0 PRK
G0 PSM
G0 PSN
G0 PSU
G0 PTJ
G0 PTS
G0 PUA
GM0 PUF
GW0 PUH
G0 PUU
G0 PVK
G0 PVW
G0 PWR
G0 PWZ
G0 PYB
GW0 PYN
G0 PYT
G0 PYZ
G0 PZA
G0 PZE
G0 PZI
GM0 RAB
G0 RAH

G0 RAP
GM0 RAY
G0 RAZ
G0 RBB
G0 RBG
G0 RCC
G0 RCK
GW0 RCZ
G0 RDI
G0 RDQ
G0 RDW
G0 REJ
G0 REM
G0 RES
G0 RFO
G0 RGB
G0 RGF
G0 RGR
G0 RGV
G0 RGY
G0 RHT
G0 RIN
G0 RIO
G0 RJP
G0 RKF
G0 RKL
G0 RKO
G0 RLR
G0 RMA
G0 RME
GI0 RMZ
G0 RNE
G0 RNG
GM0 RNR
G0 RNT
G0 RNU
G0 ROC
G0 ROG
G0 ROI
G0 ROJ
G0 ROK
G0 RPH
G0 RPP
G0 RPR
GI0 RQK
G0 RRA
GI0 RSH
G0 RSN
G0 RSP
G0 RSQ
G0 RTD
GM0 RTG
G0 RTK
G0 RTX
G0 RUE
G0 RUJ
G0 RUP
G0 RVA
G0 RVF
GW0 RVG
G0 RVJ
G0 RVL
G0 RVN
G0 RVT
G0 RVU
G0 RVY
G0 RWB
G0 RWD
G0 RWF
G0 RWG
G0 RWK
G0 RWR
G0 RXE
G0 RXG
G0 RXH
G0 RXL
G0 RXN
G0 RXT
G0 RXW
G0 RXY
G0 RYC
G0 RYF
G0 RYH
GI0 RYN
GW0 RYT
G0 RYY
G0 RZC
G0 RZE
G0 RZF
G0 RZJ
G0 RZR
G0 RZV
G0 SAA
G0 SAG
G0 SAM
G0 SAN
GI0 SAP
G0 SAQ
G0 SAS
G0 SAW
G0 SBD
G0 SBF
G0 SBN
G0 SBS
G0 SBW
G0 SCF
G0 SDB
G0 SDH
G0 SDI
G0 SDN
G0 SDO
G0 SDP
G0 SDR
G0 SDT
GM0 SDV
G0 SDY

G0 SEA
G0 SEM
G0 SEV
G0 SFD
G0 SFF
G0 SFH
G0 SFL
G0 SGB
G0 SGN
G0 SHE
G0 SHQ
G0 SHW
GM0 SIC
G0 SID
G0 SIF
GM0 SIM
G0 SIP
G0 SIT
GM0 SIZ
G0 SJI
G0 SJK
GM0 SJY
G0 SKB
G0 SKE
GW0 SKO
G0 SKP
G0 SKT
G0 SKU
G0 SLO
G0 SLS
G0 SMA
G0 SMD
G0 SMG
G0 SMI
G0 SML
G0 SMX
G0 SNC
G0 SNE
G0 SNJ
G0 SOA
G0 SOJ
G0 SOL
G0 SOM
G0 SOO
G0 SOP
G0 SOR
G0 SOT
G0 SPE
G0 SPR
G0 SQZ
GI0 SRM
GW0 SSB
GW0 SSI
G0 STE
G0 STG
G0 STN
G0 STQ
G0 STU
G0 STV
G0 SUG
G0 SUJ
G0 SUM
G0 SVG
G0 SVT
G0 SWG
G0 SWP
G0 SWQ
G0 SWR
G0 SXB
G0 SXC
G0 SXF
G0 SXH
G0 SYA
G0 SYG
G0 SYJ
G0 SYM
G0 SYO
GM0 SYU
G0 SZP
G0 SZR
G0 SZS
G0 TAP
G0 TAU
GW0 TBA
G0 TBD
G0 TBK
G0 TBN
G0 TBR
GW0 TBT
G0 TBX
G0 TCD
G0 TCX
G0 TCZ
G0 TDF
G0 TDI
G0 TEC
G0 TEQ
G0 TER
GM0 TET
G0 TEY
G0 TEZ
G0 TFH
G0 TFS
GM0 TFW
G0 TGC
G0 TGD
GW0 TGI
G0 TGJ
G0 TGT
G0 TGY
G0 TGZ
G0 THG
G0 THM
GI0 THZ
G0 TIK
G0 TIM

G0 TIQ
G0 TIY
G0 TJB
G0 TJH
G0 TJZ
G0 TKD
G0 TKO
G0 TKP
G0 TKV
G0 TLB
G0 TLC
GM0 TLD
G0 TLG
G0 TLK
G0 TLL
GM0 TLM
G0 TMB
G0 TMC
G0 TMD
G0 TMP
G0 TMX
G0 TMY
G0 TNI
G0 TNJ
G0 TNR
G0 TNT
G0 TNU
G0 TNV
G0 TOG
GW0 TOI
G0 TOR
G0 TOU
G0 TOV
G0 TPC
G0 TPS
G0 TPT
G0 TPU
G0 TPV
G0 TPX
G0 TPZ
GW0 TQM
G0 TQN
G0 TQO
G0 TRF
G0 TRT
G0 TRX
GW0 TRZ
G0 TSP
GW0 TSW
GM0 TTA
G0 TTD
G0 TTJ
G0 TTU
G0 TTV
G0 TTX
G0 TUA
GM0 TUB
G0 TUD
GM0 TUH
G0 TUK
G0 TUY
G0 TVF
G0 TVH
G0 TVI
G0 TVJ
G0 TVP
G0 TVY
G0 TWN
G0 TWW
G0 TXE
G0 TXI
G0 TYH
G0 TYW
G0 TYX
G0 TZA
G0 TZE
G0 TZK
GM0 TZN
G0 TZS
G0 UAL
GI0 UAM
G0 UAN
GM0 UAT
GI0 UAX
G0 UAZ
G0 UBC
G0 UBD
G0 UBE
G0 UBQ
G0 UBV
G0 UCF
G0 UCR
G0 UDA
G0 UDC
G0 UDK
G0 UDM
G0 UDX
G0 UEI
GM0 UEJ
G0 UEN
G0 UER
G0 UES
G0 UEX
G0 UFH
G0 UFK
G0 UGC
G0 UGE
G0 UGF
GW0 UGT
GW0 UHA
GI0 UHE
G0 UHY
G0 UII
G0 UIK
GW0 UIM
GM0 UIN

G0 UJA
G0 UJK
GM0 UKI
G0 UKN
G0 UKR
G0 ULE
GM0 ULK
G0 ULR
G0 ULV
G0 UMF
G0 UNJ
G0 UNO
G0 UNU
G0 UOA
G0 UOG
G0 UOH
G0 UON
G0 UOO
G0 UOR
G0 UPA
G0 UPC
G0 UPQ
G0 UPX
G0 UQN
G0 UQR
G0 UQX
G0 URE
G0 URG
G0 URK
G0 URM
G0 URR
G0 URW
G0 URY
G0 USD
G0 USH
G0 USL
G0 USP
G0 USR
G0 USU
GI0 USX
G0 UTL
GI0 UTS
G0 UTY
G0 UUK
G0 UVA
G0 UVB
G0 UVM
G0 UWG
GW0 UWM
G0 UWY
G0 UXA
G0 UXC
GI0 UXE
G0 UXM
G0 UYB
G0 UYL
G0 UYN
GI0 UZC
GI0 UZG
G0 UZL
G0 VAK
GW0 VAW
G0 VBA
G0 VBC
G0 VBJ
G0 VBS
G0 VCE
G0 VCG
G0 VCH
G0 VCL
G0 VCM
G0 VCP
GM0 VCQ
G0 VCS
G0 VDG
G0 VDI
G0 VEA
G0 VEO
G0 VFG
G0 VFI
G0 VFK
GM0 VFR
G0 VGF
GI0 VGQ
G0 VGS
G0 VGX
G0 VGZ
G0 VHM
G0 VHS
G0 VHZ
G0 VIH
G0 VIN
G0 VIP
GW0 VIR
G0 VIU
G0 VIW
G0 VIZ
G0 VJD
G0 VJG
G0 VJT
GI0 VJV
G0 VJW
GI0 VKP
GM0 VKQ
G0 VKT
G0 VKW
G0 VLG
G0 VLH
GW0 VLN
G0 VLO
G0 VMG
GW0 VML
G0 VMX
G0 VNG
G0 VNP
GW0 VNR

G0 VNS
G0 VNZ
GM0 VOM
G0 VOW
G0 VOY
G0 VOZ
G0 VPF
G0 VPL
G0 VPM
G0 VPR
G0 VQH
G0 VRD
G0 VRI
G0 VRN
G0 VRR
G0 VSV
G0 VTH
G0 VTR
G0 VTX
G0 VUF
G0 VUO
G0 VUV
G0 VUW
G0 VUZ
G0 VVE
G0 VVM
G0 VVN
G0 VVS
G0 VWC
GM0 VWG
G0 VWI
G0 VWL
GM0 VWR
GW0 VWS
G0 VWY
G0 VXN
G0 VXO
G0 VXU
G0 VYH
G0 VYO
G0 VYS
G0 VZG
G0 VZJ
G0 VZM
G0 VZQ
GM0 VZU
G0 VZW
G0 VZY
G0 WAF
G0 WAG
GI0 WAI
G0 WAJ
G0 WAK
G0 WAP
G0 WAU
G0 WAV
G0 WBE
G0 WBJ
G0 WBX
G0 WBY
G0 WCA
G0 WCF
G0 WCG
G0 WCP
G0 WCW
G0 WCY
G0 WDA
GM0 WDD
G0 WDM
G0 WEC
GW0 WEE
G0 WEH
G0 WEJ
G0 WEN
G0 WFJ
G0 WFS
G0 WFW
G0 WFY
G0 WGC
G0 WGF
GW0 WGK
GI0 WHA
G0 WHE
G0 WHF
G0 WHG
GI0 WHI
G0 WHJ
GM0 WHM
G0 WHP
G0 WHX
G0 WIU
G0 WIV
G0 WJF
G0 WJQ
G0 WJT
G0 WKB
G0 WKC
G0 WKD
G0 WKG
G0 WKK
GW0 WKR
G0 WKV
G0 WLB
G0 WLL
G0 WLO
G0 WMF
G0 WMM
G0 WNA
G0 WNC
G0 WNH
G0 WNO
G0 WNQ
G0 WNT
G0 WNV
G0 WOD
G0 WOE

G0 WOK
G0 WOY
G0 WPA
G0 WPB
GW0 WPG
G0 WPJ
G0 WPN
G0 WPP
G0 WPR
G0 WQB
G0 WQD
G0 WQI
G0 WQK
G0 WQL
G0 WQS
G0 WQU
G0 WQV
G0 WRB
G0 WRF
GW0 WRJ
G0 WRO
G0 WRP
G0 WSE
GM0 WSG
GD0 WSK
G0 WSM
GM0 WSO
G0 WSQ
G0 WSS
G0 WST
G0 WSV
G0 WSW
G0 WTH
GW0 WTT
G0 WTZ
G0 WUD
G0 WUE
G0 WUT
G0 WVK
G0 WVU
G0 WVX
G0 WWG
G0 WWV
G0 WWW
G0 WXB
G0 WXI
G0 WXS
G0 WXT
G0 WXU
G0 WYD
G0 WYL
G0 WZI
G0 WZR
G0 XAX
G0 XBI
G0 XBJ
G0 XBU
G0 XEL
G0 XEM
G0 XGK
G0 XJB
G0 XRO
G0 XTV
G0 XXL
G0 XYS
G0 YAK
G0 YAP
GI0 YEW
G0 YLM
G0 YNM
G0 YOY
G0 ZAA
G0 ZEN
GI0 ZGB
G0 ZGN

## G*1

G1 AAE
G1 AAS
G1 ABR
G1 ABY
G1 ACH
G1 ADD
G1 AED
G1 AEG
G1 AES
G1 AEV
G1 AGI
G1 AGT
G1 AHA
GM1 AIH
G1 AIV
G1 AJI
G1 AJQ
G1 AJR
G1 AKN
G1 AKR
GW1 AKT
G1 AKX
G1 ALT
G1 AMI
G1 AMV
G1 AMW
G1 ANT
G1 AOU
G1 AOY
G1 APV
G1 AQQ
GM1 ARG
G1 ARI
G1 ARN
G1 ARV
G1 ASE
G1 AST
G1 ATI
G1 ATU

GW1 AUJ
G1 AUN
G1 AVD
G1 AVI
G1 AVQ
G1 AYL
G1 BAD
G1 BAF
G1 BCL
G1 BCP
G1 BCR
G1 BCZ
G1 BDK
G1 BED
G1 BEI
G1 BEM
G1 BFQ
G1 BFU
G1 BFZ
G1 BGY
G1 BHM
G1 BIQ
G1 BIR
G1 BJC
G1 BJH
G1 BJJ
GW1 BKH
GW1 BLD
G1 BMF
G1 BML
G1 BND
GM1 BNH
G1 BOE
G1 BOP
G1 BOZ
G1 BPH
G1 BPR
G1 BQC
G1 BRV
G1 BSN
GM1 BSP
G1 BSQ
G1 BSR
G1 BTU
G1 BUO
G1 BVI
G1 BWT
G1 BXB
G1 BXG
G1 BYY
G1 BZB
G1 BZG
G1 CAB
G1 CAE
G1 CAJ
G1 CBD
G1 CCC
G1 CDV
G1 CEB
G1 CHB
G1 CIK
G1 CKL
G1 CLG
G1 CLH
G1 CLL
G1 CMU
G1 COR
G1 COT
G1 CPP
G1 CPQ
G1 CQB
G1 CQL
G1 CQN
G1 CQO
G1 CQV
G1 CSC
G1 CSM
GW1 CSU
G1 CTF
G1 CUX
G1 CWC
G1 CWN
G1 CWO
G1 CWP
GI1 CXA
GM1 CXD
G1 CXG
GW1 CXT
GM1 CYB
G1 DBC
G1 DBD
G1 DCG
G1 DDH
G1 DDY
G1 DEA
G1 DET
G1 DFL
G1 DHE
GW1 DIQ
G1 DIW
G1 DJA
G1 DJG
G1 DJI
G1 DKA
G1 DLC
G1 DLL
G1 DMQ
G1 DMX
G1 DOK
G1 DOS
G1 DOW
G1 DQX
G1 DRA
G1 DRF
G1 DRX
G1 DSC

G1 DST
G1 DTV
G1 DUE
G1 DUH
G1 DVW
G1 DVX
G1 DWG
G1 DWH
G1 DWN
G1 DXS
G1 DYD
G1 DYE
G1 DYX
GI1 EAG
G1 EAY
GW1 ECA
G1 ECO
G1 ECZ
G1 EEL
GW1 EEP
G1 EEQ
G1 EEU
G1 EFI
G1 EFY
G1 EGC
G1 EHN
G1 EHT
G1 EJE
G1 EJX
G1 EJY
G1 EKH
G1 EKW
G1 ELG
G1 EMD
G1 EMU
G1 ENI
G1 EQT
G1 ERJ
G1 ESV
GW1 ETH
G1 EWA
G1 EWP
GW1 EWQ
G1 EYO
G1 EZG
G1 EZV
G1 FBH
G1 FBK
G1 FCG
G1 FDB
G1 FDK
GI1 FDY
G1 FGB
G1 FGT
GM1 FJQ
G1 FMD
GM1 FML
G1 FNR
G1 FPS
G1 FPU
G1 FQE
G1 FQV
G1 FRS
G1 FSH
G1 FTF
G1 FTO
G1 FTQ
G1 FUA
GM1 FUD
G1 FXE
GM1 FXJ
G1 FYG
G1 FYX
G1 GAB
G1 GBL
G1 GBO
G1 GBQ
G1 GCT
G1 GCU
G1 GCX
GM1 GFJ
G1 GGC
G1 GHF
G1 GHM
G1 GIB
G1 GIL
G1 GJF
G1 GJM
G1 GKU
G1 GKY
G1 GLA
G1 GNQ
GW1 GOF
G1 GPL
G1 GRD
G1 GRG
G1 GRU
G1 GRW
GW1 GTI
G1 GTO
G1 GTV
G1 GUE
G1 GUI
G1 GVE
G1 GVI
G1 GWN
G1 GWR
G1 GWV
G1 GXM
G1 GZY
G1 HAG
G1 HAK
GW1 HAN
G1 HEH
G1 HFC
G1 HFD

G1 HHL
G1 HIZ
G1 HJH
G1 HJW
G1 HKE
G1 HLF
G1 HLW
G1 HMA
G1 HMO
G1 HMP
G1 HNI
G1 HOK
G1 HPD
G1 HRE
G1 HRI
G1 HRW
G1 HSN
G1 HSS
G1 HUI
G1 HUL
G1 HUP
G1 HUW
G1 HWN
G1 HXM
G1 HYD
G1 HZA
G1 HZW
G1 IAH
G1 IAR
G1 IBH
G1 IDN
G1 IFL
G1 IGK
G1 IGZ
G1 IID
G1 IIE
G1 IIN
G1 IKO
G1 ILB
G1 ILN
G1 ILW
G1 IMH
G1 IMJ
G1 IML
G1 IMN
G1 IMQ
G1 INC
G1 INF
G1 IOJ
G1 IOS
G1 IOV
G1 IQZ
G1 IRB
GI1 ISG
G1 IUN
G1 IVE
G1 IWX
G1 IXY
G1 IYM
G1 IYU
G1 IZF
G1 IZQ
G1 JAJ
G1 JAK
G1 JAN
G1 JAR
G1 JAS
G1 JBC
G1 JBD
G1 JET
GW1 JEV
G1 JFV
G1 JFW
G1 JFX
G1 JGQ
G1 JHI
G1 JHS
GM1 JJJ
G1 JME
G1 JMO
G1 JMX
G1 JPL
G1 JPV
G1 JQI
G1 JQJ
G1 JQW
G1 JRG
G1 JSA
GW1 JSH
G1 JSZ
G1 JTD
G1 JTG
G1 JTO
G1 JTY
G1 JUX
G1 JXI
GM1 JZQ
G1 KAJ
G1 KAN
G1 KAW
G1 KBU
G1 KDQ
G1 KDU
G1 KEA
G1 KEJ
G1 KEW
G1 KEX
G1 KFC
GM1 KFD
GM1 KHL
G1 KIK
G1 KMA
G1 KMO
GM1 KNP
G1 KOO
G1 KPS

G1 KQJ
G1 KTJ
G1 KTN
G1 KUI
GW1 KVD
G1 KVM
G1 KVN
G1 KWJ
G1 KXF
G1 KXW
G1 KXY
G1 KYO
G1 KZL
G1 LAJ
G1 LCX
G1 LDI
G1 LDM
G1 LFU
G1 LGA
G1 LGX
G1 LIE
G1 LIF
G1 LII
G1 LIQ
G1 LIS
G1 LIW
G1 LJA
G1 LJR
G1 LJU
G1 LJY
G1 LLB
G1 LNT
GW1 LNY
G1 LOA
G1 LOP
GW1 LOR
G1 LOT
G1 LPW
G1 LSJ
G1 LUS
G1 LUV
G1 LVN
G1 LWA
G1 LXI
G1 LXP
G1 LXR
G1 LXW
G1 MAK
G1 MAW
GM1 MBT
GW1 MCD
G1 MCP
G1 MCX
G1 MDK
G1 MEZ
GD1 MFF
G1 MFG
GM1 MFN
G1 MHU
G1 MIB
GI1 MIC
GM1 MII
GW1 MIK
G1 MIR
G1 MJD
G1 MKH
G1 MKJ
G1 MLO
G1 MLR
G1 MMS
G1 MPH
G1 MSE
G1 MSU
G1 MTC
G1 MTT
G1 MTY
G1 MUK
GW1 MVL
G1 MWI
G1 MWP
G1 MXS
G1 NAF
G1 NAG
G1 NBQ
G1 NCI
G1 NDA
G1 NDG
G1 NEJ
G1 NFT
GM1 NGH
G1 NHB
G1 NHH
G1 NHW
G1 NIB
G1 NID
G1 NIG
G1 NIM
G1 NIS
G1 NKU
GM1 NLD
G1 NLV
GM1 NLX
G1 NMH
G1 NNP
G1 NOC
G1 NOD
G1 NOL
G1 NOX
G1 NPT
G1 NPV
G1 NQG
G1 NQP
G1 NQW
G1 NRL
G1 NRT
G1 NRU

Withheld

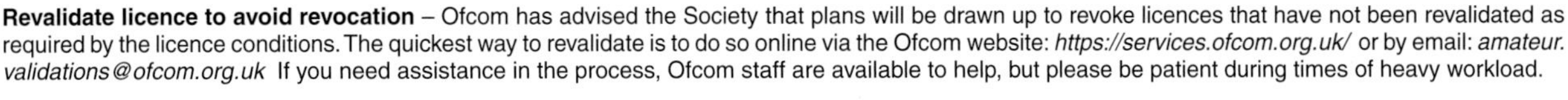
**IMPORTANT NOTE**

**Revalidate licence to avoid revocation** – Ofcom has advised the Society that plans will be drawn up to revoke licences that have not been revalidated as required by the licence conditions. The quickest way to revalidate is to do so online via the Ofcom website: *https://services.ofcom.org.uk/* or by email: *amateur.validations@ofcom.org.uk* If you need assistance in the process, Ofcom staff are available to help, but please be patient during times of heavy workload.

G1 NRV
G1 NSA
G1 NSU
G1 NTE
G1 NTH
G1 NUC
G1 NUJ
G1 NUR
G1 NUU
G1 NVQ
G1 NZB
G1 NZM
G1 NZO
G1 NZT
G1 OAC
G1 OAH
G1 OAT
G1 OAY
G1 OBD
G1 OBF
G1 OCA
G1 OCJ
G1 OCN
G1 OEN
G1 OEX
G1 OFA
G1 OFZ
G1 OGU
G1 OHN
G1 OHS
G1 OHT
G1 OHZ
G1 OIF
GM1 OIN
G1 OIP
G1 OJJ
G1 OJP
G1 OKT
G1 OLJ
G1 OMK
GM1 OMO
G1 OMQ
GI1 ONL
G1 ONN
G1 ONY
G1 OOA
G1 OOE
GW1 OPE
G1 OPH
G1 OPK
G1 ORY
G1 OUR
G1 OUT
G1 OVA
G1 OWP
G1 OXD
G1 OXL
G1 OXX
G1 OYQ
G1 OYS
G1 OYW
G1 OYY
G1 OZF
G1 OZT
G1 OZU
G1 OZZ
G1 PAC
GI1 PBO
G1 PCM
G1 PCS
G1 PCW
G1 PCX
G1 PDZ
G1 PEP
G1 PFP
G1 PFQ
G1 PGF
G1 PHL
G1 PHQ
G1 PHZ
G1 PIB
G1 PIC
G1 PIE
G1 PIR
G1 PJW
G1 PKY
G1 PKZ
G1 PLB
G1 PLM
G1 PLT
GM1 PLY
G1 PML
G1 PMN
G1 PMR
GM1 PNP
G1 POS
G1 PPA
G1 PPC
G1 PPH
G1 PPP
G1 PRA
G1 PRB
G1 PRX
G1 PTH
G1 PTO
GI1 PVO
G1 PVY
G1 PWW
G1 PXJ
G1 PYA
G1 PYT
G1 RAD
G1 RAH
G1 RBK
GM1 RBM
G1 RCL

GM1 RCP
G1 RCQ
G1 RCT
G1 REQ
G1 RFO
G1 RHV
G1 RIJ
G1 RKF
G1 RKO
G1 RLP
G1 RMH
G1 RNA
G1 ROI
G1 RRD
G1 RRM
G1 RSB
G1 RSH
G1 RSL
G1 RSZ
G1 RTG
G1 RTJ
G1 RTN
G1 RUN
G1 RUV
G1 RUX
G1 RUY
G1 RVD
G1 RVE
G1 RVU
G1 RVV
G1 RVX
G1 RWK
G1 RWS
G1 RWU
G1 RXC
G1 RXR
G1 RYO
G1 RZT
G1 RZV
G1 SAA
G1 SAO
G1 SBB
G1 SBG
GW1 SBO
G1 SCP
G1 SCX
G1 SDC
G1 SDD
G1 SDT
G1 SEH
G1 SEN
G1 SFI
G1 SFN
G1 SFO
G1 SFS
G1 SGC
G1 SGX
G1 SHC
G1 SHM
G1 SHV
G1 SJF
G1 SKL
G1 SLN
GM1 SLW
G1 SMA
G1 SMX
G1 SNH
G1 SNJ
GM1 SNL
GM1 SNW
G1 SOA
G1 SOU
G1 SPC
G1 SPO
G1 SQB
G1 SQH
G1 SQM
G1 SQY
GM1 SRX
G1 SRZ
G1 SSO
G1 SSR
G1 SST
G1 STE
G1 SUI
G1 SUT
G1 SUV
G1 SWN
G1 SWY
GW1 SXO
GM1 SXV
GM1 SXZ
GW1 SYG
G1 SYS
G1 TAQ
G1 TBA
G1 TBC
G1 TBL
G1 TCM
G1 TDB
G1 TDG
G1 TDM
G1 TED
G1 TEJ
G1 TFI
G1 TFV
G1 TGC
G1 TGJ
G1 TGM
G1 THX
GW1 TKO
G1 TLB
G1 TLF
G1 TLG
G1 TLQ
G1 TLY

G1 TMJ
G1 TNG
G1 TNL
G1 TOA
G1 TOP
GI1 TOU
G1 TOX
G1 TQC
G1 TQP
G1 TQW
G1 TRB
G1 TRD
GM1 TSE
GM1 TSJ
G1 TSQ
G1 TSY
G1 TTM
G1 TTW
G1 TTY
G1 TUK
G1 TUO
G1 TVJ
G1 TVL
G1 TXC
G1 TXD
G1 TXY
G1 TYH
G1 TYV
G1 TYY
G1 TZG
G1 TZJ
G1 UAI
G1 UAJ
G1 UAN
G1 UAS
G1 UAU
G1 UBA
G1 UBB
G1 UBW
G1 UCA
G1 UCP
G1 UCW
G1 UDA
G1 UDH
GW1 UDK
G1 UDL
G1 UDO
G1 UEB
G1 UEG
G1 UFN
G1 UFQ
G1 UGM
G1 UGY
G1 UHD
G1 UHE
G1 UHI
G1 UHQ
G1 UHU
G1 UHY
G1 UHZ
G1 UIF
G1 UJB
G1 UJO
GW1 UJU
G1 UKE
G1 UKI
G1 UKQ
G1 UKR
G1 UKU
G1 UKX
G1 UKY
G1 ULN
G1 ULS
G1 UME
GW1 UMH
G1 UML
G1 UMR
G1 UMV
G1 UMX
G1 UND
G1 UNF
G1 UNL
G1 UNP
G1 UNW
G1 UOJ
G1 UOM
G1 UOZ
G1 UPC
G1 UPH
G1 UPK
G1 UPL
G1 UPR
G1 UPU
G1 UPY
G1 UQO
G1 USC
G1 USS
GW1 UTI
G1 UVQ
G1 UVY
G1 UWF
G1 UWG
G1 UWL
G1 UWR
G1 UWW
G1 UXA
G1 UXB
G1 UXC
G1 UXG
G1 UXH
G1 UXK
G1 UXP
G1 UXQ
G1 UYA
G1 UYE
G1 UYM

G1 UYP
G1 UZL
G1 VAE
G1 VAF
G1 VBI
G1 VBR
G1 VBW
G1 VBX
G1 VCA
G1 VCD
GW1 VCN
G1 VCP
G1 VCT
G1 VDA
G1 VDB
G1 VDY
G1 VEB
G1 VEQ
G1 VFD
G1 VFE
G1 VFG
G1 VFJ
GM1 VFS
G1 VGF
G1 VGT
GM1 VGZ
G1 VHG
G1 VHQ
G1 VIE
G1 VIK
G1 VIM
G1 VJW
G1 VKP
G1 VKZ
G1 VLB
G1 VLC
G1 VLM
G1 VLR
G1 VLV
GW1 VLW
G1 VMB
G1 VMG
G1 VMH
G1 VMJ
G1 VML
G1 VMR
G1 VMS
G1 VMT
G1 VMU
G1 VMW
GW1 VMY
G1 VNA
G1 VOA
G1 VOK
G1 VPB
GM1 VPK
GM1 VPL
G1 VPP
G1 VPR
G1 VPV
G1 VPX
G1 VQN
G1 VQO
GM1 VQU
G1 VQW
G1 VRG
G1 VRH
G1 VSL
G1 VSV
G1 VSW
G1 VSX
G1 VTA
G1 VTD
G1 VTI
G1 VTK
G1 VUE
GM1 VUH
GM1 VUL
G1 VUS
GM1 VUT
GM1 VUU
G1 VUX
G1 VVP
G1 VWV
G1 VWW
G1 VXJ
G1 VXK
G1 VXL
G1 VXQ
G1 VYK
GI1 VYZ
G1 VZC
G1 VZM
G1 VZP
G1 WAT
G1 WBN
G1 WCB
G1 WCL
G1 WCO
G1 WCR
G1 WDF
G1 WDM
G1 WDV
G1 WDW
G1 WEA
G1 WEB
G1 WEU
G1 WEX
G1 WFK
G1 WFV
G1 WGH
G1 WGJ
GW1 WGV
G1 WHC
G1 WHH
G1 WHN

G1 WHP
G1 WIA
G1 WIH
GW1 WIO
G1 WJZ
G1 WKP
GM1 WKR
G1 WKV
G1 WLM
G1 WMD
G1 WNS
G1 WOB
G1 WOQ
G1 WPD
G1 WRK
G1 WRM
G1 WSA
G1 WTF
G1 WUJ
G1 WUP
G1 WVF
G1 WVN
GM1 WWG
G1 WWV
G1 WWX
G1 WXA
G1 WXB
G1 WXI
G1 WXR
GW1 WXZ
GI1 WZA
G1 XAD
G1 XAE
GI1 XBH
G1 XBU
G1 XCS
G1 XCW
G1 XCZ
G1 XDA
G1 XDC
GM1 XDH
G1 XDT
G1 XEL
G1 XEM
G1 XEN
G1 XFQ
G1 XGF
G1 XIG
GM1 XIJ
G1 XIK
G1 XIP
G1 XIX
G1 XJP
G1 XJS
G1 XJV
G1 XKM
G1 XKR
G1 XKT
G1 XLP
G1 XLX
G1 XMF
G1 XMK
G1 XMV
G1 XNS
G1 XOL
G1 XOX
G1 XPC
G1 XPR
GI1 XPV
GW1 XQT
G1 XQX
G1 XSF
G1 XSZ
G1 XTB
G1 XTP
G1 XWJ
G1 XWT
G1 XWU
G1 XYH
GM1 XYL
G1 XYU
G1 XYX
G1 XZL
G1 YAI
G1 YAR
G1 YBN
G1 YBQ
G1 YCB
G1 YDH
GW1 YDN
G1 YDT
G1 YEF
G1 YEH
G1 YEX
G1 YEY
G1 YFF
G1 YFR
G1 YFU
G1 YFV
G1 YFX
G1 YFY
G1 YGD
G1 YGJ
G1 YHR
G1 YIB
G1 YIF
G1 YIJ
G1 YIK
G1 YIO
GW1 YJE
G1 YKV
G1 YLU
GM1 YMI
G1 YMU
G1 YMW
G1 YNE

G1 YNG
G1 YNI
G1 YNL
G1 YNW
G1 YNY
G1 YNZ
GM1 YOG
G1 YOK
G1 YOM
G1 YOR
G1 YPA
G1 YPB
G1 YPG
GM1 YQH
G1 YQJ
G1 YQK
G1 YRV
G1 YRW
G1 YST
GM1 YUO
G1 YUV
G1 YVF
G1 YVR
G1 YVT
G1 YWC
G1 YWR
G1 YXF
GM1 YXM
G1 YYE
G1 YYV
G1 YZB
G1 YZN
G1 YZR
G1 YZS
G1 YZY
G1 ZAF
G1 ZAL
G1 ZAT
G1 ZCK
G1 ZCV
G1 ZCW
G1 ZDB
G1 ZDC
G1 ZDI
G1 ZDK
G1 ZDM
G1 ZEB
G1 ZEF
G1 ZEO
G1 ZEZ
G1 ZGC
G1 ZGU
GM1 ZGX
G1 ZGZ
G1 ZHS
GW1 ZHX
G1 ZIE
G1 ZIF
G1 ZIT
G1 ZIY
GW1 ZJC
G1 ZJM
GW1 ZJV
G1 ZKC
G1 ZKJ
G1 ZLG
G1 ZLH
G1 ZLM
G1 ZLU
G1 ZMV
G1 ZNE
GM1 ZNI
G1 ZNL
GW1 ZOI
G1 ZOM
G1 ZOV
G1 ZQA
G1 ZQC
G1 ZQI
G1 ZQZ
G1 ZRF
G1 ZRN
G1 ZRZ
G1 ZUG
G1 ZUK
G1 ZVB
G1 ZVL
G1 ZVN
G1 ZVQ
G1 ZWA
G1 ZWT
G1 ZWW
G1 ZXA
G1 ZXH
G1 ZXQ
GM1 ZXT
G1 ZYB
G1 ZYC
G1 ZYX
GW1 ZZX
G1 ZZY

## G*2

G2 AA
G2 ADC
G2 ASF
G2 AVV
G2 BDV
G2 BFC
G2 BGI
G2 BRH
G2 BXH
G2 DVP
G2 DXU
G2 FQS
GM2 FZ
G2 HPG
G2 HR
G2 IV
G2 MT
G2 VH

## G*3

G3 AGA
G3 AGW
G3 AHD
G3 AHW
G3 ARO
G3 ATI
G3 BCY
G3 BK
G3 BRW
GM3 BSQ
G3 BXF
G3 BYV
G3 BZW
G3 CAR
G3 CDR
G3 CEN
G3 CKR
G3 COO
G3 CPS
G3 CQQ
GM3 CSO
G3 CXX
G3 DAM
G3 DCU
G3 DIH
G3 DJR
G3 DKJ
G3 DKN
G3 DLP
G3 DNE
G3 DNK
G3 DNQ
GW3 DRK
G3 DWW
GI3 DZE
G3 EEO
G3 EFB
G3 EGY
G3 EHG
G3 EIG
G3 EJS
G3 EKL
GJ3 EML
G3 ENG
G3 EPN
G3 EWE
G3 EYB
G3 FAB
G3 FKB
G3 FRW
G3 FSO
G3 FVO
G3 GNF
GM3 GNX
G3 GRM
G3 GUL
G3 GVB
G3 GWI
G3 GZJ
GM3 HBT
G3 HCM
G3 HCN
G3 HCU
G3 HDS
G3 HEL
G3 HER
G3 HIJ
G3 HPP
G3 HSR
GM3 HZX
G3 IEW
G3 IGR
G3 II
G3 IIA
G3 IJN
G3 IMR
G3 INF
G3 ITK
G3 IVB
G3 IZV
G3 JBQ
G3 JFH
G3 JFL
G3 JJM
G3 JKN
G3 JPL
GW3 JRT
G3 JXB
G3 KCJ
G3 KEF
G3 KER
G3 KFN
G3 KGB
GW3 KLB
G3 KNE
G3 KNM
G3 KOU
G3 KPP
G3 KQD
G3 KSG
G3 KSU
GI3 KSY
G3 KUM
G3 KUS
G3 KWA
G3 KWT
G3 KXU
G3 KYG
G3 LBL

G3 LCT
G3 LDW
G3 LFZ
G3 LKO
G3 LKZ
G3 LLL
G3 LMD
G3 LMK
G3 LMO
G3 LPP
G3 LSS
G3 LWK
G3 LX
G3 LXI
G3 LYE
GW3 LYF
G3 LYK
G3 LZG
G3 MAL
G3 MAO
G3 MDC
G3 MEJ
G3 MGM
G3 MGP
G3 MGQ
G3 MGV
GW3 MHW
G3 MII
G3 MJO
G3 MMH
G3 MMP
G3 MMR
G3 MTM
G3 MUI
GD3 MXG
G3 MZF
G3 MZK
G3 NAC
G3 NAS
G3 NCM
G3 NCR
G3 NDM
G3 NGD
G3 NHO
G3 NKR
GW3 NLN
G3 NLZ
G3 NNF
G3 NNH
GM3 NQV
G3 NRB
G3 NRT
G3 NSB
G3 NSM
GW3 NTK
G3 NUE
GM3 NUF
G3 NWD
G3 NWG
G3 NYY
G3 OAG
G3 OBX
G3 ODR
G3 OFA
G3 OFL
G3 OGF
G3 OGY
G3 ORH
G3 OSB
GM3 OSW
G3 OUF
G3 OUK
G3 OVM
G3 OVQ
G3 OWF
G3 OWS
G3 OXQ
G3 OYE
GM3 OYO
G3 OYP
G3 OYS
G3 OZQ
G3 OZS
G3 PBA
G3 PBU
G3 PCN
G3 PCQ
G3 PCR
G3 PD
G3 PDG
G3 PDK
GI3 PGD
G3 PGV
G3 PGX
G3 PHA
G3 PHR
G3 PHU
G3 PJD
G3 PJI
GW3 PKM
G3 PKW
G3 PLH
G3 PLS
G3 PLU
G3 PMM
G3 PNY
G3 PPG
GM3 PQU
G3 PRN
G3 PTD
G3 PTU
G3 PUK
G3 PUV
G3 PVS
G3 PVW
G3 PVX

G3 PWH
G3 PXQ
G3 PXT
G3 PYA
G3 PYL
GM3 PYU
GJ3 RAX
G3 RAY
GW3 RBA
G3 RBG
GW3 RBM
GI3 RBX
G3 RCU
G3 RD
G3 REA
G3 REN
G3 RFF
G3 RFL
G3 RFT
G3 RHN
G3 RJQ
G3 RLB
GW3 RME
G3 RNQ
G3 ROI
G3 RR
G3 RRX
G3 RSS
G3 RSY
G3 RTU
G3 RVU
GM3 RXF
G3 RXL
G3 RXQ
G3 RZA
G3 SAZ
G3 SCB
G3 SCU
G3 SDQ
GW3 SEV
G3 SEW
G3 SIP
G3 SJZ
G3 SKC
GW3 SKP
G3 SLR
G3 SMP
G3 SMU
G3 SMX
G3 SNX
G3 SPB
G3 SPE
G3 SPX
G3 SPZ
G3 SQV
G3 SR
G3 SRO
G3 SSC
G3 SSG
G3 SSO
G3 STR
G3 STS
G3 SUF
G3 SVL
GM3 SWF
G3 SWM
G3 SYF
G3 SYX
G3 SZO
G3 SZT
G3 TAH
G3 TAM
G3 TAN
G3 TBA
G3 TCA
GM3 TCE
G3 TCR
G3 TDJ
G3 TGP
G3 TGW
G3 TIB
G3 TIC
G3 TIQ
GM3 TKV
G3 TKW
G3 TLE
G3 TMG
GW3 TMH
G3 TMZ
G3 TOE
G3 TOI
G3 TOR
G3 TOU
G3 TQM
G3 TRF
GW3 TSH
G3 TSK
G3 TTK
G3 TVQ
G3 TWE
G3 TWZ
G3 TXA
G3 TYB
G3 TYY
G3 TZH
G3 UAU
G3 UBU
G3 UCM
G3 UCV
G3 UDU
G3 UFK
G3 UFO
G3 UFP
G3 UGG
G3 UGK
G3 UGY

G3 UHR
G3 UIE
GM3 UIN
G3 UJL
G3 UKU
G3 ULF
G3 UMK
GD3 UMW
G3 UNC
GW3 UNH
G3 UNR
G3 UPK
G3 USB
G3 UUW
G3 UWI
G3 UXB
G3 VAA
G3 VBS
G3 VDX
G3 VEQ
G3 VFP
G3 VFV
GI3 VGL
G3 VGU
G3 VHB
GW3 VHG
G3 VHJ
G3 VHY
G3 VII
GW3 VLP
G3 VLT
GM3 VMB
G3 VMH
G3 VMS
G3 VNC
G3 VPD
G3 VRE
G3 VRL
G3 VSS
G3 VUF
G3 VUP
G3 VUQ
G3 VW
G3 VWF
G3 VWL
G3 VWP
GM3 VXI
GW3 VXP
G3 VXX
G3 VZZ
G3 WA
G3 WCS
G3 WFB
G3 WFK
G3 WGM
G3 WHL
G3 WHR
G3 WIC
G3 WIE
G3 WIR
G3 WKD
G3 WKU
G3 WM
G3 WMM
G3 WMR
G3 WMU
G3 WNU
G3 WOF
G3 WOX
G3 WPO
G3 WPS
G3 WQO
G3 WQP
G3 WQT
G3 WRU
G3 WRV
G3 WSN
G3 WTA
GI3 WTG
G3 WVY
G3 WVZ
G3 WW
G3 WWM
GD3 WWW
G3 WXI
G3 WXL
G3 WXX
G3 WZD
G3 WZN
G3 XAD
G3 XAJ
G3 XCR
G3 XDE
G3 XDH
G3 XDQ
G3 XDV
G3 XEB
G3 XEK
G3 XFA
G3 XFE
G3 XFV
G3 XFZ
G3 XGM
G3 XGQ
G3 XGS
G3 XHF
G3 XHY
G3 XIR
G3 XJK
G3 XJO
G3 XKQ
G3 XKR
G3 XKT
G3 XLF
G3 XLV
G3 XNH

G3 XNO
G3 XNT
G3 XPE
G3 XPO
G3 XPY
G3 XQX
G3 XRE
G3 XSA
G3 XSH
G3 XSJ
G3 XTX
G3 XUA
G3 XVF
G3 XWJ
G3 XXT
G3 XYS
G3 XYX
G3 XZS
G3 XZT
G3 YAM
G3 YAP
G3 YBD
G3 YBT
G3 YC
G3 YCA
G3 YCR
GM3 YEH
G3 YFN
GM3 YGS
G3 YIP
G3 YIS
G3 YIV
G3 YJO
G3 YJR
G3 YL
G3 YLE
G3 YLG
G3 YLO
G3 YMP
G3 YNH
G3 YNT
G3 YOK
G3 YPP
G3 YPQ
G3 YPX
G3 YQD
G3 YRE
G3 YRJ
G3 YRO
G3 YRW
G3 YRZ
G3 YSC
GW3 YSH
G3 YTH
G3 YTK
G3 YUE
GD3 YUM
G3 YUV
G3 YVF
G3 YWP
G3 YWQ
G3 YXX
G3 YXZ
G3 YYD
G3 YYH
G3 YYO
G3 YYT
G3 YZG
G3 YZJ
G3 ZAF
G3 ZCC
G3 ZCU
G3 ZDD
G3 ZDR
G3 ZDV
G3 ZEH
G3 ZEV
G3 ZFJ
G3 ZHH
G3 ZHI
G3 ZHW
G3 ZHX
GM3 ZKF
G3 ZLB
GM3 ZLC
G3 ZLW
G3 ZMC
G3 ZMU
G3 ZMY
G3 ZOD
G3 ZOE
G3 ZPF
GD3 ZPQ
G3 ZQX
G3 ZRK
G3 ZRV
G3 ZSE
GM3 ZSP
G3 ZSV
G3 ZUD
G3 ZVG
G3 ZVJ
G3 ZWI
GM3 ZXH
GW3 ZXI
GM3 ZXL
G3 ZXP
G3 ZXX
G3 ZYS
G3 ZYW
G3 ZZO

## G*4

G4 AAO
G4 ABD
G4 ABF
G4 ABP
G4 ADC
G4 ADN
G4 ADX
GM4 AFG
G4 AHS
G4 AJC
G4 AJM
G4 AKU
G4 AKZ
G4 ALG
G4 ALH
G4 ALI
GW4 AMZ
G4 ANA
G4 ANS
G4 AOG
G4 AON
G4 AOV
G4 APT
G4 APV
GW4 AQU
G4 AQV
G4 AQY
G4 ARP
G4 ASE
G4 AST
G4 ATH
G4 ATI
G4 ATX
G4 AVB
G4 AVD
G4 AVU
G4 AWC
G4 AWP
G4 AXH
G4 AXN
G4 AXP
G4 AYF
G4 AYJ
G4 AYT
G4 AYV
G4 AZF
G4 AZO
G4 AZR
G4 AZZ
G4 BAZ
G4 BBP
G4 BBS
GM4 BCK
G4 BDE
G4 BDO
G4 BDP
G4 BDY
G4 BEE
GD4 BEG
G4 BFP
G4 BFY
G4 BFZ
G4 BGQ
G4 BGX
G4 BIJ
G4 BIP
G4 BJM
G4 BKM
G4 BLF
G4 BLI
G4 BLJ
G4 BLM
G4 BLN
G4 BLQ
G4 BLX
G4 BMF
GM4 BNU
G4 BOY
G4 BPF
G4 BPG
G4 BPQ
G4 BSU
GW4 BTH
GM4 BWD
G4 BWQ
G4 BXO
G4 BXP
GW4 BZD
G4 BZO
G4 CAO
G4 CAW
G4 CBR
G4 CCX
G4 CDE
G4 CDM
G4 CDT
G4 CEH
G4 CFF
G4 CFJ
G4 CGS
G4 CIL
G4 CIN
G4 CIV
G4 CJC
G4 CJF
G4 CJS
G4 CJX
G4 CJZ
G4 CKN
G4 CKR
G4 CLK
GM4 CLQ
G4 CLT
G4 CLU
G4 CND
G4 CNE
GM4 CNF
G4 CNG
G4 COK

G4 CPK
G4 CPR
G4 CQJ
G4 CRF
G4 CRJ
GM4 CRV
G4 CSN
G4 CSS
G4 CTG
G4 CTL
G4 CTP
G4 CTR
G4 CTS
G4 CUJ
G4 CUS
G4 CVY
G4 CWS
GI4 CXH
G4 CYE
G4 CYQ
G4 CZE
G4 CZJ
G4 CZS
GM4 DAE
G4 DAR
G4 DBL
G4 DBS
G4 DCQ
G4 DCR
G4 DCS
G4 DCV
G4 DDR
G4 DED
G4 DEE
G4 DGK
GW4 DGU
G4 DHA
GM4 DHJ
G4 DHO
G4 DHP
G4 DHQ
G4 DIF
G4 DIW
GW4 DJW
G4 DKF
G4 DKK
G4 DKO
G4 DKR
G4 DLB
G4 DLX
G4 DMA
G4 DMD
G4 DME
GI4 DOH
G4 DOI
G4 DPE
G4 DPN
G4 DPX
G4 DPY
G4 DQY
G4 DRK
GI4 DRY
G4 DSB
G4 DST
G4 DTD
GM4 DTF
G4 DUC
G4 DUG
G4 DUH
G4 DUK
G4 DUZ
G4 DVF
GW4 DVQ
G4 DVS
G4 DVW
G4 DWQ
G4 DWV
G4 DXA
G4 DXG
G4 DXX
G4 DYA
G4 DYW
G4 DZE
G4 DZF
G4 DZL
G4 DZR
G4 EAA
G4 EAR
G4 ECA
G4 ECM
G4 ECY
G4 EDE
G4 EDS
G4 EEI
G4 EEM
G4 EER
G4 EFT
G4 EFW
G4 EGH
G4 EHA
G4 EHH
GW4 EHZ
G4 EIB
G4 EID
G4 EIS
G4 EIT
G4 EJB
G4 EJN
G4 EKK
G4 ELD
G4 ELN
G4 ELR
G4 EMM
GM4 EMO
G4 EMR
G4 EOZ

G4 EPJ
G4 EQB
G4 EQH
G4 EQQ
G4 ESQ
G4 ETH
G4 ETQ
G4 ETV
G4 EUE
G4 EUL
G4 EUM
G4 EUQ
G4 EWS
G4 EWX
G4 EXA
G4 EXQ
G4 EXW
G4 EYL
G4 EZR
G4 EZT
G4 EZV
G4 FBA
G4 FBC
G4 FBJ
G4 FBW
GM4 FBX
G4 FCR
G4 FDB
G4 FDL
G4 FDO
G4 FDR
G4 FFJ
G4 FHA
G4 FHI
GM4 FID
G4 FIK
G4 FIS
GM4 FIX
G4 FJQ
G4 FJR
G4 FKK
G4 FLN
G4 FLQ
G4 FLU
G4 FMG
G4 FOK
GI4 FP
G4 FPH
G4 FPK
G4 FPS
G4 FRE
G4 FRS
GW4 FRU
G4 FSC
G4 FTC
G4 FTF
G4 FTJ
G4 FTN
G4 FUB
G4 FUF
G4 FUN
G4 FUT
G4 FVG
G4 FVN
G4 FWC
G4 FWG
G4 FWX
GW4 FXB
G4 FXO
G4 FYA
G4 FYF
G4 FYX
G4 FZN
G4 GAD
GW4 GAS
G4 GAU
G4 GBG
GW4 GCB
G4 GCD
G4 GCM
G4 GCP
G4 GCW
G4 GCX
G4 GCZ
GM4 GEF
G4 GER
G4 GES
G4 GFK
G4 GFX
G4 GGS
G4 GHD
G4 GHH
G4 GHJ
G4 GHP
G4 GHU
GM4 GIF
G4 GII
G4 GJC
G4 GJL
G4 GJN
G4 GJW
G4 GKA
G4 GLT
GM4 GM
G4 GMA
G4 GME
G4 GMJ
G4 GMO
G4 GMP
G4 GMV
G4 GMY
G4 GNC
G4 GOC
G4 GOH
GW4 GOQ
G4 GOY

G4 GPK
G4 GPU
GM4 GQB
G4 GQL
G4 GRI
GW4 GRW
G4 GSQ
G4 GSV
G4 GTA
G4 GTF
G4 GTG
G4 GTI
G4 GTP
G4 GTR
G4 GTT
G4 GTW
GM4 GUF
G4 GUI
G4 GVB
G4 GVC
G4 GVD
G4 GVX
G4 GWD
G4 GWO
G4 GYB
G4 GZJ
G4 HAD
G4 HAQ
G4 HBJ
G4 HCF
G4 HCL
G4 HCW
G4 HDK
G4 HEQ
GI4 HFB
G4 HFF
G4 HFT
G4 HGI
GI4 HGQ
G4 HGU
GW4 HHD
G4 HHI
G4 HHT
G4 HHV
G4 HIP
G4 HJG
G4 HJM
G4 HKF
G4 HKK
G4 HKM
G4 HKT
G4 HLP
G4 HLR
G4 HMG
GI4 HMI
G4 HMJ
G4 HMS
G4 HNB
G4 HNH
G4 HNY
G4 HOV
GM4 HPF
G4 HPG
G4 HPI
G4 HPL
G4 HPW
G4 HRK
G4 HRM
G4 HSU
G4 HSY
G4 HTP
G4 HUK
G4 HUV
GW4 HVE
G4 HVK
GW4 HVN
G4 HVX
G4 HXD
G4 HXT
G4 HYC
G4 HYX
GW3 HZA
G4 HZB
G4 IAZ
G4 IBK
G4 ICD
G4 ICT
G4 ICV
G4 IDK
G4 IDO
G4 IDY
G4 IDZ
G4 IEJ
G4 IFB
GM4 IFC
G4 IFQ
G4 IGO
G4 IGW
G4 IHH
G4 IHL
G4 IIG
G4 IIQ
G4 IIU
G4 IJC
G4 IJE
G4 IJL
G4 IKH
G4 IKK
G4 IKP
G4 IKR
G4 IKZ
GM4 IMD
G4 IMN
G4 IMO
G4 IMZ
G4 INK

G4 INT
G4 IOX
G4 IPE
G4 IPP
G4 IPQ
G4 IPW
G4 IPZ
G4 IQE
G4 IQI
GW4 IQP
G4 IRN
G4 IRQ
GI4 IRW
G4 IRX
G4 ISI
G4 ISP
G4 ITF
G4 IUB
G4 IUG
G4 IUR
G4 IUW
G4 IVH
G4 IVK
G4 IVS
G4 IVV
G4 IWB
G4 IWJ
G4 IXN
G4 IYJ
G4 IYL
G4 IYU
G4 IZC
G4 JAG
G4 JBC
G4 JBI
G4 JBX
G4 JCI
G4 JCN
G4 JCO
G4 JEC
G4 JEG
G4 JEN
G4 JEQ
G4 JET
G4 JFB
G4 JFK
G4 JFO
G4 JFT
G4 JGB
G4 JGL
G4 JGP
G4 JHD
G4 JHF
G4 JHM
GM4 JIB
G4 JIS
G4 JJI
G4 JKD
G4 JKO
G4 JLB
G4 JLQ
G4 JLT
G4 JLU
G4 JMH
G4 JMJ
GW4 JMN
G4 JMV
G4 JNC
G4 JNN
G4 JNY
GM4 JOJ
G4 JOK
G4 JON
G4 JOP
GM4 JPI
G4 JQE
G4 JQM
G4 JQP
G4 JQT
G4 JRC
G4 JRH
G4 JRL
G4 JRN
G4 JRO
G4 JSL
GM4 JTJ
G4 JTY
G4 JUT
G4 JUY
G4 JVF
G4 JVG
G4 JWB
G4 JWT
G4 JXL
G4 JXS
G4 JXX
G4 JXY
GM4 JZJ
G4 KAG
G4 KAJ
G4 KAN
G4 KAO
G4 KAW
G4 KAY
G4 KBC
G4 KBD
G4 KCS
G4 KCW
G4 KDG
G4 KDY
G4 KEU
GW4 KEV
G4 KFE
G4 KFK
G4 KFO
GM4 KHA

G4 KHV
G4 KIJ
G4 KIL
G4 KIU
G4 KIV
GM4 KJQ
GW4 KJW
G4 KLG
G4 KLH
G4 KLL
G4 KMC
G4 KMV
G4 KND
G4 KNK
G4 KNM
G4 KNP
G4 KOB
G4 KOH
GI4 KOP
G4 KPB
G4 KPJ
G4 KPW
G4 KQU
G4 KRB
G4 KRM
G4 KRV
G4 KSD
G4 KSE
G4 KSW
G4 KTE
GD4 KTS
G4 KUI
G4 KVA
G4 KWL
G4 KXN
G4 KYA
G4 KYC
G4 KYD
G4 KYF
G4 KYN
G4 KYV
G4 KZS
G4 KZY
G4 LAC
G4 LAH
G4 LAX
G4 LBB
G4 LBC
G4 LBI
G4 LCD
GM4 LCZ
G4 LDG
G4 LDY
G4 LDZ
G4 LEB
G4 LEF
G4 LEL
G4 LFM
G4 LFU
G4 LGD
G4 LGE
G4 LGF
G4 LGZ
G4 LHN
G4 LHU
G4 LHX
G4 LIB
G4 LIP
G4 LJJ
G4 LJZ
GW4 LKI
GI4 LLX
GD4 LMJ
G4 LMS
G4 LMZ
G4 LNA
G4 LND
G4 LNF
G4 LOU
GW4 LPB
G4 LQP
G4 LQT
G4 LQY
G4 LRR
G4 LRS
G4 LRX
G4 LTF
GM4 LTJ
G4 LTX
G4 LUG
G4 LUL
G4 LUP
G4 LUZ
GI4 LWR
G4 LXE
G4 LXI
G4 LXK
GI4 LXL
G4 LXQ
G4 LXS
G4 LXT
G4 LXZ
G4 LYJ
G4 LZL
G4 LZM
G4 LZW
G4 MAD
G4 MAM
G4 MAP
G4 MAW
G4 MBN
G4 MBS
G4 MBU
G4 MBY
G4 MCL
G4 MCN
G4 MDI
G4 MDP
G4 MDX
G4 MEC
G4 MEZ
G4 MFU
G4 MGL
G4 MHN
G4 MHU
G4 MIN
G4 MIU
G4 MIY
G4 MJH
G4 MJM
G4 MJO
G4 MJS
G4 MJZ
G4 MKS
G4 MLA
G4 MLJ
G4 MLP
GI4 MNO
G4 MOF
G4 MOM
GM4 MOX
GM4 MPU
G4 MPV
G4 MPX
GM4 MQ
G4 MQE
G4 MR
G4 MRH
G4 MRM
G4 MSF
G4 MSH
G4 MSO
G4 MTQ
G4 MUF
G4 MUR
G4 MVC
G4 MVH
G4 MVK
G4 MVL
G4 MVN
G4 MVR
G4 MVU
G4 MVV
GW4 MWK
G4 MWM
G4 MWR
G4 MXA
G4 MXR
G4 MXU
G4 MYC
G4 MYI
G4 MZJ
G4 MZS
G4 NAB
G4 NAD
G4 NAR
G4 NAZ
GM4 NBZ
G4 NCE
G4 NCN
G4 NCT
G4 NDH
GW4 NDW
G4 NEC
GM4 NEV
G4 NEX
G4 NFB
GM4 NFC
G4 NFD
G4 NFJ
G4 NFU
G4 NGE
G4 NGH
G4 NHJ
G4 NHS
G4 NIC
G4 NIS
G4 NIW
G4 NJE
G4 NJH
G4 NKL
G4 NKM
G4 NKT
G4 NKV
G4 NLP
G4 NLX
G4 NLY
G4 NMI
G4 NMX
G4 NNA
GM4 NNC
G4 NNU
G4 NNW
G4 NOV
G4 NOW
G4 NPJ
G4 NPX
G4 NQF
G4 NQK
G4 NQX
G4 NRL
G4 NSG
G4 NSI
GW4 NSU
G4 NTB
GI4 NTF
G4 NTM
G4 NTN
GW4 NUL
G4 NUV
G4 NVB
G4 NVF
GW4 NVO
G4 NVR
G4 NVZ
G4 NWF
G4 NXI
GI4 NXK
G4 NXM
G4 NYC
G4 NYN
G4 NYS
G4 NZF
G4 NZV
G4 OAC
GM4 OAQ
G4 OAT
G4 OBI
G4 OBL
G4 OCO
G4 OCP
G4 ODC
GD4 OEL
G4 OFY
G4 OGD
G4 OGE
G4 OGF
G4 OGI
G4 OGP
G4 OGT
G4 OHK
G4 OHN
G4 OIP
G4 OIX
G4 OIZ
G4 OJC
G4 OJO
GW4 OJT
G4 OKK
GW4 OKT
G4 OLN
G4 OLT
G4 OLV
GI4 OMA
G4 OME
GW4 OMF
G4 OMH
G4 OMX
G4 ONE
G4 ONK
GI4 ONL
G4 ONW
G4 OOP
G4 OOR
G4 OPQ
G4 ORF
G4 ORM
G4 ORR
G4 OSW
G4 OTK
G4 OTQ
G4 OUL
G4 OVK
G4 OVP
G4 OVU
G4 OWF
G4 OWI
GW4 OWO
G4 OWW
G4 OXC
G4 OXF
GW4 OXW
G4 OXY
G4 OYD
G4 OYF
G4 OYS
G4 OZH
G4 OZP
G4 OZZ
G4 PAG
G4 PAN
G4 PAO
G4 PAP
G4 PBL
G4 PBV
G4 PC
G4 PCA
G4 PCI
G4 PCS
G4 PDF
G4 PDL
G4 PDS
G4 PDV
G4 PEB
G4 PEC
G4 PF
GW4 PFC
G4 PFD
G4 PFN
G4 PGI
G4 PGP
G4 PHJ
G4 PHW
G4 PIO
G4 PIV
G4 PIX
G4 PJA
G4 PJC
G4 PJF
G4 PJG
G4 PLC
G4 PLN
G4 PLP
G4 PM
G4 PMN
G4 PMQ
G4 PMX
G4 PNF
G4 PNJ
G4 PNW
G4 POA
GW4 POO
G4 PQA
G4 PQL
G4 PQT
GW4 PRV
G4 PSA
G4 PSB
GM4 PSF
G4 PSG
G4 PSN
G4 PTB
G4 PTI
G4 PTJ
G4 PTN
G4 PTS
G4 PTX
G4 PUF
GM4 PUJ
G4 PUR
G4 PVA
G4 PVB
GI4 PXI
G4 PXL
G4 PXO
G4 PXW
G4 PXZ
G4 PYM
G4 PZE
G4 PZK
G4 RAD
G4 RBI
G4 RBO
G4 RBV
G4 RCA
G4 REL
G4 REQ
G4 REV
G4 RFM
G4 RFQ
G4 RFX
GD4 RFZ
G4 RGK
G4 RGV
G4 RHD
G4 RHN
G4 RHO
G4 RIC
GW4 RII
GW4 RIT
GM4 RIW
G4 RJC
G4 RKJ
G4 RKY
G4 RLQ
GW4 RMI
G4 RMK
G4 RMM
G4 RNL
G4 ROE
G4 ROO
G4 RPX
G4 RQE
G4 RQY
G4 RRC
G4 RRJ
G4 RRZ
G4 RS
G4 RSA
G4 RSB
G4 RSH
G4 RSQ
G4 RSR
G4 RSY
G4 RTB
G4 RTK
G4 RUV
G4 RVB
G4 RVM
G4 RVR
G4 RWC
G4 RWO
G4 RWU
G4 RXN
GI4 RXT
GI4 RYD
GI4 RYN
G4 RYU
G4 RYX
G4 RZG
G4 RZK
G4 RZS
GW4 SAF
GW4 SAG
G4 SAP
G4 SAQ
G4 SBT
G4 SCX
G4 SDG
G4 SDM
G4 SDN
G4 SDP
G4 SDR
G4 SDW
G4 SFI
G4 SFM
G4 SFU
GI4 SFX
GM4 SGB
G4 SGM
G4 SGO
G4 SGZ
G4 SHP
G4 SHZ
G4 SIO
G4 SIU
G4 SIV
G4 SJO
G4 SJX
G4 SJY
G4 SKH
G4 SKK
G4 SKR
G4 SKS
G4 SLU
G4 SLX
G4 SMN
G4 SMY
G4 SNK
G4 SNS
G4 SNT
G4 SOX
G4 SPG
G4 SPK
G4 SPO
G4 SPX
G4 SQU
G4 SRA
G4 SRB
G4 SRG
GI4 SRK
G4 SRS
G4 SSE
G4 SSG
GM4 SSI
G4 SSK
G4 SST
G4 SSX
G4 STA
G4 STM
G4 STR
G4 SUH
G4 SUI
GD4 SVD
G4 SVF
G4 SVN
G4 SWI
G4 SWL
G4 SWW
G4 SWX
G4 SXS
G4 SYJ
G4 SZD
G4 SZM
G4 TAA
G4 TAI
G4 TAS
G4 TAX
G4 TBA
G4 TBD
G4 TBL
GJ4 TBW
G4 TCJ
G4 TCN
G4 TCU
G4 TCZ
G4 TDS
G4 TEI
G4 TET
G4 TEV
G4 TFE
G4 TFN
G4 TFR
G4 THD
G4 THJ
G4 THL
G4 THR
G4 TIL
G4 TIN
G4 TIO
G4 TJB
G4 TJE
G4 TJJ
G4 TJT
G4 TKA
G4 TKG
G4 TKI
G4 TKJ
G4 TKN
G4 TKR
G4 TKV
G4 TLP
G4 TLV
G4 TMO
G4 TMT
G4 TMW
G4 TNI
G4 TNN
G4 TNT
GM4 TNW
G4 TNX
G4 TOC
G4 TOJ
G4 TOK
G4 TPA
G4 TQF
G4 TQG
G4 TQJ
G4 TRK
G4 TRQ
G4 TSM
G4 TSR
G4 TTI
GW4 TTU
G4 TTW
GI4 TUE
G4 TUG
G4 TUS
G4 TUT
G4 TUY
G4 TUZ
G4 TVH
G4 TVK
G4 TVP
G4 TVR
GW4 TWE
G4 TWI
G4 TWJ
G4 TXB
GM4 TXP
G4 TXR
G4 TXS
GM4 TXX
GW4 TYB
G4 TZJ
G4 TZP
G4 UAO
GM4 UCD
G4 UCO
GM4 UDC
G4 UDJ
G4 UDR
G4 UEM
G4 UEQ
G4 UHC
G4 UHD
G4 UHN
GI4 UHP
G4 UIN
G4 UIX
G4 UJD
G4 UJG
G4 UKF
G4 UKR
G4 ULA
G4 ULF
GM4 ULS
GW4 UMX
G4 UNR
G4 UNT
G4 UNU
GM4 UOQ
G4 UPE
G4 UPF
G4 UPZ
G4 UQB
G4 UQC
G4 URH
G4 URT
G4 USG
G4 USI
G4 UTT
G4 UTW
G4 UVE
G4 UVP
GW4 UWD
G4 UWQ
GI4 UXR
G4 UXU
G4 UYL
G4 VAD
G4 VAJ
G4 VBB
G4 VBT
G4 VBW
G4 VCD
G4 VCH
G4 VCK
G4 VCT
G4 VDE
G4 VDU
G4 VDV
G4 VEK
G4 VFA
G4 VFM
GW4 VFS
G4 VFT
G4 VGE
G4 VGH
G4 VHF
G4 VIN
G4 VJG
G4 VJM
G4 VJU
GM4 VJW
G4 VJX
GM4 VJY
GM4 VKB
G4 VLC
G4 VLM
G4 VLO
G4 VLR
G4 VLY
G4 VOM
G4 VOX
G4 VPO
G4 VPT
G4 VQC
G4 VQO
G4 VQQ
G4 VRQ
G4 VRR
G4 VSF
G4 VSZ
G4 VTF
G4 VTH
G4 VTK
G4 VTX
G4 VUB
GI4 VUJ
G4 VUO
GW4 VUT
G4 VUX
G4 VVG
G4 VVU
G4 VXC
G4 VXQ
G4 VXY
G4 VYM
G4 VYS
G4 VYW
G4 VYX
G4 VZG
G4 VZN
G4 VZO
G4 VZV
G4 VZZ
G4 WAD
G4 WAU
G4 WBB
G4 WBM
G4 WBR
G4 WCS
G4 WDL
G4 WDU
G4 WEA
G4 WEO
G4 WFQ
GI4 WHR
G4 WHX
G4 WIF
G4 WIX
G4 WJI
G4 WJT
G4 WKA
G4 WKH
G4 WKJ
G4 WLC
G4 WMI
G4 WMT
G4 WNC
G4 WNE
G4 WNM
G4 WNO
G4 WNT
G4 WOC
G4 WOF
G4 WPD
G4 WPN
G4 WPS
G4 WQG
G4 WQI
G4 WQR
G4 WQV
G4 WQX
GI4 WRI
G4 WRT
G4 WRW
G4 WRY
G4 WRZ
G4 WSZ
G4 WTF
G4 WTV
G4 WVE
G4 WVG
G4 WVU
G4 WVX
G4 WWK
G4 WWS
GW4 WXM
G4 WXN
G4 WXP
G4 WYJ
G4 WZE
G4 WZK
G4 XAK
G4 XBT
G4 XCA
G4 XCP
G4 XCS
GM4 XDA
G4 XDD
G4 XDF
G4 XEC
G4 XEH
G4 XEM
G4 XEN
G4 XFA
G4 XFB
G4 XFD
GU4 XGU
GM4 XHJ
G4 XHS
G4 XIB
GM4 XID
G4 XIF
GW4 XIH
G4 XIV
G4 XIY
GI4 XJA
G4 XJP
G4 XJW
G4 XKH
G4 XKX
G4 XKY
GW4 XLS
G4 XMG
G4 XMW
G4 XNG
G4 XNH
G4 XNJ
G4 XOB
G4 XOF
G4 XOO
G4 XOT
G4 XOX
G4 XPC
G4 XPE
G4 XPF
G4 XPM
G4 XQL
G4 XQS
G4 XQY
G4 XRI
G4 XRS
G4 XRZ
G4 XSH
G4 XTN
G4 XUT
G4 XWG
G4 XWJ
G4 XWV
G4 XYT
G4 XZL
G4 XZU
G4 XZV
G4 YAD
G4 YBC
G4 YBI
G4 YBO
G4 YBU
G4 YCB
G4 YCC
G4 YCK
G4 YDK
G4 YDU
G4 YEA
G4 YED
G4 YEH
G4 YEN
G4 YEY
G4 YEZ
G4 YFE
G4 YFN
G4 YFP
G4 YFQ
G4 YHE
G4 YHF
G4 YHV
G4 YHW
G4 YIB
G4 YIJ
G4 YIL
G4 YJO
G4 YKC
G4 YKJ
G4 YKS
G4 YKU
G4 YLA
G4 YLB
G4 YLJ
G4 YLV
G4 YMK
GU4 YMV
GM4 YNA
GW4 YNP
G4 YOB
G4 YOE
G4 YOH
GW4 YOM
G4 YOQ
G4 YOU
G4 YOW
G4 YPX
GM4 YPY
G4 YQY
G4 YRB
G4 YRG
G4 YRK
G4 YRS
G4 YUM
G4 YUP
G4 YUQ
G4 YVZ
G4 YWL
GI4 YWW
G4 YWY
G4 YXG
G4 YXL
G4 YXT
GM4 YZI
G4 YZX
GM4 ZAA
G4 ZAJ
G4 ZAT
GW4 ZAU
G4 ZAZ
G4 ZBA
G4 ZBR
G4 ZBV
G4 ZBY
G4 ZCU
G4 ZCZ
G4 ZDK
G4 ZED
G4 ZEK
G4 ZEP
G4 ZFL
G4 ZFZ
G4 ZGH
G4 ZIA
G4 ZIO
G4 ZJF
G4 ZJN
G4 ZJW
G4 ZKK
G4 ZKU
G4 ZLH
G4 ZLL
G4 ZLZ
G4 ZMF
G4 ZMG
G4 ZMI
G4 ZOD
G4 ZOM
G4 ZOW
G4 ZQP
G4 ZRE
G4 ZRL
G4 ZRP
G4 ZSF
G4 ZSN
G4 ZTI
G4 ZUF
G4 ZUQ
G4 ZUR
G4 ZVC
G4 ZVE
GI4 ZVY
G4 ZWL
G4 ZWP
G4 ZXE
G4 ZXX
GW4 ZYG
G4 ZYJ
G4 ZZB
G4 ZZR

## G*5

G5 BMH
G5 CL
G5 ECO
G5 FS
G5 PH
G5 TA
G5 VU
G5 XW
G5 ZL

## G*6

G6 AAD
G6 AAL
G6 AAW
G6 ABF
G6 ABQ
G6 ABR
G6 ACA
G6 ACT
G6 ACY
G6 ADK
G6 ADW
GI6 AEG
G6 AEN
G6 AFY
G6 AHY
G6 AJY
G6 AKE
G6 ALI
G6 ALM
G6 AMP
G6 AMU
G6 AMY
G6 ANF
G6 ANG
G6 ANM
G6 ANN
G6 AOU
GW6 APR
G6 APT
G6 APU
G6 AQF
G6 AQK
G6 AQN
G6 AQP
G6 AQQ
G6 ARB
G6 ARN
G6 ARU
G6 ASB
G6 ASI
G6 ASP
G6 ASW
G6 ATJ
GM6 AWC
G6 AWI
GW6 AWS
GW6 AWT
G6 AXF
G6 AXM
G6 AXW
G6 AYK
GM6 AZA
G6 AZF
G6 AZJ
GI6 AZS
G6 AZU
G6 AZV
G6 BAI
GI6 BAJ
G6 BBE
G6 BCC
G6 BCT
G6 BDA
G6 BDE
G6 BDV
G6 BEX
G6 BEZ
G6 BFD
G6 BFT
G6 BFW
G6 BGW
G6 BHK
G6 BHM
G6 BHW
G6 BIR
G6 BJM
G6 BJS
G6 BKH
G6 BLL
G6 BOD
G6 BOJ
G6 BOL
G6 BPC
G6 BPM
G6 BQI
GM6 BQJ
G6 BRF
G6 BRH
G6 BRL
G6 BRO
G6 BSE
G6 BSK
G6 BSU
G6 BTL
GI6 BTN
GW6 BUB
G6 BUC
G6 BVN
G6 BWR
GW6 BWX
G6 BXA
G6 BZA
G6 BZB
G6 BZO
G6 BZX
G6 CAF
G6 CAO
G6 CAT
G6 CAU
G6 CBR
G6 CBX
G6 CCA
G6 CGB
G6 CGE
G6 CHH
G6 CHO
G6 CJF
G6 CJK
G6 CJN
G6 CKT
G6 CLE
G6 CLI
G6 CLO
G6 CMG
G6 CMK
G6 COQ
G6 COW
G6 CPC
G6 CPI
G6 CPZ
G6 CQA
G6 CQD
G6 CQM
G6 CQO
GM6 CRB
G6 CRL
G6 CRP
G6 CRV
G6 CSY
G6 CVC
G6 CVI
G6 CVL
GW6 CVX
G6 CWL
G6 CWR
G6 CWS
G6 CXB
G6 CYZ
G6 CZC
G6 CZF
G6 CZQ
G6 DAZ
G6 DBE
G6 DBT
G6 DCM
G6 DDE
G6 DDQ
G6 DDS
G6 DDX
GI6 DEH
G6 DEN
G6 DEQ
G6 DFD
G6 DFF
G6 DGT
G6 DHJ
G6 DHS
G6 DHY
G6 DIK
G6 DIV
G6 DIY
G6 DJA
G6 DJI
G6 DJT
G6 DKB
G6 DKC
G6 DMA
G6 DMR
G6 DNN
G6 DOM
G6 DOP
G6 DPC
G6 DPP
G6 DPR
G6 DQM
G6 DQN
G6 DQW
G6 DRR
G6 DSE
G6 DSM
G6 DTB
G6 DTM
G6 DTX
G6 DUG
G6 DUS
G6 DVU
G6 DVZ
G6 DWI
G6 DWX
G6 DXH
G6 DXM
G6 DYA
G6 DYC
G6 DYD
G6 DYE
G6 DYJ
G6 DYX
GI6 DZC
G6 DZD
G6 DZH
G6 DZM
G6 DZW
G6 EAL
G6 EAS
G6 EAV
G6 EBN
G6 ECF
G6 ECH
G6 EDL
G6 EEA
G6 EEY
G6 EFH
G6 EFN
G6 EFX
G6 EFY
G6 EGG
G6 EGM
G6 EGV
G6 EHB
G6 EHO
G6 EHU
G6 EIA
G6 EIG
G6 EIJ
G6 EIL
G6 EIW
G6 EJA
G6 EJG
G6 EJP
G6 EJV
GW6 EJY
G6 EKG
G6 EKJ
G6 ELL
G6 ELO
G6 ELQ
G6 ELY
G6 EMH
G6 ENC
GI6 ENG
G6 ENH
G6 EOC
G6 EOF
G6 EOL
G6 EOY
G6 EPD
G6 EQJ
G6 ERC
G6 ERV
G6 ETE
GW6 EUS
G6 EVD
G6 EVT
GM6 EWV
G6 EXF
G6 EXO
G6 EXY
G6 EYB
G6 EYX
G6 EYY
G6 EZK
G6 EZS
G6 FAB
G6 FBI
GM6 FCW
G6 FCX
G6 FDN
G6 FEO
G6 FEP
G6 FGK
G6 FGO
G6 FGP
G6 FHW
G6 FID
G6 FIQ
G6 FIR
G6 FIU
G6 FJT
G6 FKQ
G6 FLX
G6 FMZ
G6 FNE
G6 FNF
G6 FNI
G6 FNL
GM6 FOM
G6 FOP
GM6 FQJ
G6 FRT
G6 FRU
G6 FSH
G6 FSP
G6 FSS
G6 FTN
G6 FTP
G6 FUB
G6 FUW
GW6 FVK
G6 FWA
G6 FWC
GI6 FZI
G6 GAX
G6 GBH
G6 GCX
G6 GDT
G6 GEA
G6 GEI
G6 GEJ
G6 GEY
G6 GGE
G6 GGP
G6 GGU
G6 GHQ
G6 GIY
G6 GJQ
G6 GJT
G6 GLP
G6 GMC
G6 GMW
G6 GNN
G6 GOU
G6 GPS
GW6 GQP
GM6 GRE
G6 GRG
G6 GRX
G6 GTD
G6 GTN
G6 GTR
G6 GUN
G6 GUP
G6 GUR
G6 GUX
G6 GVG
GM6 GVK
G6 GWR
G6 GXH
G6 GYJ
G6 GZH
G6 HAC
G6 HAE
G6 HAG
G6 HAL
G6 HBK
G6 HBN
G6 HBX
G6 HCL
G6 HCU
G6 HDP
G6 HEW
G6 HFI
G6 HHB
G6 HHP
G6 HIL
G6 HIZ
G6 HJD
GW6 HJO
G6 HJP
G6 HJR
G6 HKQ
G6 HMM
G6 HNF
G6 HNM
G6 HNU
G6 HOL
G6 HOM
G6 HOQ
G6 HOU
G6 HPC
G6 HPO
G6 HRN
G6 HSF
G6 HSM
G6 HSV
G6 HTL
G6 HUA
GJ6 HUL
G6 HVI
G6 HVP
G6 HXK
G6 HXV
G6 HYN
G6 HYU
G6 HZW
GD6 IA
G6 IBC
G6 IBH
G6 IBI
G6 IBO
G6 IBQ
G6 ICF
G6 ICJ
G6 ICP
G6 IDI
G6 IDZ
G6 IEA
G6 IEZ
G6 IFD
G6 IFF
G6 IFX
G6 IFZ
G6 IGA
G6 IGR
G6 IHN
G6 IIG
G6 IJJ
G6 IKT
G6 ILE
G6 ILW
G6 IMF
G6 IMT
G6 IMU
G6 IPU
G6 IQA
G6 IQL
G6 IRB
G6 IRS
G6 ISC
G6 ISD
G6 ISO
GI6 ISQ
GI6 ISW
G6 ITD
G6 ITF
G6 ITH
G6 ITK
G6 IVT
G6 IZC
G6 IZI
G6 IZN
G6 IZS
G6 JAD
GM6 JDZ
G6 JEA
G6 JEG
G6 JEJ
GM6 JEQ
G6 JEV
G6 JFS
G6 JFT
G6 JGS
G6 JHD
GM6 JHE
G6 JHR
GM6 JIC
G6 JID
G6 JIE
G6 JIV
G6 JJE
G6 JKG
G6 JKQ
G6 JKT
G6 JLH
G6 JLM
G6 JMF
G6 JON
G6 JOX
G6 JPA
G6 JPJ
G6 JQV
G6 JQW
GM6 JR
G6 JRV
G6 JSB
G6 JSK
G6 JSV
G6 JTB
G6 JTN
GM6 JUU
G6 JUZ
G6 JVD
G6 JVS
G6 JVV
G6 JVY
G6 JWE
G6 JWR
G6 JWV
G6 JXA
G6 JXN
G6 JYG
G6 JYQ
G6 JZZ
G6 KAR
G6 KAX
GM6 KBG
G6 KBJ
G6 KBR
G6 KDF
G6 KDQ
GM6 KEC
G6 KFN
G6 KFW
GW6 KFY
G6 KGI
G6 KGW
GM6 KGZ
G6 KIL
G6 KJR
G6 KJU
G6 KLS
G6 KLW
G6 KMA
G6 KME
G6 KMM
G6 KMT
G6 KMV
G6 KMY
G6 KNF
G6 KNH
GW6 KNX
G6 KOA
GI6 KOI
G6 KOS
G6 KOY
G6 KOZ
G6 KPM
G6 KQI
G6 KQP
GM6 KRD
G6 KRF
G6 KSM
G6 KSN
G6 KUA
GM6 KUC
G6 KUG
G6 KUS
G6 KUT
G6 KV
G6 KWJ
G6 KWN
G6 KWO
GM6 KXO
G6 KYD
G6 KYX
G6 KZD
G6 KZJ
G6 KZT
G6 KZX
G6 KZZ
G6 LAM
G6 LBA
G6 LDO
G6 LDD
G6 LCC
G6 LDI
G6 LDK
G6 LDL
G6 LDN
G6 LDV
G6 LDZ
G6 LFN
G6 LGF
G6 LGZ
GW6 LHD
G6 LHJ
G6 LHO
G6 LHW
G6 LIQ
G6 LIS
G6 LJJ
G6 LLB
G6 LLX
G6 LMK
G6 LMZ
G6 LOH
G6 LPF
G6 LPY
G6 LPZ
G6 LQU
G6 LQW
G6 LRE
G6 LRQ
G6 LTG
G6 LTS
G6 LTT
G6 LTZ
G6 LUG
GW6 LUH
G6 LUW
G6 LUZ
G6 LVW
G6 LWY
G6 LXK
G6 LXX
G6 LYY
G6 LZU
G6 MAF
G6 MCI
G6 MCJ
G6 MDD
G6 MDT
G6 MDU
G6 MEK
GM6 MEN
G6 MFD
G6 MGB
GM6 MGS
G6 MHG
G6 MHH
G6 MII
G6 MIT
G6 MIW
G6 MIY
G6 MJG
G6 MJN
G6 MLJ
G6 MMC
G6 MNM
GW6 MNQ
G6 MOG
G6 MOY
G6 MRI
G6 MSQ
GM6 MSS
G6 MTZ
G6 MVC
G6 MVO
G6 MWJ
G6 MWP
G6 MWR
GW6 MXJ
G6 MXR
G6 MXT
G6 MYW
G6 MZK
G6 MZS
G6 MZX
G6 NCF
G6 NCN
G6 NCQ
G6 NCR
G6 NCU
G6 NDI
G6 NDS
G6 NEB
G6 NEM
G6 NER
G6 NET
G6 NFF
G6 NFT
G6 NGO
G6 NHR
G6 NIL
G6 NIS
G6 NJF
G6 NJI
G6 NJM
G6 NJP
G6 NKD
G6 NKG
G6 NKM
G6 NLL
G6 NNN
G6 NNV
G6 NOS
G6 NOT
G6 NPQ
G6 NPU
G6 NQF
G6 NQP
G6 NRC
G6 NSB
G6 NSY
G6 NTI
G6 NTK
G6 NUA
G6 NUD
G6 NUM
G6 NUN
G6 NUT
G6 NVA
G6 NWO
G6 NWR
G6 NWV
G6 NXR
G6 NZB
G6 OAJ
G6 OAY
G6 OBW
G6 OCD
G6 OCL
G6 OCQ
G6 OCT
G6 OCW
G6 OEY
G6 OFQ
G6 OGY
G6 OHB
G6 OI
G6 OIL
G6 OJB
G6 OJU
G6 OLS
G6 OMT
G6 OMY
G6 ONC
G6 ONN
G6 OOO
G6 OPB
G6 OQK
G6 ORA
G6 ORC
G6 ORN
G6 OTI
G6 OTM
G6 OTU
G6 OVN
G6 OVQ
G6 OVR
G6 OVU
G6 OVY
G6 OVZ
G6 OXV
G6 OXW
G6 OXY
G6 OYE
G6 OZV
G6 OZW
G6 PAF
G6 PAI
G6 PBS
G6 PBT
G6 PCA
G6 PCD
G6 PCU
GW6 PDH
G6 PDV
G6 PDW
GM6 PER
G6 PFE
G6 PHA
G6 PHB
G6 PHQ
G6 PJL
G6 PJZ
G6 PLN
G6 PLP
G6 PLV
G6 PMN
G6 PNB
G6 PNH
G6 PNX
GM6 POL
G6 POQ
G6 POW
G6 POY
G6 PPJ
G6 PPX
G6 PRB
G6 PRT
G6 PRV
G6 PSI
G6 PTJ
G6 PTL
G6 PTY
G6 PUB
G6 PUC
G6 PVF
G6 PVQ
G6 PWO
GW6 PWT
G6 PWY
G6 PXI
G6 PXK
G6 PXT
G6 PZP
GI6 PZW
G6 RBH
G6 RCA
GW6 RCP
G6 RDG
GM6 RFG
GM6 RFQ
GM6 RGZ
G6 RHZ
G6 RIB
G6 RIK
G6 RIO
G6 RJT
G6 RKE
G6 RMS
G6 RNE
G6 RNM
GI6 ROP
G6 RPX
G6 RQO
G6 RQT
G6 RRL
G6 RSC
GW6 RSY
GW6 RTV
G6 RUK
GW6 RUL
GW6 RUQ
G6 RWI
G6 RWT
G6 RWV
G6 RYC
G6 RZA
G6 SAF
GM6 SAG
G6 SBF
G6 SBR
GW6 SCD
G6 SCR
G6 SDJ
G6 SDP
G6 SEM
G6 SET
G6 SFN
G6 SFP
GM6 SHB
G6 SIC
G6 SIY
G6 SJF
G6 SJH
G6 SJY
G6 SKL
G6 SKO
G6 SLK
G6 SLM
G6 SLU
G6 SMZ
G6 SNJ
G6 SNL
G6 SNO
G6 SNR
G6 SPC
G6 SQC
G6 SQD
G6 SQX
G6 SRX
G6 SSF
G6 SSX
G6 SUD
G6 SVB
G6 SVN
G6 SVY
G6 SXW
GM6 SYC
G6 SYT
GW6 SZF
G6 SZK
G6 SZQ
G6 TAT
G6 TAU
G6 TBA
G6 TCO
G6 TDC
G6 TDL
G6 TDP
G6 TEC
G6 TFB
G6 TFC
G6 TFY
G6 TGD
G6 TGO
G6 TGP
GW6 TGV
G6 TGZ
G6 THB
G6 THR
G6 THX
G6 TIH
G6 TIM
G6 TIX
G6 TJS
G6 TKI
G6 TKK
GD6 TKX
G6 TLH
G6 TLR
G6 TME
GJ6 TMM
GW6 TMW
G6 TNO
G6 TNZ
G6 TOB
G6 TOZ
G6 TPK
G6 TPV
G6 TQO
GI6 TQT
G6 TQU
G6 TRD

**IMPORTANT NOTE**

**Revalidate licence to avoid revocation** – Ofcom has advised the Society that plans will be drawn up to revoke licences that have not been revalidated as required by the licence conditions. The quickest way to revalidate is to do so online via the Ofcom website: *https://services.ofcom.org.uk/* or by email: *amateur.validations@ofcom.org.uk* If you need assistance in the process, Ofcom staff are available to help, but please be patient during times of heavy workload.

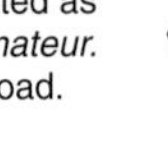


Withheld

G6 TRE
G6 TRS
G6 TRT
G6 TRU
G6 TSH
G6 TTL
G6 TTW
G6 TUH
G6 TUW
G6 TUZ
G6 TXW
G6 TXZ
G6 TYJ
G6 TYQ
G6 TYS
GW6 TZB
G6 UAD
G6 UAU
G6 UAX
G6 UAY
G6 UBB
G6 UBD
G6 UBL
G6 UBM
G6 UBP
GM6 UCJ
G6 UDE
G6 UDN
G6 UEB
G6 UEJ
G6 UEO
GM6 UFJ
G6 UFT
GW6 UGD
G6 UGL
G6 UID
G6 UJB
G6 UJH
G6 UKB
G6 UKK
G6 UKV
G6 ULP
G6 ULX
G6 UNM
G6 UOD
G6 UQB
G6 UQQ
G6 URA
G6 URB
G6 URE
G6 URY
G6 USH
G6 UT
G6 UTC
G6 UTD
G6 UTP
G6 UTQ
G6 UUM
G6 UVE
G6 UVH
G6 UWH
G6 UXD
G6 UXN
G6 UYG
G6 UZF
G6 UZH
G6 VDH
G6 VDP
GI6 VGW
G6 VIR
G6 VIW
G6 VIX
G6 VJE
G6 VKG
G6 VLB
GM6 VME
G6 VND
G6 VNH
G6 VOH
G6 VQU
G6 VQY
G6 VRD
G6 VRH
G6 VRT
G6 VTU
G6 VUI
GM6 VUL
G6 VVA
G6 VVC
G6 VVM
G6 VVT
G6 VWC
GM6 VXB
G6 VXM
G6 VXX
G6 VYS
G6 VYT
G6 WAR
G6 WBJ
G6 WBS
GM6 WBV
GW6 WCW
G6 WDV
G6 WEX
G6 WEZ
G6 WFD
G6 WFQ
GW6 WGD
G6 WGK
GW6 WGP
G6 WHI
G6 WII
G6 WIM
G6 WJP
G6 WKW
GI6 WLL
GM6 WLN
G6 WLZ
G6 WNA
G6 WNO
G6 WNW
GM6 WOE
G6 WOL
GM6 WOM
G6 WOV
G6 WOZ
G6 WQB
G6 WQU
G6 WRA
G6 WRG
GJ6 WRI
G6 WRW
G6 WSI
G6 WTA
G6 WTX
G6 WVK
G6 WVX
G6 WWO
G6 WXC
G6 WZH
G6 WZO
G6 XAC
G6 XAI
G6 XAJ
G6 XBC
G6 XBF
G6 XBW
G6 XBX
G6 XCF
G6 XCJ
G6 XDP
G6 XDQ
G6 XEG
G6 XFD
G6 XFI
G6 XFP
G6 XFW
G6 XGL
G6 XHM
G6 XIP
G6 XIQ
G6 XJU
G6 XLU
G6 XLV
G6 XMI
G6 XMO
G6 XMY
G6 XNC
G6 XOI
G6 XPL
G6 XPM
G6 XQW
G6 XRA
G6 XRQ
G6 XRX
GW6 XSP
G6 XTL
G6 XTN
G6 XTO
G6 XTR
G6 XTW
G6 XUB
G6 XVN
G6 XVO
G6 XVT
G6 XVW
G6 XWE
G6 XWN
G6 XWO
G6 XWT
G6 XWU
G6 XXW
GW6 XYE
G6 XYT
G6 XYZ
G6 XZL
G6 XZV
G6 YAF
G6 YAL
G6 YBM
G6 YBY
G6 YCK
G6 YGD
G6 YGG
G6 YGZ
G6 YHT
G6 YHU
G6 YHV
G6 YJN
G6 YKT
G6 YLJ
G6 YMP
G6 YMV
G6 YNZ
G6 YON
G6 YOS
GW6 YOV
G6 YPU
G6 YQY
G6 YRM
G6 YTC
G6 YUH
G6 YUI
G6 YVE
G6 YVX
G6 YWB
G6 YWC
G6 YWX
G6 YYX
G6 ZAB
G6 ZBB
G6 ZBI
G6 ZDQ
G6 ZFT
G6 ZFY
G6 ZGD
G6 ZIM
G6 ZIS
G6 ZKH
G6 ZLA
G6 ZLI
G6 ZMC
G6 ZMF
G6 ZMK
G6 ZMY
G6 ZMZ
G6 ZOD
G6 ZOI
GW6 ZOO
G6 ZOV
GM6 ZPD
G6 ZQE
GM6 ZQQ
G6 ZQX
G6 ZQY
G6 ZRL
G6 ZSD
G6 ZSN
G6 ZSR
G6 ZTA
G6 ZTV
G6 ZUC
G6 ZVE
G6 ZWQ
G6 ZXU
G6 ZYD
G6 ZYF
G6 ZYG
G6 ZYP
GW6 ZZP
G6 ZZX

## G*7

G7 AAC
G7 AAF
G7 AAQ
G7 AAW
G7 ACE
G7 ACU
G7 ACW
G7 AEB
G7 AED
G7 AEG
G7 AEJ
G7 AEO
G7 AEP
G7 AEU
G7 AFM
G7 AGZ
G7 AHE
G7 AHH
G7 AHJ
G7 AHQ
G7 AHS
G7 AHY
G7 AIE
G7 AII
G7 AIJ
G7 AIS
GM7 AIT
G7 AIV
G7 AJC
G7 AJM
G7 AKF
G7 AKL
G7 ALG
G7 ALN
G7 ALW
GM7 AME
G7 AMF
G7 AMG
G7 AMH
GM7 AMJ
GI7 AMK
G7 AMO
G7 AMU
G7 ANJ
G7 ANZ
G7 AOO
G7 AOP
G7 AOR
GW7 AOZ
G7 APC
G7 AQU
G7 AQX
G7 AQY
G7 ARC
G7 ARU
G7 ASJ
G7 ASX
GM7 ATQ
G7 ATV
G7 AUG
G7 AVO
G7 AXC
G7 AXF
G7 AXG
G7 AXJ
G7 AXU
G7 AXV
G7 BAP
G7 BBF
G7 BBL
G7 BBP
G7 BBQ
G7 BCB
G7 BCE
G7 BCL
G7 BDP
G7 BDQ
G7 BED
G7 BEX
G7 BFD
G7 BFK
G7 BFO
GW7 BFP
G7 BGA
G7 BHB
G7 BHN
G7 BHQ
G7 BIU
G7 BJA
G7 BJH
G7 BJJ
G7 BJQ
G7 BJU
G7 BJY
G7 BKI
G7 BKK
G7 BKU
G7 BKV
G7 BKX
G7 BLR
G7 BMA
G7 BMS
G7 BNV
G7 BPE
GW7 BQC
GM7 BRH
G7 BRV
G7 BSV
G7 BTA
G7 BUG
G7 BUQ
G7 BUT
G7 BUZ
G7 BVK
GW7 BVY
G7 BWP
G7 BZX
G7 CAK
GM7 CAN
GM7 CAQ
G7 CAR
G7 CAW
G7 CAY
G7 CBG
G7 CBL
G7 CBM
G7 CCI
G7 CCZ
G7 CDA
G7 CDK
G7 CDZ
G7 CFJ
G7 CFO
G7 CFR
G7 CFU
G7 CFV
G7 CFY
G7 CGG
G7 CGK
G7 CGV
G7 CHO
G7 CHT
G7 CHX
G7 CHY
G7 CIB
G7 CJY
G7 CKD
G7 CKF
GW7 CKR
G7 CKU
G7 CKX
G7 CLZ
G7 CMJ
G7 CMK
G7 CMS
G7 CNF
G7 CNH
G7 CNM
G7 CNQ
G7 COI
G7 COV
G7 COW
G7 COY
G7 CPB
G7 CPC
G7 CQH
G7 CRO
G7 CSE
G7 CTA
G7 CTD
G7 CTX
G7 CUG
G7 CUH
G7 CUI
G7 CVH
G7 CVJ
G7 CWA
G7 CWH
G7 CWQ
G7 CWX
G7 CWY
G7 CXR
G7 CYK
G7 CYO
G7 CYP
G7 CZH
G7 CZO
G7 CZT
G7 CZZ
G7 DAA
G7 DAK
G7 DBA
G7 DBE
G7 DBJ
GM7 DBK
G7 DBL
G7 DBM
G7 DBS
G7 DCD
G7 DCG
G7 DCO
G7 DCV
G7 DDY
G7 DEL
G7 DEM
G7 DEQ
G7 DER
G7 DET
G7 DFM
G7 DFO
G7 DGR
G7 DGU
G7 DGW
G7 DGZ
G7 DHL
G7 DIH
G7 DIQ
G7 DIS
G7 DJW
G7 DKE
G7 DKJ
GW7 DKL
G7 DKM
G7 DKQ
G7 DKW
G7 DLC
G7 DLH
G7 DLM
G7 DLS
G7 DLV
G7 DMD
G7 DMJ
G7 DMO
G7 DOH
G7 DOI
G7 DOM
G7 DOO
GW7 DOT
G7 DPE
GI7 DPP
G7 DPY
G7 DQD
G7 DQF
G7 DQG
G7 DQM
G7 DQR
G7 DRI
G7 DRK
GI7 DRS
G7 DSG
G7 DSI
G7 DSJ
G7 DSK
G7 DSX
G7 DUO
G7 DUP
G7 DUR
G7 DUW
G7 DVG
G7 DVY
G7 DWD
G7 DWK
G7 DWQ
G7 DXS
G7 DXY
GM7 DYA
G7 DYG
G7 DYI
G7 DYO
GI7 DYQ
G7 DZI
G7 DZX
G7 EAG
G7 EAP
G7 EBK
GI7 EBM
G7 EBP
G7 EBY
G7 ECK
G7 ECL
G7 ECV
G7 EDC
G7 EDM
G7 EDQ
G7 EDT
GW7 EEP
G7 EET
G7 EFC
G7 EFE
G7 EGI
G7 EGK
G7 EGN
G7 EGZ
G7 EIM
G7 EIO
G7 EIT
G7 EIX
G7 EJF
G7 EJG
G7 EKI
G7 EKY
G7 ELK
G7 ELM
G7 EMF
G7 EMK
G7 EML
G7 EOB
G7 EOL
GW7 EOO
G7 EOQ
G7 EOR
G7 EOT
G7 EOX
G7 EOZ
G7 EPG
G7 EQM
G7 EQU
G7 EQZ
G7 ERB
G7 ERM
G7 ERO
G7 ERU
G7 ERX
G7 ESB
G7 ESK
G7 ESP
GD7 ESR
G7 ETA
G7 ETI
G7 ETN
G7 ETR
G7 ETT
G7 ETU
G7 ETW
G7 ETX
G7 ETZ
G7 EUA
G7 EUH
G7 EUM
G7 EUW
G7 EVA
G7 EVE
G7 EVW
G7 EWG
GI7 EXN
G7 EYX
G7 EZG
G7 EZM
G7 EZP
GI7 EZR
G7 EZT
G7 EZW
G7 EZY
G7 FAG
G7 FAL
G7 FAN
G7 FAX
G7 FBD
G7 FCR
G7 FCT
G7 FCV
G7 FDP
G7 FDR
G7 FDT
G7 FDU
GI7 FDZ
G7 FEK
GI7 FEW
G7 FEX
G7 FEZ
G7 FFA
G7 FFE
GI7 FFL
GM7 FGF
GI7 FHC
G7 FHI
G7 FHJ
G7 FHK
G7 FHN
GI7 FHT
G7 FHW
G7 FHY
G7 FIF
G7 FIT
G7 FJA
G7 FJI
G7 FJO
G7 FKB
G7 FKE
GM7 FKH
G7 FLO
G7 FLR
G7 FLT
GI7 FLW
G7 FMM
GI7 FMN
G7 FMX
G7 FNA
G7 FNE
G7 FNS
G7 FNT
G7 FOF
GW7 FOJ
G7 FOK
G7 FPE
G7 FPL
G7 FPM
G7 FQF
G7 FQG
G7 FQH
G7 FQW
G7 FRA
G7 FRB
G7 FRI
G7 FTB
G7 FTL
G7 FTN
G7 FTP
G7 FTQ
G7 FTR
G7 FTY
G7 FUJ
G7 FVB
G7 FVD
G7 FVF
G7 FVG
G7 FVT
G7 FVV
G7 FWL
G7 FWO
G7 FWR
G7 FWT
G7 FXM
G7 FXV
G7 FYO
G7 FYR
G7 FZE
G7 FZV
GW7 FZW
G7 FZX
G7 GAD
GM7 GAF
GM7 GAM
G7 GAN
GM7 GAS
GM7 GAV
GM7 GAX
G7 GBA
G7 GBC
G7 GCK
G7 GCO
G7 GCQ
G7 GCR
GW7 GDH
G7 GDR
G7 GEQ
G7 GFS
G7 GFU
G7 GFW
G7 GGS
G7 GGU
G7 GGX
G7 GGZ
G7 GHA
G7 GHK
G7 GHT
G7 GHV
G7 GJF
G7 GJH
G7 GKO
G7 GLE
G7 GLK
G7 GLN
G7 GLO
G7 GLT
G7 GMF
G7 GMG
GI7 GMJ
G7 GMN
G7 GNB
GW7 GNF
G7 GNG
G7 GNH
GM7 GNK
G7 GNL
G7 GNM
G7 GOC
G7 GOP
G7 GPR
G7 GQF
G7 GQP
GM7 GRK
GI7 GSH
G7 GSL
G7 GSN
G7 GSP
G7 GSQ
G7 GSU
G7 GSW
G7 GTB
G7 GTO
G7 GTV
GW7 GTW
G7 GUD
G7 GUE
G7 GUF
G7 GUJ
G7 GUN
G7 GVK
G7 GVV
G7 GVZ
GI7 GWB
GW7 GWM
G7 GWR
G7 GWV
G7 GXF
G7 GXG
G7 GXK
G7 GXY
G7 GYB
G7 GYP
G7 GZE
G7 GZG
G7 GZH
G7 HAA
G7 HAH
G7 HAT
GW7 HAV
G7 HAW
G7 HCD
G7 HCE
G7 HDA
G7 HDB
G7 HDE
G7 HDI
G7 HDJ
G7 HDO
G7 HDX
G7 HEL
G7 HEM
G7 HER
G7 HEU
G7 HEX
G7 HGP
G7 HIF
G7 HIP
G7 HJY
G7 HKD
G7 HKF
G7 HKK
G7 HKM
G7 HKP
G7 HKY
G7 HLF
G7 HLH
G7 HLJ
G7 HLL
G7 HLO
G7 HMC
G7 HMJ
G7 HMS
G7 HNK
G7 HNO
G7 HNQ
G7 HNX
G7 HOB
G7 HOD
G7 HOI
G7 HOW
G7 HPB
G7 HPE
G7 HPG
G7 HPK
G7 HPM
G7 HPP
G7 HPS
GW7 HPW
G7 HQB
G7 HRU
G7 HRW
G7 HSE
G7 HSG
G7 HST
G7 HTP
G7 HTS
GI7 HUR
G7 HUW
G7 HVT
G7 HVW
G7 HWC
G7 HWD
G7 HWF
G7 HWJ
G7 HWK
G7 HWL
G7 HWT
G7 HWU
G7 HWW
G7 HXC
G7 HXF
G7 HXJ
G7 HXM
G7 HXQ
G7 HXS
G7 HXV
G7 HXX
G7 HYF
G7 HYK
G7 HZI
G7 HZX
G7 IBI
G7 IBO
G7 IBY
G7 ICH
G7 ICQ
G7 ICS
G7 ICW
G7 ICX
G7 ICY
G7 IDB
G7 IDD
G7 IDF
G7 IDJ
G7 IDL
G7 IEE
G7 IEI
G7 IFF
G7 IGB
G7 IGC
G7 IGZ
G7 IHA
G7 IHS
G7 IIM
GI7 IIU
G7 IJK
G7 IJM
G7 IJR
GW7 IJS
G7 IJT
G7 IKR
G7 IKT
G7 IKW
GD7 IKY
G7 ILK
G7 ILV
G7 IMF
G7 IND
G7 INM
G7 INN
G7 INO
G7 INP
G7 INS
G7 INV
G7 INX
G7 IOV
G7 IOW
G7 IOY
G7 IPB
G7 IPT
G7 IQJ
G7 IQT
G7 IQW
GW7 IRV
G7 ISO
G7 ISY
G7 ITD
G7 ITL
G7 IUJ
GW7 IUK
G7 IUO
G7 IUU
G7 IUV
G7 IVC
G7 IVH
G7 IVJ
G7 IVP
G7 IVR
G7 IVY
G7 IWG
G7 IWS
G7 IXA
G7 IXJ
G7 IXL
G7 IXO
G7 IXU
G7 IYB
G7 IYD
G7 IYR
G7 IYS
G7 IYT
G7 IYW
G7 IYZ
G7 IZD
G7 IZG
G7 IZH
G7 IZI
G7 IZQ
G7 IZR
G7 IZX
G7 IZZ
G7 JAB
G7 JAC
G7 JAD
G7 JAH
G7 JAK
G7 JAX
G7 JBB
G7 JBE
G7 JBQ
G7 JBT
G7 JCJ
G7 JCP
G7 JDV
G7 JDW
G7 JEW
G7 JEX
G7 JFF
G7 JFH
G7 JGD
G7 JGL
GJ7 JHF
GI7 JHJ
G7 JHL
G7 JHO
G7 JHP
GM7 JHQ
G7 JIC
G7 JID
G7 JIQ
G7 JIW
G7 JJF
G7 JJI
G7 JJZ
G7 JKB
G7 JKO
G7 JKV
G7 JKX
G7 JKZ
G7 JLI
G7 JLM
G7 JLR
G7 JLZ
G7 JMG
G7 JMI
G7 JND
G7 JNT
G7 JNY
G7 JOB
G7 JOG
G7 JOK
G7 JON
G7 JOQ
GM7 JOV
G7 JOX
GM7 JOY
G7 JPD
G7 JPH
G7 JPU
GI7 JPW
G7 JQE
GD7 JQI
GM7 JQK
G7 JQL
G7 JRV
G7 JRW
G7 JRZ
G7 JSF
G7 JSP
G7 JTQ
G7 JTS
G7 JTT
G7 JUE
G7 JUG
G7 JUS
G7 JUY
G7 JVL
G7 JVW
G7 JWB
G7 JWR
G7 JYE
G7 JYF
G7 JYM
G7 JYR
G7 JYS
G7 JYU
G7 JZD
G7 JZE
G7 JZL
G7 JZO
GM7 JZP
G7 JZR
G7 JZU
G7 JZW
G7 KAD
G7 KAH
GI7 KAW
G7 KBF
GW7 KBL
G7 KBP
G7 KBV
G7 KBY
G7 KCF
G7 KCO
G7 KCS
G7 KCW
G7 KCZ
G7 KDA
G7 KDC
G7 KDP
G7 KDZ
G7 KEH
G7 KEJ
G7 KES
G7 KET
G7 KEZ
G7 KFA
G7 KFC
G7 KFD
G7 KFH
G7 KFI
G7 KFL
G7 KFT
G7 KFW
G7 KFX
G7 KFY
G7 KGC
G7 KHN
G7 KIG
G7 KJF
G7 KJG
G7 KKH
G7 KKL
GW7 KKN
G7 KKP
G7 KKR
G7 KKS
G7 KKV
G7 KLA
G7 KLD
GM7 KLI
GI7 KMC
GI7 KMJ
G7 KMX
G7 KMZ
G7 KNE
GW7 KNF
G7 KNI
G7 KNO
G7 KNP
G7 KNV
G7 KNW
G7 KNZ
G7 KOA
G7 KOQ
G7 KPB
G7 KPN
G7 KPO
G7 KPP
G7 KPQ
G7 KPR
G7 KPT
G7 KPU
G7 KPX
G7 KPZ
G7 KQA
G7 KQC
G7 KQF
GM7 KQG
G7 KQI
G7 KQL
G7 KQM
G7 KQO
G7 KQY
G7 KSI
G7 KSW
G7 KTE
G7 KTK
G7 KTM
G7 KTN
G7 KTO
GI7 KTU
G7 KUF
G7 KUJ
G7 KUZ
G7 KVE
G7 KVK
G7 KVO
G7 KWC
G7 KWX
G7 KWZ
GM7 KXB
G7 KXD
G7 KXE
G7 KXM
G7 KXW
G7 KXX
G7 KXY
G7 KYB
G7 KYC
G7 KYN
G7 KZF
G7 KZX
G7 KZZ
G7 LAE
G7 LBS
G7 LBY
GW7 LCB
G7 LCC
G7 LCG
GW7 LCL
G7 LCM
G7 LCU
G7 LCY
GI7 LDE
GW7 LDF
G7 LEC
G7 LEE
G7 LEU
G7 LFR
G7 LFW
G7 LGD
G7 LHA
GW7 LHB
G7 LHE
GI7 LHF
GI7 LHG
G7 LHM
G7 LIN
G7 LIZ
G7 LJV
G7 LJW
G7 LJY
G7 LKB
G7 LKN
G7 LLQ
G7 LLS
G7 LMC
G7 LMF
G7 LMH
G7 LMJ
G7 LMK
G7 LMN
G7 LMP
G7 LMY
G7 LNC
G7 LOF
G7 LOH
G7 LOI
G7 LOM
GW7 LOP
G7 LOS
GI7 LOU
G7 LQE
G7 LQH
GM7 LQR
G7 LRD
G7 LRE
G7 LRF
G7 LRJ
G7 LRK
G7 LRQ
G7 LRR
G7 LRS
G7 LSH
G7 LSL
G7 LSR
G7 LSS
G7 LST
G7 LSV
G7 LSY
G7 LTQ
G7 LUC
G7 LUD
G7 LUS
G7 LUU
G7 LUV
G7 LVF
G7 LVO
GM7 LVV
G7 LVW
G7 LVX
G7 LVY
G7 LVZ
G7 LWK
GM7 LWM
G7 LWN
G7 LWT
G7 LXC
G7 LXL
G7 LXT
G7 LXX
G7 LYP
G7 LYQ
G7 LYR
G7 LZI
GW7 LZK
G7 LZW
G7 MAX
G7 MAZ
G7 MBU
G7 MCI
GM7 MCP
GJ7 MDH
G7 MDR
G7 MEI
G7 MFK
G7 MFN
G7 MFT
G7 MGK
G7 MGL
GM7 MGN
G7 MHK
G7 MHM
G7 MHN
G7 MIJ
G7 MIL
G7 MIW
G7 MIX
G7 MJA
G7 MJG
G7 MJK
G7 MJL
G7 MJN
G7 MJW
G7 MJZ
G7 MKO
G7 MKR
G7 MKS
G7 MLL
G7 MLZ
G7 MMA
G7 MMF
G7 MMM
G7 MMS
G7 MMT
G7 MNH
G7 MNM
G7 MNN
G7 MNP
G7 MNY
GM7 MOI
G7 MOT
GW7 MPW
G7 MQB
G7 MQD
G7 MQL
G7 MQT
G7 MQV
G7 MRC
G7 MRD
G7 MRG
G7 MRI
G7 MRJ
G7 MRK
G7 MRN
G7 MRP
GI7 MRT
G7 MRV
G7 MSD
GM7 MTK
G7 MUI
G7 MUL
G7 MUQ
G7 MVC
G7 MVP
G7 MVR
G7 MWF
G7 MWN
G7 MXB
G7 MXJ
G7 MYB
G7 MYL
G7 MYR
G7 MYS
G7 MYZ
G7 MZH
G7 MZN
G7 MZO
G7 MZR
G7 MZT
G7 NAD
GM7 NAG
G7 NAM
G7 NAS
GW7 NAU
G7 NBK
G7 NBN
G7 NBW
G7 NCK
G7 NCN
G7 NCO
G7 NCT
G7 NDF
G7 NDG
G7 NDH
G7 NDJ
G7 NDL
G7 NDY
G7 NEI
G7 NEL
G7 NEO
G7 NFJ
G7 NFP
G7 NFS
G7 NFU
G7 NGS
G7 NGV
G7 NGZ
G7 NHA
G7 NHG
GU7 NIE
G7 NIG
G7 NIP
GW7 NIS
G7 NJC
G7 NJF
G7 NJK
GW7 NJO
GI7 NKK
G7 NKL
GM7 NKP
G7 NKS
GM7 NKT
G7 NLB
G7 NLQ
GW7 NLX
G7 NMF
GM7 NMG
G7 NMO
G7 NNE
G7 NNF
GM7 NNG
GI7 NNY
GM7 NOA
G7 NOJ
G7 NOO
G7 NOT
G7 NOV
G7 NOX
G7 NOY
GM7 NOZ
G7 NPK
GM7 NPO
G7 NPP
G7 NPW
G7 NQH
GW7 NQI
GM7 NQL
G7 NQO
GM7 NQP
G7 NQS
G7 NRE
G7 NRH
G7 NRI
G7 NSA
GI7 NSC
G7 NSR
G7 NSU
G7 NSY
G7 NTB
G7 NTC
G7 NUK
G7 NUT
GM7 NVA
G7 NVE
G7 NVK
G7 NVO
G7 NWA
G7 NWB
GI7 NWI
G7 NYC
G7 NYK
G7 NZK
GM7 NZT
G7 OAE
G7 OAK
G7 OAU
G7 OBA
G7 OBH
GM7 OBI
G7 OBK
G7 OBO
G7 OBU
G7 OBY
G7 OBZ
G7 OCD
G7 OCG
G7 OCM
G7 OCO
G7 OCS
G7 OCT
GM7 ODA
G7 ODD
G7 ODE
G7 ODJ
G7 ODO
G7 ODQ
G7 ODX
G7 ODY
G7 OEL
G7 OEM
G7 OEP
G7 OEQ
G7 OEX
G7 OFC
G7 OFR
G7 OFX
G7 OGA
G7 OGG
G7 OGX
G7 OHA
G7 OHK
G7 OHT
G7 OIE
G7 OII
G7 OIJ
G7 OIW
G7 OJH
G7 OJN
G7 OJR
G7 OJW
GM7 OKD
G7 OKE
GM7 OKU
G7 OKZ
G7 OLD
G7 OLI
GM7 OLP
G7 OLS
G7 OLT
G7 OMT
GM7 OMZ
G7 ONG
G7 ONK
G7 ONM



GW7 ONS
G7 ONX
G7 ONY
G7 OOA
G7 OOD
G7 OOJ
G7 OOR
G7 OOW
G7 OOX
G7 OOZ
G7 OPC
G7 OPO
G7 OQH
G7 OQJ
G7 OQK
G7 OQN
G7 OQS
G7 OQV
G7 OQZ
G7 ORH
G7 ORI
G7 OSG
G7 OSL
G7 OTY
G7 OUJ
G7 OUL
G7 OVA
G7 OVC
G7 OVV
G7 OWE
G7 OWH
G7 OXM
G7 OXX
G7 OYC
G7 OYO
G7 OZF
G7 OZK
G7 OZO
G7 OZY
G7 PAB
G7 PAJ
G7 PAL
G7 PAN
G7 PAS
G7 PAV
G7 PAZ
G7 PCA
G7 PCJ
G7 PCL
G7 PCM
G7 PCN
G7 PCQ
G7 PDL
G7 PDP
G7 PDY
G7 PDZ
G7 PEA
G7 PEK
GI7 PEM
G7 PEQ
G7 PET
GW7 PEY
G7 PFB
G7 PFE
G7 PFH
G7 PFN
G7 PFO
G7 PFQ
G7 PFU
G7 PFX
G7 PGB
G7 PGS
G7 PGZ
G7 PHF
G7 PHN
G7 PHU
G7 PHZ
GM7 PIA
G7 PIM
G7 PIQ
G7 PJC
G7 PJT
G7 PJV
G7 PKF
G7 PKO
G7 PKS
G7 PKU
G7 PKV
G7 PKW
G7 PKX
G7 PLF
GM7 PLG
GM7 PLN
G7 PLX
G7 PLY
G7 PLZ
G7 PME
G7 PMJ
G7 PMS
G7 PNQ
G7 PNR
G7 PNU
G7 POG
G7 POJ
G7 POU
G7 PPB
G7 PPE
G7 PPF
G7 PPM
G7 PPT
G7 PQF
G7 PQN
GW7 PQR
G7 PQU
G7 PQV
G7 PRA
G7 PRN
G7 PRR
G7 PRY
G7 PSA
G7 PSQ
G7 PSX
G7 PTL
G7 PTQ
G7 PTU
G7 PUD
G7 PUQ
G7 PUV
G7 PVJ
G7 PVR
G7 PVS
GM7 PVV
G7 PWE
G7 PWH
G7 PXA
G7 PXB
G7 PXD
G7 PXE
G7 PXK
G7 PXP
G7 PXT
G7 PYI
G7 PZI
G7 PZS
GI7 PZV
G7 PZX
G7 RAC
G7 RAD
G7 RAR
G7 RBG
G7 RBV
G7 RCD
G7 RCQ
G7 RCX
G7 RDC
G7 RDI
G7 RDM
GM7 RDO
GW7 RDV
G7 REB
G7 RFG
G7 RFI
G7 RFR
G7 RGB
G7 RGH
G7 RGQ
GM7 RGW
G7 RID
G7 RIH
G7 RIK
GM7 RIQ
G7 RIS
GW7 RJC
G7 RJK
G7 RJL
G7 RJT
G7 RJU
G7 RJV
GM7 RJZ
G7 RKI
G7 RKK
G7 RKM
G7 RKY
G7 RLI
G7 RMB
G7 RMH
GW7 RMN
GM7 RMV
G7 RNE
G7 ROF
G7 ROS
G7 RPF
G7 RPG
G7 RPH
G7 RPO
G7 RPS
G7 RPX
G7 RQA
G7 RQB
G7 RQE
G7 RQG
GW7 RQM
GW7 RQN
G7 RQS
G7 RQT
G7 RQU
G7 RQX
G7 RRF
G7 RRI
G7 RRP
G7 RRQ
G7 RSD
G7 RSI
G7 RSO
G7 RST
G7 RSV
G7 RSZ
GW7 RTK
G7 RUA
G7 RUG
G7 RUL
G7 RUM
G7 RUP
G7 RUT
G7 RUV
G7 RVJ
G7 RVL
G7 RVS
G7 RWA
G7 RWH
G7 RWI
G7 RWJ
G7 RWO
G7 RWQ
G7 RWU
G7 RWX
G7 RXG
G7 RXH
G7 RXP
G7 RXR
G7 RXT
GI7 RXV
G7 RXY
G7 RYE
GM7 RYR
G7 RYV
G7 RZA
G7 RZD
G7 RZF
G7 RZM
G7 RZT
G7 RZU
GW7 SAB
G7 SBH
G7 SBQ
G7 SBV
G7 SBW
G7 SCC
G7 SCF
G7 SCQ
G7 SCY
G7 SDJ
G7 SDV
G7 SEH
GI7 SEI
G7 SET
G7 SFC
G7 SFQ
G7 SFR
G7 SFV
G7 SGG
G7 SGS
G7 SHG
G7 SHX
GM7 SIC
G7 SII
G7 SJV
G7 SKE
G7 SKI
G7 SKJ
G7 SLD
G7 SLK
G7 SLM
G7 SLO
G7 SLS
G7 SLT
G7 SMA
G7 SMK
G7 SMW
G7 SMY
G7 SOG
G7 SOI
G7 SOJ
G7 SOQ
G7 SPG
G7 SPI
G7 SPK
GW7 SPR
G7 SPS
G7 SPT
G7 SPU
G7 SPV
G7 SPW
G7 SQL
G7 SQR
G7 SQU
G7 SQX
G7 SRE
G7 SSE
G7 SSZ
G7 STA
G7 STB
G7 STF
G7 SUD
G7 SUH
G7 SUI
G7 SUP
G7 SUZ
G7 SVB
G7 SVG
G7 SVI
GM7 SVK
G7 SVW
G7 SWC
G7 SWP
G7 SXA
G7 SXE
G7 SXH
G7 SXQ
G7 SXR
G7 SXX
GW7 SYB
G7 SYI
G7 SYK
G7 SYL
G7 SYO
G7 SYP
G7 SYZ
G7 SZD
G7 SZJ
G7 SZN
G7 SZR
G7 SZT
GI7 SZV
G7 TAC
G7 TAG
G7 TAH
G7 TAI

G7 TAO
G7 TAR
G7 TAS
G7 TBE
G7 TBH
G7 TBO
G7 TBT
G7 TCI
G7 TCK
G7 TCL
G7 TCM
G7 TDO
G7 TDP
G7 TDW
G7 TDX
G7 TDY
G7 TDZ
GW7 TED
G7 TEF
G7 TEM
G7 TER
G7 TFD
G7 TFH
G7 TFJ
GM7 TFP
G7 TGE
G7 TGL
G7 TGO
G7 TGR
G7 TGV
G7 TGY
G7 THD
G7 THT
G7 THX
G7 TIA
G7 TIC
G7 TII
G7 TIS
G7 TIZ
G7 TJE
GW7 TJK
GW7 TJS
G7 TJW
G7 TJX
G7 TJY
G7 TKE
G7 TKJ
G7 TKK
G7 TKL
GW7 TKZ
G7 TLN
G7 TLP
G7 TMG
G7 TMJ
GM7 TMW
G7 TNA
G7 TNC
G7 TND
GW7 TNF
G7 TNL
G7 TOC
G7 TOG
G7 TOP
GI7 TOW
GI7 TOX
G7 TPI
G7 TPL
GI7 TPP
G7 TPR
G7 TPV
GI7 TPX
G7 TQG
G7 TQJ
G7 TQS
G7 TQY
G7 TQZ
G7 TRE
G7 TRJ
G7 TRN
G7 TRW
G7 TRZ
G7 TSL
G7 TSX
G7 TTR
G7 TTT
G7 TUA
G7 TUJ
G7 TUY
G7 TVB
GM7 TVD
G7 TVH
G7 TVI
G7 TVP
G7 TVX
G7 TWB
G7 TWI
G7 TWR
G7 TWX
G7 TWZ
G7 TXH
GM7 TXJ
G7 TXM
G7 TYA
G7 TYV
GM7 TYY
GW7 TZH
G7 TZY
G7 UAA
GM7 UAU
G7 UBF
G7 UBJ
G7 UBL
G7 UBT
G7 UBU
G7 UCC
G7 UCX

G7 UDH
G7 UDP
G7 UDO
G7 UDJ
G7 UDT
G7 UEB
GM7 UED
G7 UEH
G7 UEM
G7 UEO
G7 UEQ
G7 UEY
G7 UFA
G7 UFG
G7 UFJ
G7 UFP
G7 UFS
G7 UGE
G7 UGF
G7 UGG
G7 UGK
G7 UGN
G7 UGS
G7 UGT
GI7 UGV
G7 UHF
G7 UHN
G7 UHV
G7 UIE
G7 UIL
G7 UIQ
GI7 UJD
G7 UJG
G7 UJM
G7 UJN
G7 UJZ
G7 UKG
G7 UKT
G7 ULE
G7 ULT
G7 UME
G7 UMP
G7 UMR
G7 UMU
G7 UMX
G7 UND
G7 UNE
G7 UNF
GW7 UNH
G7 UNN
GI7 UNQ
G7 UOT
G7 UOV
G7 UOZ
G7 UPE
G7 UQF
G7 UQN
G7 UQP
G7 URA
G7 URB
GM7 URD
G7 URN
G7 URU
G7 UTK
G7 UTL
G7 UTN
G7 UVC
G7 UVD
G7 UVG
G7 UVH
G7 UVT
G7 UWA
G7 UWF
G7 UWT
G7 UXG
G7 UXI
G7 UXV
G7 UXW
G7 UXX
G7 UYH
G7 UZM
G7 VAC
G7 VAJ
G7 VAK
G7 VAM
G7 VAV
G7 VAX
G7 VBB
G7 VBC
G7 VBP
G7 VBQ
G7 VBR
G7 VBX
G7 VCD
G7 VCO
G7 VCX
G7 VDR
G7 VEC
G7 VEO
G7 VEQ
G7 VEZ
G7 VFD
G7 VFK
G7 VFW
G7 VFZ
G7 VGG
G7 VGP
G7 VGU
G7 VGW
G7 VGZ
G7 VHP
G7 VHS
G7 VHV
G7 VJC
G7 VJL
GI7 VJN

G7 VJR
G7 VJX
G7 VKF
G7 VKP
G7 VKQ
G7 VKU
G7 VKZ
G7 VLG
G7 VLN
GJ7 VLY
G7 VMC
G7 VMG
G7 VMI
G7 VMX
G7 VNB
G7 VNR
G7 VNU
G7 VNY
G7 VOF
G7 VPI
G7 VPK
G7 VPO
G7 VPZ
GW7 VQD
G7 VQQ
G7 VQU
G7 VQV
G7 VRC
G7 VRD
G7 VRI
G7 VRT
G7 VSH
G7 VSQ
G7 VTF
GW7 VTO
G7 VTU
G7 VTV
GI7 VUF
G7 VUR
G7 VUV
G7 VUZ
G7 VVE
G7 VVG
G7 VVU
G7 VWH
G7 VXV
G7 VXX
G7 VYC
G7 VYJ
G7 VYL
G7 VZE
G7 VZJ
G7 VZN
G7 VZT
GI7 VZW
G7 VZX
G7 WAJ
G7 WAM
G7 WAP
G7 WAY
G7 WBB
G7 WBI
GM7 WBP
G7 WBX
G7 WCA
G7 WCC
G7 WCI
G7 WCM
GM7 WCO
G7 WDK
G7 WDR
G7 WDT
GM7 WEF
G7 WEL
G7 WEX
G7 WFA
G7 WFB
G7 WFE
G7 WFY
G7 WGF
G7 WGJ
G7 WGN
G7 WHC
G7 WHF
G7 WHJ
G7 WHK
G7 WHN
G7 WIA
G7 WIB
G7 WIE
G7 WII
G7 WIJ
G7 WIP
G7 WIR
G7 WIT
GI7 WIU
G7 WIW
G7 WIZ
GM7 WJI
G7 WKE
G7 WKN
G7 WKX
G7 WKY
G7 WLI
GI7 WLS
G7 WLW
G7 WLZ
G7 WMA
G7 WMC
G7 WOT
G7 WRO
G7 WRS
G7 WST
G7 XAL
G7 XJS
G7 ZZZ

## G*8

GW8 AAG
G8 ACE
G8 ACM
G8 ACZ
G8 ADM
G8 AFC
G8 AFZ
G8 AGC
G8 AGQ
G8 AGR
G8 AIT
G8 AJF
G8 AJN
G8 ALB
G8 AMP
G8 AMV
G8 AMZ
G8 ARG
G8 ARP
GM8 ARV
G8 ASZ
G8 ATH
G8 ATO
G8 ATU
G8 ATV
G8 AUC
GM8 AUT
G8 AWO
GW8 AXK
G8 AXW
G8 AXX
G8 AXZ
G8 AZC
G8 AZI
G8 BBI
GM8 BBS
G8 BDA
G8 BEG
G8 BFF
G8 BFT
G8 BGG
G8 BJG
G8 BJY
G8 BKS
G8 BMD
GW8 BNL
G8 BOU
G8 BPR
G8 BPV
G8 BQR
G8 BTS
G8 BTT
G8 BUJ
G8 BUN
G8 BUR
G8 BUY
GW8 BVI
G8 BXU
G8 BZY
G8 CAR
G8 CBL
G8 CBM
G8 CCG
G8 CCR
G8 CDF
G8 CDJ
G8 CEE
G8 CHB
G8 CHQ
G8 CHW
G8 CIL
G8 CJS
G8 CJZ
G8 CKF
G8 CKT
G8 CMJ
G8 CNE
GW8 COE
GW8 COJ
G8 COV
G8 CQA
GW8 CQK
G8 CQP
G8 CSA
G8 CTZ
G8 CUV
G8 CVL
G8 CVY
G8 CWM
G8 CZV
G8 CZZ
G8 DAZ
G8 DBN
G8 DCZ
GM8 DGZ
G8 DID
G8 DJK
G8 DJM
G8 DJN
G8 DKC
G8 DLA
G8 DLW
G8 DMP
G8 DMS
G8 DNW
G8 DOG
G8 DQQ
G8 DQX
G8 DRL
G8 DVG
G8 DVK
G8 DVR
G8 DWA
G8 DWN
G8 DYF

G8 DYK
G8 EAM
G8 EAP
[illegible]
G8 EBJ
GW8 EDE
G8 EDJ
G8 EEZ
G8 EFP
G8 EFV
G8 EGT
G8 EHK
G8 EIA
G8 EIK
G8 EJG
G8 EKO
G8 ELA
G8 ELS
GI8 ELZ
G8 EMO
G8 EOP
G8 EPA
GJ8 EPM
G8 EQA
G8 ERJ
G8 ERT
G8 ESB
G8 ETC
G8 ETQ
G8 EUD
G8 EUK
G8 EUQ
G8 EWG
GW8 EYG
G8 EYH
G8 EZO
GU8 FBO
G8 FBZ
G8 FCW
G8 FEC
G8 FEG
G8 FEL
GI8 FGD
G8 FGF
G8 FGK
G8 FGN
G8 FHF
G8 FHN
G8 FIW
G8 FKY
G8 FPN
G8 FQE
G8 FQW
G8 FRA
G8 FRF
G8 FRL
G8 FSO
G8 FTI
G8 FTO
G8 FUP
G8 FUQ
G8 FUR
GW8 FVY
G8 FXB
GM8 FXD
G8 FYE
G8 FYH
G8 FZA
GW8 FZC
G8 GBW
G8 GCM
G8 GCU
G8 GD
G8 GDF
G8 GDS
GM8 GEC
G8 GFG
G8 GFH
G8 GFI
G8 GGG
G8 GGP
GJ8 GGY
G8 GHD
G8 GHF
G8 GHX
G8 GIJ
G8 GIY
G8 GKA
G8 GKQ
GW8 [illegible]
G8 GKU
G8 GLM
GW8 GLU
G8 GLW
G8 GMC
G8 GMD
G8 GME
G8 GNC
G8 GND
G8 GNI
G8 GNW
G8 GOI
G8 GOW
G8 GQT
G8 GRA
G8 GSQ
G8 GSR
G8 GTC
G8 GUV
G8 GVK
G8 GWD
G8 GWL
G8 GWT
G8 GXP
G8 GYZ
G8 HAJ

G8 HAK
G8 HAX
G8 HBA
G8 HBM
G8 HBS
G8 HCB
G8 HCO
G8 HCU
G8 HCY
G8 HED
G8 HEM
G8 HFI
GM8 HGT
G8 HHV
G8 HHX
G8 HIK
G8 HIM
G8 HIP
G8 HJT
G8 HLU
G8 HLW
G8 HNB
G8 HOD
G8 HOH
G8 HON
G8 HOP
G8 HOV
G8 HQJ
G8 HQV
G8 HRH
G8 HRK
G8 HRR
G8 HRZ
G8 HSO
G8 HSV
G8 HTH
G8 HTX
G8 HUA
G8 HUE
G8 HUL
G8 HVR
G8 HVY
G8 HWG
GM8 HWZ
GW8 HZW
G8 IAY
G8 IB
G8 ICJ
G8 IFR
G8 IFW
G8 IGP
G8 IGS
G8 IGV
G8 IGY
G8 IIV
G8 IJA
GI8 IJF
G8 IJN
G8 IJP
G8 IJZ
G8 IKL
G8 IKP
G8 ILF
G8 ILM
G8 ILS
G8 IMD
G8 IML
G8 IMN
G8 IMP
G8 IMR
G8 INE
G8 INI
G8 INN
G8 INP
G8 INU
G8 INV
G8 IQG
G8 IQK
G8 IQP
G8 ISX
G8 IUV
G8 IVC
G8 IVI
G8 IWN
G8 IXA
G8 IXE
G8 IYP
G8 IZZ
G8 JAO
G8 JA[illegible]
G8 JAU
GM8 JAV
G8 JA[illegible]
G8 JBE
G8 JC
G8 JCI
G8 JCW
G8 JEF
G8 JFJ
G8 JFW
G8 JGC
G8 JGH
GM8 JGI
G8 JGO
G8 JGQ
G8 JIX
GM8 JKQ
G8 JKR
G8 JLE
G8 JLZ
G8 JNF
G8 JNS
GI8 JOD
G8 JOJ
G8 JOR
G8 JPZ

G8 JTJ
G8 JTY
G8 JUF
G8 JUX
G8 JUZ
G8 JVD
G8 JXA
G8 JXJ
GM8 JYJ
G8 JZA
G8 JZE
G8 JZY
G8 KAI
G8 KBC
G8 KBI
G8 KBO
G8 KBV
G8 KBZ
G8 KCM
GM8 KCQ
GM8 KCS
G8 KCV
G8 KDG
G8 KDI
GI8 KFA
G8 KFB
G8 KGA
G8 KGH
G8 KHJ
G8 KHR
G8 KHS
G8 KHW
G8 KHZ
G8 KIO
G8 KIT
G8 KJC
G8 KJV
G8 KKS
G8 KLD
G8 KLQ
G8 KLR
G8 KMH
GJ8 KNV
G8 KPM
G8 KPS
G8 KQW
G8 KRA
G8 KSP
G8 KUE
G8 KUH
G8 KUJ
G8 KUL
G8 KUW
G8 KVM
G8 KVP
G8 KWC
G8 KWI
GM8 KWQ
G8 KWR
G8 KWX
G8 KXM
G8 KYT
G8 KYU
GI8 KZU
GI8 LBD
G8 LBV
G8 LCO
G8 LDP
G8 LEC
G8 LGI
G8 LGW
G8 LGX
G8 LIB
G8 LIL
G8 LIR
G8 LIT
G8 LJC
G8 LJO
G8 LKD
G8 LKI
G8 LKM
G8 LLB
G8 LLK
G8 LMD
G8 LMN
G8 LMS
G8 LMX
G8 LNR
G8 LNT
G8 LOF
G8 LOL
G8 LOY
GW8 L[illegible]
G8 LVI
G8 LVK
G8 LVZ
G8 LWY
G8 LXI
GW8 LXO
GM8 LYS
G8 LYT
G8 LZI
G8 LZR
G8 LZU
G8 LZV
G8 MAC
G8 MAQ
G8 MAU
G8 MBI
G8 MBT
G8 MCD
G8 MCG
G8 MCV
G8 MCZ
G8 MDA
G8 MEJ

G8 MES
G8 MHK
G8 MHR
G8 MIF
G8 MJW
G8 MKK
G8 MKM
G8 MLQ
G8 MMD
G8 MMH
G8 MML
GI8 MNK
GI8 MOA
G8 MPP
G8 MQH
G8 MQN
GW8 MRB
G8 MRJ
G8 MSA
G8 MSG
G8 MSR
G8 MSW
G8 MTM
G8 MTQ
G8 MUE
G8 MUQ
G8 MVF
G8 MVP
G8 MVV
G8 MWF
G8 MWK
G8 MWQ
G8 MXE
G8 MXZ
G8 MZF
G8 NAV
G8 NAX
G8 NBD
G8 NBR
GW8 NCF
G8 NCW
GI8 NCX
G8 NDA
G8 NDD
G8 NDL
G8 NDP
G8 NDT
G8 NEZ
G8 NFU
G8 NHN
G8 NHR
G8 NIC
GU8 NIS
G8 NJD
G8 NJJ
G8 NJW
G8 NKD
G8 NKV
G8 NKX
G8 NKY
G8 NLY
G8 NMC
G8 NMI
G8 NML
G8 NMU
G8 NMW
G8 NNT
G8 NOC
G8 NPF
G8 NPT
G8 NPV
G8 NQA
G8 NRY
GM8 NSU
G8 NSV
G8 NT
G8 NTI
G8 NTU
G8 NUC
G8 NUG
G8 NUP
G8 NVF
G8 NVJ
G8 NWR
G8 NXG
G8 NXI
G8 NXV
GW8 NYS
G8 NYW
GW8 NZH
G8 NZM
G8 NZU
G8 OBR
G8 OBV
G8 OCN
G8 ODF
G8 ODP
G8 OEF
G8 OEL
G8 OEM
G8 OER
G8 OES
G8 OFJ
GW8 OFT
GW8 OGI
G8 OGJ
G8 OGO
G8 OHL
G8 OHR
G8 OIS
G8 OLD
G8 OLT
G8 OMG
G8 OML
G8 OMU
G8 ONK

G8 ONX
GW8 OPK
G8 OPV
G8 [illegible]
G8 OQN
GM8 ORE
G8 ORV
G8 OSA
G8 OSD
G8 OSE
G8 OSN
G8 OTE
G8 OTJ
G8 OUA
G8 OUU
G8 OUX
G8 OVT
G8 OWB
G8 OWL
G8 OWX
G8 OYA
G8 OZG
G8 OZV
GM8 OZW
G8 OZZ
G8 PAD
G8 PBJ
G8 PCB
G8 PCF
G8 PEF
G8 PEO
GM8 PEV
G8 PFE
G8 PFG
G8 PFO
G8 PFS
G8 PGA
G8 PGD
GW8 PGL
G8 PHN
G8 PHP
G8 PIB
G8 PID
G8 PIZ
GM8 PKL
G8 PKY
G8 PLL
G8 PLW
G8 PLZ
G8 PME
G8 PMQ
G8 PMU
G8 PNC
G8 PND
G8 PNO
G8 PNX
G8 POC
G8 POW
G8 PPS
G8 PRF
G8 PRI
G8 PRR
G8 PSE
G8 PSP
G8 PTD
G8 PTJ
G8 PTP
G8 PTR
G8 PUO
G8 PUX
G8 PVB
G8 PVD
G8 PVJ
G8 PVT
G8 PWC
G8 PWQ
G8 PWX
G8 PWY
G8 PXB
G8 PYP
G8 PYT
G8 PYX
G8 PZJ
G8 PZL
G8 PZP
G8 PZT
G8 QR
G8 RAL
G8 RCP
G8 RDF
G8 RDH
G8 RDO
G8 RDY
G8 REU
GM8 REY
G8 RFN
G8 RHT
G8 RIA
G8 RIE
G8 RJA
GM8 RJD
GW8 RJU
G8 RJX
GI8 RKC
G8 RLJ
G8 RLZ
G8 RMM
G8 ROD
G8 ROM
G8 ROZ
G8 RPH
G8 RPK
G8 RPV
G8 RQP
G8 RQY
G8 RTC

G8 RTW
G8 RTZ
G8 RUM
G8 RVC
G8 RVX
G8 RWX
G8 RXK
G8 RXL
G8 RXM
G8 RXP
G8 RXU
G8 RZ
G8 RZA
G8 RZJ
G8 SAE
G8 SAJ
GI8 SAM
G8 SBE
G8 SBF
G8 SCH
G8 SCK
G8 SCU
G8 SDC
G8 SDT
G8 SEJ
G8 SEZ
G8 SFR
GM8 SHU
G8 SIQ
G8 SJP
G8 SKU
G8 SLN
G8 SLX
GW8 SNG
G8 SNH
G8 SOZ
G8 SQF
GM8 SQM
G8 SRD
G8 STU
G8 SUP
G8 SUT
G8 SVC
G8 SVE
G8 SVF
G8 SVL
G8 SVO
G8 SVU
G8 SWG
G8 SWZ
G8 SXC
GI8 SXN
G8 SYR
G8 SYU
G8 SYZ
G8 TA
G8 TAM
G8 TBI
G8 TBK
G8 TCM
G8 TCQ
G8 TDB
G8 TDL
G8 TDW
G8 TEE
G8 TET
G8 TFH
G8 TFI
G8 TFK
G8 TFO
G8 TGR
G8 THS
G8 TIC
G8 TIJ
G8 TIM
G8 TIS
G8 TJQ
G8 TKR
G8 TKV
G8 TLQ
G8 TLX
G8 TLZ
G8 TMA
G8 TMY
G8 TNC
G8 TOK
G8 TPA
G8 TPR
G8 TPT
G8 TPX
G8 TPY
GM8 TQY
G8 TRB
G8 T[illegible]
G8 TRS
G8 TSE
G8 TSJ
GM8 TT
G8 TTK
G8 TV
G8 TVH
G8 TVT
G8 TWH
GM8 TWJ
G8 TWL
G8 TXM
G8 TXO
G8 TYV
G8 TZT
G8 UAW
G8 UCB
G8 UCM
G8 UDH
G8 UEC
G8 UED
G8 UEM

G8 UEP
G8 UEQ
G8 UFD
G8 UFK
G8 UFN
G8 [illegible]
G8 UGB
G8 UGD
G8 UGM
G8 UGT
G8 UGZ
G8 UHN
G8 UIB
G8 UJB
G8 UJP
G8 UKN
G8 UKT
G8 ULG
G8 ULO
G8 ULV
G8 UMX
G8 UNZ
G8 UOD
G8 UQI
G8 UQQ
G8 URC
G8 URE
G8 URP
G8 USX
G8 UUA
G8 UUH
G8 UUL
G8 UVC
G8 UVE
G8 UVK
G8 UVQ
G8 UVR
G8 UVS
G8 UWF
G8 UWJ
G8 UWO
G8 UWU
G8 UWW
G8 UXE
G8 UXU
G8 UYA
G8 UYC
G8 UYN
G8 UZJ
GW8 UZL
G8 UZO
GM8 VAU
G8 VAZ
G8 VBO
G8 VCB
G8 VCW
G8 VDO
G8 VFH
G8 VFN
G8 VFS
G8 VGC
GW8 VGG
G8 VHC
G8 VIQ
GM8 VIW
G8 VJ
G8 VJH
G8 VJS
G8 VKB
G8 VKD
GM8 VKT
G8 VKU
G8 VKV
G8 VLQ
G8 VNA
G8 VNE
G8 VNU
G8 VNW
G8 VOP
GW8 VOV
G8 VPQ
G8 VPS
G8 VQM
G8 VRI
G8 VSK
G8 VTF
G8 VTN
G8 VUR
G8 VUW
G8 VVK
G8 VVQ
G8 VWN
G8 VXH
G8 VXV
G8 VXX
G8 VYD
G8 VZA
G8 VZE
G8 WAR
G8 WBB
G8 WBH
G8 WBI
G8 WCK
G8 WED
G8 WEG
G8 WEN
G8 WHS
G8 WHZ
G8 WID
G8 WIG
G8 WJZ
G8 WMQ
G8 WMZ
GW8 WND
G8 WNX
G8 WOW

G8 WOY
G8 WPD
G8 WQB
G8 WQM
G8 WQU
G8 W[illegible]
G8 WT
G8 WTR
G8 WTV
G8 WVW
G8 WWB
G8 WWS
G8 WY
G8 WYZ
G8 WZB
G8 XAC
GW8 XAH
G8 XBA
G8 XBC
G8 XBI
G8 XBP
G8 XBW
GM8 XBZ
G8 XCH
G8 XCK
G8 XFA
G8 XFP
G8 XFX
G8 XGA
G8 XGE
GM8 XGI
GM8 XGP
G8 XGR
G8 XHO
G8 XIT
G8 XIU
G8 XIX
G8 XJR
G8 XJU
G8 XKB
G8 XKR
G8 XKV
G8 XLM
G8 XLV
G8 XLY
G8 XNI
G8 XNQ
GW8 XNV
G8 XOT
G8 XPA
G8 XPF
G8 XPL
G8 XPV
G8 XQB
G8 XQF
G8 XQQ
G8 XRR
G8 XTK
G8 XTX
G8 XUJ
G8 XUX
G8 XVA
G8 XVC
G8 XVD
G8 XVV
G8 XWG
G8 XWI
GI8 XWJ
G8 XXE
G8 XXF
G8 XXK
G8 XYY
G8 XZD
G8 XZO
G8 YBL
G8 YCR
G8 YCV
G8 YDU
G8 YFG
G8 YHH
G8 YHM
G8 YHS
G8 YHW
G8 YIH
GM8 YIJ
G8 YJM
G8 YKF
G8 YKW
G8 YLB
G8 YLU
G8 YLX
G8 YMV
GW8 YNJ
G8 YNM
G8 YOJ
G8 YOV
G8 YPC
G8 YPE
G8 YPJ
G8 YQD
G8 YQK
G8 YQQ
G8 YRG
GM8 YHI
G8 YRJ
G8 YRO
G8 YSK
G8 YSO
G8 YTC
G8 YTZ
G8 YUQ
G8 YVR
G8 YXE
G8 YXL
G8 YZD
G8 YZH

G8 ZAZ
G8 ZBT
GI8 ZBW
G8 ZBX
G8 ZCQ
G8 [illegible]
G8 ZDW
G8 ZEC
G8 ZEG
G8 ZEZ
G8 ZGI
G8 ZGX
G8 ZHC
G8 ZHD
G8 ZHL
G8 ZHP
G8 ZJD
G8 ZJJ
G8 ZJP
G8 ZKK
G8 ZKV
G8 ZKY
G8 ZKZ
G8 ZMH
G8 ZMP
G8 ZNU
G8 ZNW
G8 ZOH
G8 ZPL
G8 ZPU
G8 ZQO
G8 ZQR
G8 ZRL
G8 ZSG
G8 ZSH
G8 ZSN
G8 ZTQ
G8 ZUD
G8 ZVE
G8 ZVJ
G8 ZWQ
G8 ZWS
G8 ZXJ
G8 ZXW
G8 ZYO
G8 ZZE
G8 ZZZ

## M*0

M/ K2BBC
M0 AAE
M0 AAI
M0 AAO
M0 AAX
M0 AAY
M0 ABE
M0 ABH
MM0 ABM
M0 ABR
M0 ABS
M0 ACD
M0 ACE
M0 ACF
M0 ACG
MM0 ACH
M0 ACI
M0 ACQ
M0 ADC
M0 ADH
M0 ADI
M0 ADO
M0 ADP
M0 ADS
MI0 ADX
M0 AED
M0 AEG
M0 AEH
M0 AEI
MI0 AEY
M0 AFA
M0 AFE
M0 AFG
M0 AFH
M0 AFL
M0 AFO
M0 AGC
M0 AGG
M0 AGK
MM0 AGN
M0 AHN
M0 AHP
M0 AHX
M0 AIA
M0 AII
M0 AIO
M0 AIQ
M0 AIU
M0 AJL
M0 AJM
M0 AJN
M0 AJO
M0 AJW
M0 AJY
M0 AJZ
M0 AKB
M0 AKH
M0 AKO
M0 AKT
MI0 AKU
MM0 AKW
M0 ALA
MM0 ALJ
M0 ALW
M0 AMO
M0 ANG
M0 AOP

M0 AOV
M0 APG
M0 APO
M0 APS
M0 APT
M0 A[illegible]
M0 AQC
M0 AQM
M0 AQS
M0 AQY
M0 ARF
M0 ARP
M0 ARW
M0 ASA
M0 ASH
MJ0 ASP
M0 ASW
M0 ASZ
M0 ATE
M0 ATM
M0 AUE
M0 AUI
M0 AUQ
M0 AUT
M0 AUV
M0 AVC
M0 AVD
M0 AVE
M0 AVG
M0 AWF
M0 AWR
M0 AWS
M0 AWV
M0 AWW
M0 AXB
M0 AXF
M0 AXH
M0 AXK
MW0 AXM
M0 AXP
M0 AXQ
M0 AXY
M0 AYD
M0 AYJ
M0 AYP
MI0 AYR
M0 AYW
M0 AZA
M0 AZH
M0 AZI
M0 AZM
M0 AZO
M0 AZQ
M0 BAN
M0 BAS
M0 BBA
M0 BBB
M0 BBJ
M0 BBP
M0 BBX
M0 BBY
M0 BBZ
M0 BCD
M0 BCY
M0 BDP
M0 BEG
M0 BEW
M0 BFE
MM0 BFF
M0 BFN
M0 BFU
MW0 BFY
M0 BGA
M0 BGC
M0 BGD
M0 BGL
M0 BGM
M0 BGP
MW0 BGQ
M0 BGV
M0 BHB
M0 BHL
M0 BHU
MM0 BHY
MM0 BIM
M0 BIU
M0 BIW
MM0 BJA
M0 BJH
M0 BJW
M0 BJY
M0 BKB
M0 BKQ
M0 BKH
M0 BKW
M0 BKZ
M0 BLB
M0 BLD
M0 BLE
M0 BLK
M0 BLP
M0 BLQ
M0 BMH
M0 BMK
M0 BMO
M0 BNA
M0 BNE
M0 BNH
MI0 BNJ
M0 BNK
M0 BNY
M0 BOG
M0 BOJ
M0 BOP
MI0 BOT
M0 BOV

M0 BOZ
M0 BPH
M0 BPI
M0 BPJ
M0 BPK
M0 [illegible]
MM0 BQA
MM0 BQP
M0 BQU
MM0 BQV
M0 BQW
MI0 BRC
M0 BRK
M0 BRV
M0 BSA
MW0 BSR
M0 BTQ
MW0 BUD
M0 BUI
M0 BUJ
M0 BUL
M0 BUO
M0 BUQ
M0 BUU
M0 BUX
M0 BVL
M0 BWR
MM0 BXH
M0 BXR
M0 BXT
M0 BXW
M0 BYH
M0 BYQ
M0 BYW
M0 BZG
M0 BZT
M0 BZW
M0 CAH
M0 CAL
M0 CAP
MW0 CBC
M0 CBE
MI0 CBH
M0 CBJ
M0 CBU
MM0 CBV
MD0 CCE
M0 CCO
M0 CDD
M0 CDI
M0 CDP
M0 CEE
M0 CEF
M0 CEL
M0 CEN
M0 CEP
M0 CET
M0 CEV
M0 CFI
M0 CFN
M0 CFO
M0 CFU
M0 CFV
M0 CFW
M0 CGC
M0 CGD
M0 CGI
M0 CGL
M0 CGM
MI0 CGQ
M0 CHK
M0 CHW
MW0 CHZ
M0 CID
M0 CIJ
M0 CIL
MM0 CIN
M0 CIQ
M0 CIU
M0 CIV
M0 CJW
M0 CJX
M0 CKN
M0 CKQ
MM0 CKR
M0 CLA
M0 CLF
MM0 CLX
M0 CMB
MW0 CMD
M0 CMM
M0 CMR
MM0 CMV
MM0 CMX
MI0 CMY
M0 CNQ
M0 CNS
M0 COA
M0 COG
M0 COH
M0 COK
M0 COX
M0 COY
M0 CPA
M0 CPJ
M0 CPP
M0 CPR
M0 CPX
M0 CPY
MM0 CPZ
MI0 CQA
M0 CQE
M0 CQG
M0 CQK
MW0 CQU
M0 CQY

M0 CRT
M0 CRX
M0 CSH
M0 CSI
M0 CSL
M0 CSM
M0 CSN
MM0 CSS
M0 CSX
M0 CTW
M0 CUB
M0 CUV
M0 CUX
M0 CVL
M0 CVM
M0 CVN
M0 CVX
M0 CWL
MM0 CWM
M0 CXC
M0 CXF
M0 CXG
M0 CXM
M0 CXN
MM0 CXR
M0 CXT
M0 CXX
M0 CYP
M0 CYQ
M0 CYS
M0 CYW
M0 CYZ
M0 CZI
MW0 CZJ
M0 CZV
M0 CZW
M0 CZY
M0 DAF
M0 DAJ
M0 DAO
M0 DAQ
M0 DAU
M0 DAV
M0 DBL
MI0 DBN
M0 DBP
MW0 DBQ
M0 DCJ
M0 DCK
M0 DDD
M0 DDJ
M0 DDL
M0 DDM
M0 DDN
M0 DDP
M0 DDR
M0 DDS
M0 DED
M0 DEG
M0 DEM
M0 DFC
MI0 DFG
M0 DFI
M0 DFK
M0 DFM
MM0 DFS
M0 DFT
M0 DFU
MM0 DFV
M0 DGD
M0 DGF
M0 DGG
M0 DGL
M0 DGP
M0 DGW
M0 DGY
M0 DGZ
M0 DHD
M0 DHG
MM0 DHH
MW0 DHJ
M0 DHL
M0 DHR
M0 DHZ
M0 DIA
M0 DIF
M0 DIH
M0 DIP
M0 DIR
M0 DIU
M0 DIY
M0 DIZ
M0 DJE
M0 DJJ
M0 DJK
M0 DJN
M0 DJP
M0 DJS
M0 DJU
M0 DJV
M0 DJY
MW0 DJZ
M0 DKA
M0 DKE
M0 DKH
M0 DKI
M0 DKK
M0 DKO
MM0 DKQ
M0 DKR
M0 DKW
M0 DKZ
M0 DLD
M0 DLJ
M0 DLK
M0 DLV



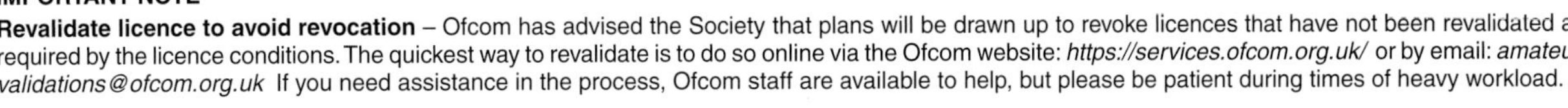

**IMPORTANT NOTE**

**Revalidate licence to avoid revocation** – Ofcom has advised the Society that plans will be drawn up to revoke licences that have not been revalidated as required by the licence conditions. The quickest way to revalidate is to do so online via the Ofcom website: *https://services.ofcom.org.uk/* or by email: *amateur.validations@ofcom.org.uk* If you need assistance in the process, Ofcom staff are available to help, but please be patient during times of heavy workload.

Withheld

MI0 DLW
M0 DMC
M0 DMG
M0 DMH
MI0 DML
M0 DMM
M0 DMP
M0 DMQ
M0 DMW
M0 DNA
M0 DNC
M0 DNE
M0 DNG
MI0 DNI
MW0 DNQ
M0 DNS
M0 DNT
MM0 DOE
M0 DOF
M0 DOG
M0 DOI
MW0 DOJ
M0 DOQ
M0 DOU
M0 DOV
M0 DPA
MI0 DPB
M0 DPC
M0 DPD
M0 DPE
M0 DPM
M0 DPN
M0 DPO
M0 DPP
M0 DPR
M0 DPU
M0 DPX
M0 DPZ
M0 DQE
M0 DQJ
MM0 DQM
M0 DQR
M0 DQS
M0 DQT
M0 DQV
M0 DQW
M0 DQX
M0 DQY
M0 DRC
M0 DRH
M0 DRP
M0 DRV
M0 DRW
M0 DRX
MM0 DSD
M0 DSH
M0 DSJ
M0 DSK
M0 DSL
M0 DSQ
M0 DST
M0 DSU
MW0 DTC
M0 DTG
M0 DTM
M0 DTP
M0 DTR
M0 DTT
M0 DTU
M0 DTV
MI0 DTX
M0 DUC
M0 DUE
M0 DUF
M0 DUG
M0 DUH
M0 DUI
M0 DUJ
M0 DUK
M0 DUL
MM0 DUS
M0 DUW
M0 DUZ
M0 DVA
MM0 DVC
M0 DVE
M0 DVO
M0 DVP
MM0 DVS
M0 DVU
M0 DVV
M0 DVY
M0 DWA
MI0 DWJ
M0 DWN
M0 DWO
M0 DWV
M0 DWY
M0 DXB
MM0 DXE
MM0 DXL
M0 DXO
M0 DXU
MD0 DXW
M0 DXY
M0 DYD
M0 DYE
M0 DYF
M0 DYI
MI0 DYJ
M0 DYK
M0 DYL
M0 DYM
M0 DYN
M0 DYY
M0 DZI

M0 DZJ
M0 DZN
M0 DZP
M0 DZQ
M0 DZR
M0 DZS
M0 DZU
M0 DZY
M0 DZZ
M0 EAA
M0 EAC
M0 EAG
M0 EAH
M0 EAJ
M0 EAP
M0 EAV
M0 EAW
M0 EBA
M0 EBB
M0 EBC
M0 EBE
M0 EBF
M0 EBH
M0 EBK
MM0 EBL
M0 EBM
M0 EBS
M0 EBW
M0 EBZ
M0 ECA
M0 ECB
M0 ECD
M0 ECE
M0 ECG
M0 ECH
M0 ECI
M0 ECO
M0 ECT
MM0 ECU
M0 ECV
M0 EDB
M0 EDC
M0 EDD
M0 EDG
M0 EDH
M0 EDI
M0 EDJ
M0 EDK
M0 EDT
M0 EDV
M0 EDW
M0 EDY
MM0 EDZ
M0 EEA
M0 EEC
M0 EED
M0 EEE
M0 EEF
M0 EEI
M0 EEJ
MM0 EEY
M0 EGH
M0 EGM
MW0 EIB
M0 EIG
M0 EIP
M0 EIQ
M0 EJN
M0 ELG
MM0 ELI
M0 ELS
M0 ELT
M0 EMS
M0 ENA
M0 ENG
M0 EOG
M0 EPV
M0 EPW
MM0 ERA
M0 ESE
M0 ESG
M0 ETM
M0 ETO
MW0 EUG
M0 EWS
M0 EXV
M0 EYB
M0 FAB
M0 FAQ
MI0 FBI
M0 FDA
M0 FDE
M0 FDF
MI0 FDS
M0 FDY
M0 FEW
M0 FFF
M0 FGH
MU0 FGM
MI0 FHB
M0 FHN
M0 FIN
M0 FMA
M0 FMC
MM0 FME
MM0 FMF
M0 FNJ
MM0 FNR
M0 FOS
M0 FOZ
M0 FRE
M0 FRZ
M0 FWD
M0 FWZ
MM0 FZQ
M0 FZY

M0 GAA
M0 GAF
M0 GAJ
M0 GAK
M0 GAM
M0 GAP
M0 GAU
M0 GAW
M0 GAZ
M0 GBE
M0 GBM
M0 GBP
M0 GBQ
MW0 GBR
M0 GBS
MM0 GBV
M0 GBX
M0 GBY
M0 GCE
M0 GCG
M0 GCJ
M0 GCK
M0 GCL
M0 GCM
M0 GCN
M0 GCO
M0 GCP
M0 GCW
M0 GCZ
M0 GDA
M0 GDB
M0 GDC
M0 GDE
M0 GDF
M0 GDK
M0 GDN
MI0 GDO
M0 GDQ
M0 GDR
M0 GDW
M0 GDY
M0 GDZ
M0 GEE
MU0 GEG
M0 GEH
M0 GEJ
M0 GER
M0 GET
M0 GEV
M0 GFB
M0 GFQ
MM0 GFS
M0 GFT
MI0 GFU
M0 GFV
M0 GFY
M0 GGA
M0 GGE
MM0 GGI
M0 GGR
M0 GGS
MI0 GGY
M0 GHB
M0 GHD
M0 GHF
M0 GHG
M0 GHH
M0 GHP
M0 GHQ
M0 GHS
M0 GHU
M0 GIC
M0 GIH
M0 GIK
M0 GIO
M0 GIR
M0 GIV
M0 GIX
M0 GJE
M0 GJF
M0 GJO
M0 GJQ
M0 GJR
M0 GJW
M0 GJY
M0 GJZ
M0 GKC
M0 GKH
MD0 GKI
M0 GKM
MW0 GKX
M0 GKY
M0 GLC
M0 GLD
M0 GLH
MM0 GLM
M0 GLN
M0 GLO
M0 GLR
MM0 GLW
M0 GLZ
M0 GMB
M0 GML
M0 GMM
M0 GMV
M0 GMX
M0 GMY
M0 GND
M0 GNE
M0 GNG
MW0 GNT
M0 GNV
MM0 GNZ
M0 GOE
M0 GOJ
MM0 GOR

M0 GPA
M0 GPI
M0 GPM
M0 GPR
MU0 GPS
M0 GPT
M0 GQA
M0 GQC
M0 GQH
M0 GQK
M0 GQL
M0 GQN
MW0 GQO
M0 GQQ
M0 GQT
M0 GQY
MM0 GRK
M0 GRL
M0 GRQ
M0 GSA
M0 GSE
M0 GSF
MM0 GSG
M0 GSJ
M0 GSM
M0 GSU
MI0 GTA
M0 GTC
M0 GTD
MM0 GTF
M0 GTK
M0 GTV
M0 GTW
M0 GTZ
M0 GUB
M0 GUI
M0 GUP
MM0 GUS
M0 GUT
M0 GVA
M0 GVB
M0 GVD
M0 GVG
M0 GVH
M0 GVO
M0 GVR
M0 GVS
M0 GVU
M0 GVV
M0 GWI
M0 GWJ
M0 GWP
M0 GWS
M0 GWW
M0 GWX
M0 GXA
M0 GXD
M0 GXF
M0 GXJ
M0 GXL
M0 GXP
M0 GXR
M0 GXS
M0 GXT
M0 GXX
M0 GXZ
M0 GYD
M0 GYE
M0 GYQ
M0 GYT
M0 GZG
M0 GZO
M0 GZP
M0 GZQ
M0 GZR
M0 GZT
M0 GZV
M0 GZY
M0 HAA
M0 HAC
M0 HAE
MM0 HAI
M0 HAJ
M0 HAK
M0 HAQ
MI0 HAV
M0 HAW
M0 HAX
M0 HBA
M0 HBB
MI0 HBD
M0 HBF
M0 HBK
M0 HBP
M0 HBQ
M0 HBR
M0 HBS
M0 HBZ
M0 HCB
MM0 HCD
MM0 HCF
M0 HCH
M0 HCJ
M0 HCL
M0 HCR
M0 HCS
M0 HCU
M0 HDB
M0 HDD
M0 HDF
M0 HDH
M0 HDL
M0 HDO
M0 HDY
M0 HEA
M0 HEC

M0 HEE
M0 HEH
M0 HEI
M0 HEK
M0 HEL
M0 HEO
M0 HEQ
M0 HER
M0 HES
M0 HEU
M0 HEV
MI0 HEZ
M0 HFD
M0 HFJ
M0 HFK
M0 HFL
M0 HFM
M0 HFN
M0 HFP
MI0 HFT
M0 HGB
M0 HGC
M0 HGE
M0 HGH
M0 HGI
M0 HGL
M0 HGO
M0 HGP
M0 HGQ
M0 HGR
M0 HGT
M0 HGU
MM0 HGW
M0 HGX
M0 HGZ
M0 HHE
M0 HHJ
M0 HHK
M0 HHL
M0 HHN
MM0 HHO
M0 HHQ
M0 HHS
M0 HHY
M0 HHZ
M0 HIB
M0 HIF
M0 HII
M0 HIK
M0 HIR
M0 HIS
MM0 HIU
MI0 HJH
MW0 HJM
M0 HJS
M0 HJT
M0 HJV
MM0 HJX
M0 HJZ
M0 HKD
M0 HKM
M0 HKN
M0 HKO
M0 HKQ
M0 HKY
M0 HKZ
M0 HLG
M0 HLL
M0 HLT
M0 HMA
M0 HMD
M0 HMG
M0 HMH
M0 HMK
M0 HML
MM0 HMM
M0 HMN
M0 HMP
M0 HMQ
M0 HMW
M0 HNB
M0 HNP
MM0 HNR
M0 HNS
M0 HNU
M0 HNV
M0 HNZ
M0 HOC
M0 HOE
M0 HOG
M0 HOO
M0 HOS
M0 HOW
M0 HOX
M0 HPA
M0 HPD
M0 HPI
M0 HPO
M0 HPY
M0 HQN
M0 HQS
MW0 HQT
M0 HQV
M0 HQW
M0 HQX
M0 HRE
M0 HRF
MM0 HRI
M0 HRJ
M0 HRK
M0 HRR
M0 HRS
M0 HRU
M0 HRX
M0 HSD
M0 HSE

M0 HSF
MM0 HSK
M0 HSL
MM0 HSM
M0 HSN
M0 HSP
MM0 HST
M0 HSY
M0 HTD
M0 HTH
M0 HTM
M0 HTN
M0 HTP
M0 HTT
M0 HTW
M0 HTZ
MM0 HUB
M0 HUC
M0 HUE
M0 HUK
M0 HUO
M0 HUP
M0 HUT
M0 HUX
M0 HVH
M0 HVJ
M0 HVT
M0 HVX
M0 HVY
M0 HWA
MI0 HWB
M0 HWE
M0 HWK
M0 HWX
M0 HWZ
M0 HXD
M0 HXJ
M0 HXL
M0 HXP
M0 HXQ
M0 HXR
M0 HXT
M0 HXU
M0 HXW
M0 HXY
MI0 HYB
M0 HYF
M0 HYO
M0 HYR
M0 HYS
M0 HYT
M0 HYU
MW0 HYV
M0 HYW
M0 HYY
M0 HYZ
M0 HZG
M0 HZQ
M0 HZS
M0 HZZ
M0 IAB
M0 IAI
MW0 IAN
M0 IAO
M0 IAP
M0 IAQ
M0 IAR
M0 IAU
MW0 IAV
M0 IAY
M0 IBA
M0 IBB
MW0 IBF
M0 IBG
M0 IBP
MW0 IBS
MW0 IBU
M0 ICC
MJ0 ICD
M0 ICF
M0 ICH
M0 ICM
M0 ICN
M0 ICR
M0 ICV
M0 ICW
MM0 ICX
M0 ICY
M0 IDA
M0 IDB
M0 IDD
M0 IDF
M0 IDH
M0 IDP
M0 IDU
M0 IEE
M0 IEI
M0 IEN
M0 IFL
M0 IFR
M0 IGA
MD0 IGD
M0 IGS
MW0 IHD
M0 IHO
M0 IHV
M0 IHZ
M0 IIC
MW0 IJD
M0 IJM
M0 IJS
MI0 IKF
MJ0 IKL
M0 IKO
M0 IKQ
M0 IKY

MJ0 ILB
MI0 ILJ
M0 ILS
M0 IMA
MM0 IMC
M0 IMZ
M0 INA
M0 INK
M0 INN
MM0 INT
M0 INV
M0 IOD
M0 ION
M0 IOR
M0 IOY
M0 IPA
M0 IPC
M0 IPU
M0 IQB
MM0 IRM
M0 IRQ
M0 ISA
M0 ISM
M0 ISO
M0 ISS
M0 ITC
M0 ITF
MM0 ITH
M0 ITM
M0 ITU
M0 IVQ
MM0 IWD
M0 IWF
MW0 IWH
M0 IXA
MM0 IXD
MI0 JAA
M0 JAB
M0 JAH
M0 JAL
MI0 JAS
M0 JAU
M0 JBA
M0 JBE
M0 JBJ
M0 JBM
M0 JBP
M0 JBR
MW0 JBX
M0 JCI
M0 JCU
M0 JCW
M0 JDC
M0 JDH
M0 JDK
M0 JDM
M0 JDN
M0 JDO
M0 JED
M0 JEE
M0 JEL
M0 JEN
M0 JES
MI0 JFC
M0 JFK
M0 JFN
M0 JFS
M0 JFZ
M0 JGD
M0 JGL
M0 JGR
M0 JGX
M0 JHA
MW0 JHE
M0 JHO
M0 JHR
M0 JIG
M0 JIN
M0 JIY
M0 JJS
M0 JKA
MM0 JKO
M0 JLA
M0 JLG
M0 JMA
M0 JMG
M0 JMM
M0 JMO
MM0 JNB
M0 JNC
M0 JNH
M0 JNI
M0 JNK
MI0 JNR
M0 JOA
M0 JOE
M0 JOF
M0 JOG
M0 JOS
M0 JPF
M0 JPH
MI0 JPJ
M0 JPX
MI0 JPZ
M0 JQZ
M0 JRB
M0 JRC
MI0 JRD
M0 JRG
M0 JRK
MM0 JRM
M0 JRS
M0 JSC
M0 JSF
MM0 JSM
M0 JSO

MI0 JST
M0 JTA
M0 JTI
M0 JTR
MM0 JTV
M0 JUD
M0 JUE
M0 JUG
MM0 JVB
MI0 JVI
M0 JVP
M0 JWB
M0 JWF
M0 JWG
M0 JWK
M0 JWS
M0 JWT
M0 JXB
MM0 JXK
M0 JXN
M0 JYK
M0 JYL
M0 JZH
M0 JZX
M0 KAF
M0 KAH
M0 KAL
M0 KAS
M0 KAT
M0 KAV
M0 KBE
M0 KBF
MM0 KBM
M0 KCB
M0 KCM
MM0 KCN
M0 KCT
M0 KDB
M0 KDD
M0 KDK
M0 KDN
MM0 KDP
M0 KEH
M0 KEI
M0 KEK
M0 KEX
M0 KFT
MM0 KGB
MM0 KGN
MW0 KIK
M0 KIS
M0 KIT
M0 KJA
M0 KJB
M0 KJP
M0 KJR
M0 KJV
M0 KJW
M0 KKH
M0 KKT
M0 KKX
M0 KLC
M0 KLF
M0 KLO
M0 KLP
M0 KMA
M0 KMF
M0 KMM
M0 KMP
M0 KMV
M0 KNA
M0 KNC
M0 KNG
M0 KNO
M0 KNT
M0 KNY
MI0 KOA
MW0 KOP
MM0 KOS
M0 KOX
M0 KPH
M0 KPL
MM0 KPZ
MM0 KQA
M0 KRB
M0 KRF
M0 KRG
M0 KRI
M0 KRJ
M0 KRT
M0 KSD
M0 KSH
M0 KSJ
M0 KSM
M0 KSP
M0 KSW
M0 KTA
MM0 KTC
M0 KTK
M0 KTM
MW0 KTX
M0 KTY
M0 KUB
MJ0 KUC
M0 KVC
M0 KVL
MU0 KWD
M0 KXO
MI0 KYE
M0 KYO
M0 KYT
M0 KYW
M0 LAC
M0 LAJ
M0 LAK

M0 LAM
M0 LAN
M0 LAP
MM0 LAX
M0 LBC
M0 LBT
M0 LBV
M0 LBZ
M0 LCB
M0 LCD
MM0 LCG
M0 LDA
M0 LDE
M0 LDM
M0 LDN
M0 LDU
M0 LDW
M0 LEV
M0 LFA
M0 LFG
M0 LFO
MI0 LFR
MI0 LGD
M0 LGH
M0 LGV
M0 LHT
M0 LIN
MM0 LIV
M0 LJB
M0 LJP
M0 LKB
M0 LKC
M0 LKQ
MI0 LLL
MI0 LLZ
M0 LMK
M0 LNZ
M0 LOM
M0 LON
M0 LOV
M0 LOY
M0 LPD
M0 LPE
MI0 LPO
M0 LRE
M0 LRI
M0 LRQ
MM0 LRT
M0 LRX
M0 LSK
M0 LTC
M0 LUG
M0 LUI
M0 LUX
M0 LWB
M0 LWO
M0 LXQ
M0 LXT
M0 LYC
M0 LYT
M0 LYX
M0 LZA
M0 LZP
M0 MAA
M0 MAD
M0 MAE
M0 MAK
M0 MAM
M0 MAS
M0 MAV
M0 MBF
M0 MBJ
MM0 MBK
MM0 MBN
M0 MBP
M0 MBQ
M0 MBU
M0 MBW
M0 MBX
M0 MBY
M0 MCD
MM0 MCF
M0 MCK
M0 MCR
MM0 MCU
M0 MCX
M0 MDA
M0 MDB
MD0 MDI
MM0 MDK
M0 MDL
MM0 MDM
M0 MDR
M0 MDW
M0 MDX
M0 MDY
M0 MEC
M0 MEE
M0 MEM
M0 MES
M0 MFD
M0 MFF
M0 MFM
M0 MGC
M0 MGD
M0 MGG
M0 MGH
M0 MGT
M0 MGY
MM0 MHA
M0 MHB
M0 MHS
MM0 MIA
M0 MIB
M0 MIN
MI0 MIO

M0 MIS
M0 MIW
MI0 MIX
M0 MJA
M0 MJC
M0 MJJ
M0 MJL
M0 MJO
MW0 MJP
M0 MJR
M0 MJU
M0 MKA
M0 MKB
M0 MKD
M0 MKI
M0 MKK
MM0 MKM
M0 MKS
M0 MKT
MM0 MKW
M0 MKX
M0 MLC
M0 MLN
M0 MLX
M0 MMA
M0 MMB
M0 MME
M0 MMH
M0 MMI
MM0 MMK
M0 MMM
M0 MMY
M0 MMZ
M0 MNK
M0 MNM
M0 MOA
M0 MOE
M0 MOF
M0 MOG
MM0 MOK
M0 MOM
MM0 MOZ
M0 MPC
M0 MPG
M0 MPH
M0 MPJ
M0 MPR
M0 MQB
M0 MQV
M0 MRA
M0 MRB
M0 MRD
M0 MRR
MM0 MRU
M0 MRZ
M0 MSD
MM0 MSK
M0 MSL
M0 MST
MM0 MSU
M0 MSW
M0 MTB
M0 MTE
M0 MTH
M0 MTL
M0 MTM
M0 MTT
M0 MTV
M0 MTY
M0 MUD
M0 MUF
M0 MUG
M0 MUJ
M0 MUS
MW0 MUT
M0 MUX
M0 MVA
M0 MWC
M0 MWD
MW0 MWE
M0 MWG
M0 MWH
M0 MWI
M0 MWM
MW0 MWV
MW0 MWZ
M0 MXL
M0 MXV
MM0 MXW
M0 MXY
M0 MYO
M0 MYR
M0 MZL
MW0 MZM
MM0 MZW
M0 NAD
M0 NAH
MI0 NAJ
MM0 NAN
M0 NAT
M0 NAV
M0 NBG
M0 NBR
M0 NBX
M0 NBY
M0 NCB
M0 NCF
M0 NCR
M0 NCT
M0 NCW
M0 NCY
M0 NDB
M0 NDN
M0 NDS
M0 NDW
M0 NEB

M0 NEE
M0 NEI
M0 NEL
M0 NES
M0 NEY
M0 NFA
MM0 NFC
M0 NFE
M0 NFN
M0 NFP
M0 NGA
M0 NGD
M0 NGE
M0 NGO
M0 NGT
M0 NHQ
M0 NIA
M0 NIK
M0 NIQ
M0 NIT
M0 NIX
M0 NJA
M0 NJB
M0 NJC
M0 NJN
M0 NKC
M0 NKD
M0 NKI
M0 NKL
M0 NKM
M0 NLB
M0 NLO
MI0 NML
M0 NMP
M0 NMQ
MW0 NMS
M0 NNN
M0 NOD
MM0 NOP
M0 NOT
M0 NOX
M0 NPE
M0 NPK
M0 NPN
M0 NQN
MM0 NRA
M0 NRD
M0 NRF
M0 NRK
M0 NRO
M0 NRT
M0 NRX
M0 NSA
M0 NST
MM0 NTD
M0 NTE
M0 NTL
M0 NTP
M0 NTS
M0 NUE
M0 NUT
M0 NVH
M0 NWB
M0 NWH
M0 NWR
M0 NXS
M0 NYE
M0 NZH
M0 OAH
M0 OAK
M0 OAM
M0 OAP
M0 OAV
M0 OBI
M0 OBS
M0 OCD
M0 OCH
M0 OCM
M0 OCN
M0 ODC
M0 ODV
M0 OEY
M0 OFA
M0 OFC
M0 OFI
M0 OFP
M0 OFR
M0 OFX
M0 OGG
MD0 OHM
MM0 OIL
M0 OJM
M0 OKE
M0 OLA
M0 OLF
M0 OLI
M0 OLT
M0 OLY
M0 OMB
M0 OMM
M0 OMN
M0 ONA
M0 ONE
M0 OOB
M0 OOM
M0 OON
M0 OOX
M0 OPR
M0 OPV
M0 OPW
MM0 OPX
M0 ORD
M0 ORO
M0 OSA
M0 OSC
M0 OSG

M0 OSK
MW0 OTG
M0 OTM
M0 OUZ
M0 OVA
M0 OVE
M0 OVL
M0 OVR
M0 OWE
MW0 OWW
M0 OXE
M0 OYG
MW0 OZZ
M0 PAB
MW0 PAD
M0 PAE
M0 PAH
M0 PAN
M0 PAP
MI0 PAS
M0 PAU
M0 PBA
M0 PBC
MW0 PBD
M0 PBK
M0 PBL
M0 PBM
M0 PBQ
M0 PBS
M0 PCC
M0 PCF
M0 PCI
M0 PCM
M0 PDJ
M0 PDM
M0 PDT
M0 PEK
M0 PEY
MI0 PFD
M0 PGA
M0 PGE
M0 PGG
M0 PGR
M0 PGT
M0 PHI
M0 PHS
MM0 PHT
M0 PHV
M0 PIH
M0 PIO
M0 PIW
M0 PJB
M0 PJH
M0 PJO
M0 PJW
M0 PKB
M0 PKD
MM0 PKM
M0 PKP
M0 PKT
M0 PKZ
M0 PLA
MW0 PLD
M0 PLF
M0 PLK
M0 PLL
M0 PLM
M0 PMD
M0 PME
MM0 PMK
M0 PMT
M0 PMZ
M0 PNP
M0 PNX
M0 POL
M0 PPC
M0 PPE
M0 PPK
M0 PPL
M0 PPY
MM0 PQN
M0 PRJ
M0 PRL
M0 PRY
MI0 PSU
M0 PSX
M0 PTB
MI0 PTC
M0 PTD
M0 PTH
MI0 PTK
M0 PTM
M0 PTU
MW0 PUP
M0 PUX
M0 PUZ
M0 PVG
M0 PVO
M0 PVX
M0 PWG
MI0 PWH
M0 PWJ
M0 PWK
M0 PWW
M0 PWZ
M0 PXB
M0 PXR
MM0 PYR
M0 PYU
M0 PZG
M0 PZT
M0 PZW
M0 RAA
M0 RAH
M0 RAJ
M0 RAO

M0 RAQ
M0 RAV
M0 RAY
M0 RBS
M0 RBW
MM0 RBZ
MW0 RCB
MM0 RCO
M0 RCQ
M0 RDE
M0 RDH
M0 RDJ
MM0 RDM
M0 REA
M0 REL
MW0 REN
M0 RER
M0 RES
M0 RET
M0 RFC
M0 RFD
M0 RFI
M0 RFL
M0 RFR
M0 RFX
M0 RGA
M0 RGJ
M0 RGT
MU0 RGU
M0 RGZ
M0 RHF
M0 RHJ
M0 RHV
M0 RHZ
M0 RIF
MM0 RIP
M0 RIT
M0 RIX
M0 RJA
M0 RJC
M0 RJD
M0 RJF
M0 RJL
M0 RJP
M0 RJV
M0 RKL
M0 RKS
M0 RLB
M0 RLH
M0 RLK
M0 RLX
M0 RMC
MM0 RMR
M0 RNA
M0 RNE
M0 RNS
M0 RNT
MM0 RNY
M0 ROA
MM0 ROB
M0 ROD
M0 ROE
M0 ROF
M0 ROI
M0 ROQ
MM0 ROS
M0 ROT
M0 ROU
M0 RPA
M0 RPC
M0 RPH
M0 RPQ
M0 RRA
M0 RRB
M0 RRO
M0 RRQ
M0 RRS
M0 RRU
M0 RRV
M0 RSL
MW0 RSS
M0 RTB
M0 RTG
M0 RTH
M0 RTI
M0 RTR
MI0 RTX
M0 RTZ
M0 RUN
M0 RUS
M0 RUT
M0 RVM
M0 RVP
M0 RVR
MM0 RVZ
M0 RXJ
M0 RXQ
M0 RXW
M0 RXY
MM0 RYE
MI0 RYL
M0 RYN
M0 RYW
M0 RYZ
MI0 RZA
M0 RZF
M0 RZK
M0 RZM
M0 RZW
M0 SAE
M0 SAF
M0 SAG
M0 SAN
M0 SAS
MW0 SAW
MM0 SBE

M0 SBG
M0 SBI
M0 SBM
M0 SBS
M0 SBU
M0 SBV
M0 SBW
M0 SCE
M0 SCH
M0 SCI
M0 SCJ
M0 SCK
M0 SCL
M0 SCM
M0 SCQ
M0 SDI
M0 SDN
M0 SDX
M0 SDZ
M0 SEE
M0 SEH
MM0 SEN
M0 SES
MM0 SFB
M0 SFU
M0 SGB
M0 SGC
M0 SGL
MM0 SGM
M0 SGN
M0 SGO
M0 SGP
MW0 SGT
M0 SHL
M0 SHX
M0 SIA
M0 SIB
M0 SIE
MW0 SIG
M0 SIM
M0 SIS
M0 SJA
M0 SJB
M0 SJC
M0 SJF
M0 SJM
M0 SJP
M0 SJS
MM0 SJT
MM0 SKB
M0 SKE
MM0 SKJ
M0 SKL
M0 SKR
M0 SKS
M0 SLA
M0 SLJ
M0 SLK
M0 SLM
M0 SLS
M0 SLT
M0 SMF
MW0 SMI
MI0 SMK
M0 SMM
M0 SMO
M0 SMR
M0 SNF
MM0 SNO
M0 SNR
MM0 SOB
M0 SOM
M0 SON
M0 SOP
M0 SPF
M0 SPG
M0 SPI
M0 SPO
M0 SPU
M0 SQL
M0 SQO
M0 SRF
M0 SRQ
M0 SRW
MM0 SRY
M0 SSC
M0 SSI
M0 STB
M0 STC
M0 STD
M0 STG
M0 STH
M0 STP
M0 STQ
M0 STR
MM0 STW
M0 STY
M0 STZ
M0 SUE
M0 SUL
M0 SUT
MM0 SVK
M0 SVM
M0 SWA
M0 SWG
M0 SWW
M0 SWX
M0 SXD
M0 SXF
M0 SXP
M0 SXR
M0 SXY
M0 SXZ
M0 SYC
M0 SYP
M0 SZA

MW0 SZE
M0 TAH
MM0 TAI
M0 TAO
M0 TAR
M0 TAU
M0 TBF
MI0 TBL
MM0 TBM
M0 TBO
MW0 TBP
M0 TBT
M0 TBU
M0 TBX
M0 TBZ
M0 TCA
M0 TCG
M0 TCH
M0 TCO
M0 TCW
M0 TDA
M0 TDH
MM0 TDI
MW0 TDR
M0 TEC
M0 TED
M0 TEE
M0 TEH
M0 TEM
M0 TEV
M0 TFD
M0 TFG
M0 TFT
M0 TGA
MM0 TGH
MI0 TGL
M0 TGR
M0 TGZ
MI0 THD
M0 THI
M0 THL
M0 THX
M0 THZ
MI0 TIB
M0 TIC
M0 TID
M0 TIG
MM0 TIS
M0 TJH
M0 TJP
MM0 TKL
M0 TKO
M0 TLA
M0 TLE
MI0 TLG
M0 TLS
M0 TLT
MI0 TME
M0 TMK
M0 TML
M0 TND
MM0 TNK
MI0 TNN
M0 TNR
M0 TNY
M0 TOA
M0 TOC
M0 TOE
M0 TOT
MW0 TOX
M0 TOY
M0 TPM
M0 TPN
M0 TPS
M0 TPX
M0 TQM
M0 TQR
M0 TRA
M0 TRJ
M0 TRQ
M0 TRX
M0 TSE
M0 TSF
MM0 TSG
M0 TSH
M0 TSK
M0 TSL
M0 TSR
MM0 TSS
M0 TST
M0 TSX
M0 TTA
M0 TTD
MM0 TTM
M0 TTN
M0 TTS
M0 TTW
MM0 TUG
M0 TUL
M0 TVE
M0 TVI
MW0 TVJ
M0 TWA
M0 TWB
M0 TWD
M0 TWE
M0 TWI
M0 TWT
M0 TXG
MI0 TXI
MM0 TXT
M0 TYA
M0 TYL
M0 TYM
M0 UAB
M0 UAF

M0 UAL
MI0 UAN
M0 UBA
M0 UCB
M0 UCQ
M0 UCW
M0 UCX
M0 UDP
MI0 UDR
M0 UEA
MM0 UED
M0 UFM
M0 UFO
MI0 UGB
M0 UGX
M0 UHF
M0 UKD
M0 UKG
M0 UKL
M0 UKR
M0 ULF
M0 ULT
M0 UNC
M0 UNG
M0 UNO
M0 URC
M0 USA
M0 USE
MI0 USH
M0 USP
M0 USR
M0 UTC
M0 UTE
M0 UVA
M0 UVK
M0 UVR
M0 UXE
M0 UXH
MM0 VAE
M0 VAF
MI0 VAX
M0 VBP
M0 VCD
M0 VCH
M0 VCM
M0 VCO
M0 VCT
M0 VCW
M0 VDC
M0 VEE
MM0 VEG
MI0 VEV
M0 VFU
M0 VGB
M0 VGK
M0 VGP
M0 VHF
MW0 VHS
M0 VIA
M0 VIC
M0 VID
M0 VIE
MI0 VIM
M0 VIP
M0 VIX
M0 VJL
M0 VKA
M0 VKD
M0 VKH
MI0 VKO
M0 VKV
M0 VMF
MM0 VMP
M0 VMT
M0 VOA
M0 VOC
MD0 VOD
M0 VOP
M0 VOR
M0 VOW
M0 VOX
M0 VPN
M0 VPT
M0 VRA
M0 VRC
M0 VRF
M0 VRH
M0 VRL
M0 VRR
M0 VSN
M0 VST
M0 VSV
M0 VTB
M0 VTF
M0 VTM
M0 VTT
M0 VTX
M0 VTZ
M0 VUC
M0 VVF
M0 VVS
M0 VWG
M0 VWH
M0 VXA
M0 VXK
MW0 VXO
M0 WAL
M0 WAN
M0 WAW
M0 WBA
M0 WBH
M0 WBN
MM0 WBT
M0 WBX
M0 WCQ
M0 WCY

M0 WDF
MM0 WDR
M0 WDX
M0 WEF
M0 WFM
M0 WER
M0 WGH
M0 WGN
M0 WGT
M0 WGZ
M0 WHD
M0 WHH
M0 WHI
M0 WHW
M0 WID
MW0 WIW
M0 WJD
M0 WJH
M0 WJJ
M0 WJN
MM0 WKC
M0 WKW
MM0 WLC
M0 WLM
M0 WLO
M0 WLR
M0 WLZ
M0 WMC
M0 WMD
M0 WNH
M0 WNP
M0 WOK
M0 WOZ
M0 WPD
M0 WPG
MW0 WPH
M0 WQX
M0 WRM
M0 WRT
M0 WSS
M0 WTB
M0 WTD
M0 WTD
M0 WTF
M0 WTR
M0 WTS
M0 WTT
M0 WUE
M0 WUT
M0 WWA
MI0 WWG
M0 WWS
M0 WWW
M0 WXB
M0 WXG
M0 WYA
M0 WZC
M0 WZD
M0 WZF
M0 WZW
M0 XAA
M0 XAE
MM0 XAF
M0 XAG
M0 XAN
M0 XAP
M0 XAS
M0 XAZ
M0 XBE
M0 XBF
M0 XBO
M0 XBP
M0 XCB
MM0 XDB
MW0 XDD
MM0 XDG
M0 XDW
MW0 XEN
MM0 XFM
M0 XFR
MI0 XFT
M0 XGP
MW0 XHO
M0 XIE
M0 XII
M0 XIM
M0 XIN
M0 XIP
M0 XIT
M0 XIX
M0 XJD
M0 XJC
M0 XKF
M0 XKR
M0 XLD
M0 XLC
MW0 XLF
M0 XLG
M0 XMM
M0 XMR
M0 XMT
M0 XNA
M0 XPU
M0 XPO
M0 XRS
M0 XSW
M0 XSX
M0 XTA
M0 XTC
M0 XTF
M0 XTH
M0 XTR
M0 XVR
M0 XVT
M0 XXA
M0 XXF

M0 XXI
M0 XXT
M0 XXY
M0 XXZ
M0 XZA
M0 XZF
M0 XZL
M0 YAA
M0 YAP
M0 YAT
M0 YAZ
M0 YBO
M0 YBX
MW0 YCC
M0 YCM
M0 YDD
M0 YEL
MM0 YEN
M0 YEW
M0 YFR
M0 YGR
M0 YHB
M0 YHC
M0 YHS
M0 YIK
MI0 YJR
M0 YME
M0 YMK
M0 YNE
M0 YNG
MI0 YOH
MM0 YOR
M0 YOU
M0 YOY
M0 YPZ
M0 YQV
M0 YRS
MM0 YSO
M0 YSU
MM0 YTM
M0 YTX
M0 YUI
MM0 YUL
M0 YYC
M0 YYZ
M0 YZZ
M0 ZAD
M0 ZAG
M0 ZAH
MI0 ZAO
M0 ZAT
M0 ZAX
M0 ZAZ
M0 ZBB
M0 ZBU
M0 ZCC
M0 ZCH
M0 ZCR
M0 ZDF
M0 ZDM
M0 ZDR
M0 ZDZ
M0 ZEP
M0 ZER
MI0 ZES
M0 ZFA
MM0 ZFR
M0 ZFX
M0 ZGD
M0 ZGN
M0 ZGY
M0 ZHA
M0 ZHN
M0 ZHX
M0 ZIE
M0 ZIX
M0 ZLB
M0 ZLR
M0 ZLW
M0 ZMF
M0 ZNH
M0 ZOC
M0 ZOD
M0 ZOL
M0 ZOS
M0 ZQI
M0 ZQR
M0 ZRA
M0 ZRN
M0 ZRQ
M0 ZRT
M0 ZSD
M0 ZSP
M0 ZSU
M0 ZTP
M0 ZTT
M0 ZTX
MM0 ZUN
M0 ZUU
M0 ZWE
M0 ZXA
M0 ZXP
M0 ZXT
MI0 ZXZ
M0 ZYZ
MM0 ZZB
M0 ZZK
M0 ZZM
M0 ZZR
M0 ZZW
M0 ZZX
M0 ZZY

### M*1

M1 AAA
M1 AAD
M1 AAG
MW1 AAK
M1 AAM
MW1 AAT
MM1 AAY
M1 ABD
M1 ABE
M1 ABK
M1 ABP
M1 ABQ
M1 ACI
M1 ACM
M1 ACS
M1 ACU
M1 ACZ
M1 ADA
M1 ADC
M1 ADE
M1 ADG
M1 ADW
M1 AEC
M1 AEM
M1 AES
M1 AFD
M1 AFM
M1 AFR
M1 AFY
M1 AGB
M1 AGF
M1 AGJ
M1 AGN
M1 AHB
M1 AHE
M1 AHO
M1 AHS
M1 AHZ
M1 AIF
M1 AIU
M1 AJB
M1 AJJ
M1 AJK
M1 AJR
M1 AJS
M1 AJV
M1 AKJ
M1 AKS
M1 ALI
MW1 ALN
M1 ALY
M1 AMH
M1 AMK
M1 AMN
M1 AMT
M1 ANH
M1 ANV
M1 ANW
M1 ANX
M1 AOH
M1 APN
M1 APU
M1 APV
M1 AQF
M1 AQH
M1 AQK
MM1 AQM
M1 AQO
M1 AQV
M1 AQW
M1 ARA
M1 ARN
M1 ARW
M1 ARZ
M1 ASF
M1 ASO
M1 ASW
M1 ASY
M1 ATF
M1 ATH
M1 ATM
M1 ATW
M1 ATZ
MM1 AUD
M1 AUS
MI1 AUU
M1 AVE
MM1 AVG
M1 AVP
MM1 AVR
M1 AWG
M1 AWH
M1 AWZ
MM1 AXA
M1 AXI
M1 AXL
M1 AXN
M1 AXO
MW1 AXV
M1 AXY
MI1 AYE
M1 AYI
M1 AYM
M1 AYX
M1 AYZ
M1 AZT
M1 AZZ
M1 BAU
M1 BAW
M1 BAZ
M1 BBM
M1 BBN
MI1 BBQ
M1 BBT
M1 BBV
M1 BCF
M1 BCJ
M1 BCV

M1 BCX
MM1 BDB
M1 BDI
M1 BDK
M1 BDM
M1 BEA
M1 BEG
M1 BEK
M1 BEY
M1 BFH
M1 BFP
M1 BFZ
M1 BGB
M1 BGC
M1 BGQ
M1 BGU
M1 BGV
M1 BHA
M1 BHD
M1 BHR
M1 BHS
M1 BIA
M1 BIJ
M1 BIN
M1 BIR
M1 BIZ
M1 BJH
M1 BJI
M1 BJL
MM1 BJO
M1 BJR
M1 BKC
MI1 BKD
MW1 BKM
M1 BKU
M1 BLM
M1 BLV
M1 BMJ
MM1 BMN
MI1 BNE
M1 BNJ
M1 BNL
M1 BNM
M1 BNP
M1 BNW
M1 BNZ
M1 BOC
M1 BOI
M1 BPJ
M1 BPT
MM1 BRH
M1 BRQ
M1 BRR
MM1 BSA
M1 BSC
M1 BSZ
M1 BTF
MM1 BTJ
M1 BTY
M1 BUD
M1 BUV
M1 BUZ
M1 BVA
MM1 BVW
M1 BWB
M1 BWO
M1 BWT
M1 BWW
M1 BWY
M1 BXE
M1 BXG
M1 BXI
M1 BXT
M1 BXW
M1 BXY
M1 BYA
M1 BYB
M1 BYN
M1 BYO
MM1 BYY
M1 BZM
M1 BZN
M1 BZV
M1 CAF
M1 CAM
M1 CAN
M1 CBG
MI1 CBP
M1 CBW
M1 CCC
M1 OOO
M1 CCP
M1 CDE
M1 CDG
M1 CDH
M1 ODM
M1 CDU
M1 CDZ
M1 CEB
M1 CEE
M1 CFH
M1 CEK
M1 CEL
M1 CEN
M1 CEU
M1 CEV
M1 CEX
M1 CFJ
M1 CFK
MW1 CFN
M1 CFV
M1 CGE
M1 CGH
M1 CGP
M1 CGX

M1 CHG
M1 CHN
M1 CHO
M1 CHV
M1 CHW
M1 CIA
M1 CID
M1 CIW
M1 CIX
M1 CIY
M1 CJK
M1 CJP
M1 CJU
M1 CJY
M1 CKC
M1 CKG
M1 CKM
M1 CLF
M1 CLK
MI1 CLN
MD1 CLV
MW1 CMC
M1 CMP
MM1 CMS
MM1 CMV
MI1 CNO
M1 CNU
M1 CNW
M1 COI
MW1 COK
M1 CPE
M1 CPH
M1 CPM
MW1 CPO
MW1 CPU
M1 CPV
M1 CPX
M1 CQB
M1 CQO
M1 CQQ
M1 CRB
M1 CRG
M1 CRN
M1 CRS
M1 CRV
M1 CRX
M1 CSB
MW1 CSK
M1 CSO
M1 CSQ
MI1 CSS
M1 CTA
MW1 CTP
M1 CUE
MM1 CUI
M1 CUJ
M1 CUW
M1 CUZ
M1 CVA
M1 CVE
M1 CVJ
M1 CVR
M1 CVZ
M1 CWM
M1 CWR
M1 CXB
MM1 CXE
M1 CXF
M1 CXJ
M1 CYE
M1 CYI
M1 CYS
M1 CYV
M1 CYW
M1 CYY
M1 CZB
M1 CZC
M1 CZR
M1 CZT
M1 CZW
M1 DAT
M1 DAZ
M1 DBB
M1 DBE
M1 DBG
M1 DBP
M1 DBS
M1 DBT
M1 DBZ
M1 DCG
M1 DOJ
M1 DCN
MI1 DCO
M1 DCQ
M1 DCU
M1 DOW
M1 DDD
M1 DDE
M1 DDK
M1 DDL
M1 DDN
M1 DDV
M1 DDX
M1 DEC
M1 DED
M1 DEI
M1 DEL
M1 DEN
M1 DFG
MW1 DFX
MW1 DGA
M1 DGC
M1 DGD
M1 DGF
M1 DGH
M1 DGN

MM1 DGO
M1 DGZ
M1 DHB
M1 DHD
M1 DHH
M1 DHL
M1 DHP
M1 DHZ
M1 DIC
M1 DID
M1 DIG
M1 DII
MI1 DIP
M1 DJE
M1 DJF
M1 DJK
M1 DKC
M1 DKU
M1 DKV
MI1 DLB
M1 DLC
M1 DLF
M1 DLJ
M1 DLL
M1 DLW
M1 DLY
M1 DMC
M1 DMJ
M1 DMQ
M1 DMS
M1 DNF
M1 DNL
M1 DNO
M1 DNS
M1 DNV
M1 DOC
MD1 DOJ
M1 DOP
M1 DOW
M1 DPB
M1 DPD
M1 DPM
M1 DPV
M1 DPZ
M1 DQA
M1 DQF
M1 DQY
M1 DRC
M1 DRN
M1 DRQ
M1 DRX
MW1 DSB
M1 DSL
M1 DSR
M1 DST
MI1 DTI
M1 DTJ
M1 DTX
MI1 DUM
M1 DUQ
M1 DUX
MM1 DUY
M1 DUZ
M1 DVM
M1 DWM
M1 DWO
M1 DWV
M1 DWX
M1 DXA
M1 DXF
M1 DXK
M1 DXP
M1 DXZ
M1 DYR
M1 DYX
MW1 DZA
M1 DZI
M1 DZK
M1 DZO
M1 DZU
M1 EAC
M1 EAG
M1 EAR
M1 EAU
M1 EAX
M1 EBM
M1 EBO
MM1 ECE
M1 ECF
M1 ECK
M1 ECN
M1 ECP
M1 ECR
M1 EDB
M1 EDG
MW1 EDN
M1 EDT
M1 EDU
M1 EDX
MI1 EEM
M1 EFU
M1 EEV
MM1 EFI
M1 EFL
M1 EFQ
M1 EFS
M1 EFY
M1 EGF
M1 EGH
M1 EGI
M1 EGJ
M1 EGO
M1 EHL
M1 EHM
M1 EHU
M1 EHY

M1 EIB
M1 EIK
M1 EIV
M1 EJC
M1 EJH
MM1 EJK
M1 EJL
M1 EJU
M1 EKG
M1 EKI
M1 EKT
M1 EKV
M1 EKX
M1 EKZ
MW1 ELC
MW1 ELF
M1 ELJ
M1 ELK
M1 ELT
M1 ELY
M1 EME
M1 EMF
M1 EMH
M1 EMI
M1 EMJ
MI1 EMT
M1 EMZ
M1 ENG
M1 ENH
M1 ENI
M1 ENO
M1 ENR
M1 ENU
M1 ENV
M1 ENW
M1 ENY
M1 EOA
M1 EOE
M1 EOG
MW1 EON
M1 EOX
M1 EPL
M1 EPM
M1 EPO
M1 EPQ
M1 EPW
M1 EPY
M1 EPZ
M1 EQP
M1 EQS
M1 ERM
M1 ERS
M1 ESA
M1 ESC
MI1 ESG
M1 ESL
M1 ESR
M1 EST
M1 ESU
M1 ETI
M1 ETU
MW1 EUA
MM1 EUI
M1 EUS
M1 EUW
M1 EUZ
M1 EVA
MI1 EVG
M1 EVK
M1 EVV
M1 EWH
M1 EWK
M1 EWZ
M1 EXA
M1 EXE
M1 EXP
M1 EXR
M1 EXX
M1 EYB
M1 EYE
M1 EYK
M1 EYW
M1 EYX
M1 EYY
M1 EZA
M1 EZF
M1 EZS
M1 EZU
M1 EZV
M1 EZW
M1 EZY
M1 FAC
M1 FAM
MM1 FAU
M1 FAV
M1 FDO
MM1 FBE
M1 FBG
MM1 FBH
M1 FBJ
M1 FBR
M1 FBT
M1 FBU
M1 FDV
M1 FBX
M1 FBY
M1 FBZ
M1 FCD
M1 FCI
M1 FCJ
M1 FCP
M1 FDB
M1 FDI
MI1 FDJ
MW1 FDQ
M1 FDS

M1 FDT
MW1 FDU
M1 FDZ
M1 FEA
M1 FER
M1 FEP
M1 FEV
M1 FFK
M1 FFQ
M1 FFZ
M1 FGA
M1 FGE
M1 FGF
M1 FGG
M1 FGI
MU1 FGK
MW1 FGL
M1 FGN
MW1 FGQ
M1 FGR
M1 FGS
M1 FGU
M1 FGY
M1 FGZ
M1 FHD
M1 FHG
M1 FHI
MM1 FHK
M1 FHM
M1 FHU
M1 FID
M1 FIF
M1 FIH
M1 FIJ
M1 FIM
M1 FIN
M1 FIO
M1 FIQ
M1 FIU
M1 FIV
M1 FIW
M1 FJE
M1 FJI
M1 FJN
M1 FJO
M1 FJU
M1 FJV
M1 FMW
M1 FXY
M1 GAZ
M1 GGY
M1 GLA
M1 GLO
M1 GMJ
MW1 GMW
M1 GNB
M1 GPD
M1 GPO
M1 GRY
M1 GWZ
M1 HAC
M1 HAT
MI1 HCO
M1 HDD
M1 HIS
M1 HOG
M1 HSB
M1 HTC
M1 HUL
M1 HVA
M1 HVD
M1 HZJ
M1 ICL
M1 IJJ
M1 ISR
M1 IVA
M1 IWB
MM1 JCM
M1 JDK
M1 JED
M1 JEN
M1 JJK
M1 JKL
M1 JMH
M1 JMR
M1 JMW
M1 JNC
M1 JPF
M1 JPM
M1 JRC
M1 JRO
M1 JRV
M1 JSC
M1 JUL
M1 JUN
M1 JWT
M1 KAB
M1 KAH
M1 KDA
M1 KEN
M1 KET
M1 KFM
M1 KGB
M1 KOO
M1 KIQ
M1 KLS
M1 KTM
M1 KTO
MW1 LCR
M1 LEE
M1 LIV
M1 LKY
M1 LNB
M1 MAR
M1 MAS
M1 MAX

MW1 MDH
M1 MHS
M1 MJT
M1 MOW
M1 MOZ
M1 MPR
M1 MSG
MW1 MWD
M1 MZX
M1 NCD
M1 NCP
M1 NDF
M1 NED
M1 NIC
M1 NIX
M1 NJJ
M1 NRT
M1 NTP
M1 NWT
M1 OAK
M1 OEY
M1 OJS
M1 OMW
M1 OOO
M1 PDJ
M1 PEP
M1 PGC
M1 PHS
M1 PJC
M1 PLA
M1 POM
M1 POP
M1 PPP
M1 PRV
M1 PSH
MW1 PSY
M1 PUB
M1 PVS
M1 RCA
M1 RCT
M1 RDB
M1 REW
M1 RGN
M1 RGZ
M1 RIP
M1 RKH
M1 RMB
MM1 RON
M1 RPW
M1 RPY
MM1 RWN
M1 SAG
M1 SBH
M1 SCH
M1 SDE
MI1 SJM
M1 SMH
M1 SMJ
M1 SPH
M1 SSY
M1 STE
MM1 STX
M1 TAY
M1 TDM
M1 TGQ
M1 TIG
MM1 TLD
MI1 TMT
M1 TNT
M1 TNY
M1 TRV
M1 TTT
M1 TTY
M1 TUT
M1 TWO
M1 TXI
M1 USA
M1 VAW
M1 VCG
M1 VFO
M1 VFR
M1 VTR
MM1 VWB
M1 WCJ
M1 WEB
M1 WML
M1 WPB
M1 WRG
MI1 WWG
MI1 XDX
M1 XCR
M1 XJC
M1 XMP
M1 XUP
M1 XWD
M1 XXX
M1 YAP
M1 YJG
M1 ZRP

### M*3

MM3 ACK
M3 ACV
M3 ADF
M3 ADV
M3 AEX
M3 AFV
M3 AFW
MM3 AFZ
MM3 AGC
M3 AGJ
M3 AGO
M3 AGQ
M3 AGS
MW3 AHK
M3 AHM
M3 AHT
MI3 AIX
M3 AJB
M3 AKD
M3 AKU
MM3 ALP
MI3 ALR
M3 ALT
M3 AMG
M3 AMN
M3 AMY
M3 ANY
M3 AOB
MI3 AOI
M3 APF
MI3 APK
M3 APL
M3 APM
M3 APP
M3 APR
M3 ARC
M3 ARV
M3 ARX
M3 ASD
M3 ASQ
M3 ASU
M3 ATA
M3 ATH
MI3 ATK
M3 ATX
M3 AUD
M3 AUE
M3 AUH
M3 AUR
M3 AUV
M3 AUW
MW3 AVB
MW3 AVC
MI3 AVG
M3 AVN
M3 AVP
MI3 AVT
M3 AWF
MM3 AWJ
MM3 AWL
M3 AWV
M3 AXJ
M3 AYG
M3 AYK
M3 AYR
M3 AZC
M3 AZD
M3 AZW
M3 BAF
M3 BAP
M3 BAV
M3 BAX
M3 BBE
M3 BBJ
M3 BBM
M3 BBT
M3 BBV
M3 BBW
M3 BCN
MI3 BCO
M3 BCT
M3 BDE
M3 BDF
M3 BDI
M3 BDL
M3 BDM
M3 BDO
M3 BDS
M3 BEH
M3 BEI
M3 BEO
M3 BEY
M3 BFA
M3 BFS
M3 BFW
M3 BGO
M3 BHN
M3 BHV
M3 BHX
M3 BIJ
M3 BIM
MI3 BIN
M3 BJC
M3 BJD
MW3 BJI
MM3 BJU
M3 BJX
M3 BKF
M3 BKL
M3 BKR
M3 BKZ
M3 BLL
M3 BLP
M3 BLS
M3 BLV
M3 BLW
M3 BMD
M3 BMY
M3 BNH
M3 BNI
M3 BNO
M3 BNP
M3 BNT
M3 BOM
M3 BOQ
M3 BOX
MW3 BOY
M3 BPB
M3 BPE
M3 BPK
MI3 BQA

M3 BQR
M3 BQU
M3 BQV
M3 BQW
M3 BRP
M3 BSA
M3 BSL
MM3 BSQ
MW3 BSX
M3 BTC
M3 BTD
M3 BTE
M3 BTV
M3 BTY
M3 BUB
M3 BUG
M3 BUK
M3 BUO
MW3 BVC
M3 BVN
M3 BVZ
M3 BWM
M3 BWO
M3 BXA
M3 BXI
MM3 BXL
MW3 BYG
M3 BYH
M3 BYY
M3 BYZ
M3 BZG
MM3 BZO
M3 CAI
M3 CBD
M3 CBI
M3 CCR
M3 CDC
M3 CDD
MI3 CDF
M3 CDH
M3 CDJ
M3 CDS
M3 CDZ
M3 CEC
M3 CED
M3 CEE
M3 CEF
M3 CEH
M3 CEI
M3 CEK
M3 CEY
M3 CFC
M3 CFE
M3 CFH
M3 CFK
M3 CFL
M3 CFN
M3 CFQ
M3 CFS
M3 CFW
M3 CFY
M3 CGB
M3 CGV
M3 CGW
M3 CGY
M3 CHA
M3 CHC
M3 CHD
M3 CHH
M3 CHO
M3 CIB
MI3 CID
M3 CIN
M3 CJQ
M3 CJV
M3 CJW
M3 CKB
M3 CKG
M3 CKS
M3 CKV
M3 CLF
M3 CLN
M3 CLR
M3 CLV
M3 CLZ
MM3 CMD
M3 CMT
M3 CNU
M3 CNX
M3 COB
M3 COH
M3 COO
MM3 COR
M3 COS
M3 COX
M3 COY
M3 CPA
M3 CPB
M3 CPJ
M3 CPP
M3 CPT
M3 CPU
M3 CPV
M3 CQE
M3 CQQ
M3 CQI
M3 CQJ
M3 CQK
M3 CQL
M3 CQM
M3 CQZ
M3 CRB
M3 CRE
M3 CRG
M3 CRJ
M3 CRU

M3 CSP
M3 CSW
MI3 CTF
M3 CTG
M3 CTH
M3 CTJ
M3 CTL
M3 CTW
M3 CUD
M3 CUG
M3 CUI
M3 CUO
M3 CUP
M3 CUQ
M3 CUS
M3 CVF
M3 CVG
M3 CVI
M3 CVT
MI3 CWB
M3 CWD
M3 CWM
M3 CWR
M3 CWS
M3 CWT
M3 CWU
MM3 CWW
M3 CXL
M3 CXS
M3 CXW
M3 CYP
M3 CYR
M3 CYY
M3 CZC
M3 CZD
M3 CZH
M3 CZO
M3 CZZ
M3 DAJ
M3 DAT
M3 DBD
M3 DBI
M3 DBM
M3 DBO
M3 DBQ
M3 DCA
MM3 DCB
M3 DCG
M3 DCO
M3 DCQ
M3 DDL
MI3 DDN
M3 DDO
M3 DDP
MM3 DDS
M3 DDV
M3 DEG
M3 DEH
M3 DEK
M3 DEO
M3 DEQ
M3 DEU
M3 DFE
M3 DFI
M3 DGC
M3 DGE
MW3 DGW
M3 DHC
MM3 DHL
M3 DHP
M3 DID
MW3 DIF
M3 DII
M3 DJD
M3 DJS
M3 DJT
M3 DKD
M3 DKF
MM3 DKQ
M3 DKR
M3 DKW
M3 DKX
M3 DKY
M3 DLA
M3 DLL
M3 DLS
M3 DLV
M3 DLY
M3 DMB
M3 DMC
MM3 DMH
M3 DMV
M3 DMX
MI3 DNK
MM3 DNL
M3 DNZ
M3 DOE
M3 DOI
M3 DOQ
M3 DOV
M3 DPD
M3 DPE
M3 DPL
M3 DPO
M3 DPR
M3 DPU
M3 DPW
M3 DQC
M3 DQF
M3 DQO
M3 DQT
MI3 DRL
M3 DRP
M3 DRQ
M3 DRT
M3 DSN

M3 DSO
M3 DTF
M3 DTS
M3 DTW
MI3 DUA
M3 DUF
M3 DUP
M3 DUV
M3 DUW
M3 DUX
M3 DUY
M3 DVE
M3 DVI
M3 DVK
M3 DVS
M3 DVX
M3 DWJ
M3 DWN
MI3 DWS
M3 DWT
M3 DWU
M3 DWY
M3 DXB
M3 DXC
M3 DXH
M3 DXJ
M3 DXP
M3 DYD
M3 DYG
M3 DYH
M3 DYI
MM3 DYJ
M3 DYP
M3 DYS
M3 DYX
M3 DZJ
M3 DZY
M3 EAC
M3 EAG
M3 EAJ
M3 EAM
M3 EAO
M3 EAV
M3 EBN
M3 ECA
M3 ECH
M3 ECL
M3 ECM
M3 ECT
M3 ECX
M3 EDB
M3 EDE
M3 EDF
M3 EDM
M3 EEA
M3 EEB
M3 EEY
M3 EFE
M3 EFF
M3 EFM
M3 EFY
M3 EFZ
M3 EGN
M3 EGO
M3 EGP
M3 EHE
M3 EHN
M3 EHO
MI3 EHQ
M3 EHW
M3 EHX
MI3 EIE
M3 EII
M3 EIR
M3 EIS
M3 EJA
M3 EJG
M3 EJH
M3 EJT
M3 EJZ
M3 EKB
M3 EKQ
MM3 EKX
M3 ELE
M3 ELH
M3 ELJ
M3 ELU
M3 EMF
M3 EMH
M3 EMP
MI3 EMT
M3 EMY
M3 ENB
MW3 END
M3 ENQ
M3 ENT
MM3 ENW
M3 ENX
M3 EOA
M3 EOG
M3 EOK
MW3 EPA
M3 EPB
M3 EPE
MI3 EPN
M3 EPP
M3 EPY
M3 EPZ
MI3 EQC
MI3 EQI
MI3 EQJ
M3 EQN
MI3 EQR
MI3 EQV
M3 EQX
M3 EQZ

M3 ERG
M3 ERK
M3 ERL
M3 ERW
M3 ESC
M3 ESH
MM3 ETD
M3 ETP
M3 ETY
M3 EUC
M3 EUO
M3 EUT
M3 EVL
M3 EVZ
M3 EWC
M3 EWJ
M3 EWL
M3 EXE
M3 EXO
M3 EXQ
M3 EXU
M3 EYE
M3 EYG
MM3 EYJ
MU3 EYU
MM3 EYV
M3 EZA
M3 EZT
M3 EZW
M3 EZX
M3 FAD
M3 FAQ
MI3 FAV
MM3 FAW
M3 FBE
M3 FBH
MI3 FBT
M3 FBZ
M3 FCC
M3 FCD
M3 FCE
M3 FCH
MW3 FCI
M3 FCJ
M3 FDD
M3 FDE
M3 FDI
M3 FDK
M3 FDP
M3 FDT
M3 FDW
M3 FDY
M3 FEE
M3 FEH
MM3 FEP
M3 FER
MM3 FEV
MI3 FEZ
M3 FFG
M3 FFW
MM3 FGD
MW3 FGE
MM3 FGJ
M3 FHN
M3 FIA
M3 FID
M3 FIF
M3 FIN
M3 FIR
M3 FIV
M3 FJU
M3 FKF
M3 FKH
M3 FKJ
M3 FKT
MW3 FKX
M3 FLF
M3 FLH
M3 FLQ
M3 FLW
M3 FLY
M3 FMA
M3 FMG
M3 FMJ
MI3 FML
M3 FMO
M3 FMT
M3 FMX
M3 FMZ
M3 FNF
M3 FNK
M3 FNP
MI3 FNU
M3 FNW
M3 FOB
M3 FOI
M3 FON
M3 FOT
M3 FOU
M3 FPP
M3 FPX
M3 FPY
M3 FQB
M3 FQE
M3 FQF
M3 FQH
M3 FQK
M3 FQL
M3 FQP
M3 FQV
M3 FQW
M3 FRA
M3 FRF
M3 FRI
M3 FRW
M3 FSA

M3 FSJ
M3 FSM
M3 FSO
MM3 FST
MM3 FTL
M3 FTO
M3 FT3
M3 FUE
M3 FUL
M3 FUU
MW3 FVF
M3 FVI
M3 FVK
MM3 FVQ
M3 FVT
M3 FVU
M3 FVV
M3 FWB
M3 FWG
MM3 FWQ
MW3 FXH
M3 FXN
M3 FXR
M3 FXS
M3 FXV
M3 FYD
M3 FYH
M3 FYO
M3 FYP
MI3 FYT
M3 FYU
M3 FYX
M3 FZG
M3 FZH
M3 FZK
MJ3 FZN
M3 FZP
M3 FZQ
M3 FZT
MW3 FZU
M3 FZW
M3 FZY
MM3 GAI
MW3 GAJ
M3 GAR
M3 GAS
M3 GAU
M3 GBG
M3 GBQ
MI3 GBU
M3 GBV
M3 GCA
M3 GCE
MW3 GCG
MW3 GCL
M3 GCO
M3 GCQ
M3 GCU
M3 GCV
MW3 GDA
M3 GDB
M3 GDJ
M3 GDP
M3 GDS
M3 GDW
MI3 GDZ
M3 GED
M3 GEG
MI3 GEO
M3 GEV
M3 GFC
M3 GFK
M3 GFN
M3 GFU
M3 GFV
M3 GFX
M3 GFY
M3 GGC
M3 GGG
M3 GGI
M3 GGK
M3 GGL
M3 GGP
M3 GGT
M3 GGU
M3 GGX
M3 GGY
M3 GGZ
M3 GHC
M3 GHJ
M3 GHP
M3 GHQ
M3 GHV
M3 GIA
MI3 GIB
MM3 GIN
M3 GIO
M3 GIP
M3 GIU
M3 GJA
M3 GJF
M3 GJJ
M3 GJL
MM3 GJP
M3 GJ3
M3 GJZ
M3 GKC
MW3 GKF
M3 GKP
M3 GKQ
M3 GKS
M3 GKV
MW3 GLB
M3 GLE
M3 GLF
M3 GLG

M3 GLU
M3 GLW
M3 GLX
MM3 GMF
M3 GMH
M3 GMN
M3 GMO
M3 GMQ
M3 GMZ
MI3 GNC
M3 GNE
MW3 GNH
M3 GNL
M3 GNO
M3 GNQ
M3 GNS
MI3 GNV
M3 GNW
M3 GNX
MI3 GOC
M3 GOF
MW3 GOG
M3 GON
M3 GOO
M3 GOP
MM3 GPF
M3 GPO
M3 GPT
M3 GPU
M3 GPZ
MW3 GQF
M3 GQG
M3 GQH
M3 GQQ
M3 GQU
M3 GRD
M3 GRE
M3 GRQ
M3 GRU
M3 GRW
M3 GRX
MI3 GRZ
M3 GSA
MM3 GSN
M3 GSS
M3 GSV
M3 GTI
M3 GTS
M3 GUA
M3 GUE
M3 GUK
M3 GUL
M3 GUP
M3 GUT
M3 GUW
M3 GVI
MW3 GVP
M3 GVR
M3 GVX
MW3 GVY
M3 GVZ
M3 GWE
M3 GWL
M3 GWM
M3 GWO
MW3 GWP
M3 GWX
M3 GWY
M3 GXC
MW3 GXF
M3 GXS
M3 GXT
M3 GXW
M3 GYG
M3 GYS
M3 GYX
M3 GZA
M3 GZB
M3 GZH
M3 HAA
M3 HAH
M3 HAO
M3 HAR
MW3 HAY
M3 HBA
M3 HBI
M3 HBO
M3 HBZ
M3 HCZ
M3 HDB
M3 HDU
M3 HDR
M3 HDT
MW3 HDY
M3 HEF
M3 HEN
M3 HEP
M3 HES
M3 HEW
M3 HEZ
MW3 HFB
MW3 HFK
M3 HFM
MW3 HFV
MJ3 HFY
M3 HGB
MW3 HGD
M3 HGF
M3 HGS
M3 HGY
M3 HHJ
M3 HHL
MW3 HHT
M3 HHW
M3 HHY
M3 HIJ

**IMPORTANT NOTE**

**Revalidate licence to avoid revocation** – Ofcom has advised the Society that plans will be drawn up to revoke licences that have not been revalidated as required by the licence conditions. The quickest way to revalidate is to do so online via the Ofcom website: *https://services.ofcom.org.uk/* or by email: *amateur.validations@ofcom.org.uk* If you need assistance in the process, Ofcom staff are available to help, but please be patient during times of heavy workload.

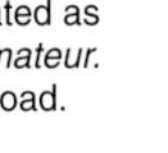

Withheld

M3 HIY
M3 HJH
M3 HJL
M3 HJR
M3 HKK
M3 HKS
M3 HKU
M3 HKX
M3 HLO
M3 HMX
M3 HNC
M3 HNH
M3 HNI
M3 HNY
M3 HNZ
M3 HOB
M3 HOF
M3 HOI
M3 HOJ
M3 HOO
M3 HOQ
M3 HOS
MM3 HOZ
M3 HPG
M3 HPJ
M3 HPL
M3 HPP
M3 HPR
MW3 HPS
M3 HPX
M3 HQG
M3 HQI
M3 HQO
M3 HRI
M3 HRP
M3 HRQ
M3 HRR
M3 HSA
MW3 HSB
M3 HSN
M3 HSP
M3 HSQ
M3 HSY
M3 HTB
M3 HTC
M3 HTD
M3 HTT
M3 HTU
M3 HTV
MM3 HTZ
M3 HUA
M3 HUC
M3 HUF
M3 HUH
M3 HUI
M3 HUJ
MM3 HVB
M3 HVF
M3 HVG
M3 HVI
M3 HVR
M3 HVT
M3 HWB
M3 HWE
M3 HWF
M3 HWG
M3 HWK
M3 HWT
M3 HWZ
M3 HXA
M3 HXE
M3 HXI
M3 HXP
MM3 HXX
M3 HYA
M3 HYC
M3 HYK
MI3 HYM
M3 HYW
MM3 HYZ
M3 HZF
M3 HZG
M3 HZV
M3 HZY
M3 IAB
MW3 IAJ
M3 IAS
M3 IAT
M3 IBB
M3 IBD
M3 IBF
M3 IBH
M3 IBL
M3 IBN
MM3 IBO
MM3 IBW
MM3 IBX
M3 ICJ
M3 ICX
M3 ICY
M3 ICZ
M3 IDL
MW3 IDP
M3 IDS
M3 IDU
M3 IEO
M3 IES
M3 IEY
M3 IFC
M3 IFH
M3 IFX
M3 IGI
M3 IGK
M3 IGL
M3 IGP
M3 IGW

M3 IGX
M3 IHE
M3 IHG
M3 IHH
M3 IHT
M3 IIB
M3 IIE
M3 IIF
M3 IIQ
M3 IIS
M3 IIY
M3 IJB
M3 IKA
MI3 IKF
M3 IKL
M3 IKX
M3 IKY
M3 ILE
M3 ILF
M3 ILI
M3 ILL
MM3 ILO
MW3 ILP
M3 ILX
MM3 IMN
M3 IMT
M3 IMU
M3 IMW
M3 INB
M3 INI
M3 INK
MM3 INV
M3 INX
M3 INZ
MM3 IOB
M3 IOE
M3 IOP
M3 IOY
MW3 IPB
M3 IPI
MW3 IPN
MM3 IPU
M3 IQH
M3 IQK
M3 IQL
M3 IQR
MW3 IRD
MM3 IRI
M3 IRK
M3 IRL
M3 IRN
M3 ISF
MM3 ISV
M3 ITE
M3 ITF
M3 ITJ
M3 ITP
M3 ITQ
M3 ITR
M3 ITX
M3 IUB
M3 IUG
M3 IUL
MM3 IUQ
M3 IUR
MM3 IUY
M3 IVE
M3 IVH
M3 IVJ
M3 IVL
M3 IVP
M3 IWH
M3 IWS
M3 IWY
M3 IXI
M3 IXS
M3 IXV
M3 IXW
M3 IXX
M3 IYC
M3 IYD
M3 IYF
MI3 IYJ
M3 IYL
M3 IYN
M3 IYS
M3 IYT
M3 IYU
M3 IYZ
MM3 IZC
M3 IZR
M3 IZS
MW3 IZX
M3 JAH
M3 JAJ
MW3 JAT
M3 JBA
M3 JBC
M3 JBD
M3 JBL
M3 JBN
M3 JBO
M3 JBR
M3 JBU
M3 JBX
M3 JCG
M3 JCJ
M3 JCM
MI3 JDC
M3 JDD
M3 JDE
M3 JDK
M3 JDV
M3 JDY
MM3 JDZ
M3 JEG

M3 JEL
M3 JEO
M3 JEU
M3 JEV
M3 JEX
M3 JEZ
MW3 JFG
M3 JFI
M3 JFJ
M3 JFM
M3 JFO
M3 JFU
M3 JGD
MI3 JHD
MI3 JHF
M3 JHI
M3 JHK
M3 JHO
M3 JHQ
M3 JHU
M3 JIB
M3 JIO
M3 JIY
M3 JJA
M3 JJF
M3 JJK
MW3 JJO
M3 JJR
M3 JJX
M3 JKC
MM3 JKF
M3 JKO
M3 JKQ
M3 JKV
M3 JLC
M3 JLG
M3 JLN
M3 JLO
M3 JLP
M3 JLU
M3 JLY
M3 JLZ
M3 JMB
M3 JME
M3 JMF
M3 JMH
M3 JMN
MI3 JNF
M3 JNO
M3 JNS
M3 JOB
M3 JOG
M3 JOK
M3 JOX
M3 JOZ
M3 JPE
M3 JPR
MW3 JPT
M3 JPW
M3 JPY
M3 JQA
M3 JQE
M3 JQF
M3 JQL
MM3 JQO
M3 JQP
M3 JQQ
M3 JQZ
M3 JRB
M3 JRP
M3 JSC
M3 JSP
M3 JSR
M3 JSU
M3 JSW
M3 JSX
M3 JSY
M3 JTD
M3 JTG
M3 JTL
M3 JTP
MW3 JTQ
M3 JTR
M3 JTW
M3 JTY
MD3 JUA
MM3 JUG
M3 JUS
M3 JUT
M3 JUU
M3 JUX
M3 JVE
M3 JVL
MI3 JVN
MM3 JVO
M3 JVU
MM3 JVZ
M3 JWH
M3 JWI
M3 JWL
M3 JWO
M3 JWU
M3 JWX
MW3 JWY
M3 JXB
M3 JXT
M3 JYJ
M3 JYQ
M3 JYU
MM3 JZB
M3 JZH
M3 JZQ
M3 JZS
MW3 JZU
MW3 JZZ
M3 KAA

M3 KAO
M3 KAZ
M3 KBA
M3 KBI
M3 KBK
M3 KBO
M3 KBU
M3 KBV
MI3 KBW
M3 KCK
M3 KCN
MI3 KCW
M3 KDB
M3 KDD
M3 KDT
M3 KEP
M3 KEQ
M3 KEX
M3 KFA
MW3 KFD
MM3 KFF
M3 KFX
M3 KFY
M3 KFZ
M3 KGA
M3 KGC
M3 KGH
M3 KGM
M3 KGR
MW3 KGX
MW3 KHD
MM3 KHN
M3 KHO
MM3 KHQ
M3 KHS
M3 KHV
MM3 KIC
M3 KIH
MM3 KII
MM3 KIJ
M3 KIK
M3 KIV
MW3 KIW
MM3 KIX
M3 KJI
M3 KJN
M3 KJO
M3 KJX
MW3 KJZ
M3 KKJ
M3 KKR
M3 KKV
MM3 KLA
M3 KLD
M3 KLG
M3 KLI
MW3 KLQ
M3 KLR
M3 KLW
MW3 KLX
M3 KMG
MI3 KMI
MI3 KMJ
M3 KMM
MW3 KMV
M3 KNG
MM3 KNI
M3 KNQ
M3 KNX
M3 KOD
MW3 KOE
M3 KOH
M3 KOO
M3 KOQ
M3 KOV
MM3 KPA
M3 KPH
M3 KPN
M3 KPR
M3 KPS
M3 KPV
M3 KPX
M3 KPY
M3 KQA
M3 KQE
MW3 KQH
M3 KQJ
M3 KQQ
M3 KQZ
M3 KRG
M3 KRJ
M3 KRU
M3 KRV
M3 KRW
M3 KSB
M3 KSF
MW3 KSL
M3 KSR
M3 KST
M3 KSU
M3 KSY
M3 KSZ
M3 KTF
M3 KTN
M3 KTQ
M3 KTU
M3 KUA
M3 KUF
M3 KUL
M3 KUR
M3 KUT
M3 KUW
M3 KVF
M3 KVS
MM3 KVX
M3 KWA

M3 KWE
M3 KWK
M3 KWN
M3 KWO
M3 KWQ
M3 KWW
M3 KXN
M3 KXO
M3 KXQ
M3 KXW
MM3 KXX
M3 KYF
MM3 KYN
M3 KYS
M3 KYT
M3 KYU
M3 KYY
M3 KZG
M3 KZL
M3 KZM
M3 KZN
M3 KZO
M3 KZU
M3 LAA
M3 LAC
M3 LAE
M3 LAH
M3 LAI
M3 LAL
M3 LAM
M3 LAR
M3 LBF
M3 LBH
M3 LBI
MW3 LBU
M3 LCA
M3 LCB
MM3 LCC
M3 LCK
M3 LCM
M3 LCQ
M3 LCV
MM3 LCX
M3 LDG
M3 LDN
M3 LDP
M3 LDV
M3 LDZ
M3 LEA
M3 LEM
M3 LEQ
M3 LEU
M3 LFA
M3 LFD
M3 LFF
M3 LFJ
M3 LFK
M3 LFM
M3 LFS
M3 LFT
M3 LFW
M3 LFX
M3 LFY
M3 LGA
M3 LGG
M3 LGP
M3 LGQ
M3 LGT
M3 LGV
M3 LHC
M3 LHD
M3 LHH
M3 LHJ
M3 LHL
MI3 LHS
M3 LHT
M3 LHY
M3 LIA
M3 LIF
M3 LII
MW3 LIJ
M3 LIQ
M3 LIR
M3 LJD
M3 LJN
M3 LJO
M3 LJU
MW3 LJV
M3 LKA
MI3 LKG
M3 LKI
M3 LKK
M3 LKL
M3 LKX
M3 LKZ
MM3 LLE
M3 LLI
M3 LLJ
M3 LLO
M3 LLR
MM3 LLY
M3 LMO
M3 LMP
MM3 LMT
M3 LMY
M3 LND
M3 LNK
M3 LNS
M3 LNV
M3 LOD
M3 LOH
M3 LOJ
M3 LOM
M3 LON
M3 LOV
M3 LOW

M3 LPD
MI3 LPO
M3 LPV
M3 LPX
M3 LPY
M3 LPZ
M3 LQG
M3 LQH
M3 LQM
M3 LQQ
MM3 LQR
MM3 LQS
M3 LQT
M3 LRA
M3 LRB
M3 LRD
M3 LRL
M3 LSD
M3 LSJ
M3 LSM
M3 LSN
M3 LSP
M3 LSQ
M3 LSY
M3 LSZ
M3 LTC
M3 LTD
M3 LTE
M3 LTF
M3 LTJ
MI3 LTK
M3 LTM
M3 LTN
MM3 LTO
M3 LTQ
M3 LTZ
M3 LUB
M3 LUE
M3 LUF
M3 LUH
M3 LUI
M3 LUJ
M3 LUM
M3 LUQ
MW3 LUR
M3 LUT
M3 LVE
M3 LVG
M3 LVH
M3 LVL
M3 LVO
MM3 LVQ
M3 LVS
M3 LVU
M3 LVV
M3 LVW
M3 LWA
M3 LWB
M3 LWD
M3 LWE
M3 LWH
M3 LWI
M3 LWL
M3 LWM
M3 LWN
M3 LWS
M3 LWY
M3 LXC
MI3 LXG
M3 LXM
MI3 LXO
MM3 LXQ
MM3 LXX
M3 LXY
MM3 LYB
MM3 LYD
M3 LYF
M3 LYJ
MI3 LYK
M3 LYO
M3 LYT
MW3 LYW
M3 LYY
M3 LZB
M3 LZE
M3 LZG
M3 LZI
M3 LZJ
M3 LZM
M3 LZN
M3 LZP
M3 LZQ
MW3 LZW
M3 LZX
MW3 LZZ
M3 MAF
M3 MAH
M3 MAI
M3 MAL
MM3 MAO
M3 MAQ
M3 MAW
M3 MAX
M3 MBN
M3 MBP
M3 MBW
M3 MCH
MW3 MCJ
M3 MCO
M3 MCP
MI3 MCZ
MW3 MDL
M3 MDQ
M3 MDT
MI3 MDZ
M3 MEQ

M3 MEX
M3 MFI
M3 MFM
M3 MFQ
M3 MFW
MM3 MGG
M3 MGM
M3 MGX
M3 MHB
M3 MHJ
M3 MHM
M3 MHO
M3 MHT
M3 MHU
MM3 MIL
M3 MIY
M3 MJB
M3 MJC
M3 MJP
MW3 MJQ
M3 MJT
MM3 MJU
M3 MJW
M3 MKL
M3 MKM
M3 MKR
M3 MKU
M3 MKX
M3 MLC
M3 MLD
M3 MLE
M3 MLN
M3 MLO
MI3 MLS
M3 MLU
M3 MLY
M3 MMH
M3 MMR
MM3 MMT
MM3 MMX
M3 MNA
M3 MNE
M3 MNF
M3 MNH
M3 MNM
M3 MNN
M3 MNO
M3 MNP
M3 MNZ
M3 MOA
M3 MOI
MM3 MOM
MI3 MOS
M3 MOU
M3 MPD
M3 MPF
M3 MPG
M3 MPN
M3 MPQ
M3 MPU
M3 MPV
MM3 MPX
M3 MPY
M3 MQD
MW3 MQE
M3 MQF
M3 MQG
M3 MQK
MW3 MQN
M3 MQQ
M3 MQS
M3 MQU
M3 MQV
M3 MQW
M3 MQZ
M3 MRP
M3 MRR
M3 MRT
M3 MSE
M3 MSI
M3 MSM
M3 MSS
M3 MTA
M3 MTG
M3 MTH
MM3 MTJ
M3 MTN
M3 MTV
M3 MTY
MM3 MUC
M3 MUH
M3 MUJ
M3 MUM
MI3 MUR
M3 MUV
M3 MUY
M3 MUZ
M3 MVA
MM3 MVC
M3 MVD
M3 MVF
M3 MVG
M3 MVH
MI3 MVP
M3 MVQ
M3 MVS
M3 MVW
M3 MVZ
M3 MWF
M3 MWK
MW3 MWL
M3 MWP
M3 MWU
M3 MWY
M3 MXB
M3 MXD

M3 MXS
MM3 MXT
M3 MXY
M3 MYA
M3 MYB
M3 MYJ
M3 MYL
M3 MYN
M3 MYO
M3 MYP
M3 MYR
M3 MYU
M3 MYV
M3 MYX
M3 MYY
M3 MZH
M3 MZI
M3 MZL
M3 MZQ
M3 MZS
M3 MZU
M3 MZY
M3 NAA
M3 NAM
M3 NBA
M3 NBC
MW3 NBF
M3 NBM
MW3 NBO
M3 NBR
M3 NBT
M3 NBW
MW3 NCF
M3 NCR
MW3 NCU
M3 NCV
M3 NCW
MI3 NCX
M3 NCY
M3 NCZ
M3 NDD
MM3 NDH
M3 NDP
M3 NDV
M3 NDY
M3 NEJ
M3 NEK
M3 NEU
M3 NEZ
M3 NFC
M3 NFD
MM3 NFI
MM3 NFX
MW3 NGI
M3 NGL
M3 NGQ
M3 NGT
M3 NGW
M3 NHF
M3 NHJ
M3 NHL
M3 NHO
M3 NHQ
M3 NHT
M3 NHX
M3 NIB
MW3 NIH
M3 NIL
M3 NIM
MI3 NIO
M3 NIU
M3 NIY
M3 NJG
M3 NJL
M3 NJT
M3 NJZ
M3 NKA
M3 NKD
M3 NKH
M3 NKI
M3 NKJ
MM3 NKM
M3 NKS
M3 NKT
MW3 NKV
M3 NKY
M3 NLB
M3 NLC
M3 NLL
M3 NLO
M3 NLS
M3 NLU
M3 NLV
M3 NMA
MW3 NMC
M3 NME
M3 NMF
M3 NMH
MM3 NMN
MM3 NMO
M3 NMQ
M3 NMT
M3 NMW
M3 NMY
M3 NMZ
M3 NNB
M3 NNC
M3 NND
M3 NNF
M3 NNK
M3 NNL
MW3 NNR
M3 NNX
M3 NOC
M3 NOG
M3 NOK

MD3 NOT
M3 NOU
M3 NPJ
M3 NPL
M3 NPP
M3 NPQ
MM3 NPV
M3 NPY
M3 NQC
M3 NQD
M3 NQF
M3 NQG
M3 NQJ
M3 NQP
M3 NQQ
M3 NQV
MW3 NQW
MW3 NQX
M3 NQZ
M3 NRD
M3 NRO
MW3 NRS
MW3 NRU
M3 NRY
MM3 NRZ
M3 NSE
M3 NSW
M3 NSY
MM3 NTF
M3 NTM
M3 NTO
M3 NTS
M3 NTU
MW3 NUA
M3 NUD
M3 NUF
M3 NUJ
M3 NUK
M3 NUR
MM3 NUS
M3 NUZ
M3 NVB
M3 NVI
M3 NVM
MM3 NVT
M3 NVU
M3 NVZ
M3 NWB
M3 NWG
MM3 NWI
M3 NWM
M3 NWN
M3 NXB
M3 NXG
MI3 NXI
M3 NXL
M3 NXM
M3 NXN
M3 NXP
M3 NXU
M3 NXV
M3 NXW
M3 NXX
M3 NYE
M3 NYJ
M3 NYK
M3 NYL
M3 NYO
MW3 NYR
M3 NYT
M3 NYU
M3 NYV
M3 NZB
M3 NZC
M3 NZD
M3 NZF
M3 NZL
M3 NZP
M3 NZQ
M3 NZS
M3 OAD
M3 OAF
M3 OAN
M3 OAO
M3 OAU
M3 OAV
M3 OBE
M3 OBH
M3 OBP
M3 OBY
M3 OCB
MI3 OCD
M3 OCF
M3 OCH
M3 OCI
M3 OCM
M3 OCT
M3 OCU
M3 ODD
MM3 ODF
M3 ODG
MM3 ODI
M3 ODJ
M3 ODQ
M3 ODS
M3 ODU
MI3 ODZ
M3 OEA
M3 OEK
M3 OEP
MD3 OET
M3 OEU
MM3 OEW
MW3 OEX
MW3 OEZ

M3 OFG
M3 OFK
M3 OFL
M3 OFM
MD3 OFP
M3 OFQ
MM3 OFT
M3 OFY
MW3 OFZ
M3 OGA
MW3 OGE
M3 OGF
M3 OGH
M3 OGI
M3 OGJ
MW3 OGO
M3 OGP
M3 OGQ
M3 OGW
M3 OGX
M3 OGZ
M3 OHA
M3 OHH
M3 OHM
MW3 OHT
M3 OHU
M3 OHV
MM3 OHW
MM3 OID
MM3 OIE
MM3 OIF
M3 OIG
M3 OIH
M3 OIJ
M3 OIO
MW3 OIP
MW3 OIQ
M3 OIR
M3 OIT
M3 OIU
M3 OIW
MW3 OJA
M3 OJB
M3 OJH
M3 OJI
M3 OJM
M3 OJQ
M3 OJT
M3 OJZ
M3 OKD
M3 OKK
M3 OKL
MW3 OKM
M3 OKN
M3 OKQ
M3 OKV
M3 OKW
M3 OLB
M3 OLG
M3 OLH
M3 OLJ
M3 OLL
MM3 OLP
M3 OLR
M3 OLS
M3 OMB
M3 OMJ
M3 OMK
M3 OMM
MD3 OMN
M3 OMO
M3 OMP
M3 OMQ
M3 OMR
M3 OMV
M3 OMW
MW3 ONA
M3 ONJ
M3 ONL
M3 ONQ
M3 ONR
M3 ONS
MW3 ONT
M3 ONU
MM3 ONW
M3 ONZ
M3 OOB
M3 OOJ
M3 OOK
M3 OON
M3 OOO
M3 OOS
M3 OOV
M3 OOW
M3 OPE
M3 OPF
M3 OPH
M3 OPJ
M3 OPQ
M3 OPR
M3 OPY
M3 OQA
M3 OQE
M3 OQF
M3 OQM
M3 OQO
M3 OQP
M3 OQU
M3 OQW
M3 OQX
M3 OQY
M3 ORA
M3 ORF
M3 ORH
M3 ORI
M3 ORJ

M3 ORW
M3 ORX
MM3 OSD
M3 OSH
M3 OSL
M3 OSM
M3 OSR
M3 OSX
MI3 OTC
M3 OTD
M3 OTH
M3 OTK
M3 OTL
M3 OTN
M3 OTO
M3 OTX
M3 OTY
M3 OUA
M3 OUB
MM3 OUD
M3 OUE
M3 OUJ
M3 OUK
M3 OUM
M3 OUO
M3 OUP
M3 OUR
M3 OUW
M3 OUX
M3 OUZ
M3 OVB
M3 OVH
M3 OVI
M3 OVL
M3 OVN
M3 OVP
M3 OVQ
M3 OVS
M3 OVU
M3 OVW
M3 OVY
M3 OWA
M3 OWC
M3 OWD
M3 OWI
M3 OWJ
M3 OWK
M3 OWL
M3 OWM
M3 OWR
M3 OWT
M3 OWV
M3 OWW
M3 OWX
MM3 OWY
M3 OXA
M3 OXF
MM3 OXG
M3 OXI
M3 OXJ
M3 OXL
M3 OXM
M3 OXS
MW3 OXT
M3 OXU
M3 OXW
M3 OXX
M3 OYA
M3 OYF
M3 OYG
M3 OYH
M3 OYI
M3 OYK
M3 OYM
M3 OYO
M3 OYT
M3 OYV
M3 OYX
M3 OYY
M3 OZA
M3 OZL
M3 OZP
M3 OZQ
M3 OZR
M3 OZS
M3 OZV
MI3 OZX
M3 OZZ
M3 PAA
M3 PAB
M3 PAJ
M3 PAK
M3 PAO
M3 PAQ
M3 PBD
M3 PBF
M3 PBH
MM3 PBJ
M3 PBM
M3 PBZ
M3 PCB
MM3 PCM
M3 PCU
M3 PCY
M3 PCZ
M3 PDA
M3 PDI
M3 PDS
M3 PEG
M3 PEI
M3 PEJ
M3 PEK
M3 PEO
M3 PEP
M3 PEW
M3 PEX

M3 PFD
M3 PFG
MM3 PFI
M3 PFO
M3 PFP
M3 PFQ
M3 PFS
M3 PFT
M3 PFV
M3 PFW
M3 PFZ
MW3 PGJ
M3 PGM
M3 PGQ
MI3 PGR
MM3 PGV
M3 PGW
M3 PGX
M3 PGZ
M3 PHB
M3 PHI
M3 PHK
M3 PHN
M3 PHV
M3 PIF
M3 PII
M3 PIJ
M3 PIR
M3 PJA
MI3 PJB
M3 PJD
M3 PJY
M3 PKA
M3 PKD
M3 PKF
M3 PKG
MI3 PKJ
MI3 PKK
M3 PKN
M3 PKV
M3 PKX
M3 PKY
M3 PLF
M3 PLG
M3 PLK
MM3 PLM
M3 PLQ
M3 PLR
MI3 PLY
M3 PMA
MW3 PMB
M3 PMC
MI3 PMD
M3 PMF
M3 PMG
M3 PMJ
M3 PMP
M3 PMQ
M3 PMV
MW3 PND
M3 PNJ
MI3 PNM
M3 PNP
M3 PNS
M3 PNU
MM3 PNX
M3 POA
M3 POG
MI3 POS
M3 POU
MI3 POY
M3 POZ
MW3 PPF
M3 PPJ
M3 PPP
M3 PPS
MM3 PPT
M3 PPV
M3 PQA
M3 PQC
M3 PQD
M3 PQH
M3 PQK
MM3 PQR
M3 PQW
M3 PQX
M3 PRC
M3 PRF
MI3 PRI
M3 PRO
MM3 PRQ
M3 PRT
M3 PRW
MW3 PRX
M3 PSK
M3 PSM
M3 PSP
M3 PSQ
M3 PST
M3 PSX
MM3 PTJ
M3 PTK
MI3 PTW
M3 PTY
M3 PUG
M3 PUK
MM3 PUM
M3 PUS
M3 PUV
M3 PUX
M3 PUY
M3 PVA
M3 PVD
MW3 PVH
M3 PVJ
M3 PVK

M3 PVL
M3 PVM
M3 PVN
M3 PVO
M3 PVS
MM3 PVW
M3 PWA
M3 PWB
M3 PWF
MW3 PWG
M3 PWI
M3 PWJ
M3 PWQ
M3 PWR
M3 PWU
M3 PWV
M3 PWX
M3 PXC
M3 PXD
M3 PXH
MM3 PXI
MM3 PXJ
M3 PXR
M3 PXS
M3 PYA
M3 PYF
M3 PYH
MW3 PYN
M3 PYP
M3 PYU
M3 PYZ
M3 PZA
M3 PZD
M3 PZE
M3 PZG
M3 PZH
M3 PZI
M3 PZM
M3 PZP
M3 PZT
M3 PZU
M3 RAO
M3 RAP
M3 RAX
M3 RBV
M3 RBZ
M3 RCA
MW3 RCF
M3 RCK
M3 RCN
MM3 RCP
M3 RCU
M3 RDE
M3 RDG
M3 RDI
M3 RDN
M3 RDT
M3 RDZ
M3 REE
M3 REF
M3 RER
M3 RES
M3 RFP
M3 RFT
M3 RFU
M3 RFY
M3 RFZ
MM3 RGI
MW3 RGM
M3 RGQ
M3 RGT
M3 RGX
M3 RGY
M3 RHD
M3 RHE
M3 RHM
M3 RHU
M3 RHV
MW3 RHY
M3 RIB
M3 RIH
MM3 RIN
M3 RIR
M3 RIY
M3 RJA
MM3 RJD
MM3 RJQ
MM3 RJR
M3 RJT
MW3 RJV
M3 RJW
M3 RJY
MM3 RKA
M3 RKM
M3 RKP
MM3 RKU
M3 RKX
M3 RLI
MM3 RLL
M3 RLQ
M3 RLR
M3 RLU
M3 RLY
M3 RLZ
M3 RMC
M3 RME
MI3 RMK
MI3 RMM
MI3 RMO
MW3 RMR
M3 RNB
M3 RND
M3 RNE
MM3 RNF
MW3 RNI
M3 RNJ

MW3 RNP
M3 RNR
M3 RNV
M3 ROE
M3 ROH
M3 ROJ
MM3 ROT
M3 RPC
M3 RPG
M3 RPL
M3 RPM
M3 RPN
M3 RPU
MW3 RPY
M3 RQF
M3 RQL
M3 RQS
M3 RQT
M3 RQU
M3 RQZ
MM3 RRD
M3 RRK
M3 RRL
MM3 RRO
M3 RRS
M3 RRX
M3 RSK
M3 RSL
M3 RSP
M3 RSQ
MM3 RSR
M3 RSW
M3 RTF
M3 RTX
M3 RUA
M3 RUD
M3 RUJ
M3 RUN
M3 RUU
MM3 RUX
M3 RVA
MM3 RVF
MM3 RVG
M3 RVL
MW3 RVU
MW3 RVW
M3 RVY
M3 RVZ
M3 RWE
MI3 RWG
M3 RWL
M3 RWT
M3 RWU
M3 RWY
MM3 RXC
M3 RXI
M3 RXR
M3 RXS
M3 RXV
M3 RXX
M3 RYH
M3 RYK
M3 RYL
M3 RYM
M3 RYQ
M3 RYV
M3 RYX
M3 RZH
M3 RZK
M3 RZR
M3 RZS
MW3 SBG
MW3 SBH
MI3 SBI
MW3 SBK
M3 SBL
M3 SBU
M3 SBV
M3 SBZ
M3 SCE
M3 SCK
M3 SCS
M3 SCT
M3 SCY
M3 SCZ
M3 SDM
MW3 SDO
M3 SDT
M3 SEH
M3 SEP
M3 SEQ
M3 SER
M3 SEW
M3 SFB
M3 SFE
M3 SFF
M3 SFH
M3 SFI
M3 SFO
M3 SFQ
M3 SFR
M3 SFU
MM3 SFV
M3 SGC
M3 SGN
M3 SGO
M3 SGY
M3 SHC
MM3 SHD
M3 SHE
M3 SHH
MI3 SHM
MW3 SIA
M3 SIE
M3 SIF
M3 SIH

M3 SIJ
M3 SIU
MM3 SIV
M3 SIW
M3 SJC
M3 SJG
M3 SJZ
MM3 SKE
M3 SKG
M3 SKO
M3 SKS
M3 SKX
M3 SLE
M3 SLR
M3 SLU
M3 SLV
M3 SMO
M3 SMS
M3 SMU
M3 SNC
M3 SND
MM3 SNM
M3 SNP
M3 SNT
M3 SNU
M3 SNV
M3 SOH
M3 SOI
M3 SOJ
MW3 SOK
M3 SOM
M3 SOO
MW3 SOR
M3 SOU
M3 SOX
M3 SPB
M3 SPD
M3 SPF
M3 SPM
M3 SPO
MW3 SPV
MI3 SPW
MW3 SPX
M3 SQB
M3 SQC
M3 SQD
MM3 SQF
M3 SQR
M3 SRE
M3 SRN
M3 SRX
M3 SRZ
M3 SSE
M3 SSN
M3 SSP
MI3 SST
M3 SSY
MM3 STB
M3 STE
M3 STF
M3 STG
MM3 STM
M3 STO
M3 STX
M3 STZ
MW3 SUA
MM3 SUH
M3 SUN
M3 SUQ
M3 SUZ
M3 SVA
M3 SVE
M3 SVG
M3 SVI
M3 SVL
M3 SVQ
M3 SVR
M3 SVS
M3 SVV
M3 SVX
M3 SWL
M3 SWN
M3 SWR
M3 SWX
M3 SXA
M3 SXD
MI3 SXG
M3 SXN
M3 SXO
M3 SYA
M3 SYE
MW3 SYG
M3 SYK
M3 SZH
M3 SZJ
M3 SZL
MM3 SZN
M3 SZP
M3 SZR
MI3 SZU
M3 SZV
MW3 SZW
M3 SZX
M3 SZZ
M3 TAI
M3 TAJ
MI3 TAO
M3 TAQ
M3 TBB
M3 TBD
M3 TBI
M3 TBN
M3 TBR
M3 TCA
M3 TCH
M3 TCK

M3 TCL
M3 TCM
M3 TCO
M3 TCV
M3 TCZ
M3 TDA
MI3 TDF
M3 TDI
MM3 TDN
M3 TDQ
M3 TDV
M3 TDW
M3 TDZ
M3 TEA
M3 TEC
MM3 TEJ
M3 TEO
M3 TER
M3 TEU
M3 TFQ
M3 TFU
M3 TGF
M3 TGG
M3 TGH
M3 TGI
M3 TGN
MW3 TGX
M3 TGY
M3 THG
M3 THL
M3 THT
MM3 THU
M3 THZ
M3 TIH
M3 TIO
MM3 TIP
M3 TIR
M3 TIW
M3 TJA
M3 TJD
M3 TJN
M3 TJP
M3 TJY
M3 TKA
M3 TKB
M3 TKC
M3 TKF
M3 TKL
M3 TKY
M3 TKZ
M3 TLA
MM3 TLE
MM3 TMD
M3 TMH
M3 TMS
MI3 TMV
M3 TNI
M3 TNR
M3 TNU
M3 TOH
M3 TOQ
M3 TOU
M3 TOX
M3 TPC
M3 TPL
MI3 TPP
M3 TPX
M3 TQC
MW3 TQE
M3 TQL
M3 TQQ
M3 TQR
MM3 TQW
M3 TQZ
M3 TRD
M3 TRN
M3 TRQ
M3 TRS
M3 TRU
M3 TSC
M3 TSK
M3 TSL
M3 TSM
M3 TSS
M3 TSU
M3 TTG
M3 TTJ
MM3 TTQ
M3 TTV
M3 TTW
M3 TUE
M3 TUK
M3 TUT
M3 TUV
M3 TVE
M3 TVS
MM3 TVX
M3 TWJ
M3 TWO
M3 TWU
M3 TWX
M3 TWZ
M3 TXB
MM3 TXD
M3 TXE
M3 TXN
M3 TXO
M3 TXX
M3 TXZ
M3 TYD
M3 TYF
M3 TYN
M3 TYY
MM3 TZA
M3 TZD
M3 TZH

M3 TZJ
M3 TZL
MI3 TZM
M3 TZT
M3 TZV
MI3 TZZ
M3 UAB
M3 UAD
MW3 UAF
M3 UAU
M3 UBA
M3 UBI
MM3 UBJ
M3 UBM
M3 UBO
M3 UBP
M3 UBV
M3 UBW
M3 UBZ
M3 UCB
M3 UCG
M3 UCK
M3 UCM
M3 UCQ
M3 UCT
M3 UCW
MW3 UDE
M3 UDH
M3 UDM
M3 UDP
M3 UEM
M3 UEO
M3 UES
M3 UFB
MW3 UFM
MW3 UFN
M3 UFO
M3 UFP
M3 UFR
MW3 UFV
M3 UGN
MI3 UGP
M3 UHA
MI3 UHD
M3 UHE
M3 UHM
M3 UHR
M3 UHU
M3 UHZ
M3 UIH
MW3 UIZ
M3 UJG
M3 UJI
M3 UJK
MM3 UJU
M3 UJW
M3 UKA
M3 UKE
M3 UKI
M3 UKL
M3 UKZ
M3 ULA
M3 ULF
MW3 ULG
M3 ULH
MI3 ULK
M3 ULR
M3 ULV
M3 UME
M3 UMG
M3 UMI
M3 UMK
M3 UMO
M3 UMP
M3 UMS
M3 UMT
MW3 UNA
M3 UNE
MW3 UNM
M3 UNS
M3 UNV
MM3 UNX
M3 UOA
M3 UOH
M3 UOI
M3 UOQ
M3 UOT
M3 UOU
MW3 UPC
M3 UPD
M3 UPE
M3 UPH
M3 UPM
M3 UPR
M3 UPU
M3 UPW
M3 UQS
M3 UQU
MI3 UQV
M3 UQX
M3 URC
MJ3 URI
M3 URJ
M3 URM
M3 URP
M3 URY
M3 USE
M3 USG
M3 USI
MW3 USL
MM3 USO
M3 USY
M3 UTA
M3 UTC
M3 UTD
MW3 UTF

M3 UTG
M3 UTL
M3 UTW
M3 UUB
M3 UUC
M3 UUI
M3 UUM
M3 UUQ
M3 UUU
M3 UVE
M3 UVI
MM3 UVL
M3 UVR
M3 UVZ
M3 UWC
M3 UWG
M3 UWK
M3 UWQ
M3 UXA
M3 UXD
M3 UXJ
M3 UXP
M3 UXT
MM3 UXV
M3 UXW
M3 UYB
M3 UYN
M3 UYR
M3 UZC
M3 UZD
M3 UZG
M3 UZR
M3 UZS
M3 UZY
M3 VAI
M3 VAL
M3 VAO
M3 VAP
M3 VAZ
M3 VBK
M3 VBO
M3 VBS
M3 VBW
M3 VBX
M3 VCH
M3 VCL
M3 VCU
M3 VDJ
M3 VDK
M3 VDR
M3 VDS
M3 VEB
M3 VED
MI3 VEI
M3 VEO
M3 VEU
M3 VEV
M3 VFA
M3 VFD
MM3 VFG
M3 VFH
M3 VFT
M3 VGC
M3 VGI
M3 VGO
M3 VGR
M3 VGU
M3 VGY
M3 VID
M3 VIE
M3 VIF
M3 VIK
M3 VIT
M3 VIX
M3 VIY
M3 VJQ
M3 VJR
M3 VJY
M3 VJZ
M3 VKC
M3 VKE
M3 VKH
M3 VKI
MM3 VKK
M3 VKW
M3 VKZ
M3 VLC
M3 VLK
M3 VLR
M3 VLW
M3 VLX
MM3 VLY
M3 VLZ
M3 VMB
M3 VMC
MW3 VMK
M3 VMO
M3 VMX
M3 VNA
M3 VNB
M3 VND
M3 VNF
M3 VNY
M3 VOD
M3 VOE
MI3 VOF
M3 VOK
M3 VOM
M3 VOO
M3 VOV
M3 VPC
M3 VPF
M3 VPI
MM3 VPU
MM3 VPZ
M3 VQB

MM3 VQO
M3 VQT
M3 VQU
M3 VQW
M3 VRF
M3 VRG
M3 VRH
MI3 VRJ
MM3 VRT
M3 VRW
M3 VSA
M3 VSD
M3 VSM
M3 VSP
M3 VSR
M3 VSV
M3 VSY
M3 VTD
M3 VTI
M3 VTM
M3 VTY
M3 VTZ
M3 VUB
M3 VUE
M3 VUF
M3 VUG
M3 VUI
M3 VUM
MM3 VVG
M3 VVM
M3 VVP
MM3 VWA
MI3 VWB
M3 VWI
M3 VWL
M3 VWN
M3 VWS
M3 VWT
M3 VWU
M3 VWX
M3 VWZ
M3 VXD
M3 VXT
M3 VXU
M3 VXW
M3 VXZ
M3 VYG
M3 VYI
M3 VYL
M3 VYW
M3 VYZ
M3 VZA
M3 VZB
M3 VZE
M3 VZG
M3 VZK
M3 VZM
M3 VZR
M3 VZT
M3 VZW
M3 VZY
M3 WAA
M3 WAE
M3 WAG
M3 WAH
MI3 WAI
M3 WAO
M3 WAQ
M3 WAR
MW3 WAU
M3 WAZ
M3 WBW
MM3 WBX
M3 WCC
M3 WCD
M3 WCF
M3 WCH
M3 WCJ
M3 WCK
M3 WCL
M3 WCT
M3 WCU
M3 WDD
MW3 WDG
M3 WDM
M3 WDQ
M3 WDS
M3 WDW
M3 WDX
M3 WEC
MM3 WED
M3 WEK
M3 WEL
M3 WEN
M3 WEO
M3 WER
M3 WEX
M3 WFD
M3 WFI
M3 WFV
M3 WFW
M3 WFX
M3 WFZ
M3 WGF
M3 WGG
M3 WGH
M3 WGN
M3 WGQ
M3 WGT
M3 WGX
M3 WHD
M3 WHE
M3 WHF
MI3 WHM
M3 WHO
M3 WIB

M3 WIE
M3 WIF
M3 WIS
M3 WIY
M3 WJB
M3 WJC
MI3 WJH
M3 WJJ
MW3 WJQ
M3 WJS
M3 WJX
M3 WKA
M3 WKD
M3 WKQ
MM3 WKS
M3 WKT
MM3 WKU
MI3 WLC
MM3 WLE
M3 WLF
M3 WLH
M3 WLJ
MM3 WLM
M3 WLN
M3 WLP
M3 WLQ
M3 WLR
M3 WLT
M3 WMA
MW3 WMF
M3 WMN
M3 WMT
M3 WMZ
M3 WND
M3 WNL
M3 WNO
M3 WNU
M3 WNW
M3 WNY
M3 WOE
M3 WOH
M3 WOJ
M3 WOM
M3 WON
M3 WOO
M3 WOU
M3 WPD
M3 WPR
M3 WPZ
MI3 WQH
M3 WQJ
M3 WQK
M3 WQP
M3 WQQ
M3 WQU
M3 WQZ
M3 WRL
MW3 WSA
MW3 WSD
MW3 WSG
M3 WST
M3 WSX
M3 WSY
M3 WTJ
M3 WTQ
MW3 WTX
M3 WTZ
M3 WUC
M3 WUF
M3 WUT
M3 WUY
M3 WVA
M3 WVE
M3 WVM
M3 WVS
M3 WVU
M3 WWB
M3 WWE
M3 WWF
MW3 WWI
MW3 WWL
M3 WWT
MI3 WWX
M3 WXK
M3 WXM
M3 WXO
M3 WXQ
M3 WXV
M3 WYD
MW3 WYE
M3 WYH
M3 WYP
M3 WYU
M3 WYX
M3 WZC
MW3 WZD
M3 WZI
MM3 WZK
M3 WZO
M3 WZQ
MD3 WZU
M3 WZX
MI3 XAA
M3 XAD
M3 XAL
M3 XAP
M3 XAQ
M3 XAS
MM3 XBA
M3 XBD
M3 XBI
M3 XBJ
M3 XBK
M3 XBP
M3 XBQ
M3 XBU

M3 XBV
M3 XBW
M3 XCC
M3 XCD
M3 XCG
M3 XCK
MM3 XCL
M3 XCM
M3 XCO
M3 XCQ
M3 XCZ
M3 XDA
M3 XDC
M3 XDN
M3 XDO
MI3 XDR
M3 XDT
M3 XDY
M3 XEC
M3 XED
M3 XEE
M3 XEH
M3 XEM
M3 XEV
M3 XEW
M3 XEZ
M3 XFE
M3 XFF
M3 XFJ
M3 XFL
M3 XFQ
M3 XFU
MW3 XFV
MM3 XFW
M3 XFY
M3 XGE
MM3 XGF
M3 XGJ
MM3 XGK
M3 XGM
M3 XGO
M3 XHA
MW3 XHD
M3 XHG
M3 XHI
M3 XHJ
M3 XHP
M3 XHR
M3 XHS
MI3 XIB
M3 XII
M3 XIR
M3 XIX
M3 XJD
M3 XJM
M3 XJN
M3 XJR
M3 XJU
M3 XJV
MI3 XJY
M3 XKB
M3 XKC
M3 XKK
M3 XKQ
M3 XKR
M3 XKS
M3 XKT
M3 XKV
M3 XKW
M3 XKZ
M3 XLA
M3 XLD
M3 XLF
M3 XLI
M3 XLN
M3 XMF
M3 XMI
M3 XML
MM3 XMM
M3 XMR
M3 XMV
M3 XMX
MM3 XND
M3 XNG
M3 XNI
M3 XNJ
M3 XNS
M3 XNY
M3 XOB
M3 XOF
M3 XOL
M3 XOM
M3 XOO
M3 XOP
MW3 XPA
M3 XPB
M3 XPC
MM3 XPE
M3 XPO
M3 XPQ
M3 XPV
M3 XPZ
M3 XQA
M3 XQD
M3 XQR
M3 XQU
M3 XRA
M3 XRB
M3 XRF
M3 XRL
M3 XRU
MM3 XSB
M3 XSH
M3 XSL
M3 XSO
M3 XSQ

M3 XSS
MM3 XSW
M3 XTB
M3 XTH
M3 XTO
M3 XTQ
M3 XTW
M3 XTX
MI3 XUB
M3 XUK
M3 XUL
M3 XUM
M3 XUQ
M3 XUZ
M3 XVA
M3 XVE
M3 XVN
M3 XVV
M3 XWA
M3 XWD
M3 XWH
M3 XWJ
M3 XWL
M3 XWT
M3 XXC
M3 XXD
M3 XXF
MI3 XXH
M3 XXN
M3 XYD
M3 XYF
M3 XYG
M3 XYL
M3 XYQ
M3 XYR
M3 XYY
M3 XZI
M3 XZL
M3 XZM
MW3 XZO
M3 XZQ
M3 XZU
MW3 XZZ
M3 YAF
M3 YAH
M3 YAR
M3 YAU
MM3 YAY
M3 YAZ
M3 YBE
MM3 YBH
M3 YBO
M3 YBP
M3 YBS
M3 YBV
MM3 YBY
MI3 YBZ
M3 YCA
M3 YCC
M3 YCF
M3 YCH
M3 YCP
MM3 YCY
M3 YDC
M3 YDE
M3 YDG
M3 YDO
MW3 YDR
M3 YEB
M3 YED
MW3 YEF
MM3 YEH
MW3 YEI
M3 YEO
M3 YEV
M3 YEX
M3 YEY
M3 YFC
M3 YFK
M3 YFO
M3 YFP
MM3 YFU
MM3 YGE
M3 YGH
M3 YGN
M3 YGX
M3 YHB
M3 YHE
M3 YHI
MW3 YHX
M3 YIA
M3 YIB
M3 YIN
MM3 YIO
M3 YIP
M3 YIW
M3 YJC
MM3 YJO
M3 YJV
MM3 YJX
M3 YKB
MI3 YKD
MI3 YKG
M3 YKJ
M3 YKQ
M3 YKU
M3 YKX
MM3 YKY
M3 YLC
M3 YLF
MW3 YLI
M3 YLY
M3 YMO
M3 YMP
M3 YMR
M3 YMT

M3 YMZ
M3 YNF
M3 YNR
M3 YNT
M3 YNU
M3 YNV
MM3 YNW
M3 YOA
M3 YOB
MW3 YOD
M3 YOJ
M3 YON
M3 YPC
M3 YPL
M3 YPM
M3 YPO
MW3 YPT
M3 YPV
M3 YPZ
M3 YQB
MM3 YQE
M3 YQF
M3 YQI
M3 YQK
MI3 YQN
MW3 YQS
MW3 YQV
MW3 YQY
M3 YRD
M3 YRG
M3 YRI
M3 YRK
MM3 YRL
MM3 YRN
M3 YRQ
M3 YRU
M3 YRY
M3 YSB
M3 YSG
M3 YSH
M3 YSI
M3 YSK
M3 YSO
M3 YSP
MI3 YSR
M3 YST
M3 YSX
M3 YTD
M3 YTJ
M3 YTK
MW3 YTO
M3 YTT
M3 YTU
M3 YTX
M3 YTY
M3 YUG
M3 YUJ
M3 YUL
MM3 YUO
M3 YUZ
MM3 YVH
M3 YVI
M3 YVL
M3 YVM
MW3 YVO
M3 YVP
MM3 YVQ
M3 YVX
M3 YWA
M3 YWB
M3 YWQ
MM3 YWS
M3 YWV
MM3 YWX
M3 YWY
M3 YXG
M3 YXI
M3 YXO
M3 YXY
M3 YYA
M3 YYC
M3 YYL
MW3 YYP
M3 YYY
M3 YZD
M3 YZF
M3 YZG
M3 YZL
M3 YZT
M3 YZY
M3 ZAD
M3 ZAG
M3 ZAH
M3 ZAJ
M3 ZAX
MM3 ZBA
M3 ZBE
M3 ZBK
M3 ZBN
M3 ZBP
M3 ZBU
M3 ZBY
MM3 ZCK
M3 ZCL
M3 ZCP
M3 ZCV
M3 ZCZ
M3 ZDD
M3 ZDO
M3 ZDP
M3 ZDY
M3 ZDZ
MM3 ZEA
MM3 ZEF
M3 ZEG
M3 ZEK

M3 ZEP
MM3 ZEU
M3 ZEZ
M3 ZFA
M3 ZFF
M3 ZFM
M3 ZFP
MD3 ZFQ
MJ3 ZFT
M3 ZFU
M3 ZFX
M3 ZGB
M3 ZGL
M3 ZGQ
MD3 ZGV
M3 ZGZ
M3 ZHA
M3 ZHK
M3 ZHN
M3 ZHO
M3 ZHR
MW3 ZHT
MM3 ZIB
M3 ZIC
M3 ZIG
MW3 ZIJ
M3 ZIP
M3 ZIQ
M3 ZIW
M3 ZJC
MM3 ZJI
M3 ZJR
M3 ZJU
M3 ZJW
M3 ZJZ
MW3 ZKB
M3 ZKC
M3 ZKG
M3 ZKH
M3 ZKK
MM3 ZKM
MM3 ZKN
M3 ZKV
MW3 ZLB
M3 ZLC
M3 ZLG
M3 ZLH
MW3 ZLM
M3 ZLN
MU3 ZLQ
M3 ZLT
M3 ZMA
M3 ZMF
M3 ZMG
M3 ZMH
M3 ZMK
M3 ZND
M3 ZNE
M3 ZNH
M3 ZNI
MM3 ZOF
M3 ZOS
MM3 ZOT
M3 ZOW
MM3 ZOX
M3 ZPA
M3 ZPF
M3 ZPH
M3 ZPI
M3 ZPK
M3 ZPN
M3 ZPP
M3 ZPQ
MM3 ZPS
M3 ZPU
M3 ZPV
M3 ZPX
M3 ZQD
M3 ZQE
M3 ZQH
M3 ZQI
M3 ZQQ
M3 ZQR
MI3 ZQT
M3 ZQY
MW3 ZQZ
M3 ZRD
M3 ZRI
M3 ZRJ
M3 ZRN
M3 ZRT
M3 ZRU
M3 ZSN
M3 ZSP
MM3 ZSR
M3 ZSS
M3 ZSX
M3 ZSZ
M3 ZTA
M3 ZTE
M3 ZTG
M3 ZTO
M3 ZTQ
M3 ZTY
M3 ZUE
M3 ZUH
M3 ZUI
M3 ZUQ
M3 ZUX
MI3 ZVM
M3 ZVN
M3 ZVO
M3 ZVP
M3 ZVQ
M3 ZVZ
M3 ZWV

M3 ZWZ
M3 ZXJ
M3 ZXM
M3 ZXP
M3 ZXS
M3 ZXV
M3 ZXW
M3 ZYA
M3 ZYB
M3 ZYC
M3 ZYD
M3 ZYH
MM3 ZYJ
M3 ZYL
M3 ZZB
M3 ZZC
M3 ZZZ

## M*5

M5 AAD
M5 AAE
MW5 AAO
M5 AAS
M5 ABP
M5 ABV
M5 ACM
M5 ACU
M5 ACZ
M5 ADU
M5 ADX
M5 ADZ
M5 AED
M5 AEJ
MM5 AEK
M5 AEX
M5 AFD
MI5 AFK
M5 AFO
M5 AFU
M5 AGC
M5 AGH
M5 AGT
M5 AHQ
M5 AHR
M5 AIF
M5 AIL
M5 AIM
M5 AJF
MM5 AJT
MM5 AJV
M5 AKA
M5 ALI
M5 AML
M5 ASF
M5 CDS
M5 CJW
M5 CSM
M5 DBH
M5 DEN
M5 DWI
MW5 EPA
M5 EVT
M5 FET
M5 FLY
M5 FRA
M5 GTI
M5 GUM
M5 GWH
M5 IPX
M5 JAB
M5 JAN
MM5 JDZ
M5 JFS
M5 JSW
M5 JTS
MI5 KAW
M5 LOE
M5 MAN
M5 MKW
M5 MOW
M5 PDL
M5 RJL
MI5 SPY
M5 STC
M5 TAW
MM5 TGW
MW5 UFO
MI5 UTC
MM5 WIG
M5 WMF
M5 WNS
M5 ZZR

## M*6

M6 AAB
M6 AAI
M6 AAN
M6 AAS
M6 AAT
MW6 AAW
M6 ABA
MW6 ABC
M6 ABH
MI6 ABI
M6 ABJ
M6 ABL
MW6 ABU
M6 ABW
M6 ABY
M6 ACG
M6 ACK
M6 ACL
M6 ACN
M6 ACO
M6 ACT

M6 ACU
M6 ADA
M6 ADH
M6 ADK
M6 ADN
MI6 ADU
M6 ADP
MM6 ADR
M6 ADW
M6 AEA
M6 AEH
M6 AEI
M6 AEO
M6 AEP
M6 AEV
M6 AFD
M6 AFE
M6 AFJ
MM6 AFO
MI6 AFP
MM6 AFR
M6 AFT
M6 AFV
M6 AFZ
MD6 AGB
M6 AGC
M6 AGE
MM6 AGM
M6 AGN
M6 AGP
M6 AGU
M6 AGW
M6 AGX
MM6 AHC
M6 AHI
M6 AHM
M6 AIG
MM6 AIX
M6 AIY
MW6 AJD
M6 AJE
M6 AJG
MM6 AJK
M6 AJR
M6 AJS
M6 AJT
M6 AJU
M6 AJV
M6 AJW
M6 AJY
M6 AJZ
M6 AKB
M6 AKD
M6 AKL
M6 AKM
M6 AKR
M6 AKS
M6 AKU
M6 AKX
M6 ALA
M6 ALF
MM6 ALN
M6 ALR
M6 ALS
M6 ALV
M6 AMB
M6 AMH
MW6 AMP
M6 AMQ
M6 AMS
M6 AMU
M6 AMY
M6 AMZ
MM6 ANA
M6 ANC
M6 ANF
M6 ANG
MW6 ANH
M6 ANJ
M6 ANK
MM6 ANL
M6 ANQ
MW6 ANU
M6 ANW
M6 ANZ
M6 AOA
M6 AOC
MM6 AOD
M6 AOI
M6 AOM
M6 AOO
M6 APE
MI6 APF
M6 APH
M6 APJ
M6 APK
M6 APL
M6 APN
M6 APP
M6 APQ
M6 APT
M6 APV
M6 APX
M6 APY
M6 APZ
M6 AQB
M6 AQC
MW6 AQF
M6 AQS
MW6 AQX
M6 AQZ
M6 ARA
M6 ARE
MM6 ARG
MI6 ARZ
M6 ASA

M6 ASB
M6 ASG
M6 ASR
M6 ATE
M6 ATN
M6 ATU
M6 ATT
M6 AUN
M6 AUR
M6 AUS
M6 AUT
M6 AUV
M6 AUW
M6 AUY
MW6 AVB
M6 AVC
M6 AVI
M6 AVP
M6 AWF
M6 AWH
M6 AWK
M6 AWM
M6 AWT
MW6 AWX
M6 AXA
M6 AXE
M6 AXK
M6 AXN
MW6 AXO
M6 AXQ
MW6 AXV
M6 AXY
M6 AXZ
M6 AYB
M6 AYF
MD6 AYO
MM6 AYT
MW6 AYX
M6 AZB
M6 AZI
M6 AZJ
M6 AZL
M6 AZW
M6 AZZ
MM6 BAE
M6 BAF
M6 BAL
MM6 BAO
M6 BAP
M6 BAT
M6 BBI
M6 BBN
M6 BBV
M6 BBW
M6 BBY
M6 BCA
M6 BCD
M6 BCE
M6 BCI
M6 BCL
M6 BCM
M6 BCO
M6 BCP
M6 BCR
M6 BCS
M6 BCT
M6 BCX
M6 BDB
M6 BDC
M6 BDE
M6 BDF
MW6 BDH
MW6 BDK
M6 BDN
M6 BDP
M6 BDT
M6 BDW
M6 BDY
M6 BDZ
M6 BEA
M6 BEO
M6 BEP
MM6 BER
M6 BET
M6 BEV
M6 BEW
M6 BEY
M6 BFA
M6 BFD
M6 BFE
M6 BFT
MI6 BFW
M6 BFY
M6 BGN
M6 BGQ
M6 BGU
M6 BGX
M6 BGY
M6 BHG
M6 BHL
M6 BHR
M6 BHV
M6 BHW
M6 BHX
M6 BIC
M6 BIK
MM6 BIL
M6 BIN
M6 BIQ
MI6 BIW
M6 BIZ
M6 BJD
M6 BJH
M6 BJR
M6 BJS
M6 BJU

M6 BJV
M6 BKB
M6 BKG
M6 BKW
M6 BKX
M6 BKY
M6 BKZ
M6 BLD
MJ6 BLJ
MM6 BLL
M6 BLR
M6 BLU
M6 BMA
MM6 BMD
M6 BMG
M6 BMH
M6 BML
M6 BMS
M6 BMU
M6 BMX
MI6 BND
MW6 BNE
MI6 BNI
M6 BNR
M6 BNZ
M6 BOH
M6 BON
M6 BOO
M6 BOU
M6 BPF
M6 BPN
M6 BPP
M6 BPT
M6 BPZ
M6 BQI
M6 BQT
M6 BQU
M6 BQX
M6 BQY
M6 BRA
M6 BRC
M6 BRD
M6 BRE
M6 BRG
M6 BRM
M6 BRW
MM6 BRX
M6 BSD
M6 BSE
M6 BSF
M6 BSH
M6 BSJ
MW6 BSM
M6 BSN
M6 BTD
M6 BTI
MI6 BTJ
M6 BTL
M6 BTO
MI6 BTU
M6 BTV
M6 BTW
M6 BTX
M6 BTZ
M6 BUK
M6 BUN
M6 BUO
MW6 BUP
M6 BUQ
M6 BUT
MW6 BUV
MW6 BUZ
M6 BVA
M6 BVB
M6 BVF
MI6 BVH
M6 BVL
M6 BVO
MW6 BVT
M6 BVY
M6 BWA
M6 BWD
M6 BWI
M6 BWM
MM6 BWS
M6 BWT
M6 BWX
M6 BXC
M6 BXE
M6 BXG
MM6 BYD
M6 BYC
M6 BYG
M6 BYI
M6 BYW
M6 BZJ
M6 BZL
M6 BZO
M6 BZR
M6 BZS
M6 BZZ
M6 CAB
M6 CAC
M6 CAL
M6 CAM
M6 CAO
MM6 CAX
M6 CBB
M6 CBK
M6 CBR
M6 CBS
MM6 CBT
M6 CBV
M6 CBX
MM6 CBZ
M6 CCC

M6 CCD
MM6 CCK
M6 CCL
M6 CCP
MI6 CCT
M6 CDA
MW6 CDF
M6 CDI
M6 CDM
M6 CDP
M6 CDS
M6 CDX
M6 CEC
M6 CEK
M6 CER
M6 CEV
M6 CFB
M6 CFC
MM6 CFF
M6 CFG
M6 CFJ
M6 CFK
M6 CFM
M6 CFP
M6 CGA
M6 CGE
MI6 CGG
MI6 CGI
M6 CGN
M6 CGP
M6 CGR
M6 CGU
M6 CHA
M6 CHE
M6 CHK
M6 CHR
M6 CIB
M6 CIC
M6 CID
M6 CIH
M6 CII
M6 CIL
M6 CIN
M6 CIO
M6 CIR
MI6 CIT
MI6 CIV
M6 CJB
M6 CJO
M6 CKA
MI6 CKB
M6 CKG
M6 CKK
M6 CKN
M6 CKS
MM6 CKW
M6 CKX
M6 CLD
M6 CLE
M6 CLH
M6 CLN
M6 CLR
M6 CLS
M6 CLU
M6 CMA
M6 CMD
M6 CME
MI6 CMH
M6 CMJ
M6 CMN
M6 CMP
M6 CMS
M6 CMV
MM6 CMX
M6 CNB
M6 CNE
M6 CNF
M6 CNH
M6 CNJ
M6 CNM
MJ6 CNU
M6 CNW
M6 COA
M6 COG
M6 COJ
M6 COO
M6 COQ
M6 COR
M6 COS
M6 COT
M6 COZ
M6 CPD
M6 CPF
MI6 CPG
M6 CPI
M6 CPL
M6 CPY
M6 CQA
M6 CQC
M6 CQI
M6 CQM
M6 CQP
M6 CQQ
M6 CQT
M6 CQW
M6 CQZ
M6 CRB
M6 CRD
M6 CRK
M6 CRL
M6 CRM
M6 CRN
MW6 CRS
MI6 CRV
M6 CRY
M6 CSB

M6 CSD
M6 CSH
M6 CSQ
M6 CST
M6 CSV
M6 CSY
M6 CTF
M6 CTK
M6 CTM
M6 CTP
M6 CTR
MW6 CUB
M6 CUG
M6 CUH
M6 CUI
M6 CUK
M6 CUO
M6 CUT
M6 CUY
M6 CVA
M6 CVD
M6 CVG
M6 CVI
M6 CVL
M6 CVR
M6 CVS
M6 CVT
M6 CVX
M6 CVZ
M6 CWE
M6 CWM
MI6 CWS
M6 CWT
MI6 CWY
M6 CXE
M6 CXG
M6 CXK
M6 CXL
M6 CXR
M6 CXT
MW6 CXX
M6 CXY
M6 CYH
M6 CYI
M6 CYJ
MW6 CYK
M6 CYZ
M6 CZE
M6 CZG
MW6 CZM
M6 CZP
M6 CZQ
M6 CZT
M6 CZX
M6 CZY
M6 CZZ
M6 DAA
M6 DAH
MM6 DAJ
M6 DAV
M6 DBC
M6 DBM
M6 DBO
M6 DBR
M6 DBY
MW6 DBZ
M6 DCC
MW6 DCF
M6 DCG
M6 DCV
M6 DCX
M6 DDA
M6 DDG
MM6 DDJ
M6 DDK
M6 DEB
M6 DEH
M6 DEN
M6 DEQ
M6 DET
M6 DEU
M6 DEW
M6 DEX
M6 DFB
M6 DFF
M6 DFN
M6 DFO
M6 DFR
M6 DFV
M6 DGD
M6 DGE
M6 DGF
MW6 DGL
M6 DGT
M6 DGV
M6 DGX
MM6 DHA
M6 DHC
M6 DHD
M6 DHF
M6 DHH
M6 DHJ
M6 DHL
M6 DHM
MW6 DHN
M6 DHO
M6 DHP
M6 DHQ
M6 DHR
MD6 DHY
M6 DIA
MW6 DID
M6 DIN
M6 DIP
M6 DIR

MM6 DIS
M6 DIW
MM6 DIX
MI6 DIY
M6 DJB
M6 DJD
M6 DJE
M6 DJF
M6 DJH
M6 DJM
M6 DJO
M6 DJR
M6 DJU
M6 DJV
M6 DKB
M6 DKD
MM6 DKE
M6 DKG
M6 DKH
M6 DKJ
M6 DKP
MI6 DKQ
M6 DKV
M6 DKZ
M6 DLE
M6 DLF
M6 DLG
M6 DLK
M6 DLN
M6 DLR
M6 DLS
MW6 DLZ
M6 DMI
M6 DMJ
MM6 DML
M6 DMM
M6 DMR
M6 DMT
M6 DMU
M6 DMV
M6 DMX
M6 DMZ
M6 DNA
MI6 DND
M6 DNE
M6 DNH
M6 DNK
M6 DNT
M6 DNU
M6 DNW
M6 DNY
M6 DOE
MI6 DOH
MI6 DOI
M6 DOK
M6 DOL
M6 DOO
M6 DOP
MW6 DOQ
MM6 DOS
MI6 DOT
MW6 DOU
M6 DOV
M6 DOY
M6 DPB
MW6 DPD
M6 DPE
M6 DPK
M6 DPM
M6 DPS
M6 DPT
M6 DQA
M6 DQB
M6 DQE
M6 DQI
M6 DQL
M6 DQM
M6 DQX
MM6 DRA
M6 DRB
M6 DRC
M6 DRD
M6 DRE
M6 DRL
M6 DRN
M6 DRQ
M6 DRS
M6 DRX
M6 DRY
M6 DSA
M6 DSD
M6 DSF
M6 DSG
M6 DSI
M6 DSK
M6 DSL
M6 DSM
MM6 DSX
M6 DSZ
M6 DTB
M6 DTI
M6 DTK
M6 DTW
M6 DTZ
M6 DUA
M6 DUC
M6 DUQ
M6 DUS
M6 DUU
M6 DVB
M6 DVD
M6 DVG
M6 DVI
MM6 DVJ
M6 DVK
M6 DVL

M6 DWB
M6 DWD
M6 DWH
M6 DWK
M6 DWL
M6 DWN
M6 DWQ
M6 DWR
M6 DWU
M6 DWX
M6 DXB
M6 DXD
M6 DXE
M6 DXM
M6 DXP
M6 DXR
MI6 DXT
M6 DXU
M6 DXX
M6 DXY
M6 DYE
M6 DYJ
M6 DYK
M6 DYQ
M6 DYS
M6 DYT
M6 DYZ
M6 DZD
M6 DZE
M6 DZG
M6 DZI
M6 DZJ
M6 DZK
M6 DZL
M6 DZM
MW6 DZO
M6 DZW
MM6 EAB
M6 EAM
M6 EAQ
M6 EAR
M6 EAU
M6 EAV
M6 EAY
M6 EBF
M6 EBH
M6 EBK
M6 EBM
M6 EBT
M6 EBV
M6 ECA
M6 ECF
M6 ECG
M6 ECI
M6 ECK
M6 ECL
M6 ECN
MM6 ECO
M6 ECQ
M6 ECS
M6 ECU
M6 ECY
M6 EDE
M6 EDF
M6 EDG
M6 EDT
M6 EDU
MM6 EDX
M6 EEE
MW6 EEI
M6 EEK
M6 EES
M6 EEV
M6 EEY
M6 EFA
M6 EFN
M6 EFO
M6 EFQ
M6 EFT
M6 EFZ
M6 EGI
M6 EGQ
M6 EGR
M6 EGY
MM6 EHG
M6 EHJ
M6 EHN
M6 EHS
M6 EHT
M6 EHZ
M6 EIQ
M6 EIT
M6 EIV
M6 EJD
M6 EJE
M6 EJH
M6 EJN
M6 EJQ
MI6 EJZ
M6 EKG
M6 EKH
M6 EKJ
M6 EKR
M6 EKU
M6 EKZ
M6 ELA
M6 ELG
M6 ELK
M6 ELM
M6 ELO
M6 ELP
M6 ELT
M6 EMB
MM6 EMD
M6 EMN



**IMPORTANT NOTE**

**Revalidate licence to avoid revocation** – Ofcom has advised the Society that plans will be drawn up to revoke licences that have not been revalidated as required by the licence conditions. The quickest way to revalidate is to do so online via the Ofcom website: *https://services.ofcom.org.uk/* or by email: *amateur.validations@ofcom.org.uk* If you need assistance in the process, Ofcom staff are available to help, but please be patient during times of heavy workload.

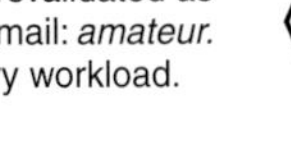

Withheld

MW6 EMR
M6 EMU
M6 ENG
M6 ENO
M6 ENT
M6 EOD
M6 EOF
MW6 EOG
M6 EOH
MW6 EOI
M6 EOJ
M6 EOL
MW6 EOO
M6 EOV
M6 EPB
M6 EPC
M6 EPG
M6 EPI
M6 EPK
M6 EPP
M6 EPR
M6 EPU
MM6 EPY
M6 EQP
M6 EQU
M6 EQX
M6 ERB
M6 ERE
M6 ERG
M6 ERJ
M6 ERK
M6 ERM
M6 ERV
M6 ERX
M6 ERY
M6 ESB
M6 ESC
M6 ESD
M6 ESE
M6 ESF
M6 ESG
M6 ESI
M6 ESK
M6 ESM
M6 ESN
M6 ESQ
M6 ESU
M6 ESX
M6 ETB
M6 ETF
M6 ETI
M6 ETK
M6 ETM
M6 ETO
M6 ETQ
M6 ETT
M6 ETV
M6 ETX
M6 EUD
M6 EUF
M6 EUG
M6 EUJ
M6 EUM
M6 EUN
M6 EUQ
M6 EUR
M6 EUS
M6 EUT
MI6 EVG
MI6 EVJ
MM6 EVN
M6 EVP
M6 EVQ
M6 EVT
M6 EVV
M6 EWB
M6 EWE
M6 EWF
M6 EWG
M6 EWM
M6 EWS
M6 EXM
M6 EYC
M6 EYE
M6 EYN
M6 EYQ
M6 EZB
M6 EZD
M6 EZH
M6 EZI
M6 EZJ
MD6 EZL
MM6 EZQ
M6 EZU
M6 EZW
M6 FAA
M6 FAD
M6 FAI
M6 FAL
M6 FAO
MW6 FAP
M6 FAQ
M6 FAR
MW6 FAT
M6 FAV
M6 FAZ
MW6 FBC
M6 FBG
M6 FBI
MI6 FBL
M6 FBX
M6 FCK
M6 FCL
M6 FCO
M6 FCP
M6 FCU

M6 FCY
M6 FDA
M6 FDB
M6 FDC
M6 FDE
M6 FDJ
M6 FDM
MM6 FDO
M6 FDP
M6 FDQ
M6 FDV
M6 FEG
MM6 FES
M6 FET
M6 FEV
M6 FEW
M6 FEY
M6 FFA
M6 FFC
M6 FFG
M6 FFH
M6 FFL
MW6 FFX
M6 FGB
MW6 FGD
M6 FGF
MM6 FGL
M6 FGO
M6 FGP
M6 FGT
M6 FGU
M6 FGV
M6 FGX
M6 FHA
M6 FHC
MI6 FHL
M6 FHT
M6 FHU
M6 FHW
MI6 FHZ
M6 FIE
M6 FIH
MW6 FIM
M6 FIP
M6 FIQ
MD6 FIS
M6 FJB
MI6 FJG
M6 FJK
M6 FJP
M6 FJR
M6 FJY
M6 FKK
M6 FKL
MM6 FKM
M6 FKT
M6 FKX
M6 FLD
M6 FLI
M6 FLM
M6 FLP
M6 FLT
M6 FLV
MM6 FMA
M6 FMB
MW6 FMD
M6 FMH
M6 FMK
M6 FMQ
M6 FMR
MW6 FMX
MW6 FNC
M6 FND
M6 FNH
M6 FNI
M6 FNJ
M6 FNN
M6 FNS
M6 FNW
M6 FOB
M6 FOH
M6 FOI
M6 FOM
M6 FON
M6 FOP
M6 FOQ
M6 FOS
MM6 FOU
M6 FOV
M6 FPA
M6 FPB
MW6 FPH
M6 FPR
M6 FPS
M6 FPT
M6 FPZ
MM6 FQB
M6 FQI
M6 FQQ
M6 FQS
M6 FQT
M6 FQV
M6 FRA
M6 FRB
M6 FRH
M6 FRL
M6 FRM
M6 FRN
MM6 FRU
M6 FRX
M6 FSF
M6 FSH
M6 FSK
MM6 FSL
MM6 FSS
M6 FSX

MM6 FSY
M6 FTB
M6 FTD
M6 FTE
M6 FTH
M6 FTJ
M6 FTL
M6 FTM
M6 FTR
MM6 FTS
M6 FUB
MW6 FUG
M6 FUI
M6 FUP
M6 FUS
M6 FUV
M6 FUY
M6 FUZ
MI6 FVF
M6 FVG
M6 FVI
M6 FVQ
M6 FVU
M6 FVX
M6 FWA
M6 FWF
M6 FWH
M6 FWI
M6 FWJ
M6 FWK
M6 FWN
M6 FWR
M6 FWU
M6 FXC
M6 FXF
M6 FXH
M6 FXK
MM6 FXN
MM6 FXP
M6 FXR
M6 FXS
M6 FXT
M6 FXV
M6 FYC
M6 FYI
M6 FYO
M6 FYS
M6 FYU
M6 FYX
M6 FYZ
M6 FZA
M6 FZH
M6 FZI
M6 FZL
M6 FZO
M6 FZP
M6 FZU
M6 GAJ
M6 GAY
M6 GAZ
M6 GBT
M6 GCA
M6 GCF
M6 GCG
M6 GCH
M6 GCR
MJ6 GCU
M6 GCZ
M6 GDF
M6 GDG
M6 GDK
M6 GDM
M6 GDR
M6 GDU
M6 GDW
M6 GDX
M6 GEB
M6 GEC
M6 GED
M6 GEH
MM6 GEI
M6 GEM
MW6 GEN
M6 GEO
MI6 GER
M6 GES
M6 GEZ
M6 GFB
M6 GFC
M6 GFD
M6 GFG
M6 GFI
M6 GFJ
M6 GFL
M6 GFT
MM6 GFU
M6 GFX
M6 GGA
MM6 GGB
M6 GGC
M6 GGK
MM6 GGP
M6 GGR
M6 GGY
M6 GHE
M6 GHG
M6 GHK
M6 GHL
M6 GHO
M6 GHU
MI6 GHZ
M6 GIE
M6 GIG
M6 GIJ
M6 GIK
M6 GIO

M6 GIR
M6 GJF
MI6 GJK
M6 GJR
M6 GJU
MM6 GJW
M6 GJZ
M6 GKA
MM6 GKD
M6 GKJ
M6 GKL
M6 GKO
M6 GKP
M6 GKR
M6 GKY
M6 GKZ
M6 GLA
M6 GLB
M6 GLE
M6 GLF
M6 GLG
M6 GLL
M6 GLQ
M6 GLR
M6 GLT
MI6 GLY
M6 GMB
M6 GME
MI6 GMG
M6 GMH
M6 GML
M6 GMN
M6 GMW
M6 GMX
M6 GMZ
M6 GNB
M6 GNE
M6 GNI
M6 GNK
M6 GNL
M6 GNQ
MI6 GNT
M6 GOH
MW6 GOJ
M6 GOK
M6 GOP
M6 GOT
M6 GOV
M6 GPH
M6 GPW
M6 GPY
M6 GQE
M6 GQL
M6 GQM
M6 GQN
MM6 GQO
M6 GQQ
M6 GRE
MM6 GRG
M6 GRI
M6 GRL
M6 GRM
M6 GRS
M6 GRW
M6 GSA
M6 GSB
M6 GSE
M6 GSI
MM6 GSM
M6 GSN
M6 GSU
M6 GSV
M6 GTA
M6 GTF
M6 GTG
MM6 GTJ
M6 GTL
MI6 GTN
M6 GTP
M6 GTX
M6 GTZ
M6 GUD
M6 GUG
M6 GUI
M6 GUK
M6 GUL
M6 GUN
M6 GUP
M6 GUQ
M6 GUR
M6 GVB
M6 GVE
M6 GVG
M6 GVH
M6 GVK
M6 GVO
M6 GVQ
M6 GVV
M6 GVX
M6 GVZ
MW6 GWP
MW6 GWR
M6 GWV
MW6 GWY
MM6 GXB
MM6 GXM
M6 GXN
M6 GXQ
M6 GXT
M6 GXX
M6 GXY
M6 GYJ
M6 GYN
M6 GYO
M6 GYW
M6 GYX

M6 GZC
M6 GZH
MM6 GZM
M6 GZP
M6 GZQ
M6 GZV
M6 HAA
M6 HAB
M6 HAI
M6 HAJ
M6 HAL
M6 HAN
MM6 HAO
M6 HAP
M6 HAQ
M6 HAW
M6 HAX
M6 HAZ
M6 HBA
M6 HBL
MW6 HBO
M6 HBW
M6 HBZ
M6 HCC
M6 HCF
MM6 HCH
M6 HCN
M6 HCX
M6 HDF
M6 HDJ
M6 HDL
M6 HDN
M6 HDQ
MW6 HDR
M6 HEC
M6 HEL
M6 HEM
M6 HES
M6 HEU
M6 HEV
M6 HFD
M6 HFH
M6 HFL
M6 HFS
M6 HFW
M6 HGB
M6 HGC
M6 HGK
M6 HGL
M6 HGP
M6 HGT
M6 HHG
M6 HHI
M6 HHS
M6 HIC
M6 HIE
M6 HIF
M6 HIH
MM6 HIN
M6 HIO
M6 HIQ
M6 HIR
MM6 HIS
M6 HIW
M6 HIX
M6 HIY
M6 HJJ
M6 HJL
M6 HJN
M6 HJO
M6 HJP
MW6 HJW
M6 HJY
M6 HKD
M6 HKE
M6 HKF
M6 HKG
M6 HKH
M6 HKM
MI6 HKP
MM6 HKR
M6 HKV
M6 HLB
M6 HLD
M6 HLH
M6 HLI
M6 HLJ
M6 HLK
M6 HLM
M6 HLN
MM6 HLO
MM6 HLV
MW6 HLX
M6 HMA
M6 HMF
M6 HMH
MW6 HMI
M6 HMJ
M6 HMN
M6 HMO
MI6 HMT
M6 HMX
M6 HMY
M6 HNE
M6 HNK
M6 HNL
M6 HNM
M6 HNP
M6 HNR
M6 HNU
MM6 HNY
M6 HOB
M6 HOC
M6 HOD
MW6 HOE
MM6 HOJ

M6 HOL
M6 HOW
M6 HOX
M6 HOZ
M6 HPA
M6 HPB
M6 HPE
M6 HPI
M6 HPJ
M6 HPM
M6 HPO
M6 HPP
M6 HPQ
MM6 HPU
M6 HPW
MI6 HQA
M6 HQF
MM6 HQG
M6 HQH
MM6 HQJ
M6 HQP
M6 HQQ
MI6 HQS
M6 HQT
MI6 HQU
M6 HQV
MM6 HQY
MI6 HRA
MW6 HRB
M6 HRH
M6 HRP
M6 HRU
M6 HRY
MM6 HSD
M6 HSJ
M6 HSN
M6 HSO
M6 HSU
M6 HSW
M6 HSZ
M6 HTA
M6 HTD
M6 HTE
M6 HTL
MW6 HTO
MM6 HTP
M6 HTR
M6 HTV
MI6 HTY
M6 HUA
M6 HUB
M6 HUD
M6 HUG
M6 HUJ
M6 HUO
M6 HUU
M6 HVB
MW6 HVC
M6 HVG
M6 HVJ
M6 HVQ
MM6 HVT
M6 HVV
M6 HVW
M6 HVZ
M6 HWA
M6 HWB
M6 HWF
M6 HWI
M6 HWK
M6 HWP
M6 HWR
MW6 HWS
M6 HWU
M6 HWY
M6 HWZ
M6 HXH
M6 HXJ
M6 HXL
M6 HXM
M6 HXP
M6 HXQ
M6 HYC
M6 HYK
M6 HYO
M6 HYP
M6 HZS
MM6 IAD
MM6 IAE
M6 IAH
M6 IAM
M6 IAP
M6 IAU
M6 IBA
M6 IBE
MM6 IBK
M6 IBM
M6 IBN
M6 IBS
M6 IBT
M6 IBU
M6 IBW
M6 IBX
M6 IBZ
M6 ICC
M6 ICE
M6 ICN
M6 ICS
MM6 ICT
M6 ICW
M6 ICY
M6 IDA
M6 IDH
M6 IDL
M6 IDT
MI6 IED

MM6 IEE
M6 IEG
M6 IES
M6 IET
M6 IFM
M6 IFR
MW6 IGB
MI6 IGD
MI6 IGL
M6 IGN
M6 IHB
MM6 IHI
M6 IIC
M6 III
M6 IIT
M6 IJA
M6 IJS
M6 IKO
M6 ILA
M6 ILD
MW6 ILE
MM6 ILL
M6 ILV
M6 ILW
M6 IMD
M6 IMG
M6 IMJ
M6 IMO
MD6 IMX
M6 INE
M6 INK
M6 INO
M6 INZ
M6 IOC
M6 IOK
M6 IOP
MM6 IOT
M6 IOU
M6 IPC
M6 IPD
MM6 IPO
M6 IPR
M6 IPS
M6 IPT
M6 IPV
M6 IPX
M6 IPZ
M6 IRG
M6 IRN
MM6 IRQ
M6 IRR
M6 IRS
M6 IRT
M6 IRV
M6 IRX
MW6 ISC
M6 ISL
M6 ISO
M6 ISS
M6 IST
MM6 ISY
M6 ITA
MM6 ITB
M6 ITE
M6 ITF
MM6 ITO
M6 ITT
M6 IUZ
M6 IVY
M6 IWC
M6 IWZ
M6 IXX
M6 IYO
M6 IZD
M6 IZS
M6 IZZ
M6 JAA
MW6 JAC
MM6 JAH
M6 JAI
M6 JAO
M6 JAT
M6 JAX
M6 JAY
M6 JBD
M6 JBF
M6 JBJ
M6 JBK
M6 JBP
M6 JBS
M6 JBW
M6 JBY
M6 JCI
M6 JCK
M6 JCO
M6 JCY
M6 JDB
M6 JDE
M6 JDG
M6 JDH
MD6 JDI
MM6 JDJ
MM6 JDO
M6 JDP
M6 JDR
M6 JDS
M6 JDT
M6 JDW
M6 JDZ
MM6 JEE
M6 JEF
M6 JEI
M6 JEN
M6 JER
MM6 JES
M6 JEU

M6 JEX
M6 JFD
M6 JFM
M6 JFQ
MI6 JFT
M6 JFX
M6 JFZ
M6 JGA
M6 JGB
M6 JGD
M6 JGF
MM6 JGL
M6 JGO
M6 JGW
M6 JGX
M6 JHA
MW6 JHM
M6 JHW
M6 JHX
M6 JHY
M6 JIA
M6 JIB
M6 JIJ
M6 JIM
M6 JIO
M6 JIP
M6 JIV
M6 JJA
M6 JJC
M6 JJP
M6 JJR
M6 JJT
M6 JKC
M6 JKE
M6 JKF
MW6 JKG
M6 JKI
M6 JKK
M6 JKL
MM6 JKO
M6 JKP
M6 JKZ
M6 JLC
M6 JLF
M6 JLG
M6 JLI
M6 JLK
M6 JLN
M6 JLO
M6 JLP
M6 JLS
M6 JMA
M6 JME
M6 JMK
M6 JMS
MI6 JMT
MI6 JMU
M6 JMY
M6 JMZ
M6 JNK
M6 JNT
M6 JNY
M6 JOB
M6 JOC
M6 JOF
M6 JOJ
M6 JOL
M6 JOM
M6 JOQ
M6 JOT
M6 JPA
M6 JPE
M6 JPG
M6 JPJ
M6 JPN
M6 JPY
M6 JQB
MW6 JQM
M6 JQN
M6 JRA
M6 JRD
M6 JRG
M6 JRM
M6 JRN
MM6 JRP
M6 JRQ
M6 JRR
M6 JRV
M6 JRX
M6 JRY
M6 JRZ
M6 JSA
M6 JSB
M6 JSC
M6 JSD
M6 JSE
M6 JSF
M6 JSG
M6 JSK
MW6 JSL
M6 JSO
M6 JSY
M6 JSZ
M6 JTB
M6 JTF
M6 JTJ
M6 JTS
M6 JTT
M6 JTX
M6 JTY
M6 JUA
M6 JUD
MW6 JUL
M6 JUN
M6 JUR
MW6 JUS

M6 JVH
M6 JVN
M6 JVR
MM6 JWA
MI6 JWB
MM6 JWM
M6 JWR
M6 JWS
M6 JWZ
M6 JXA
M6 JXB
M6 JXC
M6 JXK
M6 JXL
MM6 JXM
M6 JXP
M6 JXR
M6 JXV
M6 JXW
M6 JXZ
M6 JYA
M6 JYC
M6 JYK
MM6 JYM
M6 JZP
M6 JZX
M6 KAD
M6 KAI
MM6 KAJ
MI6 KAK
M6 KAL
M6 KAP
MW6 KAW
M6 KAY
M6 KBL
M6 KBM
M6 KBQ
M6 KBR
MM6 KBU
MI6 KBZ
M6 KCH
M6 KCN
M6 KCW
M6 KCX
MI6 KDE
M6 KDG
MW6 KDH
M6 KDI
M6 KDL
M6 KDN
M6 KDS
M6 KDT
M6 KDU
MI6 KDY
M6 KEI
M6 KEJ
M6 KEK
M6 KEN
MM6 KEO
MW6 KEW
M6 KEX
M6 KEY
M6 KEZ
M6 KFD
M6 KFL
M6 KFO
M6 KFT
M6 KFY
M6 KGG
M6 KGH
MM6 KGT
M6 KGW
MW6 KGZ
M6 KHC
M6 KHI
M6 KHO
M6 KHT
M6 KHW
MW6 KHZ
M6 KIB
M6 KIC
M6 KID
M6 KIM
M6 KIS
M6 KJC
M6 KJG
M6 KJH
M6 KJN
M6 KJO
M6 KJS
M6 KJT
M6 KJV
M6 KKC
M6 KKE
M6 KKH
MI6 KKN
M6 KKP
M6 KKS
M6 KKT
M6 KKZ
MM6 KLB
M6 KLE
M6 KLG
M6 KLJ
M6 KLO
M6 KLS
M6 KLY
M6 KMB
M6 KME
M6 KMH
M6 KMM
MM6 KMO
M6 KMT
M6 KMX
MW6 KNA
MM6 KNE

M6 KNH
M6 KNI
MM6 KNR
M6 KNX
M6 KNZ
M6 KOF
M6 KOG
M6 KOL
M6 KOO
M6 KOP
M6 KOR
M6 KOT
M6 KPB
M6 KPH
M6 KPM
M6 KPS
M6 KPY
M6 KRA
M6 KRB
M6 KRC
M6 KRG
M6 KRJ
M6 KRN
M6 KRO
M6 KRS
M6 KRT
M6 KRU
M6 KRY
M6 KSB
M6 KSK
M6 KSN
M6 KSR
M6 KTE
M6 KTF
M6 KTG
MM6 KTL
M6 KTR
M6 KUS
M6 KVC
M6 KVR
M6 KVY
M6 KWB
M6 KWD
M6 KWE
MM6 KWF
M6 KWO
M6 KWR
M6 KWS
MI6 KXT
M6 KXZ
M6 KYA
M6 KYD
M6 KYH
M6 KYM
M6 KYR
M6 KYT
M6 KYY
MM6 KZA
M6 KZO
M6 KZT
M6 KZY
MI6 LAJ
M6 LAN
MM6 LAT
M6 LAU
M6 LAW
M6 LAX
M6 LBA
M6 LBB
M6 LBF
M6 LBG
M6 LBH
M6 LBJ
M6 LBO
M6 LBZ
M6 LCA
M6 LCB
M6 LCE
MM6 LCJ
M6 LCM
M6 LCS
M6 LCZ
M6 LDA
MM6 LDC
M6 LDI
M6 LDO
M6 LDP
M6 LDT
M6 LDX
M6 LED
M6 LEI
M6 LEL
M6 LEM
M6 LEN
M6 LEO
M6 LER
M6 LES
M6 LEZ
M6 LFA
M6 LFF
M6 LFP
M6 LFV
M6 LGE
M6 LGG
M6 LGK
M6 LGN
M6 LGO
MM6 LGP
M6 LGU
M6 LGW
M6 LGY
M6 LGZ
M6 LHC
M6 LHK
M6 LHT
M6 LIA

M6 LIX
M6 LJC
M6 LJH
MM6 LJQ
M6 LJW
M6 LKB
M6 LKC
MM6 LKG
M6 LKH
M6 LKQ
M6 LKV
M6 LLA
M6 LLJ
M6 LLP
MM6 LLR
MW6 LLT
M6 LLX
M6 LMA
M6 LMD
M6 LMF
M6 LML
MW6 LMM
M6 LMN
M6 LMO
MM6 LMP
M6 LMR
M6 LMS
M6 LMX
M6 LNA
M6 LNH
M6 LNN
M6 LNT
M6 LNW
M6 LNY
MI6 LNZ
M6 LOA
M6 LOI
MM6 LOM
MM6 LOR
M6 LOU
M6 LOV
M6 LOX
M6 LPA
M6 LPJ
M6 LPL
M6 LQH
M6 LRC
M6 LRD
M6 LRE
M6 LRJ
M6 LRM
M6 LRP
M6 LRR
M6 LRS
M6 LRT
M6 LRY
M6 LSC
M6 LSD
MM6 LSF
M6 LSL
M6 LSM
M6 LSQ
M6 LSR
M6 LSS
M6 LSU
M6 LSX
M6 LSZ
M6 LTA
M6 LTE
M6 LTF
M6 LTH
M6 LTT
M6 LTY
M6 LTZ
M6 LUB
M6 LUE
MM6 LUF
M6 LUL
M6 LUN
M6 LUS
MW6 LUV
M6 LUX
M6 LVG
M6 LVT
M6 LWC
M6 LWH
MM6 LWW
MM6 LXA
M6 LXB
M6 LXG
MW6 LXL
M6 LXZ
M6 LYE
M6 LYL
M6 LYM
M6 LYR
M6 LYT
M6 LYW
M6 LYX
M6 LZE
M6 LZJ
MI6 LZL
M6 MAC
MW6 MAE
M6 MAI
M6 MAN
M6 MAQ
M6 MAR
MW6 MAU
MM6 MAV
M6 MAZ
M6 MBA
MI6 MBJ
M6 MBL
M6 MBM
M6 MBT

MW6 MBV
M6 MBW
M6 MCD
M6 MCG
M6 MCI
M6 MCN
MM6 MCV
M6 MDA
M6 MDD
M6 MDE
M6 MDF
M6 MDI
M6 MDK
M6 MDP
M6 MDQ
MW6 MDS
MI6 MDV
M6 MDW
M6 MDY
M6 MEE
M6 MEF
MM6 MEH
M6 MEI
M6 MEM
M6 MER
M6 MET
M6 MEW
M6 MEX
M6 MEY
M6 MEZ
M6 MFH
M6 MFP
M6 MFQ
MI6 MFT
M6 MFW
M6 MFX
M6 MFY
M6 MGB
M6 MGE
MI6 MGI
MM6 MGK
M6 MGL
M6 MGS
M6 MHB
M6 MHC
M6 MHG
MM6 MHP
M6 MHR
M6 MHS
M6 MHX
M6 MHZ
M6 MIJ
M6 MIW
M6 MIX
M6 MJB
M6 MJE
M6 MJH
M6 MJJ
M6 MJK
M6 MJQ
M6 MJU
M6 MJW
M6 MJX
MM6 MKA
M6 MKC
M6 MKG
MI6 MKI
M6 MKL
M6 MKP
M6 MKQ
M6 MKR
M6 MKS
M6 MKT
M6 MKZ
M6 MLB
M6 MLC
M6 MLD
MW6 MLJ
M6 MLS
M6 MLW
M6 MLZ
M6 MMA
MI6 MMD
MI6 MME
M6 MMJ
M6 MMK
M6 MMO
MM6 MMV
MD6 MMW
M6 MMZ
M6 MNA
M6 MNB
M6 MNF
M6 MNJ
M6 MNM
M6 MNN
MM6 MNY
M6 MOA
M6 MOE
M6 MOL
M6 MOR
M6 MOT
M6 MPA
M6 MPC
M6 MPG
M6 MPJ
M6 MPN
M6 MPU
M6 MQG
M6 MQZ
M6 MRA
M6 MRB
M6 MRE
M6 MRF
MI6 MRO
MM6 MRQ

M6 MRZ
M6 MSB
M6 MSD
M6 MSK
MM6 MSL
M6 MSQ
M6 MSZ
M6 MTB
MW6 MTF
MW6 MTP
M6 MTX
M6 MTY
M6 MUG
M6 MUM
MW6 MUN
M6 MUQ
MW6 MUV
M6 MUX
M6 MVC
M6 MVG
M6 MVJ
M6 MVO
MM6 MVP
M6 MVR
M6 MVS
M6 MWH
M6 MWI
M6 MWK
M6 MWL
MW6 MWM
M6 MWR
M6 MWU
M6 MWY
M6 MXI
M6 MXJ
M6 MXL
M6 MXS
M6 MXW
M6 MYA
M6 MYJ
M6 MZA
M6 MZM
M6 MZS
M6 NAE
M6 NAF
MM6 NAH
M6 NAR
M6 NBA
M6 NBB
M6 NBC
M6 NBD
M6 NBE
M6 NBF
M6 NBJ
M6 NBK
M6 NBT
M6 NBZ
MM6 NCN
M6 NCT
M6 NDA
M6 NDH
M6 NDJ
M6 NDL
M6 NDV
M6 NDW
M6 NDX
M6 NEB
M6 NEK
M6 NEP
M6 NER
MM6 NEX
M6 NFA
M6 NGC
M6 NGG
M6 NGL
M6 NGS
M6 NGT
M6 NHB
M6 NHI
M6 NIF
M6 NIH
MI6 NIR
M6 NJA
M6 NJC
M6 NJF
M6 NJR
M6 NJW
M6 NJY
M6 NKE
MW6 NKF
M6 NKI
M6 NKN
M6 NKS
MM6 NLC
M6 NLD
M6 NLH
MW6 NLU
M6 NLY
M6 NMA
MW6 NMB
M6 NMC
M6 NMK
M6 NMO
M6 NNG
M6 NNH
M6 NNL
M6 NNN
M6 NOG
M6 NOO
MM6 NOP
M6 NOV
M6 NOW
M6 NOZ
M6 NPC
M6 NPH
M6 NPK

M6 NPM
M6 NPP
M6 NPR
M6 NPS
MI6 NPT
M6 NPU
M6 NRD
M6 NRP
M6 NRZ
M6 NSA
M6 NSB
M6 NSF
MM6 NSN
MI6 NSU
M6 NSV
M6 NTC
M6 NTE
MW6 NTG
MM6 NTH
MM6 NTI
M6 NTO
M6 NTS
M6 NTU
M6 NTZ
M6 NUE
M6 NUH
M6 NUN
M6 NUT
M6 NUU
M6 NVI
M6 NVK
M6 NWD
M6 NWG
M6 NWJ
MM6 NWR
MM6 NXC
M6 NXD
M6 NXN
M6 NYC
M6 NYK
M6 NYS
MJ6 NZD
M6 OAA
MI6 OAC
M6 OAH
M6 OAK
M6 OAM
M6 OAS
M6 OAY
M6 OBA
M6 OBD
M6 OBE
MI6 OBG
MM6 OBI
M6 OBL
MM6 OBN
M6 OBT
M6 OBV
M6 OCA
MW6 OCF
M6 OCL
M6 OCM
M6 ODI
M6 ODJ
M6 ODT
MM6 ODX
M6 OEE
M6 OEL
M6 OFC
M6 OFO
M6 OFP
M6 OFR
M6 OFS
M6 OFX
M6 OGB
M6 OGG
M6 OGN
M6 OGW
M6 OGY
M6 OHB
M6 OHH
M6 OHM
M6 OHV
M6 OIK
MM6 OIL
M6 OIS
M6 OJA
M6 OJJ
M6 OJR
M6 OJW
M6 OKA
M6 OKE
MI6 OLA
M6 OLC
M6 OLG
M6 OLO
M6 OMB
M6 OMC
M6 OMD
M6 OME
M6 OMI
M6 OMM
M6 OMN
M6 OMO
M6 OMV
M6 ONC
M6 ONE
M6 ONG
M6 ONI
M6 ONK
M6 ONR
M6 ONV
M6 ONX
M6 ONY
M6 OOF
M6 OOH

M6 OON
MI6 OOR
M6 OOS
M6 OOT
M6 OPH
M6 OPP
M6 OPR
M6 OPW
M6 ORA
M6 ORD
M6 ORP
MD6 ORR
M6 ORS
M6 ORZ
M6 OSA
M6 OSB
M6 OSC
M6 OSE
M6 OSG
M6 OSH
M6 OSK
MW6 OSS
M6 OTA
M6 OTO
M6 OTZ
M6 OUE
M6 OUK
M6 OUT
M6 OVA
M6 OVO
M6 OWE
M6 OWK
M6 OWN
M6 OWR
M6 OWW
M6 OXO
M6 OYS
M6 OZD
M6 PAB
M6 PAH
MW6 PAK
M6 PAL
M6 PAQ
M6 PAR
M6 PBA
M6 PBB
M6 PBC
M6 PBG
M6 PBH
M6 PBQ
M6 PBS
M6 PCA
M6 PCG
M6 PCH
M6 PCI
M6 PCK
M6 PCO
M6 PCS
M6 PCY
M6 PDF
M6 PDH
M6 PEE
M6 PEH
M6 PEK
M6 PEV
M6 PFC
M6 PFE
M6 PFF
M6 PFQ
MW6 PFS
M6 PGE
M6 PGG
M6 PGK
M6 PGM
M6 PGO
M6 PGR
M6 PGS
MI6 PGV
M6 PHB
M6 PHD
M6 PHI
M6 PHN
M6 PHR
M6 PIE
M6 PIF
M6 PIO
M6 PIQ
M6 PIV
M6 PIX
M6 PIZ
M6 PJC
MI6 PJF
M6 PJO
M6 PJW
M6 PKA
M6 PKB
M6 PKG
M6 PKM
M6 PKP
M6 PKS
M6 PKW
M6 PKY
MI6 PKZ
M6 PLA
M6 PLD
M6 PLF
M6 PLG
M6 PLJ
M6 PLL
M6 PLM
MM6 PLU
MW6 PMC
MW6 PME
M6 PMI
M6 PMO
M6 PMS

M6 PMX
M6 PNA
M6 PNE
M6 PNM
M6 PNO
M6 PNS
MW6 POS
M6 PPG
M6 PPI
MW6 PPM
M6 PPR
M6 PPW
M6 PQY
M6 PQZ
MM6 PRG
M6 PRI
M6 PRJ
M6 PRL
M6 PRR
M6 PRX
M6 PSA
M6 PSB
M6 PSD
M6 PSE
M6 PSH
M6 PSI
M6 PSK
M6 PSP
M6 PSS
M6 PSU
M6 PSW
M6 PTA
M6 PTB
MM6 PTC
M6 PTD
M6 PTG
M6 PTL
M6 PTN
M6 PTS
M6 PTT
M6 PTW
MW6 PUB
M6 PUC
M6 PUK
M6 PUR
MI6 PVA
M6 PVG
M6 PVP
M6 PVX
MM6 PWA
M6 PWI
MI6 PWJ
M6 PWW
M6 PWX
M6 PWY
M6 PXB
MM6 PYT
M6 PYU
M6 PZL
M6 RAA
MI6 RAB
MW6 RAE
M6 RAF
M6 RAG
M6 RAI
M6 RAJ
M6 RAK
M6 RAT
M6 RAU
M6 RAY
MM6 RBC
M6 RBD
M6 RBG
M6 RBJ
MI6 RBL
M6 RBM
MM6 RBR
M6 RBS
MM6 RBX
M6 RCH
M6 RCM
M6 RCQ
M6 RCT
M6 RCX
M6 RDB
M6 RDE
M6 RDG
M6 RDH
M6 RDJ
M6 RDO
M6 RDY
M6 REA
MM6 REC
M6 REN
M6 REP
M6 RET
M6 REU
M6 REZ
MM6 RFT
MM6 RFU
M6 RFX
M6 RFY
M6 RGB
MI6 RGG
M6 RGJ
M6 RGL
M6 RGV
MW6 RHE
M6 RHG
M6 RHH
M6 RHJ
M6 RHM
M6 RHP
M6 RHS
M6 RHV
MM6 RHW

| M6 RHZ | M6 RSC | M6 SDO | M6 SMO | M6 SUS | M6 TET | M6 TNX | M6 TZW | M6 VAS | M6 VVP | M6 WLM | M6 XCV | M6 XUK | M6 YTE | M6 ZPI |
|---|---|---|---|---|---|---|---|---|---|---|---|---|---|---|
| M6 RIC | M6 RSG | M6 SDY | M6 SMP | M6 SVA | M6 TEX | M6 TOB | M6 UAA | M6 VBA | M6 VVV | M6 WMC | M6 XDD | MW6 XUL | MM6 YTT | M6 ZQQ |
| M6 RIF | MI6 RSK | M6 SED | M6 SMS | M6 SVC | M6 TFA | MI6 TOH | M6 UAC | M6 VBC | M6 VVW | M6 WMD | M6 XDH | M6 XUP | M6 YTY | M6 ZRA |
| M6 RIG | M6 RSM | M6 SEH | M6 SMT | M6 SVD | M6 TFC | M6 TOP | M6 UAM | M6 VBG | M6 VWG | M6 WMO | MI6 XDK | M6 XVE | M6 YVM | M6 ZRB |
| M6 RII | M6 RSN | M6 SEL | M6 SMU | M6 SVK | M6 TFI | M6 TOX | M6 UAU | M6 VBM | M6 VXH | MM6 WMR | M6 XDL | M6 XVZ | MM6 YWF | M6 ZSL |
| M6 RIP | M6 RSQ | MM6 SEM | M6 SMV | M6 SVO | M6 TFM | M6 TPJ | M6 UAV | M6 VBO | M6 VXL | M6 WMT | MM6 XDM | M6 XWT | M6 YWZ | M6 ZSR |
| M6 RIW | M6 RSW | MM6 SEQ | M6 SMW | M6 SVR | M6 TFR | M6 TPK | M6 UBJ | M6 VCO | M6 VXO | MM6 WMU | M6 XDP | M6 XWZ | M6 YXZ | M6 ZSU |
| M6 RIX | M6 RSX | M6 SET | M6 SMZ | M6 SVS | M6 TFS | M6 TPL | M6 UBU | M6 VCT | M6 VXX | M6 WNA | M6 XDS | M6 XYA | M6 YYC | M6 ZSY |
| M6 RJA | M6 RTB | M6 SEW | M6 SNA | M6 SVW | M6 TFU | M6 TPN | M6 UCC | M6 VDN | M6 VYZ | MM6 WOA | M6 XDY | M6 XYM | MM6 YYP | M6 ZTC |
| M6 RJC | M6 RTH | M6 SEX | M6 SNB | M6 SWJ | M6 TFX | M6 TPP | M6 UCH | M6 VDQ | M6 VZM | M6 WON | M6 XED | M6 XYS | M6 YYZ | M6 ZTE |
| M6 RJD | M6 RTJ | MW6 SFA | M6 SND | M6 SWM | MW6 TGA | M6 TPW | M6 UCK | M6 VEI | M6 VZO | M6 WOZ | M6 XEH | M6 XYX | M6 YZA | M6 ZTT |
| M6 RJG | M6 RTK | M6 SFD | M6 SNI | M6 SWR | M6 TGE | M6 TPZ | M6 UCW | M6 VER | MM6 VZT | M6 WPJ | M6 XEN | M6 XYZ | M6 YZV | M6 ZTZ |
| M6 RJI | MI6 RTL | M6 SFE | M6 SNK | MM6 SWW | M6 TGW | M6 TQM | M6 UDD | M6 VEX | MI6 WAC | M6 WPK | M6 XEO | M6 XZA | M6 YZW | M6 ZUJ |
| M6 RJT | MM6 RTO | M6 SFG | M6 SNL | M6 SWX | M6 TGX | M6 TQY | M6 UDF | MM6 VEY | M6 WAE | M6 WPM | M6 XES | M6 XZQ | M6 YZZ | M6 ZUK |
| M6 RJU | MW6 RTS | M6 SFI | M6 SNM | M6 SWY | M6 THD | M6 TRA | M6 UDI | M6 VFO | M6 WAH | M6 WPZ | M6 XET | MM6 XZX | M6 ZAE | M6 ZUM |
| M6 RJV | MM6 RTV | M6 SFN | MM6 SNT | M6 SWZ | M6 THE | M6 TRB | M6 UDP | M6 VFX | M6 WAK | M6 WQI | M6 XEW | MI6 XZZ | M6 ZAG | M6 ZUU |
| M6 RKG | M6 RTW | M6 SFO | M6 SOA | M6 SXL | M6 THF | M6 TRG | M6 UDZ | M6 VGG | MW6 WAL | M6 WQW | M6 XFD | MW6 YAA | M6 ZAI | M6 ZVV |
| M6 RKH | M6 RTY | M6 SFU | M6 SOB | M6 SXR | M6 THG | M6 TRJ | M6 UEX | M6 VGL | M6 WAO | M6 WRA | M6 XFV | MM6 YAB | M6 ZAJ | M6 ZWD |
| M6 RKJ | M6 RUD | M6 SFX | MI6 SOH | M6 SXS | M6 THH | M6 TRM | M6 UGG | M6 VGX | M6 WAR | M6 WRI | M6 XGP | M6 YAC | M6 ZAM | M6 ZXA |
| M6 RKN | M6 RUF | M6 SFY | M6 SOJ | M6 SXW | MI6 THK | M6 TRN | M6 UGV | M6 VHC | M6 WAU | M6 WRK | M6 XII | M6 YAI | M6 ZAO | MM6 ZXJ |
| MI6 RKV | M6 RUU | M6 SFZ | M6 SOK | M6 SYD | M6 THL | M6 TRY | M6 UJC | M6 VHD | M6 WAZ | M6 WRP | M6 XIM | M6 YAK | M6 ZAP | M6 ZZI |
| M6 RKY | MI6 RVC | MM6 SGI | M6 SOL | MW6 SYL | M6 THM | M6 TSB | M6 UJI | MW6 VHF | M6 WBC | M6 WRZ | M6 XIS | M6 YAL | M6 ZAT | M6 ZZO |
| M6 RKZ | M6 RVF | M6 SGK | MM6 SOM | M6 SYN | M6 THN | M6 TSD | M6 UKD | M6 VHS | M6 WBD | M6 WSA | M6 XJA | M6 YAX | MD6 ZBC | M6 ZZR |
| M6 RLF | MI6 RVK | M6 SGR | M6 SOP | M6 SYP | M6 THR | M6 TSE | M6 UKE | M6 VID | M6 WBL | MM6 WSD | M6 XJB | M6 YAZ | M6 ZBF | M6 ZZT |
| M6 RLH | MW6 RVL | MI6 SGU | M6 SOQ | M6 SYR | M6 THT | M6 TSF | M6 UKP | M6 VII | MM6 WBQ | M6 WSF | M6 XJH | M6 YBF | M6 ZBI | M6 ZZW |
| M6 RLI | M6 RVP | M6 SGW | M6 SOY | MW6 SYS | M6 THX | M6 TSL | M6 UKR | M6 VIK | MW6 WCD | MW6 WSL | M6 XJK | M6 YBG | M6 ZBK | M6 ZZZ |
| MW6 RLJ | M6 RVT | M6 SGX | M6 SPB | MW6 SZE | M6 THZ | M6 TSO | M6 UKV | MI6 VIM | M6 WCH | M6 WSN | M6 XKA | M6 YBM | M6 ZBO | |
| M6 RLM | M6 RWC | MM6 SHE | M6 SPD | M6 SZS | M6 TIB | M6 TSQ | M6 UKW | M6 VIN | M6 WCJ | M6 WSX | M6 XKX | M6 YCN | M6 ZCC | |
| M6 RLO | MW6 RWD | M6 SHF | MI6 SPE | M6 SZY | M6 TIE | M6 TSS | M6 UKY | M6 VIP | M6 WCM | M6 WTA | M6 XLA | M6 YCQ | M6 ZCJ | |
| M6 RLS | M6 RWH | M6 SHN | M6 SPF | MW6 TAB | M6 TIK | M6 TTA | M6 UKZ | M6 VIR | M6 WCO | M6 WTB | M6 XLB | M6 YDJ | M6 ZDD | |
| M6 RLY | M6 RWN | M6 SHO | M6 SPI | M6 TAC | MI6 TIM | M6 TTC | M6 ULF | M6 VIS | M6 WCS | M6 WTH | M6 XLD | M6 YDM | MI6 ZDK | |
| M6 RMB | M6 RWR | M6 SHR | M6 SPO | M6 TAE | M6 TIN | M6 TTD | M6 ULP | M6 VJB | M6 WDA | MW6 WTI | M6 XLL | M6 YDS | M6 ZDM | |
| M6 RMC | MI6 RWW | M6 SHS | M6 SPQ | MW6 TAI | MW6 TIP | M6 TTG | M6 ULT | M6 VJK | M6 WDE | M6 WTK | M6 XLV | M6 YDT | MM6 ZDR | |
| M6 RMF | M6 RXD | M6 SHT | M6 SPR | M6 TAM | MD6 TIS | M6 TTJ | M6 UMA | M6 VJM | M6 WDJ | M6 WTN | M6 XLY | M6 YDX | MM6 ZDX | |
| M6 RMH | M6 RXF | M6 SHX | M6 SPV | M6 TAN | M6 TIV | M6 TUB | M6 UMF | MW6 VJP | M6 WDK | M6 WTS | M6 XMD | M6 YEA | MW6 ZEC | |
| M6 RMM | M6 RXM | M6 SHY | M6 SPW | M6 TAP | M6 TIX | M6 TUC | MW6 UMH | M6 VJS | M6 WDP | M6 WUF | M6 XML | M6 YEC | M6 ZED | |
| MI6 RMP | M6 RYB | MM6 SIB | M6 SPX | M6 TAQ | M6 TIZ | M6 TUI | M6 UML | M6 VJT | M6 WDT | M6 WUT | M6 XMM | M6 YEM | M6 ZEY | |
| M6 RMQ | M6 RYE | M6 SIG | M6 SQA | M6 TAR | M6 TJC | M6 TUP | M6 UNE | MM6 VKO | M6 WDU | M6 WVD | M6 XMP | M6 YEP | M6 ZEZ | |
| M6 RMY | M6 RYO | M6 SIN | M6 SQM | M6 TAS | M6 TJK | M6 TUS | M6 UNK | M6 VKR | M6 WDZ | M6 WVI | MM6 XMT | M6 YER | M6 ZFS | |
| M6 RMZ | M6 RYS | MW6 SIO | M6 SQR | M6 TAV | M6 TJM | M6 TUZ | M6 UNO | M6 VKV | M6 WEA | M6 WVM | M6 XNX | M6 YEW | M6 ZGP | |
| M6 RNB | M6 RYW | M6 SIQ | MW6 SQS | M6 TBB | M6 TJR | M6 TVA | M6 UNX | M6 VLC | M6 WEC | M6 WWA | M6 XOC | M6 YFL | MM6 ZGT | |
| M6 RNG | M6 RZD | M6 SJI | M6 SQT | M6 TBE | M6 TJT | M6 TVG | MW6 UOJ | M6 VLF | M6 WEF | M6 WWP | M6 XOF | MM6 YFS | MM6 ZHC | |
| M6 RNH | M6 RZR | M6 SJL | M6 SQZ | M6 TBI | M6 TJW | M6 TVR | M6 UOK | M6 VLJ | M6 WEG | M6 WWW | M6 XOP | MW6 YGC | MW6 ZIG | |
| MW6 RNJ | M6 RZT | M6 SJM | M6 SRH | MW6 TBP | M6 TKB | M6 TVT | MM6 UOS | M6 VLM | M6 WEH | M6 WXA | M6 XOZ | MM6 YGI | MM6 ZIM | |
| M6 RNL | M6 SAB | M6 SJP | M6 SRJ | M6 TBW | M6 TKD | M6 TVZ | M6 URB | M6 VLV | M6 WEI | M6 WXG | M6 XPA | MM6 YHF | M6 ZIN | |
| M6 RNR | M6 SAJ | M6 SJS | M6 SRK | MW6 TBZ | M6 TKF | M6 TWB | M6 URF | M6 VMW | M6 WEK | MW6 WXM | MI6 XPG | M6 YHT | M6 ZIX | |
| M6 RNW | M6 SAO | M6 SJT | M6 SRP | M6 TCC | M6 TKH | M6 TWC | M6 URL | M6 VMZ | M6 WEY | M6 WXZ | M6 XPM | MW6 YJE | M6 ZJC | |
| MM6 RNY | MM6 SAP | M6 SJY | MM6 SRT | M6 TCJ | MM6 TKM | M6 TWF | M6 URU | M6 VNA | M6 WFD | M6 WYE | MW6 XPS | M6 YJP | M6 ZJO | |
| M6 ROK | M6 SAR | M6 SKA | M6 SRW | M6 TCP | M6 TKO | M6 TWG | M6 URZ | M6 VNE | M6 WGA | M6 WYL | M6 XRB | M6 YKK | MM6 ZJS | |
| MM6 ROO | M6 SAT | MM6 SKB | M6 SRX | M6 TCQ | M6 TLA | M6 TWO | M6 USB | M6 VNS | M6 WGW | M6 WYS | M6 XRE | M6 YKV | M6 ZKC | |
| M6 ROQ | MM6 SBA | MI6 SKC | M6 SRY | M6 TCR | M6 TLD | M6 TWP | M6 USL | MM6 VOE | M6 WGY | M6 WZA | M6 XRG | M6 YKZ | M6 ZKD | |
| M6 ROR | MM6 SBC | M6 SKE | M6 SSE | MW6 TCT | M6 TLE | M6 TWR | M6 USR | M6 VOJ | M6 WHB | M6 WZL | M6 XRH | M6 YLB | M6 ZKG | |
| M6 ROS | M6 SBQ | MI6 SKG | M6 SSG | M6 TCU | M6 TLW | M6 TWT | M6 UTA | M6 VON | M6 WHC | M6 WZT | M6 XRJ | M6 YMC | M6 ZKL | |
| M6 ROZ | M6 SBX | MM6 SKI | M6 SSH | M6 TCW | M6 TMB | M6 TWX | M6 UTD | M6 VOO | M6 WHE | MW6 XAH | M6 XRK | M6 YMD | M6 ZKM | |
| M6 RPA | M6 SCE | MM6 SKJ | M6 SSJ | MW6 TDG | MI6 TME | M6 TXB | M6 UTE | M6 VOP | M6 WHR | M6 XAJ | M6 XRM | M6 YMJ | MM6 ZKO | |
| M6 RPG | M6 SCJ | MI6 SKK | M6 SSK | MW6 TDH | M6 TMI | M6 TXC | M6 UTM | M6 VPC | MM6 WHY | M6 XAN | M6 XRY | M6 YNE | M6 ZKX | |
| M6 RPI | M6 SCK | M6 SKO | M6 SSL | MW6 TDN | MM6 TMN | MW6 TYD | MI6 UUC | M6 VPN | M6 WIA | M6 XAP | M6 XSG | M6 YOA | MM6 ZKZ | |
| M6 RPK | M6 SCL | M6 SKS | MW6 SSQ | MW6 TDQ | M6 TMT | M6 TYF | M6 UVR | M6 VPR | M6 WJD | M6 XAZ | M6 XSJ | M6 YOM | M6 ZLG | |
| M6 RQL | MW6 SCO | M6 SLF | M6 STB | MW6 TDT | MW6 TMW | M6 TYO | M6 UWE | M6 VPT | MI6 WJH | M6 XBF | M6 XSM | M6 YOP | MM6 ZMD | |
| MW6 RRB | M6 SCR | MW6 SLH | M6 STE | M6 TDW | M6 TMX | M6 TYP | M6 UXH | M6 VRA | M6 WJO | M6 XBG | M6 XSP | M6 YOR | MW6 ZNQ | |
| M6 RRE | M6 SCS | M6 SLQ | MM6 STG | M6 TDX | M6 TNE | M6 TYW | M6 UXP | M6 VRS | M6 WJR | M6 XBH | M6 XSW | M6 YPH | M6 ZOA | |
| M6 RRM | M6 SCT | M6 SLS | M6 STL | M6 TEA | M6 TNF | M6 TZF | M6 UXY | M6 VSP | M6 WJV | MM6 XBK | M6 XSY | M6 YPK | M6 ZOE | |
| M6 RRS | M6 SCY | MW6 SLW | M6 STO | M6 TEB | M6 TNG | M6 TZG | M6 UZI | M6 VTL | M6 WKA | M6 XBT | M6 XTC | MI6 YPT | M6 ZON | |
| M6 RRT | M6 SDB | M6 SLX | M6 STS | M6 TEC | MM6 TNK | M6 TZH | M6 VAC | M6 VTR | M6 WKM | M6 XCB | M6 XTE | M6 YRK | M6 ZOT | |
| MM6 RRX | M6 SDC | M6 SLY | MI6 STT | M6 TED | M6 TNP | M6 TZM | M6 VAD | M6 VTX | M6 WLJ | MI6 XCG | MM6 XTS | M6 YRU | M6 ZOZ | |
| M6 RRY | M6 SDD | M6 SMC | M6 STW | M6 TEH | M6 TNS | M6 TZN | MW6 VAF | M6 VVC | M6 WLK | M6 XCM | M6 XTX | M6 YSO | M6 ZPC | |
| M6 RRZ | M6 SDM | M6 SMJ | M6 STX | M6 TEI | M6 TNW | M6 TZT | M6 VAO | M6 VVM | M6 WLL | MW6 XCT | M6 XUD | M6 YSZ | MI6 ZPG | |

# Permanent Special Event Callsigns

GB2CW RSGB, 3 Abbey Court, Priory Business Park, Bedford, MK44 3WH
GB3HQ RSGB, 3 Abbey Court, Priory Business Park, Bedford, MK44 3WH
GB3RS RSGB National Radio Centre, Bletchley Park, Milton Keynes, MK3 6EB
GB3VHF RSGB, 3 Abbey Court, Priory Business Park, Bedford, MK44 3WH
GB4FUN RSGB, 3 Abbey Court, Priory Business Park, Bedford, MK44 3WH
GB4RS Radio Society President, Church Road, Conifers, Littlebourne, Kent, CT3 1UA
GB0AC Air Cadets, AR Centre, St Johns Road, Tunbridge Wells, TN4 9UU
GB0MAC Merseyside Air Cadets, Charles Gordon House, 90 (Speke) Sqdn ATC, Woolton Road, Liverpool, L19 5QN
GB0MWM Military Wireless Museum, Kidderminster, 47 Oakfield Road, Worcs, DY11 6PL
GB0SMA Stow Maries Aerodrome, Purleigh, Stow Maries Aerodrome, Chelmsford, CM3 6RN
GB0SNB Secret Nuclear Bunker, Crown Buildings, Kelvedon Hatch Nuclear Bunker, Kelvedon Hatch Lane, Kelvedon Hatch
GB0SSB Golf Bravo Zero Sierra Sierra Bravo, Troywood, Crown Buildings, Nr St. Andrews, Fife
GB0WFB WORTHING FIRE BRIGADE, ARDSHEAL ROAD, WORTHING FIRE STATION, Worthing
GB0YAM Yorkshire Air Museum, Halifax Way, Yorkshire Air Museum, Elvington, York, YO41 4AV
GB1BM Brooklands Museum, Brooklands Road, Brooklands Museum, Weybridge, KT13 0QN
GB1CHF CoalHouse Fort, Princess Margaret Road, Coalhouse Fort, East Tilbury, Essex, RM18 8PB
GB1FDC Brede Steam, Stubbs Lane, Brede Steam ARS, Brede, Nr Rye
GB1PBL Portland Bill Lighthouse, Portland, Portland Bill Lighthouse, Dorset, DT5 2JT
GB1ROC Royal Observer Corp, Portadown, 60 Derrylettiff Road, BT62
GB2CAV Cavalier, Chatham, Chatham Historic Dockyard, Chatham, ME4 4TZ
GB2CPM Chalk Pits Museum, Station Road, Amberley Museum and Heritage Centre, Amberley, Arundel, BN18 9LT
GB2CWP Christopher Whitton Panton, The airfield, East Kirkby, Lincolnshire Aviation Heritage Centre, SPILSBY, Lincolnshire, PW23 4DE
GB2EVR Eden Valley Railway, Appleby, Warcop Railway Station, Cumbria, CA16 6PR
GB2GM GOLF MIKE, Poldhu, Poldhu ARC, Mullion, Cornwall
GB2HAM Harrington Aviation Museum, Harrington Aviation Museum, Sunnyvale Farm, Lamport Road, Harrington
GB2NLO Norman Lockyer Observatory, Salicombe Hill, Norman Lockyer Observatory, Sidmouth, Simouth, EX10 0NY
GB2OWM Orkney Wireless Museum, Kiln Corner, Orkney Wireless Museum, Junction Road, Kirkwall
GB2RA Royal Artillery, Royal Arsenal, R.A. Firepower Museum, Woolwich, London
GB2RFM Redoubt Fortress Museum, Eastbourne, Royal Parade, East Sussex, BN22 7AQ
GB2RGM Royal Gunpowder Mills, Beaulieu Drive, Royal Gunpowder Mills, Waltham Abbey, Essex, EN9 1JY
GB2RHQ Regional Head Quarters, French Lane, Hack Green Secret Bunker, Nantwich, CW5 8BL
GB2RN Royal Navy, The Queens Walk, HMS Belfast, London, SE1 2JH
GB2SJ Souter Lighthouse, Coast Road, Souter Lighthouse, Sunderland, SR6 7NH
GB2SPY B2 SPY Museum, Church Whitfield, Dover Construction Club, Dover, Kent, CT16 3HZ
GB3RN Royal Navy, Newgate Lane, HMS Collingwood, Fareham, PO14 1AS
GB4MSE Morayvia Sci-Tech Experience, North Road, Morayvia Sci-Tech Experience Project, Kinloss, Moray, IV[illegible]
GB4UAS Ulster Aviation Society, Halftown Road, Long Kesh Airfield, Lisburn, BT27 5RF
GB8ROC Romeo Oscar Charlie, Kilrea, BlackRoack Road, Co. Londonderry, BT51 5XH

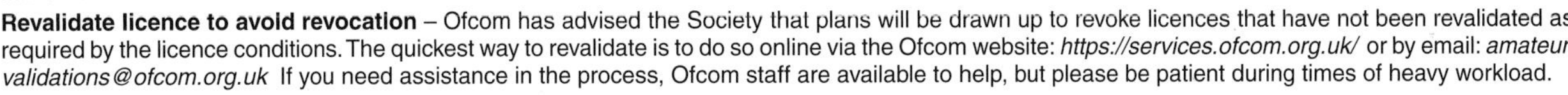

**IMPORTANT NOTE**

**Revalidate licence to avoid revocation** – Ofcom has advised the Society that plans will be drawn up to revoke licences that have not been revalidated as required by the licence conditions. The quickest way to revalidate is to do so online via the Ofcom website: *https://services.ofcom.org.uk/* or by email: *amateur.validations@ofcom.org.uk* If you need assistance in the process, Ofcom staff are available to help, but please be patient during times of heavy workload.

# Special Contest Calls

| | |
|---|---|
| G0B (M) | Scottish DX & Contest Club, M0AIK |
| G0D | QRZ Amateur Radio Grp of Sussex, G3YNN |
| G0F (M) | FUNNY CONTEST G, G0FRT |
| G0G (M) | STROMNESS ACADEMY ARC, M0SLB |
| G0T | Mr Alex Shafarenko, M0SFR |
| G0W | Mr Nigel Newby, G0VDZ |
| G1A | UK Young Contesters Group, M0UKC |
| G1C (M) | Mr John Ross, GM1BSG |
| G1J (M) | Mr Jim Martin, MM0BQI |
| G1N | Mr Gordon Brown, G3MZV |
| G1T | Sands AR Contest Group, M0SCG |
| G1X (M) | Mr J MacLean, MM0CCC |
| G2A (J) | The Jersey Amateur Radio Society, G3DVC |
| G2A (M) | NORTH OF SCOTLAND CONTEST GROUP, G2MP |
| G2B | Bittern DX Group, M0NBG |
| G2C | Christchurch ARS, G7MUD |
| G2E | Eagle Radio Group, M0ERG |
| G2F | Wisbech Amateur Radio Club, M5ARC |
| G2L | Luton VHF Group, G3SVJ |
| G2R | Stevenage & DARS |
| G2T (M) | COCKENZIE & PORT SETON ARC, M0CPS |
| G2U | Angus Wilson |
| G2V (M) | Mr CW Tran, GM3WOJ |
| G2W | Worcester Radio Amateurs Association, M0ZOO |
| G2X | HAMPSHIRE HILLTOPPERS AMATEUR RADIO CLUB, M0HHT |
| G2Y (M) | Mr James Hume, MM0DXH |
| G2Z (M) | SCOTTISH-RUSSIAN AMATEUR RADIO SOCIETY, M0DGR |
| G3A (M) | OBO STRATHMORE ARC, G3GBZ |
| G3B | Shefford ARC, G2DPQ |
| G3C (M) | Mr J Hayes, GM0NBM |
| G3F (M) | Mr SG Cooper, GM4AFF |
| G3H | Hoover GW3RDB |
| G3H (W) | HOOVER ARC, G3RDB |
| G3M | Reigate Amateur Transmitting Society |
| G3P | Mr M J Chamberlain, G3WPH |
| G3Q | Flaxton Moor Contest Group, G3QI |
| G3R | Mr Stephen Brumby, G0DCK |
| G3S | Over Wyre ARG, M0WYR |
| G3T | TORBAY ARS, G8NJA |
| G3U | Mr Brian Gale, G3UJE |
| G3V | Verulam ARC, G3VER |
| G3W (M) | WIGTOWNSHIRE ARC, G4RIV |
| G3X (M) | Mr C Penna, GM3POI |
| G3Y | Mr Ian McCarthy, G3YBY |
| G3Z | Telford & District Amateur Radio Society, G3ZME |
| G4A | Mr Justin Snow, G4TSH |
| G4C | Essex CW Club, G1FCW |
| G4E | Mr Christopher Page, G4BUE |
| G4F | Lincolnshire Poachers Contest Grp, G6LI |
| G4G | Richmond Radio Group, M0RRG |
| G4L | Mr Anthony Bettley, G4LDL |
| G4O (M) | Mr G Berrich, GM0IIO |
| G4P (W) | Pembrokeshire Radio Society, G0EJE |
| G4R | Flight Refuelling ARS, G4RFR |
| G4S | Mr NS Cawthorne, G3TXF |
| G4T | Mr Keith Maton, G6NHU |
| G4U | Iain Haywood |
| G4W (M) | Woodpecker Contest Group, M0WCG |
| G4X | Mr David Mason, G3RXP |
| G4Z | A1 Contest Group, G4ZAP |
| G5A (M) | GM DX Group |
| G5B | Five Bells Group, G4SIV |
| G5C (M) | CROCODILE ROCK AMATEUR GROUP, M0ROB |
| G5D | Tall Trees Contest Group, M0TTG |
| G5E | Mr Derek Moffatt, G3RAU |
| G5F (D) | Mr Robert Finch, GD4RFZ |
| G5K (I) | BCG (Bluehill Contest Group), M0LLZ |
| G5L | Northants Expedition Group, G0ING |
| G5M (M) | The Jaggy Thistles |
| G5M (M) | The Jaggy Thistles, M0TJT |
| G5N | Tynemouth Amateur Radio Club, G0NWM |
| G5O | Stockport Radio Society, G6UQ |
| G5R (W) | Mr Ron Stone, GW3YDX |
| G5T | The Triode Amateur Radio Group, M0TDE |
| G5U | Mr David Mason, G3RXP |
| G5V (M) | Windy Yett Contest Grp |
| G5V (M) | Windy Yett Contest Grp, G5VG |
| G5W | WPX Contest Group, G0WPX |
| G5W | Mr Donald Beattie, G3BJ |
| G5X (M) | KM Kerr |
| G5Y (Z) | Eshaness Radio Club, M0ZET |
| G6A | The North West 320 DX Club, M0AJB |
| G6K (I) | RWC Cummings |
| G6M | Mr Victor Lindgren, G4BYG |
| G6T | Mr T L Burbidge, G4MKP |
| G6W (W) | Dragon Amateur Radio Group, G4TTA |
| G6X | Mr Robert Rushbrooke, M0KLO |
| G6Z (M) | Gavin Taylor |
| G7A (M) | Kilmarnock & Loudoun ARC |
| G7A (M) | Kilmarnock & Loudoun ARC, G0ADX |
| G7M (M) | Montrose Amateur Radio Station, G3KC |
| G7O | Mr Edwin Taylor, G3SQX |
| G7R (M) | Mr Jim Fisher, GM0NAI |
| G7R (M) | Inverclyde Contest Group, M0APF |
| G7T | The Squarebashers, M0ADJ |
| G7V (M) | North of Scotland Contest Group |
| G7V (Z) | North of Scotland Contest Group |
| G7V (M) | North of Scotland Contest Group, G2MP |
| G7V (Z) | North of Scotland Contest Group, G2MP |
| G7Y | Mr Michael Clark, M0ZDZ |
| G8A | South Wirral Contest Group, G3CSA |
| G8D | Central Contest Asso. |
| G8M (M) | Mr Edmund Holt, GM0WED |
| G8M (M) | Orkney ARS, M0MWW |
| G8N | Northampton Radio Club, G8LED |
| G8P | Parallel Lines Contest Group, G4LIP |
| G8T | Black Sheep Contest & DX Group, M0BAA |
| G8W (M) | Border ARS, G0BRS |
| G8X | Mr Timothy Hugill, G4FJK |
| G9A | North Wakefield Radio Club |
| G9D | UK DX Contest Club, M0HGC |
| G9F | G4BVY |
| G9M | Thorpe Camp Museum Radio Group, M0TCM |
| G9N | STIRLING AND DISTRICT ARS, G6NX |
| G9Q (I) | Greenisland Electronics Amateur Radio Society, M0PQR |
| G9T (W) | Wrexham ARS, G4WXM |
| G9V | The Vulture Squadron C.G., M0VSQ |
| G9W | Mr Mark Haynes, M0DXR |
| G9X | Mr A Rees, MW1LCR |
| GB0RSC | Mr Nigel Cole, GW7VJK |
| GB175PO | Mr Armando Martins, M0PAM |
| GB4DM | Mr Martin Childs, G7WBX |
| GI5I | Mr Richard White, GI4DOH |
| GI5K | Mr C J Smith, INDIVIDUAL |
| GI6K | Mr Robert Cummings, GI0KOW |
| GJ2A | Jersey Amateur Radio Society, G3DVC |
| GM0B | Scottish DX Contest, M0AIK |
| GM1C | Mr John Ross, GM1BSG |
| GM1J | Mr Jim Martin, MM0BQI |
| GM1X | Mr J MacLean, MM0CCC |
| GM2A | Prof. KM Kerr, GM4YXI |
| GM2T | Cockenzie & Port Seton ARC, M0CPS |
| GM2V | Mr CW Tran, GM3WOJ |
| GM2Y | Mr James Hume, MM0DXH |
| GM3C | Mr J R Hayes, GM0NBM |
| GM3D | Mr Paul Glasper, GM0BKC |
| GM3F | Mr S G Cooper, GM4AFF |
| GM3W | Wigtownshire ARC, G4RIV |
| GM3X | Mr CS Penna, GM3POI |
| GM4O | Mr G Berrich, GM0IIO |
| GM4X | Mr Bernard McIntosh, GM4WZG |
| GM5A | GMDX Group, G8VL |
| GM5M | The Jaggy Thistles, M0TJT |
| GM5X | Mr KM Kerr, GM4YXI |
| GM6Z | Mr G A Taylor, GM0GAV |
| GM7A | Kilmarnock and Loudoun ARC, G0ADX |
| GM7R | Mr Jim Fisher, GM0NAI |
| GM7V | North of Scotland Contest Group, G2MP |
| GW0A | Mr John Barber, GW4SKA |
| GW2L | Luton VHF Group, G3SVJ |
| GW4J | Mr Stewart Rolfe, GW0ETF |
| GW4P | Pembrokeshire Radio Society, G0EJE |
| GW4W | Woodpecker Contest Group, M0WCG |
| GW5R | Mr Ron Stone, GW3YDX |
| GW6W | Dragon Amateur Radio Club, G4TTA |
| GW7X | Contest Cymru, G3XEJ |
| GW8K | Barry Amateur Radio Society, G4BRS |
| GW9X | Mr Adrian Rees, MW1LCR |
| GZ5Y | Eshaness Radio Club, M0ZET |
| M0A | Mr Christopher Plummer, G8APB |
| M0B | Hadley Wood Contest Group, G4STV |
| M0C | Chiltern DX Club |
| M0C (D) | R Barden |
| M0F (I) | Windy Hill Contest Group, M0WHG |
| M0I (M) | Island Radio Club, M0IRC |
| M0M (I) | Northern Ireland Dxers Group, M0NID |
| M0O | Scarborough Special Events Group, G7OOO |
| M0P | Dr Peter Lock, M0RYB |
| M0Q (M) | Summer Isles R.C. |
| M0R (M) | RK Towers |
| M0T | Simpson ARS, M0SRA |
| M0X | Oxo Contest Club, M0OCC |
| M0Z | Mr Conor Turton, M0GVZ |
| M1A | Martlesham DX Contest Group, G0KPW |
| M1B | Mr S B Foote, G4FOH |
| M1D (M) | Croy Hill ARC |
| M1D (M) | Croy Hill ARC, M0GFA |
| M1E | MM0GOR |
| M1M | Sheffield DX Net, M0SDC |
| M1N | Mr James Patterson, M1DST |
| M1P | Crossways Contest Group, G4HXX |
| M1T | Mr M Whyte, MM0XXW |
| M1T (M) | Mr M Whyte, MM0XXW |
| M1X (M) | M Whyte |
| M2A | De Montford University ARS, G3SDC |
| M2B | South Leeds ARS, M0TSF |
| M2C (D) | R Barden |
| M2G | Mr John Muzyka, G4RCG |
| M2H | Stratford Upon Avon & DRS, G0SOA |
| M2I (W) | Denzil Contest Grp |
| M2K (U) | Guernsey ARS, G3HFN |
| M2M | Croham Callers, G0VVX |
| M2N (M) | Mr Gordon Paterson, MM0GPZ |
| M2R (M) | Inverclyde Amateur Radio Group, G0GNK |
| M2T | Chelmsford ARS |
| M2W | Whitton ARG, G0MIN |
| M2X | Mr Stephen Wilson, G3VMW |
| M2X | Strictley Contest Club |
| M2Z | South Dorset RS, G3SDS |
| M3C | Reading University ARS, M0CYB |
| M3D | Mr DI Field, G3XTT |
| M3I | Mr Kenneth Chandler, G0ORH |
| M3M | Sheffield ARC, G3RCM |
| M3N (M) | Mr Sidney Will, GM4SID |
| M3P | Harwell ARS, G3PIA |
| M3T (M) | Lanarkshire contest Group, M0LCG |
| M3W | Hallam DX Group, M0HDG |
| M3X | Burton & District Radio Society, G3NFC |
| M4A | Cambridge University Wireless Society, G6UW |
| M4C (W) | Swansea ARS, G4CC |
| M4D | Mr John Hotchin, G8DYT |
| M4F | Finningley ARS |
| M4K (D) | The Northern Lights, G0EMG |
| M4P | M0PMV |
| M4R | Granta Contest Group, M0CAM |
| M4T | Reading & District ARC, G3ULT |
| M4U | Harwich Amateur Radio Interest Group, G0RGH |
| M4W | Swindon & District ARC, G8SRC |
| M5A | The Three A's Contest Group, G0AAA |
| M5B | Cross Border Contest Group, M0ZIP |
| M5C | Colchester Radio Amateurs, M1COL |
| M5D | Litchfield ARS, G3WAS |
| M5E | Mr O I Lundberg, G0CKV |
| M5I | Mr Joseph Williamson, GI0RQK |
| M5M | Midland ARS, G1MAR |
| M5O | Mr Peter Hobbs, G3LET |
| M5S | Manchester Wireless Society, G5MS |
| M5T | Warrington ARC, G0WRS |
| M5W | Wythall Contest Group, G0WRC |
| M5Z | Mr Kazunori Watanabe, M0CFW |
| M6C | Ripon & DARS, G4SJM |
| M6M | UK 6 metre Contest Group, G0VSM |
| M6O | Mr David Aslin, G3WGN |
| M6Q | Gravesend Radio Society, G3GRS |
| M6T | Martlesham Radio Society, G4MRS |
| M6V (D) | Mr R Ferguson, GD0IOM |
| M6V (D) | Isle of Man DX Pedition Group |
| M6W | Mr DW Watson, G0DEZ |
| M6X (I) | Co.Antrim ARS DX Group, G0XYZ |
| M6Z | Mr Brian Jones, M0STY |
| M7A | Newton-Le-Willows ARC, M0NRC |
| M7C | M5RIC |
| M7M | Medway Contest Group, M0MDX |
| M7O | Mr Simon Billingham, M0VKY |
| M7P | Mr Keith Maton, G6NHU |
| M7Q | Mr A Cook, G4PIQ |
| M7T | Mr David Wicks, G3YYD |
| M7V | Mr G McGowan, G8YTF |
| M7W | Mr John Cree, G3TBK |
| M7X | Mr Darren Collins, G0TSM |
| M7Z | Mid Beds Contest Association, G4MBC |
| M8A (Z) | Shetland Contest Group, M0ZCG |
| M8C | Cray Valley Radio Society, G3RCV |
| M8T (M) | Mr James Cameron, MM0CWJ |
| M8Y | Otley Amateur Radio Society, G3XNO |
| M8Z (M) | Mr Jason O'Neill, GM7VSB |
| M9A | Mr Stephen White, G3ZVW |
| M9C | South Cheshire ARS, G6TW |
| M9K | M0SIY |
| M9M (M) | NA FIR CHLIS ARS, M0NFC |
| M9R | Rushey Mead School AR Club, G4RMS |
| M9T | Moors Contest Club, M0MCG |
| M9W (W) | Strumble Head DX & Contest Group, M0SHL |
| M9X | Dorridge Scout Group, M0XXT |
| M9Y (D) | Manpak Club |
| M9Z | Sierra Hotel ARCG, G0OBS |
| MD1R | Mr Albert Crespo, M0FCR |
| MD2C | Mr R Barden, MD0CCE |
| MD4K | The Manx Kippers, G0EMG |
| MD6V | Isle of Man DXPedition Group |
| MI6X | Co.Antrim ARS DX Group, G0XYZ |
| MM0R | Mr R K Towers, MM0RKT |
| MM1X | Mr M Whyte, MM0XXW |
| MM2N | Mr Gordon Paterson, MM0GPZ |
| MM3N | Mr Sidney Will, GM4SID |
| MM3T | Lanarkshire Contest Group, M0LCG |
| MM3Z | Mr John Cattigan, MM0LBX |
| MM8T | Mr James Cameron, MM0CWJ |
| MM9M | NA FIR CHLIS ARS, M0NFC |
| MW2I | Denzil Contest Group, G0FRE |
| MW5R | Mr Anton Koval, MW0EDX |

# Irish Republic Callsign Listings

The following pages include a list of new callsigns and amendments notified up to July 2016.
The Irish Radio Transmitters Society is indebted to the Commission for Communications Regulation for supplying much of the data in this listing. Irish callsigns begin with EI or EJ. The EJ prefix is used instead of EI when operating from offshore islands.

## EI0

**EI0A** Limerick Radio Club, c/o Ger McNamara, Kearnmara House, Ruanard, Clonlara, Co. Clare
**EI0B** Peter Green, Castlerea, Co. Roscommon
**EI0M** Mayo Radio Experimenters Network, c/o Brendan Minish EI6IZ, Raheens, Castlebar, Co. Mayo
**EI0R** WestNet Dx Group Contests, 167 St. James's Road, Greenhills, Dublin 12, D12 W6T4
**EI0W** Dundalk Amateur Radio Society, QSL via EI7DAR
**EI0Z** Amateur Portable Group, c/o John Satelle EI6GHB
**EI0AC** IRTS AREN Co-ordinator, P.O. Box 462, Dublin 9
**EI0CF** Finbar O'Connor, Lagg Road, Malin, Co. Donegal
**EI0CI** Details withheld at licensee's request
**EI0CK** Thomas McManus, Belle Ile, Cherry Lawns, Hillcrest, Lucan, Co. Dublin
**EI0CL** Michael Higgins, Galway, Ireland
**EI0CM** Roy Edwards, 16 Tara Court, Ashbourne, Co. Meath
**EI0CP** Sean Taaffe, 'San Marino', 56 Muirhevna, Dublin Road, Dundalk, Co. Louth
**EI0CT** Details withheld at licensee's request
**EI0CY** George H. O'Reilly, 97 St. Assams Avenue, Raheny, Dublin 5
**EI0CZ** Brendan Kilmartin, "Glendale", Lisduff, Clonlara, Co. Clare
**EI0DA** Vincent Rafter, Hillquarter, Coosan, Athlone, Co. Westmeath
**EI0DB** David J. Aldridge, Emerald Cottage, Cloonfane, Charlestown, Co. Mayo
**EI0DC** Colm Ward, 7 Sydenham Terrace, Ballinacurra, Limerick
**EI0DD** Thomas Hurley, 112 Ballindrum, Athy, Co. Kildare
**EI0DG** Irvine Ferris, Kilclone, Co. Meath
**EI0DJ** Breda Condon, Golf Links Road, Castletroy, Co. Limerick
**EI0DK** John Hill, 33 Merchant Square, East Wall, Dublin 3
**EI0EC** IRTS AREN Co-ordinator, P.O. Box 462, Dublin 9
**EI0HQ** IRTS Contest Station, QSL via Dave Moore EI4BZ, Dooneen, Carrigtwohill, Co. Cork
**EI0NC** IRTS AREN Co-ordinator, P.O. Box 462, Dublin 9
**EI0PL** Papa Lima DX Group, c/o Adam Kolodziejczyk, Apartment 2 Coranna, Tipper South, Naas, Co. Kildare
**EI0CAR** Carndonagh Amateur Radio Club, c/o Peter Homer EI4JR, Tullyarb, Carndonagh, Co. Donegal
**EI0DXG** EI DX Group, c/o Patrick O'Connor, EI9HX
**EI0MAR** Howth Martello Radio Group, c/o Tony Breathnach, 159 CÈide Ard MÛr, Ard Aidhin, Baile ¡tha Cliath 5
**EI0MRG** Marconi Experimental Club, c/o Michael Higgins, Galway, Ireland
**EI0NDR** North Dublin Radio Club, c/o Tony Fay EI6EQB, 48 Ardlea Road, Artane, Dublin 5
**EI0RTS** Irish Radio Transmitters Society, P.O. Box 462, Dublin 9
**EI0TED** Craggy Island DXpedition Group, c/o Steve Wright, 14 Churchfields, Lower Salthill, Galway
**EI0IRTS** Irish Radio Transmitters Society, P.O. Box 462, Dublin 9
**EI0YOTA** Youngsters On The Air Ireland, c/o Ger McNamara, EI4GXB

## EI1

**EI1A** Olivier Vandenbalck, 4 Avenue du Cerf-Volant, 1170 Brussels, Belgium
**EI1C** Cork Radio Club, QSL via EI2KA
**EI1E** Avondhu Radio Club, c/o Gerard Scannell, 3 Kingston Close, Mitchelstown, Co. Cork P67 HR44
**EI1K** Kerry Amateur Radio Group, www.kerryamateurradiogroup.com
**EI1Y** Papa Lima DX Group (Contest Call), c/o Adam Kolodziejczyk, EI5JQ
**EI1AA** Irish Leprechaun Contest Group, QSL via EI2BB
**EI1CI** Tom Fitzgerald, Fermoy Road, Ballyhooly, Co. Cork
**EI1CK** Robert W. Semple, "Algonquin", 16 Cairn Hill, Foxrock, Dublin 18
**EI1CN** Glynn W. Langston, 3005 Mills Street, Carencro, LA 70520, USA
**EI1CW** Thomas Byrne, Address withheld at licensee's request
**EI1CX** Richard F. Chambers, 9 Shrewsbury Park, Fernhill Road, Carrigaline, Co. Cork
**EI1DD** Blackrock Radio Scouts, QSL via EI2CA
**EI1DF** Eugene Ryan, 10 Mountain View, Crinken Glen, Shankill, Co. Dublin
**EI1DQ** Patrick McGrath, 15 Castleknock Way, Laurel Lodge, Castleknock, Dublin 15
**EI1DK** William E. Boles, 205 Emmet Road, Inchicore, Dublin 8
**EI1EM** John Owen-Jones, Clonboo, Corrandulla, Co. Galway
**EI1NC** North Cork Radio Group, "Lyston", 13 Forest View, Goulds Hill, Mallow, Co. Cork
**EI1SI** Details withheld at licensee's request
**EI1UN** Irish United Nations Veterans Assoc., C/O Anthony Liddy EI9IL, 26 Sarsfield Avenue, Garryowen, Limerick
**EI1DRC** Donegal Amateur Radio Club, c/o Paul Sinclair, Rarooey, Donegal Town, Co. Donegal
**EI1GHZ** Mayo VHF Group, c/o Joe Fadden EI3IX, Knockthomas, Castlebar, Co. Mayo
**EI1KARG** Kerry Amateur Radio Group, www.kerryamateurradiogroup.com

## EI2

**EI2E** Cormac McHenry, 8 Heidelberg, Ardilea, Dublin 14
**EI2V** Air Corps Signals Amateur Radio Club, c/o Noel Daly, CIS Squadron, Air Corps, Baldonnel, Dublin 22
**EI2AF** Dermot K. Donnelly, 43 Crozon Downs, Sligo
**EI2AH** Raymond S. Jordan, 35 Bryanstown Village, Drogheda, Co. Louth
**EI2AI** Dermot J. Ryan, 75 Ballinteer Crescent, Ballinteer, Dublin 16
**EI2AR** Patrick R. McCabe, Bishop Street, Tuam, Co. Galway
**EI2AW** Anthony Condon, Golf Links Road, Castletroy, Co. Limerick
**EI2BB** James R. Bartlett, "Chickamauga", Tinnahinch, Clonaslee, Co. Laois
**EI2BY** Terenure College Radio Club, c/o George Adjaye, 245 Lower Kimmage Road, Dublin 6W
**EI2CA** Paul Martin, Beech Cottage, Gorman Lough, Stackallen, Slane, Co. Meath
**EI2CC** Donal Lonergan, "Ormond", 47 Hazelbrook Drive, Terenure, Dublin 6W
**EI2CD** Jeremy Misstear, Riva, 32 Coliemore Road, Dalkey, Co. Dublin
**EI2CF** Gerald O'Reilly, Monavally, Tralee, Co. Kerry
**EI2CH** Gerard P. Morgan, N6CBX, Bunatubber, Corrandulla, Co. Galway H91 NH9X
**EI2CI** Joseph M. Purfield, 12 Wolseley Street, South Circular Road, Dublin 8, DO8 A3K7
**EI2CJ** Patrick J. Doran, Main Street, Leighlinbridge, Co. Carlow
**EI2CL** Michael McNamara, 92 Griffith Court, Dublin 3
**EI2CM** Robert L. Williams, 32 Madrona Street, San Carlos, California 94070, USA
**EI2CN** Douglas N. Turnbull, De Forest House, Coolfore, Monasterboice, Co. Louth
**EI2CR** Sean Carvin, 43 Brackenstown Village, Swords, Co. Dublin
**EI2CV** Tony Bourke, 'Hillcrest', Spur Hill, Togher, Cork
**EI2CW** Noel McIntyre, Springhill, Ballindrait, Lifford, Co. Donegal
**EI2CY** Details withheld at licensee's request
**EI2CZ** Patrick Doyle, "Shalom", Hawthorn Drive, Hill View, Waterford
**EI2DA** Declan Howard, 11 Island View Estate, Sea Park, Malahide, Co. Dublin
**EI2DB** Nicholas Mulhall, The Gate Lodge, Athy Road, Stradbally, Co. Laois
**EI2DD** Sean M. Reilly, Cratloekeel, Cratloe, Co. Clare
**EI2DF** Patrick Rohan, 139 Ballyroan Road, Rathfarnham, Dublin 16
**EI2DI** Allan McMurty, Drumirrin, Lackaduff, Loughros Point, Ardara, Co. Donegal
**EI2DJ** Michael Wright, 5 Woodview Park, The Donahies, Dublin 13
**EI2DL** Details withheld at licensee's request
**EI2DN** A. Ryan, Balreask, Trim Road, Navan, Co. Meath
**EI2DP** Evelyn A. Robinson, 26 Frascati Park, Blackrock, Co. Dublin
**EI2DR** Details withheld at licensee's request
**EI2DT** Gerald P. McGorman, "Teaghlach", Legnakelly, Clones, Co. Monaghan
**EI2DW** Karen G. Wright, 5 Woodview Park, The Donahies, Dublin 13
**EI2EB** Harry McMullan, Mill Road, Bunclody, Co. Wexford
**EI2ED** John Hosty, 17 Ardilaun Road, Newcastle, Galway
**EI2EK** John P. McCafferty, Ballinlaw, Slieverue, Co. Kilkenny
**EI2EM** Charles Lyons, 24 Oakwood Avenue, Swords, Co. Dublin
**EI2EO** Michael F. Hayes, Curry Cummer, Tuam, Co. Galway
**EI2ET** Manfred Lauterborn, Address withheld at licensee's request
**EI2EX** Details withheld at licensee's request
**EI2FA** William G. Ryan, 115 Fortview Drive, Ballinacurra Gardens, Limerick
**EI2FD** John R. Masterson, Aughamore, Ballinaloe, Co. Longford
**EI2FE** Patrick Keelan, 59 Dean Cogan Place, Navan, Co. Meath
**EI2FG** John Hearne, 4 Delwood Grove, Ballinure, Blackrock, Cork
**EI2FJ** Michael C. Griffin, Cahirtilane, Castlemaine, Co. Kerry
**EI2FL** Details withheld at licensee's request
**EI2FM** Denis Cadogan, 53 Bancroft Park, Tallaght, Dublin 24
**EI2FN** John McGowan, 15 Lr. Kindlestown, Delgany, Co. Wicklow
**EI2FQ** Flor Lynch, 'Nephin', Coronea, Skibbereen, Co. Cork
**EI2FS** Edward Taylor, 8 Cappanoole, Estuary Drive, Mahon, Mahon, Cork A96 RF99
**EI2FT** Mary T. Lyons, 24 Oakwood Avenue, Swords, Co. Dublin
**EI2GC** Seamus McGiff, Main Street, Buttevant, Co. Cork
**EI2GI** Peter Bluett, 50 Dukesmeadow Avenue, Kilkenny
**EI2GM** John J. Hill, Decomade, Lissycasey, Co. Clare
**EI2GN** John P. Ketch, 9 Rockcliffe Terrace, Blackrock Road, Cork
**EI2GO** William A. Fahy, 11 Laureston Court, Tower, Blarney, Co. Cork
**EI2GP** Thomas Rea, Bridge Street, Headford, Co. Galway
**EI2GR** Adrian W. O'Leary, 15 Elm Drive, Shamrock Lawn, Douglas, Cork
**EI2GT** Dermot Gleeson, Aoibhneas, 7 Doon Houses, Clarina, Co. Limerick
**EI2GX** Tony Stack, 46 Beechdale, Dunboyne, Co. Meath
**EI2GY** Details withheld at licensee's request
**EI2GZ** Details withheld at licensee's request
**EI2HC** Dick Bean, K1HC, 422 Everett Street, Westwood, MA 02090-2218, USA
**EI2HD** Dermot P. Miley, 26 Slievebloom Park, Walkinstown, Dublin 12
**EI2HF** Pat McGrath, Ballintemple, Castlegar, Co. Galway
**EI2HG** Damian Commins, 5 Cruachan Park, Rahoon Road, Galway
**EI2HH** Andy Linton, Ballygorey, Mooncoin, Co. Kilkenny
**EI2HI** Hugh O'Donnell, Baurleigh, Bandon, Co. Cork
**EI2HL** Mark J. Doyle, 28 Ashton Avenue, Knocklyon, Dublin 16
**EI2HN** Patrick McGrath, Carriglawn, Ballysimon Road, Limerick
**EI2HO** John Hagin-Meade, 26 Broadford Walk, Ballinteer, Dublin 16
**EI2HP** David Waugh, 16 Seaview Avenue, Millisle BT22 2BN, Co. Down, Northern Ireland
**EI2HQ** Joseph Quigley, Newtown Kells, Co. Kilkenny
**EI2HR** David Hooper, 14 Corbally Way, Westbrook Lawns, Tallaght, Dublin 24
**EI2HS** Robert Hyde, 12 Avondale Drive, Kilcohan, Waterford
**EI2HV** Patrick K. Keogh, 27 St Joseph's Square, Fermoy, Co. Cork
**EI2HW** John M. Forristal, "Seawinds", Islandtarsney, Fennor, Tramore, Co. Waterford
**EI2HX** Patrick J. Fitzpatrick, 24 Ascail A DÛ, Yellow Batter, Drogheda, Co. Louth
**EI2HY** Anthony O'Rourke, 13 Hazel Road, Togher, Cork
**EI2IA** Eoghan " hUallach·in, 171 Martello, Port MearnÚg, Co. ¡tha Cliath
**EI2IB** Michael Kiely, Ballinrush Lower, Kilworth, Co. Cork
**EI2IF** Patrick P. Rosney, 15 Tegan Court, Mucklagh, Tullamore, Co. Offaly
**EI2IG** Ivan O'Sullivan, Rathoneigue, Bartlemy, Fermoy, Co. Cork
**EI2IH** Hugh Galt, Rathconry, Carrowkeel, Kincon, Killala, Co. Mayo
**EI2II** Enda Broderick, Cloonacastle, Kylebrack, Loughrea, Co. Galway
**EI2IJ** Mary T. Daly Scanlon, Meehan, Coosan, Athlone, Co. Westmeath
**EI2IM** Donald F. Lynch Jr., 1517 West Little Neck Rd., Virginia Beach, VA 23452, USA
**EI2IN** Brendan F. Logue, Whitecross, Julianstown, Co. Meath
**EI2IO** Barry Bridgeman, 2 Killena, Knockraha, Co. Cork
**EI2IP** Robbie Phelan, Kilmeedy West, Kinsalebeg, Via Youghal, Co. Waterford
**EI2IQ** Fr. Robert Swinburne, Salesian House, Milford Grange, Castletroy, Limerick V94 X832
**EI2IS** Dermott Finegan, Apartment 18, James McSweeney House, Berkley Street, Dublin 7
**EI2IT** Thomas Hallinan, P.O. Box 20, Cahir, Co. Tipperary
**EI2IU** Details withheld at licensee's request
**EI2IV** James Linehan, 35 Maulban, Passage West, Co. Cork
**EI2IW** Martin McElwee, 7 Columcille Close, Kerrykeel, Letterkenny, Co. Donegal
**EI2IX** Michael Kingston, 5 Merval Crescent, Clareview, Limerick
**EI2IZ** Details withheld at licensee's request
**EI2JA** Peadar Slattery, 74 Bettyglen, Raheny, Dublin 5
**EI2JB** John A. Burke, The Green Road, Rathduff, Golden, Co. Tipperary
**EI2JC** Noel Walsh, 53 Kennedy Park, Roscrea, Co. Tipperary
**EI2JD** Thomas Caffrey, The Slip, Clogherhead, Co. Louth
**EI2JE** Details withheld at licensee's request
**EI2JF** Paul Hurley, 17 Turlough Gardens, Fairview, Dublin 3
**EI2JG** John Geoghegan, 120 St James Road, Walkinstown, Dublin 12
**EI2JL** Nicholas Cummins, 25 Pinewood Park, Rathfarnham, Dublin 14
**EI2JM** Michael Goss, Ballymorris, Aughrim, Arklow, Co. Wicklow
**EI2JP** Ron Hahn, Rose Hill, Rosslare Strand, Co. Wexford
**EI2JQ** Christian Kieffer, 39 Route de Reisdorf, L-6311 Beaufort, Luxembourg
**EI2JT** Don Harney, 1 Hunter's Vale, Maryborough Woods, Douglas, Cork
**EI2JU** Maurice O'Sullivan, 49 Ardfallen Estate, Douglas Road, Cork
**EI2JV** Paul Michael Albone, 47 Wainsfort, Rochestown Road, Cork
**EI2JW** Patrick Gerard McGuinness, 9 Farmdale Road, Carshalton Beeches, Surrey SM5 3NG, England
**EI2JX** Philip Lawrence, 7 The Woodlands, Lowestoft, Suffolk NR32 5EZ, England
**EI2JY** James P. Lappin, Cashel, Ayle, Westport, Co.Mayo
**EI2JZ** Joe Guilfoyle, 58 Willow Park Crescent, Glasnevin, Dublin 11
**EI2KA** Tim McKnight, Gortadrohid, Reengaroga Island, Baltimore, Co. Cork
**EI2KB** Details withheld at licensee's request
**EI2KC** Anthony Murphy, 80 Cedarfield, Donore Road, Drogheda, Co. Louth
**EI2KD** Rodney Roulston, Court, Milford, Letterkenny, Co. Donegal



For information about amateur radio in Ireland, see: *www.irts.ie*
For information about amateur radio licensing in Ireland, see: *www.comreg.ie*

EI2KE Patrick J. McLoughlin, 102 Aglish Estate, Castlebar, Co. Mayo
EI2KF Details withheld at licensee's request
EI2KG Antonio Iafrate, Dublin Street, Monasterevin, Co. Kildare
EI2KH Seamus F. Holland, 21 Valentia Road, Drumcondra, Dublin 9
EI2KI Details withheld at licensee's request
EI2KJ Adrian T. O'Gorman, Hillside Cottage, Castlelumney, Dunleer, Co. Louth
EI2KK Details withheld at licensee's request
EI2KL Details withheld at licensee's request
EI2KM Details withheld at licensee's request
EI2MC Jack Quinn, 1764 Brandee Lane, Santa Rosa, California 95403, USA
EI2ADB Thomas J. Boland, 8 Hillsbrook Grove, Perrystown, Dublin 12
EI2AFB John Bergin, Grange, Ballyragget, Co. Kilkenny
EI2AIR Ballooning Amateur Radio Club of Ireland, c/o Aidan Murphy, P.O. Box 9927, Dunshaughlin, Co. Meath
EI2AMB D. Tocher, 2 Mount Shannon Road, Lisnagry, Co. Limerick
EI2APB Matthew Gavigan, 41 Watermeadow Drive, Old Bawn, Dublin 24
EI2BLB Noel McBrearty, Dunmore, Carrigans, Co. Donegal
EI2BRB Details withheld at licensee's request
EI2BWB William J. Dundon, Ardshanbally, Adare, Co. Limerick, V94 Y2YY
EI2CHB Eamonn O'Connor, 441 Howth Road, Raheny, Dublin 5
EI2CPB George Adjaye, 245 Lower Kimmage Road, Dublin 6W
EI2CRG Carlow & District Amateur Radio Club, c/o Pat Hutton, EI6HF, 50 Ash Grove, Tullow Road, Carlow
EI2CTB Aidan V. Grant, Hillview, Ballyknock, Tramore, Co. Waterford
EI2DDB Details withheld at licensee's request
EI2DLB Keith Chadwick, 32 Deansrath Road, Clondalkin, Dublin 22
EI2DUB Details withheld at licensee's request
EI2EBB John A. McGennis, 4 College Drive, Terenure, Dublin 6W
EI2ECB Kenneth W. Gaul, 109 Kimmage Road West, Terenure, Dublin 12
EI2EDB Colm Donelan, 11 Cloghanboy Crescent, Ballymahon Road, Athlone, Co. Westmeath
EI2EHB Maurice Wilson, 27 Vernon Gardens, Clontarf, Dublin 3
EI2EJB Wilfred J. Higgins, Ashbury, 24a Waterloo Lane, Ballsbridge, Dublin 4
EI2ELB Niall J. Coveney, Athassel, Burnaby Road, Greystones, Co. Wicklow
EI2EMB David B. Meehan, Ballincurrig, Leamlara, Midleton, Co. Cork
EI2EUB Owen J. O'Sullivan, Coole Cottage, Coole Lane, Gowlane South, Donoughmore, Co. Cork
EI2EVB Paul J Kelly, 16 Drumacrin Avenue, Bundoran, Co. Donegal
EI2FCB Terry McEvoy, The Cinema, Monasterevan, Co. Kildare
EI2FDB Sean Lynch, Killeenacoff, Westport, Co. Mayo
EI2FEB John Lyons, 24 Oakwood Avenue, Swords, Co. Dublin
EI2FJB Martin Hanley, 22 St Asicus Villas, Athlone, Co. Westmeath
EI2FRB Donogh Roche, Address withheld at licensee's request
EI2FRC Fingal Radio Club, c/o Erin's Isle GAA Club, Farnham Drive, Finglas, Dublin 11
EI2FZB Alan Geoghegan, 21 Hop Hill Vale, Tullamore, Co. Offaly
EI2GBB David Magee, 40 Bellevue Court, Father Russell Road, Dooradoyle, Limerick
EI2GCB James J. Kelly, Derrynacross, Breaffy, Castlebar, Co. Mayo
EI2GFB Declan Mayock, 36 Killyconnigan, Monaghan, Co. Monaghan
EI2GGB Gavin Heneghan, Lissarda, Co. Cork
EI2GHB Details withheld at licensee's request
EI2GKB Michael Friary, 18a Allenview Heights, Newbridge, Co. Kildare
EI2GLB Trevor Dunne, Hillcrest, Mooretown, Kildare Town, Co. Kildare
EI2GMB Details withheld at licensee's request
EI2GNB Gerard McCarthy, Boola, Glendine, Youghal, Co. Cork
EI2GPB Robert Nunn, 35 Sarsfield Terrace, Youghal, Co. Cork
EI2GQB Details withheld at licensee's request
EI2GRB Steve Sola, 10439 W. Roundelay Circle, Sun City, Arizona 85351, USA
EI2GRC GMIT Radio Club, c/o Tom Frawley EI3ER, GMIT Engineering Department, Dublin Road, Galway
EI2GSB Paul Eustace, 2 Haydens Park Walk, Lucan, Co. Dublin
EI2GTB Lydia Ritchie, 23 Derrygreenagh Park, Rochfortbridge, Co. Westmeath
EI2GUB Samuel Ritchie (Snr), 23 Derrygreenagh Park, Rochfortbridge, Co. Westmeath
EI2GXB Paudy MacKenna, Knockavota, Milltown, Killarney, Co. Kerry
EI2GYB Steven Homer, Tullyarb, Carndonagh, Co. Donegal
EI2HAB Adrian Healy, 3 Lucan Street, Castlebar, Co. Mayo
EI2HCB Desmond Sharpe, Kellystown, Slane, Co. Meath
EI2HDB Robert Murphy, Claramore, Millstreet, Co. Cork
EI2HEB Details withheld at licensee's request
EI2HFB Details withheld at licensee's request
EI2HGB Michael Keogh, Esker, Castleblakeney, Ballinasloe, Co. Galway
EI2HHB Clive Kilgallen, The Old Millhouse, Riverstown, Co. Sligo
EI2HIB Details withheld at licensee's request
EI2HJB Details withheld at licensee's request
EI2HKB Se·n McKeown, Corleackagh, Castleblayney, Co. Monaghan
EI2HLB Details withheld at licensee's request
EI2HMB Ingo Kallfass, Danziger, Strasse 10, 75015 Bretten, Germany
EI2HNB Robert Grundulis, 47 Kells Road, Crumlin, Dublin 12
EI2HOT Ballooning Amateur Radio Club of Ireland, c/o Aidan Murphy, P.O. Box 9927, Dunshaughlin, Co. Meath
EI2HPB Dave Carty, 39 Kevinsfort Heath, Strandhill Road, Sligo
EI2HQB Details withheld at licensee's request
EI2HRB Lee P. McGuire, 39 Glenrichards Cove, Pollshone, Gorey, Co. Wexford
EI2HSB Dennis Drennan, Ballyroberts, Cuffesgrange, Co. Kilkenny
EI2HTB Details withheld at licensee's request
EI2HUB Turlough McKeown, Macetown, Tara, Co. Meath
EI2HVB John King, 17 Ardcrannagh, Gortlee, Letterkenny, Co. Donegal
EI2HWB Details withheld at licensee's request
EI2HXB Tommy Browne, Bree, Malin Head, Co. Donegal
EI2HYB Declan Byrne, 1 Hillview Way, Bennettsbridge, Co. Kilkenny, R95 NX93
EI2HZB Sean Byrne, Corries Bridge, Bagenalstown, Co. Carlow, R21 RD21
EI2IHE Plassey Amateur Radio Club, c/o John Bird, Department of Maths & Statistics, University of Limerick, Limerick
EI2IPA International Police Association, c/o Jim Jeffers EI8DK
EI2MIE Dublin QRP Radio Amateur Club, Marino Institute of Education, Griffith Avenue, Dublin 9, (Correspondence c/o Ron Hall EI4AR)
EI2MRG Mayo VHF Group, c/o Joe Fadden EI3IX, Knockthomas, Castlebar, Co. Mayo
EI2NCR North County Radio Club, c/o Derek McGonagle, North Strand, Skerries, Co. Dublin
EI2PAR Phoenix Amateur Radio Club, 30 Woodview Grove, Blanchardstown, Dublin 15
EI2QRP QRP Club of Ireland, c/o William K. Ryan EI8BC
EI2RTE RTE Amateur Radio Club, c/o Michael Wright, 5 Woodview Park, The Donahies, Dublin 13
EI2SBC Shannon Basin Radio Club, c/o Brian Canning, Aughnaglace, Cloone, Co. Leitrim
EI2SDR South Dublin Radio Club, Ballyroan Community Centre, Marian Road, Rathfarnham, Dublin 14
EI2SRC Sligo Amateur Radio Club, c/o Michael B. Rooney, 84 Knocknagranny Park, Sligo
EI2TRC Thomond Radio Club, c/o William G. Ryan EI2FA, 115 Fortview Drive, Ballinacurra Gardens, Limerick
EI2WRC South Eastern Amateur Radio Group, c/o Mark Wall, 11 Kilcaragh Village, Ballygunner, Waterford
EI2WWC Wexford Wireless Club, Gorey Scout Group Hall, Gorey Sport and Leisure Centre, Esmonde Street, Gorey Co. Wexford
EI2SIRG Southern Irish Repeater Group, c/o John McCarthy EI8JA

## EI3

EI3A John Llewellyn, Pier Road, Enniscrone, Co. Sligo
EI3C Kevin Kilduff, Swellan Lower, Cavan
EI3I Josef F. Brezina, 43 Redesdale Road, Mount Merrion, Blackrock, Co. Dublin
EI3V James J. Moore, 8 Meadow Mount, Churchtown, Dublin 16
EI3Y Ian Clarke, 1 Dernan Grove, Ballina, Co. Mayo
EI3Z Shannon Basin Radio Club, QSL via EI2SBC
EJ3Z Shannon Basin Radio Club, QSL via EI2SBC
EI3AL Thomas O'Sullivan, Monavalley, Mill Road, Killarney V93 K7Y7, Co. Kerry
EI3AX William E. Curristan, Riverside House, Waterloo Place, Donegal Town
EI3BF John J. Hickey, Ard Mo Chroi, Cullen, Mallow, Co. Cork
EI3BK Jeremiah O'Sullivan, 70 Uam Var Drive, Bishopstown, Co. Cork
EI3BW Bernard Flynn, Viewmount, Clarkes Road, Gurteens, Ballina, Co. Mayo
EI3CA Brian Fox, Rosebank, Farranlea Park, Model Farm Road, Cork
EI3CD Fergal Holmes, 7 Elgin Wood, Bray, Co. Wicklow
EI3CG Rev. Alan Malone, The Square, Eyrecourt, Co. Galway
EI3CN Michael Lawlor, 16 Wainsfort Grove, Terenure, Dublin 6W
EI3CP Colum P. Clarke, Sunbury, Killough, Kilmacanogue, Co. Wicklow
EI3CW Details withheld at licensee's request
EI3CX Thomas P. Dwyer, 10 Brookhaven Park, Blanchardstown, Co. Dublin
EI3CZ Roderick Power, 50 Cabinteely Avenue, Cabinteely, Dublin 18
EI3DB Michael P. Mee, Minnistown Road, Laytown, Co. Meath
EI3DD Sean O'Rourke, Gallowstown, Roscommon, Co. Roscommon
EI3DJ Details withheld at licensee's request
EI3DN Peter O'Sullivan, Cor an Aitinn, Ardea, Tuoist, Killarney, Co. Kerry
EI3DP Jim Ryan, 11 Knockgriffin, Midleton, Co. Cork
EI3DT Sean N. Walsh, Feenagh, Quin, Co. Clare
EI3DV Gregory M. Murphy, 8 Laurelton, Bushy Park Road, Rathgar, Dublin 6
EI3DY Michael Staunton, 'Glenina', Enniskerry Road, Sandyford, Dublin 18
EI3DZ Stewart C. Crampton, 135A Ballymena Road, Doagh, Ballyclare, Co. Antrim, BT39 OTN, Northern Ireland
EI3EA Gerard A. O'Sullivan, 29 Janemount Park, Corbally, Limerick
EI3EC John K. O'Sullivan, 'Dunhallow', 4 Woodvale Road, Beaumont, Cork
EI3ED Edward H. Brooks, GD4HOX, Elmwood, Somerset Road, Douglas, Isle of Man, IM2 5AE
EI3EF Details withheld at licensee's request
EI3EG Aedan O'Meara, 42 Halldene Drive, Bishopstown, Cork
EI3EH John Kelly, Templeton Glebe, Killashee, Co. Longford
EI3EI Phyllis M. MacArthur, 11 Woodlawn Terrace, Upper Churchtown Road, Dublin 14
EI3EN Harry Houlihan, Franciscan Friary, Lady Lane, Waterford
EI3EP John Cronin, Cum, Laherdane, Ballina, Co. Mayo
EI3EQ Malcolm Bowden, Rosseragh, Ramelton, Letterkenny, Co. Donegal
EI3ER Thomas Frawley, Kiloughter, Castlegar, Galway
EI3EU Kevin B. Clancy, 11 Raheen Court, Tallaght, Dublin 24
EI3EZ Olan O'Brien, 68 Uam Var Drive, Bishopstown, Cork
EI3FA Details withheld at licensee's request
EI3FE Pierce Meagher, 158 Merrion Road, Dublin 4
EI3FF Details withheld at licensee's request
EI3FN Isaac F. Wheelock, Beech Heights, Monart, Enniscorthy, Co. Wexford
EI3FO James Fitton, Ballinara Cross, Waterfall, Cork
EI3FQ Joseph P. Walsh, 3 Grosvenor Terrace, Monkstown, Co. Dublin
EI3FR Tom Doherty, Newtown Bridge, Summerhill Road, Dunboyne, Co. Meath
EI3FU John Lawlor, Main Street, Castletownroche, Mallow, Co. Cork
EI3FV Details withheld at licensee's request
EI3FW Craig Robinson, Arrowrock Lodge Ballynary, Lough Arrow, Via Boyle, Co. Roscommon
EI3FX Harry Harty, Garryfine, Bruree, Co. Limerick
EI3FZ James Mannix, Keelclogherane, Faha, Killarney, Co. Kerry
EI3GC Details withheld at licensee's request
EI3GD Details withheld at licensee's request
EI3GE Jim Echlin, Stylebawn Cottage, Delgany, Co. Wicklow
EI3GF Michael C. Quinn, 13 Seafield, Wicklow Town, Co. Wicklow
EI3GG Gerard Elliott GI4OWA, 4 Fernbrae Gardens, Kilfennan, Derry, BT47 1XS, Northern Ireland
EI3GH Liam Field, 'Ittledy', Portrane Road, Donabate, Co. Dublin
EI3GN Timothy J. McCarthy, Ballintrim, Rostellan, Midleton, Co. Cork
EI3GO Martin J. Ffrench, Iry, Ballyfin, Portlaoise, Co. Laoise
EI3GS John P. Cowman, 29 Beaumont Lawn, Ballintemple, Cork
EI3GU Patrick A. Murtagh, 31 Seaview Park, Shankill, Dublin 18
EI3GV Brendan M. De h"ra, 7 Lakelands Lawn, Stillorgan, Blackrock, Co. Dublin
EI3GW William A. Wilson, 10 Kickham Street, Mullinahone, Thurles, Co. Tipperary
EI3GY Luke Conroy Sr., 407 Carnlough Road, Cabra West, Dublin 7
EI3GZ John J. O'Sullivan, Address withheld at licensee's request
EI3HA Anthony Casey, 4 Burma Road, Strandhill, Sligo, Co. Sligo
EI3HB Kinsale Radio Club, c/o Jeremy Sheehan, Hillcrest, Bandon Road, Kinsale, Co. Cork
EI3HE Eamonn Doyle, Kilcavan, Carnew, Co. Wicklow
EI3HF SÙnia M. Malone, 7 Clonard Drive, Dundrum, Dublin 16
EI3HG Andy Green, 44 Newtown Glen, Tramore, Co. Waterford
EI3HJ John Harte, Oldtown, Moycullen, Co. Galway
EI3HK Liam Murphy, 67 Gracepark Meadows, Drumcondra, Dublin 9
EI3HM Seosamh " hlarn·in, Na Haille Thiar, Indreabh·n, Conamara, Co. na Gaillimhe
EI3HO Details withheld at licensee's request
EI3HS Austin Grogan, 25 Townparks, Skerries, Co. Dublin
EI3HW James M. Hogan, 5 Eastlands, Tramore, Co. Waterford
EI3HX David G Lafferty, Carrickshandrum, Killygordon, Lifford, Co. Donegal
EI3HY Michael Hentschel, Ballinacrick, Kerrykeel P.O., Letterkenny, Co. Donegal
EI3HZ Noel Moore, Ennis Road, Newmarket on Fergus, Co. Clare
EI3IA David Heale, 3 Evans Wharf, Hemel Hempstead, HP3 9WU, England
EI3ID Naish Kelly, Grace Dieu, Ballyboughal, Co. Dublin
EI3IE Liam P. O'Riordan, Ballintubbrid East, Carrigtwohill, Co. Cork
EI3IF Patrick J O'Doherty, Cromogue, Dromcollogher, Charleville, Co. Cork
EI3IG Michael Everett Clarke, Ballindrimley, Castlerea, Co. Roscommon
EI3IJ Patrick C. Corkery, "Ard Alainn", 65 St. Patrick's Road, Greenhills, Dublin 12
EI3IK Brian Gundry, Beherna, Ryefield, Virginia, Co. Cavan
EI3IL Details withheld at licensee's request
EI3IM Paul Reilly, 253 Ballsgrove, Drogheda, Co. Louth
EI3IN Patrick Augustine Jennings, 43 Sutton Park, Sutton, Dublin 13
EI3IO David Court, Connogue, River Lane, Shankill, Dublin 18 D18 W2R4
EI3IP Se·n " S`illeabh·in, 14 The Crescent, Inse Bay, Laytown, Drogheda, Co. Louth
EI3IQ Michael Walsh, 437 St. John's Park, Waterford
EI3IS Enda Coffey, Corranellistrum, Rosscahill, Co. Galway
EI3IT Tony Kenny, 130 Ard O'Donnell, Letterkenny, Co. Donegal
EI3IU Jarlath J. Herron, Donegal Street, Ballybofey, Co. Donegal
EI3IV Hugh Boyle, Devlinmore, Carrigart, Letterkenny, Co. Donegal
EI3IX Joe Fadden, St. Joseph's, Knockthomas, Castlebar, Co. Mayo
EI3IY Hugh O'Leary, Stumphill, Midleton, Co. Cork
EI3IZ Terry Lynch, 77A Landscape Park, Churchtown, Dublin 14
EI3JA Peter Enright, Chapel Hill, Castleconnell, Co. Limerick
EI3JB Nicholas M. Madigan, 15 Birch Avenue, Birchwood, Waterford
EI3JD John A. Wrafter, 12 Park Court, Tara Street, Tullamore, Co. Offaly
EI3JE Neil Powell, The Bungalow, Schoolgardens, Ladysbridge, Co. Cork
EI3JF Maurice Donworth, Kilballyowen, Bruff, Co. Limerick
EI3JG Adrian Fry, 128 Sylvan Way, Sea Mills, Bristol BS9 2LU, England
EI3JK Alan Walsh, Roxborough, Ballysheedy, Co. Limerick
EI3JL David Eames, Idlewild, Farnamullan, Lisbellaw, Co. Fermanagh BT94 5EA, Northern Ireland
EI3JM John McAndrew, Ballinvoy, Aughagower, Westport, Co. Mayo
EI3JN James Lane, Avondhu, Knockaverry, Youghal, Co. Cork
EI3JO John Maloco, Department of Electronic Engineering, NUI Maynooth, Co. Kildare
EI3JP Frank Hunter, 2 Wandsworth Court, Belfast BT4 3GD, Northern Ireland
EI3JQ Dmitry Bubyagin, 13 Station Grove, Station Road, Portarlington, Co. Laois
EI3JR John Walsh, "Estuary View", Gortnaskeha, Ballybunion, Co. Kerry
EI3JS Grahame S. Duffy, 39 Cornfield Lane, Newtowncunningham, Co. Donegal
EI3JU Details withheld at licensee's request
EI3JW William R. Welch M.D., P.O. Box 425, 49 Parker Road, Osterville, MA 02655-0425, USA
EI3JX Thomas H. Perrott, Bolacreen, Gorey, Co. Wexford
EI3JY Eamonn J. G. MacIntyre, 115 Beldoo, Strabane BT82 9QL, Co. Tyrone, Northern Ireland
EI3JZ Dmitrij Pavlov, Corduff, Carrickmacross, Co. Monaghan
EI3KA John Smyth, Tiaquin, Colmanstown, Co. Galway
EI3KB Alvin Raymond Goebel, 4820 Orleans Lane North, Plymouth, MN 55442, USA
EI3KC Pat Twomey, The Rock, Carrigaline, Co. Cork
EI3KD Mark Turner, 2 The Willows, Bridgemount, Carrigaline, Co. Cork
EI3KE Seamus Campbell, Reynoldstown, Clogherhead, Co. Louth
EI3KF Andy Clements, Trim, Co. Meath
EI3KG Michal Konopka, Apt.3 Coranna, Tipper South, Naas, Co. Kildare
EI3KH Patrick Donnelly, The Laundry House, Roseberry, Newbridge, Co. Kildare

**EI3KI** Juozas Piepalius, 10 Hillsbrook Lawn, Brittas, Co. Dublin
**EI3KJ** Ivan William McCaffrey, 19250 East Palm Ln, Black Canyon City, Arizona 85324, USA
**EI3KK** Christian Hillmer, Juraring 13, 38440 Wolfsburg, Germany
**EI3KL** Gordon Harrison, 65 Trinity Gardens, Drogheda, Co. Louth
**EI3KM** Details withheld at licensee's request
**EI3SI** Larch Hill Radio Club, c/o Daniel Cussen EI9FHB, South Dublin Radio Club, Ballyroan Community Centre, Marian Road, Rathfarnham, Dublin 14
**EI3ADB** James G. Lacy, An Cuan, Fairbrook Lawn, Rathfarnham, Dublin 14
**EI3AJB** James G. Gough, 50 Culmore Point, Derry BT48 8JW, Northern Ireland
**EI3AKB** Gerard McCarthy, 140 Rusheeny Court, Rusheeny Village, Dublin 15
**EI3AYB** John G. Gilmour, Cloneen, Drumcliff, Co. Sligo
**EI3BCB** James P. Keane, Birchgrove, Ballinasloe, Co. Roscommon
**EI3BVB** Peter J. Towey, 51 Cloondara, Ballisodare, Co. Sligo
**EI3CAB** Douglas Port, 8 Betterton Drive, Sidcup, Kent DA14 4PS, England
**EI3COB** Declan Mullally, Mount Prospect, Raheen Road, Limerick
**EI3CTB** Justin Behan, 25 Birchdale Road, Kinsealy Court, Kinsealy, Co. Dublin
**EI3CUB** Tony Cronin, Chapel Road, Old Portmarnock, Co. Dublin
**EI3CVB** Laurence B. Reardon, 21 Redesdale Road, Mount Merrion, Co. Dublin
**EI3CYB** James Allen, 40 Victoria Street, South Circular Road, Dublin 8
**EI3CZB** Details withheld at licensee's request
**EI3DFB** Sarah Lee, 'Helvellyn', Knockaverry, Youghal, Co. Cork
**EI3DHB** Terence N. McCafferty, 51 Oweneragh Drive, Strabane, Co. Tyrone BT82 9DR, Northern Ireland
**EI3DIB** John A. Lofthouse, Monroe East, Ardfinnan, Clonmel, Co. Tipperary
**EI3DQB** Ray Elgy, Rooskagh East, Carrigkerry, Athea, Co. Limerick
**EI3DRG** Duhallow Repeater Group, c/o EI2HI
**EI3DWB** Finbarr Sheehy, 4 Abbey Court, Prosperous Road, Clane, Co. Kildare
**EI3DYB** P.J. Tallon, Ballinree, Mostrim, Co. Longford
**EI3DZB** Denise M. Lyons, 24 Oakwood Avenue, Swords, Co. Dublin
**EI3EBB** Alan D Foley, 23 The View, Priory Court, Watergrasshill, Co. Cork
**EI3ECB** Details withheld at licensee's request
**EI3EEB** John O'Hea, Myrtle Hill, Ballygarvan, Co. Cork
**EI3EFB** Michael A Fitzgerald, Broomfield House, Midleton, Co. Cork
**EI3EMB** Enda L Murphy, Belvedere House, Tower, Blarney, Co. Cork
**EI3ENB** Paul Norris, Clogga, Mooncoin, Co. Kilkenny
**EI3ETB** Matthew J Byrne, 60 College Rise, Newfoundwell Rd, Drogheda, Co. Louth
**EI3EVB** Garrett O'Hanlon, Ashbury House, Dunmore East, Co. Waterford
**EI3EXB** Daniel Doran, Convent View, Mooncoin, Co. Kilkenny
**EI3EZB** Brendan Meehan, 40 Holly Road, Dublin 9
**EI3FAB** Details withheld at licensee's request
**EI3FBB** Paul McEntagart, 3 Grange End, Dunshaughlin, Co. Meath
**EI3FDB** Patrick Kearney, Curraclough, Bandon, Co. Cork
**EI3FEB** Andrew McCormack, Claran, Ower P.O., Headford, Co. Galway
**EI3FFB** Eamonn Kavanagh, Ballyverane, Bansha, Co. Tipperary, E34 WE29
**EI3FKB** Paul McQuaid, 2 Brides Glen Park, Swords, Co. Dublin
**EI3FOB** Noel Lane, 53 Dundanion Road, Beaumont, Cork
**EI3FXB** Ciaran Ferry, Brinalack P.O., Gweedore, Letterkenny, Co. Donegal
**EI3GAB** Anthony R. Cummins, "Lyston", 13 Forest View, Goulds Hill, Mallow, Co. Cork
**EI3GBB** Details withheld at licensee's request
**EI3GDB** Jim Hayes, Craigue, Kildorrery, Co. Cork, P67KH36
**EI3GFB** Brendan Kehoe, Heathpark, Old Ross, Newbawn, Co. Wexford
**EI3GGB** Robert O'Sullivan, 102 Ardmore Avenue, Knocknaheeny, Cork
**EI3GHB** Allen Collinge, 2 Moyglare Village, Moyglare Road, Maynooth, Co. Kildare
**EI3GIB** Gary Renihan, Crossakiel, Kells, Co. Meath
**EI3GJB** Details withheld at licensee's request
**EI3GKB** David Reid, c/o 1172 Eyrefield Road, Curragh, Co. Kildare
**EI3GLB** Sandra Nel, Ardkirk, Castleblaney, Co. Monaghan
**EI3GOB** Mark Ford, Curkish Lane, Bailieborough, Co. Cavan
**EI3GQB** Francis Shortt, 10d Glin Park, Coolock, Dublin 17
**EI3GRB** Details withheld at licensee's request
**EI3GTB** Ciar·n Fagan, Cloonagh, Dring, Co. Longford
**EI3GUB** P·draigh M. " Murch˙, Saoirse, 40 Radhairch an Gleann, Ghr·inseach, DuglaÌs, Corcaigh
**EI3GVB** William Flynn, 141 Seapark, Malahide, Co. Dublin
**EI3GWB** Richard Bonner, Summerhill, Donegal Town, Co. Donegal
**EI3GXB** Iain Mounsey, Riverside, Coolrecuill, Tourlestrane, Tubbercurry, Co. Sligo
**EI3GYB** Michael Foertig, Valley Forge, Carrowneden, Ballyhaunis, Co. Mayo F35 H308
**EI3HAB** Details withheld at licensee's request
**EI3HBB** Details withheld at licensee's request
**EI3HCB** Jim Whitty, Wellingtonbridge, Co. Wexford
**EI3HDB** Alan Buckley, Mohurry, Kiltealy, Enniscorthy, Co. Wexford
**EI3HEB** James Dwyer, 32 Templeroan Grove, Knocklyon, Dublin 16
**EI3HFB** Ian Morrow, 90 Bracky Road, Sixmilecross, Omagh, Co. Tyrone BT 799PH, Northern Ireland
**EI3HGB** Jer Aspell, 31 Woodview Drive, Kennill Hill, Mallow, Co. Cork
**EI3HHB** Gerald Brennan, Mucklagh, Carlingford, Co. Louth
**EI3HIB** David Mee, 7 Monread Lawns, Naas, Co. Kildare
**EI3HJB** Rafal Wasowicz, 10 The Green, Tir Cluain, Midleton, Co. Cork
**EI3HKB** Bobby Wadey, 20 Clonmakate Road, Portadown, Co. Armagh BT62 1SU, Northern Ireland
**EI3HLB** Details withheld at licensee's request
**EI3HNB** Martin J. Murphy, Judesville, Convent Road, Claremorris, Co. Mayo
**EI3HOB** Details withheld at licensee's request
**EI3HPB** Graeme McCusker, 41 The Granary, Waringstown, Co. Armagh BT66 7TG, Northern Ireland
**EI3HQB** John Tubbritt, 4 Seafield, Newtown, Tramore, Co. Waterford
**EI3HRB** Joseph Desbonnet, 106 Turvisce, Doughiske, Co. Galway
**EI3HSB** Robert Leslie Whitehall, Whitywood, Irremore, Listowel, Co. Kerry
**EI3HTB** David Connolly, 21 Corran Ard, Athy, Co. Kildare
**EI3HVB** Michael Kearns, Knockina, Gorey, Co. Wexford
**EI3HWB** Aurelian V. Lazarut, 141 Branswood, Kilkenny Road, Athy, Co. Kildare
**EI3HXB** Details withheld at licensee's request
**EI3HYB** Details withheld at licensee's request
**EI3HZB** James Devereux, 11 Heather Grove, Marley Wood, Grange Road, Dublin 16
**EI3RCW** Waterford Institute of Technology Amateur Radio Club, Cork Road, Waterford

## EI4

**EI4J** Shane P. Halpin, "Dookinella", Rathmullen Road, Drogheda, Co. Louth
**EI4L** John Kelly, 25 Abbey Court, Abbey Farm, Celbridge, Co. Kildare
**EI4AB** Christopher Connolly, 29 Marian Park, Waterford
**EI4AL** Michael Burke, Baylough, Athlone, Co. Westmeath
**EI4AR** Ronald G. Hall, 32 Marino Green, Marino, Dublin 3
**EI4BB** Brendan Daly, 28 Templeogue Wood, Templeogue, Dublin 6W
**EI4BC** Alexander D. Patterson, Malvern, Delgany, Co. Wicklow
**EI4BK** T. Deegan, 27 Oakland Drive, Greystones, Ennis Road, Limerick
**EI4BS** Patrick Trant, Reencaheragh, Portmagee, Killarney, Co. Kerry
**EI4BX** Jim Bellew, Claret Rock, Upper Faughart, Dundalk, Co.Louth
**EI4BZ** David Moore, Dooneen, Carrigtwohill, Co. Cork
**EI4CE** Con O'Callaghan, 14 Lower Main Street, Newmarket, Co. Cork
**EI4CF** Rev. Fr. Niall Foley, Mullagh, Loughrea, Co. Galway
**EI4CG** Mike Babe, Lisrenny, Tallanstown, Dundalk, Co. Louth
**EI4CI** Pierce O'Brien, Clavinstown, Drumree, Co. Meath
**EI4CM** Details withheld at licensee's request
**EI4CN** Michael P. Brown, Curraghmore, Tullogher, Co. Kilkenny
**EI4CP** Jim Smith, 'Laurelin', Killincarrig, Delgany A63 V099, Co. Wicklow
**EI4CS** Rev. Bro. Francis P. Crummey, Christian Brothers, Bective Street, Kells, Co. Meath
**EI4CV** James Menton, 73 The Oaks, Avondale, Trim, Co. Meath
**EI4CW** Robert J. Brown, 10 Anchor Watch, Shore Street, Donaghdee BT21 0GA, Co. Down, Northern Ireland
**EI4CX** Details withheld at licensee's request
**EI4DC** Patrick Tuohy, "Mill View", Ballinagough, Whitegate, Co. Clare
**EI4DE** Details withheld at licensee's request
**EI4DH** Denis Rooney, Cornagilla, Manorhamilton, Co. Leitrim
**EI4DK** Details withheld at licensee's request
**EI4DM** Mike Fogarty, 11 Willow Park Place, Athlone, Co. Westmeath
**EI4DN** Joe Kirk, 23 Seatown Place, Dundalk, Co. Louth
**EI4DO** William J. Harvey, 20 Eaton Square, Terenure, Dublin 6W
**EI4DP** Details withheld at licensee's request
**EI4DQ** Tom Cocking, 'Scartlea', Saleen, Cloyne, Co. Cork P25 WP74
**EI4DT** James Cullinan, Ballyfrunk, Ballycallan, Co. Kilkenny
**EI4DU** Leslie W. Long, 79 Hawthorn Heights, Letterkenny, Co. Donegal
**EI4DW** Ken McDermott, Curraghamone, Ballybofey, Co. Donegal
**EI4DY** Andrew P. Henry, 6 Belview Court, Greenhill Road, Wicklow Town, Co. Wicklow
**EI4DZ** Noel Cameron, 16 St. Mary's Crescent, Westport, Co. Mayo
**EI4ED** Tony Whelan, 5 Allenagh, Longford, Co. Longford
**EI4EF** George F. Peterson, 152 Clonkeen Crescent, Kill-o-the-Grange, Co. Dublin
**EI4EK** William McCauley, Drumoghill, Manorcunningham, Letterkenny, Co. Donegal
**EI4EL** Christopher Flanagan, Kilnameela Cottage, Ahiohill, Enniskeane, Co. Cork
**EI4EU** Patrick J. Clifford, 75 Newark Street, Lindenhurst, NY 11757, USA
**EI4EV** Thomas Walshe, 2 Clonkeen Crescent, Dun Laoghaire, Co. Dublin
**EI4EW** William H. Kelly, 3 Clonrose Court, Ard na Greine, off Malahide Road, Dublin 13
**EI4EY** John Phelan, Racefield, Church Road, Raheen, Limerick
**EI4EZ** Details withheld at licensee's request
**EI4FD** Michael Keane, 59 Mornington Heights, Trim, Co. Meath
**EI4FE** Details withheld at licensee's request
**EI4FF** Rev. Donal Kilduff CC, Kildallan, Ballyconnell, Co. Cavan
**EI4FH** Details withheld at licensee's request
**EI4FK** Hugh McNulty, Ardi MhÛr, College Farm Road, Letterkenny, Co. Donegal
**EI4FV** Joseph F. Dillon, 5 Verbena Lawn, Sutton, Dublin 13
**EI4FX** Liam Fitzgerald, 'Melrose', Dwyers Road, Midleton, Co. Cork
**EI4GA** Paul Smith, Sheepgrange, Drogheda, Co. Louth
**EI4GB** Liam Mangan, "Mount Ievers", Sixmilebridge, Co. Clare
**EI4GC** John C. Farrell, Curravarahane, Bandon, Co. Cork
**EI4GD** Gerry Cregg, Killaraght, via Boyle, Co. Sligo
**EI4GG** Ronan O'Gorman, 7 St. Senan's Road, Lifford, Ennis, Co. Clare
**EI4GJ** John Slocum, Shanakill, Ballymacoda, Co. Cork
**EI4GK** John J. Donelan, "Inniscarra", Ballybride Road, Shankill, Dublin 18
**EI4GL** Brian O'Daly, Tralee, Co. Kerry
**EI4GO** John H. Dalton, 12 The Anchorage, Clarence Street, Dun Laoghaire, Co. Dublin
**EI4GP** Garrett Sinnott, Ballycadden, Bunclody, Co. Wexford
**EI4GQ** John D. Crichton GI4YWT, 10 Dann Drive, Lisnagelvin BT47 2HW, Londonderry, Northern Ireland
**EI4GR** Michael Dorgan, Ballinasare, Annascuil, Co. Kerry
**EI4GV** Peter Vekinis, Ballyvogue Cottage, Goleen, Co. Cork
**EI4GX** Joseph Earley, 3 Whitworth Terrace, Drumcondra, Dublin 3
**EI4GZ** James Gallagher, 58 St. Jarlath Road, Cabra, Dublin 7
**EI4HE** Robert McGrogan, 12 Castle Court, Killiney, Co. Dublin
**EI4HF** Michael J. Lee, 'Helvellyn', Knockaverry, Youghal, Co. Cork
**EI4HH** James Holohan, 7 Hilton Gardens, Balinteer Avenue, Balinteer, Dublin 16
**EI4HI** Details withheld at licensee's request
**EI4HQ** Cormac Gebruers, Springfield, Cobh, Co. Cork
**EI4HR** Ray Martin, 96 College Rise, Drogheda, Co. Louth
**EI4HT** David Ryan, 11 Knockgriffin, Midleton, Co. Cork
**EI4HV** Jimmy Hamill, 11 Windsor Avenue, Coleraine, Co. Derry BT52 2DR Northern Ireland
**EI4HW** Frankie McEvoy, 12 Rousseau Grove, Norwood, Waterford
**EI4HX** Peter B. Grant, 37 Glenmore Park, Dundalk, Co. Louth, A91 H0A6
**EI4HY** Prakash Madhavan, Beechview, Clownings, Straffan, Co. Kildare
**EI4HZ** Anne Clear, Ballyvogue Cottage, Goleen, Co. Cork
**EI4IA** Details withheld at licensee's request
**EI4IB** Nicholas Jordan, Tristernagh, Ballynacargy, Mullingar, Co. Westmeath
**EI4IF** James Duggan, Cahirkeem Strand, Eyeries, Bantry, Co. Cork
**EI4IH** Joseph O'Neill, Hermitage, Collon Road, Slane, Co. Meath
**EI4II** Bernard Gondard, Sea View, Forth Commons, Wexford
**EI4IJ** Martin Michael Scanlon, Meehan, Coosan, Athlone, Co. Westmeath
**EI4IK** Shane McKeever, Address withheld at licensee's request
**EI4IM** John W. Spendlove, Tullaghanrock, Edmonstown, Ballaghadereen, Co. Roscommon
**EI4IN** Ben Gaughran, 9 Hillside Estate, Skerries, Co. Dublin
**EI4IO** Rev. John Malcolm Drummond, 14 Bulls Head Cottages, Turton, Bolton BL7 0HS, England
**EI4IP** Sean Kennedy, 114B Newfield, Drogheda, Co. Louth
**EI4IQ** Noel Clarke, Smithstown, Maynooth, Co. Kildare
**EI4IR** P. J. O'Neill, Address withheld at licensee's request
**EI4IT** John Nee, Carrygawley, Letterkenny, Co. Donegal
**EI4IU** Kenneth O'Connell, 17 Greenmount Avenue, Ballinacurra Weston, Limerick
**EI4IV** William Horace Raitt, Main Street, Stranorlar, Ballybofey P.O., Lifford, Co. Donegal
**EI4IW** Derrick A. Hare, "Whitelodge", Mount Gabriel, Schull, Co. Cork
**EI4IY** John M. Dilks, c/o Handley Farm Bungalow, The Clays, Brant Broughton, Lincolnshire LN5 0RN, England
**EI4IZ** Details withheld at licensee's request
**EI4JA** Joseph Deery, Drum Carbit, Malin, Lifford, Co. Donegal
**EI4JD** Mark Lee, Ballydoole, Pallaskenry, Co. Limerick
**EI4JF** Anatolij Mezenkov, 7 Inchera Lawn, Mahon, Cork
**EI4JG** Details withheld at licensee's request
**EI4JK** Josef Veith, Kilbrown, Goleen, Co. Cork
**EI4JL** Michael A. Gorman, 4 The Glen, Boden Park, Rathfarnham, Dublin 16
**EI4JM** Keith Martin, "Sitka", Cronroe, Ashford, Co. Wicklow
**EI4JN** Conor O'Neill, Garranes, Templemartin, Bandon, Co. Cork
**EI4JO** William J. Concannon, Carrowbeg, Claremorris, Co. Mayo
**EI4JP** Terence Jeacock, Tirgrasau, Adfa, Newtown, Powys, SY16 3DD, Wales
**EI4JQ** William Watson, 7 Darwin Close, Lichfield, Staffordshire WS13 7ET, England
**EI4JR** Peter Homer, Tullyarb, Carndonagh, Inishowen, Co. Donegal, F93 TO26
**EI4JS** Christopher Holt, 92 Hazelwood Park, Artane, Dublin 5
**EI4JT** Wieslaw Polchlopek, 3 Elm Drive, Ennis Road, Caherdavin Lawn, Limerick
**EI4JU** Joseph O'Regan, Mein, Knocknagoshel, Tralee, Co. Kerry
**EI4JW** Stephen Blake, Seechaun, Kincon, Ballina, Co. Mayo
**EI4JX** Christian Noel Armoogum, 152 Corke Abbey, Bray, Co. Wicklow
**EI4JY** Alex Labunskij, 100 Priory Court, Eden Gate, Delgany, Co. Wicklow
**EI4JZ** Bronislaw Opach, The Bungalow, Ardsallagh, Athlone Road, Roscommon F42 ED73
**EI4KA** Details withheld at licensee's request
**EI4KB** Details withheld at licensee's request
**EI4KC** Brian Smith, 39 Wood Ville Manor, Dundalk, Co. Louth
**EI4KD** John Barry, Lackabeha, Carrigtwohill, Co. Cork
**EI4KE** Seamus Keenan, 30 Ballynabee Road, Bessbrook, Co. Armagh, BT35 7HD, Northern Ireland
**EI4KF** Erik Carling, "Lyndhurst", Tullynascreena, Dromahair, Co. Leitrim
**EI4KG** Waldemar Antonik, 52 Barnwall Court, Bremore, Balbriggan, Co. Dublin
**EI4KH** Denis O'Flaherty, Mitchelsfort, Watergrasshill, Co. Cork
**EI4KI** Robert Hammond, Co. Galway
**EI4KJ** Eoin O'Connor, 31 Fernlea, Carrigaline, Co. Cork
**EI4KK** Tomasz Bednarski, 2 Arus Bun Caise, Galway
**EI4KL** Details withheld at licensee's request
**EI4KM** Stephen Ormondroyd, Tintern Lodge, Tintern, Saltmills, New Ross, Co. Wexford, Y34 EV10
**EI4ABB** Aengus Cullinan, Address withheld at licensee's request
**EI4ACB** Mark G. Davis, 44 Keatingstown, Wicklow, Co. Wicklow
**EI4AFB** Kenneth W. McAllister, "Lermoos", Domoono Road, Malahide, Co. Dublin
**EI4AGB** Cyril Moriarty, Hillquarter, Coosan, Athlone, Co. Westmeath
**EI4AJB** John R Ashe, 49 Dean's Walk, Sleepy Valley, Richill, Co. Armagh BT61 9LD, Northern Ireland
**EI4ALE** Galway VHF Group, c/o Steve Wright, 14 Churchfields, Lower Salthill, Galway
**EI4AND** Kevin Connolly, 27 Haven Hill, Summercove, Kinsale, Co. Cork
**EI4AZB** Michael Jordan, Mullymucks, Roscommon
**EI4BBS** Ballybrack Scout Troop, Scout Hall, Church Road, Ballybrack, Co. Dublin
**EI4BGB** Brendan Finn, 6 Ashbrook Avenue Road, Dundalk, Co. Louth
**EI4BZB** John V. Goff, 14 Hillcrest Court, Lucan, Co. Dublin
**EI4CJB** Keith R. Roberts, Gwenallt, Lon Crecrist, Trearddur Bay, Holyhead LL65 2AZ, Wales
**EI4CKB** Michael Martin, Tullycanna, Ballymitty, Co. Wexford
**EI4CLB** Ronan B. O'Neill, 55 Hillside, Greystones, Co. Wicklow
**EI4CQB** Francis G. Halford, 46 The Downings, Prosperous, Naas, Co. Kildare
**EI4DBB** David M. O'Sullivan, 23 Millford, Athgarvan, Newbridge, Co. Kildare

EI4DCB Dan Gallagher, Jardim Do Sol, Inchigaggin, Carrigrohane, Co. Cork
EI4DIB Tony Allen, Address withheld at licensee's request
EI4DJB Rory A.J. Hinchy, 'Waverley', 299 Navan Road, Dublin 7
EI4DNB D.R. Rowkins, Mountpleasant, Strokestown, Co. Roscommon
EI4DOB Details withheld at licensee's request
EI4DSB John Kirwan, 101 Castlenock Park, Castlenock, Dublin 15
EI4DUB Tony O'Regan, Ballincrokig, Dublin Pike, Co. Cork
EI4DVB C.J. Colgan, 11 St. Johns Park, Moira, Craigavon BT67 0NL, Northern Ireland
EI4DWB Trevor J. Jones, St. Bridgets, Muingwee, Lyracompane, Listowel, Co. Kerry V31 X638
EI4EAB Brendan Power, 12 Faughart Road, Crumlin, Dublin 12
EI4EEB Martin Sweeney, Lisaleen, Tuam, Co. Galway
EI4EFB Thomas O'Dea, Kilcornan, Clarinbridge, Co. Galway
EI4EHB Anthony M. Rogers, Rosedale, Cove Lane, Dunmore Road, Waterford
EI4EIB Michael Sweeney, Lisaleen, Tuam, Co. Galway
EI4ELB Bartholomew O'Leary, 1 Mellows Park, Renmore, Galway
EI4EOB Nigel Barlow, Heathstown, Coralstown, Mullingar, Co. Westmeath
EI4EPB Brendan L. Kelly, 145 Swords Road, Whitehall, Dublin 9
EI4EQB Paul R. Delany, 4 St Mary's Villas, Donore, Co. Meath
EI4ERB Andrew Earley, Address withheld at licensee's request
EI4ESB Dermot Madsen, 7 Gracefield Avenue, Artane, Dublin 5
EI4EUB John F. Malone, 1 New Row, Belvedere, Mullingar, Co. Westmeath
EI4EVB Details withheld at licensee's request
EI4EYB John A. Bracken, Address withheld at licensee's request
EI4FCB DÈagl·n " Meachair, 8 Cnoc na Manach, Cill Mhant·in
EI4FDB William Keyes, 76 Captain's Avenue, Crumlin, Dublin 12
EI4FEB Details withheld at licensee's request
EI4FIB Stuart Keyes, 76 Captain's Avenue, Crumlin, Dublin 12
EI4FMB Mike J. Griffin, 9 Lissanalta Avenue, Dooradoyle, Limerick
EI4FNB Mark Kilmartin, Hollylodge, Newtown, Nurney, Co. Carlow
EI4FOB John Kenneth Collins, Marken House, Ballygaddy Road, Tuam, Co. Galway
EI4FPB Laurence Wright, 5 Woodview Park, The Donahies, Dublin 13
EI4FQB Gerard Kelly, Curranrue, Burren, Co. Clare
EI4FWB Ron Cartin, 6 Arranmore Avenue, Phibsboro, Dublin 7
EI4GAB Rolando D. Dela Cruz Jr., Valleyview, Derryduff, Coomhola, Bantry, Co. Cork
EI4GBB Nicholas James Jackman, 302 Bridgewater Quay, Arklow, Co. Wicklow Y14 FD92
EI4GDB John Joseph Aston, 3 Valley Road, Darley, Harrogate HG3 2QE, North Yorkshire, England
EI4GEB Leslie Ferguson, Rosendale, Skule, Fedamore, Co. Limerick
EI4GGB Owen O'Reilly, Bellfield, Gaybrook, Mullingar, Co. Westmeath
EI4GHB Tony Gallagher, Address withheld at licensee's request
EI4GIB Paul Whelan, Woodhaven, Allenwood South, Co. Kildare
EI4GJB Declan Sheehan, 1 Nicholas Street, Limerick
EI4GKB Steven McKenna, Cluain MhÛr, Clonmellon, Co. Westmeath
EI4GLB Terence Webb, 4 Haven Bank, Palmer Road, Rush, Co. Dublin
EI4GMB Details withheld at licensee's request
EI4GNB Details withheld at licensee's request
EI4GOB Fionnbarra Mac TrÈanfhir, Tuaim Dhubh, Buirgheas " nDrÚna, Cill Chainnigh
EI4GPB Patrick McCauley, 5 Brookmount Rise, Omagh BT78 5AL, Co. Tyrone, Northern Ireland
EI4GRC Galway Radio Experimenters Club, c/o Gerry Ormond, 4 Elm Park, Renmore, Galway
EI4GSB Michael Lockwood, 5 Marsh Street, Cleckheaton, Liversedge, West Yorkshire BD19 5BW, England
EI4GTB Details withheld at licensee's request
EI4GUB Details withheld at licensee's request
EI4GVB Michael McLaughlin, 2 Gilroy Chalets, Downings, Letterkenny, Co. Donegal
EI4GWB Paul Reilly, 11 Ballsbridge Wood, Shelbourne Road, Ballsbridge, Dublin 4
EI4GXB Ger McNamara, Kearnmara House, Ruanard, Clonlara, Co. Clare
EI4GYB Roland V. Byrne, St. Gerard's, Knockthomas, Nurney, Co. Carlow
EI4GZB Pete Taite, 20 Parkhill Court, Dublin 24
EI4HAB Krishna Kaushal Panduru, 7 Cathair Dannan, North Circular Road, Tralee, Co. Kerry
EI4HBB Barry Campbell, 3B Crewe Road, Upper Ballinderry, Lisburn, Antrim BT28 2PL, Northern Ireland
EI4HCB Thomas Nevin, Racecourse Road, Roscommon, Co. Roscommon
EI4HDB Mark Mullaney, Gransha, Clones, Monaghan
EI4HEB Details withheld at licensee's request
EI4HFB Robert Ring, 60 St. Teresa's Road, Kimmage Road West, Dublin 12
EI4HGB Details withheld at licensee's request
EI4HHB Kieran Smalle, 22 Oakfield, Ashleigh Wood, Monaleen, Co. Limerick
EI4HIB Details withheld at licensee's request
EI4HJB Details withheld at licensee's request
EI4HKB Patrick James Flood, 815 Woodlands East, Castledermot, Co. Kildare
EI4HLB Jordan Cummins, "Lyston", 13 Forest View, Goulds Hill, Mallow, Co. Cork
EI4HMB Michael Fenlon, Tinnahinch, Graiguenamanagh, Co. Kilkenny
EI4HNB Daniel McGlashin, Cranny Road, Frosses, Co. Donegal
EI4HOB Gerard Madden, 192 Ardilaun, Portmarnock, Co. Dublin, D13 FA03
EI4HPB William Greer, 11 Mullaghcarton Road, Lisburn, Co. Antrim BT28 2TE, Northern Ireland
EI4HQB Francis Kearney, 45 Sperrin Park, Omagh, Co. Tyrone BT78 5BA, Northern Ireland
EI4HRB Dermot Tynan, Somerset Cottage, Aughinish (Nr. Kinvara), Co. Clare
EI4HSB Patrick McMahon, 87 O'Neill Park, Clones, Co. Monaghan
EI4HTB John Corry, 13 Corbally Court, Naas, Co. Kildare
EI4HUB Adrian Dilo, 1 Central Park, Mullingar, Co. Westmeath
EI4HWB Details withheld at licensee's request
EI4HYB Details withheld at licensee's request
EI4HZB Details withheld at licensee's request
EI4KRC Kilkenny Radio Club, c/o Michael Drennan EI7GH, Ballyroberts, Cuffesgrange, Co. Kilkenny
EI4LRC Limerick Radio Club, c/o Anthony Condon, EI2AW, Golf Links Road, Castletroy, Co. Limerick
EI4CARA Crossakiel Amateur Radio Association, The Post Office, Crossakiel, Kells, Co. Meath

## EI5

EI5C Signals Amateur Radio Club, c/o Michael Moore, CIS Sch Combat Support College, Plunkett Barracks, Curragh Camp, Co. Kildare
EI5J Frances Taheny, Pearse Road, Sligo
EI5V Signals Amateur Radio Club, c/o Patrick Molloy, 2nd Field Signals Company, Cathal Brugha Barracks, Dublin 7
EI5AL Paul Charles Layton, 189 Old Youghal Road, Cork
EI5AR J. Lysaght, 47 Mount Farren, Assumption Road, Co. Cork
EI5BF Details withheld at licensee's request
EI5CA Tom O'Brien, 31 Radharc na Coille, Ballycasey, Shannon, Co. Clare
EI5CD Desmond J. Walsh, 17 The Rise, Owenabue Heights, Carrigaline, Co. Cork
EI5CE Details withheld at licensee's request
EI5CK Robert O'Donnell, 4 Ballyman Road, Enniskerry, Co. Wicklow
EI5CL William B. Mannion, "An Culaun", Galway Road, Tuam, Co. Galway
EI5CN James D. Chadwick, 6 The Orchard, Monkstown Valley, Monkstown, Co. Dublin
EI5DD Stephen Wright, 14 Churchfields, Lower Salthill, Galway
EI5DE Edward Tuthill, Green Briar, Loughanure, Clane, Co. Kildare
EI5DG Thomas A. McGuinn, 13 Avondale Lawn Extension, Blackrock, Co. Dublin
EI5DH Thomas Moore, Ballynacragga, Ennis Road, Newmarket-on-Fergus, Co. Clare
EI5DI Paul O'Kane, 36 Coolkill, Sandyford, Dublin 18
EI5DK Michael Kinsella, 9 Shannon Grove, Corbally, Limerick
EI5DO Albert G. Brown, 19 James Connolly Park, Clondalkin, Dublin 22
EI5DR Edward F. Kelly, Cregganavar, Breaffy, Castlebar, Co. Mayo
EI5DS Edward Collins, Ashford, Dublin Road, Malahide, Co. Dublin
EI5DT Patrick Keeney, Legandara, Lifford, Co. Donegal
EI5DW Details withheld at licensee's request
EI5DY William A. Murphy, 96 Langton Park, Newbridge, Co. Kildare
EI5DZ Details withheld at licensee's request
EI5EE J. Harvey Makin, 'Rowson Heights', Goggins Hill, Ballinhassig, Co. Cork
EI5EF Details withheld at licensee's request
EI5EH Patrick Egan, NR2N, 113A Pilatus Platz Unit #A, Freehold, NJ 07728, USA
EI5EI Gerry O'Sullivan, Cherryhound, The Ward, Co. Dublin
EI5EM Tony Breathnach, 159 CÈide Ard MÛr, Ard Aidhin, Baile ¡tha Cliath 5, DO5 PD26
EI5EN Romano Morelli, 12 Capel Street, Dublin 1
EI5ER Frank Nolan, 92 Roselawn, Tramore, Co. Waterford
EI5ET Robert H. Allen, 39 Deerpark, Ashbourne, Co. Meath
EI5EV Joseph Murphy, Carriganurra, Slieverue, Co. Kilkenny
EI5EZ Declan J. McLoughlin, 90 Broom Lane, Broom, Rotherham, S60 3EW, England
EI5FA Graham P. Clarke, 2 Chestnut Court, Collinswood, Collins Avenue, Dublin 9
EI5FC John A. Coakley, Glen-na-Smol, Galway's Place, Douglas West, Co. Cork T12 K3T6
EI5FE Brendan Somers, 37 Carricklawn, Coolcotts, Wexford
EI5FI Diarmuid J. O'Sullivan, 1 Greenlea Avenue, Terenure, Dublin 6W
EI5FK Charles Coughlan, Address withheld at licensee's request
EI5FR Details withheld at licensee's request
EI5FT Colm D. Leahy, Trabolgan, Whitegate, Midleton, Co. Cork
EI5FV John Twamley, 30 Grange Park Crescent, Raheny, Dublin 5
EI5FY William Rice, 'Oakwall', 16A Presentation Road, Galway
EI5GB Martin Whyte, 79 The Paddocks, Tipper Road, Naas, Co. Kildare
EI5GE Joseph T. Leahy, 'Galtymore', Bansha, Barnlough, Co. Tipperary
EI5GG Details withheld at licensee's request
EI5GJ Thomas Mortell, 3 The Lawn, Hayden's Park, Lucan, Co. Dublin
EI5GM Jeremy M. Sheehan, Hillcrest, Bandon Road, Kinsale, Co. Cork
EI5GN Iain Fisher, 29 Seapoint, Brittas Road, Wicklow
EI5GP Nick Greaves, Forty Shilling Cottage, Oghill, Monasterevin, Co. Kildare
EI5GS International Police Association, c/o Jim Jeffers EI8DK
EI5GT Desmond Chambers, Burrishoole, Newport, Co. Mayo
EI5GX Eamonn G. Roddy, 7 Marlborough Avenue, Derry City BT48 9BQ, Northern Ireland
EI5HB Details withheld at licensee's request
EI5HD Daniel Coughlan, 157 Shanganagh Cliffs, Shankill, Dublin 18
EI5HE Sean Corcoran, Baltydaniel East, Mallow, Co. Cork P51 X47V
EI5HG J. Eugene O'Malley, 11 Silvercourt, Tivoli, Cork
EI5HH John Madden, The Cottage, 53 Clarendon Street, Derry, Northern Ireland
EI5HI Helen O'Sullivan, 3 Kiltegan Lawn, Rochestown Road, Co. Cork
EI5HJ Gerard Molloy, Knockacarhanduff, Upperchurch, Thurles, Co. Tipperary
EI5HL John Rogers, 23 The Grange, Donore, Co. Meath
EI5HM Joseph Harding, 13 New Houses, Donecarney, Co. Meath
EI5HN Details withheld at licensee's request
EI5HP Anthony C Marston, Stormsfield, Station Road, Askeaton, Co. Limerick
EI5HQ Details withheld at licensee's request
EI5HS Patrick J. Geoghegan, Kilmolin, Enniskerry, Bray, Co. Wicklow
EI5HT Dermot Fagan, 9 Ashdale Close, Kinsealy, Co. Dublin, K67 E235
EI5HU Martin Sheridan, Blackrock Road, Dundalk, Co. Louth
EI5HV Brian Tansey, 28 Oak Park, Castle Redmond, Midleton, Co. Cork P25 H622
EI5HW Aidan Murphy, Address withheld at licensee's request
EI5HX Walter E. Roberts, 24 Leeds Road, Barwick in Elmet, Leeds LS15 4JD, England
EI5HZ Paul D. Reilly, 44 The Drive, Castletown, Celbridge, Co. Kildare
EI5IC Details withheld at licensee's request
EI5IE Patrick A. Mulreany, Tulach na GrÈine, 51 Miller Ridge Road, Wellington NV 89444, USA
EI5IF Patrick Molloy, 71 Bannow Road, Cabra West, Dublin 7
EI5IH Gerard Richardson, Woodbine Cottage, Carnin, Ballyjamesduff, Co. Cavan
EI5II Thomas Walsh, 28 Clonshaugh Park, Clonshaugh, Dublin 17
EI5IL Patrick Gaffney, 32 St Anne's Drive, Montenotte Park, Cork
EI5IN Keith Nolan, 'Laurel Grove', Loughanstown, Knockdrin, Mullingar, Co. Westmeath
EI5IO Details withheld at licensee's request
EI5IP Ari Pietikainen, Krattivuorentie 9 D 12, FIN-02320 ESPOO, Finland
EI5IQ Dermot Wall, Cabra, Dublin 7
EI5IS Pat Duffy, Gowel, Charlestown, Co. Mayo
EI5IU James Ryan, 10 MacCurtin Villas, College Road, Cork
EI5IV Stanley Raitt, Main Street, Stranorlar, Ballybofey P.O., Lifford, Co. Donegal
EI5IW Robert Mannion, Coolaght, Claremorris, Co. Mayo
EI5IX Padraic Baynes (Jnr), Derrygorman, Westport, Co. Mayo
EI5JB James McGrory, 2 Beechwood Road, Letterkenny, Co. Donegal
EI5JC John W. Ferguson, Mountargus, Redcastle, Co. Donegal
EI5JD Laurence Mahon, Leaba Sioda Park, Ballyellis, Gorey, Co. Wexford
EI5JE Peter Henshaw, 31 Slievebloom Park, Walkinstown, Dublin 12
EI5JF Andy Jay, 13 Hazelwood Avenue, Glanmire, Cork, Co. Cork
EI5JG Kevin Rock, 71 Rowan Heights, Marley's Lane, Drogheda, Co. Louth
EI5JL Details withheld at licensee's request
EI5JM Trevor E. Hamman, 4 Gallows Hill, Ennis, Co. Clare
EI5JQ Adam Kolodziejczyk, Apartment 2 Coranna, Tipper South, Naas, Co. Kildare
EI5JR Una Murray, 80 Canterbury Park, Kilfennan, Co. Derry BT47 6DU, Northern Ireland
EI5JS John Clavin, Aughamore, Kilmainhamwood, Kells, Co. Meath
EI5JV Adrian Brentnall, Gortnagrough, Ballydehob, Co. Cork
EI5JX Jeffrey Smith, 54A Blackstaff Road, Kircubbin, Co. Down, BT22 1AF, Northern Ireland
EI5JZ Details withheld at licensee's request
EI5KA Udo Lauterborn, 40 Colmcille Road, Shantalla, Galway
EI5KB Aidan J. Cornyn, 2 Drumshambo Road, Leitrim Village, Co. Leitrim
EI5KC Gerry Breen, 30 Corcoran Terrace, Kells Road, Kilkenny
EI5KD Hilary Barry, Lackabeha, Carrigtwohill, Co. Cork
EI5KE Roy Steele, 1 Ulverton Court, Ulverton Road, Dalkey, Co. Dublin
EI5KF Gerard D. Scannell, 3 Kingston Close, Mitchelstown, Co. Cork, P67 HR44
EI5KG David Sherwood, Coroin Mhuire, Maudlintown, Wexford
EI5KH Robert Brandon, 16 Woodley Road, Dun Laoghaire, Co. Dublin
EI5KI Paul Barlow, Geal·n, Durham Road, Sandymount, Dublin 4
EI5KJ Keith Crittenden, Ballybrennan, Bree, Enniscorthy, Co. Wexford
EI5KK Details withheld at licensee's request
EI5KL Details withheld at licensee's request
EI5KM Details withheld at licensee's request
EI5ACB P. Buckley, Main Street, Banteer, Co. Cork
EI5ASB T. O'Neill, 13 Pondsfield, New Ross, Co. Wexford
EI5AXB Leo Hilliard, Smeria Bridge, Duagh, Listowel, Co. Kerry
EI5AZB James Hennessy, Gotoon, Hospital, Co. Limerick
EI5BBB Cecil E. Fairman, Finn View House, Trennamullin, Ballybofey, Co. Donegal
EI5BEB Patrick J. Bowe, 197 Gloucester Avenue, Chelmsford, Essex CM2 9DX, England
EI5BGB Details withheld at licensee's request
EI5BHB Kieran O'Carroll, International Air Transport Association, 703 Waterford Way, Suite 600, Miami, Florida, 33126, USA
EI5BPB Details withheld at licensee's request
EI5BVB Peter J. Mathews, Creevelea, Emyvale, Co. Monaghan
EI5BWB John J. Crerand, College Farm Road, Letterkenny, Co. Donegal
EI5CAB Alan C. Hildebrand, 64 Carton Court, Maynooth, Co. Kildare
EI5CEB Frank Bourke, Graigue Lower, Cuffesgrange, Co. Kilkenny
EI5CHB Patrick J. Martin, 24 Tamarisk Avenue, Kilnamanagh, Tallaght, Dublin 24
EI5CLB Francis McAuley, 61 Templeroan Park, Templeogue, Dublin 16
EI5CRC Cork Radio Club, QSL via EI2KA
EI5CTB Trevor C. Plowman, 5 Mackie's Place, Dublin 2
EI5CYB Joe Ivory, Address withheld at licensee's request
EI5DDB Brendan H. Cornyn, Dowra, via Carrick-on-Shannon, Co. Leitrim
EI5DJB Lawrence Hoey, 34 The Walk, Oldtown Mill, Celbridge, Co. Kildare
EI5DLB Kevin Foley, 18 Pine Grove, Ashling Heights, Raheen, Limerick
EI5DPB Karl P. Madden, Rosgraerin House, Lagore Road, Dunshaughlin, Co. Meath
EI5DRB Donal Caulfield, 10 Beechcourt, Killiney, Co. Dublin
EI5DSB Eamonn Sheeran, 12 Blackwater Drive, Navan, Co. Meath
EI5DWB Brendan J. Flanagan, 111 Avondale Road, Killiney, Co. Dublin
EI5EAB Michael White, 6 Tormey Villas, Athlone, Co. Westmeath
EI5EBB Thomas W. Vickery, 4 Marino Street, Bantry, Co. Cork
EI5EEB Details withheld at licensee's request
EI5EIB Details withheld at licensee's request
EI5ERB John P. Ward, Station Road, Glenties, Co. Donegal
EI5ESB John Gartlan, Carrickedmond, Kilcurry, Dundalk, Co. Louth
EI5EXB Philip Bartlett, 10 Coolbane Wood, Castleconnell, Co. Limerick
EI5EYB Jeremiah C. Forde, 8 Newbrook Grove, Mullingar, Co. Westmeath
EI5FGB John Joseph Noonan, Summerfield, Youghal, Co. Cork
EI5FJB Patrick Kelleher, O'Briens Place, South Abbey, Youghal, Co. Cork

EI5FOB John Doherty, Ballybrennan, Drombanna, Co. Limerick
EI5FQB Clifford Hack, [illegible], Co. Offaly R42 Y752
EI5FSB Susan Blythe-Hess, 3 Havana Court, Eastbourne, East Sussex BN23 5UH, England
EI5FUB Gary Scott Gillies, Aughurine Cottage, Aughurine Village, Ballaghadereen, Co. Roscommon
EI5FVB Patrick B. O'Shea, Address withheld at licensee's request
EI5FXB Patrick Tuohy, Tipperary Road, Cahir, Co. Tipperary
EI5FYB Gareth Martin, "Sitka", Cronroe, Ashford, Co. Wicklow
EI5FZB David W. Douglas, Carberry Lodge, The Stocks, Athboy, Co. Meath
EI5GAB Conor Gilmer, 19 Lorcan Villas, Santry, Dublin 9
EI5GBB Noel Kelly, Munsboro, Roscommon
EI5GDB Joseph Flanagan, 54 Marren Park, Ballymote, Co. Sligo
EI5GEB P·draig Foley, Sc·th na Coille, Kilcronan, Whitechurch, Co. Cork
EI5GFB Details withheld at licensee's request
EI5GGB Michael O'Callaghan, Dromcolligher Road, Knockanglass East, Freemount, Charleville, Co. Cork
EI5GHB John Walsh, Cluain Damh, Gleann Oistìn, Baile Uì Fhiach·in, Co. Mhaigh Eo
EI5GJB Leslie McCarthy, 3 Kilmahon, Shanagarry, Co. Cork
EI5GKB Damian Boylan, Durhamstown, Bohermeen, Navan, Co. Meath
EI5GLB Victor Loughlin, 49 Clanmalire Close, Portarlington, Co. Laois
EI5GMB Details withheld at licensee's request
EI5GNB Brendan Byrne, Ballour, Ballon, Co. Carlow
EI5GOB Francis Lenane, Faha, Ring, Dungarvan, Co. Waterford
EI5GOB Francis Lenane, Faha, Ring, Dungarvan, Co. Waterford
EI5GRB Declan Fitzgerald, Colligan Mountain, Dungarvan, Co. Waterford
EI5GSB Denis Long, 2 Pairc na hAbhainn, Cloyne, Co. Cork
EI5GTB Paul Sinclair, Rarooey, Donegal Town, Co. Donegal
EI5GUB Richard Gorman, Ferefad, Longford, Co. Longford
EI5GVB Tim Collins, Cloonabricka, Ballinamore Bridge, Ballinasloe, Co. Galway
EI5GWB Patrick Butler, Saint Hilary, Lower Road, Cobh, Co. Cork
EI5GXB Laurence M. Clarke, Dromada, Ladysbridge, Co. Cork
EI5GYB Mark Mc Nicholas, Address withheld at licensee's request
EI5GZB Noel Smith, Garryricken, Windgap, Co. Kilkenny
EI5HAB Details withheld at licensee's request
EI5HBB Eoghan Kinane, Ballytarsna, Mullinavat, Co. Kilkenny
EI5HCB Details withheld at licensee's request
EI5HDB John O'Brien, Tiermoyle, Latteragh, Nenagh, Co. Tipperary, E45 WC03
EI5HEB Details withheld at licensee's request
EI5HFB Colm Brazel, 10 Dargle Wood, Knocklyon, Dublin 16
EI5HHB Details withheld at licensee's request
EI5HIB Noel O'Loughlin, Ross, Ballyfin, Co. Laois, R32 K5Y9
EI5HJB Details withheld at licensee's request
EI5HKB Joseph Cluney, Barrack Street, Passage East, Co. Waterford
EI5HLB Details withheld at licensee's request
EI5HMB Fintan Phelan, 152 Roseberry Hill, Newbridge, Co. Kildare
EI5HNB Details withheld at licensee's request
EI5HOB Kevin O'Brien, Tiermoyle, Latteragh, Nenagh, Co. Tipperary, E45 WC03
EI5HPB Cassiano Morgado de Aquino, Vantage Apartments Central, Building 4 Apartment 204, Leopardstown, Dublin 18
EI5HQB Details withheld at licensee's request
EI5HRB Ronan Daly, Lissadonna, Cloughjordan, Co. Tipperary
EI5HSB Michael McGearty, Laragh, Maynooth, Kildare
EI5HTB Lukas Nardella, Address withheld at licensee's request
EI5HUB Christopher Cummins, Suffolk Street, Kells, Co. Meath
EI5HVB Reiner Wickel, Knockroe West, Allihies, Beara, Co. Cork
EI5HWB Dereck White, Address withheld at licensee's request
EI5HXB Albert Price, Kilcolumb, Kilmaley, Ennis, Co. Clare
EI5HYB Dominic McManus, Gabalis Lodge, Barrowhouse, via Athy, Co. Kildare
EI5HZB Michael Fallon, Ballyorban, Monkstown, Co. Cork, T12 KD7P
EI5MRC Mellary Radio Club, c/o James Farrell EI8IG, Mellary Scout Group, Cappoquin, Co. Waterford
EI5TCR Tir Conaill Amateur Radio Society, c/o Danny Bonner, Cloughwally, Lettermacaward, Co. Donegal

## EI6

EI6D Leo Purcell, "Gleann-na-Greine", Naas, Co. Kildare
EI6S George McClarey, "Rosemount", Mountnugent, Co. Cavan
EI6AD David Connolly, The Pier, Ballycotton, Co. Cork
EI6AG Alexander F. Barrett, Address withheld at licensee's request
EI6AI William Long, Dunkineely, Co. Donegal
EI6AK John A. Mooney, 'The Cottage', 48 South Douglas Road, Cork
EI6AL Dave O'Connor, Silver Howe, Sydenham Mews, Corrig Avenue, Dun Laoghaire, Co. Dublin A96 RF99
EI6AU M.F. Whelan, 457 Collins Avenue, Whitehall, Dublin 9
EI6AX Bernard O'Sullivan, Cahermore, Bantry, Co. Cork
EI6BA Thomas J. Foley, 40 Hillcourt, Donnybrook, Douglas, Cork
EI6BT Jerry Cahill, Killea, Broomfield East, Midleton, Co. Cork
EI6BY Hugh McGivern, Navvy Bank, George's Quay, Dundalk, Co. Louth
EI6CB Cornelius J. Connolly, The Square, Skibbereen, Co. Cork
EI6CF Sean Grant, Clonmacken Road, Caherdavin, Limerick
EI6CJ Details withheld at licensee's request
EI6CK Guy Jean-Francois Poinboeuf, Belan, Moone, Co. Kildare
EI6CV Patrick Ronaghan, Moyne Hall, Cavan
EI6CY Patrick J. Manning, White Gables, Keevagh, Quin, Co. Clare
EI6CZ Rolande E. Hall, 'Pinewood Lodge', 16 Tullyvarraga Hill, Shannon, Co. Clare
EI6DH Colin Kennedy, 74 Dunmore Park, Ballymount, Clondalkin, Dublin 22
EI6DJ Robert Hickson, 1 Watermill Close, Oldbawn, Tallaght, Dublin 24
EI6DK Victor P. Moran, 32 Anne Devlin Avenue, Rathfarnham, Dublin 14
EI6DL Anthony Magliocca, Court Devenish, Athlone, Co. Westmeath
EI6DN John Molloy, Kilmoon, Ashbourne, Co. Meath
EI6DO A.W. Boles, [illegible] Road, [illegible], Co. Dublin
EI6DP Gerard P. FitzGerald, 96 Grianan, Westbury, Corbally, Co. Clare
EI6DT Anthony Enright, "Crossfield", Firhouse Road, Templeogue, Dublin 16
EI6DU William Guiry, Sunville, Woodpark, Castleconnell, Co. Limerick
EI6DW Patrick Lynch, Churchtown, Carndonagh, Co. Donegal
EI6DX Details withheld at licensee's request
EI6DY John Gilligan, 14 Arden Vale, Tullamore, Co. Offaly
EI6EC Fr. Michael Reaume, Marianist Community, 13 Coundon Court, Killiney, Glenageary, Co. Dublin A96 KOT9
EI6EE Peter J. Gillen, Address withheld at licensee's request
EI6EF Frank Malone, Elmhill, Grove Road, Malahide, Co. Dublin, K36 WN2
EI6EG Joseph Cosgrave, 103 McIntosh Park, Pottery Road, Dun Laoghaire, Co. Dublin
EI6EH Tom Clarke, Balrath, Kells, Co. Meath
EI6EQ Daniel Costello, Ballyfin Road, Portlaoise, Co. Laois
EI6ER Michael P. O'Sullivan, 14 Pleasant Drive, Mount Pleasant, Waterford
EI6ES Thomas Bluett, Convent Road, Clonakilty, Co. Cork
EI6ET John F. Murphy, Cooleanig, Beaufort, Killarney, Co. Kerry
EI6EV Donal " hUallach·in, 171 Martello, Port MearnÛg, Co. ¡tha Cliath
EI6EW Anthony Baker, 3 Le Hunt House, Brennanstown, Cabinteely, Co. Dublin
EI6EY Michael J. McElligott, Skehenerin, Listowel, Co. Kerry
EI6EZ Joe Martin, 13 Cherryfield View, Hartstown, Clonsilla, Dublin 15
EI6FB Niall Syms, 71 Hazelwood, Shankill, Dublin 18
EI6FE Paul Kirkby, 43 Gleann an Oir, Tullyvarragh, Shannon, Co. Clare
EI6FF John B. McClintock, Moneygreggan, Newtowncunningham, Lifford, Co. Donegal
EI6FG Anthony McGarry, Ballydoogan, Magheraboy, Sligo Town
EI6FI Details withheld at licensee's request
EI6FJ John Higgins, Apartment 4, 124 Cromwell Road, London SW7 4ET, England
EI6FL William B. Chapple, 3 Knapton Terrace, Knapton Road, Dun Laoghaire, Co. Dublin
EI6FM Dermot P. O'Connell, Ballinvarrig, Whitechurch, Co. Cork
EI6FN Patricia Keenan, West End, Bundoran, Co. Donegal
EI6FR Declan P. Craig, 167 St. James's Road, Greenhills, Dublin 12
EI6FY Richard Prendergast, Mogeely Road, Castlemartyr, Co. Cork
EI6FZ Dermot Flanagan, 132 Upr. Kilmacud Road, Stillorgan, Co. Dublin
EI6GA Brendan Lynch, 37 Hazelwood, Shankill, Dublin 18
EI6GF Michael J. McLoughlin, Spencerstown, Murrintown, Co. Wexford
EI6GG Paul Cotter, Derryvillane, Glanworth, Co. Cork
EI6GH Padraig F. Barry, 21 Belgrave Road, Monkstown, Co. Dublin
EI6GI John D. Hickey, c/o 1st Field Arty. Regt., Collins Barracks, Cork
EI6GJ Edmond Fitzgerald, 'Sea Winds', Ballinamona, Shanagarry, Midleton, Co. Cork
EI6GL Ronan W. Lynch, 37 Hazelwood, Shankill, Co. Dublin
EI6GM Karl Reddy, 18 rue des sources, L-6579 Rosport, Luxembourg
EI6GN Dan R. Nelson, 6 Streedagh Point, Grange, Co. Sligo
EI6GQ Details withheld at licensee's request
EI6GS Domhnall " Chn·imhsì, Clochbhaile, Leitir-Mhic-a-Bh·ird, D´n na nGall, Co. Dh´n na nGall
EI6GT Michael P. Larkin, 2 Manor Court, Greencastle Road, Moville, Co. Donegal
EI6GV Padraig McCormack, Main Street, Kilnaleck, Co. Cavan
EI6GX James A. Whelan, 26 Main Street, Kilcoole, Co. Wicklow
EI6GY Ian C. White, 79 Hawthorn Drive, Hillview, Waterford
EI6GZ Brendan Wall, Riverstown, Rathfeigh, Tara, Co. Meath
EI6HB Denis O'Flynn, Ladysbridge P.O., Ladysbridge, Castlemartyr, Co. Cork
EI6HD Patrick H. O'Keeffe, 8 Clonard Grove, Sandyford Road, Dublin 16
EI6HE Steve Canavan, Station Road, Corofin, Co. Clare
EI6HF Patrick N. Hutton, 50 Ash Grove, Tullow Road, Carlow
EI6HH Richard Bermingham, 17 Fremont Drive, Melbourn, Bishopstown, Cork
EI6HJ Donal Skelly, Bredagh Glen, Moville, Co. Donegal
EI6HL John Walsh, 26 Verbena Grove, Kilbarrack, Dublin 13
EI6HN Denis Bernard Rowe, 23 Glenwood Drive, Onslow Gardens, Commons Road, Cork
EI6HQ D. M. Brosnan, Killarney Road, Castleisland, Co. Kerry
EI6HS Paul Breen, 21 SlÌ an Aifrinn, Athlone, Co. Westmeath
EI6HT Melne H. Rouwhof, Tigh-na-Mara, Durrus, Co. Cork
EI6HW Noel Mulvihill, Hillquarter, Coosan, Athlone, Co. Westmeath
EI6HX Christopher Hannigan, 4 Silverhill Road, Strabane, Co. Tyrone, Northern Ireland
EI6HY Michael J. Bonar, Galwolie, Cloghan, Co. Donegal
EI6HZ Richard Coleman, 45 Carriganarra, Ballincollig, Co. Cork
EI6IA Christopher Maverley, 3 St. Stephens Place, Friar Street, Cork
EI6IB Fergus Millar, Dromore, Carrick on Shannon, Co. Leitrim
EI6IC Francis G. Fahy, Athenry Road, Loughrea, Co. Galway
EI6IF Denis Collins, Knockfeering, Mylerstown Road, Robertstown, Naas, Co. Kildare
EI6IH Jan Serridge, 21 Lissara Heights, Warrenpoint, Co. Down BT34 3PG, Northern Ireland
EI6IK Anthony Hodgins, 31 Cherry Brook Drive, Drogheda, Co. Louth
EI6IL Don Brennan, Ramsfort, Tullydonnell, Togher, Drogheda, Co. Louth, A92 XE81
EI6IM Michael J. Grifferty, Sweetwell, Swinford, Co. Mayo
EI6IN Sean McMorrow, 'San Juda', Convent Avenue, Bray, Co. Wicklow
EI6IQ David Corbett, 28 Woodview, Killygoan, Co. Monaghan
EI6IR John Francis McDonnell, Burrin, Ballyglass, Claremorris, Co. Mayo
EI6IT Leo James McGranaghan, Eastwood House, Dromboe Ave, Stranorlar, Ballybofey, Co. Donegal
EI6IU Eunan McCarron, Shannagh, Raphoe, Co. Donegal
EI6IW John Edgeworth, 48 Shelbourne Park, Limerick
EI6IZ Brendan Minish, Raheens, Castlebar, Co. Mayo
EI6JA Stephen O'Leary, Fanisk, Killeagh, Co. Cork
EI6JB Rory O'Brien, Ballinahalla, Moycullen, Co. Galway
EI6JC Michael Burke, Bodenstown, Sallins, Naas, Co. Kildare
EI6JF Mark G. Healy, 16 The Heath, Alderbrook, Ashbourne, Co. Meath
EI6JG Brian McConnell, Carheen, Tourmakeady, Claremorris, Co. Mayo
EI6JK Mark Condon, Mullaghadoey Hill, Castlerea, Co. Roscommon
EI6JR Joe Murray, 80 Canterbury Park, Kilfennan, Co. Derry BT47 6DU, Northern Ireland
EI6JU Kieran Burke, Ironpool, Kilconly, Tuam, Co. Galway
EI6JW Details withheld at licensee's request
EI6JX Details withheld at licensee's request
EI6JY Lenz Grassl, Eulenweg 2, D-84036 Landshut, Germany
EI6JZ David Gilmartin, 50 Addison Drive, Addison Park, Glasnevin, Dublin 11
EI6KA Details withheld at licensee's request
EI6KB Aengus O'Fearghail, The Reask, Dunshaughlin, Co. Meath
EI6KC Details withheld at licensee's request
EI6KD Robert Goryl, 119 Caragh Court, Naas, Co. Kildare
EI6KF Arthur E. McGahey, 4049 E. 133rd Circle, Thornton, CO 80241, USA
EI6KG Andy Ronan, 2219 West Eastwood Avenue, Chicago, IL 60625, USA
EI6KH Details withheld at licensee's request
EI6KI Laurence Frank Blatt, Hillview House, Ballynoe, Castlemahon, Co. Limerick
EI6KJ Larry Murphy, 6 Brooklawn, Rushbrooke, Cobh, Co. Cork
EI6KK Sean Rafferty, 13 St. Lukes Crescent, Milltown, Dublin 14
EI6KL Brian Whelehan, Mount Odell, Carriglea, Dungarvan, Co. Waterford, X35 AX74
EI6KM Anthony McDermott, 139 Courtown Park, Kilcock, Naas, Co. Kildare, W23 WY92
EI6ALB Leo O'Leary, 70 Mourne Road, Drimnagh, Dublin 12
EI6AMB John Melvin, Derreen Upper, Kilkerrin, Ballinasloe, Co. Galway
EI6AOB Robert Cullen, 35 Aylmer Road, Newcastle, Co. Dublin
EI6ARB John C. O'Sullivan, 142 Lr. Kilmacud Road, Stillorgan, Co. Dublin
EI6BCB Declan Kerr, Craigs Road, Dunmoe, Navan, Co. Meath
EI6BDB Martin Kerr, 8 Plunkett Hall Avenue, Dunboyne, Co. Meath
EI6BEB Details withheld at licensee's request
EI6BHB John Walsh, Birmingham Road, Tuam, Co. Galway
EI6BJB Ann-Marie Langston, 3805 Mills Street, Carencro, LA 70520, USA
EI6BOB Francis X. Gilmore, 32 Lr. Salthill, Galway
EI6BPB Sean Corry, 56 Beechwood Drive, Rathnapish, Carlow
EI6BSB John P. Kelly, 13 Clonmore Road, Mount Merrion, Co. Dublin
EI6BXB Details withheld at licensee's request
EI6CGB Details withheld at licensee's request
EI6CMB John P. Connolly, The Square, Skibbereen, Co. Cork
EI6CPB Details withheld at licensee's request
EI6CQB Patrick Hanley, 107 Applewood Heights, Greystones, Co. Wicklow
EI6CRB Kenneth Duffy, 16 Brookhaven Drive, Blanchardstown, Dublin 15, D15 WPHO
EI6CSB Desmond J. Walsh, Charleville Road, Tullamore, Co. Offaly
EI6CTB Ian Hurley, 38 Fairyfield, Parteen, Near Limerick, Co. Clare
EI6CWB Details withheld at licensee's request
EI6DCB Michael McGowan, 40A Dargle Wood, Knocklyon Road, Templeogue, Dublin 16
EI6DJB Details withheld at licensee's request
EI6DSB James A. Hyland, 224 Road 1, Balrothery Estate, Tallaght, Dublin 24
EI6DTB Pat Monahan, 7 Whitethorn Avenue, Inniscarra View, Ballincollig, Co. Cork
EI6DWB Patrick J. Haughey, Monamintra, Grantstown, Co. Waterford
EI6DYB James E. Farrell, The Dell, Ballycahane, Portlaw, Co. Waterford
EI6DZB Ann Farrell, The Dell, Ballycahane, Portlaw, Co. Waterford
EI6EEB John Lynn, 81 Pine Valley Avenue, Rathfarnham, Dublin 16
EI6EFB Brendan Beasley, 145 St Peter's Road, Walkinstown, Dublin 12
EI6EIB Philip C. McCarthy, Kilgobbin Cross, Ballinadee, Bandon, Co. Cork
EI6EQB Anthony F. Fay, 48 Ardlea Road, Artane, Dublin 5
EI6ESB Robert A. Stack, 78 Hunters Walk, Hunterswood, Ballycullen Road, Dublin 24
EI6EVB Frank Mason, 27 Dundanion Court, Blackrock Road, Blackrock, Cork
EI6EWB Shane R. Moloney, 6 Rockboro Road, Old Blackrock Road, Cork
EI6FDB Kieran Peyton, Dromedagore, Killasser, Swinford, Co. Mayo
EI6FGB Gerard Patrick Moyne, Inishoneil, Carndonagh, Co. Donegal
EI6FJB Tom·s Langan, Davitt's Terrace, Castlebar, Co. Mayo
EI6FLB Geoffrey Doyle, Gravelstown, Carlanstown, Kells, Co. Meath
EI6FYB Joe Diskin, Killinaugher, Ballyhaunis, Co. Mayo
EI6FZB John Diskin, Killinaugher, Ballyhaunis, Co. Mayo
EI6GAB Declan McGlone, 32 Shipley Mill Close, Kingsworth, Ashford, Kent TN23 3NR, England
EI6GBB Shamsudin Amkhadov, 136 Rusheeney Village, Hartstown, Dublin 15
EI6GCB Oliver Whelan, Address withheld at licensee's request
EI6GDB Bob Ashmore, Marble Hall, Ferndale Glen, Rathmichael, Dublin 18
EI6GEB Jimmy Hammond, Elm House, Tolerton, Co. Carlow
EI6GFB John Hall, Cloonaraher, Gurteen, Ballymote, Co. Sligo
EI6GGB Anthony Dolan, Newtown, Creagh, Ballinasloe, Co. Roscommon
EI6GHB John Satelle, "Shambles", Ardagh, near Kingscourt, Co. Meath
EI6GJB Michael Northey, Achill Mist House, Kilmeaney, Kilmorna, Listowel, Co. Kerry V31 VX95
EI6GLB Pawel Wieslaw Laskowski, 3 Preston Heights, Kilmeague, Naas, Co. Kildare
EI6GMB Details withheld at licensee's request
EI6GNB Marek Paczynski, 172 Drominbeg, Rhebogue, Limerick
EI6GOB Eddie McCrystal, 33 Richmond Park, Killyclogher, Omagh, Co. Tyrone BT79 7SJ, Northern Ireland
EI6GQB Details withheld at licensee's request



RSGB

**EI6GRB** Jason McGarrigle, Quay Street, Donegal Town, Co. Donegal
**EI6GSB** Conor Farrell, Boat Trench, Ardee, Co. Louth
**EI6GTB** Details withheld at licensee's request
**EI6GUB** Mark Armour, 51 Gaelcarraig Park, Newcastle, Galway
**EI6GVB** Details withheld at licensee's request
**EI6GXB** Liam Martin, 20 The Waterfront, Gort Road, Loughrea, Co. Galway
**EI6GYB** Damien Grehan, Knockmore, Ballina, Co. Mayo
**EI6GZB** Liam Kelly, Seskinryan, Bagenalstown, Co. Carlow
**EI6HAB** Details withheld at licensee's request
**EI6HBB** Patrick W Buckley, Braddan, Churchtown Road Upper, Dublin 14
**EI6HCB** Paul G Mannion, Quarry Road, Menlo, Galway
**EI6HDB** Details withheld at licensee's request
**EI6HEB** Peter Moore, 14 Trinity Square, Townsend Street, Dublin 2
**EI6HFB** Details withheld at licensee's request
**EI6HGB** Gerard Walsh, Rahan Near, Station Road, Dunkineely, Co. Donegal
**EI6HHB** Joseph Carragher, Carroweragh, Kilkshanny, Kilfinora, Co. Clare
**EI6HIB** Details withheld at licensee's request
**EI6HJB** Details withheld at licensee's request
**EI6HKB** Details withheld at licensee's request
**EI6HLB** Adil Usman, 29 Leeson Park, Ranelagh, Dublin 6
**EI6HMB** Details withheld at licensee's request
**EI6HNB** Details withheld at licensee's request
**EI6HOB** Paul Gledhill, Oak Lea, Greggane, c/o Post Office, Abbeyfeale, Co. Limerick, V94 D2CW
**EI6HPB** Details withheld at licensee's request
**EI6HQB** Fiachna MacMurch´, Log na Giumhaisì, An Rinn, Dungarvan, Co. Waterford, X35 K065
**EI6HRB** John Doherty, The Commons, Killybegs, Co. Donegal, F94 F9X6
**EI6HSB** Details withheld at licensee's request
**EI6HTB** Aaron O'Reilly, Stradbally North, Claranbridge, Co. Galway
**EI6HUB** Details withheld at licensee's request
**EI6HVB** Details withheld at licensee's request
**EI6HWB** Mark Smith, 2 Tullig, Aughatubrid, Cahersiveen, Co. Kerry, V23 V348
**EI6HXB** Details withheld at licensee's request
**EI6HYB** Details withheld at licensee's request
**EI6HZB** Details withheld at licensee's request

## EI7

**EI7M** East Cork Amateur Radio Group, John Barry, Lackabeha, Carrigtwohill, Co. Cork
**EI7T** Tipperary Amateur Radio Group, c/o Thomas Hallinan, P.O. Box 20, Cahir, Co. Tipperary
**EI7AF** Details withheld at licensee's request
**EI7AG** Stafford Charles McConnell, Killaloe, Co. Clare
**EI7AK** Details withheld at licensee's request
**EI7AU** Details withheld at licensee's request
**EI7BA** John Tait, Ballykennefick, Whitegate, Co. Cork
**EI7BR** David Fitzgerald, 36 Vardens Road, London, SW11 1RH, England
**EI7BV** John Breen, Address withheld at licensee's request
**EI7CC** Peter R. Ball, 21 Doonamana Road, Dun Laoghaire, Co. Dublin, A96 W6K3
**EI7CD** Sean Nolan, 12 Little Meadow, Pottery Road, Dun Laoghaire, Co. Dublin
**EI7CN** Patrick J. Ryan, 16 Knockiel Drive, Rathdowney, Co. Laois
**EI7CR** Michael Mullins, 41 Menin Road, Napier 4110, New Zealand
**EI7CS** Brendan Rooney, Lower Road, Glencar, via Sligo, Co. Leitrim
**EI7CV** Sean Linehan, 2 College Grove, Dunshaughlin, Co. Meath
**EI7CW** Clare Dixon, 1 Castlepoint, Crosshaven, Co. Cork
**EI7CX** Ian McStay, 37 Clonkeen Drive, Foxrock, Dublin 18
**EI7CY** Joseph P. Lawless, 45 Coolamber Drive, Rathcoole, Co. Dublin
**EI7DF** Roderick Walsh, "Duneala House", Moortown, Rathcoffey, Co. Kildare
**EI7DH** Michael Ennis, 47 Singleton Road, Scarborough, Ontario M1R 1H8, Canada
**EI7DR** Paul P. Farrell, 43 Glenmore Drive, Drogheda, Co. Louth
**EI7DV** Edward O'Loughlin, Mountrice, Monasterevin, Co. Kildare
**EI7DW** Tony O'Connor, 41 Wesley Heights, Sandyford Road, Dublin 16
**EI7EC** George Moran, 56 Rivervalley Grove, Swords, Co. Dublin
**EI7EH** Alan Shattock, The Stone House, Letternoosh, Westport Road, Clifden, Co. Galway
**EI7EJ** June Dunne, 26 Duncreggan Road, Derry BT48 0AD, Co. Derry, Northern Ireland
**EI7EK** Luigi Infante, Dublin Street, Monasterevin, Co.Kildare
**EI7EL** Thomas Lande, 15 Haldene Villas, Bishopstown, Cork
**EI7EM** John Fitzgerald, The Old Schoolhouse, Inch, Whitegate, Co. Cork
**EI7EO** Tom Kelly, Address withheld at licensee's request
**EI7EU** Thomas A. Buckley, 6 Cherry Drive, Listowel, Co. Kerry
**EI7EZ** Brendan Joyce, Pillar Park, Buncrana, Co. Donegal
**EI7FD** John M. Cashman, 'Garsview', Crush, Glanmire, Co. Cork
**EI7FE** Liam O'Brien, 33 Heywood Heights, Clonmel, Co. Tipperary
**EI7FG** Aileen Finucane, 7 Rectory Slopes, Bray, Co. Wicklow
**EI7FH** Declan Joyce, 3 Taobh Uisce, Coscorrig, Loughrea, Co. Galway
**EI7FI** Morgan H. Evans, Coolagad, Redford, Greystones, Co. Wicklow
**EI7FJ** William McLoughlin, Spencerstown, Murrintown, Co. Wexford
**EI7FL** Bernard McMahon, 34 Rose Park, Kill Avenue, Dun Laoghaire, Co. Dublin
**EI7FM** Derek Peyton, "Primrose Cottage", 1 Killincarrig Cottages, Graystones, Co. Wicklow
**EI7FN** John A. Hegarty, "The Nook", 1 Cookstown Road, Moneymore, Derry, Northern Ireland
**EI7FO** Tobias Stapleton, Swiss Cottage, Knocknagore, Crosshaven, Co. Cork
**EI7FR** Harry McGrath, WB2EZM, 300 East Overlook, Apt. 614B, Port Washington, NY 11050, USA
**EI7FV** Vincent McGettrick, "Creggan", Greenfield Road, Sutton Cross, Dublin 13
**EI7FX** Thomas Lambe, Gray Acre Road, Newtownbalregan, Dundalk, Co. Louth
**EI7FY** Patrick Geary, Kilbeg, Ladysbridge, Co. Cork
**EI7GF** Patrick M. O'Neill, 8 Abbey Drive, Rathculliheen, Ferrybank, Co. Waterford
**EI7GG** Charles Harkin, 7 Knowhead Road, Muff, Co. Donegal
**EI7GH** Details withheld at licensee's request
**EI7GK** P·draig " Meachair, 8 Cnoc na Manach, Cill Mhant·in
**EI7GL** John M. Desmond, 4 Rathmore Lawn, South Douglas Road, Cork
**EI7GM** Paul Kearney, Address withheld at licensee's request
**EI7GN** John Sherwood, Kerinstown, Killucan, Co. Westmeath
**EI7GP** Michael F. Foley, 67 Park Road, Strabane, Co. Tyrone, Northern Ireland
**EI7GU** Thomas Walk, Lamprechstr 26, P.O. Box 110355, D-63719 Achaffenburg, Germany
**EI7GV** Peter Clandillon, 108 Abbey Park, Baldoyle, Dublin 13
**EI7GW** Joseph Breen, 7 Watermill Road, Raheny, Dublin 5
**EI7GY** Joe Ryan, 34 Watson Road, Killiney, Co. Dublin, A96 EF60
**EI7HA** Charles J. Reason, 91 Fortfield Road, Terenure, Dublin 6W
**EI7HC** Details withheld at licensee's request
**EI7HE** Francis Fitzpatrick, 45 Oaklands, Arklow, Co. Wicklow
**EI7HF** Emmet A. Caulfield, 10 Beechcourt, Killiney, Co. Dublin
**EI7HG** Charles Farnan, 22 Lavarna Road, Terenure, Dublin 6W
**EI7HI** Details withheld at licensee's request
**EI7HK** Patrick J. Murray, 4 Dunboy, Brighton Road, Foxrock, Dublin 18
**EI7HM** Brendan A. Rooney, 5 Oakwood, The Paddocks, Enniscorthy, Co. Wexford
**EI7HN** Vincent J. Neff, 14 Westgate Road, Bishopstown, Cork
**EI7HO** John O'Connell, Apartment 5, Roxboro House, Bailick Road, Midleton, Co. Cork
**EI7HP** Brian Hoffmann, 1 Springmount, Tramore, Co. Waterford
**EI7HQ** Kay Eyman, WA0WOF, 321 W. 3rd Street, Ottawa, Kansas 66067, USA
**EI7HT** Tom McGrath, Rose Cottage, Piperstown, Dublin 24
**EI7HU** Michael G. McCarthy, Ballintrim, Upper Aghada, Midleton, Co. Cork
**EI7HX** Jack Fenlon, Address withheld at licensee's request
**EI7HZ** Christopher Kenneth Youens, Derrynaneane, Kilmactranny, Boyle, Co. Roscommon F52 RH56
**EI7IA** Michael Doyle, 23 Cabra Drive, North Circular Road, Dublin 7
**EI7IB** Details withheld at licensee's request
**EI7IG** John Ronan, 56 Meadowbrook, Tramore, Co. Waterford
**EI7II** Albert A. Kleyn, Clashadoo, Durrus, near Bantry, Co. Cork
**EI7IJ** Thomas Kenny, Aughnadrung, Virginia, Co. Cavan
**EI7IK** Bill Igoe, 28 The Paddocks, Naas, Co. Kildare
**EI7IL** Joseph Hernon, 13 Portersgate View, Clonsilla, Dublin 15
**EI7IN** John Pirollo, "Lismar", Graddum, Crosserlough, Co. Cavan
**EI7IO** Joe Cahill, 17 Oakley Drive, Earls Court, Waterford
**EI7IP** Charles Kinsella, Johnstown, Sea Road, Arklow, Co. Wicklow
**EI7IQ** John Corless, Coolaght, Claremorris, Co. Mayo
**EI7IR** Paraic Loughnane, Grange, Eyrecourt, Ballinasloe, Co. Galway
**EI7IS** Mark Wall, 11 Kilcaragh Village, Ballygunner, Waterford
**EI7IT** Patrick Joseph Bonner, Daleside, Callancor, Drumkeen, Convoy, Co. Donegal
**EI7IV** Christopher McGinty, Rathgooane, Dromore West, Co. Sligo
**EI7IX** Dermot Adams, Kings Hill, Westport, Co. Mayo
**EI7IY** Robert J McGowan, 42 Grange Erin, Douglas, Cork
**EI7IZ** Details withheld at licensee's request
**EI7JB** Brian E. Hoare, Drumaleague, Kilclare, Carrick on Shannon, Co. Leitrim
**EI7JC** Aidan Noone, "The Hermitage", Deerpark Road, Ravensdale, Dundalk, Co. Louth
**EI7JE** William Almeraz, KS6Y, 9265 Hillside Drive, Spring Valley, CA 91977-2147, USA
**EI7JF** Details withheld at licensee's request
**EI7JG** John Williams, 149 Viewmount Park, Dunmore Road, Waterford
**EI7JK** Details withheld at licensee's request
**EI7JL** Matthew O'Brien, The Post Office, Kiltegan, Co. Wicklow
**EI7JM** Paul Barnett, Gort Road, Gortyarn, Carndonagh, Co. Donegal
**EI7JN** Tony Byrne, 86 Slievemore Road, Drimnagh, Dublin 12
**EI7JO** John Crawford-Baker, 131 Gobbins Road, Islandmagee, Larne, Co. Antrim, BT40 3TX, Northern Ireland
**EI7JP** Denis Connolly, 43 Ashbrook, Ennis Road, Limerick
**EI7JQ** Pete Nel, Ardkirk, Castleblaney, Co. Monaghan
**EI7JR** Seamus Ryan, 6 Glencairn Lawn, The Gallops, Sandyford, Dublin 18
**EI7JS** Nigel A. Knapton, 4 Crabmill Lane, Easingwold, York YO61 3DE, England
**EI7JT** Eamonn Quinn, Tonystacken, Scotstown, Co. Monaghan
**EI7JV** Charles H. Lyons, 313 N. Tremont Dr., Greensboro, NC 27403-1546, USA
**EI7JX** Frits Remmers, Postelse Hoeflaan 95, 5042 KC-Tilburg, The Netherlands
**EI7JY** Details withheld at licensee's request
**EI7JZ** Claus Stehlik, 4 Dundanion Terrace, Blackrock Road, Blackrock, Cork
**EI7KA** Pierce Nicholas Dunphy, Brownswood, Enniscorthy, Co. Wexford
**EI7KB** Michael Eric Goulbourne, Flat 2 Orchard Mews, 15 York Road, Formby, Liverpool L37 8DN, England
**EI7KC** Details withheld at licensee's request
**EI7KD** Oleg Solovyov, Apt. 66, The Steelworks, Foley Street, Dublin 1
**EI7KE** Christoph Weritz, DL9YEL, Shelmalier Commons, Barntown, Co. Wexford
**EI7KF** Richard White, 28 Lord Warden's Parade, Bangor BT19 1YU, Co. Down, Northern Ireland
**EI7KG** John Bradley, Heatherstone House, Annaslee, Burnfoot, Co. Donegal
**EI7KH** Marek Kubita, 24 Chelmsford Road, Ranelagh, Dublin 6
**EI7KI** Matthew M. Fahy, Athina, 26 Chestnut Grove, Caherdavin Lawn, Co. Limerick
**EI7KJ** Details withheld at licensee's request
**EI7KK** John Anderson, Lisduff, Ardagh, Co. Longford
**EI7KL** John Colley, Mullaun, Cloonacool, Tubbercurry, Co. Sligo
**EI7KM** Frederick Elder, c/o Glackmore, Aught Road, Muff, Co. Donegal
**EI7AAB** Chris Yeates, 75 Georgian Village, Castleknock, Dublin 15
**EI7ALB** Simon Kenny, 1 Tullyglass Court Lr., Shannon, Co. Clare
**EI7ASB** Details withheld at licensee's request
**EI7AVB** Details withheld at licensee's request
**EI7BFB** David D. Tobin, 8 Ferndale Road, Glasnevin, Dublin 11
**EI7BMB** Tony Moore, 19 Parkhill West, Tallaght, Dublin 24
**EI7CAB** Michael J. Stack, Carrigabruse, Virginia, Co. Cavan
**EI7CDB** Details withheld at licensee's request
**EI7CEB** Fintan Sheerin, 105 Five Oaks, Dublin Road, Drogheda, Co. Louth
**EI7CGB** Details withheld at licensee's request
**EI7CHB** Derek McGonagle, North Strand, Skerries, Co. Dublin
**EI7CLB** Details withheld at licensee's request
**EI7CMB** Joe Moore, "Chez Nous", Sandyhall, Julianstown, Co. Meath
**EI7CQB** Billy O'Connor, Killarney Road, Castleisland, Co. Kerry
**EI7CSB** Michael Scannell, 2 Shrewsbury, Ballinlough, Co. Cork
**EI7CTB** Brian Menton, 190 St. Donagh's Road, Donaghmede, Dublin 13
**EI7DAB** Tom Rogers, 11 Griffith Place, Waterford City
**EI7DAR** Dundalk Amateur Radio Society, 113 Castletown Road, Dundalk, Co. Louth
**EI7DBB** Details withheld at licensee's request
**EI7DGB** Stan O'Reilly, Stradbally North, Clarinbridge, Co. Galway
**EI7DKB** Ken FitzGerald Smith, Passage West, Co. Cork
**EI7DMB** Adrian Jackson, 7 Caislean Oir, Athenry, Co. Galway
**EI7DOB** Michael Munnelly, Gibbstown, Navan, Co. Meath
**EI7DPB** Details withheld at licensee's request
**EI7DSB** Liam Rainford, 16 The Beeches, Ballina, Killaloe, Co. Tipperary
**EI7ECB** Gerard McGinley, Liskey, Ballindrait, Lifford, Co. Donegal
**EI7EEB** Peter J. Duggan, Address withheld at licensee's request
**EI7EGB** Margaret Lane, Ballincurrig, Leamlara, Midleton, Co. Cork
**EI7EHB** Details withheld at licensee's request
**EI7EIB** Andrew O Baoill, 42 Monivea Park, Galway
**EI7EJB** Joseph O'Callaghan, Main Street, Ballyclough, Mallow, Co. Cork
**EI7EQB** Fergal Purcell, 3612 Crestcreek Court, McKinney, TX 75071, USA
**EI7ERB** Desmond Murphy, Tubberfinn, Donore, Co. Meath
**EI7ESB** Sean Santry, Woodfield, Clonakilty, Co. Cork
**EI7EUB** Michael Joseph McArdle, Big Ash, Knockbridge, Dundalk, Co. Louth
**EI7EXB** Enda McDonnell, 7 Woodville, Loughrea, Co. Galway
**EI7EZB** David Elliott, 18 Deerpark Lawn, Castlenock, Dublin 15
**EI7FAB** John Browne, 31 Knockaphunty Park, Castlebar, Co. Mayo
**EI7FCB** Kim Lee Souza, 19 Pinebrook Heights, Clonsilla, Dublin 15
**EI7FEB** Michael O'Grady, 17 O'Connell Road, Tipperary Town, Co. Tipperary
**EI7FHB** Pat Walsh, 262 Glanntan, Golflinks Road, Castletroy, Co. Limerick
**EI7FQB** Liam McMahon, 4 Auburn Avenue, Dun Laoghaire, Co. Dublin
**EI7FRB** Noel Daly, 76 Castlenock Way, Laurel Lodge, Castlenock, Dublin 15
**EI7FSB** Thomas Rickard, 'Denard', Rogerstown, Southshore Road, Rush, Co. Dublin
**EI7FTB** Details withheld at licensee's request
**EI7FXB** Eoin Doherty, 8 Ozier Park Terrace, Waterford City
**EI7FYB** David Stearn, 3 Pinewood Lawn, Monang, Dungarvan, Co. Waterford
**EI7FZB** Gareth Wilmott, 12 Millbrook Court, Old Tramore Road, Waterford
**EI7GBB** Kevin Sanderson, The Shambles, Carrickboy, Co. Longford, N39 V327
**EI7GCB** Jennifer Johnson, Kilmurry House, Kilmurry, Fermoy, Co. Cork
**EI7GDB** Details withheld at licensee's request
**EI7GEB** David Morgan, 9 Lambertstown Manor, Kilmessan, Co. Meath, C15 K276
**EI7GGB** Details withheld at licensee's request
**EI7GHB** Colin Hemming, Ballyfintan, Gortymadden, Loughrea, Co. Galway
**EI7GIB** Brian Lucey, 21 Pembroke Meadows, Passage West, Co. Cork
**EI7GJB** Aidan Murtagh, Address withheld at licensee's request
**EI7GKB** George Donaldson, 9 Millrace, Bealnamulla, Athlone, Co. Roscommon
**EI7GLB** Derek Holmes, The Mews, Tinode, Blessington, Co. Wicklow
**EI7GMB** Artur Mejsak, 26 Inse Beag, Doughiska Road, Galway
**EI7GNB** Jonathan Smyth, 4 Glenullin Road, Garvagh, Coleraine, Co. Derry BT51 5DQ, Northern Ireland
**EI7GOB** Details withheld at licensee's request
**EI7GPB** Details withheld at licensee's request
**EI7GQB** Piotr Wysmyk, 16 Caisle·n "ir, Athenry, Co. Galway
**EI7GSB** Ciaran Culligan, 16 Castlepark Avenue, St. Joseph's Road, Mallow, Co. Cork
**EI7GTB** Details withheld at licensee's request
**EI7GUB** Tony Doyle, 6 Tara Hill Road, Rathfarnham, Dublin 14
**EI7GWB** Alan Keaveney, Belleville, Athenry, Co. Galway
**EI7GXB** Details withheld at licensee's request
**EI7GYB** David Malone, 3 Leinster Street East, North Strand, Dublin 3
**EI7GZB** Keith Thomas, Laurel Hill, Rathanker, Monkstown, Co. Cork
**EI7HAB** Leif Mariussen, 19 Wolverton Glen, Glenageary, Co. Dublin, A96 W9E8
**EI7HBB** LLoyd Clark, Address withheld at licensee's request
**EI7HCB** Jaroslaw Partyka, 137 Glean Allain, Tullyallen, Drogheda, Co. Louth
**EI7HEB** Keith Wallace, 69 Longwood Park, Rathfarnham, Dublin 14
**EI7HFB** David Zielinski, 8 Edward Walsh Road, Togher, Cork, Co Cork
**EI7HHB** Albert White, Address withheld at licensee's request
**EI7KRC** Kells Radio Club, Address withheld at licensee's request
**EI7MRE** Mayo Radio Experimenters Network, c/o Brendan Minish EI6IZ, Raheens, Castlebar, Co. Mayo
**EI7NET** Westnet DX Group, c/o Declan Craig, EI6FR, 167 St. James's Road, Greenhills, Dublin 12
**EI7SDX** Two Counties DX Cluster Group (Shankill), Connogue, River Lane, Shankill, Dublin 18

**EI7TRG** Tipperary Amateur Radio Group, c/o Thomas Hallinan, P.O. Box 20, Cahir, Co. Tipperary
**EI7WDX** Two Counties DX Cluster Group (Bray), c/o Hugh Forde, 4 Trinity Street, Wexford, Co. Wexford
**EJ7NET** Westnet DX Group, (see EI7NET)

## EI8

**EI8I** Larry Duggan, Cashellachan, Ballyshannon, Co. Donegal
**EI8K** Details withheld at licensee's request
**EI8U** Details withheld at licensee's request
**EI8Z** Terence J. Upton, 16 Huntstown Road, Mulhuddart, Dublin 15
**EI8AJ** John Reddington, 28 Abbey Park, Baldoyle, Dublin 13
**EI8AR** Rev. Bro. John Shortall, Benildus House, 160A Upper Kilmacud Road, Dublin 14
**EI8AT** James Stafford Murray, St Annes, Parade Ground, Arklow, Co. Wicklow
**EI8BC** William K. Ryan, 11 Wendell Avenue, Martello, Portmarnock, Co. Dublin
**EI8BD** John J. Gallagher, "Calderstones", Link Road, Brownshill, Carlow
**EI8BH** James Terence Barnes, "Whitegables", 95 Crawfordsburn Road, Bangor, Co. Down BT19 1BJ, Northern Ireland
**EI8BP** SÉamus McCague, 10 Dromartin Close, Goatstown, Dublin 14
**EI8BR** Leo J. McHugh, 1 Glenageary Woods, Upper Glenageary Road, Dun Laoghaire, Co. Dublin
**EI8BX** John J. Keely, Walshestown, Lusk, Co. Dublin
**EI8CC** Ger Gervin, Rathduff, Cullen, Co. Tipperary
**EI8CE** Aidan McGrath, "Tinhalla", Carrick-on-Suir, Co. Waterford
**EI8CG** Rev. Fr. J. Griffin, Ballard Road, Miltown Malbay, Co. Clare
**EI8CN** Pat Flynn, 127 Pine Valley Avenue, Rathfarnham, Dublin 16
**EI8CR** Bryan A. Yeomans, Bayview House, Front Strand, Youghal, Co. Cork, P36 YW88
**EI8CS** Daniel J. O'Sullivan, Derragh, Cullen, Mallow, Co. Cork
**EI8CT** Sean R. Reilly, 46 Sycamore Road, Dundrum, Dublin 16
**EI8CZ** Patrick J. O'Leary, 103 Dorney Court, Shankill, Co. Dublin
**EI8DA** Patrick Brennan, 22 Highfield, Drogheda, Co. Louth
**EI8DC** Lillian Higgins, Roevahagh, Kilcolgan, Co. Galway
**EI8DD** Tom McNamara, Annach, Rahoon Road, Galway
**EI8DH** Bernard Walsh, Otterstown, Athboy, Co. Meath
**EI8DJ** Donal Kelly, Olivette, Camden Road, Crosshaven, Co. Cork
**EI8DK** Jim Jeffers, 51 East Avenue, Park Gate, Frankfield, Douglas, Cork
**EI8DN** Liam McNulty, Town Parks, Raphoe, Co. Donegal
**EI8DQ** Aemar Higgins, 1 Cairnshill Park, Cairnshill Road, Belfast BT8 4RG, Northern Ireland
**EI8DY** Joseph Donnelly, Garryvadden Lower, Blackwater, Co. Wexford
**EI8EA** Keith Burnside, 4 Cuttles Road, Comber, Newtownards, Co. Down BT23 5YX, Northern Ireland
**EI8EB** Thomas Schewe, Bramfelder Weg 27, D-22159 Hamburg, Germany
**EI8EC** Thomas Finucane, 7 Rectory Slopes, Bray, Co. Wicklow
**EI8EE** Kevin Fitzsimons, 4 Station Road Cottages, Sutton, Dublin 13
**EI8EL** Bernard P. Curtin, 42 Lime Trees Road East, Maryborough Estate, Douglas, Cork
**EI8EM** Alan Cronin, College View, Clonroadmore, Ennis, Co. Clare
**EI8EN** Details withheld at licensee's request
**EI8EO** James R. Moloney, 17 Beechbrook, Delgany, Co. Wicklow
**EI8EQ** Ben Croly, Flat #10, Marcon Court, Triq-It, Tamar, Qawra, St. SPB, Malta
**EI8ET** Noel Grier, 9 Rushbrook, Claremorris, Co. Mayo
**EI8EU** John Sullivan, Moneymore West, Oranmore, Co. Galway
**EI8EV** James O'Hara, Binghamstown, Belmullet, Co. Mayo
**EI8EY** John Cooley, 18 McDermot Avenue, Mervue, Galway
**EI8FC** James Ryan, Tullovin Bridge, Croom, Co. Limerick V35 K289
**EI8FG** Michael Mulcahy, Reanagoshel, Newmarket, Co. Cork
**EI8FH** Malcolm Joyce, 111 Mount Prospect Drive, Clontarf, Dublin 3
**EI8FI** Kevin M. Keane, Knopogue, Ballyduff, Co. Kerry
**EI8FV** Martin Hughes, Knockane, Beechcourt, Passage West, Co. Cork
**EI8FW** Padraig Moroney, Main Street, Broadford, Co. Clare
**EI8GD** James Tighe, 1 Avondale Drive, Hanover, Co. Carlow
**EI8GE** Dr. John Malone, Greenfield House, 1 Santa Sabina Manor, Greenfield Road, Sutton, Dublin 13
**EI8GG** James Ryan, Skehanagh, Watergrass Hill, Co. Cork
**EI8GH** Thomas P. Finn, Pallas, Portlaoise, Co. Laois
**EI8GJ** Gerard McGrane, 2 Ferndale, Navan, Co. Meath
**EI8GM** Peter D. White, 109 Philomena Terrace, Irishtown, Dublin 2
**EI8GO** Thomas Molloy, 8 The Avenue, Grantstown Village, Co. Waterford
**EI8GP** Martin J. Gillespie, Carrickmagrath, Ballybofey, Co. Donegal
**EI8GQ** Garry Wilson, 5 Fernwood Crescent, Lehenaghmore, Togher, Co. Cork
**EI8GR** Aidan Riordan, 1 Portmarnock Crescent, Carrick Hill, Portmarnock, Co. Dublin
**EI8GS** Jim Barry, Windsor Hill, Glounthaune, Co. Cork
**EI8GU** Thomas Mooney, 21 Manor Court, Dunshaughlin, Co. Meath
**EI8GV** Eugene O'Connor, Ashbourne House, Kilcow, Castleisland, Co. Kerry
**EI8GW** David A. Perris, 44 Westpark, Tallaght, Dublin 24
**EI8GX** Kevin O'Sullivan, 3 Kiltegan Lawn, Rochestown Road, Cork
**EI8GY** Details withheld at licensee's request
**EI8GZ** Finbar Moloney, 21 Clonard Park, Sandyford Road, Dundrum, Dublin 16
**EI8HA** Jim Murphy, Kilcarrig, Bagenalstown, Co. Carlow
**EI8HC** Eoin Savage, Carrnahone, Beaufort, Killarney, Co. Kerry
**EI8HH** Andreas Imse, Hinter der Kirche 31, D-55129 Mainz, Germany
**EI8HJ** Ronan Coyne, East End, Inishbofin, Co. Galway
**EI8HL** Robert G. Barry, Address withheld at licensee's request
**EI8HO** Gerard Dykes, 30 St Benildus Avenue, Ballyshannon, Co. Donegal
**EI8HP** Details withheld at licensee's request
**EI8HQ** John Quinn, Newtown, Kilcolgan, Co. Galway
**EI8HR** William Power, 8 Leamy Street, Waterford City
**EI8HS** John Kelleher, 40 Rosewood Estate, Ballincollig, Co. Cork
**EI8HT** Gerald Kenneally, 23 Knockaverry, Youghal, Co. Cork
**EI8HU** Patrick J O'Mahony, Ballintotas, Castlemartyr, Co. Cork
**EI8HV** Michael Regan, 8 Glanmire Drive, Glanmire, Cork
**EI8HX** John P Maher, 61 Lower Newtown, Waterford
**EI8HY** W.P. Hughes, Address withheld at licensee's request
**EI8HZ** John Edmundson, Drumbuoy, Lifford, Co. Donegal
**EI8IC** Tim Makins, Gubnagree, Bawnboy, Co. Cavan
**EI8IE** Michael McCann, 36 Quarry Road, Cabra, Dublin 7
**EI8IF** Michael O'Connor, "Ashfield", Raheen, Golden, Co. Tipperary
**EI8IG** James A. Farrell, 12 Byrneville Estate, Dungarvan, Co. Waterford
**EI8IH** Ciaran McCarthy, 35 D`n-na-Mara Drive, Renmore, Galway
**EI8IN** Larry Hess, 3 Havana Court, Eastbourne, East Sussex BN23 5UH, England
**EI8IO** Garrett Trant, 9 Trafalgar Court, Greystones, Co. Wicklow
**EI8IP** Patrick J. O'Reilly, Glass & A.L.U. CAD Ltd, Unit 10 Kells Business Park, Kells, Co. Meath A82 VP04
**EI8IQ** Pat Whitty, Crosstown, Ballycogley, Killinick, Co. Wexford
**EI8IU** Brian Canning, Aughnaglace House, Cloone P.O., Co. Leitrim
**EI8IZ** Darragh " HÉiligh, 35 Marian Park, Drogheda, Co. Louth
**EI8JA** John A. McCarthy, 24 Sunrise Crescent, Cork Road, Waterford
**EI8JB** Charlie Carolan, Crewbawn Cottage, Drogheda Rd, Slane, Co. Meath
**EI8JC** Kohei Nishiyama, JR0BAQ, 2176, Ogawa, Maki-mura, Higashikubi-ki-gun, Niigata 943-0648, Japan
**EI8JD** Eamonn O'Brien, 94 Rory O'Connor Park, Dun Laoghaire, Co. Dublin
**EI8JE** Richard J. Cullinan, Annaghbeg, Ballyartella, Nenagh, Co. Tipperary
**EI8JF** Rodger Adair, Faugher Lower, Dunfanaghy, Co. Donegal
**EI8JG** Details withheld at licensee's request
**EI8JK** Anthony Baldwin, Rathlin, Kilcrohane, Bantry, Co. Cork
**EI8JO** David S. Warwick, 36 Annetts Hall, Borough Green, Kent TN15 8DZ, England
**EI8JQ** Paul Hegarty, 1 Ard na Boinne, Dublin Road, Trim, Co. Meath
**EI8JR** Jim Armstrong, 4 The Covert, Woodfarm Acres, Palmerstown, Dublin 20
**EI8JT** Philip Pollock, 30 Monastery Grove, Enniskerry, Co. Wicklow
**EI8JU** Daniel McFadden, 6 Doonwood, Buncrana, Co. Donegal
**EI8JV** H. A. Sinclair, 43 Edgcumbe Gardens, Belfast BT4 2EH, Northern Ireland
**EI8JW** Artur Laskowski, Address withheld at licensee's request
**EI8JX** Axel-Joachim Kaltenborn, Rinnaney, Foxford, Co. Mayo
**EI8JY** Ruthann O'Connor, Tuam Road, Kilmaine, Co. Mayo
**EI8JZ** Details withheld at licensee's request
**EI8KA** Joseph Loughrey May, 1 Ard na T·na, Castletowncooley, Rivers-town, Dundalk, Co. Louth
**EI8KC** Gerrit Verhoef, Youghal Cottage, Newtown, Nenagh, Co. Tipperary
**EI8KD** Details withheld at licensee's request
**EI8KE** Donald J. Shelton, Geashill, Co. Offaly
**EI8KF** Details withheld at licensee's request
**EI8KG** Dave Sholdice, 27 Ashton Wood, Herbert Road, Bray, Co. Wicklow
**EI8KI** Louis Ryan, 32 Broomville, Dublin Road, Portlaoise, Co. Laois
**EI8KJ** James Clarke, Kells, Co. Meath
**EI8KK** Daniel E Deck, 57 Bruach Na H'Abhainn, Quinn Road, Ennis, Co. Clare
**EI8KL** Timothy Jones, 33 Moonlaun, Tramore, Co. Waterford
**EI8KM** Details withheld at licensee's request
**EI8AAB** Michael Pettigrew, The Coach House, Collierstown, Tara, Co. Meath
**EI8BAB** Martin O'Rourke, Dromsikin, Castlebellingham, Co. Louth
**EI8BDB** Patrick J. McCann, 24 Woodview Court, Glenalbyn Road, Stillorgan, Co. Dublin
**EI8BEB** David J. Dillon, 84 Knocknaganny Park, Sligo
**EI8BHB** Details withheld at licensee's request
**EI8BLB** William Grant, Listerlin, Tullogher, Mullinavat, via Waterford, Co. Kilkenny
**EI8BRB** Ivan Sproule, Bella, Collooney, Co. Sligo
**EI8BSB** Details withheld at licensee's request
**EI8CHB** David Kearney, Ballybeg, Currow, Killarney, Co. Kerry
**EI8CLB** George D. O'Reilly, 51 Prospect View, Stocking Lane, Rathfarnham, Dublin 16
**EI8CMB** Joseph Clarke, Balrath, Kells, Co. Meath
**EI8DGB** Sean Sheehy, 48 Murmont Road, Montenotte, Cork
**EI8DRB** Gerry Kavanagh, Derry, Cregannabeg, Oranmore, Galway
**EI8DZB** Martin Burke, Moneygreggan, Newtowncunningham, Lifford, Co. Donegal
**EI8EFB** Mark Cusack, Ballisk, Donabate, Co. Dublin
**EI8EJB** Brian Whelan, 32 Commons Grove, Dromiskin, Co. Louth A91 X379
**EI8EPB** Seamus A. Ryan, Modeshill, Mullinahone, Co. Tipperary
**EI8EDB** Thomas J. Harte, Stonestown, Robinstown, Navan, Co. Meath
**EI8ETB** Thomas McCoy, 46 Ashford Court, Pinecroft, Grange, Douglas, Cork
**EI8EUB** Tony Cooke, Ohean House, Athy, Co. Kildare
**EI8EXB** Gerry Ormond, 4 Elm Park, Renmore, Galway
**EI8EYB** George W Quinlan, 16 Plunkett Terrace, Cobh, Co. Cork
**EI8FAB** Anthony Lyle Smith, Gurrane, Portmagee, Co. Kerry
**EI8FBB** John Murphy, Barranastook Upper, Old Parish, Dungarvan, Co. Waterford
**EI8FDB** Bernard Tyers, Moonvoy, Tramore, Co. Waterford
**EI8FFB** Martin G. Evans, 13 Ashley Close, Cherrymount, Waterford City
**EI8FGB** Dennis Butcher, Knockalahara, Cappoquin (via Mallow), Co. Waterford
**EI8FHB** Steve Brosnan, 4 Black Rod Close, Hayes, Middlesex UB3 4QJ, England
**EI8FIB** Details withheld at licensee's request
**EI8FKB** Seamus C. McGrory, Address withheld at licensee's request
**EI8FLB** Details withheld at licensee's request
**EI8FNB** P·draig Naughton, Villa Nova, Togher, Ballinasloe, Co. Roscommon
**EI8FOB** Michiko Nishiyama, JF0KYK, 2176, Ogawa, Maki-mura, Higashikubiki-gun, Niigata 943-0648, Japan
**EI8FQB** Thomas Phelan, Tivoli Crest, Waterford Road, Tramore, Co. Waterford
**EI8FRB** John Duggan, Carrick Road, Kilmaganny, Co. Kilkenny
**EI8FWB** Alan Bennett, Melrose, Beaumont Crescent, Ballintemple, Co. Cork
**EI8FXB** Details withheld at licensee's request
**EI8GBB** Details withheld at licensee's request
**EI8GCB** Details withheld at licensee's request
**EI8GDB** Brian Haughey, 3 Friars Tale Place, Staatsburg, NY 12580, USA
**EI8GEB** Caroline Hartigan, Doonagore, Doolin, Co. Clare
**EI8GFB** Christopher Davies, 9 Templeogue Wood, Templeogue, Dublin 6W, D6W CK49
**EI8GGB** Andrew Hill, 5 Gortnaclohy Heights, Skibbereen, Co. Cork
**EI8GIB** John O'Grady, 23 Hollywood Grove, Ballaghaderreen, Co. Roscom-mon
**EI8GJB** Details withheld at licensee's request
**EI8GKB** Details withheld at licensee's request
**EI8GLB** Stephen Webster, 14 Owendore Crescent, Rathfarnham, Dublin 14
**EI8GMB** Ron Skingley, Coomkeen, Durrus, Bantry, Co. Cork
**EI8GNB** David Casey, 4 Seven Oaks, Frankfield, Cork
**EI8GPB** Philip Hosey, 13 Glenelly Gardens, Omagh, Co. Tyrone, BT79 7XG, Northern Ireland
**EI8GQB** Olivier Vandenbalck, 4 Avenue du Cerf-Volant, 1170 Brussels, Belgium
**EI8GRB** Eamonn McGuinness, 32 St. Laurence Park, Garryowen, Co. Limerick
**EI8GSB** John Lunnon, Clonakilty, Co. Cork
**EI8GTB** Details withheld at licensee's request
**EI8GUB** Simon Barnes, 191 Marlacoo Road, Portadown, Co. Armagh, BT62 3TD, Northern Ireland
**EI8GVB** Details withheld at licensee's request
**EI8GWB** Victor Reijs, 15 Shenick Grove, Skerries, Co. Dublin
**EI8GXB** Details withheld at licensee's request
**EI8GYB** Details withheld at licensee's request
**EI8GZB** Pat Ryan, Sycamore View, Adare, Co. Limerick

## EI9

**EI9E** Network Southern Area Radio Experimenters Club, c/o John Hearne, RTE, Fr Mathew Quay, Cork
**EI9I** Emerald Isle Contest & DX Group, c/o John Corless, Coolaght, Claremorris, Co. Mayo
**EI9O** Eoin Fagan, The Gables, Listraghee, Ballinalee, Co. Longford
**EI9P** Phil Cantwell, 'Villa Maria', Manorland, Trim, Co. Meath
**EI9AD** Details withheld at licensee's request
**EI9AE** Norman Miller, G3MVV, "Oak Tree", Ashwood, Arklow, Co. Wicklow
**EI9AL** Brian Toner, 19 Springfield Drive, Dooradoyle, Limerick
**EI9BD** Jim Naughton, Crannagh, Cloghans P.O., Ballina, Co. Mayo
**EI9BW** Henry Boyle, 120 Pinebrook Road, Artane, Dublin 5
**EI9BX** William F. Hurley, 20 Chestnut Grove, Caherdavin Lawn, Limerick
**EI9BZ** Denis T. Walsh, Charleville Road, Tullamore, Co. Offaly
**EI9CC** Richard K. Wilson, 9 Navan Road, Castleknock, Dublin 15
**EI9CE** John McGorman, Teaghlach, Legnakelly, Clones, Co. Monaghan
**EI9CF** Seamus O'Dea, 27 Millmount, Mullingar, Co. Westmeath
**EI9CI** Timothy O'Mahony, Roxtown, Fedamore, Kilmallock, Co. Limerick
**EI9CJ** Tom McDermott, Rockmarshall, Jenkinstown, Dundalk, Co. Louth
**EI9CK** Rev. Fr. Nicholas O'Grady, St. Paul's Retreat, Mount Argus, Dublin 6W
**EI9CN** Larry McGriskin, Derryowen, Barna Road, Knocknacarra, Galway
**EI9CP** Peter Clancy, 'Seapoint', Fane Mouth, Blackrock, Co. Louth
**EI9CS** Donald Riordan, Lombardstown, Pallasgreen, Co. Limerick
**EI9CW** Raymond Bonar, 28 Cherry Grove, Naas, Co. Kildare
**EI9DA** Kieran Daly, Granarogue, Carrickmacross, Co. Monaghan
**EI9DB** Peter F. McGovern, Barran, Blacklion, Co. Cavan
**EI9DC** Dermot Cronin, Derhill, Church Road, Killiney, Co. Dublin
**EI9DJ** Dermot E. O'Dwyer, Dublin Road, Navan, Co. Meath
**EI9DK** Sean Dunne, Rockfield Lodge, Boicetown, Togher, Co. Louth
**EI9DL** Thomas McLoughlin, Cranley More, Mostrim, Co. Longford
**EI9DM** Raymond Long, Dunkineely, Co. Donegal
**EI9DO** Patrick Kennedy, 16 Pebble Hill, Maynooth, Co. Kildare
**EI9DR** Gerard M. O'Doherty, 22 Meadowvale Close, Raheen, Limerick
**EI9DS** James Waters, Clonsilla, Gorey, Co. Wexford
**EI9DW** Dr, David W. Hughes, 81 Cahard Road, Ballynahinch BT24 8YD, Co. Down, Northern Ireland
**EI9DZ** Gerald E. Birkhead, "Lus na SÌ", 15 CrannÛg, Mohill Road, Keshcar-rigan, via Carrick-on-Shannon, Co. Leitrim
**EI9EB** Edward McCourt, Gortglass Cottage, Gortyglas Lake, Cranny, Co. Clare
**EI9ED** Ronald F McGrane, Cavan Road, Kells, Co. Meath
**EI9EE** Michael Grady, Fycorrenagh, Cullion Road, Letterkenny, Co. Donegal
**EI9EF** Patrick J. Kenny, Slieve Ban, Malin Head, Co. Donegal
**EI9EH** Patrick Murphy, PO Box 62, Dolores 03150, Alicant, Spain
**EI9EL** Ken O'Brien, 2 Dartry Park, Dublin 6
**EI9EM** Maurice H. McFadden, 121 Greystown Avenue, Belfast, BT9 6UH, Northern Ireland
**EI9EO** Michael Mulkerrin, 315A Templeogue Road, Dublin 6W
**EI9EQ** Egidio A. Giani, Ballybrack, Kilmacthomas, Co. Waterford
**EI9ER** Dermot Ryan, 53 Culmore Road, Palmerstown, Dublin 20
**EI9ES** Peter Morrison, Address withheld at licensee's request
**EI9EW** William Furlong, Ballycoheir, New Ross, Co. Wexford
**EI9EZ** Patrick Geoghegan, Fairfield Lodge, Lios an Oir, Lismore, Co. Waterford
**EI9FB** Joseph Bannon, 2 The Marina, Deerpark, Boyle, Co. Roscommon

**EI9FD** Edward Navagh, Rathdrinagh, Beauparc, Navan, Co. Meath
**EI9FE** Mike Hoare, "Glencoe", Ballykisteen, Tipperary
**EI9FF** Ted Kennedy, 45 Erne Dale Heights, Ballyshannon, Co. Donegal
**EI9FK** William Somerville-Large, 'Vallombrosa', Thornhill Road, Bray, Co. Wicklow A98 E9F4
**EI9FM** Noel Lafferty, Carrickshandrum, Killygordon, Lifford, Co. Donegal
**EI9FN** Percy Masters, Middletown, Williamstown, Co. Galway
**EI9FP** Thomas P. Gaffney, 56 Hollybrook Road, Clontarf, Dublin 3
**EI9FV** Gerry Lawlor, Co. Dublin
**EI9FX** Mark O'Rourke, Derrycammagh, Castlebellingham, Dundalk, Co. Louth
**EI9FY** Patrick Devine, Begrath, Tullyallen, Drogheda, Co. Louth
**EI9GA** Declan Goggin, Desert, Clonakilty, Co. Cork
**EI9GB** John Doherty, 3 Rockfield Terrace, Buncrana, Co. Donegal
**EI9GH** Jacqueline Holland, Warrensbrook, Enniskeane, Co. Cork
**EI9GI** Details withheld at licensee's request
**EI9GL** Paul Healy, 20 Blackheath Drive, Clontarf, Dublin 3
**EI9GO** Eamonn G. Phelan, 14 Ursuline Crescent, Waterford
**EI9GQ** Eamon Skelton, 1 The Crescent, Estuary Drive, Blackrock, Co. Cork
**EI9GT** Peter J. McNally, 'Lorna', 57 Holmpatrick, Skerries, Co. Dublin
**EI9GW** Michael J. Flynn, Cullentragh, Mayo Abbey, Claremorris, Co. Mayo
**EI9GY** Patricia Mangan, "Mount Ievers", Sixmilebridge, Co. Clare
**EI9GZ** Gerald V. Gavin, 14 Parslicktown Avenue, Mulhuddart, Dublin 15
**EI9HC** Stephen Nolan, "Churchview", Barrack Lane, Athboy, Co. Meath
**EI9HD** James McGrory, Carrickanee, Inch Island, Co. Donegal
**EI9HG** Joe Ryan, Kylebeg, Newtown, Nenagh, Co. Tipperary
**EI9HK** Tony Clifford, 'Bofeenaun', Brookfield, Rochestown Road, Cork T12 Y28P
**EI9HL** Mark McNulty, Town Parks, Raphoe, Co. Donegal
**EI9HM** Finbarr Carroll, 19 Gurteen North, Annascaul, Co. Kerry
**EI9HO** Joseph P. Clarke, 61 Whitebridge Manor, Killarney, Co. Kerry
**EI9HQ** Declan Lennon, 45 Pearse Park, Sallynoggin, Dun Laoghaire, Co. Dublin
**EI9HR** Richard Ryan, 22 Convent Hill, Waterford
**EI9HV** Maurice Keating, 20 Kincora Avenue, Clontarf, Dublin 3
**EI9HW** John FitzGerald, Springville House, Balrath, Kells, Co. Meath
**EI9HX** Patrick G. O'Connor, Togher, Ballinasloe, Co. Roscommon
**EI9HZ** Eoin O'Cleary, Address withheld at licensee's request
**EI9IA** Bruce T. Marshall, QSL via K1AJ, PO Box 4242, Andover, MA 01810-0814, USA
**EI9IB** Mark J Boothman, Norton Place, Ballysax, The Curragh, Co. Kildare
**EI9ID** Anthony Stack, 33 Park Court, Ballyvolane, Cork
**EI9IF** Alan Dean, 50 Eden Court, Dunshaughlin, Co. Meath
**EI9IG** Sean Doran, 11 Hillbrook Estate, Tullow, Co. Carlow
**EI9IK** Dominic J. Nolan, 64 Oakleigh, Longwood, Co. Meath
**EI9IL** Anthony Liddy, 26 Sarsfield Avenue, Garryowen, Limerick
**EI9IM** Derry Lawlor, 29 Grove Park Avenue, Finglas, Dublin 11
**EI9IN** Brian Crowley, 56 Gleann Tuarigh, Youghal, Co. Cork
**EI9IO** John P. Cronin, Tinegeragh, Watergrasshill, Co. Cork
**EI9IP** George A. Carey, Carnmalin, Malin Head, Co. Donegal
**EI9IR** Michael Levin, 1701 Diana Drive, Winter Park, FL 32789, USA
**EI9IS** Michael J. Shevlin, Bomany, Letterkenny, Co. Donegal
**EI9IT** Charles McHugh, Donegal Street, Ballybofey, Lifford, Co. Donegal
**EI9IU** Valentine Duggan, Apartment 8, Cathedral Close, Cathedral Square, Waterford
**EI9IW** Details withheld at licensee's request
**EI9IX** Dick Gibson, 93 Cavan Road, Dungannon, Co. Tyrone BT71 6QN, Northern Ireland
**EI9IY** Albert Lahoud, 'Parnassus', Boskill Road, Caherconlish, Co. Limerick
**EI9IZ** Hazel Leahy, 'Galtymore', Barnlough, Bansha, Co.Tipperary, E34 WN92
**EI9JA** Padraic Baynes (Snr), Derrygorman, Westport, Co. Mayo
**EI9JB** Laxon Mack, Mullen, Frenchpark, Co. Roscommon
**EI9JF** Nicki Mullally, 1172 Eyrefield Road, Curragh, Co. Kildare
**EI9JG** Details withheld at licensee's request
**EI9JK** Anthony Cummins, 91 Killinarden Heights, Tallaght, Dublin 24
**EI9JL** John Martin Toland, 23 Convent Road, Carndonagh, Co. Donegal
**EI9JM** William Denmead, 19 City View Mews, The Downs, Banduff, Co. Cork
**EI9JO** John W. Gill, Reenard, Cahirciveen, Co. Kerry
**EI9JQ** Dez Watson, 3 Brunel Drive, Biggleswade, Bedfordshire SG18 8BT, England
**EI9JS** Dominic Curtin, Derrynabrook, Cloontia, Ballymote, Co. Sligo
**EI9JU** Gerard McLaughlin, Lisfannon, Burt, Co. Donegal
**EI9JV** Laurent Pierre Schoumacker, Coolboy Big, Letterkenny, Co. Donegal
**EI9JW** Michael Dornan, Hillsborough Parish Church, 2 Beechgrove, Ballynahinch, Co. Down BT24 8NQ, Northern Ireland
**EI9JX** Arnold Mallows, Kilmurry House, Kilmurry, Fermoy, Co. Cork
**EI9JY** Robert William Stokes, Toberroe, Glinsk, Ballymoe, Co. Roscommon
**EI9JZ** Details withheld at licensee's request
**EI9KA** Details withheld at licensee's request
**EI9KB** Billy Cullen, 35 St Manntans Road, Wicklow Town, Co. Wicklow
**EI9KC** Details withheld at licensee's request
**EI9KD** Adrian Ryan, 1 Woodhill Grove, Woodhill, Ardara, Co. Donegal
**EI9KF** Hugh Bradley, 22 Park Street, Dundalk, Co. Louth
**EI9KG** Martin Farnan, 23 Larchfield Park, Goatstown, Dublin 14
**EI9KH** Details withheld at licensee's request
**EI9KI** Details withheld at licensee's request
**EI9KJ** Tom Foote, "The Moorings", Tonabrocky, Bushy Park, Galway
**EI9KK** Edwin J Doyle, Miskaun Glebe, Aughnasheelin, Ballinamore, Co. Leitrim
**EI9KL** Jeremiah J. Kelleher, Glanogue, Drommahane, Mallow, Co. Cork
**EI9ADB** Details withheld at licensee's request
**EI9AEB** Anthony T. Cullen, 18 Knockmeenagh Road, Dublin 22
**EI9AKB** James Travers, 59 Mondalea Woods, Firhouse Road, Dublin 16
**EI9ALB** Details withheld at licensee's request
**EI9BAB** Keith Garland, 17 Seaview Wood, Shankill, Dublin 18
**EI9BFB** Details withheld at licensee's request
**EI9BJB** James Bustard, Clarcam, Donegal
**EI9BXB** Louis O'Toole, 2 Seapark, Malahide, Co. Dublin
**EI9CBB** John F. Marron, Briarless, Julianstown, Co. Meath
**EI9CDB** Finbarr McCarthy, Kilross, Mackeys Cross, Clogheen, Co. Cork
**EI9CPB** John V. Gardner, 44 Munster Street, Phibsborough, Dublin 7
**EI9CSB** Paul Thim, 1 The Orchards, Gorey, Co. Wexford, Y25 E978
**EI9CTB** Thomas J. Fahy, Rathgeal, Carragh Grove, Knocknacarra, Co. Galway
**EI9CUB** Patrick McCabe, Carrickacromin, Mountain Lodge, Cootehill, Co. Cavan
**EI9CZB** Christopher Mann, Woodford, Listowel, Co. Kerry
**EI9DDB** Derek O'Hanlon, 3 Cherry Court, Granstown Village, Waterford
**EI9DFB** Gerard Barron, Kildermody, Kilmeadon, Co. Waterford
**EI9DHB** Kieran Howley, 7 Corcoran Terrace, Kells Road, Kilkenny
**EI9DPB** David McCabe, 26 North Summer Street, North Circular Road, Dublin 1
**EI9DQB** Patrick Crawley, Townparks, Ardee, Co. Louth
**EI9DSB** Details withheld at licensee's request
**EI9DVB** Liam Brady, "Stonehaven", Eaton Brae, Shankill, Dublin 18
**EI9DZB** Stephen J. Curran, 70 Gilford Road, Sandymount, Dublin 4
**EI9EIB** Roy Tallon, 1 Blackwater Abbey, Navan, Co. Meath
**EI9EMB** Details withheld at licensee's request
**EI9EOB** Joseph Gallagher, 2 The Glen, Chapel Road, Dungloe, Co. Donegal
**EI9EQB** Bruno Nardone, 2A Oxmantown Road, Dublin 7
**EI9ESB** Details withheld at licensee's request
**EI9EYB** Revere Richardson, Cloughdoolarty North, Fedamore, Co. Limerick
**EI9FBB** David Deane, 7 Spriggs Road, Gurranabraher, Cork
**EI9FCB** Alan Lester, 28 O'Connell Avenue, Turner's Cross, Cork
**EI9FDB** W. K. Donald, 15 Kingsland Parade, Portobello, Dublin 8
**EI9FEB** Michael Watterson, 11 Laurel Park, Patrickswell, Co. Limerick
**EI9FHB** Daniel Cussen, c/o South Dublin Radio Club, Ballyroan Community Centre, Marian Road, Rathfarnham, Dublin 14
**EI9FMB** Martin Philip Hayes, Kilsteague, Annestown, Co. Waterford
**EI9FNB** John Doherty, 43 Hawthorn Heights, Letterkenny, Co. Donegal
**EI9FQB** Bernard Reilly, Dernavagy, Aughnacliffe, Co. Longford
**EI9FTB** Maurice R. O'Donovan, Address withheld at licensee's request
**EI9FVB** Declan Horan, 6 Glincool Grove, Ballincollig, Co. Cork, P31 WY91
**EI9FXB** Details withheld at licensee's request
**EI9FYB** John J. Cunnane, Terra Nova, Spencer Park, Castlebar, Co. Mayo
**EI9FZB** Details withheld at licensee's request
**EI9GAB** Krzysztof Jankowski, 3 Bayview Rise, Killiney, Co. Dublin
**EI9GBB** Details withheld at licensee's request
**EI9GDB** Niall Behan, Cois Locha, Carraghadoo, Kilcolgan, Co. Galway
**EI9GEB** Trevor Hooper, 46 St Johns Crescent, Clondalkin, Dublin 22
**EI9GGB** Michael Fitzgerald, 35B Garden City, Gorey, Co. Wexford
**EI9GHB** Brendan McNamara, Carhucore, Ogonnelloe, Scariff, Co. Clare
**EI9GIB** Details withheld at licensee's request
**EI9GJB** Details withheld at licensee's request
**EI9GKB** Details withheld at licensee's request
**EI9GLB** Jim Hall, Ballinamona, Ballycanew, Gorey, Co. Wexford
**EI9GMB** Anthony Bond, 25 Claddagh Park, Tom Bellew Avenue, Dundalk, Co. Louth
**EI9GNB** Details withheld at licensee's request
**EI9GPB** Ronan Griffin, 38 Thorncliffe Park, Orwell Road, Dublin 14
**EI9GQB** Fran Griffin, 38 Thorncliffe Park, Orwell Road, Dublin 14
**EI9GRB** Hans Krauss, Station Road, Crookstown, Co. Cork
**EI9GRC** St. Joseph's College Radio Club, Garbally College, Ballinasloe, Co. Galway
**EI9GSB** Lisa Collins, "Lyston", 13 Forest View, Goulds Hill, Mallow, Co. Cork
**EI9GTB** Brian Cowley, 14 Beaulieu Mews, Greenhills, Drogheda, Co. Louth
**EI9GUB** Details withheld at licensee's request
**EI9GVB** Details withheld at licensee's request
**EI9GWB** Arkadiusz Tokarski, 133 Lissadyra, Ballygaddy Road, Tuam, Co. Galway
**EI9GXB** Details withheld at licensee's request

# UK Surname Index

## A

Aanestad T ... G0LVA
Abberley M ... 2E0XCV
Abberley M ... M3XCV
Abberley S ... M3XIG
Abbey A ... G3OVH
Abbey J ... G7KLN
Abbishaw J ... G6CQH
Abbot A ... G4NPA
Abbot S ... G4NPB
Abbott A ... G1GOP
Abbott A ... G6GBL
Abbott C ... M6ZBM
Abbott D ... 2E0GYG
Abbott D ... G4MPT
Abbott D ... G4RFU
Abbott D ... M3DRA
Abbott D ... M6DZV
Abbott H ... G7LGY
Abbott J ... G8CWJ
Abbott J ... G8JPW
Abbott J ... M3RYT
Abbott P ... G0FUI
Abbott R ... G7RCL
Abbott S ... G1GOQ
Abbott W ... M6AYS
Abbruscato J ... G0AOH
Abdullah S ... M3UTZ
A'Bear D ... G0VZV
Abel K ... 2E0DRI
Abel K ... M6XKG
Abel K ... M6YDA
Abel L ... M3LJA
Abel M ... M3UMA
Abel R ... M6FRF
Abel R ... G4FKX
Abel S ... G7ETC
Abela C ... G0ATP
Abela K ... M6BAY
Abell B ... G1SKV
Abernethy P ... G8HGG
Abeynayake K ... G7RFS
Abid M ... M6BYQ
Abousaid R ... M0XDY
Abraham A ... 2E0UOK
Abraham A ... M3UOK
Abraham D ... G4UTC
Abraham G ... M6GEA
Abraham I ... M6IKA
Abraham J ... M6LGH
Abraham M ... G1DDK
Abraham M ... M6MDX
Abraham R ... MI3RBM
Abraham S ... M3SAR
Abraham S ... M6SAD
Abrahams A ... G8UOJ
Abram J ... G6CZZ
Abram M ... G1CZU
Abram P ... 2E0WWR
Abram W ... GI6KJC
Abrams A ... M0DOK
Abrey C ... G3RZY
Abson A ... 2E0SVZ
Abson A ... M6ZDL
Abson T ... G6MLS
Aburrow P ... G4BQY
Ace P ... GW4SPL
Acheson B ... G1JFQ
Acke P ... G3FYF
Ackerley G ... G3VUN
Ackerley J ... G8TKQ
Ackerley N ... G3RIR
Ackerman B ... G0GNP
Ackland N ... G0IIK
Ackley P ... G3LRP
Ackrill D ... G0DJA
Ackrill D ... M0HVK
Ackroyd A ... M3XPI
Ackroyd B ... G8ZVK
Ackroyd D ... G4SYV
Acott J ... G4ILH
A'Court G ... M0GAC
Acres B ... G4MXQ
Acton G ... 2E0HXX
Acton G ... M0TXX
Acton J ... G0NFH
Acton J ... G1DOX
Acton L ... M3ZRY
Adair D ... G3BVB
Adam A ... GM0VFD
Adam M ... M6CTG
Adam N ... 2E0CZU
Adam N ... M6BKS
Adam N ... GM8LFA
Adam P ... G7KXS
Adam R ... GM4ILS
Adam R ... G7JAQ
Adam S ... MM3GOI
Adam T ... GM0NBA
Adam W ... G1TJT
Adamek R ... G6IYY
Adams A ... GM0KZG
Adams A ... G3YOA
Adams A ... 2E0KMP
Adams A ... G1EGZ
Adams A ... G7OPB
Adams A ... M0ZCM
Adams A ... M6AZA
Adams B ... G0CPZ
Adams B ... G4RFV
Adams B ... MM6BDA
Adams C ... G3URL
Adams C ... G3YNC
Adams D ... G3JUU
Adams D ... G4YLK
Adams D ... G8IJG
Adams D ... G8KTV
Adams D ... M3DEA
Adams D ... G0GIE
Adams D ... 2I0NWO
Adams D ... G1TCK
Adams D ... MI0NWO
Adams D ... MI3NWO
Adams D ... G4ZEW
Adams F ... G8YTU
Adams G ... G3ICA
Adams G ... G3LEQ
Adams G ... GM8FVN
Adams H ... G6PYR
Adams I ... G4EJG
Adams J ... 2E0GXI
Adams J ... G0WSY
Adams J ... GM1JHU
Adams J ... G6AFK
Adams J ... GI7HYU
Adams J ... M6GOC
Adams J ... M6JJX
Adams J ... G4JZL
Adams J ... G8BDM
Adams J ... G3VZF
Adams J ... G7CIA
Adams K ... G4JIH
Adams K ... G7OAH
Adams L ... G4RKV
Adams M ... G0AMO
Adams M ... G1ICH
Adams M ... G4IYA
Adams M ... G6BZL
Adams M ... G6ZLJ
Adams M ... M6YBA
Adams M ... G6VMR
Adams M ... G4LOF
Adams M ... G3ZLQ
Adams P ... 2E0HUJ
Adams P ... G0HWY
Adams P ... G3UKE
Adams P ... G4XKA
Adams P ... G6LZB
Adams P ... G1YLE
Adams P ... MM5AMM
Adams P ... G0NOV
Adams P ... G7EGU
Adams R ... G0JLE
Adams R ... G0UMK
Adams R ... G8NYH
Adams R ... G8RWJ
Adams R ... G8KOD
Adams R ... MM6VGS
Adams R ... 2E0IEI
Adams R ... M6GZE
Adams R ... G6NYG
Adams S ... G0FRV
Adams S ... G0KVZ
Adams S ... G0OMM
Adams S ... GD0ULF
Adams S ... G8MZW
Adams S ... M6ZSD
Adams T ... G1ZSK
Adams T ... G4CHD
Adams T ... M3TJI
Adamson A ... G4SPV
Adamson B ... M6BMJ
Adamson D ... G6ODF
Adamson D ... 2M0OVD
Adamson D ... MM0OVD
Adamson G ... GM0RDA
Adamson L ... G6PVT
Adamson M ... G1UTQ
Adamson R ... G6KWU
Adamson R ... 2M0CLU
Adamson R ... MM0TIE
Adamson R ... MM6BKP
Adamson W ... MM0CHV
Adams-Spink G ... GD4IUM
Adaway S ... M0TNG
Adby B ... 2E0CBN
Adby B ... M0HBW
Adby R ... M6MBE
Adcock G ... G4EUK
Adcock J ... G0FWD
Adcock M ... G8CMU
Adcock M ... G8DIO
Adcock P ... G1JNR
Addicott M ... G7SSA
Addidle V ... GI3VHM
Addis B ... M0ART
Addis C ... G6DSP
Addis G ... G3TEB
Addison C ... MM6CBI
Addison J ... G6YOZ
Addison J ... M5AGI
Addison J ... MM6JEA
Addison L ... 2E0LLC
Addison L ... M6MGD
Addison M ... 2E0MCA
Addison M ... G4TQY
Addison M ... G6MJA
Addison-Lees C ... G4OHV
Addy J ... 2E0BQQ
Addy J ... M3UPK
Addy W ... G3KRX
Adedeji V ... M6MXO
Adem A ... G7EMH
Adey A ... G0HDD
Adey A ... G7BCO
Adey G ... M6GCV
Adgie G ... M6AKA
Adie W ... GM7LDU
Adiga R ... M6NVG
Adjey C ... MI3CDA
Adkin D ... M6FLH
Adkin F ... G3LAU
Adkins A ... G3MVU
Adkins A ... M3REZ
Adkins B ... M3BJV
Adkins B ... M1CQN
Adkins D ... M5AGV
Adkins G ... G1IQA
Adkins H ... MM6OCP
Adkins L ... M3LMA
Adkins R ... M3AOM
Adkins S ... M3SVN
Adlam D ... M3OWU
Adlam S ... M3HTR
Adler J ... 2E0JJA
Adler J ... M6CVB
Adlington J ... M0DVT
Adlington J ... M3LVA
Adlington M ... 2E0MAZ
Adlington R ... M0BOB
Admans M ... G8ITJ
Admans S ... M6HIP
Adrain S ... GI4SQL
Adrian G ... G7CUF
Adrian K ... MM6KHA
Adshead D ... 2E0BPX
Adshead D ... M0GUM
Adshead D ... M3VGZ
Aedy A ... G4GMT
Affleck L ... 2M0LUG
Affleck L ... MM3INY
Affolter A ... G0CKH
Afford A ... G6AFG
Afford L ... G4SHY
Afolabi M ... M6NMT
Agacy R ... G1EQM
Agacy R ... M0EQM
Agar D ... G0EBW
Agar J ... G8AZA
Agar T ... 2E0DXV
Agboma F ... 2D0CQT
Agboma F ... MD6FEA
Ager A ... 2E0GZV
Ager B ... M6BRZ
Ager D ... G7RUY
Ager J ... G1VRJ
Ager T ... G4UXJ
Aggus R ... G4CZZ
Agnew J ... GI6IRL
Agnew R ... G1ZNX
Agnew S ... G1ZIM
Agnew T ... MI6XBA
Agombar K ... G8BOJ
Ahern T ... 2E0TMA
Ahern T ... M6RMN
Aherne T ... G6UGT
Ahier P ... 2J0ODX
Ahier P ... MJ0PMA
Ahier P ... MJ6AUD
Ahmed J ... 2E0JAA
Ahmed J ... 2E0JAA
Ahmed J ... M3BGT
Ahmed M ... G0RNH
Ahmed M ... M3GKK
Ahmed M ... M3MAA
Ahmed S ... M3SCA
Aicheler B ... M0OZJ
Aiello A ... M1AEV
Aigeldinger H ... G0PTI
Aiken D ... G4CDO
Aiken J ... GM0AZU
Ailsby E ... G4MZL
Aindow J ... G4GUV
Aingo R ... G6GA
Ainger A ... G1ZYJ
Ainger L ... M6LAC
Ainley M ... [illegible]
Ainscough D ... G1OMY
Ainscow P ... G7EKG
Ainslie D ... G6FXR
Ainsworth B ... G4GPW
Ainsworth J ... G7UCB
Ainsworth R ... G1JMP
Ainsworth R ... G4UPU
Ainsworth R ... G6GEN
Ainsworth R ... M0TLC
Aird A ... GM0UGG
Aird K ... M0KBC
Aird R ... MM0RFA
Airey A ... G7MKQ
Airey L ... G3GEJ
Airey R ... G1NDV
Airey S ... M6IRU
Airs D ... G7VIB
Airs M ... G7VIA
Airs S ... 2E0ZSA
Airs S ... M3ZSI
Aisher J ... G4YPA
Aissa D ... M6FCQ
Aisthorpe-Buckley P ... G4TMF
Aitchison C ... G4JDG
Aitchison M ... G4YFK
Aitchison P ... G3LSQ
Aitchison W ... G6CZX
Aithison J ... G0JFC
Aitken B ... GM0CBA
Aitken D ... MM6HIZ
Aitken E ... 2E0PDG
Aitken E ... M3PDG
Aitken J ... GM8NFG
Aitken L ... MM3UXO
Aitken P ... M3XHM
Aitken R ... G1JVU
Aitken R ... GM4SUR
Aitken S ... GM0IOA
Aitken W ... MM0CNV
Aitkenhead D ... GM4BHU
Aitkenhead R ... GM4UQG
Aizlewood B ... M0DAY
Aizlewood J ... G0DLT
Ajeti B ... MD0HOF
Ajmani A ... M6BGB
Akam J ... G8BIH
Akanyeti O ... M3RPA
Akay T ... 2E0SWT
Akay T ... M6TAA
Akehurst M ... G3MIQ
Akena M ... MD6YOO
Akerman D ... 2E0LTX
Akerman D ... M0RPI
Akerman D ... M6EQN
Akester D ... G8UPK
Akhurst W ... G1HDK
Akiki M ... G7LBP
Akines J ... G8XXI
Akinin J ... 2E0BBX
Akinin J ... M0GWR
Akinin J ... M3XHC
Akira J ... M6FGK
Akred L ... 2E0NGC
Akred L ... M6LRA
Aksamit E ... 2E0EAU
Aksamit E ... M6AQN
Akse G ... G0UVR
Akula P ... 2M0PKA
Al Busaidi A ... M6EEF
Al Saeed M ... 2M0VXB
Al Saeed M ... MM6VXB
Alabaster K ... 2E0WEK
Alabaster K ... M6WEK
Alamudi A ... 2E0BWM
Alava Moreira H ... M0JHP
Al-Azzawi A ... M0PSI
Alban H ... MW3IXZ
Alban R ... GW3SPA
Albers F ... GM4YGN
Alberti G ... M0HKL
Albinson A ... G5AU
Albison D ... M3AUB
Albon K ... G0CVB
Alborough P ... 2E0CCR
Alborough P ... M3BQM
Albrighton I ... 2E0CGU
Albury D ... G7OZH
Alcaino Pizani M ... M0VAP
Alcaino Pizani M ... M6SGV
Alce P ... M3PBA
Alcock A ... GI6SBW
Alcock G ... G8RXY
Alcock J ... G4CJM
Alcock J ... G8XQL
Alcock N ... GI4SPU
Alconcel L ... M6NAN
Alden L ... G7VML
Alder J ... G4GMZ
Alder M ... G4UTR
Alder N ... G3DYO
ALDER S ... G0HTS
Alder S ... G0PWK
Alder W ... M3LWX
Alderman G ... G3BNE
Alderman I ... 2E0GQT
Alderman I ... M6GQT
Alderman J ... G4JBA
Alderman K ... G0HRI
Alderman T ... G4LOD
Alderman W ... G7PHH
Alders N ... 2E0MDU
Alders N ... M0NAU
Aldersey R ... G0DPE
Aldersey J ... G1AQI
Aldersley J ... M6RIA
Aldersley S ... M6SSA
Alderson B ... G3KJX
Alderson F ... G0LRU
Alderson H ... G3BKJ
Alderson H ... M0HRA
Alderson I ... M6LWA
Alderson R ... G4ZQC
Alderton A ... G7GUB
Alderton I ... G6IZA
Alderton R ... G4OQK
Aldhous L ... G4UYF
Aldous R ... G8CBU
Aldred A ... M3TAG
Aldred D ... 2E0LCA
Aldred D ... M0XCT
Aldred D ... M6DLA
Aldred D ... M6KFC
Aldred G ... G4YGD
Aldred J ... G1UEO
Aldred L ... M3XZJ
Aldred T ... M0VVM
Aldridge A ... G0CIG
Aldridge A ... G3PJQ
Aldridge D ... 2E0DMA
Aldridge D ... G3VGR
Aldridge D ... G6FOV
Aldridge G ... M0HHG
Aldridge G ... M3TYO
Aldridge I ... G4AJU
Aldridge R ... 2E0EBA
Aldridge R ... M0TMT
Aldridge R ... M6EST
Aldridge T ... G4GJR
Aldus K ... G0TAH
Aldwinckle-Day J ... M6TTM
Aldworth E ... G3FHN
Alecio A ... G0RKC
Alefs C ... G1MPG
Alejo-Blanco K ... M6KIE
Aleksander N ... M0NPA
Alesbury D ... G3HSV
Alexander C ... G0TID
Alexander C ... M3CJA
Alexander D ... G3KLH
Alexander D ... GI8ZFZ
Alexander E ... 2E0IAS
Alexander G ... G0OSD
Alexander I ... G4AKD
Alexander I ... GM7OTT
Alexander J ... GM0KWW
Alexander J ... G1HEJ
Alexander J ... GW6CJJ
Alexander J ... GM7OJJ
Alexander J ... MI0JPO
Alexander J ... M3OVX
Alexander J ... MW1BAJ
Alexander J ... M3TGJ
Alexander K ... M3EMN
Alexander L ... GM0LVK
Alexander L ... GM1MRS
Alexander M ... GM7IHJ
Alexander M ... G8YDJ
Alexander P ... GW4RXO
Alexander P ... MI0CUN
Alexander R ... 2E0WWZ
Alexander R ... G4ZPB
Alexander R ... M3WWZ
Alexander R ... GM0DEQ
Alexander S ... G3TLY
Alexander S ... 2E0ZZC
Alexander S ... M6ZZC
Alexander S ... G7ETN
Alexander T ... G1AKV
Alexander W ... 2E0DGU
Alexander W ... MD3WCA
Alexandre A ... GJ4YBM
Alexis F ... M3XFA
Alfei J ... MW3UZH
Alford C ... G6SRU
Alford J ... G4DOE
Alford J ... G4NBW
Alford R ... M3ZRA
Alford R ... M1COE
Alford R ... M3COE
Alfrey F ... 2E0MFA
Alfrey F ... M0MFA
Alfrey F ... M6IOW
Algar S ... M6BLI
Alger J ... G7NGB
Ali A ... 2E0DUQ
Ali A ... M6GNY
Ali D ... 2D0NOT
Ali D ... M6UCP
Ali J ... M6GAV
Ali M ... 2E0ALB
Ali M ... M3XMA
Alincastre J ... 2E0UTC
Alincastre J ... M0SGZ
Alincastre J ... M6BBA
Alison N ... G0RPL
Al-Kattan S ... G6SFE
Alker G ... M3MET
Alkhateb A ... M6SNP
Allaker P ... 2E0EPA
Allaker S ... M6MCG
Allan A ... M6CUW
Allan A ... 2M0HUD
Allan C ... MM5CFA
Allan C ... MM3VKP
Allan D ... G3WYP
Allan E ... G7RGO
Allan G ... GM4HYF
Allan G ... GM4SBP
Allan G ... GM6FDQ
Allan G ... G7DFX
Allan G ... GM8SNB
Allan G ... G0BGA
Allan J ... G3IJA
Allan J ... GM6UWF
Allan J ... G7IHX
Allan J ... G4LTH
Allan P ... G1NBP
Allan P ... G1NPA
Allan P ... G4NTA
Allan P ... G4RYO
Allan R ... MM3EHM
Allan R ... M0OSE
Allan R ... G3TQZ
Allan W ... GM1XIN
Allan W ... G4ODS
Allan-McWilliams C ... M6RPW
Allanson A ... M3VEW
Allanson A ... M3UEW
Allanson C ... MM1YAM
Allanson J ... M3JIA
Allanson L ... M3UXI
Allanson P ... G0CEW
Allanson S ... G0TYS
Allardyce J ... GM6PKP
Allart J ... G8SVR
Allbright A ... M0ABZ
Allbright R ... G3RCE
Allbutt S ... G3WXU
Allchin A ... G4JGH
Allchin B ... MJ1EPG
Allchin J ... G8XXJ
Allcock A ... G8ZSK
Allcock L ... 2E0MCS
Allcock L ... M6LSA
Allcock P ... G1EHB
Allcock R ... G7JWO
Allcote D ... 2D0IOM
Allcote R ... 2D0RLA
Allcote R ... MD0RLA
Alldrick A ... G3ZOY
Alley E ... G0PZT
Alley P ... G3KJW
Allen A ... G4HRH
Allen A ... G4XEJ
Allen A ... M3AYP
Allen A ... M0TCD
Allen A ... M6TTT
Allen B ... 2E0CDS
Allen B ... G7BVZ
Allen B ... M0BZE
Allen B ... MI3MWA
Allen B ... M6BRY
Allen C ... G4SYA
Allen C ... G8RBI
Allen C ... M6CAQ
Allen D ... G0BWE
Allen D ... G3SDT
Allen D ... G4CNZ
Allen D ... G6FDS
Allen D ... G6TLN
Allen D ... G7OWX
Allen D ... G8BZT
Allen D ... G8LHD
Allen D ... M0APK
Allen D ... M1ETN
Allen D ... GI1ELP
Allen D ... GI8RQI
Allen D ... M0DUU
Allen D ... M6DMP
Allen E ... G3DRN
Allen E ... G3JHP
Allen F B S C ... G6JYO
Allen G ... G0OCC
Allen G ... G3IUO
Allen G ... G3JTK
Allen G ... GI4OZJ
Allen G ... M3GAF
Allen G ... M3KKN
allen G ... M6FCV
Allen G ... G0KNX
Allen G ... G3HST
Allen G ... M6INA
Allen G ... 2E0DZJ
Allen G ... G8LZG
Allen G ... M6HFO
Allen H ... G4TLW
Allen I ... G6VGH
Allen J ... 2E0BCX
Allen J ... G1PVU
Allen J ... G4VHX
Allen J ... M0AOB
Allen J ... M0DIJ
Allen J ... M0JAE
Allen J ... M3EOX
[illegible] ... [illegible]
Allen J ... 2I0CVN
Allen J ... 2I0NGK
Allen J ... M6GXS
Allen J ... MI6TCU
Allen J ... MI6TMZ
Allen J ... G0SGV
Allen J ... G6PKM
Allen J ... 2E0DVV
Allen J ... M6EYV
Allen J ... M6HGN
Allen K ... GI4RSI
Allen K ... M0BWW
Allen K ... M0KAD
Allen K ... M6KCG
Allen K ... M0HEF
Allen L ... G0EIB
Allen L ... G3CJD
Allen L ... M0TCF
Allen M ... G1IPP
Allen M ... G4OVL
Allen M ... G6RBR
Allen M ... M3MKJ
Allen M ... 2I0CKN
Allen N ... G0RCN
Allen N ... G4IHR
Allen N ... G4MIS
Allen N ... M3YJN
Allen N ... G4VVF
Allen N ... 2E0CGK
Allen P ... G1DYT
Allen P ... M0GTE
Allen P ... M1EUR
Allen P ... M3KZS
Allen P ... G1GVJ
Allen P ... 2E0BRT
Allen P ... G0IAP
Allen P ... M3VPJ
Allen P ... M6IPJ
Allen P ... M6DBF
Allen R ... G0NBE
Allen R ... G3NVL
Allen R ... G4VLI
Allen R ... M1BOB
Allen R ... M3BGE
Allen R ... M6RBN
Allen R ... M6RFA
Allen R ... G4WOI
Allen R ... G6JTV
Allen R ... 2E0REL
Allen R ... M6REL
Allen S ... G1SCA
Allen S ... M3CGM
Allen S ... M3SUI
Allen S ... G4CYR
Allen S ... M3VHB
Allen S ... MI6CQS
Allen S ... 2E0ZYL
Allen T ... G7HKZ
Allen V ... G1HRD
Allen W ... 2E0PYA
Allen W ... G8JCN
Allen W ... M0PYA
Allenby F ... M1AEQ
Allenet R ... GJ3XZE
Allenson M ... G3TGD
Allerton Austin M ... G3ZCJ
Alley P ... 2E0PBR
Alley P ... M0PAJ
Alley P ... M6PAO
Allgar K ... M3ZKA
Allgar K ... M6KGA
Allgar M ... M1MPA
Allgood A ... M6LYY
Allgood I ... G1NOO
Allgood P ... G8RXZ
Allgood R ... G0KSD
Allgood S ... 2E0AQH
Allibone D ... G0SZT
Allies K ... G0LDU
Allin G ... G4ZUC
Allin J ... G3VLN
Allin M ... G4HWI
Allin P ... 2E0HDM
Allin P ... M3WBA
Allingham T ... MI1AVH
Allington S ... 2E0SDK
Allington S ... M3KDK
Allinson N ... G8ILB
Alliott S ... G1UQT
Allisette M ... GU4EON
Allisette R ... GU4CHY
Allison B ... MM1HMZ
Allison D ... G0JQR
Allison D ... G3IZA
Allison D ... G7BGM
Allison D ... G7REC
Allison E ... G4JNQ
Allison G ... G0OQZ
Allison G ... M0GBA
Allison G ... M1CXV
Allison J ... G4URW
Allison J ... M0JUK
Allison L ... M6GMY
Allison M ... G4JJM
Allison M ... G0UQO
Allison P ... M3UQA
Allison R ... GD4FJI
Allison S ... G6BNR
Allison T ... G0TYM
Allison V ... G3TNX
Allman B ... G6HLL
Allmark M ... [illegible]
Allnutt A ... G6DTH
Allnutt P ... G7PSS
Allott M ... 2E0GPV
Allott P ... G7HGD
Alloway C ... MW3FYA
Allport A ... G7UCN
Allport B ... G1RKD
Allsebrook D ... G1VAC
Allsop J ... G3OGX
Allsop J ... G1FEX
Allsopp D ... G0DFA
Allsopp D ... G4UHW
Allsopp J ... G4YDM
Allsopp P ... G4LCM
Allsopp T ... M6TIO
Allum J ... 2E0EOI
Allwood E ... GW4UAJ
Allwood L ... G3VQO
Allwood P ... G8XBY
Allwood R ... M3ZRV
Almas E ... 2E0EFN
Almas E ... M6KQW
Almeida M ... M3WPM
Almey C ... M0PDQ
Almond D ... M3UXU
Almond D ... M0ZFF
Almond J ... G1KST
Almond L ... 2E0ZAH
Almond S ... G7JSE
Almond T ... G4WNW
Alperowicz B ... G0LPN
Al-rawi B ... 2E0KBP
Al-rawi B ... M0KBP
Al-Shakarchi A ... 2E0OAO
Al-Shakarchi A ... M0SHN
Al-Shakarchi A ... M6SHK
Alston B ... M6GVL
Alston S ... G4VSR
Alston T ... G8RLH
Alston-Pottinger B ... G4YZF
Alston-Pottinger S ... G6YZF
Altman B ... G7MNE
Alton R ... G3ZYD
Alvey T ... M6TWA
Alward N ... 2W0NCA
Alway D ... G7PKJ
Alwyn-Clark T ... G7JRD
Amare B ... G7VXK
Ambach M ... G0NPW
Ambler A ... G3SNG
Ambler D ... G1BKL
Ambridge N ... G4FRL
Ambrose C ... G7CPQ
Ambrose P ... M6PMA
Ambrose S ... G7NBQ
Amer J ... G0ALQ
Amery C ... G8AXN
Ames C ... G4SVI
Ames J ... G8MUF
Ames J ... G4XVI
Ames M ... G1IQF
Ames S ... G4OAV
Amesbury S ... 2E0BAX
Amesbury S ... M0GIA
Ameson R ... G1URD
Amies M ... G7RMJ
Ammundsen J ... M0ATS
Amond P ... M6FYB
Amos A ... M1BEP
Amos A ... 2E0VVA
Amos A ... M0VVA
Amos A ... M6VVA
Amos C ... M3YHZ
Amos D ... M6ASH
Amos F ... G4WUM
Amos G ... M6UGA
Amos J ... G1KGU
Amos K ... G2DPQ
Amos K ... G4YRF
Amos M ... M3KUG
Amos M ... M6OMX
Amos M ... G0ACD
Amos P ... M1AJT
Amos R ... G6EJF
Amos R ... 2E0ZAF
Amos R ... M0ZAV
Amos R ... M3ZWL
Amos S ... M6CZF
Amphlett M ... 2E0MKF
Amplett M ... M0AXV
Amunyela L ... M6AWP
Ancill C ... M6CRA
Ancill S ... M6STA
Anderson A ... G0RFM
Anderson A ... 2E0BKE
Anderson A ... M0KUP
Anderson A ... M3UHX
Anderson A ... 2M0NAX
Anderson A ... MM6NAI
Anderson A ... 2M0YYU
Anderson A ... G4JR
Anderson A ... G7WJP
Anderson B ... G7UHX
Anderson B ... 2E0YND
Anderson B ... M3TND
Anderson C ... G1DQD
Anderson C ... G8OTC
Anderson C ... M1DGP
Anderson C ... M0DNO
Anderson C ... MM6CIA
Anderson C ... 2E0LTC
Anderson C ... M6RWX
Anderson C ... G4TBN
Anderson D ... 2E0LEA
Anderson D ... GW0GFN
Anderson D ... GM4JJJ
Anderson D ... GM4SQM
Anderson D ... GM6JNJ
Anderson D ... G6YBC



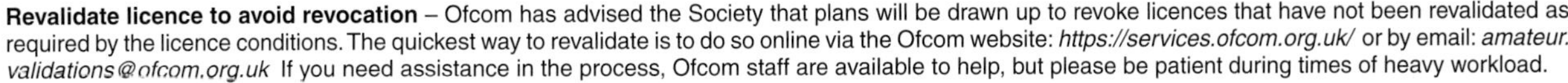
**IMPORTANT NOTE**

**Revalidate licence to avoid revocation** – Ofcom has advised the Society that plans will be drawn up to revoke licences that have not been revalidated as required by the licence conditions. The quickest way to revalidate is to do so online via the Ofcom website: *https://services.ofcom.org.uk/* or by email: *amateur.validations@ofcom.org.uk* If you need assistance in the process, Ofcom staff are available to help, but please be patient during times of heavy workload.

Anderson D GM6BIG
Anderson D GM6WTT
Anderson D M0HTX
Anderson D 2M0ONS
Anderson D MM6MMG
Anderson F G1OPZ
Anderson F GI4ZAH
Anderson G GM0VGI
Anderson G G1VGP
Anderson G G3NPA
Anderson G GI4TPI
Anderson G G4ZA
Anderson H GW1NED
Anderson I G3OAX
Anderson I G7VVO
Anderson I M1FHQ
Anderson I MM3XIA
Anderson J 2E0PSP
Anderson J G0MPP
Anderson J G1OZD
Anderson J G4DBP
Anderson J G4ZYL
Anderson J G7CHC
Anderson J 2E0MUA
Anderson J GW0AKV
Anderson J G4AUS
Anderson J G6ZCI
Anderson J M3MUA
Anderson J M6JCG
Anderson J GI6GBK
Anderson J MI0AAZ
Anderson J M1EZL
Anderson J M6ALW
Anderson J M6JCT
Anderson J M6RPP
Anderson J MM6FXQ
Anderson K M0BEO
Anderson K G1OUG
Anderson M G0GNI
Anderson M GI3WWY
Anderson M G7STL
Anderson M GI8YWE
Anderson M M3MXA
Anderson M M3WLO
Anderson M M5ACX
Anderson M 2E0CEA
Anderson M M6CEA
Anderson M G4VMA
Anderson N M1CQU
Anderson N 2E0BNA
Anderson N MM6NAD
Anderson P G7DQC
Anderson R GI0UAG
Anderson R G4AEV
Anderson R G8HVZ
Anderson R M0RWA
Anderson R GM0SCW
Anderson R M6NBN
Anderson R 2W0RGA
Anderson R MW6RGA
Anderson S 2M0SWA
Anderson S MM3NLH
Anderson S MM3SWA
Anderson S G0EAT
Anderson T G4ZBE
Anderson T 2E0IKH
Anderson W 2M0BZZ
Anderson W MM3XWS
Anderson W GM3OIV
Anderson-Mochrie I G3VCM
Anderton C G1TAR
Anderton C G0DNQ
Anderton G G6WIT
Anderton H 2E0MBA
Anderton N G0KLF
Anderton T G4YYL
Anderton W M0WHA
Ando S M6UAP
Andover A 2E0GMG
Andover A M3VGT
Andre R G0SSG
Andreang K G4GZN
Andres P G0VJN
Andress J G0JCA
Andrew A G7WKP
Andrew A 2E0CFH
Andrew A M0NRP
Andrew A M6YMA
Andrew C 2M1VXB
Andrew C G8YKE
Andrew D G4GZH
Andrew D G6TRA
Andrew D G6OUT
Andrew L G4VZS
Andrew M G8NRP
Andrew P G7VVB
Andrew P M0GPW
Andrew P G7GCW
Andrew R GM3WFJ
Andrew S G3SNA
Andrew S GM7WGM
Andrew W GM1MRY
Andrews A G1VAO
Andrews C 2E0FKM
Andrews C G0SZE
Andrews C G4RVE
Andrews C M6BIU
Andrews C M6FWB
Andrews D G4CWB
Andrews D G4EZZ
Andrews D G4NNP
Andrews D G4ZOG
Andrews D G3MXJ
Andrews G 2E0KPR
Andrews G G1XWN
Andrews G M0CWY
Andrews G M0ETA
Andrews H G7UXU
Andrews J G3APU
Andrews J G3GEF
Andrews J G3RMY
Andrews J G4IWN
Andrews K 2E0GSJ
Andrews M G0PHR
Andrews M G6IRG
Andrews M M6MIA
Andrews M M3ILY
Andrews N M3KLU
Andrews P G1KUQ
Andrews P G6MNJ
Andrews P G8VGQ
Andrews P G4XKM
Andrews P G4TOI
Andrews P M3XUR
Andrews R G1JMN
Andrews R GW1XUD
Andrews R G3UZW
Andrews R G6SRV
Andrews R M0JMN
Andrews R G6MNI
Andrews R G6TQZ
Andrews R G4BWB
Andrews S G6RPW
Andrews S M1EXO
Andrews W GW2DHM
Andronov G G6IRJ
Andronov I G1VVX
Angel J G6YJR
Angel R G4ZUP
Angell D G0BHS
Angell G G4CQA
Angell P G3ZTX
Angell R G4CCE
Anger D M0BMF
Angier T G0HMK
Angiolini J GM0GFV
Angiolini M GM1SBD
Angold P G3WTQ
Angove A G6ZWI
Angove C G4NCS
Angove S M3KGE
Angus A 2E0CWF
Angus A M0HNT
Angus A M6CXF
Angus B 2E0TPE
Angus B G6GQC
Angus G MI6CUZ
Angus H G0EVS
Angus H GM4MUZ
Angus J G1ZDR
Angus R G0CHP
Angus S MM3SYO
Angwin F G1IQE
Annan A MM1CCR
Anness P G0WCO
Annetts S G1FXL
Anostalgia M 2E0OTP
Anostalgia M M6IKI
Ansari M M6MTA
Ansdell B M6ARI
Ansell B G7BIY
Ansell B M6AII
Ansell C M1DFK
Ansell D 2E0NKC
Ansell D G3ZWY
Ansell D M3NKC
Ansell I G4YBN
Ansell L 2E0LFE
Ansell L M6KUD
Ansell P G4MZK
Ansell P G7VPU
Ansell-Wood M M6MAW
Anson M G4TQC
Anstead L G4HOU
Anstee R G6RSU
Anstie D G7UVB
Anstie D 2E0HOO
Anstis P M3VYB
Anstock D G6DZT
Anstock P G6OJZ
Antcliffe D M6DFZ
Anthoney G GM0AAX
Anthoney M MM3TYA
Anthony D 2E0BDJ
Anthony J G3KQF
Anthony J M6ILS
Anthony J G4TZE
Anthony M G4THN
Anthony M 2E0BOJ
Anthony M M3TNB
Anthony R G0DFY
Anthony T 2E0ECG
Antill S M6SRA
Antimano J 2E0JOD
Antley A G3UTG
Antrobus C G3JCJ
Anziani T G4ZWN
Ap-Dafydd P GW4SXA
Aplin P M1BFO
Apperly M G6FSU
Applebee W G4ZKJ
Appleby B M6ZAV
Appleby B M6CSJ
Appleby F G4RYH
Appleby I M6GBH
Appleby J M3YJA
Appleby M G3ZNU
Appleby N G0WIW
Appleby P G4BLS
Appleby P G8CXW
Appleby R G3INU
Appleby R M1CBO
Appleby R M6GZB
Appleby S M1CFG
Applegarth K M6KEP
Applegate A 2E0CMC
Applegate A M6BBO
Applegate S G7SYD
Appleton A G1HDO
Appleton C G4GBK
Appleton D G4KCP
Appleton J G3XPZ
Appleton J G4SVA
Appleton K G4USC
Appleton M G1EHE
Appleton P M6FJU
Appleton V G6CHC
Appleyard A G4ZBS
Appleyard A G6PVU
Appleyard P G0AMX
Appleyard S G3PND
Appleyard T G0LNV
Apps C M6CJA
Apps N G1FTK
Aquilina M GW1RQM
Arak R G4EEJ
Arakawa T G0RTA
Aram C M1EHV
Aram G G3SET
Aram S G6SRY
Arblaster C M6CBD
Arblaster J M3VYV
Arbon C 2E0NOC
Arbon M G4SAW
Arbuckle S M1AZF
Arcari D GM0ISA
Archer A M3TBG
Archer B G4KEZ
Archer C G4VFK
Archer C M6ZCA
Archer J G0EQD
Archer K G4CMZ
Archer M M6BKV
Archer P G6AKK
Archer P 2E0ZPA
Archer P M0PJA
Archer P M6PJA
Archer R G6WXS
Archer T 2E0XXT
Archer T M6TEZ
Archibald C MI5HNA
Archibald K MM6EQY
Ardern Z M3INQ
Ardley W G0BOF
Ardrey B M6BJA
Arenas Martinez L M6GHC
Arey R M0APY
Argent J 2E0VMC
Argent J M3NQK
Argent P M1EGP
Argument H G0UGI
Argyle J G8FED
Argyle K G1GEV
Argyle M MW6MBZ
Argyle W M3WBQ
Arif M M6HUP
Arigho D G3NVM
Aris F G0LYG
Arkell R G0UYP
Arkinstall E M0KZB
Arkinstall J M3JKB
Arkwright N G6BKY
Arlette D G0AEW
Arliss M G7GEI
Armand J M6DOX
Armatage G G0KGQ
Armatage S M1SAB
Armer N M0BUR
Armes W 2E0WHA
Armes W M0WIA
Armes W M3WIA
Armistead B GM4LBN
Armistead C GM4DMI
Armitage A M0GZB
Armitage C G0GFA
Armitage D G0AOU
Armitage G 2E0BRP
Armitage G M3WWU
Armitage J G0OHR
Armitage P G8SPM
Armitage P M6AYM
Armitage R G1RFC
Armitage R G3TZE
Armour M G8YKG
Armour T GM6FOT
Armson A 2E0MRA
Armson A M6OKC
Armstrong B M3NFF
Armstrong A G0FBW
Armstrong A G1GQY
Armstrong A G3YZW
Armstrong A MI6ALL
Armstrong A M6FYM
Armstrong A M6APA
Armstrong B GM0EHL
Armstrong B G0IBG
Armstrong B G3EDD
Armstrong B M5CMO
Armstrong C G1GQZ
Armstrong D G6SRT
Armstrong D G6XAT
Armstrong D G0FYQ
Armstrong E M6EKA
Armstrong F G4NHC
Armstrong F 2E0YGH
Armstrong F M6TTO
Armstrong F G3JRL
Armstrong G GI4XGO
Armstrong G M1AHU
Armstrong H M3ZYF
Armstrong J G6YPF
Armstrong J G7ROC
Armstrong J GW3EJR
Armstrong J G4VPU
Armstrong J G8MVH
Armstrong J M6DAD
Armstrong K G1AQX
Armstrong K G8OYB
Armstrong K M1DEY
Armstrong K 2E0KYI
Armstrong K M0KYI
Armstrong K M3JZA
Armstrong L G6CCN
Armstrong M G0BMQ
Armstrong M 2E0TYE
Armstrong M M6GZN
Armstrong N G6AWY
Armstrong N GM7PNX
Armstrong N M0BCI
Armstrong N M6NIK
Armstrong N 2I0OAZ
Armstrong N MI6OAZ
Armstrong P 2E0PVW
Armstrong P M6XIP
Armstrong R G0BIA
Armstrong R GM4GOW
Armstrong R G4JYT
Armstrong R G8SPE
Armstrong R M0AAM
Armstrong R MM0BPF
Armstrong R M3LCE
Armstrong R G3PGC
Armstrong R M6RMA
Armstrong S G1EZI
Armstrong S MM3WZH
Armstrong T M6HWJ
Armstrong W G3PRE
Armstrong W GI4XTC
Armstrong W G4TMX
Armstrong-Bednall A G7MLT
Armstrong-Taylor J M3TNJ
Arnell J M0DXZ
Arner C M3LYU
Arnett M M0MAH
Arnfield M M5ACS
Arnison M G4THC
Arnold A 2E0NPH
Arnold A G0PYE
Arnold A G8NPH
Arnold A M3NPH
Arnold A M6DUJ
Arnold A M6DIH
Arnold B G3FP
Arnold C G3PJC
Arnold D GM0JPG
Arnold D G7OGN
Arnold D G8YQA
Arnold D M6NDF
Arnold D 2E0CAQ
Arnold D M0LDQ
Arnold D M6SUV
Arnold G G1IDQ
Arnold G M3TGA
Arnold I M0CLD
Arnold I G4IJM
Arnold I G8HDM
Arnold J G3WXH
Arnold J G4NPH
Arnold K G0CAY
Arnold K G3XNP
Arnold K G8BAY
Arnold L G8AHE
Arnold M M3AGH
Arnold M M3EFL
Arnold P G4CFG
Arnold P G1CJI
Arnold P M6MLG
Arnold P 2E0CXH
Arnold P M6BXY
Arnold R G1VTU
Arnold R M6RXA
Arnold R G1GRB
Arnold S G1HSA
Arnold S G4FXA
Arnold V G8PWE
Arnott J G1WKK
Arnott M M3HHX
Arnott S M3LSO
Arnould R MW0RBA
Aron S MM3CKP
Arpino S M3VZP
Arris T G4OSB
Arrow I M3ZDT
Arrow J M0JWA
Arrow T 2E0GPL
Arrow T M3UAW
Arrowsmith B G0KOU
Arrowsmith F GD3HFC
Arrowsmith G G8XHU
Arrowsmith H G7KNK
Arrowsmith H M3KNK
Arrowsmith J M3LWG
Arrowsmith J G4IWA
Arrowsmith K G4VVL
Arrowsmith K GM6JOS
Arscott C G8JVA
Arscott D G6YJO
Arscott J G3VSL
Arscott P G6GFA
Arscott T 2E0TTA
Artamonovs A MM6HZO
Arter D G1NLQ
Arter J G7MEZ
Arter S 2E0DDZ
Arthur D G1KBG
Arthur D M6ETL
Arthur E M6XZK
Arthur J GM7DTC
Arthur J MM3XJA
Arthur J G1CPC
ARTHUR J GJ4JVP
Arthur M G0BVS
Arthur S M3ZSA
Arthurs D G7OGR
Artingstall B G1BBK
Artis E 2E0GVB
Artus S G0EBS
Arundale J M1CHS
Arundel C G1YNH
Arundel C G4KDX
Arundel C M3YNH
Arup P G8WSP
Asbury A MM1JAA
Asbury L M1LYN
Asbury P G7MGX
Asbury P M0PCA
Ascroft A 2E0LTV
Ascroft A M6FDT
Ash A G3PZB
Ash B G3LVL
Ash D GU1BWW
Ash J 2E0HEX
Ash J M0HEX
Ash J M6HEX
Ash N 2E0FGB
Ash N G6ASH
Ash P GU7APA
Ash S M1SGA
Ash W G8XSA
Ashall B 2E0VCB
Ashall N G6KRS
Ashbee G G6LKV
Ashbee G G6ZLS
Ashbee J G7JOW
Ashbee T G6LKW
Ashburner E G0EHV
Ashby A M6GXW
Ashby A 2E0YNT
Ashby A M6YNT
Ashby D G0BOT
Ashby D G3SUV
Ashby M G6WRB
Ashby P G1ZLC
Ashby P G6UZG
Ashby P 2E0IPX
Ashcroft B M0TER
Ashcroft B M1NIS
Ashcroft B M3BMW
Ashcroft C G1KGV
Ashcroft D G4RYI
Ashcroft D M6DON
Ashcroft G G0CGW
Ashcroft G G8AKL
Ashcroft K G3MSW
Ashcroft K M3OPV
Ashcroft N G4IRU
Ashcroft P G4CTI
Ashcroft P G8XOB
Ashcroft P G0RLS
Ashcroft S G0JYN
Ashcroft S M3RYI
Ashdown B G4KZT
Ashdown C G0FHF
Ashdown D M0AWY
Ashdown M G0BQV
Ashdown N G4NAJ
Ashdown P G6AWZ
Ashe A G8SRV
Ashe H MI6RCV
Ashe J GI8RLE
Asher G M3DTD
Asher N M6NEA
Asher P M6LTD
Ashfield N G0OYR
Ashfield S G4XMO
Ashford A G8CMD
Ashford D M6DGA
Ashford E M6TDY
Ashford G GW0RQP
Ashford H G3WGY
Ashford I G8PWE
Ashford L G6HZI
Ashford L G7SMV
Ashford P G7TKT
Ashley E G0BIN
Ashley E M3NFH
Ashley J M6JBA
Ashley N M3FMI
Ashley R 2E0RPA
Ashley R M0HNE
Ashley R M6REH
Ashley S G8OSZ
Ashley W 2E0GYN
Ashley W M3ACU
Ashlin C G4MUQ
Ashman B 2E0EWK
Ashman C M3KHJ
Ashman C M3JCA
Ashman N G3NSI
Ashman R 2E0RJA
Ashman R G4JVJ
Ashman R G6NYC
Ashman R G7BPO
Ashman R 2E0TEF
Ashman R G3TBL
Ashman R M6TEF
Ashmore A M6NDO
Ashmore D G3SXI
Ashmore J G7KVZ
Ashmore J G8GXF
Ashton A G3XAP
Ashton A M6AOS
Ashton A 2E0TCI
Ashton A M6EJA
Ashton A G1NAQ
Ashton B G0KDX
Ashton B M0GGT
Ashton D G6GEV
Ashton D M6GXJ
Ashton G G1ARD
Ashton G M0AUG
Ashton K G7IBH
Ashton K M6BJK
Ashton L G4TDQ
Ashton M M3NKZ
Ashton M 2E0XFH
Ashton M M3XFH
Ashton P G0DCS
Ashton P G4CHG
Ashton P M6HVM
Ashton P 2E0BNW
Ashton P M0WFK
Ashton P M3WFK
Ashton R G0AXR
Ashton T G4PBZ
Ashton T G8MII
Ashton W MW0DCQ
Ashton-Cox R M6YYL
Ashton-Hilton D 2E0YCD
Ashton-Hilton D M6YCT
Ashton-Jones J G7LVG
Ashwick R M3WIC
Ashwood D G3TVX
Ashworth A M3IMR
Ashworth A M3TOY
Ashworth A G4ZCG
Ashworth C MD3WXS
Ashworth C M3DIY
Ashworth C 2E0CXU
Ashworth C M6CYG
Ashworth D G0NLN
Ashworth D G3SYA
Ashworth E G0PFM
Ashworth E G6XCD
Ashworth E G7DNM
Ashworth H G3CUF
Ashworth J G6JOV
Ashworth M 2E0MPA
Ashworth M M3ZQF
Ashworth R 2W0RFX
Ashworth R MW0VGH
Ashworth R MW3RFX
Askam A 2E0GEN
Askam A M6CSX
Askam W G0VNQ
Asker B G1NJG
Asker T G1KZD
Askew D G3PCG
Askew D 2E0KCX
Askew F G4NLG
Askew G MM3NTX
Askew G GM1OPO
Askew J G3ZMK
Askew J G7TIV
Askew K G7AGA
Askew P M0IDI
Askey J G6SWJ
Askey P M6JTC
Askey W G8LXN
Askham P G6NJO
Askin C G8ZFD
Aslam A M6PZM
Aslam M G0KMK
Aslan J G6LFJ
Aslin D G3WGN
Asling N M6EMH
Asling S M3JKJ
Asling S M3JKP
Aspden E 2E0IEA
Aspden E M6IEA
Aspden K G8WZW
Aspey S G1ZEU
Aspinall D G3XIP
Aspinall J GW4ZAW
Aspinall P G4ZLJ
Aspinall S M6SEA
Aspinall S M6DYG
Aspinwall B G3MBO
Aspital D M6DJA
Aspland J G4PFZ
Asquith C G4ENB
Asquith C G8LJQ
Asquith D G4UVD
Asquith P G4ENA
Astbury C M0ISF
Astbury C M6BXV
Astbury G M3GLA
Astbury J M3JLA
Astbury R M3RDA
Astbury R M0EAF
Astbury S M0SFA
Astbury T GM0GMD
Astbury-Rollason N G4NRR
Astill W M6ELB
Astle J G8OCA
Astley A G0AJA
Astley D M0SAQ
Astley I 2E0VKN
Astley I M0IAA
Astley I M3VKN
Astley M G7CAH
Aston A G0IWD
Aston J G6EQT
Aston J G7OBF
Aston K G0WZV
Aston M G6HXB
Aston R 2E0BIO
Aston R M1AZV
Aston R G3WND
Atack K G4WAS
Atanassov K M0NKA
Atfield M 2E0ZPT
Atfield M M0MTA
Atfield M M6MEA
Athawes A G4RUZ
Atherfold J G0FZB
Atherley A G4ORW
Athersmith J M0KLJ
Athersmith P 2E0TIL
Athersmith P M6TIL
Atherton M G3ZAY
Atherton M 2E0CKL
Atherton M M0UFA
Atherton M M6BHF
Atherton M M6AXF
Atherton P G8EXS
Atherton P G7ILL
Atherton R G0JIT
Atherton T M6TCX
Atherton V MJ6VAA
Atherton W M3EUF
Atifeh H 2E0HAB
Atifeh H M0HOU
Atkin C G4YAM
Atkin F G4VHH
Atkin T M6TJA
Atkins A G1VNZ
Atkins B 2E0OBI
Atkins C G6OOT
Atkins C G7MPZ
Atkins C G8AFA
Atkins C M3KJY
Atkins D G7USV
Atkins D M1BUL
Atkins D 2E0DHA
Atkins D 2E0DZY
Atkins D M3JZL
Atkins D M6MTU
Atkins G G3HQG
Atkins G G0ATK
Atkins I G0HOX
Atkins J G0WGB
Atkins J G0SDF
Atkins K M6KFA
Atkins L M6LAA
Atkins N GI4CUV
Atkins N M0PYE
Atkins P G4CKK
Atkins P G4DOL
Atkins R G0MWH
Atkins R M6ETW
Atkins R M6ATK
Atkins T G0JJK
Atkins T G4ABN
Atkins W 2E0AJT
Atkinson A G4BYD
Atkinson A M6RNM
Atkinson B G0SWO
Atkinson B G3TEP
Atkinson B G8SRZ
Atkinson C 2E0XBM
Atkinson C G8NZD
Atkinson C M0XBM
Atkinson D G4ACL
Atkinson D G8DRE
Atkinson D G0WOU
Atkinson D M3TYV
Atkinson D M6NTZ
Atkinson D G1LZL
Atkinson G G7HEJ
Atkinson G G8GWP
Atkinson G G8RPI
Atkinson I G6REM
Atkinson J G0COI
Atkinson J G4XBU
Atkinson J M0NKY
Atkinson J G8XSD
Atkinson K G1OGH
Atkinson L G4FAA
Atkinson L M6OOB
Atkinson M G8ZVM
Atkinson P G0WXD
Atkinson P G7NRG
Atkinson P G1RRU
Atkinson P G6FOW
Atkinson R G0SWB
Atkinson R G3OPH
Atkinson R M6VMJ
Atkinson S G0IQM
Atkinson S M3SAA
ATKINSON S G3YPS
Atkinson T M1FAA
Atkiss B G3VYA
Atrill N G6GFO
Atrill N M3NJA
Attack A G7IYN
Atter K G3PTI
Atterbury B G7HCB
Atterbury R G4NQI
Attew P G0GRB
Attfield M GW6YPA
Atthill R G3KTM
Attle A G0RNY
Attlesey M M0MEA
Atton S M1DMB
Attree D G6TLP
Attree M M6MCY
Attridge I G1BPD
Attwater P 2E0JHF
Attwater P M6JHF
Attwell J G0MKE
Attwood J 2E0TWQ
Attwood J M6TWQ
Attwood P M6WYR
Attwood R G0PTA
Attwood S M0GEL
Atwell S G6IRF
Au S G0NYJ
Aubin J G0RKP
Aubrey M M6TDA
Aubury J 2E0JCA
Aubury J M6JCA
Auchterlonie L MM0LJA
Auckland N 2E0NAJ
Auckland N M0NAF
Auckland N M6NAJ
AUCKLAND S G4NSO
Aucoin S 2E0GHX
Aucote R G0BYF
Aucott G G7EKT
Audcent A G4MCE
Aughey D GI7VBS
Aughey F GI8XSB
Aughey R GI3VPV
August R MM0BDA
Augustus M 2E0EUW
Augustus M M0JEA
Augustus M M6EUW
Auker-Howlett A G0FLT
Auld C GM0SUY
Auld D 2I1ALE
Auld D GI0UTE
Auld G M1ULD
Auld G M3XLC
Auld P G1JNQ
Aulsebrook J G0CQT
Ault A M3GPP
Ault A 2E0PNR
Ault D G1ZEK
Ault J G3KTU
Aunger D G4MXP
Aunger F G6FJA
Aungiers G G0KMP
Aungiers G G4PNH
Ausher G M1BJS
Austen C 2E0TUB
Austen C M6TJN
Austen D G1EHF
Austen D M3EHF
Austen J G6UZJ
Austen K GW4GJA
Austen K M1AZO
Austen M M0AEN
Austen P 2E0WCO
Austen R 2E0TUS
Austen R M6TJO
Austen R M0PLY
Austen S G8WLB
Austen-Jones S M0CYF
Austerfield N M6EUE
Austin A 2E0HEO
Austin A G1JZK
Austin A 2E0XYX
Austin A M6TPA
Austin B G1IKV
Austin B G0GSF
Austin C M6CWK
Austin C M0NGC
Austin D 2E0AUN
Austin D G1XUW
Austin D 2E0NNX
Austin D M6NNX
Austin F G4NFA
Austin G 2E0FAQ
Austin G GD1LQV
Austin G G4DPA
Austin G G4OAU
Austin G M6GAX
Austin G G6NYH
Austin G M5GUY
Austin J G6EIZ
Austin J M3PHF
Austin J G8NHO
Austin K G6ATK
Austin K G8JSC
Austin K G7ABF
Austin L G0NMD
Austin M 2E0MWA
Austin M G0VPH
Austin M G3REM
Austin M G8JRW
Austin M M3MSN
Austin M M1NIZ
Austin M G1GDA
Austin N G6DSG
Austin N G6WAO
Austin N G7MKJ
Austin P G7BXA
Austin P G0AXQ
Austin S G1XXR
Austin T M3UCF
Austin W G7HIO
Austrin D M6DBA
Austwick J G0RIP
Austwick M M6MJA
Authers B M1EAN
Authers S G4SWA
Auty C GM4GPP
Auty E G4DYM
Auty J G3TFO
Auty S G3JRY
Auty W G0CVM
Au-Yeung L 2W0LBN
Aveling B G8VPH
Avenell M G1YOS
Averill N GI0SRP
Averill-Elias C G4VHB
Avern J G6HHE
Avery D M1CSI
Avery E G3WBB

Avery I M6IAV
Avery K G3AAF
Avery [illegible]
Avery R [illegible]
Avery S G7JKW
Avery-Hawkins S M3HAW
Aviss H G7JUD
Avon P G8LHZ
Awbery R G3YZN
Awcock N G8GWB
Axcell B M6UVB
Axe J G4EHN
Axford G G4AQZ
Axford M M1ARI
Axon A G8POS
Axon M 2E0DFZ
Axon N M3GKI
Axtell R M6RBO
Axtell T G4SVC
Axworthy M 2E0DHC
Ayer S G8WVB
Ayers D G4SXT
Ayers J G4SBN
Ayers R G4ZWB
Ayers R M1HFX
Ayers S G7OQQ
Ayers W G0NOU
Ayers-Hunt P G6NMA
Aykroyd L G0NAS
Ayland T M6CZD
Ayley R G6AKG
Ayling A M3AMA
Ayling M G4BNO
Ayling R G3YUH
Ayling S G4ASL
Aylward B G6ODE
Aylward B M3TII
Aylward G G0XAN
Aylward J G6NYF
Aylward J M3TIJ
Aylward W GI3UIH
Aylwin K G4SEK
Aynge R G6IRE
Ayre E 2E0SDZ
Ayre E M0ZGT
Ayre E M6LCG
Ayre G G0MFR
Ayre L 2E0LRA
Ayre P G4TFF
Ayres A H 2E0WYE
Ayres A H M0TAY
Ayres A H M6TAY
Ayres B GU1HTY
Ayres C 2U1EKE
Ayres C G0KTC
Ayres J G3DQT
Ayres M G4OQG
Ayres M M6MQE
Ayres N G4ADR
Ayres R GU4ASO
Ayres W GU1WJA
Ayris D G4GZA
Ayriss K G8MXV
Ayton A G4YEJ
Azizoff N 2E0NAZ
Azzaro B M6BEE
Azzaro S M6SSC
Azzopardi L 2E0KDI

## B

Babb R M6BQB
Babbage A G0WWL
Babbage A G4YKK
Babbage N G1DZB
Baber J G7ARJ
Babic D M3FSD
Babic S M3FSB
Bache A M6EDA
Bache D 2E0JTM
Bache D M6JPD
Bache J G7JMZ
Bache K M6RKB
Bache S 2E0CUY
Bache S M6ZSB
Bacheta J M3GVN
Back A M6AOB
Back C G6HCH
Back R M6RPQ
Backham R G8KOC
Backhouse A 2E0ACF
Backhouse C G7VNN
Backhouse G M3FLA
Backhouse J M3SLI
Backhouse K G4RHR
Backhouse W G4HZI
Backus J G4PFJ
Ba[illegible] P M1PJB
Bacon A M3JZP
Bacon G G7INC
Bacon J G3YLA
Bacon P 2M0DKX
Bacon P MM0PHB
Bacon P G3ZSS
Bacon P M6PJB
Bacon R G3WHJ
Bacon S M3JZO
Bacon T M0SCB
Badami R M0RXB
Badcock A G8IPQ
Badcock A M3IPQ
Badcock C G1NPI
Baddeley J G6YCO
Baddeley J G7DOY
Baddeley S M0HWM
Baddeley T M3UWY
Baddeley W G4SVR
Badger E G3OZN
Badger G G3OHC
Badger J G4YZO
Badham G M3FLK
Badham J [illegible]
Badham P G0WXJ
Badham R G4PKE
Badley P M0PIB
Badz W G4CGF
Baechle R M0LRB
Bagely P M6PGF
Bagg A 2E0IIM
Baggaley D G7MMK
Baggaley S G4RQG
Baggett P G4ORX
Baggley T 2E0OMT
Baggley T M3OMT
Baggott G G4GUW
Baggott S 2E0TVV
Baggott S M6TVV
Bagley C G4RSL
Bagley D G1NFB
Bagley G G3FHL
Bagley G G7ODV
Bagley J G4BDW
Bagley R G4HQD
Bagley W G1UOY
Bagnall J G4FSH
Bagnall L G6WXJ
Bagnall M G0ICW
Bagshaw J G1YBT
Bagshaw J G1GQB
Bagshaw K G8YTX
bagshaw P G3NEO
Baguley C G0CEJ
Baguley M G7LQD
Baguley M G8ORM
Baguley P G4HUF
Bagwell C G1UAL
Bagwell R G4HZV
Bagwell R G7WLY
Bagwell W G3YDE
Bagworth A G1YUU
Baiey R G1CJK
Baigent G M6GJB
Bailes J G7NQX
Bailey A 2E0BLY
Bailey A G3SFM
Bailey A M0KMB
Bailey A M1FMC
Bailey A M3JTU
Bailey A M3KZR
Bailey A M1AUP
Bailey A M6AIH
Bailey A MW6TES
Bailey B G1UFA
Bailey B G3MAH
Bailey C 2E0TQF
Bailey C G3TRX
Bailey C G6CQR
Bailey C M3EIJ
Bailey C M3TQF
Bailey D G3ZNR
Bailey D G4FFM
Bailey D G4GUC
Bailey D G3WNW
Bailey D G6HEF
Bailey D M6XLS
Bailey D M6AVF
Bailey D 2E0DGD
Bailey D M0HLX
Bailey D M6DGD
Bailey D M0DSY
Bailey D G0MWM
Bailey E G4NNI
Bailey E G4LUE
Bailey F M1EYH
Bailey G 2E0FMS
Bailey G G1ZTJ
Bailey G M6OYF
Bailey I M0HVO
Bailey J 2E0FTV
Bailey J G0FJB
Bailey J G3BNW
Bailey J G6JEB
Bailey J G6PCN
Bailey J G8KWV
Bailey J G8PLJ
Bailey J M3ACY
Bailey J M3UJL
Bailey J 2E0RDF
Bailey J G6KAE
Bailey J G7PYB
Bailey J M0MTW
Bailey J M0DFK
Bailey K 2E0KMB
Bailey K G0GZV
Bailey K G4DFD
Bailey K M3JLF
Bailey K M3WUX
Bailey K M3UTN
Bailey L G0IDE
Bailey L G4HME
Bailey L G4JLJ
Bailey L M3BPC
Bailey L M3IQP
Bailey L [illegible]
Bailey M G3ZEK
Bailey M G4RHB
Bailey M M1CHM
Bailey M G8OKD
Bailey M 2E0MCH
Bailey M M3TLU
Bailey M M6MBF
Bailey M M6MQB
Bailey M 2E0BQH
Bailey M M6GTR
Bailey N G4GPJ
Bailey P G7BHY
Bailey P G8JSN
Bailey P G8UXX
Bailey P M3XLB
Bailey P M6MTV
Bailey P 2E0BYG
Bailey P 2E0COI
Bailey P G3PJB
Bailey P M6BEZ
Bailey P M6EGO
Bailey R G0VFS
Bailey R G4PJL
Bailey R G4PPP
Bailey R G6WLE
Bailey R G7NFN
Bailey R G8IBE
Bailey R G0IAX
Bailey R G3WCQ
Bailey S G6ZHF
Bailey S G8KGE
Bailey S M3UZB
Bailey S G4MCQ
Bailey S G3VDK
Bailey T G0CRF
Bailey T G6CRF
Bailey W G2CHI
Bailey W G4ZXV
Bailey W G6UKN
Bailey W G7PZF
Baillie A M6FHF
Baillie A M6OPT
Baillie C 2I0TUV
Baillie C MI6TUV
Baillie J GI3XEQ
Baillie J GI4PGN
Baillie J GI7ALH
Baillie R GI4TUV
Baillie D M6DGZ
Baillie D MM6DGC
Baillie D 2M0DGB
Baillie D MM0GOG
Baillie D MM3NVD
Baillie-Searle C GD4EIP
Baillie-Searle P MD0DMV
Bailur B M0OKB
Baily A G7GPI
Baily C G7KID
Baily J G4MDG
Bain A GM1VSR
Bain A M5AIQ
Bain G G1IDV
Bain I 2E0TLY
Bain I M0KEG
Bain I M3TLY
Bain J G7CRY
Bain J MM3UMZ
Bain L 2E0LAB
Bain L M3UNY
Bain R M1RSB
Bain T 2E0EJC
BAINBRIDGE C M3FLI
Bainbridge C MM6IBB
Bainbridge J G1PGH
Bainbridge K 2E0LRD
Bainbridge K M3ZWP
Bainbridge M G4GSY
Bainbridge P M3OCA
Bainbridge P MM0PTE
Bainbridge R G0AXO
Bainbridge S M1SWB
Baine L MI5CFM
Baines A G7TDN
Baines C M6JKT
Baines C M6GII
Baines D 2E0DBS
Baines D G8OUI
Baines D M3DBS
Baines D M3JMQ
Baines D G7RCP
Baines D M3APE
Baines G 2E0SDI
Baines G G7VDT
Baines J G4TUX
Baines J G6DGQ
Baines J G7LXI
Baines M G7OWV
Baines M M0RIS
Baines P M3KUV
Baines R G1PHK
Baines R 2E0FFS
Baines R M1KIA
Baines R M3KTA
Baines R G3YBO
Baines R M6CKI
Baines S M6HOF
Baines T 2E0TUR
Baines T M3UAE
Baines T G4AEB
Baines-Jones D G1WHU
Bains D M0DKD
Bai[illegible] D G8EVT
Baird A G0PTM
Baird B MI6BAI
Baird H 2E0HNK
Baird H [illegible]
Baird J GM7FHN
Baird K MM0HSV
Baird M MM3MMB
Baird M MM3NGV
Baird N GM4JNB
Baird P G7GBN
Baird S MI0XLK
Baird S GM0FHS
Bairstow A G3NBS
Bairstow A G4RSW
Bairstow A 2E0VBM
Bairstow A M6GYA
Bais A MM6APO
Baister M G0UDZ
Bajjon A [illegible]
Bajjon A [illegible]
Bak Z 2M0DGJ
Bak Z MM0HVF
Bak Z MM6ERO
Baker A G0TQR
Baker A G3PFM
Baker A M0TBA
Baker A M1DJX
Baker A G7HQJ
Baker A G4GNX
Baker A G6TVK
Baker A G8UBD
Baker A M6AWB
Baker A G3UXY
Baker B G8XOE
Baker B GM0JRQ
Baker C G1SLA
Baker C G3HQS
Baker C G4FFN
Baker C G4HAY
Baker C G4KFJ
Baker C G4WUV
Baker C G7ADP
Baker C G7OSB
Baker C M0BHA
Baker C M3ORU
Baker C G4LDS
Baker C M6BAK
Baker C M0HXI
Baker D 2E0JTB
Baker D 2E0GDG
Baker D G3IYF
Baker D G6HNI
Baker D G8SSX
Baker D G8THH
Baker D M3KJV
Baker D 2E0MFV
Baker D M6CPT
Baker D M6MFV
Baker D 2E0GET
Baker D G0GWL
Baker D G8XCE
Baker D M6DXQ
Baker D M6JIY
Baker D M6NHD
Baker D M6XTL
Baker D M6DNN
Baker E G1POM
Baker E G2FMW
Baker E G7VLJ
Baker F 2E0FBL
Baker F M6ENH
Baker G G0CQI
Baker G G4ZRZ
Baker G M0RGB
Baker G M6IOS
Baker I GM6HFH
Baker I MW0IBZ
Baker J G1PGH
Baker J G3TMB
Baker J G4XIP
Baker J G6LIB
Baker J G7JMB
Baker J G7STT
Baker J M3MXC
Baker J G7ACG
Baker J G3YHB
Baker J G0MTQ
Baker J M0GWQ
Baker J M6BXX
Baker J M6VST
Baker J 2I0ZXD
Baker J MI0NWA
Baker J MI6NOS
Baker K 2E0SSX
Baker K G3WTV
Baker K G4RPV
Baker K G8WUO
Baker K M0SSK
Baker K M6OLP
Baker K M3ZHX
Baker K 2E0CTL
Baker K M6BGA
Baker L G4RZY
Baker L G0TEO
Baker M G0GMB
Baker M G1JUP
Baker M G3SUK
Baker M G3ZBP
Baker M G4CEI
Baker M G4HGR
Baker M G4OTL
Baker M G4SHM
Baker M G4WUW
Baker M G4YZR
Baker M 2E0SAY
Baker M G4RUR
Baker M M6SKU
Baker M M0GWB
Baker M M6AJJ
Baker M [illegible]
Baker N G4VYH
Baker N G7MZL
Baker N M6FTY
Baker N 2E0GCM
Baker N M0VCE
Baker N M6GCM
Baker P G1LUN
Baker P G4HSO
Baker P G4PBF
Baker P G6ESQ
baker P G7IPH
Baker P G8ENW
Baker P G8IXL
Baker P M3JIC
[illegible]
Baker R G1DDX
Baker R G6FVB
Baker R G8DLP
Baker R G4SFY
Baker R G0AIH
Baker R G0BLB
Baker R G4DJC
Baker R GD4HPN
Baker R MW6TIQ
Baker R G3OVD
Baker R M3RNO
Baker R G1RVC
Baker S G7HRM
Baker S GD8EXI
Baker S M6ROX
Baker T G3WNP
Baker T G4CLE
Baker W G1GRZ
Baker W G7GFH
Baker W G1ZBW
Baker W GW4RGI
Baker-Munton A G0KPY
Bakes C M6EBI
Bakrania P G0MHA
Balboa I M6BNV
Balchin B M3WHB
Balderson D M6NTY
Balderson R G6IVW
Balderston C G4XMP
Balding A G6JAS
Balding J G6TLS
Balding S G7VRK
Baldock D G6NKL
Baldock K G7HLP
Baldock K G3UID
Baldock R G0FFQ
Baldock R G0PCP
Baldry J M3OZE
Baldry K G7OZE
Baldry M G6MCB
Baldry M G7BYI
Baldwin A G7HMK
Baldwin A G6CAR
Baldwin D 2E0EKM
Baldwin D G4BFR
Baldwin D M6DBQ
Baldwin D M6MTL
Baldwin D M0ESB
Baldwin G G4KDU
Baldwin J G3UHK
Baldwin J G4KGE
Baldwin J G4RYB
Baldwin J G7HNF
Baldwin J 2W0JCV
Baldwin J MW6JCV
Baldwin L G6WDC
Baldwin L M6LBU
Baldwin L G6GTJ
Baldwin M M3FOR
Baldwin O G0RCL
Baldwin P G3LCF
Baldwin P G4CQV
Baldwin R G6ZDS
Baldwin R M6RNZ
Baldwin R M3RQQ
Baldwin S G0FRD
Baldwin S 2E0EXJ
Baldwin S M0WXY
Baldwin S M6EXJ
Baldwin T M1BAN
Baldwin T M6MZU
Bale D M3NLI
Bale I G4SOL
Bale S M6FPW
Bales D G8NIL
Balfour G GM0SHD
Balfour G GM4PYJ
Balfour K MM6ABO
Balharrie D G8FKH
Balharrie D M0DGB
Balister R G3KMA
Balkham S G7KMA
Balkham S M3AXB
Balkwell R G1WPL
Ball A G4EDK
Ball A G7HNR
Ball A G8RSE
Ball A G3UQW
Ball A M3VTH
Ball A GW0VWD
Ball A G8VWV
Ball D GW1ZLL
Ball D G7ANB
Ball D M1BCB
Ball D 2E0DHY
Ball E G4MYY
Ball E G4UOZ
Ball E G6DGW
Ball E G0PFI
Ball E MM6EBZ
Ball G G6WXI
Ball G M1EER
Ball G G4OJF
Ball H M3XNA
Ball I M3ZYI
Ball J G3UYX
Ball J G4CEY
Ball J G4KIP
Ball J G6HNR
Ball J M3VJM
Ball J 2E0BON
Ball J 2E0XZI
Ball J M6BUL
Ball K G3XVW
Ball K M6AIF
Ball [illegible]
Ball L [illegible]
Ball L G4YCE
Ball M G7MGQ
Ball M M6MBQ
Ball M G8LZK
Ball N M3EPC
Ball N G1HHU
Ball P 2E0PBL
Ball P G0URC
Ball P G3HQT
Ball P G6KIH
Ball P G4PRB
Ball R G0INZ
Ball R G4IRS
Ball R G4UXB
Ball R G6HNS
Ball R G7MYJ
Ball R 2E0PFY
Ball R M6FXU
Ball S G4BBZ
Ball S G1VGM
Ball S G2CNN
Ball S M0CTL
Ball S M3TEV
Ball T GM1AXI
Ball V G7RXI
Ball V 2E0MIB
Ball V M3VUZ
Ball W G3ZKD
Ball W G1FNN
Ballance J G1YBG
Ballance M G8AQB
Ballance P G6HEB
Ballantyne I G7JXT
Ballantyne J G8AKF
Ballantyne M G1MOV
Ballard J G0SMR
Ballard M 2E0MKB
Ballard M M0FCW
Ballard M M0KMW
Ballard N G1HOI
Ballard R G3TWB
Ballard S G8BFK
Ballard S G8OTD
Ballard S M6SNN
Ballard W G7IGV
Ballinger A M3WHA
Ballinger K G0RQQ
Ballinger L M3GVJ
Ballinger T G0RYR
Ballington P 2E0RTE
Ballington P M6RCG
Balloch I GM3UTQ
Balls D M3DEJ
Balls J G4ALC
Balls J M0CKE
Balls K M6ZTO
Balls L G3YYQ
Balls M G8HDK
Balm R M3IVX
Balme S G0GVB
Balmer A M6ZET
Balmer B G0EFG
Balmer S G4OPY
Balmford J G6DAP
Balmforth A G3RKQ
Balmforth R M6BCB
Balsdon C G1BOB
Balsdon D M6DIZ
Bamber B G4RKU
Bamber D G4ZZS
Bamber M G0SHY
Bamber M M6CEB
Bamber R G6JIY
bamber R MI1DAW
bamber R MI5DAW
Bambrey R GM7SWB
Bambridge P G7VRY
Bambrook D 2E0DAB
Bambrook D M3DAB
BAMBROOK R G8NRR
Bambrough D 2E0ODB
Bambrough D M0ZXW
Bamford C G4FQP
Bamford C M3OFN
Bamford H G0WFK
Bamford H M6HQZ
Bamford I G0BGH
Bamford J G0ORJ
Bamford W G1STO
Ban V M3VUG
Banach S M0SEB
Banahan L M3UPB
Banahan M G4BUH
Banasiak H 2E0HTB
Banasiak H M0HTB
Banasiak H M6HTX
Banaszak L [illegible]
Banbury P G8BNK
Bance B G1XFR
Bancroft D G8PPR
Bancroft D MW6BDS
Bancroft H [illegible]
Bancroft H MW0HVB
Bancroft H MW6HDB
Bancroft J G4XXP
Bancroft J 2E0JRB
BANCROFT M 2E0SAL
BANCROFT M M0MVK
BANCROFT M M3MIB
Bancroft M G8NQN
Bancroft M M3HGW
Bancroft R M3OKP
Band K G3WSB
Banda S M6URS
Bandara [illegible] G7NLF
Bandtock A [illegible]
Banester J G4CKB
Banester R G4CDN
Banfield J G7PCF
Banfield J G1SSS
Banfield R G8LQZ
Banfield R M3BBA
Banfield V GW8RLI
Bangalore S M3SSI
Bangle M G4WMP
Banham J 2E0JMB
Banham J M0UKS
Banham N G4YFV
Banham T G6WWA
Banham T M6OKM
Banister C G7TQC
Banister D G7TDQ
Banks A M1EYL
Banks A 2E0GFF
Banks A M0YMA
Banks A M6ADB
Banks B G4POR
Banks B G6JJB
Banks B M0DZG
Banks B M3UPL
Banks D G0VFB
Banks D G6KIE
Banks D M0EJB
Banks D MM0GGG
Banks D 2E0DKB
Banks D M6ICQ
Banks E M1BLW
Banks G G0PMB
Banks G GM7HNU
Banks H M0AHV
Banks J G1XLN
Banks J M3NKB
Banks J M3VRB
Banks K G0DAY
Banks M M6MTN
Banks N G1JJE
Banks P G7NTP
Banks P M6NNK
Banks R G4WND
Banks R G7UTH
Banks R G7GPJ
Banks R M6UJM
Banks S G0CLJ
Banks S G1MZG
Banks S G6VVE
Banks T 2E0BTF
Banks T M0GXE
Banks T M3XWE
Banner A G7UMW
Banner B M6BAN
Banner F 2E0BPF
Banner L G7SFS
Banner M G6YCL
Banner M 2E0GDM
Banner S G8FPG
Bannerman I MM6BSK
Bannerman J MM6BMQ
Bannerman R GM4LUD
Bannerman W G8YSJ
Bannier P GU4SXM
Bannister A G0OVQ
Bannister B MM6MCX
Bannister D G0DEB
Bannister D G1PKR
Bannister D G4FNZ
Bannister D MW6YAE
Bannister E G0RBA
Bannister G GI8SKR
Bannister M M6MYB
Bannister M MW6MGZ
Bannister P G1HNH
Bannister P G7IMQ
Bannister R 2E0ASX
Bannister R G3FVR
Bannister R G8PIN
Bannister S G4HNZ
Bannister S 2E0ZSB
Bannister S M3ZSB
Bannister T 2E0DJU
Bannister T M6EPA
Bannon P M6ZPB
Bannon P 2E0RNA
Bannon P M0XPB
Bannon P M6RNA
Bannon R M3PBB
Bano F M3UCA
Bansal A G6VSE
Bansil E M6EMA
Bansil G 2E0NNH
Bansil G M0NNH
Bansil K M3NNH
Bant L G7DCM
[illegible]
Bantock K M0MFL
Bantock K M6KBE
Bantoft A M1CCQ
Baraclough G G8DLT
Barba M M6AIL
Barber B G0EUR
Barber B G0OYP
Barber B G7TPG
Barber C G0JPQ
Barber C M3KUB
Barber C G4BGP
Barber D G0LSX
Barber D G4JBW
Barber D G7ILY
Barber D G8STE
Barber G 2E0WGB
Barber G G8CHN
Barber G G4KYO
Barber I [illegible]
Barber I [illegible]
Barber J G0LAU
Barber J GM4NNH
Barber J G7NEM
Barber J M3YXF
Barber J G3TTJ
Barber J G1JRD
Barber J GW4SKA
Barber J M0AOH
Barber M MW3SKW
Barber M M3URA
Barber M M6MMY
Barber P G6SFH
Barber R G7HYG
Barber R G8WLV
Barber R M5REG
Barber S M6POE
Barber S 2E0FTV
Barber S M6FTV
Barber S G6CJR
Barber T G3TRB
Barber V G7AZH
Barbery K 2E0ADT
Barbieri J 2E0BTV
Barbieri J M0GOQ
Barbieri J M3XWV
Barbour A M1HZZ
Barbour R MM0ASB
Barbour S 2M0APX
Barbour W G1GES
Barclay A 2M0TGN
Barclay A MM0WXE
Barclay A MM6TGN
Barclay B MM6POV
Barclay C MM0OWL
Barclay D M0BPM
Barclay D MM6PFT
Barclay E G0SBX
Barclay J MM6JJB
Barclay L G3HTF
Barclay P GM0PXV
Barclay S M6FWG
Barczynski H G1GMM
Bard A G1XKZ
Bardell D M6RZO
Bardell I 2E0GXX
Bardell I M3GXX
Bardell J G8HHR
Bardell M M0AEQ
Bardell N M3RZO
Bardell N G3WKA
Barden A M3MOP
Barden J G6UWK
Barden P M6HXA
Barden P G0JUY
Bardgett T M1AIK
Bardsley S G8YKO
Bardy A G6ODA
Bareham K G1RRR
Bareham R G8ZGY
Barfield T G0NSO
Barfoot C G1TJH
Barfoot J G8BJO
Barham C G4MYB
Barham D G0DGW
Barham T M6SKZ
Barker A G6XVZ
Barker A G8HTB
Barker A G8IYZ
Barker A M3DQJ
Barker A M3OAB
Barker A M6DOF
Barker A M3TXP
Barker A M6RIN
Barker B 2E0GCW
Barker B G4VRT
Barker B G3NIJ
Barker C G1EZJ
Barker C G7WDO
Barker C M3CIE
Barker C M6TUD
Barker C M6PKD
Barker D G4GZL
Barker D G8GBU
Barker D M3IDD
Barker D G1DOT
Barker D [illegible]
Barker D M0RRN
Barker D M6YUK
Barker F MW6EGH
Barker F GI4SXV
Barker F G4AUQ
Barker F G6YAQ
Barker F G0BKE
Barker G G6LQM
Barker G G6SFY
Barker H G4RXY
Barker H G3SYR
Barker J 2E0KIP
Barker J G0UPB
Barker J [illegible]
Barker J GM3PDX
Barker J G3WAL
Barker J G3ZLD
Barker J G4JOB
Barker J M0ABP
Barker J 2E0GOO
Barker J M6JIW
Barker J M1EAZ
Barker J 2E0BPN
Barker J M3ZVX



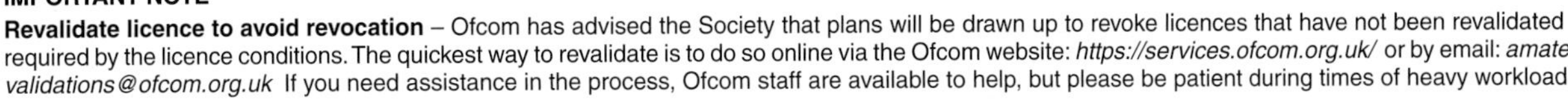
**IMPORTANT NOTE**

**Revalidate licence to avoid revocation** – Ofcom has advised the Society that plans will be drawn up to revoke licences that have not been revalidated as required by the licence conditions. The quickest way to revalidate is to do so online via the Ofcom website: *https://services.ofcom.org.uk/* or by email: *amateur.validations@ofcom.org.uk* If you need assistance in the process, Ofcom staff are available to help, but please be patient during times of heavy workload.

Barker K G0OBA
Barker K GD6MLV
Barker K G7PEX
Barker K M6THI
Barker L G1TWF
Barker L G4TJY
Barker L M6LOC
Barker M G7JXU
Barker M MU3UWX
Barker M 2E0REC
Barker M M6MCU
Barker N G4LRN
Barker N 2E0NHB
Barker N M0HZR
Barker N M6NBS
Barker P G0DZU
Barker P G1MQB
Barker P G4HPS
Barker P G8BBZ
Barker P 2E0PBY
Barker P G8DBK
Barker P G4BPV
Barker P G8ECZ
Barker R G0NUN
Barker R G0VAZ
Barker R G3UTL
Barker R GW3WVV
Barker R G4JNH
Barker S G0JUM
Barker S G0WYP
Barker S M6BZV
Barker S GW7HDS
Barker S 2E0CYU
Barker T G0HLA
Barker T G6XVY
Barker T G4AZT
Barker T M6ACD
Barker W G0MVW
Barker W G4JIQ
Barker W GI4ODT
Barker W G6PRP
Barker-Gunn J M6AIT
Barker-Mawjee S M6ZIB
Barkley D G0DPI
Barkley I G7EDK
Barkley P M6PGZ
Barkley R G1GGN
Barkley R G7IXC
Barley J M0DUT
barling R G4TGP
Barlow A G1GKR
Barlow B G0ADL
Barlow C G1ORL
Barlow C G7GYN
Barlow D 2E0OTE
Barlow D G0UJU
Barlow D M0OTE
Barlow D M3OTE
Barlow D G1MZD
Barlow D G3PLE
Barlow G M3GBA
Barlow J G3VOU
Barlow J G6SUV
Barlow J G7BJN
Barlow K G0BZU
Barlow L G8ZSM
Barlow M G6MMA
Barlow S G7OTE
Barlow S M6ICR
Barma J M6JCN
Barnaby M 2E0DJQ
Barnacle G G4ILT
Barnard A G6OEJ
Barnard C GM6GFQ
Barnard J G7RPJ
Barnard K GD4MMA
Barnard M M6WWM
Barnard N 2E0NAQ
Barnard N M6CPP
Barnard P 2E0PGB
Barnard P M0PGB
Barnard R 2E0DHE
Barnard R M0HVC
Barnard T M3UGA
Barnard W M6WTT
Barnden T M6EFE
Barnes A G0FVX
Barnes A M3AFF
Barnes B G4IJA
Barnes C 2E0GTB
Barnes C G4OVM
Barnes C G8CBB
Barnes C G4KQK
Barnes C 2E0ZHG
Barnes C M3ZHG
Barnes D G1VHN
Barnes D G3UVB
Barnes D G6IVE
Barnes D G7MGM
Barnes D G0RIF
Barnes D 2E0DBA
Barnes D M0LOW
Barnes D M3ERR
Barnes E 2E0ACV
Barnes E 2E0HMS
Barnes E G0WHL
Barnes F M1FRB
Barnes G G4SGA
Barnes G G6FPF
Barnes G G8ZNK
Barnes G M6GAB
Barnes H 2E0ACQ
Barnes I MD6HUV
Barnes J G0GPV
Barnes J G4JKB
Barnes J G7DMP
Barnes J G8AHN
Barnes J G8NSK
Barnes J G0DDT
Barnes J G3RDH
Barnes J M6KPI
Barnes J 2E0EVB
Barnes J M6EVB
Barnes K G7OOB
Barnes K G8KRB
Barnes K G8GTI
Barnes K G4MKQ
Barnes M G6BQQ
Barnes M G6OCA
Barnes M M1CQF
Barnes M M3GHE
Barnes M M3ZCN
Barnes N G1SGP
Barnes O M1DYW
Barnes P G4YNO
Barnes P M3OKC
Barnes P M0DVD
Barnes P 2E0UAR
Barnes P M6KVA
Barnes R G0DKS
Barnes R G0VFZ
Barnes R G8AZN
Barnes R 2E0OES
Barnes R G0PCE
Barnes R G8CJQ
Barnes R M0TRY
Barnes R M3XUG
Barnes R G4YGV
Barnes R G4YLI
Barnes S 2I0SAI
Barnes S MI0SAI
Barnes S MI3XUS
Barnes T M0TJB
Barnes T G4LLM
Barnes V M0BXQ
Barnes W G0RAT
Barnes W G6SJA
Barnes W G6XML
Barnes W 2E0WOS
Barnes W M0WOS
Barnes W M3WOS
Barnes-Martin B M6BAD
Barnetson I GM0ONN
Barnett A G7OLH
Barnett A M3EBA
Barnett B 2E0FZK
Barnett B G8GLP
Barnett D 2E0CDK
Barnett D G1LVH
Barnett D MW3XDB
Barnett D 2E0KDB
Barnett D G7UGR
Barnett D M6DDI
Barnett D M6KWK
Barnett D G0RHH
Barnett G M1ACA
Barnett J GI6GRV
Barnett J G7GIJ
Barnett J GM7IFX
Barnett J M0DRE
Barnett J G0JGB
Barnett J M6OLI
Barnett J G8RLN
Barnett K G7CED
Barnett K G0HOF
Barnett M 2E0EAW
Barnett M G0HCO
Barnett N G0JUR
Barnett P G4GSK
Barnett P G4TMC
Barnett R G4XCV
Barnett R G6NBL
Barnett T G4HVD
Barnett-Bone M G3VOO
Barney D G3VIC
Barnfather M M6MRM
Barnham D G0SMS
Barnish R G1OJT
Barnwell D G8KWJ
Barnwell M G4FKQ
Barnwell S G4YAA
Baroch R M3PSF
Baron C G0HXQ
Baron J G6NBF
Baron P G0WWF
Baron P G2PB
Barr A G4LLZ
Barr C G0SLU
Barr C G3PFO
Barr D GM7LOK
Barr G GI4NBO
Barr I 2I0IMB
Barr J M0EDP
Barr J GI1CET
Barr K MI3KMB
Barr L G7KOF
Barr M G4OQN
Barr M MI6WAG
Barr M GI6IBL
Barr P GI4GOV
Barr R GI4MNN
Barr S G0CLV
Barraclough D 2W0DRB
Barraclough D MW6AEM
Barraclough G G8MKG
Barraclough I G7DWY
Barraclough I M0INB
Barraclough S G0SJB
Barraclough T G3XVG
Barraclough T G3VSR
Barrasford J G6WML
Barrass B G7DGP
Barrass M G0RWW
Barratt A G1DPI
Barratt D M1NAD
Barratt G G4VEO
Barratt J G3WVQ
Barratt J 2E0JBX
Barratt R G4WJB
Barratt T M0THB
Barratt T M6THB
Barrell D G4BMC
Barrett A M6NBH
Barrett A M6ZZS
Barrett A G2BWW
Barrett A G7IIF
Barrett A G8DOR
Barrett A 2E0CKY
Barrett A M6AIV
Barrett B M1DBM
Barrett C 2E0RYP
Barrett C GW4VEI
Barrett C M6EBY
Barrett D G1ZRQ
Barrett D G3ZQH
Barrett D M0IVW
Barrett D MM0DNX
Barrett D 2E0DXJ
Barrett D M6AWE
Barrett E M6PUY
Barrett G M0BTG
Barrett G G8ZOJ
Barrett J G7OZP
Barrett J G4GQV
Barrett J G0SGF
Barrett J M6JIR
Barrett J M6ODM
Barrett J G4AMX
Barrett J 2E0GMQ
Barrett K GW4NBY
Barrett K GW0SIS
Barrett M G6VGO
Barrett P G4CTM
Barrett P G4WQU
Barrett P G6EJI
Barrett P M6PWB
Barrett P G8PYE
Barrett R G0TDE
Barrett R G3YCY
BARRETT R G4CWC
Barrett R G4ZIZ
Barrett R M1CJF
Barrett R 2E0YXB
Barrett R M3YXB
Barrett R M1AEZ
Barrett R G2VS
Barrett S G4HTZ
Barrett S M6GWU
Barrett T G8JJK
Barrett V G3KDD
Barrett V G1NXR
Barrett-Sprot A 2E0TUG
Barrett-Sprot A M6EZN
Barrick R G7FVH
Barrie G 2M0ZEB
Barrington B G3ZQW
Barrington N G4PUZ
Barrington S G1KBC
Barron A 2E0TAZ
Barron C G4JNZ
Barron K M3IPD
Barron R G7MZX
Barrott D G4GVN
Barrow D M6BUU
Barrow J 2E0JRB
Barrow K M3KBB
Barrow P M6POP
Barrow R M1EYG
Barrow R G8ILD
Barry C GM1DLS
Barry D G3ONU
Barry D G4WWP
Barry D M6BAR
Barry E G7JQT
Barry J 2E0CUJ
Barry J 2W0USN
Barry J G8SLP
Barry K M3AFR
Barry K 2E0UOJ
Barry K M3UOJ
Barry M G4MAB
Barry M M6OVR
Barry P G3KFU
Barry P G3RJS
Barry P G8OPA
Barry R G7BYK
Barry S 2W0VVO
Barry S MW0VVO
Barry S MW3VVO
Barry T G0CEM
Barsby D G6JFL
Barstow D 2E0IJK
Bartels G G8OPO
Barter A 2E0FAV
Barter A M6ZAF
Barter A G8ATD
Barter D 2E0DEV
Barter R M0ZAF
Barter R M3XJQ
Barter S G8CPZ
Barth A M0ALC
Bartha R 2E0PNP
Bartha R M6BEU
Barthel C M6BXN
Bartholomew J G8GNX
Bartholomew P G0PQW
Bartlam R G0HDF
Bartle A G0TTR
Bartle A M1IFT
Bartle M G0UJD
Bartle W 2E0FOL
Bartle W M6FOL
Bartlett A 2E0MIZ
Bartlett A M3VWR
Bartlett C M1EUX
Bartlett D M3DSA
Bartlett E M3MSQ
Bartlett J G0GZB
Bartlett J G3ZND
Bartlett M G3PXH
Bartlett M G6YCF
Bartlett M 2M1VFO
Bartlett M G7PCG
Bartlett M M6EEP
Bartlett M M6MNZ
Bartlett N M0CPU
Bartlett P MD0DLC
Bartlett R G4LJN
Bartlett S G1JBJ
Bartlett T G3ITB
Bartlett W G4KIH
Bartley B M0BAR
Bartley G M3GBB
Bartley J M6VWD
Bartley M 2E0MBB
Bartley N GM7UFO
Bartolo J GM7FGH
Barton A G6GKK
Barton A M3UVQ
Barton A M6AIA
Barton D G1JPF
Barton D G0SBB
Barton D G8YKS
Barton E M6EMM
Barton G 2E0GBN
Barton G M6GBN
Barton H M6HJB
Barton J 2M0LRO
Barton J 2E0OMV
Barton J G0RPD
Barton J G6SZB
Barton J M0OMV
Barton J 2E0JJB
Barton J M0BWC
Barton J M0JJB
Barton J M3JJB
Barton J M6BTK
Barton K G1ZFB
Barton M G8OOQ
Barton N G4BZV
Barton R G6SPN
Barton R M0DZC
Barton S G0FSB
Barton W G1ZNT
Barton W G7JIM
Bartram A G7ELS
Bartram G GM1JNS
Bartram J G0JYL
Bartrum R MW0WNL
Barugh H G1ZVE
Barusevicus G M0GEN
Barusevicus P M3CAP
Barville P G3XJS
Barwell B M6RDD
Barwell C G4ZPL
Barwell C G6UKO
Barwell F G4ZCY
Barwell F G6AKS
Barwell J G7VBF
Barwell S MW3USP
Barwick B G0HRF
Barwick J GW6MAB
Barwick M G1BPE
Barwick P G4SUO
Barwood D G7VPS
Barwood D G4EGR
Baseden Butt D 2E0XDD
Baseden Butt D M3XDD
Baseden D M3MUX
Basford D G0UGJ
Basford M M6INV
Basford R M6CKD
Basford R G3VKM
Bashford B G7PIK
Bashford R GW3PGJ
Bashford T 2E0FSZ
Bashir A G0TVM
Baskerville N G0UQQ
Baskerville S G1JTC
Baskeyfield M G7SEJ
Bason M G1ZBG
Bass A M6JFR
Bass C 2E0XDY
Bass J G0DKY
Bass J G3KMD
Bass R G4IPH
Bassett A G0LLX
Bassett C G1LRK
Bassett C M6COE
Bassett I M5AXA
Bassett K G7NBU
Bassett S M0SBY
Bassett S M6BAH
Bassett S M6SBI
Bassett-Smith I G3XPR
Bassford B G4YNX
Bassford R G2BZR
Bassford S G3YZB
Bassi C M6FMF
Bassnett J G8WMG
Basson D M1NTY
Bastable J G8FDF
Bastable R G6IUK
Basterfield A M3TQX
Bastin C M1BAS
Bastin D G4XVY
Bastin M M1AIY
Baston P G0PJA
Bastow R G3BAC
Bastow S G8HQM
Batchelor C M6BSZ
Batchelor J M1GUS
Batchelor M G0RBB
Batchelor M 2M0RRT
Batchelor M MM6BJJ
Batchelor N M3UKO
Batchelor R G1ALK
Batchelor S 2E0HLL
Batchelor S M1HLL
Batchelor T G0JUW
Batchelor V G3TTG
Bate A 2E0LMD
Bate A M6YEH
Bate C G8MCT
Bate D G4XEE
Bate D G6LQG
Bate E G3LUC
Bate F M1FJH
Bate G G0FHT
Bate H G8AMD
Bate L M6LMB
Bate S G7NZO
Bate S G4XVO
Bate T 2E0TTB
Bate T M6TTB
Bateman A G7HZU
Bateman A M0DNU
Bateman A 2E0BFF
Bateman B 2W0ZVR
Bateman B MW0ZVR
Bateman E 2E0HTK
Bateman E M3EGB
Bateman I G3ZKH
Bateman J M0GTO
Bateman M G7JPN
Bateman N G3UUB
Bateman P GW0BAH
Bateman R M5EHG
Bateman S G4ZBL
Bateman S MW6GLK
Bates A GM0KNT
Bates A G1BUJ
Bates A G1XYD
Bates A G3AZW
Bates A G6MOI
Bates C G4ORY
Bates C GM6KAY
Bates D G0LZL
Bates D G1NCK
Bates D G1OLE
Bates D M0DUY
Bates D M3ZQG
Bates D MI3SYF
Bates G 2E0GPG
Bates G M0YUG
Bates G M6GPG
Bates G G6HFF
Bates J G0BZP
Bates J G3OJK
Bates J G4PLU
Bates J G6YSB
Bates J M1FFS
Bates K G7KUU
Bates K G1XQI
Bates K G3KGV
Bates M 2E0MPB
Bates M G8FBM
Bates N M6NHK
Bates P G0IYD
Bates P G1NZD
Bates P GM4BYF
Bates P G0KWQ
Bates P G3XIA
Bates R G0IZE
Bates R M0CSQ
Bates R MI6TXS
Bates R G0YDX
Bates R 2E0SFC
Bates R M3UMX
Bates S G1IGE
Bates S G7KHF
Bates S G1YJW
Bates S 2D0SEO
Bates S MD0AAV
Bates S MD6SJX
Bates V G6MML
Bates V G6SSH
Bateson A 2E0CDE
Bateson D 2E0TBF
Bateson D M6DBE
Bateson S 2E0NMK
Bateson S M6EHF
Bateson W G4RTS
Bateup M M3ZMB
Batey A G7LBL
Batey J G1TIF
Batey W G0DEO
Bath G G3NMZ
Bath M G4EEZ
Bath R M6PUF
Bathe D G6TPO
Bathurst A G0WCB
Bathurst A G4PNB
Batiste R 2U0WGE
Batiste R MU6FXB
Batkin F G0OCR
Batley J G0IID
Batley S 2E0SUF
Batley S M0SUF
Batsman G 2D1HVL
Batsman J MD0XRA
Batson J M3IKI
Batson P M0PBX
Batt K M3XNR
Batt M G3SJI
Battelle T M6TTE
Batten I G1FVC
Batten J G8BIJ
Batten R M0DBX
Battersby M G3ZZH
Battersby R G0IMB
Battershill P G0IUH
Battle-Welch J G4NYZ
Batty A G8OZD
Batty C G1MBE
Batty D M3JYG
Batty F 2E0AFR
Batty G M0PAF
Batty G M0PPP
Batty H G2API
Batty K G6LBO
Batty P 2E0GXY
Batty P G1HMY
Batty P M3BAT
Batty R G0LBB
Bauer A G7NGQ
Bauer M M6MZB
Bauers C G4JUV
Baugh D M3KZT
Baugh J M1ERN
Baugh M G0SRR
Baugh M G8JHH
Baugh S G4AUC
Baughan A M0AVP
Baughan J 2E0VBR
Baughan J M0VBR
Baughan J M6VBR
Baulf N 2E0DIT
Baulf N M0KOT
Baulf N M6DIT
Bauly B G6SYW
Baum K G1VNS
Baum K G6ZDP
Bautista J G4JTC
Baverstock C G4WCK
Baverstock J G6ANI
Baverstock S G6OAI
Baviello E GM0MJR
Bavin J GM0GMI
Bavin S GM1TFZ
Bavister C M6WCR
Bawden A G1FJF
Bawden R M6ASJ
Bawley R G7VCN
Bax G G4SBD
Baxendale G G4IBS
Baxendale H G4OJK
Baxendale T M3VRU
Baxter A G1MPP
Baxter C G7KJR
Baxter F GM3VEY
Baxter G G8OAD
Baxter J G4NMV
Baxter J M3HBM
Baxter K GM7IRI
Baxter M 2E0CYZ
Baxter M M3MKB
Baxter M G8DHU
Baxter O G0WRQ
Baxter P G1DEZ
Baxter P G4EOW
Baxter P G4ZVN
Baxter R 2E0ENZ
Baxter R G1YZH
Baxter R G3WTB
Baxter R G7SYS
Baxter R 2M0SOE
Baxter R M0ZVF
Baxter R M6BAX
Baxter W G6SRZ
Baxter W M0WAB
Bay M M0MBO
Baybrook M G1XRM
Bayes M G4DZC
Bayley P G4ASG
Bayliff M M0XGG
Bayliffe G G6PVW
Bayling A G4AZS
Baylis A G3LCL
Baylis B G1HOJ
Baylis D G4EKQ
Baylis I MW6BUH
Baylis J G4LMA
Baylis M G1GSK
Baylis P G6NYL
Bayliss A MW6AEX
Bayliss D G4FJJ
Bayliss D G4OEF
Bayliss E M6LIB
Bayliss I G8CZQ
Bayliss J 2E0OJB
Bayliss J M3NZK
Bayliss M G1YLV
Bayliss N GD0IDU
Bayliss P G6SGE
Bayliss R M1RSJ
Bayliss T G2HHH
Bayliss T G4OPL
Bayliss Z M3LIV
Bayman M M0ZKK
Bayman M M6DSQ
Baynes N G4OFO
Baynes S G0AQA
Baynton S 2E0CQH
Baynton S M6BVM
Bays C M0HAM
Bazar O M3OXY
Bazley J G3HCT
Bazley L MD3LJB
Bazley M MD3MLB
Bazley N GD6AFB
Bazyk J G4WOB
Beach D M3DLB
Beach J GW7PMA
Beach J G1KJH
Beach K G0UBJ
Beach M G7NMT
Beach T MW3ZTB
Beacham A MM0IBX
Beacher D GM0SRQ
Beacher G 2E0GPB
Beacher G M0PPG
Beacher G M6GPB
Beachey J G1WTL
Beacon J G1JBG
Beacon P G8NAP
Beacroft W M6WBO
Beadle D G7JQZ
Beadle G G8UBF
Beadle J G7BNC
Beadle R G0KGR
Beadle R G3VQG
Beadle R G7ODG
Beadle S G0DWC
Beadman T M1BKQ
Beakhust D G3OSQ
Beakhust M G0DGF
Beal A G6CSK
Beal A M3CVO
Beal E G4HNX
Beal L G4YVJ
Beal M G0AOB
Beal S G3WZK
Beale A M1EAI
Beale A M1AZQ
Beale D G0DBX
Beale D G1KII
Beale G 2E0WGB
Beale G M0WGB
Beale G M3ORE
Beale M M5CBS
Beale T M6XTB
Beales A G6ZDV
Beales D A G3MWO
Beales S 2E0BBM
Bealey J M3PZF
Bealing L G6PJP
Beam G M0EHA
Beament R G8KSM
Beamish C M6GHY
Beamish S G7JCF
Beamond T G3VLF
Bean A M3FLB
Bean B G6AGO
Bean B G6PKS
Bean B M3KTV
Bean C G3VWD
Bean D G4OUS
Bean G G4OHF
Bean M G7UTC
Bean N G6DGR
Bean N G8NOB
Bean P M6NEC
Beane D G0TAG
Beaney L 2E0FSE
Beaney L M6FSE
Bearchell R G6YRY
Beard A G0GJR
Beard B 2E0IBB
Beard B M6OBB
Beard D G8FMX
Beard M M6PSV
Beard V M1BDY
Beardmore E G4LCL
Beardmore G 2E0LME
Beardmore G M3LME
Beardmore J G6DZX
Beardmore R M6RJB
Beards J M3LLQ
Beards P G4IZX
Beardshall R M3SNY
Beardshaw P G1UGL
Beardsley A 2E0ETE
Beardsley A M6FFF
Beardsley M M3YSN
Beardsley M G8WSB
Beardsley M M1BVT
Beardsley R 2E0REB
Beardsley R M6REB
Beardsmore J G3SIO
Beardsmore P G4IXY
Beardsmore R G0KYK
Beardwood T M0TJW
Bearne D M3GDV
Bearne R G4DUA
Bearpark T G4KFA
Beasley A G0CXJ
Beasley C G1DPJ
Beasley D G1GLN
Beasley G G3LNS
Beasley M 2E0RKM
Beasley M M6KMJ
Beasley R G1CSN
Beastall D G7VHU
Beaton A GM4BAP
Beaton M MM6WAW
Beaton W 2M0ZNQ
Beaton W MM3ZNQ
Beatrup C G7IIS
Beatrup M G7DIZ
Beattie A G0TNL
Beattie C G3YSR
Beattie D G3BJ
Beattie F MI6FEB
Beattie R G6CKW
Beattie S GI0LTT
Beattie W GM8AT
Beatty S MI6SWB
Beauchamp S G3SYD
Beaugie R G1JFT
Beaumont A M0TIU
Beaumont G G7LHV
Beaumont H M6AUH
Beaumont J G4EIM
Beaumont J G4JPB
Beaumont M G4VCX
Beaumont P G8HPJ
Beaumont R G6AXC
Beaumont R G7FNU
Beaumont R M6SUN
Beaumont S G7IBD
Beaumont T G4DVZ
Beaumont T M0CMF
Beaumont T M0URX
Beavan F G1SMJ
Beavan J M6LFG
Beaven B G4BZU
Beaver G 2E0GXB
Beaver G G4CLD
Beaver G M3GXB
Beavis D M6DDB
Beavis H M6GQW
Beavis S MW6ESA
Beavon J G3PPR
Beazley A G3WEF
Beazley S G7BIM
Bebbington R M0BWP
Beccles J M6JLZ
Beck B M0PAL
Beck D M1DIB
Beck D 2E0DTC
Beck D M6BNL
Beck D M6WBK
Beck J M6EDB
Beck N G0DMJ
Beck O M6OJB
Beck P G0VQS
Beck P M6DPO
Beck R G6NOW
Beck R M3YLB
Beck R M0DES
Beck R M0LYD
Beck R M6RJB
Beck T G8RSX
Beckers B G6EBO
Beckers R GW4VBV
Beckett B 2E0LAH
Beckett B M0SMA
Beckett B M6BEC
Beckett D G4UQW
Beckett E G4CMR
Beckett E M6GQK
Beckett G G7JHE
Beckett H M0DQK
Beckett J MI6PEJ
Beckett J G4FWA
Beckett K M3FWO
Beckett M 2E0MAF
Beckett M 2E0NRB
Beckett M M0MOS
Beckett M M3WCZ
Beckett M M6NRB
Beckett S G1JZL
Beckett T 2E0MTB
Beckham T G8MAF
Beckingham G G7GQC
Beckingham J G7JUP
Beckley D G0KCB
Beckley M G3OTR
Beckley N M1EEY
Beckly D G0PGI
Beckman H G4YNI
Beckwith A M1EEW
Beckwith L G3NFP
Beckwith P M1EBS
Beckwith P M6PEB
Be-Dard V M3VBT
Beddall M M6MYC
Beddard M M1EGX
Beddington G G6LZM
Beddoe K G3YOM
Beddow D G6CBB
Beddow G G0CYO
Beddows J G3NQK
Bedell M G0IDL
Bedell S 2E0GOC
Bedford A M1DQU
Bedford A 2E0YDA
Bedford A M3VTV
Bedford D G6YRV
Bedford D M6CNN
Bedford D G3ZVH
Bedford D M6VTT
Bedford J G0ETI
bedford K 2E0ODL
bedford K M3ODL
Bedford L 2E0LJB
Bedford L M3ODK
Bedford M G0DXK
Bedford M G4AEE
Bedford P G7BSL
Bedford R 2E0IJQ
Bedford R G0CUX
Bedford R M6IJQ
Bednarski D M0HIO
Bedson D G1KVI
Bedwell P G3KCD
Bedwell R G1VBB

| | |
|---|---|
| Bedworth D | M6HIB |
| Bee D | 2E0ILN |
| [illegible] | M0ILN |
| Bee D | M0ILN |
| Bee G | G6UCO |
| Bee L | M6KCF |
| Bee R | G6PXN |
| Bee R | G3SZS |
| Beebe R | GU4YOX |
| Beeby I | G1VFW |
| Beech C | MW1BLE |
| Beech D | 2E0XCB |
| Beech D | G8JMP |
| Beech E | M6EKB |
| Beech G | G0KBN |
| Beech J | M3MKV |
| Beech J | G8SEQ |
| Beech K | M6KDB |
| Beech K | G4WRB |
| Beech N | M3NMB |
| BEECH R | G1HUM |
| Beech R | G4ZIS |
| Beecham C | G0CMU |
| Beecham J | G1LAP |
| Beecham P | G6PZ |
| Beecham R | G4OQV |
| Beecher A | G4MMG |
| Beecher C | G8XXM |
| Beecher T | G1XBE |
| Beechill E | 2E0JCK |
| Beechill E | M0JCK |
| Beeching A | G0WIX |
| Beeching T | G7FHV |
| Beecroft J | M3KOA |
| Beecroft M | M6CVN |
| Beecroft R | G6SDY |
| Beed B | M0PDB |
| Beedan D | GD0PDN |
| Beedham S | 2E0NSQ |
| Beedham S | M3NSQ |
| Beedie C | MM6CWB |
| Beedie J | MM6JNB |
| Beedle L | GW0VMS |
| Beedle T | GW0TOM |
| Beehlar J | G3ZCT |
| Beeney J | M6BNY |
| Beeney J | G7VAE |
| Beeney M | G7VAD |
| Beer A | G0GPO |
| Beer D | M3HWX |
| Beer H | M6XHF |
| Beer M | G3OGZ |
| Beer M | GW1DPL |
| Beer N | G0NIN |
| Beer R | G0FAE |
| Beer R | G1UTF |
| Beer S | 2W0RFT |
| Beer S | MW3IQY |
| Beer S | MW0CWF |
| Beer S | MW1COY |
| Beer S | MW6VNP |
| Beers A | M3AKR |
| Beesley C | G4OUG |
| Beesley F | G8CZE |
| Beesley G | M0GEB |
| Beesley M | G4RUN |
| Beesley P | GW1HNG |
| Beesley P | G4PGG |
| Beesley S | G7PHB |
| Beesley-Reynolds C | G0UFP |
| Beeson M | G7PIJ |
| Beeson P | G1CNN |
| Beeson S | MM6SIM |
| Beestin B | G8STR |
| Beeston A | G8WSQ |
| Beeston C | G4YCG |
| Beeston P | G0OCY |
| Beet D | M0WAM |
| Beetlestone M | G8YQC |
| Beeton D | G1RLD |
| Beeton D | MM6RLD |
| Beever E | 2E0CFV |
| Beever P | G6HNP |
| Beevers M | G8SPD |
| Beevers P | G3JFW |
| Beevers T | M1EWD |
| Beezer J | M3JFB |
| Beezley C | G4FEA |
| Begg C | G4MOF |
| Begg D | GM3YXJ |
| Begg D | GM6GFL |
| Begg P | M6UTY |
| Beggs M | GI4TMB |
| Beggs N | G4SPS |
| Beglin A | G4RCY |
| Behal V | [illegible] |
| Behan C | M1CDT |
| Behan T | G0NON |
| Behan T | M6TBM |
| Behane H | [illegible] |
| Beharrell M | M6HLL |
| Behrooz-kafshdooz M | M3SDQ |
| Beier H | M3HJB |
| Beier P | 2E0PKB |
| Beier P | M0EVP |
| Beier P | M3DYF |
| Beier P | M3PLB |
| Beighton T | M6SAI |
| Beilby W | G1OSG |
| Beir E | G1KOG |
| Beirne M | G0DCO |
| Beith J | 2E0HAW |
| Beith J | G8CON |
| Beith M | GM0OXS |
| Beith N | M3NBB |
| Beith S | G1ZVC |
| Bekara M | M3XSZ |
| Bekenn W | G0WIU |
| Belcher D | G0MCU |
| Belcher [illegible] | [illegible] |
| Belcher J | M1BFX |
| Belcher M | G1SSL |
| Belcher M | M3SSL |
| Belcher R | GW3XIS |
| Belcher R | GW4PCJ |
| Belete M | M1DYH |
| Belfield J | G0NNZ |
| Belgium G | 2E0HCC |
| Belham C | M3NDJ |
| Bell A | 2E0ANQ |
| Bell A | G1AQP |
| Bell A | G1YVZ |
| Bell A | G4IAB |
| Bell A | GW4JJW |
| Bell A | G6AXO |
| Bell A | M0WHY |
| Bell A | M3JBZ |
| Bell A | GM4MPY |
| Bell A | G4MHQ |
| Bell A | 2E0BEE |
| Bell A | G4YWX |
| Bell A | 2I0LOR |
| Bell A | G1BDQ |
| Bell A | MI6BAJ |
| Bell A | M6GUM |
| Bell B | G0NUD |
| Bell B | G0HQT |
| Bell C | G0PXQ |
| Bell C | G8AKC |
| Bell C | G3NIE |
| Bell C | M0RTM |
| Bell C | G1ZSG |
| Bell C | 2E0BPP |
| Bell C | M6DTY |
| Bell C | M3PKH |
| Bell C | G3RWP |
| Bell C | G8PY |
| Bell C | M3XYH |
| Bell D | G4NLL |
| Bell D | G4TYN |
| Bell D | GM6TMH |
| Bell D | MI3BYJ |
| Bell D | M3GHD |
| Bell D | M3UTQ |
| Bell D | G8CZG |
| Bell D | M6FHB |
| Bell D | G6AFS |
| Bell D | MM6NAB |
| Bell F | G1TIH |
| Bell G | G3NPT |
| Bell G | M1GDB |
| Bell G | G4PMY |
| Bell G | G6LLD |
| Bell G | 2E0GBE |
| Bell G | M6GHM |
| Bell H | G3MAZ |
| Bell H | M6HPG |
| Bell I | GI6PLO |
| Bell I | GW8ZIL |
| Bell J | GM0EDR |
| Bell J | G4FOL |
| Bell J | G4LSA |
| Bell J | GI7TMQ |
| Bell J | M0CON |
| Bell J | M0GFN |
| Bell J | 2E0BXD |
| Bell J | M0PSB |
| Bell J | M6JOE |
| Bell J | 2E0GBK |
| Bell J | G0CMM |
| Bell J | GM0JMO |
| Bell J | GM4SLY |
| Bell J | M0HBN |
| Bell J | M6BMB |
| Bell J | G7CND |
| Bell J | M3FJB |
| Bell J | 2E0ASU |
| Bell K | M0KFB |
| Bell K | M3GUG |
| Bell K | 2I0LRN |
| Bell K | M0BCK |
| Bell K | MI6LRN |
| Bell K | M3YLV |
| Bell L | 2E0LBI |
| Bell L | M6LFB |
| Bell L | M6LBL |
| Bell M | G0ECM |
| Bell M | G1WXK |
| Bell M | [illegible] |
| Bell M | M3HVH |
| Bell M | G4CXT |
| Bell M | G4PPE |
| Bell M | G6UGW |
| Bell M | M6MBX |
| Bell M | M6MVT |
| Bell M | [illegible] |
| Bell M | M0TEB |
| Bell M | M6DTF |
| Bell M | M6MII |
| Bell M | [illegible] |
| Bell M | G8TCH |
| Bell N | GI4OHW |
| Bell N | G4WLJ |
| Bell P | G0DWR |
| Bell P | G0GWI |
| Bell P | G4KVR |
| Bell P | M3IXM |
| Bell P | 2E0HCT |
| Bell P | MI0CIB |
| Bell P | 2I0PBZ |
| Bell P | MI6PBZ |
| Bell R | G0WYY |
| Bell R | G1ILO |
| Bell R | G3RTB |
| Bell R | [illegible] |
| Bell R | [illegible] |
| Bell R | GM8HEG |
| Bell R | M0DRO |
| Bell R | G4VNA |
| Bell R | M6EFY |
| Bell R | G1MWS |
| Bell R | G7IAS |
| Bell R | M0GMG |
| Bell S | G1EWH |
| Bell S | G7JTK |
| Bell S | M1DFC |
| Bell S | M3LEB |
| Bell S | M0SVB |
| Bell T | G0LEF |
| Bell T | MM6TGB |
| Bell T | G1FRD |
| Bell T | M6APS |
| Bell W | G0JAL |
| Bell W | G1ZTB |
| Bell W | G4WMB |
| Bellaby M | G7BGY |
| Bellamy A | M0AQP |
| Bellamy B | G7IIO |
| Bellamy F | G8MOF |
| Bellamy J | G3TRD |
| Bellamy M | G1TPC |
| Bellamy M | M3BFB |
| bellamy N | M0NIC |
| Bellamy R | G0NRB |
| Bellamy R | G4JZV |
| Bellamy R | G0BIQ |
| Bellamy W | M6WRB |
| Bellas M | G0MDV |
| Bellas M | G1FNP |
| Bellenot R | G0TEL |
| Bellerby R | G3ZYE |
| Bellfield A | G4GLN |
| Bellhouse I | 2E0ICB |
| Belling J | G0RHO |
| Bellinger D | G7DRR |
| Bellingham D | G8JBT |
| Bellis C | M3WOC |
| Bellis G | G7PCX |
| Bellis G | M3PCX |
| Bellis J | 2E0OSG |
| Bellis J | GD0TFO |
| Bellis S | 2M0YIC |
| Bellis S | MM6XSB |
| Bellis W | 2E0HPZ |
| Bell-Stephens D | M6MKD |
| Belshaw M | G6PJD |
| Belshaw T | G3UTS |
| Belshaw W | MI0CZF |
| Belson G | M0HLO |
| Belt G | G0SCV |
| Belt G | G4ZPO |
| Belt G | G7KXT |
| Belton D | 2E0DWB |
| Belton D | M3ZNR |
| Belton T | M0DTB |
| Beltrami P | 2E0ZGX |
| Beltrami P | M3ZGX |
| Beltrami X | M3ZHC |
| Benbow J | 2E0NWB |
| Benbow J | M3TWV |
| Benbow J | M6NWB |
| Benbow R | G4YDI |
| Bence G | G4RXF |
| Bence J | 2M0JSB |
| Bence J | MM3JSB |
| Bence S | MM3JLS |
| Bendall A | G0EXA |
| Bendall D | G1RAX |
| Bendelow D | M3ZSY |
| Bendelow T | M3ZSV |
| Bender M | G4WQL |
| Bendermacher P | GM0PEX |
| Bendoris D | MM3YUY |
| Bendrey D | G7BYN |
| benes N | M3PHQ |
| Benfield A | G0NZE |
| Benfold K | G1NQN |
| Benford G | 2E0IGM |
| Benford G | M6IGM |
| Bengey P | 2E0PPD |
| Benham A | G8FSL |
| Benham D | G3YJP |
| Beniston M | [illegible] |
| Benitez R | M0DHP |
| Benjamin A | M1AEJ |
| Benjamin H | G0MND |
| Benjamin T | G7KRE |
| Benko R | MI0HHU |
| Benko R | MI6BZC |
| Benn A | G1WWH |
| Benn D | G4GOP |
| Bennellick T | 2E0HNN |
| Bennellick V | G3WPN |
| Bennet C | M1CAH |
| Bennet J | M0IIE |
| Bennett A | G0AHB |
| Bennett A | G1JHX |
| Bennett A | G4KNX |
| Bennett A | G4VVM |
| Bennett A | G6YCG |
| Bennett A | G6YTW |
| Bennett A | G7WCB |
| Bennett A | M0CFB |
| Bennett A | M1LEO |
| Bennett A | M3AGB |
| Bennett A | G0OPI |
| Bennett A | 2E0BVK |
| Bennett A | G0HSA |
| Bennett A | G6ORS |
| Bennett C | G4VXN |
| Bennett C | G6WXN |
| Bennett C | G7JDB |
| Bennett C | M0RBI |
| Bennett C | [illegible] |
| Bennett D | 2E0FME |
| Bennett D | G0VPS |
| Bennett D | G0WQQ |
| Bennett D | G4GLH |
| Bennett D | M0SKA |
| Bennett D | M3FME |
| Bennett D | 2E0DJY |
| Bennett E | GI0OHG |
| Bennett E | G3ZJO |
| Bennett E | G1HFT |
| Bennett E | G3THT |
| Bennett G | G0SHU |
| Bennett G | G1FRD |
| Bennett G | G6WNB |
| Bennett G | G3CYL |
| Bennett H | M6BUA |
| Bennett I | G6TVJ |
| Bennett I | M3XGB |
| Bennett I | G6HNJ |
| Bennett I | M0IGB |
| Bennett I | G4PYG |
| Bennett J | 2E0DTN |
| Bennett J | G3KLC |
| Bennett J | G3PVG |
| Bennett J | M3EJX |
| Bennett J | M3LTW |
| Bennett J | M3CDY |
| Bennett K | G4WDZ |
| Bennett K | G8AVV |
| Bennett K | G0DUK |
| Bennett K | M1CVK |
| Bennett L | G4HLN |
| Bennett M | G3UKL |
| Bennett M | G4HUO |
| Bennett M | G4XPU |
| Bennett M | G6XIR |
| Bennett M | G7PSV |
| Bennett M | G8NLK |
| Bennett M | MW6CIY |
| Bennett N | M0CAA |
| Bennett N | G4IGG |
| Bennett N | M0GTQ |
| Bennett N | 2E0FGQ |
| Bennett N | M6NJB |
| Bennett N | 2E0NIB |
| Bennett N | M6NIB |
| Bennett P | 2W0ACD |
| Bennett P | 2E0BHA |
| Bennett P | GW0KTE |
| Bennett P | G1YQI |
| Bennett P | G3VDU |
| Bennett P | G4NMY |
| Bennett P | G7RHT |
| Bennett P | M3PDD |
| Bennett P | M3SHZ |
| Bennett P | G6LLF |
| Bennett P | G7PEO |
| Bennett P | G8DDN |
| Bennett R | G4DIY |
| Bennett R | G6GLT |
| Bennett R | M0EAZ |
| Bennett R | M0RPB |
| Bennett R | G4GSS |
| Bennett R | GW6ZGY |
| Bennett R | G7NLZ |
| Bennett R | G2ARV |
| Bennett R | G4KZQ |
| Bennett R | G0FMJ |
| Bennett S | 2E0SWB |
| Bennett S | G0XAI |
| Bennett S | G3TMX |
| Bennett S | M3SDB |
| Bennett S | M6SBE |
| Bennett S | M6CBP |
| Bennett T | G1URZ |
| Bennett T | G7MST |
| Bennett W | G1VSD |
| Bennett W | G6RBO |
| Bennett W | M1DWW |
| Bennett W | G4WWB |
| Bennett-Blacklock M | 2E0ESC |
| Bennett-Blacklock M | M6MBB |
| Bennetts M | 2E0MEI |
| Bennetts M | M6DZB |
| Bennewitz F | G6RPH |
| Bennie S | GM4PTQ |
| Bennington S | G1KSI |
| Bennion R | G7UQL |
| Bennison G | [illegible] |
| Bennison W | 2E0WDC |
| Bennison W | M0WBS |
| Bennison W | M6WDS |
| Benns A | M0AIS |
| Benoy J | G8PSC |
| Bensley A | G3PTZ |
| Benson A | [illegible] |
| Benson A | G3XDM |
| Benson A | G0MZZ |
| Benson B | G8VNF |
| Benson C | G3MUX |
| Benson F | GM8EKF |
| Benson G | 2E0LGB |
| Benson G | M0LGB |
| Benson J | MM0CWB |
| Benson J | GI0NYI |
| Benson J | M3JBB |
| Benson Jnr B | M3XRQ |
| Benson M | 2E0XOJ |
| Benson M | M3XOJ |
| Benson P | G0PVF |
| Benson P | G8CSQ |
| Benson P | G0SPA |
| Benson S | 2E0BJG |
| Benson T | 2E0XTV |
| Benson T | M3XTV |
| [illegible] | [illegible] |
| Benstead F | G4YZD |
| Benstead S | G3ZAE |
| Benstock A | G1HHT |
| Bent A | M3LUZ |
| Bent A | G7IER |
| Bent D | G7FCC |
| Bent J | G0FSM |
| Bent P | G6ZLD |
| Bentham B | M0BMB |
| Bentham B | M6OBO |
| Bentham B | M6XXB |
| Bentham J | M3SGZ |
| Bentham R | G4SHC |
| Benting Y | GM7DMN |
| Bentley A | M6FUD |
| Bentley B | G1ELQ |
| Bentley B | G4GQS |
| Bentley D | G6LKZ |
| Bentley D | G4RVK |
| Bentley D | M6DBG |
| Bentley D | G8MMG |
| Bentley D | MW6EPX |
| Bentley G | G3XHB |
| Bentley J | 2E0DIH |
| Bentley J | G4JIX |
| Bentley L | M3LYC |
| Bentley M | G0CBW |
| Bentley M | G0MVR |
| Bentley M | G6AKN |
| Bentley M | 2E0DYJ |
| Bentley M | M6EZC |
| Bentley P | G3VUD |
| Bentley P | G4BIM |
| Bentley P | G6BQM |
| Bentley P | M6PHK |
| Benton D | G7TLC |
| Benton D | G6JYR |
| Benton H | M3HCB |
| Benton J | G8UXW |
| Benton K | GU0NHD |
| Benton N | G4KBS |
| Benton R | G4WKW |
| Benton W | G1OIO |
| Benwell D | M3YSV |
| Benwell G | G4BHP |
| Benyon R | G4KSK |
| Benzie E | G1PCN |
| Beresford A | M6EOS |
| Beresford C | 2E0CJB |
| Beresford C | G8DXT |
| Beresford C | M0TVA |
| Beresford C | M3ZAW |
| Beresford D | M6DEI |
| Beresford I | 2E0OEZ |
| Beresford I | M3SPA |
| Beresford J | G6PCX |
| Beresford P | M3AZH |
| Beresford P | M3KXE |
| Berg J | G4LON |
| BERGERET G | M0TUN |
| Bergin P | M0AAC |
| Bergman A | G4NBN |
| Bergstrom-Allen N | G0AWH |
| Berisford T | M6TKR |
| Berkeley A | G1COX |
| Berkeley R | G0WUO |
| Berkerey A | G1HGT |
| Berks S | G8ZGM |
| Berlyn M | M6TTW |
| Bernard A | MM3NGJ |
| Bernard A | MM0OVV |
| Bernard M | G6JRL |
| Bernard M | G8GJU |
| Bernard M | G4AKQ |
| Bernasinski M | M6COI |
| Berrall N | 2W0MSL |
| Berrall N | MW6NAG |
| Berridge J | G0OXX |
| Berridge R | 2E0AAK |
| Berrie N | M0NQB |
| Berriman A | G8JAB |
| Berrisford J | 2E0RLG |
| Berrisford J | M6RLG |
| Berrisford M | 2E0MFI |
| Berrisford M | M6SKQ |
| Berrisford R | G0UQI |
| Berrisford T | G0UPV |
| Berrow A | G4MVB |
| Berrow S | M6ZBE |
| Borrow S | M6SVV |
| Berry A | 2E0JBJ |
| Berry A | G4VRM |
| Berry A | G6SLG |
| Berry A | G7VPN |
| Berry A | M3BJB |
| Berry A | 2M0OXX |
| Berry A | MM0OXX |
| Berry A | MM6OXX |
| Berry B | G6FFR |
| Berry C | 2E0BKK |
| Berry C | G6RIM |
| Berry C | M1ENJ |
| Berry C | M3ENJ |
| Berry C | M1EWT |
| Berry C | M0GXV |
| Berry D | G0FQO |
| Berry D | M3KQT |
| Berry D | M6PIC |
| Berry D | G4DFB |
| Berry G | 2E0ZTD |
| Berry G | M3ZTD |
| Berry I | M3IMB |
| Berry I | G1VIR |
| Berry J | G7LCK |
| Berry J | [illegible] |
| Berry J | M1ENK |
| Berry J | [illegible] |
| Berry J | M6GCE |
| Berry J | M6JBB |
| Berry L | M3JQJ |
| Berry M | G7NTS |
| Berry M | G8BHX |
| Berry M | M6CCZ |
| Berry M | G1LWX |
| Berry N | M6FNB |
| Berry N | G0EVG |
| Berry P | G4ZJC |
| Berry P | G6SOY |
| Berry R | GM6JJN |
| Berry R | G6NBK |
| Berry S | G4IWR |
| Berry S | G4LRT |
| Berry S | M3BER |
| Berry S | M3SLB |
| Berry S | G0KIK |
| Berry S | M6SLB |
| Berry T | M3TDB |
| Berry V | G0EVE |
| Berry W | G1RHE |
| Bertola G | 2E0GSB |
| Bertola G | M3VDF |
| Bertoneri A | M6DZN |
| Bertos G | GW4YVN |
| Bertram E | MM3YPH |
| Bertram J | G7HOV |
| Bertram J | GM0GMN |
| Bertram P | GJ8PVL |
| Bertram R | 2M0KLL |
| Bertram R | MM0LBF |
| Bertram R | MM3LWJ |
| Berwick P | G4BOP |
| Berwick P | G6CKZ |
| Berwick R | M6UDX |
| Besley P | G8WUS |
| Bessant G | G8WMK |
| Bessell R | GD0WHV |
| Bessell-Baldwin Y | M1EAJ |
| Bessent S | G1JJA |
| Best A | M6WUB |
| Best C | M1HZR |
| Best C | G0ATZ |
| Best D | G6TRN |
| Best D | M1EBD |
| Best D | 2I0DTE |
| Best D | G3SQO |
| Best D | MI0OBC |
| Best D | M3VWO |
| Best D | MI6DTE |
| Best G | MI0ASV |
| Best J | M1EBC |
| Best J | M6JOU |
| Best J | 2E0YQT |
| Best J | M3YQT |
| Best K | M1KSB |
| Best L | G3THM |
| Best L | M3DKO |
| Best M | G0NAU |
| Best M | G0IKR |
| Best P | MD0PMN |
| Best P | G8CQH |
| Best R | GI3VAF |
| Best R | M6RSI |
| Best S | 2E0XBE |
| Best S | M3XBE |
| Best V | G3HGD |
| Best W | G0SCY |
| Beston W | M6BWB |
| Bestwick B | 2E0FGY |
| Bestwick B | M6FGY |
| Beswarick E | G6BTP |
| Beswick B | G1OHD |
| Beswick K | M6VCU |
| Beswick T | M1FGM |
| Bethell J | 2E0JBI |
| Bethell J | M6JBG |
| Bethell L | M6BRB |
| Bethell N | M0HVS |
| Bethell P | G0RRX |
| Bethell P | G6NBE |
| Bethell S | M3HBP |
| Bethune N | GM4IUS |
| [illegible] | [illegible] |
| Bette- Bennett J | GJ7FGS |
| Betteridge J | 2E0JNH |
| Betteridge J | M6DDY |
| Bettie D | G6DYR |
| Bettles E | G3KXE |
| Bettles J | M0VIU |
| Bettley A | [illegible] |
| Bettley G | G8KWD |
| Bettridge N | M6NCB |
| Betts A | G4PPH |
| Betts A | M1AIA |
| Betts A | G0VLC |
| Betts J | G4TTM |
| Betts I | M0CBT |
| Betts L | 2E0UNY |
| Betts L | M6UNY |
| Betts M | G4EAK |
| Betts M | G4FFW |
| Betts N | 2E0NVB |
| Betts P | G0PAB |
| Betts R | G0TRB |
| Betts T | G6XMB |
| Bettyes G | G7NDQ |
| Bevan A | G3GZZ |
| Bevan A | G4FAV |
| Bevan A | M3ZCA |
| Bevan D | G4DMR |
| Bevan D | G4WPO |
| Bevan D | G4XUV |
| Bevan I | G4JGQ |
| Bevan J | [illegible] |
| Bevan J | M3JLB |
| Bevan J | M6VAZ |
| Bevan J | G4OJL |
| Bevan K | G3XKB |
| Bevan K | MW6ARX |
| Bevan M | G0BRL |
| Bevan N | G1HAX |
| Bevan P | MW1DCI |
| Bevan R | G3XPA |
| Bevan T | G4TWW |
| Bevan T | M0ACV |
| Beveridge J | G8TLT |
| Beverstock K | G3YZZ |
| Bevington A | G7UNB |
| Bevington A | G8IYH |
| Bevington B | G0KJG |
| Bevington P | G4ZUI |
| Bevins A | G7CVZ |
| Bevins A | M0DGA |
| Bewick B | M6VFN |
| Bewley J | G1SQI |
| Bewley K | G0JGV |
| Bewley L | M3IOT |
| Bewley M | G7LLD |
| Bexley A | GW4JKK |
| Bexon I | G4OCS |
| Bexon N | 2E0GAC |
| Bexon T | M6TJB |
| Beynon C | GW3WSU |
| Beynon D | G7VGB |
| Beynon J | G4ILL |
| Beynon M | MW0BEY |
| Beynon M | GW4GSH |
| Beynon-Fisher G | M6GCB |
| Beyoglu M | 2E0ZUT |
| Beyoglu M | M6ZUT |
| Bhakoo A | 2E0CRC |
| Bhakoo A | M0KBA |
| Bhakoo A | M6CCR |
| Bharrich S | M3RHK |
| Bhart J | M3LQC |
| Bhatia M | M3PQI |
| Bhogal J | G7LOE |
| Biadon M | 2E0OKS |
| Biadon M | M6BHM |
| Bianchini K | 2E0ETV |
| Bianchini K | M0ZRX |
| Bianchini K | M6FMZ |
| Bias E | 2E0EBG |
| Bias E | M0MEB |
| Bias E | M6EBB |
| Bibb C | G0EAN |
| Bibb M | G7PWK |
| Bibby F | G4PDD |
| Bibby J | G6JBY |
| Bibby J | 2E0JBY |
| Bibby J | M6SRI |
| Bibby J | G6EJT |
| Bibby M | G3NJY |
| Bibby R | G1PIX |
| Bibby R | M0PIX |
| Bichard A | GU4XGB |
| Bichard G | M0GXB |
| Bichard L | 2U0LRB |
| Bick J | 2E0CYP |
| Bicker R | GI0WJI |
| Bickers J | 2E0JMY |
| Bickers J | M0JIB |
| Bickersteth P | G8RBS |
| Bickerton M | G7BXL |
| Bickham W | G3TJH |
| Bickle D | MM1FHZ |
| Bickley P | M3WOL |
| Bickley R | G0EBK |
| Bickley R | G4MZQ |
| Bicknell C | G6FDX |
| Bicknell E | M3YAO |
| Bicknell J | M6FLR |
| Bicknell-Thompson R | G1OGV |
| Bicknell-Thompson R | M3OGV |
| Biddiscombe M | G3YKZ |
| Biddiscombe S | G8CNF |
| Biddle P | G4LUX |
| Biddle R | G0ROY |
| Biddle T | G6LPS |
| Biddlecombe K | G1YVI |
| Biddlecombe R | G4THV |
| Biddles C | G6EJU |
| Biddles M | [illegible] |
| Biddulph G | G3XHP |
| Biddulph M | G7CGB |
| Bide S | 2E0EAI |
| Bide S | M6SMB |
| Bidwell A | M3HPZ |
| Bidwell J | MW0OUC |
| Bidwell J | MW3OUC |
| Bidwell J | MW0SGX |
| Bidwell J | MW3SGX |
| Bidwell M | 2E0KJI |
| Bidwell M | M0KJI |
| Bidwell P | G6DAU |
| Bieber D | G4AIR |
| Bieber S | G7RNB |
| Bielawski E | G6JTX |
| Bielen J | M0VWW |
| Bienko P | M0MNV |
| Biernacki W | M0ULC |
| Bierney T | G8IWF |
| Bierton K | G6KHD |
| Bietz C | 2E0YOK |
| Bietz C | M6OKY |
| Biggadike P | G8JAN |
| Bigger J | G7HDW |
| Biggin A | G7DGE |
| Biggin J | [illegible] |
| [illegible] | [illegible] |
| Biggin P | M0PBN |
| Biggs J | G0FGC |
| Biggs J | G7BRP |
| Biggs K | G1PGD |
| Biggs K | G8TVM |
| Biggs K | M1CAX |
| Biggs P | M6PHA |
| Biggs T | G6GJN |
| Biginton D | G0WYG |
| Biginton D | G1WYG |
| Bignell M | G1MZM |
| Bignell M | G1ZLD |
| Bigsby P | 2E0PBF |
| Bigsby P | M6PAF |
| Bigwood J | G8KNN |
| Bigwood P | G3WYW |
| Bilke F | G3UZD |
| Bilke K | G8VIV |
| Bilkey M | G8MNC |
| Bill D | M0ZLL |
| Billam J | G7KWA |
| Billinge D | M3DBX |
| Billingham A | 2E0EOD |
| Billingham A | M6LJB |
| Billingham H | G8IPF |
| Billingham J | G4AGQ |
| Billingham J | G8AAC |
| Billingham N | G6JJA |
| Billingham N | M0EBI |
| Billingham P | 2E0MIY |
| Billingham P | M6MIY |
| Billingham S | 2E0BJY |
| Billingham S | M0VKY |
| Billingham S | M3NFB |
| Billings A | 2E0PBH |
| Billings A | M3PHZ |
| Billington C | G0TNO |
| Billington G | 2E0PNC |
| Billington G | G3EAE |
| Billington G | M3LJI |
| Billington Z | M1BFI |
| Bills A | G3KZG |
| Billson C | G8HVF |
| Bilmen J | G7RFZ |
| Bilsel T | M6BSL |
| Bilsland R | G7WIG |
| Bilson D | M0IKE |
| Bilson G | 2E0FDF |
| Bilson G | M0BAY |
| Bilson J | M0JOY |
| Bilson L | M0BMY |
| Biltcliffe M | G8EKN |
| Biltcliffe R | G2BSJ |
| Bilton A | G8STM |
| Bilton D | M6WET |
| Bilton F | G6FFQ |
| Bilverstone M | 2E0YXO |
| Binding I | G4RVG |
| Bindley K | 2E0BQC |
| Bindley K | M3XEA |
| Bindon D | G3RSU |
| Bindon G | G1VVE |
| Biner P | G4FSE |
| Bines D | M0RHG |
| Binfield J | M3UVJ |
| Bingham A | G0JBH |
| Bingham C | 2E0FTE |
| Bingham D | 2E0XRM |
| Bingham D | M0XRM |
| Bingham J | GI4TAJ |
| Bingham J | 2I0EJT |
| Bingham J | MI6AJA |
| Bingham J | 2E0TAG |
| Bingham J | M3TQU |
| Bingham M | G8FDR |
| Bingham M | G1PGX |
| Bingham P | 2I0PBM |
| Bingham P | MI3GJI |
| Bingham P | G0OUN |
| Bingham P | MM3WUI |
| Bingham W | G3NYB |
| Bingham W | G4WUS |
| Bingley A | G1SIU |
| Binks M | G0LJF |
| Binnall D | M6MAD |
| Binnell C | [illegible] |
| Binnell J | M3DVU |
| Binnington F | G0SCM |
| Binnion S | MW0FIY |
| Binns D | G0DOW |
| Binns E | G4OSO |
| Binns G | G0FOU |
| Binns J | [illegible] |
| Binns K | G0NJG |
| Binns L | G3RGN |
| Binns M | G0TVU |
| Binns M | G4USP |
| Binns S | M0STF |
| Binns T | G1SRA |
| Binny I | [illegible] |
| Binswanger P | M0DXF |
| Biorka Z | MM0GQF |
| Biram D | G6TVA |
| Birbeck S | M0CHJ |
| Birch A | G4NXG |
| Birch A | 2E0LFI |
| Birch A | 2E0YGB |
| Birch A | G1VJY |
| Birch A | M0YGB |
| Birch A | M3YGB |

**IMPORTANT NOTE**

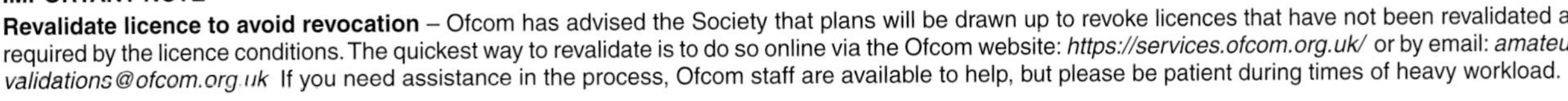
**Revalidate licence to avoid revocation** – Ofcom has advised the Society that plans will be drawn up to revoke licences that have not been revalidated as required by the licence conditions. The quickest way to revalidate is to do so online via the Ofcom website: *https://services.ofcom.org.uk/* or by email: *amateur.validations@ofcom.org.uk* If you need assistance in the process, Ofcom staff are available to help, but please be patient during times of heavy workload.

Birch B G0CGT
Birch B G7CMI
Birch D G0ORM
Birch D GW4HDZ
Birch D M0DLP
Birch D G0GKH
Birch D G0UDB
Birch E G0WTO
Birch E G8HLQ
BIRCH G G0GWY
Birch G M6WGB
Birch G G6KTR
Birch J G0PZW
Birch J G4GGZ
Birch J G7FUW
Birch J G7GLH
Birch J M3YJB
Birch J 2E0KRK
Birch J M6KRK
Birch K M1AMZ
Birch M G3KMO
Birch R 2E0ARJ
Birch R G5HI
Birch S G0AXS
Birch S M3PSB
Birchall D G4MAU
Birchall P 2D0GBM
Birchall P MD0VMD
Birchall P MD3VMD
Birchall R G3KMV
Birchall S M1EJS
Birchall T G4ERQ
Birchenough A GD0KEO
Bircumshaw J M3LDI
Bird A G6HOC
Bird B G7WDD
Bird B G8DZN
Bird D G4DHY
Bird D G4JIK
Bird D G4OOY
Bird D G8XOC
Bird D G6EJD
Bird G G3GDB
Bird G G4FRI
Bird J G0PIS
Bird J G0UEC
Bird J G1ALA
Bird J G3GIH
Bird J G4CEK
Bird J G4CPN
Bird K G3ZAW
Bird K G4JED
Bird K M6VRR
Bird K 2I1AXH
Bird M G4KFB
Bird M M1DRM
Bird N G8URZ
Bird N G6LQI
Bird P G1EHM
Bird P M1EGG
Bird P M1EBI
Bird R 2E0ZAP
Bird R G7URS
Bird R G8DUF
Bird R G8ZVS
Bird R M1CBI
Bird R M3RJB
Bird R M3ROQ
Bird R M6CEN
Bird R G4ALY
Bird R GU4XIT
Bird S 2E0BKV
Bird S GI6EBX
Bird S M1ERH
Bird S M3OZI
Bird S G0UFE
Bird S M6ZZB
Bird W G4JAV
Bird W GI4XIR
Birdsall L M3SVO
Birk G G0DKX
Birkbeck D G6HHK
Birkbeck J G3IGV
Birkby G G1TTB
Birkby G M0SSD
Birkenshaw D M1EZR
Birkenshaw S G0TVL
Birkett A M0HFA
Birkett I GM0FTX
Birkett J G8OPP
Birkett M G3OBZ
Birkett R G0UQC
Birkhead A 2E0ZCM
Birkhead A M3BCM
Birkhead G G4KOQ
Birkill S G8AKQ
Birkin C M0NOI
Birkitt R M3SFK
Birkmyre H G1HAB
Birkmyre J G1DNA
Birks D G4SFD
Birks E G3XXO
Birney C MI3JXG
Birnie N 2E0NBZ
Birnie N M3NBZ
Birse I 2M0GMB
Birse J G4ZVD
Birt A G3NR
Birt D G7FEP
Birt E G7SAI
Birt N G7UOQ
Birt S G3ZNG
Birtchnell C G1DNO
Birtwhistle S G7TYO
Birtwistle A G4YYD
Birtwistle E G7SBJ
Birtwistle F 2E0RGB
Birtwistle J GM3UQU
Bisbey D 2E0UVP
Bisbey D M0HPB
Bisbey D M6BIS
Bischoff J 2E0JXB
Bischtschuk A G6HBF
Biscombe P G0IXV
Bishop B G1LGY
Bishop D G8DHA
Bishop D 2E0DCJ
Bishop D 2E0VNK
Bishop D M0VNK
Bishop D M3VNK
Bishop D M6CQE
Bishop E G0KZA
Bishop E M3PNI
Bishop G G0WUG
Bishop G G6MAA
Bishop I G7GFP
Bishop I G8GYL
Bishop J G1TFM
Bishop J G7VPQ
Bishop J G8OYY
Bishop K G1ATL
Bishop K G3LQB
Bishop L M6CVF
Bishop M G0GQT
Bishop M G8YZF
Bishop M M0EJW
Bishop N G4LQE
Bishop N M6XWD
Bishop P G1TQY
Bishop P MW3HSZ
Bishop P M3LGJ
Bishop P M6API
Bishop R GD0SQL
Bishop R G3GGG
Bishop R G4PNI
Bishop R G7GCB
Bishop R MW6RBZ
Bishop R M6GZD
Bishop S G6XCK
Bishop S G1XUU
Bishop T M3PTB
Bishop T M6TKG
Bishton S M6BUW
Bisiker N 2E0NFB
Bisiker N M0NFB
Bisiker N M6ATF
Bisley B G3OFI
Biss R G7SUS
Biss T M6BGI
Bisseker R G3SRQ
Bisset D M0KBT
Bisset R MM6FXW
Bisson D 2E0DHB
Bisson J G0MHF
Bisson R MJ0RGR
Biyikli D M0TUR
Blabey K GM0FZM
Black A G6YTV
Black C GD3SKH
Black C G4LOJ
Black C GI4MBQ
Black D 2E0AQI
Black D M0DAN
Black D M1DAN
Black E M6CIG
Black G G3XPQ
Black G MM0CBL
Black G 2I0GKB
Black G MI6GKB
Black J G0UKA
Black J MI3JTB
Black J 2M0GRK
Black J G0FQV
Black J G6TVR
Black K M6KJB
Black M GM0PIV
Black M G4HJY
Black M GW8GOC
Black M M1BSV
Black M GI4OYG
Black N G4NOC
Black N G4RYS
Black P 2D0DRM
Black P M0CNE
Black S 2E0DOF
Black S M3SVB
Black S M6NTT
Black S G4PSS
Black T GI4IKF
Black W GI0LXN
Blackburn A M0BLZ
Blackburn A G0REP
Blackburn A M6EGN
Blackburn C 2E0GOM
Blackburn D G1JBE
Blackburn E M0AQH
Blackburn G G7IWK
Blackburn H 2D0HEB
Blackburn H MD0HEB
Blackburn H G7NDO
Blackburn J G4ACI
Blackburn J G4EAB
Blackburn J M3PKQ
Blackburn J G7JDB
Blackburn K G0LUU
Blackburn K G3VGK
Blackburn K G6HNQ
Blackburn M GD7IEH
Blackburn M G7UKR
Blackburn N G8CHA
Blackburn O 2E0OPB
Blackburn O M3OPB
Blackburn R G0DAG
Blackburn S M6LBE
Blackburn S M6WSU
Blackburn T G6AKX
Blackburne C M0CVC
Blackburne P M6PAI
Blacker R G4GBE
Blacker R M0PES
Blackett J G4IOG
Blackett K G4IGU
Blackett P G7BDK
Blackett R M3WPN
Blackham B 2E0IJO
Blackham B M3BBL
Blackham K G7LOZ
Blackham S 2E0FEM
Blackham S M6FEM
Blackhorse-Hull D M6DAX
Blackie D 2E0WQK
Blackie D M0WQK
Blackie J GM8TCG
Blackie P 2E0JQK
Blackie P M3NCH
Blacklaw P GM7DPI
blacklock M G7VZM
Blacklock O M6VCC
Blackman C 2E0EOK
Blackman D G6TNR
Blackman I G4OCQ
Blackman J G1AQV
Blackman K G8ERV
Blackman M G1YJH
Blackman P M1FJP
Blackman R M6RSB
Blackman-Wells D G1WXC
Blackmoor C G0HAW
Blackmoor G G0BLT
Blackmore D M3GZD
Blackmore D G6LPV
Blackmore I M3UJM
Blackmore M G0VBK
Blackmore M M6BJL
Blackmore N G8ARH
Blackmore P M6BJP
blackmore R MW3VNZ
Blackmore T G1XXV
Blackmun C M3ZCB
Blackmur C M1BIK
Blacksell G G6CLA
Blackshaw J G8NPR
Blackstock I MM6FPY
Blackstone B M0LWM
Blackwell A G1WJO
Blackwell A M6FKN
Blackwell B G3YVW
Blackwell C G6BXO
Blackwell D G4PWP
Blackwell D G4SMD
Blackwell D G4UFX
Blackwell D G4XOH
Blackwell J M6NMS
Blackwell J G7VWN
Blackwell M G3VUH
Blackwell R GM4PMK
Blackwell R G8ZPO
Blackwell R 2E0FUD
Blackwell R M0NIL
Blackwell R M6DQC
Blackwell S M3UFX
Blackwell-Chambers K M1CQP
Blackwood G G0JAQ
Bladen J G4FZA
Blades G G6OTL
Blades J G0AAU
Blades J G4ZFX
Blades J MM3DHG
Blades S M6RIJ
Bladon A G3GZX
Blagburn D G6DAD
Blagden P 2E0GXK
Blagden P M6GXK
Blagg K G1UGG
Blagg M 2E0BSN
Blagg M M3ZEY
Blaikie S 2E0SRB
Blaikie S M0HBJ
Blain B G4WJQ
Blain C M0XEY
Blain F G3JLN
Blain J G7LSF
Blain J G4SKU
Blain K G7TEP
Blain R G3NTI
Blair A G6KXW
Blair A G8GYN
Blair D G8VU
Blair J MM3DKA
Blair M G8WSS
Blair M M0DYO
Blair N G7ASZ
Blair P G3LTF
Blair R M6IIG
Blair T G3YTG
Blake A G1BCB
Blake A 2E0NCF
Blake A M0SNT
Blake A M6FBK
Blake B G0XAP
Blake B G1XOT
Blake C M0AIJ
Blake C MI1EQI
Blake D 2E0BLN
Blake D G2FT
Blake D M0WBD
Blake D M3MEO
Blake D 2E0EAV
Blake D M0NMI
Blake D M6EIW
Blake E M3XOT
Blake F G8AAF
Blake G GJ0NTD
Blake G G4WMF
Blake G G8GNZ
Blake H G3OFW
Blake K 2I0BIQ
Blake M G7ART
Blake M M3AUK
Blake P G8ZFX
Blake P M6BLK
Blake R M3GYH
Blake S G8MAG
Blake S G8RMI
Blake W G0BHT
Blakeley D G3KZN
Blakeley G G4CBM
Blakeley M G4YKX
Blakely M M3DZC
Blakeman T G6QN
Blakemore A G8FFN
Blakemore D G8LGT
Blakemore G G7AMD
Blakemore P G7ACR
Blakemore R G1IKF
Blakeney A M1BXC
Blakeney P G8BLB
Blakeston A G7MTG
Blakeway R G1PXM
Blakey J G3LRI
Blakley S MI0AAW
Blamey J 2E0NNQ
Blamey J M3NNQ
Blamey K G4KKB
Blamire A 2E0GTZ
Blamire A M6AXB
Blamires A 2E0MBQ
Blamires A M6MRT
Blampied D G4GJO
Blampied N M6GSY
Blampied P GU7DSB
Blanch K 2E0KJB
Blanch K M3BKV
Blanchard A G8ZSP
Blanchard J G4VVS
Blanchard K 2E0FAN
Blanchard L G7OHW
Blanchard T M6TFB
Blanchard W G3JKV
Blanchard W G3LHB
Blanche J 2E0EDW
Blanche J 2E0EDV
Blanchflower W MI5WJB
Bland A G6SLH
Bland B M6EOB
Bland C G3DAE
Bland D G8KIK
Bland D M1YOW
Bland G M3GTB
Bland J G1WSC
Bland J G0HRO
Bland M G4WPE
Bland N 2E0NGB
Bland N M0JEC
Bland N M3NBL
Bland P G7MNQ
Bland P M3HGP
Bland P G8MBK
Bland S M6DLU
Blandford R G1PCA
Blandford R 2D0RSB
Blandford R M0RBG
Blandford S G7GSF
Blane R M0THN
Blaney R M0RJB
Blaney S M6SBL
Blankley T G8SBJ
Blankley W G8CMK
Blanning R G0NZU
Blanshard K M0KPB
Blasdell R G3ZNW
Blatch K J M6LFQ
Blaxall G G7LIT
BLAXLAND F G4WSH
Blaxland J 2W0JYI
Blaxland J MW6EDQ
Blay J M6UTI
Blay P G6AFL
Blay W G0GSK
Blaylock J 2E0MUW
Blayney R GW4NPC
Bleach R G7VWC
Bleaney J G4UOO
Bleaney R 2E0IHB
Bleaney R M0RBK
Bleaney R M6AKY
Blears R M1RWB
Blease P G6NBP
Bledowski J GM4YWU
Bleek N G4WKT
Bleiker P G4KVX
Blemings R G4YYH
Blenkinsop M M3WUS
Blest T G6OEI
Blew S G8JXK
Blewett M G4VRN
Blewitt C G4JXJ
Blewitt R G7UHY
Blewitt R G7HJG
Blewitt S M1CVH
Blezard C G4RNC
Blezard J 2E0SLK
Blezard J M6BLZ
Blichfeldt J G0OEQ
Blick J MM6KSJ
Bligh F G0MTA
Bligh J 2U0BGE
Bligh J MU0ZVV
Blight B G8AAD
Blight D G0PGL
Blight J G4SOF
Bligh-Wall J 2E0CXN
Bligh-Wall J M6DVA
Bligh-Wall S M6HRE
Blinco T G8KNJ
Blinkhorn S G1XGP
Bliss D G8OMQ
Bliss F G3IFB
Bliss I M1CSU
Bliss M G4AQS
Bliss P M3PKM
Blissett A G4OPD
Blizzard P G0WLG
Block C 2E0DYT
Block C M1CVF
Block J G7PSF
Blockley M G1FYU
Blomeley G G0NWJ
Blomfield M MW6MTG
Bloodworth A G1IQG
Bloodworth J G4VWO
Bloodworth R G4VWP
Bloom S G0BEE
Bloomer B G4KES
Bloomfield B G4OKB
Bloomfield D G0KUC
Bloomfield D G4ATL
Bloomfield G GD4PXY
Bloomfield H M1BYT
Bloomfield K G1FVU
Bloomfield R MM0AOL
Bloomfield S M6FUU
Bloomfield T MM3GYU
Bloor A M3YIL
Bloor G G3UD
Bloor J M1BPK
Bloor P G8WSY
Bloor R G7RSM
Bloor V G4VJB
Bloore M 2E0HRH
Bloore M M0PNC
Bloore M M6JMB
Blore G G8VEE
Blore H G0EMB
Blore R M3SAI
Blore T G0TGB
Bloss M M1ATU
Blott R G7OXV
Blount A M0SIR
Blount C 2E0IET
Blount C G0DJL
Blount C M0IET
Blount C M3IET
Blount H M6JNH
Blount M M3IEV
Blount S M3SGF
Blow J M6WOB
Blow R M6RNC
Blower G G4MQR
Blower J M6JBX
Blower J 2E0YFZ
Blower J M3YFZ
Blower R G4KMK
Blower W G1RJD
Blowers J G4DWU
Bloxam R M1DHO
Bloxam T GW3LJS
Bloxham J G8WKE
Bloxham M G1KXQ
Bloxidge C G8EFU
Bloxidge C M0DBM
Bloxsome D M6DIB
Bloy C M0DQO
Bloyce G G0UHU
Bluck M G6DUT
Bluer E M5ACT
Bluer H G3UUZ
Bluff S 2E0DUJ
Blumson S G0WNB
Blundell D GW0TPL
Blundell E G3RXI
Blundell I M3LBJ
Blundell J G4UQS
Blundell J M0JME
Blundell J M6JLB
Blundell M G6ITW
Blundell P 2E0WFD
Blundell S G3MLQ
Blunden M G3PFH
Blunden P G7NSK
Blunden R MW6VLR
Blundson T G7NTA
Blunt D G4PFU
Blunt G G7JCQ
Blunt P G0UXG
Blunt R G1SGA
Blunt S G8CRB
Bluthner P M0ASO
Blyth A GM4TAL
Blyth D M0GRF
Blyth P G1LOK
Blyth P 2E0CEH
Blythe A MI3IGO
Blythe D G3KCT
BLYTHE D M1DIN
Blythe H M6SQI
Blythe M G4HFO
Blythe N G0SQI
Blythe P M0PSH
Blythe R G4MWH
Blythe S G6KIV
Blythe W G0PPH
Blything J G1LEX
Boag A 2E0PPK
Boag A M3GKG
Boag A M3YII
Boag J 2M0MGY
Boag j M3UZW
Boag K GI4TPY
Boag N G1UZW
Boakes G G8PPQ
Boal C M6HFK
Boalch D G0BEP
Boaler K M6CUU
Boaler P G4YDO
Boam G M6YAD
Boam P G8MZZ
Boan R M6MNQ
Boar O 2E0OBB
Boar O M3OBB
Board C 2E0FWD
Board C M3EYZ
Board M G8GUS
Boardall R G8AJZ
Boardman A G7ROM
Boardman D 2D0CVB
Boardman D M3ZKU
Boardman F G3XUB
Boardman G G4CDI
Boardman J G8RIM
Boardman R 2E0CME
Boardman R M6BBU
Boardman T M1TOM
Boarer C M3RBQ
Boast G M3UGB
Boast J G3WDL
Boast P M6PEG
Boast P G3XBI
Boato A M0THO
Bobby M G0WZZ
Bobby S 2E0JYN
Bobby S M3WZZ
Boberschmidt L M0LEO
Bock O M0TAO
Bocock C G0SXU
Bocutt A M6ABB
Bodaly G G0PYI
Boddington L G4JDC
Boddy H G4MGP
Boddy I 2E0JIA
Boddy I M6IJB
Boddy M M6EJY
Boddy M M6MXB
Boddy R G4MGQ
Bodecott M G1ZDX
Boden B G0HGI
Boden D 2M0BET
Boden D MM3BOJ
Boden D MM0LGR
Boden D M6LUC
Boden J G3LXE
Boden J G4CDZ
Boden P G6MOD
Boden S G4XCK
Bodenham D G6XKV
Bodie I G8END
Bodie J M3HBS
Bodie L M3ZKO
Bodill M G6IBN
Bodily T M0DOD
Bodle J G7WHX
Bodle J M5ABJ
Bodle M G0HCI
Bodley C 2W1EIN
Bodley G M0RZC
Bodman D G1UGV
Bodman D G4UJJ
Bodnar L 2E0TOY
Bodnar L M0XER
Body B G0RIZ
Body D G0HTJ
Body J G6FPC
Boehme L MM0DWF
Boele F M0CZX
Boev V 2E0VLB
Bogg G M6XGB
Bohan H GM0FIQ
Boids J M3XJO
Boittier R G8CUA
Bokor R G1CLT
Bolam S G0OKF
Bolan G M6GTB
Boland C G0RBM
Boland D 2E0FXN
Boland J G0NEO
Boland M 2E0GMO
Boland M M0MAO
Boland M M6GMO
Bold R G4CNW
Bolderson P G1OUX
Bolderston B GU1HYN
Bolderstone T 2E0TMB
Boldireff Strzeminski A M0HWJ
Bolger V M1AOD
Bolingford D G6AAR
Bolla A 2E0BMB
Bolland J G6OEM
Bolsover J G7IHD
Bolt B G0FGE
Bolt C G3NN
Bolt D G1HHC
Bolt E G1HHD
Bolt N MI0RUC
Bolt N MI3NRB
Bolter J G3NNY
Bolton A G1EAB
Bolton A G3BMI
Bolton A M0ZEN
Bolton A G4VSQ
Bolton A M3VBZ
Bolton C MM6LFN
Bolton C 2E0HST
Bolton C M0WFX
Bolton D GW8UQC
Bolton D M0ETQ
Bolton D M6APU
Bolton D G6SSQ
Bolton J G3YYG
Bolton J G4WEL
Bolton J M0ETP
Bolton J M1FHJ
Bolton J G6LZZ
Bolton L 2E0NVK
Bolton L M3NVK
Bolton M G0OVV
Bolton M GM0DBW
Bolton M 2E0XHG
Bolton N G3HMV
Bolton P G3CVK
Bolton P G4CXE
Bolton P M0HDP
Bolton P 2E0ROP
Bolton P M0NOC
Bolton R GI4BBE
Bolton R G4SSJ
Bolton R M6TXZ
Bolton S G0DTW
Bolton S M0KDA
Bolton T G4OUM
Bolwell D G3JCM
Bona R G3SGX
Bonar J 2E0ONV
Bonar J M3ONV
Bonar R G1ONV
Bond A 2E0MEO
Bond A M3UXL
Bond A M0AZZ
Bond B G3ZKE
Bond C 2M1IGQ
Bond C G1ODQ
Bond C G8BLP
Bond C 2E0CPB
Bond C M3YPB
Bond D M0NZR
Bond D M6XIT
Bond J G7IIH
Bond J G8APY
Bond J G8RMP
Bond J M3JCQ
Bond M GM4HYR
Bond M G8XSU
Bond M G1ALL
Bond N G3IHX
Bond N M3ZYT
Bond P M0PBZ
Bond R G1ANK
Bond R 2E0EAF
Bond R M3UHY
Bond S G7CAF
Bond S M1CZL
Bond S G4XZK
Bondar P G6FTE
Bonds P G6GUH
Bondy D G4NRT
Bone D G4MOI
Bone H G3EHQ
Bone J G7OOE
Bone K G8UXY
Bone K G6RTN
Bone N G7UUR
Bone R M0UOO
Bone R M3UOO
Bone S G4PPJ
Bone W G7MWB
Bones D G4RSD
Bones K GI4ERM
Bones K GI8KEP
Bones W G4CFP
Bonfield D G4JXK
Bonfield D G7UDM
Bonham S G7APL
Boniface A G4XSA
Boniface O G4DSC
Boniface S M6SJO
Bonnar B MI1BNO
Bonner A G2BHY
Bonner A M1DPJ
Bonner C G3TGF
Bonner J G0GKP
Bonnett D M1AMJ
Bonnett J G6BNJ
Bonney A G0KTD
Bonney D M6VCB
Bonney J M3UBF
Bonney S G7VGA
Bonnor H G4LYL
Bonsall C G3UWR
Bonser M G7KXN
Bonser W G0SNB
Bonsey P M0BSC
Bonson P G4FUY
Bontoft I G4ELW
Boocock A G4REG
Boocock F GD4IRP
Booer A G3ZYR
Booker P 2E0FMG
Booker P M6FMG
Booker R 2E0RBZ
Booker R M6RAZ
Booker W M5ASK
Bookham D G8JNI
Bookham J 2E0JLB
Bookham J M6OBJ
Bookham M 2E0DFF
Bookham M M6TXT
Bookham N 2E0INV
Bookham N M6IVS
Booley I M0CQD
Boom A M0AWB
Boon B G4TRE
Boon C G8FGZ
Boon I G4TRF
Boon M 2E0CUU
Boon M M6CHI
Boon T M0AFC
Boone J M0YOJ
Boonham A G6DAQ
Boor A G3OS
Boor G G8NWC
Boorman L G3DT
Boorman P G0JBA
Boorman T GW1FBL
Boot A G1OQW
Boot D G1UCC
Boot D M3NGZ
Boot J M3GKX
Boote K G1SQC
Boote S G6BER
Booth A M1AYC
Booth A M0SMG
Booth A M6LYS
Booth A G4YQQ
Booth A MM6FYW
Booth A M6BVX
Booth A 2E0ZXV
Booth C 2E0FWD
Booth C G3HKA
Booth C G8BPS
Booth D G6FWK
Booth D 2E0DBN
Booth D M6DMB
Booth E M3KXS
Booth G G8DZJ
Booth I G7HRP
Booth I 2W0IDT
Booth I MW0IDT
Booth J M1BAC
Booth J G6ZHB
Booth K M6KAV
Booth L G4XDT
Booth L 2E0XKD
Booth L M0XKD
Booth L M3XKD
Booth M G4DCF
Booth M G4RSF
Booth M G8EGM
Booth M G8HKS
Booth M M0HKS
Booth M G3YNO
Booth N G1MOB
Booth N M6XVC
Booth N 2E0UHF
Booth N M0RAR
Booth N M0CVO
Booth N M6NKB
Booth O M6OLL
Booth P M1AYA
Booth P M1CRP
Booth P M1PKB
Booth P M3BVA
Booth P M5ADW
Booth P M0PWB
Booth P G4PAA
Booth P 2E0DLA
Booth P M6DLB
Booth R G0TTL
Booth R G1WQH
Booth R GI4GCN
Booth R G4RSG
Booth R G8BIW
Booth R M1RIC
Booth R 2E0RBU
Booth S G4DFS
Booth T G4YTD
Booth T G8DPH
Booth W G7JQF
Booth-Isherwood N G4SNN
Boothman K M0KAI
Boothman M M0AOZ
Boothroyd J G0MTJ
Boothroyd K G1FYS
Boots A M6BCC
Bootyman T G8TKY
Bootz C M0ELO
Borda A M3ALB
Bore J M5TAM
Bore P M3AGA
Boreham P M3YJU
Borer M G0NYM
Borer P M0PFW
Borg J G0AIS
Borkowski C G4ILP
Borland A G4DBD
Borland B GM1FBM
Borland J GM8ZEJ
Borland W G3EFS
Borley D G4CAX
Borradaile J G8CAU
Borrell N G8LIE
Borrell S 2E0DDU
Borrell S M0XLB
Borrell S M6SEY
Borrett D M6DDV
Borrett G G1NLS
Borrett J M6JGU
Borrett N G8NXB
Borrett P G3XTC
Borrow M G6GKL

Borrow R M1CXY
Borrowdale G G0OFR
Burrows D GM0LLJ
Borszlak K M0KEJ
Borthwick J M6FBO
Borthwick K M3SII
Borthwick M GM0HLK
Borthwick Q G6DLM
Bortowski J G0FPY
Boruch R M0LKS
Bosanquet-Bryant N G6TNQ
Bosanquet-Bryant P G6DLZ
Bosher T GU0UVH
Boskett S GI7IFW
Bosley E M6BOZ
Bosley M G0FGS
Bosnyak G 2E0GCD
Boss I G1SNQ
Boss R G4ZDE
Bosson B 2E0GWR
Bosson B M0SWL
Bossons C M3WXI
Bostock A M0LLW
Bostock M 2E0DTF
Bostock M M0WCA
Bostock M M6TGU
Bostock R M3BZQ
Boston A GI8VTK
Boston C M3YCB
Boston L G1ONK
Boston S G3MYY
Boswell G G0TMW
Boswell M M6BBT
Boswell P GM4AEK
Boswell P G8JVW
Bosworth D G4NAC
Bosworth I G8PKG
bosworth J 2E0JAQ
Bosworth J 2E0CZJ
Bosworth J G8BAV
bosworth J M3JTO
Bosworth W M0DVG
Botham B G0BTB
Botham I 2E0CBL
Botham I M3ZMX
Botham M M6FTK
Botherway A G0AJB
Botley A M3PVP
Bott B 2E0BRI
Bott K M6KRV
Bott T GW6NLO
Bott V G4AFZ
Botterell D 2E0HHB
Botterill G 2E0GAZ
Botterill S G7LFL
Bottle J M1BWJ
Bottom J G3SDG
Bottomley A M6GYT
Bottomley D G3GAQ
Bottomley E G0HNP
Bottomley H G4WCY
Bottomley H G6JRM
Bottomley H G8BCL
Bottomley J G7RKE
Bottomley J G3YFP
Bottomley J G3TQQ
Bottomley K G8MJF
Bottomley M G0WNJ
Bottomley-Mason J G1CFZ
Bottoms C G4PIP
Bottrell A M1AZJ
Bottrill A M0DIN
Bottrill A M3ADB
Boucher A M6ARB
Boucher K G4KBA
Boucher T G3OLB
Boucher W G1GBC
Bouchier G M6GBF
Boudier C MJ3CMB
Boudier S MJ3GBJ
Boughton C G0VUC
Boughton D G7PER
Bougourd R MU3MNG
Bougourd T M3IEU
Bould D M1ECV
Bould D M3ECV
Boull G G4NVH
Boull J M3NVH
Boull L M3NLW
Boult B G7HMQ
Boult J G4LOM
Boultbee G G3VHI
Boulter C G1UXY
Boulter K MW0KAB
Boulton C G4IOS
Boulton K M3OMX
Boulton R G4VXV
Boulton R M6YHR
Boultwood P 2E0PTS
Boultwood P M0PTQ
Boultwood P M6PGB
Bound H M6FNM
Boundey G G0VII
Bounds A G3KDF
Bounds M G6DAC
Boundy S G0EDH
Bounford M G8GLD
Bourdon S M3TWB
Bourhill G 2M0BEB
Bourhill G MM0FZV
Bourhill K MM3ZVY
Bouris K M3HXW
Bourke C G4UOR
Bourn L M6LCP
Bourn R G7CUL
Bourne B G4DLC
Bourne B M0BIK
Bourne C G0RFF
Bourne C G1EJD
Bourne C G4LTK
Bourne D M6KLA
Bourne F G3YJQ
Bourne G 2E0GHB
Bourne G M0EAM
Bourne J G3VYD
Bourne J G6JJK
Bourne L M3YJY
Bourne M G7MTV
Bourne P G4WUX
Bourne S M0CYD
Bourne S M6HVX
Bourne T M3PIQ
Bourner H G3NCB
Bourner N G4JYU
Bourner N G8MLB
Bourner T MW6RGK
Bournes A G6YTY
Bousfield M G0AFU
Bousfield T G0SJU
Boutell C G7HMN
Bover M GW8KCY
Bovey C G0AFT
Bow K G7HIC
Bowden B G0ENN
Bowden C GM4MFB
Bowden C G7IPX
BOWDEN C G7LKZ
Bowden C M6YAH
Bowden C G3OCB
Bowden D G1MPT
Bowden D M0WOB
Bowden G G4JBY
Bowden K G4SBE
Bowden P G4MKW
Bowden R G4JOU
Bowden R G7FKZ
Bowden R G3IXZ
Bowden S G4TDP
Bowden T M0TRB
Bowden W G0MAL
Bowditch A G4WSB
Bowdler K G0KMB
Bowe P G1MVI
Bowell E G0OTE
Bowell M M0MBB
Bowell R G3LRL
Bowen B 2E0BTB
Bowen C M1BSI
Bowen D 2E0ZJA
Bowen D M0UAA
Bowen D M3YVK
Bowen D G6ATS
Bowen G M3JPG
Bowen I G0JJY
Bowen J GW3TSQ
Bowen J GW6MMM
Bowen J G8DET
Bowen M MW3CHZ
Bowen M GW3KGI
Bowen N M6BWN
Bowen P G3TZL
Bowen P 2E0PNN
Bowen P M0PNN
Bowen P M6FSW
Bowen R 2E0YLL
Bowen R M6ZOL
Bowen R M6HXI
Bowen S 2W1SRB
Bowen T G8JOY
Bowen T G4JKQ
Bowen T MW6BOW
Bowen V 2E0VRD
Bowen V M6VRD
Bowen W G4AAU
Bower A G3MKU
Bower C G0VRH
Bower D 2E0YDB
Bower D M0SDB
Bower D M6YDB
Bower J G0KED
Bower L G4HKY
Bower M 2E0HXT
Bower M M3HXT
Bower P G0LNK
Bower P G3OFT
Bower R G4DLG
Bowering A G0JKU
Bowerman S G0OGM
Bowers A G7NIZ
Bowers A MM3MZX
Bowers B G0VAX
Bowers C G4YSH
Bowers D G7SYU
Bowers D M0DKV
Bowers D G4AVC
Bowers J G6BIM
Bowers J G8FWF
Bowers J G0NLL
Bowers J G0FAD
Bowers M G7NOR
Bowers D G0MOX
Bowers-Edgley D M6WOT
Bowes A G4XTW
Bowes B G6CRG
Bowes C 2E0UBT
Bowes C M3UBT
Bowes J G4MB
Bowes J G4ZFY
Bowes J M1DPY
Bowes J MM0OKG
Bowes P M6PRB
Bowhay G G3ZYL
Bowhill A G4FTI
Bowhill F G1SSZ
Bowie A GM7BOZ
Bowker A M0AGJ
Bowker A G7RHU
Bowker B G7MRO
Bowker D M1EKH
Bowker J G1ZWQ
Bowker N M6NIC
Bowkett S 2E0SMB
Bowkett S M3FIK
Bowkett V 2E0VJB
Bowkett V M3VJJ
Bowlas D G6VGT
Bowler A G6ZTF
Bowler C M6FZK
Bowler D M3BFY
Bowler R G3ZHV
Bowles A M6KDQ
Bowles D G0CRK
Bowles D G1CFK
Bowles D G6IBD
Bowles D M0DAB
Bowles F GM4KAV
Bowles J G0GMJ
Bowles P G3ECM
Bowles P M1CGR
Bowles R M6BLX
Bowles S G0UWF
Bowles W G0KCZ
Bowley A G6CHI
Bowley C G4OQL
Bowley C 2E0CAL
Bowley J 2E0DAJ
Bowley J M3VAQ
Bowlzer A M6CHQ
Bowmaker A G0EBP
Bowmaker A G0REV
Bowman A G4NJN
Bowman A G8ZIA
Bowman A M6EJV
Bowman B MM3OMI
Bowman C 2E0PMC
Bowman C M0NLP
Bowman C M6BFG
Bowman D G0BOH
Bowman D G0MRF
Bowman I G7ESY
Bowman L 2W0WFB
Bowman L MW0WFB
Bowman L MW6WFB
Bowman M GM4LVW
Bowman M M6FTW
Bowman P M6TBH
Bowman R 2E0RBA
Bowman R M0RAZ
Bowman R G4DDV
Bowman S G7ESX
Bown C M5SAV
Bown D M3JOS
Bown M 2E0SEA
Bown M M0ESU
Bown N 2E0CGW
Bown N M0NIB
Bown N M6BWN
Bown P M0ONY
Bown R G8JVM
Bown T 2E0NAG
Bown T M3NWY
Bowron A M3EVI
Bowron P G6UCT
Bowry N GM6OUL
Bowskill C M6FZR
Bowskill T M3EAT
Bowyer A G1KQN
Bowyer A G4OYH
Bowyer D G3NHB
Bowyer D M1AEI
Bowyer K 2E0FDS
Box A G0HAD
Box F G8RWM
Box M M6RZA
Box M G3RZG
Box S G1PUZ
Boxall B M6FQJ
Boxx P M6MRP
Boyce A G8SGM
Boyce C G4XWP
Boyce E G1LAW
Boyce G M1AHN
Boyce J GI0TJJ
Boyce M GM0RRK
Boyce P GW4NHB
Boyce R G6PXQ
Boyce R G8TUU
Boyce S G3VBV
Boycott T G1ODJ
Boyd A 2I0FXP
Boyd A G4LRP
Boyd A MI6EXP
Boyd A GM4UQD
Boyd D MI6GFQ
Boyd E M5PYE
Boyd I GI7GKC
Boyd J GM7RAK
Boyd J M3UKJ
Boyd J G7NPT
Boyd K GI4SLQ
Boyd M M3YUA
Boyd M M6OYD
Boyd N G0UGD
Boyd P MI3CXM
Boyd P GM6PCW
Boyd S G0AOQ
Boyd S 2M0XBD
Boyd S MM0SBO
Boyd S MM6XBD
Boyd S 2M0DKU
Boyd S MM6SUB
Boyd S MM6FZD
Boyd T 2I0TAA
Boyd T MI3VHW
Boyd W G4BID
Boyd W GI4ZOS
Boydell J G3TAX
Boydon E G1SJG
Boydon M G1UDT
Boyes A 2E0MLX
Boyes A G0PWO
Boyes D 2E0KVJ
Boyes D M6KVJ
Boyes G MI6GBC
Boyes J G4YJX
Boyes M 2E0FKT
Boyes R M6BZK
Boylan A MI3LFE
Boylan I M6OAU
Boyle A M0CXX
Boyle C 2E0DKE
Boyle C M6FCM
Boyle J 2M1EZA
Boyle K 2I0KB3
Boyle K MI6KOB
Boyle M M6MQL
Boyle P G0NVT
Boyle R GI0VTS
Boyles S 2E0LTD
Boyne A G3YVH
Boyns R M1FCG
Boys C G8BCO
Boys C M0FRS
Boyter G MM3ZJY
Boyton J G0HKE
Bozac P G0GOE
Bozikis P 2E0PNB
Bozikis P M0PNB
Bozikis P M6TNB
Bozman B G6OUJ
Bpophy I G1OBA
Brabazon R M6BQS
Brabbins S G6KJT
Bracci M M0GSC
Brace D M0DOA
Brace H M3DIM
Brace J GW3JBZ
Brace R G3WOO
Bracegirdle R G4UGB
Bracewell C G4HVF
Bracey A G7VOK
Bracey E G4UTS
Bracey R G4BZI
Bracey R G6IUD
Bracher C G4SXR
Bracher D 2E0CMZ
Bracken R G6EQZ
Brackenridge J G1YKL
Brackpool A M6BQV
Brackstone A M0SAB
Brackstone L G7UFF
Brackstone L M0NRY
Brackstone L M3LEN
Brackstone P G6VCF
Bradberry D G4OZM
Bradburn J M3LIB
Bradbury A G6YPK
Bradbury G G6DAO
Bradbury I G7ADF
Bradbury J 2E0JAB
Bradbury J 2E0HHG
Bradbury J G6LFG
Bradbury M G6VGS
Bradbury M M0MCT
Bradbury M M6TBO
Bradbury P G4EXK
Bradbury P G4SCY
Bradbury P G7MKG
Bradbury S G0URT
Bradbury-Harrison R G4GOT
Bradd K M0KWB
Braddock C M6NCL
Brade A M0XBR
Brade J G4CDH
Brade R G3VIR
Brade W G8ESW
Bradfield H G0VTL
Bradfield P G1GSN
Bradfield R G4XIM
Bradford B M3NXU
Bradford I G4ZQV
Bradford K M6KCE
Bradford M G3HLI
Bradford M G8FXC
Bradford S M1AJU
Bradley A M6ZIP
Bradley A GI4YPS
Bradley A G1NBT
Bradley C G0RFS
Bradley C G1FYF
Bradley C M3UOB
Bradley C M6CGB
Bradley C 2E0VRX
Bradley C M6BEQ
Bradley D 2E0IPA
Bradley D G6HV
Bradley D G7SVM
Bradley D M0BQC
Bradley D 2E0DTB
Bradley D G3VUY
Bradley D MI3XWW
Bradley D MI3YYT
Bradley D G0PBU
Bradley E G3UYY
Bradley G G0IQQ
Bradley G G0JQP
Bradley G G0FJZ
Bradley I G8UCC
Bradley J G3MAV
Bradley J G6FIH
Bradley J M3SLD
Bradley J M0CRU
Bradley J G6SOX
Bradley K M3KHE
Bradley L G4RZD
Bradley L MI3YXX
Bradley L 2M0LWB
Bradley L MM0VSU
Bradley L MM6LWB
Bradley M M6CNA
Bradley M 2E0BYW
Bradley M 2E0MDR
Bradley M M6MVB
Bradley M GI4MQA
Bradley M G6ASJ
Bradley M 2E0AKW
Bradley N G4IWO
Bradley O 2E0OLI
Bradley P G4BZE
Bradley P M3LCW
Bradley P G3UJO
Bradley P G7VVK
Bradley R G1DQL
Bradley R MI1EZZ
Bradley R MI3STW
Bradley R M6EUX
Bradley R G0OKD
Bradley S G1GFC
Bradley S G8HYM
Bradley S M6PBP
Bradley S 2E0CSY
Bradley S M6BYY
Bradley S M6SRE
Bradley S 2E0CND
Bradley S M3CND
Bradley T M3WEA
Bradley T G1KBE
Bradley T G4TDU
Bradley W MI3IOH
Bradley W 2I0WBU
Bradley W 2E0WJB
Bradley W 2I0WWB
Bradley W MI0WWB
Bradley W M6EHV
Bradley W MI6USC
Bradnam S G3XXX
Bradnock J G0IZQ
Bradshaw A G1BDU
Bradshaw A M3WVT
Bradshaw A 2E0CKC
Bradshaw A M6APB
Bradshaw B G0LVJ
Bradshaw B G6CJT
Bradshaw D G3SUX
Bradshaw D M3NAO
Bradshaw E 2E0CXP
Bradshaw E G1YLM
Bradshaw G G7WGO
Bradshaw G M3WVF
Bradshaw L G0MRL
Bradshaw M 2E0WVD
Bradshaw M M3WVD
Bradshaw M M3LBQ
Bradshaw M 2M0FLG
Bradshaw M MM6FLG
Bradshaw N G6UWI
Bradshaw N M3UNI
Bradshaw P G4CTE
Bradshaw R G4PDE
Bradshaw R 2E0ETN
Bradshaw R G7CQG
Bradshaw R M0TKT
Bradshaw R M6ETN
Bradshaw S G4OUK
Bradshaw S G8GMB
Bradshaw S M3GMB
Bradshaw S G3WEJ
Bradshaw W MI6WDB
Bradshaw X MM6SHB
Bradwell M 2E0EDB
Bradwell R G0FGP
Brady D 2E0GRG
Brady E M6ECB
Brady G G0UFI
Brady I 2E0FCO
Brady J MM0BIR
Brady K G0MKN
Brady K GI6RBD
Brady L M3YPS
Brady M 2E0GGG
Brady M G1XGN
Brady M M0RKB
Brady M M3LIU
Brady M M0EED
Brady T G8HEB
Braeman A M6RFE
Braeman N G4FUP
Bragg M G1BDY
Braggs H G4JLX
Braham H G3OJS
Braham R GW4RYA
Braidwood P M1TCP
Braidwood S G3WKE
Brailey K G6FXH
Brailsford C G7FHA
Brain A 2E0HDF
Brain C G4GUO
Brain D G7RUJ
Brain D G3XMC
Brain M G0VNB
Brain T M3EGM
Braisby C 2E0XSG
Braisby C M0XSG
Braisher K M6AZR
Braithwaite D M6YLD
Braithwaite H M6FUJ
Braithwaite I G4COI
Braithwaite J G3PWK
Braithwaite R G3DAQ
Bramall J G4HUD
Brambley J G0JWD
Brame D M3YDV
Bramham G 2E0GEG
Bramham G M3VBP
Bramley A G4NDU
Bramley A G4OYO
Bramley D G0CYR
Bramley J M3UHC
Bramley L M0LHB
Bramley R G4NDT
Brammeld H G0HNQ
Brammer A G6RFR
Bramwell W G6KTC
Bran K G3UDC
Branagan G G6PKV
Branagh D GI7JEM
Branagh J GI3YRL
Branagh K GI4SBA
Branch C G0UUR
Branch C M6EEH
Branch H G7RVY
Branch M G1OWI
Branch Y G7AKP
Brand A 2E0HWK
Brand A M0HKW
Brand A M3UFS
Brand J G3SSN
Brand J G0MGN
Brand R G6CEZ
Brand R G0SJR
Brand V G3JNB
Brander B GM7OWU
Brandhuber J G4PDY
Brandon B G4TDU
Brandon D G4UXD
Brandon M M3RVX
Brandon S 2E0HEV
Braniff B GI1XTK
Brannan W MM6DBN
Brannon D G0AVE
Brannon P G4IQJ
Bransby D G7RTJ
Branson D 2M0TAS
Branson D MM3UBD
Branson J 2M0GYM
Branson J MM3UBR
Branton A G8VUS
Branton S G6XWK
Brasenell D G6ZLY
Brash A M6LOB
Brash C MM3ZSK
Brash H GM3RVL
Brash P GM4PGM
Brasher M 2E0EVP
Brasher M M6ATJ
Brashill M G2DPA
Brashill M G6YAS
Brashill M M5EXY
Brashill S M3HRT
Brasier A G3XMP
Brass G MM6GYR
Brass M G4YMB
Brassington M G4UMG
Brassington P G4MDM
Brassington R G6YCN
Bratchley J M6BED
Brattle D M5DHB
Braund G G3ZPI
Braunstein S M6NYM
Bravery R G3SKI
Brawn D G4XJE
Brawn J 2E0CES
Brawn J M0HVR
Bray A MW0CIS
Bray B G0ELB
Bray C G3RWQ
Bray D 2W0SRD
Bray D G4MKI
Bray E 2E0DTO
Bray E G8OEJ
Bray E M0HFF
Bray G GD3SRN
Bray M G0JLP
Bray M M0AGR
Bray M 2E0MDN
Bray M M0MBZ
Bray M M6MEJ
Bray P M6HLE
Bray R G3KEL
Bray R G8LPI
Bray R M3ONK
Bray R G0OYZ
Bray T G7UBQ
Bray T G0JNZ
Bray V MW0CIT
Bray W M0WHB
Bray W M3BXC
Braybrooke P G6XVQ
Braybrooke S G1EOH
Brayley I M6AKE
Brayshaw C M0VOL
Brayshaw J G0FNS
Brayshaw P G7GNU
Brazenall P G0NXD
Brazier G GM8VAM
Brazier K G7AFT
Brazier K M6KRZ
Brazier L G4ZBC
Brazier L M6LGB
Brazier P G0OCW
Brazier P G6JFN
Brazier R G0OSG
Brazington K G4IZV
Brazington K G8UXX
Brazinskas M M6LTU
Breaden B G6OUM
Breadon L GI7NMK
Breakspear R G1DOL
Breakwell K G6JJG
Breakwell M MW3JQC
Brealy D G3TSE
Breame F G8ISI
Brearley L 2E0DXZ
Brearley L M6GQY
Brearley M 2E0GKB
Brebner D G6HOB
Breck P G7EWS
Breckell N G7GOK
Breckell P M3HNV
Breckons C G6XWD
Breckons N 2E0BVR
Breed D M0YDB
Breedon K G1HMZ
Breeds A G8ZVX
Breen A MD6MGP
Breen D G0FQI
Breen J MM6JBN
Breen S G7HZQ
Breese P G7PZQ
Breese S M3WMU
Breet B M1ABV
Breeze A G0UUZ
Breeze R G8RTB
Breffit M M3FRX
Breimann K MM6IKB
Breimann M MM6IMB
Breingan J G0RRO
Brelsford I G7CRA
Bremner H GM3SER
Bremner J 2E0MDV
Brenchley S 2E0DOZ
Brend A G4PXE
Brendish E M0CMQ
Brenig-Jones M G3ZEQ
Brennan C MI3OYP
Brennan G MI1EIH
Brennan G MI3BPS
Brennan G GI4GOL
Brennan J MW3GVU
Brennan J M6JEJ
Brennan K GI6OCC
Brennan M MW3KCL
Brennan P MW0CFQ
Brennan R MI3CQO
Brennan T GI7ILA
Brennan W GI3FTT
BRENT N 2E0RAO
Brent P G4LEG
Brentnall A G0THW
Brenton P G8EHD
Brereton M G8ALE
Brereton N M0FOG
Breslin G GI4ZLD
Breslin M 2E0ZMB
Breslin M M3SQG
Breslin P M3LMB
Brett A 2E0HXD
brett C 2E0BEI
Brett G G1VVM
Brett J G0TFP
Brett J G6NBI
Brett K M6WKB
Brett M 2E0HWV
Brett M G7AHR
Brett M 2E0LTJ
Brett M G4XVM
Brett R M3WPC
Brett S G4COT
Brett S M6TVB
Brett V M6WVB
Brettle P GW7JHK
Brew R M3RTI
Breward J M6XBR
Brewer A G1GMV
Brewer D M3YNX
Brewer K G4WYO
Brewer P G4OSJ
Brewer S MW0ATK
Brewerton D M0EZP
Brewster C G6HNV
Brewster H M6BQD
Brewster J 2E0BLQ
Brewster M M3YLZ
Briant C M6CGV
Briant R 2E0CPD
Briant R M0RBV
Briant R M6HXI
Briant S M6STF
Brice C 2E0AUV
Brice B G0LUI
Brice J 2E0HTM
Brice K M3LHQ
Brice T M3UIL
Brice-Stevens P G0WAT
Bricknall K G1DZY
Bricknell I M0KRW
Bricknell R G0SCT
Brickstock E G3IVQ
Brickwood J G1TTG
Brickwood N G6DAI
Brickwood N G7UFW
Bridge G 2E0GAO
Bridge G M0YYA
Bridge G M6GEF
Bridge J G0RGC
Bridge J G3ZVQ
Bridge J M6JTU
Bridge M M0CEB
Bridge M G2VC
Bridge H G4WMV
Bridge S M3MSB
Bridgehouse J M6LEA
Bridgehouse J G1VTP
Bridgehouse M 2E0MCK
Bridgehouse M M0MBM
Bridgehouse M M3UXF
Bridgeland A M0DUQ
Bridgeland M 2E0MAB
Bridgeland M M3MBC
Bridgeman P G3SUY
Bridgeman S M3WBS
Bridgen D G3VCX
Bridgen W G4XSB
Bridger M 2E0MCN
Bridger M M6MPB
Bridger P M3BKU
Bridger S M1GIZ
Bridges C GM4NGJ
Bridges C 2E0OBC
Bridges C M6OBC
Bridges D M6NET
Bridges J M3VPD
Bridges M G0LLC
Bridges M M6MRC
bridges N 2E0EOD
Bridges P GM3OBG
Bridges P G6DLJ
Bridges S M3IOX
Bridges S M6SJA
Bridgland C G8KUA
Bridgland R G4ALZ
Bridgland-Taylor T G0JWJ
Bridgman J 2E0ADJ
Bridgnell D G4VPJ
Bridgwater R G8RJB
Bridle D M1EZX
Bridle G G1GSY
Bridle K G1DZZ
Bridle N M5NEV
Bridle P G6CHD
Bridson J 2E0IZR
Bridson J G0IZR
Bridson J M3RXT
Bridson R G3VEB
Brien T G1CYY
Brier R M0RBB
Brierley A G8TIU
Brierley D G3YGJ
Brierley D 2E0PGB
Brierley D M3PGB
Brierley G 2E0CCG
Brierley G M3LPK
Brierley G 2E0ALJ
Brierley G M0OYZ
Brierley G M3SVZ
Brierley G G4SRP
Brierley N G2DNJ
Brierley R M3LPI
Brierley R 2E0RUZ
Brierley R M0RUZ
Brierley R M6RUZ
Brierley T 2E0TAB
Brierley T M3ATB
Briers A G0KZT
Briers J MW3FPF
Brigden N G1UBV
Briggs A 2E0ZRL
Briggs C G8EXN
Briggs C M0EJF
Briggs D G0BDM
Briggs D G0JJR
Briggs D G0LQC
Briggs D G1XVR
Briggs D G7VSE
Briggs D M6FQK
Briggs D G8OUM
Briggs E G3IJU
Briggs E M6ENF
Briggs G M6GAD
Briggs G M6AXU
Briggs J G7LYH
Briggs J G7PIR
Briggs J G6GWP
Briggs J M6JDM
Briggs L M6LAQ
Briggs L M6LIN
Briggs M G4SMB
Briggs N 2E0LEF
Briggs N M0VSP
Briggs N M6APG
Briggs R G4AMY
Briggs R G4YJB
Briggs S G7VZK
Briggs S M6SNJ
Briggs S G3FBT
Briggs S M3LQL
Brigham G G3LEO
Brigham I M0LKL
Brigham L M6LLB
Brigham M M0RNG
Bright A G1BQG
Bright A M0TTB
Bright A M6BBC
Bright D M3YKN
Bright J G4HJI
Bright K M0KBB
Bright K G4KEB
Brightman G G7DPF
Brightman M G4NHD
Brightmore S M6SBM



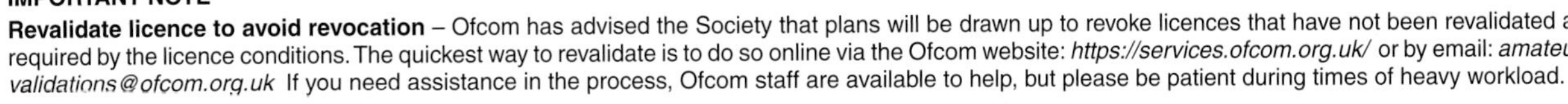
**IMPORTANT NOTE**

**Revalidate licence to avoid revocation** – Ofcom has advised the Society that plans will be drawn up to revoke licences that have not been revalidated as required by the licence conditions. The quickest way to revalidate is to do so online via the Ofcom website: *https://services.ofcom.org.uk/* or by email: *amateur.validations@ofcom.org.uk* If you need assistance in the process, Ofcom staff are available to help, but please be patient during times of heavy workload.

Brighton A M6LXM
Brighton A G7BQY
Brighton A M6USP
Brighton D G4ISK
Brighton M G6KAI
Brightwell R G0TBU
Brightwell V 2E0CQP
Brignall R M6RCB
Brigstocke C G0MOI
Briley T G0EIH
Brimble-Brice A M3GUJ
Brimecombe J G3GUX
Brimley M G0WSB
Brims A 2E0FIZ
Brind C G0CDS
Brind G G4CMU
Brindle G G3VXE
Brindle H G3YZY
Brindle J G0DVT
Brindle P M3PMO
Brindley E G4CZH
Brindley M G6VTX
Brindley P G0HEV
Brindley R M6RQB
Brindley W G0MVT
Brink C 2E0DTL
Brink C M0LTD
Brinkley R G0UPD
Brinkworth N G3UFB
Brinnen D G7BUK
Brinnen J M3FEY
Brinnen M 2E0GBH
Brinnen M M0DZA
Brinnen N G3VDV
Brinton A G7DIR
Brion C G4BTN
Brion S M1BLJ
Brisbar G G3LWU
Brisco L M3YVW
Briscoe A 2E0NTA
Briscoe A M0XAL
Briscoe A M6NTA
Briscoe J GM8AOB
Briscoe M G0WJC
Briscoe S G1YHV
BRISLEY R 2E0AKU
BRISLEY R M3KQY
Brislin A G6WPO
Bristeir R G0ADG
Brister J G6AK
Brister J G8IXX
Bristow A G8YBH
Bristow B G4KBB
Bristow C GI3PSQ
Bristow C G7TGB
Bristow D G0EOH
bristow J M6HQB
Bristow K G3XCY
Bristow M M3TXM
Bristow N 2E0EEM
Bristow R M6RBA
Britain K 2E0VAA
Britain K G8EMY
Brito da Silva A M6WDN
Britt R 2E0BCQ
Britt R G8WTM
Britt R M0KXK
Brittain D G0UQB
Brittain J G1LEO
Brittain M M3XMB
Brittain R G6IUF
Brittain R M3RZB
Brittan H G8FGQ
Britten J M0ALD
Britten S M0SRB
Britten-Jones W MW1WYN
Britt-Hazard S 2E0COL
Brittle M MW6CEX
Brittleton G M6GLN
Britton A GM1JKJ
Britton A MM0MGB
Britton C G4IQO
Britton D G0SCK
Britton J G7MHL
Britton L G0BNG
Britton P M3XTR
britton R G8FUO
Britton R MD3OED
Britton R GW6DGU
Britton V M3VKB
Broad B G6LZX
Broad B M0HBC
Broad C G7OQL
Broad D G1PVT
Broad D M3EUU
Broad G G6AXE
Broad M G1BDP
Broad M G6MNN
Broad P G0SWU
Broad S GD1AQY
Broadbent A G3PYH
Broadbent C G1RPP
Broadbent C G7JUV
Broadbent C M3EFW
Broadbent G M3UYH
Broadbent J G4MBK
Broadbent N G8CDG
Broadbent S M3PQT
Broadbere P M3PRU
Broadbridge R M0RWB
Broadbridge R M1DYS
Broadfoot J G7GEA
Broadhead P G7OLC
Broadhead S G0HQU
Broadhurst G G0BVC
Broadhurst G G0SVP
Broadhurst H M6HBE
Broadhurst J M3VBI
Broadhurst L M6LOD
Broadhurst M M6VEC
Broadhurst S M3VRP
Bradley J G3WBP
Broadley P G0HSN
Broadley R M1VRC
Broadway M G4GFI
Broadway M G6LLG
Broadwith R M6RCA
Brock A G4BOZ
Brock C G6DGV
Brock C G3ISB
Brock C 2E0DRT
Brock C M6MUD
Brock G G1ZBH
Brock G G8WRY
Brock W G8MPM
Brockbank C G3RCD
Brockett J G4KXP
Brockett P G1LSB
Brockie E GM4EHB
Brocklebank J M0BIC
Brocklehurst D G4VDB
Brocklehurst R M6ILB
Brocklehurst S G6VMV
Brocklesby J M3JBE
Brockman T G4TPH
Brockway J G4OUH
Brockway M G4OUI
Brockwell J G8HYK
Broderick S 2E0CVM
Broderick S M6XAG
Brodie D G4ZDT
Brodie E G6HTB
Brodie I M6VGR
Brodie J G3RYH
Brodie R GM7IHR
Brodie R M0GJL
Brodie S G0PSY
Brodie S G7SNR
Brodie T 2E0HSL
Brodribb B G1EHS
Brodribb P G3ONL
Brodrick C M0BKF
Brodrick R G4LAF
Brodrick T G1DOJ
Brodzky J G3HQX
Brogan C MM3UMY
Brogan J G3NIN
Brogan M G8JHE
Brogan V MI6TYR
Brokenshaw R M5RIC
Brolan P G1MPU
Broll A M3NWD
Broll S M6FKB
Bromage M G7ABZ
Bromfield A G6JJI
Bromfield D G4KFI
Bromfield G G0MIT
Bromfield G G7ACK
Bromfield M M3MUI
Bromfield P M6PDZ
Bromiley J G2ANC
Bromley G G4NID
Bromley G G4UTN
Bromley K M3KPB
Bromley M M0CLM
Bromley N GW3IVR
Bromley P G6MMJ
bromley P M6HMM
Bromley R G1TJR
Bromley T G1WPR
Bromsgrove B G0AJL
Bromsgrove T G4STZ
Bronze A M3ARB
Brook A GD0JKA
Brook A G7WGP
Brook D 2E0TTS
Brook D M3TTS
Brook E 2E0GNU
Brook E M3XGI
Brook J G7VDH
Brook J G7VWW
Brook J M0GOO
Brook K G4ENR
Brook N G0KIY
Brook T G3WBQ
Brooke A G6YCE
Brooke D G0TRH
Brooke D G7SCO
Brooke M G4SKO
Brooke M G8HXR
Brooke P M1BIB
Brooke S G1GKK
Brooker A G4WGZ
Brooker M G0WVM
Brooker P G3WXC
Brooker R M0UYR
Brooker S M6SUI
Brooker V G7NHY
Brooker-Carey A G3OGH
Brooker-Evans D M3ZZI
Brookes A G1ZPO
Brookes A G7HWM
Brookes A G6UCW
Brookes C G1XMP
Brookes D M6BNP
Brookes D M3WXP
Brookes E MW0TAF
Brookes G G3NSO
Brookes G G6SYX
Brookes G GD4IZL
Brookes G GD8PPU
Brookes G 2E0GDB
Brookes J M6SXD
Brookes J M3ZHI
Brookes K G1NOS
Brookes K G0UZE
Brookes K G7RTX
Brookes L G0HPG
Brookes L M6TDB
Brookes P G0HPH
Brookes R G4PUM
Brookes R G8DZW
Brookes R G6YCM
Brookes R M6RFK
Brookes S G1IGC
Brookes S G1WLU
Brookes S M6XTD
Brookes T G0JRZ
Brookes T M0NTA
Brookes T M3CIO
Brookfield B G4ZDD
Brookfield D G1PJL
Brook-Foster D G3NRU
Brookhouse L M6HGA
Brookhouse L M6HGD
Brooking P G4SHH
Brooks A G3IDB
Brooks A G4KIQ
Brooks A G4VAV
Brooks A G7MKP
Brooks A M3FNA
Brooks A M3XAN
Brooks A M1AFQ
Brooks A M0LAY
Brooks C 2E0CIO
Brooks C G1IGA
Brooks C M3YSL
Brooks C M6HTW
Brooks D 2E0IWD
Brooks D G1PGN
Brooks D M0BNZ
Brooks D M0IWA
Brooks D M3DIW
Brooks D G0VIE
Brooks D G4IAR
Brooks G G1YMN
Brooks G G3LVB
Brooks G G4KGF
Brooks G GM4NHX
Brooks G 2E0GMS
Brooks G M6GHB
Brooks I G0AAY
Brooks I G1RVK
Brooks I G6IAN
Brooks J G0GDJ
Brooks J M0HJO
Brooks J M0HSH
Brooks J G4IAQ
Brooks K G0SPH
Brooks K M3KJB
Brooks M G8OXE
brooks M M1CKU
Brooks M M0CDF
Brooks M 2E0XUL
BROOKS N G0RNB
Brooks N G4BMH
Brooks O G1OLY
Brooks P G0SIA
Brooks P G4NZQ
Brooks P G6PQP
Brooks P G1OJQ
Brooks P G6FAX
Brooks R G3FGP
Brooks R G3YLL
Brooks R G7JGZ
Brooks R MM3GEW
Brooks R M6RSS
Brooks S M3ISH
Brooks T M6DUW
Brooks V G8DPE
Brooks W G4DTT
Brooksbank E GW6YUC
Brookson C G4GBA
Broom A 2M0KZI
Broom D 2E0HCB
Broom R G4LFE
Broom S M0ZTE
Brooman M G0LGE
Broome J G0CCF
Broome J G7LUB
Broomfield C G8CAA
Broomfield J G0JRV
Broomhall I G0JYZ
Brophy E MM6NED
Brophy M G1MGZ
Broscomb K M6BYH
Brosnan S G1JLM
Brosnan T GD4DRO
Bross O 2E0NAD
Bross O M0JRX
Brotherhood A M0TAB
Brotherhood A M1CJT
Brotherhood G 2E0GBV
Brotherhood G M0HOB
Brotherhood G M6CIU
Brothers P G6AXH
Brotherton P G7SCN
Brotherton R M1AXX
Brotherton R G7JSC
Brothwell I G4EAN
Brothwood R G6ICH
Brough B G1BCE
Brough B GD4PTV
Brough C M6PTV
Brough D 2E0KEZ
Brough D G3HUR
Brough D M3KEZ
Brough D 2E0DYB
Brough D GW7EMO
Brough D M0SWC
Brough D M6DYB
Brough J M3NNY
Brough M 2E0BYM
Brough M M6ZRO
Brough M G0TUO
Brough R M6MRX
Brough S G1UUL
Broughton A G1EAE
Broughton C G1RSK
Broughton D G1IEX
Broughton G G7TZV
Broughton J G4ZSV
Broughton P G3ZJF
Broughton P M3KQB
Broughton R G0MRB
Broughton R G4LZK
Broughton R G6YTR
Broughton R M6GLD
Browell A M3BPL
Brower J G4JLV
Brown A 2M0DXC
Brown A 2E0GJC
Brown A 2E0IJY
brown A 2E0TWB
Brown A G0AFJ
Brown A G0FKI
Brown A G0GWD
Brown A G4CCB
Brown A G4GPV
Brown A G4VSB
Brown A G4WSI
Brown A G4WWG
Brown A G4ZIL
Brown A G6RFS
Brown A G6UZR
Brown A G7ERI
Brown A G7KMW
Brown A GI7URC
Brown A G8GLV
Brown A G8INO
Brown A G8NPP
Brown A MM0ALY
Brown A MW0WEE
Brown A MI3CGU
Brown A M0HUA
Brown A MM6DBJ
Brown A M6JUX
Brown A G1ALD
Brown A GM7VCV
Brown A G8LTN
Brown A M3ZYV
Brown A M6AFB
Brown A G0NBD
Brown A G1XYS
BROWN A GI0KDH
Brown A M6AJB
Brown A M6CYY
Brown A M6NEG
Brown A 2E0CMH
Brown A M0HLR
Brown A M3AJV
Brown A M6TEY
Brown A MM6OOP
Brown B G3JFD
Brown B G3PZU
Brown B M3DNC
Brown B M6SIE
Brown B M6DUH
Brown B M3HOV
Brown C G0JRM
Brown C G0MGU
Brown C G1UKG
Brown C G3CEI
Brown C G4LIL
Brown C G6ZWC
Brown C G7VNO
Brown C G8ORR
Brown C M0TES
Brown C MI3CGT
Brown C M3FVJ
Brown C G7SOH
Brown C G4CLB
Brown C G7GJZ
Brown C GM4WEW
Brown C G0DWE
Brown C G1HHB
Brown C G4NQJ
Brown C MM3GVE
Brown C M6VWE
Brown C GM0RLZ
Brown C M6AVQ
Brown D 2E0RFQ
Brown D 2E0GLS
Brown D G0EMF
Brown D G0HBB
Brown D G0LYX
Brown D G0NWV
Brown D G3UOC
Brown D G3YGD
Brown D G4LNM
Brown D G4MUI
Brown D GM4RWE
Brown D GM6JFP
Brown D GM6JUA
Brown D G6LQP
Brown D G6MMG
Brown D GM6RAK
Brown D G6UZT
Brown D G7NTO
Brown D G7OVE
Brown D G8ECI
Brown D GM8FFH
Brown D G8GDH
Brown D MM0DBC
Brown D M0DND
Brown D MI3DBB
Brown D M3LQP
Brown D M3RFQ
Brown D MM3ROV
Brown D MM3UDB
Brown D M6DTJ
Brown D G4UNS
Brown D M0DCB
Brown D MM0ZBD
Brown D M3TYQ
Brown D G6PRL
Brown D M6CEQ
Brown D G4YZA
Brown E G1YYD
Brown E G4STK
Brown E M1CLW
Brown E M1WVS
Brown E M3TOT
Brown E G3ZIE
Brown E 2E0KKC
Brown F G8FPW
Brown G 2E0SKA
Brown G G0JSL
Brown G G0PWU
Brown G G2BJK
Brown G G4VRX
Brown G M3BMV
Brown G G4WUA
Brown G GW0PUP
Brown G 2E0JBF
Brown G M6BWE
Brown G G3MZV
Brown G 2E0YZM
Brown G GD0KWM
Brown G MM6EEQ
Brown G M6YZM
Brown H G0EMS
Brown H MM6HMB
Brown I GM0ILB
Brown I G0NPO
Brown I G3TVU
Brown I G4XTG
Brown I G4YSN
Brown I G7UWV
Brown I G8RBK
Brown I MM6IAI
Brown I G3TLH
Brown I G7PLP
Brown J 2E0AQE
Brown J 2M0JHN
Brown J 2E0BCF
Brown J 2E0CEZ
Brown J G0DPX
Brown J GD0ELY
Brown J G0JSM
Brown J G0OGG
Brown J G0PIA
Brown J G0UFY
Brown J G1MBG
Brown J G1XMI
Brown J G3ECP
Brown J G4EJE
Brown J G4NUK
Brown J GM4VLX
Brown J GM4ZIT
Brown J G6ETC
Brown J G6PCP
Brown J G6YPJ
Brown J G7FGD
Brown J G8FZW
Brown J M0DQB
Brown J M0JTB
Brown J M0XWD
Brown J M1EHZ
Brown J M1JKB
Brown J MI3FSR
Brown J M3LIW
Brown J G0PBF
Brown J GM6VRC
Brown J GM7SPA
Brown J G1HHQ
Brown J M6PBV
Brown J M6JYB
Brown J G6ZGU
Brown J M3RGG
Brown J M6JRB
Brown J 2M0GCF
Brown J G0PJU
Brown J G4UBB
Brown J GM7HHB
Brown J GW8EHQ
Brown J MM0GCF
Brown J MM0JMB
Brown J M6CJU
Brown J G3JOE
Brown J M6PEI
Brown K G0IBS
Brown K G0IKB
Brown K G3XQE
Brown K G4ZNY
Brown K G6NKI
Brown K G7EXO
Brown K G8IRL
Brown K 2E0IHO
Brown K 2E0KLV
Brown K MM0TCP
Brown K M3IHO
Brown K M3VAF
Brown K M3ZVS
Brown K MM0GHT
Brown K MM1XJS
Brown K MM3KNY
Brown L G3JBF
Brown L G4ZQS
Brown L G7LSB
Brown L 2W0GYB
Brown L MW6GYB
Brown L G7LQO
Brown L 2E0LBB
Brown L 2E0LFK
Brown L M6BTA
Brown M G0JLS
Brown M G0JXI
Brown M G0MNH
Brown M GM0SGH
Brown M G3UDP
Brown M G3ZXM
Brown M G4ZPN
Brown M G6SHQ
Brown M M3UAK
Brown M M3WGM
Brown M G3OVE
Brown M M0DBA
Brown M G3HUD
Brown M G0DCG
Brown M M0BJR
Brown M M0MFB
Brown M 2J0SZI
Brown M G3ZPW
Brown M G4RAA
Brown M G6UKC
Brown M M0APC
Brown M MJ3SZI
Brown M M3VCK
Brown N 2E0DNB
Brown N 2E0BVQ
Brown N GM4VHZ
Brown N G5NB
Brown N G6JIR
Brown N G8NCK
Brown N G8NVB
Brown N M3DNB
Brown N M6NVB
Brown N 2E0NLB
Brown N M6NDB
Brown N M3YYR
Brown N M6NKY
Brown N G7MVU
Brown P 2E0GUV
brown P 2E0PLA
Brown P G0DGU
Brown P G3PZL
Brown P G3WUZ
Brown P G4IDV
Brown P G4WWY
Brown P G6LPX
Brown P G6PJC
Brown P G7ILD
Brown P G8GLB
Brown P G8OEK
Brown P M3VWG
Brown P M5BTB
Brown P G3WRI
Brown P G4AJE
Brown P G8OXD
Brown P M3WEF
Brown P M6PZB
Brown P MI6PGI
Brown P M3NVC
Brown P 2E0FDT
Brown R 2E0RCB
Brown R G0NGG
Brown R G0PBW
Brown R G1LCY
Brown R G1NLZ
Brown R G1SCO
Brown R G1ZOB
Brown R G3LQP
Brown R G3SCZ
Brown R G3TOF
Brown R G3WPT
Brown R G3ZNT
Brown R GI4BXB
Brown R G4LZT
Brown R G4QA
Brown R G7BZM
Brown R G7LYB
Brown R G7OJO
Brown R G7PKD
Brown R G7SYW
Brown R G8CXV
Brown R G8OKE
Brown R M0CJJ
Brown R M3LFP
Brown R 2E0UDE
Brown R M0RKY
Brown R M6RLB
Brown R GI6IVJ
Brown R M0AHZ
Brown R 2E0ZRB
Brown R M3ZRB
Brown R G3RXO
Brown R G7SPZ
Brown R M3WVJ
Brown S 2M0MSB
Brown S 2E0SHE
Brown S G1ZBB
Brown S G4ATU
Brown S G4ELI
Brown S G4NYL
Brown S G4PRF
Brown S G6CRD
Brown S G8LZY
Brown S M0DAH
Brown S M3LDF
Brown S MM3MGK
Brown S M3SHB
Brown S M3SKB
Brown S M6SDS
Brown S G1UEV
Brown S G6UKM
Brown S M6EXF
Brown S M6EYH
Brown S M6ISB
Brown S 2E0CGG
Brown S M0REZ
Brown S M6TDM
Brown T G0NSA
Brown T G2HMK
Brown T G7FJK
Brown T G7NUG
Brown T G7TQE
Brown T MM0TGB
Brown T M6AMX
Brown T M6MDB
Brown W G0HZD
Brown W GM0IYA
Brown W G3FBU
Brown W G4MGK
Brown W GD4XTT
Brown W M0WBB
Brown W MI3DZD
Brown W G1BBY
Brown W G3NQX
Brown W G3NTM
Brownbill J 2E0JBG
Brownbill J M6AVN
Browne A G8YKM
Browne D 2E0DMB
Browne D G7HCQ
Browne D G7PRH
Browne D MI3CST
Browne D G4XKF
Browne G GI0MXT
browne G G3MMJ
Browne G M1CQS
Browne I 2E0ZVG
Browne J G3XZG
Browne J M3MGQ
Browne J 2M0CQI
Browne J MM6RKT
Browne K M3SHQ
Browne K M6FMJ
Browne K 2E0KTX
Browne L G0VCD
Browne M G8AOG
Browne M 2E0BXV
Browne P 2E0PFB
Browne P 2E0PMB
Browne P M0HWQ
Browne P M6PBK
Browne R G0NUU
Browne R 2E0ATY
Browne R M0MYC
Browne R M3YEM
Browne S G0GWA
Browne S G4SJU
Browne T 2I0TDL
Browne T MI0HXB
Brownen T 2E0XOL
Brownen T M0XOL
Brownen T M6XON
Brownett G G4FAZ
Browning A M3TUO
Browning D G0KKC
Browning D G3UEY
Browning G M6YEQ
Browning M G0SMB
Browning R GI0MQN
Browning R G4BEB
Browning R G4UMW
Browning R M1RKB
Browning S G8CSK
Browning S G8WUR
Brownlees J GI6VCG
Brownley A M3BUH
Brownlie I GM4EGD
Brownlow M G4LCU
Brownsea D 2E0ZDB
Brownsea D M0ZDB
Brownsea D M6ZDB
Brownsea J M6ZJB
Brownsett J G6PAA
Brownsett P G7RDA
Broxtom M MW0CHI
Broxtom M MW1BXX
broxup W G4OPN
Broyd M M6MLO
Broyles H M0CPE
BRUCE A G4NSM
Bruce A G6OCO
Bruce A GM7JGH
Bruce A M1DFO
Bruce A G8EEK
Bruce B 2E0FPO
Bruce B 2I0LPG
Bruce C M6LNE
Bruce D 2E0HYE
Bruce D G3IGZ
Bruce D M1DNA
Bruce D MM1EYZ
Bruce D M3HYE
Bruce D M0LNE
Bruce E M3UXE
Bruce G MM3GLH
Bruce J 2M0AYU
Bruce J GM1IEL
Bruce J GI4SJB
Bruce M G3TUY
Bruce M 2E0FSM
Bruce M M3UXM
Bruce M M6RFP
Bruce M 2E0ERE
Bruce M M0ITI
Bruce M M6HNZ
Bruce P G1FJH
Bruce P M3ZIN
Bruce P G4WPB
Bruce R GM8LON
Bruce R M6GZO
Bruce T G6IAT
Bruce-Smith M G8OBK
Bruckner R G6LPT
Bruckshaw D G1JRF
Brudenell D G4BUL
Bruetsch N M6NBU
Bruines N MW3IFZ
Brumby E GM8YKT
Brumby P G0RRM
Brundle K M6EHY
Brundle M G6EBL
Brundrett P 2E0HPB
Brundrett T M0SIN
Brundrett T M1SIN
Brundrit M M6BRN
Brunning A G4PTW
Brunning K 2E0HKB
Brunning K M0HKB
Bruno-Gaston I M6IBG
Brunsdon M G6UZO
Brunsdon M M3BRU
Brunsdon M MM0KCS
Brunsdon M MM3MDB
Brunt D G6KTE
Brunt D G7UWS
Brunt F M0FCB
Brunt F M6MNV
Brunt I M6IZA
Brunt J G0TPY
Bruntlett K G4MXI
Bruntnell V G7GEU
Brunton A G4HTG
Brunton D G1XWX
Brunton D M0DSB
Brunton M GM8GDN
Brunton P G0RBV
Bruring A G7MJD
Brusch D G0SCQ
Brush C MI6LIT
Brush N GW0MOQ
Brushwood P G4VIQ
Brutnall G G4PAV
Bruton A 2E0TCB
Bruton A M6OJC
Bruty T M3EWU
Bruyneel M M6YWO
Bryan A M3HIT
Bryan C G4EHG
Bryan C G6OAN
Bryan C G6TPI
Bryan C G7ANK
bryan D G4JCL
Bryan D M1CYX
Bryan D M3SKC
Bryan G G8MCA
Bryan H G6TMN
Bryan M G4DTB
Bryan M GW6ATT
Bryan M M1TVR
Bryan M M3MTR
Bryan M 2E0SNE
Bryan M M6FIT
Bryan P G1XLE
Bryan R G0NLA
Bryan R M0HRT
Bryan S G7VHG
Bryan S 2E0TIV
Bryans D MI3LXN
Bryans M M3UMB
Bryant A G1FTV
Bryant A G3NVB
Bryant D G1JFU
Bryant D G7JLS
Bryant D G0NES
Bryant G G0TZV
Bryant G G6NBM
Bryant H G4TTG
Bryant I M0EBO
Bryant I M1CWB
Bryant J G1XAM
Bryant J G8KMM
Bryant J G4LTR
Bryant J 2J0YAY
Bryant J G4CLF
Bryant J MJ6DEY
Bryant J G8XIJ
Bryant K G0BSK
Bryant K G7JLT
Bryant M M3ZMZ
Bryant M 2E0MWB
Bryant M M0UFC
Bryant M M6MWB
Bryant M M6FEL
Bryant M G6NLP
Bryant M G6TRM
Bryant P G0JWV
Bryant P G8ZFI
Bryant R G4UBM
Bryant R G8KWN
Bryant R G3WBC
Bryant S G3YSX
Bryant T G3SB
Bryce A MI1FAR
Bryce A MM0CDW
Bryce G GM3JOB
BRYCE G G7LGI
Bryce G G8LQN
Bryce J G4LWY
Bryden J G4HNQ
Bryder J G0JOP
Brydges J 2W0BGS
Brydges J MW3ONG
Brydon I G0PMZ

Bryson C 2M0CBC
Bryson C MM3ZQW
Brzenczek P G0TTN
Bubb A M0ZAB
Bubb G G7KNS
Bubez J G0SOF
Bubloz G G4FNL
Buchan A G6ZOJ
Buchan A GM0EFH
Buchan D M6EZX
Buchan N G0MAR
Buchan P G3INR
Buchan R G7PQM
Buchan R 2M0DHI
Buchan R MM6DHI
Buchanan G GM3KCY
Buchanan I MI3FPB
Buchanan J GM4VGR
Buchanan M MI3MBM
Buchanan M 2E0MBT
Buchanan M M6MKB
Buchanan N G0KUF
Buchanan S GM0MIS
Buchanan S MI3CGZ
Buchanan W GM1EAH
Buchan-Terrey R 2E0IGN
Buchan-Terrey R M6GKQ
Buchner I 2E0ICB
Buchner I M0ICB
Buck D G4HAS
Buck F M0FGB
Buck F M1WDX
Buck G G0RFQ
Buck J G7RWF
Buck J M6YJB
Buck M GW4NHH
Buck M G6YCI
Buck M G6KGR
Buck P G8AUL
Buck P G3LWT
Buck R G4RXQ
Buck T G6VXL
Buck T M0TBJ
Buck T M6WGC
Buckby R G3VGW
Buckenham H G3PGN
Buckerfield D M1GHT
Buckett W G3ODO
Buckie I G6WXK
Buckingham D G0ENJ
Buckingham J G4BCS
Buckingham P G6ECS
Buckingham S G7TZN
Buckland A 2E0EBR
Buckland A M6EBR
Buckland C G8IYJ
Buckland D G3JKM
Buckland J G4SLL
Buckland M 2E0BUK
Buckland M G0XAE
Buckland R 2E0RCL
Buckland R M0RBX
Buckland R M6DBI
Buckland-Hoby M 2E0GMD
Buckle D G0FUR
Buckle I G0MIF
Buckle J G1LES
Buckle M G4RMV
Buckle P G1VBA
Buckle R M0AZS
Buckle R G6HGM
Buckle T G0EWV
Buckler J G0HNI
Buckley A G4STW
Buckley A M3TLP
Buckley A 2E0IBU
Buckley A M6WBM
Buckley A M1BFF
Buckley A M6BSU
Buckley C G4HHO
Buckley C G7VKJ
Buckley D G0EQV
Buckley D G7ORT
Buckley D 2E0WBE
Buckley D M6WBE
Buckley E G0CVC
Buckley E M1BQO
Buckley G G0PNG
Buckley G M6GDB
Buckley G M6GAT
Buckley H 2E0SBN
Buckley H M6HBB
Buckley J G4HGL
Buckley J M0CCU
Buckley L G3PTX
Buckley M G7RME
Buckley M M1CCF
Buckley M M3ACF
Buckley P G0APB
Buckley P M3VJX
Buckley D G1GIE
Buckley S 2E0JIR
Buckley S M3JIR
Buckley S M6SJB
Buckley S M1EMD
Buckley-Brown M G1RAF
Buckman A M3AEE
Buckman R G3CZL
Buckmaster P G1DQF
Bucknell A M6BDI
BUCKNELL D G8ZPH
Bucknell T G4AFS
Bucknell W G0ULQ
Buckton I 2E0CNX
Buckton I M3ITL
Buckwell G G0MBU
Buckwell R G0MBV
Buckwell R M3VEJ
Buczkowski W M6WWB
Budas [illegible] [illegible]
Budas [illegible] [illegible]
Budas C MM3CVB
Budas M 2M1MIC
Budas M GM4VTB
Budas S 2M1SJB
Budas V GM3VTB
Budd C G0LOJ
Budd C G4NBG
Budd D G6DAH
Budd D G6NLS
Budd K 2E0DBS
Budd M G7CSS
Budd M G7JZS
Budd M M0MPB
Budd S M0AIB
Budden G G3WZP
Budden J G0PCW
Budding A 2W0BOK
Budge G GW0MGQ
Budgen P GM7MAG
Budina H GI1LBI
Buer P M1BZJ
Bues M G3OPB
Bues M G8AAI
Buffham I G3TMA
Bufton N G0MNO
Bufton N G1HNF
Bufton R 2E0CIP
Bugg M G6HNN
Bugg T G6XMM
Buggs D G1YRF
Buggs D G3DBJ
Buggs L 2E0HBF
Buick P 2E0HKK
Buik D G0DAB
Buist M 2E0OOM
Buist M M6MAY
Bukin G G0HPS
Bulbrook J G6MMB
Bulger G G3WIP
Bull A G3ICB
Bull A G1MLK
Bull B 2E0UTT
Bull B M0UTT
Bull B M6EHX
Bull C G6TLX
Bull D G6PJE
Bull G G7HMB
Bull G 2E0NTC
Bull G M0NTC
Bull J M0JPB
Bull J M3BUA
Bull J G8WVH
Bull K 2E0KAB
Bull K M0KAB
Bull M 2E0ESO
Bull M G4MIK
Bull M M0TSM
Bull M M3ULL
Bull M GM1MLY
Bull P M1CVB
Bull P M3BTZ
Bull P 2E0CRH
Bull P M0HKV
Bull P M6BSQ
Bull S M6SIH
Bull W 2E0WDB
Bull W M3IPT
Bullard A 2E0EEU
Bullard A M6EEU
Bullard D G7TWA
Bullen G G7NJW
Bullen L 2E0GKR
Bullen L M0VKR
Bullen L M3OLZ
Bullen R M3PPZ
Bullen R 2E0BHH
Bullen R M0KRP
Buller B G4JUM
Bullett V G3EAO
Bulleyment G G3XIV
Bullimore R G4VOT
Bullions M M6NEY
Bulloch A MM1LBA
Bullock A G8MKQ
Bullock A M0DDE
Bullock D G0UWO
Bullock J M6BCJ
Bullock J M6OAT
Bullock N G3TAQ
Bullock P G4WUO
Bullock R G1FJI
Bullock R G0EML
Bullock R G4XRV
Bullock S M6PRV
Bullock W G4CVG
Bullough N M0CDN
bullough P G1CWD
Bulman A 2E0CIO
Bulman A M0HIA
Bulman A M6AZU
Bulman J G1PWM
Bulmer H G4FZS
Bulmer M G6MNB
Bulmer P G0TTS
Bulmer R G8RFV
Bulpin J GW0BNN
Bultitude D G7SVT
Bultitude I G6KHW
Bumford J G0GTN
Bumstead M 2E0DKT
Bumstead M M0IAX
Bumstead M M6FKY
Bumstead S M6FLJ
Bunce F G0GAJ
Bunce M [illegible]
Bunce P [illegible]
Bunce R M1EBN
Bunce R M3DXR
Bunce S 2E0XXB
Bunce S M3XXB
Bunce T G4RRR
Bunce T M3TVN
Bundell R G0KQR
Bundle N G4LRV
Bundy M G4WVD
Bunker R M0BZZ
Bunkum C G1JXS
BUNN D G3OEQ
Bunn G G7MLC
Bunn M G6ITU
Bunn P G6UZL
Bunney D G7NDI
Bunney R G8ZMM
Bunting A G1PSW
Bunting A G8JCS
Bunting D M0BMT
Bunting G MI0GPB
Bunting G 2E0GWB
Bunting G M0HLP
Bunting G M6GWB
Bunting I 2E0IWB
Bunting I M6IWB
Bunting M G1EAJ
Bunting M M6BJM
Bunting P M6YSB
Bunting R M3NGU
Bunting S M0BPQ
Bunworth S M6TPO
Bunyan A G3XLR
Bunyan E G6OCM
Burbage B G7AQL
Burbanks J G3SJJ
Burbeck M M6AIZ
Burbeck P G0OMH
Burbeck R G4NOB
Burbidge L 2E0VLL
Burbidge L M3VLL
Burbidge M G0VPU
Burbidge T G4MKP
Burbridge C M3VCB
Burch G G7VQI
Burch H 2E0WAF
Burch H M3WPK
Burch J G4TEN
Burchell A M1BPN
Burchell C G3NKQ
Burchell D G0ANV
Burchell G G7VXS
Burchell J G8JVV
Burchell P G0VSO
Burchell R G0WBV
Burchell S G1DAZ
Burchmore A G4BWV
Burchmore M G0ARQ
Burcombe T M3MZP
Burdass J G6NLN
Burden B M1VHF
Burden B M3WOK
Burden C M6COY
Burden C G4AXC
Burden D 2E0DJB
Burden D M0LDI
Burden D M3WLV
Burden I G0RRI
Burden P G1WFO
Burden P G3UBX
Burden W G3XCJ
Burdess R G4OFU
Burdett C M6ZCB
Burdett D G7MFW
Burdett J G1KNA
Burdett J G0SUT
Burdett J 2E0JAX
Burdett J M0CSE
Burdett J M6JHB
Burdett R G0DLB
Burdis B G7OXN
Burdis B G0WZB
Burdis S M6WZB
Burdon I G1TKQ
Burdsall D 2E0TLX
Burdsall D M0TLX
Burdsall D M3TLX
Burfield A 2E0KFM
Burfield A M0KVR
Burfield A M3GTV
Burfield M M1MPB
Burfield P G6GLH
Burfoot P G8GGM
Burg R G8WSC
Burge A GI3OTU
Burge A G6ALB
Burge D MW0ALG
Burge J G3VHN
Burge N G0BTQ
Burgess A 2E0STT
Burgess A 2W0FAR
Burgess A MW6FAN
Burgess A G4GLV
Burgess B M6DKU
Burgess C G8EWL
Burgess C M1FFX
Burgess C G0RKE
Burgess D 2E0BLA
Burgess D G0HGV
Burgess D G4IYS
Burgess D M3BXZ
Burgess D MM6HHB
Burgess F MI6RLU
Burgess G G4HDY
Burgess I G0TAZ
Burgess J G0KKP
Burgess J G4BNP
Burgess K G1HFS
Burgess K MW6BAU
Burgess L M6LAG
Burgess M G6UCQ
Burgess M MM0MLB
Burgess M G7HID
Burgess N M0COI
Burgess O G6SSM
Burgess P 2E0BZT
Burgess P G4BCH
Burgess P G6OUO
Burgess P M3ZPB
Burgess P 2E0PBT
Burgess P G3VPT
Burgess P M0CCQ
Burgess P M0PBT
Burgess P M6PBT
Burgess R G3RXG
Burgess R G8KJT
Burgess R G0XGM
Burgess S 2E0GJD
Burgess S G3JHH
Burgess S G4CQO
Burgess S G4TRM
Burgess S G7CPN
Burgess S M0YCQ
Burgess S G4NMS
Burgess S M6FSJ
Burgess T G8VGU
Burgin D G7MII
Burgin D M1GTI
Burgin K G4CRG
Burgoine S G7GBE
Burgoyne J MM0JOK
Burgoyne J MM3EJB
Burhouse G G4MVA
Burin W G8PDE
Burke A G6BEN
Burke B G4HIY
Burke C M0USK
Burke C M3USK
Burke D 2I0JXO
Burke D MI3JXO
Burke J G1JAB
Burke J G3MVX
Burke J GI4RVF
BURKE J GM4TNP
Burke L M3ZNZ
Burke L M6LTB
Burke M GM6SZJ
Burke M M0AMS
Burke P 2E0FAC
Burke P M0PEB
Burke P M3SQO
Burke R G4ZAQ
Burke S G6TVP
Burke S M5ZZZ
Burke T G3UPM
Burke T G4UGR
BURKILL N M0BZU
Burkin D M0VMC
Burkinshaw T M6TBX
Burkitt A G1VLD
Burkitt A M6CEU
Burkitt C G3PZE
Burkitt N G4FXT
Burleigh D G4WIZ
Burleton A 2E0GHD
Burleton A M0GZL
Burleton A M6GHD
Burling M G1HND
Burling R G0CMB
Burling S 2E0FLW
Burlington G 2E0JYX
Burlington G G4HXQ
Burlong R M3WOI
Burman B G4PYI
Burman D M3XQQ
Burman J M3YUK
Burman R G4RSN
Burn C G8WYW
Burn D GI0AQD
Burn G G7HLW
Burn H G0RKN
Burnage S M6SCB
Burnand P G0VET
Burnand S M3EVM
Burndred E G0KBJ
Burnell K M1DZI
Burnell K M3ILA
Burnell M G8OUI
Burness K 2E0KBN
Burness K M6KBS
Burnet C GM4TVB
Burnet J G1LDN
Burnet N 2E0IFC
Burnet N M6NLB
Burnet P G3YQJ
Burnet R G0EUC
Burnet W G3WZA
Burnett A G0IFC
Burnett A G4LSU
Burnett A G1SPU
Burnett C G6GCI
Burnett D G0MXB
Burnett D M3CCQ
Burnett F G4CYB
Burnett F G3RSM
Burnett G G6IBP
Burnett G MM0YET
Burnett H 2E0IYY
Burnett J G3OLW
Burnett J G4NDP
Burnett J G6GCJ
Burnett J 2E0OAK
Burnett J M6JBU
Burnett M M3CTB
Burnett M M6MBU
Burnett P G1DAT
Burnett P G4BLL
Burnett P G7SKA
Burnett R G6YTX
Burnett S GM4LHW
Burnett T 2U0TKB
Burnett T MM0TOB
Burnett T 2M0DTP
Burnett T MM6FYF
Burnett-Provan A 2E0VAV
Burnham C 2E0TBS
Burnham C M6CXB
Burnham H M0WYE
Burnham J G3UJK
Burnham J M6AOJ
Burnham T M3TBU
Burnie J G3ZXO
Burningham R G8UNP
Burningham Z M6BWL
Burnitt K G8HI
Burnley P G8ZMQ
Burns A MM0CXA
Burns A 2E0BVD
Burns A M6ALB
Burns A GW0UXJ
Burns B MI0TGO
Burns C M6YOB
Burns D GI0UXD
Burns G M6GTV
Burns G MM6GLO
Burns I G0AFH
Burns J G4ZXJ
Burns J M3AQP
Burns M MW6RJX
Burns N GM4DIN
Burns P GI4NLQ
Burns P M3DYR
Burns P MI3PDL
Burns R 2E0HJS
Burns R G7FMB
Burns R G3OOU
Burns S M0WUS
Burns S M3AUP
Burns S MM6AKQ
Burns T G0LVX
Burnside B G7CRV
Burnside K GI4IYO
Burnside R GI4RXS
Burnside S 2M0CNR
Burnside S MM0PDD
Burnside S MM6BKQ
Burr B M6CXC
Burr C G3VYX
Burr J G4TEU
Burr M M6MBR
Burrell D G4KTR
Burrell E G3LPU
Burrell J G8OZH
Burrell M M6BRL
Burrell R M0GAG
Burrett G G0KDW
Burridge A M6GUZ
Burridge C M6JSX
Burridge N G8NDR
Burridge W M6MUF
Burrill J M1BMW
Burrow F G8BME
Burrow J G0NYD
Burrow P G0GGE
BURROW R M3WTD
Burrows A G1HHS
Burrows B 2M0XFM
Burrows B MM3XFM
Burrows C G0GFI
Burrows C G8FFU
Burrows D G6LLL
Burrows D M3DBU
Burrows D 2E0BUZ
Burrows D M3ZUB
Burrows E M3EBU
Burrows G M3ZLA
Burrows J G0GPK
Burrows J G1LCS
Burrows J G3SUI
Burrows J G7RTN
Burrows J M6JAB
Burrows J G6GFC
Burrows M 2E0SBM
Burrows M 2E0FTI
Burrows M M6JSJ
Burrows P G1CZN
Burrows P M0KOI
Burrows P M6PGA
Burrows R 2E0BMJ
Burrows R G6DUN
Burrows R G6WEH
Burrows R G8BYI
Burrows S 2E0SHB
Burrows S G6MMD
Burrows S M3NXE
Burrows S M5BXB
Burrows T G6EJM
Burrows W G0ANE
Burrows-Ellis R M1DUD
Burrows-Ellis R M5AEF
Bursnall M G3NUB
Burston K G4XKK
Burstow M M3UQE
Burt A 2W1CAU
Burt A G6WXM
Burt B MM0GLX
Burt C G0PVV
Burt D 2E0GXI
Burt D G0DAX
Burt D M0CXI
Burt G GM3OXX
Burt J G7WLO
Burt K M6KBH
Burt M G7MHB
Burt M G8RGU
Burt P G3NBQ
Burt P 2E0BZS
Burt P M6NLO
Burt R G1YKZ
Burt S MM3YCI
Burt W M0AHT
Burtenshaw P G4ZWE
Burton A G1CWJ
Burton A M0DBY
Burton A G4VUA
Burton A M0GAV
Burton A M0LAZ
Burton A 2E0ERP
Burton A M0TNC
Burton A M6SQL
Burton B G1PHU
Burton C G8EGL
Burton C G1XET
Burton D G0GKY
Burton D G0SFV
Burton D G8LKS
Burton D MW6AHV
Burton G M0CHO
Burton G M0HCI
Burton G M1EXS
Burton H G0WWQ
Burton J G0CAP
Burton J G4TQE
Burton J G4ZHX
Burton J G6IVP
Burton K G4JRW
Burton K G6SSN
Burton K M6FXO
Burton M M0LHA
Burton N 2E0VTR
Burton N G6BBH
Burton P M1EQD
Burton P G3GEX
Burton P G3ZPB
Burton P G4CUR
Burton R G1XEH
Burton R G4KPX
Burton R M0BCZ
Burton R M6GTU
Burton R M0RHD
Burton R 2E0ROD
Burton R M0RHB
Burton S 2E0ENN
Burton S M6ENN
Burton S M6SQC
Burton W G4CWA
Burtsal S M0SBT
Buruma S M6EHK
Bury E G8VFL
Bury P G1LNQ
Burzynski M M0GQU
Busby A G4LFT
Busby G G4ORB
Busby P G0JCD
Busch M M0UKM
Buschl B G8KAS
Bush A G1EAM
Bush B 2E0DLO
Bush B G4WEY
Bush B M0LBB
Bush D G4SVS
Bush D M3YDB
Bush I M1IRB
Bush J 2E0DWM
Bush M G3LZM
Bush P M0AWH
Bush P M1PAB
Bush P G0OWC
Bush R G6PKY
Bush R 2E0CIR
Bush R G7RVH
Bush T M3KUQ
Bushby D 2E0EFQ
Bushby D MM0GKT
Bushby J G3WLV
Bushell A G0NUR
Bushell K 2E0BWK
Bushell K M0KVD
Bushell P G6HQ
Bucholl R G0HTM
Bushell R G4UNM
Bushell R G8KAE
Bushnall S M6OYZ
Bushnell I M6ICB
Bushnell M G0TEI
Bushnell M G4AZH
Buskin B M3HLX
busley N M6UUE
Buss D G1IBU
Buss F M0FLD
Buss W M6WJB
Bussell D G4EOT
Bussey C G1EXM
Bussey S G1EXK
Busson M G8MER
Bustard W GI0UVD
Buswell C M6WEJ
Butchart D 2E0ULH
Butcher A G8ZFL
Butcher A GM0ROU
Butcher A M6BDO
Butcher C M0XED
Butcher G G4RLA
Butcher G G0JFD
Butcher H GM0SUF
Butcher H M6GXP
Butcher J G3LAS
Butcher J G4GWJ
Butcher J M0JHB
Butcher J M6SHQ
Butcher J G4HKZ
Butcher K G4NLO
Butcher M G1SLG
Butcher P G4PMZ
Butcher P G3UDH
Butcher P G4GXB
Butcher R G3UDI
Butcher S G1MMA
Butchers M M3JBM
Butchers W G8XST
Butkus A 2E0WUF
Butland R G6RKJ
Butland R M3AZP
Butler A 2E0SSG
Butler A G0JCG
Butler A G1EAN
Butler A G7JWH
Butler A M0EVI
Butler A M3SSG
Butler A M6BSI
Butler A MM0DBR
Butler A G6MNL
Butler B M6BAM
Butler C G6ZGI
Butler C M6BOY
Butler D G0PFO
Butler D G4ASR
Butler D G4ZMP
Butler D G8JMK
Butler D MW3TUB
Butler D 2E0IJK
Butler D M0TFY
Butler E GI0JQQ
Butler E M6TPR
Butler F G0LWI
Butler F G8EQY
Butler G G4JVA
Butler G G6ZHJ
Butler G M0BDS
Butler H G4JSW
Butler I 2E0ISO
Butler I M3ISO
Butler J G8RKO
Butler J M3JAB
Butler J G0NRK
Butler J GM3ZMA
Butler J GD0NFN
Butler J G4JOW
Butler L M6SLL
Butler M G0LMD
Butler M G4OCR
Butler M G8KTX
Butler M M3MGO
Butler M M6GXV
Butler M M1MRB
Butler M G0GXZ
Butler M G0HCX
Butler M GW0MNP
Butler M G4UXC
Butler P G4URM
Butler P G1GTA
Butler P G6AXK
Butler P G6OHK
Butler P G7KMO
Butler P M6AOL
Butler R G0VKL
Butler R G4JXC
Butler R G1XYG
Butler S G6VUE
Butler S GI7ISX
Butler S MI3ISX
Butler S MW6SWN
Butler S M6JFN
Butler S G6XMA
Butler T G6VVZ
Butler-Roskilly A G3ZRJ
Butlin R G3YKS
Butson I G4HKC
Butt E G6BZW
Butt J G0JJG
Butt K M6BPH
Butt L G4KON
Butterfield D 2E0LDV
Butterfield D M3MZG
Butterfield C G4FPE
Butterfield G M1EEN
Butterfield P G4AAQ
Butterfield R G3VEV
Butterfield R G4UOI
Butterfield R M6AHL
Butters B GW0UAY
Butters C M3ZST
Butters J G6FBA
Butters J M6DEZ
Butterwick J G0AOO
Butterworth A G7DPW
Butterworth B G8FOT
Butterworth D G7JMU
Butterworth D G7VDQ
Butterworth G M6GDV
Butterworth J G0LHN
Butterworth J G0UOS
Butterworth P G0GPH
Butterworth R G0FYH
Butterworth R G4IJB
Butterworth R G6FDU
Buttery C G1KAK
Buttery J G7OPJ
Buttery N M3YNR
Buttery P M3DRW
Buttle D M0XDD
Button A G3YSK
Button C G0RPY
Button C 2E0ZCB
Button C M0YCB
Button J G8JMB
Button J M5AJB
Button J M0JFB
Button L G1LNR
Button S 2E0SNS
Button S M6SNS
Buttress H G3VHL
Buttress P G8NAM
Buttriss L M3BFU
Buxey K 2E0KBX
Buxton A GW8NBI
Buxton C G1KGQ
Buxton G 2E0GCB
Buxton G M3GKH
Buxton G G4VXG
Buxton J G7JSB
Buxton M 2E0CCK
Buxton M M0TTK
Buxton M M3MTB
Buxton P M1BLX
Buxton S G6NLU
Buzzing B G4BYM
Buzzing P G4EFS
Byard M 2E0MHM
Byard M M6MLQ
Byard R M6RBU
Byars G G8ZCV
Byatt M M1AHF
Bye A G3TCI
Bye E G8FAX
Byers D G4IZU
Byers D G6OCB
Byers T 2E0HYE
Byers T M0HYE
Byers T M6BYE
Byers W G3XIU
Byfield D M1FFA
Byford R G4MKR
Byford S G6UZM
Bygate R G1UUT
Bygrave R G0KNJ
Bygrave R G8WRV
Bygrave S M1SSB
Byles M G6UWS
Byne D G3MRQ
Byng D M6OAQ
Bynorth M G8SCI
Byrd M M6FNF
Byrne A G6SOZ
Byrne A G7EZE
Byrne B G0NSW
Byrne B G6HCI
Byrne B MI1AYL
Byrne C 2E0UAM
Byrne C G6JJF
Byrne C M3UAM
Byrne D G0LVN
Byrne E G1RUZ
Byrne E G8LF
Byrne E M6VWP
Byrne J 2E0JBS
Byrne J 2E0RFU
Byrne J M0RFU
Byrne J M6BYR
Byrne L M6UCY
Byrne M G3RYZ
Byrne M M6KEB
Byrne P M0BAI
Byrne P M6HNO
Byrnes M M6BKA
Byrom D G6LQE
Byron N G0PZM
Bysshe P G3WYK
Bystryakov G 2E0LSE
Bystryakov G M0SME
Bywater M M0DFF
Bywater S 2E0BCW
Bywater S M0EBN

## C

Cabban E G0ETU
Cabban M G0XDI
Cabban P G4OST
Cable R G4WSL
Cadd R G0DFV
Caddick J G4TCP
Caddick J G4VZL
Caddick J G8NWS
Caddick J M3NVS
Caddis J GM4UYK
Caddy R G0VBM
Caddy S 2E0FKH
Cade C G0DMB
Cade S G8SQY
Cadet D 2E0PKL
Cadet D M6PKL
Cadey A 2E0CPB
Cadey G G0CEY
Cadier A 2E0GTI
Cadman C G1DDI
Cadman D G3RLO
Cadman D G6XGF
Cadman M M1MAL
Cadman M 2E0ECM
Cadman M M6EVF
Cadman P G4JCP
Cadman S G1DIG

UK Surnames

**IMPORTANT NOTE**

**Revalidate licence to avoid revocation** – Ofcom has advised the Society that plans will be drawn up to revoke licences that have not been revalidated as required by the licence conditions. The quickest way to revalidate is to do so online via the Ofcom website: *https://services.ofcom.org.uk/* or by email: *amateur.validations@ofcom.org.uk* If you need assistance in the process, Ofcom staff are available to help, but please be patient during times of heavy workload.

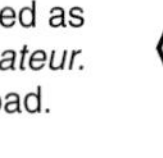

Cadman T G0HFE
Cadogan C G3XWB
Cadwaladr G M5CAD
Cadwallader R G1WMS
Cady D G0HGO
Cafe J M0JLE
Caffrey M 2E0MWC
Cafolla D GI4DOM
Cage I G4CTZ
Cahalan A MI6CAD
Cahill B M3FHV
Cahill D G3TGE
Cahill G 2E0TED
Cahill G 2E0ITE
Cahill G G4GXW
Cahill G M3WHY
Cahill P G0LBZ
Cahill T G0UXN
Caiden D MM0WST
Caiden D MM3PYX
Cain A M3GZT
Cain A G3YCX
Cain C G7KZG
Cain D 2D0YLX
Cain D MD3YLX
Cain D MD6IKR
Cain G G3DVF
Cain J G4GDF
Cain M M6FSU
Cain P G8IZW
Cain P M1DPU
Cain W M6WDV
Caine A M3NVA
Caine C G4IWS
Caine L G8HFL
Caine R G0LGW
Caine S G6NPW
Caine S M1ENX
Caine W G6JNZ
Caine W M3WTC
Caines C G1OQF
Cains R G7GLW
Cainsford-Betty C G8OHP
Caira R G4URD
Cairney J 2M0BYT
Cairney J MM3RGZ
Cairney T G1DPT
Cairns A MM3ODV
Cairns A MM6TUG
Cairns A M6VBS
Cairns B 2E0FAU
Cairns B M3FYV
Cairns E G0INA
Cairns I MI6ICD
Cairns J G3ITT
Cairns J 2E0YJL
Cairns J GI0NNK
Cairns J M3YJL
Cairns P 2E0LKC
Cairns P M6LCH
Cairns P M6OBY
Cairns Thomas C M6HJU
Cairns Thomas G M6HJQ
Cairns W GM7OPN
Caithness J M3ZUW
Caithness W MM3LQK
Cake A G1RUL
Calder G 2E0NIF
Calder G M0NIF
Calder I GM0WNS
Calder J G7HSN
Calder J G6EAM
Calder N GM0ERB
Calderbank K M3KEY
Calderley N M6FFV
Calderwood D GI4VHO
Calderwood L MI6UBE
Caldicott B G7NME
Caldicott D M6DLJ
Caldwell A 2M0TXY
Caldwell A GM0NGJ
Caldwell D M3EPQ
Caldwell G G6VVG
Caldwell P G4PAC
Caldwell S MM6SWC
Caley D G0PTL
Caley M G8DMT
Caligari E G0MRM
Caligari E G6TKY
Calkin G G4RTO
Calkin R G4MNT
Call L 2E0CNO
Call L M6HCI
Callaghan A M6GEY
Callaghan G G4SPE
Callaghan H G4OQH
Callaghan J G8VFM
Callaghan J M6YEL
Callaghan J GM4ZNS
Callaghan J G4GNO
Callaghan P 2E0BOF
Callaghan P G4WHL
Callaghan P G4WHM
Callaghan P M3LRF
Callaghan T G0DVG
Callaghan T GM6WTH
Callan N GW7HVA
Callanan T G0HNO
Callanan W MM0DBF
Callaway B G4FIG
Callegari A G3OMD
Callegari C G8CRC
Calleja R M6RAR
Callicott C G4DJJ
Callis J 2E0BQE
Callis J M3XCU
Callis M 2E0XMC
Callis M M6XMC
Callister W MD3ZHD
Callow M 2E0RMT
Callow M M6RKC
Callow N G0TUP
Calloway H G6CSR
Callum J G3ZMO
Calpin P G6HAT
Calter P G0RMD
Calthorpe E G0HXL
Calver R G4JGX
Calver S M1WHO
Calver S M3EMS
Calver S G6TXV
Calvert A G7MZW
Calvert A 2E0VXT
Calvert A M0XAJ
Calvert A M6VRT
Calvert B G7ITZ
Calvert C 2E0WCC
Calvert C M6LEJ
Calvert E G4EIC
Calvert I G0PCM
Calvert J MI3BKA
Calvert M M3XFC
Calvert R G0DSO
Calvert R G6BXR
Calvert R M6HXK
Calvert-Toulmin B G4YZH
Calvin A GI4XFE
Calvin A MI6RHL
Calvin D GI7WLA
Calvi-Parisetti P MM0TWX
Camac D G1IOO
Camber L G0XAM
Cambridge W G1CRT
Came T M3XSR
Cameron B MM3BUZ
Cameron B 2E0MCM
Cameron C G1EHX
Cameron C GM0HIG
Cameron D G1BZR
Cameron E MM0BIX
Cameron H G4VIS
Cameron H GM0GSG
Cameron J G4XNF
Cameron J G4YIM
Cameron J MM0CWJ
Cameron J 2E0EEB
Cameron J M6EEB
Cameron R GM4OHY
Cameron R M1CAO
Camley R GM4ZGU
Camm A G1UBL
Cammies S G3VNI
Cammish A 2E0OKZ
Cammish A M3OKZ
Cammish M G1FJJ
Camp A M6CDQ
Camp D 2E0TNE
Camp J 2E0FOX
Camp M G3MFK
Camp N G7KFQ
Camp N M3KFQ
Campanario D 2E0CIA
Campanario D M0ASR
Campbell A GM0WNR
Campbell A GM1JNC
Campbell A GM3NKG
Campbell A G6DTT
Campbell A M3JNJ
Campbell A M6LFM
Campbell A M3WKC
Campbell A GM8ICC
Campbell A 2M0CMA
Campbell A MM6AMA
Campbell B GM0FQQ
Campbell B MI0MSB
Campbell B M6POB
Campbell C GM1LUZ
Campbell C MM5AGM
Campbell C MI6CWC
Campbell C G7KDM
Campbell C M0LUS
Campbell C M6AIN
Campbell D GI3TAC
Campbell D G4ELG
Campbell D GI8PGJ
Campbell D M3MVO
Campbell D G0MWV
Campbell D 2M0DAC
Campbell D GW6VTZ
Campbell Davis T G3YMM
Campbell E G8PHS
Campbell G GM3HNE
Campbell G 2E0ZGA
Campbell G M0ZAM
Campbell G M3ZGA
Campbell G MM6GOR
Campbell H MM6HMC
Campbell H MM6HSC
Campbell I 2E0IAN
Campbell I G3URK
Campbell I GM4FFP
Campbell I G6ICC
Campbell I GM6TIB
Campbell I MM6DTV
Campbell J GM1TCP
Campbell J GM1YKE
Campbell J GM4RUP
Campbell J GI6UFU
Campbell J G7OKR
Campbell J GI8TME
Campbell J MM1JAC
Campbell J MM3OQV
Campbell J M3PCQ
Campbell J MI1DJW
Campbell J MI5AHG
Campbell J G4IUA
Campbell J GM4LHJ
Campbell J G4SGJ
Campbell J G7PUA
Campbell J MW6BVG
Campbell K G4NOX
Campbell K MI3DOD
Campbell M M0EQY
Campbell M M3HXS
Campbell O 2E0OPC
Campbell O M6OPC
Campbell P G7LSG
Campbell Q G4OEU
Campbell R 2E0RMC
Campbell R G6ABO
Campbell R GM6OQN
Campbell R M0HNL
Campbell R M0HME
Campbell R GM4CMI
Campbell S G3HDM
Campbell S GM4OSS
Campbell S MM0CFE
Campbell S MM3VBF
Campbell S 2M0BZB
Campbell T GI1GKI
Campbell T G8IWR
Campbell T MM0TCQ
Campbell T M6YTC
Campbell T MI5TCC
Campbell W GI4NKY
Campbell W 2I0WAS
Campbell W MI0WJC
Campbell W MI6WRM
Campbell-Black R 2E0YLP
Campbell-Black R M6YLP
Campigli P MW6SAW
Campion J 2E0VWL
Campion P G1OGB
Campion S M0KUR
Camplin P M3KZV
Camsey C M3OBU
Canavan W 2E0ECI
Canavan W M6CIE
Cane C M0COQ
Cane D M6DTC
Cane R G4KRH
canham D G8YKY
Cann R G4WHK
Cann T G4CTC
Cannam P M3IRP
Cannam S M3IRQ
Cannell J G7OAI
Cannell R G0RIC
Canning A 2E0BOT
Canning A G2NF
Canning C M6CES
Canning D GI0VIF
Canning J G1OMI
Canning J M0OJG
Canning R G0ARF
Cannings D G4DWC
Cannings M G1AZC
Cannon A 2E0DZP
Cannon B G8DIU
Cannon C G0VIA
Cannon D G1OKF
Cannon D G4IWQ
Cannon D G4ZHE
Cannon D GD4MCR
Cannon D M6DAL
Cannon G 2E0IKW
Cannon G M0IKW
Cannon P G8EAJ
Cannon R GW1LKG
Cannon R G8OTG
Cannon R 2E0EOZ
Cannon S 2E0HWQ
Cannon T G0HNL
Cannon T G6YLW
Cannon T G7MUT
Cannon T G0VQR
Canny P GI4MAZ
Cansfield D G0LDJ
Cant B M0IOW
Cant G G0CQB
Cant J M6DIO
Canterbury I 2W0BGQ
Canterbury I MW3IUS
Cantwell J G1ZRS
Cao T M3ZNO
Cape S M0LDX
Capel A G4ROX
Capell V G3JOR
Capewell P G0MQC
Capindale J G0CSV
Capon D M0GVI
Capon G M1CLO
Capon I G0KRL
Capon I G7APP
Capon J G8WTN
Capon N G1URJ
Capon R M6RRR
Capovila J M6FWO
Capp D G3CPT
Capper M M0MSC
Cappleman F 2E0BAJ
Cappleman F M0EAK
Capron A G8RWU
Capstick D 2E0JHD
Capstick D M0IKT
Capstick D M6JHD
Capstick E GM7OAF
Capstick E M0RCD
Capstick M G4RCD
Capstick M G4HOL
Carberry L M0HOK
Carberry P M6JED
Carberry T G4SBW
Carby I GM4XXO
Carby I G4YCS
Card M G7PGH
Cardell D G3ZUC
Carden A 2E0POB
Carden A M0HXG
Carden A M6BWQ
Carden D G3RIK
Carden J G0CWH
Carder E 2E0HLF
Carder R G7SRK
Carding D G6SGD
Cardno W GM0NRT
Cardwell C G4COV
Cardwell H M5AMV
Cardwell J 2E0UBW
CARDWELL P G3FXI
Cardwell P G7UEX
Care C G3WDM
Care M GD4MEX
Care W G7RIO
Carena M G8NIK
Carey D M6DMW
Carey D M6JHO
Carey J G8HSI
Carey J M6EMO
Carey M GW4VSE
Carey M M0POP
Carey N 2E0NAM
Carey N M0NAW
Carey N M6NKC
Carey P G3UXH
Carey S G4MJW
Carey W M6WDQ
Carfoot J M1JAN
Carfoot J M3GNB
Carfoot S G4JOT
Cargill D G0EPE
Cargill W GM0UZV
Carhart T G4KYE
Carins J G8YMD
Carins J M1BKI
Cariss R G7ACD
Carless G 2E0ZXG
Carless G M0ZXG
Carless G M3ZXG
Carless W M3WDC
Carlile A G0MNI
Carlile L G7GLL
Carlin G 2E0GVC
Carlin G 2E0GVD
Carlin J GM0NAE
Carlin M 2E0MCW
Carlin R MI3TXI
Carlin S 2I0SEC
Carlin S MI0NLY
Carlin S MI6EEC
Carline J G4JJY
Carlisle G G3SRJ
Carlisle S M6EQZ
Carlisle T GI8KYI
Carlisle T MI1EVD
Carlsen D G3XRC
Carlson R G6NPC
Carlson T 2E0HHE
Carlton A G7KBD
Carman A M6ADT
Carman D M0DRK
Carman J G1URW
Carmichael D 2E0DKA
Carmichael D M6EWN
Carmichael K GM0MNW
Carmichael M MM3JJI
Carmichael R GM0HYW
Carmichael R MI6RYC
Carmichael W G4BQR
Carn J M6WAQ
Carnall A G6FXZ
Carnegie P GM1CMF
Carnegie P GM8JCF
Carney A 2E0YDT
Carney G G4DRZ
Carney I G8JHM
Carney M 2E0MCL
Carney M M3LML
Carney R G6VTH
Carp J G3NHS
Carp O M0SMZ
Carpenter A GM0SNT
Carpenter A G3YWW
Carpenter C G1NQB
Carpenter D 2W1IHN
Carpenter D G6WAS
Carpenter D 2E0PDQ
Carpenter G M1GPC
Carpenter H G7DEH
Carpenter J M6JFF
Carpenter J G1YDQ
Carpenter J M0LJC
Carpenter L G4CNH
Carpenter P M0ZVB
Carpenter R 2E0DDP
Carpenter S G3ZQF
Carpenter S G8HUF
Carpenter S 2E0OZI
Carpenter S M6OZI
Carpenter S 2E0SJQ
Carpenter S M6SCX
Carpenter S M6SVJ
Carpenter T 2E0CFE
Carpenter T 2E0DYX
Carpenter-Beale O 2E0OCB
Carpenter-Beale O M3VXO
Carr A G6AFT
Carr B G4LOY
Carr B G4UHQ
Carr C G7UVY
Carr D 2E0HPD
Carr D G6PZS
Carr D G8MLW
Carr D G8NRU
Carr D 2E0KAC
Carr D 2E01DLR
Carr D M3KAC
Carr D M3UEX
Carr G G3KEK
Carr I GM6SEV
Carr K G1WUH
Carr L 2E0FNJ
Carr M 2E0BQK
Carr M 2E0CTZ
Carr M M3VOU
Carr M M6FBB
Carr N G0JHC
Carr N G1SHU
Carr N G6ZGO
Carr N M3OYU
Carr P G4CMC
Carr P M6HYT
Carr R M6FZA
Carr R M6EAT
Carr S 2E0HPC
Carr W M0BSB
Carr W M3WMC
Carragher J M3ESN
Carre P GU4XEA
Carrel H MJ3HWC
Carress I G8EAH
Carress W GI0WYK
Carrett D G4OPK
Carrick D G1VIG
Carrick G G3MRV
Carrick Smith J G6CFA
Carrick-Smith L G2FVL
Carrig T G8UYW
Carrigan S G4OUJ
Carrington C G4PDU
Carrington E G0RCF
Carrington J G0VNW
Carrington J M0NTK
Carrington R G0RRZ
Carrington R G6EAH
Carrington S M1ERU
Carroll D M6RLN
Carroll J G0KQP
Carroll J G3NFW
Carroll J 2E0JFC
Carroll J M6JFC
Carroll K MM0KRC
Carroll K MM3RCX
Carroll M M6MEC
Carroll P MW6PLC
Carroll R M1DBC
Carroll R G7MQW
Carroll S MI6COM
Carroll T G0HCD
Carroll T G7NHD
Carroll T M3TCX
Carroll W GM0PKP
Carrott D G0PMM
Carruthers D G0MGT
Carruthers D G6RVZ
Carruthers G GW4HGJ
Carruthers J G0NPQ
Carruthers M MI6ENR
Carruthers P G8MNL
Carruthers S M1ERP
Carruthers S M3MIR
Carruthers T G4XLA
Carrythers H M0HXK
Carskake D 2E0BMF
Carslake D G7JDI
Carslake R G4UMJ
Carslaw A GM0BCY
Carson B GM0GCO
Carson C G7OVK
CARSON D G4IHO
Carson J G3OXK
Carson K 2E0BTK
Carson N M3NYG
Carson P G8GGO
Carter A 2E0GHE
Carter A G6XPZ
Carter A G7IJD
Carter A M3EYH
Carter A M6CBO
Carter A M6ZAY
Carter B G0TPR
Carter B G1AVA
Carter B G8ADD
Carter B M5BAZ
Carter B G0TXA
Carter C 2E0FQE
Carter C G4TYA
Carter C GM1KWX
Carter C G1PCR
Carter C M6CYC
Carter D G3VYW
Carter D G4DJD
Carter D G8PGO
Carter D G8SOI
Carter D M1DDR
Carter D M6BSY
Carter D 2E0VGC
Carter D G6BIU
Carter D G8STJ
Carter D M0UAC
Carter D M6DKT
Carter D M6GBU
Carter E G8EFK
Carter E M3HQV
Carter F M3IBS
Carter G G0ISE
Carter G G8GYK
Carter G 2E0HTC
Carter G M6AUE
Carter G G6XXL
Carter G M6KGV
Carter H M3EAE
Carter I G0GRI
Carter I G6ZXN
Carter J G0GIL
Carter J G0LHZ
Carter J G4GEY
Carter J G4LPY
Carter J G7JTH
Carter J M1EFP
Carter K 2E0EBX
Carter K M6KGR
Carter L G6HCF
Carter L G7DNV
Carter L G8WW
Carter L MW0AMJ
Carter M G1UUS
Carter M G1VRP
Carter M G6CHJ
Carter M G6ETX
Carter M G6JNV
Carter M G7AJT
Carter M M6DYX
Carter M G7HNL
Carter M M3XYT
Carter M 2E0MPC
Carter M G0NEV
Carter M M6DNS
Carter M GW1DPV
Carter N G6SZP
Carter N M6SXA
Carter N M6NAC
Carter P G6FTJ
Carter P M6AFS
Carter P M6PBE
Carter P G6CLK
Carter R 2E0BJT
Carter R G4NDM
Carter R G4VSO
Carter R G6XQB
Carter R M0AIY
Carter R M3LBT
Carter R G0ZEP
Carter R G8WSW
Carter S G4FYB
Carter S G6JNW
Carter S G6PPY
Carter S MI6CXU
Carter S G7PZM
Carter T G1HJP
Carter V G8YQH
Carter W G6WSX
Cartin A MI0GRN
Cartin C MI3XEY
Cartin D MI0UTY
Cartledge A MM1WKD
Cartledge B G1JYB
Cartledge K M3YOG
Cartlidge D M0DWM
Cartlidge J GM0URU
Cartlidge S G0MJG
Cartmel C G4EST
Cartmell I M3LIU
Cartmell M GM8LFI
Cartmell P G6MED
Cartmell R G1LRM
Cartwright A 2E0AZK
Cartwright A 2E0FPK
Cartwright A G7MNS
Cartwright A M1DAM
Cartwright B 2E0GOP
Cartwright C M6CSE
Cartwright C M3XWN
Cartwright C M1EOZ
Cartwright D G7ODR
Cartwright H M3DZK
Cartwright J GW0KRQ
Cartwright J GW1WZI
Cartwright J G7GAP
Cartwright J 2E0RXN
Cartwright J M6RXN
Cartwright K 2E0COJ
Cartwright K M0KCC
Cartwright K M6BXO
Cartwright L G0UVL
Cartwright L G6KJY
Cartwright M G4DVM
Cartwright M G8WSV
Cartwright N G4DKX
Cartwright P G4YWA
Cartwright P 2E0AYS
Cartwright P G1GAN
Cartwright R M1NUS
Cartwright T G8EVD
Cartwright W G0DQO
Cartwright W G6IJQ
Cartwright-Proctor S M6SIW
Carty W 2E0BFN
Carty W M3JIT
Carvell C G0UBK
Carvell M G6DNH
Carvell M 2E0ZBZ
Carvell M M0ZBZ
Carvell M M3ZBZ
Carvell R G0PSG
Carvell T G0TFV
Carver C G4EYO
Carver G 2E0LHS
Carver G M6FKO
Carver J G0IZC
Carver M G4NCP
Carvill E G8RHU
Carvill G GI1ANG
Carvill J 2E0JEM
Carvill J M3JEM
Carvill T MI6TOC
Carville R G1XZQ
Carvin S G0DRD
Carwardine M M6MGC
Carwood S GD6SPI
Carwood W G0WVT
Casambros V M6BOK
Cascarino C M3KZB
Cascino S G6ZNW
Case D G4BTI
Case I G1SBW
Case L G8BJQ
Case M M3UMM
Case S M6AZF
Case S M0ALH
Casemore P G3SGF
Casewell I G8DUO
Casey D G4MVE
Casey G M6GSC
Casey I G7DNG
Casey M 2E0YES
Casey M M0TLY
Casey M M3GBC
Casey S G4JQK
Casey S G7VOQ
Casey S M6BWK
Cash C G0SMO
Cash D G7MEG
Cash D M3MEG
Cash G 2E0LGZ
Cash G M3LGX
Cash G G7LNO
Cash J G7AZA
Cash M M3LGY
Cash S M3FPU
Cash S M6SRC
Cash T M6FZX
Cashman M 2E0CGR
Cashmore V G4MOK
Caslake S G0DIR
Casling L M3LSF
Casling J G3MWZ
Caspell C G0IAC
Casper C G7CYN
Caspersz A GD1TMW
Cass G G4HEV
Cass T 2E0LCN
Cassar A MW6GSL
Cassar V G7CXO
Cassell R G8RFC
Cassell R M3AUF
Cassell T G8UVF
Cassells P 2E0TXP
Cassells P M0TXP
Cassells P M6TXP
Cassere D G3TAF
Cassidy E M0ECC
Cassidy F G4HBI
Cassidy F GM6JEP
Cassidy G MI6GPV
Cassidy J G4XXT
Cassidy M M0CAS
Cassidy P M0DDI
Cassidy R G6YTO
Cassidy S M0DOM
Cassidy T MM0BQJ
Cassidy T GM4RKM
Cassidy W G0WUI
Cassling R G4RPC
Casson A M1RDX
Casson A M6JUB
Cast D G4XQD
Castanheira M 2E0CAO
Castell A M6UUU
Castelow J G6MHR
Castle B G3ZJX
Castle C 2E0CTC
Castle C 2E0BSC
Castle C G7HFP
Castle D G1PGJ
Castle E GM0WRH
Castle I G6DUI
Castle M M6KQL
Castle P G7PTX
Castle P 2E0PWC
Castle P M6PWC
Castle W G6OQJ
Castledine N G4ALB
Castley K G0FDJ
Castley K G6NVC
Castro M M6AAL
Caswell J G7JTV
Caswell J 2E0EPM
Caswell J M6JLL
Caswell J GW7HQL
Caswell J 2E0TUF
Caswell J M3TUF
Caswell P G6PPV
Caswell R G4GMN
Catchpole D G0PFN
Catchpole M 2E0CSJ
Catchpoole B G1WMV
Catchpoole B M0TAD
Cater A G0EVX
Cater C M1DAU
Cater D G4WHZ
Cater G 2E0DNU
Cater G M6DNV
Cater J 2E0JWC
Cater J M0KCA
Cater J M6JWC
Cater T 2E0UOG
Cater T M3UOG
Cates L G4AVE
Catherall L G0MMT
Catherwood D G3ZRN
Catleugh D G8YDR
Catley C M6CAE
Catlin C 2E0XSD
Catlin C M0XSD
Catlin C M6XSD
Catling G G8ABX
Catling P G4FVA
Catlow R G7ICE
Catney R MI3PMR
Catney W 2E0IPC
Cato R G8YLA
Cato R M0YLA
Caton A G0OQP
Caton R G0LHD
Cattar B G4AED
Cattanach A MM0DRA
Cattanach P G7RGA
Cattani A G4RMJ
Cattel C 2E0LEC
Cattel P M3PSC
Cattell A M6YPA
Cattell A 2E0UDA
Cattell A M0UDA
Cattell A M6OUD
Cattell D G1KDO
Cattell M M1AED
Catterall A MM6BZQ
Catterall D G0NZR
Catterall L 2E0GBI
Catterall L M6LSB
Catterall M M3UOM
Catterall P M6PFZ
Catterall P G4OBK
Cattermole A M3EBZ
Cattermole D 2E0EVE
Cattermole D 2E0IXX
Cattermole D M3VDT
Cattermole K M3KIT
Cattermole L M3LDC
Cattermole P 2E0MIG
Cattermole P M0SDY
Cattermole P M3MIG
Cattermole T G6DNA
Catterson J G0GIF
Cattigan J 2M0MAV
Cattigan J MM0LBX
Cattigan J MM6LAD
Cattle G 2W0KMU
Cattle G 2E0ENN
Cattle G MW3KMU
Cattle S G1XSA
Cattley T G0CWZ
Catton P 2E0FYL
Catton T 2E0TJC
Cattrall C G4UGK
Catts A GU3LPV
Caudy C GW1KQV
Caulfield A M3ZYR
Caulfield J M6KWT
Caulfield Kerney F G6VUF
Caulfield N M0NDC
Caulton M G8XXU
Caunce K G0PTT
Caunt E G7IID
Causby C M1TET
Causer J M6GZG
Causer W G1ZTK
Cavalcante Pinheiro Filho C 2E0LUK
Cavalcante Pinheiro Filho C M6IVE
cavanagh M M6ODL
Cavanagh T M6CUP
Cave B 2E0GBJ
Cave B M6TXI
Cave C G1HXZ
Cave C G6ABP
Cave C G7VNC
Cave D M0LOU
Cave H G4RLX
Cave M M3OCP
Cave M M3ODN
Cave R M3SLF
Cave R M1LOU
Cave S M6BFX
Cave T G0JRX
Cavendish R G8XTD
Caveney D MM6DJC
Cavie G M0JJH
Cawkwell G G3ULD
Cawley H M1ARS
Cawley J G8FAW
Cawley R G8LPC
Cawood M G8IJT
Caws A M3ECU
Cawser D G0FSF
Cawser J G1NAU
Cawson F G2ART
Cawte E M1DTS
Cawthorne N G3TXF
Cawthorne S G4KOO
Cawthron V G4ODG
Cayford B M3UHO
Cecil W GM3KHH
Cedar B G8BMQ
Celyn R 2E0RHO
Celyn R M3LNH
Centanni J G0CKE

Ceresole J G8BSD
Cerveny M M0KBW
[illegible] G6SPG
[illegible] [illegible]
[illegible] [illegible]
Chacko S 2M0DSY
Chacko S MM6OAI
Chadbund P G6JXC
Chadburn C G7MGV
Chadburn C M0BCH
Chaddock A G3RTM
Chadwick A G3AB
Chadwick A G7KDJ
Chadwick D M6GOX
Chadwick D G0SKK
Chadwick F G0HFC
Chadwick G G6CSN
Chadwick J M0ANQ
Chadwick K G4EVC
Chadwick K G8XPB
Chadwick K M0TMO
Chadwick L G0OJB
chadwick L G0TTO
Chadwick L M6LEU
Chadwick N G3VMK
Chadwick P 2E0DHO
Chadwick P M6EWO
Chadwick P G3RZP
Chadwick R 2E0RCC
Chadwick R M0RCC
Chadwick R G8FCT
Chadwick S G0NRW
Chadwick T G4ZVU
Chadwick T G0CSU
Chadwick W G1IBS
Chady V M6VCA
Chaffey G 2E0GSC
Chaffey G M0OHI
Chaffey G M3OHI
Chaffey M M6MCL
Chak C M0VVR
Chaldecott J M0CZQ
Chalk A 2E0HNH
Chalk A M0BXJ
Chalk R M3RHI
Chalker R G1JRR
Chalkley M M6KIF
Chalkley P G1FGC
Challacombe N G0LGG
Challans P 2E0PGC
Challans P M6PGQ
Challen A G6DTW
Challen P G1EBW
Challen S G1EBX
Challender C G8PHQ
Challenger J G7PIB
Challinor A G1ELX
Challinor C G0EXD
Challinor J M3XGD
Challinor P G4YVD
Challis A G0VRE
Challis A G8VEM
Challis D M0ZLI
Challis J GM1USN
Challis J MM3USN
Challis L G0MJJ
Challis L G8SKG
Challis M G6BTR
Challis N 2E0MTX
Challis N M0WBG
Challis N M6MBP
Challis S G6AUD
Challis S 2E0CYP
Challis S M0HSU
Challis S M6DVW
Challoner K M3JNR
Challoner S G6VUX
Chalmers A GM4TEF
Chalmers B M0NBA
Chalmers C G6ETZ
Chalmers C M6CDC
Chalmers D G0IYE
Chalmers D G3WQG
Chalmers G G0ALW
Chalmers J G8VPO
Chalmers J 2E0CZJ
Chalmers J M0JBZ
Chalmers J M6CZJ
Chalmers M G0ALX
Chaloner D G3VSJ
Chaloner G M0GWC
Chaloner M G0UGR
[illegible] [illegible]
Chamberlain A G0AKC
Chamberlain A G4ROA
Chamberlain A G7VGN
Chamberlain A M6SCQ
Chamberlain A G7CFC
Chamberlain B G0WLS
Chamberlain B M3XBC
Chamberlain B M3YGU
Chamberlain D G6NUI
Chamberlain D G7EPE
Chamberlain D M6DRT
Chamberlain F G3XBN
Chamberlain G G4NPD
Chamberlain I G7TIE
Chamberlain M G3WPH
Chamberlain M 2E0XFF
Chamberlain M M0ZSS
Chamberlain M M6ESS
Chamberlain N 2E0EDG
Chamberlain P G4XHF
Chamberlain P M6BHU
Chamberlain P 2E0BTX
Chamberlain P M6GBR

Chamberlain R G3VYU
Chamberlain S G0JQZ
Chamberlain W G6FSG
Chamberlin S G0UYA
[illegible] [illegible]
Chambers A 2E0HKC
Chambers A M6HKC
Chambers B G4TAM
Chambers B G8AGN
Chambers B M6BNC
Chambers C G1AYU
Chambers C G1TQU
Chambers C G8MVJ
Chambers C M1ONE
Chambers C M3ONE
Chambers C 2E0CCA
Chambers D GI4OYI
Chambers D G4SYT
Chambers D M1AMI
Chambers D M3DFC
Chambers E G0HKC
Chambers G M3UBG
Chambers H M3HMC
Chambers I 2E0WEF
Chambers J M1ACC
Chambers J M1BHN
Chambers J M6BUC
Chambers J M6BSS
Chambers J K GI4OQY
Chambers K G0LUP
Chambers K GI0USS
Chambers K G8CTB
Chambers M G1AJE
Chambers M G8ZLT
Chambers N GD0IRM
Chambers N M6NEV
Chambers P G1LQC
Chambers P G7HIT
Chambers P 2E0PTA
Chambers P GD4ZUU
Chambers P M3PTA
Chambers R G0IIP
Chambers R GI4IOO
Chambers R G8BCA
Chambers R G8EZZ
Chambers R G4BAQ
Chambers R M6TGP
Chambers S G6VTE
Chambers S G6WYD
Chambers S M1ERV
Chambers S G8IHT
Chambers T G0KQK
Chambers T M6TFP
Chamings P G6USO
Champ C G1VSO
Champion A G7LBH
Champion A 2E0TWK
Champion A M0WSE
Champion G M3NGC
Champion I G3PUX
Champion M M3MGP
Champion M G1MPD
Champion P G3KKZ
Champion R G6KHG
Champion S G8LOF
Champness M 2E0IWT
Champness M M3IWT
Chan C G6BUP
Chan H M3OHY
Chan L MM0LMC
Chan T M3YKO
Chan W GM0PLH
CHANCE A 2E0EBZ
Chance A G1HYX
CHANCE A M3OJJ
Chance C G7EYS
Chance E M6NEM
Chance F M3FNC
Chance W M0GRO
Chance W M6WJC
Chancellor M MI0BDZ
Chance-Read J G4BOU
Chandler B MW3WXN
Chandler B M6GNC
Chandler J G0NVM
Chandler K G0ORH
Chandler K G7HNN
Chandler N G1UFS
Chandler P G3PID
Chandler P M1FRJ
Chandler R G4XUZ
Chandler R 2E0CXL
Chandler R M3UUZ
Chandler S M3KQC
Chandler S G3UDD
Chandler S 2E0SCQ
Chandler S M6SCV
Chandler W G0TAA
Chandler W G0VBG
Chandler W GW4LWD
Chandless L GD0FLH
Chandoo A 2E0ILM
Chaney P M3IJO
Chaney R G1CQA
Chang D M6CDG
Chang L MW6HKL
Channon J G8IKK
Chanter M 2E0MGC
Chanter M M0WMB
Chanter M M6CQG
Chantler J M3CAE
Chantler R M1CCL
Chantry G G0VEU
CHAPATON D G0CWL
Chapelle J M6EVC

Chaplin A 2E0CVC
Chaplin A M6CSZ
Chaplin A 2E0BJB
Chaplin A M0GGZ
Chaplin A [illegible]
Chaplin D G4CEM
Chaplin G GD8MAA
Chaplin J M3YTZ
Chaplin K G4NWO
Chaplin N 2E0LOE
Chaplin N M6LOE
Chaplin T G1UGH
Chaplin-Madden J M6CHJ
Chapman A GW3RIY
Chapman A 2E0TEE
Chapman A G4FZC
Chapman A M0YEE
Chapman A M6AXI
Chapman A G0NEF
Chapman A 2E0UWT
Chapman A M3UWT
Chapman A 2E0BQW
Chapman A M3SVP
Chapman B G0KKR
Chapman B G4UIP
Chapman B MW0CWS
Chapman B MW3DWZ
Chapman C G4LZF
Chapman C G8BUV
Chapman C M1AQP
Chapman D G3NGK
Chapman D GM4NVI
Chapman D G4SJV
Chapman D GI6DNI
Chapman D G4PPN
Chapman D M6FDG
Chapman G M6UDS
Chapman G G7RMG
Chapman G M0RMG
Chapman G M3RMG
Chapman H GW0TSL
Chapman H G8VEN
Chapman H G3NZL
Chapman I G0MTK
Chapman I G8VZS
Chapman J 2E0CIQ
Chapman J G4CCI
Chapman J G4FIT
Chapman J GI4LVC
Chapman J G7GJY
Chapman J G7UUP
Chapman J M3IWR
Chapman J G6PGT
Chapman J M6JGQ
Chapman K G4YZN
Chapman K G7CSJ
Chapman K M6KWC
Chapman L G4ZID
Chapman L M6LKF
Chapman M 2E0ECN
Chapman M G0BQI
Chapman M G0EMT
Chapman M G4NGF
Chapman M G4ZKE
Chapman M G8XIN
Chapman M M1ABM
Chapman M M6EXD
Chapman M M0MCH
Chapman N M1DTC
Chapman N M6NEF
Chapman N M3SNN
Chapman P G0SOX
Chapman P G1PAT
Chapman P G3JSR
Chapman P G4HUH
Chapman P G4JCG
Chapman P G4VBS
Chapman P G5VH
Chapman P M3IYX
Chapman P G0NMP
Chapman P G7DXN
Chapman P M6PGY
Chapman R G0NLG
Chapman R GM4IGS
Chapman R G7KEA
Chapman R G7PEU
Chapman R G3XPC
Chapman S G4UDD
Chapman S M1CKO
Chapman S M6CWI
Chapman S M6VPX
Chapman T G0MKA
Chapman T G4GWU
Chapman T G0TIO
Chapman T M3MXX
Chapman T M3OTQ
Chapman T 2E0OOE
Chapman T G0OOD
Chapman T M6BGZ
Chapman T M6XAF
Chapman T [illegible]
Chapman T M6FXG
Chapman T 2E0CHL
Chapman T M6ASP
Chapman V G0KAI
Chapman W G4FDA
Chappell A G3HVJ
Chappell B G0LPV
Chappell C G4MEK
Chappell H G6PPU
Chappell J G0JZT
Chappell J G6KJK
Chappell J G0BWB
Chappell J G1VUP
Chappell K G1DOA
Chappell K G1OPG
Chappell M G0GQX

Chappell R G6XPY
Chappell R M0RCI
Chappell S G3XRJ
Chappell S G4ZTQ
[illegible] [illegible]
Chapple C G4KRX
Chapple M G8XQS
Chapple T G3SGZ
Chappuis K M6KLC
Charbit A M3AJC
Charbonneau R 2E0RGC
Chard P G6NPZ
Charge R M3DTC
Charles A 2E0DQT
Charles A G4ORE
Charles C M6EVI
Charles D M1CZO
Charles H G7VZI
Charles J G3KVG
Charles J M1BTR
Charles S G3XYA
Charles S 2E0SSC
Charles S M6SJC
Charlesworth G M0HZA
Charlesworth H G4FMQ
Charlesworth M G0WWZ
Charlesworth R G4UNL
Charlotte N M6NDC
Charlton A G4HZP
Charlton A G6NUZ
Charlton A M0NUZ
Charlton D 2E0FSX
Charlton E G8UFO
Charlton G G1DKV
Charlton J 2E0EEK
Charlton J G3VRF
Charlton J M0KEE
Charlton M G0RMC
Charlton M G6ZQS
Charlton M G6OXZ
Charlton P G0UKL
Charlton R G1EBZ
Charlton R G3CPC
Charlton R G8SAN
Charlton R G7DAR
Charlton W M6BYL
Charlwood M G7GFQ
Charman G G0OMN
Charman M GW4UEJ
Charman T G6HBJ
Charman W M1BSB
Charnley F G4RPW
Charnley F G7CLR
Charnley J 2E0AFI
Charnley P G6SLO
Charnock D G0BCU
Charnock G M1FDK
Charnock J G4WXX
Charter J 2E0FZC
Charter J M3FZC
Charteris R G0NFO
Charteris R G1ZNV
Charters S G7JWW
Charters-Reid J M6ZEN
Chase B G4MYE
Chase C G7MTJ
Chaston R G1HOP
Chatel A M1BYG
Chater-Lea D G4EPX
Chatfield G G0LEH
Chatt L M6LMC
Chattenton K G4KIR
Chatterton D G8YXQ
Chatterton D 2E0FNY
Chatterton D M6FNY
Chatterton J 2E0JBL
Chatterton J M0OAA
Chatterton J M6BPQ
Chatzikos D 2E0BZO
Chatzikos D M3XPY
Chau J M3UFE
Chau S M6FBY
Chaudhry M G4UTE
Chauhan B 2E0KTV
Chauhan B M0KTV
Chauhan B M6BCH
Chaundy J 2E0JLC
Chave W M3GSR
Chawdhry H 2E0TRY
Chawdhry H M6HXC
Chawner D G0NTG
Chawner M G0SPC
Chaytor R G7SNJ
[illegible] [illegible]
Cheadle D M6NUG
Cheadle E G3NUG
Cheasley A G8DYP
Cheater G G4FUA
Cheatle C G3MYC
Chebsey D 2E0DEC
[illegible] M6DDC
Checketts A 2E0ITC
Chedzoy M G8UIG
Cheek P M6PCD
Cheema M G1WVM
Choor A G1IOP
Cheer E G8JEM
Cheeran G 2E0GLT
Cheeran G M0HPF
Cheers K G6TBV
Cheese J M1TCQ
Cheese N G1TZC
Cheeseman A G1VUP
Cheeseman D MW6ONZ
Cheeseman G G0DCN
Cheeseman I G1DOG

Cheeseman P G4XST
Cheeseman T M1TCI
Cheeseworth W G3HTT
Cheesewright N G7QJC
Cheesewright N M0MAW
Cheesley R M0LIV
Cheesman H 2E0HWC
Cheesman H M6HWC
Cheesman P G3KDE
Cheetham A G4AFI
Cheetham D 2E0IDM
Cheetham D G0HEX
Cheetham E G6PLT
Cheetham G G7UVF
Cheetham G G0EHK
Cheetham K G4RWD
Cheetham M MM6MYK
Cheetham R G0FLQ
Cheetham R G0NYR
Cheetham R G3JZT
Chell E MW3JVH
Chell M 2W0NRA
Chell P M6PDC
Chell R G0MKL
Chell R G6YIS
Chen C M3HZC
Chen X M3JYW
Chenery A G6PLU
Chenery D G0PPI
Chenery D G7IEY
Chenery G G4BVI
Cheney C G3RSE
Cheng A M6ACC
Cheng J M6CSW
Cheng T M6KKL
Chennells J G4CIJ
Chenoweth D G1HYQ
Cheriton D G6VGZ
Cherrie H GM0NHT
Cherrington D G4WRX
Cherrington D G4FIF
Cherry A GM0TEA
Cherry C 2E0BUJ
Cherry C M3WJV
Cherry M M6ARU
Cherry J M6EDI
Cherry M G1JYH
Cherry S G3SJK
Cheseldine P G8HMZ
Cheshire A G4EEL
Cheshire P M0GMO
Chesney W GI4DCC
Chester B G4DXB
Chester M M0MFC
Chester P G3YPD
Chester P 2E0CMQ
Chester P M6BCQ
Chester S G4DRA
Chesterman J G6HRA
Chesters E GM7FLZ
Chesters G M0YIJ
Chesters K G4NXW
Chesters L M0ELC
Chesters L M3LIN
Chesters M GM0OYU
Chesterton W G0AIZ
Chesterton W G0WHC
Chesworth G GM7OKX
Chesworth W M6NEW
Chetcuti J GW3PYX
Chettle S G8ATB
Chetwynd J G1RCW
Cheung A M6HFJ
Cheung H M6WUN
Cheverall C M3KRS
Cheverall R M0DTA
Chew C M3YGO
Chew C M6FBR
Chew G G7SQH
Chew R M0EBD
Chewter W G0IQK
Chibnell - Smith T G7IFR
Chick A G6ABM
Chick H M3BXG
Chick I M0IAT
Chick J G4NWJ
Chick R M0FAK
Chicken E G3BIK
Chiddick J G6PWJ
Chiddick R G8AUN
Chidgey C G3YLV
Chidgey D M6CJE
Chidgey M M3YHV
[illegible] [illegible]
Chidlow F G3WCM
Chidwick A G4XDW
Chidzey R G3IOM
Chiesa M M6BGG
Chiew S M3YYV
Chilcott M G7KII
Chilcott R G4RRA
CHILD J M3WXX
Child J G4PIR
Child P G6WVS
Childs A M0CBK
Childs C M1CSE
Childs D G7VGY
Childs G G3XEW
Childs G M3MEI
Childs J G4RUS
Childs L M6DNR
Chilinski A G0UOB
Chilton C 2E0RIZ
Chilton E G1GTF
Chilton F G7IZW
Chilton G G1JPS

Chilton N G2DJM
Chilton R G3IKQ
Chilvers B G0CUL
Chilvers D 2E0WPZ
[illegible] [illegible]
[illegible] [illegible]
Chin A GI0XAC
Chin I GI0TWX
Chin J G7NIL
Chinnappa G M6OIC
Chinnery J G0ONS
Chinnock J GM8RSC
Chinnock J MW0AGE
Chinnock P G0LCD
Chipman P G0GEF
Chippendale D G1ECC
Chipper H G0UWI
Chipperfield M M6VFC
Chipperfield R 2E0EVJ
Chipperfield R 2E0EVK
Chipperfield R M0VFC
Chipperfield S M6HDU
Chipperfield T G3VFC
Chisholm D MM6YHO
chisholm J G7KGP
chisholm J M5TTT
Chisholm M M0JJE
Chisholm N M3NEC
Chisholm R M3GTG
Chisholm S MM6EVS
Chisholm T G0RTC
Chislett C G0CAM
Chislett D G4XDU
Chisman J G4ADS
Chittenden K G8VAF
Chitty M G1SMB
Chitty P G8LGS
Chiu G M3ZJJ
Chiu Y M3UYK
Chivers D G3XNX
Chivers J G0AKK
Chivers J G4DJP
Chivers M 2E0JTN
Chivers M M0JTN
Chivers M M6JTN
Chmielewski J G4GQA
Chng S 2E0YIP
Chng S M0LAH
Chng S M6OAL
Cholerton F G1PIY
Chomer J G0ODM
Chong P 2E0CQV
Chong P M0HJA
Chong Y M6CYL
Choong L G0MFY
Chorley A G4HSN
Chorley A G4BKH
Chorley B G0RWA
Chorley D G6AYD
Chorley H G0PWV
Chorley P G7XPC
Chorley P G4YMG
Chorlton F 2E0LED
Chorlton W 2E0ROI
Chorlton W 2E0WRC
Chorlton W M3ROI
Chouings M G3XAW
Choules S G6PWF
Chown J G0BQE
Chown J M6JGV
Chown M G0AYA
Chown R M3RWV
Chown W G8FAD
Choy T M0TCC
Christensen J M6JEL
Christian J G8CRV
Christian R G3GKS
Christie C GM1TGY
Christie D MI3DCM
Christie D GI0ISQ
Christie D MI0BYR
Christie G GI4VWC
Christie G MM6YGT
Christie H M6HJC
Christie J GI7RAM
Christie J GM1ASY
Christie S MI0SAM
Christie S G3WHB
Christie T G7JHC
Christie T GM6RQW
christie powell C 2W0BBT
Christieson E 2E0GIZ
Christieson M M0FCD
[illegible] M0OYO
Christlo W G0KNL
Christmas E G0JOS
Christmas M GM1SRR
Christmas S G0BEC
Christmas T G1IUT
Christodoulou P G0VBI
Christodoulou S M0DVK
Christofi G G0JKZ
Christofi S 2E0YXZ
Christofi S M6YYY
Christofi X M6SNO
Christoforou J M6AXP
Christopher C G4PZJ
Christopher H G3VBI
Christopher N M6DYY
Christopher R G0ISI
Christopher T M6TLC
Christopoulos P 2E0AUU
Christy J G1ECI
Chronopoulos P M0PCX
Chroston R MM0ZRC
Chrumka D M6DOM

Chruscinski A 2E0EIX
Chruscinski A M3OQG
Chrysostomou P G6JEU
Chrzanowski M G7IAM
Chu KY M3XQY
[illegible] [illegible]
Chubb A M6VSA
Chubb D G4LUY
Chubb F G3ARE
Chuck S M6TTR
Chudasama A M6EGJ
Chung L M1CTO
Church A 2E0FRF
Church D 2E0FRK
Church D M0DSZ
Church I G1HQH
Church J G4INI
Church J MM6HBY
Church K G8XIR
Church K G1HYU
Church M G4ENZ
Church Q MI6UVS
Church R G3KJC
Church S G1LIK
Church S M3KUY
Church S M6FEX
Churcher M 2E0MJC
Churchill C G8OTH
Churchill E G1PBX
Churchill I G8XIM
Churchill J G4TEK
Churchill J G7PWJ
Churchill J 2E0EGP
Churchill J M0ARQ
Churchill J M6EGP
Churchill J G7ONI
Churchill K M1CZA
Churchill L M6FWP
Churchill R M6RTC
Churchill S MI3FCA
Churchley A G4EAQ
Churchman H G0CTH
Churchman M G1SVN
Churchward C M1DXL
Churchyard E G3TVR
Churms M G4NQL
Chuter B G8CVV
Chuter C M1FCC
Chuter S 2E0BUI
Cianni M 2E0SES
Cianni M M0OER
Cianni M M6LFT
Ciechan M M0JCZ
Cilia R G1PLV
Ciotti C M3BDC
Ciotti P G3XBZ
Civil G G0LKY
Civil R G6ION
Civita L G0USA
Civita L G7ALC
Clabon I G4FVB
Clabon I G0OFN
Clacher N G4XQF
Clack A G6WYE
Clack M 2E0KZC
Clack M M0KZC
Clack M M3KZC
Clack S G4YEK
Clack S G6NUX
Clague R G6WGY
Clamp G G3YTX
Clamp K G3ZOW
Clamp R G0IMD
Clampin G G0YKC
Clampitt A G0GWH
Clancy J G4TKS
Clancy M G1MWT
Clapp D M1ANL
Clapp D M0GMT
Clapp D M1RAD
Clapp R M0BCC
Clapp T 2E0LDZ
Clapp T M0TDZ
Clapp T M6LDZ
Clapp W G0GDT
Clapperton M G0OYA
Clare F G3NWS
Clare F G8GT
Clare J M6MOM
Clare J M6JCC
Clare M G7TLD
Clare N 2E0NGG
Clare N M3NGG
Clare N G6LNF
Clare R M0REC
Clare R M6BEK
Clare R 2E0WTY
Clare R M0WTY
Clare R M3WTY
Clare-Noon R G4EJQ
Clarey J M0CGR
Clarey S M0CJZ
Claridge M G0UWU
Claridge R G8GYM
Claridge R G7DBT
Claridge V G0WUW
Clark A G3BII
Clark A GM3DIN
Clark A G3UBY
Clark A GW4SKP
Clark A G4WIY
Clark A G7ABR
Clark A MM0DHQ
Clark A M0DPV
Clark A M0PDF
Clark A M3DXD

Clark A G7KYJ
Clark A M0XYZ
Clark A M6ACX
Clark A G8EVI
Clark A M0NLR
Clark A MM0CYO
Clark A M3YTQ
Clark A M0HQE
Clark A M6LLD
Clark B G1ECE
Clark B G3HGL
Clark B G3VCL
Clark B G4KZI
Clark B MM3BSC
Clark C 2M1DZS
Clark C G1GQJ
Clark C G8BKQ
Clark C M0PDE
Clark C G1GTH
Clark D 2E0HGX
Clark D GW0KQV
Clark D GW0KWA
Clark D G0WDC
Clark D G6FTH
Clark D G7BIK
Clark D GM7OSQ
Clark D G7UDE
Clark D M3HGX
Clark D 2E0DCL
Clark D G0HVB
Clark D GM1MSO
Clark D G8GPF
Clark D M6KSC
Clark D 2W0FCF
Clark D MW6XXC
Clark D G4ELJ
Clark E 2M1EUV
Clark E 2M1GLD
Clark E G0WIC
Clark E G0WNF
Clark E G3PNU
Clark F 2E0FWC
Clark G 2E0KLT
Clark G G0MGC
Clark G G0TYQ
Clark G G3ZHU
Clark G G6AGA
Clark G M0KLT
Clark G G0BKR
Clark G 2E0EHT
Clark G M0IUM
Clark G M6MPL
Clark H G3YOY
Clark H 2E0HNC
Clark H M6FED
Clark I G1XTD
Clark I G6UIF
Clark I M6ARR
Clark I G7NIB
Clark I M6AVY
Clark J 2E0JHC
Clark J 2E0JPC
Clark J 2W1FGR
Clark J GM0LBN
Clark J G1OOB
Clark J G4DBE
Clark J G6IIZ
Clark J MM0CNF
Clark J M0IDC
Clark J M3JHC
Clark J M3JNQ
Clark J G4NFV
Clark J M6ZXI
Clark J 2E0JZC
Clark J MI6JJZ
Clark J M6JZC
Clark K G4MPQ
Clark K MW3HZB
Clark K MM3KVN
Clark K G0CQO
Clark L 2E0GSL
Clark L M0LAE
Clark L M6LXC
Clark L G6LIK
Clark L M6FYD
Clark L M6CBM
Clark L MM6LIL
Clark L 2E0BPT
Clark L M3VYM
Clark L M6LLC
Clark L MM6LCL
Clark M GM6AES
Clark M GM6OFO
Clark M G8VLP
Clark M G1JBM
Clark M M6JHC
Clark M G0EHR
Clark M M1EPX
Clark M M6CLK
Clark M M6ZDZ
Clark M M6SAC
Clark N G0LGT
Clark N MM0HZO
Clark N G6YMA
Clark N M6NWC
Clark N M0BNC
Clark P G0DTI
Clark P G0TBO
Clark P G0VCY
Clark P G0VDT
Clark P G4FUG
Clark P G6RCD
Clark P G7WKH
Clark P M0PWC
Clark P M6HPY
Clark P G4PGS



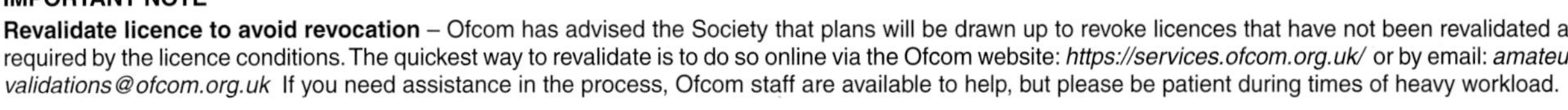
**IMPORTANT NOTE**

**Revalidate licence to avoid revocation** – Ofcom has advised the Society that plans will be drawn up to revoke licences that have not been revalidated as required by the licence conditions. The quickest way to revalidate is to do so online via the Ofcom website: *https://services.ofcom.org.uk/* or by email: *amateur.validations@ofcom.org.uk* If you need assistance in the process, Ofcom staff are available to help, but please be patient during times of heavy workload.

Clark P G8BZR
Clark P M1EBY
Clark P M6GRO
Clark R G0CXV
Clark R G0IGA
Clark R G0MOU
Clark R G0PBR
Clark R GM1BXI
Clark R G8GQJ
Clark R MM0DTW
Clark R M0DZT
Clark R M3DUL
Clark R M5ABT
Clark R G3LFE
Clark R G4DDP
Clark R G8BXC
Clark R G0FXR
Clark R M6LLE
Clark S G4XGR
Clark S M1SAC
Clark S MM3VSC
Clark S M0JPP
Clark S MM3WVN
Clark S G8HNA
Clark S M6CSA
Clark S M6CLA
Clark T G4ZLU
Clark T 2E0GYW
Clark T GM0WFA
Clark V GM3WSR
Clark W G0WTC
Clark W GM1MSN
Clark W GM4XND
Clarke A G4NFY
Clarke A M1ZXZ
Clarke A M3DSI
Clarke A G7CST
Clarke A 2E0GCI
Clarke A M0KIN
Clarke A M3LAP
Clarke A G1GTQ
Clarke A G6XXN
Clarke A M3XCA
Clarke A M6DEF
Clarke B G0MXX
Clarke B G4ICB
Clarke B GW6RDV
Clarke B G8MUV
Clarke B G0ECZ
Clarke C G1ATG
Clarke C G4XCX
Clarke C G6BWA
Clarke C G3SQU
Clarke C G4LPW
Clarke D G0IDD
Clarke D GW0MVS
Clarke D G3KYZ
Clarke D G6XWY
Clarke D G7KAO
Clarke D G8RFP
Clarke D G8TEB
Clarke D M1EJE
Clarke D M3YCD
Clarke D M3LBP
Clarke D M3UPO
Clarke E G3UYD
Clarke E M6VZF
Clarke F G3WII
Clarke F G7BXG
Clarke G G0FWX
Clarke G G4LKM
Clarke G G8XXV
Clarke G 2E0DAQ
Clarke G M3VWD
Clarke G M3XGL
Clarke G M6GRC
Clarke G 2I0GIF
Clarke G G1DNZ
Clarke G G3YTW
Clarke G M6DCD
Clarke G MI6GIF
Clarke H GW0PYU
Clarke I G1CPX
Clarke I 2E0DUM
Clarke I M6IVN
Clarke J G0JDL
Clarke J G0OBE
Clarke J G1HIU
Clarke J G3OWQ
Clarke J GI4GUH
Clarke J G6IIP
Clarke J M3HXH
Clarke J M0JHC
Clarke J GI4HCN
Clarke J 2E0CWQ
Clarke J G1CEU
Clarke J M1EJG
Clarke J M6JDU
Clarke J M6KQJ
Clarke J M6XBB
Clarke K G0JKC
Clarke K MW6WLB
Clarke K M6KBV
Clarke K 2E0NCN
Clarke L G1LQB
Clarke L G7RSE
Clarke M 2E0NOW
Clarke M G3CQL
Clarke M G7DUK
Clarke M G7RHD
Clarke M G8GGS
Clarke M M0HKG
Clarke M M1BNI
Clarke M M3HHB
Clarke M M3JEP
Clarke M M3LQE
Clarke M G8FHI
Clarke M 2E0MFF
Clarke M M6FFY
Clarke M 2E0MEV
Clarke M G0VAE
Clarke M M0CRA
Clarke M M0DEQ
Clarke M MI5MTC
Clarke M M6MHV
Clarke N G0CAS
Clarke N G5VO
Clarke N M0WKR
Clarke N GM8PZR
Clarke P G7FML
Clarke P G7WJZ
Clarke P MI0TRC
Clarke P M1AHJ
Clarke P MI3DFR
Clarke P G1GTR
Clarke P G3LST
Clarke P G8EDH
Clarke R G0IBI
Clarke R G1GTS
Clarke R G8UNO
Clarke R M6EBL
Clarke S GI7MDJ
Clarke S G8LXY
Clarke S M3OUV
Clarke S M3ULO
Clarke S M6AKO
Clarke S M6RPZ
Clarke S G3ODX
Clarke S M6CJV
Clarke S M6DSE
Clarke S M6MDM
Clarke S G7UID
Clarke S M3UZL
Clarke T 2E0KNA
Clarke T G4RKD
Clarke T M0WET
Clarke T M6FLX
Clarke T M6TCZ
Clarke W 2E0IHB
Clarke W M1AJQ
Clarke W M3WAC
Clarke W M6WMH
Clarke W G0NGE
Clark-McIntyre J 2E0JCM
Clark-McIntyre J M6AEK
Clarkson A G2CJK
Clarkson C GM3ZEU
Clarkson D G4JLP
Clarkson F G7TWU
Clarkson G G7ITM
Clarkson J 2E0PDP
Clarkson J G8XJT
Clarkson J M3PDP
Clarkson M G2FLW
Clarkson R G4IHY
Clarkson S M3GWC
Clarkson W G4XYR
Clasper R GM0LOT
Claughton J G0MPM
Clavey D 2E0CQC
Clavey D M0OBY
Clavey D M6RGC
Claxton J G4SCM
Claxton M G7TYT
Claxton R G0TOT
Clay A M0AXJ
Clay A M6XAB
Clay B G6ECN
Clay E G0BTT
Clay J M6JNC
Clay J M6JRC
Clay M G4PJP
Clay P M0PCK
Clay R G0VJI
Clay R G1DNY
Clay R M0VZT
Clay S 2E0BST
Clay S M0SKC
Clay S M3OAJ
Clay T G3TSV
Clay-Burley M M3OVF
Clayden M G6MIC
Claydon B M6BJT
Claydon C G4ITC
Claydon C GM4ZJI
Claydon F G7SMT
Claydon J 2E0WSM
Claydon M G6ABJ
Claydon R M3LQO
Claydon T G8RZL
Clayphon A G6IJK
Claypole C M6GFA
Clayton A G4SEG
Clayton A 2E0PIX
Clayton A M6LOK
Clayton A G7HZZ
Clayton A M0ASC
Clayton B 2E0PEC
Clayton B M6PEC
Clayton B 2E0EBD
Clayton B M0IBR
Clayton B M6EBO
Clayton C G3VCY
Clayton C G4UUU
Clayton D M6WSK
Clayton D G4OCF
Clayton E G0KZH
Clayton G G0VXD
Clayton G 2E0GVC
Clayton G M6GVC
Clayton H M5OOO
Clayton I G6DHW
Clayton J G0JRC
Clayton J G0KJC
Clayton J G4PDQ
Clayton M G8VMF
Clayton M M3MRC
Clayton M M6YEB
Clayton M M6ZMC
Clayton N G0GVS
Clayton N G0IFS
Clayton N M3BYX
Clayton P G4ANQ
Clayton R G0DAM
Clayton R G3ZJT
Clayton R G8SDU
Clayton R 2E0OOO
Clayton R G4SSH
Clayton R G7ROY
Clayton S G7FTM
Clayton T 2E0TCP
Clayton T G0TKJ
Clayton T G1YTX
Clayton T M6BLG
Clayton W M0WHC
Claytonsmith F G3JKS
Claytonsmith M G4JKS
Cleak J G4JBQ
Cleak L G8SBK
Cleal A M3RNX
Cleal D M0DLX
Cleall P G8AFN
Clear R G4FFX
Clearn A MI6EQD
Cleary G MM3NYY
Cleary P G3WZE
Cleary R 2E0DTK
Cleaton J G4GHA
Cleave A G8XQN
Cleaver A G0LCB
Cleaver D G0JVF
Cleaver J M3SJQ
Cleaver N G4HOW
Cleaver R GW8KSL
Clee J G8EYQ
Clee M GW6OVD
Cleeter J M6JAZ
Cleeton G G3LBS
Cleeve J G3JVC
Clegg D 2E0MIB
Clegg G G1LCN
Clegg G G4FIQ
Clegg G GI7PUG
Clegg J G3UYG
Clegg J G4BIC
Clegg P G3MGH
Clegg P G3TTB
Cleghorn T G4ZTC
Cleghorn T G0VYB
Cleland C GI0DHW
Cleland M 2M0MMB
Cleland M MM3BQK
Clelland A G3UUQ
Clem G G7IUE
Clemence J G3YHK
Clemens D G3VXM
Clemens P G4RAR
Clement D M6LPS
Clement F M6FRC
Clement I M6IHC
Clements A G0PIK
Clements A 2E0CUB
Clements A M6BRQ
Clements B M6MEO
Clements B 2E0CBB
Clements D G6KTG
Clements F 2E0FRC
Clements F M0TAQ
Clements G M3RRU
Clements G M6GCC
Clements J G3SGS
Clements M G7IPA
Clements M M0MDC
Clements P GM6LDG
Clements R G0FGW
Clements S G1YBB
Clements S G6NQB
Clements T G0IKP
Clements T G0STF
Clements W G6RMJ
Clementson E MI5ALO
Clementson G G4TIM
Clemmetsen A G3VZJ
Clemons A G4ZRY
Clench D G0BVV
Clennell G G1LMZ
Clennell G M0GEC
Clenshaw D M0COJ
Cleveland R G0AKO
Cleverley C M3YHU
Cleverley J M3HWS
Cleverley M GW6MIH
Cleverley M G7PNF
Cleverley R G0GJE
Cleverley R G4DMC
Cleverley R GW4RKZ
Clewer A G4CTA
Clewer D 2E0CLW
Clewer D M0VZS
Clewer D M3VZS
Clewes A 2E0RAL
Clewes B G7GXR
Clewley I G7KAK
Clews A G1ASU
Clews D G6GLZ
Clews M 2E0MFC
Clews M G1VVF
Clews M M3CER
Clews R G0SMZ
Cliff D G1IGW
Cliff N G1VIT
Cliff R GM0SXP
Cliff R GM8DFC
Cliffe D G8EQC
Cliffe D G0JWE
Cliffe G M3GHO
Cliffe S G1PDS
Clifford B G7GFR
CLIFFORD B M6UBC
Clifford C G4IIC
Clifford G G4SHK
Clifford J 2E0BHF
Clifford J G4BVE
Clifford J G8DHT
Clifford M MM6MJC
Clifford M MW6BZP
Clifford M 2E0RBH
Clifford M M3VYS
Clifford P M0PTR
Clifford R G8VUU
Clifford-Smith R M1FJA
Clift R G4VRJ
Clift K G0ECL
Clift L M3NVO
Clift-Jones A G4YCJ
Clifton A G4JCZ
Clifton B G7OWQ
Clifton C 2E0HJE
Clifton D G3WOK
Clifton D G8PIY
Clifton F M6FPF
Clifton G G7NZM
Clifton J G3XYE
Clifton J G6LAE
Clifton R G0MYC
Clifton S G4WBT
Clinch G G3RPD
Clinchant A 2E0BQY
Cline D G4VGY
Cline S G4MDZ
Clingan J G3TNI
Clink S GM1VBE
Clint T G8TMD
Clinton G G3OYT
Clinton R G0BUX
Clinton W G8KZN
Clode C G0CJC
Clohessy D G4BMO
Cloke F G0EON
Cloke R G1EXU
Cloke T G0JYX
Close B G0VLQ
Close C G1HQE
Close D G1VKC
Close D G3MFG
Close D 2E0DNO
Close D M6DCO
Close D G6ECT
Cloude E G7FAQ
Clough B G6YUX
Clough D G8CEP
Clough D M6DAC
Clough F G1GCF
Clough H M6ARJ
Clough H M3TOJ
Clough J GM0MDD
Clough M 2E0TNV
Clough M M0TNV
Clough M M3TNV
Clough P M6CLO
Clough R M6RYN
Clough T G4PHR
Clough W G3USW
Cloutman R G6PWL
Clover I G4UGD
Clover J M6OSF
Clover R G0RMU
Clow R 2E0EBU
Clow R M6FZT
Clowes B G4HBZ
Clowes J M3LWP
Clowes M G0FOC
Clowes P G6KKN
Clowes W G3ICZ
Clubley D G6XXJ
Clubley R G8PLO
Cluer G G4AVV
Clues B G0KOY
Cluee B G0LXG
Cluley G G7FGR
Cluley J G4YIG
Clune P G0LUL
Clunnie A M3RYZ
Clutson R G0WHO
Clutterbuck D M0AVZ
Clutterbuck P G4FTQ
Clutton M G4OAB
Clutton M G4VQH
Clyne J M6HTM
Clyne N G8LIU
Cmoch S G1DPW
Coad J G6IZQ
Coad K M3IZQ
Coad M M0XMC
Coady J G7INY
Coady J G8PZS
Coady N M6NCX
Coate D G1YHE
Coates A G3TVV
Coates A G6NPE
Coates A G7DJN
Coates A G0EWT
Coates B G0SHM
Coates C 2E0KCC
Coates C G1VQI
Coates D G1HEN
Coates E MI3SRG
Coates G G0COA
Coates I GM1MSS
Coates J G1WJG
Coates J G6PJT
Coates K G3IGU
Coates P G4VMC
Coates P G6PVC
Coates R G4DIH
Coates R M6RCD
Coates S M6VGE
Coates W G1OFX
Coates W MI3BLN
Coathup A M1KMC
Coatman R G1ZME
Coaton A G4TFO
Coats A 2E0WBX
Coats A M6WLR
Cobain P MI6MIB
Cobb A G0WJK
Cobb A MI0BTK
Cobb B G1DBH
Cobb E M0GUH
Cobb F G3YIH
Cobb J G3VHS
Cobb J M1FFV
Cobb J M0JBF
Cobb J M6JVB
Cobb L G3UI
Cobb R 2E0RBP
Cobb R M6RPC
Cobb R MW0COB
Cobbold D 2E0PUG
Cobbold D M6CHL
Cobbold J 2E0COB
Cobbold J M3OSF
Coben S G1KGO
Cobern R M6ONO
Cobley J G4RMD
Cobley J GM0DWH
Cobley P M3THN
Coburn M G4DUL
Coburn P G0NOP
Coburn S G0NOO
Coburn S G7COB
Coburn W 2E0CPS
Coburn W M0YAH
Coburn W M6ZYX
Cochrane A G4UIO
Cochrane D G8IHF
Cochrane G 2E0GDO
Cochrane G 2M0GCS
Cochrane G MM0GHM
Cochrane G MM3GDC
Cochrane H GM0LYH
Cochrane J MM3VEE
Cochrane L MM0LEN
Cochrane P G3RVC
Cochrane R G8DRQ
Cochrane R M0RAC
Cockayne D M1DGE
Cockayne P G7GJA
Cockayne R 2E0RPC
Cockayne R M3XCF
Cockayne T 2M0OML
Cockayne T MM0OML
Cockayne T MM3OML
COCKBILL R G0KTP
Cockburn A M0DAS
Cockburn A 2E0NFC
Cockburn A M6NFC
Cockburn A M6MNC
Cockburn D G0VWQ
Cockburn D M0SED
Cockburn D 2M0YFR
Cockburn D MM3YFR
Cockburn G 2E0GTE
Cockburn G M0TYN
Cockburn G M6GAK
Cockburn K G0SKJ
Cockburn K M6SLU
cockburn S G0VXB
Cockcroft G G6WEI
Cocker A G7LFM
Cocker D G4RYE
Cockerell G M6EXZ
Cockerell W G0LKI
Cockerill A G4UTV
Cockerill B M0CNH
Cockerill E G4GOZ
Cockett A 2E0BWT
Cockett A M6AEW
Cockett E M6ELY
Cockfield B G0KRK
Cocking N G8PNM
Cocking R G1BEB
Cockings D G3WF
Cockings R M0CKS
Cockle A M6ZXQ
Cockman A G4UTF
Cockman P G4UAI
Cockman R G1HNN
Cockram A M0BJE
Cockram O G8OWZ
Cockram P G8CUG
Cockram T G8GFZ
Cockrill J G4CZB
Cockroft J G1MSR
Cockroft R M0PIE
Cocks D M6EPN
Cocks J M0BHV
Cocks K GM6AXZ
Cocks S G4ZUL
Cocks T 2E0CNU
Cocks T M6BOA
Cockshaw W G4NBH
Cockshoot I 2E0COV
Cockshoot S G1WWR
Codd D MW6COD
Codd M G3NNA
Coddington P M1BKL
Codling M G7FEA
Codling T G3UPI
Codrai G M6GYC
Codrai R 2E0RXC
Codrai R M6RCS
Codrai T G3WQY
Cody M M6NAM
Coe A G1MKV
Coe A G4PNL
Coe A G6UHS
Coe D G7SDC
Coe D 2E0DDC
Coe D M3XOE
Coe E G0COE
Coe K G7VUU
Coe K M0AZW
Coe M M0DMD
Coe S M0BHO
Coenraats P G0CEC
Coffey D G0BPM
Coffey D G1SFU
Coffey E M6IZY
Coffin B G7EPX
Coffin G G3XFN
Coffin S G7JJW
Cogan T G7USJ
Cogdon W M6WLC
Cogger G G4YRT
Coggins D G4NBP
Coggon G G4FYE
Coggon I 2E0IMC
Coghill F GM0PXR
Coghlan A M0JJC
Coghlan A M6HAU
Coghlan H 2E0HMC
Coghlan H M6DYU
Coghlan P G6IOB
Coghlan T M6EUC
Cogle I MM3WJZ
Cogman J M6KJR
Cogzell R G3OTY
Cohen A G6YIP
Cohen E 2M0IEC
Cohen E MM3IEC
Cohen I G0EFL
Cohen M G8NNX
Cohen M G4XWA
Cohen M M6NBX
Cohen P GM3LKY
Coiley M M6MCJ
Coisson D M0CUH
Cokayne F G0VFU
Coker C G4FCN
Coker C G8GCS
Coker G G6CLD
Coker M G1LIG
Colbeck P G4PHK
Colbert E M6OAB
Colborn I MM6JDN
Colbourn N G0HOD
Colbourne J 2E0TWT
Colclough C G1VDP
Colclough G 2E0IIJ
Colclough G G6ALN
Colclough H M3GLC
Colclough R G7FSA
Colcombe A 2E0VFA
Colcombe A M6VFA
Coldbeck D G6ABG
Colderwood J 2E0BMI
Colderwood J M3RQY
Coldham F G1GGK
Coldham G M6HHA
Coldicott M G8DKV
Coldicott P G1UNN
Cole A G4XAE
COLE A G8TVU
Cole A G7VGJ
Cole A 2E0WKG
Cole A M6WKG
Cole B G0WLC
Cole B G7WWW
Cole B G3PQJ
Cole C G8PJQ
Cole C G8FIG
Cole C 2E0XDM
Cole D M6UDC
Cole D G3RCQ
Cole D M3DLC
Cole D 2E0DJC
Cole D M6DGJ
Cole E 2E0EJA
Cole E G7MRH
Cole F G4HSX
Cole G G1ODK
Cole H G3VJE
Cole I G6GBT
Cole K M0KJC
Cole K M3KJC
Cole K M6FNU
Cole K M6KCC
Cole L G8HIG
Cole M G0DKR
Cole M G4ZRC
Cole M G6FPH
Cole M M6OKR
Cole M G1DNT
Cole M M3MIX
Cole M M3UGH
Cole N G7SSN
Cole N GW7VJK
Cole N MW3LDY
Cole P G0XAO
Cole P G6JVP
Cole P G7SSQ
Cole P G3JFS
Cole P M1PAC
Cole R 2E0AYT
Cole R G7MPH
Cole R 2E0RDJ
Cole R M6RDC
Cole R G8ZJK
Cole R G3REB
Cole S MI6SCC
Cole S G3YOL
Cole S G4BLE
Cole S M0TEY
Cole S G0KEY
Cole S G3ZDG
Cole T G7EEG
Cole V GW7HOM
Cole W G0KFW
Cole W G4JUW
Cole W M3HXM
Cole W G0PDA
Cole Z M6IAC
Colebrook C G8TJR
Colebrook R G8YBO
Coleby E M0ERN
Coleby E M1ETS
Coleby E M5ERN
Colegate J G3RVX
Colegate P G3RVY
Coleman A G0PBO
Coleman A M6XTP
Coleman A M6PDD
Coleman B G4NNS
Coleman C G7KSH
Coleman C G8GGR
Coleman D M6DAZ
Coleman D G0UTR
Coleman G G3ZEZ
Coleman G G4IVT
Coleman G G7PYT
Coleman G G7RLX
Coleman G 2E0GPC
Coleman G M0YIG
Coleman G M6GPC
Coleman G M6GIL
Coleman G 2E0IIT
Coleman I G1YVV
Coleman I M6DGN
Coleman I G4GBT
Coleman I M6WWR
Coleman J G0KGI
Coleman J M1FJD
Coleman J G3YGB
Coleman J MI6FRQ
Coleman L G0WJA
Coleman L G6KGA
Coleman M 2W0JYC
Coleman M G5BH
Coleman M G6TID
Coleman M MW0JYC
Coleman M 2E0ZMZ
Coleman M M6MFJ
Coleman P M3ZUA
Coleman R G0BXP
Coleman R G1JKP
Coleman R G4LQX
Coleman R M3BQN
Coleman R 2E0RBO
Coleman R M3YPW
Coleman S 2E0USC
Coleman S G4YFB
Coleman S M3USC
Coles A G3SAQ
Coles B G0EHW
Coles C G8CJT
Coles C M6WFL
Coles D G3NYD
Coles D 2E0DQL
Coles D G4NZY
Coles D G7GZC
Coles D M6VTG
Coles E 2E0DIH
Coles E 2E0BBY
Coles F G7TKB
Coles F M3KFR
Coles G G0CCA
Coles G G8PBY
Coles J G4IJU
Coles J M1FIR
Coles J G1ANA
COLES M 2E0GZZ
Coles M G0AJX
COLES M M3MPC
Coles M G0VXC
Coles M M0CIE
Coles P G0XAF
Coles R M0XDL
Coles R M6DLO
Coles R M6SKY
Coles S M5AJZ
Coles S G0BKU
Coles V G4EGN
Coley A G1HQG
Coley R M0RLC
Colgan N GI1FSJ
Colhoun C GI0BZM
Collar M GM8AGM
Collar P G3TGN
Colledge S G0AOC
Colledge-Wiggins J M6CUN
Collerton K G4BDC
Colles J M0BVT
Colless C 2E0IHI
Colless C M0RTW
Colless C M6FHI
Collett A G0JSU
Collett A G4NBS
Collett C G1PZP
Collett J GI3AMY
Collett M G7GCI
Collett P G4LYC
Collett R G4AUN
Collett R G4LRQ
Collett S M3HEI
Collette R G3CUR
Colley E 2E0HZV
Colley E M3EGC
Colley J GI4XJC
Colley M G1JGE
Colley R G1DPX
colley S G1ILC
Collick A G0BXU
Collick P G1LOW
Collie F G4YQJ
Collie G MM5TUW
Collier A G6WXZ
Collier A G8WZJ
Collier A G6IZK
Collier D G6WGZ
Collier D M6HKS
Collier D M6PHJ
Collier G GM0LOD
Collier G G8PQZ
Collier G M0GMC
Collier H G4JXI
Collier J M6EAE
Collier J M6JDC
Collier J G0OSU
Collier J M6HUN
Collier M 2E0ZJB
Collier M M0ZJB
Collier M M3ZJB
Collier P G4ZPC
Collier P 2E0DYV
Collier P M6GQZ
Collier Q G3WRR
Collier S G3LLD
Collier T G7UKA
Collier W G0AYF
Collier W G0TGU
Collier Webb N GJ0VJP
Colligan A MI3MRG
Colligan T G8UCK
Collinge B G7NYD
Collingham P 2E0YZZ
Collings K M6FKE
Collings L G7AVZ
Collings P G7VKK
Collings S G4SAC
Collings S G4YFF
Collings S G4SGI
Collings T G6GCM
Collingswood M M3NCO
Collingwood J G0VBX
Collins A 2E0AMB
Collins A G4SOR
Collins A M3YDA
Collins A G0MUN
Collins A 2E0BPN
Collins A M1BPD
Collins A M6LDY
Collins A M6SEK
Collins A 2E0DIQ
Collins A M6DIQ
Collins B G0FMB
Collins B G3MXA
Collins B MW0GBW
Collins B 2E0PKY
Collins B M6MRL
Collins C 2E0KXD
Collins C G0JAP
Collins C G0JBC
Collins C G3WEQ
Collins C G7PWS
Collins C M0IAM
Collins C M3KXD
Collins C MM6CLC
Collins C M6MBG
Collins C G6FYA
Collins D 2E0LXD
Collins D G1XZG
Collins D G4NHW
Collins D M3JLW
Collins D G0TSM
Collins D GW6XGA
Collins D M6DHW
Collins D G0DJC
Collins D G8WZK
Collins D M1DPX
Collins D MM6GDY
Collins E 2E0SNU
Collins E G4UPS
Collins F G4UTG
Collins G G6EMB
Collins G G6XUJ
Collins G M3XXG
Collins I G8ZVZ
Collins I M6LVC
Collins J 2E0AGQ
Collins J G0EPV
Collins J G0ILH
Collins J G0KFM
Collins J G0OKL
Collins J G8JHG
Collins J G8TSC
Collins L G0IPB
Collins L G4LWC

Collins L MW3UIN
Collins M G0GCA
Collins M G4WOJ
Collins M G7JDN
[illegible]
Collins M M3VMA
Collins M M0XVI
Collins M G4XKZ
Collins M G6TVX
Collins M G7FTA
Collins M M3MEB
Collins M M6HFF
Collins M M1IKE
Collins N M3NPC
Collins N G3SPN
Collins N M6NHN
Collins P G4GHZ
Collins P G7PNP
Collins P G8ZWA
Collins P M0PDC
Collins P 2E0PCI
Collins P M0BSW
Collins P M3XPF
Collins P M5PWR
Collins R G0VTA
Collins R G1BEC
Collins R G1DNP
Collins R G2AQJ
Collins R G3ROC
Collins R G3TRC
Collins R G4PCE
Collins R G6VUG
Collins R M1FJL
Collins S G0WDQ
Collins S M0BTO
Collins S M0KLA
Collins S M6CQK
Collins S G8HAM
Collins S M6SLO
Collins S 2E0HHK
Collins S M6ZZQ
Collins T G1PNC
Collins T M0TCT
Collins T M6DUT
Collins T G6ALJ
Collins V 2E0EQI
Collins V G1THA
Collins W GI6JXG
Collins W G7MDY
Collinson A G4BNS
Collinson D G0OLO
Collinson E G0VWX
Collinson H G6WZM
Collinson J M1CWO
Collinson L 2E0EBX
Collinson L M6LEC
Collinson R G7SZO
Collinson T G1TDN
Collinson T G4RVO
Collis A G3NWH
Collis Bird N G0WZK
Collis G 2E0GRY
Collis G G4YBA
Collis G M0GRY
Collis M G7AGC
Collis M M6MBI
Collis R G8TZU
Collis R M0LET
Collis S 2E0STA
Collis S M5SLC
Collison A M6CFY
Collister P G4DLY
Colliton J G0IIE
Collopy M G3XVC
Collyer M M6EKM
Colman A 2E0GWC
Colman A M6CKQ
Colman D G6VLV
Colman K 2E0UJD
Colman K G8RZZ
Colman K M0UJD
Colman K M3UJD
Colman M M6MSC
Colman R G1YJJ
Colman R M0RBC
Colman S M3ZSC
Colman S M3VUH
Colman T M3PLU
Colman-Whaley R M6SBK
Colmer E G4KMF
Colpman M 2E0MUZ
Colpman M M0SWT
Colpman M M6MUZ
Colquhoun F M6UMM
Colquhoun W M6RPI
Colson G M6BPC
Colson J G4XMY
Colson R G4CYN
Colson R M6EJI
Collard D G3SYM
Coltman H G3PVJ
Colton A G0UYE
Colton D G7SQY
Columbine C G4ANU
Columbine J 2E0CEU
Colver D M0WMX
Colville A G7UOU
Colville H G8LKW
Colvin I 2E0BSX
Colwell P M1BDS
Colwill J G0DQH
Colyer P 2E0PCZ
Colyer P M0PCZ
Colyer P M6PCZ
Comben K 2E0KWC
Comben K M3YKC

Comben M M3YMC
Comben P G7AAR
Comber A M6AAK
Comber M M0CXB
[illegible]
Comer G G6USG
Comerford A 2E0OXF
Comerford A M6CFW
Comerford J G0ENT
Comis A 2E0BHB
Comis P 2E0BHC
Comis S G4XDM
Comley-Ross P M6AZM
Commander C G8PDM
Common B G4SUA
Compagno M M3UTV
Comper S M6WPN
Compton A G1SHH
Compton A G4IDW
Compton C G1CHV
Compton C G7RJO
Compton J G4COM
Compton L G0HKI
Compton R G1ZPU
Compton R M0ZPU
Comrie C MM6CNC
Conaghan M MI0HOZ
Conaghan O MI6LJO
Conboy G M6LUI
Concannon W G4EDM
Conce C G7SHI
CONDE D M0CQR
Conduit C G4KCZ
Conduit P M6PHC
Cone R G7JBZ
Coneley L 2E0ATB
Coneley T G0SFG
Conibear I G4TAH
Conlan T 2M1BYW
Conlin T G1EWM
Conlon A 2M0RCZ
Conlon A MM0POD
Conlon A MM3RCZ
Conlon J G4RKB
Conlon J GI7NEB
Conlon K G0RLO
Conlon L G0BBE
Conlon M G0ZMC
Conlon M G7TMC
Conlon S M3KEC
Conlon T MI0GCV
Conlon T MI3OFX
Conneely R G0HKB
Connell A 2M0GYN
Connell A MM6EBD
Connell J G0DRJ
Connell L G0DGB
Connell M G8HDL
Connell M M3JGU
Connell T G8EXQ
Connelly A M6ACY
Connelly J MM0SIL
Connelly J GM3RGU
Connelly P M3KUS
Connelly S M0EEL
Connery L GW4ZBN
Connett J M6XJC
Connett R G0WSC
Connolly A 2I0EBB
Connolly A MI6EDP
Connolly D G3KHK
Connolly D M0HVN
Connolly D M6GUV
Connolly E GI6FXY
Connolly G MI3ZJN
Connolly J 2E0JCY
Connolly J M3JCY
Connolly M G0NKC
Connolly P G0CUN
Connolly P MI0BOK
Connolly R G4LLN
Connolly R GI7IVX
Connolly R 2E0TSP
Connolly R M6RPV
Connolly S 2E0CCB
Connolly S M6TYE
Connolly T G7BYV
Connolly W GM4ZET
Connon P MM6MLT
Connor D G7LKI
Connor D M6JTD
Connor E G4DDB
Connor F G4FMI
Connor I G7PHD
Connor J G1OQV
Connor N M6CUE
Connor P M3ONM
Connor P G0XTE
Connor S G6WAU
Connor S M3PZO
Connors I 2E0OBX
Connors I M6MAO
Connors P G4PLZ
Conrad H G7SPP
Consitt D G4HUT
Consolante R G1FEW
Constable A G4IXQ
Constable B G4HPD
Constable C M1EXL
Constable C M3WOW
Constable D M6TRH
Constable I M6CIF
Constable J 2E0ATF
Constable J 2E0GOM
Constable J M5FUN
Constable J M0DEX

Constable M G4FMC
Constable S G7GOA
Constance J G0VGD
Constance J G7PKP
[illegible]
Constantine C M6PVC
Constantine E 2E0BVJ
Constantine L G6CQG
Constantine M 2E0BOO
Constantine M G0MIC
Constantine R G1WRE
Constantine R G3UGF
Convery F GI3ZTL
Convery G MI3GEI
Convery W G6WQN
Conway A G3WQL
Conway B G6BLC
Conway D M0YDC
Conway J M0DLG
Conway J MM1EVJ
Conway J M3XWC
Conway P G3UFI
Conway P G4LAN
Conway R G7HRH
Conway S MM3LWT
Conway S M6HZZ
Conway T MI3LZA
Cooch J G4MRX
Coogan J M3XWF
Coogan P MI6PAT
Coogan P G6FFU
Cook A G4PIQ
Cook A G6HIA
Cook A G6HQX
Cook A G7IEO
Cook A M3NGP
Cook A M6APC
Cook A M6GUF
Cook A 2E0DLF
Cook A M6EPF
Cook A G0IKQ
Cook B G0RIX
Cook B G4BWJ
Cook B G7TYP
Cook B G8PHG
Cook B M3NGN
Cook B 2E0KXX
Cook B M6KXX
Cook C 2E0AXZ
Cook C G0GMC
Cook C GM0OMC
Cook C M0CPC
Cook C M6CPC
Cook C G3OTH
Cook C M0VHR
Cook D G6XXB
Cook D M0DNJ
Cook D M6YDC
Cook D 2E0MJZ
Cook D 2E0EAD
Cook D 2E0TGV
Cook D M3ZCU
Cook D 2E0DCX
Cook D M0VAD
Cook E G1HYT
Cook E G0VZE
Cook E G3NAV
Cook E M3EDC
Cook F G4MTW
Cook G G4CUI
Cook G GW4RKX
Cook G M0GCC
Cook G G8TEC
Cook H G1JRL
Cook H G4SYD
Cook I M6GYS
Cook J G0AFQ
Cook J G0DPC
Cook J GW1AXU
Cook J G1AZA
Cook J G3OPW
Cook J G4OYC
Cook J G6GCK
Cook J G7VUL
Cook J G8LBG
Cook J M1JEC
Cook J 2E0GMM
Cook J 2E0WHH
Cook J M0AXL
Cook J M3OEG
Cook J M3WHH
Cook J M3SUJ
Cook J 2E0JTQ
Cook J M0JTQ
Cook J MW0JIJ
Cook J M6JTQ
Cook K 2E0IMQ
Cook K M6KBD
Cook L G8PWT
Cook L MW3AFX
Cook L 2E0DTX
Cook L M0LDZ
Cook L M6XEL
Cook M 2E0CHT
Cook M 2E0MJC
Cook M G1ELZ
Cook M G7RIA
Cook M M3PVC
Cook M G0PXE
Cook M M6ERU
Cook M 2M0MFK
Cook M MM6FIF
Cook M G1NQH
Cook M G0TPO
Cook M MW6KCQ
Cook M M6KYC

Cook M M6SXM
Cook N 2E0NJC
Cook N G0MQV
Cook N M3NJC
[illegible]
Cook N G7JVE
Cook N M6MGN
Cook O 2E0GBD
Cook O M6FMO
Cook P G0PEQ
Cook P G4NCA
Cook P G6GFG
Cook P M3BWF
Cook P G8XOM
Cook R 2E0SOX
Cook R 2E0YWN
Cook R G0JVK
Cook R GM1MYR
Cook R GM4BYT
Cook R G4XHE
Cook R G6WYF
Cook R M0ALK
Cook R M0KAA
Cook R 2E0RHC
Cook R M6RHC
Cook R M6VVE
Cook R M6YBD
Cook R G7IQD
Cook R M0TMB
Cook R MM3ZDI
Cook S G1HOL
Cook S G6BPK
Cook S M6NRX
Cook S G8CYE
Cook S M6EPS
Cook S M6MBC
Cook S 2E0ETP
Cook S G3YBR
Cook T M6FJT
Cook T G8CWE
Cook W G0KQX
Cooke A G3IFX
Cooke A G4UCT
Cooke A 2E0CKE
Cooke A M6SXO
Cooke B 2E0BZJ
Cooke B G0FCO
Cooke B G0ION
Cooke B GW6MHV
Cooke B M3CEB
Cooke C M6DPA
Cooke D 2I0BSA
Cooke D G8RFZ
Cooke D G8TDP
Cooke D MI3ATT
Cooke D M1CSL
Cooke D 2E0ZDC
Cooke D M0ZDC
Cooke D M6ZDC
Cooke E G0HMO
Cooke E GW7MMH
Cooke E M0ATT
Cooke G G4NIX
Cooke G G4WDC
Cooke G G6LFT
Cooke G M0CUQ
Cooke G M0TGV
Cooke G M6GCJ
Cooke I M0HTA
Cooke J G6TYB
Cooke J 2E0XAO
Cooke J M3UJZ
Cooke J GM8OTI
Cooke J M3UOD
Cooke K G6AYH
Cooke L G4FMJ
Cooke M G4DYC
Cooke M G6YMD
Cooke P G6ETL
Cooke P M6AUA
Cooke P M6CLX
Cooke R M0RDC
Cooke R M3RCD
Cooke R G7PMB
Cooke R G8ETR
Cooke R GM1MQA
Cooke R G3LDI
Cooke S G4NIY
Cooke S M6GGL
Cooke S G1MPW
Cookman L M3BSM
Cooknell D G3DPM
Cooknell F G2CO
Cooksey J M1CXX
Cooksey C M6XVS
Cookson D G0DRM
Cookson J G6FTR
Cookson J G4XWD
Cookson R G4KEC
Cookson R G7CUA
Coole D G8OFR
Cooley M G3XOC
Cooling T G4XMQ
Coolledge G M0ANU
[illegible]
Coombe J M6OSO
Coombe R G4DSQ
Coomber A M6VAJ
Coomber D G8MJX
Coomber D M0UXB
Coomber G G4YOF
Coomber G G0NBI
Coomber P G8LOP
Coombes A 2E0LAX
Coombes A M6AWC
Coombes D G7BHR

Coombes D M1VPN
Coombes F 2M0BIN
Coombes F MM3VVZ
Coombes G MW3GHC
[illegible]
Coombes J G1GDM
Coombes P G6AXY
Coombes P 2E0PCO
Coombes P M6PCX
Coombes R G4IGL
Coombes R G3ZNH
Coombes R G4ZEJ
Coombes R G6RBZ
Coombes R M6AGI
Coombes S M6AFW
Coombes S 2E0SUZ
Coombes S M0SUZ
Coombes S M6SUU
Coombes W G4ERV
Coombs B 2E0CZO
Coombs M G3VTO
Coombs M G4YBB
Coombs P G4XYI
Coombs R M0RTV
Coombs S M5SUE
Coombs S 2E0GMW
Coombs T G1IWE
Cooney S M6FBV
Cooney W MI3IRY
Coonick C M1EZB
Coop A M3XFZ
Cooper A 2E0TRK
Cooper A 2E0BRT
Cooper A G0CQD
Cooper A G0OWJ
Cooper A G1OOG
Cooper A G3TKD
Cooper A G4VMY
Cooper A M3ZAV
Cooper A G4YSG
Cooper A G4ZBZ
Cooper B 2E0OOC
Cooper B G0HCT
Cooper B G6GNO
Cooper B M3OOC
Cooper B G0TUM
Cooper B G4RKO
Cooper B M6TLT
Cooper B G8UOL
Cooper C GM3JKC
Cooper C G8MHD
Cooper C M3LYA
Cooper C G1SDJ
Cooper C 2E0FCZ
Cooper C G4XTX
Cooper C M6FCZ
Cooper D G0HVP
Cooper D G1EXV
Cooper D G4NZX
Cooper D G6DRC
Cooper D G7RXZ
Cooper D GM8SAP
Cooper D G8UWL
Cooper D G0KYR
Cooper D G6MIF
Cooper D M6KPX
Cooper D 2E0OQH
Cooper D M3OQH
Cooper D 2E0WTG
Cooper D M0WTG
Cooper D M3WTG
Cooper E G3GEG
Cooper E G4VAS
Cooper E G3YBA
Cooper F G0BYU
Cooper F G4ZWI
Cooper F M6IOG
Cooper G 2E0TGC
Cooper G G0ESH
Cooper G G0KVK
Cooper G G1PII
Cooper G G3HJP
Cooper G G4OQX
Cooper G G4XXS
Cooper G M3TGC
Cooper G G1JKO
Cooper G 2E0GAU
Cooper G M0OKK
Cooper G M6OKK
Cooper G G4LFA
Cooper H G0NWC
Cooper H G3SWW
Cooper H M6HPC
Cooper H G0BLQ
Cooper I G0RMS
Cooper I G3XVV
Cooper I G7AON
Cooper I G8GLC
Cooper I M6ATQ
Cooper J G4GKU
Cooper J G4RAC
Cooper J G7VIE
Cooper J G8WUU
Cooper J M3FWT
Cooper J M3VJC
Cooper J MW6GFQ
Cooper J GW0ACH
Cooper J G3XEV
Cooper K M6KPC
Cooper M G0KOI
Cooper M G1ANI
Cooper M G1VUG
Cooper M G6JEY
Cooper M G6MGH
Cooper M G8MQF
Cooper M M3JZX

Cooper M M6MCB
Cooper M M6WCT
Cooper M G0SVZ
Cooper M M0MBG
Cooper M M6RCV
Cooper N G6PPK
Cooper N G8NYZ
Cooper N G1ZLY
Cooper P G0KDA
Cooper P G0VUM
Cooper P G1JKN
Cooper P G8ZPE
Cooper P MM3PDC
Cooper P M3RPS
Cooper P M6EYG
Cooper P GU0SUP
Cooper R G1DBZ
Cooper R G4GPB
Cooper R G4TEW
Cooper R G4TSB
Cooper R G4WFR
Cooper R G6IIU
Cooper R M6RHY
Cooper R M6FSC
Cooper R M6LOL
Cooper S G0EPI
Cooper S G0WZC
Cooper S G3YTI
Cooper S GM4AFF
Cooper S G7RXX
Cooper S MI0DDW
Cooper S M1AVM
Cooper S M3EWW
Cooper S G6MGA
Cooper T G4CBY
Cooper T G4XOP
Cooper T G6XWZ
Cooper T M6TIY
Cooper T 2E0ZRT
Cooper T M0TAK
Cooper T M0ZRR
Cooper T M6ZRT
Cooper T MW6CTD
Cooper W GW0OTY
Cooper W G4CIA
Cooper W G4WJM
Cooper W G7FDD
Cooper W G0KDL
Cooper W M0TAP
Cooper W M3XYI
Cooper W M5BIL
Cooper-Hutley S M3EHJ
Coore C MM3LWZ
Coote A 2E0FQR
Coote A M6FQR
Coote D M6DBH
Coote G G1IPU
Coote J G0THV
coote M G7MTE
Coote M M6UCS
Coote N GM8TVV
Coote P M6PDK
Coots H G7WJE
Cope D M3YCO
Cope D G8RYE
Cope J G4DYJ
Cope K M3LKO
Cope R G4KMX
Cope S G4VAX
Copeland D G0JXJ
Copeland J G6NPJ
Copeland L M6BHQ
Copeland N GI8RPT
Copeland P G0FJS
Copeland P G7CEY
Copeman C M3TUD
Copeman M M3HQU
Copeman P G8SWW
Copeman T G4NVL
Copland A GM1SXX
Copland I G1HJL
Copland J M6JFB
Copley L M1EZH
Copley L M3EZH
Copley S G0UEM
Coppen D G8DEL
Coppenhall D 2E0OSY
Coppenhall D M6OSY
Copper E M3ILJ
Copperwaite A G8AQO
Copperwaite A M0DDC
Coppin P 2E0IMJ
Coppin P M0YMJ
Coppin P M3IMJ
Copping I M0BNF
Copping O M6NCD
Copplestone J G0RFM
Copse A 2E0HDG
Copse A M6CDO
Copse M M0HQC
Copsey A G1OWJ
Copsey C G8ULQ
Copsey D 2E0XDC
Copsey D M0XDC
Copsey D M3NUU
Copsey M G8UUV
Copsey R G4TDF
Copson J G3TUL
Copus R M6EML
Corallini A G0DCW
Corallini D G4YOS
Coram B G0MPA
Coram H 2E0HOO
Coram H M3UNF
Corben J G4EXT
Corbett A G0WFV

Corbett A G7RAF
Corbett A M6HVI
Corbett D M3KRR
Corbett G G0JZO
Corbett M G4MPW
[illegible]
Corbett M M6XRA
Corbett M G4BIY
Corbett M M3UJE
Corbett M G8VBW
Corbett N MI6NCC
Corbett R 2E0HTU
Corbett S M0CRN
Corbett W 2E0BFD
Corbett W M3FDO
Corbidge J G8SUQ
Corbin P M3WPJ
Corbishley C M3NAL
Corbishley P GM8NAL
Corbishley S M6ZOM
Corcoran J GW7MLN
Corcoran M GI4TTL
Corcoran T 2E0TCO
Corcoran T M6TKC
Corcoran T 2E0TYC
Corcoran T M6TYC
Corcoran V M6VCK
Cordell N M1NCC
Corden R 2E0ATV
Corden S M6SHC
Corder D G6SPB
Corderoy C GI4CZW
Corderoy J GD6LFD
Cordes D M0YVG
Cordial J G6OFM
Cordier A M1CXK
Cordingley I G4AQR
Cordner S 2E0HAJ
Cordner S M0HET
Cordrey P G0UTB
Cordwell A G0NFY
Core A G0AGC
Corfield D G7NBJ
Corfield J G4ZKG
Corfield R M0AYY
Corke M G8RJQ
Corker A G4NJI
Corker B G8FBQ
Corker T 2E0WNT
Corker T M3WNT
Corkett G M6PWE
Corkey R MM0RBN
Corkill D G4DJK
Corkish D 2D0RGW
Corkish D GD0IFU
Corkish W GD0HEY
Corless A M6XSS
Corless D M0HGA
Corlett M MD3LWQ
Corlett W GD1GHK
Corley P M0PGC
Cormack D G4VZR
Cormack J GM4JUE
Cormack M MM6DRJ
Cormack M MM6WKC
Cormack R MM6JUE
Cornall B M1AXP
Cornall J G1HJO
Cornelius A M6TAF
Cornell C G8GWK
Cornell J G6ZKM
Cornell P G4TBI
Cornell P G8CQG
Cornell R G0GOH
Cornell R G7HBO
Corneloues G M0CGE
Corner C G1MHA
Corner D G3ZXF
Cornes F M6FCC
Cornes I G4OUT
Cornes K G4THY
Cornes P G6YLZ
Corney R 2E0RFM
Corney R M6RFM
Cornish G 2E0GCY
Cornish G M6GCK
Cornish J M3UKX
Cornish R G0DOK
Cornish W G7OKI
Cornish W M3WCM
Cornmell K 2E0KCO
Cornmell K M0KCO
Cornmell K M3KCO
Cornthwaite R M0BZS
Cornwall B G3ZFX
Cornwall R 2E0RWC
Cornwall R M3RWC
Cornwell R G1WEF
Cornwell S 2E0SDC
Corp A G1ZIM
Corps A G7CDO
Corr L MI6LFK
Corr M M6EJJ
Corr R G4GXM
Corrieri R MM3YVU
Corrigan J MM6BIY
Corrigan M 2E0MYT
Corrigan M GW8LKX
Corrigan M M6CHC
Corrigan S G1TSV
Corsi D G6FED
Corson S G1GJD
Corstorphine S MM3KZD
Cort-Wright P G3SEM
Cory C G3MEV
Cosens N 2E0NAC
Cosens P 2E0MEL

Cosens P 2E0PEC
Cosford H G8ACL
Cosgrif C G0DZC
Cosgrove C 2M0CLN
Cosgrove C M0FYH
[illegible]
cosgrove J 2M0BMK
Cosgrove J G4UOW
Cosgrove J MI0JPC
Cosgrove T MI3GWQ
Cosgrove W G1EXR
Cosham A M6AJC
Cosham G G6AUE
Cosham I 2E0IDC
Cosham I M6IDC
Cosic A M3WAP
Cossar D GM3WIL
Cossey A G7WEW
Cossey D M1AEH
Cossey K 2E0KGC
Cossey K M0GTL
Cossey K M6KGC
Cosson J 2E0BIS
Cost A M6ABR
Costa F M0HOJ
Costa G MM0GJC
Costello C G1LQM
Costello M 2E0AXB
Costello M G1TQT
Costello P G1IAV
Costello R G6BUU
Coster J GM3SHR
Costford I 2I1GIH
Costford T MM0BHX
Costigan P G1DAX
Coston R M1ALF
Cothey D M3IDA
Coton D G8KDD
Coton I G8XCJ
Coton K M6KCT
Cott J GD6SSV
Cottage D G7UWL
Cottam D G0HVN
Cottam G M1AQY
Cottam P 2E0DYM
Cottee B G7SMQ
Cotter A MW6ALC
Cotter A MW6RAW
Cotterell R 2W0RDD
Cotterell R MW3DQB
Cotterell T M1CJX
Cotterill K G7NHQ
Cotterill S M0ANK
Cotterill S G6AYE
Cottham T G4KTB
Cottier J G8KRV
Cottington R M1CMW
Cottis S G4KMH
Cottis S G8TFR
Cottle A 2E0DAR
Cottle D G4XWQ
Cottle E G7OOF
Cottom I M3ULM
Cotton A G0JHQ
Cotton B G3VXY
Cotton C M6UBI
Cotton D G1ATA
Cotton D GW7EHD
Cotton D M0HIQ
Cotton E M3HME
Cotton F G0LFI
Cotton L G7CBY
Cotton M 2E0ZAC
Cotton M G4HBY
cotton M G7MVE
Cotton P M3KAQ
Cotton P M6EMF
Cotton S G7CBZ
Cottrell A M6AMC
Cottrell D G7GLQ
Cottrell G G3XGC
Cottrell J G1OKY
Cottrell M M1MAD
Cottrell N G3JFR
Cottrell R G3VOS
Cottrell S G3SHY
Cotz D M6DCA
Coubrough D M3CAD
Coubrough J 2M0JBZ
Coubrough J MM6JCU
Couch M GW4ZQY
Couch P G8UHJ
Couch P M6PCP
Couch R G6ERI
Couchman B G4IYC
Couchman H G8WHB
Couchman M M0SAC
Couchman V M3XVC
Couchy D G8BJA
Coughlin A GW3TOB
Coughtrie J GM0RUW
Coull P G3XVY
Coulsey P M6COU
Coulson B G4NZZ
Coulson D G0OAP
Coulson I GM0KKE
Coulson I G8HUG
Coulson I M0GZC
Coulson I M3ZWM
Coulson M M3AHO
Coulson S 2E0VBB
Coulstock B G6TXY
Coulston S M0BXC
Coultas M G0SLP
Coulter D G4MTR
Coulter D G8ORO



**IMPORTANT NOTE**

**Revalidate licence to avoid revocation** – Ofcom has advised the Society that plans will be drawn up to revoke licences that have not been revalidated as required by the licence conditions. The quickest way to revalidate is to do so online via the Ofcom website: *https://services.ofcom.org.uk/* or by email: *amateur.validations@ofcom.org.uk* If you need assistance in the process, Ofcom staff are available to help, but please be patient during times of heavy workload.

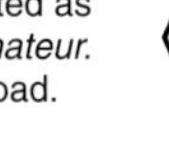

Coulter M G0MEY
Coulter P MI3FBW
Coulter R GI8VKA
Coulthard A M6ANP
coulthard S M0CJK
Coulthard W G0FPO
Coulthart D G8VBX
Coupar D GM3YVX
Coupe B G4RHZ
Coupe D G4ZML
Coupe D M0SRO
Coupe D G7DGF
Coupe E G1WYB
Coupe G M0RMF
Coupe J G0ASH
Coupe M G1EUT
Coupe N G3KBS
Coupe P G8UKO
Coupe S M6KFF
Couper T 2M0LNR
Couper T MM6TSC
Courcoux M G3EBP
Courree S M6ROU
Course A G4HND
Court A M6GAG
Court D G4RPA
Court D G3SDL
Court H G1UHO
Court K G4GSZ
Court M G4AKB
Court P G1ROK
Court S 2E0CAW
Court S M0SCO
Court S M3CAW
Court T G4IAG
Court T 2E0YOP
Court T M3YOP
Courtenay B G7UFI
Courteney M G4VWE
Courtier-Dutton D G3FPQ
Courtnell E G7CQZ
Courtney B M0GFF
Courtney D GI8PDK
Courtney G G1LHE
Courtney-Crowe S G0LFP
Courtney-Crowe S G7DEI
Couse D G8ILW
Couse W M0ABQ
Couser D MI3OIB
Cousins B MI6GPZ
Cousins C 2E0PWF
Cousins C M6EGV
Cousins M G0FND
Cousins P G1LVR
Cousins P G4NJJ
Cousins S M6STC
Couston C 2E0TSM
Couston C M0XSM
Couston C M3UTK
Coutts A GM0MZD
Coutts A GM4PWR
Coutts D GM3VTH
Couzens G G3NTA
Couzens G M6CYN
Couzins J G1KMJ
Covel J M5SJS
Covell-London V G0APV
Covell-London V G6TXQ
Coveney L M6FPD
Coveney R M6FPE
Coventon E G4LHY
Coventry D M0GNJ
Coventry N M0NAU
Coverdale C G3YWL
Coverdale M G4LTI
Coverley C M3XYU
Covey T G4ARO
Covill B M6KWW
Cowan A M6DOB
Cowan A 2E0KRB
Cowan A GM0UDL
Cowan A M0SOT
Cowan C GM0UIG
Cowan C MM6CWN
Cowan D M6DKC
Cowan G GM7GXI
Cowan J GM7OIN
Cowan J GM0HNJ
Cowan M M3XYC
Cowan P M3JFA
Cowan R GM4SRL
Coward A M6YTU
Coward C G3YTU
Coward J M6BIX
Coward M G3NFJ
Coward M G3PRH
Coward M G7JKD
Coward R G4XKR
Cowdell K G8BDZ
Cowdell S G4XED
Cowdery R G3UKB
Cowdrey D M6CFE
Cowee J G6FYC
Cowell B G6USL
Cowell J M3TZI
Cowell K G0OKV
Cowell M G1XNX
Cowell W G0OPL
Cowell W GM8LQL
Cowen C 2E0TBZ
Cowen C M6GLX
Cowgill S 2E0RSG
Cowhey M 2E0MAO
Cowie D GM8KSJ
Cowie D 2M0DWC
Cowie D MM3YXN
Cowie I MM0CVH
Cowie S G0EZB
Cowin P M3ITH
Cowles J MW6BVR
Cowles R M0TIX
Cowles R M3DCS
Cowley A G3FCM
Cowley A M6NAQ
Cowley B G4WDH
Cowley C M6EZY
Cowley D G0DAC
Cowley G M3YGC
Cowley J 2E0AIM
Cowley L GW6VBR
Cowley L M6LAL
Cowley M G0GAG
Cowley M G7RTQ
Cowley N G6MFU
Cowley R G8FTE
Cowlin F G8JSE
Cowling D 2E0DGC
Cowling D M0HDV
Cowling D M6GDC
Cowling D MW6TDC
Cowling G G0FRX
Cowling J G3HWM
Cowling P G8BMZ
Cowling R G8ZWF
Cowlishaw R 2E0HCT
Cowlishaw R M3BEK
Cowman D M3MEU
Cowper P M6POM
Cowperthwaite E G4UJI
Cowperthwaite R M6LPO
Cowsill A G4MBA
Cox A G0OBG
Cox A 2E0COX
Cox A G4BSV
Cox A G4PCB
Cox A M3MIO
Cox A M6COX
Cox B G4KWX
Cox B M0BVW
Cox B G0DMH
Cox B 2E0CKR
Cox B M0HFZ
Cox B M6BKU
Cox C M6ZXL
Cox D G0NYS
Cox D G3KHZ
Cox D G8DUI
Cox D M3MYM
Cox D M0CIK
Cox D GI4OHH
Cox D 2E0WWS
Cox D M3YXW
Cox D M0DTK
Cox F G1OPW
Cox G M3RLX
Cox H G0OBH
Cox H G8HOU
Cox I GM6JNQ
Cox I M6YFX
Cox J M3DNX
Cox J M6CWO
Cox K G7JXB
Cox L G1JGF
Cox M G1XWM
Cox M M0XOC
Cox M M0GQB
Cox M M6MBY
Cox M M0TAM
Cox M M3IQS
Cox M M6MOX
Cox N G0JZA
Cox P G0UQY
Cox P G7MGT
Cox P G4BUB
Cox P M6FZB
Cox P M6XPC
Cox R G0BAG
Cox R G0FOK
Cox R G1MKE
Cox R G1OVO
Cox R G3LQJ
Cox R G4AEL
Cox R M3UQY
Cox R G3PLP
Cox S G0JGF
Cox S G1DRY
Cox S G7EKJ
Cox S M0DUP
Cox S M3SVC
Cox T G0PXP
Cox T GI7JUH
Coxhead M 2E0AFL
Coxhill D G8CXT
Coxon D G0GHM
Coxon D G0DTC
Coxon J M0DHE
Coxon J M6JCX
Coxon K G0HDV
Coxon L GM1MUY
Coxon P M6PQD
Coxon R M1CCY
Coxon S M6SPG
Coy D G3PYI
Coy R M6CGX
Coyle A G0HTX
Coyle D GI4DGI
Coyle D M6DGY
Coyle E GI4EPK
Coyle G M3NSM
Coyle M GI0NWN
Coyle R GM4ZMK
Coyne B G3DCO
Coyne J G1UFH
Coyne J G6GKP
Coyne J G4ODV
Coyne L M3WRZ
Cozens M G0JDQ
Cozens M M6ZOC
Crabb H 2E0HUM
Crabb H M6YES
Crabb I M3IAC
Crabb J M6EZP
Crabb N 2E0NKI
Crabb N M6TKE
Crabb R G4GHI
Crabb S 2E0TZY
Crabb S M0TZY
Crabb S M6TZY
Crabbe G 2I0GLC
Crabbe G MI6GLC
Crabbe J G3WFM
Crabbe L G3CON
Crabtree A G6YLX
Crabtree L M3YIC
Crabtree M G1NAA
Crabtree P G7TOO
Crabtree P M0PJC
Crabtree S M0CRZ
Crabtree S M3SZC
Cracklow J G4YMF
Cracknell D 2E0WHF
CRACKNELL D M0XXO
Cracknell D M6WHF
Cracknell M G0CPU
Cracknell P G0KDT
Cracknell V G4KPZ
Craddock A M0WSB
Craddock J G4YDT
Craddock T M3LDM
Crafer A G8VJY
Craft M G4NSN
Craft P GM0RKU
Cragg A G8YKV
Craggs A GW8ARC
Craggs D G3RYP
Craggs R M1BCZ
Craggs S G0BAU
Craib G GM1YGW
Craib J GM1LKD
Craig A M0GLJ
Craig C GM0EWU
Craig D G4YYC
Craig D G2HIX
Craig D G6KGU
Craig D GI8LCJ
Craig D M3URX
Craig G 2M0BJU
Craig G G4IWD
Craig G MM0BAG
Craig G MM0GON
Craig G MM3OZU
Craig G 2I0GFO
Craig G MI6GFO
Craig I G6VGV
Craig I M6DXH
Craig I 2E0ACA
Craig J GM3HVK
Craig J GI6OQL
Craig J G3SGR
Craig J G7GMB
Craig J MM0DVZ
Craig K G3ZSX
Craig L 2I0LNZ
Craig M G6NCE
Craig M GM6BEY
Craig N MI3FGK
Craig P G7BOH
Craig P G7ULL
Craig P 2E0PRC
Craig P M6SFC
Craig P 2E0DLA
Craig R G0MCT
Craig S GI8WHP
Craig S GI4SSF
Craig T M6LTC
Craigen W G4GTX
Craine J GD3XNU
Craioveanu G M0HZF
Crake D G6DTN
Crake D M0DFA
Crake M G4HUQ
Crake P G0OVA
Craker D G0CEP
Cram D GM3NIG
Cramond J GM4NHI
Cramp A G7HSA
Cramp K M3GQI
Cramp N G1HYA
Crampton M G8DLX
Crampton N M0FZW
Crampton S GI3XDD
Crampton T M6FZW
Cranage J G8RHC
Cranage M G8OFA
Crane C G1UKW
Crane D M1ESI
Crane D MM6SEI
Crane E 2E0IGA
Crane G 2E0ZRQ
Crane G M3ZRQ
Crane G M3GRY
Crane K G1VGA
Crane K M3ESK
Crane L M3OIC
Crane M G0FLU
Crane M G0KHZ
Crane P G7IPI
Crane S G0KUY
Crane S G0CUH
Crane S M6HTB
Crane T G7WHZ
Craner K 2E0HOL
Craner M 2E0MNC
Craney B 2I0CEI
Craney B MI0HHV
Craney B MI3YVB
Cranfield J G8FWK
Crangle J M0BLI
Crank C M6GMP
Crank J 2E0JAC
Crank J M0JBC
Crank K M1KCB
Crankshaw P GM7VXR
Cranston A M6CBQ
Cranston J 2E0CRN
Cranston J M6SRF
Cranston P MI1ERL
Cranston R M3WBN
Cranwell M G0PNA
Cranwell R M0BCF
Crashley D M6AIU
Crask S G7AHP
Craske S G3ZLS
Craswell C 2E0VHF
Cratchley G G4ILI
Crathorne R G8TBW
Craven A 2E0GIK
Craven A 2E0HAP
Craven A M6AVO
Craven C G6HIB
Craven D 2E0KDV
Craven D M0KDV
Craven D M3KDV
Craven F G4LAW
Craven H M5COL
Craven K G4LKP
Craven M 2E0MXC
Craven M M0MXC
Craven M M6MXC
Craven P M1PVC
Craw P G3CCX
Crawford A MI3EGD
Crawford A M3VNM
Crawford A 2E0VEX
Crawford A M0MKV
Crawford A MM6ACM
Crawford A M6LHA
Crawford C G0GKI
Crawford E G4GUQ
Crawford G G6IOE
Crawford J GI4PGH
Crawford J M1TES
Crawford J G0VNY
crawford M G3KDU
Crawford N MI0ABN
Crawford N MI3EOD
Crawford N G7SQM
Crawford R GM0KEQ
Crawford R GM6NUL
Crawford S M3RTP
Crawford S M6DXC
Crawford S G6HCQ
Crawford T GI4OPH
Crawford T 2I0EPC
Crawford T MI6CMU
Crawford-Baker J GI0HWO
Crawford-Baker M MI3JCB
Crawley D G8OPC
Crawley K G1BRP
Crawley M 2E0MDC
Crawley M M3YMG
Crawley T MD6FGZ
Crawshaw B M0DPJ
Crawshaw C GW1YHL
Crawshaw G G0ENF
Crawshaw M G8CME
Crawshaw M G4BLH
Crawshaw P G3UFV
Crawshaw S 2E0CTU
Crawshaw T M6HFX
Craxton R G3IKL
Creaser J 2E0JXX
Creaser J M0XXJ
Creaser J M6JXX
Creasey J G4FSD
Creaseyy J G4RIP
Creasy E G4FBI
Creber D MW1AAH
Creber S M1SHA
Credland J G8CSR
Cree A M3DVQ
Cree E M0TBK
Cree E M3TBK
Cree S G3TBK
Creed G M3GHF
Creed J M3ZFB
Creedy M M6MBH
Creek A G7BNL
Creek D G0GEZ
Creek L M0NEV
Creese F M3FSC
Creese J M6GSQ
Cregan C MM0CGZ
Cregg G M6GJC
Creighton D M0MED
Creighton K G8ONS
Creighton P G4TGQ
Creissen P G0UOQ
Crellin B G8NNA
Crellin J G0JSZ
Crellin S G1TIQ
Crellin T M1XRC
Crerar R M3XLK
Crespel P G0NSG
Crespo A M0FCR
Cressey D G4MQM
Cressey J G0VBN
Cressey M M5DJC
Cressey P G0SXW
Cresswell A G7ADS
Cresswell A M0BLR
Cresswell C M3KOL
Cresswell H G4ZCJ
Cresswell J G4AMF
Cresswell M 2E0MJX
Cresswell R 2E0VKK
Cresswell R M0VKK
Cresswell R M6VKK
Cresswell S G6VWV
Creswick M G0SVJ
Crew T 2E0CTQ
Crewe D G6DUH
Crewe J 2E0JCO
Crewe J M0JCE
Crewe J M6JCE
Crewe P M0KRU
Crewe R M3OUF
Crewe R G0WJU
Crichton A MM6HIA
Crichton J MM1DPH
Crichton J GI4YWT
Crichton S M3UTJ
Crichton V MI3YWT
Crick C G4CJR
Crick M G1YSA
Crickett A G4WIP
Crickett A G6AFX
Cridland N G4FTK
Cridland R G3ZGP
Cridland R G7LAS
Crighton A GM7ALI
Crighton B M5BJC
Crighton S M3HKH
Crill J 2E0CZD
Crill J M6CQF
Crimes M G0RCY
Crimlisk A G0BLW
Crinson D G7DSV
Cripps B G6RQZ
Cripps B M6BLC
Cripps G G7WJW
Cripps J G3XWL
Cripps N M6TZD
Crisp A G7DFW
Crisp D G4OAE
Crisp G G7MLX
Crisp J G3HXN
Crisp T M3PVV
Crissell R G1DXH
Cristofoletti A M0XTX
Critchley A G3SXC
Critchley C G0GUN
Critchley J G1DSA
Critchley P G1OJB
Critchlow A G1LVV
Croasdale C M6AQI
Croasdale I M1EHI
Crocker J G4BQB
Crocker J G4RGO
Crocker K G1ZSE
Crocker R G0KAU
Crocker T 2E0EXO
Crocker T M6EXO
Crockett H G8ACA
Crockett J MM1FHL
Crockett T G8NFP
Crockford F G6YUY
Crockford M 2E0OZE
Crockford M M0VOZ
Crockford M M6OZZ
Crockford P G8IOA
Crockford R 2E0BZX
Crockford R M6RFC
Crockford S GM1KBZ
Croft A G0SOY
Croft A G6NUS
Croft A G6XEX
Croft A G8CJM
Croft D 2E0JBC
Croft D M0YBC
Croft D M3CSZ
Croft I G7GZK
Croft J G3SXA
croft N 2E0NCC
croft N M3NCT
Croft P G6MID
Croft P G0WSP
Croft R G7VRX
Crofts F G4FLM
Crofts M G4JAQ
Crofts P G4ZHD
Crofts R G7RVW
Croker J G3WCL
Cromack A MM0BQL
Cromack H GM0FGI
Cromack H GM8SBH
Cromack J G6YLV
Cromack W 2E0WJC
Cromack W M3KFO
Cromar L G6MZF
Crombie M M3KLB
Crombie S M6SBJ
Crome D GI0WCE
Crompton D G4IAD
Crompton F G1BOO
Crompton L G6MCC
Crompton L GW6UHY
Crompton L MW0CLU
Cromwell E M3NRW
Cronin K 2E0HGM
Cronin K M1ECB
Cronin V M3VSL
Cronk A G6TXP
Cronshaw R G7VNK
Crook A G1GFD
Crook A G4AYS
Crook D G6ZQU
Crook D M6GKV
Crook M G1ZRP
Crook M G4THA
Crook N G4OKA
Crook N G4ZLP
Crook P G7AES
Crook R G6SZS
Crook R M3TLW
Crook T G6GFR
Crookall E G7BBN
Crookbain J G6SPH
Crooke D GM0RHP
Crookes A G8UCN
Crookes K G0NHO
Crookes K G7DNR
Crookes P 2E0HGF
Crookes T G4RBU
Crooks A G7EFL
Crooks N M0ANP
Crooks R G4RRU
Crooks S G0LMA
Crooks W GW4XWC
Croome G G8GCK
Croot D G7UVL
Croot S G0WXH
Cropley L G0DFC
Cropp H M6YTI
Crosby C G0EPU
Crosby E M6EJC
Crosby J 2E0RDU
Crosby J G0POU
Crosby K M6KCD
Crosby-Clarke D G4SWM
Crosfill M G7TJD
Crosland A M0HXA
Crosland D M3ZNC
Crosland P G6JNS
Crosland T G4PNK
Cross A G3WEA
Cross A G4WGE
Cross A 2E0CTA
Cross A G4GNU
Cross A M0ONZ
Cross A M6ORC
Cross B G3ZBZ
Cross B 2E0DKY
Cross B M6SFQ
Cross B G8EOV
Cross C 2E0HAW
Cross C M3CJX
Cross D G0KLQ
Cross D G7TIB
Cross D M6DDT
Cross E G4ZEG
Cross E M3ENO
Cross E 2E0EAW
Cross E M6AZH
Cross G G8MHE
Cross G G8URI
Cross G M3ZXH
Cross I G6ZNT
Cross I G7UVN
Cross J G4GOR
Cross K M3XFN
Cross M G3OKU
Cross M M1BXD
Cross M M1EJQ
Cross M 2E0VWG
Cross M M6EGX
Cross N G8JEI
Cross O G4DFI
Cross P G0GHH
Cross P G3OZD
Cross P G3RWI
Cross P G7MWH
Cross R G6LFQ
Cross R 2I0XDR
Cross R G0TRW
Cross S 2E0CRX
Cross S G0TPJ
Cross S G6NGM
Cross S M3LBR
Cross S M3NBQ
Cross T GW4LFW
Cross W G0ELZ
Crossfield J G2DML
Crossfield J G3SEQ
Crosskey S G4XDE
Crossland G G4CWG
Crossland H G4SBG
Crossland J G8PWU
Crossland S G6OFV
Crossley B M1CHF
Crossley D G6BIT
Crossley E 2E0LIS
Crossley E M0LIS
Crossley H M3EBG
Crossley M G8UWM
Crossley M M1CVL
Crossley N G0WPC
Crossley P G8THR
Crossley P M3PCC
Crossley P M6JPC
Crossley R G4CZP
Crossley S M3HHC
Crossman K M0BYZ
Crossman M G8OQR
Crosson Smith S G4LMX
Crosswell B M3PNB
Croston A M6FEJ
CROSTON S M6VIE
Crosweller P M6MTT
Crothers G GI4EXI
Crothers V MI0VAC
Crouch A G1JKL
Crouch D 2D0FGW
Crouch D MD6FGW
Crouch H G0KVC
Crouch P G4ZGC
Crouch S M3JLX
Crouch S M6EYB
Crouch T M6TZR
Croucher A G6ERJ
Croucher C G4BLD
Croucher P G4YPC
Croucher R G0KVF
Croucher R G0NRJ
Croucher S M3YSC
Crough D G8JFL
crow B GW4LFV
Crow B G4UYJ
Crow C GW6IOA
Crow J G1YNQ
Crow J G8GQG
Crow L G0TNH
Crow M M1BAA
Crow P M6BKJ
Crow R G0CRO
Crowder J MJ6ADQ
Crowder K G8JNZ
Crowe D G1OYG
Crowe D G8WTB
Crowe D M6WCU
Crowe J G6ZXO
Crowe P G4MQP
Crowe R G3RVA
Crowe T G1ANF
Crowe-Haylett B G0OKK
Crowhurst B M6CZC
Crowhurst C G7IGR
Crowhurst J M3AJA
Crowley G MM5AHO
Crowley M M3JMU
Crowley N M3NDC
Crowley P G0GPF
Crowley R GW4UGI
Crowly M G0OJC
Crowson I M6IRC
Crowther A G6ZOB
Crowther A GD0MWL
Crowther A M3VDA
Crowther B G1HYG
Crowther B G6VRF
Crowther I G3KLF
Crowther J M3KBG
Crowther J G3KMM
Crowther R M3RYY
Crowther S G6XXE
Crowther T MM6HFC
Crowther-Watson M G3IAR
Crowton G G6WIG
Croxford C G4YCW
Croxford I G3OIC
Croxford M G7HQH
Croxford P 2E0PSC
Croxford P M0PSC
Croxford P M6PSC
Croxford Simmons M 2E0BSJ
Croxford Simmons M M3LCS
Croydon A G4VTC
Crozier A GI7VXC
Crozier B 2I0BYL
Crozier B MI0GQG
Crozier B MI3LXJ
Crozier J G8NZO
Crozier J MI6BFJ
Crozier M 2I0FUT
Crozier M MI0GQI
Crozier M MI3GMI
Crozier R G0ZAP
Crozier T 2I0TPC
Crozier T MI6TPC
Crozier W MI3DNN
Crucefix D G0HTF
Cruddas C G0LKA
Crudgington J M6PUD
Crudgington N M6HJI
Cruickshank D GM6EUC
Cruise A M1DYO
Cruise S M0CUT
Crump O M3OCR
Crump P M0DNY
Cruse Howse R 2E0RIW
Cruse Q G3BV
Cruse T G0JUE
Crust P G3XYC
Crutchley M G3MRZ
Crutchley S G7HRF
crute A G8TEF
Crute T G4DGB
Cryan J 2E0HHE
Crymble N GI4MCH
Csapo P G6RFU
Cubberley F G3OCW
Cubitt A M0CCZ
Cubitt C G4MQK
Cubitt O G0TGQ
Cucchiara A M6FKF
Cuddington R G4MQL
Cuddy J M1AGY
Cudlip A G0TCQ
Cudworth A M6UAS
Cuff D G0YBU
Cuff M 2E0OEV
Cuff M M3OEV
Culak J M0ITY
Cull D M6OBK
Cull G MM3FJA
Cull J G8ALR
Cull J M0BVO
Cull T G1SUM
Cullen A G0DJX
Cullen B 2E0CXM
Cullen D GW7ORB
Cullen J M3WSE
Cullen K M6WHT
Cullen P G4KTZ
Cullen P M6PCW
Culling J G0SNF
Culling M G8UCP
Cullingworth C GM1XLH
Cullingworth S G7CCL
Cullis N G1JGD
Cullis R M0AZN
Cullum K 2E0HKC
Cullum K M0HKC
Cullup A G4FCI
Cullup T G0RCH
Cully N 2I0DYA
Cully N MI6FHJ
Culpan R G4GND
Culpan S G4GPZ
Culshaw J M6JXN
Culshaw L G0LGC
Culshaw S M0SCU
Culverwell P M6AZG
Cumberland S M6SCM
Cumbers P M6PCC
Cumiskey D GW0UXX
Cumiskey P G1OMX
Cumming A GM4HMN
Cumming A MM0GTU
Cumming H GM0HSC
Cumming J GM0GUJ
Cumming J G3VQY
Cumming R G3NOI
Cumming T M3PBR
Cummings A 2I0GWA
Cummings B 2E0HQV
Cummings B M6LHH
Cummings C G4BOH
Cummings G 2E0YSU
Cummings G M3YSU
Cummings G M3YAS
Cummings N G0IMQ
Cummings P M3RJI
Cummings R M3RGC
Cummings R GI0KOW
Cummings T M6OEN
Cummins G M6GFM
Cummins J G0OTJ
Cummins J 2E0DNS
Cummins J G4ERR
Cummins J M6BSX
Cundall C 2E0MDE
Cundall P G7LNT
Cunliffe A G4EII
Cunliffe A G6ERK
Cunliffe J G3ZOC
Cunliffe J G6LNV
Cunliffe N G4OWS
Cunliffe P G1KKH
Cunliffe R GI0HVJ
Cunliffe R G3BZB
Cunliffe R 2E0JET
Cunnah G G3OFP
Cunningham A G0PET
Cunningham A 2E0CBF
Cunningham A M0GAH
Cunningham A 2M0NLA
Cunningham A GM0NWI
Cunningham B G1DNK
Cunningham B 2E0CKT
Cunningham C MI3SXQ
Cunningham D MM3HTY
Cunningham D G0EQE
Cunningham D MM6AUI
Cunningham G G0OEM
Cunningham G M6FNL
Cunningham J 2E0XKC
Cunningham J GI1BSJ
Cunningham J G8EAN
Cunningham J M3UDK
Cunningham J M1EUF
cunningham J G0LIQ
Cunningham K GI6VCL
Cunningham K M6BZF
Cunningham K G0VKC
Cunningham K MM3GZG
Cunningham L M3VVB
Cunningham M G6IOM
Cunningham M GI7UPQ
Cunningham N 2M0NCM
Cunningham N MM3NCM
Cunningham N M6GBM
Cunningham P G0NXH
Cunningham R M0RTC
Cunningham R 2D0BCR
Cunningham R MD0MAN
Cunningham V G4FDF
Cunnington A G7VKB
Cunnington P G8LHW
Cupit D M3BIR
Cupples C GI4EQN
Cupples K GM1MMK
Curant R M6GKH
Curd A M6DHB
Curd E G7DPZ
Curley A G6SDE
Curley B G4ARX
Curling E 2E0AKY

Curling E M0LUV
Curling E M3PQS
Curlis A 2M0AKS
Curlis A [illegible]
[illegible]
Curno R M3KNF
Curnock I MW6IGC
Curnow B G3UKI
Curnow E G7CVA
Curnow J G0ETQ
Curnow P M0PMC
Curnow P M3ROU
Curnow T 2E0TCV
Curphey W G3AGC
Curr J GM7AOM
Curran A G4UMM
Curran D GM7SWX
Curran J MM6BXH
Curran M G4YTT
Curran N M0VOM
Curran P G6TLB
Curran T GM1TCN
Curran W GM1KCH
Curran-Bilbie P G6TLA
Currant R 2E0HEB
Currell A G4VBX
Currell I G3WBA
Currell R G4EIK
Currey B G3LYZ
Currey C G0FKJ
Currey N G0COX
Currie A 2E0MDK
Currie A M0HHX
Currie A M6IDE
Currie A MM6OOK
Currie B GM7HQW
Currie C G1IHS
Currie E MM0EFW
Currie G GM7CPJ
Currie H MI3MPL
Currie I 2M0TGD
Currie I MM6TGD
Currie K G4EDN
Currie S G1VPH
Currie S GI3NYJ
Currie T MI0GLG
Currie T MI3VFZ
Currie W 2E0DPX
Currigan P G6IIN
Curry A M6HBK
Curry A M3XAC
Curry C M0OJC
Curry D M6REO
Curry F M3FEC
Curry G GI6ATZ
Curry J G3UVU
Curry N M6NCA
Curry S G7HHI
Curry W 2M0IQU
Curry W MM3IQU
Curry W 2I0ETB
Curry W MI6ETE
Curry-Peace W M1DCK
Curson C GD4SCG
Curson D G1HYC
Curtis A G4JYH
Curtis B G0KEK
Curtis C G0JRR
Curtis C G3MKV
Curtis C G1TRI
Curtis C M1TAZ
Curtis D 2E0DDL
Curtis D M0LDV
Curtis D M6DXN
Curtis F G3SVK
Curtis G M6UKA
Curtis G M3IZD
Curtis J G0HVA
Curtis J G0JWY
Curtis J G0SEC
Curtis J G4VMW
Curtis J G4WDA
Curtis J MM0MBC
Curtis L G8ZWC
Curtis L 2E0WRY
Curtis L M6WRY
Curtis M M3MGZ
Curtis M M1CMN
Curtis M G4ZKH
Curtis N 2E0CMN
Curtis R G0CYC
Curtis S 2E0CJF
Curtis S G0XAK
Curtis S M0HOY
Curtis W G1HHW
Curtis W G6KTB
Curtis W M1BMD
Curtis-Smith G GI8UUN
Curtress I M3ZZX
Curwell D G8OLY
Curwen J G4ZLZ
Curwen J M6JHN
Curwen L G6THP
Curwen M M6MWC
Curwen R G3PDC
Curzon J G0MBD
Curzon Q G0BVW
Curzon R G0HBA
Curzon R G1AUY
Cushing B M0DJQ
Cushing M M6AAV
Cushion C G3RHW
Cushman D G8MZY
Cushnahan J GI4ELQ
Cusiter G GM4FVS
Cuskin G G7JTI

Cussen A 2E0CUS
Cussick K MM0TMG
Custura A 2M0NSA
Cuthbush R G4ADK
[illegible]
Cutcliffe C MW3EGZ
Cuthbert C G4FJT
Cuthbert J G3YYZ
Cuthbert J GI4OYL
Cuthbert P G0AHR
Cuthbertson A G6JRS
Cuthill J G0MMC
Cutland A GJ7RWT
Cutler C M3EUP
Cutler P G3MXF
Cutmore N G6ALG
Cutter A G0MBB
Cutter D G3UNA
Cutter D G6CP
Cutter D M6TAG
Cutts A G6BUV
Cutts B G1FNS
Cutts D G4FGC
Cutts D G4YJQ
Cutts D 2E0EBV
Cutts D G4FAW
Cutts G M0TAZ
Cutts H G3RDP
Cutts R G3RJM
Czajkowski R G6ATW
Czarnota S G7LNI
Czernuszka N M0NCZ
Czerski M M0RBD

## D

Da Dalt F M6FWZ
Dabbs S G4GFN
Dabbs T G7JYQ
Dabell P G0WOP
Dabhi H G0KSN
Dabinett D G4DEP
Daborn A G6CMF
Dacey W G1OLQ
Dackham J G1BBT
Dadak H G0MMQ
Dadd C G1HZN
Daddy P G0PSL
Dade A M6ADD
Dade D G3XCT
Dadge R M3WRM
Dadswell J G0BKP
Dafter R G4TRD
Dagger N M6NPD
Daglish R M1DZR
Dagnall A G0MNY
Dahalay A M0CAZ
Dailey A GM0REZ
Dailey K G0RVH
Dailey M 2E0MDA
Dailey M M6ARF
Daily C G1PXH
Daines P G6PHJ
Dainton S M6SHZ
Dainty M G6JAM
Daish M G8UYY
Daisley G G4UMP
Dake B M3LMD
Dakin N G4ZSO
Dakin P G4PRD
Dalby K M6KEH
Dalby P 2E0DOA
Dalby T M6TJX
Dale A 2E0LSX
Dale A M0LSX
Dale A M3LSX
Dale C G0DOA
Dale C M0CVJ
Dale C M1CTB
Dale D G8MOG
Dale E G4KTW
Dale E G1VKT
Dale G G3PZF
Dale G M6ACP
Dale G G3MFH
Dale J 2E0JDA
Dale J 2E0KNE
Dale J G2DSY
Dale J M0JDA
Dale J M3NKW
Dale J M6JDA
Dale J 2E0WJI
Dale K G0JPC
Dale K 2E0WKV
Dale K M3WKV
Dale M G6ABU
Dale M M3UYY
Dale R 2E0BGZ
Dale R M0RGD
Dale R M3IIA
Dale S G8CKV
Dale S 2E0NGF
Dale S M6NGF
Dale-Green D G4GLP
Dales J M6PPY
Daley B [illegible]
Daley G G0DAL
Daley W 2E0WLD
Daley W M6WLD
Dalgety J MM6JAE
Dalgliesh A M6CGZ
Dalgliesh J 2E0JCD
Dalgliesh J M0JCD
Dallas D MI3WDO
Dallas P 2I0PAC
Dallas P MI3UIA
Dallas P MI6PFD
Dallas S 2I0SMY

Dallas S MI0SMY
Dallas S MI6SJD
Dallas T M6WTD
Dallas W [illegible]
[illegible]
Dallaway D G6RNO
Dallaway G G8MHT
Dallen N M3IDB
Dalley E G7LWH
Dalley M G4VYI
Dalley P G6JAC
Dalley R M6BTB
Dalley S G0IBR
Dallimore B M0YDX
Dallimore B M3GBD
Dallimore R M0IDX
Dalling D GW4PHT
Dally H 2M0ZET
Dally H MM3ZET
Dally K G4FZR
Dally M G4PCD
Dalton D G4ZTY
Dalton E G3ZLJ
Dalton E 2E0ELD
Dalton E M3ELD
Dalton H G4GZK
Dalton I M1FBN
Dalton I G6CMB
Dalton K G4UEF
Dalton P MW6WDX
Dalton R G3PWS
Dalton R M6EVY
Dalton S M0HOH
Dalton W GI0VLE
Dalton-Kirby A G4DDS
Daly B M0BSD
Daly J M6WCQ
Daly K G0AUX
Daly W G0SXA
Daly W M5CAB
Dalzell A 2E0AZZ
Dalzell A 2E0UJE
Dalzell D 2E0FIV
Dalzell W M5WGD
Dalziel C GM8LBC
Dalziel M M6MXD
Dalziel R M6HMZ
Damm R M1CDQ
Damon L G3CPG
Danby A G0KGA
Danby C G0DWV
Dance A G7KWN
Dance D G4CXP
Dance K M6KPD
Dance M G3IPP
Dance R G8XKD
Dancer K MW0CCK
Dancer K MW0CCL
Dancer M MW3DEM
Dancock W G0LJK
Dancy R G3JRD
Dando J M6PHP
Dane P G4PWA
Danfer M M0BCT
Dangerfield A G1XAL
Daniel A G4PND
Daniel B G7DZY
Daniel C G7NPL
Daniel I MW6IBD
Daniel P MW6GKU
Daniel P M3ZGF
Daniel R G4RUW
Daniell R M6XRD
Daniells C MJ3KBQ
Daniells J MJ3JBQ
Daniells M MJ3MBQ
Daniells P GJ4CBQ
Daniels A G7LCS
Daniels B G0IIB
Daniels B G6DAN
Daniels C M6JCB
Daniels D GW0SRF
Daniels D G4XHP
Daniels D M6DDO
Daniels D M6DEE
Daniels D M6WIR
Daniels E 2E0LLD
Daniels E M6LDD
Daniels F G1ORN
Daniels G G0DGH
Daniels I G4VTD
Daniels I G4VUR
Daniels J M6XAA
Daniels J 2E0SPN
Daniels J M6HKA
Daniels K G6WHY
Daniels K M0AKR
Daniels K M6KDD
Daniels M G4IFJ
Daniels M G8WKK
Daniels M M3ZZA
Daniels N G6WYH
Daniels P G1USW
Daniels R G0GZL
Daniels S G8DIM
Daniels S M1BOU
Daniels S G6UIM
Daniels S M6YSD
Daniels T G7KTP
Daniels W G4HUG
Danks C G7GEP
Danks E G8BKL
Danks K G0DBI
Dann A MD3TLG
Dann C G1AXW
Dann J G3PYO

Dann M G3NHE
Dann P G7PJW
Dann P M0PGD
Dann P 2E0ICC
[illegible]
Dannatt M G8MCY
Dannatt-Brader H G3MBD
Danner J G7LAW
Dansey T G0BIX
Danton A G0SGP
Danton J GM6NYT
Danvers P M3PGD
Daramy J G0EWI
Darby B G1XBL
Darby B G6ALW
Darby C M0OTT
Darby C M6XDR
Darby D G4SRV
Darby D 2E0TVZ
Darby D G6AGN
Darby D M3TVZ
Darby G G7GJU
Darby G G7RTC
Darby G G4TVC
Darby J G1CKF
Darby N M3NJD
Darby R G0UIS
Darby R G7TZQ
Darby R M0VBD
Darby R M3VBD
Darby S 2E0DAY
Darby S M3SJD
Darbyshire J G4GIS
Darbyshire K G8EZU
Darbyshire M G7ING
Darbyshire W G3FFR
D'Arcy M M6EKD
Darcy P G4YBP
Dare B 2E0RQK
Dare B G3JFT
Dare B M0RQK
Dare E G7FBE
Dare K G6JUI
Darke P G1DXD
Darkes D G4CYG
Darkin M G3KTH
Darley J G1AOE
Darley J M6HFP
Darley M M6VOR
Darling D 2E0LDD
Darling D M3LDD
Darling J G1DSB
Darling M G0NUH
Darling W G1DNI
Darlington A G1WYD
Darlington A M6XAL
Darlington C M0DOL
Darlington J G8VNL
Darlington P M0XPD
Darlington P M6PAD
Darlington S G1UTP
Darlow C M0RSF
Darmont J M6EIM
Darragh A G8KWP
Darragh J M6JXD
Darragh P G3MNV
Darrah T 2I0HRV
Darrah T MI0HRV
Darrah T MI6AOX
Darrell G G0CTG
Darrigan S M1AOR
Darroch R GM0OVD
Dart D 2E0ORI
Dart D M0ORI
Dart D M6ORI
Dart P G0VZX
Darton K G4LSE
Darton L 2E0DEP
Darton M 2E0FDU
Darvill W M6FEI
Darwen M M6AVD
Darwent R G0UHF
Darwin J M6JID
Darwin K M6KSD
Darwood D G3YKO
Das D G7ONV
Das Neves Pedro C G6FYE
Dash D 2E0SLD
Dash D M3SLO
Dasilva-Hill G G1HER
Dasilva-Hill K G0KEN
Daskalov T M0NDZ
Datchanamourty S MI3RIV
Date A G0FQU
Daubaris P 2E0POU
D'Aubray-Dutler J M6FMU
D'Aubray-Butler J M3LLB
Daulman A G4KQL
Daum A G3GWE
Davage G M6GBD
Davenport A M3NXA
Davenport J G4OQU
Davenport K G8INC
Davenport K G0MKD
Davenport M 2E0HFS
Davenport M G0AXE
Davenport M G6NZA
Davenport M M3BKI
Davenport N G4EOX
Davenport P G0GLQ
Davenport R 2E0INC
Davenport R G3ROD
Davenport R GW4ANK
Davenport T G1HRM
Davenport T M3BKJ
Davey A G6WLX

Davey A M1BFY
Davey B G4ITG
Davey B G0LCP
Davey E G1JBT
[illegible]
Davey G G8TBT
Davey G G8VJW
Davey H G0CLD
Davey M 2E0XXX
Davey M GW4FZM
Davey M M3LRK
Davey M 2E0DWZ
Davey M M6RKD
Davey P G4VLW
Davey R G8MRI
Davey S 2I0SMD
Davey S M1CYT
Davey S M6EEO
Davey W G0THI
David A GW4OH
David A MW0JZE
David E G4LQI
David E G7EOE
David M G4MEM
David P MW6PGD
David S GW0NCU
David S 2E0SDD
David W GW4WMD
David Z MW3ZAQ
Davidge J M6CCJ
Davidson A G0UVQ
Davidson A G4CDG
Davidson A G4PSU
Davidson B G1YXG
Davidson B GI4BTG
Davidson D GM4ZGV
Davidson D MM3YOL
Davidson E MM3ERD
Davidson F 2M0LAS
Davidson F MM3GTF
Davidson G G1LOL
Davidson I 2M0IWD
Davidson I G4KDW
Davidson I GM8CZU
Davidson I G8YBR
Davidson J GI3FJX
Davidson J GM3UAG
Davidson J GM4BVZ
Davidson J G6CTA
Davidson J G1PGV
Davidson J M3PSU
Davidson K 2E0KTD
Davidson K M3KTD
Davidson M 2E0MAD
Davidson M G0GVF
Davidson M 2E0BEG
Davidson M G3YSM
Davidson M M0XMD
Davidson M M3LCI
Davidson N 2M0BPV
Davidson N MM3WMQ
Davidson P G0DOR
Davidson P G0PTE
Davidson R G4DAT
Davidson R GM4DOF
Davidson R G1XJM
Davidson R G1AVB
Davidson R M3RWD
Davidson S MM6VTS
Davidson T GI8ITD
Davidson T M0MTD
Davidson W GM4NXT
Davidson W MM3ECO
Davidson W GM4XFU
Davidson W G4YDD
Davie E G2XG
Davie M G0NJS
Davie P GM0VEK
Davies A 2E0GTT
Davies A G0EZU
Davies A G0TZY
Davies A G3IIV
Davies A G3PBI
Davies A GW4BIS
Davies A G4NUZ
Davies A G6IDO
Davies A G7NIH
Davies A G8VPX
Davies A M0AMP
Davies A MW0ARD
Davies A MW0DYO
Davies A M0RSY
Davies A M3RPX
Davies A M3SHX
Davies A 2W0VAY
Davies A G0HDB
Davies A MW3VAY
Davies A G7RZW
Davies A M6EFJ
Davies A M6FUE
Davies A M0JIP
Davies A 2E0CVF
Davies A 2E0LAR
Davies A G7DWI
Davies A M0NAZ
Davies A M6ASC
Davies A M6KBY
Davies A M6SYX
Davies A MW6OCT
Davies A 2W0SCL
Davies A GW0WXO
Davies A G4EVR
Davies A M3SMD
Davies A MW3ZDQ
Davies A G6ILH
Davies B G0VGP

Davies B G3PHL
Davies B G4UCE
Davies B G8UTK
Davies B 2E0DHK
Davies B M0HYN
Davies B M0DUL
Davies B 2E0MIV
Davies B G1AZE
Davies B G1DEO
Davies B G3OYU
Davies B G4KAZ
Davies B G8SXD
Davies B MW0DBV
Davies B M6MIV
Davies C G0HRQ
Davies C GI3HNM
Davies C G3JAU
DAVIES C GW3YBN
DAVIES C G4JIV
Davies C G4VFE
Davies C G4VWS
Davies C GM6YQA
Davies C G8IEW
Davies C GM8NBV
Davies C MW0WZX
Davies C M1ELR
Davies C M6MRS
Davies C G3OAJ
Davies C G7GZB
Davies C GW8DUY
Davies C G4FVP
Davies C G7ENR
Davies C 2W0CED
Davies C MW6CAN
Davies C M6CEF
Davies C M6DCJ
Davies C M6NFN
Davies C 2E0CSD
Davies C M G7HAE
Davies C M M3HAE
Davies D 2E0TDI
Davies D G0BVU
Davies D GW0CYG
Davies D GW0NKJ
Davies D G0TLT
Davies D GW0VEW
Davies D G1CDH
Davies D GW3YUC
Davies D G4CEN
Davies D GW4MVY
Davies D GW4RML
Davies D G6GCO
Davies D GW7SXN
Davies D G0LLG
Davies D M3ZFO
Davies D M6DCN
Davies D 2E0DDG
Davies D G0FEU
Davies D G4WVK
Davies D G4YER
Davies D G6MYY
Davies D G7CKS
Davies D GW8ICS
Davies D M3AWI
Davies D 2E0BAF
Davies D 2E0DUL
Davies D M0WYN
Davies D M6DUL
Davies DA G4EYJ
Davies E G4TAU
Davies E G4XVV
Davies E G8SSY
Davies E M3TTA
Davies E M6ELX
Davies E GW6PXW
Davies F G1ISK
Davies F M6EAX
Davies G 2W0APT
Davies G G0CIT
Davies G G4VEW
Davies G G6SBD
Davies G G7DHQ
Davies G G7MJS
Davies G G7SDM
Davies G GW7VST
Davies G MW1GLD
Davies G 2E0VVL
Davies G M3SWJ
Davies G M6CQX
Davies G GW0NDZ
Davies G M3UAR
Davies G M6GKN
Davies G G4KTQ
Davies G 2E0GAG
Davies G M0XGD
Davies H G4DZH
Davies H G6DOQ
Davies H G7KZN
Davies H M0ARL
Davies I G3KZR
Davies I G6BUW
Davies I MW3IHD
Davies I M3ROX
Davies I GI7FHU
Davies I G3IZD
Davies I 2W0ICD
Davies J 2W0BTE
Davies J GW0DZL
Davies J G0ETM
Davies J G0ISY
Davies J G0PTU
Davies J G0PVR
Davies J G0VXK
Davies J G1YJR
Davies J G1ZFD
Davies J G3KZE

Davies J G3LJD
Davies J G3PAG
Davies J G3YJD
Davies J G4XQH
Davies J G4YDJ
[illegible]
Davies J G7OAA
Davies J G8SXA
Davies J M0AXW
Davies J G6JWD
Davies J MW0ZAP
Davies J MM0JMI
Davies J M0GOX
Davies J MW3UYJ
Davies J GW0VMZ
Davies J G4UGQ
Davies J G8WNK
Davies J M1AGP
Davies J M6JKD
Davies J MW6MWG
Davies J 2E0FKA
Davies J 2E0JMD
Davies J G7KMP
Davies J G8RDJ
Davies J M0LDY
Davies J M3FKA
Davies J M6JBO
Davies J MW6RXZ
Davies K G1YHI
Davies K G4CGR
Davies K GW8NAC
Davies K M1DFM
Davies K M3DFM
Davies K M0IBH
Davies K M6HHD
Davies K MM0KCD
Davies K M6KIP
Davies K 2E0VKB
Davies K M3KJD
Davies K M6WKW
Davies L G0RDV
Davies L GM1YPJ
Davies L M6LEE
Davies L GW3FSP
Davies L M6PDU
Davies M 2E0EYC
Davies M 2E0MAR
Davies M G0VPC
Davies M G0VWB
Davies M G1CGJ
Davies M G1YDA
Davies M G4HFS
Davies M G6IPN
Davies M GW7TNS
Davies M G8JTL
Davies M M3MKD
Davies M M3POW
Davies M MW3UBY
Davies M 2E0MTD
Davies M M3UPX
Davies M G4CGH
DAVIES M G4GNY
Davies M MW0CND
Davies M G0WZY
Davies M G8ZWN
DAVIES M M6HKI
Davies M M6INM
Davies N 2E0GWI
Davies N G0AXI
Davies N G7UNU
Davies N M6NND
Davies N M6NTW
Davies N M3TQN
Davies N G8IBR
Davies N M6NSB
Davies O G4VPF
Davies P G0PBL
Davies P G1XCB
Davies P G4EYX
Davies P G8HBQ
Davies P M0CYG
Davies P MW1AUV
Davies P M3OZN
Davies P M3XSD
Davies P 2E0BZC
Davies P 2E0KPD
Davies P G8ZPD
Davies P M0ZPA
Davies P M0ZPD
Davies P M6PMD
Davies P 2E0BFG
Davies P 2E0CNA
Davies P G0KQA
Davies P G4MIO
Davies P M0PJD
Davies P M0PSD
Davies P M1EOK
Davies P M3LLZ
Davies P M6BMI
Davies P M6PDJ
Davies P G0SXY
Davies P M3RNG
Davies R [illegible]
Davies R G0UTZ
Davies R G1EHI
Davies R G1HWP
Davies R G1UQF
Davies R G1ZEX
Davies R G3TAZ
Davies R GJ4BCC
Davies R G4NDL
Davies R M0ANV
Davies R M1FFY
Davies R G7HLZ
Davies R M3HLZ
Davies R M6RSD
Davies R G0MJP

Davies R G3ZWP
Davies R M0USB
Davies R M3WZJ
Davies R M6RRD
Davies R G6VNC
[illegible]
Davies R G0WJX
Davies R G4PDK
Davies R MW6RYD
Davies S GW0AJI
Davies S G1BLV
Davies S G1EAV
Davies S G4KNZ
Davies S G6ZTR
Davies S M0SJD
Davies S M0SPD
Davies S M1EZE
Davies S M1FWD
Davies S M6SMD
Davies S MW6SCD
Davies S 2W0TKS
Davies S MW0KST
Davies S MW6KSL
Davies S G1XST
Davies T 2W1TBD
Davies T G0JLI
Davies T GW3YAF
Davies T GW4ADL
Davies T G4ZKW
Davies T M0AQF
Davies T MW0TJD
Davies T MW3BGP
Davies T M0AYB
Davies T 2E0DTD
Davies T M6TPD
Davies V M6RVA
Davies W 2E0WAY
Davies W G4HLO
Davies W G4YWD
Davies W M3WDZ
Davies W M6SZP
Davies W M6WOD
Davies-Bolton J G4XPP
Davies-Jones A G0EXU
Davin A G7NKI
Davin S 2E0AKJ
Davin S M3OQL
Davis A GW0SQT
Davis A G1NZK
Davis A G3FSA
Davis A G3MGL
Davis A G3VTR
Davis A G4TFM
Davis A G6ALZ
Davis A G6JGT
Davis A G7BDR
Davis A GW7RRS
Davis A G8GZW
Davis A 2E0JUW
Davis A M3JUW
Davis A 2E0CQX
Davis A M1DHA
Davis A M6MCZ
Davis A G6NYR
Davis A GM1HRY
Davis A G0TPE
Davis A 2E0CFY
Davis A MM6DOC
Davis B G0XIT
Davis B GI6EWO
Davis B M1EYQ
Davis B G3UJB
Davis C G0EGR
Davis C G1JRP
Davis C G1LLA
Davis C G3SZR
Davis C G3TNQ
Davis C G3VMU
Davis C G3ZFC
Davis C G7KIF
Davis C G7VTL
Davis C M3UBH
Davis C G7VQM
Davis C M6FOW
Davis C 2W0BOX
Davis C G0JII
Davis C MW0GOV
Davis C MW3SQA
Davis D G0JXQ
Davis D G4AQK
Davis D G4GJE
Davis D G8UCR
Davis D G3SVI
Davis D M3LBG
Davis D M3XZR
Davis D GM4VVY
Davis E G1ZCS
Davis E G0VQY
Davis E G8FFA
Davis E G7WBR
Davis F G6EDC
[illegible]
Davis G G6MJW
Davis G MM6YLO
DAVIS G G0TBW
Davis G G0FBH
Davis G G3ICO
Davis G G4MFX
Davis H G3IUZ
Davis H MI3LNC
Davis H G0SNQ
Davis I G4LPL
Davis I G8XCL
Davis I M0INY
Davis J 2E0AFA
Davis J G0JMD



**IMPORTANT NOTE**

**Revalidate licence to avoid revocation** – Ofcom has advised the Society that plans will be drawn up to revoke licences that have not been revalidated as required by the licence conditions. The quickest way to revalidate is to do so online via the Ofcom website: *https://services.ofcom.org.uk/* or by email: *amateur.validations@ofcom.org.uk* If you need assistance in the process, Ofcom staff are available to help, but please be patient during times of heavy workload.

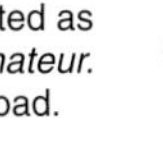

Davis J G0RVI
Davis J G3PAQ
Davis J G3WTD
Davis J G6VUJ
Davis J G6XGJ
Davis J M0VCA
Davis J M6DQN
Davis J G7IIQ
Davis J G4HGK
Davis J G6DID
Davis J M0AXN
Davis K G1FJS
Davis K M6CFI
Davis K G7FSC
Davis L M3LDE
Davis M G0BGB
Davis M G1EKC
Davis M G1UCN
Davis M G1UUZ
Davis M G4LFG
Davis M G4NXS
Davis M G4PQW
Davis M G6KYE
Davis M G7WKW
Davis M M0MJD
Davis M M3IQN
Davis M M3UPQ
Davis M G6USR
Davis M G0ROT
Davis M G4KRT
Davis M G7MBH
Davis M G8VRW
Davis M G8KMR
Davis N M1ADZ
Davis N 2I0HSL
Davis N MI6HSL
Davis O MW6AHT
Davis P G7DIL
Davis P G7KNU
Davis P M0DYQ
Davis P 2M0GFC
Davis P G0RIU
Davis P MM0NQY
Davis P MI6PDY
Davis P MM6PTE
Davis R 2E0REG
Davis R G0MEO
Davis R G1HXR
Davis R G1UFJ
Davis R G3TDL
Davis R M6NYL
Davis R G0WMU
Davis R G6PKG
Davis R G1UNQ
Davis S G3KVR
Davis S G6YPY
Davis S G7AUR
Davis S G8BFA
Davis S M3RTR
Davis S G4SJD
Davis T G7DJL
Davis T M6TJD
Davis W G7DUE
Davis W MW1BDV
Davis W M6WRD
Davis-Edmonds L 2E0HEO
Davis-Edmonds L M6HEO
Davison B G0VEI
Davison C G0JGI
Davison D GD3VFX
Davison D M1DYL
Davison F 2E0CTD
Davison G M6GPE
Davison H G3TVW
Davison H M1DYK
Davison I G0LNX
Davison J G1SBN
Davison J G3JKD
Davison K M0BJT
Davison K M6KJD
Davison L 2E0DZC
Davison L M6LFR
Davison M G3ZUB
Davison M M6GMD
Davison M M3ZME
Davison M M3UKK
Davison N M0NMD
Davison P M3DGP
Davison R G3NVX
Davison R 2E0GDA
Davison R M0WML
Davison R M6AQU
Davison R M6GRJ
Davison S 2E0HTS
Davison S M0YKS
Davison S 2I0TRM
Davison S MI0LRC
Davison S MI3RYD
Davison W MM6PDX
Davy D G6EWP
Davy M G6SWZ
Davy-Jones J G0USE
Daw A G1DSF
Daw B G7KEE
Daw R G0MCE
Dawber S GW6OGD
Dawe A G8WMF
Dawe B G0PWC
Dawe C G8TCP
Dawe D G8JBV
Dawe I G3SPI
Dawes A G6VZF
Dawes G G7VJI
Dawes G M0AEP
Dawes L M1LSD
Dawes M G8USA
Dawes P 2E0GBF
Dawes P M6AHZ
Dawes R G3SEN
Dawkes D G0ICJ
Dawkins G G8KQZ
Dawkins M G8XPD
Dawkins R G1EAX
Dawkins R G4FRH
Dawkins W M0HZW
Daws A G4PVX
Daws J M6KET
Dawson A G0NEU
Dawson B G8OKS
Dawson B M3FCO
Dawson B G4OQZ
Dawson C G8BRD
Dawson C 2E0BRF
Dawson C M0GMK
Dawson C M6CJD
Dawson D M3WHL
Dawson D G0ELJ
Dawson D G1NEV
Dawson D G7UZN
Dawson F M0DRB
Dawson F G1HCM
Dawson G G0VBT
Dawson G M0GFM
Dawson H G0OQZ
Dawson J G4XGT
Dawson K G3KWO
Dawson K G8OSX
Dawson K M0KEV
Dawson L M6LND
Dawson M 2E0MIN
Dawson M G0GYJ
Dawson M G0VYC
Dawson M G4HWJ
Dawson M M1FFM
Dawson M G1MHN
Dawson M G3TCL
Dawson M G8IC
Dawson N G7GGA
Dawson P 2E0PJD
Dawson P G6YIU
Dawson P G7TGN
Dawson P G8WVO
Dawson P M3IRU
Dawson P M3TNL
Dawson R G0EYH
Dawson R G0XAB
Dawson R GI6CMA
Dawson R G6TJE
Dawson R G4ELA
Dawson R G0PEV
Dawson R M6VED
Dawson S GI4OVN
Dawson S M3UWR
Dawson S G0HXU
Dawson T M5AJK
Dawson W M6FDI
Daswell S G1EBV
Day A G4RIM
Day A G6OXQ
Day A G7CGT
Day A M6FVE
Day A M1AMA
Day A M1BFR
Day B G3WIS
Day C M3KXI
Day C G4MAS
Day C M1EAK
Day D 2E0OZO
Day D G4TDI
Day E G4OBV
Day E G8SRN
Day F G4PZU
Day G G1PHA
Day G MW0GCD
Day I 2E0OKK
Day I M0IRD
Day J 2E0XXX
Day J G7NNM
Day J MW0JAN
Day J M0JBD
Day J M6GVI
Day K G3LDJ
Day K G8LCM
Day L G4DIV
Day L G4OUZ
Day M G4ZKI
Day M M0MXX
Day M 2W0CJI
Day M MW0HKA
Day N G4OBT
Day P G4KYY
Day P M3ZYY
Day P 2W0PCD
Day P MW6PEL
Day P G3PHO
day P M6HTQ
Day R G0AOM
Day R G1CEO
Day R G8GRB
Day R M1AKN
Day R G2FQZ
Day R M6TBV
Day S 2E0HHY
Day S G1FVS
Day S G8PAN
Day S M0GLS
Day S M3FUQ
Day S M6KMC
Day T G0SLI
Day T G0VSM
Day T G3ZYY
Day W G6FZV
Daymond D G7AVB
Daymond G M6FHR
Daymond M M3ZYQ
Daymond P G0LRJ
Daynes G G7BQU
Daynes R G0MJB
De Araujo A M0AVF
De Bank J G1XHA
De Banks M G7ITO
De Bass F G4LXD
de Basto P M6PDB
De Broise A M0NOA
De Buriatte A G0PYF
De Camps P M0PAC
De Chastelain S M6SBU
De Fraine D G8IRC
De Frece J G0TRN
De Freitas D 2M0NIT
De Freitas D MM6TNO
De Havilland R G1OHV
De Ieso R M0DSO
De Jong M M0HCE
De La Haye D 2E0DDH
De La Haye D G6BAL
De La Haye D M0MBD
De La Haye D M6DLH
De La Haye K GU0RAG
De La Hunty K M6FQE
De La Rue R G4FJC
de Lacy C G0TSQ
De Maillet A M0ZZA
De Mengel P MW0GDM
De Muth R G4JRD
De Peyer O 2E0LVR
De Peyer O M0LVR
De Peyer O M6ODP
De Putron T GU3LYC
De Renzi J G1PUY
De Rouffignac M G8OPE
De Savigny-Bower R M0RDB
De Silva D G7AGI
De Silva M G0WMD
De Ste Croix G G1HCJ
De Ste Croix R G1HCI
De Vries A M3FTZ
De Vries M GM0TQB
De Vries R 2E0SDP
De Vries R M6RDV
De Vries S M0ZJV
de Young M M3MDY
De Zeeuw H MM6HDZ
Deacon A G6DOW
Deacon C G1LUX
Deacon C G4IFX
Deacon D G7IRK
Deacon G M0UWD
Deacon J 2E0JIM
Deacon J G1TRL
Deacon J 2E0XAZ
Deacon M M0CSU
Deacon R G1DLA
Deacon S G0HLS
Deacon S G6JAK
Deacon T G0AHI
Deak B G4PNP
Deakes P G1HRJ
Deakin A G0EOM
Deakin A G1ENR
Deakin B M3YPI
Deakin C G0HRX
deakin D M0CAX
Deakin D M3FFU
Deakin R G0HYR
Deakin R G6IPQ
Deakin R M0BIH
Deakin S G0JYF
Dealey G 2E0GDY
Dealey G M3GDY
Deamer C GD3NDC
Dean A G1RSF
Dean A M6BRU
Dean B G4KCD
Dean C M3TLN
Dean D 2E0DRE
Dean D G3JSK
Dean D M3POH
Dean D G7SXG
Dean E M0DHM
Dean G 2E0GDZ
Dean G M0XAC
Dean G M6GDZ
Dean G G1IWH
Dean J 2E0NRQ
Dean J G1AVC
Dean J GM4FBP
Dean J G6NEA
Dean J M3MIJ
Dean J G8JXG
Dean J M3TNO
Dean L M0DBD
Dean L G7ELE
Dean M G4RSX
Dean M 2W0JMK
Dean M G0CQU
Dean N G0OQS
Dean P G0IAI
Dean P G0UFN
Dean P G1FVS
Dean R G1AOF
Dean R G7UEL
Dean R M0COC
Dean R M6RQD
Dean R 2E0VPO
Dean R M6YDW
Dean S G0WFE
Dean S M6SDQ
Dean T M6XBE
Dean W G4MFD
Deane A M5ADE
Deane D G3ZOI
Deane D GM4VZY
Deans K 2E0KMD
Deans K M6KMD
Deans R G0CFI
Dear N G4DFQ
Dear P G8TIO
Dearden F G4IYP
Dearden G G1YPR
Dearden J M3YEZ
Dearden J M0TOR
Dearden R M3UGZ
Dearing B G4HMM
Dearing M G0KBP
Dearman A G1NEB
Dearman D M0GTS
Dearsley R G1EMW
Deary M 2E0OAP
Deary M M6OAP
Deary M M0AIC
Deas G GM7SCJ
Death G G1BEK
Deaves A M6DGR
Debenham B 2E0RMD
Debenham B M3OGM
Debenham S 2E0PFF
Dec S M3FLP
De-Cogan D 2E0FIJ
De-Cogan D M6FIJ
Dedier J M0JDD
Dedman R G4DFY
Dee P M3ZLI
Dee R G7DIG
Deefholts B G8JDN
Deegan K G0UDG
Deehan M GI0HDO
Deeley E M0EMD
Deeley M 2E0MGX
Deeley M M0WFM
Deeming A M6TNY
Deeprose P M6PWD
Deeprose R G4VBK
Deery A M6DOJ
Deery J 2E0CWK
Deery N 2E0NAP
Deffee P G8PJD
Degerdon M G1DSG
Degg R G0JOD
Deglos M M1BBR
Deighton D G8RBV
Deighton S G8NRC
Dekkers P 2E0FOD
Dekkers P M0SSJ
Dekkers P M6SWL
Del Monte H G3UPD
Delacassa D G0NPF
Delafield H M0CIW
Delaforce G M6VKZ
Delamare R G6ALR
Delaney F G4GKT
Delaney J G1GBF
Delaney J M0JSD
Delaney M M6EGW
Delaney P G0HPQ
Delaney P G1WAS
Delaney P G8KZG
Delaney S M3MXI
Delaney W G3FOQ
Delasalle E 2E0EHD
Delasalle E M6EHD
Dell C M3OSS
Dell D G3PQF
Dell D G4WLA
Dell L G6DOZ
Dell R 2E0RFD
Dell R M0RDZ
Dell R M3URD
Dellbridge G G0PMF
Dellbridge R G0PMG
Deller R M6LRF
Dellett D MI0MSM
Dellett P GI0STC
Delrosa R 2E0KLD
Deluce P M6TCB
Delve R G1FZR
Delves A G0HXM
Delves C 2E0JDD
Delves J M3LZK
Delves J G3VHH
Delves R G7MJJ
Delwiche A G7VIO
Demczur W MM6HEZ
Demirkaya L M6DEM
Dempsey L G6PXX
Dempster B G1RVF
Dempster C M6FOD
Dempster I G4RCZ
Dempster K G0ENO
Dempster P G0GBC
Dempster R 2M1GFG
Dempster S M6EWK
Dempster W GM0MDX
Denby G MD0ALQ
Denby J G3TSA
Denby M G4GYL
Denby T M1TAD
Dence R G4DAM
Dench E G4NPE
Dench M G1TWS
Denecker V G0LMX
Denehy P M3MPT
Denford D G0HSH
Denford J G0GFK
Dengate M M6SXT
Denham C G4VLL
Denham G G1SWU
Denham L 2E0BXQ
Denham L M6LJD
Denham M 2E0MVD
Denham M M6MCO
Denham P 2E0PRD
Denham P M0PRD
Denham P M3NQI
Denham R M6FTP
Denham T M1TDD
Denim R 2E0JOC
Denim R M3JOC
Dening A G4JBH
Denison A G4ICF
Denison C G8KZY
Denison G G6JAP
Denison I G6XSC
Denison M G1JLB
Denker N G7EKL
Denley R G4HRG
Denman C G7SYT
Denman C M6NOJ
Denman E G0PXT
Denman E M3EKA
Denman G M6EIZ
Denman J G3OND
Denman S 2E0SPD
Denman S M0SZD
Denman S M6AXD
Denmead H M3GYA
Denmead J M5AFH
Dennehy M G0IZK
Dennett C GM7MYF
Dennett D G1LZZ
Dennett E G0EDM
Denney I G3JPZ
Denney S G3CIM
Denney T G3VLD
Dennick G G4MFK
Dennis A G0PZX
Dennis A M3ANW
Dennis B G4UTM
Dennis C GD7ELF
Dennis C 2E0XJP
Dennis C M6XJP
Dennis C 2E0ECD
Dennis C M6ECD
Dennis G 2E0PLE
Dennis J G7RAG
Dennis J G4ZOQ
Dennis K G7PHT
Dennis M GW3KZO
Dennis M G6KIW
Dennis M GM7RNJ
Dennis M G8BTY
Dennis M M3RHJ
Dennis M M6XAQ
Dennis P G1SQG
Dennis R G0OFA
Dennis R 2E0FFJ
Dennis R M3RDV
Dennison I M3YZU
Dennison M M3MJD
Dennison W G6USX
Denniss J G0NMJ
Denniss K G7DEE
Denny A G0JNJ
Denny C MM6CHN
Denny J M3JLD
Denny T M6TCD
Denny W 2E0WCD
Denon M M3ZIO
Denscombe C G4HAC
Densem J G4KJV
Densham B G8MQY
Densham C 2E0CRD
Densham C M3DGD
Densham M G0FPU
Densham T G4GOG
Dent A M1ELM
Dent J G1YFE
Dent J G1CSO
Dent M G6PHF
Dent P 2E0PBN
Dent P M6BPS
Dent R G4TJS
Dent R G7CSL
Denton B G4GAT
Denton C G8TUN
Denton D G0FUE
Denton E G3WLO
Denton G G8VAT
Denton H G1KEV
Denton J G7CWM
Denton J G8EGE
Denton N G7AIB
Denton R G8WKX
Denton R G7TQT
Denton R G4YRZ
Denton-Powell C G0MRR
Denut M M0SBD
Denyer A G4MLG
Denyer D G7UEC
Denyer L G0HPN
Denyer-Green B M0HBM
Denyer-Green B M6AGO
Depledge J G6LKJ
Depledge N G0TSR
Depledge T 2E0FRW
Depledge T G1VIY
Deravi F G6WDS
Derbidge K G7NCE
Derbin-Sykes B 2E0BDS
Derbyshire H M3OZC
Derbyshire N G1CKV
Derbyshire T G6USU
Derham D G3EXL
Derham R 2E0LSV
Derham R M0LSV
Dermondy I M6DER
Dermont A G8BGT
Dermont A M5AGY
Derner S M6XCC
Derrick A G4DEQ
Derrick H G7JQW
Derrick J G4PKM
Derricott R G4VPE
Derringer M M3VFC
Derry F M0SFD
Dervin C G4RBZ
Derwin S M1AFU
Derzsi Z 2E0DZS
Derzsi Z M0MBA
Derzsi Z M6DZS
Desbois R 2E0AWI
Desborough C G3NNG
Desborough J G1DLB
Desoer A M0CGA
Dessau N G6ZIO
Destoop P M6ONL
De-Thabrew N 2E0NFD
Deutsch M G3VJG
Deverell D G6USZ
Devereux E G3SED
Devereux E G4PYS
Devereux G GM7NHU
Deverill M GM6VHA
Deville P G4KWM
Deville S G6TJC
Devine A GM6BAO
Devine E G7RIJ
Devine M G1HZJ
Devine N G3DFY
Devine N M0AZV
Devine P G0HJX
Devine P G8LGE
Devine T GI4XGQ
Devlin B GM0EGI
Devlin C M6PDT
Devlin D G7RMD
Devlin M MI6DVN
Devlin P 2E0PFD
Devlin P G1SMP
Devlin P M0PDV
Devlin P M6PDE
Devlin T G4SPW
Devonshire J GW4LFF
Devos R M6SHH
Devos S M0VOS
Devos S M6VOS
Dew J G3FPY
Dew J G4UXG
Dewar D G1UCZ
Dewar I GM1BLX
Dewberry C M3CVD
Dewberry O G7VFC
Dewes R M3RDV
Dewey T G8RBU
Dewey V M0CVA
Dewhirst J M6NIN
Dewhurst C G4KLD
Dewhurst K G0CHL
Dewhurst W G6RZG
Dewick P G6PHC
Dewing K G0ULG
Dewis R 2E0BRL
Dewis R M3UYO
Dews F G3HPD
Dews K G3PMW
Dewsbery R G7PZB
Dewsbury D 2E0DLD
Dewsbury D M3PIY
De-Wynter M G1XGM
Dexter B 2E0ZEV
Dexter B M3ZEV
Dexter J M0HDX
Dexter L M3LXR
Dexter M M6USA
Dexter N M6NXZ
Dey B M0PRT
Dey M G8SNV
Dhami M G4NRI
Dhami R G4PNQ
Dharas M G1ZHL
Dheerendra P M6PRD
Dhillon A M6ASD
Dhuglas D GM4ELV
Di Domenico A M1FHT
Di Duca A G1EBB
Di Genova G 2E0GDG
Diacon G G8EWT
Diamond A GM0IKY
Diamond D G3UEE
Diaper C G1IBJ
Diaper G G1IUW
Diaper R 2E0DLR
Dibben K M3DIB
Dibben S M6SYG
Dibbins A G7OTQ
Dibden F G4SHO
Dibsdall M G8MTI
Dick A GM0IRZ
Dick A GM6KKP
Dick D G6PWQ
Dick P GM4DTH
Dick P GM8HHC
Dick R M6RKK
Dick W GM8MMW
Dickason A G0EKD
Dicken P G1PCD
Dicken P G8UUR
Dickens K G4OCH
Dickens P G6PHH
Dickenson B G7MIF
Dickenson J M6MGT
Dickenson L MM3FMY
Dickenson M 2E0NUQ
Dickenson M M3NUQ
Dicker R 2E0TAQ
Dicker R M3UUF
Dickerson C G0KFT
Dickerson R G8RLF
Dickerson R M3NGF
Dickerson R M6EJX
Dickeson M 2M0BPM
Dickeson M MM0LER
Dickeson M MM3WLA
Dickey R MI1BRS
Dickie G G7JDE
Dickie R GM0IPW
Dickinson A G0GTI
Dickinson A G3XJR
Dickinson A M3ISN
Dickinson A M0TTL
Dickinson B G1LZF
Dickinson D M3IBJ
Dickinson H M3CMI
Dickinson I G3OUI
Dickinson I G8ZSZ
Dickinson J G0WUV
Dickinson M G7DRT
Dickinson M M6GMM
Dickinson N G0FZA
Dickinson P G0VEX
Dickinson P G7RKU
Dickinson P G8KEO
Dickinson P G6DIC
Dickinson S M3ULT
Dickman P M0DFQ
Dicks R G0TES
Dicks T G7BLK
Dickson A G8DJF
Dickson A GM0KVE
Dickson A 2E0DWA
Dickson A M6DWA
Dickson C G8AVB
Dickson D MM0CDK
Dickson D M0PJX
Dickson J G0IGJ
Dickson J G4UPR
Dickson K G8XGB
Dickson K M0XGB
Dickson K MI3KDR
Dickson P M0HNF
Dickson R G1WFU
Dickson R M0DNR
Dickson T GM3DIE
Dickson W 2E0PBU
Dickson W GI0HSB
Dickson W M6PBO
Didcott C 2E0SBJ
Didmom B G4RIS
Dietsch J M3OOP
Diez J M6JGY
Digby A G0JLX
Digby C G4NEO
Digby D G8DHQ
Digby M 2W0TCM
Digby M MW0XTZ
Digby P G4NKX
Diggins G M3VUN
Diggins P G8MZD
Dighton A 2W0PPL
Dighton A MW6AZP
Di-Giulio B M6EMX
Digman E G3BVA
Digman T 2E0TDO
Digman T M0ZTD
Digman T M6CWX
Dignall K 2E0EBL
Dignall K M3DIG
Dignan J M0JJD
Dignum B G0DDE
Dilks J G7OVS
Dilks J G7JGI
Dillane L M6TVI
Dilley N G8YBT
Dillon C G3WCD
Dillon J GM3YQK
Dillon S MI3WQC
Dillon S M6VTC
Dillon W G0JJQ
Dillon W G6UIT
Dillow E G1HRL
Dillow K M6KDW
Dilworth A 2E0TDE
Dilworth I G3WRT
Dilworth J M6MVD
Dimambro D G4KTP
Dimambro M 2E0HDB
Dimambro M M1MOB
Dimambro P 2E0HDC
Dimambro P M1TXT
Dimbleby N G0TRE
Diment I GM8ZKN
Diment J G4LTC
Dimes P 2E0GPD
Dimes P M6OBZ
Dimitrov G M6ZES
Dimmick A GM0USI
Dimmick A GM4FVQ
dimmick M M6HQL
Dimmick P 2E0LFM
Dimmock A G1SUK
Dimmock A M5ADD
Dimmock F G0CFD
Dimmock J G8AZR
Dimmock M G8GIL
Dimmock R G1HIJ
Dimonaco N M6IJD
Dimopoulos S M6SDZ
Dinally M 2E0DOQ
Dineley H MM3GQY
Dinger F GM0CSZ
Dingle A M6AMD
Dingle B G4ITV
Dingle N G1XNI
Dingle R G0OCB
Dingle R G7VAY
Dingley W G0UCS
Dingmar S M3YGS
Dingwall A 2E0PDO
Dingwall A M0TUV
Dingwall A M6PDO
Dingwall F G0SGX
Dingwall J G4ILW
Dinning C MM3YWZ
Dinning J MM3LZD
Dinning R GM0GOV
Dinsbier J 2E0JGD
Diplock A G4NRV
Diplock J GW3UZS
Diplock J M3VXK
Diplock O G3NXK
Dipper A G0NKK
Diprose M G4AKA
Disley R G3KQY
Disney R G0HNZ
Disney W M3YWD
Diss D G4YAX
Diss P G6NEK
Dissanayake M G0CUA
Distin R G6IPH
Diston J M3DVM
Ditchburn S M6BNU
Ditchfield C G0JQX
Ditchfield T M0CEC
Divall C G8MCC
Divall J G4MDC
Dix C 2E0GKP
Dix D GD4JZS
Dix D GD8LZE
Dix M 2E0IAC
Dix S 2E0TOT
Dix S M6TOT
Dix W G4ZEU
Dix W 2E0WCX
Dix W M6WCX
Dixey D M1BPY
Dixey M G4OSU
Dixey P G6JLI
Dixon A 2E0AVP
Dixon A G4IVU
Dixon A G6GBU
Dixon A M3IYG
Dixon A G8BQF
Dixon A G6JAL
Dixon A M0ACC
Dixon B 2E0BPG
Dixon B G0GZT
Dixon B G1UWV
Dixon B G1YXA
Dixon B M3XFT
Dixon B G0TOK
Dixon D G0AYD
Dixon D G0BXV
Dixon D G7BPG
Dixon F G7OJU
Dixon G G6SXN
Dixon G MW6NGE
Dixon H M3HQQ
Dixon I G1BEJ
Dixon I G4BVY
Dixon J 2E0RJD
Dixon J G0NYE
Dixon J G6YIQ
Dixon J G8VNX
Dixon J M0JFD
Dixon J M1EMR
Dixon J M3HIP
Dixon J M3LKD
Dixon J M5AIO
Dixon J GW4YLF
Dixon J G6KWH
Dixon K G4XKD
Dixon L M0LED
Dixon L M3HJD
Dixon L G3XXQ
Dixon M G1TUU
Dixon M G7EYL
Dixon M 2E0MLD
Dixon M 2E0HJD
Dixon M G3LHU
Dixon M G8IQX
Dixon M M0PVA
Dixon N M6NJD
Dixon O M3OWN
Dixon P G0HHA
Dixon P G4JBR
Dixon P G6FTL
Dixon P G7JRK
Dixon P M6FDL
Dixon R 2E0DMI
Dixon R G3SNT
Dixon R G4YAV
Dixon R GW7DTB
Dixon R M1RST
Dixon R GI0BFO
Dixon R GM3ZDH
Dixon S G7UCL
Dixon S M3IVY
Dixon S G3RE
Dixon S G4IYK
Dixon T G4NHL

Dixon T 2E0TDD
Dixon T M6NIT
Dixon W G3XIH
Dixon Z [illegible]
Djali P G4ITF
D'Mellow D M0TTF
Doak T 2W0GER
Dobbs G G3RJV
Dobbs G G4LAY
Dobbs J G0OWH
Dobbs S M0DOB
Dobby I GW4LDP
Dobbyn A G0IAD
Dobdinson R G3RGD
Dobie A M6SNY
Dobie A M6GYP
Dobie S MM3KLO
Dobinson C G4YAK
Dobinson L 2E0LDQ
Dobinson L M6LDQ
Dobson D G7PHY
Dobson D G0IDE
Dobson D M6FLL
Dobson E M3WCS
Dobson G M0CJZ
Dobson G G8OFQ
Dobson G 2E0PAT
Dobson G M6EWD
Dobson I G6LNL
Dobson I G4XVW
Dobson J G8CVF
Dobson J M3JDN
Dobson J G4UNO
Dobson J G6WJD
Dobson J G8JOX
Dobson K G6XXY
Dobson K M3XXY
Dobson K 2E0FJD
Dobson K M6FJD
Dobson L M3ZPY
Dobson M M3MDN
Dobson N M6KMI
Dobson P G6ABA
Dobson R G3JDD
Dobson R G4OBX
Dobson R M3EWZ
Dobson R MI6TND
Dobson S G7PHW
Docherty A 2E0HCG
Docherty A G1TTX
Docherty A GM4FXL
Docherty C MM6ZCD
Docherty H G4ZJO
Docherty J M3VRY
Docherty J MM3ZQP
Dock J MM3LVT
Dock J 2M0DOC
Dockar D G4IDD
Docker M G3OOW
Dockerill A M3TDT
Dockerty M G7WFD
Dockery D G4IBH
Dockery S MI0UST
Dockray E G7OQB
Dockray K G0EMM
Dockray R M6RMD
Dockray S M6XSF
Dodd A G0SXK
Dodd A 2E0DOD
Dodd A M0PAI
Dodd A M3TUL
Dodd A G4YTY
Dodd D G4DKZ
Dodd D G6DOX
Dodd D GD3RFK
Dodd E M1FET
Dodd G G2DBH
Dodd I G1HZI
Dodd J G1FNU
Dodd J G4FJB
Dodd K G6GWY
Dodd L G7VTC
Dodd M GD4RFK
Dodd P G6SFW
Dodd R G0UDO
Dodd R G1BWG
Dodd S G0CIM
Dodd S G1NST
Dodd S M6ISD
Dodd T MM6TWZ
Dodd T M6REO
Dodds A 2E0TAM
Dodds A GM4YMI
Dodds D G0XXQ
Dodds B G3YRH
Dodds C G8YOT
Dodds C M6NJZ
Dodds J 2E0JAM
Dodds J 2E0HXN
Dodds J G7UTG
Dodds J M0UTG
Dodds K MD6TDU
Dodds P M3VPX
Dodds R G7IYX
Dodds R M0HGV
Dodds S G7TFL
Dodds T G0NJZ
Dodge B G3PCX
Dodge J G6ILN
Dodge N M6NHS
Dodge P G0DBY
Dodgson C 2E0CAD
Dodgson M G0EKX
Dodman K G8HRF
Dodman L 2E0LOZ
Dodman P G7RBQ

Dodshon C M0BYU
Dodson A G3MGU
Dodson B M6DJJ
Dodson D [illegible]
Dodson L G0IKE
Dodson L G6HVQ
Dodson M G6RII
Dodson M G7MAB
Dodson R G4RNK
Dodsworth G G0PHS
Dodsworth J M3EYK
Dodsworth M G0URF
Dodwell G G4CFS
Doe A M0HKK
Doe A G4NJR
Doe N G8TBU
Doermann D G7IXG
Doggett D M3NKX
Doherty D GI6AUI
Doherty D MI0BBF
Doherty D GW4HZH
Doherty F MI3MFD
Doherty J GI4AXV
Doherty J GI4XJD
Doherty J G7HIK
Doherty J MI3XGR
Doherty K GI4TED
Doherty M GI4TAV
Doherty M M6XBQ
Doherty N GM0PWS
Doherty R MI3CGA
Doherty S MW0SGD
Doherty T GI0OTC
Doig A GM1JTK
Doig G G1DBI
Doig M G4CQZ
Doig M GM7GOE
Dokic A M0KVA
Dokic A M3VAR
Dolan D M3HUY
Dolan J G3PRW
Dolan M G3KZU
Dolby A G6KJE
Dolby J G3PDD
Dollery C G3GAF
Dollery P G4TNB
Dollimore P G1LLQ
Dolling D G0FVH
Dolling P G4LQZ
Dolman H G7VGX
Dolman L G4EXN
Dolman R M3HVE
Dolphin C M6TGL
Dolphin D G0AQF
Dolton R G3HTO
Domachowski P G7JSQ
Doman C G4EZQ
Doman H M3TGT
Dominy A M3DZT
Dominy R G8YXZ
Dommett T G1MOK
Domville R G6RKS
Donachie F G4XWT
Donachie G G1LMQ
Donachie I G0CWD
Donachie I G0WJZ
Donachie P MM6EFH
Donaghy E G7WAA
Donaghy E M0CLI
Donaghy P 2E0PXD
Donaghy P M0PXD
Donaghy P M6PXD
Donaghy P MI6FHG
Donaghy R MM6AAR
Donald A GM1ROX
Donald A 2M0CXG
Donald A MM6CXD
Donald C GM6MJY
Donald C G0OEB
Donald J M0CMT
Donald J M3SDK
Donald L G1WWY
Donald N G0UZW
Donald S G6EDD
Donald V G7RXE
Donald W G7ENQ
Donaldson A GM0DGK
Donaldson A GM7CQQ
Donaldson G M3ZGD
Donaldson I G4SZA
Donaldson J MI3RST
Donaldson J GM3ZSH
Donaldson L GM4AJR
Donaldson M G1HZI
Donaldson R M0ZEM
Donaldson T MM6HNS
Donaldson W GI4GYM
Donaldson-Badger P G4OZN
Donath S M0IAG
Donati D G8BAD
Donbavand E G6BIX
Donders B G3UXJ
Donegan T M3LBD
Donin J G4YJD
Donkin B G7OUZ
Donley T G0SQX
DONN B G3XSN
Donn G G4IHS
Donn I G6FUT
Donnachie B M0DMX
Donnachie F G1RON
Donnachie M MM0EFJ
Donnan J GM0WUX
Donnan R MI6CMQ
Donne C G3YKK
Donnell J MW0AJH

Donnell W MI3TUS
Donnelly A GI4NSV
Donnelly B 2M0BRD
[illegible]
Donnelly D M0DYW
Donnelly D M3NOF
Donnelly D M6WHN
Donnelly H 2E0BCF
Donnelly H M0GNM
Donnelly H M3IHV
Donnelly I 2E0XOD
Donnelly I M3XOD
Donnelly J GI8JRE
donnelly J MM1CXO
Donnelly J 2E0BZI
Donnelly M GI3NSV
Donnelly M G4SNW
Donnelly P MI6BJZ
Donnelly P GI4VCZ
Donnelly S MI0TBN
Donnelly W 2E0RPO
Donnelly W M6RPO
Donnet R GM7KXJ
Donnett J G0MVY
Donnithorne J G8MKX
Donno R G3YBK
Donoghue P G8DNP
Donohoe D M0DTI
Donoughue G G4THX
Donovan B GW1OZW
Donovan C M6EWL
Donovan F G4ALD
Donovan H M6HZD
Donovan J G4GHK
Donovan J M6JII
Dons C GM4YMM
Dons E GM0AXY
Doogan M MI6HGS
Dooks B G0RHI
Doolan C MM3YIG
Doolan R MM3YIQ
Doole C M3OLM
Dooley G G7IIZ
Dooley J M3DOO
Dooley R GW0AZW
Dooley S M6LAF
Dooley T M3TPD
Doorbar K M3FSY
Doores J G0WEY
Doorey S G6MYZ
Doran C G3VZH
Doran E M1CYK
Doran P GI4WYE
Doran R G4LPZ
Doran R G4VRC
Dore A M1DCY
Dore J G3XPK
Dore R G7GAH
Dore T M1DCX
Dore W GW4CGE
Dorey B G7OYX
Doris C MI0AHH
Doris J MI0AHI
Doris P MI0BML
Dorling A G8CJL
Dorling R G0PJZ
Dorman A GD0AMD
Dorman A MD0GLK
Dorman H MD0HHH
Dorman I MD6IZI
Dorman J GD1LVY
Dorman K MI6LLZ
Dorman P M6GFV
Dormer C G6EWQ
Dormer D M6MOU
Dormer M G7PFL
Dornan M GI4SOY
Dornan S GI3TNK
Dornan S GI7GXZ
Dornford-Smith A GI0NQC
Dorning J G0NEB
Doroba D 2E0BFT
Doroba D M3AGE
Dorrian A GU7NHX
Dorrian C GI7VGR
Dorrian P MI6LDI
Dorrian T MI3VTJ
Dorricott B G4SDL
Dorricott T 2M0TXA
Dorrington M 2E0FTL
Dorrington N G4VAF
Dorrington P G8MMF
Dorrington S G1KGL
Dorrington-Ward S G8LOJ
Dorris P MI0HPE
Dosher J G8MGK
Dossa A 2E0SSA
Dossa A M0SAO
Dossa A M6AHD
Dossett P 2E0PAU
Dossett P M3YXU
Doswell A G7UQA
Doswell J G3VYE
Doswell K G1OET
Doubleday G G6BZQ
Douce I G0JDD
Douch G 2E0BKJ
Douch G M3UGJ
Douch J 2E0UUW
Douch J M3UUW
Dougan M GI0DWN
Dougan M MM6DDX
Dougherty G M6KVL
Dougherty P M0DXQ
Doughty A G6ZQJ
Doughty G G1VQB

Doughty G M1REC
Doughty P G3TKK
doughty R G1FRM
[illegible]
Doughty R M3XRO
Doughty R G1AUC
Douglas A G0I IDJ
Douglas A GM4HVM
Douglas A G4WDO
Douglas A M0HNG
Douglas A M6CUV
Douglas A 2E0IBI
Douglas A M6IBI
Douglas A MM6CCY
Douglas C G0OGW
Douglas C GD3ZEX
Douglas C G8LDJ
Douglas G G3SZY
Douglas I G3NID
Douglas I GM4FGS
Douglas I MM0FFC
Douglas I M0IMD
Douglas J G4DVG
Douglas J M3JID
Douglas J M3XYO
Douglas J G8GFW
Douglas L GI0RJO
Douglas M 2E0ECZ
Douglas M M0ZCE
Douglas M M3ECZ
Douglas M M3HYD
Douglas M 2M1HSG
Douglas M MM3HSG
Douglas M M6KKY
Douglas N G4SHJ
Douglas N M1EMP
Douglas P GI8KFG
Douglas R G0FUH
Douglas S G6JLL
Douglas T M6TDD
Douglas W G4NTW
Douglas W M3WLD
Douglass M G6MKD
Doull J GM6TJD
Doull W MM3YBD
Douthart S GI8WIU
Douthwaite J G6OQO
Dovaston N G4ODE
Dove D G3DOV
Dove J G8KHF
Dover G G4AFJ
Dover J G0KOH
Dover P G6FWO
Dover R MM6DVR
Dover W 2E0PGP
Dover W M6KIK
DOVETON S G0JNR
Dow A M3OBD
Dow J 2M0DQY
Dow J MM0SNK
Dow J MM6DQY
Dowd R G0RPO
Dowdall C GI1LGM
Dowdall J GW8HDH
Dowdell R G8RLD
Dowdeswell J 2E0JRD
Dowdeswell J M0JDL
Dowdeswell J M3YJD
Dowdeswell R G1WIW
Dowding M GU0PSP
Dowell A G6TJK
Dowell C G0TJQ
Dower K G0BDJ
Dower M G6XSB
Dower P 2M0UTH
Dower P MM3UTH
Dowers C GM8ZGC
Dowey J GI0HHE
Dowie A M0AUS
Dowie J G8UZV
Dowkes W G1SZT
Dowkes W M0YAL
Dowler N G8LID
Dowler P G6EAR
Dowles P G3VNP
Dowley B 2E0NCI
Dowley B M6NCI
Dowling A G3GUE
Dowling A MI6GGF
Dowling B G4IJV
Dowling J GD0TFG
Dowling P 2E0PMD
Dowling R G3NKH
Dowling R G3WMT
Dowman E G4GXY
Down A G4KQD
Down C G8MXW
Down E GW0DDK
Down G G6ZTP
Down M G4ALR
Down N G3SRX
Down S [illegible]
Down S G3USE
Down S G8HKF
Down T G1GYT
Downe S 2E0HVW
Downe S M6DYO
Downer B G3ZQI
Downer D M0GQJ
Downes B M0OIC
Downes D G0KRD
Downes I G6IPC
Downes J 2E0JYM
Downes J G8SCG
Downes L 2E0SIS
Downes L G4TMD

Downes M M3MHD
Downes V 2E0DUP
Downes V M6VAH
[illegible]
Downey M G4AHJ
Downham R G6TJJ
Downie I G8XFY
Downing A G0JQS
Downing A G3ZES
Downing A G8UWI
Downing A M0ZZI
Downing A MW3NXD
Downing M G7WIY
Downing T G3MXH
Downs G G3UCK
Downs G GI4MTZ
Downs M M6FOE
Downs N G0DND
Downs R G3OEB
Downton B G3ZQR
Dowse G G6HHH
Dowse I G0DYW
Dowse J G0GKN
Dowsett A G8VNN
Dowsett N G8PUY
Dowsett P G0HWS
Dowsett R G3RSV
Dowson J G8BTU
Dowson J MM3VIS
Dowson K G4EQS
Dowson M G8CJA
Dowson R G4TDG
Dowthwaite R G8WUY
Doxey G M6CPR
Doxey J G0WOC
Doxey J M0BZQ
Doy J G7OCH
Doyle A G0FAU
Doyle A 2E0YLE
Doyle A M6TBC
Doyle B G6IPB
Doyle B M0GPP
Doyle C MI6CAK
Doyle D G0DYG
Doyle E M0BSP
Doyle F M3HIM
Doyle H 2E0FRC
Doyle J 2E0FZY
Doyle J G3DID
Doyle J M0BSQ
Doyle J MW3GQE
Doyle J GW4FOI
Doyle J MI0JPD
Doyle J M3IJD
Doyle M G7JZJ
Doyle P G1MBN
Doyle P M6PJD
Doyle R G0JDE
Doyle S G1YLB
Doyle T G6LXE
Drabble R G1SLE
Drage A G1AZD
Drage P G7DMK
Drage R G0GOB
Drain R GI4POC
Drake D G0EZJ
Drake D M0DAZ
Drake D M6DHK
Drake G G0MFT
Drake I G3WFL
Drake J M0REV
Drake J M6EGT
Drake M G6JAR
Drake P 2E0FFW
Drake R M3KFH
Drake R M3VDY
Drake-Brockman L GM4UOD
Drake-Brockman S 2E0CHQ
Drake-Brockman S M6AUB
Drakeford K G4VSJ
Drakeley J G4FGF
Drakeley J G8GRC
Drakeley S M6HEB
Dransfield A 2E0SHW
Dransfield I M3UJT
Dransfield P M6DGP
Dransfield S G7UJT
Dransfield W G4DCY
Draper D G4JUI
Draper D M0AME
Draper J G6BAM
Draper J G8BLD
Draper M G4BSA
Draper M 2E0DRA
Draper M M0IKD
Draper M M0IKD
Draper N [illegible]
Draper N M3NBD
Draper P G0LUI
Draper P G1TKY
Dray G M3OAC
Dray J [illegible]
Dray L G3SHD
Draycott D G0LNN
Draycott M G3XTQ
Draycott P G0OXD
Draycott R G4LFS
Draycott S GM4JFH
Drayton C G1IBX
Drea J M6JQE
Drea W G0CBU
Dredge I G4DIE
Dredge W G4HJF
Dreiling G G7KBI
Drennan M GM7PZH
Dresser A G7MFP

Dresser P G4VOK
Dresser R G0FDA
Dresser R G6KVA
Dreszer M M0GNB
Drever M G0WKA
Drew A 2E0CDV
Drew B G0MBG
Drew D M5ABH
Drew L M6LFD
Drew M G0HWK
Drew P G1OPV
Drew P MW0DAR
Drew P GW6IPR
Drew R G6JYX
Drew R G8URU
Drew R G8EJC
Drew T G3MD
Drewe C G8MYJ
Drewitt T G0BMU
Drewry J G1TAU
Dring L G7HHM
Dring M M3ZLO
Dring R G0BXG
Drinkall M G6XGK
Drinkell J 2E0JLD
Drinkell J M0JLD
Drinkwater E G6SXD
Drinkwater J G0JYD
Drinkwater L G3RHR
Drinkwater M M5AEC
Drinkwater R G3ZIY
Driscoll A 2E0RFI
Driscoll A M6UTC
Driscoll D G1MVT
Driscoll E M6VAM
Driscoll J GI0TDP
Driscoll M 2E0YAW
Driscoll M M6MZZ
Driscoll N M6MKN
Driver A G1VWU
Driver A 2E0EAY
Driver C G6CMD
Driver D G7VQC
Driver E M3EYY
Driver E G1VMX
Driver G G4ZSM
Driver J M3JXX
Driver P 2E0DQD
Driver P M6FHP
Drohan G G6DIE
Drohan M G4NQM
Dronfield M G1WWI
Dronfield P G4RNA
Drouet C G8GJW
Drought D M6DRO
Druce B G3ZGT
Druce K M3ZIZ
Druitt S G8HTZ
Drumm N M5GVY
Drummond A GM6KAM
Drummond A G7PSU
Drummond G MI6GDD
Drummond J G0RGO
Drummond J M0CQW
Drummond J M1AZY
Drummond J MM6JSN
Drummond R 2M0DRO
Drummond R MM6RDU
Drummond T G0ILT
Drury A G4FZP
Drury B 2E0WKZ
Drury B M3WKZ
Drury C M0NVJ
Drury H GD4HMD
Drury I G0FXQ
Drury M 2E0GMD
Drury M M6CHS
Drury N G4SCO
Drury N G6YIK
Drury P G6UST
Drury S G4ZPQ
Drury S G6ALU
Drury S G7RLK
Dryburgh G GM7CNW
Dryburgh R M3RXD
Dryden J G4DSN
Dryden P M3PTR
Drysdale I GM3TYS
Drysder D 2M0DRY
Drysder D MM6BRV
D'Souza D M6BTH
D'Sylva S M1ESW
Du Chemin A GJ1KCB
Du Feu G G0NUO
Du Heaume J [illegible]
Du Plessis P G7FZN
Du Pré N M0NPD
Dubbins E G8OCM
Duberley C G7ABE
Dublon C GW0KZE
Dublon G M0AKD
[illegible]
Duce M G4BQF
Duchar M 2E0LOW
Duchar M M6MDU
[illegible]
Duck A G0DQT
Duck I G3DTX
Duckett S G6UFL
Duckfield K GW0BZA
Duckles C G6KIA
Duckles P 2E0SXJ
Duckles P G1SXJ
Duckles P M3SXJ
Duckles S G6UQA
Duckling S G7TAJ

Duckworth A G4BG
Duckworth C G0PUY
Duckworth C 2E0DWK
Duckworth D GM4UGN
Duckworth R G0WMZ
Duclos G [illegible]
Duddin R G8PTF
Dudding P M6KRL
Duddridge J G4NVM
Dudhill B G4NMP
Dudkowski D G4SLW
Dudley A G0HLI
Dudley E G0NED
Dudley G G0DGQ
Dudley I G3YRP
Dudley J G4RMQ
Dudley M M3YMD
Dudley M M0GHA
Dudley S M3NXK
Dudley S M3YSD
Dudman N G8GGW
Dudman P G6GTS
Duell A G1XPF
Duell D M6DUE
Duell E M6FPU
Duerden M G1KWK
Duesbury H G8YYA
Duesbury P G6KIB
Duff D GM4UGF
Duff M G4CGM
Duff W G0GWS
Duffain J 2E0GFX
Duffain J M6VGA
Duffell B G3VGZ
Duffell B G8AOE
Duffett-Smith P G3XJE
Duffie I 2E0CGP
Duffie I M0HBU
Duffie I M6AKW
Duffield A M3YTE
Duffield A G0BCO
Duffield C M3VTA
Duffield E 2E0TJU
Duffield E M0TJU
Duffield E M3TJU
Duffield J M3MKO
Duffield P M6GIP
Duffield S G7CBW
Duffield W M3WDY
Duffield-Dyche A M1DQH
Duffill C M3PHX
Duffill L G0UBY
Duffill L G6CTE
Duffin I G7HXI
Duffner W G6KGG
Duffy A GI6ZIR
Duffy A G1JCW
Duffy B G1VKJ
Duffy B M0BJD
Duffy C G0PVP
Duffy C G1NFE
Duffy C G4MPO
Duffy G MI0BTM
Duffy G 2E0XRG
Duffy G M3XRG
Duffy J GM0LIM
Duffy J G8RQF
Duffy J M6BNB
Duffy J G1YJQ
Duffy L M6DUF
Duffy P G4NPG
Duffy R M0REX
Duffy S M3IHA
Duffy V G4CJP
Duffy W MI3CXD
Dufour I G3PWB
Dufton W G3WUH
Dugdale M M1BUG
Duggan A G0LAX
Duggan G G6VNI
Duggan H G0MSY
Duggan I 2E0EAV
Duggan J G1XVM
Duggan K M6PUT
Duggan P G7POL
Duggan-Keen J G1FWE
Duggins A G4SNO
Duguid S GM4WFV
Duguid W GM4RLV
Duke G GM4GKH
Duke N 2E0FAS
Duke N M3CYW
Duke P G4YJK
Duke R G8GZV
Dukes D [illegible]
Dukes D M0DWR
Dukes J 2E0JBD
Dukes J M3JND
Dukes K 2E0KRD
Dukes K M3KRD
Dukesell D G0RKT
[illegible] G7NNZ
Duley P G6LXF
Duley R G6TYF
Dumbill K G8JYV
[illegible]
Dumitrescu C M0HCC
Dumitrescu I M6IOA
Dummer I M0DVF
Dumpleton M 2E0NCG
Dumpleton M M0NCG
Dumpleton M M3NCG
Dumpleton P 2E0YYD
Dumpleton P M0XDX
Dunbar I GM0KDP
Dunbar J MM6DYM

Dunbar M G6RRY
Dunbar P GW3WCA
Dunbar S MM3UPY
Duncan A GM8SVR
Duncan A GM4ZUK
Duncan B [illegible]
Duncan C 2M0RRO
Duncan C MM3LKV
Duncan C GM0EKM
Duncan C GM6MUZ
Duncan C MM6YUJ
Duncan D M3ZUG
Duncan E MM0HKU
Duncan G GM4ZEX
Duncan G MM0GGD
Duncan G MM1DSD
Duncan J 2E0BBY
Duncan J G3KUD
Duncan J MM0HDW
Duncan P G3TKA
Duncan P G4HIX
Duncan R MM0RDD
Duncan R MM0UDI
Duncan R M1DIR
Duncan R 2M1HRS
Duncan R M3ZUU
Duncan S G0OFT
duncan S M6HRQ
Duncan W MI6XKE
Duncombe H G3XJN
Duncombe R MW0AIE
Dundas J GM0OPS
Dunford D G0PTK
Dunford D M3CPN
Dunford P G3YXW
Dungan R G8LSI
Dungavel E MM6CPK
Dungey K M3ZKD
Dungey T 2W0DUN
Dunglinson J G4CGW
Dunham A G6OHM
Dunham A M3FTV
Dunham L G6SXB
Dunham M G1OQX
Dunham P G8JIU
Dunham R G3ZSQ
Dunhill J G1DEP
Dunkerley S G4GFZ
Dunkin J M3UIP
Dunkley J G0ITL
Dunkley R M3LYG
Dunlop A G7IET
Dunlop C G6LKH
Dunlop D GI7TTO
Dunlop J GM0WDF
Dunlop J GI3YDM
Dunlop J MM0XEA
Dunlop J MI3PUH
Dunlop J MI3VFF
Dunlop P G7NHR
Dunlop R GM3NMN
Dunlop S M1AZI
Dunmore F 2E0FEC
Dunn A MM3USV
Dunn B G4FQW
Dunn B G3XTR
Dunn C GM4OSV
Dunn C G8BEK
Dunn C M0BEK
Dunn C G0AMW
Dunn C M3ULI
Dunn C G4KVI
Dunn D G3SCD
Dunn D G3XRM
Dunn D G4NKU
Dunn D G8RZN
Dunn D M3XNV
Dunn D G8KOV
Dunn E G1UBT
Dunn F G4KPV
Dunn F M3FED
Dunn G G0PGW
Dunn G G4DYH
Dunn G G4MZI
Dunn G G6RIQ
Dunn G G8TZW
Dunn J GM0VIV
Dunn J G1HEX
Dunn J M3WZY
Dunn J M6IPH
Dunn J M6MPF
Dunn J G7HGB
Dunn K G1NDK
Dunn K G1RBH
Dunn L [illegible]
Dunn M G7ELA
Dunn M [illegible]
Dunn M M3GFH
Dunn N G4UWG
Dunn P G0NLQ
Dunn P M0TWO
Dunn P M6COR
Dunn R G4WBG
Dunn R G7LQK
Dunn R G8GDI
Dunn R [illegible]
Dunn R M6DYV
Dunn R M6SRN
Dunn R G4PMW
Dunn S G4KCR
Dunn S GM4PMH
Dunn S M3KWS
Dunn T M3GFE
Dunn W G0MKC
Dunn Y G6FPN
Dunnachie N GM6VVX



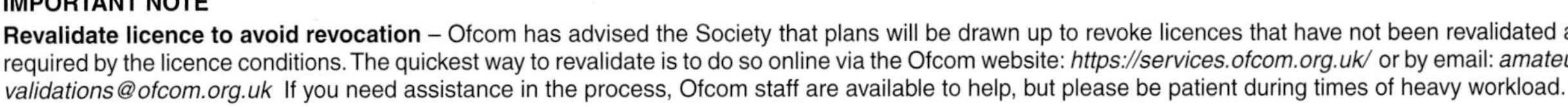
**IMPORTANT NOTE**

**Revalidate licence to avoid revocation** – Ofcom has advised the Society that plans will be drawn up to revoke licences that have not been revalidated as required by the licence conditions. The quickest way to revalidate is to do so online via the Ofcom website: *https://services.ofcom.org.uk/* or by email: *amateur.validations@ofcom.org.uk* If you need assistance in the process, Ofcom staff are available to help, but please be patient during times of heavy workload.

Dunnaker R M6KNY
Dunne C MD0EGL
Dunne D 2E0DLZ
Dunne D M0WOW
Dunne D M3LHF
Dunne D M3OFD
Dunne E G6CTH
Dunne G 2E0AWU
Dunne G M0GID
Dunne G M3GID
Dunne H G0EFN
Dunne J G0UPG
Dunne K M3VSB
Dunne M GI4MJD
Dunne M GI8YBU
Dunne P 2E0HFS
Dunne R G0MGM
Dunne S M3OFC
Dunne T 2E0HRS
Dunne T M3NBH
Dunne Z 2E0ZLD
Dunne Z M0ZOE
Dunne Z M3LHI
Dunnett A GM6PYD
Dunnett J G4RGA
Dunnicliffe P 2E0TXL
Dunnicliffe P M0TXL
Dunnicliffe P M6FIL
Dunning J M3FQM
Dunning M GD0HYM
Dunning P G4DEA
Dunning S M3FQN
Dunnington J GM4EIW
Dunsmore A MM0DLH
Dunsmore S G0JEU
Dunstan C 2E0RBR
Dunstan C M3WOY
Dunstan D M3RRN
Dunstan E M6EME
Dunstan K G4WKD
Dunstan K M0RSJ
Dunstan M G0VBR
Dunstan R G4MFQ
Dunstan W G0CSY
Dunstan W M3XOA
Dunster A M1DFB
Dunster A M3DFB
Dunster G G0MFQ
Dunthorne M 2E0FQO
Dunthorne P G4NTT
Dunwell J G1PCG
Dunwell K G4WLG
Dunwoody R 2I0RHN
Dunwoody R MI6UDR
Dunworth I G4SNL
Dupont J MM3YEC
Dupree B G4INB
Duque M M6FWQ
Durant C G4CEX
Durant T M3OOA
Durant T M3OVA
Durban J G6LNU
Durban M 2E0LNU
Durban M M3LNU
Durbin P G6PHM
Durbridge A G1OSO
Durbridge C G4EML
Durdin J G7KCC
Durell C G3PNT
Durell R G3LRX
Durey E G4MOV
Durey J M6JEK
Durey M G1NMN
Durham D G3SIR
Durham G M6GGD
Durham P G3ZOX
Durham S M0DYR
Durkin E 2E0EDT
Durkin E M6FFE
Durkin J M6GQV
Durkin M G6SGV
Durnall C G4DHL
Durno G G7NIW
Durrand J GM4XLN
Durrans J G4AYQ
Durrant A 2E0GFB
Durrant A M6RWV
Durrant B G4LVD
Durrant B G8NZB
Durrant B G0SIU
Durrant C M6AIB
Durrant D G6DOR
Durrant G 2E0EGI
Durrant J G0SWF
Durrant J G8DCD
Durrant J G8UZW
Durrant K G4UBC
Durrant M 2E0HAU
Durrant M M3DUR
Durrant P G1UNB
Durrant W 2E0CGH
Durrant W M6WFA
Durrell J G0HJZ
Durston-Wyatt J G6MBD
Dussart J G4DGQ
Dutfield-Cooke K 2E0CRB
Dutfield-Cooke K M0HYK
Duthie G G7GEF
Duthie R 2E0RJZ
Duthie R M6RJZ
Dutson E M6FYK
Dutson K G0LOH
Dutton A G1KLK
Dutton A G3TIE
Dutton C 2E0UXS
Dutton C M0UXS
Dutton C M3UXS
Dutton D M3ASI
Dutton J G0ITM
Dutton J M3XQY
Dutton L G6HFK
Dutton P MM0PHD
Dutton R M0RXD
Dutton R M3XQV
Duxbury J G1BBC
Duxbury J G6LNS
Duxbury M M6LJA
Dwight D GD1DQQ
Dwight J G1DWT
Dwyer A G0HBU
Dwyer C G3FMR
Dwyer D G1PWU
Dwyer I M3AFS
Dwyer N M3XXE
Dwyer S G1CNI
Dwyer S G6RXP
Dyce A G8TAQ
Dyde D M0DEK
Dye D G3WPG
Dye J MM1MOY
Dyer A G4ODI
Dyer B G3HNC
Dyer C G7CJD
Dyer C MM3UIX
Dyer C M6UTT
Dyer D G4DNX
Dyer D G4WUK
Dyer D GW8VCA
Dyer D G4CXQ
Dyer G G0NUS
Dyer G G8VRV
Dyer J 2E0LJD
Dyer J M6EFV
Dyer K G0KWD
Dyer K GW0RHC
Dyer N M6KKJ
Dyer P G1FZL
Dyer P G4POU
Dyer P M0PDY
Dyer P M3PDY
Dyer R G1XIE
Dyer R M1AFF
Dyer T M6TJI
Dyer W GM7JUX
Dyke J G4NVA
Dyke M G8EPC
Dyke P G0LUC
Dyke P G2PA
Dykes A GW1MNC
Dykes A M1CWV
Dykes J G0NDU
Dykes R M0IAZ
Dykes S 2E0HRB
Dymond G G4VQL
Dymond L G4PEK
Dymott A G4OAG
Dynes J GI6FTM
Dynes P GI3OZW
Dyson A G0HUW
Dyson C G0WAD
Dyson D G7DHD
Dyson G M0AOG
Dyson G M6HXO
Dyson G 2E0BJX
Dyson G M0GPG
Dyson G M3UQJ
Dyson H G4JLO
Dyson J G7RXJ
Dyson J M6HWL
Dyson R 2E0KAF
Dyson R M3KRY
Dyson S M0DNW
Dyson T G0RLL
Dyson T G1RIV
Dyson-Bawley G M0BGT
Dzundza C M6KLN

## E

Eacott-Palfrey D M3IHX
Eade J G8AJP
Eade M 2E0ESU
Eades A GD1VIO
Eades A M3KSS
Eades G G1LTE
Eades J M6LID
Eades M 2E0EKQ
Eadie J G8GZX
Eadon T M3TAE
Eady D M1AUO
Eady J G8ZIH
Eager H M6CHZ
Eager R G8KDU
Eagle G G4UTX
Eagle J M3FNY
Eaglen C M0TCE
Eagles V G8MCR
Eagleton B M0AFQ
Eagling C G6PMD
Eales M G1YEU
Eames B G3SBF
Eames B 2E0CMD
Eames B M0HFY
Eames B M6CBY
Eames D MI0DWE
Eames J 2E0PSD
Eames S G1LTV
Eamus I G3KLT
Eardley A G3UXO
Eardley G G6LMJ
Eardley J 2E0RFK
Eardley R M3RFK
Earl K G8VJU
Earl N G8DWF
Earl O M6FMI
Earl P G8LHF
Earl S M3OJW
Earland E G1OEF
Earland R G3AJK
Earland R M6BHP
Earle J G6VLY
Earle R G4FTA
Earle S G7FQP
Early R G7KJA
Earnshaw C G3DMO
Earnshaw D G4MZF
Earnshaw D G0FNJ
Earnshaw H M6YLH
Earnshaw J 2E0ZZZ
Earnshaw J G4YSS
Earnshaw J M0JFE
Earnshaw J M1NNN
Earnshaw L M3IPY
Earnshaw P G0UUU
Earnshaw R 2E0RJE
Earnshaw R M3URV
Earnshaw T M3TCY
Earp A GW7IBT
Earp C M1FHC
Earp D G4SGG
Earp K G0MZJ
Earp R M1FHB
Earp T M1FHA
Earwicker K 2E0HJE
Earwicker K M6VLB
Earye J M6DQD
Easden D 2E0VRE
Easden D M6DTA
Easdon B G0RZI
Easdown J G4HIZ
Easdown J 2E0HIZ
Easdown J M3SIZ
Easdown W 2E0WEL
Easdown W M0HIZ
Easdown W M6SIZ
Easey B G1JGM
Easey J G4XBE
Easom A G4OPI
Eason G M6UKT
Eason J G3RYQ
East B GM3NNZ
East D M6OXF
East D G4PWM
East J G0HIC
East J G0OQX
East J G7HSS
East K G4UNF
East K M6YYD
East R M1AEO
East R 2E0RHE
East R G0GEB
East R M0TBW
East R M6CNY
East S G4BMU
Easteal J G4DNH
Easter S G7URR
Easterbrook J M3VMU
Easterling D M0JXM
Eastgate G G4SNV
Eastham F G7PZE
Eastham I G7KXV
Eastham J G1BWI
Easthope R M1DUO
Eastick B G1AVF
Easting R G7NZV
Eastlake D M0SUG
Eastland A M1CJE
Eastland C G4IID
Eastman B M6PCV
Eastman M G0DDZ
Eastment J GW4LXO
Easton A M1AUH
Easton B G4XMA
Easton C G0FSG
Easton D G7AWK
Easton G M6KVV
Easton W GM8OAH
Eastope D M1ERO
Eastty K G3LVP
Eastwell A M3KQP
Eastwood D M0RWD
Eastwood H G0BGV
Eastwood J G8ONP
Eastwood K M1ADP
Eastwood S G7POT
Eate D M0HZU
Eaton A G8YEF
Eaton A M6BFK
Eaton D GU8ITE
Eaton J 2E0IHW
Eaton J M3GIY
Eaton J M0JRE
Eaton K GW1FKY
Eaton N G8DHV
Eaton P G0MBY
Eaton P G4EDW
Eaton P G1BWH
Eaton R 2E0MPE
Eaton R M6MYN
Eaton W G3TAO
Eaton-Watts D M0AVH
Eatough S G4GVQ
Eatwell C G3AJW
Eatwell W G7LZY
Eaves A G6FZW
Eaves T G4GUY
Eavis F M3FAY
Eavis M G0AKI
Eavis M G3VHF
Ebbetts R G7JWV
Ebborn K G8NGE
Ebbs D M6PDN
Ebbs R G0MIE
Ebdon R M6RXE
Eborall M G0JUQ
Ebsworth P G8CKB
Eccles A G3KNZ
Eccles C G8NMK
Eccles D M0ODS
Eccles F GI3TIJ
Eccles G G1JNG
Eccles J G7LII
Eccles M G3PPE
Eccles N M0NEC
Eccles P G4CLJ
Eccles R G8MQX
Eccleshare G G6VRI
Eccleston G G4OTS
Eccleston K M3UQP
Eccleston M M6CLQ
Ecclestone G M1AXE
Eckersall G G4HFG
Eckersley J G1RRE
Eckersley P 2E0PMZ
Eckersley P M3PDE
Eckhoff M G4HLT
Eckley D G3UFQ
Eddleston H G0FZG
Eddy B M0UOK
Eddy G G7PVE
Eddy R M3URE
Eddyvean M M0ACU
Ede R M6BVZ
Eden G G1HXT
Eden G G0BCX
Eden H G6VCR
Eden J G0EXN
Eden J G6NFJ
Eden L G1WPH
Eden P M3CBC
Eden W M6WME
Edgar C 2E0XYZ
Edgar C M5XYZ
Edgar D M0EEK
Edgar G GI4MCW
Edgar G M1GDE
Edgar G M3GDE
Edgar G M6EIX
Edgar J G4FVM
Edgar R G0KYS
Edgar R G1ANV
Edgar S M3XNZ
Edgcombe A G6TOI
Edge A M6YYD
Edge A G6XFU
Edge D M3DAE
Edge D M3FSE
Edge D M0MNO
Edge D 2E0MIX
Edge D M6DEG
Edge M G7GCU
Edge M 2E0ZYK
Edge M M3ZYK
Edge R G4EMD
Edgecock A G4AZD
Edgecock J G1NZH
Edgecombe R G8BXD
Edgecumbe L G0NXI
Edgeley J G4NWW
Edgeley R G8KZO
Edgington I G7HSW
Edgington J G0KTV
Edginton R G3AGF
Edgley B G7KEI
Edgson C 2E0CLE
Edgson J M6EXI
Edib M G3YTY
Edinborough R G0BAJ
Edinburgh G G3SDY
Edinburgh P G1DEN
Edis C M1IDE
Edis C M3EDI
Edis M G4RPT
Edis R G6XYF
Edlin G G7EIK
Edmands K M6KKD
Edmett K G1ITJ
EDMOND M 2E0BRY
EDMOND M M3MYW
Edmonds A G0HIF
Edmonds A 2E0DWU
Edmonds A M6OAX
Edmonds D G0NAX
Edmonds D G8EWN
Edmonds E G6ILU
Edmonds G G6HIG
Edmonds G G8VNO
Edmonds L M0HPZ
Edmonds L M6DYI
Edmonds M 2E0MSE
Edmonds M M0MSE
Edmonds M 2E0SER
Edmonds M M6MPZ
Edmonds P G4OFN
Edmonds R G0KVB
Edmonds R G4HHX
Edmonds R M6EOQ
Edmonds S G7ORV
Edmondson B G3HCZ
Edmondson D G0RGL
Edmondson D G7KDH
Edmondson D 2E0HDE
Edmondson D M0MDE
Edmondson D M6HDE
Edmondson J M3JEE
Edmondson J G6KKA
Edmondson P G7KDG
Edmondson R G3YEC
Edmondson R M0ZLP
Edmondson R G6BPN
Edmondson S G4CIC
Edmondson W GI0THO
Edmondson W G1WWW
Edmunds C G3MJW
Edmunds E M0BTP
Edmunds G G0KTL
Edmunds J 2E0VAF
Edmunds J M3VJE
Edmunds J M5JON
Edmunds K G8NZC
Edmunds L G0GZN
Edmunds M G0ISO
Edmunds M G0JKI
Edney D M3AUC
Edney R G8EWP
Edridge J M6WBA
Edson S G6NZG
Edward B G6ILX
Edwardes J G0WHN
Edwards A 2E0LGA
Edwards A G0HDG
edwards A G0HIR
Edwards A G0NWS
Edwards A G0SUA
Edwards A G1IDF
Edwards A G1NTN
Edwards A G1ZNK
Edwards A G3KGN
Edwards A G4GBI
Edwards A G4ZON
Edwards A G6DXD
Edwards A G7GGJ
Edwards A G7HHU
Edwards A G7JLC
Edwards A GI8UCS
Edwards A M0BAO
Edwards A G0HWU
Edwards A M6AOW
Edwards A M3CBW
Edwards A MW6BHO
Edwards B G3WCE
Edwards B M3JYE
Edwards B G3YVA
Edwards B G6HIE
Edwards B M3HIE
Edwards C G4GLG
Edwards C G6CTY
Edwards C G6YXY
Edwards C G7UDJ
Edwards C GM7UPD
Edwards C G8WVZ
Edwards C M3LVL
Edwards C M6RCP
Edwards C M6HCA
Edwards C G7MJP
Edwards C MM3YLP
Edwards D 2W0AZP
Edwards D 2E0YDK
Edwards D 2E0HGR
Edwards D GW0CVY
Edwards D GW0TMU
Edwards D G1YZF
Edwards D G4HXC
Edwards D G4KMJ
Edwards D G4TIQ
Edwards D G4ZWR
edwards D G7RAU
Edwards D M0ECF
Edwards D M1ARL
Edwards D M3DLE
Edwards D G0OWP
Edwards D G8BFV
Edwards D G8NEO
Edwards D M0CNP
Edwards D M6BVW
Edwards D M3MRU
Edwards E G3RUE
Edwards E GM7UAC
Edwards E G8HLJ
Edwards E GW8LJJ
Edwards E MW0SKD
Edwards E M3HOU
Edwards E 2W0WCQ
Edwards E MW6WCQ
Edwards G GW0LTC
Edwards G G1ENA
Edwards G GW2ABJ
Edwards G GM7WFT
Edwards G M0EDS
Edwards G 2E0HVN
Edwards G G0XDL
Edwards G M6HVN
Edwards G MW6MMU
Edwards G 2E0DFN
Edwards G M0HTG
Edwards G M6EIK
Edwards H G4RUT
Edwards I G4WRK
Edwards J G0JSE
Edwards J G0WGH
Edwards J G1LCC
Edwards J G3TCV
Edwards J G4KFY
Edwards J G4NFE
Edwards J G7PEB
Edwards J G7SSD
Edwards J GM8HBB
Edwards J M1CRL
Edwards J M3OEN
Edwards J M6OOJ
Edwards J 2E0RIS
Edwards J G0MJZ
Edwards J G0ONY
Edwards J M0JAX
Edwards J M1DHI
Edwards J M3XHL
Edwards J M3YQL
Edwards J M6BCY
Edwards K GW1EWW
Edwards K GW4LWL
Edwards K G4RYQ
Edwards K G6BWE
Edwards K G8GRL
Edwards K G0IVI
Edwards K G6GSF
Edwards K G8LPN
Edwards K M6YNY
Edwards L 2E0EET
Edwards L 2E0LGE
Edwards L G0GLI
Edwards L G6KHM
Edwards L M6GVT
Edwards M 2I0UCS
Edwards M 2E0WVM
Edwards M G4GDY
Edwards M G4KHY
Edwards M G4LHL
Edwards M GW4SJO
Edwards M MI0GJN
Edwards M M1EOR
Edwards M M1EQV
Edwards M M3ISY
Edwards M MI3UCS
Edwards M G3WYT
Edwards M MI0OIM
Edwards M M3MJE
Edwards M G1HLQ
Edwards M G8CPF
Edwards M G0JIW
Edwards M M3OAL
Edwards M G4BZM
Edwards N 2E0DQM
Edwards N G3XZB
Edwards N G4FCB
Edwards N 2E0CVL
Edwards N M3NAE
Edwards N M3ZNV
Edwards N M6BFZ
Edwards N M3WNM
Edwards N M6KTK
Edwards P 2E0HRC
Edwards P G0NFI
Edwards P G4ESL
Edwards P G4IKJ
Edwards P G4WFW
Edwards P G7TWJ
Edwards P M3HRC
Edwards P M6AKF
Edwards P 2E0PBE
Edwards P 2E0FJZ
Edwards P G8ARR
Edwards P M6SVM
Edwards R G0FIU
Edwards R G0GWE
Edwards R G0NQK
Edwards R G4BBY
Edwards R G4TSD
Edwards R G6OHR
Edwards R M0DTD
Edwards R MW3FTY
Edwards R 2D0EJR
Edwards R MD0RJE
Edwards R MD6RJE
Edwards R M3RCE
Edwards R 2E0LQW
Edwards R M0RGE
Edwards R M3LQW
Edwards R 2E0EAX
Edwards S 2E0IMS
Edwards S GM0SCA
Edwards S G4GPY
Edwards S GW4TVE
Edwards S G6EDJ
Edwards S G8GEF
Edwards S M0AFD
Edwards S M0CKI
Edwards S M3SBE
Edwards S 2E0SMZ
Edwards S GW0CNJ
Edwards T 2W0DLE
Edwards T G0PZS
Edwards T G0PWW
Edwards T 2E0CEY
Edwards T M6JET
edwards V M1BQW
Edwards V G8NGZ
Edwards W G6ZUE
Edworthy R G3URU
Edy G G4AXD
Eeles R G0SWC
Eeles R G1WWB
Egan A M6ABE
Egan C M3CVL
Egan D G0BHR
Egan D G0KZM
Egan D GW4XKE
Egan G G0NWE
Egan M M0JSE
Egan M M6TLX
Egan P G1WTX
Egan R G4XOM
Egan R M3YIY
Egerton M M3MVK
Egerton M G6UAW
Egerton S G4BWX
Egerton W GM8RBR
Eggett D G7RAJ
Egginton N M3ZXE
Egglestone B 2E0EYS
Eggleton D M0DCW
Eggleton M M3BAA
Eggleton M G4IZZ
Eggleton P G3XMQ
Eggleton S M1CSZ
Eggleton S M1WAZ
Eggs E G6CTV
Egleton D G7FFB
Egleton J 2E0JEE
Eglington R 2E0OKY
Eglington R M3OKY
Eglinton S M1CQC
Ehlen T G8NXA
Ehm H M0ZAE
Ehm H M3ZIE
Ehrenfried M G8JNJ
Eibl-Kaye G M6IBL
Eilec M M0MEI
Eizzard A M3ULU
Ejugue D M1DYE
Eke G G0TKK
Eklund G M3NQY
Ekpe U M3RZU
El Khalidi A MD0MAR
Elam D GM6WMA
Elbourn M 2E0EFA
Elcoate A G0RTH
Elcoate B G7KCN
Elcock D M6KOH
Elcock T G1KKD
Elcombe C G7VXQ
Elcombe P 2E0FQB
Elden R G8VNP
Elden S M0AKK
Elder F GI4AHD
Elder I MM0DIS
Elder M GI8RPP
Elder W MM6GID
Elderbrant B M3UYU
Elderfield J M3ZAI
Eldredge S G1FQD
Eldridge E G0HZQ
Eldridge G GI0POB
Eldridge K G6SHS
Eldridge S G8IZY
Eldridge S M3KSP
Eldridge W M0CBD
Element M G0EBD
Elena A M6LXE
Eley C G3LRA
Eley J G4VCY
Eley J G3LMR
Eley P G6VCY
Elford A G0JZW
Elford J G6YIJ
Elford M M0ULD
Elford R G0XAY
Elgar J M6JTE
Elger R M6RKE
Elgin K GI7SOB
Elgy R G8EZT
Elias A 2W0DSP
Elias D MW3WLS
Elias D M6CYT
Elias R M3VRD
Elias S 2E0SCB
Elias S M6DBB
Elkington D G0PAN
Elkington M 2E0BVT
Elkington M M0GUC
Elkington M M3UPF
Elkins P G8MCW
Ellacott D G3XOB
Ellam T G4HUA
Ellams L G0TJR
Ellard D G1ZJK
Ellaway M M3EHK
Ellefsen A G3FJO
Ellens D M6KZR
Elleray H G1KCW
Ellerby M G1HSH
Ellerington A M3VON
Ellerington L M3ZLE
Ellershaw D G1KLZ
Ellerton B M6BTE
Ellerton J G3NCN
Ellery B G0COJ
Ellery C G4EAS
Ellery E G3YIN
Ellery J 2E0ZBE
Ellery J M6SFC
Ellery R 2E0JJR
Ellery R M3PXK
Ellin S G6SWW
Ellingworth D G6ZDE
Ellinor T G4DFA
Ellinor T G8GYX
Ellins D G0GGL
Elliot C M3NVV
Elliot D MM3SWW
Elliot I G3HMB
Elliot J G3KIQ
Elliot J M6GNX
Elliot N GM0ELL
Elliot P G3MFO
Elliot R G3ZRA
Elliott A G3ZOG
Elliott A G6GEK
Elliott A M0AOJ
Elliott A 2E0AZS
Elliott A M0VED
Elliott A MD6SNV
Elliott B GM1RRJ
Elliott B GM4JTA
Elliott B G4MEO
Elliott B G1SLI
Elliott B G6AMF
Elliott C G4UJW
Elliott C G1NZL
Elliott D G0AWW
Elliott D G1YAS
Elliott D G4DUT
Elliott D G4ZOY
Elliott D G8GMA
Elliott D 2I0EAS
Elliott D M3XNN
Elliott D MI3YUT
Elliott E 2E0BVS
Elliott E G1NZP
Elliott E G3YGC
Elliott G G0KQO
Elliott G G3FMO
Elliott G GI4OWA
Elliott G G4UMT
Elliott G G8MOZ
Elliott H G4GSO
Elliott J 2E0RAD
Elliott J G0KYH
Elliott J G8CGW
Elliott J MI0JTE
Elliott J MI3YJE
Elliott J 2I0KBB
Elliott J MI6KBB
Elliott K GM4NTX
Elliott L G4OGB
Elliott L G7NZU
Elliott M G1DQU
Elliott M M0CJS
Elliott M M0MAL
Elliott M G8SAR
Elliott M MI6MQX
Elliott M G4VEC
Elliott M M0TMM
Elliott P G4MQS
Elliott P G0TXL
Elliott P M3PSE
Elliott R G4ERX
Elliott R G7JLK
Elliott R G7UFT
Elliott R GW8VFQ
Elliott R M0BTX
Elliott R MI3TRR
Elliott R G4NJK
Elliott R G8DOY
Elliott S G1IKT
Elliott S M1HJE
Elliott S M3EDS
Elliott S M1SJE
Elliott S 2E0MYX
Elliott S 2E0SBX
Elliott S G0WEX
Elliott S M6MYH
Elliott S M6SME
Elliott S M0SEL
Elliott S M3XHV
Elliott T MI3CQB
Elliott T G0EHX
Elliott V G0IGB
ELLIOTT W GI4OYM
Elliott-West P G0KFY
Ellis A G0NXS
Ellis A M0AWE
Ellis A G7HEP
Ellis B G3VXF
Ellis B G8NVC
Ellis C GD7ESU
Ellis C M6KIT
Ellis C M5AGG
Ellis C G6DXC
Ellis D G4FBB
Ellis D G6GUC
Ellis D M1BXO
Ellis D G4AJY
Ellis D G4RAB
Ellis E GD3LSF
Ellis G M0GME
Ellis G 2E0DTR
Ellis G M3VQJ
Ellis H G7BUN
Ellis H G7FKS
Ellis J G0ELO
Ellis J G0MVU
Ellis J G4ABE
Ellis J G6LXL
Ellis J G7PFY
Ellis J M3YMI
Ellis K G0DPJ
Ellis K G8HGM
Ellis M G4GDL
Ellis M G4ROM
Ellis M G4UDE
Ellis M G4VXB
Ellis M G7UKF
Ellis M M1BXJ
Ellis M M3HNE
Ellis M M5AKT
Ellis M 2E0MCQ
Ellis M M6MTW
Ellis N G6RIC
Ellis N G1OIS
Ellis N G6MKJ
Ellis P 2E0PUS
Ellis P G0ANL
Ellis P G0SSX
Ellis P G0UVG
Ellis P G0VMQ
Ellis P G3YAS
Ellis P G4AEM
Ellis P G8REF
Ellis P M0GIE
Ellis P M3LQA
Ellis P G1ILF

Ellis R G0AGB
Ellis R G1ZRE
Ellis R [illegible]
Ellis R [illegible]
Ellis S G1YMP
Ellis S M6HCE
Ellis T 2E0TJE
Ellis T G7AJG
Ellis T G7PST
Ellis T G8HIO
Ellis T G8SVT
Ellis T 2E0TRE
Ellis T M3TNN
Ellis W 2E0WJE
Ellis W G0UUX
Ellis W G4DQZ
Ellis W M3DPY
Ellis W G0MMY
Ellison A M0GNC
Ellison B G0SIW
Ellison B G1JBW
Ellison C G4WYF
Ellison C G8RBW
Ellison C 2M0CBE
Ellison C MM6CBE
Ellison D G0JTA
Ellison D G6BXS
Ellison D M3DYY
Ellison E M3XEJ
Ellison G G3LZN
Ellison G G8OOF
Ellison J G1XXF
Ellison J G2PK
Ellison J G0KPA
Ellison K M1CPB
Ellison K M6ELL
Ellison M G8RRS
Ellison P G0WIG
Ellison R M6RDQ
Ellison S G7APS
Ellison S M3MLK
Elliston E G6ILT
Elliston M GU7CNI
Ellner J M0BYY
Ellsmore J G1BEG
Ellsom A M6AAE
Ellwood J GW0RLQ
Ellwood M M0YIM
Ellwood P G0RHF
Ellwood P M3IYE
Ellwood-Thompson R G1HIN
Elmer M M3XME
Elmer P G4LSQ
Elmes A G7VNG
Elmore P 2E0PJE
Elmore P M6PJE
Elmore S G4MBL
Elms P G0IJU
Elms R GW0TVX
Elms R G8FAR
Elmy D M6DME
Eloie S M6SLZ
Elphick C G7HFL
Elphick L M6COP
Elsdon J G4RLS
Elsdon S G4TUH
Else A G4RPD
Else R G1BFK
Elsey J M6JGN
Elsey M G4YME
Elsey P 2E0CQS
Elsey P 2E0ECV
Elsey P M0WXU
Elsey P M6BZD
Elsigan G M0XVL
Elsley E G3YUQ
Elsley M G6IEE
Elsmore J 2E0JWE
Elsmore J M3YYW
Elsmore J 2E0JPE
Elsmore J M6SAX
Elsom C G1POC
Elsom P G1OPD
Elston I G3YRX
Elston L M3WIV
Elston W 2E0AAZ
Elstone J M0VPC
Elstub P 2E0KLS
Elsworth D G0ADJ
Elsworth G M3MII
Elsworth J M0CDY
Elsworth K G6XYD
Elsworth-Wilson E M3GCD
Elsworthy G 2E0GCE
Elsworthy G M3HWY
Eltham D M6DNB
Elton W GW3RIH
Elvers M M1BGS
Elvin D G0INO
Elvin I [illegible]
Elvins I G3WUG
Elwell D G4MUS
Elwell P G8PIP
Elwell-Sutton S GM0ETP
Elwin A M6BFU
Elwood D M1MDE
Elwood P G0POC
Elworthy S GW1ZNC
Ely B G3TGB
Ely K M0KAE
Emans M M6OOX
Emanuel D MW3CNL
Emanuel L M6IOI
Emanuel S GW0VFF
Emary B GW1DQV
Emary S M0NDL

Ember C 2E0JES
Ember C M0JXS
Ember C M6JIC
[illegible]
Emberton R G7IEB
Emblem-English T G7ITX
Emblen C G0FQZ
Emblen K G0VCJ
Embleton A G3BNF
Embleton A G4MBH
Embling M G0BSA
Embrey A G3KNG
Embrey D M0GTN
Emeney T G3RIM
Emeny R G4JAC
Emerson B 2I0OTC
Emerson B MI6GIN
Emerson D G3SYS
Emerson H GI8RLG
Emerson J G0BAN
Emerson R G6NEZ
Emerton S 2E0HFA
Emery A G3YWG
Emery D G4FEB
Emery F G4ASI
Emery J M3XFB
Emery J 2E0DZT
Emery J M6DZT
Emery M G8IRM
Emery M G8OHS
Emery R 2E0RNE
Emery R G0WDT
Emery R G3FYX
Emery R M6RGU
Emery R M6WSR
Emery T G4ZRF
Emery-Ford K 2E0KVF
Emery-Ford K M0KVF
Emery-Ford K M6KVF
Emlay K M3XGC
Emlyn-Jones S GW4BKG
Emm M G0TQZ
Emmans P G8XTR
Emmanuel C G0DXF
Emmerson A G8PTH
Emmerson A 2E0IQX
Emmerson A M0IQX
Emmerson A 2E0UTX
Emmerson A M0UTA
Emmerson B M6BJE
Emmerson G G8PNN
Emmerson G M1GPE
Emmerson L M1EBH
Emmerson M G3OQD
Emmerton P G4IOV
Emmett D G3TMR
Emmett M M6GEQ
Emmett S G7RGG
Emmett S M0TZT
Emmett T 2E0HMB
Emmott J G3ANG
Emmott M M3XVJ
Emmott R GM7TTU
Emms S G0RDU
Emons E GD1MYM
Empringham P G6GZS
Emsden G M0TPH
Emson N G4UXO
Endacott A G3TLK
Endean D 2E0CZN
Endean D M6DLW
Endean D G0UBG
Endean M G7VQL
Endean P G1BZM
Enderby D GM0FMW
Enderby J M0DMI
Endersby R G1AZJ
Endicott J G1WZG
Endicott J G3UWH
Endicott J G6FBJ
Enever J G7EPM
Enfield P M3LXP
England A GW4OEJ
England A M3LHW
England A M0WXF
England E MW0HVL
England J G4XYY
England J M6JKM
England P G1MYQ
England P [illegible]
England P MI3ISC
England R G4REH
England S M3SEE
England T G8GJV
Engledow F M0WVG
English C G6ZOE
English C MM6CPE
English D G6LXP
English G G8NQK
English J M6UKG
English R G4NVV
English V G1OAU
Ennion D M0JXE
Ennion D M3JXF
Ennis J G3XWA
Ennis M G7VBD
Ennis R MI3CSS
Enoch D G3KLZ
Enrico P M0CNZ
Enright K M0DZM
Enright R G0RJE
Entwisle S 2D0ZYX
Entwisle S GD0HGG
Entwisle S MD6HGG
Entwistle A G4LVI
Entwistle E M0AQE

Entwistle M G6VHE
Entwistle N G0BRM
Entwistle Y G1SXY
Epps H M0ADN
Epton D G4LNB
Epton W G3NGJ
Erber M G7HJQ
Erbes E M0HDK
Erents S G3SSW
Erinjeri J 2E0WRK
Erkiert P G4BKS
Erlank A M6LXX
Ernest A GW3LQE
Ernster O M6AOP
Ernster P G4NYY
Erratt C G0MXY
Erridge F M1GFE
Erridge H G6CUA
Errington G M0GAE
Errington J G8FWA
Errington S G0UUF
Errock G G3HCO
Erskine N GW0GDI
Erskine W GM1BTL
Erwood A GM7AFE
Escott P M6FYE
Escreet B G4SPC
Escreet J M0BTN
Escreet P G1FGI
Esdale D G0RMX
Esdale D M6FUR
Eselgroth R G0HRK
Eskelson A G0POY
Esp S M3HRE
Espey A MD3MAN
Espey J MD3GAB
Espie A 2M0GGY
Espie S MM3FOE
Esposito S G4MGG
Esser M G6MZN
Essery D M3DSE
Essex A G8WCX
Essex J G8JCL
Estevez D M0HXM
Estibeiro J G0EUL
Etchells C 2E0CTE
Etchells C M6CCE
Etchells J G1LWE
Etchells R 2E0FPA
Etchells R M0FPA
Etchells R M3FPA
Etches C M6CRE
Etheridge A G0HXF
Etheridge J G4DKP
Etheridge S 2E0SPA
Etheridge S M6SPA
Etherington B G1ISP
Etherington D M6NWO
Etherington W G0JZU
Etherton-Scott W G1HLS
Etwell K G1UNU
Eunson P GM8YEC
Eustace A M0RON
Eustice N M0NJE
Evans A G0CWG
Evans A G0PBP
Evans A GW1TJK
Evans A GW3XXB
Evans A G3ZZX
Evans A G4BPE
Evans A M1ASV
Evans A M3ISG
Evans A M6MFN
Evans A M3UEG
Evans A G4HDR
Evans A M6DGQ
Evans A M1VIP
Evans A G3PXI
Evans A MW1CFE
Evans A M0BLM
Evans B G7IJY
Evans B G0GOO
Evans B GW8YLK
Evans B 2E0FDD
Evans B G0EHE
Evans B G1ILG
Evans B G6YXT
Evans B M3ILG
Evans B M6FDD
Evans C GW0IRP
Evans C G3XKE
Evans C G4EYA
Evans C G4NHE
Evans C G6CKE
Evans C GW6IZZ
Evans C M0DAX
Evans C MW0ICE
Evans C MW3CCE
Evans C M3IQJ
Evans C M3KCC
Evans C [illegible]
Evans C G3LUO
Evans C GM4FVO
Evans C M1BUU
Evans D 2E0SPB
Evans D 2E0AMW
Evans D 2E0EQE
Evans D G1GBV
Evans D G1JRU
Evans D GW1XFB
Evans D G3IVK
Evans D G3YNK
Evans D G6DPE
Evans D G6EDF
Evans D G6EDM
Evans D G7UCZ

Evans D G7WLC
Evans D M1FFE
Evans D M3AXW
Evans D [illegible]
Evans D [illegible]
Evans D M3JKI
Evans D G0EVA
Evans D GW1ANW
Evans D G4AMJ
Evans D G4EQR
Evans D G4GTE
Evans D G7CJS
Evans D G7RAB
Evans D MW6AGS
Evans D MW6DAI
Evans D MW6LIW
Evans D M6VDE
Evans D G8POI
Evans D M6VAK
Evans E 2E0IFW
Evans E G1PDA
Evans F G8AWM
Evans G G0HDC
Evans G G0HOP
Evans G GW0HPC
Evans G G1WKZ
Evans G G3ZZV
Evans G G4IQK
Evans G G6AUK
Evans G G8NVZ
Evans G G8XAS
Evans G GW1CIY
Evans G M6YPG
Evans G G6MKQ
Evans G 2E0CPF
Evans G M6BQF
Evans G GW1JVB
Evans G G4JDE
Evans G G1YJB
Evans G G8WXU
Evans G M0ODX
Evans G M3ADX
Evans G G0LJI
Evans G G7NOI
Evans G M3IAP
Evans G M3UST
Evans H M3HJE
Evans H GI0AZB
Evans H M6HSE
Evans I G1BTN
Evans I G1OBC
Evans I G6JAF
Evans J 2W0YFC
Evans J G3VDB
Evans J GM3VJY
Evans J G4NXB
Evans J G6LMI
Evans J G6NNO
Evans J G8AGJ
Evans J G8ITI
Evans J G8TWR
Evans J M3LSU
Evans J M3OXV
Evans J M3PPY
Evans J MI6JHE
Evans J G7CEC
Evans J M3UWZ
Evans J M3ZBF
Evans J 2I0JIE
Evans J MI6JIE
Evans J M3TVJ
Evans J MW3ZTH
Evans J MW6JDY
Evans J G3LWR
Evans K G8IOS
Evans K M0AQQ
Evans K G3VKW
Evans K G3VUR
Evans K M6CYU
Evans L G7IGU
Evans L M0BZN
Evans L M0BAV
Evans L M6PZZ
Evans L M0PXI
Evans L G7JIB
Evans L M6HVD
Evans M GI0AYG
Evans M G0GQK
Evans M G0KFV
Evans M G1HSF
Evans M G4MMH
Evans M GW4TPG
Evans M G4VLN
Evans M G8HVT
Evans M MW0CNA
Evans M M0KWV
Evans M M0TTU
Evans M M6MPE
Evans M GW3UCJ
Evans M 2E0EKT
Evans M M0UDW
Evans M G7EBI
Evans N 2E0ESJ
Evans N G0AOE
Evans N [illegible]
Evans N M3HNM
Evans N M3SGG
Evans N G1HSG
Evans N G6EWX
Evans N M6NDE
Evans N M0NDE
Evans N GI4BDR
Evans N G0MMB
Evans P G0RBJ
Evans P GW4KCQ
Evans P G4RAV

Evans P G4RUJ
Evans P G7REH
Evans P G8JHO
Evans P G8SYS
[illegible]
Evans P MW0UXH
Evans P M0ILT
Evans P M3MMP
Evans P 2E0HPL
Evans P G4BKI
Evans P M6MSN
Evans P M3ZPE
Evans P 2E0CTF
Evans P M0HNJ
Evans P M6PRE
Evans R 2E0AWW
Evans R G0RPX
Evans R G0VCW
Evans R GW0WLQ
Evans R G3LQC
Evans R G3NNB
Evans R G3PRO
Evans R G3VHE
Evans R G4CXK
Evans R G4GEZ
Evans R G4XAT
Evans R G7ODB
Evans R G7VWG
Evans R G8FOZ
Evans R G8KHV
Evans R M0CVT
evans R M0SPB
Evans R M3TVC
Evans R G0DHB
Evans R G0CDV
Evans R G4AGE
Evans R G6PMC
Evans R G8AVC
Evans R M3ENE
Evans R G1YQM
Evans R G0HLB
Evans R M0RPE
Evans R G0NDB
Evans R G8XTO
Evans S 2E0PPY
Evans S G4XEL
Evans S G6AHX
Evans S G7CVF
Evans S GW7GWO
Evans S G8KQV
Evans S M1SAM
Evans S M3KFE
Evans S M3YRC
Evans S G1WLD
Evans S M3IWX
Evans S G1EKM
Evans S M6ZSA
Evans S G0EVJ
Evans S G4ZVQ
Evans S M6SFP
Evans T G1ISN
Evans T M3ENF
Evans T M5TLE
Evans T M6LDW
Evans T 2I0TME
Evans T 2W0TAI
Evans W G0OBB
Evans W G4EQM
Evans W GW4LKS
Evans W G4POG
Evans W GW4PWZ
Evans W M6BNX
Evans W G6ILY
Eve C GJ7AOG
Eveleigh D G1VYM
Evely N G0CEL
Evelyn B M6BZE
Evenden C M6EKX
Evenden R M3RPE
Evennett C G1OJL
Evennett S M6KSS
Evennett T G0LGF
Evens R GW3YQH
Everall M G6FTA
Everall M M3FTA
Everard A G0ARZ
Everard A G4IEC
Everard D G7CEW
Everard J G0SZO
Everard K G0NKZ
Everard P G4UNW
Everard P G7GFX
Evered B M6EVE
Evered R M6TRE
Everest G G3XUP
Everest G G1TZX
Everest R G4EOE
Everett A M6ASE
Everett J G0HJK
Everett M G3LFR
Everett P G7SRV
Everett P G3SFE
Everett R G1DWC
Everett S G6MKO
Everett C G7GRR
Everett S M6SAE
Everett W G0ODK
Everingham J G4TRN
Everington K G6YII
Everist J G4CVC
Everitt A G0LMO
Everitt B G6OQV
Everitt M G6KRN
Everitt M G8OYQ
Everitt M 2E0CER
Everitt R G1JZT

Everitt R G4ZFE
Everitt S G0WXA
Everley C G4PPK
Everley M G6GUD
[illegible]
Everson J [illegible]
Everson H G6DIO
Everson S 2E0BSK
Everson S M3WHV
Everton L G4NBI
Everton P M6GOE
Everton W G7GJT
Eves A G0JJM
Eves D G0RNP
Eves T G6DIM
Evetts D G3RKP
Evetts N M6UBS
Evill J G6HJV
Evison R G7JFU
Evrall J G3SZE
Ewald B G0NXL
Ewan J G0UEO
Ewart B G8ZML
Ewart B 2M0WLX
Ewart B MM6OWL
Ewart C MM0ABJ
Ewart D MM6PYX
Ewen I M3IJE
Ewen J G6RGA
Ewence A M3OKE
Ewen-Smith B G3URZ
Ewen-Smith J G4JKA
Ewer H M3VYE
Ewer N G3ZUS
Ewing A MM6YEZ
Ewing C G4DBR
Ewing E 2M0UEA
Ewing E MM3UEA
Ewing H G7GRB
Ewing J 2E0FKJ
Ewing P GM1FNX
Ewing P GM0WEZ
Ewing R 2M0BSE
Ewing R MM6SOR
Ewington P G8FFZ
Ewington P M3AEA
Excell A G4DII
Exell S G1VNU
Exley H G4FOT
Exton M G7NUM
Extrance R M6BPO
Eycott G 2E0FEC
Eycott G M3YOW
Eyers M G0IHI
Eyers S G1VNM
Eyes J G7OMN
Eyes R G8RON
Eyles A M0AYV
Eyles C G3SJH
Eyles S G8NXE
Eyre A 2E0EIV
Eyre A M0PUD
Eyre B G3TPP
Eyre C 2E0CJD
Eyre C M3SQT
Eyre D G0TFD
Eyre D G0UQZ
Eyre D G1UUJ
Eyre J M6PPS
Eyre J G7TZZ
Eyre P G8ZIY
Eyre R 2E0GIG
Eyre R M0XAI
Eyre R M1FLY
Eyre S M0HUG
Eyre S 2E0SLE
Eyres J M1EGN
Ezard C M6GZC
Ezard D M6SDE
Ezard S 2E0SEZ
Ezard S M3SEZ
Ezra R G3KOJ

## F

Fabian S M1EDA
Fabrega P M6NBQ
Fabris S M0YUX
Faccenda S MM3ZSF
Facer D G1OJD
Fadil M G4CCA
Fagan A M3HEO
Fagan D G4OTU
Fagan D M3UDF
Fagan W G7IUI
Fagence H G1DHB
Fagg D 2E0DJF
Fagg D M3PQB
Fagg M G4DDY
Fagg P G4CCY
Fagg R M3ZEJ
Fahey N [illegible]
Fahey N M6BLQ
Faichney K G4ZJE
Fails V GI4WWF
Fair R MM6SUR
Fairbairn I MM0IMF
Fairbairn M M6SEU
Fairbairn N 2E0VDC
Fairbairn N M0KID
Fairbairn N M6VDC
Fairbank D G0AII
Fairbotham K G0UDP
Fairbourn P M6PZF
Fairbourn S 2E0VAU
Fairbourn S M0TTE
Fairbourn S M6VAU

Fairbrass G G0UIF
Fairbrass G G8LUV
Fairchild D G0DDF
Fairclough A M6TXA
Fairey A [illegible]
[illegible]
Fairfax J G4KVT
Fairfull R MM0HDA
Fairgrieve A GW3OQK
Fairgrieve O GD0PKW
Fairhall J M6JOO
Fairholm R G1UUV
Fairholm R GM4VBE
Fairhurst A G6ZIY
Fairhurst D G4GTS
Fairhurst G G4WGF
Fairhurst J G6DBJ
Fairhurst N M6NPF
Fairhurst P G0KLU
Fairhurst V M0COV
Fairman F G0NLV
Fairmington P G3PLJ
Fairweather S G6BEL
Faithfull B G3KOS
Faithfull J G0DCU
Faithfull M G6UBH
Faithfull R M0ALF
Faiz N G7VLF
Fakes R M6RQE
Falconer A 2M0ECK
Falconer A MM3ZQX
Falconer K GM3YOI
Falconer S G7GUO
Falding G G3LGF
Falkner D G8LKQ
Falkner D M0DRF
Falkowski M M0MTF
Fallaize C MU0FAL
Falle P GJ8PCY
Fallick T G4FYI
Fallon J G3SGV
Fallon O 2E0WCL
Fallon O M6WCL
Fallon P G8RIB
Fallon P 2I0OTW
Fallon P MI6OTW
Fallows A G4UCL
Fallows G G1BMB
Fallows I G0UAA
Fallows J GI4OWB
Fallows R 2E0BHB
Fallows S M3RFR
Falomir Montanes E M6EFM
Faloon K GM0WPU
Faloon K GM8JWQ
Falstein D G6BAT
Fambely P G0BHP
Fancourt J G3HEE
Fang F 2E0CFG
Fang F M6AOT
Fang X M3XZF
Fanning Ba Bd Cf P G6UFI
Fanning P G1RDU
Fanning T 2E0EPQ
Fantham A G3TGL
Fantom G M6CUQ
Faragher L G7AYI
Faram A G1XIO
Fard D M3NDF
Fardell P G0LQU
Fardoe G G1NXB
Fare G G3OGQ
Farey B G8GHR
Farey R G6LLP
Farhall R M6EWI
Farina J 2E0BLC
Faris T M0CDB
Farley C G4WVF
Farley C G6GEX
Farley C M0GEX
Farley H G4SNI
Farley P G3TDF
Farley R G4LOG
Farline G G6AIB
Farlow C G0RUZ
Farman P G7SCE
Farmar A G1GZI
Farmar A M3GZI
Farmer A G0PDP
Farmer A G4MDR
Farmer C M6CKF
Farmer G G8XQZ
Farmer K G4VJT
Farmer M G7MRF
Farmer M G6BD
Farmer M G3VAO
Farmer N G7CTE
Farmer P G1UOD
Farmer R G1KEB
Farmer S G7BSP
Farmer T G4RBH
Farmer T [illegible]
Farmer T G6PPA
Farn D G4HRY
Farnbank G M6EIU
Farnborough A G0WDT
Farndon C G0IYW
Farnell P G6CNL
Farnell S 2E0SJF
Farnell S M0SLF
Farnham D G0BPL
Farnham M M6MSF
Farnie G G4FXM
Farnington K M6KFE
Farnley R G0KTR
Farnworth M GM0TOF

Farnworth M G8YLM
Farquhar A GM0JOV
Farr D G4WUB
Farr G G1ILH
Farr G G0UTC
[illegible]
Farr K G6WFM
Farr M G4CAJ
Farr R G4DVX
Farr S G4STE
Farr T M3ZLY
Farrall R G6POZ
Farrance K G0AKF
Farrance R G3TRH
Farrant D 2E0DRJ
Farrant D M0NDT
Farrant J G8UDS
Farrant J M0GJD
Farrant L M0LTW
Farrant M G0HEQ
Farrant S G1JCT
Farrar A 2E0BMY
Farrar A M0GVX
Farrar A M3PGK
Farrar E G7LOV
Farrar G 2E0GAR
Farrar G M3PGU
Farrar J G3UCQ
Farrar J G4KXF
Farrar K G1HZR
Farrar P G8SJA
Farrar P G0JMZ
Farrar R M3RYR
Farrar S M3TCT
Farraway M 2E0HAL
Farrell A G8HTO
Farrell G G8PYD
Farrell J GI4JOR
Farrell L 2E0LTF
Farrell M G0PFT
Farrell M G8ASG
Farrell M G0WFD
Farren P M6EEN
Farrer F 2E0OVR
Farrer F M6PDV
Farrer G MM6GHF
Farrer J M0DCZ
Farrer J G3XHZ
Farrer J M3VMQ
Farrer R 2M0SRF
Farrer R MM0VTV
Farrer R MM3SRF
Farrington B M0CFZ
Farrington J G0VAV
Farrington N M6ZNF
Farrington R M3ZID
Farrington-Smith N M6FNX
Farrow A G0TAM
Farrow B G3FIR
Farrow J G7SXJ
Farrow N M3XEX
Farrow P G1HTN
Farrow S G0PCD
Farthing K M0CLO
Farthing S G0XAR
Fasham M G7ITS
Fatu R M0HZH
Fauchon P G0JSP
Faul D M1FBS
Faulconbridge J G0GWN
Faulkner A G1SCV
Faulkner A G8WXV
Faulkner A G8EYP
Faulkner A M6AND
Faulkner C 2E0BIM
Faulkner C M6CRF
Faulkner D G4DWF
Faulkner D G4ODF
Faulkner G G6FPO
Faulkner J G0CYX
Faulkner J M3RPQ
Faulkner J M6JCF
Faulkner K G6YXV
Faulkner K G4EQZ
Faulkner L M6LEF
Faulkner M 2E0MJF
Faulkner M 2E0RNF
Faulkner M M6OEM
Faulkner P 2E0PCF
Faulkner P G8TMJ
Faulkner P M6PCF
Faulkner P [illegible]
Faulkner R G0CAL
Faulkner R M6FIZ
Faulkner R G1OXT
Faulkner S G0CQS
Faulkner S [illegible]
Faulkner S 2E0SCF
Faulkner S M0TGT
Faulkner S M6SCF
Faulkner W M5WJF
Faulkner-Court J G0CDO
Faulks J 2E0RNX
Faulks J M6RNX
Fautley J 2E0XEN
Fautley J M0NGY
Fautley J M3XEN
Fautley P G0ASG
Fautley R G3ASG
Faversham R G0ISM
Fawcett A G0VGN
Fawcett A GM0PMO
Fawcett B G4FEJ
Fawcett C G8YIG
Fawcett D 2E0FFZ
Fawcett D M3AOC

**IMPORTANT NOTE**

**Revalidate licence to avoid revocation** – Ofcom has advised the Society that plans will be drawn up to revoke licences that have not been revalidated as required by the licence conditions. The quickest way to revalidate is to do so online via the Ofcom website: *https://services.ofcom.org.uk/* or by email: *amateur.validations@ofcom.org.uk* If you need assistance in the process, Ofcom staff are available to help, but please be patient during times of heavy workload.

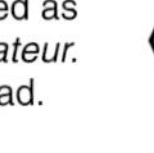

Fawcett S G0BVA
Fawcett S M6SPP
Fawcett S G7VMO
Fawdon C G8GBP
Fawke R G4YGY
Fawkes B G3VQW
Fawkes C G4UDG
Fawkes P G4MOC
Fawkes R G0HUV
Fawley S M6FAW
Fay C G4SBB
Fay C G4UDB
Fay I G1HLT
Fay J M6ELS
Fay K G0AMZ
Fay P G8NDV
Fayers D G3YKC
Fazey J G6XYO
Fear C M6FUA
Fearby W M6HEW
Fearis M GI4ISH
Fearn D M3DAF
Fearn J M6JMF
Fearn J G4EHX
Fearnhead G M6BCW
Fearnhead S M6WLA
Fearnley A G1SEF
Fearnley D G3XYI
Fearnley R M3FRJ
Fearns J G7RFH
Fearnside G G0BWY
Fearon M 2E0LJS
Fearon M M3YBT
Fearsaor-Hughes R MM0TIR
Feary G G4MDH
Feasey M M0ALX
Feast J G6SEF
Feast M 2E0MFS
Feast M M6SMX
Feast Z G6SEE
Feather C G1ZCC
Feather J M3JDF
Feather W M0PWF
Featherby D 2E0DNF
Featherby D M0ZEB
Featherby D M6DNF
Featherstone B M6USG
Featherstone C M6FJC
Featherstone D G4JFD
Featherstone J G4OVO
Featherstone M M6BLF
Featherstone P G3XOP
Featherstone R G4MWJ
Featherstone S G6IPW
Feaviour T G0CFT
Feay J G1KFG
Feay K G1GZK
Feeley D G6DIZ
Feeley J G4MRB
Feeney S G0LNO
Feetenby D 2E0FSV
Featham N G7WDM
Fegen A GM0LYT
Feilen R GM7WED
Feist A G3PMV
Fekete R M6FEK
Feldman S G3GBN
Felds R M5DLA
Felgate J G4YVK
Felix F MD0TAT
Fell J G0API
Fell K M0KRL
Fell K M3NBI
Fell M G4TRP
Fell S G7MHO
Fellingham N M0WNF
Fellingham N M3WNF
Fellingham P G7FJC
Fellows A G7UGC
FELLOWS B G0ONH
fellows D 2E0KPP
Fellows D G8FBF
Fellows E G8ENA
Fellows E M3FEL
Fellows G G7GHP
Fellows J G8EQI
Fellows L G4HCZ
Fellows M G8GKR
Fellows P M6EHW
Fellows S M3FGU
Fellows T G7KYF
Fells C G0LUW
Felstead H M0HJF
Feltham J M6JFH
Feltham M M6AWG
Felton E M6ERF
Felton G G0FEM
Felton G G8FDI
Felton M G1LSN
Felton P G1PMF
Felton R G0ENB
Felton S G1DAK
Fenelon M G4WJE
Fenlon E G1YZT
Fenn J G4UJV
Fenn P G6FVF
Fenn R M3WXD
Fenn S G8ZTB
Fenna D M0DSF
Fennah A G1JNI
Fennah F G0EHS
Fennell A G0JIF
Fennell J G4HOY
Fennell J G8RPD
Fennell P G0FPM
Fennelly D G7ITW
Fenner N G1NMP
Fenney P 2E0BZU
Fenney P M6PMF
Fensch N G8FYX
Fensom J M6FKH
Fensome E G6ZJD
Fentham A G3TON
Fentiman T M3VWF
Fenton C G0NAR
Fenton D G4MIH
Fenton J 2E0DUS
Fenton J M6YEG
Fenton R G3NQF
Fenton S MM6CTH
Fenton W G3ZJP
Fenton-Coopland C G8EUV
Fenton-Coopland J G4ENC
Fenwick B G8BTC
Fenwick F G0AOJ
Fenwick J G3VCK
Fenwick S G3AIO
Fenwick T G4MXX
Ferdenzi S G8GNO
Fereday B G4TDO
Fereday B G6MZT
Fereday M G3VOW
Fereday M G8IMH
Ferenc A 2E0BWY
Ferenc M M3ZLV
Ferentiuk M G1CIA
Ferguson A GM7IIL
Ferguson A 2W0GAY
Ferguson A MW0SIP
Ferguson C MM6GIW
Ferguson C 2E0BSI
Ferguson C M6CHF
Ferguson C GM4HNK
Ferguson D GM3YMX
Ferguson D G6RXY
Ferguson D M6MQK
Ferguson E G6JJT
Ferguson G MI6GAU
Ferguson G 2E0SEW
Ferguson G M3YGF
Ferguson I GM0ILQ
Ferguson I M6IAW
Ferguson J GM4ARJ
Ferguson J GI4GPC
Ferguson J GI4RYP
Ferguson J G8STW
Ferguson J M3HLV
Ferguson K G0HYP
Ferguson L M6LDF
Ferguson M G0JPU
Ferguson M M3IFE
Ferguson N G0BPK
Ferguson P G7TUH
Ferguson R GD0IOM
Ferguson R GM3YTS
Ferguson R GD4GNH
Ferguson R G4KQO
Ferguson R GM5CX
Ferguson R 2E0VDE
Ferguson R M0RBF
Ferguson R M3VDE
Ferguson S GI6NFK
Ferguson S M3VYT
Ferguson S MM6SFF
Ferguson S M6XAK
Ferguson T GM1OST
Ferguson T M3TBF
Ferguson T 2I0TJF
Ferguson T MI6TJF
Ferguson W 2M0AAQ
Ferguson W GM4AGL
Ferguson W GM6VCV
Ferguson W MM6WBP
Fergusson J M3XCJ
Fergusson S G0DYH
Ferigan D G3ZYV
Fermor G G8RCZ
Fern M G6HWR
FERN R M3PQQ
Fern T G4LOH
Fernandes E G4FIH
Fernandez A G4XPT
Fernandez R G3NMT
Fernando G 2E0GXF
Fernando G M6GXF
Fernant D G4PGO
Ferne M G6GGZ
Ferneyhough A G0EVH
Fernie C G0RDB
Fernie D G4EYB
Fernie P 2E0PFA
Fernie P M3SOV
Fernihough R G1GAT
Ferns D G7ABQ
Ferns G M6GSF
ferrans J 2M0OOT
ferrans J MM3OOT
Ferrari P G4IHI
Ferreira P M6PQF
Ferreira R M6ENW
Ferrie T GM4KHI
Ferrier J G0ATW
Ferries A MM3XAF
Ferrin J GI4JIW
Ferrington D 2E0XDF
Ferrington D M0XDF
Ferrington J M3XJF
Ferrington N 2E0XNF
Ferrioli A 2E0FJA
Ferrioli A M0IDM
Ferrioli A M6FJA
Ferris A GM0LKT
Ferris B G3HKF
Ferris B G7HAR
Ferris D M0BDL
Ferris E G8YXR
Ferris I GI8YYM
Ferris J 2E0BCE
Ferris M G1JTM
Ferris P G0LLE
Ferris R GI0OUM
Ferris R G0VZB
Ferris S G0RMB
Ferrol J 2E0IBT
Ferrol J M3IBT
Ferrol J 2E0KFT
Ferrol J M0JKF
Ferrol J M3KFT
Ferrow D M3DFW
Ferry B M6JIK
Ferry J M6ZEF
Ferryman R G4BBH
Ferula O MM1BJG
Fesel J M3TXU
Few K G8JDQ
Fewings G G7SUM
Fewkes A G4WAF
Fey A 2W0FEY
Fey A MW6PEY
Fey E M0FEY
Fey M 2E0FSY
Fiander D G4APS
Fiddy R M6ABQ
Fidler R G4EXZ
Fidler R 2E0CVJ
Fidler R M0ZRF
Fido B 2E0CZB
Fidoe J G0LCV
Fiedler C G0GYP
Field B G3TTY
Field D G3WNV
Field D G3XTT
Field D M6AEN
Field D MM3TLQ
Field E 2E0WWF
Field E M6XTT
Field F G0CEN
Field J 2E0DLS
Field J G7JSH
Field K G0KFF
Field N G4LQF
Field N M3XFD
Field R G0OOU
Field R G4JQJ
Field R M1CQR
Field R G0MQD
Field R G8LGM
Field S G4SYR
Fielden C M6AER
Fielder A G0CPT
Fielder F G6XFR
Fielding A G4CPM
Fielding A G7CMP
Fielding B M1EUL
Fielding D M3VAT
Fielding D G8KHU
Fielding E G3WDN
Fielding G G1DFP
Fielding G 2E0NON
Fielding G G7HUC
Fielding G M0PYG
Fielding G M6PYG
Fielding J G4VIL
Fielding M G3JPO
Fielding M G4IOJ
Fielding P G4TMA
Fields B G4XDJ
Fields I G7IGF
Fieldsend D G6HCW
Fieldsend N G1BWZ
Fifield B G7MFH
Filby D M3LOT
Filby D 2E0DCN
Filby R G0HJR
Fildes G G4ZYY
Fildes G M6GMF
File M M6MFF
Fill D G3UBB
Fill N M6NYA
Fillingham D G4OVR
Final M G4TOO
Finbow C M6CCF
Finbow N G7OBX
Finbow P G0DEH
Finch A 2E0FCH
Finch A M6FCH
Finch A G3XQM
Finch B GM6YGW
Finch D 2E0EAZ
Finch F MW3FJW
Finch G G6OHQ
Finch I 2E0HIL
Finch J G3VNU
Finch J G4YVB
Finch J G6PQI
FINCH K G3XIQ
Finch L GW8PSJ
Finch M G0VXJ
Finch R G3JYS
Finch R G4AJO
Finch R G4DDM
Finch R 2E0POQ
Finch R G1NZQ
Finch R M0POQ
Finch R M3POQ
Finch R G0IFL
Finch R 2E0DNH
Finch R M6RFZ
Finch S G0HMG
Finch S 2E0SAF
Finch S M3UCU
Finch T G1LWF
Fincham P M0BTY
Fincher M G6SWO
Fincher P G1ANQ
Findlay A G0FIN
Findlay A G8GLZ
Findlay D M0CXL
Findlay I GM7VYR
Findlay I MM3VYR
Findlay R G3PRF
Findlay T GM4DOZ
Findlay W GM4ZNC
Findler B M0GBF
Findon G G3TQF
Findon M M6MVF
Fineman N G4JHU
Fineman R G4BOL
Fineron P GM8PEB
Fingerhut G G0ENW
Fingerhut-Holland H MW0ECY
Fingerhut-Holland H MW3NTE
Finlay A GM3NEQ
Finlay B G4USK
Finlay F MI3TFF
Finlay N M3LZR
Finlay T GI6BVQ
Finlayson A MM0BPV
Finlayson I GM0WZO
Finlayson J GM4IXH
Finlayson L M6NFO
Finlayson S M6MFS
Finlayson V G7IBX
Finn A M3FIZ
Finn A M3ZUZ
Finn A 2E0GCL
Finn A M6GBV
Finn K M0KVN
Finn K M1KVN
Finn M 2E0MAQ
Finn M M0MPF
Finn M M3XOH
Finnegan C M0RAB
Finnegan J GI4FFL
Finnegan M MI3CQX
Finnegan S G7EYE
Finnemore D G3WJJ
Finneran T G0PKN
Finnesey E G0CTC
Finney G 2E0TPP
Finney G M3GVT
Finney G 2E0YYT
Finney G M6YYT
Finney G 2E0YYF
Finney G M6YYF
Finney J M3UJF
Finney M G6NZO
Finney P G4WBF
Finnie P MM3YFT
Finnigan M G4UJF
Finnis C 2E0CSF
Finnis C M3CRL
Finnis C M3CSF
Finnis R G0HDY
Finon A G7DIB
Fiorentini M G7HUC
Firby I G7VCG
Firby I M3IAF
Firby J G3UOI
Firks D G0SYT
Firmin G GM4GQM
Firmin J G4GUS
Firmin P G0FUU
Firmstone J G6DBP
Firmstone P G0MDQ
Firth A G0FUY
Firth A GM6WOF
Firth A G7UKK
Firth A G4ETD
Firth B G4KCT
Firth D G3WLT
Firth D 2E0GLL
Firth D G7VBU
Firth D M6TIU
Firth E G6XSK
Firth G G3MFJ
Firth H GM4ZZH
Firth I M6IDF
Firth J G1TMF
Firth J G4BHL
Firth J GM8YRE
Firth J M6JDF
Firth J 2E0UAE
Firth J M3ULS
Firth K G7WGL
Firth M G3ZJV
Firth M G4JMT
Firth M G6EDU
Firth M G7PMX
Firth M M3MLF
Firth M 2E0FIR
Firth M M0MJF
Firth M M6FIR
Firth R G6IMH
Firth R G1USK
Firth S G8SFI
Firth T G1FFH
Firth T M0JQK
Fish A G4GPL
Fish C M6XCF
Fish E M6LFZ
Fish F G4YFD
Fish H M3SCF
Fish J GM0ANG
Fish J G4NRP
Fish J GM3NYG
Fish K G1ZVZ
Fish L G0LQN
Fish L M6LWF
Fish W G4FBO
Fishbourne E M1EIR
Fisher A G8TAU
Fisher A M3PFF
Fisher A M6CCH
Fisher A G3WSD
Fisher A G6CUK
Fisher A M6VEN
Fisher B G8TWT
Fisher C 2W0MXT
Fisher C MW0MXT
Fisher D 2E0DFT
Fisher D G4EXU
Fisher D G8ELH
Fisher D G8UZY
Fisher D G0DNF
Fisher D G6FSK
Fisher D G6YQJ
Fisher D MM0WXD
Fisher E G0FFN
Fisher E M6YLI
Fisher F G3LXJ
Fisher G G0DSU
Fisher G G0WTL
Fisher G G4MJX
Fisher G G4OZQ
Fisher G G6UTK
Fisher H GM0EFQ
Fisher J G0IVZ
Fisher J G4XQZ
Fisher J G6GLW
Fisher J GM0NAI
Fisher J G1KIT
Fisher J M1BRY
Fisher K G4ITR
Fisher K G6LMR
Fisher K G8VOB
Fisher L 2E0KYQ
Fisher L G7LOA
Fisher M G3UBI
Fisher M G4WYW
Fisher M G4OKH
Fisher M M3XBF
Fisher M M3UDD
Fisher M G6VTA
Fisher M M6FIG
Fisher N M0FUN
Fisher N M6FCD
Fisher P G0LFV
Fisher P G4GHQ
Fisher P G7WER
Fisher P M0BDH
Fisher P M0FIS
Fisher P G4PPL
Fisher P G6MKZ
Fisher P G1VVH
Fisher P M6MRN
Fisher R G0UEB
Fisher R G4KBK
Fisher R G4KJU
Fisher R M5RMF
Fisher R G1XMH
Fisher R G3PWJ
Fisher R M6KCY
Fisher S G1SMY
Fisher S G1YTV
Fisher S G6YXO
Fisher S M6SAF
Fisher S M3ZFS
Fisher S G3YWU
Fishlock D G6WOT
Fishlock T G1BFF
Fishwick A M1CGI
Fishwick R G6FUY
Fishwick S M1CGJ
Fisk G M6SHD
Fisk J M3PFN
Fisk P 2E0CPJ
Fisk P M3PFK
Fisk R G4CPV
Fisk R G7AVU
Fisk S G0UWW
Fisk T 2E0TFX
Fisk T M0TFX
Fisk T M3TFX
Fiske S M6BRP
Fisken D G8BDU
Fitch G G4WKZ
Fitches H G7BCW
Fitchett M M0BMX
Fitt G G4ILN
Fitton A M6CQD
Fitton F M6KKF
Fitton G G4MDT
Fitton S G0WWE
Fitzakerley B M3PTV
Fitzgerald A G0MEN
Fitzgerald D G4CKS
Fitzgerald D M3DPF
Fitzgerald E G3XUX
Fitzgerald J G4JGF
Fitzgerald J G8XTJ
Fitzgerald J GI6UFO
Fitzgerald M M6MFZ
Fitzgerald R G0HBW
Fitzgerald R M3RJF
FitzGerald T G4UQE
Fitzgerald W GW4YMJ
Fitzgerald-O'Connor B 2E0TBW
Fitzgerald-O'Connor B M0LGN
Fitzgerald-O'Connor B M6CXP
Fitzgibbons M G8VHB
Fitzherbert H G4GAP
Fitzhugh S G6PHT
Fitzjohn M M0HCT
Fitzjohn S G7TBU
Fitzmaurice A G0KXW
Fitzpatrick B MW1BNY
Fitzpatrick D M0CHL
Fitzpatrick G G1CGU
Fitzpatrick J G7DFP
Fitzpatrick J M3LRU
Fitzpatrick J M6GQH
Fitzpatrick M M0DZL
Fitzpatrick P M0CLK
Fitzpatrick S G1YMC
Fitzpatrick S G7SJD
Fitzpatrick V M3SKU
Fitzpatrick-Browne L G8RBX
Fitzsimmons B MM6SGZ
Fitzsimmons B G0GGM
FitzSimons J G4JFS
Fitzsimons J G6WEW
Fitzsimons R GI0HHZ
Fitzsimons W GI4OXO
Fitzsimons W MM3RTH
Fitzwater J G4HVO
Fitzwater L G6GGY
Flack B G4AMP
Flack G 2E0BKD
Flack G M3FHO
Flack M G7VCE
Flaherty D M3DUO
Flaherty I G4RPJ
Flanagan B MI3EZK
Flanagan C G7NRO
Flanagan D G4ZAR
Flanagan D G6KXP
Flanagan G G7FPR
Flanagan M 2E0IKA
Flanagan M M3ZVU
Flanagan P 2E0DGQ
Flanagan P M6EOC
Flanagan S M1SMF
Flanagan S MI3EAQ
Flanagan T MI0TLF
Flanders R G4PEA
Flanner F G3AVE
Flannigan E G4SZB
Flannigan J 2M0DKV
Flannigan J MM6EGC
Flather A G3XMK
Flatman C G4WTD
Flatman M G0FDT
Flatman N G0EBQ
Flatt P M6PFL
Flatters D G7ENC
Flatters D G7IQO
Flattley J G1TNK
Flavell P M1PVF
Flawn L M3HKV
Flaws M MM0HUQ
Fleet C G0PVN
Fleet M G1NYP
Fleet R G1FZS
Fleet T G0FSL
Fleetham S G8NSO
Fleetwood D G0IXZ
Fleetwood D G1ZDF
Fleetwood D G6WPR
Fleetwood D M3ZDF
Fleetwood J G0UJP
Fleetwood M G0TMF
Fleetwood M G3XEF
Flello D G4WOS
Fleming A M0DWU
Fleming A 2E0TCC
Fleming A M6AMO
Fleming D 2E0CDO
Fleming G G0IIG
Fleming G G0ODY
Fleming G G1MHF
Fleming G 2E0GPF
Fleming G M0HPP
Fleming G M6GPF
Fleming G MM6HQC
Fleming H M3OHZ
Fleming M G8PTL
Fleming M 2E0HZM
Fleming S GI7FCM
Fleming S 2M0SEF
Fleming V 2E0VKY
Fleming V M6VKY
Fleming W GM8JGB
Fleming W MM0WLL
Flemming A M3MDS
Flemming I G3ZDQ
Fletcher A G4AVF
Fletcher A G4SCS
Fletcher A MM0GYG
Fletcher A M6ZAB
Fletcher A 2E0UKX
Fletcher A G8EHS
Fletcher B G0SWH
Fletcher B G4MFW
Fletcher B G4CWL
Fletcher C G3DXZ
Fletcher C M6FOK
Fletcher D G0IDF
Fletcher D GU4WRP
Fletcher D G8CJH
Fletcher D G8HTF
Fletcher D M0GYC
Fletcher D 2E0FTM
Fletcher D M6TMF
Fletcher D M6TBS
Fletcher E G0UAY
Fletcher G G7TEG
Fletcher H G3TVM
Fletcher H MU3ZHF
Fletcher I M1DQE
Fletcher I M0IAF
Fletcher J 2E0JOE
Fletcher J GM0LYO
Fletcher J G4EDD
Fletcher J G4EGB
Fletcher J M3JOW
Fletcher J 2E0KBA
Fletcher J M3UJP
Fletcher J M6NOH
Fletcher J M3OMU
Fletcher J G4EDX
Fletcher J M1EUN
Fletcher J M3LBZ
Fletcher K M6KNS
Fletcher L G4SXH
Fletcher L M6AIQ
Fletcher M G0NVS
Fletcher M G6EDT
Fletcher M G8JYS
Fletcher M M6RTN
Fletcher M 2E0CWB
Fletcher M M0TXK
Fletcher M M6CQH
Fletcher N G4YNK
Fletcher N M3NVF
Fletcher P G6NZL
Fletcher P M1PAF
Fletcher P 2E0ECM
Fletcher P M6DPF
Fletcher R G6EDR
Fletcher R G6RIJ
Fletcher R G7EXD
Fletcher R MM3ZRF
Fletcher S 2E0GMV
Fletcher S G0SZJ
Fletcher S G4SFQ
Fletcher S G7GVJ
Fletcher S G4GXL
Fletcher S G4RFC
Fletcher T M6HNW
Fletcher T M6TFE
Fletcher T MW6TFL
Fletcher T M6STY
Fletcher V G0NEC
Fletcher W G3NXT
Fletcher-Cowen E G6YQI
Flett A GM0HTT
Flett G GM0HQG
Flett J GM4MKU
Flevill R M6CRR
Flew G M6KGF
Flewitt M G1WSW
Flicos P G3ZWD
Flikkema E 2E0FLI
Flikkema E M6FNA
Flindall W G4IPN
Flindell W G0FAA
Flinn B G6DPH
Flinn G 2E0CSN
Flinn G M6BYZ
Flint A G4VLV
Flint E M0RSG
Flint G G6WOI
Flintoft A M6HGY
Flis L MM0IPD
Flitterman D G0PZC
Flood A G0SNZ
Flood G G0APY
Flood P M6ACB
Flood W G8XKH
Flook P M3CFU
Florence K M6ZBS
Florentin J G8AVQ
Flores-Watson J M3JSF
Flounders B G1VJB
Flower H M3HAZ
Flower J G1LAO
Flowers J G0JLF
Flowers S M6SBR
Flowers V G8QM
Floyd A G3PNQ
Floyd P 2I0LDC
Floyd P MI6BII
Floyd S GM3KXQ
Fludgate W G1YYU
Flunder D M6DAF
Flute M MM6MQA
Flux C G7IVF
Flynn B GM8BJF
Flynn C G7EOG
Flynn C MM6BCF
Flynn D M3WLG
Flynn L G4UWP
Flynn M 2E0OZH
Flynn M 2E0RZH
Flynn M M0OZH
Flynn M M3OZH
Flynn M MM6MWF
Flynn R GW1XXL
Flynn T M6BYM
Fochtmann M G4BHJ
Foden C GM4LNN
Foden M G3UPA
Foden R G6NFR
Foden W G7BAB
Fogarty C GI1RSR
Fogarty E M6EBE
Fogerty M M6JVW
Fogg J G8UZZ
Fogg M G0RVK
FOGG R G3ZCD
Foggin N G8ETI
Foister M 2E0FWM
Foley A 2E0LLI
Foley A M6FIV
Foley C G0XCF
Foley D 2I0YDF
Foley D MI0YAM
Foley D MI3YDF
Foley D 2E0CTU
Foley D M6JFE
Foley M GI3SOO
Foley M 2I0YMF
Foley M MI0YMF
Foley M MI3YMF
Foley P GI8YGG
Foley R MI3NSF
Foley T G4FYO
Foley T G7MTF
Folgate R G3KDY
Folkerd C 2E0DTE
Folkerd D 2E0DTF
Folkes M G7RQD
Folland J G7CUO
Folland M G7PVF
Follant J GW0CBL
Follett J GW1MBV
Fone C G4LAI
Fone D GW8VRS
Fong N M3TNK
Fooks G G4UTY
Fooks S M3ONH
Foord D M3UBS
Foord E G0RJJ
Foot H M3OPW
Foot M G4WHY
Foot N G4WHO
Foot S MJ6SSF
Foote A 2E0FAE
Foote A M6ASF
Foote B G0SKC
Foote C G8IPN
Foote N GI4MNF
Foote R G4GQP
Foote S G0TWI
Foote S G1EUM
Foote S G4FOH
Footring L G4SDI
Footring M 2E0AIK
Forber A M6BWV
Forber C 2E0FSH
Forber C M0ZLK
Forber S M6YZX
Forbes A G4UST
Forbes A GM7RJG
Forbes A 2E0OJE
Forbes D M6ZDF
Forbes J 2E0FTC
Forbes J M6FSO
Forbes L MM3VSX
Forbes N M3NXF
Forbes P G0NZJ
Forbes P M6PXF
Forbes S G6YQT
Forbes S M3SXF
Forbes S G6YIE
Forbes T G0JWO
Forbes W G0NNN
Ford A G6FPP
Ford A G6HBQ
Ford A GM4UEH
Ford A M0TET
Ford B G0EYF
Ford B M1BSO
Ford C G4ZVS
Ford C M6FEQ
Ford D G0IDP
Ford D G4UKP
Ford D G7POS
Ford D G6PHU
Ford E G3DEY
FORD G G0MHC
Ford G G0PEJ
Ford I G1GZM
Ford J G0HJD
Ford J G6LFW
Ford J M0DJI
Ford K G1BBI
Ford K G1SLU
Ford K G4ZTG
Ford K G7CNZ
Ford M M1BSP
Ford M G4KBP
Ford M M3MBG
Ford N 2M0CWV
Ford N MM6DKN
Ford P 2I0HWW
Ford P G0RHK
Ford P G3VRU
Ford P G4WFL
Ford P MI3PCF
Ford P G3WDE
Ford R G0DRS
Ford R G0ENY
Ford R G4GTD
Ford R M0AFY
Ford R M3RFW
Ford R G0FFF
Ford R M6RBV
Ford R M6XRF
Ford S G4BEM
Ford S G4DPV
Ford T GW7ESF
Ford T M6TLF
Ford W G0TGP
Forde E MI6GHA
Forde G G0VXY
Forde L G4ZLF
Forde W 2I0JFO

UK Surnames

Forde W M6JFO
Forder H 2E0ABE
[illegible] M6RLA
Forder M [illegible]
Forder M G7KGH
Fordham M G0LQV
Fordham S G4HOP
Fordyce A G1WLW
Fordyce G 2E0YBR
Fordyce G M6BIH
Foreman C G1IMI
Foreman D G4NEE
Foreman G 2E0IIC
Foreman K MM0GNH
Foreman M G4HCI
Foreman M G7LSZ
Foreman M G4HBT
Foreman T M1CWG
Fores I M6MUJ
Forester I 2E0FOR
Forester I M6FOR
Forhead B G7TUQ
Forknell M G7JXF
Forrest A M1KAZ
Forrest A 2E0WET
Forrest A M0CVU
Forrest A M6PRO
Forrest B 2E0RZX
Forrest B M0RZX
Forrest B M6RZX
Forrest D GM7UTD
Forrest D M0ACM
Forrest J G3VVW
Forrest J G4WJV
Forrest K G1BTI
Forrest P M6XWB
Forrest S M3YSF
Forrester A M6CYE
Forrester A M6WIM
Forrester T G4WIM
Forrester W G3XAN
Forrester W MM0HSB
Forrest-Mcneill S MM0SFM
Forrest-Webb R 2E0AFP
Forryan A G4OKD
Forsey D G1DAV
Forsey J G4ETS
Forsey M G8PWK
Forshaw D M6TFV
Forshaw J M6MZN
Forshaw P G0JJI
Forshaw P G4HSS
Forshaw S M6MVN
Forshaw S M6GOS
Forss R 2E0IXI
Forss R M6IXI
Forss T M6MOC
Forster A G0UYG
Forster A G1YDD
Forster A G4NNJ
Forster C G4FFU
Forster C G8LCC
Forster D G0NUL
Forster D G0VSZ
Forster D G3KZZ
Forster D G3NWY
Forster D G7VZL
Forster D M1DKZ
Forster D 2E0BLZ
Forster D M3SZS
Forster G G0CPD
Forster G G0VSY
Forster G G7VTE
Forster G MM0GFR
Forster I G4FFV
Forster J G1SKQ
Forster K G8RDA
Forster K M3KEF
Forster M 2E0GQU
Forster M M1BXM
Forster M M1MSF
Forster P G0CBN
Forster P G3VWQ
Forster R G3SMN
Forster R G8EIE
Forster T G7WGE
Forster W G6TOC
Forster-Pearson R G6GDD
Forsyth A G6BJB
Forsyth C 2M0CLF
Forsyth C G0TQS
Forsyth D 2E0AJK
Forsyth D MM6DHF
Forsyth E G6KHR
Forsyth I GM3PUY
Forsyth J 2M0JCF
Forsyth J M0LTK
Forsyth J G0PAU
Forsyth J GM4OOU
Forsyth M G1OWD
Forsyth S [illegible]
Forsyth W GM1OVJ
Forsythe A M6AKK
Forsythe I MI6ILF
Forsythe J GI0PGC
Fort P G3ZCX
Fortescue M G7IMH
Fortescue R G1HTO
Forth D G1CDY
Fortnum C G1JGR
Fortt S G0OBT
Fortune D GM7SKB
Fortune R GM4WTK
Forward D G7MSS
Forward D 2E0EOW
Forward J G3HTA

Fosbraey R G7BWV
Fosbrook C G0KCF
[illegible] G0URO
[illegible] G7UEN
Foss P G0JYE
Foss R G8YLR
Fossey J G0TUX
Foster A 2E0HBY
Foster A M6FOZ
Foster A 2E0RJU
Foster A G6VHG
Foster B G4XRM
Foster C M6CJF
Foster C G7BJB
Foster C GW6MKR
Foster C G1NXI
Foster D G3KQR
Foster D G4IPI
Foster D G7FPZ
Foster D G7URL
Foster D G4PHP
Foster D G6UTF
Foster D M0NZL
Foster E M3HCA
Foster G G1DRG
Foster G M6GZR
Foster G M3VBV
Foster G 2E0GBB
Foster G M6AQO
Foster H M3HFO
Foster H 2E0EZY
Foster I G8CAM
Foster J G0MYH
Foster J G1HTL
Foster J G1PJK
Foster J G4GSL
Foster J GM4PWQ
Foster J G8TML
Foster J G8WSZ
Foster J 2E0FZJ
Foster J M0JHF
Foster J M6FZJ
Foster J 2W1DGM
Foster J G0FZZ
Foster J G3TKB
Foster J 2E0RQN
Foster J M0RQN
Foster J M6RQN
Foster K G8GIH
Foster K M6KPF
Foster K 2E0BXM
Foster K M3VHH
Foster K M1CGM
Foster M G1SWE
Foster M G8AMG
Foster M G3VOF
Foster M G4KLE
Foster M G1TZZ
Foster M M6MFO
Foster N G0ULA
Foster N G0NBJ
Foster N M6BIB
Foster N 2E0NAF
Foster N M3NAF
Foster P G0OSV
Foster P G1PAF
Foster P G4TQZ
Foster P G7TMO
Foster P G1LOV
Foster R G0GRU
Foster R G1NJI
Foster R G0WEA
Foster R G0LCO
Foster R G1SIX
Foster S G4EEF
Foster S G4MPK
Foster S M6SFT
Foster T G7TMF
Foster T G3PQP
Foster W 2E0ELT
Foster W M0KLK
Foster W M6WJF
Fosters A GM3OXA
Fothergill B G8KPD
Fothergill J G8RER
Fotheringham S M6SUF
Fouche D G0DWF
Fougere T G4MHK
Foulds A M3LQB
Foulds D G4GFE
Foulds J M3JNT
Foulds P G1SRD
Foulds R 2M0MTQ
Foulds [illegible] MM0MTQ
Foulds I G3GFC
Foulger K M6LCC
Foulkes C M6CDL
Foulkes F G3KOM
Foulkes L M6LLL
Foulkes P G6YXW
Foulkes [illegible]
Foulser S G6UTL
Foulser S G8VQA
Found A G7BYG
Fountain G G1ODZ
Fountain N G7ELC
Fountaine A G7AFL
Fountaine S G1CBY
Fouracres S G8MBE
Fower A 2E0HRN
Fower C M3HRN
Fower D M0FAZ
Fowle C G7HKT
Fowle G 2E0AMW
Fowle G M0CJC
Fowler A G0GFP

Fowler B G7ODZ
Fowler D 2E0WVF
Fowler D G4YWG
Fowler [illegible] M0TDI
Fowler [illegible] G4MDN
Fowler G G8XKI
Fowler G G6NZN
Fowler H G4RXH
Fowler I G8UFX
Fowler J G4VHG
Fowler J G7KLZ
Fowler J M1RJJ
Fowler K G7NVB
Fowler M G8XZQ
Fowler M G8XTU
Fowler N MM0CJH
Fowler P G4TQO
Fowler R G3IQF
Fowler R M3RYN
Fowler R M6TCM
fowler R G0JJP
Fowler S G0NKU
Fowler S G0TGM
Fowler W G0JHJ
Fowler W M3YCN
Fowles D G8VJR
Fowles D M3GEZ
Fowles G G3XGV
Fox A M0MBI
Fox B G7PVG
Fox B G0OJR
Fox C G3PKL
Fox C G7UEV
Fox C 2E0PXY
Fox C M3PXY
Fox C 2E0PXZ
Fox C M0PXZ
Fox C M3PXZ
Fox C G0WYR
Fox C 2E0PIT
Fox C M6FVW
Fox D G4SSZ
Fox D G8YLS
Fox D 2E0BVY
Fox I G4UDF
Fox J G3KHR
Fox J M1BMV
Fox J G0OAZ
Fox K M3VEX
Fox M G0EYT
Fox M G4YUN
Fox N G6YQN
Fox N M3GFO
Fox N M6HOV
Fox P G2YT
Fox P G8HAV
Fox R M3MER
Fox R G0UOI
Fox S G4FAB
Fox S G4GVV
Fox S M6SJF
Fox S M3STQ
Fox W M3LQF
Foxall B G0PCF
Foxall D M0JDE
Foxall S G7EKW
Foxley R G8YRF
Foxon A MD3JGS
Fox-Roberts P GI0USQ
Foxton T G0KOE
foy D MI6TRF
Foy D G4WCO
Foy F 2E0FBC
Foy F M6FOY
Foy J G0JAF
Foy J 2E0DNF
Foy M G0SKI
Foyen N M1BUP
Foyston D 2E0DID
Foyston D M3UAX
Fradgley K G0WBA
Fradley D M0DJF
Fradley J M6JJF
Fradley M M3MIQ
Fradley S 2M0BKL
Fradley S MM3ZXL
Fradley T G6NLQ
Fraley D G1JYR
Frame J MM1JWF
Frame K G6IZU
Frame M 2E0MAL
Frame M M3MCF
Frame M GM3ZWG
Frampton D [illegible]
Frampton G G3LRH
Frampton J [illegible]
Frampton J G6CUE
Frampton J G6NNK
Frampton P M0CNX
Frampton P M3RGN
Frampton [illegible] 2E0SID
Frampton S M0ZID
Frampton S M6SCA
France A G0VUH
France D G1IQB
France J G3KAF
France L GW3PEX
France M 2E0APJ
France P M3EFC
France P M6OOO
France R G3ZOU
France S M3IRF
France S M3YKA
Francis B GW1KQY
Francis B MW3BMF
Francis D MM0DYX

Francis D M3UCH
Francis I 2E0IWF
Francis I M6IWF
Francis [illegible]
Francis J [illegible]
Francis J G4XVE
Francis J G8ZTD
Francis J M3JRF
Francis J 2E0LEM
Francis K 2E0KLF
Francis K M0JVC
Francis K M6KLF
Francis M G0GBY
Francis M G3LOV
Francis M M3TOI
Francis P M1FFC
Francis P G7MMW
Francis R G0EYP
Francis R G0TJT
Francis R G4EIE
Francis R M1DJN
Francis S M0SGF
Francis S M6RCN
Francks K M0BFB
Frank A G0TDY
Frank S G7SOP
Frankcom K G3OCA
Frankham L G4GFJ
Frankland M 2E0IAX
Franklin A G6GAF
Franklin A M1FEW
Franklin C G1FLI
Franklin C G6XSL
Franklin D 2E0FRE
Franklin D G8VFI
Franklin D G0MQL
Franklin I 2E0SQN
Franklin I M6SWS
Franklin J G1XAJ
Franklin K M3ZKF
Franklin K G0NUA
Franklin K G3JKF
Franklin L G3LWF
Franklin L G7VSN
Franklin M G1WSM
Franklin M G0MNN
Franklin M G3VYI
Franklin P M0PJF
Franklin P G1FOA
Franklin R G3ITH
franklin R GM4XHH
Franklin S G3XVH
Franks C G4PSI
Franks J G3SQQ
Franks J 2E0RNO
Franks J M0VIT
Franks J M6JPF
Franks M G4MKF
Frankum S G0WZH
Fraser A 2M0AVL
Fraser A GM3AXX
Fraser A G8MHO
Fraser A GM8NET
Fraser A MM0DFZ
Fraser A M0NER
Fraser A MM6EPV
Fraser C GM0HBF
Fraser C MM0IRC
Fraser D G4SXG
Fraser D GM6JRX
Fraser D MM6FYR
Fraser I G3TVT
Fraser I GM8MHU
Fraser I M0FRH
Fraser I M6GBK
Fraser J GM0OKJ
Fraser J GM4WJA
Fraser K G4FMA
Fraser P G3HZT
Fraser P MM1BJZ
Fraser R M1DHM
Fraser R M3WWH
Fraser R 2M0VUV
Fraser R MM0VUV
Fraser R MM6VUV
Fraser R GM0NTL
Fraser S G1FBQ
Fraser-Hopewell C 2M0FRA
Fraser-Hopewell C MM6CFH
Fraser-Smith K M6KFS
[illegible] GM0MVQ
Fray H GW6RCK
Fray T M0FRA
[illegible]
Frayne G M6DXV
Frazer A G0DMU
Frazer G GI4SJQ
Frazer S 2I0KPA
Frazer S MI0KPA
Frear M G0MEF
Frearson J GM1MZZ
Frearson J G3OVK
Frearson T G0COY
Freckelton C M6FXX
Frederick D G4YXM
Frederick G G4NZO
Frederick N G6YXX
Free J M3WID
Free M G0OGE
Free S M3SFZ
Freeborough D G1JLE
Freeburn G MM6HOO
Freeburn R GI6WHZ
Freedman A M3VSH
Freedman D G3VSH

Freedman M 2E0EPD
Freedman P 2E0EPE
Freeland J GM4SZG
[illegible] MM1CMU
[illegible]
Freeland R MM3BHD
Freelove J 2W0JPF
Freelove J MW3JPF
Freeman A G0PIT
Freeman A M0HAZ
Freeman B G3ITF
Freeman B 2E0STP
Freeman C G6HWT
Freeman D G0LTP
Freeman D G7RTR
Freeman D M6TRI
Freeman G G4FCC
Freeman G M0GEF
Freeman G MI3GFA
Freeman J G4MGX
Freeman J G4XQQ
Freeman J M0EJG
Freeman J 2E0BPE
Freeman J M0GJU
Freeman J M3XQZ
Freeman J G1OHU
Freeman J G6KRG
Freeman M G8UST
Freeman M G8YYW
Freeman N M3IEP
Freeman P 2E0BXF
Freeman P G4DMS
Freeman P 2E0PEF
Freeman P M0HQO
Freeman R G3TCZ
Freeman R G4SDJ
Freeman R 2E0URF
Freeman R M1AYU
Freeman R M3URF
Freeman S G3LQR
Freer J GM7LJE
Freer K G8RJF
Freer K M6KJQ
Freer R G0PBY
Freeston D G4DBF
Freestone J GD4TEM
Freestone M GD8UOZ
Freestone P G4PFL
Freeth A M6MKH
Freeth G G4HFQ
French A GM3XQP
French C G8ZAJ
French D G3TIK
French D G4PKO
French I G0WGV
French J M3FRE
French J G7HEZ
French J 2M0JSF
French J G4IET
French L G8VJP
French M G8WSH
French M G3ZXD
French P GM0PKF
French P G8FUJ
French P M3GWZ
French R M3JIV
French S G6SWT
French S G6ZTZ
Frencham R M0GKJ
Frend P GI0FZT
Frenzel C G4SNJ
Freshwater A M6BRK
Freshwater R G6CNK
Frettsome C G1ECS
Frettsome R G4WPW
Fretwell G G2CFC
Fretwell P G4UFC
Frew G MM3IKS
Frew J GM3YLD
Frew J MM6JFU
Frew R G3SEF
Frewen W G0FLA
Frey S GM8ZQY
Friar R M6DWY
Friberg H M3RPK
Fricker R M3ULN
Fricker T G0NQN
Friday N G8FRY
Friedman M G1END
Friel A G1GWE
Friel C [illegible]
Friend A G1EUN
Friend J G4HLI
Friend [illegible] G7VPD
Friend S M6TCI
Frier R M1RHW
Friis A G4LNC
Fripp G M3UOY
Fripp R 2E0DGT
Fripp R M6DPN
Frisby J G8DJU
Frisby S [illegible]
Frisholm G M6GUS
Friston R G0FMI
Friswell A M3XXU
Frith B M6SGN
Frith I G4GIH
Frith W G3FRE
Frizell J G1WSE
Frizzell C 2E0CAE
Frizzell C M3XGU
frizzell S G7PUW
Froggatt A G1MZP
Froggatt B G1KVP
Froggatt B M6BJF
Froggatt D G8CXA

Froggatt L 2E0EFT
Froggatt N 2E0WEE
Froggatt N M6FRO
Froggatt P 2E0FRO
[illegible] G1HTM
Froggatt S G1LMN
Froggatt S 2E0VET
Froggatt S M6ARC
Froggatt T G7JGF
Froggatt T G0EOI
Froggatt S M6SKF
Frohnsdorff M M3MFF
Froley B GW4ZVO
Fronters A M0HKH
Froom A M0ORF
Frosdick M G6VZG
Frost A G0MVM
Frost A G8UDV
Frost A G8YWJ
Frost A M0MYJ
Frost C G0KEB
Frost C G1XRO
Frost C M3VME
Frost D G7VQE
Frost D MM6EQM
Frost D MM3MMO
Frost F M1DYD
Frost J G0DCR
Frost J G4SYL
Frost M G0IDH
Frost P G3VYK
Frost P 2E0LGO
Frost P M6ZLD
Frost P G4NLD
Frost P M0HJE
Frost P M6AQG
Frost R G3SZF
FROST R G3UUV
Frost R G1SZK
Frost R M6OST
Frost S G4VNM
Frost S M6SIF
Frost S G0NDA
Frostick V M0PPO
Froud J G3YHH
Froude N M6NCF
Frow R G0VAI
Frowd K G4ZCM
Fry A G0VXX
Fry A G4WBV
Fry C G3NDI
Fry D G0BBJ
Fry D G4JSZ
Fry D G1MJI
Fry E GM6LKQ
Fry J 2E0RUG
Fry J M6TRT
Fry J G4UNX
Fry M M3MEF
Fry M G0GLU
Fry M G7NDS
Fry M G6BXT
Fry P G4AKG
Fry P G4SBF
Fry P G0FUS
Fry P M3WZR
Fry P M0GUO
Fry R MW0FRY
Fry R M6RBF
Fry S GW6UFH
Fry S M6SFV
Fry S M0SWF
Fry W G1YRR
Fryer A M3MBF
Fryer B M6KIL
Fryer C 2E0FRY
Fryer C M6FRY
Fryer C M6PCE
Fryer D G4KEG
Fryer D G6DBQ
Fryer D G1OQG
Fryer D M0COM
Fryer G 2E0CNY
Fryer N MJ1CNB
Frykman G G0GNF
Frylinck R M3RHO
Ftaiha I M0HJW
Ftaiha I M6CJX
Fudge H G3DZS
Fudge J M6AYQ
Fugard A 2E0HWD
Fugard A M0INF
Fugard A M6ESJ
Fuge A G4DEU
Fuggle H [illegible]
[illegible] G4IIY
Fujita K [illegible]
Fujita K M0OCX
Fukuda R M0CFF
Fulbrook M 2E0WZT
Fulbrook M M0WZT
[illegible] M6WZT
Fulbrook P 2E0EAN
Fulbrook P M0PNA
Fulbrook P M6ABN
Fulcher K [illegible]
Fuldner N M0IJX
Fullagar R 2E0RPF
Fullagar R M0RPF
Fullagar R M3RPF
Fuller A G4OYN
Fuller A M3YIE
Fuller D G1CQR
Fuller D G4IQB
Fuller D G4XPY
Fuller E M0BXB

Fuller F G4GCJ
Fuller G G3TFF
Fuller G G4VFH
Fuller G MM3HQL
[illegible]
Fuller J G4WPI
Fuller J 2E0IAD
Fuller J M3PDK
Fuller J M0GPD
Fuller K G0VOB
Fuller L G0ULN
Fuller L G7MYY
Fuller L 2E0PDU
Fuller L M0PDU
Fuller L M3PDU
Fuller M M6FUL
Fuller N G8MVS
Fuller P G0PVQ
Fuller P 2E0PFX
Fuller P G4FNJ
Fuller P M0PFX
Fuller P M6PFX
Fuller R G0EBG
Fuller R G0HFK
Fuller R G6PWS
Fuller R G6YQU
Fuller R G8CEZ
Fuller R M3TXV
Fuller S M3HTA
Fuller T M3ZIM
Fuller W 2E0WRF
Fullerton B 2M0YIO
Fullerton I G6ILZ
Fullerton W MM3AQM
Fullick D M3URO
Fullwood A M0TMF
Fullwood A M1TMF
Fulton A 2E0PUB
Fulton A M6ACF
Fulton D GI4OUN
Fulton J 2M0AWY
Fulton J G4XFC
Fulton S MI3OUN
Fulton W 2M0MOK
Fulton W MM6WUL
Fung C M3YCK
Fung K Y M3YKF
Funnell A M1ECI
Funnell C G7IZV
Funnell M G3YQW
Funnell P G4BWN
Funnell P G8AFI
Furby M M6JIS
Furlong C M6CAF
Furlong D MD3PYB
Furlong G M6GAF
Furlong K M6BPB
Furlong K M6KRF
Furmage G GM0NAQ
Furminger S G7TYH
Furmston A G0TCW
Furneaux N M6NFX
Furness D M3AHL
Furness P G0WGL
Furness R GD4IHC
Furness R G4JEY
Furness R GM3RUI
Furness T G4WNG
Furness W G3SMM
Furniss H M6HFG
Furniss R G7FBY
Furnival M G8ETP
Furnivall M M6FAY
Furr L M6IIL
Fury E M3EBF
Fury O M3OMF
Furze W G0UIQ
Fusniak R G3TFX
Fussey R GD3BQE
Futcher H G8BTL
Futter O G8PIO
Fyall G GM0NXO
Fye R 2E0FYE
Fye R M3ZFY
Fyfe C G8UUW
Fyfe F G8CIJ
Fyfe G MM3RHT
Fyfe J MM3FYF
fyfe J 2M0FYF
fyfe J MM0JRF
Fyffe A GM4ENF
Fyrth J G0DPS
Fyrth J G4PTM
Fysh A M6AJF
Fysh J [illegible]
Fysh M M6MJF
Fyson J [illegible]
Fyson J G1YBK
Fyvie G MM6ZAZ

## G

[illegible] G7LNY
Gabbitas R G3KHU
Gabel P G1TAI
Gabell F G4LDJ
[illegible] M0XJT
Gabriel N G4JUZ
Gabriel P G4IKI
Gabriel S G3HCQ
Gabriel T 2E0BJV
Gabriel T M3PXF
Gadalla N M3KRQ
Gadd C G4JZZ
Gadd T 2E0GSF
Gadeberg E G0CMT
Gadney R G7KIV
Gadsby S G1OFW

Gadsden D G4NXV
Gadsden P G3MTP
Gaffney D M6DXG
Gaffney E G4KRJ
[illegible]
Gaffney J G4UAA
Gage B G4PGA
Gagen P G4OTC
Gagg J G4XRB
Gagliardi P 2E0KPC
Gagliardi P M0KPC
Gagliardi P M3GAG
Gagnon A G4DGW
Gailer J G3RTD
Gain C G7VZR
Gain C M0VZR
Gain W G1IFW
Gainey P G0DZM
Gainford E M3HFA
Gainswin S G1GMG
Gainza M M6MFG
Gaisford R MM0FWG
Gait P G6XQO
Gajewski C 2M0AYZ
Galbraith A MM3FET
Galbraith A MI6DAY
Galbraith C MM3GBL
Galbraith G G1NOR
Galbraith G M0ADR
Galbraith J G7EVI
Galbraith N M3ZNG
Galbraith T MM3TWG
Gale B M3WYJ
Gale B G3UJE
Gale C M3FTK
Gale D G1GVM
Gale D G8WZR
Gale E G4TMV
Gale G M1ATB
Gale J G3LLK
Gale J G8UIL
Gale K M0KEL
Gale L M3NHU
Gale M GM4PXB
Gale P G3OJG
Gale P 2E0CHN
Gale R M3NHV
Gale S M0AZR
Gale T 2E0BCK
Gale T G1XDV
Gale T G8JUS
Gale T M0BUT
Gale T M0EAQ
Gale Z M6ZAQ
Galea C G0LPP
Galea M G7PDO
Galer P M0PJG
Gall D G0AOK
Gall I GM8BNH
Gall R GM4UFD
Gallacher C G4JCX
Gallacher G 2E0GVS
Gallacher J GM8FHK
Gallacher K M1EAW
Gallacher W G0IRY
Gallagher A M6GJA
Gallagher A G0TOE
Gallagher A G6YDP
Gallagher C G4ZTW
Gallagher D MI6ODG
Gallagher J 2E0DDJ
Gallagher K M3PFL
Gallagher K G0KEV
Gallagher M G8OKN
Gallagher R M6RIQ
Gallagher S MM6KCB
Gallagher T G8EDN
Gallagher V M6VJG
Gallamore G G8BRU
Gallear B G8VPR
Gallery K M6KPG
Galley G G0JZL
Gallichan A G7SRL
Gallichan A M0ASD
Gallichan D MJ1COO
Gallienne J GU4WMG
Gallimore D M1CZZ
Gallimore F G4PKX
Galliver G M0SOX
Gallon M G3KTT
Gallop A M3RKV
Gallop D [illegible]
Gallop J G3YIW
Gallop K M6BLM
[illegible]
Galloway C G3RNV
Galloway J MM0JOQ
Galloway M 2E0YZX
Galloway M M6TMG
Galloway S M3BZC
Galloway T M3YPU
[illegible] GW6NNF
Galsworthy B GW0UIZ
Galt T 2M0AZW
Galvin A G0IKD
[illegible] G8KYO
Gamble B M6GHQ
Gamble C M6NQT
Gamble E G0ASZ
Gamble K M0KIG
Gamble P GI7FJY
Gamble P G1SGZ
Gamble R GM3UYR
Gamble W MI6WKN
Gamblen J G4ERS
Gambles R G6DBU

UK Surnames

**IMPORTANT NOTE**

**Revalidate licence to avoid revocation** – Ofcom has advised the Society that plans will be drawn up to revoke licences that have not been revalidated as required by the licence conditions. The quickest way to revalidate is to do so online via the Ofcom website: *https://services.ofcom.org.uk/* or by email: *amateur.validations@ofcom.org.uk* If you need assistance in the process, Ofcom staff are available to help, but please be patient during times of heavy workload.

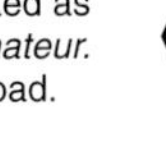

Game P G6TQF
Gammage R G6BVR
Gammage T G4TAG
Gammage T G3YOV
Gammans D G7EQR
Gammer G M3LHG
Gammer J M3LHE
Gammon A G1EUQ
Gammon I G4SCV
Gamulea Schwartz N M6CKY
Gandy R GM0MNV
Gane G G6PTX
Gane P GM4SUF
Gange L M0GEK
Ganley P G4YWJ
Gannaway J G3YGF
Ganner D G0DPG
Ganner D M3DPG
Ganson J MM0BTD
Gant R G0LXP
Ganti S M6XTF
Gao C M3YSZ
Gapper A G8PQA
Garbett C G0RCX
Garbett S G4XDX
Garbutt B G0LPX
Garbutt D G7GAK
Garbutt K G0PBA
Garbutt M G0OEJ
Garbutt N G4PJJ
Garcia C G8DWW
Garcia-Quismondo T 2E0TGQ
Garcia-Quismondo T M0MEW
Garcia-Rodriguez J G6BNW
Garczynski S G0RFA
Garde P G6MCE
Garden G G4LJR
Garden L G0EJI
Gardener J G1ECV
Gardiner D G8BAS
Gardiner D M3DGJ
Gardiner G G3WEB
Gardiner G G4GRP
Gardiner G 2I0DKQ
Gardiner G MI6JGK
Gardiner I G3PHD
Gardiner J M0CSR
Gardiner J 2E0MGT
Gardiner J M6GYD
Gardiner J M3XCB
Gardiner K G8NFD
Gardiner M G4MVX
Gardiner P G1MRI
Gardiner S GM7GWW
Gardiner T GI1CKU
Gardner A G0OZB
Gardner A G4OKC
Gardner A G7DHG
Gardner A MM1GAR
Gardner A G0NTH
Gardner A M6SEO
Gardner B M3KTT
Gardner C G6IRP
Gardner C MI1DQB
Gardner C MI3FBX
Gardner D G1LTL
Gardner D M0RFK
Gardner D M6DFG
Gardner D GM4JZB
Gardner D M6OZB
Gardner E 2E0HLH
Gardner G G0HEM
Gardner G G1CFJ
Gardner G G4ZEN
Gardner G G7RRO
Gardner I G3CDM
Gardner J GM3VPN
Gardner J 2E0WHD
Gardner J M6SPT
Gardner J G7RUX
Gardner J GU7CQN
Gardner J MU0CHN
Gardner K G0OKX
Gardner K G0VFL
Gardner K G7SZG
Gardner M M1EXW
Gardner N G0CQZ
Gardner P G1DCU
Gardner P G1PWY
Gardner P G0IPE
Gardner R G3CGE
Gardner R M3SPY
Gardner S G4PSP
Gardner S M3XZB
Gardner T G3YVZ
Gardner W G1HMW
Gare C G3WOS
Garfirth S G6ANR
Garfitt M G2CKR
Garforth A G3IGC
Garforth I M6IGH
Garland B GW0LDZ
Garland C G3RJT
Garland D M3UAG
Garland H G0NFJ
Garland M G1JGS
Garland P G6MCX
Garland R 2E0RUS
Garland R M3SQS
Garland S G8GHL
Garlick G G1BJZ
Garlick M G4YNG
Garlick M M1AUR
Garlick S G7API
Garlick S G7HVF
Garman A M6AAX
Garman G 2E0YTF
Garman G M3YTF
Garmany H GM0GYQ
Garn A 2E0YAX
Garn A M6YAF
Garner A MW6HLU
Garner A M0DJA
Garner A M1DJA
Garner A G0DPY
Garner C 2E0DEP
Garner C G8RFY
Garner C M3WUB
Garner D GW8DYR
Garner D M0GYS
Garner G G6XQP
Garner G GW8NXK
Garner G G1OWK
Garner J 2E0HQH
Garner J G1ACA
Garner J G2BGG
Garner J G4ZKQ
Garner J M3VOL
Garner J G3ZJG
Garner J M3WUJ
Garner M G1WUC
Garner P GW0INN
Garner R G7HHL
Garner R G8LAN
Garner T G1ULP
Garner T G3XZY
Garner W G3UHT
Garnett D M0PWT
Garnett J G1YUX
Garnett L 2E0BWA
Garnett L M6SVF
Garnett P G4LZJ
Garnett W M3WLL
Garnett-Frizelle R G7ORW
Garnham C G6MCG
Garnham D M6DNX
Garnham J G0PMX
Garnham T G1IVK
Garrard D G8EOM
Garrard I M0IDG
Garrard I M6IDG
Garrard J 2E0CGI
Garratt C M6EZZ
Garratt D G0FXL
Garratt D G1IFX
Garratt D G1WSD
Garratt D M6GTD
Garratt F G4HOM
Garratt I M3ZWG
Garraway P 2E0PCC
Garraway P M6EHP
Garrett B G6ENO
Garrett D G8MZA
Garrett F M1RES
Garrett F G3MVZ
Garrett G G3IJW
Garrett J G1PJR
Garrett J G3RHP
Garrett M M0MJG
Garrett P M6GZY
Garrett P M0HJQ
Garrett R G0MUR
Garrett S G4EVN
Garrett S 2E0SAA
Garrett S M3RRZ
Garrett T M6GJY
Garrington D GM3RFA
Garrington M GM7SPB
Garrod L M6RVE
Garrod N G0OQK
Garrott E G0LMJ
Garry B M3VBG
Garry E M6DFI
Garry M 2E0PDL
Garry M G0VYT
Garry M M0LOB
Garry M M3UVT
Garry N M3NSG
Garside K G4LGH
Garston P G4SII
Gartell P M1AXD
Garters J G8JLD
Garthwaite A M0RTL
Garthwaite A M1OXR
Garthwaite P G3OXR
garthwaite S M3ITM
garthwaite S M3SPG
Gartland J G8JMG
Garton B M6BSG
Garton M G8WJY
Garton M M0CZE
Gartshore D 2M0DSG
Gartshore D MM0TDB
Gartside J G8NQI
Garvey R M6REE
Garvey T G0BML
Garvey T M3OQQ
Garwell S G1PNB
Garwood D G8MGQ
Garwood J 2E0VRT
Garwood J M6MPK
Garwood M G4DLD
Garwood S G7USP
Gascoigne B G8PXU
Gascoigne D G4OSY
Gascoigne J 2E0XLX
Gascoigne J M6XLX
Gascoigne O M6ORO
Gascoigne P G4IMB
Gascoyne K MD6KFH
Gascoyne M 2E0ICT
Gascoyne M M0ICT
Gash G M0GUD
Gash P G7AOA
Gash R 2E0BBQ
Gash S M6SLG
Gaskell C M1XCG
Gaskell D G0REL
Gaskell D M6DWG
Gaskell E G0RJX
Gaskell P G4MWO
Gaskell R 2E0ADA
Gaskell R G0RKG
Gasken R GM4RXD
Gaskin A M6EXN
Gaskin C G7MNZ
Gaskin J M3XJG
Gaskin J 2E0JKG
Gaskin J M0JKG
Gaskin P G8AYY
Gaskin P MM1FEO
Gaspar L M3BPF
Gasper M G8EUE
Gass J G4NGL
Gass P G4XZC
Gasser D G4KWY
Gasser W M0GPK
Gaston A G0HPK
Gaston C G4KEI
Gaston M M3MCG
Gaston-Johnston L 2E0LGJ
Gateley A G1NAN
Gatenby F M3IFG
Gater K G7STC
Gater M G4ICC
Gates A G0ARV
Gates B M6GXH
Gateson S M6CGF
Gathergood M G0OIE
Gathergood R G4LUA
Gatrell A G4SVB
Gau S 2W0PCT
Gau S MW0PCT
Gau S MW6PCT
Gaudie B MM3LLU
Gaughan J GM4FEO
Gaukroger C G7CLO
Gauld A G0KFG
Gauld G M0EAL
Gault A GI6PYP
Gault A MI3TUZ
Gault C MI6TZP
Gault J GM0LPB
Gault R 2I0RPM
Gaunt B G0LPG
Gaunt C G7BRZ
Gaunt E GM7UXH
Gaunt G G4IJO
Gaunt J G0JWG
Gaunt K G3ADZ
Gaunt K G7CIY
Gaunt L G4MLV
Gaunt M M3XAK
Gauntlett B G4LYU
Gauntlett G G3VLL
Gaur P 2E0JPR
Gaur P M0JAI
Gaur P M6ETU
Gauson J MM0CAE
Gaut D 2E0UDG
Gaut D M3UDG
Gaut J G0CCV
Gautier-Lynham P G5BCO
Gautrey N G6GGW
Gavin C G0POG
Gavin J G0XAS
Gavin P 2E0PTG
Gavin P M0URL
Gavins F M3XMS
Gaw S GM4XWL
gawan R M0AKQ
Gawn T G8UKI
Gawne A GD7LAV
Gawthorpe B G7RDP
Gawthorpe E G8FEK
Gawthrope A G0RVM
Gay F G3CFV
Gay R 2E0BSR
Gaylard M G6IUQ
Gayler M G4SDZ
Gayne A G7KPF
Gaynor L M3WVX
Gayther B G4PUX
Gayther J G0TPD
Gayton R G8ATC
Gazi A MW3PGN
Gazinski P MM3DVD
Geall B 2E0DFA
Gealy R G3PTG
Gear F 2E0CCY
Gear F M3NWH
Gear L M3LTG
Gearey J MW0COZ
Gearey S 2E0SMG
Gearey S M0SGE
Gearey S M3OIA
Gearing R G1XNC
Geary A MI3LWU
Geary S G4JZA
Gebbie P G8YQN
Gebhardt K G7AUF
Geddes D G4RKR
Geddes R G8GGI
Gedvilas E G8XVJ
Gedvilas E M0SDA
Gedvygas T M6BOT
Gee A G1IMM
Gee A G1AGA
Gee B G0VRU
Gee B G3LDG
Gee B M6GEE
Gee C G0CKM
Gee C G4ZUN
Gee D G4KYX
Gee D G7NAP
Gee D M6DFC
Gee D G1AGB
Gee G M6FVC
GEE J G4BAV
Gee J G7GEE
Gee M G0BDR
Gee M G6EXG
Gee R 2D0BOR
Gee R M3SVJ
Gee W M6WSM
Geen K G4LRB
Geer J G8WJB
Geere D G3UON
Geeson B G4KHJ
Geeson P G8DIY
Geeson R G3NJX
Gegg D 2E0DMJ
Gehammar A G4YHN
Geier P 2E0FZO
Geier P M1CKA
Geiger P G4CNI
Geisau J M0JVG
Geldart D G8BRK
Geldart M 2M0MGM
Geldart M MM3RLG
Geldart T G4PXR
Gelder J M6JKW
Gell B G6XSS
Gell F G3JTO
GELL J G4EAX
Gell N G7PYQ
Gell P G0NIK
Gellatly J G3ZVV
GEMMELL A GM0DVO
Gemmell D 2E0EBW
Gemmell D M3RBF
Gemmell F G4GGR
GEMMELL H GM8ZAK
Gemmell K MW6KEL
Gemmell T G4RPO
Gemmill A G1DFR
Gener A 2E0DBP
Genes T G4POP
Genes T G6CNQ
Genge N M6NRG
Genon H GW4JPJ
Gent C G4AKE
Gent D G3MON
Gent M G0RUV
Gent N GM4HQU
Gent S G1VQV
Gentile G M0HUW
Gentle D G4RVL
Gentles J GM4WZP
Gentry M G6DXP
George A M6EEW
George A GM0EIT
George B G3ZOH
George B G7TFU
George C G7MYN
George C M0MYN
George D GW1OUP
George D 2E0BHQ
George D G8TUH
George D M3SXZ
George D M3YUB
George D MM3TQI
George D GM0DQV
George G GM8FFK
George G G0OTF
George K G3XPJ
George K M1EKK
George K M6DVF
George M G0NFL
George M G0VKF
George M G3XYG
George M M6GXZ
George P GW1YHA
George P M3WVB
George R 2W0TRD
George R G4VTU
George R G8RDB
George R G6AII
George S M6WWE
George T G3NJG
George T G4AMT
George T GW4ZRW
George T M6UFF
George W G6TQC
George-Powell M G3NNO
Geraghty D G4UIT
Geraghty J G0PJG
Gerard C G4PKW
Gerard D G7IMT
Gerard H G0AED
Gerard K GM4TPX
Gerhardi V G6GDI
Gericke C G4MKX
German R G3OZT
German R G8VCL
Germaney D G7HYS
Gerrard A G4TFU
Gerrard A G6WZY
Gerrard C G3LSJ
Gerrard C G4AXL
Gerrard C GD4OEA
Gerrard C MM3YBG
Gerrard I MM3IAG
Gerrard I G0EFZ
Gerrard K 2E0HUQ
Gerrard M 2M0RND
Gerrard M G6OKC
Gerrard M MM0ROV
Gerrard M MM3MVY
Gerrard P G7JJC
Gerrard R G3LAZ
Gerrard R G7HVO
Gerrard R GM1STW
Gerrard T M3TUC
Gerrard T G7MGC
Gerrard W G4ZRB
Gerrity J G4OQP
Gervais D G7JHV
Gething D G3XZK
Gething L M3LAG
Getty M GI4DNW
Getty R GI6JRY
Ghafoor A M6ZAD
Ghani N G3WZH
Ghassempoory M GW7JDX
Ghetti G G1CPD
Ghillyer K G4YGZ
Ghost M M6HST
Giacani E G1KSW
Giannakopoulos G M3YGT
Gibb D G6BME
Gibb I 2I0GHY
Gibb I MI0IIG
Gibb I MI3GHY
Gibb J G0GVT
Gibb J M1CNH
Gibb M GM0GIB
Gibbard J G1BES
Gibbings A G3FDW
Gibbings A G6SWD
Gibbins A M6TXX
Gibbins K M6KGE
Gibbon D G4DTQ
Gibbon J G1GVP
Gibbon J G3XAG
Gibbon S 2W0BUQ
Gibbon S MW3ZWO
Gibbons B M3ZBG
Gibbons D G0DUM
Gibbons D M0GIB
Gibbons E G7BWE
Gibbons F G0MPR
GIBBONS G G1PEU
Gibbons H G6CVY
Gibbons J 2E0JGG
Gibbons L M6AQY
Gibbons M M0MSG
Gibbons P G8ZHN
Gibbons R G0FOT
Gibbons R M6RRA
Gibbons R GW0AIY
Gibbs A 2W0WMB
Gibbs A G0RGP
Gibbs A G0RSY
Gibbs A G0VSB
Gibbs A G1SPJ
Gibbs A G3PHG
Gibbs A MM1DQV
Gibbs B G3MBN
Gibbs B M3JVR
Gibbs C G8GHH
Gibbs C M0GRW
Gibbs D G7IJI
Gibbs E M6HMD
Gibbs F G7UUB
Gibbs F M3FTI
Gibbs G G3AAZ
Gibbs G M6GFF
Gibbs G M6EHO
Gibbs I G4GWB
Gibbs I G8NYJ
Gibbs J G0NGQ
Gibbs J G3LIO
Gibbs J G3ZZZ
Gibbs J G4MGY
Gibbs J G4UQR
Gibbs J G8JZO
Gibbs K 2E0KHG
Gibbs K M6KAO
Gibbs L G1NMQ
Gibbs L 2E0ZPN
Gibbs M G3PSR
Gibbs P G4DFG
Gibbs P 2E0MRQ
Gibbs P M3ZJG
Gibbs R G0RBQ
Gibbs R G0UBA
Gibbs R G7IWW
Gibbs R G8BNB
Gibbs R 2E0GOQ
Gibbs R M6GIB
Gibbs S G1JQK
Gibbs S M6SGD
Gibbs S G4YYR
Gibbs S 2E0SGG
Gibbs S M6SDG
Gibbs W G4EBO
Gibson A G0TFT
Gibson A GI0TJV
Gibson A G7TGF
Gibson A G0RCI
Gibson A G1TKE
Gibson B G0ADU
Gibson B GD4UKD
Gibson B G6WZZ
Gibson C GM0UKZ
Gibson C G6VMB
Gibson C 2E0TYT
Gibson C M6GBI
Gibson C M6MSJ
Gibson C M6KAT
Gibson C 2E0FKU
Gibson C M0PSK
Gibson C M3AIE
Gibson C 2E0SYY
Gibson C M0SYY
Gibson C M6OFF
Gibson D G4LXA
Gibson D G6ZUO
gibson D M6HNN
Gibson D 2E0DEK
Gibson D M6ATP
Gibson D GI3OQR
Gibson E G0JUI
Gibson E M6ERZ
Gibson F M1BQS
Gibson G G3ZFZ
Gibson H G2BUP
Gibson H G6AIG
Gibson I GI4MDD
Gibson I M0AYU
Gibson J 2E0AWR
Gibson J G6CNW
Gibson J 2E0ZSJ
GIBSON J G3WYN
Gibson J M0ZSJ
Gibson J M3ZSJ
Gibson K GI7JAM
GIBSON K G4MIV
Gibson K MM3GIR
Gibson L G3RCX
Gibson L G6UMN
Gibson L G8VML
Gibson M GI7JEB
Gibson M MM0YMG
Gibson M M6VKG
Gibson M G1EUU
Gibson M G7VKG
Gibson M G8PFR
Gibson M M3YNS
Gibson N M3HUX
Gibson P G0EOK
Gibson P GM0GYN
Gibson P GW1ENG
Gibson P G1LDC
Gibson P G1WVS
Gibson P G7LJA
Gibson P 2E0TZO
Gibson P G1ZQN
Gibson P G4UVB
Gibson P G6EXC
Gibson P M0TZO
Gibson P M1DSE
Gibson P M3TZO
Gibson P MI3CAB
Gibson R G4WGR
Gibson R GI3NPP
Gibson S M6SRG
Gibson S 2M0TOR
Gibson-Ford K G6XDY
Giddens D G3IKB
Giddings B G1JLG
Giddings J G3XLB
Giddings N G8VZR
Giddings R GW0RCG
Gidman G M3EOT
Giering D M3NRV
Giffard J M3WNV
Giffen I GM4MIG
Giffin M M0DWW
Gifford A G8EZD
Gifford A G7BPX
Gifford B G0MPZ
Gifford G G1ACB
Gifford G G4PFK
Gifford M M1DLE
Gifford P G3AWP
Gifford R G7SGM
Gilbert A G0OMD
Gilbert A M0APH
Gilbert A G4ENW
Gilbert B G6IUS
Gilbert B G7OYF
Gilbert B G8EWF
Gilbert B G0BOO
Gilbert C MW3JEK
Gilbert C 2E0DLG
Gilbert D G0NFA
Gilbert D G3OYL
Gilbert D M6DZA
Gilbert D G1MHP
Gilbert E G3YBE
Gilbert E G8TMM
Gilbert I G0FNF
Gilbert J G0OFD
Gilbert J M1JES
Gilbert J M3CQP
Gilbert J M6JGZ
Gilbert J G0XZT
Gilbert J G7TVQ
Gilbert L G1CJC
Gilbert M MM6MJG
Gilbert P G7PXR
Gilbert R 2E0RBI
Gilbert R G0ROB
Gilbert R G3UEZ
Gilbert R G3YVI
Gilbert R M0RJT
Gilbert R M1DMN
Gilbert R M3RBI
Gilbert S M6GCO
Gilbert S G4UCJ
Gilberts A 2E0TCG
Gilberts A M6TCG
Gilbertson A G8UDZ
Gilbertson G G4SGU
Gilbey A G4YTG
Gilbey D G1DEQ
Gilbey D G1ITL
Gilbody C GI4XFS
Gilbody H GI4WVN
Gilbody J M6OCD
Gilboy N G0WPM
Gilbraith B G7MXQ
Gilby P G8AFU
Gilchrist D G1JRW
Gilchrist M MM6WGT
Gilchrist R G0TUE
Gilchrist R MM6CHV
Gilchrist S M3HJF
Gildersleve I G3YAR
Giles B M0BHG
Giles D G7OXA
Giles D M1CVM
Giles G G1ECY
Giles G G6VRU
Giles M G7KNM
Giles N G7GPL
Giles O 2E0OCG
Giles P M0DKU
Giles R G0VWV
Giles R G4LBH
Giles R M3ZAE
Giles S G4FDI
Giles T G1UPX
Giles T G4CDY
Giles-Holmes M G4IML
Gilfillan A G0FVI
Gilham A M6TID
Gilham D G7LNM
Gilham R G6OKB
Gilham S M6TSI
Gilhespy J M6KAS
Gilhooly D M3FMP
Gilhooly F GM0AXX
Gill A G0SNK
Gill A G8UXD
Gill A M6RTR
Gill C G0CCB
Gill C G3XKL
Gill C G8EEM
Gill D GW4YCO
Gill D G6IIK
Gill D M6ZRZ
Gill D G0FTR
Gill D GM0HLV
Gill J G0RQG
Gill J G3UAE
Gill J G6AIK
Gill K G1DRI
Gill K G4YDX
Gill K G7OMF
Gill L M3LFG
Gill M 2M0ALS
Gill M G0WAN
Gill M G3VJX
Gill M M5AGZ
Gill M G0KVS
Gill P G0WID
Gill P GD3YTE
Gill P G4IEV
Gill R 2E0CXP
Gill R 2E0GHR
Gill R 2E0AOK
Gill R G3CXP
Gill R G3NKJ
Gill R G3ROQ
Gill R G4KOY
Gill R G7BHG
Gill R M1CXP
Gill R M3GHR
Gill R G8DSU
Gill S 2M0SGQ
Gill S MM0SGQ
Gill S MM3SGQ
Gill T G7SLZ
Gill T G8IBO
Gill T M3SLZ
Gill W G0PPK
Gill W G1PFZ
Gillain L G4YEO
Gillam G G3ZHA
Gillard A G1CGP
Gillard A G1BXX
Gillard A M6TXU
Gillard A 2E0SSD
Gillard A M6YPX
Gillard B G4VVP
Gillard M M6MGX
Gillard S 2E0CCL
Gillard S M0GYP
Gillard S M6SGL
Gillatt D 2E0CIK
Gilleard T G8ZLF
Gillen K MM0CXB
Gillen K G0VRS
Gillen P G4GVW
Gillen S G0WYZ
Giller J G1JGT
Gillespie A G6MCN
Gillespie C MM3FGI
Gillespie D GI0OZQ
Gillespie F G7THI
Gillespie F GI7UPU
Gillespie H MI3DKN
Gillespie I MM0AMY
Gillespie I 2I0FBY
Gillespie W MM3FUG
Gillet E GW7PFK
Gillett A M3LUU
Gillett B G1SWI
Gillett B M6BTM
Gillett D G3WAG
Gillett S M6SGG
Gillham F G0KUU
Gilliatt R G8BGL
Gillies D GM1PKN
Gillies D GM4HSR
Gillies D MM0AMW
Gillies D MM6ZDG
Gillies G G7VAU
Gillies N G7BCK
Gillies S MM1FZR
Gilligan B M3ASZ
Gilligan D G1OGY
Gilliland B 2I0RGD
Gilliland B MI6BGD
Gilliland F GI1MJJ
Gilliland P G8SGF
Gilling R G0AHV
Gillingham D 2E0GBG
Gillingham D M6DMG
Gillingham I G0FLB
Gillingham M 2E0MJG
Gillingham M M0NCN
Gillingham M M3NCN
Gilliver J G6JPG
Gillman J G7OMQ
Gillmore D G0DER
Gillon A G1GWJ
Gillott D G4TMZ
Gillott J G6YOR
Gillott M G1KPZ
Gillott W G7BZE
Gillson I G1MZW
Gilman A G4GFD
Gilmore A G1ZHD
Gilmore D GI4KIX
Gilmore I G4FOS
Gilmore J M0CJN
Gilmore P M1EMU
Gilmore R 2I0RWG
Gilmore R MI0RGX
Gilmore R MI6BJG
Gilmore S MI0BAT
Gilmore T G0UXR
Gilmour D MM3UKG
Gilmour G 2M0GIL
Gilmour G MM0HLQ
Gilmour J GM8EJS
Gilmour J M6OTK
Gilmour S 2I0SNG
Gilmour S MI0SNG
Gilmour T GM1VZG
Gilmour T M6BQN
Gilowski E G7VHZ
Gilpin J M6CFZ
Gilpin J G6REA
Gilpin K M6FDU
Gilroy W G6YIW
Gilruth J GM1VWA
Gilruth S MM6SGQ
Gilson P G3WSZ
Giltrow N G8GJG
Gilzean I G8ZRD
Gimber G G7PMV
Gimbert B M6OUI
Ginever J 2E0JHG
Ginever J M0JHG
Gingell R G6BUY
Ginger N G1IFV
Ginger W M1BSY
Ginn R G7CQA
Ginsberg D GD0JGX
Ginsburg B G1INI
Ginsburg R G1INJ
Ginty M M6ICK
Giovinazzo J M3ZLL
Gipp D G4OCU
Girard J 2E0MBG
Girard J M6PWT
Girdwood S M6ELJ
Girgis M G1JRX
Girling C G8XCR
Girling D G4POT
Girling P 2E0ALD
Girling R G4FCD
Girt J G1HSL
Gisby A G0UZD
Gissing W 2E0IVO
Gissing W M0IVO
Gissing W M3IVO
Gittens A 2E0HHA
Gittings B M3VAG
Gittins E G6YDT
Gittins E M3YDT
Gittoes T G7JYJ
Giudice G G6TQH
Giudice M M3KSI
Giuliani G G0WMX
Giuliano N 2E0MNG
Giuliano N M6MNG
Given D 2I0ITY
Given D MI0TUB
Given J M6JBZ
Given P G3ZDK
Given T MM3PHC
Givens A GM3YOR
Givens R M0HQM
Gizzi F G6XQR
Glacken K 2M0WRX
Glacken K MM0KJG
Glacken K MM3KJG
Gladden R G4IRY
Gladders M 2E0GLA
Gladders M M3NUH
Gladding N MW6NLG

Gladman A ... 2E0KIT
Gladman A ... M6ADG
Gladman C ... M0AWN
[illegible]
Gladman T ... M6AMI
Gladman T ... M1CEA
Gladwin D ... G4WLV
Gladwish D ... G6HVX
Glaisher P ... G4RWW
Glaister B ... MD3OKH
Glaister C ... MD6ZEE
Glaister H ... GD0BCN
Glaister K ... MD3OKG
Glandfield P ... M0PSE
Glanville J ... G3TZG
Glanville S ... G0OGL
Glaser A ... G3ZEN
Glasgow A ... MI6HGI
Glasgow G ... 2E0COD
Glasgow R ... GM4UYZ
Glashan A ... GM4JCM
Glasper M ... MM6GYL
Glasper P ... GM0BKC
Glasper P ... MM6BKC
Glass C ... 2E0DPL
Glass C ... M0HYH
Glass D ... G7PUK
Glass J ... G4OJG
Glass S ... 2E0JBK
Glass S ... G4HSK
Glass S ... M3JBK
Glasscock L ... G6IGK
Glasscott E ... G3TSF
Glaysher K ... 2E0KBG
Glaysher K ... M6KBG
Glazebrook D ... M3DMG
Glazebrook K ... G0DPO
Glazier M ... G1HSI
Glazzard N ... M6ZGZ
Glazzard S ... G1XYO
Gleadall S ... M0BBW
Gleadell D ... M1RIG
Gleaden N ... M3ZXA
Gleave A ... M6GEG
Gleave B ... G1JPT
Gleave D ... G6JPT
Gleave R ... 2E0GGI
Gleave R ... M6CPX
Gleave W ... G8YWK
Gledhill K ... M3KKG
Gledhill O ... M6BFV
Gledhill P ... G7BHE
Gledhill R ... G3ZYN
Gledhill V ... G4JIN
Gleed B ... G0IOU
Gleed L ... M3YJJ
Gleek D ... GD0MOX
Gleek V ... G4WIS
Gleeson J ... M3MDI
Gleeson J ... M0JPG
Gleeson J ... M3ZBX
Gleeson R ... G6TCD
Glen A ... G0OQR
Glen M ... G0DQS
Glen S ... MM6AXT
Glenday T ... M6ERS
Glendinning H ... M6ZTA
Glendinning K ... GM4EZJ
Glendinning M ... G7GIS
Glendinning S ... GI7JKM
Glenn C ... G4ZCR
Glenn C ... MI6CGQ
Glenn D ... M3HBX
Glenn I ... G8MEX
Glenn J ... G7RIE
Glenn M ... M6NMG
Glenn W ... GI4KUM
Glennie H ... MM6HIG
Glennon J ... GM0ZAM
Glennon M ... G4JVZ
Glew W ... G4NEG
Glicklich J ... 2E0IRN
Glicklich J ... M6NAV
Gliddon B ... G4NGB
Glitsun C ... M3WXW
Gloess C ... M6CRG
Gloistein M ... GM0HCQ
Glossop I ... G7ILG
Glossop R ... M3GUF
Glotham G ... G0CLX
Glover A ... 2E0GNN
Glover A ... M3GNN
Glover C ... G6OKA
Glover D ... G6AZP
Glover D ... G6RMA
Glover D ... 2E0HNF
Glover D ... M6DPG
Glover D ... M6RRS
Glover F ... G6HWA
Glover H ... M1HLG
Glover J ... [illegible]
Glover J ... G3FIC
Glover J ... G4TOX
Glover J ... G8YEJ
Glover J ... [illegible]
Glover J ... 2E0JGE
Glover J ... M6JGE
Glover K ... G7HHN
Glover K ... M1DRB
Glover M ... G0BKJ
Glover M ... G0ISK
Glover M ... M3MMG
Glover M ... G6JPR
Glover M ... G0PDM
Glover N ... G6XDZ
Glover P ... M0CNL

Glover R ... M0BJX
Glover R ... G8IUC
Glover R ... G0WGP
[illegible]
Glover T ... G6MSC
Glover W ... G4BQW
Glover W ... M0AAN
Glydon P ... G4XML
Glyn R ... G0JAG
Glynn A ... G4WZS
Glynn J ... M6UAF
Glynn M ... G3AAS
Glynn R ... 2E0TTW
Glynn R ... M6EJF
Glynn R ... G0NFR
Goacher D ... G3LLZ
Goacher J ... G1WLX
Goacher J ... M0GOA
Goad G ... G0SIO
Goadby C ... G8HVV
Goadby D ... G8BZN
Goadby P ... G3MCP
Goan S ... G1AUU
Goatman A ... G1SEW
Goben P ... G4BVV
Gobey A ... 2E0YSP
Goble C ... G4OZX
Gocher R ... M6BMV
Godber P ... G4YTF
Godbold P ... G4UDU
Godbold S ... G0NRX
Goddard A ... G1STK
Goddard A ... G7TEA
Goddard A ... M0TEA
Goddard B ... G7IYI
Goddard C ... G4XAN
Goddard F ... G4OVS
Goddard G ... G0GNW
Goddard G ... G6DDU
Goddard G ... M3UBB
Goddard J ... G0JOM
Goddard J ... M6TYK
Goddard K ... M6KGK
Goddard M ... G0OJU
Goddard M ... G6MCY
Goddard N ... G0OAS
Goddard N ... G3UXR
Goddard P ... 2E0PFG
Goddard P ... M3VFE
Goddard R ... M0AZB
Goddard T ... 2E0TUK
Goddard T ... M6LUA
Godden C ... G4BXI
Godden D ... M0DAG
Godden I ... G4CZX
Godden L ... G1HSJ
Godden M ... G0ACQ
Godden N ... G7GSC
Godding C ... M6CPA
Godding D ... G0VHK
Godding J ... 2E0CAJ
Godding J ... M1JON
Godding J ... M3JDG
Godfrey B ... G0OVC
Godfrey C ... M3ZFI
Godfrey D ... G0KIU
Godfrey D ... M6MTC
Godfrey J ... M3YNE
Godfrey J ... 2E0HVZ
Godfrey J ... M0WWD
Godfrey J ... 2E0PSW
Godfrey J ... M0JOH
Godfrey J ... M0PSW
Godfrey J ... M3PPG
Godfrey J ... 2E0JSG
Godfrey K ... GW3VEW
Godfrey P ... G4BAN
Godfrey P ... G8JBD
Godfrey P ... M3CZX
Godfrey S ... G7AJR
Godfrey T ... M3YRS
Godfrey T ... M6VPZ
Godley A ... G7IWV
Godlieb E ... G4XRG
Godlington I ... G7BJE
Godney A ... M0CBG
Godolphin A ... 2E0EIO
Godolphin P ... 2E0AOK
Godolphin P ... G4XTA
Godrich S ... G7OOT
[illegible]
Godsiff A ... G4NMT
Godward C ... G1GDJ
Godwin A ... G0WYN
Godwin C ... 2E0BSW
Godwin D ... G7NJG
Godwin E ... G0PCB
Godwin J ... G0NHG
Godwin K ... G0PCA
Godwin N ... G6XQT
Godwin P ... G4III
Godwin R ... G3VDH
Godwin W ... G8NXY
Godzisz H ... M6DOG
Goff I ... M3LNY
Goff M ... M0MLG
Goff M ... M3MLG
Goff R ... G1EDA
Goff R ... G4FON
Goffin L ... G0VDR
Goffin L ... G7RPK
Goh T ... M6TSG
Gohill D ... G7PJG
Gohl M ... G7BAC
Golaszewski H ... MW3WWO
Gold A ... G3SKR

Gold J ... G1PFY
Gold M ... G1VNV
Goldbey J ... G4DUW
Goldby M ... 2E0DUW
Golden A ... M6INA
Golder P ... G0NGA
Goldfinch J ... M3PQG
Goldie A ... GM0DEX
Goldie J ... G0UVT
Golding A ... G3UKD
Golding B ... G6AUR
Golding Brown A ... G1OPJ
Golding D ... G8LNC
Golding D ... 2E0BPL
Golding D ... M3WDV
Golding F ... G8MIF
Golding H ... G4SPT
Golding H ... G7UOD
Golding J ... G7VHX
Golding J ... M1DNQ
Golding M ... M6MGU
Golding N ... G6RIG
Golding P ... G1KLW
Golding R ... G3VZG
Golding R ... G7BQA
Goldingay C ... G4DFC
Goldman M ... G4LCB
Golds P ... G1ZZC
Goldsbrough D ... M3ASC
Goldsbrough J ... M0BYV
Goldsbrough S ... M6DIF
Goldsmith A ... 2E0GOL
Goldsmith A ... M0NKR
Goldsmith A ... M3UYG
Goldsmith C ... M6FQH
Goldsmith F ... 2D0FHG
Goldsmith J ... G0DLW
Goldsmith J ... G4KTX
Goldsmith J ... M6JAU
Goldsmith L ... 2E0LJG
Goldsmith L ... M0LJD
Goldsmith L ... M6LJD
Goldsmith P ... G8ISM
Goldsmith T ... M0GOL
Goldspink A ... G0GHE
Goldstein A ... MD6FQF
Goldstraw A ... G0PSH
Goldstraw J ... MM1EDY
Goldstraw W ... G0DTQ
Goldsworthy K ... M6BFL
Goldsworthy T ... G4BHD
Goldthorpe D ... 2E0IAZ
Goldthorpe D ... M6IAZ
Goldthorpe M ... M6GOL
Goldthorpe P ... G0AQS
Goldup R ... M6FHS
Goligher R ... GI4LIF
Golightly G ... G0IGH
Golightly J ... G0IGC
Golightly J ... G6KEH
Golik D ... M6PNG
Golland B ... G7GMR
Golley C ... G4JYF
Gollins T ... M6TPG
Golsby R ... M0DOH
Gomez C ... G6BYF
Gomez Lozano E ... 2E0DKG
Gomez Lozano E ... M0KLB
Gomez Lozano E ... M6ATD
Gonczarow P ... 2E0GON
Gonczarow P ... M6PGW
Gonsalves T ... G0OYJ
Gonzales N ... 2E0NJE
Gonzalez N ... M3NFG
Gooch J ... M6ZZK
Gooch L ... G6PTI
Gooch M ... M3WLX
Gooch R ... M0DWG
Gooch R ... M3HZK
Good P ... G7JME
Good P ... G7HCL
Good P ... M0EBG
Good T ... G7MIT
Good W ... G7VQO
Goodacre K ... M0BQH
Goodale J ... M0HAU
Goodall A ... G4RFP
Goodall A ... G7MAV
Goodall B ... 2E0JRC
Goodall B ... G8BUB
Goodall D ... 2E0DDG
Goodall D ... G0GYH
Goodall D ... G6VZS
Goodall F ... 2E0BKY
Goodall F ... M3IQI
Goodall J ... G0SKR
Goodall J ... G0BOR
Goodall M ... G0MGI
Goodall N ... GM7GNO
Goodall P ... M3XHK
Goodall D ... C0ONQ
Goodall R ... GM0UGZ
Goodall R ... 2E0ITN
Goodall R ... GW0RVR
Goodall R ... M6ITN
Goodall S ... 2E0FTA
Goodayle P ... G7IOI
Goodayle P ... M3PJG
Goodburn P ... M3OVC
Goodby L ... M3CPY
Goodchild A ... M3FSS
Goodchild K ... G1FBU
Goodchild K ... G6EFO
Goodchild P ... M3PGO
Goodchild P ... G4CLL
Goodchild R ... G8FTW

Goode A ... G4TXE
Goode A ... G8LCI
Goode C ... G4HHY
Goode D ... G0HCP
Goode M ... G4SMA
Goode N ... M6NGD
Goode P ... G7HOE
Goodearl G ... 2E0FPI
Goodearl M ... G4XNO
Gooden R ... G0CWW
Goodenough A ... 2E0MKC
Goodenough P ... G3TJS
Goodenough T ... M6FXL
Gooderham R ... M1RJG
Goodes F ... G7IBN
Goodes M ... G1LHD
Goodey J ... G6XSY
Goodey M ... G0GJV
Goodfellow D ... MM1CFC
Goodfellow D ... 2E0XQK
Goodfellow D ... M3XQK
Goodfellow P ... G4KUQ
Goodfellow P ... G8SHR
Goodfellow W ... 2M0WDG
Goodfellow W ... MM6FXZ
Goodfield G ... GW4CNL
Goodger F ... G0GOX
Goodger P ... G0BAI
Goodhall P ... 2E0SQL
Goodhall P ... M3PHP
Goodhand C ... M6CMG
Goodhand C ... G6MWD
Goodhand C ... M5ADA
Goodhand P ... M6WZK
Goodhew B ... G8ONY
Goodier B ... G1SYV
Goodier B ... G0PRM
Goodier G ... G6PMW
Goodier I ... G0UWK
Goodier J ... G4KUC
Goodier R ... M6RCF
Goodings A ... G1GER
Goodings J ... G8VAD
Goodison D ... G0LUH
Goodland T ... GM4LER
Goodliffe J ... G7PLE
Goodman B ... M6GBO
Goodman D ... GI4JFP
Goodman D ... M6OLY
Goodman E ... G4LEM
Goodman I ... G8TWZ
Goodman I ... M6IVR
Goodman J ... M3VLB
Goodman J ... 2E0EHF
Goodman J ... G3WOA
Goodman J ... G4PIJ
Goodman J ... M0RVJ
Goodman J ... M6GIS
Goodman M ... 2E0FBE
Goodman M ... G4UQA
Goodman M ... M6FTF
Goodman N ... G4XFG
Goodman P ... G4LKT
Goodman P ... G7ALR
Goodman R ... G3KOB
Goodrich G ... G4NLA
Goodrich M ... G3YAD
Goodrich M ... M0AOM
Goodrick M ... G3ROP
Goodridge A ... G7WDN
Goodridge G ... G1KTS
Goodridge L ... G7STD
Goodridge M ... GW0VND
Goodridge P ... 2E0XKO
Goodridge P ... M0XKO
Goodridge P ... M3XKO
Goodridge S ... 2E0SJG
Goodrum M ... G3ZQU
Goodson J ... GW6VET
Goodson J ... M6EKS
Goodson P ... G4PCF
Goodway D ... G8DAM
Goodwill D ... G1VAB
Goodwin A ... M3FDV
Goodwin C ... G3WTT
Goodwin D ... G0LJW
Goodwin D ... G8KSC
goodwin D ... M0BDU
Goodwin D ... MM3WKF
Goodwin D ... M6WDG
Goodwin D ... G4SOT
Goodwin E ... G0MSW
Goodwin E ... G1XHR
Goodwin E ... 2E0PAX
Goodwin E ... M6UVF
Goodwin F ... G0SPL
Goodwin F ... G8AGB
Goodwin G ... G0IHA
Goodwin G ... G0LNB
Goodwin H ... G1NSG
Goodwin J ... G6CNX
Goodwin J ... [illegible]
Goodwin J ... G0PRF
Goodwin J ... M6JGG
Goodwin L ... G6ANO
Goodwin M ... G3WKH
Goodwin M ... G7NBE
Goodwin M ... M3BBB
Goodwin M ... G8VAN
Goodwin M ... M3XEF
Goodwin P ... 2E0CPQ
Goodwin S ... G0RYM
Goodwin S ... G1VWP
Goodwin S ... M0SJG
Goodwin T ... G0HOA
Goodwin T ... G7TBM

Goodwin W ... 2E0IWG
Goodwins R ... GM6HVY
Goody B ... G7VGC
Goody J ... M1JOS
Goody R ... 2E0RCF
Goody R ... M3OBO
Goody T ... GM0MXZ
Goodyear B ... G6AUP
Goodyear J ... 2E0CGS
Goodyear J ... M3ETH
Goodyer G ... G6NMQ
Goodyer K ... M6KAG
Goodyer T ... G0ATG
Goold A ... 2E0GLD
Goold A ... M3ORB
Goolden J ... M6GOO
Goolding B ... G0UVN
Goom M ... M1ACN
Goozee H ... G6GFJ
Gopan R ... 2E0XUU
Gopan R ... M0XUU
Gopan R ... M3XUU
Gopikrishna Y ... MD3YGK
Gorbutt T ... M3ZHV
Gorczynski J ... MM0LGS
Gordon A ... 2M0SOP
Gordon A ... GM3GKJ
Gordon A ... G4BCT
Gordon A ... GM4PCT
Gordon A ... G3XOI
Gordon A ... G4TTB
Gordon A ... M3RAA
Gordon A ... G7GJV
Gordon A ... G6RXQ
Gordon C ... M0IIM
Gordon C ... M6ENU
Gordon D ... G0MRD
Gordon D ... G6ENT
Gordon D ... G6ENN
Gordon D ... M3UKY
Gordon E ... G4VMU
Gordon F ... 2M0COT
Gordon F ... GM3ALZ
Gordon F ... MM0ODL
Gordon F ... MM6ALZ
Gordon G ... GM6KFO
Gordon G ... M6GKG
Gordon H ... GM7KRQ
Gordon I ... G6ENU
Gordon I ... G8IFT
Gordon J ... G4LIA
Gordon J ... G7WGI
Gordon J ... M3WQN
Gordon J ... G4XUI
Gordon K ... G4AQJ
Gordon K ... M3KGG
Gordon K ... M3KSG
Gordon M ... G4BRW
Gordon N ... GW1AUT
Gordon P ... GD3GCE
Gordon R ... G3UHJ
Gordon R ... G4HCG
Gordon S ... G6ENS
Gordon S ... GW7GKN
Gordon S ... GI7GRY
Gordon S ... 2E0SEB
Gordon S ... M0STT
Gordon S ... 2E0MLG
Gordon S ... M6GXU
Gordon T ... G6MWB
Gordon W ... G3SEG
Gordon W ... GI4DXK
Gordon-Laycock W ... G3XYD
Gordon-Smith D ... G3UUR
Gore J ... M6AGZ
Gore R ... M6AEZ
Gore S ... M1BZF
Gore S ... 2I0VTZ
Gorecki L ... M6CMB
Gore-Thorne G ... M1EHF
Gore-Thorne G ... M3ANH
Gorlinski J ... M0RMT
Gorman A ... MM6EOU
Gorman B ... G1VTQ
Gorman C ... MI6CBG
Gorman F ... MM6FTG
Gorman G ... 2I0LJW
Gorman G ... MI3XOI
Gorman J ... M3WJN
Gorman P ... M6PMG
Gormley V ... G4GVG
Gornall A ... G6IVB
Gornall D ... M3FSQ
Gornall L ... GI1BZT
Gornall R ... G7DME
Gorny V ... G4HHL
Gorringe K ... 2E0DWW
Gorse C ... 2E0HPI
Gorse C ... M6HPL
Gorse D ... G7HRQ
Gorsuch I ... G1AOX
Gorton A ... [illegible]
Gorton J ... GD4JMG
Gorton J ... G4UTJ
Gorton R ... G3NIQ
Gorton H ... G6VPH
Gorwills S ... G8NKN
Gosal-Tooby E ... M6WND
Gosbee K ... G6NTW
Gosby S ... G8OVZ
Gosden A ... G7GDC
Gosi M ... 2E0DZN
Gosi M ... M6HMU
Goslin M ... M6FWT
Gosling B ... G6KVI
Gosling D ... M3LOE

Gosling J ... M3MFZ
Gosling K ... G1LOE
Gosling M ... M3VDL
Gosling M ... M3MWG
Gosling P ... 2E0BHU
Gosnell P ... G0PLC
Gosney K ... M6EZV
Gospel D ... G6IGO
Goss G ... G6DEA
Goss G ... G0RKS
Goss K ... M6EIF
Goss M ... M3FDM
Gostelow P ... 2W0HOH
Gostelow P ... MW6HOH
Gostick P ... G0SJV
Gostling N ... 2E0GOS
Gostling N ... M3GOZ
Gotch D ... G8FTX
Gotch M ... G0IMG
Gott G ... G3MUO
Gott G ... G6PFJ
Gott I ... G1KKS
Gott R ... M6RGF
Gotts M ... G7HHT
Goudie W ... GM4WXQ
Gough C ... G1URR
Gough C ... 2E0CBG
Gough D ... GD0DZH
Gough D ... 2E0DWG
Gough D ... M6GVJ
Gough E ... G4ZEY
Gough I ... 2E0IMG
Gough J ... GI0TQD
GOUGH J ... GW3WXA
Gough J ... M1CWA
Gough K ... GW1SRB
Gough L ... G0CWO
Gough L ... GI1BEU
Gough M ... M3NTI
Gough M ... M3NTQ
Gough N ... G3AWK
Gough N ... G7BMP
Gough P ... M0PET
Gough P ... 2W0CDZ
Gough P ... MW6PDG
Gough R ... G0BET
Gough R ... G8RUX
Gough W ... M6WNE
Goulbourn G ... G3TBG
Goulbourne D ... G4EHK
Goulbourne D ... G0NPK
Goulbourne M ... M0COO
Gould A ... G4UAM
Gould A ... M6EQV
Gould B ... G0FZE
Gould B ... M3MNQ
Gould D ... G3UEG
Gould D ... M0DJD
Gould E ... G7KGI
Gould G ... M0AZJ
Gould J ... GI3SUM
Gould J ... G3WKL
Gould J ... G4POD
Gould J ... M1DPO
Gould J ... G6JPQ
Gould M ... G4OKE
Gould M ... M1AZA
Gould M ... M3PIK
Gould P ... G4FVZ
Gould P ... G6DBY
Gould P ... G7PAG
Gould P ... M6FEU
Gould P ... 2E0EDY
Gould R ... M0CFR
Gould R ... G8PSO
Gould S ... G0VJJ
Goulden D ... G1JZU
Goulden J ... G0TBM
Goulding C ... 2E0CAR
Goulding J ... 2E0BDV
Goulding J ... M3MVM
Goulding L ... G4EPW
Gould-Martin J ... M6JGM
Gouldstone R ... G3TAG
Goulsbra D ... G4UHZ
Goulty C ... M3MFG
Goulty J ... G8BBV
Gourlay I ... MM0HRL
Gourlay M ... 2M0MMG
Gourlay M ... MM0MMG
Gourlay M ... MM3IQD
Gourley D ... G0MJY
Gourley G ... G0OZJ
Gourley J ... M6JEV
Govan C ... M0CUG
Govan I ... 2E0BZV
Govan I ... M3ZNT
Gove C ... GM4LCJ
Govier J ... G6YJD
Gow A ... 2E0IFM
Gow A ... 2M0DKF
Gow A ... MM0JWT
Gow B ... 2E0IJL
Gow H ... G3LAG
Gow K ... 2E0IFN
Gow P ... G7VHJ
Gow R ... 2E0KID
Gowans A ... M3JXY
Gowans N ... M6GOW
Gow-Barber A ... M6EEG
Gowen P ... G3IOR
Gower B ... G8NNP
Gower D ... G1ROH
Gower D ... G4GRJ
Gower G ... G0VAY
Gower I ... G8VHG

Gower J ... M0MGF
Gower R ... M5RHG
Gower R ... M6RQB
Gowers D ... G0IZV
Gowers H ... 2E0UVB
Gowers H ... M0UVB
Gowers R ... M6UVB
Gowing C ... 2E0GOW
Gowing C ... M0GOW
Gowing H ... G3BNP
Gowland G ... G1GCY
Gowler R ... 2E0NFK
Gowler R ... M6WNW
Goy A ... G4HJD
Goy S ... G1PLU
Goyder D ... G2SZ
Gozzard C ... 2E0CEO
Gozzard C ... M0CDG
Gozzard C ... M6GOZ
Gozzard D ... M6DIG
Grabham E ... M6TVH
Grabianski A ... M1SKI
Grace A ... M3GRF
Grace A ... G0OFZ
Grace D ... GW4OUU
Grace D ... G3VM
Grace D ... M6SDT
Grace G ... 2E0TPN
Grace J ... G3VVR
Grace P ... G4MPG
Grace P ... G8XXZ
Grace P ... MW0DCT
Grace-Bolton S ... M6XNU
Gracey V ... GI3WEM
Gracie J ... MM3JGR
Gracie P ... GD1FBS
Gradwell W ... M3YWG
Graffham K ... 2E0HAQ
Graffham K ... M0GWE
Graffham K ... M3KEV
Graffham M ... G6AUO
Graffham O ... G1JDE
Gragon G ... 2E0BAB
Graham A ... 2W0OAG
Graham A ... G7OCC
Graham A ... MW0YAG
Graham A ... MM3DFG
Graham A ... MW6YAG
Graham A ... M6HWW
Graham A ... 2M0PMR
Graham A ... M3YJQ
Graham A ... G3TXL
Graham A ... M6EJR
Graham B ... G4LGB
Graham C ... G3XIG
Graham C ... MM3DZW
Graham C ... M3PGY
Graham C ... 2M0YCG
Graham C ... MM3YCG
Graham D ... GW0IXQ
Graham D ... G0VGJ
Graham D ... G6XSZ
Graham D ... G7SKV
Graham E ... 2E0THS
Graham E ... G0TVV
Graham G ... GM4ZHL
Graham G ... MI6MSG
Graham H ... G1AUR
Graham H ... GI6JPO
Graham I ... M3MZO
Graham I ... M6IBF
Graham J ... G0BVG
Graham J ... G0GHB
Graham J ... GM1XPE
Graham J ... G3HDT
Graham J ... G4LIC
Graham J ... G6HFW
Graham J ... GM7UVS
Graham J ... GW6CNS
Graham K ... G7IVG
Graham K ... G8ZWU
Graham K ... M0KGM
Graham K ... G7ELH
Graham K ... G0AVB
Graham L ... M6XLG
Graham M ... G6EXE
Graham M ... MI3OHF
Graham M ... MI6TMR
Graham M ... G3XMG
Graham N ... GI3RXV
Graham N ... M3UNG
Graham N ... G8ENY
Graham P ... 2E0PGA
Graham P ... G0PJY
Graham P ... G6JZE
Graham P ... 2E0RAK
Graham P ... M0PGX
Graham P ... M0FIL
Graham R ... C0ESW
Graham R ... G3OAY
Graham R ... G7PTH
Graham R ... MI0CRY
Graham S ... GM4ZFS
Graham S ... GM7DRY
Graham S ... G0THS
Graham S ... G0TII
Graham S ... M0BCV
Graham T ... GM7UFN
Graham T ... G8OBT
Graham T ... M3TCG
Graham T ... M3TXL
Graham W ... GM1BNS
Graham W ... GM1SYC
Graham W ... MI0JAY
Graham Z ... MM3ZDG
Grahame I ... 2E0IBS

Grainge B ... G3JPM
Grainger B ... G4TOG
Grainger B ... M6TGC
Grainger D ... G4UQM
Grainger J ... G0JZL
Grainger J ... M3UVA
Grainger I ... M3RLG
Grainger M ... M1DMX
Grainger M ... GI7UCS
Grainger M ... G0VQB
Grainger P ... 2E0PGR
Grainger P ... G0SLN
Grainger P ... G4XWR
Grainger P ... M0PEG
Grainger R ... M6RRG
Grainger S ... 2E0BDO
Grainger S ... M3CCJ
Grainger S ... M3KZW
Grainger S ... M3NFQ
Grainger T ... G4GGL
Granatt M ... M0RYK
Granby P ... GW4OKF
Grandfield H ... G0DOU
Grandshaw R ... G7WBA
Grane J ... G6MCQ
Grange M ... G4JJX
Grange T ... G7MXL
Grannell P ... G4TQB
Grannon A ... M3KUH
Granshaw A ... G6AZR
Grant A ... G1VAG
Grant A ... G4KPD
Grant A ... M0VAG
Grant A ... M6LCV
Grant C ... G0AHO
Grant D ... G4UAY
Grant D ... G8GJQ
Grant E ... MM6EER
Grant F ... GM0PKQ
Grant G ... GM3UKG
Grant G ... M0TTO
Grant G ... MM0CUG
Grant I ... G4MWG
Grant J ... G3SQN
Grant J ... G3TYA
Grant J ... M6GRT
Grant J ... M0CDQ
GRANT J ... G4OVT
Grant L ... G3TJU
Grant L ... MM6EEX
Grant M ... GM0JVC
Grant M ... GW7AFC
Grant P ... G6MRW
Grant P ... G4YDW
Grant P ... G4OEB
Grant P ... G0UGW
Grant P ... G1NKT
Grant R ... GM4DQJ
Grant S ... 2E0SJG
Grant S ... GM4YHS
Grant S ... GM7KHA
Grant S ... GW3UWL
Grant S ... G6ENR
Grant T ... M0TDG
Grant T ... G4VKJ
Grantham D ... M3YRB
Grantham E ... G7AKV
Grantham G ... G8PQB
Grantham M ... 2E0MLV
Granville M ... GI8AFS
Graph A ... G4YGB
Graseley F ... M0CKX
Grassby N ... G4CPY
Grassi M ... G0PRH
Grattan H ... G3XKS
Grattan K ... GD4RGR
Gratton E ... G4OCN
Gratton G ... G1HCU
Gratton T ... G3TGG
Gratton W ... M3YPG
Graupner L ... GM0GNY
Gravel L ... M3ZPO
Gravell A ... GW8VUV
Gravell A ... 2E0TPG
Gravell A ... M0TPG
Gravell A ... M6TBG
Graven N ... 2E0NDG
Graven N ... M6NDG
Gravenor W ... G0KVM
Graver A ... G8SEY
Graver J ... G7VYT
Graves A ... G1KHU
Graves D ... G0AYP
Graves J ... G0KSJ
Graves J ... G8NHM
Graves J ... G7CJO
Graves L ... G4BCP
Graves M ... G7RUN
Gravoc N ... GM1WKH
Gravestock W ... M1EGZ
Gray A ... 2W1CNN
Gray A ... GM0ART
Gray A ... G0MYA
Gray A ... GW1BCI
Gray A ... G1XGW
Gray A ... GW1GNI
Gray A ... G4UFV
Gray A ... G7BWF
Gray A ... G7GRQ
Gray A ... G4DJX
Gray A ... MM6THU
Gray A ... M6XAO
Gray B ... G0GRR
Gray B ... G4TFB
Gray B ... G6WME
Gray B ... MU0EDN

UK Surnames

**IMPORTANT NOTE**

**Revalidate licence to avoid revocation** – Ofcom has advised the Society that plans will be drawn up to revoke licences that have not been revalidated as required by the licence conditions. The quickest way to revalidate is to do so online via the Ofcom website: *https://services.ofcom.org.uk/* or by email: *amateur.validations@ofcom.org.uk* If you need assistance in the process, Ofcom staff are available to help, but please be patient during times of heavy workload.

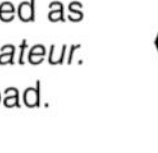

| | |
|---|---|
| Gray C | 2E0ECL |
| Gray C | M0ZNP |
| Gray D | G0FLX |
| Gray D | G1JDF |
| Gray D | GJ3XOJ |
| Gray D | G3YPL |
| Gray D | G4FQV |
| Gray D | G4TFC |
| Gray D | M0DLL |
| Gray D | M3TYW |
| Gray D | G7CNC |
| Gray D | G0LEA |
| Gray D | M6GSZ |
| Gray D | 2E0BSQ |
| Gray D | M0GMD |
| Gray D | M6EMG |
| Gray E | G0CTZ |
| Gray G | G1HMT |
| Gray G | G1KQU |
| Gray H | G1MNP |
| Gray I | G0SNU |
| Gray I | MW0CXW |
| Gray I | G3VAJ |
| Gray J | G0ASL |
| gray J | GM3PLO |
| Gray J | GM3LRG |
| Gray J | GW6ZUS |
| Gray J | GM7PBB |
| Gray K | G0LFE |
| Gray K | MM0AWJ |
| Gray K | G1SCB |
| Gray L | G1HTT |
| Gray L | G3FTK |
| Gray L | M3OHR |
| Gray L | G8OUG |
| Gray M | 2E0FJL |
| Gray M | G1HNU |
| Gray M | G6CKY |
| Gray M | M0ABK |
| Gray M | 2E0MKG |
| Gray M | M0MKG |
| Gray M | M3CZE |
| Gray M | M6MGY |
| Gray M | G0OXY |
| Gray M | G4EPU |
| Gray P | G1FLL |
| Gray P | G0HYT |
| Gray P | GM0FWY |
| Gray R | 2E0RFG |
| Gray R | G0DOB |
| Gray R | GW1NWF |
| Gray R | G4AWO |
| Gray R | G6SVV |
| Gray R | G7HIX |
| Gray R | M3RFG |
| Gray S | 2W1EYZ |
| Gray S | G0ASK |
| Gray S | G0RKD |
| Gray S | G6XQX |
| Gray S | G7LHS |
| Gray S | MM6SGF |
| Gray S | M6SII |
| Gray S | M0MPY |
| Gray T | G6VEG |
| Gray W | MM1DTN |
| Gray W | 2E0WEG |
| Graydon D | G1EDE |
| Grayer G | G3NAQ |
| Gray-Jones R | GM6URP |
| Gray-Jones S | 2M0URP |
| Gray-Jones S | MM6URP |
| Grayshon P | G1AOR |
| Grayson D | G1UDE |
| Grayson D | 2E0IUK |
| Grayson D | M0IUK |
| Grayson D | M6IUK |
| Grayson E | G6OJX |
| Grayson I | G3RYK |
| Grayson M | G4OTE |
| Grayson R | MW3URG |
| Grayson T | M6CZW |
| Grayson V | GW0HYH |
| Grayson V | GW8JDB |
| Gray-Thompson S | GM7WHQ |
| Gready J | GJ6ENP |
| Greany J | G3OWX |
| Greatbatch A | G7NHZ |
| Greatbatch D | G4KCU |
| Greathead A | G3ZID |
| GREATHEAD I | 2E0CEG |
| Greatorex D | GM0AEG |
| Greatorex E | 2E0LIZ |
| Greatorex E | 2E0LIZ |
| Greatorex E | M3ZIL |
| Greatorex K | G0THF |
| Greatorex M | M1LLL |
| Greatorex P | G4FEM |
| Greatorex S | 2E0SAZ |
| Greatorex S | M3SAZ |
| Greatrex A | GW4OQB |
| Greatrex M | GW4HDB |
| Greatrix B | G4ICZ |
| Greatrix G | G7HNM |
| Greatrix J | M3MIS |
| Greatrix R | M3RHG |
| Greatwood V | 2E0HRD |
| Greatwood V | M0VHG |
| Greatwood V | M6HRD |
| Greaves A | G3JOX |
| Greaves B | M6SIR |
| Greaves C | M0CTQ |
| Greaves D | 2E0IDG |
| Greaves D | M6DYA |
| Greaves F | G7POA |
| Greaves G | M6GVS |
| Greaves J | 2E0JMG |
| Greaves J | G3UXM |
| Greaves K | G0PVE |
| Greaves N | G4VET |
| Greaves R | G4JVV |
| Greaves R | 2E0ZRG |
| Greaves R | M3SGV |
| Greaves R | G0MOH |
| Greaves S | M3XZS |
| Greaves S | 2E0XAY |
| Greaves S | M6SEV |
| Grech P | M1HMP |
| Grech-Cini W | G1PZA |
| Greco V | M3JQN |
| Greed P | G1HDG |
| Greed P | G3MQD |
| Greed P | G3PZV |
| Greed W | G0MZQ |
| Greed W | G4GBX |
| Green A | 2E0RAG |
| Green A | G0CRE |
| Green A | G3PBR |
| Green A | GW4JGU |
| Green A | G6BFM |
| Green A | G7HSB |
| Green A | G8BNG |
| Green A | M0RTE |
| Green A | M3FZS |
| Green A | M3LEX |
| Green A | M3TIY |
| Green A | M3XDQ |
| Green A | G3UZF |
| Green A | G4ZWO |
| Green A | M1AEG |
| Green A | M6RNV |
| Green A | G0OJJ |
| Green A | G3TRL |
| Green A | M6TBK |
| Green A | M6YNN |
| Green B | G1EVI |
| Green B | G3KCB |
| Green B | G4HYZ |
| Green B | G6IRU |
| Green B | M3EVJ |
| Green C | 2E0CVG |
| Green C | GM4VUG |
| Green C | M3FWJ |
| Green C | M6AYU |
| Green C | GW1WTZ |
| Green C | G4SAJ |
| Green C | G3OPX |
| Green C | G8VJO |
| Green D | 2E0BQO |
| Green D | 2E0HKY |
| Green D | G0LJG |
| Green D | G4ABY |
| Green D | G4ZFV |
| Green D | G6XEB |
| Green D | G8HPV |
| Green D | M0HPV |
| Green D | MM0VWR |
| Green D | M1EGW |
| Green D | M1FAJ |
| Green D | M3DVG |
| Green D | M3IQF |
| Green D | M3SBS |
| Green D | M0DRG |
| Green D | 2W1FUD |
| Green D | G1VLU |
| Green D | G4OTV |
| Green D | G6ZBT |
| Green D | MM3WFU |
| Green D | M6MDG |
| Green D | G7SZW |
| Green E | G4EZM |
| Green E | G0ATS |
| Green F | G3GMY |
| Green G | MW6GAA |
| Green H | G3AMH |
| Green H | G4VAG |
| Green I | G7CWI |
| Green I | G1IXF |
| Green J | 2E0FPW |
| Green J | 2E0PMT |
| Green J | GW0SXZ |
| Green J | G1DVU |
| Green J | G1GWO |
| Green J | G3PYF |
| Green J | G3ZNV |
| Green J | G4UPI |
| Green J | G8MKW |
| Green J | G8SIM |
| Green J | M0BMD |
| Green J | G0IIF |
| Green J | G3WVR |
| Green J | M0ACN |
| Green J | M1JHG |
| Green J | G4RRH |
| Green K | G0PHP |
| Green K | G0SEW |
| Green K | G4CYC |
| Green K | M1ACL |
| Green K | 2E0KGJ |
| Green L | G6BZG |
| Green L | G6DPL |
| Green M | GJ1EXC |
| Green M | G6MDC |
| Green M | G6PVA |
| Green M | G7PFI |
| Green M | G4PMG |
| Green M | M3ICH |
| Green M | MI6XAM |
| Green M | M6CEY |
| Green M | G1HYO |
| Green M | G7DYD |
| Green M | G8NCS |
| Green M | M3YFN |
| Green M | 2E0BLL |
| Green N | 2E0PPK |
| Green N | M0GWK |
| Green N | M3PPK |
| Green P | G7LGS |
| Green P | 2E0PCG |
| Green P | G0ABI |
| Green P | G0ELM |
| Green P | G3EWM |
| Green P | G4LWF |
| Green P | G4MEB |
| Green P | G4PHL |
| Green P | G4VZT |
| Green P | G6VTN |
| Green P | GM7LAC |
| Green P | G7ONR |
| Green P | G7VTS |
| Green P | 2E0PWG |
| Green P | G8LQM |
| Green P | M6HIM |
| Green P | 2E0PNG |
| Green P | M6PWG |
| Green R | G0DCF |
| Green R | G1BHV |
| Green R | G3ENO |
| Green R | G3TRG |
| Green R | G4JII |
| Green R | G4UDV |
| Green R | G8MIH |
| Green R | G6WUD |
| Green R | G1AUQ |
| Green R | G6PAJ |
| Green R | G6ULJ |
| Green R | MW6RDL |
| Green R | M0HJY |
| Green R | M6BRJ |
| Green R | M0RJG |
| Green S | G4DNA |
| Green S | G4EKM |
| Green S | G4HDE |
| Green S | GM4JXP |
| Green S | G4YZM |
| Green S | M0ZBT |
| Green S | G1INK |
| Green S | G6JPM |
| Green S | MM6SGO |
| Green T | G3GLL |
| Green T | G7AJS |
| Green T | M0DXN |
| Green T | M0TGX |
| Green T | M3PTG |
| Green T | M5TMG |
| Green T | G7AHB |
| Green T | M6TRP |
| Green TWO T | G4ZBG |
| Green V | G1IXE |
| Green W | G4DMB |
| Green W | G4KGX |
| Green W | M5AGW |
| Green X | G0BMG |
| Greenacre D | G0TGR |
| Greenacre J | G7NLP |
| Greenall A | G8WBU |
| Greenall I | 2E0EHB |
| Greenall J | G8OWS |
| Greenall T | M3TPG |
| Greenall W | M0GRE |
| Greenall W | M1AIX |
| Greenan P | GI3RNO |
| Greenaway B | G3THQ |
| Greenaway T | M3XTG |
| Greenbank A | G3ZVM |
| Greenbank A | G4VIO |
| Greenbeck B | G8LIP |
| Greenberg B | M6BDG |
| Greendale S | G8OUS |
| Greene C | M3EYR |
| Greene J | G0SZG |
| Greene N | M3ETQ |
| Greenfield J | G0VPZ |
| Greenfield J | M6GHR |
| Greenfield K | M6KIG |
| Greenfield M | G8MTB |
| Greenfield S | G1TWH |
| Greenfield S | M3TAN |
| Greengrass R | G4NRG |
| Greenhalgh A | 2E0SUD |
| Greenhalgh D | G0IWN |
| Greenhalgh D | G0KDB |
| Greenhalgh E | G0AQI |
| Greenhalgh G | G0MOF |
| Greenhalgh H | 2E0WOZ |
| Greenhalgh J | M6JJG |
| Greenhalgh P | G3XGE |
| Greenhalgh P | M3UZA |
| Greenheart S | 2E0SYE |
| Greenhough A | 2E0UIQ |
| Greenhough A | M3NMU |
| Greenhough J | G4UIQ |
| Greenhough J | G6NMU |
| Greenhough R | G4KMW |
| Greenhow M | M3MOV |
| Greenland A | G7JGQ |
| Greenland A | M3JGQ |
| Greenland C | G4SEZ |
| Greenland D | 2E0DPO |
| Greenland D | M0HZM |
| Greenland D | M6FLS |
| Greenland S | 2M0SCG |
| Greenland S | MM6SCG |
| Greenleaf A | G6IGU |
| Greenlees G | GM4NSL |
| Greenley J | G6OJV |
| Greenough F | G4EHY |
| Greenough K | G8BEQ |
| Greenshields I | G4FSU |
| Greenshields K | M6KYB |
| Greensmith G | G8JQS |
| Greensmith M | 2E0MGA |
| Greensmith M | M6MJI |
| Greensted N | G8DBU |
| Greenstreet N | G4BOJ |
| Greenway D | M0BJO |
| Greenway P | GM0PRO |
| Greenway P | M3PHG |
| Greenway-Brown J | M0JGB |
| Greenway-Brown V | 2E0VGB |
| Greenwood A | G0IPN |
| Greenwood A | G1IYA |
| Greenwood C | G4TFT |
| Greenwood C | M0HHF |
| Greenwood C | 2E0CNQ |
| Greenwood C | M0HRM |
| Greenwood C | M6CSG |
| Greenwood D | G7JXQ |
| Greenwood G | G3OAR |
| Greenwood G | G4LIX |
| Greenwood G | MM6WSG |
| Greenwood I | GI0AIJ |
| Greenwood J | G0KNH |
| Greenwood J | G3ZJY |
| Greenwood J | M6DKO |
| Greenwood J | G3KRZ |
| Greenwood J | M6KBX |
| Greenwood M | G3YPE |
| Greenwood M | M6MIG |
| Greenwood P | 2E0ONE |
| Greenwood R | G4UFZ |
| Greenwood R | G4YOR |
| Greenwood R | M6MPY |
| Greenwood S | G2BUJ |
| Greenwood S | G4TWG |
| Greenwood S | G4KAM |
| Greenwood T | G4AYR |
| Greep S | G4EET |
| Greer A | GI7INR |
| Greer B | G0JEE |
| Greer D | G4EEH |
| Greer J | MM6JOX |
| Greer M | G4UJO |
| Greer R | 2E0BFF |
| Greer R | MI6TLG |
| Greer T | GI4TGR |
| Greetham P | G1XLL |
| Greeves B | G4BJO |
| Greevy J | G6JVA |
| Gregg D | MI0IRZ |
| Gregg G | 2I0GSG |
| Gregg G | MI6GSG |
| Gregg J | G4MAK |
| Gregg M | G0KNN |
| Gregg N | 2E0KPI |
| Gregg P | G0AHM |
| Gregg T | G1WQU |
| Gregor G | G4OWH |
| Gregorian N | G0ATO |
| Gregory A | G1ZDT |
| Gregory A | G7AQF |
| Gregory A | G7AYP |
| Gregory A | G4KJS |
| Gregory A | M3IRR |
| Gregory C | 2E0ZCG |
| Gregory C | G3JXC |
| Gregory C | M3ZCG |
| Gregory C | G8GYY |
| Gregory D | G0SLV |
| Gregory H | G3VDF |
| Gregory H | M3FXQ |
| Gregory J | G4PFO |
| Gregory J | G8IHA |
| Gregory J | M3KXF |
| Gregory J | M3NGM |
| Gregory K | G3WEU |
| Gregory L | M6LIE |
| Gregory L | M6FXY |
| Gregory M | G0JYQ |
| Gregory M | G4LCH |
| Gregory M | G4HGM |
| Gregory M | G1HRH |
| Gregory M | 2E0MKE |
| Gregory M | M6MKE |
| Gregory N | G0HIK |
| Gregory P | G0BHH |
| Gregory P | 2E0CYO |
| Gregory P | G6KZI |
| Gregory R | GW8FNO |
| Gregory R | M3MGU |
| Gregory R | G4FQT |
| Gregory S | 2E0SOT |
| Gregory S | G7ESI |
| Gregory S | M3LFH |
| Gregory S | M3SOT |
| Gregory S | 2E0SGY |
| Gregory S | M6SGA |
| Gregson C | M0DBT |
| Gregson D | G6IGV |
| Gregson D | M0CZU |
| Gregson M | M0HTV |
| Gregson N | 2E0NSG |
| Greig A | G0CLM |
| Greig A | 2E0SKK |
| Greig A | M0RND |
| Greig A | M6AGG |
| Greig A | MM3PLC |
| Greig C | MM0AOQ |
| Greig D | G8KFJ |
| Greig J | GM0REW |
| Greig M | 2M0MMM |
| Gresswell C | G0WFH |
| Gresswell D | G3PWY |
| Gresty S | G0FRB |
| Gretton A | M6HIK |
| Grevatt D | G7AIF |
| Grevatt J | G8VCH |
| Grevett D | G0BCW |
| Grey D | G0JCP |
| Grey K | G3WNR |
| Grey M | G1IVI |
| Grey N | 2E0NGR |
| Grey N | M6NJG |
| Grey P | GW1PAV |
| Grey R | G1MTA |
| Greywolf D | G7JVN |
| Gribben D | M3YGL |
| Gribben K | 2E0KTG |
| Gribben K | M0XDJ |
| Gribben Z | M3ZLP |
| Grice B | 2E0BCD |
| Grice M | 2E0FNG |
| Grice M | M0ZAI |
| Grice M | M6FNG |
| Grice N | G0MKP |
| Grice P | G4INA |
| Grice T | GM4PSL |
| Gridley A | G4XPJ |
| Gridley M | M3SZK |
| Grierson C | GM4YLN |
| Grierson M | G3TSO |
| Grieve G | G3MSM |
| Grieve J | 2M1JKG |
| Grieve J | G6JHG |
| Grieve J | GM1MWK |
| Grieve J | GM0HTH |
| Grieve J | GM0OTI |
| Grieve J | GM8RTI |
| Grieve L | GM0NTI |
| Griffey P | 2W0YYP |
| Griffin B | G0CGE |
| Griffin B | G0HEA |
| Griffin B | G3XRK |
| Griffin D | M3ZCM |
| Griffin D | 2E0BKQ |
| Griffin D | M0ZDG |
| Griffin D | M3UYC |
| Griffin F | G1DDF |
| Griffin G | G0AQH |
| Griffin I | G4RZZ |
| Griffin J | G6XQY |
| Griffin J | 2E0ARA |
| Griffin J | G6ZUV |
| Griffin J | G8CLW |
| Griffin J | M0CDL |
| Griffin M | GW8THM |
| Griffin M | M1AEP |
| Griffin M | M0AGY |
| Griffin M | G3IIN |
| Griffin N | 2E0NEV |
| Griffin N | MI6NCG |
| Griffin P | G4IFU |
| Griffin P | G1MJA |
| Griffin R | G4XFZ |
| Griffin R | GI7BET |
| Griffin R | M0EZO |
| Griffin S | G1OHH |
| Griffin S | G6DBZ |
| Griffin W | M0BRU |
| Griffith C | G1PLE |
| Griffith D | G0OAB |
| Griffith D | M6WYN |
| Griffith J | G0UKF |
| Griffith J | G1DCI |
| Griffith P | M6PSG |
| Griffith R | G0GYI |
| Griffith R | G8LEM |
| Griffith S | G0WWP |
| Griffiths A | G6YWL |
| Griffiths A | G8VAE |
| Griffiths A | G8LJY |
| Griffiths A | M6RDX |
| Griffiths A | MW3ZKX |
| Griffiths B | G6NTY |
| Griffiths C | G7UHT |
| Griffiths C | MW6ULX |
| Griffiths C | M3CMG |
| Griffiths C | M0XDK |
| Griffiths D | 2W0DMG |
| Griffiths D | G0CEQ |
| Griffiths D | G4DMG |
| Griffiths D | GD4PKV |
| Griffiths D | G6RAO |
| Griffiths D | M3BLR |
| Griffiths D | M3RWR |
| Griffiths D | 2E0DTG |
| Griffiths D | G3RDQ |
| Griffiths D | GW3XHG |
| Griffiths D | M6DTG |
| Griffiths D | M6WYG |
| Griffiths D | G7UMF |
| Griffiths E | MM3GOT |
| Griffiths F | G3MED |
| Griffiths G | 2E0AWT |
| Griffiths G | GW0PDB |
| Griffiths G | GW3LHK |
| Griffiths G | G3ZIL |
| Griffiths G | G7IZA |
| Griffiths G | G8FXA |
| Griffiths G | M0GIQ |
| Griffiths G | M1DJC |
| Griffiths G | G3STG |
| Griffiths G | G0KQS |
| Griffiths H | GW0OUH |
| Griffiths H | G7MQQ |
| Griffiths H | MW6HLG |
| Griffiths I | M3HAJ |
| Griffiths J | G6RUO |
| Griffiths J | G7CBU |
| Griffiths J | M1DGX |
| Griffiths J | M6BQP |
| Griffiths J | 2E0JCT |
| Griffiths J | G8JWP |
| Griffiths J | M6BQL |
| Griffiths J | M6JGI |
| Griffiths J | G4OSX |
| Griffiths J | 2W0JAI |
| Griffiths K | G0EVP |
| Griffiths K | M6YKI |
| Griffiths L | GW6PFK |
| Griffiths L | M5LMG |
| Griffiths L | M3XKN |
| Griffiths L | M0AQZ |
| Griffiths L | M1CGF |
| Griffiths M | G1ZHN |
| Griffiths M | G3WLG |
| Griffiths M | G6KIZ |
| Griffiths M | 2E0MDG |
| Griffiths M | M0MDT |
| Griffiths M | M3UXR |
| Griffiths M | G6IVC |
| Griffiths M | MW1AND |
| Griffiths N | G7UUG |
| Griffiths N | 2E0KAX |
| Griffiths N | G0WPO |
| Griffiths N | M6NAO |
| Griffiths N | M6NGR |
| Griffiths O | M0OPG |
| Griffiths P | G8FGY |
| Griffiths P | M1FJF |
| Griffiths P | M6PNC |
| Griffiths P | G8PNE |
| Griffiths R | G0AUF |
| Griffiths R | GW0IXK |
| Griffiths R | G3JTQ |
| Griffiths R | G6JVB |
| Griffiths R | M1CAR |
| Griffiths R | M3KBZ |
| Griffiths R | G0TMA |
| Griffiths R | M6RGI |
| Griffiths R | GW0DIV |
| Griffiths R | M3UEC |
| Griffiths R | G7NHB |
| Griffiths S | GW2CGF |
| Griffiths S | G4SWN |
| Griffiths S | G4SWO |
| Griffiths S | MW3GPG |
| Griffiths S | M3YWF |
| Griffiths S | M6HTF |
| Griffiths T | G3NPZ |
| Griffiths T | M3GRI |
| Griffiths T | G4GRN |
| Griffiths T | G7PEE |
| Griffiths W | 2E0WFG |
| Griffiths W | G4IEU |
| Griffiths W | G6CVW |
| Griffiths W | M3WFG |
| Griffiths W | GW0UHJ |
| Griffiths W | GW7PIN |
| Griffiths W | MD1DXW |
| Griffiths W | MD3DXW |
| Grigg G | G0FOY |
| Griggs A | G4KMB |
| Griggs D | G0IPT |
| Griggs M | G4YJN |
| Griggs P | G4KPE |
| Griggs S | M0ASJ |
| Grigor A | G7TPW |
| Grigsby N | M0NRG |
| Grigson P | G0TLE |
| Grimbleby S | M3FVA |
| Grimbleby T | G4MPL |
| Grime A | G7EQK |
| Grime K | G4PMV |
| Grime M | G4PSE |
| Grimes A | G0GVZ |
| Grimes B | G0LGZ |
| Grimes M | G6NMK |
| Grimes S | G1VIS |
| Grimley F | M6TJY |
| Grimley R | G6GSG |
| Grimmer V | M3VLN |
| Grimmett A | G6HEE |
| Grimshaw D | M0BXM |
| Grimshaw G | G3PWN |
| Grimshaw G | G3TQX |
| Grimshaw G | 2D0CGB |
| Grimshaw G | MD3VUQ |
| Grimsley M | M6JMJ |
| Grindel E | G4VGM |
| Grindell D | G3RLL |
| Grindle M | G1ZKE |
| Grindrod A | G4EXF |
| Grindrod A | G8WCT |
| Grindrod D | 2E0SHG |
| Grindrod M | G4EDY |
| Grindrod M | G8ILN |
| Grinling A | G4FZF |
| Grint E | G4XLY |
| Grint J | 2E0PPO |
| Grint J | M6RNS |
| Grint W | G1FGK |
| Grinter H | G0LCC |
| Grinter H | G1OYH |
| Gripp R | 2E0RCU |
| Gripp R | M0RVC |
| Gripp R | M6AVK |
| Grisley A | G4DNJ |
| Grisley E | GU7TQX |
| Gritton M | M0UXO |
| Grizzell C | M1AFZ |
| Groat A | 2E0AZU |
| Groat A | M3ICN |
| Groat N | 2E0BQX |
| Groat N | M3MYT |
| Groeber N | G1FDD |
| Groeger J | G4XXW |
| Grogan W | G4APP |
| Gromadzki J | M0ULR |
| Gromadzki J | M3ULQ |
| Gromen-Hayes A | M3NGY |
| Groom C | G1GDT |
| Groom D | M3IZI |
| Groom I | G0FCA |
| Groom J | G0CMW |
| Groom K | G0VTV |
| Groom M | G3WME |
| Groom P | G4FIE |
| Groom P | G6YWN |
| Groom R | G4RKP |
| Groome B | G1WPG |
| Grosjean P | G4SFS |
| Grossart C | GM0KLO |
| Grossmith E | G3WOH |
| Grosvenor A | 2E0BDB |
| Grosvenor J | 2E0BSU |
| Grosvenor J | M3NLF |
| Grounds A | M6EWT |
| Grounds R | G8CMO |
| Groundwater P | MM6PHG |
| Grounsell I | G1UUF |
| Grout K | M1CML |
| Grout R | M0RWG |
| Grout R | M6RMG |
| Grove A | M0GAD |
| Grove G | G3PXU |
| Grove M | G0GUG |
| Grove S | G4BSM |
| Grover A | G6DBX |
| Grover K | G3KIP |
| Grover K | G7THK |
| Grover L | M6LGI |
| Groves C | G6FPQ |
| Groves C | M0FPQ |
| Groves C | M6CJG |
| Groves F | G0AUB |
| Groves G | 2E0GJE |
| Groves G | M6GJE |
| Groves J | G0VQA |
| Groves K | G3HDF |
| Groves L | M6LYN |
| Groves M | M6DXL |
| Groves P | G3MWQ |
| Groves P | G7UWP |
| GROVES S | G1IUZ |
| Groves S | MM0XPZ |
| Groves S | MM3SWG |
| Groves T | M3CAA |
| Groves T | G4KUJ |
| Groves W | 2E0NUC |
| Groves W | M3NUC |
| Grubb J | 2E0IJF |
| Grubb R | G3FNL |
| Grubb S | M6HPT |
| Grudgings N | M6BFS |
| Gruffydd L | G4CFC |
| Grundey J | GM7ESM |
| Grundey J | MM3IOM |
| Grundy A | G7DEC |
| Grundy A | G7OIR |
| Grundy A | M0ADY |
| Grundy D | 2E0DFB |
| Grundy D | M3WXU |
| Grundy G | G4YJW |
| Grundy G | G3XEC |
| Grundy J | G7ULN |
| Grunewald U | G0BBB |
| Grunwald R | G4EOC |
| Grylls B | G4ZCN |
| Grzybek T | MI6AHR |
| Grzywaczewski J | MW0TDQ |
| Gubbins R | G4BKQ |
| Gubric J | M0GGH |
| Gudgeon G | 2E0IID |
| Gudgeon J | G4MDU |
| Gudgeon L | M3LJF |
| Gudgeon R | M6ROJ |
| Gudgeon V | M6VHA |
| Gudgin G | G6KGK |
| Gudgin G | G6KGL |
| Gudziunas E | M0LYQ |
| Guerrero C | G0GSX |
| Guess J | 2E0JWG |
| Guess J | M0JXG |
| Guess J | M6JHG |
| Guess R | M3HDL |
| Guest A | M3IXY |
| Guest D | GM3TFY |
| Guest G | M6GGV |
| Guest I | G7OIT |
| Guest J | G4GON |
| Guest J | G4YVA |
| Guest M | G0VYN |
| Guest P | M6PXG |
| Guest R | G4RWG |
| Guest S | G6BSP |
| Guest S | 2E0TJI |
| Guest S | M6ETR |
| Guffick I | G7JXY |
| Guffogg J | G4UAL |
| Guilbert P | GU0DXX |
| Guild G | G0VQL |
| Guilford J | M1DYC |
| Guilfoyle W | G3UBC |
| Guinan C | G0FZF |
| Guinan G | 2E0IIV |
| Guinan G | M0UTH |
| Guite J | G4FNP |
| Gulley J | GW4TQD |
| Gullick D | G0DGA |
| Gullick J | G0DGE |
| Gulliford B | G6ANV |
| Gulliford G | G0DVP |
| Gulliford G | G4SQI |
| Gulliford K | MW6ECR |
| Gulliver R | G4EDQ |
| Gully D | G4YOC |
| Gully P | G2BQP |
| Gulyas I | GW0CKL |
| Gumb J | G4RDC |
| Gumb R | G0CTS |
| Gumbrell G | G1AUM |
| Gumbrell M | G6CVV |
| Gunby M | G4HOD |
| Gundry B | G4SBU |
| Gundry G | G1LGB |
| Gundry G | M0CLG |
| Gunia J | G7HIJ |
| Gunn A | M3CCB |
| Gunn C | G8VCJ |
| Gunn D | 2E0BQF |
| Gunn D | M0OFL |
| Gunn D | M3UFG |
| Gunn F | G8PZX |
| Gunn G | G8ACT |
| Gunn G | M0CBA |
| Gunn J | G4FCT |
| Gunn J | G3JSG |
| Gunn M | M6GKT |
| Gunn N | G8IFF |
| Gunn S | G6JRZ |
| Gunn S | M1SUE |
| Gunn W | GM6JHH |
| Gunnell N | M1NMG |
| Gunner R | G7ORG |
| Gunning D | 2E0ALZ |
| Gunson R | M6FYN |
| Gunstone K | M0AGA |
| Guppy J | G8FET |
| Guppy P | G0BCH |
| Guppy R | G0TNM |
| Gupta M | M6ZSH |
| Gupta P | M6POG |
| Gurbutt A | G0NQA |
| Gurney I | G7RPP |
| Gurney L | G4LBJ |
| Gurney P | M0CVR |
| Gurney P | M1GUR |
| Gurney S | G0ETZ |
| Gurney T | M6ENB |
| Gurney-Smith F | G0JIV |
| Gurnhill A | G0ROW |
| Gurowich R | M5ENM |
| Gurr K | G8FRS |
| Gurr M | G0MAZ |
| Gurr M | G7BVH |
| Gurton I | G0CPN |
| Gurton I | G8ASP |
| Guscott R | M5GUS |
| Guthrie R | MM3RIX |
| Gutten N | G6IGW |
| Gutteridge B | 2E0MZB |
| Gutteridge B | M6BAA |
| Gutteridge J | G3PEZ |
| Gutteridge N | G8BHE |
| Gutteridge P | M3NVR |
| Guttridge J | G3JQS |
| Guttridge P | G3TCU |
| Guttridge R | G4YTV |
| Guy B | G8IEV |
| Guy C | G6MRY |
| Guy C | G4DDI |
| Guy D | G0ANP |
| Guy E | G4RMG |
| Guy H | G4TFW |
| Guy M | G0VNJ |
| Guy M | G4OJD |
| Guy N | G6YEA |
| Guy P | G6IVD |
| Guy T | 2E0TCF |
| Guy T | M0TGY |
| Guy T | M6TEG |
| Guy W | G4ZSD |
| Guy W | MM6VLG |
| Guymer C | G1ZRT |
| Gwilliam E | G7EXX |
| Gwilliam J | MW0GWL |
| Gwilliam S | G8XGG |
| Gwillym R | M6NGU |
| Gwillym S | M6GWL |
| Gwilt J | GI4MUE |
| Gwynn J | G0JDC |
| Gwynn R | G8KFD |
| Gwynne A | GW3LNR |
| Gwynne G | G8VCI |
| Gwynne R | G4CKT |
| Gwyther T | MW3TJG |
| Gynane M | G7AWW |
| Gyngell C | M6SJQ |
| Gyngell P | M0PLG |
| Gynn R | G3SBP |
| Gyton J | M3GQB |
| Gyurgyak L | G0BMP |

## H

| | |
|---|---|
| Haase R | G4VHE |
| Habens I | G3WXG |
| Haberman A | G0NMB |
| Hack M | G8SLU |
| Hack M | M6CXZ |
| Hack P | M6FID |
| Hacker C | M3HVY |
| Hacker J | G1DER |
| Hacker R | 2E0GDO |
| Hacker T | G1EDH |
| Hackett D | G0GBV |
| Hackett D | G0WDK |
| Hackett J | G6VZZ |
| Hackett M | M3ZQC |
| Hackett P | G3WCJ |

| | |
|---|---|
| Hackett R | G4LAJ |
| Hackett S | G8VVG |
| Hackett T | M0ABA |
| Hackford M | G0IBZ |
| Hacking I | G3VDU |
| Hackling D | M3XIM |
| Hackman A | M3UJQ |
| Hackney C | G4VMD |
| Hackney M | G6UPH |
| Hackney S | 2E0HQW |
| Hackwell K | G4VFR |
| Hackwill T | G4MUT |
| Hackwood S | M3SDJ |
| Hadaway D | G4MCM |
| Hadden A | MW0BBL |
| Hadden D | M0CES |
| Hadden R | GI6EGE |
| Haddleton J | M0OBU |
| Haddleton M | M6PTF |
| Haddock A | G6VBQ |
| Haddock C | G3UZM |
| Haddock F | M1FRH |
| Haddock W | 2E0CDQ |
| Haddock W | M6AFY |
| Haddon C | 2E0DSX |
| Haddon J | G4FCA |
| Haddon M | MJ3HMA |
| Haddon M | G4ZIY |
| Haddow J | M6CNZ |
| Haddrell C | G4OPO |
| Haden B | G1NTP |
| Haden H | G4GWF |
| Haden S | G1PRF |
| Hadfield A | M3JGX |
| Hadfield A | M6AIO |
| Hadfield K | M6KLH |
| Hadfield M | 2E0FIF |
| Hadfield M | M0MKH |
| Hadfield P | M1TZR |
| Hadfield R | G4ANN |
| Hadfield R | G6OVX |
| Hadfield S | M3UDU |
| Hadjidakis D | G1GIJ |
| Hadjigeorgiou C | G7DWM |
| Hadjioannou J | G8GXS |
| Hadland R | GW8WXP |
| Hadler P | G4CZU |
| Hadley B | G0JKY |
| Hadley B | G6JKY |
| Hadley J | 2E0UWK |
| Hadley J | M6UWK |
| Hadley K | M6ODE |
| Hadley M | M6HGJ |
| Hadley N | G4BSW |
| Hadley P | M6EVK |
| Hadley P | MM6FSV |
| Hadley R | G8GKH |
| Hadley S | M6SUE |
| Hadley T | G4NNO |
| Hadnum M | G4NGD |
| Haener E | M0HXS |
| Haestier D | M3HXG |
| Haffenden P | G4SVQ |
| Hagan C | GI6IES |
| Hagan D | 2E0SCE |
| Hagan D | M6BMN |
| Hagan E | GI0EUG |
| Hagan L | M6LHG |
| Hagan M | M3IBE |
| Hagan V | GI6NNP |
| Hagemans K | M0HZN |
| Hagen J | G0LPQ |
| Haggart J | G3JQL |
| Haggas G | M3VDV |
| Hagger D | G1DEU |
| Hagger L | G6HSW |
| Hagland A | G0MSA |
| Hagland A | M0WAJ |
| Hague D | 2M0AKI |
| Hague D | MM3OHD |
| Hague J | GM3JIJ |
| Hague J | G6HSR |
| Hague M | G0PYD |
| Hague N | 2M0NIA |
| Hague N | MM6NIA |
| Hague P | G0HTG |
| Hague R | G3ZQV |
| Hague R | G4XOU |
| Hague T | M0AFJ |
| Hagues J | G1JAH |
| Hagues R | G7GAB |
| Hahn M | G4JHB |
| Hahn S | G5EDQ |
| Haig J | G4VXU |
| Haig J | M0XGK |
| Haigh A | G4PLE |
| Haigh A | G6UMU |
| Haigh D | M0OMD |
| Haigh D | G7UQA |
| Haigh N | 2E0EOP |
| Haigh N | M3PAX |
| Haigh S | G0VOK |
| Haigh S | G3SJW |
| Haigh S | G7TOA |
| Haigh S | 2E0FGM |
| Haigh S | M0HAG |
| Haighton F | G4INU |
| Hail K | MM3XKH |
| Hail L | MM6LAH |
| Haile J | G8ADC |
| Hailes N | G4OTB |
| Haills A | G6YJH |
| Hailstone B | G4BEO |
| Hailstone C | M6CDH |
| Hailstone L | MW3LHA |
| Haime G | M6EKF |
| Haimes J | 2E0JEH |
| Haines A | GM7VLC |
| Haines B | G1IYB |
| Haines B | GD8FAT |
| HAINES C | GU3KQ |
| Haines C | M6GGU |
| Haines D | M3FHP |
| Haines G | M1BTA |
| Haines G | M3GWH |
| Haines G | G4SXY |
| Haines J | G6WNG |
| Hainesborough M | G0PYV |
| Haining R | G4ZDY |
| Hainsworth D | G3ZNK |
| Hainsworth G | G4JFC |
| Hainsworth P | G4JVD |
| Hair J | G0LSU |
| Haith P | G0KUD |
| Hakes J | G4KWJ |
| Hakes J | G0HSK |
| Hakes K | G4KWK |
| Haladij M | M0DAE |
| Halbert C | G7RWC |
| Halbertsma S | G7CRK |
| Halcrow A | 2M0BDT |
| Halcrow A | MM0LSM |
| Halcrow C | MM6ZBG |
| Haldane C | 2M0SSO |
| Haldane C | MM0SSG |
| Haldane C | MM3YNP |
| Halden M | G1SES |
| Hale A | G4ZKT |
| Hale C | G3SFB |
| Hale C | M0ORR |
| Hale D | M0DJH |
| Hale J | G3FTH |
| Hale K | 2E0KHI |
| Hale K | G0AKH |
| Hale M | GW0POA |
| Hale M | G4EQK |
| Hale M | G6PAO |
| Hale P | G2HS |
| Hale R | G0ISG |
| Hale S | G6PAP |
| Hales K | G0RQF |
| Hales R | G4XGI |
| Hales R | M3RPH |
| Hales R | M3VZN |
| Hales T | M6GXC |
| Halewood S | M6SUH |
| Haley A | G4MMT |
| Haley L | M3ZZJ |
| Haley M | GD1XMA |
| Haley T | 2E0TMY |
| Haley T | M0TTH |
| Haley T | M3TTH |
| Halford E | M6ERH |
| Halford J | G4EWZ |
| Halford J | M6OSW |
| Halford M | G8XJL |
| Halford N | M3VNH |
| Halford R | 2E0FVJ |
| Halford R | G8UJQ |
| Halifax D | G6NRH |
| Hall A | G1MGF |
| Hall A | G4CRN |
| Hall A | G8DLH |
| Hall A | G8YNG |
| Hall A | M3GQP |
| Hall A | GW7RKC |
| Hall A | M0BWU |
| Hall A | M6ALY |
| Hall A | M6AHH |
| Hall A | G0ROV |
| Hall B | G0EDC |
| Hall B | M0LPF |
| Hall B | M1STI |
| Hall B | M3KQR |
| Hall B | M3KUE |
| Hall B | 2E0BQL |
| Hall B | M3MWQ |
| Hall B | M6WBG |
| Hall B | M6BHJ |
| Hall B | 2E0BKB |
| Hall B | G6MWM |
| Hall B | M3ZYO |
| Hall C | G1ISX |
| Hall C | G4FGW |
| Hall C | G4HJB |
| Hall C | GM4JPZ |
| Hall C | G6HTH |
| Hall C | G6LWA |
| Hall C | G8EUF |
| Hall C | 2E0HTE |
| Hall C | M6DYN |
| Hall C | GM1EMX |
| Hall C | MM0TWK |
| Hall C | M3UJN |
| Hall C | 2E0HAC |
| Hall C | G4SET |
| Hall C | M3XHU |
| Hall D | G0KCC |
| Hall D | G0SYS |
| Hall D | G4TID |
| Hall D | G6FZC |
| Hall D | G0TOZ |
| Hall D | G7EDF |
| Hall D | G7NOF |
| Hall D | G8VZT |
| Hall D | M0HDJ |
| Hall D | M0KWK |
| Hall D | M1CDL |
| Hall D | M5AFG |
| Hall E | G4RGP |
| Hall E | G6EXN |
| Hall E | M3OLN |
| Hall F | G8UQV |
| Hall F | G8IQA |
| Hall G | G1KOH |
| Hall G | G3RZJ |
| Hall G | G4XZI |
| Hall G | G6VPJ |
| Hall G | M0HFR |
| Hall G | MM1DEE |
| Hall H | 2E0HLH |
| Hall I | G3IWH |
| Hall J | 2E0BPW |
| Hall J | G0HKK |
| Hall J | G0ODQ |
| Hall J | G1ZHZ |
| Hall J | G3FJL |
| Hall J | G3WLD |
| Hall J | G4LGX |
| Hall J | G4WTZ |
| Hall J | M0JNX |
| Hall J | M3KYK |
| Hall J | M3XDV |
| Hall J | M6FLY |
| Hall J | G0DZZ |
| Hall J | G4BOQ |
| Hall K | 2E0KDH |
| Hall K | GM1FGZ |
| Hall K | G4OSK |
| Hall K | M6KDJ |
| Hall K | M6KRH |
| Hall L | 2E0HPM |
| Hall L | G6XAR |
| Hall L | M1DNC |
| Hall L | M6FKA |
| Hall L | M6RKL |
| Hall L | G4IGC |
| Hall M | 2E0GYD |
| Hall M | G0BQK |
| Hall M | G0ODN |
| Hall M | G1IMD |
| Hall M | G3USC |
| Hall M | GM3VQQ |
| Hall M | G4GSB |
| Hall M | G6CHT |
| Hall M | G7NFO |
| Hall M | G8YNH |
| Hall M | M3BNU |
| Hall M | M3EBK |
| Hall M | M3LQX |
| Hall M | 2E0WCE |
| Hall M | M6MHW |
| Hall M | GM8IEM |
| Hall M | 2E0CDU |
| Hall M | 2E0MMH |
| Hall M | M0GVY |
| Hall M | M3MKH |
| Hall M | M6DBX |
| Hall M | M6FFB |
| Hall N | G0DWJ |
| Hall N | G4DQL |
| Hall N | M3NJH |
| Hall N | M1DKK |
| Hall N | 2E0TBT |
| Hall O | 2E0SWE |
| Hall O | M6BIV |
| Hall P | 2E0OQZ |
| Hall P | G4AQA |
| Hall P | G4YEE |
| Hall P | G6SDI |
| Hall P | M0BJS |
| Hall P | M0GMQ |
| Hall P | M3OQZ |
| Hall P | M6PKT |
| Hall P | G0ELC |
| Hall P | G1JMD |
| Hall P | G1UKZ |
| Hall P | M6PGU |
| Hall R | G3IKX |
| Hall R | G3XIY |
| Hall R | G4DVJ |
| Hall R | G4PGD |
| Hall R | G4UGU |
| Hall R | G4WGA |
| Hall R | G4XDV |
| Hall R | G7UPP |
| Hall R | MW0AMI |
| Hall R | M1ANR |
| Hall R | M3OCQ |
| Hall R | GM0OGN |
| Hall R | MM3BRR |
| Hall R | M6FXE |
| Hall R | M6PYP |
| Hall R | 2E0RBY |
| Hall R | M0RBY |
| Hall R | M6RGH |
| Hall R | G0GEA |
| Hall R | G4CFV |
| Hall R | G1RLA |
| Hall S | 2E0EVX |
| Hall S | 2E0GTE |
| Hall S | 2E0HKS |
| Hall S | G1RJN |
| Hall S | G1RSC |
| Hall S | G3NQA |
| Hall S | G4VEG |
| Hall S | G6VWH |
| Hall S | M0CUF |
| Hall S | M3AEZ |
| Hall S | M3USH |
| Hall S | M6AUC |
| Hall S | M3USJ |
| Hall S | G4USP |
| Hall S | G8CCF |
| Hall S | G8DGC |
| Hall S | M0SGH |
| Hall S | M3HAL |
| Hall S | M6SCH |
| Hall S | M5AFY |
| Hall T | G1TOB |
| Hall T | G8DIQ |
| Hall T | G8HFW |
| Hall T | M3MHZ |
| Hall T | 2E0BSM |
| Hall T | M0GPH |
| Hall T | M3YFI |
| Hall W | G3RMX |
| Hall W | G8KSA |
| Hall W | M3IWO |
| Hall W | G4WMH |
| Hallam J | G0DME |
| Hallam J | M3YSA |
| Hallam P | GI4GVS |
| Hallam R | G4HLB |
| Hallam R | G8JGU |
| Hallam S | G7BMM |
| Hallam S | G1SIM |
| Hallam S | 2E0LGH |
| Hallam S | M3LGH |
| Hallam T | G6ADD |
| Hallas P | 2E0HLP |
| Hallas P | M3HLP |
| Hallas P | 2E0ICE |
| Hallatt I | 2E0KZL |
| Hallatt I | M3LYP |
| Hallet M | G3MDR |
| Hallett G | G4RQP |
| Hallett J | G6CAC |
| Hallett M | MW3FVH |
| Hallett P | G7FXY |
| Hallett R | 2W1FVH |
| Hallewell P | M0TWG |
| Halley H | MM6HAH |
| Halliburton M | M6RKA |
| Halliday A | G0IZN |
| Halliday C | G0FNA |
| Halliday D | G7CFS |
| Halliday J | G1TVW |
| Halliday J | G4HMX |
| Halliday J | G6TCV |
| Halliday J | G7SWZ |
| Halliday M | M6BLN |
| Halliday S | 2E0XAR |
| Halliday S | M6TOK |
| Halligan T | GM0WPW |
| Hallin J | G7DQA |
| Halliwell B | 2E0HJB |
| Halliwell D | G8OSJ |
| Halliwell I | M0ALT |
| Halliwell R | M0RMH |
| Halliwell W | G0LKJ |
| Hall-Osman R | G1OFL |
| Halloway C | M6CEH |
| Halloway C | M6NRF |
| Halloway K | M6KIR |
| Halloway L | 2E0LJH |
| Halloway L | M6ODF |
| Halloway M | G1NRF |
| Halloway P | M0PKH |
| Hallows B | G3XPI |
| Halls C | G1KBF |
| Halls C | M3VZZ |
| Halls D | G4LVG |
| Halls G | M3GHA |
| Halls K | G8VJG |
| Hallson P | M0PLP |
| Hallsworth D | M6DLL |
| Hallsworth D | G4VOV |
| Hallsworth F | M0AFF |
| Hallsworth M | G1GYC |
| Hallsworth R | G6XTD |
| Hallsworth S | 2E0FAT |
| Hallwood N | 2E0GRM |
| Hallybone H | G8IBL |
| Halman J | G3JNP |
| Halmshaw B | G0JSF |
| Halpin P | 2E0JEN |
| Halpin P | M0EEH |
| Halsall J | M3XWR |
| Halsall J | 2E0GLR |
| Halsall J | M1DIE |
| Halsall P | G0HXD |
| Halsall R | M3DAV |
| Halse G | G3GRV |
| Halsey D | G7LAN |
| Halsey I | G1JBZ |
| Halsey S | M6WSH |
| Ham A | GW8YTO |
| Ham A | GW1UOV |
| Ham B | G0NPV |
| Ham D | G3ZUO |
| Ham N | 2E0BUF |
| Ham N | M0VVQ |
| Hambidge A | M0DZU |
| Hamblen E | G3MZA |
| Hambleton L | MI6VAI |
| Hambleton S | G7ETS |
| Hamblett P | G0TKT |
| Hamblett P | G8AAI |
| Hamblett R | G0DNU |
| Hamblin R | G0TKZ |
| Hambly D | G0PCT |
| Hambly P | G0TUJ |
| Hambly R | 2E0KAK |
| Hambly R | M0HAF |
| Hambly R | M6AGK |
| Hambrook J | GM1RJS |
| Hamer H | G1JHP |
| Hamer J | G3LMQ |
| Hamer K | G0FSR |
| Hamer K | M0ASU |
| Hamer S | M6MNT |
| Hames D | G7FCL |
| Hamill A | MI0HMY |
| Hamill E | MI6EQC |
| Hamill I | G3SMF |
| Hamill J | GI4ORI |
| Hamill S | M6BSB |
| Hamill W | GI4KTZ |
| Hamilton A | GI0TJJ |
| Hamilton A | GI4HVI |
| Hamilton A | G4ERD |
| Hamilton A | 2E0KDF |
| Hamilton A | MI6AGV |
| Hamilton A | M6BVE |
| Hamilton A | 2M0VTB |
| Hamilton A | GM4FH |
| Hamilton A | MM3VTB |
| Hamilton B | GI3VYY |
| Hamilton B | GM6RQU |
| Hamilton C | M1ECQ |
| Hamilton C | G3XWV |
| Hamilton D | G1LGW |
| Hamilton D | G4GLC |
| Hamilton D | G4TKW |
| Hamilton D | G8XJK |
| Hamilton D | MI0DWD |
| Hamilton D | M1DWT |
| Hamilton D | MI3BIE |
| Hamilton D | MI3CKF |
| Hamilton D | MI6PIR |
| Hamilton D | GM4CAM |
| Hamilton G | 2E0SBS |
| Hamilton G | M3UZZ |
| Hamilton H | GM4JRF |
| Hamilton H | 2E0FSN |
| Hamilton H | M0SBR |
| Hamilton H | M6HGH |
| Hamilton I | M3FLL |
| Hamilton J | G4JSD |
| Hamilton K | M6KAZ |
| Hamilton K | M6MOQ |
| Hamilton K | MM6KER |
| Hamilton L | GM3ITN |
| Hamilton M | G3KEV |
| Hamilton M | GM3TAL |
| Hamilton N | GM0ARY |
| Hamilton N | G1DRR |
| Hamilton N | G4TXG |
| Hamilton S | G4UIY |
| Hamilton S | M3MZN |
| Hamilton S | M6LLO |
| Hamilton T | G0HIN |
| Hamilton T | M3AHR |
| Hamilton T | M6ALM |
| Hamilton V | 2M0PID |
| Hamilton V | MM0PID |
| Hamilton V | MM3VAH |
| Hamilton W | M1CYN |
| Hamilton W | 2I0WSH |
| Hamilton W | G0CXF |
| Hamilton-Cooper S | G1MVQ |
| Hamilton-Sturdy W | GI1TRZ |
| Hamilton-Sturdy W | MI0CNI |
| Hamlet W | M6HQN |
| Hamlett J | G3EOO |
| Hamlin G | M1BDJ |
| Hamlyn K | G4WQB |
| Hamlyn S | GW7IPS |
| Hamm A | G4EBI |
| Hammersley G | G3ZCL |
| Hammersley M | M3OEO |
| Hammersley P | G7IXP |
| Hammett A | G3VWK |
| Hammett C | 2E0ADQ |
| Hammett C | M3CKH |
| Hammett H | G3OVX |
| Hammett R | G0DNV |
| Hammett T | M0TRV |
| Hammond A | G3PGA |
| Hammond A | G6GZZ |
| Hammond A | G7AJJ |
| Hammond A | M3EOZ |
| Hammond B | G4SIJ |
| Hammond B | M6GAS |
| Hammond D | G3KAR |
| Hammond E | M6GVU |
| Hammond G | 2E0FSR |
| Hammond G | G1NPC |
| Hammond G | G8YJS |
| Hammond G | G7TPH |
| Hammond I | G8ZYM |
| Hammond J | G0FLP |
| Hammond J | 2E0GUA |
| Hammond J | 2E0HMD |
| Hammond J | M6GUA |
| Hammond M | 2E0HYX |
| Hammond M | G4HIE |
| Hammond N | M6NKH |
| Hammond P | G1EDK |
| Hammond P | G6XEF |
| Hammond P | G7CNX |
| Hammond P | G8SHC |
| Hammond R | G1NFN |
| Hammond R | G8VSX |
| Hammond R | G4DRW |
| HAMMOND R | G4FKR |
| Hammond R | M6HXN |
| Hammond R | 2E0COM |
| Hammond R | G0RAH |
| Hammond R | M6WKD |
| Hammond S | GM6PNJ |
| Hammond S | M6DPI |
| Hammond S | M6EIC |
| Hammond W | G4SOB |
| Hammond W | G6BRD |
| Hammond W | M1HQX |
| Hammonds S | G4KUR |
| Hammonds V | 2E0ACG |
| Hamnett P | M0PLH |
| Hamon A | GU4WTN |
| Hampshire J | G6ZUZ |
| Hampshire S | M6SMG |
| Hampson A | MW6HPN |
| Hampson A | M6IZE |
| Hampson B | M3HUV |
| hampson D | M6HOQ |
| Hampson E | G8RXB |
| Hampson G | G0JIL |
| Hampson G | G7KSP |
| Hampson G | G4IXW |
| Hampson I | G1DFT |
| Hampson J | G4LPO |
| Hampson K | G3WFW |
| Hampson K | G4DUX |
| Hampson M | G8YBZ |
| Hampson R | 2E0GPT |
| Hampson R | M6GPT |
| Hampson R | G4MGH |
| Hampson S | 2E0ZSH |
| Hampson S | M3ZSH |
| Hampson T | G6DEG |
| Hampton A | G7EWA |
| Hampton D | G3UHU |
| Hampton D | 2E0UBU |
| Hampton D | M0WRW |
| Hampton M | M5MDH |
| Hampton P | M3LVX |
| Hampton P | 2E0CZZ |
| Hampton P | M6CZS |
| Hampton P | G4ADJ |
| Hampton R | G3BPP |
| Hamstead R | G4RTH |
| Hanaghan M | G8HPW |
| Hanburger M | 2E0DEU |
| Hance G | M6GNJ |
| Hance S | M6GNS |
| Hance W | MW3OLX |
| Hancock A | G3SLS |
| Hancock A | GM4IEF |
| Hancock A | G8HPS |
| Hancock A | GW8BWX |
| Hancock D | G3OMY |
| Hancock D | M6DPH |
| Hancock D | G8IYD |
| Hancock D | M0BVM |
| Hancock G | G4HEW |
| Hancock G | M1EGC |
| Hancock G | G0UNC |
| Hancock I | G6BJG |
| Hancock J | G8YNI |
| Hancock J | G4TAK |
| Hancock J | G6ISM |
| Hancock J | M6WMV |
| Hancock K | G0MGJ |
| Hancock K | G4KIY |
| Hancock L | M6LRH |
| Hancock M | G8EYY |
| Hancock M | M6KSW |
| Hancock N | M6SSY |
| Hancock P | G0EFS |
| Hancock P | G6ISG |
| Hancock P | GU3WOW |
| Hancock R | GW4SCK |
| Hancock R | G6BBD |
| Hancock R | M6RFF |
| Hancock R | G4BBT |
| Hancock S | G1CIM |
| Hancock S | G3GBD |
| Hancock S | M0CGS |
| Hancock S | GU6RWD |
| Hancock S | G1XRJ |
| Hancocks N | G4XTF |
| Hancocks N | G7SBP |
| Hancox J | M3AVF |
| Hancox J | M6JHV |
| Hancox R | G7UWG |
| Hancox W | G7JZK |
| Hand A | M3WCI |
| Hand B | M3YHH |
| Hand F | MI6UIM |
| Hand G | M3YHF |
| Hand I | G1DRP |
| Hand K | G8NTJ |
| Hand L | M6LAV |
| Hand M | G0WCI |
| Hand R | G7TPH |
| Hand S | MI6SFH |
| Hand W | M6TCL |
| Handcocks A | G6UMT |
| Handford T | G4YTH |
| Handforth N | M0AUF |
| Handley A | G6VYC |
| Handley B | 2E0JRZ |
| Handley B | M0NOK |
| Handley C | G6KTK |
| Handley F | 2E0IVT |
| Handley I | 2E0RPD |
| Handley I | M0RPD |
| Handley I | M3RPD |
| Handley J | M3JDJ |
| Handley J | G4DNE |
| Handley L | M3AAQ |
| Handley P | 2E0CZS |
| Handley P | M0PVI |
| Handley P | M3PVI |
| Handley R | G3GJQ |
| Handley S | G6APX |
| Handrick H | 2E0IOS |
| Hands C | G4KQV |
| Hands D | G1PGS |
| Hands G | G0FBG |
| Hands R | G4EJU |
| Handscombe F | G4BWP |
| Handscombe J | G7DNT |
| Handscombe K | M0XON |
| Handstock R | G4TOY |
| Handy E | G4RTI |
| Handy F | G4XNE |
| Handy J | G6UVU |
| Handy P | M3JQV |
| Handyside R | GM0NDO |
| Hanfrey P | M1BOD |
| Hankin J | G4NEH |
| Hankin J | M3DLT |
| Hankin R | M1ACO |
| Hanking N | G0DSX |
| Hankins A | 2E0KCF |
| Hankins G | G8EMX |
| Hankins T | G7RVI |
| Hankinson P | G8PAL |
| Hanks C | G0MWX |
| Hanley A | MW6AQJ |
| HANLEY D | G3YVY |
| Hanley J | M3WVI |
| Hanley J | 2M0OSC |
| Hanman P | 2E0CJS |
| Hanman P | M0PZR |
| Hanman P | M6ANI |
| Hanmer P | G0WMY |
| Hanmore F | M3FLC |
| Hann J | 2E0EBO |
| Hann J | M3EBO |
| Hann R | 2E0RPH |
| Hanna A | 2E0IWZ |
| Hanna A | GI0SMU |
| Hanna A | M0IWZ |
| Hanna A | M3IWZ |
| Hanna A | MI6OKS |
| Hanna L | GI3WHA |
| Hannaby E | G0AEX |
| Hannaford G | G1VQG |
| Hannaford J | G7KCE |
| Hannaford M | 2E0HBG |
| Hannah C | G4LOP |
| Hannah D | MM6DAF |
| Hannah S | 2E0SIX |
| Hannah W | MM6CCO |
| Hannam G | M0DGJ |
| Hannam M | G1EUZ |
| Hannam P | G6TSJ |
| Hannam P | G3ZNB |
| Hannan G | G1JVF |
| Hannan P | G1XVT |
| Hannan R | G4RQJ |
| Hannant D | G4YDQ |
| Hannant R | M6TCE |
| Hannell C | G0INQ |
| Hannemann R | G7GQH |
| Hanner A | M0FVV |
| Hanney J | G8BDF |
| Hanney N | G4YPE |
| Hanney R | M5BUF |
| Hannigan B | MI3BXQ |
| Hannigan C | GI0VJE |
| Hannigan K | MI3JVV |
| Hannigan P | MI3JVJ |
| Hannigan R | M0DER |
| Hannington D | 2E0DCH |
| Hannington P | G6UVS |
| Hannon A | 2E0PUE |
| Hannon A | M3PUE |
| Hannon D | M3KVG |
| Hanraads M | G0ILZ |
| Hanrahan R | G6BJQ |
| Hanratty T | G0JRT |
| Hanscombe S | G6NRM |
| Hansell B | G0BXH |
| hansell G | M6YGL |
| Hansell J | G4COS |
| Hansen A | G3VLJ |
| Hansen L | G7MWI |
| Hansford C | G7VCJ |
| Hansford D | M6DPY |
| Hansford R | M0RSH |
| Hansford V | G4WTX |
| Hansley A | G0JRN |
| Hansom J | G3UUF |
| Hanson A | M6AXH |
| Hanson A | G4LVV |
| Hanson A | 2E0BUD |
| Hanson A | M3AKH |
| Hanson D | G7MJX |
| Hanson D | G8KOM |
| Hanson F | G3RBD |
| Hanson G | G0UUS |
| Hanson G | G4CPA |
| Hanson J | G8AYJ |
| Hanson N | G6VNO |
| Hanson N | M6NNC |
| Hanson P | G4IFR |
| Hanson P | M0VPH |
| Hanson P | G0NVY |
| Hanson R | 2E0RLR |
| Hanson R | M6RLR |
| Hanson S | M6SBN |
| Hanson-Brown A | M1JAK |
| Hanson-Collins N | 2E0XUZ |
| Hanton J | G4MCA |
| Hanwell D | 2E0NNE |
| Hanwell D | M6NNE |
| Harada A | G4INX |
| Harber B | G8DKK |
| Harber J | G3JWH |
| Harber P | G4LGY |
| Harber R | G6LHQ |
| Harber T | G4SZS |
| Harbinson P | 2E0FFL |
| Harbison A | GI0SRL |
| Harbison R | GI3PDN |
| Harbottle J | G0OWV |
| Harbour D | G0EID |
| Harbour K | 2E0WKH |
| Harbour K | M6HAR |
| Harbour S | G3PQB |
| HARBRON D | 2E0JDH |
| HARBRON D | M0GDV |
| HARBRON D | M0HJD |
| Harbron D | G7PHG |
| Harbron J | M3UJH |
| Harbron M | M3XQB |
| Harcourt G | G4EBE |
| Harcourt R | G7MUB |
| Harcourt S | 2E0SMA |
| Harcourt S | M6WXN |
| Hard C | G4ITJ |
| Hardacre G | GM1PZT |
| Hardacre P | M0KXQ |
| Hardacre P | M6TEE |
| Hardaker I | G1YBA |
| Hardaker M | G4HJH |
| Hardcastle A | G0STK |
| Hardcastle D | G3DGH |
| Hardcastle J | G0JUA |
| Hardcastle J | G3JIR |
| Hardcastle P | G7SLP |
| Harden A | 2E0XAH |
| Harden C | M0BIJ |
| Harden D | 2E0OXO |
| Harden G | M1EKM |
| Harden K | M3KAK |
| Harden R | G4DUB |
| Harders J | G0VNK |
| Hardes S | G7ICV |
| Hardie A | G0VUL |
| Hardie C | G3YQP |
| Hardie D | G4ITY |
| Hardie N | G7GRH |
| Hardie R | G4SNU |
| Hardie R | G7JVK |
| Hardie W | GM6RGY |
| Hardill G | 2E0FSK |
| Hardill G | M6GAH |
| Hardiman R | G1RPV |
| Harding A | G0HVT |
| Harding A | G0VHQ |
| Harding A | G1JHM |
| Harding A | M3YKH |
| Harding C | 2E0CQM |
| Harding C | G6XAK |
| Harding C | G7UWW |
| Harding C | M0XCH |
| Harding C | M3CWV |
| Harding C | M3MVN |
| Harding C | G1XOZ |
| Harding D | G0DQI |
| Harding D | G3YHG |
| Harding D | M1BOL |
| Harding D | G1PMA |
| Harding D | G6YWU |
| Harding D | M0LSN |
| Harding E | 2E0AUQ |
| Harding E | G6UMH |
| Harding H | G0HZY |
| Harding H | M6HAY |
| Harding J | G0ABW |
| Harding J | G3TZU |
| Harding J | G3XFL |
| Harding J | G4PFR |
| Harding J | M6JDL |
| Harding K | G7FFI |
| Harding K | M0UND |
| Harding M | G7DHJ |
| Harding P | G4WUQ |
| Harding P | G6YOP |
| Harding P | GW4LJS |
| Harding R | G1XNN |
| Harding R | G3RJH |
| Harding R | G4SBM |
| Harding R | G6BJL |
| Harding S | G4JGS |
| Harding S | G4PYU |
| Harding S | G7RTA |
| Harding S | G6XAN |
| Harding S | G4OBN |
| Harding T | G3RGQ |
| Harding T | M3UUN |
| Harding V | G0KCE |
| Hardinges D | GD7HSO |
| Hardingham J | 2E0JDE |
| Hardingham J | M6JFV |
| Hardingham M | M0ZMX |
| Hardman A | 2E0IHS |
| Hardman A | M3AHZ |
| Hardman D | G0VLV |
| Hardman J | M0IGY |
| Hardman J | M1DMH |
| Hardman M | G0PXQ |
| Hardman R | G3LGV |
| Hardman R | G6LWC |
| Hardstone J | G3TFR |
| Hardwick D | 2E0ZDH |
| Hardwick D | M0ZDH |
| Hardwick F | G1DFV |
| Hardwick G | G8ERQ |
| Hardwick J | G4ALA |
| Hardwick M | 2E0JMF |
| Hardwick P | M0PDH |
| Hardwick P | M3EMX |
| Hardwick R | G1NUH |
| Hardwick T | 2D0IMN |
| Hardwick T | MD3FMN |
| Hardy A | G1BTF |
| Hardy A | 2E0HDY |
| Hardy A | M6TGH |
| Hardy B | 2E0ZAI |
| Hardy B | M6MGJ |
| Hardy C | G1FMT |

**IMPORTANT NOTE**

**Revalidate licence to avoid revocation** – Ofcom has advised the Society that plans will be drawn up to revoke licences that have not been revalidated as required by the licence conditions. The quickest way to revalidate is to do so online via the Ofcom website: *https://services.ofcom.org.uk/* or by email: *amateur.validations@ofcom.org.uk* If you need assistance in the process, Ofcom staff are available to help, but please be patient during times of heavy workload.

Hardy C M0ATZ
Hardy D G4BXH
Hardy D G8AJA
Hardy D M1PAH
Hardy D 2E0HBD
Hardy D G8ROU
Hardy D M6HDY
Hardy D M6USE
Hardy G G4HBL
Hardy J G3KND
Hardy J G8OCE
Hardy J 2E0JUL
Hardy J M0JHD
Hardy J M3VNG
Hardy K 2E0XBK
Hardy K G8LEG
Hardy K G0HPV
Hardy L M6CKE
Hardy M M3WPV
Hardy M G4FHQ
Hardy M M0FCG
Hardy M M1KEY
Hardy P M3HTO
Hardy P M6BMK
Hardy P G3VNH
Hardy R G3TIX
Hardy R M3MXM
Hardy S M6VKH
Hare A MW6ATC
Hare C 2E0SKL
Hare C M6LSK
Hare D G3NHV
Hare J G1EXG
Hare T M6HTJ
Hares G G6IXM
Harfield I G8OUH
Hargan R GI3TME
Hargate R M3RNU
Hargeaves R M3IHN
Hargrave A G0AXA
Hargraves J G8ZTF
Hargreaves A G4SYW
Hargreaves A 2E0TRH
Hargreaves A M6ARH
Hargreaves C G6NRL
Hargreaves D M6ATZ
hargreaves J G1EDM
Hargreaves K M1KWH
Hargreaves L M6LEH
Hargreaves P M6BXD
Hargreaves R G3OHH
Hargreaves R G4IVO
Hargreaves R G4YVQ
Hargreaves S 2M0GZA
Hargreaves S MM0GZA
Hargreaves S MM3YPN
Hargreaves S G1GGI
Hargreaves W G0PLG
Hargrove D M6TTV
Haria H GD4KEP
Hark R 2E0RSS
Hark R M0RKD
Harker R G4CMK
Harkess A MM3MFN
Harkess R GM3THI
Harkin G MI3TJV
Harkins P G1CZH
Harkishin H M6HRV
Harkness D GM4NNK
Harkness I G0UED
Harkness R 2E0RHT
Harkness R M6RHT
Harkness S M3SMY
Harknett J G3TVH
Harknett R G3KZC
Harland A G6BBG
Harland E G3VPF
Harland L G6AIU
Harland M M6AWJ
Harley C M3KBF
Harley D M6HDA
Harley I G6BJJ
Harley I M0IFH
Harley J M6JER
Harley K 2E0RGK
Harley K M3RGK
Harley N M3NHD
Harley P G4UDH
Harley P M3UPJ
Harling I G7HFS
Harling J G8RGN
Harling P G4LDD
Harlow C M3CWH
Harlow J G3SHL
Harlow P M3VLH
Harlow R M0REH
Harlow R 2E0PKS
Harlow R M6GOU
Harlow T M3TFW
Harman A G8LTY
Harman C 2E0KDG
Harman C GD4JOO
Harman E 2E0EDE
Harman E M0HIE
Harman E M6EDY
Harman G 2E0IRX
Harman J G8GIU
Harman J G8ZTG
Harman L M3URH
Harman M G3NZP
Harman N 2E0ABQ
Harman P G1FWR
Harman P G4XGD
Harman R G0BAP
Harman R G0BSH
Harman R G7GQO
Harman R G8XHN
Harman R MM0BRG
Harmer C M0HMR
Harmer D G4AOL
Harmer-Knight R G4LBT
Harmsworth H G3ACQ
Harness P G8LDW
Harness P G7LSP
Harnett R G4WXJ
Harney M G4YMH
Harper A G7VIY
Harper A G0PQG
Harper A G1BFG
Harper B GM0JFL
Harper B G0TOS
Harper B G8VAR
Harper B M1CEM
Harper Bill J G3IZM
Harper C GM0JFK
Harper C G6GJD
Harper C G8YVQ
Harper C G7DIO
Harper C M6EVD
Harper C MI6UNC
Harper C M6VVB
Harper E GI0AZA
Harper E G0WXL
Harper G 2E0AIL
Harper G G4IVZ
Harper G G4WHA
Harper G G6SMW
Harper G G4NVN
Harper J G0MJX
Harper J G3RLJ
Harper J GM8KIQ
Harper K G0EKH
Harper K 2M0ONW
Harper L G0CJA
Harper L G4FNC
Harper M G7SRH
Harper N G3ZCV
Harper N G6RZY
Harper P G4MZM
Harper P 2E0HUU
Harper P 2E0PDZ
Harper P M0PDZ
Harper P M6EZR
Harper R 2E0AZM
Harper R G3KSF
Harper R GW4VGB
Harper R M3GIE
Harper R G0KPU
Harper S M6XST
Harper T G0KIN
Harpham D G1EZU
Harpur P G3ZWF
Harrad B G3NPS
Harrad B G8LDV
Harradine A M0CRP
Harrall R 2E0AVK
Harrap C G1SWX
Harrap C M3CPH
Harratt B 2E0CIY
Harratt J 2E0GSW
Harrell M M3NPI
Harries B GW7FXX
Harries D G4WYN
Harries D GW0MWN
Harries I GW7SXU
Harries M G4PXF
Harries R G4PWF
Harries W 2W0BFV
Harries W MW3WFH
Harrigan J GI4JRA
Harrigan R MM3EKL
Harriman R 2E0ERG
Harriman R M6RAH
Harrington A G1XLW
Harrington A G7JJD
Harrington D G3VQM
Harrington E M6UTB
Harrington J M6ALU
Harrington J G7CLM
Harrington J MM3AWC
Harrington M G8WWW
Harrington M G0ISP
Harrington M GM7BNF
Harrington V M6VDH
Harriott D G3IIO
Harriott J 2E0JDO
Harriott M M1RKY
Harris A 2E0BAY
Harris A G3EWF
Harris A G8SYC
Harris A G8ZNB
Harris A M3IQQ
Harris A M3LMV
Harris A M6WWZ
Harris A G6FEM
Harris A G4SJI
Harris A G6YMI
Harris B 2E0KPT
Harris B G0DRH
Harris B G0NPN
Harris B G3GTF
Harris B G3HZR
Harris B G3XGY
Harris B G1JMC
Harris C G7UDX
Harris C G4TBO
Harris C M6LEY
Harris C M3PYT
Harris C M6XYL
Harris D G0CER
Harris D GW0ONU
Harris D GM0TPI
Harris D G1POK
Harris D G2BOF
Harris D G6DEV
Harris D G6FEI
Harris D G6VSG
Harris D G7MXT
Harris D M3XDH
Harris D M6GYV
Harris D 2E0MVT
Harris D M6ZDA
Harris D 2E0DJH
Harris D 2E0JIG
Harris D G7JAS
Harris D G8INA
Harris D M0RNI
Harris D M3ZUF
Harris D M6DAW
Harris D M6IFO
Harris D GW7VBE
Harris D G6XTK
Harris D G3XBX
Harris E G3LQO
Harris E M6EHC
Harris F M3IDC
Harris F G1JHN
Harris G G7NJB
Harris G G3TPQ
Harris G G4KWZ
Harris G M6GNH
Harris G M6ZTS
Harris I G3WAE
Harris I G4IDH
Harris I G6PFX
Harris I M3IRH
Harris J 2E0JSH
Harris J G1OEP
Harris J G1TQR
Harris J G3PFJ
Harris J GW3RYE
Harris J G4DRV
Harris J G4GOA
Harris J G4VMO
Harris J M6JEH
Harris J 2E0BVZ
Harris J G4PFT
Harris J M0GUR
Harris J M3YGV
Harris J M6JHS
Harris J G0OZO
Harris J G3LWM
Harris J G4VYE
Harris J M6AWU
Harris J M6GHV
Harris K G0IEE
Harris K G1HWA
Harris K G6YLR
Harris K G6ZVB
Harris K G8MWN
Harris K G1FMU
Harris K G1IOT
Harris K G8LGP
Harris K M6EIH
Harris L G0ULH
Harris M G0UUG
Harris M G3YIA
Harris M G6DEP
Harris M M1CAY
Harris M 2E0EYL
Harris M M6ARK
Harris M 2E0VNL
Harris M M6VNL
Harris M G3VUI
Harris M M3YOQ
Harris M MD6MIK
Harris M 2E0DNC
Harris M G0HOC
harris M 2E0EKR
Harris N G0GZU
Harris N G3TXC
Harris N G6FYL
Harris N M0FSH
harris N M0GRB
harris N M3EKR
Harris N G4GQE
Harris N 2E0DZV
Harris N G1IUA
Harris P 2E0FAH
Harris P G0POM
Harris P G0WUA
Harris P G3OBV
Harris P G4SPZ
Harris P G4ZOB
Harris P M3AZE
Harris P M3FPH
Harris P M3HVN
Harris P G0LSE
Harris P M6MPQ
Harris P 2E0HNJ
Harris P G4BDQ
Harris P M6LCC
Harris R G0MOW
Harris R G0NCL
Harris R G0SBC
Harris R G3OTK
Harris R GD4FFA
Harris R G4GIY
Harris R G8BIR
Harris R GW8DUP
Harris R G8MZR
Harris R G6XTJ
Harris R 2E0TUW
Harris R G0KIA
Harris R G6IQF
Harris R G8OCF
Harris R M0TUW
Harris R M6RYL
Harris R G1AFW
Harris R M0GSO
Harris R M6FEH
Harris R G3ZFR
Harris R M6RIZ
Harris R M6FUT
Harris S G0SJH
Harris S G4ZGZ
Harris S G6SVL
Harris S M3RZF
Harris S M6SUX
Harris S 2E0ZST
Harris S M6ZST
Harris S M0WFO
Harris T 2E0TEH
Harris T M6BHA
Harris T G2KF
Harris T M3XOV
Harris U G4RDA
Harrison A GM0GON
Harrison A G0PXD
Harrison A G1NRM
Harrison A G6PXJ
Harrison A M3IFA
Harrison A G0BAX
Harrison A G7SEG
Harrison B 2E0BWH
Harrison B G0ILD
Harrison B G0SNS
HARRISON B G1YUB
Harrison B G7PQD
Harrison B M0WSR
HARRISON B M3ALX
Harrison B M3WSR
Harrison B M3XBH
Harrison B M6CWJ
Harrison C G4MZU
Harrison C G8ZIC
Harrison C G4TTS
Harrison C G8KRG
Harrison C GW0TWR
Harrison D G0FXD
Harrison D G1MDE
Harrison D G4PGQ
Harrison D G4ROR
Harrison D G6WVM
Harrison D G7NQZ
Harrison D M0EAU
Harrison D 2E0KQV
Harrison D M6EXQ
Harrison D G1SPA
Harrison D M6DVH
Harrison E 2E0DLD
Harrison F G3XII
Harrison F G4MJT
Harrison G GD0NRN
Harrison G G2BHG
Harrison G G7JWI
Harrison H G0IVX
Harrison H G1IOT
Harrison I M6IFH
Harrison I G4JPX
Harrison J G4CLP
Harrison J G4HCB
Harrison J GD4XWF
Harrison J G8XXA
Harrison J M6JRH
Harrison J GM0NTR
Harrison J M6EXS
Harrison J M3CWC
Harrison J M6WWJ
Harrison J M6GKM
Harrison K G0KBZ
Harrison K G1NDL
Harrison K M3KVH
Harrison L G4RDY
Harrison M G0IYK
Harrison M G3HKH
Harrison M G3HUB
Harrison M G3USF
Harrison M G6DPS
Harrison M G6YWV
Harrison M M3EVV
Harrison M M6KMS
Harrison M G6NVT
Harrison M M6DGG
Harrison M 2E0EAO
Harrison M M0ROK
Harrison M M6BBM
Harrison M M6VOY
Harrison M 2E0RXT
Harrison M M6YBT
Harrison M M6CHX
Harrison M GM1THR
Harrison P G0MYM
Harrison P G1PQK
Harrison P G3WAB
Harrison P G6NVS
Harrison P G8MJH
Harrison P G8OZY
Harrison P M3AQG
Harrison P G4VAM
Harrison P G8ADY
Harrison P G4ZWX
Harrison R G0SHJ
Harrison R G3TMQ
Harrison R G3VPR
Harrison R G4LMF
Harrison R G8HGN
Harrison R G8PFZ
Harrison R G8WNQ
Harrison R M3ZRH
Harrison R G1YXH
Harrison R GD4VBA
Harrison R 2E0RJH
Harrison R 2E0HDF
Harrison R G4UJS
Harrison R M0RJX
Harrison R M6RJH
Harrison R M0BTZ
Harrison R M1BHP
Harrison S 2E0NKF
Harrison S G0AJF
Harrison S G0TOQ
Harrison S G0UTN
Harrison S G1KSC
Harrison S G7FPS
Harrison S M1EQW
Harrison S M1SJH
Harrison S M3NKF
Harrison S M3WPH
Harrison S M6SNQ
Harrison S G4JJS
Harrison S M0IBW
Harrison S M6SPH
Harrison T 2E0TFH
Harrison T GM3NHQ
Harrison T 2E0OWH
Harrison W G1ACV
Harrison W G3SNH
Harrison W G7PMK
Harrison W M0CPL
Harrison-Webb G M6GHW
Harris-Pugh L M6FKU
Harriss S G0RTI
Harrisson C G3OPJ
Harrod D M6NUR
Harrod R G4JUR
Harrold C G4RRN
Harrold G G7EOA
Harrold M G3VXA
Harrold M M1WEH
Harrop A 2E0MND
Harrop A M3MND
Harrop B M6FUO
Harrop D G6YLQ
Harrop D G0FUO
Harrop D G4BMQ
Harrop I G7EGQ
Harrop I M3IJH
Harrop J M3TMX
Harry P G4BOF
Harsley J M3MKY
Harste R G8LAB
Harston T GW8WRC
Hart A MM0DHY
Hart A G8VTV
Hart B G7VHO
Hart B G8UVZ
Hart D GI4DAV
Hart D G4YGH
Hart D G1DRW
Hart D G1JEA
Hart D G4XXK
Hart D M6DNZ
Hart E G0SRZ
Hart F MW3BOC
Hart G M0GRH
Hart G 2E0GPH
Hart G M0PGH
Hart G M6GHT
Hart H G4TWK
Hart J G3ZGA
Hart J G4SQV
Hart J G7VHN
Hart J G8GIK
Hart J G7RKW
Hart J M6ECZ
Hart J M6JLW
Hart J G4POF
Hart J M0EAD
Hart J M6MVK
Hart K M3JZD
Hart L G6RSI
Hart L G8ZVI
Hart M G6REC
Hart P 2E0DEN
Hart P G0THD
Hart P G3SJX
Hart P G4JSM
Hart P G4XVP
Hart P G7HUK
Hart P M6PHT
Hart R 2M0RUP
Hart R G0NNE
Hart R M3DTH
Hart R 2E0RIH
Hart R M6JGH
Hart R 2E0RJO
Hart R M0RCN
Hart R M6RJO
Hart T G3VFO
Hart T G4KPF
Hart T M0MEY
Hartas P G6HTA
Harte M MI3DLO
Harte R G4VBI
Hartgroves S G4OIA
Hartigan C G7KLP
Hartigan J G0UWH
Hartin J GI0EWP
Hartland A G8WOX
Hartland B M6DQH
Hartland R G8IVO
Hartle R M3XAV
Hartless D M3ASN
Hartless G 2E0PGI
Hartless G M0PGI
Hartless G M6BHH
Hartless I M6RDW
Hartley A G6URF
Hartley A G8PRH
Hartley A M3UHB
Hartley B G0GGT
Hartley C G3VJV
Hartley C G7BBD
Hartley D M1BHP
Hartley D M3UQZ
Hartley D 2E0FUH
Hartley E G1CSY
Hartley G M6BNJ
Hartley J 2E0FUJ
Hartley J G4ING
Hartley J G3WGQ
Hartley K G7CKQ
Hartley L M6CHG
Hartley M M6XCZ
Hartley N G7SZF
Hartley P 2E0HZY
Hartley P G0SKN
Hartley P M6PLI
Hartley R G0NCY
Hartley S G0FUW
Hartley W 2E0DYP
Hartley W M6WRH
Hartnell D M6YWX
Hartridge J G1WID
Hartropp D M6FHD
Hartshorn C M6ROW
Hartshorn D G7SFF
Hartshorn G G1YPT
Hartshorn V G3VZO
Hartshorne J M3TEY
Hartt C G8ZEV
Hartung M G3NQN
Hartwell D G0VOV
Hartwell H GW0PNC
Hartwell J G3UOJ
Harvey A G1XFX
Harvey A G7JTF
Harvey A GI7VIW
Harvey A M3ZNJ
Harvey A M6HLA
Harvey A G6JCK
Harvey A MW3UYX
Harvey B G8RIW
Harvey C G3XWM
Harvey C G3ZIO
Harvey C MM6VRH
Harvey C G6LVT
Harvey C M6VIG
Harvey D G1GYF
Harvey D G3XBY
Harvey D G4BJN
Harvey D G6DJH
Harvey D M6HHQ
Harvey E M3BFJ
Harvey E M3DLZ
Harvey F G4YOA
Harvey F G7FEE
Harvey F M6FFW
Harvey G 2E0CLG
Harvey H M1EMX
Harvey I G4COR
Harvey I G7BJG
Harvey J 2W0KAY
Harvey J MW0KAY
Harvey J M3LDJ
Harvey J M3BDQ
Harvey K G1WMN
Harvey K M1KDH
Harvey L G3MJN
Harvey L G4OOX
Harvey M 2E0GRW
Harvey N GM0NLU
Harvey N G4WTQ
HARVEY O G8KQH
Harvey P G1OPT
Harvey P 2E0SNA
Harvey P G0RLA
Harvey P G1ZTN
Harvey P G1ZZA
Harvey P M0SET
Harvey P M3YRO
Harvey R G0FQD
Harvey R G3IJV
Harvey R G3YHM
Harvey R GU4JHH
Harvey R G4YKZ
Harvey R M3PIL
Harvey R 2E0RVE
Harvey R M6RVY
Harvey S G0OYF
Harvey S G6KHN
Harvey S G6SVJ
Harvey S MM0SJH
Harvey S 2E0SBK
Harvey S M6SLA
Harvey S 2M0SMM
Harvey S MM0HUF
Harvey S MM6CXA
Harvey S G7IHN
Harvey T M6MHE
Harvey T G6APW
Harvey W GM4VYQ
Harvey-Ross M M6ETZ
Harvie D MM6FFQ
Harwood A G4HHZ
Harwood A G4PQU
Harwood C M6HVP
Harwood D 2E0YYG
Harwood D M6YYG
Harwood D 2E0VNO
Harwood D M0VNO
Harwood D M3VNO
Harwood F 2E0TPD
Harwood F M0VLC
Harwood F M3WUV
Harwood I G8LHT
Harwood J G3WLY
Harwood M M0XAT
Harwood R G8GJM
Harwood R M6HUQ
Harwood R M6OWT
Harwood R G0HRT
Harwood S G4OWT
Harwood S G6CWH
Haselden F G1CCW
Haselden O M0DYA
Haseldine M M3LFZ
Haseldine S G8EBM
Hashim M M6MHM
Haskett E G4OZG
Haskins P M1FZL
Hasko S G8RSA
Haslam A M6GHI
Haslam A G7NZR
Haslam J G4KBQ
Haslam J M3GHI
Haslam M G3UJU
Haslam P G4WMA
Haslam T G3XSI
Haslam-Brunt P 2E0PHB
Haslehurst A G6JCV
HASLEHURST D G4VLK
Haslehurst H G6JCT
Haslem G MI6ECC
Haslewood C G7TCW
Haslip G G8DWX
Haslip J G8ZAT
Hasman C G7KFN
Hasman I G3XFU
Hasman I G7KFM
Hassall G G0WHD
Hassall K GW1PND
Hassall S 2E0SCX
Hassall S M6ALQ
Hassall T 2E0AHU
Hassall T G7KLT
Hassan S M0SMH
Hassel H GM4SSA
Hassell Bennett R G3WJN
Hassell R G4NBQ
Hassmann P G4REX
Haste J G8UVY
Hasted E G3BHF
Hasted T G7WKC
Hastie J G4IRV
Hastie K G4DKH
Hastilow P G4NHA
Hastings D G6WMG
Hastings D M3AZF
Hastings J G6EQS
Hastings K 2E0KEI
Hastings R 2E0BOB
Hastings R M3RDH
Hastings T G0CFN
Hastry B G1OND
Hastry J 2E0HEF
Hastry M 2E0GMO
Hastry M M1GMO
Hastwell C M1DZP
Hastwell C M3CSH
haswell A G6JSI
Haswell C MM1AEL
Haswell G G7MWU
Haswell M G4KAX
Haswell T M6GTQ
Hatch D M6XYH
Hatch E G3ISD
Hatch J G3OOL
Hatch M G8EXK
Hatch R G0UHS
Hatch R G6CGQ
Hatcher D G7NGF
Hatcher P 2E0DRC
Hatcher R 2E0ALW
Hatcher R G7BKN
Hatcher R M5ALC
Hatchman D M6HWM
Hateley A M6DYR
Hateley P 2E0OTB
Hateley P M6AQD
Hately J GM4EQY
Hatfield D G4WBP
Hatfield J G6JCM
Hatfield J G7EAT
Hatfield R G8NVT
Hatfield R MW0HAT
Hatfull F 2E0CNP
Hatfull F M6FBH
Hatfull M 2E0CNS
Hatfull M M6MSH
Hathaway D G1EPD
Hathaway R G3JHI
Hathaway R 2E0BII
Hathaway S GW7FBV
Hathaway T G4WVH
Hatherall E G6BXU
Hatley B G4JPE
Hatt G 2E0CHU
Hatt G M0HEJ
Hatt G M6ATL
Hatt J G1CBS
Hatt M G3YTR
Hattam M G4KGA
Hatter C M3FMV
Hatter P G6PYL
Hatter P M3ODC
Hattersley A G0CWS
Hattersley M G1EVA
Hattie W GM4LYV
Hatton D G1BYQ
Hatton D G4VWI
Hatton D G6VBK
Hatton J 2E0KBE
Hatton J GM4RJX
Hatton J 2E0JOH
Hatton J M0YJO
Hatton J M6LDL
Hatton K G3VBA
Hatton L G4WPG
Hatton M M3RIE
Hatton R G0VVP
Hatton R 2E0HAT
Hatton R M6HAT
Hattrick W G0PZB
Hatwood M G4VLU
Haughey J M1EVN
Haughey M M3GIK
Haughey P MI3NCC
Haughie G M1AEB
Haughton D M0CTN
Hauser M 2E0XOR
Hauser M M0XOR
Hauser M M6XOR
Hausler M G1NHG
Hauton H G0TNS
Hauton J G7ERS
Hauton J M0ERS
Hauxwell R 2E0DRX
Havard F G4VEY
Havard M G4USN
Havart R M0HJI
Havell G G7IHV
HAVENHAND B G3OOP
Haver C G8VVC
Haver C G1IXV
Haver I G6VEY
Havercroft C G8CTX
Haverson R G4DHT
Havran V G4VGN
Haw A G7KHT
Haw C M0KOG
Hawbrook J M6FHE
Hawes A G1JPP
Hawes B G3YLY
Hawes B G8ZRG
Hawes B G0KML
Hawes B 2E0BMH
Hawes B G2KQ
Hawes B M3SSO
Hawes C G0UCH
Hawes D M6NJK
Hawes I G1IYE
Hawes J G4JHP
Hawes J G0DWD
Hawes J G8CQX
Hawes K M0KLH
Hawes M G7TAF
Hawes P G4CKW
Hawes P M6MOB
Hawes R M6PZA
Hawes R M6UES
Hawes T 2M0TIP
Hawes T MM6TYM
Hawke K 2E0BIM
Hawke R G4FPG
Hawken D G6NQO
Hawken D M3APQ
Hawker G G1REL
Hawker M M0RKA
Hawker M M6WKI
Hawkes A M0ZDU
Hawkes C G0BVQ
Hawkes C G7RSA
Hawkes D G4FOR
Hawkes D M1BBS
Hawkes D M6YPD
Hawkes J 2E0BNH
Hawkes J M0XFX
Hawkes J M3SKV
Hawkes P M6FUN
Hawkes R G6FEJ
Hawkes R M6RQH
Hawkes S M6SNH
Hawkes W 2I0WMH
Hawkes W MI6WAF
Hawkesford T M6TVW
Hawkings J G4GVE
Hawkings K G4RXB
Hawkins A 2E0ELK
Hawkins A G4GKK
Hawkins A G4TCC
Hawkins B M3TLJ
Hawkins B G4YBH
Hawkins C G0CEU
Hawkins C G3VLC
Hawkins C G4RND
Hawkins C M3HCG
Hawkins C G6FGC
Hawkins D G6XEL
Hawkins D G7AAS
Hawkins D G7LEL
Hawkins D G8KNF
Hawkins E GW4ZEA
Hawkins G G0PSK
Hawkins G G1YOA
Hawkins G G4ATQ
Hawkins I G3PNO
Hawkins I G0TAI
Hawkins J G8CPN
Hawkins J M3JKZ
Hawkins J MW6LCX
Hawkins J MM0HVA
Hawkins J MM6HBF
Hawkins J M6JBT
Hawkins K M6BJO
Hawkins L M0LDH
Hawkins M G7OKF
Hawkins M G0HZL

Hawkins N M6HSP
Hawkins N M6IGS
Hawkins P G4JAA
Hawkins P G4KHU
Hawkins P M0VEC
Hawkins P G3YUD
Hawkins R 2E0XWD
Hawkins R M0BDJ
Hawkins R M3LVF
Hawkins R G4VWF
Hawkins R M0RHS
Hawkins R M0RDY
Hawkins S 2E0RHS
Hawkins S M0ODE
Hawkridge A G0OWI
Hawkridge C G4RBC
Hawkridge C G4UKA
Hawkridge M M6OMH
Hawkridge W G4MXM
Hawkshaw M G7AQA
Hawksworth A 2E0ECO
Hawksworth A M6EXC
Hawksworth G G3JQC
Hawley D G6AUY
Hawley J G7TYB
Hawley J M6HJD
Hawley S GM8ZKU
Haworth A G0PYW
Haworth A G7PZU
Haworth B G1ZED
Haworth K 2E0XTC
Haworth K M0TNX
Haworth L M3UOL
Haworth M M3HQN
Haworth P G6OWI
Hawrylyshen A M0WTH
Haws C 2E0HWG
Haws C M6HWG
Hawthorn J 2E0ESX
Hawthorn J M6LOS
Hawthorn R G4TCA
Hawthorn W MM3RHA
Hawthorne D 2I0FPT
Hawthorne P MI0EDF
Hawthorn-Slater G G7KIO
Hawtree R G0TST
Hawxby A G1KAO
Haxton A GM0ARH
Hay B M3XEO
Hay B G4KRO
Hay C M3YQR
Hay C MM6CHY
HAY D G0TGW
Hay E MM3BHG
Hay J G1GYH
Hay J M0TGC
Hay J M6LIV
Hay K M6AEB
Hay N G3ZWR
Hay N G8GWM
Hay R 2M0RMH
Hay R MM3NXY
Hay R 2M0REH
Hay R MM0TQH
Hay R MM3TQH
Hay S G3HYH
Hay S 2M0HAY
Hay S MM0HAY
Hay W GM6AOJ
Hayati F M6HSF
Haycock E M3HEJ
Haycock S 2E0VZL
Haycock S M3VZL
Hayden E G4LZU
Hayden S M6SJH
Hayden T G0HCN
Hayden T M6ZAH
Haydn Smith P M0GKP
Haydock A G0SVQ
Haydock S G4KZW
Haydon A G7MID
Haydon D G4NLH
Haydon H 2E0ZEN
Haydon M G1KVO
Haydon P 2E0RGY
Haydon S G4WBI
Haydu Jones T G3OAD
Haye C G0YOU
Haye T G7DNF
Hayers G 2E0ELI
Hayers G M6WJL
Hayes A 2I0BAD
Hayes A M3INJ
Hayes B G4WXD
Hayes B 2E0BNB
Hayes B M0HAS
Hayes B M3JZM
Hayes C G7ACN
Hayes C G7CIH
Hayes C G1AUI
Hayes C 2E0XCH
Hayes C M3HSM
Hayes C M6BSV
Hayes D 2E0LDY
Hayes D M0GDX
Hayes D G4AKY
Hayes E M3INH
Hayes F 2E0ABL
Hayes G GM7IHZ
Hayes J G3FPH
Hayes J GM0NBM
Hayes L 2W0LEN
Hayes M G4BMD
Hayes P G3POQ
Hayes P M0BDW
Hayes P 2E0SCH

Hayes P M0PIT
Hayes P M6ETC
Hayes R M0RKH
Hayes R M6RIH
Hayes S M0CNY
Haygarth C G1CPO
Haygarth P 2E0HYG
Hayhurst A G6UPM
Hayhurst J M0DBB
Hayhurst M G0ECB
Hayhurst T G6UPL
Hayler P G1ITE
Hayler P M5ITE
Haylett B G8MTA
Haylett B G8VQE
Haylock P G7KJW
Haylor A M6HND
HAYLOR P G6DRN
Hayman J G6BVS
Haymes M G1KJQ
Haynes A 2M0SYL
Haynes A G4URA
Haynes A M6CRX
Haynes B G0IOE
Haynes B G6XTG
Haynes B MM6BFH
Haynes C 2E0EBL
Haynes C M6EVX
Haynes C 2E0HCW
Haynes C M6HCW
Haynes D M0CVZ
Haynes G G4FLY
Haynes G M6GVP
Haynes J M6HBV
Haynes K G3WRO
Haynes L G0BAW
Haynes L MM0BCR
Haynes M M0DXR
Haynes M 2E0MHR
Haynes M G6NVU
Haynes N G0VQO
Haynes P G8MGZ
Haynes P M6PBU
Haynes R 2E0RAH
Haynes R 2M0RYL
Haynes R M6RBH
Haynes S G0CLG
Haynes S G7HRZ
Haynes S M1SAZ
Haynes S MM6PDA
Haynes W G3STT
Hayselden R G7USI
Haystead J 2E0YLH
Hayter D G3OCI
Hayter G G0AZD
Hayter R G8BFO
Hayter S G6RAQ
Haythornwhite K M6VGU
Hayward A 2E0BBI
Hayward A G1GYQ
Hayward A M0NOS
Hayward A M1ETM
Hayward B G0KIC
Hayward B G0HHW
Hayward C M1EQO
Hayward D G2UH
Hayward D M1CNN
Hayward D M1GDH
Hayward G 2E0GRA
Hayward J M0BBE
Hayward J M0GXH
Hayward L M0SEC
Hayward L M6EHE
Hayward M G3LGA
Hayward M M6MRH
Hayward M G6JSF
Hayward P G0PSD
Hayward P G0TNG
Hayward P G3JMX
Hayward P G4PGB
Hayward P 2E0NPP
Hayward P M3ZJF
Hayward P M3ZMV
Hayward R G4OPR
Hayward R G4RTY
Hayward R G0RZG
Hayward R 2E0MIT
Hayward S M6LBR
Hayward S 2W0OZO
Hayward S MW6OZO
Hayward T G3HHD
Hayward T M3ZNX
Hayward W G4LJT
Hayward W G7NBD
Haywood A G7CRQ
Haywood I G4SGX
Haywood J G7KPM
Haywood J G7SSW
Haywood J M6DVZ
Haywood K G8HXE
Haywood K M3KMH
Haywood K 2E0KSH
Haywood K M6KSH
Haywood L G8KJJ
Haywood M M0BZY
Haywood P 2E0PCH
Haywood P M0VVZ
Haywood R G6GES
Haywood S G7CXT
Haywood S G8TFB
Haywood S G8UZQ
Haywood-Samuel M 2W0FOG
Hayzen D G7LXH
Hazel D M0AUA
Hazel R M6LDG
Hazel T 2E0TPH

Hazel T M3TPH
Hazell C G6CWF
Hazell E G0HAU
Hazell J 2E0JRH
Hazell J G7VWM
Hazell M G1EDP
Hazell N MD6HSV
Hazel-McGown M MM0ZIF
Hazeltine J 2E0JDK
Hazeltine J M6JHZ
Hazelwood P G4KZB
Hazle C MM3HJC
Hazlehurst N M6NTN
Hazlett C MI3EGJ
Hazlett J MI3RYJ
Hazlewood B G4DYU
Hazlewood G 2E0HAZ
Hazlewood G M3GHH
Hazlewood K M6RRW
Hazlewood R G0CGQ
Hazzard D G4HUM
Hazzledine M G0MRY
Head A G3YPY
Head A M3LKU
Head D G7PNE
Head G G4EBY
Head G G8JBP
Head J G0JCH
Head J G4VUD
Head M G1WLO
Head M M6MXH
Head N G4ZAL
Head N G7LUO
Head N M0DBO
Head P G4FYY
Head P G4LKW
Head R G8GBM
Head R M0RHE
Head R M3KFK
Head R G4XBW
Head R G8IMJ
Head S G8KOS
Head T G1KIZ
Heading C M1FDO
Headland D G8UJF
Headland R G4XRX
Headridge G MM6EOZ
Heagren E 2E0IJU
Heagren J 2E0HEA
Heagren J M3FQG
Heald F G4YDB
Heald G G0VLJ
Heald J G0UEA
Heale D G6HGE
Healen J G8EHF
Heales M G7OMA
Healey B M6GMQ
Healey C G0NCS
Healey D M3CVM
Healey J G7ANQ
Healey J 2E0IAY
Healey P 2E0IAZ
Healey P G0OON
Healey P 2E0IAE
Healey R G8TQP
Healey T G7BNS
Healy L M1AAC
Healy P G8HZQ
Healy P G1MKS
Heaney B G8IIS
Heaney G GI7TGJ
Heaney J G6RUY
Heaney L M6FCW
Heap A G1OCS
Heaps C G4NOR
Heard C G7HBV
Heard D G7YCK
Heard G M1BOA
Heard G M6GGH
Heard S M3VUY
Heard S 2E0EQY
Heard S M0SWH
Heard S M3EQY
Hearn A G3UEQ
Hearn C G3UIB
Hearn D G1VAY
Hearn E M6ECH
Hearn G G3REU
Hearn M G4AMI
Hearn N G4XTR
Hearn S G7UUW
Hearne A GJ8CEY
Hearne F M3HYO
Hearne M GW1YFP
Hearne R GW0NKH
Hearnshaw P M6AXW
Hearsey M G8ATK
Hearson P G3SIU
Hearsum D G2HLP
Heartfield T G7URM
Hearty A MM6AHB
Heasley J GI4GID
Heasman A G8GJO
Heasman D GM4LMG
Heasman N G4XDK
Heater C G6TSX
Heath A G8BHY
Heath A G4GDR
Heath B GU7MXZ
Heath B 2E0OBS
Heath B M6CAH
Heath C G8VVB
Heath D M6XEP
Heath E 2E0EAZ
Heath E M6ETH
Heath J G4IRB
Heath M G0BAO
Heath P G0CZY

Heath R G4HWN
Heath R G6NQY
Heath R MM1DZW
Heath R MM3RGH
Heath R G3LUV
Heath R G4VER
Heath R G6FQL
Heath R G4XIZ
Heath S G6MTE
Heath T 2E0MFD
Heath T M6HEA
Heath T M6TIW
Heath-Anderson A G1DIO
Heath-Coleman E G6YWZ
Heathcote D 2E0TAK
Heathcote D G0AOD
Heathcote D M3TBZ
Heathcote M M0LCK
Heathcote P M1FIB
Heather A G0PQA
Heather D G6SNA
Heathershaw J G4CHH
Heathfield H M6HPH
Heathfield J G1YKI
Heathfield K G3SDO
Heatley R G0OPV
Heaton D G1WZK
Heaton D MD6RHN
Heaton J G1YYH
Heaton K G0KNY
Heaton R G3JIS
Heaton R G3UGX
Heaton S 2E0BLK
Heaton S M3NUL
Heaton-Bentley M 2E0MKH
Heaton-Bentley M M6KHB
Heavens C G4AMD
Heaviside J G3NYX
Heaviside K G0NPG
Heaysman A G1IBP
Hebb D 2E0FHZ
Hebborn J GM4TJL
Hebden A G8BYB
Hebden C G3GRQ
Hebden C 2E0GRR
Hebden D G0MMX
Hebdige R G1IOR
Hebel S G4UCU
Hebenton C 2M0MBE
Hebenton C MM3VPK
Hebenton D GM4API
Hector G G4TVD
Hector H MI3IXE
Hedderley R G8JNR
Hedge B G6JCY
Hedgecock S 2E0OVF
Hedgecock S M0SHQ
Hedgecock S M6AEE
Hedges A G6PYM
Hedges J G7ANQ
Hedges M G0JHK
Hedges M 2E0IAE
Hedges M G7EDZ
Hedges R M1AVW
Hedges S G3DBV
Hedges S G7UXK
Hedicker P G1SED
Hedicker S G8XJO
Hedison P G4NLC
Hedley D G0OYN
Hedley E 2E0AOS
Hedley E 2E0EYF
Hedley G G0SOF
Hedley G G8BDQ
Hedley J GM8TXC
Hedley M MW6HED
Hedley P G6PTT
Hedley R G8MFF
Heed B G8JAW
Heel M G0GZR
Heeley A G3PFT
Heeley R G4EPD
Heeley R G8TXA
Heeley T G3SWU
Heenan L 2E0BHC
Heenan M 2E0SFS
Heenan M M6PZW
Heerma Van Voss S GM0UMJ
Heesom R G0VYK
Heffer G G4NSJ
Heffer R M6HFR
Hefferman W G8XJN
Hefford A M6ASQ
Hefford A G1NEN
Hegarty D M0HTL
Hegarty J GI4NFW
Hegarty S 2E0SZH
Hegarty S M3AMB
Hegarty S 2E0VWW
Hegarty S M3VWW
Heggan G MI3GSW
Heggie A GM0NAZ
Heigh A 2W1FJG
Heighton M G6DET
Heilbron S G3MIP
Hein J GM1YME
Heiney P 2E0DFO
Heiney P M0HWV
Heinonen J M3VHE
Heirene B 2E0HAS
Heirene B M3BCH
Hekman P M0XPA
Hele Kergozou de la Boessiere J M6JHK
Heley M M1EJX
Helgesen K G7CVY

Helie I GM7VHQ
Heller L G1HSM
Hellewell A G4PNT
Hellier A G4ZZD
Helliwell P G7SME
Hellon M M6THC
Hellowell J M3JQS
Helm A G4BCX
Helm G G6VEZ
Helm P G6AMX
Helm P G8AEN
Helman B G4TIC
Helsdon B G6XRE
Hembrow C M6PXL
Hemenway A G1VIZ
Hemesley R MM1RAH
Heming R 2E0VJH
Hemingway D G8EGG
Hemingway N G7BPM
Hemmens H GW4KUS
Hemming A G1MJO
Hemming J G0UYT
Hemming M 2E0RKX
Hemming M M0RKX
Hemming M M6RKX
Hemming R MW6FIX
Hemmings C G8JKB
Hemmings D M3MBH
Hemmings K M3XTP
Hemmings M 2E0TSC
Hemmings M G0NCQ
Hemmings M M0CMH
Hemmings M M3YCM
Hemmings N G4NQQ
Hemmings R G3VCT
Hemmings R GM0TVT
Hemmings S M0EDA
Hemmings S M3LPF
Hemmings S M6SKH
Hemmins D G6DRP
Hemphill P G0HUG
Hempsall K G8SMH
Hempsall T M6TJJ
Hempstead R G8TOI
Hemsil K G7TLK
Hemsley P M3TFK
Hemstock M 2E0HFW
Hemstock M M3VRV
Hemsworth S G0JTT
Henderson A G0EJF
Henderson A G1RRW
Henderson A GM1WYV
Henderson A 2E0ZED
Henderson A M0ZDD
Henderson A M3ZED
Henderson A G3SXH
Henderson A MW3UIQ
Henderson B G0EET
Henderson C MM3CLA
henderson D G0PVT
Henderson D G7FKP
Henderson D G8YQO
Henderson D G4NTC
Henderson D MW6KDA
Henderson D MM6AKZ
Henderson E G3LYD
Henderson G M6GJH
Henderson G GM6JKU
Henderson G GM3RTJ
Henderson I G1WEV
Henderson J G0KKH
Henderson J GM4HKV
Henderson J G1AOZ
Henderson J M1DSU
Henderson J 2I0JCH
Henderson J GI7SBF
Henderson J MI6EAS
Henderson J 2M0BUY
Henderson J MM3YMN
Henderson K GM4AOR
Henderson K MM6DSC
Henderson K G6SVH
Henderson K M3SVH
Henderson M MM3UET
Henderson N M6ELH
Henderson P GM4VRE
Henderson P M3FUR
Henderson P MW3UIY
Henderson P GI4KBW
Henderson P M0GTR
Henderson P M3SKQ
Henderson R 2W0AUC
Henderson R G0LNA
Henderson R GM0UET
Henderson R GM4DTJ
Henderson R M0GTJ
Henderson R G3NAN
Henderson R G3ZEM
Henderson R 2E0RFH
Henderson R M3RFH
Henderson S G0EES
Henderson S G1BNV
Henderson S GI4OUP
Henderson S MM0SAH
Henderson T GI7GVI
Henderson T G1PJB
Henderson W 2E0WDH
Henderson W GM0VIT
Henderson W G3KOZ
Henderson W M3WDH
Hendon D G8DPQ
Hendricks D M6GLS
Hendricks G G1OKV
Hendricks K M6EGL
Hendron J GI8DHW
Hendry B 2E0ODT

Hendry B MM0WAX
Hendry C G0ODR
Hendry G M0GZS
Hendry I MM1DVG
Hendry J 2E0OFF
Hendry K G0BBN
Hendry K G8TSP
Hendry L 2E0OCL
Hendry L M0OCL
Hendry P M6DNJ
Hendry R GM6OLM
Hendry R M6RHO
Hendry S G0ASN
Hendy P G1OFY
Hendy R MI3CQR
Hendy S 2E0ZEH
Hendy S M0ZEH
Hendy S M3ZEH
Henery R G7KIW
Henk A G4XVF
Henley D G3OQO
Henley L M3MRS
Henman M G6UVN
Henman M M6GOG
Henn D G8LIY
Henne G G1XUE
Henneman R G6JKV
Hennessey M 2E0SPG
Hennessey M M0MBV
Hennessey M M6LWT
Hennessey P M3XPH
Hennessy P G4PWY
HENNEY K 2E0DMQ
Hennigan T G4GNW
Henniker C M6KWL
Henning A G0DPQ
Henningway J M6CWU
Henretty D G0KOF
Henry A MI6DOD
Henry D GM7VZV
Henry D MM0AOF
Henry J GI0DVU
Henry J MM3HQC
Henry K G4CFB
Henry M 2M0SEG
Henry M MM6MFA
Henry P GW8SZC
Henry P 2E0SQK
Henry P M6HSB
Henry R GW4JQQ
Henry R M0RHH
Henry R MM3RHH
Henry R MM5AON
Henry S G7VDJ
Henry S G4GNS
Henry S MI3VXI
Hensby D G8TKD
Hensby D 2E0TKD
Hensby G 2E0JEF
Henshall A 2E0KLR
Henshall G M0RGC
Henshall R G0VJR
Henshall R G4NIL
Henshall R M0CVK
Henshall R M3AVZ
Henshaw A G3VQR
Henshaw B M6HEN
Henshaw G G6CDT
Henshaw J G0GBN
Henshaw J G6AOF
Henshaw M G8PQN
Hensman A M3TXH
Hensman D 2E0PLY
Hensman R M3DIS
Henson M 2E0HGY
Henson P G4IIH
Henson P M0BPY
Henson R M0ROY
Henson S G6IXS
Henson S 2E0KNC
Henson W G6JCI
Henson W M6SIK
Henstock A G4XIN
Henstock G G0HEN
Henstridge G M3UED
Henstridge G G4UED
Henville J G6FVD
Henville R G3TPH
Henwood P G3RWF
Hepburn B G8BGI
Hepburn G MM6KVI
Hepburn J 2E0SCO
Hepburn J M3WLU
Hepburn M GM0OSX
Hepburn R MI3OZI
Hepburn R GM7DAJ
Hepburn R MM6RHQ
Hepke K G4NJB
Heppenstall B G4CWU
Heppenstall L M6LHE
Hepple T G4UNI
Hepplestone D G4KUL
Heptonstall C G1HEP
Hepworth A G4VOG
Hepworth C G1DVO
Hepworth D G4LKX
Hepworth D GM4ZDH
Hepworth D G7ABT
Hepworth F G4SAV
Hepworth G G1LFI
Hepworth H M0EBV
Hepworth I 2M0BVN
Hepworth I MM0IHE
Hepworth I MM3XIZ
Hepworth S G0KMN
Herant J M3ZJH

Herbert D G0VXE
Herbert D G4PPS
Herbert D MW0CUA
Herbert G G0NXA
Herbert J G4UIH
Herbert M M3WQA
Herbert M MW6XDA
Herbert P M5ABN
Herbert V G0ATB
Herbison T MI0IOU
Hercombe B G4JCH
Herd A M6DHT
Herd C 2E0CVD
Herd C M6TBR
Herd D G4SOZ
Herd D MM3UDV
Herd R MM0BGO
Herd T G0TJY
Herdman M M3BLO
Heredge P G4PWI
Herf L G0EOP
Heritage A G4EOG
Heritage C G6UVO
Heritage F M0AEU
Heritage H GM0PUN
Heritage M M3UYF
Herke D G8IBC
Herlingshaw C M6CLI
Herman Cranmer P G4TFP
Herman S M0SSH
Hermes W G3YHC
Hern A G1UFX
Herod P G8TQI
Herod S G8EAX
Heron H G1HTF
Heron K M6KTY
Heron L M6LCY
Heron M M1KDP
Heron R G8KFN
Herpe M M0AEZ
Herrett C G4LSV
Herrick B M6AST
Herridge D G1VSE
Herridge D M1DNZ
Herridge M 2M0ORK
Herridge M MM0MDH
Herridge M MM6ORK
Herries J G0MKY
Herring A GM0CQQ
Herring P G4FQF
Herring P G8WYI
Herring R G4BNE
Herringshaw R G4HTH
Herrington J G0VJH
Herrmann P G0SVB
Herron I G0VHB
Herron P 2E0CVZ
Herron P M6CVE
Herron W GM4LHQ
Hersey P G4UDW
Hersey R G8ENY
Hersom J G7BYE
Hersom M 2D0CBT
Hersom M MD6MHA
Herwig C G4YVE
Heselton K M1EMC
Heselwood R G4UYI
Hesketh D G6MTB
Hesketh J G1OLM
Hesketh J G1JZX
Hesketh K M0AQK
Hesketh P G4LIG
Hesketh W MI5EEM
Heslop R G3KMQ
Heslop R 2E0HES
Heslop R M6GEV
Hess J G8CCO
Hessom D G4GFM
Hester P G7IVW
Hetherington C G8XVO
Hetherington C G4ZOH
Hetherington F GM3UCN
Hetherington J G4YGU
Hetherington K G1YUN
Hetherington L G8GRD
Hetherington M 2E0MWH
Hetherington R GM1OVW
Hetherington R MI6LLI
Hetherington S G7DMH
Hetherington W G7DMG
Hetherington W M5WAH
Hettiarratchi D G1VOP
Heward A G0PVY
Heward G G4BJM
Hewart J 2E0JEH
Hewart J M6FHI
Hewat A G0NTH
Hewer I G4CLR
Hewes A G1VCZ
Hewes C G6SFF
Hewes J G6AOA
Hewett A G0KJD
Hewett A M0CFH
Hewett C M3UCJ
Hewett J G4SVE
Hewett L G4PBE
Hewett N G0DDV
Hewett P G1ENP
Hewett R M0LEP
Hewgill N M1CGO
Hewick S 2E0HDX
Hewins F G1HFW
Hewins M G3WKI
Hewis G 2E0GRF
Hewis G M0HWS
Hewis G M3ZJV

Hewitt A G3SVD
Hewitt A G6PFN
Hewitt A GI8LUR
Hewitt B G0AHG
Hewitt B G8IPY
Hewitt C G0PAE
Hewitt C G7LPF
Hewitt D GM0CAD
Hewitt D G8ZRE
Hewitt E G0PVJ
Hewitt F G6JCX
Hewitt G G6IXN
Hewitt I G4SVL
Hewitt I G8SNF
Hewitt I M3SNF
Hewitt J G0CUO
Hewitt J GM3ZOT
Hewitt J G7IFM
Hewitt J G7TUV
Hewitt J 2E0JMH
Hewitt J 2E0DSU
Hewitt J M0SLH
Hewitt J M6JMH
Hewitt J G0BNZ
Hewitt J M3JGH
Hewitt J M6FBQ
Hewitt K G6DER
Hewitt L G0PVO
Hewitt L G2HNI
Hewitt M G0SQS
Hewitt M G4AYO
Hewitt M G4NCU
Hewitt M G7JVC
Hewitt M G7TKM
Hewitt M G7TKP
Hewitt M M3SVD
Hewitt M G7IIN
Hewitt N G8JFT
Hewitt N M3UKN
Hewitt N 2E0NDH
Hewitt P G8JCV
Hewitt P M3FNM
Hewitt R G6TTD
Hewitt R G4MHJ
Hewitt S GI4MDO
Hewitt S M3SJH
Hewitt S GI7PIZ
Hewitt T G7WEM
Hewitt T M3TDH
Hewitt W G1EDT
Hewitt W G3YFD
Hewitt W G4WEP
Hewlett C GM1PFU
Hewlett J M1AYN
Hewlett J 2D0JTY
Hewlett P M1BKE
Hewlett T M6TCH
Hewson B G0TVO
Hewson D G6YXB
Hewson M M3GFZ
Hewson P M3TSI
Hewson R G4KFF
Hewson R G4NJA
Hewson R 2E0MMX
Hewson R M6MMI
Hextall C 2M1HVR
Hey C MW6HEY
Hey P G4JHS
Heyburn J MI6JVC
Heyes A G1VSK
Heyes A G3ZHE
Heyes A M3LFU
Heyes G M0JIL
Heyes M G0EIM
Heyes P M6ZJT
Heymans C G8CMP
Heyne N 2E0AVZ
Heyne N M0EDQ
Heyne N M3BPZ
Heys J G3BDQ
Heys J G1XZV
Heys S M6TKU
Heywood C G7FVR
Heywood F G4IDT
Heywood J G4IAL
Heywood M 2E0MLS
Heywood M M0ICK
Heywood N M6GZK
Heywood R G1MET
Heywood S M3ITZ
Heywood-Bell T G7RNC
Heyworth A G0VAH
Hibberd A G8AQN
HIBBERD D G0RFO
Hibberd D G0ODH
Hibberd D MM3ZRZ
Hibberd F M0BVV
Hibberd G G3PYP
Hibberd G G4OTX
Hibberd J G3YMU
Hibberd W M0TYW
Hibbert C G1AIA
Hibbert J G3YCV
Hibbert J G6UMX
Hibbert J G8LZO
Hibbert M M1DEJ
Hibbett F G8LAY
Hibbin D G4AOP
Hibbin D G8CPK
Hibbitt M G4IKL
Hibbitt M G3ULN
Hichisson K M1EPK
Hicken C M0DMY
Hicken M G6ZVD
Hickey A 2E0CKS
Hickey A G7AAI

UK Surnames

**IMPORTANT NOTE**

**Revalidate licence to avoid revocation** – Ofcom has advised the Society that plans will be drawn up to revoke licences that have not been revalidated as required by the licence conditions. The quickest way to revalidate is to do so online via the Ofcom website: *https://services.ofcom.org.uk/* or by email: *amateur.validations@ofcom.org.uk* If you need assistance in the process, Ofcom staff are available to help, but please be patient during times of heavy workload.

Hickey A M6HKY
Hickey D MI6DCH
Hickey J G1JMH
HICKEY J M3HJJ
Hickey J G7HIQ
Hickey M 2E0MGH
Hickey M 2E0SYN
Hickey M M6CXO
Hickey P G3WDX
Hickey P G6LVI
Hickey R G6LVJ
Hickford G M0BLS
Hickford G M0GRA
Hickford M 2E0ORT
Hickford M M0HEY
Hickford M 2E0MJH
Hickford M M0MJH
Hickford M M3MJH
Hickin A G3PXL
Hickinbottom I M6IMH
Hickling P 2E0LAQ
Hickling P M3LQL
Hickling T G7HBU
Hickman A G0IAS
Hickman B G6PZH
Hickman E G7JWJ
Hickman G G8KPV
Hickman J G8JWE
Hickman J 2E0BHU
Hickman J M3NKN
Hickman M G3VGE
Hickman M G7BGZ
Hickman M G8XUN
Hickman M M0MUZ
Hickman M M0HVV
Hickmott R G8MFV
Hicks A G0TXY
Hicks A G1DSM
Hicks A G1EDU
Hicks A M3HWV
Hicks A M6OLT
Hicks A M6ATH
Hicks A G8JVI
Hicks C G0WTB
Hicks C M6CWP
Hicks C M0HLV
HICKS D G8EPR
Hicks D M0HLD
Hicks G M0NBK
Hicks J G4VTM
Hicks J G8NMT
Hicks J G6XRF
Hicks L G4GMS
Hicks L M0GWZ
Hicks P G4DVP
Hicks P G4KCX
Hicks R 2E0HVT
Hicks R M3RJH
Hicks R 2E0MEX
Hicks R M6HCS
Hicks S 2E0HVU
Hicks S G6DYK
Hicks W 2W1GAC
Hicks W G1FBZ
Hicks-Arnold E G0CDZ
Hickson N 2E0NRH
Hickson N M0NRH
Hickson N M3ILM
Hickson T M3TSA
Hickton D G6DRH
Hide R G0LFF
Higbee D G0SQH
Higbee K G6VPK
Higgin M G0ECG
Higginbotham J GM4GXR
Higgins A M6DBV
Higgins A G8GLY
Higgins A GM0PTY
Higgins B 2E0HIG
Higgins B M6BGH
Higgins C G3NRQ
Higgins C MM3SYU
Higgins C M1WIN
Higgins C M6TOL
Higgins D G3XVR
Higgins D G8XMH
Higgins D MM0HGN
Higgins E 2M0EDY
Higgins F G3YNG
Higgins G G3UBD
Higgins J G4UAF
Higgins J MI1CSA
Higgins J M3JER
Higgins J GM3ZXG
Higgins J G6VLT
Higgins JP P G6NVW
Higgins K G1HRV
Higgins L MM3HKE
Higgins M GI3YMT
Higgins M MM0DUN
Higgins M G4ZVL
Higgins M G7CBI
Higgins M G8BUF
Higgins N G4ZQL
Higgins N G7VIK
Higgins P M3YWM
Higgins P MM6PRH
Higgins P M3LPU
Higgins R 2E0RZL
Higgins R M6RZL
Higgins S 2M0GTR
Higgins T M3HIG
Higgins T M3JCE
Higgins W G1MCW
Higginson D GM8JET
Higginson J MI3FXE
Higginson J M6JPH
Higginson P G8IZR
Higginson P 2E0DXO
Higginson P G7EZH
Higginson P M6PJH
Higginson T M0WNV
Higginson T M1GXL
Higginson W G4EGG
Higgs A 2E0HGE
Higgs A G6XAG
Higgs D M0NTH
Higgs G G4AWG
Higgs G G6BSS
Higgs P G4IGF
Higgs R G0IBE
Higgs S 2E0GJO
Higgs S G8YTR
Higgs T G4TUA
Higgs W M5AFE
Higham A G6ZBV
Higham A M3ZBV
Higham A M6ALH
Higham A M6AJH
Higham M 2E0VAM
Higham M M3XMH
Higham P G8JLM
Higham R M3IUX
Higham R G3VDS
Higham-Hook N M3NAH
Highams T M1ABY
Highfield A 2E0MLA
Highfield A M3HHN
Highfield A 2E0LZB
Highfield A M6HGF
Highfield N 2E0LSI
Highfield N M0LSI
Highfield N M6EPZ
Highley K G4MOL
Higlett K G1XKJ
Higlett M G6WTM
Higlett S G8ZYT
Higson J G4NTY
Higton A M0HLF
Higton B GM0VBE
Higton G G8KGK
Higton G M3KGK
Higton M G8YNK
Higton R G3XXR
Higton S M6YNK
Hilbery N G8LPA
Hilbery N M0LPA
Hilbourne A G6NVY
Hildebrand P M1APH
Hildebrand R G6EAZ
Hildich M G4BWR
Hildreth J G6NWS
Hiles R G4SXZ
Hiley B 2E0MDT
Hiley B M6CJQ
Hill A 2E0HKR
Hill A 2E0ZTG
Hill A G0JTP
Hill A G0XBA
Hill A G4BQJ
Hill A G4VPZ
Hill A G8GZN
Hill A M0ZTG
Hill A M3NHW
Hill A M3OTG
Hill A 2E0MFT
Hill A M3MFT
Hill A M6BOQ
Hill A G4PYQ
Hill A G7KSE
Hill A M6RDA
Hill A G8MGP
Hill B M3BJH
Hill B M3PUU
Hill B MI5HIL
Hill B G2BAR
Hill B G0PGZ
Hill B M6CMF
Hill C G2KG
Hill C M6KIX
Hill C 2E0CHH
Hill C G7VLH
Hill C M0LLO
Hill C M6CJH
Hill C G8RAX
Hill D 2E0DLH
Hill D G3ZCH
Hill D G3ZRB
Hill D G6OLJ
Hill D G8EEA
Hill D M0HIL
Hill D M3LRI
Hill D 2E0UGO
Hill D M3UGO
Hill D 2E0CCN
Hill D G0SNG
Hill D M0AZK
Hill D M0BWY
Hill D M1AFX
Hill D G4LKU
Hill D G8BEH
Hill E G4HZE
Hill E G6OWB
Hill E MI0TBE
Hill F G3YWH
Hill F G6APQ
Hill G G0NPC
Hill G G6EOA
Hill G G0ODA
Hill H G1EML
Hill I G4SWR
hill I M6CEZ
Hill I M3TBQ
Hill J G3JIP
Hill J G3XYH
Hill J G4CFH
Hill J G4NBR
Hill J G7CLY
Hill J G8FWH
Hill J G8YNP
Hill J M3JCH
Hill J 2E0JHT
Hill J M0TJC
Hill J M0XOX
Hill J M6JHT
Hill J 2D0JEA
Hill J MD3WFJ
Hill J M6SJE
Hill J G8HUY
Hill K G0VSL
Hill K G1EMM
Hill K G3CSY
Hill K G3XGU
Hill L G6ISY
Hill L MW3PKC
Hill L M0ARM
Hill M 2E0KFX
Hill M G0BEV
Hill M G0JAC
Hill M G4KUY
Hill M G4MJF
Hill M G4SUE
Hill M 2E0KNM
Hill M 2E0PIA
Hill M M6KNM
Hill M G7GOV
Hill M M6MCH
Hill N G4UKO
Hill N G7BVS
Hill N M3MNV
Hill P G0DRX
Hill P G1NRX
Hill P G4FRM
Hill P G4IOA
Hill P G6PNG
Hill P G6PNO
Hill P MI3KIL
Hill P 2E0PGH
Hill P M6PGH
Hill P G0REO
Hill P M1PJH
Hill R G0FJE
Hill R G0IUW
Hill R G3YTN
Hill R G4DLT
Hill R G6XEN
Hill R G8LEB
Hill R G8THE
Hill R M0HXE
Hill R M0RFH
Hill R M3XYZ
Hill R G0IMV
Hill R G3SMZ
Hill R G6TSL
Hill R GM0FTG
Hill R J G1YEW
Hill S 2E0SLH
Hill S G0JJV
Hill S G4XRO
Hill S G6FPX
Hill S G6PFP
Hill S M6VOL
hill S M6CAS
Hill S 2E0PEP
Hill S G1HRU
Hill S M3VXY
Hill S M6PEP
Hill S GD0OUD
Hill S GD7HTG
Hill S 2E0DPK
Hill T G0SPF
Hill T GM4NWK
Hill T G4YFC
Hill T G7VJY
Hill V GW4HDF
Hill W M1BKF
Hill Z M6ZYH
Hill-Adams G 2E0GHA
Hill-Adams G M0HAG
Hill-Adams G M3ZWJ
Hillam M M6AFH
Hillan R G8YVS
Hillard B G4PDG
Hillary M M3CZB
Hillary M M5KZI
Hillbeck A M3HQW
Hillbeck G M3UQF
Hillbeck K M3UFU
Hillcox C M6LZY
Hilleard D G4CQM
Hilliard J G4YRX
Hilliard R G7LPK
Hilliar-Mills M M3ZHU
Hillier C 2E0IDC
Hillier L M6AFC
Hillier M G0DDW
Hillier M G0MQM
Hillier P G1MUQ
Hillier P G6AIO
Hillier R M6RNI
Hillier T M6VBD
Hilling J G3HRX
Hillman B G0UXO
Hillman C G1OCH
Hillman C G7UWR
Hillman D M3EFQ
Hillman G G3OKH
Hillman M G1MPI
Hillman M 2E0VKZ
Hillman M M0TVV
Hillman S M6LHB
Hills A G1RCE
Hills B G8AOO
Hills C G8KKH
Hills D G6MFM
Hills D G6PYF
Hills F G0KUQ
Hills J G4YRA
Hills J M3WRN
Hills M GW1VMA
Hills M M6ZPE
Hills R G3HRH
Hills W G4LGU
Hill-Smith C G7RKX
Hillson B M6OPS
Hillum R G6PAE
Hillyer D G0RQO
Hilsley R G3PBT
Hilton A G8FME
Hilton B M6MYT
Hilton D G6VXE
Hilton G G4NKW
Hilton J G0IPC
Hilton J G1HAC
Hilton J G3JMZ
Hilton J M3HAC
Hilton J G0RJL
Hilton J GM1ZVJ
Hilton K M3RFI
Hilton M M1DHG
Hilton N G7KMD
Hilton N M0NMH
Hilton N M3MQH
Hilton P M3HSR
Hilton P G7MYT
Hilton R G1LKH
Hilton R G7DLD
Hilton R G4WZI
Hilton V 2E0AWZ
Hilton W G7JZI
Hilton W G0NRI
Hilton-Jones D G4YTL
Himmo O G1KGA
Himsworth S M6RTA
Hinchcliffe J MM6JXH
Hinchliffe B M3FJC
Hinchliffe J G0JTL
Hinchliffe M G0LQK
Hinchliffe N G7SVU
Hinchliffe P M3ILV
Hinchliffe R G8PDP
Hinckley A M3YYE
Hincks R M3ZTL
Hind D G3VNG
Hind D M6DDU
Hind D 2E0DSH
Hind D M0ZRD
Hind D M3VDH
Hind G G1FMV
Hind J G6BMG
Hind P M3VTX
Hind W G8EDS
Hinde J G4PED
Hinde P 2E0BFZ
Hinde R 2E0SPZ
Hinderwell N G8IFN
Hindl N M6LIH
Hindle A M3OPS
Hindle C M5CJH
Hindle D 2E0GHK
Hindle D M6SBV
Hindle H G3WBG
Hindle N G7VEX
Hindle P G0LYC
Hindle S M0ETY
Hindle S G8NVS
Hindley F GM3VBY
Hindley M G4VHM
Hindmarsh C G7RKJ
Hindmarsh J G8XGS
Hindmarsh S 2E0SNM
Hindmarsh S M6SMA
Hinds C M6FJE
Hinds G G1GKH
Hinds J 2E0EOU
Hinds J M0EOU
Hinds M M6MOH
Hinds S M3SPJ
Hindson T M1ETT
Hine D G3VFF
Hine D G4DIG
Hine M G6VFA
Hine N 2E0PNA
Hine R G8AQH
Hines D M1CXW
Hines G M3GTT
Hines G G6SAQ
Hines R M3LMQ
Hing J M3MQB
Hingley N G4JSV
Hingley-Hickson J M6FRR
Hingston B G6UPR
Hinks P G4LTK
Hinksman V G8JSR
Hinsley G M3JIL
Hinson D G0JKH
Hinson P GW8PFT
Hinton C G7JRM
Hinton G G7NCW
Hinton J G0OUK
Hinton J G4EZE
Hinton K G7MAT
Hinton S M6RFN
Hipkin I GM7TWM
Hipkin S G0OFF
Hipkiss A M0ZDO
Hipwell J G0LQM
Hipwell W G3HTX
Hipwood T G8OEU
Hird D G4RVH
Hirons J G6TGJ
Hirons P G1CEI
Hirst A M3IAO
Hirst A G8BRF
Hirst A G3TOW
Hirst C G0EPY
Hirst D M1AGH
Hirst D M1ANQ
Hirst G 2E0RKR
Hirst G M0RKR
Hirst G M6RKR
Hirst I M0CQH
Hirst J 2E0OVL
Hirst M M3NGO
Hirst M M0OLG
Hirst N 2M0YAG
Hirst N MM0GTB
Hirst N MM3YMQ
Hirst P G0GXF
Hirst P G0PHI
Hirst P G7NDC
Hirst R G4APO
Hirst R G4XJG
Hirst R M3BVP
Hirst R 2E0DDD
Hirst W G6TSM
Hiscock J G4RLM
Hiscoe G G4GYF
Hislop J G7OHO
Hislop R G0UEH
Hitch E G8ZYH
Hitch N G8ZYI
Hitcham C G1MPL
Hitchcott A 2E0DHZ
Hitchcott A M3UNH
Hitchen B G1VZW
Hitchen L G7OET
Hitchens A 2E0HGE
Hitchens A M3HGE
Hitchens J 2E0CFP
Hitchens J M6AFN
Hitches S G7HFE
Hitchins B G4CTU
Hitchins D G4GMB
Hitchins E M6EDH
Hitchins R M0DYV
Hitchman M G3HAN
Hives M M3NYQ
Hixson N M0ALB
Hizzey P G6YLO
Ho C G7TJV
Ho F M3VHO
Ho J M3ZNL
Ho W M3VHC
Hoad R G1ZMG
Hoare B G0VEC
Hoare B 2E0FKS
Hoare B M3FKS
Hoare C G4AJA
Hoare D G1DFW
Hoare D 2E0DBH
Hoare D M3ZFN
Hoare H G4PJD
Hoare J M6HHH
Hoare J G6XRH
Hoare M G4NBC
Hoare M 2E0CEU
Hoare N M0NEH
Hoare N M6NEH
Hoare R G0DUF
Hoath P G7JBW
Hoban J G0EVT
Hoban J G3EGC
Hobbs A G1TQN
Hobbs A G3OJX
Hobbs A M3MVR
Hobbs B G1KJX
Hobbs D GW0LYF
Hobbs D G8PVG
Hobbs G G1JRZ
Hobbs G G0SCX
Hobbs I M6ILH
Hobbs J 2E0HEP
Hobbs J G1FTD
Hobbs J M0KDT
Hobbs J M6HQD
Hobbs J 2E0CTW
Hobbs J M0ZGB
Hobbs J M6HBS
Hobbs K M6FLQ
Hobbs K G7NCV
Hobbs L G7IYH
Hobbs M G6YLN
Hobbs N G7LXV
Hobbs N G7IYG
Hobbs P G3LET
Hobbs R G6IRY
Hobbs R M6RGQ
Hobbs R M3OUU
Hobbs R G7KHZ
Hobbs S G6XRI
Hobbs S G7EXZ
Hobbs T M3TBH
Hobday S G3SKV
Hobden D GM3XMY
Hobden S G0VAJ
Hobin J G3OJ
Hobin J G3XIX
Hobkirk A G4VZH
Hobley J G4FBQ
Hoblin R G6AOH
Hobro D G4IDF
Hobson A M3ZZN
Hobson C G0MDK
Hobson C G7JGE
Hobson G M5EAY
Hobson J 2E0FIE
Hobson J M6MMF
Hobson K G1YPQ
Hobson L G0CUI
Hobson M GM8KPH
Hobson P G6OLU
Hobson P G4HOJ
Hobson-Smith T M6WPI
Hockedy J M3IKD
Hockenhull N M0NAB
Hockey J G4YJH
Hockey P GW7VOO
Hockin D G4UGT
Hockin M M3UMJ
Hocking A G0FHX
Hocking D G4FSS
Hocking J G0FHY
Hocking J 2E0WSZ
Hocking J M6WNB
Hocking M M0OMA
Hocking P G8ZDS
Hocking V 2E0AZJ
Hocking V M0VMH
Hocking V M3FXX
Hockley A G4ZUW
Hockley J G0ANW
Hockley J G0XYL
Hodby T M0CHU
Hodder J MW6HBM
Hodder K M3DOA
Hodder R G7IFD
Hodder S 2E0CRV
Hodder S M6MSU
Hoddinott L M6LDB
Hodds R G0KLG
Hoddy M G0JXX
Hodge A G4UPY
Hodge D G4FFS
Hodge F 2E0EPO
Hodge J G8CCD
Hodge K G4NEI
Hodge K GM3JIG
Hodge L 2E0ESK
Hodge M G7OLG
Hodge P M0WRI
Hodge R G3YJN
Hodge R G4MMI
Hodge R GM4PPT
Hodge S M6FXA
Hodges B G4JKF
Hodges C 2E0RBH
Hodges D G6IXH
Hodges E G4YUO
Hodges G GD3KRT
Hodges G MW6TJE
Hodges G G7OEY
Hodges G M0DBG
Hodges J G0WRN
Hodges J G7DLE
Hodges K G6TDG
Hodges L G0EJV
Hodges M G4OPE
Hodges M G6WZN
Hodges M G4PJD
Hodges M M6DJQ
Hodges P 2E0XYA
Hodges R G0NFZ
Hodges R G4DDH
Hodges R G6MFB
Hodges R G0RYL
Hodges S G7BMT
Hodges Z 2E0CJZ
Hodgetts B G4YQG
Hodgetts C 2E0OGY
Hodgetts C M6CVH
Hodgetts G G7FXZ
Hodgetts J G2FXZ
Hodgetts S G4FAE
Hodgetts S G1LMU
Hodgetts T G4AYD
Hodgetts T G4USQ
Hodgkins I G6AJC
Hodgkins J G3JZP
Hodgkins R M1ALM
Hodgkinson A 2E0TBV
Hodgkinson A G1SAM
Hodgkinson A G3LLJ
Hodgkinson B G7BOY
Hodgkinson B M6FHM
Hodgkinson D G0EZL
Hodgkinson G GI7TPO
Hodgkinson G G3SUN
Hodgkinson M G4HCC
Hodgkinson P G4IPB
Hodgkinson P M3XUJ
Hodgkinson P G8ZYR
Hodgkinson R 2E0RSH
Hodgkinson R G3ZMM
Hodgkinson R M0ZMM
Hodgkinson R M3RSH
Hodgkiss I G0FMP
Hodgkiss P G3WTY
Hodgson A G0ADO
Hodgson A 2E0GIX
Hodgson A M0OTH
Hodgson A M6GIX
Hodgson A G6KSK
Hodgson A 2E0WHU
Hodgson A M6WHU
Hodgson B 2E0JHO
Hodgson B G1HAH
Hodgson B M6DKF
Hodgson C G0JYJ
Hodgson C G0OBN
Hodgson D M0AAF
Hodgson D 2E0WBQ
Hodgson D M6DTH
Hodgson D M6EHU
Hodgson E G0SBO
Hodgson E G3RAR
Hodgson G G0BZH
Hodgson G G8UBN
Hodgson J G3YKB
Hodgson J G8TYX
Hodgson J M3TVK
Hodgson M M0NOE
Hodgson M M1MIC
Hodgson N G3WAH
Hodgson P G0JYC
Hodgson P G0CLT
Hodgson P M6PID
Hodgson R G3DUW
Hodgson R G4KBH
Hodgson S G0LII
Hodgson W M3HPN
Hodkin A G1YLG
Hodkinson J G6COB
Hodkinson K G4MDE
Hodkinson S G7CGN
Hodkinson S M0SPH
Hodnett J G8NPD
Hodson A M3HOD
Hodson B M0OAB
Hodson H G7TLL
Hodson J 2E0RMM
Hodson J M6RML
Hodson J MW6HDV
Hodson K G8RWZ
Hodson M G6VUN
hodson M M1APQ
Hodson P G8RBY
Hodson P 2E0DGV
Hodson R 2E0HPD
Hodson R M0IFT
Hodson R M6HPD
Hodson S G1XTA
Hodson T M6MNS
Hodson W G6GKG
Hoe A M3HBG
Hoe A M3HOE
Hoe P M3FFO
Hoey G G4JCF
Hoey I MI6AJO
Hoey J GM0ARD
Hoey M G1HSO
Hoey R GI8SJS
Hoey R M3AKE
Hoey T GI0BJH
Hoffman R G0CBO
Hofman G M6GTH
Hofman J M6AEL
Hofstedt A M6GYI
Hogan C M6ZZE
Hogan C M6CRH
Hogan D G3PUZ
Hogan G M6WXY
Hogan J G4RLR
Hogan M G4RMN
Hogan R G4VCQ
Hogan R M6RKI
Hogan W G6VFB
Hogarth H GM8CSE
Hogarth S G7EWL
Hogben G G0JUL
Hogben R M3RGD
Hogg A G0AUR
Hogg A M6IGZ
Hogg A GM8LKL
Hogg C G3NRZ
Hogg D G7KIT
Hogg D G4CAF
Hogg E GM8NZL
Hogg G M0GDH
Hogg H G3NGX
Hogg J G1YHP
Hogg J G2OG
Hogg J G3NUA
Hogg J M6DOW
Hogg J 2M0DTZ
Hogg J MM0GKN
Hogg K G0UZJ
Hogg M G7SYJ
Hogg R G3ZCY
Hogg S G0RMJ
Hogg S 2E0SMX
Hogg S M6TDO
Hogg W 2D0WFH
Hogg W MD6WHG
Hoggan A G8ASX
Hoggan J 2E0AVT
Hoggan M M3NYA
Hoggan T M3OII
Hoggard P M1BLO
Hoggard R G7PUL
Hoggarth J G8WSU
Hoggett K G8RHM
Hogwood R M3BFG
Hoile A M6EVL
Holbrook A 2E0GTA
Holbrook A M0CST
Holbrook J G6VWF
Holbrook J M6JJH
Holbrook M 2E0MAH
Holbrook M MM0MBH
Holbrook V G6CWW
Holbrook-Bull M 2E0ZXR
Holbrook-Bull M M6OOD
Holbrough M G0IKI
Holburn D G3XZP
Holdaway P G0DCP
Holdbrook D M6HKU
Holden A M0DOW
Holden A M3DPX
Holden A M6GND
Holden A G7SMN
Holden B G4SXE
Holden B M3HZP
Holden C M6CPH
Holden C G7BLD
Holden D G3WUN
Holden D G8DPW
Holden D M3KUO
Holden D 2E0VXI
Holden D M6VXI
Holden E G3MAJ
Holden G G0OYI
Holden J G8ZGS
Holden M M3HZM
Holden P G0JSG
Holden R G8SKA
Holden S G3VEK
Holden V 2W0CBJ
Holder A G4ZBH
Holder A M3UEY
Holder L MW0LCH
Holderness B MM6EJO
Holderness C G6LVM
Holderness L GW0CCO
Holderness R G3XDA
Holdford C G7KXZ
Holdford I 2E0DUU
Holdford I M6IJH
holding E G0DVS
Holding J G4ASP
Holding M M6BJC
Holding M G0OHD
Holding W G6PUV
Holdridge N M3KKA
Holdroyd D M0CYU
Holdsworth A G8OO
Holdsworth D G1VGO
Holdsworth D G6COG
Holdsworth D M3XJE
Holdsworth J G4XFF
Holdsworth M G0FOH
Holdsworth M G7NFG
Holdsworth M G8HZI
Holdt C M3ZON
Holdt K M3NUO
Holdup A G6GEP
Holdup A M0GEP
Holdway A G0KQT
Holdway K G8DYI
Hole I G0OUG
Hole J M6JDQ
Hole M G0GJC
Hole R G0IKC
Holford G MM6GBP
Holgate B MM6HGM
Holgate M M3OHC
Holgate N MM6NWH
Holgate R G6XTT
Holgate S G8YTP
Holker P G3OIP
Holland A M0PAR
Holland A M3SUK
Holland A G0DIA
Holland A G4VFL
Holland A M6RIY
Holland B M3BXN
Holland B M0DLT
Holland B 2E0BSG
Holland B M0GOB
Holland B M6ARS
Holland C G0FYP
Holland Carter J G3OWB
Holland D G3WFT
Holland D G4LDT
Holland D G8WTZ
Holland D M0HVQ
Holland D M6BOP
HOLLAND D 2E0BCB
HOLLAND D M0GDT
Holland D M0RLO
Holland F GI0AIQ
Holland G G0VHJ
Holland G G7HJD
Holland G G7UWO
Holland G 2E0HCL
Holland G M6WID
Holland J G3GHS
Holland J G4COQ
Holland J G8TTJ
Holland J M6EZF
Holland K G3MCD
Holland M G3IWM
Holland M G7VRJ
Holland N M1NPH
Holland P GI0EWE
Holland P G3XLI
Holland P G4TCB
Holland P G6CHX
Holland P G6IFE
Holland P M0SRN
Holland P G3TZO
Holland P 2E0CIG
Holland P M0JAD
Holland P M6BBG
Holland R G4DTZ
Holland R G1AUH
Holland R M6YWG
Holland S G7VOH
Holland S G8ANT

Holland T G0VWT
Holland T M6ISG
Hollands P G1PXW
Hollas P M0GHY
Hollebon U G4ULL
Hollerbach J G8ZEM
Holles A G7FSD
Holley C M3CVH
Holley K G4IMU
Holley M G1ACY
Holley M G4EHQ
Holley M G3CIL
Holley P G6NWC
Holley S G8WBO
Hollick R G6ZAX
Holliday M G4DCK
Holliday T G7KWQ
Hollidge G G6BOF
Holliman P G4EAZ
Hollinger W GI7DBZ
Hollinghurst M G4NOE
Hollings A 2E0KOS
Hollings A M0IPS
Hollingsbee I G3TDT
Hollingshurst S G3GEV
Hollingsworth D G3VKU
Hollingsworth K G7AFO
Hollingworth L G4NCV
Hollinrake D M3KXV
Hollins C M5CHH
Hollinshead N G6IFS
Hollis A 2E0HRJ
Hollis A M3XNU
Hollis B M3DWA
Hollis D M6EPH
Hollis F G4LNG
Hollis J G8ION
Hollis P G7BMC
Hollis P G8BPY
Hollis P M3BPY
Hollis P M6HLS
Hollis S G7OYP
Hollis S M6SBY
Hollister C G4SQQ
Hollingworth B G6GGV
Holloman N M6NFH
Hollow D G4HRE
Hollow R G4SOK
Holloway A G7MWJ
Holloway B G0GFR
Holloway C G7FKJ
Holloway D 2W0MMD
Holloway D G3BBX
Holloway D G6UPQ
Holloway J G4LFQ
Holloway J M0DTJ
Holloway K M3IFK
Holloway K M6GYQ
Holloway K 2E0SLS
Holloway K M0SHK
Holloway K M6KAH
Holloway M G4YRY
Holloway P M0DZH
Holloway P M6ZXB
Holloway R G3HUA
Hollowell E GW0GUY
Hollowood J G7LMI
Holman D 2E0DAI
Holman D 2E0KUC
Holman D M0ORS
Holman D M0YDH
Holman D M6WOW
Holman E G8WMC
Holman G G6TGE
Holman G G8YNF
Holman H G7HJJ
Holman I G7VYQ
Holman K 2E0KNH
Holman K M0KNH
Holman K M6MGF
Holman P M6EMP
Holman P M3TXQ
Holman T M6GGZ
Holmden H G4KCC
Holme E G4YYB
Holme J M3BBC
Holme L M3LMH
Holmes A G4CRW
Holmes A MI3RLA
Holmes A M3XAH
Holmes A 2E0WSX
Holmes A M0HIX
Holmes A M6ARP
Holmes A 2E0BEP
Holmes A M0EDE
Holmes A G4ION
Holmes A G6VTR
Holmes A G8LVM
Holmes B M3BAH
Holmes C G0JHG
Holmes C G0LJH
[illegible]
Holmes C 2E0GAV
Holmes C M6AEG
Holmes C M6BQK
Holmes C M0JPT
Holmes D G3JSV
Holmes D G4FZZ
Holmes D G4KIZ
Holmes D G4ZAO
Holmes D G8LVL
Holmes D G4XEO
Holmes D G8ZXZ
Holmes D 2E0TQB
Holmes D M3TQB
Holmes D 2E0CAU

Holmes De Wyvill Sinclair H 2E0IIG
Holmes E G3ALK
Holmes E M3EMA
Holmes L MI5AJI
Holmes E G4TLY
Holmes G G0PZD
Holmes G G6ENZ
Holmes G M0GCH
Holmes H G4TWT
Holmes J G3PKQ
Holmes J G8STY
Holmes J M6AAO
Holmes J G0BNF
Holmes J G3UEU
Holmes J G3WSW
Holmes J G4VMG
Holmes K G6BTX
Holmes K G6IRW
Holmes L M3LDH
Holmes M 2E0ADN
Holmes M 2W1CLC
Holmes M G0TYN
Holmes M G6AIZ
Holmes M G8MMN
Holmes M MM3YMM
Holmes M 2E0ZLO
Holmes M M0ZLE
Holmes M M6ZAC
Holmes M M6FZS
Holmes N M6NRH
Holmes P 2E0PLH
Holmes P G0MAY
Holmes P 2E0JZU
Holmes P 2E0PFH
Holmes P 2M0PXH
Holmes P 2E0ZSE
Holmes P M0SEV
Holmes P MM3ZNN
Holmes P M6JZU
Holmes P M6PFH
Holmes P M6ZSE
Holmes P 2M0SWM
Holmes P 2E0PJH
Holmes P M0IRK
Holmes R 2E0DHG
Holmes R G7SVQ
Holmes R GM0WRU
Holmes S G0HBO
Holmes S G4TWS
Holmes S G7ECE
Holmes S M3WSH
Holmes S M6ACZ
Holmes V G8GYP
Holmes W G6SPQ
Holmquest P 2E0WPH
Holmquest P M0PMH
Holmquest P M6PPH
Holmshaw R G0KMF
Holmwood R G8PRK
Holohan A G7OAV
Holohan D MD6LET
Holroyd A G6PFZ
Holroyd M 2E0MJH
Holroyd M M3CTT
Holroyd T G3YHD
Holroyd W G7JGW
Holroyd W M3JGW
Holstead J G3OZC
Holt A 2E0EMX
Holt A M6AGH
Holt A M6EXG
Holt A M0MVM
Holt B M0GDJ
Holt C G6IRX
Holt D G4LRD
Holt D G1BTV
Holt E GM0WED
Holt F G0WVY
Holt F M0GED
Holt G G3PTS
Holt G M6WYD
Holt K G3IBQ
Holt M 2E0COM
Holt P G1FOE
Holt P G4AIB
Holt P G4LPP
Holt P G8YJQ
Holt R GM1BNP
Holt R M0AIT
Holt R G3TTU
Holt R G8NOF
Holt S G0GEU
Holt S MM3SUS
Holt S M3SZY
Holt C 2E0CCH
Holt S M0MRF
Holt S M6SGH
Holt T MM3TLH
Holt W G7DHM
Holtam M G1EDX
[illegible]
Holtham M G0EIG
Holtham P G3ZXY
Holtham R G4EKS
Holtom R MM0LVV
Holton J G3PSC
Holton J G8OGR
Holton P 2E0WAK
Holton P M0WKO
Holton P M3WYQ
Holyer L M1CJB
Holyhead R G6FGA
Holyoake D M3TOF
Holyoake R G4WAY
Holyoake V G4OJJ

Holyoake V G7OJY
Holzapfel A M0ATD
Homan J G6ZEN
Homans S G4BNM
Home R M0ETE
Homer A G1IHJ
Homer B G8IWX
Homer B M3IUZ
Homer M G1YMH
Homer M G6AIQ
Homer P M6WCC
Homer S G6HFZ
Homer S MI0KQU
Homsey J M3HOM
Hone D M5AJP
Hone J M6AFF
Honey D M0DHO
Honey P 2E0VLT
Honey P M0PHO
Honey P M3VLT
Honey W G6GAB
honeyball J G1LMS
Honeybone P G7AVF
Honeybourne R M3XMZ
Honeyman R GM0NUI
Honeywell M G0ABB
Honeywell P G7FCJ
Honour P G4IHX
Honywood S M3YRV
Hood A M6MNO
Hood A GM7GDE
Hood A GM4UXX
Hood B M3OEE
Hood B M6SSW
HOOD D G4PNC
Hood J G3MZI
Hood J MM0AKM
HOOD J M3JLH
Hood J GM4COX
Hood R G4BIA
Hood R GM8MNM
Hood T GM4LRU
Hoodless D G8ZAU
Hook C G4FJW
Hook D M0REB
Hook D M5ADF
Hook D M6BXF
Hook L G3PPO
Hook R G3ZLM
Hooker D G8IIK
Hooker G G4OEM
Hooker T G8OBB
Hookham R G4GWV
Hookham R M0RBH
Hooks E GI0GDF
Hooks M GW3MKT
Hooks M G7IXM
Hooks M 2M0AIE
Hooles M G3LGR
Hooper A G1JMF
Hooper B G0FZI
Hooper B M6SYH
Hooper D G0CYU
Hooper D G4FVW
Hooper D 2W0DNV
Hooper D MW6CTS
Hooper D M0ALO
Hooper G M6GRH
Hooper I 2E0HPR
Hooper I M6HPR
Hooper I G4PRQ
Hooper J G3XEI
Hooper M G7UUK
Hooper M M6BEM
Hooper P G3KSP
Hooper P M3KFL
Hooper R G4LXR
Hooper R G6HTS
Hooper R M5YEX
Hooper T M3ZAY
Hoose D G4BSD
Hoose J G0NYK
Hooton D G6VFC
Hope B G4EXE
Hope B G4ZPP
Hope G 2E0YQC
Hope G G1DLJ
Hope G G6YLD
Hope G M3TGO
Hope H 2E0CUO
Hope H M0HYG
Hope H M6CNL
Hope I 2E0IJH
Hope J G3RXM
Hope J G7JDI
Hope J G7KYG
Hope J 2E0JWH
Hope R G0DRQ
Hope R G0PNB
Hope R G6VMF
Hope R G6YEY
[illegible]
Hope R 2E0BXK
Hope R GW8TVX
Hope R M6HOP
Hope R G6LVN
Hope T 2E0DUZ
Hopewell J G4JPQ
Hopewell M 2E0THZ
Hopewell M M0XMH
Hopewell M M6MGH
HOPEWELL P G4DCI
hopgood L 2E0DYY
Hopkin O MW6ORH
Hopkins A G4YKW
Hopkins B 2E0FOX

Hopkins B G7KDR
Hopkins B M0MPS
Hopkins B MW6KBK
Hopkins C 2E0GOE
Hopkins C G7EOC
Hopkins C M6WUH
Hopkins D G0WOZ
Hopkins G M0GHO
Hopkins G M1DCF
Hopkins H 2E0HAN
Hopkins H M3UHH
Hopkins I G4WUH
Hopkins J G1NAP
Hopkins J GM1XJE
Hopkins J G4LPT
Hopkins J M1PXB
Hopkins J 2E0TXJ
Hopkins J M6TXJ
Hopkins J G7LXY
Hopkins M G0OEI
Hopkins M G3FHG
Hopkins M M3PBQ
Hopkins M M1CTG
Hopkins N G1IME
Hopkins O 2E0OKH
Hopkins O M6OKH
Hopkins P G0LIW
Hopkins P M3SUW
Hopkins R G1APL
Hopkins R GW4NOS
Hopkins R G4ZUE
Hopkins R M6HPK
Hopkins R 2E0BZH
Hopkins R M1BHO
Hopkins R G3IWW
Hopkins S 2E0DGM
Hopkins S M6HBT
Hopkins S M6SDH
Hopkins S G1IHL
Hopkins T G1HBC
Hopkins T GW4EVL
Hopkins T G8TYY
Hopkins T M6TGY
Hopkins V 2E0COV
Hopkins V M0TAV
Hopkins W M3YOX
Hopkinson A G0SIY
Hopkinson A G0VKI
Hopkinson A G1BCN
Hopkinson A G8OJQ
Hopkinson A M0RAL
Hopkinson J G6VPL
Hopkinson M G0AUH
Hopkinson M GM4JDK
Hopkinson P GM4EBX
Hopkinson P G6GWU
Hopkinson P 2E0UHJ
Hopkinson P M3UHJ
Hopkinson T G0AUG
Hopley A G7WBU
Hopley C G8ICT
Hopley I MM1ABA
Hopley S G4YTK
Hoppe D G0NVC
Hoppe M G0GLA
Hoppe R M0WHP
Hopper A 2E0NNB
Hopper A M0NNB
Hopper A M6NNB
Hopper B G0CRN
Hopper C M6DWI
Hopper D G0NFG
Hopper J G8XML
Hopper S G4HS
Hopps K M0BLN
Hopson A M0HOP
Hopson A M1HOP
Hopson L G6SKF
Hopton D G1NYJ
Hopton J GW3WMP
Hopton S G0VZT
Hopwood F G8BIA
Hopwood J G0EDT
Hopwood P G3UKH
Hopwood R 2E0RGR
Horabin T G0LNW
Horbaczewskyj P G4ZXO
Horder D G7HDR
Hordley T G8BXQ
Hore C G6GWX
Hore K G0UHD
Hore R G8LTC
Horgan R M3WKK
Horley J G4KMF
Horn C 2E0OOF
Horn E MM0HFU
Horn J 2E0FLN
Horn J M0NUX
Horn J M6FLN
Horn L G6DYU
Horn L M3DYU
[illegible]
Horn M 2E0WGF
Horn M M6GQJ
Horn P G3GGH
Horn T G8MTF
Horn W G4OPD
Hornal D MM6DDN
Hornbuckle R G3ZFF
Hornby A G1HBD
Hornby A M6PLS
Hornby D G0WCK
Hornby D G7UCP
Hornby E G7CYQ
Hornby J G4RAK
Hornby K M0GOI

Hornby L M3WWD
Hornby R G8UWD
Hornby R M0RWH
Hornby W 2E0ARS
Horne A G4CKQ
Horne A G0TPH
Horne A G4QJV
Horne B G4NLB
Horne C M3CGH
Horne D G0JUT
Horne H G0MEX
Horne K GM3YBQ
Horne K GM4FSF
Horne L GW0JTE
Horne L G6DDR
Horne M GM7UQM
Horne P G3JRH
Horne R G0JVN
Horne R M3FCS
Horne R G4HBA
Horne R G8KRT
Horne S M0WDZ
Horner D M3DRH
Horner D 2E0DAH
Horner D M0TEG
Horner D M3PYO
Horner J G6RUP
Horner J M6JSH
Horner R 2E0RIN
Horner R M3YRH
Horner S G8YNE
Horner S 2I0LLG
Horner S MI0LLG
Horner S MI6LLG
Horning B GM4TOE
Hornsby A G0KTX
Hornsby J G0AJH
Horobin P G6KJH
Horoszko M G4BDX
Horrabin C G3SBI
Horridge C 2E0FAM
Horridge C M6FLC
Horrobin T G3WDD
Horrocks B G8OUT
Horrox P 2E0YMT
Horrox P M6UMC
Horry J M6JKH
Horry J 2E0CBU
Horry J M6MOW
Horry R M1GOH
Horsburgh D G3UOM
Horsburgh G GM4XHV
Horsburgh J GM1RDG
Horsefield I G4OPP
Horseman L G4IPF
Horsewell W M0WAH
Horsey B G3TTP
Horsfall A G7DCT
Horsfall A G4CBW
Horsfall F G7MUH
Horsfall G G3GKG
Horsfall G G8CWQ
Horsfall J G0TOB
Horsfall M G7DMS
HORSFALL R G7KEK
Horsfield A M5ALU
Horsfield E G4UFR
Horsfield J G0UBM
Horsfield J M0DKX
Horsfield K G1HIP
Horsfield M G6LWK
Horsfield M G1HIO
Horsford R G8LKK
Horsley D 2E0DQQ
Horsley D M3DQQ
Horsley J M6JIH
Horsley K 2E0IGG
Horsman A G0MBA
Horsman B G4MZC
Horsman D G4TNE
Horsman R G3XOD
Horsoo T 2E0BQB
Horsoo T M3VPB
Horsten T 2E0ETT
Horsten T M0TRN
Horsten T M6TTH
Horswell C G8SSP
Horton A 2E0TAT
Horton A G0GWP
Horton A G4BOV
Horton A M6HTN
Horton B G0HNG
Horton B M3OEM
Horton D 2E0EOQ
Horton D M3EOQ
Horton D G0RFY
Horton D G3RZF
Horton D M3GYH
Horton D M6YIN
Horton K G0AEL
Horton K G4TZT
[illegible]
Horton L G4GRM
Horton L M3GYI
Horton M G4DMH
Horton M M0MEH
Horton M M0MTH
Horton N M0ZEY
Horton P G7IOO
Horton R G4AOJ
Horton R G3XWH
Horton R G7ACO
Horvath R 2E0CYL
Horwell M M6KGL
Horwood D G7HGQ
Horwood D M1LRX

Horwood D M3LRX
Horwood G G6IFR
Horwood W G7OCQ
Hose A M6DQW
Hosea J MM0JWH
Hosegood C G7IRU
Hosegood C M3CAQ
Hosey P 2I0MSO
Hosey P MI0MSO
Hosfield J G0SVK
Hosker N G8UEK
Hosker N M3UEK
Hoskin I G4GDU
Hoskin M MM3HAF
Hoskin M G6GGN
Hosking A MD0CFM
Hosking C G0PGB
Hosking J G0JTF
Hosking J G8DEX
Hoskins A M3OOU
Hoskins A M6DMY
Hoskins A M0DOP
Hoskins A M3DMY
Hoskins J G4XYM
Hoskins J M3NDO
Hoskins M G6ZVL
Hoskins S M1AJM
Hossack W GM3UBJ
Hossell M 2E0SSY
Hossell M M6MHO
Hossle H G1ZSZ
Hostekens M G1EHU
Hotchen K G6VVL
Hotchin J G4ATA
Hotchin J G8DYT
Hotchin M 2E0FVS
Hotchin M M0HOM
Hotchkiss G G4BEQ
Hotham S G8FAS
Hotson N G0TDG
Houbart D M1DNJ
Hough B G0KKF
Hough J G0KRR
Hough J G8JQG
Hough K 2E0HOK
Hough K M0KOH
Hough K M6HOK
Hough R G3XTN
Hough W G0IHF
Hought C GI4SFZ
Houghton A G0WMB
Houghton A G0WWH
Houghton C G4JNE
Houghton C M0DFL
Houghton C M3CJH
Houghton D G0MXW
Houghton D G0TQG
Houghton D G3UPY
Houghton D M0FCI
Houghton F G3VZM
Houghton H G8XMO
Houghton I M1BUQ
Houghton I G0ADT
Houghton J G1KEP
Houghton J G4NIA
Houghton J M0PAY
Houghton K G7PYW
Houghton M M6MMH
Houghton N G0GHD
Houghton N 2E0TWZ
Houghton P M1ELQ
Houghton P M3PXL
Houghton P 2E0GKA
Houghton P M0XCX
Houghton P M6WKL
Houghton R G4HTX
Houghton R G3SCL
Houghton R M6CZR
Houghton S G6TGQ
Houghton S G7TAV
Houghton T G1KPI
Houghton V G7ERH
Hould N M0CFD
Houlden C 2E0DQZ
Houldershaw H G7GZV
Houlding S G0BYA
Houldridge D M0EFE
Houldridge I G6GKT
Houlihane T G7BIX
Hoult D G8FPA
Hoult M M6XMH
Houltby A G3ZSF
Houltby C G3UHS
Houlton G M3ZGH
Houlton R M6HOU
Houlton S G3GRL
Hounslow D M3HCP
Hounslow N G4TMY
Hourican T GI1RAA
Hourston R G7GLZ
Housden D M3UVX
[illegible]
House K 2W0KPH
House K MW6CDV
House S M6DUK
Houseago P G8SGB
Housego J M0TIF
Houssart S 2E0HOU
Houssart S M0OSY
Houssart S M6EYD
Houston C MM0UKW
Houston C MM6CLM
Houston D M6DYH
Houston G MI6XGN
Houston J GM4OPU
Houston J MI5AMO

Houston K G8KCH
Houston R MM3KQI
How C G4JCA
How D G0PAR
Howard A MW6XRO
Howard A 2E0SEP
Howard A M0FRG
Howard A M6FUQ
Howard A M6NWI
Howard B G0LJD
Howard B G1SVP
Howard B MW1DFQ
Howard B M3WKM
Howard C G0OWR
Howard C M3XCH
Howard C 2E0TUT
Howard C M6OND
Howard C GD0HWA
Howard C GD8ANU
Howard C G7UBP
Howard C 2E0OCH
Howard C M0UCH
Howard D 2E0WJR
Howard D G4IZA
Howard D G7GQX
Howard E G8KPE
Howard G M6FKR
Howard H M0BGR
Howard I G8WAW
Howard J G0KKU
Howard J G4EVI
Howard J G0BUV
Howard J G0UFZ
Howard J G6YLA
Howard J G0NJO
Howard K G1TTC
Howard K G7VLD
Howard K M3KTH
Howard K G0FEQ
Howard M G1OSJ
Howard M GW7HGU
Howard M G7SOE
Howard M G8RES
Howard M M1CNE
Howard M G4OVV
Howard P G0WPH
Howard P G1OFG
Howard P G4JKC
Howard P G4UMB
Howard P G7FPW
Howard P G0AFN
Howard R 2E0WBW
Howard R M6WWD
Howard R GM4EGX
Howard S G1VON
Howard S G7DNQ
Howard S 2E0KFO
Howard S M3WVC
Howard T G7LBM
Howard W G6TGM
Howarth A G0RHY
Howarth B G0YRT
Howarth C G7SOV
Howarth D 2E0MEH
Howarth D M0TCB
Howarth D M6AXC
Howarth G G0DRL
Howarth G G4YVO
Howarth J G0AEZ
Howarth J G1VKG
Howarth J G4WNI
Howarth J G7NOQ
Howarth J M3MOB
Howarth K G6ZVO
Howarth L G0GEH
Howarth N G4YAC
Howarth N GD1HVL
Howarth P G7ULM
Howarth P GM3YAC
Howarth R G4GIV
Howarth R G7SYY
Howarth S G4YGA
Howarth T G4BKF
Howarth W M3GSQ
Howat A G7LZB
Howchen T G1FBW
Howcroft S G1EEA
howcroft S G6IFQ
Howden A M0MCA
Howden D M6WCZ
Howden M M0JKP
Howe A M3HJG
Howe A M1BCU
Howe C G4RTJ
Howe C M6OCH
Howe F M5AFI
Howe G 2E0IGN
Howe G G8XDU
Howe G M6DTO
Howe J G3NXZ
Howe K G4KDH
Howe N M0NIG
Howe P G4CHL
Howe P G4IIO
Howe R GW3PLB
Howe I G0ETP
Howe T G1AJC
Howe T G1VID
Howe T G3VID
Howe W MW0WIL
Howell A G4ABL
Howell A G7SWH
Howell A G0VJM
Howell B G4LOI
Howell C G0VWF
Howell C G4LJU

Howell D GW3WCV
Howell G G4JVT
Howell G G6KVK
Howell G M5KVK
Howell J G4BXZ
Howell J GM4ZQH
Howell J M0AMX
Howell J M3XOU
Howell J G0ICB
Howell K G4YKE
Howell L M5AKW
Howell M G0ZMH
Howell M G1UBH
Howell P G0NAP
Howell P G3ZTZ
Howell P M1PGH
Howell P M3PGH
Howell P M0DCV
Howell R G0HZE
Howell S M6NGA
Howell T GW6FBV
Howell W G0TMK
Howell W G8NTG
Howell-Jones A M0THJ
Howell-Pryce J G4ZHI
Howells A G1APQ
Howells A G8UEI
Howells C G4DXP
Howells D 2E0GPG
Howells E M0YVK
Howells E G0GAL
Howells G G8SXI
Howells G GW1FKL
Howells J G6PUR
Howells J GW4BUZ
Howells M G7IAT
Howells N MW0JLN
Howells P M3YCT
Howells R G4FFY
Howells R G8GWX
Howes A G0HOV
Howes C G0MVV
Howes C G6BAY
Howes D G4KQH
Howes J G0NMS
Howes M M3SEY
Howes M M6WXP
Howes P GM4TTC
Howes R G7PJD
Howes R M0DTH
Howes R G4OWY
Howes R G6AUW
Howes S G6RHA
Howett C G4ILR
Howett P G4MD
Howgate A G7WHM
Howgego R G4DTC
Howie D G0GSJ
Howie E 2E0IEK
Howie J GM4DIJ
Howie J GM7JGR
howie W 2E0WBH
Howie W GM7IUF
Howison J MM6RBE
Howitt R M6IPA
Howker F M1FDH
Howkins M G4CPG
Howland A G4ZKS
Howland M G4MIX
Howland P G0WJN
Howlett A G1HBE
Howlett B G0VZA
Howlett D M0VTG
Howlett D M3VTG
Howlett D G8FIF
Howlett J 2E0JZK
Howlett J M0JZK
Howlett J M6JZK
Howlett R G1OSE
Howlett R G6RHB
Howlett S M1SRH
Howman A G0FVF
Howorth D G4IFT
Howorth N G4LNE
Howroyd S M3OOH
Howse A M1CBZ
Howse D M1CBY
Howse G G6YLB
Howse M 2E0MLH
Howse M G7PTV
Howse R G1RYY
Howsen A M0VLA
Howsham I G0CVN
Howson P GM8GAX
Howson R M6YDD
Howson T M6TEV
Howton D G4PWV
Hoy E M3ESH
Hoy H G4ZWM
Hoy J G6NQU
Hoy M M3HOY
Hoy P M3VEN
Hoy S M3XIL
Hoyland J G0HOJ
Hoyland M M6MLH
Hoyle C M3HSE
Hoyle D G4HGN
Hoyle G M0BAU
Hoyle N M3PNH
Hoyle P G6KJA
Hoyle P M6BYP
Hoyle S G7MAJ
Hoyle S 2E0YYK
Hoyle S M6YYK
Hoyle T 2E0TTG
Hoyle T M6TMH

UK Surnames

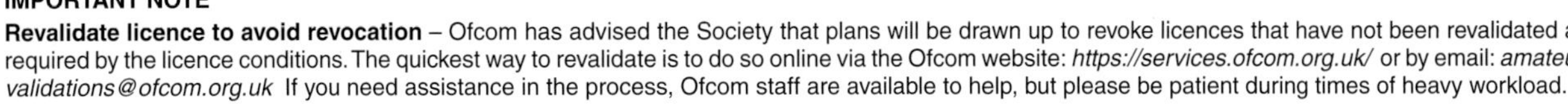
**IMPORTANT NOTE**

**Revalidate licence to avoid revocation** – Ofcom has advised the Society that plans will be drawn up to revoke licences that have not been revalidated as required by the licence conditions. The quickest way to revalidate is to do so online via the Ofcom website: *https://services.ofcom.org.uk/* or by email: *amateur.validations@ofcom.org.uk* If you need assistance in the process, Ofcom staff are available to help, but please be patient during times of heavy workload.

Hoyle W G0DRV
Hoyle-Jackson I MW3TKI
Hoyte A M6LLH
Hristov A M0HOV
Hruszka F M0ZKA
Hruza P G4SGQ
Hrycan J M0BEX
Hrynkiewicz I 2E0OUK
Huband P 2E0PPH
Hubball M G1JHL
Hubball S M3SRH
Hubbard A M3HUB
Hubbard A G7TTY
Hubbard A GW6JSO
Hubbard A MW0GUV
Hubbard B G4KEX
Hubbard D M3HZH
Hubbard D M6HOT
Hubbard F G0IBY
Hubbard J G4PWG
Hubbard J M0CQV
Hubbard J G0RGG
Hubbard M G3OVL
Hubbard M 2E0HUB
Hubbard M M6MDH
Hubbard M M6YFT
Hubbard N G8OSH
Hubbard N M3NHZ
Hubbard P G4DKB
Hubber G G3NVJ
Hubberstey P G7TCB
Hubert P G3YWM
Hubert R G6ZEQ
Hubner J G4YHG
Huby A M6PNT
Hucker W MW3WRH
Huckle A M6STH
Huckle D M3PIH
Huckle H M3TXS
Hudd J M3YVY
Huddart L M6LKE
Huddart M G4OQR
Hudders A M3UEP
Hudders H G0THH
Huddleson B MI6HBI
Huddleston A G3WZZ
Huddleston D G0SGT
Huddleston T GI8HUD
Huddleston W M0CLR
Huddlestone J G1UJX
Hudgell G G6DYM
Hudman R G7OBC
Hudsmith G G4LTM
Hudson A G8ZEE
Hudson A MI6RFD
Hudson B G0TUI
Hudson B M3ZBH
Hudson B 2E0HUT
Hudson C G1XYF
Hudson D G4WOE
Hudson D G4XUW
Hudson E MI6EFD
Hudson G 2E0GAH
Hudson G GM4SVM
Hudson I G8BQT
Hudson J G0DBC
Hudson J G4NS
Hudson J M0FJM
Hudson J G3PEW
Hudson J M0CMW
Hudson J MW6JCH
Hudson J G4ABQ
Hudson K G0OBP
Hudson K M3KAN
Hudson L 2E0LMH
Hudson L M0LMH
Hudson L M3YUV
Hudson M G7VDS
Hudson M G1WMJ
Hudson N M0OJO
Hudson P 2E0BDB
Hudson P G3UMM
Hudson P G4ANV
Hudson P G4VAH
Hudson P G7UCB
Hudson R G0TUU
Hudson R G4JFN
Hudson R G4YUL
Hudson R G8MSM
Hudson R G1IVL
Hudson R G7BIV
Hudson R MM6NRE
Hudson R G1XZW
Hudson S G4CEL
Hudson S G7MIE
Hudson W G0CSW
Hudspeth J M3JAZ
Hudspeth K G0ARK
Hueck A G6MWS
Huelin R 2E0HRJ
Huelin R M0HRP
Huelin R M6HRJ
Huelin S 2J0DXA
Huelin S MJ0ULE
Huff A G4WYI
Huffadine R G3VXH
Hufschmied S G6EAX
Huggett L G3ZAL
Huggett M G0LEJ
Huggett M 2E0WVE
Huggett M M6WVE
Huggins J G0DIP
Huggins J G0DZX
Huggins M G6XRK
Huggins N G8TOP
Hughes A 2E0SOX
Hughes A G0LBA
Hughes A G1SPM
Hughes A G1TFB
Hughes A G1ZSY
Hughes A GW3YGH
Hughes A G4KOR
Hughes A G4ZOF
Hughes A G7UUH
Hughes A M3XAO
Hughes A M6COV
Hughes A 2E0KHA
Hughes A G0WMH
Hughes A M3KHA
Hughes B G0TCI
Hughes B G4CUQ
Hughes B G4XVR
Hughes B G7NNR
Hughes B G4SGE
Hughes B MM3XOK
Hughes C G0FBT
Hughes C G6YMH
Hughes C M0HRD
Hughes C G0DQW
Hughes C G6DYW
Hughes C G7IAK
Hughes C M6AVA
Hughes C M6GZW
Hughes C M3FTW
Hughes C M0GNF
Hughes D G0ODX
Hughes D G0RQH
Hughes D G4PDR
Hughes D G7EWH
Hughes D G7VOA
Hughes D G8PMJ
Hughes D G8WPL
Hughes D M0CCX
Hughes D M0KDH
Hughes D MD1DNT
Hughes D M3ODH
Hughes D 2E0GUH
Hughes D M6GUH
Hughes D G0HIZ
Hughes D G4JNS
Hughes D G7MEU
Hughes D M3LFC
Hughes D G7LFC
Hughes D M6DRH
hughes E G0LEI
Hughes E M3TSE
Hughes E MM3WQO
Hughes G G0DQY
Hughes G G0GJH
Hughes G G0MYD
Hughes G M6HRK
Hughes H 2E0CJN
Hughes H G0ECJ
Hughes H G6KAV
Hughes H G8XEN
Hughes H M6HRI
Hughes H 2E0BXG
Hughes H G4HSC
Hughes H M6HJT
Hughes H 2E0GMZ
Hughes H M0GMZ
Hughes J G0JQK
Hughes J G1FBI
Hughes J G3RRM
Hughes J G4KGT
Hughes J G7CJC
Hughes J G7UIZ
Hughes J G8DML
Hughes J MM6HUE
Hughes J 2E0DNR
Hughes J M6TWH
Hughes J GW1CHS
Hughes J M0HHI
Hughes J M3YTQ
Hughes J M6ZJH
Hughes J M6JLH
Hughes K 2E0HNB
Hughes K G1RAO
Hughes K GW3SUH
Hughes K G4CJT
Hughes K G4XVS
Hughes K M3CBN
Hughes K MW1FJK
Hughes K 2M1EKI
Hughes L M6HUZ
Hughes M G0PJM
Hughes M G1HBF
Hughes M G1KAS
Hughes M G3KBH
Hughes M G3VFZ
Hughes M GM4ISM
Hughes M G4MEH
Hughes M G8OCO
Hughes M M0DMZ
Hughes M M3WMH
Hughes M M6HMG
Hughes M 2E0DVU
Hughes M 2E0KAG
Hughes M 2E0MBO
Hughes M M3YZV
Hughes M M6DVU
Hughes M M6HMP
Hughes N 2W0BMM
Hughes N GI4SZP
Hughes N G0EFP
Hughes P G0ABE
Hughes P G0BXC
Hughes P GI0VHG
Hughes P G4VQI
Hughes P G8AOZ
Hughes P G8IPT
Hughes P G3OSR
Hughes P G0PWH
Hughes P G1ACD
Hughes P 2E0DPH
Hughes P GM1ZCD
Hughes P G7VEH
Hughes P M6PHL
Hughes P G0KHQ
Hughes R 2E0LRP
Hughes R G0AGJ
Hughes R G0FTI
Hughes R G0RBH
Hughes R G4PBD
Hughes R G4PLH
Hughes R G6AZX
Hughes R G8JBQ
Hughes R M3HUG
Hughes R M6BUB
Hughes R M6RGZ
Hughes R M6GBY
Hughes R 2E0GAL
Hughes R M6RVH
Hughes R M3RYG
Hughes S M3SAB
Hughes S M3SHJ
Hughes S M3YOZ
Hughes S M6SWH
Hughes S G0WBL
Hughes S G6ZVU
Hughes T G4KST
Hughes T G4DSO
Hughes T M6BST
Hughes V G0CVH
Hughes V M0BET
Hughes V MI1AUI
Hughes W G0LYJ
Hughes W G0MYN
Hughes W G4LVY
Hughes W G6JXS
Hughes W M3PTQ
Hughes W G6IRZ
Hughes-Burton D M3CYQ
Hughes-Burton R M6RHD
Hughes-Burton S M3CYU
Hughesdon R G0NCX
Hughes-Lai E G7MME
Hugill T G4FJK
Hugman D G8JMY
Huish M G3VRV
Hulands M G4BHT
Hulbert C M0PZC
Hulbert K G6BJR
Hulett J G3OCH
Hulin D GM6HGF
Hulin D MM0XPG
Hull A G0KYD
Hull B GD0SFI
Hull B M6BVP
Hull E 2E0EFH
Hull E M0TFH
Hull E M3UYQ
Hull F GI4WME
Hull J M1DMT
Hull K 2E0LKH
Hull N G6ZVV
Hull P G1NTK
Hull P G4DCP
Hull P M0GFX
Hull R M0AUW
Hull S 2E0CJJ
Hulme A G0CDY
Hulme A M6ZEB
Hulme E G3BQT
Hulme G G0DLK
Hulme J M1APC
Hulme K M0GYN
Hulme M 2E0LME
Hulme M G3WI
Hulme P G0NPE
Hulme R 2W0HUL
Hulme R MW3CBS
Hulme S G3SRM
Hulme T M3ZHW
Hulmes T GW7KYT
Hulok A M6SDF
Hulse L M6LHN
Hulse P G7KAT
Hulstone B M6GXG
Hultquist B M0HIG
Human M G1MPC
Humberstone I 2E0DXB
Humberstone I M3AJD
Humberstone S G0ORK
Humble K M6OKH
Humble R G8DXO
Humble T 2E0APN
Humby P G6AUS
Hume C M3PAE
Hume D M3YDH
Hume J G1XAS
Hume J 2M0GUL
Hume J MM0DXH
Hume J MM3JHS
Hume J M3YJH
Hume P MI0PCJ
Humes I GI6DRK
Hume-Spry P G0FOE
Humm A G4CWE
Humm D 2E0DFL
Humm D M6EHI
Humm S G8CHO
Hummerstone L M0VRT
Hummerstone P M3KYD
Humpage R M3LCZ
Humphrey B M6BEH
Humphrey D 2E0DGH
Humphrey D 2E0DMY
Humphrey D M6EYL
Humphrey G G0MAH
Humphrey G G3BMQ
Humphrey G M0OTL
Humphrey J G7VEH
Humphrey J M3TBW
Humphrey L 2E0AJP
Humphrey L G6BFP
Humphrey L G8MHA
Humphrey M G0SWY
Humphrey N G3VHW
Humphrey P M0EAS
Humphrey R 2M0RWH
Humphrey V G8CPQ
Humphreys A G6WFW
Humphreys A GI8WJN
Humphreys C M0ASE
Humphreys D G4GHR
Humphreys D G4XDG
Humphreys D 2W0TRS
Humphreys D MW6DBU
Humphreys F G0CWU
Humphreys J G1LSX
Humphreys K 2E0YMR
Humphreys K M6YMR
Humphreys L G0JBM
Humphreys M G0LHL
Humphreys N 2E0DEM
Humphreys P G8WBP
Humphreys P M6PIH
Humphreys P 2E0CBA
Humphreys P G0UTU
Humphreys P G6YMS
Humphreys P M6HRC
Humphreys R M3FZE
Humphreys S 2E0SJH
Humphreys S M3KYG
Humphreys W G7KON
Humphries A M6CEO
Humphries B G0AGU
Humphries C 2E0NEC
Humphries C M6WUG
Humphries D G4ETG
Humphries D M6HCD
Humphries E G0EZT
Humphries F G3DCE
Humphries I M6LVJ
Humphries J G0EAW
Humphries J G7MKB
Humphries J G7FFZ
Humphries M G3LRQ
Humphries M 2E0MIH
Humphries M M6MHU
Humphries M G8PAB
Humphries T 2E0TNC
Humphries T M6BCK
Humphries T 2E0BIC
Humphries T G0OLS
Humphries T M3BIC
Humphries W G0OOI
Humphries W G7IHL
Humphries W M0IHL
Humphris C M0DXJ
Humphris F M0BJN
Humphriss A 2E0NDZ
Humphriss A M3NDZ
Humphrys R G6ULD
Humpoletz J G3ITL
Humpston E G8RZS
Humpston G G4GYO
Hundy P G7OPD
Hunnisett I GD0RNF
Hunsdale D G4ZOI
Hunsley A 2M0LBH
Hunsley A MM6RHI
Hunt A G0NFV
Hunt A G1YLJ
Hunt A G4TYT
Hunt A G7TMH
Hunt A G8GHB
Hunt A G6ISB
Hunt A M3GNY
Hunt A M6UFC
Hunt A M3HVS
Hunt A M6GJX
Hunt A G6MEW
Hunt A M6ESR
Hunt A G6NRK
Hunt B G4SJN
Hunt B M1EJI
Hunt B M3BSH
Hunt B G4WZZ
Hunt B G8BII
Hunt C G1HSP
Hunt C G6VZU
Hunt C G7KIL
Hunt C G8MYG
Hunt C G8VXR
Hunt C M5IAN
Hunt C G6WVO
Hunt C M6CBH
Hunt D G4VNE
Hunt D M1CTK
Hunt D M6DZZ
Hunt D GD6MFR
Hunt D G7SBK
Hunt E 2E0EUE
Hunt G G7RFM
Hunt G MM1FHR
Hunt G M6AET
Hunt I G1MXM
Hunt J G2FSR
Hunt J G3LPN
Hunt J G3PVU
Hunt J G3RJE
Hunt J G6NNS
Hunt J G6XRL
Hunt J G8JPA
Hunt J M3AJS
Hunt J M3PUQ
Hunt J M3VUA
Hunt J G7GFM
Hunt J 2E0CFQ
Hunt J G1GDS
Hunt J G6IFV
Hunt J M0HCP
Hunt J M0HUN
Hunt J 2E0JGH
Hunt J M0JGH
Hunt K 2E0WBZ
Hunt K M6WBZ
Hunt K 2E0SYD
Hunt K M0BPC
Hunt K M3IIP
Hunt K M6FHO
Hunt L G8RKH
Hunt M G1ADB
Hunt M GM6AJA
Hunt M M3MSH
Hunt M GW6VBN
Hunt M M0JZT
Hunt M M6MNH
Hunt M M6HNT
Hunt N M6EHH
Hunt P G1ZHM
Hunt P G3TVL
Hunt P G8CRZ
Hunt P 2E0PAH
Hunt P G8HJK
Hunt P G4AEO
Hunt R G0CLR
Hunt R G4MIJ
Hunt R G4XAB
Hunt R M0RBT
Hunt R M0RHI
Hunt R M3RMH
Hunt R M0AKJ
Hunt R M6RPH
Hunt R G4PRL
Hunt S G1VUK
Hunt S G3YQ
Hunt S GM4YDC
Hunt S G6YBV
Hunt S M1FJB
Hunt S G3TXQ
Hunt S M1FIG
Hunt T M6THV
Hunt T G4DWM
Hunt T M6TGI
Hunt T M6THJ
Hunt W G0SZX
Hunt W M6WDH
Hunter A G6LBQ
Hunter A 2E0ZOZ
Hunter A M6AVM
Hunter A 2I0POD
Hunter A 2E0CUC
Hunter C MM6PHO
Hunter C M3PTX
Hunter D M3CGI
Hunter D M3CSR
Hunter D M6DEC
Hunter E 2E0FWN
Hunter F GI4NKB
Hunter F GI6DEY
Hunter G M3XMU
Hunter G M3VGH
Hunter G G1OSL
Hunter G GM3ULP
Hunter G G8WWD
Hunter G G6YFF
Hunter H GM0LOO
Hunter J GM0EEH
Hunter J G3KAK
Hunter J GW4GZX
Hunter J GM7AYW
Hunter J GI7JLD
Hunter J G7VQX
Hunter J M6JAM
Hunter J M6BHN
Hunter K G7PQB
Hunter K GM0JHE
Hunter K G4ENJ
Hunter M MI0CLP
Hunter M 2E0YYY
Hunter M M6MMM
Hunter N G8DQN
Hunter P 2E0PHX
Hunter P M6PHX
Hunter P G0GSZ
Hunter P MM6GQP
Hunter R M6RGO
Hunter R 2E0FEO
Hunter R M0TLO
Hunter R M6FEO
Hunter R G7FVA
Hunter R MI3VWH
Hunter S G7UAK
Hunter S 2E0GDH
Hunter S M0SXH
Hunter S M6EMW
Hunter S GM3WZV
Hunter T MM3XIW
Hunter T M6ZXC
Hunter W GM3HUN
Hunter W GM7BAS
Hunter W G8CQV
Hunting J GM8DPV
Huntington J M3FVC
Huntley M G6GGT
Huntley P G1ZQG
Huntley R G1VWZ
Hunton I G0EOX
Hunton W G7HAF
Huntriss T 2E0FMB
Huntsman M G4MZN
Huntsman R G3KBR
Hupton D G0OMF
Hurd A G7PUP
Hurkett J M3IIV
Hurlbutt J 2E0DCV
Hurlbutt J M6CLG
Hurley C M6UXB
Hurley C MW6RPS
Hurley P G6LWD
Hurley R G0CGH
Hurley S 2E0OWL
Hurley T G7KRH
Hurlock N M6NHX
Hurnandies J G8KAM
Hurp P G1RBY
Hurrell B G6UPI
Hurrell D G3JMK
hurrell G G7GGH
Hurrell M G4UGV
Hurrell R G4HSM
Hurrell S M3YYS
Hurren B M6HRN
Hurren M M3MHN
Hurren T M0THA
Hurst A G7VGO
Hurst B G4XIL
Hurst C G4BJB
Hurst C G7MER
Hurst G G0RTZ
Hurst G G4MIT
Hurst G M0XID
Hurst I 2E0CQM
Hurst J G4EFY
Hurst J M1EDO
Hurst J M6BUF
Hurst K G7ULS
Hurst M G0CCM
Hurst M G4ASZ
Hurst M M6HDD
Hurst M M3YMH
Hurst N M3SDN
Hurst P G3PCT
Hurst P G0CCN
Hurst R G7HIU
Hurst S M0ARZ
Hurst S G8LIK
Hurst T G0GEV
Hurt A G0OJY
Hurt R G0HDS
Hurton T G6SJG
Husband M M1CYM
Husk J G8GLI
Hussain A M3YHJ
Hussain M M6HUS
Hussain U M6UZH
Hussey D G0MIB
Hussey R GM1FGN
Hutasuhut H M6HBG
Hutchens J GM0SYY
Hutchens K M6NDN
Hutchings D G6YMU
Hutchings F G8SXJ
Hutchings M G4ZIW
Hutchings M G8EAD
Hutchings P G8LRD
Hutchings R G4MEA
Hutchings S G0DVE
Hutchings T G4POB
Hutchings W G0UUA
Hutchins J G6ISA
Hutchins K G0BMN
Hutchins P M6PTX
Hutchins W G0HVC
Hutchinson A M1VGH
Hutchinson B G3VGH
Hutchinson D G8SQH
Hutchinson D G1XGE
Hutchinson D M3UBL
Hutchinson E G0CEB
Hutchinson G GI8MIV
Hutchinson G M0OTA
Hutchinson G M6GIH
Hutchinson I 2E0IHH
Hutchinson I M6IHH
Hutchinson J MM6JHH
Hutchinson J 2M0CPV
Hutchinson J MM0JHL
Hutchinson K G8UPF
Hutchinson L M0AIX
Hutchinson P G0OXT
Hutchinson S G0HYS
Hutchinson S GM3WPA
Hutchinson S G7KBZ
Hutchinson S G7REJ
Hutchinson S M3MOQ
Hutchinson S M3JSH
Hutchinson S M3XHN
Hutchinson T G3SQW
Hutchinson T GI4YRP
Hutchinson T M6TAH
Hutchinson V GI0EJU
Hutchinson W GI4FUM
Hutchison A 2M0ZAX
Hutchison A MM0KFX
Hutchison A MM6HUT
Hutchison D GM7RPT
Hutchman W GI0WAH
Hutley A M0SPS
Hutley K G0VDP
Hutley O M3OIL
Hutson D G0PNE
Hutson J MM0RAG
Hutt G G0HZC
Hutt J M3JEH
Hutt K G0TSH
Hutt L G7AYO
Hutton A M5CBR
Hutton A GM4ZRH
Hutton C G8CJD
Hutton C MM6EQW
Hutton D M3PDH
Hutton G 2I0SXM
Hutton G G7OVB
Hutton G MI3SXM
Hutton G M6DKS
Hutton H G6WJW
Hutton J M3JNY
Hutton J MM6JOH
Hutton K G4VSY
Hutton L G3CLW
Hutton N M6NHZ
Hutton R G3WYH
HUTTON R G8FPU
Hutton R M0ATB
Hutton R MM3WVP
Hutton R 2E0KCP
Hutton R 2M0RDH
Hutton R M0KCP
Hutton R MM0RBR
Hutton R M6KDP
Hutton R MM6ROH
Hutton S MI0CCG
Hutton S MI0LJM
Hutton T G0HUT
Hutton T GI1ACN
Hutton T G1WDQ
Huxham K G8MLI
Huxham M G4UXP
Huxley H 2E0HUX
Huxley R G1BNX
Huxley W G3RIB
Huyton S G8XDV
Hyams P GW4OZU
Hyams R G4NYA
Hyatt C G0FBC
Hyde A GM1CHT
Hyde A G4ILF
Hyde A G8VQX
Hyde B 2E0AES
HYDE B GM0MHE
Hyde B G0SFA
Hyde C G7JKH
Hyde D 2E0IFW
Hyde D G0PDH
Hyde D G1OKI
Hyde D 2E0DKZ
Hyde D M0HLI
Hyde D M0DCH
Hyde E G0RFG
Hyde F M0HYD
Hyde G M0CTP
Hyde I M6LBI
Hyde J G0GDS
Hyde J G1NML
Hyde J M0NYX
Hyde J M3ZOG
Hyde J M6XAD
Hyde K 2E0ISK
Hyde M 2E0HJO
Hyde M G4IHZ
Hyde N GM4LTL
Hyde P 2E0BFW
Hyde P G4CSD
Hyde P G7KIQ
Hyde P G1EFO
Hyde P M6BBS
Hyde R G3ZDW
Hyde T G6KHA
Hyde T G7ANO
Hyde-Dryden S M0SHD
Hyder T G7JEJ
Hydes A G3XSV
Hydes D M6NLF
Hydes N G4OPB
Hydes R M0TDM
Hyett M M6MAH
Hyett R G4MBJ
Hyland-Davis G M3YGD
Hyland-Davis S M6MWA
Hylton A M3XJH
Hylton H 2E0TBR
Hylton H M6AUL
Hyman H G3IZQ
Hyman M M0MBS
Hyman S G4CCT
Hyndman A G7GTG
Hyndman E MI0DNB
Hyndman J MI0DMT
Hynes B GM0CQV
Hynes I G4UCV
Hynes P G8TEL
Hynes T G4DUV
Hyslop A GM0UDY
Hyslop K G1NPJ

# I

Ian B G1RLK
I'Anson P G3YFV
I'Anson R G4CGG
I'Anson S M3XSI
I'Anson-Holton J G0BST
Iason J M6EIJ
Iball K M3KML
Ibbetson A G3XAQ
Ibbett M M0MPI
Ibbitson H 2E0MYB
Ibbitson H M0MYB
Ibbitson T G0VTI
Ibbitson T G1RHW
Ibbotson A G1OKB
Ibbotson A M6UAJ
Ibbotson R G0DVY
Ibbotson S 2E0SJI
Ibrahim A 2E0GUC
Ibrahim D G7SCU
Icke K G4XQG
Icke S G4ZWY
Ickringill J G3HHU
Igali L M3IGA
Igglesden H G4TPW
Iggulden A M6MMX
Igo P GI4NJQ
Ikeda-Chew C M3ZGO
Ikin E G4VPC
Ikin K G1RWX
Ikonomou V G0VAS
Iles C G0VOE
Iles P 2W0PCE
Iles S 2E0OWC
Iles S M6OWC
Ilett A M0BSV
Ilie H 2W0IMD
Ilie H MW0UNU
Ilie H MW6KPA
Iliffe J G8SVN
Iljin P M0HMZ
Illidge P 2E0PAE
Illidge P M0MCI
Illidge S G4HBS
Illman D G7CVM
Illman R M1DKW
Illman S G6VF
Illsley J G2FHF
Illsley S G0BES
Illston M G1YYP
Ilott J G4KWW
Ilott T G4EOJ
Ilsley D G4ZVW
Ilsley M M6MTI
Ilston J G0JAN
Ilston J G1OXO
Imamura S MM0JVQ
Imianowski A G4KKU
Imms S M0IMM
Imperato J GW6ITB
Impey M M6IRJ
Import S G7ITT
Ince A G4YUK
Ince A GM0BZS
Ince D G4UPO
Ince T M6TII
Inch W M1ESH
Ind P 2E0IND
Ind P M0IND
Ind P M6IND
Ingamells C 2E0ZTL
Ingamells C M6CAI
Ingamells R G8HZJ
Ingate R M3CAM
Ingerslev L G4RZC
Ingham B M3BRT
Ingham D G4RPL
Ingham D G7LTR
Ingham D M3LTR
Ingham G G4VSV
Ingham J G3RMQ
Ingham K M3NSJ
Ingham P G6HDD
Ingham R M0GRI
Ingham W G4DWO
Ingle N G3ZNE
Ingle P G6HYI
Ingle P G7CUP
Ingle R G6YFH
Ingle R G0MQI
Ingle T G7MZE
Ingleby T M6CBA
Ingledew S 2E0CZC
Ingledew S M6GEP
Ingles C 2D0CFZ
Ingles C MD0PWI
Inglis A G6OAS
Inglis F GM3NLB
Inglis G MM6SZA
Inglis N 2E0NAS
Inglis N M6NSJ
Inglis N MM0GHN
Inglis R MM0RSI
Inglis R MM6CXJ
Inglis W MM6WAI
Ingmire G G7BNW
Ingram A G1OYM
Ingram C G4ZSY
Ingram C MM6OMT
Ingram C G3TDX
Ingram F M3YWI
Ingram J G7BSK
Ingram J G4FDS
Ingram K M1DIM
Ingram K G7SVF
Ingram K M0GAQ
Ingram M G7VOX
Ingram P G3GYC
Ingram P G4OZL
Ingram P GM4UPN
Ingram R G3YIY
Ingram S M3LUP
Ingram W G4PEF
Ingram W G4ZKM
Ingrey D 2E0CGV
Ingrey D M0HBV
Ingrey D M6ALT
Inman K G6GEL
Inman R G3MYG
Inman S 2E0KBJ
Inman S M3SBJ

Innes D ... GM7GVD
Innes D ... MM3FYN
Innes G ... G4YDH
Innes J ... MM0JXI
Innes I ... MI6JBI
Innes J ... GI1JQP
Inness M ... G6GMF
Inns D ... GM0RTY
Inns S ... G8YDE
Instone A ... 2E0HUN
Instone A ... M6IAQ
Instone G ... G6KAW
Instone P ... G4SOA
Instone S ... G0NPL
Inwood C ... M3FQT
Inwood D ... M6INW
Inwood J ... 2E0KUF
Inwood J ... M6DIV
Ioannou J ... M3JMI
Ion D ... 2E0DLP
Ion D ... M0OAC
Iona X ... G4RTC
Iqbal M ... M0MIQ
Iqbal M ... M6BQZ
Iredale G ... 2E0GSR
Iredale G ... M6EZA
Iredale P ... G8HCZ
Ireland C ... G0SSN
Ireland E ... 2M0EGI
Ireland E ... MM0WEI
Ireland E ... MM3WEI
Ireland G ... G3CCL
Ireland J ... G0AQB
Ireland L ... GW0JHH
Ireland M ... 2E0CIV
Ireland M ... M0MIE
Ireland M ... M3YMY
Ireland P ... G0TGX
Ireland R ... G3YXQ
Ireland S ... G3ZZD
Ireland T ... M3PVB
Iremonger I ... M6IFI
Ireson A ... G8GOM
Ireson M ... G3OKB
Ireson R ... G6HGG
Irish C ... G6PDE
Irish R ... G4LUF
Irlam I ... G7WLL
IRONMONGER C ... G6HYF
Irons C ... M6ICA
Irons K ... G1FVA
Irvin H ... G1DEX
Irvin K ... M6KRI
Irvine A ... GM7OAW
Irvine A ... MM3OAW
Irvine A ... G0TRU
Irvine G ... 2M0GLI
Irvine G ... MM6GLI
Irvine N ... 2E0NBE
Irvine N ... M0NCE
Irvine N ... M6FUF
Irvine R ... MM1RIK
Irvine R ... G7KHV
Irvine R ... MM0XRI
Irvine R ... MM6XRI
Irving A ... GM8WGU
Irving D ... M0HRZ
Irving G ... 2M0ZBF
Irving G ... MM6EZO
Irving I ... G0BAQ
Irving M ... G3ZHY
Irving R ... 2E0LDF
Irving R ... G1UCR
Irwin A ... M6EKY
Irwin B ... G4DSR
Irwin B ... G4XZF
Irwin H ... GI0LTF
Irwin K ... M6KWI
Irwin N ... G0SSO
Irwin P ... GI6FEN
Irwin P ... G0RYQ
Irwin R ... 2E0YPG
Irwin R ... G8MFR
Irwin R ... G7TUP
Irwin T ... GM4VIK
Irwin W ... MM0BMA
Isaac D ... M0RMS
Isaac E ... 2E0BYJ
Isaac R ... G0HAE
Isaac Sneath L ... 2E0LUY
Isaac Sneath L ... M0LUY
Isaacs A ... G3SGL
Isaacs C ... G3PIY
Isaacs J ... M0CBI
Isard-Brown J ... M3JLE
Isbell M ... 2E0MGI
Isbell M ... M0MGI
Isbell M ... M6EKQ
Isham C ... GD3OFC
Isherwood C ... M3CFI
Isherwood J ... 2E0JIF
Isherwood J ... M0OND
Isherwood J ... M6GIM
Isherwood M ... G4VSS
Ishmael D ... G7CWE
Islam M ... M0EDV
Isles J ... G3NEH
Isles J ... M6ZZA
Ismay A ... 2I0IZI
Ismay A ... MI6AHO
Isom D ... G0DAI
Isom T ... G4RLC
Ison A ... G7BNM
Ison C ... G4DAP
Ison G ... M3FRU
Issac B ... 2E0DJZ
Issac B ... M6FFO
Issatt S ... 2E0DBZ
Issatt S ... M3DBZ
Issatt S ... M3SRI
[illegible] ... G7NOE
Isted M ... G0MYP
Ivanov I ... M0YGM
Ivermee C ... M0WYM
Ivers C ... M3IHS
Ives A ... G4DDT
Ives C ... 2E0CAG
Ives E ... G7CTG
Ives J ... M3JGI
Ives P ... G3NIW
Ives R ... G3MSL
Iveson D ... 2E0DKI
Iveson D ... M0GWH
Ivison W ... M3IVI
Ivory J ... M0MDG
Izzard J ... 2E0BUR
Izzard M ... G8XWR
Izzard P ... 2E0CCF

## J

Jack A ... G0VQW
Jack C ... M6YAY
Jack J ... G2AJW
Jackaman K ... G1HBR
Jackett R ... G0JSJ
Jacketts A ... G8UBP
Jacklin C ... M3DIT
Jacklin D ... GW3VNZ
Jacklin H ... G1XAA
Jacklin J ... G1JZN
Jacklin R ... M1REJ
Jacklin S ... M3JUO
Jackman D ... G4WPT
Jackman H ... G4YHK
Jackman P ... G1TST
Jacks G ... M6GGJ
Jackson A ... 2E0KSO
Jackson A ... G0UEU
Jackson A ... G7SVE
Jackson A ... MM0MUL
Jackson A ... M0RCL
Jackson A ... M1AAS
Jackson A ... G3DTP
Jackson A ... G4FNK
Jackson A ... GI4TCR
Jackson A ... G8JAC
Jackson A ... G1KLP
Jackson B ... G0GII
Jackson B ... G0NYL
Jackson B ... G4UFQ
Jackson B ... M0IGR
Jackson B ... G4MKT
Jackson B ... G8FFM
Jackson B ... M6WIC
Jackson B ... M6EXX
Jackson B ... G1IUL
Jackson C ... 2E0HVM
Jackson C ... G7UPN
Jackson C ... M6CRJ
Jackson C ... G0EUD
Jackson C ... 2E0ANH
Jackson C ... M0XCJ
Jackson D ... G0GQP
Jackson D ... G1FQI
Jackson D ... G1KYN
Jackson D ... G4ESY
Jackson D ... GI4NAE
Jackson D ... G4PYH
Jackson D ... 2E0YDJ
Jackson D ... M0YDJ
Jackson D ... G0EGG
Jackson D ... G6ZET
Jackson D ... G0WJD
Jackson E ... G6WJX
Jackson F ... G4XHC
Jackson F ... G6FTB
Jackson F ... M0JSZ
Jackson F ... G3ZMX
JACKSON G ... G4CKH
Jackson G ... M3FLJ
Jackson G ... M3GPJ
Jackson G ... M3MQJ
Jackson G ... G1HWH
Jackson G ... M3HWH
Jackson H ... 2E0HJJ
Jackson I ... G3OHX
Jackson I ... G3TYP
Jackson I ... G8AWH
Jackson I ... M6IGJ
Jackson J ... 2E0AWK
Jackson J ... 2E0IAB
Jackson J ... G0FVS
Jackson J ... G0NDD
Jackson J ... G1XQP
Jackson J ... GM3DDL
Jackson J ... G4CSV
Jackson J ... G4OPV
Jackson J ... [illegible]
Jackson J ... G8WWO
Jackson J ... MM3OFE
Jackson J ... 2E0DHF
Jackson J ... M6FHM
Jackson J ... M6JJL
Jackson J ... M3SXU
Jackson K ... G4KXG
Jackson K ... G8RAV
Jackson K ... 2E0CPK
Jackson K ... M0HHC
Jackson K ... M6BKL
Jackson K ... G4NEJ
Jackson K ... M0XLT
Jackson L ... G0NPJ
Jackson L ... G4HZJ
Jackson M ... G0BYK
Jackson M ... G0FDE
Jackson M ... G1LLZ
Jackson M ... M6EGG
Jackson M ... M6MHJ
Jackson M ... G0JVR
Jackson M ... G0SMJ
Jackson M ... M0ACK
Jackson M ... M6EXR
Jackson O ... G4VYU
Jackson P ... 2M0WTE
Jackson P ... 2E0ANG
Jackson P ... G1GJT
Jackson P ... G3KNU
Jackson P ... G3WQ
Jackson P ... G4WBH
Jackson P ... G7AMW
Jackson P ... 2E0JXN
Jackson P ... M1AQJ
Jackson P ... M3JXN
Jackson P ... M6PNJ
Jackson P ... G0PPQ
Jackson R ... G1ZGF
Jackson R ... G6VVS
Jackson R ... M0BLO
Jackson R ... G7ANA
Jackson S ... G0UGY
JACKSON S ... G1PEE
Jackson S ... G3ZBB
Jackson S ... G6ONW
Jackson S ... G8SZZ
Jackson S ... M3IPM
Jackson S ... M6SOG
Jackson S ... M6SJJ
Jackson S ... M6SVN
Jackson S ... 2E0JCX
Jackson S ... M6FDS
Jackson S ... 2E0STX
Jackson S ... M0WTX
Jackson S ... M6WTX
Jackson T ... G7HRJ
Jackson T ... G4HYY
Jackson V ... GW4UPG
Jackson W ... G0DLL
Jackson W ... GI4TCS
Jackson W ... G7VPL
Jackson W ... 2M0WBJ
Jackson W ... G8OQV
Jackson W ... MM0HYM
Jackson W ... MM6WBJ
Jacob C ... G3VPG
Jacob C ... G1DEY
Jacob F ... GW0OET
Jacob S ... G0HSV
Jacobi P ... G8ZEK
Jacobs A ... G0PAD
Jacobs C ... G4LEP
Jacobs C ... G8WAV
Jacobs C ... 2E0TRW
Jacobs C ... M0KTT
Jacobs C ... M6BSA
Jacobs D ... 2E0VCC
Jacobs D ... 2E0DFJ
Jacobs D ... M6RFJ
Jacobs J ... G1GFW
Jacobs J ... G4VYA
Jacobs N ... 2E0RKO
Jacobs R ... G8KNU
Jacobs R ... M3KRE
Jacobs R ... G7ONE
Jacobs S ... G3SUS
Jacobs T ... G7EIE
Jacobsen M ... G1AJD
Jacot G ... G0SHN
Jacovides N ... M0DEA
Jacquemai E ... G4YLG
Jacques A ... G8IWQ
Jacques D ... G7VTW
Jacques J ... G4AXF
Jacques J ... G1NQO
Jacques P ... G6YMY
Jacques S ... G0PSJ
Jagdev P ... G7IVN
Jaggard W ... G1OIK
Jaggard W ... M3OIK
Jagger D ... G3KAJ
Jagger T ... M6AYC
Jaggs P ... G7IQM
Jago C ... 2E0PAA
Jago P ... G8ECR
Jakins A ... G7GWA
Jakins E M ... G8HKP
Jakowuik A ... G7GPG
Jakstas A ... 2E0LTU
Jakstas A ... M0PZD
Jakstas A ... M6IZP
Jakubowski B ... M6BIH
Jakubowski M ... M0TNB
Jakusz-Gostomski A ... M5ACR
James A ... G1CHN
James A ... G1VOC
James A ... G3KZT
James A ... G3RUV
James A ... G4IVD
James A ... G6FVJ
James A ... G7FFT
James A ... G7VIR
James A ... G7WDS
James A ... 2E0OSO
James A ... M6FQM
James A ... G0KLK
James B ... GW4TFX
James B ... M3NNZ
James B ... G4PCK
James B ... GW0WGW
James C ... G0SDD
James C ... G6MUJ
James C ... MD3XUA
James C ... G7SUT
James C ... G7DDV
James C ... G0CLY
James D ... 2W0DJS
James D ... G0FNM
James D ... G0FCQ
James D ... G0OYO
James D ... G4DOC
James D ... GM4UTK
James D ... GW6DFX
James D ... MM1GBS
James D ... MW3NUX
James D ... 2E0CNB
James D ... GW6JSJ
James D ... G8XUL
James D ... M0BCN
James D ... M0JAP
James D ... M6CCM
James D ... M6DDQ
James D ... M6IAT
James D ... 2E0TLG
James D ... M6PDP
James E ... G3KQG
James E ... G4PQM
James F ... 2E0GTW
James F ... G0LOF
James F ... G4YVF
James G ... 2E0PSN
James G ... M6PSN
James H ... G3HZP
James H ... G3UPZ
James I ... G4OQJ
James I ... M1TSU
James I ... M0CJY
james J ... GW0KPD
James J ... G7NLY
James J ... MW3ASG
James K ... 2E0ESM
James K ... G0BFK
James K ... G8EZR
James K ... G4XQA
James K ... G8PGH
James L ... G2DD
James L ... G4JJR
James M ... 2E0RNM
James M ... G1FOF
James M ... G6BZE
James M ... G7SFL
James M ... G7UTS
James M ... MI0MGJ
James M ... M3EZY
James M ... M3RNM
James M ... GW4YID
James M ... G0USF
James M ... 2E0PBV
James M ... G3OZK
James M ... GW8PKV
James M ... M6PDL
James M ... G6IXE
James M ... M1FGH
James N ... MW6HFY
James N ... G1NIC
James O ... 2E0FQY
James P ... G1HPU
James P ... G6PZF
James P ... 2E0ITU
James P ... G0IQI
James P ... M6BGM
James P ... M6ITU
James P ... G6NSQ
James P ... MI6PCJ
James R ... G0LIA
James R ... G0REA
James R ... G3AHE
James R ... G3ZSZ
James R ... GM4CXM
James R ... G4TOT
James R ... M3SNL
James R ... 2E0ONO
James R ... M0JJA
James R ... M6REJ
James R ... 2M0RJJ
James R ... MM0RJJ
James R ... MM6GHX
James S ... G3NZW
James S ... 2E0HAV
James S ... M0ZEL
James S ... M6DQZ
James S ... M6XDX
James S ... 2E0BQH
James S ... M0SGJ
James S ... M3FGG
James S ... M3NOY
James T ... G0VAU
James T ... G4EWW
James T ... M0TMJ
James T ... M6TMP
James W ... M0GIP
James W ... M1PAS
Jameson A ... G0VTJ
Jameson B ... G6VPN
Jameson C ... G0UFU
Jameson C ... M3KCA
Jameson G ... G7OMO
Jameson I ... 2E0IGQ
Jameson M ... M0MIK
Jameson N ... 2I0NDJ
Jameson N ... MI0NDK
Jameson P ... M1CIS
Jameson R ... MI6WRT
Jameson T ... GM3VYJ
James-Robertson R ... G4ELL
Jamieson B ... M3VQS
Jamieson C ... M3HPF
Jamieson D ... MM0EAI
Jamieson G ... 2M0GEJ
Jamieson G ... MM0TGG
Jamieson G ... MM6XGJ
Jamieson J ... GM1PUA
Jamieson M ... G0RHU
Jamieson M ... GI4VIZ
Jamieson M ... 2M0CTI
Jamieson M ... MM6CNV
Jamieson N ... G8LCP
Jamieson N ... M0BUG
Jamieson R ... GM1JPJ
Jamieson R ... G6RVH
Jamieson R ... M3RGJ
Jamieson W ... GM7ANE
Jamieson-Colville A ... M3WFO
Jamil M ... G0RAN
Jamison A ... 2I0BID
Jamison A ... MI0GTI
Jamison A ... MI3OEQ
Jamison W ... G3XGH
Jandrell A ... G4HWH
Janes H ... G0TIG
Janes M ... 2E0BBS
Janes P ... G1SXU
Janes R ... G0JNA
Janes R ... GW4IUN
Janezko R ... M6ZTB
Jannetta L ... G8YSH
Janowicz D ... 2E0SPH
Janowicz D ... M0ZPL
Janowicz D ... M6FRJ
January M ... M3JNX
Jappy D ... MM6HHJ
Jaques M ... G6WKZ
Jardine A ... GM7SAK
Jardine A ... MM0SAK
Jardine D ... G0FDV
Jardine J ... G3XJY
Jardine J ... G7UYJ
Jarman G ... G0GBS
Jarman L ... M3LAJ
Jarman M ... G7VYN
Jarman S ... G1YGP
Jarrard D ... G0DEP
Jarratt B ... G1ATY
Jarratt K ... G6WLA
Jarrett A ... G1EFP
Jarrett A ... G4FRZ
Jarrett D ... G4DCJ
Jarrett G ... G8NDB
Jarrett G ... G4PPV
Jarrett M ... 2E0MRJ
Jarrett P ... G1UZC
Jarrett P ... G4CDJ
Jarrett P ... G8FTP
Jarrett S ... M1BWN
Jarvice D ... 2E0DNJ
Jarvice D ... M6TAJ
Jarvie J ... G3XTI
Jarvie W ... GM8IIH
Jarvill W ... G7SYC
Jarvis A ... GD0NTA
jarvis A ... M6HVO
Jarvis B ... GD0NTB
Jarvis B ... M6DTL
Jarvis B ... M1CMR
Jarvis C ... G8YJT
Jarvis C ... M0YJT
Jarvis D ... G4CEU
Jarvis D ... G4TFZ
Jarvis D ... M3OMZ
Jarvis G ... G0WGJ
Jarvis H ... G7EXT
Jarvis I ... G6CYO
Jarvis I ... M6BIJ
Jarvis J ... G3SUG
Jarvis J ... G4NEY
Jarvis N ... G4SPD
Jarvis P ... G3OWJ
Jarvis P ... 2E0XPP
Jarvis P ... M0BHN
Jarvis P ... M3XPP
Jarvis R ... G4OBA
Jarvis R ... G7OHM
Jarvis R ... G8UBU
Jarvis R ... G4JPA
Jarvis S ... G4TIA
Jarvis S ... M3HIO
Jarvis S ... M0BJL
Jarvis S ... M3UFJ
Jarvis T ... G6XTZ
Jarvis W ... G6SND
Jasper D ... M6XMA
Jasper G ... G8TWA
Jasper R ... G0CIH
Javes S ... G1BGC
Jawor S ... M3UGV
Jay A ... G8JAY
Jay A ... M3YAJ
Jay B ... G8TEO
Jay C ... G4TDR
Jay M ... M3NEE
Jay P ... 2E0PJY
Jay P ... M6RGY
Jayne B ... G0DPL
Jayne C ... M1FAT
Jaynes D ... M3RUW
Jays P ... M3IEA
Jeacock T ... G0EZY
Jeacock T ... G8CDV
Jeans C ... G4IPJ
Jeans W ... G3GAA
Jeary C ... M6NWP
Jeavons A ... G3OZL
Jeavons P ... G8XUM
Jebb J ... G8YDC
Jebbett M ... M0MRJ
Jebbett T ... G0GPZ
Jeckells G ... M0BHW
Jedryka A ... 2E0PLK
Jedrzejczak A ... M6DLK
Jedryka B ... M6JBL
Jeeves M ... G6CBY
Jeeves R ... G7EPR
Jeffcoate S ... G7IMB
Jefferies D ... G6GPR
Jefferies G ... M1ASR
Jefferies J ... 2E0HUR
Jefferies J ... M6HUR
Jefferies L ... G0BXZ
Jefferies N ... G8RNV
Jefferies R ... G0BQG
Jefferies R ... M0AMF
Jefferies T ... G1JPK
Jefferies W ... 2E0WEJ
Jefferies W ... M0GBC
Jefferiss A ... M1HVJ
Jefferiss R ... M3MVJ
Jeffers J ... G0UNB
Jeffers M ... G1VDC
Jefferson D ... G6PZE
Jefferson D ... M6MDJ
Jefferson I ... G1JCC
Jefferson I ... G4IXT
Jefferson I ... M0JCC
Jefferson K ... M3KEJ
Jefferson M ... G8WYB
Jefferson P ... M1AZM
Jefferson R ... G8UZM
Jefferson S ... M0LAA
Jefferson S ... M1DOA
Jefferson T ... G4IAJ
Jefferson W ... M6WEN
Jeffery A ... G8SIG
Jeffery A ... M3ZTT
Jeffery C ... G3JLK
Jeffery D ... G4HWM
Jeffery D ... M6EDJ
Jeffery E ... G4UUJ
Jeffery G ... G7BRM
Jeffery H ... 2E0IAK
Jeffery K ... 2E0CQL
Jeffery K ... M0KAO
Jeffery K ... M6BQW
Jeffery M ... G6JVK
Jeffery M ... M0MAT
Jeffery M ... 2E0GDB
Jeffery M ... M3MAR
Jeffery N ... M3NLJ
Jeffery R ... G0GZI
Jeffery R ... G0MID
Jeffery R ... G6DSA
Jeffery R ... G6RBM
Jeffery R ... MW0DYZ
Jeffery S ... G7DSA
Jeffery S ... M6SBB
Jeffery S ... G1FAA
Jefferys A ... GU4RUK
Jefferys D ... G6IWZ
Jeffery-Wright H ... G6VDY
Jefford A ... G8GON
Jefford M ... G8UWE
Jefford P ... M1AFP
Jefford P ... M3ADL
Jefford T ... G7HIY
Jeffrey C ... MI6CEJ
Jeffrey J ... G3RDF
Jeffrey M ... 2E0MCJ
Jeffrey M ... M6ZEJ
Jeffrey R ... 2E0SCO
Jeffrey S ... GM0UYS
Jeffreys S ... G0IDB
Jeffries L ... M0AUY
Jeffries P ... G8TNH
Jeffries R ... G4KAR
Jeffryes J ... M6JGT
Jeffs M ... G7LJN
Jeffs R ... M0RMJ
Jeffs R ... M1KEJ
Jeffs W ... G3OAF
Jelfs M ... G8KNS
Jelley G ... G7DFV
Jelley G ... M3DFV
Jempson B ... G4OVW
Jenkin C ... M6RIV
Jenkin I ... M3NFE
Jenkin M ... G0DRN
Jenkings C ... M6CKT
Jenkins A ... M1LSG
Jenkins A ... G7CYD
Jenkins A ... 2M0LEX
Jenkins B ... G0VGC
Jenkins B ... G7MOY
Jenkins B ... G7PKG
Jenkins B ... G8KLE
Jenkins C ... 2E0INC
Jenkins C ... M0UWT
Jenkins C ... M6CJN
Jenkins C ... M6ERA
jenkins C ... M3LHU
Jenkins C ... G0VRX
Jenkins C ... M0XFU
Jenkins C ... M6XFU
Jenkins C ... G6JPC
Jenkins D ... 2E0DOJ
Jenkins D ... GW0PUM
Jenkins D ... G4TND
Jenkins D ... MW0LMW
Jenkins D ... MW3DIL
Jenkins D ... 2E0DSJ
Jenkins D ... M6JWL
Jenkins D ... GM8MNR
Jenkins D ... MW6HDT
Jenkins E ... 2E0SWB
Jenkins G ... G6GAC
Jenkins J ... 2E0JBJ
Jenkins J ... GW0ADO
Jenkins J ... G0KHY
Jenkins J ... G4LJW
Jenkins J ... M3JZV
Jenkins J ... MW0SGR
Jenkins J ... M0JRJ
Jenkins J ... G7CZF
Jenkins K ... 2E0ADO
Jenkins L ... 2E0AFS
Jenkins L ... MW3LEW
Jenkins M ... G0SBK
Jenkins M ... M3MSX
Jenkins M ... GW7EYP
Jenkins M ... GW3OMN
Jenkins N ... G7RQV
jenkins P ... G0ECK
Jenkins P ... G7HTU
Jenkins P ... GW8HWL
Jenkins P ... M3USW
Jenkins P ... M6PLE
Jenkins P ... GW4MII
Jenkins P ... M6EWC
Jenkins R ... GI4RVT
Jenkins R ... GW4UYT
Jenkins R ... GW6VFH
Jenkins R ... G8HDP
Jenkins R ... M3MYQ
Jenkins R ... G0WTK
Jenkins R ... G8TBF
Jenkins R ... G6CYR
Jenkins S ... 2E0GYC
Jenkins S ... MM3SWU
Jenkins T ... G6AJG
Jenkins W ... MM0WKJ
Jenkins W ... G4USW
Jenkinson B ... G7BBJ
Jenkinson B ... M3BJJ
Jenkinson D ... 2E0APW
Jenkinson G ... M1CYJ
Jenkinson G ... M3CYJ
Jenkinson H ... G0FAW
Jenkinson J ... G0HGM
Jenkinson J ... G8CVS
Jenkinson J ... M3JMJ
Jenkinson K ... G0LDY
Jenkinson M ... G0RGE
Jenkinson N ... 2I0SFA
Jenkinson N ... MI6PUX
jenkinson P ... G0VGK
Jenkinson R ... 2E0SWF
Jenkinson R ... M3SWF
Jenkinson R ... G4SEF
Jenkinson S ... G6HBZ
Jenkinson W ... G0OJT
Jenkins-Powell C ... G7MFR
Jenks D ... G1RFQ
Jenner A ... G0GQH
Jenner A ... G7KNA
Jenner D ... M3TWP
Jenner G ... G3KIW
Jenner J ... G0PEG
Jenner L ... M3JCT
Jenner S ... 2E0AXN
Jenner S ... M3DSU
Jennings A ... M3FPG
Jennings B ... G4TLM
Jennings B ... M3FBG
Jennings E ... M3XWY
Jennings F ... G0MXE
Jennings G ... G0MJL
Jennings J ... G0EOG
Jennings J ... G4VOZ
Jennings J ... G6URK
Jennings J ... G8LM
Jennings L ... M3LJZ
Jennings M ... G6ZEW
Jennings M ... MW3CXA
Jennings M ... 2E0XXM
Jennings M ... M3XXM
Jennings P ... M0CPD
Jennings P ... G7FMF
Jennings P ... G3RMJ
Jennings R ... G0JHD
Jennings R ... G3NXV
Jennings R ... G3SOE
Jennings R ... GI4WRJ
Jennings R ... GI4NFH
Jennings R ... M0GNK
Jennings S ... GI4RKC
Jennings T ... 2E0FNZ
Jennings Z ... M3ZBJ
Jensen F ... G1HQQ
Jensen J ... MW0ANX
Jensen I ... M0RXA
Jensen P ... G4GZT
Jensen P ... G7BTP
Jensen P ... M0EPJ
Jenson B ... 2E0BUO
Jenson B ... M0LBJ
Jenson B ... M3EOL
Jephcott A ... G6JPE
Jephcott R ... G0XNN
Jepsen R ... M0RPJ
Jepson A ... G0JYS
Jepson A ... 2E0USA
Jepson A ... M6HWE
Jepson D ... M6JEP
Jepson L ... 2E0LDJ
Jepson L ... M0NEX
Jepson L ... M6LDJ
Jepson W ... M0BLU
Jeram H ... M6HYJ
Jermany C ... G1EBP
Jermy R ... G0UVP
Jermyn A ... M3KPQ
Jerome C ... M6CBJ
[illegible] ... M0DFJ
Jerrome-Jones M ... GD4WDY
Jervis C ... G8JBC
Jervis S ... M3OLE
Jesinger D ... M0CJG
Jess A ... MI3MDV
Jesson L ... M3AQJ
Jesson T ... G1KFB
Jessop A ... 2E0JRA
Jessop A ... M0JRA
Jessop A ... M3JRA
Jessop K ... G3TAA
Jessop K ... M3HVU
Jessop M ... M0HFO
Jessop M ... M6HJE
Jessop P ... G8KGV
Jessup G ... G4BKB
Jessup G ... G7AYA
Jessup L ... M0LDJ
Jessup P ... G7VQW
Jewell B ... G7WKV
Jewell B ... M0BRB
Jewell C ... G7UHS
Jewell C ... 2E0CHZ
Jewell C ... M6RAL
Jewell C ... M0DWK
Jewell D ... G0LUA
Jewell E ... G4ELM
Jewell K ... M6KBJ
Jewell L ... G4OCX
Jewell S ... G4DDK
Jewitt D ... M3XHQ
Jewitt N ... 2E0BQM
Jewitt N ... M0NCK
Jewitt N ... M3VWY
Jewitt P ... G4ENL
Jewkes M ... G0TWH
Jewson D ... G3XFB
Jex A ... G0OOR
Jex L ... 2E0YAO
Jillings G ... G3WMJ
Jinks F ... G3XVQ
Jinks J ... G1WRU
Jinks M ... G0GIT
Jobber A ... G6EXU
Jobber R ... M6RRJ
Jobbins P ... G8OQG
Jobbins R ... G1DFZ
Jobbling A ... M6BFI
Jobes R ... G4XYP
Jobling C ... G4YHP
Jobson A ... G7SUQ
Jocys J ... G4WQD
Jodrell D ... M3DMJ
Jodrell M ... M6MHD
JODRELL P ... M3OXN
Johannessen P ... M3XNT
Johansen D ... G8TSG
Johanssen N ... G1ZPA
Johansson P ... G1NGN
John A ... GW8MFQ
John A ... M6FCN
John C ... M0AHJ
John C ... M6BBR
JOHN D ... G3WCB
John D ... MW0SBJ
John E ... G3SEJ
John G ... GW1EOI
John G ... G8ZRN
John R ... G4DDD
John S ... G0OJS
John W ... GW1YKT
Johns A ... G4TVJ
Johns A ... G6HAA
Johns C ... M3VAE
Johns D ... G0VNO
Johns D ... GW4XES
Johns D ... MW6LSW
Johns E ... G0RWI
Johns F ... GW0CXK
Johns I ... M6IJJ
Johns J ... GW4ZBU
Johns M ... 2E0SDO
Johns M ... M0MDJ
Johns M ... M6TBU
Johns M ... MW6TTS
Johns M ... G0GZM
Johns N ... M0ASI
Johns R ... G4FVX
Johns R ... G4NXD
Johns R ... 2W0RTJ
Johns R ... MW0RLJ
Johns R ... M0AAC
Johns W ... G7MTT
Johns W ... MW6WCI
Johnsen H ... GM0DHZ
Johnson T ... M0TFU
Johnson A ... G0PUK
Johnson A ... G3WHJ
Johnson A ... G4EFH
Johnson A ... G4FWH
Johnson A ... G4UFG
Johnson A ... G6BLK
Johnson A ... G8JVU
Johnson A ... M0AVY
Johnson A ... MM1CKW
Johnson A ... M6OAJ
Johnson A ... G6FYR
Johnson A ... M6EMQ
Johnson A ... M6AVJ
Johnson A ... G4RQK

**IMPORTANT NOTE**

**Revalidate licence to avoid revocation** – Ofcom has advised the Society that plans will be drawn up to revoke licences that have not been revalidated as required by the licence conditions. The quickest way to revalidate is to do so online via the Ofcom website: *https://services.ofcom.org.uk/* or by email: *amateur.validations@ofcom.org.uk* If you need assistance in the process, Ofcom staff are available to help, but please be patient during times of heavy workload.

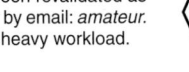

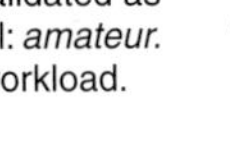

UK Surnames

Johnson A G0AUV
Johnson B 2E0BJA
Johnson B G3LOX
Johnson B G6MYO
Johnson B M0BOI
Johnson B M0CTI
Johnson B M1ABC
Johnson B M3IZP
Johnson B M3NAR
Johnson B G6URM
Johnson B 2E0XBN
Johnson B G0GAQ
Johnson B G3XIB
Johnson B M0XBN
Johnson B M6XBM
Johnson C G0BZN
Johnson C G4BFT
Johnson C G4PYD
Johnson C M3FYR
Johnson C M6VDR
Johnson C M3VWP
Johnson C M6LVE
Johnson C G7RBL
Johnson C M6ITI
Johnson C M6AMN
Johnson C G6LRT
Johnson C 2E0WDY
Johnson C M6NCU
Johnson C M6WDY
Johnson C 2E0CTJ
Johnson C G6PHX
Johnson C M6CTJ
Johnson D 2E0APX
Johnson D G0NGI
Johnson D G1PNL
Johnson D G3GAH
Johnson D G3HLG
Johnson D G6RTY
Johnson D G8AZM
Johnson D MI3AVJ
Johnson D M3JQG
Johnson D M3ROF
Johnson D M6NNU
Johnson D G0IES
Johnson D G3MPN
Johnson D G4DHF
Johnson D G4DPZ
Johnson D G7TZX
johnson D MW6HUY
Johnson D G4UIA
Johnson D M6DSJ
Johnson E G4LUW
Johnson E M6DVS
Johnson F M0BAL
Johnson F M6FJJ
Johnson F M6FKJ
Johnson F G0GSR
Johnson G G0SSK
Johnson G G4ZWA
Johnson G G7ATW
Johnson G MW0MEX
Johnson G M3GCJ
Johnson G 2E0GJJ
Johnson G M0GJJ
Johnson G M3HFU
Johnson G M6GEJ
Johnson G 2E0MZL
Johnson G M6GDJ
Johnson G M6MZL
Johnson G G8YYL
Johnson H G0KGE
Johnson H G1PHJ
Johnson H G1YMV
Johnson H G4XNK
Johnson I G1DGW
Johnson I M1CZI
Johnson I 2E0OCM
Johnson I G0SUQ
Johnson I M6HDO
Johnson J 2E0HOF
Johnson J G0AHJ
Johnson J G0FDS
Johnson J G0VIJ
Johnson J G4XTE
Johnson J G7AOQ
Johnson J M0BRT
Johnson J M3EGY
Johnson J M3FDB
Johnson J M3JQY
Johnson J M3MRJ
Johnson J 2E0CXD
Johnson J 2E0JTW
Johnson J M0HOQ
Johnson J M6BPK
Johnson J M6EAO
Johnson J M6JEY
Johnson J G7UAT
Johnson J 2E0JAJ
Johnson J G1JER
Johnson J G6VZM
Johnson J G0KSC
Johnson K 2E0KAJ
Johnson K G4FRF
Johnson K G7NHV
Johnson K G8NTD
Johnson K G4WBO
Johnson K M0DZB
Johnson K 2U1EKH
Johnson K G1NRN
Johnson K G6ZEY
Johnson K 2E0OAH
Johnson K M6OZY
Johnson L M3DKC
Johnson L M6CDY
Johnson L G4XMX
Johnson L M6LUK

Johnson M G0GCK
Johnson M G1RJA
Johnson M G3YQZ
Johnson M G6THC
Johnson M G8KEJ
Johnson M G8UJV
Johnson M M0MMT
Johnson M M3JZN
Johnson M M3KCU
Johnson M M6MIE
Johnson M G3ZUI
Johnson M M6MVA
Johnson M G4ZRT
Johnson M M6WWF
Johnson M M1MAJ
Johnson M M6HFV
Johnson M M3EGV
Johnson M 2E0MKJ
Johnson M G0BPU
Johnson M G6ONV
Johnson M G8MYF
Johnson M M5MUF
Johnson N G0SNX
Johnson N G7UIO
Johnson N 2E0NRJ
Johnson N M0NRJ
Johnson N M6NRJ
Johnson N G4AJQ
Johnson N G8HQO
Johnson N M3NRJ
Johnson P 2E0VEF
Johnson P 2E0EXA
Johnson P G0DIG
Johnson P G0LEU
Johnson P G0PPJ
Johnson P G0UMV
Johnson P G1FHH
Johnson P G1JST
Johnson P G3UMV
Johnson P G4EMV
Johnson P G4LXC
Johnson P G4TLO
Johnson P G4UCX
Johnson P G4UMV
Johnson P G4YFX
Johnson P G4ZSX
Johnson P G6DFC
Johnson P G7RFD
Johnson P G8BFC
Johnson P G8JYX
Johnson P G8LVC
Johnson P M0ALE
Johnson P M3PJJ
Johnson P G4RMT
Johnson P M6NNJ
Johnson P G7DDF
Johnson P G4TCT
Johnson P G4TMI
Johnson P G8KB
Johnson R 2E0BGV
Johnson R 2E0RLJ
Johnson R 2E0HAF
Johnson R G0BCZ
Johnson R G0GFC
Johnson R G0MAT
Johnson R G1ZUC
Johnson R G3VZT
Johnson R G3XOV
Johnson R G7COA
Johnson R GM7CZC
Johnson R G7HHK
Johnson R G7RFE
Johnson R M0AKE
Johnson R M0RAU
Johnson R M3FKV
Johnson R M3KOF
Johnson R GW0BCL
Johnson R 2E0MPB
Johnson R GM7NHS
Johnson R M6CNI
Johnson R 2W0BJR
Johnson R G7JHW
Johnson R MW3OSI
Johnson R M5ROB
Johnson R G4BWF
Johnson R M3YNM
Johnson S G0USJ
Johnson S G7OIB
Johnson S 2E0SEJ
Johnson S M0SBK
Johnson S M6SGY
Johnson S M3YAW
Johnson S M6EKO
Johnson T G0ABM
Johnson T G0WBR
Johnson T M1FCH
Johnson T M3FFA
Johnson T M3HXQ
Johnson T 2E0ITJ
Johnson T M0HEW
Johnson T M6TJZ
Johnson W G0AXZ
Johnson W G4ZXZ
Johnson W M0ABW
Johnson W M6PAG
Johnson W 2E0EMF
Johnson W 2E0WAJ
Johnson W M3JWQ
Johnson W M3FFE
Johnston A G6YNA
Johnston A G8ROG
Johnston A 2E0RHI
Johnston A M1SKY
Johnston A M3YWC
Johnston B MM3RBJ
Johnston C M3NJQ

Johnston C MI6FNO
Johnston D GI6RMO
Johnston D 2E0DWJ
Johnston D M6DWJ
Johnston E MD6ABT
Johnston E M6EAA
Johnston G GD7OZA
Johnston G G0RDN
Johnston J G3PHJ
Johnston J MI3JRJ
Johnston J MI6MNL
Johnston J GM4ENP
Johnston J G4OCC
Johnston K GM1KKI
Johnston K G4BCB
Johnston L G8DQF
Johnston M GM0MRJ
Johnston M MI3ZMJ
Johnston P G0FVN
Johnston P G6HGI
Johnston R GM3XUW
Johnston R G3YEK
Johnston R GI6WFX
Johnston R G7MHF
Johnston S GI4IBV
Johnston S G4WZM
Johnston S M6TZE
Johnston S 2E0BXW
Johnston S M6SPJ
Johnston T 2M0TJO
Johnston T MM0HZI
Johnston T MM6FXM
Johnston V G3KXV
Johnston V G3NUL
Johnston V G1PKS
Johnston W GI4GST
Johnstone C G8XDR
Johnstone D 2E0DLJ
Johnstone D GM1JZM
Johnstone D GM4EVS
Johnstone H G1RRG
Johnstone I MM0IRJ
Johnstone J G0CBC
Johnstone P GM0KMJ
Johnstone R GM0WWX
Johnstone R GM1YGV
Johnstone R MM6RXJ
Johnstone S GI0HHV
Johnstone W GM0BPF
Johnstone W M0BCE
Johnston-Stuart C M0GBH
Johnston-Stuart R M6RBY
Joiner D GM8BZP
Joiner M G3ZYZ
Joiner W G4OAX
Joll J G0TQT
Jolley D G1IQU
Jolley F G4XHZ
Jolley G G4BUF
Jolley M G3URN
Jolliffe E G3IMX
Jolliffe R G3ZGC
Jolliffe V M3ZGC
Jolly A G1WRF
Jolly I G4BTW
Jolly M G1OZV
Jolly P G6FEQ
Jolly R M3RXG
Joly G G6DFY
Jonas C 2E0ZCJ
Jonas C M0ZCJ
Jonas R G6EPL
Jones A G0JFA
Jones A G0MKW
Jones A G0PZN
Jones A G0YSS
Jones A G1URF
Jones A G1YBI
Jones A G3SGA
Jones A G4NHF
Jones A G4ORJ
Jones A GW4RUX
Jones A GW4TFS
Jones A G4VMZ
Jones A G6NSK
Jones A G7APQ
Jones A G7LNP
Jones A G7MIN
Jones A G7UEK
Jones A G8EKZ
Jones A M0EUS
Jones A MW0SWB
Jones A M1AWS
Jones A MW1SAS
Jones A M3FMK
Jones A M3KKX
Jones A M3MIN
Jones A M6ABZ
Jones A M6BGJ
Jones A M6TDI
Jones A G0PJC
Jones A G6VFI
Jones A 2W0EUO
Jones A 2W0LFY
Jones A M0JSA
Jones A MW3XVR
Jones A M6ATG
Jones A MW6EUO
Jones A M6GXL
Jones A MW6SBT
Jones A 2W0CZP
Jones A M6ALJ
Jones A MW6DVQ
Jones A G4TDZ
Jones A M6ALX
Jones A G7HCN

Jones A MW6ICU
Jones A G3CTZ
Jones A G1EBT
Jones A GW4VPX
Jones A 2E0LLA
Jones A G4RWY
Jones A M0KLW
Jones A M3UBE
Jones A M6GBA
Jones A MW6GNA
Jones A G0UIW
Jones A G4ICU
Jones A M1COJ
Jones A G7VHD
Jones B G1BDI
Jones B G1FUJ
Jones B G1VJN
Jones B G3GCW
Jones B GW3WRE
Jones B G4ISQ
Jones B GW4OPW
Jones B G4YFZ
Jones B G7IFI
Jones B G7KNN
Jones B G8EXJ
Jones B G8OYT
Jones B MW0CTX
Jones B M3BSJ
Jones B M3GEN
Jones B M1DIL
Jones B G6RIZ
Jones B M3TFG
Jones B 2E0FSG
Jones B M6AMJ
Jones B M6JHU
Jones B 2E0FJZ
Jones B G0UKB
Jones B G1BDH
Jones B G4PBY
Jones B GW6ZYI
Jones B MW3SNH
Jones C G0IQN
Jones C G0LSJ
Jones C G1PKO
Jones C GW3NKM
Jones C G4DHV
Jones C G6OAV
Jones C G6ZEZ
Jones C G7SPM
Jones C G8BWP
Jones C G8JRL
Jones C MW1AFW
Jones C MW1BTM
Jones C M3NZN
Jones C M3TOR
Jones C M3WDB
Jones C G0DYL
Jones C G6MYL
Jones C 2E0LJC
Jones C GW0JCB
Jones C M6CLJ
Jones C G0FTU
Jones C 2M0BRH
Jones C 2E0CCJ
Jones C G0FIJ
Jones C G8GFB
Jones C MM0GPL
Jones C MM3FPI
Jones C M6FMW
Jones C G6ERZ
Jones C 2E0CLJ
Jones C M6EAD
Jones C MW6GMJ
Jones C M6WCG
Jones C 2M0YCJ
Jones C MM0YCJ
Jones C MM6YCJ
Jones D 2E0DDJ
Jones D G0DSR
Jones D G0EYW
Jones D G0IBW
Jones D G0UDJ
Jones D G0WVV
Jones D G1XJN
Jones D GI3KVD
Jones D GW3XYW
Jones D G4EIN
Jones D G4FAH
Jones D G4FQR
Jones D G4GDS
Jones D G4LXH
Jones D G4RVJ
Jones D G4SXD
Jones D GW4XMU
Jones D G6PGG
Jones D GW6REF
Jones D G6WAG
Jones D G7VSJ
Jones D G8IMX
Jones D G8JPJ
Jones D GW8XMW
Jones D MW1DUJ
Jones D M1FAX
Jones D M3PMU
Jones D M3YHC
Jones D M6DQQ
Jones D MW6DRV
Jones D M6WCN
Jones D G7SSB
Jones D M3PVX
Jones D MW6SEP
Jones D M6HXB
Jones D 2E0BWV
Jones D 2E0CPO
Jones D 2E0DYN
Jones D 2E0KBO

Jones D 2E0VCD
Jones D 2E0ZZF
Jones D 2E0IIE
Jones D G1PUK
Jones D G3VGD
Jones D G4FAQ
jones D G4GRU
Jones D G6KFR
Jones D G7FNQ
Jones D G8KKD
Jones D G8PBX
Jones D M0BER
Jones D M0IBT
Jones D M0ICZ
Jones D M3AZR
Jones D M3DAO
Jones D M3ITT
Jones D MW3XVB
Jones D M3YXC
Jones D M3ZVD
Jones D M6ADU
Jones D MW6AUX
Jones D M6DJX
Jones D M6FWX
Jones D M6OCO
Jones D M6SKV
Jones D M6TEL
Jones D G3UVR
Jones D G0REE
Jones D G7HCC
Jones D M3DEI
Jones D M1IHM
Jones E 2E0AKK
Jones E 2E0AEJ
Jones E GW0MLN
Jones E G0TMJ
Jones E G1HBV
Jones E G1VRA
Jones E G3EUE
Jones E G3FJI
Jones E G3JQK
Jones E G3ZLX
Jones E G4JPP
Jones E G4YML
Jones E G7NFM
Jones E G7TSB
Jones E G8XMU
Jones E M0AEV
Jones E M0ERJ
Jones E M0GJV
Jones E M3GTH
Jones E MW3VWE
Jones E G8CDC
Jones E 2W0ZZU
Jones E MW3ZZU
Jones E G0RLV
Jones E G7UNV
Jones F G1HBW
Jones G GW0DDL
Jones G G0KAM
Jones G G1PQY
Jones G G3VAS
Jones G G3VKV
Jones G G3VSB
Jones G G4OEX
Jones G G6PVK
Jones G G6STS
Jones G G7CMM
Jones G G7TIM
Jones G M0GRJ
Jones G MW0GTY
Jones G MW3GDL
Jones G M3IHB
Jones G M3LJX
Jones G M3XRY
Jones G 2E0GGO
Jones G 2E0TAO
Jones G M0GUF
Jones G M3UQI
Jones G G4TPV
Jones G M6GCQ
Jones G G1FEO
Jones G 2W0ZBC
Jones G M1DBF
Jones G GW0ANA
Jones G M0AJX
Jones G 2E0RSV
Jones G G4DPH
Jones G GW4UCK
Jones G M0RSV
Jones G M3ZZV
Jones G M6YZF
Jones G G6TGR
Jones H GW0LLD
Jones H GW0SAJ
Jones H G1NCN
Jones H G4XZJ
Jones H G6YFL
Jones H M0ZAY
Jones H MD6YBE
Jones H M0HSJ
Jones H G4VQS
Jones I G0GPS
Jones I GW1IQS
Jones I G3TLP
Jones I G4JVC
Jones I GW7SDE
Jones I G8WKH
Jones I MW0CAB
Jones I 2E0CLZ
Jones I M0IAJ
Jones I M6IAJ
Jones I 2E0WNA
Jones I G1OXJ
Jones I G4MLW
Jones I M6FZM

Jones I 2W0IUN
Jones I G4FQU
Jones I MW0IUN
Jones I MW6IUN
Jones I 2E0DIU
Jones I M6EOY
Jones J 2W0BLA
Jones J 2E0JOX
Jones J 2E0OVT
Jones J 2E0EEP
Jones J G0BPQ
Jones J G0HQK
Jones J GW0KJZ
Jones J G0OFB
Jones J GW0PND
Jones J G3JTJ
Jones J G3NPJ
Jones J G3TQL
Jones J G3UED
Jones J G3WVL
Jones J G4KYK
Jones J G4LPU
Jones J G4TLL
Jones J GD4WOW
Jones J G4XTU
Jones J G6NSG
Jones J G8AZT
Jones J M0CCN
Jones J M0DKP
Jones J M0DOC
Jones J MI0JPL
Jones J M0OMI
Jones J MW3JTJ
Jones J M3KUK
Jones J M3NMM
Jones J M3NVL
Jones J M3OVT
Jones J G1RAG
Jones J M5AFV
Jones J M3MOF
Jones J M0HZT
Jones J 2E0CYE
Jones J 2E0MJJ
Jones J 2E0EQR
Jones J G1UVN
Jones J G3TMP
Jones J G4PKP
Jones J GW7NJT
Jones J M0CIH
Jones J M3TJJ
Jones J M3XRI
Jones J M3YQU
Jones J M6CKM
Jones J M6DYF
Jones J M6OHN
Jones J 2E0XJJ
Jones J M3ZKY
Jones J MW6VBE
Jones K 2E0HOT
Jones K 2E0HUC
Jones K 2E0HXT
Jones K G0IUI
Jones K G0KIR
Jones K G0RJA
Jones K GW1BDF
Jones K G3RRN
Jones K G4AHO
Jones K G4FPY
Jones K G4SGV
Jones K GW4SUD
Jones K G4UXL
Jones K G8CZM
Jones K GW8YYF
Jones K M3HYV
Jones K M3UWI
Jones K M3VKJ
Jones K 2E0GWK
Jones K M6GWK
Jones K G6OWS
Jones K M3UXK
Jones K G3PSZ
Jones K M0AET
Jones K M0DLR
Jones K M6AGQ
Jones K MM6FNQ
Jones K G7TSO
Jones K M6KTJ
Jones K M6LVK
Jones K M3VVA
Jones K 2E0RTQ
Jones K M0RTQ
Jones K M6VUL
Jones K M6WJJ
Jones L G4KLT
Jones L G8FUB
Jones L M0LEE
Jones L M1LMJ
Jones L M3MZT
Jones L M6LJJ
Jones L 2W1HFZ
Jones L 2E0DUW
Jones L M6GGO
Jones L G7NIR
Jones L GW4RDW
Jones L M6SLJ
Jones L 2E0YBL
Jones L M6YAT
Jones L M3YKT
Jones L M0ACL
Jones L G4WKQ
Jones L M6FQW
Jones L 2E0LCJ
Jones L M0PWY
Jones L M3VLJ
Jones M 2E0FGT
Jones M 2E0MAJ

Jones M 2E0GFW
Jones M GW0ADY
Jones M G3WTZ
Jones M G4MFN
Jones M G4NBM
Jones M G4NKC
Jones M G4NNL
Jones M G4TIF
Jones M G4WTA
Jones M G4WVC
Jones M G6DFZ
Jones M G6ITJ
Jones M G6OAU
Jones M G7BBY
Jones M GW7ODP
Jones M G7UWI
Jones M G8RRN
Jones M M0MAJ
Jones M M1CGB
Jones M M1CQX
Jones M M3BFX
Jones M M3MMJ
Jones M M3MYZ
Jones M M3OFJ
Jones M G0BCP
Jones M G0MIX
Jones M GW6VRN
Jones M M6DJY
Jones M G0MGX
Jones M G0MLJ
Jones M G0WVL
Jones M G1YWY
Jones M M0UTD
Jones M M1DQI
Jones M M6CMZ
Jones M M6DYD
Jones M M1DOU
Jones M M3FJE
Jones M G1SGG
Jones M 2E0MIJ
Jones M 2E0MLJ
Jones M G6GOS
Jones M G7JCD
Jones M M0VLN
Jones M M6MTJ
Jones M M6PMH
Jones M M6TMJ
Jones M G6PYI
Jones M MW6CGH
Jones N G0NJJ
Jones N GW0VQZ
Jones N G6HYP
Jones N G8IWO
Jones N MJ0AQJ
Jones N 2E0NJO
Jones N M6NAT
Jones N M0NBJ
Jones N M6NEZ
Jones N M6XSI
Jones N 2E0NTJ
Jones N G4XXJ
Jones N MW6NAX
Jones N M6NHJ
Jones N M6NTJ
Jones N M3SLQ
Jones N M6NJT
Jones O GW1BDG
Jones O GW3DRV
Jones O M6OCJ
Jones P GW0JCT
Jones P G0TIS
Jones P G0VHL
Jones P G1ZEA
Jones P G3ESY
Jones P G3YLV
Jones P G4ERF
Jones P G4GNK
Jones P GW4ZCL
Jones P G6AJK
Jones P G6IIM
Jones P GW7ASL
Jones P G7RKT
Jones P G7TZD
Jones P M3WAY
Jones P M6EGD
Jones P 2E0PBP
Jones P M0ICO
Jones P M3PJI
Jones P M3XQJ
Jones P M6DPJ
Jones P M6FYT
Jones P M6KTM
Jones P M6XXD
Jones P 2W0LMM
Jones P G1GKV
Jones P G4DXO
Jones P G4UKU
Jones P M6DXZ
Jones P M6GPO
Jones P 2E0ZDX
Jones P G1PVN
Jones P M6DCB
Jones P 2E0CQG
Jones P 2W0PJJ
Jones P GW4HAT
Jones P G4UGW
Jones P M0HIW
Jones P MW0PJJ
Jones P MW3TWQ
Jones P MW6PJJ
Jones R GW0BYZ
Jones R G0EUV
Jones R G0EUZ
Jones R GW0KLY
Jones R GW0LOI
Jones R G0UCI

Jones R G1LRU
Jones R G1OIB
Jones R G1WVW
Jones R G3NKL
Jones R G3YIQ
Jones R G4AIJ
Jones R G4KQQ
Jones R G4RIU
Jones R G4SAS
Jones R G4SWH
Jones R G6DSD
Jones R G6LRU
Jones R G6TGW
Jones R G7CMF
Jones R G7KAX
Jones R G8MBQ
Jones R M0CRJ
Jones R M0REJ
Jones R M0SPM
Jones R M1ROD
Jones R M3JUM
Jones R M3KCG
Jones R M3REJ
Jones R G3MDK
Jones R G7DUY
Jones R 2W0BPJ
Jones R G0MHY
Jones R GM1MYF
Jones R G4NMU
Jones R MW3WFF
Jones R M6BYN
Jones R M6DGU
Jones R M6JFS
Jones R 2E0RBX
JONES R G4FCV
Jones R G7VJU
Jones R 2E0XKL
Jones R G0BGR
Jones R G7RGR
Jones R M0XKL
Jones R M3XKL
Jones R G0RMG
Jones R G4OEK
Jones R G7SWW
Jones R M0ANC
Jones S 2E0IBA
Jones S 2W0LDX
Jones S GW0COU
Jones S GW0UHX
Jones S G1YKY
Jones S G3LXB
Jones S G4AXW
Jones S G4DSF
Jones S G4GNV
Jones S G4KQP
Jones S G4ZAX
Jones S G6FES
Jones S G7SRJ
Jones S MW0AWO
Jones S M0BWH
Jones S M0SJJ
Jones S M0SZQ
Jones S M3LXA
Jones S M3RVN
Jones S MW3SNJ
Jones S M6SUJ
Jones S M6DUX
Jones S M6ISJ
Jones S M6SID
Jones S G4MQQ
Jones S G8RYX
jones S M6ERP
Jones S G1KNX
Jones S M6SHJ
Jones S 2E0NLW
Jones S GW0GEI
Jones S M6DYW
Jones S M6NLW
Jones S M6JUG
Jones T 2E0TJX
Jones T G0IVT
Jones T G0IYV
Jones T G0RBW
Jones T G1HPV
Jones T G1MGN
Jones T G3UUL
Jones T G4TNF
Jones T G4VEQ
Jones T G7IRD
Jones T G7VMQ
Jones T G8DVF
Jones T GW8WEY
Jones T MW0CLT
Jones T MW0TCJ
Jones T M1DTO
Jones T M1FAF
Jones T M3OTB
Jones T M3TRJ
Jones T M3YBK
Jones T G4RQQ
Jones T G6MUQ
Jones T M0MLV
Jones T M3TJO
Jones T GW7JUB
Jones T G0HGN
Jones T G0JAI
Jones T G6OAW
Jones T M6TEJ
Jones T G4IPR
Jones T M6TEP
Jones T G1HPS
Jones T G0WGM
Jones T MW6LDS
Jones V G6VIC
Jones W G0GOR
Jones W G0KZW

Jones W G0NIX
Jones W G1DLP
Jones W GD4XOD
Jones W [illegible]
Jones W [illegible]
Jones W G7MUX
Jones W G7WHP
Jones W G8DSG
Jones W M0LTO
Jones W M1WEJ
Jones W M3VVJ
Jones W M3XZV
Jones W M6GWO
Jones W 2E0WBO
Jones W 2E0YAV
Jones W GW0IVG
Jones W M0YAV
Jones W M6GZI
Jones W M6WRJ
Jones W M6WTG
Jones W M6YCR
Jones-Martin S M6SSO
Jones-Robinson J M6EFG
Jonusas M M6LYP
Jopling J G0CXX
Jopson B G0UKP
Jopson J 2E0KCL
Jopson J M6KCL
Jopson L G6QA
Jordan A M0GJX
Jordan A G0HAS
Jordan B M6FBJ
Jordan B G7SNT
Jordan B G4EWJ
Jordan C GM7GUL
Jordan C G0PZO
Jordan C G0NGN
Jordan D G1TNR
Jordan D M1FHX
Jordan F G4BZR
Jordan H G0NWL
Jordan I G4GET
Jordan J 2E0HGJ
Jordan J 2E0IBT
Jordan J G7HLV
Jordan J M0GCU
Jordan J M0JWJ
Jordan K G7SNP
Jordan K 2E0CJB
Jordan K GD0KQE
Jordan L G4KJP
Jordan L G6GXE
Jordan M G1FVH
Jordan M G4VAO
Jordan M G7EVQ
Jordan M G7NKE
Jordan M M0OGS
Jordan M G4ASQ
Jordan R 2E0URJ
Jordan S 2E0GMA
Jordan S M3STJ
Jordan W MI3WJO
Jorgensen D 2W1ETN
Jorquera P G1EYS
Joseph L MM6BPY
Joseph M MW3MKN
Josephs R M1BYQ
Josey C 2W0CDJ
Josey C MW0HMV
Joshi M M0SHI
Josi M M0HSX
Josko S G4ZVZ
Joslin J G3NPY
Jouhal S M1JHL
Jovanovic A M6MQH
Jowett D G8FJR
Jowett J G3CFR
Jowett K M1KDJ
Jowett R 2M0WWM
Jowett R MM0WWM
Jowett R MM3WWM
Joy A G8ZEW
Joy B G0DJV
Joy G M3TQG
Joy M G7CWT
Joy R G0LSG
Joyce A G6REG
Joyce C G1NSQ
Joyce D G0FGJ
Joyce E G8ELG
Joyce G M0EDP
Joyce J M6OFM
Joyce J MM6IAB
Joyce L G0SZK
Joyce P 2E0OFM
Joyce P M0OFM
Joyce P M6TOG
Joyce R G3WLM
Joyce W 2E0CCW
Joyce W M6WGJ
[illegible]
[illegible]
Joyner M 2E0BTG
Joyner M GM3EIY
[illegible]
Joynes A 2W1EUR
Joynes H G7SNF
Joynson E M6OCU
Joynson M G1VTE
Joynson P 2E0ENZ
Joynson-Ellis M 2E0GDF
Joynson-Ellis M M3ZVT
Joynt J G4WVY
Jubb A G3PMR
Jubb E M6EMJ
Jubb S G6PDJ

Juby M G8RML
Judd B G0RZM
Judd G G4ULQ
Judd I G8IDJ
[illegible]
[illegible]
Judge A G0NCW
Judge A G0PQF
Judge A G1ARU
Judge D M3IWK
Judge D 2E0RBC
Judge H G7IWU
Judkins P G3OMJ
Judson A M3OAT
Judson R G0RHJ
Juett D G1TPV
Juffs M G4NVY
Juggins J G7VFA
Juhe M 2E0MJD
Juhe M M0XJP
Jukes G MW6XGD
Jukes I GW1MNU
Jukes J M0JMJ
Jukes W 2E0AVD
Jukna S M0UNN
Jul-christensen F G4MJC
Julian H G3UFX
Julian P G7PRO
Julians M G6ZBO
Julius V MM6VAB
Jump D 2E0BKT
Jump H G7RGV
Juner K GM7GIF
Juniper M M6MZJ
Jupp B G0SDE
Jupp D G6ITM
Jupp M G1HWY
Jurczyszyn M M3ZGI
Jurgaitis G M0NSP
Jurisic M M6XCA
Jurkiewicz M M1EOP
Jusko M M0GLV
Just B M3HML
Just C G8SZG
Justice J G6ZFA
Justice M G1JMK
Justin A GD8DAI
Justin B G8JMO
Justin L 2D0XLJ
Justin L MD3XLJ
Justin P GD4AZL
Justin S GD6XUD
Juszczak K 2E0KKJ
Juszczak K M6MKJ
Jutting I M3ZKL
Juvonen M 2E0TTI
Juvonen M M0MTI
Juvonen M M6TTI

## K

Kaby J M3XBO
Kacprzyk J MI6VOZ
Kaczmarek J G7GBJ
Kahlbau L M0LKD
Kaighin J MD6AGF
Kaine J G4RPK
Kaiser R G4MFE
Kakoutas C M0WCK
Kalas P G3VCN
Kalawsky R G7NII
Kaliski M G0ULI
Kalogerakis B 2E0VBK
Kalogerakis B M6RBK
Kaluarachchi A M6NDI
Kamal S 2E0SSK
Kamal S M3UON
Kamm D M0BKV
Kane C MM0ABB
Kane D MI3FOJ
Kane G MI3FKI
Kane I 2E0VAW
Kane J GM0ODB
Kane J MI3VQH
Kane R GM4UWN
Kane T MM0KNE
Kane W GI7MBP
Kanelis M G8ZFQ
Kapadia S 2D0LHR
Kapfunde G M6GKC
Kapoutsis C G6URT
[illegible] M0RVF
Karaalp L M6EGM
Karaalp O M6ENJ
Karande A G1TNP
[illegible] G6PGQ
Karbhari I M0IAK
Karchev L M0GYU
Karklins E GW6KRK
Karkoszka J G0RLY
Karlstad J M0JMS
[illegible] G1SDK
[illegible]
Karpinski R 2E0WNI
Karpinski R M0WNI
Karpinski R M3WNI
Karpuk H M6HDK
Karpukhina S M6KSV
Karthauser L 2E0LBK
Karthauser L M0LBK
Karthauser L M3LBK
Karuppannan Rajan P M6PKR
Kashkoush E MI3DDK
Kashkoush M MI0MRV
Kashkoush M MI3TXT
Kasprzyk M M0HHP
Kassai M M1MPK
Kasser J G3ZCZ

Kassier H 2E0OMI
Kassier H M0HTK
Kassier H M6DZH
Kalaria D M6END
[illegible]
[illegible]
Katon Y G0GRV
Katz J 2E0BTU
Katz J M3YGR
Katz W G8MZQ
Katzmann M G4NYV
Kavanagh B G1RVT
Kavanagh G M3VXG
Kavanagh J 2I0OJK
Kavanagh J MI6OJK
Kavanagh J M3JMK
Kavanagh M GM4VKI
Kavanagh P M3VWJ
Kavanagh R M0RJK
Kawayida D M6EOM
Kay A G1RUG
Kay A G4SPY
Kay A G8UWG
Kay A M0CJO
Kay B G4VBJ
Kay C G4TIH
Kay D G0MXH
Kay G G4WMY
Kay H G0FAB
Kay J G6OBG
Kay J M0RYA
Kay J 2E0JEK
Kay J M6BMZ
Kay L G0KBS
Kay L 2E0ETC
Kay L M3ZHZ
Kay M G1EOJ
Kay M MW3OEJ
Kay M G7VAS
Kay P G0KUX
Kay P G1YJI
Kay P G3KOD
Kay P 2E0PBO
Kay P M6PPK
Kay R G0LHV
Kay R G0MZP
Kay R G3NSW
Kay R G3OQF
Kay S G3OMA
Kay S M3XGY
Kay S M6SPK
Kay T M0TKA
Kay W GM8UGO
Kaye A G1YIL
Kaye D G6PGM
Kaye D M0CTF
Kaye E 2E0EDS
Kaye E M6EDS
Kaye M G3WPQ
Kaye M GM7GTX
Kaye N M3NGK
Kaye R G0JEA
Kaylor D G4OBB
Kay-Newman D M0AOD
Kazlauskaite I M6BZY
Kaznowski M G6OBA
Keable D G1ICA
Keable T M6TKX
Keal B G4HDU
Kealey M 2E0YJY
Kealey M M6MAJ
Keane P G4DUQ
Keane P M0GVN
Keane P M6CTY
Keane T MW6TRK
Kear C G4JHQ
Kearey N G8EYM
Kearley S M0SJK
Kearnes R G7HJK
Kearney F 2I0FPK
Kearney F MI0PPA
Kearney F MI6FPK
Kearney J M3JNB
Kearney M 2E0ANN
Kearney R G3XOK
Kearns A G3AQF
Kearns D G0HVS
Kearns D G8TTI
Kearns K G8LYV
Kearns M G4XAR
Kearns T G4GIX
Kearsley T G4WFT
Keasley P G0JXR
Keast O 2M0OLK
Keast R 2E0EJK
Keast R M6CUC
Keates D G0DJK
Keates R G8LIX
Keating M G4WVM
Keating T 2E0HOS
Keatley A M6ASK
Keats T GM4CCN
[illegible]
Keay G G4ELC
Keay R MM3OGS
Keay T M6TSM
Kebbell M G4UWF
Keddie A M0KED
Keddie D GM1RGM
Keddie J G7LUN
Kedward S 2W0KED
Kedward S MW0JZM
Kedward S MW6KED
Keeble A G0MJC
Keeble A G4RUI
Keeble Buckle F G8SVZ
Keeble C G3TUU

Keeble D G7WAF
Keeble E G1EOK
Keeble G G6CDU
Keeble J M3NKP
Keeble K [illegible]
[illegible]
Keeble S M6BKT
Keech A G4PPW
Keech N G7VDV
Keechan B G0GFE
Keefe J 2E0EWL
Keefe R G4SIS
Keegan G G6DGK
Keegan J 2E0DZV
Keegan J M3JMK
Keelan M M3KVW
Keeler D G3KXI
Keeler E G4FPM
Keeler J G4EZN
Keeler M M6KEE
Keeler P M1CNG
Keeler T M1BCY
Keeley C M3RSN
Keeley J G0MFB
Keeley R G8IMM
Keeley H 2E0RAA
Keeley V M0WQR
Keeley W G6RAV
Keeley-Osgood R G0GIA
Keeling B G4EUW
Keeling J G0WQC
Keeling M 2E0GRT
Keeling T G6OBD
Keely D G0OGI
Keen A G7CSX
Keen A G7DWN
Keen C G8BYC
Keen D G3JVN
Keen D G7DQZ
Keen J G7PZT
Keen K G6RXV
Keen P G0VGY
Keen S G4PWS
Keen S 2E0WIZ
Keen S M6WIZ
Keenan A GM4SFA
Keenan J GI4WAH
Keenan J GM0WFB
Keenan K G0XKK
Keenan M GI4SZW
Keenan P GI0PTQ
Keenan S GD4MDY
Keene G G7EVF
Keene S 2E0GTD
Keens C G8KYK
Keens C G8NDN
Keep R G7SPE
Keep T 2E0KEP
Keep T M0KEP
Keep T M6GYY
Keepin K G7UMS
Keeping C G4PTF
Keeping M G8AVZ
Keery T MI0CGV
Kees D M0DKS
Keeton L G0JNT
Keevil S M3SKY
Keeys W G7LAX
Keig J MD3LJS
Keighley L G3FLV
Keighley P G0KPH
Keightley M G8BLK
Keightley N G0BNR
Keightley S G3ZZL
Keightley S M0BII
Keiller B 2M0BDR
Keiller B MM0ZAL
Keilty M 2E0AAN
Keilty M M6AOQ
Keir A G4KZO
Keitch K G0IFA
Keith D G3RQF
Keith E M6HGX
Keith V M6HGU
Keith-Hill R M6RMK
Kelday J G1TBE
Keleher J G8PUN
Kelk J G4JMP
Kelk V G1FDL
Kell G 2E0GAK
Kell P G7GCF
Kell S G4KEL
Kelland C G0JEK
Kelland C G6LRY
Kelland O M6OKI
Kellaway G G0ITE
Kellaway G G6AY
Kellaway H GW3CBA
Kelle A G4AYB
Kelleher A G0WXE
Kellett D G8ZZB
Kellett M [illegible]
Kellett M G8HMJ
Kellett T G3EGF
Kelleway M G4XIU
Kelley C [illegible]
Kelley C G1YIE
Kelley K M6LDK
Kellingley P G7HOK
Kellner B M3PBK
Kellow N M3EWY
Kellow T G3ZHK
Kelly A G4LVK
Kelly A G6MUX
Kelly A M6YLE
Kelly A 2I0TAN

Kelly A G4TYD
Kelly A MI0TBD
Kelly A MI3ZYU
Kelly B 2E0RBK
[illegible]
[illegible]
Kelly C MD3CPK
Kelly C 2I0CKB
Kelly D G4HXN
Kelly D 2M0FSB
Kelly D MD3YUQ
Kelly D MM6FSB
Kelly D M6HFN
Kelly E G0RJG
Kelly F G4TFV
Kelly F MD3KSN
Kelly G G6OWT
Kelly G GM8MST
Kelly G M3OOE
Kelly G MM3XGP
kelly G G4FQN
Kelly G GD4NTR
Kelly G MI6YPY
Kelly I M0PCB
Kelly J 2E0KGQ
Kelly J 2E0NHC
Kelly J G0MRK
Kelly J G1OGC
Kelly J GM3TCW
Kelly J M3IDF
Kelly J M3KBE
Kelly J M3KGQ
Kelly J M3KPL
Kelly J M3NHE
Kelly J M3YZH
Kelly J GM0SYV
Kelly J MM0TBH
Kelly J G3YGG
Kelly J M3XHB
Kelly J M3XZD
Kelly J M6FLK
Kelly J G2HFR
Kelly K G3VKF
Kelly L G7OPY
Kelly L 2E0VHV
Kelly L M3VHV
Kelly M G0WON
Kelly M G1NMR
Kelly M GI4MEQ
Kelly M G4XGP
Kelly N G4UUF
Kelly P 2E0KGP
Kelly P G3SDH
Kelly P GI4BGB
Kelly P G4ENK
Kelly P G6ITO
Kelly P GW7TZG
Kelly P M0DGT
Kelly P M0KGP
Kelly P M3KGP
Kelly P 2E0PCA
Kelly P M0NED
Kelly P M6GVA
Kelly R GM0GRD
Kelly R G1PVR
Kelly R MM0ODI
Kelly R 2E0REK
Kelly R M0REK
Kelly R M6REK
Kelly S G6HDF
Kelly S GD7DUZ
Kelly S MD3DUZ
Kelly S MI3IYP
Kelly S M6JJE
Kelly S M6GJG
Kelly T 2E0TDJ
Kelly T 2E0ZMI
Kelly T M6GJI
Kelly T M6ZMI
Kelsall G G7UTR
Kelsall K G0YKK
Kelsall P G0CYB
Kelsall R G0BDW
Kelsall R G1VBL
Kelsall R G8JDD
Kelsey B G1AFK
Kelsey L 2E0LRK
Kelsey L M3NQL
Kelsey P G1MLV
Kelsey-Stead W G0COQ
Kelso J 2I0EOS
Kelso J MI6WJK
Kelso R M3RJK
Kember P G7EWY
Kemble M G0JMK
Kemble P 2E0CWN
Kemble P G3UYK
Kemish R M6OSL
Kemmis P G4MGI
Kemp A 2E0SVV
Kemp B 2E0BRK
Kemp B [illegible]
Kemp C G7VCM
Kemp C M6NSC
Kemp D G0POQ
Kemp D [illegible]
Kemp D 2E0RVV
Kemp D M6GUW
Kemp E M0APW
Kemp E M0XEK
Kemp F G4JEO
Kemp G G0EIR
Kemp G G4WGK
Kemp G G4DSA
Kemp G G1PVA
Kemp I G4AEG

Kemp J G0WJH
Kemp J GW8SBN
Kemp J M0AAR
Kemp J M6GDA
[illegible]
[illegible]
Kemp P G4XBJ
Kemp R GM0KQB
Kemp R G4VUW
Kemp R M3MXO
Kemp S G7IMY
Kemp S M0DYB
Kemp T G6JDC
Kemp T 2E0TEZ
Kemp T M0TAJ
Kemp T M3TQJ
Kemp W GM8WWY
Kemplay A M3NMK
Kemplen D G1NSV
Kempson C G3JJT
Kempson H G3BHM
Kempson M M3OFV
Kempster J M5AEO
Kempton A G1BYS
Kempton R G6CYT
Kempton R M6XCO
Kemsley A G4OPT
Kenchington J MW3IKC
Kendal A G6WJJ
Kendal W G4MKD
Kendall A G4LEN
Kendall G G0SSV
Kendall G 2E0GPK
Kendall G M6GPK
Kendall I G6ARO
Kendall J G8JKC
Kendall J G1FKM
Kendall M M6MKK
Kendall M G0EMK
Kendall N G4JNK
Kendall N 2E0NJK
Kendall N M3PXQ
Kendall P M0EJL
Kendall R G8BNE
Kendall R M6GVY
Kendall S G0MOM
Kendall T G4LDB
Kendal-Ward N G7SZB
Kendrick A G3RDW
Kendrick A M3LVY
Kendrick A M6DXA
Kendrick B MW3FFL
Kendrick C G0STW
Kendrick I M6EGB
Kendrick J M0PLN
Kendrick M G6CYU
Kendrick R M0DOY
Kendrick S 2E0SJK
Kendrick S M0KWS
Kendrick S M6SAK
Kendrick S 2E0SMK
Kendrick S M6SMK
Kenealy G G1ZKZ
Kenington P G1ISR
Kennard D G0KEI
Kennard H G4TLE
Kennard J G6BDH
Kennaugh A GD6HCB
Kennaugh J MD3KCT
Kenneally B G6AJT
Kennedy B G3ZUL
Kennedy C MM0KNN
Kennedy C M6CMK
Kennedy D M5DNK
Kennedy D MM6YRO
Kennedy D G3VZE
Kennedy D G6DHI
Kennedy E GM0VMV
Kennedy F M0FSK
Kennedy F 2E0HQA
Kennedy F M3BOB
Kennedy G G0IEQ
Kennedy G G3OGK
Kennedy G 2E0GRX
Kennedy G MM3CNS
Kennedy H 2E0GJG
Kennedy H M3HAT
Kennedy H 2E0RPY
Kennedy H M0HWY
Kennedy H M6NSE
Kennedy I G4DND
Kennedy J [illegible]
Kennedy L G0TUV
Kennedy L G4TEP
Kennedy M G1YHG
Kennedy M G0ADG
Kennedy N GM0MTZ
Kennedy P 2E0BDC
Kennedy P G0EWD
Kennedy P G1CCX
Kennedy P G1CLJ
Kennedy P [illegible]
[illegible]
Kennedy R GI1WYZ
Kennedy R G7RKO
Kennedy R [illegible]
Kennedy R 2M0IOB
Kennedy R MM3WWV
Kennedy R G8UMB
Kennedy S G0FCU
Kennedy S G0LRI
Kennedy T G0PLK
Kennedy T G0PLK
Kennedy T GI7HVC
Kennedy V MW6BTC
Kennedy W G3MCX

Kennedy W 2E0BYY
Kennedy-Bright R M1DQG
Kennedy-Brown G M6BVK
Kennett D G7CMB
Kennett M [illegible]
[illegible]
[illegible]
Kenny B M3CBH
Kenny C G3LTK
Kenny M M6VVT
Kenny M 2E0MNJ
Kenny R G1NNF
Kensall M G7FMI
Kensall R 2E0CPF
Kensett G M3RHR
Kent A G3DSZ
Kent A G3THG
Kent A G4ZSC
Kent A G6EXZ
Kent A 2E0YZQ
Kent A M0XAK
Kent A M6ALK
Kent A G8PBH
Kent B G6MUW
Kent B G6EXX
Kent C 2E0UAB
Kent C M6CBC
Kent D G7LIW
Kent E G1HCC
Kent G G0VNA
Kent I M6FUM
Kent J G8OQC
Kent J 2E0KJJ
Kent K G1SCQ
Kent M G3PSS
Kent M G4SUK
Kent M G8PHM
Kent M M6MKO
Kent M M6OSI
Kent N 2E0NKT
Kent N M6TRC
Kent P 2E0BBO
Kent R G0ROS
Kent R G4POY
Kent R G3KCF
Kent S M3SMN
Kent W G3YCN
Kent W M3HAU
Kentell M M0GVE
Kentfield H G1UID
Kentish P M6XMS
Kentish R GW4JRK
Kenton K 2E0PNX
Kenton K M3PKU
Kenton L M3OBL
Kenton R M3LBC
Kent-Woolsey P G6ULS
Kenward R G6WTD
kenway W M6BBB
Kenworthy N G0WEF
Kenyon A GW4DOO
Kenyon C 2E0NPS
Kenyon D 2E0INT
Kenyon G 2E0OFF
Kenyon G M6GRK
Kenyon I G4VAP
KENYON J M6WKH
Kenyon N G4AYU
Kenyon P G8CAK
Kenyon S G8YHF
Kenyon T G6CIA
Kenyon W G1FDN
Kenyon W G6TTX
Kenyon W G0EIZ
Kenyon-Brodie P G1WQL
Kenzie B G4PDI
Keogh A MM6KFJ
Keohane A G3XUC
Keon A G0DLF
Ker J G0ICP
Kerby T 2M0TXK
Kerby T MM0TKE
Kerby T MM6TXK
Kergozou R M6CHD
Kerins J GM1NEW
Kermode B G8ESK
Kernaghan H GI3USK
Kernahan N G1EKU
Kerner N M3NPK
Kernick A M3REM
Kernohan H GI0JEV
Kernohan J [illegible]
Kernohan W GI4OGQ
Kerr A GM3KBP
Kerr A GI4IVI
Kerr A [illegible]
Kerr A G6QOG
Kerr A M1BWS
Kerr A 2E0DCY
Kerr A M0KRR
Kerr A M6DCY
Kerr A 2E0WWA
Kerr A [illegible]
Kerr A M3ELT
Kerr D GM0DBK
Kerr G M3IME
Kerr J G6RLG
Kerr J 2E0IGY
Kerr J MM6XXV
Kerr K GM4YXI
Kerr L M6LCK
Kerr P G6OBB
Kerr P M6BUX
Kerr R GI0RUC
Kerr R GM4FDT
Kerr R M3RBT

Kerr S G7KLJ
Kerridge B M3ZXB
Kerridge C M3UQB
KERRIDGE D G7JXL
Kerridge K 2E0AXA
Kerridge R G1FPK
Kerrison A [illegible]
Kerrison B G7GAZ
Kerrison C M0DLY
Kerrison S M3SJK
Kerr-Munslow A M1FGO
Kerry B M3XIY
Kerry M G1SXT
Kerry M G4BMK
Kerry N M3ZOH
kerry N M6HNG
Kerry P G3SFK
Kerry P G6LSD
KERSEY A G0IBN
Kersey B M3LPN
Kersey E G4LBU
Kershaw C G8WFP
Kershaw D M6EOR
Kershaw E G7SEU
Kershaw P G6EYA
Kershaw R G4PJE
Kerslake S M3XWX
Kerslake W M0BSH
Kerstein N G4BPN
Kerswill L 2E0ICM
Kerswill R 2E0KER
Kerswill R M0KER
Kerswill R M3WXY
Kerton A G4CJV
Kerton P G0EOZ
Kerton P 2E0INT
Kerton P M6CTA
Kessel M G4WJX
Kesterton E G6HGK
Keston D G8FMC
Keszei C 2E0EMP
Keszei C M6LCU
Ketley H G1JGY
Kett A G8VLL
Kett K G4NVP
Kett L 2E0LKE
Kett L M0LKE
Kett L M6LKT
Kettle A G6ARM
Kettle K M3UBU
Kettley A G8HTN
Kevern M M1BKS
Kewell T G0JDM
Kewn J G7TBJ
Key B G6PGO
Key C G8VVR
Key J G4SXX
Key M G3WHG
Key M G4JQF
Key S G6LSB
Key S M6SKX
Key S 2E0RTM
Keys D GI0LDI
Keyser I G3ROO
Keyte B G3SIA
Keyte M 2E0MKX
Keyte M M0MJK
Keyte P G1AFJ
Keyte R G3UPS
Keyworth A G4MWL
Khachaturian A G7GJN
Khalaf M G4KRD
Khan D M3GKB
Khan J M6EQS
Khan K M6KDK
Khandros S G0CVI
Khayer H M6HKX
Kibble N M6NOK
Kibblewhite K G0HER
Kicman A M0TRE
Kidane B M1CIM
Kidd A G3JRS
Kidd C G3YTQ
Kidd D G3XNK
Kidd M G6JVO
Kidd P M3PMK
Kidd R M0EVK
Kidd S M3SKD
Kiddell L 2E0KPL
Kiddell L M6LPK
Kidder G G3NZO
Kiddle A [illegible]
Kidger C G0LOL
Kidman M G7LAF
Kidner M 2E0KDR
Kidner M M3RTU
Kidwell T [illegible]
Kiel S 2E0WEO
Kiel S M6WEO
Kielthy M G7CIK
Kiely D G0RBD
Kiely J M6MLF
Kiely W [illegible]
Kiely T 2E0FVK
Kier E G1DTS
Kierman J G1HQW
Kiernan P 2E0HAJ
Kiernan S G4ROI
Kiff H G0DOC
Kift G GW0LDQ
Kightley A G3MZZ
Kightly T M6TFK
Kijak R MD0RKI
Kijewski R M6BIA
Kilbey G G7MLW
Kilburn D M0CKG



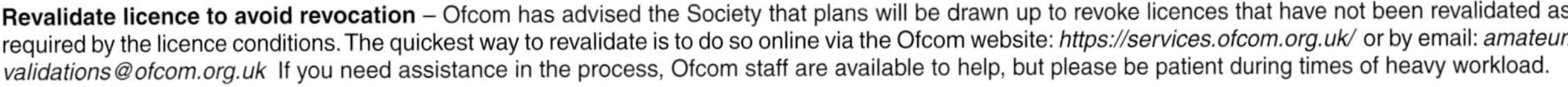
**IMPORTANT NOTE**

**Revalidate licence to avoid revocation** – Ofcom has advised the Society that plans will be drawn up to revoke licences that have not been revalidated as required by the licence conditions. The quickest way to revalidate is to do so online via the Ofcom website: *https://services.ofcom.org.uk/* or by email: *amateur.validations@ofcom.org.uk* If you need assistance in the process, Ofcom staff are available to help, but please be patient during times of heavy workload.

Kilday M MM3MPK
Kiley P M6PEN
Kiley S 2E0ZIP
Kiley S M3XHH
Kilfeather T 2E0DTQ
Kilfeather T M6TPI
Kilgore R GI0WYO
Kilgore, MBE T MI1FCB
Kilgour D MM3YDK
Kilgour G GM7KVU
Kilkenny I G7FEG
Kilkenny I M3ATC
Kilkenny M 2E0MKK
Kilkenny M G1IUF
Kilkenny M M3MKK
Kill C G1OQI
Killeen J G3KPV
Killeen W G1VHW
Killen R M0GTH
Killian J M3JSK
Killian S 2E0JSK
Killick K G0SNM
Killing A M1DHC
Killingback K M6KPK
Killingley D M6HKJ
Killington R M0VPG
Killman J 2E0CTO
Killman J M6KAX
Killner R M6RJK
Killoran M M6MNP
Kilminster J G7TPB
Kilmister J G0CQJ
Kilmister S G6MOT
Kilner E G1KZA
Kilpatrick H GM1ASA
Kilroy J G4PBC
Kilroy T 2E0UCV
Kilroy T M3UCV
Kilvington C G8EPH
Kim S M3VIR
Kim T M0GCB
Kimbell G G3TCT
Kimber B G1NVO
Kimber B G1SGS
Kimber D G8HQP
Kimber K G4ZXF
Kimber M M6UAD
Kimber N M0KBH
Kimber P G1HSX
Kimber R G1BZW
Kimber S G6ARR
Kimberlee P 2E0DBL
Kimberlee P M0PYT
Kimberlee T M6GAL
Kimberley R G8AVK
Kimblin S G6OBE
Kimm A G4YMQ
Kimm T 2E0HLU
Kimmitt M G4GOO
Kimoto Y G0TDX
Kimpton J G4WAO
Kimura K M6HQW
Kin R G7BOB
Kincaid A 2I0IPB
Kincaid A GI4TOR
Kincaid A MI3NUW
Kinch J 2E0FSF
Kinch R 2E0FSG
Kind N 2E0ALA
Kind N M3NRK
Kinder M G0CZD
Kinder M G0KYB
Kinder R M6ISH
King A GI6BDI
King A G6KTX
King A G7ANY
King A M0ANL
King A M1LOL
King A 2E0BTO
King A 2E0KVR
king A GM0OTU
King A G6UQC
King A M3YFJ
King A M6KNG
King A 2E0XEE
King A G0DDJ
King A G0IAG
King A M0AYX
King B G3SGK
King B G8ARA
King B G8CHC
King C G1UTN
King C G6PGN
King C G7JAO
King C G7LQN
King C G7NBL
King C MM0HOL
King C M3IJT
King C 2E0OAA
King C M6SRO
King C G6MYT
King C 2E0BME
King C GW1VRW
King C M3UWV
King C M6XFM
King D 2E0FUZ
King D G0RWJ
King D GM1FPD
King D G1PWF
King D G4BBQ
King D G4EOD
King D G6KWA
King D G8GFY
King D GM8XUK
King D M0HVD
King D G0DMK
King D G4JKE
King D GM6OOA
King E GM1ATW
King E G4HWC
King G G1HXN
King G G3XTH
King G G4FTL
King G G8BJB
King G M0BHR
King G G1VQH
King G M0BHP
King G G3XSD
King H G0KMV
King H G8DXV
King H G3ASE
King H MM6HSK
King I G6VIK
King I G8VTX
King I M3ZKI
King I 2E0IDK
King I M0IDK
King I M6IDK
King J GD0JIM
King J G1XFE
King J G4AND
King J GD6JIM
King J G7VTT
King J 2E0JJK
King J M0JJK
King J M6JAK
King J 2E0JAF
King J G0RSA
King J M0JAF
King J M3YZJ
King J 2E0TFE
King J M0BXG
King J M0ORC
King J M3TFE
King K G3RGE
King K M3KVJ
King K M3MNY
King L GD0VTC
King L M0LHK
King L 2E0USQ
King L M3USQ
King M G1XPW
King M G3XKD
King M G4GLI
King M 2E0GBO
King M G7SFD
King M M6MUK
King M 2E0DTV
King M M0REM
King M M6MKY
King M G0RAM
King N G3DNS
King N G8NGM
King N M3NLA
King N G0OHK
King P G0OIK
King P G1KFQ
King P G1RGG
King P G2RSA
King P G4GFY
King P G4JXE
King P M0DVR
King P G1XXW
King P G3WKP
King P G6BOK
King P G7IFL
King P G8KJP
King P M6ROI
King P MW0DHF
King P M3IAQ
King R G0AKL
King R GI0OHU
King R G0TDV
King R G0THQ
King R G0VSS
King R G4NIZ
King R GW4XJK
King R G6XWM
King R G7AJX
King R GM7BOW
King R G7IMV
King R MW0RPK
King R M5DIK
King R M1REK
King R M6RLK
King R G6CYA
King R M3RAK
King R G8CHK
King R G4VXD
King S M0KRA
King S G1LHQ
King S M0KKB
King S M6ZEL
King T 2E0TKV
King T G1FDO
King T M3TKV
King T M3ZTK
King T G1SWK
King T G4IVL
King V M3ZVK
King W G0WIY
Kingdon A G4MUA
Kingdon G G4RTX
Kingdon G G3VZF
Kingdon M G4XYB
Kinger M G4UMS
King-Evans R M6RUE
Kinghan A GI4OZI
Kinghorn H G4EOM
Kingon-Rouse N MM1EYI
Kings A G0HBD
Kingsley N G3RCB
Kingsley-Lewis N G7PSK
Kingsley-Williams P G1PKW
Kingston P M6BCN
Kingstone G 2E0RZM
Kingstone G M3RZM
Kingstone S 2E0FAJ
Kinley L 2E0LEN
Kinley P M1DBA
Kinnersley A G7OSO
Kinnersley A 2M0EWY
Kinnersley A MM6EWY
Kinney G M1EIW
Kinney P MI6PHH
Kinney V MI6GZF
Kinrade D GD4EBA
Kinrade R G6POC
Kinsella B G7KIN
Kinsella J G0UGX
Kinsella M G1RNL
Kinselley N G1BYT
Kinsey D M3KIN
Kinsey J M6JVK
Kinsey M 2E0LCO
Kinsey P G6ONZ
Kinson A G0KOC
Kinton S G6PGP
Kipling L M1KIP
Kipling M 2E0MLK
Kipling M M0MLK
Kipling M M6FYL
Kipp M G4FBK
Kipping M G1GAR
Kirby A G1HOU
Kirby A M6JKA
Kirby A M3OYR
Kirby C M0LEY
Kirby D GW0PLP
Kirby G G0PXM
Kirby J G3JYG
Kirby L M3KRB
Kirby L G4CRT
Kirby M G1EEO
Kirby M M0MWK
Kirby M M3MJY
Kirby M M3KBY
Kirby P 2E0TMD
Kirby P 2E0YPJ
Kirby P G3XUD
Kirby P M0YPJ
Kirby P M3YPJ
Kirby P M6TMD
Kirby P G8HQW
Kirby P 2E0PZK
Kirby P M0OPK
Kirby P M6DGI
Kirby R M3KKZ
Kirby R M6RHK
Kirby R G3VQS
Kirby S 2E0GUN
Kirby S M1ECH
Kirby T G4VXE
Kirk D G7RIU
Kirk D G4NXA
Kirk G G4FKG
Kirk G M6KVK
Kirk H G1ADE
Kirk I G6URR
Kirk J 2E0JFK
Kirk J G3ZDF
Kirk L GI4MMJ
Kirk M G0JPZ
Kirk M G1LBK
Kirk M G4UCZ
Kirk M M0MEL
Kirk P G0IYU
Kirk R M3RNY
Kirk T G3OMK
Kirk W G0FTK
Kirkbride S MM3ZCS
Kirkbright S G7VLB
Kirkby D G8WRB
Kirkcaldy L 2E0VIA
Kirkcaldy L M0VSD
Kirkcaldy L M6VIA
Kirkden D 2E0ZDE
Kirkden D M0ZDE
Kirkden D M6DES
Kirkden P 2E0CPZ
Kirkden P M0ZPK
Kirkden P M6PKU
Kirkham A G7MQF
Kirkham A G7TMM
Kirkham D G8SFU
Kirkham J G7RWW
KIRKHAM J M1DGK
Kirkham J M3JVK
Kirkham M G8RHN
Kirkham P G6CYV
Kirkland A G4HDO
Kirkland C G1VIN
Kirkland C M0ATY
Kirkland D MW0KAK
Kirkland G GM0KDO
Kirkland K GM4HCE
Kirkman C G1JVH
Kirkman M M0KIR
Kirkman N G7KIE
Kirkpatrick A MI6OWD
Kirkpatrick B G8OKR
Kirkpatrick G M0DHN
Kirkpatrick J MM3JKX
Kirkpatrick L M6LJK
Kirkpatrick S MU6WZY
Kirkpatrick S 2E0BOY
Kirkpatrick S 2E0CAF
Kirkup P G0RTU
Kirkwood D MI0OPM
Kirkwood D MI1OPM
Kirkwood D G3YQO
Kirkwood J G7RAE
Kirkwood R G4RIK
Kirkwood S G4ZDU
Kirkwood S G1LAT
Kirkwood S MI3TEM
Kirsch R G8CJG
Kirsop P GM4WCE
Kirtley N G3RQR
Kirton D G4EHR
Kirton J G8WWJ
Kirton O M6IRK
Kisiel J M0AZG
Kissack A GD0TEP
Kissack A MD6ZKK
Kissack B G3MTD
Kissack R MD6WFK
Kisselev K M0BDQ
Kissin A 2E0YAP
Kissin E 2E0ERK
Kitchen A G7COD
Kitchen B G4GHB
Kitchen D G3KVP
kitchen I G0WZM
Kitchen S G7UEJ
Kitchen S M6URG
Kitchener C G4LYB
Kitchener C G8IMI
Kitchener J G7UYW
Kitchener M G6LSC
Kitchener R G4IKQ
Kitchener S G6DZJ
Kitching A G0DHS
Kitching L G3LEK
Kitching R G3XVS
Kitching W G4FBZ
Kiteley D G7SCZ
Kiteley M G7WDC
Kitney A M3WFE
Kitson D G3TRK
Kitson G G4ZAD
Kitson K G0BWQ
Kitson M G1MBM
Kitson P G7GXE
Kitt M M0WZM
Kitt N M0FVD
Kittle A G4GWT
Kitto J M6JKX
Kitto T M3UWM
Kittrick A G0EVR
Kittrick E G0FEV
Klee S M6GQS
Klein H G0VKS
Klein J G4DFJ
Kliffen J G0ACA
Klima R M0GOY
Klimaszewski M M6EMK
Kluger-Langer N 2E0BCC
Klunder J G7KTQ
Klymenko S M6EKC
Knaggs C G0LYZ
Knapp G G3NMJ
Knapp K M6CEI
Knapp R M0BEV
Knapp W M0WBK
Knapp W M3WBK
Knappett P M0PJK
Knapton M M6UNI
Knapton N G1JKE
Knapton N M3JKE
Knatchbull H G4VLP
Kneale D MD3UMN
Kneale F G4DBG
Kneale H G7MIP
Kneale J GD0BFN
Kneale S M6CUD
Kneebone A G6CEP
Kneebone P G6FQP
Kneebone P G7EKD
Kneeshaw S 2E0KBL
Kneeshaw S M6SKW
Knell A G0BNE
Knell M M3LIX
Kneller B G8BAQ
Kneller S 2E0ZSK
Kneller S M6ZSK
Knibb E G4ZJR
Knibbs D GW0IRC
Knibbs K G1GHG
Knight A G2HKQ
Knight A M3UOZ
Knight A M3YBR
Knight A M3ZBS
Knight B G7DMZ
Knight B G7MSC
Knight C GM0CFK
Knight C G8IPK
Knight C M0EAY
Knight D G6MTG
Knight D MW0KAK
Knight E G7UBB
Knight G G3XRD
Knight G M0GJK
Knight G 2E0BUP
Knight G GM8FFX
Knight H 2E0MGB
Knight H M0XLX
Knight J G0CYI
Knight J G3JRK
Knight J G3TPB
Knight J M6JWK
Knight J M0NDU
Knight J M6JFK
Knight J G1PPK
Knight J M6JPK
Knight K G1ASD
Knight K G4DFZ
Knight K G8MWE
Knight L 2E0BUN
Knight M G4BNW
Knight M G7OZU
Knight M 2E0BFH
Knight N G8VTY
Knight P 2E0PKK
Knight P G4CEC
Knight P G6EPN
Knight P G8BAK
Knight P G4BMM
Knight R G8ZRQ
Knight R M3EGU
Knight R G7CSM
Knight R G3VUK
Knight R 2E0KWG
Knight S G0BGI
Knight S M3ZJK
Knight S G1UBN
Knight T G2FUU
Knight T G4FHK
Knight T G8AOI
Knight T M0ALR
Knight T M3YZE
Knight W G7HMW
Knight W M6ADI
Knighton D M6DEK
Knighton H M3HMK
Knighton J G1DVH
Knighton R G0GER
Knighton R G4NYB
Knights A 2E0BFM
Knights D G0MPK
Knights J M6IVI
Knights K G7VNP
Knights M G3TQY
Knights M M6BID
Knitter A M3IKJ
Kniveton K G4IDU
Knock G G4FTX
Knock R G8SNQ
Knott C G3WMX
Knott D M0DKT
Knott D M3DKT
Knott D G3UYL
Knott E G1XCY
Knott J G6POE
Knott M G0ICE
Knott M G0WCR
Knott P GI1KGZ
Knott S 2E0SGK
Knott S M0SGK
Knott S 2E0CVQ
Knott S M6CTN
Knowler A G7AJK
Knowler D G4SFB
Knowler M G3YIB
Knowler W G7AZM
Knowles A M3VCO
Knowles A G3LUA
Knowles A G0AUE
Knowles C M6ITD
Knowles D G3UVA
Knowles F G3AKI
Knowles G 2E0KNL
Knowles G M6KNL
Knowles G M6SOZ
Knowles H M6HKW
Knowles J G3RPA
Knowles J G4OFP
Knowles J M3JLK
Knowles J M3WJK
Knowles J G0KWE
Knowles K G0IUV
Knowles L G1ZOY
Knowles M M3MRK
Knowles M 2E0MTH
Knowles M M6FTT
Knowles M 2E0EVA
Knowles M 2E0MYK
Knowles M M0MYK
Knowles M M6CET
Knowles M M6EVA
Knowles N G6SDG
Knowles P G4YPK
Knowles P G7DEY
Knowles P G4XDP
Knowles R G0LZX
Knowles R G1VWL
Knowles R G4HQA
Knowles S G3UFY
Knowlson C 2E0GES
Knowlson C G0OPG
Knowlson C G7IOB
Knowlson M G7KHE
Knox G G0IGK
Knox G G4CPD
Knox P G7LNK
Knox P M6PFY
Knox R G1RYM
Knox R GI3WEL
Knox T MM6KNO
Knox T M0TRF
Knox W G7BIL
Kober G M6GRZ
Kobiela G G6XRY
Koch A G1DGY
Koch D M3BJZ
Koeller A M5AGB
Koenig J G1XRQ
Kok A M1EYT
Kok D M1EYU
Koker P G3YPU
Kokinis G M6HSY
Kolbe G G4LPJ
Kolderman C M0PIA
Kolesnik K 2E0KJK
Kolesnik K M0KJK
Kolesnik K M6KJK
Kolonko E GM0HQF
Komenan K M6KKG
Kong W M3OHQ
Konos W GI4TUJ
Konowicz R G0YYY
Konstantynowicz M M6ERI
Konstas I MM0HLU
Kooner J M3EWV
Koopman D G1TLH
Koops J G0VZO
Koops J G7SKL
Korben A M6FZE
Kornreich J G5CDC
Kosarzecki J MM6KOS
Kossick A 2E0KVA
Kosteletos T M6TKK
Kostryca D M0AYF
Kotarba K M0EAE
Kotowicz A G6UZA
Kottis E M3VKT
Koval A M0EDX
Kowalczyk D G8NTZ
Kowalczyk Z G4GCU
Kowalski G G0RDG
Kowalski R M1AKV
Kowcun C G6VWI
Kozakowski M M0CWT
Koziolek R M6KOZ
Kozlowski K 2E0KFK
Kozlowski K M6KFK
Kozlowski S M6HNH
Kozminski J G8CPB
Kraft E G4FTP
Kratzer C G0SLY
Krauskopf D MM6SJK
KRAVCHENKO V G0KBO
Kraven I G4JIJ
Krawczyk M 2M0KAU
Krawczyk M MM6KAU
Krawczyk R MM1FAS
Kreissl A M3VOJ
Kressman R G3SIT
Kreuchen K G0PER
Krinks J M6JKR
KRISHNAMURTHY T M6FDW
Krol J 2E0FTX
Krol J M6HRT
Kroon F M0HXC
Krzeminski P M0SYJ
Krzymuski J G4DQW
Kuba A M6LPF
Kubecki S M6SQJ
Kugler R G8VQS
Kuik M G1SHI
Kuipers J G4GUX
Kulikovsky C 2E0HEM
Kulpinski K M0HQB
Kurczab M M0HVI
Kurdi L M6GZX
Kurian P G4HYT
Kurn D M6DRK
Kurnatowski A G4XTK
Kurtz E 2E0EDA
Kusin M GM0PDQ
Kusin V GM4HCO
Kuss C G0DZI
Kuss C G6DZI
Kuttikkate F 2E0KFB
Kuttikkate F M6KFB
Kveksas D M6URA
Kveksas M 2E0VBT
Kveksas M M0VBT
Kveksas M M6KVS
Kvilums A M1LCL
Kvilums A M3CTO
Kwan H M3KHT
Kwiatkowski T 2E0XMT
Kwiatkowski-Zelazny M M6MIZ
Kwok T M0WSC
Kyle A G8TBV
Kyle A M0KYL
Kyle D MI1BLZ
Kyle E 2I0BKI
Kyle E MI0WGW
Kyle E MI3UHI
Kyle J G6PGJ
Kyle J GM4CHX
Kyle N GI4CQL
Kynaston B M1DJO
Kynaston J G1ZLB
Kynaston J G4AHZ
Kypriadis A M1EEZ
Kyriacou K G0ELU
Kyriacou M M3TYG
Kyriacou T 2E0PLC
Kyriacou T M0TKS
Kyriacou T M3WSV
Kyriakides J G3WZO
Kyte A M6TLK
Kyte G G4RPP
Kyte P 2W0LLY

## L

La Pierre P G3STP
La Traille L G7LPM
Labourn M M0LAB
Labron P G0DWO
Labrosse F 2E0FFL
Labrosse F M0LBR
Labrosse F M6FLA
Lacaman D MW6BGO
Lace B M3YBL
Lace S M1MCW
Lacey C G0IMX
Lacey D G4JBE
Lacey E 2E0HXR
Lacey F G4NSW
Lacey J G3GLB
Lacey M M6MHL
Lacey R G8XNC
Lacey R MW0XMI
Lachs M M6OBH
Lack M G7KYL
Lacken R G4DKM
Lacy A G4AUD
Lacy K G3ZON
Lacy S G4TMR
Laczko I G0BQQ
Ladd A M6FAK
Laddiman S M0BLH
Ladell A M3SPL
Ladley C G1PRS
Ladley T 2E0BAE
Ladner L G1IVO
Laepong S MM1COS
Lafferty C G4KDS
Lafferty P GM0HWB
Laffin J 2E0DIA
Lagan D M0LBD
Lagar R G0RFT
Lai C M0HFQ
Lai C M6BHK
Lai C M6LCT
Lai O M6LDH
Lai P Y M3YUH
Lai Y M6JYL
Laidler I M0RZE
Laight R M0PML
Lailvaux G 2M0GRE
Lailvaux G MM6YPN
Lainchbury J G1IJJ
Lainchbury J G4XIQ
Laing J G6KLQ
Laing P M3PAU
Laird C G4UCY
Laird F G8TLU
Laird P MM0AUP
Laister A 2E0TZM
Laister A M3TMZ
Lait P G0IFQ
Laity A M0LAT
Lake A G8MAY
Lake B M1BPU
Lake C G7KTL
Lake D G3ZCA
Lake E 2E0WXM
Lake F G4YXS
Lake J 2E0GBM
Lake K G8ZIP
Lake K G8FVE
Lake M M3ZJE
Lake M M6MLK
Lake N 2E0NLK
Lake N M3NLK
Lake P G8ZZL
Lake R G8MFH
Lake S G1WFG
Lake S 2E0GCO
Lake S M6EXY
Lake S M6SLK
Lake T 2E0TOM
Lake T M3TJL
Lake T M6TJL
Laker B M3XQX
Laker J M1BIY
Laker P G1INB
Lakey B G1FWZ
Lakhaney H GM0DXI
Lalji S MD6SHL
Lam A 2E0LAM
Lam C M0WAI
Lam T G0REU
Lam W G4KPG
Lam W M6HUH
Lam Y M3YLM
Lamb D G1IZA
Lamb D GD0ACK
Lamb F G4GDO
Lamb G GI0LAM
Lamb G G7GMU
Lamb J G0WXG
Lamb J G1FQX
Lamb J G1IVP
Lamb J 2E0LFX
Lamb J M6LFX
Lamb K G4BUW
Lamb M G0DSP
Lamb M G1LFX
Lamb M G7PSL
Lamb N G1BMN
Lamb P 2E0KZJ
Lamb P G3VRW
Lamb P M3KZJ
LAMB R G4TVX
Lamb V G7AIY
Lambarth R G8HAU
Lambe J G7MNK
Lambe R MI6CVO
Lambert A G8HER
Lambert A G8TNU
Lambert A M5ADL
Lambert A G0UIX
Lambert A G7JAF
Lambert A G8OXS
Lambert A M6AFA
Lambert B G4YJA
Lambert C G3TA
Lambert C G4PCN
Lambert C G7VNL
Lambert D G1ABA
Lambert D G4POI
Lambert D G8HGL
Lambert E G4AYL
Lambert E GD3FKI
Lambert G G0IXT
Lambert G M3UEQ
Lambert Hurley J 2E0XIK
Lambert Hurley J G3MOT
Lambert Hurley J M0XIK
Lambert Hurley J M3XIK
Lambert I G4LWG
Lambert I M6IAL
Lambert J 2E0JRL
Lambert J G3FNZ
Lambert J G8OQT
Lambert J M6BXA
Lambert J M6POA
Lambert J M6TNU
Lambert K G7BVL
Lambert K 2E0CJF
Lambert K M6AHU
Lambert L G1XEP
Lambert M G3XGZ
Lambert M M3MTE
Lambert N G7HCO
Lambert N M3JOA
Lambert N 2E0HOQ
Lambert N G7BFH
Lambert P G0TGK
Lambert P G1LSZ
Lambert P G3CYX
Lambert P G8XNO
Lambert P M6PCL
Lambert R G6GND
Lambert T G8EZL
Lamberton R G4LAM
Lambeth C G1YDI
Lambeth M G2AIW
Lamble D 2E0NYF
Lamble D M6NYF
LAMBLE J G6NGA
Lambley G M1CNI
Lambley R G8LAM
Lambourne R G1FXC
Lambrianou J G0PRT
Lamden D G7RJW
Lamerick M G6KTN
Lamerton N M6XNL
Lamford K G6ODT
Laming G G4JBD
Lamkin R G8WDX
Lamming P G3NXL
Lamont A G4KDE
Lamont A M6HKQ
Lamont D G3WPV
Lamont J G4MLQ
Lamont T M3YRJ
Lampard S G6AJX
Lampard S M6SJZ
Lampett M G7DQQ
Lampson M M0GZF
L'Amy M GJ6WMZ
Lanaway S M3SYN
Lancashire W G4SSL
Lancaster A G0JCC
Lancaster B G6YCW
Lancaster D M6LNC
Lancaster P G6ZOL
Lancaster P G7NVI
Lancaster P M1EKU
Lancaster P M3EKU
Lancaster R M3RZW
Lancastle M G7VYZ
Lancefield A M3UZK
Land A 2E0GNW
Land A M6GNW
Landen-Turner G G0OXA
Lander A 2E0WAU
Lander A M6VAL
Lander D G4LQL
Lander G G3OOH
Lander T M6TBL
Landers L GM5CGA
Landless A M0WOJ
Landless C M3YCZ
Landless J 2E0LEG
Landless J M6EMC
Landless J 2E0JLX
Landless J M3XJL
Landon E G3MHT
Landon E M3XAU
Landon K G3IXI
Landon M M6CCA
Landor C GJ4YLP
Landor E G0IPO
Landor P GJ3AME
Landragin M 2E0RIA
Landragin M M6WIF
Landragin R 2E0RHL
Landragin R M6SOO
Landragin R 2E0WOB
Landragin R M6ZUB
Landricombe L G0KYE
Lane A G4CQW
Lane A M0ALN
Lane B M3ZVF
Lane C M6BKF
Lane D G3VOM
Lane D G6AVY
Lane D M0CMI
Lane D M0DRL
Lane D G6FCS
Lane D M6DCL
Lane E M0NOV

| Name | Callsign |
|---|---|
| Lane G | M3YNK |
| Lane G | M3ZGY |
| [illegible] | [illegible] |
| Lane J | G8XNA |
| Lane K | M6VAG |
| Lane K | 2M0KTL |
| Lane K | MM0KTL |
| Lane M | G3VOV |
| Lane M | G6YGV |
| Lane M | M1EZG |
| Lane M | G0SHC |
| Lane M | G1SCT |
| Lane N | M3KSE |
| Lane N | GI6FOR |
| Lane P | G1MNX |
| Lane P | MW0BWM |
| Lane P | M3PWL |
| Lane P | 2E0OTY |
| Lane P | M6ODY |
| Lane R | G4AWU |
| Lane R | 2E0BWL |
| Lane R | M0RWL |
| Lane R | M3USZ |
| Lane S | GW4PRP |
| Lane V | G4FRA |
| Lane-Wells L | M6WLY |
| Laney J | M3EQQ |
| Laney R | G4RAE |
| Lanfear A | G4CQI |
| Lang A | MM0THE |
| Lang D | G6NAG |
| Lang G | G0OBQ |
| Lang G | G1BGK |
| Lang M | G4DVK |
| Lang N | M6NGW |
| Lang R | M3GZU |
| Lang R | M6RLX |
| Langabeer P | M0GQV |
| Langan J | G0KKO |
| Langdale S | G1JYK |
| Langdon B | G4PUD |
| Langdon C | G0VNH |
| Langdon C | M0ZOM |
| Langdon D | G6CKM |
| Langdon G | 2E0CDK |
| Langdon G | M3AIL |
| Langdon J | M6FZY |
| Langdon K | M3KWL |
| Langdon M | G4WHV |
| Langdon P | G0EXB |
| Langdon R | G4RHL |
| Langdon R | G8GZR |
| Langdon T | G3MHV |
| Lange A | GJ7DTA |
| Langer R | G4RQF |
| Langfield P | G7EBF |
| Langford A | G0PQY |
| Langford A | G4ARY |
| Langford A | G4IQW |
| Langford B | G4GWP |
| Langford G | G0MKU |
| Langford G | G4LNZ |
| Langford G | M1BGL |
| Langford P | G1MRE |
| Langford P | G8ZDT |
| Langford P | M3LQJ |
| Langford R | G4RLT |
| Langford R | G4FAD |
| Langford R | M6RIK |
| Langford T | G4VHL |
| Langford-Brown A | G3NUN |
| Langham C | G7URT |
| Langham C | G8XWH |
| Langham M | 2E0HDW |
| Langham S | G0DMN |
| Langham T | G4ZDF |
| Langhamer K | M1CYL |
| Langley A | G3KHQ |
| Langley C | G3XGK |
| Langley D | M3KDM |
| Langley S | 2E0HOK |
| Langlois E | GJ6TPD |
| Langlois L | 2J0COQ |
| Langlois L | MJ0LEL |
| Langlois L | MJ6BDJ |
| Langlois S | GJ4ODX |
| Langmaid C | 2E0BQU |
| Langmaid C | G3RAM |
| Langmaid C | M6CSL |
| Langmead A | G1OOE |
| Langmead D | 2E0HAM |
| Langmead D | M3GTA |
| Langmuir I | G7NAO |
| Langmuir R | M3VSO |
| Langridge D | G0CYC |
| Langridge D | M3IBZ |
| Langsley H | G6PMF |
| Langson S | G1LMT |
| Langstaff G | G6SMI |
| Langston S | G1JNX |
| Langton A | GM4HTU |
| Langton R | G6PSQ |
| Langton S | M3MUU |
| Langton S | G6ART |
| Langtree I | G0KBA |
| Langwade M | G3VET |
| Lanham A | G0TVC |
| Lanham K | G1XFL |
| Lanham M | M6TTF |
| Laniosh B | G4NZK |
| Lankester M | G8ATL |
| Lankshear D | G3TJP |
| Lankshear K | G6CXN |
| Lannon R | GW6ZHM |
| Lanoe G | MU6GBG |
| Lansdell S | 2E0EDD |
| Lansdown P | M6PVL |
| Lansdown S | G4SDT |
| Lansdowne J | G0HFA |
| Lansley N | [illegible] |
| Lapham-Crozier D | 2E0OPS |
| Lappage D | M6DHG |
| Lappin D | MI6YSW |
| Lappin J | GI0OND |
| Lapthorn R | G3XBM |
| Lapwood R | M1RJL |
| Larbalestier P | G4TEB |
| Larcombe G | G6AJV |
| Larcombe J | M6ABG |
| Larcombe M | 2E0BRD |
| Larcombe M | G7OIA |
| Larcombe M | G7ZMS |
| Larden J | M0EBT |
| Larden S | M1EVF |
| Large G | 2E0IHT |
| Large J | G1ZIV |
| Large J | G7MLO |
| Large K | M1KGL |
| Large L | G4CYZ |
| Large R | M1AGK |
| Large T | M3XFG |
| Largent S | G4VBQ |
| Larimer R | GI7THY |
| Lark A | 2E0XDX |
| Lark A | M0XAB |
| Lark A | M6LRK |
| Larke R | GI6BDN |
| Larkin A | G1JNY |
| Larkin D | M3KAX |
| Larkins J | G8WKT |
| Larkins L | 2E0LAL |
| Larkins L | M3FZJ |
| Larkins S | M0SBF |
| Larman B | M3IWJ |
| Larman T | 2E0TWL |
| Larman T | M0TWL |
| Larman T | 2E0TWW |
| Larman T | M0TWW |
| Larner C | 2E0MOB |
| Larner C | M0RMO |
| Larner C | M6RMO |
| Larner P | M3VUV |
| Larrigan G | M3MMN |
| Larsen D | G4FMY |
| Larsen N | G7RXO |
| Larsen T | G7PHI |
| Larson J | G3NBL |
| Larson N | G7RXB |
| Larssen J | G6TBJ |
| Larter R | G4YFI |
| Lasbury R | 2W0LAZ |
| Lasbury R | MW3RVP |
| Lasbury W | G6DQT |
| Lascelles P | G3YLW |
| Lash P | G6KSO |
| Lashbrook A | M6AJL |
| Lasher N | G6HIU |
| Lashley J | 2E0ODF |
| Laskey T | G7SYE |
| Laslett R | G8KQA |
| Lassemillante J | G0JOC |
| Lassester N | M0IST |
| Lassman A | G7PHE |
| Last D | GM4THP |
| Last D | G8XKT |
| Last D | G6LEU |
| Last D | M5ABC |
| Last E | G7HMF |
| Last J | G3MZY |
| Last P | G3THS |
| Last R | G0FCH |
| Last R | 2E0CWI |
| Last R | M6CHU |
| LAST S | G8ETN |
| Laszkiewicz A | G1GHY |
| Laszkiewicz A | G1PSS |
| Laszkiewicz L | G3KAU |
| Laszlo J | M6NOA |
| Latham A | G0DIM |
| Latham A | G7VFQ |
| Latham D | G6HXL |
| Latham F | G1DMW |
| Latham K | M6CSS |
| Latham P | 2E0SFB |
| Latham P | M6SFB |
| Latham R | M0ADW |
| Latham S | G7NUN |
| Latham S | G0NQG |
| Latimer D | G3VUS |
| Latimer S | M6OOL |
| Latimer-Sufit M | M5AHF |
| Latta B | G7KIS |
| Latter E | G0UMI |
| Latter S | M6SGC |
| Lattin G | G6ZHO |
| Lattka C | G0PPI |
| Latto A | [illegible] |
| Latto D | 2M0DSL |
| Latto D | MM0KDY |
| Lau H | MW6LHW |
| Laud N | G0MMU |
| Lauder D | G0SNO |
| Laugher S | G7LYN |
| Laughlan A | G1YCM |
| Laughlin R | M3WMO |
| Laughton DA | G1CUG |
| Laughton K | G1ESW |
| Laurence A | G3GGI |
| Laurence A | G7PSZ |
| Laurence A | M3PSZ |
| Laurence S | M6BWW |
| Laurie M | M1EJJ |
| Laurie W | MM0WAK |
| Lauritson S | G8ZEX |
| Lauxman McCorkell J | MM6VZA |
| [illegible] | [illegible] |
| Lavelle K | M6KML |
| Lavelle L | G3SSZ |
| Lavender D | M3ZYM |
| Laver R | G8GDC |
| Laver R | M6DYP |
| Laverick J | G4PFE |
| Laverick R | 2E0DMZ |
| Laverick R | M3PFE |
| Laverie J | 2E0JWL |
| Laverock P | G8GHQ |
| Laverty R | 2I0BTT |
| Laverty S | GI3RQU |
| Lavery C | M6MAG |
| Lavery J | M6EAJ |
| Lavery T | GI8OLH |
| Lavin H | G4HAP |
| Lavin K | G7LMR |
| Lavis C | G6HIQ |
| Lavis J | G4LYG |
| Law A | G4ANE |
| Law A | M6WMZ |
| Law C | M0OKT |
| Law D | G0GUD |
| Law E | G3WGE |
| Law E | G4CVW |
| Law G | M3GHL |
| Law J | G8ILG |
| Law J | M3GAE |
| Law K | G4WMZ |
| Law K | G4ALF |
| Law M | G6OKU |
| Law N | G1OCR |
| Law N | G0ADA |
| Law R | GI1PVE |
| Law R | G8VLR |
| Law R | M0DNK |
| Law R | MM6RAM |
| Law T | G6TQL |
| Law Y | MD6KEU |
| Lawes A | G1KKF |
| Lawes D | G6CLU |
| Lawes G | G3PLT |
| Lawes N | G8ZHR |
| Lawford P | G6PIM |
| Lawford S | M3XRH |
| Lawford T | G4KGY |
| Lawler A | M6LWP |
| Lawless E | 2E0IJM |
| Lawless P | GM4PGV |
| Lawless T | GM6JOD |
| Lawley D | G4BUO |
| Lawley E | G4ERL |
| Lawley E | G3MMX |
| Lawley E | G8ADX |
| Lawlor G | M6FTO |
| Lawlor P | GM3MUA |
| Lawman S | G0UIH |
| Lawrance A | G3RZV |
| Lawrance B | G4ZYO |
| Lawrance G | G4EOB |
| Lawrance M | M1DBK |
| Lawrence A | M6MJL |
| Lawrence A | G6JGP |
| Lawrence A | M3SQH |
| Lawrence A | M6AFL |
| Lawrence A | M6IBO |
| Lawrence A | M6HJG |
| Lawrence B | G0SHO |
| Lawrence B | G4LSL |
| Lawrence C | G1JHB |
| Lawrence C | M0XPL |
| Lawrence C | G1KAT |
| Lawrence D | G0PGK |
| Lawrence D | G4FYT |
| Lawrence D | G6HXR |
| Lawrence D | G6XAW |
| Lawrence D | M0BPS |
| Lawrence D | M3MHZ |
| Lawrence D | G0RTQ |
| Lawrence D | G4VKC |
| Lawrence E | M3EJL |
| Lawrence G | G0XGL |
| Lawrence G | G6XAV |
| Lawrence H | G3YYE |
| Lawrence J | G3JGA |
| Lawrence J | G3PQY |
| Lawrence J | G3WEZ |
| Lawrence J | G4VII |
| Lawrence J | G4VYN |
| Lawrence J | G3MEY |
| Lawrence J | G0KYL |
| Lawrence K | G6FAH |
| Lawrence K | G8SSE |
| Lawrence L | 2E0HLP |
| Lawrence L | G0HRD |
| Lawrence L | M3ULE |
| Lawrence M | G0SZN |
| Lawrence M | M0WMT |
| Lawrence M | 2E0XVX |
| Lawrence M | M6CFX |
| Lawrence M | M0XVX |
| Lawrence P | G1MUT |
| Lawrence P | GM7JYW |
| Lawrence P | G4XAL |
| Lawrence P | M0MRQ |
| Lawrence R | G1YCR |
| Lawrence R | G4LDA |
| Lawrence R | G8AWB |
| Lawrence R | G0RQZ |
| Lawrence R | MI6HLY |
| Lawrence R | M6TLB |
| Lawrence R | M0CWC |
| Lawrence S | G0SCL |
| Lawrence S | G4EOF |
| Lawrence S | G7PMU |
| Lawrence T | 2E0HGU |
| Lawrence T | MM0TDI |
| Lawrence W | G8WRI |
| Lawrenson C | GM3SJY |
| Lawrie G | MM3GPL |
| Lawrie R | GM7SFE |
| Lawrie S | GM0VWZ |
| Lawrie W | GM4VZI |
| Laws J | M6HWN |
| Laws K | G0CZR |
| Lawson A | M6FJQ |
| Lawson D | G0FBM |
| Lawson D | 2E0HMM |
| Lawson D | M3UCO |
| Lawson D | 2E0NCB |
| Lawson D | M0GGK |
| Lawson D | M3NCB |
| Lawson D | M6SJN |
| Lawson F | G1AET |
| Lawson J | G0GVA |
| Lawson J | G3WSV |
| Lawson J | M6PPD |
| Lawson K | G4AAH |
| Lawson K | G7FLS |
| Lawson M | M3CSM |
| Lawson M | G0FCJ |
| Lawson M | 2M0TZB |
| Lawson M | MM6NLP |
| Lawson M | M1MCL |
| Lawson M | G4YQA |
| Lawson M | M6HNI |
| Lawson P | G4FCL |
| Lawson R | G1RLB |
| Lawson S | G0TBC |
| Lawson S | G6WPL |
| Lawson W | 2M0WLA |
| Lawson W | MM0MLD |
| Lawson W | MM6YLA |
| Lawson-reay J | G8WFS |
| Lawton B | G0HWV |
| Lawton D | G0ANO |
| Lawton D | G6NSZ |
| Lawton D | G8ANO |
| Lawton G | G1XKD |
| Lawton G | G7EVY |
| Lawton I | 2E0DQU |
| Lawton I | M6IAX |
| Lawton J | G0LHU |
| Lawton J | G4ZOC |
| Lawton J | G4YAW |
| Lawton K | G4YQW |
| Lawton K | G1AYP |
| Lawton K | M3KAL |
| Lawton M | M6MEP |
| Lawton P | G4XZG |
| Lawton P | G7IXH |
| Lawton R | M3YJG |
| Lawton S | M3LNM |
| Lawton S | MW3SLL |
| Lax D | G4AHN |
| Lax K | G3VVL |
| Laxton A | M6XAY |
| Laxton B | G8IJE |
| Laxton P | G3SVZ |
| Lay L | 2E0LBL |
| Lay L | M0LBY |
| Lay L | M3RQW |
| Lay N | M3VUC |
| Lay N | G8FXG |
| Lay S | 2E0LAY |
| Lay S | M6LAY |
| Laycock F | M1BWH |
| Laycock G | G3WZW |
| Laycock G | G3XWN |
| Laycock G | G7ADW |
| Laycock R | G1IMY |
| Layden J | M6SDL |
| Laye R | M6RZW |
| Layland A | G4NUS |
| Layland L | 2E0COA |
| Layland L | M0LGL |
| Layland L | M6EEL |
| Layland L | M6LTL |
| Layland N | M6CFN |
| Layne D | G0AHK |
| Layphries T | G0XTL |
| Layton A | G6BTC |
| Layton A | G8YYX |
| Layton D | G1XJZ |
| Layton J | G4AAL |
| Layton J | M0ADA |
| Layton M | M3UML |
| Lazzari J | G8NAI |
| Le Boutillier K | GU0EFD |
| Le Boutillier K | MU3EFB |
| Le Boutillier P | GU3UOQ |
| Le Carpentier B | GW4SUN |
| Le Couteur Bisson A | G4WZH |
| Le Feuvre P | [illegible] |
| Le Feuvre R | [illegible] |
| Le Good G | G4GUN |
| Le Grosley N | G4SFV |
| [illegible] | [illegible] |
| Le Grove D | M0TIN |
| Le Grys B | G3GOT |
| Le Jehan C | GJ0FYB |
| Le Lievre B | GU4LJC |
| Le Mare D | 2E0BWP |
| Le Mare D | M6DLM |
| Le Moine M | 2E0MQC |
| Le Moine M | M6MQC |
| Le Page B | 2E0CKO |
| Le Page J | 2U1FNQ |
| Le Page L | GU4SYQ |
| Le Page N | GU4NYT |
| Le Piez R | M0RLP |
| Le Poer Trench Brown S | G7DTG |
| Le Prevost A | MD3LEP |
| Le Quesne F | GJ4HSW |
| Le Roux J | M0ICI |
| Le Serve W | M0WLS |
| Le Serve W | M6WLS |
| Le Vallois P | 2E0PLV |
| Le Vallois P | M0PLV |
| Le Vallois P | M6PLV |
| Le Ves Conte M | GW6YCT |
| Le Vine D | G1FKP |
| Lea P | G8DLZ |
| Lea R | G3DEN |
| Lea T | M0RIA |
| Lea V | 2E0VJL |
| Lea V | M3VJL |
| Lea W | G8NQY |
| Leach A | G1VAJ |
| Leach A | G3WIW |
| Leach A | G4WOQ |
| Leach H | G4ANE |
| Leach H | M6RXL |
| Leach I | G1NRY |
| Leach J | G0FVO |
| Leach J | G7OXK |
| Leach J | M6LPN |
| Leach K | M1CDX |
| Leach L | GW8JLY |
| Leach L | G1WZO |
| Leach M | G4HGV |
| Leach M | 2E0MTL |
| Leach M | M0TGF |
| Leach M | M6NXY |
| Leach P | G0OSP |
| Leach P | G7GBZ |
| Leach P | G8NSS |
| Leach R | M1RAL |
| Leach T | G6YCV |
| Leach V | G4GFT |
| Leach W | G8LYG |
| Leack M | G6ZHL |
| Leadbeater R | G8DVW |
| Leadbetter G | MM1DEA |
| Leadbetter S | 2E0BYH |
| Leadbetter S | M3XUE |
| Leader J | G0KLJ |
| Leader-Chew T | G6LEB |
| Leadill P | M1EHD |
| Leah R | G0TZD |
| Leahy T | G7ELZ |
| Leak H | G0RJT |
| Leak M | G4VNC |
| Leak S | G6IQC |
| Leak S | G0TGH |
| Leake A | G0NAA |
| Leake G | M3EZB |
| Leake G | G8SLE |
| Leake M | M3WBM |
| Leake W | M6WDL |
| Leaker S | M1BUX |
| Leal C | G3ISX |
| Leaman J | M3IBY |
| Lean G | G3WJG |
| Leaney N | G1JKF |
| Lear A | M6TLL |
| Lear J | M3TQD |
| Lear V | G3TKN |
| Learmonth I | 2M0STB |
| Learmonth I | MM0ROR |
| Learmonth I | MM6MVQ |
| Learoyd C | G6SKS |
| Learoyd W | G6SKT |
| Leary D | G8JKV |
| Leary J | G4YSB |
| Leary J | G6ZOT |
| Leary T | GW4WPA |
| Leask D | GM7VPT |
| Leask E | GM6UNQ |
| Leask R | G3XNG |
| Leask S | 2E0EEO |
| Leask S | M0IOI |
| Leask S | M6NOT |
| Leat C | G4SAB |
| Leather P | G4JYK |
| Leatherbarrow G | G4KQC |
| Leatherbarrow J | G6DDC |
| leatherbarrow T | M6HTU |
| [illegible] | [illegible] |
| Leathes S | M3EFV |
| Leathley-Andrew R | G8GMU |
| Leaver H | M3KAU |
| Leaver I | G7DBO |
| Leaver T | M3BRO |
| Leavesley J | M3ZJM |
| Leavey A | G0SCG |
| Leavold L | G0AJJ |
| Leavold R | G6VOV |
| Leavold T | M0TJL |
| Leaworthy I | [illegible] |
| Leaworthy T | M3JQX |
| Leazell S | M3UJJ |
| Lebaldi A | M3KNV |
| Lebaldi L | M3KHM |
| Lebaldi M | M3KIQ |
| Le-Brun P | G0HHN |
| Lecaille J | M3RWK |
| Leckey B | M6AYG |
| Leckie D | GM4NFI |
| Leckie T | GM0HZO |
| Ledbury P | G4GGH |
| Leddington N | M3NFL |
| Leddington W | G8PTS |
| Ledeux N | G8ZTM |
| Ledger A | G4PYA |
| Ledger A | G6LBR |
| Ledger A | M6BMF |
| Ledger C | G3UBL |
| Ledger D | 2E0ILH |
| Ledger J | 2E0DBO |
| Ledger J | M0IDX |
| Ledger M | G8HXD |
| Ledson B | 2E0CWO |
| Ledson B | M6CWD |
| Ledson K | M6MFU |
| Lee A | 2E0KDE |
| Lee A | G1VLA |
| Lee A | M0BYB |
| Lee A | M0KDE |
| Lee A | 2E0TMX |
| Lee A | M0LKY |
| Lee A | M6LKY |
| Lee A | G0EDK |
| Lee B | G0TWT |
| Lee B | G1IIX |
| Lee B | G7FTF |
| Lee B | G8DOW |
| Lee B | G4VDJ |
| Lee B | MW6GQD |
| Lee C | 2E0CLL |
| Lee C | 2E0GAD |
| Lee C | 2E0IIA |
| Lee C | G0BNY |
| Lee C | G0FZO |
| Lee C | G4GQY |
| Lee C | G7VEL |
| Lee C | M0GVT |
| Lee C | M6AVH |
| Lee C | G3GXG |
| Lee D | G0ROX |
| Lee D | G1BGO |
| Lee D | G1OTA |
| Lee D | G6YGP |
| Lee D | G8ZZK |
| Lee D | M0WDL |
| Lee D | M1BRZ |
| Lee D | MW1MFY |
| Lee D | M6VKE |
| Lee D | 2M0RRS |
| Lee D | G0AXC |
| Lee D | G6XUV |
| Lee D | G0DJU |
| Lee D | G3OMZ |
| Lee D | G4UHJ |
| Lee D | G4NIF |
| Lee D | G8PWA |
| Lee E | M0BQD |
| Lee F | M1BZI |
| Lee G | G4XXI |
| Lee G | G2ARY |
| Lee H | G6FLY |
| Lee H | MW3UAA |
| Lee H | M3YHL |
| Lee I | MM0IEL |
| Lee J | G0INT |
| Lee J | G1HLV |
| Lee J | G4JTK |
| Lee J | G4LHE |
| Lee J | G4TTJ |
| Lee J | G7AGO |
| Lee J | GM7ORX |
| Lee J | G8LII |
| Lee J | G4AEH |
| Lee J | M6EIY |
| Lee J | G0OPA |
| Lee J | G4EQJ |
| Lee J | G4TSN |
| Lee K | G4JXU |
| Lee K | GM6KDB |
| Lee K | M3KBL |
| Lee L | G3MCE |
| Lee L | 2E0EMB |
| Lee L | M6AKJ |
| Lee M | 2E0MJL |
| Lee M | 2E0XBX |
| Lee M | G0LXV |
| Lee M | G3VYF |
| Lee M | G6FKN |
| Lee M | G6OPK |
| Lee M | M0JWL |
| Lee M | M0TSN |
| Lee M | M1BHC |
| Lee M | M3MJL |
| Lee M | M3TSN |
| Lee M | MW0MJB |
| Lee M | G8BGM |
| Lee M | M0HHR |
| Lee M | M6LGQ |
| Lee M | M0YMN |
| Lee N | G4OYR |
| Lee N | G4WKY |
| Lee P | 2E0GJE |
| Lee P | G0DLP |
| Lee P | G0MFH |
| Lee P | G0MHK |
| [illegible] | [illegible] |
| Lee P | G3ZKO |
| Lee P | G4UVX |
| Lee P | G8DWP |
| Lee P | [illegible] |
| Lee P | M0TRC |
| Lee P | G4GEW |
| Lee R | G0BWK |
| Lee R | G8CQR |
| Lee R | M6RPL |
| Lee R | G0LEE |
| Lee R | M0GXW |
| Lee R | M6RLC |
| Lee S | G1RWR |
| Lee S | M3GZE |
| Lee S | M6YOG |
| Lee S | G7GCD |
| Lee S | G0JPF |
| Lee S | M6SLE |
| [illegible] | [illegible] |
| Lee T | M0BQB |
| Lee T | M3YTL |
| Lee T | 2E0KIL |
| Lee T | M3HNK |
| Lee T | M0GHK |
| Lee V | M1DOO |
| Lee V | M6VIV |
| Lee W | G0ESU |
| Lee W | GW3MFY |
| Lee W | G3MHR |
| Leece T | 2D0TRL |
| Leech D | G7DIU |
| Leech D | MM0LOZ |
| Leech D | M1FFF |
| Leech M | G0KCD |
| Leech M | 2E0LAD |
| Leech M | M3UXG |
| Leech N | M6NAK |
| Leech P | MM3PSL |
| Leech P | G0GQV |
| Leech S | G4WEE |
| Leeder D | G8VLS |
| Leeder G | M1CBV |
| Leedham G | M6GCN |
| Leedham Hawkes G | M6CDE |
| Leedham J | G0BKA |
| Leedham V | G0BKB |
| LEEDS C | M1ACT |
| Leeds R | G4RZN |
| Leek C | M6CTV |
| Leek J | G8NST |
| Leek L | G0NOB |
| Lee-Koo C | M3YGY |
| Leeman P | G1ZBA |
| Leeman S | MM6SRL |
| Leeman T | G0MLM |
| Lee-Marr C | MM3XFP |
| Leeming A | G4LLQ |
| Lee-Ray S | 2E0CDI |
| Lees A | G1UTJ |
| Lees A | G1YUS |
| Lees B | G0JBO |
| Lees C | G4WXG |
| Lees D | G4SVV |
| Lees E | G1UZS |
| Lees F | 2E0YAR |
| Lees F | M3ZVI |
| Lees G | G0MGL |
| Lees G | G6CXM |
| Lees G | M6GCD |
| Lees J | G0WLJ |
| Lees K | M6HML |
| Lees L | G1DMH |
| Lees M | G4KGL |
| Lees M | G8IAN |
| Lees M | M3MYK |
| Lees N | G0ONW |
| Lees N | M3NCL |
| Lees P | G1JSK |
| Lees P | G4WIG |
| Lees R | G0PFE |
| Lees R | G8DQE |
| Lees R | G8EBT |
| Lees R | G1GLS |
| Lees T | G0ICG |
| Leese C | M5BAD |
| Leese D | G6ZJM |
| Leese D | M3KWZ |
| Leese P | G6UXX |
| Lees R | M3ZBL |
| leesley G | G4WEC |
| Lees-Oakes R | G1JAA |
| Leeson B | M6RSL |
| Leetch J | MI3OHP |
| Leetham P | G4YVV |
| Lefever J | G4CAZ |
| Lefevre S | M1AHA |
| Lefton M | M6BPL |
| Legate C | G1HZD |
| Legg A | G6JWO |
| Legg D | G8WUF |
| Legg G | G0EEF |
| Legg K | 2E0CWZ |
| Legg K | M0KEB |
| Legg K | M6FGG |
| Legg P | M1IOW |
| Legg S | 2E0SLJ |
| Legg S | M6SDI |
| Leggat J | GM0LOK |
| Leggat S | GM7CPY |
| Legge D | G7TQA |
| Legge V | G4ZXP |
| Legge W | GM0OSJ |
| Leggett A | G4KNN |
| Leggett A | M0NWK |
| Leggett D | M3NOS |
| Leggett C | [illegible] |
| Leggett J | G1JVL |
| Leggett J | M6LGT |
| Leggett K | G4JKZ |
| Leggett M | [illegible] |
| Leggett M | M6JAL |
| Leggett S | M3UFA |
| Legood A | GM8IXZ |
| Legrain J | 2E0CKQ |
| Legrain J | M0HDU |
| Legrain J | M6BKO |
| lehane J | M0CEW |
| Leicester P | M0FOX |
| Leigh A | G0FBX |
| Leigh A | G4DEN |
| Leigh A | G4JZC |
| Leigh A | G8SXQ |
| Leigh C | G0VVR |
| Leigh D | M0GYI |
| [illegible] | [illegible] |
| Leigh R | G0HKN |
| Leighs A | G4SXK |
| Leighton A | GJ6SNQ |
| Leighton C | G4VBM |
| Leighton F | GJ4WRR |
| leighton J | 2E0YDX |
| Leighton J | G4RQS |
| Leighton L | G6HXW |
| Leighton M | G3UKM |
| Leighton S | 2M0CSZ |
| Leighton S | MM6CNO |
| Leiper A | MM3TEQ |
| Leiper D | MM0OZY |
| Leiper G | GM4VQY |
| Leitch I | G0PAI |
| Leitch J | 2M0JLS |
| Leitch J | MM6MCT |
| Leitch M | G6FIV |
| Leitch P | GI1EOS |
| Leitch S | MI3SFL |
| Leitch W | GI6HKE |
| Leith J | GM6GJW |
| Leith J | MM6JLX |
| Leith S | GM7LWA |
| Lekesys J | G4BYW |
| Lelliott M | G4DNK |
| Lelliott M | G8DDH |
| Lelliott R | G8XNB |
| Leman C | G7TSP |
| Leman R | G8CDD |
| LeMasonry P | M0AZP |
| Lemay J | G4ZTR |
| Lemer M | M3MML |
| Lemin M | G4UUB |
| Lemmon L | 2E0DVF |
| Lemmon L | M6KOA |
| Lemon J | G4FYJ |
| Lemon N | MW0CIA |
| Lempriere D | G4SXQ |
| Len S | M0GKA |
| Leng P | 2E0PFL |
| Leng P | M6DLQ |
| Lenicker R | 2E0LTZ |
| Lenicker R | M6MTZ |
| Lenihan G | G8YQU |
| Lennard D | G4ZHK |
| Lennard J | G1HXP |
| Lennard P | G3VPS |
| Lennon A | GI4MUN |
| Lennon A | M6AML |
| Lennon J | G4CRM |
| Lennon M | M0ABI |
| Lennon P | G4CMP |
| Lennox B | G0ZIG |
| Lennox C | G4LXU |
| Lennox E | G4JKY |
| Lennox J | G3NSP |
| Lennox M | G6VZB |
| Lennox V | G7PMF |
| Lenthall R | G0RUS |
| Lenton A | G8UUG |
| Lenton S | M3UUG |
| Leo S | M0HID |
| Leonard C | G6FHK |
| Leonard C | G4ERO |
| Leonard D | G0ORT |
| Leonard D | M6EQR |
| Leonard G | G4DDN |
| Leonard G | G1GNX |
| Leonard H | 2W0HRL |
| Leonard H | MW3LLV |
| Leonard I | G4UKV |
| Leonard J | M6JFW |
| Leonard M | G4YJM |
| Leonard M | 2E0XYM |
| Leonard M | M3XYM |
| Leonard P | GI0CAH |
| Leonard S | G6GTH |
| Leonard W | 2I0WGL |
| Leonard W | MI0WGL |
| Leonard W | MI6WGL |
| Leong H | G7KRT |
| Leong R | G6ODU |
| Le-Petit M | M6MLP |
| Leppard V | M0VLL |
| Lerner D | G4XWZ |
| Lerpiniere E | M0EOL |
| Lertruengpanya T | 2E0SNX |
| Lertruengpanya T | M0SNX |
| Lertruengpanya T | M6SNX |
| Lesiecki P | M0PLO |
| Leslie D | G6CBL |
| Leslie H | MD3LEG |
| Leslie J | GM6UNL |
| Leslie P | G1BGM |
| Leslie R | G4VHK |
| Leslie S | 2D0STL |
| Leslie-Reed P | [illegible] |
| Lesniowski M | M0MIR |
| Lester B | M1EWP |
| Lester C | 2E0CSI |
| Lester C | M0GGL |
| Lester C | M3JYP |
| Lester D | 2E0LPD |
| Lester D | M0WYH |
| Lester D | M6LPD |
| Lester E | G6NAP |
| Lester G | G4JBF |
| Lester P | G8ZHS |
| Lester R | G8UBJ |
| Lester R | 2E0WUN |

UK Surnames

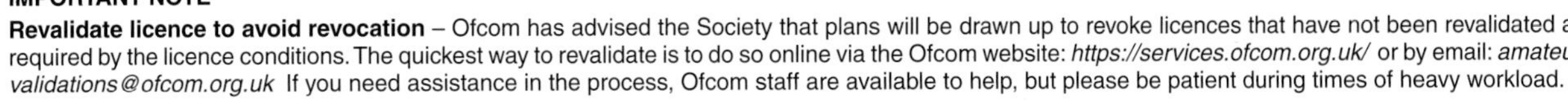
**IMPORTANT NOTE**

**Revalidate licence to avoid revocation** – Ofcom has advised the Society that plans will be drawn up to revoke licences that have not been revalidated as required by the licence conditions. The quickest way to revalidate is to do so online via the Ofcom website: *https://services.ofcom.org.uk/* or by email: *amateur.validations@ofcom.org.uk* If you need assistance in the process, Ofcom staff are available to help, but please be patient during times of heavy workload.

Lester R M3RVQ
Lester S M6SAL
Letchford A M6KYE
Lethbridge L G3SXE
Lether A G4UQI
Letters P GM1VYF
Letters P MI0PJL
Lettington P 2E0PLL
Lettington P M6BBZ
Letts J G3URQ
Letts R G0LGA
Leung F M3XUF
Leung L M3SVF
Lever D G7VWA
Lever I G8CPJ
Lever J G4TEZ
Leverett W M1WTL
Leverett W M3WTL
Leveridge M 2E0LEV
Leverington P G8WWI
Leverington R M6BWF
Leverton D M3HGT
Leverton G M3XKY
Levesley J G0HJL
Leveton B G8UJO
Levetsky P M0PMM
Levett A G8HCJ
Levett B G3TXH
Levett C G4WYL
Levett D 2E0EBK
Levett D M6EUH
Levett J G3VTL
Levett M G4TSQ
Levey D 2D0UAY
Levey D MD3XSG
Levi R G3NQT
Levick W G7OLF
Levie B G8VZY
Levine G M0PAX
Levingston C G0XBQ
Levinson-Withall M 2E0BLG
Leviston C 2E0CPG
Leviston C M0KPW
Leviston C M6AYY
Leviston J G3NFB
Levitt A G4EFX
Levitt K G2FSJ
Levitt P G4HAI
Levitt P G8ITG
Levring E G0VZL
Levsen S 2E0SXF
Levy A G7UET
Levy A M5ALG
Levy D 2E0CVW
Levy D M0HWD
Levy D M6CWG
Levy M G4ACU
Levy M G8LHI
Levy P 2E0PLX
Lewczenko L M6LAI
Lewicky P M6CAJ
Lewin B 2E0DOW
Lewin B M0CEO
Lewin B M6LFE
Lewin C 2E0GHX
Lewin D G4PKT
Lewin G G0NEN
Lewin G 2M0DTJ
Lewin G MM6CYQ
Lewin P M3PLN
Lewin P M3RKZ
Lewin R G8OWA
Lewing D G8MWD
Lewington A M3VGF
Lewington S MM6DXJ
Lewinton-Smith K M3XZT
Lewis A G0TNQ
Lewis A G1ZSF
Lewis A G3XHM
Lewis A G4ELK
Lewis A G8YCI
Lewis A M1DKA
Lewis A 2W0RMY
Lewis A MW6ADZ
Lewis A G0JTU
Lewis A M6RTZ
Lewis A G7BSF
Lewis A 2E0PAL
Lewis A 2E0DQP
Lewis A M6DQP
Lewis A 2E0LAA
Lewis A G3FHT
Lewis A G6FIT
Lewis B G4SJH
Lewis B G4VPS
Lewis B M3BYL
Lewis B 2E0DDK
Lewis B 2E0VUK
Lewis B M3VUK
Lewis B M6DUY
Lewis C 2M0BMN
Lewis C 2E0CGL
Lewis C G3NHL
Lewis C G3YTL
Lewis C G7UTY
Lewis C M0EGN
Lewis C M3EYS
Lewis C M3LCL
Lewis C MM3NNO
Lewis C 2E0CWC
Lewis C M0SER
Lewis C M3ZCW
Lewis C G4CYY
Lewis C M0BEA
Lewis D G1FMW
Lewis D G1YCN

Lewis D G4CLC
Lewis D G4KPH
Lewis D G6SLY
Lewis D G4HBK
Lewis D GW7VSO
Lewis D G1AEQ
Lewis D M3DEE
Lewis D G4NIP
Lewis G G1TTK
Lewis G G4NKR
Lewis G G7KWP
Lewis G M3FSU
Lewis G 2E0RGL
Lewis G M0GYY
Lewis G M3TYC
Lewis G M0BCJ
Lewis G M6CYM
Lewis G M3WXG
Lewis G 2E0MET
Lewis I M1BUJ
Lewis I G7KGV
Lewis J G0CKU
Lewis J G0UUM
Lewis J G0UUN
Lewis J G1EEZ
Lewis J G3TCY
Lewis J G7KWO
Lewis J MM0CKF
Lewis J M0CMS
Lewis J M1BGK
Lewis J M3XKJ
Lewis J M3YMX
Lewis J MW0POB
Lewis J GW0RAD
Lewis J G3MYI
Lewis J MW6JFL
Lewis J G6KAY
Lewis J M6TKI
Lewis K GI1KDS
Lewis K MW0KDL
Lewis K MW3DJZ
Lewis K M6RLZ
Lewis L G7LBD
Lewis L 2E0LJL
Lewis L M0LJL
Lewis L M6GMU
Lewis L GW7KDU
Lewis M 2W0BGI
Lewis M 2I0LXW
Lewis M 2E0CRA
Lewis M GW0JDY
Lewis M G1LQH
Lewis M G1MDS
Lewis M G4CSE
Lewis M G4ZAC
Lewis M G7DGC
Lewis M MW0BRO
Lewis M M1CZF
Lewis M M3HBC
Lewis M MI3LXW
Lewis M MW3NUP
Lewis M 2E0MBK
Lewis M GW7KDU
Lewis M M3YZO
Lewis M M6MEL
Lewis M G0TWL
Lewis M G1UXW
Lewis M M6FEF
Lewis M M6RLA
Lewis M 2E0UAO
Lewis M M3UAO
Lewis M M6OML
Lewis N G4EPM
Lewis N MW0JGE
Lewis N M0NUG
Lewis N MW3ORY
Lewis N G1UQK
Lewis P 2E0BIY
Lewis P G3RQX
Lewis P G8MAV
Lewis P M3KIZ
Lewis P 2E0DGL
Lewis P G4APL
Lewis P MI1AIB
Lewis P M3IZB
Lewis P M6PFK
Lewis P G3WBI
Lewis P G4VFG
Lewis P G4ZFP
Lewis P G6NSU
Lewis P 2E0PGL
Lewis P M6PGL
Lewis R G1OXF
Lewis R GW3FZV
Lewis R G3YCO
Lewis R G4UIL
Lewis R G6EOK
Lewis R G6UNN
Lewis R G6WKI
Lewis R M6RLE
Lewis R GW0WRI
Lewis R G8RAU
Lewis S G4PLK
Lewis S MW3LGS
Lewis S M3XQO
Lewis S GM6BNS
Lewis S 2I0SHZ
Lewis S MI6MOI
Lewis T MW3DJV
Lewis T M6SNU
Lewis T M0ZUS
Lewis T M6TAL
Lewis T GW0KJT
Lewis V G4DQP
Lewis W GW3RTA
Lewis W 2W0RKF
Lewis W M3BJL

Lewis W MW6RKF
Lewkowicz S G1CMZ
Lexton H G4JDT
Lexton S G1UGO
Ley R 2E0GKM
Ley R M0GKG
Ley R M3RMV
Leybourne P MM5PSL
Leyland D 2E0DWF
Leyland D M6WDF
Leyland T G6NAJ
Leyshon P MW3ZSM
Li B M6LBQ
Li J M6KLK
Li J 2E0HNF
Li K G0TOY
Li L M3OGC
Li S M6LXH
Li Song Y 2E0YUN
Li Song Y M0LSY
Lickley A G7POV
Lickley A G1CMC
Lidbetter P G4NER
Liddard D M1DKL
Liddell A G7JWE
Liddell J MM6AHY
Liddell S GM4BGS
Liddiard N G1MJV
Liddle D GM1PSZ
Liddle G GM4OAS
Liddle K G7TRG
Liddle N G8KWH
Lidster A M6AYV
Lidster M M6PDI
Lidstone G G4OQ
Lie F M3MIU
Liepziger R G1OCK
Liffchak L G6ODW
Lifton A G0PEH
Liggins A G6NNA
Liggins T G0FYU
Light R G4NXI
Light S M0SCA
Light S M6SLD
Lightbody A G4CTY
Lightbody D MM6BOD
Lightbown E 2E0JQI
Lightfoot A G4ONJ
Lightfoot G G0WTI
Lightfoot J M3LJJ
Lightfoot J M5AGS
Lightfoot J M6UKB
Lightfoot N 2E0NOD
Lightfoot N M3NOD
Lightfoot P G0OFW
Lightfoot P G6ZHU
Lightfoot R G3MZN
Lightly A G6UXY
Lightly A G3VFL
Lightly J M3FTJ
Lihou N G8OVO
Likitplug P M6BTF
Liles S M3VXH
Lill N G4YWR
Lilley D G4IAU
Lilley G G4TYO
Lilley N G0TMZ
Lilley P 2E0DLT
Lilley R 2E0ISQ
Lilley R M0ISQ
Lilley R M3ISQ
Lilley S M6SIY
Lillingstone K G4SOP
Lillis A M6FSN
Lillis C G1VZB
Lilly A G8VGI
Lillywhite A G8MEM
Lima Barbosa M MW6DFX
Lima Matos R 2E0RUI
Lima Matos R M0RLM
Limb D G7MDT
Limb J G8FXX
Limb R M0FXX
Limb R G4PVY
Limbert I 2E0CYS
Limbert K M3KAY
Limbert K M3CYS
Limbert R G4IUP
Limbert T M3TNY
Limebear R G3RWL
Limehouse R G3WTN
Linacre T G4VKV
Linahan K MW6KLL
Lincoln A G4CIG
Lincoln E G3OAL
Lincoln J GM0JOL
Lincoln J GM8DFX
Linda M G4GTH
Lindell S G4IEH
Linden A M3RZL
Linden R GD4RFI
Linden S G0DUA
Lindenbergh M G1DKI
Lindgren B G0BMZ
Lindgren V G4BYG
Lindley A M0DEL
Lindley B G4ITQ
Lindley C M6WNC
Lindley D GW0BNO
Lindley J G8MQK
Lindley K G8NDK
Lindley M G0UNY
Lindley P 2E0GBP
Lindley R G7IAE
Lindon R M3TGE
Lindop B G3WUA

Lindop D G6XDN
Lindsay A G4NRD
Lindsay C G3KTZ
Lindsay C G4VJI
Lindsay C M3KDY
Lindsay C GM1MLW
Lindsay D G0PLZ
Lindsay D GM4HQF
Lindsay D G7CXM
Lindsay D MM0GDL
Lindsay G GM0BEL
Lindsay G G0KGL
Lindsay G G4KOT
Lindsay G G8BZL
Lindsay I GM0TWB
Lindsay I 2M0VVS
Lindsay I GM8LYQ
Lindsay I MM0IAL
Lindsay I MM3VVS
Lindsay I M0UWS
Lindsay J 2E0FCE
Lindsay J GM4ZRX
Lindsay K G0JWL
Lindsay K M0BPP
Lindsay M G0IYY
Lindsay P G4CLA
Lindsay R GI4AIO
Lindsay R GM4HUX
Lindsay S G0KDS
Lindsay-Smith R G3YBS
Lindsay-Smith S G1KXP
Lindsey P G3UDV
Line D G4ZRM
Line S G3XYO
Lineham P G8XQA
Linehan B G0UCK
Lines A M6AZD
Lines D M3TGZ
Lines G G7ISR
Lines J G6BEB
Lines J G6XBG
Lines M G1SEO
Lines P M3PML
Lines P GW4KOE
Lines R G8LQP
Linfield G M3GML
Linfoot A G1UVK
Linfoot J G0CPP
Linford J G3WGV
Linford R G3YQF
Ling A G7TXU
Ling H G4CCH
Ling H MM6HBR
Ling M M6LXP
Ling P 2E0HXZ
Ling R 2E0HXY
Ling S 2M0SPL
Ling S MM0SPL
Ling S MM3PZJ
Ling W 2E0EJD
Lingard D G0CLH
Lingard E G3WNQ
Linge N G6BVF
Lingham K M3KLY
Linham T M0BRH
Linkins B G8TXT
Linksted S GM7KMM
Linley E M0CDU
Linn B G1JTZ
Linn F MM6AUJ
Linnell D G0MJK
Linnett D M6KKM
Linnett G M6GFZ
Linney A G7RNX
Linney K G3UDA
Linney S G8GWR
Linney S M1AVV
Linsdall D G8TEQ
Linsley K G0PXF
Lintern D GW4VEB
Linton D GI0ITJ
Linton M G0IYM
Linton R G8XMZ
Linton S MI3SIL
Linton S G8JUT
Linton S M3YNN
Lintott R G0KDR
Linzey R G6KXB
Lionheart W M0WLH
Lipkowitz I MM3SDP
Lipman N G6ASA
Lippett A M1LIP
Lippett I 2E0ICI
Lipscomb R G0JBP
Lipton A M6FQO
Liptrott S G0HKZ
Liptrott S G4EGY
Lishman C 2E0BBL
Lishman C M3HJV
Lishman T G0WVD
Lisi D M3UHV
Lisi D M3YZK
Lisi S M6SLI
Lisle D G7VWO
Lisle M M6HUX
Lisle M GM4FPZ
Lisle R G0SLR
Lisney D G0FVT
List H 2E0OIN
List H M3OIN
List J 2E0BHJ
List J G8WLY
List J M3KYH
Lister A M3DGN
Lister D GW0ZDL
Lister H G7AJN

Lister H M3OCL
Lister J G7COG
Lister M G1JUI
Lister M G8FCQ
Lister M G8NYM
Lister P M3PLI
Lister R G8IXP
Lister T G0IEH
Liston-Brown I G8GSL
Litchfield D G0NQV
Litchman M G0TOC
Litchman M G7KJV
Litobarski S G6FDP
Litten H M3YAI
Little A G4PSO
Little A G7EYV
Little A M6CXW
Little A 2E0BVO
Little A M3YIV
Little B M6BDL
Little B GI4PID
Little C M3SXK
Little D G7PQP
Little D M3DFL
Little E M3LEK
Little E M3SSU
Little F G0HJB
Little J GM3KEZ
Little J M1SHE
Little J M3HSS
Little N GW3YVN
Little P G0WYV
Little R G0MZK
Little R G0NDC
Little S M0ABT
Little S G7KSQ
Little W 2E0WML
Little W M6FJV
Littleboy N G8YEQ
Littlechild B M6XXI
Littlechild G M6EYW
Littlechild P G0HWT
Littlechild S M6ZOG
Littlefair J MM6JCL
Littlefield R G4GYJ
Littleford A M3IXJ
Littleproud B G3AMK
Littler A G0LRM
Littler C G4YKH
Littler D G4RIE
Littler J G4HPH
Littler J G0LVY
Littler J G6TPG
Littler T MI3TBL
Littlewood D G6DCT
Littlewood D M0RCU
Littlewood J G6VKP
Littlewood J 2U0EJL
Littlewood J MU6GCI
Littlewood R G4YET
Littlewood R G3XLX
Littlewood S M3KYV
Littlewood V M3XQG
Littman C G8JKD
Litt-Wilson M M3MLI
Liu H M6LLW
Liu P M6XYY
Liu S M3WTO
Liu T 2E0BOV
Liu X M3UCL
Lively G G3KII
Livermore E G7LQP
Liversidge C M3UEJ
Liversidge D G4JMY
Liversidge G G0KYJ
Liversidge R M0CBN
Livesey B M6AEQ
Livesey C G0LJC
Livesey E M3IJF
Livesey G M6CGL
Livesey I G8NXJ
Livesey I M3NXJ
Livesey M M1MKL
Livesey M M6MLY
Livesey W G6MIU
Livesley A G8JAI
Livesley J G4CML
Livesley J G4YAB
Livings D M3TLO
Livingston G G3LAI
Livingston H GM0FTH
Livingston J G4FDD
Livingston M G6PIB
Livingstone C MM6CTL
Livingstone D GI0DQJ
Livingstone D G0PDE
Livingstone D MM6GHH
Livingstone J 2M0IRC
Livingstone J MM3RQC
Livingstone N 2E0EHB
Livingstone N M0NPL
Livingstone N M6EHB
Livingstone-Lawn C MW3UFH
Livsey D G4BQH
Livsey M 2E0AZB
Livsey P 2E0AZA
Lixenberg J G3MOL
Lles S G6YFG
Llewellyn C GW0RTP
Llewellyn F M3ETB
Llewellyn G MW3EPK
Llewellyn J G4JTM
Llewellyn M G0IVB
Llewellyn S G0ESO
Llewellyn S 2E0LAC
Llewellyn S M6SAA

Llewellyn T GM4RKH
Llewellyn W M1CHU
Llewelyn D G8VXU
Llewelyn G G3RXD
Lloyd A G1TLW
Lloyd A M3JAL
Lloyd A MW3VGJ
Lloyd A M6KPL
Lloyd B 2E0BGL
Lloyd B G1NNA
Lloyd B G6CLW
Lloyd B G8WPA
Lloyd D G1HRA
Lloyd D G6CXO
Lloyd D G6MJB
Lloyd D G7HII
Lloyd D G4HJT
Lloyd D G6CLX
Lloyd E G0HEJ
Lloyd F M1BQT
Lloyd G G1NNB
Lloyd G G4FNO
Lloyd G M3UDA
Lloyd G 2E0DYD
Lloyd G M6EKI
Lloyd H MW6FTC
Lloyd J G4NDD
Lloyd J G4OLS
Lloyd K 2E0AIT
Lloyd K G3XTP
LLoyd K G7BIQ
Lloyd K M0KLL
Lloyd L G0NTT
Lloyd M G1PKP
Lloyd M G3KLY
Lloyd M M0MVL
Lloyd P 2E0PGL
Lloyd P G1BLQ
Lloyd P GI8YJV
lloyd P M3FFV
Lloyd P 2E0LUD
Lloyd P M6PPZ
Lloyd R G4IQA
Lloyd R G1OKP
Lloyd S GW0JQT
Lloyd S GW1OII
Lloyd S G8IAM
Lloyd S MW0BBU
Lloyd T G6UED
Lloyd W 2E0WJL
Lloyd W 2W0SKG
Lloyd-Jones D G7SLJ
Lloyd-Owen J 2E0AMB
Lo S 2E0DUF
Lo S M6FYJ
Loach E G4ZXS
Loach F G4JUD
Loach G G4PVZ
Loach M G8UDJ
Loader A M6HXU
Loader E G6HXU
Loader J 2E0JKL
Loader M G8ONR
Loader P M6MGQ
Loader R M3LOA
Loades P M6ETG
Loake G G0GBI
Loane S 2E0IAS
Loasby A M3CQT
Lobb M G4EKV
Lobban F G4JQW
Lobo-Kazinczi P M6PAE
Loch P G0TCF
Lock A G3NWL
Lock A M0FBM
Lock C G4OJU
Lock D G7JTR
Lock D G7UZO
Lock D M6HKB
Lock D 2E0DGL
Lock D M6ENV
Lock M G4ZZP
Lock N M0CBF
Lock P G4STB
Lock P G7JUR
Lock P G8ZTN
Lock P M0RYB
Lock S G7CTT
Lock T G8ZZG
Locke D GW3TKG
Locke D G3LKV
Locke J G8VQQ
Locke J M3JHL
Locke M M6OTT
Locke R 2E0FBA
Locker J G7MIZ
Lockerbie C MM6EUA
Lockett D G7JTD
Lockett G M6ZRL
Lockett N G4EMB
Lockett T 2E0HDZ
Lockett T M3TLM
Lockey F G0RXQ
Lockhart B MI3WBL
Lockhart B G6UNX
Lockhart G G3NQZ
Lockitt M G7RXW
Lockley B M3SBT
Lockley M G4WAM
Lockley P M0DYH
Lockstone K M0KIL
Lockstone K M6LKI
Lockwood A G1YOF
Lockwood D G3ZTR
Lockwood D G8UAI
Lockwood D G4CLI

Lockwood J G3XLL
Lockwood M G6SCG
Lockwood M G1XCC
Lockwood P G8SLB
Lockwood P M3PGL
Lockwood P M6HCU
Lockwood P G7ELV
Lockwood T G4HZN
Lockyear C 2E0CNL
Lockyear C M3VNL
Lockyear E G8KZJ
Lockyear J 2E0JAO
Lockyear R M3EFX
Lockyer D 2E0DHD
Lockyer D M6ERQ
Lockyer I 2E0IAJ
Lockyer I M3INL
Lockyer J G0MQB
Lockyer J G0MUZ
Lockyer W MW6WRC
Locock D G7CQB
Loda P M0WPL
Lodeweegs W M1BDO
Lodge A M3IZN
Lodge A G4MGW
Lodge C G4IIK
Lodge D G8TOT
lodge H G7PPS
Lodge P G0TIA
Lodge S G0KWG
Lody B G0PCZ
Lofthouse N G1DJQ
Lofthouse S M3NWK
Lofthouse S 2E0RCI
Loftus C G4IFI
Logan A 2E0OSX
Logan A M0OSX
Logan A 2I0LXS
Logan A MI6CVW
Logan B G0VDN
Logan D G8OTZ
Logan D 2E0DOZ
Logan D M6DOZ
Logan G G0AVU
Logan J M3ZJL
Logan J M0OOT
Logan P MI3LDO
Logan R G4WXF
Logan T GM3VBT
Logan W G4EZF
Logsdon M G8FZI
Logue L MI3LGL
Logue R GI8ZDB
Loker T 2E0BVU
Loker T M0GPE
Loker T M3VUX
Lokuge R G0TAO
Lomas A G0VVG
Lomas A M1MLM
Lomas A M3EUW
Lomas D G4XOW
Lomas E G0KZO
Lomas G G4SYC
Lomas J G0UTC
Lomas P 2E0DNB
Lomas R G0OMZ
Lomas S G1NUN
Lomax M M0YYV
Lomax R 2E0MAN
Lomax R M3CUR
Lombao C 2E0DLV
Lombao C M0HQU
Lombao C M6DPU
Lomond H M0HMO
London D G0VGB
Londors A 2E0OKC
Loney S G1XBR
Long A G1SJT
Long A G3XDL
Long A GM4BRM
Long A G6CXI
Long B G0KUZ
Long B G4ULI
Long D 2E0TKF
Long F G3TUF
Long H G6LVB
Long I M6FBT
Long J G4IQZ
Long J G4HAG
Long J M1ENA
Long K G7JZY
Long L G1MGI
Long L M3UCY
Long M G6TPE
Long M GM7FTK
Long M G4BIN
Long N 2E0NJC
Long N M3VQZ
Long P G0PHE
Long P M3BRL
Long R M6RFL
Long S 2E0BST
Long S M3UEL
Long V G4DUM
Long W GM0VCN
Longbottom M 2E0AYK
Longbottom M M0EHL
Longhurst D G7UEI
Longhurst J G3VLH
Longhurst P G3ZVI
Longhurst P G7WBM
Longley M G6GVL
Longmore T 2E0LQR
Longmore T M0LQR
Longmore T M6TML
Longson D M6FWW
Longson M G0NMY

Longstaff B G1NTI
Longstaff D G4WCD
Longstaff J G3VJR
Longstaff P G6UAJ
Longton J G4WFE
Longuet A G8ZGQ
Lonie Jr C 2E0DCF
Lonie Jr C M0WCL
Lonie Jr C M6LJR
Lonnen M GM1FTG
Lonnon B G3ZUM
Lonsdale C G0FJH
Lonsdale I 2E0LON
Lonsdale I M0ZWT
Lonsdale I M6DFT
Lonsdale P M0LHR
Lonsdale P M1BZZ
Lonsdale S G7FFW
Looker R M3RHL
Looker S 2E0XSL
Loomes P 2E0HDU
Loomes P M3CKD
Loon D G1PMJ
Loose J G4BJS
Loose P M3LCU
Loosemore C G3WVM
Loraine T G0LOC
Loram B G0ILK
Lord A G0PCK
Lord A G1YEZ
Lord A G4KHT
Lord A G4PXC
Lord A G7RFA
Lord A GM7DAP
Lord A G7HNG
Lord A G8DQZ
Lord A M0DQZ
Lord B G7UCT
Lord C G3YUU
Lord E G3ZAU
Lord J G4XRK
Lord J 2E0SCD
Lord J M6FPC
Lord K 2E0GKL
Lord M G0PAS
Lord P G8PGI
Lord R GW3NCT
Lord S 2E0SXY
Lord S M3SIM
Lorenowicz M MM0WRO
Lorenzen J G7ORS
Lorimer C MM3FKO
Lorimer M M3EVC
Lorimer T GM0GTL
Lorimer T GM0IWX
Lorn T M0GZW
Lorne A M6HNX
Lorton J G4EKL
Lorton J G4KPI
Losardo N G1CTQ
Loseby R M6RBQ
Lote C 2E0CDL
Lote C M0RJK
Lote C M3KDL
Loten P G8KFK
Lott A G1AEU
Lott C M3HCL
Lott K 2E0KAL
Lott K M0KAR
Lott R G4FXE
Lotz J G3VUL
Loucks W G4MNI
Loud C M6HCG
Lough P G1BJK
Loughlin J G4DKQ
Loughlin J 2W0CJG
Loughlin J MW3YHW
Loughran B M3YAL
loughran C 2E0BEH
loughran C M0GGO
loughran C M3JOM
Loughran E GI4WNH
Loughran N M3OYQ
Loughran V GI6TBC
Loughren D MM6DUV
Loughren J MM6GON
Loughrey N GM4TTD
Loughrey N GI6JJR
Loukes A 2E0RTY
Loukes A M0JZG
Loukes A M6RTU
Loukes R G7MZA
Lovato T M6RPJ
Lovatt B G7LIE
Lovatt B M0CAD
Lovatt M G0JCN
Lovatt M 2E0LTT
Lovatt M M0LTT
Lovatt M M6MLA
Lovatt T G1XII
Love A G0EJR
Love D G1LFR
Love D G4RBQ
Love D G7ORK
Love D G0DRA
Love G G6XPF
Love J G0EPL
Love K G0GPB
Love K G1ESX
Love M 2E0BOM
Love M M3OVE
Love P 2E0DFQ
Love P G0KOK
Love P G6USA
Love R M0KSA
Love W G0NXQ

UK Surnames

Loveday G … G3IUV
Loveday J … GM0APN
Loveday J … M0IWF
[illegible] … G3LLG
Loveden P … M3TKU
Lovegreen A … GM4FLX
Lovegrove G … G7KLV
Lovejoy D … M6IRL
Lovejoy M … G3IXN
Lovelady P … G3LNL
Loveland L … G3KZX
Loveland P … G4SYB
Loveland R … G0PVI
Loveland R … G2ARU
Loveland T … M6FAJ
Lovell C … G3JUW
Lovell D … M3LCF
Lovell J … G8JHL
Lovell J … G3JKL
Lovell K … M3YRZ
Lovell K … M3YWO
Lovell L … M6TVC
Lovell R … G4FHN
Lovell R … G8LMC
Lovell S … G1BLJ
Lovell S … G8XPZ
Lovell T … M0ZLF
Lovell T … M6KLD
Lovell T … G6RNA
Lovelock J … 2E0LLO
Lovelock J … M0SOU
Lovelock J … M6LZP
Lovelock P … M5ABR
Lovelock T … G8DFU
Lovely N … G4RGE
Loveridge A … G0JGH
Loveridge J … M3YFX
Loveridge N … 2E0NRL
Loveridge R … GU1IIW
Lovering M … M6EWH
Lovesey R … 2E0RJL
Lovesey R … M0RBL
Lovesey R … M3LYQ
Lovesey S … G0DBM
Lovett C … G1MMN
Lovett P … G6HXZ
Loveys D … M6DFL
Lovius B … M3XVU
Low D … 2E0BZI
Low G … 2E0BZH
Low G … G0RLF
Low G … GM4LXM
Low G … GM4UZR
Low J … G0IDJ
Low M … G1ZZL
Low R … G0ECU
Low S … GM4KGZ
Low S … M0SJL
Lowcock I … M1DWQ
Lowcock K … 2E0VEK
Lowcock K … M3VYD
Lowden A … G3FIA
Lowder W … G0KYM
Lowe A … G1INA
Lowe A … G7DTS
Lowe A … M3VUP
Lowe A … M6YVN
Lowe B … G0RDR
Lowe B … G4PPC
Lowe B … G7PPC
Lowe B … G1UGJ
Lowe C … G0UMY
Lowe C … M6FFI
Lowe C … G1IVG
Lowe D … G3YER
Lowe D … M3IXO
Lowe D … G1IVF
Lowe E … M3WTP
Lowe G … M3FRT
Lowe G … M3LBN
Lowe G … G0NRA
Lowe I … 2E0BEV
Lowe I … G0PDZ
Lowe J … G1WHY
Lowe J … G3XZX
Lowe J … M3SJL
Lowe K … GW1BAI
Lowe K … M3KOJ
Lowe K … G4NZN
Lowe M … G0KKV
Lowe M … G0NZA
Lowe M … G4YCD
Lowe M … M3YVG
Lowe M … G7SRI
Lowe M … G1NNU
Lowe M … G1ZQV
Lowe M … G6JKF
Lowe M … M3YVF
Lowe P … G1EVR
Lowe P … M6RHV
LOWE R … G0FRL
Lowe R … G4ZAM
Lowe S … G0OYQ
Lowe S … M0DIL
Lowe S … M1CKK
Lowe S … M3SML
Lowe S … M3SWS
Lowe S … M0XZX
Lowe S … M6XQX
Lowe T … G3XWO
Lowe V … G1IND
Lowe V … M3IND
Lowe W … G0MLC
Lowe W … G4ONG
Lowenthal J … M6LOW
Lowery A … 2E0LXA
Lowery A … M6AXL
Lowes C … MW6HYS
Lowes G … G4NJW
Lowin M … M6MOK
Lowis R … M3HEL
Lown K … G4PTE
Lown S … G8WLL
Lowndes F … M6FML
Lowrie P … G6TBE
Lowrie P … GI7JYK
Lowrie P … MI5JYK
Lowson R … G2HKG
Lowther K … MW3IQE
Lowthian A … G6SQL
Lowthian H … M3WBR
Loxley D … M3LOX
Loxley W … G4LMV
Loxton P … M3OUH
Loyd A … 2E0LAV
Loyd A … M0XYX
Loyd S … G7TXR
Lubrani A … G6SBG
Luby M … M0DGK
Lucas A … G4LVA
Lucas B … G0TAR
Lucas C … G8ZBC
Lucas C … M6NDK
Lucas D … G0BIE
Lucas D … G4MAG
Lucas D … G8XND
Lucas D … G6AJW
Lucas D … G8DKI
Lucas D … G8XDD
Lucas F … G7IZM
Lucas G … GM4EJI
Lucas G … G0MJO
Lucas J … G8CXF
Lucas J … G8FRI
Lucas J … MM6MML
Lucas J … M6BIE
Lucas K … M0CUI
Lucas M … 2E0HAC
Lucas M … G7GRU
Lucas M … M3IXF
Lucas M … M3ZOO
Lucas M … M6MEN
Lucas P … MM3DDQ
Lucas S … M6EBX
Lucas S … 2E0JKB
Lucas S … M0JKB
Lucas S … M6JKB
Lucas S … G8APZ
Lucas S … G8PUB
Lucas W … G0TBS
Lucas Z … M6ZLA
Lucas-Davis E … G4YAS
Luchi E … MI6EAC
Luck L … G8UKH
Luckett C … 2E0CPL
Luckett C … M0UGR
Luckett C … M3UGR
Luckett E … M6ELZ
Luckett N … M6ETS
Luckett R … G6JAY
Luckhaus S … G4VGL
Lucking C … MM6PND
Lucking I … GD8RNM
Lucock B … G0LCJ
Lucock V … M6HJZ
Lucocq E … MW0AQT
Lucyk R … G8PXA
Ludar-Smith G … G8ZIW
Ludders P … M3LUD
Ludgate K … G1MSY
Ludlam S … M0SYM
Ludlow D … G4ETX
Ludlow G … 2E0GDL
Ludlow G … G8WWC
Ludlow J … GW3ZTH
Ludlow V … G3JLZ
Ludwell R … G3ZZQ
Ludziss S … M6LSV
luetchford B … M3GKE
Luff D … M1DRL
Luft P … G3SEZ
Lugard C … G4LZE
Lugg F … M0BHQ
Lugmayer E … G3KIJ
Lugsden J … M6BXL
Luhman G … G0ETA
Lui D … M6LST
Lukasz Z … M0LTP
Luke B … GW3XJC
Luke D … MW1BUN
Luke K … GW0HNE
Luker R … G1HKR
Lumb J … G0ARU
Lumb P … M6GSP
Lumbard S … G4YIF
Lumley D … 2E0EJU
Lumley T … G1SBZ
Lumm J … M6JHL
Lummis K … G6JZV
Lumsdon G … G0GKO
Lundberg O … G0CKV
Lundean B … G3ZHT
Lundegard T … G3GJW
Lunden N … M6HEJ
Lunn A … G4JAX
Lunn A … M3NUE
Lunn A … 2E0ZGL
Lunn A … G0EYE
Lunn A … G6UNU
Lunn B … 2E0VGK
Lunn B … M3VGK
Lunn D … G7TNO
Lunn P … M6UHT
Lunney K … MI1FHE
Lunnon C … GI7UBY
Lunnon R … G1MIY
Lunson [illegible] … [illegible]
Lunt G … G0EDF
Lunt M … G8ALD
Luper M … 2E0EFO
Lupton B … M0RXZ
Lupton F … G0MDR
Lupton I … G1ROD
Lupton K … G6ZHS
Lupton R … G1ARH
Lupton S … G7UOH
Lupton T … 2E0KEA
Lupton T … M3THY
Lupton W … 2E0EGU
Lurcook D … G4ERW
Luscombe D … G8JWC
Luscombe G … M6LUZ
Luscombe P … G0TDQ
Luscombe R … 2J0RZD
Luscombe R … MJ0RZD
Luscombe R … MJ3RZD
Lusted J … 2E0LSS
Lusty D … 2E0ERJ
Lusty E … G3YIE
Lusty G … G3DWI
Lusty R … M0AKS
Lutas P … G6VDK
Lutley A … 2E0DKP
Lutley A … M0LUT
Lutley A … M6LUT
Lutman P … G8YPN
Lutte N … M0NNL
Lutte N … M3YQQ
Luttrell M … M6MDL
Lux L … G4OHA
Luxton J … G0BYL
Luxton J … G6XYL
Luxton K … G0AYM
Luxton M … M3IEG
Luxton M … 2E0MCB
Luxton M … M3BTN
Ly H … 2E0HVL
Ly H … M6HVL
Lyall D … M6YDG
Lyall J … M6JLY
Lyall P … G8FRH
Lyall V … M6WOV
Lycett A … M6CGY
Lycett C … M0HCZ
Lycett C … M6HUC
Lycett D … G6BYL
Lycett J … G0MSZ
Lycett S … G1YFD
Lycett S … M6SPL
Lydall H … GM4DIZ
Lyddall R … 2E0DPZ
Lyddall R … M6LYD
Lydford A … M3NNA
Lye B … M6TER
Lye F … M1LYE
Lyes D … M6TFD
Lyford B … G0BPA
Lyford R … M0BBV
Lyle C … MW6OZF
Lyle G … GI0IUP
Lyman A … M6NPL
Lymer A … GM0DHD
Lynas D … MI0MEV
Lynch C … M6ZCL
Lynch D … G6VJM
Lynch D … M6GIQ
Lynch D … M3UUO
Lynch D … 2E0EVZ
Lynch D … M6EVZ
Lynch G … G0RTN
Lynch J … G0NXX
Lynch J … M1NXX
Lynch J … M5DZH
Lynch J … 2E0TBI
Lynch J … G3VBU
Lynch J … M0WAU
Lynch J … M6JPL
Lynch K … 2E0DUA
Lynch K … M6WDI
Lynch M … G4HKS
Lynch R … 2M1EJI
Lynch R … 2E0FIT
Lynch R … M0NVQ
Lynch R … MM1AWV
Lynch R … M6RCL
Lynch S … G0IKE
Lynch S … G8TRU
Lynch V … 2E0ZVL
Lynch V … M0LCR
Lynch V … M6ZVL
Lyne C … 2E0TCN
Lyne C … M0TCN
Lyne C … M0TIM
Lyne J … G0PEW
Lyne R … G0JTD
Lyne T … M0BPU
Lynn A … 2E0LYN
Lynn E … MI5LYN
Lynn J … 2E0JRN
Lynn J … M0JRL
Lynn M … M0LYI
Lynn M … G1KOT
Lyon C … G3EIZ
lyon C … M0ETS
Lyon D … M6XMX
Lyon G … G0TRY
Lyon G … 2E0CHY
Lyon G … M6AMV
Lyon J … M6SPC
Lyon M … G0WOA
Lyon O … M6LYO
Lyon P … G0DIC
[illegible] … 2E0XKX
Lyon S … M6HUI
Lyon-McKeil D … M0GVK
Lyons A … M0DRI
Lyons B … G4SOQ
Lyons B … G1ISS
Lyons D … 2E0DDO
Lyons D … M6CMI
Lyons E … GI6DCX
Lyons J … G0MLL
Lyons J … GW0WGN
Lyons J … G6NTE
Lyons N … G4JWV
Lyons R … G4ZZL
Lyons T … GI0PVG
Lyons T … GI7GHC
Lythaby A … G6KLF
Lythall A … M3FJD
Lythall D … M3FFK
Lythall R … G0TLA
Lythall R … G7KMF
Lythgoe G … M6DRG
Lythgoe J … G8CLY
Lytollis S … M3TFM
Lyttle A … GM4VGU
Lyttle G … MI3VCI
Lyttle P … G1FOM
Lyttle R … MI6RBP

## M

Ma G … M1GGG
Ma J … M3ZCJ
Maas R … 2E0IXH
Maas R … M3IXH
Maby C … G4NAQ
Mac Gregor Of Stirling M … G4EZG
MacAdam G … M5AGR
Macalister I … MM0TFU
Macarthur I … G3NUQ
Macaulay A … G7KLS
Macaulay B … G4ABX
MacAulay N … MM3FGH
MacAuley G … M3AQN
Macauley S … M0PMR
MacBeth M … G1MAC
Macbeth R … G8VLY
Macconnell D … GM0VYY
Maccormick P … G0VRY
MacDiarmid I … G6BGH
MacDiarmid W … GM4YMD
Macdonald A … G3NPM
MacDonald A … 2E0EHV
MacDonald A … M0VLT
MacDonald A … M6UAX
MacDonald A … 2M0CGE
Macdonald A … GM4TRH
Macdonald A … MM3FGL
Macdonald C … G4ZGM
Macdonald C … MM6CEM
Macdonald D … GM0BNQ
Macdonald D … GM0MGO
Macdonald D … GM8CIF
Macdonald D … GM0FSY
Macdonald E … G4TTY
Macdonald F … G1ABQ
Macdonald F … GI4VRF
Macdonald G … G4AZG
Macdonald I … 2M0IBW
Macdonald I … G1ABM
Macdonald I … GM7JED
Macdonald I … MM0SUS
MacDonald I … 2M0IAH
MacDonald I … 2E0IAL
MacDonald I … G8AVM
Macdonald I … M0IAD
MacDonald I … M6ITM
MacDonald I … M6LBK
MacDonald I … MM6UEN
Macdonald J … G1OTZ
MacDonald J … GM4XZN
Macdonald J … G8ZTR
MacDonald J … MM0IEJ
Macdonald J … MM0BED
Macdonald J … 2M0APB
MacDonald J … GM7REY
MacDonald K … MM5DOQ
Macdonald L … MM6GPD
Macdonald M … G0VME
Macdonald N … GM4BVU
Macdonald P … G0BOQ
MacDonald R … 2E0AIZ
MacDonald R … M6AGT
MacDonald S … 2M0VKO
MacDonald S … G0ELN
Macdonald S … GM4GUL
MacDonald S … MM3VKO
Macdonald S … 2E0FTH
Macdonald S … M3SZM
Macdonald S … G4AQB
MacDonald S … 2E0PSZ
MacDonald S … M0PSZ
MacDonald S … M6PSZ
MacDonald T … 2M0GOE
MacDonald T … MM3GOE
MacDonell M … 2E0XCO
MacDonell M … M0XCO
MacDonell M … M6LCF
MacDougall A … MM6AHN
Macdougall H … GM4FBU
MacDougall-Smith R … M6TNH
Macduff R … GM4AWB
Mace F … M6FMS
Mace J … G3ZTU
Macey C … G6SGM
Macey D … G6STD
Macey J … M3TKQ
Macey T … G8JIS
Macfadyen A … G3ZHZ
Macfarlane A … GM0MAC
Macfarlane I … 2E0TSO
Macfarlane I … M0TSA
MacFarlane J … 2I0PPW
MacFarlane J … MI0PPW
MacFarlane J … MI6MCF
Macfarlane N … GM7NYB
Macfarlane S … GM4ZWJ
Macfie R … G4FAX
Macgillivray K … GM4XKP
MacGra E … MI6BLW
macgra E … MI6DCE
Macgregor A … G3HMG
Macgregor A … G3LYG
MacGregor A … M3MXP
MacGregor D … G4IDJ
MacGregor D … M3SCQ
Macgregor G … GM7GBD
MacGurk E … 2E0EMG
MacGurk E … M0MAQ
MacGurk E … M3VTQ
Machalski J … M1FBW
Macham J … G0GOI
MacHardie I … G3YMV
Machen K … 2E0KMA
Machen K … M6KMA
Machin A … G6LFA
Machin F … G3LUZ
Machin G … M3ZGM
Machin J … M1JWM
Machin R … G6EOO
Machniak F … G4ZPH
Machon D … 2W0DRK
Machon D … MW0HXX
Machon D … MW6DRM
MacInnes A … MM3BCA
MacInnes A … MM6INN
Macintyre A … MM6HNQ
MacIntyre E … GI4DYE
Maciver A … G7LKR
Maciver D … G1SJU
MacIver I … G8FLS
mack G … M0CUS
Mack J … MM6NUP
Mackay A … M6AVR
Mackay A … G4OLK
Mackay A … M3UVV
Mackay C … 2E0MDC
Mackay C … GM0KVD
Mackay C … MM0BAC
Mackay C … M0TVL
Mackay D … GM1RLV
Mackay F … GM0WXX
Mackay G … M6RYK
Mackay J … GI4OCK
Mackay P … 2E0NDW
Mackay R … M1ACK
Mackay R … 2M0WIC
Mackay R … MM6NBI
Mackean R … GM4HAO
Macken D … G4DQA
Mackendrick A … M6NBL
Mackenny D … G0LJJ
MacKenzie A … G4NUO
Mackenzie A … M3ZTW
Mackenzie A … 2I0RMK
Mackenzie A … MI0RMK
Mackenzie A … GM0DKK
Mackenzie B … G0RWL
MacKenzie C … MM6FSQ
MacKenzie D … 2M0DNM
Mackenzie D … GM0TNK
Mackenzie D … GM4HJQ
MacKenzie G … GM4NUN
Mackenzie I … GM7GAE
MacKenzie J … 2M1ECF
Mackenzie J … GM0HJU
Mackenzie J … G6SLZ
Mackenzie J … MM6JCZ
Mackenzie J … 2I0TJK
Mackenzie J … MI0JBK
Mackenzie J … MI3TJK
Mackenzie K … 2M0UAL
Mackenzie K … M6KWA
Mackenzie K … MM6ULL
Mackenzie L … GM0TGG
Mackenzie M … GM4MFO
MacKenzie M … 2E0DNX
Mackenzie N … MM6BIO
Mackenzie N … GM3WIJ
Mackenzie N … GM4GTV
Mackenzie P … GM0HNV
Mackenzie S … G4XIW
Mackenzie S … MM6CFA
Mackenzie T … 2E0TMC
Mackenzie T … M6TIH
Mackenzie W … G0FSH
MacKenzie W … 2M0WMJ
Mackett J … M3BBF
Mackey R … G3SEY
Mackie G … G4NNB
Mackie P … MM0PWM
Mackie S … 2M1ETM
Mackie S … MM0DAA
Mackie W … G4AIE
MacKimm P … G8HDS
Mackimm S … M3SMK
Mackinlay A … G6IHW
Mackinnon D … GM0ADF
Mackinnon D … G1TFY
MacKinnon I … MM3IMK
Mackinnon J … GM0EUM
Mackinnon J … GM4EKC
Mackinnon M … MM0UIG
MacKinnon M … GM4AJV
Mackinnon [illegible] … [illegible]
MacKinnon N … G4WAZ
Mackinnon W … GM7GIO
Mackintosh C … M3YNC
Mackintosh K … MM0GKB
Mackinven P … G4TFH
Macklin B … G4BHE
MacKmin M … G6PUE
Macknish J … G0TZR
MacKrell G … GW3KAX
Mackrell R … G8ZGF
Maclaine E … GI4IWP
Maclaine T … MI6UTV
Maclean A … 2E0TDT
Maclean A … M6GKK
MacLean C … 2M0JAL
MacLean C … MM3AQW
Maclean D … GM4ODW
Maclean H … G0FEP
Maclean I … G0SRO
MacLean J … MM0CCC
Maclean J … MM6OEC
Maclean J … M6OOZ
Maclean K … GM6BGL
Maclean K … GM7JFN
MacLennan A … MM6IAY
Maclennan C … G1HKF
Maclennan D … G3KGM
Maclennan I … GM0DRU
Maclennan I … GM6BGJ
Maclennan M … GM4TJD
MacLennan S … GM4UMA
MacLeod A … 2M1GYX
Macleod A … MM0IOB
Macleod C … 2E0PDH
Macleod C … M6AOK
MacLeod C … GM6TYX
Macleod G … MM3SCO
Macleod J … GM0CBQ
MacLeod J … GM7DXT
MacLeod J … MM6SML
MacLeod M … 2M1EOV
Macleod M … MM1EWA
Macleod N … GM4DHN
Macleod R … GM4EWL
Macleod R … GM8DRA
MacLeod R … GM4DZX
Macleod R … M1ABT
MacLiesh C … G1FKS
MacIver D … GM6OHF
MacLucas N … MM0NDM
Macmanus E … G4VVE
Macmanus K … 2E0VVE
Macmillan B … M0SOE
Macmillan D … G3GNA
Macmillan D … MM3HMM
Macmillan S … GM7OMU
Macmillen P … G6SIX
MacMurray S … M6SQO
Macnab C … 2M1LPT
Macnab I … MM6PKC
Macnamara D … G1DON
MacNaught G … G3WOV
MacNaughton D … MM0DOT
Macnauton A … 2E0TMN
Macnauton A … M3HCE
Macnauton K … M3KMN
MacNee C … MM6MNE
Macolive P … G0LQW
Macphee J … GM3VNW
Macpherson A … G3WLA
Macpherson C … GM0EWX
Macpherson D … GM7OBM
Macpherson I … G3RXU
Macpherson I … G4ZPZ
Macpherson S … 2E0SUE
Macpherson S … MM0STU
Macrae A … M3XQM
Macrae K … MM6KHM
Macrae M … M6KZU
Macrides A … M0FXB
Macrides L … 2E0ZDJ
Macrobbie W … GM1BQP
MacRory R … GI4JTS
Macvean L … G4LES
MacWalter A … G3TSZ
Madagan P … G3RQZ
Madden A … M6BWG
Madden B … GI0RWO
Madden D … GM1OCZ
Madden D … G4ZPI
Madden J … GI0PXS
Madden J … GI6KCX
Madden P … GW8HYT
Madden V … MM0CXZ
Madden V … MI3UIV
Maddex C … [illegible]
Maddex E … 2E0YPK
Maddex V … G0RAS
Maddex V … G8YPK
Maddison D … 2E0DOM
Maddison F … 2E0IFA
Maddison M … G4HIC
Maddison R … G0GNE
Maddison R … G6RLM
Maddock R … MM3PFA
Maddock R … M0PNZ
Maddock R … M3PNZ
Maddox B … M3TID
Maddox P … G6PHZ
Maddox W … M0WCM
Maddy K … M6KZM
Madely F … G0VBP
Madge A … M3TQA
Madge C … G8OXU
Madore D … G1HRQ
Madsen R … MW3ZSO
Mady E … 2E0ZHN
Maennel P … M0HAN
Magee A … GI0BFD
Magee D … G4HAJ
Magee M … MM3UTU
Magee P … MI6OYB
MaGee S … G7VFJ
Magee S … 2M0HYZ
Magee T … GI0BEB
Magee W … GI3XLK
Mageehan D … M1CZY
Mageehan D … M3CZY
Magfhogartai L … GW0HGP
Maggs B … 2E0BJM
Maggs B … M0NKS
Maggs C … 2W0PHP
Maggs C … MW3XZP
Maggs P … G0OVY
Magill B … G3RMF
Magill H … GI0LGV
Magill I … GI4HCX
Magill S … GI0LWO
Magnall M … M6ABV
Magness I … M0HUI
Magnus A … G3ZFT
Magnus-Watson P … G6FCJ
Magnuszewski E … G6UAP
Magowan D … GI7OOM
Magri R … G0NCH
Magrys S … 2E0IQT
Maguire A … 2E0ZMO
Maguire A … M3ZMO
Maguire B … MI6FQY
Maguire D … GI0BQX
Maguire G … 2E0JED
Maguire J … 2E0XTM
Maguire J … GI4UHA
Maguire J … M3EEJ
Maguire J … M3XTM
Maguire J … GM0PQV
Maguire P … MI6PMM
Maguire R … MI1RAY
Maguire S … MI6SMM
Maguire T … G0NHP
Magwood R … GW1XZI
Mahany B … G4XAG
Mahany D … G4XAH
Maheshwarappa M … M6MFM
Mahon J … G0VVQ
Mahon R … M0RSM
Mahoney G … GW6NHB
Mahoney I … M6EIS
Mahoney J … G4KKT
Mahoney J … G1BAL
Mahoney J … M0MHY
Mahoney J … G6FCL
Mahoney K … M1KEV
Mahoney M … M0MAI
Mahoney P … GD8UMY
Mahoney R … M3EGF
Mahoney W … G3TZM
Mahony C … G6INI
Mahood K … G0OXV
Mahood P … GM8LYO
mahorney R … M0WGA
Mahrer P … GJ0KYZ
Maiden A … 2E0TZD
Maiden A … M6ZTD
Maiden J … G8STI
MAIDMENT J … G8FWE
Maile P … MI0BME
Main A … GM0XAV
Main A … MM6BXQ
main J … M3LPR
Main R … GM1BNA
Main T … GM4DCL
Maines J … G8RAC
Mainhood D … G3HZW
Mainwaring A … G4ZXQ
Mainwaring D … 2E0ZMM
Mainwaring D … M3ZMM
Mainwaring G … MW3GTM
Mainwaring M … 2W0BKA
Mainwaring M … MW0HAB
Mainzer S … M3KWF
Mair C … GM7NUQ
Mair J … GM0JFH
Mair K … 2M0KVM
Mair K … MM6KCM
Mairos A … G4YPI
Mairs H … MI3BRX
Maisey P … G4XPV
Maish A … G4ADM
Maitland A … MM3WNH
Maitland G … G3XYB
Maitland R … [illegible]
Majhail M … 2E0MMJ
Majhail M … M0MMJ
Majhail M … M6MMQ
Majoch A … M0VQP
Majoch A … M3VQP
Majoch E … M6HDG
Major A … G4LOB
Major B … G6VNW
Major C … G7UYB
Major G … M0BZO
Major M … GU7PVI
Major N … G0HFL
Makeham B … G4BQC
Makepeace R … G0TRJ
Makin J … G1ARF
Makin M … M6KIN
Makins S … M3ZXC
Makkinje J … M1DBL
Maksimavicius V … MI0GOZ
Malbon A … G8MIA
Malcher A … GD4TFM
Malcolm A … G8DEC
Malcolm C … G3UYN
Malcolm D … M6MLV
Malcolm I … GM3VWY
Malcolm R … GM4SXJ
Malcolm-Brown W … 2E0SUU
Malcolm-Brown W … M0SUU
Malcolm-Brown W … M6WDS
Malcom J … GD6SYB
Male B … M3OQJ
Male R … GM1JWJ
Malekout D … G6EEF
Males J … G4DEW
Males S … G0EVZ
Maley B … G7JWX
Maley I … M6IMA
Malhi A … G1TBN
Malia J … M0JAQ
Malia J … M6DAM
Malik A … 2E0HAQ
Malik T … M6BLS
Malin A … M6ATM
Malin W … M3DYL
Malins D … MW6DZR
Mallaband T … M0TPJ
Mallalieu Howard D … G6LVS
Mallender R … M3AZS
Mallet D … G4HIF
Mallett A … M1CVT
Mallett D … G3HUL
Mallett P … G0BEJ
Mallett R … G0NJD
Mallett T … G6LCL
Malley J … M0MLY
Mallichan J … 2E0IJQ
Mallichan J … G6SXC
Mallichan J … M0FMY
Mallin A … G1KVQ
Mallin D … G1MOW
Mallin H … M1ABG
Mallinson B … G4MYW
Mallinson G … G3NAK
Mallinson G … M0DVW
Mallinson M … G0LDB
Mallinson R … G0EQI
Mallinson S … M3SGI
Mallinson S … M3UUS
Mallolm G … MM3LWO
Mallory C … 2E0VCM
Mallory C … G6MYH
Mallory C … M3VCM
Mallows A … G4FUZ
Mallows C … G8NTY
Mallows F … G1GYJ
Mallows J … M1WRX
Mallows K … M1KTY
Mallows V … G3TSM
Malme P … G4PQP
Malone J … GM7DZK
Malone M … G7WBZ
Malone M … M6TGF
Malone P … M3PHJ
Maloney E … G7NEE
Maloney J … M3AHQ
Maloney R … G0OKN
Maloney S … G0NYZ
Malpas H … M3KQW
Malpas R … GW0DKG
Malpass A … G7CAG
Malpass P … G7FFM
Malpass S … G0OGS
maltas I … M1DHJ
Maltby C … G2HLB
Maltby I … M6PBY
Maltby J … G4FGY
Maltby R … G8RDG
Maltby R … M3ECF
Maltman M … MM3NPG
Malyon M … G6NHA
Malyon N … 2E0DVX
Malyon N … M6DVX
Mamijs G … MI6GTY
Man H … G6KNU
Man J … G0NSI
Man L … G8RTK
Man L … M6LCQ
Manchester G … M3PRS
Manchett B … G4NZC
Mandal S … M1BJB
Mandall C … G1ARL
Mander A … G8HPY
MANDER G … G6YHE
Mander R … G4DYY
Mander V … G8HOS
Manderson L … G4IXL
Mandeville L … [illegible]
Mandville C … M6RCE
Manekshaw M … GM4UKG
Manford P … G8XGK
Mangan J … GI0DPV
Mangan M … G1FON
Mangan S … G1WQN
Mangles T … G4STP
Mankin A … GM7BDD
Manklow C … G4CGV
Manktelow K … GD3SKZ
Manley F … M1CNJ
Manley M … G4SMM
Manley M … 2E0RMR



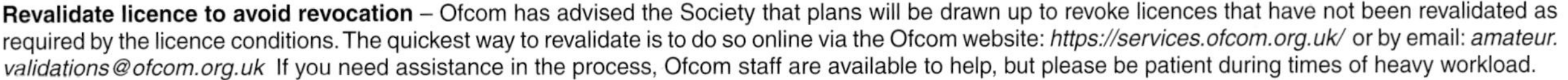
**IMPORTANT NOTE**
**Revalidate licence to avoid revocation** – Ofcom has advised the Society that plans will be drawn up to revoke licences that have not been revalidated as required by the licence conditions. The quickest way to revalidate is to do so online via the Ofcom website: *https://services.ofcom.org.uk/* or by email: *amateur.validations@ofcom.org.uk* If you need assistance in the process, Ofcom staff are available to help, but please be patient during times of heavy workload.

Manley S M3FTE
Mann A G4CMY
Mann C 2E0BFX
Mann C G6TFP
Mann D M0DLM
Mann E M6ANN
Mann E G8EKH
Mann F GM0KUP
Mann G G0VHH
Mann J G7KYH
Mann J MM0JOM
Mann K G7OGO
Mann L M6LGM
Mann N G1BXL
Mann N MM3TPF
Mann P G6WWV
Mann P G7OJA
Mann P M0BRE
Mann R G1GKF
Mann S G8PUK
Mann S G6XID
Mann T G7MSK
Mann T G7IYM
Mannerfelt W M0FCA
Mannering P 2E0BAH
Mannifield R 2M0CFA
Mannifield R MM0GXY
Mannifield R MM3RKF
Manning A G1XBX
Manning A G1ZBJ
Manning A M0UCK
Manning B 2E0HSD
Manning C GW1SMT
Manning C M6BTQ
Manning D G0KBM
Manning D 2E0DSK
Manning D M6DCQ
Manning F G0IQH
Manning G GD4GLM
Manning J G6KWZ
Manning J M3XNO
Manning M G0IGM
Manning P G1LKJ
Manning P G6ZPL
Manning P M3LKJ
Manning R G8UBX
Manning R G8VLZ
Manning R M6ZPL
Manning R G8TAY
Manning R M6BOB
Manning S G1IRG
Mannion D M3HZE
Mannion R G3XFD
Mannix D G3XER
Mannix E G8OXI
Mannock R M0RFM
Manos K M0HIH
Mansel R G6AWO
Mansell D G8SFT
Mansell G M0CRO
MANSELL J G7GEL
Mansell M G4MAN
Mansell P GM3VKN
Mansell R G3TZD
Mansell R M0RXV
Mansell R G0OVK
Manser M M3RIU
Manser P G8NQC
Manser R G3LFV
Manser R G1DMR
Manser R M6RFV
Manser R M0GIF
Mansfield A G4YIZ
Mansfield A 2E0GGM
Mansfield A M0YGG
Mansfield A M6GGM
Mansfield C G0EFA
Mansfield D M6ANE
Mansfield G M6MFD
Mansfield L G0WQY
Mansfield L G8MXT
Mansfield M M0AWD
Mansfield N G6ZPV
Mansfield P 2E0PMV
Mansfield P M0PMV
Mansfield P M6PMZ
Mansfield P 2E0PBJ
Mansfield P M6PSM
Mansfield P M3GPM
Mansfield R 2E0RMZ
Mansfield R M0RMZ
Mansfield S M6XSR
Mansfield T G4KAU
Mansley J G1FKT
Manson A MM6BOS
Manson I GM1PSU
Manson R G1PWS
Mantell I G4NRY
Mantell S G4NRX
Manthorp W G7BLT
Mantle G G0OYY
Mantle G G7ACJ
Mantle R G8ZAD
Manton R G4ILQ
Mantovani G G4ZVB
Manwaring J G7UXQ
Mape G G7PXS
Mapeley D M0BZK
Maple P G0BPZ
Mapp B G3NVP
Mappin D G4EDR
Mappin D G8HWQ
Mapson M G3UUI
Maqbool S M6SKL
Maqbool S M6BMY
Marbus W G0ERL
March D G4DYT
March E G8EOJ
March N M3NKO
March P G7MFX
March P 2E0PDM
March P M6MTR
March P G8FMT
March P M0FMT
March S G4TRI
Marchant B G0SMH
Marchant C MW6FYY
Marchant D G8JQV
Marchant P G3UWM
Marchant P G4OIM
Marchant P 2E0DBI
Marchant P M0WAF
Marchant P M3UEF
Marchant R G3TAJ
Marchant R G6YHF
Marchington A G7AXL
Marchington R G4MRQ
Marchini S G4TOZ
Marcomini S 2E0MJO
Marcot S M1ENE
Marcus C M6NMW
Marcus E GM4YRE
Marcus J M1CXA
Marden D M6FJM
Marden S M6HKO
Mardle D G6WCX
Mardle P G0HWI
Mardlin D GM6JBF
Mardlin D 2E0DWM
Mardlin D M3TXG
Mardo A G0OED
Marflow C G3VWA
Margaswamy A M3WJY
Margetts C G7VJM
Margetts C M0BQE
Margetts E GM4BOA
Margetts I M3IMM
Margetts M M6MGM
Margetts R G8XRG
Margolis L G3UML
Margolis R G4TTZ
Margrave F G4VRG
Margrave S M6ENZ
Marinho A G0XBG
Marino M 2M0WMU
Mariott S M3PRY
Maris A G3XDK
Marjoram D G0JVT
Marjoram D M6TOR
Mark W MI3UHL
Markeson G G0IOK
Markettos A 2E0BUV
Markey B G0NMH
Markey D M6SUP
Markey G M0GGM
Markey S MM3FMB
Markfort R G8TQJ
Markham G G0OZR
Markham J G6INF
Markham K M6GNM
Markham P G0OSO
Markham P G0GYY
Markham R G0KYN
Markham R M0AGT
Markley N G7HUG
Marks A G7DPV
Marks A 2E0BLJ
Marks A M3SWQ
Marks B 2E0WHB
Marks B M6UHN
Marks C G4ZPJ
Marks J G6GVH
Marks J G7FPJ
Marks J G8WGN
Marks J 2E0DQO
Marks J M6EZS
Marks P G8UGS
Marks P G8VZZ
Marks R G4ZFC
Marks V G6IHO
Markwick S G0MUC
Marland J M3YTA
Marlborough A M3RLB
Marles S M3NAQ
Marlett J M0JGM
Marley G G0VFV
Marley G M0MGK
Marley N 2E0CYC
Marley N M0HQP
Marley N M6BOE
Marlow A 2E0PMX
Marlow A M3PMX
Marlow C G6NVJ
Marlow C M6BUY
Marlow D G4OMV
Marlow E 2E0HBT
Marlow J G0LDR
Marlow J G1ORP
Marlow J M3ORP
Marlow K G7AFQ
Marlow M G0DUS
Marlow P G8BTV
Marlow P M6MQP
Marlow S G7ITU
Marney P 2E0PDM
Marobin L G0OOS
Marquardt L G1VDW
Marques Gomes A MM0KZA
Marquis I G0FPV
Marquiss R G8SQP
Marr J 2E0ZML
Marr J G4WUI
Marr S 2E0GEB
Marr S M6HEE
Marrable P G4DTM
Marrai F M1CPD
Marren C M1TRC
Marriott A GM0GFL
Marriott A G1EFF
Marriott A G3VWC
Marriott A G7GTH
Marriott A M6BYS
Marriott C G4RCP
Marriott D G6IQP
Marriott G G4ZSP
Marriott H G4ERT
Marriott J G8BFH
Marriott J G8VWU
Marriott J M6OJM
Marriott J M6JJM
Marriott K M5ADM
Marriott K G6NHY
Marriott L 2E0LJK
Marriott L M0LJK
Marriott L M6LJM
Marriott L G4FFE
Marriott M G0OPC
Marriott N MM5YLO
Marriott P GM0VXA
Marriott P M3GJN
Marriott P GW0RUD
Marriott S M0SHM
Marriott-Levett J M6TXH
Marrison A M3UVY
Marron J G7WEN
Marron J M0ASN
Marrows A G4REC
Marrs M M6EVU
Mars A M6TUT
Marsden A G3NTD
Marsden A GM4FEI
Marsden A M6AXX
Marsden D G0DUG
Marsden D G1ZQE
Marsden D G3ZHP
Marsden D G4RMC
Marsden G M0FDX
Marsden G 2E0GMA
Marsden G M6GMA
Marsden J G0RRR
Marsden J G7KKW
Marsden K G7CLX
Marsden M G4AXX
Marsden M G8BQH
Marsden N G4BQN
Marsden P GD3LGQ
Marsden W G0PDK
Marsden W G4YJY
Marsden W G8CET
Marsh A M6JWF
Marsh B G1JOA
Marsh B M6JUC
Marsh C G0PXB
Marsh C G1PHV
Marsh C G8IYN
Marsh C GW8REV
Marsh C 2E0CRS
Marsh C G4HKQ
Marsh C M6CPM
Marsh D 2E0ICU
MARSH D M3FWR
Marsh D M3TIZ
Marsh E 2E0EJM
Marsh E G1XKY
Marsh E G3ILE
Marsh E M3OFS
Marsh G M3XUW
Marsh G GM4EKI
Marsh I 2E0IAG
Marsh I G4EXD
Marsh I G7FUV
Marsh I M0UAT
Marsh I M3ZBQ
Marsh J GD0IPK
Marsh J G3SYG
Marsh J G7BRB
Marsh J M3YEJ
Marsh J 2E0MIU
Marsh J M0JFM
Marsh J M6MIU
Marsh J 2M0DOL
Marsh J MM6MIS
Marsh K G7JUC
Marsh K G8NEI
Marsh K M3JUC
Marsh K M3KJM
Marsh K M0AID
Marsh L G6DKF
Marsh L M3XYN
Marsh M 2E0MJP
Marsh M G4GGC
Marsh M 2E0MEZ
Marsh M M6DZP
Marsh N G1SGR
Marsh N G1ZBL
Marsh N M3NJM
Marsh P 2E0PIP
Marsh P G4WFZ
Marsh P M3OUI
Marsh P M0EYT
Marsh R G4GRZ
Marsh R G4YZK
Marsh R G8TYH
Marsh R M3YBJ
Marsh S G1BNG
Marsh S M0CUD
Marsh S G4BWG
Marsh S M6MSO
Marshall A GW0FJQ
Marshall A M1PMR
Marshall A G6MHF
Marshall A G6LUU
Marshall A M6SFR
Marshall A 2E0TBD
Marshall A M3UIS
Marshall A G3MJM
Marshall B G1DVD
Marshall B G4BJF
Marshall C 2E0ETH
Marshall C G4IOK
Marshall C M0NAK
Marshall C M6BNQ
Marshall C G7JJJ
Marshall D 2E0IBJ
Marshall D G0NQW
Marshall D M0DMA
Marshall D MM1ELE
Marshall D 2E0IUI
Marshall D G8MGD
Marshall D M0AXE
Marshall D M3ZUD
Marshall E G4PPB
Marshall F G1MVG
Marshall F G2XQ
Marshall G G0FEJ
Marshall G G1WRH
Marshall G G3HWS
Marshall G G3ZDU
Marshall G GM4GVJ
Marshall G G6HLR
Marshall G 2E0BNI
Marshall G M3ORZ
Marshall G G6MLH
Marshall H G0JMN
Marshall H G6JTW
Marshall I 2E0AOZ
Marshall J G0NPU
Marshall J GM1FAF
Marshall J G4KFP
Marshall J G4MFV
Marshall J G4MHF
Marshall J G8ZXT
Marshall J G3RKH
Marshall J G8MGO
Marshall J M3RWI
Marshall J M6FRG
Marshall K G0OGJ
Marshall K G3ZTI
Marshall K G4IIB
Marshall K G4LNQ
Marshall K G7NDB
Marshall K G7TLR
Marshall K M3YHD
Marshall L G1EPF
Marshall L M3LHM
Marshall L M3ZQV
Marshall N G0XTM
Marshall N M0BVY
Marshall P G8MOL
Marshall P M1BBB
Marshall P M1BJC
Marshall P G0MLF
Marshall P G8SGH
Marshall P M6CCV
Marshall P M6EHR
Marshall P G0BBK
Marshall R 2E0CAW
Marshall R G0PFY
Marshall R GM3RXZ
Marshall R GM4GIO
Marshall R GD4KEW
Marshall R G7PHL
Marshall R G8GFA
Marshall R G8HLE
Marshall R G3SBA
Marshall R G4ERP
Marshall S G0UAI
Marshall S M3YSM
Marshall S G0JKJ
Marshall S G1RYQ
Marshall S G3YXH
Marshall S M0SKM
Marshall S G6NHG
Marshall T 2E0GZT
Marshall T M6GZT
Marshall W G4IOD
Marshall W M6WCV
Marshallsay M G8DYG
Marshland B M0SWD
Marshman J G4LIO
Marshman M G6GYF
Marsland D M3JIH
Marsland L G0DBE
Marsland T M6TMM
Marsters D G1KIJ
Marston A G1FTH
Marston A G7VDU
Marston A G4UIF
Marston B 2E0BCM
Marston B M3VDU
Marston D MW0CRI
Marston F G0ZIP
Marston F G7VIP
Marston M G7EFG
Marston M M3PZC
Marston M M1EZT
Marston N G0ASM
Marston R GW1WRV
Mart P G8NHD
Marten T G0AOA
Marter J 2E0JAC
Marter J M3JRM
Marter M 2E0MJM
Marter M M3MJM
Martich K GW4YKM
Martin A G4HBV
Martin A G7JRU
Martin A GM7ONJ
Martin A G8ZPW
Martin A M0BMZ
Martin A M6DWO
Martin A M6ROF
Martin A MM6AAA
Martin A 2W0KOP
Martin B M6BMW
Martin C G0RYP
Martin C G0UAC
Martin C G0XAU
Martin C G7KNQ
MARTIN C M0MTN
Martin C M3CIJ
Martin C MM3EDW
Martin C M6CGM
Martin C 2E0PPM
Martin C M0ECM
Martin C M6CJM
Martin C M6ROC
Martin C 2E0LIT
Martin C M0LIT
Martin C M1EIU
Martin C G7GGF
MARTIN D 2M1EDM
Martin D G1IJQ
Martin D G1YEV
Martin D G3RUZ
Martin D G4DBZ
Martin D G4RST
Martin D G6HKL
Martin D G6OZH
Martin D G7AEY
Martin D G7HOL
Martin D G8IOJ
Martin D M6DXS
Martin D 2E0DCM
Martin D G0OKA
Martin D MI0AIH
Martin D M0EMM
Martin D M6CPJ
Martin D G0ORO
Martin D 2M1EBJ
Martin D 2E0WOW
Martin D M6PGN
Martin D G3ODC
Martin De La Fuente M M0HAO
Martin E G0LAA
Martin E G0TKR
Martin E G0WYU
Martin E M6NMD
Martin E M6LIZ
Martin E G0UBO
Martin F M3LYZ
Martin G GM3NVQ
Martin G GI3XCZ
Martin G G4KOU
Martin G G6GQF
Martin G G0HXR
Martin G 2E0JAN
Martin G M3OUL
Martin G 2E0GDM
Martin H G3JDO
Martin H G3VES
Martin H G3XCD
Martin H G6JTI
Martin H M0CNW
Martin H M6MRG
Martin H GM4UYE
Martin I G6CKL
Martin I G6ZNO
Martin I G6DQO
Martin I G8RVZ
Martin J 2E0JJM
Martin J G1KIB
Martin J G4GWE
Martin J GD4RAG
Martin J G4TMQ
Martin J G4ULM
Martin J G6HIV
Martin J G7PDR
Martin J G8JGM
Martin J G8XLB
Martin J MI3BYQ
Martin J M3JJM
Martin J M3JMY
Martin J 2E0XAV
Martin J M3XSU
Martin J M6GYM
Martin J MM0BQI
Martin J 2I0OMA
Martin J 2E0VTK
Martin J G4ECE
Martin J G4NLJ
Martin J G7VEY
Martin J M0VTK
Martin J M3VTK
Martin J M3ZHQ
Martin J M6CEG
Martin J MI6MPH
Martin J MI6MRJ
Martin K 2W0EGK
Martin K G4UBK
Martin K MM0KJM
Martin K MI3BUT
Martin K M3KVL
Martin K G0GFQ
Martin K G1MIE
Martin K G0PQO
Martin K G7CUY
Martin L GM0FVJ
Martin L M3ZHM
Martin L 2E0LGG
Martin L M6LDM
Martin M G0JCZ
Martin M M0LAW
Martin M M3MWM
Martin M M6GQX
Martin M 2E0GWM
Martin M MM3YTB
Martin M M6FDN
Martin M M6GWM
Martin N G6HVE
Martin N M1BTO
Martin N G6NHK
Martin P G0BUW
Martin P G3PSU
Martin P G3VLW
Martin P G4KHK
Martin P G6EON
Martin P G6GVM
Martin P GW7LDP
Martin P G7NJM
Martin P G8DZC
Martin P G8LZS
Martin P G8YPL
Martin P M0NJM
Martin P G0NXY
Martin P G4AZC
Martin P M0KDX
Martin P 2E0CYS
Martin P 2E0PME
Martin P G0CTR
Martin P G4ISJ
Martin P M0NWI
Martin P M3OHN
Martin P MW6PBF
Martin P M6PWK
Martin P M6VDX
Martin P M6BGF
Martin P MU6CPV
Martin P M6NIE
Martin R GW0RWM
Martin R G0UNW
Martin R G0WKL
Martin R G6CKK
Martin R G7IED
Martin R G7SDG
Martin R G8EBQ
Martin R MI3AIN
Martin R G7TIN
Martin R G8VQN
Martin R M0LLY
Martin R G3WKH
Martin R M6MRV
Martin R G6NAV
Martin S 2E0BKP
Martin S 2E0TVR
Martin S 2E0BKP
Martin S G0JPI
Martin S G1KNI
Martin S M3PRN
Martin S 2M0TIA
Martin S MM0TIA
Martin S M0GOT
Martin S M3NMJ
Martin S G6WHH
Martin S G7HON
Martin S M0PDP
Martin S M6OBS
Martin S M3AIS
Martin S G8OGP
Martin T 2E0LYR
Martin T G1CKJ
Martin T GI4JWW
Martin T G7CEN
Martin T M0CLH
Martin T M3LYR
Martin T M3TYM
Martin W M0CZN
Martin W 2I0EKN
Martin W M6BQO
Martindale A G3MYA
Martindale J G4VPA
Martindale P G7OHD
Martindale W M6DQT
Martinelli V G0NNT
Martinez J G3PLX
Martinez K M6KMZ
Martins A M0PAM
Martins S M3CDE
MARTLAND A G1AHM
Martland D G0PXL
Martlew B G1NXS
Martorano G G4NNZ
Martyn A 2E0CAK
Martyn A M6ARM
Martyn Clark J G6KTO
Martyn J G7JAN
Martyn L G7PUZ
Martyr D 2E0FYC
Martyr S 2E0SAM
Marvelley S GW4TGA
Marvill J G0UFW
Marwick D GM1RQD
Marwood A G8SSL
Mascall H G7LNY
Mase B G3GLA
Maskell K G4YTU
Maskew R G4TDW
Maskill R G4PYR
Maskort G G0DKO
Maskrey S G6FDK
Maskrey S M3DVA
Maslen I G4BYR
Maslen W GM4LPG
Maslin A G6VJK
Mason A G1NKN
Mason A M1MTV
Mason A M3IUK
Mason A M3NNI
Mason A G8FSV
Mason A GW0TKX
Mason A M6VAP
Mason A G6JSR
Mason B M3OJY
Mason C 2E0BWN
Mason C GD4UZE
Mason C M3EHH
Mason C G0HRJ
Mason D G0PBH
Mason D G0SOV
Mason D 2E0ZDM
Mason D G3RXP
Mason D G3USD
Mason D G3ZPR
Mason D G6EWK
Mason D G4NDC
Mason D M6DTS
Mason E G0ASP
Mason E G0PXX
Mason F G3TWN
Mason G G0OSC
Mason G G3YJG
Mason G G4IWF
Mason G G4PTK
Mason G G4VCA
Mason G G7VIL
Mason G M6SPU
Mason H G4AOA
Mason H MM0HLN
Mason I G0EWZ
Mason J 2E0FNQ
Mason J GW0VSW
Mason J G8NWL
Mason J M3KJK
Mason J M3ZHY
Mason J M6JMN
Mason J 2E0CVP
Mason J G8YFK
Mason J M6CYO
Mason J G3ZJZ
Mason J G7PBV
Mason J MM3VVI
Mason K G4JIO
Mason K M1NEW
Mason K M6KLM
Mason L G1CBL
Mason L M1AJA
Mason L G4HTD
Mason L M6LHM
Mason L 2E0LDM
Mason L M3YAX
Mason L M3YOH
Mason L M6LBM
Mason M G0DHL
Mason M 2E0JVP
Mason M MM6AHX
Mason M M6CDR
Mason N M6NDM
Mason P G0VAG
Mason P G1RCI
Mason P G3NNN
Mason P M0BCW
Mason P M1BTU
Mason P G0LFQ
Mason P 2E0ZAE
Mason P M3ZYE
Mason P M6PCM
Mason R 2E0RAM
Mason R G0AQU
Mason R G1GKA
Mason R G4SIE
Mason R G6HKS
Mason R G6LEK
Mason R M3MXF
Mason R M3YHQ
Mason R G4YPG
Mason R G3TDM
Mason R G4GVR
Mason R M6RKM
Mason S G3CKE
Mason S G4EWT
Mason S G4KNR
Mason S G6WPE
Mason S M3LIY
Mason S 2E0CMF
Mason S M0SDM
Mason S M6AZT
Mason T MM3VYY
Mason T M3TEG
Mason V GM4GGF
Mason W G8ZXY
Mason W M0AGW
Massen M G4YDZ
Massey B G0UNG
Massey C 2E0DDB
Massey C G3HDX
Massey D G8LRS
Massey E G8XQH
Massey H GI7FNP
Massey I M0DZW
Massey J G6YCZ
Massey J G6LBE
Massey P M6PVM
Massey T M6CEE
Masshedar R M0HPL
Massheder J GM6JGH
Massheder P G0EIF
Massie B MM0GDI
Massie M M6PBR
Massimiani C M6WEU
Massimino R 2E0BOI
Massimino R M3UWW
Massimo C M3CSI
Massolt P M0HQA
Masson A GM3PSP
Masson T G7SJK
Masterman M G7PTA
Masterman S G4YEI
Masters C 2E0HBJ
Masters C G3XED
Masters E G0KRT
Masters F 2E0FMY
Masters G G7TEZ
Masters G M3GCM
Masters J G3MBM
Masters J M6UEB
Masters M M6MUB
Masterson M G4GGT
Masterson M MI6MXZ
Masterson T MI6RXC
Masterton J G8FUL
Matcham C G0ALB
Mate G M3BSF
Maternaghan R GI1KHF
Mather A G0OHY
Mather A G1ZOQ
Mather A G3ZIK
Mather A MM6EOE
Mather C G1PCU
Mather E 2E0HEG
Mather F M6FEE
Mather G G8DHE
Mather G M6GMR
Mather J M6CFR
Mather L G8OKI
Mather P 2E0PMM
Mather P 2E0GHZ
Mather R 2E0RSB
Mather R G4ZDG
Mather R 2E0SOZ
Mather S GM4OGM
Mather S 2E0BUE
Mather S M3URT
Mathers D G4PQB
Mathers F G3JBJ
Mathers G MM1CAC
Mathers J GM4EAW
Mathers J GI7TEB
Mathers M G4ODD
Matheson A G3ZYP
Matheson A G8BZJ
Matheson A M0GQR
Matheson C MM6CMY
Matheson D 2E0YZC
Matheson D M3YZC
Matheson J 2E0JDM
Matheson P GW0HCB
Mathew J G7OXY
Mathew S M6VWN
Mathews P G0BLM
Mathews P G4WJH
Mathewson C 2E0MTC
Mathewson C M3MTC
Mathewson D 2E0IOG
Mathias A GW0TLJ
Mathias E M1CNS
Mathias W GW8CNS
Mathieson E M0HBH
Mathieson-Dodd D M6DMD
Mathlin L 2E0DEH
Mathlin L M6LMJ
Matias J M1ELS
Matkin P G7ASY
Matley S 2E0SIM
Matley S M3ALZ
Maton D M6DHU
Maton K G6NHU
Maton S G0KXL
Maton S M6SHU
Mattacks C G3KQQ
Mattacks H G3EKJ
Mattey N G6LUY
Matthes N G8YVM
Matthes N M0NAM
Matthew J G6NID
Matthew K G0WYS
Matthew P 2E0MPX
Matthew P M0PXM
Matthew P M6KMG
Matthew-Brown P M3FWA
MATTHEWMAN A GD4GWQ
MATTHEWMAN C GD4FWQ
Matthewman L M3XLM
Matthewman V 2D0VMN
Matthewman V MD3VMN
Matthews A G0SLZ
Matthews A G1AQF
Matthews A G3UNM
Matthews A G3VFB
Matthews A G4DGF
Matthews A M3MCA
MATTHEWS B G4UZF
Matthews B 2E0EAA
Matthews B M0SAA
Matthews B M3ZPW
Matthews B G0JWF
Matthews C 2E0NYM
Matthews C M3CRD
Matthews C G1LQT
Matthews D GW0CKX
Matthews D G0ICD
Matthews D G0OWE
Matthews D G3ZZP
Matthews D M3NFZ
Matthews D M3XDM
Matthews D M6DCU
Matthews E G3NPL
Matthews G GM0UUB
Matthews G G4LLI

| Name | Callsign |
|---|---|
| Matthews G | 2E0GPX |
| Matthews H | G8WUM |
| Matthews H | M1FOO |
| Matthews H | M3TOB |
| MATTHEWS H | G4SJJ |
| Matthews J | 2E0JCM |
| Matthews J | G0UWD |
| Matthews J | G3HUX |
| Matthews J | G6ASK |
| Matthews J | G7RCW |
| Matthews J | M3SRV |
| Matthews J | G3WZT |
| Matthews J | M6JMX |
| Matthews K | G7UUD |
| Matthews K | 2E0ENP |
| Matthews K | M6ENP |
| Matthews K | 2E0KIM |
| Matthews L | G0OXP |
| Matthews L | G3LHS |
| Matthews L | M6AQT |
| Matthews M | 2E0BYC |
| Matthews M | G3ZBF |
| Matthews M | G6OYF |
| Matthews M | G6USD |
| Matthews M | G8SZR |
| Matthews M | G6MTY |
| Matthews M | G8OLP |
| Matthews N | G8GTZ |
| Matthews P | G4AWZ |
| Matthews P | G7CDI |
| Matthews P | G7UUA |
| Matthews P | G8SEV |
| Matthews P | M0LBM |
| Matthews P | M0PHM |
| Matthews P | M1EJO |
| Matthews P | G6HMA |
| Matthews P | M6BTY |
| Matthews R | G0ECI |
| Matthews R | G2CD |
| Matthews R | G3SAH |
| Matthews R | G4KGO |
| Matthews R | MM6RDM |
| Matthews R | M1CUX |
| Matthews S | G1BHB |
| Matthews S | G1IAB |
| Matthews S | G8DHF |
| Matthews S | M3JLV |
| Matthews S | G1VKN |
| Matthews S | G4VJN |
| Matthews S | M6FVA |
| Matthews T | G3RGC |
| Matthews W | GW4YWM |
| Matthewson J | M0HXH |
| Matthewson P | M6XWG |
| Matthiae D | G0VVZ |
| Matthias M | M3IGZ |
| Mattiello M | M0KNX |
| Mattingley-Scott M | M0MMS |
| Mattinson H | G1CUC |
| Mattinson S | G0GBR |
| Mattison C | G0SLB |
| Mattison J | G0PTG |
| Mattock J | G4VBO |
| Mattocks J | G4TEQ |
| Mattos P | G8LOU |
| Matts G | 2E0LDE |
| Matwiejczyk E | M6ZYK |
| Matynka A | M0RXM |
| Maude A | G4JBK |
| Maude J | G4PVS |
| Maude M | G1UPP |
| Maudsley J | 2E0PXP |
| Maudsley J | M0PXP |
| Maudsley J | M6PXP |
| Maudsley W | G4ROU |
| Maufe A | G8WBK |
| Maughan I | G7LET |
| Maughan J | G3ZMG |
| Maughan S | G0FFB |
| Maule J | G7LVM |
| Maule R | G3OEF |
| Maund V | G8CZP |
| Maunder A | G7IWA |
| Maunder J | G4RNR |
| Maunder R | 2E0CXI |
| Maunder T | G0IOI |
| Maver P | GM0VYL |
| Mavin D | G4OIV |
| Mavin M | G0KGY |
| Maw J | G0DIT |
| Mawby M | G4CQN |
| Mawdsley D | G1LFM |
| Mawdsley D | G3POG |
| Mawdsley P | G7VQR |
| Mawhinney D | GI4KJO |
| Mawhinney D | GI6BNI |
| Mawhinney J | GI8GZM |
| Mawhinney S | MI6HCP |
| Mawn B | G0BKD |
| Mawson A | G0JVI |
| Mawson A | G0UXI |
| Mawson A | 2E0CIN |
| Mawson A | M6GAM |
| Mawson D | G0TKG |
| Mawson D | M1ALA |
| Mawson H | G1VVT |
| Mawson J | GM7KZL |
| Maxey M | G8CTJ |
| Maxfield D | G3ZRQ |
| Maxfield V | G8NWM |
| Maxted D | GW7DVJ |
| Maxted K | GM4JMU |
| Maxwell A | M3PCW |
| Maxwell B | G1GWF |
| Maxwell C | 2M0ELP |
| Maxwell C | MM0ELP |
| Maxwell D | GM0ELP |
| Maxwell J | G7DXC |
| Maxwell J | G3BJD |
| Maxwell M | G8YOY |
| Maxwell N | GW3UMD |
| Maxwell R | G8MKI |
| Maxwell S | G6IHD |
| Maxwell S | GM7UJO |
| Maxwell T | GM0TKC |
| Maxwell W | M0HLW |
| Maxworthy J | G8WQC |
| May A | M1LXM |
| May A | M1BSX |
| May B | 2E0TFI |
| May C | 2E0OON |
| May C | M6ZCM |
| May C | MI6OFN |
| May D | 2E0DDZ |
| May D | M6DAR |
| May D | GW4REI |
| May F | G8AAR |
| May G | G4RZF |
| May G | M0AHF |
| May G | G8TIX |
| May G | G1UTS |
| May G | M0GMA |
| May H | G7VDN |
| May J | G4JFG |
| May J | MI0AQX |
| May J | MI0BES |
| May J | G0CDB |
| May J | G4YMZ |
| May J | 2M0VXL |
| May J | MM3VXL |
| May K | G4APB |
| May L | G4HHS |
| May M | G0UPO |
| May N | G0MIG |
| May N | G4FWN |
| May P | GM7KTY |
| May P | M3ZPM |
| May P | M6YHQ |
| May R | G4YLQ |
| May R | G8IBP |
| May R | M6HRX |
| May R | G7KFZ |
| May R | MM3RXM |
| May R | M6RIU |
| May T | MM1EGS |
| Mayall A | G6OJK |
| Mayall J | G3VPH |
| Mayall N | M3VTN |
| mayall W | M6HME |
| Maybin P | 2E0NLM |
| Maybin P | M3NLM |
| Maycey W | G4YIH |
| Maycock B | G3JWQ |
| Maycroft D | M6DDM |
| Maydew D | 2E0MAY |
| Maydew D | M0DCM |
| Mayell G | M6GPM |
| Mayer P | G0SOK |
| Mayer P | G0KKL |
| Mayer S | G6TEL |
| Mayer S | G1NTR |
| Mayer T | G8KST |
| Mayers A | G6ZHY |
| Mayers D | G6VKL |
| Mayes A | G4ZQJ |
| Mayes A | G1GXX |
| Mayes A | 2E0HIQ |
| Mayes D | G3MMA |
| Mayes J | G1PQT |
| Mayes K | G0EBL |
| Mayes L | G8XLA |
| Mayes N | G0JUK |
| Mayfield J | G0LXX |
| Mayfield R | M0RWM |
| Mayfield T | G4YQD |
| Mayfield T | G7BHU |
| May-Golding J | G7BPZ |
| Mayhew A | G8TQK |
| Mayhew L | G8RDK |
| Maylin A | M1DKP |
| Mayman A | G6JFU |
| Mayman C | G2ABR |
| Maynard B | G6LUO |
| Maynard G | G3XZJ |
| Maynard J | M0DZV |
| Maynard M | G0OIX |
| Maynard P | G0TCP |
| Maynard R | G4YRM |
| Maynard V | M3VBM |
| Mayne G | G4IPV |
| Mayne K | GI4FLG |
| Mayne P | M0CRD |
| Mayo C | G0KVR |
| Mayo D | G4MUL |
| Mayo G | G4EUF |
| Mayo J | M0JEM |
| Mayo J | G1WMK |
| Mayock D | MI0YCK |
| Mayor H | G4MGB |
| Mayor S | 2E0TFT |
| Mayor S | M6JOG |
| Mayson B | G1LHL |
| Mayson M | 2E0YSO |
| Maytum R | G7BLJ |
| Mazura I | G7LAL |
| Mc Dermott J | G4NTL |
| Mc Dermott J | G8ZXQ |
| Mc Ewen A | GM3PGY |
| Mc Geown T | GI0MSG |
| Mc Geown T | GI7IEZ |
| Mc Glynn J | G0HOB |
| Mc Gregor G | MM3GQT |
| Mc Namee M | GI7JGT |
| Mcadam S | G4VBD |
| McAdam W | G0EYL |
| Mcadams C | GJ7DNI |
| McAfee G | GI7SLN |
| McAleer P | MI6IRE |
| McAleer R | G0VLF |
| McAleer W | GI3MMF |
| McAlister P | G3YFK |
| McAlister R | GI0DFD |
| McAllister J | GM1AYT |
| McAllister K | G7OOH |
| McAllister M | MM0OVK |
| McAllister N | GM0FSW |
| McAllister R | M3NJK |
| Mcallister-Bowditch A | 2E0SCS |
| Mcallister-Bowditch A | M6WSB |
| McAlonan D | GM4SFT |
| McAlpin D | G8UPI |
| McAlpine A | MM3DZG |
| McAlpine D | GI6NTP |
| McAlpine N | G8OSG |
| McAlpine P | GI3WFP |
| McAndrew B | M3ILZ |
| Mcandrew I | G4DSI |
| Mcanespie B | GI0JRD |
| Mcara C | G1EFG |
| McArdle J | 2M0OIC |
| McArdle J | GM6OFB |
| McArdle J | MM6MCA |
| McAreavey W | G7JSV |
| McArthur D | MM1AHL |
| McArthur D | MM3DIZ |
| McArthur D | 2E0DMD |
| McArthur M | M6MVV |
| McArthur P | MM3HYG |
| McArthur S | G0CBI |
| McArthur T | MM6TAT |
| Mcateer P | GI7IRJ |
| McAteer S | GI0NOX |
| McAulay I | GM6CZM |
| McAulay J | GM6NIC |
| McAuley D | MI3DMM |
| McAuley J | MI3UMC |
| McAuley K | MI0CRQ |
| Mcauley P | GI4JIC |
| McAuley R | MI6DUP |
| McAuley S | 2I1SWD |
| McAuley S | MI1DOG |
| McAuley S | MI3SWD |
| McAuslan D | G3ZMH |
| McAuslan D | 2E0CEK |
| McAuslan D | M0MDO |
| McAuslan D | M6DFM |
| McAvoy G | G4NFT |
| McAvoy I | G0RPA |
| McAvoy J | GM3RPM |
| McBain W | M3PYI |
| McBain W | MM6HQI |
| McBirnie A | 2E0FZK |
| McBlain W | MM3UVF |
| McBride A | MI3WQT |
| McBride A | GM0IMW |
| McBride J | GI1CAI |
| McBride J | MI0MCB |
| McBride J | 2I0DJM |
| McBride J | MI3DJM |
| McBride J | 2E0BVE |
| McBride K | MM3YQX |
| McBride M | M1DCV |
| McBride P | G6DQK |
| McBride P | MM1DHU |
| McBride T | 2E0VWX |
| Mcbride T | M3VNQ |
| Mcbride T | MM6ATU |
| McBride W | GI0JFF |
| McBrien G | G6SHD |
| McBrien H | M6HRM |
| McBurney J | G4AUR |
| McBurney R | GI3HJH |
| McBurney W | GM0TTY |
| McCabe A | GI0VWU |
| McCabe B | MM6HQK |
| McCabe D | G8JXP |
| McCabe I | G0FYD |
| McCabe J | GM4RPE |
| McCabe J | GI0KUH |
| McCabe M | MM0GUW |
| Mccabe S | GI4TAP |
| McCaffery G | 2E0DJJ |
| McCaffery G | M0KCF |
| McCaffery G | M6GPM |
| McCaffery K | G7FRW |
| McCaffrey B | GM8EZS |
| McCaig A | 2M0LAW |
| McCaig A | MM3SUV |
| McCalden A | G8ZMC |
| McCaldin A | GI0HXH |
| Mccaldon P | G0DPK |
| McCall A | 2M0WXS |
| McCall A | MM0WXO |
| McCall A | 2E0MLL |
| McCall A | M6CLL |
| Mccall C | G6VGA |
| McCall D | GM0NIA |
| McCall J | GM3HGA |
| Mccall J | G3ZBS |
| Mccall M | M6MPM |
| McCall P | G4RGF |
| McCall T | 2M0BCL |
| McCall T | MM0GKU |
| Mccallan M | GI4RYL |
| Mccallion A | GI4VKS |
| McCallum A | G6NHL |
| McCallum A | M6ADM |
| McCallum D | G6CWZ |
| McCallum G | GM3UCI |
| McCallum J | M3MCU |
| McCallum J | G4YMC |
| McCallum M | G0GBU |
| McCallum M | MM3VRI |
| McCallum M | MM3UCI |
| McCallum R | G4VNG |
| McCallum T | MM6FEX |
| Mccallum W | GM0POD |
| McCalmont B | MI3EOH |
| McCammick C | MI6CAV |
| Mccance S | MM1SYD |
| McCandlish W | G1FMX |
| McCann A | G3AZI |
| McCann A | G3PS |
| McCann A | G4GAF |
| McCann A | 2I0ROC |
| McCann A | MI6DED |
| Mccann B | M3HFX |
| McCann C D | GI4XAA |
| McCann D | MM1DMU |
| McCann G | MI6NOE |
| Mccann I | G4RFJ |
| Mccann J | GI3YBZ |
| Mccann J | GI8BNC |
| McCann K | 2E0DRU |
| McCann K | G6YYN |
| McCann N | MI6NVM |
| McCann P | MM3YUS |
| McCann R | G8OKB |
| Mccann S | G4HFZ |
| McCann T | M3FBJ |
| McCarrison J | MI0ABD |
| McCart R | GM0JBE |
| McCartan S | MI6XOX |
| McCarthy A | M3BAL |
| McCarthy C | 2E0MAC |
| McCarthy C | G3XVL |
| McCarthy D | 2E0DYK |
| McCarthy D | M6XBJ |
| McCarthy G | M6LHV |
| McCarthy H | M6HHM |
| McCarthy I | G3YBY |
| Mccarthy K | M3GGA |
| McCarthy K | M6AQR |
| Mccarthy M | G0KCH |
| McCarthy P | M1AWT |
| McCarthy P | M3SJV |
| Mccarthy T | G1HWK |
| McCartney B | G4DYO |
| McCartney C | MI6LFU |
| Mccartney D | G4TXA |
| Mccartney G | G4VYR |
| McCartney J | 2D0JBE |
| Mccartney M | G7IFU |
| Mccartney R | G4BDJ |
| Mccartney W | G3SOA |
| McCarty D | M0FGA |
| McCash J | MM3CBO |
| McCaughey G | MI6SIS |
| Mccaughey W | GI4XJJ |
| McCaulay P | G7VTN |
| Mccauley D | G1SPQ |
| McCauley P | MI1FCQ |
| McCausland B | GI0KPF |
| McCausland J | GI0IVJ |
| McCausland P | MI3PJM |
| McCaw J | 2I1JMC |
| McCaw J | MI0JML |
| McCaw J | MI3JMC |
| McCay L | MM5AII |
| McCay M | MI3NVX |
| McClean S | GI0PFL |
| Mcclean S | GI4ESI |
| Mcclean S | MI0RSO |
| Mcclelland A | GM0BFW |
| McClelland A | G0DKN |
| McClelland C | MI0MCC |
| McClelland D | MM3OCY |
| McClelland I | M5AKZ |
| McClelland P | G1OVY |
| McClements A | MM6FPI |
| McClements E | MI3IFO |
| McClements R | G4CID |
| Mccleverty A | GW0VEM |
| McClew D | G7IJL |
| McClintock C | MM1AUF |
| McClintock G | GM8YUI |
| McClintock M | MM1AUG |
| Mcclintock W | G3VPK |
| Mcclintock W | GI4MAJ |
| Mcclory B | G3UJA |
| McCloskey D | 2I0DMC |
| McCloskey D | MI3RUV |
| McCloskey J | MI6JMC |
| McCloud C | G4EFB |
| McCloy M | MI3LVZ |
| Mccloy S | G1SMC |
| McCluney D | GI4MVQ |
| McClung K | MM6AAM |
| McClure C | GM0HJV |
| McClure D | MM3OIT |
| McClure J | MM6JAN |
| McClure J | M6JSM |
| McClure K | GM1XDK |
| McClure M | MI3MMC |
| McClure V W | G8WCQ |
| Mcclurg C | GI4ISR |
| McClurg D | M6GRQ |
| McCluskey P | 2M0BEL |
| McCluskey P | MM3KSV |
| McCluskey R | G0AXJ |
| McCluskey S | M6AXJ |
| McClymont C | 2M0NTY |
| Mccoll A | G7BZU |
| McColl J | 2E0BHN |
| McColl J | M3RKJ |
| McColl J | M0ZYT |
| McColl J | M3XWB |
| McCollam P | GI7FQD |
| McCollin J | G7JHM |
| McCollum C | M3PLV |
| McCollum J | MI3XUC |
| Mccolm R | M3ORV |
| Mccomb G | GI4CZO |
| McComb J | G6ZIC |
| McCombe B | G3ZJW |
| McCombe S | G4WLP |
| McConkey A | M6SDN |
| McConnachie A | G7RRJ |
| McConnachie C | MM6IMP |
| McConnell B | MI3VEQ |
| McConnell C | 2I0BAC |
| McConnell C | MI3ILL |
| Mcconnell G | G0DIB |
| McConnell J | M3DBF |
| McConnell L | G6AMV |
| McConnell M | MM3KVY |
| McConnell S | GI4MBM |
| Mcconnell T | MI3IIH |
| McConnell T | 2M0TGM |
| McConnell T | MM6FPX |
| McConnell T | M0THM |
| Mcconnell W | GI6PAZ |
| McConochie A | MM3TAV |
| McConville D | GI6FQT |
| McCoo W | 2E0WNM |
| McCoo W | M6WNM |
| Mccook E | MM3PXO |
| Mccord C | MI3CIZ |
| McCord D | MI3DSM |
| McCorkell S | MM6TZX |
| Mccormack A | GM1AHF |
| McCormack J | GI4CSO |
| McCormack M | M6MJM |
| Mccormack N | GM1AHG |
| McCormack P | M0BKD |
| McCormack S | M6GQB |
| McCormick A | MM6HQE |
| McCormick C | 2I0NTH |
| McCormick C | MI6BZI |
| McCormick D | G0VCB |
| McCormick G | 2I0GYL |
| McCormick G | MI6YLG |
| Mccormick I | G0JJF |
| McCormick J | GI7FCP |
| McCormick K | MM0WRX |
| McCormick M | M3FLZ |
| McCormick R | G4ZCA |
| McCormick S | 2M0STV |
| Mccormick S | G7KOS |
| McCormick S | MI0SRM |
| McCormick S | GI7TFK |
| McCormick S | 2I0YLT |
| McCormick S | MI0YLT |
| McCormick S | MI6YLT |
| McCormick T | G1JHG |
| McCormick W | MI6WAM |
| McCormick W | 2I0WMC |
| McCormick W | GI6EJW |
| McCormick W | MI6BMC |
| McCormick W | MI6WDM |
| Mccormick X | GI7PWQ |
| McCosh B | MM0MSH |
| McCosh J | M6NBG |
| McCourt B | G8ZJH |
| McCourt I | M3YGZ |
| McCourt M | 2I0EQC |
| McCourty G | M0SOA |
| McCown J | M0JWM |
| Mccoy L | G1KON |
| McCoye R | G0OCF |
| Mccracken R | GM4CEA |
| McCracken R | G6EIH |
| McCracken S | GI4LGP |
| McCrae D | M3UIK |
| McCrae D | M6DMC |
| McCrae J | MM6JJQ |
| McCrandles D J | GM4SNZ |
| McCrea J | GI4MRN |
| Mccrea R | GI3WBR |
| McCreadie A | 2E0GKD |
| McCreadie R | G0FGX |
| McCready A | GI7TVV |
| McCrimmon T | G4LQM |
| Mccron J | M3DHW |
| Mccron C | M3DHV |
| McCrory P | GI7RTB |
| McCrum I | MI5AFL |
| McCrum M | 2I1HNZ |
| McCrystal A | 2E0ZOR |
| Mccrystal E | GI7FHZ |
| Mccrystal E | MI3FHZ |
| McCrystal K | MI3KRL |
| McCuaig I | MM3IMC |
| McCubbin R | G4OLA |
| Mccudden A | GM4DLU |
| McCue J | M3HHJ |
| McCue W | G6HKN |
| McCue W | 2M0BIL |
| McCue W | MM0ELF |
| McCue W | MM3FCG |
| McCuish K | MM6EBJ |
| McCullagh J | GI4BWM |
| McCullagh J | GI4UDI |
| McCullagh J | GI7JKA |
| McCullagh P | MI6BVN |
| Mccullagh S | GI6EEH |
| McCullagh S | GI6NAQ |
| Mcculloch A | G4HTL |
| McCulloch I | G1RYS |
| Mcculloch J | G0IUM |
| McCulloch J | M0GOF |
| McCulloch J | GM1SRP |
| McCulloch M | GM1XEB |
| McCulloch S | G4TPO |
| McCullough A | MI0TNZ |
| McCullough D | MI1VOX |
| McCullough G | MI3N3R |
| McCullough J | GI4SFE |
| McCullough L | GI4RMA |
| McCullough T | MI3MRF |
| McCullough T | GI3SCM |
| Mccullough W | GI4BQI |
| McCully N | MI0DGX |
| Mccurdy A | MM3MHQ |
| McCurdy S | 2M0BOS |
| Mccurrach R | G4ASF |
| McCurrie P | G4ADP |
| McCurry J | GI4SZU |
| McCurry R | GI4OCL |
| McCurry T | G3VSK |
| McCusker A | MI6XEM |
| McCusker G | 2I0WGM |
| McCusker G | MI0WGM |
| McCusker G | MI6WGM |
| McCutcheon G | GI1WLJ |
| Mccutcheon J | M0JMC |
| McCutcheon M | GI6MTL |
| McCutcheon R | G1WVZ |
| McCutcheon T | G0HPL |
| Mcdade A | M1CDP |
| McDaid P | MI1DRP |
| McDaid P | MI3AGR |
| McDaid R | MI6RAC |
| Mcdermid A | MM1BJP |
| McDermid A | M6MLN |
| McDermid I | MM3ISA |
| Mcdermott D | G4HXU |
| McDermott F | M3VST |
| McDermott J | M6JEQ |
| Mcdermott J | GM6LEZ |
| Mcdermott M | GM0WIB |
| McDermott M | G6NAD |
| McDermott M | G7UHW |
| Mcdermott R | 2E0AKL |
| Mcdermott R | M0GFO |
| McDermott R | M3NLX |
| McDermott R | G6TDR |
| McDermott-Roe A | G8UJS |
| McDiarmid D | G3FMU |
| McDicken W | GM4XMD |
| McDonald A | MM6FVD |
| McDonald B | 2I0MBI |
| McDonald B | MI3SYI |
| McDonald D | G4SMQ |
| McDonald D | G0AQZ |
| McDonald D | M3SQE |
| McDonald D | M6MKU |
| McDonald D | 2I0YLT |
| McDonald D W | G0OTT |
| McDonald E | M0ACB |
| McDonald G | GM4LZO |
| McDonald H | GM1VFR |
| Mcdonald I | MM1DAK |
| Mcdonald J | GI0THR |
| McDonald J | GM1VYG |
| McDonald J | MI6JMD |
| McDonald J | G8PJC |
| McDonald J | M6HBU |
| McDonald K | 2I0KEW |
| McDonald K | MI3IHZ |
| McDonald M | 2M0MCD |
| McDonald M | MM3NZX |
| McDonald M | M6SOV |
| Mcdonald N | MM6NAA |
| Mcdonald P | GI0SFT |
| Mcdonald P | GI7KVR |
| Mcdonald P | GI0USW |
| McDonald P | MM6MOY |
| Mcdonald R | G0IZP |
| McDonald R | G3DCZ |
| McDonald R | MM6FCA |
| McDonald R | MI6RME |
| McDonald R | MM3FZI |
| McDonald S | G0IKN |
| McDonald S | MW3EPJ |
| McDonald T | G7OGT |
| McDonald W | 2I0WBD |
| McDonald W | MI6WAB |
| McDonnell A | MI3WTT |
| McDonnell B | G6MAC |
| McDonnell C | M0EAO |
| McDonnell D | 2I0EQR |
| McDonnell D | G0RZB |
| McDonnell J | GI8ZHW |
| McDonnell L | 2E0LKM |
| McDonnell L | M6LKM |
| McDonnell P | G1FIM |
| McDonnell T | 2I0EQS |
| McDonnell T | GI4WTT |
| McDonnell T | MI3EQS |
| McDonough P | M3VPY |
| McDonough P | 2E0VRQ |
| McDonough P | M0TVR |
| McDonough P | M3VPM |
| McDougall A | GM0AYT |
| McDougall A | GM0LAH |
| McDougall D | M3DLJ |
| McDougall K | G1NPN |
| McDougall M | M6MAM |
| McDougall N | M3VVQ |
| Mcdowall J | G7ICD |
| McDowall J | MM6JTG |
| McDowall W | GW6ZMN |
| McDowell A | MI3LXZ |
| McDowell A | G0KOO |
| Mcdowell G | GI3XDX |
| McDowell H | MI6XOD |
| McDowell M | GI4FNU |
| McDowell R | GI6IHM |
| McElhatton K | GI3NFM |
| McElmurray S | MI6SFK |
| Mcelroy A | 2E0IDM |
| McElroy D | GI0MSH |
| Mcelroy P | G4DHW |
| McElvanna J | GI4OVE |
| Mcelvenney J | G3LLV |
| McElwee T | 2E0TMH |
| McElwee T | M0TMP |
| McElwee T | M6VMC |
| McEnteggart I | G8GQF |
| McErlean H | 2I0BFB |
| McErlean H | MI0HMC |
| McErlean H | MI3LQN |
| McErlean J | 2I0LPO |
| McErlean M | 2I0WAI |
| McErlean M | MI3UKW |
| McEvoy A | G0SSL |
| McEvoy D | M3VJI |
| Mcevoy H | M6HFM |
| McEwan C | GM0LVL |
| McEwan D | M0MCE |
| McEwan I | GM0IMZ |
| McEwan I | MM6SHM |
| Mcewan R | G4CHM |
| Mcewan R | GM4VWV |
| McEwan W | MM6HRZ |
| McEwen A | G4RQW |
| McEwen A | M1CDV |
| McEwen A | 2E0YEW |
| McEwen A | M6PRA |
| McEwen C | G3VKQ |
| McEwen G | M6MCE |
| McEwen I | M6KTC |
| McEwen P | M3CTN |
| McEwen P | G1GYM |
| Mcewen P | G4PUQ |
| McEwen R | M0GYM |
| McEwen W | GI0SZH |
| McFadden K | M6AIP |
| McFadden M | GI3VCI |
| McFadden P | 2I0VOF |
| McFadden P | MI6VOF |
| McFadden P | 2E0PGM |
| McFadden P | M0PAO |
| McFadden P | M6AZS |
| McFadden R | MI0ENR |
| Mcfadyen A | G0JZE |
| Mcfadyen J | GM0KBR |
| McFadyen J | G4JMM |
| McFall M | GI4NXJ |
| McFalls E | M3OUS |
| McFarland E | G3GMM |
| McFarland M | G0GLX |
| McFarland P | G7BZY |
| McFarland V | GI4RNP |
| McFarlane B | M3KAE |
| McFarlane E | G8KKN |
| McFarlane G | MM6GBX |
| McFarlane J | G3JUX |
| McFarlane J | M6XJM |
| Mcfarlane K | G3ICG |
| McFarlane T | MM3OJE |
| McFaul W | GI4FHB |
| Mcferran D | GM0OPX |
| McFetridge N | G8FIE |
| McForsyth M | GM4LDX |
| McGahon P | G1IKG |
| McGann A | M3HSI |
| McGann G | M3AYS |
| McGarrigle D | M6DTM |
| McGarrigle I | G4JIU |
| McGarry B | GI4HDJ |
| McGarry G | GI3ECQ |
| McGarry P | G1NEG |
| McGarry P | M3NEG |
| McGarvey A | 2I0TXM |
| McGarvey A | MI0TXM |
| McGarvey A | MI6XMG |
| McGarvey P | G0IVD |
| McGarvie D | G0END |
| McGaughey L | 2E0CJA |
| Mcgee C | G4CPL |
| McGee G | G3MDM |
| Mcgee J | G4HBR |
| Mcgennity B | G4DBM |
| McGeouch G | MM3SNB |
| McGeough G | M3GMG |
| McGeough K | G7EKM |
| McGhie A | G0THF |
| McGhie A | G3AUE |
| Mcghie D | G6CZO |
| Mcgifford J | GM0KUJ |
| McGill A | GM4XOI |
| McGill G | MI3GJG |
| McGill J | 2E0MCG |
| McGill K | M6KAM |
| Mcgill W | 2E0BMG |
| McGill W | GM0DXB |
| Mcgill W | M3LFO |
| McGillewie P | M0ZPM |
| McGillian J | GI4NNM |
| McGilp F | G0URB |
| McGinty J | MM0ZZO |
| McGinty J | GM4GZO |
| McGivern P | G4JXH |
| Mcglasson D | G6NVF |
| McGleenan D | M6CTO |
| McGlen E | G6INK |
| McGlone D | 2E0FAA |
| McGlone D | M0TMX |
| McGlone D | M3FAA |
| McGlone D | 2I0CFW |
| McGlone D | MI3DWQ |
| McGlynn B | 2E0CQN |
| McGlynn B | M6SBO |
| Mcglynn I | 2M0VLF |
| McGlynn R | MM6ZYZ |
| McGoff A | 2E0TGF |
| McGoff A | M3BIB |
| Mcgoldrick H | MI0CAC |
| McGoldrick J | 2I0ESA |
| McGoldrick J | MI6EQA |
| McGoldrick J | 2I0EIG |
| McGoldrick J | MI6XOS |
| McGoldrick J | MI0JAT |
| McGoldrick M | 2E0MEL |
| McGoldrick P | G6AAC |
| McGonigall C | M6RRU |
| McGonigle N | MI3NMG |
| McGougan C | MM3OJV |
| McGowan B | M6SMQ |
| Mcgowan C | MM0CWI |
| McGowan C | G0KPE |
| McGowan C | MM0NDX |
| Mcgowan D | G1NAT |
| McGowan E | MI3JVX |
| Mcgowan G | G4FGJ |
| McGowan G | G8YTF |
| McGowan G | M0VAA |
| Mcgowan I | GM1RIG |
| McGowan I | G8OFZ |
| McGowan J | GM0FSV |
| McGowan J | M6GKI |
| McGowan J | M0MAC |
| McGowan J | M1CUC |
| McGowan M | 2E0MFG |
| McGowan M | M6TTP |
| McGowan P | G7VDD |
| McGowan P | GM1COF |
| McGowan R | GM0DUX |
| McGowan V | MM6MVM |
| McGowan V | M6VMG |
| McGrath D | M3TYX |
| McGrath N | G7AQK |
| Mcgrath P | GD0BCJ |
| McGrath P | 2E0MPG |
| McGrath P | M6MAK |
| Mcgreevy E | GM0LKS |
| McGregor A | M3TLT |
| McGregor J | GM4LGM |
| McGregor M | M0VVT |
| McGregor R | M3ZRM |
| Mcgregor S | GM4ZOA |
| McGregor S | M0SMC |
| McGreish A | M3YXJ |
| McGreish G | M3YXK |
| McGroarty N | M6NYZ |
| McGrogan L | G4UUE |
| McGrorty M | 2M0FXX |
| McGrorty M | MM6AJI |
| McGrory M | MI0GRG |
| McGrory S | G4GNP |
| McGuckian S | M6MKX |
| Mcguckin K | GI0RDM |
| Mcguffie W | G7BYU |
| McGuigan S | G8MFI |
| Mcguigan T | M0AGV |
| McGuigan T | G7CIT |
| McGuinness A | MI0KMJ |
| McGuinness P | G8ZXU |
| McGuinness P | M6EQF |
| McGuinness P G | G4FDN |
| McGuinness P G | G8PAT |
| Mcguinness S | M3RUO |
| McGuire C | G4WFF |
| Mcguire J | G3LNW |
| Mcguire L | G7NBG |
| Mcguire L | M3NBG |
| Mcguire T | G0AVH |
| McGuire W | G0TYC |
| McGurk B | 2E0VCE |
| Mcgurk I | 2M0LNF |
| Mcgurk I | MM3LNF |
| McGurk J | 2M0TOK |
| McGurk T | MM6RCK |
| McHale J | G0HEW |
| McHale J | G6LFC |
| Mchale M | G4PWD |
| McHardy A | G6AWP |
| Mchardy D | MM3SQM |
| McHugh B | G3THF |
| Mchugh B | G6WIO |
| McHugh D | M6HCB |
| McHugh M | 2E0MBV |
| McHugh M | M0WFU |
| McHugh W | GI4SRQ |
| McIlroy G | G0PNR |
| McIlroy H | GI4CYU |
| Mcilroy J | G4EQX |
| Mcilveen J | GI0OON |
| Mcilvenna S | GI7MDP |
| McIlwaine S | M6LOF |
| McIlwee A | GI1GME |
| McIlwraith J | GM4ZTO |
| McInally M | M0AJC |
| McInerney T | G0VYQ |
| McInnes A | M6CYD |
| McInnes B | G7SKW |
| McInnes C | GM4YHO |
| Mcinnes G | M0FOR |
| McInnes K | M0KSR |
| Mcinnes K | GI6KDN |
| Mcinnes T | G7CXB |
| McInnes-Boylan J | 2E0JMC |
| McInnes-Boylan J | M0OBZ |
| McInnes-Boylan J | M6ZFE |
| McIntier S | GM1LXM |
| Mcintosh A | G0JBV |
| McIntosh B | GM4WZG |



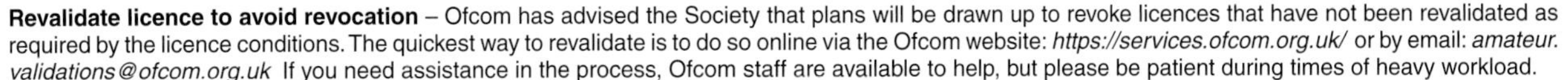

**IMPORTANT NOTE**

**Revalidate licence to avoid revocation** – Ofcom has advised the Society that plans will be drawn up to revoke licences that have not been revalidated as required by the licence conditions. The quickest way to revalidate is to do so online via the Ofcom website: *https://services.ofcom.org.uk/* or by email: *amateur.validations@ofcom.org.uk* If you need assistance in the process, Ofcom staff are available to help, but please be patient during times of heavy workload.

McIntosh D M0DWX
McIntosh S 2M0ROT
McIntosh S MM6ROT
Mcintosh W GM0OTS
McIntyre A GM7BYB
Mcintyre B G7KBE
McIntyre C MI3FPN
McIntyre C MM6AQM
McIntyre J GM4ARU
McIntyre M GI3YDH
McIntyre M M6MMC
Mcintyre N 2E0NMC
Mcintyre N M0NMC
Mcintyre N M3HXZ
McIntyre N MM3UNQ
McIntyre T 2E0TAX
Mcintyre-Stewart S 2M1HKA
McIver C G7MYO
McIver E G8SDN
McIver I G0BKN
McIver M G8XEF
McIver S MM3GQR
McKae W G4ILA
McKain T M3TMM
Mckane K G7NHW
McKavanagh J GI4SIP
Mckay A G0HLU
McKay A GM3OZB
McKay A 2I0NGM
McKay A MI6NGM
McKay D MM5AJW
McKay D G0NUT
McKay D G1JWG
McKay G GM3SPT
McKay G GM4YWS
McKay G G8MOK
McKay H 2M0HZL
McKay H MM0HZL
McKay J G4HOK
Mckay J MM0DQP
McKay M MM0DUR
McKay M MM0LGT
Mckay P G3WQU
McKay P MM3VXP
McKay P MM3PDM
McKay R 2I0ZFZ
McKay R MI6FNZ
McKay S G6XJJ
McKay S MI6MCK
McKean I 2E0CQB
McKean I M6CQB
McKean M M6OUS
McKechnie A G4XFV
McKechnie J MM0JMK
McKee D GI0GPG
McKee M GI4MXV
McKee N MI0NOR
Mckee P GI0MHB
Mckee R T G4LJK
Mckee T G8FWD
McKee T 2I0TCJ
McKee T MI0TBV
McKee T MI6TFG
McKeen B GI6DKQ
McKeen M MI3WMK
McKeever G MI3NWU
McKeever J GI7TDA
McKeever P MI3UIM
Mckeever R G0HEF
Mckeever W GI0DSG
Mckellow P G8XGO
McKelvie G G7USC
McKendley V MW6MIQ
McKenna A G7NGX
McKenna J M3FTP
McKenna J G4UWM
McKenna M 2E0HAH
McKenna M M0MBT
McKenna M M6MRK
McKenna M 2E0ENQ
McKenna M M6ENQ
McKenna T M6WFE
McKenzie A M0ELA
McKenzie A G0SMN
McKenzie C G8LQO
McKenzie C M0CSO
McKenzie D MM0DMK
Mckenzie J G0CNU
McKenzie J MM3JPS
Mckenzie K MM6BNO
Mckenzie L M3PZN
McKenzie M G8RWN
Mckenzie O MM6BNN
McKenzie P G8PHB
McKenzie P MM3DYT
McKenzie R G0FFK
Mckenzie R M1XZG
McKenzie R M3PZL
McKenzie R GM0AUL
McKenzie S 2M0RCD
McKenzie S MM0HCO
McKenzie S MM6AOH
Mckenzie W GM6UCN
McKeown F G4WXI
McKeown I MI3SXI
McKeown J M3KVK
McKeown K M3KUZ
McKeown P MI3EYB
McKeown T MW6TPM
McKeown W GI1RBI
McKeracher F G3RSI
Mckersie R GI0BFA
Mckie A M6MOJ
McKie D G4ZPW
Mckie J M6MKM
McKie R 2E0TKE

McKie R M3TKE
Mckillop C MM3IBM
McKillop J G8IEI
McKillop W G8CIT
McKimm R GI3UPG
McKinlay C MM6CMM
McKinlay G GM0IVQ
Mckinlay R G4FDU
McKinley A MI6HAF
McKinley C MI6ZXT
McKinley G MI6GMK
McKinley T M6DRI
McKinney C M0KIB
McKinney D GI4MXW
McKinney J M0BPO
McKinney M GI4MAC
Mckinney R GI4LQU
McKinney W J M GI3TZB
McKinnon J G8JIT
McKinnon J MM6LPT
McKinnon S 2M0OXQ
McKinnon S MM0PAZ
McKinnon S MM3OXQ
McKinnon S G0TBI
McKinnon Z MM6SWT
McKinty R GI3GTR
Mckittrick N MI3CBJ
McKnight E 2I0EBS
McKnight E MI6EBS
McKnight P G0RTM
McKnight R 2E0GYC
McKnight R G2YC
McKnight R M3TEI
McKnight T G8YPH
Mckone D G4DRX
McKone M M3YRX
McKown L G3WXN
McKune I G1OXQ
McLachlan A M6ASM
Mclachlan D G1BGF
McLachlan D G4KOW
Mclachlan I MM6DAT
Mclachlan K MM6DWC
Mclachlan R G3OQT
Mclachlan S MM3KMX
McLaren B GM4SVW
McLaren D G0OLD
McLaren K B GM0EMC
Mclaren N G4OAR
McLaren N GM0HZI
Mclaren P GM6TUE
McLaren P 2M0ZBH
McLaren P MM0ZBH
McLaren P 2E0PJM
McLaren P M6BFM
McLaren R GM4FQG
McLaren W G1GSJ
McLarnon P MI6PML
Mclauchlan M MM0PMW
Mclauchlan M MM1DPC
McLaughlan J GM7CZU
McLaughlin A M3JSM
Mclaughlin D GM0PHG
Mclaughlin D G4RGH
McLaughlin D MI6XRC
Mclaughlin I G1IKH
McLaughlin J M0JBW
McLaughlin K M3JKM
McLaughlin L M3LMC
McLaughlin M MI0DVH
McLaughlin S 2M0FDZ
McLaughlin S MM3BPR
McLaughlin S M3EOY
McLaughlin S M3SPR
McLaverty K 2I0IDJ
McLaverty K MI6IDJ
Mclay D G8FVC
Mclean A M6XZY
McLean A 2E0UTD
McLean A M3OEH
McLean C G6FOH
McLean C M1ANC
McLean C 2E0ESN
McLean D G6GXS
McLean D MM3WHS
McLean D GM3SUZ
McLean D M6GPU
McLean E GM4EWM
McLean J G0BMH
McLean J M6HUW
McLean L GM4LFK
McLean N 2E0VOD
McLean N M3UFW
McLean W GM4REF
Mcleary I MM1CPP
McLeary M MM3MTM
McLellan A MM3DCN
McLellan R G6XKO
McLelland C MI0JZZ
Mclelland J 2M0RMP
Mclelland J MM6WBU
Mcleman L GM1JDJ
Mcleman S MM3AWD
McLenaghan I G8HPF
McLennan A G7DCF
McLennan C M6NCE
McLennan S GM0ERV
McLennan S G7JNS
McLeod G 2M0DQN
Mcleod J G4MWW
McLeod K M6KRM
McLeod P G0LJP
McLeod R M6ALP
McLeod T GM4TRZ
McLeod-Stangroom F GM6CRX
McLernan G MI3GHW

Mclernon A MI3CON
Mclintock R G1TGZ
McLlroy K A GI4JJF
Mclocklin A G7RHI
Mcloughlin D G4XEI
Mcloughlin F G1GAD
Mcloughlin F M3FRD
McLoughlin J GD0KPN
McLoughlin M G6PZN
McLoughlin N M3NEI
McLoughlin R GI8TAX
McLoughlin S 2E0DQQ
McLoughlin S M3GNM
McLuckie A MM0CKK
McLuckie J GM0TKE
Mcluckie S GM4HML
McLusky E 2E0AGI
McLusky E G1ZBO
McLusky E M5ACF
McLusky P 2E0BDV
Mcmackin A G4NHQ
McMahon B GI4KEQ
Mcmahon C G6FCI
Mcmahon E M3YWJ
McMahon I M3WUN
McMahon J G4YZP
McMahon J G4KVD
McMahon J M6AUO
McMahon K G8JJR
Mcmahon L G8SQK
Mcmahon N M1ECW
Mcmahon P MI3NJU
McMahon P MI1ASN
McMahon R G0CBP
McMahon R G7KJP
McManus B G8FDE
McManus C G8FFC
Mcmanus D GI1SZC
McManus P GM7AAJ
Mcmanus T G0CNV
McMartin J MM3RCR
McMaster J GM0TBH
McMaster R GI1TFC
McMaster R GI7NOW
McMath A MM6INC
McMath M M6DPQ
McMaw G GI4JYJ
Mcmillan A G4GZM
McMillan A G1JXG
Mcmillan A G4SSO
Mcmillan A GM0GDD
McMillan D MM6DSD
McMillan G 2E0FLN
Mcmillan P M0PJM
Mcmillan R G3IUC
McMillan R GM3SAE
McMillan R GM8JUY
Mcmillan R GM6JIL
McMillan S MM6STM
McMillan T G1CCM
McMillan W GM4WMM
McMillen A MI6HAD
McMillen W GI4RCK
McMinn A GM6VYY
McMinn D GM4XJY
Mcminn M M3MFR
McMinn R G6REY
McMinn W GM6VYZ
McMonigle J MM3OIZ
McMorland J 2M0TTF
McMorland J MM0GUE
McMorland J MM6AYR
McMorrin D M6FDX
McMorrow E MW6EXV
McMullan I GI0RDJ
McMullan J M0JMY
McMullan P G4RXK
McMullan-Bell P M6PMB
McMullen A M1CTM
McMullen J 2E0TDR
McMullen J M6FVH
McMullen S G0WEV
Mcmullen W MI6WMN
McMullin A G8TQH
McMullon A G0SDJ
Mcmurray E MW3NBN
McMurray H GW4MTD
Mcmurray J GW6UAS
McMurtrie S 2E0MJS
McMurtrie S M0SSM
McMurtrie S M6MCM
McMurtry A GI3MBB
McNab D G0PRY
McNab W G0WMN
McNally P MM6HFZ
McNally S M0CNN
McNally W G3YDO
McNamara P G8FMZ
Mcnamara T G8SFQ
McNamara T M6TDV
McNamara W G0AWG
Mcnaught J G3UJZ
McNaught R GM6MHC
McNaughter T 2I0IMO
McNaughter T MI3IMO
McNaughton C M6MKF
McNaughton D M G8KOF
McNeany M 2E0HLW
McNeice A GI4OTG
McNeil A MM3SAK
McNeil A 2E0CFD
McNeil G G3PLY
Mcneil N G1MTP
McNeil R G0RON
McNeil S MM1EHO
McNeill A 2M0TXR

McNeill A MM6TRX
McNeill E MM0DSM
McNeill E MM3DEC
McNeill H M6HJM
McNeill J G6LCS
McNerlin A GI7JRG
Mcnerlin J GI4EBS
McNicholl D 2E0VTA
McNicholl D M0VTA
Mcnicol A GM0MIW
McNiece N M6NTM
McNiel D G4LQW
Mcniff J GM4HRJ
McNinch M GI6JGB
McNulty A MI3TMN
McNulty C 2E0CDM
McNulty C M3VZC
Mcnulty N GM0DVH
McParlane J G8VOY
McPartland A GI6FHD
McPartland C G4EVP
McPhail A G4WYH
McPhail M M1DAB
McPheat E G4OEC
Mcphedran A GM3GTQ
McPhee J 2M1HIN
McPhee M 2E0FBA
Mcphee S G0MWS
McPherson A GM7FPN
McPherson E MM0EMC
McPherson E M0GBZ
McPherson I M3YYM
McPherson I G4EWV
McPherson M G0GYO
McPherson P G3TEL
McPhillips H M6TGR
McPhillips J MM0SEK
McPolin T MI6TOA
McQuade P G8AKP
McQuaid S GI4LDN
McQuaid T GI1OMD
McQuail P G8DCJ
McQuarrie A G0FMN
McQueen C G7TIK
McQueen J GM1MON
Mcqueen S GM8MRW
Mcquillan R G0HVX
McQuillian S MM0BSM
Mcquire L G0ETV
McQuirk D M6DMQ
McRae B MM3DQV
McRae K MM3DQX
McRitchie G M6HWX
McRobie J M6BPV
McShane J G1NTL
Mcshea B G7MCS
Mcshea P G1YQN
McSherry B MM0SMB
McSherry J G4VIA
McSherry M 2E0MAX
McSherry M M0WCR
Mcsoley J G1YAH
McSorley J GD4LHT
Mcspadden D M3ETI
Mcspirit P M3OWQ
Mcsweeney J GI4CFQ
McSweeney M M6GVD
McTaggart A 2W0MCT
McTaggart A MW3KNE
Mctaggart A MM0CJT
McTaggart D GM0EEG
McTaggart D 2E0CEN
McTaggart D M3ZEI
McTaggart P G6BJO
McTait R G2BKZ
McVay J M3NFA
McVeagh N M6NCM
McVeigh P G4HTW
Mcveigh S MI0SMV
McVey P G3GMC
McVicar J G6MCV
McVittie G MM1DWU
McVittie J G GM0NBG
McVittie M G1UIO
McWatters A G3ZRL
Mcwhinnie C G0MQW
Mcwhinnie M GM0DYU
McWhirter R GI6CXD
McWilliam A G7UZG
McWilliam A GM4MAI
McWilliam N M0RAT
McWilliams L GM0LWD
McWilliams M MI6MYW
McWilliams R GI8LTB
McWilliams T MI6UBT
Md Ali A M0PGL
Mdlongwa Q M3TZS
Meachen E G3SFV
Meachen M M0HLZ
Meacock D G0UFC
Mead A G4KQE
Mead A G6EEE
Mead A G8HUS
Mead A G4DOA
Mead C G1YRM
Mead D G8WQZ
Mead D MW0MWL
Mead D M6BDM
Mead J G6IGY
Mead J G6TDW
Mead L G3ZGQ
Mead L M6EXL
Mead M GW3TZT
Mead P 2E0HGG
Meade I GJ7DNJ
Meadley R M0DPQ

Meadowcroft M 2E0MOZ
Meadowcroft M M6MAA
Meadowcroft M M6MOZ
Meadows C G6KMQ
Meadows C G4KWH
Meadows D G4TGB
Meadows M G4GUG
Meadows M G7BPQ
Meadows P M0PGM
Meads M G3WOT
Meads V G7HLU
Meadwell S G6LEI
Meagher M 2I0ZXM
Meagher M MI6ZTM
Meaker D G6FFB
Meaker J G3RKM
Meaker T M6DQS
Meakin C M1DJB
Meakin C 2E0CTM
Meakin C M3UJY
Meakin D M6OCR
Meakin D M0AVL
Meakin J M0BBK
Meakin N 2E0NHM
Meakin S G6XBD
Meakin T M6TRV
Meakins D G4SCJ
Meakins D G6NHV
Meakins R G8HKN
Meal R MW0RGM
Meale L G4MNA
Meanley G G0GRM
Means G G1EFK
Meanwell P 2E0FGA
Meanwell P M6FGA
Mear D M3LFV
Meares G G0JGM
Mears A 2E0HYM
Mears A M6HYM
Mears C G0AFR
Mears D GW4ONI
Mears G 2E0ICT
Mears G M3PIA
Measom J G6IHU
Measom P G0ZZZ
Measures L G7GKQ
Mecca K M6BOI
Medcalf A G1IPE
Medcalf A G0FRO
Medcalf M G1EFL
Medcalf M M0VAM
Medcalf M M3VAM
Medcalf R G4RUA
Medcalf R G8XDL
Medcraft R GD3JVM
Meddings D 2E0EIU
Meddings J G4DGM
Medhurst A 2E0MED
Medhurst A M0EBQ
Medland A G4WAV
Medland J M6ZIY
Medland R 2E0RCM
Medland R M0RBM
Medley D G0OYX
Medley K G4WUG
Medley L 2E0LEY
Medlicott G M0GGU
Medway R M6DFY
Medza L M6LUM
Mee B G7EXH
Mee C G4JQV
Mee H G5MY
Mee J M6DXI
Mee M G7NFY
Mee M G7UQJ
Mee S G4CTV
Meech R 2E0IJZ
Meech R M0IJZ
Meech R M3ANX
Meecham W G0GAP
Meehan D 2E0UDM
Meehan D M6UDM
Meehan M 2E0MXM
Meehan M M0HHA
Meehan M M6MXM
Meehan R M6RJM
Meek A 2E0LGV
Meek E M3LMU
Meek G 2E0GQB
Meek I MW3NHC
Meek J G6BGY
Meek J 2E0JLM
Meek J M0XJM
Meek L 2E0STK
Meekers E G4SNR
Meekins A M6AHP
Meekins D M6RNF
Meeks J M3HLA
Meerman M G7DQE
Meerman M M0MPM
Meerman P 2E0PHM
Meerman P M0PAQ
Meerman P M6PHM
Meers H G3RTY
Megone R 2E0FTH
Megson G M6CXI
Mehaffey T MI6MTO
Mehmet E 2E0BWD
Mehmet E M6OJO
Meigh S G6ING
Meijer J M0JPM
Meikle H GM0DQC
Meikle L G3TTI
Meiring P G0BSX
Meisenbach W M0CKU
Mekka R G4AWY

Melaku E M1DYI
Melbourne M G0UYQ
Melbourne M G8EHX
Melbourne P G8GML
Meldrum D G3XCO
Meldrum J G8PSS
Melham A G7VNM
Melhuish F M1ART
Melhuish H M6HMK
Melhuish S G4TJC
Melia A G3NYK
Melia C M0GIZ
Melia G G0OPM
Melia R G0CMR
Melia T M6TVM
Mellett G G4MVS
Mellett P G3PIJ
Mellin D G4NKP
Melling B 2E0FJP
Melling P M6ESO
Melling S G8FUH
Mellings C G7KYW
Mellings D G0WJS
Mellings F G0SOG
Mellish M M3FKN
Mellish P M3IEW
Mellor A M3PQL
Mellor C M0HTS
Mellor D G7IRS
Mellor D 2E0DBM
Mellor D M3UDZ
Mellor D G4EWK
Mellor D G0GXT
Mellor I G7EMZ
Mellor K G4RTA
Mellor P G4BIK
Mellor R G0EHO
Mellor R G0GCJ
Mellor S G4TZG
Mellor S 2E0GUY
Mellor S M3XXS
Mellors G G1KNK
Mellors P G0RCP
Melman E 2E0EWM
Melman E M3ZEW
Melody M 2W0HPM
Melton J G0ORX
Melton R G0OJP
Melton S 2E0WYZ
Melton S M6AQL
Melville D G0IOP
Melville I G4EZP
Melville J GM6HLT
Melvin C G4VCB
Melvin J G3LIV
Melvin K GM3ZBR
Melvin S G8UEE
Melvin T GM8MJV
Membury G G8DJW
Membury G M3DOR
Memory D G1IAQ
Menday J 2E0AOG
Menday R G4CXL
Menday V G8HCL
Mendham B M0PVP
Mendoza G G4EUC
Mendum K G8RPA
Menguy J G4VDX
Menhinick J G6RTE
Mennell J M6OAV
Menown P GI4FZD
Menzel K G0FIT
Menzies A MI6MHI
Menzies D MM3HKG
Menzies G G4IQV
Menzies I GM1FSU
Menzies R GM0FHJ
Menzies T GM1GEQ
Mepham A G4CBZ
Mepham A M6TOY
Mepham F G4CDL
Mepham R G4TPJ
Mercer A G4CZK
Mercer D G3YHQ
Mercer D G1TPA
Mercer E MI3IRV
Mercer I G3ZER
Mercer J 2I0HRM
Mercer J MI6MIH
Mercer L 2E0GNI
Mercer N G3UVQ
Mercer P G0TPM
Mercer R G1ZSV
Mercer T M6XTM
Merchant F G7CWN
Merchant J M3HXF
Merckel P MM1CIR
Meredith B M0BBM
Meredith B G4EBG
Meredith C M6MUP
Meredith G G0KXV
Meredith J 2E0FNX
Meredith J M3BTJ
Meredith J G8CZJ
Meredith K M3FIM
Meredith L M6LLM
Meredith M G6EEU
Meredith P M3YAV
Meredith T 2E0TDM
Mereuta V M6VLD
Merifield J G0FKY
Merison S M3FUD
Merrell B G8PUH
Merrell C G8OIV
Merrell R G4GUJ
Merrick D G8VVM

Merrick J 2E0EVM
Merrick L G0VIG
Merrick P M1PTE
Merrick-Jenkins R GW8JJZ
Merridale D 2D0HFT
Merridale D M0ODM
Merridale D M6HFT
Merridale S M6FFT
Merrifield R M6FNE
Merrifield S G7SFI
Merrifield S 2W0BJE
Merrifield S MW3RXK
Merrilees C GM3EOB
Merrill A 2E0GRA
Merrill I M1EBW
Merrils G G0UQF
Merriman P G3YJE
Merriman R G8RIS
Merrin L G0WAX
Merrin M G0WGA
Merrington D G6HKP
Merrington E 2E0FKD
Merrington G G1IVV
Merrington G M3IVV
Merrington H 2E0HEE
Merrington J G7NCP
Merritt F G8LWO
Merritt J 2E0MUT
Merritt J M0HSZ
Merritt J M6ZBA
Merrix A G0TEM
Merry D G1NTV
Merry R G8YDB
Merrylees A G0DCI
Mersi B G7TKG
Merz W G1DTE
Mesbah A 2E0KAR
Mesbah A M0NPT
Mesny H GJ3LFJ
Messenger D M3WMV
Messenger J 2E0XOT
Messenger J M0XOT
Messenger N M3MES
Messer R G3KIL
Messingham R M3TQP
Messner A MM0GNX
Mestel D M3XZG
Meszaros A M6ECJ
Metcalf A G8ETU
Metcalf A M6FKW
Metcalf E G4YDE
Metcalf H G6VXR
Metcalf K G4YSP
Metcalf K G8OWO
Metcalf W G8XLE
Metcalfe B M0AMB
Metcalfe D M6NUF
Metcalfe E G4XLC
Metcalfe I G6TUG
Metcalfe J G0VQJ
Metcalfe J G4YOV
Metcalfe K 2E0WTD
Metcalfe K M0NEG
Metcalfe K M6AAP
Metcalfe L 2E0YSF
Metcalfe L M6IEO
Metcalfe P G6BKL
Metcalfe R 2E0PIW
Metcalfe R M0FTL
Metcalfe R M3HLD
Metcalfe R M6CSC
Metcalfe R GW1LRN
Metcalfe S 2E0CMT
Metcalfe T G1SVI
Metcalfe T G7LTW
Metcalfe W G6HKY
Methven E G1SLP
Metselaar a MD6EZM
Metson H M3ZZH
Mettam R G6XUX
Metters P 2E0PEM
Metters P M6PEM
Mew J 2E0FDM
Mew R G6HKF
Mewis R G1NBO
Meyer B G8YEP
Meyer G M3JPM
Meyer G M1BSM
Meyer K M3WOQ
Meyer M 2E0APZ
Meyer M M6YPZ
Meyers H G3CMU
Meynell S M0AIR
M'Garry-Durrant A G8PL
Miariti A M6EFL
Miazek J M6JIF
Micallef L 2E0LCM
Micallef L M0LCM
Michael D GM0KCY
Michael D G0TSU
Michael D M3XCX
Michael E GM0PKX
Michael K G6OJN
Michael S M6FJW
Michaelis S M6AUZ
Michaelson K G3RDG
Michalak M M3VIB
Michalowski L M0HCM
Michalowski M M6MPP
Michie G GM0IXO
Michie L GM7PXJ
Michilin G M0KUY
Mickleburgh R G4BRF
Micklewright R G3MYM
Middleditch M 2E0MJM
Middleditch M M3OWO

Middlehurst K M6KDM
Middlehurst P G1DVA
Middleton A G1DTF
Middleton A G8WPF
Middleton A M6FRW
Middleton A M6MYD
Middleton D G3VSV
Middleton E G7MOH
Middleton G G4SQK
Middleton G G6EER
Middleton H G0OSR
Middleton I G4BNX
Middleton J G6MGZ
Middleton J 2E0CON
Middleton J G4TXO
Middleton J M0RHQ
Middleton J M3WUH
Middleton J M6BAW
Middleton K G4EJH
Middleton K M3MID
Middleton L G6OPD
Middleton M G2FXV
Middleton P G1CFA
Middleton R G7RDJ
Middleton S G7VYY
Middleton T 2E0HNS
Middleton V G1LFD
Midgeley R G8TVZ
Midgley J G3SAO
Midmore J G3ZXW
Midwood J G7PTD
Midwood P M0SSF
Midworth N G0WTA
Miers F M0PPM
Mieske K 2E0HJA
Mifflin F M0FWM
Mifsud D 2M0MIF
Mifsud D MM6DCM
Mikicki K 2I0HBO
Mikicki K MI0HZD
Mikicki K MI6EMI
Mikolka L 2E0SVK
Mikolka L M0LMI
Mikolka L M6LMI
Milano M M6MLM
Milbourne J M6MIL
Milburn C G1CHM
Milburn C G7NLA
Milburn C G7FDW
Milburn J G8YUJ
Milburn L 2E0ZLM
Milburn L M3RXO
Mildenhall T 2E0BHZ
Milenkovic R G4TNJ
Miles A G1YDG
Miles A M3NRI
Miles A 2M0LFS
Miles A MM0TMZ
Miles A MM6LFS
Miles B M6BCU
Miles C G4TYR
Miles C M0DFH
Miles D G7LED
Miles D M6KVG
Miles E 2E0HKM
Miles E G0RLH
Miles F G3VBE
Miles G G3NIR
Miles G G3TOV
Miles G G7CZL
Miles G M6FCG
Miles I 2W0IWM
Miles I MW6IWM
Miles J 2E0JOF
Miles J M3JOF
Miles K MW6ENY
Miles L G4LNR
Miles L G4TWP
Miles L G6LWZ
Miles N M6NRM
Miles N M6NWM
Miles P GW3KDB
Miles P G1GOY
Miles R G1WUM
Miles R G6FTY
Miles R M3WUM
Miles R G4XXH
Miles R 2W0RAD
Miles R GM4CAQ
Miles R MW0PIC
Miles R MW3ZLX
Miles T GW3NXR
Miles-Williams W G0LZG
Milford L M0LAS
Milioto M M6HWH
Millar A G6JLU
Millar B GI0RYU
Millar C 2E0CMR
Millar C M6CMR
Millar D GM3JJQ
Millar F GM8ZTV
Millar F MM3MMI
Millar G GI0KVQ
Millar G GM4FSB
Millar I G1KMS
Millar I M0GHR
Millar J 2M1IBX
Millar J M6BUG
Millar J 2I0JOS
Millar J MI6JOS
Millar L GM7MBB
Millar P MI3PQM
Millar R G3WLW
Millar S M0AOK
Millar W GI4UPC
Millar W GI6CAG

| Name | Callsign |
|---|---|
| Millar W | MI6NUM |
| Millard A | 2E0TNE |
| Millard B | M1CYP |
| Millard B | M6RRI |
| Millard D | M3FVE |
| Millard D | G8NEY |
| Millard D | M0GHZ |
| Millard D | 2E0MRD |
| Millard D | M6DEL |
| Millard P | 2E0NEY |
| Millard S | M6SSM |
| Millard S | 2E0SUS |
| Millard V | M3VFM |
| Millbank F | G1XWZ |
| Millen D | G7RUC |
| Millen M | M0MLM |
| Millen R | M6GUC |
| Miller A | G1BAQ |
| Miller A | GM1FAI |
| Miller A | GM4ACM |
| Miller A | G7PLV |
| Miller A | G8YAS |
| Miller A | G0KTU |
| Miller A | MM6FDU |
| Miller A | M6TXM |
| Miller B | G7BQT |
| Miller B | G8INL |
| Miller B | M1HHL |
| Miller B | M3DBG |
| Miller C | G6YHL |
| Miller C | G8KVO |
| Miller D | G6JMC |
| Miller D | G6LEY |
| Miller D | G6LYM |
| Miller D | G7TIY |
| Miller D | G4JHI |
| Miller D | G6AWF |
| Miller D | M6NLR |
| Miller D | M0DEN |
| Miller D | GM4MPR |
| Miller E | M3TFZ |
| Miller G | G1DHM |
| Miller G | G6UCI |
| Miller G | G8YYC |
| Miller G | G6WWY |
| Miller G | M3WWY |
| Miller G | G8JIP |
| Miller I | GM4JAE |
| Miller I | G7FFV |
| Miller I | M3OAS |
| Miller I | G3NXX |
| Miller J | G1DKY |
| Miller J | G1OBM |
| Miller J | G3LRU |
| Miller J | G3RUH |
| Miller J | G4CGL |
| Miller J | G6XII |
| Miller J | M0CMP |
| Miller J | 2E0PJJ |
| Miller J | M3MJJ |
| Miller J | G6CLP |
| Miller J | G6SGW |
| Miller J | M6KVD |
| Miller K | G0BWJ |
| Miller K | M1AYR |
| Miller K | M1DJG |
| Miller K | M3DBA |
| Miller K | M3DJG |
| Miller L | G3YEQ |
| Miller M | G0RMO |
| Miller M | 2E0CEM |
| Miller M | G7EGX |
| Miller M | M0ILM |
| Miller N | G3MVV |
| Miller N | G6CDW |
| Miller N | G6BCL |
| Miller N | M3EUM |
| Miller P | GW1PKM |
| Miller P | G4AAW |
| Miller P | G4REE |
| Miller P | 2E0TWP |
| Miller P | M6PJM |
| Miller P | G6ANA |
| Miller R | G4DSY |
| Miller R | G4UKX |
| Miller R | M0BUA |
| Miller R | 2E0RJX |
| Miller R | GI3TJM |
| Miller R | M0RMI |
| Miller R | M6EYJ |
| Miller R | G4FDP |
| Miller R | G8NSX |
| Miller S | M6FLW |
| Miller S | G8JMS |
| Miller T | G1CKR |
| Miller T | G1HWO |
| Miller T | G4BYE |
| Miller T | G6HLU |
| Miller T | G6YHK |
| Miller T | G8XQD |
| Miller W | G3OTW |
| Miller W | GM3PMB |
| Millerchip P | G7NBZ |
| Millerchip R | M0BEO |
| Millership E | G7FND |
| Millett R | G0HBJ |
| Millican B | G4OFA |
| Millichamp T | 2E0TDV |
| Millichip J | M3IJV |
| Milligan D | 2I0WDD |
| Milligan D | MI6WDD |
| Milligan G | G1CPU |
| Milligan J | MI6EJK |
| Milligan W | GM4TPQ |
| Milliken R | G8LGU |
| Millin A | M6AKT |
| Millin D | G4LXY |
| Millington D | G6GSI |
| Millington R | G7MVO |
| Millington J | 2E0GVJ |
| Millington P | M6PTM |
| Millington R | 2E0RJM |
| Millington R | M0RJM |
| Millington R | M3RZV |
| Millington W | 2E0TGB |
| Millington W | M0WBF |
| Millington W | M6AZY |
| Million J | M3ZAL |
| Millis P | M3KXZ |
| Millman I | 2E0MIL |
| Millman I | M0IHM |
| Millman R | G3PJY |
| Mills A | G4GPR |
| Mills A | G8JRZ |
| mills A | M3OLW |
| Mills A | G1PMK |
| Mills A | G3TRR |
| Mills A | M6CWF |
| Mills B | G3VMP |
| Mills B | G7THJ |
| Mills B | M3THJ |
| Mills C | 2E0GLY |
| Mills C | G7EVT |
| Mills C | M1EGV |
| Mills C | M3YLN |
| Mills C | G4BYZ |
| Mills D | G1AFI |
| Mills D | G1UEA |
| Mills D | G7UVW |
| Mills D | M0HCN |
| Mills D | G4PDM |
| Mills D | M6TVX |
| Mills D | M6WBX |
| Mills E | 2E0GGL |
| Mills E | 2E0GHI |
| Mills E | G4DGL |
| Mills E | G4XXB |
| Mills E | M3LVR |
| Mills F | G4XCY |
| Mills F | G4XXA |
| Mills G | G3TWY |
| Mills G | G6ZMG |
| Mills G | G0SSC |
| Mills J | G0FUV |
| Mills J | GI0GQG |
| Mills J | G3VUO |
| Mills J | G4GHM |
| Mills J | G4NOY |
| Mills J | G6LWT |
| Mills J | G6XLC |
| Mills J | G7BDS |
| Mills J | G7BRJ |
| Mills J | G8DBP |
| Mills J | G4XRJ |
| Mills J | MM3CWO |
| Mills J | M6HRO |
| Mills J | GW8HWS |
| Mills K | 2E0KMI |
| Mills K | G4ZOK |
| Mills K | M0KMI |
| Mills K | G0GIN |
| Mills K | M6HBD |
| Mills K | G4DIS |
| Mills K | 2E0BQP |
| Mills K | M0KVM |
| Mills K | M3UOF |
| Mills L | G1HWR |
| Mills M | G0HMD |
| Mills M | G6SYA |
| Mills M | G8CXZ |
| Mills M | 2E0OLF |
| Mills M | M0MLZ |
| Mills M | M6OLF |
| Mills M | G3TEV |
| Mills N | G8HUR |
| Mills P | M6FEC |
| Mills P | M6PBM |
| Mills R | G3ZAV |
| Mills R | G4LPD |
| Mills R | G4MZB |
| Mills R | G8UKY |
| Mills R | G8XRL |
| Mills R | G4ZES |
| Mills R | M3TET |
| Mills S | G7KJT |
| Mills S | M3LXF |
| Mills T | G3UNS |
| Mills T | G3YAI |
| Mills T | G4RQA |
| Mills V | G7AZT |
| Millson V | M6VCM |
| Millsott G | M1CSG |
| Millward A | G7WBW |
| Millward B | G0MDN |
| Millward K | M6KCA |
| Millward M | G0LNT |
| Millward M | M3NEW |
| Millward P | 2E0CWQ |
| Milne A | GM4BFX |
| Milne A | G8LHP |
| Milne A | M3DKK |
| Milne B | G4HIV |
| Milne C | 2E0IMM |
| Milne C | M6IMM |
| Milne D | G6VMI |
| Milne D | GM6WLJ |
| Milne D | G0WXZ |
| Milne G | MM0GXQ |
| Milne G | GM1ZQF |
| Milne G | GM4BLO |
| Milne G | MM5ISS |
| Milne J | 2E0BYF |
| Milne J | GM7AUW |
| Milne J | MM3AUX |
| Milne J | M3RYF |
| Milne M | 2E0MAA |
| Milne M | M0MAF |
| Milne M | M3RXQ |
| Milne R | G6LKG |
| Milne R | G3AKN |
| Milne R | M0DDK |
| Milne S | GM4KOI |
| Milne S | 2M0OAB |
| Milne S | MM0PSM |
| Milne T | 2E0AVA |
| Milne T | G4CMG |
| Milner C | G3ZJK |
| Milner E | 2E0ERM |
| Milner G | G8NWK |
| Milner G | M6FQL |
| Milner J | G7VUP |
| Milner J | M3MRM |
| Milner J | M6MQN |
| Milner P | G1HWJ |
| Milner R | G8RIK |
| Milner S | 2E0EUR |
| Milner S | M6EKP |
| Milner Smith A | M6NXA |
| Milns M | G7LTO |
| Milone K | M0KOO |
| Milosevic J | G1ZFX |
| Milsom G | 2E0BKT |
| Milsom G | M0GGW |
| Milsom G | M3TUU |
| Milsom H | G1WYP |
| Milsom P | G4GSA |
| Milsom S | G8SFA |
| Milton C | G6XIF |
| Milton R | GW0NIY |
| Milton R | 2E0REY |
| Milton R | M3UVC |
| Milton-Eldridge A | M3PQU |
| Mina M | M3BLX |
| Minard C | GW0WUL |
| Minaudo J | G1EOA |
| Minchin P | G6NHW |
| Mindel S | G6NVI |
| Minett A | G3SPP |
| Minett D | G3WPP |
| Minihane K | G1OCY |
| Minihane M | G0UHQ |
| Minish B | G7CUW |
| Minks B | M3KQD |
| Minnock J | 2E0ATH |
| Minnock S | 2E0ABW |
| Minshull M | G1XYV |
| Minshull M | M3MGJ |
| Minter R | G0HVO |
| Mintern K | G8ATP |
| Minton A | MW6NYE |
| Minton B | G0BKH |
| Minton D | G1MCG |
| Miocinovic F | M6MIO |
| Mir D | M0DSX |
| Mir H | M6MIR |
| Mirams F | G6YDO |
| Mirams H | G8UTW |
| Mirams J | G6KJM |
| Miranda R | 2E0RVI |
| Miranda R | M0RVI |
| Miranda R | M6RVI |
| Mirchev M | M0HZV |
| Mirfield J | M6XNO |
| Mirjalili Mohanna S | M6HOM |
| Mirtle P | GM1YFO |
| Miskimmin G | GI6SFO |
| Miskimmin R | GI0DUP |
| Missenden K | 2E0KMZ |
| Missenden K | M6EQE |
| Misson P | MM6PRZ |
| Mister C | G0DAZ |
| Mistofsky M | GM4KLO |
| Misztela J | M6FPP |
| Mitchard D | M6DWM |
| Mitchell A | G0EEJ |
| Mitchell A | G3YJZ |
| Mitchell A | G4BZJ |
| Mitchell A | G4ICE |
| Mitchell A | G6EIO |
| Mitchell A | M6AJM |
| Mitchell A | G1AHS |
| Mitchell A | G0TDK |
| Mitchell B | G0NXN |
| Mitchell B | G3HJK |
| Mitchell B | G4MLI |
| Mitchell B | GI8TWB |
| Mitchell B | M0MIT |
| Mitchell C | G0PKM |
| Mitchell C | G0SKA |
| Mitchell C | M1ELB |
| Mitchell D | G6WNX |
| Mitchell D | G7FGZ |
| Mitchell D | G7MZK |
| Mitchell D | M6MIN |
| Mitchell D | MM0CZH |
| Mitchell D | M6DNM |
| Mitchell D | G2DLX |
| Mitchell E | G1POD |
| Mitchell E | MM1BMK |
| Mitchell F | G3UZX |
| Mitchell G | G7BJR |
| Mitchell G | M6HFI |
| Mitchell G | G4AUL |
| Mitchell G | G7FMJ |
| Mitchell H | G2AMG |
| Mitchell H | G6IHG |
| Mitchell Hynd L | 2M0UMH |
| Mitchell Hynd L | MM0UMH |
| Mitchell Hynd L | MM6LMH |
| Mitchell I | G0BUK |
| Mitchell I | G7MZJ |
| Mitchell I | M3VOH |
| Mitchell I | G4NSD |
| Mitchell I | GW4XAZ |
| Mitchell I | 2E0ITM |
| Mitchell J | G0JPM |
| Mitchell J | GD0PLQ |
| Mitchell J | G3KEQ |
| Mitchell J | G4TPR |
| Mitchell J | G6OGZ |
| Mitchell J | M0JLM |
| Mitchell J | M1JDW |
| Mitchell J | 2E0RMJ |
| Mitchell J | M6JEZ |
| Mitchell J | GW0GRQ |
| Mitchell J | G6DQB |
| Mitchell J | M3WZN |
| Mitchell J | M3XVQ |
| Mitchell K | G7NHL |
| Mitchell K | 2E0XXK |
| Mitchell K | M0XXK |
| Mitchell K | M3XXK |
| Mitchell K | 2I0KXM |
| Mitchell K | MI6KXM |
| Mitchell K | M6KJM |
| Mitchell K | MM6KGM |
| Mitchell L | MM6BOQ |
| Mitchell M | 2M0BEC |
| Mitchell M | MM3IOF |
| Mitchell M | G1MDQ |
| Mitchell N | GI0LRZ |
| Mitchell N | G7TUG |
| Mitchell N | M3WMI |
| Mitchell P | 2E0PWM |
| Mitchell P | G1PJM |
| Mitchell P | M3PWM |
| Mitchell P | G4XYK |
| Mitchell P | G8BRG |
| Mitchell R | G1IZN |
| Mitchell R | G3YBM |
| Mitchell R | GM4HJK |
| Mitchell R | G4KVC |
| Mitchell R | G7IOC |
| Mitchell R | MM0JUL |
| Mitchell R | M3IOC |
| Mitchell R | M6RMU |
| Mitchell R | G8REO |
| MITCHELL S | G0JNQ |
| Mitchell S | M0HQL |
| Mitchell T | G0GLH |
| Mitchell T | G1JZY |
| Mitchell T | G3LMX |
| Mitchell T | 2E0RWF |
| Mitchell T | M0TMH |
| Mitchell T | GM4OHT |
| Mitchell W | G4IBI |
| Mitchell W | M3MUQ |
| Mitchell W | M1EBU |
| Mitchell-Watson C | 2E0HLO |
| Mitchell-Watson C | M3CAZ |
| Mitchell-Watson W | G0LUM |
| Mitchelson N | G0IYQ |
| Mitchener G | M0GIM |
| Mitchener J | G0DVJ |
| Mitchinson D | G4KGN |
| Mitchinson J | G7GLA |
| Mitchinson N | G0FTJ |
| Miyake S | M0BJJ |
| Mloduchowski T | 2E0MTM |
| Mloduchowski T | M6MQT |
| Moakes S | 2E0BDI |
| Moakes S | M3LQD |
| Moan T | G4MSJ |
| Moar J | GM4EFR |
| Moat R | G1PQX |
| Moate K | G0CRL |
| Moate R | G0CHG |
| Mobberley P | G8UEF |
| Mobbs A | G8EEY |
| Mobbs A | M0RGF |
| Mobbs A | M0RGY |
| Mobbs C | G8UHW |
| Mobbs D | G1IPD |
| Mobbs D | G4MEE |
| Mobbs J | G0ONG |
| Mobbs J | M3YEU |
| Mobbs J | M6JMG |
| Mobley D | G1AHQ |
| Mobley L | M3WGK |
| Mobley S | G7JVF |
| Mocatta F | M3MOC |
| Mock A | M0OCK |
| Mock A | M6OCK |
| Mock C | G6FLU |
| Mock J | M3NYX |
| Mockford A | M0GKD |
| Mockford A | G8ZGK |
| Mockford C | G7APM |
| Mockridge N | G6EED |
| Mocroft W | M5LLT |
| Modha J | M3FAE |
| Modi N | 2E0NBM |
| Moe A | M0MRM |
| Moerman A | MM6BMP |
| Moerman B | MM6AUG |
| Moffat D | 2E0DMX |
| Moffat G | MM0AWU |
| Moffat I | G0OZS |
| Moffat J | MM6JPM |
| Moffat K | GM0CFW |
| Moffat M | 2E0IKM |
| Moffat M | M6IKM |
| Moffat N | M1ZXG |
| Moffat S | G4LHF |
| Moffat T | 2M0MOF |
| Moffat T | MM6TCS |
| Moffatt A | M0FAT |
| Moffatt D | M0JIW |
| Moffatt D | G3RAU |
| Moffatt G | G4ZRA |
| Moffitt C | M6MOF |
| Moffitt P | M6PDM |
| Moffitt T | GI4KQA |
| Moger D | G8ZRU |
| Moger D | 2E0DBQ |
| Moger D | M0ZYD |
| Moger D | M6AYK |
| Mogford K | 2E0KEQ |
| Mogford K | M0KEQ |
| Mogford K | M6KEQ |
| Mogford P | GW4PCO |
| Moggeridge J | G1YHB |
| Moggeridge M | M6XTR |
| Moggridge A | G0OEA |
| Mohammed J | M3UOC |
| Mohammed K | G0OVE |
| Mohammed Shafi M | 2E0MSM |
| Mohammed Shafi M | M6AQE |
| Moir C | 2M0HCF |
| Moir G | 2M0BXN |
| Moir G | MM3YQO |
| Moir J | 2M0YOY |
| Moir J | MM6YOY |
| Moir N | GM7RVR |
| Moissejev S | 2E0KGB |
| Moissejev S | M0TLN |
| Moissejev S | M3WNX |
| Moisy S | G2FXJ |
| Mok W | M3YHM |
| Mokes C | 2E0SAU |
| Mokes C | M3TKP |
| Mold J | G1GMX |
| Mold R | 2E0ABD |
| Mold R | G7PTZ |
| Moldoveanu R | 2E0TRX |
| Moldoveanu R | M6REX |
| Mole C | M0LAL |
| Mole C | M6CFT |
| Mole F | M0GHW |
| Mole J | M3DFU |
| Mole S | G1ERU |
| Mole S | M3SMM |
| Mole S | G1CBK |
| Mole T | G8WBL |
| Molendijk G | 2E0XNL |
| Molendijk G | M6GFH |
| Moles R | 2E0MOL |
| Moles R | M3YRM |
| Molinghen J | G8ZOO |
| Moll R | M5REV |
| Mollard G | M0MOL |
| Mollart J | G4IMV |
| Moller C | G3XPW |
| Moller G | GW4WFM |
| Moller J | GI6JMD |
| Molloy D | M0HGD |
| Molloy M | G0PGQ |
| Molloy M | G4DWR |
| Molloy S | MI3BRJ |
| Molloy W | 2E0DKM |
| Molloy W | M6DKM |
| Molnar K | M0SLC |
| Molyneaux G | M0LYN |
| Molyneux C | G6YHP |
| Molyneux J | G6DCH |
| Molyneux K | M0ABF |
| Monaghan J | MI6JBH |
| Monaghan K | MM6BQH |
| Monaghan K | 2E0TWO |
| Monaghan K | M3HKM |
| Monaghan N | M6RKQ |
| Monaghan S | 2M0OVV |
| Monaghan S | MM3OVV |
| Monaghan T | G1VJQ |
| Monahan J | 2E0JPM |
| Monahan J | M6JMO |
| Monahan R | GM1FLQ |
| Moncaster T | 2E0TTM |
| Moncaster T | M0TMA |
| Monckton C | G7APO |
| Money D | G3HKD |
| money D | M6HJH |
| Money D | M6HDM |
| Money E | M6HVY |
| Money M | M6PBE |
| Money P | G4UVA |
| Money T | 2E0AOL |
| Money T | M0AKY |
| Monk A | GM7LFT |
| Monk D | G6EET |
| Monk R | G0HUQ |
| Monk R | M3MON |
| Monk S | G0EVI |
| Monk W | G1TBI |
| Monks G | G4AYH |
| Monks I | G0NKM |
| Monks J | M0AGL |
| Monksummers B | M3UZE |
| Monksummers C | G0ZFF |
| Monksummers R | G6TER |
| Monnery D | G4VFC |
| Monnington D | M3AXX |
| Monro M | G8DLL |
| Monshall L | M6BLO |
| Montague C | 2M0PBC |
| Montague R | G7CHB |
| Montague R | M3RHP |
| Montanana N | G8RWG |
| Monte J | G8LCS |
| Monteith C | G0CXY |
| Monteith G | MI6EAF |
| Monteith J | MI3JTM |
| Montford R | G6RPD |
| Montford W | G3JZI |
| Montgomery A | MM0CTQ |
| Montgomery C | G0AWM |
| Montgomery F | MI1FRM |
| Montgomery J | G0SUH |
| Montgomery K | G8ECG |
| MONTGOMERY P | G1KKA |
| Montgomery P | M6MON |
| Montgomery R | 2I0RBV |
| Montgomery S | 2I0SHM |
| Montgomery W | MI6FAU |
| Monument N | G0RWQ |
| Monument N | G4BTX |
| Moodie D | GM4FOZ |
| Moodie D | GM8KJO |
| Moodie R | MM0AMV |
| Moodie W | G6EEB |
| Moodie W | MM6WMM |
| Moody A | MW6ICM |
| Moody B | G0UGA |
| Moody D | G0HVQ |
| Moody E | M0AXX |
| MOODY F | G4HVW |
| Moody G | M0BEJ |
| Moody G | G4LIM |
| Moody I | G6FDO |
| Moody J | M6BGC |
| Moody J | 2E0JGM |
| Moody K | G4NRZ |
| Moody K | M3RUK |
| Moody M | M0LMO |
| Moody R | M0GYA |
| Mookerjee S | M0OKS |
| Moon A | G3RGB |
| Moon E | M3MZR |
| Moon G | 2E0GTM |
| Moon G | M6CWA |
| Moon H | G6KQJ |
| Moon J | G4WAX |
| Moon R | G0CLF |
| Mooneapillay J | 2E0NCY |
| Mooneapillay J | M6NCY |
| Mooney C | 2I0NAT |
| Mooney C | MI3WFT |
| Mooney D | MI6XBL |
| Mooney D | 2E0DCZ |
| Mooney D | M6DCZ |
| Mooney E | GI4EQA |
| Mooney J | G0ANM |
| Mooney J | G6ZMO |
| Mooney J | MI3WVL |
| Mooney J | M6CNX |
| Mooney L | M6VHM |
| Mooney N | M0NFI |
| Mooney N | M6NFI |
| Mooney P | G4FWK |
| Mooney W | G3VZU |
| Mooney W | G4LOR |
| Moonie C | MI6MHK |
| moorcroft K | G7OWB |
| Moorcroft S | M6FAE |
| Moore A | G0EAM |
| Moore A | G1KXZ |
| Moore A | G1PRP |
| Moore A | GD3GBG |
| Moore A | G3VSU |
| Moore A | G4BRL |
| Moore A | G4RHX |
| Moore A | M1AIM |
| Moore A | M1CFZ |
| Moore A | G4AFX |
| Moore A | G0RFE |
| Moore B | GI8DGB |
| Moore B | MI0BMM |
| Moore B | MW6VOW |
| Moore C | G4ZOX |
| Moore C | G6CZS |
| Moore C | G7DAH |
| Moore C | M1ALR |
| Moore C | G6EYI |
| Moore C | M6FSG |
| Moore C | M6OZT |
| Moore D | G0DHM |
| Moore D | GM0GRL |
| Moore D | GI0TSA |
| Moore D | G1THG |
| Moore D | G3LSA |
| Moore D | G6DOH |
| Moore D | M0RIC |
| Moore D | M0TYG |
| Moore D | 2E0XLG |
| Moore D | GM4HPK |
| Moore D | MM3RDP |
| Moore D | G0FZH |
| Moore E | GD1RHT |
| Moore E | 2E0EDM |
| Moore E | M0TEK |
| Moore E | 2E0ICY |
| Moore E | M0RZY |
| Moore E | M6ORE |
| Moore F | G3VST |
| Moore G | G0IOF |
| Moore G | G6JBL |
| Moore G | G7NVZ |
| Moore G | M0ORE |
| Moore G | 2E0CWM |
| Moore G | GI4YDP |
| Moore G | M0GUU |
| Moore G | M6GAC |
| Moore G | G4BNT |
| Moore H | 2E0HTR |
| Moore H | M6LRV |
| Moore I | G0CAX |
| Moore I | GM4KLN |
| Moore I | M0WIZ |
| Moore I | M6TCF |
| Moore J | G0AXB |
| Moore J | G1RYI |
| Moore J | G7RMY |
| Moore J | M3JMX |
| Moore J | M6HJV |
| Moore J | M3NNJ |
| Moore J | 2E0ZAL |
| Moore J | GM0EWW |
| Moore J | G7TUM |
| Moore J | G8GEV |
| Moore J | GW8GQE |
| Moore J | M6BLV |
| Moore J | M6HBC |
| Moore K | G1PRE |
| Moore K | G7HCT |
| Moore K | M3WGB |
| Moore L | G3VFH |
| Moore M | G3NRM |
| moore M | M6UBR |
| Moore M | 2W1GKJ |
| Moore M | M6HCO |
| Moore N | 2E0FET |
| Moore N | GI7CMC |
| Moore P | GM0DNH |
| Moore P | GI8UIU |
| Moore P | G8MKS |
| Moore P | M6PAJ |
| Moore P | 2I0ETW |
| Moore P | 2E0FYQ |
| Moore P | G3TUW |
| Moore P | MI0HWG |
| Moore P | M6FYQ |
| Moore P | MI6NID |
| Moore P | 2E0TZZ |
| Moore P | GI0VAB |
| Moore P | M0TZZ |
| Moore P | M6TZZ |
| Moore R | 2E0ETB |
| Moore R | G0KPG |
| Moore R | GI3PLL |
| Moore R | G3YOC |
| Moore R | G3YPM |
| Moore R | G3YUX |
| Moore R | G4LWU |
| Moore R | G8AXR |
| Moore R | MM1FJM |
| Moore R | M3RMX |
| Moore R | G4CEJ |
| Moore R | G1IWT |
| Moore R | G4KRF |
| Moore R | M6RVM |
| Moore R | M6ACV |
| Moore S | 2E0SJM |
| Moore S | G0GTV |
| Moore S | G1NNN |
| Moore S | G6FKP |
| Moore S | GI8YTH |
| Moore S | MM6TXN |
| Moore S | M6TZU |
| Moore S | M3KIU |
| Moore S | M6FOT |
| Moore T | 2E0ZTM |
| Moore T | G0CXD |
| Moore T | G3AJD |
| Moore T | MI3TJM |
| Moore T | GW1JOV |
| Moore W | G1BKZ |
| Moore W | G8DTT |
| Moore W | GW4THK |
| Moore W | G7AIR |
| Moore W | G7EUG |
| Moore X | G4DML |
| Moorecroft B | G4UDY |
| Moorecroft M | G1WXT |
| Moores R | G3GZT |
| Moorey B | G7KVT |
| Moorey D | M0BYI |
| Moorfield G | G3ZBG |
| Moorhead T | G0VMT |
| Moorhouse G | 2E0ITV |
| Moorhouse G | M3CKU |
| Moorhouse J | G3VQF |
| Moorhouse J | M6ZYQ |
| Moorhouse M | G7RAI |
| Moppett C | M6CSM |
| Moppett J | 2E0JBM |
| Moppett J | M3SWV |
| Moppett S | M1DNE |
| Moran D | M3UDC |
| Moran D | 2E0DMM |
| Moran D | 2E0YJF |
| Moran D | M3YJF |
| Moran J | 2E0HLA |
| Moran J | M0BGU |
| Moran J | G3XUM |
| Moran K | G4MOH |
| Moran M | M0RAN |
| Moran P | G0JIB |
| Moran P | G0WER |
| Moran P | MM0CEZ |
| Moran V | MM3VHM |
| Morbey D | M3YQM |
| Morcom C | G8VEI |
| Mordas E | G4TJM |
| Mordaunt R | M3DFS |
| Morden W | G1OHL |
| Mordue R | G1SDN |
| More J | GM6SDV |
| More K | M0CKM |
| Moreau P | G0FWU |
| Morecroft A | G4DFP |
| Morecroft J | G8OOC |
| Morehen A | M6UPS |
| Moreland D | G7FGA |
| Moreland L | M3FCR |
| Moreman E | G2BTZ |
| Moreno J | MM0IBO |
| Moreton C | 2E0RED |
| Moreton C | M0LUK |
| Moreton C | M6USK |
| Moreton J | M3DVB |
| Moreton K | G1VBY |
| Moreton R | M0EWW |
| Moreton W | G7VYW |
| Morey K | G4NCD |
| Morey R | G4NAK |
| Morgan A | 2E0XAM |
| Morgan A | GW0HCK |
| Morgan A | G0MWY |
| Morgan A | GW1SGE |
| Morgan A | G6JGF |
| Morgan A | G8WWM |
| Morgan A | M0XAM |
| Morgan A | M3XAJ |
| Morgan A | M3XAM |
| Morgan A | M5ZAP |
| Morgan A | M6ZIZ |
| Morgan A | G1MME |
| Morgan A | GD1MIP |
| Morgan A | G6PPD |
| Morgan B | G0LXR |
| Morgan B | GW0GQC |
| Morgan B | MW6ZBR |
| Morgan C | G3ZTJ |
| Morgan C | G4OLY |
| Morgan C | G8HCW |
| Morgan C | G8LPX |
| Morgan C | G8NBF |
| Morgan C | M0XTD |
| Morgan C | GW3RYR |
| Morgan C | M0PCH |
| Morgan C | M6CON |
| Morgan D | 2E0MYH |
| Morgan D | 2E0ZRM |
| Morgan D | 2E0HRM |
| Morgan D | GW0POZ |
| Morgan D | G4NSA |
| Morgan D | G6MLI |
| Morgan D | G7BWO |
| Morgan D | M3HRM |
| Morgan D | 2E0CXV |
| Morgan D | G8LMI |
| Morgan D | M1FDN |
| Morgan D | M6DCI |
| Morgan D | M6DNP |
| Morgan D | G3ZKN |
| Morgan E | G1BGJ |
| Morgan E | M3WQV |
| Morgan E | G4AIU |
| Morgan F | 2E0FNB |
| Morgan F | GW8THL |
| Morgan G | G3FNO |
| Morgan G | G3SNR |
| Morgan G | G3ROG |
| Morgan G | G3POM |
| Morgan H | G1FWF |
| Morgan H | GI4ZTU |
| Morgan H | M6FSM |
| Morgan I | GM3OZJ |
| Morgan I | G6VKS |
| Morgan I | M6IJM |
| Morgan J | G0KPQ |
| Morgan J | G0PXO |
| Morgan J | G3HAA |
| Morgan J | G3ZHL |
| Morgan J | G8SYV |
| Morgan J | M1CIE |
| Morgan J | M6JTM |
| Morgan J | MI3FCK |
| Morgan J | 2E0TOF |
| Morgan J | G1PLJ |
| Morgan J | G3YIK |
| Morgan J | GW4UEP |
| Morgan J | G8OHH |
| Morgan J | M0HCW |
| Morgan J | M0TOF |
| Morgan J | M6BVV |
| Morgan J | M6TOF |
| Morgan Jones I | M0IMJ |
| Morgan K | G1KHM |
| Morgan K | G3NWX |
| Morgan K | M3LFL |
| Morgan K | M3VQD |
| Morgan L | GM0ATQ |
| Morgan L | G0WDG |
| Morgan L | G4MSW |
| Morgan M | G4WLK |
| Morgan M | G6RGT |
| Morgan M | G6DQA |
| Morgan P | G1DSJ |
| Morgan P | GW1RZE |
| Morgan P | G7SKF |
| Morgan P | M0AXZ |
| Morgan P | M6WFC |
| MORGAN P | G0KYX |
| Morgan P | MD6IOM |
| Morgan R | 2E0ROB |
| Morgan R | G6NTQ |
| Morgan R | M0TTT |
| Morgan R | M3VKM |
| Morgan R | MW6AIC |
| Morgan R | 2E0CZT |
| Morgan R | M0RHO |
| Morgan R | M6RFG |
| Morgan R | G0BMB |
| Morgan R | GW8VGB |
| Morgan R | M6PTO |
| Morgan S | GW0SQY |
| Morgan S | G4SUS |



**IMPORTANT NOTE**

**Revalidate licence to avoid revocation** – Ofcom has advised the Society that plans will be drawn up to revoke licences that have not been revalidated as required by the licence conditions. The quickest way to revalidate is to do so online via the Ofcom website: *https://services.ofcom.org.uk/* or by email: *amateur.validations@ofcom.org.uk* If you need assistance in the process, Ofcom staff are available to help, but please be patient during times of heavy workload.

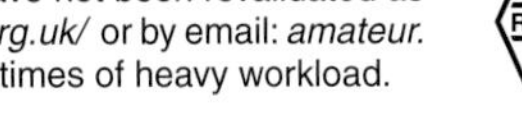

| Name | Callsign |
|---|---|
| Morgan S | G8VCU |
| Morgan S | MM3OYB |
| Morgan S | M6BNF |
| Morgan S | 2E0YAB |
| Morgan S | M3YDS |
| Morgan S | MW6CBL |
| Morgan S | M6VIO |
| Morgan T | GW4SML |
| Morgan T | G3XMM |
| Morgan T | G0CAJ |
| Morgan T | GD3UAS |
| Morgan W | M3WAF |
| Morgan-Lucas M | M6GLU |
| Morgans C | G4LUO |
| Morgan's R | M3TPU |
| Mori A | 2E0JVM |
| Moriarty A | G4TRY |
| Moriarty M | M6PIA |
| Moriarty R | G0IFN |
| Moriarty S | M1AOG |
| Moring S | G6NAX |
| Morison I | G0DMU |
| Moritz J | M0BMU |
| Morley A | G7OZI |
| Morley A | G7FNN |
| Morley C | M0HRH |
| Morley D | M3RVM |
| Morley D | M6MJV |
| Morley E | G7WGZ |
| Morley E | M0ACA |
| Morley H | GW4IUK |
| Morley I | G8VPD |
| Morley J | G4FSQ |
| Morley M | G1PHN |
| Morley M | G4BQA |
| Morley M | G4OII |
| Morley N | G8OZT |
| Morley O | G7WHA |
| Morley P | G4FIV |
| Morley R | G4BNL |
| Morley R | M3FEA |
| Morley R | G8PRN |
| Morley R | M3CZL |
| Morley S | G0MCV |
| Morley S | G7OZJ |
| Morley T | G4DUJ |
| Morley T | G4WAG |
| Morley T | G6MAD |
| Morling A | G6CDV |
| Morling P | M3NNM |
| Morne A | G6EYS |
| Morphett C | G1VIF |
| Morphew D | M6SLC |
| Morphew N | M3JBF |
| Morrall R | G8ZHA |
| Morrell A | M0YDK |
| Morrell S | M6SAV |
| Morrell W | M6ULZ |
| Morrell-Cross L | G7HEY |
| Morrell-Cross L | M3TIE |
| Morrell-Tourle B | G7HLG |
| Morrey S | G0MOR |
| Morrice J | G0OAJ |
| Morrin B | G0FYW |
| Morris A | 2E0ICK |
| Morris A | 2E0IUH |
| Morris A | 2E0HDE |
| Morris A | G0BRQ |
| Morris A | G0UOZ |
| Morris A | GM1RKI |
| Morris A | G1VKB |
| Morris A | G7TIW |
| Morris A | M0CWX |
| Morris A | M0HDE |
| Morris A | M3IUH |
| Morris A | M6WEE |
| Morris A | G4ENS |
| Morris A | M6MTK |
| Morris A | G4KKS |
| Morris A | G6ZPR |
| Morris A | G6VQN |
| Morris A | G6KLC |
| Morris B | G3VBG |
| Morris B | G4KSQ |
| Morris B | G4XXF |
| Morris B | G6VIF |
| Morris B | M6BMO |
| Morris C | 2E0AZK |
| Morris C | G0CUZ |
| Morris C | G6FOF |
| Morris C | 2E0TYG |
| Morris C | M0TBB |
| Morris C | M6TYG |
| Morris C | 2M0DIB |
| Morris C | MM6NSM |
| Morris C | M6CSI |
| Morris D | G0CGS |
| Morris D | G0VID |
| Morris D | G3REW |
| Morris D | G3WFH |
| Morris D | G4GVZ |
| Morris D | G4HMR |
| Morris D | G6RHN |
| Morris D | M0ADG |
| Morris D | M3BXE |
| Morris D | GM3YEW |
| Morris D | G6CKJ |
| Morris D | G7DOA |
| Morris D | MW0DVM |
| Morris D | M3DKG |
| Morris D | GW6FNB |
| Morris G | G1ATZ |
| Morris G | G1GCJ |
| Morris G | G3ATZ |
| Morris G | G4CEP |
| Morris G | G4YWN |
| Morris G | G7VND |
| Morris G | 2E0EPL |
| Morris G | M6GPN |
| Morris G | G6KQD |
| Morris G | M6FMN |
| Morris G | G3SGC |
| Morris G | M0AXO |
| Morris G | M6GJM |
| Morris H | G3MUJ |
| Morris H | M0VCC |
| Morris H | G4BWL |
| Morris I | M3TYI |
| Morris I | G7UAV |
| Morris J | G0EYA |
| Morris J | G3XHW |
| Morris J | G4ANB |
| Morris J | G4LMK |
| Morris J | G6PEP |
| Morris J | M3POP |
| Morris J | M3VOZ |
| Morris J | 2M0FLJ |
| Morris J | MM0TSB |
| Morris J | MM3SQJ |
| Morris J | G8ENS |
| Morris J | 2E0BYO |
| Morris J | G0DVB |
| Morris J | G4BXS |
| Morris J | G6DDF |
| Morris J | GM8ZFW |
| Morris J | M3YWU |
| Morris J | M6RQM |
| Morris K | 2E0OPU |
| Morris K | G1BSY |
| Morris K | M3OPU |
| Morris K | G1MMD |
| Morris K | M6BXW |
| Morris K | M0WIK |
| Morris K | M6KEM |
| Morris L | G8PMR |
| Morris M | G1MFK |
| Morris M | G4LDW |
| Morris M | G6NVH |
| Morris M | G7CSI |
| Morris M | G7DNX |
| Morris M | M3XEL |
| Morris M | G4WFC |
| Morris M | M3FBR |
| Morris N | G8KOQ |
| Morris N | 2M0MHN |
| Morris N | MM0NHM |
| Morris N | MM6NHM |
| Morris N | M6NJM |
| Morris N | G1ISY |
| Morris O | 2E0OZY |
| Morris O | M3OTM |
| Morris O | M3OZY |
| Morris P | 2E0EXL |
| Morris P | G1IAG |
| Morris P | G4XNR |
| Morris P | G7GJO |
| Morris P | M0ZPG |
| Morris P | M3PFM |
| Morris P | 2E0PCQ |
| Morris P | M3UNT |
| Morris P | M6PCQ |
| Morris P | G7NRR |
| Morris P | G6EES |
| Morris R | 2E0DZJ |
| Morris R | G0JZH |
| Morris R | G0OKI |
| Morris R | G0SVU |
| Morris R | G0VZH |
| Morris R | G3MHY |
| Morris R | GM4RRP |
| Morris R | G6XLB |
| Morris R | GW7EXQ |
| Morris S | G4FQM |
| Morris S | G8KIZ |
| Morris S | G7MFE |
| Morris S | 2E0SPM |
| Morris S | M0SXM |
| Morris S | M6FEN |
| Morris S | M6SPM |
| Morris S | M6VLS |
| Morris S | M1XTN |
| Morris S | M6EXA |
| Morris T | GW6JFV |
| Morris V | G0HTO |
| Morris V | G4PLY |
| Morris W | GW1NBW |
| Morris W | M6FST |
| Morris-Jones D | M6DOA |
| Morrison A | 2E0BDD |
| Morrison A | G3KGA |
| Morrison A | GM4HQZ |
| Morrison A | MM6FGR |
| Morrison A | MM0IOL |
| Morrison C | GI4FUE |
| Morrison C | M1PTT |
| Morrison D | G4JHW |
| Morrison D | GM0LZE |
| Morrison D | GM1BAN |
| Morrison D | MM3YHA |
| Morrison G | MM6LRX |
| Morrison I | G3ZWM |
| Morrison I | GM4MIM |
| Morrison J | G0ICT |
| Morrison J | G0JAM |
| Morrison J | 2I0MMA |
| Morrison J | MI6MUC |
| Morrison J | 2E0ABN |
| Morrison J | 2I0TXB |
| Morrison K | GM6KTP |
| Morrison K | G7CFW |
| Morrison K | G7OAX |
| Morrison K | M1VHT |
| Morrison L | G6OFD |
| Morrison L | GM7ADU |
| Morrison M | GM7ADY |
| Morrison M | MM3YUU |
| Morrison P | G3ZDT |
| Morrison P | G0VHT |
| Morrison R | G6CKN |
| Morrison R | GM7PKT |
| Morrison R | MM3YMU |
| Morrison S | MM6IVP |
| Morrison W | G8LQB |
| morrison W | MM3KVV |
| Morrison-Bates W | M6RGX |
| Morrison-Smith D | G7VBY |
| Morris-Roe G | M6GKF |
| Morriss A | G4GEN |
| Morrissey A | GJ3YLI |
| Morrissey B | G4YK |
| Morrissey M | G3HUK |
| Morrow A | MI6MFR |
| Morrow C | M6MXX |
| Morrow D | M0HZP |
| Morrow H | GI4KSH |
| Morrow I | MI1CCU |
| Morrow N | MI6CCU |
| Morrow S | 2I0GCN |
| Morrow S | MI0ULK |
| Morse G | G4FRB |
| Morse G | M0RCE |
| Morse J | G0URB |
| Morse M | M3MMZ |
| Morse N | 2E0NKM |
| Morse N | M6NKM |
| Morse V | G8IK |
| Morse W | MW3AQI |
| Morson-Pate F | M6FMP |
| Morstatt J | G7AMQ |
| Mort A | G4AXY |
| Mort D | M0TCL |
| Mortiboy P | M3XTK |
| Mortimer C | G4OXR |
| Mortimer C | G0WBC |
| Mortimer E | M6CDU |
| Mortimer F | G0ULS |
| Mortimer G | G3DQG |
| Mortimer J | 2E0CXE |
| Mortimer J | M0JAM |
| Mortimer R | G4AUG |
| Mortimer T | G2JL |
| Mortimore R | GW4BVJ |
| Morton A | G7PZL |
| Morton A | GM8BJJ |
| Morton A | M6SFJ |
| Morton B | G4HWA |
| Morton C | GM3RUP |
| Morton C | 2I0NIE |
| Morton C | MI3NIE |
| Morton D | G4LOX |
| Morton D | G6EYJ |
| Morton D | M6EBP |
| Morton D | G1IAD |
| Morton E | G4CDC |
| Morton G | G0IGU |
| Morton I | G4KNT |
| Morton J | GM1GDO |
| Morton J | M6TZO |
| Morton K | M3PSI |
| Morton M | 2E0XIS |
| Morton N | G0SSE |
| Morton R | G0FTO |
| Morton R | G7WJJ |
| Morton S | M6XEE |
| Morton T | G4HZT |
| Morton T | GM4RTN |
| Morton T | G7TPD |
| Morton T | M3VQL |
| Morton-Thurtle P | G8UIV |
| Morys P | G4HQX |
| Mosby J | 2E0CRI |
| Moscrop T | M3LXU |
| Mosedale N | G8TBL |
| Mosedale R | G4PJK |
| Moseley A | G1ERY |
| Moseley A | G7VOT |
| Moseley B | M3BJW |
| Moseley C | M3LLT |
| Moseley J | G0NYH |
| Moseley J | G1SOY |
| Moseley J | G2CIW |
| Moseley M | G6UAN |
| Moseley M | M6WDC |
| Moseley R | G7DRG |
| Moser G | G3HMR |
| Moses D | 2E0EHP |
| Moses D | M3EHP |
| Moses R | G6HKZ |
| Mosley H | G1KHH |
| Mosner G | M0NAX |
| Moss A | G0MJU |
| Moss A | G0ORY |
| Moss A | G3URJ |
| Moss A | G4VVT |
| Moss A | G7WAQ |
| moss A | M6HXE |
| Moss A | M0CZA |
| Moss A | G6DDA |
| Moss A | M1BYH |
| Moss B | G0FYE |
| Moss C | M0CFX |
| Moss C | M6CJW |
| Moss C | G3TXK |
| Moss C | G8IEZ |
| Moss F | G1TTL |
| Moss G | G0LCT |
| Moss G | M0GQE |
| Moss G | 2E0YGS |
| Moss G | M6GMS |
| Moss H | G4OYT |
| Moss J | G0KTW |
| Moss J | G1KIW |
| Moss M | G0EJD |
| Moss M | G3RZI |
| Moss M | G8NVX |
| Moss M | 2E0NKR |
| Moss M | M6MYR |
| Moss P | G0AHE |
| Moss P | G0UYF |
| Moss P | G1NCD |
| Moss P | G3ZHJ |
| Moss P | G4BUP |
| Moss P | G7JUJ |
| Moss P | M3OSP |
| Moss P | G3SQA |
| Moss R | G7MWC |
| Moss R | M0SSY |
| Moss R | G7UVO |
| Moss S | M6TQW |
| Moss T | M3HVP |
| Mossman J | G7JKK |
| Mossman J | M0SSE |
| Mossop F | G0DUB |
| Mossop J | G1GWS |
| Mossop L | M0LSA |
| Moston I | G6GQG |
| Mostyn D | M3DRM |
| Motala S | G7RCK |
| Moth I | G4MBD |
| Moth W | G7IRH |
| MOTHEW A | G0LWM |
| MOTHEW A | G7EEE |
| Mott A | G6EYD |
| Mott G | G1FRL |
| Mott K | G0HRR |
| Mott R | G0ECX |
| Mottart E | GM0FHD |
| Mottart E | GM0JLJ |
| Mottershead R | M3VKF |
| Mott-Gotobed C | G7LJB |
| Mott-Gotobed C | G4ODM |
| Mott-Gotobed J | G6JDP |
| Mottram J | 2E0SEY |
| Mottram J | M6LLU |
| Mottram J | G8MUX |
| Mouland D | M6DGM |
| Mould A | 2E0ESQ |
| Mould A | G7FAZ |
| Mould G | G7VCT |
| Mould J | M6JLM |
| Moulder A | G0PBN |
| Moulder K | 2E0BZA |
| Moulder K | M3MXV |
| Moulding C | G4HYG |
| Moulding C | 2E0KGV |
| Moulding C | M0KGV |
| Moulding C | M3KGV |
| Moulding D | M3OJP |
| Moulding F | G6KUJ |
| Moule I | M3XRR |
| Moule J | G4ZKD |
| Moule P | 2E0BWU |
| Moule P | M3OJN |
| Moules A | G1ERZ |
| Moulsdale F | M6FME |
| Moulson D | G1AWK |
| Moult J | G0BGY |
| Moulton B | G7BKL |
| Moulton P | G0CHY |
| Mount B | G4JFX |
| Mount C | G4YJU |
| Mount J | M6FHQ |
| Mount K | G7DOF |
| Mount R | G7DOE |
| Mountain C | G1ESC |
| Mountain D | G1WIS |
| Mountain D | G4KSA |
| Mountain G | 2E0UGM |
| Mountain G | M6UGM |
| Mountain J | G6JMB |
| Mountain J | G6YDN |
| Mountain L | G7LZM |
| Mounter D | G4GRT |
| Mountford B | M6TGV |
| Mountford D | M3IRS |
| Mountford K | M0CDZ |
| Mountford K | M3XKM |
| Mountford N | M6FMY |
| Mountford W | M1ARH |
| Mountifield A | G4CJO |
| Mouradian S | M6CYP |
| Mourant A | GJ7HTV |
| Mourant A | MJ0BJU |
| Mourelatos G | M6OPA |
| Moverly C | M6MOV |
| Mowat J | GM7RDY |
| Mowbray I | G7TET |
| Mowbray J | G0EVF |
| Mowbray T | G3VUE |
| Mowlam J | M3WJU |
| Moxey J | G3MOE |
| Moxham P | G4GJU |
| Moxham R | G7AXN |
| Moxham R | M0RCM |
| Moxley-Wyles J | M3JMW |
| Moxon R | G1UIB |
| Moy F | M1CPC |
| Moye J | G7TAT |
| Moye P | 2E0CQR |
| Moye P | M6BTT |
| Moye T | M6NUL |
| Moyle H | G8SWC |
| Moyle J | G1AWJ |
| Moyle R | G0UWB |
| Moyler B | G3LTM |
| Moys R | 2E0MOY |
| Moys R | M0YSR |
| Moys R | M6FFR |
| Moyse B | G4IJR |
| Moyse J | G7FLI |
| Moyses S | G0GJL |
| Moysey K | 2E0KMF |
| Mozolowski M | GM4HJO |
| Mrzyglod M | G7TNQ |
| Muchamore K | G0AKM |
| Muchowski J | 2M0WSK |
| Muchowski J | MM0WSK |
| Mucklow M | G4FIA |
| Mudd A | G0HOI |
| Mudd D | 2I0CGZ |
| Mudd D | MI6AOZ |
| Mudd S | G0FNB |
| Muddimer C | G0PAO |
| Mudge B | G3MDD |
| Mugele S | G1RXV |
| Mugford R | GW3SFQ |
| Muggeridge G | G1AEX |
| Muggleton R | G0HZK |
| Muir A | G0IJI |
| Muir C | GM0UOU |
| Muir D | GM4XNQ |
| Muir G | MM0GEO |
| Muir I | GM0OQV |
| Muir J | G6CKH |
| Muir M | G3WOM |
| Muir N | GM1PHD |
| Muir R | G3LHN |
| Muir R | GM4BQD |
| Muir R | M6UIR |
| Muir S | 2M0CVK |
| Muircroft E | 2E0AGV |
| Muircroft G | 2E0BKF |
| Muirhead D | M6DDH |
| Muizelaar B | M0CBP |
| Mukans C | MW3VNR |
| Mulcahy A | G3LBM |
| Mulcahy A | M6EIR |
| Mulcuck T | 2W0BAO |
| Mulder B | G7WBY |
| Muldoon A | MW6FPL |
| Muldowney M | M3KLF |
| Mulford L | G0RHB |
| Mulhern J | MM6MHN |
| Mulheron J | GM3ZNC |
| Mulheron P | GM0VXQ |
| Mulholland A | MI0BPB |
| Mulholland B | M6HOS |
| Mulholland R | GM0ATA |
| Mulholland T | 2I0TJM |
| Mulholland T | MI0TJM |
| Mulholland T | MI3FSX |
| Mullan D | GI6EIR |
| Mullan K | MI0KAM |
| Mullan S | MI6JAD |
| Mullaney D | G0WJJ |
| Mullaney K | G4FIN |
| Mullaney M | MI6CAY |
| Mullaney T | M0TEZ |
| Mullany J | G4GIG |
| Mullard D | M6EYZ |
| Mullard R | G4BQV |
| Mullarkey J | 2E0BFA |
| Mullarkey J | M3MVV |
| Mullarkey W | G3RSW |
| Mulleady B | GM0KWL |
| Mullen A | M3UFF |
| Mullen J | GM0CNP |
| Mullen P | 2E0XPM |
| Mullen P | M0XPM |
| Mullen P | 2E0CJK |
| Mullen P | M0PMJ |
| Mullen P | M6PMJ |
| Mullen R | M0DIQ |
| Mullen R | M5RAG |
| Mullen R | 2E0IIP |
| Mullen S | G7NJX |
| Mullenger D | G8SXB |
| Muller C | G4NUF |
| Muller C | G6JZW |
| Muller H | G8KIW |
| Mulley B | 2E0BLP |
| Mulligan D | MI6MQF |
| Mulligan E | MI0EFM |
| Mulligan S | G4CBA |
| Mulligan S | G6VPU |
| Mulligan S | MI0JLC |
| Mullin A | M0HKE |
| Mullin E | G1KSK |
| Mullin J | G0TEV |
| Mullin L | G6DQU |
| Mullinder A | M6AAD |
| Mullineaux C | G8KED |
| Mullineaux P | G3XEN |
| Mulliner D | M1AEK |
| Mullins D | G3RGM |
| Mullins J | M0MRY |
| Mullins P | G6MQD |
| Mullins S | G7GGT |
| Mullis M | M3BOU |
| Mullis P | G7MTX |
| Mullock D | G7GFC |
| Mullock E | G4OCJ |
| Mullord A | G7VFU |
| Mulraney T | M3TMY |
| Mulryan N | M1CGQ |
| Mulvana D | M1AKH |
| Mulvaney A | G6MAJ |
| Mulvany C | G0JRY |
| Mulye J | G0VEH |
| Mumby J | 2E0FQT |
| Mumford A | G4RXG |
| Mumford M | MM3MEH |
| Mumford W | G7LFZ |
| Muncey N | G7VPA |
| Munday A | G8YOX |
| Munday D | 2E0FRB |
| Munday D | M3FRB |
| Munday J | G7VBL |
| Munday R | G1DYR |
| Munday S | G8JNO |
| Mundell T | M6AUP |
| Munden G | G3NIL |
| Munden R | G7TUS |
| Mundiya T | G4RFF |
| Mundy A | 2E0LWR |
| Mundy F | G3XSZ |
| Mundy J | G4XYS |
| Mundy M | G0GNV |
| Mundy M | G1TDL |
| Mundy P | G6UXW |
| Mundy R | G8POP |
| Mundy R | M6KBT |
| Munford A | 2E0MUN |
| Munford A | M6APM |
| Munir A | G0MNA |
| Munir R | M6RJQ |
| Munir U | 2E0UMR |
| Munir U | M6UMR |
| Munn E | M3EVB |
| Munn G | MM3ZGK |
| Munn M | G1JCL |
| Munn M | M0DRQ |
| Munnery N | G6PBG |
| Munns R | G6FVZ |
| Munro A | MM0BGW |
| Munro C | MM0MRO |
| Munro D | GM4XWS |
| Munro D | M5ALS |
| Munro D | M0KGK |
| Munro E | 2M0EMM |
| Munro E | MM0MUN |
| Munro H | GM0MCJ |
| Munro I | GM4GVK |
| Munro I | GM4VXM |
| Munro I | G8NLF |
| Munro J | G0MIZ |
| Munro J | MM3XUX |
| Munro N | G6YYQ |
| Munro N | GM4KGK |
| Munro R | GM3PIL |
| Munro R | MM3YOC |
| Munro S | GM0UKD |
| Munro T | 2E0BZE |
| Munro T | M3YVD |
| Munro-Smith B | M0WEB |
| Munson P | M0KEF |
| Munson P | M6PRM |
| Munt P | G3KTA |
| Munt R | G0WSH |
| Munt R | G7WSH |
| Munton A | M3EVP |
| Munton C | G4XMS |
| Munton D | G3ORD |
| Munton J | M6JMM |
| Murakami M | G0NJP |
| Murch G | 2E0GMU |
| Murch J | G1MDJ |
| Murchie P | G4FSG |
| Murden D | G3MWM |
| Murdie J | MI0NYC |
| Murdoch A | 2E0VXR |
| Murdoch A | M0YFT |
| Murdoch J | GM1VFQ |
| Murdoch J | GM3JMM |
| Murdoch J | G3YSD |
| Murdoch M | M6TOV |
| Murdoch P | G0TWO |
| Murdoch S | G4VHP |
| Murdoch S | GI7ULG |
| Murdoch W | 2E0WUL |
| Murdoch-McKay S | M6SIU |
| Murfin A | G7HMZ |
| Murfitt B | M1JLM |
| Murfitt D | M3YDY |
| Murfitt P | G3YQL |
| Muriel R | G3ZDM |
| Murkin R | G0GUV |
| Murly C | 2E0CRM |
| Murly G | G6YHW |
| Murphy A | 2E0CCI |
| Murphy A | GI1RXM |
| Murphy A | GI6ANC |
| Murphy A | MI0GHI |
| Murphy A | M3HTG |
| Murphy A | M3VZU |
| Murphy A | G4ASM |
| Murphy B | GW6MLL |
| Murphy C | G4IBM |
| Murphy C | G1SVJ |
| Murphy C | M0HLS |
| Murphy C | M6BFO |
| Murphy D | 2M0TSR |
| Murphy D | 2E0HSB |
| Murphy D | M6ACE |
| Murphy E | GM3SBC |
| Murphy E | G6NTM |
| Murphy E | G0VVT |
| Murphy I | G8PBI |
| Murphy J | G0UDI |
| Murphy J | GI7DWF |
| Murphy J | G8RSE |
| Murphy J | MI3BSN |
| Murphy J | M6SUC |
| Murphy J | G1BQH |
| Murphy J | MI1CTQ |
| Murphy K | GI4SZQ |
| Murphy K | G8RIC |
| Murphy K | M5KJM |
| Murphy K | G4XBG |
| Murphy L | 2E0AZL |
| Murphy L | G7RUQ |
| Murphy M | G0CDQ |
| Murphy M | GI0CTI |
| Murphy M | G0LQI |
| Murphy N | M0CJE |
| Murphy O | M6OPM |
| Murphy P | GI4OMK |
| Murphy P | M6EEZ |
| Murphy P | M0MDP |
| Murphy R | G3NRX |
| Murphy R | G7AEQ |
| Murphy R | M3CEV |
| Murphy R | 2E0RHM |
| Murphy R | M3UVM |
| Murphy R | 2I0RGM |
| Murphy R | MI0RYM |
| Murphy R | MI6RGM |
| Murphy S | G6PMJ |
| Murphy T | G4WJW |
| Murphy T | M1TEM |
| Murphy W | MW6BMM |
| Murray A | GM0BPT |
| Murray A | GM3DOD |
| Murray A | GM4FIZ |
| Murray A | GM4EJX |
| Murray A | 2M0JAT |
| Murray B | MM0ERK |
| Murray B | M6MRY |
| Murray B | G4VVK |
| Murray C | G0MIA |
| Murray C | GM4EAU |
| Murray C | MI3CEM |
| Murray C | 2E0BNC |
| Murray C | 2E0LMG |
| Murray D | G0NCO |
| Murray D | G8GTR |
| Murray D | G1GPE |
| Murray D | M6GYZ |
| Murray D | M0RFY |
| Murray E | 2I0FSL |
| Murray E | MI0XAX |
| Murray E | MI6TUX |
| Murray F | G3GML |
| Murray G | GI1JXE |
| Murray G | G4RDG |
| Murray G | GM7OQE |
| Murray G | G7SKH |
| Murray G | G8XEC |
| Murray G | G8PXO |
| Murray G | 2M0MUR |
| Murray G | MM0MUR |
| Murray G | MM6MUR |
| Murray H | GM7MWL |
| Murray H | M6ADJ |
| Murray H | G3NBY |
| Murray H | G1PVD |
| Murray J | GM4PUS |
| Murray J | MM1BFE |
| Murray J | M6MFB |
| Murray J | 2E0BPS |
| Murray J | 2E0COF |
| Murray J | M0JCM |
| Murray J | M3XGZ |
| Murray J | M3YZN |
| Murray K | GI0JRI |
| Murray K | G0PLB |
| Murray K | GI0STM |
| Murray L | MI3XRT |
| Murray M | GM0CMO |
| Murray M | M3OHO |
| Murray M | G8IBK |
| murray M | M6HVR |
| Murray M | M3IKE |
| Murray N | G7ECQ |
| Murray P | M3FSV |
| Murray P | 2E0GEM |
| Murray P | M3XPM |
| Murray R | G7DDR |
| Murray R | G7IWZ |
| Murray R | MM3DHE |
| Murray R | MM0RRM |
| Murray R | M6EXT |
| Murray R | 2M0DSU |
| Murray R | MM3YKR |
| Murray S | G1FHI |
| Murray S | MI3SVM |
| Murray S | 2I0SGM |
| Murray S | G3ZQL |
| Murray S | MI0SAP |
| Murray S | MI3SEO |
| Murray S | MM3WNP |
| Murray S | M6BQQ |
| Murray T | GM6RGD |
| Murray U | MI0UNA |
| Murray U | MI1UNA |
| Murray W | G1CRN |
| Murray W | MI0WJM |
| Murray W | MI1JOE |
| Murray-Shelley R | G4KCV |
| Murrell D | G0KQY |
| Murrell M | M3MPM |
| Murrell P | G0DUH |
| Murt C | G0MWW |
| Murtha P | G3WZG |
| Murthwaite P | M3PFY |
| Murton V | G0MQK |
| Musgrave J | G6KBS |
| Musgrave M | M3SZD |
| Musgrave M | G0ESL |
| Musgrave M | G4NVT |
| Musgrave S | M0HOI |
| Mussard M | G7TOU |
| Mussell B | G4CXJ |
| Mussell S | G7SZA |
| Mussell T | 2E0TRA |
| Mussell T | M3OGU |
| Musselle E | 2E0DKJ |
| Musselwhite P | G7RGJ |
| Musselwhite P | M0HAL |
| Musson C | G1UAZ |
| Musson E | G1WTB |
| Mustafa H | MW3HAS |
| Mustard A | GM3NCO |
| Mustard E | G0BLU |
| Mustchin P | G0LFH |
| Musther A | G7UZY |
| Musto R | G4MHE |
| Mustoe H | 2E0GSX |
| Muswell R | MD6MED |
| Mutavdzic P | M3PXT |
| Mutch G | M0HWT |
| Mutimer A | G6YYU |
| Mutkin M | 2E0CQO |
| Mutkin M | M0MBR |
| Mutkin M | M6BTG |
| Mutlow D | M3MFU |
| Mutter P | G8XDM |
| Mutter P | M0BEH |
| Mutton M | G0DEF |
| Mutton M | M3TMQ |
| Mutton R | G3EVT |
| Mutton R | G4ACZ |
| Muxlow E | 2E0GCB |
| Muxlow P | G8HDJ |
| Muzyka J | G4RCG |
| Muzyka P | G6LCP |
| Myall G | M3RVK |
| Myall J | G8OHG |
| Myall S | M1DOT |
| Myciunka S | M6WAA |
| Mycock B | 2E0BMV |
| Mycock R | G0DTT |
| Myers A | G1TLC |
| Myers A | G7PTT |
| Myers D | M0MYE |
| Myers G | M3NIZ |
| Myers G | M0PLT |
| Myers J | M6OLS |
| Myers P | G3UWT |
| Myers R | G8ZRM |
| Myers R | G8LUL |
| Myerscough R | G0SIQ |
| Myford I | G4HTO |
| Myint K | G0DSQ |
| Myland A | G0JIS |
| Myland S | 2E0YLY |
| Myland S | M0YLY |
| Mylchreest C | G1EVV |
| Myler-Cook B | 2E0CDF |
| Myler-Cook B | M6AEC |
| Mynett A | G3HBW |
| Mynn M | 2E0PWL |
| Mynn M | M0PWL |
| Mynn M | M6PWL |
| Mynors T | M3LAQ |
| Myszka J | MW0FDG |
| Myszka P | M6GNU |

## N

| Name | Callsign |
|---|---|
| Nadin T | M1EOU |
| Nagle M | G4RZI |
| Nagy R | 2E0NAR |
| Nagy R | M6NRA |
| Naik P | 2E0CXE |
| Naik P | M0HQR |
| Naik P | M6CKU |
| Nailer A | G4CFY |
| Nairn M | G4SUU |
| Nairne P | G1CPA |
| Nairne P | M0AFZ |
| Naish N | G4MHS |
| Najman R | G0NXM |
| Nakagawa T | 2E0PIO |
| Nakagawa T | M0JPN |
| Nakagawa T | M3PIO |
| Nakajima C | 2E0CHA |
| Nall A | M0NDO |
| Nally J | G8KCB |
| Nancarrow D | G3RID |
| Nance A | G0KKS |
| Nandalan S | M0SBH |
| Napier A | GM1TBW |
| Napieralski M | M0LPK |
| Napp P | G1USF |
| Napper D | M1BXU |
| Napper R | G4FXU |
| Nappin D | G3MLS |
| Narayanankutty M | M6KNB |
| Narinian V | M0GCX |
| Narroway C | G6FAF |
| Nash A | G4ZHZ |
| Nash D | G0NZT |
| Nash G | G4XBD |
| Nash J | G1FSX |
| Nash J | G4SFP |
| Nash J | M6KDF |
| Nash L | G0DSN |
| Nash M | G0IHC |
| Nash M | M3ZMN |
| Nash M | G0LCH |
| Nash M | G8FNH |
| Nash N | 2E0NDR |
| Nash N | M0NGL |
| Nash N | M3NDR |
| Nash N | M3UFK |
| Nash P | 2W0PCN |
| Nash P | M0ATL |
| Nash R | G0DVL |
| Nash R | G1BSZ |

Nash R G4GEE
Nash S G0UQT
Nash S M0GBK
Nash S M6NAS
Nash S 2E0SJN
Nash S 2I0STN
Nash S MI0WWF
Nash S MI6STN
Nash T M6FKS
Nassau M 2E0MTT
Nassau M M0NJX
Nassau M M6NJX
Nathan D M6FTN
Nathan P 2E0PQR
Nation F M3HER
Nattrass G G0OGD
Nattress J 2M0NGO
Nattress J MM3XLO
Naughton J G1UWQ
Naughton J GM4WQH
Naughton P GM4WGC
Naughton W M6WJN
Navier B G0OVT
Naylor B G3SHF
Naylor B M6ODA
Naylor C G7PCW
Naylor C M6NZN
Naylor D G0JEC
Naylor D G3YXS
Naylor D G4JWA
Naylor D G3SKN
Naylor E G6UEH
Naylor I G8YAT
Naylor J G0NIQ
Naylor J GW3SCX
Naylor J G4KLX
Naylor J G7ROI
Naylor J M3HVW
Naylor J M0NTT
Naylor K G7ILP
Naylor K M0EBU
Naylor L 2E0NYX
Naylor L M3LNR
Naylor M G4CDF
Naylor M M3MGN
Naylor P G1JJK
Naylor P M6PAN
Naylor R G1MJT
Naylor R G4MSY
Naylor R M1CAE
Naylor W G7WBJ
Nazer D M3ZER
Neachell S 2E0SAN
Neades P G1YFC
Neal B 2E0TCU
Neal B M6WCA
Neal D G0SPK
Neal D G7ENA
Neal D G4ESG
Neal H M1CXN
Neal J G4NQC
Neal J 2E0MPJ
Neal J M3WQX
Neal J M0HXO
Neal M G7NJP
Neal P 2E0TJP
Neal P M3XDZ
Neal R G0IEB
Neal R G0KVG
Neal T G0CQH
Neal T G0NMC
Neale A M3UAJ
Neale A M6OAN
Neale B G4FBN
Neale C G3TMU
Neale D G8WBN
Neale D G7DHW
Neale D G1AJK
Neale H G3REH
Neale J G7OJZ
Neale J M0CHD
Neale M G0HLW
Neale M M6MJN
Neale P G0VVK
Neale P G3UHN
Neale R G4OIE
Neale S G1TUL
Neale S M3SHN
Neale S M6OKO
Neale S M0JTN
Neale S M6SXN
Neale W G0WAB
Neale-Gardner H M1EYA
Neary B G3VHZ
Neary C G0TMH
Neary I G3NTF
Neary J G0NAJ
Neary J G4ZQM
Neary J GM0XFK
Neary Y M3YVE
Neate B G1XKQ
Neate J G1WUU
Neaves A G4BKA
[illegible] 2E0AWC
Needham A M3YXL
Needham J MM6JNN
Needham P G8RQN
Needham P M3UPN
Needham P G4MSI
Needham R G0WXC
Needham W GW6AMK
Needle G M6WOE
Needs R G4NUG
Neely D G6ZSF
Neenan A G0MPO
Neenan J M3DIU

Nehan A G4HUE
Nehmzow C M0ASG
Nehmzow H M3OQD
Nehmzow U M0ASF
Neighbour J M3JFF
Neil C G0MVC
Neil E 2E0BGP
Neil G GM1VDZ
Neil P M6PNL
Neil S G6AEB
Neill B GI7UQW
Neill D 2I0FPB
Neill D MI6IPB
Neill D 2E0DNL
Neill F G6NJT
Neill J GI0CDM
Neill R MI0RJN
Neill R MI3PDN
Neilson D G4ZNZ
Neilson R GM0EFT
Neish W MM6LAO
Nelhams-Wright K M6YMY
Nell A G3VYS
Nellis J GM4PLI
Nelmes B G4FQH
Nelmes C M0GSI
Nelmes C M3GSI
Nelmes J M3PNV
Nelmes M M3XOR
Nelson A G8OFX
Nelson A 2E0AJQ
Nelson A GM4IIR
Nelson A G8KVN
Nelson A M0APN
Nelson A M6RWJ
Nelson B GJ4KBM
Nelson C 2E0NEL
Nelson C M3WFY
Nelson D G7GUK
Nelson D M0BAJ
Nelson D M0HBT
Nelson D M6BDQ
Nelson G M3UKH
Nelson G MI6GDN
Nelson I M6IDP
Nelson J G4FRX
Nelson J MW3HNP
Nelson J G4KLA
Nelson K M3OVM
Nelson M M6AWO
Nelson P G1TUZ
Nelson P G4XAW
Nelson P G8GOO
Nelson R G0CJO
Nelson R G0TZO
Nelson R G1CKT
Nelson R G3ZLF
Nelson R G7TMR
Nelson R G8BBK
Nelson R M3ENS
Nelson R 2I0RMD
Nelson R MI0RMD
Nelson R MI6RAV
Nelson R 2I0EGN
Nelson R G1NGE
Nelson S G6BBK
Nelson S M6SPN
Nelson T 2I0RVH
Nelson T MI0RVH
Nelson T MI6TNI
Nelson W 2I0XAN
Nelson W GI4PXM
Nelson W MI6WAN
Nelson-Jones L G4JDW
Nenova J M3CGJ
Nerurkar V GM0VOL
Nesbitt A G0CWQ
Nesbitt A 2E0AKR
Nesbitt A M0GSV
Nesbitt A M3MIH
Nesbitt A G3SGY
NESBITT E GI0MSI
Nesbitt G M1BXF
Nesbitt W GI3TZX
Nesling S 2E0XXO
Nesling S M6UGN
Ness A M3BIO
Ness J G8XKW
Nethercot P G3HXK
[illegible]
Nethercott J M0EEP
Nethercott J M1EEP
Nethercott J M0TWM
Netherton S G8WBG
Netherway R G0PDV
Netting S M0SPN
Nettleship M G4VLZ
Nettleton R G3YED
Neufeld R M0NAI
Neumann A M0DEJ
Neumunn D 2E0WCM
Neumunn D M6MLL
Nevard E M3EWN
Nevard P 2E0IJN
Nevill A G7JDH
Neville D 2E0BCJ
Neville D M6DJN
Neville J G1HGB
Neville K G0EUJ
Neville M G4JUK
Neville P G4NZB
Neville R 2E0DZL
Neville R G3KWB
Neville R GW8RAS
Neville R M3RKN
Nevin J G4ZHG

Nevins K 2E0KLN
Nevins K M0KLN
Nevison A G4OSH
New A M6BNW
New E G0BSJ
Newberry P G0VQD
Newbold I G8KSZ
Newbold M G0VVA
Newbold M G1MJN
Newbold S G1RDX
Newbould A 2E0VRC
Newbould A M6VRC
Newbould K G4VRW
Newbould K M6LHD
Newbury C G8JGE
Newbury L G6CKD
Newbury M M6MMN
Newbury S G8RIR
Newby E M0TMC
Newby G G1ZBU
Newby J M3OPM
Newby L M3FOD
Newby N M3XNB
Newby N G0VDZ
Newby P G4AVL
Newby-Robson C G1YVS
Newcombe B MM0NEW
Newcombe J G6AML
Newell A G7CFX
Newell A M0NAP
Newell A M1DSV
Newell C G8IQF
Newell D G0AKS
Newell D M3BIK
Newell J 2E0HFN
Newell J G4UHK
Newell J M3JWN
Newell M G1HGD
Newell M G3IUE
Newell M G1AYH
Newell M G8KOE
Newell M 2E0MTN
Newell M M0VTR
Newell M M3BPN
Newell N GI3YMY
Newell P M0CEX
Newell R G6XKX
Newell S G1EFS
Newell V M6VCS
Newell Z M6ZLN
Newey D G3NDN
Newey J G6RJE
Newey M G4SND
Newey W M6WNN
Newgas D M0NEU
Newham J 2E0HHU
Newham M M6TTN
Newham R G7SEK
Newhouse S 2E0NWY
Newhouse S M0NWY
Newhouse S M6NWY
Newing D M0BHH
Newland J G1ETD
Newland O G7SNB
Newland T G0TJN
Newlands A GM4TXN
Newlands A GM4VAY
Newlands M G4FYG
Newman A G4AVX
Newman A G7PCV
Newman B G3MMN
Newman C G0VXS
Newman C G4JCJ
Newman C MM1LJB
Newman C G4BVM
Newman D M0CQQ
Newman D M3DZN
Newman D 2E0HBE
Newman D G1NNR
Newman D GW4CGZ
Newman G G4YCP
Newman G M3IVN
Newman J GW0FVC
Newman J G0PGT
Newman J G0VDU
Newman J G4MGO
Newman J M0OOD
Newman J M3JRN
Newman J G6XBS
Newman K G4CZA
Newman L G3UUU
Newman M G3UWZ
Newman N M0SHF
Newman N 2E0NBG
Newman N 2E0CDD
Newman P G4KIM
Newman P G8UDI
Newman R G3VZL
Newman R GW4BCF
Newman R G7SAX
Newman R M0CBM
Newman R G4JIE
Newman T 2E0HSA
Newman T M3TEN
Newman T M0AQW
Newman T M0TNE
Newman V G0EFI
Newman V G7AIC
Newnam K 2E0FDJ
Newnham G G7EOH
Newnham M G6NZ
Newns A M0SEA
Newport M G1GZG
Newport M G8VCN
Newport R G0UUI
Newport S G4DEV

Newsam C M6REF
Newsham E G4ERN
Newsome A MW0XDN
Newsome D M6FGC
Newsome J M0CCD
Newsome P M0HIM
Newsome R G1MSD
Newstead G G8EKG
Newstead R G3CWI
Newstead R M6AJX
Newstead S G7PRC
Newstead T G0LQX
Newth M 2E0HQP
Newth P 2E0WTH
Newth P M6PCN
Newton A 2E0WPW
Newton A G0LPU
Newton A G7HAS
Newton A M3TJQ
Newton B G1HGC
Newton C 2E0WEC
Newton C G0FGZ
Newton C M3CSK
Newton C M6EIP
Newton D M3HJN
Newton G G4UOS
Newton G G4VUP
Newton G G7SMH
Newton G M3GCN
Newton G M6KRR
Newton G GM4LSD
Newton I 2E0DXL
Newton I M6HBQ
Newton J G0BZJ
Newton J M0YGW
Newton J G6KQS
Newton L M6NUW
Newton M G3UKW
Newton M G7SBZ
Newton M M3LSE
Newton M 2E0AKO
Newton M G7IDH
Newton M M0MRK
Newton O M3ORN
Newton P 2E0AKN
Newton P GM0EZR
Newton P G7OLW
Newton P M0GSN
Newton P M3LQV
Newton P G7PRI
Newton R G7LDD
Newton R G8YAU
Newton R G0EWH
Newton R 2E0RDN
Newton R M6RFQ
Newton T 2E0EZL
Newton T M3LSK
Newton T 2E0ZKD
Newton V 2E0EMH
Newton W G7GMZ
Newton W M0WFN
Newton-Goverd D MW0CJB
Ney M G7AZW
Neyland T G3RPL
Ng C M6ILC
Ngai S M3WSW
Ngan N M6GZJ
Ngi H M3YEE
Nguyen T M0TMN
Nguyen T M3TNM
Nias J G3VRB
Niblock A GI7USA
Nice I G6RUM
Nice P G8IER
Nichol J G0EUN
Nichol R G0FVD
Nicholas A G3ZUE
Nicholas C G0LZV
Nicholas D G0WWA
Nicholas G G8MMM
Nicholas G G7EVG
Nicholas J G3OIN
Nicholas J G8HMV
Nicholas J M3JLI
Nicholas J M6JNQ
Nicholas J MW6EEJ
Nicholas L M6LBC
Nicholas M M0AHS
Nicholas M 2E0TCQ
Nicholas S G0NYY
Nicholas S G6ADU
Nicholas T GW4RVA
Nicholas T G7VOI
Nicholas W M0HHC
Nicholas-Letch J G3PRU
Nicholl A MI3ASH
Nicholl A MI3BEG
Nicholl C 2E0DKW
Nicholl C M6DKW
Nicholl F GM0CVD
Nicholl I 2I0FLO
Nicholl J GM4WZD
Nicholl M MI3TLV
Nicholl N MI3NEN
Nicholl R MI3RNN
Nicholl S MI3STY
Nicholl W MI0RIB
Nicholls A G0HEL
Nicholls A M3INC
Nicholls A M6YAW
Nicholls A G0LUB
Nicholls B G6BLU
Nicholls B M3VSF
Nicholls B G8VCO
Nicholls B G7FWD
Nicholls C 2E0CBZ

Nicholls C M3ZVH
Nicholls D G3VMZ
Nicholls D GI3ZVZ
Nicholls D G6AEC
Nicholls D G6RQA
Nicholls D G7IBU
Nicholls D G0GTC
Nicholls D M0TBQ
Nicholls G G1PPX
Nicholls I M3IGN
Nicholls I 2E0CXK
Nicholls I M6CVQ
Nicholls J G0KXG
Nicholls J M3BMQ
Nicholls J 2E0JJN
Nicholls J M0JJN
Nicholls J M6JJN
Nicholls J M0PDG
Nicholls M G3LQX
Nicholls M G4HIA
Nicholls M 2E0XRZ
Nicholls M M0XRZ
Nicholls N G4TCK
Nicholls P G0ONA
Nicholls P G0VXV
Nicholls P G1BGH
Nicholls P G7RIB
Nicholls P M3PMN
Nicholls P 2E0PVN
Nicholls P M0PVN
Nicholls P M6PVN
Nicholls P G0VMK
Nicholls P G7MKV
Nicholls R G4ZHY
Nicholls R G8TTP
Nicholls R G8GRO
Nicholls R M6NLN
Nicholls S G0JFM
Nicholls S G1YOU
Nicholls S M6TBN
Nichols A G1XJT
Nichols C G0FZD
Nichols D G3BCE
Nichols E M6ELE
Nichols J M1DZM
Nichols K G4FNI
Nichols L G3GMW
Nichols S M6PUP
Nichols S G0KYA
Nichols T M0TAN
Nichols W M0WPN
Nicholson A 2E0LIV
Nicholson A G8FLV
Nicholson A M3MZC
Nicholson A M3NPA
Nicholson A M0GEU
Nicholson A M3NRQ
Nicholson A M6EQG
Nicholson B G0MZF
Nicholson D 2E0DEE
Nicholson D G4GGE
Nicholson D MM3SYB
Nicholson D G8GKX
Nicholson E M3ZOU
Nicholson F G4AFR
Nicholson G MM3XUY
Nicholson I 2E0ZMR
Nicholson I M3ZMR
Nicholson J 2E0JJN
Nicholson J M1JJN
Nicholson J M3JJN
Nicholson J M3NPE
Nicholson J G1MOZ
Nicholson J M6JBE
Nicholson J G1OSI
Nicholson K G1TRF
Nicholson K 2E0CJM
Nicholson K M0KAN
Nicholson K M6AVZ
Nicholson K M6CYV
Nicholson M G1LEH
Nicholson M G8YVP
Nicholson M G4FRW
Nicholson M G8NYK
Nicholson P G1EFT
Nicholson P G3MYZ
Nicholson P G8UXL
Nicholson R G4SQG
Nicholson R G8ZOV
Nicholson R M3MZA
Nicholson R 2E0DSV
[illegible]
Nicholson R M6FIW
Nicholson T G1VRC
Nickalls P G0AQA
Nicklin L M3MSC
Nickson K G1OTI
Nicol A G8BXA
Nicol J MM3JIN
Nicol J MM0SMD
Nicol P G8VXY
Nicol R G4YMY
Nicol R G1NZZ
Nicol S MM6EHL
Nicole D M0CYJ
Nicoletti I G7ODN
Nicoll L MM6LON
Nicoll R 2M0HMN
Nicoll R MM6RCO
Nicoll S 2M0SMN
Nicoll S MM6SMN
Nicoll W G3KWN
Nicolle S G4MUV
Nicolson D GM4RGU
Nicolson D M1DAS
Nicolson D G4WJG
Nicolson G G0FBS

Nicolson I MM6YYB
Nicolson J G3KGT
Nicolson J GM8WJK
Nicolson P M5PIP
Niel P M0AZC
Nielsen B MM3UDL
Nieto L OM1DVA
Niewiadomski B M0NIE
Niewiadomski P 2E0HGO
Niewiadomski P M6HGO
Niggemann D 2E0DTN
Niggemann D M0KRD
Niggemann D M6KRD
Nightingale A G3ZPU
Nightingale E G3UUG
Nightingale J G8AEU
Nightingale M G8VQK
Nightingale P G8XQI
Nightingale R G7MNO
Nightingale S G6XLR
Nightingale S G8NEL
Nikitits A 2E0IOZ
Nikolaidis V 2E0VNN
Nikolaidis V M3VNN
Nikolic P G6NJR
Nilan K M6KMN
Nilan P G0LVG
Niles C G0SWL
Nilon L M3TLL
Nilski Z G3OKD
Nilson M G8EKD
Nilsson M M6SWE
Niman J G8GAJ
Niman M G6JYB
Nimash T 2E0WXT
Nimmo M GM8JVZ
Niner T G1RFS
Nisar F M6NIS
Nisbet A 2E0COH
Nisbet A M0HLY
Nisbet A M6SLN
Nisbet A MW0MWA
Nishio H M0HNO
Niven C MM0IBL
Niven I G6EGY
Niven M G4AHT
Nix D G4TBK
Nixon A G4DPO
Nixon A G1EFU
Nixon B G6EGU
Nixon C G7LYL
Nixon D MM0RLN
Nixon J M0CLL
Nixon J M1EHB
Nixon J M6ZZD
Nixon L G6YZU
Nixon M G4YSO
Nixon P G4BBI
Nixon R G0XVC
Nixon R 2E0SLN
Noake R M3VRN
Noakes D G6IYD
Noakes M G4JZQ
Noakes M M3WUW
Noakes P M6PJN
Noakes R 2E0HWU
Noakes R 2E0KES
Noakes R M6RNO
Nobbs K G0KSS
Nobes J M6JNE
NOBLE A G7BSO
Noble B G4JAJ
Noble F 2E0FRG
Noble F M0NBL
Noble F M3YHY
Noble G M1BAV
Noble G M6GJN
Noble G G8LDU
Noble J G0EVD
Noble J G3NHF
Noble M G4BVP
Noble P 2E0PFO
Noble P M0PFO
Noble P M6LFO
Noble P M6PFO
Noble R G4UYR
Noble S GI4SAM
Noble S M0BVU
Noble W MW6MWN
Noblet M 2E0ILO
[illegible] M6ILO
Nocera S G1VNE
Nock A G0BFZ
Nock B G4BXD
Nock D G0CVA
Nock D G7KHW
Nock D M6GBB
Nock G G0EJO
Nock M G7WHU
Nock R 2E0RWN
Nock R M0RWN
Nock R M3RWN
Nock R G4LWN
Nock R G8UID
Nock R G1YRQ
Noda N M0RJZ
Noden C G0JPB
Noden J G8IOK
Noe D M0NDJ
Noe D M3NOE
Noel D M3DVN
Noel H M3HLN
Noel J M5DRW
Noel L 2E0FPM
Noel L M3FPM
Nofrerias Mondejar J M6JNM

Noke S G4MOE
Nokes A G4WZA
Nokes A M3ULZ
Nokes D G3JVR
Nokes R G6JTK
Nolan A M0DZX
Nolan B G0NRW
Nolan C G0MIJ
Nolan D 2E0EUI
Nolan D M6EUI
Nolan G G6ZQA
Nolan J G7LUR
Nolan K G3ZIV
Nolan K GI7NET
Nolan M 2E0MPN
Nolan M M3PQV
Nolan P M6HQR
Nolan R G3KWK
Nolan T M0TFN
Nolan V G8XEI
Nolan W G8PZI
Noller S M6GUJ
Nolson R G0PMU
Noon A 2E0GJJ
Noon C 2E0SBF
Noon D GI3MMG
Noon G G1BYJ
Noon J M6AVV
Noon M M3MJN
Noon R 2E0DIN
Noon R M3YLU
Noon W 2M0WFN
Noon W MM6AHJ
Noonan G 2M1IBH
Norbury M GW4GLU
Norbury T G1FWU
Norcliffe B G8DTS
Norcott B G7GES
Norden A 2E0LFT
Norden A M0WWV
Norden A M6WKZ
Nordgren L M0GHE
Norfolk A G6DBC
Norfolk D G4AWM
Norfolk I 2E0IDN
Norfolk I M6IDN
Norfolk M M6MNX
Norgrove G M0WOD
Noriega A M6KYO
Norman A G0GGB
Norman A M3NOR
Norman C G4MHR
Norman C 2E0CXJ
Norman C M6CQV
Norman D M3IDK
Norman F G0NCT
Norman I G6UEG
Norman J G0AZR
Norman J G0LRE
Norman L MM1FHO
Norman M 2E0BEV
Norman M G4MGN
Norman M G4XDL
Norman M M6MDN
Norman P G0WMC
Norman P G1DUI
Norman P G6UEI
Norman P M3FCN
Norman P M3PHR
Norman P M0PRN
Norman P MW6PMN
Norman P M6YDN
Norman P M6ZUF
Norman P 2E0BJP
Norman R G0AFY
Norman R G3NAI
Norman R G7IQZ
norman S 2E0BCO
norman S M0FZX
Norman S M1ABU
Norman S M3WZV
Norman S M0MVB
Norman T M0CNU
Norman W G8VCQ
Normandale S G6LVG
Norminton S G1BMP
Norridge D G0BPX
Norrie A M0GPU
Norrie A GM1MQE
Norrie J GM8UMN
Norrington J 2E0NOZ
Norrington J M0NOZ
Norrington T M3TPN
Norris B 2E0CML
Norris B G1UAH
Norris C 2E0GDA
Norris C 2E0CMK
Norris C M0IOT
Norris C M6CPN
Norris D G4TUP
Norris D G7VDI
Norris D M3OEB
Norris G G1RDZ
Norris G M6GNN
Norris G G6DCS
Norris J G1IWC
Norris J G4MLR
Norris J G0DDU
Norris K G0WZG
Norris M 2E0BVH
Norris M M0MEN
Norris M M3YIX
Norris N M0NTN
Norris P G4VUN
Norris P G3XVN
Norris S M3HJJ

Norris S M3ZIX
Norris S G4KSR
Norris T G6FKY
Norsworthy N M6NMN
North B M3PKL
North B 2E0BWQ
North B M6SIX
North C G4GBP
North M G4GMK
North M G8HKK
North P G8VVP
North R G4KHR
North R G8TZN
North R M3PKE
North R M0GDU
North R G3YYK
North S G4JRJ
Northall A M6AQQ
Northall J 2E0FTW
Northcote H M3TXF
Northcote W 2E0WHN
Northcote W M3WHN
Northcott C 2M0FFY
Northcott C MM6HCK
Northcott R G1YPM
Northeast D G7LWY
Northeast S G7UPL
Northey M G8JCD
Northfield J G7ELX
Northmore P G4KMM
Northover P 2E0PJN
Northover P M3PJN
Northover T G0TOD
Northover W 2E0WWN
Northover W M3WWN
Northrop C G6MQG
Northrop K M6XKN
Northway B G0BNJ
Northway R 2E0RCN
Northwood C 2E0NTW
Northwood C M3NTW
Northwood D G4VQJ
Northwood E G3WRL
Northwood M M6MUT
Norton A G1ABJ
Norton A M6SEE
Norton A 2E0DGR
Norton A M6DZX
Norton C G4FTG
Norton C M3XCN
Norton C GM0AZV
Norton D G0UHI
Norton D G7OCY
Norton F G0GJX
Norton F G6IHB
Norton J G4TLS
Norton J G0GNU
Norton J G3PLW
norton L G4JNW
Norton L MI3WLW
Norton M G8HJH
Norton M M3PAI
Norton R G1JOD
Norville A 2E0CFL
Norwood C M6BVC
Norwood J G1CMH
Norwood R M3FHI
Norwood W M3WUK
Noszkay T G0TCH
Nothard J G0NVD
Notman M G0JBS
Notschild A G3RSF
Nott T 2E0TVM
Nottage A G0XBV
Nottingham M G1XIY
Nowak P M6PNR
Nowell C M3CEN
Nowell J G4FUO
Nowikow C G8BUI
Noy J G8VPE
Noyce W G0OBM
Nudd E M6NME
Nugorski A M6ALO
Numata K M0HTQ
Nunn A 2E0KDA
Nunn A G8AXO
Nunn C G8NEH
Nunn D G3JMJ
Nunn J G0MZN
Nunn L G6YZB
Nunn P G8UPJ
Nunn R G1UUI
Nunneley A G0RIT
Nunnen C M6CBN
Nunns V G0NKO
Nur A M6SOU
Nurse J G3UUC
Nurse M 2E0MPN
Nurse M G4ZFT
Nurse P G0IFT
Nursey S M0AJI
Nussey A G7VIV
Nutboom T M0YBU
Nuthall D G4ZOU
Nutkins C G1EFX
Nutkins R G0HFT
Nutt A M6AIW
Nutt A 2E0PYM
Nutt D GM3YDN
Nutt I M0ECQ
Nutt M 2E0RNU
Nutt M M0RNU
Nutt M M6RNU
Nutt P G4WLI
Nutt P G8IKW
Nutt R G4LIJ

UK Surnames

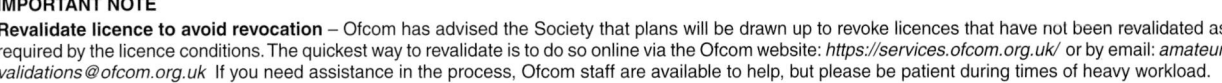

**IMPORTANT NOTE**

**Revalidate licence to avoid revocation** – Ofcom has advised the Society that plans will be drawn up to revoke licences that have not been revalidated as required by the licence conditions. The quickest way to revalidate is to do so online via the Ofcom website: *https://services.ofcom.org.uk/* or by email: *amateur.validations@ofcom.org.uk* If you need assistance in the process, Ofcom staff are available to help, but please be patient during times of heavy workload.

Nutt S 2E0YUD
Nutt S G3OCR
Nutt S G8ILJ
Nutt S M3YUD
Nuttall B 2E0OYG
Nuttall D G4VFQ
Nuttall D G4ZST
Nuttall F 2E0FNB
Nuttall G G8XRS
Nuttall J M1CRQ
Nyakabau S M6EFU
Nye C G0EKN
Nye D 2E0ACK
Nye E 2E0JOY
Nye R G3LPY
Nye T G8ZBN
Nyman M G4OMP
Nyquist K M0GJA

## O

O Broin C M0ZCO
Oag V G6XDK
Oakden R G6DDP
Oakes A G6AHO
Oakes D 2E0COG
Oakes D G0FRN
Oakes E G7JWL
Oakes G G3WRK
Oakes L G8FOY
Oakes S G4VXX
Oakes W G1YQY
Oakey W M3UFL
Oakley A G4JMO
Oakley A G4HYD
Oakley B G1DUJ
Oakley B M3UBK
Oakley B G4PBJ
Oakley C G3IPD
Oakley C G4YGE
Oakley D G0DFI
Oakley D G8LSH
Oakley E G2AZM
Oakley E M1BWR
Oakley J G4VQZ
Oakley J 2E0BXA
Oakley J M6JON
Oakley P G0BVD
Oakley R G1DDR
Oakley R G8GRT
Oakley S M3LEY
Oakley W M3HYQ
Oakton F G0IWF
Oates D G0DXO
Oates D MM3PEY
Oates G G8YAZ
Oates J GM0VIY
Oates J G3LZI
Oates M M6FVO
Oates P G7VNE
Oates-Miller C G1JOW
Oatey A M0AVN
Oatey A M3NSB
Oatway G G8BQK
Obermaier G G4LRH
Obey G G0MSF
O'Boyle G GI6NDM
O'Brian J 2E0CIT
O'Brian J M6BDD
O'Brien A G6AOB
O'Brien A MI6OMA
Obrien B G0UCT
O'Brien B G0JNV
O'Brien B M6OBR
O'Brien D M1OBR
Obrien E G3WIO
O'Brien J M0JOB
O'Brien J G7URJ
O'Brien J G4OGL
O'Brien J M6AAJ
O'Brien K M6KYN
O'Brien L M6LCI
O'Brien M M3FAK
O'Brien M 2E0ABY
O'Brien M M6XMN
O'Brien P MI3POB
O'Brien R G3ZCI
O'Brien S G8NLS
O'Brien T G0PIL
O'Brien-Bird N M0NGB
O'Buitigh D GI7OMY
O'Callaghan M GD1RYF
O'Callaghan R G1EPL
Ochiai Y M0DCP
Ochot A MD0GFK
Ockenden M G3MHF
Ockendon C G3TPO
Ockett G 2E0AAC
OConnell A 2E0IAO
OConnell A M6XAC
O'Connell A 2E0LGW
O'Connell C GI1RXL
O'Connell J G0PZF
O'Connell M G4ZIU
O'Connor D GM1ROM
O'Connor I M0IOC
Oconnor J G4GSM
O'Connor J MM3FJX
O'Connor L M6PXT
Oconnor M M1KES
O'Connor M 2E0OIL
O'Connor M M0MKO
O'Connor M M6OHL
O'Connor P GD7FSJ
O'Connor P G4SFG
O'Connor S G1YKK
O'Connor W G8DXI
Oczerklewicz R M6RKO
Odam M G7LVN
Odd H G7JYG
Oddie D 2E0DAO
Oddie D M3NZR
Oddy G G8BVR
O'Dea P G4XPI
Odegaard P G0DJT
O'Dell J G4RCF
Odell L G4ROC
O'Dell M G0AUJ
Odell R M0RJO
Odell S M0CUY
Odle P 2E0HRY
Odle P M3HRY
Odle R 2E0RIO
Odle R M3ANE
Odlum K G4WBW
O'Donnell D GM7BRL
ODonnell J MM3TNF
O'Donnell J G7VCF
O'Donnell J M6GHJ
O'Donnell M G8CCV
O'Donnell S G4MSV
O'Donnell S G4RVP
O'Donnell S M6AFX
O'Donoghue B G0KQH
O'Donoghue M M6RST
ODonoghue S M3UII
O'Donovan A G8NKM
O'Donovan M 2E0IRE
O'Donovan M M6LFW
O'Driscoll J M6JOD
O'Driscoll M M6MOD
O'Dwyer M M1CKZ
O'Farrell J G4YIS
Offer C M1MHZ
Offer I G4FDX
Offer R G1ZJP
Officer G G1NVL
Offler F GM3YAO
Offord R G0JAR
Offord R G3ZYX
Offord R M3ZBR
O'Flaherty L GI0UYY
O'Flaherty L GI1YEA
O'Flanagan D M3HTF
OGarr D G1CHE
Ogbua E M6IDM
Ogburn A M3ELP
Ogden A G1FZV
Ogden A G6JZN
Ogden B G0GML
Ogden F G4JST
Ogden G G6XKY
Ogden J G0TRK
Ogden J M3ZOI
Ogden J M1ZEM
Ogden R G1ZKN
Ogden S M1BSF
Ogden T G0UCA
Ogg D M0OGY
Ogg G GM1BOT
Ogidih C M6GGX
Ogiela S M1CJN
Ogier J G0SUU
Ogilvie I G6VDX
Ogle A M3NUM
Ogle G G8RAK
Ogle M G1DLH
Ogle P M6PFU
Oglesby J M3XPW
Oglesby R G6FNJ
O'Gorman J G0RGU
O'Hagan J G4PFY
O'Hagan J G6HIX
O'Hagan N MI3NOH
O'Hagan N G8NZK
O'Hale D 2I0DBK
O'Halloran M M6PSY
O'Hara A G0KEX
Ohara G G8GUH
O'Hara J GI0EFW
O'Hara J G4PCR
O'Hara J G6WSZ
O'Hara K 2E0MPO
O'Hara N M3NJO
O'Hare C GM0DYF
O'Hare H G0PUQ
Ohare J GM3WNB
O'Hare P MI6POH
O'Hea A M1EWM
O'Hea J M1JIM
Ohennessy C GM4VVX
Ohta M G0CEO
Oka T M0CER
O'Kane D MM3VZI
Okane O GI6OJC
O'Kane P G3OTV
Okanoue I M0CMZ
Okas J G3YGE
Okeefe S 2E0GDF
O'Keeffe A M3ZUC
O'Keeffe R G0FFL
O'Keeffe-Wilson G G4MIA
Okelly P G3PKY
Okorji I M3SQX
Okubo A G0LHB
Olbrien J G4ZKR
Old A M3LTT
Old J M3FOK
Old S M3OLD
Old S M0OLD
Oldfield B 2E0DTS
Oldfield J G4VJL
Oldfield J M6JWQ
Oldfield M G8TJI
Oldfield S G1BHF
Oldford B G6UDX
Oldham J M6GHN
Oldham K G8LDB
Oldham N M6RHR
Oldham P G0NVO
Oldham P MM1HWB
Olding C MW3HAQ
Oldman J 2E0EAL
Oldman J M0GZM
Oldman J M6JRO
Oldrid N M3YNY
Oldroyd J G6EZI
Oldroyd R G4JFV
Olds A G3KFP
O'Leary M M0JOL
Olesen G GM3MQO
Olesen M MM0MLO
Oliphant D M6JAF
Oliphant J G7OWP
Oliphant S G7IIC
Olivant M G7MSG
Olive J G7GNS
Olive J M3BUU
Olive R 2E0HQO
Olive R M6HQO
Oliveira A 2E0VIS
Oliveira A M6GIC
Oliver A G4FGO
Oliver B G8DFI
Oliver C M3CGO
Oliver D G4HPT
Oliver D G4XQX
Oliver D G6UQO
Oliver D G7BRF
Oliver D 2E0DJI
Oliver D M0JJM
Oliver D M6DJI
Oliver D G4KJK
Oliver D G8YFH
Oliver D M3VDO
Oliver E G1AIA
Oliver G G0BJR
Oliver G G7CCS
Oliver G M0GOP
Oliver J GI1VMF
Oliver J G4HMC
Oliver J M6JOR
Oliver J G7PBK
Oliver J G6CND
Oliver K G4SLG
Oliver K G7GZJ
Oliver K M6BGL
Oliver M M6MQD
Oliver M G4HOC
Oliver M G7TXW
Oliver M GW4NOO
Oliver P G1JDO
Oliver P G7CGC
Oliver P M3OLQ
Oliver P M0AWP
Oliver P GD4TSO
Oliver P G4ABT
Oliver R G3NDS
Oliver R MW3UDO
Oliver R M6RLV
Oliver R M6RSO
Oliver S G0NLJ
Oliver S G1CGD
Oliver T 2E0VAI
Oliver T M0VHC
Oliver T M6TOM
Oliver W 2E0DNI
Oliver W M0NVY
Oliver W M6KGB
Ollerhead D G4JMF
Ollerhead D G8GIZ
Ollerton D G6HIO
Ollerton M G7GMD
Olliffe P G0CEI
Olliver R G6VDW
O'Loughlin M 2E0MZM
O'Loughlin M M6MJO
Olsen G M1BEX
Olsen J G4ZBK
Olsen T GM0EQW
Olsen T M0HXN
Olson P G7FMW
Olson P M0AJJ
Olver E M3ZZE
Olver I G1LLU
Olver J M6TLS
O'Mahoney J M0DIG
O'Mahony N M6NOM
Omalley J G0OJG
O'Malley J M3YOT
O'Malley S G7ANV
Omar A M0WLY
Omar A M6DLY
Omar G M3HVO
Omara J G0PPX
O'Meara J G8TBB
O'Neal E M3EMO
Oneil T GM4PRO
O'Neill A MM6ARN
O'Neill C G0CGD
ONeill G G0LUY
O'Neill G 2I0CLS
O'Neill G MI6RAS
O'Neill G G7RES
O'Neill J GI4XKI
O'Neill J G7GAG
O'Neill J GM7VSB
O'Neill J M1AOX
O'Neill L M6WYY
O'Neill M G8WEM
O'Neill M M3KKO
Oneill P G3SPO
O'Neill P G0FCG
O'Neill S 2E0SMO
O'Neill T G4AHC
O'Neill T G3ZGI
O'Nion J G0SMM
Onion P G0DZB
O'Nion P G0PSO
Onione D G8WGQ
Onions G G4JVH
Onions G G8FCO
O'Nions J G1LHV
Onions L G6ZSG
Onions N G8HUT
Onions N M1DXO
Onions S G0RNX
Ootam S G8SJO
Openshaw F G4TWB
Ophert I MI3RXU
Opie C 2E0CSO
Opie C M6BKE
Opie P G4MKG
Opitz H G0KEC
Opitz H G1XDJ
Oram E G1NUO
Oram M M3DPB
Oram M G6GIF
Oram M M1ALT
Oram P GM8ZOW
Oram R G0FXI
Oram S 2E0TMC
Oram S M6TMC
Orange C 2E0CAA
Orange G G0DLQ
Orange J M3MRN
Orange S M0DLB
Orbell M 2E0OBL
Orbell M M0OBL
Orchard A G1OEB
Orchard A G6RXK
Orchard F G4GBC
Orchard G G0CNK
Orchard G G6JKK
Orchard J 2W0KJO
Orchard J MW6OKJ
Orchard K G3TTC
ORCHARD K M6HVK
Orchard L G7MXN
Orchard N 2W0HDC
Orchard N MW0WSD
Orchard P G0EYZ
Orchard R G7ARP
Orchard R G8HZN
Orchard R 2W0RMO
Orchard R MW6TJH
Orchel H G7TFG
Orchiston A G7SYQ
Ord P G7CLG
Ordish C M6CRO
Ore A M1CQL
Ore N M1CQK
O'Regan B G8NMH
O'Reilly A MI6AOR
O'Reilly A M6MXA
O'Reilly G 2I0GCC
O'Reilly G MI6GAQ
O'Reilly J 2E0FFM
O'Reilly J M6JTO
O'Reilly K GI7UIP
O'Reilly K GW0KIG
O'Reilly K G6INM
O'Reilly M GW7EMV
O'Reilly W G4UHT
Orfanidis K M0KYR
Orford G G4FRO
Orford J G3PBF
Orgee A G1ICQ
Orgel H G8DUT
Orgill D G0ABP
Orgill N G6LBG
O'Riordan S 2D0NSY
O'Riordan S MD6MPS
Orlebar G G7EBL
Orlebar G M5GJO
Orme D M6KCI
Orme J M0HXN
Ormerod D G3SLT
Ormerod J M3GDX
Ormerod P M1DQX
Ormerod S G0WCS
Ormiston T 2M0NMD
Ormond R G0EDO
Ormondroyd S G8RHQ
Ormsby-Rymer J G1JOR
Ormston J M6HTK
Ornstein S G8PPA
O'Rourke J G7NJZ
O'Rourke M M0ARY
O'Rourke P G7SCX
O'Rourke W MM6WOC
Orpen M 2E0ORP
Orr C GM1CCN
Orr F GI4XHO
Orr J G3KYE
Orr J G3VKB
Orr J MI0BOU
Orr J M0JOR
Orr M 2E0AVB
Orr N GI6GAG
Orr N MI3VFJ
Orr P M1DPQ
Orrells J G6DQY
Orrin J G3BBK
Ortega Navarro M M0MIG
Orton A G0GQI
Orton R G3VGX
Orton R 2E0BIB
Orton R M3GPN
Orwin P M6KZE
O'Ryan L 2E0CAT
O'Ryan P G8WWF
OSBAND M M6AZQ
Osborn A G4OZY
Osborn B G0BLO
Osborn C G4UXV
Osborn C M0CSC
Osborn C G3XIZ
Osborn D GD4HOZ
Osborn K G7GSD
Osborn M GM8ZMF
Osborn S G8JZT
Osborne A G3SLI
Osborne B G0MPJ
Osborne B GW0SZU
Osborne B G1BHG
Osborne C 2E0OZY
Osborne C M0OZI
Osborne C M6CQO
Osborne D GW1SVV
Osborne D M0RCY
Osborne D M6CKZ
Osborne D M6DSO
Osborne D M6DYO
Osborne I G0FIW
Osborne I G0KKT
Osborne J G3HMO
Osborne J G4GSC
Osborne J G4URS
Osborne J G8WBY
Osborne K GM6KOR
Osborne K G7NMI
Osborne M 2E0KZM
Osborne M G1BUV
Osborne M G3YGM
Osborne M M0KZM
Osborne M G4XOL
Osborne M M6OZM
Osborne N G4JEI
Osborne N M6NSO
Osborne P G4RPF
Osborne P G7HBN
Osborne P M6POZ
Osborne R G1JWD
Osborne R G4BVT
Osborne T M1BSU
Osborn-Jones C G4CSI
Osbourn C G0FRU
Osbourne D M3EKY
Oseland P 2W0PWO
Oseland P MW0GYV
Oseland P MW6PWO
Osgathorpe R G4VWB
O'Shanohun S G4FWT
O'shaughnessy A G0ITO
O'shaughnessy A G6ZMX
O'Shea B G6LHG
Oshea D M6EPD
O'Shea J 2E0JSO
O'Shea J G4GDP
O'Shea J M3JSO
O'Shea P M3AQK
Oskis D G0DOM
Osmand D M3ZDK
Osment M G8AIP
Osmond A G8PPN
Osmond A M3SAO
Osmond A G5DJW
Osprey G MI6HWV
Ostapiuk J 2E0DZQ
Ostapiuk J M6HSI
Ostatek A M6KCP
Ostley T M0CZP
O'Sullivan F M1FIP
O'Sullivan J G4PPG
O'Sullivan J G4VYJ
O'Sullivan N M3NON
O'Sullivan P G1IKL
O'Sullivan S G8VPG
Oswald C G0GBW
Oswald D GM3COQ
Oswald R G7PIP
O'Tani H G8OTA
Othen N G7UNW
Otley J G4CYA
O'Toole J G7UYT
O'Toole J M0HEM
O'Toole M G8ZME
Ott F G0SQP
Otter B G3TOA
Otter G M0GPO
Otter J G3KNJ
Otterson M G8VQJ
Otterson W GI4GPA
Otterwell P 2E0VBX
Otterwell P M6VBX
Ottewell P G3ZRE
Ottley R G3WIA
Ottolini R G6INO
Ottway R G4ULZ
Oubridge B G7AUE
Oubridge M G1SYU
Oughtibridge G G4EIL
Oughton A 2E0FMW
Oughton A G6VAL
Oughton B G4AEZ
Oultram D G6GYC
Oura I G0WAW
Oura M G7STQ
Ousbey C G0CHO
Outen S GW0DWQ
Outhwaite R GU3YVV
Outram W M6WIL
Outterside S G0EOY
Ovenden C G7FJU
Ovenden N M1CRF
OVERALL S 2E0HFH
OVERALL S M3HFH
Overbury F G7DPR
Overell P G4FXI
Overend M G0PMP
Overend S G4BVS
Overland J 2E0FPU
Overson C 2E0AJX
Overthrow M 2M0MMO
Overthrow M MM0MOB
Overton C 2E0CMO
Overton D G4EWE
Overton G G4BJD
Overton J G4GUA
Overton M G1JDP
Overton P GM0MHD
Overton P MM1MHD
Overton P M3UXN
Overy G G1AJN
Overy R 2E0ERD
Overy R M6ERD
Ovey P G6SEK
Owen A G4POW
Owen A MW0BEL
Owen A M3OCS
Owen A M6CYS
Owen A M6MRU
Owen A G7HEK
Owen A G0MTP
Owen B G1EWY
Owen B G8IXK
Owen C G3PIO
Owen C G4ITP
Owen C M0DNF
Owen C M0WEN
Owen C M3ZCO
Owen D G0OES
Owen D G1CIV
Owen D G1OXB
Owen D G3MCA
Owen D G4XNV
Owen D G6TFV
Owen D G8XRW
Owen D GW6LHF
Owen D 2E0GIA
Owen D G6NNB
Owen D M6GIA
Owen E M5DJO
Owen G 2E0GVO
Owen G G0KPV
Owen G 2E0PCL
Owen G M0YGJ
Owen G M6PPT
Owen H G1PIH
Owen H G4OVH
Owen I M1ADV
Owen J G8MLK
Owen J M5JAO
Owen J 2E0BLT
Owen J 2E0VTT
Owen J G4VWL
Owen J M3VDN
Owen J M6GTT
Owen J M6JPO
Owen K M3KMO
Owen K M6KVO
Owen L G0AZE
Owen L GW0TSE
Owen L G1DBL
Owen M G3ZML
Owen M G6UZY
Owen M GW0PZZ
Owen M G4JSX
Owen M G4YTA
Owen N G8JGL
Owen N M6NJO
Owen O M3WSC
Owen P G6ZSH
Owen P G8UUS
Owen P M3HIX
Owen P M6PEO
Owen P M6PTR
Owen P GW0KAX
Owen R G6DDO
Owen R G7PNG
Owen R 2E0RDQ
Owen R M6BVD
Owen S M3ULB
Owen S 2E0OSH
Owen S M6PZP
Owen S 2W0SWO
Owen S MW3SWO
Owen T G4PSH
Owen V G3JGE
Owen W 2E0BKZ
Owen W M0GMN
Owen W M3UGX
Owen-Jones J G3VLO
Owens A G0AIO
Owens C G1ZHI
Owens C MM3ZTS
Owens D MW3DTO
Owens G G0FQP
Owens G M6GJO
Owens J G4GDM
Owens J MW6JJO
Owens M M0ESZ
Owens P G0HAK
Owens P G1PWO
Owens R GW0OPP
Owens R G8IIC
Owens R 2E0EID
Owens T M3TSO
Owens W G4GJS
Owings D 2E0EFC
Owings D M6ZKA
Oxberry L M6DJK
Oxborrow A 2E0XAE
Oxborrow A M6OXY
Oxenham S M3SQQ
Oxford K M6KTO
Oxlade A 2E0BKO
Oxlade A M3RGU
Oxlade R G7RFX
Oxlade R G7RGU
Oxlade-Gotobed A M1CQI
Oxley D G0BJL
Oxley I G7EJO
Oxley M 2E0XLY
Oxley M M0XLY
Oxley M M6LXY
Oxley R G3WWI
Oxley R 2E0VRR
Oxley R M6RRV
Oxley W G4CRB
OXTOBY H GI0JHR
Ozanne D GU3UMX
Ozwell K 2E0OZW
Ozwell K M3HVV

## P

Pace J M6DQF
Pacheco III D M0BWS
Pacitti-Lamb M M6MLX
Pack S G7WIQ
Packard K G0MLO
Packer J 2E0JIX
Packer J M6JIX
Packer J G3NRD
Packer K G7WCN
Packer M G4FFC
Packham D M6ZAW
Packham T M3TDP
Packington B G4LTS
Packman C G6XDI
Packman R 2E0CPR
Packman R M6BZH
Packman V G4VYC
Padbury R G4GAB
Paddison E M3XUV
Paddock A G0CRB
Paddock G G3ZGY
Paddock R M6BZG
Paddon D 2E0FPP
Paddon L M3LDQ
Paden R M0RSP
Padfield D G6HOR
Padgett B G4VWT
Padgett M G0TPN
Padgham S G1GLG
Padley D G8TBY
Padley N G7PWA
Padmore R G4MLB
Paduch P M3FZB
Paffett S 2E0EEK
Paffey A 2E0PAN
Paffey A M3PQE
Paffey E M3PQF
Pafrey S M0ORN
Paganuzzi R G4YEG
Pagden T M0PAG
Pagdin N M0PDX
Page A G3IVP
Page A G3UUM
Page A G7ILI
Page A G7LPT
Page A M3LRN
Page A G6UYJ
Page A 2E0CLY
Page A M0JLY
Page A M6BBP
Page A M6WTL
Page B 2W0BTP
Page B G3KLK
Page B G4YFT
Page B M3FRS
Page B M6ULY
Page C G0TRM
Page C G0WJV
Page C MM3UQT
Page C G4BUE
Page D 2E0LGT
Page D G0TIJ
Page D G7WHI
Page D M3VQV
Page D G0NID
Page G G1EGB
Page G G8ZOY
Page I 2E0FJV
Page I G4WIR
Page J G6JMO
Page J G6XDG
Page J M0BUE
Page J M0JDP
Page J M1CEY
Page J 2I0JPP
Page K G4RNZ
Page K M3KZP
Page N 2M0KPE
Page N G0PAG
Page N MM3XNP
Page P G7FIA
Page P G0EMR
Page P MI0UYD
Page P G4FTY
Page R G4KKR
Page R G4YAN
Page R G6TUS
Page R M3RJP
Page R G7IFJ
Page R M6RKP
Page R G8FSJ
Page R G4DYR
Page S M6SDX
Page T M6VQV
Page W GI6UUT
Page-Jones M G7VZY
Page-Jones R G3JWI
Paget L MM3ONX
Paget L GM0ONX
Paget M MM6PGT
Paget S G7EEN
Pagett J G4YTJ
Paice D G3MXK
Paim E G0UUT
Pain K M3WVG
Pain P 2E0DZR
Pain P M6CKV
Pain R G6NAL
Pain R G4PZL
Paine D G0DAV
Paine D G1CVR
Paine D G3ASX
Paine D G0HID
Paine J M6GJL
Paine J G8BUZ
Paine J M6IEM
Painter A G3BPF
Painter C M0GHX
Painter D G0LEV
Painter D G4PNX
Painter I M6ZBD
Painter J M3YLQ
Painter K G4SZO
Painter L G0KQI
Painter N G0FFA
Painter P G4PVP
Painter P G3TEX
Painter S M6SAH
Painting P G3OUC
Painting R G4FAP
Painting S G0LTX
Painting S M1ABX
Painting S M3IEF
Painton R G4YSZ
Painz W G0WFP
Pairman A GM3UA
Paisnel N GJ1YOT
Paland C MJ1CYD
Paland C MJ3CDP
Palawinna C 2E0ZCP
Palawinna C M6ZCP
Paley A G3XAX
Paley F G7MTI
Palfreeman A G0VLK
Palfreman D G4XXZ
Palfrey A G8IMZ
Palfrey J G3MXP
Palfrey J G4CYI
Palfrey L G4CQX
Palgrave Brown I G4IXD
Palik S M0DBH
Palin B G4AHK
Palin N 2E0NKP
Palin N M6NKP
Palin P GM8TTD
Palin P MM0CYR
Palin S M3SBP
Paling P G4RWV
Palir L 2E0YVR
Palir L M3YVR
Palk S G0OUO
Pallant K G0OSI
Pallett S G4JDP
Pallett S G8OZQ
Pallister D G0KVO
Pallister R M6KKA
Pallot T MJ6THP
Palmer A G1YXJ
Palmer A G4VDF
Palmer A G6BHI
Palmer A G8VUK
Palmer A 2E0CNW
Palmer A 2E0GDY
Palmer A M0STL
Palmer B G1BHO
Palmer B G4ECO
Palmer C M3CLP
Palmer C G4FMO
Palmer D 2E0ADR
Palmer D GM0AEY
Palmer D G0LUK
Palmer D G1DHQ
Palmer D G1ICX
Palmer D G3ZDY
Palmer D G4LYD
Palmer D G4PFX
Palmer D GW4XMV
Palmer D G6BHH
Palmer D G6CMV
Palmer D G6TAH
Palmer D 2E0EOL
Palmer D G7URP
Palmer D M3DCP
Palmer D M6KVM
Palmer E G7MGW
Palmer G G0FZC
Palmer G G1JDT
Palmer H G4PFW
Palmer I G7SNC
Palmer J G4BEI
Palmer J M3RHB
Palmer J MW6BZX
Palmer J M6JHP
Palmer K G7IZE
Palmer L G6FKW

Palmer L M6FCX
Palmer M G1OQM
Palmer M G3KQP
Palmer M [illegible]
Palmer M G8PHJ
Palmer M M3NCE
Palmer M M5HOT
Palmer M 2E0ZBB
Palmer M G0OIW
Palmer M M6ZBB
Palmer M 2E0MXP
Palmer M M1EHW
Palmer M M1MDP
Palmer M M6DLP
Palmer N G4GCI
Palmer O M3OLI
Palmer P M6BHZ
Palmer R G1RFB
Palmer R G4FDK
Palmer R M0ANB
Palmer R M3DPQ
Palmer R G7OPG
Palmer R M6RLP
Palmer S 2E0SPY
Palmer S G7UIU
Palmer S M0GSP
Palmer S GM0EQS
Palmer S M6MSP
Palmer T 2E0MKI
Palmer T M0IKM
Palmer T M3IKM
Palmer V G1VNH
Palo E 2E0EDP
Palo E M6ESZ
Paloschi J G0WZD
Pamment A G0APP
Pampling A G3RSP
Panaitescu C M6ZGR
Panayiotou E M6KYS
Panayiotou P M3FTU
Panczel S 2E0AVS
Panczel S M3AJN
Panesar T 2E0CCQ
Pang K G7RHM
Pankhurst S 2E0VAO
Pannell A G0AHA
Pantall S G1PUU
Panting R G4ELY
Panton A G6UGG
Panton B G1TWY
Panton D MM0MPA
Panton G M3ZNF
Panton M M1COQ
Pantony D G3KXB
Pantony S 2E0ZFX
Pantony S M0HUD
Pantony S M6ZFX
Pantrey D G7LMT
Papadopoulos G G0PIU
Papadopoulos T M0SVA
Papaioannou C G7TWW
Papanikolaou V 2E0VAS
Papanikolaou V M0VAS
Papanikolaou V M3VAS
Papazoglou L M0BFV
Pape J G0NYQ
Papper P M3PCP
Papworth A G3WUW
Papworth G G8AUJ
Papworth G M3GCP
Papworth H G6XKE
Papworth J G6LUM
Papworth J M3JVP
Papworth S M1WDK
Papworth S M3FKW
Paradas J 2E0WFC
Paradas J M0WTC
Paradas J M3WFC
Paradi J 2E0JGP
Paradi J M6JGP
Parbery I M6OAF
Parcell A G8BIX
Pardington I G7TOF
Pardivalla A G7PPL
Pardoe A GM0HUO
Pardoe G M6GLZ
Pardoe R G0MHZ
Pardy F G3DZJ
Parfett J G0LAD
Parfitt A G6XKF
Parfitt H 2E0XAI
Parfitt H M3XAI
Parfitt J 2E0SKZ
Parfitt J M3SKZ
Parfitt J 2E0XPJ
Parfitt J M0XPJ
Parfitt J M3XPJ
Parfitt R M3BOR
Parfitt C M6AOF
Parfitt T 2E0XPT
Parfitt T M6XPT
Parfitt T G6DFR
Parfrey J 2E0UKD
Parfrey J G8JTU
Parfrey J M3UKD
Pargeter A G1ODT
Pargeter G G7KTH
Parham S G8IEA
Paricsi A M6ATI
Paris M M3NIT
Paris R 2E0EPT
Parish J G0RHV
Parish J G4UAU
Parish L M3UJX
Parish W M0IBN
Park A G3NZV

Park B G0AJZ
Park B G0TME
Park C GM7NVG
[illegible] 2E0XPK
Park C M3XPK
Park D G3PSV
Park D MM6DWP
Park E G3ZVS
Park G G8KJK
Park H GM4NUU
Park H G4UME
Park J GM0OFM
Park J GM4XJF
Park J M3TGD
Park J 2E0FJP
Park J M3FJP
Park P G4KVK
Park R GM0MXP
Parke M MI6WBN
Parke M MI6GNP
Parker A 2E0SGI
Parker A G3TXE
Parker A G4DUE
Parker A G8IWB
Parker A G8VMQ
Parker A 2E0CSV
Parker A M0HRW
Parker A M6BXS
Parker A G6DFV
Parker A G0KKD
Parker A M6GDP
Parker A G3KAG
Parker B G0FLG
Parker B G0USM
Parker B G3KOQ
Parker B M1EKA
Parker C G0OKT
Parker C G1PAK
Parker C G4CAY
Parker C G4OOI
Parker C M1FII
Parker C M3TBP
Parker C 2E0XCP
Parker C M0ZCP
Parker C M3XCP
Parker C G6BDM
Parker D GM0AIR
Parker D G7OGL
Parker D G8OMB
Parker D M5DAP
Parker D 2E0DFP
Parker D M0GYO
Parker D M6DFP
Parker D 2E0YAL
Parker D M0YOL
Parker D M3TBV
Parker D G4DZU
Parker E 2E0CWX
Parker E G3HEH
Parker E G6BOQ
Parker G G4EMK
Parker G G4JTP
Parker G G4ONV
Parker G M0GLP
Parker G M6GLP
Parker H G8GUN
Parker H M0AYO
Parker I G4IWV
Parker I G6PMO
Parker I G4YUZ
Parker I M0AAP
Parker J 2E0CWP
Parker J G0PLX
Parker J G3FKY
Parker J G3OLX
Parker J G8ILU
Parker J M3LDL
Parker J G3IWV
Parker J G0UID
Parker J G6EHJ
Parker J M1DUC
Parker K G3PKR
Parker K G7MOO
Parker K G8SYA
Parker K G8HTA
Parker K M6KJP
Parker L G6ITV
Parker L G5LP
Parker M G0UMP
Parker M G4IUF
Parker M G6KXD
Parker M G6NWK
Parker M M6DZY
Parker M 2E0MJE
Parker M M6IDB
Parker M 2E0LRG
Parker M G4YFU
Parker M G8PXI
Parker M M0SAZ
Parker M M6LRG
Parker N G1AJS
Parker N G7VQA
Parker N G8CYA
Parker O M0LQA
Parker P G1VVW
Parker P G3AVN
Parker P GD4UQO
Parker P M6PMP
Parker P G3UAP
Parker R G3JPG
Parker R G4OLP
Parker R G4ZBO
Parker R G8HNM
Parker R G8KFF
Parker R M6YAR
Parker R 2E0MNY

Parker R M6FQU
Parker R 2E0VHA
Parker R M3VHA
Parker R M6MLI
Parker R G4MUJ
Parker R M6PPP
Parker S GW0LYK
Parker S G1ZBY
Parker S G4OOH
Parker S G6MJM
Parker S G8TLC
Parker S G4HQH
Parker S M3MEE
Parker S M6CPO
Parker S 2E0SJP
Parker S M6MJP
Parker S G0BAY
Parker S 2E0PKR
Parker T M0BGE
Parker T M3ZYX
Parker T G0TIW
Parker T M0TVR
Parker-Larkin C G8UVG
Parkes A G7AXW
Parkes A M3KPU
Parkes A M3OYS
Parkes B G1WTY
Parkes C M6MFC
Parkes D M6FXJ
Parkes H G3NZS
Parkes J 2E0JPX
Parkes J M6ELC
Parkes J G6BYK
Parkes M M3RET
Parkes M M6MMB
Parkes R G3REP
Parkes R G7MFO
Parkes W GI8AIR
Parkhouse A 2E0UFM
Parkhouse A M6YON
Parkhouse D M3YPX
Parkhouse R M3ECS
Parkhurst A G6UQZ
Parkhurst G G3TOZ
Parkhurst R G1NYS
Parkin A G1NVN
Parkin A G1OER
Parkin D G8GOT
Parkin G G0ISJ
Parkin J G4KZV
Parkin L G3UVY
Parkin M G0UQV
Parkin M M3OAM
Parkin M G0JMI
Parkin P 2E0PBK
Parkin P G7BUF
Parkin P M6PBL
Parkin R G1SGM
Parkin R G7RUS
Parkin R G7VEF
Parkin S G0EVN
Parkin S G4PYV
Parkin T G6SKK
Parkin W G7EBX
Parkins G M3GLM
Parkins R G0GGA
Parkins S G1GBH
Parkinson A M1FJJ
Parkinson C G8SPP
Parkinson C M0AFW
Parkinson D G7LKY
Parkinson D M3TFB
Parkinson D M6DPP
Parkinson D 2I0SJV
Parkinson D MI6SJV
Parkinson F 2M0HDA
Parkinson F G0SBP
Parkinson G M6ANV
Parkinson H M3WIJ
Parkinson I G3YRQ
Parkinson M M3VXN
Parkinson N G3KFB
Parkinson R 2E0RFE
Parkinson R M0GDP
Parkman A M3AYJ
Parkman L M3OJX
Parks G M0GRV
Parks M G4UPD
Parkyn M G7EYM
Parmenter J G0TIL
Parnell C G0HFX
Parnell C G8VDQ
Parnell J G3WJP
Parnell M G8JPV
Parnell N G4ZXI
Parnell-Brookes A M6BPA
Parnham M M6GCX
Parr A G7HCJ
Parr A G8CVQ
Parr B G4TML
Parr C G7MTA
Parr C M0DDB
Parr D G6OSO
Parr D G8DFY
Parr E G0EMX
Parr E G6CGO
Parr E M1FHP
Parr F G4CWV
Parr G G4AXU
Parr J G6WVL
Parr M G1FLV
Parr N G6MRN
Parradine A M3NYI
Parradine F G0FDP
Parrett J 2E0BQV
Parrett J M0GQD

Parrett J M3UQH
Parrett W G7VGK
Parris J 2E0CRI
Parris J M6JSP
[illegible]
Parris S M3IQC
Parrish A M3MHL
Parrish C G4RPI
Parrish E 2E0NSR
Parrish E M6NSR
Parrish H G7PBH
Parrish L 2E0IHF
Parrish P G7MWS
Parrish R 2E0CEJ
Parrish R G0FKK
Parrish R M3MIP
Parrish R G0DMV
Parrish S G0IVP
Parrish V 2E0AEQ
Parris-Hughes S M6COH
Parrott A G1PBY
Parrott D G4LOD
Parrott G G3WHU
Parrott H G6XKK
Parrott J M3PAP
Parrott J M3HQP
Parrott L G0AMU
Parrott R G3HAL
Parrott R M6ROA
Parrott W G8WLD
Parry A G3YYR
Parry A G4XDB
Parry A G4SZI
Parry C 2E0TLC
Parry C M0ZPZ
Parry C M6CZV
Parry D G8MFU
Parry D M6SOF
Parry E G4EVD
Parry E M6EYU
Parry E M6FQA
Parry G G7OSR
Parry J G7NFT
Parry J G3VVC
Parry J G8BMH
Parry J GJ8RRP
Parry K 2E0TAR
Parry K M6KAA
Parry L G8AMK
Parry N 2E0CQJ
Parry N M0NAG
Parry N M6NAL
Parry P M0DLZ
Parry P G1HGY
Parry R G6IYP
Parry R M3IIJ
Parry R G1LBU
Parry R G4VCL
Parry S MW3SFP
Parsley G G4EMH
Parsloe D M6WTM
Parslow D G4XBI
Parslow J GD4UHB
Parsons A G8CAH
Parsons A G8DTA
Parsons A G8RVO
Parsons A G6XPO
Parsons A 2W0VOC
Parsons A MW6VOC
Parsons B GW0KZK
Parsons B G7FIJ
Parsons B G4YJS
Parsons B M6BPH
Parsons C G7KQN
Parsons D G0LWN
Parsons D G0TCL
Parsons D G1OOZ
Parsons D G3SLJ
Parsons D G4BZB
Parsons D G7VTQ
Parsons D M3PBP
Parsons E G3TMD
Parsons F M0FCP
Parsons G G0AOL
Parsons G G1WXU
Parsons G G8APL
Parsons G M0AHY
Parsons G 2E0XDZ
Parsons G M6XDZ
Parsons I M6EAN
Parsons J G1AJU
Parsons J G8MBJ
Parsons J M0ARO
Parsons J M0CUJ
Parsons J M6PRT
Parsons L M0CCV
Parsons L M0LPC
Parsons L G0ULP
Parsons M G4ORP
Parsons P GW4VRQ
Parsons P MW0PJP
Parsons R G0TLQ
Parsons R G3RBP
Parsons R M3ECW
Parsons R M6HJP
Parsons S G4DTW
Parsons S MM3PPA
Parsons S G4NZG
Parsons S G4UHM
Parsons T 2E0TRP
Parsons V M1AKL
Partington D G6EHL
Partington F G4BZP
Partington F G8DTM
Partington J M0YOT
Partington K G6PBI

Partington R M0CIC
Partner A G3HKT
Partner M M6XPB
Partner R G6XPB
Parton A G1PPZ
Parton B 2E0PTI
Parton B M3PTI
Parton P G6JMG
Parton V 2E0BLD
Parton V M3PXU
Partridge B M6GTQ
Partridge C G0GWG
Partridge C G8AUU
Partridge D G1LMW
Partridge D M0LMS
Partridge G G6HZK
Partridge H G7DTV
Partridge H M6HPS
Partridge I G3PRR
Partridge J G0VCV
Partridge M G1KYK
Partridge N G7PLS
Partridge P G1FLW
Partridge R G8KHI
Partridge W 2E0EUY
Partyka M M6VPS
Parvin D 2E0DJP
Parvin D M0PRV
Parvin D M6DJP
Pascal R GM8LNH
Pascall H M6HBP
Paschalis L 2W0FCM
Pascoe A M0CLE
Pascoe E G4DKD
Pascoe G G8LXS
Pascoe J G4ELZ
Pascoe K G8ZQM
Pascoe K M3ZQM
Pascoe N G3IOI
Pascoe R G0BPS
Pascoe S G8SGI
Pascoe S 2E0JBW
Pascoe W M6WTP
Pasek M G4BZS
Pasfield J G1JCP
Pash B G4VLH
Pashler T MI6ODN
Pashley I M6GOB
Pashley R G8OFN
Pashley V M6PAX
Pashley S M6SSP
Pasika D M6DAP
Pasika R 2E0ROM
Pasika R M0ROM
Pasika R M3XYA
Pask D M0TGM
Paskin P G6AFA
Paskins J G4MVP
Pasquet P G4RRA
Pass I 2E0ICP
Pass I M0ICP
Pass I M3XKP
Pass I M6LHP
Pass R G4MGV
Passam L M6LJP
Passam M G8FHC
Passam N M0DSR
Passey A 2E0BTR
Passey A M3SFN
Passey C GW1ABX
Passey C M6ENI
Passey D 2E0BYA
Passey D M0MYA
Passey D M3BYA
Passfield J G1JIJ
Passmore A G3UCF
Passmore B G0BWP
Passmore B G4XEF
Passmore G GW4HGS
Passmore H G0LHX
Passmore P MW6FGQ
Passmore S M0BKL
Paster R 2E0RTP
Paster R M0RTP
Paster R M6RTP
Pastwik D M3ZHP
Patatu T M6GPI
Patchett B G4VBP
Patchett S M3WAL
Patching K G6GPV
Patching W G8WYS
Pate C MM3YQP
Patel A GD7DNP
Patel J G7OMI
Patel N M3INP
Patel N 2D0GEE
Patel N MD6OMG
Patel D G7VYB
Pateman A M3KZI
Pateman S G8WPO
Paterson A G0HAL
Paterson A GM0FQD
Paterson A G8CQQ
Paterson A M3NPZ
Paterson B [illegible]
Paterson D G8CZJ
Paterson D MM3XDP
Paterson D M6NYY
Paterson D 2M1ENI
Paterson D 2M1ENK
Paterson E 2I0OHE
Paterson E MI3OHE
Paterson G GM3KJZ
Paterson G 2E0EMI
Paterson G MM0GPZ
Paterson H MM3LOF

Paterson I G4TZM
Paterson I 2E0WGO
Paterson I M0WGO
Paterson I M3WGO
Paterson J GM3COB
Paterson K MM0DHO
Paterson M MI3ZMP
Paterson M MM3LNT
Paterson N GM8IID
Paterson P G1JDQ
Paterson P GM8PLR
Paterson R G1VIW
Paterson R MM3UOR
Paterson S 2M0WTT
Paterson S 2M0FSP
Paterson S MM6FSP
Paterson T MM6REV
Paterson W 2E0WJP
Paterson W M6WJP
Patient J 2E0JPU
Patient J M0GGX
Patient J M3JPU
Patis A G8UFF
Patman F G6KSR
Patman K G8MEE
Patmore A M6FDK
Patmore P G3WYD
Paton A MM0DVB
Paton C GM7TFN
Paton C MM3CJP
Paton G M6NCR
Paton J G6RAZ
Patrick A G6XOX
Patrick A MM3WWQ
Patrick A 2E0TPA
Patrick A M0TPA
Patrick A M6TPX
Patrick D M1FCF
Patrick D G8KAP
Patrick G 2E0EKB
Patrick G M0EKB
Patrick I M1MAB
Patrick I M6EON
Patrick J 2E0DAO
Patrick M G1LWL
Patrick R G1GWX
Patrick R M3KPD
Patrick S G1VVB
Patrick-Gleed B M6BPG
Patrick-Gleed G M3YJP
Patrick-Gleed J M3YJM
Patrovits F M3BMH
Pattemore J G4NHO
Patten G G1MHB
Patten J G3PIN
Pattenden B G1TLA
Patterson B G3KYP
Patterson A GM4MCV
Patterson A G6AGP
Patterson A G8FXL
Patterson A G8GIG
Patterson A M0CCS
Patterson A MI0BDX
Patterson E G0SLK
Patterson G G4XUG
Patterson G M3INO
Patterson I 2E0BYN
Patterson I M6ILP
Patterson I MM6IPP
patterson J G7SUV
Patterson J G7TFX
Patterson J G4UIR
Patterson J M1JPS
Patterson K GI0OEH
Patterson K GI4BJK
Patterson M GW0TWF
Patterson N G7OFU
Patterson P 2E0CZI
Patterson P M6DQV
Patterson S G1CLD
Patterson S G4UNG
Patterson W GM4YRO
Patterson W G3YDT
Patterson W M6WAP
Pattinson F G7ANH
Pattinson J 2E0FUR
Pattinson J M6PWM
Pattinson N G0JVU
Pattinson R G3KVX
Pattinson W G0JCK
Pattison M G1PZD
Pattison P G6KFZ
Pattison W M6GMT
Pattison-Turner A M6ZPT
Pattman M G0SDW
Patton A G1BHB
Patton C MI3FHM
Patton L M6POW
Patton T GI4OKU
Patton T GI4OKU
Patty J GI3SG
PAUL A [illegible]
Paul A GM6IWF
Paul D GI4NGP
Paul D G6YFZ
Paul D G1NRE
Paul D G4DZJ
Paul E G3CUY
Paul J 2E0AVQ
Paul J M3AQF
Paul J 2E0AKQ
Paul J M0JCH
Paul J M3NNV
Paul M GU6XCM
Paul R M6EQL
Paul S MM3HLG
Pauley D M0DNZ

Pauley M G1MKP
Pauley O G0BGX
Pauley R G8VDJ
Pauley W MM3OZW
Paulick A M0ISL
[illegible]
Pauline J M6EVZ
Pauling F M6ENC
Paulizky A M3NLN
Paull N GW1CUQ
Pavelin E M6OSM
Pavelin P G4WWH
Pavey C G3JHU
Pavey C M6KCU
PAVIA J G7RBR
Pavier C G0BYX
Pavis D GI0BDZ
Pawlak A 2E0CKM
Pawlak A M6AVU
Pawley K M0PAW
Pawlik L G0LSP
Pawson I G0FCT
Pawson L M6XLP
Paxman N 2E0NPX
Paxman N M6NPX
Paxman R M6AVS
Paxton A G4BIZ
Paxton A M3MQI
Paxton A 2E0TAC
Paxton A M3TPI
Paxton J G0FXK
Paxton S G6VLC
Pay D G4WTU
Pay K G1EGE
Payas L G0SEU
Payas S G4YNH
Payea D G7AFZ
Payne A G3RBJ
Payne B G4CJY
Payne B G8UXB
Payne C G6KVE
Payne C G8AJM
Payne C G4OWL
Payne D 2E0DBP
Payne D GI0AWK
Payne D G1UYZ
Payne D G4UUG
Payne D G8OBP
Payne D M3AYQ
Payne E G6UQI
Payne F M6ATY
Payne G G4EUG
Payne G G7BUS
Payne G M6DPL
Payne H G7LQY
Payne J G6CNF
Payne J G8XZB
Payne J M3JFP
Payne J M6JTP
Payne J G7VZD
Payne K G8FYK
Payne K M6WFY
Payne K MD6IUH
Payne K M0CPF
Payne L G0NNU
Payne L G4MWX
Payne M G1PER
Payne M G3ZRM
Payne M G4ITX
Payne M 2E0RDP
Payne M M6MIP
Payne N M3NEP
Payne N 2E0ZZT
Payne N M3HUS
Payne P M6NZL
Payne P M6PFB
Payne R GM4AWA
Payne R G4SUX
Payne R G8UPD
Payne R M5ACJ
Payne R M6RAP
Payne R 2E0BLR
Payne S G0UEK
Payne S G7RHE
Payne S M0HBI
Payne T M3TEP
Payne W M0WAZ
Paynter D G0FJR
Payton P G6ZSU
Payton P G7COP
Peabody K M3IYO
Peace C M6PNX
Peace J M3LEF
Peacey K GW0VMD
Peach A 2E0AUM
Peach D G3VXO
Peach D G7OBF
Peach G G7SLL
Peach M G7OBD
Peach P G3GOS
Peachey A G0BXJ
Peachey D C0L6I
Peachey D G1UWD
Peachey D G6IOW
Peachey M GM1UWE
Peacock B M6BGP
Peacock C GW6CZE
Peacock C G1VTN
Peacock D G3NOP
Peacock D G0BJI
Peacock D M6DPV
Peacock E G7JVJ
Peacock G G3NHP
Peacock G M0BLV
Peacock J M6JRL
Peacock K G7OQT

Peacock M G1NQU
Peacock N G4KIU
Peacock R M3RBP
Peacock R M6PIT
Peacock R M3XRV
Peacock S 2E0AVM
Peain J 2E0XDI
Peain J M3XDI
Peak W G4VDH
Peake A GW3SRG
Peake A MW3KHH
Peake A G6MJQ
Peake C G4UDN
Peake C G0NZI
Peake J GW1WWE
Peake J G4LTZ
Peake M G3UIJ
Peake P G8FVM
Peake S M3BCW
Peake V G4GEP
Peakman K M6OYA
Pealing J G3WCU
Pearce A G6IOX
Pearce A GM7VLZ
Pearce A G8GOR
Pearce A M3AEJ
Pearce A G0AZQ
Pearce B G7ACA
Pearce B M0CMN
Pearce B M3XEG
Pearce D G1PBF
Pearce D G6WMU
Pearce D G7GQD
Pearce D M0CNG
Pearce F G1ONQ
Pearce G G4LSX
Pearce G 2M0YEQ
Pearce G MM0YEQ
Pearce G MM3YEQ
Pearce H G8XNH
Pearce H G4BXC
Pearce I GM7STI
Pearce J G1OOU
Pearce J G3MEC
Pearce J G8IWI
Pearce J G6FRB
Pearce L MW3WLB
Pearce M 2E0MEP
Pearce M G4KQY
Pearce M G6GVO
Pearce M G7SNX
Pearce M M3RIP
Pearce M 2E0XMP
Pearce M M0GYH
Pearce M M0MIM
Pearce P G0DIS
Pearce P G0IMA
Pearce P 2E0FAN
Pearce P M6PGP
Pearce R G0GHT
Pearce R G0LTV
Pearce R G4VKE
Pearce R G7UTT
Pearce R G8TLP
Pearce R M0BJP
Pearce R G0SYF
Pearce S M6SQF
Pearce S 2E0STI
Pearce S M3XXL
Pearce T G0CXW
Pearce W G8OJR
Pearcey C 2E0KFR
Pearcey C M6KDR
Peard R G0SON
Pearl B G4FCX
Pearless C G3PGK
Pearless S G6AZG
Pearman C M3WYZ
Pearn B M6BLP
Pearn G G6NKS
Pears I M6ICP
Pears J G0FSP
Pears J M0COT
Pears R G4TRV
Pears V G1OHX
Pearsall T G4OWK
Pearsall T 2E0CII
Pearsall T M0PQI
Pearsall T M6AQK
Pearse C G4WGD
Pearse J M6JMP
Pearse L G7LMX
Pearse L M3LMX
Pearsey R G6RBP
Pearson A GM1MLS
Pearson A M6DCP
Pearson B G4CVS
PEARSON B M3PWZ
Pearson B M6BXP
Pearson C M3CDI
Pearson C G5VZ
Pearson C M0JHQ
Pearson D G1OJO
Pearson D GM3TLA
Pearson D GU3ZOM
Pearson D G4EIC
Pearson D GW7RQI
Pearson D 2E0DMU
Pearson D G4PMA
Pearson D G4UFS
Pearson D G7AKM
Pearson D G7GHE
Pearson D M0HZK
Pearson D M6AGY
Pearson D M6FHY
Pearson D M3WQG

UK Surnames

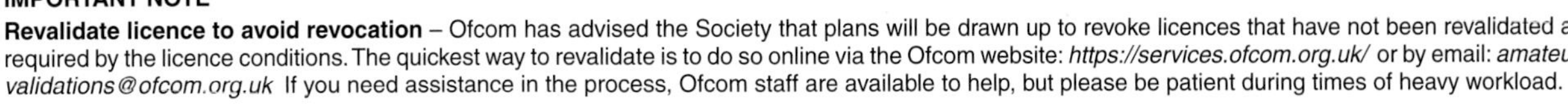
**IMPORTANT NOTE**

**Revalidate licence to avoid revocation** – Ofcom has advised the Society that plans will be drawn up to revoke licences that have not been revalidated as required by the licence conditions. The quickest way to revalidate is to do so online via the Ofcom website: *https://services.ofcom.org.uk/* or by email: *amateur.validations@ofcom.org.uk* If you need assistance in the process, Ofcom staff are available to help, but please be patient during times of heavy workload.

Pearson F G4JVM
Pearson F G7FSH
Pearson G GM8BHR
Pearson G 2E0GPE
Pearson G M0PEA
Pearson G M6GHP
Pearson H G0SET
Pearson I 2E0FJN
Pearson J G0VTM
Pearson J G1WXF
Pearson J G4KDM
Pearson J G8XUH
Pearson J M0CAV
Pearson J M6JCD
Pearson J G1FTU
Pearson J G7UBO
Pearson K G0CRX
Pearson K MI6EDK
Pearson L 2E0LUL
Pearson L M3JRQ
Pearson L G3VNT
Pearson M G1GBI
Pearson M G1VAL
Pearson M M0AAK
Pearson N M3NLP
Pearson N M3XTT
Pearson P MI3PPI
Pearson P G0MTB
Pearson P M6PPJ
Pearson P G4GXI
Pearson P G0UIB
Pearson R G3XUH
Pearson R G4GOX
Pearson R M0CTM
Pearson R G0PEY
Pearson R G3XWU
Pearson R G4FHU
Pearson R M6RGP
Pearson S 2E0ASW
Pearson S 2E0BNO
Pearson S 2E0GYH
Pearson S G0NLX
Pearson S G4AGH
Pearson S M0CLW
Pearson S M6HVF
Pearson T G0DXT
Pearson T MI6OTP
Pearsons C G0DWM
Peart B G4RKF
Peart C G1FHK
Peart R G0FHK
Peasey E G0UGM
Peat D G1GMH
Peat D G0RDP
Peat S G0CBT
Peatman W G3EJH
Peberdy E G0DKV
Peberdy H M6FHV
Pebody B G0MWU
Pechey D G8NMO
Pechey W G4CUE
Peck B M3HGL
Peck D G8RKG
Peck D G8LAU
Peck E 2E0GIH
Peck E M3GIH
Peck G G3ZVN
Peck G G4OIG
Peck G G8CXK
Peck J G3YCE
Peck M M3THE
Peck T 2E0JSJ
Peck T M6TDF
Peck W G6OTS
Pecknett R G4KUX
Pedder C G3VBL
Pedder D G3LFX
Peden J G3ZQQ
Peden J G8DQP
Peden J M6EYK
Pedley C G3YHN
Pedley D G8EMA
Pedley J G1UBC
Pedley J G7TUD
Pedley J MW6PED
Pedley J G1YBM
Pedley M 2E0AYW
Pedley M M3NMP
Pedley P G1RLR
Pedreschi M G3MZX
pedro R GD7HIH
Peech R M0RAP
Peel C G4NFL
Peel E G8CCQ
Peel J G6TAI
Peel K 2E0BNV
Peel K M3PXE
Peel R G4TKP
Peel R G8NEF
Peel R M3IOK
Peel S 2E0PEL
Peel S 2W0SCP
Peel S MW0GIN
Peel S M0SMP
Peel T G4KWE
Peel W G4URV
Peel W M6ARD
Peeling R G6MXE
Peerless J G3JPJ
Peerless J MD3THQ
Peers K G1NVY
Peers M G6FMS
Peers M G8HYP
Peet A G7TOY
Peet J G4PXJ
Peet M G3ZJJ

Peeters R GU0VPA
Pegg A M0JAG
Pegg D GM7FLG
Pegg F G0HXC
Pegg S M6KSP
Peggram J 2E0JKP
Peggram R G7RJH
Pegler A G3ENI
Pegrum C 2E0CMP
Pegrum C M0NAY
Pegrum C M6BOX
Pegrum C GM8DKG
Pegrum J G3XCK
Peiperl M GD0UHK
Peirce J G0MQE
Peirson J G3UYC
Pelham J 2E0CEP
Pelham J M0HBX
Pelham J M3SDV
Pelham T 2E0CNV
Pell A M1AQX
Pell A M0DME
Pell C G3WLH
Pell J G6CMX
Pell M M0BSL
Pellatt M G4APG
Pellatt R G4LJI
Pellett R G3RZC
Pelling R G0BUJ
Pelling R G1GFZ
Pellowe D G4IGZ
Pellowe V G4VAL
Pells D G4DTP
Pemberton A G0NGH
Pemberton A MM6PEA
Pemberton B M3ZYW
Pemberton C G4GNG
Pemberton D G6HZJ
Pemberton D G7VGL
Pemberton G G1BVV
Pemberton G G7NEH
Pemberton I G0EHA
Pemberton J G4WXO
Pemberton K M3ZKE
Pemberton M G6DAY
Pemberton M G4DDL
Penaluna K G6BEH
Penaluna R 2E0RNP
Penberthy R G3ZFP
Pendle A 2E0REN
Pendlebury S M6SPZ
Pendleton J G0SOU
Pendleton T G4IRH
Pendrick G M5GAC
Penfold A G0BEX
Penfold A M3VPA
Penfold C 2E0AYX
Penfold C M0OSM
Penfold C G7GSX
Penfold C M3CFP
Penfold H M6HIL
Penfold P 2E0RAJ
Penfold P M3MFE
Penfold R G1COW
Penfold R G1OKD
Pengilly J M6JYS
Penistone N G0FLI
Penketh N G3RYY
Penman R G4BYB
Penman T MW6WFJ
Penn A GI8LFY
Penn E G7PEN
Penn M G7JYY
Penn R GI4GEL
Penna C GM3POI
Penna T MM3POI
Penn-Bixby P 2D0IGL
Pennell A G0NSK
Pennell L G3KME
Pennell L G8PMA
Pennell S M6HUF
Pennell T M6NSK
Pennells A G6TGB
Penney A G4WEM
Penney C G7IAU
Penney G G4ECF
Penney I G1JUO
Penney J G0LAG
Pennington A 2E0PDQ
Pennington J G0OGX
Pennington J G3RTP
Pennington J G7TYJ
Pennington M M6MCP
Pennington P G4EGQ
Pennington R G1NVS
Pennison D 2E0DDF
Pennison D M6DFD
Pennock R G0FLV
Penny D G0AIL
Penny D G3PEN
Penny M G7DTR
Penny P G6TAF
Penny R 2E0RWP
Penny W M3UVK
Pennycook R G1PGQ
Pennykid R MM3LGU
Penprase V M3RNK
Penrose D G1CWZ
Penrose R M3SPP
Pentecost K G7BPN
Pentecost P GW0LNM
Pentecost S G6YNT
Pentin D G1ZEW
Pentland J GM0KLP
Pentland W GM0CWR
Pentler M MM3IIG

Pentney K M3AKW
Pentney L M3LCP
Pentney R M1AOB
Pentz H M3YHP
Pentz T M3UTP
Penver R G8VBK
Penycate P 2E0FVL
Peperell A G3TOP
Pepler A M6PNP
Pepper D G0SLJ
Pepper D G8HYU
Pepper E G3YWA
Pepper J GD0CHQ
Pepper J G4EPA
Pepper J M1AOL
Pepper J G6ZSQ
Pepper M G4WEH
Pepper R G4MVZ
Pepper S G6EZH
Pepper Z M6ZLP
Peppiatt M G7TKW
Percival D G8FXU
Percival G GM0CFC
Percival J 2E0JCP
Percival J M3XLR
Percival K G7GFK
Percival L M3JWV
Percival R G7SGO
Percival R G1EQU
Percival R G4VCJ
Percival R G6CGY
Percival-Alwyn L M6RPY
Percy H G0IBT
Percy K 2E0CID
Percy K M6BCZ
Percy M G8NWZ
Peregrine S M3JZE
Perek A M0UNJ
Perera I G4FJX
Perera P G0USK
Perera P G4AJG
Perez J G4SVY
Perez-Mendez E M3DDZ
Perfect A 2E0PFT
Periam C M3VLG
Perkin F G1TAZ
Perkin K M1ABF
Perkin R M0RMP
Perkins A G0FGR
Perkins A G7RBB
Perkins A M0PER
Perkins A G1BNE
Perkins A G6KDY
Perkins C M3FAC
Perkins C G4JSP
Perkins C G4VFX
Perkins D G4VZK
Perkins D M3TIQ
Perkins G 2E0GPS
Perkins H G3NMH
Perkins I M6IWP
Perkins J 2E0AZQ
Perkins J GW3PPB
Perkins J G0NJT
Perkins M M6CCX
Perkins N G4EJW
Perkins N 2E0PMA
Perkins P G7DQL
Perkins P G3OUV
Perkins R MM6AGL
Perkins S G4FPV
Perkins S G8LLS
Perkis R G5UI
Perks A GM1UIR
Perks A MM3DMZ
Perks D M3JYO
Perks E G3VWX
Perks G G6VAA
Perks I G0JRE
Perks R G0LBQ
Perks R G6EKM
Perks R G4ICI
Perocevic M M3USB
Perovic V 2E0DSZ
Perovic V M0HZC
Perrat R G1XLT
Perreas P M0LPT
Perrett A G4ZDR
Perrett B 2W0BDW
Perrett J G6KPD
Perrett L G4XWE
Perrett M G0ESY
Perrett M G8LCE
Perrett M GW1UHF
Perretta P G3SEA
Perrin B G3NRH
Perrin B G7CIV
Perrin B MD3MCB
Perrin C M6PEZ
Perrin D M0RBU
Perrin R G4AFY
Perrin R G4DFT
Perring S 2E0SXP
Perring S M3SXP
Perrins G M1BIX
Perrins P G4NQW
Perrott C G6ANJ
Perrott N G4TAW
Perrow A M0HSS
Perry A G0AYY
Perry A G3ATX
Perry A G7LPO
Perry B G1XVL
Perry B G4ZUD
Perry C GM0EFD
Perry D G1CQG

Perry D G4YVM
Perry D M6GYU
Perry E GM0EFC
Perry G G1IDJ
Perry G G1UUK
Perry G G4OED
Perry G M0GTP
Perry G G0GEP
Perry G G1LTG
Perry H G3KXS
Perry I G7UVV
Perry J G3OHS
Perry J G4ERH
Perry J G4FJP
Perry J 2E0JNP
Perry J G0GOD
Perry J M0JNP
Perry J M6GSW
Perry K G1WGM
Perry K 2E0FOU
Perry L M6PLZ
Perry L M6TBJ
Perry M G0WIT
Perry M G8AKX
Perry M G8NYD
Perry M MD0IOM
Perry N G6DQZ
Perry O G4ASX
Perry O G0DRR
Perry R G3ZRP
Perry R G6LUJ
Perry R G8CVP
Perry R G6YNL
Perry R G6KXN
Perry S MD3PER
Perry S M6SLR
Perryman B M3CMM
Perryman E G1GKW
Perryman H G4KKJ
Perryman K GW4RKI
Person A G0TJP
Pert A GM7HIR
Pert J GM0HIM
Perver R GI8YJD
Perzyna R G8ITB
Pesarini G G0VPV
Pesch G M0DXM
Pescod C G4BMW
Pescod R G3UCD
Pesendorfer M 2E0DTY
Pesendorfer M M0OTO
Peskett G G3LOF
Pesticcio D MW6ETY
Petchey M 2E0RFF
Petchey M M6EBC
Petch-Harrison J M0PRF
Peter D 2E0BZB
Peterkin I G3WDU
Peters A G1AIO
Peters A G1UTZ
Peters C G0NDF
Peters C M0DMJ
Peters D 2E0FZU
Peters D G6GAW
Peters D M0DWP
Peters D 2E0BMJ
Peters F M0MLH
Peters G G0EGQ
Peters G G8COR
Peters I G0NWY
Peters I M5PLY
Peters J G1BAX
Peters J G3VMJ
Peters J M3YXE
Peters K M1AHR
Peters M 2E0WMD
Peters M M3ESQ
Peters M 2E0MRP
Peters M M6MAP
Peters M G4EFE
Peters M M1EXQ
Peters N M3HGM
Peters O M0DBJ
Peters T MW0CGP
Peters T MW3VMY
petersen P M6LPP
Peterson G G8SUG
Peterson M 2E0GYB
Peterson M M0GYB
Peterson M M3GYB
Peterson P G0HUF
Peterson R M0RCP
Peterson T M3OOL
Peterson T M0KCE
Peterson W G4EZU
Pethen M G6MDF
Pether J G4JGG
Pether J G4YKG
Petherick K G0ODU
Petifer B G0SYR
Petifer B G8DTQ
Petit-Brown N M6LPB
Petraitis S G4IFM
Petre K G4NMK
Petri D G8CCJ
Petri G G0PXA
Petri R G0OAT
Petrie A G4DJZ
Petrie A M6ACJ
Petrie C 2E0SCR
Petrie C M6CJP
Petrie D GW1MVZ
Petrie I 2E0HIT
Petrie I GM4EHP
Petrie I M6HIT
Petrie J GM3FDN

Petrie R G0SLL
Petrie R G0UMS
Petropouleas I M3WUQ
Pett R G3SHK
Pettefar N M0NJP
Pettefar N M3VOW
Petters J G3YPZ
Pettett R G7TKI
Pettett W GM1BVT
Pettican D G1WSF
Pettican G G4MYQ
Pettifer J G7SJP
Pettifor J G8LQF
Pettigrew A G0FPT
Pettigrew D GM0EWF
Pettigrew M G0WLR
Pettigrew R M6RJJ
Pettinger I G4YTM
Pettipher S MW0DXX
Pettis T 2E0CXQ
Pettit B 2E0FDG
Pettit B M6XBP
Pettit G G1VSH
Pettit M G3UCW
Pettit N G6LOJ
Pettit S 2E0TVD
Pettit S M0HQH
Pettit S M6SCP
Pettitt C G0EYO
Pettitt C G6SL
Pettitt D M0DNP
Pettitt N M6NHP
Pettitt S M0MOI
Petts A G3PXF
Petts A M3DAM
Petts R M3RYA
Petty D G8ULM
Petty D G8VSV
Petty P G4CVD
Pevy A G4XYW
Pevy W G4CWP
Phaff J G4KDN
Phair W 2I0LJQ
Phair W MI3LJQ
Phanco G GM4KHE
Pharaoh M G3LCH
Pharoah M M3UDJ
Phelan A M1FIE
Phelan G G8EPS
Phelps D M6EXH
Phelps J G4RMX
Phelps J G6DFM
Phelps M G1KGE
Phelps P G6UDF
Phelps P G6IOV
Phibbs R MU6RAN
Philip D M1ALX
Philip E GM4JLZ
Philipp B G4PUP
Philipps S G0VRV
Philipson J G4LBQ
Phillips A M0BZC
Phillips A MW0EAT
Phillips A M3PQN
Phillips A 2E0FAP
Phillips A M0HAP
Phillips A G0JUV
Phillips A M6FQP
Phillips B G0GJG
Phillips B G0GRO
Phillips B G4DUF
Phillips B M0DHI
Phillips B G4ZLN
Phillips C 2E0DEO
Phillips C G4LXJ
Phillips C GM4WUR
Phillips C 2M0YSR
Phillips C MM0YAB
Phillips C MM6YAP
Phillips C G3RLA
Phillips D 2E0ZDA
Phillips D G0WAE
Phillips D GW4KQ
Phillips D G4SDO
Phillips D G8AAE
Phillips D GW8HF
Phillips D GM0OLF
Phillips D GW7CYT
Phillips D G8CXI
Phillips D G1EGI
Phillips D G0YAE
Phillips D G7SXB
Phillips E 2E0AYO
Phillips E G8MVY
Phillips F M6BZW
Phillips G GI4AFH
Phillips G G4KWO
Phillips G GW6UGC
Phillips G M0HWO
Phillips G G0KRB
Phillips G G3XTZ
Phillips G G7HHW
Phillips H G1MUM
Phillips H 2M0HJP
Phillips H MM0VPF
Phillips H MM6VPF
Phillips H M6BXM
Phillips I G7CQK
Phillips I 2E0DII
Phillips I M6IAO
Phillips J G1DSZ
Phillips J GW3VOL
Phillips J G4JKH
Phillips J G8POE
Phillips J M0ASL
Phillips J M0WBC

Phillips J MD3EVY
Phillips J MI0MAP
Phillips J M6HXF
Phillips J 2E0GHP
Phillips J G3LDC
Phillips J G3PXX
Phillips J G6ZCR
Phillips J M3JAP
Phillips J M3JKT
Phillips J M3UJO
Phillips K 2E0GOK
Phillips K G8WNB
Phillips K M0MCW
Phillips K G1POJ
Phillips K G8BOQ
Phillips K G6CXV
Phillips K M6FGI
Phillips K M6MZY
Phillips L G1FNA
Phillips L G8NBO
Phillips L M6LSP
Phillips M G0CBB
Phillips M G0MBZ
Phillips M G4CIO
Phillips M G4GTZ
Phillips M M0CUP
Phillips M M3DCL
Phillips M MW3IWC
Phillips M 2E0GPA
Phillips M G1JXL
Phillips M G7LTT
Phillips M M6GPA
Phillips M M6NVA
Phillips M 2E0NZD
Phillips M G6WPJ
Phillips M M0MRP
Phillips M M6CNS
Phillips M M6MPX
Phillips M 2E0CPE
Phillips M M0HPW
Phillips N GM6PGV
Phillips N G7POC
Phillips N 2E0BPU
Phillips N M3FUB
Phillips P GW4MIP
Phillips P M0ZED
Phillips P G4TPK
Phillips P G7KBR
Phillips P M6CNQ
Phillips P MW6TDE
Phillips R GM0VRP
Phillips R G1GGB
Phillips R G4IQQ
Phillips R G4SBS
Phillips R G7FKF
Phillips R G7IZC
PHILLIPS R G6OIH
Phillips R M0PHE
Phillips R M3OSQ
Phillips R M6FEP
Phillips R MM6RSP
Phillips S G0LOE
Phillips S G4EYR
Phillips S G6UDI
Phillips S M3TRO
Phillips S 2E0MMZ
Phillips S 2W0CYF
Phillips S MW6BRF
Phillips T GI8JPF
Phillips W M0WDP
Phillips W M3WMP
Phillips W 2E0WPS
Phillips W M0WPS
Phillips W M6WPS
Phillipson B G7OCK
Phillipson C G8IJC
Phillipson J G4BEZ
Phillipson R G7OOI
Phillpott A G4RJZ
Phillpott A G4IMP
Phillpott I 2E0UHS
Phillpott I M3IMP
Phillpott N M3PNA
Philp A GM3WAP
Philpot C G6KBC
Philpot J 2E0BFP
Philpot R G4TTQ
Philpot S 2E0EFG
Philpott A G1JYZ
Philpott C G4YNC
Philpott M 2E0DFI
Philpott M M6EHA
Philpott R G0AUW
Philpott R G3VCH
Philpott T 2W1JOL
Philpott T M3YTP
Philps J 2E0JND
Philps J M6JND
Phin R G7CRN
Phipps E M6EFF
Phipps G G3IPG
Phipps J G4HVG
Phipps J 2E0JAW
Phipps P 2E0PDB
Phipps P M6PEF
Phipps R G7AYE
Phipps R G4DIC
Phipps S M6SGP
Phiri E 2E0EPR
Phiri E M6ESP
Phizacklea K M3HVL
Phoenix A G7CBR
Phythian S M6STP
Piatkowski M M6JYI
Pibworth D G4KWT
Piccavey S 2E0BYI

Pick D G3YXM
Pick J M3YXM
Pick M M3MVI
Pickard A GW4OES
Pickard B G4TFD
Pickard D GD8EUH
Pickard J M3BVX
Pickard W G8TRR
Pickavance A M6TKP
Pickerill E G0IUK
Pickering A M3GMY
Pickering A G0GKK
Pickering B 2E0BRU
Pickering C G1AJT
pickering C G8DHJ
pickering D G7KSV
Pickering D M1EJD
Pickering D G7UAY
Pickering E G3LPS
Pickering G G4RAY
Pickering J G0WZJ
Pickering J M3JPI
Pickering M M6LHF
Pickering P G3ORP
Pickering R G0BCF
Pickering R G3SLK
Pickering R G3UWP
Pickersgill A M3ESG
Pickersgill F G3XXN
Pickett A 2E0TDP
Pickett A M0TDP
Pickett A M3IXK
Pickett C M5LRO
Pickett R G7BZC
Pickford C M3TNE
Pickford D G8TNE
Pickin S M1SRP
Pickles A G1AWU
Pickles A 2E0CKI
Pickles A M6BCG
Pickles C G7KPS
Pickles D G3XVA
Pickles J G4CWM
Pickles K G4XZM
Pickles R G3VCA
Pickrell B G8ARM
Pickstone R G6NCL
Pickstone S G1FLX
Pickup B M0RNR
Pickup E G4KWF
Pickup J G4RYT
Pickup P M0BXF
Pickwoad R M6WFR
Pickworth M G4VMI
Picot A G6LGR
Picton A G4OJV
Pidd C M3CMP
Piddington G 2E0GCP
Piddock C G1WTH
Pidgeon A G6CBP
Pidgeon D G0MRP
Pidgeon J M1AVB
Pidwell K 2E0KRN
Piecha M G4RWF
Pierce A G0RVE
Pierce A GM4TRS
Pierce A G0GZF
Pierce D G6VEI
Pierce E M6ADF
Pierce G M6BYT
Pierce S 2E0PSK
Piercy L M6PRC
Piercy M G4VEA
Pieroni E G0ITZ
Pierson H G3MXV
Pierson K G1APU
Pierson P G3MVM
Pieters C G6KRY
Piggin D M1BTI
Piggott B G0NSH
Piggott J G4LMY
Piggott J GW7WFI
Piggott K G4XKV
Pigott H G0MYL
Pike A G7FFS
Pike B M6BAZ
Pike B 2E0PIK
Pike B M0PIK
Pike B M3ZUY
Pike D G3VMI
Pike D G4UVV
Pike D 2E0PAO
Pike D M6PIK
Pike G M6XXX
Pike H M6SNG
Pike M M0RAX
Pike M M1ELN
Pike M M6MGV
Pike P GW7MMG
Pike R M3YNI
Pike R G7JLF
Pike S G4ZLK
Pike S G7RPW
Pilbeam S G6AOS
Pilcher B M1DMR
Pile J GW4EPF
Pile K G0JXP
Pile P G7ZZY
Pilgrim R M3BBS
Pilkington A G0KAB
Pilkington A 2E0GAP
Pilkington A M6AJP
Pilkington B G0OCK
Pilkington D G6VJP
Pilkington D 2E0MLE
Pilkington D M6MLE

Pilkington F G0OCL
Pilkington R M6CXH
Pilkington S G3NNT
Pill A G0NRZ
Pill M G4MQB
Pilling F G4HWK
Pilling J G4DPU
Pilling N G4LYE
Pillinger S G6DDJ
Pilot M GW1DTA
Pilton I 2E0LCE
Pilton I M6IPL
Pimblett P G0TPP
Pimblott P G3XVP
Pimlott J G8IDE
Pinborough R M0CEU
Pinch T G4ETP
Pinchen R M1AOU
Pinchin H G3VPE
Pinchin R G6FDD
Pinder A G8XZC
Pinder C 2E0VCP
Pinder C M6VCP
Pinder J M6JPI
Pinder M G4STI
Pine C G4OIW
Pine R G0RWT
Pine R G3RRP
Pingel D M0OPY
pink A G3RMZ
Pink A 2E0CZF
Pink D G6EGO
Pink D G7OPI
Pink J G8MM
Pink T M3TEL
Pinkard I G6XKJ
Pinkard K G1LED
Pinkerton J GI6GIE
Pinkerton R GI0NCA
Pinkhardt W M0SMS
Pinkney A G4VOU
Pinkney L 2E0LBJ
Pinkney L M6LEQ
Pinkney M G6UGA
Pinkney N G6FJG
Pinkney S G7ACM
Pinkowski L 2M0FTA
Pinkowski L MM6FTA
Pinna J G7GLS
Pinnell M G4VQT
Pinnell W G3XWK
Pinnock A G0PIN
Pinnock D G3HVA
Pinson J G6JJP
Pinto L G1LRR
Pinvisase E M3WVY
Piotrowski W G0BOE
Pipe I 2E0IZP
Pipe I M0IRP
Pipe I M6IRP
Piper A M6BNA
Piper B G1KRX
Piper D G4JSQ
Piper D G4LMN
Piper E G0BUZ
Piper F M0APZ
Piper K G0CHE
Piper M G0HOQ
Piper M G6UEV
Piper P G0DDY
Piper R G1VBO
Piper R G3MEH
Piper R G6XCO
Piper R MW6CLP
Piper S G1YRE
Pipes M G4DKV
Pipkin J G7FEF
Pipkin N 2E0JDP
Pipkin N M0NJJ
Pipkin N M6NJP
Pires M M3EZJ
Pirie C GM0TCU
Pirie J MM0JGP
Pirie P 2M0PTE
Pirie P MM3UDI
Pirrazzo P M0BMR
Pirrie T M0BBT
Pitcher J G3JNI
Pitcher M G3SNP
Pitchford C G7RLV
Pitchford J G1ZPQ
Pitchford S M3FNR
Pitchford D M3DPP
Pitfield I G6YFY
Pitfield J G0EZI
Pither J G0JME
Pithers J M6UKJ
Pitkethley A 2M0XXP
Pitkethley A MM0XXP
Pitkethley A MM3XXP
Pitkin D GW0PNI
Pitkin I G4KJD
Pitkin J GW7WCR
Pitman D MW6DSP
Pitman E G0BOC
Pitman K G0FLD
Pitman R M3WSO
Pitt A G4YIC
Pitt C G1FWY
Pitt G G8YWL
Pitt J G3VRY
Pitt K MW3MOJ
Pitt M G4KPM
Pitt N G1GFA
Pitt N MW6NPW
Pitt W G4IES

UK Surnames

| Name | Callsign |
|---|---|
| Pitt W | G8MXR |
| Pitt W | 2E0SKA |
| Pittard H | M6HMA |
| Pittaway D | [illegible] |
| Pitter R | G0CXU |
| Pitts G | G0EGT |
| Pitts J | G4SHN |
| Pitts M | G6SMJ |
| Pitts P | G3GYE |
| Pitts R | G8SDE |
| Pitty J | G4PEO |
| Pivac M | G0JML |
| Place T | G7JWD |
| Place T | G6DRG |
| Placidi M | 2E0BJL |
| Plackett B | 2E0NLY |
| Plackett B | M6RBB |
| Plaice A | G4MKE |
| Plail A | G8GRQ |
| Plain I | M3GZP |
| Planck K | GI4NKK |
| Plant A | G1IAL |
| Plant A | G0MBL |
| Plant A | G3NXC |
| Plant C | M6LZT |
| Plant D | G1HLP |
| Plant D | G3JPU |
| Plant E | G7PKY |
| Plant G | M0EUI |
| Plant G | G4YYG |
| Plant J | M0JCL |
| Plant J | M1JCL |
| Plant J | M3RZG |
| Plant K | G3NIC |
| Plant M | G1JHD |
| Plant N | G4WUJ |
| Plant P | G4YYP |
| Plant R | G7SEO |
| Plant R | G6FHR |
| Plant S | M1BKW |
| Plant S | M3AXT |
| Plaskitt M | G0MAF |
| Plaster M | G3OJL |
| Plastow A | M6AKP |
| plastow B | G4DRU |
| Plater D | G4MZY |
| Plater R | M3LRP |
| Platt A | M6KMK |
| Platt D | 2E0MXW |
| Platt D | G8BMG |
| Platt D | M0SAD |
| Platt D | M3MXW |
| Platt D | G3JNJ |
| Platt G | G6SNI |
| Platt M | G4XUM |
| Platt R | G8ROS |
| Platt T | G7DRD |
| Platten M | G1BOX |
| Platts F | G0OLL |
| Platts G | G8DJT |
| Platts R | G8OZP |
| Platts S | 2E0PWK |
| Platts S | G0NXT |
| Platts S | M3PWK |
| Player C | G8FFF |
| Player R | G7PTB |
| Playford C | G6HOS |
| Playford K | G6MRP |
| Playle P | G4SGN |
| Pledger P | G4TZO |
| Plenderleith J | G3OOK |
| Pleshkevich V | G0WKW |
| Plested J | G4GYS |
| Plested M | GM6IYJ |
| Plested R | G6AFE |
| Plewa L | G6FJE |
| Plews S | M1FBI |
| Pleydell P | G7UBX |
| Plimmer R | GW3UEP |
| Plitsch A | M1ATP |
| Plowman J | G3AST |
| Plowman M | MW6FMV |
| Plowman S | M6SEB |
| Plowright J | G7CSF |
| Plows J | M3XJP |
| Plows S | M0SDP |
| Pluck R | G7JTB |
| Pluck T | G7OZQ |
| Plucknett G | G4FKA |
| Plucknett W | G8HGP |
| Pluckrose B | G4VOW |
| Plumb C | M3CKO |
| Plumb N | G0PDV |
| Plumb S | M3CKM |
| Plume S | M6SWP |
| Plumley J | GW4SLI |
| Plumley R | M6RUH |
| Plummer A | G3YMZ |
| Plummer C | G8APB |
| Plummer C | [illegible] |
| Plummer D | 2M0CXI |
| Plummer D | MM0HVW |
| Plummer D | MM6DVO |
| Plummer M | G3MDI |
| Plummer M | M0ZQB |
| Plummer P | 2W0EPE |
| plummer P | MW0ABV |
| Plummer P | MW6EPE |
| Plummer S | M6BRH |
| Plummer T | M3TFP |
| Plumridge D | G3KMG |
| Plumridge K | G0MMA |
| Plumridge K | G4BYY |
| Plumtree B | G6MQU |
| Plumtree R | G0RKQ |
| Plumtree S | G3OSP |
| Plunkett G | MI6HDH |
| Plunkett N | 2E0NDP |
| [illegible] | [illegible] |
| Plunkett N | M6NDP |
| Plunkett P | G8BQZ |
| Pluright D | 2E0KHO |
| Pluright D | M0KHO |
| Poate N | M3NAT |
| Pochat D | M6POD |
| Pochat O | M6OPO |
| Pochat R | M6ROP |
| Pochojka S | M0EAB |
| Pocock C | M0AYS |
| Pocock G | GM1SQZ |
| Pocock R | G0JZY |
| Pocock R | G8MKO |
| Pocock S | G0CPV |
| Pocock S | G4GTU |
| Podmore B | M0EOT |
| Podmore B | G3INQ |
| Podmore G | G7VAG |
| Podmore G | G8BCF |
| Podvoiskis J | G0NPI |
| Poel W | G8CYK |
| Poffley R | G6CZB |
| Pogorzelski A | G1XVW |
| Pogson T | G0PLD |
| Pointon A | 2E0COY |
| Pointon D | M3TLZ |
| Pointon G | G6MQI |
| Pointon K | G4IBN |
| Pokusinski Z | G4JQU |
| Polain T | G7TBW |
| Polakovs A | 2E0LBA |
| Polakovs A | M6LYA |
| Poland J | M3UFY |
| Polesel A | M0POA |
| Polgreen B | MW0CNB |
| Poll D | G8IKA |
| Pollak S | M0KGA |
| Pollard A | 2M1HTR |
| Pollard A | G0NHZ |
| Pollard A | G0LXW |
| Pollard A | G6LGW |
| Pollard D | 2E0CSG |
| Pollard D | 2E0XAA |
| Pollard D | 2W1EKR |
| Pollard D | G0UWX |
| Pollard E | GM3KJI |
| Pollard G | GM7RYK |
| Pollard G | GW8DOA |
| Pollard G | M3PGI |
| Pollard I | M0FLC |
| Pollard I | M0IMP |
| Pollard J | G7BYS |
| Pollard J | M0JDB |
| Pollard K | G4KGP |
| Pollard L | M1LAP |
| Pollard L | M3YEA |
| Pollard M | G8BWA |
| Pollard N | M5AOI |
| Pollard N | MM6NCP |
| Pollard T | M3OSU |
| Pollard-Wilkins B | G8DXU |
| Pollett T | 2E0TPY |
| Pollett T | M6TPY |
| Polley D | G0JPY |
| Polley D | G7VZQ |
| Polley D | 2E0EML |
| Polley D | M6CSK |
| Polley T | G1YDJ |
| Pollington D | 2E0TRF |
| Pollock A | GM0PEI |
| Pollock C | G0PIY |
| Pollock J | G4CGO |
| Pollock M | GI6IOU |
| Pollock M | G8KMP |
| Pollock M | MI3PMW |
| Pollock R | GI0WZW |
| Pollock T | GI4PQV |
| Pollock W | GI3NVW |
| Polson J | GI6ROI |
| Pomeroy D | M6MOP |
| Pomery D | 2E0CYI |
| Pomfret A | G3LZZ |
| Pomfret D | G7SPL |
| Pomfret I | G6MHO |
| Pomfret M | 2E0MNP |
| Pomfret M | M6INT |
| Pomfret M | G6LCX |
| Pomfrett C | 2E0FUH |
| Pomfrett C | M0EEG |
| Pomfrett C | M3FUH |
| Pomfrett V | M3NRX |
| Pomfrey-Jones A | 2E0LPJ |
| Pomfrey-Jones A | M3XIP |
| Pomphrett C | G8PIC |
| Pomphrey D | MM3LFI |
| Pomroy G | G0ILI |
| Pomroy M | G1ZUH |
| Pond C | G0VDO |
| Pond D | G0IRH |
| Pond W | G1YJL |
| Ponder S | M0CAY |
| Ponsford M | M0GWD |
| Pont B | G8SUV |
| Pont N | G8SUW |
| Pontiero A | GM4UBF |
| Ponton J | G0RWU |
| Pook H | 2E0FIQ |
| Pool D | G6MAY |
| Pool G | G0DKQ |
| Poole A | G8VZB |
| Poole A | G1KXJ |
| Poole B | G3MRC |
| Poole B | G4UJL |
| Poole C | M6XCP |
| Poole C | M6HQX |
| Poole D | M0DXP |
| Poole D | G7NZZ |
| Poole E | G8GHP |
| Poole I | G1RIX |
| Poole I | G3YWX |
| Poole J | GJ1TJP |
| Poole J | M0JMP |
| Poole J | M6HKZ |
| Poole J | M6JFP |
| Poole M | 2E0VWK |
| Poole M | G7UJY |
| Poole M | M0VWK |
| Poole M | M1HFM |
| Poole M | M3VWK |
| Poole M | M6CLY |
| Poole M | M6MWP |
| Poole P | GD3ENV |
| Poole R | M0GWM |
| Poole T | G3VMT |
| Pooler C | 2E0ACW |
| Pooley D | G7UWZ |
| Pooley L | G7KLR |
| Pooley N | G1VII |
| Pooley T | G0RFL |
| Poolman C | M3GEK |
| Poore P | M0BSJ |
| Poore R | G4OHC |
| Poore V | G3ZSB |
| Poots D | MI0SRR |
| Poots D | MI3AJK |
| Popa F | M6PFV |
| Popa O | 2E0OVI |
| Popa O | M0OVI |
| Popa O | M3XBN |
| Pope C | G4CMM |
| Pope D | M1BRU |
| Pope G | G3ASV |
| Pope G | G4XRD |
| Pope G | G3ZOL |
| Pope I | G4GUE |
| Pope K | G4TNA |
| Pope M | MW0HUU |
| Pope M | M6SCI |
| Pope N | G0GSN |
| Pope N | G4AXA |
| Pope N | G7BNI |
| Pope P | 2E0OSE |
| Pope R | G4HXH |
| Pope S | MW3NQH |
| Pope W | G4TIG |
| Popek S | G4BJP |
| Popely D | G6GYM |
| Popgueorguiev I | M0INP |
| Popham E | G4TBF |
| Pople G | G4AVJ |
| Popovic I | M6FHN |
| Popple J | 2E0SDY |
| Popple J | M0SND |
| Popplewell A | G1PCQ |
| Popplewell E | G3IQX |
| Popplewell J | G8OYF |
| Porch A | G8AOK |
| Porcher R | M6HCY |
| Poriyath J | 2E0DVY |
| Poriyath J | M6FVM |
| Port D | G6INU |
| Portch D | G0KPZ |
| Porteious J | M3JDX |
| Porteious M | M3MEP |
| Porteous A | G7GDV |
| Porter A | G0BZW |
| Porter A | G1POR |
| Porter A | M6HTC |
| Porter B | G3YJA |
| Porter B | M3XCS |
| Porter B | M6NTL |
| Porter C | G0AYC |
| Porter C | GM0TAE |
| Porter D | 2E0JHP |
| Porter D | G0GBP |
| Porter D | G1JZZ |
| Porter D | G4NFO |
| Porter D | G1ULG |
| Porter D | G4OYX |
| Porter D | M3HRV |
| Porter E | M6EMY |
| Porter F | G0OQE |
| Porter G | G4GOJ |
| Porter G | G4TYM |
| Porter G | M3GAP |
| Porter G | M0NAA |
| Porter H | GI0IVO |
| Porter I | G6IYS |
| Porter J | GI0GGY |
| Porter J | GM4XRP |
| Porter J | 2E0SCN |
| Porter J | M6INX |
| Porter J | G4OHJ |
| Porter J | G3YZR |
| Porter J | G4AGN |
| Porter J | M3UZV |
| Porter J | M6LMG |
| Porter K | G4WEN |
| Porter K | G6UCY |
| Porter K | G7OKV |
| Porter K | M6KDO |
| Porter L | M0LGP |
| Porter L | G7IUB |
| Porter M | 2E0MMP |
| Porter M | 2E0CAQ |
| Porter M | G4TJK |
| Porter M | G8BVL |
| Porter M | G7PHC |
| Porter M | 2E0XTL |
| Porter M | G8XYJ |
| Porter M | M3XTL |
| Porter M | G7CVC |
| Porter M | G4OKS |
| Porter N | G0IRK |
| Porter N | [illegible] |
| Porter N | G4AFA |
| Porter P | 2E0CPV |
| Porter R | G3NII |
| Porter R | G3VXK |
| Porter R | G4VRP |
| Porter R | G8OLL |
| Porter S | G0UOV |
| Porter S | G4FGR |
| Porter S | G4NHP |
| Porter S | G8GTD |
| Porter S | M6SDP |
| Porter S | G7LWU |
| Porter T | M3TGP |
| Porter T | M6TLP |
| Porter W | M6CBU |
| Porteus D | M0DQL |
| Porteus R | G6YAR |
| Portlock G | G1FKJ |
| Portnoy D | G0FSJ |
| POSTANS A | G6OIF |
| Postle P | G1DXQ |
| Pothecary M | G4FUU |
| Pottage J | G1XVF |
| Potten D | G7EAQ |
| Potter A | G0JIR |
| Potter A | G1JHY |
| Potter A | GM7GLJ |
| POTTER B | M6BJN |
| Potter C | M3YUR |
| Potter C | M0DDT |
| Potter D | 2E0VBY |
| Potter D | G0DMP |
| Potter D | G1OCL |
| Potter D | M0VBY |
| Potter D | M3VBY |
| Potter D | M3VIG |
| Potter D | M6PAY |
| Potter E | M6HNJ |
| Potter G | G0EOF |
| Potter H | G8LSA |
| Potter H | G8RRR |
| Potter J | G7WGA |
| Potter J | G8JMU |
| Potter J | G4PFF |
| Potter J | M6FRZ |
| Potter L | 2E0LAI |
| Potter L | G3ESK |
| Potter N | G7GEX |
| Potter P | M6UAE |
| Potter R | M0NDY |
| Potter R | M3XRP |
| Potter S | G1JHZ |
| Potter S | 2E0TKY |
| Potter W | G1XCK |
| Pottinger J | G1FIZ |
| Potts A | G1TDV |
| Potts A | 2E0DRZ |
| Potts A | M6DRZ |
| Potts A | M6DPX |
| Potts D | G6KBD |
| Potts D | G8JPU |
| Potts D | M0HKP |
| Potts D | M6HKN |
| Potts D | M3HWP |
| Potts D | M6DWT |
| Potts I | GI7AUY |
| Potts J | GI6EGJ |
| Potts J | GM4IYZ |
| Potts J | M6JPS |
| Potts K | G6HGR |
| Potts M | MM0GUX |
| Potts R | G4PEN |
| Potts R | M6BYV |
| Potts S | G0AYC |
| Potts S | M6LAP |
| Potts V | 2E0GDK |
| Pougher T | G0MKZ |
| Poulet D | G7NUC |
| Poulson C | M0JKQ |
| Poulson C | M3ZOP |
| Poulson C | G4PZN |
| Poulson R | M6JKQ |
| Poulter D | M6HCV |
| Poulter J | M6HCT |
| Poulter M | M1CLI |
| Poulter M | M6GCP |
| Poulter R | G0CHJ |
| Poulter S | G0PNT |
| Poulton D | M0RVX |
| Poulton D | G7NVE |
| Poulton D | M6DDF |
| Poulton G | G4HNG |
| Poulton H | G4WKB |
| Poulton K | 2E0LDN |
| Poulton K | M6XAX |
| Poulton P | M0BVZ |
| Pounder A | [illegible] |
| Pounder D | MM6VDP |
| Pounder F | G0GRZ |
| Pounder R | 2E0RJP |
| Pounder R | [illegible] |
| Pounder R | M0OZD |
| Pounder S | M1EKD |
| Pounder S | M5LAR |
| Pountain S | G0SMP |
| Pountain S | G7KRZ |
| Poupard A | G6MXV |
| Povall J | M6ZJP |
| Povey E | G6CHA |
| Povey N | G1JLX |
| Povey T | G8SQA |
| Povey W | G8NXQ |
| Povoas L | G4FZL |
| Powe I | M3NZZ |
| Powell A | G0PFZ |
| Powell A | [illegible] |
| Powell A | G0UKG |
| Powell A | G1VXX |
| Powell A | G7LOY |
| Powell A | G8IPA |
| Powell A | M3YLO |
| Powell A | M6MFK |
| Powell A | 2E0JDF |
| Powell A | M0ABG |
| Powell A | M0LAO |
| Powell A | M6EKK |
| Powell B | G4DIA |
| Powell C | G3YNJ |
| Powell C | M6VQC |
| Powell C | G6NWN |
| Powell C | 2W0XBC |
| Powell C | M3ZWK |
| Powell D | 2E0DMP |
| Powell D | 2E0VDP |
| Powell D | G4FJH |
| Powell D | G4JVX |
| Powell D | G4TWC |
| Powell D | M3VDP |
| Powell D | G1CTO |
| Powell D | G3XLW |
| Powell D | G6EZY |
| Powell D | M6YDV |
| Powell E | G8JOC |
| Powell E | 2E0EOK |
| Powell E | M6OEP |
| Powell G | G8XYU |
| Powell G | M0AKA |
| Powell G | M1EDF |
| Powell G | M6GPP |
| Powell H | G1MRP |
| Powell H | M0HPS |
| Powell I | M0LLS |
| Powell J | G3ZOL |
| Powell J | G4MZZ |
| Powell J | M6JPX |
| Powell J | MW6JNP |
| Powell J | 2E0CYU |
| Powell J | 2E0JEZ |
| Powell J | M0JEZ |
| Powell J | M0JLP |
| Powell J | M3YSQ |
| Powell J | M6EGE |
| Powell K | G0PPM |
| Powell K | G1JAG |
| Powell K | M3JIJ |
| Powell K | G1NCG |
| Powell L | 2W0BBO |
| Powell L | MW0GCS |
| Powell M | G8DXH |
| Powell M | M3LWV |
| Powell M | M3MTP |
| Powell M | 2E0SKD |
| Powell M | M6SKD |
| Powell M | M3RQJ |
| Powell N | G8OXG |
| Powell N | M6KMF |
| Powell N | M3NHP |
| Powell P | G1PYJ |
| Powell R | G1HLY |
| Powell R | G3OGP |
| Powell R | G4URP |
| Powell R | G4VOY |
| Powell R | G6CCQ |
| Powell R | G7MTW |
| Powell R | M0RPO |
| Powell R | G6JME |
| Powell R | M3RZP |
| Powell R | 2E0DCK |
| Powell R | M1CFW |
| Powell R | M6CSF |
| Powell R | G0AOZ |
| Powell R | G3RUJ |
| Powell S | 2E0SNP |
| Powell S | G4WYC |
| Powell S | M6FHX |
| Powell S | G3WRA |
| Powell S | M6SGJ |
| Powell T | G1OVG |
| Powell V | M3YCJ |
| Powell W | MW0WRP |
| Powell Y | G0OGP |
| Power C | G6UGE |
| Power C | M6TSP |
| Power D | G3SCJ |
| Power J | GM0KTO |
| Power J | G1WVK |
| Power J | 2E0EZK |
| Power J | M6FZK |
| Power J | MW6JFI |
| Power J | M6CJI |
| Power P | G8BVB |
| Power P | 2E0ULC |
| Power S | G0DQQ |
| Power S | G7CXU |
| Power W | G0EON |
| Power W | 2E0WGP |
| Power W | M3WZG |
| Powers A | G0JNK |
| Powers A | 2E0DXV |
| Powers M | G4XHX |
| Powers P | G8YSA |
| Powers R | G8CKN |
| Powis C | M3CIP |
| Powis D | 2E0BEW |
| Powis D | G4HUP |
| Powis D | M3IKR |
| Powis D | 2E0CFX |
| Powis D | M0WIS |
| Powis D | M3ZIR |
| Powis M | M6BQC |
| Powis N | 2E0OCV |
| Powis N | M6OCV |
| Powles C | G4UBQ |
| Powlesland C | G8CQZ |
| Pownall K | M6DFU |
| Pownall R | M6FTU |
| Pownall R | G0ISH |
| Powney A | G0TUN |
| Powney A | G1ZAK |
| Powrie G | G4PNV |
| Powrie M | G4EVZ |
| Poxon A | M1EYO |
| Poxon J | G4UPA |
| Poynter A | G0AKR |
| Poynter F | G0PFJ |
| Poyser I | G6NWN |
| Poyser S | 2E0SAK |
| Poyser S | M0GSR |
| Poyser S | M3OYC |
| Pragnell J | G1EPO |
| Prakash K | M1DFW |
| Prall V | G7FFC |
| Prandoczky A | M0CSF |
| Prangnell R | M3RGP |
| Prank V | G8VOC |
| Prasad G | M3XMP |
| Pratchett M | M3NTR |
| Prater G | G4TZK |
| Prater G | G4AEI |
| Prater I | M0WLF |
| Prater W | G0EOL |
| Pratley D | G6INV |
| Pratley G | G0OBJ |
| Pratley T | G7JRP |
| Pratt A | G0SHP |
| Pratt A | G3YSQ |
| Pratt A | GM7FWA |
| Pratt D | G8KPY |
| Pratt D | G3KEP |
| Pratt D | G4DMP |
| Pratt D | M3DBY |
| Pratt D | G1VSM |
| Pratt E | G4MID |
| Pratt E | GM7KBK |
| Pratt G | M1ADK |
| Pratt J | M3PQJ |
| Pratt K | MU3KBP |
| Pratt M | G4RHY |
| Pratt M | G7MRY |
| Pratt M | M6CNP |
| Pratt N | G0PXK |
| Pratt N | G3RTO |
| Pratt R | G0WWU |
| Pratt R | G1WQC |
| Pratt R | G3RGP |
| Pratt R | M0CGO |
| Pratt R | G4DDX |
| Pratt S | G7SFJ |
| Pratt S | G8VHO |
| Pratt S | G7MMJ |
| Praveen A | M6ITY |
| Precious P | G7PKH |
| Preda E | 2E0DWV |
| Preda E | M6FVZ |
| Preece A | G3TCO |
| Preece A | M0CXO |
| Preece B | 2W0JMX |
| Preece C | 2E0YDM |
| Preece C | M0YZF |
| Preece C | M6CAP |
| Preece D | G4TCO |
| Preece F | 2E0AXI |
| Preece F | G4FGP |
| Preece G | M3UAV |
| Preece J | G1EGK |
| Preece J | G1UMS |
| Preece J | M0JOD |
| Preece J | 2E0AVX |
| Preece J | M0JFP |
| Preece J | M3VNI |
| Preece J | M6JCP |
| Preece L | G1KUG |
| Preece M | M0BQF |
| Preece R | 2E0VAR |
| Preece S | 2E0CIM |
| Preece S | M6YVR |
| Preece T | G0IAH |
| Preedy A | G3LNP |
| Preedy G | M1BQM |
| Preen J | 2E0HLS |
| Preen K | M3KCP |
| Prendergast M | G0RDD |
| Prendiville D | G0LON |
| Prenter A | MI1BOE |
| Prentice A | 2M0WWX |
| Prentice A | MM6CQU |
| Prentice G | G8VHN |
| Prentice K | M6KFP |
| Prentice M | G7SWR |
| Prentice M | 2E0MEK |
| Prentice M | [illegible] |
| Prentice N | MI6NTP |
| Prentice T | G8INZ |
| Prescott G | G4OMG |
| Prescott G | 2E0GWP |
| Prescott L | G4XCQ |
| Prescott M | G4WYZ |
| Prescott N | M3RCC |
| Prescott P | G1WXW |
| Prescott S | M0TUT |
| Preskey B | M6FBP |
| Presland J | G3IUY |
| Press I | 2E0GOZ |
| Pressley A | G4BXQ |
| Prestage B | G0WCH |
| Prestidge M | G2BXP |
| Preston B | G4KJA |
| Preston C | G7TGG |
| Preston D | G3KFS |
| Preston D | [illegible] |
| Preston D | [illegible] |
| Preston D | M0KXY |
| Preston J | 2E0JRP |
| Preston J | 2E0JSM |
| Preston J | M6FFJ |
| Preston J | M6BEJ |
| Preston L | M0BSI |
| Preston M | M3LTH |
| Preston M | G0THY |
| Preston P | G7LUK |
| Preston P | M0GOH |
| Preston P | M0HZE |
| Preston R | G0RAX |
| Preston R | G1EGL |
| Preston R | G6YGB |
| Preston S | G0RYW |
| Preston S | G1VDO |
| Preston S | G7TCQ |
| Preston S | 2E0BHY |
| Preston W | G3EFL |
| Prestwood M | G3PDH |
| Pretty R | G8ZXL |
| Prettyjohns K | G4KSU |
| Preval N | G1ECK |
| Preval N | M6NES |
| Prew R | G8EPQ |
| Priamo R | 2E0ISZ |
| Priamo V | G6ZCS |
| Priborsky F | M1FEM |
| Price A | 2E0GUN |
| Price A | G0ARP |
| Price A | GW1ERA |
| Price A | G1OXH |
| Price A | G4TLT |
| Price A | G6TIQ |
| Price A | G6UDG |
| Price A | G8KBG |
| Price A | G8PGF |
| Price A | GW8YJN |
| Price A | M0MTX |
| Price A | M1CZH |
| Price A | M3ORL |
| Price A | M6EKT |
| Price A | GM0IDV |
| Price A | M3VII |
| Price A | 2J1EDR |
| Price B | GW4DVB |
| Price C | 2E0CJP |
| Price C | 2E0DCA |
| Price C | G6PEG |
| Price C | M0DMF |
| Price C | M3FHQ |
| Price C | M3UKB |
| Price C | G7ECA |
| Price C | 2E0HSB |
| Price C | G0FZP |
| Price C | G1RDJ |
| Price C | G0LZJ |
| Price C | G7GVP |
| Price C | G8KTE |
| Price C | M0GNL |
| Price D | GW0SRE |
| Price D | G1ODB |
| Price D | G3RLF |
| Price D | G4CQT |
| Price D | G7CUB |
| Price D | G7TIX |
| Price D | 2E0DVP |
| Price D | G3LYU |
| Price D | G4BIX |
| Price D | G4MSA |
| Price D | M0PRI |
| Price D | M6DVP |
| Price D | G0ADZ |
| Price D | G7WGD |
| Price E | 2W0CKV |
| Price E | M6LBV |
| Price E | M6ETP |
| Price F | G8WSF |
| Price F | 2E0CEF |
| Price F | M0HCA |
| Price G | G0NVV |
| Price G | G3MPP |
| Price G | M1BBU |
| Price G | G4UZG |
| Price I | GW0OSB |
| Price J | G0ILC |
| Price J | G1DKK |
| Price J | G3RNP |
| Price J | G4OIK |
| Price J | G4OIL |
| Price J | G6IQY |
| Price J | G7JCX |
| Price J | G7NGI |
| Price J | M0JEP |
| Price J | M3UEE |
| Price J | [illegible] |
| Price M | 2E0LPR |
| Price M | 2E0NSS |
| Price M | G0JXG |
| Price M | GW0UKC |
| Price M | G1XFO |
| Price M | GW4EGS |
| Price M | G7PRW |
| Price M | M3MTL |
| Price M | M3MXH |
| Price M | M3NSS |
| Price M | MW0SBX |
| Price M | MW3THI |
| Price M | M3VGX |
| Price N | G7EWX |
| Price N | M3XNK |
| Price N | M6NEL |
| Price N | 2M0NOP |
| Price O | 2E0NWE |
| [illegible] | [illegible] |
| Price O | M6YNL |
| Price P | G6GYN |
| Price P | G8NOP |
| Price P | M3PLP |
| Price P | G6OSR |
| Price P | MW0LEA |
| Price P | GW0TDA |
| Price P | MW6PLP |
| Price R | GM0DWY |
| Price R | G0VMW |
| Price R | GW3SYL |
| Price R | G3VTD |
| Price R | GW4PCX |
| Price R | G6JJX |
| Price R | G8DTF |
| Price R | M0EGA |
| Price R | M6JAG |
| Price R | M6RPX |
| Price R | M6JIC |
| Price R | MW0DSV |
| Price R | G4EVX |
| Price S | G4BWE |
| Price S | G8VHL |
| Price S | G1KQP |
| Price S | MW6SDV |
| Price S | M6SXP |
| Price T | G6WUR |
| Price T | 2E0DUB |
| Price T | M6DUB |
| Price T | MW3TPJ |
| Price W | GW0TTF |
| Price-Gore C | G6GDR |
| Prichard A | G0CPA |
| Prichard B | G4BQS |
| Prichard J | M3CSN |
| Prichard S | M3VZH |
| Prickett W | G4NCF |
| Priddy A | G4UBI |
| Pridham R | G4BVB |
| Pridmore M | 2E0HJZ |
| Priest G | M6GJP |
| Priest I | M3YVJ |
| Priest M | 2E0EBQ |
| Priest M | M6GWQ |
| Priestley A | G8MWX |
| Priestley B | G7JWQ |
| Priestley D | 2E0DME |
| Priestley G | G6RXF |
| Priestley G | G7JZM |
| Priestley G | 2E0GRP |
| Priestley G | M0GRP |
| Priestley G | M6GRP |
| Priestley M | G7HEN |
| Priestley N | G0CBK |
| Priestley R | G0IHO |
| Priestley S | M0SGS |
| Priestman J | 2E0JOP |
| Priestman J | M3ZLJ |
| Priestnall F | G6LHA |
| Priestner D | M3CRV |
| Prietzel E | GM4ZCV |
| Primmer J | G8NYC |
| Prin O | 2E0WAA |
| Prin O | M0WAG |
| Prin O | M3WPI |
| Prince A | M1GAP |
| Prince A | M1TAP |
| Prince D | M3TYZ |
| Prince D | G6MPX |
| Prince E | G3KPU |
| Prince G | G7CDU |
| Prince I | G6EZG |
| Prince J | 2E0AGE |
| Prince K | G0NUP |
| Prince K | G4TQL |
| Prince M | GW7EUL |
| Prince M | M6EQB |
| Prince N | G3VSI |
| Prince P | G8XXC |
| Prince R | G3CAJ |
| Prince T | 2E0TPR |
| Prince T | M6PCU |
| Pring L | M3LOI |
| Pringle A | GM3MAS |
| Pringle C | G1SVQ |
| Pringle C | G6BTB |
| Pringle N | M0ADB |
| Pringle R | 2E0CKW |
| Pringle R | M6BLH |
| Prinnett S | M3BHK |
| Prior A | G4UWW |
| Prior C | G3DTU |
| Prior C | 2E0CKP |
| Prior C | M6CKP |
| Prior D | M0HPT |
| Prior G | 2M0GAN |
| Prior I | M3FGO |
| Prior J | G1DIQ |
| Prior K | G4TRW |
| Prior K | G8XHK |
| Prior K | M0CKR |
| Prior L | G8IXC |
| Prior M | G1VOB |
| Prior P | M1CAK |
| Prior R | G0ALJ |
| Prior R | G3MTG |
| Prior R | G0WWR |
| Prior S | G4SJP |
| Prior S | G8KQB |
| Prior S | M6ESY |
| Prisk S | G7DWO |

UK Surnames

**IMPORTANT NOTE**

**Revalidate licence to avoid revocation** – Ofcom has advised the Society that plans will be drawn up to revoke licences that have not been revalidated as required by the licence conditions. The quickest way to revalidate is to do so online via the Ofcom website: *https://services.ofcom.org.uk/* or by email: *amateur.validations@ofcom.org.uk* If you need assistance in the process, Ofcom staff are available to help, but please be patient during times of heavy workload.

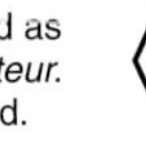

Pritchard A 2I0UAD
Pritchard A G0RNC
Pritchard A G0RUY
Pritchard A MW3XIJ
Pritchard A MI6UAB
Pritchard A G3ODB
Pritchard A MW0HGM
Pritchard B G8MCJ
Pritchard C G7JAE
Pritchard C M6CEP
Pritchard C M6CSP
pritchard D 2E0DEQ
Pritchard D G3JSJ
Pritchard D M3DOM
Pritchard D G4ULP
Pritchard E G0BJZ
Pritchard E G4USX
Pritchard E M6ZTP
Pritchard G G4ZGP
Pritchard G M6URC
Pritchard I G1XYN
Pritchard J GW3WLN
Pritchard J GM4DMQ
Pritchard J G7JXD
Pritchard J 2E0JCN
Pritchard J M0JWP
Pritchard J M6CTB
Pritchard K M6KWP
Pritchard K G3WVG
Pritchard K MI6KRP
Pritchard L G0RQS
Pritchard L G0TRI
Pritchard L G3IUW
Pritchard M G3VNQ
Pritchard M G7EVP
Pritchard N G8AYM
Pritchard P G6MPT
Pritchard P G8LLD
Pritchard P 2E0TCK
Pritchard P 2E0VCU
Pritchard P M6GSH
Pritchard P M6PIP
Pritchard P M6TCK
Pritchard P G1NPP
Pritchard P MW3WQE
Pritchard P M6PYO
Pritchard R G0AZX
Pritchard R G0UDH
Pritchard R G1BHQ
Pritchard R GI0PNP
Pritchard S M0EDR
Pritchard T G0GSM
Pritchett L G1YPZ
Pritt D G8TZE
Privett R G4MZV
Probert C G7ULC
Probert D 2W1CPS
Probert M GW4HXO
Probert P M3ZCI
Probert P G1PJP
Probert S M6HRG
Probst A M3NYF
Probst P 2E0PHL
Probst P M3SKN
Probyn G G7WIC
Procter B G8AWN
Procter B M3AWN
Procter J G0FQN
Procter K G3EPO
Procter M M1ECT
Procter N G0IZL
Procter P 2E0HQY
Procter R G0ENA
Procter R G4LAK
Procter S G6PDM
Procter S G8PDY
Proctor A M3FDQ
Proctor A G1VVL
Proctor C G1XLG
Proctor C G8MBM
Proctor C G8XLG
Proctor D G0EDE
Proctor D G7RYN
Proctor D M3HVX
Proctor D G4JYW
Proctor D M0IOK
Proctor J MI6GOA
Proctor J G0WSA
Proctor R G4PZW
Proctor R M1ATI
Prodger J M6CZI
Proffitt J G6RJH
Profitt R G4HRU
Prosser A 2U0DWD
Prosser A MU0WLV
Prosser A MU6GXE
Prosser D GU0BDI
Prosser G G7EJN
PROSSER K G8TRO
Prosser M G6TUD
Prosser N G1DYQ
Prosser O M0WIN
Prosser P G4TJA
Prosser R G1RAP
Prosser R G4BUS
Prosser R G4OJP
Prosser S G6HZH
Prothero E G0RNO
Protheroe-Thomas W GW4TGL
Proud D G0AWR
Proud L G0LLP
Proudfoot J G4ISS
Proudler A G4OMI
Proudlove P G6NAH
Proudman J 2E0LLM
Proudman J M3LLM
Proudman R M6FBM
Prouse D G8TYD
Prouse W G4SGW
Prout C G7NKU
Prout C M3NKU
Prout D G0LTE
Prout D G3VCV
Prout D 2E0DWP
Prout D M3BRV
Prout J M6SSX
Provan S GW0UQH
Provins R G0RGJ
Provis D MW3USS
Provis P 2W0XMG
Provis P MW0XMG
Provis P MW3XMG
Prowse C G7MDV
Prowse G G7ERQ
Prowse P G4UON
Pruden J G0GEL
Prunty B GI0OUZ
Pryce S G0EIY
Pryde J GM8JJN
Pryde R GM3LGU
Pryer J B A G0LAZ
Pryer S 2E0XBT
Pryer S M6SKP
Pryke G G6INX
Pryke I 2E0CQQ
Pryke I M0IAH
Pryke J G8BXH
Pryke S M3VCP
Pryke S 2E0IAF
Pryke S M0VCP
Pryor D M3HZA
Pryor R G8XRP
Przybyla A G0EPP
Przybylski D M6HHO
Puddifoot J G8JJF
Puddy J M6GXD
Pudsey F 2M0BUT
Pudsey F MM0WAP
Pudsey F MM3RQP
Puffett A G0JVH
Puffett F G0MTW
Pugh A G8ASD
Pugh B MW3VFB
Pugh C M1BIL
Pugh D 2E0DPU
Pugh D M0ZXY
Pugh D M6ZXY
Pugh D G0SJS
Pugh J G0IKZ
Pugh J G3UBH
Pugh K GW3SIK
Pugh K 2E0WMG
Pugh K GM7DHA
Pugh L G1OAX
Pugh M G4VPD
Pugh M GW0KKT
Pugh M M6DUD
Pugh N G1GZZ
Pugh P MW0PRP
Pugh P G7RYM
Pugh R GM0OTB
Pugh S G1UFM
Pugh V G2CQX
Pugsley C M6FWL
Pugsley R M0BWV
Pulev N MM6SGE
Pulford C 2E0TBX
Pulford C M6CRP
Pulford J G4RQO
Pulfrey B G4VIM
Pullan K G8NZR
Pullan M M0FEU
Pullen A G8BXJ
Pullen D 2E0PUL
Pullen D M6PUL
Pullen F G4XXX
Pullen J G4TGE
Pullen M M6ICH
Pullen P G7UZI
Pullen R G0OII
Pullen R GW1MAX
Pulley P G6XLG
Pullin D GW4IUL
Pulling M M6KDV
Pulling R G3REV
Pullinger P G8JCB
Pulman A MI6WOF
Pumford-Green J GM4SLV
Puncer M G0BLV
Punch S M1EVZ
Puncher B G0IJK
Pung C G6RVP
Punjabi P 2E0PMP
Punjabi P M6PRP
Punshon K G4APJ
Punter G G8IIG
Punter R M0PUN
Purbrick D M0SCP
Purbrick R G4JDH
Purcell A G0KFS
Purcell C G0DOG
Purcell C GW0WLN
Purcell J G0WUU
Purcell M M1DAP
Purcell T G0DOE
Purcell T M3WCO
Purcell T M3TPY
Purcer K G6HZG
Purchase B G3FWD
Purchase C G8JCC
Purchon G G6CFC
Purchon J G4EMQ
Purchon N G0WKM
Purdy H M6HEP
Purdy I G4ULN
Purdy J G6JIF
Purdy P G6VAD
Purdy R G6HZX
Purdy R G4RBP
Purdy S M5TNT
Purdy V G3JVP
Purgal-Woods J M3ZVW
Purkins N G1EIB
Purkis I MM6IJP
Purkiss B G7SFY
Purkiss C M3NMR
Purkiss N M3VIA
Purkiss S M3XSP
Purnell A M6KTP
Purnell M G4XIX
Purnell M G6INW
Purrier G M1AVU
Purse B G1RNV
Purseglove A G0JDG
Purser D M3RVJ
Purser M G0NEM
Purser S G4SHF
Purser W G2AXO
Pursglove A M0TRP
Purtell J GM0IST
Purtell N GM0NYP
Purves H M0CBK
Purves R GM4IKT
Purvess J G0FWP
Purvis B M6RTX
Purvis E G7NQR
Purvis E M0HMS
Pusey J M3IWG
Pusey M G8DTE
Putnam R G0ILN
Puttick M G3LIK
Puttick N G4PGW
Puttock E M6BBE
Puttock J G0BWV
Puttock J G6WPK
Puttock K 2E0BQZ
Puttock K M0PUT
Puttock K M3UNB
Puzey C M6PZY
Pybus A G8EZG
Pybus B G0WKZ
Pybus C G8TBX
Pybus S M1SPY
Pye D G1UCT
Pye D 2E0DPY
Pye D M6PYE
Pye J M1PYE
Pye J M6YPE
Pye K M0CFT
Pye M M6MAT
Pye R G4IUH
Pye R G8AAT
Pye S M3YXQ
Pykett D G0BEN
Pykett D G0IIQ
Pykett R G7BGT
Pyman R M1AEX
Pymm J G4JPK
Pymm J M1TAT
Pynappels S 2I0SEH
Pynappels S MI0PYN
Pynappels S MI6PYN
Pyne A G4UNH
Pyner R 2E0ZPY
Pyner R M6ZPY
Pyrah M G6ENA
Pyrah R G6UDA

## Q

Qassim M 2E0UKM
Qassim M M3UKU
Quade S G6WYQ
Quaintance R G0DIZ
Quaite G GI4MHD
Quantrill G G6WFS
Quantrill L G0DTP
Quantrill S M0DCU
Quarman K G8CBE
Quarmby J G3XDY
QUARTERMAINE J 2E0FUQ
Quarterman G G3NHX
Quarton C G5KC
Quarton S G2YL
Quayle F GD4DPK
Quayle G G4NAV
Quayle G GW0CYK
Quee M G3ZWW
Queeley C MW0COD
Queen G 2M0XZX
Quemby M M6LAM
Quest A G4UZN
Quest N G7VVL
Quested P G0DRT
Quick D 2E0ITI
Quick P M1BWZ
Quick R G7AZV
Quick S G1MMT
Quicke M G0MCQ
Quickfall P G4RLU
Quigg D MI3INS
Quigg H M0HQJ
Quigg J GI6VWS
Quigg J MI3JDQ
Quigg S 2I0BHT
Quigg S MI0GGB
Quigg S MI3IHY
Quigley D G3PRI
Quigley J MI6WPP
Quigley K M0HWH
Quillien K 2M0TLE
Quilter G M3ULX
Quin H GM0DXE
Quin T 2I0WAH
Quin T MI3YBI
Quince A G0UBX
Quiney R G6EHG
Quiney T 2E0VXX
Quiney T M0VXX
Quiney T M3VXX
Quinlan K G7DZD
Quinn A MI0DBK
Quinn A G1KQZ
Quinn E M3EWQ
Quinn H GI0WLW
Quinn J GI4PCQ
Quinn J GM6LIN
Quinn J 2I0JSQ
Quinn J MI6JSQ
Quinn N G0BAF
Quinn N GI4KJC
Quinn P GI7FCW
Quinn P MM0PJQ
Quinn P MI0CRR
Quinn S G0SLQ
Quinn S 2E0YEP
Quinn S M0MMR
Quinn T GM0GYM
Quinn T M1ZZY
Quinn T M0CSD
Quinnear D G0POK
Quinnell I MM0URN
Quinney D M6HTG
Quinney E 2D0EDQ
Quinnin C G0ECQ
Quint G G7TTH
Quinton A M3UAQ
Quirk P G0VMF
Quirk P G1RQI
Quy A G0FEO

## R

Rabbett M GI0IOT
Rabbitt I G1BHR
Rabbitt J M3BQT
Rabbitt O M3RZI
Rabbitt W G0PZP
Rabbitts J M3ZOR
Rabbitts J GM8LFB
Rabbitts T G8HUH
Rabe C G6NLC
Rabey B G8NYR
Rabey C G0RMR
Rabjohns J G3YBG
Rabl M M6EEA
Rabone D G4FNR
Rabone L M3XBL
Rabone P M3XPL
Rabson I G8LKB
Rabson J G3PAI
Rabstaff G G3ZVT
Raby A M3XYK
Raby A GW4ZVV
Raby B G8GTV
Raby F G8RF
Raby J G6DGX
Race M G1DCX
Racher P G6MQJ
Rack M G1KBL
Rackett J G3EZB
Rackham A 2E0DGN
Rackham M G4JKV
Rackham P G3IRQ
Radcliffe A GD3FXN
Radcliffe A MD3BZA
Radcliffe C G4SRF
Radcliffe L G8XUE
Radcliffe P G0FNP
Radcliffe R GW0GJD
Raddy S G6YNV
Rademaker A M0LTN
Radford A M0BAE
Radford C G7VJQ
Radford C M6NEO
Radford D G3KMY
Radford G M3NIF
Radford I M6TDK
Radford J G0PCY
Radford K G3SZU
Radford M M3XFI
Radford P 2E0PJR
Radford P G0ADW
Radford P M0LZM
Radford S GW0CYK
Radford S M6SBD
Radivan G G8OFI
Radley A G0TTM
Radley C M6RCY
Radley D G4ABI
Radley L G4JDS
Radley R 2E0RJR
Radley R M6FFZ
Radmall P M3PRZ
Radtke J G7LKC
Radulescu G 2E0UKK
Radulescu G M6UKX
Radulescu G 2E0HTV
Radulov M 2E0YRM
Radulov M M0YRM
Radulov M M6LZM
Rae A GM4ENN
Rae G G4MLO
Rae I GD8BUE
Rae J G0FQF
Rae J MM3OJR
Raeburn R M0HNX
Raehse Felstead J 2E0HAF
Raeper A 2E0TQS
Rafferty B GI1RIB
Rafferty C MI6EJT
Rafferty F G0HMS
Rafferty J G4YVX
Rafferty K MM3KYO
Rafferty S GI0EJT
Rafferty S G6NXP
Rafferty S G7DMQ
Rafferty W MI6RUC
Rafferty-Floyd C MI6ECV
Raffield K G3CQU
Raffill J 2E0XJR
Rafique M M3FJQ
Rafter R G1XNK
Rafter S 2E0ZAJ
Rafter S M3OCJ
Ragg W G4ORS
Rai M M6AXM
Railton C G3YQV
Railton L G1OKK
Raimbach D G3ZWK
Rainbow M G6OTP
Rainbow W M6HPX
Raine A GM8VYZ
Raine A M6FDR
Raine C GM8MNG
Raine D G4ZHA
Raine D 2E0MTY
Raine E M6WEP
Raine I GI8TSI
Raine J M6JPR
Raine K G4VRB
Raine M M6IRM
Raine P M6PCR
Raine R G4RXR
Raine R 2E0SPR
Raine R M6PRN
Raine S M6SBF
Rainer D G4VTQ
Rainer M G6TIU
Rainer P 2E0AYY
Rainer P M0RNP
Rainey E 2I0EAR
Rainey E MI6HGV
Rainey J GI4YCZ
Rainey M GM0KMA
Rainey P 2E0RNS
Rainey P M3RNS
Rainey R GI0RBS
Rainy Brown G G1NAB
Raisey-Skeats S GM1GCB
Raistrick A G7IKS
Raistrick K M0TKD
Rajagopal B 2E0DNE
Rajanayagam D 2E0TAV
Rajanayagam D M0RVT
Rajanayagam D M6RPD
Ralls A G3PDP
Ralls G 2W0HFU
Ralls G MW6GRB
Ralph A G8XLH
Ralph A M6APR
Ralph C 2M0INS
Ralph C MM0INS
Ralph C MM6INS
Ralph D G1UCO
Ralph D G4UZC
Ralph K 2E0CXO
Ralph K 2E0IKM
Ralph N M3KQS
Ralph R G4KSG
Ralph S G7EIA
Ralph T M6TON
Ralph T M5ATR
Ralphson C 2E0CWR
Ralphson C M6CWR
Ralston P G6CIP
Ram S 2M0NAN
Ram S MM0YSK
Ram S MM6CZH
Ramachandran J M6JIN
Ramm A G7OSJ
Ramm D G7OSK
Ramm R MW0AIZ
Ramplin R G0NGW
Rampton D G4ZEL
Ramsay A GM6BML
Ramsay D G0RWY
Ramsay E GM7AUX
Ramsay J GM3OQI
Ramsay J G3OZZ
Ramsay J G4NCZ
Ramsay J G8LOZ
Ramsay J M3JLR
Ramsay R G4BBJ
Ramsay R M1CPL
Ramsay S 2M0LAO
Ramsay S MM6LAK
RAMSAY W M0WZZ
Ramsbottom A M6FFD
Ramsbottom M M6LIS
Ramsdale C 2E0FZM
Ramsdale C M3FZM
Ramsdale E G3GWC
Ramsdale K G4YKR
Ramsden A G0VDQ
Ramsden B G4VGF
Ramsden C G4YPV
Ramsden D G6XJT
Ramsden H G4YKB
Ramsden S 2E0XJT
Ramsden S M0CCA
Ramsden T G6AYS
Ramsden T M3KMT
Ramsell D 2E0CGJ
Ramsell D M0KAJ
Ramsell W 2E0PBB
Ramsey A G4MQF
Ramsey C G8VZD
Ramsey D G3UAA
Ramsey D G6SNN
Ramsey F G4CNX
Ramsey G G4KMP
Ramsey M M6BBH
Ramsey N M0VNR
Ramsey N M6EUL
Ramsey P GM4WNQ
Ramsey P M3PRA
Ramsey R G7ONL
Ramshaw R G3RLD
Ramskill M M1AJG
Rance J 2E0JAR
Rance T G4HTB
Rand A 2E0FBM
Rand A M0HYJ
Rand S GD7RVP
Randall A M6JHI
Randall B 2E0IKB
Randall C G4RBR
Randall J G0DAF
Randall J G3OAZ
Randall J G6WAY
Randall J M6JMR
Randall K G3RFH
Randall M MW5MWR
Randall P G1SJZ
Randall P G3NJV
Randall P G8HXT
Randall P M3RMZ
Randall P MW0PDR
Randall P MW0PRC
Randall S 2E0SPR
Randall S G4GGX
Randall S M3PYJ
Randall S G4XYG
Randall W M3WTR
Randall-Cook P G8WGD
Randell B 2E0IAT
Randell P G6HKH
Randerson P M6FVJ
Randle I G6TJY
Randle J 2E0CLI
Randles A M0GWF
Randles D G7JHZ
Randles D 2E0DDR
Randles D G4AFT
Randles D M6UAR
Randolph W G0KCC
Rands A G6PEH
Ranger J M3ECQ
Ranger K G0KJK
Ranger M G6GWE
Rank J G4NSE
Rankin B MM6YUP
Rankin C MM6XRS
Rankin D GM0NTY
Rankin D G3YQB
Rankin G G4FVL
Rankin J GM1BKR
Rankin J GM4VDG
Rankin J 2M0JCG
Rankin J MM6AVE
Rankin K 2E0FDY
Rankine K G4VZW
Ransford D G4NNY
Ransom E M3PYS
Ransom R G6VBE
Ransom R 2E0IGU
Ransom S G6MIS
Ransom T G4FET
Ransome J MM6CCW
Ransome R G0NFE
Ransome R G7MPF
Ranson C G8LBS
Ranson D G0IWI
Ranson J G7NND
Ranson R G3ZTB
Rantin D 2I0FEX
Rantin D MI3FEX
Rantin E 2I0JAP
Rantin E MI0KAG
Rantin R 2I0RRE
Rantin R MI0RRE
Rantin R MI3RRE
Rantin S MI3WBU
Rao R 2E0KRR
Rao R M6RRK
Raper K G8IKG
Raphael C M0XCR
Rapson C G0OZM
Rapson N G0RJI
Rapson N 2E0RNW
Rapson N M6WNR
Rasbarry D M6EDO
Rasell M M6EYA
Rashleigh K G8ORX
Rasmussen P M6PLR
Rasmussen S G1MVF
Rasooli Nia H M6DLD
Raspin C G4KUE
Ratcliff A M1ALE
Ratcliff B M3NHA
Ratcliff L G4GYP
Ratcliff S M6NHA
Ratcliffe C G1MXO
Ratcliffe K G6VGN
Ratcliffe M M6ZVD
Ratcliffe P 2E0PFR
Ratcliffe P M6TFF
Ratcliffe R G4LRY
Ratcliffe R G4ACY
Ratcliffe S G0BYQ
Ratcliffe S G0IUA
Rath P M6MLU
Rathbone M G3ZII
Rathbone N G4KZU
Rathore Z 2E0JSR
Ratigan J G7DAL
Rattenbury J G1HQN
Rattenbury P G6OIA
Ratter J GM0JDB
Rattigan J G0GJM
Rattley S M6HJF
Rattray C GM3HYX
Rattray E MM6FKD
Rattray F G4SPR
Rattray H GI0MSK
Rattray W GM4NDV
Rauch C G8GKL
Ravelini J G1UPT
Ravelini T G1HIG
Raven A M6FBZ
Raven P G4KLM
Raven R G4CFW
Raven T G4ARI
Ravenhill-Lloyd B M6EYI
Ravenscroft P G0ULO
Ravenscroft R G0RAV
Raven-Vause F M6FRV
Ravilious N M0ARH
Rawcliffe S G4YXU
Rawdon A G6GHE
Rawley J M6FVR
Rawlin C 2E0CMZ
Rawlin C M0PEM
Rawlin C M6GRV
Rawlings A M0ANS
Rawlings F G4JOD
Rawlings G G8CUN
Rawlings J G1CUM
Rawlings J G4ULG
Rawlings K M0KIM
Rawlings L G3FET
Rawlings M M3TYU
Rawlingson I G1XZA
Rawlingson J G1XZB
Rawlins A G1CGH
Rawlins A M6OTL
Rawlins F G0OFX
Rawlinson A 2E0ZJQ
Rawlinson A M0ZJQ
Rawlinson A M3ZJQ
Rawlinson C M6YLC
Rawlinson J G4XET
Rawlinson J 2E0ZJO
Rawlinson J M0ZJO
Rawlinson J M3ZJO
Rawlinson P M0PBR
Rawlinson R M0RDR
Rawlinson T G0RNA
Rawson A G1MCI
Rawson J G0BQO
Rawson M M6OSX
Rawson M G0DJM
Rawson M M6IKY
Rawson R G1SQW
Rawson R 2E0RCR
Rawson R M0RCX
Rawson S G0MAA
Rawson W GI3XHL
Raxter D G7WEB
Raxworthy A 2E0CNM
Raxworthy K G7COQ
Raxworthy P G0AVP
Raxworthy P G6XHJ
Ray B M5BRY
Ray I G4VZB
Ray J G8DZH
Ray K G1GXB
Ray M G4XBF
Ray P M0TUK
Ray R G3ZHS
Ray R G8CUB
Ray S GD1GXC
Ray S 2E0RSD
Ray S M6BRR
Ray S 2E0SBH
Ray S M6SJR
Ray T G8DEJ
Ray W G4HES
Raybould D G7HJX
Raybould J G4PQI
Raybould L M1BZG
Raybould P G8CGM
Raybould W G4DFE
Raybould W G8FUI
Raybould Z M6ZBL
Rayland E G4ASK
Rayment C 2E0VAC
Rayment C M0ZAQ
Rayment C M3YCR
Rayment R G0TTI
Raymer B G6FDI
Raymond C GW0LIK
Raymond J GW4ZYV
Raymond T M6HVU
Rayne D 2E0DRN
Rayne J 2M0CNZ
Rayne J MM0JRR
Rayne J MM6BTR
Rayner A 2E0BLT
Rayner A G1KPV
Rayner A M6IMI
Rayner A M6AGR
Rayner B G1VZT
Rayner D G0AFP
Rayner D G3NYR
Rayner D G4XNP
Rayner D M3WPP
Rayner R G6WAN
Rayner S 2E0HGR
Raynes J G0BWG
Raynor K G8GXN
Raynor K M6KTI
Raynor L G0NFB
Raynor M 2E0VAT
Raynor M M0ZIM
Raynor M M6DEV
Raynor M 2E0MLF
Raynor P G6EUF
Raynor R GM7MWX
Raynor-Smith R G0VYU
Rayson J 2E0APY
Rayson P G1AJZ
Rayson P G8YJZ
Razey R G0ADH
Razzell A G0PBM
Rdwards S M3SMZ
Rea G MI0CXE
Rea M G6IWT
Rea M 2E0MEQ
Rea M M6MHQ
Rea P GW6OMV
Rea R G0HQX
Rea S MI3SRL
Rea W 2I0WRR
Rea W MI3REA
Reacher D G7VTH
Read A G0GMS
Read A G1AEA
Read A G3SXR
Read A G6CYE
Read A G6VQC
Read B 2E0AGQ
Read B G3JDT
Read B M0EWG
Read C G0SGM
Read C G4NPY
Read C G4TZA
Read D G7CQX
Read D M3FOV
Read D M3WFB
Read D G0OIF
Read G M3TCU
Read G G4UPB
Read J G4THU
Read J M3JRR
Read J 2E0EYP
Read J M0HGY
Read J M3EYP
Read J MM6BHS
Read M M6TRD
Read M 2E0EGO
Read N G6WHS
Read N M5DND
Read P G1ZPJ
Read P G4YAR
Read P G6ZZE
Read P M3XVZ
Read R G4OVJ
Read R G8CCN
Read R M1EWF
Read R M5BMW
Read T M1EYP
Read T M1BIG
Reade J G4OHQ
Reade P G8VUM
Reade W G4XTO
Reader A M3YFM
Reader A G0CRJ
Reader D M6DJL
Reader K G3LUH
Reader P G3TSR
Reader T G6UML
Readhead V G0EGW
Reading D G6VJR
Reading G G1IFH
Reading N G4WQS
Readman M G3YTZ
Readman R 2E0WLK
Readman R M0WLK
Readman R M3WLY
Reale A G8MOS
Rean A G4VXH
Reaney C G0ROD
Reaney M 2E0DPF
Reaney M M6TVY
Reanney R G0AKU
Reap S G8RJM
Rear S M1SEM
Reardon D G4EJK
Reason A G8WPV
Reason A M0HNH
Reason G G4EBF
Reason G 2W0GJR
Reason G MW6WUK
Reason J 2W0PEG
Reason J MW6RJY
Reason K M3TSV
Reavell M M3MHR
Reavill K G7USM
Reavill M 2E0CHX
Reay A 2E0UKA
Reay A M3XJX
Reay D G8UHO
Reay G G0GOZ
Reay P G6TAK
Rebisz S M6RSJ
Recardo A G0LFZ
Recardo D G0LFY
Recht S M3ZFH
Redall P G4FZV
Reddall L M3YXS

Reddaway J ........ G6MRO
Reddecliffe G ........ G7UBK
[illegible] F ........ M6FRD
Reddon T ........ G1AHW
Redding A ........ G0COM
Redding C ........ G6CSL
Redding D ........ M1BAD
Redding G ........ G0GGQ
Redding R ........ G3VMR
Reddington G ........ G6ZDB
Reddington T ........ M3TDM
Reddish A ........ G6MGQ
Reddish N ........ G0ORE
Reddish N ........ G8XUB
Reddish T ........ G0PLA
Reddish T ........ G7HSL
Redfearn J ........ M3VHU
Redfern B ........ G0CGI
Redfern C ........ G4CZR
Redfern D ........ M3TMI
Redfern I ........ G8DUW
Redfern J ........ M6JRE
Redfern J ........ 2E0BMP
Redfern M ........ M3UWU
Redfern P ........ G4CLN
Redfern P ........ 2E0HYI
Redford J ........ G3SXP
Redford K ........ G4FRR
Redford S ........ M6ZEP
Redgewell G ........ G4TTN
redgrave J ........ M6HWD
Redhead D ........ M6TYT
Redhead G ........ G4KXW
Redhead J ........ 2E0JDI
Redhead J ........ M0JLR
Redhead J ........ M6XJR
Redman A ........ G4KUF
Redman C ........ G4IDR
Redman D ........ M3UFT
Redman J ........ G4KXL
Redman M ........ M0BBH
Redman M ........ M0GYL
Redmayne C ........ G4GLW
Redmayne D ........ M3EVF
Redmill D ........ G6ZTM
Redmond C ........ 2E0CRU
Redmond C ........ M6ASV
Redmond E ........ M6CKO
Redmond G ........ G0TXU
Redmond J ........ MM3XXI
Redmond K ........ G6FKE
Redmond R ........ 2E0LLE
Redmond W ........ GI0LMR
Redmore M ........ M6JBV
Redpath J ........ G7CEB
Redpath R ........ G6AHR
Redpath W ........ GM4JEM
Redrup J ........ M3MNR
Redshaw A ........ G6DVE
Redstall M ........ M0MWR
Redway S ........ G4TRA
Redwood C ........ G6MXL
Reece G ........ G4YGL
Reece G ........ G6HTT
Reece K ........ G8UYB
Reed A ........ 2E0IJS
Reed A ........ G4HCD
Reed A ........ G4HKR
Reed A ........ M3XMO
Reed A ........ 2E0GKY
Reed A ........ M3GKY
Reed A ........ G6NLD
Reed C ........ G1RPO
Reed C ........ G6WKN
Reed C ........ M0MFP
Reed D ........ G7UCO
Reed D ........ G8NYB
Reed D ........ G4YDR
Reed D ........ G4ZCW
Reed D ........ G6RCY
Reed G ........ G4YBD
Reed J ........ G0JRB
Reed J ........ G0NOH
Reed J ........ G3ZMD
Reed J ........ G4ZTA
Reed J ........ M0JDR
Reed J ........ M1DDW
Reed M ........ M3FZV
Reed M ........ G4BZF
Reed M ........ 2E0CZC
Reed N ........ G0GUT
Reed P ........ G4BVH
Reed P ........ M3OPG
Reed Q ........ G0IHF
Reed R ........ G1IYO
Reed R ........ G3ZIG
Reed S ........ G0AEV
Reed S ........ G4VQR
Reed S ........ MW3ESE
Reed S ........ G1FCU
Reed S ........ G7MMC
Reed S ........ MW3ESF
Reed T ........ M0MSF
Reed T ........ M1DDY
Reed T ........ M3ZTR
Reed W ........ G0WAL
Reed W ........ MW0BYS
Reeds G ........ M1CKJ
Reeds G ........ M3ACA
Reeds M ........ G0HOT
Reekie D ........ G4UHR
Reekie H ........ GM8HSY
Reeks M ........ 2E0MGR
Reeks M ........ M6MGR
Reeley A ........ G4OIN
Reeman K ........ G8RAN
Rees A ........ MM6IAR
Rees A ........ G3IQY
Rees C ........ GU3TUX
Rees C ........ MW6GOY
Rees D ........ [illegible]
Rees D ........ M1AHM
Rees D ........ M3ARM
Rees D ........ M6PFA
Rees D ........ MW6BHD
Rees E ........ MW6EDR
Rees F ........ GW0JRF
Rees G ........ MW3MHG
Rees G ........ M3OTS
Rees I ........ G8JWD
Rees L ........ GW1ADY
Rees M ........ G3KTI
Rees M ........ G3SRR
Rees P ........ MW0PJR
Rees P ........ 2E0NVS
Rees P ........ M0NVS
Rees R ........ GW0GPQ
Rees R ........ GW0RTR
Rees R ........ G1EPR
Rees R ........ GW4PUC
Rees R ........ GW6HUD
Rees R ........ G8ZAX
Rees R ........ GW0KFL
Rees R ........ GW0DIX
Rees T ........ G6NXL
Rees T ........ 2W0TRR
Rees T ........ MM0AXR
Rees V ........ GU8FSU
Rees W ........ GW0NLB
REES W ........ GW6NXH
Rees W ........ M0WPR
Reeson M ........ G8OOS
Reeve A ........ G6ZNJ
Reeve A ........ G7KEP
Reeve D ........ G6TIW
Reeve D ........ M6KBA
Reeve D ........ G4NRW
Reeve E ........ 2E0MZZ
Reeve E ........ M6ELR
Reeve G ........ G1HHO
Reeve H ........ G4ACJ
Reeve I ........ 2E0IDR
Reeve I ........ M0IDR
Reeve I ........ M6IDR
Reeve J ........ G8ATS
Reeve J ........ M6MES
Reeve M ........ G0GXU
Reeve M ........ G3WGH
Reeve M ........ M0NRW
Reeve M ........ M0PTO
Reeve N ........ M6NDR
Reeve N ........ M6NTR
Reeve P ........ G4GTN
Reeve V ........ M6FTX
Reeves A ........ G4ZFQ
Reeves A ........ G7DTT
Reeves C ........ G8FVI
Reeves D ........ G1PAD
Reeves D ........ 2E0DKR
Reeves D ........ M6DKR
Reeves D ........ MW6DDR
Reeves H ........ M6EGF
Reeves J ........ G0MVX
Reeves J ........ G4SJM
Reeves J ........ GW4SRE
Reeves K ........ G3WZI
Reeves K ........ M6AYW
Reeves L ........ M3VPP
Reeves M ........ G6YNW
Reeves M ........ M6UKM
Reeves N ........ 2E0GWD
Reeves P ........ G8GJA
Reeves P ........ M6PFR
Reeves R ........ G1TWW
Reeves R ........ M0ROJ
Reeves R ........ M0RRR
Reeves R ........ G8VOI
Reeves R ........ M3VOI
Reeves R ........ M6RGR
Reeves T ........ G0GHL
Reeves W ........ 2E0WBR
Reeves W ........ M6WBV
Reffell G ........ G0EEA
Regan A ........ G6FOI
Regan C ........ M0HGG
Regan E ........ GI7CTW
Regan F ........ G3RMD
Regan J ........ MW6AOY
Regan M ........ G4UQY
Regan M ........ M3UTY
Regan S ........ G7FIK
Regnart A ........ G8YFA
Reich J ........ M6FUH
Reich R ........ G8WKA
Reichenfeld I ........ M0RGI
Reichmann R ........ G1WKO
Reid A ........ G3AIF
Reid A ........ MM0TXO
Reid A ........ G6BOP
Reid A ........ MI6SQN
Reid A ........ GI7IMU
Reid B ........ GI1CUH
Reid B ........ GM7JDS
Reid Bamford T ........ MM6EXU
Reid C ........ 2E0SDM
Reid C ........ M6REI
Reid C ........ G0VQM
Reid C ........ G8NMM
Reid C ........ M3YRR
Reid D ........ G0BZF
Reid D ........ M1DNY
Reid D ........ MM6BXR
Reid D ........ MI1AWM
Reid D ........ GI8SKN
Reid D ........ MI0SDR
Reid G ........ G4OIS
[illegible] ........ [illegible]
Reid I ........ GM0MYV
Reid J ........ GM4LQR
Reid J ........ GM7KPE
Reid J ........ G7THZ
Reid J ........ M3FPS
Reid J ........ M0BDF
Reid J ........ MM0XCP
Reid J ........ GM3NUU
Reid K ........ G2UT
Reid L ........ G8NKJ
Reid M ........ 2E0DYL
Reid M ........ MM6OMJ
Reid M ........ 2M0MRO
Reid M ........ M6ENX
Reid P ........ GM0FCI
Reid P ........ 2I0PRL
Reid P ........ MI6FIU
Reid R ........ GI3UBA
Reid R ........ GI4XFX
Reid R ........ MI6RCR
Reid R ........ G1CNZ
Reid S ........ G6KVG
Reid S ........ MM3MXN
Reid T ........ G0UFL
Reid T ........ MI1DEZ
Reid W ........ G4HEJ
Reidy P ........ G6INA
Reigate S ........ G1IEY
Reilly A ........ G4BVW
Reilly A ........ G6KOE
Reilly A ........ M0ATV
Reilly D ........ G0RIE
Reilly G ........ 2E0CYD
Reilly G ........ 2E0HVI
Reilly G ........ MI3GER
Reilly J ........ GM3HOM
Reilly J ........ M0JJR
Reilly J ........ MI6OPD
Reilly M ........ G6WGM
Reilly M ........ 2E0MTR
Reilly M ........ M3XZY
Reilly N ........ G0VOK
Reilly P ........ G4VSX
Reilly P ........ G8BVU
Reilly P ........ G4VYL
Reilly R ........ MI3CCN
Reilly R ........ MI3PPD
Reilly R ........ M6VKN
Reilly T ........ MM3TRZ
Reilly-Cooper A ........ M3YET
Rekers G ........ G0UQO
Remedios Y ........ GD4UDT
Remnant A ........ 2E0ZUX
Remnant D ........ G7LXP
Remnant D ........ M0SAT
Remnant M ........ G4YYM
remnant R ........ 2E0VOK
remnant R ........ M0VOK
Remsbury A ........ G3OLH
Ren H ........ M6LHS
Ren X ........ M3YXR
Renaut J ........ G8DJL
Rendall D ........ GM0MHS
Rendall M ........ G0VXW
Rendell B ........ 2E0IJR
Rendell F ........ G4HXK
Rendell K ........ M3AIG
Rendell P ........ G6TJZ
Rendell W ........ M6WRE
Render D ........ M3TVV
Renggli G ........ G0GID
Rengifo C ........ 2E0ICI
Renmans J ........ M0ZOV
Renmans L ........ M3LPJ
Renner B ........ G4YEF
Rennick P ........ G4IZJ
Rennie A ........ GM1VXE
Rennie C ........ 2E0OLO
Rennie C ........ M0OLO
Rennie D ........ 2E0HGG
Rennie E ........ G4ZMS
Rennie G ........ GM3OUU
Rennie J ........ GM4LFL
Rennie J ........ GM4UWX
Rennie S ........ MM3XLQ
Rennison A ........ G3GSL
Rennison K ........ M3KYQ
Rennison P ........ G1YZI
Rennison R ........ G7KUR
Rennison H ........ G1YZJ
Rennolds R ........ G0BXS
Renny A ........ M6FCI
Renny N ........ M6RNN
Renouf C ........ M3WXH
Renowden B ........ M3KRZ
Renshaw D ........ [illegible]
Renshaw D ........ M6DCK
Renshaw G ........ M3ZGR
Renshaw P ........ G8VOH
Renshaw H ........ 2M0CEL
Renshaw R ........ MM0RJR
Renshaw R ........ MM6AHK
Renshaw T ........ G0WOM
Renton A ........ G7VVX
Renton G ........ G7DEU
Renton K ........ G6JMJ
Renvoize P ........ G0JKM
Renwick A ........ G0PMW
Reoch G ........ G4LHA
Repton C ........ M0GUL
Restall D ........ G4FCU
Restall J ........ 2E0CCC
Restall J ........ M3FEG
Retford L ........ M3ZGU
[illegible] J ........ G6LCU
[illegible] R ........ G6CXY
Reveley S ........ M6XIE
Revell C ........ 2E0EYC
Revell J ........ M6BPX
Revell J ........ M6JAR
Revell M ........ M6SXC
Revell N ........ 2E0IAI
Revell P ........ G3IOB
Revell S ........ G3PMJ
Revill R ........ G4GKZ
Rewaj F ........ G4TAZ
Reynard B ........ M3IOQ
Reynard T ........ G7HJT
Reynolds A ........ G4XYC
Reynolds A ........ M3FYM
Reynolds A ........ M3HYI
Reynolds B ........ M3BJR
Reynolds B ........ G3ONR
Reynolds C ........ G3JPT
Reynolds C ........ M3YXV
Reynolds C ........ G3GJA
Reynolds C ........ G8EQZ
Reynolds D ........ G7DPU
Reynolds E ........ G6DKE
Reynolds E ........ M1BDD
Reynolds E ........ G4YFF
Reynolds G ........ G1XIV
Reynolds G ........ G7BZQ
Reynolds G ........ M0ZRG
Reynolds G ........ M6ZRG
Reynolds I ........ 2E0AWM
Reynolds I ........ G0UVE
Reynolds I ........ G4RRD
Reynolds J ........ 2E0GQD
Reynolds J ........ G3RSD
Reynolds J ........ G6WEL
Reynolds J ........ M3GQD
Reynolds J ........ M3IZW
Reynolds J ........ M3NOJ
Reynolds J ........ 2E0RNJ
Reynolds J ........ G0UWV
Reynolds J ........ M0RJH
Reynolds J ........ M6JNR
Reynolds J ........ 2E0BXZ
Reynolds J ........ M0JCR
Reynolds J ........ M6JCR
Reynolds K ........ 2E0IDH
Reynolds K ........ G7ESO
Reynolds K ........ M3IDH
Reynolds L ........ G8LCK
Reynolds L ........ 2E0LDR
Reynolds L ........ M0LDR
Reynolds L ........ M3LRZ
Reynolds M ........ G0AOS
Reynolds M ........ G4KXO
Reynolds M ........ G4ZMR
Reynolds M ........ G6NLZ
Reynolds M ........ M3GUQ
Reynolds M ........ 2E0CJO
Reynolds M ........ M0LEH
Reynolds M ........ MM6HRF
Reynolds M ........ M6ITX
Reynolds P ........ GD4YGM
Reynolds P ........ G3NOA
Reynolds R ........ G3AVL
Reynolds R ........ G3IDW
Reynolds R ........ G4OMS
Reynolds R ........ M6RZZ
Reynolds R ........ G0UFF
Reynolds S ........ G6AHN
Reynolds S ........ G8NVH
Reynolds S ........ M3IWN
Reynolds S ........ GM4CAB
Reynolds S ........ G7BNB
Reynolds T ........ G7IRW
Reynolds T ........ G0UFD
Reynolds W ........ M6VIX
Reywer G ........ 2E0GRI
Reywer G ........ M0IAS
Reywer G ........ M3XNC
Rhead M ........ M1DXQ
Rhenius C ........ G1VJG
Rhenius S ........ 2E0AAF
Rhenius S ........ 2E0FLR
Rhenius S ........ M6DTR
Rhind-Tutt M ........ G4BSK
Rhodes A ........ G1XWD
Rhodes B ........ 2E0FEG
Rhodes B ........ G4ZVP
Rhodes B ........ M6VSB
Rhodes C ........ M0CJD
Rhodes C ........ M6ZCR
Rhodes D ........ G6LST
Rhodes D ........ M6DSR
Rhodes E ........ M3PVQ
Rhodes E ........ 2E0PVQ
Rhodes E ........ G4WNF
Rhodes G ........ G7CUF
Rhodes G ........ M5AFX
Rhodes J ........ G7BRU
Rhodes J ........ M0JUK
Rhodes J ........ M6JRJ
Rhodes K ........ 2E0FEF
Rhodes K ........ G1CFB
Rhodes L ........ M6LXR
Rhodes M ........ G0SVD
Rhodes M ........ G0VPW
Rhodes M ........ M1AHY
Rhodes M ........ G3XZO
Rhodes M ........ G1EPS
Rhodes P ........ G4DCM
Rhodes P ........ G6PAR
Rhodes P ........ G3XJP
Rhodes S ........ G1YHN
Rhymes H ........ M3HAD
Rhys J ........ G3YMN
Rhys M ........ G8KTC
Ribeiro A ........ M0HNC
Ribton M ........ M1ETC
Ricalton A ........ G6OTV
Ricalton A ........ G6OTW
Ricalton W ........ G4ADD
Rice A ........ G8GYH
RICE F ........ G7LPP
Rice J ........ 2I0BPO
Rice J ........ G8UEY
Rice J ........ MI0JAR
Rice J ........ MI3FPE
Rice J ........ M3NEA
Rice J ........ MI6AFM
Rice L ........ G3TVC
Rice N ........ 2E0LTR
Rice P ........ G3WUB
Rice P ........ G6AYU
Rice P ........ MI3EZF
Rice P ........ 2M0CEX
Rice P ........ MM0VPR
Rice P ........ MM6ANB
Rice S ........ G0SGR
Rice S ........ G4NEA
Rice S ........ M6ZLC
Rice V ........ G6UER
Rich A ........ G1XWK
Rich C ........ M6SWD
Rich D ........ 2E0DWR
Rich I ........ M6IMR
Rich L ........ M6LHR
Rich P ........ G4DYG
Rich R ........ 2E0RXR
Rich R ........ M6RXR
Richards A ........ G4RYK
Richards A ........ G7KWF
Richards A ........ G7RHF
Richards A ........ M3YNJ
Richards A ........ GW3SFC
Richards A ........ M1BCR
Richards A ........ 2E0DUO
Richards A ........ M6DUO
Richards B ........ 2E0GMM
Richards B ........ G7VBN
Richards B ........ M6STI
Richards B ........ G4TLR
Richards C ........ 2E0CBU
Richards C ........ M3KRN
Richards D ........ G3XXM
Richards D ........ G7EUT
Richards D ........ G7MBY
Richards D ........ M6CUL
Richards D ........ 2E0BTW
Richards D ........ 2I0DHR
Richards D ........ 2E0DOX
Richards D ........ G1XGZ
Richards D ........ G6CYF
Richards D ........ G6UYM
Richards D ........ M0GOK
Richards D ........ MI3DHR
Richards D ........ M3XRK
Richards D ........ M6ADX
Richards F ........ G4TJU
Richards G ........ G7PFT
Richards G ........ G0PWX
Richards G ........ G1WJK
Richards G ........ MW3BBQ
Richards G ........ M3GPR
Richards G ........ M3UYL
Richards H ........ GW4DWN
Richards J ........ G0ALC
RICHARDS J ........ G0KJM
Richards J ........ GD3BPG
Richards J ........ G6BWJ
Richards J ........ G6WKO
Richards J ........ M6CTZ
Richards J ........ M6ZXZ
Richards J ........ 2E0DZH
Richards J ........ G0AQR
Richards J ........ G6TZE
Richards J ........ MW6IAG
Richards K ........ GW0HYU
Richards K ........ G3HSU
Richards K ........ G4UZO
Richards K ........ 2E0CUL
Richards K ........ M0KUL
Richards L ........ M6BJB
Richards L ........ G1JBB
Richards M ........ G0EIQ
Richards M ........ G1RMN
Richards M ........ GW4APF
Richards M ........ G6FJI
Richards M ........ G6WCI
Richards M ........ G7VAB
Richards M ........ MM1DXU
Richards M ........ M3MQP
Richards M ........ G8CDA
Richards M ........ M6ZMR
Richards M ........ [illegible]
Richards N ........ G8TFY
Richards O ........ MW3EQE
Richards P ........ G1DUQ
Richards P ........ G8ASC
Richards P ........ M5DUO
Richards P ........ G0AVD
Richards P ........ M6PVR
Richards P ........ 2E0CYV
Richards P ........ M6DYC
Richards P ........ G0DCZ
Richards R ........ 2E0CAH
Richards R ........ G4IRD
Richards R ........ GW3CR
Richards R ........ M0DQN
Richards S ........ 2E0PAV
Richards S ........ G4LTY
Richards S ........ M0PAV
Richards S ........ M6FBW
Richards S ........ G0XHF
[illegible] ........ [illegible]
Richards S ........ M03ER
Richards S ........ G4HPE
Richards T ........ G0SII
Richards V ........ G0GAW
Richardson A ........ 2E0GHF
Richardson A ........ G3VDZ
Richardson A ........ G4CGD
Richardson A ........ G4WMQ
Richardson A ........ G4ZEB
Richardson A ........ M0ONQ
Richardson A ........ G6YGH
Richardson A ........ M6RHX
Richardson B ........ G0WVE
Richardson C ........ G0AHD
Richardson C ........ G0FXT
Richardson C ........ G3NAE
Richardson C ........ G4LQG
Richardson C ........ G4OCZ
Richardson C ........ G6VKI
Richardson C ........ M3MFX
Richardson C ........ 2E0ZYG
Richardson C ........ G4DPD
Richardson C ........ M6ZYG
Richardson D ........ G0AWZ
Richardson D ........ G4TST
Richardson D ........ G4ZGQ
Richardson D ........ G6MGN
Richardson D ........ M6NNY
Richardson D ........ 2E0DPR
Richardson D ........ M0SMN
Richardson D ........ M6CGK
Richardson D ........ G4GED
Richardson D ........ G4YTQ
Richardson F ........ M1BFV
Richardson G ........ G1SHN
Richardson G ........ GI1VAZ
Richardson G ........ G4RVY
Richardson G ........ G7GGG
Richardson G ........ G6NLX
Richardson G ........ G7HCR
Richardson G ........ 2E0ZNZ
Richardson G ........ M0ZNZ
Richardson G ........ M6ZNZ
Richardson H ........ G0CZU
Richardson H ........ GD0JWR
Richardson H ........ M6HRS
Richardson H ........ M0HPR
Richardson I ........ G1HQK
Richardson I ........ G3XLP
Richardson J ........ G0WYQ
Richardson J ........ G6JGR
Richardson J ........ G7CUD
Richardson J ........ G8IAJ
Richardson J ........ M3IRM
Richardson J ........ M3JHR
Richardson J ........ M5ASR
Richardson J ........ M6KNC
Richardson J ........ 2W0REJ
Richardson J ........ M0JWR
Richardson J ........ MW0WRY
Richardson J ........ M5JWR
Richardson J ........ G8KJI
Richardson K ........ G3SPV
Richardson K ........ G7AAY
Richardson K ........ G8RFF
Richardson K ........ G0PEK
Richardson L ........ G3FWU
Richardson L ........ M6HLR
Richardson L ........ M3UHQ
Richardson M ........ 2E0DLL
Richardson M ........ G8IUM
Richardson M ........ M0ARX
Richardson M ........ G6ZTH
Richardson M ........ G4XMR
Richardson M ........ M6WRG
Richardson M ........ M3TIK
Richardson N ........ G4SFH
Richardson N ........ G3ZPL
Richardson N ........ M6PQR
Richardson P ........ G4IBZ
Richardson P ........ G8GBE
Richardson P ........ G8NGJ
Richardson P ........ 2E0MNU
Richardson P ........ G4UBR
Richardson P ........ G8ETV
Richardson P ........ M0MNU
Richardson P ........ M3MNU
Richardson P ........ M3RTT
Richardson P ........ G8MLA
Richardson P ........ M3MLA
Richardson R ........ 2E0CJC
Richardson R ........ G3WRD
Richardson R ........ G4MQW
Richardson R ........ G4PKZ
Richardson R ........ G0BER
Richardson R ........ G3XMB
Richardson S ........ G4URG
Richardson S ........ GW7NJQ
Richardson S ........ G8XXG
Richardson S ........ M0TQZ
Richardson S ........ G7MQP
Richardson S ........ GM8NYV
Richardson S ........ M0SLP
Richardson S ........ G4WCP
Richardson S ........ GW0AWT
Riche F ........ M6VOX
Richens J ........ G1MHZ
Richens K ........ G3ZGU
Richer M ........ G3UYE
Riches A ........ M3HPO
Riches A ........ M3UAR
Riches C ........ G3UBP
Riches D ........ G0XEG
Riches D ........ [illegible]
Riches D ........ M1BZK
Riches D ........ M6LGF
Riches F ........ 2E0FSX
Riches F ........ M6FNR
Riches M ........ G1TXO
Riches M ........ M6OTH
Riches R ........ M0HWL
Riches S ........ G7KWS
Richey R ........ MI6REM
Richings L ........ MM0HQI
Richley T ........ 2D0HBT
Richley T ........ MD3HBT
Richman C ........ GM7GTS
Richman D ........ M6EFR
Richmond A ........ 2E0AXD
Richmond A ........ GU3ONJ
Richmond C ........ G0TOO
Richmond C ........ G3WPR
Richmond D ........ M3BMI
Richmond I ........ G8CLJ
Richmond J ........ G1KFH
Richmond L ........ 2E0AXE
Richmond P ........ G8GVV
Richmond P ........ M0HBL
Richmond S ........ G1DCY
Richmond-Hardy J ........ G8BHC
Richter B ........ M0HDN
Richter H ........ G4OAI
Richtering B ........ M0CZR
Richtering S ........ M6RSR
Rick A ........ G4NTJ
Rickaby M ........ 2E0MBW
Rickard D ........ G4KNI
Rickard K ........ M3AUL
Rickard T ........ G8WQT
Rickard-Worth M ........ G0SPZ
Rickerby C ........ G1NWA
Rickerby J ........ MM6EVO
Rickers D ........ G3HEU
Ricketts M ........ M6RHU
Ricketts R ........ G3VGY
Ricketts R ........ G7AGG
Rickman L ........ 2E0LGR
Rickman S ........ G0JSR
Rickward A ........ GD4ZZN
Rickward M ........ GD3ZZN
Rickwood A ........ M6RKU
Rickwood C ........ M3OUG
Rickwood J ........ G3JJR
Rickwood P ........ 2E0PPR
Riddell A ........ G4BFC
Riddell A ........ GM6AQB
Riddell G ........ GM7IKB
Riddell P ........ G7HIN
Riddell S ........ GM3YCB
Ridden D ........ G6GXG
Riddick A ........ 2E0TEN
Riddick A ........ M0RDI
Riddick A ........ M6AMR
Riddick D ........ G0LZW
Riddick D ........ M3VZQ
Riddick K ........ M6KCR
Riddick M ........ M6URR
Riddington R ........ G4IHT
Riddiough P ........ MM6OIR
Riddiough R ........ GM4SQO
Riddle C ........ GW0DGJ
Riddle D ........ G4BMR
Riddle P ........ M3MUO
Riddle P ........ 2M0DGI
Riddle P ........ MM0ZOG
Riddle P ........ MM6BFR
Riddle S ........ 2E0EYI
Rideout A ........ G1AKD
Rideout J ........ 2E0PUN
Rideout J ........ M3PUN
Rideout R ........ G4OVI
Rider B ........ G1KTY
Rider R ........ 2E0LNG
Rider R ........ M6RAO
Ridgard S ........ G1JIG
Ridge R ........ 2E0RER
Ridge R ........ M0RRX
Ridge R ........ M6RER
Ridge W ........ M6RGE
Ridgeon P ........ G8URG
Ridgeon P ........ M0CTC
Ridgeon S ........ M0STV
Ridgers P ........ 2E0DZF
Ridgers P ........ M3GOI
Ridgeway C ........ G7FZB
Ridgway D ........ 2E0GSM
Ridgway S ........ G3TZO
Ridgwell K ........ G6XHI
Riding C ........ G8JIE
Riding K ........ M6KMR
Riding [illegible] ........ G6WGF
Riding P ........ M6SQD
Ridings I ........ M0IPR
Ridley A ........ G3ZLR
Ridley C ........ G1DLK
Ridley C ........ G8GKC
Ridley C ........ G0NLM
Ridley D ........ G6OSH
Ridley G ........ 2E0DVI
Ridley G ........ M6GGI
Ridley J ........ 2E0SHA
Ridley J ........ M6JLR
Ridley P ........ M1PTR
Ridley P ........ 2E0PPA
Ridley P ........ M6CKL
Ridley P ........ G7NYF
Ridley S ........ 2E0MEG
Ridley S ........ M3TTK
Ridout J ........ M0BWB
Ridpath M ........ M3XSV
Ridsdale I ........ 2E0MFM
Ridsdale I ........ M6SUK
Riebold P ........ G8XHD
Rieger-Ridd N ........ G7MOK
Rigazzi-Tarling N ........ M0BSF
Rigby B ........ G8UVN
Rigby C ........ G1CFG
Rigby D ........ 2E0DEX
Rigby D ........ G4KXV
Rigby E ........ G6GVZ
Rigby G ........ M3JIU
Rigby G ........ 2E0UNI
Rigby G ........ M0UNI
Rigby G ........ G8MPG
Rigby G ........ G3KTJ
Rigby H ........ G0MEQ
Rigby J ........ G8XLI
Rigby J ........ G8XNL
Rigby K ........ G3TRE
Rigby M ........ M0TGW
Rigby M ........ M3VCZ
Rigby M ........ G4FUI
Rigby M ........ G8LBT
Rigby N ........ G7IFO
Rigby P ........ G0PXI
Rigby P ........ G8LMF
Rigby R ........ 2E0DFV
Rigby S ........ M1SWR
Rigden G ........ M3SNX
Rigden P ........ 2E0GTB
Rigden R ........ 2E0RBG
Rigden R ........ M6KXS
Rigelsford K ........ G3XXC
Rigg A ........ M0RIG
Rigg P ........ G0EUP
Rigg P ........ G0TVB
Riggott P ........ G4XGN
Riggs A ........ G7UUL
Riggs C ........ G3XAS
Riggs D ........ M3MGD
Riggs J ........ G4KDK
Riggs J ........ G8PVR
Riggs S ........ GW6PDR
Rigler A ........ 2E0WSR
Rigler A ........ M6DUN
Rignall M ........ G3OYX
Riley A ........ G4DQG
Riley A ........ G7CQW
Riley A ........ 2E0WBL
Riley A ........ M6MEQ
Riley B ........ G3STJ
Riley C ........ G4JQX
Riley C ........ 2E0DBY
Riley C ........ M3ZWI
Riley C ........ M6GSD
Riley D ........ M0DFX
Riley D ........ M1ARF
Riley D ........ M3DGR
Riley D ........ M1DUA
Riley D ........ M6JDD
Riley I ........ G0UZF
Riley J ........ G0RPG
Riley J ........ G3TZA
Riley J ........ G3ZKG
Riley J ........ G7BRA
Riley J ........ M1BSN
Riley J ........ G1MSK
Riley J ........ G6OIB
Riley J ........ G6UKQ
Riley J ........ 2E0JRR
Riley K ........ 2E0CKH
Riley K ........ G7UKN
Riley K ........ G8XLZ
Riley M ........ G4CSZ
Riley M ........ M1BOP
Riley M ........ M5BOP
Riley M ........ 2E0BME
Riley M ........ M6MDR
Riley P ........ G0KTT
Riley P ........ M6GPR
Riley R ........ G4NQZ
Riley R ........ G4VPI
Riley S ........ G0MUH
Riley S ........ MM6FSR
Riley T ........ G8UQR
Riley W ........ G0HBN
Riley W ........ G0CZL
Riley-Kydd D ........ 2E0PWR
Riley-Kydd D ........ M3VSG
Riley-Marsland A ........ M3JII
Riley-Moxon C ........ G7CER
Riman D ........ 2E0SCC
Riman D ........ M6HJR
Rimell I ........ M3NXH
Rimell S ........ G1AIG
Rimington J ........ G6IFH
Rimington I ........ 2E0LCR
Rimington L ........ M6IQN
Rimmer A ........ GM4XRY
Rimmer A ........ M6EKV
Rimmer B ........ G0JCO
Rimmer D ........ 2E0DHQ
Rimmer D ........ M6EKW
Rimmer D ........ G4VYP
Rimmer E ........ 2E0GLR
Rimmer J ........ M6JHQ
Rimmer L ........ G0BRZ
Rimmer M ........ M3UGK
Rimmer M ........ M3WFL
Rimmer P ........ G4YVI
Rimmer R ........ G3RQS



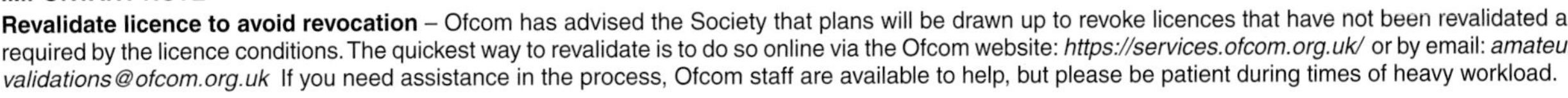
**IMPORTANT NOTE**

**Revalidate licence to avoid revocation** – Ofcom has advised the Society that plans will be drawn up to revoke licences that have not been revalidated as required by the licence conditions. The quickest way to revalidate is to do so online via the Ofcom website: *https://services.ofcom.org.uk/* or by email: *amateur.validations@ofcom.org.uk* If you need assistance in the process, Ofcom staff are available to help, but please be patient during times of heavy workload.

UK Surnames

| Name | Callsign |
|---|---|
| Rimmer R | GD3YEO |
| Rimmer R | G8CFD |
| Rimmer R | 2E0BMO |
| Rimmer R | M3RXH |
| Rimmer W | G8EYA |
| Rimmington P | 2E0CFK |
| Rimmington P | M6PER |
| Ring C | 2E0RTC |
| Ring C | M6AIE |
| Ring G | G4YSE |
| Ring S | M0SVR |
| Ringrose R | G4KYU |
| Ripley A | M0DCD |
| Ripley D | 2E0FVY |
| Ripley S | G4AVK |
| Ripley Z | M3ZAZ |
| Rippengill G | G4AGY |
| Rippin M | 2E0DMH |
| Rippin R | M0AFV |
| Rippin R | M3HZN |
| Rippington M | G0IEV |
| Rippon E | M0EPR |
| Riseborough D | G7UFV |
| Rish B | G0SFP |
| Rish M | MD3AAI |
| Rispin B | G3VLR |
| Ristic B | G0DHG |
| Ritchie A | GM3WYL |
| Ritchie A | M3FMQ |
| Ritchie A | G4VMX |
| Ritchie B | G6UHL |
| Ritchie D | MI3CIW |
| Ritchie J | GM0GYT |
| Ritchie J | GM0RSI |
| Ritchie L | MM6SXY |
| Ritchie M | G6NLG |
| Ritchie M | GM8ADK |
| Ritchie R | G8UYF |
| Ritchie T | G0GSL |
| Ritchie W | G0KJJ |
| Ritchie W | GM3IWX |
| Ritchley B | M1EGM |
| Ritchley P | M1EGL |
| Ritossa R | M0ITA |
| Ritson D | G7DBV |
| Ritson J | M6JIQ |
| Ritson J | 2E0JGR |
| Ritson K | G0PKR |
| Ritson M | G0VRT |
| Ritson M | G0MIK |
| Ritson V | G7FAD |
| Rittman W | 2E0GYZ |
| Rivers A | M0RDA |
| Rivers B | G1FHO |
| Rivers B | G4WGX |
| Rivers I | G6FDG |
| Rivers J | G0LZF |
| Rivers J | G0GCQ |
| Rivers J | G7MAR |
| Rivers M | G0VIM |
| Rivers M | G0DYM |
| Rivers P | G7PSC |
| Rivers P | G4XEX |
| Rivett I | G8WPU |
| Rivett K | M6BFF |
| Rivett N | G3OXS |
| Rivron D | M0KXD |
| Rix L | G3XJW |
| Rix R | G6UEQ |
| Rixon A | G4JIR |
| Rixon A | M1EIO |
| Rixon B | G3IOJ |
| Rixon E | GD6XHG |
| Rixon J | G0COH |
| Rixon P | M3AAS |
| Rixon R | G6NXM |
| Rizzo C | G7ISD |
| Roa Vicens J | M6RVJ |
| Roach J | G4ETO |
| Roach J | G4WXZ |
| Roach M | G3TWJ |
| Roadnight D | GD7IEF |
| Roaf B | M3KKS |
| Roan K | G7KDQ |
| Robb A | MI3SXR |
| Robb G | GM8KXF |
| Robb J | 2E0JCZ |
| Robb J | G4WMN |
| Robb J | M0JRZ |
| Robb J | M6FJL |
| Robb R | MM3VNU |
| Robb W | MM6BLY |
| Robbins B | G0CFM |
| Robbins G | MM3IQA |
| Robbins G J | 2M0AGS |
| Robbins G J | GM3KMF |
| Robbins J | M6DMN |
| Robbins K | G3PFV |
| Robbins M | G4TXD |
| Robbins R | G0MDM |
| Robbins S | M0WPA |
| Robelou M | G8YUR |
| Roberson N | G0UMM |
| Robert D | 2U0EFR |
| Robert D | MU0EFR |
| Robert D | MU6GFR |
| Roberts A | G0FMX |
| Roberts A | G1YPH |
| Roberts A | G4DTU |
| Roberts A | G4XBZ |
| Roberts A | G6AZE |
| Roberts A | G6BMP |
| Roberts A | G7BAV |
| Roberts A | M0CRH |
| Roberts A | M0TNT |
| Roberts A | M0TRO |
| Roberts A | M1CJZ |
| Roberts A | M3GVF |
| Roberts A | M6KUT |
| Roberts A | G7JDA |
| Roberts A | M0GYK |
| Roberts A | M1FJG |
| Roberts A | G0HKF |
| Roberts A | G4ZIB |
| Roberts B | G0HGC |
| Roberts B | G4DBQ |
| Roberts B | G4VYG |
| Roberts B | G4ZCP |
| Roberts B | 2E0ORT |
| Roberts B | M6YKS |
| Roberts B | M3VQN |
| Roberts B | M6BDR |
| Roberts C | G0FBO |
| Roberts C | G1CXQ |
| Roberts C | G4COJ |
| Roberts C | G4ZFJ |
| Roberts C | M6CRZ |
| Roberts C | G0CNG |
| Roberts C | G4EVA |
| Roberts C | G4YJT |
| Roberts C | M1FCV |
| Roberts D | 2E0BPV |
| Roberts D | G0DRO |
| Roberts D | G0GHG |
| Roberts D | G0GWC |
| Roberts D | G0RKB |
| Roberts D | G0TFI |
| Roberts D | G0VOG |
| Roberts D | G0WMW |
| Roberts D | G1DUS |
| Roberts D | G1LAN |
| Roberts D | G1UCI |
| Roberts D | G1ZMJ |
| Roberts D | G3UBV |
| Roberts D | G4GSR |
| Roberts D | G4MFI |
| Roberts D | G6FKR |
| Roberts D | GI7DIT |
| Roberts D | G8KBB |
| Roberts D | G8UDG |
| Roberts D | M3LBX |
| Roberts D | M6TJQ |
| Roberts D | M6BAG |
| Roberts D | 2E0DRQ |
| Roberts D | G0JBT |
| Roberts D | G7MYM |
| Roberts D | G8NZN |
| Roberts D | M6EEM |
| Roberts D | M6GWC |
| Roberts E | G0TKL |
| Roberts E | G1AJV |
| Roberts E | G7TVL |
| Roberts E | M0DOR |
| Roberts E | M3EWR |
| Roberts E | M3GIF |
| Roberts F | G4UYM |
| Roberts F | M6CCI |
| Roberts G | G0HUK |
| Roberts G | GM0RYA |
| Roberts G | G1KNQ |
| Roberts G | G1MPR |
| Roberts G | G8DHI |
| Roberts G | G4JXN |
| Roberts G | M0LPL |
| Roberts G | M3HAM |
| Roberts G | M0ATI |
| Roberts G | G7RLZ |
| Roberts H | G6NLE |
| Roberts H | M6HRR |
| Roberts H | G6JWL |
| Roberts H | M3UUE |
| Roberts H | M3HGR |
| Roberts I | G0IJY |
| Roberts I | G4ASH |
| Roberts I | G6CYH |
| Roberts I | G0IYX |
| Roberts I | 2E0VTV |
| Roberts I | M6ITV |
| Roberts J | 2E0FQC |
| Roberts J | G0LLL |
| Roberts J | G0PVU |
| Roberts J | G0ZPV |
| Roberts J | G4UQK |
| Roberts J | G6OFZ |
| Roberts J | G6OIX |
| Roberts J | G6OIY |
| Roberts J | G6OSJ |
| Roberts J | M3CDL |
| roberts J | M3CMX |
| Roberts J | M6JJI |
| Roberts J | M3UJR |
| Roberts J | 2E0BVB |
| Roberts J | 2E0LZM |
| Roberts J | G6DFH |
| Roberts J | G8FDJ |
| Roberts J | M1VCD |
| Roberts J | M3XID |
| Roberts J | M6CTI |
| Roberts J | M6RIO |
| Roberts J | M6VTA |
| Roberts J | G4ZMM |
| Roberts K | 2E0CWS |
| Roberts K | G1HHM |
| Roberts K | G8VDP |
| Roberts K | M0KDR |
| Roberts K | M3CBX |
| ROBERTS K | M3NQN |
| Roberts K | G1NGL |
| Roberts L | G0MMW |
| Roberts L | G0OLR |
| Roberts L | G4PIA |
| Roberts M | G0HUN |
| Roberts M | G0HZB |
| Roberts M | G0RUH |
| Roberts M | G1ZLA |
| Roberts M | M3MBR |
| Roberts M | M3XAR |
| Roberts M | 2E0DVW |
| Roberts M | G4AWK |
| Roberts M | G4HFI |
| Roberts M | 2E0FVV |
| Roberts M | M6FVV |
| Roberts M | 2D0VYN |
| Roberts M | G0OBV |
| Roberts M | GW7TJM |
| Roberts M | MW6FQD |
| Roberts M | M6VMR |
| Roberts M | MD6VYN |
| Roberts M | M3YHN |
| Roberts M | M6FQC |
| Roberts M | MW6FGJ |
| Roberts N | G6TZO |
| Roberts N | 2E0ACS |
| Roberts N | G6MQP |
| Roberts N | 2E0BJS |
| Roberts N | GU4IJF |
| Roberts N | G4KZZ |
| Roberts N | M0RBJ |
| Roberts N | M3VFP |
| Roberts P | G4KKN |
| Roberts P | G7SJS |
| Roberts P | G0OER |
| Roberts P | M1DBW |
| Roberts P | M3PBW |
| Roberts P | M3PSR |
| Roberts P | 2E0MFN |
| Roberts P | 2E0PCU |
| Roberts P | G0JPE |
| Roberts P | G4DJB |
| Roberts P | G6KVR |
| Roberts P | M6MNK |
| Roberts P | M6PDR |
| Roberts P | M6PFG |
| Roberts R | 2E0AZW |
| Roberts R | G0WCU |
| Roberts R | G4IGT |
| Roberts R | G4TYH |
| Roberts R | G6ZDH |
| Roberts R | G7DSQ |
| Roberts R | G7HGI |
| Roberts R | G8VUN |
| Roberts R | G8ZKG |
| Roberts R | M0CSK |
| Roberts R | G7SZZ |
| Roberts R | M6BFP |
| Roberts S | 2E0SJR |
| Roberts S | G0TFX |
| Roberts S | GD1SWF |
| Roberts S | G1YRY |
| Roberts S | GM4HWS |
| Roberts S | G4OSR |
| Roberts S | GW6BMR |
| Roberts S | G6ZMD |
| Roberts S | M0SJR |
| Roberts S | M3ITI |
| Roberts S | M6SHV |
| Roberts S | M6SGM |
| Roberts S | M3UWF |
| Roberts S | M0XSR |
| Roberts T | G6IDW |
| Roberts T | G7RNF |
| Roberts T | G0ESD |
| Roberts T | M6AEJ |
| Roberts T | M1AUW |
| Roberts T | G0GJN |
| Roberts V | GM1ZNR |
| Roberts V | GW7PBP |
| Roberts V | M3VYN |
| Roberts W | G0CGA |
| Roberts W | G3LOE |
| Roberts W | G7DDQ |
| Roberts W | G3GXQ |
| Roberts W | G0HIJ |
| Roberts W | G1UCG |
| Roberts W | G7LXB |
| Roberts W | M0HHM |
| Roberts W R | 2E0HIO |
| Robertshaw B | M3CBV |
| Robertshaw N | G0NHM |
| Robertson A | G4IAO |
| Robertson A | G8RHZ |
| Robertson A | GM3ZXB |
| Robertson A | MM6ATR |
| Robertson B | GM4OIJ |
| Robertson B | G4POL |
| Robertson B | G4RJO |
| Robertson C | M3TIF |
| Robertson C | GM0HBK |
| Robertson D | 2E0DAR |
| Robertson D | 2M0RTD |
| Robertson D | G0HYN |
| Robertson D | GM0MWJ |
| Robertson D | GM1TDT |
| Robertson D | M0RGO |
| Robertson D | MM0RTD |
| Robertson D | M3PWO |
| Robertson D | GI7UGP |
| Robertson G | M0BHK |
| Robertson G | MM3MRX |
| Robertson G | GM6AOR |
| Robertson G | G6KQN |
| Robertson H | G0GLJ |
| Robertson H | GM3RQQ |
| Robertson J | G0KJU |
| Robertson J | GM0KYU |
| Robertson J | G0PJV |
| Robertson J | G1DUT |
| Robertson J | GM7REG |
| Robertson J | 2M0JVR |
| Robertson J | GM6GPH |
| Robertson J | GM4OFI |
| Robertson J | MM5AES |
| Robertson K | GM1PWL |
| Robertson K | M3KVR |
| Robertson K | G1JIE |
| Robertson M | 2E0FIE |
| Robertson M | G0WTR |
| Robertson M | GM7IBM |
| Robertson M | G7VCK |
| Robertson M | MM6MJR |
| Robertson M | G6CMS |
| Robertson M | G1HDX |
| Robertson M | G0IOR |
| Robertson M | G3USX |
| Robertson M | MM3VNT |
| Robertson N | G0ORG |
| Robertson N | GM8EUG |
| Robertson N | 2E0NIX |
| Robertson N | M0WWH |
| Robertson N | M6GWW |
| Robertson N | M6RJN |
| Robertson P | G0RQN |
| Robertson P | GM4RAH |
| Robertson P | M3BPG |
| Robertson P | MM6BBK |
| Robertson R | GM0CEA |
| Robertson R | 2E0BJK |
| Robertson S | GI7VCR |
| Robertson S | G8OIY |
| Robertson S | M0CAR |
| Robertson S | MM5FWD |
| Robertson S | 2M0IMP |
| Robertson S | MM3ZYS |
| Robertson S | M6BLE |
| Robertson T | M6TDR |
| Robertson W | GM1SZM |
| Robertson-Mudie F | GM0HMM |
| Robey A | G4RFN |
| Robilliard R | GU2RS |
| Robins A | G1UUP |
| Robins B | 2E0SLR |
| Robins D | 2E0ODS |
| Robins D | M6FNV |
| Robins D | G1KTZ |
| Robins D | G6FUD |
| Robins F | G3GVM |
| Robins M | G4FYQ |
| Robins M | M0HBE |
| Robins M | M3TZN |
| Robinson A | G0GAR |
| Robinson A | G0WEK |
| Robinson A | G1FMA |
| Robinson A | G1KSE |
| Robinson A | G1ZAY |
| Robinson A | G3XEY |
| Robinson A | G3ZYQ |
| Robinson A | M1CLX |
| Robinson A | M3CWZ |
| Robinson A | M3OPX |
| Robinson A | G0VPO |
| Robinson A | G1OJS |
| Robinson A | G8BOB |
| Robinson A | G3TQA |
| Robinson B | G3ZEB |
| Robinson B | G4YIA |
| Robinson B | M6BEN |
| Robinson B | G4NOL |
| Robinson B | 2E0REU |
| Robinson B | G6ZTD |
| Robinson B | M6PFP |
| Robinson C | 2E0RON |
| Robinson C | G0RDC |
| Robinson C | G1KJG |
| Robinson C | M0BXD |
| Robinson C | M6CSR |
| Robinson C | M1BMR |
| robinson C | M6HLW |
| Robinson C | G7ONB |
| Robinson C | GI7PJU |
| Robinson C | G0OXL |
| Robinson C | M0MRC |
| Robinson C | M1CNP |
| Robinson C | 2I0EIR |
| Robinson C | MI0HCK |
| Robinson C | MI6GRF |
| Robinson C | M6FKP |
| Robinson D | GW0CKK |
| Robinson D | G0NXF |
| Robinson D | G1HQJ |
| Robinson D | G1JIR |
| Robinson D | G3MQR |
| Robinson D | G3UQR |
| Robinson D | GJ3YHU |
| Robinson D | GI4OSG |
| Robinson D | G6AYX |
| Robinson D | GI7MDK |
| Robinson D | G8RFL |
| Robinson D | 2E0HEF |
| Robinson D | 2E0XAG |
| Robinson D | G7EHY |
| Robinson D | M0LMN |
| Robinson D | M3WQF |
| Robinson D | M3XIE |
| Robinson D | M6DBP |
| Robinson D | M6HEF |
| Robinson D | G0MQH |
| Robinson D | G1NYZ |
| Robinson D | 2E0JBQ |
| Robinson D | M6PYD |
| Robinson E | G1AIB |
| Robinson E | G4MET |
| Robinson E | GI0ZER |
| Robinson E | G3JGP |
| Robinson E | G6OSK |
| Robinson F | G3TPV |
| Robinson F | G4EEQ |
| Robinson F | G8REQ |
| Robinson F | G0EVO |
| Robinson G | 2M1FQI |
| Robinson G | G0LNS |
| Robinson G | GI4CSP |
| Robinson G | G7BUR |
| Robinson G | MM3FQI |
| Robinson G | M3MQY |
| Robinson G | G4AKW |
| Robinson G | M6GAI |
| robinson G | M6BUS |
| Robinson G | G4ZMH |
| Robinson G | 2M0ESL |
| Robinson G | MM0OMG |
| Robinson G | MM6ESL |
| Robinson G | G4KXU |
| Robinson H | G3VVE |
| Robinson H | G7IHE |
| Robinson H | M0BWQ |
| Robinson I | M0EGV |
| Robinson I | GW1AWH |
| Robinson I | G8WGE |
| Robinson I | M6ISR |
| Robinson J | 2E0AHK |
| Robinson J | GD3VHK |
| Robinson J | G4AZX |
| Robinson J | G4TSV |
| Robinson J | G6CIO |
| Robinson J | G7SCL |
| Robinson J | G8ADA |
| Robinson J | G8PQJ |
| Robinson J | GM8RPE |
| Robinson J | MW3KHY |
| Robinson J | M3OZB |
| Robinson J | G6GCY |
| Robinson J | M3UBQ |
| Robinson J | M3HND |
| Robinson K | G4KKZ |
| Robinson K | G4LVN |
| Robinson K | 2E0JLK |
| Robinson K | M6RRC |
| Robinson K | G4CYO |
| Robinson K | G0LCE |
| Robinson L | G0LHR |
| Robinson L | M0LEZ |
| Robinson L | MM1ATR |
| Robinson L | GI4NSS |
| Robinson L | M3RQR |
| Robinson L | MI6LCR |
| Robinson M | G3ZZM |
| Robinson M | G4UUA |
| Robinson M | GI7NFB |
| Robinson M | G7PQL |
| Robinson M | 2E0OMG |
| Robinson M | M3XZN |
| Robinson M | M5TLA |
| Robinson M | M6UEA |
| Robinson N | GI4PES |
| Robinson N | MI3NPR |
| Robinson N | G0FOB |
| Robinson N | G4OIR |
| Robinson N | M6AZX |
| Robinson P | 2I1EXU |
| Robinson P | GD0DVC |
| Robinson P | GI0MTE |
| Robinson P | G0VZI |
| Robinson P | G3MGJ |
| Robinson P | G3MRX |
| Robinson P | G4CHI |
| Robinson P | G4IZH |
| Robinson P | G4UEA |
| Robinson P | G4ZKN |
| Robinson P | G6WZE |
| Robinson P | M3HVA |
| Robinson P | M3PNY |
| Robinson P | G0FSD |
| Robinson P | G8FNG |
| Robinson P | 2E0PYR |
| Robinson P | M6PYR |
| Robinson P | G0EYR |
| Robinson P | G0MQJ |
| Robinson P | G1NCO |
| Robinson P | M3ZNY |
| Robinson P | 2E0PRO |
| Robinson P | G0SDM |
| Robinson P | M3AIR |
| Robinson R | G1AIF |
| Robinson R | G1FXT |
| Robinson R | G1JUD |
| ROBINSON R | G6YGJ |
| Robinson R | G8BWH |
| Robinson R | MI0RSN |
| Robinson R | MI3FEO |
| Robinson R | MM3WVQ |
| Robinson R | 2E0KCZ |
| Robinson R | M0KCZ |
| Robinson R | M3KXG |
| Robinson R | M6KCZ |
| Robinson S | G0AIM |
| Robinson S | G0BQW |
| Robinson S | GI0URI |
| Robinson S | G4WNV |
| Robinson S | G6FKS |
| Robinson S | M3NXZ |
| Robinson S | G4NSH |
| Robinson S | M6SXI |
| Robinson S | G8POO |
| Robinson S | M5POO |
| Robinson S | M1BJE |
| Robinson S | M6SRB |
| Robinson S | GW0WMT |
| Robinson S | M3JNU |
| Robinson T | G3KOA |
| Robinson T | GI4RFH |
| Robinson T | G4URX |
| Robinson T | M0TWR |
| Robinson T | GM3WUX |
| Robinson T | G8JDC |
| Robinson T | M6TNR |
| Robinson T | M6SKT |
| Robinson V | G4JTR |
| Robinson W | G7OES |
| Roblin M | MW6MSE |
| Robnett A | M0HYL |
| Robottom-Scott S | 2E0ZSR |
| Robottom-Scott S | M6GYF |
| Robson A | G0DJO |
| Robson A | G4THI |
| Robson A | M6FBF |
| Robson A | MM6EWR |
| Robson A | GM8YIK |
| Robson C | G1COY |
| Robson D | GM8ZKF |
| Robson D | MW3FTC |
| Robson D | GM6CMQ |
| robson D | M6BIG |
| Robson J | G0NHJ |
| Robson J | G1SIO |
| Robson J | GW1TFL |
| Robson J | G3HMQ |
| Robson J | G1YUL |
| Robson J | M0BYM |
| Robson J | G4GBY |
| Robson K | G0NHK |
| Robson K | G1JLP |
| Robson M | GM8RGO |
| Robson P | G3NZK |
| Robson R | G3IDY |
| Robson S | MW3FTB |
| Robson T | G0VWW |
| Robson W | GM8IIO |
| Roby G | G0CBD |
| Roche A | G7WKG |
| Roche K | G8GOS |
| Roche M | G3YJS |
| Roche M | M3TKN |
| Rochester K | G0LCS |
| Rochester K | G8VR |
| Rochester M | GM6ZCX |
| Rochester M | GM6ZCY |
| Rochford T | G7RDQ |
| Rock A | 2E0RLY |
| Rock T | G1HMI |
| Rocke M | G1YNJ |
| Rockett E | G3SXQ |
| Rockliffe D | G7UMY |
| Rocks B | MI6IBR |
| Rod J | G1RSE |
| Rodd E | G1LTC |
| Rodda J | G4VTN |
| Rodda S | G4PEM |
| Roddam J | M3BTI |
| Roddan S | G3CSC |
| Rodden K | G0ASI |
| Roddis G | M0YLG |
| Roddy E | G0MPT |
| Roddy T | G6OTQ |
| Roderick D | M6DRR |
| Rodger A | 2E0HXC |
| Rodger B | MM3ATI |
| Rodger C | 2M0CRR |
| Rodger C | MM3LYH |
| Rodger D | 2M0DJT |
| Rodger J | 2M0FYG |
| Rodger L | M1LES |
| Rodger P | M1PAM |
| Rodgers A | M0VAT |
| Rodgers A | G4RVS |
| Rodgers A | MM3VYA |
| Rodgers C | MM3YIH |
| Rodgers C | M0BIN |
| Rodgers G | GI6ATD |
| Rodgers I | G7RBC |
| Rodgers J | G1KOD |
| Rodgers J | M6EII |
| Rodgers L | M6RCI |
| Rodgers M | G0GDL |
| Rodgers M | G7BQS |
| Rodgers M | GD7BOJ |
| Rodgers M | MD0BJM |
| Rodgers P | M6MPR |
| Rodgers R | G1UKH |
| Rodgers S | M6EVR |
| Rodgers S | M6PJX |
| Rodgers T | 2E0IIL |
| Rodgers W | G1UOR |
| Rodinson J | G7UZA |
| Rodley J | 2E0JMR |
| Rodley N | M0BNR |
| Rodley P | M0OXZ |
| Rodley R | G1TUS |
| Rodman D | G0TVW |
| Rodmell G | M0GWA |
| Rodmell P | G3ZRS |
| Rodrigues L | M3ZLR |
| Rodriguez B | M3OXB |
| Rodriguez Cemillan J | M0GXK |
| Rodway C | M0PHP |
| Rodway C | M3RLO |
| Rodway C | M3UHP |
| Rodway J | G4FRK |
| Rodway P | G3UFF |
| Rodwell G | MD6GRR |
| Rodwell J | G6IWU |
| Rodwell S | MM3VSU |
| Roe B | G4LVR |
| Roe B | G6XFB |
| Roe D | G1GNP |
| Roe F | GM0ALS |
| Roe I | G1ROE |
| Roe J | G4IMS |
| Roe M | M0GXM |
| Roe N | G1DHY |
| Roe N | G4ACW |
| Roe P | M6ROE |
| Roe W | G6BPY |
| Roebuck A | M1ROE |
| Roebuck D | G0LJM |
| Roebuck D | G8EMH |
| Roebuck I | G0SVX |
| Roebuck J | M0CRW |
| Roebuck K | G7IOF |
| Roebuck M | 2E0BJQ |
| Roebuck R | M3RPR |
| Roeschlaub R | G1NBY |
| Roets V | M6ENE |
| Roeves T | G3RKF |
| Roff B | G0RVS |
| Roff G | G3TJI |
| Roff J | 2E0EMK |
| Roffey A | G1MMZ |
| Roffey G | G7BIP |
| Rofix L | 2E0LFR |
| Rofix O | 2E0OLE |
| Rogalewski R | G8VOQ |
| Rogalski W | M0OSH |
| Rogan S | M3XXO |
| Rogers A | G0KHH |
| Rogers A | G3SYZ |
| Rogers A | G6HTZ |
| Rogers A | G7AZC |
| Rogers A | M3UKR |
| Rogers B | G6ZTL |
| Rogers B | G1XRF |
| Rogers B | G6YGI |
| Rogers C | G0CNL |
| Rogers C | G3KEG |
| Rogers C | G1WAE |
| Rogers C | MM0CZK |
| Rogers C | M6CTT |
| ROGERS C | G1YAB |
| ROGERS C | G4YUG |
| Rogers D | G1LLW |
| Rogers D | G1PIF |
| Rogers D | G1UVJ |
| Rogers D | G3UOO |
| Rogers D | G4XJL |
| Rogers D | G6WZP |
| rogers D | M3XRD |
| Rogers D | M6DRP |
| Rogers D | 2E0MQA |
| Rogers D | M3MQR |
| Rogers D | M5SSB |
| Rogers D | M6MND |
| Rogers D | M6VEL |
| Rogers D | 2E0CZR |
| Rogers D | M6BYF |
| Rogers E | G0CLC |
| Rogers E | G1OMZ |
| Rogers E | M3GEX |
| Rogers E | M3HFT |
| Rogers F | G0FZN |
| Rogers F | G0NXE |
| Rogers G | G0RJV |
| Rogers G | G3TFL |
| Rogers G | G6OMN |
| Rogers G | M1DYU |
| Rogers G | M3IRJ |
| Rogers G | M6GSR |
| Rogers G | G8ABB |
| Rogers G | M3TQY |
| Rogers H | G1ATQ |
| Rogers H | G3NHR |
| Rogers H | M6HCR |
| Rogers I | G6LGM |
| Rogers I | G0BON |
| Rogers J | G0LAK |
| Rogers J | G1PJZ |
| Rogers J | G3PQA |
| Rogers J | G4ATR |
| Rogers J | M0JAV |
| Rogers Jones D | G0KLT |
| Rogers Jones M | M3KLT |
| Rogers K | G4TZL |
| Rogers K | G8ZUF |
| Rogers L | G1LAR |
| Rogers M | 2E0GXE |
| Rogers M | 2E0GXV |
| Rogers M | 2E0HSR |
| Rogers M | 2E0BPY |
| Rogers M | G4RGB |
| Rogers M | M3UIJ |
| Rogers M | G0UPY |
| Rogers N | G0GGG |
| Rogers N | G0JZF |
| Rogers N | G6BHE |
| Rogers N | M3MRA |
| Rogers N | 2M0UGL |
| Rogers N | MM0SEY |
| Rogers N | MM3UGL |
| Rogers O | G1FXD |
| Rogers O | M3XBS |
| Rogers P | G0GXH |
| Rogers P | G0SNW |
| Rogers P | G6FMN |
| Rogers P | M0BVI |
| Rogers P | M1PUW |
| Rogers P | M3ZPR |
| Rogers P | M1ALO |
| Rogers P | 2E0CBO |
| Rogers P | G8GXO |
| Rogers P | M0HHD |
| Rogers P | M3YQC |
| Rogers R | G2FGT |
| Rogers R | G3CWH |
| Rogers R | G8PPF |
| Rogers R | G8TLH |
| Rogers R | G6UNR |
| Rogers R | M0BBR |
| Rogers R | G6KKW |
| Rogers S | G1ERF |
| Rogers S | G4HER |
| Rogers S | G4TAO |
| Rogers S | G4UUH |
| Rogers S | 2E0CYX |
| Rogers S | M6SKR |
| Rogers S | M6ZRJ |
| Rogers S | G1WFS |
| Rogers T | G0MND |
| Rogers T | G4YSQ |
| Rogers T | G6IMW |
| Rogers T | G6TZT |
| Rogers T | M3IEM |
| Rogers T | M3MGI |
| Rogers T | M0BNB |
| Rogers T | MM6IHQ |
| Rogers T | 2E0BTQ |
| Rogers T | M6TLR |
| Rogers W | GM0CII |
| Rogerson J | G0GEQ |
| Rogerson R | G3IUJ |
| Rogerson R | 2E0RDG |
| Roguszczak D | M0GQS |
| Rohrer C | G7VJE |
| Rohrlach L | G1XPB |
| Rohsler N | G4ZTM |
| Roissetter J | M0GOM |
| Roithmeir L | MU0GSY |
| Rolf D | M0DMR |
| Rolf G | G4RTW |
| Rolfe D | M3YFY |
| Rolfe H | G0BSF |
| rolfe J | G4UET |
| Rolfe S | G0ETF |
| Rolinson C | G7DDN |
| Rolinson L | M6FAB |
| Rolland A | M3YHR |
| Rolland G | G3USR |
| Rollason A | G6DHD |
| Rollason J | G3WCO |
| Rollason M | G4PAH |
| Rolley J | G1NZN |
| Rolley R | G8IKH |
| Rollin P | G4AFU |
| Rollings A | G0AGV |
| Rollings D | G0AJT |
| Rollings M | G4XLG |
| Rollins W | G1WJR |
| Rollinson P | M1DCE |
| Rollinson R | M6ROL |
| Rollinson S | M1COV |
| Rollitt A | G6HTY |
| Rollitt-Smith A | 2E01BGN |
| Rollo D | GM3GRG |
| Rolls A | G8UHK |
| Rolls M | 2E0MPR |
| Rolls M | M3JZI |
| Rolph C | G7HXW |
| Rolph D | G3DXD |
| Rolph D | G0UYC |
| Rolph J | M3SOY |
| Rolph M | G0HZA |
| Rolph M | M0MCY |
| Rolt L | M6MLR |
| Romang K | G4SKN |
| Romang K | M3KIR |
| Romanis J | M0BQZ |
| Romano A | GW1PFK |
| Romanov A | M3ZTN |
| Romanov A | M0XBI |
| Rome J | G8BPH |
| Romer H | G3CIK |
| Rominger O | M6OVI |
| Romocea C | M6ROM |
| Romocea C | M0OXD |
| Romocea C | M3OXD |
| Ronan M | 2E0KNB |
| Ronan M | M0KNB |
| Roney V | G4CJK |
| Ronnie A | G6WAZ |
| Roobottom E | G0WUM |
| Rood T | M6OTR |
| Rook K | G4XCB |
| Rooke D | G0OEW |
| Rooke J | G4AP |
| Rooke L | G4DDE |
| Rooker R | MW3NKG |
| Rooker S | G4UUI |
| Rooks L | G3PUO |
| Room R | M0RCR |
| Roomes D | G0RQL |
| Rooney A | M3NBK |
| Rooney C | G3VCR |
| Rooney D | MM6DAQ |
| Rooney J | GM1TDU |
| Rooney K | M3MQM |
| Rootes D | M3FGR |
| Roots B | M6YOU |
| Roots R | G1FHR |
| Roots T | G1PPB |
| Rope S | 2E0AWG |
| Rope S | M0EBJ |
| Roper B | G6WMT |
| Roper C | G4FPB |
| Roper C | G4TIV |
| Roper D | G4ZFD |
| Roper D | M0DAD |
| Roper G | G0AEP |
| Roper G | G8AKM |
| Roper I | M3ILR |
| Roper I | 2E0FBY |
| Roper I | G6XHK |

Roper L G0LYN
[illegible]
Roper M G1WTS
Roper M 2E0DXW
Roper M M6GHS
Roper M 2E0RPR
Roper M M0RPR
Roper M M6RPR
Roper M G8NTQ
Roper R G1SVD
Roper S G7MUE
Roper S M0AEC
Roper T M6ASO
Ropinski G M6RJF
Ropinski R 2E0ICU
Ropinski R G4IBW
Ropper I GM0UHC
Rosa J M0WPT
Rosa J M6EPT
Rosamond P G4LHI
Rosbottom S G0JEH
Roscoe D M3JQW
Roscoe G M0BYT
Roscoe J MW6JTR
Roscoe N G0RXA
Roscoe R G4MYZ
Rose A 2W1SDR
Rose A G0TNY
Rose A G1IFF
Rose A GM3WED
Rose A G4LXV
Rose A G6VEN
ROSE A M3KOU
Rose A G4AUE
Rose C GM1BUY
Rose C G4OOJ
Rose C G7WCF
Rose C G8MKE
Rose C M3OYJ
Rose C M6EXE
Rose C 2E0CZM
Rose C M0HTF
Rose C M6DXK
Rose D G0PYS
Rose D G1KKE
Rose D G7BPF
Rose D G7MLK
Rose D M0BAW
Rose D M3IAA
Rose E G1CPM
Rose G M3FKM
Rose G G4EDH
Rose H M0HOT
Rose I G6IDU
Rose I M0DKJ
Rose I G0HDZ
Rose J G0IEW
Rose J G3OGE
Rose J M6RJL
Rose K G0DZV
Rose L GD4KAB
Rose M G3VPA
Rose M G3XKU
Rose M G8PAG
Rose M M3FKL
Rose M M3RLM
Rose M 2E0XMK
Rose M M0XMK
Rose M M3XMK
Rose M M6HVH
Rose P G0BCS
Rose P GM3ZZA
Rose P G4STO
Rose P G7TDD
Rose P G4NXR
Rose P G4TRR
Rose P M6DTX
Rose R G1FMC
Rose R M6EIE
Rose S G4HDD
Rose S G7PMG
Rose T G4OXD
Rose T G8PUR
Rose U M0BQO
Rose W 2E0AXL
Roseaman D G8FLL
Rosema K 2E0KRX
Rosema K M0KRX
Rosema K M6KRX
Rosen P G3YFW
Rosenberg I GD4XEW
Rosenbrand M M0GLT
Rosenschein A 2E0TTT
Rosenschein D M6GGG
Roser G G7LJQ
Rose-Round R M0BOL
Rosevear D G4RLN
Rosevear I G3GKC
Rosewall [illegible] G3OOY
Rosewarn D [illegible]
Rosewarn D M0CIF
Rosewell M M6HNA
[illegible] M6SNZ
Rosier A G1FLY
Rosindale J G0CUO
Roskruge N G1WLN
Ross A GM3VVF
Ross A GM7ORJ
Ross A MM3SRK
Ross A GM4PMT
Ross A 2M0WTN
Ross A MM6PSX
Ross A 2E0BWT
Ross B G0PJF
Ross C 2E0ROS
Ross C G4UFP

Ross C MM6KIW
Ross C GM8HBY
Ross D G0RIW
Ross D G1DBR
Ross D G1ULR
Ross D G4LOO
Ross D GM4PVQ
Ross D GI4SNA
Ross D G7NXV
Ross D M3FFI
Ross D MM5ALX
Ross D M5FOX
Ross D MM6WCK
Ross D GM0IQD
Ross D G0VHO
Ross E G3TJC
Ross E 2E0EYE
Ross F G1MYO
Ross G MM6ELU
Ross H M3KOR
Ross H MM6HLZ
Ross I 2E0IVR
Ross I GM0IJR
Ross I GI0UTV
Ross I G1WZB
Ross I G4DPF
Ross I 2E0IMZ
Ross I GM0TGE
Ross I M6IMZ
Ross J G0LBO
Ross J G3WWG
Ross J M0AWI
Ross J M3OSA
Ross J GM1BSG
Ross J MM6DBT
Ross J M6JHR
Ross K 2M0AKT
Ross L 2E0KOR
Ross L 2E0MFC
Ross L M3JQM
Ross M M3MRO
Ross M M6MFL
Ross M M6FFK
Ross N M6NVR
Ross R G1BIM
Ross R MI1RDR
Ross S G6HUH
Ross S M6EKL
Ross S M6LWR
Ross T GM4YWI
Ross W M0WMR
Ross W G7PRB
Rossant A M3YAD
Rosselle J GM3AEI
Rosser A M6NON
Rosser A M3XYW
Rosser C 2W0BED
Rosser D 2W0DTR
Rosser S M3KHK
Rosser S 2E0KHK
Rosser W G8UHM
Ross-fraser W G1YED
Rossi S G0SHT
Rossiter D G7VVF
Rossiter J 2E0AFC
Rossiter P G4CMX
Rossmann W GM0BTK
Rostant N 2E0TRI
Rostant N M6NRO
Rotari N M0GRT
Rote M M3DDY
Rotgans D G6RRS
Roth B M1CKQ
Roth G M3UJV
Rothera G MM6PKO
Rotheram I M0ODD
Rotheram I M6VMP
Rotherham G MM3EXW
Rothery R G1UVD
Rothery R G6RFL
Rothery R M6SCN
Rothon R MM6FVK
Rothwell A G3ZED
Rothwell D G4YPH
Rothwell Hughes N M0ECX
Rothwell M G4KFT
Rothwell P G7AGB
Rothwell R G7JDK
Rothwell W G0VDE
Rouget P G4RDM
Rough J G7GKX
Roughley J M3RRJ
Roughton E [illegible]
Roullier F G7TAX
Rounce R G4UKZ
Round D G0BTA
Round G M6PYH
Round J G6LDA
Round M G0WDU
Round T [illegible]
Rourke C G8YHY
Rous R G0WUZ
Rouse A G8EWC
Rouse C G4ESU
Rouse D G4TZQ
Rouse D M3RQO
Rouse D G1XDK
Rouse G G4MUP
Rouse G G7PVU
Rouse J M3KFP
Rouse K M6HJA
Rouse L G6IFN
Rouse M G1YKX
Rouse R M1ENZ
Rouse S 2E0XZZ
Rouse S M6KON

Rout D G8PAI
Rout M M3RMD
[illegible]
Routledge R M0VKX
Rovardi P G4HSB
Rowan A 2E0HLO
Rowan A M6LHO
Rowan F G4RWN
Rowan F G8OMW
Rowan M G6SDW
Rowan N 2E0RWN
Rowan-Jenkins A MI6XEX
Rowat M G6CGI
Rowberry E G1YPU
Rowberry G M0TXD
Rowbotham F G7SWE
Rowbotham J G4AQT
Rowbotham M G1GQQ
Rowbottom A M0KTR
Rowcroft N G4NIV
Rowden J M0HFH
Rowe A 2E0BMW
Rowe A G6AVP
Rowe A G7HDC
Rowe A M0PUB
Rowe A M3OVO
Rowe A 2E0AFT
Rowe C G4MAR
Rowe C M6CYA
Rowe C MW6LTR
Rowe D M6PSR
Rowe G M3UUL
Rowe G G0UAD
Rowe I G4WWL
Rowe J M0BKK
Rowe J MW1FGV
Rowe J MW3FGV
Rowe J M3JRI
Rowe J M1FBL
Rowe J 2E0KFH
Rowe J M6CDK
Rowe J M6DUM
Rowe M G7BLX
Rowe M G8JVE
Rowe M M3MXZ
Rowe M G1TLE
Rowe P G0BKQ
Rowe P G1DIA
Rowe P G6WKQ
Rowe R 2E0BJD
Rowe R M3MQX
Rowe R 2M0RDR
Rowe R MM6RDR
Rowe S M6SSS
Rowe T G4XQB
Rowe T G8NNU
Rowell A G7HDZ
Rowell A G8WMW
Rowell B G0MRZ
Rowell B G4IOR
Rowell E G1JIH
Rowell M G1BPS
rowland A G1LEL
Rowland A G4OJQ
Rowland B 2E0BAD
Rowland G G8PKJ
Rowland I M3HNQ
Rowland K M6LEX
Rowland M G4YUA
ROWLAND M G4FTW
Rowland M M6FQX
Rowland N M3NIC
Rowland P G1AKE
Rowland P G4YSJ
Rowland R G0FXC
Rowland S M6TVE
Rowlands A 2E0CYM
Rowlands A G6DUC
Rowlands A M3MWO
Rowlands D M3ZRK
Rowlands E G8BCI
Rowlands G M0BTU
Rowlands G G8TSV
Rowlands G M3MWE
Rowlands I M0SSB
Rowlands J G0JVY
Rowlands J G1IPY
Rowlands J G4OJS
Rowlands L 2E0LXR
Rowlands L MW0EOP
Rowlands M G8HUV
Rowlands P M3OJS
Rowlands R G4HKX
Rowlands R G4UHS
Rowlands I 2E0G3W
Rowlands T M0RMY
Rowlands T M3RAU
Rowlands T M6CTX
Rowlands T M6WVC
Rowlandson D G7IHL
Rowlandson S G1FIP
Rowland-Stuart A 2E0CWW
Rowland-Stuart A M6WKF
Rowland-Stuart J M6RRF
Rowles B G1ZDU
Rowles C M6EBW
Rowles J G4ZUH
Rowles M GW4WWN
Rowles R GW4FOM
Rowlett R G0EYM
Rowley A G0TML
Rowley A G8MYK
Rowley B 2E0CPI
Rowley C G0ORL
Rowley D G7WGK
Rowley G G4WKG

Rowley G G8BAG
Rowley H G1AKA
Rowley J G8VSR
Rowley J G3KAL
Rowley K M3TKW
Rowley L 2E0GYZ
Rowley L M3AUN
Rowley M M1EBK
Rowley M MD0GCI
Rowley P M1IRM
Rowley W M1BQZ
Rowling J G1ZUZ
Rowney K MW3UEU
Rowney R G4KFL
Rowntree F G8UYK
Rowntree G G7DKY
Roworth A M0OWO
Roworth L M6LDR
Rowsby A G4CQS
Rowsby S G8GEB
Rowse D G3LOD
Rowse R G3RTR
Rowsell C G4NHN
Rowsell F G8PQH
Rowsell J 2E0CIX
Rowsell K M1FAY
Rowsell P G1ULQ
Rowsell R G4SIF
Rowson-Brown A M6CIP
Rowthorn E G1OOS
Roxbrough H 2E0HRY
Roxbrough H M3UQG
Roxburgh A G7NOS
Roxburgh D GI8YJF
Roxby T G1LPS
Roy M G4PPU
Roy S G1IDE
Roy S M0SAR
Roy T GM7VQB
Roy W GM6NXN
Royan N 2E0CLX
Royce B M3IFJ
Royce K 2E0KNT
Royce K M0KWR
Royce K M3KNT
Roychoudhuri R M1EHJ
Roycroft R G1NXV
Royds A M6ODS
Royle G G4FAS
Royle J G3NOX
Royle N G7CAA
Royle T G0JNE
Roynon M M0MTR
Roys M M6NOL
Royston A GW8PBM
Rozday J G0DOZ
Roze J G0VMP
Rozenek P M0PPR
Rozentals L G4YGJ
Rozier M M6FGZ
Rozier T G1NYI
Rozier T M0TCR
Rozier T M1TCR
Ruane T M1CZX
Ruaux A 2E0HGT
Ruaux R G3RMK
Rubery B M6PMR
Rubins R GD4HAB
Ruchomski K MM6TEQ
Ruck D G6RFH
Ruck J MW3JDA
Rucklidge P GM4KKV
Rudcenko S G0KBL
Rudd B G3ELS
Rudd J 2E0DJR
Rudd J M3UYV
Rudd M G4LOA
Rudd P GM0CQL
Rudd P GU1MUP
Rudd P M0PCR
Rudd R 2E0RUD
Rudd R G4HWF
Rudd R M3PBE
Rudd R M3RKR
Rudd T G0AOY
Ruddell A G4BGH
Ruddell A MI6AJN
Ruddell F GI4MWA
Ruddell J MI6JOY
Ruddell T 2I0TJR
Ruddell T MI3TJR
Rudderham T G7HJR
Ruddick T 2E0TXG
Ruddick T M3TXA
Ruddle J GW0WLI
Ruddock J GD7KWT
Ruddock L G0UHM
Ruddy M 2I0MRY
Ruddy M MI6MSR
Ruddy O M6HDI
Ruder P [illegible]
Rudge C G0WYF
Rudge C G6LAW
Rudge G G0CDP
Rudge L G8PKB
Rudge L M6VTB
Rudge M G4IDC
Rudge S G7RKQ
Rudgewick-Brown N G4LLL
Rudkin B G4MEF
Rudley G M6GAN
Rudling A G7DYB
Rudling D 2E0CSK
Rudling D M0HPU
Rudling D M6BZA
Rudnicki A M1FCX

Rudwick P G3RDR
Rufes J 2E0GMN
Rufes J M0ZEE
Rufus J [illegible]
Ruff J G3MOA
Ruff M 2E0GPF
Ruff S GI0IQA
Ruffell S G7JIF
Ruffle P G0PJR
Ruffle S G4EAG
Rugen D 2E0OLG
Rugen D M3BHP
Rugen G G4OVD
Ruiz K G4SGF
Rule C G1ZPC
Rule C M3RUL
Rule E G3FEW
Rule G GM7OCU
Rule L 2E0FDK
Rule M M1HGV
Rule T MW3LZC
Rumball C M6PEX
Rumbelow A G3KKC
Rumbelow L M6RUM
Rumbelow M G1BHS
Rumbelow R G8UYL
Rumble N G4OEH
Rumble T G0DIU
Rumble T G8ETD
Rumbol N G4WAK
Rumbold A G3ORX
Rumbold D G4RYV
Rumbold T G6AYY
Rumboll R MJ6RBI
Rumens D G4BOO
Rumens M G4UHU
Rumney A G6UYN
Rump M G8VIC
Rumsam P G0VYX
Rumsby A M6XAI
Rumsey G M0GCR
Runcie G MM6ZZG
Rundle A M0ATA
Rundle C M6CJR
Rundle G M0BKG
Runyard D M0PRA
Runyard D M6DPR
Ruocco A 2E0UVO
Ruocco A M3UVO
Ruocco P M3YPR
Rusbridge D G1UQC
Rusby G G0SPB
Rusby I G1DAE
Rush B M6CIS
Rush D G4CRE
Rush M 2I0EIB
Rush M MI6RSH
Rush S 2E0IKV
Rush S M6VKA
Rush W G8IIZ
Rushbrooke M MI6AUU
Rushby P 2E0BIU
Rushby P M0ICA
Rushby P M3PYR
Rushforth L G4RIQ
Rushlow R M0TEX
Rushmer D GM7OGS
Rushton A GW0UZK
Rushton B G7IJW
Rushton C G0PMY
Rushton J 2E0BEA
Rushton J G7BRS
Rushton J G1AII
Rushton J G0WZL
Rushton K G1NXT
Rushton M M6LJT
Rushton M G4SBQ
Rushton R G4YRR
Rushton R M6TTK
Rushworth P G8AHR
Russ A G7FRH
Russ G G7KOI
Russ M G8VQH
Russell - Bishop A 2E0RBA
Russell A G3OMT
Russell A G4WEV
Russell A G7AQV
Russell A G3MFL
Russell A M6PUS
Russell A G0VRM
Russell B G8KFS
Russell C G0NSL
Russell C 2E0CDR
Russell C M3RZF
Russell D G0VKY
Russell D G3XLN
Russell D G8RCO
Russell D M0AXC
Russell D M1DLM
Russell D M5IGE
Russell D [illegible]
Russell D M0GPJ
Russell D M1AXM
Russell E G0SEY
Russell F 2E0AJD
Russell F G0IZJ
Russell F G1YEP
Russell F G4WVP
Russell F G7LUL
Russell G G0CAK
Russell G 2M1DHG
Russell G G6GMH
Russell H 2E0HBB
Russell H M6HRL
Russell I G1HPZ
Russell I G6PSO

Russell J G0LVR
Russell J G6AVS
Russell J GM7FIS
[illegible]
Russell J M6HFU
Russell J G4SEU
Russell J G6LUK
Russell J 2E0CGT
Russell J M6JBM
RUSSELL K G4RZQ
Russell K G8MHI
Russell L M1AHT
russell M G0UKM
Russell M G0WLD
Russell M G4ZVX
Russell M 2E0RCT
Russell M M0ROC
Russell M M3RCT
Russell M M6MXR
Russell P 2M0ZEE
Russell P MM6PJR
Russell P G7AHO
Russell P M0CDC
Russell P G6UYK
Russell R G3KLD
Russell R GI4JLF
Russell R G4TOH
Russell R G0RUT
Russell R MI6VJR
Russell R 2E0DFM
Russell R G4BAU
Russell R M6RMR
Russell R MM6CHM
Russell S M1ARX
Russell S M3YZA
Russell S M1CNX
Russell S M3GPX
Russell S 2M0SRX
Russell S MM6SNR
Russell T M6LIK
Russell T G4ZRV
Russell W G4EDZ
Russell-Smith D G0UMD
Russoff M G3OMR
Russon A G0GUC
Russon D G7PUN
Russon J G3APL
Russon S 2E0BGM
Russon S M3ISI
Rust D G7MOD
Ruston A M1AZG
Rusu M 2E0LGL
Rusu M M6LGL
Ruth D G4TAT
Rutherford A G0DHI
Rutherford D GD7KUG
Rutherford G G7SCT
Rutherford G M0SCT
Rutherford G G3YTC
Rutherford J G7FFR
Rutherford J M1JWR
Rutherford L M1EIZ
Rutherford R G0MWZ
Rutherford R M0DCC
Rutherford T G3TEC
Rutkowski B M6VOG
Rutkowski J G4XAU
Rutland M G0VIX
Rutt H G0DHR
Rutt M G6YOG
Rutt S G0MSR
Rutt S 2E0RIO
Rutt S G3BOK
Rutt S M3UMR
Rutt T 2E0HYC
Rutt T M0HYC
Rutt W G0DEJ
Ruttenberg M G7TWC
Rutter A G8HCK
Rutter D M6XPN
Rutter J M6EIA
Rutter K G1EYD
Rutter K G0FXS
Rutter S G7VAH
Rutter-DaCosta M M6MRR
Ruud S G0TSJ
Ruzgar G MM3RUZ
Ryall A G7KRY
Ryall M 2E0LBE
Ryall M M6BVU
Ryall P M3TCR
Ryalls C GW0KLN
Ryalls C G8HRA
Ryalls C M0CIR
Ryalls P MW3YBB
Ryan A 2E0ICO
Ryan A G6AQL
Ryan A G6LUD
Ryan A M3IRX
Ryan A G3VJN
Ryan D GM7ODP
Ryan D [illegible]
Ryan D M0GIW
Ryan D MM6MNW
Ryan H G0JYH
Ryan J G1EIG
Ryan J G3NAW
Ryan J GI0EEO
Ryan M G7EQO
Ryan P M3XPR
Ryan P M6PTY
Rycott C M3ZFL
Rycroft D G4OKO
Rycroft P G8FFW
Ryder A M3TSG
Ryder D G4HHM

Ryder G M0BNO
Ryder M G0CDA
ryder P G0TSG
Ryder P M0FCT
Ryder S [illegible]
Ryder T G1MMI
Ryder T [illegible]
Rye M G4WTE
Ryland A G0PTR
Rylatt R G3VXJ
Rylett I G6NLW
Ryley J G3KQV
Rymel S M6SSR
Rymer J M0JHM
Rymsza J G1YFA

## S

Saagi L G6TMQ
Sabbatella Riccardi E M3UQD
Saben P 2E0GNN
Saben P M3PHS
Sabido R GW4LKE
Sabin D G0MLE
Sables M G7NTY
Sables P G0TZC
Sables P G4MRU
Sach M G8KDF
Sacharewicz R G1XVY
Sadanandam A M3OME
Sadauskas D M0HMJ
Saddington A M3UFZ
Saddington S G2FXQ
Saddler M MM6MSA
Saddler R M6UPE
Sadka A M0LEB
Sadler A GD7NBF
Sadler A G7MUY
Sadler C M3IZV
Sadler D G0MWA
Sadler F M0CVS
Sadler G M3GCS
Sadler J 2E0JAZ
Sadler J M0JAZ
Sadler J M6JRS
Sadler L M0YZV
Sadler M M0BHE
Sadler M M0KMT
Sadler M 2E0MEY
Sadler M M6CKR
Sadler P M6PAA
Sadler R G4BMP
Sadler R G4FAJ
Sagar A G1FUG
Sagar J GW3ARS
Sage J G7JYL
Sage P GW0UHO
Sage R M6FWV
Sage R M6FVS
Sager D G7GJI
Sager D G8QZ
Sager D M3DPS
Sager J G8ONH
Saggers C G0OSK
Saggerson T G4WSE
Saich B G1PRZ
Saiger J M1DAH
Saiger S M0BOH
Sainsbury P G1BFS
Sainsbury R G0OSW
Saint W M6AUM
Sait A G0OIV
Salaman R G4JSE
Salata A G4PLX
Sale A G0GJW
Sale T G8RDN
SALEK J M0ICL
Sales B G6SGY
Sales J 2E0TTL
Sales J M0VLF
Salisbury A G4RYJ
Salisbury A G8KSF
Salisbury G G0KTY
Salisbury S M6AOF
Salisbury W G8KSE
Salkeld A MM0WXT
Sallis T G6FFH
Salmon C G0REB
Salmon C G6ZGB
Salmon D G4KCN
Salmon J 2E0RMI
Salmon J M3RMI
Salmon M G3XVV
Salmon P G7NFF
Salmon S G7DIE
Salmon S [illegible]
Salomon P G3XQO
Salsbury D M0AVA
Salt A G0HEE
Salt A 2D0UTL
Salt B G0OZP
Salt B G3VVG
Salt D [illegible]
Salt F G7AJE
Salt K G7DXQ
Salt L M6UKO
Salt M [illegible]
Salt M M3DLU
Salt P GW0MBN
Salt R G0DJQ
Salt W M0CBQ
Salter A G6MMR
Salter B G8POQ
Salter D G1ERM
Salter D G8UIO
Salter D M0GNP
Salter D M0SAL
Salter G 2E0GWS

Salter G M6GWS
Salter H M6BIM
Salter H M6TFT
Salter J 2E0EER
Salter J [illegible]
Salter J M6FLU
Salter K G4FKU
Salter N 2E0GKF
Salthouse J G8LDC
Saltmarsh A M1DJP
Saltmarsh G MW6CDZ
Saltmarsh K 2W0CWL
Saltmarsh K MW0HUY
Saltmarsh K MW6CWL
Saltmer M G7RRY
Saltmer S G8SGX
Salvesen T GM3ODP
Salzman M G1ZSR
Samber D G1EIH
Samborskyy M M0ERY
Sambrook G G4ROK
Samet F G0EVU
Sammons A G0EAG
Sammons R G6XZP
Sammut A M3UEN
Sammut L M3ZLS
SAMOUELLE A G0BOM
Sampathkumar S M0DRS
Sampathkumar S M3VNS
Samphire R 2E0LGE
Samphire R M0LGE
Samphire R M6LGA
Sampson A GM4BVD
Sampson C G0VZK
Sampson C M1BEW
Sampson D M1FCE
Sampson D M1DOZ
Sampson G G4SRX
Sampson K G1MXC
Sampson L G3JSU
Sampson M M6SAM
Sampson N MM1FHS
Sampson P G1BGQ
Sampson P G4OVF
Sampson P 2E0LLW
Samson M G6XMT
Samson P G8IVB
Samuel D GW4JDZ
Samuel M G3ZRR
Samuels A G0ABN
Samuels B G0FJA
Samuels E G6NIX
Samuels H M3NIA
Samuels J G0FYL
Samuels J G4FOB
Samuels P G4VSD
Samways A G6LPC
Samways R G0GGN
Samwells H G7MSH
SANCHEZ-GARCI F G1TDO
Sanchez-Garci K M3TYS
Sancto V G0LFM
Sandall B G3LGK
Sandall J G1DCZ
Sandaver E G4KIT
Sandell D M1DHY
Sandell G 2E0GNS
Sandell G M3HNU
Sandell P G4HQB
Sander J GI4BUJ
Sanders C G4KCM
Sanders C G6KQK
Sanders D G0KRY
Sanders D M6YJD
Sanders G G0MTF
Sanders H G3IGH
Sanders H M1DKY
Sanders H G3CRH
Sanders J 2E0AWE
Sanders J G1LMI
Sanders J 2E0DVN
Sanders J M6BNM
Sanders J M6JYZ
Sanders J M1AWN
Sanders L G2HCA
Sanders M G8HST
Sanders M G8LES
Sanders M G3RWV
Sanders N GM6WTP
Sanders R G0MUK
Sanders R M1SAN
Sanders S [illegible]
Sanders S M0CDS
Sanders S M1APB
Sanders S M3JIK
Sanders T G1ZWH
Sanders-Hewett H M6RXS
Sanderson A M1EKL
Sanderson A 2E0SDA
Sanderson B M6JNX
Sanderson C 2E0BRG
Sanderson D [illegible]
Sanderson E [illegible]
Sanderson I G8DTX
Sanderson I M6RED
Sanderson J [illegible]
Sanderson J G7POW
Sanderson J G4MQV
Sanderson J G8UYM
Sanderson J 2E0JCS
Sanderson J M3JCS
Sanderson K G7MSF
Sanderson K G4KCF
Sanderson M 2E0IEO
Sanderson M M0IEO
Sanderson M M1AQI

UK Surnames

**IMPORTANT NOTE**

**Revalidate licence to avoid revocation** – Ofcom has advised the Society that plans will be drawn up to revoke licences that have not been revalidated as required by the licence conditions. The quickest way to revalidate is to do so online via the Ofcom website: *https://services.ofcom.org.uk/* or by email: *amateur.validations@ofcom.org.uk* If you need assistance in the process, Ofcom staff are available to help, but please be patient during times of heavy workload.

Sanderson M M6MEV
Sanderson N G8GFF
Sanderson N M3AAY
Sanderson P M6CXS
Sanderson R 2E0FMC
Sanderson R M1TLK
Sanderson R M3TLK
Sanderson S 2E0BNE
Sanderson S M6NFR
Sanderson T G4OEY
Sandever D G7MQU
Sandford J G6FEX
Sandford L G6DKM
Sandford S M3SQV
Sandham M G7VLA
Sandham P GW0VST
Sandham T 2E0VVJ
Sandham T M6TOD
Sandiford P G3STF
Sandilands D MM3XFX
Sandilands E GM0FSZ
Sandilands S G6XZO
Sandilands T G0HBV
Sandle W G0SBZ
Sandler M G6MVS
Sandoz M G3GBS
Sands A G0AOX
Sands A G1FPC
Sands K MW1COB
Sands M 2E0VHC
Sands M G3IIW
Sands N 2I0GLY
Sands S G0AOW
Sandwell R M6NPN
Sandy D G4RSP
Sandy N M6NJS
Sandys J G0RGX
Sandys J G8DXZ
Sanford D GW8NCN
Sanger W G7FOT
Sangster G GM4OBD
Sangster J M0JYM
Sangster R G8DCX
Sankey B G7RWY
Sankey D M6FDF
Sankey J G0SGI
Sankey W M6WRS
Sansom A G4KIF
Sansom I G7UPZ
Sansom J M0JAO
Sansom J M3XTE
Sansom K M0MMC
Sansom M G0POT
Sansom R M6EYO
Sansom V M3VJS
Sansum J G8GHT
Sant D G0SBA
Santagata A G7ANG
Santer C G6JVT
Santer E G8SLM
Santer E G8YBY
Santer L G3PVB
Santhosh V MD6EGA
Santillo A G0XAW
Santos A G4PMJ
Sanvoisin N G8TPF
Sapstead I 2E0CIR
Sapstead I M3IJS
Sapsworth D G3YMW
Sargent G G8ZRV
Sargeant A G0KZN
Sargeant C G8FKF
Sargeant L M3LEL
Sargeant M M3ZKT
sargeant P 2E0DZO
sargeant P M6HNV
Sargeant W 2E0WDS
Sargeant W M3UEZ
Sargent A G0KVA
Sargent A G8HXW
Sargent A M3HYF
Sargent B G0HTL
Sargent C 2E0HTV
Sargent C M3SGE
Sargent D G4JYE
Sargent J G8CKS
Sargent J G8KIH
Sargent L 2E0DLK
Sargent L M0HVE
Sargent M G4SCB
Sargent P G8RYO
Sargent T G4BFS
Sargent W G0CTF
Sarll P 2E0BGX
Sarll P M3HZO
Sarosi T M6PFM
Sarratt P M3FVX
Sarre R GU4HUY
Sartorius A M3VOA
Sartorius C G4PSR
Sartorius M G3GXG
Sarwar S M3WRK
Sasse H G8TYF
Sate A G4LSK
Sati H M6HSX
Sati Z M6ZIA
Satterthwaite R G6BMY
Saueressig J G0ADK
Saul A G7DIW
Saul D G4EKZ
Saul M G8XGT
Saul P G8EUX
Saunby M G4NFP
Saundercock V M0AVS
Saunders A G1OZB
Saunders A G3VLB
Saunders A G4LLW
Saunders A G7IRF
Saunders A G7NRS
Saunders A G7VEE
Saunders A G8PPD
Saunders A M1BAI
Saunders B GW1SPW
Saunders B M3GXI
Saunders B 2E0CPQ
Saunders B M0XBS
Saunders B M6BPJ
Saunders C G0TVS
Saunders C GW0VRL
Saunders C G4KPS
Saunders C G4ZCS
Saunders C G7KGR
Saunders C M6SAU
Saunders D G0WFT
Saunders D G3OWE
Saunders D G4HZR
Saunders D M3BLF
Saunders D GM7RBW
Saunders D 2E0OAI
Saunders D M6DWS
Saunders G G3OYN
Saunders G G7UZX
Saunders G MM3IIT
Saunders G M6IFT
Saunders H G1OUA
Saunders J G3KCV
Saunders J G4XSI
Saunders J 2E0DQX
Saunders J M6JEO
Saunders J M6ODD
Saunders J G3OLU
Saunders K G0OYC
Saunders K G8SFM
Saunders M 2E0BDP
Saunders M 2E0FZH
Saunders M G3ORV
Saunders M G6IWC
Saunders M G0DAU
Saunders M M6RUN
Saunders N G1PJV
Saunders N GM6GMZ
Saunders N 2E0CSQ
Saunders P G4AQE
Saunders P G6HGU
Saunders P 2E0SPU
Saunders P M3SPU
Saunders R 2E0BJD
Saunders R G0ERY
Saunders R G1LQX
Saunders R G6VAX
Saunders R G7HHQ
Saunders R G6OUA
Saunders R M6REQ
Saunders S G6IKC
Saunders S 2E0CLH
Saunders S M6BGE
Saunders S GM7MTQ
Saunders S MM3MTQ
Saunders T M6DFJ
Saunders W G6IAJ
Saunderson T MI6WTZ
Savage A MM6YND
Savage A 2I0DUR
Savage A MI0OBR
Savage A MI6DUR
Savage B 2E0DCD
Savage B M0LMB
Savage B M6BOF
Savage C G7UII
Savage D 2E0IOU
Savage J G1AEB
Savage J G6VBD
Savage J 2E0VDM
Savage J M0VDM
Savage J M6BOJ
Savage J 2E0SVG
Savage J M6SVG
Savage M M3EQW
Savage M M0IBQ
Savage P G7LTG
Savage P M6YAS
Savage R G3LTX
Savage S 2E0BAP
Savage S MI6HLC
Savastano S M3ZII
Saveall J G1FEJ
Savegar J G4ZSG
Saveker C G8AMU
Saverton R G6DKS
Savery R M3SZO
Savigar R G6IBW
Saville A 2E0HSP
Saville A M3CMK
Saville G G3DNN
Saville G G1RNZ
Saville K M6RWS
Saville N M6MOG
Saville S M6FFM
Savin C G0IYT
Savin G G4OYP
Savin M G7KWD
Savin N G0SVN
Savin R G4XYN
Saving P M3PPR
Savory A M3SBA
Savory W GM4HUL
Saw R G0CJQ
Saward R M1DKF
Sawday K G8YZA
Sawdy J G3XJM
Sawford G G0GRS
Sawford J GD8CAB
Sawford L G6APD
Sawford M MW3PJU
Sawkins L M6LSJ
Sawkins R G0GWM
Sawyer A 2E0ALL
Sawyer C G0WWO
Sawyer C G1OLT
Sawyer D 2E0FXD
Sawyer D M6FXD
Sawyer D M6DXW
Sawyer D G4DAY
Sawyer G M1EZC
Sawyer G 2E0SSL
Sawyer G M6GSJ
Sawyer J M6DVY
Sawyer J 2E0VJO
Sawyer J M0JSX
Sawyer J M3VJO
Sawyer P G7LTP
Sawyer S 2E0NHR
Sawyer W 2E0LIW
Sawyer W M6LWS
Sawyers B G6ZKU
Sawyers K M3CFM
Sawyers K M3MDK
Sawyers M G4VHI
Sawyers P 2E0FHO
Sawyers S M5FAB
Saxby K G7JVO
Saxon D 2E0DKD
Saxon E GM8YRX
Saxon M G8HSS
Saxton C M6CVK
Saxton M M0AJD
Saxton R G4RRX
Saxton T G3LJR
Sayegh M G0UCC
Sayer E G6MVW
Sayer J G6XCC
Sayer R G0BCT
Sayer R G8IJS
Sayers A 2E0TGS
Sayers A G6KSV
Sayers A M3TGS
Sayers B G7AZJ
Sayers M 2E0CEQ
Sayers M G1VJE
Sayers P G7MJI
Sayers R G8IYK
Sayers T M3XWM
Sayle N M6XLR
Sayles C 2E0CVV
Sayles D G7WGX
Sayles D M0SAY
Sayles P M6PNY
Sayner T G0VWP
Sayre R 2E0RYS
Sayre R M0RYS
Sayre R M6RSY
Saywell M M6SSB
Scaife R G3RSB
Scaife R G7PAF
Scales T M6BHT
Scally J MM3VFK
Scambell L M6LAS
Scambell M M6FWE
Scandrett A G4KFC
Scanlain S GM1VAD
Scanlan J MM6CJC
Scanlin G GW0NXW
Scanlon T G7VJH
Scannell B 2E0DPD
Scannell B M6FCR
Scannell J 2E0HPF
Scannell S G3ZSU
Scaplehorn D G0NQJ
Scarce R G7RMQ
Scarcliffe J M6ZWP
Scarfe R G4TUK
Scarfe T G6JDW
Scargill D G0UND
Scargill G G8UAF
Scargill R G4MBE
Scarisbrick A G7ELG
Scarlett A M3NFU
Scarlett G G0SUI
Scarlett M G4CAK
Scarlett P M6PEQ
Scarlett R G4HZF
Scarlett W G3RXS
Scarr A G0LWU
Scarr G M3OBM
Scarr J G6LBL
Scarr M M3UAY
Scarr M G1XSV
Scarratt P M3XPS
Scarratt R G0WRE
Scarsbrook A G4AIW
Scarsbrook B G7TNT
Scarth A G4IZQ
Scates C G1NWG
Scattergood G GM6AAJ
Scattergood G MM0BSX
Sceal J G4ZBW
Schafers G GM1MXE
Schall N M3OJK
Schamp E G1MZH
Scheffer A 2U1EJF
scheffer J GU0GUX
Scherrer J G4ACP
Schiffeldrin G G4LYM
Schiffman C G8TND
Schlatter P G0NJQ
Schleswick J 2E0CTK
Schleswick J M6CDT
Schmidt A M3ICA
Schmidt H M0GPY
Schmidt K 2E0HLM
Schmidt K M0HLM
Schnaar H 2E0IGJ
Schneider K G1JEH
Schneider R G8FQN
Schnurr L G0AAN
Schoales R G4NYW
Schoenmaker P 2E0LGS
Schoenmaker P M6LUG
Schofield A G6APE
Schofield A M0PXY
Schofield A M6PXY
Schofield C 2E0DRG
Schofield C M0ICS
Schofield C M6GRD
Schofield C GM7KUN
Schofield D G7OKY
Schofield D GM8PAH
Schofield D M0DMS
Schofield D G4OJI
Schofield H G6JVX
Schofield I M0CGF
Schofield J 2E0BCS
Schofield J M3JUF
Schofield J M6PUN
Schofield K M6OJI
Schofield L M6EWU
Schofield L M3VSZ
Schofield N G4CAT
Schofield P G7VME
Schofield S G4PJT
Schofield S G4VMF
Schofield T G4ZLI
Schofield W G0BAK
Scholefield P G4TSF
Scholefield R 2E0DOG
Scholefield R M0UEZ
Scholes A G6OGT
Scholes D G7NHC
Scholes E G4KHG
Scholes G G8OHC
Scholey C G0CJV
Scholey G M6HWT
Scholey I M6WRX
Scholey R M0HDR
Scholey R M6EUU
Scholte B G1SIG
Scholz J M3YPD
Schonborn M M3IDJ
Schonborn S M3SPQ
Schoof G G1SWH
Schoof G M0AWX
Schoolar J G4LYY
Schoth B G3WJM
Schrager G G3XKY
Schranz P G7VGT
Schrier S G7KQT
Schröder H M0IBK
Schuler A 2E0BPI
Schuler A M3WEQ
Schultz V M6VSC
Schulz J G1JJQ
Schuy L 2E0CKX
Schwabe T M0TMS
Schwingen Z M6POC
Scivetti A G1PXQ
Sclater A G0JVL
Scleparis E G4YBX
Scobbie D GM6KDD
Scobie M M0IME
Scoble J M6PHY
Scoles W G1YMA
Scollay K MM6ZRX
Scopes T M6TSR
Scotcher D 2E0OTI
Scotcher D M3OTI
Scotching S 2E0FEB
Scotching S M3SQP
Scothern A 2E0HVB
Scothern D G1YFQ
Scothern J G6XYR
Scothorn G M0BJK
Scotland R M1ESV
Scotney J G6BKD
Scotney P G8BKD
Scotson S 2E0SWS
Scotson S M3NUI
Scott A G1TWT
Scott A G6MDS
Scott A G6NIZ
Scott A G6UGZ
Scott A G7MWM
Scott A G7RSK
Scott A G7TXF
Scott A G8BDX
Scott A M6GQU
Scott A M6HJK
Scott A G4TNU
Scott A MW6ADS
Scott B M6XBS
Scott B M6HKT
Scott B M6BSO
Scott B G6UHD
Scott C GM7CPL
Scott C M6ZEA
Scott D 2E0WTZ
Scott D G0KJP
Scott D G6HPK
Scott D G6LDP
Scott D M3ZDS
Scott D G8UAE
Scott D M6VBZ
Scott D G6MVR
Scott D G4OWQ
Scott D M3UWE
Scott E GM8RMR
Scott E M6EJS
Scott E M3WYV
Scott F G6UDB
Scott G 2E0GXX
Scott G M1FNE
Scott G G8TRY
Scott I M3LGI
Scott I M3OLF
Scott I M6IAS
Scott J G0HGH
Scott J G0JFP
Scott J GM3KJE
Scott J G7HOT
Scott J GM7UJJ
Scott J G8LCA
Scott J MM3WZL
Scott J GM0WRR
Scott J G1OWZ
Scott J G1UDR
Scott J GM4WZL
Scott J G4ZTF
Scott J M0RDX
Scott J M3SGJ
Scott J M3XAW
Scott K G0MCP
Scott K G4FOY
Scott K M0BPX
Scott K M1CYR
Scott K 2E0FHR
Scott K MM0KSS
Scott L MI3RIL
Scott L MM6GKE
Scott L 2E0PYN
Scott L M6AVG
Scott M G3LYP
Scott M G6XZA
Scott M M0AVU
Scott M 2M0INE
Scott M MM6PAU
Scott M G7IRN
Scott M M6IOL
Scott M M6MSI
Scott M MM6OAG
Scott N G6FLE
Scott P GM0VOU
Scott P G1RVP
Scott P G7DDD
Scott P M5SRE
Scott R 2E0DFE
Scott R 2E0GEZ
Scott R G0IAL
Scott R GM1VBD
Scott R G4HWW
Scott R G4KVQ
Scott R G4NLI
Scott R M0CRY
Scott R MM1RMS
Scott R M3BMU
Scott R MI3TRY
Scott R MI5RJS
Scott R M6EWJ
Scott S 2E0BNJ
Scott S G1PJI
Scott S 2D0DJX
Scott S MD0ZDJ
Scott S MD6XDJ
Scott S MM6TMS
Scott T G1GXF
Scott T G7VTJ
Scott T M0TJS
Scott T M0VTJ
Scott T M3BYS
Scott T MI3IAI
Scott T M6BDX
Scott W 2E0CIX
Scott W G0RWS
Scott W G1ZVO
Scott W G3FUJ
Scott W GM4YAU
Scott W GI7AXB
Scott W G7UNZ
Scott W M3OAG
Scott W 2M0CTB
Scott W MM6VWT
Scott Whittle H 2E0HSW
Scott Whittle H M0HSW
Scott-Brown J M3FNO
Scott-Dickinson P G0JRH
Scotter J G0MST
Scott-Green A G4GWR
Scott-Green K G4GUK
Scott-Martin M M3YYO
Scotton M 2E0BRA
Scott-Telford H G7UTB
Scovell P G4JDF
Scrase C G8SHF
Scrase C G4MRD
Scrase S G8MHN
Screen A M6EYF
Scrimshaw P M0HSG
Scriven J G4YNU
scrivener W G0BQC
Scrivens B M3NMX
Scrivens C M6SOT
Scrivens J G4IWU
Scrivens P G0HHL
Scrivens R G3LNM
Scroggins K G1TQH
Scroggs B 2E0YBS
Scroggs B G0BDS
Scroggs G G1IGN
Scrogie N G4CPQ
Scrutton A G1KCS
Scrutton A M3GIX
Scullion G MI6GGN
Scullion M 2M0OMS
Scullion M MM0OMS
Scullion R M6LVA
Scully J G3GUR
Scully J M0WYB
Scully K M3BWZ
Sculthorpe G G6TVI
Scutt P G3IBI
Seabourne F G0VQT
Seabridge C G0VRK
Seabridge D 2E0FBS
Seabridge R G4SEA
Seabright I G0WYT
Seabright V M6ILR
Seabrook A G3ZQB
Seabrook D G4LJG
Seabrook D G7WCG
Seabrook I M0CIY
Seabrook M M6HHE
Seabrook P M3PJS
Seabrook P M3PZX
Seabrook S M0ECS
Seaford P G8XTW
Seager C M0DPS
Seager J G6BHB
Seager J G0UCP
Seago A G4KDL
Seagrave M MW3YNA
Seal G G4RLO
Seal J G3UJG
Seal R G0RMP
Seal R MW3KNR
Seal S G1CUZ
Sealey D M6UUA
Sealey K M3KJS
Sealey P G6MOZ
Seals A G0TZP
Sealy R G4SRD
Seaman G G1MAV
Seaman I M3RZN
Seaman J GI4XAP
Seaman J M6MBK
Seaman P G1TIJ
Seaman P G3OTN
Seaman P M3GFW
Seaman V G4RAP
Sear L G3PPT
Searl M G6SBN
Searle A G3BDT
Searle A G6XYS
Searle C G1XYR
Searle D MM0HQD
Searle E G3VMY
Searle G G4GRR
Searle H M6HAS
Searle J G0LXC
Searle J G4ZAS
Searle J 2E0BXX
Searle J M6SJU
Searle R M3EMU
Searle S G0OWK
Searle T G7BAE
Searle T M6VXT
Searle T G4BM
Searles P G4PAS
Searles-Bryant S M6WTF
Searley M G0AVJ
Sears G G4WXK
Sears K G1RHB
Sears N G3YEG
Sears R G0AHL
Seath N G4FI
Seath N G7FUM
Seath S M3SNZ
Seaton J M6CFD
Seaton J 2E0JGS
Seaton J M0JGS
Seaton J M6JGS
Seaton L M6EOT
Seaton P M3UKF
Seaton P GM0AKJ
Seaton R G6VQW
Seatory H G1XWS
Seaward M M0SMJ
Seaward R G0TCA
Seccombe D G1DDS
Seddon D G4VCO
Seddon D M1BRX
Seddon F G3RPO
Seddon F G4WFK
Seddon H G1VYA
Seddon J G4GMI
Seddon J G4IJD
Seddon J G6LDY
Seddon K G0KHA
Seddon R G4VXW
Seddon R G6HPL
Seddon S M6SDU
Sedgbeer A G0MAS
Sedge R G7GRJ
Sedgebeer N MW3NCS
sedgebeer S GW3RVG
Sedgebeer V GW4TVU
Sedgley V G3YIC
Sedgwick L G8XOV
Sedgwick N G4HDL
Sedman A G3LAA
Seear J 2E0JCR
Seeby D 2E0IHJ
Seed B G7TPS
Seed T M0FGC
Seedhouse D 2E0DPS
Seedig A 2E0SEE
Seedig A M6AZE
Seedle B G3UIT
Seeds A G8DOH
Seeley R 2E0RDE
Seelig A 2E0DOP
Seelig A M6ENK
Seeney C G6LDO
Sefton J M3JAC
Sefton M G0CYL
Sefton P M0PCS
Sefton W G7PVZ
Segal D G1DMS
Segal L G6XLL
Segar A GM0SXO
Seggar S G0NQY
Segrove J M3PMI
Seidner H G0ROA
Seitz P G7HTN
Seivwright D MM6AIK
Sejwacz A G7MFA
Sejwacz J M3HGZ
Sejwalz D M3HHQ
Selby I G4DRI
Selby P M6HAC
Seldon C G3SCY
Seldon H G7NIA
Seldon H M0BWN
Self A M6GBQ
Selfridge H 2I0RHQ
Selfridge J GI1JHQ
Sell D G1JQR
Sell F G0GGH
Sell G M0CKL
Sell J M6BLB
Sellar L G8XMS
Selleck K G3SNU
Sellen D G3YAJ
Sellers A G8TZJ
Sellers B 2W0PEH
Sellers B MW0PEH
Sellers B MW3PEH
Sellers J G3VDE
Selley M G1THJ
Selley P G7FJZ
Sellick A G7LPB
Sellick J G8LCZ
Sellick P M6HEG
Sellick S GM1LXA
Sellman R G4CRK
Sellman S M0AYA
Sellors A 2E0BYK
Sellors C M3KUM
Sellwood D G4CGA
Selman L G0LYQ
Selmes A G4HGH
Selmes A G4KLF
Selvey A M3VEY
Selvey M M3UXX
Selway L G7POI
Selwood P G3YDY
Selwood P G8BGV
Selwood P M6KEG
Selwood R G7AUU
Selwood R M3AHJ
Selwyn J G7UDU
Selwyn M G3TLD
Selwyn S G7GZU
Selwyn-Smith C M0CGW
Semark A G4OJN
Semark A G8LUP
Semmens N G3PUQ
Semple A 2E0HTY
Semple G MM6GGE
Semple J GI3OYG
Semple M MI3CIV
Semple R 2E0HUB
Senft R G0AMP
Sengupta P G1YJY
Senior A M3VFL
Senior B G8YGT
Senior C M0GLQ
Senior D M6NCD
Senior D G4MIB
Senior G G3HRU
Senior K 2E0XDJ
Senior M GM3PAK
Senior M GM1DCB
Senior M G4EFO
Senior N GM1ZOX
senior P G0CSK
Senior P G4JNL
senior P G6ZFX
Senior R G4IEZ
Senior S M3YYG
Senitt J G1PKV
Senter I G6RDD
Seo R M0SEO
Sephton A G3IJL
Sephton I M3AOQ
Sephton P M0KDM
Sephton P M6MRD
Sercombe A G7RBS
Sergeant A GM4HRL
Sergeant D G3YMC
Sergeant F G4HTE
Sergeant J 2E0EXY
Sergeant J G8YCP
Sergeant J M3BAN
Sergent P G4ONF
Serlin J G3TLU
Sermons A G7LCD
Sermons B G8IUD
Sermons C G1IUD
Serplus C GI7LCQ
Serridge J GI0VVC
Service N GM0TXJ
Serwa E G8ZIK
Sessions G 2E0FWX
Sethuraman S M3VRL
Seton M G4PLV
Setter B G4YAQ
Setter P 2E0SET
Setter P M6NFW
Setterfield D G1SYZ
Setterfield D G6LLU
Setterfield W G2IF
Severe L M0WAE
Severn D G8LNG
Severn M MI3SEV
Severn P G1NHX
Severs M G6UJR
Seward B M6BTS
Sewell B M6BSC
Sewell D G4FVK
Sewell D 2E0DWS
Sewell D M0WCS
Sewell D M6BUI
Sewell J M3XFS
Sewell J MD0SEW
Sewell K 2E0KSW
Sewell K M6KVN
Sewell M G7PBO
Sewell M M3TWY
Sewell P 2E0PES
Sewell P M6DIU
Sewell S G4VCE
Sewell S M3JWZ
Sex M G4AZQ
Sexton C G0TEB
Sexton D M0IZS
Sexton J G6ZWZ
Sexton R G4IZS
Sexton T G1KUN
Seymour A M6AEU
Seymour C M6HNF
Seymour D G7VCY
Seymour E M6ANM
Seymour J G0CWK
Seymour J M6ANO
Seymour P M3SRY
Seymour R G1JOO
Seymour T G4ZYZ
Seymour-Smith G G6REW
Shackleton A G1UVE
Shackleton B 2E0YKS
Shackleton B M3YKS
Shackleton C M5AEH
Shackleton D M1DHT
Shackleton E G0EQH
Shackleton H 2E0HFV
Shackleton L M6YOS
Shackleton T G6PSZ
Shackley N G0OSX
Shadbolt P G6LBJ
Shadbolt R M3GZW
Shaddick E G7EIS
Shaddick R M1EPN
Shades J GM0EPO
Shadwell W G6EUY
Shafarenko A M0SFR
Shafto K GW8KEV
Shail K G8AXV
Shailes S 2E0SBD
Shailes S M6MAS
Shajan N 2E0FMX
Shajan N M6CZB
Shakeshaft S G4YNS
Shakespeare F G3HKN
Shakespeare G 2E0VCY
Shakespeare G M3VCY
Shakespeare S M6LIP
Shalders A G6MPN
Shallcross L M3VYK
Shalley C G0FDD
Shaman T G4DXT
Shambhu S M6IAF
Shambrook P G7AKI
Shambrook W G8SUJ
Shams-Nia R G7JUZ
Shams-Nia R M3JUZ
Shanahan J M6CYX
Shand G G7DZR
Shand L 2E0WSS
Shand L M3WSS
Shand R GM4RAI
Shane B G1JOL
Shane C M0NTY
Shankland J MM1JAS
Shanklin E G0UWA
Shanks C G4MFR
Shannon A G3KKJ
Shannon D M6EFC
Shannon J M3XPN
Shannon M G6NXV
Shapero P G4NCK
Shapland R G4ATB
Sharam A 2E0CJV
Sharam A M6AYL
Shard R G4XIE
Shardlow J G4EYM
Shardlow M G3SZJ
Share J G3OKA
Sharif A 2E0MEU
Sharif A M6MEU
Sharif C M6CSU
Sharkey J G1BBH
Sharma A M6EGU
Sharman A G7GPU
Sharman A G0UWS
Sharman B G0FBL
Sharman I M6RDF
Sharman J G3YCH
Sharman J G4TKO
Sharman K G0TQC



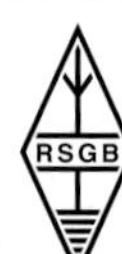

Sharman K G7VJA
[illegible]
Sharman M M6UWS
Sharman M M6JAV
Sharman R G1NGR
Sharon C G4YUF
Sharp A G4CWY
Sharp A G6AJS
Sharp A G7PKK
Sharp A 2E0SHP
Sharp A G0JIA
Sharp A M6SHP
Sharp C G7BBU
Sharp C MM6KLT
Sharp D M0XDS
Sharp D M3CJI
Sharp D G8FWC
Sharp E G3RDN
Sharp G M0RED
Sharp G G8ISE
Sharp I G7MZY
Sharp J G0JYK
Sharp J G1NWQ
Sharp J GM4DEX
Sharp J M3ABQ
Sharp M G0DWZ
Sharp M M6WFH
Sharp O MM6FJS
Sharp P G1KNU
Sharp R G1GGT
Sharp R G6RMV
Sharp R G4VNR
sharp R G4MNB
Sharpe C G7WCP
Sharpe C G0WYI
Sharpe D G8HUO
Sharpe D G8NIE
Sharpe D G3ZUN
Sharpe D M6ILM
Sharpe G G8RFW
Sharpe I M6RTG
Sharpe J G1WWP
Sharpe J G1ZGH
Sharpe K 2E0KAS
Sharpe K M0KAU
Sharpe K M6KAN
Sharpe M G0BUB
Sharpe N M3NPS
Sharpe P G6KCG
Sharpe P G8RRC
Sharpe R G4FKY
Sharpe R G4LQH
Sharpe R M3DZQ
Sharpe R G6IIF
Sharpe W GI4ILZ
Sharpen D 2E0DCS
Sharpen P G3XXG
Sharples A MM6YQP
Sharples D G4ZMB
Sharples G G3YYC
Sharples G G7GTU
Sharples I 2E0IDA
Sharples J M5KEN
Sharples K G7LPW
Sharples N G4NFS
Sharples R G3PKD
Sharples S G4UNE
Sharples W G4MPI
Sharps L G4OHP
Sharrad J M6IFF
Sharred D G3NKC
Sharred S G4JGV
Sharrock P M3PPQ
Sharrott I G0CND
Shasby L 2E0WIE
Shasby M 2E0ETA
Shasby M M0GLI
Shasby M M3RYO
Shatford J 2D0LBG
Shatford J MD0JNS
Shatford J MD6JNS
Shattock A G4RTP
Shaughnessy T M3JKG
Shaul S 2E0SIY
Shaul S M0SIY
Shaul S M3SIY
Shave W G8OYL
Shaves P M3UMV
Shaw A G6CMN
Shaw A G6RDO
Shaw A M0GNA
Shaw A M0GOQ
Shaw A GW1ALV
Shaw A G0IPG
Shaw A 2E0GDT
Shaw A G0HSD
Shaw A G6OTZ
Shaw A M0GGQ
Shaw A M0UKO
Shaw A M6UKH
Shaw A G1KXX
Shaw B G1IQK
Shaw B G6HFS
Shaw C G4EKJ
Shaw C G6EUI
Shaw C M6EOA
Shaw C G7TJQ
Shaw C 2E0JWS
Shaw C G6XZC
Shaw C G8SMZ
Shaw C M6EMS
Shaw D G0PUD
Shaw D G6AVN
Shaw D G6OTE
Shaw D G8MDG
Shaw D G8RJO
Shaw D M0DSI
Shaw D M3FVW
Shaw D M3LYX
Shaw D M3RJO
Shaw D G0TXJ
Shaw D M5TXJ
Shaw D G3PCL
Shaw E G7DWH
Shaw E M6GUU
Shaw F G6NIW
Shaw F GM0CDC
Shaw F G3NXS
Shaw G G8ZUI
Shaw G M0AQA
Shaw G MM3GSL
Shaw G M6GJS
Shaw G 2I0GTO
Shaw G MI3GTO
Shaw I G4MWD
Shaw J 2E0WIG
Shaw J 2E0FIX
Shaw J GM0EDQ
Shaw J G0ERI
Shaw J G3CAZ
Shaw J G3YFE
Shaw J G4ETI
Shaw J G4RAJ
Shaw J M6JKS
Shaw J G1XFM
Shaw J 2E0PWD
Shaw J G3ZKZ
Shaw J G4NKI
Shaw J M6JEM
Shaw K 2E0BRQ
Shaw K M3SHW
Shaw K M6KHS
Shaw L G8YMU
Shaw L M6LOZ
Shaw L M6LSH
Shaw M 2E0WPE
Shaw M M3HXO
Shaw M M6SHA
Shaw M G3UYB
Shaw M G4EKW
Shaw N G6CFU
Shaw N M3OFA
Shaw P G0KHK
Shaw P G0UFV
Shaw P G8UHT
Shaw P M0EHS
Shaw P M0PSS
Shaw P M3WDK
Shaw P G4MSP
Shaw P M6DKL
Shaw P 2E0PRS
Shaw P G1YSR
Shaw P M0YES
Shaw P 2E0PSR
Shaw P G6PMR
Shaw P M0NAL
Shaw P M6GAE
Shaw S MM0SVE
Shaw S M3GUO
Shaw S 2E0SCV
Shaw S M6BKI
Shaw S M6PHF
Shaw V G4SOC
Shaw W GD0STR
Shaw-Ashton I G1ZAG
Shayler P G6TSF
Shayler P M3TSF
Shea D G1CAY
Shead A G6RGI
Shead F G4VVQ
Shead G G4WXT
Shearan C 2E0HNX
Sheard A G0CEF
Sheard A M0UUU
Sheard P G3YCJ
Shearer A G1TAY
Shearer A GM3SWK
Shearer B MM1HMV
Shearer C G4IEG
Shearer M G4KVU
Shearer M MM3MHS
Shearer S G6DWS
Sheargold P M3VOB
Shearing C G0KVX
Shearing N G8XGW
Shearing R GW6AYR
Shearing S M1ACJ
Shearman A MM3OYL
Shearman M M1BGF
Shearme J G2SH
Shears E M1EYS
Shears J 2E0DUH
Shears J M6FKG
Shears L G1RIR
Shears N M6GWI
Shears R G8IWT
Shears R G8KW
Shears R G8SHE
Shearson L M6HTZ
Sheath C M6UWO
Sheath S M3MUB
Sheather E G0UOD
Sheehan D M6DSH
Sheehan J G4GWI
Sheehan K M1EZK
Sheen N G0JLL
Sheen R M6VRO
Sheer H G4YAZ
Sheffield I GM3VEI
Sheffield P 2E0PSO
Shekhdar H 2E0MCC
Sheldon A G0VSP
Sheldon K G4NIJ
Sheldon K M3OSW
[illegible]
Sheldon W G0ZDJ
Sheldrake J G0IVV
Shelford R G6BRV
Shelford R G7HMI
Shell M GM0IET
Shellam A M3FYZ
Shelley B MW0RLD
Shelley G G7MSQ
Shelley M M6HXD
Shelley M GW3XJQ
Shelley N G4JYP
Shelley R G3NZY
Shelley R G8VVY
Shelley S M3NSH
Shelley T G0MFV
Shelley W 2E0SHY
Shelley W M6ACA
Shelswell A M0BPW
Shelton B G0MEE
Shelton C G4PCP
Shelton G G0MMD
Shelton M 2E0SFT
Shemeld D G8YXI
Shemming H M1DGY
Shemwell J M3OPN
Shenfield S G6VQV
Shenstone S M5BFL
Shenton K 2E0KMS
Shenton R M0SRJ
Shenton S G4VFJ
Shephard B 2E0UBN
Shephard B M0ORY
Shephard B M6UBN
SHEPHARD D 2E0GKE
Shephard D M6DCS
Shephard J M0TYK
Shephard S M6SSD
Shepherd A 2E0LSB
Shepherd A G3RKK
Shepherd A GM8BSQ
Shepherd A M3OLU
Shepherd A M6OOM
Shepherd B M0ARK
Shepherd B M3FZO
Shepherd C M6BPU
Shepherd C G0COL
Shepherd D G3LCS
Shepherd G G8BKH
Shepherd G M6PGC
Shepherd H G0KMW
Shepherd I G4LJF
Shepherd I G4EVK
Shepherd I G8ZZW
Shepherd J G0NQI
Shepherd J G0SEB
Shepherd J G1SDH
Shepherd J G8GET
Shepherd J M3SEJ
Shepherd J M6BUE
Shepherd K G0ADC
Shepherd L M0DDU
Shepherd M GM0PRQ
Shepherd M M6GMV
Shepherd M 2E0CNE
Shepherd M M0HGS
Shepherd M M6AWL
Shepherd N 2E0GPY
Shepherd N G8ADZ
Shepherd N G4AWW
Shepherd N G8EIN
Shepherd P G4ZMN
Shepherd P G7DXV
Shepherd R G8WCH
Shepherd R 2E0GWX
Shepherd S M0SHP
Shepherd T M1BHZ
Shepherd T GM4BUA
Shepley M M6MJZ
Sheppard A 2E0YAP
Sheppard A M3YAP
Sheppard B G4CVF
Sheppard C G6VSY
Sheppard C MI6LGX
Sheppard C G7BNO
Sheppard D M0DXS
Sheppard E 2E0FTF
Sheppard G 2E0ENF
Sheppard G M6BUJ
Sheppard I M0GXO
Sheppard J G4WOD
Sheppard K G7BNN
Sheppard M M3KSK
Sheppard M M6NHT
Sheppard M 2E0CPC
Sheppard P G4EJP
Sheppard P G6UTT
Sheppard P M3PSD
Sheppard R G8CQC
Sheppard S M6STJ
Sheppard T M3GQM
Sheppard V M6VNT
Sheppeck J G0AMY
Shepperd W G3WMA
Shepperley R M1MRS
Shepperley R M3SMR
Sheratte H G0KVU
Sherbourne J M0CVB
Sherburn P G7PBC
Sherdley P G3LPL
Sherer A 2E0OZQ
Sherer A G3TEU
Sherer I G8RQH
Shergold J G6LVC
Shergold J G8WQW
Shergold K G8RCE
[illegible]
Sheridan A G4FUW
Sheridan C MM0FCM
Sheridan D G7RYL
Sheridan N 2E0NJS
Sheridan N M0NJS
Sheridan P G4SHB
Sheridan T M0GCA
Sheriden N G0LWG
Sheriff M GM4ZVF
Shering G G8VHX
Sherlock M G4ZYN
Sherlock R G6RQJ
Sherman A 2E0GEV
Sherman A M3YFH
Sherman B G6EPU
Sherman K G7VFL
Sherman T G8XAJ
Sherrard R GI3VAW
Sherratt A G4TGM
Sherratt A G7ETM
Sherratt L G6IWD
Sherratt M G7MFZ
Sherratt P 2E0OBI
Sherratt P 2E0GSC
Sherratt P M3UHN
Sherratt S G8FAK
Sherrey M 2E0FKV
Sherrey M M0SKV
Sherrey M M6FKV
Sherriff B G0RNV
Sherriff T G0CHV
Sherriffs W GM8GHV
Sherry G M0RRY
Sherry J GM0AZC
Shersby B G0ISL
Shersby J G3TVD
Sherwin A M6EQJ
Sherwin D 2E0DHS
Sherwin D M3DHS
Sherwin E 2E0EWS
Sherwin K G0WGI
Sherwin S M0BKN
Sherwood B G0GVX
Sherwood C 2E0HQJ
Sherwood C G4TMG
Sherwood L G1ZST
Sherwood N G0CMK
Sherwood P G1KSH
Sherwood P G4AUY
Sherwood P MW6BEG
Sherwood R G8GEE
Shettler J 2D0RNR
Shettler J M6JCS
Shevchenko T M6THS
Shewan I GM4DZM
Shewan J G3UZB
Sheward L G6UUQ
Shewring M G8SIT
Shewry M 2M0SWY
Shibata M M0WXO
Shield E G8GVN
Shield M GM0PRQ
Shield P G8BXM
Shields A G0WYM
Shields B G3OIH
Shields B 2E0DCV
Shields B G7SJX
Shields B M3SJX
Shields D 2E0JWJ
Shields D G8XUW
Shields D M3JWJ
Shields F M3LZL
Shields J MM3DOP
Shields M M3YBU
Shields P G7IIB
Shields P M0USY
Shields P G4PJS
Shields S M6MJS
Shillabeer A M3SHI
Shillabeer S M6SHI
Shilliday A 2I0CBV
Shilliday A MI3YTH
Shillington D G8OKZ
Shillito P G8GVW
Shillitto H G0CSS
Shilson J G1IHY
Shingler D G4TCX
Shingler J M3FIP
Shingler M M0KDU
Shingles O G1MUC
Shinn F G0JKG
Shipham M M3XHT
Shipley S G0PQX
Shipman H 2E0RDZ
Shipman H M0RCZ
Shipman R M6RFS
Shipp A M0CFG
Shippen D G6LDM
Shipperley A G0CTD
Shipperley G G1EYG
Shipney B MM0DNH
Shippey R 2E0HAS
Shippey H M3RBX
Shipton B G7FEQ
Shipton D 2E0BTZ
Shipton E G0DSJ
Shipton R G4KYI
Shiradski C G6APH
Shiradski I M6MID
Shireby R G0OJF
Shires D M3KRX
Shires I M6IAN
Shires K G1SPX
Shires P M6MEB
Shirlaw P 2E0BRJ
Shirley C M3CGS
Shirley D G0VXX
Shirley D MM1AFT
Shirley E M3RVK
Shirley F M3FES
Shirley R G7MGG
Shirley V G1OYF
Shirley V G0ORC
Shirras S G0BRX
Shirtliff P G8MED
Shirville G G3VZV
Shirvington R G0ICC
Shirvington S G1VXS
Shitov A M6EUV
Shiu W M3OSY
Shockness M M6RUK
Shoesmith C G0SLH
Shone A G4VWX
Shone A M5ALA
Shone D G4FBG
Shone L G4XME
Shone S M6HCJ
Shonfield P G1OAN
Shons A G1PPD
Shoosmith M G1IJM
Shooter R MD6RAQ
Shopland M 2E0CLD
Shopland M M0OGI
Shore D G0EOJ
Shore M G0NNO
Shore M M6KES
Shore P G8IYE
Shore R G8RAJ
Shorey S G3ZPS
Shorland M G3WIK
Short A G2DGB
Short B G3YEU
Short C 2D0SCJ
Short C M3SCJ
Short E M6ZEK
Short H M3WQL
Short J G4NCJ
Short J G4SMT
Short J G1DJI
Short L 2E0JDC
Short P G0UFB
Short P G4WJJ
Short P G8RFD
Short P M6EDM
Short R G3GNR
Short R G8OFO
short S M6HYX
Short T G6FFL
Short W G0LWE
Short W G3BEX
Shortall M M6CFL
Shorten D G7SRB
Shorter M G8YMN
Shorthouse J G4FPA
Shorthouse J M0AGU
Shortland M G0EFO
Shortland P G0AGD
Shortland S 2E0SSJ
Shortreed M 2E0RSM
Shotter J G0UZY
Shotter M G0CWX
Shoubridge D M3LUO
Shoubridge R M3TFS
Shoulder R G8FJG
Showell J M3FLV
Shoyer M 2E0CCW
Shrago M G1WVD
Shread M G6TAN
Shreeves S 2E0GCH
Shreeves S M3GCH
Shrewsbury A G3KAN
Shrimpling B G7VKY
Shrubsall J G7GSR
Shrubsall W G0UKJ
Shufflebotham J 2E0CGX
Shufflebotham J M3NBU
Shulver P G8TXK
Shulver R M6HQM
Shurety C G7NIX
Shurley M M0TVG
Shurmer J G7BZR
Shurmer P M3ZPZ
Shute D G3GGN
Shute J G3GTA
Shuttlewood P G0INJ
Shuttleworth C G4LKZ
Shuttleworth D 2E0LXF
Shuttleworth D M0PSY
Shuttleworth D M3NPO
Shuttleworth D M6ELN
Shuttleworth K M0KHS
Shuttleworth M M0EBR
Shuttleworth P G6EPX
Shuttleworth R M3EXY
Siarey G G1FJO
Sibert P GW8SBO
Sibley A 2E0XVZ
Sibley A G0DDW
Sibley A M0GVQ
Sibley A M3LLN
Sibley C M6KSA
Sibley D 2E0SIB
Sibley D M6SBZ
Sibley D G8MAR
Sidaway K M6KAQ
Siddall A 2E0LJR
Siddall A M6ZZV
Siddall B G7USQ
Siddall J 2E0IBP
Siddall M M6MBO
Siddall R G4DIT
Siddle A G7TBX
[illegible]
Siddle J M6EJP
Siddle M M3YDI
Siddle-Ward A M3DEL
Siddons A G4ZDQ
Siddons C G1SID
Side J 2E0ALH
Side J M3IYY
Sides M 2E0VMR
Sidey C M0PIP
Sidhu-Brar A 2E0SHV
Sidhu-Brar A M0SHV
Sidnell J G0DFF
Sidney C G0LXI
Sidney C G4RJY
Sidwell P M0VEY
Sidwell R G1SKW
Sidwell R M6RDZ
Sidzhimov F M0SFI
Siebert A M3FJN
Siebert H G3BFL
Siebert L M3FJR
Siebert R M3HAK
Siemieniago A G4LZZ
Sieroslawski A G4XKC
Sierota A G8LVF
Siertsema W G3KCZ
Sievert R G1GMF
Sifford G G4TGG
Silburn D M0RSA
Silcock R 2E0CZK
Silcock R M6CYF
Silcocks B G7NQJ
Silcocks R G1CZW
Silcox D GW6RWJ
Silcox R 2E0BLW
Silcox R M0LHS
Silcox R M3VHI
Silk S 2E0GTF
Silkstone D 2E0WFM
Silkstone D M0VGA
Silkstone D M6WFM
Sillars D G4IKY
Sillars E G8FMA
Sillence A G4MYS
Sillence C G6WZD
Sillito R G0NXC
Sillitoe J G4MTG
Sillitoe J G8TXL
Sills J 2I0VFO
Sills J MI0GTM
Sills J MI3OZK
Silva O 2E0OPO
Silva O M6DKY
Silver B G8JLB
Silver M 2E0GSK
Silver M M0GSK
Silver M M6GSK
Silver S M3PFU
Silvera R G4WDS
Silvers S M3KKB
Silvers T G6OYV
Silversides R M3SHK
Silverson D G2DPY
Silverton D G7KME
Silvester W G4DAQ
Silveston D G3GDH
Silveston N M3ILB
Silvey C G1VJJ
Silvey R G4YQL
Sim A 2M0VNW
Sim A MM3VNW
Sim G G0TLN
Sim J G8VIB
Sim J M6HHF
Sim K M0BKS
Sim K 2E0KVK
Sim K M0KVK
Sim K M6EYR
Simarpi J G6SIM
Simcock C M1SIM
Simcock D M1ANT
Simcock G M3VNP
Simcox P M6PAS
Sime P GM8YGI
Simes S G6DKK
Simister B G4OOC
Simkin C G0REC
Simkins C G8FKP
Simkins C G8IYS
Simkins J 2E0OIJ
Simkins J M6SIJ
Simkins M M6HOI
Simkins M G7OBS
Simkins M M3OBS
Simlat J M1DHW
Simmen P M6EHQ
Simmens M G4YQP
Simmons A GM1HNZ
Simmonds A G0HND
Simmonds B 2E0BRS
Simmonds B M3PYW
Simmonds D G3JKB
Simmonds J G1YFI
Simmonds J G3GLX
Simmonds J G4XNW
Simmonds J G6MVQ
Simmonds K G0PIB
Simmonds K G6TWR
Simmonds N 2E0TVS
Simmonds N M0KZP
Simmonds N M6TVS
Simmonds N G4MPH
Simmonds P M6WAX
Simmonds P M6PLH
Simmonds R G0RSS
Simmonds R M3NOA
Simmonds S M4GJY
Simmonds S G6WLM
Simmons A G1TID
Simmons B G6SDC
Simmons B G8XJB
Simmons D G1MAL
Simmons D M3DWV
Simmons D 2E0AAI
Simmons D 2D0EDL
Simmons D M0RIU
Simmons D M6KND
Simmons D M6LDE
Simmons D G1UKS
Simmons G G0VXM
Simmons H G0XAZ
Simmons J G8SXU
Simmons J M3KPF
Simmons J G6MPE
Simmons K M3RKE
Simmons L M3SIS
Simmons M G0LRW
Simmons N GI7GJX
Simmons P GW0VYF
Simmons R M3ZWH
Simmons S G4HXY
Simmons T G6YRJ
Simmons T G8JFX
Simmons V G6TSC
Simms B G8GYV
Simms D G4FQZ
Simms P G1CQF
Simms P G6HPE
simms R 2E0RSI
simms R M3NNG
Simms R M3RLS
Simms S M3SXV
Simon E GM4GZW
Simon G GU7CMH
Simon J M0KWP
Simon M MM3NQT
Simon P G4JTX
Simonds A G8KUV
Simons C G0KEE
Simons D G7DKB
Simons J G3ZVK
Simons J G8HWI
Simons L 2E0ABT
Simons L M6LTM
Simons M M6JTL
Simons P G4CCZ
Simonsohn M 2E0HVE
Simonsohn M M6HVE
Simpkins C G3MCL
Simpkins D G6VOE
Simpkins G G8EOH
Simpson A M3ARS
Simpson A G3UMF
Simpson A 2I0MVP
Simpson A MI0MVP
Simpson A G6IKH
Simpson B G3PEK
Simpson B G8PZF
Simpson B G1KWA
Simpson C G7LCW
Simpson C M0DMB
Simpson D GM3LVA
Simpson D G4BZL
Simpson D GI4PRH
Simpson D G6NGN
Simpson D G6PII
Simpson D G8NDF
Simpson D G8OCS
Simpson D M6WAS
Simpson D M3UIC
Simpson D MM6ACI
Simpson D MI6DVM
Simpson D 2E0CIJ
Simpson D M0HDT
Simpson D M6AZV
Simpson D G6ZGF
Simpson E G0LES
Simpson E 2I0PRM
Simpson E MI0PRM
Simpson E MI3WWJ
Simpson G G4LWQ
Simpson G M3BTG
Simpson H GM1FTZ
Simpson I GM3YND
SIMPSON I GM4MBG
Simpson I G7HGF
Simpson J G0BQP
Simpson J GI0WYB
Simpson J G4GFV
Simpson J GM4YZT
Simpson J G4BUI
Simpson K G0BUC
Simpson K GW4JGW
Simpson K G4YPQ
Simpson K G8XBV
Simpson L M3WSQ
Simpson M 2E0GLT
Simpson M G0UHF
Simpson M G3UVM
Simpson M G8BOI
Simpson M M0MSS
Simpson M M1APX
Simpson M 2E0UHL
Simpson M G1WRD
Simpson M M6BQE
Simpson P G3GGK
Simpson P G3XQZ
Simpson P G7SEY
Simpson P G8FQS
Simpson P G0RUR
Simpson P G1KGC
Simpson P 2E0PSX
Simpson P M0HIU
Simpson P M0PIU
Simpson P M6PXS
Simpson R G0KDG
Simpson R G3OMS
Simpson R G3UWE
Simpson R GM7NZI
Simpson R 2E0RGS
Simpson R G1DGL
Simpson R G4NOP
Simpson R M0RBQ
Simpson R M6AOV
Simpson R 2E0KDT
Simpson R MI6GNF
Simpson R M6KVT
Simpson S 2E0FON
Simpson S G4ZSS
Simpson S M1BVX
Simpson S MM3PXG
Simpson S G4SGD
Simpson T G3NSF
Simpson V GI4BDL
Simpson V GM6UJG
Simpson-Fraser A GM6FLL
Sims A G4VHV
Sims A M3APA
Sims B 2E0BXS
Sims B M0GQP
Sims B M3YZI
Sims C 2E0NII
Sims C G1ILY
Sims C M3NII
Sims C M6UFO
Sims E G6TVD
Sims G G4GNQ
Sims H G0LJS
Sims J G1UEQ
Sims M 2E0MSI
Sims M M0IMS
Sims M M6IMS
Sims M G7RBA
Sims M 2E0UGF
Sims M G7LCV
Sims M M0WHO
Sims M M3UGF
Sims R G4VNS
Sims R G4CVX
Sims S G8NFZ
Sims S M0ZZE
Sinclair A G0AMS
Sinclair A GM7HUD
Sinclair A GD3TNS
Sinclair B G0MBS
Sinclair B G1GRM
Sinclair C M6CLZ
Sinclair D G4BFV
Sinclair D G8PGE
Sinclair E 2E0HZI
Sinclair G 2M0GXZ
Sinclair G GW6APK
Sinclair G MM0KGS
Sinclair G MM6GZS
Sinclair G M6LAZ
Sinclair H GI4GOS
Sinclair I G1GIA
Sinclair I GM1GXH
Sinclair I 2M1ELU
Sinclair J GM4YAA
Sinclair J M6SNC
Sinclair J MM3YSJ
Sinclair J MI0ZSC
Sinclair M G4BVF
Sinclair N 2M1AVZ
Sinclair N 2E0CNH
Sinclair N 2E0TLD
Sinclair P GI0UQK
Sinclair P G3UCA
Sinclair S G4EKF
Sinclaire M M6CAG
Sindall R 2E0ETD
Sindall R M6ETD
Sinderbury S M6UFR
Singam K MD1FFR
Singer H G0IAA
Singer I G7VDA
Singer I M0BMJ
Singer M 2E0MWN
Singer M M0FJN
Singer M G7HYM
Singer R M6HAK
Singfield C M6NUC
Singfield M M6CZA
Singh A M6DFQ
Singh D M6ING
Singh R M3URS
Singh-Gill M G4ZZV
Singlehurst-Ward I M0ISW
Singleton B G0IVO
Singleton C G0FOQ
Singleton D GI3UZJ
Singleton E M6FCT
Singleton G M6NFF
Singleton J G4WJR
Singleton J G6EQB
Singleton K G7MCK
Singleton P 2E0EBN
Singleton P M6EBN
Singleton R G7OXP
Sinkinson E G4KFH
Sinton D 2I0VOQ
Sinton M MI3XIU
Sinton R GI3ONF

UK Surnames

**IMPORTANT NOTE**

**Revalidate licence to avoid revocation** – Ofcom has advised the Society that plans will be drawn up to revoke licences that have not been revalidated as required by the licence conditions. The quickest way to revalidate is to do so online via the Ofcom website: *https://services.ofcom.org.uk/* or by email: *amateur.validations@ofcom.org.uk* If you need assistance in the process, Ofcom staff are available to help, but please be patient during times of heavy workload.

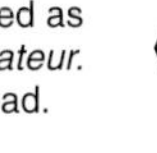

Sircombe I G0FCM
Sirignano B G4FZG
Sirkett N G1FPP
Sirley I G4OLZ
Sisley M G8ZID
Sismey T G4XQV
Sissens S 2E0CZG
Sissens S M0ZSM
Sissens S M6GEK
Sissons C G4PXN
Sivapragasam J GD4MSE
Sivapragasam N GD4IQD
Sivaraman Valsala S M6JYO
Sives A GM6OGN
Sives M MM6SIV
Siviter D 2E0IHE
Siviter D G7EWD
Siviter F M3FXU
Siviter G M1FFG
Siviter K M3LDX
Sivyer A M3ZNM
Sivyer F G6SQS
Sixsmith P G0JFE
Sizer P GW6OTD
Sizmur S G6YRI
Skakle B GM1INS
Skarzynski A 2E0SKI
Skarzynski A M0JEK
Skarzynski A M6ABS
Skates D M6DAU
Skea E MM0CMO
Skea G 2E0GKS
Skea G M0GUK
Skea G M6GKS
Skeels W G4HOI
Skeggs I M6FZC
Skelcher C G3YHF
Skelhorn H G8BPU
Skells R G0XTA
Skells R G8YXJ
Skelton B G4VSK
Skelton M M1EZJ
Skelton R GW7NVM
Skeoch I GM7NAA
Skerratt S 2M0PCW
Skerratt S MM6HSS
Skerritt P G6ZYZ
Skerritt T MW6TJS
Skerry K M3YWH
Skertchly J G6JPS
Skerton T MD6FFP
Sketcher B G0PVB
Sketchley J G4DCE
Skewes M 2E0CWE
SKIDMORE B GW4ERB
Skidmore D G7VJD
Skidmore J G8ZCJ
Skidmore K M1DDI
Skidmore P 2D0SRP
Skidmore P MD0SRP
Skidmore P MD6PRS
Skidmore S GM7RQK
Skidmore W G6CSC
Skillen J GI4TSK
Skillings C G0FKG
Skillington F G4DFU
Skilton D G1KPU
Skingley R G6IWK
Skingley R M6RLQ
Skinley M M0MAX
Skinner A G0BIR
Skinner A M3JGJ
Skinner A MD6CIX
Skinner C M3VIO
Skinner D 2E0BBG
Skinner D M3IXU
Skinner E M3MHP
Skinner F G0EHQ
Skinner I 2E0BOG
Skinner J G7VSM
Skinner J M3JFS
Skinner J 2E0ZOM
Skinner J M3ZZS
Skinner L G4VEH
Skinner L MM0ACR
Skinner M 2E0MES
Skinner M M0SEM
Skinner M M1AGR
Skinner M M3ELS
Skinner M M3MSP
Skinner M M6MQS
Skinner P G7MLJ
Skinner P M6BPE
Skinner R G6EQF
Skinner R G6JQD
Skinner R MM0ACT
Skinner R 2E0KBK
Skinner R M0KWN
Skinner R M6KWN
Skinner S M3ZUS
Skinner T M3MIV
Skinner T M6BUR
Skipper W G0ZAT
Skipworth D G2FFD
Skirving S 2E0SIA
Skirving S M6ARL
Skittrall J 2E0BLI
Skittrall J M0SKI
Skittrall J M3XSK
SKIVINGTON P G4UUM
Skolar P G4EYV
Skolik A M0SKO
Skorupinski L G1GRN
Skoyles E G7NFW
Skoyles R G3MTJ
Skrobanski Z G3XDZ

Skrzypecki A 2E0SKR
Skulski G G6TNE
Skupski C 2E0VMA
Skupski C M6VMA
Skupski G G0VMA
Skuse K G0PQD
Skye D G3PLR
Skye D G8EPK
Skyner M G4GHT
Slack G G4AKR
Slack S 2E0ERV
Slack T G4ANW
Slade A G0IJN
Slade D G4YTB
Slade D 2E0DSB
Slade D M6DFS
Slade J G6WWM
Slade J M6JSV
Slade K 2E0RZB
Slade K M6RZB
Slade M G4ONS
Slade M M0CUK
Slade P G4DPP
Slade P M0PKV
Slade R G6PCC
Sladen P 2E0PHS
Sladen P M6EET
Sladen P G8BTD
Slaney A G3YZT
Slaney I G0RTF
Slaney J G0DMA
Slapper S 2E0SST
Slapper S M6SST
Slark P G1JMV
Slater A G0BDN
Slater B G6XYX
Slater C G6EUG
Slater D 2E0HIP
Slater D M0HLB
Slater D M6KOM
Slater F G1ONH
Slater G G1OSH
Slater G M6BZN
Slater J G0WCJ
Slater J G6FIO
Slater J G7FEL
Slater J G6EUO
Slater J M6FJF
Slater J G1UGB
Slater K G1LWY
Slater K M6BZM
Slater M G3NML
Slater M M0SSO
Slater N G4WLE
Slater P G1GRP
Slater P G1NRK
Slater P G0PGS
Slater R M0BOQ
Slater S M6RJR
Slater S G0PQB
Slater W G0UUC
Slater W M3XYX
Slator R G4MSN
Slatter B G4DF
Slatter D G0CJG
Slatter D G6WZC
Slatter G GW0KQU
Slatter R G0RGW
Slatter R G8EXF
Slattery J 2E0JSN
Slaughter A G0KMC
Slaughter K G6TKB
Slaughter S G1BYO
Slaven K MM6SLV
Sleat W GM3FJA
Slee D 2E0IVY
Slee D M6TUR
Slee M M3JJU
Sleeman G G7PGY
Sleigh A G1OSA
Sleigh R G0KBK
Sleight C G6EVC
Sleight J G3OJI
Slessor G GM1THS
Slevin E 2E0ESS
Slevin E M0TAA
Slevin E M3XQW
Slight D GM8CFS
Slight P G7PMQ
Slim A M6ADY
Slim J M6KYK
Slim R G4TJI
Sliman J G0XJS
Slimmon R G0WAY
Slingsby A G3ZWN
Slingsby C G8ADH
Slingsby G G0IWJ
Slinn A G7KTR
Sliwinski S G4ILX
Sloan A MM3LDR
Sloan C MI3FOL
Sloan C MM6LGS
Sloan D MI3IFI
Sloan J M0AEK
Sloan K 2E0VBN
Sloan K M3VBN
Sloan P GM6PLG
Sloan T GI4AHP
Sloane T G6AAB
Slobin J 2E0OTT
Slocombe D G8KOL
Slone R G1FVP
Slotwinski K M6KAR
Slough J M0BKA
Slup P M0GZE
Slyfield R G0ISC

Smale J M6JZY
Smale S M0FBB
Smale S G1ORT
Smales G M1CCG
Small A M0DRN
Small B G0CHB
Small C G3PFX
Small C 2E0CDS
Small C M0MTS
Small D GM6BGQ
Small D MM3NMI
Small D G7WBH
Small G G6RTD
Small J G4BEU
Small M G4DVI
Small S G4HJE
Small W G0GHO
Smallbone B M3UCZ
Smalley J G7OOS
Smallman R M3NOM
Smallwood G G4TJN
Smallwood H 2E0ACR
Smallwood I G0ANK
Smallwood J G7KJD
Smallwood M G0GXX
Smallwood R G6POO
Smallwood R G8DGR
Smallwood T G7LOG
Smallwoods C MI3TWM
Smart A G0UYH
Smart A G8ZZY
Smart A GM0RML
Smart A M6MYS
Smart C G8OCV
Smart G M1GEO
Smart J G0SRY
Smart J G8NTS
Smart J G0CTP
Smart K 2E0KCB
Smart K M3TKO
Smart L GW0LBI
Smart L 2E0POI
Smart L M6POI
Smart M G3UXQ
Smart P G4SDU
Smart P 2E0PSM
Smart P M0ZMB
Smart P M6AVL
Smart R G6BHA
Smart R G6DWO
Smart S MM3SES
Smeaton E M0AGS
Smedley D M3EKZ
Smedley F G3VJJ
Smedley S M6BEB
Smeed A 2E0FIA
Smethers R G3NLY
Smethurst D MW3YBX
Smethurst E M6FVL
Smethurst J G0FMO
Smethurst J G3UGC
Smiley P MI0PJS
Smillie D GM4FKD
Smillie G G6TRY
Smit P G0LIY
Smith A 2E0AXT
Smith A 2E0LRJ
Smith A 2E0TAU
Smith A 2M1HCP
Smith A 2M1HCQ
Smith A GM0FQS
Smith A GW0NHE
Smith A G0RYV
Smith A GW0TXP
Smith A G0UAS
Smith A GI0VIB
Smith A GU1DWO
Smith A G1JVY
Smith A G1LDY
Smith A G1VNL
Smith A G1YMJ
Smith A G3KVT
Smith A G3MPB
Smith A G3WPD
Smith A G4AUB
Smith A G4MDJ
Smith A G4MTH
Smith A G4OEP
Smith A G4XEZ
Smith A G4ZSA
Smith A G6AGY
Smith A G6FNQ
Smith A G6YRC
Smith A GM6YRH
Smith A G7ANX
Smith A GM7CTV
Smith A G7FWE
Smith A G7IZU
Smith A G7VIG
Smith A G8KDM
Smith A GI8WBZ
Smith A MI0AWL
Smith A M0HLA
Smith A M0TEF
Smith A M0VIG
Smith A M0XVX
Smith A M3AWS
Smith A MM3BWV
Smith A M3LJK
Smith A M3LRJ
Smith A M6TBD
Smith A 2E0FML
Smith A 2E0TTY
Smith A M6ADE
Smith A M6NOD
Smith A 2E0GQD

Smith A G0WAS
Smith A G3LHI
Smith A M3ARU
Smith A M3XZK
Smith A M6BBQ
Smith A M6LBY
Smith A 2E0TPO
Smith A M0KFO
Smith A MM0MOT
Smith A M6KPO
Smith A GM4KQS
Smith A G4PVC
Smith A G1DIK
Smith A M6TPE
Smith A 2E0VKG
Smith A 2E0CSD
Smith A G1HOD
Smith A G1WGO
Smith A M0CTR
Smith A M0VKG
Smith A M3WUA
Smith A M6BZT
Smith A M6RPU
Smith A M6TOW
Smith A M6XAS
Smith A 2E0DQB
Smith A G4FAI
Smith A G7PAK
Smith A M3NWZ
Smith A M6FKQ
Smith A M3BKB
Smith A M6EPO
Smith B G0DAH
Smith B G0OZL
Smith B G0TPB
Smith B G0UYV
Smith B G1GDR
Smith B G1JOJ
Smith B G1TJW
Smith B G4BYL
Smith B G4EQC
Smith B G4ETN
Smith B G4IAT
Smith B G4MHX
Smith B GM4RAZ
Smith B G4UJP
Smith B G6TET
Smith B G6WWS
Smith B G7DOS
Smith B G8UYR
Smith B M1DXB
Smith B M3AOP
Smith B MD3BCS
Smith B M0NAE
Smith B M6AKI
Smith B 2E0EXX
Smith B 2E0PXW
Smith B M3PXW
Smith B M6BSP
Smith B 2E0DIX
Smith B M3XHY
Smith B G0IER
Smith B G1EIO
Smith B G1XKL
Smith B G3WCY
Smith B G4THF
Smith B GM8DAB
Smith B M6GIY
Smith B M6XAV
Smith B 2E0GCC
Smith C 2E0EBR
Smith C 2E0IGK
Smith C G0IHU
Smith C G0JTN
Smith C G0LIN
Smith C G0VJC
Smith C G1DIM
Smith C G1FEF
Smith C G3MPF
Smith C G3UFS
Smith C G4NUX
Smith C G4VCP
Smith C G4WLS
Smith C G4XBS
Smith C G6DFB
Smith C G7JXJ
Smith C G8KVU
Smith C G8LMW
Smith C M0PBD
Smith C M3CIS
Smith C M3HQB
Smith C M3JHV
Smith C M1DOS
Smith C G0PGA
Smith C 2W0CYY
Smith C MW6CVC
Smith C G4VQP
Smith C 2E0CNG
Smith C 2E0MZU
Smith C G1III
Smith C G6XZM
Smith C M3WYR
Smith C M6AWN
Smith C M6PXK
Smith C M6TAK
Smith C G4FZH
Smith C M3WRQ
Smith C M6SMF
Smith C 2E0NOK
Smith C G1PGI
Smith C GW3MOV
Smith C G6THM
Smith C G7SWV
Smith C G8AWI
Smith C M0GLL
Smith C M0SPC

Smith C MM6CBH
Smith C M6CGS
Smith C M6FCE
Smith C MM3TVQ
Smith D 2E0HID
Smith D G0GQY
Smith D GD0JCF
Smith D G0JPL
Smith D G0JYU
Smith D G0PXY
Smith D G0SXM
Smith D GD1DES
Smith D G1OPA
Smith D GW3TMS
Smith D G3UGJ
Smith D G3XPD
Smith D G4COE
Smith D G4EQE
Smith D GM4MPC
Smith D G4PBO
Smith D GM4PKJ
Smith D G4UQU
Smith D G4XUR
Smith D G6FIL
Smith D G6NZW
Smith D G6SBI
Smith D G6ZFI
Smith D G7CLH
Smith D G7GMQ
Smith D G7MQE
Smith D G7NKH
Smith D G8DVN
Smith D G8IDL
Smith D G8KTG
Smith D G8OMC
Smith D MM1BJT
Smith D M1DLG
Smith D MD3OAK
Smith D M3OVZ
Smith D MI3VPO
Smith D 2E0DGS
Smith D 2E0DNX
Smith D 2E0ELO
Smith D M0VIR
Smith D M3ZZD
Smith D M6GBZ
Smith D M6LFC
Smith D G8OUY
Smith D 2E0DFS
Smith D 2E0FBD
Smith D 2E0FWY
Smith D 2E0SBZ
Smith D G0KCN
Smith D G0WVW
Smith D GM3PML
Smith D G4DAX
Smith D GM6KEV
Smith D G7CEA
Smith D G7JRC
Smith D M0DSS
Smith D MM0HVU
Smith D M0SBZ
Smith D M0STI
Smith D M3CBY
Smith D M3UQL
Smith D M3XTA
Smith D M6DBL
Smith D M6DJS
Smith D M6EYM
Smith D M6FBD
Smith D M6FWY
Smith D M6PLB
Smith D G7AHT
Smith D G4TBG
Smith D M6MDC
Smith D G8YMT
smith D M6UVD
Smith D GM0EEY
Smith D G1ZJQ
Smith D M6FIO
Smith D M0BLF
Smith D M6DLC
Smith D M0KPK
Smith E G0BAM
Smith E G0BKL
Smith E G0VYV
Smith E G0XBO
Smith E G3BPQ
Smith E G3LHG
Smith E G4MJU
Smith E GI4MRZ
Smith E G4VHQ
Smith E GW4VTG
Smith E G4YZC
Smith E GM8GJI
Smith E M0BUF
Smith E M3IKV
Smith E M3KIO
Smith E M3KVU
Smith E M3LKE
Smith E MM3LYS
Smith E G8HWJ
Smith E G1SKR
Smith E 2E0BRC
Smith E M3TXR
Smith F 2E0TLB
Smith F G4GOM
Smith F M3OAX
Smith F M3SFJ
Smith F G1IZB
Smith F G1NKV
Smith F G3YGL
Smith F G7FCU
Smith G 2E0HAL
Smith G 2M0RZE
Smith G 2E0GNK

Smith G G0CUB
Smith G G0HPA
Smith G G0OLZ
Smith G G0PGJ
Smith G G1BQI
Smith G G1EIP
Smith G GM1KUI
Smith G G3GGU
Smith G G4AJJ
Smith G G4EBK
Smith G GW4EUA
Smith G GM4GZD
Smith G G4PWB
Smith G G4RVZ
Smith G GI4XFN
Smith G GM4XUS
Smith G GW6TEO
Smith G G6WRY
Smith G G6XMU
Smith G G7DOW
Smith G G7IMZ
Smith G G7IXK
Smith G G7LTU
Smith G G7RRD
Smith G G8DST
Smith G GW8JWL
Smith G G8RTN
Smith G M0AMM
Smith G MM0GSS
Smith G M3HBB
Smith G MM3JRK
Smith G M3ZZW
Smith G 2E0GSA
Smith G M6GSS
Smith G G6FLQ
Smith G G6AGZ
Smith G G8AOJ
Smith G GM4XGY
Smith G G0BZC
Smith G G4ZUS
Smith G G6WLP
Smith G M3YQH
Smith G 2E0JOG
Smith G G3PLN
Smith G G3SNO
Smith G G4NMD
Smith G G6APJ
Smith G G6STF
Smith G M3UQO
Smith G M6GCS
Smith G MW6JBQ
Smith G M6PPL
Smith G G3ZZI
Smith H G0IZB
Smith H G0LQT
Smith H GM1CQC
Smith H G3IVF
Smith H G4KNQ
Smith H GI4PMP
Smith H 2E0HZS
Smith H M3RQG
Smith H M6HZL
Smith H M3WHX
Smith H G3YSN
Smith I GM1FEM
Smith I G4FCY
Smith I G4GDX
Smith I G4WZQ
Smith I G6RHV
Smith I G6WJM
Smith I G7TJZ
Smith I G8RYL
Smith I 2M0DIF
Smith I MM6DHZ
Smith I G1EYJ
Smith I G1RPE
Smith I G7HMA
Smith I M6EOX
Smith I M6JSR
Smith I M6RVN
Smith II M M0HMF
Smith J 2E0JPS
Smith J 2E0UYB
Smith J 2E0IHK
Smith J G0CCQ
Smith J GW0DRI
Smith J G0FKF
Smith J G0JJS
Smith J G0NZN
Smith J G0TQP
Smith J G0UAO
Smith J G1BRD
Smith J G1KCR
Smith J G1OAW
Smith J G1VTS
Smith J G3SLX
Smith J G3SMV
Smith J G3WTS
Smith J G3YWS
Smith J G3ZQC
Smith J G4KJJ
Smith J G4PET
Smith J GM4UYP
Smith J G4XJS
Smith J G4YLT
Smith J G4ZMA
Smith J G6FLH
Smith J G6SGZ
Smith J G7AXM
Smith J G7MWW
Smith J GM7VFR
Smith J G8XER
Smith J G8ZQB
Smith J M1ESD
Smith J M1JWS
Smith J M3AIZ

Smith J M3BXX
Smith J M3CIG
Smith J M3IHU
Smith J M3JDS
Smith J M3JOJ
Smith J 2E0JME
Smith J 2E0TBH
Smith J G0OFE
Smith J G0OIY
Smith J G0WWT
Smith J G6AAK
Smith J G7NTG
Smith J G7PFG
Smith J MM0CJF
Smith J M3SZT
Smith J M6AEY
Smith J M6JTI
Smith J MI0JSJ
Smith J M3YVZ
Smith J 2M0VPU
Smith J M6BRO
Smith J M6UHF
Smith J MI0AEX
Smith J 2E0XVF
Smith J M0XVF
Smith J M3XVF
Smith J 2W0BVS
Smith J 2M0HJS
Smith J GI0USC
Smith J G1IHE
Smith J G3JZF
Smith J GW3ZJS
Smith J GM4EOU
Smith J G4VEL
Smith J G4WNU
Smith J G5GX
Smith J G6FLR
Smith J G6NJE
Smith J G6VAR
Smith J G7MUN
Smith J G7UGW
Smith J M3SMI
Smith J MM3UDQ
Smith J M6CVJ
Smith J M6PPV
Smith J G1BXQ
Smith J M6JAJ
Smith J MW6XAE
Smith J M6PHZ
Smith J 2E0CJZ
Smith J M6AMG
Smith J M0GFD
Smith J M3XXA
Smith Jones A G0OIU
Smith K 2E0IJL
Smith K G0DHT
Smith K G3JCR
Smith K G3JIX
Smith K G4KEN
Smith K G4PEU
Smith K G4YYE
Smith K G8CBO
Smith K G8DNL
Smith K M0KMS
Smith K M6WLF
Smith K G8SUM
Smith K 2E0SBC
Smith K M0SBC
Smith K M0YAC
Smith K G8DBO
Smith L G0RNM
Smith L G0RPF
Smith L G0VPT
Smith L G1AEJ
Smith L M0LRS
smith L M3VCV
Smith L M6LJS
Smith L M6LHJ
Smith L G4CFK
Smith L G7GNA
Smith L GW4VNK
Smith L M6FFU
Smith M 2M1IIW
Smith M G0CES
Smith M GW0DIQ
Smith M G0DPT
Smith M G0FVU
Smith M G0FXY
Smith M G1DFF
Smith M G1MBW
Smith M G1WRO
Smith M G3RMN
Smith M G3TRV
Smith M GM3WHT
Smith M G3WXM
Smith M G4BTE
Smith M G4HMA
Smith M G4MFS
Smith M G6OES
Smith M G6WLQ
Smith M G6YZR
Smith M G6ZGA
Smith M G7SDD
Smith M G7SDQ
Smith M G7VGH
Smith M G8EWD
Smith M G8GAT
Smith M G8NAG
Smith M G8YVC
Smith M GM8ZEQ
Smith M G8ZUU
Smith M M0DZD
Smith M M0MGS
Smith M 2E0SZZ
Smith M G3UAF
Smith M G4FQI

Smith M M0XMS
Smith M M6SZZ
Smith M 2E0YME
Smith M G1KLI
Smith M M0INI
Smith M M0VPK
Smith M M6INI
Smith M M6YME
Smith M G0BIW
Smith M G0CHC
Smith M G0TVD
Smith M M0MWS
Smith M M3YGQ
Smith M M6YWA
Smith M 2E0MDZ
Smith M M6MDZ
Smith M M6MTS
Smith M 2E0DDE
Smith M 2E0GTT
Smith M G4DWX
Smith M G7TKO
Smith M G8EGU
Smith M M0MSX
Smith M M0ZVX
Smith M M3PYD
Smith M M6DDE
Smith M M6IKE
Smith M M6MBS
Smith M M6MSM
Smith M M6WOK
Smith M G4OKM
Smith M G6NIO
Smith M G8EII
Smith M M0MTJ
Smith M M6FWS
Smith N G0ILA
Smith N G0UQJ
Smith N G3YII
Smith N G4EQD
Smith N G7TBF
Smith N 2E0NSC
Smith N G4DBN
Smith N M0LAF
Smith N M0NAS
Smith N M3NHS
Smith N GW7AUQ
Smith N MD6NSS
Smith N 2W0OCF
Smith N G0NIG
Smith N MM0BUH
Smith N M0GZH
Smith N M3SUF
Smith O M6OCS
Smith P 2M0PSA
Smith P 2E0DEA
Smith P GM0ATL
Smith P G0BZX
Smith P G0CQR
Smith P G0EOS
Smith P G0JPJ
Smith P G0JTM
Smith P GD1HIA
Smith P G1KEI
Smith P G1LTI
Smith P G1OYZ
Smith P G1SNI
Smith P G1VBP
Smith P GW1XBG
Smith P G2DPL
Smith P G3OZP
Smith P G3PPU
Smith P G3PZZ
Smith P G3UPW
Smith P G3YWF
Smith P G3YWT
Smith P G4BJG
Smith P G4EES
Smith P G4JNU
Smith P G4ZWQ
Smith P G6WBG
Smith P G6XND
Smith P G7HQF
Smith P G7JGY
Smith P G7PNM
Smith P G7PQX
Smith P G7UWE
Smith P G8CYL
Smith P G8IAR
Smith P G8JSL
Smith P G8OLK
smith P M0BIT
Smith P M0CPT
Smith P M0PGS
Smith P MM0PSA
Smith P M0ZCW
Smith P M1FIL
Smith P M3EVR
Smith P G0VMR
Smith P 2E0EME
Smith P 2W0GNG
Smith P 2W0YBZ
Smith P G3WPB
Smith P G7OFI
Smith P M0GVL
Smith P M0KAP
Smith P MW0YBZ
Smith P M3MXJ
Smith P M6EYS
Smith P M6MJD
Smith P MW6PAC
Smith P MW6YBZ
Smith P M6YPS
Smith P G0BQB
Smith P M6SUM
Smith P 2W0RRY
Smith P G3TJE

| Name | Callsign |
|---|---|
| Smith P | G8JZI |
| Smith P | MM3ERP |
| Smith P | [illegible] |
| Smith P | M6DPL |
| Smith P | MW6ZAN |
| Smith P | G0MEA |
| Smith P | G4LWB |
| Smith P | G0DUI |
| Smith R | G0BWU |
| Smith R | G0ERS |
| Smith R | G0HJM |
| Smith R | G0JXO |
| Smith R | GM0KDC |
| Smith R | G0LLB |
| Smith R | G0LZY |
| Smith R | G0RRC |
| Smith R | G0VKX |
| Smith R | G1AEF |
| Smith R | G1EIR |
| Smith R | G1POV |
| Smith R | G3LVW |
| Smith R | G3VKT |
| Smith R | G3ZEJ |
| Smith R | G4BZG |
| Smith R | GM4IOB |
| Smith R | G4KZK |
| Smith R | G4NJT |
| Smith R | G4NME |
| Smith R | G4TBJ |
| Smith R | G4TTX |
| Smith R | G4WSF |
| Smith R | G4YUI |
| Smith R | G4ZXA |
| Smith R | G6EQI |
| Smith R | G6RNF |
| Smith R | G6SNV |
| Smith R | G6TFJ |
| Smith R | G6ZKX |
| Smith R | GM7GKT |
| Smith R | G7NEG |
| Smith R | G7TDR |
| Smith R | G7VYI |
| Smith R | G8HMA |
| Smith R | G8YZC |
| Smith R | M0RBE |
| Smith R | M1DEG |
| Smith R | M3AJU |
| Smith R | M3HJW |
| Smith R | M3KDO |
| Smith R | M3RDS |
| Smith R | GM4RGS |
| Smith R | G4IYE |
| Smith R | 2E0AVW |
| Smith R | G1MNY |
| Smith R | GW4MTE |
| Smith R | GW8XJC |
| Smith R | M1BNG |
| Smith R | M6EUZ |
| Smith R | 2E0ERT |
| Smith R | M0XOM |
| Smith R | M6ROB |
| Smith R | 2E0CVY |
| Smith R | 2D0VJK |
| Smith R | G0AYQ |
| Smith R | G0LEY |
| Smith R | G4GDG |
| Smith R | G4LZP |
| Smith R | G4NCI |
| Smith R | G6BWM |
| Smith R | G7JTZ |
| Smith R | M0NAQ |
| Smith R | M0ROO |
| Smith R | M3JTZ |
| Smith R | M3UHG |
| Smith R | M3YYQ |
| Smith R | M6CFQ |
| Smith R | M6TMA |
| Smith R | 2E0CBQ |
| Smith R | 2E0CLP |
| Smith R | 2E0IOS |
| Smith R | 2E0UTD |
| Smith R | M6BHB |
| Smith R | G6NHO |
| Smith R | G3SVW |
| Smith R | G4DXW |
| Smith S | 2E0MVH |
| Smith S | 2W0VAG |
| Smith S | G0BQZ |
| Smith S | G0RDZ |
| Smith S | G1WYC |
| Smith S | G3ROW |
| Smith S | G3WMY |
| Smith S | GM4LUS |
| smith S | G4SSV |
| Smith S | G6GJY |
| Smith S | G6OUX |
| Smith S | G6STJ |
| Smith S | G7JFM |
| Smith S | G7JKY |
| Smith S | G7KHL |
| Smith S | [illegible] |
| Smith S | G7OKT |
| Smith S | G8AZR |
| Smith S | M0ROT |
| Smith S | M0GMS |
| Smith S | MW0TBI |
| Smith S | M0YLS |
| Smith S | M3FRQ |
| Smith S | M3LKY |
| Smith S | MW3VVW |
| Smith S | M6OMZ |
| Smith S | G0MXU |
| Smith S | M1FMJ |
| Smith S | M0SRS |
| Smith S | M0SCS |
| Smith S | M1SCS |
| Smith S | G4LSG |
| Smith S | 2E0JTI |
| [illegible] | [illegible] |
| Smith S | G0TDJ |
| Smith S | [illegible] |
| Smith S | M3JTI |
| Smith S | M6UEH |
| Smith S | 2E0TSD |
| Smith S | M0TSD |
| Smith S | M0BFT |
| Smith S | M0ZAR |
| Smith T | G0OIS |
| Smith T | G1AYI |
| Smith T | G1GHU |
| Smith T | G1VYB |
| Smith T | G4ZZZ |
| Smith T | G6AQI |
| Smith T | G6MPK |
| Smith T | G7RVT |
| Smith T | G8MQT |
| Smith T | M0TFS |
| Smith T | M3OFU |
| Smith T | G4DKC |
| Smith T | M6GMC |
| Smith T | M1DDB |
| Smith T | M6WDO |
| Smith V | G1JJR |
| Smith V | G6RYW |
| Smith V | G6ZKZ |
| Smith V | MM3VIV |
| Smith W | G1JMW |
| Smith W | G3MKE |
| Smith W | GW4OXL |
| Smith W | G4OZC |
| Smith W | G4ROH |
| Smith W | G4TCM |
| Smith W | G6MPJ |
| Smith W | 2E0EGR |
| Smith W | GM0ENQ |
| Smith W | GD0PLR |
| Smith W | G3KJS |
| Smith W | M0WRS |
| Smith W | M0XGR |
| Smithen C | M6LWM |
| Smithers B | 2E0CEB |
| Smithers B | M6HVS |
| Smithers C | G4CWH |
| Smithers R | G4HSD |
| Smithers T | G0KTN |
| Smith-Gauvin S | GJ0JSY |
| Smithies A | M0SAV |
| Smithies D | G6RCX |
| Smithies S | G3OKS |
| Smiths P | 2E0KBD |
| Smiths P | M0KBD |
| Smithson A | G1VCU |
| Smithson C | G8TTU |
| Smithyes K | G0MDJ |
| Smoker J | G6PLF |
| Smoker M | G1ZAW |
| Smolkovic K | M1CVU |
| Smout D | 2E0MIS |
| Smout D | M0LIJ |
| Smout D | M6UDY |
| Smout J | M1JSS |
| Smy J | GM4ILE |
| Smye J | G7SSG |
| Smyth A | GI3POS |
| Smyth A | G3XNE |
| Smyth A | M6TOE |
| Smyth I | G0VSG |
| Smyth I | MI3RXF |
| Smyth J | GI4LZS |
| Smyth J | MI3UIW |
| Smyth J | 2I0BIR |
| Smyth J | MI3RIF |
| Smyth J | M6JJD |
| Smyth K | G3UTA |
| Smyth M | GI0TMS |
| Smyth M | G3YFM |
| Smyth M | M0MGA |
| Smyth M | G0BXM |
| Smyth P | G0MQU |
| Smyth P | M3ZPJ |
| Smyth R | GI4CBG |
| Smyth R | M6BHI |
| Smyth W | GI7KHR |
| Smyth W | G6MDM |
| Smyth W | GI8RNG |
| Smythe D | GI1VPA |
| Smythe T | M6ENM |
| Snaden H | G4YNV |
| Snape D | G4GWG |
| Snape E | G0BBI |
| Snape G | M6HOG |
| Snape J | M3ZZQ |
| Snape K | G3UPN |
| Snape N | M3NEL |
| Snape R | 2W0XDT |
| Snape R | MW0XDT |
| Snape T | G7LPN |
| Snary R | G4OBE |
| Sneap C | 2E0IHY |
| Sneap M | G3ZYC |
| Sneath A | G8YMW |
| Sneath R | G8XUX |
| Sneddon A | 2W0MJA |
| Sneddon A | MW0MUM |
| Sneddon A | MW6MWS |
| Sneddon G | MM6ZGS |
| Sneddon J | MW0EQL |
| Sneddon R | M6RGS |
| Snelgrove A | G7VON |
| Snelgrove J | GM4OFC |
| Snelgrove J | G7VOM |
| Snell B | G0TSB |
| Snell C | G8LVW |
| Snell D | M1DGS |
| [illegible] | G6PRF |
| Snell J | G0RDO |
| [illegible] | [illegible] |
| Snellgrove G | G0LTV |
| Snellin K | G6DNL |
| Snelling A | M6SEN |
| Snelling R | G7BSC |
| Snelling-Nash C | M6CSN |
| Snelson A | 2E0WAV |
| Snelson A | M0WAV |
| Snelson A | M6WAV |
| Snelson P | M0PWS |
| Snelson S | 2E0MHE |
| Snelson S | M3SZQ |
| Sniadowski J | G4BRH |
| Sniezek P | 2D0CTX |
| Sniezek P | MD0TCX |
| Sniezek P | MD6CTU |
| Sniezko-Blocki M | G0MCM |
| Snitch P | G1BYP |
| Snook P | 2E0XHL |
| Snook P | M0SPJ |
| Snook P | M3XZH |
| Snook T | G6AHC |
| Snow A | G4FLS |
| Snow A | G8YMR |
| Snow C | G1XVD |
| Snow C | M3YPP |
| Snow D | M1AMB |
| Snow D | G3PRL |
| Snow J | G4TSH |
| Snow M | G6MBL |
| Snow M | M6OAE |
| Snow P | G1DMN |
| Snow R | GD0BSP |
| Snow W | G0CYD |
| Snowdon C | M6GZL |
| Snowden D | M6SNW |
| Snowden J | M6NIV |
| Snowden L | G0UQI |
| Snowden M | M3RCQ |
| Snowden N | G1XJO |
| Snowden R | M3SNO |
| Snowden W | M3WPS |
| Snowdon D | M3ECD |
| Snowdon E | M1FCZ |
| Snowdon O | M0XOS |
| Snowling J | G1DYN |
| Soaft I | G4TGV |
| Soakell J | M0AYC |
| Soames C | G0TZZ |
| Soames D | 2E0USV |
| Soames D | M0USV |
| Soames D | M6USV |
| Soames E | M6RIE |
| Soane A | M0ABY |
| Soane M | G0XBC |
| Soane N | M6WYX |
| Soar R | M6ZZY |
| Soars A | G0ALI |
| Soars A | G4VCN |
| Soars C | G6VAW |
| Soars I | G3HGI |
| Soars M | G4TCI |
| Sobanski J | G6KOB |
| Sobey D | 2E0DAX |
| Sobey D | M6BIF |
| Sobey R | G4AER |
| Soble A | G1ZND |
| Soble M | G8UVU |
| Sobye P | G0PNM |
| Sockett A | G0KXZ |
| Sockett A | M6CLW |
| Soden G | G7LPV |
| Soffe W | 2E0SOF |
| Soffe W | M3JZT |
| Softley M | G7JDQ |
| Softley R | G0JUN |
| Sohal G | G6DWM |
| Sohst R | G6RVS |
| Sohst R | M3RVS |
| Sojkowski D | G6DJS |
| Solanki S | M6SNE |
| Soldan L | M0SDU |
| Sole C | M3ICF |
| Sole D | M6DGS |
| Sole G | 2E0GAF |
| Sole G | M6GDS |
| Sole M | G0PFA |
| Sole M | G4UQF |
| Solkow G | G6WFF |
| Sollazzo G | G1WNZ |
| Sollis J | MW6JGC |
| Solly I | M0CAG |
| Solomon J | G6KNK |
| Solomon J | M3YFL |
| Solomon S | M6GLJ |
| Solomons M | GD8DKW |
| Solomons R | G6CRR |
| Soltysik A | G4KWQ |
| Soltysik J | M1FJC |
| Soltysik N | M3STR |
| Somerfield P | [illegible] |
| Somers G | G7VTV |
| Somers O | G6RZJ |
| Somerville A | MM1ICE |
| Somerville A | M6BGS |
| Somerville C | MM3EYM |
| Somerville C | G6IVY |
| Somerville C | M6XBV |
| Somerville G | M3ZGS |
| Somerville G | MM6GBS |
| Somerville R | MM3EYN |
| Somerville R | M6RMS |
| Somerville Roberts B | 2E0KCN |
| Somerville Roberts B | M6NQR |
| Somerville Roberts R | M0RJS |
| Somerville W | G4WJS |
| Sommerfield D | G4TAY |
| Sommers C | M6ZAU |
| Sondhis J | G4EIV |
| Sonley J | G3XZV |
| Sonnet B | MM6VFL |
| Sood A | 2E0TKX |
| Sood A | M0TKX |
| Sood A | M6TEK |
| Soper A | M3LLK |
| Soper M | G0VVY |
| Soper M | M3VVY |
| Sorab A | G6VXZ |
| Sorab P | G3NDO |
| Sorbie T | GM3MXN |
| Sorensen T | G8MFO |
| Sorger B | G4FBY |
| Sorockyj S | G0LCG |
| Sorrell I | G7HQC |
| Soundy C | G8EDQ |
| Sousa P | M6HSA |
| Souter E | GM8PIV |
| South G | G4YAP |
| South S | M6SFM |
| South W | GW0HNT |
| South W | G7MND |
| Southall C | G8UEZ |
| Southall G | G7WLV |
| Southall G | 2E0LMR |
| Southall G | M6CNG |
| Southall H | M3JXI |
| Southall J | G4DMT |
| Southall M | 2E0MWS |
| Southall M | G3WWS |
| Southall R | G4NLK |
| Southby P | G8BAJ |
| Southerington R | G0IOZ |
| Southern B | 2E0DDN |
| Southern B | M6DNQ |
| Southern G | G3RWW |
| Southern P | M3CSV |
| Southern R | G3RST |
| Southern R | G4WAP |
| Southern R | G6BDY |
| Southern S | 2E0FUN |
| Southern S | M0RXX |
| Southern S | M3VHZ |
| Southernwood D | M6SVY |
| Southey C | M6SQK |
| Southey D | G0EYX |
| Southgate F | 2E0CMY |
| Southgate F | M6FAS |
| Southgate H | 2E0EAS |
| Southgate K | G3XSC |
| Southgate M | 2E0ZAU |
| Southgate M | M6ZAX |
| Southgate P | G6ILD |
| Southon E | G0MRA |
| Southon N | G8XYA |
| Southward D | G8YGM |
| Southwell A | 2E0SBB |
| Southwell A | M0SBB |
| Southwell A | M6AES |
| Southwell E | G4PXH |
| Southwell G | G4HVR |
| Southwell S | G0LFN |
| Southwell T | G4FEU |
| Southworth J | GM1KZG |
| Southworth R | M3EIA |
| Southworth S | M1SWS |
| Southworth W | G0VYP |
| Soutter L | G4XHK |
| Soutter V | G0OXW |
| Sowden G | G3WGZ |
| Sowden G | M0STS |
| Sowden G | M3BVL |
| Sowden R | M6RIS |
| Sowerbutts J | G1BQQ |
| Sowerby B | G4MXY |
| Sowman J | M0JSN |
| Sowter B | G3NAP |
| Sowter G | G8ULJ |
| Sowter K | M3UXH |
| Spacagna T | G8OQP |
| Spacek L | M0LFS |
| Spacey J | G0RII |
| Spacey M | G1NWZ |
| Spacey R | G0LUD |
| Spackman C | G3GYQ |
| Spafford M | G0TJG |
| Spain A | 2E0HPT |
| Spain T | G3NSS |
| Spalding D | G1JQH |
| Spalding I | G4RYM |
| Spalding R | M3YUN |
| Spalding-Reffold G | M6GVR |
| Spanner E | G0DFU |
| Spanton E | G8JTG |
| Sparey D | G0DLJ |
| Spark G | G7FQY |
| Sparke D | G6NZY |
| Sparke S | [illegible] |
| Sparke P | G7TNU |
| Sparkes D | G0UJI |
| Sparkes S | M0DFD |
| Sparks B | GM4JYB |
| Sparks C | M1BQE |
| Sparks C | M6FDY |
| Sparks F | G7JVQ |
| Sparks J | G0SPX |
| Sparks K | M1BQD |
| Sparks K | M0SKY |
| Sparks N | G4ZUX |
| Sparks R | G0TEI |
| Sparks S | 2E0MUD |
| Sparks S | M1BQC |
| Sparrey D | [illegible] |
| Sparrey N | G0IMK |
| Sparrey T | G0CAT |
| Sparrow B | G0DZY |
| Sparrow C | G1PSL |
| Sparrow D | G8XOR |
| Sparrow M | M6EGS |
| Sparrow N | G7LNU |
| Sparry R | G3BJC |
| Spashett P | G1IHB |
| Spaven G | G1HQO |
| Spavins B | G7STG |
| Spaxman A | 2E0BLF |
| Spaxman A | M3UXC |
| Spaxman B | M6BFC |
| Speak D | G6MZW |
| Speak N | G4XNS |
| Speak R | M1RJS |
| Speak T | G6TEX |
| Speak T | M3OVG |
| Speak W | M3OUQ |
| Speake J | G3URX |
| Speakman A | M1APL |
| Speakman B | G3UBS |
| Speakman J | G4TUM |
| Speakman J | G1SUH |
| Spear L | 2E0GRL |
| Spear N | G4RWI |
| Spearing J | G0BUD |
| Spearing T | G7DVO |
| Spearman D | G0IMU |
| Spears A | 2E0SJA |
| Speechley D | G4UVJ |
| Speed D | M3KVD |
| Speed H | G3BMO |
| Speed P | G4YVW |
| Speed R | M0DEP |
| Speers J | G0BVM |
| Speight A | G3PYW |
| Speight A | M6SAS |
| Speight M | G8ZNL |
| Speight N | 2E0BGC |
| Speight N | M3MUP |
| Speight T | G0TEE |
| Speirs D | 2M0DOI |
| Speirs D | MM6GFG |
| Speirs G | G0AGN |
| Speirs G | G1DAU |
| Speller E | G3LZR |
| Speller J | G0EBI |
| Speller J | GJ3YLN |
| Speller N | 2E0FGH |
| Speller N | M6FGH |
| Spelman P | G4EKD |
| Spence C | 2E0CAS |
| Spence C | M0AVW |
| Spence D | GM4MTI |
| Spence D | M0TTY |
| Spence J | G0FPI |
| Spence J | G4BSS |
| Spence M | G4SOH |
| Spence P | G1UDS |
| Spence R | MI3LMR |
| Spence R | GM7RDH |
| Spence R | G8WUG |
| Spence S | M6FOX |
| Spence S | MM0DGI |
| Spence T | M0DCG |
| Spence W | MI3WES |
| Spenceley G | G4DQN |
| Spenceley N | G8JUG |
| Spencer A | G3PMO |
| Spencer A | M3HEC |
| Spencer B | G0AYI |
| Spencer B | G0BDE |
| Spencer B | G1HII |
| Spencer B | G7UTE |
| Spencer B | G4YNM |
| Spencer B | G4GVI |
| Spencer C | G2HBA |
| Spencer C | G6RHK |
| Spencer C | M3OOY |
| Spencer C | 2E0FCS |
| Spencer C | G6UFV |
| Spencer C | M6FKZ |
| Spencer Chapman J | G3WUK |
| Spencer D | G3LGW |
| Spencer D | G8YZY |
| Spencer D | M1SKA |
| Spencer D | G0HKW |
| Spencer D | M6ENS |
| Spencer F | G0DMS |
| Spencer F | M3OOX |
| Spencer G | G3ZIN |
| Spencer G | G4DRR |
| Spencer G | G4OLO |
| Spencer G | G4TVT |
| Spencer G | G8FOL |
| Spencer G | M3VVH |
| Spencer G | G4CXW |
| Spencer G | M3ZVB |
| Spencer H | G4FZC |
| Spencer I | G3ULO |
| Spencer J | G0UNE |
| Spencer J | G3WTO |
| Spencer J | G8UMA |
| Spencer J | M3DPI |
| Spencer J | M3OPC |
| Spencer K | M3OPD |
| Spencer M | G4PPR |
| Spencer M | G8UML |
| Spencer M | G4OYZ |
| Spencer M | G3UOD |
| Spencer N | M6NJE |
| Spencer P | G0LYR |
| Spencer P | G3PSW |
| [illegible] | G7DUR |
| Spencer R | G1UNG |
| Spencer R | [illegible] |
| Spencer R | GM8CEA |
| Spencer R | G0PMI |
| Spencer R | M0ANO |
| Spencer R | MJ6ORG |
| Spencer S | 2E0SJS |
| Spencer S | G3ILO |
| Spencer S | M0URJ |
| Spencer S | M3MFS |
| Spencer S | MM6HTS |
| Spencer T | MM0GAI |
| Spencer W | G0ERW |
| Spencer W | G0TUJ |
| Spender D | G4DHU |
| Spendlove D | G4CMH |
| Spendlove J | G4DXY |
| Spensley R | M0DWQ |
| Sperry J | G4CQH |
| Spevack O | M1DDF |
| Spevack R | 2E0RJS |
| Spevack S | M0CDJ |
| Spice S | 2E0IEE |
| Spicer C | G4ZWD |
| Spicer C | M6COL |
| Spicer D | G1ORK |
| Spicer E | M0MNG |
| Spicer G | 2W0ZWR |
| Spicer G | MW0ZWR |
| Spicer G | MW3ZWR |
| Spicer G | 2E0CVX |
| Spicer J | G1IQN |
| Spicer J | G3DNH |
| Spicer J | G8LJU |
| Spicer K | G3RPB |
| Spicer K | G6KQ |
| Spicer M | M3MSY |
| Spicer P | G0PJS |
| Spicer S | M3XMJ |
| Spicer T | G8IQT |
| Spickernell F | M6FCS |
| Spiers A | 2E0ICW |
| Spiers G | 2E0TIG |
| Spiers G | M6TIG |
| Spiers J | MM1DSX |
| Spiers K | G1ALU |
| Spiers M | M6CPZ |
| Spiers R | M3TUQ |
| Spiers W | MM3UHK |
| Spillett J | G1WSN |
| Spillett J | M3WSN |
| Spillett M | G4UAW |
| Spilling R | G0DEE |
| Spilman P | 2E0PJS |
| Spilman W | G3OTD |
| Spindler I | G1NKF |
| Spink B | GM0KZX |
| Spink G | G3WUI |
| Spink J | G1BUQ |
| Spink J | G1SBK |
| Spink R | GM0WRV |
| Spink S | G0MUJ |
| Spinks A | M3BVY |
| Spinks D | M0IRS |
| Spinks G | G1IGP |
| Spinks G | G8BVY |
| Spinks I | G8XOU |
| Spinks I | M0XOU |
| Spinks L | G3YOQ |
| Spinks M | G1BQR |
| Spinks M | M6GEY |
| Spinner R | M6OYU |
| Spinney G | G0IFF |
| Spires C | G7HVL |
| Spires C | G7VHC |
| Spires E | G7PXX |
| Spirrell R | G7PBT |
| Spiteri J | G4SOM |
| Spittle M | G6NJL |
| Spittlehouse A | G7IMD |
| Splaine J | G4EPH |
| Spoard J | G0GZW |
| Spong L | M6LSG |
| Spong P | MW6PWS |
| Spooner D | G0LWL |
| Spooner D | G0OEK |
| Spooner D | G3WDS |
| Spooner D | G6TVC |
| Spooner D | 2E0YWP |
| Spooner D | M3YWP |
| Spooner J | G8OPI |
| Spooner K | G3TNY |
| Spooner M | G4PFG |
| Spooner N | M3YIF |
| Spooner P | G1JIW |
| Spooner P | G4HFU |
| Spooner R | 2E0YTT |
| Spooner R | M3LPQ |
| Spooner S | M6SFS |
| Sporton D | M3LGF |
| Sporton J | M0AUK |
| Spowart P | M3FUV |
| Spragg K | G8WKZ |
| Spragg R | G4HPY |
| Spragg S | G0RQI |
| Spratley J | M6JFJ |
| Spratley K | G0CJD |
| Spratt A | G4IJJ |
| Spratt H | G0JJE |
| Spray R | G7LKV |
| Spreadbury R | G3XLG |
| Spridgen J | G4LQJ |
| Spriggs G | G3PFE |
| Spriggs J | G6AHV |
| Spriggs L | 2E0LKS |
| Spriggs L | [illegible] |
| Spriggs L | M6LKS |
| [illegible] | [illegible] |
| Spring N | G1LUC |
| Springall P | G1WXS |
| Springall R | 2E0ERS |
| Springall R | M6ERN |
| Springate F | G3BWV |
| Springate P | M1JJS |
| Springett J | M0BTR |
| Springett M | M3KER |
| Sprint S | G7EVR |
| Sproates K | M3XSA |
| Sproson D | G8VBI |
| Sproston D | 2E0DCP |
| Sproston D | M6DEO |
| Sprott A | MW6GIU |
| Sproul G | 2M0SAX |
| Sproul G | MM0SAX |
| Spruce G | G6ZYX |
| Spry A | G0KDY |
| Spry J | G1TDP |
| Spry M | G1DKE |
| Spry P | M6CJJ |
| Spurgeon J | G4LKD |
| Spurgeon M | G0MEC |
| Spurgeon P | M1EOV |
| Spurgeon R | G3OMB |
| Spurling A | M6GVM |
| Spurr E | 2E0LMS |
| Spurr E | M3LKM |
| Spurr G | G0PFH |
| Spurr P | G3YYN |
| Spurr S | M1CIG |
| Spurway O | 2E0ZXX |
| Spurway O | M6ZXX |
| Squance B | M3BAS |
| Squance E H | GI4JTF |
| Squance M | G6IBU |
| Squance M | G3HTB |
| Squibb G | G4MPA |
| Squibb N | G4HZX |
| Squire C | G7VUH |
| Squire D | G6TAP |
| Squire G | M1BXQ |
| Squire J | G0AGZ |
| Squire O | 2E0OJS |
| Squire S | G3EBV |
| Squire W | G4NMF |
| Squires C | G3XCS |
| Squires D | G4DAC |
| Squires J | G0UQU |
| Squires K | M3WQI |
| Squires P | G3OIF |
| Squires R | G7OFM |
| Squires R | 2E0CJJ |
| Srinivasan C | M3HSV |
| St George S | M6KLR |
| St John D | G3PQD |
| St John-Murphy T | G0HTK |
| St Leger J | G3VDL |
| St Quintin D | G4PCZ |
| St Quinton E | 2E0STQ |
| St Quinton E | M0LIE |
| St Quinton E | M6BZU |
| Staal L | G1BPU |
| Stabbins L | G0OCT |
| Stabbins M | G0OCS |
| Stabler A | 2E0DHW |
| Stabler A | M0ZLH |
| Stabler A | M6AWS |
| Stables E | G3ODD |
| Stables J | G3ZIJ |
| Stables M | M3AWQ |
| Stacey A | G3BXS |
| Stacey A | M3KVC |
| Stacey C | G8YVW |
| Stacey C | M6AFG |
| Stacey J | 2E0YAS |
| Stacey J | G0VPJ |
| Stacey J | G8BXO |
| Stacey J | M6YTS |
| Stacey J | M6NIL |
| Stacey R | G4XQW |
| Stacey S | M0ACI |
| Stack T | 2E0KKO |
| Stack T | M6KKO |
| Stackhouse N | G1SCL |
| Staddon B | G6UMH |
| Staddon H | G6STI |
| Staddon K | G7VNQ |
| Staerck C | 2E01EXP |
| Staerck P | 2E0SKY |
| Staff C | G7LXA |
| Staff C | M0LXA |
| Stafford A | G3NYZ |
| Stafford A | G4VPM |
| Stafford A | G1ALR |
| Stafford J | G1YIQ |
| Stafford J | GI7KEC |
| Stafford J | G7TSQ |
| Stafford P | G0MWU |
| Stafford R | [illegible] |
| Stafford R | M0STA |
| Stageman J | G4WVT |
| Stagg B | G3ZQJ |
| Stagg C | G1YQL |
| Stagg V | G3IKN |
| Stagles A | G3RBY |
| Stain C | G0IUN |
| Stainer E | G4EJS |
| Staines O | M0WAS |
| Staines S | G7WAS |
| Stainforth -Small D | G7CUU |
| Stainforth W | G0VHI |
| Stainforth W | G7GHI |
| Stainsby F | G0NVA |
| Stainton D | G0MDI |
| Stainton G | G1MQQ |
| Stainton J | 2E0PAD |
| Stainton J | M3WUG |
| Stainton P | G4WMO |
| Stainton P | G6ZRV |
| Staite C | M3IKN |
| Staite P | 2E0MID |
| Staite P | M0MID |
| Staite P | M6VBF |
| Staley M | M6GIZ |
| Stalker A | G1BPV |
| Stalker T | G7TZU |
| Stallard C | M6GDO |
| Stallard J | 2E0JRS |
| Stallard J | M6YRS |
| Stallard W | G0NHB |
| Stalley D | M1DUB |
| Stalley K | G4HHA |
| Stallibrass D | M6DDP |
| Stallibrass P | M0PDA |
| Stallon D | G0PKJ |
| Stallworthy S | G1LZS |
| Stamford J | 2E0XFR |
| Stamford M | G4IIA |
| Stamford M | G8NOD |
| Stamford R | G6PCE |
| Stammers K | G0SXG |
| Stamp G | G8NNS |
| Stamp L | G8TFW |
| Stamp T | 2E0TSA |
| Stamp T | M6TSA |
| Stamper H | G3KYM |
| Stamper L | 2E0DNW |
| Stamper L | M6CVP |
| Stamps V | G0BEZ |
| Stanbridge M | G3RHU |
| Stanbury J | G0BBO |
| Stancer C | G1RVH |
| Stancey G | G3MCK |
| Stancliffe K | G0FKS |
| Standen B | G0GUW |
| Standen D | G6TFE |
| Standen M | G0JMS |
| Standen P | G6FJL |
| Standige M | G3MBU |
| Standing M | G1HIB |
| Standing W | G8YGK |
| Standley L | G0FCZ |
| Standley P | G1GSB |
| Standley P | G8RW |
| Standley P | M0XPS |
| Standring T | G1HGF |
| Stanfield B | M3OHJ |
| Stanford D | G8XYQ |
| Stanford E | MI0ALS |
| Stanford J | M0GLF |
| Stanford M | G8VHK |
| Stanford P | M3TNW |
| Stanford R | G1HDR |
| Stanford T | M3TWS |
| Stanhope G | G6AVT |
| Stanhope P | GM1PST |
| Stanhope P | M3JIE |
| Stanhope S | M3GHS |
| Staniewicz G | G8HPN |
| Staniforth A | G4ZDX |
| Staniforth A | G8HVX |
| Staniforth B | G7AJP |
| Staniforth R | G3EGV |
| Staniforth S | G8RSV |
| Staniland M | G6ZKS |
| Staniland R | M0RDS |
| Stanleigh R | G4DHK |
| Stanleigh R | G8HNS |
| Stanley A | G0TFC |
| Stanley A | MM6BIP |
| Stanley A | M0AST |
| Stanley C | 2E0GYO |
| Stanley C | GM1MCN |
| Stanley C | G6UXU |
| Stanley D | MI0BSU |
| Stanley D | M3DEB |
| Stanley D | GW0FJP |
| Stanley D | G1ORG |
| Stanley D | 2E0NCE |
| Stanley D | M0LMR |
| Stanley D | M0EDQ |
| Stanley G | 2E0XGS |
| Stanley G | M6XGS |
| Stanley G | G8GYI |
| Stanley G | M0LUD |
| Stanley H | M6HNS |
| Stanley I | G6MDH |
| Stanley K | G6CPE |
| Stanley K | M3KMS |
| Stanley M | G1UVI |
| Stanley M | M0CHS |
| Stanley M | M0AOT |
| Stanley N | M6BBL |
| Stanley P D | G3RSN |
| Stanley R | [illegible] |
| Stanley R | G7OED |
| Stanley R | 2E0ACZ |
| Stanley R | G0SSZ |
| Stanley S | G1EIV |
| Stanley S | M1DQQ |
| Stanley T | G4TXK |
| Stanley V | M6FYV |
| Stanley W | GI0OHT |
| Stanmore A | M3WDU |
| Stanmore K | G0IYO |



**IMPORTANT NOTE**

**Revalidate licence to avoid revocation** – Ofcom has advised the Society that plans will be drawn up to revoke licences that have not been revalidated as required by the licence conditions. The quickest way to revalidate is to do so online via the Ofcom website: *https://services.ofcom.org.uk/* or by email: *amateur.validations@ofcom.org.uk* If you need assistance in the process, Ofcom staff are available to help, but please be patient during times of heavy workload.

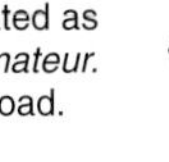

Stanners D G3HEJ
Stanners N G8AAU
Stansfield A G6IIA
Stansfield A M6URM
Stansfield D G0EVV
Stansfield D M6DPZ
Stansfield J G1ASN
Stansfield R G3UAX
Stansfield R M1CBU
Stansfield T G4LPS
Stansfield T M3KCQ
Stant L 2E0LTS
Stant L M0LTS
Stant L M6LTS
Stanton B G7HSY
Stanton D G0FWF
Stanton D G0OLX
Stanton D G8CUX
Stanton G G3SCV
Stanton J G6XYU
Stanton J M3JJS
Stanton M G0VJK
Stanton M G4CCQ
Stanton R G3XKG
Stanton R G4KDI
Stanton R G4KFZ
Stanton S G4YJP
Stanton S G6JRE
Stanway G G3INP
Stanway J MM3CGC
Stanway M G4NSZ
Stanway M G8NSZ
Stanway R M3NSZ
Stanway T G4DVA
Stanyer K G0UKK
Stapleford J G1YQU
Staplehurst J G6REH
Staples D G1PNX
Staples H M3GGN
Staples R MW6BQA
Stapleton C G6OCE
Stapleton D M6HDS
Stapleton F GM0GKF
Stapleton J GM0GKR
Stapleton L GM0GEE
Stapleton M G7DWU
Stapylton G M6ECM
Starbuck J M0HYA
Stark A GM0BUE
Stark A G8OIJ
Stark R G8SIE
Starkey B G4YOZ
Starkey E G6OPV
Starkey F G8TJG
Starkey G G4ZBF
Starkey G G6POP
Starkey K G0LDP
Starkey M G4SCL
Starkey W M6AHF
Starkie D G4AKC
Starley P G4RCR
Starling B M6RND
Starling D M6DAS
Starling G G3ALG
Starling G 2E0GST
Starling G M0GSZ
Starling G M6GST
Starling J G3WJS
Starling P G8EBX
Starling P G8SED
Starling R 2E0RFS
Starmer D G4TGW
Starnes K G7MOW
Starr B G6ZRS
Starr C M1JCS
Starr R M6RDK
Starrett L 2E0CYT
Starrett L M6CWZ
Start B G1GLZ
Startup L G0IXP
Staruszkiewicz J GM4LCP
Stasuik M G1UDX
Staszewski B M6XSZ
State P M6EJB
Statham A M1BDL
Statham B G0HSW
Statham C G0HME
Statham J 2E0SRJ
Statham J M0MLE
Statham J M6RJS
Statham M M3YMS
Statham M 2E0MSA
Statham M M6MCS
Stather K G0CWP
Staton B G6UJI
Staton M G4BGT
Staton M 2E0BEF
staton P G4FXY
Staton S G1XRE
Staveley R G8JWK
Stayt C G0LUN
Steabler G M5ACD
Stead D G7MSN
Stead M M3XBZ
Stead P M3PUL
Steadman A M5AEE
Steadman B G8FSN
Steadman F GW6AAG
Steadman M G7JUN
Steadman P G6LMB
Steam P M6PDS
Steans R G0KDI
Stearn G G4PHC
Stearn J G7DKZ
Stearn R G1SOG
Stearn R M3SOG
Stears P G4LMM
Stebbings E G6ZKY
Stebbings M G8BIG
Steddy S GW0WDV
Stedman J M0JAJ
Steed F G8FMI
Steed P G0VEP
Steed R G3YUJ
Steed S G0IQC
Steeden C G4DJY
Steedman G M1GCS
Steedman G M0BGS
Steel A G7AMS
Steel A M0GKR
Steel G G0RDY
Steel G G3TOJ
Steel J G3VJI
Steel J G7TUK
Steel J M0ZAK
Steel J M6PHU
Steel L G2BBI
Steel R G0HLL
Steel R G1OAE
Steel R GM4WWU
Steel R G1TPO
Steel T M6TRS
Steele A 2E0TON
Steele A G1HEA
Steele A G1ZNZ
Steele A GM7IEU
Steele B G0BDK
Steele C M6EFW
Steele D G0VDV
Steele D G8UPO
Steele D M3TVD
Steele G 2M0ANE
Steele G GM0WUR
Steele G GW3SIY
Steele G MM6COF
Steele H MM0HSA
Steele H MM3YHS
Steele I G1RIX
Steele J 2I0BXJ
Steele J GM0WUQ
Steele J GM6BRU
Steele J GM7VDM
Steele J M3FWS
Steele J MI3LZF
Steele K M3RCI
Steele L M3LLC
Steele L G8MEH
Steele M M6MOS
Steele M MI6MSV
Steele P G4PMS
Steele R G6LPB
Steele R 2E0DVK
Steele R G6TVB
Steele R 2E0BXE
Steele T GW0AVW
Steele T 2E0RBN
Steele V GI1WGK
Steele W GM0WUP
Steele W GM7VDL
Steele W MM3WUP
Steen B G4UFU
Steenson K GI8XSY
Steenvoorden L G0NBC
Steeper T G7JFI
Steeples M M3LGM
Steer A 2E0TJS
Steer C MW3YCL
Steer D G4XFM
Steer L 2E0LEO
Steer L M3NCP
Steer W G8CYG
Steers H M0HUS
Steggles I M6CMO
Stein O MM3ONI
Steingold W M6WCE
Steinhoefel C G0SYP
Stelfox A G6EKS
Stellar T G6RCT
Stellig R G4CK
Stelmasiak J G7DRX
Stemp H G0DJS
Stemp N G7KAV
Stenbacka C G0NGD
Stenhouse R 2E0CYE
Stennett R G0HFN
Stennett W G1SOX
Stenning M G7WBO
Stenning P G4JA
Stephen C G6ZFU
Stephen D MM0AOY
Stephen G 2M1HLE
Stephen J GM0CHM
Stephen R G0MYX
Stephen R G3ZEF
Stephens A G6GIU
Stephens A G8VVZ
Stephens C G3MGS
Stephens D GW3PYD
Stephens D G4ANY
Stephens D G4CMQ
Stephens D M6JEG
Stephens G M3OBQ
Stephens G M6GFP
Stephens G G7SHW
Stephens G GW0DXZ
Stephens G M1GRA
Stephens H G1EIX
Stephens H GW4ZRK
Stephens I M3UHS
Stephens J GW4UVC
Stephens J G7WJK
Stephens J G1VVU
Stephens M M1MBZ
Stephens M M3MBZ
Stephens M M0HZX
Stephens N G1FND
Stephens N 2E0NFS
Stephens P G0BDB
Stephens P G8GTU
Stephens P M0PHL
Stephens R G0NPA
Stephens R G8XEU
Stephens S G1LGJ
Stephenson A M0MEG
Stephenson B GM0IPV
Stephenson B G4VRU
Stephenson C G4DCD
Stephenson C G8JZX
Stephenson D G3KUL
Stephenson D 2E0KPX
Stephenson G G0TNC
Stephenson J 2E0SCM
Stephenson J G1ZFG
Stephenson J G6CPF
Stephenson J M6SBW
Stephenson J M3VXM
Stephenson L M3LSS
Stephenson M G0JHU
Stephenson M M3ITU
Stephenson M G8JXS
Stephenson S G8HBZ
Stephenson S MI1EIE
Stephenson W G3AQB
Steponitis B G6ONI
Sterland C G1FYE
Sterry A G8DVS
Sterry H G1AVW
Sterry R G4BLT
Steuerwald J MI6GQI
Steuwe C M0YCS
Steven A GM4IPK
Steven G MM1DQW
Steven J GM8EXU
Steven J M3XUO
Steven R 2E0GXU
Stevens A G0NBP
Stevens A G6FIN
Stevens A GM7PSH
Stevens A G8NVI
Stevens B G0DZQ
Stevens B G0WZX
Stevens B G6STE
Stevens B G8KKA
Stevens B G8YUP
Stevens B M3YUP
Stevens C 2E0HVZ
Stevens C G4LAU
Stevens C M6NGO
Stevens C M6YHS
Stevens F G3URV
Stevens F G4BHC
Stevens F G7LEB
Stevens F M3ZJD
Stevens G 2E0GLS
Stevens G M6GPS
Stevens G G1ZEC
Stevens H GW6WOB
Stevens H MW0JAW
Stevens J G4EPC
Stevens J GM4XAV
Stevens J G7WFQ
Stevens J M1DRZ
Stevens J 2E0JCQ
Stevens J M0JCQ
Stevens J M6JCQ
Stevens J M3LUA
Stevens K 2E0KSG
Stevens K M6KSG
Stevens K G4BVK
Stevens L M6NXT
stevens L MW6HLQ
Stevens M G0GKL
Stevens M G0KAS
Stevens M G0SWW
Stevens M G0UUP
Stevens M G3CPN
Stevens M G4CFZ
Stevens M G7MES
Stevens M G8CUL
Stevens M G8YEN
Stevens M M0CUL
Stevens M M6LFJ
Stevens M G7SFA
Stevens M M6GPJ
Stevens N G6RZR
Stevens P G4EOR
Stevens P G7LRB
Stevens P M0PST
Stevens P G8TMQ
Stevens P G3SES
Stevens R G0JBZ
Stevens R GM1ADI
Stevens R G1ERQ
Stevens R G3MAI
Stevens R G3TDH
Stevens R G3TVI
Stevens R M6RNQ
Stevens R 2E0XRS
Stevens R M3YFW
Stevens S G0WAM
Stevens S G1PQO
Stevens S M6SQB
Stevens T 2E0ASF
Stevens T 2E0GQN
Stevens T G3VIX
Stevens W G0LYI
Stevenson A G4ZAI
Stevenson A G6LSW
Stevenson A M3IDQ
Stevenson A G4TXL
Stevenson A MM3XMT
Stevenson A M0UGH
Stevenson A M3AHS
Stevenson C MM0DXC
Stevenson D GM6FIK
Stevenson D G8NXS
Stevenson D MM0RAM
Stevenson E 2I0RIR
Stevenson E MI6RIR
Stevenson F 2E0EVH
Stevenson F G4IOE
Stevenson F M0FJS
Stevenson G MW6FKC
Stevenson I G3YNU
Stevenson I MI3LXE
Stevenson I 2E0GBA
Stevenson I M3XNM
Stevenson J GI0IJB
Stevenson J GM0JVV
Stevenson J G4JCS
Stevenson J G0EJQ
Stevenson J M0JCS
Stevenson J M6ICO
Stevenson J GI0SSA
Stevenson M G1GPM
Stevenson M G8ALS
Stevenson M 2E0MAZ
Stevenson M M0MAZ
Stevenson M M3XIO
Stevenson M GI4XSF
Stevenson P G1AEI
Stevenson P G7KDI
Stevenson P M5DAD
Stevenson P G8YMM
Stevenson P 2E0PGS
Stevenson P M3PGS
Stevenson P G7TRB
Stevenson R MM3PEV
Stevenson R 2E0RMS
Stevenson R 2E0RMS
Stevenson R M0TDC
Stevenson R M3RMS
Stevenson W GM4WTS
Steventon J M6ECX
Steventon M G4GWH
Stew J 2E0JRS
Steward A G0CHR
Steward D G4OLU
Steward I G3ZRG
Steward R G7JSW
Steward R M1DNG
Stewardson P G0EVM
Stewart A GI0PCU
Stewart A GM4TOQ
Stewart A GM6UHC
Stewart A GM6YRN
Stewart A MM0GFP
Stewart A MI6POF
Stewart A MM0CZM
Stewart A MM6DQG
Stewart A G G4BRB
Stewart C MM0CTU
Stewart C MI3PZV
Stewart C GM0TFE
Stewart C MM0GNS
Stewart C MM6CCS
Stewart C MW6PNZ
Stewart D 2M1EPV
Stewart D G0LEP
Stewart D G0RHG
Stewart D G3TIR
Stewart D G4UVG
Stewart D MM0CTT
Stewart D MI1FIS
Stewart D MM3TNG
Stewart D 2E0DMS
Stewart D M3ZTU
Stewart D M6KDC
Stewart D G4UEO
Stewart D M1FEK
Stewart G G6HEJ
Stewart G M6SON
Stewart I G1IMS
Stewart J GM1MDO
Stewart J G6BWN
Stewart J G6YRB
Stewart J G7NJD
Stewart J MI0AFT
Stewart J MM0DMU
Stewart J MI1DPL
Stewart J 2E0CUV
Stewart J M6GRU
Stewart J M6JJS
Stewart J 2M0JOK
Stewart J G0CPR
Stewart J MM0ZXI
Stewart J M1BQF
Stewart J MM6JOK
Stewart K G0TWV
Stewart K M3HGH
Stewart L 2I0LBS
Stewart L G8TLL
Stewart L MI0LBS
Stewart M G1EIZ
Stewart M G4RNW
Stewart M G6OOK
Stewart M MM3WYM
Stewart N MM3WYI
Stewart N 2M0KPZ
Stewart N MM6KLZ
Stewart N GM1CNH
Stewart N MM0BMG
Stewart P G7EAH
Stewart R GI7PJF
Stewart R M6FAX
Stewart R G4PBP
Stewart R G8BHH
Stewart S GI7MWA
Stewart S M3YSS
Stewart S M0SSR
Stewart T GM0BKX
Stewart T 2E0TDS
Stewart T M6TDS
Stewart V GM3OWU
Stewart W GI0PJH
Stewart W G3RLT
Stewart W GI4EIZ
Stewart W GI4MYT
Stewart W G7PHR
Stewart W G8MNO
Stewart W GM8YRT
Stewart W 2E0DSS
Stewart W M6DSS
Stewart W M0BAP
Stewart W M0WUL
Stewart-Roberts H M6HSR
Stewart-Whyte B M3WSU
Stickland A G4LUN
Stickley C MD6LXW
Stiddard S M6SBS
Stiff B G1NEZ
Stiles E G0BHK
Stiles M M6EQQ
Stiles R G3UEN
Stilgoe D G1NYN
Stilgoe G G0NRF
Stilgoe T G0MLH
Stiling W G6MBH
Still A G4KZX
Still D G8FVJ
Still R G8PVK
Stiller C G0AUI
Stillman M 2E0SCK
Stillman M M0HJB
Stillman M M6BXJ
Stillwell D G8SIU
Stillwell S G7MHV
Stilwell P 2E0POZ
Stimpson A G1BJE
Stimpson D G7GRO
Stimpson J G4YSF
Stimpson J M3RUR
Stimpson P G7HGT
Stimson D G3THC
Stimson K MW3BOP
Stinchcombe I G4DMM
Stinson D M3ZUL
Stinson D 2E0VAN
Stinson D M0YEP
Stinson D M3VXC
Stinson R GI0UJG
Stinton D G0VBZ
Stinton H GM6IDF
Stinton I G6MQK
Stinton M M3JZK
Stinton P G4TPS
Stirk A 2E0UPU
Stirk A M0UPU
Stirk A M6UPU
Stirk B G6EQD
Stirland A G3LZC
Stirling C GM8MOI
Stirling J GM3UWX
Stirling J GM0SFQ
Stirling M M3ESS
Stirling N M3NST
Stirling W GM4DGT
Stirrup L G7KZY
Stirrup R MI3UFD
Stirrup T G6LOC
Stirzaker I M0VPE
Stisted F M6FHH
Stitt J GI7WCS
Stitt M M6GOF
Stitt T GI4RXM
Stoaling A GU7OYU
Stoate R G6VPW
Stobbs S G4OOK
Stobbs W M6RWB
Stock J G7NAI
Stock L MI6LSY
Stock W G4XYH
Stockbridge P G4OTI
Stockdale C 2I0TUI
Stockdale C MI0HRO
Stockdale M G0MKK
Stockdale P G7TBC
Stockdale R G8JRN
Stocker C G8FQZ
Stocker D 2E0DSI
Stocker D M6DQO
Stocker G G4DZK
Stocker K G1SVL
Stocker K MD3KHW
Stocker M M3MSJ
Stocker S M0SDS
Stocker V M0VCS
Stockill T G4GPQ
Stocking J G4IWI
Stocking M M0CTK
Stocking P M0GSX
Stockley A G8ELP
Stockley D G4ELP
Stockley G M0IDL
Stockley J G3FMW
Stockley J G8MNY
Stockley J G4KOK
Stockley K G4UQN
Stockley K M3TNH
Stockley L G3EKE
Stockley P M1BYI
Stockley R G7MVY
Stockley W G6RGN
Stocks A G0OPQ
Stocks D G4VUM
Stocks D G0VTU
Stocks G M3LUW
Stocks J M1CZM
Stocks J M3CZM
Stocks J G0BFJ
Stocks P M0OWS
Stocks T G4ZGE
Stockton A G1PKG
Stockton D GM4ZNX
Stockton D M6CNK
Stockton D M6DMA
Stockton J G6IKU
Stockton M G4SZX
Stockton M M0AAD
Stockwell A M1SJA
Stockwell C MU6STK
Stockwell L M1DPE
Stockwell N G0RIK
Stockwell P G6KFD
Stockwell R MU0FBO
Stoddart D G4VMB
Stoddart G 2M0FSF
Stoddart G MM0GTG
Stoddart J G1JMS
Stoddart R M0BYJ
Stoddart R M0BZA
Stoddon R G4DLP
Stoelwinder G G6IJW
Stoeteknuel H MM0XAU
Stogdale H G4FEQ
Stoker A G8BPW
Stoker D G1GEY
Stoker G M0EUK
Stoker N M0NRS
Stokes A G3ZRH
Stokes B M3IIN
Stokes C G1ZQO
Stokes C G6MVF
Stokes D G4HTY
Stokes G G0FVB
Stokes G M0BVQ
Stokes G M3ZJS
Stokes G G0BYH
Stokes H G0VNE
Stokes H GI7IPO
Stokes J G0IIA
Stokes J G6VFO
Stokes J M6KFG
Stokes K G1FSE
Stokes K M3YBN
Stokes M G3ZXZ
Stokes M 2E0BKN
Stokes M M0MAY
Stokes M M3RPZ
Stokes M G4HCY
Stokes N M6NIQ
STOKES P M6HXG
Stokes R G4TDV
Stokes R M0RWS
Stokes R M3ZRS
Stokes R M6BJQ
Stokes S G1BKB
Stokes T G0LNE
Stokes T G0WTD
Stokes T G1ASG
Stokes T G7NER
Stokes T 2E0EBP
Stokes V M3NXQ
Stokes-Herbst P G4VZC
Stokoe J M0HTU
Stoll A M6GNV
Stollard G M0GES
Stolting D GM0SEF
Stone A G1JWO
Stone A G3UIS
Stone A G4OJR
Stone A G4VSI
Stone A G8VEQ
Stone A G1NMW
Stone A MI3VXQ
Stone B G3JFC
Stone B G6SRE
Stone B M6BSW
Stone B G4BPJ
Stone C G7BYW
Stone C M0ONS
Stone D 2E0DPG
Stone D G8NGF
Stone D M3AYL
Stone D G8FNR
Stone E 2E0LZE
Stone F 2E0BUM
Stone F M3NWQ
Stone F G6VEJ
Stone G G1CWW
Stone G 2E0STO
Stone I M6ISA
Stone J G0DFE
Stone J MM0AJQ
Stone J M3HNL
Stone J G4ITB
Stone J G6OOH
Stone K 2E0KEG
Stone K G4UBT
Stone L M0EPX
Stone L M1LTS
Stone M G0OXZ
Stone N M6LPI
Stone P G3LWD
Stone P 2E0ZKT
Stone P M3UER
Stone P 2E0PAB
Stone P G7EVC
Stone R G1EJA
stone R G4TVW
Stone R G7WFZ
Stone R G8SEE
Stone R G3YDX
Stone T G1SUP
Stone W M3WMS
Stonebridge P G8ZQA
Stonebridge P M3ZQA
Stoneham M G4RVV
Stonehouse A G6CCB
Stonehouse J GW0JDS
Stonehouse W G3HPC
Stoneley B G0RLB
Stoneley B M3SBB
Stoneman R G6FAL
Stoneman W G6UNA
Stoner B G6ZRO
Stoner B M3ZRO
Stones G G7KYX
Stones I GM0SZA
Stones J M6JSS
Stones S G0NER
Stoney D G8PTN
Stoney S M6SES
Stooke A M6KCS
Stooke E M3ZRP
Stoole D G4RIB
Stopford A M6DZU
Stopford J G8UWS
Stopforth W M3FNH
Stopforth W G4LBX
Stoppard J 2E0BNK
Stoppard J M0ZIG
Stoppard J M3SRQ
Stoppard R G3XAZ
Stopper M M6BXB
Storace-Rutter W G0WLF
Storer A 2E0AWS
Storer A M3BSI
Storer D 2E0CXA
Storer D M6BKH
Storeton-West B G8JUK
Storeton-West T G8BTX
Storey A G3TEH
Storey A M6FVN
Storey B G4ZGG
Storey C G3HTC
Storey D G8CYX
Storey G G8LIH
Storey J G1JON
Storey J G8SH
Storey J 2E0DUD
Storey J M3YVT
Storey N G8YCQ
Storey P G4OIQ
Storey P G3YBH
Storey P G3ZGG
Storey S 2E0PSH
Storey S M0HTI
Storey S M6DVV
Storey T G8HZS
Storie H MM0GWO
Stork F G3TEE
Storkey B 2E0EWN
Storkey M G0MVE
Stormes E 2E0EDI
Stormes R G7COU
Stormont A G3ZPM
Stormont W G0EGC
Storr C 2E0CIK
Storrie C G7JBD
Storry A 2E0FRQ
Storry B G3WSM
Storry J G4OID
Storry L M0LSS
Stoten C M6LSE
Stothard A G0JNG
Stott A M6CUX
Stott B G1WAP
Stott D M6XTY
Stott G G3ZRY
Stott J G0CJX
Stott L M3OBN
Stott M G1YRJ
Stott M G0NEE
Stott P 2E0BUJ
Stott P G0HEU
Stott S G4SFJ
Stout R GM1CEJ
Stout R G0NRM
Stow G G4MCU
Stow K M6KGS
Stow T G0SWS
Stowe K M6BYA
Stowell J G0THJ
Stracey B G7MKF
Stracey M G0UBL
Stracey P G4HEC
Strachan A GW4IFE
Strachan A GW7HFZ
Strachan A GM7NNS
Strachan I 2E0ARV
Strachan I GM4FLP
Strachan M 2M0LEW
Strachan M MM0LWS
Strachan M MM6LEW
Strachan W GM3ZRT
Strachan-Buckley P M0NGI
Strachan-Buckley P M6NGI
Stradling M M6OMS
Strafford R G3MRT
Straker B M6FVP
Straker R G8BVQ
Straker R G8ELW
Strand T G6FNY
Strandberg J M0HZY
Strang G GM0KZJ
Strange A M1BEQ
Strange A M6EDN
Strange C 2E0FBK
Strange C M6YHV
Strange D G1ERS
Strange G G8IWJ
Strange I G8RNU
Strange J M0VRS
Strange M G8HHO
Strange M 2E0MBS
Strange M M0MSZ
Strange M M6MMS
Strange S 2E0DLO
Strange S M0SYS
Strange S M6EYP
Strangeway R G4FOW
Stratford B M3CEZ
Stratford C M6CPS
Stratford G 2E0UKG
Stratford G M3UKP
Stratford R M0HMB
Stratford S 2E0SMS
Stratford S M3SYL
Stratfull J G3IJS
Strathdee B GM4REN
Strathdee K GM0LDX
Stratton H G3HCS
Stratton H G8HCS
Stratton J G4AHM
Stratton J G7HMU
Stratton R G3XKV
Stratton R MM3YGI
Straughan J G7SCV
Straughan M 2E0NER
Straughan M M6FMC
Straughan R G6VED
Stravens R G1IHA
Straw A M1MOD
Strawbridge J M3WNZ
Strawbridge M M0RTY
Streatfield P M3ZUT
Street A G0JYI
Street A G4KSY
Street A 2E0FTQ
Street A M6FTQ
Street C G8XET
Street C M3LVK
Street G 2E0GRS
Street G M3FIH
Street G 2E0CZW
Street G M6SGS
Street I MM0DEC
Street J 2E0CQZ
Street J M0JSR
Street J M6BXK
Street M G0WFL
Street M G4NXL
Street M M6UGX
Street M 2E0XMS
Street M M0LRO
Street M M6MSX
Street M G3JKX
Street P G1PHS
Street P G8ZES
Street P M0GAN
Street R G3TJA
Street S G3VFU
Street S 2M0ZFG
Street S MM0ZFG
Street S MM6ZFG
Street T G0IFD
Streeter B 2E0BSF
Streeter D G1OOW
Streeter M M6MST
Streluk S G0PFQ
Stretch S 2E0CBH
Stretch V G4LVO
Stretton K M3KPG
Stretton P M3ZZL
Strevens A G0MGH
Strevens C G4ZHT
Strickland F M6CXN
Strickland J G0GDV
Strickland M G6WSF
Strickland M 2E0ZMS
Strickland M M0ZMS
Strickland M M3ZMS
Strickland P 2E0CPX
Strickland P M6SOE
Strickland T G4EOA
Stride A G7MYI
Stride A M6CFV
Stringer C G6IDG
Stringer D 2E0DAL
Stringer D M3ZCX
Stringer G G7CJG
Stringer J G0LNI
Stringer J G3TYO
Stringer J 2E0FPV
Stringer J 2E0SLM
Stringer J M3YHG
Stringer J M6TDJ
Stringer K G0COG
Stringer L G4GZG
stringer R M6HPF
Stringfellow P M3ECJ
Stringfellow R G4TZR
Striplin P G1FXS
Stripp A G7VEI
Strobel D G0EVY
Strode E G6XYV
Stromstedt A M0GZX

Stronach J GI3LQY
Stronach P MM3PTS
Strong A M6AHS
Strong C GM1MIG
Strong M G1UMY
Strong N G0CWA
Strong P G0DIH
Strong R M6SRD
Strong R 2E0RES
Strong R M6RES
Stroud C G1BJN
Stroud M G8IMS
Stroud M 2E0HOG
Stroud M M6STR
Stroud R GM4PSJ
Stroud R G7OMM
Stroud R G7WJV
Strowger M M3GZJ
Strudwick A 2E0LHC
Strudwick M G8TPP
Strudwick P G8CDB
Struthers J GM8CVN
Strutt B G1XZX
Strutt J G4BON
Strutt J G4XTS
Stuart A G3ZHB
Stuart C M1APS
Stuart D M1DOR
Stuart D M6ILY
Stuart E MM3IZO
Stuart F G8FEZ
Stuart G GM0VFY
Stuart H MM0HAR
Stuart I G8DOB
Stuart J GM4SFW
Stuart J G8LWC
Stuart J GM7LSI
Stuart M G6BQC
Stuart M MM0MNS
Stuart P G0JCY
Stuart P 2E0CWE
Stuart P M6CIM
Stuart R 2M1DZX
Stuart S M3MBI
Stuart S G4XSG
Stuart W G0TKF
Stuart W M0LOG
Stubbs A M6TGS
Stubbs C G1IZD
Stubbs C M0LZZ
Stubbs D G4EFD
Stubbs D M6DST
Stubbs E M6EMZ
Stubbs J G8GUJ
Stubbs L G4DBX
Stubbs M G8IMB
Stubbs N G1JJT
Stubbs N M3XNX
Stubbs P G7VCP
Stubbs R G8XLL
Stuckey E GW0DLA
Stuckey I G6TEQ
Stuckey M G0SQK
Stuckey R GW1NZF
Studd G G0WMG
Studd J G1MGU
Studdart A 2E0DPI
Studdart A M6AFK
Studdart C M6CLF
Studdart H G7AAU
Studdart P G0PNO
Studdart S G7AAV
Studeny R M6RIL
Stump D G7CRM
Stumpf W G4PUO
Stunden P G0MTY
Sturgeon C 2M0LXX
Sturgeon C G4NOT
Sturgeon J 2E0JON
Sturgess A G1LTK
Sturgess A M3HEV
Sturgess B MW6BSR
Sturgess P G0HZG
Sturman A G1CDQ
Sturman A G4HIQ
Sturman B M6AWI
Sturman J G7EQG
Sturmey T G4MJI
Sturrock F MM6CFS
Sturrock I GM4DEK
Sturrock J GM0COK
Sturt A G8SIK
Sturt M M3IUC
Styler S G8IHC
Styles R G3NSD
Styles J M0JPS
Styles J GW4HZM
Styles M 2E0SYS
Styles P G4PXX
Styles R 2E0TOP
Styles R 2E0RKS
Styles R M0ZRS
Styles R M6AYP
Styles R 2E0TFO
Styles R M0TFO
Styles R M3TFO
Styles R 2E0XCM
Styles R M0LEX
Styne N G6OVA
Stypka M MM6BYJ
Su H M3MBV
Suart I GM4AUP
Such K 2E0UVZ
Such K M0HBO
Sucharyna Thomas L M3DXL
Suckling A G0WLX

Suckling C G3WDG
Suckling P G4KGC
Suckling R G6MVN
[illegible] [illegible]
Suddaby M M6WMW
Suddell C 2E0PML
Suddell C M0VUE
Suddell C M6EGK
Suddes D G0HKQ
Suddes K M1DYP
Suffling R 2E0BQN
Suffling R M3XLY
Suffolk N G1ZYN
Sugden G G1OAZ
Sugden J G6GNE
Sugden R G0GLZ
Sugg K G8TTX
Sugg N 2W0DTM
Sugg N MW0KIJ
Suggate G G3NPI
sugihara S 2E0WGI
sugihara S M0WGI
sugihara S M3WGI
Sugrue J G6TRX
Suleyman M M1MUS
Sulieman A M3YKI
Sullivan A G6TSE
Sullivan D 2E0DSX
Sullivan D M0XDA
Sullivan D M6NSX
Sullivan G M6RFI
Sullivan J G4HLA
Sullivan K G3KYF
Sullivan K MI6KZS
Sullivan L G8GVZ
Sullivan M M0GKW
Sullivan R M6SUL
Sullivan S G0TJE
Sully E G0VIQ
Summerfield C 2W0DSO
Summerfield C MW6CQL
Summerfield O M6OSU
Summerfield R G0LDO
Summerfield T M1ASS
Summerhill J 2E0BYQ
Summerhill J M0MMO
Summerhill J M6BGT
Summerhill K G4SUM
Summers A G4MYU
Summers A G4KNO
Summers B G8GQS
Summers C G0GSH
Summers E G1TYP
Summers H G0UPL
Summers J MM0JBS
Summers J M5JWS
Summers J 2E0OEM
Summers J M6WDW
Summers K MW3LSL
Summers K M0CWZ
Summers M 2E0CFI
Summers M M0HAH
Summers M M6CFO
Summers M M6PDQ
Summers N G3RKJ
Summers R G0LTO
Summers S M6SPS
Summers T M1ALU
Summers T M6SEC
Summerwill D M0DWS
Sumner A 2E0AED
Sumner A M3NER
Sumner D G3PVH
Sumner D G6HPT
Sumner D M1SUM
Sumner I G3VPX
Sumner K G0DTI
Sumner K 2E0KNF
Sumner M G6AHD
Sumner S G6BRW
SUN W M3UDS
Sunderland D G6FHM
Sunderland D M0FHM
Sunderland G G8UTH
Sunderland J G3TQC
Sunderland M MW6CTE
Sunderland M G1BQV
Sunley F G0LEL
Sunley N 2E0NCS
Sunley N M6NCS
Sunouchi T G0WND
Sunter G G4VOB
Sunter R G6HMN
Suresh S M3EQL
Surgey D G6MBF
Surgey T G3ZGT
Surman D M0BWF
Surman J G7CKP
Surooprajally A G0EAU
Surplice M M0AZE
Surrage R G0MUQ
Surrey T M6REY
Surtees B G1YNO
Susa P M0SPK
Suslowicz G G8KGS
Sutcliffe C G3MZC
Sutcliffe C G6MBV
Sutcliffe G G8TXW
Sutcliffe J G0TFK
Sutcliffe M G3VAK
Sutcliffe M G8VSI
Sutcliffe P G4YYO
Sutcliffe S G0MPI
Suter G 2E0BHP
Suter G M0GJS
Suter J M3UJS

Suter S G0GXI
Sutherland B M0CVP
Sutherland C M3SUT
[illegible] [illegible]
Sutherland D 2E0FOK
Sutherland D M6FOJ
Sutherland G GM0UPE
Sutherland G MM0GXU
Sutherland G G0WCZ
Sutherland J GD1MZJ
Sutherland J G4LOV
Sutherland J G7USG
Sutherland M G0AIG
Sutherland N M6AXS
Sutherland P G7RVC
Sutherland R GM7BCC
Sutherland R MM3BCC
Sutherland S G0TTE
Suttenwood D G4PBR
Suttenwood R G6KNM
Suttie D 2M0XDS
Suttie G MM3XGS
Suttie A 2E0SUX
Suttle A M0SGA
Sutton A G3XSR
Sutton A G6VDA
Sutton A 2E0MUS
Sutton A M6MUS
Sutton A GM6ZAK
Sutton A G0XPD
Sutton B M0TUX
Sutton C G3UHV
Sutton C M3YCS
Sutton D G0IPH
Sutton D G3ZAJ
Sutton D G0WTM
Sutton D G7SSJ
Sutton D G8VRN
Sutton E G7GQM
Sutton G G4EVW
Sutton J G0CHN
Sutton J G0TDM
Sutton J G3TVY
Sutton J G6JUP
Sutton J G7FMQ
Sutton J G7GQL
Sutton J G6RTG
Sutton M G1PEK
Sutton M G8KEA
Sutton M M6FIB
Sutton R 2E0DUI
Sutton R G0ODI
Sutton R M6FLZ
Sutton R G1WVV
Sutton R M3ZCE
Sutton R 2E0SUT
Sutton R M6RDS
Sutton T 2E0TGG
Sutton T M6TGG
Sutton T M3CCS
Sutton W GM3YKP
Sutton W G3FWI
Sutty M M3KWR
Suyat B M1CQM
Svanda Z M6ZPS
Swaby D M3DVH
Swaby M G4XOE
Swaddle H G0GXO
Swail W GI4YPR
Swain A G3KWY
Swain A G8VYO
Swain A M3WAV
Swain A M6KLX
Swain C 2E0BIT
Swain D G3KMS
Swain E 2E0CKQ
Swain G 2E0IDE
Swain G 2E0BFJ
Swain G M3JYH
Swain G G0JON
Swain J M0JAK
Swain J M6NRS
Swain M G6RHJ
SWAIN M G8MMP
Swain M G7VNJ
Swain N G4CBS
Swain P G4GXQ
Swain S G4LPF
Swain S G0FYX
Swaine F G1FWS
Swainson D G3OXN
Swainson P M6DNL
Swale D G8ETS
Swales A G1DWU
Swales J M6ITQ
Swallow A G8NOS
Swallow J G7PDH
Swallow N G0LFA
Swallow P G8EZE
Swan A 2E0DVD
Swan A M0EUY
Swan G GW4FXF
Swan G G8ASJ
Swan I G3MNS
Swan M 2E0SKE
Swan M M6FGE
Swan N G1BLU
Swan P G0WIE
Swan P M0DYG
Swan S M3EUY
Swancutt S M1CSC
Swanepoel E M0ESW
Swanepoel P M3PSS
Swann D 2E0AII
Swann D M3KVI
Swann G G0WSD

Swann J M6MYL
Swann J M0BUY
Swann M M0XWS
Swann N G1TFX
Swann N M1NAS
Swann P M3SOQ
Swann R G6GQI
Swannick P M6PHS
Swansbury P M0GIY
Swanson D G3OSI
Swanson J G3NPC
Swanson J G4CBD
Swanson P G3NNV
Swanston M M6RKS
Swanton P M1EFT
Swanwick J G7UVP
Swanwick M G6CZD
Swanwick V G7TMU
Swarbrick P G3ZGN
Swarbrook D G4TGS
Swarbrook P G8YMS
Swartz J G0UCE
Swatton J G0PYJ
Swaysland G G4DNE
Swearman A GD7KAM
Sweatman J M0JDS
Sweatman P M0GMU
Sweeney A MI0HPX
Sweeney B M3SYY
Sweeney B G1ETQ
Sweeney C MW6CJS
Sweeney D M0KFU
Sweeney E MW6ESW
Sweeney M M0SWE
Sweeney P G4VUI
Sweeney P MI6PFI
Sweeney R G0PMS
Sweeny P G3SXT
Sweet A 2E0UZK
Sweet A M6UZK
Sweet B M6BAS
Sweet D G7FMV
Sweet G M3YXH
Sweet H G1KQE
Sweet M G1INU
Sweet R G6STK
Sweetapple A G4FVU
Sweeting P G0GMA
Sweeting P G6MVD
Sweetingham F G4ROS
Sweetland D G8LMY
Sweetland D M5LMY
Sweetlove S M1EIJ
Sweetman B G3ROM
Sweetman E G3UAZ
Sweetman M G1YLN
Swetmore R G3VTE
Swietlik R M3RCS
Swiffen J M0JSP
Swiffin A GM8OEG
Swift A G1PRW
Swift A G6WGA
Swift D G7EEJ
Swift D M0AVK
Swift D 2E0DNC
Swift D M0SFT
Swift D M6DNC
Swift G M0ONI
Swift J G3CTP
Swift J M3MKZ
Swift J 2D0BWJ
Swift J M0UTX
Swift J M3YLK
Swift J M3YLL
Swift M G4MJA
Swift M M3MSZ
Swift M G6NJJ
Swift N G7ENS
Swift N 2E0NPN
Swift N M3SQZ
Swift P G0MIH
Swift-Hook J G1DFI
Swinbank P GW8AHB
Swinbank R M0DTS
Swinbourne S G0LJV
Swinburne R M0WSN
Swindale L 2E0AFN
Swindells G 2E0PDX
Swindells G G3THV
Swindells I 2E0SWZ
Swindells I M0SWZ
Swindells M M6ZLM
Swinden J G0NSZ
Swinden S M5CKN
Swingewood J G4CVU
Swingewood P 2E0CVU
Swingewood P M0TVU
Swingler A G0SIE
SWINNERTON R G4MXE
Swinney R G0FRZ
Swinyard P M6PHE
Sword A M0SWO
Sword G G6ILC
Swynford P M3PUB
Swynford P G0PUB
Swynford R M3HBX
Swynford-Lain R G7AOK
Sycamore A G3IVC
Sycamore M M6CHP
Sycamore P 2E0PBS
Sygerycz A M0SYG
Sygerycz A M1SYG
Sykes B G2HCG
Sykes C G3XJZ
Sykes D 2E0MYE
Sykes D G0JOX

Sykes D M3MYE
Sykes E G6HYJ
Sykes J G3GGR
Sykes J G0CML
Sykes K G6NXW
Sykes M G4GAK
Sykes M M0MJC
Sykes P M3BEE
Sykes P M3PWS
Sykes P 2E0DYQ
Sykes P M6SYK
Sykes R G3NFV
Sykes R G7ROP
Sykes R M0HMX
Sykes S M1EGD
Sykes W G0RIJ
Sylvester D G3RED
Sylvester K M1ERY
Sym A G6JSN
Syme G GM3EFH
Symes J G3LNN
Symes N M6CWV
Symington R GI6TFF
Symmonds M 2E0RTN
Symmonds M M0RUM
Symmonds M M6KBI
Symonds A G8DQK
symonds D G1VHY
Symonds D M3IFB
Symonds G 2E0DRY
Symonds G M6BBF
Symonds G G3XPT
Symonds J G0RPU
Symonds J 2E0JSS
Symonds K M3KLS
Symonds K G4NUJ
Symonds L M6UDA
Symonds P M1PFS
Symonds P M6BBD
Symonds S M0PDL
Symonds S M3SYZ
Symonds T M3AYC
Symons C 2E0SYM
Symons J M0GUN
Symons L G3UFJ
Symons M G3ZPJ
Symons T MW0CNC
Symons T 2E0ZSY
Symons T M3SYV
Symons W 2E0WVS
Syms C G8VYP
Szabo G M6YUL
Szabo I M6HET
Szabo P M6PSO
Szendzielarz V G0ONF
Szewczyk P M0HFB
Szikszai J M6CVV
Szikszai Z M6CVU
Sznober Z 2E0ZGS
Szondy L G4OJW

## T

Taaffe S G4DQT
Tabberer H G3YVK
Tabberer M 2E0KGD
Tabberer M M6KGD
Tabberer P G8JJP
Taber K G7FTD
Taberer D M1CWD
Taberner F G0UGS
Taberner J G6BWO
Taberner J M6UNA
Tabor K M6HSG
Tabor R G7EUB
Tachibana M M0DHU
Tackley K G3BRQ
Tacon C M6FZQ
Tadesse K G7VYF
Taft J G0MSS
Taft M G7IMR
Tagg G 2E0KGT
Tagg G M0ICG
Tagg G M6GTK
Tagg P G8PIQ
Tagg R G1BCU
Taggart D 2E0TAG
Taggart D GI0OXK
Taggart J M3HSC
Taggart J GI4GNT
Taggart N M3NGE
Taggart W G4BKR
Taggerty P GM0IJD
Taha H G1MLC
Tahla M G7TTX
Tailford D M3OTP
Tait A G0CTD
Tait A GM4LBE
Tait C GI0TSS
Tait C MM3LZU
Tait C 2M0YAF
Tait D G3AFB
Tait J G8MGF
Tait K MM3IDH
Tait M M0CSB
Tait R 2M0CTN
Tait S M6SRS
Tait V GI4LKG
Tait W G4EHD
Talaber D GD0CAG
Talabi A M1ETW
Talbot A M6TBT
Talbot A G4JNT
Talbot B G1BIN
Talbot C G0EYU
Talbot D G0JHT
Talbot D G6PBN
Talbot G M3GLT

Talbot G G3VAL
Talbot I 2E0ZXQ
Talbot I M0ZXQ
Talbot I M3ZXQ
Talbot J M0JTT
Talbot L M3LIA
Talbot M 2E0MGL
Talbot M M3MGL
Talbot N G2DZH
Talbot N G4EBQ
Talbot P G8BAZ
Talbot P M1DGQ
Talbot R G0SBV
Talbot R G3ORK
Talbot S G1TOL
Talbot-humphries T M6ZGY
Talbot-Jones J M0JTJ
Talbott M G4NWM
Talbott R G4LRO
Talbott W G4NWN
Taljaard G 2E0GUT
Taljaard G M6GUT
Talkowski L G0SDQ
Tallentire J G4AFE
Tallis T G4UUQ
Talmage F G0LVF
Talplacido A MM6EWX
Tam N M3TND
Tam W M3TWK
Tame C M3XAG
Tames S G8LCL
Tamir G M0HMI
Tamkin C G3EWT
Tamlin J G3LCY
Tamlyn J G7WLM
Tamplin A G4XCE
Tamplin P G0VGR
Tams D M6XDW
Tams R G1OQB
Tams R G4TCG
Tan C M1BBH
Tan J M0JPT
Tan P M0HUH
Tandy C G6FGJ
Tandy G G8VKO
Tandy M G0MJT
Tandy P G1VOQ
Tanfield R G6CVE
Tang H M0THY
Tang H M6THY
Tanji M M6UED
Tankard V G1KSN
Tankaria D G6ESM
Tann M G6ZGC
Tannahill G G6JQE
Tannahill R GM0ERT
Tanner B G6HUI
Tanner C 2E0GLV
Tanner C M0LLK
Tanner C M6GLV
Tanner D M0DAC
Tanner D G4FLR
Tanner F GI0ZAK
Tanner J G8AER
Tanner J M6GTC
Tanner N 2W0PEE
Tanner N MW6NRT
Tanner P G4VTO
Tanner S G7NIU
Tanner S M3VTO
Tanseli A 2E0SIS
Tanseli A M6WYZ
Tansley N G1IRQ
Tanswell D G6LAU
Tant R G4WNP
Taperell M M0BEM
Taplin A M3IVD
Taplin M G3WAM
Taplin N M0BPN
Tapp R G1XXH
Tapping C G0LXB
Tarbatt D G7SKR
Tarbett K G0CKI
Tarbuck D G6IKM
Targett D G8FDZ
Targonski R G0BZT
Tarling M M6MIT
Tarling R M0TRK
Tarling R M6AIJ
Tarmey P G6AZL
Tarnowski D MI0SLE
Tarpey B M3RLT
Tarr D G3OUA
Tarr G G8YGO
Tarr G M0DCY
Tarr M GM0VKN
Tarr R G3PUR
Tarr R M6RYA
Tarran C G8DXF
Tarrant C G0KRH
Tarrant D M0GUJ
Tarrant J 2E0DFG
Tarrant J M0HWP
Tarrant J M6EQK
Tarrant M 2E0CYZ
Tarrant M M6ATZ
Tarry D G4FKP
Tarry D G0WFO
Tarry G G4WGU
Tarry K G0FIC
Tart E 2E0GQW
Tart E M0GQW
Tart E M3GQW
Tartt N M1NTV
Tarver F G8AIM
Tarver J M3JHT

Tasker G 2E0OFK
Tasker K M6KTZ
Tasker P G6VBJ
Tasker S M0SMT
Tate A M3UUX
Tate C M0SUN
Tate C M6IBH
Tate D G6ZFO
Tate D M6OTB
Tate H M3HMT
Tate J G3LGT
Tate M G3MHX
Tate M G4YGQ
Tate M M3RGE
Tate T M6TST
Tateishi N M0NAO
Tatem S G1SPT
Tatham L G1IEB
Tatham P M3ZPT
Tatlow C G0FCB
Tatlow D M1BEO
Tatlow F M1ERF
Tatlow K G4XLO
Tatlow M G6NGF
Tatman V G4IMH
Tatnall B G4ODA
Tattersall D M3TLD
Tattersall P G4SYG
Tattersall R 2E0DPQ
Tattersall R M0MLJ
Tattersall W G4WHT
Tatterson G G0ETL
Tatterton A G6TWX
Tatton R GM3SRV
Tattum S M3OEF
Tatum W G4OJB
Taute L M3XKI
Tavender A M0BVN
Tavener J G4XCM
Tavener R M0CSV
Taverner S G7VDX
Tawn J G4EMT
Tawney A GD1USI
Tayler B G4KUK
Tayler J M6JLA
Tayler S 2E0SVT
Tayler S M6SVT
Tayler-Grint M M6FSD
Taylerson D G3PPC
Taylforth C G7WEK
Taylforth S G7SNQ
Taylor A 2E0ALF
Taylor A 2E0TAY
Taylor A 2E0GNR
Taylor A G0TPA
Taylor A G1THW
Taylor A G3NYE
Taylor A GW3TMJ
Taylor A GM4HBQ
Taylor A G4PVN
Taylor A G4RZE
Taylor A G4SSC
Taylor A G4ZII
Taylor A G6XNI
Taylor A G7NKV
Taylor A G1MSA
Taylor A G4XTZ
Taylor A M0AUR
Taylor A 2E0VAJ
Taylor A M0WNT
Taylor A M6ANT
Taylor A 2E0HTM
Taylor A M6AYN
Taylor A MW6NHC
Taylor A 2E0GPU
Taylor A 2E0TMF
Taylor A G7EKC
Taylor A G8MXQ
Taylor A M3EKC
Taylor A M3YEK
Taylor A M6AMT
Taylor A G0LAN
Taylor A G4ALT
Taylor A M6AAG
Taylor A 2E0DIG
Taylor A M6KIA
Taylor B G0EIA
Taylor B G0TPG
Taylor B G1HJD
Taylor B G3TIN
Taylor B G3ZAG
Taylor B G6TOX
Taylor B G7CKL
Taylor B G8TPC
Taylor B 2E0JLR
Taylor B M6FCF
Taylor B M0NSI
Taylor B M0GPN
Taylor B M3LOY
Taylor C G0GYA
Taylor C GM0IMH
Taylor C G3USA
Taylor C G4OTD
Taylor C G7MCT
Taylor C G8DXM
Taylor C M0ROT
Taylor C MM0TYR
Taylor C M1CBT
Taylor C M6CAR
Taylor C 2E0PNK
Taylor C M6PNK
Taylor C M6THQ
Taylor C 2E0DOE
Taylor C G1SCN
Taylor C M6EYX
Taylor C G1LBM

Taylor C 2E0CDT
Taylor C G7VTR
Taylor C M0HLC
Taylor C M3ZRR
Taylor D 2E0HNI
Taylor D 2E0EXI
Taylor D 2E0HUF
Taylor D G0AUN
Taylor D G0GBL
Taylor D G0LSK
Taylor D G0VZZ
Taylor D G1VDE
Taylor D G1YSX
Taylor D G3IZF
Taylor D G4HRB
Taylor D GM4RZW
Taylor D G6CIF
Taylor D G6EWJ
Taylor D GM6JWH
Taylor D G7RWN
Taylor D G8APW
Taylor D G8SWK
Taylor D MM0DRT
Taylor D M3CUU
Taylor D M3OHX
Taylor D G0EGH
Taylor D 2E0KZH
Taylor D 2E0LBZ
Taylor D G1TBT
Taylor D G4EBT
Taylor D GD4FMB
Taylor D GM8FMR
Taylor D M0DCS
Taylor D M0KZH
Taylor D M6BPD
Taylor D M3CYM
Taylor D 2E0DLR
Taylor D G4NXP
Taylor D M6EQT
Taylor D G4PQX
Taylor D G0VLI
Taylor D 2E0DZZ
Taylor D M0GZZ
Taylor D M3UQN
Taylor E 2E0CAT
Taylor E 2E0EAT
Taylor E G6FKB
Taylor E G7RZN
Taylor E M0ITV
Taylor E MI0MFI
Taylor E M6CKJ
Taylor E MI6EAI
Taylor E G3SQX
Taylor E M6EOW
Taylor E M0LCA
Taylor F 2E0EBJ
Taylor F G0SCU
Taylor F G1JMY
Taylor F G4DUO
Taylor F G8NSD
Taylor G 2M0DES
Taylor G 2E0GTL
Taylor G 2E0IGW
Taylor G G0FHO
Taylor G G0MJA
Taylor G G0UNF
Taylor G G1AZZ
Taylor G G1SDX
Taylor G G4JZF
Taylor G G4KPU
Taylor G G4PJY
Taylor G G6DNV
Taylor G G6JBQ
Taylor G G7CFT
Taylor G G7MHQ
Taylor G G7UJC
Taylor G G7VDK
Taylor G G8ZFU
Taylor G M3PNF
Taylor G MM3XUI
Taylor G 2W1EAN
Taylor G M0TLR
Taylor G M3UGT
Taylor G G6APB
Taylor G M3AHU
Taylor G MM6XGT
Taylor G M6GCT
Taylor G 2E0VPT
Taylor G G0PUW
Taylor G G7GWT
Taylor G M3VPT
Taylor H GM0CNW
Taylor H G3KFG
Taylor H G8GAR
Taylor H M0XXX
Taylor H M3YJT
Taylor I G3ORG
Taylor I G6VJA
Taylor I 2E0EDX
Taylor I G4LXX
Taylor I M3WUO
Taylor I M5CYM
Taylor I M6BJI
Taylor J 2E0JPT
Taylor J G0OJW
Taylor J G0TAO
Taylor J G4DFX
Taylor J G4KKG
Taylor J G4REU
Taylor J G4UVF
Taylor J G4VHJ
Taylor J G4VTA
Taylor J G4ZVK
Taylor J G6CVB
Taylor J G7PTM
Taylor J G8HTM

**IMPORTANT NOTE**

**Revalidate licence to avoid revocation** – Ofcom has advised the Society that plans will be drawn up to revoke licences that have not been revalidated as required by the licence conditions. The quickest way to revalidate is to do so online via the Ofcom website: *https://services.ofcom.org.uk/* or by email: *amateur.validations@ofcom.org.uk* If you need assistance in the process, Ofcom staff are available to help, but please be patient during times of heavy workload.

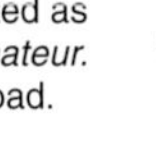

Taylor J M0BNP
Taylor J M0CYE
Taylor J M1CLZ
Taylor J M3FGO
Taylor J M3IDY
Taylor J M3JST
Taylor J M3KEW
Taylor J M3MQC
Taylor J MD3ONP
Taylor J G0RFN
Taylor J M6FAF
Taylor J G8ADQ
Taylor J G8TXX
Taylor J 2E0CWJ
Taylor J 2E0HLF
Taylor J G4ETM
Taylor J G6VJC
Taylor J G6XD
Taylor J G6XNJ
Taylor J G6YQW
Taylor J G8PUE
Taylor J M0HTE
Taylor J M3ZMU
Taylor J M6CNR
Taylor J M6HLF
Taylor K 2E0KRT
Taylor K 2E0HFX
Taylor K G0RLN
Taylor K G0TQV
Taylor K G1KKJ
Taylor K G3LME
Taylor K G3NNW
Taylor K G4GAI
Taylor K G4YQK
Taylor K G8TOX
Taylor K M0BBQ
Taylor K MM0KTE
Taylor K M3KRP
Taylor K G4ZTZ
Taylor K M6KTX
Taylor K G8EVR
Taylor K M6KEV
Taylor K M1SLH
Taylor K M6TQF
Taylor L 2E0HEK
Taylor L G0EBF
Taylor L G4ZJD
Taylor L G6FKL
Taylor L M0CMK
Taylor L M0JLT
Taylor L 2E0LKT
Taylor L G1XIH
Taylor L G2ALN
Taylor L M0LKT
Taylor L M3TUJ
Taylor L G0TJC
Taylor L G1LEN
Taylor L M3EXJ
Taylor L G1STQ
Taylor L M6CZU
Taylor L GD4ZAB
Taylor M G0EAE
Taylor M G0LGJ
Taylor M G1GSG
Taylor M G1NDQ
Taylor M G3UCT
Taylor M G3YSG
Taylor M G4OBC
Taylor M G4OMT
Taylor M G4VSW
Taylor M G4XDC
Taylor M G4ZIF
Taylor M G6AGR
Taylor M G7UXR
Taylor M G8BMP
Taylor M M0BZR
Taylor M M0LCY
Taylor M M1BVP
Taylor M M1MOG
Taylor M M3AGI
Taylor M G4YMT
Taylor M G8YUO
Taylor M GM4XRT
Taylor M M6MDT
Taylor M M6HMS
Taylor M M6PTP
Taylor M G1JPI
Taylor M M6MIC
Taylor N G0UOM
Taylor N G0WYA
Taylor N G3BHA
Taylor N G3TOQ
Taylor N G4HLX
Taylor N G4UFJ
Taylor N G7DMX
Taylor N M0CEQ
Taylor N M1DSQ
Taylor N G0RUF
Taylor N G7RMX
Taylor N 2E0NCK
Taylor N M6NCK
Taylor N MJ3IOJ
Taylor P G0GFZ
Taylor P G0ILO
Taylor P G0TCE
Taylor P G0VSJ
Taylor P G0WRK
Taylor P G1FET
Taylor P G1LZH
Taylor P G3RRG
Taylor P G4LIR
Taylor P G4MBZ
Taylor P G4VVD
Taylor P G7SLY
Taylor P G8VSH
Taylor P MD0DPG
Taylor P M3GSM
Taylor P M3HPT
Taylor P GW0MXG
Taylor P G7TEO
Taylor P G4OHB
Taylor P G8OYM
Taylor P G8WAL
Taylor P M0HQQ
Taylor P MM3OQR
Taylor P 2E0DKO
Taylor P 2E0CJD
Taylor P G8BCG
Taylor P M6IQO
Taylor P 2E0DSQ
Taylor P 2E0PAF
Taylor P G4KIN
Taylor P G8TMR
Taylor P M0VSE
Taylor P M6ESV
Taylor P MW6FNT
Taylor R G0HYG
Taylor R G0LQO
Taylor R G0RLT
Taylor R G1XRT
Taylor R GJ3ECC
Taylor R G3KAP
Taylor R G3LDY
Taylor R G4BEV
Taylor R G4CGU
Taylor R G4CQQ
Taylor R G4FDG
Taylor R G4GXO
Taylor R G6KLH
Taylor R G6RKF
Taylor R G7VSL
Taylor R G8JZZ
Taylor R M0AKZ
Taylor R MJ0JER
Taylor R M3AVU
Taylor R M3BBY
Taylor R M3KKF
Taylor R 2E0TVX
Taylor R M0TVX
Taylor R M3TCD
Taylor R GW3WWH
Taylor R M1AGA
Taylor R MW6HGZ
Taylor R 2E0RNT
Taylor R 2E0TUH
Taylor R M0HJL
Taylor R M6TUH
Taylor R G4ROB
Taylor R G6XNQ
Taylor R G8OST
Taylor R G8RYK
Taylor R M3LDT
Taylor R M6EAP
Taylor R MM6RCZ
Taylor R G1IPI
Taylor R G1RET
Taylor R G3JAL
Taylor R G4BEL
Taylor R M0BDB
Taylor R M1ANK
Taylor R M0LRD
Taylor R M3RRV
Taylor R G0JSA
Taylor R MI6TFN
Taylor S 2E0CHV
Taylor S G0INK
Taylor S G1JEZ
Taylor S G1NTX
Taylor S G6KPW
Taylor S G7VGM
Taylor S M1CVX
Taylor S M3SET
Taylor S M3SVT
Taylor S M3OTR
Taylor S M6FSA
Taylor S G0GQO
Taylor S G4CKX
Taylor S M6EUK
Taylor S M1FAI
Taylor S G4EDG
Taylor S G6LPG
Taylor S MM1FDF
Taylor S M3CFJ
Taylor S MM6TIR
Taylor T GM0GHN
Taylor T G0UCX
Taylor T G8DQD
Taylor T G0PSE
Taylor T M6TPT
Taylor T M6AUK
Taylor V G0AQT
Taylor W GM0CZM
Taylor W G0KWF
Taylor W G0RUX
Taylor W G3YLU
Taylor W G3YTT
Taylor W G4DLZ
Taylor W GM4OEZ
Taylor W G6NWT
Taylor W G6SKM
Taylor W G8HPL
Taylor W G6NGV
Taylor W G4YPF
Taylor W 2E0BDQ
Taylor W 2E0BXY
Taylor W 2E0EZX
Taylor W G4WGT
Taylor W M0WJT
Taylor W M3KJE
Taylor W M3ZFK
Taylor W G1MHM
Taylor-Cram J 2E0COC
Taylor-Mccormick S M3NPX
Taylor-Toms S M3SKT
Taynton B GM1PEL
Taynton E GM8DKB
Teague C G7DUI
Teague R G7VGE
Teale G 2E0GFE
Teale G M6GFE
Teanby P G4GZC
Tear S G4YYI
Teasdale B G0NSP
Teasdale D 2E0DBZ
Teasdale D M3HSJ
Teasdale J G0BVO
Teather P G4PLT
Tebay D G0MHN
Tebboth I G6FIP
Tebbutt P G4ONZ
Tebbutt R G7KJX
Tecklenberg D G0XAH
Tecklenberg G 2E0TEK
Tedbury W G4PQS
Tee A G8XIY
Tee W G4BYO
Teed A 2E0LWT
Teed A M3RMU
Teed J G3WWT
Teesdale R GW4URB
Teffera A M1DYF
Tegg P G8RYJ
Teklehaimanot N M1DYG
Telco P M1PWT
Telecki J M0PLX
Telfer J 2E0VHZ
Telfer J M6VHZ
Telford A G8IRN
Telford B G6TNI
Telford D G1FBE
Telford D M3YTV
Telford J G1REO
Telford M G6OET
Telford N G7IOI
Telford S G7HUJ
Telford W G6DJY
Temblett R M0AKF
Temperley H M1DXN
Tempest P G4HJS
Temple A MI3KKP
Temple D G0OYS
Temple E M3WDN
Temple P GM0EDJ
Temple P M3VGP
Temple-Heald J G6BBN
Templeman M G0TCY
Templeman M G7HVN
Templeton B MM6BXT
Templeton I GM0JQE
Tempo R M6RTQ
Tench J G4ZAY
Tennant A G4VOJ
Tennant D G1ONJ
Tennant G G8VSN
Tennant J G7ION
Tennant N M6NVT
Tennant P G0OPT
Tennent I M1IAN
Tennison W M6FKI
Tenwolde R G4WOL
Teperek R GM4HVS
Tepper G M0CTJ
Termie K M1BMC
Terraneau H G2FYO
Terras R GM1VJD
Terrell D MW6DJT
Terrell I G4JXZ
Terrell N MW6NFG
Terrell P MW6TRL
Terris I GM0VXZ
Terry A G4PZV
Terry B G6PEA
Terry C G7HUO
Terry H G0IGT
Terry L G7MZS
Terry M GW8TBG
Terry M G7HUP
Terry N G4IPM
Terry N G4YTI
Terry P M1CCN
Terry R M6HSQ
Terry S 2E0KUH
Terry S M6KUH
Terry S G8OCT
Terry V M6OGK
Terry W G4LUT
Tester M G4TBM
Tester P G8CIG
Teszner P M3YBF
Tetchner R G8VKI
Tetley G M3DKZ
Tetley M 2E0MRM
Tetley M G3RIX
Tetlow J G4SFN
Tett A G6WDR
Tett F G0OID
Tew B G3WFF
Tew G G7WFK
Tew G G8GZC
Tew G M6TAD
Tew G G4RFA
Tew M 2E0BZM
Tew M M3ZMQ
Tew R G4JDO
Tew R G6DKI
Tew R M1CNL
Tewkesbury P G8IUG
Tewsley E M0CYM
Thacker A M6DGO
Thacker C G6MEI
Thacker G M0DEF
Thacker I G6UOH
Thacker J 2E0FOW
Thacker K M6THA
Thacker L 2E0IBN
Thacker M G4FJF
Thacker P G4HSZ
Thackeray P G8YZL
Thackeray R G8FXN
Thackery A G4BBL
Thackery M M6TMY
Thackray A G1YTG
Thackray P 2E0DLX
Thackrey N GM0VPG
Thain D MM0DAT
Thain N M3MRQ
Thain P G7MOB
Thaiss I G7JXX
Thake N M3SQI
Tham K G0CRU
Thane E G0TED
Tharanee K 2E0KRS
Tharme B G4KVL
Tharp V G0CQY
Thatcher B G0JTR
Thatcher D 2E0CWJ
Thatcher R 2E0NDT
Thatcher R M6NDT
Thatcher-Sharp G 2W0YAE
Thatcher-Sharp G MW0YAE
Thatcher-Sharp G MW3YAE
Thayne M G3GMS
Thayne R G7ZRT
Theaker B G7RTO
Theaker S 2E0SDT
Theaker S M0SDT
Theakston R G0UKO
Thearle P M6AUQ
Theasby G G8BMI
Theedom B G8LYW
Theedom J G6XNK
Theobald J G3EQM
Theobald K M6NST
Theobold A M6TFO
Theobold L M6LFL
Theodorson E G8SWL
Theodorson J G4MTP
Thexton J G3URE
Thiebaut M MM6PIB
Thiele W M0IBD
Third E G4NZE
Thirkell E GM4FQE
Thirlaway S M0BZB
THIRLWALL S 2E0AAO
THIRLWALL S M6BWZ
Thirlwell T G0VFW
Thirsk J G4PAT
Thirst P G1ANZ
Thirst T G4CTT
Thistlethwaite B M6BMT
Thistlethwaite K M6KTT
Thom D G3NKS
Thom J GM3WJE
Thom W M3WDT
Thomalla D G7GGM
Thomas A G0TCV
Thomas A G0VSK
Thomas A G1DKX
Thomas A G1PYY
Thomas A G4UAT
Thomas A G7KDN
Thomas A 2W0KPN
Thomas A MW6KPN
Thomas A M6HUK
Thomas A G7OQO
Thomas A M6TDZ
Thomas A G0SFJ
Thomas A M0HIY
Thomas A M0RHR
Thomas B G0FHL
Thomas B G0KOJ
Thomas B G0NNR
Thomas B G0PID
Thomas B G3JLQ
Thomas B G4EIY
Thomas B G4KGU
Thomas B M1CCA
Thomas B M6GTO
Thomas B G1ZZG
Thomas C 2W0HAC
Thomas C G1KTW
Thomas C G1UTC
Thomas C G1YLQ
Thomas C G3YGR
Thomas C G4IZI
Thomas C G4OEQ
Thomas C G7MQC
Thomas C M1ZZA
Thomas C M3PYY
Thomas C G4ZCT
Thomas C G6BRY
Thomas C G8VBE
Thomas C G3PSM
Thomas C M6CRT
Thomas D 2E0OBZ
Thomas D G0BJK
Thomas D G0HAY
Thomas D G0VQX
Thomas D GW0WGE
Thomas D GW4AZI
Thomas D G4HHJ
Thomas D G4IKX
Thomas D GW4KYT
Thomas D G4OGW
Thomas D GW4XUE
Thomas D G6VAZ
Thomas D G7TIR
Thomas D MW0DFN
Thomas D MW0HYP
Thomas D M0RLI
Thomas D M3OBZ
Thomas D M6ZXR
Thomas D 2W0VEH
Thomas D GW0BBC
Thomas D GW3RWX
Thomas D MW0RUH
Thomas D M1AKT
Thomas D MW3RUH
Thomas D MW3VEH
Thomas D M6DEJ
Thomas D M6FXI
Thomas D GW1HEV
Thomas D M6GQF
Thomas E G8TTE
Thomas E M0BTI
Thomas E 2E0MTE
Thomas E M6MTE
Thomas E 2W0BKM
Thomas E MW3VFN
Thomas F G1WNL
Thomas F G3JRE
Thomas G G0SUB
Thomas G G1GRT
Thomas G G3JKE
Thomas G MW0AMN
Thomas G MW0AMQ
Thomas G MW0GWT
Thomas G M6EYT
Thomas G M3KKQ
Thomas G 2E0XUH
Thomas G M0XUH
Thomas G M3XUH
Thomas G G4AWJ
Thomas G M6BNH
Thomas G M6GDT
Thomas G G4JYL
Thomas H G0NPM
Thomas H GW4SYO
Thomas H MW0ATG
Thomas H M3HET
Thomas H M6POQ
Thomas H MW6HTT
Thomas I G1BXT
Thomas I G4PLL
Thomas I G8ZLL
Thomas I M0IRT
Thomas J G0RLI
Thomas J G0SDZ
Thomas J G1GST
Thomas J G3RXA
Thomas J G4DVV
Thomas J G4MYN
Thomas J GW4PAF
Thomas J G4ZBQ
Thomas J G7NFR
Thomas J M0JTH
Thomas J 2E0BVG
Thomas J M6JTH
Thomas J M6EBG
Thomas J G0GIH
Thomas K G4VYF
Thomas K G8IFH
Thomas L 2W0LLT
Thomas L G4YTN
Thomas L GW4ZXG
Thomas L MW3OLT
Thomas L M6LGD
Thomas L GM0TKB
Thomas L MW0LEW
Thomas L MW0COF
Thomas L M6LBT
Thomas M G0VQG
Thomas M G6BBR
Thomas M G6OIO
Thomas M M0ARV
Thomas M M0AZT
Thomas M G0WXN
Thomas M M0BIZ
Thomas M G0WKH
Thomas M M6YYM
Thomas M G4FCZ
Thomas N 2U0NPT
Thomas N G4LGO
Thomas N MU3NTH
Thomas N MW6FGN
Thomas N 2W0VDW
Thomas N G4MRL
Thomas N G4WZT
Thomas O 2E0OLT
Thomas O M0OLE
Thomas O M6LRO
Thomas P G1FFO
Thomas P G6ZZF
Thomas P G8KTA
Thomas P M6ORB
Thomas P M6PLO
Thomas P G3LZO
Thomas R G0DKF
Thomas R G0RCU
Thomas R GW0SLC
Thomas R G3VNB
Thomas R G4GCQ
Thomas R G4JHA
Thomas R GW4TVQ
Thomas R GW6AGS
Thomas R G6AGT
Thomas R G7MEA
Thomas R G7SBO
Thomas R G8IAK
Thomas R G8IWE
Thomas R MW0COE
Thomas R M3JUY
Thomas R M3TMR
Thomas R G4RWR
Thomas R 2W1ETW
Thomas R G4JJP
Thomas R M6RXT
Thomas R G6JBN
Thomas R G8JQW
Thomas R G4WJO
Thomas S 2E0UPT
Thomas S G0GVE
Thomas S G0PGX
Thomas S G6EQL
Thomas S M1ACQ
Thomas S M3UPT
Thomas S MM0HZJ
Thomas S M1ACB
Thomas T G0ABT
Thomas T G7UBD
Thomas T M6SBH
Thomas T G0JDO
Thomas T M0XGT
Thomas T MW3VJN
Thomas V G4UWR
Thomas V G7EWK
Thomas V G1GFF
Thomas W G3ORN
Thomas W G4AEP
Thomas W G4DQQ
Thomas W GW4WYX
Thomas W GW7VSF
Thomas W M0WAY
Thomas-Jones J M3WCE
Thomasson A GM0GAT
Thomasson P M6PTH
Thompson A G4TRU
Thompson A G7PFD
Thompson A M1BOZ
Thompson A M3FIX
Thompson A M3IDO
Thompson A MI3MOT
Thompson A M6KSI
Thompson A G6HUN
Thompson A MM0EQE
Thompson A MM1EQE
Thompson A M6EBA
Thompson A G4CXZ
Thompson A M3XSN
Thompson A M6TLY
Thompson A G8KSX
Thompson B G0KUR
Thompson B G1YAE
Thompson B G4KAL
Thompson B G4LKF
Thompson B G4LNT
Thompson B G8JJE
Thompson B G6TXB
Thompson B G6GYV
Thompson C G0GBE
Thompson C G1SJB
Thompson C G6FLW
Thompson C G6VKA
Thompson C M0DGU
Thompson C MM0ZCT
Thompson C M1CEC
Thompson C G0KLA
Thompson C G7BPI
Thompson C M6KEF
Thompson C MI6TIJ
Thompson D G4UPK
Thompson D G4ZPM
Thompson D GI8JOA
Thompson D M3JJT
Thompson D 2E0DTH
Thompson D G3OXG
Thompson D G7WAW
Thompson D M6WBH
Thompson E G3VZR
Thompson E M1DPI
Thompson F GW0UZX
Thompson F G4PMB
Thompson F G6EZR
Thompson G GI1SYM
Thompson G G3RCZ
Thompson G G3XBH
Thompson G M3GTQ
Thompson G M3XYP
Thompson G M0CGT
Thompson G M5GHT
Thompson G G0WTW
Thompson H G6AVL
Thompson I M3NSX
Thompson J 2E0EMN
Thompson J G0FGG
Thompson J GI0RBC
Thompson J G0SBY
Thompson J G0WQW
Thompson J G0XVS
Thompson J G3OKT
Thompson J G4OVG
Thompson J G4THG
Thompson J G4YES
Thompson J G6HUO
Thompson J GI6UUC
Thompson J M1APF
Thompson J 2I0RZT
Thompson J MI0GIJ
Thompson J MI3RZT
Thompson J 2E0TJT
Thompson J M3YZQ
Thompson J G8OEO
Thompson J G8UCV
Thompson J M3LXV
Thompson J M6EVM
Thompson J M6JPT
Thompson J M6JWT
Thompson J M3XQT
Thompson J M6LQO
Thompson J G1ODN
Thompson K G3AMF
Thompson K G3VSE
thompson K G7ECG
Thompson K M6KTH
Thompson K M6TKA
Thompson K G1PJO
Thompson L G3VYZ
Thompson L G4WZU
Thompson L G8ARF
Thompson L M0CIP
Thompson L M6LAE
Thompson L M6MYF
Thompson L M3XWK
Thompson L MI6PBI
Thompson M 2E0ZMT
Thompson M GD3EFD
Thompson M G7HYZ
Thompson M M0VES
Thompson M M0ZMT
Thompson M M3NLQ
Thompson M M3ZMT
Thompson M M6GYH
Thompson M 2E0WPT
Thompson M GD3JIU
Thompson M G6HUP
Thompson M G7THF
Thompson M M3RLH
Thompson M M6MPT
Thompson N G7UZS
Thompson N M3NTJ
Thompson N M6NBY
Thompson N G4CLY
Thompson N G1KMN
Thompson P G0TLU
Thompson P G1MIL
Thompson P G1ZUB
Thompson P G7OYD
Thompson P G8CSY
Thompson P G8DDY
Thompson P G8TNB
Thompson P G7AYL
Thompson P 2E0PGT
Thompson P 2E0PWP
Thompson P M0CPW
Thompson P M0PAA
Thompson P M3UDN
Thompson P G3JTT
Thompson P G8ZLN
Thompson R 2E0RPT
Thompson R G3TKF
Thompson R GM4YPL
Thompson R G6BIA
Thompson R G7MRZ
Thompson R G7RXK
Thompson R G8GDZ
Thompson R G8ZFT
Thompson R MI3TPR
Thompson R G0UAK
Thompson R GI3UEX
Thompson R M6BQJ
Thompson S 2E0SLT
Thompson S G0SEN
Thompson S G4RCH
Thompson S GI7DZE
Thompson S M0DPF
Thompson S G6VLK
Thompson S G7IFB
Thompson S M3REX
Thompson S 2E0WDM
Thompson S G8TNA
Thompson S M3VZV
Thompson S M0OMT
Thompson S M3SRT
Thompson T 2E0ADP
Thompson T G6EQP
Thompson T G8HNT
Thompson T M3EQP
Thompson T 2E0DHJ
Thompson T M6EXB
Thompson T M0ALZ
Thompson T G4AVN
Thompson T MI0MOD
Thompson T GI4SIZ
Thompson W 2I0TLT
Thompson W G3YAG
Thompson W GI4UUC
Thompson W MI0TIP
Thomsen B G0GSY
Thomson A 2M0CSX
Thomson A GM3AHR
Thomson A GM3YXY
Thomson A G4AHW
Thomson A G4OLF
Thomson A MM0KZJ
Thomson A M1CWW
Thomson A MM3CSX
Thomson A M6LMW
Thomson A MM6YST
Thomson A M1DRK
Thomson B GM8PSV
Thomson B MM0GTX
Thomson B 2E0GXE
Thomson B M6GCY
Thomson C G3PEM
Thomson C G6AVK
Thomson C MM6CIJ
Thomson D G3RGS
Thomson D MM0EAX
Thomson D M6TFQ
Thomson D M6DVE
Thomson D G3WBS
Thomson D G7DST
Thomson F 2M0BMF
Thomson I G1OZR
Thomson I GM0URD
Thomson J GM4CXF
Thomson J G8GUX
Thomson J GM8ZJS
Thomson J M0FWO
Thomson J GM0AXM
Thomson J MM3JGT
Thomson M G8SYD
Thomson M MM6CKC
Thomson M GM4JEJ
Thomson M 2M1GBG
Thomson N 2E0NEO
Thomson N M0NEO
Thomson N M3VRX
Thomson P GM1XEA
Thomson P GM0OPK
Thomson R GM0KDF
Thomson R MM0KOZ
Thomson R 2I0SEK
Thomson R MI0PLC
Thomson R MI3SEK
Thomson R GM3OBC
Thomson R GM8IOL
Thomson R G4LMW
Thomson R G8XCW
Thomson R G7RFO
Thomson R MM6RGT
Thomson R MM6TSN
Thomson S G4SJL
Thomson W M6BLT
Thomson W MM6LBS
Thomson-Best L MM3UUT
Thong R M1RTT
Thorburn B M0BDE
Thores O GM4JKT
Thorley D M1AGE
Thorley E G0JBR
Thorley J G0SPY
thorley K M1CFA
Thorley M G6KDU
Thorley P 2E0TDH
Thorley P M6TDP
Thorley R G7MGA
Thorman T G0TRD
Thorn C G6OEW
Thorn C G3STZ
Thorn D 2E0BHD
Thorn J G6XZS
Thornber B G0LVT
Thornber J G6GNC
Thornber S G6SGA
Thorndike D G4OMN
Thorndike P G0XOX
Thorndyke C G0MEZ
Thorndyke J G0MEV
Thorne A G6KPX
Thorne C G4ZTS
Thorne G G0JJD
Thorne G G0TIP
Thorne G G3UIF
Thorne G M3TGL
Thorne J 2W1YEG
Thorne J M0CWN
Thorne J MW3YEG
Thorne K G3NBZ
Thorne M G3ZUT
Thorne N G4RIH
Thorne N M3CPX
Thorne P M1EUM
Thorne R GW1SGH
Thorne S M3UVS
Thorne S 2E0WSW
Thorne S M3WZS
Thorne T G4KQZ
Thorne T MM0TJR
Thornes E M6ERC
Thornett A M6THO
Thornett R M6TYB
Thornhill J 2E0JTH
Thornhill J M3ZTX
Thornley C G6CVD
Thornley J G1NUS
Thornley N G7UGY
Thornley P M6ELW
Thornsby G 2E0BRC
Thornsby M G6XGT
Thornsby P G6SUR
Thornton A G0JVW
Thornton A M0TIW
Thornton A M1EAB
Thornton B G1IIO
Thornton C G7PWI
Thornton D M0FCY
Thornton D G1ASR
Thornton G G8EKW
Thornton G G4YJC
Thornton H GM3PKV
Thornton I MW6IMT
Thornton J G1LML
Thornton J G3YDL
Thornton L 2E0LTH
Thornton L M3SNQ
Thornton M G1XUH
Thornton M G1SYP
Thornton M M0CCW
Thornton M M3VPN
Thornton P G6NGR
Thornton R G4JCY
Thornton R G3WKW
Thornton S M3EYW
Thornton S M3YDL
Thornton T G1CNV
Thornton W G0NSC
Thornton W M3EYX

Thornton-Evison P G8PTY
Thorogood B M0TBR
Thorogood S MM3SXI
Thorp A G7MHD
Thorp D G4RCR
Thorp I G6WBT
Thorp M G3PQM
Thorpe A G3UHX
Thorpe C G0SAY
Thorpe D G1LPQ
Thorpe D G4RNT
Thorpe D M6RUG
Thorpe D G4FKI
Thorpe G M3TMG
Thorpe G 2E0BLX
Thorpe G M0GKO
Thorpe G M3RVE
Thorpe J G0OKZ
Thorpe J M6OWM
Thorpe M G1ICI
Thorpe R M3BJE
Thorpe R M0HXF
Thorpe S G4NST
Thorpe S M0TGM
Thorpe T G7NID
Thorpe-Morgan C M6BJX
Thow W GM4GNR
Thoyts M G0TXN
Threadingham E M0CXY
Threakall P 2E0MBR
Threakall P M0PTG
Threakall P M6FLE
Threapleton W G4PEL
Threlfall T GW4TGT
Threlfall-Rogers S G8UIW
Thresher E M6GQG
Thresher J 2E0YOM
Thresher J M0YOM
Thresher J M3YOM
Thresher S M6GXR
Throne J GI0AYB
Throne J MI3XYB
Throne N MI3NYB
Throne R MI3XEB
Throup C M3XSY
Throup J M3ZGT
Thrower G M3YSW
Thrower K M6KPT
Thrower N G3YSW
Thulborn C M3DLP
Thurbon A G4GZO
Thurgood A G4LNY
Thurlow A G3WBN
Thurlow B G6OVC
Thurman J M6FII
Thurman P G1BLB
Thurman-Newell D M6DTN
Thursfield N G6HUR
Thwaites B G3CVI
Thwaites J G8PWO
Thwaytes D G1GDB
Thwaytes P G4WOH
Thyer M 2E0BBZ
Thynne A G7IKG
Tibbert B G3RKZ
Tibbett G 2E0AFH
Tibbetts S G7VJT
Tibbits M M3CET
Ticehurst G M3KRO
Ticehurst T M3TJT
Tickell H G4BJJ
Tickell W G8EMB
Tickle I G4ZJH
Tickle R M0PSR
Tickle R M5RPT
Tickle S G1MTJ
Tickner M M3XHZ
Tidder A G8CHI
Tideswell I G7HKQ
Tideswell N 2E0BHS
Tideswell N M3XKF
Tideswell S G7ORN
Tidey I G1AML
Tidman M M0DWB
Tidmarsh J M6JST
Tidmarsh M G7JMQ
Tidmarsh S G4VMM
Tidnam R G4JOI
Tidswell A G6WQJ
Tidswell D 2E0BOZ
Tidswell D M3WUE
Tidswell R G8KGR
Tidwell C G0DAF
Tier P M0GIU
Tierney J G4JYQ
Tierney J MI6RKD
Tiesdell-Smith T M0TBS
Tietz P G0AEU
Tiffany B G3TXX
Tilbee A G4HXE
Tildesley J M0DIT
Tiling A M6DZF
Till D M6PVZ
Till W G8DDV
Tiller G G7UHF
Tiller J G4AZU
Tiller R G8XYR
Tillett G G6MDN
TILLETT O G3TPJ
Tilley D M3IHR
Tilley F G1XJK
Tilley J G7KRI
Tilley P G6BQE
Tilley R G6HMV
Tilley T G1BBA
Tilley T GW4YJI
Tillin J G8GJC
Tilling G M6GVN
Tillotson H G0XIL
Tillotson P M0AMX
Tillson G G3TJX
Tillson J M3TIL
Tilly K M6KTA
Tilly S 2E0AOU
Tilly S G1BNN
Tiltman D 2W1CYC
Timbrell D M6DTT
Timbrell G G4STH
Timbrell G G6SIG
Timbrell H G4YLO
Timbrell L G6GOX
Timlett M G1NVV
Timlett P G0LWC
Timm C M3BVM
Timmins A MM6BGV
Timmins L 2E0LST
Timmins L M3YLT
Timmins P G0NDV
Timmis J M1CMM
Timmons R M6KVX
Timms A G0FKW
Timms A M6EIB
Timms B G7FLX
Timms M 2E0UAC
Timms M G4ZYH
Timms R G8VBC
Timms S G7OST
Timms T G7DTK
Timms W G3BWI
Timson D G6FMF
Tindale C G1HEZ
Tindall N G7UMA
Tindill C G8LNQ
Tingay R M0DLE
Tingay R G3OPG
Tingay S M3OAQ
Tink A G7DRU
Tinker C GW1GSW
Tinker D M1FJQ
Tinker J G3PKC
Tinkler D M6GSX
Tinkler P M1EDW
Tinley D G7DJT
Tinn D 2E0TIN
Tinn D M0XXL
Tinn D M3ZFW
Tinnion W M0DXT
Tinsell-Stanton M 2E0DGG
Tinsell-Stanton M M0MCO
Tinsell-Stanton M M3TIC
Tinsley J G0FZU
TINSLEY S G0TDC
Tinson R G3XPM
Tinton B G3SWC
Tipler A M6BHC
Tipler G G8CBA
Tipp C G1ZDY
Tipper A G4KXR
Tipper A G8THZ
Tipper B G3WWL
Tipper M G6AFZ
Tipper T G1MZT
Tippett D MW6NAZ
Tipping J MI6FWD
Tipping N G4WED
Tipping T G0WHZ
Tisdale J G4NRA
Titcombe M M0MLT
Titcombe R M0RHT
Tite C G0EYG
Tite J G0HMF
Tither P G1MVE
Titheridge C G4DYI
Titheridge D G4EZX
Titherington P G4RWM
Titley A G8RKX
Titmarsh B G8SAU
Titmarsh B G7UAH
Titmarsh K 2E0LSR
Titmarsh K M6EMT
Titmarsh R 2E0RBQ
Titmarsh R M6RMT
Titmus A 2E0DEG
Titmus A M0MRI
Titmus A M6ELD
Titmuss R G0AWI
Titmuss R G6MZV
Tittensor M G4EKG
Tittensor P G4PVM
Titterington G G4ZDN
Titterington R G3ORY
Titterington S G7LKL
Titterton J M1AZB
Tivey P 2E0TOX
Toal P M6BME
Toas A G6TRW
Tobias J MM1ANP
Tobin A M0BAM
Tobin J G6PBO
Tobin C M6CKH
Toby C G4TZF
Todd A GM7DXE
Todd A M6TCY
Todd A M6CXV
Todd C M6AAU
Todd C G4ABM
Todd D GW7OIK
Todd D M0DWT
Todd G G8VYQ
Todd G MI6AFI
Todd G 2I0GWX
Todd H G0HTD
Todd J G1HEY
Todd J G4XLM
Todd J M0DDK
Todd J M0NFF
Todd M G1HJS
Todd P M3VLO
Todd R 2I0RGT
Todd R G3WNC
Todd R G4ULD
Todd R GI7AQO
Todd R M1TOD
Todd R M6RTI
Todd R 2I0RVT
Todd R MI6RAD
Todd R GI0STS
Todd S M6SEG
Todd S MW3ZWS
Todd W MI6WEZ
Todd-White B G3OJZ
Todman K 2E0XKT
Todman K M0KJT
Todman K M6XKT
Todman L M0HKI
Todorovic I M3KKI
Todorovic S G7RUR
Tofts P G3WBK
Tofts R 2E0CUW
Tofts R M0HNK
Tofts R M6XRT
Togwell B G4FTZ
Toh H G0CCG
Toher M 2E0ZOT
Toher M M0XEE
Toher M M3VFS
Toher S M3ZVA
Toher W 2E0UCH
Toher W M0VLI
Toher W M3VLI
Tohill J GM0FKP
Tointon M 2E0UEL
Tointon M M0HTY
Tointon M M6DWV
Tokely A M0LTA
Tokley E G3TFV
Tokley K M3HPY
Tokley R G4MDB
Tokley S M3ZAN
Tolcher P 2E0PLR
Tolcher P M0LAI
Tolcher P M6LBN
Tolhurst S M3GUH
Toll P G7BGO
Toller D 2E0DWE
Toller D M6GJD
Tollerfield R G3SQD
Tolman R 2E0UBH
Tolman R M6UBH
Tolman W G4RWK
Tolmie M M6MWT
Tolputt S G8ILN
Tolson J G0AHU
Tomalin A G3PTB
Tomalin R MI0RPT
Tombe L 2M0BIK
Tombs C GW4MOG
Tombs D G8VYT
Tombs J G7HDU
Tomczyk W M0HJN
Tomes P G3YPT
Tometzki E G0TIZ
Tomkins A M6HLT
Tomkins G G1IEP
Tomkins G M5IEP
Tomkins G M6GTE
Tomkins L 2E0LES
Tomkins P G1LBH
tomkins S G1UHB
Tomkins T G1VXY
Tomkins V G4KEE
Tomkinson C G4GEO
Tomkinson R M0RCT
Tomkinson S G1UYT
Tomlin D G7RYA
Tomlin E G3YFU
Tomlin P M1PGT
Tomlin R 2E0GXS
Tomlins A G4ZXB
Tomlins J M1CBH
Tomlinson A M6RUP
Tomlinson B [illegible]
Tomlinson B M6ZBT
Tomlinson C M3GAV
Tomlinson D M0DSN
Tomlinson D M6EPQ
Tomlinson G G1EJK
Tomlinson G G0DJQ
Tomlinson G M0BOC
Tomlinson H G3UHW
Tomlinson H G7PWU
Tomlinson J G0DFO
Tomlinson J G3MGX
Tomlinson J M3LNN
Tomlinson J M3ZFJ
Tomlinson K G8YCK
Tomlinson P 2E0PJI
Tomlinson P G4CBL
Tomlinson P G6TKW
Tomlinson P MW0CDO
Tomlinson P M0EME
Tomlinson P M0PJT
Tomlinson R G4TGJ
Tomlinson S G3KVJ
Tomlinson S 2E0TOL
Tomlinson S M0STO
Tomlinson S M6GRX
Tomlinson S M0VFR
Tomlinson V M3VRA
Tommasini W M0GPV
Tommey D G1MDG
Tommey J G0FHT
Tompkins B M6BKK
Tompkins C M6DLV
Tompkins J 2E0JCE
Tompkins M M3FHK
TOMPKINS R G8LDY
Tompsett S G8LYB
Tompson A G8LSS
Toms P G4XMZ
Tomschey S G8ZJO
Tomsett D G1ORB
Tomson A G0RRV
Tomson I G0LOZ
Ton W MM3ICD
Toner A 2I0EBF
Toner A MI6EAZ
Toner C M3MYI
Toner K M3MYG
Toner P MM6RPN
Tong C G7DSU
Tong T M6TTY
Tonge A G0BVT
Tonge G G4IDG
Tonge H M3JPP
Tonge J G7JRT
Tonge K 2E0NCO
Tonge K M6NCO
Tonge P G6YSZ
Tongs R 2E0RFL
Tongs R M6HBN
Tonkin B G7DUC
Tonkin M GW7TZI
Tonkin M M6FWM
Tonks J 2E0AIT
Tonks J G8ZZT
Tonks W G0PWQ
Tonkyn J M6FJI
Tonner K G1RFX
Toogood C G1MCY
Toohey E G6ZFK
Toohey K M3ZWQ
Toolan J G1AVZ
Tooley F G3HPB
Tooley I G7EHS
Tooley M G4BBD
Tooley R M6XRX
Toombes N M0CJM
Toombs D G8FXM
Toombs D M3FXM
Toomer A M1ETX
Toomer G M6GWT
Toomer W 2E0ZPR
Toomey-Langford F MW6FIN
Toon B G1OYU
Toon D G4IXF
Toon I G4RJG
Toon I G4YIT
Toon J G0FNH
Toon T M3CCF
Toop G G0ULZ
Toop K G7PCE
Tooth R G4RZR
Tootill D M0DUV
Tooze L G7VCB
Topham D GM3WKB
Topham J G0RCJ
Topham M M6DTP
Topham P G8KDO
Topley J G4PTZ
Topley R 2E0RDI
Topley R M6OPL
Topliss R G0OTH
Topping A 2E0ECG
Topping C GM6HGW
Topping D G3HYG
Topping J G6OER
Topping J M6JIT
Topping N 2E0NWT
Topping R G8SOU
Topping S G0MTD
Topple M 2E0GUI
Topple M M6EIO
Topsfield A G4NBF
Topsfield A M0TOP
Tordoff D 2E0ZWA
Tordoff D M0WMO
Torrance Smith R G4SWO
Torley M 2I0MMT
Torley M MI6MMT
Torr R GD0KSL
Torrance D 2E0GBU
Torrance D M0TOZ
Torrance P G4HAK
Torring J G6TKH
Torrington M M0SDG
Torry J M0ATQ
Torry P G3SMT
Torunski H G0GLG
Tory P G3VMQ
Tose P M0CLN
Tosh J 2E0HOT
Tosh W MI0DLF
Tostevin P G8MAD
Tostevin S MU6GUE
Toth A M6CWQ
Totten J G7LWF
Totterdell A M3HEE
Totterdell B M3XBT
Totterdell N G4FAL
Tottle J G6LDW
Totty C 2J1CWH
Totty C GJ7UIT
Totty J 2J1CWG
Tough I G0IHK
Toulson T G6FGL
Tourish D MM6DCT
Tourish R 2M0ADT
Tourish R MM6RBT
Tournant J G0DEU
Tournier J G3INZ
Tout M G4EUR
Tovey M GW4XSX
Towell G MM6HXY
Towell P G0AYX
Towers D G1GTK
Towers D A G8SZX
Towers E G4HYI
Towers K G0CEV
Towers M G7AFW
Towers R MM0RKT
Towle C G0WKJ
Towle C M0AJT
Towle J G0WPF
Towle J G4PJZ
Towler G G4NGS
Towler K M0KPT
Towler L G7JJP
Towler L M3LPT
Towler R G0UKS
Towler S G0VTD
Townend D G4GNA
Townend D G6VBA
Townend D G0RQX
Townend G G4SDX
Townend N G7JLO
Towner J 2E0KWT
Towner J M3YWR
Townley E GI4XFY
Townley E M3HTE
Townrow P G6LTB
Towns C G8BKE
Towns J M1ADX
Towns K GM6IKN
Townsend A M3VBH
Townsend A G4NMA
Townsend B M3CGP
Townsend C G8PUT
Townsend C M3PUT
Townsend C 2E0TIF
Townsend C M6TIF
Townsend C G0WVA
Townsend D G7NIN
Townsend E G8DGH
Townsend E G6XNN
Townsend G G3LSX
Townsend G G8ANN
Townsend G 2E0GGT
Townsend G M3GGE
Townsend I M6ISZ
Townsend J M0JCT
Townsend J M3JUL
Townsend J M3WTU
Townsend J G8JAD
Townsend J M0CLJ
Townsend J M6GQR
Townsend K M6HHC
Townsend M 2E0END
Townsend M M0TXS
Townsend M G4EQL
Townsend R G6CIE
Townsend R M1ENQ
Townsend R M6RTM
Townsend R M3ZYN
Townsend S MW0AEL
Townsend S M6SLT
Townsend S M6DSV
Townsend S M6ABX
Townsend W MW0AXA
Townshend B 2E0BFS
Townsley S G7SPN
Townson D 2E0GDD
Townson I M1BGY
Townson M 2E0EGV
Townson T M3DXI
Toyne D 2E0DAT
Toyne D M3NKL
Toynton A M6RDP
Tozer J G3XLZ
Tozer M M6MQM
Tozer S 2W0ZAA
Tozer S MW3USX
Tozer T 2E0TZR
Tracey A G4PZX
Tracey E G4YBI
Tracey J G6GTB
Tracey W GM4UBJ
Tracy J G1TUI
Trahearn M G0OYM
Trahearn S MW0VRQ
Trahearn-O'Brien J MW6JZZ
Trail I M1FFY
Traill K G4XUJ
Traill P MM3DHN
Traill T GM0LUF
Train G G4LEX
Trainer D G3UPJ
Trainer F G7RYW
Trainor J G1LMC
Tran A M3TFA
Tran C GM3WOJ
Trangmar N 2E0TUX
Trangmar N M0YMM
Trangmar N M3POV
Trangmar R G6ORE
Tranter A 2E0LJT
Tranter A M0LJT
Tranter A M6CQJ
Tranter J G4XFT
Tranter J M3MIF
Tranter J M6EDW
Tranter P M6EDV
Trathen M M6MQJ
Tratt L [illegible]
Tr[illegible] G3WYW
Travers J GW4UVN
Travers P G1HEW
Travett R G8XYS
Travis C 2E0CAJ
Travis C M0TAL
Travis C M6CMT
Travis D G4GXD
Travis J G4GWX
Travis M G1YCK
Travis M M0MPT
Travis P 2E0GXL
Travis R GD4DNP
Trayhurn B 2E0BNT
Trayhurn B M0ZOR
Trayhurn B M3RZJ
Trayner C G4OKW
Traynor A G0FCX
Traynor A MM3KDN
Traynor B 2I0MFJ
Traynor D G1URQ
Traynor J MI0JBT
Treacher D 2E0HBS
Treacher M G0ODS
Treacher R 2E0RCV
Treacher R M0MCV
Treacher R M3RCV
Treacher S 2E0CVN
Treadwell P G7PCT
Treanor L MI6LOT
Treanor N G8WAJ
Treasure K G0SYI
Treble L MM6LJE
Treble R MM6IIO
Treen D G0RDT
Trefry J G0TLZ
Tregay A G7AYQ
Tregear D M3UVD
Trehearne O M6OJT
Treherne R M3XIV
Trelease R M0IRI
Tremain G G6BBM
Tremain K 2E0SPT
Trembath P M0SBL
Tremble C G0PKE
Tremble L M3ELV
Tremelling R MW1FGB
Trench K M1EWV
Trenchard M GU6JQF
Trend A G4BER
Trend L 2E0LEZ
Trend L M6LTO
Trend V M0AEJ
Trengove T G1SZD
Trent A G0VGT
Trent J G8YMZ
Trent R M6ROY
Trepess P G4HBD
Trethewey T G4YXJ
Trett A G6XNP
Trett E G0BZV
Trett J G6JTT
Trett R G8JWT
Trevan R M0VEC
Trevarrow J M3HZW
Trevelyan D M3YLR
Trevett J G4GKX
Trevor B G6SUK
Trevor S M6ZSV
Trew C M3JXV
Trew T G8JXV
Trewin N M1EQA
Tribe H M1BFG
Tribe J G0IUY
Tribe M G1RPT
Tribe M G7NJI
Tribe S G0IEY
Tribute D G1OEQ
Trice J G0RGM
Trick A M6TBQ
Trick M M3TME
Tricker S G0AZP
Tricker S G6YSQ
Trickett D G8TIA
Trickett J G4JMC
Trickey E G4DCX
Trickey G G4ODW
Trickey R M6RUS
Tricklebank J M0VGG
Trigell A G3JAF
Trigg F M0ZEQ
Trigg K G7OJX
Trigg M 2E0ARG
Trigg S M6SUZ
Trim B G8YRL
Trim G G4XSC
Trim M G1USZ
Trim R 2E0RAI
Trim R M6RLT
Trimble A M1ERD
Trimble S M1ERA
Trimmer B G0FHC
Trimmer J M6HTI
Trimmer P GW4KCY
Trinder K GM7DFI
Trinder W G4UCC
Tring A G3WYB
Tring A 2E0BWI
Tring A M0YRG
Tringale L G0OLT
Tringham D G0BSD
Tripathy J MI3ZJT
Tripney R 2M0MMF
Tripney R MM0RDT
Tripp A G6XTC
Tr[illegible] G1RUO
Tripp J G3YWO
Tristram P G0UAV
Trivett B G0TTQ
Trivett M M6MAX
Trnjakov U M6UKI
Trofimiuk W M6WIK
Trohear A M6RFO
Trolan M G6SHF
Trolan V G1STP
Troll P G7PKQ
Trollope D M6ALE
Trollope N G4FAT
Tromans D G4CGB
Tromans K G0LBT
Tromans M 2E0TRO
Tromans M M3UWB
Troop D G8TGD
Troop N G4IQR
Tropman D G7UHG
Troscheit S 2M0BYK
Troscheit S MM0TRS
Troscheit S MM6THO
Troth M M3UHW
Troth M 2E0MJT
Troth P 2E0TTH
Troth P M6PMT
Trotman D G4WPR
Trotman N G8SYE
trott A G7MIS
Trott R G3YQN
Trott S 2E0TSJ
Trott S M3TSJ
Trott S G8ZOE
Trotter D M6HAM
Trotter J G4HPX
Trotter S M3RZY
Trotter S G6KVY
Trotter W G0EGE
Troughton E G0VNV
Troughton I 2E0CUK
Troughton I G8CRH
Troughton I M6TRQ
Troughton J M6CUR
Troughton R GD4AZJ
Troughton R G3ZSJ
Trousdale A G4LXW
Trouse G M1EQB
Trow R 2E0CVE
Trowell E G2HKU
Trowsdale R G6HMG
Trowse A M6CHT
Troy A G4KRN
Troy M G0WFQ
Truberg P GW4JOG
Truckel G G0LXL
Trudgen A G4YAF
Trudgeon T G0ENZ
Trudgett A M3ZUJ
Trudgian D 2E0BQJ
Trudgian D M0TGN
Trudgian D M6DAN
Trudgill G G7MVX
Trudgill R G4VLA
Trudgill S G2HFP
Trueblood M M1CEW
Truelove R M3UZN
Trueman A G1IZH
Trueman E M3EJR
Trueman R M6RWT
Truitt P G4WQO
Truman A M0DNV
Truman G 2E0TRU
Truman G M6TRU
Truman J M6TWI
Truman M G8SLC
Trundle M G3TCG
Trunks J M6JLT
Trunley H G3RPZ
Truran P G4BPR
Truran R GW0MAV
Truscott D M6DVT
Trusler A G0FIG
Truslove I G6UX
Trussler R GM0CSN
Trusson C G3RVM
Trybulski J G7SRZ
Tr[illegible] G7CKD
Trzeciak-Hicks R 2E0ROY
Trzepietowski A M0HWI
Tsakonas N M6ZBP
Tsang E M0VAN
TSE R M3TZE
Tse W MW0HGK
Tsimperidis I M0ISQ
Tsioumparakis K M1KOS
Tsoi C M6ZCT
Tsuzuki S M0BLD
Tubb D G1SNO
Tubbs J G0DWS
Tubey C G1ZMW
Tubis C G8HJD
Tubman E G4SIL
Tuck G G4RNI
Tuck P G4TKF
Tuck P G8YNC
Tucker A G4PCW
Tucker A G6LPD
Tucker B M0APL
Tucker B 2E0TUC
Tucker B M3TUH
Tucker C G3TXZ
Tucker C G4RTV
Tucker C G4DCH
Tucker D G6HMX
tucker D M0CPB
Tucker D M1AXG
Tucker D G1KHS
Tucker G MW1EAA
Tucker G MW3RRW
Tucker J G7LNG
Tucker K G1HEQ
Tucker L G8LNU
Tucker M 2E0IBN
Tucker M M6CCN
Tucker N G7URW
Tucker P G4DWZ
Tucker R G3KTC
Tucker R G6AVI
Tucker S G1FFR
Tucker S M3TUW
Tuckett R G8TQV
Tuckett R M0TQV
Tuckley C G8TMV
Tuddenham E G3XFF
Tudge A M1CRZ
Tudor E G3INY
Tudor G M6SMR
Tudor M G0LZI
Tudor S M6SQE
Tuer K G0VRZ
Tuff H G8HYL
Tuff V G7PYR
Tuffill B M0FFS
Tuffill M M6ANY
Tuffin R M6CJK
Tuffin S G4YAL
Tuffrey B G0LHM
Tuffrey M G8LHQ
Tuffs M M3VTP
Tuffs P G4HEB
Tufnail B G0HOS
Tufnell B 2E0DLC
Tufnell B M6DLX
Tugman C G4VUF
Tugwell B G0LTD
Tugwell E G0FIP
Tugwell P M6PST
Tuke J GM3BST
Tulk M 2E0HSJ
Tulk R G1GXQ
Tullett M 2E0EJX
Tullock G G1IBF
Tullock P M0DPH
Tully C G4CYF
Tully C G7ESE
Tully K G8XAX
Tully P M1CCX
Tully W G0ANX
Tumilty S MI6TUM
Tunbridge C G7LIK
Tunbridge D G1SNU
Tunbridge R G1FXX
Tungate A G0JBJ
Tunna C G4VLT
Tunney A 2E0TUN
Tunney A M6TUN
Tunnicliffe D G4BCA
Tunstall E G3MSO
Tunstall L 2E0LEE
Tunstall L 2E0HXM
Tunstall M G0SKM
Tunstall W 2E0BVJ
Tunstall W M6BWC
Tunstell T M3WHG
Tupman K G0BTU
Tuppeny G G4LOE
Turbefield J G0HBX
Turbert T G4TOM
Turford D M6GGT
Turford S 2E0SUE
Turford S M3PUZ
Turk E G7BQM
Turk I M1DJS
Turk J M1CIJ
Turk P G3PQC
Turkington N GI0BEY
Turkington W MI6WPT
Turkington W 2I0WBF
Turkington W MI6WBT
Turland N G7LNV
Turley B G6GUT
Turley J M3ZWX
Turley P G1CQT
Turley R G0IYJ
TURLEY R G4TSA
Turley S G7JYZ
Turley T G0TYZ
Turlington R G8ATE
Turnbull B G0OIM
Turnbull D 2E0CNN
Turnbull D M6BKM
Turnbull H G3KDW
Turnbull I G6KMG
Turnbull J G6TNK
Turnbull J G7NHE
Turnbull J M3ABY
Turnbull M G4ILM
Turnbull M G7PWL
Turnbull N MM6NOR
Turnbull R GM0VUY
Turnbull S MM0TBY
Turnbull S MM3TBY
Turnbull S MM3XSF
Turnbull V GM7AHA
Turnbull W G3SBT
Turnell K G4HLW
Turner A G0FMU
Turner A G4CPE

UK Surnames

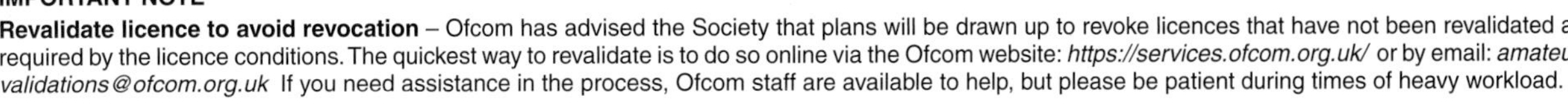
**IMPORTANT NOTE**

**Revalidate licence to avoid revocation** – Ofcom has advised the Society that plans will be drawn up to revoke licences that have not been revalidated as required by the licence conditions. The quickest way to revalidate is to do so online via the Ofcom website: *https://services.ofcom.org.uk/* or by email: *amateur.validations@ofcom.org.uk* If you need assistance in the process, Ofcom staff are available to help, but please be patient during times of heavy workload.

Turner A G4OWN
Turner A G4RUL
Turner A G4TXV
Turner A G4UUT
Turner A G4XBC
Turner A G4XQE
Turner A G7PYV
Turner A 2E0TAJ
Turner A 2E0RKB
Turner A G1GTM
Turner A M0MOR
Turner B G3RLE
Turner B G3YNF
Turner B G4WZN
Turner B G8CEX
Turner B M6SQU
Turner C G4HKP
Turner C G4SEP
Turner C M3PNO
Turner C M3CLT
Turner C G0VUT
Turner C G7MXM
Turner C G7TZO
Turner C M0BVD
Turner C G1WZM
Turner C G3VTT
Turner D G1IIY
Turner D G3SBM
Turner D G3WEI
Turner D G4PST
Turner D G4SWY
Turner D GW4WMK
Turner D M0DJT
Turner D M3FPZ
Turner D G0OZG
Turner D G0VVF
Turner D G4KEY
turner D G1SKE
Turner D 2E0DEZ
Turner E G1MSB
Turner E G1XES
Turner E G4IRG
Turner E G7NCG
Turner E M0BRM
Turner E G7DBN
Turner E M6FPQ
Turner F G8NFM
Turner G 2E0BVQ
Turner G G4SQJ
Turner G G7OBP
Turner G G1VLS
Turner G M1DHV
Turner H G4YRH
Turner H M6AKG
Turner I G4NHR
Turner I 2E0IMT
Turner I M0IMT
Turner I M3OND
Turner J G1IEO
Turner J G4FZY
Turner J GI6KBX
Turner J G7HKU
Turner J G8NDE
Turner J G8PFL
Turner J M3WRA
Turner J M0NWT
Turner J M6NWT
Turner J 2E0BMD
Turner J 2E0JKT
Turner J G0KFO
Turner J G3UST
Turner J G6MEH
Turner J G7EJK
Turner J M0JVT
Turner J M0UCD
Turner J M6TEM
Turner K 2E0GJP
Turner K G1SLO
Turner K G4GZB
Turner K GW6WEU
Turner K G7RYO
Turner K G8GIF
Turner K G8JLA
Turner K M0MRT
Turner K M6KTN
Turner L G4DLA
Turner L 2E0BLS
Turner L M3XLS
Turner M GJ0PDJ
Turner M G3VYN
Turner M G3ZMN
Turner M G6ZFZ
Turner M M0XYL
Turner M M3BIZ
Turner M M3ISJ
Turner M G7UGA
Turner N G0PPY
Turner N G4DDZ
Turner N GU7NCZ
Turner N M3NQS
Turner P G0UCN
Turner P G1CKY
Turner P G3TIG
Turner P G7BXJ
Turner P MM3TUR
Turner P G4HKB
Turner P 2E0NAI
Turner P M6CIK
Turner P M3FPT
Turner R G0GME
Turner R G0LXF
Turner R G3SMD
Turner R G6COZ
Turner R M0CJI
Turner R M3RTE
Turner R GM7SAQ
Turner R M6ORM
Turner S G1LCE
Turner S G3IST
Turner S G3UJI
Turner S G6FJO
Turner S M0BZX
Turner S M1DSZ
Turner S M3MXG
Turner S 2E0BQG
Turner S M0GRR
Turner S M6FJZ
Turner S M6TAZ
Turner S G0HCR
Turner T G7HRL
Turner T M3ZGG
Turner T GM4YAT
Turner T M6AFU
Turner V G0RJC
Turner W G4LML
Turner W GI4LZR
Turner W GW6MNC
Turner-Hicks P G0IXS
Turner-Smith F G3VKI
Turner-Smith S G8YCL
Turney D M0DKN
Turnham N G6PSA
Turnham P G4VLS
Turowski M M3TKT
Turpie R 2M0RTA
Turpie R MM0RTT
Turpie R MM6RBT
Turquand A G1FXB
Turrell N M6NTB
Turtle W MI3LRR
Turton A G0LGO
Turton C M0GVZ
Turton D G0VMN
Turton D G1JVM
Turton D M0JVM
Turton E G6EWH
Turton J G7OCX
Turton J M3JCU
Turton R 2E0AWZ
Turton W G4MSG
Turvey B G3UMT
Turvey K G6BGA
Turvey M G4URN
Tusler J 2E0BOD
Tuson I G0MQR
Tust M G4LUQ
Tust R G8IEL
Tust R M3IEL
Tuthill P G1BMT
Tutt B G4ZZK
Tutt G 2E0AOF
Tutt I G0PEC
Tutt I G0RNN
Tutt M G8LLJ
Tuttle M G0TMT
Tutty B M0DCO
Tuvey C G4IXB
Twaddle D MI6VBB
Twagirayezu E M6EDZ
Twaites P M6PTZ
tweddle S 2E0SJT
tweddle S M0GCT
Tweed J M6TWD
Tweed N M3ZAM
Tweed S GW7HJN
Tweedie D G4GQZ
Tweedie M M0MFT
Tweedie S GI4RXX
Tweedie S M6TWE
Tweedie T GI7RAH
Tweedy G G0LOP
Tweedy J G7MPJ
Tweedy W GI4FGH
Tween M M0EBX
Twells M G7DWX
Twells M M0DEV
Twemlow W 2E0WJT
Twemlow W M0TWJ
Tweney P G1RCX
Twibell G G8FKL
Twidale D G1FEP
Twigg G M3VUO
Twigg M G4ZXT
Twiggs R G4HIN
Twiss G G3RUG
Twist B G3NFY
Twist R G4AEY
Twitchen M M6NDY
Twort A 2M0OSK
Twort A MM3OSK
Twort K G8CHY
Twose R M3CVW
Twyford A G8TSZ
Twyman D G6LJR
Twyman R G1ZQR
Twynam S M3CUH
Tyblewski J G4BZA
Tybora G G1HEU
Tydeman M M3YZP
Tye A G8ZZV
Tye D G3ZIB
Tyerman D G3KCG
Tyerman J G7EJH
Tyerman N M0MJT
Tyers D G0FJD
Tyers J M1JTA
Tylee J G4REK
Tyler A G1YAF
Tyler A G1GKN
Tyler A M0BAH
Tyler B G8EQO
Tyler C GM8MYO
Tyler D G4BWO
Tyler D G8LWA
Tyler D G4PIE
Tyler G 2E0BTS
Tyler G M6GJT
Tyler J G8XZX
Tyler J G4GCL
Tyler K G4CRP
Tyler M 2E0TYL
Tyler M G4NUB
Tyler M M3TYL
Tyler R G0CFB
Tyler S G1ZAR
Tyler S G4UDZ
Tyler S M6AZK
Tyler-Moore M M6MTM
Tynan M M6TYN
Tynan W G3SJR
Tyrell P G8RSK
Tyreman H G3SLL
Tyreman J G6TVE
Tyrer T G6TKV
Tyrrell J G3OBL
Tyrrell M G6GAK
Tyrrell S M6TYS
Tyrwhitt-Drake A M0TDK
Tysiorowski J G4GLQ
Tysoe D M0CAJ
Tysoe K G1DYL
Tyson D G4ERY
Tyson D 2E0AUI
Tyson J G6MDG
Tyson M G4YTC
Tyson P G0RLJ
Tyson R G4NFR
Tyson R G8LGY

## U

Ubonis N M0NPQ
Udall J M0APD
Udall R G2HKS
Udall R M3XUT
Ukommi U M3UIU
Ul Haq Z M3ZIA
Ullersperger J 2E0JLN
Ullersperger J M6JIL
Ullman F G4KQT
Ulvenmoe L M6GBL
Ulyatt A 2E0ULY
Underdown D G3MBK
Underhay B G1YES
Underhill C G0MMI
Underhill E G4UVW
Underhill E 2E0EUN
Underhill E M0TAX
Underhill M G3LHZ
Underhill P M3UYW
Underhill T G4MWP
Underwood A GW0AJU
Underwood B G4BUD
Underwood C G1WTW
Underwood D MW0DRU
Underwood D G4RCJ
Underwood D M0UGD
Underwood D M3UGD
Underwood F G0DEK
Underwood I M0PPS
Underwood J 2E0IJX
Underwood J G0PGY
Underwood J GW1IIZ
Underwood J M1EUE
Underwood J M3EUE
Underwood J 2E0JCU
Underwood J M6YYU
Underwood K G3SDW
Underwood M M6WHZ
Underwood N G4LDR
Underwood N G4WRD
Underwood R G6MAW
Underwood V GW0HYL
Unstead P G0TNP
Unsworth A G8BCJ
Unsworth C G7GJM
Unsworth M G7LVS
Unsworth P M6UNS
Unsworth R G3PJW
Unsworth R M6AYI
Unsworth R G3WPF
Unsworth R M6BVQ
Unsworth S G0HWX
Unsworth T G8FZT
Unwin A M6IER
Unwin C M3VQQ
Unwin D G8ZUZ
Unwin D G0FMT
Unwin J G4JHN
Unwin J 2E0NUL
Unwin J M6WIX
Unwin P G4HDS
Upchurch A G7JZC
Upcott A GW6HMJ
Uphill C MW6CRU
Uphill M 2W0UAA
Uphill M MW0MAU
Uprichard K MW6KCJ
Upstone J GW4MOZ
Upstone T G4DPT
Upton B M3RBU
Upton D G1UDW
Upton D G8CTR
Upton G G4SJG
Upton J G4RSS
UPTON P G7CCV
Upton R G7AYB
Upton R G0TLS
Urban G M6LKL
Urban M M6URX
Uren J M0JIU
Urlings L MM0WRL
Urquhart A G1NCM
Urquhart D G4DR
Urquhart H GM0UTD
Urquhart S M3URQ
Ursell C G1JVO
Ursell F G1JVN
Urwin C G3OPE
Urwin R G6RRJ
Usher A G4HZW
Usher D G6ZKC
USOV I M6USO
Utley A G4YXB
Utley D G0PAZ
Utting A G0AUT
Utting A G1WZQ
Utting C G1BIU
Utting N GJ7LJJ
Uttley D G8MEC
Uttley P 2E0PHU
Uttridge P G1EQF

## V

Vaci M MM6MVI
Vadgama H G8XNN
Vage R 2I0BSH
Vage R MI0TFK
Vailati Facchini G MD6GVF
Vaile A 2E0UAK
Vaile A M6YAV
Vainas D 2E0DIM
Vainas D M0VAI
Vainas D M6DIM
Vaisey L G1LOU
Valdez E M0HUV
Vale D MD0HJR
Vale D M3XOQ
Vale F G3CWT
Vale R G3YHI
Vale R 2E0VMV
Vale R M0VMV
Vale R M3VMV
Vale T G0LUQ
Valente M G4EBN
Valenti G G6XJN
Valenti M G6XGV
Valentine A G6MAT
Valentine G G0UVX
Valentine M G4ANP
Valentine N G3KWJ
Valentine S G4WPH
valentine S MW3UNZ
Valentine S G4RWS
Valentine T GM1XHZ
Valerio P GW4KTT
Valkov I M0IVE
Valle Espin J M3IDW
Valle Espin J 2E0CUA
Valle Espin J M0ABO
Valleley P G0LQD
Valler G G6FKA
Vallis P G0RAL
Vallis R G0COC
Vallow P G4FSK
Valori S G4VUV
Valvona N M6MGA
Valvona S 2E0EBN
Van Aswegen W G0EMV
Van Beers L G7RLO
Van Breemen G M0HDQ
van Cleak R G6OPY
Van De Vondel J M6CXM
Van Den Bergh M M3SQU
Van Den Bergh T M3SUY
Van Den Bergh V G6DIF
Van Den Bogaerde R M3RNL
Van Den Bosch T M6FPV
Van Den Langenberg F M3FAL
Van Der Elsen J 2E0UPA
Van Der Elsen J M0UPA
Van Der Elsen J M6ULA
Van Der Linde I 2E0RYX
Van Der Linde I M6LOG
Van Der Steeg M G7BEP
Van Driel H G5MUN
Van Dyke J GM0RYD
Van Falier P G1VAN
Van Haaren D G4XVH
Van Klinkenberg P G7IYF
Van Praag S G4YFS
Van Schie H M3ELN
Van Stigt N G4RWA
Van Wezel L M0LVW
Van Zuilen C M5TWO
Vanbeck D G1GBX
Vanbesien J M6VBJ
Van-Boques-Tal J M6AWY
Vance C G7HPI
Vance D GI3XZM
Vance I G3WMS
Vance K 2E0HMQ
Van-Den-Bergh W G6YTZ
Van-Den-Langenberg N M3OPA
Vander Byl W G8UTQ
Vanderahe A M6GNO
Vandervord J 2E0HPS
Van-der-Wijst R 2E0ROV
Van-der-Wijst R M0MCP
Van-der-Wijst R M6ROV
Vanderydt T MM0RAI
Vane B G4SEJ
Vane S M6LRL
Vane-Stobbs R G4EDC
VanKassel S G4VWG
Vann M G3RLV
Vano I M0GFZ
Vansittart R G6GHP
Vanson B G0MBP
Vanstone D M0JMV
Vanstone S M0SGV
Varasani R M6RVR
Vardy M 2E0IQO
Vardy M 2E0MKV
Vardy M M6MKV
Varga A G6TEB
Varga L M6FPN
Varghese C MW0EYE
Varley A G0PNQ
Varley A M1WWW
Varley E M6EMV
Varley F G2FCP
Varley J M3WEJ
Varley M M6MPV
Varley R G8ODK
Varley R G7CKG
Varnals K G1UAY
Varnes N G4YXX
Varney T G4RLP
Varnham J G7NQU
Varoudis T M6VAR
Varty A G6CQC
Vasarhelyi A 2E0JVA
Vasarhelyi A M6VAV
Vasek J G0OHW
Vasey D M3ZDV
Vasey J G1JDV
Vaslet B G6GXZ
Vaslet C G8OID
Vaslet S G8LTD
Vasper R G3VIY
Vassie A M6VAY
Vassie D M6VAW
Vaughan A G8WQE
Vaughan B G0VJB
Vaughan B G3MCV
Vaughan B G3TGO
Vaughan B MI1CCT
Vaughan C G1BHW
Vaughan D G0JLV
Vaughan D G4FBV
Vaughan D M3DPV
Vaughan D G8ZZS
Vaughan E 2E0CFM
Vaughan J G3XLE
Vaughan J GW7NGU
Vaughan J G3DQY
Vaughan K 2W0HAK
Vaughan M M0VAU
Vaughan M M3MHV
Vaughan M M6MIF
Vaughan P G6KPJ
Vaughan P G8EJQ
Vaughan P G4HNU
Vaughan R G4VYK
Vaughan R G3FRV
Vaughan R G4WGB
Vaughan S G4WXC
Vaughan T MI3CCT
Vaughton G G0HRH
Vause R MM1CLR
Vaux S M3SOF
Vavasour E G8EHM
Veal A G0LTB
Veal J MM3ITA
Veal M M3VRM
Veale D G0XVL
Veale F G4LEA
Veale G M6GAR
Veale P G3ZXV
Veall E G6HMS
Veary M M0CMC
Vecenans E M6RDI
Vecenans H 2E0GCG
Vecenans H M6GBE
Veitch C G4LEV
Veitch J G4MRK
Veitch P G7VJG
Vemic B M0HWW
Venables A G1RFI
Venables C M1EBL
Veness J G4PWE
Venison R G6HMF
Venn C G6NDA
Venn K G0GFD
Venn T G3RPV
Vennard J MM0JJV
Vennard R GM0SEI
Vennard R GM7KSA
Venner S G0TAN
Venries H MM6HAV
Venugopalan G 2E0VGV
Venugopalan G M0VGV
Venugopalan G M6VGV
Venugopalan V M6BWH
Venus H G0VKH
Verardi M M0LEF
Verduyn J G5BBL
Verghese L 2E0LGV
Verity J G3ONV
Verity M GM0RMV
Verity S M6BHE
Verma R 2E0DBT
Vernalls T G6IMS
Vernau C M6CLV
Vernon A G4ZMU
Vernon C 2E0CAV
Vernon C G0TQJ
Vernon C G4CTD
Vernon C G4HIW
Vernon C G8PEN
Vernon C M0TJV
Vernon C M3VEM
Vernon M G6HVA
Vernon P MM3NJV
Vernon-Jones W G0VFE
Verrall K 2M0SBP
Verrall K MM3YBQ
Verrall M 2E0BAK
Verrall M G8TGB
Verrall R G0VEB
Verrechia M 2E0CIH
Verrechia M M3MJV
Verth S MM3WBV
Vesma V G8GYB
Vesma V M0NZA
Vesma V M6DDD
Vials R 2W0IVZ
Vicarage R G6BHY
Vicary H G0RJN
Vichitcheep N M3NUB
Vick A M3KHC
Vickers A G3HFM
Vickers A G8EQB
Vickers B M1BHE
Vickers B M6NBO
Vickers D 2E0VKS
Vickers D G4SEQ
Vickers D M0VKS
Vickers D M3VKS
Vickers I 2E0ZIV
Vickers I M3ZIV
Vickers J G3ORI
Vickers J G8BHK
Vickers K G3YKI
Vickers K G3RHQ
Vickers L M6LGV
Vickers P G1EYT
Vickers P G4RHJ
Vickers P M6AWQ
Vickers R G6DVP
Vickers R M6FQN
Vickers R M6RDN
Vickers S 2E0SRV
Vickers S M0SSV
Vickers S M6SRV
Vickers W G4PHB
Vickerstaff J G4EIG
Vickerstaff R 2E0VKQ
Vickerstaff R M1ADT
Vickerstaff R M3ADT
Vickerstaff R M3AKQ
Vickery B GW4CSY
Vickery C G4YCV
Victory C G8YOE
Videan E G4LWV
Vigors R M3LQY
Vile B 2E0LYD
Vile B M0IML
Vile B M6BDV
Villena Bota J G4VPL
Villette P M6PAV
Villiers G M6AZN
Vince C G0TLI
Vince P G8ZZR
Vince R G8DRK
Vincelli D 2E0DVM
Vincelli D M6FCJ
Vincent A 2E0GBN
Vincent A M0URF
Vincent A G7FUQ
Vincent A 2W0VKA
Vincent A MW3VKA
Vincent B G3SXV
Vincent B G4FVV
Vincent C M0VIN
Vincent C 2E0PCR
Vincent C M6PCB
Vincent C 2E0AEC
Vincent D G3OKY
Vincent D G7TZB
Vincent D M3TZB
Vincent D M6DHV
Vincent I G4MLY
Vincent J G1PVZ
Vincent J G8GWJ
Vincent M G3UKV
Vincent M G8UKV
vincent M M3PEQ
Vincent R G4FRV
Vincent R M0RDV
Vincent R M6PUQ
Vincent R G7GIG
Vincent T MW3IPK
Vincent T M3UKV
Vincent-Squibb C M3CVS
Vincent-Squibb G G0BJA
Vincent-Squibb R G0BSN
Vincz A M0GLU
Vine A M0GJH
Vine C G3XXF
Vine G G3KLV
Vine G G0HCY
Vine J G8NWI
Vine J 2E0RMV
Vine J M6RMV
Viner M G4CJJ
Vines C MM6KCV
Viney D M0CME
Viney J G3ZIC
Viney M G0ANN
Viney S M0SJV
Vining A G0TFR
Vinnicombe S G6UXM
Vinnicombe S M0BWL
Vinnicombe W M6VWA
Vinson J G4YGP
Vinters A G0WFG
Vinters J G7NSN
Vinton J GM0EMQ
Vipond P G1RLT
Vipond P M3PJV
Virdee J M3EKP
Virgo S M6JJV
Virtue M G1NIT
Viswambaran J M6OPJ
Vitiello A M6VIT
Vitiello C G8EDX
Vivash D G6BJY
Vivian A M6AKV
Vivian D G3VOB
Vivian J G4PBN
Vivian R G1NFQ
Vize W G6REQ
Vizor J G8CPA
Vizoso A G3XVB
Vlismas T GW0TMV
Voak D M6GYE
Vodden B GW3WBU
Voges R G7ILX
Voisey R G8IDK
Voisey T G1ZFF
Voke C M3SYW
Volante L G0MTN
Volck R GW3RKV
Vollbrecht J M0HVM
Voller G G3JUL
Voller K GW0CMI
Voller K G0VZN
Voller T G8ILP
Volpe A G0BZB
Von Bergmann T M6LVX
Von Fircks N G4YFJ
Vosper D M5AKY
Voss J G0FMG
Voss M G8ERA
Vousden J G8WIR
Vowles A MW6GIV
Vowles R G1WVM
Vranic N G8PLI
Vrentzos E M6VRE
Vye J G0ANS
Vyner S M6VBP
Vyvyan H G1YTL
Vzor S 2E0NYC
Vzor S M0NYP
Vzor S M6STU

## W

Wachs N M6EPL
Wackett C G0GVN
Waddell A GM4LAO
Waddell B GM4XQJ
Waddell J GW0GUA
Waddilove A G4KYH
Waddingham B M6ZBW
Waddingham I M6FJO
Waddingham R G0HPM
Waddington A M0WAD
Waddington A M5AMN
Waddington A M1DJI
Waddington C GM4FYH
Waddington G MM1DJJ
Waddington J M3FKK
Waddington P G0FWA
Waddington P M6WAD
Waddington V G4JSS
Waddington W M1MIJ
Waddoups A G1KOR
Waddy J 2E0JVV
Waddy J M0JVV
Waddy J M6JVV
Wade A G1PPO
Wade A G4AJW
Wade A G7EVK
Wade A M3PPO
Wade C 2E0CDW
Wade C M0KCW
Wade C M6CDW
Wade D G4UGM
Wade G M6GKW
Wade I G4VKX
Wade I G3NRW
Wade J G0IRI
Wade J M6JWP
Wade J M6JWX
Wade K G7GYR
Wade M M0CZC
Wade M M0EDU
Wade N G4XOJ
Wade P 2E0PHW
Wade P G4TCE
Wade P G7JVB
Wade P M0BZV
Wade P M6WCF
Wade R G7UXD
Wade R G6ZWM
Wade R G8GSU
Wade S M0SVV
Wade T G4IDL
Wade W G4FCF
Wadeson J M3FQX
Wadey D M3DSW
Wadhams P G0VRW
Wadley P GU4YBW
Wadsworth B G3OLP
Wadsworth C 2E0LLL
Wadsworth G M6STV
Wadsworth S M3FQA
Wadsworth T G4SNQ
Wadwell G G4ORU
Wagenaar A G0ALR
Wager D G7VBJ
Wager G M6GCW
Wager I G8MKN
Wager M G7RAZ
Wager T G3TPI
Wager-Bradley C G1AHT
Wager-Bradley C G3NEP
Wagg H G6RYM
Wagg T G8VEZ
Waghorne D G6DPW
Waghorne K G4SSP
Waghorne R M3WRJ
Wagiel B M0GPX
Wagner E 2E0NQU
Wagner E M0NQU
Wagner E M3NQU
Wagner F G0IRQ
Wagstaff A G1KBJ
Wagstaff B M0KEY
Wagstaff D 2E0WAG
Wagstaff P M0WEC
Wagstaff S M6EWV
Wai Ming T M0VSR
Waight K G8EOZ
Waight R G0PNN
Wain D G7SCP
Wainman C G4KBI
Wainwright A M1AOF
Wainwright B G0HDP
Wainwright C G7JRJ
Wainwright C G4AMN
Wainwright D M6PBX
Wainwright J G6PBW
Wainwright J G8AVO
Wainwright J M3JWW
Wainwright K G7KBH
Wainwright P G7LSD
Wainwright P M1CBC
Wainwright R G3YMH
Wainwright R M6TIC
Wainwright R M3XRW
Wainwright S G4DFO
Waistell M M0PBO
Waite A M1EEQ
Waite C M6CIW
Waite D M0GPJ
Waite F M1CDJ
Waite J G0SLW
Waite L G6RFJ
Waite M GW7PQS
Waite N G3KOX
Waitt R G6LJE
Wake A G3GIB
Wake J G6VPV
Wake J 2E0JPA
Wake J M0JPA
Wake R G7BXS
Wake R M3BXS
Wakefield D GM0KTH
Wakefield D M0DPW
Wakefield D M3ZLW
Wakefield H G6CTP
Wakefield J 2E0GHZ
Wakefield J M0XIG
Wakefield P M1DCH
Wakeford D G0DUN
Wakelam P M0CKC
Wakelam R M6RJW
Wakeley R G3YEP
Wakeling A G7MFY
Wakeling T G7UQQ
Wakely A G4ZMY
Wakely M G3NWW
Wakeman A G0IUD
Wakenell J G8UGL
Wakenell J M0UGL
Walch J G3RVI
Walch N M3NSO
Walcot S M3VQG
Walcott C M0FZU
Walczak J 2E0PLS
Walczak J M0PLS
Walczak J M6BXZ
Walden B M3WTB
Walden C G6AHH
Walden M G0IJZ
Walder-Davis I G0KCA
Waldie A G3NOC
Waldman M MW6HDP
Waldock S M6ATA
Waldron C G3ZZU
Waldron C M6CAW
Waldron C GM0NHL
Waldron D G4LUB
Waldron G G1MXD
Waldron J M6VEG
Waldron M M0BLT
Waldron M M1AEA
Waldron P G6ZAM
Waldron S MM3EJV
Waldron W G0FGO
Waldron W G0OAW
Waldron W G0UKT
Waldron W G4YGT
Wale B M6XMB
Wale C M0KHA
Wale D G0EWR
Wale G M6PEU
Wale G W G4BCG
Wale L M6BWJ
Wale P M3TGK
Wales G G7GUG
Walford J GM3POT
Walford P G0RFX
Walford T G3PCJ
Wali Zangana E M6EWZ
Walker A 2E0GGT
Walker A G0FGA
Walker A G1EJQ

UK Surnames

Walker A G1VYS
Walker A G3MPW
Walker A [illegible]
Walker A [illegible]
Walker A G4ORQ
Walker A G4UWS
Walker A G4XDR
Walker A M0HFX
Walker A M0WFA
Walker A M3LTV
Walker A 2E0WNW
Walker A G7RCU
Walker A G8UKZ
Walker A M6DQK
Walker A M6LBX
Walker A G4RNX
Walker B G0LCU
Walker B G7KRM
Walker B 2E0RCO
Walker B G4PCL
Walker B G6WZL
Walker B M3VXB
Walker B 2E0VJX
Walker B M0VJX
Walker B M6VJX
Walker B G6LQR
Walker B G0HDI
Walker B G0OMB
Walker B G1FNF
Walker C 2E0FWA
Walker C G3USO
Walker C G3VTS
Walker C G4WHN
Walker C M1MST
Walker C G1ETZ
Walker C G6BHX
Walker D 2E0DHX
Walker D G1PPU
Walker D G3ULL
Walker D G4DCW
Walker D G4DEM
Walker D G6ZAF
Walker D G8UCY
Walker D M6FSZ
Walker D 2E0DBB
Walker D 2E0FLA
Walker D G3BLS
Walker D M3WZF
Walker D M6DAG
Walker D M6YLR
Walker D G6JTD
Walker E G0FNM
Walker E GM7RMF
Walker E M3OSC
Walker E G0KAQ
Walker F 2E0MQT
Walker F G3JWN
Walker G G0PXZ
Walker G G4DAF
Walker G G6GLB
Walker G G6JDF
Walker G G6SKR
Walker G GI7RCH
Walker G GM8YUM
Walker G M6GBJ
Walker G M0GSL
Walker G M6GEU
Walker G M0OAT
Walker G M6WKR
Walker G G1ULB
Walker H 2E0KOI
Walker H G1NWH
Walker H G4XWN
Walker I 2E0IMW
Walker I G0KAK
Walker I G3RJF
Walker I G3VNY
Walker I G8ILZ
Walker I M0IMW
Walker I M6IMW
Walker I G6OXN
Walker J 2M0BDN
Walker J GM0DJG
Walker J G0WMJ
Walker J G2AXQ
Walker J G3RDZ
Walker J G4SSW
Walker J G6FYU
Walker J G6VIN
Walker J G7DXX
Walker J G8GIN
Walker J G8KKU
Walker J M3GTK
Walker J G0LGB
Walker J M6HMW
Walker J 2E0GZF
Walker J G0FRY
Walker J GM4FAU
Walker J G4FHF
Walker J 2E0XBG
Walker J M6INW
Walker J M6SFW
Walker J 2E0JMW
Walker K G4AFS
Walker K G6IMJ
Walker K G8DIR
Walker K M3BXY
Walker K GM0TCC
Walker L 2E0LJW
Walker L G1CBB
Walker L M3BAO
Walker L M6LVW
Walker L G4ULT
Walker L 2E0LPW
Walker L M0LPW
Walker L M6LPW

Walker M G1VIP
Walker M G4IJI
Walker M G8TRO
Walker M [illegible]
Walker M M0IAW
Walker M M3ZAA
Walker M M6MGW
Walker M G6POV
Walker M MM6WMK
Walker M 2E0VVC
Walker M M0VVC
Walker M M6BJY
Walker M M6TTL
Walker M G4OGZ
Walker M G6UOX
Walker N G8AYC
Walker N M0NOW
Walker N M1SPW
Walker N M6FZV
Walker P 2E0GXQ
Walker P G0CPJ
Walker P G0RDX
Walker P G4DBY
Walker P G4HHH
Walker P G6NDH
Walker P G8HMG
Walker P M0BRI
Walker P G0MMH
Walker P G4PLW
Walker P G4RRM
Walker P G6KUI
Walker P M0AFR
Walker R G0OUJ
Walker R G0RAE
Walker R G3XYJ
Walker R G3YKW
Walker R G3ZJQ
Walker R G4FNG
Walker R G7IVU
walker R G7NDT
Walker R G7SLV
Walker R G7SMZ
Walker R G7VUB
Walker R G8ERN
Walker R M3LDS
Walker R 2M0RWZ
Walker R M0BPT
Walker R M0WBR
Walker R M6LRU
Walker R MM6RWZ
Walker R G0UTP
Walker R 2E0RAF
Walker R G0TAK
Walker S G1PPQ
Walker S M3TRP
Walker S M3SWK
Walker S 2E0DBQ
Walker S G7OAJ
Walker S M0SBA
Walker S M6BKN
Walker T G0TWE
Walker T 2E0MKT
Walker T M0TSW
Walker T M6TMK
Walker V M3XVW
Walker V M6VVN
Walker W 2E0BCG
Walker W GM3LAW
Walker W G3RNX
Walker W G6WKU
Walker W G7KTD
Walker W M0DYU
Walker W G3HTJ
Walker W M0WTW
Walker-Kier S G7KYD
Walker-Riley C M6UTP
Walkley J G0BMT
Walkling P G8RNT
Walkling P M5PSW
Walkup C G0LQZ
Wall B G1EUA
Wall C G6SQT
Wall C M3VIW
Wall C GM1KHU
Wall D M3JZF
Wall D M6ETA
Wall E G0LGK
Wall F G0CKD
Wall G G7PMW
wall G M6ASZ
Wall H G4CBT
Wall J G3GQK
Wall K G0LRK
Wall M 2E0KVB
Wall M M6WMA
Wall N M3YXT
Wall N G8GCO
Wall R M3YVN
Wall S 2E0YYZ
Wall S M6SNF
Wall T M3IHQ
Wallace A GI6JOP
Wallace A M6WDR
Wallace A 2E0TGL
Wallace A M6SOC
Wallace B G7MVN
Wallace C G8PTW
Wallace C M0HCV
Wallace D G3OCP
Wallace E GM4TFJ
Wallace E GM4XLU
Wallace E MM3TWW
Wallace G GM0DNG
Wallace G GM4MSL
Wallace G MM3WGW
Wallace J 2E0HOS

Wallace J M6JJW
Wallace K G3LQW
Wallace K G1GZS
Wallace L M1ECM
Wallace M G8AXA
Wallace M G8RFE
Wallace M MD3OIS
Wallace M M6MAL
Wallace N G4SYF
Wallace N MD0FIX
Wallace N G0UIP
Wallace P G0WFM
Wallace P G1OAR
Wallace P M0OAR
Wallace R GM0AOF
Wallace R G4MXF
Wallace R G7JJX
Wallace R GM0MZH
Wallace S G6HSI
Wallace U GM1LTM
Wallace V 2E0VSW
Wallace V M3XQE
Wallbank A G4CIZ
Wallbank R G0XAT
Waller A 2E0DNK
Waller A M3BVQ
Waller A M6SWI
Waller D G0FPN
Waller D G3SUL
Waller E M6EPW
Waller F 2E0FWR
Waller F M0WFR
Waller F M6FRK
Waller G G0CKA
Waller I G4TQT
Waller J M0ANH
Waller J G3DAV
Waller L M6BNT
Waller M G0PJO
Waller N M3LBM
Waller P 2E0DRB
Waller R G0CJZ
Waller R G0DHA
Waller R G6HGD
Waller R G7IBF
Waller R G6ICZ
Waller S 2E0SAZ
Waller S M3OWZ
Waller T G3KBI
Waller W M1WAW
Wallett J G4KNS
Walley B G7EHU
Walling D G1DXM
Wallington H G8IJM
Wallis A G0FQA
Wallis A G4DEO
Wallis A G4LDC
Wallis A G4TQS
Wallis A MM0XTW
Wallis D G3KXF
Wallis G G4DQB
Wallis I G0PLS
Wallis J G0OGB
Wallis J GW4TJQ
Wallis J M0CAN
Wallis J M6JRW
Wallis K G1UYW
Wallis M G0CRD
Wallis M G4EIA
Wallis P G7CRU
Wallis P M6PAW
Wallis R G6OCF
Wallis R M1CJM
Wallis S G3YPK
Wallis T G6BWK
Wallman A G4XBX
Wallman A 2E0RPE
Wallman A M6RPE
Walls A 2E0WWT
Walls A M6WWT
Walls D M0DIW
Walls M M6CIQ
Walls S G4HNW
Wallstone D M3FLE
Wallstone P M3LBY
Wallwork A G4BOB
Wallwork C G6AHK
Wallwork C G8DBH
Wallwork C M6XVJ
Walmsley C 2E0WMY
Walmsley G G4IXE
Walmsley J G7FNM
Walmsley M G8RIP
Walmsley M G0VOF
Walmsley P G0NGK
Walmsley R 2M0RDK
Walmsley R MM6RWA
Walmsley S 2W0NQE
Walmsley S MW0GEI
Walmsley S MW3NQE
Walpole B G4INF
Walpole D G0OOB
Walpole M G8VMZ
Walpole R G3VYG
Walrond S [illegible]
Walsh A G0YCZ
Walsh A G6HSG
Walsh A G7VZS
Walsh B 2E0WJB
Walsh C G7IAW
Walsh C M3LYV
Walsh C M6CMW
Walsh C M6XTK
Walsh D M0WDU
Walsh E M6GOI
Walsh G G6ORO

Walsh H M0BEQ
Walsh I GM4OLH
Walsh I G6IML
Walsh J [illegible]
Walsh J [illegible]
Walsh J G0JJW
Walsh J M0EIW
Walsh J M6GOQ
Walsh J G6BUH
Walsh K G0CAE
Walsh K M3SBQ
Walsh M 2E0AYQ
Walsh M 2E0HXP
Walsh P G3PJV
Walsh P G7HKN
Walsh P 2E0KCW
Walsh P M0PKE
Walsh P M6BWP
Walsh P M6PHW
Walsh R G4ZNK
Walsh R M1DGL
Walsh R M3KXB
Walsh S 2E0OKP
Walsh T G4FMM
Walsh T 2E0WTQ
Walsh T M0HNN
Walsh T M6CPQ
Walsh V M6BFB
Walster B GJ3SND
Walstra B M0WAO
Walsworth S G0TAL
Walter P GM8SNE
Walter P G6BRP
Walter R G1RLF
Walters A M6MEK
Walters B GW3XHD
Walters B G4UNJ
Walters D G0ENU
Walters D M0DEW
Walters D G4DFV
Walters G G7ISE
Walters J G6RKG
Walters J M6DLI
Walters K G0GZO
Walters K G0MLB
Walters M G3JVL
Walters M 2E0MDH
Walters M M6BWO
Walters M M6BGW
Walters M 2E0WLY
Walters M M6WOL
Walters P G0GYU
Walters P G3THW
Walters P G8JGF
Walters S MW1DTT
Walters S 2E0GRH
Walters S G7VFY
Walters S M6RRH
Waltham A G4UAV
Walther W 2E0WOL
Walther W M3XEI
Walton A G0RAR
Walton A G3ZKQ
Walton A M6EQO
Walton A 2E0TAL
Walton A M3ZXN
Walton B M3UMD
Walton B G7DGD
Walton C G6FXE
Walton C G6TNA
Walton C M3CZW
Walton C M3ZGE
Walton E M6TFZ
Walton E G4FSN
Walton F G4XWM
Walton G M6DUI
Walton G M3YFG
Walton G G3XBE
Walton I G8PYU
Walton J G1CSA
Walton J G4KKO
Walton J G4FRD
Walton J G6FLK
Walton K G7HQY
Walton K G7PMO
Walton L G1PSH
Walton M G0AZM
Walton M M6MNI
Walton P G1IEC
Walton P G4WAL
Walton P M3GHG
Walton P M1BNH
Walton P 2E0DHV
Walton R G4XUA
Walton R G7SSK
Walton T M6BQR
Walukiewicz I G8IUP
Wand E G4ZOR
Wandless K G7VKA
Wane G G7ODM
Wane J G0OFY
Wanford A G6LTN
Wanford A M0BSZ
Wang A M3YCU
Wang D M0BPW
Wang C 2I0CYW
Wang C MI3ZCY
Wang J 2E0JQW
Wang J M0JQW
Wang J M6JQW
Wang R M3RUI
Wang Y M6IDO
Wankling C G6DOF
Wanklyn D 2W1AOD
Wann G G0DNI
Want R G4VXP

Wantling C G3TNE
Waples M G6CPX
Waples M G0VRF
[illegible]
[illegible]
Warburton D GW7HOC
Warburton D MW5HOC
Warburton D GM0LVI
Warburton D G6LKB
Warburton G M1GWA
Warburton J G0KHJ
Warburton P G8UGK
Warburton P MW6YDP
Warburton W G0WSI
Ward A G0PZU
Ward A G3PZX
Ward A M3TOE
Ward A GI4VQK
Ward A 2E0WAE
Ward A G8KGG
Ward A M3GOV
Ward A M6PUG
Ward B G1ZWB
Ward B G3SZV
Ward B G3XKH
Ward B G8BRL
Ward C 2E0FHQ
Ward C G1WFA
Ward C G3TAI
Ward C G4HON
Ward C G8EPZ
Ward C G8SMA
ward C M6HPV
Ward C M6ZPH
Ward C G6YVJ
Ward D G1FVE
Ward D G1HFH
Ward D G1MTU
Ward D G3ZLE
Ward D G4NNX
Ward D G4PGJ
Ward D G8KBH
Ward D M0CZT
Ward D G7CWO
Ward D G4AOQ
Ward D M6CND
Ward D M6DWW
Ward D G0MDO
Ward E G0CMQ
Ward G G0HRL
Ward G G3MZB
Ward G G6TDJ
Ward G G7TCD
Ward G M3IHC
Ward G GI3ZCK
Ward I 2E0UUA
Ward J G0SUL
Ward J G0WQA
Ward J G3PNP
Ward J G4KXK
Ward J G6MMT
Ward J G8AWY
Ward J G8YOK
Ward J G7SNW
Ward J M6JCJ
Ward J M3WOX
Ward K G1ITV
Ward K G4TYP
Ward K G6YSN
Ward K M3KMW
Ward K G4RJD
Ward L G0PRI
Ward L GW1YBF
Ward L G4EPL
Ward L G1RZJ
Ward L M3XVK
Ward M 2E0CTT
Ward M G0KDQ
Ward M G4GHL
Ward M G4MNP
Ward M G4ZJP
Ward M G4ZXN
Ward M G8XEZ
Ward M G1JPC
Ward M M6MVW
Ward M M6ACQ
Ward M 2E0SAT
Ward M M3OYZ
Ward M M3VYF
Ward M M6AIR
Ward M G3KZD
Ward N G0ASX
Ward N G4OOQ
Ward N G7OVM
Ward N GW3OPC
Ward O M6OLE
Ward P G1BZD
Ward P G4GYI
Ward P G6BBI
Ward P M3PWE
Ward P G1VHC
Ward R G1TYU
Ward R G1VOY
Ward R G2BSW
Ward R [illegible]
Ward R G7SGK
Ward R G8LGA
Ward R M3HGO
Ward R G7BJD
Ward R M6WGD
Ward R G5NF
Ward S G0RBI
Ward S G1XVC
Ward S G4VWA
Ward S G6BCM
Ward S G6ZAL

Ward S G7NDN
Ward S 2E0SXC
Ward S M6DBS
Ward S [illegible]
Ward S M6FYW
Ward S G0JKE
Ward S M6AKH
Ward S G8TNS
Ward T G0SKD
Ward T 2E0BWX
Ward T G7WAE
Ward T M0NTI
Ward T M6TWW
Ward T M6VFR
Ward T 2E0HYK
Ward T 2E0CNC
Ward T M0GOC
Ward T M6TKW
Ward T 2E0UAV
Ward T M0UAV
Ward T M6TYL
Ward V G0FEI
Ward W GM0KTJ
Ward W G4NRE
Ward W G6EHE
Wardale C M5RFD
Wardale G M0ANN
Wardale P G0EPR
Warde S M6RTE
Wardell C M0CUU
Wardell R G7TOB
Warden J GM0JZV
Warden N M6NBW
Warden P M1EMO
Wardill C G7CYF
Wardlaw D G3RYW
Wardle A M6XAW
Wardle C G0PQR
Wardle C G1LKK
Wardle E G4XZA
Wardle G G1UDB
Wardle J G3MAU
Wardle J G7CJW
Wardle J G8UTY
Wardle J 2D0JKW
Wardle J G4CVA
Wardle L M3MSU
Wardley S G6JEF
Wardman D 2E0HWN
Wardman D M3HWN
Wardman G GW0PSV
Wardy G G4YFO
Ware C G4LIY
Ware M G4BJT
Ware M M6RKW
Ware R G7PRD
Wareham G MM3ZKQ
Wareham R G0GXS
Wareham S M6LNX
Wareham-Kirk A M6EZT
Wareing M G8IUQ
Warhurst C M6CWW
Warhurst C M0CQF
Waring A G6SRJ
Waring D M6DWF
Waring G M0AYI
Waring G M6GFY
Waring J 2E0JCB
Waring J M6JTW
Waring L GI3WUO
Waring M G8YPQ
Waring N M1LMO
waring P M3NXC
Waring W G0OYL
Waring W G3GGS
Waring W G8MRN
Wark D GM3FRU
Warke H GI6GAQ
Warman B G0EFY
Warman I M6IFW
Warman P G0ODP
Warmington T MI3KFI
Warnaby B G1GTP
Warncken D GJ4YMX
Warne R G0UHG
Warnecke I 2E0DUE
Warnecke I M6WFI
Warner A G3XUF
Warner A GM7LTX
Warner A M0GWV
Warner B M3YBW
Warner D M6BAC
Warner D G4AFQ
Warner D G4OER
Warner D M0CKP
Warner D GM7PXL
Warner E G4GSG
Warner F G4EWI
Warner J G6LTR
Warner M GW0KYY
Warner N M0NWW
Warner R G3SAR
Warner R G0KIF
Warner R G8HOI
Warner S G7DCJ
Warner S G8AQP
Warner S M3OER
Warner T M0FRG
Warnes A M6AWA
Warnes G G4YAH
Warnes K G0OBK
Warnock A G7FAS
Warnock G GI0VGL
Warnock G M6GGW
Warnock N GW4JCK
Warr A M3XNE
Warr C G0AWM

Warr D G4RQI
Warr M G0CWF
Warr T MM6TEW
[illegible]
Warren A [illegible]
Warren B G4ANZ
Warren C 2E0FNY
Warren D G0FKX
Warren D M5RST
Warren D M3YKZ
Warren E G8HSR
Warren G G4BYS
Warren J G0DCJ
Warren J G4LJY
Warren J M3JHW
Warren L M6GQA
Warren M 2E0FDD
Warren M G1JWL
Warren M G8MLD
Warren M G8VZI
Warren R M1RGW
Warren R G8WAP
Warren S 2E0NBR
Warren S G7LIH
Warren S M0SCW
Warren S M3WSI
Warren S G8EXZ
Warren S 2E0KPO
Warren S M0KPO
Warren S M3KPO
Warren V G0CMP
Warren W G0PNF
Warrender R G8ASW
Warrener P G4HOF
Warrillow S M0CYT
Warrilow A G4GXZ
Warrilow I G7ILS
Warriner K G8GEA
Warriner M G0TTG
Warriner M G8HGI
Warriner P G0AOP
Warriner P MI6PWR
Warrington E G4EMW
Warrington J G1KCU
Warrington J G8AKE
Warry R M1RMW
Warwick A G0IZI
Warwick C GW8DSO
Warwick C 2E0BZG
Warwick C M0HUZ
Warwick C M3ZIH
Warwick D G4EEV
Warwick D G8IKS
Warwick D M0WAR
Warwick F GI8MOV
Warwick G G6ROS
Warwick J G0CDR
Warwick P 2E0PWI
Warwicker W 2E0ZWW
Warwicker W M0ZWW
Warwicker W M3ZWW
Warwick-Oliver E G3YGA
Washbourne T MI3BCR
Washbrook E M6AUF
Washbrook P M6TAO
Washby A G6ZJI
Washby J G1KNZ
Washington A G0PTD
Washington D M6XJS
Washington J G4VUH
Waspe D G4HQM
Wassell I G4KDR
Wastie M M0BRP
Watanabe S M0SMW
Watch M G8MRN
Waterall L 2M1ANY
Waterfall G M6GWF
Waterfall M G8NXD
Waterfield D G0WEO
Waterfield J M0BAZ
Waterhouse D G4TDB
Waterhouse D M6WSC
Waterhouse I M1APT
Waterhouse J G0JHW
Waterhouse J M3CHU
Waterhouse K 2E0KVE
Waterhouse K M3GUM
Waterhouse R M6KBO
Waterloo B G6HGX
Waterloo B M0DDA
Waterloo B M3ADJ
Waterman N G7RZQ
Waterman R G4KRW
Waters A G1BET
Waters A M3BET
Waters B G4GIM
Waters C G0BRH
Waters C G3TSS
Waters C M6GZU
Waters D M0AFX
Waters D M3BOV
Waters D 2E0NBC
Waters D G8LHS
Waters D G1TIK
Waters F G0UQP
Waters G MW0SWH
Waters G M6OLD
Waters I G3KKD
Waters J G1VVY
Waters J M1BMQ
Waters K G6GVF
Waters N 2E0EHY
Waters N 2E0NRW
Waters N M6NRW
Waters P G3OJV
Waters P G0LRP

Waters R G0RSW
Waters R M1DVO
Waters R M3MWT
Waters R G0AXD
Waters S M0DOY
Waters T G0GQJ
Waters T M0OAD
Waters T GW4IMC
Waters W G3OYB
Waterson K G1VGI
Waterson R M6FZF
Waterson W M6OOW
Waterton N G6RPK
Waterton W G7EFV
Waterworth C 2E0ZRX
Waterworth C G6AHF
Waterworth C M3ZRX
Waterworth D G4HNF
Wathen-Blower D G0DWB
Watkin A G4GSD
Watkin B M3WTN
Watkin Ba Hnd A G7NNA
Watkin S G8HRW
Watkins B M6BVS
Watkins C 2E0WGC
Watkins C M0WGC
Watkins C M6CGW
Watkins G M0TTX
Watkins J G3EHW
Watkins J G4VMR
Watkins J G8NCU
Watkins J M3YWE
Watkins K G3AIK
Watkins K G8IXN
Watkins K G0ERF
Watkins M G0NBB
Watkins M G1PRH
Watkins M M0CPN
Watkins M G0MBW
Watkins M 2E0DHX
Watkins P M0HBY
Watkins T GW0JTJ
Watkins T G4VSL
Watkins T M3CWA
Watkins T M6SKN
Watkins Y M0YNK
Watkins Y M6YAN
Watkins-Field J 2E0CIT
Watkinson A M3HIN
Watkinson C 2M0BZL
Watkinson C MM0XPT
Watkinson C MM6CRW
Watkinson D G6JYN
Watkinson J G3EEH
Watkinson K G0ECN
Watkinson S G0LBE
Watkiss D M6CDN
Watling B G4ZPA
Watling I G4NYD
Watling M G3PCW
Watling N G0UML
Watling P M3OJU
Watling S 2E0FIB
Watling S M3FIB
Watmough A 2E0WAT
Watmough A M3AJW
Watmough B M0CVG
Watmough D G0LRO
Watmough J G7NFK
Watmough M M3NFK
Watmough M M0RMN
Watney C G7BKJ
Watson A 2E0GNU
Watson A G1FPY
Watson A G1KAG
Watson A G3DCV
Watson A MM6AON
Watson A D G4DZS
Watson B G1HFY
Watson B G8UDA
Watson B G0UTM
Watson B GM3LLP
Watson B G0RDH
Watson C GW0PCJ
Watson C GM1CCI
Watson C G7NEC
Watson C MM6CEW
Watson C M6YOT
Watson C G7JSS
Watson D 2E0IXC
Watson D G1FEV
Watson D G1TBK
Watson D M3EVE
Watson D M3IXC
Watson D M3IXD
Watson D M0WDJ
Watson D G3YXO
Watson D G0DEZ
Watson E M0CSP
Watson E M3FIY
Watson E M6SJW
Watson E M3ZWF
Watson F G3HRE
Watson F G4MSQ
Watson F G6PBQ
Watson G G0HVH
Watson G G0RIY
Watson G G3XGD
Watson G G8UHV
Watson G M1FER
Watson G M3GJW
Watson G M6GAW
Watson G 2E0DEI
Watson G GW4EVJ
Watson G 2E0GWE
Watson G M6GZA

UK Surnames

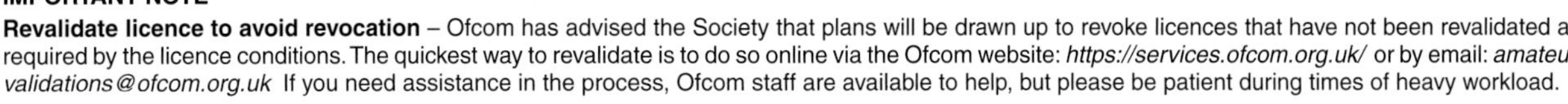

**IMPORTANT NOTE**

**Revalidate licence to avoid revocation** – Ofcom has advised the Society that plans will be drawn up to revoke licences that have not been revalidated as required by the licence conditions. The quickest way to revalidate is to do so online via the Ofcom website: *https://services.ofcom.org.uk/* or by email: *amateur.validations@ofcom.org.uk* If you need assistance in the process, Ofcom staff are available to help, but please be patient during times of heavy workload.

Watson G 2E0GFW
Watson G M6GFW
Watson H M1ELW
Watson I 2M0ISA
Watson I 2M0CFB
Watson I MM0GYX
Watson J 2E0BUA
Watson J G4ZKA
Watson J G6BHS
Watson J M3FIW
Watson J M3PVU
Watson J GM1OQT
Watson J G7DAZ
Watson J MM0TPD
Watson J M3APO
Watson J M6JIZ
Watson J M6JGJ
Watson K G4MOT
Watson K M1FEQ
Watson L M1FES
Watson L G7DXB
Watson M G1OOJ
Watson M G3JME
Watson M G3WMQ
Watson M G6VIQ
Watson M G8SFF
Watson M G4WNZ
Watson M G7PTC
Watson M M3XQL
Watson N G4LCE
Watson P G3GJJ
Watson P G3PEJ
Watson P G8CRM
Watson P M3IUV
watson P MW3WJP
Watson P 2E0PAK
Watson P M0PKW
Watson P M6WAT
Watson P M6TPV
Watson R 2E0OYN
Watson R G0MKG
Watson R GM0NJL
Watson R G7EHR
Watson R M0RKW
Watson R M3OYN
Watson R 2E0PYC
Watson R M0XAW
Watson R M6BNG
Watson R G4WQT
Watson R G4CVM
Watson R GM5BDW
Watson S 2E0TSU
Watson S G8ZMG
Watson S M1GSM
Watson S M3JQT
Watson S G1KWF
Watson S M6CFU
Watson S G3IEJ
Watson T G0NPP
Watson T M6CZL
Watson T G0JAA
Watson T G4JYG
Watson V G7AWG
Watson W GM4YWV
Watson W G4EHT
Watt A G3NXO
Watt A G3WZJ
Watt A G3ZBU
Watt D GI7GUT
Watt D M0MMX
Watt E GM0IOY
Watt F M3EUR
Watt G G1GID
Watt G M3GCR
Watt H M0DEY
Watt I GM4ZRR
Watt J MI0OBE
Watt J MM6AFQ
Watt K MW3ZKW
Watt K MI6KJW
Watt R G1ELE
Watters D GM8NVE
Watterson P G1AAC
Watton D G4EFG
Watts A M0MZX
Watts A G4UFK
Watts A G7MJV
Watts A M6HAE
Watts A M6KWH
Watts B G0SXN
Watts B G4ZSZ
Watts B M0BWO
Watts C G8SEK
Watts C M0GMW
Watts C M3YCV
Watts C G0NVX
Watts C G7PVL
Watts C G4KLB
Watts C G8SWO
Watts D G6ZZS
Watts D G7BME
Watts D M6TJG
Watts D G3XWD
Watts G G0EVW
Watts H G0BBV
Watts H G0XBL
Watts H M6ARY
Watts J G0BCR
Watts J G3ZFV
Watts J M3JWM
Watts J M6WAJ
Watts J M3VQI
Watts J 2E0NLP
Watts J GW8KXW
Watts J M6ATS
Watts K G7JJG
Watts M G0IAK
Watts M G7KSS
Watts M M0DDV
Watts M M0DFY
Watts M M6WMJ
Watts M G1WHT
Watts M M6MWW
Watts N G1ZUS
Watts N M3ZWB
Watts R GM4ZZW
Watts R G6YTB
Watts R G7NLJ
Watts R G4VIF
Watts S G3XXH
Watts S G6YSL
Watts S G1GUI
Watts S M3YPA
Watts T G4ZZY
Watts T M5ECX
Watts T G3PIZ
Watts W G4VIW
Watts-Read R G7VFX
Watwood K M3KEL
Waud M G6IDL
Waud N G1OVH
Waudby A M3YUC
Waudby S 2E0SCA
Waugh B 2M1DZW
Waugh D GI3OBO
Waugh G M0DQH
Waugh H GM7PPN
Waugh I G0WIZ
Waugh J M3OYW
Waugh W G5BW
Way C G3LWJ
Way D G0WTG
Wayer P G6YSO
Wayer P M3GVC
Waygood P G4YXR
Waygood R G4OXK
Waylett N G3YQG
Wayman J G0COZ
Wayman J G4DRS
Wayman L M3WEZ
Wayman S G4JQL
Waymark B M6WAY
Wayne R G0NDY
Waywell G G8UQY
Weal C M6BGR
Weale C G8DKD
Weale G GW3LEW
Weale G G4ACS
Weale N M6ZOO
Wear D G4DPJ
Wearing C GM8GIQ
Wearing J G8BVF
Wearing R G1HPB
Wears N M0NJW
Weatherall A 2E0SOB
Weatherall A 2E0TWA
Weatherall A M6WEV
Weatherall P G3MLO
Weatherer J G4RJF
Weatherhead H G7CNP
Weatherhead H M3HUW
Weatherill D G0LCX
Weatherley C G1HKU
Weatherley M G1SAK
Weatherley T G3WDI
Weatherspoon W G4NSC
Weatherup R M0RSW
Weaver A G7JMW
Weaver A M3XJK
Weaver C G1YGY
Weaver C G6LYD
Weaver C G3IZW
Weaver D G6BWP
Weaver G G4MUW
Weaver J G2HNA
Weaver K G1SKI
Weaver M G0MJF
Weaver M G4GMW
Weaver M 2E0WMP
Weaver M M6WMP
Weaver P G4JMB
Weaver P G7VCZ
Weaver P 2E0SDQ
Weaver P M3GDK
Weaver P G4ROV
Weaver R GW3KXX
Weaver R 2E0CUE
Weaver R M0HNI
Weaver R M6RCW
Weaver R 2E0ZBW
Weaver R G0UQE
Weaver R M3ZBW
Weaver W M3PSO
Weaving P G0PAW
Weaving R G3NBN
Webb A 2E0HQZ
Webb A G0MSO
Webb A G6UXG
Webb A G6XJF
Webb A M6AMW
Webb A M1ANN
Webb A 2E0BNF
Webb A M0GRU
Webb A M3OIV
Webb A G4LYF
Webb A M1MVX
Webb A M3PYG
Webb B G1NFO
Webb B M0WBJ
Webb C 2E0WEB
Webb C G4FWM
Webb C G4JFF
Webb C G4NYJ
Webb C G6NVD
Webb C G6OXI
Webb C M3KHI
Webb C M6JJK
Webb D 2E0EOO
Webb D 2E0GXH
Webb D G4YHQ
Webb D G6OXJ
Webb D G8ZJE
Webb D M3ZZF
Webb D 2E0DIP
Webb D G0SSY
Webb D G6JOR
Webb D M6DWE
Webb D G0WRL
Webb D 2E0DDW
Webb D G1KTF
Webb D M0GUZ
Webb D M6DDW
Webb F G4ETZ
Webb F G7OKO
Webb I G6TNW
Webb J G6SCM
webb J M6HMQ
Webb J G0RYO
Webb J G8RDP
Webb J M3RAE
Webb J G7EPN
Webb K 2E0HYT
Webb K M6GWZ
Webb L M6LWJ
Webb M G0UGQ
Webb M G4SHA
Webb M G8VZJ
Webb M GD6ICR
Webb N G1AMN
Webb P G0KUE
Webb P G6LMC
Webb P 2E0AQU
Webb P G1RLI
Webb P M0BMN
Webb P M3HCW
Webb P M3UNL
Webb P G7NBI
Webb P M0CKA
Webb P M3XZE
Webb R G0WEB
Webb R G1SIP
Webb R G3NDK
Webb R G8VBA
Webb R M3RXW
Webb R G1WYA
Webb R M6EIG
Webb R G6TXH
webb S 2M0IOK
Webb S G0AEN
Webb S G3TPW
Webb S G4GHO
Webb T G8TVC
Webb T G1OEM
Webb T M6TGQ
Webb W G1BWP
Webb W GM3NGW
Webb W GW4ZUA
Webb W MW6WFF
Webber A M6FMT
Webber A G1PRM
Webber B G1ABW
Webber C G4NGR
Webber D G0PBS
Webber D MD6SFL
Webber D G3LHJ
Webber H G4ZYR
Webber J G8DNH
Webber J M6HGE
Webber J M6WBR
Webber J G6FGV
Webber J M0CZB
Webber K G1WQY
Webber M G6VKX
Webber P G8KLC
Webber P G0HJW
Webber S 2E0WRS
Webber S M0ZZT
Webber S M3DHA
Webber S M3WRS
Webber W G7DRO
Weber M M0AGP
Weber R M6YRW
Webley F M3ROW
Webley V G0RKV
Websdale J 2E0JPW
Webster A G1EWC
Webster A G3JQ
Webster A G1UAF
Webster B G7WJC
Webster C G3TBJ
Webster C MM3ZUP
Webster D 2E0HER
Webster D G7AYS
Webster D G7VSG
Webster D G8MYV
Webster D M0BDD
Webster E G0VDJ
Webster F G0VBQ
Webster F G3YON
Webster F G6ZJK
Webster G G3VOT
Webster G G6JDH
Webster G GJ7RIY
Webster G M6GAO
Webster G 2E0CSH
Webster G G6OBJ
Webster G M0OSB
Webster G M6BYU
Webster H 2E0GYR
Webster J G0DRE
Webster J G3YOO
Webster J GM7KIY
Webster J MM6WEB
Webster J 2E0JWW
Webster J M3NVG
Webster K G3LLE
Webster K G6DMM
Webster K G7DWV
Webster K G8VMP
Webster M M3MDW
Webster N G2DWB
Webster N GM4RXW
Webster P G1AAK
Webster P G1YMR
Webster P G3ZTV
Webster P G4EVE
Webster P G8FJA
Webster P M3NPW
Webster P 2E0IFF
Webster P GM7EEY
Webster P M3HJU
Webster P M3IFF
Webster P G3WEG
Webster R GM0DZW
Webster R GM3MOR
Webster R G4EGM
Webster R M0DWZ
Webster S G0NWF
Webster S GM1WMU
Webster T G4ZVA
Webster T 2E0WYT
Webster T M0WYT
Webster T M6WYT
Webster W G0RSV
Weddell A G1JFF
Weddell J M1DTU
Wedderburn J G4JOV
Wedgbury C G4FNQ
Wedgbury N G6CUQ
Wedge A 2E0IED
Wedge A M0IED
Wedge J M0WEV
Wedgwood A G0TJD
Wedgwood A G6HFB
Wedgwood B G0SBU
Wedlock T MI0BVG
Weeden A G4MIE
Weeden B G2FSH
Weedon R G0IWW
Weeds C G1BZE
Weekes R G6FGW
Weeks G G8WAM
Weeks I G7DAB
Weeks K G0OBO
Weeks L M6LCW
Weeks S M6GLM
Wegg G G0LPT
Weigh J G6LXW
Weight D 2E0DSW
Weight D M0MRL
Weight D M6DSW
Weight R G1OTN
Weightman N MW6TLN
Weightman R 2E0KNU
Weightman R M6KNU
Weightman S 2E0SML
Weightman S M6SAN
Weiner J G3YIF
Weinstock J G0CCJ
Weir C G4MWQ
Weir E G6CEM
Weir G 2M0WEV
Weir G MM3WEV
Weir G G6NOL
Weir J MM6WER
Weir K G4SEW
Weir M 2E0NVP
Weir M M3NVP
Weir M G1IRX
Weir R 2E0DMW
Weir R M3DMW
Weir S GM3SAN
Weir W GI7UDV
Weir W MI5AFM
Weiss L G6OBO
Weiss S G6EFE
Welbourn R G0VJY
Welburn I G4EMA
Welburn W G7CAS
Welch A 2E0XEA
Welch A M6CDB
Welch D G0ATD
Welch D G7WBE
Welch D M1ELI
Welch G G8EBD
Welch G MM5AHM
Welch G 2E0XDG
Welch G M0XZG
Welch G M6XDG
Welch J M6EKE
Welch J 2E0XJL
Welch J M6LRW
Welch L G4JSK
Welch R G0LQF
Welch R G1OGR
Welch R G3OFX
Welch T G0TAX
Welch W 2E0TDF
Welch W M0TDF
Welch W M3XTF
Weldon A G4OMZ
Weldon P G4DNI
Welford D M6CAT
Welford I G4RKK
Welford J G3WOD
Welford J M0JLW
Welford P G4YKQ
Welford W G4RKL
Welger S G7MGY
Welland A G0OIQ
Welland B M1UKC
Welland M M0WAQ
Wellard J M3YAA
Wellard J G6ZAA
Wellard J M0ZAA
Wellbeloved L G1NBU
Wellbeloved R G3LMH
Wellborn S M6XGF
Wellburn R M3XPU
Weller A G6ORJ
Weller A M0OAL
Weller A GM8BSU
Weller B M6WHO
Weller C G4ONH
Weller D M3TLB
Weller D GM7RYT
Weller J G0GNA
Weller K G3WBL
Weller M GI4IZF
Weller M M3MZW
Weller P GM3XOQ
Wellings D G4KLJ
Wellings S G0CXO
Wellington G M6ORT
Wellington L 2E0LSW
Wellington W M0WJW
Wellman A M1BPS
Wellon S G6DMG
Wells A 2E0CFT
Wells A G4ERZ
Wells B G0JEZ
Wells C G6XOG
Wells C G7NZY
Wells C G8FMD
Wells C M1FEX
Wells D M0CEM
Wells D 2E0DWT
Wells D G0GPE
Wells D G7IJC
Wells D M0KWY
Wells D M0WEL
Wells D M6DBK
Wells D M6WEL
Wells D M6JXF
Wells E G7FED
Wells E G3XCE
Wells F M3TZF
Wells G GM4YGS
Wells G M6GWE
Wells G M3WGZ
Wells J 2E0ZBD
Wells J G0IWB
Wells J G3IZG
Wells J GW8KZA
Wells J M3FGX
Wells J 2E0GYP
Wells J M6JAS
Wells J G4BSC
Wells J M6LFY
Wells J G8YFP
Wells K G0WAC
Wells K 2E0DVO
Wells K M0KAW
Wells K M6FDH
Wells K M6KFW
Wells L M6EBU
Wells M 2E0WHO
Wells M G1CHQ
Wells M G3XBW
Wells M G4JES
Wells M G7TNZ
Wells M 2E0NUG
Wells M M0VPL
wells M M6HWO
Wells N 2W0CGM
Wells P G0JEW
Wells P G4PAI
Wells P M0FSN
Wells P M0VGC
Wells P G4RKN
Wells R G0ITS
Wells R G8BNR
Wells R M0RWW
Wells R M6REW
Wells R M3RDW
Wells R 2E0RXW
Wells R M0RCK
Wells R M6ELI
Wells S M1AFV
Wells S G7KRB
Wells T G1AWD
Wells T M0BRL
Wellsbury N G8TPM
Wellspring M G8AWE
Wellstead G M6ODC
Wellsted R M0RNW
Wellsted S M6MMR
Welsby J G0AJW
Welsh C 2E0IDI
Welsh C 2M0CRQ
Welsh C MM0RKN
Welsh C MM6CRQ
Welsh J G0NVZ
Welsh J GI4JXM
Welsh J MM0AYE
Welsh J M6JSW
Welsh N M3NCD
Welsh R G4NEQ
Welsh R G8MIN
Welsh R MM0RWJ
Welthy B G7BND
Welthy B M3BAW
Welton S G7BXU
Wemyss D GM7FYB
Wendes R G7RCC
Wendon B G6JMX
Wengraf I 2E0BBA
Wenham A G3ZXA
Wenlock M M6MGO
Wenlock R M6MGG
Wenman K G1YTO
Wenn C G8YAE
wenseth F 2M0EFI
wenseth F MM0EFI
wenseth F MM3LKR
Wensley S G4SVG
Wentworth A G7PYN
Wentworth D G0NLT
Wentworth P G0DZA
Werba P G7FXO
Werndle L 2E0ZWE
Werner K G7RTI
Werner S M0ASY
Wernham J MD3WBC
Wernham L MD3LPW
Wernham M MD3UGY
Wernham R GD3MBC
Wernham T G0SBH
Werrell B MW6OCC
Werrett V M0VAW
Wersby S G7TAE
Wertheim R M3OAZ
Weseley A G3XBQ
Wesil D G0RPJ
Wesley P M0COP
Wesselby A M6GJV
Wesson G M1EZD
Wesson R G3MXZ
West A G6YBN
West A G7LNB
West A 2E0FMA
West A M6TAU
West B G4STD
West C 2E0CHW
West C G8DYA
West C M3VCW
West D M3TEE
West D G1PUO
West D GW3TYI
West D G3KTP
West F G1LVZ
West F G1SCY
West G G0JAJ
West G G3YZK
West G G7HZS
West G M1AUY
West G G0FAS
West G G8POK
West J G0KIM
West J G1JKX
West J G4AOS
West J G4OGG
West J G8MIW
West J G0DBD
West J G4LRG
West K M3LVM
West L G0GNQ
West L M6YBB
West M G1AAR
West M G1YQP
West M G4EJM
West M G0UCD
West M 2E0KIA
West M G7UUC
West M M6AYJ
West Of Stow J G8CJW
West P G4LLG
West P G6RHL
West P G4ADG
West R G3SHX
West R G3VSQ
West R G4ZIH
West R G8DVU
West R 2E0WST
West R M0VGC
West R M6WST
West S M1FEZ
West W G6TWD
Westaby G G0ECS
Westall F G4MUU
Westall S G6LXU
Westall S M0CKO
Westbrook B G4WBA
Westbrook E G6FGY
Westbrook G M6GUY
Westbrook T G7NKJ
West-Bulford C G8JXU
Westbury D G3OXL
Westbury P G0UAP
Westbury P G1JAL
Westbury T G3TBW
Westby D G4UHI
Westby I M6ERW
Westcott C 2E0EOF
Westcott E G4ONC
Westcott R MW0EAN
Westell S G3YFG
Westerman A GM8ZCS
WESTERMAN J G4OOB
Western F M0BZI
Western R G3SXW
Westgate D G6WSN
West-Knights L G8FBK
Westlake A G1DFM
Westlake B G1ODD
Westlake D G0AIX
Westlake J G8SBS
Westlake R G8MVC
Westlake S G8SQZ
Westlake W G8MWW
Westlake W 2E0NWR
Westlake W M0RHW
Westland A M1VPL
Westland D M0DJW
Westland M GM0UGH
Westley K G4WEZ
Westley R G0NNG
Westley S 2E0FXP
Westmeckett R G4MRW
Weston B 2E0WBT
Weston B M3WBT
Weston C GM3VAP
Weston D G0WWD
Weston E G0VPY
Weston G G0RSU
Weston G G6ZGK
Weston G G6LJC
Weston G M6GWG
Weston G 2E0DMV
Weston G M6GVW
Weston I G4UAQ
Weston J 2E0WEZ
Weston J M3RMQ
Weston J G3LYW
Weston K M3KGJ
Weston K M3XLW
weston K G0BKW
Weston K G7IHP
Weston L M3ULW
Weston M G7AFV
Weston P 2E0AED
Weston P M6DKK
Weston R G0LIB
Weston R G4XZS
Weston R G6RRV
Weston R G3RGJ
Weston S 2E0WES
Weston S G7VSW
Weston T 2E0FBN
Weston T M6FYG
Weston W GW4VKG
Westripp P G0SLD
Westwater M G4KAT
Westwell H G3CTQ
Westwell P G4HLF
Westwell P 2E0ZSU
Westwell P M3ZSU
Westwood C G3VFD
Westwood D M0DSC
Westwood D G4NPN
Westwood E M6EAL
Westwood G 2E0FDP
Westwood G G3VWJ
Westwood H G4NRF
Westwood I M6WBF
Westwood I G3VPQ
Westwood J G1NWM
Westwood L 2E0PPZ
Wetherall E GW7VMT
Wetherell K G6IMN
Wetherill T M6FTI
Wetherill D M6HKK
Wetton B MM6BQG
Wetton D M6DCW
Wetton R G8PHV
Wevill K G4UKW
Whadcoat A 2E0WBS
Whadcoat A M6WHA
Whalan P M3XEQ
Whale G M6OVE
Whale M G7MNG
Whaling G G0PPR
Whall B 2E0VBW
Whall B M0VBW
Whall C M3WCX
Whall G 2E0XGW
Whall G M0XGW
Whall G M3XGW
Whall S 2E0XSW
Whall S M0VSW
Whall S M3VSW
Whalley B 2E0AIY
Whalley D G4EIX
Whalley J 2E0FAB
Whalley J M6JWO
WHALLEY M G1CIT
Whalley P G3YXN
Whalley R 2E0RWB
Whalley R M6RWP
Whalley R 2E0REM
Whalley R M6AVT
Whalley S G4DVN
Whalley S M0SGW
Whan D G4OXU
Wharfley D 2E0NJJ
Wharfley D M0GIG
Wharfley D M3NJJ
Wharton A M0DNN
Wharton E G4NUY
Wharton G G0NLO
Wharton J G7LLY
Wharton J M1BSE
Wharton J G0AZH
Wharton M M1DGW
Wharton M M6MKW
Wharton R G0IEN
Wharton T G0AZG
Whateley R G1TPN
Whateley T G3NMW
Whatley F G3JOT
Whatley M G7FZJ
Whatley M G7IBL
Whatley P G1VDT
Whatley R G1JXA
Whatling B G0BRW
Whatling M M6WYW
Whatmore A G4UVZ
Whatmore W G4BJX
Whatmough D G4ORV
Whatmough G M6GDQ
Whatmough R 2E0JKR
Whattam R G8ACQ
Wheal M 2E0MWW
Wheat B 2E0HXB
Wheat B M3HXB
Wheatley A G1RBA
Wheatley G G4HNJ
Wheatley J G0JSC
Wheatley J G8JUC
Wheatley K M0KHZ
Wheatley L G1XPD
Wheatley M G0JJO
Wheatley P G0AAT
Wheatley R M3WRB
Wheatley R G7TZW
Wheatley S G8KNC
Wheatley T M6TRW
Wheatley-Hince C M0KKA
Wheddon R M6RFR
Wheeldon B G1CYQ
Wheeldon J M0JHW
Wheeldon N G0IMP
Wheeldon R G7NRV
Wheeldon S 2E0LDS
Wheeldon S M0SNW
Wheeldon S M6CHO
Wheele A G3AKJ
Wheeler A G4NWS
Wheeler A M1CQT
Wheeler A M3NZG
Wheeler A M6KEA
Wheeler A 2E0DJV
Wheeler A M3ZQN
Wheeler B M0EXM
Wheeler C G0TOX
Wheeler C G6YAH
Wheeler C M3BXH
Wheeler D M6DJW
Wheeler E M0EMW
Wheeler G M6NGK
Wheeler G G1ODE
Wheeler G G8SGP
Wheeler J G7CRR
Wheeler J G8EMU
Wheeler J M3IEQ
Wheeler J G7WBL
Wheeler J GM0UYZ
Wheeler J M0JFW
Wheeler J M6WJM
Wheeler J G0IUE
Wheeler K 2E0KTW
Wheeler K M0AXG
Wheeler K GW0SXS
Wheeler L G0EFR
Wheeler M G6DOD
Wheeler M G5FM
Wheeler M M3UIF
Wheeler M M6XBX
Wheeler O G0NCE
Wheeler P G4CDX
Wheeler P G4PFA
Wheeler P G8LSC
Wheeler R G3MGW
Wheeler R G6GOW
Wheeler R G8HLH
Wheeler R M6CGJ
Wheeler R G6UXE
Wheeler S M0HRY
Wheeler T G7MIM
Wheeler W G0BNW
Wheelhouse G M3GWW
Whelan B 2E0TOG
Whelan B M6FAC
Whelan C G0WFF
Whelan D 2E0LOG
Whelan D M0LAG
Whelan E 2E0CQQ
Whelan J G6NOI
whelan J G7HHZ
Whelan K G3KRW
Whelan M M6AOG
Whelan N G7COC
Whelan R G3PJT
Whelan T G1BIF
Whelan T M6XAT
Wheldon C M0DEO
Whenham G G3TFA
Wherrett C G4IIX
Whetstone G G1SXB
Whetstone J G4OUB
Whetton R G4XKL
Wheway J G4JJQ
Wheway J M3JVW
Whibley A G0JKP
Whiffen J G7FQE
Whiffin I G0CTQ
Whiffin J M3DCJ
Whiffing P M0PGW
Whiffing P M1DPW
Whiles G G6LYE
Whiley S G6ORM
Whillier R G3WUL
Whillock A G4ZLX
Whincup D G1EYY
Whinney K M1ALH
Whipp A G1ZBP
Whish D 2E0CEE
Whistance K G1UFL
Whiston G G8RCL

Whitaker A G3RKL
Whitaker A G3UOS
Whitaker D M6DWZ
Whitaker [illegible] [illegible]
Whitaker M 2E0BGJ
Whitaker M G4DNG
Whitaker M M0NXP
Whitaker S M3SJM
Whitbourn S G0SWE
Whitbread H G1MOS
Whitbread K G3XDU
Whitburn A M3EAI
Whitby C G1DJU
Whitby E G6XOD
Whitby F G6XOE
Whitby P M1ATJ
Whitby R G8MEI
Whitby R G8ENB
Whitcher A G7JUL
Whitcher W 2E0BNZ
Whitcher W M3WBI
Whitchurch K G6NQM
Whitchurch M M1DTG
Whitchurch P G3SWH
Whitcomb J GM7DLY
Whitcombe D MW6DGW
Whitcombe W G6SIQ
White A 2E0RWT
White A G0IWZ
White A G0OHA
White A GM3HEN
White A G4IOQ
White A G4IPY
White A G4YBG
White A GM6JOA
White A G6OLV
White A GW6VKY
White A G6YBH
White A G7AUP
White A G7JNM
White A G7JVG
White A G7OAS
White A G8YUK
White A M0DJB
White A GM0KAZ
White A M6HFQ
White A M0OOO
White A M3VQF
White A M6WTE
White A G1WRY
White B 2E0BJW
White B M1FFP
White B M6BJW
White C G4FKE
White C G4JBL
White C G4ROP
White C G4TXF
White C G6AOV
White C G7NJE
White C G4VFU
White D G1VUY
White D M1AIN
White D 2W0UZO
White D MW0UZO
White D MW3UZO
White D G3ZPA
White D M0PLR
White D M0TOG
White D M3XMQ
White D M0GNU
White D M0YDW
White D M3YDW
White D 2E0RRF
White D G3OHL
White E G8XUU
White E M6LBW
White E 2E0ODO
White E M3ODO
White E G4JIG
White F G4FLW
White F G8CYT
White F MI3FSW
White G 2E0XGA
White G G1RTW
White G G8APM
White G MM3KUU
White G M3XGA
White G G7WID
White G G0GLW
White G GM7NPR
White G G8EZV
White G MM3YTI
White H G0FRM
White H G0WUH
White H G3NKW
White H G4LFB
White H MI3HSW
White I G3SEK
White J G4MNE
White J GM4RDI
White J G0LJF
White J G7BWI
White J G8DX
White J M0PRO
White J G7LFQ
White J M6ITL
White J M6JCR
White J M1AIS
White J G4BCZ
White J G0RNS
White J M3YXP
White J MM6JAW
White J G8CCL
White J M3CCY
White K G0RSL
White K M3HSH

White K 2E0KHW
White K G7TRM
White K G4KTU
White K M0KMW
White L G0JAO
White L G1YXY
White M G0MTI
White M G1KKG
White M G1MSG
White M G4HZG
White M G6GXY
White M G8FXV
White M G8IQC
White M M1ECM
White M M1EZP
White M 2E0DZE
White M M6MWD
White M 2E0RGO
White M M0VNG
White M G3WOE
White M G4YRV
White M M6FCT
White N 2E0NSW
White N M0ITX
White N M0WNW
White N M3NAW
White N MM3OIX
White N M6NAW
White O M1CWY
White P G0BHA
White P G0DDA
White P G1LGQ
White P G3WJI
White P G4VQF
White P G6CJB
White P G6IQI
White P G6OZT
White P G8DOF
White P M0BXU
White P 2I0DHC
White P 2E0IPW
White P G0WHY
White P G7ULJ
White P MI6PBW
White P MW6PNW
White P M6PRW
White R G0DQB
White R GI0RYK
White R G1SAJ
White R G4PGY
White R G8UAD
White R M3WYT
White R G6NFE
White R M0RFW
White R G0AGO
White R G8SPC
White R G4ZJK
White S G1EYZ
White S G4XXD
White S G6TYT
White S M0TTI
White S M3ZSW
White S G1MTB
White S G3ZVW
White T G4YQS
White T G6VAE
White T G8JHA
White T M3GGV
White T G0BXL
White T GI7THH
White T M6WTW
White V G7BCI
White W G4TFI
White W G8GHK
White W 2E0WJA
White W M0HHW
White W M6BVI
Whitear G M3CZJ
Whiteford I MM3TWA
Whitehall J G0VAD
Whitehead A M6AHW
Whitehead B G8XPQ
Whitehead C G0PHD
Whitehead C M1AMW
Whitehead C G1XNG
Whitehead C G3XAC
Whitehead D G3FDZ
Whitehead D G4NRH
Whitehead D M3FBN
Whitehead G G0GHW
Whitehead G 2E0JII
Whitehead G G3YLJ
Whitehead G M3YLJ
Whitehead I G1DTQ
Whitehead J 2E0JGW
Whitehead J M6WBB
Whitehead J M6ARQ
Whitehead L G6XJC
Whitehead M G0UXI
Whitehead M M0DXV
Whitehead M M0VFU
Whitehead M GM0PHW
Whitehead P G4SCE
Whitehead P G8HEU
Whitehead P M6JPW
Whitehead R G4GWZ
Whitehead R M6BEX
Whitehead R M3RNW
Whitehead R M6YLJ
Whitehead R G3ZUK
Whitehead S 2E0SDW
Whitehead S G3RDA
Whitehead S G8FGB
Whitehead S M0SJW
Whitehead S M3NHI
Whitehead S G6UUR

Whitehead S G8LKA
Whitehead T G6HRX
Whitehead T GM7FDS
Whitehouse A G4DTK
Whitehouse A M0BHM
Whitehouse B G1LWH
Whitehouse B G0FGK
Whitehouse D 2E0OPM
Whitehouse D G4OSI
Whitehouse D M0WRD
Whitehouse D M6WHS
Whitehouse D G0ALA
Whitehouse E 2E0TBE
Whitehouse E M0VAH
Whitehouse E M6EAW
Whitehouse G G7LUF
Whitehouse J G3UEK
Whitehouse J G6LJU
Whitehouse K G3OHN
Whitehouse R G3YJW
Whitehouse R G6BCG
Whitehouse R G8FMW
Whitehouse R M3ZAR
Whitehouse S GW7RRM
Whitehouse S G8HIQ
Whitehouse S M1AGW
Whitehurst F G6VSQ
Whitehurst G M3NYM
Whitehurst J G6CUT
Whitelam A M3WJA
Whitelaw C M0RKF
Whitelaw K M6BEL
Whitelegg L G0CCU
Whiteley B 2E0RCA
Whiteley B M0KLM
Whiteley C G3NYS
Whiteley I 2E0TFM
Whiteley I M6KDZ
Whiteley I 2E0IZW
Whiteley I M6IZW
Whiteley K M3NVE
Whiteley M GM0SVS
Whiteley M G6JTC
Whiteley P M0AFS
Whiteley P M0PDW
Whiteley P 2E0ZFV
Whiteley P M3ZFV
Whiteley R G4EUJ
Whiteley S G2DAN
Whitelock D M3KYZ
Whitelock D M6NBV
Whitelock P G0GMY
Whitelock-Wainwright D M0CHR
Whiteman P G8NPZ
Whiteman S M1SCW
Whitemore B 2E0GWF
Whiten E 2E0RKK
Whiten E M0GNO
Whiten E M3RKK
Whitenstall R G7AIH
Whiteoak H M5AIB
Whiteside J 2E0HPJ
Whiteside J M0HPJ
Whiteside J M6JWW
Whiteside L G0MEW
Whiteside L G7TVT
Whiteside N G4HUN
Whiteside O G7FTS
Whiteside R M3OGD
Whiteside W G8MGG
Whiteway G GW4VWY
Whitewood E M6EWW
Whitfield D G8VMY
Whitfield E G8HYI
Whitfield H G0GBQ
Whitfield H G6AUC
Whitfield L M0ZMO
Whitfield M G0EBZ
Whitfield M G4MPJ
Whitfield M M0MJW
Whitfield P G0VWE
Whitfield R 2E0RTW
Whitfield R G3KML
Whitfield S G3IMW
Whitfield S MJ0SIT
Whitford P G3MME
Whitford-Robson J M3YYJ
Whitgreave A G6SKP
Whitham E G6CPY
Whitham E G8ZEI
Whitham G G0FPA
Whitham N G4SEN
Whithorn S G3NAY
Whiting B G1DYV
Whiting C G3VMV
Whiting D 2E0PCV
Whiting G G3MMS
Whiting G GM7MZZ
Whiting J G4RKG
Whiting J G6JUT
Whiting N G1BRK
Whiting P G4YQC
Whiting P G7OPS
Whiting R G8NIU
Whiting S M3GUU
Whiting T M1ADN
Whitington J G1KHM
Whitley D 2E0ZDW
Whitley D M0WIT
Whitley D M6DFW
Whitley M G7ONF
Whitley R G4GJI
Whitling P G0NEP
Whitlock A G0VFM
Whitlock A G8ALQ
Whitlock I M6BEI

Whitlock M G8EZB
Whitlock S M3KRM
Whitmarsh C G0FDZ
Whitmarsh K M1KRW
Whitmore A M0PYF
Whitmore I G4WIA
Whitmore J M6JOW
Whitmore K G0WKN
Whitmore P M6AHA
Whitnear S G0BPR
Whitney G G8RSI
Whittaker A 2E0CXF
Whittaker A G8BFM
Whittaker A G4NGV
Whittaker B G3LUW
Whittaker C G0UXF
Whittaker C GJ7SLU
whittaker D 2E0EHM
Whittaker D G3XAB
Whittaker D G4FSJ
Whittaker D M6OXB
Whittaker E G0IZJ
Whittaker F G1BAA
Whittaker F G4IAY
Whittaker I G6ICV
Whittaker J G6PRA
Whittaker J G3UK
Whittaker J M0AAS
Whittaker K M0KFW
Whittaker N 2E0NYE
Whittaker N G3KLN
Whittaker P 2E0FLD
Whittaker P G1ZDG
Whittaker S M6WTR
Whittaker S 2E0SCW
Whittaker S M0WSW
Whittaker S M6SCW
Whittaker T G1JKV
Whittaker T G3JNM
Whittall P M3LTP
Whittam T G1HKR
Whittam T M3GDI
Whittam T M3TAW
Whitten M M3PPU
Whitten P G4PRW
Whitten S MI3SLT
Whitter I G3NSL
Whittering R G3URA
Whittick J M3ZIF
Whitticombe A GW4ODN
Whitticombe M 2E0MGW
Whittingham A G0CBJ
Whittingham N G4ISU
Whittingham P G1XAP
Whittingham S G4CLG
Whittington B G7PMY
Whittington J G3SHZ
Whittington J M3VJW
Whittington M G1VGK
Whittington P G8WHD
Whittington R G3UQD
Whittle A 2E0DKU
Whittle B G3YBU
Whittle D G6ESK
Whittle D G8SYM
Whittle D M1EQN
Whittle E M0DKL
Whittle G 2E0GLW
Whittle G 2E0MRG
Whittle G M3IIW
Whittle G G6GVR
Whittle H G7RFT
Whittle K MD6KBW
Whittle M G1EQJ
Whittle P G4BBU
Whittle R M3RXP
Whittles B G4KUD
Whittlestone P G3OAH
Whittock B G7IYA
Whittock J G4VKO
Whitton D G4YEB
Whitton D M3WTA
Whitton G M6HGW
Whitton J 2E0JCC
Whitton J M6JBR
Whitton K G1RWT
Whitton N M0CQN
Whitty B G3HWX
Whitty D G4FEV
Whitwam A G0TLP
Whitwell R G4FRI
Whitwell S M6PII
Whitworth A G1YWN
Whitworth D 2E0AZF
Whitworth E G4TUO
Whitworth I G8JHC
Whitworth J 2E0DXK
Whitworth J M6JWG
Whitworth P M6PWH
Whitworth P G7FXW
Whomoo J G3WMD
Whotton M M6MPW
Whtye A GM7KVB
Whyatt A G8NQO
Whyatt A M3NQO
Whyatt D 2E0DBW
Whyatt D M3ZDW
Whyatt M GM4SKB
Whyatt W 2E0WIL
Whyatt W M3WPW
Whyborn D G4KIK
Whyborn N G4JNX
Whyle B G4YIV
Whyman A 2E0DET
Whyman A M3DVP
Whyman A M6WCY

Whysall D G6XJD
Whysall J G6XJE
Whysall P G6LYA
Whucker R G0DNY
Whyte I G0CPF
Whyte K MM3NWF
Whyte M MM0XXW
Whyte T MM6DZC
Whytock J G7TOZ
Wibberley K G8PEA
Wibberley M M0GVW
Wiblin D G6XJI
Wiblin M G4WZJ
Wickenden C G4ICH
Wickenden R GM6IQH
Wickenden R M0FZR
Wickens D G6WZA
Wickens F G3NXN
Wickens M M6TSZ
Wickers P G7EQX
Wickham A G3IAZ
Wickham M G4IGK
Wickham M G8DGW
Wickham R G4JQB
Wicks A G8BSP
Wicks D M3DFP
Wicks D M3OPT
Wicks G G0OIR
Wicks G M6SWG
Wicks R G1RJW
Wicks R G0JEQ
Wicks T G6FIB
Widders R GD3LFD
Widdett S G4DFN
Widdows M G7BWW
Widdowson A G3PET
Widdowson B M3OHL
Widdowson J M0HQZ
Widdowson L M3ICO
Widdowson S M3IXT
Widdowson S G7ULW
Widdowson S 2E0YNI
Widdowson T M6ETJ
Wideman I M6IGK
Widger P G0HNW
Wieckowski A M0GPQ
Wiegold B GW6WTK
Wienrich C G0REN
Wierdis A M0HJJ
Wiese A GI7GSB
Wiewiorka J GM3YKA
Wigg M GM0OAA
Wigg S G1KQH
Wigg W G5OW
Wiggans W MW0VWC
Wiggins B 2E0KBF
Wiggins B M6KBF
Wiggins D MI3WDI
Wiggins G G4XMJ
Wiggins J G8KUZ
Wiggins M 2E0MKW
Wiggins M G7BJC
Wiggins R M3RCW
Wiggins T M6ZLL
Wigginton C G6DOI
Wiggs D G4FYM
Wightman I M6BPM
Wightman T 2E0TIM
Wightman T M1JCB
Wightman W 2E0DRW
Wightman W M6KWG
Wigley P G4RVU
Wigmore R G4NOU
Wignall E G4EAJ
Wignall H GM0TFQ
Wignall K G0PFU
Wilberforce P GM4AXS
Wilberforce R M3CJE
Wilburn L M6DMO
Wilby E G4EHJ
Wilby P G3YRU
Wilcock B M6FGM
Wilcock G M3UGW
Wilcock P G4VPW
Wilcockson A 2E0IJE
Wilcockson C 2E0IJD
Wilcockson E M0HWN
Wilcockson M G7KYI
Wilcockson R G1VPE
Wilcox A G1FPZ
Wilcox B G6TRO
Wilcox C G4HQC
Wilcox D G3JSA
Wilcox D M0DAW
Wilcox H M6FWC
Wilcox J G4GBW
Wilcox J G0GIV
Wilcox P M6IDD
Wilcox R M0GTT
Wilcoxen Z M6ZJW
Wild A 2E0CHK
Wild B 2E0WXD
Wild B M0WLD
Wild B M6WXD
Wild C GU6TKE
Wild C 2E0TWI
Wild D 2E0SDJ
Wild D M0SDJ
Wild G G3RFN
Wild I G7SWQ
Wild J G4LJB
Wild J G6IMQ
Wild J M0JVW
Wild L G4OAN
Wild P GU6NCZ
Wild R M3JGN

Wild R G6FJP
Wilday A G4TUF
Wilday R G6SOA
Wilde B G3VWH
Wilde B M3UIB
Wilde D G0UOO
Wilde D M0DPW
Wilde G G4JXR
Wilde J G0FOI
Wilde M M3VSQ
Wilde P G3XDS
Wilden G G6WIL
Wilder L GD4SYI
Wilders S G3ZEO
Wilderspin C 2E0DAK
Wilding J G4NLW
Wilding N G0CQC
Wildman D G8HZL
Wildman G M0GIL
Wildman L G0NUZ
Wildman S M0SHY
Wildridge D MM1CHQ
Wildsmith J G6CUY
Wildsmith J 2E0ZXJ
Wildsmith J M0ZXJ
Wildsmith J M3XSJ
Wileman R G0FPZ
Wiles C G0KNW
Wiles D G1EUD
Wiles J G4WQZ
Wiles J G8IRS
Wiles K M6KRW
Wiles P G7EYR
Wiles S G4GDC
Wiles T G4HWV
Wiley P G8ZFS
Wilinski R MM6RWI
Wilkerson G G8BPN
Wilkes A G4NTV
Wilkes A G7KJE
Wilkes A G4HYW
Wilkes B 2E0IKB
Wilkes B 2E0RDX
Wilkes B G4RWQ
Wilkes C 2E0DKS
Wilkes D G0TUC
Wilkes D GM4LIS
Wilkes G G4SEL
Wilkes I M6IRW
Wilkes J 2E0EXK
Wilkes J G7PHK
Wilkes J M3EXK
Wilkes L G3KJK
Wilkes L M3UPP
Wilkes M M6WOM
Wilkes P 2E0VTS
Wilkes P M0VTS
Wilkes P M3VTS
Wilkes R G0DUQ
Wilkes R G0OWU
Wilkes R G4TQR
Wilkes R M0MWT
Wilkes R 2E0HQD
Wilkes R M3HQD
Wilkes S M6EIL
Wilkes S G7LEX
Wilkie A G0BLS
Wilkie C G0CBM
Wilkie G GM0RMT
Wilkie J GM0OFL
Wilkie M G1UKA
Wilkie T 2E0FYZ
Wilkin P MI6PHQ
Wilkins A GW8VFF
Wilkins A G8KSH
Wilkins A M3KSH
Wilkins A M6WKS
Wilkins C G0MMJ
Wilkins D G5HY
Wilkins D G6DMF
Wilkins D G7JAV
Wilkins D G6CVP
Wilkins G G0TSK
Wilkins G 2E0JDL
Wilkins G M6FFN
Wilkins H G3TBF
Wilkins I GM6OXL
Wilkins J G1WUY
Wilkins J G6DIQ
Wilkins J M6DMF
Wilkins M M3CUK
Wilkins M 2E0BXC
Wilkins M M3UMW
Wilkins N G6KBZ
Wilkins P G4LRL
Wilkins R G2ALM
Wilkins R G7OXH
Wilkins R G8NHG
Wilkins S G0NIF
Wilkins S G7MDM
Wilkins T GM6KON
Wilkinson A G1DIR
Wilkinson A G3RHZ
Wilkinson A G4PVU
Wilkinson A G6RIY
Wilkinson A G7SKX
Wilkinson A M3REP
Wilkinson A M6ATW
Wilkinson A M6NIX
Wilkinson B G0GBG
Wilkinson B G0PXH
Wilkinson C G0NQE
Wilkinson C G0OSA
Wilkinson C M6FUW
Wilkinson C M6WKY
Wilkinson C G7LAK

Wilkinson C GD8GRE
Wilkinson C 2E0CVE
Wilkinson C G4UXH
Wilkinson D 2E0SVW
Wilkinson D 2E0ESW
Wilkinson D G0DWO
Wilkinson D G4KNV
Wilkinson D M1BTD
Wilkinson D M3BHI
Wilkinson D M3YDJ
Wilkinson D G7MCE
Wilkinson D M6EYY
Wilkinson E G4XHT
Wilkinson E M6GJQ
Wilkinson G 2E0OGZ
Wilkinson G G3XDP
Wilkinson I G8VUG
Wilkinson I G4RJA
Wilkinson J G1LKL
Wilkinson J GI1XIB
Wilkinson J G4HGT
Wilkinson J G4WIL
Wilkinson J G8MSY
Wilkinson J M1ZAR
Wilkinson J 2I0WKE
Wilkinson J G6NQL
Wilkinson J M6OAO
Wilkinson J MI6YAM
Wilkinson L M6NNA
Wilkinson M M0BZH
Wilkinson M 2E0MDJ
Wilkinson M M6LKA
Wilkinson M G0HCE
Wilkinson N G4HVT
Wilkinson N G4HCK
Wilkinson P G4MWF
Wilkinson P G8SAX
Wilkinson P G7LPD
Wilkinson R G0VXG
Wilkinson R G3KWW
Wilkinson R G4JUH
Wilkinson R G4OKY
Wilkinson R G6GVI
Wilkinson R G6LDJ
Wilkinson R G3VVT
Wilkinson R M3RIA
Wilkinson R MI6WKE
Wilkinson S G4IQF
Wilkinson S M3IAE
Wilkinson W G0ESA
Wilkinson W G3WJH
Wilkinson W G3XJI
Wilkinson W G4MSK
Wilkinson W M3PBU
Wilks D G3VCG
Wilks G G8DVJ
Wilks K G8MVD
Wilks K M0DPY
Wilks M M6WWK
Will S GM4SID
Willan W G7IKM
Willans T G1OUY
Willard G 2E0GGW
Willard G M0HHB
Willard H G1HML
Willard K G6WHT
Willats J G6YPM
Willby J 2E0JWY
Willby J M0WBY
Willby J M6JWY
Willcocks P G4BWY
Willcocks P G8AIE
Willets K M3ZXX
Willets V G8OJK
Willetts A G4LTT
Willetts A G8DEM
Willetts C M1XXT
Willetts D M6SCU
Willetts G G0AAM
Willetts G G0DBJ
Willetts J 2E0GEL
Willetts J M6UHU
Willetts P G0KYG
Willetts P G6YAK
Willetts P M6SRZ
Willey D G1WVO
Willey D M3WVO
Willey L G4ABW
Willford T G8OPX
Willgoss M G4XRR
Williams A GW0FYO
Williams A G0GST
Williams A G0MJV
Williams A G0NTJ
Williams A G0RDS
Williams A G1YWI
Williams A G3MHD
Williams A G3ZKI
Williams A G4CHJ
Williams A G4PQY
Williams A G6OLY
Williams A G7JLG
Williams A G8CKJ
Williams A G8FBW
Williams A GW8LRO
Williams A MW0RDF
Williams A M1DKM
Williams A M1DYJ
Williams A MM3GPB
Williams A M3JJH
Williams A M3NFJ
Williams A M3YOO
Williams A 2E0WCB
Williams A G3ATI
Williams A G6EUT
Williams A M3UTM

Williams A MW6FAM
Williams A M6FQZ
Williams A M6SGT
Williams A M6WCR
Williams A M6BRT
Williams A M0EWQ
Williams A 2E0UPH
Williams A M0UPH
Williams A M6UPH
Williams A 2W0DQT
Williams A MW6EFK
Williams A GM7SXI
Williams A M0CNM
Williams A 2E0NDY
Williams A G3MCB
Williams A M0JOO
Williams A M6ASW
Williams A MW6AWV
Williams A M3XIH
Williams B G0KGT
Williams B GW1VAW
Williams B G4RZM
Williams B GM6OSZ
Williams B M3BWT
Williams B M3WBJ
Williams B G7FSI
Williams B GW0GHF
Williams B G1ORS
Williams B M3ORT
Williams B M6PGX
Williams C G0ESK
Williams C G4EYT
Williams C G4FXQ
Williams C G4GKY
Williams C G6AMW
Williams C G6DOK
Williams C G8HJG
Williams C G8SFD
Williams C MW3TAF
Williams C M6FJX
Williams C MM3GOX
Williams C MW6ZOD
Williams C 2E0XVT
Williams C 2E0ZPM
Williams C G7NBP
Williams C G8HJF
Williams C M0YVT
Williams C M3GQS
Williams C M6XVT
Williams C M6ZPM
Williams C 2I0SUB
Williams C M6FZN
Williams D 2E0AZU
Williams D G0ESI
Williams D G0JMR
Williams D G0LSQ
Williams D G0ODE
Williams D G0PWA
Williams D G1LTH
Williams D G3CCO
Williams D G3XJA
Williams D GW4BNJ
Williams D G4CVN
Williams D G4LPA
Williams D GW4WQC
Williams D G6TOY
Williams D G7GQW
Williams D G7IRP
Williams D G7PMI
Williams D G7PYH
Williams D G7TXX
Williams D MW0ATR
Williams D M1EPI
Williams D MW3HBF
Williams D M0WHR
Williams D M3WHR
Williams D 2E0DTW
Williams D M6RPM
Williams D M5ADI
Williams D G6ONE
Williams D G4BII
Williams D G4UNB
Williams D G4UUW
Williams D G7LPZ
Williams D M3ZRW
Williams D MW6AHQ
Williams D M6DRW
Williams D M6LNS
Williams D M6VCN
Williams D 2E0CFB
Williams D 2E0DER
Williams D M0MON
Williams D M3UUY
Williams D M0UAS
Williams D M0PSJ
Williams D 2E0DAW
Williams D M3VIJ
Williams D G3ORL
Williams E 2E0ECW
Williams E 2E0AZA
Williams F G4JZR
Williams E G4LHR
Williams E G4NUA
Williams E G4VHS
Williams E G6VYK
Williams E M0ECZ
Williams F MM3GOY
Williams E M3JQK
Williams E G0ULL
Williams E G0WMQ
Williams E M1EWJ
Williams E M6CLB
Williams E M0ATX
Williams E M6WLE
Williams E G0AGL
Williams E M0TEN

**IMPORTANT NOTE**

**Revalidate licence to avoid revocation** – Ofcom has advised the Society that plans will be drawn up to revoke licences that have not been revalidated as required by the licence conditions. The quickest way to revalidate is to do so online via the Ofcom website: *https://services.ofcom.org.uk/* or by email: *amateur.validations@ofcom.org.uk* If you need assistance in the process, Ofcom staff are available to help, but please be patient during times of heavy workload.

Williams F G1XWO
Williams F M6FNK
Williams G 2E0DIG
Williams G G0HHD
Williams G G1AGM
Williams G G1MCT
Williams G GD1RGT
Williams G G1TBX
Williams G G1YHJ
Williams G G4DXN
Williams G G4LZQ
Williams G G4MTF
Williams G GW4VMT
Williams G G8YPV
Williams G M6ATX
Williams G M6GZZ
Williams G G6JUQ
Williams G 2E0WKT
Williams G M0WKT
Williams G M6WKT
Williams G 2W0CLT
Williams G GM4FGL
Williams G MW6BOC
Williams G M3SZF
Williams G M6GFS
Williams G M6GLW
Williams G G7ADH
Williams G 2E0GIW
Williams G G4LCF
Williams G M6PAM
Williams G GW7AOE
Williams G G2DLK
Williams G G4FKH
Williams H G3ROS
Williams H G3WZS
Williams H G4WNA
Williams H G4MOP
Williams H G1JXX
Williams H GW1MTH
Williams H G7BNZ
Williams I G4MEI
Williams I 2E0TLM
Williams I G0PEF
Williams I G5WQ
Williams I M0BCG
Williams I M0TLM
Williams I M6IEW
Williams I M6TLM
Williams I 2E0GAQ
Williams I M6KTS
Williams I G4TUD
Williams I M0GWY
Williams J 2E0JMW
Williams J G0GXQ
Williams J G0KZI
Williams J G0SXE
Williams J G0TFL
Williams J G1FSF
Williams J GW1VRR
Williams J GW3SSK
Williams J G4WLT
Williams J G4WVB
williams J G6JQH
Williams J G6WVD
Williams J G7GGN
Williams J G7JHX
Williams J G8LGC
Williams J G8RHP
Williams J G8VSF
Williams J G8XJE
Williams J M0AMZ
Williams J M0JDW
Williams J MM3GMP
Williams J M6MNU
Williams J 2E0SDV
Williams J M6GDI
Williams J 2E0YJW
Williams J M0YJW
Williams J MM3NRX
Williams J M3YJW
Williams J M3VBL
Williams J G0VMC
Williams J 2E0GMT
Williams J G0DQM
Williams J G0DSK
Williams J G0MNC
Williams J GW3KCQ
Williams J G4DYK
WILLIAMS J G4RRL
WILLIAMS J G4TSG
Williams J GD6OXG
Williams J GD8NTR
Williams J G8VKS
Williams J 2E0CAP
Williams J M3XIF
Williams J M6MZI
Williams J M6YJK
Williams K 2E0AYI
Williams K GW0KWO
Williams K GW0RNK
Williams K G6BHQ
Williams K G6XCV
Williams K G7NHF
Williams K M0BAK
Williams K M3CLO
Williams K M0XKW
Williams K GW4WOV
Williams K G8RDT
Williams K M6CYB
Williams L G0ASQ
Williams L M3NZV
Williams L MW3OEC
Williams L M6ARV
Williams L G0FLW
Williams L MW0HNM
Williams M 2E0JRW
Williams M 2E0NAY
Williams M G0BEU
Williams M G0JMW
williams M GW0OUV
Williams M G0TRP
Williams M G0WIL
Williams M G3LCQ
Williams M G4ZMW
Williams M G7WEP
Williams M G8POL
Williams M G8SGV
Williams M G8ULL
Williams M M0VET
Williams M M1TAF
Williams M MW3MRL
Williams M M3MWV
Williams M M3NCQ
Williams M M3PXP
Williams M M6DBD
Williams M MW6SEF
Williams M GM1HJX
Williams M M3VWM
Williams M M3WZP
Williams M G4GRS
Williams M G4PGX
Williams M G8MIC
Williams M M6MHY
Williams M 2E0FCD
Williams M MW3VNV
Williams M MW1FEU
Williams M 2E0FCC
Williams M 2E0BFC
Williams M G6BMZ
Williams M M0GKV
Williams M M0MRW
Williams M M3LMZ
Williams M M6HSH
Williams N GW0JFQ
Williams N GW1YXR
Williams N G3BYG
Williams N M0VKC
Williams N MW3XNW
Williams N 2E0TNB
Williams N M6BYX
williams N M6HWQ
Williams N MW6FHK
Williams N G0RPM
Williams N M6FRP
Williams N G7MRL
Williams N M0CRM
Williams N M0HJG
Williams N M6NBP
Williams O G0PHY
Williams O 2E0BOC
Williams O GW0IXM
Williams O G8ITX
Williams O M0GMH
Williams O M3BOO
Williams P 2E0DIV
Williams P G0IQP
Williams P G0IRT
Williams P G0LJB
Williams P G0OJX
Williams P G1BRF
Williams P G1CJJ
Williams P GW3NUO
Williams P G3XXE
Williams P G4LIQ
Williams P G4NPU
Williams P G6IEI
Williams P G7MYD
Williams P M0BCL
Williams P M1BPW
Williams P M3VOY
WILLIAMS P G3YZQ
Williams P GM4VXA
Williams P G8INS
Williams P M6IJW
Williams P M6UKC
Williams P 2E0ELE
Williams P G3XRI
Williams P M0RGN
Williams P M6MTD
Williams P G6EUR
Williams P G8RDQ
Williams R 2E0BLG
Williams R 2E0YAD
Williams R G1ELJ
Williams R G1PRL
Williams R G1WWA
Williams R G3TVN
Williams R GW4BGD
Williams R G4NLU
Williams R G4NYK
Williams R G4PMM
Williams R G4RWH
Williams R G6ZJN
Williams R G7DEG
Williams R G7NGN
Williams R G8CMG
Williams R G8JSF
Williams R G8VVX
Williams R G8YPR
Williams R G8YRW
Williams R M0RLW
Williams R MW3MVT
Williams R M3NDU
Williams R M3NHN
Williams R M3PNR
Williams R G0OOF
Williams R M3TXJ
Williams R MW6HEI
Williams R M6RZP
Williams R M6RIT
Williams R 2E0RBW
Williams R G0PEB
Williams R GD4AGM
Williams R G7ATJ
Williams R G8MBU
Williams R M1BGT
Williams R M6COB
Williams R M6NLA
Williams R GW4HSH
Williams R M0RMW
Williams R GW7UXY
Williams R G7LND
Williams R 2E0HAS
Williams R M3VUJ
Williams R M6RUT
Williams S 2E0SLP
Williams S 2W0SNW
Williams S 2E0HHL
Williams S GW0OFH
Williams S G0VNI
Williams S G0VOJ
Williams S G4OGO
Williams S G4RIO
Williams S GW4UIE
Williams S G4ZDP
Williams S G8SUN
Williams S M0EDO
Williams S M1ECY
Williams S MW3SNW
Williams S M3FNT
Williams S 2W1AID
Williams S G6LJX
Williams S 2W0BMR
Williams S MW3PBV
Williams S 2E0SAW
Williams S M0WSA
Williams S GW0RHE
Williams S GW6CUR
Williams S M6CUS
Williams S G7OQG
Williams S MW6CCG
Williams S M6KBC
Williams T GW0HXS
Williams T G1EWE
Williams T G1ITS
Williams T G1KOX
Williams T G1XHO
Williams T G3XLS
Williams T G6KBQ
Williams T M6EXW
Williams T M6ASI
Williams T 2E0TMO
Williams T M6TMO
Williams T G4YVY
Williams T M3EHA
Williams V 2E0VPW
Williams V G6XNU
Williams V M6VPW
Williams V 2E0VKW
Williams V M6VKW
Williams W G0IQZ
Williams W G0JMJ
Williams W G1FWC
Williams W GW4PEX
Williams W G4RCM
Williams W G4ZYM
Williams W G7AQD
Williams W GW8TGS
Williams-Davies M G6UWW
Williamson A GI0NWG
Williamson A GM0RAO
Williamson A G0WUY
Williamson A G6FVM
Williamson A M6VET
Williamson A 2E0SND
Williamson A M3REQ
Williamson B G7UYI
Williamson B M3WXB
Williamson B M6BAQ
Williamson C G4IEB
Williamson C M3JYZ
Williamson C 2M0BYI
Williamson C MM0MYL
Williamson C MM3YUX
Williamson D G0EGP
Williamson D M6LRB
Williamson D MD6TSW
Williamson D MM3YUW
Williamson E M6EJW
Williamson F G4VDZ
Williamson G GM0BUI
Williamson G 2E0IJW
Williamson G M3OQK
Williamson I 2E0INW
Williamson I G0EQC
Williamson J GI3JOZ
Williamson J G4NTG
Williamson J G4XJN
Williamson J G6WYL
williamson J M0JWE
Williamson J MM6YCB
Williamson J M6FZG
Williamson K G8IJI
Williamson K M6BQM
Williamson L G8RSQ
Williamson M G1OQO
Williamson M G8ATG
williamson M M3HQS
Williamson M G1IJC
Williamson P G4WUU
Williamson P M6EFP
Williamson R 2E0RWX
Williamson R G0MTT
Williamson R G4NPT
Williamson R G8WRL
Williamson R M1DXG
Williamson R G4GGI
Williamson R M6AXR
Williamson S M3SCX
Williamson S G3WGU
Williamson T 2E0TAW
Williamson T GW8UAP
Williamson T M3MNT
Williamson V G6HSC
Williamson W G0WDW
Williamson W G3RUO
Williamson W GM8MMA
Williamson-Brown E G7BUL
Willicombe D G0DEC
Willies C G6DFA
Willies D G3HRK
Willimot C M3CMW
Willimot D 2E0IGI
Willingham D 2E0SLO
Willingham D M6WSW
Willingham J 2E0IHO
Willingham J M1FFN
Willingham P G3YPW
Willis A 2E0ETJ
Willis A G0IOO
Willis A G0VKE
Willis A G4XNA
Willis A G6PFF
Willis A M6OAW
Willis A M3USF
Willis B GW4ZUJ
Willis C G0ADB
Willis C G7JIN
Willis D GW0JDW
Willis D G4NMC
Willis D G6BXV
Willis D G7KRO
Willis D M6UZZ
Willis D 2W0XTP
Willis D MW0TMI
Willis F M0AZY
Willis H G3ZPK
Willis J G4DOQ
Willis J G4MES
Willis M M6HIJ
Willis M 2E0MWJ
Willis P G0IYS
Willis P G3GLW
Willis P M6PDW
Willis R G6XCU
Willis R G6NPP
Willis R M6TGK
Willis S G7PDU
Willis S M3GKJ
Willis T M3TGW
Willis T 2E0CJQ
Willis T M6TWL
Willis T G7NCD
Willis W G6GJV
Williscroft R G4GYA
Willison M M6MRW
Willkins R G0LPF
Willmer I G4TRG
Willmore A 2E0MOR
Willmot N G3ZHC
Willmott M M6RMW
Willmott P G6KCV
Willmott R G6HSD
Willmott S M3WNC
willmott W G7VQJ
Willott H M0OPS
Willoughby A M3CDV
Willoughby C G4GKC
Willoughby K M3WYA
Willoughby M G7UQV
Willoughby S M0SDW
Willoughby W G1HFK
WILLOX E M0BKX
Wills B GM0SDS
Wills D G3XKX
Wills D G3NMX
Wills E 2E0AAX
Wills G G3YYW
Wills J G4AXO
Wills J MD6HVA
Wills K M6MCR
Wills M G0TNF
Wills M G8RJZ
Wills M G3OIL
Wills M M3XGV
Wills N G7PQW
Wills R G0BDP
Wills R G7VFE
Wills R G3ZZS
Wills S 2E0LMK
Wills S M6SAQ
Wills S M3SJW
Wills T G8PZD
Wills-Browne N G1VAA
Willsher A 2E0VYW
Willsher A M0VYW
Willson A G3WNS
Willson B G0RNQ
Willson B G8AKU
Willson C M6NRC
Willson D M3ITK
Willson G M0AQO
Willson K 2E0KWW
Willson K M0KWW
Willson K M6LUD
Willson M G8JRF
Willson P G6GTC
Wilman G 2E0CSU
Wilmot M G0TAT
Wilmot P M6BTP
Wilmot R 2E0UWI
Wilmot R G0LHE
Wilmot R G4PEY
Wilmot R M3UWJ
Wilmot R G0JHL
Wilmot R G3RRI
Wilmot T MI6TAW
Wilmott B G8TGH
Wilmott C M0OXO
Wilmott C M3ZYZ
Wilmott F G8JFC
Wilmott L M3ZLZ
Wilmshurst H G1BIA
Wilmshurst M G1JZG
Wilmshurst T G3IBY
Wilsdon P G1ZTG
Wilsher C M6GNG
Wilsher M M6OCB
Wilsher M M6BNK
Wilson A G0IRJ
Wilson A G1OGE
Wilson A GM6KPL
Wilson A G6NDJ
Wilson A G6OZU
Wilson A G6ZAC
Wilson A G7UUT
Wilson A GM8NVG
Wilson A M1FBF
Wilson A MM3JFW
Wilson A G3JHS
Wilson A 2I0IYH
Wilson A G4BYI
Wilson A MI3IYH
Wilson A 2E0GZY
Wilson A M1EPR
Wilson A M3EPR
Wilson A GM6UHE
Wilson A 2E0MXR
Wilson A 2E0TCX
Wilson A G6NQQ
Wilson A M6AWZ
Wilson A M6IGW
Wilson A M6VWW
Wilson A G0UGO
Wilson A M6HLP
Wilson A 2E0TVW
Wilson A G0HWP
Wilson A G3MAE
Wilson A G6EUU
Wilson A M6AVW
wilson A 2E0HYD
Wilson B GW4KFD
Wilson B G8PK
Wilson B M1VLS
Wilson B G0JOG
Wilson B 2E0YZA
Wilson B M6HBH
Wilson B G0KFQ
Wilson B G6HWI
Wilson B M0MCN
Wilson C G0AFZ
Wilson C G0HQN
Wilson C G0VAR
Wilson C G4VVZ
Wilson C G6UVB
Wilson C G8MEA
Wilson C M0YZA
Wilson C M1OCN
Wilson C M5IMI
Wilson C G7PAY
Wilson C 2E0EXC
Wilson C G8ZCK
Wilson C 2E0RCW
Wilson C M6ECW
Wilson C M3WCR
Wilson C 2E0CXW
Wilson C G3VCQ
Wilson C G4AZM
Wilson C M6AYH
Wilson C 2E0FLF
Wilson C M6FLF
Wilson D GM0BRJ
Wilson D G1HKS
Wilson D GI3LEG
Wilson D G3OST
Wilson D G4AWF
Wilson D G4OKZ
Wilson D G4OMD
Wilson D G7LBO
Wilson D G7RKV
Wilson D G7VEB
Wilson D G7WGY
Wilson D G8YPY
Wilson D M0OBW
Wilson D M3DJW
Wilson D M3WGY
Wilson D 2I0FNN
Wilson D MI6FIK
Wilson D M3YXD
Wilson D M6HDW
Wilson D 2E0DAA
Wilson D G4JKR
Wilson D G4OLL
Wilson D 2E0DLM
Wilson D M0BFA
Wilson D G0RPW
Wilson E G0ECW
Wilson E G3OJQ
Wilson E MM3TOV
Wilson E 2E0ENW
Wilson F G0OHF
Wilson F G7LHT
Wilson F G3YQA
Wilson G 2E0APG
Wilson G M3BGN
Wilson G M6GGQ
Wilson G M6LMT
Wilson H G3ZIF
Wilson H G4KTG
Wilson H M3HWW
Wilson H G6OBT
Wilson I G0SPQ
Wilson I G1KVW
Wilson I G1XOG
Wilson I G3YUZ
Wilson I G4BGW
Wilson I G4JPG
Wilson I G6YAI
Wilson I G1JXP
Wilson I G4UPX
Wilson I G4ZCD
Wilson J 2M1EDT
Wilson J G0SDL
Wilson J G3CYU
Wilson J G4KOJ
Wilson J G4SMX
Wilson J G6PTF
Wilson J G6WQH
Wilson J G6YVS
Wilson J GM7MMI
Wilson J MM0DXD
Wilson J M0JRW
WILSON J M1TUG
Wilson J M3PHO
Wilson J MM3VEG
Wilson J M3ZTP
Wilson J 2E0TBL
Wilson J 2E0WLN
Wilson J G8ULH
Wilson J M6EUP
Wilson J M6JWD
Wilson J 2E0YOZ
Wilson J M6HOY
Wilson J 2E0HFA
Wilson J M0JKN
Wilson J M6HFA
Wilson J G3PCY
Wilson J G3UUT
Wilson J G7BHW
Wilson J M3XQH
Wilson J MM6JMI
Wilson J M3KUN
Wilson K G0WHQ
Wilson K G4FPO
Wilson K G4SMK
Wilson K G4WHF
Wilson K M1CNY
Wilson K M0TVT
Wilson K G4WGN
Wilson K G6LTK
Wilson L G4TDC
Wilson L M3LRW
Wilson L MW3VEL
Wilson L 2E0KOM
Wilson L M0KOM
Wilson M G0TFU
Wilson M G1AAP
Wilson M G1CSS
Wilson M G4GOU
Wilson M G4RQL
Wilson M G4SYE
Wilson M G6CQB
Wilson M G6LJH
Wilson M M1SNM
Wilson M M3WIX
Wilson M M6GUO
Wilson M M6ZWB
Wilson M M3MST
Wilson M M6MCW
Wilson N M3MUF
Wilson O M6EUB
Wilson O M6ORW
Wilson P G0ELX
Wilson P G0VCA
Wilson P G1ELK
Wilson P G4CPW
Wilson P G4DUI
Wilson P G6GTZ
Wilson P G7EOK
Wilson P G8KEK
Wilson P M1AKF
Wilson P 2E0DGA
Wilson P G0TIX
Wilson P G1PQJ
Wilson P M0VRW
Wilson P M6EOK
Wilson P G0NGP
Wilson P G1RCN
Wilson P G6MQY
Wilson P G7VIH
Wilson P M3VTL
Wilson P MI6RFW
Wilson P MI6SEZ
Wilson P G0ULM
Wilson R 2E0HKE
Wilson R G0FEK
Wilson R G0VHY
Wilson R G1BAB
Wilson R G1EUF
Wilson R G4AVS
Wilson R GM4BIT
Wilson R G4HIH
Wilson R G4NZU
Wilson R G4TYW
Wilson R GD4ZPR
Wilson R M0BOM
Wilson R GI0BRO
Wilson R M3FWU
Wilson R 2E0RRC
Wilson R M6SDW
Wilson R M6DII
Wilson R M6WIB
Wilson R G1RKJ
Wilson R G3YZO
Wilson R GI4SZY
Wilson R MM3UOE
Wilson R MM6HEQ
Wilson R M6RTD
Wilson S 2E0DMU
Wilson S G4HNO
Wilson S M3NOW
Wilson S M6XBO
Wilson S M3VCQ
Wilson S M6FAH
Wilson S M6NSW
Wilson S G6BOX
Wilson S M0BOX
Wilson S G3VMW
Wilson S 2M0RVF
Wilson S M6RNE
Wilson S 2M0WJS
Wilson S MM6WJS
Wilson T GM0FNE
Wilson T GM4DPC
Wilson T GI4PBS
Wilson T GI4PBT
Wilson T G6MQZ
Wilson T GI4VJZ
Wilson T G1OAM
Wilson W M3AYT
Wilson W MM3ENP
Wilson W M0BWI
Wilson W MI0DNM
Wilson W MI6WJW
Wilson W MI6WPW
Wilson Y MI3MIE
Wilson-Shah C M0PKL
Wilton K M3UCC
Wilton P M1CNK
Wilton P M1PKW
Wilton R G0CIX
Wilton R G3ZHO
Wilton V G0ORV
Wiltshire D G3LKW
Wiltshire D G6ZZR
Wiltshire J M1CXI
Wiltshire L G6GCW
Wiltshire L G0IAY
Wiltshire N G6AQW
Wiltshire R G1UMY
Wiltshire R G7SFM
Wiltshire R M0CKV
Wiltshire T G8AKA
Wimbury W G4TGK
Wimlett G G8GLS
Wimpenny J GW0LVH
Winch C M3WKL
Winch M 2E0ENG
Winch M M3ORQ
Winch T 2E0HWJ
Winchester A GM0SSQ
Winchester G GM4CUX
Winchester G M6CCB
Windass K M6ELF
Windass T M6YEY
Windebank J G0KJN
WINDER R G1HFA
Windle A G8VG
Windle A M6HUI
Windle M G0SJP
Windle T M6XTA
Windsor A G1OQU
Windsor C 2E0BSS
Windsor C M3OYE
Windsor P GM4OCA
Windsor S GM6KMK
Windus D M3TPZ
Winfield C 2E0FSH
Winfield D G6XOR
Winfield G G4MDK
Winfield K G0TZM
Winfield N G0UWO
Winfield P G0WRT
Winford S G8CRX
Wing G G4AUV
Wing M G1NJV
Wing N 2E0JRT
Wing N M6JRT
Wing S 2E0XLM
Wing S M0SCX
Wingfield C 2E0WIN
Wingfield I G4IHM
Wingfield J G4VLF
Wingfield J G6JQX
Wingrove M GD6VIO
Winiberg M G8GFS
Winkler A G0FJJ
Winkley A 2E0XAW
Winkley A M6ARW
Winkley D G1DYC
Winkley P M6CPW
Winkup R G4ZLT
Winkworth R G3BGF
Winkworth R G4AZA
Winlove-Smith S G7BPR
Winnan C M6CTW
Winnard K GW3TKH
Winnett G M6ZYE
Winnett P G4RSU
Winning C G4YWZ
Winning W G1NBK
Winship R 2E0CCG
Winship T G0DSB
Winslow B G3TYG
Winson J M3IZH
Winson J M3IZJ
Winson J M3VHQ
Winson K M3IZM
Winstanley C G4CDR
Winstanley D 2E0PKU
Winstanley E G3SUA
Winstanley R M6CDL
winston G G8OPY
Winstone J M6TNC
Winter D GW7CSK
Winter G G3XCW
Winter H GW6HRL
Winter I G3KJN
Winter J G1AMS
Winter J M3WCQ
Winter K G8WCA
Winter M G3OHP
Winter N G7EED
Winter S G0NVJ
Winter T 2E0HLC
Winter T M6AXG
Winter T G4AOK
Winterbottom A G6ZJV
Winterbottom G 2E0GEF
Winterbottom G M3YHT
Winterbottom P M6PKN
Winterbourne J GM1KWG
Winterburn D G3DQQ
Winterburn J G4NSB
Winterburn R G1NSB
Winterflood C G4NNN
Winter-Kaines M G4LAP
Winters C G4MQG
Winters D G6EZM
Winters J G4SGY
Winters L G4IPL
Winters R G3IPL
Winters S G4XUQ
Winters W G0PQI
Winterton P G1BMW
Winthrop R M0PTT
Wintle J G0UJW
Winton A GM3MWX
Winton A GM6XW
Winton D G1ICK
Winton D GM7RXL
Winton K 2E0OJD
Winton K M0MCL
Winton K M6OJD
Winton T G7RLQ
Winton V G4JUN
Winward K 2E0JKD
Winwood H G4GPF
Winwood K M0RSD
Winwood K M0WYZ
Winwood P G8KIG
Winwood T 2E0HGA
Winwood T M3HGA
Winyard G G4XWW
Winyard T M0TPW
Wirthner M G4RDH
Wisbey A G4ECS
Wisbey D G0RIQ
Wisbey G G7IRG
Wise C G1YXT
Wise M M6KTW
Wise M G0GPX
Wise P M1DVV
Wise R G1IJY
Wise S G1FHY
Wiseman A G7CIQ
Wiseman A M0CGB
Wiseman B G6KQZ
Wiseman C G0RDK
Wiseman C G1PUV
Wiseman D G6UPA
Wiseman D G7SUA
Wiseman F G3GRY
Wiseman G G7JXR
Wiseman I G1LSK
Wiseman J G8BPQ
Wiseman N M0NIW
Wiseman R M5WIZ
Wiseman R G3PXV
Wiseman T G3DEJ
Wishart D MM3WJD
Wishart D MM5DWW
Wishart I MM0GSW
Wishart J 2E0JWP
Wishart J M6CHH
Wishart R G0FDH
Wisher S G8CYW
Wiskow D 2E0WIS
Wiskow D M6WIS
Wiskow J M6JLE
Wislocki T G4JRY
Wissun C G7FPU
Witchard M G8YOC
Witchell J 2E0JFW
Witchell J G4OTJ
Witchell J M3PIW
Withall J M6KAE
Withall P G4ZSW
Witham W M3WWR
Wither A G0ADP
Witherall D G7UAL
Withers A G1AAG
Withers A M0BTL
Withers B M6OMR
Withers D G0HUD
Withers G G0VAP
Withers H M1AMP
Withers J M3WIT
Withers R M6TXR
Withers S G0MBQ
Withers T G3HGE
Witherspoon J G8DSM
Withey M G0KZD
Withnall S M0GPC
Withnell S G0AIN
Witley P G0KHF
Witney R G4ICP
Witt B M6WIT

Witt D M6XDO
Witt J G8ISJ
Witter A G0VAM
[illegible]
[illegible]
Witts D [illegible]
Witts D MW6DYL
Witts J G6BBW
Witts P MW0PVW
Wixon A G4PNM
Wnekowski L M6WEW
Woan G MM0WOA
Wogden B M6KXQ
Wogden M G4KXQ
Wohlgemuth J 2E0DRV
Wohlgemuth M M1EXJ
Wojcik C 2E0GJT
Wojcik M M0GHC
Wolf C G6LSO
Wolfe B G3MTR
Wolfe D G1EUG
Wolfe L M3VNX
Wolfe P G4EGU
Wolfenden E GW7OJT
Wolfson L G4VUK
Wolfson M M0MUC
Wolk R G7SUU
Wollaston M M6CDD
Wollaston R G4IVB
Wollen I G3UZI
Wollen P M6GNR
Wolohan J M3JBW
Woloszun G MW0GRZ
Wolstencroft A 2E0IGP
Wolstencroft C GM3MTW
Wolstenholme L G0RDF
Wolten R G3ZIM
Wolverson A M1EDL
Womack D G4TNY
Wong A 2E0TBQ
Wong C M3WCY
Wong H M6HYW
Wong J 2E0TJG
Wong L G7EPL
Wong N 2E0NWA
Wong N M6NWA
Wood A G1FJD
Wood A GM3LIW
Wood A G3RDC
Wood A G3VJM
Wood A G6EVX
Wood A G6VIY
Wood A G7HFW
Wood A G7MEE
Wood A M1BNK
Wood A G4LED
Wood A G0ENV
Wood A G4EEE
Wood A 2E0LSL
Wood A G7NTI
Wood A M3CNC
Wood A GM7KFS
Wood B G4FPI
Wood B G4RDS
Wood B G4RFO
Wood B G6GOV
Wood B G0OQQ
Wood C G1ILJ
Wood C G3PZN
Wood C G3TAW
Wood C G4XOG
Wood C GM6NOO
Wood C GD6TWF
Wood C G7FKX
Wood C G7PIG
Wood C G8JSM
Wood C MM3WWP
Wood C G8PSZ
Wood C M1DVJ
Wood C G1ZOS
Wood C M6DAK
Wood C G0EZX
Wood D G0ABV
Wood D G0WZA
Wood D G3HKO
Wood D G4GDT
Wood D G4KAE
Wood D G7BNK
Wood D M1CVG
Wood D G4JEF
Wood D GD0VIK
Wood D GM1PUR
Wood D M6HUM
Wood D G1LNA
Wood D G4CQR
Wood D G1TIW
Wood D MM0ALM
Wood D M3WOD
Wood D G4ZJL
Wood E G0JLU
Wood E G4PKF
Wood E G4YZP
Wood E G6LOR
Wood E MD3EEW
Wood F 2E0XYT
Wood F G1GET
Wood F G4IVC
Wood F G8NSE
Wood F G1DDA
Wood G G3VIP
Wood G G4XCR
Wood G G6YVD
Wood G G8NRF
Wood H G7NAL
Wood H M0DHX
Wood I G8MBV

Wood J G0KUI
Wood J G0MPQ
Wood J G3VG
[illegible]
Wood J G8MTV
Wood J [illegible]
Wood J M6ONS
Wood J G8GHO
Wood J G0PSI
Wood J G3YQC
Wood J G4AL
Wood J G7NNU
Wood J 2E0FRI
Wood J M1CUY
Wood J M6DFE
Wood J M6CYW
Wood K 2E0EQQ
Wood K G4JOA
Wood K G4YKV
Wood K G7BCS
Wood K M3AXZ
Wood K M3XST
Wood L M6KCK
Wood M G0ITU
Wood M G0SYQ
Wood M G1KVR
Wood M G4VIT
Wood M G7KPH
Wood M G4HLZ
Wood M M0AYG
Wood M G7HMV
Wood M M0KSG
Wood M M6EZE
Wood N G4UFL
Wood N G4OXG
Wood N M6ERR
Wood O M6HMV
Wood O M6OLW
Wood P G0FQC
Wood P GM0HWQ
Wood P G4UIW
Wood P G6ZFG
Wood P M6WVH
Wood P G0PJI
Wood P G8POG
Wood R G1PJT
Wood R G4UDK
Wood R G4XRW
Wood R G6GHU
Wood R G6GVS
Wood R G6RZS
Wood R G8GUA
Wood R M1DLR
Wood R G1RZZ
Wood R M3RQB
Wood R 2E0RLW
Wood R G4PEW
Wood R M0UKA
Wood R M6RLW
Wood R G0LZD
Wood R G7UUN
Wood R G8MFM
Wood R G3WEW
Wood R MM6CEL
Wood S G0BJD
Wood S G6GVU
Wood S GW6LSL
Wood S G8LWQ
Wood S G8WHR
Wood S G8YIN
Wood S G8ZQG
Wood S G1HFE
Wood S M6XER
Wood S G8JBM
Wood T G3UNI
Wood T G4KFS
Wood T G7VJJ
Wood T M3KPZ
Wood T 2E0TWS
Wood T 2E0WTA
Wood T M0TWS
Wood V G1NVE
Wood W G4VNX
Wood W G7OLU
Wood Y 2E0EAK
Woodard J G8YOG
Woodard K M6KWX
Woodberry R G0IEO
Woodbine D 2E0CIS
Woodbine D M6BAV
Woodbridge C M0GMI
Woodburn P 2E0EET
Woodburn P M0PWD
Woodburn P M6PEW
Woodbury G G0WIS
Woodcock B G1OOM
Woodcock B G4CIB
Woodcock J G0WBI
Woodcock L G4RHK
Woodcock P M3KQF
Woodcock R G4GZU
Woodcock R G7FDA
Woodcock S 2E0BBN
Woodcroft D M1DMG
Wooden E G3PRQ
Woodfield B G3REL
Woodfield P G3SFU
Woodfield S M6GYK
Woodfin H 2E0DQJ
Woodfin H M6EVH
Woodfin P 2E0YPW
Woodfin P M0YPW
Woodfin P M6YPW
Woodford A 2M0CPN
Woodford A MM0ZAW
Woodford A MM6ACW

Woodford D MM6ZDW
Woodford G G0KNM
Woodford I G1SHQ
[illegible]
Woodford R [illegible]
[illegible]
Woodford S M0BBO
Woodgate R G6RJW
Woodhall D G3ZGZ
Woodhall F G0MZY
Woodhams D 2E0WPD
Woodhams D M6TCN
Woodhams F M0RIK
Woodhams K 2E0RRL
Woodhams K M6RRL
Woodhead B G0RGN
Woodhead P G8NJI
Woodhead S G4YAJ
Woodhead W M6ZEW
Wood-Hill G G4TWH
Woodhouse A G1WFJ
Woodhouse A M6WJA
Woodhouse A 2E0WAP
Woodhouse A M6APW
Woodhouse B G1GXW
Woodhouse D 2E0UIP
Woodhouse D M6RWE
Woodhouse D G3TWX
Woodhouse E G6ORL
Woodhouse E G7FBT
Woodhouse G G0PHC
Woodhouse H G3MFW
Woodhouse J G6GPF
Woodhouse J G7VLL
Woodhouse M G7GQB
Woodhouse M M5AJO
Woodhouse P M0VCR
Woodhouse T M0TAW
Woodhouse T M3OGL
Wooding D M3CQW
Wooding M G6IQM
Wooding P G0VQY
Wooding T G0VUN
Woodington M 2W0TMB
Woodington M GW4KAC
Woodland A G4KVP
Woodland F G1ZYS
Woodland J G4KHQ
Woodland J 2E0TEW
Woodland J M0JSW
Woodland J M6BUD
Woodland M G7KZJ
Woodland R 2W0JJW
Woodland R MW6RGW
Woodley C G3XPU
Woodley G G1OWM
Woodley N G6KPT
Woodley R G1HKT
Woodley R M3HKT
Woodley T 2W0DAP
Woodley T MW6THW
Woodlock C G0UDE
Woodman D G4ULV
Woodman G G8XAO
Woodman J M0NYW
Woodman J M1PRO
Woodman R G4DSD
Woodmass P M6PFW
Woodmore S M0GXN
Woodnutt D G0RAU
Woodnutt H G6IYA
Woodnutt J G6XDB
Woodridge D M6DBW
Woodroffe B M0TDW
Woodroof J 2E0WSJ
Woodroof J M3WSJ
Woodrow R M6RFB
Woodruff C M3LNQ
Woodruff J 2E0BGO
Woodruff J M3OFB
Woodruff M M3JSQ
Woodruff M 2E0CYR
Woodruff M M6DTU
Woodruff S M6WFG
Woodruffe D G1AGK
Woodrup A 2E0RAF
Woods A 2E0CDZ
Woods A G0FCV
Woods A G7SQW
Woods A M3NZW
Woods B G7MNT
Woods B G8TQZ
Woods B M6PVV
Woods B 2E0SSN
Woods B M0AKI
Woods B M0SSN
Woods B M6SSN
Woods D G0NIL
Woods D G3ONI
Woods D G7CCH
Woods D MM3XDW
Woods D G6LXV
Woods D M3TFI
Woods F G1HKM
Woods F 2E0GBT
Woods G G0AXU
Woods G G0TJI
Woods G G3LPT
Woods G GW4JPC
Woods G M3WHQ
Woods I M3FOS
Woods I MM3XVD
Woods J G0MPW
Woods J G1EUH
Woods J G3OQC
Woods J G4RLL
Woods J G7HQP

Woods J M0PUC
Woods J MI1CUS
Woods J MI6SPY
Woods J G7MRY
Woods J 2M0EAC
[illegible]
Woods J 2E0JDA
Woods J 2E0YRW
Woods J M3FOQ
Woods J M3YRW
Woods J MM6ZUY
Woods K E G8GVL
Woods L 2E0KLW
Woods L M6KLW
Woods M M1PLC
Woods O M3WRO
Woods P GM4JLD
Woods P G8HHZ
Woods P G3HLN
Woods P 2M1IBE
Woods P MM0IBE
Woods P GM0LIR
Woods R G8XAN
Woods S M3VPQ
Woods S M6SWO
Woods T G8STF
Woods T M1BDH
Woods T M3BDH
Woods T MM0WCT
Woods W GW0EWY
Woods W MW6WHW
Woodsford A M3LHX
Woodstock N M3ZWN
Woodward A G1AAL
Woodward A 2E0FKZ
Woodward A M3SFC
Woodward B G0KDD
Woodward B G0CKF
Woodward D M3HPM
Woodward E M1BMU
Woodward E 2E0EJW
Woodward E M6EVW
Woodward G G8RLW
Woodward G M3OQI
Woodward H GW4JUC
Woodward I G1IAW
Woodward I G6OSV
Woodward J G7UCG
Woodward K G6AAZ
Woodward P 2E0PAW
Woodward P G4JWL
Woodward R G0MWE
Woodward R G8GRS
Woodward S M3NTZ
Woodward T G3XAU
Woodworth G G4ZAG
Woodworth R M3XOW
Woodyard P 2E0PAP
Woodyard P G3ETP
Woodyard P M3UGQ
Wooff C G3SPJ
Wooffindin K G4NCB
Wookey D G6UXK
Wookey P G6ESJ
Woolard N G1ZFS
Wooldridge J 2E0CBH
Wooldridge J M6WOO
Wooldridge M M1MPW
Wooldridge P G1HKP
Wooldridge T 2E0BSB
Wooldridge T M1CND
Wooldridge T M3TPW
Woolen D GW3SRF
Woolfall D G0CCT
Woolfenden E G7LDR
Woolfenden G G0TUW
Woolford A G3SNN
Woolgar D G4UII
Woolgar J GW0FZY
Woolgar R M3OTU
Woolgar S G4UIH
Woolger D 2E0FSI
Woolger D M6FSI
Woolger M M6MPO
Woolhouse P G7PWV
Woollams D G4ZSR
Woollams G G4YZL
Woollard A G6TWA
Woollard E G3YZV
Woollard H 2E0BND
Woollard M G7USX
Woollard R G0TUL
Woollard R G8RCK
Woollen E M0AKI
Woollen I 2E0WDI
Woollen J M0AVQ
Woollen J M3ONB
Woollen W G4MFP
Woollen W G8GDQ
Wooller L G8GEZ
Woollett B 2E0XBW
Woollett B M0XBW
Woollett B M6XBW
Woolley C G7HLD
Woolley C GD3ZZF
Woolley D G8AMJ
Woolley E 2E0CLM
Woolley H M0HRW
Woolley L 2E0LOL
Woolley L M3MQA
Woolley M 2E0MAP
Woolley M M3LXS
Woolley M M3PKZ
Woolley P 2E0PEW
Woolley P M0PEW
Woolley P M3WPO
Woolley R G4HIJ

Woolley R G4OZD
Woolley R G6VGC
Woolley R MD1RPC
Woolley R G8VHI
Woolley R 2E0MWT
[illegible]
Woolley S M6YLE
Woolley S M1CRE
Woollin M G4ADE
Woolliss J G4NPS
Woollons J G1OSP
Woollven K G8CLK
Woolmer D G1FFU
Woolnough B G6LKA
Woolnough B M5ADQ
Woolridge L M6LGJ
Woolridge M G1JFL
Woolridge R G7LNJ
Woolridge R M3LNJ
Woolrych H G4TIX
Woolrych H G6EVY
Woolsey K M6KNW
Woolterton G G8TAE
Woolterton M G4TAD
Woolvin J M0JPW
Woolvin T M6TGJ
Woomans I G4NCY
Wooster D G1SHT
Wooster S G7VSP
Wootten T M0FFX
Wootton A G0BXD
Wootton N G7LPY
Wootton N M3ENY
Wootton P G1EQL
Wootton P 2E0FYI
Wootton T G7GDA
Wootton W G0FBQ
Worden H G0KXD
Wordley C G4LAE
Wordsworth D G0MYP
Wordsworth R G0GPR
Wordsworth T G7OFV
Worgan H GW1XJJ
Worger S 2E0SNJ
Worger S M6SWA
Workman D G0OHJ
Workman E G3RUD
Worledge P G1AAH
Worley H G4WZB
Worley J G4TYY
Worley K G8TEK
Worley M M6BHY
Worlledge A M3UNR
Worlledge C M1ANO
Worlledge P M1MUM
Worlledge P M0BHJ
Wormald C G0PJW
Wormald D G3GGL
Wormald T M0TRW
Wormall D G1CDN
Wormall R G1CDO
Wormwell B G3WGK
Worner S G6WVR
Wornham J GD4RVQ
Worrall F M6TFW
Worrall G M3MOH
Worrall K G3MCC
Worrall K M1BVI
Worrall M M1AWC
Worrall S G7RBT
Worrall T G0TCJ
Worsdale I G4RUE
WORSDALE P G0CEG
Worsell R G4CUG
Worsfold A G8XCY
Worsfold C G0HDH
Worsfold M G0VPX
Worsfold M G4PRJ
Worsley A M1LAN
Worsley J M6JFA
Worsley K G1FOW
Worsley R G0MYR
Worsnop J G0SNV
Worsnop J G7DFC
Worsnop J G4BAO
Worth D M6FPG
Worth R G4ZQF
Wortham C G4AGC
Worthing J M1BCM
Worthington B M3DXN
Worthington C M1VSR
Worthington E M3XMY
Worthington E M3ZSD
Worthington I G8ZSD
Worthington J M0AVQ
Worthington J M0VVV
Worthington T GM4PXG
Worthington W G3ZBM
Worthy I G6POJ
Worton K M6DTD
Worton L M0LNX
Worvell B G1NWT
Worvill M G4CHH
Woudstra M G8FEJ
Wozniak J G6WPL
Wrack K M0BMW
Wragg A G4ZNI
Wragg K G0FEZ
Wragg M G0HMX
Wragg P G8ITU
Wragg S G7LPE
Wragge S G1XOW
Wraight D M0DWC
Wraight D M3DWD
Wraight J G0DHJ
Wraith I G7GHH

Wraith J M6XJW
Wraith R M6NFL
Wraith S M6SHW
Wrampling R G7TFA
Wratten D G6XJB
[illegible]
Wratton T GM4CAU
Wray B M0WRA
Wray H M0AGO
Wray M G6DIR
Wray M G7KJO
Wray M G4SOI
Wren C G8UCZ
Wren C G0KLD
Wren G G3XQJ
Wren G M6GSO
Wren G M6BOR
Wren J G3IRA
Wren M M0MLW
Wrench W G7AKJ
Wren-Hilton M M6WRN
Wresdell J 2E0ADL
Wresdell J G3XYF
Wressell D M0WDG
Wright A G0LIS
Wright A GI0VGV
Wright A G0WTF
Wright A GM3IBU
Wright A G4EPN
Wright A G4OJY
Wright A M6AWR
Wright A G0KRU
Wright A G7ARK
Wright A M6AQA
Wright A 2E0XRX
Wright A M0TEI
Wright A M6BBX
Wright A M3XWP
Wright B G0XAD
Wright B G7UWB
Wright B G4HJW
Wright B M6YNA
Wright C G0EEZ
Wright C G3YFL
Wright C GM4HWO
Wright C G6ZWL
Wright C G7ODT
Wright C G7UWC
Wright C 2E0CJW
Wright C M6CJT
Wright C 2E0YRK
Wright C G7FYG
Wright C M6CHW
Wright C M0EMR
Wright C M6EED
Wright D G0BNU
Wright D G0GJA
Wright D G0MTV
Wright D G3UUY
Wright D G3VBQ
Wright D G3WTR
Wright D G6ORH
Wright D G7LVE
Wright D M1BZR
Wright D M1ECC
Wright D G7TRL
Wright D M3TZX
Wright D 2E0KWM
Wright D G0HBZ
Wright D G0OIX
Wright D G3XOU
Wright D G4BKE
Wright D G8BKG
Wright D G8EQD
Wright D M0DFW
Wright D M0EQD
Wright D M0KWM
Wright D M6KWM
Wright E G0SVH
Wright E G1VQK
Wright E M6HEK
Wright F G1DFN
Wright G G3AER
Wright G G8KPG
Wright G M1AUZ
Wright G MI3CCA
Wright G M6AGA
Wright G M6GBW
Wright G 2E0CCF
Wright G M3TZQ
Wright G G4FUJ
Wright H G4KOV
Wright H GI6GNA
Wright H M3XHW
Wright H G0EKK
Wright I G4HRQ
Wright I G6JRI
Wright I M6IWA
Wright J 2E0LLX
Wright J 2E0JGW
Wright J G0ANH
Wright J G0KTS
Wright J G1LUF
Wright J G3VPW
Wright J G6KNE
Wright J G0TNQ
Wright J GM7CPR
Wright J M3LLX
Wright J M3LPE
Wright J M3OFH
Wright J 2E0JCW
Wright J M6JCW
Wright J M6OWA
Wright J G4ZQT
Wright J 2M0GEK
Wright J G0OWA

Wright J G3SZG
Wright J G4DMF
Wright J G6POI
Wright J M0DSW
Wright J MM6CJZ
[illegible]
Wright K G4FYN
Wright K GI4KCO
Wright K G4ZTD
Wright K G6MAM
Wright K M0KHW
Wright K M1EVH
Wright K M6NMR
Wright L G6ZJS
Wright L G0PBJ
Wright L GW8UAM
Wright M 2E0TTZ
Wright M G0VSH
Wright M G0VWH
Wright M G6ELG
Wright M G8ZLU
Wright M M3MEW
Wright M M3OTZ
Wright M M3UNK
Wright M G1EUI
Wright M G4RLF
Wright M G1VPS
Wright M G4GXN
Wright M G8NWU
Wright N G6JDO
Wright P G0WXF
Wright P G1VBQ
Wright P G3JDM
Wright P G4CGP
Wright P G6PBZ
Wright P G8GYS
Wright P G8IOW
Wright P G8JQH
Wright P M0CVW
Wright P 2E0CKJ
Wright P M3PWW
Wright P 2E0WRI
Wright P G4MHA
Wright P M6PXW
Wright P 2E0BMU
Wright Q 2E0DQG
Wright Q M0OAE
Wright R 2E0GNE
Wright R G0EEN
Wright R G1SWZ
Wright R G3TOY
Wright R G3WZR
Wright R M3NVQ
Wright R M3WYF
Wright R G0WET
Wright R G8SWM
Wright S G0PWL
Wright S G4CPC
Wright S G4GFC
Wright S G4LBY
Wright S G6OBU
Wright S G7ARF
Wright S M3LXH
Wright S M3XYJ
Wright S G6YRK
Wright S M0IGG
Wright S MM3TZP
Wright T 2E0LCW
Wright T G0LZS
Wright T G0UNK
Wright T G3KVE
Wright T G6NFB
Wright T 2E0CTR
Wright T G0DRW
Wright T M3PMY
Wright V M3WGV
Wright W GM3UCH
Wright W G4BNK
Wright W G7EPY
Wright W G0FAH
Wright W M6WIQ
Wrighton H G7KGD
Wright-Williams N G3UTE
Wrigley D M5BGR
Wrigley D G6GXK
Wrigley J GD7DPG
Wrigley W GD7ARS
Wring D G4WRQ
Writer E G1AAQ
Wrobel J M6WRO
Wroblewski R GM8YAQ
Wroe D P G0MXD
Wroe G G4IUJ
Wroe P G0KXY
Wroe R G1WTN
Wroe R G6TAS
Wuille J G3SZM
Wunderlich W G4FXR
Wyard A G7FSR
Wyatt A G8LSD
Wyatt A G1CFE
Wyatt B G1CWQ
Wyatt C G6MIT
Wyatt D G0VAL
Wyatt G GW3NDB
Wyatt G GW0ADA
Wyatt J G1ZFI
Wyatt M G0CNA
Wyatt M G7STM
Wyatt M 2E0MSW
Wyatt M M6MSW
Wyatt P G7WDG
Wyatt R G6MQN
Wyatt R M0CIO
Wyatt R 2E0RDW
Wyatt R M0RDW

Wyatt S M0OXR
Wyer W G4CVO
Wyeth K G6CUV
Wyeth R M0RAW
Wylde S M3WYL
[illegible]
Wyles P M3IPZ
Wyles S G6VJU
Wylie A M0DBI
Wylie B G7RQO
Wylie G GM0GMO
Wylie J G4LYX
wylie N MI3OHG
WYLIE P G0GZE
Wylie P M3GCT
Wylie P MI3TKK
Wylie R MI0RJW
Wylie R MI3MJI
Wylie S MI3CBL
Wylie T GM4FDM
Wylie T MI0TMW
Wynes G G3TLV
Wynford-Thomas D G3YQM
Wynn B G8TB
Wynn C 2E0IFL
Wynn M 2E0BGQ
Wynn R G4BNB
Wynn V G7DRW
Wynne C 2E0BTA
Wynne C M3LXK
Wynne G M6GWN
Wynne H GM6AQR
Wynne J 2E0BTD
Wynne J M3LXB
Wynne J M6DSU
Wynne R G0JSO
Wynne-Jones T G6ZFV
Wynters D G6KCJ
Wyse A G3IWE
Wysocki N G6CPO
Wyspianski A G1AWF

## X

Xian M M0TDD
Xu B M0XUB
Xu B M6BXU

## Y

Yakub D 2E0DIL
Yakub D M0YKB
Yakub D M6DIL
Yale J G3ZTY
Yallop A G3SVQ
Yallop M G4YNT
Yam J G6REV
Yamamoto T M0OJX
Yan J M0JSH
Yao Z 2E0FBH
Yao Z M0HZB
Yap A M6AYE
Yapp S M3SAY
Yardley J G3HFZ
Yardley K M3YYK
Yardley P G0INS
Yardley S M6HGR
Yardley T M3YDM
Yarker A G3TAY
Yarnall J M1AUN
Yarnall J G1JLQ
Yarnold A M3ZBI
Yarnold P G7DSO
Yarnold R GW6DOC
Yarrow D GD6TDX
Yarrow M 2M0MJY
Yarrow M MM0MJY
Yarrow M MM6MJY
Yarrow N GM4PJR
Yarrow R 2E0AAJ
Yarrow R M6MPD
Yarwood B M3ZQJ
Yarwood P M0PJY
Yates A G6CPS
Yates A G8RAO
Yates A M0DVQ
Yates A M3FLU
Yates A 2E0DMI
Yates A M6BGK
Yates A M3BDA
Yates A G6LUF
Yates A MM6NRQ
Yates B G7UOS
Yates B G0DIZ
Yates B G4TVN
Yates D G0PBE
Yates D G3PGQ
Yates D 2E0DIJ
Yates D M0DIJ
Yates J G0NNF
Yates J G1UZD
Yates J G3IDC
Yates J G3MNJ
Yates J G1SQA
Yates Jones H G7RGI
Yates K M3IVA
Yates K G3XGW
Yates K G1HGA
Yates L M3NMV
Yates M M6ALG
Yates N M3GDQ
Yates P G0NPY
Yates P G7BZD
Yates R G8ZUL
Yates R M6RGD
Yates S GW0HNS
Yates S G7ENM
Yates S M3LHZ
Yates S G7ETK

UK Surnames

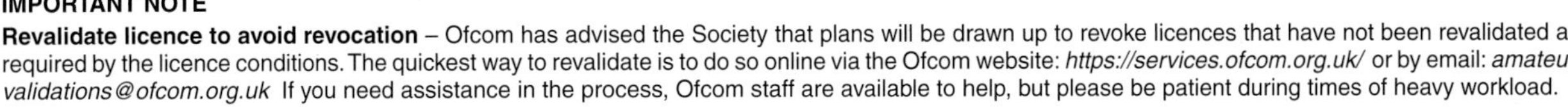
**IMPORTANT NOTE**

**Revalidate licence to avoid revocation** – Ofcom has advised the Society that plans will be drawn up to revoke licences that have not been revalidated as required by the licence conditions. The quickest way to revalidate is to do so online via the Ofcom website: *https://services.ofcom.org.uk/* or by email: *amateur.validations@ofcom.org.uk* If you need assistance in the process, Ofcom staff are available to help, but please be patient during times of heavy workload.

| Name | Callsign |
|---|---|
| Yates T | G3RWE |
| Yates T | G6LUE |
| Yates T | MM6TFY |
| Yaxley P | G4YLW |
| Yaxley R | G3YHO |
| Yea P | G0WKU |
| Yeaman D | G4ASY |
| Yeandel J | G4OOL |
| Yearl H | G0WKI |
| Yearley S | M0SJY |
| Yearp A | M0DOS |
| Yearsley K | M0KGY |
| Yearsley P | G1UTM |
| Yeates D | G4FND |
| Yeates K | G0TTW |
| Yeates P | G1XXE |
| Yeatman D | G1AGW |
| Yeatman P | G0PSF |
| Yeend J | G3CGD |
| Yeldham H | G6XOU |
| Yeldham S | M3LFO |
| Yendell S | M1AWX |
| Yendole J | M6JXY |
| Yeo A | M6YEO |
| Yeo D | G0IAE |
| Yeo I | G0PCQ |
| Yeo K | G7UCR |
| Yeoman D | G4SQA |
| Yeoman R | GM3ZGH |
| Yeomans A | M0HXV |
| Yeomans M | G4YTO |
| Yeomans M | M1FCW |
| Yeomans T | G4FOD |
| Yerrell R | M6YRL |
| Yetton T | G1SOB |
| Yiangou A | M0VKJ |
| Yilmaz A | G3PRK |
| Yip T | M3YIT |
| Yirrell M | G6RXD |
| Yohn C | MM6CJY |
| Yohn E | M6AHE |
| Yohn S | G7OEW |
| York A | G4EQP |
| York E | M6GKX |
| York E | G8HOR |
| York G | G8MXD |
| York G | G8UUC |
| York J | G3KJY |
| York M | G1BKI |
| York M | GW0NKG |
| York R | G1USV |
| York R | G6ZGH |
| York S | G8JUV |
| York S | 2E0SBL |
| York S | M6SAY |
| Yorke A | 2E0BAV |
| Yorke A | G7KUM |
| Yorke A | M3YOE |
| Yorke D | G4JLG |
| Yorke M | G4ASW |
| Yorke N | 2E0NEI |
| Yorke N | M0NKE |
| Yorke N | M6NEI |
| Yorke P | G6WBX |
| Yorke T | G1JWY |
| York-Jones P | G8CYU |
| Yorkston A | M3XAY |
| Yotov V | M6AQV |
| Youd J | G0JBY |
| Youd N | G1AAD |
| Youd W | M6LNK |
| Youde J | G0GUF |
| Youell R | G4VME |
| Youlden M | M0CLB |
| Youll G | 2E0CZA |
| Youll G | M6DQR |
| Young A | G1AJY |
| Young A | G3YBP |
| Young A | G4BVG |
| Young A | M0FYA |
| Young A | M0IKB |
| Young A | MM0LUP |
| Young A | 2E0GRN |
| Young A | M6GRN |
| Young A | MM0GYD |
| Young A | MW6AYA |
| Young A | G6NFC |
| Young A | 2M0BUX |
| Young A | MM0GSQ |
| Young A | MM6ART |
| Young A | M6YCA |
| Young B | GW6TYO |
| Young B | MI3JQD |
| Young B | G0SCI |
| Young B | M6YBV |
| Young B | M3SBY |
| Young C | 2E0BAU |
| Young C | G4CCC |
| Young C | G8KHH |
| Young C | MW0KRS |
| Young C | M3IJZ |
| Young C | M5KHH |
| Young C | M6KOI |
| Young C | M6HCZ |
| Young C | MM6YNG |
| Young D | GM0DYD |
| Young D | G1IDZ |
| Young D | G1NSD |
| Young D | G4RQU |
| Young D | G4ZHN |
| Young D | G8TVW |
| Young D | G8VXB |
| Young D | M0AOA |
| Young D | 2W1ACM |
| Young D | 2E0DYG |
| Young D | 2E0LZT |
| Young D | M0YAY |
| Young D | M3DAY |
| Young D | M3LZT |
| Young D | M6YGD |
| Young D | MM6ZDY |
| Young D | MM3UOS |
| Young D | G8ZQJ |
| Young E | 2E0FRY |
| Young E | G0SQE |
| Young E | M6EDD |
| YOUNG F | G7NBV |
| Young G | G8MGE |
| Young G | M3HAI |
| Young G | MW6CUA |
| Young H | G3LCI |
| Young H | G4JTO |
| Young I | G7III |
| Young J | G0BIV |
| Young J | G3UIK |
| Young J | G4PPZ |
| Young J | GM4RCN |
| young J | G6LYJ |
| Young J | G6MMS |
| Young J | 2E0JYA |
| Young J | G3KLP |
| Young J | MI0RJY |
| Young J | MM6RLL |
| Young J | MW3UZP |
| Young J | G1NIV |
| Young J | G4KZD |
| Young K | G3HUO |
| Young K | G3ZCG |
| Young K | G6UXF |
| Young K | 2E0GDN |
| Young K | 2E0KLY |
| Young K | M3SJY |
| Young K | M6ITW |
| Young K | 2E0JKY |
| Young K | M6FVT |
| Young K | M6JKY |
| Young L | G7APU |
| Young M | G4KPL |
| Young M | M3YOU |
| Young M | M6YSM |
| Young M | M6ESH |
| Young M | MW6FBA |
| Young N | M3NJY |
| Young N | M0NFY |
| Young P | GM0GBH |
| Young P | G1MRX |
| Young P | G6AHE |
| Young P | G7SQC |
| Young P | G0HWC |
| Young R | GM0GRW |
| Young R | G0TUS |
| Young R | G1OIZ |
| Young R | G4NQS |
| YOUNG R | G4XYD |
| Young R | G6CIT |
| Young R | G6FVL |
| Young R | GM7ITG |
| Young R | GI7PBQ |
| Young R | G7RNQ |
| Young R | M3RDY |
| Young R | G6JOL |
| Young R | M6KKI |
| Young R | G8XFK |
| Young R | 2E0JPD |
| Young R | M3ZNP |
| Young R | M6FFS |
| Young S | M0BLY |
| Young S | M0GHV |
| Young S | 2M0SRY |
| Young S | MM6DKI |
| Young S | 2M0VFV |
| Young S | MM6SMY |
| Young T | G1FXM |
| Young W | G4DTL |
| Young W | G4SJT |
| Young W | G7BEJ |
| Young W | G3NZR |
| Young W | MM3VYU |
| Younge E | G3IVH |
| Younger A | G0RMN |
| Younger K | GM3OIB |
| Younger K | M6KJY |
| Youngman-Smith N | G1ZRR |
| Youngs D | G3JIE |
| Youngs S | G1KYV |
| Youster B | G8PRP |
| Yovchev I | 2D0PBW |
| Yovchev I | MD6IVO |
| Yoxall C | M6YOX |
| Yoxall G | M0CYX |
| Yu B | M6YZH |
| Yu J | G3ZQT |
| Yuen P | M3OIY |
| Yuill G | M3VDZ |
| Yuill S | G7UHL |
| Yukawa T | G0WWM |
| Yung C | M0GZK |
| Yung W | M3XOZ |
| Yunnie C | M6TUK |

## Z

| Name | Callsign |
|---|---|
| Zabalujevs A | M3YSY |
| Zak K | G0OQI |
| Zakharov V | G0KGD |
| Zakrzewski A | MI0HYQ |
| Zakrzewski R | M0MVO |
| Zalicks L | G4YOT |
| Zammit C | G3WXD |
| Zammit C | G8DUV |
| Zanek P | M0PPZ |
| Zara P | G1UFT |
| Zarattini M | G4HJL |
| Zarucki B | M0DGQ |
| Zarucki C | M3ZCR |
| Zaulincy Adams A | M6TZA |
| Zaza G | M0GXC |
| Zdziech C | 2E0HWI |
| Zdziech C | M0CXQ |
| Zeal C | G4BGM |
| Zeal R | G7PRK |
| Zeller A | 2E0ZLA |
| Zeller A | M6ZLR |
| Zemlicka J | M0GBO |
| Zennadi A | M6EJG |
| Zerafa A | G8CKK |
| Zerafa A | M0BUV |
| Zerafa R | M6EZG |
| Zhan Y | M6BZB |
| Zhang P | M6ZPZ |
| Zhang X | M6FIA |
| Zieba E | M6DZQ |
| Zieba R | M3RWZ |
| Zielinski R | G0ELG |
| Ziemacki A | G1GBR |
| Zimmermann P | G0UPS |
| Zimnowlocki A | MM3DFZ |
| Zissler J | G0CCS |
| Zlobinski M | M0GNY |
| Zlotnicki R | M6ZXO |
| Zmajkovic Z | M6DJG |
| Zollman P | G4DSE |
| Zorzi G | M0HEP |
| Zubrzycki D | 2E0DMN |
| Zubrzycki D | M0ZUB |
| Zubrzycki D | M6ATB |
| Zubrzycki M | G4PYW |
| Zulkfli N | M6ZAS |
| Zygadllo D | M3ZYG |
| Zywicki L | M0ICJ |

# Postcode Index

## AB

(Aberdeen)

| Postcode | Callsign |
|---|---|
| AB1 6JU | GM4VRE |
| AB10 6DT | MM6WEB |
| AB10 6ED | 2M0EMM |
| AB10 6ED | MM0MUN |
| AB10 6JD | GM4HTU |
| AB10 6QH | GM0NRT |
| AB10 6QQ | GM7VZV |
| AB10 6QQ | MM0AOF |
| AB10 6QW | GM4JLZ |
| AB10 6RA | GM1MCN |
| AB10 6SB | GM6GFQ |
| AB10 7BS | MM3OMI |
| AB10 7JE | GM3KJE |
| AB11 7DG | MM3ZRZ |
| AB11 7SF | MM6APO |
| AB11 7TZ | GM4AJR |
| AB11 7WD | 2M0RND |
| AB11 7WD | MM0ROV |
| AB11 7WD | MM3MVY |
| AB11 7WD | MM3YBG |
| AB11 8FX | 2M0AKS |
| AB11 8FX | MM0GDG |
| AB11 8FX | MM3JJC |
| AB11 9JD | MM3JPS |
| AB12 3DZ | 2M1IIW |
| AB12 3JJ | MM1ABA |
| AB12 3NG | GM1THS |
| AB12 3PH | MM6MOY |
| AB12 3RL | GM3NUU |
| AB12 3RN | GM8FFX |
| AB12 3RW | MM0KSS |
| AB12 3SH | GM7BYB |
| AB12 3SU | MM0GQF |
| AB12 4LZ | MM6BXQ |
| AB12 4NY | GM4RAZ |
| AB12 4NY | GM4RGS |
| AB12 4QA | MM0AOY |
| AB12 4QX | 2M0MJY |
| AB12 4QX | MM0MJY |
| AB12 4QX | MM6MJY |
| AB12 4TF | GM3ZEU |
| AB12 4XL | GM8GDN |
| AB12 4XT | 2M0DWC |
| AB12 4XT | MM3YXN |
| AB12 4XT | MM3ZGK |
| AB12 5AB | GM0DBW |
| AB12 5DQ | GM4UWN |
| AB12 5LJ | MM0DBR |
| AB12 5QT | 2M1HRS |
| AB12 5XT | GM0VFY |
| AB13 0EF | GM7WGM |
| AB13 0ER | GM0OPX |
| AB13 0JB | GM3TLA |
| AB14 0LN | GM4ZUK |
| AB14 0NX | GM4TEF |
| AB14 0TU | GM0CQV |
| AB14 0UE | GM4ZGV |
| AB15 4BP | GM1MYF |
| AB15 4UH | GM8ADK |
| AB15 5EX | GM4FBP |
| AB15 6AN | GM3OUU |
| AB15 6BH | 2M1HTR |
| AB15 6DU | 2M0GEJ |
| AB15 6DU | MM0TGG |
| AB15 6DU | MM6XGJ |
| AB15 7QA | GM4EKC |
| AB15 7QN | GM4BAP |
| AB15 7SX | GM3VEY |
| AB15 7XP | GM7RNJ |
| AB15 7YB | 2M0AVL |
| AB15 7YB | MM0DFZ |
| AB15 8LQ | GM3WIJ |
| AB15 8LQ | GM4GTV |
| AB15 8LT | GM3HGA |
| AB15 8RL | GM1STW |
| AB15 8SF | GM3BSQ |
| AB15 8SF | GM4GVK |
| AB15 9AR | GM4NVI |
| AB15 9JX | GM4IXH |
| AB15 9PL | GM8MFZ |
| AB15 9QF | GM3VAP |
| AB15 9QQ | GM3TYS |
| AB15 9HE | MM6IAI |
| AB15 9RH | GM4EGX |
| AB16 5QG | MM3RBJ |
| AB16 5[illegible] | 2M0[illegible]YL |
| AB16 5RP | 2M0SYI |
| AB16 5RP | MM0BCR |
| AB16 5RR | MM6AIK |
| AB16 5SR | MM3KQI |
| AB16 6FN | GM6JBF |
| AB16 6FU | GM0MYQ |
| AB16 6NX | GM7KBK |
| AB16 6PD | GM8MHU |
| AB16 6WD | MM0CVH |
| AB16 6WF | GM4EMX |
| AB16 6XY | MM6CIA |
| AB16 7DQ | 2M0FYG |
| AB16 7SL | 2M0PKA |
| AB2 3YS | GM7HNU |
| AB21 0AH | MM6TIR |
| AB21 0JZ | GM4NNK |
| AB21 0LH | MM0ALY |
| AB21 0NG | GM8BZP |
| AB21 0QF | 2M0MRO |
| AB21 0RG | GM6UHC |
| AB21 0RJ | GM8GHV |
| AB21 0RS | GM1BNP |
| AB21 0TW | GM0IPV |
| AB21 0TW | GM4SID |
| AB21 0TY | MM1FDF |
| AB21 0TZ | GM6GMZ |
| AB21 7DD | MM6HRF |
| AB21 7FN | GM3WSR |
| AB21 9HJ | MM0CZH |
| AB21 9HS | GM0PTY |
| AB21 9JJ | 2M0NIA |
| AB21 9JJ | MM0CUG |
| AB21 9JJ | MM6NIA |
| AB21 9PQ | MM0DOT |
| AB21 9QS | 2M0BPV |
| AB21 9QS | MM3WMQ |
| AB21 9QX | GM7NHS |
| AB21 9RH | GM0SZA |
| AB214NR | MM3SWW |
| AB22 8LJ | GM7MWL |
| AB22 8LY | GM7ITG |
| AB22 8TW | GM7MMI |
| AB22 8WG | GM0MCJ |
| AB22 8WL | MM1EYZ |
| AB22 8WY | GM0JOV |
| AB22 8XB | GM6MJY |
| AB22 8YD | MM6FSR |
| AB22 8ZB | MM5ISS |
| AB22 8ZH | 2M0DHI |
| AB22 8ZH | MM6DHI |
| AB23 8BD | GM8BNH |
| AB23 8EH | GM1FSU |
| AB23 8PW | 2M0RMH |
| AB23 8PW | MM3NXY |
| AB23 8QD | GM4YWV |
| AB23 8QN | GM1LKD |
| AB23 8TS | GM4THP |
| AB23 8UT | 2M1ENI |
| AB23 8WZ | MM6AGL |
| AB23 8XY | GM7SPB |
| AB23 8YG | MM6JPM |
| AB24 1WT | 2M0NSA |
| AB24 1WT | 2M0STB |
| AB24 1WT | GM4BFX |
| AB24 1WT | MM0ROR |
| AB24 1WT | MM6MVQ |
| AB24 2RP | GM0AUL |
| AB24 2XS | MM6BNN |
| AB24 2XS | MM6BNO |
| AB24 3JL | GM8BSU |
| AB24 3PA | MM6EUA |
| AB24 3TS | MM6CJY |
| AB24 4HX | GM4CAU |
| AB24 4NH | GM4TVB |
| AB24 4NJ | GM6SDV |
| AB24 4NW | GM7RXL |
| AB24 5BE | MM0TJR |
| AB24 5EP | 2M0URP |
| AB24 5EP | GM6URP |
| AB24 5EP | MM6URP |
| AB24 5EP | MM3OQR |
| AB25 1DQ | GM4HQF |
| AB25 2YF | MM0DBF |
| AB3 1NX | GM7IHR |
| AB30 1XZ | GM4YRE |
| AB30 1YS | MM6LVV |
| AB31 4AL | MM0IBX |
| AB31 4EN | GM6EUC |
| AB31 4HG | GM0TCU |
| AB31 4LS | GM8GJI |
| AB31 4QA | MM3XAF |
| AB31 4QL | 2M0BZL |
| AB31 4QL | MM0XPT |
| AB31 4QL | MM6CRW |
| AB31 4RY | GM0OSXQ |
| AB31 5AU | GM4MFB |
| AB31 5HA | MM0ACR |
| AB31 5HA | MM0ACT |
| AB31 5TS | MM6YQP |
| AB31 5UY | 2M0DJT |
| AB31 5XA | GM0PKQ |
| AB31 5XW | MM0DRT |
| AB31 5YG | 2M0CXI |
| AB31 5YG | MM0HVW |
| AB31 5YG | MM6DVO |
| AB31 5ZF | GM7BDD |
| AB31 6HH | GM0AZV |
| AB31 6DT | GM0RAO |
| AB31 6NL | GM6NXN |
| AB32 6WS | GM8BSQ |
| AB32 6WW | GM1XEA |
| AB32 6XH | MM0JOM |
| AB32 6XY | GM0GAT |
| AB32 6YE | MM3HLG |
| AB32 7EQ | GM4NHI |
| AB33 8BH | GM8ZFW |
| AB33 8ER | 2M0PTE |
| AB33 8ER | MM3UDI |
| AB33 8HQ | GM0MHD |
| AB33 8HQ | MM1MHD |
| AB33 8JU | GM0OSJ |
| AB33 8NX | MM6MYK |
| AB33 8QA | MM6ELU |
| AB33 8QQ | MM6IJP |
| AB33 8QW | MM5CFA |
| AB33 8UB | GM7IEU |
| AB34 4TA | GM8AT |
| AB34 4UH | MM3VNU |
| AB34 4YG | GM3GG |
| AB34 4YG | GM4HWS |
| AB34 5JF | GM4JXP |
| AB34 5JZ | GM0KDP |
| AB34 5PQ | 2M0EFI |
| AB34 5PQ | MM0EFI |
| AB34 5PQ | MM3LKR |
| AB35 5QH | GM4RLV |
| AB35 5RX | MM6HDZ |
| AB35 5SF | GM8FVN |
| AB35 5UT | GM0DZW |
| AB35 5ZQ | MM6CHN |
| AB36 8US | 2M0COT |
| AB36 8US | GM3ALZ |
| AB36 8US | MM0ODL |
| AB36 8US | MM6ALZ |
| AB37 9BG | MM6IHQ |
| AB37 9ET | MM3WFU |
| AB37 9HW | MM3WQO |
| AB38 7BD | MM3YDK |
| AB38 9NU | GM1TGY |
| AB38 9SQ | MM3XMT |
| AB39 2AD | GM1JPJ |
| AB39 2BZ | 2M0DRY |
| AB39 2BZ | MM6BRV |
| AB39 2EG | GM4PVQ |
| AB39 2GF | 2M0BXN |
| AB39 2GF | GM4HVS |
| AB39 2GF | MM3YQO |
| AB39 2JA | MM0CFE |
| AB39 2LU | GM4VQY |
| AB39 2PL | MM6MLT |
| AB39 3PF | GM4KOI |
| AB39 3PF | GM6OSZ |
| AB39 3QG | MM6TFY |
| AB39 3RB | GM3PML |
| AB39 3SY | MM0ALM |
| AB39 3UL | 2M0MGM |
| AB39 3UL | MM3RLG |
| AB4 5SE | GM1VSR |
| AB41 6BJ | GM4MBG |
| AB41 6RT | MM6JAN |
| AB41 7DF | GM4LTL |
| AB41 7DH | GM7LDU |
| AB41 7DS | GM7KRQ |
| AB41 7JY | GM1WKH |
| AB41 7NZ | MM3OZW |
| AB41 7PH | MM6DXJ |
| AB41 8BA | GM0PKF |
| AB41 8BH | GM4FVS |
| AB41 8QS | GM3UAG |
| AB41 8QW | GM4GNR |
| AB41 8UJ | GM7LAC |
| AB41 8UJ | MM6BOS |
| AB41 8YH | GM2MP |
| AB41 8YH | GM4YXI |
| AB41 9HF | GM0JEF |
| AB41 9JB | GM6AOR |
| AB41 9LW | GM4OBD |
| AB41 9NF | GM0MIS |
| AB42 0LN | MM0CWI |
| AB42 0NG | GM7LJE |
| AB42 0NY | GM1TBW |
| AB42 0PZ | MM3TAV |
| AB42 0RE | MM1BMK |
| AB42 0RE | MM6BOQ |
| AB42 0TB | GM8AGM |
| AB42 0TQ | MM6FYR |
| AB42 1GS | MM3AWD |
| AB42 1HB | MM0CJF |
| AB42 1HL | GM1JNS |
| AB42 1NX | GM4UFD |
| AB42 1RD | GM6KAM |
| AB42 1UQ | GM4PXB |
| AB42 2HW | GM1KBZ |
| AB42 2UF | GM7FYB |
| AB42 3AT | MM3PDM |
| AB42 0AY | GM1CCB |
| AB42 3BP | GM6WTT |
| AB42 3DD | 2M0CBE |
| AB42 3DD | MM6CBE |
| AB42 3DN | 2M0BVN |
| AB42 3DN | MM0IHE |
| AB42 3DN | MM3XIZ |
| AB42 [illegible]LA | MM[illegible] |
| AB42 4HA | MM6ZAZ |
| AB42 4JU | GM1XIN |
| AB42 4JU | GM6FDQ |
| AB42 4JU | GM8PLR |
| AB42 4NL | GM4EHP |
| AB42 4RD | MM6IAR |
| AB42 5AY | GM7OJJ |
| AB42 5AY | GM7OTT |
| AB42 5BL | MM0XTW |
| AB42 5DE | GM3ZMA |
| AB42 5DG | MM3WKF |
| AB42 5EH | MM6OMT |
| AB42 5ES | GM0RSI |
| AB42 5GT | MM6VTS |
| AB42 5HR | GM1KUI |
| AB42 5LR | MM6BQH |
| AB42 5RR | MM6SLV |
| AB42 5WE | MM0AOQ |
| AB42 7TB | GM8ZKU |
| AB42 8HX | GM7NNS |
| AB423HZ | GM0DYU |
| AB43 6NA | MM0HZJ |
| AB43 6NE | GM1KZG |
| AB43 6NN | MM0CAE |
| AB43 6NQ | GM7OGS |
| AB43 6SY | MM3FYN |
| AB43 7ED | MM0JGP |
| AB43 7JS | GM4PMH |
| AB43 7JT | MM3LNT |
| AB43 7JT | MM3LOF |
| AB43 7NW | MM1DAK |
| AB43 8WA | 2M1VXB |
| AB43 9DH | MM1CAC |
| AB43 9NL | GM1INS |
| AB43 9PU | GM3ZOT |
| AB44 1RP | GM3UBJ |
| AB44 1SL | GM8DAB |
| AB44 1YA | GM8SVB |
| AB45 1DB | GM1HRY |
| AB45 1DB | GM4VVY |
| AB45 1DZ | GM4TOE |
| AB45 1HS | MM0HKU |
| AB45 2BQ | GM0FHD |
| AB45 2BQ | GM0JLJ |
| AB45 2JR | GM1CCI |
| AB45 2JT | MM0BUH |
| AB45 2PJ | GM4PSJ |
| AB45 3BR | GM6JOA |
| AB45 3RB | GM0WIB |
| AB45 3UD | GM1ROX |
| AB5 2BJ | GM1VBD |
| AB51 0ES | GM4DZM |
| AB51 0JT | GM0APN |
| AB51 0PJ | MM6THU |
| AB51 0SN | GM0PKX |
| AB51 0XA | GM0TGE |
| AB51 0XA | MM0DTL |
| AB51 0XX | MM6ACM |
| AB51 3UA | MM6NRQ |
| AB51 3UB | GM7NUQ |
| AB51 3WJ | GM0TFQ |
| AB51 4FL | GM0VGI |
| AB51 4RQ | GM4DIN |
| AB51 4TB | GM1RDG |
| AB51 5BY | MM6FVD |
| AB51 5HE | GM7UPD |
| AB51 5JE | GM4YAU |
| AB51 5QT | GM3XOQ |
| AB51 5QZ | GM0FIQ |
| AB51 5RH | GM1FSZ |
| AB51 7HH | MM3MXN |
| AB51 7QP | 2M0GMB |
| AB51 8TB | MM6TSN |
| AB51 8TQ | GM1MKC |
| AB51 8TS | MM6JEA |
| AB52 6HX | MM5FWD |
| AB52 6JY | GM1KHU |
| AB52 6JY | MM6FYW |
| AB52 6NU | MM6SRL |
| AB52 6PD | MM0TXO |
| AB52 6UN | MM3ERP |
| AB52 6ZD | GM0UYS |
| AB53 4GN | GM3OXX |
| AB53 4GS | MM0UDI |
| AB53 4UD | GM0EMQ |
| AB53 5QP | GM4OCA |
| AB53 5QP | GM6KMK |
| AB53 5RN | GM0EQS |
| AB53 5TQ | GM7CPJ |
| AB53 6SL | GM4TRS |
| AB53 6TE | GM0NGJ |
| AB53 6UP | MM6ACI |
| AB53 8ED | GM1LXA |
| AB53 8HD | GM0BNQ |
| AB53 8LT | GM4ZEX |
| AB54 4GD | MM3VSX |
| AB54 4GD | MM6SIM |
| AB54 4JB | GM8MNM |
| AB54 4NN | MM3VSU |
| AB54 4NN | MM6EQM |
| AB54 4PF | 2M0IOK |
| AB54 4XS | MM6DZC |
| AB54 5NY | GM0WPU |
| AB54 5NY | GM8JWQ |
| AB54 6AT | GM6KDB |
| AB5[illegible] | GM1XLH |
| AB54 7LQ | MM3TPF |
| AB54 7SY | GM8REG |
| AB54 7XR | GM4NXT |
| AB55 4AH | MM6NAD |
| AB55 4AZ | MM0CMD |
| AB55 4AZ | MM3JIN |
| AB55 4EF | GM7TWM |
| AB55 5AG | GM0EIT |
| AB55 5AP | GM0LVK |
| AB55 5AP | GM1MRS |
| AB55 5EB | MM1DVC |
| AB55 5ED | MM1DTN |
| AB55 5ED | MM3TQI |
| AB55 5EG | GM7DXT |
| AB55 5EH | GM4FGL |
| AB55 5EQ | MM0DVB |
| AB55 5FX | MM3AWC |
| AB55 6LP | GM1HNZ |
| AB55 6LQ | MM0IEL |
| AB55 6QU | GM4WJA |
| AB55 6UR | GM8PSV |
| AB56 1DD | GM3UKG |
| AB56 4LD | 2M0HJS |
| AB56 4LD | MM3UDQ |
| AB56 4NB | MM3TLQ |
| AB56 4NR | GM4UWX |
| AB56 4PS | 2M0FSP |
| AB56 4PS | MM6FSP |
| AB56 4QA | GM4OEZ |
| AB56 4QW | GM4PMT |
| AB56 4SD | GM8JCF |
| AB56 5AL | GM1KWG |
| AB56 5BS | MM6IMF |
| AB56 5BW | GM0ARY |
| AB56 5EP | GM3KHH |
| AB56 5YD | GM4YGS |
| AL1 1DW | 2E1BDB |
| AL1 1UA | G0NVZ |
| AL1 2PG | G6BDW |
| AL1 2PP | M6NGO |
| AL1 2QS | G6CMD |
| AL1 3AG | G8VMP |
| AL1 4ES | M6XAL |
| AL1 4PZ | G3XXF |
| AL1 4RD | G3YCY |
| AL1 4SN | M6GJQ |
| AL1 4TT | G0MVY |
| AL1 4UX | M6LOW |
| AL1 4XZ | G0LQU |
| AL1 5AE | M6ADI |
| AL1 5BX | 2E0MRM |
| AL1 5ES | M6JGT |
| AL1 5EX | G4BIX |
| AL1 5LF | G7ARF |
| AL1 5NJ | G4DDV |
| AL1 5NS | G4COL |
| AL1 5NS | M0GTJ |
| AL1 5PZ | M6RNF |
| AL1 5QB | M0CZX |
| AL10 0UB | G1ROH |
| AL10 8DQ | G4RMD |
| AL10 8HE | G6INV |
| AL10 8QF | G1IHE |
| AL10 9GW | M0TMS |
| AL10 9LE | G3VIX |
| AL10 9LJ | G0SVD |
| AL10 9LR | M6IDO |
| AL10 9NT | M1JES |
| AL10 9PB | G4SNI |
| AL10 9PE | 2E0BXF |
| AL10 9RW | G3NQN |
| AL10 9SB | G6UBH |
| AL2 1HL | M1FCX |
| AL2 2AH | G1ARL |
| AL2 2AR | G8GYK |
| AL2 2QL | G8UCP |
| AL2 3DD | G7GJN |
| AL2 3ND | G0BEE |
| AL2 3PU | G1WVD |
| AL2 3SE | G7SDG |
| AL2 3SJ | G0HTL |
| AL2 3SJ | G0KVA |
| AL2 3SL | G4GOU |
| AL2 3SR | G1IHS |
| AL2 3ST | G6CKW |
| AL2 3TD | G8KGG |
| AL3 4AS | G4WMB |
| AL3 4HH | G3GEX |
| AL3 4JZ | G4WSL |
| AL3 4NJ | G4HVG |
| AL3 4TL | G6PWL |
| AL3 5RE | G3PZF |
| AL3 5TD | 2E0RTY |
| AL3 5TD | M0JZG |
| AL3 5TD | M6RTU |
| AL3 5TU | 2E1HHE |
| AL3 6HP | G8SGF |
| AL3 6LR | 2E0BIY |
| AL3 6LR | M3KIZ |
| AL3 7EW | 2E0CIM |
| AL3 7EW | M6YVR |
| AL3 7HD | 2E1EWN |
| AL3 7JB | G0EHO |
| AL3 7NL | G4YSJ |
| AL3 7PF | G6USO |
| AL3 [illegible] | [illegible] |
| AL3 8EE | G8KLC |
| AL3 8HW | G0OIK |
| AL3 8HW | M3MNY |
| AL3 8JN | G3LFV |
| AL3 8JN | M6RFV |
| AL4 0AZ | M6FJI |
| AL4 0DH | G0LZW |
| AL4 0DH | G8BNR |
| AL4 0DN | G8MEA |
| AL4 0DR | 2E1IDI |
| AL4 0EX | 2E0WAP |
| AL4 0EX | M6APW |
| AL4 0EY | M6GBF |
| AL4 0GE | G6PFP |
| AL4 0JG | G4HHJ |
| AL4 0NS | M6RTZ |
| AL4 0NW | M6TYT |
| AL4 0QA | G6DYM |
| AL4 0QR | G0MDR |
| AL4 0QS | 2E0KHA |
| AL4 0QS | M3KHA |
| AL4 0UP | G8HZQ |
| AL4 0XA | G4XJS |
| AL4 8HN | G8HXT |
| AL4 8JD | G4WNP |
| AL4 8PE | G4BOU |
| AL4 8PR | G0HGO |
| AL4 8RY | G6AHE |
| AL4 8TP | G0XBC |
| AL4 8TP | M0ABY |
| AL4 9AF | G4ZRA |
| AL4 9AP | M6POG |
| AL4 9BB | 2E0SKI |
| AL4 9BB | M0JEK |
| AL4 9BB | M6ABS |
| AL4 9DW | G4IXY |
| AL4 9DZ | M6XJM |
| AL4 9EH | G6PWS |
| AL4 9HD | M6APR |
| AL4 9JU | G4BEO |
| AL4 9JX | G4AUE |
| AL4 9LS | M6RTX |
| AL4 9LX | M6SQO |
| AL4 9NZ | G4CZA |
| AL4 9PJ | G0GJC |
| AL4 9PS | G7IFJ |
| AL4 9PS | M6MCY |
| AL4 9PW | G7PKH |
| AL4 9QG | G4BPR |
| AL4 9QH | M6BVS |
| AL4 9QN | M6ACD |
| AL4 9SJ | G3XYH |
| AL4 9TG | G4DJX |
| AL4 9TG | M0SCY |
| AL4 9TG | M6MXD |
| AL4 9UY | M6GDB |
| AL4 9XB | G4ZES |
| AL4 9XJ | M6VBS |
| AL4 9XQ | 2E0IOZ |
| AL5 1EF | 2E1BZB |
| AL5 1EZ | G7HDR |
| AL5 1HD | G4DOC |
| AL5 1JQ | G3UHN |
| AL5 1LL | G1XEH |
| AL5 1RF | 2E0IHB |
| AL5 1RF | G8CLJ |
| AL5 1RF | M0RBK |
| AL5 1RF | M6AKY |
| AL5 1RJ | G3UXO |
| AL5 1SR | G8CLY |
| AL5 1TD | G8KEK |
| AL5 2AL | G0DFF |
| AL5 2BE | 2E0EEM |
| AL5 2HJ | M0AYS |
| AL5 2QJ | G4GEZ |
| AL5 3AS | G1BRP |
| AL5 3LJ | 2E0JSG |
| AL5 3NX | G4OGZ |
| AL5 4BT | G1ZYJ |
| AL5 4DB | G0CPN |
| AL5 4DB | G8ASP |
| AL5 4HE | G3SBA |
| AL5 4HN | G4KCN |
| AL5 4LP | G3YWA |
| AL5 4TE | M5WIZ |
| AL5 5BH | 2E1HEB |
| AL5 5BT | G4OAV |
| AL5 5DW | G3MSW |
| AL5 5DW | G8XOB |
| AL5 5HR | G4RIK |
| AL5 5LY | 2E0SUU |
| AL5 5LY | M0SUU |
| AL5 5LY | M6WDS |
| AL5 5PW | G7GCI |
| AL5 5QS | G1ZGF |
| AL5 5QU | M0MLT |
| AL5 5SJ | G7EFL |
| AL5 5ST | M1CBV |
| AL5 5SU | G0RVH |
| AL5 5SU | G3SDG |
| AL6 0DB | G1MMZ |
| AL6 0DH | G0URC |
| AL6 0DZ | G0NYY |
| AL6 0EN | G3AQF |
| AL6 0PS | G7UOU |
| AL6 0RU | G8TRU |
| AL6 0RW | G3KOX |
| AL6 9BB | G6KJA |
| AL6 9NW | G3XDM |
| AL6 9PX | G0WTT |
| AL6 9TF | G4SZI |
| AL7 1BY | G4TUH |
| AL7 1DU | G0IIG |
| AL7 1DW | 2E0XYT |
| AL7 1NL | G7BDS |
| AL7 1SD | G0WAT |
| AL7 2BW | G6BUP |
| AL7 2HF | G6POE |
| AL7 2PZ | M0HSQ |
| AL7 2QJ | G5KW |
| AL7 2QJ | G8FXM |
| AL7 2QJ | M3FXM |
| AL7 3LR | G0NJS |
| AL7 3RJ | G8TVW |
| AL7 3RL | G0MKN |
| AL7 3TD | G1PGJ |
| AL7 3TE | G3TNE |
| AL7 3TN | G0AII |
| AL7 3TW | G0HVX |
| AL7 3XN | M6HSE |
| AL7 4AG | 2E1PDM |
| AL7 4AJ | G4LFB |
| AL7 4AQ | G0JJM |
| AL7 4JT | G3HLN |
| AL7 4RL | G3ZQI |
| AL7 4TG | M1FEK |
| AL7 4TX | 2E0TDS |
| AL7 4TX | M6TDS |
| AL8 6HT | G4JVA |
| AL8 6RF | G4LWV |
| AL8 6RH | G1LLW |
| AL8 6TA | G0CYC |
| AL8 6WW | 2E0HOG |
| AL8 6WW | M6STR |
| AL8 7AW | M6GCV |
| AL8 7BA | M1PWT |
| AL8 7BG | G4MDC |
| AL8 7DQ | G0UBO |
| AL8 7EE | M0GJK |
| AL8 7EN | G0VDT |
| AL8 7HX | M0JLW |
| AL8 7HZ | G4POB |
| AL8 7LF | G6CXY |
| AL8 7LH | G4DRI |
| AL8 7NN | G8TVZ |
| AL8 7QY | M1CYT |
| AL8 7QY | M1DYP |
| AL8 7SG | G4TIM |
| AL8 7SN | G6FVF |
| AL8 7ST | G8CYX |
| AL8 7TJ | M6SQK |
| AL9 5AS | G1EKM |
| AL9 7AY | M0BMU |
| AL95HQ | G7CMB |

## B

(Birmingham)

| Postcode | Callsign |
|---|---|
| B1 1LS | M6UED |
| B1 1UJ | G3OTY |
| B1 1UJ | G4ACF |
| B10 0LE | 2E0XTV |
| B10 0LE | M3XTV |
| B10 9TA | G7LOZ |
| B10 9TB | G6HLR |
| B11 3QJ | G8EMX |
| B11 4NJ | G0IVD |
| B12 9ER | G0NHP |
| B12 9JQ | G1YIL |
| B13 0BA | G6DSD |
| B13 0BH | G7UGC |
| B13 0BH | G8ENA |
| B13 0BH | G8GKR |
| B13 0DL | G4KZV |
| B13 0DL | G4OHM |
| B13 0EJ | G4PUD |
| B13 0JL | M3JPP |
| B13 0NR | G6DRN |
| B13 0PR | G4GIG |
| B13 0RQ | G4BJS |
| B13 8JZ | G0EBW |
| B13 8NZ | M0AEJ |
| B13 9AD | M6AKA |
| B13 9EN | G4DFO |
| B13 9QR | 2E0BTU |
| B13 9QR | M3YGR |
| B13 9TY | G3YHF |
| B14 4DS | G4ETZ |
| B14 4LS | G1POR |
| B14 4NX | M6GYH |
| B14 4TE | G8MYK |
| B14 4TG | G1LZH |
| B14 4TH | 2E0MEX |
| B14 4TH | M6HCS |
| B14 4TJ | 2E0MSE |
| B14 4TJ | M0MSE |
| B14 4TW | M3LTH |
| B14 5AF | M6RZA |
| B14 5BT | 2E0LPD |
| B14 5BT | M0WYH |
| B14 5BT | M6LRD |
| B14 5DS | 2E0NYC |
| B14 5DS | M0NYP |
| B14 5DS | M6STU |
| B14 5DU | M[illegible] |
| B14 6DU | M6WJC |
| B14 5EF | G3PQP |
| B14 5HD | G4MTG |
| B14 5HD | G8TXL |
| B14 5LX | M1CPC |
| B14 5QD | M6AOJ |
| B14 5SN | G7PIR |
| B14 5XX | 2E0SDD |
| B14 6AD | G8YQC |
| B14 6DE | G2BBC |
| B14 6DE | G3YXM |
| B14 6DE | M3YXM |
| B14 6HH | G4ZVS |
| B14 6LD | M6STJ |
| B14 6TN | G1FDD |
| B14 7AS | G4JGH |
| B14 7DB | M0DGQ |
| B14 7DB | M3ZCR |
| B15 2AA | G7ARP |
| B15 2DF | M6HSF |
| B15 2RU | M6SNZ |
| B15 2TT | G7AFQ |
| B15 2XA | G3UOA |
| B15 2XA | G8CQH |
| B15 3JB | G3NMW |
| B15 3RL | G4NMC |
| B16 0EF | G1JDE |
| B16 0EF | G6AUO |
| B16 0LL | G6ZDS |
| B16 9EY | G1XFL |
| B17 0AT | G3SJH |
| B17 0JX | 2E0GAZ |
| B17 0NW | G0KOI |
| B17 0QT | G4NZY |
| B17 8BN | G4HWN |
| B17 8BP | 2E0KUF |
| B17 8BP | M6DIV |
| B17 8PT | M0BVY |
| B17 8PY | G4KXV |
| B17 8QH | 2E1BHB |
| B17 8QH | G4XDM |
| B17 9EB | M0BZY |
| B17 9JU | G4LQF |
| B17 9JX | M6EFM |
| B17 9LE | G4YDO |
| B17 9QX | G1KUQ |
| B17 9RE | M5PIP |
| B17 9SD | G6PGO |
| B17 9TZ | 2E1DLA |
| B18 4NE | G1UBT |
| B20 1AJ | M3LIY |
| B20 1EQ | M3LTV |
| B20 2AQ | G6EEB |
| B20 2AS | M3OCQ |
| B20 2HN | G6KHN |
| B20 2NX | G7WKG |
| B21 8AU | 2E0HBD |
| B21 8AU | M6HDY |
| B21 9NU | G1AJK |
| B21 9QD | M3EWV |
| B23 5DF | 2E0FOX |
| B23 5DY | M0OSH |
| B23 5LE | G0PCP |
| B23 5RG | G7RYA |
| B23 5RH | G4ZSM |
| B23 5XE | G8LIX |
| B23 5XR | G8NZO |
| B23 6BT | G0RAR |
| B23 6NN | G6IHW |
| B23 6PJ | M3PKQ |
| B23 7DG | M6MVF |
| B23 7JJ | 2E0LPJ |
| B23 7JJ | M3XIP |
| B23 7JR | G0VBQ |
| B23 7LT | G6JAC |
| B23 7PB | G0LNE |
| B23 7XL | G7EKJ |
| B24 0SN | M0KRW |
| B24 0TQ | 2E0GDM |
| B24 0TQ | G4JDC |
| B24 8LX | G4ZCR |
| B24 9DN | G4CVW |
| B24 9LX | G6GBT |
| B24 9NE | M6CKT |
| B24 9NF | G8YJT |
| B24 9NF | M0YJT |
| B24 9RJ | M3UTV |
| B24 9RX | M6ZOG |
| B25 8EJ | 2E0DJV |
| B25 8EJ | M3ZQN |
| B25 8JF | G8AFI |
| B25 8LQ | G8XXZ |
| B25 8NZ | G7PSC |
| B25 8TJ | M6AUO |
| B25 8XS | M6GYK |
| B26 1DY | G3NXC |
| B26 1LD | M0OCH |
| B26 1PL | G8AYY |
| B26 1PR | G3KYE |
| B26 1TT | G6FGW |
| B26 2AF | G3JTT |
| B26 2AW | G1VSD |
| B26 [illegible] | G6KHM |
| B26 2HW | 2E0MCH |
| B26 2HW | M6MBF |
| B26 2NY | M6DEG |
| B26 2PP | M0GNP |
| B26 2PP | M0SAL |
| B26 2UL | G1CQF |
| B26 3BX | M0WSN |
| B26 3EL | G7HNR |
| B26 3HY | G8DHI |
| B26 3LA | 2E0EUN |
| B26 3LA | M0TAX |
| B26 3LS | G8KGS |
| B26 3PW | G4NPG |
| B26 3ST | G0LBQ |
| B27 6BB | G1OOW |
| B27 6EH | M6XBR |
| B27 6QD | G4EKQ |
| B27 6RL | 2E0AXB |
| B27 7AJ | 2E0YOM |
| B27 7AJ | M0YOM |
| B27 7AJ | M3YOM |
| B27 7HJ | G4JGV |
| B27 7LA | M6CYY |
| B27 7PB | 2E0DHZ |
| B27 7PB | M3UNH |
| B27 7RU | G6HFZ |
| B28 0DQ | M6CUP |
| B28 0HH | G1BLK |
| B28 0HR | M6EQS |
| B28 0LA | G8ADD |
| B28 0QX | G6KVR |
| B28 0TW | G3RGD |
| B28 0TZ | M6EXQ |
| B28 0UT | 2E0DTV |
| B28 0UT | M0REM |
| B28 8DX | M0PGD |
| B28 8EZ | G3VPE |
| B28 8LZ | G7IJI |
| B28 8PF | 2E0TBR |
| B28 8PF | M6AUL |
| B28 9EQ | G3KLD |
| B28 9NT | G1SWU |
| B29 4AG | G8GDZ |
| B29 4AH | G3ZKQ |
| B29 4LT | 2E1GYZ |
| B29 4LT | M1EBK |
| B29 4NH | G4KLA |
| B29 4RD | G7EJN |
| B29 5DG | 2E0MBR |
| B29 5DG | M0PTG |
| B29 5DG | M6FLE |
| B29 5DR | M3ONM |
| B29 5LE | G1LWH |
| B29 5NX | G0HDF |
| B29 6NH | M0JZT |
| B29 6PX | M0BCN |
| B29 6RP | M3LXV |
| B29 7QJ | G4RCZ |
| B29 7QQ | M6RDN |
| B29 7SG | G4NCY |
| B30 1DR | G4NPA |
| B30 1DR | G4NPB |
| B30 1LY | G8ORR |
| B30 1NG | G0BWP |
| B30 1PU | 2E0GAH |
| B30 1RF | M1EAN |
| B30 1RQ | G7BZM |
| B30 1TH | G8DOY |
| B30 1UZ | G6SRU |
| B30 2BA | G4YUI |
| B30 2DY | G0UYT |
| B30 2EB | M6HIP |
| B30 2LS | M3VCB |
| B30 2RB | M6TVI |
| B30 2SH | G4OMP |
| B30 2UY | M6FRF |
| B30 3BZ | G1EAX |
| B30 3QN | G0MQC |
| B30 3QR | M6SYG |
| B30 3RP | M6BJN |
| B31 %ny | M6HME |
| B31 1AL | G6VGT |
| B31 1DX | G4RPV |
| B31 1NE | G7DNX |
| B31 1UJ | M0SIR |
| B31 1UQ | 2E0DOQ |
| B31 1XH | 2E0AZL |
| B31 1XH | M3HTG |
| B31 2AX | G8XIZ |
| B31 2BJ | G8ZXY |
| B31 2EB | G1FVC |
| B31 2EE | M0RTE |
| B31 2EJ | G4YTO |
| B31 2FG | G0FPN |
| B31 2FS | G1YLQ |
| B31 2FS | G8VBE |
| B31 2FT | G8AHE |
| B31 2HD | G6IRJ |
| B31 2JG | G3XVW |
| B31 2LY | G1JRF |
| B31 2PP | G8VNL |
| B31 2QB | G8KAE |
| B31 2QZ | G2AZM |
| B31 2RP | M0VAT |
| B31 3BT | M3XOQ |
| B31 3HD | G[illegible]LKW |
| B31 3HU | 2E0PGT |
| B31 3LA | M6GWG |
| B31 3NL | M6FRP |
| B31 3QY | G1JON |
| B31 3QY | G3OHM |
| B31 3QY | G8SH |
| B31 3ST | G1HOL |
| B31 3SY | M3XHM |
| B31 3XE | M6FBO |
| B31 4AJ | 2E0PVN |
| B31 4AJ | M0PVN |
| B31 4AJ | M6PVN |
| B31 4AU | G7FUW |
| B31 4BS | M0BAZ |
| B31 4ES | G0JGH |
| B31 4HD | G4XLO |
| B31 4LJ | 2E0JWY |
| B31 4LJ | M0WBY |
| B31 4LJ | M6JWY |
| B31 4LN | M0ALH |
| B31 4SS | G6TDG |
| B31 4SU | 2E0INC |
| B31 4SU | M0LWT |
| B31 4SU | M6ERA |
| B31 4TA | M3UTM |
| B31 4TQ | G8MIT |
| B31 5AE | M6DJW |
| B31 5AU | 2E1BHF |
| B31 5DP | 2E1FZY |
| B31 5DP | M0BSP |
| B31 5DP | M0BSQ |
| B31 5HH | 2E0UGO |
| B31 5HH | M3UGO |
| B31 5HT | 2E0GCB |
| B31 5HT | M3GKH |
| B31 5HT | M3HIO |
| B31 5JB | M6UEB |
| B31 5NT | M3HNE |
| B31 5NY | 2E0REN |
| B31 5QB | 2E0NCO |
| B31 5QB | M6NCO |
| B31 5RD | G6KAE |
| B32 1AY | M6ZBL |
| B32 1DR | G4SWA |
| B32 1EG | G4XUA |
| B32 1EL | G4CGU |
| B32 1HE | G4XEJ |
| B32 1JA | 2E0KEG |
| B32 1LB | G1MAR |
| B32 1LB | G3MAR |
| B32 1LB | G8BHE |
| B32 1LB | G8OHM |
| B32 1NB | G4LLW |
| B32 1PG | 2E1AIT |
| B32 1PG | G0PWQ |
| B32 1QT | G0LGO |
| B32 2AA | G7TPB |
| B32 2BA | G4JCZ |
| B32 2BT | M6OOO |
| B32 2LY | M6HSQ |
| B32 2LZ | 2E0PPH |
| B32 2NW | G0PPX |
| B32 2SB | G1NHX |
| B32 2TT | G4ZTM |
| B32 2UX | G4CKK |
| B32 3AL | G1BXL |
| B32 3EJ | M3UZV |
| B32 3NL | 2E0CLI |
| B32 3NL | M6EYH |
| B32 3TA | G1PKG |
| B32 4LW | 2E0DLF |
| B32 4LW | M6EPF |
| B33 0NR | G4SAS |
| B33 0RL | G7DQQ |
| B33 0YL | G6BNO |
| B33 0YP | G4FGF |
| B33 8AB | M0DZO |
| B33 8JP | 2E0MJX |
| B33 8SL | 2E0UOK |
| B33 8SL | M3UOK |
| B33 9BU | G7FPZ |
| B33 9DL | G8KCB |
| B33 9NZ | G6SFW |
| B34 6JL | G4LTT |
| B34 6PZ | G6NOW |
| B34 6TL | G6FKW |
| B34 7AY | 2E1SJG |
| B34 7AY | M0CDQ |
| B34 7AY | M6LCV |
| B34 7BU | M0BEM |
| B34 7LP | G4ZPI |
| B34 7QX | G0EOS |
| B34 7RS | G4LAJ |
| B34 7RU | 2E0VJX |
| B34 7RU | M0VJX |
| B34 7RU | M6VJX |
| B35 7PD | G8TBW |
| B35 7PF | M3VVH |
| B36 0AD | G6WYQ |
| B36 0BX | G6VMR |
| B36 0EH | M1NPH |
| B36 0HH | M3VVA |
| B36 0HS | G0UUG |
| B36 0JT | 2E0DEZ |
| B36 0LQ | G0OKU |
| B36 0PH | [illegible] |
| B36 0RT | G0ELJ |
| B36 0UH | G7UEJ |
| B36 8RW | G7LKI |
| B36 8HA | G0OTF |
| B36 8QG | G3NQA |
| B36 9DY | M3VVB |
| B36 9HX | G4SGA |
| B36 9JD | G4IMB |
| B36 9JD | G8PXU |
| B36 9JD | M3SZY |
| B36 9LL | G6VXL |
| B36 9LL | G6YCI |
| B36 9LL | G7OZE |
| B36 9LL | M3OZE |

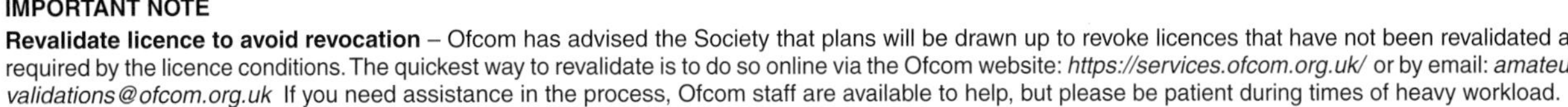

**IMPORTANT NOTE**

**Revalidate licence to avoid revocation** – Ofcom has advised the Society that plans will be drawn up to revoke licences that have not been revalidated as required by the licence conditions. The quickest way to revalidate is to do so online via the Ofcom website: *https://services.ofcom.org.uk/* or by email: *amateur.validations@ofcom.org.uk* If you need assistance in the process, Ofcom staff are available to help, but please be patient during times of heavy workload.

| Postcode | Callsign |
|---|---|
| B36 9SN | G3DID |
| B36 9ST | G4LRN |
| B36 9TB | G4HAI |
| B36 9TB | G8ITG |
| B36 9TS | M3JYA |
| B36 9TW | G1GFW |
| B36 9TY | G4XPV |
| B37 5HS | M3CIP |
| B37 5HX | G0FCQ |
| B37 6DL | G4SPY |
| B37 6SB | G6APQ |
| B37 6TN | M3LDT |
| B37 7EE | G4HPT |
| B37 7HS | G4TDF |
| B37 7JJ | G0MLH |
| B37 7PA | M3XJP |
| B37 7RD | G1PRF |
| B38 0AB | G1BUJ |
| B38 0AL | M0BTO |
| B38 0DN | G3YKO |
| B38 0EP | G0MTN |
| B38 0EP | G0WRC |
| B38 0EP | G7WAC |
| B38 8AJ | G1UOD |
| B38 8DA | G8NLK |
| B38 8DT | G4AEG |
| B38 8LB | G0MKU |
| B38 8LS | G1XPD |
| B38 8PH | G4KRT |
| B38 8PW | G3TGL |
| B38 8PW | G8XUN |
| B38 8TH | G0AHC |
| B38 9AG | 2E0SRV |
| B38 9AG | M0SSV |
| B38 9AG | M6SNN |
| B38 9AG | M6SRV |
| B38 9HS | M6SYH |
| B38 9LA | G0BOT |
| B38 9LW | M6DCW |
| B38 9PA | G0WYT |
| B38 9QR | M6MNZ |
| B38 9QT | G1MZM |
| B38 9RY | G0OKI |
| B38 9UW | M3VXB |
| B42 1EU | G7TVL |
| B42 1HF | G0FBQ |
| B42 1HG | M0NKA |
| B42 1LP | G1HUM |
| B42 1LY | G4WSI |
| B42 1LY | G6HPE |
| B42 1PL | G0DOG |
| B42 1PZ | G6UGA |
| B42 1RT | 2E0SWS |
| B42 1RT | M3NUI |
| B42 1RY | G1EDT |
| B42 2BX | G1MOV |
| B42 2DT | G4NTV |
| B42 2EA | M3EYK |
| B42 2EH | M3PHX |
| B42 2HF | 2E0WPS |
| B42 2HF | M0WPS |
| B42 2HF | M6WPS |
| B42 2HJ | G0GPF |
| B42 2HT | G8ASW |
| B42 2JA | M3TET |
| B42 2JW | G8SFQ |
| B42 2LX | M0GYA |
| B42 2NU | G1LTE |
| B42 2PQ | G6NHW |
| B42 2QB | G6UGZ |
| B42 2QJ | G4HOM |
| B42 2RL | G6YQW |
| B42 2SQ | G4NBW |
| B43 5AG | M6KDF |
| B43 5EH | M0ANK |
| B43 5HH | 2E1ILM |
| B43 5JR | G7IKG |
| B43 5LN | M3OHX |
| B43 5LY | G0NOV |
| B43 5ND | 2E0XAW |
| B43 5ND | M6ARW |
| B43 5PG | M6GGX |
| B43 6BB | G3FIA |
| B43 6HX | G0TEM |
| B43 6JR | G1MRP |
| B43 6NA | M3EMO |
| B43 6QE | G6YSB |
| B43 7HG | G6LPB |
| B43 7HX | 2E0NVP |
| B43 7HX | M3NVP |
| B43 7JW | M3NVR |
| B43 7PG | G4PFK |
| B43 7PQ | 2E0CVU |
| B43 7PQ | M0TVU |
| B44 0AL | G6WOI |
| B44 0AY | 2E0VMV |
| B44 0AY | M0VMV |
| B44 0AY | M3VMV |
| B44 0LB | G1BJK |
| B44 0LF | G0FOC |
| B44 0NF | G7PZQ |
| B44 0NF | M3WMU |
| B44 0SN | M0WET |
| B44 8AB | M6KTX |
| B44 8EN | M1CSC |
| B44 8ER | G7CWO |
| B44 8JB | G4DIP |
| B44 8LQ | G6WIG |
| B44 8RG | 2E0ELD |
| B44 8RG | M3ELD |
| B44 8RL | G0WMU |
| B44 8RS | G7OPD |
| B44 8SW | 2E0GOM |
| B44 8SW | G8CHA |
| B44 9BY | G4RZD |
| B44 9NY | G8TZW |
| B44 9QH | G7ORT |
| B44 9RP | G6NZW |
| B44 9RR | G3URV |
| B44 9SS | G4BTK |
| B45 0DA | 2E1IAT |
| B45 0EN | 2E0SVZ |
| B45 0EN | M6ZDL |
| B45 0JB | G8VXY |
| B45 0JS | M6LDU |
| B45 0JY | G0HPG |
| B45 0JY | G0HPH |
| B45 0JY | M6BNP |
| B45 0JY | M6SXD |
| B45 0NB | G6ZSG |
| B45 8EH | G0NJT |
| B45 8GU | 2E0AAI |
| B45 8GU | M6KND |
| B45 8HP | G4TSB |
| B45 8NG | M0VGG |
| B45 8NZ | G7UMY |
| B45 8QU | 2E0VUK |
| B45 8QU | M3VUK |
| B45 8SJ | M3JYE |
| B45 8TQ | 2E1ANG |
| B45 8TQ | 2E1ANH |
| B45 9DA | G1XKY |
| B45 9EU | 2E0AQE |
| B45 9EU | M0JTB |
| B45 9HW | G6YXW |
| B45 9HY | G7RLV |
| B45 9LR | G4YTJ |
| B45 9LW | M0BYJ |
| B45 9RU | M3KCQ |
| B45 9SZ | G0HVN |
| B45 9SZ | G3IUB |
| B45 9SZ | G8IUB |
| B45 9TT | M6FXJ |
| B45 9UH | M6SQI |
| B45 9WA | M6ILR |
| B45 9XB | G8IFT |
| B45 9XB | M6GKG |
| B46 1EP | G4OPE |
| B46 1EP | G4XIQ |
| B46 1HL | G6NTY |
| B46 1PD | G3UPA |
| B46 1RU | G8MWE |
| B46 1SA | 2E0MEY |
| B46 1SA | M6CKR |
| B46 1SN | G0CKE |
| B46 1SN | G3ZUM |
| B46 1TW | G6VNC |
| B46 1UF | G1KAT |
| B46 3EH | G0BUV |
| B46 3LZ | G4YQD |
| B46 3NE | G4FTY |
| B47 5DX | 2E0KHG |
| B47 5DX | M6KAO |
| B47 5EN | G3OIC |
| B47 5HJ | G8TWZ |
| B47 5HY | G4VPD |
| B47 5NR | 2E0HPD |
| B47 5NR | M0IFT |
| B47 5NR | M6HPD |
| B47 5PX | G6JYO |
| B47 5QE | G0NES |
| B47 6HP | G1PYJ |
| B47 6HP | G4OJL |
| B47 6LX | G0ICJ |
| B47 6LX | G1WAC |
| B47 6LX | G4WAC |
| B48 7NE | G1PHN |
| B48 7TB | G0BNZ |
| B48 7TB | G4SVL |
| B48 7TL | G8RIM |
| B49 5DD | G1YFD |
| B49 5DD | G4GYI |
| B49 5EG | 2E0BOT |
| B49 5EG | M0OJG |
| B49 5HA | G0UGY |
| B49 5HY | G6MOZ |
| B49 5LJ | G0UMV |
| B49 5LJ | G3UMV |
| B49 5LJ | G4UMV |
| B49 5LJ | G6LRT |
| B49 6AP | M1EZD |
| B49 6DR | G1IKL |
| B49 6HQ | G3USA |
| B49 6LF | G3EVT |
| B49 6LF | G4ACZ |
| B49 6LX | G1WLU |
| B49 6LX | G1ZUU |
| B49 6QY | G6BAY |
| B5 7NE | G6DWS |
| B50 4AN | G4OHJ |
| B50 4AN | M3MEB |
| B50 4AP | G0VBZ |
| B50 4AR | G3VRF |
| B50 4NP | 2E1MGB |
| B50 4NP | M0XLX |
| B50 4NX | M0BHM |
| B50 4QE | M0MWK |
| B6 4PP | M6MRX |
| B6 5HW | 2E0DHY |
| B60 1AL | G1BIU |
| B60 1AW | G1GDR |
| B60 1BH | G3MWQ |
| B60 1BN | M0FRA |
| B60 1BP | G1FPY |
| B60 1DG | G8DEC |
| B60 1DS | G4LVK |
| B60 1DY | G4OJS |
| B60 1DY | G4WZA |
| B60 1DZ | G6NQB |
| B60 1HE | G0KHK |
| B60 1HN | G4BBU |
| B60 1HN | M3RXP |
| B60 1HW | G4AHK |
| B60 2DB | 2E1GNE |
| B60 2EB | G0BLT |
| B60 2EB | G0HAW |
| B60 2LH | M0WOD |
| B60 2LJ | G4TJS |
| B60 3AY | G0EVY |
| B60 3EP | G4UIW |
| B60 3GP | M6OBK |
| B60 3HB | G0BIR |
| B60 3HB | G0EHQ |
| B60 4EB | 2E1GXI |
| B60 4EB | G1TCK |
| B60 4EB | G6NYG |
| B60 4EB | M3CBC |
| B60 4LH | M6NAN |
| B60 4LS | G6OPY |
| B60 4NF | G4FNQ |
| B61 0DX | G6OIF |
| B61 0EL | G4AHO |
| B61 0ER | G6EET |
| B61 0ER | M1CLZ |
| B61 0JP | M6SOZ |
| B61 0JT | G4LRL |
| B61 0LG | M1ESD |
| B61 0LQ | G0AQF |
| B61 0LU | G4NZK |
| B61 0PA | G8XUW |
| B61 0PB | G0NBI |
| B61 0PB | G2FKZ |
| B61 0PB | M0RSE |
| B61 0PB | M6VAJ |
| B61 0SE | G1LMU |
| B61 0TT | M3KGO |
| B61 7BE | G1IEC |
| B61 7DA | M6KLR |
| B61 7EB | G4CQS |
| B61 7EB | G8GEB |
| B61 7JG | G0BGA |
| B61 7PR | M0LLS |
| B61 8HY | M3NFL |
| B61 8HY | M3SKU |
| B61 8NQ | G0SUQ |
| B61 8PE | G4OTC |
| B61 8PN | 2E0MJT |
| B61 8PN | M3UHW |
| B61 8UA | G4PKW |
| B61 9BH | 2E0MIS |
| B61 9BH | M0LIJ |
| B61 9BH | M1JSS |
| B61 9BH | M6UDY |
| B61 9JN | M3XNE |
| B61 9JT | G4CAF |
| B61 9JW | M3AXT |
| B61 9LH | M3OJS |
| B62 0HU | G1WTH |
| B62 0HW | 2E0UAE |
| B62 0HW | M3ULS |
| B62 0JL | M0JFW |
| B62 0LN | G4OHB |
| B62 0NE | G0EVM |
| B62 8EU | G7DIZ |
| B62 8EU | G7IIS |
| B62 8EZ | M0VCR |
| B62 8JS | G2FXZ |
| B62 8JS | G6GQJ |
| B62 8LJ | M3SAR |
| B62 8LJ | M6IKA |
| B62 8LJ | M6MDX |
| B62 8LJ | M6SAD |
| B62 8LR | 2E0KPP |
| B62 8LU | M6EFJ |
| B62 8NQ | M3YWM |
| B62 8SH | G6IJQ |
| B62 8TB | G4YAH |
| B62 8TH | G1IAL |
| B62 9AW | G3VBV |
| B62 9DX | 2E0DSS |
| B62 9DX | M6DSS |
| B62 9HJ | M3CPY |
| B62 9NQ | G4VPZ |
| B62 9NR | G4DPZ |
| B62 9PP | G1EBB |
| B62 9QJ | 2E0MKB |
| B62 9QJ | M0FCW |
| B62 9QJ | M0KMW |
| B62 9QW | G4JSV |
| B62 9RF | G4PVP |
| B62 9TF | 2E0SLJ |
| B63 1BB | G1FET |
| B63 1BZ | G4OQX |
| B63 1DU | G8TXA |
| B63 1EQ | 2E1WIN |
| B63 1EQ | G1PPZ |
| B63 1HD | M6RKB |
| B63 1JQ | G4LWF |
| B63 1JQ | G4MEB |
| B63 1JY | G4AYK |
| B63 1JY | G4NQW |
| B63 2AY | G7VJT |
| B63 2BD | G7VIE |
| B63 2DW | G6XYO |
| B63 2JA | G7FIJ |
| B63 2JJ | G0GUG |
| B63 2PP | G4EHR |
| B63 2PR | M3XMJ |
| B63 2PY | G1VQV |
| B63 2SY | G8TIA |
| B63 2SZ | G8ZPD |
| B63 2SZ | M0ZPD |
| B63 2TB | G6LDA |
| B63 2UY | G1KII |
| B63 2XH | M0YIM |
| B63 3DN | G8XSA |
| B63 3JX | G4UMY |
| B63 3QR | G0OUJ |
| B63 3QR | G4FAV |
| B63 3QR | M3EFL |
| B63 3RZ | M3FFA |
| B63 3RZ | M3FFE |
| B63 3ST | M6EUX |
| B63 3TJ | G1BEG |
| B63 3TP | 2E0CUY |
| B63 3TP | 2E0JTM |
| B63 3TP | G7JMZ |
| B63 3TP | M6EDA |
| B63 3TP | M6JPD |
| B63 3TP | M6ZSB |
| B63 3TZ | M6ARK |
| B63 4BX | M6EMS |
| B63 4HD | M0CTK |
| B63 4HG | G6ALW |
| B63 4HQ | G1VOB |
| B63 4HQ | G3DTU |
| B63 4PB | G0NLA |
| B63 4QG | G1BRF |
| B63 4RN | M1BZG |
| B63 4TJ | M1EVF |
| B63 4XN | G7TUV |
| B64 5LA | G8MUV |
| B64 6DU | G0PPJ |
| B64 6EA | M0EHS |
| B64 6QX | G0BHR |
| B64 6RB | G4JFF |
| B64 7EZ | G3VPX |
| B64 7EZ | G7BRA |
| B64 7HH | 2E0FGT |
| B64 7HJ | G6NNO |
| B64 7LE | G4ISQ |
| B65 0AG | M6NEL |
| B65 0EZ | M6CPJ |
| B65 0HF | G6LOR |
| B65 0LY | M3FGU |
| B65 0NP | G0WYA |
| B65 0QE | M1FFG |
| B65 0QE | M3FXU |
| B65 0RL | G0KNM |
| B65 0RL | G1SHQ |
| B65 0RR | 2E0JAM |
| B65 0RR | 2E0LEE |
| B65 0RR | 2E1HXM |
| B65 0RR | 2E1HXN |
| B65 8DT | G4YIV |
| B65 8DW | M3VBI |
| B65 8DW | M3VRP |
| B65 8HT | 2E0KAS |
| B65 8HT | M0KAU |
| B65 8HT | M6KAN |
| B65 8HU | M6NDM |
| B65 8NX | M0GSX |
| B65 8PB | M6OAQ |
| B65 9BQ | G1OOZ |
| B65 9DZ | G0PAI |
| B65 9DZ | G4WBC |
| B65 9DZ | M3XNK |
| B65 9HZ | G0OYF |
| B65 9JN | G0KJG |
| B65 9JX | 2E0HAZ |
| B65 9JX | M3GHH |
| B65 9LG | G0MJT |
| B65 9NJ | M0IMM |
| B65 9NT | G3TRG |
| B65 9NT | G8TRG |
| B65 9SD | G4FJJ |
| B67 5AY | G1JVM |
| B67 5AY | M0JVM |
| B67 5DD | G0CDP |
| B67 5DH | G1UCO |
| B67 5PD | G7UNB |
| B67 6DA | M6DKK |
| B67 6HU | M0BBW |
| B67 6LA | G1UUZ |
| B67 6PR | G1FFO |
| B67 6QS | M3JJH |
| B67 6QX | G0TVR |
| B67 6QX | M3DVU |
| B67 7BX | G4KVC |
| B68 0BJ | M0MEL |
| B68 0NA | G8DEM |
| B68 0NE | G4TCC |
| B68 0NU | G8PTF |
| B68 0PU | 2E0PYA |
| B68 0PU | G4RWY |
| B68 0PU | M0PYA |
| B68 0SW | 2E0XYX |
| B68 0SW | M6TPA |
| B68 8AQ | G4SFG |
| B68 8BE | M1AED |
| B68 8HY | M3EFQ |
| B68 8LT | G4OJJ |
| B68 8LT | M3TOF |
| B68 8NG | G0UFJ |
| B68 8PP | M1WEH |
| B68 8PR | G4OYT |
| B68 8PT | G2BXP |
| B68 8QH | G3NAI |
| B68 9DP | G1SAN |
| B68 9DP | G1XFO |
| B68 9DP | G6IQY |
| B68 9DU | M3XVZ |
| B68 9ES | G8MKE |
| B68 9LU | G4VRX |
| B68 9PW | G0NNF |
| B68 9PW | G8RAO |
| B68 9TB | G7RTQ |
| B68 9UL | G0WXA |
| B69 1BA | 2E0TAM |
| B69 1BA | G0NJZ |
| B69 1BA | G7IYX |
| B69 1BA | M6NJZ |
| B69 1BU | G7CBW |
| B69 1JU | 2E0TAT |
| B69 1NP | G4JVH |
| B69 1NP | G8FCO |
| B69 1NT | G3NZS |
| B69 1PW | G4YFT |
| B69 1QA | G1WHY |
| B69 1SE | G6EOR |
| B69 1SW | 2E0SOX |
| B69 1TP | G8SPU |
| B69 1UD | G8TFB |
| B69 2DZ | 2E0VGC |
| B69 2DZ | M6DKT |
| B69 4QA | G7LOE |
| B69 6RN | G3VJX |
| B70 0SL | G4DVM |
| B70 6DX | 2E0WBZ |
| B70 6DX | M6WBZ |
| B70 6RQ | 2E1LJW |
| B70 6RQ | M6LRU |
| B70 7ES | 2E0TVV |
| B70 7ES | M6TVV |
| B70 7EW | G6ZYZ |
| B70 7EW | G7SRG |
| B70 7LS | G1TBK |
| B70 8JY | 2E0KVE |
| B70 8JY | M3GUM |
| B70 8LD | G0BZP |
| B70 8PL | M3LMD |
| B70 8QR | M3LLC |
| B70 9ES | 2E0ZGL |
| B70 9ES | M3NUE |
| B71 1DQ | M6BSI |
| B71 1DX | M3DIU |
| B71 1NJ | G0WCK |
| B71 1RU | G4PMW |
| B71 1RU | G6FPN |
| B71 2AA | 2E0SJS |
| B71 2AA | M3MFS |
| B71 2DY | G7RCP |
| B71 2DY | M3APE |
| B71 2PB | M6ZGY |
| B71 2QJ | 2E0BMU |
| B71 2QU | 2E0SWB |
| B71 2QW | 2E0RWN |
| B71 2QW | G4LWN |
| B71 2QW | G8UID |
| B71 2QW | M0RWN |
| B71 2QW | M3RWN |
| B71 3BT | G6VFO |
| B71 3DA | M6EED |
| B71 3EE | M1DMT |
| B71 3HX | G6MPT |
| B71 3LA | 2E0LIT |
| B71 3LA | 2E0XAV |
| B71 3LA | M0LIT |
| B71 3LA | M6GYM |
| B71 3LX | G0RCX |
| B71 3NE | 2E1IHE |
| B71 3NF | G0HKK |
| B71 4BA | M6KNY |
| B71 4HN | 2E0BCD |
| B71 4LQ | M3OFU |
| B71 4LR | M1GAP |
| B71 4LR | M3TYZ |
| B72 1AG | G1AAL |
| B72 1HB | G7ETS |
| B72 1HP | G3PLP |
| B72 1JP | G0UQJ |
| B72 1JU | G1NFN |
| B72 1YE | G4TLR |
| B72 1YF | G0VXK |
| B72 1YZ | G8NFD |
| B73 5AR | G7FBY |
| B73 5EA | G4TYR |
| B73 5EH | M3FSU |
| B73 5EL | G8EZB |
| B73 5JY | G1KYK |
| B73 5LD | M3TGE |
| B73 5LF | G0VZL |
| B73 5LT | G0GEP |
| B73 5LT | G1LTG |
| B73 5SP | G7EKW |
| B73 6NZ | G4OFN |
| B73 6NZ | G7ORV |
| B73 6PG | G8AMD |
| B73 6QA | G4LBT |
| B73 6QR | G4NBI |
| B73 6UQ | G1NPA |
| B74 2BS | G6HOC |
| B74 2BU | M3IZN |
| B74 2DA | 2E1EQE |
| B74 2DG | M0CUD |
| B74 2EA | G7JJW |
| B74 2JE | G8KFF |
| B74 2LA | G4DDD |
| B74 2PS | G3MCB |
| B74 2QA | G3JZF |
| B74 2TB | G4ABW |
| B74 3AA | G3RDW |
| B74 3HF | G8YNI |
| B74 3JU | G6VPL |
| B74 3LR | G6RBO |
| B74 3NP | G1MGZ |
| B74 3NP | G1OBA |
| B74 3PG | G3XFN |
| B74 3PG | G4IWF |
| B74 3PQ | G7OBC |
| B74 3QE | G0PJG |
| B74 3QE | M0AZK |
| B74 3QP | G1EYT |
| B74 3RF | 2E0KBF |
| B74 3RF | M6KBF |
| B74 3TU | G1XDV |
| B74 3UD | M3XCN |
| B74 4BL | M3YOH |
| B74 4DQ | G1AZE |
| B74 4DY | G6REY |
| B74 4HL | G8EFU |
| B74 4HL | M0DBM |
| B74 4JF | G4YZO |
| B74 4JS | G3WWL |
| B74 4LU | G4ORY |
| B74 4SG | G8XMH |
| B74 4SJ | G7NJW |
| B74 4UG | 2E0MAJ |
| B74 4UG | M0MAJ |
| B74 4UG | M3OFJ |
| B74 4XG | G8OHS |
| B74 4XR | G1JXX |
| B74 4YD | G1GFA |
| B74 4YD | G7SER |
| B75 5AQ | M0AZE |
| B75 5LD | G1NSG |
| B75 5LH | G0KLK |
| B75 5PQ | G6VIY |
| B75 5TJ | G7VBJ |
| B75 6AU | G0EVH |
| B75 6AX | M3FQX |
| B75 6DB | G6HNS |
| B75 6DG | 2E0RFQ |
| B75 6DG | M3RFQ |
| B75 6DH | G7UCG |
| B75 6DW | G3AVE |
| B75 6EN | 2E0GTL |
| B75 6EN | G1AZZ |
| B75 6EN | G3MYC |
| B75 6EN | M3PNF |
| B75 6SN | G8TBB |
| B75 7AA | G6AGO |
| B75 7AA | G6DGR |
| B75 7BL | G6BBR |
| B75 7LQ | M6FOT |
| B75 7ND | G1VVF |
| B75 7TH | G6HSR |
| B75 7UU | M6LZY |
| B76 1EA | G7IXH |
| B76 1FN | M3WAP |
| B76 1HU | G8VUU |
| B76 1HY | M3HOD |
| B76 1JQ | G1BUQ |
| B76 1JR | G1XKN |
| B76 1LZ | G8IOS |
| B76 1PJ | G6KPX |
| B76 1QZ | G8ERN |
| B76 1XR | G6UED |
| B76 1YD | G0HID |
| B76 1YE | 2E0CCB |
| B76 1YE | M6TYE |
| B76 1YR | 2E1DSU |
| B76 2PT | G3VNY |
| B76 2QH | M1EGX |
| B76 2RP | G1OKK |
| B76 2SY | G1FQX |
| B76 2TG | M6MKH |
| B76 9AP | G0GOD |
| B77 1AB | 2E0HZS |
| B77 1BT | G0WKI |
| B77 1BY | G6EOO |
| B77 1DF | M6BTN |
| B77 1HB | G4GYA |
| B77 1JD | G4MFN |
| B77 1NY | 2E0CCC |
| B77 1NY | M3FEG |
| B77 1PE | G1MGN |
| B77 1QR | G4ORW |
| B77 1QT | G6RZY |
| B77 2BE | G4JAV |
| B77 2EG | M6IZY |
| B77 2HQ | G3YUX |
| B77 2LD | G0GUD |
| B77 2LP | M6FAH |
| B77 2NA | G0PFQ |
| B77 2RH | G8MKT |
| B77 2RS | M3JWN |
| B77 2RY | G8SCG |
| B77 2TZ | 2E0JDO |
| B77 3BH | 2E0DPS |
| B77 3BT | G4LZV |
| B77 3BT | G4NWM |
| B77 3BT | G4NWN |
| B77 3BT | G8UXX |
| B77 3DG | M6TFW |
| B77 3HZ | 2E0NTA |
| B77 3HZ | M0XAL |
| B77 3HZ | M6NTA |
| B77 3JH | G4NWO |
| B77 3JW | M5ACJ |
| B77 3JZ | G8RFL |
| B77 3LA | G6LPX |
| B77 3LH | M1FEQ |
| B77 3NB | M3FLB |
| B77 3PE | 2E0BME |
| B77 3PE | M3UWV |
| B77 3PP | M1FER |
| B77 3PP | M1FES |
| B77 3PR | M6AQA |
| B77 3QG | G7HJG |
| B77 3QG | M1CVH |
| B77 4AA | G4INA |
| B77 4AQ | G0SKK |
| B77 4BZ | G0LHR |
| B77 4DF | M6APU |
| B77 4DL | G6NHG |
| B77 4EJ | G0KFF |
| B77 4EJ | G4ZPJ |
| B77 4EP | G4AUS |
| B77 4EP | G6ZCI |
| B77 4EU | 2E0JRW |
| B77 4EU | G0LTR |
| B77 4EU | M0WHR |
| B77 4EU | M3WHR |
| B77 4EZ | M3FLZ |
| B77 4HT | G0IKQ |
| B77 4JJ | 2E0LCR |
| B77 4JJ | M6ION |
| B77 4JL | G0TRB |
| B77 4JW | M3ASN |
| B77 4JZ | M6HMS |
| B77 4LD | M6SUX |
| B77 4LU | M3RLT |
| B77 4NA | G1UZW |
| B77 4NA | M3UZW |
| B77 4NA | M3YII |
| B77 4QY | 2E0EHB |
| B77 4QY | G3YTT |
| B77 4QY | M0NPL |
| B77 4QY | M6EHB |
| B77 5BB | G0VTU |
| B77 5EY | G0FXL |
| B77 5GG | M6DSO |
| B77 5JD | 2E1FDY |
| B77 5JE | 2E0ZAP |
| B77 5JE | M3RJB |
| B77 5LD | G0HAY |
| B77 5PQ | G8VPH |
| B77 5QE | M6GCJ |
| B77 5QF | G6MOD |
| B77 5TD | G6FBH |
| B78 1BQ | M6OYF |
| B78 1DA | M6TWD |
| B78 1DE | G7ODM |
| B78 1JS | G0FFA |
| B78 1JY | M3ZQG |
| B78 1LS | 2E0LLE |
| B78 1LS | M6CKO |
| B78 1LW | M6AMG |
| B78 1NJ | G8NAP |
| B78 1NU | G6DYR |
| B78 1QZ | G7UKF |
| B78 1RN | G6MUW |
| B78 1SY | G2BZR |
| B78 2AW | G8ACA |
| B78 2EP | G4NRY |
| B78 2ER | 2E0JBM |
| B78 2ER | M1DNE |
| B78 2ER | M3SWV |
| B78 2ER | M6CSM |
| B78 2ET | M6FAJ |
| B78 2JR | M3OHJ |
| B78 2JU | G0FEO |
| B78 2LA | G0GEF |
| B78 2LG | 2E0GEV |
| B78 2LG | M3YFH |
| B78 2LX | G6VWI |
| B78 3DE | G3LGW |
| B78 3HQ | G8MCT |
| B78 3HS | M6DEJ |
| B78 3HW | G8OSX |
| B78 3NW | M6CRJ |
| B78 3QD | G0NXY |
| B78 3RA | G4NRX |
| B78 3SW | G7UHG |
| B78 3SZ | M1APQ |
| B78 3TJ | G4ICI |
| B78 3YA | G4SBS |
| B79 0DJ | G6VBD |
| B79 0HR | M6JMN |
| B79 0JR | M6JKT |
| B79 0LD | M1EDF |
| B79 7BE | G0BFC |
| B79 7BQ | G0FGK |
| B79 7SQ | M6EEP |
| B79 7UU | M6IWP |
| B79 8BE | M6NIN |
| B79 8BZ | 2E1SHE |
| B79 8DE | G8KRV |
| B79 8DN | M6LFD |
| B79 8EJ | G8NCK |
| B79 8EY | 2E0TLM |
| B79 8EY | 2E1JMW |
| B79 8EY | M0TLM |
| B79 8EY | M6TLM |
| B79 8JB | G6DIF |
| B79 8JB | M3SQU |
| B79 8NB | M3SQV |
| B79 8NF | 2E0FAJ |
| B79 8PE | 2E0TSD |
| B79 8PE | M0TSD |
| B79 8QB | G7WFK |
| B79 8UA | M3SUY |
| B79 9BP | G3ENO |
| B79 9HP | G8RVZ |
| B79 9JA | 2E0CLP |
| B79 9JA | M6BHB |
| B79 9JJ | M6GJP |
| B8 1PS | M0HMZ |
| B8 2AU | G3RSC |
| B8 2AU | M1CQN |
| B8 2AU | M3ASZ |
| B8 2AU | M3BJV |
| B8 2AU | M3LMA |
| B8 2AU | M3SVN |
| B8 2EA | G7GGT |
| B8 2LB | M6TOL |
| B8 2PD | G0NFR |
| B8 3LL | M3AOM |
| B8 3LL | M3REZ |
| B80 7BZ | M3RKE |
| B80 7HD | G1YFA |
| B80 7HN | G4YJC |
| B80 7JJ | G7APL |
| B80 7LX | G4LMF |
| B80 7PG | G4STE |
| B80 7PL | M3VDL |
| B80 7RD | G6FDO |
| B80 7RR | G3CON |
| B80 7SH | G0WDU |
| B9 5NG | G7LTG |
| B9 5RY | G7LXV |
| B90 1DS | G7DDN |
| B90 1DS | M6FAB |
| B90 1LF | G4AQJ |
| B90 1RW | G7JHX |
| B90 2BB | G3XBY |
| B90 2BQ | G7GFP |
| B90 2BU | M6FMI |
| B90 2DR | G7WBJ |
| B90 2EJ | G1MJO |
| B90 2HB | G4ZVZ |
| B90 2HS | M6HVF |
| B90 2HY | G4RTI |
| B90 2LN | G7IMQ |
| B90 2PR | G7IMR |
| B90 2PU | M3UKX |
| B90 2QF | M0GWM |
| B90 2QW | G6AQW |
| B90 2TA | M0DOH |
| B90 3DF | G0IHU |
| B90 3DQ | M3IBS |
| B90 3HX | G4KOR |
| B90 3JE | M1DJG |
| B90 3JE | M3DJG |
| B90 3JF | M3FAE |
| B90 3JR | G8ZUL |
| B90 3JZ | G6FIO |
| B90 3LG | G4TBJ |
| B90 3LJ | 2E0DBL |
| B90 3LJ | M0PYT |
| B90 3LJ | M6GAL |
| B90 3PL | G8XQA |
| B90 3RE | G6IHB |
| B90 3SA | G3MRZ |
| B90 3SA | G8SAN |
| B90 4BU | G6BWT |
| B90 4BU | M3DOX |
| B90 4BX | G1ZQE |
| B90 4PH | G4BQW |
| B90 4PH | M6AXW |
| B90 4PN | G6HNR |
| B90 4QR | G0NFZ |
| B90 4RN | 2E0UDM |
| B90 4RN | M6UDM |
| B90 4RU | G6UUR |
| B91 1BS | G6VUJ |
| B91 1DD | G4MPG |
| B91 1DQ | G7KZV |
| B91 1DQ | G7MAB |
| B91 1DX | G4AMI |
| B91 1DY | M3ZAR |
| B91 1LN | 2E1GYC |
| B91 1LN | 2E1GYD |
| B91 1LN | G4WMH |
| B91 1PR | G4LOE |
| B91 1QB | G1FMW |
| B91 1TJ | G3LUA |
| B91 1TQ | G8GBM |
| B91 1TS | M0DWX |
| B91 1TZ | M0AEC |
| B91 2EG | G3DEJ |
| B91 2EH | G4EET |
| B91 2JH | G6VLK |
| B91 2PL | G6JEB |
| B91 3GA | G1RBX |
| B91 3JY | G1BHB |
| B91 3LL | G6KMQ |
| B91 3LL | M3YWD |
| B91 3ND | G4KSG |
| B91 3PW | G1ZLC |
| B91 3UD | G3WZI |
| B91 3XR | G4NYG |
| B92 0AS | 2E0MSW |
| B92 0AS | M6MSW |
| B92 0BS | G8SAN |
| B92 7BU | 2E0XMS |
| B92 7BU | M0LRO |
| B92 7BU | M6MKY |
| B92 7BU | M6MSX |
| B92 7DF | 2E0ZSR |
| B92 7DF | M6GYF |
| B92 7EE | G4YZA |
| B92 7ES | G8GWR |
| B92 7EY | G6BJG |
| B92 7HB | G6DFH |
| B92 7HD | G6PVA |
| B92 7HE | G1STK |
| B92 7HE | G4MCY |
| B92 7HH | G4FJB |
| B92 7JA | G8YMU |
| B92 7JB | G0HLW |
| B92 7JF | G4LCH |
| B92 7JH | G4VMO |
| B92 7JY | M3IKN |
| B92 7NU | G4KQV |
| B92 7PN | M1LCL |
| B92 7PN | M3CTO |
| B92 7QL | G8HQP |
| B92 7RU | M6RJB |
| B92 8AL | G6EDF |
| B92 8DB | M3XCJ |
| B92 8DP | G7LED |
| B92 8DX | G4CEX |
| B92 8EE | G3GEI |
| B92 8EE | G4BBT |
| B92 8NB | G4MVB |
| B92 8NN | G4PCE |
| B92 8QA | M1CXA |
| B92 8QS | G8ULQ |
| B92 9BJ | G1ASU |
| B92 9DB | G1FUG |
| B92 9DQ | G0PHR |
| B92 9HH | G7OJO |
| B92 9LQ | 2E0SSL |
| B92 9LQ | M6GSJ |
| B92 9ND | G4KWO |
| B92 9NW | G7JVF |
| B92 9PT | G7GQX |
| B92 9QH | G4EIG |
| B93 0HX | G0IZQ |
| B93 0PT | G3OIF |
| B93 8AW | M0LYQ |
| B93 8DN | G4RWG |
| B93 8DN | M6TTT |
| B93 8NN | G7FFS |
| B93 8QP | G3UFQ |
| B93 8RA | G8IK |
| B93 8RN | G1DWU |
| B93 8RN | M6ITQ |
| B93 9AW | G6VKS |
| B93 9EQ | G3KEK |
| B93 9JL | G4MAU |
| B93 9LA | G3TZM |
| B93 9LG | M0RXV |
| B93 9LQ | G1JYR |
| B93 9PA | G1VIW |
| B93 9PP | G4CVM |
| B94 5DP | G3GBS |
| B94 5EB | G8NOF |
| B94 5LP | G7OKF |
| B94 5LP | M3JKZ |
| B94 5RZ | G6HSD |
| B94 6LE | G4KRO |
| B94 6LN | M0HRT |
| B94 6QH | M3KKO |
| B94 6QY | 2E0RNJ |
| B94 6QY | M0RJH |
| B94 6QY | M6JNR |
| B95 5BA | G0EPL |
| B95 5LR | G0CRB |
| B95 5NN | G4PIP |
| B95 6AB | 2E1FRC |
| B95 6BH | G3UOC |
| B95 6BH | G4CGR |
| B95 6JA | G3XTI |
| B96 6BB | G7VTR |
| B96 6DY | M6KET |
| B96 6EB | M0CUS |
| B96 6ED | G1DCY |
| B96 6ED | G3RZI |
| B96 6LT | G6CUQ |
| B96 6NG | 2E0EGP |
| B96 6NG | M6EGP |
| B97 4JL | G0TOX |
| B97 4LX | G3WF |
| B97 4NP | G3KWK |
| B97 4PN | G8MGK |
| B97 4RL | G8DSM |
| B97 4SP | M0JEM |
| B97 5AA | G2FXJ |
| B97 5AY | G3TNI |
| B97 5DF | G4ZWR |
| B97 5EN | M1ANN |
| B97 5EP | G4SGV |
| B97 5FP | G4NTG |
| B97 5JA | G1JJA |
| B97 5LX | G0TPG |
| B97 5NG | G0ORE |
| B97 5NG | G8XUB |
| B97 5NW | G0RMG |
| B97 5NW | M6IOI |
| B97 5PZ | G1XVY |
| B97 5QT | G1TQU |
| B97 5RX | G0EYO |
| B97 5RX | G6SL |
| B97 5TB | G4NRP |
| B97 5UA | G3OXL |
| B97 5UN | G4SRV |
| B97 5WB | G4NYZ |
| B97 5XS | G0NRF |
| B97 5YL | G3SAH |
| B97 6EL | G1NEB |
| B97 6EN | G4WXG |
| B97 6JY | M3PSF |
| B97 6LP | M3JFS |
| B97 6NJ | G2HHH |
| B97 6NS | G0TNH |
| B97 6PG | M1ERO |
| B97 6PH | G1AJV |
| B97 6TS | G0CLM |
| B97 6UF | G7RBQ |
| B98 0AG | G7GFM |
| B98 0AS | G4MUV |
| B98 0BH | G4SWR |
| B98 0BJ | 2E1FPM |
| B98 0BJ | M3DVN |
| B98 0BJ | M3FPM |
| B98 0BJ | M3HLN |
| B98 0BJ | M5DRW |
| B98 0BX | G1AVW |
| B98 0EH | M3TXQ |
| B98 0EW | M6MBH |
| B98 0EY | 2E0PWM |
| B98 0EY | M3PWM |
| B98 0HJ | G3CKE |
| B98 0HJ | G6WPE |
| B98 0JQ | G0PLA |
| B98 0JQ | G7HSL |
| B98 0LA | M3VNM |
| B98 0NA | G1DOA |
| B98 0NF | G1RAG |
| B98 0NF | M5AFV |
| B98 0NL | 2E0WTH |
| B98 0NL | M6PCN |
| B98 0PJ | G0CAX |
| B98 0PP | 2E0ULC |
| B98 0QT | G7AZH |
| B98 0SB | G4DFC |
| B98 0TQ | G6DVP |
| B98 7EJ | G0UWH |
| B98 7NH | G1DHM |
| B98 7NR | G0NVV |
| B98 7PD | G6FVL |
| B98 7PL | M6AGA |
| B98 7PW | G3TBW |
| B98 7QA | G3WJN |
| B98 7QF | 2E0ETH |
| B98 7QF | M0NAK |
| B98 7QF | M6BNQ |
| B98 7RF | M3EFX |
| B98 7SZ | G6OUX |
| B98 7TL | G1RFI |
| B98 7UT | G4UDK |
| B98 7XE | G0UQE |
| B98 7XT | G1MZT |
| B98 7YD | G6HCW |
| B98 7YL | G4LDB |
| B98 7YU | G3TQD |
| B98 7YZ | M0AMF |
| B98 8DJ | G0NSW |
| B98 8HT | 2E0HOL |
| B98 8HT | 2E0MNC |
| B98 8JS | M6MWD |
| B98 8PL | G0KWQ |
| B98 8QL | G0LFY |
| B98 8QL | G0LFZ |
| B98 8RD | 2E0DUO |
| B98 8RD | G1DUO |
| B98 8RD | M5DUO |
| B98 8RD | M6DUO |
| B98 8RL | G7KMW |
| B98 8RW | G6VUN |
| B98 8SQ | G4KNX |
| B98 9EH | G3KFS |
| B98 9JH | G6ERJ |
| B98 9JX | G4TRI |
| B98 9JX | G8EOJ |
| B98 9LE | M6DWY |

## BA

### (Bath)

| Postcode | Callsign |
|---|---|
| BA1 1SR | M6XTM |
| BA1 2BL | G4YNM |
| BA1 2TD | 2E0DUF |
| BA1 2TD | M6FYJ |
| BA1 2UU | G0EJR |
| BA1 2XH | M3XUR |
| BA1 3HH | M0RCR |
| BA1 3LU | G3LYW |
| BA1 3LU | G4LYW |
| BA1 3PE | G7KCC |
| BA1 3PY | G3TKF |
| BA1 3PY | G4HTV |
| BA1 3RB | G6MZW |
| BA1 4DZ | G3UMM |
| BA1 4NQ | 2E0PDZ |
| BA1 4NQ | M0PDZ |
| BA1 4NQ | M6EZR |
| BA1 4NR | G1BCB |
| BA1 5JU | M3TCG |
| BA1 5SW | G6UTK |
| BA1 5SY | G1ZUC |
| BA1 5TB | M0VNR |
| BA1 5TB | M6EUL |
| BA1 5TW | G6UZG |
| BA1 6EF | M3GZP |
| BA1 6EF | M6GPN |
| BA1 6JR | G1MDC |
| BA1 6NA | G8DRK |
| BA1 6NP | G7MZY |
| BA1 6NZ | M6GTQ |
| BA1 6QN | G7VSJ |
| BA1 6QW | M3ZYQ |
| BA1 7BA | G0WZY |
| BA1 7SB | G1OPW |
| BA1 7TJ | G3RVX |
| BA1 7TT | G4JQB |
| BA1 8AD | G0WAW |
| BA1 8AD | G7STQ |
| BA10 0BS | G3KZR |
| BA10 0HH | G0WRL |
| BA10 0JD | G7VNC |
| BA10 0RJ | G4WFX |
| BA11 1AQ | G3XBW |
| BA11 1RR | M3IKD |
| BA11 2BD | G1JPK |
| BA11 2DZ | 2E0BHQ |
| BA11 2DZ | M3SXZ |
| BA11 2LA | G8BDU |
| BA11 2NR | G0JDM |
| BA11 2QD | M0IBW |
| BA11 2TN | G4OWH |
| BA11 2UR | M6RZW |
| BA11 3DP | G8VGI |
| BA11 3DY | M3BPG |
| BA11 3LR | M0EDA |
| BA11 3LR | M3LPF |
| BA11 4AB | M6KMK |
| BA11 4JA | G4XAG |
| BA11 4JA | G4XAH |
| BA11 4JB | M0HBH |
| BA11 4NR | G7VQX |
| BA11 5AR | M6HBV |
| BA11 6PR | G1JAL |

BA12 0AE G7AZV
BA12 0ES 2E0JZK
BA12 0ES M0JZK
BA12 0ES M6JZK
BA12 0JW G0BGI
BA12 0JW G4ZUP
BA12 0PP M6ZPH
BA12 0PW G7FHA
BA12 0RN G0AYD
BA12 6JX G4SSP
BA12 6LR G4KDK
BA12 7AE 2E0TMX
BA12 7AG G7VHC
BA12 7AP G3NFJ
BA12 7BB M6FRP
BA12 7HE M0RSG
BA12 7PA G1ACY
BA12 8BU M3HNL
BA12 8EB G0GGG
BA12 8EZ G4ILF
BA12 8HY G3ZNH
BA12 8LL M1DIR
BA12 8LY G7COA
BA12 8NE M6MKD
BA12 8NW M0BYL
BA12 8NW M0HZT
BA12 8TB M6CJN
BA12 8TF G4YMG
BA12 9DU M0SJG
BA12 9EF M1SRP
BA12 9LH G7DDV
BA12 9LY M1FMJ
BA12 9PN G3MHV
BA12 9PN G4WHV
BA12 9PN M0ZYF
BA12 9PT 2E0PYC
BA12 9PT M0XAW
BA12 9PT M6BNG
BA13 3AQ M1BPW
BA13 3AU M6PNP
BA13 3ES G4IQZ
BA13 3GS 2E0WZT
BA13 3GS M0WZT
BA13 3GS M3WZT
BA13 3HL 2E0JWJ
BA13 3HL M3JWJ
BA13 3HN G7DTG
BA13 3HP G0VFS
BA13 3HQ G4YSE
BA13 3HQ M6SBS
BA13 3JW G7LND
BA13 3LQ M0ADW
BA13 3PN 2E0EBO
BA13 3PN M3EBO
BA13 3QL M5EAY
BA13 3RW G6ZDE
BA13 3SH G1WFO
BA13 3UE G6ESJ
BA13 3UH G0GHW
BA13 3XF 2E0DNC
BA13 3XF M0SFT
BA13 3XF M6DNC
BA13 3XG M0XBI
BA13 4AT G4YXS
BA13 4BH G3WZH
BA13 4EA G0PVN
BA13 4LG G4JQN
BA13 4NX G4JQX
BA13 4NY G4CLC
BA13 4NY M1FHB
BA13 4TH G0JYL
BA14 0DT M6TEP
BA14 0HD G0HAS
BA14 0HG G0RKB
BA14 0HG G0RTO
BA14 0HS M1WAZ
BA14 0LH G4KHK
BA14 0LH G6CKL
BA14 0LJ G7FXY
BA14 0QL M6OMR
BA14 0QP M3DHV
BA14 0QP M3DHW
BA14 0QS G7OIB
BA14 0RE G7NGI
BA14 0RX G7KNU
BA14 0TD G1UGV
BA14 0TD M6LEX
BA14 0TE M0WIZ
BA14 0TX G0KCZ
BA14 0UJ G6RAE
BA14 0UX G6HQX
BA14 6EH M1CKJ
BA14 6EH M0ACA
BA14 6EZ G0HEL
BA14 6FG M0LAZ
BA14 6JG M0PUB
BA14 6JQ G0LJG
BA14 6JZ G3VMZ
BA14 6NA G4UJJ
BA14 6QP G7JQW
BA14 6SA G4SPE
BA14 7BN 2E1EYI
BA14 7HD G7PEE
BA14 7LF G5RBI
BA14 7PE G8GUA
BA14 7PG 2E0BUA
BA14 7PG G1HFY
BA14 7PG M3PVU
BA14 7PH G4GFJ
BA14 7PR G0VYU
BA14 7PZ G7EPX
BA14 7RS G3AZW
BA14 7UN G0KHQ
BA14 7UY M6OSI
BA14 8QP G1TST
BA14 8RY G0GKH
BA14 8WD 2E0DUQ
BA14 8WD M6GNY
BA14 9BQ G0WPL
BA14 9ED M3FMI
BA14 9EG G0JYF
BA14 9GG M6ALE
BA14 9HH G0BQG
BA14 9HL G0BNG
BA14 9HS M6DNO
BA14 9JZ M3WAY
BA14 9LQ G3PYF
BA14 9PH G7PQW
BA14 9PW G0HFX
BA14 9PW G8BYI
BA14 9RB G0TOE
BA14 9RB G6YDP
BA14 9TB G0EUR
BA15 1AX G0IAK
BA15 1HS G7RGI
BA15 1HZ G7KBD
BA15 1JE G4BMR
BA15 1JF G4YPE
BA15 1LL G8MGQ
BA15 1LX 2E0MSI
BA15 1LX M0IMS
BA15 1LX M6IMS
BA15 1RJ G0OFT
BA15 1SE G4VVZ
BA15 1SE G4ZAP
BA15 1SJ M3OKE
BA15 1TB G3GKC
BA15 1TB M3VAT
BA15 1TJ G3BBX
BA15 1UD G7SPM
BA15 2BH G1LCN
BA15 2BX G1FJS
BA15 2DL G0GRI
BA15 2DL G2BQY
BA15 2DL G6ZXN
BA15 2DL M1BQY
BA15 2HG G4LYG
BA15 2HL M0AXW
BA15 2SB G3YIQ
BA16 0BY G4PLY
BA16 0HX G7SDD
BA16 0HX M0BRH
BA16 0HY G4JBW
BA16 0RL G4LWQ
BA16 0RY G1PVU
BA16 0SA G1ZTM
BA16 0TE G1FGK
BA16 0TE G4XLY
BA16 9PE G7NCE
BA16 9PF G0BNF
BA16 9QN G6YZF
BA16 9QQ G6BJQ
BA16 9RJ G1OOB
BA2 0DH G0LTE
BA2 0DH G8LJY
BA2 0DZ G4BHP
BA2 0EA G4OMG
BA2 0HB G7IRF
BA2 0HB M1LRX
BA2 0HB M3LRX
BA2 0JF G6YB
BA2 0PE M6VIG
BA2 0PS G4CBS
BA2 1AE G4YTN
BA2 1DY M6FZF
BA2 1DY M6OOW
BA2 1HS G6VIF
BA2 1NW G4YCE
BA2 1NW G7ANB
BA2 1NW M6YDV
BA2 1PY 2E0CXK
BA2 1PY M6CVQ
BA2 2DL 2E0ICP
BA2 2DL M0ICP
BA2 2DL M3XKP
BA2 2DR G4CVD
BA2 2LZ G0KTN
BA2 2NG M1ZZY
BA2 2NG M3BFY
BA2 2PG G8CJT
BA2 2PS 2E0PBL
BA2 2QB G6ZKU
BA2 2TB 2E0TFO
BA2 2TB M0TFO
BA2 2TB M3TFO
BA2 2UD G1EEZ
BA2 3AE M6KDC
BA2 3BS G0FUW
BA2 3BS M3MOB
BA2 3JL G8XZB
BA2 3JW 2E0PPD
BA2 3PP 2E0EOL
BA2 3PP G4EJS
BA2 3PP M6KVM
BA2 4DH M6FWZ
BA2 4DW 2E0CTK
BA2 4DW M6CDT
BA2 4HS G3NAW
BA2 4LP G0CQC
BA2 4RJ G4BBD
BA2 5AL G7NFN
BA2 5JE G8FRI
BA2 5NF G4NDT
BA2 5NF G4NDU
BA2 5PL M3ZJE
BA2 5PU 2E0MIZ
BA2 5PU M3VWR
BA2 6AL G4GON
BA2 6DE 2E1CZO
BA2 6DE G3VTO
BA2 6DE M5SUE
BA2 6DF G3YWG
BA2 6PJ G7AYL
BA2 6UE G6MBF
BA2 6XG M6RTQ
BA2 7AY M0HFO
BA2 7BA G4FEA
BA2 7GR G3TTJ
BA2 7HN 2E1DAR
BA2 7LA G1OPZ
BA2 7NR G2BUP
BA2 7NZ G7MRY
BA2 7SS M6KDJ
BA2 7SS M6TJD
BA2 8AB G0SDQ
BA2 8AF G0LIB
BA2 8EF G1WFU
BA2 8EQ M0WYB
BA2 8HT M0UAS
BA2 8HT M6PSJ
BA2 8SA G4PVX
BA2 8TG M6FIZ
BA2 8TY M6IPJ
BA2 9AF M6DBW
BA2 9DZ M1TAP
BA20 2AZ G7SFY
BA20 2BD G4JBH
BA20 2BD M3VBH
BA20 2DB G8AFN
BA20 2EH G1FZL
BA20 2EH G4ERN
BA20 2PD G4GNV
BA21 3AH G0UMS
BA21 3BT G7LNJ
BA21 3BT M3LNJ
BA21 3EY G3CFV
BA21 3JB G7AIB
BA21 3LF M6FMT
BA21 3NN M6GYZ
BA21 3QP M0ZFF
BA21 3QU M6WMV
BA21 3SQ M0BAO
BA21 3TB G0TIJ
BA21 3TE 2E0KNH
BA21 3TE M0KNH
BA21 3TE M6MGF
BA21 3TW G7MSK
BA21 4AW G7GGJ
BA21 4BA G8WBT
BA21 4BD G6GLZ
BA21 4DD G4EVI
BA21 4HF G7EAQ
BA21 4JF M0WOB
BA21 4JF M3VQF
BA21 4NN G3KCV
BA21 4NX 2E0MJM
BA21 4NX M3OWO
BA21 4PG G4KKG
BA21 4RJ M0BHO
BA21 5DG G0HEQ
BA21 5FP M3SJW
BA21 5FQ M0SCA
BA21 5HA G6FHR
BA21 5JB G0AIL
BA21 5JE G3MYM
BA21 5NY G7AIC
BA21 5RJ M6RIV
BA21 5SH G3ZLQ
BA21 5SP G3CMH
BA21 5SP G3ICO
BA21 5SU G6IUQ
BA21 5XA G3OBL
BA21 5XQ G6LLP
BA22 7AG G8DXO
BA22 7NQ M1FFO
BA22 7QZ G8EGG
BA22 8BW G0LNI
BA22 8HF G4SMD
BA22 8JY 2E0FFW
BA22 8NS G1XNK
BA22 8NY G0WRK
BA22 8RB G7WBE
BA22 8RB M3ARS
BA22 8SG M0DZH
BA22 8UR G3YPL
BA22 9BT M3OQJ
BA22 9BT M6KMW
BA22 9EN G3AST
BA22 9HF 2E0BHH
BA22 9HF M0KRP
BA22 9LF G0EON
BA22 9LY G4BMO
BA22 9QW G1GAN
BA22 9RR G2AMG
BA3 2AS 2E0WXD
BA3 2AS M0WLD
BA3 2AS M6WXD
BA3 2AX G8HKP
BA3 2JH G4KVI
BA3 2JH M3FED
BA3 2NB G7IYA
BA3 2PR G0JLF
BA3 2PR G1IHI
BA3 2RH G4ZNK
BA3 2RZ G6VJP
BA3 2SD 2E0DFG
BA3 2SD M0HWP
BA3 2SD M6EQK
BA3 2SL 2E1DOZ
BA3 3NB M3XSA
BA3 3PD G8XTO
BA3 3XL G4ZEU
BA3 4AN G1ORL
BA3 4BB G6EYI
BA3 4BR G3RHU
BA3 4BR G4PSP
BA3 4BZ G0TJP
BA3 4FX G0HKR
BA3 4GT G4VVP
BA3 4HA M6BEB
BA3 4HG G7SSA
BA3 4HT 2E0CSE
BA3 4HT M6XAF
BA3 4JL G4MKE
BA3 4QH G1ORN
BA3 4QH G8WKK
BA3 4QH G8WKL
BA3 4SS G6KPD
BA3 5AL 2E1FPV
BA3 5PG M0SBY
BA3 5PG M6SBI
BA3 5PJ G7VCY
BA3 5PP G7KEP
BA3 5PR G8FWF
BA3 5PT G0HIC
BA3 5UY G8AVO
BA34SF M3WNC
BA4 4AZ M1ANL
BA4 4JP M6EVE
BA4 4JP M6TRE
BA4 4PS M6AXR
BA4 4PS 2E0TUW
BA4 4PS M0TUW
BA4 4QN G8EDQ
BA4 4SN G0WKM
BA4 4TL G4YJH
BA4 4TR M6EPO
BA4 5DH M1ERH
BA4 5GL M6DGA
BA4 5GL M6TDY
BA4 5JW M3BVK
BA4 5JW M3CGS
BA4 5JW M3FES
BA4 5JX 2E0JFW
BA4 5JX M3PIW
BA4 5LE M0PRF
BA4 5LG G7FPW
BA4 5LX M6YWA
BA4 5PX G4YJX
BA4 5QE 2E0ORI
BA4 5QE M0ORI
BA4 5QE M6ORI
BA4 5UD 2E0FGQ
BA4 5UD M6NJB
BA4 5UR G0BKU
BA4 5XR G8KKA
BA4 5XW M6RZZ
BA4 5YG G6RRY
BA4 6BB G7ORK
BA4 6JG G7SDQ
BA4 6NG G4STH
BA4 6NG G4YLO
BA4 6NG G6SIG
BA4 6PN M3ULZ
BA4 6SG M3RNW
BA4 6TS G7LMX
BA4 6TS M3LMX
BA5 1AH M5ENM
BA5 1DG G3OJL
BA5 1JA M1FJL
BA5 1SA G2BJK
BA5 1UD G1WQU
BA5 2BX G7LVN
BA5 2DG G8BFV
BA5 2EN G4KQQ
BA5 2ER G0FZI
BA5 2FE M3MYM
BA5 2FN G3IJU
BA5 2GA M6BJK
BA5 2HS G6SIM
BA5 2HZ G4ZDR
BA5 2JE M6TPV
BA5 2JU G3ZJF
BA5 2QL G4FSU
BA5 2QL M6KYB
BA5 2UY G4JJP
BA5 2XG G3IUZ
BA5 3BA G7LXA
BA5 3BA M0LXA
BA5 3ED G4SFS
BA5 3FG G0AQA
BA5 3HY G4XWE
BA5 3JS G7NTS
BA6 8AW G6UVO
BA6 8EJ G0MYH
BA6 8EJ G1XRM
BA6 8HG M0BYZ
BA6 9AN G0GJX
BA6 9JW M3MMZ
BA6 9PA M6MQB
BA6 9PG M6FCT
BA6 9PH G5FM
BA6 9PH M0IOA
BA6 9SH M6AWR
BA6 9TQ G4FOB
BA7 7EH M1FFP
BA7 7FF G1YWY
BA7 7HE G0ANP
BA7 7JY G6ZJK
BA7 7LA G7PBT
BA7 7LA G8YEO
BA7 7LT G7VCJ
BA8 0BP G0ENW
BA8 0DN G6NDA
BA8 0ED M0GOT
BA8 0HJ 2E0RNO
BA8 0HJ M0VIT
BA8 0HJ M6JPF
BA8 0HR M1EYT
BA8 0JG G4KHU
BA8 0JH G7UWR
BA8 0JP G4WJW
BA8 0LH G3RCF
BA9 8AL G6UZH
BA9 8LY 2E0CYC
BA9 8LY M0HQP
BA9 8LY M6BOE
BA9 9BS G8GJA
BA9 9BZ G4YXX
BA9 9EJ M0FZR
BA9 9LS G4BIN
BA9 9LT G7SSG
BA9 9NL G8SEK
BA9 9RB G6IRX
BA9 9SB G6RCT

## BB (Blackburn)

BB1 1TW M0BXM
BB1 2AS 2E0MOZ
BB1 2AS M6MOZ
BB1 2DR G3YWH
BB1 2ER 2E0MQC
BB1 2ER M6HIU
BB1 2ER M6MQC
BB1 2HB G0PXH
BB1 2HY G4XHZ
BB1 2JQ G0DTI
BB1 2NN G4FSD
BB1 3HL M6TVM
BB1 3LP G1DJQ
BB1 4BH G7WJC
BB1 4EE M3TIQ
BB1 4ES G6SUV
BB1 4JX G7WBZ
BB1 4ND G7IJL
BB1 4NL M6MFL
BB1 4NP G6SZS
BB1 4PD 2E0BGO
BB1 4PD M3OFB
BB1 5HQ G4CDR
BB1 5QU G0MTY
BB1 6LF 2E0DIL
BB1 6LF M0YKB
BB1 6LF M6DIL
BB1 7EX G4BJJ
BB1 7EX G8EMB
BB1 8HU 2E0KLW
BB1 8HU M6KLW
BB1 8HU M6PVV
BB1 8LP G1BBC
BB1 8NS 2E0BPP
BB1 8QZ M0SKA
BB1 8RB M6WKH
BB1 9DP G3SQO
BB1 9HH G1YJW
BB1 9HX G1MXO
BB1 9NE G0DAG
BB1 9NF 2E0PME
BB1 9NF M0NWI
BB1 9NF M6PWK
BB1 9NH M3JRI
BB1 9PW G0GSN
BB1 9QT G7MOB
BB1 9QY G0HTD
BB1 9RR M6OLI
BB1 9SA G0VOF
BB10 1BA G0KMK
BB10 1EU G4EOX
BB10 1EU G6FTB
BB10 1HU 2E1FUH
BB10 1HU M3LNN
BB10 1HZ M3LIU
BB10 1JA 2E0DWK
BB10 2HB G8YYX
BB10 2JT G0ECG
BB10 2LG M3OAT
BB10 2NP M3CFJ
BB10 2QP G8FLS
BB10 3AG G4YMQ
BB10 3BS M3NSJ
BB10 3DS G3KJY
BB10 3EG M6NFE
BB10 3JF G6OHK
BB10 3JF M3JAB
BB10 3JG G0ZGF
BB10 3NG G4OKZ
BB10 3NU G6SYI
BB10 3PQ M0OQD
BB10 3QH G3XAB
BB10 4AJ G6PPY
BB10 4ET G1CVR
BB10 4HX 2E0LFI
BB10 4JA G3MBU
BB10 4LA G8BEK
BB10 4LA M0BEK
BB10 4LB G0MLL
BB10 4LZ M0LGP
BB10 4PD 2E0GON
BB10 4RD M6RGW
BB10 4QH G0OCK
BB10 4QH G0OCL
BB10 4TE G4JSK
BB11 1UG G0FNJ
BB11 3LH G1HZL
BB11 3NX M3HLV
BB11 3PR M0AIS
BB11 4BJ M3RFR
BB11 4BP G8XCE
BB11 4DN G0UAA
BB11 4NP M6VGU
BB11 4QG M3PFK
BB11 5EA 2E0CSG
BB11 5HP G7RRS
BB11 5HR M6HUF
BB11 5HX M3KVL
BB11 5RB G4OPN
BB12 0DF M3HLA
BB12 0EF G0DZC
BB12 0EF G1ZBP
BB12 0JG G0MUH
BB12 0JJ G6YRC
BB12 6AA G8IUQ
BB12 6DT 2E0CKT
BB12 6JT G6PRA
BB12 6NG G6RIM
BB12 6NJ 2E0TUN
BB12 6NJ G6IFV
BB12 6NJ M6TUN
BB12 6NQ G3YGC
BB12 6NZ G7DMS
BB12 6PW M6JQE
BB12 7AU 2E0BLZ
BB12 7AU M3SZS
BB12 7DB G4RQA
BB12 7DP M6CXW
BB12 7HT G3ROS
BB12 7HY G0DLT
BB12 7QG G3XAC
BB12 7QH G4NYL
BB12 8AF G7KTD
BB12 8DR 2E0SBD
BB12 8DR 2E1CXF
BB12 8DR M6MAS
BB12 8JB M0MMX
BB12 8RP G1JCW
BB12 8RP 2E0FHR
BB12 8RP M6SIR
BB12 8SH M0NWT
BB12 8SH M6NWT
BB12 9EE M6TCM
BB12 9LR G3YGD
BB12 9QA G0RTU
BB18 5BS G1BDY
BB18 5ED G7PZU
BB18 5JB G4GOZ
BB18 5LB M1DHA
BB18 5LB M6RQD
BB18 5LQ 2E0BYM
BB18 5LQ 2E0CKS
BB18 5LQ M6HKY
BB18 5NH G7WAW
BB18 5NW G4LWG
BB18 5PD G7SNQ
BB18 5PD G7WEK
BB18 5PR M6EYY
BB18 6DD G6FEQ
BB18 6LX 2E0ZTM
BB18 6NA M3VPQ
BB18 6NA M3VQD
BB18 6WA G1VAL
BB2 2LT M6MAA
BB2 2NQ G1WYB
BB2 2PT M6KXQ
BB2 2TX G0TPE
BB2 3EH G4HYT
BB2 3HU M6FFD
BB2 3HU M6LIS
BB2 3JS 2E0SCW
BB2 3JS M0WSW
BB2 3JS M6SCM
BB2 3JS M6SCW
BB2 3JZ 2E0JQI
BB2 3LQ M3WOX
BB2 3NZ G1BWI
BB2 3ST M0CAS
BB2 3TP M6LBE
BB2 3UE M6MMH
BB2 4AU 2E0MPO
BB2 4AU M3NJO
BB2 4EU 2E0MFC
BB2 4EU M3JQM
BB2 4FU M6XBO
BB2 4HS M6HUI
BB2 4NG G6ZAL
BB2 4NQ G0VNV
BB2 4PQ G6OWI
BB2 4QJ G4UCC
BB2 4QT 2E0MIV
BB2 4QT M6MIV
BB2 4RQ M0MCW
BB2 4TD G7PZT
BB2 4TW G7VPL
BB2 4TY 2E0THH
BB2 4TY M6ARH
BB2 5AH M0BHP
BB2 5DT G7VIY
BB2 5EJ 2E0BFN
BB2 5EJ G6ZKZ
BB2 5EJ M3JIT
BB2 5EQ M3ZWF
BB2 5ER G7DEC
BB2 5LE G3UUM
BB2 5NN G3OZC
BB2 6ET G7VNK
BB2 6HR M6VWD
BB2 6LW G4TMY
BB2 7DP 2E0JCM
BB2 7DP M6AEK
BB2 7DS G3JXC
BB2 7ED G3SBI
BB2 7EP G3LPS
BB2 7HA G6ILH
BB2 7PA G4HUQ
BB2 7PN G1JHP
BB2 7QS G8CFD
BB3 0AJ G6UXU
BB3 0AQ 2E1HXP
BB3 0AQ G3PJY
BB3 0AY G1VLS
BB3 0EH G8WZW
BB3 0HW G0YLM
BB3 0HZ G4JBY
BB3 0JB G6WGA
BB3 0JW G4IAT
BB3 0LU G6ETP
BB3 0QT G1ZEX
BB3 0QT G1ZFD
BB3 0QY G7DNM
BB3 0RG G1FDN
BB3 1EF M1EXS
BB3 1LQ G1JBE
BB3 1NP M3OYR
BB3 1NS G3KWO
BB3 2BS G4JGF
BB3 2DT G1BKZ
BB3 2HP G6CIE
BB3 2JH G6FKR
BB3 2LG M6AVD
BB3 2LW G6RXP
BB3 2PB M6LJA
BB3 2SA G7VZS
BB3 2SF G1XNX
BB3 2SQ G0NPJ
BB3 2SQ G1ECC
BB3 2SQ G4PSE
BB3 2SS G4IBS
BB3 2ST G1ZAG
BB3 2TR M6HEN
BB3 3AG G0BPQ
BB3 3AG M0DOC
BB3 3DR M6PPJ
BB3 3EB 2E0FRC
BB3 3EB M0TAQ
BB3 3EB M6FRC
BB3 3GZ G0KXD
BB3 3JH G3YTI
BB3 3PY G4UQK
BB3 3QA G8TJG
BB4 4AJ G4LPO
BB4 4AN G0OCW
BB4 4DZ G7JXD
BB4 4EA M3HZE
BB4 4EE G4NGV
BB4 4FN G7BXL
BB4 4HN M6FBR
BB4 4PB G0KIM
BB4 5BQ G0LRR
BB4 5BQ G4TWG
BB4 5DA G6NID
BB4 5EF M3ILV
BB4 5NA G1BPS
BB4 5NG G0PGW
BB4 5NW M6PRV
BB4 5TE G8FGB
BB4 6AY G7IAW
BB4 6BE G4ZLJ
BB4 6DS G8JCN
BB4 6EE G3XWB
BB4 6QN G7WGO
BB4 6RX G4UUE
BB4 6TH G6GLT
BB4 7HN G0RFY
BB4 7JA G0MEX
BB4 7JA M0AKS
BB4 7JZ M6KKI
BB4 7TH G7CZL
BB4 7TT G6VZZ
BB4 8AG M3HJG
BB4 8JG M3FHK
BB4 8LY 2E0ETA
BB4 8LY 2E0WIE
BB4 8LY M0GLI
BB4 8LY M3RYO
BB4 8PY G4VVK
BB4 8QH G0VYX
BB4 8QL G0ZAP
BB4 8TZ 2E0HEF
BB4 8TZ M0LMN
BB4 8TZ M6HEF
BB4 8UW G0FCA
BB4 9BT 2E0PLA
BB4 9HG G6MBV
BB4 9HG G6TKB
BB4 9JE G8GNO
BB4 9PX G0FQF
BB4 9SD G4MKT
BB4 9TG M0DDC
BB4 9TQ G4LNE
BB5 0EX M0ANL
BB5 0FN G0JSA
BB5 0HQ G0SVP
BB5 0JJ M0BZS
BB5 0NA 2E0BBL
BB5 0NA M3HJV
BB5 0ND M3GMY
BB5 0NT G7MYJ
BB5 0SB 2E0LGV
BB5 0SQ G3ILE
BB5 1SL G4ZMB
BB5 2AF 2E0HTR
BB5 2AF M6DQT
BB5 2AF M6LRV
BB5 2AS G6MKQ
BB5 2JD 2E0PCV
BB5 2JD M3LYV
BB5 2LE G6IVE
BB5 2NF G1OPV
BB5 2NY G1ZED
BB5 2NY M3UOL
BB5 2PA G1BQQ
BB5 2PF G4DPU
BB5 2PR G6EYS
BB5 2QP G6YJM
BB5 2QS M0CHJ
BB5 2TQ G7PQL
BB5 3AT G0DMH
BB5 3AT M0HFI
BB5 3AT M0HIP
BB5 3AT M6HUW
BB5 3BL M0CZU
BB5 3EY M0NFI
BB5 3EY M6NFI
BB5 3LH G4GQP
BB5 3TA G0VGN
BB5 4AG M6AOS
BB5 4AL G7DHD
BB5 4AR 2E0LLX
BB5 4AR G0UNK
BB5 4AR M3LLX
BB5 4AR M3UNK
BB5 4BZ M0JMY
BB5 4DX G4GHK
BB5 4HL G0SVH
BB5 4HL M3LPE
BB5 4JQ G4GQV
BB5 4JR G7ILY
BB5 4LY M1CEM
BB5 4PH G7WJK
BB5 5AU G2CJK
BB5 5DH G0GBU
BB5 5GG G1CTQ
BB5 5GH G3RFN
BB5 5LA G4XEI
BB5 5NQ G4FQW
BB5 5QD G3KEG
BB5 5QH G6ZFO
BB5 5RB G0LQO
BB5 5RB M1CVX
BB5 5RU G8JIE
BB5 5RX G3PUO
BB5 5XF M0AFC
BB5 5XG G6LVM
BB5 5XP G4FRF
BB5 6BD G4BYL
BB5 6BJ G1BQI
BB5 6BS G4GLW
BB5 6HP M6LJT
BB5 6JG G1VVM
BB5 6JQ G0CYR
BB5 6NB M3BLO
BB5 6PL G0TZY
BB5 6SY M6CND
BB5 6TD G8RBV
BB6 7AS G0PXI
BB6 7HU G8UQY
BB6 7JE G6PXX
BB6 7JS G0TPP
BB6 7JS G8MEC
BB6 7JU G8VJO
BB6 7LP M6WGB
BB6 7NF M0CRH
BB6 7NL G6LXU
BB6 7NL M0CKO
BB6 7PH G3SXC
BB6 7RN G4EXK
BB6 7ST M6EIA
BB6 8AA G8CZG
BB6 8BN G8LTC
BB6 8BQ G6HSG
BB6 8BT 2E0VXR
BB6 8BT M0YFT
BB6 8ET G4VEY
BB6 8HB G3DMO
BB7 1DG M0BXC
BB7 1LA M1DGL
BB7 1ND G0AZH
BB7 1PD G7RDP
BB7 2HG G6PBQ
BB7 2LD 2E0HJD
BB7 2LD M0PVA
BB7 2LD M3HJD
BB7 2LS G8XSU
BB7 2NX M0BXF
BB7 2QD G4IJD
BB7 2QH M1BOZ
BB7 2QN G0RKP
BB7 2QW G4LKZ
BB7 3DA G0AZG
BB7 3DA G0NLO
BB7 3JD G4UMB
BB7 3JW 2E0EAW
BB7 3JW M0AZH
BB7 3LB G4WJG
BB7 4EE G1LDD
BB7 4QT G6RZG
BB7 4RR M0DFH
BB7 4RS G2FLW
BB7 9BJ G3YFG
BB7 9JN G4XSG
BB7 9JW M1CLX
BB7 9TJ 2E0EXY
BB7 9XR G8IEZ
BB8 0ND G6RDM
BB8 0TX G6BXS
BB8 7AA M3TBP
BB8 7HW G1WAP
BB8 7JY G3ZCX
BB8 7NH G4HXL
BB8 8BF G0JMR
BB8 8BW G6HMN
BB8 8DP G0OKV
BB8 8DT G3KLN
BB8 8JX G4DUI
BB8 8NR G4JMO
BB8 8QS G0IQM
BB8 8SA G0RFQ
BB8 8TB G0HBN
BB8 9AA M1EAZ
BB8 9AZ M0FIL
BB8 9DF G6DFV
BB8 9ED 2E0WLN
BB8 9ED M6EUP
BB8 9PE G0UGM
BB8 9QJ G4RTS
BB8 9QR G0VQJ
BB8 9RS G7VZM
BB9 0AP G0TFK
BB9 0EG G4ZFD
BB9 0EZ G0BQC
BB9 0HZ G4ZLU
BB9 0LE G0AQZ
BB9 0QD G0BPR
BB9 0RP G4MJX
BB9 0SB G4UEA
BB9 0SJ 2E0HAQ
BB9 0ST G4MLB
BB9 0TW M1BSV
BB9 0YH G4KFF
BB9 5BN M0ETY
BB9 5DR 2E0DLZ
BB9 5DR 2E0ZLD
BB9 5DR G0ECB
BB9 5DR M0WOW
BB9 5DR M0ZOE
BB9 5DR M3LHF
BB9 5DR M3LHI
BB9 5DR M3OFC
BB9 5DR M3OFD
BB9 5DR M3VSB
BB9 5DT G4HCC
BB9 5EW G4BLH
BB9 5EW G8CME
BB9 5HB 2E1CTU
BB9 5HG 2E0MUN
BB9 5HG M6APM
BB9 5LA G0WZL
BB9 5RS G4MYU
BB9 5RX M6AZG
BB9 6BQ G0LLL
BB9 6EX G6JAL
BB9 6HE 2E0PIO
BB9 6HE M0JPN
BB9 6HE M3PIO
BB9 6LD G4WZM
BB9 6LZ G4SEG
BB9 6PT G4UCU
BB9 7BD G4TSV
BB9 7BD G4UUA
BB9 7ET 2E0RWB
BB9 7ET M6RWP
BB9 7RA G0ABV
BB9 8AB G2IF
BB9 8EE G6RUY
BB9 8QP G0GGT
BB9 8SA M6MAT
BB9 8SD G0JAF
BB9 9HR G0DFO
BB9 9JA G6PVU
BB9 9LL M6ENS
BB9 9LN G3TRK
BB9 9RH 2E0VEK
BB9 9RH M3VYD
BB9 9RR G4KTR

## BD (Bradford)

BD10 0LP G4ONZ
BD10 0NX M0AMM
BD10 0RJ G4RSF
BD10 0RJ G4RSG
BD10 8PU G1CHQ
BD10 8QX G1CFZ
BD10 8RZ M3UAV
BD10 9JW G0LJM
BD11 1HE G4TCT
BD11 1HR G4YJM
BD11 1HW M3SUT
BD11 1JL G3YXH
BD11 1LU M6MAW
BD11 2EE G8CHN
BD11 2EF M3LHW
BD11 2ET M0YDD
BD11 2JE M0BCE
BD11 2JY M6UBR
BD11 2NN G0FOI
BD11 2NN G3SVC
BD11 2PG G4UNJ
BD12 0JQ G6BIU
BD12 0PL G6CJT
BD12 0UX G0WJC
BD12 8DN G0BVQ
BD12 8PT M6JKW
BD12 9DA 2E0YZQ
BD12 9DA M0XAK
BD12 9DA M6ALK
BD12 9JE G0DFU
BD12 9LS G7HSS
BD12 9NR G0LGB
BD13 1LD 2E0WBQ
BD13 1LD M6EHU
BD13 1LP G4XGN
BD13 1NE G7NEC
BD13 1PL G4NTA
BD13 2AE M6MCJ
BD13 2BE M6EJX
BD13 2EP G8NWK
BD13 2FF M0AAM
BD13 2HQ 2E0FKU
BD13 2HQ M3AIE
BD13 2JN G0KYM
BD13 2LJ G4UKW
BD13 2QA M3JNR
BD13 2SA M6PFL
BD13 3AT M6EHQ
BD13 3BE G0SNV
BD13 3BE G7DFC
BD13 3BG M0NAP
BD13 3DQ G8FJR
BD13 3EZ G1GQB
BD13 3PQ G4JHS
BD13 4EY G0IBS
BD13 4LZ G4WFC
BD13 5AE G1HJL
BD13 5AE M0BLZ
BD13 5JF G8VSI
BD14 6AB M1EKK
BD14 6BL M0AIY
BD14 6JZ G3HMV
BD14 6PJ G1IEP
BD14 6PJ G1UHB
BD14 6PJ M5IEP
BD14 6RY G6CMN
BD15 0BE G3UCK
BD15 0HB G7HEN
BD15 0HH G4EZX
BD15 7AU G3NYR
BD15 7AU G4TIV
BD15 7PR G1OLQ
BD15 7QZ G1FKM
BD15 9BD G0IFT
BD15 9BD G4ZFT
BD15 9BT G6OSJ
BD15 9LD 2E0KEI
BD15 9LP G7EVR
BD16 1AD 2E0PJH
BD16 1AD M0IRK
BD16 1BD M0EAD
BD16 1DA G8ESK
BD16 1LN G3UOI
BD16 1LZ G0CEF
BD16 1PU M3HHC
BD16 1QB M0CRD
BD16 1QD G7CKG
BD16 1RB G4SMK
BD16 2BN M6HNO
BD16 2DY G1CWD
BD16 2QD M0AIX
BD16 2SR 2E0ERT
BD16 2SR M0XOM
BD16 2SR M6ROB
BD16 3BX G3RXS
BD16 3DF G6KJT
BD16 3DH G4YCP
BD16 3DY G0MDO
BD16 3LG G8UPK
BD16 3NE G0SUI
BD16 3PL G3NN
BD16 3QN 2E0PMM
BD16 4DR G3TXX
BD16 4DX M0MSS
BD16 4DX M1APX
BD16 4ED G8NTZ
BD16 4EE G8PZF
BD16 4LB G7UTR
BD16 4QD G4YTI
BD16 4RN G0HUK
BD16 4RW G4YWR
BD17 5DW G8FTX
BD17 5HS G1KAS
BD17 5NR G0RJC
BD17 5NR G7HKU
BD17 5RS G0YKK
BD17 5TJ G8JTD
BD17 5TJ M1RAL
BD17 6DR G0FVO
BD17 6NN 2E0PAK
BD17 6NN M0PKW
BD17 6NN M6WAT
BD17 6RT M6NRS
BD17 7PQ 2E0HTS
BD17 7PQ M0YKS
BD17 7PQ M3TLL
BD18 1EH 2E0CDR
BD18 1EH M3RZE
BD18 1HL 2E0PLK
BD18 1HL M0PLK
BD18 1NB G3TJC
BD18 2EY G1XJO
BD18 2EY M0GZL
BD18 2JB 2E1GES
BD18 2JD G7KHE
BD18 2LT G0OEJ
BD18 2LT G0PBA
BD18 2NT G0GME
BD18 2NT M3VFU
BD18 3DE G1IYA
BD18 3JD G1LZF
BD18 4AR G8ZMG
BD18 4AW G0VJB
BD18 4DY G8ORM
BD18 4EJ G0PCM
BD18 4EW G0MPP
BD18 4JZ M0TAN
BD19 3BY G0DPX
BD19 3DG G1SKE
BD19 3EJ 2E0ZDW
BD19 3EJ M0WIT
BD19 3EJ M6DFW
BD19 3PX M3ZPT
BD19 4NX M3TAN
BD19 4RU G0MZZ



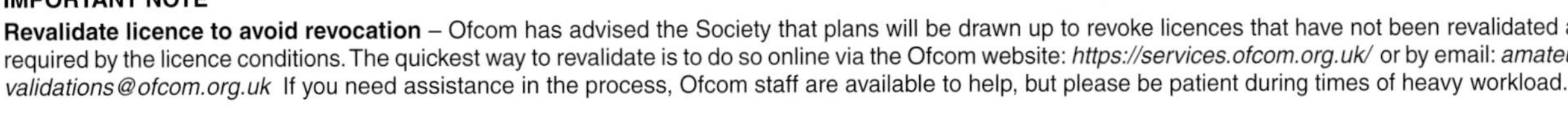

**IMPORTANT NOTE**

**Revalidate licence to avoid revocation** – Ofcom has advised the Society that plans will be drawn up to revoke licences that have not been revalidated as required by the licence conditions. The quickest way to revalidate is to do so online via the Ofcom website: *https://services.ofcom.org.uk/* or by email: *amateur.validations@ofcom.org.uk* If you need assistance in the process, Ofcom staff are available to help, but please be patient during times of heavy workload.

BD19 4SB G3WYP
BD19 5BW G1XCC
BD19 5JH G0IQQ
BD19 6DQ 2E0MBQ
BD19 6DQ M6MRT
BD19 6JH G4OSO
BD19 6JH G4OSP
BD19 6LJ G3RGN
BD2 1BQ G8AQH
BD2 2LS G4XYR
BD2 2NE G6YGJ
BD2 3AL 2E0LTZ
BD2 3AL M6MTZ
BD2 3HD G0PVB
BD2 3RB G1OCS
BD2 4HX G8MVD
BD2 4HY G4TFD
BD2 4SA G6PAR
BD2 4SG G1FVS
BD20 0LD 2E1HRC
BD20 0LD M3HOU
BD20 0LD M3HRC
BD20 0LQ 2E0SJK
BD20 0LQ M0KWS
BD20 0LQ M0PLN
BD20 0LQ M6DXA
BD20 0LQ M6SAK
BD20 0ND G6IPB
BD20 0QG M3VGH
BD20 0QQ G0JFC
BD20 5AN G3WZZ
BD20 5DB G0HOT
BD20 5EJ 2E0LXA
BD20 5EJ M6AXL
BD20 5JT G7BUR
BD20 5NW G4AFS
BD20 5TD M1EKD
BD20 5TD M5LAR
BD20 6AY G1HEA
BD20 6SP 2E0DVI
BD20 6SP M6GGI
BD20 6SZ G1OTA
BD20 6SZ G7KDH
BD20 7BH M0PWF
BD20 7DN G4CPA
BD20 7PN M0AEZ
BD20 7RW G0STK
BD20 8LT M6GZN
BD20 9AU 2E0JCB
BD20 9AU M6DWF
BD20 9AU M6JTW
BD20 9LL G0LVT
BD20 9NR G4YXB
BD21 1HY G3VDK
BD21 2QJ G0OPT
BD21 2QJ G1ONJ
BD21 2RX M0DIT
BD21 3HY M0KSC
BD21 3HY M1PAC
BD21 4NP G8VPX
BD21 4TA 2E1HFV
BD21 4TA G7HLV
BD21 4TD M0CEX
BD21 4TD M1DSV
BD21 4TF G0AJF
BD21 4TL G0BWY
BD21 4YG G0MJB
BD21 4YG G7BQU
BD21 5BD M0CAN
BD22 0AP G7SFJ
BD22 0HA G3TFF
BD22 0NU M0HSS
BD22 0QY G4AEE
BD22 6DD G0MEA
BD22 6EU M6AXH
BD22 6FF G4IUP
BD22 6HF M6KCP
BD22 6HQ M6DER
BD22 6QT 2E0RIZ
BD22 7AB 2E0CQN
BD22 7AB M6SBO
BD22 7AU M6MLV
BD22 7BP G1AVA
BD22 7DH 2E1FON
BD22 7DH M1BVX
BD22 7DN G0MMC
BD22 7EX G4TFT
BD22 7NQ G0VXS
BD22 7PD G0GVE
BD22 7QS G3UBD
BD22 7RH G0BZH
BD22 7SH G4XNF
BD22 8BJ G0RLY
BD22 8BJ G0TSJ
BD22 8HG 2E0MFI
BD22 8HG M6SKQ
BD22 8JY M6JLT
BD22 8PL G1NQU
BD22 8PR 2E0DOG
BD22 8PR M0UEZ
BD22 9DL M6UEA
BD22 9EB M1BUU
BD22 9LE G1SRA
BD22 9LE G7KRG
BD22 9SS G3OPW
BD23 1BB G0XDL
BD23 1BP G0RAX
BD23 1NS 2E0VRX
BD23 1NS M6BEQ
BD23 1QN G6AML
BD23 1TL 2E0RCI
BD23 2BZ M6ZEA
BD23 2ET 2E0PHL
BD23 2ET M3NYF
BD23 2ET M3SKN
BD23 2LE M3CWC
BD23 2PY 2E0JBC
BD23 2PY M0YBC
BD23 2PY M3CSZ
BD23 2QE M0NTT
BD23 2RR G0UCD
BD23 2SP M3SBY
BD23 3DW G8LKQ
BD23 3DW M0DRF
BD23 3DW M0YCG
BD23 3NT G7COD
BD23 3RY M0XLT
BD23 3SE G7CYN
BD23 3TH 2E0XLG
BD23 3TP G1LNQ
BD23 3TT M0BCQ
BD23 4HJ 2E0PIW
BD23 4HJ M0FTL
BD23 4HJ M3HLD
BD23 4JE G6LYE
BD23 4JW G6RIY
BD23 4LB G6LFC
BD23 4PQ 2E0HRD
BD23 4PQ M0VHG
BD23 4PQ M6HRD
BD23 5EN G6LHA
BD23 5EZ M1DEY
BD23 6SD G4YQA
BD24 0AG G4GLC
BD24 0AH 2E0PFL
BD24 0AH M6DLQ
BD24 0DP 2E0PXP
BD24 0DP M0PXP
BD24 0DP M6PXP
BD24 0EF G7VVB
BD24 0HG G4HOK
BD24 0LH G4ZVD
BD24 9DA 2E0RSM
BD24 9JP M5CMO
BD24 9QR M0ANP
BD3 7BU M3IAO
BD3 7BY G7FDW
BD3 9JT G0GNU
BD3 9NU G0JTP
BD4 0JJ 2E1HZM
BD4 0QH G4HAG
BD4 0QU M6KEM
BD4 0SJ G1ANA
BD4 0SW G8PPR
BD4 6BY G0BZV
BD4 6JY M3TZS
BD4 6PJ M0CXL
BD4 7JD 2E1EGV
BD4 7JD M1BGY
BD4 7JJ 2E1GDD
BD4 7JJ M3DXI
BD4 7JT 2E1GVC
BD4 7JT 2E1GVD
BD4 8EN M0LMO
BD4 8PB 2E0RAS
BD4 8PB M3RSX
BD4 9JH G1XGM
BD4 9JJ 2E0NOK
BD4 9LX G4KZW
BD5 8NX G4JRW
BD5 9AN M3XWN
BD5 9HB G0ITU
BD6 1ET 2E0JPA
BD6 1ET G2UG
BD6 1ET G6FDG
BD6 1ET M0JPA
BD6 1HL M6BRS
BD6 1PQ G0FUY
BD6 1QU G0IQH
BD6 1RP G4RFO
BD6 1RP M0AMB
BD6 1UL M6FKP
BD6 1UU G4YOR
BD6 2BT G6RFL
BD6 2LP G3TIX
BD6 3DJ G4UNH
BD6 3JQ G0VVK
BD6 3RR G7MAJ
BD6 3SW G7JZM
BD6 3SW M3XKY
BD6 3XE G0BTA
BD6 3YN M3AHL
BD7 2LX G7RTA
BD7 4AS M3RQO
BD7 4BG 2E1HEM
BD7 4DB G6CGI
BD8 0AA G0OWI
BD8 0EN M0GEN
BD8 0EN M3CAP
BD8 7BH 2E0GYW
BD8 9EX G7MMJ
BD9 4AX G8ZMQ
BD9 4ES M3FTU
BD9 4HG M6OMH
BD9 5EX 2E1FBY
BD9 5PA M1DJI
BD9 6DQ G0TVM
BD9 6EX G1KWK
BD9 6EZ G7PHW
BD9 6EZ G7PHY
BD9 6PU M3XJE

## BH (Bournemouth)

BH1 2PG G3PUZ
BH1 3DX G4LXY
BH1 3PH G4FUP
BH1 3SS G4UGV
BH1 3TE G3XFD
BH1 4DH 2E1CEQ
BH1 4JH 2E0FXD
BH1 4JH M6FXD
BH1 4PH G0OFE
BH10 4DT M6GFS
BH10 4EE G1INB
BH10 4EY G7IWW
BH10 4HP G3WZP
BH10 4HP M6LXE
BH10 5AW G4ZLT
BH10 5EP G1OQM
BH10 5JT G8CYT
BH10 5JT M0GIU
BH10 5LF G8UAD
BH10 5LG G4OEB
BH10 6BG 2E0WCX
BH10 6BG M6WCX
BH10 6DS G7MTT
BH10 7AA G1XES
BH10 7EU G3ZWD
BH10 7HW M0KHA
BH10 7HW M6XMB
BH11 8AW G8SLE
BH11 8BN M3MYZ
BH11 8BT M6OCS
BH11 8DF M1ELM
BH11 8EE M0VCA
BH11 8EQ 2E0RAI
BH11 8EQ M6RLT
BH11 8NN G1WAW
BH11 8PS G4DKM
BH11 8RB M6CJW
BH11 8RH G3XBZ
BH11 8RH M3BDC
BH11 8RT G3YWG
BH11 8SL M3ZWN
BH11 9DY M0AUY
BH11 9EQ G7TKG
BH11 9EU 2E1FTA
BH11 9EU G0SKR
BH11 9HJ M3ZKE
BH11 9JB G1PJB
BH11 9JD G7VJJ
BH11 9JW G0PRS
BH11 9JW G4WCK
BH11 9LJ G4XOH
BH11 9PG 2E0SPM
BH11 9PG M0SXM
BH11 9PG M6SPM
BH11 9SJ G7JNS
BH12 1BW M6TGZ
BH12 1JA 2E0WEG
BH12 1NS G1MXD
BH12 1NS M6VEG
BH12 1QE G0UCX
BH12 2EZ M6SBL
BH12 2HE G0ROZ
BH12 2HG 2E1BVY
BH12 2HT 2E0TMN
BH12 2HT M3HCE
BH12 2HT M3KMN
BH12 2JB G4RFR
BH12 2JQ M0MMC
BH12 2LQ M3SUK
BH12 3AQ M6ANV
BH12 3AW M0NMO
BH12 3BT M6GBH
BH12 3DA G4EOE
BH12 3HB G0CRY
BH12 3HE G6AHC
BH12 3HF M3KCA
BH12 3JW G7CAA
BH12 3LP G0GZN
BH12 3LP G0ISO
BH12 4BD G1UIO
BH12 4BS 2E0TPY
BH12 4BS M6TPY
BH12 4DH G0AEP
BH12 4EH M6SJZ
BH12 4FD M6MCL
BH12 4JH M6EKB
BH12 4JQ G4JQW
BH12 4JR 2E1HMB
BH12 4JR G4PPD
BH12 4LD M0GUH
BH12 4LT G0ISC
BH12 5AY M3TFW
BH12 5ET M3OEF
BH13 6EF G0MJO
BH13 7BE G4PVY
BH14 0DB G6EJH
BH14 0LL G6CGQ
BH14 0PA M1PJH
BH14 0PF G8YYA
BH14 0PJ G3RRP
BH14 0PP G3MXF
BH14 0QS G3WJJ
BH14 8AD G6ABA
BH14 8EG G3RGJ
BH14 8EY G8GAT
BH14 8NW G8IWE
BH14 8QY G3LUH
BH14 8SD M0BVI
BH14 8UE G0KKL
BH14 9AU G1NNR
BH14 9BS 2E0MVT
BH14 9BS M6ZDA
BH14 9EL G1TEX
BH14 9HU M6BLC
BH14 9JH 2E0DUD
BH14 9JH M3YVT
BH14 9LW G4UWS
BH14 9QP G0KCC
BH15 1EH G3YEK
BH15 1JY G8FDZ
BH15 1RD M3YYO
BH15 1TU G4UTG
BH15 1UA M0LEZ
BH15 1UP M0GPV
BH15 1UR 2E0DMY
BH15 1UR M6EYL
BH15 1XL G7FKS
BH15 2BZ G1YHE
BH15 2DB G0CBP
BH15 2DW G6OAI
BH15 2ED M0CZQ
BH15 2EF M6CDL
BH15 2ES G4CQW
BH15 2ES M1EZG
BH15 2EX G0JII
BH15 2EX G0KZD
BH15 2HQ G6NIO
BH15 2JQ G3WCU
BH15 3AG M0IDL
BH15 3AQ G3NIL
BH15 3EE G6TEL
BH15 3ET G3HUO
BH15 3ET G3UEE
BH15 3LG M6TDD
BH15 3NE G3FTK
BH15 3NR G3JTK
BH15 3QZ G3RZV
BH15 4DQ G3PLR
BH15 4DQ G8EPK
BH15 4HP G1UEQ
BH15 4HP G4CVX
BH15 4JD G1FPZ
BH15 4JF G6OFM
BH15 4JS G0JJI
BH15 4JX M6BUW
BH15 4JZ G7TZB
BH15 4JZ M3TZB
BH15 4LT G0KQH
BH15 4QX G1MMT
BH16 5BF M6PHY
BH16 5BZ G0UPG
BH16 5ED G8MCW
BH16 5EJ M0ATB
BH16 5EQ G1TQN
BH16 5LA G2HKQ
BH16 5LS G0RPA
BH16 5LS G4NFT
BH16 5LY M6EKE
BH16 5NS G8DLT
BH16 5PB G0JBZ
BH16 5PP G3VOB
BH16 5QT G6AKG
BH16 5RA G0IWZ
BH16 5RT G6DUN
BH16 5RX G7MUT
BH16 6AP G0ICG
BH16 6BW G7WBU
BH16 6DP G0FCV
BH16 6DP M3WHQ
BH16 6DT M3EZB
BH16 6EQ G0FRR
BH16 6EQ G3PFM
BH16 6EQ G6SFR
BH16 6EQ M1DKY
BH16 6ET G3PKL
BH16 6HJ G1JUI
BH16 6JH G1YHI
BH16 6NB G4MPQ
BH16 6NB M6BEE
BH17 7AH G3ZPR
BH17 7AH G4PRS
BH17 7AJ G0WTG
BH17 7AX 2E0CHN
BH17 7DW 2E1HBJ
BH17 7DW G7TEZ
BH17 7DW M3GCM
BH17 7EH 2E0VXT
BH17 7EH M0AIJ
BH17 7EH M0XAJ
BH17 7EH M6VRT
BH17 7EU G3MEC
BH17 7HA 2E1DBS
BH17 7SB G6MXL
BH17 7UP G0FVH
BH17 7UP M3RLX
BH17 7XT 2E1FDK
BH17 7YJ 2E0FDK
BH17 8BU 2E0EXO
BH17 8BU G1ZSE
BH17 8BU M6EXO
BH17 8DB G1HBF
BH17 8QP 2E1EOW
BH17 8SQ G7MZX
BH17 8SR 2E0AOL
BH17 8SR M0AKY
BH17 8SU G0RVI
BH17 9AH M1EEQ
BH17 9AS 2E0TCF
BH17 9AS M0TGY
BH17 9BF G4IBM
BH17 9BF G8RSE
BH17 9EE M6LDF
BH17 9WE G7GNU
BH18 8AE M0HZU
BH18 8DY 2E0UEL
BH18 8DY M0HTY
BH18 8DY M6DWV
BH18 8HZ M3WBM
BH18 8JS 2E0AMW
BH18 8JS M0CJC
BH18 8LN M3FSB
BH18 8LN M3FSD
BH18 8ND G6AOF
BH18 9AE G4BYO
BH18 9AE M0PTR
BH18 9AJ G1YHJ
BH18 9BD G3PSV
BH18 9DB G7OAX
BH18 9DF G8HNA
BH18 9DZ G7RDT
BH18 9DZ G8TNU
BH18 9DZ M5ADL
BH18 9DZ M6IAL
BH18 9DZ M6TNU
BH18 9EB G4MMT
BH18 9ED G8JMB
BH18 9EX 2E0MIL
BH18 9EX M0IHM
BH18 9HQ G4BKE
BH18 9HQ M3WYF
BH18 9HY G6PIM
BH18 9JG G3SGX
BH18 9JG M0BHH
BH18 9LB M6TJL
BH18 9ND G0WKH
BH18 9NR G0BRQ
BH18 9QZ 2E1DFZ
BH19 G4FDS
BH19 1HY G1HEJ
BH19 1HY G4ZPB
BH19 1LG G7FXO
BH19 1LL G3HFZ
BH19 1NF G6CAC
BH19 1PQ M3SGZ
BH19 1QU G7OVM
BH19 2DE G3MBM
BH19 2LB G4ZYY
BH19 2PZ G7OYX
BH19 2RG M1DCH
BH19 2RU 2E0KJK
BH19 2RU M0KJK
BH19 2RU M6KJK
BH19 2SL G0TOT
BH2 5JT G8CJH
BH2 5NR G4FNI
BH2 5RA M3XZR
BH2 5RA M3XZS
BH20 4AQ G0WFE
BH20 4DR G7VXQ
BH20 4EL G4GHA
BH20 4HA G1SPQ
BH20 4HY G0BAO
BH20 4PT G1RRW
BH20 4QX G0TGP
BH20 4SD M3KQP
BH20 4SG G1HXR
BH20 4SJ M6JHS
BH20 5RN G1LZZ
BH20 5RQ G6HCH
BH20 5RY G1EQJ
BH20 6AA M6STC
BH20 6DY M6KAT
BH20 6DY M6NEV
BH20 6EQ G7MND
BH20 6EY M0CWC
BH20 6HB G7EWY
BH20 6NN M0HXA
BH20 7BD 2E0XAM
BH20 7BD M0XAM
BH20 7BD M3XAM
BH20 7BE G1KKG
BH20 7BX G1AAH
BH20 7JF G7BRU
BH20 7LU M6BEE
BH20 7LU M6SSC
BH20 7NH M6NBL
BH20 7PA G6XEL
BH20 7PE 2E0JPD
BH20 7PE M3ZNP
BH20 7QA 2E0CTE
BH20 7QA M6CCE
BH21 1JJ G3VQR
BH21 1PJ M0JLE
BH21 1PJ M0TAK
BH21 1PL G4RLM
BH21 1QT G6NZN
BH21 1RG G6EOK
BH21 1SB G8CEZ
BH21 1SR G6JKK
BH21 1SX M0CGT
BH21 1TB G6SHS
BH21 1TP G0GOX
BH21 1TU G4WEY
BH21 1TZ M3ONK
BH21 1UQ G8MXW
BH21 1UT G0HOA
BH21 1XH 2E0JWS
BH21 2AU M1ETT
BH21 2BL G0EEF
BH21 2BL M0MJS
BH21 2DH G0UPS
BH21 2EA G0MTV
BH21 2HR G4NHE
BH21 2JB G7MQU
BH21 2JU G0DVE
BH21 2JZ G4RFV
BH21 2LA M1BAI
BH21 2LA M6REQ
BH21 2LD G3BHM
BH21 2LE G0MDK
BH21 2NG G8YLR
BH21 2NZ G0WAL
BH21 2PQ G3ZXW
BH21 2QB M1BSB
BH21 2QB 2E0AKK
BH21 2QB M0GJV
BH21 2QB M3GTH
BH21 2SF G0SKN
BH21 2ST G8CHI
BH21 2UW G0UGS
BH21 2UW G4YTA
BH21 3AA G4LBH
BH21 3AF M6CVF
BH21 3BG G3RJH
BH21 3BJ G4VZT
BH21 3DS G1VIP
BH21 3DS G8UCY
BH21 3EZ G0API
BH21 3EZ G3ZTY
BH21 3EZ G6NLC
BH21 3EZ G7MHO
BH21 3EZ G7PRO
BH21 3HL G0PRH
BH21 3HW M0EYT
BH21 3JD 2E0WCL
BH21 3JD M6WCL
BH21 3LG G0FKY
BH21 3LW M0FBB
BH21 3LZ G1YHV
BH21 3NH G7AUF
BH21 3NN 2E0BNA
BH21 3PN 2E0ZXR
BH21 3PN M6OOD
BH21 3PU G8NVC
BH21 3PZ M1EZL
BH21 3QG M3YZP
BH21 3RT 2E0DIN
BH21 3RT M3YLU
BH21 3SB G0ODP
BH21 3SL G8KNS
BH21 3SP G0KFM
BH21 3SP G3NUN
BH21 3TB M1PMR
BH21 3TF G7AGB
BH21 3TL M0MRP
BH21 3TW M0GUJ
BH21 3UA G8CRV
BH21 3UW 2E0CMT
BH21 4AE M6JWQ
BH21 4BH G4WUK
BH21 4DS M6FRV
BH21 4HD G8ZEK
BH21 5AF G6CHX
BH21 5BH M3RLB
BH21 5ET G4GXZ
BH21 5LZ G3XAS
BH21 5RD G7MYI
BH21 6RG M0SPC
BH21 6RR M3XQM
BH21 6SG G4MHF
BH21 7AN G3OAF
BH21 7JL G8YZL
BH21 7JZ M0MAI
BH22 0BQ G4PIJ
BH22 0BQ M3VLB
BH22 0BW G1GRB
BH22 0BZ G1JBM
BH22 0DW G8TEO
BH22 0EE G0KLQ
BH22 0EX M6ODC
BH22 0EY 2E1FSZ
BH22 0HN G4FOL
BH22 0JE 2E0LMS
BH22 0JE M3LKM
BH22 0LG G4VLP
BH22 8AJ M6MKO
BH22 8AL G0UJI
BH22 8BA G4VHI
BH22 8BY G1BRS
BH22 8BY G2BRS
BH22 8BY G3CPN
BH22 8JX M3FSC
BH22 8PA 2E1FSY
BH22 8PS M0WPR
BH22 8QG G8DJL
BH22 8QT M6FCD
BH22 8QW G8TSZ
BH22 8RP G8UJS
BH22 8RS G0AEL
BH22 8RW M6KDV
BH22 8RZ 2E0ADN
BH22 8RZ 2E0HJE
BH22 8RZ G8OFQ
BH22 8RZ M6VLB
BH22 8SB G3ZCL
BH22 8SF M0EWG
BH22 8SF M1EWF
BH22 8SF M5BMW
BH22 8UR 2E0DQX
BH22 8UR M6JEO
BH22 8XA 2E0YMR
BH22 8XA M5ACX
BH22 8XA M6YMR
BH22 8XN M0DOS
BH22 9AY M6GRO
BH22 9EQ G8YHF
BH22 9ES G4LJN
BH22 9HN G1RSE
BH22 9HP G4ULQ
BH22 9HW G4CFW
BH22 9JE G3WPB
BH22 9JP G0NLM
BH22 9JP G8DQE
BH22 9JQ G0CDY
BH22 9LA G4PAI
BH22 9LH M6PEB
BH22 9LJ G3WEG
BH22 9LY M3PIL
BH22 9NB G8XKD
BH22 9QB M3HJN
BH22 9QB M3ORN
BH22 9QD M6INM
BH22 9QE M0DOY
BH22 9QQ G1DNK
BH22 9SD G4ACJ
BH22 9SG G4XEE
BH23 1AS G8UFF
BH23 1JU 2E0IYY
BH23 1QL 2E1CDS
BH23 1QL G0RCN
BH23 1QX 2E0XAZ
BH23 1QX M6VZT
BH23 2AF G8ZPW
BH23 2AG G4HEC
BH23 2DF G4XWT
BH23 2DY G1NMR
BH23 2EE G0WXZ
BH23 2EH M3VRM
BH23 2HJ M0CDY
BH23 2LJ G0LOW
BH23 2LL 2E1FTI
BH23 2NG G3UZD
BH23 2NG M3LAQ
BH23 2NP M1EGM
BH23 2RP G3YPT
BH23 2SP G3UXR
BH23 2TW G6MYT
BH23 3BJ G4SEU
BH23 3BJ G6LUK
BH23 3BL 2E0KFR
BH23 3BL M6KDR
BH23 3EW G6SPB
BH23 3JA G0AHK
BH23 3JA G0MUD
BH23 3JA G7MUD
BH23 3NB 2E0CSF
BH23 3NB M3CRL
BH23 3NB M3CSF
BH23 3NS 2E1DHX
BH23 3QN G4MUU
BH23 3RE M6CAQ
BH23 3SN G4SNR
BH23 4BP G6FAH
BH23 4DD G0NIQ
BH23 4ED G0TVS
BH23 4QT 2E1MGH
BH23 4SL M6ILB
BH23 5AH 2E0CPS
BH23 5AH 2E0REK
BH23 5AH M0REK
BH23 5AH M0YAH
BH23 5AH M6REK
BH23 5AH M6ZYX
BH23 5BA G8GZN
BH23 5BE G1FLY
BH23 5DB G7UWZ
BH23 5DF G3NDS
BH23 5DX G1HQG
BH23 5HD 2E0DBB
BH23 5HD M6DAG
BH23 5PN G3YNC
BH23 5QD M3EBA
BH23 5QH G6YPJ
BH23 5SG M6BCU
BH23 6AW M6NDF
BH23 6BE G0IXT
BH23 6LP G7EUB
BH23 7BE G1LLA
BH23 7LD G0IKN
BH23 7LE M0DEP
BH23 7NJ G4TKP
BH23 7NT M6PDQ
BH23 7NW G4FND
BH23 8BS G0JIL
BH23 8BS G8YBZ
BH23 8BT G4OXK
BH23 8BU G1BLO
BH23 8DU G3PIY
BH23 8DU G3SGL
BH23 8HG G1IOP
BH23 8HG G8JEM
BH23 8NA G0HJL
BH24 1AJ G3JAZ
BH24 1FA 2E0KCL
BH24 1FA M6KCL
BH24 1LS G1RRR
BH24 1LS G1SVJ
BH24 1NX M3ZER
BH24 1PH G4ZLI
BH24 1PQ G4ZPW
BH24 1PU G8LHZ
BH24 1PX G8BDF
BH24 1QB M0TRF
BH24 1QU G3OMY
BH24 1QX G6PBO
BH24 1UY M1EHF
BH24 1UY M3ANH
BH24 1XD G6PLU
BH24 1XL G1LOU
BH24 1XX 2E0OEV
BH24 1XX M3OEV
BH24 2BH 2E1HGE
BH24 2BL G1MTA
BH24 2BT M6DTY
BH24 2HH G6EDU
BH24 2HH G6IMH
BH24 2HS 2E0DWS
BH24 2HS M0WCS
BH24 2HS M6BUI
BH24 2HY G0TKU
BH24 2HY G5HY
BH24 2JA G0SQH
BH24 2JP G0FUR
BH24 2JP G1VBA
BH24 2PE M6HFV
BH24 2PF G8SXJ
BH24 2PH G3YNJ
BH24 2QJ G3NWL
BH24 2QS M6DOB
BH24 2QW M5AGG
BH24 3BE M0HVO
BH24 3BP 2E0RFG
BH24 3BP M3RFG
BH24 3DT G3ZCI
BH24 3ER G1OCH
BH24 3HT G0MYL
BH24 3NQ G1WSN
BH24 3NQ M3WSN
BH24 4EL G0DBI
BH25 5AE M3WRN
BH25 5AY M6ITL
BH25 5BQ M0JRE
BH25 5ED M0AYC
BH25 5EJ M0EVI
BH25 5JF M0DBX
BH25 5JG G7EPE
BH25 5JP 2E1GQN
BH25 5JP 2E1HAW
BH25 5JP 2E1HFA
BH25 5JP G1ZEC
BH25 5JP G1ZVC
BH25 5JP G8CON
BH25 5JP M3NBB
BH25 5NA G0RIX
BH25 5NB G0TMZ
BH25 5NL M6CFY
BH25 5NQ 2E1IHW
BH25 5NQ G4MVP
BH25 5PE G1HHO
BH25 5PW G0HOQ
BH25 5SD 2E1DUW
BH25 5SD G4DUW
BH25 5XP G1IAV
BH25 6AB G3DCO
BH25 6BN G7KTL
BH25 6ES G4ODM
BH25 6ES G6JDP
BH25 6EY G4HFQ
BH25 6NW 2E0SEW
BH25 6NW 2E0VDE
BH25 6NW M0NFR
BH25 6NW M0RBF
BH25 6NW M3VDE
BH25 6NW M3YGF
BH25 6NZ G6LVC
BH25 6NZ G8WQW
BH25 6QE G1UWV
BH25 6SL G3UEZ
BH25 7DF M0ORE
BH25 7DQ G6IZQ
BH25 7DQ M3IZQ
BH25 7EP G7ILD
BH25 7HR 2E1IGY
BH25 7HR G8BKE
BH25 7JT M1HHL
BH25 7LU G4BJB
BH25 7LU G7MER
BH3 7AF G7HEY
BH3 7AF M3TIE
BH3 7AG M0MEY
BH3 7DG G3YUZ
BH3 7DY M1GPE
BH31 6BS G4GTH
BH31 6DH G1BEC
BH31 6DH M3JLW
BH31 6DX M6DYY
BH31 6EY G0SWF
BH31 6HJ G6VRU
BH31 6HX G8NAG
BH31 6JA M6FFM
BH31 6JJ 2E0EHV
BH31 6JJ 2E0PSZ
BH31 6JJ G4WHO
BH31 6JJ G4WHY
BH31 6JJ M0PSZ
BH31 6JJ M0VLT
BH31 6JJ M6PSZ
BH31 6JJ M6UAX
BH31 6JP 2E1HVM
BH31 6JY M0CPU
BH31 6LB 2E0RXR
BH31 6LB M6RXR
BH31 6LQ G4TMF
BH31 6QB G0VPJ
BH31 6QG G3JSJ
BH31 6XA G6SRV
BH31 7LE G3XYE
BH31 7PD G4HNJ
BH31 7PG G0DEJ
BH4 8AA M0GES
BH4 8AL M0DZT
BH4 8AL M3DUL
BH4 8AT G1HDO
BH4 8BX G7FPR
BH5 1LY G8PVK
BH5 2DJ G1PFY
BH5 2DU G4TBI
BH5 2DU G8CQG
BH5 2HT M6FCN
BH6 3DU G3WAL
BH6 3HJ G8ARA
BH6 3LU G7SUM
BH6 3ND M0MPS
BH6 3NN G4YNO
BH6 3PN G4CWY
BH6 3PZ M3ZNZ
BH6 4AE G6EZM
BH6 4DT G3KTU
BH6 4DT G4YRY
BH6 4DX G3ZHY
BH6 4DX G4RGP
BH6 4DX M3GPP
BH6 4HU G3IQX
BH6 4NB G8CRZ
BH6 5BB M6HRV
BH6 5JL G8GLY
BH6 5LD G6INW
BH6 5NW G4JYH
BH6 5PB M3XCH
BH6 5PY G8ASX
BH7 6LF 2E0SUZ
BH7 6LF G4ERV
BH7 6LF M0SUZ
BH7 6LF M6VUZ
BH7 7AA G4ELC
BH7 7AS G4BNO
BH7 7BD G0OPI
BH7 7BD G4JDW
BH8 0BT 2E0RIS
BH8 0BT M0JAX
BH8 0BT M6BCY
BH8 0DS M1EGL
BH8 0JP 2E0DFA
BH8 0JQ 2E1ICM
BH8 0NL G4STB
BH8 8NP G8FXL
BH8 8NX G0GID
BH8 8PS 2E0BKE
BH8 8PS M0KUP
BH8 8PS M3UHX
BH8 8SF M5BUF
BH8 8SR G1BNG
BH8 8SR G4WFZ
BH8 8ST M3MNR
BH8 8UA M1ERU
BH8 9AE G8OWZ
BH8 9HW G0EJO
BH8 9HY G0EGR
BH8 9PJ G7HLG
BH8 9RQ G7VJI
BH8 9SG G8MQT
BH8 9UD G3IWV
BH9 1AN M6SNQ
BH9 1DB G4KLB
BH9 1DB G8SWO
BH9 1LH G7ADH
BH9 1SH 2E0ZDB
BH9 1SH M0ZDB
BH9 1SH M6ZDB
BH9 1SH M6ZJB
BH9 1SJ M6BPV
BH9 1TX G8RAJ
BH9 2BZ G4HBD
BH9 2JE G3JAU
BH9 2JQ G6CML
BH9 2ND G3AWP
BH9 2QJ G0UAP
BH9 2UD G6ZAX
BH9 2UJ M6KIF
BH9 3AJ M0POP
BH9 3BB M3HDV
BH9 3BX M6FCI
BH9 3EH M3VUN
BH9 3LP G7VIK
BH9 3LP M3JER
BH9 3LW G1HRH
BH9 3NS G0URB
BH9 3PR M0RWB
BH9 3PR M1DYS
BH9 3PZ M1PKB
BH9 3PZ M3BVA
BH9 3SB G3BHA
BISHOP AUCKLAND
G4ZOF

## BL (Bolton)

BL0 0DP G4PAS
BL0 0LD G4YVO
BL0 0LD G6ZVO
BL0 0QA 2E1EPD
BL0 0QA 2E1EPE
BL0 0QA G3VSH
BL0 0QA M3VSH
BL0 0RY G4PVS
BL0 9BR M3IMR
BL0 9JX G7CER
BL0 9ND G3XPZ
BL0 9ND G4SVA
BL0 9QG G6SHD
BL0 9RE G6JJF
BL0 9UF G0UKM
BL0 9UT G4APJ
BL0 9XE G6CHC
BL0 9YN G0JOG
BL1 2JP G0WWH
BL1 2JX 2E0NNX
BL1 2JX M6NNX
BL1 3DJ 2E0FNB
BL1 3EZ G8JIT
BL1 3LD G0IZJ
BL1 3LP M6GXG
BL1 3PE G6VFB
BL1 3RH M3YWG
BL1 4HW G7ROM
BL1 4LZ G6TWX
BL1 4NJ G6YEA
BL1 4PA G0VUX
BL1 4RQ G0BWC
BL1 4RQ G6GVI
BL1 4RQ G6MRY
BL1 4SA G4NTC
BL1 4UA G0FRL
BL1 5GS 2E1BKT
BL1 5UN 2E1IAB
BL1 6AA G0PVP
BL1 6AH M3ZHP
BL1 6AJ M6BRP
BL1 6BL G1KQZ
BL1 6DA 2E1HLA
BL1 6DA M0BGU
BL1 6HZ M0CSP
BL1 6LU G7NRS
BL1 6LU M6ODD
BL1 6NT M3TEY
BL1 6RR G7KON
BL1 7RJ G6ZBV
BL1 7RJ M3ZBV
BL1 8PA G4RWS
BL1 8RW G1PSL
BL1 8SD G4YNK
BL1 8SF M1AWC
BL2 1AD G7DAZ
BL2 1PA G1OUG
BL2 2ET M6LUM
BL2 2SS M3DPI
BL2 2TA G0JFE
BL2 2TA G0TCY
BL2 2TA G1IOO
BL2 2UR G1JMV
BL2 3AY G1YQI
BL2 3AY G4XPU
BL2 3AY G6CVY
BL2 3BW M3UNT
BL2 3JJ M0ZVF
BL2 3LH G6HFF
BL2 3LH G6MML
BL2 3NG G4CMH
BL2 3NR M3RNX
BL2 3NS G0HYG
BL2 3PU G6OCA
BL2 3QN G6MVR
BL2 4AQ G1RRE
BL2 4DU G1ZWQ
BL2 4ET G0EWV
BL2 4LZ 2E0MLK
BL2 4LZ M0MLK
BL2 4LZ M6FYL
BL2 4NU 2E0JAC
BL2 4NU M0JBC
BL2 5BY G1ZBW
BL2 5LY G0GRX
BL2 5LY G4MUQ
BL2 5NH M6BNB
BL2 5NL G7HVO
BL2 5NP G0MYP
BL2 5NS G4TZG
BL2 5QU G6ERK
BL2 5RJ G0KEX
BL2 6BJ G1VFW
BL2 6EU G1EVR
BL2 6JJ G3TVT
BL2 6LN 2E0MYX
BL2 6LN M6MYH
BL2 6LQ M3SVT
BL2 6LT G4AQB
BL2 6NX G8YOY
BL2 6PE M0ANQ
BL2 6RQ G4CIC
BL2 6UD M3OPN
BL2 6UE M6MFU
BL2 6US G3ZIK
BL26HB M3ZPJ
BL3 1BA 2E0BVJ
BL3 1BA 2E0VBQ
BL3 1BA M0TXR
BL3 1BA M3VPM
BL3 1BA M3VXP
BL3 1BA M6BWC
BL3 1JU G4YKB
BL3 1NX G1BIF
BL3 1PN G6VPH
BL3 1PW G1EDM
BL3 1QE M3KKZ
BL3 1RG 2E0DAH
BL3 1RG 2E0RIN
BL3 1RG M0TEG
BL3 1RG M3PYO
BL3 1RG M3YRH
BL3 1RN G8ROS
BL3 1TE G8RGN
BL3 1TR M3YTA
BL3 1UB G7GLS
BL3 1UD G6CTH
BL3 3AU G7MGC
BL3 3ER 2E0IRN
BL3 3ER M6NAV
BL3 3JD G1YYH
BL3 3JH G6DEJ
BL3 3LG M1FJJ
BL3 3RB 2E0NSG
BL3 4DB G1ITV
BL3 4JU M6SXI
BL3 4LF G4TNA
BL3 4NR G8XIJ
BL3 4PB G4GIV
BL3 4PH 2E0JSR
BL3 4PP G4DFP
BL3 4QR G7RZW
BL3 4QZ G1AEQ
BL3 4QZ G1ONE
BL3 4QZ M3DEE
BL3 4SA G0BMS
BL3 4SA M3LUP
BL3 5HX G4PMV
BL3 5NU G0KAB
BL3 5PJ 2E0CAV
BL3 5PJ M0TJV
BL3 5PQ G7SKV
BL3 6QP M6EGJ
BL3 6RA M6HFN
BL3 6SJ M1FAF
BL3 6YH 2E0SIA
BL3 6YH M6ARL
BL4 0AU M6REZ
BL4 0AY M1TOM
BL4 0DS G6KQN
BL4 0HQ G1RWX
BL4 0QH 2E1HDF
BL4 0RG 2E1AMW

BL4 0RQ G0RIP
BL4 7DF M3KVR
BL4 7HS M0JDE
BL4 7LG G0IUA
BL4 7QD M0OWS
[illegible]
BL4 7TH M6KIN
BL4 8BR 2E0JMC
BL4 8BR M0NWC
BL4 8BR M0OBZ
BL4 8BR M6ZFE
BL4 8NR G1YCK
BL4 8NT G6QA
BL4 8SB M6OSX
BL4 9HT G3XUM
BL4 9LX G0KHJ
BL4 9LX M6LEE
BL4 9PT G4HYG
BL4 9QW G0TBW
BL5 1BA G1JZX
BL5 1DL M0YOT
BL5 2AF M3BBC
BL5 2AF M3LMH
BL5 2DT G6SSH
BL5 2EG 2E0XGS
BL5 2EG M6XGS
BL5 2EU G7BKL
BL5 2GR G6OBE
BL5 2HR G2ANC
BL5 2LE G0AIM
BL5 2LY G6GVR
BL5 2NT M6TCB
BL5 2PJ G7CGN
BL5 2PJ M0SPH
BL5 2RT M6MYT
BL5 2SD G6UVS
BL5 2SL M6ABV
BL5 3AX 2E1CKH
BL5 3DN G6GYC
BL5 3DP G0LBE
BL5 3HQ G4RIE
BL5 3SF M0HZX
BL5 3UZ G8IZR
BL5 3YB G4EGG
BL5 3ZD 2E0JBQ
BL5 3ZD M6PYD
BL6 4AZ G4PDD
BL6 4BB M0WHB
BL6 4BB M3BXC
BL6 4FA G4IAD
BL6 4HX G0MXH
BL6 4JF G1AQP
BL6 4LG G8WLL
BL6 4PJ G3JNM
BL6 4RQ G1VHW
BL6 5BG G0MRL
BL6 5BG M3LBQ
BL6 5QX 2E0EBJ
BL6 5RH G7CIQ
BL6 5SZ G1EFP
BL6 5TE G1VON
BL6 5TR 2E0DDN
BL6 5TR M6DNQ
BL6 5UG G4CFP
BL6 5UG G4WBO
BL6 6DJ M6CRR
BL6 6JE G0UXF
BL6 6NR G6YQI
BL6 6PZ G7GFK
BL6 6QG G3KMS
BL6 7AF M6MAM
BL6 7BE G0WBT
BL6 7ED G0KGI
BL6 7HZ G7ETK
BL6 7JU G4FSN
BL7 0DE M6OOM
BL7 0HS G0RGO
BL7 0HS G0TWD
BL7 0HS G7PSU
BL7 0PA G3HCZ
BL7 0PW G4XTG
BL7 9BJ G7CDU
BL7 9BN G1HSG
BL7 9DX G8UQV
BL7 9LL G6MAJ
BL7 9QE G3REM
BL7 9RF G3EGC
BL7 9RR M0HDQ
BL7 9SP 2E0LXF
BL7 9SP M0PSY
BL7 9SP M3NPO
BL8 1BJ G0HUQ
BL8 1DW G6DEG
BL8 1JB G0TGH
BL8 1RT G7SPL
BL8 1UE G1RAO
BL8 1UU M3KKI
BL8 1XX G6BIT
BL8 2BQ G0BZU
BL8 2HD G4QSY
BL8 2QG M3IXO
BL8 2RE G7NLP
BL8 2TS G6DGV
BL8 2TY M0MSC
BL8 2UL 2E0DQJ
BL8 2UL M6EVH
BL8 3BL G0VYP
BL8 3DB G6PZE
BL8 3DB G8AEN
BL8 3DB G8EZU
BL8 3DY 2E0DNB
BL8 3DY M3DNB
BL8 3DY M3DNC
BL8 3EZ G0LVN
BL8 3JA G0WVY
BL8 3JE G3DQQ
BL8 4BG G3BQT
BL8 4EP M0BSB
BL9 0BS G2DPL
BL9 0BS G3BRS
[illegible]
BL0 0BS G8NRS
BL9 5DL G1PKO
BL9 5DL G4KLT
BL9 5DQ G0GPH
BL9 5JG G3UGC
BL9 6JH 2E0PDG
BL9 6JH M3PDG
BL9 6JQ M6MDL
BL9 6NE M3HGH
BL9 6PP G4YYD
BL9 6RN G1YUS
BL9 6RN G4UYJ
BL9 6SJ G7LQY
BL9 7BT G4ZOI
BL9 7BU G4KQZ
BL9 7EL G8AJZ
BL9 7HB M6GJX
BL9 7QA 2E0LOE
BL9 7QA G1AKE
BL9 7QA M6LOE
BL9 7SG G7JUL
BL9 8DN G3TNQ
BL9 8HN G6MHO
BL9 8PD G1ADB
BL9 8PD G7DAL
BL9 9BS M3TIF
BL9 9DQ G0LVX
BL9 9ET M6DSV
BL9 9EY M1EKU
BL9 9EY M3EKU
BL9 9HS G6AMX
BL9 9HS M3SXU
BL9 9HZ G7TYO
BL9 9NZ G6CLX
BL9 9QB 2E0UTD
BL9 9QB M3OEH
BL9 9QE M1AGA
BL9 9UD G0RAN

## BN (Brighton)

BN1 3LS G6LFJ
BN1 3PS G3XCT
BN1 3RU G6YRJ
BN1 4AF 2E0CNW
BN1 4AF M0STL
BN1 4SG G7SRZ
BN1 5AG G8KTG
BN1 5EL G3TDL
BN1 5EP G4CFB
BN1 5EP M6ULZ
BN1 5FA G0CJZ
BN1 5FN G4FNL
BN1 5GB G3UWZ
BN1 5HH G3CUY
BN1 5NH G6CXO
BN1 5PQ M6YTI
BN1 6EB G3WBK
BN1 6GB G4HWF
BN1 6RZ 2E0WMY
BN1 6SB G4UNX
BN1 6WG G7KIL
BN1 6WL G4GDT
BN1 7BG M6KPC
BN1 7EG G3GZT
BN1 7EJ G6LMB
BN1 7FB G7UCO
BN1 7GH G6JNW
BN1 7HP G3EWT
BN1 7HX G7UDE
BN1 8AH G0SFV
BN1 8DP M6DXC
BN1 8HA G7PYT
BN1 8HF G0WCZ
BN1 8HF G8GEZ
BN1 8HJ G4BVH
BN1 8LF M6FMJ
BN1 8LP M3HXG
BN1 8NE G3WXG
BN1 8NL G2FSR
BN1 8NL G8VYP
BN1 8NL M1CTK
BN1 8NP G8VIC
BN1 8QW G8TTP
BN1 8RE G1QQD
BN1 8RH G1OYG
BN1 8SH G4WCP
BN1 8TQ G4ZWD
BN1 8UE M6FXI
BN1 8WE G3XBN
BN1 8XQ G0CKA
BN1 8YG G1GDJ
BN1 9ED M6EWC
BN10 7EF G0MIB
BN10 7JA G1KZI
BN10 7LA M3RTR
BN10 7LS M0HEP
BN10 7NS 2E0JAW
BN10 7PP G7UBB
BN10 7PS G0RNS
BN10 7RL M3UZN
BN10 7RS G4ASX
BN10 8AB M3CEZ
BN10 8DP M6SSX
BN10 8DS G4XLM
BN10 8DZ 2E0XCO
BN10 8DZ M0XCO
BN10 8DZ M6LCF
BN10 8EQ G8MSM
BN10 8HX G8KHH
BN10 8HX M5KHH
BN10 8JB G0KVF
[illegible]
BN10 8NS G7RUQ
BN10 8NS M6JDL
BN10 8PJ G8ZTG
BN10 8QD G3VQM
BN10 8RP G4PNP
BN10 8SA M0AZT
BN10 8TE M6CWK
BN10 8TH M3ZZA
BN10 8TJ G0AWG
BN11 1UG 2E0WGB
BN11 1UG M0WGB
BN11 1UG M3ORE
BN11 2DW 2E0LJK
BN11 2DW G4FFE
BN11 2DW M0LJK
BN11 2DW M6LJM
BN11 2JG M3ZNM
BN11 2LT G4DYI
BN11 2PN M6ARQ
BN11 2QE M0GEK
BN11 2SS M3ZAM
BN11 3LE M6FWS
BN11 3LL M3FTZ
BN11 3NE M6DTX
BN11 3RB G1IJY
BN11 3RU G8RKG
BN11 4EP M6RTC
BN11 4JB G0HKN
BN11 5EF G3SPN
BN11 5HB G6JEY
BN11 5QS G8RJB
BN11 5QY G8MAD
BN11 5RX 2E0OJD
BN11 5RX M0MCL
BN11 5RX M6OJD
BN12 4BH 2E1CPP
BN12 4BH M0CQQ
BN12 4HG G3YZT
BN12 4HR M0APW
BN12 4LD G3SZM
BN12 4LH G4TWK
BN12 4PN G8ZNK
BN12 4PR M6RBN
BN12 4PU M3RIP
BN12 4QA G4TRP
BN12 4TD M1KSB
BN12 4TQ 2E0VBY
BN12 4TQ M0VBY
BN12 4TQ M3VBY
BN12 4UB 2E0SRB
BN12 4UB M0HBJ
BN12 4UY M0GLQ
BN12 5BP G3LTK
BN12 5BQ G8RHU
BN12 5DJ G1DKX
BN12 5JF G7RQD
BN12 5JG M6CCI
BN12 5JW M6CCJ
BN12 5PL G4GUO
BN12 5QA G4KHM
BN12 5RD G0PBV
BN12 6AB G6JVT
BN12 6AJ 2E0DEG
BN12 6AJ M0MRI
BN12 6AJ M6ELD
BN12 6DY 2E0CHK
BN12 6EY G4UWM
BN12 6GA M6JVK
BN12 6HR G4FZF
BN12 6JE M6FQN
BN12 6LA G4GOT
BN12 6LH 2E0GPX
BN12 6LH M6DCU
BN12 6QA G3HPB
BN12 6QP M6GMQ
BN12 6QS G8MMF
BN13 1AE G4HNU
BN13 1BH M3NHU
BN13 1BH M3NHV
BN13 1DA G1OLE
BN13 1DA G3YSW
BN13 1DA M0REG
BN13 1DA M3YSW
BN13 1DG G0MVP
BN13 1DX 2E0EKB
BN13 1DX M0EKB
BN13 1DZ G4CZX
BN13 1ET M0HVD
BN13 1JS G6HXW
BN13 1LX G3SKI
BN13 1PQ G6BWP
BN13 1QX M6DMO
BN13 2AU G1YVI
BN13 2BE M1DZP
BN13 2BF M3CSH
BN13 2BH G7GOA
BN13 2DH G1KIZ
BN13 2DH G4LKW
BN13 2DT 2E0FEM
BN13 2DT 2E0KDT
BN13 2DT M6FEM
BN13 2DT M6KVT
BN13 2EN 2E0TGQ
BN13 2EN M0MEW
BN13 2HE M3MTE
BN13 2LP M3GXI
BN13 2NT G3UEQ
BN13 2PW G4BAQ
BN13 2QX G4OHC
BN13 2QY G3YHM
BN13 2SU G7EWA
BN13 2TT 2E0LDY
BN13 3AG G0JXX
BN13 3AG G4GFN
[illegible]
BN13 3AT G3XIA
BN13 3BH G3SXE
BN13 3BH G4DXO
BN13 3BZ G0SBB
BN13 3DG G4ALZ
BN13 3DH 2E0IAL
BN13 3DH M0IAD
BN13 3DH M6LBK
BN13 3DS G7EQR
BN13 3HB G6FCS
BN13 3HG G4RRU
BN13 3HH G6FJL
BN13 3HZ M6EEA
BN13 3JD G8YGK
BN13 3JG G4XHE
BN13 3JU 2E0XBT
BN13 3JU M6SKP
BN13 3LG G4FPM
BN13 3PG G7NUE
BN13 3QG 2E0BHP
BN13 3QG M0GJS
BN13 3QG M3UJS
BN13 3QQ M6OZM
BN14 0AH G3OUA
BN14 0EY 2E1AMB
BN14 0EZ G7FCU
BN14 0HR G3KGA
BN14 0HR G7GSD
BN14 0TQ 2E0FSI
BN14 0TQ M6FSI
BN14 7AE M6JMX
BN14 7BL M0AIB
BN14 7DB M6FMN
BN14 7EE G0ELN
BN14 7EG 2E0DSK
BN14 7EG G4SVQ
BN14 7EG M6DCQ
BN14 7EJ M0GMT
BN14 7EJ M1RAD
BN14 7HE G4NSJ
BN14 7HH G8MIH
BN14 7HW G8ZWC
BN14 7NE G7JRM
BN14 7PE G1KKH
BN14 7PY G6MBL
BN14 7QU 2E1EOO
BN14 7QU M3YEA
BN14 7QW 2E1HEG
BN14 7QW G8DHE
BN14 7QY M0PCR
BN14 7SB G0SWH
BN14 8AH G7PIK
BN14 8AZ G6MMJ
BN14 8DG G3UQD
BN14 8EE G1ZZC
BN14 8EL G6KHG
BN14 8ET G6GAW
BN14 9AU G2ALM
BN14 9BB M3VEM
BN14 9DG M3CWH
BN14 9DS M0OAL
BN14 9EA M3ZUJ
BN14 9JE G8RDK
BN14 9JT M0GHO
BN14 9NJ G3VXJ
BN14 9RJ G3UDH
BN15 0AE G4GPW
BN15 0DJ G0WAM
BN15 0DJ G7NLZ
BN15 0DY G4JEI
BN15 0HU G1EOM
BN15 0LP M6AJC
BN15 0LU M3YFY
BN15 0ND G7FGZ
BN15 0NG G4ZZS
BN15 0NN G8XEU
BN15 0NZ G4OEH
BN15 0QE M0KAA
BN15 0QL G1WSM
BN15 0QX G0VLC
BN15 8AS G3JSU
BN15 8DE G1UGB
BN15 8DE M1DNZ
BN15 8DF M6RFF
BN15 8EN G3IUE
BN15 8EQ M6BJI
BN15 8JN G1OCL
BN15 8LF 2E0FVV
BN15 8LF M6FVV
BN15 8LN G0UZD
BN15 8LW G4JEY
BN15 8LX 2E0EGO
BN15 8LZ G1BGC
BN15 8LZ G6BWN
BN15 8NY G0WMG
BN15 8PA G4RTX
BN15 8QD 2E0LNG
BN15 8QD M6RAO
BN15 8RL G0TLU
BN15 8RL G4GQR
BN15 8RL M0AAN
BN15 9BS G7NVI
BN15 9BT G4KIT
BN15 9BU M0EAU
BN15 9DD G1JPI
BN15 9PZ G3UFS
BN15 9QD G1DLB
BN15 9QL M0THT
BN15 9QL M3CCF
BN15 9RQ G4FIG
BN15 9SU G0AUE
BN15 9UF M0DLE
BN16 1AJ G4HNX
BN16 1AJ M1BXO
[illegible]
BN16 1DT G7KAV
BN16 1DU G8JVE
BN16 1HB G6MIC
BN16 1HE G7IWZ
BN16 1HF G0LOF
BN16 1JS M0KEL
BN16 1QS M0MNG
BN16 1QS M3DNX
BN16 1QU G4KOU
BN16 2AB G8FNH
BN16 2EF G4TMG
BN16 2HH G3ZBG
BN16 2LF G3ZZL
BN16 2NY G0BHK
BN16 2QU M6MPF
BN16 2RU G3CCX
BN16 2UB M3RBQ
BN16 3HW M0HJF
BN16 3JP M0RJK
BN16 3NP G8XDV
BN16 3NT M6FGK
BN16 3PA M1BSP
BN16 3QU G6WHH
BN16 3RZ G4EJG
BN16 4AF G8MFO
BN16 4FH M1DPY
BN16 4HE G4RVE
BN16 4HF G7AIF
BN16 4HF G8VCH
BN16 4JS G8TIU
BN17 5BX M0IMP
BN17 5BX M3PGI
BN17 5DB M3RPQ
BN17 5DP G1WXC
BN17 5DS M6SBD
BN17 5DW G6TXY
BN17 5EY G0AMP
BN17 5HE G7HNF
BN17 5HN M0FRH
BN17 5HS 2E0KZM
BN17 5HS M0KZM
BN17 5NU 2E0VKZ
BN17 5NU M0TVV
BN17 5PW 2E0GKL
BN17 5PW M6GUU
BN17 5PY G3FET
BN17 5QQ 2E0BDQ
BN17 5QQ M0WJT
BN17 5QQ M3KJE
BN17 6AS G7KEA
BN17 6AU 2E0KCB
BN17 6AU M3TKO
BN17 6BG G1AFK
BN17 6JN G0NGP
BN17 6LT G0MOU
BN17 6ND G4ZKE
BN17 6NN G1VYM
BN17 6PA G4ZFV
BN17 6QD M0RDV
BN17 6QX G4RBC
BN17 6RG G1YEW
BN17 6RG M6MCH
BN17 6RY G8DOW
BN17 6UP G3ZTZ
BN17 6UT M3CZJ
BN17 7AT G0SPZ
BN17 7BW M3EPQ
BN17 7DF G1MOW
BN17 7HB 2E0TCB
BN17 7HB M6OJC
BN17 7HN 2E0CQZ
BN17 7HN M0JSR
BN17 7HN M6BXK
BN17 7HY M6DXI
BN17 7JS G3GGN
BN17 7ND 2E0RCN
BN17 7NE M1BSO
BN17 7NE M3UJX
BN18 0AH M6RKK
BN18 0DJ G0JZU
BN18 0EG 2E0RKO
BN18 0FD 2E0MKJ
BN18 0JA G8DIU
BN18 0JF G4TSO
BN18 0JE G4WYL
BN18 0JH G0LPF
BN18 0ND G3VZJ
BN18 0PA G8KFJ
BN18 0PW G7JXB
BN18 0SD G6FKY
BN18 0SH G4ITQ
BN18 0TG M0CKM
BN18 9AP G4LAP
BN2 0FZ M6CGL
BN2 0HP M1ATU
BN2 0JH G4SLW
BN2 0JL G3XDK
BN2 0JJ G4AOG
BN2 1HE 2E0CKX
BN2 1JF 2E0DIH
BN2 1JZ G7WBO
BN2 1LE 2E0CWW
BN2 1LE M6PRF
BN2 1LE M6WKF
BN2 1LL M6NBP
BN2 2EX G4AGQ
BN2 2EX G8AAC
BN2 3BG G7MRH
BN2 3DA G6NCE
BN2 3EQ M1CKQ
BN2 3RH G1EXG
BN2 4BA M6STF
BN2 4LT G1KKF
BN2 4LU G4YLW
[illegible]
BN2 4TD G4YLG
BN2 5DA G0VAJ
BN2 5JA M3TGZ
BN2 5JS G8JFT
BN2 5PD M0IVW
BN2 5PU G0JKP
BN2 5PY G0SJP
BN2 5TD M0DKD
BN2 5UP G3UON
BN2 6BE G6JNV
BN2 6DJ G1SBZ
BN2 6PF G4NQC
BN2 6RG G0ODI
BN2 6RH M3KXZ
BN2 6SF M0OTA
BN2 6SF M6GIH
BN2 6UA M6HWM
BN2 6UE 2E0ABD
BN2 6UE G7PTZ
BN2 7DP G3ITF
BN2 7GG G8ZRM
BN2 7GL G4CBZ
BN2 7HA G4GDP
BN2 7HH G1RSC
BN2 7HH M0ATQ
BN2 8AF G2CHI
BN2 8AS M6MOH
BN2 8DH G8LGS
BN2 8DJ G0ENJ
BN2 8EZ G7NKU
BN2 8EZ M3NKU
BN2 8FQ G0ECW
BN2 8FR M0EBI
BN2 8PF M3DVB
BN2 8PH G6CIT
BN2 9YD G7FJC
BN2 9YE G4MFR
BN20 0AS G4CUG
BN20 0DJ M6GND
BN20 0EU G8GEA
BN20 7DG G7PKP
BN20 7EU G6TIQ
BN20 7HD G4DRV
BN20 7HS G8CRC
BN20 7JA M0NZL
BN20 7QE G4NYA
BN20 7TZ 2E1OLI
BN20 7XR G0UGD
BN20 8AU M0JAO
BN20 8AU M3XTE
BN20 8DA G6JME
BN20 8DT G3YXW
BN20 8HY G4MHK
BN20 8LP G4YJW
BN20 8LP G7ULN
BN20 8NT M6BVP
BN20 8SD G1KSN
BN20 8UG G1AKV
BN20 8XD G0USA
BN20 8XD G7ALC
BN20 9DJ 2E1HKY
BN20 9DJ M3DVG
BN20 9DP 2E1LGV
BN20 9DY M1EHJ
BN20 9HX M6YNN
BN20 9JA G1RMN
BN20 9JA G8CVV
BN20 9NJ G0EDS
BN20 9NJ G3ZQB
BN20 9NS G1UUP
BN20 9PP G7VAE
BN20 9QD G3ZHZ
BN20 9QJ 2E0VAJ
BN20 9QJ G1FNF
BN20 9QJ M0WNT
BN20 9QJ M6ANT
BN20 9RD G4UPY
BN20 9RJ G7GSC
BN20 9SD G4WPT
BN20 9SQ M6EYO
BN21 1QZ G8NFZ
BN21 1SH M3SPP
BN21 1UD G7GMB
BN21 1UJ G6GVM
BN21 1UJ G6ZNO
BN21 1UJ M3OYU
BN21 2AH 2E0VHC
BN21 2AH M3WUK
BN21 2BX M3TBW
BN21 2EW M6AUK
BN21 2HH G4WLV
BN21 2LQ G1SSS
BN21 2LQ G8LQZ
BN21 2LQ M3DDA
BN21 2LU G3MHF
BN21 2SD G0GZB
BN21 2TL M0LDH
BN21 2UR G1VGK
BN21 2UR G7JUD
BN21 3DD G3UCW
BN21 3QD G0RSG
BN21 3QD M0THM
BN21 4HR G6KJM
BN22 0AB G8NVB
BN22 0AJ G0JPY
BN22 0JP G4ZWE
BN22 0TT G4YGJ
BN22 0UT G7DWI
BN22 0UX 2E0DIG
BN22 0UX M6KIA
BN22 0UZ G1MNX
BN22 0XH G4NBQ
BN22 0XH G7UWG
BN22 7BT M6LGJ
[illegible]
BN22 7JF M6CHG
BN22 7JL M3EVM
BN22 7NZ G8DXU
BN22 7SZ M3YHR
BN22 8EH G0EBF
BN22 8EH G7PGH
BN22 8LW 2E1PJJ
BN22 8LW M3MJJ
BN22 8RS G7MJI
BN22 8SX M6JYS
BN22 8TQ 2E0DMV
BN22 8TQ M6GVW
BN22 8TY M0AEK
BN22 8UH G8HGM
BN22 8UQ G4SHM
BN22 8UW G7SUV
BN22 9EZ G6RGA
BN22 9EZ M3IJE
BN22 9HP G0DAX
BN22 9HQ 2E0JRP
BN22 9JL 2E1IKB
BN22 9QG G4RUL
BN22 9RH G7JVO
BN22 9RN M1DRB
BN22 9RR M3ENE
BN22 9RR M3ENF
BN22 9SF 2E0SPU
BN22 9SF G0UAI
BN22 9SF M3SPU
BN23 5AJ M3HIP
BN23 5BN G1SGR
BN23 5BN M3NJM
BN23 5PG G4XXM
BN23 5PL G3IAZ
BN23 5TH G0FIP
BN23 5TS G0LTD
BN23 5UB 2E0WGO
BN23 5UB M0WGO
BN23 5UB M3WGO
BN23 5UH G8CCO
BN23 6AB G0OZG
BN23 6AF G4EKS
BN23 6AL G6GVU
BN23 6HQ M3UDS
BN23 6JP G7TSP
BN23 6LN G0EYE
BN23 6LN G6UNU
BN23 6TQ G7PMY
BN23 7BE G6GVL
BN23 7BH G1LEO
BN23 7BT M0DUP
BN23 7DS M0CHO
BN23 7NY G1FSF
BN23 7PF G4PRJ
BN23 7PF G8XCY
BN23 7QT G4ZQS
Bn23 7RW M6DPL
BN23 7TP G8ANT
BN23 8AL M1APT
BN23 8AL M1HLG
BN23 8AX M0DJQ
BN23 8BR M3GZQ
BN23 8BX M0PHL
BN23 8DA G1FHY
BN23 8DX G0IGU
BN23 8FB G0GFP
BN23 8JD M3GUQ
BN23 8JQ 2E0SCR
BN23 8JQ M6CJP
BN23 8NS M6JFH
BN24 5AW M3TZE
BN24 5DT M0RFW
BN24 5DY M1ABM
BN24 5EZ M3ZBG
BN24 5HW M0CUY
BN24 5HW M0RJO
BN24 5NL M3RVM
BN24 5NN M6COH
BN24 5QX G4PGA
BN24 6DE G1NEZ
BN24 6DE G6GVS
BN24 6LX M0KIL
BN24 6LX M6LKI
BN24 6NH M6RFR
BN24 6RT M6DQH
BN24 6RW G0RPJ
BN24 6SL G0HVO
BN245LB 2E0BHL
BN245LB M3UYO
BN25 1DG 2E0MEU
BN25 1DG G6MAR
BN25 1DG M6MEU
BN25 1ES M6MTT
BN25 1QF G8DWX
BN25 1SP G6FYA
BN25 1SW G3DQT
BN25 2EB G1OIO
BN25 2EW G3IKF
BN25 2EW G6GAF
BN25 2HA G0GJH
BN25 2HZ G8WHD
BN25 2JZ G4BMK
BN25 2NE G4MGN
BN25 2NZ G6OHR
BN25 2RT M0RKF
BN25 2RT M6BEL
BN25 2RU G0KDD
BN25 2RU M0ABT
BN25 2TX G1VAY
BN25 2UL G3AGF
BN25 2UN 2E1IIJ
BN25 2UN M3GLC
BN25 2XW G7AFZ
BN25 3AD G3KXF
[illegible]
BN25 3EZ G8XXJ
BN25 3HH 2E0BXE
BN25 3HH 2E0RBN
BN25 3HW G4BQA
BN25 3JB 2E0PQR
BN25 3JF 2E1MPN
BN25 3JR G6FLH
BN25 3LB M1EFP
BN25 3QY M6GWE
BN25 3ST M3TPY
BN25 3ST M3WCO
BN25 3TN G0IOF
BN25 3TU G6HIE
BN25 3TU M3HIE
BN25 3UE 2E0DMX
BN25 4AL 2E1SAM
BN25 4LZ G1BAB
BN25 4NG 2E0MWC
BN25 4NU G8KUV
BN25 4PD 2E0MEZ
BN25 4PD M6DZP
BN25 4QL G7TOI
BN25 4QL M3PJG
BN25 4QR G1KVR
BN25 4QT 2E1JAC
BN25 4QT 2E1MJM
BN25 4QT M3JRM
BN25 4QT M3MJM
BN26 5HT M6HCI
BN26 5NS G3CMU
BN26 5PA M6GVU
BN26 5QA M6VAW
BN26 5QA M6VAY
BN26 6AL G0UOI
BN26 6DA G4BJP
BN26 6DQ M3VEJ
BN26 6EE M3OWF
BN26 6EL 2E0DWW
BN26 6PF M6NHN
BN27 1DS 2E0MED
BN27 1DS M0EBQ
BN27 1NA M1BSY
BN27 1NS G8URG
BN27 1NS M0CTC
BN27 1NY M3UWF
BN27 1PJ G4ZKQ
bn27 1qn M6MOV
BN27 1RL 2E0BVZ
BN27 1RL M0GUR
BN27 1RL M3YGV
BN27 1SP 2E0KFX
BN27 1SR G4RWM
BN27 1SU M6KPD
BN27 1TQ G4TWP
BN27 1TU G6YFF
BN27 1TW G6JTK
BN27 1UD 2E0RPT
BN27 1UX M6CQX
BN27 2BT G8JLD
BN27 2JY 2E0CVF
BN27 2JY M0NAZ
BN27 2JY M6SYX
BN27 2LX G1GLG
BN27 2PE M3LHU
BN27 2PE M3YNE
BN27 3AH G1RTW
BN27 3AQ M6EHY
BN27 3AT M3TEG
BN27 3AZ M6FJM
BN27 3AZ M6PDI
BN27 3BZ G3JYG
BN27 3DJ G1KAR
BN27 3DJ G4BLS
BN27 3DJ G7ECU
BN27 3DR G0UZY
BN27 3EG 2E0ERF
BN27 3EG M6BUJ
BN27 3FZ M6HKO
BN27 3HU G0RFF
BN27 3LJ G8MAF
BN27 3LS G1ATL
BN27 3ND 2E0BXG
BN27 3ND M0HJL
BN27 3ND M6HJT
BN27 3NP G4KAR
BN27 3NP G8TNH
BN27 3NX M1DMR
BN27 3QG 2E1GPE
BN27 3RA M3YDA
BN27 3TG G3NZO
BN27 3TL G4LYU
BN27 3TP M0AWY
BN27 4BG G3RXO
BN27 4DB M6VOY
BN27 4JL G0CRD
BN27 4NH 2E1BUR
BN27 4RA 2E1GTT
BN27 4TY G4HFJ
BN27 4UT 2E0ACA
BN27 4UT G3SGR
BN3 1DL M6MIZ
BN3 1PP G4HZR
BN3 1RL M1BIL
BN3 1RN G3LAZ
BN3 1TB G2NM
BN3 1TB G4JZC
BN3 2BP G0CIM
BN3 2FF M0DTB
BN3 2LH G6ONI
BN3 2RA M6HQM
BN3 2RT M0EEK
BN3 3DD M0GWZ
BN3 3RF G3LRU
BN3 3WJ G4GUX
[illegible]
BN3 5AF M0LAQ
BN3 5BE G3VBE
BN3 5DF G3MOL
BN3 5HP G7WHX
BN3 5HP M5ABJ
BN3 5ND G3ECM
BN3 5NY 2E0BHS
BN3 5NY M3XKF
BN3 5RG G7VWC
BN3 5SQ G6JAY
BN3 6BJ G8BZL
BN3 6FZ M6ETL
BN3 6HP G1VUP
BN3 6HP G3WOR
BN3 6NT G3HZT
BN3 6PJ G4DGW
BN3 6PL G3VLC
BN3 6UJ M6CJK
BN3 6WQ G0IGA
BN3 7EJ M6HZZ
BN3 7FW G7MGG
BN3 7GX G1GSK
BN3 7QB M0LHB
BN3 7QX G3LCF
BN3 8BA G4ORP
BN3 8EE G1JYH
BN3 8EQ G1KGL
BN3 8EQ G6GPF
BN3 8FE G0FUI
BN3 8LQ G6FFH
BN3 8PP G8IQX
BN41 1GG G7UTH
BN41 1PT M6VIV
BN41 1SA M6CWV
BN41 1SJ G1FZS
BN41 1SJ M3LPN
BN41 1SW M6VMR
BN41 2DN G4XKF
BN41 2DN M6FTX
BN41 2HN G1GID
BN41 2LE 2E0LDD
BN41 2LE M3LDD
BN41 2LS G7OBD
BN41 2LS G7OBE
BN41 2PN M5AJZ
BN41 2RD G0NDF
BN41 2UZ M6RSS
BN41 2WU M3OHN
BN41 2YB G1ELE
BN41 2YF G0XAN
BN41 2YF M3TII
BN41 2YF M3TIJ
BN41 2YG G0HKI
BN41 2YG G0SLW
BN41 2YL M6JOO
BN42 4AT M6BBQ
BN42 4GD M6TKK
BN42 4HG G8OYM
BN42 4LA G4YRT
BN42 4NE G0DWZ
BN42 4NH M3HXF
BN42 4QS M3XEX
BN42 4RR M6ITD
BN43 5AA M1IRB
BN43 5AN G4ZHK
BN43 5AY M3GQB
BN43 5ES G8AIP
BN43 5GD G4XUZ
BN43 5HL G4OCZ
BN43 5JH M3ITK
BN43 5LG G7VML
BN43 5LN G0FJJ
BN43 5LW G1RDU
BN43 5NE G6YIK
BN43 5NN M3RQR
BN43 5NQ G4EUK
BN43 5TH G0FIG
BN43 5WY G4JBA
BN43 6BH G4GNX
BN43 6BL G0PAW
BN43 6GH G7GMD
BN43 6GP G0FZB
BN43 6GR G0AQH
BN43 6HF G2DPY
BN43 6HY G3LLD
BN43 6LN M6MKN
BN43 6LN M0VAM
BN43 6NN G0SIU
BN43 6PF G6WDR
BN43 6PJ 2E0CCA
BN43 6RA G0SON
BN43 6RP G6ENT
BN43 6TP 2E0IDC
BN43 6TP G3XOI
BN43 6TP G6ENS
BN43 6TP G7GFX
BN43 6TP M6IDC
BN43 6YB G0IFO
BN44 3AG M1PKW
BN44 3DU M6BQV
BN44 3FJ M5NEV
BN44 3FP G3REP
BN44 3GD G4GZH
BN44 3HF M0FVV
BN44 3LJ 2E0FSE
BN44 3LJ M6FSE
BN44 3LN G6AIK
BN44 3LR G0BAF
BN44 3RH M0YDC
BN44 3RQ G1HWY
BN44 3RZ G0ANS
BN44 3TB G4RDH
BN44 3TF G1AND
BN44 3WE G3EUE
[illegible]
BN44 3WH G0NRX
BN44 3WH G1WOR
BN44 3WH G4UDU
BN44 3WH M3UJJ
BN44 3WH M3ULO
BN44 3WH M3UPO
BN44 3WJ G0UQY
BN44 3WJ G7MGT
BN44 3WJ M3UQY
BN5 9DG G8RDG
BN5 9SB G8XNB
BN5 9UX G3WMY
BN5 9YU G4XFV
BN6 8BB G4VTD
BN6 8BJ G8YKV
BN6 8BJ M6ZRJ
BN6 8BP M3ZMB
BN6 8DD G3XTH
BN6 8HR G6MJW
BN6 8HR G6YPY
BN6 8JL G4HS
BN6 8NB G0DKY
BN6 8NS G7VNG
BN6 8NU G8XYR
BN6 8PD G0GZE
BN6 8PD G4LYX
BN6 8PP M6UVD
BN6 8PR G0GMC
BN6 8PR G1ZMS
BN6 8PR G3ZMS
BN6 8SB G4CYZ
BN6 8YA G1JNX
BN6 9BZ G3SGF
BN6 9LU M0BKX
BN6 9RZ G3NYX
BN6 9TR M3NIF
BN6 9UB G8KNC
BN7 1BT M0GKP
BN7 1EN G0VPX
BN7 1HY M1ALO
BN7 1LT 2E1CCI
BN7 1LT G4XBG
BN7 1NP G3IIO
BN7 1QG G4OWT
BN7 1QG G6CWH
BN7 1QG M6OWT
BN7 1SP M6AIT
BN7 2BE G6RXK
BN7 2DL G1WSE
BN7 2DL M1CNS
BN7 2EJ G4TBM
BN7 2ET G7MOW
BN7 2HY G4PZU
BN7 2NS G0XAI
BN7 2SH G6GOS
BN7 2TX G1MXC
BN7 2UB G4YGE
BN7 3BQ 2E0DEV
BN7 3BQ 2E0FAV
BN7 3BQ M0ZAF
BN7 3BQ M3XJQ
BN7 3BQ M6ZAF
BN7 3JL G8BYC
BN8 4DS G1RKR
BN8 4GD 2E1BWT
BN8 4HP G7JKW
BN8 4JY M3BPC
BN8 4NA G5RV
BN8 4NA G6DGK
BN8 4PG 2E1JBJ
BN8 4PG G8JBJ
BN8 4PG M3BER
BN8 4PG M3BJB
BN8 4PG M3IMB
BN8 4PG M3SLB
BN8 4PS G2FQZ
BN8 4QX G0XAM
BN8 4RH M6HSR
BN8 5ET M6SCU
BN8 5HX G3YXO
BN8 5JD G7SMT
BN8 5LJ M6JNH
BN8 5PA M3IQG
BN8 6DR 2E1GJG
BN8 6DR 2E1HQA
BN8 6DR M3BOB
BN8 6DR M3IAT
BN8 6EQ M6CNZ
BN8 6HW G4SVB
BN8 6LS G4MJC
BN8 6PS G3EKJ
BN9 0JU G8ZVX
BN9 0ND G4KZX
BN9 0PS G8NQK
BN9 0RD G1YRA
BN9 9AH G1AAR
BN9 9DJ G8KAS
BN9 9DY M6ACR
BN9 9ER 2E0CAE
BN9 9ER G7PUW
BN9 9ER M3XGU
BN9 9EZ M1UKC
BN9 9LJ G6FYR
BN9 9NR M3VQI
BN9 9QB G7WHU
BN9 9SP M1DNC

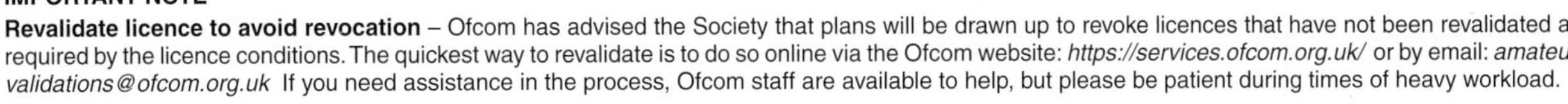

**IMPORTANT NOTE**

**Revalidate licence to avoid revocation** – Ofcom has advised the Society that plans will be drawn up to revoke licences that have not been revalidated as required by the licence conditions. The quickest way to revalidate is to do so online via the Ofcom website: *https://services.ofcom.org.uk/* or by email: *amateur.validations@ofcom.org.uk* If you need assistance in the process, Ofcom staff are available to help, but please be patient during times of heavy workload.

## BR (Bromley)

BR1 2AF G0MTA
BR1 2AT G4XGP
BR1 2EY 2E0MTT
BR1 2EY M0NJX
BR1 2EY M6NJX
BR1 2JY M6BPI
BR1 2NF G0JBP
BR1 2PR M3EFW
BR1 2UD 2E0JXB
BR1 3BL G4NCS
BR1 3DU 2E0TFE
BR1 3DU M0ORC
BR1 3DU M3TFE
BR1 3EA G6YYQ
BR1 3PU G4FXR
BR1 4DB G4OIM
BR1 4LP G8OTG
BR1 4QN M3TZX
BR1 4QY G1GYF
BR1 4SD M6RXT
BR1 5BT 2E1TAG
BR1 5BT G4AHT
BR1 5EA 2E0RTM
BR1 5EG G7CRK
BR1 5EW G0UBA
BR1 5LR M3RHB
BR1 5LR M6BHZ
BR1 5NA G0WYG
BR1 5NA G1WYG
BR2 0DN M0HOV
BR2 0EE G3IUJ
BR2 0LS G1ROK
BR2 0NW G8LHI
BR2 0RW G3IZA
BR2 0TN 2E0PAF
BR2 0UA G3EFS
BR2 6AJ M6TSR
BR2 6BF G0REE
BR2 6BF G7HCC
BR2 6DG G6YRY
BR2 6HL M6HJI
BR2 7AL M3UMB
BR2 7DY G3XDL
BR2 7DY G8ZYH
BR2 7EH M3FLV
BR2 7HE G3TSC
BR2 7HE G4XAT
BR2 7HE M1TSC
BR2 7HE M3LSU
BR2 7HE M3TVC
BR2 7HR G8MVS
BR2 7HU G4TYT
BR2 7HX G1OJO
BR2 7JA G4WJQ
BR2 7JS G4IPF
BR2 7LX G8SXU
BR2 7PT M6VJG
BR2 7QJ G8BVQ
BR2 7QN G3OQD
BR2 7QT G3NGK
BR2 7QT G8BUV
BR2 8EY G4MSK
BR2 8FG G3WOE
BR2 8LT G1BYS
BR2 8PF G1ZDR
BR2 8PP G3ZJX
BR2 8QQ G8ITB
BR2 8QQ M0XBY
BR2 9JZ M3LBP
BR3 1LE G7RUS
BR3 1NB M3UNP
BR3 1NN M6AWG
BR3 1NY G4HZX
BR3 1RE G0ELU
BR3 3AY G7TOU
BR3 3HG 2E0NCY
BR3 3HG M6NCY
BR3 3JW G7EWS
BR3 3PL G3OKY
BR3 3QA G7LSZ
BR3 3RL G7KWO
BR3 3SB G7CXO
BR3 4AE G7HYM
BR3 4AJ G4VJT
BR3 4JT M6DOM
BR3 4PZ M6FZC
BR3 4RU G8NKM
BR3 4SD G3BWV
BR3 4SS G3SZR
BR3 4XS G4NPD
BR3 5DB G1SLO
BR3 5DB G6MEH
BR3 6LJ G4RTV
BR3 6LX M0ZVB
BR3 6SN M6BXL
BR4 0HA G8VZS
BR4 0HB M1ENE
BR4 0QH G1PGS
BR4 0TA G4IKL
BR4 9AH G3ZQF
BR4 9DZ M6GJA
BR4 9JA M1LES
BR4 9JA M1PAM
BR4 9JG G8POI
BR4 9NA G6CVP
BR5 1AB G6MYZ
BR5 1AL 2E0BBG
BR5 1AL M3IXU
BR5 1BZ G8IVB
BR5 1DN M6OLP
BR5 1EY G0ZZZ
BR5 1EZ G3NAT
BR5 1EZ G4WGZ
BR5 1EZ M0UYR
BR5 1JF M6ELY
BR5 1JF M6JEM
BR5 1NF G7URM
BR5 2DL G8KPY
BR5 2EH G1FAA
BR5 2EP M6GRH
BR5 2JF G1BPD
BR5 2JR G7RUY
BR5 2LZ G6LGR
BR5 2NT G7RFM
BR5 2NY G1HIG
BR5 2NY G1UPT
BR5 2SB G7CEW
BR5 3AN G7BIP
BR5 3AT G0OMH
BR5 3AT M6AIZ
BR5 3LF G4DOE
BR5 3LG G6DEV
BR5 3LQ M3TKT
BR5 3LY G7FRH
BR5 4EQ G6XYD
BR5 4JP G1EYS
BR5 4LA G1VOJ
BR5 4LU G3BVA
BR5 4NS G0LXF
BR5 4PF G0JPM
BR5 4PF M1AHJ
BR5 4PN G4KMF
BR6 0AN G0SIO
BR6 0AQ G7PJD
BR6 0BH G1VII
BR6 0BT G7HOL
BR6 0EJ G0TZR
BR6 0EQ G4NMT
BR6 0ER G8LSC
BR6 0QE G1PJI
BR6 0QJ G0VPC
BR6 0TD G8LSI
BR6 6AY G3YQN
BR6 6BE G7DRW
BR6 6HW M1DPO
BR6 6JF M3XFB
BR6 6JU G3SIU
BR6 7QG M0GZI
BR6 7QJ M3XEF
BR6 7RT G7RGJ
BR6 7RT M0HAL
BR6 7SD G3VFD
BR6 7TD 2E0CGG
BR6 7TD M0REZ
BR6 7TD M6TDM
BR6 8BL G8KZJ
BR6 8DJ 2E0BWI
BR6 8DJ M0YRG
BR6 8EN M0LEP
BR6 8HJ G3ZOH
BR6 8HP G3MCA
BR6 8JB G6GKL
BR6 8JP G4TQO
BR6 9AA 2E0AYQ
BR6 9BS G0BZX
BR6 9BZ G8HKN
BR6 9EW M0HRY
BR6 9JG M6GTE
BR6 9LE M3HVE
BR6 9LL 2E0PAT
BR6 9LL M6EWD
BR6 9LY G1JTX
BR6 9NF G8IKA
BR6 9PN G1EOH
BR6 9PN G6XVQ
BR6 9QA G0KDV
BR6 9QA G8AXA
BR6 9RS G7AQK
BR6 9RS M3TYX
BR7 5JD G3VEK
BR7 5RN M6MMC
BR7 6AG G7CXU
BR7 6JD G7ULL
BR7 6JR M0GXN
BR7 6LA G0GZV
BR7 6SJ M1FAA
BR8 7TW G7AKM
BR8 7JA G1PUO
BR8 7JY G4IKQ
BR8 7LS M1LYE
BR8 7QT M3GML
BR8 7QZ M3KGG
BR8 7QZ M3KSG
BR8 7RJ G7LIT
BR8 7SB G3IKQ
BR8 7SE G6SYA
BR8 7UR G7PEB
BR8 8AQ G3PJB
BR8 8BH M6GSR
BR8 8DQ 2E1DQQ
BR8 8DQ M3DQQ
BR8 8DU M0HUN
BR8 8JL M0RAW
BR8 8LE M6RSL
BR8 8LP G8VJG
BR8 8NZ G4AGC
BR8 8TN M1AEP

## BS (Bristol)

BS1 4AU M0HKH
BS1 5JR G4FAZ
BS1 6EH M3NWQ
BS1 6EH M3NYG
BS10 5HH M6GXD
BS10 5PN 2E0GHD
BS10 5PN M0GZL
BS10 5PN M6GHD
BS10 5SR M3HXQ
BS10 6BP G0PID
BS10 6DT M3DWD
BS10 6JN G0JKU
BS10 7PH G0HNO
BS10 7TE G4SQQ
BS10 7XE G1EIR
BS11 0EF G4HHL
BS11 0NL M6KKJ
BS11 0QX M6LDB
BS11 0RW M6EVR
BS11 9JD M0HXI
BS11 9RT G0OKL
BS11 9RY M0GMW
BS11 9SR G1PBX
BS11 9SR G8OTH
BS11 9TX G3YHV
BS11 9TX G4AHG
BS11 9TX M3YHV
BS11 9UF G0XAK
BS11 9XB 2E0JCC
BS11 9XB G1RWT
BS11 9XB M6JBR
BS12 3LQ G4KXO
BS13 0HS M0WLF
BS13 0JG G8VUM
BS13 0NL M0PCH
BS13 0QA G7VOK
BS13 7BW G0RCU
BS13 7DB G3XED
BS13 7DF G3WCL
BS13 7LY G6FMN
BS13 7ND G7UWP
BS13 7SA G7NQJ
BS13 8AQ G6SDW
BS13 8DB M0AKF
BS13 8EF 2E0WSW
BS13 8EF M3VIU
BS13 8EZ M3WZS
BS13 8HN 2E0VTT
BS13 8HN M6GTT
BS13 8HQ G1AVB
BS13 8HQ M3RWD
BS13 8HU G8GRS
BS13 8PY M1KEV
BS13 8SA M3NRJ
BS13 9DR M6LWR
BS13 9HS M1BOB
BS13 9QQ 2E0SZH
BS13 9QQ M3AMB
BS13 9RB G0CEM
BS13 9RL M3HBP
BS14 0DS G4OJU
BS14 0EG G4YZR
BS14 0EH M0DIL
BS14 0EH M3SWS
BS14 0HH G4JFX
BS14 0NN 2E0JIL
BS14 0NN G3YLJ
BS14 0NN M0DXV
BS14 0NN M3YLJ
BS14 0NN M6BEX
BS14 0NN M6YLJ
BS14 0PB G4EPH
BS14 0PE G0DRX
BS14 0PZ G4WUB
BS14 0QG G0CCA
BS14 0RX M6MIT
BS14 8BQ M6DYW
BS14 8DY G4RAB
BS14 8EY M6PVL
BS14 8LQ G0JLI
BS14 8PZ G0KDS
BS14 8SS G7PVG
BS14 8SZ G1SLU
BS14 8TT M0CRO
BS14 9BD 2E0HCC
BS14 9BT M6MOD
BS14 9ED G4XML
BS14 9EU 2E0WXT
BS14 9LW G4NFS
BS14 9ND M3CFU
BS14 9NL G3KUL
BS14 9SF M3VRN
BS14 9TB M6FZM
BS15 1DE M0RRN
BS15 1HF G4CQI
BS15 1TA G0VBK
BS15 1UW M0EAE
BS15 1XB G0TDV
BS15 3BY G3KZC
BS15 3HH G0WPH
BS15 3JR G7MWJ
BS15 3JX G8DBP
BS15 3JZ G7IRP
BS15 3PW 2E0BYQ
BS15 3PW M0MMO
BS15 3PW M6BGT
BS15 3RA M1MVX
BS15 3RB G8VFL
BS15 3TB G0JYX
BS15 3TJ G4EQP
BS15 4BL G0GOR
BS15 4BQ G4ULV
BS15 4DR G4VJL
BS15 4HN G6NQM
BS15 4HT G7BME
BS15 4LT G7IPH
BS15 4PA G6XEB
BS15 4QB M3VAE
BS15 4QJ G4UBI
BS15 4RT G1FUJ
BS15 8AA G8DQD
BS15 8DQ G4OPO
BS15 8ES G3KAC
BS15 8ES G4BVK
BS15 8NT G4WOD
BS15 8NZ G1DFM
BS15 8NZ G1ODD
BS15 8QR M1BEP
BS15 9NN G4DCX
BS15 9PU M1MAD
BS15 9QJ G1XYR
BS15 9QP G1IHL
BS15 9SH M0VVM
BS15 9UE G8ZEX
BS15 9ZE G0CFM
BS16 1DQ G3EWF
BS16 1FB G3TEX
BS16 1FD M0DEY
BS16 1GE M6VSC
BS16 1WH 2E0EUW
BS16 1WH M0JEA
BS16 1WH M6EUW
BS16 1WL G3UHK
BS16 2AH M5AJO
BS16 2BU M6TNH
BS16 2LH M6NPD
BS16 2SW G4CXW
BS16 2UB G3RUJ
BS16 2UD G4TVD
BS16 2UF G8JUT
BS16 3DR G1IXE
BS16 3DR G1IXF
BS16 3NG G1FWF
BS16 3QY G7NZZ
BS16 3TL M0KBB
BS16 4JA M6RFO
BS16 4JD G7UUC
BS16 4PQ M0HDJ
BS16 4SQ M1BGK
BS16 5AA G7HVL
BS16 5BL G0PDV
BS16 5DE G7BLT
BS16 5HB M0MAT
BS16 5LE G3IZM
BS16 5QS G4DEU
BS16 5RU G4FHN
BS16 5RU G8LMC
BS16 5TN G4UGU
BS16 5UP G1LBM
BS16 6EG 2E0BLW
BS16 6EG M0LHS
BS16 6EG M3VHI
BS16 6HN G6CWF
BS16 6JG G4OEQ
BS16 6PN M6GFM
BS16 6PT G7IHL
BS16 6PT M0IHL
BS16 6QR G4BWJ
BS16 7BP G2BAR
BS16 7BP G4VKO
BS16 7BW M1MBZ
BS16 7BW M3MBZ
BS16 7DN G0FMJ
BS16 7HB G4JQK
BS16 7JL G6AWZ
BS16 9AG 2E0TBS
BS16 9AG M6CXB
BS16 9AZ G0KGL
BS16 9BQ G8HSR
BS16 9BZ G0XAF
BS16 9DR G0NQJ
BS16 9EA M6KGF
BS16 9EY 2E0DSB
BS16 9EY M6DFS
BS16 9HN G4RXF
BS16 9LB 2E0BFF
BS16 9LF M6JYZ
BS16 9NH G7CWN
BS16 9QF G7DSQ
BS17 1RH G4YCD
BS2 9TB G1ODE
BS2 9UB G3TVV
BS20 0DE G7IEB
BS20 0EQ G0SNP
BS20 0EY M0BJS
BS20 0JR G4VYK
BS20 0JX G4XDP
BS20 0JX G4YSZ
BS20 0LF G6TSE
BS20 0QB G4WRQ
BS20 6BB M6BGW
BS20 6JR G7AQF
BS20 6JX G4DRZ
BS20 6LD G8EZR
BS20 6LU G4YNV
BS20 6PF G4UGT
BS20 6QY G4USQ
BS20 6RG G7UTS
BS20 6SR G0RWI
BS20 6YT G6EQZ
BS20 7AD 2E0RGL
BS20 7AD M0GYY
BS20 7AD M3TYC
BS20 7AD M6TKI
BS20 7DH M6TLS
BS20 7DY G4EJH
BS20 7FG G0XZT
BS20 7FW G1UPP
BS20 7JH G0JZY
BS20 7RF M6MSY
BS20 7TQ G6ETL
BS20 8AX G4NXI
BS20 8DD G0FKJ
BS20 8JQ G4KPM
BS20 8JQ G6ZPV
BS20 8LG G4NVV
BS20 8LG G6ZOE
BS20 8RQ G4VEH
BS20 8RW G1SEW
BS21 5AQ G0ELO
BS21 5AQ M6TJA
BS21 5HB G4ONS
BS21 5HN G7ENR
BS21 6AY G6IQF
BS21 6EH M3YMX
BS21 6HR M1EPX
BS21 6JJ G1CZW
BS21 6JU G1DAX
BS21 6LE G4IWQ
BS21 6LH G0CJC
BS21 6LL 2E0LXR
BS21 6LQ G0IMA
BS21 6LQ G1YDJ
BS21 6NZ G4OCF
BS21 6RD G0AZE
BS21 6SN 2E0LZT
BS21 6SN M3LZT
BS21 6UQ 2E0IVO
BS21 6UQ M0IVO
BS21 6UQ M3IVO
BS21 6YW G1YTG
BS21 7AJ G4ULP
BS21 7DY G0AKS
BS21 7DY G1EFS
BS21 7EX G8SPC
BS21 7LU G0IFF
BS21 7LU G3WBA
BS21 7PQ G8LAN
BS21 7RL G4BBL
BS21 7TN G6BGY
BS21 7UP G6AEC
BS21 7US G7RKT
BS21 7XY 2E0CFY
BS21 7YJ M3GYA
BS21 7YJ M5AFH
BS21 7YN G8OUG
BS22 6BZ G0JGM
BS22 6DJ G1FWZ
BS22 6DQ G6AMV
BS22 6EN G8AVK
BS22 6NY 2E0HOQ
BS22 6NY M3UEQ
BS22 6RA 2E0DAR
BS22 6RA M0RGO
BS22 6RA M3PWO
BS22 6RL G0ATD
BS22 6SF G0VJM
BS22 6XP G8PRP
BS22 7FA G1OOJ
BS22 7FW G6OPD
BS22 7JX 2E0CXM
BS22 7LU M1EYL
BS22 7PF G8VZI
BS22 7PG M1EZB
BS22 7TS 2E0WSM
BS22 7YB G0IHE
BS22 8AD G1VJE
BS22 8AD M0ONS
BS22 8DD 2E0UWI
BS22 8DD G7UWI
BS22 8DD M3UWI
BS22 8DD M3UWJ
BS22 8DD M6ABZ
BS22 8JX G7HYS
BS22 8LN M6SXP
BS22 8PS 2E0BAP
BS22 8QH M1BQD
BS22 8SE G0SVA
BS22 8SE M1JMB
BS22 8TX G8GYV
BS22 8XJ M1NEW
BS22 8XR 2E0AUV
BS22 8XR 2E1HTM
BS22 8XS M3KUS
BS22 9AL G4WAZ
BS22 9AQ G7OPJ
BS22 9AY G0WMW
BS22 9BD G1XRO
BS22 9BD G4JWV
BS22 9HG M6JXY
BS22 9HT G4ZUX
BS22 9HT M1BQC
BS22 9HT M1BQE
BS22 9HU G3PLT
BS22 9LG M0FXB
BS22 9LW G4DUQ
BS22 9QS G0ADW
BS22 9SP G3GMC
BS22 9TB 2E0RCO
BS22 9UL 2E0JCO
BS22 9UL M0JCE
BS22 9UL M6JCE
BS22 9UU M0CAZ
BS22 9YD G7HVN
BS23 1RQ G3PLJ
BS23 2BH G3LJD
BS23 2HE G3RNX
BS23 2JR G3TJE
BS23 2QA M6SEO
BS23 2SY G6LQI
BS23 2UA G4UPR
BS23 2US M3HPZ
BS23 3BS G1SWX
BS23 3BS M3CPH
BS23 3BX G0TCH
BS23 3DE G1CMZ
BS23 3DF G4CXQ
BS23 3DF G8WSM
BS23 3EE M6MHY
BS23 3JE 2E0JQF
BS23 3LY M3NFJ
BS23 3PQ M6MJN
BS23 3RR G4RNZ
BS23 3RT M3NVC
BS23 3SH G3JLK
BS23 3TU G8RJO
BS23 3TU M3RJO
BS23 3XQ G4CDI
BS23 3XX M3LJX
BS23 4DB G7RUJ
BS23 4DH 2E1KIP
BS23 4HR M6EDH
BS23 4HU G6YEK
BS23 4JR G0PZB
BS23 4JY G8JNO
BS23 4JZ G0VAZ
BS23 4LQ M6CTZ
BS23 4QS G4NMV
BS23 4RF M6MWW
BS23 4YH G0IYE
BS24 0AB G0TAT
BS24 0AN M3TQX
BS24 0HY G6PZ
BS24 6TH M6OCD
BS24 7AS G7BRX
BS24 7BZ G4VMI
BS24 7DZ M3FIX
BS24 7EF G7USJ
BS24 7FH M6PAA
BS24 7GT G0WWE
BS24 7HS G8JMK
BS24 7SB M1EPN
BS24 7TJ 2E0ZDJ
BS24 8AD G6FEX
BS24 8AQ M3PLV
BS24 8BH G4HJF
BS24 8DS G3LAA
BS24 8PA G8WAL
BS24 8RZ G6XNQ
BS24 9BU M6PCL
BS24 9DY G7UUK
BS24 9EB M3UPP
BS24 9EH G0BMT
BS24 9JF G6EQI
BS24 9JW G4PWP
BS24 9LH G3WXH
BS24 9LW G3RUD
BS24 9NJ M1BDO
BS24 9RH M6IJM
BS24 9TJ G0DKM
BS25 1AL G1HHQ
BS25 1AR 2E1EMH
BS25 1AT G7SMH
BS25 1BA 2E0PMX
BS25 1BA M3PMX
BS25 1HB G3GTA
BS25 1HD G7PRI
BS25 1HL G0CHJ
BS25 1HQ G8TTX
BS25 1JE 2E0DSJ
BS25 1JE M6JWL
BS25 1JX 2E0GHA
BS25 1JX M0HAG
BS25 1JX M3ZWJ
BS25 1NB G1CGJ
BS25 1NH G3YOL
BS25 1SA G3RXG
BS25 1TG G4XYH
BS25 1TR G4RCY
BS25 1UE G7DRO
BS25 5PE M0GAC
BS25 5PR G0LCC
BS25 5PR G1OYH
BS26 2AX G4MCE
BS26 2EH G3XLX
BS26 2QW G4YKG
BS26 2RE G4ODI
BS26 2XZ G4KMB
BS27 3HS G8OGP
BS27 3HZ G0VSS
BS27 3JH 2E0VHA
BS27 3JH M3VHA
BS27 3JN M6DBP
BS27 3LE 2E0AFT
BS27 3NY G6OSR
BS27 3PB M6LSJ
BS27 3TH M6CHD
BS27 3TH M6JHK
BS27 3UB G0VVR
BS27 3XN G4KAE
BS28 4BZ G3RBJ
BS28 4HL G4FXM
BS28 4SW G0HVB
BS29 6AZ G3KVR
BS29 6BE G4OKO
BS29 6DG M6BJL
BS29 6DG M6BJP
BS29 6EA M3NKB
BS29 6JD 2E1ADP
BS3 1DP M1CWA
BS3 1LE G4LJW
BS3 1NJ M6IGH
BS3 1XB G0VJN
BS3 2BP G0FXI
BS3 2LZ M1PUW
BS3 3DY G7PXR
BS3 3EA G4XED
BS3 3EA G8BDZ
BS3 3JF M6GZU
BS3 3LU M6SSB
BS3 3PW G1HYQ
BS3 4LG M0SYJ
BS3 4NZ M6HLE
BS3 5BT G8SUV
BS3 5EG G4EGR
BS3 5ES G4OJI
BS3 5ES M6OJI
BS3 5LN G3IUO
BS3 5LR M6HQB
BS30 5JH M3OPG
BS30 5PP G4VRP
BS30 5PP G6AUR
BS30 5PW G7VVO
BS30 6DY G4SVS
BS30 6EZ G8VZB
BS30 6JA G0VDJ
BS30 6RH G0JJS
BS30 7BS G1ABA
BS30 8EJ G4MQF
BS30 8UT G4MCQ
BS30 8UT G8KGE
BS30 8YB 2E0RBG
BS30 8YB M6KXS
BS30 8YD G4FJH
BS30 8YL G1AIG
BS30 9DU G4YOC
BS30 9PX G8ZFL
BS30 9UE G2BTZ
BS30 9XB G4SNU
BS30 9XB G7JVK
BS30 9YQ G4EXZ
BS31 1AS 2E0DUH
BS31 1AS M6FKG
BS31 1BA M6SKN
BS31 1DB G1PCA
BS31 1JX 2E1GOZ
BS31 1QE G0JZH
BS31 1QF M0HBT
BS31 1QF M6BDQ
BS31 1QW G7CBI
BS31 1XD 2E1CCG
BS31 1XD G0DSB
BS31 1XG G3XAW
BS31 2BU G8OTA
BS31 2EQ G7PHE
BS31 2LJ M6GDP
BS31 2NL M3PKI
BS31 2NL M3PKL
BS31 2TR M6DYA
BS31 2TU M6TII
BS31 3DR G3VJJ
BS31 3DR M1EVP
BS31 3DU G7TFU
BS31 3DX G8VPG
BS31 3LA G6AYY
BS32 0AP G6TJZ
BS32 0BH G4GOA
BS32 0DA G7IYM
BS32 0DW G1YHN
BS32 0DZ G4GSA
BS32 0EG G0LOJ
BS32 0HB M1CTJ
BS32 4HH 2E1CYP
BS32 4LQ G4ABC
BS32 4LQ G4OST
BS32 8AF M0ZLI
BS32 8AS G4FUA
BS32 8AS M0GBH
BS32 8AS M6RBY
BS32 8AU M0JOB
BS32 8BB G4BOL
BS32 8BP G0RVM
BS32 8DP M0SUG
BS32 9AR G0RAT
BS32 9BF G4HBT
BS34 5AU G0HTS
BS34 5BG 2E0CNV
BS34 5BY M6IAF
BS34 5ER G8SYC
BS34 5HH G0UMP
BS34 5HX G4IYE
BS34 5LF M6DJJ
BS34 5NP G0NFH
BS34 5NP G1DOX
BS34 5PW G3ZZU
BS34 5PY G6ZRS
BS34 5RN G0ECM
BS34 5SA G6RUP
BS34 6EB G4DVV
BS34 6EF G1YXA
BS34 6PE M6GLW
BS34 7HZ G8RFD
BS34 7LJ G7MNO
BS34 7LU M3WLO
BS34 7RD G4YQQ
BS34 8GD G6TVJ
BS34 8GD G7RHT
BS34 8GD G8EKZ
BS34 8NG G3VM
BS34 8NG M6SDT
BS34 8NJ 2E0CCL
BS34 8NJ M0GYP
BS34 8NJ M6SGL
BS34 8PJ G0SXU
BS34 8QB G1NQB
BS34 8QH G8IMB
BS34 8RZ G4LAW
BS34 8TG M0PLO
BS34 8UD G4AEL
BS34 8XA 2E0PGS
BS34 8XA G8YMM
BS34 8XA M0NBC
BS34 8XA M3PGS
BS34 8XN G4CJV
BS35 1AY G1USV
BS35 1DP G4YZD
BS35 1JF G3XNN
BS35 1JH G4AGH
BS35 1JJ 2E0RES
BS35 1JJ M6RES
BS35 1SR G8AZT
BS35 1SX M0HFH
BS35 1SX M3EQQ
BS35 1TB G0JPU
BS35 1UE 2E0DFV
BS35 1UL 2E0UAR
BS35 1UL G4EDQ
BS35 1UL M6KVA
BS35 2DN G8MXD
BS35 2EJ G7NVZ
BS35 2EW G6RAZ
BS35 2JE G4ZOG
BS35 2LL M6CPA
BS35 2LX G1PXW
BS35 2QX G0RYM
BS35 2YD G6OLJ
BS35 2YE G3XIY
BS35 3JG G0WMB
BS35 3JL G0GZW
BS35 3JL G1FND
BS35 3LQ G4UGO
BS35 3LZ G4GMW
BS35 3NJ G0WOI
BS35 3NJ G4RAE
BS35 3PR G3HTJ
BS35 3RW G1JHD
BS35 3SB G6YNL
BS35 3TA G0NBP
BS35 3TL G6FFL
BS35 4AQ G0EBZ
BS35 4DX G0MGC
BS35 4DZ G4YHG
BS35 4HJ G0CYD
BS35 4LZ M0LDG
BS35 4LZ M5JON
BS35 4PF G0WRN
BS35 5RE G3ZUT
BS36 1BY G4PHK
BS36 1EP G3FYX
BS36 1HQ M6EZX
BS36 1NA G3GBD
BS36 2AT M3PSO
BS36 2BQ G8PQA
BS36 2EN G0LXC
BS36 2FD G4DEM
BS36 2HL G0NQG
BS36 2HT M6RXA
BS36 2NA G4BWB
BS36 2NB G4FKA
BS36 2NQ G4RKG
BS36 2RL M0GTT
BS37 4BS M0NIC
BS37 4DJ M6EIY
BS37 4EG G0IUD
BS37 4EY G3YAD
BS37 4GB G4FBK
BS37 4JX G7IBF
BS37 4LL 2E0JEZ
BS37 4LL M0JEZ
BS37 4LL M3YSQ
BS37 4LR G4IGU
BS37 4LS M3ZGM
BS37 4PF G6FFB
BS37 4PN M6ZAW
BS37 5DY M0TPW
BS37 5EX G1WVM
BS37 5PJ G6VEJ
BS37 5TF G0JZF
BS37 5UR M0SVR
BS37 5XQ M6JBE
BS37 6DQ G7GNS
BS37 6DQ M3BUU
BS37 6HE G0LXL
BS37 6JB 2E0PSM
BS37 6JB M0ZMB
BS37 6JB M6AVL
BS37 6JU G4FPI
BS37 6LA G1ZFF
BS37 6LD M6FUA
BS37 6NJ M6THQ
BS37 6XA G0UYH
BS37 6XB G7BYN
BS37 6XF M0EEP
BS37 6XF M1EEP
BS37 6XJ G1JOR
BS37 6XQ M6VSA
BS37 7AH M6DTT
BS37 7LL 2E0RKM
BS37 7LL G7TZO
BS37 7LL M6KMJ
BS37 8SA G7CJD
BS37 8UA G0JTR
BS37 8YE G0JMD
BS37 8YE G7NAR
BS37 8YW G6HKZ
BS37 9XN M0XXX
BS37 9XN M0YYY
BS39 4BH G4SND
BS39 4NT G1ZFG
BS39 5ED M3NEE
BS39 5PB G1FZV
BS39 5PL M6CBO
BS39 5PL M6PBE
BS39 5RF M6ODE
BS39 5RJ G0BLB
BS39 5SA G4OTJ
BS39 5UP G0KMV
BS39 5UT G0JCC
BS39 6ES 2E0FKV
BS39 6ES M0SKV
BS39 6ES M6FKV
BS39 6TW 2E0VHV
BS39 6TW M3VHV
BS39 6UD M5AKY
BS39 6YG G0FZP
BS39 7LU G0RWT
BS39 7LW M1ALR
BS39 7PN G1OWD
BS39 7PX G1DOJ
BS39 7PX G4LAF
BS39 7QB G7AEQ
BS39 7RP M0ALZ
BS39 7RP M0MSA
BS39 7SE 2E0DMA
BS39 7XA G8OEU
BS39 7YX G1GFD
BS4 1BN G4YTH
BS4 1JT M6VPZ
BS4 1PL G7LPP
BS4 1QE M0RCE
BS4 1RG 2E0JUW
BS4 1RG M3JUW
BS4 1RN G4RZY
BS4 1UQ G7BYK
BS4 2DL G4KUQ
BS4 2DL G8SHR
BS4 2DX G6HMV
BS4 2LJ 2E0MXR
BS4 2LJ M6VWW
BS4 2PB M6WOM
BS4 2QP G4KVT
BS4 2UP G0EXU
BS4 2UW 2E0WCB
BS4 2UW G4RAT
BS4 2UW M6WCB
BS4 2XN M3KPZ
BS4 3EY G0CCU
BS4 3JD M5AXA
BS4 3JF G6LPG
BS4 3QP G4KKU
BS4 4AF G1XXE
BS4 4BN 2E0JPH
BS4 4HN G4XCB
BS4 4JL G3XOD
BS4 4JT G6VPW
BS4 4LP G1XZV
BS4 4NA G4LEA
BS4 4NX G7AES
BS4 4QX M1ACQ
BS4 4RA G4GTD
BS4 4RT G1AIB
BS4 5AE G6PHM
BS4 5DJ G0LTB
BS4 5DZ M6FWL
BS4 5HQ M6EAN
BS4 5JA G0SYI
BS40 5EB G3XSV
BS40 5EG 2E0DDB
BS40 5HD M6DCP
BS40 5HD M6JCD
BS40 5LD 2E1DTN
BS40 5LS G4AYD
BS40 5LS G4YQG
BS40 5QG M0FCP
BS40 6AD G4MKX
BS40 6AP G1KTY
BS40 6BZ G0PCQ
BS40 6HF G3XMC
BS40 6JE G3SDH
BS40 7TL G3ZMH
BS40 8BG M0TTI
BS40 8SS G8ZOO
BS40 9YF M6ATI
BS41 8JA 2E0EEK
BS41 8JA M0KEE
BS41 8JU M6NFX
BS41 9AQ G3KOS
BS41 9AZ 2E0EKT
BS41 9FE G7HMQ
BS41 9NF G3YQV
BS48 1BB G8XXG
BS48 1HR M6USP
BS48 1JD G0CCB
BS48 1JL G4LSX
BS48 1JQ G4WAW
BS48 1JQ G7KNA
BS48 1LT 2E0GWF
BS48 1PS G8DEX
BS48 1QA G3NBN
BS48 1QF G1VBB
BS48 2AG G6ATS
BS48 2BH M0MYJ
BS48 2BN M0HCT
BS48 2DS G6WLX
BS48 2DZ G1ODB
BS48 2JN G7PVL
BS48 2QH M6ORO
BS48 2XD G0LHD
BS48 3BG G3ATX
BS48 3NE 2E0HTB
BS48 3NE M0HTB
BS48 3NE M6HTX
BS48 3NR G7UTT
BS48 3PB G3PRH
BS48 3PR G8DWW
BS48 3RX G4DEQ
BS48 4RA G0GHM
BS48 4RT G0CEN
BS48 4SX M6TAF
BS48 4YD 2E0DBW
BS48 4YD 2E0WIL
BS48 4YD G8NQO
BS48 4YD M3NQO
BS48 4YD M3WPW
BS48 4YD M3ZDW
BS48 4YH G8HVT
BS49 4AJ G6XID
BS49 4AS G0LCX
BS49 4DA G4DPH
BS49 4EB G4FZV
BS49 4EB G6GVH
BS49 4HF M6WSH
BS49 4HP G4CMC
BS49 4JU M6AEZ
BS49 4LE 2E0ZBB
BS49 4LE M6ZBB
BS49 4LN G6ANJ
BS49 4LS G4BWR
BS49 4NS M6VTC
BS49 4RB M0SWH
BS49 4RG G6BZW
BS49 5BN 2E0YCD
BS49 5BN M3FPZ
BS49 5BN M6YCT
BS49 5ES G4DYM
BS49 5ES G4FDK
BS49 5EX M0DWC
BS49 5HA 2E0JHT
BS49 5HA M0XOX
BS49 5HA M6JHT
BS49 5HB G0ALI
BS49 5HB G4TCI
BS49 5HL G4REH
BS49 5HQ G3SWH
BS5 0DL G4BWO
BS5 0SE G1JOO
BS5 6PY M6ISZ
BS5 6RJ G4TPV
BS5 6SY G8BIR
BS5 6TN M0SVA
BS5 7BQ G4GGE
BS5 7BT G1HPZ
BS5 7EJ G0SCK
BS5 7NE G0GJN
BS5 7QZ M3WMC
BS5 7RL 2E0KAC
BS5 7RL M3KAC
BS5 7SP 2E0TUK
BS5 7SP M6LUA
BS5 7SP M6TYK
BS5 7XG G4ZMW
BS5 8DX G3ZKI
BS5 8DX G4JPS
BS5 8DX G8XAA
BS5 8HF G7GLQ
BS5 8JU 2E0LJT
BS5 8JU M0LJT
BS5 8JU M6CQJ
BS5 8LN 2E0AKJ
BS5 8LN G4ZBL
BS5 8LN M3OQL
BS5 8RH G4EIA
BS5 8ST G8CKK
BS5 8ST M0BUV
BS5 8ST M6EZG
BS5 8SZ G0JLE
BS5 8TA G6YFG
BS5 8TW G0KWF
BS5 9HN M3UIC
BS5 9HN M6JWF
BS5 9JT M3MCU
BS5 9QN G4TAH
BS6 5AH 2E0PFB
BS6 5AH M0HWQ
BS6 5BQ M1BBR
BS6 5HX G8NNU
BS6 6BE G3ORV
BS6 6LH G7NJX
BS6 6NS G4FVX
BS6 6PD G4TRN
BS6 7LG G4LOX
BS6 7SU G1UGO
BS6 7XS G8UXB
BS7 0AQ G8OQG
BS7 0HS G0NZT
BS7 0LP G6YCG
BS7 0RG G0NZU
BS7 0RH G6BZG
BS7 0RH G8ZRN
BS7 0RP G8VYT
BS7 0RT G3XPJ
BS7 0SA G7UHS
BS7 0SG G7ISR
BS7 0SR G7DRU
BS7 0TT G4CSE
BS7 0UA 2E0ZST
BS7 0UA M6ZST
BS7 0UH G3FPY
BS7 0UQ G3IUV
BS7 0US G4CGF
BS7 8EX G3MGS
BS7 8JJ G8MHD
BS7 8LT G7AGI
BS7 8LU G0KKC
BS7 8QG G4YQH
BS7 8RN G4NBG
BS7 9AG M6FUN
BS7 9DW M0HCV
BS7 9ST G4ROX
BS7 9UH 2E0DQQ
BS7 9UH M3GNM
BS7 9XS 2E0RBK
BS7 9YW G7PKJ
BS8 2HF G6GGN
BS8 2QD M6GOQ
BS8 3EG G7RLQ
BS8 3GE G3YER
BS8 3PE G6DUC
BS8 4DL G8OOQ
BS9 1DR G4NYK
BS9 1NG G8HJD
BS9 1QP G4FRO
BS9 1SN G0SDW
BS9 2BA 2E0KMI

BS9 2BA M0KMI
BS9 2BW G3TCO
[illegible]
BS9 2LN G4NAQ
BS9 2LU G4WBV
BS9 2PU M6PAJ
BS9 2QP G0KJM
BS9 2QQ G8XIM
BS9 2QR G4ZBQ
BS9 2QT G0OER
BS9 2QU G8PSO
BS9 3AY 2E0CES
BS9 3AY M0HVR
BS9 3DQ G3XOB
BS9 3RN M0AWH
BS9 3SX G8GRD
BS9 3UU G4HCB
BS9 3UW G3OWX
BS9 4BU G3SJI
BS9 4EL G2BQP
BS9 4QW G8FNR
BS9 4RH G0CJG
BS9 4RS G4KSR
BS9 4TF G4OEP
BS99 5LG G0TEI
BS99 5LG G1IBO

## BT

(Belfast)

BT10 0AN GI0JRI
BT10 0AS GI6JOP
BT10 0BS 2I0FSL
BT10 0BS MI0XAX
BT10 0BS MI6TUX
BT10 0JQ GI0JRD
BT10 0QE MI0GPB
BT11 8BP GI0DPV
BT11 8LL MI6PAT
BT11 8LP GI7OMY
BT11 8PP MI6LDI
BT11 9EA 2I0OJK
BT11 9EA MI6OJK
BT11 9ED 2I0TJM
BT11 9ED MI0TJM
BT11 9ED MI3FSX
BT11 9GF GI3UIH
BT11 9LU MI6CVO
BT119ER MI1DAW
BT119ER MI5DAW
BT12 4NH 2I0GCC
BT12 4NH MI6GAQ
BT12 4QB GI7PJU
BT12 5NR MI1RDR
BT12 5NR MI3BRX
BT12 6GF GI0VWU
BT12 6JS MI0SAM
BT12 6JT MI6AHR
BT12 6NH MI6TOA
BT12 7JD 2I0KBS
BT12 7JD MI6KOB
BT13 1BP MI6DCH
BT13 2DR GI4XFN
BT13 2SB GI4GOL
BT13 3DZ GI0CDM
BT13 3LG GI4CFQ
BT13 3LR GI4PXM
BT13 3LR MI0MFI
BT13 3LR MI6EAI
BT13 3PS MI3RLA
BT13 3XN GI0ZAK
BT14 6BT MI0HPE
BT14 6ED MI6WDB
BT14 6NZ MI6SWB
BT14 6PA 2I0CVR
BT14 6PA MI6TMZ
BT14 6RZ GI4OZI
BT14 6SL MI6MSG
BT14 6TE GI4IKF
BT14 7BF MI6RBP
BT14 7GD GI7TPO
BT14 7NF MI3MPL
BT14 7NX MI6MSR
BT14 8FP MI0BDZ
BT14 8HD GI4EIZ
BT14 8JX GI4CUV
BT14 8JY GI7MBP
BT14 8PP MI3GEI
BT14 8RE 2I0ZFZ
BT14 8RE MI6FNZ
BT15 3ER MI1EZZ
BT15 3ER MI3STW
BT15 3FY GI0RWO
BT15 3LA MI5HNA
BT15 3NP MI0PCJ
BT15 3NT MI3VXQ
BT15 3QP GI0MXT
BT15 3QS MI6HAD
BT15 4EP GI1KDS
BT15 4GR GI4MNN
BT15 4JU GI7IFW
BT15 5AJ 2I0EBB
BT15 5AJ MI0EDP
BT15 5EP MI1VOX
BT15 5GA MI6FRQ
BT15 5GL GI3RQU
BT16 1JD 2I0DHR
BT16 1JD MI3DHR
BT16 1SL MI5AMO
BT16 1UU GI4SAM
BT16 1XD GI4NBO
BT16 1XU GI8SJS
BT16 2AB MI6JJZ
BT16 2BB GI4OCL
BT16 2BE 2I0TUV
BT16 2BF GI4TUV
BT16 2BF MI6TUV
BT16 2HB MI3MDV
BT16 2NT GI7KHR
BT16 2NU MI5CFM
BT16 2PQ MI3CST
BT16 2SQ MI6PDY
BT17 0AF GI1JQP
BT17 0JG GI4CSO
BT17 0NZ MI6BFJ
BT17 9EU GI4RNP
BT17 9HD MI6SIS
BT17 9PW GI7ULG
BT17 9PY GI4KCO
BT17 9PZ MI6OYB
BT17 9QY 2I0DHC
BT17 9QY MI6PBW
BT18 0BZ GI4OVN
BT18 0DY GI3GTR
BT18 0HG GI0TDP
BT18 0HH GI4JTF
BT18 0PJ GI4ZTU
BT18 0PL GI3USK
BT18 9EL GI3WUO
BT18 9EU GI0USQ
BT18 9NB GI8YJV
BT18 9NX 2I0VOQ
BT18 9NX MI3XIU
BT18 9QB 2I0LJQ
BT18 9QB MI3LJQ
BT19 1AA GI4MUE
BT19 1BJ MI3SEV
BT19 1DQ GI6ATD
BT19 1EQ GI4POC
BT19 1FG MI3CCA
BT19 1GH MI3LXZ
BT19 1HG MI3LXN
BT19 1HQ MI0CBX
BT19 1HQ MI3OHF
BT19 1LB GI6UFU
BT19 1NG GI3TAC
BT19 1NS MI0NYC
BT19 1NT GI4OCK
BT19 1QL MI6OWD
BT19 1RQ 2I0DUR
BT19 1RQ MI0OBR
BT19 1RQ MI6DUR
BT19 1RQ MI6HLC
BT19 1SR MI1EVD
BT19 1YE GI4NSS
BT19 1YG GI6IHM
BT19 1YN MI3JSH
BT19 1YU GI3XRQ
BT19 1YU MI3FSW
BT19 1YU MI3HSW
BT19 1ZQ MI3FBX
BT19 6AE GI3TZB
BT19 6AF GI4TPY
BT19 6AF GI7DIT
BT19 6AR GI3OTU
BT19 6AY GI1VPA
BT19 6BA GI3MBB
BT19 6DG GI6BDN
BT19 6DJ GI0VTS
BT19 6DQ 2I0UAD
BT19 6DQ MI6UAB
BT19 6DT GI7VIW
BT19 6DU GI3YMY
BT19 6EB MI6CBG
BT19 6FN GI3MMG
BT19 6HW GI4TCR
BT19 6HY GI6KJC
BT19 6JF GI6JGB
BT19 6LF 2I0LJW
BT19 6LF MI3XOI
BT19 6LX MI3SRG
BT19 6LZ MI3PPI
BT19 6NJ 2I0POD
BT19 6NR GI7FOD
BT19 6NX GI1SZC
BT19 6SD GI3VAF
BT19 6SD MI6HBI
BT19 6XG GI6BNI
BT19 6ZB GI6JMD
BT19 6ZH MI3FSR
BT19 6ZW GI6PLO
BT19 7FE 2I0DTE
BT19 7FE MI0OBC
BT19 7FE MI6DTE
BT19 7GB GI7GXZ
BT19 7HR GI4MRZ
BT19 7QB GI3WFP
BT19 7RB GI4WYE
BT20 0BF MI0RSO
BT20 3EP GI4LZS
BT20 3ER GI4JJF
BT20 3HA GI0HSB
BT20 3JD GI0FOD
BT20 3JF MI3VFJ
BT20 3PP GI3TZX
BT20 3PU GI0JFF
BT20 3TP GI7IFR
BT20 4DF GI0WPV
BT20 4HS GI6IVJ
BT20 4JD GI0FZT
BT20 4NP GI4OPH
BT20 4NS MI6EAC
BT20 4PE GI0HHV
BT20 4PP 2I0LNZ
BT20 4PT GI4SPU
BT20 4PT GI6SBW
BT20 4PX GI7ISX
BT20 4PX MI3ISX
BT20 4PY MI3BLN
BT20 4RQ GI3SUM
BT20 4RS GI4TMB
BT20 4TG GI8YJD
BT20 4TX GI4MRN
BT20 4US GI0BEY
BT20 4UX GI7TVV
BT20 5LT MI3CBJ
BT20 5NT MI3FEO
BT20 5NU GI8ZHW
BT20 5PN GI3UBA
BT20 5RF GI4FLG
BT20 5RQ GI4TPI
BT21 0AW MI3CBL
BT21 0BN GI1SYM
BT21 0BQ GI7FNP
BT21 0DR 2I1HNZ
BT21 0ES GI8VTK
BT21 0EY MI6KRP
BT21 0EZ 2I0ETW
BT21 0EZ GI0SSA
BT21 0EZ MI0HWG
BT21 0EZ MI1EIH
BT21 0EZ MI3BPS
BT21 0EZ MI3CQO
BT21 0EZ MI6NID
BT21 0GA GI4BXB
BT21 0HL GI1WGK
BT21 0HZ MI6CUZ
BT21 0LN GI3UPG
BT21 0NS GI7PIZ
BT21 0PU MI3FVW
BT21 0PY GI3TJM
BT21 0SH GI4PQV
BT21 0SH GI6IOU
BT22 1AF MI0AEX
BT22 1AF MI0JSJ
BT22 1AJ MI0BSU
BT22 1AU MI3MJI
BT22 1BW GI0PNP
BT22 1DZ MI0SRM
BT22 1HP GI4XSF
BT22 1JX GI0UAG
BT22 1LL GI4SJB
BT22 1LL GI4TTL
BT22 1ND GI4NAE
BT22 1NE GI6IES
BT22 1NQ 2I0GTO
BT22 1NQ GI3UZJ
BT22 1NQ MI3GTO
BT22 1QT MI0GLG
BT22 1QT MI3VFZ
BT22 1RB GI4YPR
BT22 2BG GI7NMK
BT22 2BN GI3OBO
BT22 2HY GI4PBS
BT22 2HY GI4PBT
BT22 2JQ MI0RPT
BT22 2LA GI4HCX
BT22 2LB MI6WPT
BT22 2PE MI6VAI
BT22 2QF GI4PGN
BT22 2RP MI0GUQ
BT22 2TH GI4NKK
BT22 2TR GI4EQN
BT22 2TZ 2I0FPB
BT22 2TZ GI7VCR
BT22 2TZ MI6IPB
BT23 3BN GI4AFH
BT23 4AN MI0JPC
BT23 4BN 2I0BXJ
BT23 4BN MI3LZF
BT23 4LY GI7ALH
BT23 4NA MI6OPT
BT23 4ND MI3WJO
BT23 4NT MI3ZMJ
BT23 4PD MI0RSN
BT23 4RB MI0UST
BT23 4SQ GI4MCW
BT23 4TN MI6BAI
BT23 4TP GI4JTS
BT23 4TQ GI4PGH
BT23 4UW GI0PFL
BT23 5EW GI6EGE
BT23 5HA GI7FGQ
BT23 5HE GI0SMU
BT23 5HR GI4GST
BT23 5LJ GI4TSK
BT23 5LD MI1FAR
BT23 5LN GI6CAG
BT23 5LP GI4MBM
BT23 5LI GI6VLY
BT23 5QP GI3NYJ
BT23 5RJ GI4OSG
BT23 5RN GI0EEO
BT23 5TQ GI4OWB
BT23 5TZ 2I0WWB
BT23 5TZ MI0WWB
BT23 6TZ MI6USC
BT23 5XA MI3SXI
BT23 5YX GI4IYO
BT23 6BB GI4MOA
BT23 6DE MI3VFF
BT23 6EN GI3XZM
BT23 6ES GI3VHM
BT23 6HB GI4MAC
BT23 6RZ MI3ZMP
BT23 7AD GI7UGP
BT23 7AF GI7HVC
BT23 7AN 2I0KBB
BT23 7AN MI6KBB
BT23 7AR GI0LTT
BT23 7BW MI0ABD
BT23 7BZ GI0OHT
BT23 7ED MI6CMQ
BT23 7QP GI4OYI
BT23 7RF GI0WYK
BT23 8NE GI0AQD
BT23 8NN MI6MOJ
BT23 8QS GI3AMY
BT23 8QT GI0HHZ
BT23 8RF MI3PMR
BT23 8RS MI0BVG
BT23 8RT GI0DUP
BT23 8SN MI5ALO
BT23 8TE GI3WHA
BT23 8UA MI3LVZ
BT23 8YE 2I0WMH
BT23 8YE MI6WAF
BT24 7BE GI4BJK
BT24 7BS MI6MPH
BT24 7DP MI3EZF
BT24 7FQ GI0VIF
BT24 7HU MI3NSF
BT24 7PR 2I0LXS
BT24 7PR MI6CVW
BT24 8GA GI4JOR
BT24 8HU GI8TME
BT24 8HW GI4MUN
BT24 8LB GI4TUJ
BT24 8LF GI4XLB
BT24 8LF GI6ATZ
BT24 8NQ GI4SOY
BT24 8PT GI0WJI
BT24 8QD GI4MHD
BT24 8QQ GI4AXV
BT24 8QZ GI7MDP
BT24 8QZ MI3NJU
BT24 8UN GI6DCX
BT24 8YS GI4SZP
BT25 1BD 2I0YMF
BT25 1BD GI0DVU
BT25 1BD MI0YMF
BT25 1BD MI3YMF
BT25 1BF GI7VXC
BT25 1DD MI0BAT
BT25 1LL 2I0GIF
BT25 1LL MI6GIF
BT25 1NP MI0SRR
BT25 1NP MI3AJK
BT25 1PN 2I0WKE
BT25 1PN MI6WKE
BT25 1PN MI6YAM
BT25 1RT MI6MQX
BT25 2AF MI1CUS
BT25 2AN 2I0EIG
BT25 2AN 2I0ESA
BT25 2AN MI6EQA
BT25 2AN MI6XOS
BT25 2EQ GI0UXD
BT25 2HN GI6SFO
BT25 2NT GI0TSS
BT25 2PN MI0BTK
BT25 2QJ MI0MAP
BT26 6BH GI4KSO
BT26 6BL MI0DGX
BT26 6BS 2I0VOF
BT26 6BS MI6POF
BT26 6BS MI6VOF
BT26 6BX GI4OZJ
BT26 6DJ 2I0LPG
BT26 6EG MI3VPO
BT26 6ES GI3VPV
BT26 6HL MI6TNZ
BT26 6HQ MI1BRS
BT26 6HU GI7NFB
BT26 6LJ GI4TCS
BT26 6NS GI6UUT
BT26 6PW GI4RXM
BT27 2HW MI3CSS
BT27 4BH GI7TFK
BT27 4BS GI4NFH
BT27 4DA GI4ERM
BT27 4DA GI8KEP
BT27 4EW GI6CMA
BT27 4JA GI6WFX
BT27 4NU MI3VEQ
BT27 4PL 2I0SHM
BT27 4QA GI4NLQ
BT27 4QX GI4NKY
BT27 4RY MI6GTY
BT27 4YD MI6TIJ
BT27 5BF GI4AHP
BT27 5BT GI6FOR
BT27 5BY 2I0GKB
BT27 5BY MI6GKB
BT27 5BZ MI0CZF
BT27 5DA MI6GGF
BT27 5DB GI4XIR
BT27 5HJ MI6HLY
BT27 5LF MI0OPM
BT27 5LF MI1OPM
BT27 5LF MI3TEM
BT27 5LQ GI7PBQ
BT27 5LR GI1GKI
BT27 5LW MI1FRM
BT27 5PD GI0LWO
BT27 5PD GI8TSI
BT27 5RF 2I1AXH
BT27 5RQ 2I0EIH
BT27 5RQ MI0HCK
BT27 5RQ MI6GRF
BT27 6UU 2I0ITY
BT27 6UU MI0TUB
BT27 6YW MI1ERL
BT28 1EX GI0RDJ
BT28 1HE GI0GDF
BT28 1LD GI0RYK
BT28 1PZ MI0JLC
BT28 1QD MI3OHG
BT28 1SQ GI0UVD
BT28 1YJ GI0PGC
BT28 2DN GI0TJV
BT28 2DR GI3LQV
BT28 2DW MI0DME
BT28 2ER MI6TMR
BT28 2EY MI6PGI
BT28 2GZ 2I0JCH
BT28 2GZ GI7SBF
BT28 2GZ MI6EAS
BT28 2HU MI3SYF
BT28 2HX MI3MOT
BT28 2LH GI8DGB
BT28 2PL MI0MSB
BT28 2QF MI0MOD
BT28 2TE MI6TLG
BT28 2UN MI3NSR
BT28 2XU GI4PES
BT28 3BT GI4DNW
BT28 3DS GI4XTC
BT28 3HD MI3MRG
BT28 3HG GI4LKG
BT28 3HN 2I0PRL
BT28 3HN MI6FIU
BT28 3HQ MI6MNL
BT28 3HS GI7GKC
BT28 3JH GI4MEQ
BT28 3JP GI0TJJ
BT28 3LL GI8YWE
BT28 3LP GI6GNA
BT28 3PD GI3ECQ
BT28 3QB GI4RKC
BT28 3QX MI0ASV
BT28 3QY MI0ABN
BT28 3QY MI3EOD
BT28 3RE GI4SNA
BT28 3RR GI0DVU
BT28 3RR GI4GTY
BT28 3TD GI0NQC
BT28 3TQ GI1YSG
BT28 3TR GI7GSB
BT28 3UT GI8UUN
BT28 3YB MI3TBL
BT29 4DJ GI1WLJ
BT29 4JL 2I0EQC
BT29 4JL 2I0EQR
BT29 4JL 2I0EQS
BT29 4JL GI4WTT
BT29 4JL MI3EQS
BT29 4JL MI3WTT
BT29 4JQ GI0AIJ
BT29 4JW MI5AJH
BT29 4QU 2I0OAZ
BT29 4QU MI6OAZ
BT29 4RH GI7IRJ
BT29 4SG GI7UBY
BT29 4TF MI6CXU
BT29 4WT GI7PJF
BT29 4YA MI1DJW
BT29 4YA MI6LFU
BT30 6JL MI6GPV
BT30 6NS GI1XTK
BT30 6PZ GI4NJQ
BT30 6ST MI6XOX
BT30 7AZ MI3LXE
BT30 7DA GI8UIU
BT30 7LY GI4RMA
BT30 7NU MI5AFL
BT30 7RJ 2I0OTW
BT30 7RJ MI6OTW
BT30 7SQ MI6MQF
BT30 9BS 2I0TXM
BT30 9BS MI0HNQ
BT30 9BS MI0TXM
BT30 9BS MI6XMG
BT30 9BU GI4MMJ
BT30 9BW MI3AIN
BT30 9HW MI6TND
BT30 9PD GI3HNM
BT30 9PX GI1RXM
BT31 9SJ GI0LAM
BT32 3AW MI0DNM
BT32 3RD 2I0MMA
BT32 3RD 2I0TXB
BT32 3RD GI0USS
BT32 3RD MI6MUC
BT32 3TL 2I0NGM
BT32 3TL MI6NGM
BT32 3TZ MI6MXZ
BT32 3TZ MI6RXC
BT32 3UT MI6LIT
BT32 3YA GI0WZW
BT32 4AU MI6PEJ
BT32 4BN GI7INR
BT32 4HF GI0HXH
BT32 4JL 2I0BSH
BT32 4JL MI0TFK
BT32 4LF GI3WEM
BT32 4NA MI6GPZ
BT32 4NU MI6EFD
BT32 4PE MI6RFD
BT32 4PT MI6MTO
BT32 4PY GI4GUH
BT32 4PZ 2I0SXM
BT32 4PZ MI3SXM
BT32 4RA GI8RQI
BT32 4RE GI0UQK
BT32 5JF MI0CGV
BT32 5NN GI4GPC
BT32 5PS 2I0PBM
BT32 5PS MI3GJI
BT32 5RD GI1JXE
BT33 0DT MI6AGV
BT33 0HW MI1FCB
BT33 0NQ GI3FJX
BT33 0WJ MI6PIR
BT34 1HL MI3PMW
BT34 1JG 2I0BIR
[illegible]
BT34 1JW 2I0SFH
BT34 1JW MI0PYN
BT34 1JW MI6PYN
BT34 1NZ GI0LRZ
BT34 1PW MI6TUM
BT34 2BQ MI0AQX
BT34 2BQ MI0BES
BT34 2JB MI5AHG
BT34 2NA GI6JJR
BT34 2ND GI0THR
BT34 2NY GI0WAH
BT34 2PG GI0UYY
BT34 2PG GI1YEA
BT34 2PJ GI8ZFZ
BT34 2QP 2I0MMT
BT34 2QP MI6MMT
BT34 3AN MI6AOR
BT34 3BJ GI6RBD
BT34 3DX GI8YJF
BT34 3HL GI4MBQ
BT34 3JZ MI6GMK
BT34 3NL MI3EOH
BT34 3PG GI0VVC
BT34 3SA GI1RAA
BT34 3SR MI0DDW
BT34 4AQ GI4BBE
BT34 4DA GI7IVX
BT34 4JJ GI6EBX
BT34 4LP GI4XFY
BT34 4NU GI3HJH
BT34 4XW MI3SXQ
BT34 5DE MI6WMN
BT34 5DQ MI6UNC
BT34 5EL 2I1ALE
BT34 5EL GI0UTE
BT34 5LS 2I0SEC
BT34 5LS MI0NLY
BT34 5LS MI6EEC
BT34 5TJ GI4WAH
BT35 0PJ GI0VGL
BT35 6BZ 2I0ZXM
BT35 6BZ MI6ZTM
BT35 6DD 2I0DBK
BT35 6EH GI6FXY
BT35 6LF MI3RIV
BT35 6NA GI4OVE
BT35 6NS 2I0FEX
BT35 6NS 2I0JAP
BT35 6NS 2I0RRE
BT35 6NS MI0KAG
BT35 6NS MI0RRE
BT35 6NS MI3FEX
BT35 6NS MI3RRE
BT35 6NS MI3WBU
BT35 6SD 2I0BYL
BT35 6SD 2I0FUT
BT35 6SD MI0GQG
BT35 6SD MI0GQI
BT35 6SD MI3GMI
BT35 6SD MI3LXJ
BT35 6TW 2I0MBI
BT35 6TW MI3SYI
BT35 7AA 2I0EIB
BT35 7AA MI6RSH
BT35 7GA MI6CQS
BT35 7HD GI4SZW
BT35 8AP MI6KJW
BT35 8PP MI6DVN
BT35 8PW GI1ANG
BT35 8PW MI6TOC
BT35 8XA MI3CQX
BT35 9RR MI3CQR
BT35 9RR MI3NCC
BT35 9TX GI0CTI
BT36 4QT GI4OXO
BT36 4TP GI4BUJ
BT36 4WL GI4BTG
BT36 4WQ GI4SQL
BT36 4ZW MI6MRI
BT36 5BX MI6PBI
BT36 5FG MI1AUI
BT36 5GD GI7AQO
BT36 5GZ GI4RXX
BT36 5JT 2I0WRR
BT36 5JT MI3REA
BT36 5JZ GI1WYZ
BT36 5LA 2I0BID
BT36 5LA MI0GTI
BT36 5LA MI3OEQ
BT36 5NF GI7JYK
BT36 5NF MI5JYK
BT36 5NW MI6PFI
BT36 5SJ GI4XJJ
BT36 5ST GI4KEQ
BT36 5ST MI1ASN
BT36 5WR GI4JIW
BT36 5WZ 2I0TCJ
BT36 5WZ MI0TBV
BT36 5WZ MI6TFG
BT36 6BA GI4OTG
BT36 6DD 2I0EPC
BT36 6BD MI6CMU
BT36 6LA GI4RFH
BT36 6LE MI3DNN
BT36 6LJ 2I0CLS
BT36 6LJ MI6RAS
BT36 6LJ GI4KSH
BT36 6QQ GI0BEB
BT36 6SP GI8RPT
BT36 6TZ GI8LUR
BT36 6UA MI1BOE
BT36 6UN GI4JWW
BT36 7HA GI0USW
BT36 7SU GI4GID
BT36 7TG 2I0TUI
BT36 7TG MI0HRO
BT36 7YP GI0JFF
BT37 0AZ GI4MCH
BT37 0EL GI6KVS
BT37 0GH MI6LLZ
BT37 0HH MI3OIB
BT37 0LH MI0GPF
BT37 0LH MI0XLK
BT37 0LX GI4XJC
BT37 0NL MI0OIM
BT37 0QL 2I0SUB
BT37 0RR GI0IQA
BT37 0TD GI0HHE
BT37 0UL GI4KBW
BT37 0XH GI4SFE
BT37 0XL GI0BJH
BT37 0XL GI0XYZ
BT37 0XY GI4VWC
BT37 0ZA 2I0LBS
BT37 0ZA MI0LBS
BT37 0ZY 2I0LOR
BT37 0ZY MI6BAJ
BT37 9JZ MI3FPN
BT37 9PD GI4RYL
BT37 9SH GI6ANC
BT37 9SQ MI0TBE
BT37 9TJ GI8UCS
BT38 7BL 2I0CYW
BT38 7BL MI3ZCY
BT38 7DJ MI3CGT
BT38 7DJ MI3CGU
BT38 7EH GI0USC
BT38 7EP MI3KIL
BT38 7HG GI6DKQ
BT38 7HG MI0PQR
BT38 7HN MI3JTB
BT38 7HQ MI3WMK
BT38 7JT GI8LCJ
BT38 7LD MI0KMJ
BT38 7LL MI6GZD
BT38 7LZ 2I1EXU
BT38 7NG GI0LIX
BT38 7NG GI3YRL
BT38 7NG GI4SBA
BT38 7QD GI4DAV
BT38 7RB GI6ROI
BT38 7RL MI6HDH
BT38 7RT MI3FCK
BT38 7RU MI0ZSC
BT38 7UE 2I0UCS
BT38 7UE MI0GJN
BT38 7UE MI0PJL
BT38 7UE MI3UCS
BT38 7XU 2I0TLT
BT38 7XU MI0TIP
BT38 8BF GI7VGR
BT38 8BY GI0IBC
BT38 8BY GI4GVS
BT38 8DP MI3VQH
BT38 8EW 2I0HRV
BT38 8EW MI0HRV
BT38 8EW MI6AOX
BT38 8FB GI4RVF
BT38 8FG GI4OYG
BT38 8GP MI3RLD
BT38 8GQ 2I0SMD
BT38 8GQ MI0AFT
BT38 8HA MI6HWV
BT38 8HZ GI4GCN
BT38 8JT MI6RLU
BT38 8NA GI8KYI
BT38 8NE 2I0IYH
BT38 8NE MI3IYH
BT38 8NN 2I0CEI
BT38 8NN MI0HHV
BT38 8NN MI3YVB
BT38 8SN GI0IJB
BT38 8ST MI6XEX
BT38 8SY GI6EJW
BT38 8TX GI0RUC
BT38 9AP GI0DFD
BT38 9BB GI4FUE
BT38 9DL GI8SKN
BT38 9DL MI6EXU
BT38 9EA GI6TFF
BT38 9EG GI8WDZ
BT38 9EG MI0AWL
BT38 9GZ GI0PCU
BT38 9HF GI8KFG
BT38 9JD GI7NOW
BT38 9LF GI4IZF
BT38 9ND GI6GRV
BT38 9NT MI6LGX
BT38 9NZ GI4XHO
BT38 9RL GI1TFC
BT38 9SJ MI6RUC
BT38 9SU MI0JPL
BT38 9XA GI7JEM
BT39 0BW GI0SRL
BT39 0DW GI8PDN
BT39 0HL 2I0RWG
BT39 0HL MI0RGX
BT39 0HL MI6BJG
BT39 0HS MI6WPP
BT39 0HZ GI3ZVZ
BT39 0JP GI4KQA
BT39 0PH GI4SZY
BT39 0QB GI4PRH
BT39 0QB GI6PAZ
BT39 0RJ GI4JXM
BT39 0SB GI4BWM
BT39 0SB GI7JKA
BT39 0SQ 2I0BAC
BT39 0SQ MI3IIH
BT39 0QQ MI0III
BT39 0TN GI3XDD
BT39 0TQ GI0LGV
BT39 9DD MI3FXE
BT39 9FP GI7AXB
BT39 9GN GI7FCM
BT39 9HE GI1CKU
BT39 9HT GI7GUT
BT39 9HZ GI4TAJ
BT39 9HZ GI4XGO
BT39 9HZ MI6ALL
BT39 9JS GI7DBZ
BT39 9PJ MI3TKK
BT39 9PS GI7GVI
BT39 9QW GI4JYJ
BT39 9QW GI4UUC
BT39 9QW GI6UUC
BT39 9RZ 2I0LXW
BT39 9RZ MI3LXW
BT39 9RZ MI6XAM
BT39 9SB MI6PHQ
BT39 9TS GI8VKA
BT39 9UZ 2I0RBV
BT39 9WE GI4NXJ
BT4 1LJ MI3IAI
BT4 1NA GI7PUG
BT4 1ND GI0RBC
BT4 1PR MI6SQN
BT4 1QT GI7RAM
BT4 1QU GI4IOO
BT4 2BL GI4JLF
BT4 2BY GI3TNK
BT4 2DX GI4LGP
BT4 2EH GI4GOS
BT4 2HH GI7RAH
BT4 2HS GI4FZD
BT4 2HT GI6FEN
BT4 2JX MI0RJW
BT4 2JZ GI4ILZ
BT4 2PA GI4WRJ
BT4 2RB 2I0OHE
BT4 2RB MI3OHE
BT4 2RH MI6SEZ
BT4 3BW MI6MIB
BT4 3DE GI0VAB
BT4 3DJ GI6DRK
BT4 3GD GI2BX
BT4 3GD GI4NKB
BT4 3GD GI6DEY
BT40 1EB GI8WHP
BT40 1ET 2I0GFO
BT40 1ET MI6GFO
BT40 1EW MI3BKA
BT40 1NE 2I0LRN
BT40 1NE 2I0PBZ
BT40 1NE MI6LRN
BT40 1NE MI6PBZ
BT40 1QL GI6EWO
BT40 1SE 2I0RPM
BT40 1SE MI3TUZ
BT40 1SE MI6TZP
BT40 1TE MI6WTZ
BT40 1TU GI1TRZ
BT40 1TU MI0CNI
BT40 1UB GI6VCG
BT40 1UL GI7DZE
BT40 2AJ 2I0TPC
BT40 2AJ MI6TPC
BT40 2DF GI4MXV
BT40 2EG GI4MVQ
BT40 2EJ MI6XBA
BT40 2ER MI3CIV
BT40 2ER MI3CIW
BT40 2HX GI7GJX
BT40 2JE MI3CIZ
BT40 2JH 2I0JFO
BT40 2JH MI6GHA
BT40 2JH MI6JFO
BT40 2JR MI0BOU
BT40 2PH GI4MAJ
BT40 2QA GI6DNI
BT40 2TL MI6TXS
BT40 2TU MI6ICD
BT40 2WF 2I0TAA
BT40 2WF MI3VHW
BT40 3BJ GI7USA
BT40 3DU MI0ONL
BT40 3HL GI7MDK
BT40 3JG GI4UPC
BT40 3NF 2I0XDR
BT40 3NT MI3UMC
BT40 3SD GI1OET
BT40 3SD MI6WAG
BT40 3SQ GI4RVT
BT40 3TT GI4MTZ
BT40 3TX GI0HWO
BT40 3TX GI7WCS
BT40 3TX GI0IZD
BT40 3TX MI3JCB
BT40 3UG GI0TWX
BT40 3UQ GI0XAC
BT41 1AY MI3MRF
BT41 1HH GI3HNO
BT41 1HP MI0GFE
BT41 1LN 2I0RZT
BT41 1LN MI0GIJ
BT41 1LN MI3RZT
BT41 1QF MI3CXM
BT41 2AT GI6KCX
BT41 2DP MI3TUS
BT41 2DR GI0MQN
BT41 2EU GI6OQL
BT41 2EY GI1EOS
BT41 2EY MI3SFL
BT41 2HJ GI3YDM
BT41 2JD MI6CAD
BT41 2JH GI0NNK
BT41 2PN MI0DFO
BT41 2QT GI4PID
BT41 2RF MI6DUP
BT41 2TG MI6SPY
BT41 2TR MI1DPL
BT41 2TS GI7IPO
BT41 3BA MI6IBR
BT41 3BJ GI4VJZ
BT41 3DX GI8TWB
BT41 3NH 2I0LPO
BT41 3RT GI4DCC
BT41 4EG MI6PML
BT41 4FG MI0HHU
BT41 4FG MI6BZC
BT41 4HD GI4KUM
BT41 4HD MI3NOH
BT41 4ND 2I0IDJ
BT41 4ND MI6IDJ
BT41 4NH MI6NUM
BT41 4NP GI0VLE
BT41 4NS MI6FNO
BT41 4SB GI4FUM
BT41 4SB GI4SIW
BT41 4SB GI8MIV
BT42 1AX GI4SFZ
BT42 1AX MI3FHM
BT42 1DE GI3FFF
BT42 1DE GI4HCN
BT42 1GW GI1XIB
BT42 1JA GI0JEV
BT42 1JL MI3AVJ
BT42 1JN 2I0JOS
BT42 1JN MI6JOS
BT42 1JW MI0CUN
BT42 1LY 2I0STN
BT42 1LY MI0WWF
BT42 1LY MI6STN
BT42 1NZ MI3PZV
BT42 1PU GI6KBX
BT42 1QP GI7HYU
BT42 1RW 2I0EOS
BT42 1RW MI6WJK
BT42 1RX MI3PQM
BT42 2AU GI0AYG
BT42 2BJ GI4OYL
BT42 2DG GI4TOR
BT42 2DJ MI6GDD
BT42 2DQ GI4OGQ
BT42 2DU GI1GME
BT42 2LR MI6AJO
BT42 2QH GI0THO
BT42 2QQ 2I0WDD
BT42 2QQ MI6WDD
BT42 2RG MI6KZS
BT42 2RJ MI1DEZ
BT42 2RP GI4KUZ
BT42 2RZ GI8MOV
BT42 3BE MI6GZF
BT42 3DF MI3DSM
BT42 3JE 2I0RGT
BT42 3LD GI0OHG
BT42 3LH MI6XGN
BT42 3LH MI6XOD
BT42 4AP 2I0GLC
BT42 4AP MI6GLC
BT42 4DH MI3LNC
BT42 4JG MI0TMW
BT42 4LB MI6UVS
BT42 4NJ MI3LRR
BT421BZ MI6ODN
BT43 5HE GI4ESI
BT43 5NP GI0RBO
BT43 5PR GI3XDX
BT43 5PY GI0ITJ
BT43 6DT 2I1JMC
BT43 6DT MI0JML
BT43 6DT MI3JMC
BT43 6ET MI6PHH
BT43 6JG MI3OHP
BT43 6NF GI6OJC
BT43 6PB GI4NNM
BT43 6QE 2I0IPB
BT43 6QE MI3NUW
BT43 6SX MI0IOU
BT43 6TA GI0LMR
BT43 6TL GI1KHF
BT44 0NS GI4PCQ
BT44 0QZ GI1LCM
BT44 8AD GI4XFX
BT44 8AR GI6CXD
BT44 8EF GI1BSJ
BT44 8EF GI6VCL
BT44 8EF GI7UPQ
BT44 8HH GI4RXS
BT44 8JB 2I0RVT
BT44 8JB MI6RAD
BT44 8NZ 2I0WAI
BT44 8NZ MI3UKW
BT44 8QH MI3DZD
BT44 9BN GI0WYB
BT44 9DL GI4SZU
BT44 9DT 2I0RHQ
BT44 9HZ MI0PJS
BT44 9JJ MI3XUC
BT44 9NA MI3SIL
BT44 9PE 2I0DYA
BT44 9PE MI6FHJ
BT44 9QA MI0BBF
BT44 9RH MI3UHL
BT45 5DR GI4LVC
BT45 5JF GI0EUG
BT45 5LQ MI6VOZ
BT45 5LY GI7CTW
BT45 5NU 2I0BFR
BT45 5NU MI0HMC
BT45 5NU MI3LQN
BT45 5QA GI0SRP
BT45 5QG GI4NGP
BT45 5RP MI0UTY
BT45 6BS MI0GOZ
BT45 6DN MI0GRN
BT45 6DN MI3XEY
BT45 6DS MI3UTY
BT45 6EX GI7MWA
BT45 6HW GI1BZT
BT45 6ND MI3UFD
BT45 6NH GI4WNH
BT45 6PF MI0JPD
BT45 6PU GI0PJH
BT45 6PY GI4EQA
BT45 7DT 2I0CKB
BT45 7LY GI4BGB
BT45 7NU GI3TIJ
BT45 7QF GI4NFW
BT45 7TW GI7JKM
BT45 7XS 2I0WBF
BT45 7XS MI6WBT
BT45 8AL GI3RXV
BT45 8BZ GI4BDR
BT45 8JE MI0TLF
BT45 8JE MI3EAQ
BT45 8NQ GI6OCC
BT45 8NQ MI3OYP
BT45 8PJ MI3OLM
BT46 5JR GI3ZTL
BT46 5NX MI3EYB
BT46 5TU MI3MIE
BT47 2AF 2I0RGM
BT47 2AF MI0RYM
BT47 2AF MI6RGM
BT47 2BY GI7FJY
BT47 2ES GI8AFS
BT47 2HA MI3TXI
BT47 2HH GI4YPS
BT47 2HW GI4YWT
BT47 2HW MI0KQU
BT47 2HW MI3YWT
BT47 2LD MI0SDR
BT47 2LD MI3TPR
BT47 2LD MI6RCR
BT47 2NL GI3FTT
BT47 2QB MI3JVX
BT47 2RD GI6AUI
BT47 2RD MI5TCC
BT47 2RY GI0AYB
BT47 2RY MI3NYB
BT47 2RY MI3XEB
BT47 2RY MI3XYB
BT47 2SL MI3LGL
BT47 3AQ MI3UIM
BT47 3DD 2I0MRY
BT47 3HP MI3TRY
BT47 3JW GI0NCA
BT47 3NY MI6FAU
BT47 3PR GI6BVQ
BT47 3PW GI0LDI
BT47 3PW MI3MFD
BT47 3SF MI3STY
BT47 3TE MI3CCN
BT47 3TE MI3GER
BT47 3TE MI3PPD
BT47 3TE MI6OPD
BT47 3TR MI3FGK
BT47 3TW MI3EGD
BT47 3YE 2I0TDL
BT47 3YE MI0HXB
BT47 4AD MI3OZT
BT47 4AQ GI4AHD
BT47 4AQ GI8RPP
BT47 4AS MI3TFF
BT47 4BX MI3RIL
BT47 4BX MI5RJS
BT47 4DA MI3KKP
BT47 4GA 2I0VTZ
BT47 4HQ MI3EGJ
BT47 4JN MI3RXF
BT47 4PJ GI0EFW
BT47 4PN 2I0JIE
BT47 4PN GI0AZA
BT47 4PN GI0AZB
BT47 4PN MI6JHE
BT47 4PN MI6JIE
BT47 4PP 2I0DMC
BT47 4PP MI3RUV
BT47 4ST MI6CEJ
BT47 4TJ GI4HDJ
BT47 4TT 2I0EGN
BT47 5HE GI0RJO
BT47 5JP GI0AWK
BT47 5PS MI3TJM
BT47 5QY GI0IOT
BT47 5SZ GI8ZDB
BT47 5WN MI6HAF
BT47 5WN MI6ZXT
BT47 5XS GI4OWA
BT47 5YD GI4FHB
BT47 6DU MI0UNA
BT47 6DU MI0WJM
BT47 6DU MI1JOE
BT47 6DU MI1UNA
BT47 6DU MI3CEM
BT47 6DU MI3SVM
BT47 6DU MI3XRT
BT47 6FE GI0TSA



**IMPORTANT NOTE**

**Revalidate licence to avoid revocation** – Ofcom has advised the Society that plans will be drawn up to revoke licences that have not been revalidated as required by the licence conditions. The quickest way to revalidate is to do so online via the Ofcom website: *https://services.ofcom.org.uk/* or by email: *amateur.validations@ofcom.org.uk* If you need assistance in the process, Ofcom staff are available to help, but please be patient during times of heavy workload.

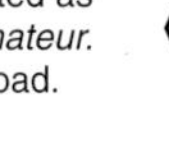

BT47 6FE MI3TWM
BT47 6HY MI6CGQ
BT47 6NF MI6GNP
BT47 6SE GI0WYO
BT47 6UG GI0IUP
BT47 6UW MI3IFO
BT47 6XA MI3DBB
BT47 6XL MI3CXD
BT47 6YB MI6SCC
BT47 6YU MI3WQC
BT48 0AD GI4MJD
BT48 0AD GI8YBU
BT48 0AD MI0DVH
BT48 0AU GI3TJJ
BT48 0BY GI0NWN
BT48 0JX MI3DKN
BT48 0QA GI4ZLD
BT48 0RL MI3CGA
BT48 0RS GI0HDO
BT48 6NS MI6WBN
BT48 7ER GI0PXS
BT48 7JN GI4DGI
BT48 7UA MI3GJG
BT48 8AQ GI0GGY
BT48 8BA 2I0HWW
BT48 8BA MI3PCF
BT48 8GQ GI3TME
BT48 8HN MI3IYP
BT48 8HX 2I0NAT
BT48 8HX MI3WFT
BT48 8HX MI3WVL
BT48 8HX MI6XBL
BT48 8JD GI0SFT
BT48 8JD GI7KVR
BT48 8JW GI0TQD
BT48 8JW GI1BEU
BT48 8JW GI4EXI
BT48 8NT GI7UPU
BT48 8PF GI4EPK
BT48 8RT MI6LLS
BT48 8SU MI6BLW
BT48 8SU MI6DCE
BT48 8TH GI0DSG
BT48 9DU GI0OZQ
BT48 9LA GI1PVE
BT49 0AE GI4EBS
BT49 0AP MI6WJW
BT49 0AP MI6WPW
BT49 0AS MI3VTJ
BT49 0AT MI6MCK
BT49 0BB MI3YXX
BT49 0BD GI0RBS
BT49 0BF 2I0NWO
BT49 0BF MI0NWO
BT49 0BF MI3NWO
BT49 0BF MI6EJK
BT49 0BH MI0JBT
BT49 0DH GI1ELP
BT49 0DQ MI3JTM
BT49 0EQ GI44OKU
BT49 0EQ GI4OKU
BT49 0HS MI3XWW
BT49 0HS MI3YYT
BT49 0NF 2I0KEW
BT49 0NF MI3IHZ
BT49 0NT GI4PMP
BT49 0NY GI4MAZ
BT49 0QB MI6JAD
BT49 0QF GI0BZM
BT49 0QF GI3KVD
BT49 0QF MI3UIW
BT49 0RG 2I0WBD
BT49 0RG MI6WAB
BT49 0RH MI6WKN
BT49 0RW MI3MBM
BT49 0SF GI3VAW
BT49 0SH MI0DWD
BT49 0SH MI3BIE
BT49 0UF 2I0BHT
BT49 0UF MI0GGB
BT49 0UF MI0GKL
BT49 0UF MI3IHY
BT49 9AD GI4YDP
BT49 9BQ GI7JRG
BT49 9BS GI6VWS
BT49 9BS MI3INS
BT49 9BS MI3JDQ
BT49 9EN MI3XGR
BT49 9EW MI3RYJ
BT49 9HQ GI4GNT
BT49 9HT 2I0BTT
BT49 9LR GI4SIZ
BT49 9LY GI4ODT
BT49 9PG 2I0SHZ
BT49 9PG MI1AIB
BT49 9PG MI6MOI
BT49 0AT MI3UIV
BT5 4FE 2I0WSH
BT5 4FT 2I0HRM
BT5 4FT MI6MIH
BT5 4NT GI7RCH
BT5 5DU 2I0HBO
BT5 5DU MI0HZD
BT5 5DU MI6EMI
BT5 5HS GI6BDI
BT5 5LT 2I0CGZ
BT5 5LT MI6AOZ
BT5 5NT GI3KYP
BT5 5NT GI8YTH
BT5 6AD GI0UTV
BT5 6AL GI7IMU
BT5 6BT GI4FNU
BT5 6DD GI7UQW
BT5 6ED MI1DQB
BT5 6FN GI7URC
BT5 6GE MI6WEZ
BT5 6NG GI0BRO
BT5 6NG GI0UEG
BT5 6PU GI0WCE
BT5 6PU GI3XLK
BT5 7AP GI6HKE
BT5 7DH GI4CBG
BT5 7EH MI6UBT
BT5 7EQ GI3MMF
BT5 7EQ GI6YM
BT5 7EY GI0OUM
BT5 7EZ GI4SSF
BT5 7HL GI8YYM
BT5 7HL MI6GPQ
BT5 7HW GI4ZOS
BT5 7JP GI7MDJ
BT5 7LX GI4VRF
BT5 7LY MI0ALS
BT5 7LZ GI0BFO
BT5 7NR GI0BDZ
BT5 7PS GI4MYT
BT5 7RL MI3SXR
BT51 3AY MI3FOL
BT51 3PZ MI0JZZ
BT51 3QZ GI4OHH
BT51 3RA MI0RTY
BT51 3RD GI8LTB
BT51 3RE GI4GPA
BT51 3RW MI6XRC
BT51 3RZ GI4VIZ
BT51 3SN MI6JMC
BT51 3TN 2I0SEK
BT51 3TN MI0PLC
BT51 3TN MI3SEK
BT51 4AR MI1BNO
BT51 4BD GI7LCQ
BT51 4DN 2I0GWX
BT51 4LZ GI7SOB
BT51 4NB 2I0GCN
BT51 4NB MI0ULK
BT51 4NB MI6MFR
BT51 4NE GI1LBI
BT51 4NG MI3VXI
BT51 4RN GI4ZAH
BT51 4SD GI3PLL
BT51 4SZ MI3CON
BT51 4TS GI4HVI
BT51 4US GI8OLH
BT51 5BJ MI1CSA
BT51 5JP MI0BYR
BT51 5RZ MI0HYQ
BT51 5RZ MI0SLE
BT51 5SB MI0DMT
BT51 5SB MI0DNB
BT51 5SB MI0MSM
BT51 5TA GI6JXG
BT51 5YR GI6GBK
BT51 5YR MI0AAZ
BT52 1EN 2I0TAN
BT52 1EN MI0TBD
BT52 1EN MI3ZYU
BT52 1EW GI4JFP
BT52 1JR GI7TEB
BT52 1JW GI6IBL
BT52 1LJ MI6RME
BT52 1NG GI1RXL
BT52 1QH GI1JHQ
BT52 1SS GI3JOZ
BT52 1TL GI4WWF
BT52 1TU MI3GFA
BT52 1TW GI6EIR
BT52 1TY MI3BRJ
BT52 1UJ 2I0PAC
BT52 1UJ MI3UIA
BT52 1UJ MI6PFD
BT52 1WN MI0CIB
BT52 1WN MI3BYJ
BT52 2BL MI3VCI
BT52 2DR GI4NRQ
BT52 2DR GI4ORI
BT52 2DY MI6BEF
BT52 2ES MI0RJN
BT52 2EU MI0HMY
BT52 2EX MI3NWU
BT52 2HD MI0HPX
BT52 2JL 2I0WAS
BT52 2JL MI0WJC
BT52 2JL MI6WRM
BT52 2ND 2I0ETB
BT52 2ND MI6ETE
BT52 2NY GI7TTO
BT52 2QA MI3CKF
BT52 2QE 2I0IMO
BT52 2QE MI3IMO
BT52 2QF MI6CWC
BT52 2QF MI6OKS
BT521SP 2I0BKI
BT521SP MI0WGW
BT521SP MI3UHI
BT53 6BZ MI0JAY
BT53 6EQ MI0ENR
BT53 6ES 2I0BPO
BT53 6ES MI0JAR
BT53 6ES MI3FPE
BT53 6EX MI0JPO
BT53 6HT MI6RYC
BT53 6JG MI3MMC
BT53 6NT 2I0TJF
BT53 6NT MI6TJF
BT53 6PT MI6GAU
BT53 6QB GI6GIE
BT53 6QF GI3OYG
BT53 7AH GI6GAQ
BT53 7AQ MI3FOJ
BT53 7AS 2I0EAR
BT53 7AS MI6HGV
BT53 7BB MI3DCM
BT53 7BE MI0MCC
BT53 7BX MI0VAC
BT53 7BZ 2I0BAD
BT53 7DE MI0GHI
BT53 7PT MI6LFK
BT53 7QL GI8DHW
BT53 7QS MI0OBE
BT53 8AB MI6MSV
BT53 8DL MI3LMR
BT53 8JT GI4JRA
BT53 8NL MI0JTE
BT53 8NL MI3CQB
BT53 8NL MI3TRR
BT53 8NY MI6GGN
BT53 8QQ MI0MCB
BT54 6AN MI3YJE
BT54 6BW MI3BUT
BT54 6BW MI3BYQ
BT54 6DQ 2I1SWD
BT54 6DQ MI0CRQ
BT54 6DQ MI1DOG
BT54 6DQ MI3DMM
BT54 6DQ MI3SWD
BT54 6DS GI8WIU
BT54 6DZ 2I0RMK
BT54 6DZ 2I0TJK
BT54 6DZ MI0JBK
BT54 6DZ MI0RMK
BT54 6DZ MI3TJK
BT54 6HS MI6WOF
BT54 6JH MI3WES
BT54 6JL GI0JQQ
BT54 6LE MI3WDO
BT54 6LP MI3POB
BT54 6PF MI0MRV
BT54 6PF MI3DDK
BT54 6PF MI3TXT
BT54 6QZ MI0CRR
BT54 6QZ MI0MRG
BT54 6RT MI1BLZ
BT55 7DL MI3CDA
BT55 7EG GI7JLD
BT55 7HA MI0GVC
BT55 7HA MI0KAM
BT55 7HF 2I0EBF
BT55 7HF MI6EAZ
BT56 8AS GI4OYM
BT56 8GP GI4CZO
BT56 8HN GI7TMQ
BT56 8JN GI4OQY
BT56 8NJ GI7SLN
BT56 8QA GI8PGJ
BT56 8SP GI0OTC
BT56 8ST MI3DOD
BT56 8SY MI3CAB
BT57 8AE MI0GBU
BT57 8AE MI0RUC
BT57 8AE MI3NRB
BT57 8RB MI6TAW
BT57 8RY MI6VBB
BT57 8SD 2I0LLG
BT57 8SD MI0LLG
BT57 8SD MI6LLG
BT57 8UX GI0ISQ
BT57 8WH MI0CCG
BT57 8WH MI0LJM
BT57 8YX GI8AIR
BT6 0AD MI3PUH
BT6 0AE GI7CMC
BT6 0AE MI1BRA
BT6 0DG GI4IBV
BT6 0DZ GI7KEC
BT6 0ER GI0DHW
BT6 0FN MI0AAW
BT6 0NA GI3XEQ
BT6 0NH GI4FGH
BT6 0NR GI3YDH
BT6 8BH GI4RCK
BT6 8NL GI3LEG
BT6 9DS 2I0IZI
BT6 9DS MI6AHO
BT6 9DW GI3XHL
BT6 9GJ GI3UEX
BT6 9JF GI4LZR
BT6 9PJ GI6NTP
BT6 9RB MI5WJB
BT6 9RH 2I0OTC
BT6 9RH MI6GIN
BT6 9RP GI0URN
BT6 9RP MI0NOR
BT6 9RX GI4GOV
BT6 9SF GI4MDD
BT6 9SF GI7JAM
BT6 9SR MI3KDR
BT60 1DS GI4XFE
BT60 1HP GI4RYP
BT60 1JA MI3LWU
BT60 1JU MI1EQI
BT60 1LE 2I0KXM
BT60 1LE MI6KXM
BT60 1LP GI4SRQ
BT60 1LQ MI3WLW
BT60 1LZ MI0DBK
BT60 1NT GI8OCR
BT60 1QR GI6TBC
BT60 1TW GI7WLA
BT60 1TW MI6RHL
BT60 2BH 2I0FPT
BT60 2BN GI0DWN
BT60 2FF MI0YCK
BT60 2GP MI3ZJN
BT60 2GP MI6LOT
BT60 2JF GI0MSI
BT60 2JQ MI3RST
BT60 2NA 2I0GWA
BT60 2NA GI0KOW
BT60 2UP MI3WBL
BT60 3AA GI0MSH
BT60 3JS 2I0CBV
BT60 3JS GI0NWG
BT60 3JS MI3YTH
BT60 3TQ GI0WLW
BT60 3TS GI0VHG
BT60 4AS GI6NNP
BT60 4BL GI0GPG
BT60 4BU GI1VMF
BT60 4BU MI3LFE
BT60 4DZ 2I0RHN
BT60 4DZ MI6UDR
BT60 4NZ 2I0KPA
BT60 4NZ MI0KPA
BT60 4NZ MI0WHG
BT61 7DF GI8RLG
BT61 7JB GI8RNG
BT61 7JD GI0NYI
BT61 7PE MI6DOD
BT61 7QU GI0LTF
BT61 7SA GI0MSG
BT61 7SA GI7IEZ
BT61 8BU MI0CLP
BT61 8EZ GI3NSV
BT61 8EZ GI4NSV
BT61 8JD GI6NFK
BT61 8NP GI0MTE
BT61 8NR MI0GRG
BT61 8NX GI0OND
BT61 8NX MI0VFW
BT61 8NX MI6YSW
BT61 8RW GI4SYM
BT61 9BB GI0MSK
BT61 9DT MI6JVC
BT61 9HA GI7DWF
BT61 9HA MI3BSN
BT61 9JX GI4FFL
BT61 9LD GI0ADD
BT61 9LD GI8RLE
BT61 9LT GI7GRY
BT61 9RW 2I0NTH
BT61 9RW MI6BZI
BT62 1EH GI0KUH
BT62 1EW MI3IRV
BT62 1EY GI0RYU
BT62 1JN GI7FCP
BT62 1JX GI0AIQ
BT62 1RN GI0UJG
BT62 2BB GI6NGF
BT62 2BB MI6NVM
BT62 2BL MI0SAP
BT62 2BL MI3SEO
BT62 2DD 2I0MVP
BT62 2DD 2I0PRM
BT62 2DD MI0MVP
BT62 2DD MI0OCG
BT62 2DD MI0PRM
BT62 2DD MI3WWJ
BT62 2DD MI6MHI
BT62 2DD MI6MHK
BT62 2EJ MI0IRZ
BT62 2LZ MI6TFN
BT62 2NE MI3VWH
BT62 2NF MI6ECC
BT62 2NJ GI3ONF
BT62 2NJ GI4IVI
BT62 3BN 2I0EJT
BT62 3BN MI6AJA
BT62 3ED GI7UDV
BT62 3ED MI5AFM
BT62 3EH MI3IFI
BT62 3HY MI6CAV
BT62 3JB MI6LSY
BT62 3JF MI6NOE
BT62 3QH GI4SJQ
BT62 3QU GI8TAX
BT62 3QX GI0DQJ
BT62 3RN GI4BQI
BT62 3TD 2I0SAI
BT62 3TD 2I0WAH
BT62 3TD MI0NID
BT62 3TD MI0SAI
BT62 3TD MI3XUS
BT62 3TD MI3YBI
BT62 3TF 2I0EBS
BT62 3TF MI6EBS
BT62 4HL MI0TRC
BT62 4HL MI3DFR
BT62 4HP GI4MDO
BT62 4HX MI6FWD
BT62 4JP 2I0RIR
BT62 4JP MI6RIR
BT622BL 2I0SGM
BT63 5AR 2I0JPP
BT63 5AR GI7VBS
BT63 5AR GI8XSB
BT63 5AR MI0UYD
BT63 5DQ GI1ACN
BT63 5HR 2I0TME
BT63 5JS GI1RSR
BT63 5LT GI6FTM
BT63 5PF GI0NOX
BT63 5PQ MI6ILF
BT63 5RJ MI3KFI
BT63 5RS GI7TGJ
BT63 5RX MI3ISC
BT63 5SW GI4YCZ
BT63 5SZ 2I0MFJ
BT63 5SZ MI6RCV
BT63 5TL GI6EEH
BT63 5TY MI6FHG
BT63 5UU 2I0TJR
BT63 5UU MI3TJR
BT63 5XT MI3IRY
BT63 5YD GI0PVG
BT63 5YD GI6FHD
BT63 5YD GI6NAQ
BT63 5YD GI7GHC
BT63 5YH GI4MXW
BT63 5YJ GI8JOA
BT63 6AT MI1AYL
BT63 6BB 2I0BIQ
BT63 6BH GI3WWY
BT63 6EU GI7FHU
BT63 6FA GI6MTL
BT63 6LY 2I0SFA
BT63 6LY MI6PUX
BT63 6NB MI6DVM
BT63 6NX MI6GOA
BT64 3AN MI3PDN
BT65 4AB GI6GAG
BT65 4AL GI0STS
BT65 4AT 2I0ZXD
BT65 4AT MI0NWA
BT65 4AT MI6NCG
BT65 4AT MI6NOS
BT65 5AE MI6TRF
BT65 5DH GI0SZH
BT65 5JF MI3LZA
BT66 6LD GI0MHB
BT66 6PS MI0TBN
BT66 6QW GI6FQT
BT66 7AN 2I0TRM
BT66 7AN MI0LRC
BT66 7AN MI3RYD
BT66 7DJ 2I0JSQ
BT66 7DJ MI6JSQ
BT66 7EB GI3POS
BT66 7HD GI8GZM
BT66 7HH GI4BDL
BT66 7LE MI1FIS
BT66 7LP 2I0RGD
BT66 7LP MI6BGD
BT66 7NE GI4GEL
BT66 7NE GI8LFY
BT66 7PD MI3RBM
BT66 7PP MI0BPB
BT66 7PQ GI4YRP
BT66 7QY GI0OHU
BT66 7RU GI3WEL
BT66 7SY MI6HGS
BT66 7TG 2I0GLY
BT66 7TG 2I0WGM
BT66 7TG MI0WGM
BT66 7TG MI6WGM
BT66 7TG MI6XEM
BT66 7TY MI0MEV
BT66 7UA GI4ISR
BT66 7UT 2I0ROC
BT66 7UT MI6DED
BT66 8DF 2I0SNG
BT66 8DF MI0SNG
BT66 8JP MI0EDF
BT66 8JZ 2I0HSL
BT66 8JZ MI6HSL
BT66 8LE GI3SCM
BT66 8LW GI7OOM
BT66 8QT MI3WQT
BT66 8RH GI0OUZ
BT67 0AP MI3GSW
BT67 0AU 2I0FNN
BT67 0AU MI6FIK
BT67 0ED MI0BMM
BT67 0FA MI0SMV
BT67 0FQ MI0GCV
BT67 0FQ MI3OFX
BT67 0GB GI4MWA
BT67 0GB MI6AJN
BT67 0GB MI6JOY
BT67 0GJ 2I0YDF
BT67 0GJ MI0YAM
BT67 0GJ MI3YDF
BT67 0HP GI4AIO
BT67 0HW MI6GBC
BT67 0JB GI4MNF
BT67 0JR MI0TGO
BT67 0JZ 2I0FBY
BT67 0LB GI0OEH
BT67 0LB MI6SJV
BT67 0LN MI3UGI
BT67 0NL GI1FSJ
BT67 0NZ 2I0GYL
BT67 0NZ 2I0YLT
BT67 0NZ GI7PWQ
BT67 0NZ MI0YLT
BT67 0NZ MI6YLG
BT67 0NZ MI6YLT
BT67 0QT GI0VGV
BT67 0RN MI5HIL
BT67 0SS 2I0GSG
BT67 0SS MI6GSG
BT67 0SZ GI6IRL
BT67 0UT 2I0DKQ
BT67 0UT MI6JGK
BT67 9BS MI3PJM
BT67 9HN MI5LYN
BT67 9JN GI4ELQ
BT7 1BL MI3BCR
BT7 2GF GI8HUD
BT7 2GJ MI0BOK
BT7 3HA GI6KDN
BT70 1SZ GI3OZW
BT70 1TH GI4XKI
BT70 1UH GI0RDM
BT70 2EL GI4VCZ
BT70 3BU GI7FCW
BT70 3DT GI0ZER
BT70 3DY GI7TDA
BT70 3DY MI6LCR
BT70 3JY GI4XAA
BT70 3LU GI4LDN
BT70 3PU GI7GHC
BT71 4AA MI6FQY
BT71 4AG GI7NEB
BT71 4AJ GI7RTB
BT71 4DW MI3NPR
BT71 5GA MI0CAC
BT71 5JL 2I0SMY
BT71 5JL MI0SMY
BT71 5JL MI6SJD
BT71 6DD GI0URI
BT71 6DR GI3NFM
BT71 6FF MI6HCP
BT71 6HX MI6EQD
BT71 6LG MI3FKI
BT71 6PS GI4CSP
BT71 6PW 2I0EXP
BT71 6PW MI6EXP
BT71 6QG GI3NPP
BT71 6QN GI3OQR
BT71 6TN GI0IVJ
BT71 7AY MI3IOH
BT71 7BH GI0KPF
BT71 7BN MI6EQC
BT71 7ER GI0KVQ
BT71 7ER MI0AVI
BT71 7ET MI6MYW
BT71 7HT GI0VIB
BT71 7JN GI8ITD
BT71 7NY 2I0WBU
BT71 7PY MI0CXE
BT71 7QF GI4TED
BT71 7QF GI4XJD
BT71 7SQ GI4SLQ
BT71 7SY GI4CYU
BT74 4DY 2I0NIE
BT74 4DY MI3NIE
BT74 4RB MI3TMN
BT74 5ND GI7NET
BT74 5NQ GI4CZW
BT74 6AZ MI6FEB
BT74 6BQ MI3GHW
BT74 6EP GI0BQX
BT74 6ET MI0BTM
BT74 6EU GI0BFD
BT74 6HD MI1FHE
BT74 6JJ MI6REM
BT74 6JP 2I0EAS
BT74 6JP MI3YUT
BT74 6JW MI6JBH
BT74 7JN GI4UHA
BT75 0LG MI6LLI
BT76 0HJ GI0EWE
BT76 0XE GI4CQL
BT78 1ES MI6AUU
BT78 1QS 2I0NMA
BT78 1QS MI3MWA
BT78 1QZ GI4OHW
BT78 2DD MI6BJZ
BT78 2DD MI6BKD
BT78 2DL GI0LXN
BT78 2EH GI0KDH
BT78 2PN MI6UTV
BT78 2PP GI7THY
BT78 3AW 2I0CFW
BT78 3AW MI3DWQ
BT78 3HL MI6WRT
BT78 3LR GI0OXK
BT78 4HX GI1CAI
BT78 4HX MI6OMA
BT78 4JT GI4KJC
BT78 4JZ GI0HVJ
BT78 4LG GI0TMS
BT78 5AL MI1FCQ
BT78 5BA 2I0FPK
BT78 5BA MI0PPA
BT78 5BA MI6FPK
BT78 5BB MI3KRL
BT78 5HG MI6PCJ
BT78 5JE MI6OTP
BT78 5RS MI3TJV
BT79 0DW 2I0JXO
BT79 0DW MI3JXO
BT79 0DZ 2I0OMA
BT79 0DZ MI6MRJ
BT79 0EB GI4RSI
BT79 0HF GI3NVW
BT79 0HX MI1CTQ
BT79 0JX GI0STC
BT79 0PG MI6ODG
BT79 0PH GI0GQG
BT79 0PL GI4UDI
BT79 0RZ GI6EGJ
BT79 0WA MI6BVN
BT79 0XS GI3XCZ
BT79 7BN GI7UCS
BT79 7FB 2I0LDC
BT79 7FB MI6BII
BT79 7FB MI6ECV
BT79 7FB MI6EJT
BT79 7JJ GI4LIF
BT79 7JX MI6PWR
BT79 7LB MI6EDK
BT79 7PQ GI0EJT
BT79 7PQ GI1RIB
BT79 7RA 2I0DJM
BT79 7RA GI4SXV
BT79 7RA MI3DJM
BT79 7SJ GI7AUY
BT79 7SJ GI7FHZ
BT79 7SJ MI3FHZ
BT79 7SJ MI6TYR
BT79 7TJ GI0EJU
BT79 7WJ GI3YBZ
BT79 7WJ GI8BNC
BT79 7XG 2I0MSO
BT79 7XG MI0MSO
BT79 9AY MI3DLO
BT79 9NH MI6SFK
BT79 9PH MI1CCU
BT79 9PH MI6CCU
BT79 9QA GI4IWP
BT8 6HU GI0BFA
BT8 6JX MI5EEM
BT8 6RG GI3YMT
BT8 6SB GI4KIX
BT8 6WL GI4JIC
BT8 7DB GI4WME
BT8 7DN GI4XAP
BT8 7RJ GI1VAZ
BT8 8BQ GI4TAP
BT8 8HQ GI4TAV
BT8 8JZ GI4TGR
BT8 8LX GI8PDK
BT8 8NL MI0BDX
BT8 8PH MI3ZJT
BT8 8TG GI4SIP
BT80 0AA MI6ENR
BT80 0JL MI6CAK
BT80 8AS MI0RJY
BT80 8BW GI8XSY
BT80 8JH MI0AHH
BT80 8LJ GI7BET
BT80 8PL GI3SOO
BT80 8PL GI8YGG
BT80 8PW MI0AHI
BT80 8PW MI0BML
BT80 8PZ MI6HGI
BT80 8QE GI6JRY
BT80 8QJ GI0JHR
BT80 8RL MI1AVH
BT80 8RS GI6NDM
BT80 8XW GI8JYA
BT80 9LB GI6WHZ
BT80 9RN MI0AIH
BT81 7DF MI6DAY
BT81 7DR MI3SLT
BT81 7NA MI3NVX
BT81 7QF 2I0NDJ
BT81 7QF MI0NDK
BT81 7QY MI0JAT
BT81 7TJ MI3CCT
BT81 7TJ MI3FCA
BT81 7WD MI3JQD
BT82 0AE GI0VJE
BT82 0AE MI3BXQ
BT82 0AE MI3JVJ
BT82 0AE MI3JVV
BT82 0AE MI5ALJ
BT82 0AJ MI3EZK
BT82 0BD GI4OKX
BT82 0DB GI4OUP
BT82 0DN MI3RXU
BT82 0DP GI3CFH
BT82 0DP GI4OUN
BT82 0DP MI3OUN
BT82 0DW MI0RIB
BT82 0DW MI3ASH
BT82 0DW MI3BEG
BT82 0DW MI3RNN
BT82 0DZ MI3CGZ
BT82 0DZ MI3FPB
BT82 0DZ MI3IGO
BT82 0GZ GI0EWP
BT82 0HN MI3NEN
BT82 0HN MI3TLV
BT82 0LZ 2I0BSA
BT82 0LZ 2I0FLO
BT82 0LZ MI3ATT
BT82 0PD MI6EAF
BT82 0PD MI6JMD
BT82 0PD MI6VJR
BT82 0PP GI4XGQ
BT82 0SF MI3NMG
BT82 8EL GI7THH
BT82 8LD GI4VQK
BT82 8LS MI0HOZ
BT82 8LS MI6LJO
BT82 8NW GI4VKS
BT82 8QQ MI6XKE
BT82 9AJ 2I0VFO
BT82 9AJ MI0GTM
BT82 9AJ MI3OZK
BT82 9DT 2I0IMB
BT82 9DT MI3KMB
BT82 9LA MI1AWM
BT82 9LS MI0HRG
BT82 9LS MI1CCT
BT82 9PN MI1DRP
BT82 9PN MI3AGR
BT82 9QL GI4DYE
BT82 9QS MI3PDL
BT82 9RQ MI6POH
BT82 9TL MI6RAC
BT9 5AA GI3ZCK
BT9 5AQ GI1KGZ
BT9 5EL MI0EFM
BT9 5FQ MI6IRE
BT9 5GZ MI3WDI
BT9 5PG GI1MJJ
BT9 5PQ GI4LQU
BT9 5QL GI4OMK
BT9 5QL GI4VIP
BT9 5QP GI4WVN
BT9 5QP GI4XFS
BT9 6HA GI3VYY
BT9 6JX MI3GWQ
BT9 6LJ GI8SKR
BT9 6NF MI6GQI
BT9 6PJ GI7JUH
BT9 6QR GI3SG
BT9 6QX 2I0NGK
BT9 6QX MI6TCO
BT9 6TJ GI3PSQ
BT9 6UH GI3VCI
BT9 7AX GI4ISH
BT9 7JD GI4DOM
BT92 0EQ MI6AFI
BT92 0EQ MI6AFM
BT92 0FS MI1RAY
BT92 0FS MI6PMM
BT92 0FS MI6SMM
BT92 0FW MI3LDO
BT92 0HJ 2I0XAN
BT92 0HJ MI6WAN
BT92 2DL GI0LEC
BT92 2DL GI6ZXD
BT92 2EW 2I0WGL
BT92 2EW MI0WGL
BT92 2EW MI6WGL
BT92 3BU GI0CAH
BT92 6NB 2I0PPW
BT92 6NB MI0PPW
BT92 6NB MI6MCF
BT92 7AR MI3JRJ
BT92 7DD 2I0WMC
BT92 7DD MI6BMC
BT92 7DD MI6SFH
BT92 7DD MI6UIM
BT92 7FN 2I0RMD
BT92 7FN MI0RMD
BT92 7FN MI6GDN
BT92 7FN MI6RAV
BT92 7GG MI6WAM
BT92 7HP MI6COM
BT92 7PP MI6NCC
BT92 7PR MI6WDM
BT92 7PT 2I0RVH
BT92 7PT MI0RVH
BT92 7PT MI6TNI
BT92 7QY GI7JGT
BT92 8AE MI6CAY
BT93 0FQ GI1OMD
BT93 1UH GI4SZQ
BT93 3FT GI0PTQ
BT93 6FR 2I1GIH
BT93 7DN GI6PYP
BT93 8DF GI3WBR
BT94 1FJ MI3JXG
BT94 1RN MI5MTC
BT94 2AS MI6YPY
BT94 2FU GI7UIP
BT94 2JA GI1RBI
BT94 3DB 2I0GHY
BT94 3DB MI0IIG
BT94 3DB MI3GHY
BT94 3EH GI8WJN
BT94 3EQ 2I0EKN
BT94 3FJ MI6NTP
BT94 4QX MI6OFN
BT94 5BX GI6RMO
BT94 5EA MI0DWE
BT94 5EW GI4VHO
BT94 5EW MI6UBE
BT94 5GL GI6JPO
BT94 5GL MI0RCF
BT94 5HJ GI6UFO

## CA

(Carlisle)

CA1 1HQ M0AOH
CA1 1LX G4RQW
CA1 2DL M6RYK
CA1 2FH M6LCK
CA1 2QE M6ELH
CA1 2QJ G3XWA
CA1 2SZ G3WTO
CA1 2TA M0SHD
CA1 2WF M3OLQ
CA1 3AS M3TUC
CA1 3HF G4KVQ
CA1 3HT G1COY
CA1 3JQ M3ZYF
CA1 3QB G1FBE
CA1 3TH G7DXB
CA10 1AJ G7UNZ
CA10 1BA G1AEI
CA10 1BX G4LQM
CA10 1DT G4WKG
CA10 1EW M3OKP
CA10 1PD G0YYG
CA10 1QR G4EXD
CA10 1QX G4GXO
CA10 1TA G4XET
CA10 2AW G4UWG
CA10 2LL G0VRZ
CA10 3AL G0TNF
CA10 3HD G3WGV
CA10 3JE G7VVX
CA10 3LT M3ITH
CA10 3NL M6MTD
CA10 3PF G0VMP
CA10 3PF G7DEG
CA10 3RZ G8OZT
CA10 3TW G0NAS
CA10 3UE G3LRA
CA10 3XH 2E0CXU
CA10 3XH M6CYG
CA11 0DT G1XTD
CA11 0XB G0AUG
CA11 0XB G0AUH
CA11 7HZ G4PZN
CA11 7JY G0MDV
CA11 7JY G1FNP
CA11 7UW G0TDM
CA11 7UW G7GQL
CA11 7UW G7GQM
CA11 8AX G0OFR
CA11 8BJ G3YEP
CA11 8EH G3ZSK
CA11 8EH G4GLQ
CA11 8PN M6LKE
CA11 8SD G1AQV
CA11 8TW G4GQZ
CA11 8UF G4FUI
CA11 8UF G8LBT
CA11 9AQ M6NNY
CA11 9HN G1VKG
CA11 9PA M5TNT
CA11 9PD M6FWW
CA11 9PF M3WBR
CA11 9SJ M6CTO
CA11 9SS G6XWD
CA11 9TR G4YLI
CA12 4AP G3ZED
CA12 4AZ M0AOT
CA12 4HS G0UQC
CA12 4NN G4ERL
CA12 4SX G0FZG
CA12 5LQ M6LAL
CA12 5SA G0END
CA13 0AR 2E1APW
CA13 0LA M0GCU
CA13 0QR G1GWE
CA13 0TJ M0AYB
CA13 0TL M6RTN
CA13 9EQ G7OHM
CA13 9JD 2E0LDF
CA13 9UA G4NEH
CA13 9UA M1ACO
CA13 9UQ G0JSO
CA13 9UQ M0WLA
CA14 1AA G3WCM
CA14 1DT 2E0OMT
CA14 1DT M3OMT
CA14 1DW M0XAT
CA14 1EE G1OAE
CA14 1HE G3RRI
CA14 1LL G0HCE
CA14 1LP G3WJH
CA14 1PY G0EMM
CA14 1PY M6RMD
CA14 1PY M6XSF
CA14 1QW 2E0IKM
CA14 1QW M6IKM
CA14 1XJ 2E0MAX
CA14 1XJ G4ZFX
CA14 1XJ M0WCR
CA14 2PT 2E0IFF
CA14 2PT M3IFF
CA14 2QY 2E0DNW
CA14 2QY M6CVP
CA14 2TP G6CSL
CA14 2UG G0OMB
CA14 2XZ M3HWP
CA14 3EH 2E0BEG
CA14 3EH M0XMD
CA14 3EH M3LCI
CA14 3EY G7NMB
CA14 3HR G6TAK
CA14 3PE M1RIC
CA14 4HY G1LEH
CA14 4JA M3LDL
CA14 4LS G1LZL
CA14 4NU G1GTA
CA14 4NX G4CJP
CA14 4PU G6PTF
CA14 4UJ 2E0BKP
CA14 4UJ 2E1BKP
CA14 4UJ G0ORO
CA14 4UJ M0EVG
CA14 4UJ M3JMY
CA14 5BQ M6LIV
CA14 5HP G0DPE
CA14 5JA G1AQI
CA14 5LZ M3SDN
CA14 5QR 2E0NWT
CA14 5QR G0MRT
CA14 5QR M0WRC
CA14 5UJ G1FYE
CA15 6BQ G0NNN
CA15 6HR M3XVK
CA15 6HT M6MSP
CA15 6PB M6GXW
CA15 7BU G0TII
CA15 7BU M0BCV
CA15 7DB M3UVA
CA15 7DX 2E0GSR
CA15 7DX M6EZA
CA15 7EG G0MWE
CA15 7ER M1EBC
CA15 7ER M1EBD
CA15 7JU M3SDK
CA15 7JW M1CIE
CA15 7QN G0ORM
CA15 7RG M0GDH
CA15 7SS G0MTQ
CA15 8RX 2E0ENP
CA15 8RX M6ENP
CA15 8ST M0VLA
CA16 6BD G0KDB
CA16 6BD G0VGJ
CA16 6DA G7ITT
CA16 6JZ 2E0AOK
CA16 6JZ 2E1EIO
CA16 6JZ G4XTA
CA16 6PX G0ANT
CA16 6PX G8TXJ
CA16 6PX M5TXJ
CA16 6QE G0OPM
CA16 6RD G0NYQ
CA16 6SP M0JKQ
CA16 6SP M3ZOP
CA16 6SP M6JKQ
CA16 6TA G7THI
CA16 6TU G7NRG
CA16 6XU G1AOZ
CA17 4BZ G8OCA
CA17 4EX G7BNB
CA17 4HR G4SPR
CA17 4NQ G4TUA
CA18 1SL 2E0DAL
CA18 1SL M3ZCX
CA18 1SW G1KGU
CA19 1UD M1PAF
CA19 1XH G3PNU
CA2 4EX 2E0BKV
CA2 4EX M3OZI
CA2 4LD M1LSD
CA2 4PZ G0NUD
CA2 5RG M0PTT
CA2 5UH 2E0TMS
CA2 5UH M6TIH
CA2 6AG M0ADG
CA2 6DH M5FAB
CA2 6JJ G8CAU
CA2 6LZ G4FQM
CA2 6LZ G8KIZ
CA2 7DA M6COB
CA2 7LD M3VZQ
CA2 7LU G4ARS
CA2 7LU G6LSO
CA2 7LU M0SOL
CA2 7RD G0JXO
CA2 7TB M3IRH
CA2 7XD G8GUH
CA2 7XY M3BSF
CA20 1AY G0SVK
CA20 1EF M3IHX
CA20 1LB M3MLI
CA20 1PZ G1HZJ
CA21 2YY G1MPG
CA22 2AT 2E0CBB
CA22 2DN M0KHZ
CA22 2DT M0HCI
CA22 2EH G4VFL
CA22 2NA G1GWF
CA22 2NA G3BJD
CA22 2QQ G1HDX
CA22 2RL M0EJB
CA22 2SG M5GVY
CA22 2UD 2E0XUH
CA22 2UD M0XUH
CA22 2UD M3XUH
CA23 3DY G1LMW
CA23 3DY M0LMS
CA24 3JL G3YSD
CA24 3JQ M6PHF
CA25 5EU G4RVH
CA25 5LR M0AIW
CA26 3AT 2E0WEO
CA26 3AT M6WEO
CA26 3QJ M6RMS
CA26 3QQ G3XIU
CA26 3QS G0RZI
CA26 3SA 2E0XSD
CA26 3SA M0XSD
CA26 3SA M6EPW
CA26 3SA M6XSD
CA26 3XN M1DZR
CA27 0AF 2E0DNX
CA27 0AN G4DSI
CA27 0AR G7KSE
CA27 0AR M0LCW
CA27 0AS G0ESI
CA27 0EG G3WIN
CA27 0EG G4PEW
CA28 6AQ M3USF
CA28 6EF M0SHP
CA28 6JZ G6WLP
CA28 6PE M6FYM
CA28 6SG M0FWM
CA28 6SW G7RKV
CA28 6SX G8IMM
CA28 6XL 2E0MIX
CA28 6XL M6DEG
CA28 7AW G7SGO
CA28 7XG G4VIA
CA28 8EP G0LJB
CA28 8HS G0HYP
CA28 8JP G7MRL
CA28 8JP M0BEE
CA28 8JP M0CRM
CA28 8PX M3YRR
CA28 8PZ G0DIG
CA28 8XJ G4OVF
CA28 8YG M0SKY
CA28 9LT G4MTR
CA28 9LT G8ORO
CA28 9PB M3AQP
CA28 9TG G0CZU
CA28 9UG 2E0BEE
CA3 0AW G0KEE
CA3 0AY M6HRT
CA3 0DB M1ETM
CA3 0LF G0TFC
CA3 0NT M6MOS
CA3 0NU G1FVA

| | |
|---|---|
| CA3 0NW | G0OGG |
| CA3 0PL | M6YJD |
| CA3 0RE | M1ERP |
| [illegible] | [illegible] |
| CA3 9SN | G0DII |
| CA4 0RN | G6DOX |
| CA4 8AT | G4WOQ |
| CA4 8BD | G8UTQ |
| CA4 8ER | M3OLF |
| CA4 8LD | G4BUI |
| CA4 8PL | G4ISS |
| CA4 9QR | G4IIB |
| CA4 9QY | M3FZV |
| CA4 9QY | M3ZTR |
| CA4 9RR | 2E0TXG |
| CA4 9RR | M3TXA |
| CA4 9TP | M3WGV |
| CA5 6BE | G1OFY |
| CA5 6NS | G0NVA |
| CA5 6NS | G7EKM |
| CA5 6NS | M3GMG |
| CA5 6PB | M3OLN |
| CA5 6QW | G4SCE |
| CA5 7AX | 2E0MXP |
| CA5 7AX | M6DLP |
| CA5 7AX | M6RLP |
| CA5 7DP | G8KAP |
| CA5 7NS | M3DLJ |
| CA5 7NW | G7IVG |
| CA5 7NW | M3VYT |
| CA6 4PE | M6DXH |
| CA6 4PZ | G0OYI |
| CA6 5RT | G2HKG |
| CA6 5SS | G4LIL |
| CA6 5TS | M1JWR |
| CA6 5UJ | M1JON |
| CA6 5UJ | M3CFM |
| CA6 5UJ | M3JDG |
| CA6 5UJ | M3MDK |
| CA6 5XX | M3FBJ |
| CA6 5XX | M3HFX |
| CA6 6LH | G4HZP |
| CA6 6NT | G8URU |
| CA6 6PA | M3WGD |
| CA7 0BZ | G3KEL |
| CA7 0DF | 2E0COB |
| CA7 0DF | M3OSC |
| CA7 0DF | M3OSF |
| CA7 1JF | G8DVW |
| CA7 2HJ | G1GDB |
| CA7 3DJ | M0DXT |
| CA7 3QX | G0ZMH |
| CA7 3QX | M0AMX |
| CA7 4EU | M6KWI |
| CA7 4JJ | G0TNL |
| CA7 4QH | G7KWQ |
| CA7 5AN | G0TRW |
| CA7 5AS | M0PMR |
| CA7 5AS | M3AQN |
| CA7 5BH | M6HWW |
| CA7 5ET | 2E0JGR |
| CA7 8AU | G0HNQ |
| CA7 9JL | 2E0PGR |
| CA7 9JL | M0PEG |
| CA7 9QW | M3TEE |
| CA7 9RE | M3UIK |
| CA8 1BN | M6BJW |
| CA8 1EX | G0LEJ |
| CA8 1LE | G6KXD |
| CA8 1PL | M3TFM |
| CA8 1SR | G0JVU |
| CA8 1SR | G7ANH |
| CA8 1TY | M3XUO |
| CA8 2BB | G4IIY |
| CA8 2NF | G0FHF |
| CA8 2NJ | G3OHL |
| CA8 7AL | G4RHC |
| CA8 7LA | M1DIN |
| CA8 9AA | G8IBK |
| CA8 9HF | M6AVE |
| CA8 9JA | G7NZR |
| CA8 9LE | G7JKD |
| CA9 3EA | 2E0ACE |
| CA9 3LH | G1RLT |
| CA9 3LH | M3PJV |
| CA9 3LZ | G0NXS |

## CB (Cambridge)

| | |
|---|---|
| CB1 1LE | M0DVK |
| CB1 2DW | 2E0BOV |
| CB1 2HA | M3XZG |
| CB1 2LG | G6TIU |
| CB1 2LP | G3ROP |
| CB1 2NG | 2E0BPE |
| CB1 2NG | M0GJU |
| CB1 2NG | M3XQZ |
| CB1 2QF | M6OSF |
| CB1 3AJ | M3KWR |
| CB1 3EF | [illegible] |
| CB1 3EG | G3TFX |
| CB1 3HH | 2E0VTA |
| CB1 3HH | M0VTA |
| CB1 3LR | M0GXM |
| CB1 3PT | 2E0CDV |
| CB1 3PW | G0IJZ |
| CB1 3QL | M0GLT |
| CB1 3RS | M6EIS |
| CB1 3RY | G4EZN |
| CB1 3RY | G6UW |
| CB1 3UB | G0SFA |
| CB1 3UB | G1OKI |
| CB1 3UF | G1KUN |
| CB1 6HS | G0DEU |
| CB1 7TZ | G1MMF |
| CB1 7UP | M6GXC |
| [illegible] | [illegible] |
| CB1 7UR | G3INR |
| CB1 8QR | G6ZEY |
| CB1 8RG | G8OUS |
| CB1 8RJ | G3KBR |
| CB1 8RX | G0XAS |
| CB1 8SG | G4EAG |
| CB1 8SH | G4GLI |
| CB1 8YU | G3BMI |
| CB1 9AL | M3UBS |
| CB1 9AZ | G6EUI |
| CB1 9GH | 2E0BLI |
| CB1 9GH | M0SKI |
| CB1 9GH | M3XSK |
| CB1 9JA | G0PNQ |
| CB1 9JS | M1DUO |
| CB1 9LN | G3UPJ |
| CB1 9YL | G7VCE |
| CB10 1NR | M3IAQ |
| CB10 1QD | M0ECC |
| CB10 1SS | 2E0XOR |
| CB10 1SS | M0XOR |
| CB10 1SS | M6XOR |
| CB10 2AH | G1WSF |
| CB10 2AJ | G4YHN |
| CB10 2AP | G4UHR |
| CB10 2DP | G4AXX |
| CB10 2DP | M0CAM |
| CB10 2EB | G6YBH |
| CB10 2HG | G8SZR |
| CB10 2LL | G8VAE |
| CB10 2RW | G4FBY |
| CB10 2TZ | G0TCI |
| CB11 3BJ | G8XLB |
| CB11 3DN | G0GWM |
| CB11 3ED | G0ICB |
| CB11 3ER | G0JML |
| CB11 3HD | G0IMG |
| CB11 3HY | M0SDW |
| CB11 3LH | G4DUV |
| CB11 3RU | G3WMS |
| CB11 3SL | G7PNP |
| CB11 3SP | G4HSN |
| CB11 4AU | M1EMO |
| CB11 4BD | M3NQS |
| CB11 4DE | G0CLJ |
| CB11 4DG | 2E0GUA |
| CB11 4DG | M6GUA |
| CB11 4DJ | 2E1HLH |
| CB11 4DJ | G0HEM |
| CB11 4DJ | G1LTL |
| CB11 4DS | M6BLB |
| CB11 4DW | 2E0NOZ |
| CB11 4DW | M0NOZ |
| CB11 4DW | M3TPN |
| CB11 4JB | G8ULM |
| CB11 4JB | M6GUB |
| CB11 4JT | G0FWX |
| CB11 4JU | G3YZO |
| CB11 4LA | G1PXH |
| CB11 4LX | G0OJR |
| CB11 4SJ | G3TXC |
| CB11 4SJ | G6FEM |
| CB11 4TH | G8EBX |
| CB11 4UU | G3NOX |
| CB11 4XG | G3XHZ |
| CB2 0QZ | G7PLE |
| CB2 1ST | M0XUB |
| CB2 1ST | M6BXU |
| CB2 1TA | M3HZC |
| CB2 1TJ | M0FFX |
| CB2 1TP | M0ERY |
| CB2 1TQ | 2E0DSZ |
| CB2 1TQ | 2E0TRY |
| CB2 1TQ | M0HZC |
| CB2 1TQ | M0WCK |
| CB2 1TQ | M6HXC |
| CB2 3BU | M3VUS |
| CB2 3HU | M0RSW |
| CB2 3JU | M0TMN |
| CB2 3JU | M3TNM |
| CB2 6SP | M0DCV |
| CB2 6SP | M5AKW |
| CB2 8DT | 2E0CQV |
| CB2 8DT | M0HJA |
| CB2 8PH | 2E0XRX |
| CB2 8PH | M0TEI |
| CB2 8PH | M6BBX |
| CB2 9AQ | M3VFC |
| CB2 9HR | G3NHB |
| CB2 9HS | G1YPU |
| CB2 9JH | 2E0BUV |
| CB2 9JR | M6RLN |
| CB2 9JT | 2E0TTM |
| CB2 9JT | M0TMA |
| CB2 9JW | [illegible] |
| CB21 4AD | G4PDI |
| CB21 4EG | G4STV |
| CB21 4EG | G6BOX |
| CB21 4EG | M0BOX |
| CB21 4EU | M0RSP |
| CB21 4JQ | 2E1HLF |
| CB21 4JQ | G7SRK |
| CB21 4JT | G1VCU |
| CB21 4LY | G1CQA |
| CB21 4RS | M3ZCA |
| CB21 4RY | G0CHC |
| CB21 5BG | M6FQO |
| CB21 5BT | G4YJK |
| CB21 5DT | G8RYL |
| CB21 5DU | M3IZV |
| CB21 5EY | G4VTU |
| CB21 5JG | G3EDD |
| CB21 5JY | G4HJW |
| CB21 5LB | G0UDI |
| [illegible] | [illegible] |
| CB21 5LU | M0HBU |
| CB21 5LY | G4FTW |
| CB21 5PE | G3ZUK |
| CB21 6BU | G4MZN |
| CB22 3BN | G1GFC |
| CB22 3BT | 2E0WWA |
| CB22 3BT | M6ZBQ |
| CB22 3DH | G4WEZ |
| CB22 3DU | G7KRT |
| CB22 3DU | G8TMV |
| CB22 3DW | M1AQP |
| CB22 3DW | M1CKO |
| CB22 3HZ | G3WFF |
| CB22 3LP | M1DBK |
| CB22 3SP | G6ASH |
| CB22 3UR | M6BSL |
| CB22 4LX | G6BBN |
| CB22 4NE | G4LCE |
| CB22 4NG | G0PGQ |
| CB22 4PA | M0ZZC |
| CB22 4QU | G0VVZ |
| CB22 4RT | G6WJX |
| CB22 4RZ | G8VQN |
| CB22 4XT | M6HPT |
| CB22 5AE | G8MEI |
| CB22 5AW | G0ETP |
| CB22 5BP | M1KTA |
| CB22 5BP | M3KTA |
| CB22 5BX | G3WLD |
| CB22 5BX | M0LAT |
| CB22 5EG | G3UUT |
| CB22 5EG | G5PI |
| CB22 5JF | G7UUT |
| CB22 5JJ | G6WKQ |
| CB22 5JW | G3TVM |
| CB22 5JZ | M0GCX |
| CB22 5LG | G1IMM |
| CB22 5LW | M6FCG |
| CB22 6RT | G6UFL |
| CB22 6RY | M6LBW |
| CB22 7QF | M1HJE |
| CB22 7RP | M3IQN |
| CB22 7RZ | G8SHC |
| CB23 1LS | G6KWA |
| CB23 2NG | G6FSV |
| CB23 2NL | G4VHK |
| CB23 2RE | G3TAG |
| CB23 2RE | G4VYG |
| CB23 2RE | M3CMX |
| CB23 2RY | G7SPZ |
| CB23 2SJ | G7ASZ |
| CB23 2TH | G4IPM |
| CB23 2TU | G0BOM |
| CB23 3XP | 2E0GGW |
| CB23 3XP | M0HHB |
| CB23 4JQ | G4UGQ |
| CB23 4JQ | G4VWS |
| CB23 5DG | 2E0SDC |
| CB23 5ED | G0ULA |
| CB23 5FN | M6CWJ |
| CB23 5HY | G0RQN |
| CB23 6AB | 2E0CIH |
| CB23 6AB | M3MJV |
| CB23 6AT | G0REU |
| CB23 6BZ | 2E0PWL |
| CB23 6BZ | M0PWL |
| CB23 6BZ | M6PWL |
| CB23 6ER | M0DES |
| CB23 6FN | M3WCX |
| CB23 6GR | G4HUN |
| CB23 7AE | G8MEX |
| CB23 7AZ | G8GEV |
| CB23 7BU | G4DGL |
| CB23 7DY | G3PJT |
| CB23 7NX | G3GGK |
| CB23 7PG | M0YCS |
| CB23 7PH | 2E0VBK |
| CB23 7PH | M6RBK |
| CB23 7PS | G3RUH |
| CB23 7PW | G3XZP |
| CB23 7QD | 2E0NHB |
| CB23 7QD | M0HZR |
| CB23 7QD | M6NBS |
| CB23 7QL | G4JZQ |
| CB23 7RY | G8EYP |
| CB23 7UZ | G7GCW |
| CB23 7XA | 2E0BKD |
| CB23 7XA | M3FHO |
| CB23 7XJ | G4NBS |
| CB23 7XJ | G4NXO |
| CB23 7XL | G3YMW |
| CB23 7XR | G6HFS |
| CB23 7XR | G6NIW |
| CB23 7XS | G4YAC |
| [illegible] | [illegible] |
| CB23 8BW | 2E0JJA |
| CB23 8BW | M6CVB |
| CB23 8EP | 2E1AEJ |
| CB23 8EP | G0NJJ |
| CB23 8EQ | G3LOD |
| CB23 8HH | G6KIB |
| CB23 8SF | G0KRB |
| CB23 8SJ | G8RPI |
| CB23 8SU | G6VTN |
| CB23 8SZ | G6TUS |
| CB23 8TD | G6HSW |
| CB23 8TD | M1MPW |
| CB23 8TQ | M6TAG |
| CB23 8TR | G1ZCC |
| CB24 3AG | M6RLV |
| CB24 3AY | G8NPR |
| CB24 3EH | G7LBL |
| CB24 3XA | 2E0RJZ |
| [illegible] | [illegible] |
| CB24 4RT | G7CRO |
| CB24 4RT | G0JDQ |
| CB24 4SR | G3XIP |
| CB24 4TR | 2E0BPX |
| CB24 4TR | M0GUM |
| CB24 4TR | M3VGZ |
| CB24 5HY | G1JZN |
| CB24 5HY | G1XAA |
| CB24 5JT | G3RGQ |
| CB24 5JU | G4NEY |
| CB24 5JX | G4XXX |
| CB24 5LS | G6XRY |
| CB24 5NJ | G8GJU |
| CB24 5PZ | G4BHJ |
| CB24 5TZ | G1MKP |
| CB24 5UR | G0BOE |
| CB24 5UR | G1JKF |
| CB24 5UU | G8FTE |
| CB24 5UX | G8IWQ |
| CB24 6BS | G0OIR |
| CB24 6DA | G1NWM |
| CB24 6DA | G8CRB |
| CB24 6DT | G4LNY |
| CB24 6EB | G0WZD |
| CB24 6ET | 2E0HWK |
| CB24 6ET | M0HKW |
| CB24 6ET | M3UFS |
| CB24 8AW | G0AWD |
| CB24 8QG | G1KIJ |
| CB24 8QR | G0GKP |
| CB24 8QR | G2XV |
| CB24 8QR | G8EVY |
| CB24 8SD | G0EUZ |
| CB24 8TB | M0BLF |
| CB24 8TG | G1DEQ |
| CB24 8TQ | G3XXX |
| CB24 8TR | M0VFC |
| CB24 8TS | M1EMG |
| CB24 8TX | G3XPW |
| CB24 8UJ | M6ESO |
| CB24 8UL | M0VMC |
| CB24 8UN | G4SUM |
| CB24 8XT | M6GCO |
| CB24 8XY | M1XZG |
| CB24 9JL | G8XPQ |
| CB24 9JU | G0THI |
| CB24 9PE | G7KDG |
| CB24 9PP | G8HSS |
| CB24 9PP | M0AQF |
| CB25 0BA | G4PZW |
| CB25 0BB | M3MKZ |
| CB25 0DT | G8NEO |
| CB25 0DT | M0CNP |
| CB25 0DW | M3HWW |
| CB25 0EL | G8DQF |
| CB25 0ER | M0DRK |
| CB25 0ES | G7SCO |
| CB25 0ES | M1BIB |
| CB25 0HL | G3XBM |
| CB25 0HT | G8BXA |
| CB25 0JU | G4FVA |
| CB25 0JX | G0SBK |
| CB25 0LD | G6ALB |
| CB25 9AD | G3KKD |
| CB25 9BQ | G3PTQ |
| CB25 9FA | G3NIE |
| CB25 9HB | G7BNM |
| CB25 9HW | G4VME |
| CB25 9NA | 2E0TWK |
| CB25 9NA | G8LOF |
| CB25 9NA | M0WSE |
| CB25 9NB | G8XLE |
| CB25 9NE | G8ZRD |
| CB25 9PF | 2E0SDY |
| CB25 9PF | M0SND |
| CB25 9PX | G3EEZ |
| CB25 9PX | G4BAO |
| CB3 0BN | 2E0TJG |
| CB3 0BN | M0HUH |
| CB3 0BN | M0JSH |
| CB3 0DG | M0HFQ |
| CB3 0DG | M6BHK |
| CB3 0DS | M0XOS |
| CB3 0GP | G0VDE |
| CB3 0HX | G4XIL |
| CB3 0JA | G4LEN |
| CB3 0LT | M1MAJ |
| CB3 0LT | M3ZCB |
| CB3 0NP | G0ANV |
| CB3 0NP | G1DAZ |
| CB3 0NS | 2E1FLN |
| CB3 0PE | M0HOT |
| CB3 0PP | G0ECL |
| CB3 6GA | G3HZP |
| CB3 9BB | 2E0YIP |
| [illegible] | [illegible] |
| CB3 9BB | M0LAH |
| CB3 9BB | M6OAL |
| CB3 9BB | M6TSG |
| CB3 9DT | [illegible] |
| CB3 0LQ | G3MRX |
| CB4 1DG | G3VGX |
| CB4 1ET | G6AIG |
| CB4 1FD | G0RDB |
| CB4 1LW | G4DNG |
| CB4 1NP | G6FKS |
| CB4 1QG | M6GVY |
| CB4 1SQ | G3ZAY |
| CB4 1SX | M6GKW |
| CB4 1TT | G3RQS |
| CB4 1TZ | G1DVD |
| CB4 1TZ | G8CTX |
| CB4 1XG | G3URX |
| CB4 1YZ | G8NNA |
| [illegible] | [illegible] |
| CB4 2AS | G7AZJ |
| [illegible] | [illegible] |
| CB4 2BJ | G8UDV |
| CB4 2DB | G3RSE |
| CB4 2DB | G8UDG |
| CB4 2HJ | G7RTR |
| CB4 2HT | G1VJG |
| CB4 2LD | G7KHL |
| CB4 2NP | G6KSV |
| CB4 2NS | M6EBG |
| CB4 2PB | 2E0LCM |
| CB4 2PB | M0LCM |
| CB4 2QR | M5CHH |
| CB4 2RA | G7DWU |
| CB4 2SH | G7NZY |
| CB4 2TA | M0HOJ |
| CB4 2UF | G4NUA |
| CB4 2UH | G8XYS |
| CB4 2UN | G1AQX |
| CB4 3AA | 2E0YUN |
| CB4 3AA | M0LSY |
| CB4 3EL | M6VWA |
| CB4 3ND | G3HRE |
| CB4 3PA | G4AJW |
| CB4 3PA | G8KNN |
| CB4 3PB | G8JKV |
| CB4 5NB | G0DAH |
| CB5 0JF | G1YAS |
| CB5 8QU | 2E0DVY |
| CB5 8QU | M6FVM |
| CB5 8QU | M6GQF |
| CB5 8SS | 2E0IHI |
| CB5 8SS | M0RTW |
| CB5 8SS | M6FHI |
| CB5 8SS | M6ZZA |
| CB5 8TE | M6RMA |
| CB5 8UB | M6AWR |
| CB5 8US | M6CFX |
| CB5 8XJ | M6FPQ |
| CB6 1AB | G0KFQ |
| CB6 1DA | G4BUH |
| CB6 1DA | M3UPB |
| CB6 1DD | G7VGH |
| CB6 1DJ | G0OKK |
| CB6 1DP | G8GLB |
| CB6 1DZ | G6YTO |
| CB6 1FH | G7HMI |
| CB6 1HT | G3YZB |
| CB6 1HU | G3KKC |
| CB6 1JD | M1BFY |
| CB6 1JJ | 2E0DPH |
| CB6 1JJ | M6PHL |
| CB6 1JS | G6YQJ |
| CB6 1JW | G0MGI |
| CB6 1RJ | M6DXL |
| CB6 1RQ | G7BNL |
| CB6 1SB | G4HWJ |
| CB6 1TL | 2E0RDF |
| CB6 1TL | M0MTW |
| CB6 1TL | M6DFK |
| CB6 1TS | M0TJC |
| CB6 2AQ | G6OIB |
| CB6 2ED | G3XIQ |
| CB6 2EQ | G1HMT |
| CB6 2HH | G1DUI |
| CB6 2HS | G1PEK |
| CB6 2NW | G4FIT |
| CB6 2PD | G0PPL |
| CB6 2QB | M0TVT |
| CB6 2QB | M6TRH |
| CB6 2QF | M0DBJ |
| CB6 2RL | 2E0DNF |
| CB6 2RL | M0ZEB |
| CB6 2RL | M6DNF |
| CB6 2RX | G3IUY |
| CB6 2ST | G4WMZ |
| CB6 2TQ | M1ZEM |
| CB6 2TU | 2E0FBE |
| CB6 2TU | M6FTF |
| CB6 2UH | M0BWH |
| CB6 3AL | G0ICE |
| CB6 3AL | G1XCY |
| CB6 3DA | G4GXY |
| CB6 3DN | G1CQR |
| CB6 3PP | G4GND |
| CB6 3PW | G0PPI |
| CB6 3PW | G4KNS |
| CB6 3PW | G7IEY |
| CB6 3ST | G4BEL |
| CB6 3UE | 2E1NPH |
| CB6 3UE | G0PYE |
| CB6 3UE | G4NPH |
| CB6 3UE | G8NPH |
| CB6 3UE | M3NPH |
| CB6 3UE | M3TGA |
| CB6 3UG | M6LFQ |
| CB6 3XE | G1VLD |
| CB7 4BU | G1UAF |
| CB7 4HU | G6IOB |
| [illegible] | [illegible] |
| CB7 4TH | G4KPX |
| CB7 4UN | G0LQV |
| CB7 4YJ | G1AAQ |
| CB7 5AL | M0TIX |
| CB7 5AL | M3DCS |
| CB7 5BJ | G3AMF |
| CB7 5BL | G8LWQ |
| CB7 5DH | G6BUU |
| CB7 5DS | G7BRP |
| CB7 5EF | 2E0GXB |
| CB7 5EF | M3GXB |
| CB7 5EL | G8PAG |
| CB7 5EX | 2E0VAS |
| CB7 5EX | M0VAS |
| CB7 5EX | M3VAS |
| CB7 5FF | G8PK |
| [illegible] | [illegible] |
| CB7 5JD | M6DZV |
| CB7 5JU | M6IKY |
| CB7 5NR | G4UJV |
| CB7 5QX | G8WAV |
| CB7 5RU | G0PIK |
| CB7 5SH | G0TLQ |
| CB7 5TN | G4ERO |
| CB7 5TW | G3FCM |
| CB7 5UU | G0EIM |
| CB7 5XE | G7FMV |
| CB7 5XH | M0GUO |
| CB8 0BY | 2E0BLR |
| CB8 0DJ | G6GQI |
| CB8 0DP | G0TEV |
| CB8 0QD | 2E0NEV |
| CB8 0SE | G3LUN |
| CB8 0SE | G8IDL |
| CB8 0SQ | G1ACB |
| CB8 7AZ | G8WUR |
| CB8 7BQ | G0SPK |
| CB8 7DU | G6JJK |
| CB8 7HQ | M6RPY |
| CB8 7RT | M3VSF |
| CB8 7SD | G6BBK |
| CB8 7SD | G8BBK |
| CB8 7YP | G1NHG |
| CB8 8AH | G0KOJ |
| CB8 8DT | G4DSN |
| CB8 8FZ | G1JLE |
| CB8 8GY | M0AOZ |
| CB8 8HU | G7VGN |
| CB8 8RL | M6AWY |
| CB8 8SW | M3YDY |
| CB8 8TZ | G3YXN |
| CB8 8UW | G3UJG |
| CB8 8YG | G8SVZ |
| CB8 8YH | 2E0TUG |
| CB8 8YH | M6EZN |
| CB8 8YW | M6JDC |
| CB8 9DF | M6GUF |
| CB8 9DG | G3NUL |
| CB8 9DG | G3SZY |
| CB8 9DQ | G0ADZ |
| CB8 9DY | G1EYZ |
| CB8 9NB | G0FRU |
| CB8 9NR | G4IQR |
| CB8 9PD | M1BJE |
| CB8 9RX | G4HVV |
| CB8 9RX | G8HVV |
| CB8 9RY | G6JDO |
| CB8 9SA | G7VNP |
| CB8 9SA | M1BXF |
| CB9 0AP | G7MVU |
| CB9 0AU | M1DQU |
| CB9 0EQ | G8LHD |
| CB9 0JR | G7EQG |
| CB9 0PH | M3DMG |
| CB9 7ED | G4PQY |
| CB9 7EE | G7LVM |
| CB9 7EW | G7MQC |
| CB9 7FU | G1WZM |
| CB9 7GH | M0JDD |
| CB9 7GH | M1RTT |
| CB9 7NA | G6MQN |
| CB9 7NN | G4KVU |
| CB9 7PT | M6JFF |
| CB9 7RF | M3ZKI |
| CB9 7SG | M6EZT |
| CB9 7WD | G0OJB |
| CB9 7WG | G8UBX |
| CB9 7XW | G1DNO |
| CB9 7YE | G8OST |
| CB9 8EG | G1BBA |
| CB9 8EH | M6BEJ |
| CB9 8LX | M5PYE |
| CB9 9DD | 2E0NOW |
| CB9 9DD | M0HKG |
| CB9 9DD | M3HHB |
| CB9 9DD | M3XCA |
| CB9 9LE | G4LYF |
| CB9 9LY | M6SBJ |
| CB9 9ND | 2E0PCI |
| CB9 9QL | G4LYB |
| CB9 9QL | G8IMI |

## CF (Cardiff)

| | |
|---|---|
| CF1 8BZ | GW3KXX |
| CF11 7BG | 2W0CYF |
| CF11 7BG | GW8WEY |
| CF11 7BG | MW6BRF |
| CF11 7DQ | MW6DZR |
| CF11 0AJ | [illegible] |
| CF11 8AJ | MW3LLV |
| CF11 9JR | 2W0TKS |
| CF11 9JR | [illegible] |
| CF11 9JR | MW6KSL |
| CF11 9NQ | MW0CCK |
| CF14 0NU | GW0LOI |
| CF14 0NZ | MW6NHC |
| CF14 0RF | GW8VRS |
| CF14 0RW | MW3KNR |
| CF14 1EH | MW6CJS |
| CF14 1EH | MW6ESW |
| CF14 1HP | MW6WFJ |
| CF14 1NQ | GW4LFW |
| CF14 1UE | GW4VGB |
| CF14 1UH | GW6RCK |
| CF14 1UN | GW4TJQ |
| CF14 2BR | 2W0LAZ |
| CF14 2BR | MW3RVP |
| CF14 2EH | GW4NHB |
| [illegible] | [illegible] |
| CF14 2HP | 2W0CDZ |
| CF14 2HP | MW6PDG |
| CF14 2LR | GW7ASL |
| CF14 2RY | 2W0PCN |
| CF14 2TE | GW4NHH |
| CF14 3DU | GW0KRQ |
| CF14 3DU | GW1WZI |
| CF14 3PL | MW3FPF |
| CF14 3QA | GW8NCN |
| CF14 4AQ | 2W1CYC |
| CF14 4DJ | GW7EMV |
| CF14 4HN | MW3XIJ |
| CF14 4HS | GW4GZX |
| CF14 4JQ | GW8LKX |
| CF14 4NS | GW0GDI |
| CF14 4SB | GW1AWH |
| CF14 4SP | GW8ARC |
| CF14 5HJ | GW4JOG |
| CF14 5HQ | MW0SGR |
| CF14 5JW | MW3HZB |
| CF14 5LR | MW3WWO |
| CF14 5PW | MW3DEM |
| CF14 5QA | GW6UAS |
| CF14 5QA | MW3NBN |
| CF14 6AE | GW4LXO |
| CF14 6AE | MW6MBZ |
| CF14 6AQ | MW3CII |
| CF14 6BJ | GW1BCI |
| CF14 6DF | GW6DFX |
| CF14 6JZ | GW0BBC |
| CF14 6JZ | GW3RWX |
| CF14 6NS | MW6FKC |
| CF14 6PE | GW0WMT |
| CF14 6QH | MW3PJU |
| CF14 6RL | GW6ZHM |
| CF14 6SS | GW6ITB |
| CF14 6SW | GW3WLN |
| CF14 7EH | GW3UZL |
| CF14 7TP | GW0TPL |
| CF14 9AF | GW4LWL |
| CF14 9AH | GW7EYP |
| CF14 9LA | GW3TKH |
| CF15 7PT | MW6XDA |
| CF15 7SE | MW3ZKX |
| CF15 8DD | GW6MNC |
| CF15 8FE | GW6VTZ |
| CF15 8RA | GW8FNO |
| CF15 8RE | GW0HCB |
| CF15 9JP | GW1CUQ |
| CF15 9NJ | MW6HYS |
| CF15 9NN | MW0ZAP |
| CF15 9NZ | GW7HOM |
| CF15 9PB | GW3TOB |
| CF15 9QR | GW3VOL |
| CF15 9SJ | GW4LFV |
| CF15 9SJ | GW6IOA |
| CF15 9TJ | GW8YYF |
| CF2 6JN | GW3UMD |
| CF23 5BY | GW8EHQ |
| CF23 5EZ | GW0WLI |
| CF23 5HE | MW6EPX |
| CF23 5NN | GW4ZAW |
| CF23 5SB | MW3PGN |
| CF23 6EJ | GW6MHV |
| CF23 6EQ | GW3XXB |
| CF23 6HN | GW4JJV |
| CF23 6HN | GW4JJW |
| CF23 6LF | GW6AGS |
| CF23 6QU | GW0MLN |
| CF23 6SZ | 2W0EUO |
| CF23 6SZ | MW6EUO |
| CF23 7DA | GW3MKT |
| CF23 7ER | 2W0BDW |
| CF23 7HN | GW0VMD |
| CF23 7HQ | GW4KOE |
| CF23 8XJ | GW0HYU |
| CF23 9AZ | GW7JDX |
| CF23 9BJ | MW6SAW |
| CF23 9DW | 2W0FCF |
| CF23 9DW | MW6XXC |
| CF23 9HX | MW0AQT |
| CF23 9NR | GW6CUR |
| CF23 9NR | GW8NP |
| CF24 0HS | GW0DGJ |
| CF24 1RW | GW7HOC |
| CF24 1RW | MW3YKL |
| CF24 1RW | MW5HOC |
| CF24 1RW | MW6YDP |
| CF24 2JT | MW3ASG |
| CF24 2NW | GW7EMO |
| CF24 3JD | GW1SMT |
| CF24 4DZ | GW8AHB |
| CF24 4LS | MW3KHH |
| CF24 4RU | GW1XZI |
| CF24 4RU | [illegible] |
| CF24 4RU | GW7BTC |
| CF3 0BU | 2W0JYI |
| CF3 0BU | MW6EDQ |
| CF3 0BY | GW7KDU |
| CF3 0DF | MW6FPL |
| CF3 0DQ | GW0FVC |
| CF3 0DU | MW6ATC |
| CF3 0NB | 2W0OCF |
| CF3 0PL | MW0RPK |
| CF3 0PS | MW6RJX |
| CF3 1RE | MW3IHD |
| CF3 1TA | MW6JHJ |
| CF3 1TA | MW6KCQ |
| CF3 2TX | MW6ETY |
| CF3 2TY | MW3AQI |
| CF3 3AG | MW3UFH |
| CF3 3HQ | GW4IUN |
| CF3 3LX | 2W0BJE |
| CF3 3LX | [illegible] |
| [illegible] | [illegible] |
| CF3 5QE | MW6TRK |
| CF3 5SY | MW6OZF |
| CF3 5UD | MW6GLK |
| CF3 7LL | GW4OES |
| CF31 1HD | 2W1GKJ |
| CF31 1PF | GW0SXZ |
| CF31 1PQ | GW0PYU |
| CF31 1RL | 2W0ZAA |
| CF31 1RL | MW3USX |
| CF31 2AT | MW3KHY |
| CF31 2BY | MW6PJP |
| CF31 2EA | GW0RVR |
| CF31 2ED | GW0FJP |
| CF31 2JA | MW3THI |
| CF31 2JG | MW3VEL |
| CF31 2JR | MW0AXA |
| CF31 2LA | GW1ERA |
| CF31 2LD | MW0SWR |
| CF31 2LJ | GW6HMJ |
| CF31 2ND | GW4MTE |
| CF31 2ND | GW8XJC |
| CF31 2NE | GW0NKG |
| CF31 2NG | GW3ZTH |
| CF31 2PF | GW1ZNC |
| CF31 2PR | MW0LCH |
| CF31 2QF | MW6AHT |
| CF31 3BA | GW7RRS |
| CF31 3DA | MW0GRZ |
| CF31 3LN | GW4SUD |
| CF31 3LN | GW8ZIL |
| CF31 4BB | GW0SZU |
| CF31 4BF | 2W0TRR |
| CF31 4EY | GW3SRF |
| CF31 4LU | MW3OEJ |
| CF31 4NR | GW1KQV |
| CF31 4PJ | MW0FDG |
| CF31 4PY | MW0AEL |
| CF31 4QX | GW1FKY |
| CF31 4QY | 2W0GNG |
| CF31 4QY | MW6PAC |
| CF31 4SJ | 2W0PWO |
| CF31 4SJ | MW0GYV |
| CF31 4SJ | MW6PWO |
| CF31 4SR | GW6MAB |
| CF31 5BA | GW6VET |
| CF31 5BT | GW7DVJ |
| CF31 5FH | GW4SML |
| CF32 0BJ | GW0MNP |
| CF32 0HL | GW0TTF |
| CF32 0LH | GW1XJJ |
| CF32 0PJ | GW4ZEA |
| CF32 0SL | GW4SKP |
| CF32 7BB | 2W0PEH |
| CF32 7BB | MW0PEH |
| CF32 7BB | MW3PEH |
| CF32 7EL | MW0TDQ |
| CF32 7SP | MW6PGD |
| CF32 8AH | GW0UZK |
| CF32 8AH | GW7VBE |
| CF32 8BL | MW1CFE |
| CF32 8HY | GW8BWX |
| CF32 8LE | MW6MWG |
| CF32 8SA | MW0DCQ |
| CF32 8SD | GW0GQC |
| CF32 8SE | MW3BBQ |
| CF32 8UU | MW3SLL |
| CF32 8YB | GW1OUP |
| CF32 8YB | GW1YHA |
| CF32 8YB | MW3DIL |
| CF32 8YB | MW3HVE |
| CF32 9AF | GW4VKG |
| CF32 9AQ | GW7TNS |
| CF32 9DH | GW3SSK |
| CF32 9DH | MW3MRL |
| CF32 9LU | GW0ACH |
| CF32 9LX | MW6NGE |
| CF32 9NH | 2W0SNW |
| CF32 9NH | MW3SNW |
| CF32 9RJ | MW0WNL |
| CF32 9SB | GW0KAX |
| CF32 9SB | GW4WVO |
| CF32 9SX | MW6CLP |
| CF32 9UE | MW3BOC |
| CF32 9UE | MW3UBY |
| CF32 9UG | GW4THK |
| CF32 9UR | 2W0WCQ |
| CF32 9UR | MW6WCQ |
| CF32 9YU | GW0CNC |
| CF32 9YN | GW4SKA |
| CF32 9YW | GW4LNP |
| CF32 9YW | GW7KYT |
| CF33 4AL | GW0GUA |
| CF33 4BE | MW6PNW |
| CF33 4EU | MW3OEC |
| CF33 4LT | [illegible] |
| CF33 4LT | MW0COF |
| CF33 4RQ | GW4IFE |
| CF33 6AP | MW0AGE |
| CF33 6AS | MW6XGD |
| CF33 6EG | GW4YMJ |
| CF33 6EU | GW0RTP |
| CF33 6HT | GW6FNB |
| CF33 6HY | MW0PDR |
| CF33 6HY | MW0PRC |
| CF34 0AB | GW0CYK |
| CF34 0BB | GW3WRE |
| CF34 0BD | GW0VJS |
| CF34 0BD | GW4VSE |
| CF34 0DE | MW1WYN |
| CF34 0DP | GW6OMV |
| CF34 0EA | 2W0DRB |
| CF34 0EA | MW6AEM |
| CF34 0HY | MW3MVT |
| CF34 0LU | [illegible] |
| [illegible] | [illegible] |
| CF34 0NT | GW0TOM |
| CF34 0NT | GW0VMS |
| CF34 0RU | MW0COD |
| CF34 0SW | 2W0JAI |
| CF34 0SW | MW6DSP |
| CF34 0TG | MW6RPS |
| CF34 0UE | MW6TDL |
| CF34 0UF | 2W0DRK |
| CF34 0UF | 2W0IUN |
| CF34 0UF | MW0HXX |
| CF34 0UF | MW0IUN |
| CF34 0UF | MW6DRM |
| CF34 0UF | MW6IUN |
| CF34 0UR | GW0IRP |
| CF34 0UW | 2W0LFY |
| CF34 0UW | 2W0LLY |
| CF34 0UW | MW6SBT |
| CF34 0YG | MW0GWT |
| CF34 9AH | MW6GMJ |
| CF34 9AH | MW6KBK |
| CF34 9DU | 2W1AOD |
| CF34 9HP | GW3XJC |
| CF34 9JL | 2W0FAR |
| CF34 9JL | MW6BAU |
| CF34 9JL | MW6FAN |
| CF34 9ST | MW0AWO |
| CF34 9TB | 2W0DMG |
| CF34 9TB | MW3GPG |
| CF340TB | MW3NCS |
| CF35 5AE | GW3MFY |
| CF35 5AL | GW0SIS |
| CF35 5AL | GW4NBY |
| CF35 5BH | GW4ZXG |
| CF35 5EG | MW1MFY |
| CF35 5HX | MW1BUN |
| CF35 5NA | GW6ZGY |
| CF35 5NA | MW0WIL |
| CF35 5NF | MW0GCD |
| CF35 5NF | MW0JAN |
| CF35 5NF | MW0PSG |
| CF35 5PW | GW0PND |
| CF35 5QA | GW6CZE |
| CF35 5RH | GW7NJT |
| CF35 6ED | GW8HWL |
| CF35 6FA | GW8TGS |
| CF35 6HH | GW3SYL |
| CF35 6JR | GW0KFL |
| CF35 6PD | GW3FSP |
| CF35 6SD | GW3RVG |
| CF35 6TF | MW0JLN |
| CF35 6TY | GW1YBF |
| CF35 6YA | 2W1FUD |
| CF35 6YD | MW3NXD |
| CF35 6YD | MW3WQE |
| CF35 6YF | MW5MWR |
| CF35 6YG | 2W0DJS |
| CF35 6YG | MW3NUX |
| CF36 3HR | GW0BAH |
| CF36 3HW | MW6NLG |
| CF36 3QH | GW0VST |
| CF36 3SW | GW6JFV |
| CF36 3YS | GW0BZA |
| CF36 5AJ | GW7DTB |
| CF36 5AN | GW3RIY |
| CF36 5AU | GW4BKG |
| CF36 5EE | GW4BCF |
| CF36 5HN | GW1MNC |
| CF36 5LP | GW4ZVO |
| CF36 5SG | GW3ARS |
| CF37 1EN | MW6GKU |
| CF37 1HT | GW0ONU |
| CF37 1HY | GW0SQY |
| CF37 1NH | GW8HF |
| CF37 1NW | GW0RQP |
| CF37 1SS | GW1FKL |
| CF37 1XE | GW0KLN |
| CF37 2DA | GW4ZVV |
| CF37 2HF | GW3YBN |
| CF37 2LG | 2W0BAO |
| CF37 2SE | 2W0CJI |
| CF37 2SE | 2W0PCD |
| CF37 2SE | MW0HKA |
| CF37 2SE | MW6PEL |
| CF37 3BW | GW8LRO |
| CF37 3DF | GW0SLC |
| CF37 3EF | GW8ICS |
| CF37 3EU | GW0JHH |
| CF37 3HU | 2W0KAY |
| CF37 3HU | MW0KAY |
| CF37 4DP | MW0ATG |
| CF37 4NR | MW3HBF |
| CF37 5DA | MW0MWL |
| CF37 5EP | MW6ZBR |
| CF37 5ER | MW6ARX |
| [illegible] | [illegible] |
| CF37 5HW | 2W0BMR |
| CF37 5HW | MW3PBV |
| CF38 1AU | GW0MOQ |
| CF38 1BW | GW1KQT |
| CF38 1DY | GW4CNL |
| CF38 1DW | MW3ZSO |
| CF38 1LY | 2W0CBJ |
| CF38 1RP | GW0WLN |
| CF38 1RY | GW8RLI |
| CF38 1UF | GW4NPC |
| CF38 2AL | MW3CHZ |
| CF38 2EN | 2W0KPH |
| CF38 2EN | MW6CDV |
| CF38 2LB | 2W0OAG |
| CF38 2LB | MW0YAG |
| CF38 2LB | MW6YAG |
| CF38 2LN | 2W0CWL |
| CF38 2LN | MW0HUY |
| [illegible] | [illegible] |
| CF38 2LN | MW6CWL |
| [illegible] | [illegible] |
| CF38 2NT | MW6MGZ |
| CF38 2NT | MW6YAE |
| CF39 0EG | MW3ZSM |
| CF39 0ET | MW0DVM |
| CF39 8AL | GW4RKZ |
| CF39 8AL | GW6MIH |
| CF39 8DF | MW6HDV |
| CF39 8DU | GW7CSK |
| CF39 8PL | GW0UHJ |
| CF39 8PL | GW7PIN |
| CF39 8PP | GW0WLQ |
| CF39 8TW | 2W0GER |
| CF39 8UG | 2W0BKA |
| CF39 8UG | MW0HAB |
| CF39 9EL | MW6BUH |
| CF39 9HT | MW0GWL |
| CF39 9UY | GW0KLY |
| CF4 5SR | GW8NNF |
| CF4 6HD | GW0NIY |
| CF4 6HY | GW1YKT |
| CF4 7PD | GW4WQC |
| CF40 1BW | MW6BZP |
| CF40 1BY | GW7RKC |
| CF40 1DE | 2W1CAU |
| CF40 1DQ | GW1BDF |
| CF40 1DQ | GW1BDG |
| CF40 1DW | MW0SBJ |
| CF40 1EN | MW6SDV |
| CF40 1EQ | MW0SKD |
| CF40 1EU | 2W0BED |
| CF40 1HF | MW0SWB |
| CF40 1HP | 2W0DLE |
| CF40 1HR | GW2FOF |
| CF40 1HR | GW4BUZ |
| CF40 1JD | MW0MWA |
| CF40 1JT | GW0KQU |
| CF40 1LX | GW6ZYI |
| CF40 1LX | MW0WRQ |
| CF40 1LX | MW0WVR |
| CF40 1NG | MW6FIN |
| CF40 1PJ | MW0PRP |
| CF40 1QY | GW4YCO |
| CF40 1TA | 2W0SCL |
| CF40 1TA | MW3ZDQ |
| CF40 2AF | MW6MMU |
| CF40 2AH | MW6CGH |
| CF40 2BY | 2W0VAG |
| CF40 2BY | MW0TBI |
| CF40 2BY | MW3VVW |
| CF40 2DA | GW1XXL |
| CF40 2DH | GW4UVC |
| CF40 2DZ | MW3FVH |
| CF40 2HD | MW6WQE |
| CF40 2HJ | MW6AEX |
| CF40 2HN | GW4SYO |
| CF40 2JD | GW0LYF |
| CF40 2LY | GW0OET |
| CF40 2NX | 2W1FVH |
| CF40 2RE | MW6KDA |
| CF40 2RY | MW3YBB |
| CF40 2UN | MW3MKN |
| CF41 7HF | GW3XHG |
| CF41 7NJ | GW0DLA |
| CF41 7PW | MW6CUA |
| CF41 7PW | MW6FBA |
| CF41 7TW | GW0FJQ |
| CF41 7UD | GW1NZF |
| CF41 7UE | MW1COB |
| CF41 7UG | GW4NOS |
| CF42 5AH | MW6TPM |
| CF42 5AH | MW6DJT |
| CF42 5AR | MW6NFG |
| CF42 5AR | MW6TRL |
| CF42 5DR | GW4XAZ |
| CF42 5DT | 2W0YAE |
| CF42 5DT | MW0YAE |
| CF42 5DT | MW3YAE |
| CF42 5HL | MW3TWQ |
| CF42 6AW | GW7CYT |
| CF42 6DF | GW0NDZ |
| CF42 6DT | 2W0VDW |
| CF42 6DT | 2W0VEH |
| CF42 6DT | MW3VEH |
| CF42 6ED | MW0TJD |
| CF42 6LB | 2W0TRS |
| CF42 6LB | MW6DBU |
| CF42 6RL | GW1VAW |
| CF42 6RL | MW6AOY |
| CF42 6SN | 2W0PHP |
| CF42 6SN | MW3XZP |
| CF42 6TF | GW8UAP |
| CF43 3LF | GW7VSO |
| CF43 3LJ | MW1BTM |
| CF43 3NW | [illegible] |
| CF43 3NW | MW3YHW |
| CF43 3RN | MW0DRU |
| CF43 4EU | GW3SUH |
| CF43 4HE | [illegible] |
| CF43 4HE | MW3IPK |
| CF43 4HE | MW3VKA |
| CF43 4HT | GW4YLF |
| CF43 4PB | MW0HVL |
| CF43 4RG | MW6KEL |
| CF44 0BQ | 2W0TMB |
| CF44 0BQ | MW6KAC |
| CF44 0DJ | 2W0ZVR |
| CF44 0DJ | MW0ZVR |
| CF44 0EJ | 2W0XMG |
| CF44 0EJ | MW0XMG |

Postcode

**IMPORTANT NOTE**

**Revalidate licence to avoid revocation** – Ofcom has advised the Society that plans will be drawn up to revoke licences that have not been revalidated as required by the licence conditions. The quickest way to revalidate is to do so online via the Ofcom website: *https://services.ofcom.org.uk/* or by email: *amateur.validations@ofcom.org.uk* If you need assistance in the process, Ofcom staff are available to help, but please be patient during times of heavy workload.

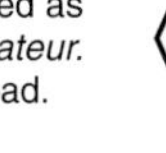

CF44 0EJ MW3USS
CF44 0EJ MW3XMG
CF44 0ER GW1HNG
CF44 0HR GW0UKC
CF44 0HY 2W0PJJ
CF44 0HY MW0HPH
CF44 0HY MW0PJJ
CF44 0HY MW6GNA
CF44 0HY MW6PJJ
CF44 0LL GW3PEX
CF44 0PB GW3SFC
CF44 0PF GW4UAJ
CF44 0PY GW4SUN
CF44 0RG GW4SCK
CF44 0TH GW4ZUA
CF44 0TP MW6AUX
CF44 0TT MW6WRC
CF44 0TU GW6MKR
CF44 6AH MW6LTR
CF44 6DD GW0OUH
CF44 6EN 2W0DTR
CF44 6LD GW3ASW
CF44 6LD GW4RUX
CF44 6LD MW0TTR
CF44 6LD MW6OCC
CF44 6NG 2W0XTP
CF44 6NG MW0TMI
CF44 6NL 2W0KOP
CF44 6PP GW6PXW
CF44 6RY MW6ALC
CF44 6RY MW6RAW
CF44 6SH 2W1HFZ
CF44 6YG MW1AND
CF44 7HA MW6LDS
CF44 7HE MW6TIQ
CF44 7NF MW6MIQ
CF44 7PP MW6LCX
CF44 8AH MW6GOY
CF44 8BW GW6YCT
CF44 8EW MW6BQA
CF44 8EW MW6LIW
CF44 8HA GW4OPW
CF44 8HL GW4SXA
CF44 8LA MW6DRV
CF44 8LF 2W0PPL
CF44 8LF MW6AZP
CF44 8LF MW6HEY
CF44 8LF MW6TJE
CF44 8TH MW0BWM
CF44 8TL MW3EQE
CF44 8UB GW0KZE
CF44 8YB GW4KCY
CF44 9DJ GW4BGD
CF44 9EW 2W0KED
CF44 9EW MW0JZM
CF44 9EW MW6KED
CF44 9JT 2W0DTM
CF44 9JT MW0KIJ
CF44 9JU MW3FJW
CF44 9LY GW0UHX
CF44 9LY GW1SVV
CF44 9ND GW4VEI
CF44 9NG MW0DAR
CF44 9RU 2W0WFB
CF44 9RU MW0WFB
CF44 9RU MW6WFB
CF44 9SA MW6BHD
CF44 9TA GW6OVD
CF44 9TX 2W0DSO
CF44 9TX MW6CQL
CF44 9YJ 2W0XBC
CF45 3AB GW0WXO
CF45 3BS GW0UHO
CF45 3BS MW3VGJ
CF45 3BY MW6BVG
CF45 3EF MW6KCJ
CF45 3NB MW6JGC
CF45 3ND GW4LWD
CF45 3NU 2W0SKG
CF45 3YA GW0VQZ
CF45 3YW GW4ZUJ
CF45 3YW MW6ENY
CF45 4BG 2W0RAD
CF45 4BG MW0PIC
CF45 4BG MW3ZLX
CF45 4BL 2W0IWM
CF45 4BL MW6IWM
CF45 4BX MW3IQE
CF45 4DZ MW6JFL
CF45 4NU MW1DTT
CF45 4SY GW7HVA
CF45 4TP 2W0KMU
CF45 4TP MW3KMU
CF46 5HG GW0JTJ
CF46 5HQ GW1IQS
CF46 5LA GW7SXN
CF46 5LB GW0OPP
CF46 5LN GW7HDS
CF46 5NJ GW4LKE
CF46 5RL 2W0BGS
CF46 5RL MW3ONG
CF46 6AB GW0PSV
CF46 6AJ MW6VBE
CF46 6HJ MW3NKG
CF46 6HQ 2W0AZP
CF46 6HQ MW3XVB
CF46 6LF MW3MOJ
CF46 6NS GW0OUV
CF46 6RY MW6EGH
CF46 6SD MW6DAI
CF47 0NB MW0DYS
CF47 0TA MW0KDL
CF47 0TB GW0UZX
CF47 0TG MW6CIY
CF47 0UX GW4PWZ
CF47 0YU 2W1SRB
CF47 8HJ GW3RYR
CF47 8HJ GW6BMR
CF47 8NJ GW0VMZ
CF47 8RH MW3YCL
CF47 8UN GW1MTH
CF47 9DA 2W0MJA
CF47 9DA MW0DOZ
CF47 9DA MW0EQL
CF47 9DA MW0MUM
CF47 9DA MW6MWS
CF47 9SD 2W0HFU
CF47 9SD MW6GRB
CF47 9SN MW6JNP
CF47 9TP GW3RTA
CF47 9YH 2W0VAY
CF47 9YH MW3VAY
CF47 9YP GW3PYD
CF48 1EE GW1WRV
CF48 1EN GW0UXJ
CF48 1EN MW0IBI
CF48 1HH MW6KLL
CF48 1HW GW6AYR
CF48 1JS GW4GJA
CF48 1JU 2W1CLC
CF48 1LU MW3BOP
CF48 1PH GW4CGZ
CF48 1TP GW6VRN
CF48 1YY GW0KIG
CF48 1YY MW3IXZ
CF48 2BB GW3LNR
CF48 2DG MW3TPJ
CF48 2RT 2W0BPJ
CF48 2RT MW3WFF
CF48 2SA MW3AFX
CF48 3HE GW0BYZ
CF48 3NY 2W0BOK
CF48 3NY GW3YUC
CF48 3PE MW3DTO
CF48 3PE MW6JJO
CF48 3PU GW0JTE
CF48 4BQ MW6HKL
CF48 4EE GW3RDB
CF48 4EE GW4ZRW
CF48 4HD GW4BIS
CF48 4HJ MW6MWN
CF48 4RN GW0WUL
CF5 1DH 2W0FCM
CF5 1GH GW1VRR
CF5 1GH GW1YXR
CF5 1HL MW3MHG
CF5 1NH MW6CTE
CF5 1RF GW8HWS
CF5 2AU GW8REV
CF5 2DH 2W0BTP
CF5 2JF 2W0USN
CF5 2JN GW2DHM
CF5 2JR MW0CCL
CF5 2QG MW6GFQ
CF5 2QP GW8RAS
CF5 2QS GW8THM
CF5 2QT GW3WCV
CF5 2RJ GW8JLY
CF5 3EH GW6RDV
CF5 3EW 2W0UZO
CF5 3EW MW0UZO
CF5 3EW MW3UZO
CF5 3RN GW0AJI
CF5 3SS GW6NHB
CF5 4AP GW4KQ
CF5 4FG GW8VFF
CF5 4FH MW0DYZ
CF5 4HJ MW3VNZ
CF5 4LN 2W0ZBC
CF5 4PG MW3HAS
CF5 4PR 2W0LEN
CF5 4QW MW3KCL
CF5 4QY MW6AYA
CF5 5AP MW6ICU
CF5 5AQ GW4FOM
CF5 5BZ 2W0GJR
CF5 5BZ MW6WUK
CF5 5DA 2W0SWO
CF5 5DA MW3SWO
CF5 5DD GW7ESF
CF5 5JS 2W0PEG
CF5 5JS MW6RJY
CF5 5NW MW3LGS
CF5 5PB 2W0YFC
CF5 5PL MW6BSR
CF5 5QD MW6EEJ
CF5 5SB MW3HSZ
CF5 5SB MW6AWV
CF5 5SB MW6RBZ
CF5 6AF MW6SEF
CF5 6BQ GW1CHS
CF5 6DA GW0WDV
CF5 6EP GW1OZW
CF5 6LQ GW3RIH
CF5 6SB 2W0PCT
CF5 6SB MW0PCT
CF5 6SB MW6PCT
CF5 6SS GW4CSY
CF5 6TN GW4JCK
CF61 1GX GW8MFQ
CF61 1TA GW6GSR
CF61 1YA GW8KEV
CF61 1YD GW4LFF
CF61 2AT 2W0IDT
CF61 2AT MW0IDT
CF61 2GS GW4MOG
CF61 2SG GW7HFZ
CF61 2UF GW3NCT
CF61 2UG GW0CNJ
CF61 3XR GW3FYA
CF61 5BD MW6ONZ
CF62 3DZ GW4ZCL
CF62 3EA MW6ZMN
CF62 3FT GW8YTO
CF62 3HL GW3TZT
CF62 3HQ GW0LTC
CF62 3LB MW6AQD
CF62 3LQ GW6OGD
CF62 4JD MW6ADS
CF62 4PD 2W0VOC
CF62 4PD MW6VOC
CF62 4PG MW6ORH
CF62 6RA GW1JVB
CF62 6RA GW1XFB
CF62 6RA GW8LJJ
CF62 7HX GW1VRW
CF62 7TG GW1ZLL
CF62 8BJ GW1XUD
CF62 8BL GW6CNS
CF62 8BR GW8KZA
CF62 8BU MW3DJV
CF62 8BU MW3DJZ
CF62 8BY GW0JCB
CF62 8BY GW0NXW
CF62 8BY MW3JDA
CF62 8HF MW1AUS
CF62 8ND GW4BRS
CF62 8ND GW6BRC
CF62 8ND MW0VRQ
CF62 8ND MW6JZZ
CF62 9DE MW6BVR
CF62 9DR GW7PFK
CF62 9DW 2W0NCA
CF62 9HD GW0SQT
CF62 9HJ GW3WSU
CF62 9HJ GW4GSH
CF62 9SW GW3CBA
CF62 9TE MW6FYY
CF62 9TG MW0RUH
CF62 9TG MW3RUH
CF63 1DW GW7AUQ
CF63 1FP MW0CWF
CF63 1FP MW1COY
CF63 1QJ GW7HQL
CF63 2FE MW3WRH
CF63 2HY 2W1SDR
CF63 3QD GW4ANK
CF63 3RE GW1ABX
CF63 4EF GW6ATT
CF63 4EF GW7VST
CF63 4HW GW0TSL
CF63 4JE 2W1ETN
CF63 4NP MW6ZAN
CF63 4PQ GW0GRQ
CF64 1DD GW3VKL
CF64 1DD MW0DHF
CF64 1TN GW3MOV
CF64 1TR GW1LKG
CF64 1WW GW3LQE
CF64 2JS MW3LZC
CF64 2PG GW0GHF
CF64 2QZ GW3WBU
CF64 2RX GW0PZZ
CF64 2RX MW3WXN
CF64 2SA GW0OTY
CF64 2TZ GW4UGI
CF64 3BE GW3UWL
CF64 3DD GW3SPA
CF64 3DH MW1EAA
CF64 3DH MW3RRW
CF64 3EN GW3FZV
CF64 3HF GW3VNZ
CF64 3HH GW1MBV
CF64 3HY MW0HAT
CF64 3JD GW4IUL
CF64 3PP MW3NHC
CF64 3QX 2W1YEG
CF64 3QX MW3YEG
CF64 3RB GW3PYX
CF64 4DJ GW8PBM
CF64 4LR GW3SFQ
CF64 4PS GW1BAI
CF64 4PZ MW6WLB
CF64 4TG GW4XKE
CF64 5BR GW3IVR
CF64 5BW GW1DQV
CF64 5GA GW4EUA
CF64 5GA GW8JWL
CF64 5JW 2W0ICD
CF64 5QD MW0GBW
CF64 5QE MW6FTC
CF64 5SX MW3UDO
CF64 5UG GW0CKK
CF71 7EX GW4MOZ
CF71 7HG 2W0DUN
CF71 7LX GW0ANA
CF71 7QR GW7NJQ
CF72 8BG 2W0YYP
CF72 8DA GW0DIV
CF72 8DA MW6HLG
CF72 8DY 2W0CGM
CF72 8HR GW4VMT
CF72 9BP 2W0HPM
CF72 9DB GW4HDZ
CF72 9ER GW1DPV
CF72 9FZ GW6PFK
CF72 9HU MW6WCI
CF72 9JB MW6LSW
CF72 9JD 2W1ACM
CF72 9LT 2W0RTJ
CF72 9LT MW6TTS
CF72 9ND GW7VSF
CF72 9QQ MW6PLC
CF72 9SB MW0PVW
CF72 9SB MW6DYL
CF72 9SJ GW0VWD
CF72 9SL MW6WDX
CF72 9SZ MW6TES
CF72 9TF GW7VOO
CF72 9TX GW4DVB
CF8 4AU GW4ZBN
CF8 4DL GW1SRB
CF8 8NA 2W0IVZ
CF81 8JS GW0LBI
CF81 8JX 2W1EKR
CF81 8LU GW7WFI
CF81 8NB GW0WGW
CF81 8NS MW6WFF
CF81 8NT 2W0WMB
CF81 8NW MW1AFW
CF81 8PP 2W0EGK
CF81 8PP MW3EGZ
CF81 8QB MW3PKC
CF81 8TG MW0TAF
CF81 8US MW3FFL
CF81 8UU 2W0HAK
CF81 9AB GW8GOC
CF81 9DZ MW3IWC
CF81 9LQ 2W1JOL
CF81 9PT MW3XNW
CF81 9RS MW6KAB
CF82 7DB 2W0UAA
CF82 7DB MW0MAU
CF82 7HF MW3ZTH
CF82 7HL MW3CXA
CF82 7JB MW0ATK
CF82 7JP MW0ICE
CF82 7JP MW3CCE
CF82 7NP GW4TPG
CF82 7QQ 2W1EUR
CF82 7RH GW0PCJ
CF82 7RZ GW6JSJ
CF82 8BR MW1FEU
CF82 8EG MW0NQH
CF82 8ET GW0TKX
CF82 8EW MW6EOP
CF82 8FN MW0XMI
CF82 8LA GW4XMU
CF83 1AN MW6CRU
CF83 1JW MW6EXV
CF83 1LA MW3ZTB
CF83 1NQ MW6FNT
CF83 1RF 2W0RDD
CF83 1RF MW3DQB
CF83 1RU MW6JBQ
CF83 1SP GW4HZM
CF83 1SQ GW6IPR
CF83 1SZ GW0IRC
CF83 2AN MW6NYE
CF83 2EU MW6HPN
CF83 2JU MW0MEX
CF83 2LA 2W0PCE
CF83 2LL MW3USP
CF83 2NE 2W0CYY
CF83 2NE 2W0YBZ
CF83 2NE MW0YBZ
CF83 2NE MW6CVC
CF83 2NE MW6YBZ
CF83 2NJ 2W0RFT
CF83 2NJ MW3IQY
CF83 2NR GW4YKM
CF83 2NZ GW0MXG
CF83 2RA MW3HAQ
CF83 2RX GW8UAM
CF83 2SP GW0AVW
CF83 2SR MW6SEP
CF83 2TX GW0OSB
CF83 2TX GW1MAX
CF83 2UN MW3TJG
CF83 2UW GW4HDF
CF83 3BT 2W0FEY
CF83 3BT MW6PEY
CF83 3BU GW6WTK
CF83 3FB 2W0MSL
CF83 3FB MW6NAG
CF83 3FT GW0KWO
CF83 3HF GW4VEB
CF83 3JH MW6HGZ
CF83 3JT GW0MAV
CF83 3QB MW6HFY
CF83 3RT GW6DOC
CF83 4BG 2W0JCV
CF83 4BG MW6JCV
CF83 4BN GW4SLI
CF83 4DD MW3WLB
CF83 4DH MW6ECR
CF83 4EQ GW0CKL
CF83 4FD MW6GSL
CF83 4GG GW0PUP
CF83 4HJ 2W0BGQ
CF83 4HJ MW3IUS
CF83 4HN 2W0BUQ
CF83 4HN MW3ZWO
CF83 8EP MW6IAG
CF83 8TA MW6PBF
CF83 8TT 2W0NQE
CF83 8TT MW0GEI
CF83 8TT MW3NQE

## CH (Chester)

CH1 1QP G8GWX
CH1 3BZ M6VBJ
CH1 3HG G8WLD
CH1 3ND 2E0BSR
CH1 4AN 2E0CCK
CH1 4AN G3SES
CH1 4AN M0GYF
CH1 4AN M0TTK
CH1 4AN M3MTB
CH1 4BE 2E0IFW
CH1 4BE M0JXE
CH1 4BE M3JXE
CH1 5AF G6LRU
CH1 5AZ G0ULM
CH1 5DP 2E1CRA
CH1 5DZ M6BNK
CH1 5DZ M6GNG
CH1 5JQ G1URW
CH1 5NW G7VPQ
CH1 5QX M1EQD
CH1 5RU M0ELC
CH1 5RU M3LIN
CH1 5SW M6DOA
CH1 5SY M1DAP
CH1 6JS G4SMM
CH1 6LU G4WXO
CH2 1BB 2E0UPT
CH2 1BB M3UPT
CH2 1HF G6VGV
CH2 1HT G8BMH
CH2 1JG M0NUG
CH2 1LX G0NLQ
CH2 1LX G7HGB
CH2 1NF G7GFC
CH2 1NL G3ZVH
CH2 1NW M0AUA
CH2 1PF G0HEJ
CH2 1RD G8RRS
CH2 2AQ G3KJS
CH2 2BX G0OIV
CH2 2DH M3UZA
CH2 2DP M6LOB
CH2 2EA G1NCD
CH2 2HL M3IHA
CH2 2LA G0ROY
CH2 2PB M6VHM
CH2 2PL G0NMD
CH2 3EW 2E0GPE
CH2 3EW M0PEA
CH2 3EW M6GHP
CH2 3EW M6LVA
CH2 3HT M3OME
CH2 3JT M0CVP
CH2 3PZ G8UEK
CH2 3PZ M3UEK
CH2 3RE G6SKP
CH2 4EA M1RWB
CH2 4RP M3HWV
CH3 5DY G1YNJ
CH3 5HA G1LML
CH3 5HD M3OAS
CH3 5HE 2E0WVD
CH3 5HE M3WVD
CH3 5HE M3WVF
CH3 5HE M3WVT
CH3 5HF M0PWS
CH3 5JB G7HEK
CH3 5JF 2E0OBZ
CH3 5JF M3OBZ
CH3 5LA G8KKN
CH3 5LE G1XPW
CH3 5LU M3JAL
CH3 5LY G1CZU
CH3 5LY G7BQY
CH3 5PT G8ZRE
CH3 5RR G4JMF
CH3 5RR G8GIZ
CH3 5RR M3XSD
CH3 5XN G1DIA
CH3 6PE G0VAH
CH3 6QX M3CZX
CH3 7EJ G8KWP
CH3 7HE G7USC
CH3 7NT G0DUB
CH3 7NT G0RCW
CH3 7NT M0LSA
CH3 7QD G6FDK
CH3 7QD M3DVA
CH3 7QQ 2E0IBA
CH3 7QQ M3OCS
CH3 8EQ G6VUX
CH3 8LP G7NEH
CH3 8LR G0DYL
CH3 8LR G6MYL
CH3 9BE G7EEN
CH3 9EH M6NUG
CH3 9JH 2E0TPA
CH3 9JH M0TPA
CH3 9JH M6TPX
CH4 0AH 2E0LUD
CH4 0AH M6PPZ
CH4 0AL M0CSO
CH4 0AQ G6FKB
CH4 0FT M6AQT
CH4 0HN G3RTR
CH4 0JD M6RDX
CH4 0LF M6WCF
CH4 0NH G0EDC
CH4 0NZ 2E0BTA
CH4 0NZ 2E0BTD
CH4 0NZ M3LXB
CH4 0NZ M3LXK
CH4 0PT G8VNN
CH4 0PU M3LBC
CH4 0QE M6AKF
CH4 0QJ G4FGC
CH4 0QT G6OKC
CH4 0RJ M3KXV
CH4 0RL G7TZX
CH4 0RN G1LDY
CH4 0SQ G8ZBC
CH4 0SZ G0ISE
CH4 7LE G6XKJ
CH4 7LU G8VAD
CH4 7LZ G1LED
CH4 7NF M1BZI
CH4 8BB G7KLN
CH4 8BG M3EVF
CH4 8BG M6NCF
CH4 8BY M3IUV
CH4 8DE M1FJP
CH4 8JG G8SLP
CH4 8LB G4UXD
CH4 8PA G8DOF
CH4 8SJ 2E0TWI
CH4 8SS M3JUY
CH4 9DG G4YNS
CH4 9NG G0CTH
CH4 9NG G1SVN
CH4 9NN G0UVT
CH4 9NN G0UWA
CH4 9NU G6YCW
CH41 0AX M1AAS
CH41 3LF 2E0LGZ
CH41 4FF G4NCI
CH41 8BZ M6THC
CH41 8HJ 2E0MTH
CH41 8HJ M6FTT
CH42 0LB G8UWG
CH42 1LW M1AOR
CH42 2BU G6RYM
CH42 3YE 2E0OQH
CH42 3YE M3OQH
CH42 3YH 2E1SOX
CH42 3YH M3SHJ
CH42 4PD G3OTW
CH42 4QX M0BZZ
CH42 5LH M5HFJ
CH42 5LR G0PKE
CH42 5LR M3ELV
CH42 6QY 2E0YJY
CH42 6QY G4EWJ
CH42 6QY M6MAJ
CH42 7JA G4WVC
CH42 7LB 2E1GMA
CH42 7LB G0PZO
CH42 8JZ M0HGG
CH42 8QA G6JZE
CH42 9LJ G0KXL
CH42 9NZ G6RVH
CH42 9NZ M3RGJ
CH42 9QA 2E1PGB
CH42 9QA M3PGB
CH43 0RQ M3PRU
CH43 0TT G4OAR
CH43 0TT G6HGI
CH43 0XB G1MHF
CH43 1TE G0PIT
CH43 2HS G6XHF
CH43 2HS M6CUL
CH43 2HS M6STI
CH43 2JL M3LXH
CH43 2JP G7SFM
CH43 2JP M0CKV
CH43 2LZ G0LZI
CH43 2NQ M1ARH
CH43 3AS 2E0GLR
CH43 3BB M3XCB
CH43 3DW 2E0ENZ
CH43 4UL G6VDX
CH43 4XR 2E0MFN
CH43 4XR M6MNK
CH43 5RP M0GNA
CH43 5XD G7NOS
CH43 6TA G4ASM
CH43 6UT G0UCP
CH43 6UY G0MHF
CH43 7NX G4MUP
CH43 7QP G3ZIC
CH43 7SE G0WMY
CH43 7SQ G6NFB
CH43 7YQ G6EWK
CH43 8SU G8FPU
CH43 9SW G3NMT
CH43 9UN G6FUY
CH43 9UQ M0HRZ
CH43 9YT M3LGX
CH43 9YT M3LGY
CH44 0AQ G1RCI
CH44 0BQ G4NXG
CH44 0BQ M0AJB
CH44 0BQ M3PSB
CH44 0BQ M3YJB
CH44 0EB 2E1GXL
CH44 0EL G1GBH
CH44 1AJ M6UHF
CH44 1AT M0AXJ
CH44 1DW G7VQM
CH44 2EY G1BYI
CH44 3AF G1PMJ
CH44 3DA G1MCR
CH44 3DA G8RXB
CH44 3DJ 2E0AAN
CH44 3DJ M6AOQ
CH44 3DZ G8NNS
CH44 3DZ G8TFW
CH44 3EE G0JQK
CH44 4DL G0DPO
CH44 5RQ G4ZCA
CH44 5RQ G7KOS
CH44 5XB 2E0DHQ
CH44 5XB M6EKW
CH44 6PR M6GSP
CH44 7BJ 2E1TAB
CH44 7BJ M3ATB
CH44 8AB G7OKR
CH44 8AE G4AHC
CH44 9AA G6LWC
CH44 9BR G4PGO
CH44 9DN 2E0WAE
CH44 9DN M6PUG
CH44 9DX G4KVP
CH44 9DZ G1NXB
CH44 9EB 2E0PCU
CH44 9EB M6PFG
CH45 0LG M6AAC
CH45 1HA G0NBD
CH45 1HD G6ABO
CH45 1JG G0WMQ
CH45 1JT G1JEZ
CH45 2LZ G4KRF
CH45 2NG M0NBJ
CH45 2NN M6EXS
CH45 2PE G8STF
CH45 4PN M6BTM
CH45 5AY G4ENK
CH45 5DB G4FPB
CH45 5EJ 2E0PSO
CH45 5HB G0AWW
CH45 5HE G1PDA
CH45 5HN G1FEP
CH45 6TD G3SEJ
CH45 6TD G8STD
CH45 6TE G3XCD
CH45 6UQ G6IHD
CH45 7LQ 2E0PJD
CH45 7LQ M3IRU
CH45 7NZ G3VEB
CH45 7QF G8TRY
CH45 7QF G8WDC
CH45 7QG G4VQS
CH45 7QQ G4KOY
CH45 8JN M3DBX
CH45 8LH M6SIP
CH45 8LR M6PMA
CH45 8NX G8JUV
CH45 8NX G8UUC
CH45 9JA G1EYJ
CH45 9LJ G6IIN
CH46 0QT G3PEZ
CH46 0RS M6DXZ
CH46 0TP 2E0RAA
CH46 0TU G3LCI
CH46 1QH G0PFT
CH46 1QZ G7NHR
CH46 1RQ M3PUL
CH46 1RW 2E0NLP
CH46 1RW M6ATS
CH46 1RX M3HQP
CH46 1RX M3PAP
CH46 2QR G6HBF
CH46 2QZ G0WUO
CH46 2SA 2E0WBL
CH46 2SA M6MEQ
CH46 2SB G0PVY
CH46 3RR M6BXW
CH46 3RX 2E0AZK
CH46 5NE M6GAW
CH46 6AS 2E0DEQ
CH46 6AX G3YHB
CH46 6BJ G6USU
CH46 6DH G4WUA
CH46 6DH M0MTC
CH46 6DS M6TGL
CH46 6EL 2E0DLV
CH46 6EN 2E0BDO
CH46 6EN M3KZW
CH46 6ET M1ELR
CH46 6FL G4VWL
CH46 6HB G4UJP
CH46 6HB G6WWS
CH46 6HE 2E0MTX
CH46 6HE M0WBG
CH46 6HE M6MBP
CH46 6HQ G3XJZ
CH46 7SJ G0CLV
CH46 7TS G0TYN
CH46 7TS M3LDH
CH46 7UP G7FND
CH46 7UT G0DNQ
CH46 8TT M0SPD
CH46 8TU 2E0CKL
CH46 8TU M0UFA
CH46 8TU M6BHF
CH46 8TX G0EQE
CH46 9PF G0LHN
CH46 9PS 2E1RSB
CH46 9RW M6GZI
CH46 9RY M3VYK
CH47 0LB G3OKA
CH47 0LQ G4ELA
CH47 0NF G7IIF
CH47 2AE 2E0LGH
CH47 2AE M3LGH
CH47 2AG 2E0YOZ
CH47 2AG M6HOY
CH47 2AZ G4RVZ
CH47 3AG M6RKQ
CH47 3AJ M3FLL
CH47 3DD G8THZ
CH47 7DA G1ZZA
CH47 9RW 2E0CDM
CH47 9RW M3VZC
CH47 9RZ M6GWU
CH48 0QQ G4DLY
CH48 3HE M0BFV
CH48 4EL M6HFM
CH48 5DG 2E0ILO
CH48 5DG M6ILO
CH48 5EG M6HJM
CH48 5HQ G8ODK
CH48 5HZ M6PLE
CH48 6DA G0VAX
CH48 6DJ G1UHO
CH48 6ET M3IYO
CH48 7EU M0SJK
CH48 7EX G3OVE
CH48 7HR G8LEM
CH48 9UF G7NIR
CH48 9UL G0SSG
CH48 9UX G0GSF
CH48 9XP 2E0ETD
CH48 9XP M6ETD
CH48 9XS M5AOI
CH49 0TD G4BKF
CH49 1RT 2E0TFM
CH49 1RT M6KDZ
CH49 1SS G6PRE
CH49 2RJ G8UZZ
CH49 2RQ G6KBQ
CH49 2RZ G8OBT
CH49 3AG G3YSM
CH49 3AW G3MAJ
CH49 3NG G0MQJ
CH49 3NP G8ZJH
CH49 3QH G0LPU
CH49 4GD G3NWR
CH49 4GD G8WWD
CH49 4GY G3VUY
CH49 4PF G0TPJ
CH49 6JD G7MIZ
CH49 6LA G4MIA
CH49 6NL M0DWU
CH49 6NQ G0CCT
CH49 6PN 2E1JEF
CH49 7LH M3TDM
CH49 7NJ 2E0BJW
CH49 7NT M6DYG
CH49 8EE G7GRU
CH49 8HQ M1MPB
CH49 8JP 2E0SQK
CH49 8JP M6HSB
CH5 1EZ G7GKX
CH5 1HH G4KDI
CH5 1HY G6KWU
CH5 1NU G1ATZ
CH5 1NU G3ATZ
CH5 1XU M3JIJ
CH5 2BR M3RPX
CH5 2DP G7SBJ
CH5 3AL G0PBJ
CH5 3EF G4MVA
CH5 3EX G0TCL
CH5 3HS G4EYO
CH5 3JG G4SII
CH5 3RW G0DRS
CH5 4AL G1RVC
CH5 4AU M3XZV
CH5 4BY M6PAI
CH5 4GG G6POO
CH5 4HW G4ZAG
CH5 4JP G0DSP
CH5 4JP G1LFX
CH5 4JW G1LEL
CH5 4LG G3KNZ
CH5 4LG M3JLK
CH5 4LG M3MPN
CH5 4QA M3FLI
CH5 4QN M0CRA
CH5 4QP G7UVO
CH5 4RE G4RQS
CH5 4RE G4VBM
CH5 4SH G6NVJ
CH5 4SN G0HRG
CH5 4SN G7AAU
CH5 4SN G7AAV
CH5 4SN M6CLF
CH5 4TF M3SUF
CH5 4TN G0UEO
CH5 4WP G0SXE
CH5 4WQ 2E0DPI
CH5 4WQ M6AFK
CH6 5AN G0UGQ
CH6 5HL G8JRL
CH6 5HU G0UIP
CH6 5HU M0GPP
CH6 5JP G6LMI
CH6 5JU M1CRE
CH6 5LN M1LSG
CH6 5LP 2E0CCG
CH6 5LT M1RES
CH6 5ND G6XQX
CH6 5QA G4ZAR
CH6 5QB G4VFE
CH6 5RQ M0GRJ
CH6 5RX G4RYJ
CH6 5SB G0GZR
CH6 5YE G7CMM
CH6 6DP G6WGY
CH6 6JE G0POG
CH6 6JS 2E1RSS
CH6 6JS M0RKD
CH6 6LS G7CCR
CH60 0BP G3WIO
CH60 1UH G6LKG
CH60 1YD G6AGP
CH60 2SG G0MQR
CH60 2UA G3NPJ
CH60 5RY G4EIC
CH60 5RZ M3YKZ
CH60 6RD M3REX
CH60 6TE G7FVR
CH60 7RA G3UVR
CH60 7RA G4MGR
CH60 7RA M0ATZ
CH60 9JT G3RLA
CH61 0HD G4DBE
CH61 3UY 2E0EFC
CH61 3UY M6ZKA
CH61 3XH G7AAI
CH61 3XS G0WAB
CH61 4UA G1RIX
CH61 4YS M6EKV
CH61 5UG G6KIV
CH61 6UP G1KJX
CH61 6UZ G4NOY
CH61 6XT G8DTS
CH61 6YH G8HLJ
CH61 8SU G4ILA
CH61 8SX G8REQ
CH61 9NT M6YIN
CH61 9PR M0CNN
CH61 9PY 2E0PYR
CH61 9PY M6PYR
CH61 9QA 2E0XGA
CH61 9QA M3XGA
CH62 0BD G3ODC
CH62 1BJ M0BQH
CH62 1DY M3YWJ
CH62 3LH M0HVS
CH62 5DB M3TFI
CH62 5DB M3TXU
CH62 6AN G6HKY
CH62 6AW G7IGU
CH62 6BE G6PYL
CH62 6BE M3FMV
CH62 6BE M3ODC
CH62 6BR G8WHR
CH62 6BR G8WSR
CH62 6BU G0GBN
CH62 6DN M6BNW
CH62 6DT G8IWJ
CH62 6DX G8LAU
CH62 6EA M0WAD
CH62 7BA 2E0HLP
CH62 7BA M3HLP
CH62 7EE 2E0DUP
CH62 7EE M6VAH
CH62 7EX G4YWD
CH62 7HA M0AUF
CH62 7JY G3NNV
CH62 8BB G6AVY
CH62 8BB M0CMI
CH62 8BN M6GHM
CH62 8BR G3YGL
CH62 8DG G1WAS
CH62 8DJ G8CVF
CH62 8DJ M6RTR
CH62 8DL G4UCE
CH62 8EB G1NPN
CH62 8EQ G3TRR
CH62 8ET 2E0ICE
CH62 8EX G0PZD
CH62 9AA G3AVL
CH62 9DF G6GFG
CH63 0JA G6TPE
CH63 0JJ G8TSG
CH63 0JW G7NHD
CH63 0LA G8EYA
CH63 0QF M0SMC
CH63 2JU G6NOI
CH63 2LW M6CXH
CH63 2NH M6AQQ
CH63 2NW G8UCV
CH63 4JS G1KEP
CH63 5JR G4NSA
CH63 5JR M6HAK
CH63 5LH G7EED
CH63 5PA G0OXA
CH63 7LR 2E0OWL
CH63 7QU M6XBJ
CH63 8LB G7TIY
CH63 9AH G6BWO
CH63 9LP G0PXO
CH63 9LU M6LSP
CH63 9NG M0HWO
CH63 9YJ M1BZZ
CH64 0UZ G4EEQ
CH64 0XB G1DEV
CH64 0XP G1NUH
CH64 1RZ G4FVZ
CH64 1TN G4JZR
CH64 2XG M0PER
CH64 2XQ G8OKZ
CH64 3RS M6PTH
CH64 3TJ 2E0IGQ
CH64 4AN G6HQ
CH64 4AR G8OJQ
CH64 4AR M0RAL
CH64 4AT G4HGL
CH64 4BJ G3NTI
CH64 4DZ G3PXX
CH64 5TJ 2E0CPB
CH64 5TJ M3YPB
CH64 6QB G0KQS
CH64 6RB 2E0MLD
CH64 6RB G6FTL
CH64 6SF G4ELK
CH64 7TR G6ADO
CH64 7TR G8MMM
CH64 7TR M3JLI
CH64 9SW G7DEY
CH65 0DJ G0OJC
CH65 2BD 2E0GWI
CH65 2BE G1AOF
CH65 2BE M0COC
CH65 5AD M1SUM
CH65 5BB 2E0PXW
CH65 5BB M3PXW
CH65 5DG M6FEK
CH65 5DG M6RBF
CH65 5DX G0HFE
CH65 6PB G7JKH
CH65 6RW 2E0RFF
CH65 6RW G0RGG
CH65 6RW M6EBC
CH65 6SB G3YCO
CH65 6TD G6ALN
CH65 7AQ G4JTK
CH65 7AZ M3LCL
CH65 7BW G0WVW
CH65 7DX M6BLV
CH65 7DZ 2E1LES
CH65 8EP 2E0IHO
CH65 8EP M3IHO
CH65 9DZ G0PJY
CH65 9EN G0CTR
CH65 9EY G8RVY
CH65 9JQ M6LAF
CH66 1JF G1PRL
CH66 1JG M3UHB
CH66 1JJ G3VZM
CH66 1JT G0VHJ
CH66 1JW G0JZJ
CH66 1JX M3AIS
CH66 1NY M3UQZ
CH66 1QS G0CGD
CH66 1QT G5MUN
CH66 1RW G0AKU
CH66 1TY G4STZ
CH66 2BG M0BAG
CH66 2BJ G4DBG
CH66 2BL G4YBJ
CH66 2GU G8XPB
CH66 2GX G6IIM
CH66 2LH G1GWS
CH66 2PA G0IEQ
CH66 2PZ G1URQ
CH66 2SX G4ZKG
CH66 2WL M3LYU
CH66 2YJ G3CSA
CH66 2YJ G4WSE
CH66 2YL G0JCG
CH66 3LL G0CDJ
CH66 3NN M6TXH
CH66 3PF G7GQW
CH66 3PH G4UXL
CH66 3QB M0AMS
CH66 3SQ G3TXH
CH66 4NA G7NHV
CH66 4NN G0KKO
CH66 4PD G0RCY
CH66 4QY M1ZXZ
CH66 4SG G8IPT
CH66 4TW G4UPB
CH66 4UF M6HCJ
CH66 5NB M3YQH
CH66 5PB M1AOX
CH7 1HH 2E0GDA
CH7 1HH M0WML
CH7 1HH M6AQU
CH7 1LD G0TCV
CH7 1SU G7MVG
CH7 1TH M3RAA
CH7 1UY M3WMH
CH7 2AG G3UOO
CH7 2AG G4HER
CH7 2AX G7MHB
CH7 2BS 2E1HUX
CH7 2BS G3RIB
CH7 2GE G0EGQ
CH7 2HQ G7AIY
CH7 2JA G0GXQ
CH7 2JL G0FHL
CH7 2JS M6BPH
CH7 2LJ 2E0NAD
CH7 2LJ M0JRX
CH7 2PA G3GAT
CH7 2PB 2E0IBM
CH7 2QP M0HLW
CH7 3BL G3UVA
CH7 3BR G7BOY
CH7 3DU 2E0DHD
CH7 3DU M6ERQ
CH7 3EL G1OTI
CH7 3ER 2E0VMC
CH7 3ER M3NQK
CH7 3ET G4GTE
CH7 3HN G4HBZ
CH7 3JU M6BNH
CH7 3LH 2E0RSV
CH7 3LH M0RSV
CH7 3LH M6YZF
CH7 3NB M0RHT
CH7 3NG G0PJA
CH7 3QA G0OHJ
CH7 3QF M6FEF
CH7 4NS M3JNX
CH7 4PY G4GDM
CH7 4QD G1PKW
CH7 4SN G6UTF
CH7 4SS G3TMP
CH7 4TU G4SDO
CH7 5AE G1FWE
CH7 5DZ G8EQI
CH7 5EL G7VFJ
CH7 5RF G8NZN
CH7 6DU G6RNV
CH7 6DY G3ITT
CH7 6ED G7NFM
CH7 6EF G4EVX
CH7 6HT M6ROX
CH7 6JF G3FPH
CH7 6NH G3TOW
CH7 6PN M6VDE
CH7 6PY G1HFW
CH7 6RT 2E0VAC
CH7 6RT M0ZAQ

[illegible]
[illegible]
CH7 6SL G3IVK
CH7 6SW G4BTW
CH7 6TD G4NEI
CH7 6TF G0PZU
CH7 6TF G4RAF
CH7 6TP G7MGW
CH7 6TR M0CVW
CH7 6TU M3KML
CH7 6US G6FKP
CH7 6UZ G3LWU
CH7 6WD M0RPE
CH7 6YP G3KJN
CH7 6YU G4GSG
CH8 7AP M6KOI
CH8 7AU M3UDA
CH8 7DF G0PFZ
CH8 7DR G0HKQ
CH8 7PJ G1CGD
CH8 7QR G4MOK
CH8 7SJ G4TIZ
CH8 7UG G6WQJ
CH8 7UJ G6VZB
CH8 7UR G7KIS
CH8 7XG G4VUH
CH8 8AX G0UDJ
CH8 8BU M0HCC
CH8 8BU M6IOA
CH8 8DL G4JUN
CH8 8DU M3PZC
CH8 8ES G0FEU
CH8 8HA M0CIH
CH8 8HE G1PJL
CH8 8JY 2E0GXI
CH8 8JY M6GXI
CH8 8LQ G8RAK
CH8 8NG M3CBX
CH8 8PR M0RMS
CH8 8QY G6VEI
CH8 9BZ M0LCK
CH8 9HJ M0ATT
CH8 9HY G6RCX
CH8 9JB M1ARM
CH8 9JB M3ARM
CH8 9JQ G7KBI
CH8 9NZ M3NGN
CH8 9NZ M3NGP
CH8 9PQ M3LBX

## CM (Chelmsford)

CM0 7AL 2E0XDM
CM0 7AL M6UDC
CM0 7AY M1CKZ
CM0 7BA M3TCR
CM0 7HJ 2E0CUS
CM0 7NF G4JDH
CM0 7QE G6VQV
CM0 7QT G7JDR
CM0 7QT M0RBI
CM0 7RD 2E1EDV
CM0 7RD 2E1EDW
CM0 7TH G0ATK
CM0 8BT G1IGN
CM0 8DP G7RGR
CM0 8EH M0CZC
CM0 8EY G4POP
CM0 8EY G6CNQ
CM0 8HS G8CCL
CM0 8HS M3CCY
CM0 8LX G8PUY
CM0 8PZ G6EUW
CM0 8QH G7OBX
CM0 8RB G1LUX
CM1 1QB G7KOI
CM1 1QD G0UIB
CM1 1TN G1FOA
CM1 1TP G6XHI
CM1 2AR G8UVY
CM1 2AT 2E0TXL
CM1 2AT M0TXL
CM1 2AT M6FIL
CM1 2DZ M1EZE
CM1 2JA G3XHC
CM1 2JA G4YTG
CM1 2NH 2E1HBB
CM1 2NJ G7RFT
CM1 2NQ G0NVM
CM1 2PR G1BNE
CM1 2PT G3SLT
CM1 2QZ 2E0CDX
CM1 2QZ M6MAO
CM1 2RR G0NAX
CM1 2SE G7GBZ
CM1 2SX M6WTF
CM1 2SY 2E0FSX
CM1 2SY M6FNR
CM1 3BJ G4IVU
CM1 3EA G4IMC
CM1 3GW M3ZJD
CM1 3HS M6PVM
CM1 3JB M0HBY
CM1 3JZ G4RAP
CM1 3NB 2E0XBK
CM1 3NU G8GWB
CM1 4DA M3DIM
CM1 4DG 2E0JCS
CM1 4DG G1CSR
CM1 4DG G3CSR
CM1 4DG G8GFF
CM1 4DG M3AAY
CM1 4DG M3JCS
CM1 4DU G1IBP
CM1 4DU G4TRF
CM1 4EF G3VCG
CM1 4EJ G8AEU
CM1 4HB G4GHO
CM1 4HD G8CDB
CM1 4HJ G1NZD
CM1 4JQ 2E0KFM
CM1 4JQ M0KVR
CM1 4JQ M3GTV
CM1 4JT 2E1IAI
CM1 4JY G4FKH
CM1 4JY G4TOO
CM1 4NU M0CSV
CM1 4NW M0ROO
CM1 4SJ G6ZWC
CM1 4UN G4HSK
CM1 4UQ G6DJS
CM1 4XU G8SGM
CM1 4XZ M0BQC
CM1 4YA G4ADG
CM1 4YG M3YNM
CM1 4YR M6HIJ
CM1 6BF 2E0KIL
CM1 6BF M3HNK
CM1 6FD G0LLB
CM1 6HY G4JJH
CM1 6JF 2E0RMI
CM1 6JF M3RMI
CM1 6JX G4DJC
CM1 6QR G3VSI
CM1 6QT M6VGE
CM1 6TX G7RCL
CM1 6UJ M3NOJ
CM1 6UJ M6ERR
CM1 6UW G8AAE
CM1 6UX G1EFL
CM1 6UX M0VAM
CM1 6UX M3VAM
CM1 6XY 2E0HDW
CM1 6XY G1EXU
CM1 6YP G8PFL
CM1 7BU M6NOM
CM1 7PG G8ETN
CM1 7PJ G7PEC
CM1 7PJ G7UHE
CM1 7PJ M0EHF
CM1 7RD M3GUO
CM1 7RX 2E0BWU
CM1 7RX M3OJN
CM1 7RY G6GOV
CM11 1AU G4EVZ
CM11 1BU M0JDR
CM11 1ET G0ORT
CM11 1ET G4HTG
CM11 1HB G0NGD
CM11 1JF G4EOW
CM11 1LH G0PXA
CM11 1QN G8CCN
CM11 1RR M6GSQ
CM11 2HU M3ZTW
CM11 2LL G1DHB
CM11 2NX G0TST
CM11 2NZ M0BZC
CM11 2PD G8URI
CM11 2QN G3UTC
CM11 2QN G4SIS
CM11 2RQ G8HGN
CM11 2RY G6KVY
CM11 2TS M6ISG
CM11 2XA G3MVV
CM11 2XE G4XTS
CM11 2XU G3OCI
CM11 2YX M6EZP
CM12 0HH G3TTB
CM12 0HL G4DKB
CM12 0JF G7PMI
CM12 0NX G4LPS
CM12 0PF G8GZV
CM12 0PY M3MQB
CM12 0SU G1UKA
CM12 0UD M6FVR
CM12 0UG M3CAD
CM12 0YE M0DKS
CM12 9DS M0HSX
CM12 9JF G8WXU
CM12 9JJ G4JQU
CM12 0JL M0ODX
CM12 9JL M3ADX
CM12 9JN G1DSG
CM12 9LX G0DQN
CM12 9NT M1EMU
CM12 9PG G4LFT
CM129LH M3VUC
CM13 1BT G3JWI
CM13 1JD M6WSR
CM13 1JX 2E0JRT
CM13 1JX M6JRT
CM13 1LW G6YUY
CM13 1RH G0UBL
CM13 1SJ G0PNO
CM13 2HZ G8NYD
CM13 2HZ M6PLZ
CM13 2LA G7HRL
CM13 2NF G8CUB
CM13 2TJ G4TTQ
CM13 2UF G4WYI
CM13 3AL G0FMU
CM13 3DD M0OGS
CM13 3DZ G0LKY
CM13 3TR 2E0RKB
CM14 4UU G7HFL
CM14 5AZ G8APZ
CM14 5AZ G8PUB
CM14 5BN G6PZF
CM14 5DB M0BAW
CM14 5DG G4XFZ
CM14 5HA 2E0OOW
[illegible]
CM14 5HA M6BWN
CM14 5JR 2E0DTD
CM14 5JR M6TPD
CM14 5NS M0GNC
CM14 5WT M6HHM
CM15 0BQ G4SYR
CM15 0NB G8DXV
CM15 0PP 2E0XDC
CM15 0PP M0XDC
CM15 0PP M3NUU
CM15 0PY G4ZON
CM15 0QX G8SPP
CM15 8BW G3ZRH
CM15 8JL G4NGS
CM15 8PF G4YVW
CM15 8SA G8DWP
CM15 9ND M0MPY
CM15 9ND M0ZNP
CM15 9QA M3NRW
CM15 9QS G4BLD
CM15 9SG 2E0KWG
CM16 5AT M6TMP
CM16 5DL G4LGY
CM16 5DP G6MBD
CM16 5EW G8RIR
CM16 6BP M3WEA
CM16 6EF M5AGV
CM16 6EW G0PYV
CM16 6FH M6OBH
CM16 6HA G6OIX
CM16 6JQ M3EYY
CM16 6JQ M3JXX
CM16 6LF G4WWY
CM16 6PJ M3WYT
CM16 6RR M6ARR
CM16 7BB G1WXS
CM16 7ET G8DZH
CM16 7HT M3GKJ
CM16 7JX G4ACL
CM17 0AE G8BUI
CM17 0EX M3MER
CM17 0HD G3UUY
CM17 0HN G8CUA
CM17 0HQ G8MHO
CM17 0JY G0BXL
CM17 0LH G3UEG
CM17 0LL G7CLM
CM17 0LQ G0CPU
CM17 0QR G8DNP
CM17 0SB G7ULS
CM17 9EU M0OAB
CM17 9EU M6OAB
CM17 9HL M6PVZ
CM17 9PL G0AWY
CM17 9PL G6MZV
CM17 9PR G4XUQ
CM17 9PZ M6PTV
CM17 9PZ M6ZRO
CM17 9QG G0KNJ
CM18 6BL 2E0SIJ
CM18 6BL G6BUT
CM18 6BL G7OBS
CM18 6BL M3OBS
CM18 6BL M6HOI
CM18 6BL M6SIJ
CM18 6EB G7IZW
CM18 6ES G0HDZ
CM18 6EY G3RYK
CM18 6HN M6BUY
CM18 6QE M0JPG
CM18 6QE M3ZBX
CM18 6QG M6GQW
CM18 6QN G8VQE
CM18 6QT M6FKI
CM18 6QY 2E0AUU
CM18 6ST G8POE
CM18 6TG M6HPX
CM18 6XL 2E0TCI
CM18 6XL M6EJA
CM18 7BY G3WRO
CM18 7EH G3TOF
CM18 7HD M6SKZ
CM18 7PG G4JQU
CM18 7QD G0HRR
CM18 7QD G1FHL
CM18 7QZ 2E0VAO
CM18 7RE M3YND
CM18 7RF G1OSI
CM18 7RJ 2E0VLT
CM18 7RJ M0PHO
CM18 7RJ M3VLT
CM18 7SY G1FIM
CM19 4DB M3GOV
CM19 4HE M6FXL
CM19 4NJ G1SFU
CM19 4NW G3GWE
CM19 4RD G7RFZ
CM19 4RH G0AGO
CM19 5HG G8TPO
CM19 5LJ G0MGU
CM19 5NZ G7AZA
CM19 5RN G0REF
CM2 0ES G1JIJ
CM2 0FJ G8NWL
CM2 0RX 2E0DNU
CM2 0RX M6DNV
CM2 0SA G8MYJ
CM2 0TY G4PVM
CM2 0UA G6BZQ
CM2 6AP 2E1GQB
CM2 6AQ G4MDB
CM2 6AQ M3HPY
CM2 6AQ M3ZAN
CM2 6AZ G8DFT
CM2 6BA G7RVY
CM2 6DH G6CNK
CM2 6DN M0MVO
CM2 6DX M1SNM
CM2 6EB G1INA
CM2 6EW 2E0SES
CM2 6EW M0OER
CM2 6EW M6LFT
CM2 6HA G4JDS
CM2 6LG M3CAM
CM2 6QE G6WFM
CM2 6QL G4KGL
CM2 6QN G4LAE
CM2 6UH G4LQE
CM2 6XN M3LOT
CM2 7AU G6JYB
CM2 7AX M3IWO
CM2 7BU G3CVI
CM2 7BU G3VMJ
CM2 7DD G4OLU
CM2 7JL G1NLZ
CM2 7JN G3PMW
CM2 7LT 2E0CPQ
CM2 7LT M0XBS
CM2 7LT M6BPJ
CM2 7PP 2E0IWT
CM2 7PP M3IWT
CM2 7SF G3JOX
CM2 7TD G3FMO
CM2 8AH 2E0WHB
CM2 8AH M6UHN
CM2 8AL M6DNR
CM2 8HY G7KLV
CM2 8NF G0RLI
CM2 8NT G3YDY
CM2 8NT G8BGV
CM2 8NY G0BDS
CM2 8PD G8GNZ
CM2 8PT G0KOU
CM2 8QB M6IPA
CM2 8RQ M6ERF
CM2 8RR M3OQI
CM2 8XJ G3VUO
CM2 8YA G3WGE
CM2 8YY G6FJE
CM2 8YY G6ZNJ
CM2 9AG M6TTR
CM2 9BD 2E1RMD
CM2 9BD M3OGM
CM2 9BZ G8VU
CM2 9DD G3PEM
CM2 9DX G1MVI
CM2 9DZ G3UCD
CM2 9DZ G4BMW
CM2 9FZ M6HNA
CM2 9GJ 2E1GHE
CM2 9HA G0POK
CM2 9JR 2E1BFF
CM2 9LF G1NNN
CM2 9LN 2E0WDY
CM2 9LN M6WDY
CM2 9LW G0SQP
CM2 9ND G4RPF
CM2 9NY 2E1IDH
CM2 9NY M3IDH
CM2 9PW G0SXK
CM2 9RD M6SPS
CM2 9RE G3PBT
CM2 9SN G0NEM
CM2 9SN G8VYK
CM2 9XQ G6ZGI
CM20 1SW M0SLC
CM20 2JG M0TJB
CM20 2PA M3WMV
CM20 2PA M6BUE
CM20 2PQ G3RSP
CM20 2PZ G4GGX
CM20 3BW M3APA
CM20 3DP G8CHO
CM20 3DW 2E0JOP
CM20 3DW M3ZLJ
CM20 3DY G4MQG
CM20 3EF G6UXX
CM20 3EY G7UFF
CM20 3EY M0NNY
CM20 3EY M3LEN
CM20 3HB G4YBN
CM20 3HL M6FQX
CM20 3JU G3VAS
CM20 3LE G8XGO
CM20 3LY 2E0MBD
CM20 3LY M3NBD
CM20 3RH M0DXR
CM20 3RH M6YGL
CM21 0AU G4PGB
CM21 0BT G8WUO
CM21 0DZ G1XRT
CM21 0ER G4SNL
CM21 0[illegible] G1ENG
CM21 9ED G7NRQ
CM21 9HG G8FUL
CM21 9NF G0UEU
CM21 9NN G3NWX
CM21 9NT G3WOO
CM22 6LG G6IWT
CM22 6NT M3HKV
CM22 6QX M6UKT
CM22 6SW G4PFJ
CM22 7AJ G7VAB
CM22 7AS G8BPH
CM22 7JR G8ACT
CM22 7LT G4PJD
CM22 7NB G6ART
CM22 7PA M1SGA
CM22 7RA M6DJQ
CM22 7RT M6TTP
CM22 7SJ G4WWP
CM22 7SQ G7VRX
CM22 7TP G7VJE
CM23 1BD M0DXZ
CM23 1HX G6ECS
CM23 2BL G3WYD
CM23 2BQ G0CGH
CM23 2HU M0GCA
CM23 2HX G4CWH
CM23 2QP G4EVR
CM23 2QU G8XDU
CM23 3EU G7CQX
CM23 3EX G3KCG
CM23 3EX M0AQO
CM23 3JN G7TET
CM23 3LR M6IFW
CM23 3NG M0AZR
CM23 3NP G4SUA
CM23 3NQ G4WPO
CM23 3PU 2E1WGB
CM23 3PU M5REG
CM23 3PX G1NRK
CM23 3PX G1OSH
CM23 3QH M3XBC
CM23 3QN M6BUA
CM23 3RJ G3PRI
CM23 3YW 2E0WAG
CM23 4AD G0PQF
CM23 4AD G1ARU
CM23 4AD G5ZG
CM23 4BL G4SNJ
CM23 4DU G0PLS
CM23 4EY G0GZU
CM23 4HU G0XAD
CM23 4JL G3SUS
CM23 4JP G4OBV
CM23 4LL G6LIB
CM23 4LL G8THH
CM23 4PD G6YMU
CM23 5AP G3ZXF
CM23 5DE G8ZPH
CM23 5DS G0PZJ
CM23 5DS G4HXH
CM23 5HN M0ITY
CM23 5HY G4GIS
CM23 5JJ G7WJV
CM23 5LS G4LSE
CM23 5NU G8SEY
CM23 5QH G0ODE
CM24 8JP 2E0ZCJ
CM24 8JP M0ZCJ
CM24 8LR G3PEW
CM3 1AS G6ANO
CM3 1DA G1UZC
CM3 1DA M1BWN
CM3 1NP G8LVW
CM3 1NR G6YJH
CM3 1QS G3WSV
CM3 1RB G6OTE
CM3 1RT G1GKN
CM3 1RT G6EEE
CM3 2AY G4VVQ
CM3 2DF G8RSX
CM3 2JA M3AAS
CM3 2JE 2E0NOD
CM3 2JE M3NOD
CM3 2LB G8LYV
CM3 2LJ G4OAU
CM3 2LP 2E0DEO
CM3 2LU M0VRS
CM3 2QS G4KTX
CM3 2SG M0DTJ
CM3 3AD G6CMS
CM3 3BY G4DBM
CM3 3EF M3MFX
CM3 3JJ 2E0BOB
CM3 4DB G1OGY
CM3 4DP G3JCM
CM3 4HT G7KSQ
CM3 4HT M6BDL
CM3 4LS G3PJC
CM3 4PG M0DTA
CM3 4PG M3KRS
CM3 4PH G7UVP
CM3 4PN M6TTW
CM3 4PS G0MWT
CM3 4PQ G0TNM
CM3 4RB M6BAK
CM3 4RP G4TNB
CM3 4RQ G3SUY
CM3 4SD G8CYK
CM3 4XL M3DRH
CM3 4XN 2E0PXD
CM3 4XN M0PXD
CM3 4XN M6PXD
CM3 5FP M0DVD
CM3 5FS 2E1EQY
CM3 5FS G7HRV
CM3 5FS M3EQY
CM3 5GH G4GYJ
CM3 5GL G8DXH
CM3 5GY G3STP
CM3 5JJ M0AMZ
CM3 5NJ G0AUR
CM3 5NL 2E0RFH
CM3 5NL M3RFH
CM3 5NY G7VEE
CM3 5PF G7NBR
CM3 5PQ G3XMB
CM3 5SB G0BCW
CM3 5WH G8NAM
CM3 5YG G4DUJ
CM3 5ZT G7BHE
CM3 5ZU 2E0JPU
[illegible]
CM3 5ZU M3JPU
CM3 6AQ G4TWL
CM3 6BX G3MAH
CM3 6BY G0IJN
CM3 6BY G0UTT
CM3 6DD G0LWL
CM3 6DF G4THV
CM3 6DP G7EHS
CM3 6DQ G6TXV
CM3 6DT G8WRB
CM3 6DU G3ZES
CM3 6EJ 2E0GRA
CM3 6EP 2E0IWZ
CM3 6EP M0IWZ
CM3 6EP M3IWZ
CM3 6EP M6LLH
CM3 6EU G1SNU
CM3 6EU G7LIK
CM3 6HU G7JYD
CM3 6JE G3HMQ
CM3 6LF G6AHD
CM3 6NF G0HWI
CM3 6NF G4BUP
CM3 6NZ 2E0DJQ
CM3 6RE G4XBE
CM3 6TP G7TAV
CM3 7AH M1GUS
CM3 7BD M3FHI
CM3 8AP M6GFF
CM3 8AR 2E0DNS
CM3 8AR M6BSX
CM3 8BG 2E0PGB
CM3 8BG M0PGB
CM3 8EL G3XYI
CM3 8EW G6BCL
CM3 8HJ G8CRX
CM3 8UT M3USH
CM3 8XA G4OGL
CM3 8XA G8EGE
CM4 0AL G3FJO
CM4 0HA G8YQO
CM4 0LJ G7RCW
CM4 0RS G8GIG
CM4 9DT G8LAB
CM4 9QA G7WKP
CM5 0BL M6MNH
CM5 0ES M6BUC
CM5 9BG G4KVR
CM5 9BG G6AXO
CM5 9HW M3YIF
CM5 9LJ G8SWM
CM5 9QL G4GZG
CM5 9QN G6NAX
CM5 9QW G4TOX
CM6 1AJ G7JXL
CM6 1BP M6WWM
CM6 1BY G7DPR
CM6 1BY M6DWO
CM6 1PQ G3WCO
CM6 1SD M0TRE
CM6 1WT G8OPI
CM6 1WU M6DVS
CM6 1XQ G8XGT
CM6 1XQ M6BCW
CM6 2AP M5BIL
CM6 2BS M1SUE
CM6 2DL G0PTG
CM6 2JR 2E0MAA
CM6 2JR M0MAF
CM6 2JR M3RXQ
CM6 2PD M6DVY
CM6 2PU G1ILG
CM6 2PU M3ILG
CM6 2SQ M3ZIF
CM6 3AW G6WYH
CM6 3AX G7TKI
CM6 3BP M0DUT
CM6 3LU M6LMC
CM6 3PJ G8HUT
CM7 1AB M3TPU
CM7 1AL G0PYW
CM7 1AR M6WTT
CM7 1DL M3RTI
CM7 1DR G1JSK
CM7 1EB G4IVX
CM7 1XF G6HSC
CM7 2RZ G6XCU
CM7 2TA M6RNZ
CM7 2WB G6OIY
CM7 3JR M3TXH
CM7 3JY M6RAP
CM7 3LG G4ICP
CM7 3NW M6MNO
CM7 3PE G3TGR
CM7 3TY M6ORM
CM7 4BY G8PLO
CM7 4JZ G4ONH
CM7 4LE G4YQL
CM7 4QG M6GGH
CM7 5EG G3WVR
CM7 5HN G4YWN
CM7 5HU G6DIO
CM7 5LW G8DJO
CM7 5LW M0IAE
CM7 5PY G0EMK
CM7 5PY G3XG
CM7 5PY M0CLO
CM7 5SD 2E0TKY
CM7 5SH G0OSI
CM7 5UE G7IRK
CM7 5UQ G1XWN
CM7 9DZ M3OIL
CM7 9ES 2E0NPX
CM7 9ES M6NPX
CM7 9LL G0DFC
CM7 9LL G1YYY
CM7 9LL G8XVO
CM7 9LR G3ITL
CM7 9NF G0DCR
CM7 9NJ G1WRH
CM7 9RH M6POC
CM7 9TX 2E0WAV
CM7 9TX M0WAV
CM7 9TX M6WAV
CM7 9UG 2E1HYX
CM7 9UR G1TZZ
CM77 6BY G4WXT
CM77 6DE G7SLJ
CM77 6RE M0SJY
CM77 7JW G6EJF
CM77 7PX G8ZIW
CM77 7UA G6IOW
CM77 8ER G3ZWW
CM77 8HN G7HHN
CM8 1DR G4VOT
CM8 1GF 2E0DSX
CM8 1GF M0XDA
CM8 1GF M6NSX
CM8 1HP M6PXT
CM8 1HR G6CSK
CM8 1JB 2E0IRX
CM8 1JB 2E0KDG
CM8 1QJ G0FWA
CM8 1QR G3MD
CM8 1SZ 2E0CSL
CM8 1SZ M0GGL
CM8 1SZ M3JYP
CM8 1TJ M3RYT
CM8 1XY G8NGM
CM8 2ES M6KRW
CM8 2LH G6GXG
CM8 2LJ G8WSC
CM8 2LL G0DJK
CM8 2NP 2E0NRH
CM8 2NP M0NRH
CM8 2NP M3ILM
CM8 2NP M3TSA
CM8 2NY M0KGK
CM8 2PA M6ENM
CM8 2PE G1SDN
CM8 2PL M6VIO
CM8 2PT 2E0XIS
CM8 2SZ G1NNB
CM8 2SZ G7BIQ
CM8 2SZ G7LEY
CM8 2UQ G7BND
CM8 2UQ M3BAW
CM8 2XG G3KXI
CM8 2XL M6LIK
CM8 2XQ G8CMO
CM8 3JR G1UZD
CM8 3LN G3LVL
CM8 3LT G7NAI
CM8 3LZ G3VYD
CM8 3NS G3XAX
CM8 3NZ M0PWB
CM8 3PH G3XVV
CM8 3QY 2E1IIV
CM8 3QY M0UTH
CM8 3RZ 2E0RNS
CM8 3RZ M3RNS
CM8 3SN M0TCT
CM8 3SP 2E1HGG
CM8 3SP G4KQE
CM8 3SP G8WOZ
CM8 3SP G8WQZ
CM9 4AB G7WAQ
CM9 4BL G6RKF
CM9 4BW M0RAX
CM9 4LF M1OCN
CM9 4PB G1WFA
CM9 4RQ G0AAN
CM9 4SE 2E0WAA
CM9 4SE M0WAG
CM9 4SE M3WPI
CM9 4US G1YKZ
CM9 4YN G0MBY
CM9 4YN G8DHV
CM9 4YX G8JLM
CM9 5DE G3VNP
CM9 5EE G1BQG
CM9 5HF 2E0KCP
CM9 5HF M0KCP
CM9 5HF M6KDP
CM9 5HY M6ROW
CM9 5JJ G8GLP
CM9 5JQ M6BZW
CM9 6AF M0ORF
CM9 6BE 2E0OVF
CM9 6BE M0SHQ
CM9 6BE M0ACE
CM9 6DJ G4HRB
CM9 6EW G1NMN
CM9 6EW G1MOV
CM9 6JF G7RFC
CM9 6JF G7SRC
CM9 6JF G8PYD
CM9 6JF M1BDR
CM9 6JH G6LJR
CM9 6LR G6PCP
CM9 6QP G3UYC
CM9 6UE M1EVZ
CM9 6UH G4MNT
CM9 6UQ G8WTM
CM9 8AY M1GFE
CM9 8BW G0UHD
CM9 8BY M1DOZ
CM9 8EH M3YWO
CM9 8HA M6FAP
CM9 8LL G0VDV
CM9 8PB G4EMB
CM9 8PD G6OXQ
CM9 8PJ G0VDP
CM9 8PR G3ZOX
CM9 8PX G0YAE
CM9 8QA G0ASN
CM9 8RJ G6CYE
CM9 8SB G3GLL
CM9 8SR G4IIH
CM9 8UA G1FXM
CM9 8UD G1FWY
CM9 8UF G6IFH
CM9 8UN 2E1DET
CM9 8UN M3DVP
CM9 8UN M6JEK
CM9 8XB G0IBN

## CO (Colchester)

CO1 1PR G0VNY
CO1 1UR 2E0BYF
CO1 1UR M3BYF
CO1 2GQ 2E1FRI
CO1 2GQ M1CUY
CO1 2HD G7LYH
CO1 2JJ M6AWP
CO1 2LF 2E0FTQ
CO1 2LF M6FTQ
CO1 2NA G0DZB
CO1 2RZ M1CYK
CO1 2UT M1FJD
CO1 2WA M0WNF
CO1 2WA M3WNF
CO10 0DJ G0PAO
CO10 0DU M3LEL
CO10 0EP G0VAS
CO10 0FH G4LSK
CO10 0JN G4GGC
CO10 0NQ G1DEU
CO10 0NT M6ICH
CO10 0QP G8AAR
CO10 0RJ 2E0REL
CO10 0RJ M6REL
CO10 0RN G6RKG
CO10 0RT G7PCG
CO10 0YE 2E0BYC
CO10 0YF G8LTY
CO10 0YU G4JAA
CO10 1JB G6DKE
CO10 1PJ G7HMF
CO10 1QX G7MLK
CO10 1YU G3PZU
CO10 2BU G7NZV
CO10 2PP G7UTC
CO10 2TP G1IUW
CO10 2TU M3TFP
CO10 5JA G1YJL
CO10 5JN 2E0TBZ
CO10 5JN M6GLX
CO10 6JB G1PMF
CO10 7JX G4DHU
CO10 7LS G1EUF
CO10 7LX M0JMV
CO10 7PN M0COJ
CO10 7PQ G4FBQ
CO10 7PS M0AUR
CO10 7PS M0CEQ
CO10 7RL G3YAI
CO10 7RS G4SWQ
CO10 7RW G3WRD
CO10 7SB G0HUG
CO10 7SJ G0PGS
CO10 7SJ G7KSH
CO10 8EQ G4HGH
CO10 8JP G6FCL
CO10 8JS G8GRL
CO10 8LQ 2E1MJH
CO10 8LQ 2E1ORT
CO10 8LQ G7SRA
CO10 8LQ M0MJH
CO10 8LQ M3MJH
CO10 8NN G8CDG
CO10 8PD G6UQZ
CO10 8PJ 2E0XRS
CO10 8PJ M3YFW
CO10 8QE G0STW
CO10 9DD G7LQP
CO10 9ET G1TZC
CO10 9QD G1TWY
CO10 9QD M1COQ
CO10 9RU G0SNU
CO10 9SD M3NIT
CO11 1NS G1XMH
CO11 1NN G1NNA
CO11 2AL G0NXH
CO11 2AL G4JVM
CO11 2BU 2E0KLF
CO11 2BU M0JVC
CO11 2BU M6KLF
CO11 2DT G7EPM
CO11 2HE M0BGE
CO11 2HE M6EWJ
CO11 2HE M6GLP
CO11 2HX G3YAJ
CO11 2HZ 2E1RFS
CO11 2JU G4UWW
CO11 2LH M0AFX
CO11 2QP M1DVO
CO11 2RB G7VFC
CO11 2SL G4HWK
CO11 2SP G6UWK
CO12 3BD G4WHK
CO12 3DS M3DZT
[illegible]
CO12 3JZ M0NWW
CO12 3LD M6WAS
CO12 3PL G8XAX
CO12 3PP M0DSY
CO12 3PP M6DSY
CO12 4BW G3UJB
CO12 4DW M0RIC
CO12 4DW M0TYG
CO12 4EQ G1VCZ
CO12 4HR G7HMN
CO12 4JU M6BLO
CO12 4LA M0BVQ
CO12 4NT M6RHK
CO12 4TS G0STW
CO12 4UH G1JRD
CO12 5AD M0CGE
CO12 5BQ G0RDX
CO12 5BS G2ABR
CO12 5DH G4JIG
CO12 5EE G4YJQ
CO12 5EJ G3YYZ
CO12 5HE G4GBY
CO12 5JF G4EYE
CO12 5LS G7VHF
CO12 5LS G8PL
CO12 5NE G8PAI
CO12 5NN G4TZM
CO13 0AN G6CDU
CO13 0AQ G0KCD
CO13 0BG G6PTI
CO13 0BH G0NLG
CO13 0BQ M0CKB
CO13 0ET G1WJR
CO13 0EW M0LWM
CO13 0JL G0MJA
CO13 0LQ G6NHU
CO13 0LQ M6DHU
CO13 0LQ M6SHU
CO13 0QH G3OMB
CO13 0SF G4OAX
CO13 0SW M3NGC
CO13 0TG G4ZDY
CO13 0TH M3LIX
CO13 0TQ M6FCV
CO13 0TW M6HQL
CO13 9DW G4PBD
CO13 9HR G3OJS
CO13 9JE G6WHT
CO13 9JZ M6BDO
CO13 9NR M6PEU
CO13 9PW 2E0HUM
CO13 9PW M6YES
CO14 8BL G0WWR
CO14 8DJ G0GYY
CO14 8DF G8HSI
CO14 8EW G0OFF
CO14 8LE G6GJV
CO14 8LT M6ALY
CO14 8NF M3MGP
CO14 8PX G3VLD
CO14 8RE G1ISS
CO14 8RG G3CO
CO14 8RG G6XOU
CO14 8RG M1COL
CO14 8RG M3LFO
CO14 8RL G3TUU
CO14 8RL G7BWF
CO14 8RR 2E1GXH
CO14 8RR G8ZJE
CO14 8SP M3ZLY
CO14 8SS G0KKT
CO14 8SS G0OIF
CO14 8UD G1HZD
CO14 8UE 2E1FPU
CO15 1EJ G3XRN
CO15 1EJ G4AQZ
CO15 1HB 2E1HGU
CO15 1HB 2E1HLP
CO15 1HB G7PMU
CO15 1LA G6DLZ
CO15 1NQ 2E1HHB
CO15 1TX 2E1HAQ
CO15 1TX M0GWE
CO15 1TX M3KEV
CO15 1UW G6EAH
CO15 1UX M6TBK
CO15 2DR M6MMN
CO15 2DD G6AGN
CO15 2JN G6WIL
CO15 2JW M6EVD
CO15 2NU 2E1IGU
CO15 2PA G6JIR
CO15 2PE G4PQU
CO15 2QB 2E0DKM
CO15 2QB M6DKM
CO15 2QT G8DMT
CO15 2RW 2E0CZM
CO15 2RW M0HTF
CO15 2RW M6DXK
CO15 2SF M6KCT
CO15 3AR G0UHU
CO15 3DW 2E0IKH
CO15 3HS G6CYA
CO15 3JR M0CHL
CO15 3LA G1YJJ
CO15 3LA M0RBC
CO15 3LA M3ZSC
CO15 3NF G7AJG
CO15 3SB M0DEX
CO15 3SE G6XJC
CO15 3SR G0DGH
CO15 4EU M6NBH
CO15 4JE G3ZEZ
CO15 4ND G0LOT
CO15 4NN G7UHY
CO15 4PA G4AEB
CO15 4PJ G8IOI
CO15 4RJ G4ZXS
CO15 4RJ G8ATL
CO15 4RU G0VAL
CO15 4RZ G0HWC
CO15 4TP M0CNL
CO15 5DJ G3LZR
CO15 5LA G4ZJK
CO15 5NA G0SMJ
CO15 5PL 2E0COI
CO15 5PL M6BEZ
CO15 5PZ G3JNI
CO15 5RH 2E1HIO
CO15 5SR G7KME
CO15 5XH M3TLO
CO15 5XY M0DNO
CO15 6DP G6CTY
CO15 6EN M6FRS
CO16 0AX G0UBJ
CO16 0BU M0PJK
CO16 0EG 2E0CWZ
CO16 0EG M0KEB
CO16 0EG M6FGG
CO16 0EP G4CYF
CO16 0HT G1JCP
CO16 7HF G0MBA
CO16 7HF G0PKT
CO16 8BN G4RUJ
CO16 8BZ 2E1GYN
CO16 8BZ M3ACU
CO16 8DB G4WHZ
CO16 8DG G8LID
CO16 8DP M6TMY
CO16 8EP G1FWR
CO16 8ET M3VZN
CO16 8EX M0ITX
CO16 8FE G7HJK
CO16 8FQ G0DZZ
CO16 8HB G0HCY
CO16 8HH M6SDI
CO16 8PH M1EAK
CO16 8RQ G0CKD
CO16 8RQ G0RIU
CO16 8RQ G7IIQ
CO16 8US G4FJT
CO16 8YD G6TNQ
CO16 8YE G0GEQ
CO16 9AN G3VMP
CO16 9EN G1IPU
CO16 9ES G3PQM
CO16 9NE G8CVP
CO16 9PS G0FIW
CO16 9PU G1HBW
CO16 9PU G8JPJ
CO16 9QP G0EBI
CO16 9RS G7BVZ
CO167DX 2E1PAW
CO2 0JE G4YJN
CO2 0NQ G4YOF
CO2 0PR G7WLC
CO2 7EN G6DNA
CO2 7LQ G4NNN
CO2 7LU 2E0JJN
CO2 7LU M0JJN
CO2 7LU M6JJN
CO2 7PE M3WWR
CO2 7RH G7OWB
CO2 7UG G7EQX
CO2 7UX G6FBB
CO2 8AJ 2E0GUI
CO2 8AJ M6EIO
CO2 8AR G6XWY
CO2 8BH G1XAJ
CO2 8BP G7PST
CO2 8EG M3EUM
CO2 8GY M3RPA
CO2 8LP 2E0DBY
CO2 8LP M6GSD
CO2 8LT M6KCU
CO2 8NS G7TAT
CO2 8NS M6NUL
CO2 8NZ G8UBU
CO2 8PA G7UFT
CO2 8QP G7RWF
CO2 8QY 2E0LSI
CO2 8QY M0LSI
CO2 8QY M6EPZ
CO2 8SJ G4DIS
CO2 8UB G7ELH
CO2 8UD G6FLW
CO2 8UD M0VLL
CO2 8UJ M6EBW
CO2 8WW G6RVP
CO2 9DT M0KTR
CO2 9DZ M0KAP
CO2 9HN M3WMO
CO2 9JR M1CZY
CO2 9JR M0OZY
CO2 0NE M6ASK
CO3 0HP G4CTA
CO3 0HP G6WYL
CO3 0HY G8VJR
CO3 0NP G7CGC
CO3 0QZ G0OSR
CO3 0RZ M1BPY
CO3 0YA M0BCK
CO3 0YJ G0NHB
CO3 3BP M6HWX
CO3 3HE M0LFS

Postcode

**IMPORTANT NOTE**

**Revalidate licence to avoid revocation** – Ofcom has advised the Society that plans will be drawn up to revoke licences that have not been revalidated as required by the licence conditions. The quickest way to revalidate is to do so online via the Ofcom website: *https://services.ofcom.org.uk/* or by email: *amateur.validations@ofcom.org.uk* If you need assistance in the process, Ofcom staff are available to help, but please be patient during times of heavy workload.

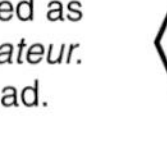

CO3 3QE M6ALU
CO3 3RS G0KOY
CO3 3SJ G1TWF
CO3 4JJ G0WZV
CO3 4JP G4JKC
CO3 4LT G4UTJ
CO3 4NQ G6DFZ
CO3 4NU M6CRX
CO3 4PS 2E0JBK
CO3 4PS M3JBK
CO3 4QN G4SOB
CO3 9AQ G6RDD
CO3 9DP G0HHC
CO3 9TR M1XXT
CO4 0AJ G0OWK
CO4 0DB M6GVI
CO4 0EA G6IIF
CO4 0HX G3MQR
CO4 0LD 2E0BMG
CO4 0LD M3LFQ
CO4 0LP G4OYH
CO4 0PJ G4AUG
CO4 0PW G8RFC
CO4 0PW M0YNK
CO4 0PW M6YAN
CO4 0QJ G6AQI
CO4 3AS G8SGB
CO4 3EY G4HKC
CO4 3FD G4HKB
CO4 3FD M6EWT
CO4 3FE G0VYC
CO4 3FN G4RKB
CO4 3FP G8DRE
CO4 3HB G4JHP
CO4 3JA M0ASF
CO4 3JA M0ASG
CO4 3JA M3OQD
CO4 3JP G0PFM
CO4 3JR M1AMI
CO4 3JR M3DFC
CO4 3LU G6KTR
CO4 3LX G7EHY
CO4 3NF G0VAE
CO4 3NF M0DEQ
CO4 3NF M3ZZN
CO4 3NL G1DTE
CO4 3QA G7NID
CO4 3SQ M0UOE
CO4 3SQ M3PQI
CO4 3SQ M3TFA
CO4 3UT 2E0CEA
CO4 3UT M6CEA
CO4 3YA G6CLA
CO4 3YH G0SLB
CO4 3YP G3ZOL
CO4 5AD G1VDC
CO4 5BD M3KMS
CO4 5DU 2E0TFH
CO4 5JL G6ZVV
CO4 5JT G8MKN
CO4 5JX 2E1DXB
CO4 5JX M3AJD
CO4 5PE G0AHE
CO4 5PY G8WBY
CO4 5RN M6XEL
CO4 5YA M3YAL
CO4 5ZR M6HWH
CO4 9RE M3JBE
CO4 9RJ M1BQZ
CO4 9RR M6HJA
CO4 9SL 2E1EWK
CO4 9ST 2E0GMM
CO4 9ST M3OEG
CO4 9WQ G7BIV
CO4 9YD 2E0IGM
CO4 9YD M6IGM
CO4 9YU G1GWO
CO5 0AA G6XIF
CO5 0AE G6KVI
CO5 0AY G4TFI
CO5 0BN G7JMW
CO5 0DT G4KTB
CO5 0EF G4TFP
CO5 0EF G7NMI
CO5 0EN M0JJH
CO5 0HJ M1ASV
CO5 0HN M0HIG
CO5 0JG M6CHP
CO5 0JG M6FSC
CO5 0JU G4ZZL
CO5 0LQ M0ECL
CO5 0LR G0IZK
CO5 0QD G3GML
CO5 0QT G6DBY
CO5 0RP G6YWU
CO5 7AS G4PZX
CO5 7ET G6KNM
CO5 7HF G3KHK
CO5 7LB G4YK
CO5 7LG M6LKF
CO5 7LJ G4PBR
CO5 8BX M1RJS
CO5 8DR G8SOI
CO5 8GG M5RHG
CO5 8GJ G8IFN
CO5 8HB G7GCB
CO5 8JU M3BSH
CO5 8LN G0JDE
CO5 8ND G4ZOR
CO5 8PX M1EVN
CO5 8QP G0APP
CO5 8RY G6OOT
CO5 8SL M1CAY
CO5 8SS M0PUC
CO5 9AG G8ADZ
CO5 9BB 2E0BXC
CO5 9BB M3UMW
CO5 9TD G3NXK
CO5 9TD G4KXF
CO5 9TH G4PYG
CO5 9TU M3VXK
CO6 1BP G0NMB
CO6 1DB G3YEC
CO6 1NY G1GMM
CO6 1SG G0WJD
CO6 1SY M6YUM
CO6 1UL G7VGC
CO6 1XG G1IUD
CO6 1XG G6LJF
CO6 1XG G8IUD
CO6 1YN 2E0JPR
CO6 1YN M0JAI
CO6 1YN M6ETU
CO6 1YP M1ESH
CO6 2DZ G8LHF
CO6 2NG M1DWW
CO6 2NH 2E0VAN
CO6 2NH M0YEP
CO6 2NH M3VXC
CO6 2NX 2E0COM
CO6 2NX M6WKD
CO6 2PF G7LNU
CO6 3BT G3SUV
CO6 3DB G0VHF
CO6 3DB G4ZTR
CO6 3DB M1CRO
CO6 3DX G8VAF
CO6 3EP G8YMZ
CO6 3HY 2E0FBA
CO6 3LX 2E0CGV
CO6 3LX M0HBV
CO6 3LX M6ALT
CO6 3NJ 2E0WDM
CO6 3NJ M3VZV
CO6 3RY 2E0ORP
CO6 4BJ G6XMM
CO6 4LT G4DKX
CO6 4PQ G7DTT
CO6 4QH G3LST
CO6 4UN G1AWK
CO6 5AG M0EJG
CO6 5AR G0IBZ
CO7 0AQ 2E0KKC
CO7 0DJ 2E0XMK
CO7 0DJ M0XMK
CO7 0DJ M3XMK
CO7 0DU G6AEB
CO7 0HE 2E0DPZ
CO7 0HE M6LYD
CO7 0LA G4IZX
CO7 0LB G4CIA
CO7 0NA G1XUU
CO7 0NN G4EUW
CO7 0NZ G6PMD
CO7 0PE G3ELS
CO7 0PR G3MGW
CO7 0QR G4JAC
CO7 0RH G7KRO
CO7 0RP G7TBU
CO7 0RS G0SOX
CO7 0SJ G1BFF
CO7 6ES G4PZL
CO7 6QY G1VZT
CO7 6RF G0VEI
CO7 6SJ G8CJD
CO7 6TP G1CHN
CO7 6TX G6VJK
CO7 6UX M6AZX
CO7 7AS M0PDE
CO7 7AS M0PDF
CO7 7EG G7USX
CO7 7GA G3YVW
CO7 7QB M6OAT
CO7 7RY G8GML
CO7 7SD 2E0BZM
CO7 7SD M3ZMQ
CO7 7SY G6IGU
CO7 7UB M6ATQ
CO7 7UQ G4EOR
CO7 8DD 2E0DDU
CO7 8DD M0XLB
CO7 8DD M6SEY
CO7 8DH G0GYI
CO7 8HS 2E0WMG
CO7 8HS G8EWC
CO7 8JA G4AZD
CO7 8JB G4URA
CO7 8JH G1TWH
CO7 8JH M0CHS
CO7 8JH M3DEB
CO7 8LH M6BAR
CO7 8LJ G4ZKS
CO7 8NN G0PJZ
CO7 9EH G0BEP
CO7 9JB G7IFB
CO7 9LD G0GGM
CO7 9LG 2E0AIK
CO7 9LG G4SDI
CO7 9LS 2E0PXY
CO7 9LS 2E0PXZ
CO7 9LS M0PXZ
CO7 9LS M3PXY
CO7 9LS M3PXZ
CO7 9NH G4FTP
CO7 9QH M1BQF
CO7 9QQ G4NXR
CO7 9QZ G7TWA
CO7 9RP M3XGB
CO7 9SD M1DYW
CO8 5BN G6WPJ
CO9 1AS G1ZLD
CO9 1BJ M1CJF
CO9 1DX M6JKA
CO9 1ED G6LMC
CO9 1EH G7UUA
CO9 1EH G7UUD
CO9 1JU G4YAX
CO9 1JU G6NEK
CO9 1LB G0GWN
CO9 1NH 2E0PAB
CO9 1NY G4FJC
CO9 1NY G8WPO
CO9 1PD G4BCV
CO9 1PD G7ORE
CO9 1PD M0NAS
CO9 1PD M3JDS
CO9 1PY M3PSI
CO9 1TD G1CCW
CO9 1UB G6OXJ
CO9 1XE M0GKW
CO9 1YB M3IRX
CO9 2BE G4TEB
CO9 2HF G0GII
CO9 2HJ M3JEP
CO9 2HN G6IAJ
CO9 2NW G7UVV
CO9 2SH M6DNX
CO9 2TA G6LHG
CO9 2TB G0CQI
CO9 2TB G1GRZ
CO9 2TF G3MMA
CO9 2UA 2E0MKW
CO9 3AZ G0LZY
CO9 3HX G0GZF
CO9 3NX G1OFX
CO9 3QJ 2E0TNC
CO9 3QJ M6BCK
CO9 3QN G4WVH
CO9 3RN M5AJB
CO9 4AB 2E0AAF
CO9 4AB 2E0FLR
CO9 4AB M6DTR
CO9 4JG G3PGN
CO9 4LN G6WHY
CO9 4LX G8JVV
CO9 4PR 2E0CNE
CO9 4PR M0HGS
CO9 4PR M6AWL
CO9 4QJ M6NCU
CO9 4QQ M6GEY
CO9 4QQ M6YEL
COWBRI DGE
G7KQN

## CR

## (Croydon)

CR0 0AA G0DDT
CR0 0BL G7VTH
CR0 0DN M3XNR
CR0 0JA M6NFL
CR0 0JE G0VQM
CR0 0JL G7GCF
CR0 0NU G0HSH
CR0 0PF M0RXZ
CR0 0RP M6HXD
CR0 1HB G7DCF
CR0 1JS G7SFA
CR0 1PJ G8TBL
CR0 1XL G1ALR
CR0 2BB 2E0WBO
CR0 2BB M6YCR
CR0 2HX G1OIS
CR0 2LP G4NHA
CR0 2LW M6EEW
CR0 2PF 2E0BPU
CR0 2PF M3FUB
CR0 2PZ G1ZDY
CR0 3AD G4MZK
CR0 3JF G0GFY
CR0 3NF G1IEY
CR0 3NZ G4WAY
CR0 3QP G8GYM
CR0 3SW G6FGY
CR0 4DN M6LHV
CR0 4EG G4DAF
CR0 4HG G7LTP
CR0 4JD M6BGG
CR0 4NN M6ZGR
CR0 4PS G3FP
CR0 4PU G4FFX
CR0 4QH G6MUQ
CR0 4QH M3TJO
CR0 5BA G3WBN
CR0 5HX G4STD
CR0 5PS G6DAY
CR0 5QA M0CCF
CR0 5QA M0VOG
CR0 5QA M1ACF
CR0 5QA M1CCF
CR0 5QA M3ACF
CR0 5ST G4KQO
CR0 6AP M3WXH
CR0 6AQ M6CGF
CR0 6DE G4DPO
CR0 6XR M6ILH
CR0 7EB G4AVV
CR0 7HY G4RWW
CR0 7PP M0WTH
CR0 7PR G3ZIO
CR0 7QP M0CVZ
CR0 7SH M3WRO
CR0 8HJ M3OSA
CR0 8HX M6VZF
CR0 8LG G7ODG
CR0 8PN G3RMN
CR0 8SB G3ALG
CR0 8YQ M3YRV
CR0 9DG G7MSF
CR0 9HE G4UIO
CR0 9LN G4FUU
CR2 0BL M3YDJ
CR2 0DZ G0HWY
CR2 0EF G3WMT
CR2 0LB G4SXY
CR2 6EE G0VVX
CR2 6EE G3ZRR
CR2 6NE G1WFG
CR2 7EF G0DJT
CR2 7ER G8MNY
CR2 7GE G3TCZ
CR2 7HH G3MCX
CR2 7JJ G4DAC
CR2 7LJ G0IZB
CR2 7LT G6ODE
CR2 7RE G4GFC
CR2 7SD G4ANQ
CR2 8AR 2E0DCZ
CR2 8AR M6CNX
CR2 8AR M6DCZ
CR2 8HR 2E1DRV
CR2 8PH G4LZE
CR2 8RA G3ZYZ
CR2 8SL M3YGO
CR2 8SL M3ZGO
CR2 9DU M1ZAR
CR2 9HY M6API
CR2 9JR G0TCE
CR2 9JY G8DNL
CR2 9LN G8IYS
CR3 0AJ G7VFX
CR3 0PE G0UQO
CR3 5BG G4BWG
CR3 5BG M6XLS
CR3 5DL G1RCE
CR3 5EL 2E1PAL
CR3 5EL G0SCR
CR3 5EL G4APL
CR3 5EL G7BSF
CR3 5EL M3BYL
CR3 5JN G6LTK
CR3 5LJ G4LXR
CR3 5LN 2E0AYT
CR3 5LX M6MQS
CR3 5QH G8LMI
CR3 5RB G7BWE
CR3 5RT G0SYR
CR3 5RT G8DTQ
CR3 5SD G0OLX
CR3 5SD G8CUX
CR3 5SH M3SOF
CR3 5TG G4ECS
CR3 5ZU G7ONI
CR3 6AD G8PRK
CR3 6BA G3ODX
CR3 6DQ G6CDW
CR3 6HN G4DTC
CR3 6NJ M3YTG
CR3 6QX G0TNM
CR3 7DL G8CPB
CR3 7EH G8HDP
CR36SA G7NGB
CR4 1JF M3WVG
CR4 1LF G0WCR
CR4 1NY G7JMQ
CR4 1XG 2E0TSA
CR4 1XG M6TSA
CR4 1XJ M3MZP
CR4 1XN G1IZN
CR4 2GA M6XCA
CR4 2LF G0LUL
CR4 2LT M3URZ
CR4 3DZ G0VTM
CR4 3JS G4RBH
CR4 3LL 2E1GOE
CR4 3LW G1KGO
CR4 3RQ G7OLH
CR4 3RS 2E0YGS
CR4 3RS M6GMS
CR4 4HD G1SCH
CR4 4LZ G1FOF
CR5 1AT M6XZK
CR5 1BB G3MZA
CR5 1DF G8EIN
CR5 1DH G4GEW
CR5 1HR G0KZT
CR5 1HR G4FUR
CR5 1HR M1FUR
CR5 1JS G1KGA
CR5 1NF M1DOR
CR5 1NL G8EDN
CR5 1QP G3CQU
CR5 1QS G0GZM
CR5 1RF G1TLW
CR5 2BL G4CJR
CR5 2EG G7OKY
CR5 2EJ G4FVL
CR5 2JF G8GAR
CR5 2LD G8RWG
CR5 2LF G3ZPB
CR5 3BP G0KUE
CR5 3DD G0FUH
CR5 3DE G2LW
CR5 3DE G3OOU
CR5 3PH G6UVB
CR5 3SJ M6EBE
CR5 3SX G3GWC
CR6 9EP 2E0UGM
CR6 9EP M6UGM
CR6 9HZ G3WZK
CR6 9JQ G6YOG
CR6 9JQ M6WEJ
CR6 9LB G4LLM
CR6 9LP G8JAC
CR6 9LU G3KKZ
CR6 9NZ M0IMJ
CR6 9TD 2E0FJA
CR6 9TD M0IDM
CR6 9TD M6FJA
CR6 9TL M6HJV
CR7 6BR G6MFM
CR7 6BX G1POM
CR7 6EB M0HNC
CR7 7AF G3UFY
CR7 7BG M6JEJ
CR7 7EN G0ELG
CR7 7HQ G4KRD
CR7 7JE G3JAL
CR7 7NP G1HER
CR7 7NQ G7HLW
CR7 7RE G0TDG
CR7 8DF G0IOO
CR7 8NY G4REK
CR7 8RP G1OTN
CR8 1BB G0DLP
CR8 1JA G7JAQ
CR8 1JB G3ZMN
CR8 1JL G0UCT
CR8 1JQ G7PWV
CR8 2DY 2E0BCF
CR8 2DY M0GNM
CR8 2DY M3IHV
CR8 2HQ G8FOT
CR8 2LR G6GFJ
CR8 3AQ M6URA
CR8 3EJ G4AOJ
CR8 3PE G0UQO
CR8 4DH G3TWJ
CR8 4DN G8TB
CR8 4JB G8XHD
CR8 4NG G4CDY
CR8 5DG M0CUP
CR8 5DG M3IDY
CR8 5GE 2E0LTU
CR8 5GE M0PZD
CR8 5GE M6IZP
CR8 5JJ G3ZXV

## CT

## (Canterbury)

CT1 1NR M3KRE
CT1 1PZ 2E1EHM
CT1 1QG G7RBB
CT1 1SJ M3AJA
CT1 1TS G6EGU
CT1 1TS G6YZU
CT1 1WX M0HWJ
CT1 1YG 2E0WPJ
CT1 1YH M0XED
CT1 2AA G1AUI
CT1 3JL G4KGY
CT1 3LD G3MMJ
CT1 3QW G7TDD
CT1 3UP 2E0DBP
CT1 3UP M3AYQ
CT10 1DR 2E0RDP
CT10 1DR G8FME
CT10 1DR M6WFY
CT10 1HN M3TDP
CT10 1HR M0ATS
CT10 1PG 2E1EKM
CT10 1QN 2E0PGC
CT10 1QN M6PGQ
CT10 1QT M6TVB
CT10 1RP G0LGW
CT10 1SR G6SFC
CT10 1TL G4GAT
CT10 2DT M0ASC
CT10 2EW M6YZH
CT10 2HA M0AQA
CT10 2HG G8DZN
CT10 2HL M6ZPE
CT10 2HY M3UVJ
CT10 2HY M6DWM
CT10 2JG G4AQE
CT10 2JL G8XJE
CT10 2ND G7FNU
CT10 2NG G0LEU
CT10 2PE 2E1HPT
CT10 2PL 2E0CPZ
CT10 2PL M0TFC
CT10 2PL M0ZPK
CT10 2PL M6PKU
CT10 2RU G3LHI
CT10 2SD G0RJJ
CT10 2SS M6XAY
CT10 2TX G0KCA
CT10 2UF M6TRP
CT10 2XN G7OHO
CT10 2XN M0DLI
CT10 2XU 2E0NJJ
CT10 2XU M0GIG
CT10 2XU M3NJJ
CT10 3AA G1VID
CT10 3AH G4PTE
CT10 3AZ G0VUT
CT10 3DE M3JKJ
CT10 3DE M3JKP
CT10 3DL M3MQP
CT10 3DR G1KMJ
CT10 3ES G4GAP
CT10 3EY G0GUW
CT10 3HN G8WQT
CT10 3LS M3LQC
CT10 3NE G0RXU
CT10 3QP G4VBI
CT10 3SD G4VPI
CT11 0BH M3UCF
CT11 0DF M6DJA
CT11 0ED G8FSV
CT11 0HT M3WOD
CT11 0JJ M3XJZ
CT11 0LL G3LHS
CT11 0LP M0CAG
CT11 0LQ M1CCL
CT11 0ND G0MBL
CT11 0PB 2E1CPF
CT11 0PB G7FMI
CT11 0PF 2E0CVJ
CT11 0PF M0ZRF
CT11 0PL G7SYE
CT11 0PQ G3WEB
CT11 0PX G0UAK
CT11 0QP G3VSU
CT11 0RN G1YZT
CT11 0RR G7DNQ
CT11 0RY G0AHA
CT11 7EF G6ENA
CT11 7HS M6HFJ
CT11 7HT G4ADS
CT11 7JU G0DVS
CT11 7LH G4IOA
CT11 7LP 2E0XDY
CT11 7NW G7ORS
CT11 8DD M6FFU
CT11 8JP M3TWY
CT11 8QA 2E1GDO
CT11 9DE G3TVD
CT11 9LP M6AVQ
CT11 9LU G0JIF
CT11 9PB G4HAK
CT11 9PW M6MFF
CT11 9QZ G4BXI
CT12 4AG 2E1HNF
CT12 4BE G6UXG
CT12 4BG G7PHD
CT12 4EL M3UGH
CT12 4EL M6RVR
CT12 4EP 2E0MRJ
CT12 4HR G6OKB
CT12 5AW G0LFM
CT12 5ED G3YCV
CT12 5ED M1DEJ
CT12 5EP G0WWP
CT12 5LX G7IFU
CT12 6DD 2E0IAJ
CT12 6DD M3INL
CT12 6DE G0TFB
CT12 6DW M6GQK
CT12 6DX 2E1HPS
CT12 6DX G7ITS
CT12 6EZ G3ZBF
CT12 6JQ G7SFD
CT12 6NX M0DIJ
CT12 6NX M3EOX
CT12 6NZ G1ODK
CT12 6SW G8UHJ
CT12 6UG M3TNH
CT13 0AQ 2E1ABW
CT13 0AQ 2E1ATH
CT13 0AS 2E1GTF
CT13 0BE M6GJY
CT13 0BG G7BPZ
CT13 0BG G7BQA
CT13 0BH G0DSK
CT13 0DW M3SOQ
CT13 0EU G4RXG
CT13 0PE G0JBA
CT13 9AS G4KPF
CT13 9JB 2E0ALH
CT13 9JB M3IYY
CT13 9JB M3KRM
CT13 9JB M3VKJ
CT13 9JE G3NCB
CT13 9JE G4JYU
CT13 9JE G8MLB
CT13 9JF G0BVA
CT13 9JF G1PJR
CT13 9NY G0MAZ
CT13 9NY G7BVH
CT13 9QA 2E0BZV
CT13 9QA M0GUG
CT13 9QA M3ZNT
CT14 0BT G0LGK
CT14 0HJ 2E0AWU
CT14 0HJ M0GID
CT14 0HJ M3GID
CT14 0JF 2E0RMV
CT14 0JF M6RMV
CT14 0JH G4WOS
CT14 0LA G4SUS
CT14 6PP 2E0LJH
CT14 6PP M0PKH
CT14 6PP M6CEH
CT14 6PP M6ODF
CT14 6QH G0HCX
CT14 7BB G1VWP
CT14 7BL M3BWZ
CT14 7EZ G0RDN
CT14 7PX M3OZC
CT14 7QB 2E0AAK
CT14 7QB G0OXX
CT14 7SA 2E0PBY
CT14 7SY M1AJU
CT14 8BT G8YXR
CT14 8DD G4YWA
CT14 8EB G0DQI
CT14 8JW G1IUA
CT14 8JW G7MSC
CT14 8JW M6ZTS
CT14 9AA M3WGZ
CT14 9AT 2E1ACK
CT14 9DQ G3MZI
CT14 9EE G1LLU
CT14 9EF 2E0JEN
CT14 9EF G8SOU
CT14 9EF M0EEH
CT14 9EW G3FBU
CT14 9HW M6DWI
CT14 9JF G3VIR
CT14 9JF M3VIR
CT14 9JR M1DNA
CT14 9LS G7HIX
CT14 9NB G0DUK
CT14 9NJ G0SRR
CT14 9NL M0ARX
CT14 9NP G4RXH
CT14 9QN M0SOE
CT14 9RG M1DEG
CT14 9RG M3AJU
CT14 9SS 2E0SMG
CT14 9SS M0SGE
CT14 9SS M3OIA
CT14 9TW G7MSS
CT15 4BT G6IOM
CT15 4HE M0HHI
CT15 4HH M0WPL
CT15 4HR G6BNW
CT15 4HX M6KWW
CT15 4HZ M3OVC
CT15 5BY G7FCL
CT15 5JD G8SMZ
CT15 5JW G6HIG
CT15 6AH G8YMD
CT15 6AH M1BKI
CT15 6BS M3VFM
CT15 6ED G6PSQ
CT15 6EJ G4OPR
CT15 6HL 2E1BTG
CT15 6HL M6LYO
CT15 6JL G6WPK
CT15 7BH G8UWS
CT15 7ES G4OJG
CT15 7HE G8MBV
CT15 7HJ G0OJU
CT15 7HJ G8UMB
CT15 7HR G3ZHU
CT15 7JN G4MIX
CT15 7JS G4AWW
CT15 7JY G4GLG
CT15 7LJ G0AXD
CT15 7LP G4VRB
CT16 1EX 2E0CUV
CT16 1PX G4DDB
CT16 1TH 2E0VDC
CT16 1TH M0KID
CT16 1TH M6VDC
CT16 2AR G1PQX
CT16 2AX G4BBH
CT16 2DF M3WSU
CT16 2EP M6JSW
CT16 2JW 2E1ABN
CT16 2JW G0ICT
CT16 2NQ 2E1NFD
CT16 2SG 2E0LYD
CT16 2SG M0IML
CT16 2SG M6BDV
CT16 3BA M6FEN
CT16 3DU M6LWM
CT16 3EE M1EJO
CT16 3EJ G4HHX
CT16 3HA G4RLX
CT16 3HZ G0ROO
CT16 3HZ G3ROO
CT16 3LQ G4UHT
CT16 3LT G7JOW
CT16 3ND M6HKQ
CT16 3NP G4WHQ
CT17 0BD 2E0BEH
CT17 0BD M0GGO
CT17 0BD M3JOM
CT17 0BS G8IYN
CT17 0DY M6BAQ
CT17 0ER G4WIY
CT17 0NP 2E1ABY
CT17 0NT G8YNG
CT17 0NX G0ADK
CT17 0PL G0SET
CT17 0PR G7BRM
CT17 0PS G0KOK
CT17 0PS G3YMD
CT17 0PS G6USA
CT17 0RX M1CJB
CT17 0SF G4FJF
CT17 9ES M3IWX
CT17 9JZ G6TRM
CT17 9LL 2E0CPL
CT17 9LL M0UGR
CT17 9LL M3UGR
CT17 9LQ G4FXE
CT17 9PN G4SMX
CT17 9QF M6ELZ
CT17 9QQ 2E0BHJ
CT17 9QQ 2E0OIN
CT17 9QQ G8WLY
CT17 9QQ M3KYH
CT17 9QQ M3OIN
CT17 9RB 2E0FQR
CT17 9RB M6FQR
CT17 9RQ G0CCG
CT17 9TX G0TBS
CT18 7AP G1URG
CT18 7AP G4IMP
CT18 7AP G4RJZ
CT18 7AP M3PNA
CT18 7BS 2E0UGF
CT18 7BS M0WHO
CT18 7BS M3UGF
CT18 7BS M6UFO
CT18 7DS G6RRS
CT18 7DS M0KBC
CT18 7DZ G4TGP
CT18 7EH M1CXI
CT18 7FA M6AKH
CT18 7FN G0BPS
CT18 7GZ G4MDZ
CT18 7HB 2E0NCI
CT18 7HB M6NCI
CT18 7JS G0IXV
CT18 7LB G3OJZ
CT18 7LL 2E1DYT
CT18 7LL M1CVF
CT18 7LQ G0VHL
CT18 7LW 2E0WUN
CT18 7LW M3RVQ
CT18 7LZ M0DRN
CT18 7NT M6PWE
CT18 7QL G4DUE
CT18 7TG M3FAK
CT18 8BU G1UFJ
CT18 8BY G8KSD
CT18 8BY M0CFH
CT18 8DF 2E0JKG
CT18 8DF M0JKG
CT18 8DF M3XJG
CT18 8DS G0VAI
CT18 8LX M3PMN
CT19 4AF G1RQI
CT19 4BX M3RWI
CT19 4EQ G0NOH
CT19 4HN M6LCI
CT19 4JA 2E0DJF
CT19 4JA M3PQB
CT19 4JA M3ZEJ
CT19 4JS M6BOK
CT19 4JT 2E0MTR
CT19 4QG M6GXS
CT19 4QH G0VWB
CT19 5AT G8YXQ
CT19 5BZ M1CMN
CT19 5DD M3HOM
CT19 5EL M3JBF
CT19 5HG G4IVL
CT19 5HG G7IWV
CT19 5JF G6GXE
CT19 5JH M6AFC
CT19 5JH M6RNI
CT19 5LS 2E0NEC
CT19 5LS M6WUG
CT19 5NA G8SUQ
CT19 5NB G4EQJ
CT19 5NE 2E0VYW
CT19 5NE M0VYW
CT19 5NR M1DFB
CT19 5NR M3DFB
CT19 5PW M1AZO
CT19 5QS G6AUK
CT19 5QY M0TUX
CT19 5RU M6RUS
CT19 5SH G8YNH
CT19 5SH M3BNU
CT19 5TA 2E0GJJ
CT19 5TA M0GJJ
CT19 5TA M3HFU
CT19 5TA M3IZP
CT19 5UB G8EKD
CT19 6BA G3XVY
CT19 6EY M6GXL
CT19 6LH 2E0DTS
CT19 6NE 2E0ALL
CT19 6PR G4MHS
CT194JW M6AZQ
CT2 0EA G0GPO
CT2 0HR G8WMK
CT2 0HR M3CSR
CT2 0HT 2E0EKR
CT2 0HT M0GRB
CT2 0HT M3EKR
CT2 0HT M6JEH
CT2 0HX G1LHL
CT2 0JZ G7NIN
CT2 0LA M1DFM
CT2 0LA M3DFM
CT2 0LA M3TTA
CT2 0LL M6DYV
CT2 0QH G3FMU
CT2 7AH 2E0OZH
CT2 7AH 2E0RZH
CT2 7AH M0OZH
CT2 7AH M3OZH
CT2 7AH M3ZIZ
CT2 7BX 2E0MNY
CT2 7BX M6FQU
CT2 7HH 2E0DFQ
CT2 7JZ G3HTT
CT2 7NA M0KKB
CT2 7NT G6WDS
CT2 7QX G1KSH
CT2 7TA M3DKO
CT2 7TF M0DCO
CT2 7TJ 2E0BEI
CT2 7UJ G0VZZ
CT2 8AY G0IEV
CT2 8PN M6JFP
CT2 8PQ G8UIV
CT2 9BL G4PCN
CT2 9BP M0SGF
CT2 9BP M6RCN
CT2 9BU M3WTU
CT2 9DB 2E0IMJ
CT2 9DB M0YMJ
CT2 9DB M3IMJ
CT2 9DL G8BRD
CT2 9JX 2E0XXK
CT2 9JX G2DLX
CT2 9JX M0XXK
CT2 9JX M3XXK
CT2 9NW M6GUC
CT20 1DA G6ZNW
CT20 1DF G6WGM
CT20 1HY G7HIY
CT20 2LU 2E0DLR
CT20 2LU M6EQT
CT20 2PQ G8IWX
CT20 2QT G7SXJ
CT20 2RP M6ECZ
CT20 2SL M6GXJ
CT20 2TY G0SLJ
CT20 2TY G2FA
CT20 3BE M0DRO
CT20 3EJ M6ASH
CT20 3LA 2E0AYY
CT20 3LA M0RNP
CT20 3LA M3BJL
CT20 3LH 2E0RDE
CT20 3NJ M3MSN
CT20 3QJ 2E1GTI
CT20 3SA G4EGQ
CT20 3TA G3XHW
CT20 3TL M0GCB
CT21 4EA G7SZO
CT21 4JP G3LWD
CT21 4JQ G3ZHT
CT21 4QA G4EDZ
CT21 4SE G4SSZ
CT21 5BT G3VYW
CT21 5RY M3PLI
CT3 1BJ G1WMJ
CT3 1ED 2E0TWA
CT3 1ED M6WEV
CT3 1JX G3JIX
CT3 1LN G0HRS
CT3 1LN G3TAJ
CT3 1LY G4FLR
CT3 1PQ M6FHQ
CT3 1SY G3KFG
CT3 1TZ G3KTZ
CT3 1UA G3RWF
CT3 1UH G4KJS
CT3 2BH 2E0VWW
CT3 2BH M3VWW
CT3 2JF G3XAQ
CT3 2LP G1MXM
CT3 2NH G8DBU
CT3 3AQ G6EPL
CT3 3HZ G6WSF
CT3 4DB G4NEE
CT3 4DS G0TXA
CT3 4HL 2E1JEH
CT3 4JN G7VQW
CT3 4LD M3OUU
CT3 4LN G4PLS
CT4 5EX G4ZHN
CT4 5JX M6WBA
CT4 5PN M6GDA
CT4 6DN G3MLO
CT4 6HX G4AZG
CT4 6JB G6GES
CT4 6JQ G7LNB
CT4 6NP G4WDI
CT4 6RT G4BQS
CT4 7AH G4ZIF
CT4 7ND G0VRW
CT4 7ND G7MPV
CT4 7NN G3XAQ
CT5 1EL G0IAA
CT5 1JQ M3PPU
CT5 1JQ M3RZY
CT5 1JZ 2E0GTB
CT5 1NS G4SIL
CT5 1QF G4RIS
CT5 1QF M3BQN
CT5 2DH G4LQI
CT5 2DH G7EOE
CT5 2DS G8BNK
CT5 2DY G8WIR
CT5 2LA 2E0TDO
CT5 2LA M0ZTD
CT5 2LA M6CWX
CT5 2LB G8NXQ
CT5 2LE G4LTS
CT5 2LE M6BIE
CT5 2NW G4PYA
CT5 2PH G0FAE
CT5 2PY G7JWX
CT5 3EJ G3KXB
CT5 3JZ G0IFS
CT5 3PJ G0AUW
CT5 3QD G4SEJ
CT5 3RF G8YMN
CT5 4BZ G0BEX
CT5 4DN G4GRJ
CT5 4DT G4WZZ
CT5 4DT G4XAB
CT5 4EL G0PPY
CT5 4EQ M1DHI
CT5 4HX G4CZU
CT5 4LA G1PVA
CT5 4LA M3MRS
CT5 4LY G7UWW
CT5 4LY M6HAY
CT5 4NY G0NBB
CT5 4NY G1PRH
CT5 4TE M3PNB
CT6 5QQ M6NSD
CT6 6BH G1KQE
CT6 6DX G7EIA
CT6 6EQ 2E0TDJ
CT6 6EQ M6GJI
CT6 6ES G3NIR
CT6 6GZ G4ELP
CT6 6GZ G4ICM
CT6 6GZ G8ELP
CT6 6HG G0ANK
CT6 6HU G4ZIH
CT6 6JA G4WZQ
CT6 6JU G4JMP
CT6 6NT G0ILO
CT6 6PP G8PPQ
CT6 6RE 2E0AIT
CT6 6RF G0HVC
CT6 6RF M1DHY
CT6 6SB G0LXB
CT6 6SB G0OXE
CT6 6SB G8ESW
CT6 6SE G2FMW
CT6 6SQ 2E1AWS
CT6 6SQ M0PAM
CT6 6SR G4SUK
CT6 6SS M3BBY
CT6 6UQ M3UCJ
CT6 7AY M6UVB
CT6 7DW G4EVD
CT6 7EE M1FZL
CT6 7EQ G8FEZ
CT6 7EW G1TQH
ct6 7hb G0ALJ
CT6 7HG G6CVD
CT6 7LE G1EDK
CT6 7NA M3BOV
CT6 7PY G4DBW
CT6 7QA M3CUH
CT6 7QD G0KFO
CT6 7RS G0IHI
CT6 7RS G1VNM
CT6 7RS G4MKI
CT6 7TA M6BNY
CT6 7TB M3PSE
CT6 7UD G4VMZ
CT6 7UW G0ETI
CT6 7XB M3SEY
CT6 7XF 2E0MDC
CT6 7XF G1DSZ
CT6 7XF M3YMG
CT6 7XG G6VRI
CT6 8AD G8KOL
CT6 8AE G4XOE
CT6 8AN G7MIF
CT6 8HG G4URD
CT6 8HX G4ZZK
CT6 8HY G1SDH
CT6 8JA M0HWM
CT6 8JS G6PKS
CT6 8JS M3KTV
CT6 8LN G6RMA
CT6 8LP 2E0ATZ
CT6 8LS G0LAA
CT6 8LT G0HMK
CT6 8LZ G7LFQ
CT6 8QW G8KDU
CT6 8QW M6CHZ
CT6 8RH M6BAC
CT6 8RX G1DKY
CT6 8SD G4RKV
CT6 8TU G0NFG
CT7 0BU M0PJG
CT7 0EL G0ANW
CT7 0HD M6EAJ
CT7 0PY G1HWH
CT7 0PY M1AIS
CT7 0PY M3HWH
CT7 9AJ M6AYS
CT7 9AS G1UEA
CT7 9AZ G0INT
CT7 9BN G8GJQ
CT7 9DF G0CIX
CT7 9ED G1JEH
CT7 9NA 2E1EHY
CT7 9QD G3OND
CT7 9QH G0CTQ
CT7 9QN G0MQB
CT7 9QN M1TCI
CT7 9RS M6LPP
CT7 9SR G3VID
CT7 9TY M3TGJ
CT7 9XE M6SIH
CT8 8AN G4SBD
CT8 8AN G4SEK
CT8 8AP 2E1DDZ
CT8 8AP G1NLQ
CT8 8AP G7MEZ
CT8 8BP M6KCC
CT8 8BT G4KPV
CT8 8BX 2E1CQP
CT8 8BX G0TBU
CT8 8HR 2E0CNU
CT8 8HR M6BOA
CT8 8LW G3PNT
CT8 8RJ 2E1CPB
CT8 8RJ G0CEY
CT88AH G6KKW
CT9 1LY M6UFF
CT9 1NS M6HCZ
CT9 1TR G4YGB
CT9 2AN G3WZG
CT9 2EJ M0AJC
CT9 2EN G3YEQ
CT9 2EN M0MVS

CT9 2NH M6EBL
CT9 2PS G0GNQ
[illegible]
CT9 2SW G0PDZ
CT9 2TD M6DGJ
CT9 3BH 2E0JFK
CT9 3DU G3YWF
CT9 3EF 2E1COV
CT9 3ES G8WMW
CT9 3HD G3SVJ
CT9 3HD G8ATD
CT9 3JB G0DFI
CT9 3JB G1DUJ
CT9 3NN M1BBS
CT9 3PT M0RST
CT9 3PT M1CBT
CT9 3PX G0LGE
CT9 3PX G6KNU
CT9 3PX M6LCQ
CT9 3RX G0CHN
CT9 3RY G6KQS
CT9 3SA G1WWR
CT9 3SL 2E1PDQ
CT9 3XJ G0NEP
CT9 3XT 2E0EMB
CT9 3XT M6AKJ
CT9 4DH G3YUH
CT9 4HA G3SHX
CT9 4LL G6ILU
CT9 4NE G7NOR
CT9 5DT G7EYE
CT9 5HD G6GHP
CT9 5HT G3TRX
CT9 5JA G4BSW
CT9 5JZ G0TBO
CT9 5LN G1AOC
CT9 5NE 2E1HQH
CT9 5NG 2E1HUC
CT9 5NP 2E0DUE
CT9 5NP M6WFI
CT9 5NT 2E1DNC
CT9 5PA G7MKF
CT9 5PF 2E0ZDE
CT9 5PF M0ZDE
CT9 5PF M6DES
CT9 5PN M6FZB
CT9 5QA G8DHJ
CT9 5QB G3ZZZ
CT9 5QB G8GHH
CT9 5UN G0WVA

## CV

(Coventry)

CV1 1FZ G1XLL
CV1 2AA G6AJC
CV1 2AL M6FKW
CV1 2AR M6JGY
CV1 2JQ G1HTT
CV1 4DJ G3YGB
CV1 4DJ G8GGR
CV1 4EB M3YVZ
CV1 4HL M1DCV
CV1 5EA G6XMT
CV1 5GU M6GKX
CV10 0BA G1JWY
CV10 0BA G1WRN
CV10 0BL M6AIU
CV10 0BX G4UQU
CV10 0BY M0CJS
CV10 0DF G0MDN
CV10 0DW G0VZO
CV10 0DW G3VDU
CV10 0DW G7SKL
CV10 0DZ G4NCV
CV10 0HP G8VHI
CV10 0HR G1VVL
CV10 0HR M6SIW
CV10 0HY 2E0IMG
CV10 0LB G0FBG
CV10 0LY M3TID
CV10 0NH M3FRU
CV10 0NL G7ANQ
CV10 0SL G0LLP
CV10 0SN G4KQL
[illegible]
CV10 7AW G1FTH
CV10 7BA M6UHT
CV10 7DE G0ZCQ
CV10 7EE G4ZUE
CV10 7ES G4AEH
CV10 7ET G3ZOY
CV10 7HG G7USI
CV10 7HP G4YOZ
CV10 7LF 2E0NEI
CV10 7LF M0NKE
CV10 7LF M6NFI
CV10 7PW G3YTW
CV10 7SW G0NZI
CV10 7SW M3RCW
CV10 8BD G6DDA
CV10 8DL M3VCO
CV10 8EG G7TYJ
CV10 8HL M3CBW
CV10 8HL M3CCB
CV10 8JQ G8ITJ
CV10 8NH G4NHF
CV10 8NU G4BBQ
CV10 8NW M6FSJ
CV10 8PA G0REO
CV10 8PT G7FSC
CV10 8PT M6CFI
CV10 8QA M6DZY
CV10 9AR 2E0NAG
CV10 9AR M3NWY
CV10 9DQ M6HHV
[illegible]
CV10 9DY M3ETQ
CV10 9DY M3EYR
CV10 9HU 2E0CJF
CV10 9HU M6AHU
CV10 9JH G7JXF
CV10 9NG G1VTU
CV10 9PB G6OQV
CV10 9PQ M6JPH
CV10 9PQ M6PJH
CV10 9QJ G8OMB
CV10 9QJ M0NDA
CV10 9QP G7SSD
CV10 9SD G1FLW
CV10 9SG G7CEN
CV11 4NP G4SHY
CV11 4RE M6KVG
CV11 4XG G7LSD
CV11 4XQ G4APS
CV11 5HP G8AWI
CV11 5LU G1EXV
CV11 5PD G6MTG
CV11 5RL G1VDP
CV11 5RL M0ABU
CV11 5RR M0GNB
CV11 5UB G1JVL
CV11 6AJ M6RVM
CV11 6BQ G4LMK
CV11 6DY M6EMF
CV11 6DY M3KWL
CV11 6DZ G4NQZ
CV11 6EE G0SBC
CV11 6EZ G8GXN
CV11 6FF G6SFY
CV11 6GA G4ASQ
CV11 6HN G3SLK
CV11 6JA G4HXC
CV11 6UU G0NCQ
CV12 0AZ G0OBE
CV12 0DE G0JUN
CV12 0EL G8GMU
CV12 0JH M0HSJ
CV12 0LD M6NKH
CV12 0LY G0LDP
CV12 0NS M3HBM
CV12 8AL M3HSH
CV12 8BE G0IYW
CV12 8BQ M3ZNV
CV12 8DA G1LBH
CV12 8DG G1OJD
CV12 8DU M1BNI
CV12 8JG M3JOS
CV12 8QU M6ENZ
CV12 8SG G1ZSR
CV12 9BX G0BVS
CV12 9DZ 2E0BGL
CV12 9EL M1JAK
CV12 9JB G0BIN
CV12 9ND G3YQZ
CV12 9PR M3SYH
CV12 9PS G1ORG
CV12 9SG M6TXA
CV13 0AU M3MFU
CV13 0LQ G3JLQ
CV13 0PR G7DDD
CV13 6BB G1SWI
CV13 6BB G7EKD
CV13 6BZ M0SDP
CV13 6HN G0OEM
CV13 6NL M1BKQ
CV2 1AF G0BJA
CV2 1EF G1NAT
CV2 1FP M6VOG
CV2 1FY 2E0TJE
CV2 1NN M6DTL
CV2 1PF G1HOP
CV2 1PP M3XXU
CV2 1SA G1AVF
CV2 2JD G6FXE
CV2 2JG G4HRY
CV2 2JS G4WKB
CV2 2JS G7DBO
CV2 2NE G6FYL
CV2 2QF G7PEN
CV2 3BH G1BNX
CV2 3DX G7JSQ
CV2 3EQ G1MSA
CV2 3HR M0HWI
CV2 3LW G8WW
CV2 3NQ G8JEC
CV2 4DS 2E0JFC
CV2 4DC M6JFC
CV2 4EU G0MTP
CV2 4GL G0UGX
CV2 4HT M6BIR
[illegible]
CV2 5BH G8SEQ
CV2 5BH M3MKV
CV2 5EH G4GEE
CV2 5FZ 2E0FNG
CV2 5FZ M0ZAI
CV2 5FZ M6FNG
CV2 5FZ M6JOD
CV2 5GJ G6WVM
CV2 5GL G8TGD
CV2 5LG M0COV
CV2 5LL G8ZLL
CV2 5NQ G1OPA
CV2 5NU G4ZXN
CV21 1BS M3JUO
CV21 1DE M6EEN
CV21 1HW G3OQO
CV21 1HW M0RTM
CV21 1JB G6GND
CV21 1JN 2E0HCW
CV21 1JN M6HCW
[illegible]
CV21 1NJ G1VAJ
CV21 1PG G0ZMC
CV21 1PG G7TMC
CV21 1PZ 2E0PPK
CV21 1PZ M0GWK
CV21 1PZ M3PPK
CV21 2QU G1FLI
CV21 2SZ 2E0SND
CV21 2SZ M3REQ
CV21 2TE G1ICA
CV21 3AB M0BBH
CV21 3BD G7BNI
CV21 3LH G8VTY
CV21 3NQ 2E1LOZ
CV21 3QH G3IKL
CV21 3SZ G0COY
CV21 4AJ G4IZI
CV21 4AN M6MIP
CV21 4BP G8EYY
CV21 4EF M0DEF
CV21 4HG G7KRE
CV21 4HJ M0HVV
CV21 4JY G7NFO
CV21 4LA M3CVS
CV21 4LT G7SRL
CV21 4LT M0ASD
CV22 5BG G0UFW
CV22 5ET G7APD
CV22 5ET G8LYB
CV22 5HN G0DLB
CV22 5HN G0RLV
CV22 5JN M0CNE
CV22 5NY M3SWK
CV22 5QJ G8AQN
CV22 5RG G4GZS
CV22 5RW G8FFZ
CV22 5RW M3AEA
CV22 5RW M6AEA
CV22 6BG M3VIG
CV22 6BG M3XRP
CV22 6EH G1BIM
CV22 6HB G3OBV
CV22 6HQ G0HWX
CV22 6JF M6JXF
CV22 6LG G4ACY
CV22 6NS M0ANH
CV22 6PG M6CIF
CV22 6PG M6LIB
CV22 6PX G8IHF
CV22 7AP G6CYT
CV22 7AQ G1GXW
CV22 7BZ G8DLX
CV22 7EG G7RGO
CV22 7EW G4SSW
CV22 7FJ G0FIN
CV22 7GN M6EIE
CV22 7HY 2E0RUG
CV22 7HY M6TRT
CV22 7PT G7GAB
CV22 7PT G7LIH
CV22 7RE G4AUB
CV22 7RG G8BLK
CV22 7TL M1AXE
CV22 7TN G1EBW
CV22 7TT G7JBW
CV23 0DB G1BIN
CV23 0DE G1UUV
CV23 0EE G3TQF
CV23 0JA G4JSX
CV23 0NN G4CFG
CV23 0NZ G0GZL
CV23 0PE G0GSA
CV23 0PH G0TZH
CV23 0PH G4AZH
CV23 0SP G7DDF
CV23 0SS 2E0NWY
CV23 0SS M0NWY
CV23 0SS M6NWY
CV23 0TS M0CNH
CV23 0UH G2LO
CV23 0UH G4IXT
CV23 8BH M6TQF
CV23 8DN G3ZJK
CV23 8HF M0GVW
[illegible]
CV23 8UF G6IQM
CV23 9EZ G4ZHY
CV23 9FN M0TXM
CV23 9QT M6HGU
CV23 9QT M6HGX
CV3 1AN M0BVV
CV3 1AU G0DKR
CV3 1EQ G0RDU
CV3 1FF M3JMI
CV3 1GG M6DZF
CV3 1GW G8MJX
CV3 1LY 2E0PES
CV3 1LY M6DIU
CV3 1NZ G1HGD
CV3 2AJ G4GDY
CV3 2AX G8YKY
CV3 2HA G3HLI
CV3 2HF G8OHG
CV3 2LZ 2E0PDP
CV3 2LZ M3PDP
CV3 2NQ G4ZDG
CV3 2NQ M6FKA
CV3 2NW G1COV
CV3 2NW G8HPV
CV3 2NW M0HPV
CV3 2PE G4ZCP
CV3 2QG G0LXI
CV3 3EQ G7MNT
CV3 3HH G1AMS
CV3 3HH M3WCQ
[illegible]
CV3 4BQ M1FCM
CV3 4FS G7LXY
CV3 5AG G0JKJ
CV3 5AU G4ROA
CV3 5DA M0IRT
CV3 5DS G6CTC
CV3 5DS G8ISJ
CV3 5HH M0BBQ
CV3 5JT G3LMQ
CV3 5LF G3TZG
CV3 5NG M6PKR
CV3 5PJ 2E0BLX
CV3 5PJ M0GKO
CV3 5PJ M3RVE
CV3 6BT G1EOK
CV3 6BT M3BMQ
CV3 6BT M3XGV
CV3 6EB G7RWY
CV3 6EH G3DNH
CV3 6EW M0BVW
CV3 6GH M0ACN
CV3 6JB M0AZY
CV3 6JG M6JYI
CV3 6JH G6RZS
CV3 6NY 2E0ROB
CV3 6NY M0TTT
CV3 6NY M5ZAP
CV31 1DQ G4DXN
CV31 1GE M0JJR
CV31 1LA M5ALS
CV31 1RJ G3XKE
CV31 1TA G4KTP
CV31 1WA 2E0CWC
CV31 1WA M0SER
CV31 1WA M3ZCW
CV31 2JX M6EFW
CV31 2JZ G0GIL
CV31 2PB G0RTI
CV31 2RB G4GEP
CV31 2TQ G3SPL
CV31 3AN M6NOD
CV31 3BD M1ERJ
CV31 3EB M6OPM
CV32 5AJ G4TYA
CV32 5JG G4PKT
CV32 5LU G6ISG
CV32 5NU G0KPH
CV32 6BA G8UIO
CV32 6BA G8WRG
CV32 6ES M6AVR
CV32 6HE G0BNE
CV32 6HY 2E1FQY
CV32 6LR G7DIW
CV32 6PB G3ZCG
CV32 6PE G8FOZ
CV32 6PW G3YJA
CV32 6QG G0LXG
CV32 6QR G8CDC
CV32 7DY M1AUK
CV32 7DZ G1YMN
CV32 7EY G3UDN
CV32 7EY G4CYG
CV32 7HT G6GSI
CV32 7JD G8AIM
CV32 7JD M3JHT
CV32 7JT G3HTB
CV32 7RX G0HCR
CV32 7SN G0DWJ
CV32 7TF G6NMA
CV33 9AQ G6VQW
CV33 9HG G1EAN
CV33 9LN G3YJN
CV33 9PX M1CZL
CV33 9QL G1PVR
CV33 9RN G1APQ
CV33 9RZ G0ASX
CV33 9TE G4USP
CV33 9TE M3EBK
CV33 9TH G1FEO
CV34 4SS G6BVR
CV34 5BX G0FGC
CV34 5FE G8GEE
CV34 5HX G6VGZ
[illegible]
CV34 5NL G7RMW
CV34 5NL G8XDL
[illegible]
CV34 5PU G0OYR
CV34 5RQ G4TIF
CV34 5SN G8YPK
CV34 5TG G1HRL
CV34 5TS G4JFS
CV34 5UJ G6YLB
CV34 5UL M0GWB
[illegible]
CV34 5XD G1WXF
CV34 5YQ M0LCA
CV34 6AR G8OVO
CV34 6BS M6AGT
CV34 6FQ G1ABM
CV34 6JQ 2E0DQG
CV34 6JQ M0OAE
CV34 6PF G3VRW
CV34 6QA G3TTC
CV34 6QJ G0GRM
CV34 6QX M3RYZ
CV34 6QY G3PXU
CV34 6XB G1XWM
CV35 0HH M0AYA
CV35 0LU G3VYE
CV35 0RE G1YYP
CV35 0SB G6NGF
CV35 0SS G8GDC
CV35 0UU M0UXA
[illegible]
CV35 7DX G4HCR
CV35 7HF G1WUM
CV35 7HF M3WUM
CV35 7TL G8NTQ
CV35 7TQ G1OUY
CV35 7UA G0OMN
CV35 8BN G3OAY
CV35 8DS 2E0TPO
CV35 8DS M0KFO
CV35 8DS M6KPO
CV35 8ED G1UTJ
CV35 8ED G6CXM
CV35 8JX M0CLM
CV35 8QP G7GUG
CV35 8QP M0XTD
CV35 8SZ G4BUD
CV35 8TY G6XCC
CV35 9BG G6ESK
CV35 9EY G3XMQ
CV35 9HQ G0IOD
CV35 9LY G6VGH
CV35 9PW M6MUT
CV36 4EH M0HGA
CV36 4FD G0CHO
CV36 4FD G1SWR
CV36 4HS G1RLD
CV36 4NG G8PYU
CV36 4NR G1AWJ
CV36 4PG G0CXJ
CV36 5DA G8HGG
CV37 0AP G1HPB
CV37 0DN G8JXP
CV37 0DZ G3YED
CV37 0JD M6BAH
CV37 0PP G4RYM
CV37 0PP G6MMD
CV37 0TT G1UNN
CV37 6DD G0TVC
CV37 6ST G6SGE
CV37 6TD G0REP
CV37 6TF G0JJY
CV37 6XG G0EDT
CV37 7DD G0CCJ
CV37 7EX 2E0BAV
CV37 7EX M3YOE
CV37 7HH 2E0JBG
CV37 7HH M6AVN
CV37 7JF G4IQW
CV37 7JS G8FRS
CV37 7PL G1EIX
CV37 7TL G3XZO
CV37 8EU M3XWC
CV37 8RA 2E1DMH
CV37 8XY G3XGV
CV37 9AD G3GMJ
CV37 9ED 2E0NDZ
CV37 9ED M3NDZ
CV37 9JN G0JUQ
CV37 9JU G6FIP
CV37 9LU G4GTS
CV37 9PN G7BRF
CV37 9QL G0BKA
CV37 9QL G0BKB
CV37 9QW G8OCS
CV37 9ST 2E1FCC
CV37 9ST 2E1FCD
CV37 9ST G8RDT
CV37 9XN 2E0BZA
CV37 9XN M3MXV
CV37 9XR G0WKT
CV4 7AY M6WYW
CV4 7DU G1KCR
CV4 8HF G0KNX
CV4 9AA G0OTT
CV4 9AN M6ZPT
CV4 9AU G6UCT
CV4 9AZ G8BZT
CV4 9BW G6PSO
CV4 9DJ M0PML
CV4 9EU 2E0ZWW
CV4 9EU M0ZWW
CV4 9EU M3ZWW
CV4 9EZ G4VCX
CV4 9JH G1YFL
CV4 9LA G6WLM
CV4 9RN G1IUL
CV4 9TD G6JPR
CV47 0ES M3BOU
CV47 0HW G0OKN
CV47 1EX G1XLG
CV47 1QQ M3LQJ
CV47 1NF G3TFA
CV47 1NQ G4MQW
CV47 1NQ G8DER
CV47 1PU G0AFZ
CV47 2AT G0CDO
CV47 2AT G0SDX
CV47 2BN G3ZST
CV47 2DF M6KWH
CV47 2FJ G4EWW
CV47 2QS G0GNF
CV47 2RA M6GVP
CV47 2UF 2E0ZPM
CV47 2UF M6ZPM
CV47 2XY G4YIE
CV47 7RT G4NXA
CV47 8LG M0ZOM
CV47 8LG M6FZY
CV47 8LX G3SNO
CV47 8NE G1KCS
CV47 8NL M6VPX
CV47 8NN G3OJI
CV47 8NN G6EVC
[illegible]
CV5 6AL G1HKP
[illegible]
CV5 6AY G3VWD
CV5 6BB G4RUN
CV5 6DJ G3PUR
CV5 6DR G3WCQ
CV5 6FF G0EMX
CV5 6GH 2E1IAX
CV5 7AD G8BHY
CV5 7AH G6MQP
CV5 7BY G8WYI
CV5 7DA M6BHT
CV5 7DP M0LEE
CV5 7EH G3MAU
CV5 7FH G1KFQ
CV5 7GP G1PPB
CV5 7HP G7DTK
CV5 7JU G3NAP
CV5 7LJ G3UKD
CV5 7ND 2E0UAC
CV5 7NE G8AHR
CV5 7NR G4HOC
CV5 7QE G4MWP
CV5 7QE G4UVW
CV5 8BD 2E0UWK
CV5 8BD M6HGJ
CV5 8BD M6UWK
CV5 8DA M0XKW
CV5 8DR 2E0RRC
CV5 8DR M6SDW
CV5 8FL G4DRU
CV5 8LE G8KVU
CV5 8LG G7III
CV5 8LS G4VCY
CV5 8LS G6VCY
CV5 9AS G0ENY
CV5 9EA M0NOI
CV5 9EN G8ALS
CV5 9HL G3JYS
CV5 9JT G4BCG
CV5 9JT G8GET
CV5 9JU G3XQE
CV5 9JU G6OES
CV5 9JY G3NAY
CV5 9LN G4STI
CV5 9LQ G8RHM
CV5 9NA G4JDO
CV5 9NS G4KUE
CV6 1BE G7LAW
CV6 1BY G0EWZ
CV6 1LF G8KTX
CV6 1NW G6KHA
CV6 1PY M0YGM
CV6 2GQ G1UWQ
CV6 2HB G8FFN
CV6 2HB G8LGT
CV6 2NN G4KZU
CV6 2PG M3ORU
CV6 3GF G1ZNK
CV6 3GJ M3ECW
CV6 3GL G6IHO
CV6 3GS G7IZV
CV6 4EY G4LTK
CV6 4FH G3CWH
CV6 4FZ M6HKW
CV6 4FZ M6JMJ
CV6 4HL M3VYN
CV6 5EH 2E0COV
CV6 5EH G3VLG
CV6 5EH M0HIN
CV6 5EH M0TAV
CV6 6BD M6JAV
CV6 6DS G1CPU
CV6 6ES M3TSE
CV6 6FA G1NML
CV6 6GY G4OQV
CV6 6JZ G6ZTF
CV6 6NB M3JSF
CV6 6QA 2E0CJW
CV6 6QA M6CJT
CV6 6RE G0DCZ
CV6 7EU 2E0TSP
CV6 7EU M6RPV
CV6 7JJ M6EXH
CV6 7LB M3LVP
CV7 7AB G4LMV
CV7 7FX G4IEV
CV7 7FX G6SCM
CV7 7LU G0EPV
CV7 7LW G8XUU
CV7 7NB 2E0TUX
CV7 7NB M0YMM
CV7 7NB M3POV
CV7 7NQ 2E1ACG
CV7 7NQ G4KUR
CV7 7NQ G7UBX
CV7 7QF G3UKM
CV7 7QF M6JGV
CV7 7QL M0CAR
CV7 7QN G1UUG
CV7 7QW G4JYF
CV7 7QW M3HYF
CV7 8AQ M3UYF
CV7 8AU G4IAG
CV7 8BF G4RWH
CV7 8BY M6DAD
CV7 8ER G3SZE
CV7 8HL M0BVX
CV7 8HL M0BVZ
CV7 8HY G6GYN
CV7 8JJ G3SPY
CV7 8JJ G3UVW
CV7 8JJ G3ZFR
CV7 8JX M3CVO
CV7 8LB M0NIF
CV7 8LB M0SQC
[illegible]
CV7 8PP G1UUK
CV7 8PR G6KXN
CV7 8PX M0MRW
CV7 9AQ M0BVU
CV7 9AU G6VJR
CV7 9AU G7AMQ
CV7 9BJ G4LML
CV7 9BX M6TNY
CV7 9FH M0DVG
CV7 9JE G1BXQ
CV7 9LR G0AIZ
CV7 9LT G1OPJ
CV7 9NJ G4WXK
CV7 9NQ 2E0OCV
CV7 9NQ M6OCV
CV8 1AH G7RYO
CV8 1BE G1UUT
CV8 1DG G6XVZ
CV8 1DJ M0AZJ
CV8 1EZ G0KPG
CV8 1FL G4GJY
CV8 1FX M6FXE
CV8 1GH 2E0FKA
CV8 1GH M0LDY
CV8 1GH M3FKA
CV8 1HX G6GQF
CV8 1JD M6OAO
CV8 1NR M3RNG
CV8 1PL G3VTL
CV8 1RB G8CYA
CV8 1SF M0URX
CV8 2AF M0BNC
CV8 2BB M0JDB
CV8 2BE G6XRI
CV8 2BJ G4DF
CV8 2BX G4YMT
CV8 2DS G0INS
CV8 2DU G0DKV
CV8 2EE G0JMW
CV8 2FE G0PYD
CV8 2JZ M6ENC
CV8 2SW 2E0PCH
CV8 2SW M0VVZ
CV8 2XH G7JYY
CV8 3ET G6WTD
CV8 3FR G0HOV
CV8 3HH M0JCT
CV8 3JF G3JZL
CV9 1HN G1SIM
CV9 1HP G8LQF
CV9 1NZ G4IWA
CV9 1PS G0LKA
CV9 2AW G0KKV
CV9 2BD M6TJQ
CV9 2DA G4XDE
CV9 2ET G4YNX
CV9 2EZ M1CNL
CV9 2EZ M6TAD
CV9 2HB G4BKA
CV9 2RP G6SGY
CV9 3BP M3GTQ
CV9 3DP G3NXV
CV9 3EH 2E0NDW
CV9 3EJ G0VZK
CV9 3EX G0HUW
CV9 3EX M6HLT
CV9 3HJ M6KDO
CV9 3LT G4KMX
CV9 3PJ G8NDB
CV9 3RE G2FXQ

## CW

(Crewe)

CW1 3AX G8DTT
CW1 3AX M6FUJ
CW1 3EG M0MTF
CW1 3EQ 2E0NKP
CW1 3EQ M6NKP
CW1 3JN 2E1EHP
CW1 3JN M3EHP
CW1 3LE 2E0BDP
[illegible]
CW1 3NW 2E0FQC
CW1 3NW M6FQC
CW1 3PJ 2E0AZZ
CW1 3PJ 2E1FIV
CW1 3PJ 2E1UJE
CW1 3PJ M5WGD
CW1 3RY G0HIZ
CW1 3SG 2E0KNC
CW1 3XJ G4OUK
CW1 3XJ G6WZY
CW1 3XN M0MBG
CW1 3XN M6BCV
CW1 3YN M6EXA
CW1 4AL 2E0RMS
CW1 4AI 2E1RMS
CW1 4AL M0TDC
CW1 4AL M3AHS
CW1 4AL M3RMS
CW1 4AU M3KQW
CW1 4BS G7BMT
CW1 4DA 2E1RAO
CW1 4DE M3DMJ
CW1 4DE M6MHD
CW1 4DF 2E0NPP
CW1 4DF M3ZJF
CW1 4DF M3ZMV
CW1 4DU M3USB
CW1 4DY G0TWH
CW1 4ES G1DBI
CW1 4HB 2E0FTV
CW1 4HD M0FTY
CW1 4HR M6POE
CW1 4HL M3ISG
CW1 4HZ 2E0HST
CW1 4HZ M0KDA
CW1 4HZ M0WFX
CW1 4JR 2E0FFM
CW1 4JR M6JTO
CW1 4JZ M3VFL
CW1 4NW M6RGF
CW1 4PS M6HRE
CW1 4PX M6JUG
CW1 4RA G4DBX
CW1 4RP M0UTD
CW1 4RS 2E0CKM
CW1 4RS M6AVU
CW1 4TD M6TGS
CW1 4TT G6INM
CW1 4TX M6CSI
CW1 5FU G0CZD
CW1 5LF G7UUG
CW1 5PB 2E0DCP
CW1 5PB M6DEO
CW1 5PX 2E0CPE
CW1 5PX M0HPW
CW1 5QE G6OBT
CW1 5RA G0CGQ
CW1 5SY G3ZFC
CW1 5YF G0AVE
CW1 5YQ M0CJK
CW1 6ES M3HQS
CW1 6HD G4DQG
CW1 6HN 2E0RAL
CW10 0DG M3LVY
CW10 0DL M0BMR
CW10 0EA M6DMA
CW10 0EZ 2E0IEA
CW10 0EZ M6IEA
CW10 0HR G3MCC
CW10 0PF 2E0CDQ
CW10 0PF M6AFY
CW10 0PJ G4ZKR
CW10 9AU 2E1GDA
CW10 9BY G0NTG
CW10 9HE G7APM
CW10 9HG G0GKN
CW10 9HL 2E0DWF
CW10 9HL M6WDF
CW10 9NW M3KKQ
CW11 1BL M3NNZ
CW11 1BN M0CJI
CW11 1BS G6MPN
CW11 1BZ G6ZGB
CW11 1EB G0BRZ
CW11 1RS M3TAE
CW11 1RW 2E0ZAC
CW11 1RW M0BZI
CW11 1SG G7OOI
CW11 1SQ G4MXE
CW11 1WN M3LIV
CW11 1XP 2E0FUH
CW11 1XP M0EEG
CW11 1XP M3FUH
CW11 1XP M3NBX
CW11 1XU M3SIS
CW11 3AS G0JNE
CW11 3AU 2E0JZC
CW11 3AU M6JZC
CW11 3BL 2E0CXN
CW11 3BL M6DVA
CW11 3BU 2E0OSY
CW11 3BU G4OHP
CW11 3BU G6IIU
CW11 3BU M6OSY
CW11 3FZ G1XRJ
CW11 3JF G7HOA
CW11 3JF M0OBW
CW11 3JF M1CNY
CW11 3QL G4UCT
CW11 3RD G6BMG
CW11 4HL G4VIF
CW11 4HS G7SHI
CW11 4HZ 2E0WIS
CW11 4HZ M6JLE
CW11 4HZ M6WIS
CW11 4PN G7CTE
CW11 4PQ G6VPJ
CW11 4RE G4PMY
CW11 4RZ G1CNN
CW11 4SP G3HUD
CW11 4SP G3JOE
CW11 4TN G6UKN
CW12 1AG 2E0GRS
CW12 1AG M3FIH
CW12 1DT M6JGG
CW12 1HL G3WNW
CW12 1LT M6PYF
CW12 1NU G0REB
CW12 1NU G4LTZ
CW12 1NY G7PZF
CW12 1QW M6CLX
CW12 1SD G7VTQ
CW12 1SE G4JYK
CW12 1SE G4NVN
CW12 1SH G6DZX
CW12 2BH M6EGB
CW12 2BS 2E0SOT
CW12 2BS M3SOT
CW12 2BY M6JKR
CW12 2EP M3XKE
CW12 2EQ M0OWO
CW12 2HN G3WRK
CW12 2HQ G7HIO
CW12 2JJ G7RMJ
[illegible]
CW12 2LL G1DUT
CW12 3AU G3LLJ
CW12 3BH G1PAD
CW12 3BN G7FJK
CW12 3DB M0AVQ
CW12 3DE G4GMZ
CW12 3DZ M3KWZ
CW12 3EE M6NND
CW12 3EP 2E1CPQ
CW12 3FF M6GCD
CW12 3HH M6IGJ
CW12 3HP G1EGE
CW12 3HQ M6DEI
CW12 3HS G4UJO
CW12 3JA M6ZOM
CW12 3JJ M6BSO
CW12 3JP G7NCD
CW12 3JY G8JPU
CW12 3JY M3ZBL
CW12 3LD M0TCL
CW12 3LL G6VMV
CW12 3LZ M6ZJT
CW12 3PL G1LCE
CW12 3PL G4FSH
CW12 3QG G0DOC
CW12 3QG M6ZYZ
CW12 3RB G3ZAU
CW12 3RH G4TAG
CW12 3RX M6AMX
CW12 3SW M6RDA
CW12 3SW M6SJE
CW12 3TD 2E1CAT
CW12 3TD G8WWF
CW12 3TU G4BER
CW12 3TU M6SBY
CW12 3TW M1EER
CW12 3TW M6VCB
CW12 3TY G8RXY
CW12 4BS 2E0CJD
CW12 4BS M3SQT
CW12 4DY G4WTZ
CW12 4EQ M6WLE
CW12 4FR G4HMX
CW12 4FR M6EAL
CW12 4HL G8UVZ
CW12 4JL M0VVG
CW12 4JL M1LOL
CW12 4JL M1REK
CW12 4JN M0GVE
CW12 4JR M3ZQB
CW12 4NE G0UXO
CW12 4PH G0CQY
CW12 4PR G3TVX
CW12 4PR M6XEP
CW12 4PS M3AVZ
CW12 4PY G7CPN
CW12 4QH G4IWV
CW12 4QR M6EHC
CW12 4QU G6XYV
CW12 4QX G6MAT
CW12 4RF G1VIY
CW12 4RS G6WXI
CW12 4SN M3HXM
CW12 4TQ G4BOH
CW12 4UX M6MNU
CW2 5AW G0NWF
CW2 5AW G4BZI
CW2 5AZ 2E1GDB
CW2 5AZ G8SIG
CW2 5AZ M3MAR
CW2 5AZ M3ZTT
CW2 5BJ G0TMK
CW2 5EZ M6EFP
CW2 5GJ M6RPW
CW2 5GP G4VXX
CW2 5GS G0FAU
CW2 5JN G3XHP
CW2 5LX G0GRO
CW2 5NR 2E0RVE
CW2 5NR M6RVY
CW2 5NS 2E0ECO
CW2 5NS M6EXC
CW2 5NZ G4XUM
CW2 6HN 2E0ZHG
CW2 6HN M3ZHG
CW2 6JB G8DHQ
CW2 6RD G0DRM
CW2 6SQ M0XPA
CW2 6TG G1LSN
CW2 6TJ G6IGW
CW2 6TT M6WAP
CW2 6TX M0VFR
CW2 6TX M0YVX
CW2 6TX M6ZBT
CW2 6UH M6DBQ
[illegible]
CW2 7DH M6HUG
CW2 7HD M3KKF
CW2 7JY M3KPF
CW2 7LE M0KRV
CW2 7NY 2E0DYB
CW2 7NY M0SWC
CW2 7NY M6DYB
CW2 8AP G0MMH
CW2 8AP G4RRM
CW2 8AP G6TW
CW2 8BD M1LEO
CW2 8BD M3AGB
CW2 8BN G4YFO
CW2 8BT G7IFI
CW2 8DS G1GRM
CW2 8EX G7GZZ
CW2 8HG G4MAG
CW2 8HJ G4LVR
CW2 8HP G0OIE
CW2 8HS G3ZBM
[illegible]
CW2 8NS M3EUU
CW2 8PB G7PWU
CW2 8PL M3NVF
CW2 8QA 2E0JDD
CW2 8QA M3LZK
CW2 8RE M0RCT
CW2 8RS G7OMN
CW2 8SH G6LVT
CW2 8SZ G8BTD
CW3 0AH 2E0KZL
CW3 0AH M3LYP
CW3 0BA G3YMU
CW3 0JB G0NBH
CW3 0JB G1XHR
CW3 9BS M3KEF
CW3 9DN G8FNG
CW3 9EG G4YPQ
CW3 9EL G3JJR
CW3 9FU G1VTN
CW3 9SZ M3YHQ
CW3 9SZ M3ZHY
CW4 7AS G3GMM
CW4 7BT G0CTP
CW4 7DR M3IFE
CW4 7LA G1DBL
CW4 8AS M3HYV
CW4 8DT G0CSY
CW4 8HX M0SUN
CW4 8JB G4NVA
CW4 8JG M0ESR
CW4 8JG M0ZED
CW4 8JG M6DNM
CW4 8JG M6NVA
CW4 8JU G4ZVA
CW4 8LJ M0YIJ
CW4 8LL 2E0OCG
CW4 8PG M3JWZ
CW4 8PP 2E0SKA
CW4 8PP M3BMV
CW5 5JE M6LPB
CW5 5LA 2E0SEY
CW5 5LA M6LLU
CW5 5QJ M0ELO
CW5 5TX M6GWN
CW5 6AL G3KMV
CW5 6AL G8UHT
CW5 6DX G4XNR
CW5 6ED G0RBA
CW5 6HJ G4PCR
CW5 6JJ G6SNI
CW5 6NL G0RBW
CW5 6QF G0VJR
CW5 7BS G4ZMR
CW5 7EJ G3PFE
CW5 7ER G0OEW
CW5 7FL G4SHC
CW5 7GU 2E0JCD
CW5 7GU M0JCD
CW5 7GU M0MHC
CW5 7HY G4DBD
CW5 7NN G1BDQ
CW5 7NX G4MHX
CW5 7PN G4PUM
CW5 7QD G4PNC
CW5 7QX G8ATB
CW5 8AQ M3FJN
CW5 8AQ M3FJR
CW5 8AQ M3HAK
CW5 8BG G6PGG
CW5 8HH G3TLV
CW5 8HW G3HEH
CW5 8JD G0ZPV
CW6 0AB G0TQJ
CW6 0BA 2E0RJM
CW6 0BA M0RJM
CW6 0BA M3RZV
CW6 0HQ G3TRL
CW6 0NL G4VPC
CW6 0PU G3NHP
CW6 0QR M3SHK
CW6 0SD G7VOI
CW6 0SX G3VUN
CW6 9AJ M0SOA
CW6 9DN G4UUD
CW6 9BP M1EIU
CW6 9JT G4KZO
CW6 9QX G4XMX
CW6 9RT M6AYI
CW6 9TF G0MMT
CW6 9TZ M6ENX
CW7 1DZ M3RVX
CW7 1EU G1ILH
CW7 1FE M3JIC
CW7 1HA M0RDR
CW7 1LU G0GZI
CW7 1LU G6DSA
CW7 1LU G7DSA
CW7 1LU M0NLJ
CW7 1NA G7CRG
CW7 1NA G8HAV
CW7 1NA G8ZTT
CW7 1NA M0TPC
CW7 1NE G7RYN
CW7 1NE M1ATI
CW7 1NJ M6CAG
CW7 1QL M1ARS
CW7 1QZ M3HGP
CW7 1RE G4TEZ
CW7 1SW 2E0RNI



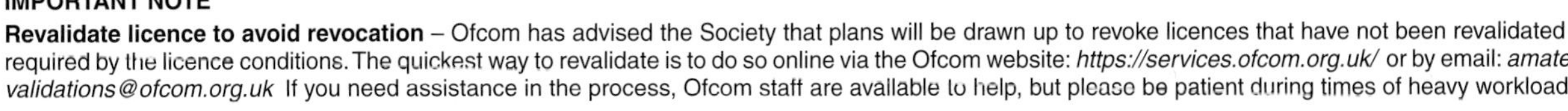
**IMPORTANT NOTE**

**Revalidate licence to avoid revocation** – Ofcom has advised the Society that plans will be drawn up to revoke licences that have not been revalidated as required by the licence conditions. The quickest way to revalidate is to do so online via the Ofcom website: *https://services.ofcom.org.uk/* or by email: *amateur.validations@ofcom.org.uk* If you need assistance in the process, Ofcom staff are available to help, but please be patient during times of heavy workload.

CW7 1SW G4JVX
CW7 1SW M3CUU
CW7 2ED 2E1HEV
CW7 2EG M3EHA
CW7 2JE M3PPQ
CW7 2LE 2E0FDS
CW7 2LJ M1DJP
CW7 2LL M6TKR
CW7 2NE M6GLN
CW7 2QD G0EOL
CW7 2UE M3XFD
CW7 2UW G6WEL
CW7 3EN G0ADU
CW7 3EN G6WZZ
CW7 3JB 2E0RJD
CW7 3JB M0JFD
CW7 3JB M3LKD
CW7 3JU M1BXM
CW7 3LE G1DBR
CW7 3NG G8UYB
CW7 4AJ G7GRO
CW7 4DP M3YUA
CW8 1BW G0VOK
CW8 1HX 2E0HGE
CW8 1HX M3HGE
CW8 1JR M3YET
CW8 1LA 2E0SDK
CW8 1LA M3KDK
CW8 1NB G6HZJ
CW8 1PL G6HXU
CW8 1PY G3ZTT
CW8 1PY G4XUV
CW8 1RD G1GYJ
CW8 2BQ G0JIT
CW8 2JB G3YWU
CW8 2JW G4MIH
CW8 2PB G1YJR
CW8 2PL M6GMP
CW8 2PP G6AFG
CW8 2TA G4VGF
CW8 2XJ G4JYP
CW8 2XN G6DQO
CW8 3DX M1ALA
CW8 3ED G4YTB
CW8 3ED G6WWM
CW8 3EZ M0XEY
CW8 3HD G4YVI
CW8 3HZ G7HFW
CW8 3JD G7VDQ
CW8 3LP M3KJB
CW8 3PN G4KCZ
CW8 3PN G6NBP
CW8 3PT G4IAB
CW8 3RH G6LCS
CW8 4AA G4XDG
CW8 4BA G0DQQ
CW8 4DF G0HXD
CW8 4EH M6SLB
CW8 4HR G6GAK
CW8 4JX G4ROK
CW8 4LL M3KOL
CW8 4PU G0BKH
CW8 4PU G1MCG
CW8 4XA G8NCS
CW9 5JL 2E0DHJ
CW9 5JL M6EXB
CW9 5LJ G0DND
CW9 5LJ M3CRV
CW9 5PX M3KKN
CW9 5PZ G1LDC
CW9 5QR G4OJF
CW9 6DA G8BJA
CW9 6DR G4XQB
CW9 6DS G1MVE
CW9 6EB G1URR
CW9 6ED M3GYH
CW9 6EZ G6LDM
CW9 6EZ M3OUQ
CW9 6EZ M3OVG
CW9 6HJ G0THJ
CW9 6LN 2E0XOJ
CW9 6LN M3XOJ
CW9 6PP G8XMZ
CW9 6PX G8VNX
CW9 7AN M6RJJ
CW9 7AR 2E0DHV
CW9 7AR M6EQO
CW9 7AR M6TFZ
CW9 7AS G0SPH
CW9 7AS M3ISH
CW9 7BF M3GHG
CW9 7EP M3LZR
CW9 7JA G0OGJ
CW9 7JB 2E1GYG
CW9 7JB G1GOP
CW9 7JB G1GOQ
CW9 7JQ M3XRQ
CW9 7LD M3CGI
CW9 7QD M3XQV
CW9 7QD M3XQY
CW9 7QH 2E0HGX
CW9 7QH G4PKX
CW9 7QH M3HGX
CW9 7QJ M0CRP
CW9 7TY M6EWU
CW9 8AR G6LJX
CW9 8BN M3NDJ
CW9 8DB M3MVK
CW9 8GF G7KTH
CW9 8GG G6ZTT
CW9 8GG G7LQD
CW9 8GP G4MUA
CW9 8PB 2E0STX
CW9 8PB M0WTX
CW9 8PB M6EAX
CW9 8PB M6MHJ
CW9 8PB M6WTX
CW9 8QA G0LBO
CW9 8QQ G4CAX
CW9 8RQ M0FCT

## DA

## (Dartford)

DA1 1ND G4ZHX
DA1 1PL G7GRB
DA1 1QT M3YJU
DA1 1TR G6UML
DA1 1XB G4APG
DA1 1YY M3ALB
DA1 2QL G7VCM
DA1 2QN M1CXK
DA1 2RZ G6MVN
DA1 3AJ M1CYN
DA1 3AN G3ZPS
DA1 3BA G7KAO
DA1 3BP G0RJL
DA1 3DE M6CAF
DA1 3DE M6GAF
DA1 3DE M6KRF
DA1 3EE M6BPC
DA1 3JU G4XKZ
DA1 3JX M0TGV
DA1 3NX G1RCV
DA1 3NX G3RCV
DA1 4PZ M6KVL
DA1 4RY G8ZHR
DA1 4SN G7WLL
DA1 4SU G4APB
DA1 5HT G6CMB
DA1 5JW G6TSC
DA1 5JW G8XJB
DA1 5NF M3YZN
DA10 0BX M6WJM
DA10 0LT G0NZR
DA10 0LU M6WCQ
DA11 0PH G4WTE
DA11 7AQ G7MIE
DA11 7BN G8JAD
DA11 7BN M1ENQ
DA11 7BN M3JUL
DA11 7EB 2E0IJH
DA11 7EZ G0DWS
DA11 7LB G8IEA
DA11 7LG G0RNP
DA11 7NY M3UUZ
DA11 7PP G6LUD
DA11 7QG G3NPS
DA11 7QG G8LDV
DA11 8BF M0HZF
DA11 8ET G0MMD
DA11 8NN G0MLF
DA11 8NQ G4ALD
DA11 8PL M1CXY
DA11 9DU M6KIG
DA11 9DX G1YXY
DA11 9LW 2E0YRM
DA11 9LW M0YRM
DA11 9LW M6LZM
DA12 1LL G6LUY
DA12 2LP G7EOC
DA12 2NN G4CCE
DA12 4AR M3TJI
DA12 4BJ G3DCV
DA12 4EL 2E0OOM
DA12 4EL M6MAY
DA12 4HD G7CJS
DA12 4HJ G3RE
DA12 4HJ G4IYK
DA12 4LQ G3KZN
DA12 4NA G4BBJ
DA12 4NJ G8XIR
DA12 5BD G1PHJ
DA12 5BD G1PNL
DA12 5DN G1OFL
DA13 0BW G3KBH
DA13 0EA G0OAT
DA13 0HH 2E0IAD
DA13 0LS G3BAC
DA13 0QA G0RJN
DA13 0SQ G4YGU
DA13 0SU G1AZJ
DA13 0TQ M6DDF
DA13 0TX 2E0RCL
DA13 0TX M0RBX
DA13 0TX M6DBI
DA13 0UD G0AFH
DA13 0UD G0FBB
DA13 9DS G3PBF
DA13 9EJ G0SSN
DA13 9EJ G6CBY
DA13 9JR M6HHF
DA14 4AW 2E0SKL
DA14 4AW M6LSK
DA14 4BE G8MCA
DA14 4ET G4YMF
DA14 4LJ G7EDA
DA14 4PS G6INU
DA14 4QU G0BAX
DA14 4RH M3JLV
DA14 5NF 2E0FBH
DA14 5NF G4FAA
DA14 5NF M0HZB
DA14 5NG G6GTC
DA14 6JQ G3BNE
DA14 6JT G8XIN
DA14 6SG M1BXU
DA15 7NL G4ILH
DA15 8AT 2E0SDW
DA15 8AT M0SJW
DA15 8AT M3NHI
DA15 8ER G1TXO
DA15 8JN G8JTG
DA15 8JT G1HYT
DA15 8RX M0DDV
DA15 8SZ G8KAM
DA15 8TA G1PRZ
DA15 9DZ 2E0PBF
DA15 9DZ G8OPA
DA15 9DZ M6PAF
DA15 9JJ G0JBT
DA16 1DE M1TAD
DA16 1EJ G0KPZ
DA16 2AW G8SWC
DA16 2BN G6NRH
DA16 2BP G0KTS
DA16 2HJ G2DZH
DA16 2HX G8GKC
DA16 2LH G1OAZ
DA16 2QD G0GKI
DA16 2RU G6ODW
DA16 3AW G0LZF
DA16 3AW G1FHO
DA17 5BG G8MIF
DA17 5BG M0JHP
DA17 5BT G7UHW
DA17 5EW G3CUR
DA17 6HB G4WJH
DA17 6JE M0PBR
DA17 6LP G0VWF
DA2 6DN G1JNQ
DA2 6HD G0IPC
DA2 6HE M6NEY
DA2 6HE M6THS
DA2 6HZ G1ZAW
DA2 6JS G4OFU
DA2 6JX G6ZAY
DA2 6LB G7FIA
DA2 6LQ M6MRV
DA2 6NB G4CVC
DA2 7ES G4IQQ
DA2 7HX M1CWG
DA2 7LP G0HRD
DA2 7NW G0DVL
DA2 7PB G0KUU
DA2 7QQ G4LGU
DA2 7RL M6CMO
DA2 7RL M6WLF
DA2 7SP G3XVC
DA2 7WB G1EWE
DA2 8BZ M0DPS
DA3 7HD G3FGP
DA3 7JR G0LJC
DA3 7NS G6XND
DA3 8DD G1FJJ
DA3 8EU G4NRV
DA3 8LG G8RW
DA3 8LG M0XPS
DA3 8LN 2E0RBI
DA3 8LN M0RJT
DA3 8LN M3RBI
DA4 0BE 2E0IAG
DA4 0BE G7FUV
DA4 0BE M0UAT
DA4 0BE M3YBJ
DA4 0BE M3ZBQ
DA4 0DF 2E1FPI
DA4 0HA G0WMC
DA4 0HA M1ABU
DA4 9DQ 2E0WAF
DA4 9DQ G0ARQ
DA4 9DQ G4BWV
DA4 9DQ M3WPK
DA4 9EW M3OKC
DA4 9EX G6YLD
DA4 9HS M3YOU
DA5 1LX G6IRP
DA5 1NW G4KUL
DA5 2ER G3XEW
DA5 2ES G4CW
DA5 2HN M3VSO
DA5 3AH G0FDZ
DA5 3BE G1FNN
DA5 3BT G6YZB
DA6 7LA M1DYD
DA6 7PA G4MB
DA6 8HU G4RPP
DA6 8JS G1MAV
DA7 4AJ M6ASM
DA7 4EP G4DDP
DA7 4EP G8BXC
DA7 4JL G4XKV
DA7 4JZ G4KOW
DA7 4LF G4DFI
DA7 4LY M3JPG
DA7 4PG G6CUE
DA7 4PQ 2E0WUF
DA7 4QD M6BZG
DA7 4RL G0FAS
DA7 4ST G8CXI
DA7 4TZ M0HFR
DA7 4UE M1CZA
DA7 4UX G7MJJ
DA7 5BT G7UUB
DA7 5BT M3FTI
DA7 5DG G3IJW
DA7 5DG G6ENO
DA7 5DZ G3KHR
DA7 5HG M3USW
DA7 5JU G0PXT
DA7 5JU M3KVJ
DA7 5NP G8JZT
DA7 5QD M6LXR
DA7 5SL G4MDG
DA7 5SL G7GPI
DA7 6AF G4DUM
DA7 6NL G0UAO
DA7 6QU G4GZN
DA7 6SG M6BLT
DA8 1DZ 2E0GPG
DA8 1DZ M0YUG
DA8 1DZ M6GPG
DA8 1LU G4LSU
DA8 1NL G3BHF
DA8 1NN G4MHJ
DA8 1NN M1AJT
DA8 1QW G3GEV
DA8 2AQ M1CXN
DA8 2DR 2E0XUZ
DA8 2JD M3VNX
DA8 2JG 2E0WTZ
DA8 2JG M0CRY
DA8 2JH G4EGU
DA8 3BA M0LHA
DA8 3BL G3NRZ
DA8 3NG 2E0JET
DA8 3QT M0SHI
DA9 9PG G8MHI

## DD

## (Dundee)

DD1 2AP MM0GDI
DD1 2JA 2M0KAU
DD1 2JA MM6KAU
DD1 4LY GM0TOF
DD10 0BS MM0HZO
DD10 0DJ GM4AFF
DD10 0DJ MM3PDC
DD10 0DN MM6MJC
DD10 0HW MM6ZZG
DD10 0HX GM4PKJ
DD10 0HX GM8ZEQ
DD10 0RB MM1DZW
DD10 0RB MM3RGH
DD10 0RY MM3NWF
DD10 0SB GM4TXN
DD10 0SL GM0ENQ
DD10 0SR GM1MLS
DD10 0SW GM3YAO
DD10 0TG GM1TDU
DD10 0TT GM0WRV
DD10 8AZ MM1COS
DD10 8EX GM4UTK
DD10 8PW MM3DFG
DD10 8TW GM3COQ
DD10 9AQ MM6XGT
DD10 9DD GM5CGA
DD10 9EJ GM0ARH
DD10 9RR MM0BIX
DD10 9TS MM3HAF
DD11 2DR GM4YWU
DD11 2LZ GM4SXJ
DD11 2NX GM0MIW
DD11 3EF 2M0GXZ
DD11 3EF MM0KGS
DD11 3EF MM6GZS
DD11 4BH MM3NJV
DD11 4EZ 2M0DOI
DD11 4EZ MM6GFG
DD11 4RA GM4YWS
DD11 4RH GM0WNS
DD11 4SR GM1XHZ
DD11 4SX GM0SHD
DD11 4SX GM3GBZ
DD11 4UX MM6IBB
DD11 5RH MM6CWB
DD11 5RH MM6JNB
DD11 5SS GM1MON
DD11 5ST 2M0MAV
DD11 5ST MM0LBX
DD11 5ST MM6LAD
DD11 5SY GM3OBG
DD2 1NY GM0IRZ
DD2 1NY GM6KKP
DD2 1PJ GM8JVZ
DD2 1QN MM6DBJ
DD2 1TU GM0ROU
DD2 2BU GM8UMN
DD2 2BU MM3XGS
DD2 2EZ 2M0LEW
DD2 2EZ MM0LWS
DD2 2EZ MM6LEW
DD2 2NH MM6AKQ
DD2 2PW GM3OQI
DD2 2PW MM0AKX
DD2 2RZ 2M0MFK
DD2 2RZ MM6FIF
DD2 2SZ 2M0RRS
DD2 3AF MM6RAM
DD2 3BF MM6YYB
DD2 3JR GM3WPA
DD2 4HP MM6BXH
DD2 4PT MM3KDN
DD2 4TT 2M0XDS
DD2 4TW GM8LON
DD2 4UA GM4JCM
DD2 4XB GM4VXM
DD2 5AH 2M0SEF
DD2 5DL GM0KKE
DD2 5PX GM4LUD
DD2 5PZ 2M0INE
DD2 5PZ MM6PAU
DD2 5QJ GM1CMF
DD2 5QN GM0TAY
DD2 5QN GM4FLP
DD2 5RB GM3LIW
DD3 0BN GM4ZFS
DD3 0JR MM3NPG
DD3 0LT MM6AAR
DD3 0PH GM4API
DD3 0QN 2M0CVK
DD3 6AQ MM0DTW
DD3 6DD GM3ZXB
DD3 6HR 2M0RRT
DD3 6HR MM6BJJ
DD3 8AF 2M0XZX
DD3 8JW GM1BOT
DD3 8PY GM5BDW
DD3 9LH 2M0MGY
DD3 9NZ MM3NRX
DD3 9RG GM0FSW
DD4 0AD GM1KCH
DD4 0NR MM0DXD
DD4 0NR MM3JFW
DD4 0PP GM7ONJ
DD4 0PW GM0GYQ
DD4 0QU GM0NLU
DD4 0TA GM3WJE
DD4 0TN 2M0RTA
DD4 0TN MM0RTT
DD4 0TN MM6RBT
DD4 0UJ GM7OWU
DD4 0XJ MM6SGQ
DD4 7EL GM0PIV
DD4 7LH MM3HKG
DD4 8AP MM6GID
DD4 8RP MM6DDX
DD4 9DB GM3ZTP
DD4 9DZ MM3PXG
DD4 9EX GM1VWA
DD4 9HQ GM0DNH
DD4 9LP GM0ISA
DD4 9ND MM0TMG
DD5 1QT GM3NHQ
DD5 1QW GM1JTK
DD5 1QW GM7GOE
DD5 2EU GM4CAB
DD5 2RE GM1MUY
DD5 2RJ GM6BML
DD5 3AT GM0URU
DD5 3BN GM0RKU
DD5 3BT 2M0DOL
DD5 3BT MM6MIS
DD5 3EF GM0CQL
DD5 3JG GM4JPZ
DD5 3LE GM3YVX
DD5 3PD GM8OEG
DD5 3QN GM4UGF
DD5 3RE MM3NLH
DD5 3WN MM0DUN
DD5 3WN MM3HKE
DD5 3WN MM3SYU
DD5 4HG GM7AUX
DD5 4HS MM6KLT
DD5 4HT GM4MUZ
DD5 4LD MM1CFC
DD5 4SW GM0VOL
DD5 4TS MM0DRA
DD6 8AP GM0TGG
DD6 8DT GM4FSB
DD6 8HL GM3MOR
DD6 8HL GM4RXW
DD6 8JH MM0BTD
DD6 8LA MM3GPF
DD6 8LF GM0CNW
DD6 8ND GM8RPE
DD6 8NP GM7IIL
DD6 8PJ MM0HGN
DD6 9DL GM3EFH
DD6 9JD 2M1IBX
DD6 9JZ GM7DPI
DD6 9LG GM4RDI
DD6 9NE MM3FZI
DD6 9NX GM4LOH
DD7 6DQ MM1ELE
DD7 6DT MM3IZO
DD7 6EX GM4FEI
DD7 6HT MM1EDY
DD7 6HW GM6RAK
DD7 6JY 2M0NGO
DD7 6JY MM3XLO
DD7 7BR MM0GGD
DD7 7BR MM1DSD
DD7 7QF MM0EQE
DD7 7QF MM1EQE
DD7 7QQ MM3XFP
DD7 7SQ GM4JEJ
DD7 7TB GM0EFT
DD8 1AD GM3KEZ
DD8 1DQ GM1JWJ
DD8 1EP MM0CXA
DD8 1JR GM0DGK
DD8 1UE MM6BHS
DD8 1UF GM3ZBR
DD8 1UF GM4GOW
DD8 1UN GM0BTK
DD8 2JN GM0OLF
DD8 2PQ GM8RSC
DD8 2SP GM0TWB
DD8 2UZ GM4RWE
DD8 3AG GM4XKP
DD8 3EY GM6AAJ
DD8 3EY MM0BSX
DD8 4BJ MM3PFA
DD8 4PG GM8LYO
DD8 5ab MM3KVV
DD8 5DB GM0CDC
DD8 5LD GM4VHS
DD8 5NT GM0BKC
DD8 5NT GM4AAF
DD8 5NT MM6BKC
DD8 5NT MM6GYL
DD9 6JD GM4LFL
DD9 6RQ GM0JZV
DD9 6SD GM7DAP
DD9 6UH GM3KC
DD9 6UH MM0ERK
DD9 7QE MM0BDA
DD9 7SZ GM6RGD

## DE

## (Derby)

DE1 1DU M6DEZ
DE1 3RW M1CWY
DE11 0DS G1PKR
DE11 0EB G3OMT
DE11 0JR G4WPE
DE11 0JR G6SLH
DE11 0LY M0APK
DE11 0LZ G6DCS
DE11 0NB G1PKV
DE11 0NB M3ASI
DE11 0RX G4EWK
DE11 0RX M1FBF
DE11 0SL M6CGM
DE11 0SP 2E1FLD
DE11 0SR M3THN
DE11 0TG G7EJO
DE11 0UU G8UZQ
DE11 0UW G0OAP
DE11 0UW M3XUJ
DE11 7BT M6GPR
DE11 7EG G0VVF
DE11 7EZ M0URJ
DE11 7HG M6MAX
DE11 7QU G6NZL
DE11 7QU M6YAD
DE11 8AA 2E0EWL
DE11 8AA G4RJO
DE11 8BD G8VBC
DE11 8BG G0WHC
DE11 8BX M6RVN
DE11 8FS M6MLG
DE11 8HD 2E0XPK
DE11 8HD M3XPK
DE11 8LX G1ZMW
DE11 9AF M6XPC
DE11 9AQ M6EZF
DE11 9AT G7AYB
DE11 9BQ G7LAS
DE11 9BW 2E0FGM
DE11 9BW M6HAG
DE11 9DZ G1OWK
DE11 9EW G0KVB
DE11 9FB 2E0BGV
DE11 9FB G7JHW
DE11 9FB M3FKV
DE11 9FB M3JQY
DE11 9FH G1SNQ
DE11 9FH G4ZDE
DE11 9HD G8FVC
DE11 9PH M1BFV
DE11 9QN M6GHV
DE11 9QU M6CDD
DE11 9QU M6DYP
DE12 6BL G4AYS
DE12 6DN M6GLJ
DE12 6EX 2E0DAQ
DE12 6EX M3VWD
DE12 6HH M6ICA
DE12 6HQ G1YBM
DE12 6JJ M3KBY
DE12 6JJ M3KRB
DE12 6JL M3XWP
DE12 6LU G0SRC
DE12 6LU G4CRT
DE12 6PL M6XAQ
DE12 6PZ M3UDF
DE12 7AS G1OAX
DE12 7BD G1MVQ
DE12 7EG G6JTV
DE12 7JG M3XUT
DE12 7JZ G7MGX
DE12 7JZ M0PCA
DE12 7JZ M1LYN
DE12 7NA G7KVZ
DE12 7PW G6MYO
DE12 7QG G7EIK
DE12 7QU G0HYR
DE12 8DL M0APD
DE12 8ES G1VQH
DE12 8JW G7EHU
DE12 8LB G0OES
DE12 8NJ G4HIW
DE13 0AD G7JEJ
DE13 0AL G6NSQ
DE13 0BT M6RFF
DE13 0EE G4YFZ
DE13 0EE G4GKZ
DE13 0FD M6BSW
DE13 0FN M6IGZ
DE13 0FQ M6OML
DE13 0GZ 2E0GPC
DE13 0GZ M0YIG
DE13 0GZ M6GPC
DE13 0HA M3BJE
DE13 0HZ 2E0HRH
DE13 0HZ M0PNC
DE13 0HZ M6JMB
DE13 0HZ M6KSW
DE13 0JD G4RAC
DE13 0JH 2E0PKS
DE13 0JH M6GOU
DE13 0JP G1ATA
DE13 0LD G1OYZ
DE13 0LD G1OZD
DE13 0LR M1CBO
DE13 0ND M1CDT
DE13 0NN M3GMD
DE13 0NT G6TOC
DE13 0NT M3RCW
DE13 0NU G4XKL
DE13 0NY G4IPV
DE13 0RD M3FKN
DE13 0RD M3IEW
DE13 0RF G4HBY
DE13 0SQ M3HRV
DE13 0UU G4VPF
DE13 0WJ G1PRW
DE13 7AJ G3CRH
DE13 7EZ G1DQU
DE13 7EZ G4UJW
DE13 7HP G6NAJ
DE13 8AB G7VSM
DE13 8DS G7KIF
DE13 8DU G0FVX
DE13 8JR G4RWD
DE13 8NT G4CHI
DE13 8PT G4ICZ
DE13 8SW G7ORN
DE13 9AA G6EIH
DE13 9AB G8VBA
DE13 9AR G0JLU
DE13 9AR G4LIR
DE13 9BJ G0FNH
DE13 9BJ G1OYU
DE13 9DE G4SXE
DE13 9DS G4KBP
DE13 9DY 2E1IIL
DE13 9EG G0SHO
DE13 9JR G0AOD
DE13 9JR M3RHK
DE13 9LE 2E1FZK
DE13 9LE G3BZB
DE13 9NH G8OZP
DE13 9QD G0JEE
DE13 9RT G3TOY
DE14 1BS G6TJK
DE14 1BW M3WJY
DE14 1QS M6FDY
DE14 2EX 2E0SMX
DE14 2EX M6TDO
DE14 2HJ G8UUR
DE14 2NB G8WSY
DE14 2NP G1NBO
DE14 2SN M3KQR
DE14 2SW M3VLH
DE14 3BY M0HFA
DE14 3DA G6FLY
DE14 3DY M3AUF
DE14 3DZ G8HZI
DE14 3FQ 2E0GSB
DE14 3FQ M3VDF
DE14 3GJ M3OTS
DE14 3HX G1SQA
DE14 3HX G7ENM
DE14 3JE 2E0CGU
DE14 3JF G6AZL
DE14 3LR G1BEJ
DE14 3SB 2E0JBY
DE14 3SB M6SRI
DE14 3TX 2E0KLR
DE15 0AB G4DIH
DE15 0AD G0WRM
DE15 0AD G4HZG
DE15 0AD M6JGR
DE15 0AR G6OVA
DE15 0BW 2E0CGJ
DE15 0BW M0KAJ
DE15 0BY G0VUC
DE15 0EJ G0IDD
DE15 0PS G3HKN
DE15 0PT G0CQS
DE15 0QJ G4PGJ
DE15 0QW G0TZM
DE15 0SL G4ELG
DE15 0TT G6JPQ
DE15 9BJ M0AGS
DE15 9BN 2E0ZAI
DE15 9BN M6MGJ
DE15 9EB G0HJR
DE15 9EW 2E0KPO
DE15 9EW 2E0NBR
DE15 9EW M0KPO
DE15 9EW M0SCW
DE15 9EW M3KPO
DE15 9EY G3CWT
DE15 9EY G4PGX
DE15 9FR G6BRY
DE15 9QN G3NFC
DE15 9QN G8EKG
DE15 9QN M6AJX
DE15 9QW M3XXG
DE15 9RL 2E0TOX
DE21 2DG G4TAY
DE21 2DH M6TDA
DE21 2HY 2E0ZCG
DE21 2HY M1BUQ
DE21 2HY M3ZCG
DE21 2JP 2E0WCE
DE21 2JP M6MHW
DE21 2UG 2E0RCB
DE21 2UG M3LFP
DE21 4DX M0NMD
DE21 4GA 2E0TGV
DE21 4GB M3DGN
DE21 4HQ G3HCS
DE21 4HQ G8HCS
DE21 4JL G1VQB
DE21 4JY M6FFI
DE21 4LG 2E0TCU
DE21 4LG M6WCA
DE21 4LY 2E0JNH
DE21 4LY M6DDY
DE21 4QS M0HDX
DE21 4RF G6VFA
DE21 5AP G4AOA
DE21 5AR 2E1ICU
DE21 5AR G4IBW
DE21 5AR M6RJF
DE21 5AX G4FQZ
DE21 5BN G4RAR
DE21 5BS G0NPU
DE21 6DZ M3JVW
DE21 6FW M0CNY
DE21 6FX G4TBK
DE21 6FZ M1CWV
DE21 6LG M6YDC
DE21 6LW M6GDL
DE21 6NX G1NZN
DE21 6NX G7AFW
DE21 6PH G1DHY
DE21 6PL G1UJX
DE21 6RG G7MUY
DE21 6SH G4EYN
DE21 6SJ 2E0EBQ
DE21 6SJ M6GWQ
DE21 6TR G1XFE
DE21 6TZ 2E0BFA
DE21 6TZ M3MVV
DE21 6UR 2E0OES
DE21 6UR M0TRY
DE21 6UR M3XUG
DE21 6WX G8BFA
DE21 7AT M0KEF
DE21 7AT M6PRM
DE21 7DS M5GAC
DE21 7DZ G4UMT
DE21 7EA G3PRF
DE21 7EN G3OCA
DE21 7EN G3ZBI
DE21 7ES G3JFT
DE21 7FX M6MGT
DE21 7GL G1UYZ
DE21 7GT M0BJT
DE21 7JW G6WYD
DE21 7JZ G3OXN
DE21 7LR 2E1GOC
DE21 7LR G0IDL
DE21 7LX G6JPS
DE21 7QA G6FVJ
DE21 7RG 2E0SYD
DE21 7RG M3IIP
DE21 7TP M3NTZ
DE22 1EQ G4OMT
DE22 1EY G8RBK
DE22 1GE M3AWQ
DE22 2BJ G3SZJ
DE22 2BJ G4EYM
DE22 2EH G6NLE
DE22 2GN 2E1DIA
DE22 2GW G6DNH
DE22 2HA 2E1FBS
DE22 2HF G1BEJ
DE22 2HF G4FPY
DE22 2HG G0RII
DE22 2HH M3WMS
DE22 2JN G0GDL
DE22 2LN 2E0GWB
DE22 2LN G3YOO
DE22 2LN M0HLP
DE22 2LN M6BJM
DE22 2LN M6GWB
DE22 2NA G6WXJ
DE22 2NB M6ZTB
DE22 2SN M3ULN
DE22 2TA G4VLJ
DE22 2UB G4VWB
DE22 2XP G4LOF
DE22 3AS G0VVA
DE22 3EF G6EBL
DE22 3GL G7ODZ
DE22 3JT M6PTY
DE22 3RG G8PXA
DE22 3UE G4CZH
DE22 4AQ M6JDM
DE22 4AY 2E0RLG
DE22 4AY M6RLG
DE22 4BW G3SMV
DE22 4FL G7RWW
DE22 4HN G3PWS
DE22 4JU 2E1ADJ
DE22 4JX 2E1EDD
DE22 4LQ M6ZLL
DE23 1BZ G4CMZ
DE23 1DN 2E0KDH
DE23 1GA G6OEI
DE23 1GN 2E0MNG
DE23 1GN M6MNG
DE23 1PR M3RCI
DE23 1RH G6XOR
DE23 1WN G0IOZ
DE23 2UB G4EJE
DE23 3RL G4COR
DE23 3XU M1MPK
DE23 4BE M6ZAV
DE23 4BJ M3JBZ
DE23 4EJ M6GNV
DE23 6BL G7SJS
DE23 6BW 2E1DBT
DE23 6DA G7GXE
DE23 6DL G3URU
DE23 6EB M6JDF
DE23 6FB 2E0OJB
DE23 6FB M3NZK
DE23 6FT 2E0KRS
DE23 6FZ M0PHE
DE23 6GA M0HLS
DE23 6GA M6BFO
DE23 6GT G3ERD
DE23 6GT G3KQF
DE23 6HB G0WQY
DE23 6HB G8MXT
DE23 6JD G8TNE
DE23 6JD M3JBB
DE23 6JD M3TNE
DE23 6JY M0DVQ
DE23 6NF M0OTS
DE23 6NF M5EHG
DE23 6NQ G6CWW
DE23 6PF G1VAB
DE23 6PS 2E1CZJ
DE23 6PS G8BAV
DE23 6QW G7SWW
DE23 6TE M0GPK
DE23 6TG G1DDI
DE23 6TG G3RLO
DE23 6TG G6XGF
DE23 6XL G3FUJ
DE23 8AX 2E0DLP
DE23 8AX M0OAC
DE23 8BP 2E0GCG
DE23 8BP M6GBE
DE23 8BP M6RDI
DE23 8LH M6EPJ
DE23 8PR M6USA
DE24 0AQ G8YCK
DE24 0AU G1CUH
DE24 0BP G7LPV
DE24 0BU 2E0RUZ
DE24 0BU M0RUZ
DE24 0BU M6RUZ
DE24 0DJ G0PRY
DE24 0ER G0PHC
DE24 0FJ G6SOX
DE24 0FT G0DKA
DE24 0GH G1HEU
DE24 0GQ G4FAE
DE24 0HZ G6KUI
DE24 0JU G4YVV
DE24 0LF 2E0VRR
DE24 0LF M6RRV
DE24 0LH G0BIE
DE24 0LP G7SZG
DE24 0PP G4AKE
DE24 0PX G8BFC
DE24 0TA G4LNQ
DE24 0UQ 2E1HVZ
DE24 0UY M1EDL
DE24 1AE 2E1GDF
DE24 3AA G3WSM
DE24 3AG M1WTL
DE24 3AG M3WTL
DE24 3BH 2E0WHO
DE24 3BH G0WAC
DE24 3BQ G1VAC
DE24 3BS G1LKK
DE24 3DZ G0IWF
DE24 3EL G8VBW
DE24 3ES G6XZO
DE24 3HE G0CPO
DE24 3HH M1JKB
DE24 3JA G4PTZ
DE24 5AF 2E0WBX
DE24 5AF M6WLR
DE24 8AT 2E0JWW
DE24 8AT M3NVG
DE24 8AZ G7BJB
DE24 8BD G4PDE
DE24 8DA M3PRS
DE24 8DU M6SPK
DE24 8EX M6PKT
DE24 8GT G0RUR
DE24 8JR M6WET
DE24 8NF G0TJT
DE24 8NP G6UGT
DE24 8NS M6TLK
DE24 8RN G4SNN
DE24 8WD G7VNO
DE24 9BT G1IWT
DE24 9DL M6TIO
DE24 9GY 2E0FOR
DE24 9GY M6FOR
DE24 9HE M6HFQ
DE24 9LZ M1AJQ
DE24 9NG G6CXI
DE24 9PD G4YYI
DE24 9PF 2E0RTE
DE24 9PF M6RCG
DE24 9PT G6XEN
DE24 9QT M6KKM
DE3 0DF G4HJL
DE3 0DU M3JIK
DE3 0ED M0WSB
DE3 0EE G4YIZ
DE3 0PA G7BHY
DE3 0PH M3LFG
DE3 0QQ G4ZHD
DE3 0QT G4MBH
DE3 0RD G0MSS
DE3 0RD G7SVQ
DE3 0RG M1FAJ
DE3 0RR G3TOV
DE3 0RR G8OWA
DE3 0RS M3JKG
DE3 0TB M3JXI
DE3 4LQ G7BJC
DE3 9AH G8GBU
DE3 9AH G8KGC
DE3 9AQ G6MWS

DE3 9BD M1MTV
DE3 9FJ G4KRW
DE3 9FL G7WGP
DE3 9FP G6WZE
DE3 9FT G8YNK
DE3 9FY G1UZS
DE3 9GH G7GJI
DE3 9GH G8QZ
DE3 9GH M3DPS
DE3 9GQ G0DMK
DE3 9GT G0CRN
DE3 9HB G4KOJ
DE3 9HZ G7GEX
DE3 9JT G1KKD
DE3 9JW G2DAN
DE3 9LN G4BKO
DE3 9LU M3JZE
DE4 2AH M3FUV
DE4 2GG G7VWN
DE4 2GL G0KKD
DE4 2GX M0MIK
DE4 2HP G0OTJ
DE4 2JJ G1CJC
DE4 2JW G4XTK
DE4 2PW G6HWR
DE4 3BA 2E0VKS
DE4 3BA M0VKS
DE4 3BA M3VKS
DE4 3BT G4ZEY
DE4 3EB M6PBU
DE4 3EP G8LXN
DE4 3EU G7NBJ
DE4 3EW 2E1AWZ
DE4 3GP G7NLJ
DE4 3HE G3RBP
DE4 3QD M3TMI
DE4 3QU G0FSB
DE4 3QY 2E0TVX
DE4 3QY M0TVX
DE4 3QY M3TCD
DE4 3TE G0BJD
DE4 4AD G4UIQ
DE4 4AD G6NMU
DE4 4EJ 2E0UIQ
DE4 4EJ M3NMU
DE4 4EX M0CVS
DE4 4FF G6PBN
DE4 4FR G1SGP
DE4 4NG M6RJR
DE4 4PG G0MMO
DE4 5AR M6BOR
DE4 5BH G8DLP
DE4 5DG G4HTW
DE4 5DJ G0FVU
DE4 5EG G3VLF
DE4 5FF G8RNU
DE4 5GU M1CFZ
DE4 5HN G4UUQ
DE4 5LA G0OWC
DE4 5LT M3OGD
DE45 1BH G4JIK
DE45 1DD G6EAZ
DE45 1DX G7ABR
DE45 1DX M3DXD
DE45 1FG M0PKV
DE45 1JB G4GWX
DE45 1NJ G3VOT
DE45 1QQ G3SPV
DE45 1RE M6WIL
DE45 1SG G6GMH
DE45 1TP G6CSC
DE5 3AS G8ZYC
DE5 3BU M6AEY
DE5 3EP G4MSG
DE5 3FG G0WWO
DE5 3FG G1OLT
DE5 3GH G4TWW
DE5 3HD G1YPT
DE5 3JL 2E1HFN
DE5 3LJ G0NYM
DE5 3PJ 2E0GRM
DE5 3PY G4GGL
DE5 3RE G3NJX
DE5 3RR G3ZYC
DE5 3RY M6RWE
DE5 8JG G0NNU
DE5 8PQ G4DXY
DE5 8RF 2E1AZA
DE5 9QG M3FYM
DE5 9QN G8HNT
DE5 9RB G8ZIY
DE5 9SP G3YQL
DE55 1BL G1IGE
DE55 1BU G7BGT
DE55 1BW G3MHR
DE55 1DJ 2E0MFS
DE55 1DJ M6SMX
DE55 1ES G8DXT
DE55 1LG M6HQN
DE55 1LN G6KPW
DE55 1RL 2E0IPA
DE55 2AJ G7DGF
DE55 2AJ M0GNU
DE55 2AS G1UAZ
DE55 2BJ G0DGQ
DE55 2BR M5RST
DE55 2DN G4CHM
DE55 2EJ G1BFK
DE55 2EP M0DAG
DE55 2ER M0HXO
DE55 2HE 2E0CND
DE55 2HE M3CND
DE55 2HS G8VSN
DE55 2JD G7SJD
DE55 2LD G0GHD

DE55 3AP G6TRQ
DE55 4AW M3JAF
[illegible] [illegible]
DE55 4AF G4KBI
DE55 4AG G1NWH
DE55 4AG G7ANO
DE55 4BL G1OGC
DE55 4BP G3JWQ
DE55 4EX M3TFK
DE55 4JD M0DEN
DE55 4JT M6JSR
DE55 4LA 2E1DBQ
DE55 4LA M0SBA
DE55 4LH M1GHT
DE55 4LT G4OHV
DE55 4NF G6TLA
DE55 4PB M6PDK
DE55 5AJ M0TRP
DE55 5HD G4AMF
DE55 5HT G7BHW
DE55 5HU M3EAT
DE55 5JJ M3EUW
DE55 5JL G6LSD
DE55 5JL M5SRE
DE55 5JN M1DDI
DE55 5LL G4VNG
DE55 5LQ G4PCL
DE55 5LT M3VLO
DE55 5LU G0DXT
DE55 5NA G4THI
DE55 5NR M6HFR
DE55 5QH G4EKD
DE55 5SH G8PWA
DE55 5TD 2E0WYT
DE55 5TD M0WYT
DE55 5TD M6WYT
DE55 5TT G6ZDB
DE55 5TY G6AZE
DE55 5TY M0BAY
DE55 5TY M0BMY
DE55 6BT 2E0RNU
DE55 6BT M0RNU
DE55 6BT M6RNU
DE55 6DX M0SMG
DE55 6EH G3LGK
DE55 6GW M6KLX
DE55 6JG G1AII
DE55 6JW G4DTP
DE55 6LD M3ZHW
DE55 6LH G7GLL
DE55 7DG G4NID
DE55 7DG G4UTN
DE55 7EF G0VBX
DE55 7ER G1XJM
DE55 7FL M3TXV
DE55 7HT G7SMZ
DE55 7JN G4LSV
DE55 7JP G3GRL
DE55 7JX 2E0SBX
DE55 7JX M6HMW
DE55 7JX M6NEO
DE55 7JX M6SME
DE55 7JY G7TYP
DE55 7JY M3SUJ
DE55 7LP G4EWZ
DE55 7NH 2E0CEM
DE55 7NH M0ILM
DE56 0EA G1RYQ
DE56 0EY G1BZE
DE56 0HD M0WMX
DE56 0HL M3FXQ
DE56 0HN G0ORC
DE56 0PF 2E0CTM
DE56 0PF G7EMZ
DE56 0PF M3UJY
DE56 0PG G4ANV
DE56 0PQ G0RLJ
DE56 0PY G8IHA
DE56 0QH G7HHZ
DE56 0QR G4RVL
DE56 0SH M6GWF
DE56 0TG G4NXL
DE56 0UB G6SKK
DE56 1BX G8VYO
DE56 1EE G0MGX
DE56 1EJ G3ZYD
DE56 1FR M6GKM
DE56 1GH 2E0YPG
DE56 1HH G8GJC
DE56 1JG G4VSI
DE56 1LX 2E1HES
DE56 1NE G0ROD
DE56 1PD G0TTY
DE56 1RE G1GMX
DE56 1RE M6FED
DE56 1UT M1EAI
DE56 1UZ G8FPA
DE56 2AL G3PDD
DE56 2BA G4UBR
DE56 2EF 2E0TDD
DE56 2EF M6NIT
DE56 2EU G8HOU
DE56 2FW M0DMF
DE56 2GG G2DJ
DE56 2GG G3VGW
DE56 2GR G0UOV
DE56 2GR G8GTD
DE56 2GS G0JNK
DE56 2LH 2E0FGY
DE56 2LH G0CXD
DE56 2LH M6FGY
DE56 2TR M0NOV
DE56 2UW G0AEU
DE56 4BH G3SNR
DE56 4DF G4RVU
DE56 4DP G4DJP
[illegible] [illegible]
[illegible] [illegible]
DE56 4FX G0TQC
DE561FP G6GDD
DE6 1AT G4HIJ
DE6 1BJ G1NDL
DE6 1BR G3TVU
DE6 1DF G1MUT
DE6 1EX G0DLS
DE6 1HR G8PKM
DE6 1LZ G0BPZ
DE6 1QB G6YWV
DE6 2AS G0LAU
DE6 2EH G3KAG
DE6 2GY G8NOP
DE6 2LP G0CQP
DE6 2LP G7KRM
DE6 3AE G3OQT
DE6 3AG M3IJF
DE6 3BZ G7EJK
DE6 3EN G4DKV
DE6 3FE G6RBM
DE6 4JS G3UBS
DE6 4LP M0BAJ
DE6 4NG G7DDR
DE6 4NJ G7EUT
DE6 4NN G3IVF
DE6 4PF G8EBM
DE6 5BE M3MQI
DE6 5HP 2E0MGA
DE6 5HP M6MJI
DE6 5JZ G6YWN
DE6 5NY G3LKV
DE6 6NB G4MIT
DE65 5DL G1PUU
DE65 5ER G4PYI
DE65 5FD 2E1BUM
DE65 5FX 2E0JWC
DE65 5FX G0WOC
DE65 5FX M0KCA
DE65 5FX M6JWC
DE65 5FY G1JJT
DE65 5HP G7BJE
DE65 5HQ M6YNK
DE65 5JG M3LFV
DE65 5LP G4RJA
DE65 5QQ G1TKY
DE65 6DW G7ILL
DE65 6EB G0HLI
DE65 6ES G6CZD
DE65 6FX M3XHN
DE65 6GP G0MRY
DE65 6GR G0DMS
DE65 6GR G8VAN
DE65 6GT G8WSQ
DE65 6HD G4PWY
DE65 6HL G0SUT
DE65 6HT 2E0DWE
DE65 6HT M6GJD
DE65 6HY G3ZIF
DE65 6JP M6KMC
DE65 6LA G1NUN
DE65 6WA G0HHL
DE7 4DA G1ZHZ
DE7 4DF G4AIB
DE7 4DF M0AIT
DE7 4DL G6CDT
DE7 4EH 2E0GBU
DE7 4EH M6TOZ
DE7 4EZ M6HNI
DE7 4HD M3IZH
DE7 4HD M3IZM
DE7 4HQ M0CQF
DE7 4JY G3XRD
DE7 4LE 2E1GOK
DE7 4LZ G0KBN
DE7 4NE M0BCJ
DE7 4PW G0ITL
DE7 5AP 2E0CTC
DE7 5AT G0LUI
DE7 5AT M6LUC
DE7 5EF G3IFX
DE7 5EX G4XCK
DE7 5NZ M0BCC
DE7 5PP M0GLJ
DE7 5RB M0BQT
DE7 6AU G4UPX
DE7 6AW G3RKZ
DE7 6AX 2E0VGK
DE7 6AX M3VGK
DE7 6DD G1GNP
DE7 6DE G8DKV
DE7 6DX G4OUR
DE7 6EG G4NOB
DE7 6EZ G3NNN
DE7 6FR G4LPF
DE7 6GB 2E1GMV
DE7 6GB G4DAM
DE7 6GB G4EDD
DE7 6GU G8KSW
DE7 6HW G0HJF
DE7 6LW 2E0TDT
DE7 6LW M6GKK
DE7 6PE M3ZYV
DE7 6XB G7UUP
DE7 8AL 2E1BFP
DE7 8AW G8WRY
DE7 8EY G1DMH
DE7 8NL G0JVB
DE7 8PW 2E0CMR
DE7 8PW 2E0FSH
DE7 8PW M6CMR
DE7 8PX G6KBC
DE7 8QP G0RTQ
DE7 8QP M3SQH
DE7 8SJ 2E0DTF
DE7 9QJ M0WCX
[illegible] [illegible]
DE7 9HE G1YEV
DE7 9HJ M3AQK
DE7 9HJ M6EPD
DE7 9JW G3ZDK
DE7 9LF M1DYC
DE72 2AU G0JKC
DE72 2AU G0NHR
DE72 2BA G0PBM
DE72 2BJ G4NYB
DE72 2DF G4RBZ
DE72 2DQ 2E0GEN
DE72 2DQ M6CSX
DE72 2GL 2E0AVW
DE72 2GL M0LJC
DE72 2GS G3ZOW
DE72 3AQ 2E0RRF
DE72 3DF M6KMI
DE72 3DZ G4EHX
DE72 3EE G4EAX
DE72 3GN G8KEA
DE72 3HR G7TLL
DE72 3JJ M0SDS
DE72 3LN G8YZC
DE72 3NP G0DPQ
DE72 3PN G4OQL
DE72 3PN G6CHI
DE72 3RP G1DGY
DE72 3RR G0JWE
DE72 3TE G1SGZ
DE73 1AG M3IHU
DE73 1BW G3NYZ
DE73 1BX G0MZJ
DE73 5AS 2E1BHC
DE73 5BX M1BGT
DE73 5SJ G1SPT
DE73 5SU G7CYF
DE73 6PB 2E0PFT
DE73 6PF G7DZR
DE73 6PS G6OCE
DE73 6RD G4LPZ
DE73 6RP G4XYD
DE73 6ST G4DUT
DE73 6XH G1PER
DE73 7GY G0JCK
DE73 7HD M3IIN
DE73 7JJ G3XER
DE73 8AG G4FQI
DE73 8AG G8DBO
DE73 8BD G8MZA
DE73 8BX G8NFM
DE73 8EB 2E0GRL
DE73 8FG M3ZZL
DE73 8GA 2E0WRI
DE73 8GA M6PXW
DE73 8JH M3HBT
DE74 2DN G0SOU
DE74 2DN G4IRH
DE74 2DS G0WUG
DE74 2DT G0RDY
DE74 2EJ G4TCA
DE74 2EL G6GBU
DE74 2GQ G3JRS
DE74 2JG 2E0MPJ
DE74 2JG M3WQX
DE74 2LF G0EKK
DE74 2LG G7FLS
DE74 2NG 2E1CPI
DE74 2NG G8BAG
DE74 2NJ G4DTZ
DE74 2PG 2E0RJU
DE74 2PJ G1ZEK
DE74 2PQ G3KTP
DE74 2QQ G8PTW
DE74 2RX G4UVG
DE74 2SR G7PNG
DE74 2XA G0VNQ
DE74 2XB G1LCR
DE74 2XB G1SPA
DE75 7AN G7LGY
DE75 7BN 2E1FAT
DE75 7BN G6XTD
DE75 7FW G7MLT
DE75 7GQ G4DMF
[illegible] [illegible]
DE75 7HF G7JJC
DE75 7HG G3LZC
DE75 7JX M1EUR
DE75 7LY G8IOW
DE75 7NE M6BIA
DE75 7NJ G4GBC
DE75 7NJ M3MBI
DE75 7PQ G0FEZ
DE75 7PQ G4TYY
DE75 7PQ G4ZNI
DE75 7PQ M0HFF
DE75 7PZ G1GKK
DE75 7QB G8UCC
DE75 7TY M3UFX
DE75 7UN 2E0CID
DE75 7UN M6BCZ

## DG

## (Dumfries)

DG1 1BT G1OXQ
DG1 1PP G4JKB
DG1 1QN G1JVU
DG1 1RB M5AKZ
DG1 1RT G7TUD
DG1 1RZ G2AJW
DG1 1UZ 2E0RDG
DG1 3HE G4JR
DG1 3HF G0EFU
[illegible] [illegible]
DG1 3LH M6BXF
DG1 3RJ G3OXK
DG1 3SL M6BNT
DG1 4BU G4TNJ
DG1 4DN G0WIZ
DG1 4DW G4RJF
DG1 4EW G0BQQ
DG1 4HN M6EIX
DG1 4HU G8VBX
DG1 4LU M3OYW
DG1 4NA M3AYS
DG10 9JU G3OFT
DG11 1NF M3MID
DG11 1RL 2E0NJS
DG11 1RL M0NJS
DG11 1RL M6LTC
DG11 1SA G8UPI
DG11 2BA M0WHA
DG11 3AE G3UHT
DG11 3AT G0WMH
DG11 3BA G1VLA
DG11 3DU G0CBC
DG11 3JH G0BVG
DG12 5QL G3YLU
DG12 6BU G6LYJ
DG12 6HX G4WHA
DG12 6HX G6SMW
DG12 6QX G0FQV
DG12 6QX G6TVR
DG12 6SZ G0PMW
DG13 0AX G4BDJ
DG13 0HJ G6NJL
DG13 0JW G1GES
DG14 0RZ G6LJE
DG14 0UY G1CUC
DG14 0XY M6AEB
DG16 5JS G3MRV
DG16 5LB G3KAK
DG2 0AJ G4RPO
DG2 0NB M3MFR
DG2 0QX G0RJG
DG2 0QX G1EOA
DG2 7AS 2E0EBW
DG2 7AS M3RBF
DG2 7DH G1SVQ
DG2 7LW G6PFJ
DG2 7NS 2E0RHT
DG2 7NS M6RHT
DG2 7PL G8XKW
DG2 8DA M6BLE
DG2 8QR 2E1SCO
DG2 9BP G4XUJ
DG2 9EE 2E0NEO
DG2 9EE M0NEO
DG2 9EE M3VRX
DG2 9NR G0AVB
DG2 9NU M3ELT
DG3 5AN M3BWT
DG4 6JZ M0CIK
DG4 6LD G4MAY
DG4 6LD G8ZXQ
DG5 4DX 2E0HOS
DG5 4DX M6JJW
DG5 4EB G6VVG
DG5 4EE G4BRB
DG5 4EN G3LJR
DG5 4FA G4HAA
DG5 4FA M1BHO
DG5 4FA M3PBQ
DG5 4GY M6RJN
DG5 4QN M6JFB
DG6 4HL G4DSO
DG6 4LH M3ZJL
DG6 4PE G4KOO
DG6 4QD M3SLD
DG6 4RN G6RXQ
DG6 4TT M3UJQ
DG6 4XT M0KBT
DG7 1AH 2E0WUL
DG7 1BL M6SNY
DG7 1EG M6ACV
DG7 1GB G6ZLY
DG7 1HF M6GYP
DG7 1HH G0BWU
DG7 1HR M6IAW
DG7 1HS M6ACY
DG7 1HX G1BZR
DG7 1HX G4VIS
DG7 1LG M6COW
DG7 1LQ M0FNL
DG7 1LW G4TPR
DG7 1LW G6WNX
DG7 1TN M6DPV
DG7 2EA M0GOF
DG7 2JE M6DMC
DG7 2NG M3WDT
DG7 2PW G8CJG
DG7 2PZ G4WFE
DG7 3BF G3JKS
DG7 3JW M6KCR
DG7 3JW M6URR
DG7 3LD M6DYH
DG7 3UR M6DPQ
DG8 0HQ M6DGZ
DG8 0LD 2E0NIX
DG8 0LD M0WWH
DG8 0LD M6GWW
DG8 0PN G6FXZ
DG8 6AB G6AQL
DG8 6BH G3MZX
DG8 6HF M6MNQ
DG8 6NU M6PTR
[illegible] [illegible]
DG8 7AF G7ZYE
DG8 7HN G4JNW
DG8 7HU 2E0TKE
DG8 7HU M3TKE
DG8 8BA G4DUX
DG8 8DQ 2E0NSW
DG8 8DQ G3SEK
DG8 8DQ G5RP
DG8 8DQ M0WNW
DG8 8DU G4FZH
DG8 8LD G0TFF
DG8 8LD G3AUE
DG8 8NE 2E0GDF
DG8 8NE M3ZVT
DG8 8NZ G0ECU
DG8 8NZ G0IDJ
DG8 8PP G6FSG
DG8 9AB 2E0HEO
DG8 9AB M6HEO
DG8 9DT G4GDF
DG8 9EE G8AVM
DG8 9HX 2E0GKD
DG8 9JA M6KTY
DG8 9JG G4XAW
DG8 9LJ G7NFF
DG8 9LL M6FKB
DG8 9LT 2E0JWL
DG9 0AU G6WAZ
DG9 0DX G0AGV
DG9 0DX G0HPL
DG9 0HL 2E0BOY
DG9 0LJ G0UWV
DG9 0LN G0AJT
DG9 0LW G0VYB
DG9 0RY G4ZIL
DG9 7BX G1FMX
DG9 7HR G0CXY
DG9 7LL M3NWZ
DG9 7TA G0AGN
DG9 8BQ G4LPT
DG9 8DB G3TGG
DG9 8ET M3MZO
DG9 8EY G6KXP
DG9 8HH 2E0WJP
DG9 8HH M6WJP
DG9 8RD M6DGY
DG9 8RD M6RHO
DG9 8TD G4YYF
DG9 9AW 2E0WML
DG9 9AW M6FJV
DG9 9BZ G0HPK
DG9 9BZ G4RIV
DG9 9BZ M3MCG
DG9 9EX G1ZIV
DG9 9HF M6DPO
DG9 9NE G4DLG
DG9 9PS M6WGD

## DH

## (Durham)

DH1 1HA G0PMX
DH1 1JE M6KTA
DH1 2BB 2E0UDG
DH1 2BB M3UDG
DH1 2DF G4KOT
DH1 2JR 2E0OGZ
DH1 2JU G7BYE
DH1 4EN G6RMJ
DH1 4NB G7VIL
DH1 4QX G8AVV
DH1 5DQ G4EBN
DH1 5EH G7DGD
DH1 5FN G8PSS
DH1 5HP G7VJH
DH1 5HT G1SLP
DH1 5HZ M0CKC
DH1 5JN G0DGB
DH1 5JN G0DRJ
DH1 5LA M6NHD
DH1 5LN M6EUB
DH1 5NL G3JQL
DH1 5PB M6RHX
DH1 5PN M0HAM
DH1 5PT 2E0UTC
DH1 5PT M0SGZ
DH1 5PT M6BBA
DH1 5PU G4PFE
DH1 5RJ G4LOM
DH1 5ZG G0VLF
DH1 5ZG G1EUZ
DH2 1JJ G3ZIJ
DH2 1NR G1NCL
DH2 1QH G4OCQ
DH2 1QY M1ELW
DH2 1RS M6BHL
DH2 1TA 2E0DEK
DH2 1TA M6ATP
DH2 1UF G4PSS
DH2 2BY 2E0DGQ
DH2 2BY M6EOC
DH2 2LD G7TFX
DH2 2LX G4ODE
DH2 2NX 2E0LOW
DH2 2NX M6MDU
DH2 2PB M6WYG
DH2 2SD G7OES
DH2 3DD G8YWK
DH2 3EA G6LCL
DH2 3EU G8APW
DH2 3HY G7DPZ
DH2 3JS G4MJA
DH2 3LU G0LEA
DH2 3LX M3NJD
DH2 3TH G4UIA
[illegible] [illegible]
DH3 1HP G1JGE
DH3 1HP G1KYN
DH3 1RW G4YMC
DH3 2EY G0UEC
DH3 2JG G0CGW
DH3 2JL M0CBP
DH3 2PT M0CSF
DH3 3ED M0BPM
DH3 3PR G4OSK
DH3 3PR M3GQP
DH3 4HN G1JDP
DH3 4HU G1REO
DH3 4LU G6GLR
DH3 4LU G8FBQ
DH4 4BW M0FCY
DH4 4BW M0HRA
DH4 4EF M1ADK
DH4 4HN M6CLO
DH4 4HN M6GYS
DH4 4HU 2E0BMW
DH4 4HU M3OVO
DH4 4PE M3EKP
DH4 4XR 2E0GRI
DH4 4XR M0IAS
DH4 4XR M3XNC
DH4 5BB M0EXM
DH4 5ND M3JEE
DH4 5NR M6HMD
DH4 5NR M6SGD
DH4 6EA G0AAU
DH4 6JG G0FDA
DH4 6JG G4VOK
DH4 6JG G6KVA
DH4 6LA M3IVI
DH4 6PA G3NMD
DH4 6PA M0RZE
DH4 6QQ G0FRZ
DH4 7FB 2E0ODB
DH4 7FB M0ZXW
DH4 7HP 2E0JDH
DH4 7HP M0GDV
DH4 7HP M0HJD
DH4 7LP G0FBL
DH4 7RD G0OLO
DH4 7RZ M1FBI
DH4 7TD G0EHV
DH5 0BH M0ETP
DH5 0BH M0ETQ
Dh5 0dq 2E0AAC
DH5 0DQ M6ZRL
DH5 0EF G6VMF
DH5 0ES M6TKX
DH5 0NR M0EGA
DH5 0NW M3BFG
DH5 0SE G6BIM
DH5 8DP 2E0HNJ
DH5 8EX G4WJV
DH5 8GB M0RNR
DH5 8HN M3JMQ
DH5 8HR G0CKI
DH5 9EL M0AHT
DH5 9LB 2E0IAS
DH5 9LJ M3KUV
DH5 9LX G7GJV
DH5 9QX 2E0DLA
DH5 9QX M6DLB
DH5 9RD M6FMS
DH5 9RR G4HBR
DH6 1DA G0KGQ
DH6 1LB 2E0MBB
DH6 1LS G6WML
DH6 1PA G4SYD
DH6 1QB G6CQH
DH6 1RH G4RHL
DH6 1RJ G0WPM
DH6 2NF G0WIC
DH6 2QD M0IOC
DH6 2RG M6BLK
DH6 2TR G0PXQ
DH6 2YJ 2E0PUL
DH6 2YJ M6PUL
DH6 3DB M3ZAL
DH6 4BE 2E0NMK
DH6 4DL [illegible]
DH6 4DG G7DIG
DH6 4DN G7MVX
DH6 4NN G4WUI
DH6 4NP G0DUG
DH6 4QW M6CAT
DH6 5ES G6LLD
DH6 5QA G7NFG
DH6 5QB G0SLP
DH61LB M3GBB
DH7 0BQ G6CQC
DH7 0HL G0OGB
DH7 0QA 2E0MPG
DH7 0QA M6MAK
DH7 6JU G0EMF
DH7 6LZ M6PNL
DH7 6NE 2E0WET
DH7 6NE M6PRO
DH7 6SG M0CDC
DH7 6TF G0CUO
DH7 6TW G3SNT
DH7 6UZ G0VHB
DH7 6XF G4NTW
DH7 7AN M6ZSV
DH7 7DP M3ZGE
DH7 7FD G1RVT
DH7 7NN G7MBH
DH7 7RL M0BAR
DH7 8PQ G7UKA
DH7 8RY M0DOW
DH7 8RY M3UPX
[illegible] [illegible]
DH7 8TG G1WRF
DH7 9AG M6FSZ
DH7 9AU G0MNH
DH7 9FR 2E0JVA
DH7 9FR M6VAV
DH7 9JQ 2E0LLC
DH7 9JQ M6KAM
DH7 9JQ M6MGD
DH7 9LY G7TLR
DH7 9NF M6ESH
DH7 9TB M0NIW
DH7 9UW G6XCO
DH7 9YA G6PTT
DH8 0AW G0OGD
DH8 0RF G0XVC
DH8 5JF G1EJQ
DH8 5LS G0JRT
DH8 5LS G4PFQ
DH8 5PA G0OHA
DH8 6AF M6FOX
DH8 6EZ M0HTV
DH8 6HZ 2E1EPA
DH8 6JG M3UJN
DH8 6JT M6NHK
DH8 6JZ 2E0GBB
DH8 6JZ M6AQO
DH8 6TL G0GKK
DH8 7AE G0SBU
DH8 7AE M6GMD
DH8 7DB M6ICP
DH8 7DT M6WKY
DH8 7DY G3UTS
DH8 7EH M0MFL
DH8 7EH M6KBE
DH8 7JB 2E0CIN
DH8 7JB M6GAM
DH8 7JE M0WDJ
DH8 7QT G1NVN
DH8 7SB G8IAJ
DH8 7SJ G3EGF
DH8 7UJ M6XJC
DH8 8DD G7UXQ
DH8 9AP G3KMG
DH8 9JU G1BET
DH8 9JU M3BET
DH8 9RF G8KZN
DH9 0QD G1NTI
DH9 0QP G1HYG
DH9 0QR G7LQK
DH9 0QU G1HEZ
DH9 6BE G1HEX
DH9 6ET 2E0DYD
DH9 6ET M6EKI
DH9 6HX 2E0GBN
DH9 6HX M6GBN
DH9 6JD M3TSO
DH9 6JU M0BZX
DH9 6XE G7OVB
DH9 7AJ G6FLK
DH9 7PH M5ABT
DH9 7QE 2E0MAL
DH9 7QE M3MCF
DH9 7TR G1AIF
DH9 7TZ M0AYI
DH9 8DG G1KRX
DH9 8DG G6UEV
DH9 8EQ G4WAB
DH9 8EQ G7GJU
DH9 8QY G1OGH
DH9 8UH G4MHA
DH9 9HB G0KYL
DH9 9LZ G0LEI
DH9 9PA G0IVX
DH9 9PA M6JIK
DH9 9PT G0VII
DH9 9PT M3KQB
DH9 9UH M0CLD

## DL

## (Darlington)

[illegible] [illegible]
DL1 1HG G8BWH
DL1 1JG M6AAJ
DL1 2DU G0SBP
DL1 2DX G0FGG
DL1 2HY G0LUY
DL1 2TA G6VPV
DL1 2TA G8XQS
DL1 2TU G4NYJ
DL1 3HH G6SPQ
DL1 3HZ 2E0MRQ
DL1 3HZ M3ZJG
DL1 3ND M0WLY
DL1 3ND M6DLY
DL1 4EG 2E0NYM
DL1 4EG M3CRD
DL1 4LG G8OWS
DL1 5DF G0TJC
DL1 5DF G1LEN
DL1 5DX G4AUN
DL1 5EA G6XCD
DL1 5LH G0JHD
DL1 5LX 2E0LRP
DL1 5LX M3HUG
DL10 4ET M6ABR
DL10 4PQ G3XHB
DL10 4PQ G8YPN
DL10 4PS G4DWM
DL10 4PS M6JJI
DL10 4QY M0OJC
[illegible] [illegible]
[illegible] [illegible]
DL10 5AY M0DIN
DL10 5AY M3ADB
DL10 5BJ 2E0BNC
DL10 5BJ M3WUS
DL10 5DB G3RLV
DL10 5DJ G0MQV
DL10 5PE M0BYU
DL10 6DX G6RNR
DL10 6EE M0JWR
DL10 6EE M5JWR
DL10 6SB 2E1GUN
DL10 6SB G8HQW
DL10 6SB M0RRG
DL10 6SB M1ECH
DL10 7AE G8ZNB
DL10 7AG G4VGY
DL10 7BG M3NPZ
DL10 7BQ G0RYS
DL10 7DL G4DBY
DL10 7DQ G6CUK
DL10 7DX M3VON
DL10 7JE 2E0WGC
DL10 7JE M0WGC
DL10 7JE M6CGW
DL10 7JP G7LLY
DL10 7LQ G6XJJ
DL10 7RT 2E0BEA
DL10 7SP M1EAB
DL10 7SS M3UBU
DL11 6HH G0JTA
DL11 6PY M6LAA
DL11 6QX G8TOQ
DL11 7AJ M3CDI
DL11 7NH M6SPP
DL11 7RD G3XWV
DL12 0AH 2E0BWP
DL12 0AH M6DLM
DL12 0AQ 2E0BWN
DL12 0AQ M3EHH
DL12 0JD M6EVM
DL12 0QU G0ADO
DL12 0RP G4IPB
DL12 0ST G7LIE
DL12 0ST M0CAD
DL12 0UU G4AFE
DL12 8EB G0NRK
DL12 8EB G0PPS
DL12 8EB G8PRU
DL12 8LF 2E0BFM
DL12 8LQ 2E0BPF
DL13 1JD G6FZC
DL13 2SY M6MYB
DL13 2XY G6OTL
DL13 3BW M6HVU
DL13 4DS 2E0AIM
DL13 4DU G4TMI
DL13 4ED M6KWT
DL13 4EE M6CZW
DL13 4JS G4OJY
DL13 4LW G4LDT
DL13 4LW G8WTZ
DL13 5NF G0LLC
DL13 5RH G4KUX
DL14 0DH G0OCB
DL14 0DH G1XNI
DL14 0DH M6AMD
DL14 0RN M3WIT
DL14 0RP 2E1EYS
DL14 0RX G0EQD
DL14 0TG G1ASR
DL14 6AR M3RLO
DL14 6ET M3UOM
DL14 6TU M0BLN
DL14 6UB G6VTR
DL14 6UB M3BAH
DL14 6UB M3EMA
DL14 7DS G0KIC
DL14 7GH M3FSV
DL14 7LZ M0ESZ
DL14 8AE G0GWL
DL14 9EJ G3PYZ
DL14 9EJ G4CCF
DL14 9EJ G4LRG
DL14 9LF 2E0GWE
DL14 9LF M6GZA
DL14 9LG G4TTF
DL14 9LG M0ACV
DL14 0LQ G1YNO
DL14 9PY G0JQP
DL14 0PY M3UOB
DL14 9RY 2E0LDQ
DL14 9RY M6LDQ
DL15 0BJ 2E0BJA
DL15 0BJ M0BPC
DL15 0BJ M3NAR
DL15 0DA G4STP
DL15 0DS M3XWM
DL15 0HN G6LQR
DL15 0LQ 2E0CEG
DL15 0LQ G1YUB
DL15 0LQ M3ALX
DL15 0NS M0MEG
DL15 0PP 2E0CWQ
DL15 0PP M6JDU
DL15 0QL 2E0LIW
DL15 0QL M6LWS
DL15 8NN G4FZS
DL15 8QU G4PSI
DL15 9AE M6RTG
DL15 9DB G7OCK
DL15 9DY G7PTT
DL15 9QX G0HCD
DL15 9SE G4VIO
DL15 9SF M1GSM
[illegible] [illegible]
DL16 6DW G7ESY
DL16 6ER M3AHR
DL16 6HP G7ESX
DL16 6LY G0TSR
DL16 6NB G0RNY
DL16 6RN 2E0XVF
DL16 6RN G0KNN
DL16 6RN M0XVF
DL16 6RN M3XVF
DL16 6TG M0PJM
DL16 6TT M6SCB
DL16 6TZ G7MKQ
DL16 7BA G0PWK
DL16 7DZ G8NSX
DL16 7HF G7VYZ
DL16 7HN G1LPS
DL16 7HU G7KJR
DL16 7HZ G0JEC
DL16 7QS G0BVO
DL16 7TA M3AHO
DL16 7TU 2E0HDF
DL16 7XJ M5CJH
DL17 0RW G0BNY
DL17 0RW G4CAY
DL17 8DA G4MQV
DL17 8DA G8UYM
DL17 8DL G4JIX
DL17 8NG G4XHP
DL17 8QE G1LGQ
DL17 8SX 2E0DXV
DL2 1AU G0AIH
DL2 1HF G3SGY
DL2 1JG 2E0UCH
DL2 1JG 2E0ZOT
DL2 1JG M0VLI
DL2 1JG M0XEE
DL2 1JG M3VFS
DL2 1JG M3VLI
DL2 1JG M3ZVA
DL2 1LF M3PGL
DL2 2AL 2E1DPK
DL2 2AL G8HUY
DL2 2LN M3KRX
DL2 2PX G4NLL
DL2 2SL G8MTV
DL2 2SQ M1SAZ
DL2 2YE G3EKT
DL2 2YE G4HIY
DL2 3JJ G4MIJ
DL2 3NS 2E1HHL
DL2 3NS M1DYJ
DL2 3RY G3GJJ
DL2 3SX M0OAT
DL2 3SX M6WKR
DL3 0AH G3UHJ
DL3 0AL 2E0PCQ
DL3 0AL M6PCQ
DL3 0EJ M6ZET
DL3 0HX G8HZS
DL3 0JY G2HMK
DL3 0NW G4YFS
DL3 6ES G7TCD
DL3 6HN G4JIR
DL3 7HP G1DMN
DL3 7SJ G1BIA
DL3 8BH G4OXU
DL3 8BH G8YBO
DL3 8HU G6KMG
DL3 8HY G4FVP
DL3 8HY M0NFD
DL3 8LD G0ARZ
DL3 8NH M3ONH
DL3 9AT 2E0GBA
DL3 9AT M3XNM
DL3 9DT G6BCG
DL3 9PB G3CDM
DL3 9PF G0MSZ
DL3 9SA G0VQG
DL3 9SR M3MCA
DL3 9XL G0LUW
DL3 9XS G8TJR
DL4 1AP G7BPR
DL4 1BH G3LUC
DL4 1EB [illegible]
DL4 1EB M0SDE
DL4 1FL G4KEL
DL4 1QA G1DFN
DL4 2AZ M3NFU
DL4 2DN 2E0SUX
DL4 2DN M0SGA
DL4 2LE 2E0XLY
DL4 2LE M0XLY
DL4 2LE M6LXY
DL4 2LE M6MYF
DL4 2LH M0MYE
DL5 4HX M6ZAY
DL5 4PF M6WIT
DL5 4PF M0XDQ
DL5 5HZ 2E0DLL
DL5 6RF G3OAL
DL5 6RU G3JKD
DL5 7BD M0PST
DL5 7BQ M0WRI
DL5 7DR G1AZC
DL5 7DX G8LCP
DL5 7DX M0BUG
DL5 7DY 2E0RRL
DL5 7DY 2E0WPD
DL5 7DY M6RRL
DL5 7DY M6TCN
DL5 7LN G3FBT
DL5 7NA M3UVY
DL5 7NP M3VRY
DL5 7PQ G1ZEU
[illegible] [illegible]
DL0 1DZ G0GBG
DL6 1ED G0JQA
DL6 1ED G3KJX
DL6 1EE G8FLV
DL6 1HP G1XRE
DL6 1QQ G0LEL
DL6 1QT G8IHT
DL6 1RQ G0GCK
DL6 1SJ G7NKJ
DL6 2AA G7HHK
DL6 2BD G1JER
DL6 2BE G3MAE
DL6 2LB 2E0FBC
DL6 2LB M6FOY
DL6 2RD G0WWF
DL6 2RD G2PB
DL6 2SN G1HLV
DL6 3QA M0BBK
DL7 0HL 2E0DNH
DL7 0HL M6RFZ
DL7 0QR 2E0HWN
DL7 0QR M3HWN
DL7 8AD M3XYJ
DL7 8BN 2E1CQQ
DL7 8BN G7COC
DL7 8HT M6HGY
DL7 8JF M0MFT
DL8 1PX G3JZP
DL8 1SX G7HSN
DL8 2AN G7VAU
DL8 2DR G3GEJ
DL8 2HT G1WWH
DL8 2JE G4AFU
DL8 2LF G3LEO
DL8 2NH M3ZSY
DL8 2PZ M6PHV
DL8 2QF G4NUY
DL8 3HN G8AMJ
DL8 3RH G3ZMO
DL8 4LU M3LOA
DL8 4NA G1BLV
DL8 4QN 2E1AZW
DL8 4QN G0OLR
DL8 5EL M1FGH
DL8 5HU G0BAQ
DL8 5JQ G3XPQ
DL8 5JS 2E0CQX
DL8 5JS M6MCZ
DL8 5QN G8LNQ
DL8 5QN M0KXD
DL9 3BW M6RKU
DL9 3NJ 2E1COG
DL9 3NJ G1YQY
DL9 3NP M6FGM
DL9 3RA G7WBW
DL9 4HX M0DIG
DL9 4PG 2E0TCN
DL9 4PG M0TCN
DL9 4PG M6TIA
DL9 4XA M0HCM

## DN

## (Doncaster)

DN1 2NP G0UQU
DN1 2PZ G4OEM
DN1 2QU G1PBY
DN10 4BE G4OZN
DN10 4BU G0OKZ
DN10 4DP G8EGL
DN10 4EF M1FGM
DN10 4EN G6KGG
DN10 4HT M1WDK
DN10 4LG 2E0CTO
DN10 4LG M6KAX
DN10 4NG G0EJD
DN10 5BS G6GWP
DN10 6HE G8JFC
DN10 6NW M0LLW
DN10 6QF M1CTB
DN10 6RT M3ZFS
DN10 6SW 2E0XAH
[illegible] [illegible]
DN11 0BT M0BZQ
DN11 0FR G4ZLP
DN11 0JG M0YDD
DN11 0JP 2E0DMJ
DN11 0JP 2E0KPI
DN11 0LT G0VHI
DN11 0LT G7GHI
DN11 0NQ G0PXX
DN11 0NQ G6USL
DN11 0PU M3EGM
DN11 0QF M0CTN
DN11 0QF M3CTN
DN11 0QX M0CTQ
DN11 0RW M3EYW
DN11 0RW M3EYX
DN11 0RY G7JDH
DN11 0RZ M3BZQ
DN11 0SB 2E1IOS
DN11 0SB 2E1RAD
DN11 0SH M3WXB
DN11 0TD G4WZI
DN11 0TS M3ZLS
DN11 0UP G4NQM
DN11 0UP G6DIE



**IMPORTANT NOTE**

**Revalidate licence to avoid revocation** – Ofcom has advised the Society that plans will be drawn up to revoke licences that have not been revalidated as required by the licence conditions. The quickest way to revalidate is to do so online via the Ofcom website: *https://services.ofcom.org.uk/* or by email: *amateur.validations@ofcom.org.uk* If you need assistance in the process, Ofcom staff are available to help, but please be patient during times of heavy workload.

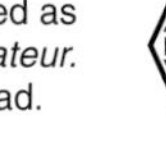

DN11 0UR G8HIQ
DN11 0YD M0DMZ
DN11 8DT M6SDP
DN11 8ET G7DUI
DN11 8HP G4ZVP
DN11 8HT G0FVD
DN11 8HW 2E0ZCB
DN11 8HW M0YCB
DN11 8HZ M6ROE
DN11 8LL G4BZG
DN11 8LU M3IOK
DN11 8QP M6FYD
DN11 8QU M6TDR
DN11 8SN M6ENE
DN11 9BW G0TTL
DN11 9ET G6HMS
DN11 9EW 2E0DFT
DN11 9PW M0BOI
DN11 9UE M6BGP
DN11 9UH G7EDF
DN11 9UL M3XPN
DN12 1BD 2E0DKS
DN12 1BD 2E0IKB
DN12 1ED 2E0SEJ
DN12 1ED M0SBK
DN12 1ED M6SGY
DN12 1NQ M3SMD
DN12 1NW 2E0OXO
DN12 1NW M3KAK
DN12 2AA 2E0CRS
DN12 2AA M6CPM
DN12 2JG 2E0XPM
DN12 2JG M0XPM
DN12 2JW G3NXZ
DN12 3DF G3WBG
DN12 3DG M6EMJ
DN12 3LB 2E0RDX
DN12 4EL M0FAT
DN12 4EN M0TOR
DN12 4LR M3LTG
DN12 4SB G8LGC
DN12 4SB M0DID
DN14 0AT 2E0DXZ
DN14 0AT M6GQY
DN14 0BU M0ROW
DN14 0JX G4FPO
DN14 0LN G4III
DN14 0NW G6LUE
DN14 0PD G8DML
DN14 0PG M0JKP
DN14 0PX G8VAT
DN14 0QY G4AFZ
DN14 0RD G8FCT
DN14 0SB G7OYD
DN14 5HQ M3HVU
DN14 5JB M6JIR
DN14 5NY G4GNP
DN14 6DH G0PQX
DN14 6DY M3BRW
DN14 6EB M6BPX
DN14 6EL M3NVE
DN14 6HW G8FWC
DN14 6JX G3LYZ
DN14 6NA G0GLZ
DN14 6QW G8VHL
DN14 6RR M6MRU
DN14 6SH 2E0CAD
DN14 7EN G3ZGT
DN14 7FD G4NLG
DN14 7HD G0KHZ
DN14 7HE G4DBN
DN14 7JL G1RFC
DN14 7NT M6BIB
DN14 7TP M3KQS
DN14 8DW M6AFH
DN14 8ET G0FRX
DN14 8EX G4BDX
DN14 8HL G4LKD
DN14 8HL M0NTG
DN14 8LR G4GRP
DN14 8RP G6PXJ
DN14 8RW G7UJT
DN14 8RW M3UJT
DN14 9JH M3SRI
DN14 9NG 2E0HCT
DN14 9NG M3GUG
DN14 9NG M3PKH
DN14 9QZ G0UYF
DN15 0AD G7MTJ
DN15 0BE M3OZB
DN15 6SN M6FJW
DN15 7AT G0RMD
DN15 7BT G0OQX
DN15 7EH G1HKM
DN15 7EN 2E0XNL
DN15 7EN M6GFH
DN15 7LE M6TAZ
DN15 7PJ G0DLL
DN15 7PJ M6BYQ
DN15 8AB G3KFB
DN15 8AU G0EFY
DN15 8BP G0RKQ
DN15 8LA M6NTM
DN15 8NS 2E1FVS
DN15 8NS M0HOM
DN15 8PA M0RMT
DN15 8PE M0PDC
DN15 8TS G8NQN
DN15 8UG G6JVO
DN15 9AG G0DVG
DN15 9EW G4CDC
DN15 9EX M3IPM
DN15 9HH G0JRR
DN15 9NQ M0GBB
DN15 9NQ M0OGY
DN15 9NR G6XPT
DN15 9NW 2E0AWT
DN15 9NW M0GIQ
DN15 9QG M6MCC
DN15 9QY G8TLL
DN15 9RT G6CMX
DN15 9RU 2E0FTW
DN15 9RU M1DOS
DN15 9UY G0UZJ
DN158LL M3VTP
DN16 1EB G4JJY
DN16 1EY G0GWY
DN16 1HL G4HFZ
DN16 1HT 2E1MIN
DN16 1HT M1FFM
DN16 1NA G0OKF
DN16 1NE 2E0GBD
DN16 1NE M6FMO
DN16 1QH 2E0FQT
DN16 1RW 2E0MBO
DN16 1RW M6HMG
DN16 1RW M6HMP
DN16 2AJ 2E0PFY
DN16 2AJ M6FXU
DN16 2BE G4JRY
DN16 2LQ G1YNQ
DN16 2LR 2E0CQD
DN16 2PA M6PHS
DN16 2RP M6GRC
DN16 3DW M0KBH
DN16 3EN G8GIH
DN16 3HQ M6KNS
DN16 3LQ G0TNS
DN16 3PE G6AOA
DN16 3PH G0PQY
DN16 3SA G8LIP
DN16 3SW G4NJA
DN17 1AF G4CBY
DN17 1AS G0RUH
DN17 1DJ G1CDY
DN17 1DU G1ZRT
DN17 1SA 2E0DGC
DN17 1SA G0UTM
DN17 1SA M0HDV
DN17 1SA M6GDC
DN17 1SW 2E1CZB
DN17 1TS 2E1GXY
DN17 1TS M3BAT
DN17 1YB 2E0CZA
DN17 1YB M6DQR
DN17 1YL 2E1HHA
DN17 2BD G8YVC
DN17 2EQ G0NYL
DN17 2EQ G3KNU
DN17 2GE G4STW
DN17 2LW 2E0RNP
DN17 2NQ G7LAL
DN17 2TP G0ECS
DN17 3HZ G7URL
DN17 3PB M0ACB
DN17 3RA G6PAJ
DN17 3SB G0JEU
DN17 3SB G4ZGE
DN17 3ST G0JRB
DN17 3SZ G0PAB
DN17 3TT G4EQD
DN17 4AT G6DBC
DN17 4DG G4FZR
DN17 4ET G8MOF
DN17 4HU G8ZGY
DN17 4LH G4LSL
DN17 4NB M3ZPO
DN17 4PP G4VOV
DN17 4PQ M1LMO
DN17 4RZ G0NVD
DN18 5BS 2E0GDN
DN18 5BS M0IIE
DN18 5BS M6ITW
DN18 5DP G1ZQG
DN18 5DZ G3RHQ
DN18 5HH G0HMD
DN18 5LE M0AEP
DN18 5LN 2E0RWX
DN18 5NH G4IWR
DN18 5QA G0PHP
DN18 5TD G8RQH
DN18 5TW M3USJ
DN18 6AE G4TGE
DN19 7AU G0KAT
DN19 7EE G4JPK
DN19 7EE M1TAT
DN19 7EY G1CUM
DN19 7QB G7PXX
DN19 7QG M3BEE
DN19 7QG M3PWS
DN2 4BB M1BUX
DN2 4DA M3KJY
DN2 4GD M6GPO
DN2 4JN M6RBQ
DN2 4JS M3LKY
DN2 4LF M3WPV
DN2 4QA G0IDZ
DN2 4RF M6DJL
DN2 4RJ 2E0PBP
DN2 4RJ M6DPQ
DN2 4RW M6AMC
DN2 5EU G6OSK
DN2 5HN G4PJL
DN2 5RF G3XGU
DN2 6AN G4LKX
DN2 6DF G8EOV
DN2 6DT G4ZGM
DN2 6EN M6PGZ
DN2 6HF G0VID
DN2 6HF G7GJO
DN2 6HF M0BDL
DN2 6JH M0FCI
DN2 6LG M1CHM
DN2 6LN 2E0ECG
DN2 6PA M6OAE
DN20 0AU M3UPN
DN20 0BB G3MHT
DN20 0HW G0HDV
DN20 0NN G8YAU
DN20 0PP G0AOJ
DN20 0SE 2E0SXF
DN20 8DY M3CVD
DN20 8PW M0ICJ
DN20 8QR M6MOF
DN20 8SG M6PDM
DN20 8SH G0CMK
DN20 8SW M6HFF
DN20 9BE G3MPW
DN20 9DG G4ROC
DN20 9FN M3YUN
DN20 9HU G8YYW
DN20 9JG G6GNE
DN21 1BQ M3VYF
DN21 1BT G3SCJ
DN21 1BX 2E1EKQ
DN21 1DD G1HKF
DN21 1DH G7AVU
DN21 1NA G7OLF
DN21 1PA G7GMR
DN21 1QZ M1EBN
DN21 1QZ M3DXR
DN21 1RG M0AYF
DN21 1SS 2E0RKS
DN21 1SS M0ZRS
DN21 1SS M6AYP
DN21 1TT G8VRN
DN21 1WB G7MDM
DN21 1XQ M3LAJ
DN21 2HQ G6FOW
DN21 2JA G3YPS
DN21 2JZ G3UHS
DN21 2RY M6JVW
DN21 2TH M3TNL
DN21 2TU G6YVS
DN21 3DL 2E0FAM
DN21 3DL G1CIM
DN21 3DL M0CGS
DN21 3DL M6FLC
DN21 3GT G0BIA
DN21 3GT G3PSG
DN21 3JY G4XHC
DN21 3QZ G4YET
DN21 3RU G7WGY
DN21 3RU M3FWU
DN21 3RU M6WGY
DN21 3SL G7JQT
DN21 3SR G3TMD
DN21 3TL M3GGN
DN21 3TZ G0PZW
DN21 4AF G4GZA
DN21 4BA 2E0BCX
DN21 4BA G3VBI
DN21 4BA M0JAE
DN21 4BB G4SXZ
DN21 4BY M6LHJ
DN21 4DF M3FWD
DN21 4ER G4MBK
DN21 4FE M6CGX
DN21 4NE 2E1HXY
DN21 4NE 2E1HXZ
DN21 4NE G4CCH
DN21 4NF G0MPK
DN21 4SW G6PHC
DN21 5BN M0KEY
DN21 5BU 2E0SIY
DN21 5BU M0SIY
DN21 5BU M3SIY
DN21 5BZ G3RAU
DN21 5DN 2E0LQR
DN21 5DN M0LQR
DN21 5DN M6TML
DN21 5LB G8SFF
DN21 5LY G7IYQ
DN21 5PE G4JTX
DN21 5PT G6WVO
DN21 5QU G4VHH
DN21 5TQ G3SET
DN21 5UT G0SQS
DN21 5UT G7JVC
DN21 5XJ M6LHE
DN21 5XP 2E0CJK
DN21 5XP M0PMJ
DN21 5XP M6LIE
DN21 5XP M6PMJ
DN22 0AS G7SRI
DN22 0BY G6NRL
DN22 0FL G7KYL
DN22 0HU M3BAL
DN22 0JX G0TGB
DN22 0LG G4SOI
DN22 0LN G1EWH
DN22 6LR 2E1FRW
DN22 6NH G0MYA
DN22 6NW G4XTU
DN22 6QH G0CEB
DN22 6QU 2E0LDV
DN22 6QU M3MZG
DN22 6RT G7IBD
DN22 6SF G4YWX
DN22 6SQ M0PAL
DN22 6UB G0HLU
DN22 6UF G3DXZ
DN22 7AD 2E1HYI
DN22 7BT G6GKK
DN22 7DX M6XAK
DN22 7LJ G3NTD
DN22 7QW M6JMF
DN22 7TL 2E0SNS
DN22 7TL M6SNS
DN22 7UW G4WBH
DN22 7XA M6MEK
DN22 8AJ G0IAS
DN22 8AY G4OCU
DN22 8NL G4VEO
DN22 8PQ G0NPG
DN22 8RS G4AWU
DN22 9LJ G3OS
DN22 9NQ G0VUL
DN3 1AN G1MCT
DN3 1AW M3XYN
DN3 1BJ M6GWT
DN3 1BS G4NZX
DN3 1DP G0UZW
DN3 1DP G7RXE
DN3 1DS G3UWT
DN3 1JU G4KKJ
DN3 1LY G0EZY
DN3 1LY G8CDV
DN3 1NY G4DDS
DN3 1NY M6DMO
DN3 2AR M6RCD
DN3 2AZ 2E0PKB
DN3 2AZ G7MEX
DN3 2AZ M0BOH
DN3 2AZ M0EVE
DN3 2AZ M0MEO
DN3 2AZ M1DAH
DN3 2AZ M3DYF
DN3 2AZ M3HJB
DN3 2AZ M3PLB
DN3 2AZ M3WQI
DN3 2DE G1IND
DN3 2DE M3IND
DN3 2DJ M6JIF
DN3 2EP M6WCZ
DN3 2ES M3MSC
DN3 2HE M3YJG
DN3 2HN G4PNT
DN3 2PE G4MDE
DN3 2PQ G7GEI
DN3 3AJ G7JTF
DN3 3HE G7GZK
DN3 3HG M3PTV
DN3 3HS G0GOH
DN3 3PB G0WKA
DN31 2DN G6NDH
DN31 2NB G1WSC
DN31 2PW M3YAV
DN32 0HN 2E0OZW
DN32 0HN M3HVV
DN32 0JQ M6PFU
DN32 0NG G8BMZ
DN32 0NH M6EJP
DN32 0NH M6LZX
DN32 0NQ M3WJA
DN32 7JE G1HCI
DN32 7JE G1HCJ
DN32 7PL G0CJV
DN32 7SB G7HJR
DN32 8BA G4ZRV
DN32 8DR M6RGU
DN32 8DZ G7VKY
DN32 8HR M6CBP
DN32 8HR M6SBE
DN32 8NU M6PRC
DN32 8NX M0DDY
DN32 8QS G7LIW
DN32 9DZ G7NCV
DN32 9HJ M6NBO
DN32 9JQ M6HBC
DN32 9LB M6WMJ
DN32 9NP M6EPS
DN32 9NU M3XPU
DN32 9PQ M1BYQ
DN32 9PY G1KBL
DN32 9QG M0DMY
DN33 1BG G4HZF
DN33 1DE G3DAE
DN33 1DN M6NGA
DN33 1EZ G8XXI
DN33 1JB M0AJT
DN33 1LU 2E0KQV
DN33 1NL M3UIP
DN33 1RF M6KEC
DN33 1SB M0SDG
DN33 2BB G0KUD
DN33 2EA G4NPS
DN33 2HF G3DSZ
DN33 2JS G1HZN
DN33 2LG G6EXZ
DN33 2NW G1FVE
DN33 2PL G0CGZ
DN33 2PL G0IOR
DN33 3BL G0CSV
DN33 3BS G1BKJ
DN33 3BW G4VIM
DN33 3EE G3PLN
DN33 3JP G0JNT
DN33 3JT G3SWU
DN33 3JY M6MJA
DN33 3NG G4MSY
DN33 3RA M6MJL
DN34 4AD 2E1FHZ
DN34 4DP 2E0OZQ
DN34 4NJ M6AVY
DN34 4PN G0IGC
DN34 4PN G0IGH
DN34 4PW M0FIS
DN34 4QH G0HXL
DN34 4SG M5ACD
DN34 5DB G0RUS
DN34 5JP 2E0GCI
DN34 5JP M0KIN
DN34 5JP M3LAP
DN34 5LX M1NCC
DN34 5PE G6RIQ
DN34 5RB 2E0GUH
DN34 5RB M6GUH
DN34 5RE G7EOG
DN34 5TG 2E1HIL
DN34 5UQ G4JMY
DN34 5UQ G4VUP
DN35 0JQ G8ULH
DN35 0NN M1BYI
DN35 0QW G3RSD
DN35 0SE M0CDU
DN35 0SF G4YSO
DN35 7AP M3TYW
DN35 7BB M6XPN
DN35 7DX G0LFA
DN35 7DX M6LAV
DN35 7DX M6TCL
DN35 7JJ 2E0YQT
DN35 7JJ M3YQT
DN35 7NG G4YVJ
DN35 7NP 2E0GRN
DN35 7NP 2E0JKY
DN35 7NP M6FFS
DN35 7NP M6GRN
DN35 7NP M6JKY
DN35 7SD G4ZVX
DN35 7UR M0KWK
DN35 8EA M0BZU
DN35 8HY M6PXK
DN35 8HZ G0MNI
DN35 8LE M6SNF
DN35 8PD G0GSY
DN35 8PL G6VJA
DN35 8PR G7FFW
DN35 8PX M6RKP
DN35 8QN G0PBR
DN35 8QN M6RED
DN35 9DE G4SRF
DN35 9PF M6NEA
DN35 9PP G7CHB
DN35 9PP G7PSV
DN35 9PP M3RHP
DN36 4DF G8ZLF
DN36 4DS G0BBT
DN36 4EA G7DEU
DN36 4HE M0SMT
DN36 4JZ G3YQ
DN36 4LG 2E0GNW
DN36 4LG M6GNW
DN36 4LJ G1JZL
DN36 4NS 2E0WNW
DN36 4NS M6DQK
DN36 4PX G8RHN
DN36 4RB M0ANU
DN36 4ST G6CCB
DN36 4TT G7IMZ
DN36 4UF G3ZSF
DN36 4WH M3UII
DN36 5AQ G3VIP
DN36 5BD G4HOF
DN36 5BG 2E1CAW
DN36 5BG G0NQW
DN36 5BH 2E0CTZ
DN36 5BH M6FBB
DN36 5BQ G4PYD
DN36 5DS G4GAB
DN36 5JE G4OII
DN36 5LP M0DER
DN36 5LS G4XFG
DN36 5NF M1CYP
DN36 5NJ G4SEP
DN36 5PG M6CVK
DN36 5PL G4GOJ
DN36 5QS G4LOY
DN36 5SQ M6FJU
DN36 5TP G7GHK
DN37 0EE M1CDQ
DN37 0JR G3TVC
DN37 0NQ M0JJE
DN37 0NR G0HOI
DN37 0QE 2E0VBM
DN37 0QE M6GYA
DN37 0UA G4WOH
DN37 0UG 2E0PBJ
DN37 0UG M6PSM
DN37 0UX M0CBQ
DN37 7BA G0IIQ
DN37 7DB G0ATW
DN37 7DF M6DCS
DN37 7DF M6SSD
DN37 7ER M6SKT
DN37 7ER M6TNR
DN37 8LB G0OZO
DN37 9DU M6MNI
DN37 9EE M0COI
DN37 9HE G0MAF
DN37 9QJ G3CNX
DN37 9QJ G4EBK
DN37 9QJ G4GZ
DN37 9QL G3RGC
DN37 9QZ 2E0SFT
DN37 9RH G1WHT
DN37 9RN G8RIW
DN37 9RX G4DXB
DN37 9RZ G1BRB
DN38 6AP 2E1GCB
DN38 6AP G8HDJ
DN38 6AU G7TRG
DN38 6BD G4FHF
DN38 6DX M6FOK
DN38 6HF G7IJC
DN39 6RB 2E0PSX
DN39 6RB M0PXS
DN39 6RB M6PXS
DN39 6UQ G0FEQ
DN4 0BT G8XKI
DN4 0EP 2E0SOF
DN4 0EP M3JZT
DN4 0UD 2E0NDH
DN4 5EE G8XTU
DN4 5EG G6RQJ
DN4 5EQ 2E11BGN
DN4 5EY M1EUF
DN4 5QF G0CYB
DN4 6AD M3OYZ
DN4 6DQ G3NYB
DN4 6DX G3AAF
DN4 6DX G7PUL
DN4 6EZ 2E0IMC
DN4 6EZ G4FYE
DN4 6HE G1HZR
DN4 6HX G6ZKS
DN4 6LF G0LOL
DN4 6NX 2E0EAT
DN4 6NX M0ITV
DN4 6PD 2E1DBZ
DN4 6PD M3DBZ
DN4 6QB G0GTI
DN4 6RE M3CMP
DN4 6SN G8GHB
DN4 6SQ G4OEK
DN4 6TB M3WSI
DN4 6TU G0DAM
DN4 6TU G3TCQ
DN4 6TU M6WSK
DN4 6UR M3KFH
DN4 6UT G7KEK
DN4 7EG G7CDI
DN4 7HU G3VHZ
DN4 7JH G7KMO
DN4 7JQ G7TMO
DN4 7JY G0NMJ
DN4 7JY G4DMH
DN4 7RB M0MCH
DN4 8QS M6TKA
DN4 8SB M6MDT
DN4 8SX G8RFZ
DN4 8TN G6MWD
DN4 8TN M5ADA
DN4 9AF M6MKD
DN4 9AJ G3OFP
DN4 9AR G3TTU
DN4 9DA M0KAI
DN4 9DB M0AGJ
DN4 9DQ G0SSC
DN4 9EL M3KEW
DN4 9LA G0LJV
DN4 9QR G0LHM
DN40 1AR G7BTP
DN40 1DW G4YAM
DN40 1EE 2E1HWU
DN40 1HH G0NBC
DN40 1LT 2E0CMZ
DN40 1LT M0PEM
DN40 1LT M6GRV
DN40 1NR G2HLB
DN40 1PT G0KTX
DN40 1PW G0DOB
DN40 1RB G7PLS
DN40 2AS G7GOV
DN40 2AZ G8XFY
DN40 2DQ G0HDS
DN40 2EQ G7BRZ
DN40 2EQ M0ZLF
DN40 2EQ M6KLD
DN40 2JH G3VCX
DN40 2SG G1RSK
DN40 3AX M0PLY
DN40 3DA G4YTQ
DN40 3PS M3KYV
DN40 3PT G3YKK
DN41 7RD G4YHP
DN41 7RE G1LWL
DN41 7SR G7RLK
DN41 8AN G8FKF
DN41 8AP G0LWG
DN41 8ED G1OPD
DN41 8EL M0MAL
DN41 8ER G6PEH
DN41 8HE G4KAL
DN41 8HG G6SXN
DN5 0DW M6SRB
DN5 0ED 2E1HKM
DN5 0ED M0RTQ
DN5 0EG M0BDU
DN5 0LH G7VYY
DN5 0NT G0IBG
DN5 0PQ M3GFO
DN5 0RA M6BMW
DN5 7BL G7ANA
DN5 7EN G4OVS
DN5 7LH G3VLL
DN5 7LW G0UQQ
DN5 7QF G8LHT
DN5 7RE G7ONV
DN5 7RY G1SCO
DN5 7SD 2E0HNH
DN5 7TE G0GNP
DN5 7UH M6TIW
DN5 8EB M6RUM
DN5 8EH G8JJR
DN5 8ER G6NBF
DN5 8HZ G7BGZ
DN5 8NQ G0DQB
DN5 8PQ M6LUZ
DN5 8QA G0PXE
DN5 8QN G4ANP
DN5 8QP G3IGU
DN5 8QW G0HLJ
DN5 8RR G6GNO
DN5 9DY M3KKA
DN5 9HD G4KBS
DN5 9HD G7ACO
DN5 9QX M6PBY
DN5 9QZ 2E0BQR
DN5 9QZ M0SGJ
DN5 9QZ M3FGG
DN5 9ST G4PCD
DN6 0AR G6GYV
DN6 0EZ G0TSH
DN6 0EZ G7AYO
DN6 0LU 2E0BSJ
DN6 0LU M3LCS
DN6 7AX G7CTG
DN6 7BT M0RWA
DN6 7EA G4JII
DN6 7EE G0GBQ
DN6 7JL M3LQB
DN6 7JQ M3GJZ
DN6 7JQ M3YEZ
DN6 7JZ G6HWI
DN6 7JZ G6NSZ
DN6 7LX M3WXP
DN6 7NW M3JVK
DN6 7QQ 2E0AWZ
DN6 7QQ 2E0AXA
DN6 7QQ M6DGG
DN6 8AR G1ILF
DN6 8EB G3UWR
DN6 8EH G1JTZ
DN6 8JL M0CCD
DN6 8JQ G4MRU
DN6 8PD G0AHV
DN6 8RJ G8WCX
DN6 9BY G3WZW
DN6 9HJ G7UBO
DN6 9HX G7VMO
DN6 9JX G0DQM
DN69JU 2E0UBT
DN69JU M3UBT
DN7 4AH M0DLY
DN7 4AH M3GCN
DN7 4AH M3SJK
DN7 4AS M1EJX
DN7 4BS G3VZE
DN7 4BT M6MXH
DN7 4BU G1HOU
DN7 4HB G0TJR
DN7 4HB M3NVS
DN7 4HS 2E1SKA
DN7 4LZ M6WDG
DN7 5BS G8TVU
DN7 5JF M1ALU
DN7 5PE G4VVE
DN7 5TQ G4ALF
DN7 6AB G3OZD
DN7 6AH G0LCT
DN7 6BP M0YRF
DN7 6EH M6BTP
DN7 6ER G8PRH
DN7 6JX M3IKJ
DN7 6PF G7SKW
DN7 6PF M0KSR
DN7 6QQ G7BPF
DN8 4AX M0GIW
DN8 4NG M3RRV
DN8 4NU G0BKD
DN8 4PR M6NTZ
DN8 4QN 2E0RHC
DN8 4QN M6RHC
DN8 4QN M6VVE
DN8 4QR G3WDL
DN8 4QR M3UGB
DN8 4QT G0LNT
DN8 4SF M1EMR
DN8 4ST 2E0VVE
DN8 4ST 2E1GLT
DN8 5BS G4HZN
DN8 5EA G7MGQ
DN8 5EL M6WCV
DN8 5JG G3KPU
DN8 5QS G6VYC
DN8 5TL G4SZX
DN8 5YW G7MZW
DN9 1DP G4WJE
DN9 1DY M6GCW
DN9 1HA G7ABT
DN9 1HR 2E0RTQ
DN9 1HR M0RTQ
DN9 1HR M6VUL
DN9 1JL G4HOY
DN9 1JL M0XRC
DN9 1JS G4FUH
DN9 1JS G4GZB
DN9 1LB 2E0SWF
DN9 1LB G1IAB
DN9 1LB M3PWL
DN9 1LB M3SWF
DN9 1LJ G7IMD
DN9 1LL G4NBM
DN9 1LR G4GZC
DN9 1LT 2E0CMY
DN9 1LT M3HGW
DN9 1LT M6FAS
DN9 1MB G4NEG
DN9 1NW 2E0EDI
DN9 1NW G7COU
DN9 1PG G0OPA
DN9 1RG G4ZWQ
DN9 1RG G6ONE
DN9 1RT G8EVI
DN9 1RT M0NLR
DN9 1SD G8NDK
DN9 2DG G4YUK
DN9 2FB G4OGB
DN9 2HP M3HGA
DN9 2HX G8BPS
DN9 2JQ G0UQV
DN9 2JX 2E1PMT
DN9 2JX G0TTR
DN9 2JX M1IFT
DN9 2LH G6BWA
DN9 2LR G8BAQ
DN9 2NX G4XRO
DN9 3AE 2E0GWP
DN9 3AE M3RCC
DN9 3AJ G4RHY
DN9 3AJ M3RYN
DN9 3BA M0CAX
DN9 3ED 2E0MRG
DN9 3EG M3XPW
DN9 3JR G4RHZ
DN9 3LN M0CJZ
DN9 3PE G0RON
DN9 3PQ G0TUU

## DT
## (Dorset)

DT1 1LH M0AFR
DT1 1LL G1DZZ
DT1 1NJ M3LDS
DT1 1NZ G1GMV
DT1 2BA G2DGB
DT1 2BS M0EJW
DT1 2BY G6KXW
DT1 2DP G8DJW
DT1 2DP M3DOR
DT1 2DY M6DDB
DT1 2EF G4CFY
DT1 2EL M0MZX
DT1 2JJ 2E0VWK
DT1 2JJ M0VWK
DT1 2JJ M3VWK
DT1 2LZ G3YUD
DT1 2NL M6POW
DT1 2NR G4EAS
DT1 2PE G7JIM
DT1 2PE G8SDS
DT1 2PF G4ZPO
DT1 2PS G3XIG
DT1 2SB G1RUL
DT1 3RJ G6PBG
DT1 3RR M1CSG
DT1 3SU M1ALT
DT1 3WP G4EDK
DT10 1DD G4JBL
DT10 1EW M6FKJ
DT10 1EZ 2E0MSA
DT10 1EZ M6MCS
DT10 1HL M3EWU
DT10 1HQ G4CEL
DT10 1LQ G8MNO
DT10 1NA M6KTJ
DT10 1NY G4RKV
DT10 1PS G1SDJ
DT10 2AQ G0SEN
DT10 2BE G3RTD
DT10 2DL 2E0SDP
DT10 2DL M6RDV
DT10 2DT M3PAI
DT10 2HF G7GPJ
DT11 0AA G3ZDQ
DT11 0AY 2E0RHS
DT11 0AY M0ODE
DT11 0AY M0RHS
DT11 0AY M6NSC
DT11 0DQ G6XAG
DT11 0EF G8RGU
DT11 0HF G7RMG
DT11 0HF M0RMG
DT11 0HF M3RMG
DT11 0HU G0NXQ
DT11 0JG G3VOO
DT11 0RY M6GZY
DT11 0SN M3GNY
DT11 0SS G6DAI
DT11 7DE G3KWN
DT11 7DL G1XCK
DT11 7GF 2E0NUQ
DT11 7GF M3NUQ
DT11 7HA M0UND
DT11 7HB 2E0DKR
DT11 7HB M6DKR
DT11 7HH G4GKX
DT11 7LW G3TPH
DT11 7LX 2E1EJC
DT11 7LZ G7PCE
DT11 7NG G4VJI
DT11 7PA G7JTB
DT11 7RF M6UKA
DT11 7RT M1LXM
DT11 7RU G3XGY
DT11 7SS G0AUK
DT11 7UQ M0GKD
DT11 7UT G0ULZ
DT11 7XG G3PCW
DT11 7XU M0WAM
DT11 8BP M6FKH
DT11 8JD 2E1ZPR
DT11 8JY G4TEN
DT11 8RH G3RGE
DT11 8SD M6ECX
DT11 8TA G7NUN
DT11 9BU G4TRM
DT11 9DW G1YHG
DT11 9HG G6JSN
DT11 9NN G8BXQ
DT11 9PH G4ZLX
DT11 9PS 2E0HVE
DT11 9PS M6HVE
DT11 9QP G6PJD
DT2 0BP G0OJP
DT2 0ER G0LOH
DT2 0JJ G3LOX
DT2 0JX G6BSP
DT2 7BB G4LOA
DT2 7HA G8GYL
DT2 7HP G8BAZ
DT2 7HT M6WYN
DT2 7HX M1NAD
DT2 7QS M3OPW
DT2 8AE G3KKJ
DT2 8BG G4DRS
DT2 8BG G4JQL
DT2 8BG M3WEZ
DT2 8BQ G1HYX
DT2 8EB G4PQX
DT2 8EW M1EZX
DT2 8FJ 2E0PWP
DT2 8FJ M0PAA
DT2 8FJ M3UDN
DT2 8HS M0TRW
DT2 8JW G1WPG
DT2 8PE G0HVA
DT2 8PP 2E0SDZ
DT2 8PP G4TFF
DT2 8PP M0ZGT
DT2 8PP M6LCG
DT2 8QJ G0FIT
DT2 8QR G0PZN
DT2 8QR G4SXD
DT2 8TS G1RPT
DT2 8TX G7TNO
DT2 9DD G4AXU
DT2 9DS M6BPO
DT2 9ES G3IVC
DT2 9HZ 2E1HSA
DT2 9HZ G0PGT
DT2 9HZ M3TEN
DT2 9JN G0KYH
DT2 9JT G8DTE
DT2 9JU G6DOR
DT2 9LT G1LUC
DT2 9QB G0TZO
DT2 9QB M0HUV
DT2 9QX G3ZGN
DT2 9RG G4GUV
DT2 9RY G4JQS
DT2 9TW M0JOL
DT3 4AX M6NSO
DT3 4BA 2E1LEC
DT3 4BA 2E1NRQ
DT3 4BA M3PSC
DT3 4BB G3SBT
DT3 4BL G8HCJ
DT3 4HG G3VPF
DT3 4JP G7LPN
DT3 4LD G0ZQE
DT3 4LD G4PTU
DT3 4LD M6DUK
DT3 4NZ M6WFL
DT3 5AE G4JMM
DT3 5AG G8AVG
DT3 5BG G3ZGP
DT3 5BP G3RAF
DT3 5BP G6FAD
DT3 5BP M0MJG
DT3 5HE G3OWE
DT3 5HF 2E0SEA
DT3 5HF M0ESU
DT3 5JS M0MCE
DT3 5LF G4XSC
DT3 5NG G3YWW
DT3 5PB G4RSL
DT3 5PF G3SDO
DT3 5QA G3RZG
DT3 5RP M6GKF
DT3 5RT M6EKF
DT3 5SA M0ACC
DT3 6AA M0XDL
DT3 6AA M6DLO
DT3 6AS G4TRW
DT3 6BG G0LQI
DT3 6DE G0NEV
DT3 6JW G1KDO
DT3 6LE G0FBS
DT3 6LF G0BMQ
DT3 6LF G3JRL
DT3 6LZ G6GYG
DT3 6NE G3EGV
DT3 6NH G3MSM
DT3 6NL G2CO
DT3 6PD G3XCY
DT3 6PT G3LAG
DT3 6PT G6LBJ
DT3 6PT M0BQO
DT3 6QL G1HTO
DT3 6RB M3UYL
DT3 6RD G0SEC
DT3 6RH G6XSK
DT3 6RP G3PGK
DT4 0AB G8VCN
DT4 0AJ M6SHW
DT4 0AS 2E1DNX
DT4 0AS G0RWL
DT4 0BA M3NMJ
DT4 0BE G0VLJ
DT4 0DS G1YHB
DT4 0ET G4ZXP
DT4 0EZ 2E1OZY
DT4 0EZ M0WIK
DT4 0EZ M3OZY
DT4 0FE G0EVW
DT4 0FE G3SDS
DT4 0FE G8WQ
DT4 0JE M5RIC
DT4 0JX G0ROX
DT4 0NA G4OWY
DT4 0NA G6AUW
DT4 0NJ G0ECX
DT4 0QF M3LUU
DT4 0QL 2E0PBS
DT4 0QY G1CAN
DT4 0SA 2E1TNE
DT4 7DY M1HFX
DT4 7HQ M3WJU
DT4 7JH G3YIC
DT4 7LX M3ZUZ
DT4 7PS G4SDJ
DT4 8RF G3UHU
DT4 8RJ G4XRR
DT4 8SG G0LVF
DT4 8SQ 2E0TPH
DT4 8SQ M3TPH
DT4 8SQ M3YTQ
DT4 8TX M3KIO
DT4 9AL G4FYT
DT4 9AL M3MHZ
DT4 9AU M0SGV
DT4 9BH G0HOS
DT4 9EZ 2E1HFS
DT4 9HJ G7OLW
DT4 9JU 2E1HSJ
DT4 9LE G8ZVS
DT4 9LQ M3MUF
DT4 9LU G2XQ
DT4 9NU G4BEI
DT4 9RL G8VKI
DT4 9RN G3VYX
DT4 9RR G7DOW
DT4 9RX M3XPI
DT4 9SA G4MFS
DT4 9SG G6LNF
DT4 9SP M6KPL
DT4 9UE G1BBT
DT4 9UU M6YOT
DT5 1AS 2E1RBH
DT5 1AS G0RYL
DT5 1BD G4WWH
DT5 1JP M3KKS
DT5 1NH G0PNG
DT5 2AA 2E1DSX
DT5 2AA G4DOL
DT5 2AA G4ZIY
DT5 2AB G0BZW
DT5 2AR G4YAA
DT5 2AY G0ACQ
DT5 2DE G0CAE
DT5 2DJ G2FHF
DT5 2DP M3ORV
DT5 2EA G4RAK
DT5 2EE G4CFZ
DT5 2HJ 2E1DQZ
DT5 2JG G7NIU
DT5 2JX M3VWP
DT5 2JZ 2E1NRL
DT5 2JZ G7EIS
DT6 3DR G8MCC
DT6 3EB G7KLZ
DT6 3PW G0JUV
DT6 3RW M6KLC
DT6 3UB G4PPZ
DT6 4AN G3PXH
DT6 4DW G8JCC
DT6 4EH M0CYG
DT6 4ES G0EZJ
DT6 4ET G4MZL
DT6 4HX G4JFG
DT6 4NQ G4XMZ
DT6 4NU M6CIS
DT6 4NY G6LEU
DT6 4NY M5ABC
DT6 4QL M0AXC
DT6 4QN G8IUN
DT6 4RG G4JWA
DT6 4RY M3BPF
DT6 4RY M3YAD
DT6 5HX G3MTP
DT6 5LS G6RMV
DT6 5QY G4UHU
DT6 5RF G4MZY
DT6 5RF G7BYG
DT6 6AL M3FRE
DT6 6BB G4HNG
DT6 6DF G0HET
DT6 6DF G1EFX
DT6 6DU G3ZUE
DT6 6JE G6VSE
DT6 6LR M0DRB
DT6 6PE G8ZTN
DT6 6SA G1EFG
DT7 3DT G8XBY
DT7 3EZ M0TGC
DT7 3SL G1GTM
DT7 3SX G0AYY
DT7 3UR G1ENA
DT8 3BQ G1SCN
DT8 3ES G6BTP
DT8 3HD M6BRZ
DT8 3HU G8JBQ
DT8 3JT G8LOJ

DT8 3PJ G1XHA
[illegible] G1PAC
DT8 3HF G4BQN
[illegible]
DT9 3RT G7AUU
DT9 3RT M3AHJ
DT9 4AT G3OPJ
DT9 4BJ G0FAA
DT9 4BP G7VQE
DT9 4EP G4DCH
DT9 4SZ M1BUP
DT9 5BD M0ARO
DT9 5BX M6URM
DT9 5ED G7UPP
DT9 5FD G4INX
DT9 5PD G8CIG
DT9 6AF M3DIB
DT9 6AR M6TER
DT9 6BG G7KBE
DT9 6BU G4GHI
DT9 6EJ M1FFA
DT9 6HL G3UCT
DT9 6HX M6DGM
DT9 6NQ G8AFA
DT9 6RG G2UH

## DY

(Dudley)

DY1 1SL G8ZZT
DY1 2AZ G0NXD
DY1 2DX G4JCP
DY1 2EF M0MGS
DY1 2ET 2E0NZD
DY1 2EU M3LWG
DY1 2GG G0OIY
DY1 2NL M6ROF
DY1 2NN M6LID
DY1 2NZ 2E0LSB
DY1 2NZ M3OLU
DY1 2PL M6SKU
DY1 2QZ G1BTV
DY1 2RT G7EDZ
DY1 2RX M6EVI
DY1 2SL G6VJC
DY1 2SN G4XME
DY1 2SN G6PPA
DY1 2UN G6ZSU
DY1 2UW M6IRW
DY1 3DF G6IRG
DY1 3ED G7OLC
DY1 3JA M6TSZ
DY1 3LE G0OWU
DY1 3NX M3XOV
DY1 3TN G7WGD
DY1 4DZ G4ZLK
DY1 4EP M6RNN
DY1 4NE 2E0COJ
DY1 4NE M0KCC
DY1 4NE M6BXO
DY10 1LQ M3NGK
DY10 1LR G0NFO
DY10 1LR G1ZNV
DY10 1SE 2E0DUB
DY10 1SE M6DUB
DY10 1SS G3MRC
DY10 1SS M0JMP
DY10 1UE M6GZB
DY10 1XJ 2E0KKO
DY10 1XJ M1DTO
DY10 1XJ M6KKO
DY10 1YH G7ABZ
DY10 2BZ G0MBG
DY10 2EZ M6RLZ
DY10 2HB G0MJX
DY10 2HD M0UCK
DY10 2LU G8DKD
DY10 2QY 2E0JGS
DY10 2QY M0JGS
DY10 2QY M6JGS
DY10 2RH G7EZE
DY10 2RY M3XNV
DY10 2ST G0ISG
DY10 2TH G1XJZ
DY10 2TP G4XCX
DY10 2UN G6BAM
[illegible]
DY10 2XG G4EFS
DY10 2XT G3SZG
DY10 2YB G0MJY
DY10 2YB G0WVT
DY10 3AA G6KRC
DY10 3AA G8WQX
DY10 3AG 2E1DLR
DY10 3AQ G4TXV
DY10 3BH G4OBC
DY10 3DG G4ILQ
DY10 3DT 2E0DGG
DY10 3DT M0MCO
DY10 3DT M3TIC
DY10 3EQ M0TTX
DY10 3FY M6GEU
DY10 3JG G0RQO
DY10 3JZ G7RDQ
DY10 3LH G7ILG
DY10 3ND 2E1IFW
DY10 3ND M0NYX
DY10 3ND M1EJG
DY10 3ND M3ZOG
DY10 3QR G8OXG
DY10 3QS G0EYW
DY10 3QS G4AFY
DY10 3QZ G6DQZ
DY10 3RY M6MCB
DY10 3TL G8TPM
DY10 3UA M6DTD
[illegible]
DY10 3YL G4OIK
DY10 3XP M3WVJ
DY10 3XS M6BXP
DY10 4HG G4ZLN
DY10 4JT 2E0WMP
DY10 4JT M6WMP
DY10 4NE M6GVN
DY10 4RS G4ROJ
DY11 5DL G8AKX
DY11 5DW G0FJD
DY11 5DY G6ORM
DY11 5EB G8UZV
DY11 5HJ G6BDY
DY11 5JN G0GPS
DY11 5JP G1AJU
DY11 5LB 2E0VEX
DY11 5LB M0MKV
DY11 5LB M6LHA
DY11 5LU G8TLH
DY11 5LZ G0VBP
DY11 5ST G8UEF
DY11 5TZ G4ZIB
DY11 5UA G0HRL
DY11 5UJ M0BRU
DY11 6AA G3TAW
DY11 6AU G4XWD
DY11 6AU M3XHH
DY11 6BX G4CTU
DY11 6BX G4GXP
DY11 6DQ G4PRD
DY11 6DQ M6NRH
DY11 6EE M6CEY
DY11 6JU G0EOF
DY11 6LF G3GGL
DY11 6LX G3ZUL
DY11 6PL G4BXD
DY11 6PZ G0HTF
DY11 6RL G3ZQQ
DY11 6RL G8DQP
DY11 6TE G8GYI
DY11 6TP M6RFN
DY11 6UG M6LVC
DY11 7BY G7DCJ
DY11 7DR G8SKA
DY11 7EW G0GFE
DY11 7HD G4ALT
DY11 7LA G4NPN
DY11 7PE M6RMW
DY11 7XS G1UNU
DY11 7XU G1JRX
DY12 1BY M6BGZ
DY12 1DB G0UDI
DY12 1DD G1OZB
DY12 1JH M0PGS
DY12 1TR G0ESH
DY12 2BP M6RTD
DY12 2HD G8EPR
DY12 2HT G4OIL
DY12 2HX G0TKT
DY12 2HX G8AAL
DY12 2JX G0LOZ
DY12 2PU G0KRC
DY12 2PU G4SPZ
DY12 2QG G4OCH
DY12 2UZ G7IQD
DY12 3AA G4SNO
DY12 3AE M0ABF
DY13 0AR G4DKP
DY13 0EB G0TUW
DY13 0EL G7ESI
DY13 0EQ G7JWL
DY13 0EW G4OPV
DY13 0HE G7PLP
DY13 0HJ G7KZJ
DY13 0HJ G8RCE
DY13 0JG M3XRR
DY13 0JG M6JEV
DY13 0JR G0TLP
DY13 0JT G0IBT
DY13 0LL G1LBK
DY13 0LT M6CAA
DY13 0NU G7SCZ
DY13 0NU G7WDC
DY13 0NY G0PMF
[illegible] G8SRD
DY13 0RU M3MGO
DY13 0RX G0PJM
DY13 8EL 2E0MOI
DY13 8EL G3GJL
DY13 8EL M3YRM
DY13 8JB M3YCT
DY13 8JG G7WFQ
DY13 8JG M1DRZ
DY13 8LP G6CBB
DY13 8LR G3VHL
DY13 8NT G6EHG
DY13 8NX M0LNX
DY13 8QB M6XWD
DY13 8QD G0PMG
DY13 8QP G0PWC
DY13 8RA G0FSH
DY13 8RZ G0LJK
DY13 8SR G8DUW
DY13 8SU G6CPO
DY13 8TA 2E0TEH
DY13 8TA M6BHA
DY13 8TF G0MKL
DY13 8TF G6YIS
DY13 8TH 2E1CVE
DY13 8UQ G7KPF
DY13 8XW G0MAL
DY13 9EU G8BKL
DY13 9LR G7SAI
DY13 9ND G7RBT
[illegible]
DY13 9NZ M0VXX
DY13 9NZ M3VXX
DY13 9PB G7IEO
DY13 9RY G7RXX
DY13 9RZ G6PZH
DY13 9SJ M6RQH
DY14 0QB G8IYE
DY14 0RG M6WME
DY14 0UA G4EQK
DY14 8AG G0EZT
DY14 8EB G8DFI
DY14 8EZ G4DFN
DY14 8JJ G1DFR
DY14 8LS G1WWA
DY14 8QQ G4EMD
DY14 8UF 2E0IZP
DY14 8UF M0IRP
DY14 8UF M6IRP
DY14 9DB G4ZKD
DY14 9HP G0EXB
DY14 9HX G0IMK
DY14 9LH 2E0BTR
DY14 9LH M3SFN
DY14 9LJ G3WBL
DY14 9NR G3WPQ
DY14 9TJ G8ETR
DY14 9YR G7FMF
DY2 0BN M0MWT
DY2 0DG G7WLV
DY2 0EE G0AMW
DY2 0EE G0BFZ
DY2 0EF M1RSJ
DY2 0HZ G4WBP
DY2 0JY M3AYP
DY2 0NW G1HRU
DY2 0NW G7ROP
DY2 0NW M0AKZ
DY2 0RZ M3UWU
DY2 7RA G4GSB
DY2 7RA G7CRS
DY2 7TQ G7AXW
DY2 7TS G6NTQ
DY2 8BB 2E0SLR
DY2 8BY M6WDH
DY2 8DH 2E0MTB
DY2 8DH M3FWO
DY2 8DY G6DFB
DY2 8ER G0GUC
DY2 8EY 2E0PPY
DY2 8HP 2E0NTC
DY2 8HP M0NTC
DY2 8LE M6NJO
DY2 8XT M6WFC
DY2 9EU G7IZM
DY2 9EZ M3RWR
DY2 9HE G8TEK
DY2 9HR M3YZA
DY2 9JN M3NCL
DY2 9JT M0GQE
DY2 9LA M0HRC
DY2 9LN M0LPF
DY3 1AL G6DKM
DY3 1LB G0BZT
DY3 1LF 2E0VPO
DY3 1LF M6YDW
DY3 1PD G1GST
DY3 1RF G0DPJ
DY3 1RL G4ICU
DY3 1TG G3XEV
DY3 1UU G7RXZ
DY3 1XQ M6PGF
DY3 1XW G0CSW
DY3 1YA M6FBM
DY3 2AU M6DTS
DY3 2AX 2E0GOS
DY3 2AX M3GOZ
DY3 2BB G4DFE
DY3 2BB G8FUI
DY3 2DQ G2HDF
DY3 2DQ M0BLT
DY3 2DQ M1AEA
DY3 2DQ M5HDF
DY3 2HJ M3IDD
DY3 2JF G7GEU
DY3 2JU G0CUZ
DY3 2LR G8WSB
DY3 2LR M1BVT
DY3 2PS G4HCZ
DY3 2QQ M0SJV
DY3 2RH G4NME
DY3 2RX G4IFM
DY3 2SH M6ARI
DY3 2UN G0LQF
DY3 2UU M3FGO
DY3 3AS G7GEL
DY3 3BH M1AZG
DY3 3DH G0GIE
DY3 3EF G8BOP
DY3 3LB G0OCR
DY3 3LH G0DAQ
DY3 0LN G7JIN
DY3 3PH M6ZSD
DY3 3QJ 2E0IMT
DY3 3QJ M0IMT
DY3 3QJ M3OND
DY3 3TH M0KSA
DY3 3YQ G3IIV
DY3 4NQ G0TMF
DY3 4NQ G1ZDF
DY3 4NQ M3ZDF
DY3 4NT G1BLJ
DY3 4RA G8KBG
DY4 0AF M6GZG
DY4 0AJ G6VAA
DY4 0AQ G7VUR
DY4 0SU M3HAJ
DY4 0TU [illegible]
DY4 4AU G3ZQS
DY4 4AU M0BPT
DY4 7PU 2E0SMZ
DY4 7PU M3OAL
DY4 7TP 2E0XEA
DY4 7TP M6CDB
DY4 8HZ G7JCD
DY4 8JX M6TGJ
DY4 9BJ M6NGD
DY4 9PQ G4OUH
DY4 9PQ G4OUI
DY4 9QS M3SMZ
DY4 9SR G6NJR
DY5 1DB G1YMH
DY5 1EQ G4PQI
DY5 2AY G6BEB
DY5 2BS M0DZD
DY5 2BU G1DIG
DY5 2DA G4NUS
DY5 2EL G7EVI
DY5 2HB 2E0NJO
DY5 2HB M6NAT
DY5 2HT G4YVA
DY5 2LQ G1YRQ
DY5 2LY G8JTL
DY5 2NG M1CWD
DY5 2PH G7FAZ
DY5 2QF M6CFE
DY5 2QG G4FAH
DY5 2SZ M6XVS
DY5 2XH M6ZXH
DY5 2XY G6HZK
DY5 3BU M3LTP
DY5 3DP G4TGM
DY5 3DP G6IWD
DY5 3JE G6LUM
DY5 3JE G6XKE
DY5 3NT M3IFK
DY5 3NY G4TDB
DY5 3NZ M1CGB
DY5 3PE G4DYU
DY5 3QH M6NJS
DY5 3RQ M6NAK
DY5 3YY G4NLW
DY5 4AW 2E0MLA
DY5 4AW M3HHN
DY5 4EF G0JKY
DY5 4EF G6JKY
DY5 4EF M6SUE
DY5 4EQ M0RSD
DY5 4EQ M0WYZ
DY5 4EQ M3FNC
DY5 4EQ M6LIP
DY5 4EQ M6PMR
DY5 4EX G0OWJ
DY5 4HA M3LKO
DY5 4JF 2E0TSC
DY5 4JJ M6PTP
DY5 4LB M3LTW
DY5 4LG M3CGM
DY5 4ND 2E0VCD
DY5 4ND M6DJX
DY5 4QL G8WSF
DY5 4QN G8TMQ
DY6 0BB 2E0KAB
DY6 0BB M0KAB
DY6 0HL G3SIO
DY6 0HL M1CBI
DY6 0HL M1DRM
DY6 0HL M1EGG
DY6 0HX G8NWS
DY6 0JG G4CGB
DY6 0LG M6EMM
DY6 0LG M6HJB
DY6 0LL G4CVU
DY6 0LN G0SRY
DY6 7AA G1MTU
DY6 7HG 2E1FPW
DY6 7QE 2E0ARA
DY6 7QE G0IPX
DY6 7QE G6ZUV
DY6 7QE M0CDL
[illegible]
DY6 7RQ G4LVA
DY6 8BT G1WFJ
DY6 8DJ 2E0PPZ
DY6 8EE G6PYI
DY6 8HH G7HVF
DY6 8JU M6YLI
DY6 8LL G4IEB
DY6 8LW 2E0BLT
DY6 8LW M3VDN
DY6 8PD G8YZF
DY6 8NQ 2E0NLY
DY6 8RZ G8KPG
DY6 8SP G4YBT
[illegible] G6CTR
DY6 8XP M3YVJ
DY6 9AD M6KEE
DY6 9DX G3ITH
DY6 9DY M3ZWH
DY6 9ET G4KEB
DY6 9EY 2E0KPD
DY6 9HB G0HCO
DY6 9HB G0RHH
DY6 9PR M6RIA
DY6 9RB M6YNA
DY6 9RE G3APL
DY6 9RJ G6PSZ
DY6 9RJ M6AKG
DY6 9RP G4VNE
DY6 9SS G4ITF
DY6 9TQ G1WJK
[illegible] G4CYB
DY7 6AD G1WDS
DY7 6BN G0TBI
DY7 6BN M0VMW
DY7 6BZ M6IGS
DY7 6DR G4ACS
DY7 6DX G3TDC
DY7 6ED G4IES
DY7 6ED G8MXR
DY7 6EE G0TDR
DY7 6HF G0AOW
DY7 6HF G4WRA
DY7 6HN G0ONH
DY7 6HW G3KZG
DY7 6LG M6EVK
DY7 6LS G6SRS
DY7 6LS G8UAE
DY7 6RR G0DBJ
DY7 6RW G0AOX
DY7 6SP G7FXZ
DY8 1AJ G4XNW
DY8 1AX M3UFF
DY8 1BJ G7HEZ
DY8 1EX G3IVQ
DY8 1HD G4MD
DY8 1NW M3ZWX
DY8 1QX M6FBY
DY8 2BE M1DBC
DY8 2BE M5LLT
DY8 2DE G8XNL
DY8 2DE M6KIP
DY8 2HL G3ORI
DY8 2LH G4OLL
DY8 3AB M3UUE
DY8 3AX G4DYK
DY8 3BD G1RDJ
DY8 3EF G6BDH
DY8 3EH 2E0ITC
DY8 3EP G0EZX
DY8 3ER M6KUT
DY8 3JE G3NDN
DY8 3JH 2E0KIA
DY8 3JH G7DTV
DY8 3JR G7JWJ
DY8 3NA M6AHW
DY8 3NG G0UUZ
DY8 3PH G4HWH
DY8 3PJ G4FYQ
DY8 3RN M6PDC
DY8 3RP G0MWS
DY8 3SY M3TFG
DY8 3UF G8OKB
DY8 3UZ G7FMQ
DY8 3XT G4CVK
DY8 3XT G4WAO
DY8 3XU M3ZJM
DY8 4BW M0KMT
DY8 4HX M1CZH
DY8 4JE M6SHA
DY8 4JQ M3PML
DY8 4QF G6WMR
DY8 4QF G8RLN
DY8 4QS G0FLW
DY8 4RN G4OLS
DY8 4SF G8PIP
DY8 4XS G0ASK
DY8 4XS G0ASL
DY8 4XS G4TFB
DY8 4XS G4TFC
DY8 4XU G0NLT
DY8 4XX G1IHJ
DY8 4YW M6ATW
DY8 5AU G0WBA
DY8 5JB M6HIB
DY8 5JJ G4WIG
DY8 5LR G1VHN
DY8 5LU 2E0ECM
DY8 5LU M6EVF
DY8 5PH G0KVK
DY8 5PU G0OYY
DY8 5RF M6DIH
DY8 5YJ G0OVV
DY9 0JH G7RLX
DY9 0LX G6YAK
[illegible]
DY9 0RE G4VPE
DY9 0RT G6DDO
DY9 0ED G7VFL
DY9 0UP G0TZV
DY9 0YB 2E0ZDH
DY9 0YB G4IIC
DY9 0YB M0ZDH
DY9 0YE 2E0ZXJ
DY9 0YE G3PWJ
DY9 0YE M0ZXJ
DY9 0YE M3XSJ
DY9 0YH G1NZZ
DY9 7AZ G0WWA
DY9 7DT G7JDK
DY9 7EW M6JBT
DY9 7LA M1BQM
DY9 7PS G3XOV
DY9 8DE G1OHV
DY9 8SG G8JBP
DY9 8SG M3LKU
DY9 8YG M3KJM
DY9 9AE G0HTJ
DY9 9AE M3FFV
DY9 9AH G1DCU
DY9 9AH M3KTT
DY9 9BZ M6PMG
DY9 9DE G7EZH
DY9 9DL G7OJZ
DY9 9DL M0CHD
[illegible]
DY9 9EJ G0EWH
DY9 9EJ G0KZM
DY9 9EL G4XOM
DY9 9HH G7JYZ
DY9 9HT G6RSI

## E1

(East London)

E1 0EE G6USX
E1 0EE M3FLC
E1 2AG 2E0MTM
E1 2AG M6MQT
E1 2QR 2E0SAW
E1 2QR M0WSA
E1 2QR M6MFZ
E1 8PW M0HZY
E10 5QD 2E0OON
E10 6DX 2E0WJB
E10 6DX M6EHV
E10 6DX M6SRE
E10 6EE G4VXP
E10 6JD M3SJV
E10 6JL 2E0COL
E10 6QT G8VZD
E10 7EB G0TVV
E10 7JS M0GZK
E10 7LQ M0GHV
E11 1BA G7JRJ
E11 1JX G0VEH
E11 1JX G4ONP
E11 2JD G4GTZ
E11 2LD M3HTF
E11 2PP G4GGT
E11 2QQ M0CMQ
E11 2SH 2E0BYO
E11 2SH M3YWU
E11 4AA G7LAF
E11 4HY G1JPS
E12 5AX 2E0TMD
E12 5AX M6TMD
E12 5DZ G1SEF
E12 5EF G6RVZ
E12 5HD 2E1FZC
E12 5HD M3FZC
E12 5NB 2E0EWM
E12 5NB M3ZEW
E12 5PQ M3FNM
E12 6AA M0EDP
E12 6RE M3GVN
E14 0FN M6EBI
E14 0SL M3YSM
E14 3AJ G8XCJ
E14 3UA G8IWR
E14 6HP G6DFY
E14 6HU M1HMP
E14 6LS G1GLN
E14 7DX G8BBC
E14 7DX M5AEO
E14 7JX G0BEZ
E14 7TE 2E0VGV
E14 7TE M0VGV
E14 7TE M6VGV
E14 8BG G0UPL
E14 8EY G0KBO
E14 8LH G4LXH
E14 8PH M0VWW
E14 8SR G1RYS
E15 3AL M0PSS
E15 4LP M6PDS
E16 1LQ 2E0NLM
E16 1LQ M3NLM
E16 1QW M6AUM
E16 2QW M6PFV
E16 2QY G1NPJ
E16 3JB 2E0XMT
E16 3JY M6WHT
E16 3NJ M3TGL
E16 3NJ M3UVS
E16 3QL 2E0CXO
E16 3RR G6XDN
E16 3RY M0GYU
E16 3TA G0MXE
[illegible] M6WTR
E16 4NA M6KCD
E17 3RJ G3PKQ
E17 4BD G1HLS
E17 4ES G8JDN
E17 4HH G8PWK
E17 4PR M0TNL
E17 4QY G4IWD
E17 5AZ M1ADZ
E17 5BL M6JGN
E17 5EY G8KNF
E17 5HN G4TLT
E17 5RG 2E0PNB
E17 5RG M0PNB
E17 5RG M6TNB
E17 6NN M6RTE
E17 7ER M1EUE
E17 7HD M3EUE
E17 7HD M6EKY
E17 8AF M1CQM
E17 9AZ M0MLH
E17 9EH G8BVY
E18 1AP G0FMB
E18 1BN G8ZHS
E18 1DG 2E0RCW
E18 1DG M6ECW
E18 2AB G3XXC
E18 2DA G4PSR
E18 2DB M6IBG
E18 2HB 2E0LGS
[illegible]
E18 2NJ G6SPN
E18 2PL G0KHX
E18 2PZ G6XBD
E18 2QA G4LKT
E1W 2BG M0IBD
E1W 2QW G1ETD
E1W 2UT M6UBC
E2 0HE G6EUU
E2 8LP 2E1HBT
E2 9LY G7RGA
E3 2BB G0NXM
E3 3EG M0CVA
E3 4DB 2E0MSM
E3 4DB M6AQE
E3 4GS 2E0BCK
E3 4GS M0EAQ
E3 4HT M0GFD
E3 4HT M3XXA
E4 6DR G6EXN
E4 6EG M6ITY
E4 6HX M6NJT
E4 6NA G4CGA
E4 6RP M6DZQ
E4 6RT G4RMQ
E4 6UF M6DQF
E4 7BP G3JPG
E4 7DT 2E0EAL
E4 7DT M0GZM
E4 7DT M6JRO
E4 7DX G8IUC
E4 7EG 2E0JGE
E4 7EG M6JGE
E4 7HN G2XG
E4 7JG M1BLJ
E4 7LG G8JYX
E4 7ND G6XRF
E4 7NR G7ITX
E4 7PZ G0DPT
E4 7PZ G1MBW
E4 7PZ M3OAX
E4 7RA 2E0TKX
E4 7RA M0TKX
E4 7RA M6TEK
E4 8BG M3WPC
E4 8DZ G4NEL
E4 8DZ G8MJH
E4 8HD G4RSX
E4 9DU G0LWN
E4 9HE G6SKR
E4 9JA 2E0LDN
E4 9JA M6XAX
E4 9NW G8XAO
E4 9RE 2E0MMZ
E4 9RE G5YC
E4 9RE M0BPQ
E4 9SJ G7IQM
E5 0LY M6BKF
E5 0RG G4SHO
E6 2BS G1SJU
E6 2DQ G4MVX
E6 3BW M0GTO
E6 3DH G1ERS
E6 3RY G0JYJ
E6 5XX 2E0TBW
E6 5XX M0LGN
E6 5XX M6CXP
E6 6BB M6LEG
E6 6HG 2E0CKR
E6 6HG M0HFZ
E6 6HG M6BKU
E7 0DN M0GHE
E7 0HN G7KWS
E7 0JS G0LLE
E7 0JT M6SQD
E8 2AD M0DYW
E8 2ET M6LOC
E8 3BT M6TZD
E8 3LE G0IFD
E8 4DA 2E0ZSK
E8 4DA M6ZSK
E8 4LN G4UAV
E9 7EH G3SJW
E9 7HU G3WUB

## EC

(East Central London)

EC1M 7AJ 2E0KRX
EC1M 7AJ M0KRX
EC1M 7AJ M6KRX
EC1R 5XB G3WIP
EC1V 7DX G0MZF
EC1V 7NS M6BTH
EC1Y 8NL G0JKM
EC1Y 8TB G7CUF
EC4V 3EJ G0PCE
EC4Y 7EX G8PBK

## EH

(Edinburgh)

EH1 2HR GM7ORX
EH10 4EQ GM1THR
EH10 4JH GM7SWB
EH10 5AG 2M0PCW
EH10 5AG MM6HSS
EH10 5BJ 2M0YCJ
EH10 5BJ MM0YCJ
EH10 5BJ MM6YCJ
EH10 5DS GM4DTH
[illegible]
EH10 5JY 2M0CFA
[illegible]
EH10 5JY MM3RKF
EH10 5LW MM0AKM
EH10 5NU MM3FKO
EH10 5QH MM0JVQ
EH10 5SJ GM7GNO
EH10 5SL GM0VMV
EH10 5SN GM7RMF
EH10 6ER GM8OTI
EH10 6HN GM8BJF
EH10 6HN MM6MWF
EH10 6NN GM4ZOA
EH10 6PS MM0YMG
EH10 6PU GM3YMX
EH10 6XE MM0KCS
EH10 6XE MM3MDB
EH10 7AQ GM3PSP
EH10 7BU GM3HAM
EH10 7BU GM4BYF
EH10 7BU GM6KAY
EH10 7DF GM4DMI
EH10 7DF GM4LBN
EH10 7DG GM7REG
EH10 7HY GM7WFT
EH11 1HB MM0HVA
EH11 1HB MM6HBF
EH11 1JZ GM8LYQ
EH11 1LR MM6EVO
EH11 1RP GM0AXY
EH11 1RP GM4YMM
EH11 1SS GM0KZJ
EH11 1TN MM3IIG
EH11 1TT MM0ZBD
EH11 3GZ GM4AJV
EH11 3PF GM4DIZ
EH11 3SY MM3UDV
EH12 5AY GM6AXZ
EH12 5HG 2M0CLF
EH12 5HG 2M0JCF
EH12 5LG GM7JGR
EH12 5NG GM3HNE
EH12 5RF GM4HYR
EH12 5RP GM3HUN
EH12 6HN MM3DVD
EH12 6NB GM0BPT
EH12 6NB GM0CMO
EH12 6UH GM0CBQ
EH12 6XB GM4DIJ
EH12 7EB GM7UJJ
EH12 7EW GM1CNH
EH12 7EW MM0BMG
EH12 7HP MM6CFA
EH12 7JW 2M0CBC
EH12 7JW MM3ZQW
EH12 7PD GM1PEL
EH12 7PU GM8CSE
EH12 8DH GM0PXV
EH12 8DL GM0SSQ
EH12 8DL GM3VJY
EH12 8DL GM4CUX
EH12 8EE GM8DKB
EH12 8JH GM4YWI
EH12 8LD 2M0DGJ
EH12 8LD MM0HVF
EH12 8LD MM6ERO
EH12 8NQ GM7VLZ
EH12 8QA GM4ZRR
EH12 8QU MM0DMU
EH12 8QW GM3RVL
EH12 8SP MM0XRI
EH12 8SP MM6XRI
EH12 8SY GM1CHT
EH12 8TB GM4DHN
EH12 8UW MM3LWZ
EH12 8XW GM3XUW
EH12 9JF MM3GOI
EH13 0AF 2M0RMN
EH13 0AF 2M0SMN
EH13 0AF MM6LON
EH13 0AF MM6RCO
EH13 0AF MM6SMN
EH13 0EW GM4ZQH
EH13 0JH MM3UKG
EH13 0NA 2M0WRX
EH13 0NA [illegible]
EH13 0NA MM3KJG
EH13 0RA GM8CVN
EH13 9AG GM8IIO
EH13 9BE MM0WXT
EH13 9BL MM6KOS
EH13 9DB GM7RBW
EH13 9LF GM0FMW
EH14 1EF GM3SHV
EH14 1ET GM8PAH
EH14 1HE MM1ICE
EH14 1HQ GM4DZW
EH14 1HW GM4EZJ
EH14 1JW GM6NIA
EH14 1NJ GM0VXB
EH14 1NJ MM6VXB
EH14 1UH MM0AWU
EH14 2LB GM0AXX
EH14 3DZ GM4YLN
EH14 3EE MM6MSA
EH14 5DW GM3OWU
EH14 5DX GM6MUZ
EH14 5EZ GM4XZN
EH14 5JN GM8LKL
EH14 5JN GM8NZL
EH14 5PP GM4REN
EH14 5QH MM6RCK
EH14 5RN MM6LAO
EH14 5RR MM0WAK
EH14 5SA GM7RYT
EH14 5SX [illegible]
[illegible]
EH14 5SY GM1FEM
EH14 5SY MM0PSA
EH14 7ED GM1GDO
EH14 7EJ GM4EAU
EH14 7HD GM0CFW
EH14 7HS GM4RAH
EH14 7JE GM1JKJ
EH14 7JE MM0MGB
EH15 1JP MM6BMQ
EH15 1NB GM8BHR
EH15 1NH GM7CZC
EH15 1RA GM3DIE
EH15 1SG GM0VIV
EH15 2BD GM1HJX
EH15 2HE 2M0INS
EH15 2HE MM0INS
EH15 2HE MM6INS
EH15 2PX 2M0TXK
EH15 2PX 2M0XBD
EH15 2PX MM0SBO
EH15 2PX MM0TKE
EH15 2PX MM6TXK
EH15 2PX MM6XBD
EH15 3QG GM4BHU
EH16 4ED MM6JAE
EH16 4LX MM3VVI
EH16 4TG GM7IFX
EH16 4UH 2M0WMU
EH16 4UH MM6TXN
EH16 5DD 2M1GFG
EH16 5EE 2M0BUX
EH16 5EE MM0GSQ
EH16 5EE MM6ART
EH16 5EE MM6SGE
EH16 5UL GM4WCE
EH16 5UW GM8LFI
EH16 6HF GM4UPN
EH16 6YE GM4HWO
EH17 7RP 2M0FFY
EH17 7RP MM6HCK
EH17 7SE 2M0SRF
EH17 7SE MM0VTV
EH17 7SE MM3SRF
EH17 7SE MM6GHF
EH17 7SQ MM6FJD
EH17 8AW GM4DTJ
EH17 8DX MM3MTM
EH17 8SR GM6OUL
EH17 8UF 2M0DTP
EH17 8UF MM6FYF
EH18 1AB GM4DCL
EH18 1LN GM6SEV
EH19 2EH GM0OTU
EH19 2EU GM8HSY
EH19 2LD GM4SUR
EH19 2PQ GM0FQS
EH19 3JG 2M0SRY
EH19 3JG MM6DKI
EH19 3LQ GM4GGF
EH2 3NS MM3LGU
EH20 9EE GM3BST
EH20 9JT 2M0SWA
EH20 9JT MM3SWA
EH20 9LU GM7LTX
EH20 9PA GM6JGH
EH20 9RR GM0PQV
EH20 9SN GM6WMA
EH21 6AR GM7VLC
EH21 6EX GM0VYL
EH21 6RR MM0JXI
EH21 6SB GM6KFO
EH21 6TU 2M1DZW
EH21 6TU GM7PPN
EH21 7RD GM0EFQ
EH21 8AA 2M0BEB
EH21 8AA MM0FZV
EH21 8AA MM3ZVY
EH21 8JT GM1FAF
EH21 8JT GM4GVJ
EH22 1EE GM4CMI
EH22 2LW MM0CHV
EH22 2NS GM1CCN
EH22 2NZ GM0LOD
[illegible]
EH22 3BN GM4HQU
EH22 3DB GM0SUF
EH22 3DR 2M0KPZ
EH22 3DR MM6KLZ
EH22 3LT GM0TFE
EH22 4BW MM1CMU
EH22 4BW MM3ALG
EH22 4BW MM3BHD
EH22 4QD 2M0YSR
EH22 4QD MM0YAD
EH22 4QD MM6YAP
EH22 4SJ MM0GFP
EH22 5AX MM3KVY
EH22 5BY 2M0YAG
EH22 5BY MM0GTB
EH22 5BY MM3YMQ
EH22 5DT GM7DRY
EH22 5DT MM3ZDG
EH22 5ER 2M1BYW
EH22 5HX GM0PKP
EH23 4BT GM0KDC
EH23 4EQ MM0BAC
EH23 4EU MM3DHN
EH23 4GL GM7UFO
EH23 4JF 2M1HIN
EH23 4NT GM0IMZ
EH23 4SQ MM3CGC
EH25 9NP GM4LHW
EH25 9NR GM4DEK
[illegible]
EH25 9SR GM6BBU
EH25 9TD MM0AJQ
EH26 0EL GM0MJR
EH26 0LL 2M0CEX
EH26 0LL MM0VPR
EH26 0LL MM6ANB
EH26 8BW MM6DWC
EH26 8DF GM1GEQ
EH26 8DG GM3YXJ
EH26 8EF GM0LUF
EH26 8HH MM6GBS
EH26 9AN GM1OVW
EH26 9EE GM6GFL
EH26 9HS GM1PSZ
EH27 8AN GM7KZL
EH27 8AU 2M0DIF
EH27 8AU MM6DHZ
EH27 8BH GM7HHB
EH27 8BH MM3RIX
EH27 8BS MM1RAH
EH27 8DQ GM1ROM
EH28 8PD GM4XHQ
EH28 8PD MM0DLH
EH29 9AN GM1YFO
EH29 9DU MM3TUR
EH29 9DX MM0SFM
EH29 9DZ GM6BEY
EH3 5LP MM0WXD
EH3 5NA GM8EKF
EH3 5PD GM7PNX
EH3 5QP MM0DNH
EH3 6HB GM6KOR
EH3 6RP 2M0BDN
EH3 7TZ GM6CZM
EH3 7TZ GM6NIC
EH30 9PL GM7RAK
EH30 9PQ GM4XWL
EH30 9UP GM8ZMF
EH30 9XG GM0SUY
EH32 0AN GM4LRU
EH32 0AQ 2M0WLA
EH32 0AQ MM0MLD
EH32 0AQ MM6YLA
EH32 0BB 2M1IBH
EH32 0BP GM0TCC
EH32 0BY MM3FJX
EH32 0EE GM4UYZ
EH32 0EE MM0CPS
EH32 0HZ 2M1HVR
EH32 0LQ GM8ZJS
EH32 0PG 2M1DZS
EH32 0PG 2M1EUV
EH32 0PG 2M1GLD
EH32 0SY 2M0BUY
EH32 0SY MM3YMN
EH32 0TA GM4TAL
EH32 0TN GM4IKT
EH32 0TU MM0AMV
EH32 9AT MM0CGZ
EH32 9FE GM0FZM
EH32 9GF MM6SGF
EH32 9HF MM3MFN
EH32 9PX GM7PSH
EH32 9RD 2M1IBE
EH32 9RD MM0IBE
EH33 1HU MM3XDW
EH33 1QE 2M0EAC
EH33 1QE MM3EWI
EH33 1QE MM3XVD
EH33 1QJ MM6JLX
EH33 2AL GM0NTL
EH33 2AL MM0WCG
EH33 2AR MM3ITA
EH33 2AS MM0CCC
EH33 2EB MM0DXC
EH33 2EE 2M1ETM
EH33 2HR MM6AUJ
EH33 2PL MM1ATY
EH33 2QT MM6ZCD
EH34 5BB GM7MAG
EH34 5EH GM3GKJ
EH37 5QB 2M0KLL
EH37 5QB MM0LBF
EH37 5QB MM3ISA
EH37 5QB MM3LWJ
EH37 5QD 2M0GRE
EH37 5QD 2M0GZA
EH37 5QD MM0GZA
EH37 5QD MM3YPN
EH37 5QD MM6YPN
EH37 5QS 2M0DXC
EH37 5RA 2M0PMR
EH37 5RL MM3EYM
EH37 5RL MM3EYN
EH37 6RS GM8MNG
EH37 5SQ GM3OIB
EH37 5UJ MM3DZW
EH37 5UP GM8MJV
EH38 5YD 2M0NOP
EH39 4BP GM0EHL
EH39 4NH GM0WPW
EH39 4PS GM4KGZ
EH39 5JE MM0JMI
EH4 1NH 2M0GUL
EH4 1NH MM0DXH
EH4 1NH MM3JHS
EH4 1NT GM6WTP
EH4 1PJ GM8SOK
EH4 2AE MM0WST
EH4 2AE MM3PYX

Postcode

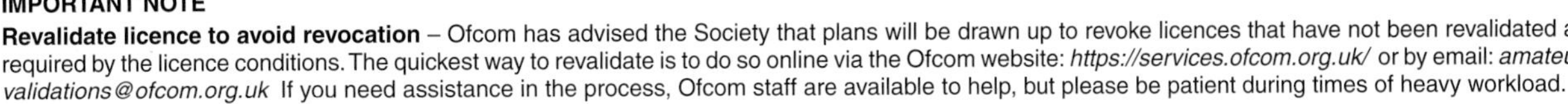

**IMPORTANT NOTE**

**Revalidate licence to avoid revocation** – Ofcom has advised the Society that plans will be drawn up to revoke licences that have not been revalidated as required by the licence conditions. The quickest way to revalidate is to do so online via the Ofcom website: *https://services.ofcom.org.uk/* or by email: *amateur.validations@ofcom.org.uk* If you need assistance in the process, Ofcom staff are available to help, but please be patient during times of heavy workload.

EH4 2AU GM4SVM
EH4 2EF GM4GIO
EH4 2JU GM4DMQ
EH4 2ND MM0MRO
EH4 2PE GM0HWQ
EH4 2TT GM1CQC
EH4 2UE GM1PHD
EH4 3JU GM8MST
EH4 3TP GM0AXM
EH4 5AW GM0RMV
EH4 5BF GM4HCE
EH4 5BQ GM7IRI
EH4 5DT GM4WZP
EH4 5JA GM3SBC
EH4 5PY GM6PYD
EH4 6EB GM6UNQ
EH4 6EY MM6FXW
EH4 6JJ MM6TEQ
EH4 7AH GM3KBP
EH4 7BJ GM7KTY
EH4 7BJ GM7PZH
EH4 7DF GM0WFB
EH4 7HN GM1UWE
EH4 7RQ MM3BUZ
EH4 8BA MM6NOR
EH4 8DL MM3ICD
EH4 8HH MM6JXH
EH40 3BH GM4HJQ
EH40 3EB GM8PEB
EH41 3BE 2M0WDG
EH41 3BE MM6FXZ
EH41 3LS MM1DSX
EH41 4QA 2M0TJO
EH41 4QA MM0HZI
EH41 4QA MM6FXM
EH41 4RU GM3LGU
EH42 1AY GM4FQG
EH42 1BA GM0JPG
EH42 1GJ 2M0FLJ
EH42 1GJ MM3SQJ
EH42 1NW GM6KEV
EH42 1QT MM6GDY
EH42 1RT GM1SYC
EH42 1UF MM3KZD
EH42 1YR GM3VEI
EH44 6JT MM0ABJ
EH44 6LZ 2M0NMD
EH44 6NJ 2M0HJP
EH44 6NJ MM0VPF
EH44 6NJ MM6VPF
EH45 8DH 2M0LAO
EH45 8DH MM6LAK
EH45 8HB GM6JFP
EH45 8JE GM1RKI
EH45 8NU GM0UTD
EH45 8PP GM4WTK
EH45 8PW GM4LDX
EH45 8QZ MM6DAT
EH45 9DB GM1MSS
EH45 9ER GM8CFS
EH45 9HX GM8NAL
EH45 9LX GM1OPO
EH45 9LX MM3NTX
EH46 7BA GM4OGM
EH46 7BD GM8ZCS
EH46 7BE 2M0SPL
EH46 7BE MM0SPL
EH46 7BE MM3PZJ
EH46 7BE MM6HBR
EH46 7BH GM4ZDH
EH47 0BH 2M0WEV
EH47 0BH MM3WEV
EH47 0JQ 2M0HZL
EH47 0JQ MM0HZL
EH47 0LH MM6TMS
EH47 0NL MM0SMB
EH47 0SE MM0HSR
EH47 0SE MM0PTE
EH47 7QJ MM6CMM
EH47 8AT MM6DSC
EH47 8EW MM0WKJ
EH47 9AZ MM6DGC
EH47 9BW 2M0DGB
EH47 9BW MM0GOG
EH47 9BW MM3NVD
EH48 1AS GM8ZKN
EH48 1DD GM0KBR
EH48 1DF GM4LHJ
EH48 1DU MM0LGS
EH48 2AF MM6GBX
EH48 2AU 2M0RDK
EH48 2AU MM6RWA
EH48 2BD MM6DDN
EH48 2BU MM3UPY
EH48 2GZ GM1WMU
EH48 2LT MM0IEJ
EH48 2PB MM5AES
EH48 2RG MM6RBE
EH48 2RR MM0DIS
EH48 2TD GM1PSU
EH48 3BU MM6RCZ
EH48 3HB MM6STM
EH48 3HG MM0HVU
EH48 3JB GM0DXB
EH48 3JU GM0EWF
EH48 3LA MM6EOU
EH48 3PP MM5AHO
EH48 4BB MM3HTY
EH48 4BB MM6INC
EH48 4DP 2M0XFM
EH48 4DP MM3XFM
EH48 4HG 2M0CTB
EH48 4HG GM0EDQ
EH48 4HG MM6VWT
EH48 4LD GM4LZO
EH48 4NU GM4BQD
EH48 4NW MM1DQV
EH482HH GM8IOL
EH49 6BP 2M0MIF
EH49 6BP MM6DCM
EH49 6BS GM3PDX
EH49 6DD GM1FAI
EH49 6DQ MM6HMC
EH49 6HA GM4LUS
EH49 6HA GM4XUS
EH49 6LW GM7VPT
EH49 6SD 2M0GAN
EH49 6SF MM3TZP
EH49 6SH GM0MWJ
EH49 7AP GM4CAQ
EH49 7BP GM4YPL
EH49 7BP MM0ZCT
EH49 7BS GM0FTH
EH49 7JU GM3VTH
EH49 7LD GM0RUW
EH49 7LN 2M0CQI
EH49 7LN MM6RKT
EH49 7ND 2M0APX
EH49 7ND MM0ASB
EH49 7QE MM6AHX
EH49 7RJ MM0GTU
EH49 7SR MM6EOZ
EH5 1EX MM0CJT
EH5 1EX MM0XXW
EH5 1FD MM6HQE
EH5 1FT GM8YIK
EH5 2JJ GM0CII
EH5 2NJ GM7UXH
EH5 3AR GM0TEA
EH5 3NH GM4AOR
EH51 0BQ 2M0IRC
EH51 0BQ MM3RQC
EH51 0DD GM0CQQ
EH51 0NX 2M0DRO
EH51 0NX MM6RDU
EH51 9ER GM0IKY
EH51 9PE MM0FWG
EH51 9QD GM4HRL
EH52 5BX MM6HSC
EH52 5HQ 2M1GBG
EH52 5NS 2M0MMG
EH52 5NS MM0MMG
EH52 5NS MM3IQD
EH52 5PN GM4VDG
EH52 6FB 2M0ZEB
EH52 6HA MM0FFC
EH52 6LY GM0UHC
EH52 6PL GM0DHD
EH52 6QB GM3OIV
EH52 6TH MM0IRJ
EH52 6UW MM3MRX
EH52 6XT MM6EUV
EH53 0EG GM7GIF
EH53 0NG MM0PWM
EH53 0NT 2M0CRR
EH53 0NT MM3LYH
EH53 0SJ GM0ERT
EH53 0TE GM7RYK
EH54 5AD GM0NAZ
EH54 5JP GM6OGN
EH54 5JP MM6SIV
EH54 5LE MM6FFQ
EH54 5LP GM0REZ
EH54 5LP MM0MOC
EH54 5NA 2M0FTA
EH54 5NA GM0ALS
EH54 5NA MM6FTA
EH54 6BG MM6WAW
EH54 6DF MM0DMK
EH54 6EE 2M0AWY
EH54 6HE 2M0DIB
EH54 6HE 2M0MHN
EH54 6HE MM0NHM
EH54 6HE MM6NHM
EH54 6HE MM6NSM
EH54 6JD 2M0GGY
EH54 6JD MM3FOE
EH54 6JJ GM0CBA
EH54 6LP 2M0RWH
EH54 6LT GM7GTS
EH54 6PG GM4UQD
EH54 6PG MM6FVK
EH54 6PQ MM6JNN
EH54 6RJ GM4HML
EH54 6RP GM3UCN
EH54 6TB MM6HAV
EH54 6UX MM1EYI
EH54 7AB MM6JMI
EH54 7BP 2M0TZB
EH54 7BP MM6NLP
EH54 7BZ GM7AHA
EH54 7DW MM3KLO
EH54 8EN 2M0DQY
EH54 8EN MM0SNK
EH54 8EN MM6DQY
EH54 8EW MM0HOL
EH54 8EW MM6HSK
EH54 8HQ MM0GEO
EH54 8HW GM8MYO
EH54 8JB GM1PZT
EH54 8JG 2M0LAS
EH54 8JG MM3GTF
EH54 8JN 2M0OSC
EH54 8JN MM0HSA
EH54 8JN MM3YHS
EH54 8LA MM6NRE
EH54 8NS MM3MPK
EH54 8NX MM6JCZ
EH54 8QP MM6EQY
EH54 8RW MM3GBL
EH54 8RW MM3TWG
EH54 8TA MM6XRS
EH54 9AA MM0DEC
EH54 9AA MM0NEW
EH54 9BZ MM3DYT
EH54 9DT GM0LOT
EH54 9DW 2M0CNZ
EH54 9DW MM0JRR
EH54 9DW MM6BTR
EH54 9EE GM7HUD
EH54 9JE GM3ZZA
EH54 9JG 2M1AVZ
EH54 9LD MM3IDR
EH55 8EW MM6XXV
EH55 8LW GM7IKB
EH55 8NL 2M0CPV
EH55 8NL MM0JHL
EH55 8SU MM3YTI
EH6 4DU 2M0BUT
EH6 4DU MM0WAP
EH6 4DU MM3RQP
EH6 4EZ MM3EXW
EH6 4NY GM0HSC
EH6 4RU GM7GAE
EH6 4TE GM6HVY
EH6 4TR GM4HQZ
EH6 5JB 2M0TIP
EH6 5JB MM6TYM
EH6 7BZ MM6HQI
EH6 7HN 2M0PID
EH6 7HN 2M0VTB
EH6 7HN GM4FH
EH6 7HN MM0PID
EH6 7HN MM3VAH
EH6 7HN MM3VTB
EH6 7JH GM4IUS
EH6 7QQ MM6AUG
EH6 8AF GM4MCV
EH6 8BA MM0PHD
EH6 8QD MM6BYJ
EH6 8QP GM7AUW
EH6 8QP MM3AUX
EH65RY MM3WBV
EH7 4HN 2M0TOK
EH7 4NJ GM3XQP
EH7 5JA GM1YME
EH7 5NG 2M0AYZ
EH7 5SA 2M0LBH
EH7 5SA MM6RHI
EH7 5TT GM7IHZ
EH7 5UF GM0HLK
EH7 6AU GM0LVL
EH7 6TH MM1FHL
EH8 7DW MM0BQI
EH8 7JL GM4JEM
EH8 7SP 2M0ECK
EH8 7SP MM3ZQX
EH8 8DZ MM6NCP
EH8 8HA MM0HLU
EH8 9HX MM6HZO
EH9 1AY MM0HQI
EH9 1PL MM0RHL
EH9 1QG GM7OCU
EH9 1UF GM0PVQ
EH9 2EH GM4XNQ
EH9 2EL GM3FRU
EH9 2JL GM4HAO
EH9 2JY 2M1GYX
EH9 2NZ MM4GZW
EH9 2NZ MM3NQT
EH9 3AR MM0GYG

## EN

## (Enfield)

EN1 1BS M6PMK
EN1 1NE 2E0DGD
EN1 1NE M0HLX
EN1 1NE M6DGD
EN1 2BZ G4OBE
EN1 2JU G3YJZ
EN1 3AE G0NEU
EN1 3DE G1GTS
EN1 3DW G4RTC
EN1 3NQ G3RWL
EN1 3NT G8PSF
EN1 3NU G8TWR
EN1 3RG M6WWE
EN1 3UP 2E0DHE
EN1 3UP M0HVC
EN1 3UT M5AJK
EN1 3UU G0NQV
EN1 4AD M1EJQ
EN1 4BD G3ATC
EN1 4BD M5AGW
EN1 4DY 2E0RHL
EN1 4DY 2E0RIA
EN1 4DY 2E0WOB
EN1 4DY M6SOO
EN1 4DY M6WIF
EN1 4DY M6ZUB
EN1 4HR G4BUB
EN1 4PS G4IPR
EN1 4SY M6SGC
EN1 4TX G7MNK
EN10 6EE G6XAR
EN10 6HD G0HBL
EN10 6JA G4UOI
EN10 6JN G6DMF
EN10 6JN M6DMF
EN10 6JN M6WKS
EN10 6LD G4GAK
EN10 6NY G4XVY
EN10 6QE 2E1IBS
EN107NR M6CKJ
EN11 0BB G1SDK
EN11 0JS G8HTA
EN11 0LP M3MRJ
EN11 8RW G0JXR
EN11 9DG G1MRE
EN11 9DG G4ARY
EN11 9DL M6EHD
EN11 9DX 2E0CHW
EN11 9DX M3VCW
EN11 9EP G3VSJ
EN11 9JR G1PZP
EN11 9JS G4YUZ
EN11 9NR G1CAY
EN11 9QH M0LSE
EN11 9QH M0XDS
EN11 9QN G7HYG
EN11 9QS 2E0RNT
EN11 0QH M6MZJ
EN2 0DZ M0NOE
EN2 6NQ G6ODA
EN2 7BY G4XIN
EN2 7BY M0LTO
EN2 7DB M1MUS
EN2 7EL G8AQO
EN2 7EL M0DDC
EN2 7EL M0HCY
EN2 7EN G8MOL
EN2 7PD M3FDM
EN2 8EN M6ZIP
EN2 8FG 2E0KRR
EN2 8FG M6RRK
EN2 8HS G4KZD
EN2 8HS G4ODR
EN2 8ND M6VNT
EN2 8QL G7MNE
EN3 4EF G0HAK
EN3 4EF G1PWO
EN3 4QE G3RKJ
EN3 5SE M0DJI
EN3 6EB M3VYE
EN3 6HE 2E1FQE
EN3 6NT G7VVF
EN3 7AD 2E0ZUT
EN3 7AD M6ZUT
EN3 7HA G0TSN
EN3 7HA G3JNJ
EN3 7JF G4GJO
EN3 7QA 2E1HVI
EN3 7RW M3XFA
EN4 0BB G6DLM
EN4 0EX G3UJV
EN4 0EX G3VER
EN4 0EX G4VER
EN4 0EX G8VER
EN4 8DQ G0CEQ
EN4 8EA 2E1DLS
EN4 8NJ G3LSX
EN4 8PU G8DVJ
EN4 8TU G8RPA
EN4 8UX M6NJH
EN4 9AS G7MGM
EN4 9DS G0LCS
EN4 9DS G8VR
EN4 9PA G8OQC
EN4 9QT G3GMY
EN4 9RL G1MPU
EN4 9TX G0NSI
EN5 1LQ G8PNE
EN5 1LX M6RJL
EN5 2BP G8FVE
EN5 2BW G1ALA
EN5 2DB M6DUT
EN5 2DP M6VCA
EN5 2EX M6SNG
EN5 2JL G0SNO
EN5 2LE G4BWY
EN5 2LE G8AIE
EN5 2LS G4GPR
EN5 2NQ G3IQY
EN5 2QU 2E1JOD
EN5 3BB G0TOY
EN5 3BP 2E1HAS
EN5 3BP M3BCH
EN5 3JR G6NVI
EN5 4JG M6FUT
EN5 5QH G3YFW
EN5 5RH G8TAU
EN5 5TN G4WRD
EN6 1QW G3RTE
EN6 1QW G6AY
EN6 1XW G1END
EN6 2BD G3WEA
EN6 2BE M6EDB
EN6 2BQ G4EZU
EN6 3AB G6LKA
EN6 3AB M5ADQ
EN6 3AJ M0ATL
EN6 3DZ G7BHR
EN6 3HG G4HCY
EN6 4DZ G0ICK
EN6 4DZ G4JIJ
EN6 4JE G4ZSC
EN6 5HE G4HTE
EN6 5HU G3WFM
EN6 5HU G4ECT
EN6 5LX G8ZCK
EN6 5NB G7RHI
EN7 5EW G3PID
EN7 5JT G3SSZ
EN7 5NL G3MVM
EN7 6AN 2E1HRJ
EN7 6AN G7OYP
EN7 6AN M6HLS
EN7 6AQ 2E0ROM
EN7 6AQ M0ROM
EN7 6AQ M3XYA
EN7 6AQ M6DAP
EN7 6AS M6NWC
EN7 6DF G8KKH
EN7 6HB 2E1JGM
EN7 6HU G1AUU
EN7 6LG M1GDB
EN7 6SB 2E0VVA
EN7 6SB M0VVA
EN7 6SB M6OMX
EN7 6SB M6UGA
EN7 6SB M6VVA
EN7 6TF G8XTR
EN8 0BW 2E0SSA
EN8 0BW M0SAO
EN8 0BW M6AHD
EN8 0EW M5BXB
EN8 0HL M0XID
EN8 0JL M6KYS
EN8 0QU M6GDT
EN8 0QU M6JTH
EN8 0RF G0KLU
EN8 7JJ G4GRN
EN8 8DA 2E1EGI
EN8 8LZ G3XNP
EN8 8LZ G8BAY
EN8 8NB G3RYQ
EN8 8NH G4SGN
EN8 8PS M6GKV
EN8 8UX G8VLP
EN8 9AP G6YNT
EN8 9BZ G0PXY
EN8 9JJ G6XBS
EN8 9QY G0NLV
EN8 9RQ G4WIP
EN8 9RQ G6AFX
EN9 1PS M3UBK
EN9 1SZ M3TEP
EN9 2AL G4XVM
EN9 2DD G2FUU
EN9 2LB G3HYG
EN9 3DS 2E0DDF
EN9 3DS M6DFD
EN9 3HP 2E0AWC
EN9 3LD G4MOI
EN9 3NS M3HSM

## EX

## (Exeter)

EX1 1SL G0SZX
EX1 2BG G8NEI
EX1 2SR G7EQO
EX1 3AH G4ARE
EX1 3AH M0GJX
EX1 3AH M0GMO
EX1 3EQ 2E0DHX
EX1 3GA G3WVM
EX1 3JQ 2E0CRD
EX1 3JQ M3DGD
EX1 3NP M3DDY
EX1 3NT 2E0GHR
EX1 3NT M3GHR
EX1 3RA G6HUO
EX1 3XP G4KXR
EX10 0AY M6FSM
EX10 0ER G6IDG
EX10 0LR M3YFN
EX10 9BW 2E0ZZT
EX10 9BW M3HUS
EX10 9ED G3UZF
EX10 9EW G8GTV
EX10 9EX 2E0BZS
EX10 9EX M6NLO
EX10 9FJ G3ZWY
EX10 9JA G3DCE
EX10 9JJ G4UBB
EX10 9LF M6VQC
EX10 9LS G1SED
EX10 9LS G6RUM
EX10 9NY G6BHY
EX10 9SU M6YWX
EX10 9TJ G0AXC
EX10 9TJ G6XUV
EX10 9TJ G7AGO
EX10 9TJ M0KSO
EX11 1AR G4BNP
EX11 1BX G1OEF
EX11 1BY G8LAY
EX11 1DT G0WGH
EX11 1EN G7DYB
EX11 1EP G7NBZ
EX11 1JA G8NGE
EX11 1JA M6OSM
EX11 1JJ 2E1VAR
EX11 1PF G0WYF
EX11 1PF G6LAW
EX11 1RL 2E0KVJ
EX11 1RL M6KVJ
EX11 1SY G8NVT
EX11 1TD G4UUH
EX11 1TD M6MTI
EX11 1UY G4BOP
EX11 1XH G4WLP
EX11 1XP G3ZEJ
EX12 2AD G6WZP
EX12 2DB G6JJA
EX12 2DJ G4XXK
EX12 2EQ 2E0EBL
EX12 2EQ 2E0MHR
EX12 2EQ M6EVX
EX12 2EQ M6HUQ
EX12 2EQ M6HVP
EX12 2HH G2BSW
EX12 2HN G8VXU
EX12 2NJ G7VCB
EX12 2NT G8WCQ
EX12 2PD G1NFQ
EX12 2SB G0CTF
EX12 2SS M3ECS
EX12 2TP M3GPR
EX12 2TU G3LVB
EX12 3BD M6TAO
EX12 3BL 2E0TWO
EX12 3BL M3HKM
EX12 3DD 2E0ZCM
EX12 3DD M3BCM
EX12 3LT M6ZYE
EX12 4AG M1AVB
EX13 5BL G1KNU
EX13 5BS G4VHG
EX13 5LE M1AXM
EX13 5RW G0AOS
EX13 5SQ G6WWY
EX13 5SQ M3WWY
EX13 5SY G3ZVW
EX13 5SZ G7CMP
EX13 5TD G4KAM
EX13 7AF G8CYG
EX13 7LU G3GOS
EX13 7PB G0HRH
EX13 7RW G4WNU
EX13 7SQ M3YKH
EX13 7ST G3CFR
EX13 8AP G3CYX
EX13 8AQ G4FVU
EX13 8TT G0GHH
EX13 8TT G8CXA
EX14 1AX G4PQS
EX14 1AX G7RMX
EX14 1JB G3GRQ
EX14 1LT M0EZO
EX14 1QZ G0BCO
EX14 1QZ M3VTA
EX14 1QZ M3WDY
EX14 2DF M3ZDS
EX14 2GP 2E0SJD
EX14 2GP G3USE
EX14 2GP G8HKF
EX14 2TT G0KJJ
EX14 2YP G0WWL
EX14 3EX G0JSC
EX14 3EX G0JST
EX14 3EX G7JST
EX14 3NL G4ENJ
EX14 3PY M6HHQ
EX14 3UP G4TRU
EX14 3WA M3FTE
EX14 4QS G3OLB
EX14 4QZ G4BXS
EX14 4RE G4TIG
EX14 4UL M1VPN
EX14 4XH G8SHF
EX14 9AL G0TEB
EX14 9SA G0CWK
EX14 9TA G0BVC
EX14 9TQ 2E0OSE
EX15 1BQ G4TLL
EX15 1DW G4HMA
EX15 1EH 2E0PAO
EX15 1EH M6PIK
EX15 1FJ 2E1FWX
EX15 1GH 2E0TIV
EX15 1QD M0AKJ
EX15 1SS G0SXN
EX15 1TE G1XKZ
EX15 1UD G0KOF
EX15 2LW M0CVR
EX15 2PZ 2E0CZE
EX15 2RN G8HPS
EX15 2RN G8IYD
EX15 2RN M0BVM
EX15 2SH G0IFA
EX15 2TT G6VDA
EX15 3AU G6TJJ
EX15 3BA G3WNV
EX15 3DN G0HPS
EX15 3DS G0OED
EX15 3HJ M6GXR
EX15 3LH G4SCV
EX15 3QG M3YNI
EX15 3RD M6FJE
EX15 3RY G4UPS
EX15 3SE G6FVD
EX15 3XA G0TAZ
EX16 4AW 2E0IFC
EX16 4AW M6NLB
EX16 4BE G6XWM
EX16 4BN G6SWD
EX16 4BW G7JGZ
EX16 4ER G4XXD
EX16 4ET G4ZAL
EX16 4LN M3IEG
EX16 4PY G8MFF
EX16 4QR M0PUN
EX16 5AF M6JSX
EX16 5DP G0MAS
EX16 5JE M3TME
EX16 6AR G7VZD
EX16 6BU M6CPZ
EX16 6EB 2E0CZC
EX16 6EB M6GEP
EX16 6EE G0IFC
EX16 6EE G4TSW
EX16 6HE 2E0MEL
EX16 6JT G0JCH
EX16 6RB G6BYG
EX16 6RG 2E0AYI
EX16 6RG M3CLO
EX16 6RJ G4YCV
EX16 6TQ M0HMJ
EX16 7BS G0GLQ
EX16 7ED G4FJK
EX16 7RE M3ZJK
EX16 8AZ M0GMS
EX16 8LQ G3ZRJ
EX16 8NZ G6HV
EX16 8PP 2E0KCF
EX16 9AY G8MBE
EX16 9BT G8NGJ
EX16 9DW G6YWZ
EX16 9JQ G6ASK
EX16 9NN 2E0RNW
EX16 9NN M6WNR
EX16 9PJ G3HXK
EX17 2DH G4YAQ
EX17 2DH G6TWD
EX17 2DN G1JXA
EX17 2EJ G0VVY
EX17 2EJ M3LLK
EX17 2EJ M3VVY
EX17 2EP G3HTA
EX17 3HE M3MEF
EX17 3JN G4WJJ
EX17 3NB 2E0NRJ
EX17 3NB M0NRJ
EX17 3NB M6NRJ
EX17 4PL M0PXI
EX17 4QX M0GEB
EX17 4RU M3TBU
EX17 5AL G3WGN
EX17 5AL G4RRA
EX17 5NA G0JUM
EX17 6DH M1AEI
EX17 6PA M6TVW
EX18 7BQ M6DVT
EX18 7BR M6FEJ
EX18 7BU M3NRW
EX18 7DD G7NTY
EX18 7QX G0ABI
EX18 7SN M1DAS
EX19 8DP G8JRW
EX19 8JB 2E0SAT
EX19 8JB M6AIR
EX19 8JW M6BPE
EX19 8JX G3LBM
EX19 8QU G4RCB
EX19 8SL G3ZQR
EX2 4DG G8LKA
EX2 4JS G3AJK
EX2 4LR M6EUV
EX2 4SJ G6ZTP
EX2 5DX M6OOZ
EX2 5EP 2E0SJM
EX2 5ES 2E0RQK
EX2 5ES M0RQK
EX2 5JT M0REB
EX2 5JX M6SQB
EX2 5NH G6BJL
EX2 5QJ G7WLM
EX2 5QN G8RCZ
EX2 6AN G0BNW
EX2 6EA G7UIU
EX2 6EG M0CPF
EX2 6JJ G4BQH
EX2 6LE M3RPS
EX2 6LS M3YDV
EX2 7AQ M0TCF
EX2 7BD M3IEA
EX2 7BQ M0NJE
EX2 8GZ 2E0CZD
EX2 8GZ M6CQF
EX2 8LA M6AJL
EX2 8XN 2E0ECW
EX2 8XN 2E0HXT
EX2 8XN G0JQS
EX2 8XN M3HXT
EX2 9AH 2E1GZZ
EX2 9AH M3MPC
EX20 1EA G3PSZ
EX20 1EA G7VMQ
EX20 1HZ G7CIH
EX20 1PL M6BPU
EX20 1PR G0FJA
EX20 1PR G0FYL
EX20 1QQ G4NJK
EX20 1QX G8JZZ
EX20 2HX G3YGA
EX20 2JF G3VDL
EX20 2NG G1BOB
EX20 3AJ G4NFP
EX20 3EZ G0FUV
EX20 3EZ G6LWT
EX20 3LU G4BTN
EX20 3NX 2E0CER
EX20 3QW G0GHT
EX20 3RF G0DRC
EX20 3RF M0DWR
EX20 4DQ G4XIX
EX20 4LL G0IVO
EX21 5BP G4FIV
EX21 5BQ G0OAW
EX21 5EF G0ASG
EX21 5HA G0LSI
EX21 5LP G0DOW
EX21 5LT G3WIM
EX21 5LT G8UIL
EX21 5LT M3WYJ
EX21 5PX M3AFF
EX21 5TX M6KPI
EX21 5UD G0IKC
EX21 5UF G3BYG
EX22 6DA 2E1GQU
EX22 6DA G7VTE
EX22 6DA M1BNR
EX22 6HB G0CLC
EX22 6QN G8MWW
EX22 6RS G1GZI
EX22 6RS M3GZI
EX22 6TB 2E0MYB
EX22 6TB M0MYB
EX22 6UU G4WWR
EX22 6UU G6WWR
EX22 6UU M0BKV
EX22 6XX G3VDH
EX22 7AA G6LAU
EX22 7BQ G3JME
EX22 7BQ M0OZJ
EX22 7DL G0HFN
EX22 7DL G0RYO
EX22 7DU M6VKZ
EX22 7ED G6REW
EX22 7EW G4CQM
EX22 7NB M1AOB
EX22 7NB M6DEM
EX22 7NY G0RQL
EX22 7NY G1DIF
EX22 7NY M0OMC
EX22 7RN G0KDY
EX22 7RT G6ILD
EX22 7SL G0AOB
EX22 7YG G3ZGI
EX23 0AH M6JBI
EX23 0AJ M0LSS
EX23 0BN G0OKN
EX23 0DT G4FWT
EX23 0ES G0TDE
EX23 0HY G4NCJ
EX23 8AE G8XZX
EX23 8AP G1IQN
EX23 8BD G3XNE
EX23 8EN G4KHY
EX23 8EU 2E0FWC
EX23 8EU G4YTC
EX23 8FF G8YWL
EX23 8HZ G3KMQ
EX23 8JB G8YRW
EX23 8LN M6EJV
EX23 8LR G6MAA
EX23 8LR G8DHA
EX23 8NA G0DBD
EX23 8PD G8ULJ
EX23 8PG G1RLB
EX23 8QF M0MJW
EX23 8RY G4IDU
EX23 8SA M6JIC
EX23 8SB G7TRM
EX23 9AF M0SMJ
EX23 9AG G1IJQ
EX23 9AZ G4OIA
EX23 9BB G4NLH
EX23 9BJ G4ASF
EX23 9BP 2E0RCM
EX23 9BP M0RBM
EX23 9ES M6CHW
EX23 9JA G0WTI
EX23 9JA G8MEH
EX23 9JR G4OJQ
EX23 9LE 2E1EOQ
EX23 9LE M3EOQ
EX24 6JU G6LUJ
EX24 6PN G0GIN
EX3 0DY G1DKE
EX3 0DY M6AZM
EX3 0JW M6RVD
EX3 0LF G3EQM
EX3 0LG G3IMW
EX3 0NA G4ETO
EX3 0PE G8ATC
EX31 1PT M1FWD
EX31 1QA G3XYG
EX31 1QF G3AKJ
EX31 1QX G4RVJ
EX31 1RZ G0OQZ
EX31 2BH G6RNT
EX31 2HL M6DHW
EX31 2HY M0DJH
EX31 2JT M6DBV
EX31 3AP G8NBO
EX31 3BS G3PGA
EX31 3BS G6GZZ
EX31 4HY G6TAP
EX31 4JD M6RZI
EX31 4JQ G4RZI
EX31 4JR G0BUJ
EX31 4LP G0RVS
EX31 4NH G4RDA
EX31 4PR G4YCW
EX32 0DF G4SGW
EX32 0JN G4SJP
EX32 0JN G8KQB
EX32 0ND M3UKF
EX32 0NJ G0WUA
EX32 0PY M6GLD
EX32 7AS G3ZFV
EX32 7AY G7FJZ
EX32 7EN M3VJM
EX32 7FA 2E0TGC
EX32 7FA M3TGC
EX32 7HN G6USZ
EX32 7HW G0YEF
EX32 8DN G4TNE
EX32 8EJ G0DNV
EX32 8ET M6FEQ
EX32 8JR M6SWG
EX32 8LA M3HJF
EX32 8NW G0NPV
EX32 8PS M3MSU
EX32 8QH G4KXQ
EX32 9BG G8VCQ
EX32 9BW G4XWQ
EX32 9DG M1CVK
EX32 9DP M3XZT
EX32 9JX M3BOR
EX33 1BB G4XZF
EX33 1DH G0DUH
EX33 1LL G1MJI
EX33 1NN 2E0HVZ
EX33 1NN M0WWD
EX33 1PP 2E0FEO
EX33 1PP M0TLO
EX33 1PP M6FEO
EX33 2EH G0SHP
EX33 2EL G3MTD
EX33 2EZ M3IXF
EX33 2HL G8EZZ
EX33 2LT G0LKI
EX33 2PF 2E0IXX
EX34 0HQ M0AME
EX34 7BT G1FEJ
EX34 7BX 2E1NJC
EX34 7BX G4NGB
EX34 7BX M3NJC
EX34 8DT G0HGM
EX34 8DW G8RPD
EX34 8DZ 2E1GBN
EX34 8HS 2E0PBV
EX34 8HS M6PDL
EX34 8NH G4RWK
EX34 9HP G3JUW
EX34 9LG G1GBC
EX34 9LL M6IAT
EX34 9LN G0DIZ
EX34 9LN M3MRU
EX34 9LS 2E0JQK
EX34 9LS 2E0WQK
EX34 9LS M0WQK
EX34 9LS M3NCH
EX34 9LS M6TFF
EX34 9TA 2E1BTK
EX34 9TA G1JWO
EX35 6EW M6BNA
EX36 3HJ G8VEQ
EX36 3HL 2E0NAY
EX36 3HL G0EOP
EX36 3HL G4XUG
EX36 3HL M0VET
EX36 3RD G6TWR
EX36 4AL G0HIW
EX36 4BH G4JBR
EX36 4DB 2E0GPT
EX36 4DB M6GPT
EX36 4DB M6IZE
EX36 4EW G4RVG
EX36 4EY M6BPB
EX36 4HJ G8BXO
EX36 4HJ G8SSS
EX36 4JU G1KBF
EX36 4PX G4YUO
EX37 9HY 2E0CNB
EX37 9HY M0JAP
EX37 9JG G1EEO
EX37 9QE G8RIS
EX37 9ST 2E0DYM
EX38 7BE G4INI
EX38 7NB G3DZS
EX38 7NZ M1AEO
EX38 8AL G0OKA
EX38 8AS G4UFK
EX38 8BY G0BVD
EX38 8EE G6KEH
EX38 8EX G0CVI
EX38 8HZ G8UZY
EX38 8NR G8TBU
EX39 1BD G4NWJ
EX39 1BS G0XCF
EX39 1BS M6XHF
EX39 1DF G6YSL
EX39 1HE M0AWI
EX39 1HS G4TCK
EX39 1LS M3KZS
EX39 1LT G0AYM
EX39 1LT G0BYL
EX39 1LT G6XYL
EX39 1NW G4FRA
EX39 1NX G0OXW
EX39 1NX G4XHK
EX39 1PA G1ZUS
EX39 1PE G0PGK
EX39 1PW G7MWI
EX39 1QY 2E0TUC
EX39 1QY M3TUH
EX39 1RD G4SRP
EX39 1SG G0UNB
EX39 1UW G6WZN
EX39 1XE G4CHD
EX39 2DN M3YBN
EX39 2EA 2E0AJX
EX39 2LL G1ZTJ
EX39 2RR G3MNV
EX39 3BN G8SFD
EX39 3DF M0BRB
EX39 3DJ G3YGJ
EX39 3EQ M3KWS
EX39 3ET G4SOF
EX39 3LZ G2FKO
EX39 3LZ G3JKL
EX39 3NE G4DXT
EX39 3PH G4CZK
EX39 3PW G3ZLS
EX39 3QZ G1ZVZ
EX39 3RE G6LPV
EX39 3RW G1SVP
EX39 3RW G3YIN
EX39 3SB G4PMB
EX39 3SB G6EZR
EX39 4AP G3YBP
EX39 4AP G4SJT
EX39 4BE 2E0KMG
EX39 4BE M6KBD
EX39 4BX M3VUY
EX39 4HA 2E0AZJ
EX39 4HA M0VMH
EX39 4HA M3FXX
EX39 4HG G0GFK
EX39 4LZ G0NKK
EX39 4QR G0VWQ
EX39 4RS G3APU
EX39 5DJ G4YBB
EX39 5HN G0JRZ
EX39 5HY G3RSF
EX39 5JA M3VJW
EX39 5JP 2E0RGK
EX39 5JP M3RGK
EX39 5PL G3VYS
EX39 5PL G8FXG
EX39 5RH G8VNO
EX39 5SF M6ENI
EX39 5SH G8CPN
EX39 5ST M3WLL
EX39 5XY M6CCH
EX39 6BQ M3VYB
EX39 6HE G0VXB
EX39 6JA G7VJA
EX4 0AP M0THJ
EX4 1NH 2E0CNL
EX4 1NH M3VNL
EX4 1NJ G0FGE
EX4 1NX G4EDG
EX4 1PQ M3GSR
EX4 1QW M0KFW
EX4 1TA G3YBG
EX4 2AP G7BAE
EX4 2AQ G4CPN
EX4 2DF G3YRX
EX4 2EF M0UWD
EX4 2EG M1BAS
EX4 2EU G1YPM
EX4 2JP G4PCB
EX4 2JP G8NZB
EX4 2LR G4FYJ
EX4 4SJ G0TQS
EX4 4SJ G4BPV
EX4 5AJ G4PXE
EX4 5AN 2E0VIA
EX4 5AN M0VSD
EX4 5AN M6VIA
EX4 5DN G4SJU
EX4 5EJ G8SPM
EX4 6EY G3SXH
EX4 6HG G6IKC
EX4 6ND M0BBV
EX4 6NG G1VNH
EX4 7DY 2E0WRS
EX4 7DY M0ZZT
EX4 7DY M3WRS
EX4 7EA G3YBK
EX4 8BE G6FXH
EX4 8DU M6GNR
EX4 8HB G3ZVI
EX4 8QD G0BTQ
EX4 9DY G4KEE
EX4 9ES 2E0HVW
EX4 9ES M6BYO
EX4 9HP M0CQW
EX4 9HP M1AZY
EX4 9JT G3TLH
EX5 1EE G3XYO
EX5 1HG M6LNK
EX5 1HP M0DKV
EX5 1JD G3VRV
EX5 1LL G4TDV
EX5 1PP G4INF
EX5 1QB G4HHY
EX5 1QS M6UJM
EX5 1RA M3HXS
EX5 2BY G3RUV
EX5 2EX G7NBU
EX5 2TJ G7VNQ
EX5 3DX G4AP
EX5 4BB G3ORN
EX5 4EP G7ANX
EX5 4HT M0WPN
EX5 4QP M6MGX
EX5 5JB M6JGJ
EX5 5NB M6OBY
EX6 6AP 2E0KMP
EX6 6AP M0ZCM
EX6 6AP M6AZA
EX6 6ET G1PHU
EX6 7AD G4WTU
EX6 7SR G4BZE
EX6 7TZ G3KQG
EX6 7XS G4WVT
EX6 8AH G7SYW
EX6 8SD G7UEK
EX7 0AB G0KYS
EX7 0AD M6RDP
EX7 0AN G4UVE
EX7 0AN M3GSI
EX7 0AN M6GSI
EX7 0BP G0JIS
EX7 0BZ M6KPK

EX7 0EA G4ONV
EX7 0FJ M3DOM
EX7 0HB G4KPP
[illegible]
EX7 0JZ G1ZVO
EX7 0LD G7URW
EX7 0NA G4VXH
EX7 0NW G1THJ
EX7 0PD G8NVS
EX7 0QL G3AVN
EX7 0RP G1MVT
EX7 9BS G3WEJ
EX7 9BU G7CNP
EX7 9BU M3HUW
EX7 9DY G0WGJ
EX7 9DZ G3WWT
EX7 9FA G0BBO
EX7 9HP M6AYJ
EX7 9JB G0CAS
EX7 9NQ G4WLA
EX8 1DT G4SET
EX8 1EH G0RKC
EX8 1QN G4EBO
EX8 1RX M3GUJ
EX8 2JW M3LPU
EX8 2JZ G0KNY
EX8 2PU M3LHQ
EX8 2PU M3UIL
EX8 2QR G4VQL
EX8 2SG G1YOF
EX8 3DT G7BWO
EX8 3EJ G1GZG
EX8 3EJ G1MQC
EX8 3HY G0XRC
EX8 3HY G4YRM
EX8 3JX G4UUW
EX8 3LA G4DPJ
EX8 3NB G1ILY
EX8 3NE G3UZM
EX8 3NL G3GNR
EX8 4DN M1CML
EX8 4ET M1DTC
EX8 4EU G8EKW
EX8 4EY G3ZDG
EX8 4JE M3LMQ
EX8 4LA G0NGH
EX8 4LF G8GON
EX8 4LF G8UWE
EX8 4LU M6NHX
EX8 4NA M6AFF
EX8 4PJ G8CQR
EX8 5AP G0OJX
EX8 5EE G0ETZ
EX8 5JD M6AFA
EX8 5PN G0GHO
EX8 5PQ G0AVJ
EX8 5SP G0XJS
EX9 6AG G3LDY
EX9 6JF G3EPO
EX9 6JN G6UEI
EX9 7AS G4WJZ
EX9 7BP G4JBF
EX9 7JE G7AKJ
EX9 7JL G4ZQF

## FK

## (Falkirk)

FK1 1RL GM0TXJ
FK1 2AG MM1CKW
FK1 2AP GM6PKP
FK1 2BX MM0RSI
FK1 2BX MM6WAI
FK1 2EA 2M0RZE
FK1 2EA MM0GSS
FK1 2EA MM3JRK
FK1 2RP GM6JUA
FK1 2RP MM0DBC
FK1 2TB GM1BVA
FK1 3BA MM6BOD
FK1 4BJ 2M0MMF
FK1 4BJ MM0RDT
FK1 4JF GM0KWL
FK1 4LY MM6EJO
FK1 4PA MM0OZY
FK1 4PB GM6RGY
FK1 4PB MM1CLB
FK1 4QB GM4SNZ
FK1 5BQ 2M1ENK
FK1 5DS GM0KUP
FK1 5HU MM0RAM
FK1 5JX GM1LUZ
FK1 5JX MM5AGM
FK1 5NZ MM0KCD
FK1 5UE GM4MIG
FK10 1DD GM0WWX
FK10 1NQ GM4DCT
FK10 2BN GM0UUR
FK10 2DT GM8FMR
FK10 2EG GM0UGH
FK10 2ER 2M0RCI
FK10 2ER MM0GKU
FK10 2JG GM0LWD
FK10 2JU GM4WQH
FK10 2LW MM3ZTS
FK10 2QU 2M0CSZ
FK10 2QU MM6CNO
FK10 2RX GM4ARJ
FK10 2TT 2M0TIA
FK10 2TT 2M0WSK
FK10 2TT MM0TIA
FK10 2TT MM0WSK
FK10 3AP GM6VCV
FK10 3AW GM0FRC
FK10 3AW GM8GAX
FK10 3BG GM0WUU
[illegible]
FK10 3FG MM6HIA
FK10 3JN GM8JJN
FK10 3PD GM0PEI
FK11 7DG GM4FXL
FK12 5AL GM4HUX
FK12 5BL MM6DWP
FK12 5DS GM0NAQ
FK12 5HU GM1PST
FK12 5JU GM1BVT
FK12 5NN GM4XHV
FK13 6HB GM0KTL
FK13 6HB MM0KTL
FK13 6HF GM0TTY
FK13 6HT GM8CIF
FK13 6NT MM0GNX
FK14 7BA MM0GJC
FK14 7BH MM6TWZ
FK14 7JZ GM1FTG
FK14 7LQ MM0BIR
FK14 7NT MM6FKD
FK15 0DF GM0DUX
FK15 0DU 2M0SWM
FK15 0EB GM3YTS
FK15 0EB GM5CX
FK15 0EB GM8VL
FK15 0HQ GM1FSQ
FK15 9AR 2M1IGQ
FK15 9AT 2M0PXH
FK15 9AT MM0AXR
FK15 9AT MM3ZNN
FK15 9ES GM6UCN
FK15 9HB GM0GRL
FK15 9JL GM4VZY
FK15 9NA GM0GMD
FK15 9PX GM0MXP
FK15 9RA GM1RIG
FK17 8LA GM8SAP
FK2 0BJ GM7GLJ
FK2 0EJ MM3KUU
FK2 0FN 2M0DKX
FK2 0FN MM0PRB
FK2 0GU GM7KMM
FK2 0HB MM0VWR
FK2 0LP MM0DSM
FK2 0LP MM3DEC
FK2 0LY MM3OIX
FK2 0NB MM6LRX
FK2 0QW MM3ZQP
FK2 0RE MM0MPA
FK2 0SU GM4MIM
FK2 0TF GM0KLO
FK2 0TJ GM6KDD
FK2 0UP GM7DZK
FK2 0UX GM3JJQ
FK2 0XT 2M0BKL
FK2 0XT MM3ZXL
FK2 7BJ MM3YQP
FK2 7BQ MM6DBT
FK2 7NH MM6KER
FK2 8AQ MM3ZSF
FK2 8DX GM0FTG
FK2 8EE GM0ZAM
FK2 8GD GM7UJO
FK2 8GE GM8VYZ
FK2 8JP GM1IEL
FK2 8NP GM0HZI
FK2 8PP 2M0TGD
FK2 8PP MM0OKG
FK2 8PP MM6TGD
FK2 8RB GM3VQQ
FK2 9JR 2M0TXR
FK2 9JR MM6TRX
FK2 9QQ GM4XQJ
FK2 9UF GM0PGD
FK2 9UT GM0NJL
FK20 8RQ GM0EWW
FK3 0BH GM7NAA
FK3 0BU GM3FDN
FK3 0DE GM3IWX
FK3 0JJ 2M0OAB
FK3 0JJ MM0PSM
FK3 8HJ MM6RDM
FK3 8LZ 2M0JLS
FK3 8LZ MM6MCT
FK3 8NT MM0CNV
FK3 8NZ GM7GTX
FK3 8TG 2M1ANY
FK3 8YG MM6HHJ
FK3 9AT GM3YKA
FK3 9JD GM6JWH
FK3 9JD MM0GXU
FK3 9JN 2M0WWX
FK3 9JN MM6CQU
FK39EQ MM3TEQ
FK4 1BJ GM0DKK
FK4 1GD GM6MHC
FK4 1HE GM0AEY
FK4 1HX MM3VEG
FK4 1HY GM0IYA
FK4 1TP MM1BFE
FK5 3JU GM3UCH
FK5 3LH GM4EJX
FK5 4DD 2M0OXX
FK5 4DD MM0OXX
FK5 4DD MM6OXX
FK5 4DX MM0NDX
FK5 4LT GM0KMJ
FK5 4QF MM0SAJ
FK5 4UF GM0OKJ
FK6 5AG 2M0VUV
FK6 5AG MM0VUV
FK6 5AG MM6VUV
FK6 5EG GM4YRR
FK6 5HY GM0AZG
[illegible]
FK6 5LP MM6TSC
FK6 6QJ 2M0VVS
FK6 6QJ MM0IAL
FK6 6QJ MM3VVS
FK6 6QJ MM6HXY
FK6 6RB GM4NTX
FK7 0DT GM0VRP
FK7 0LH GM4YMD
FK7 0LJ GM4VGR
FK7 0NP GM0WUQ
FK7 0NP GM7VDM
FK7 0NQ 2M0ANE
FK7 0NQ 2M0BYT
FK7 0NQ GM0WUP
FK7 0NQ GM0WUR
FK7 0NQ GM7VDL
FK7 0NQ MM3RGZ
FK7 0NQ MM3WUP
FK7 0QE GM6UNL
FK7 0QP GM4UYE
FK7 0QP GM6NX
FK7 0QW MM6PRZ
FK7 7BS MM6YRO
FK7 7QZ 2M0GEK
FK7 7QZ MM6CJZ
FK7 7RR GM6PNJ
FK7 7UL GM0NWI
FK7 7XA GM0WFR
FK7 8NQ GM8MOI
FK7 8PX GM1BSG
FK7 8PX MM6HLZ
FK7 8UF MM3RCR
FK7 9BY GM7EHN
FK7 9LP GM1SBD
FK8 1EP GM0RMT
FK8 1SA MM0JMK
FK8 2JE MM3RHA
FK8 2JX GM0KQB
FK8 3JY GM0HZO
FK8 3PD 2M0CNR
FK8 3PD MM0PDD
FK8 3PD MM6BKQ
FK8 3PW GM0EGI
FK8 3PW MM0AIK
FK8 3SZ MM0WCT
FK9 4HH MM6PEA
FK9 4PS GM4PJR
FK9 4SA GM0SVS
FK9 4SA GM1XJE
FK9 5AD 2M0FXX
FK9 5AD MM5AMM
FK9 5AD MM6AJI
FK9 5BE MM0BSM
FK9 5EU GM4PYJ
FK9 5LR GM4RCN
FK9 5LR GM4RTN
FK9 5LS GM0FSV
FK9 5PX MM1CCR

## FY

## (Fylde)

FY1 2AQ G1UGG
FY1 2BN M3SBT
FY1 2DW M6WHO
FY1 2JS 2E0MHE
FY1 2JS G3CTQ
FY1 2JS M3SZQ
FY1 2NY M6JIT
FY1 2PW G1NGR
FY1 2QJ M6HVH
FY1 2QL G7HEJ
FY1 2RW G1YMR
FY1 3NN G1TIH
FY1 3RB G1MET
FY1 4BN M6GBL
FY1 4DZ G3IZG
FY1 4JQ 2E0DAX
FY1 4JQ M6BIF
FY1 4LD M6HVD
FY1 4QP G0UHS
FY1 5HP G4GOR
FY1 5JB M6IAJ
FY1 5NA G1OLM
FY1 5NJ G4PNI
FY1 5NJ M3PNI
FY1 6QJ G0HZL
FY1 6DW M3CLT
FY1 6DZ G7FIK
FY1 6JP G4EYX
FY1 6LP G0NBW
FY1 6LY M1EOZ
FY1 6NW G4ZPN
FY1 6PR M6GQA
FY1 6RR M6VVN
FY2 0AW G7MXN
FY2 0BW 2E0OVO
FY2 0DZ 2E0UZK
FY2 0DZ M6UZK
FY2 0EN G1BFG
FY2 0LW M0TUT
FY2 0PD M3BAA
FY2 0QG G0SLV
FY2 0QL G3XGZ
FY2 0RJ M6XTF
FY2 0SD G1ZRE
FY2 0SR G6WWV
FY2 0TR G3WGU
FY2 0TR G8ATG
FY2 0TU G4WYF
FY2 0UG M1ALF
FY2 0WQ G7POL
[illegible]
FY2 0XH M3RKN
FY2 9AQ G4TUM
FY2 9EN M6XAS
FY2 9EP M6HXO
FY2 9EQ G6XNU
FY2 9JN G1KJQ
FY2 9QT G3LLE
FY2 9QW G1NQN
FY2 9TA M6TTE
FY2 9TX M6HUZ
FY3 0AL G3UIT
FY3 0BU G3VFV
FY3 7BQ M6DFU
FY3 7BQ M6FTU
FY3 7JN G8CSY
FY3 7PW G7EPY
FY3 7PW M0BOL
FY3 7QS M3KDY
FY3 7RR M3MEW
FY3 7RS G1CFE
FY3 7UB M3VKF
FY3 7UE M6BQP
FY3 7UJ G1CWQ
FY3 7UJ M3UUO
FY3 8DP G4OMS
FY3 8JA G6FCI
FY3 8JJ M6XMA
FY3 8JU 2E0MLL
FY3 8JU M6CLL
FY3 8PF 2E0EZY
FY3 9AS M0DQL
FY3 9TN G4IGZ
FY3 9TN G4VAL
FY4 1JJ M6IVI
FY4 1JP M3MGZ
FY4 1LB G4MPT
FY4 1PW G0LLX
FY4 1QG G1EEA
FY4 1QR G4AKC
FY4 1QS G4EZM
FY4 1SX G6IRU
FY4 2BT G4XPI
FY4 2BW G4JAJ
FY4 2DH G7PTM
FY4 2DZ G1EOJ
FY4 2NF M1CDX
FY4 3JQ G6IGV
FY4 3LH G6BXO
FY4 3LY M3RPR
FY4 3NH G0FYD
FY4 3PQ M6DSR
FY4 3QA G6BGH
FY4 3QA M3BPL
FY4 3QQ G4YLQ
FY4 3QQ M6LLO
FY4 4AF M6MOK
FY4 4AX G1NQO
FY4 4AZ M6XNU
FY4 4JJ G1HSO
FY4 4NA G6GJD
FY4 4NY 2E0DYY
FY4 4QS G4YVQ
FY4 4RQ G4XQF
FY4 4TF M6BJC
FY4 4UD G4RXK
FY4 4UZ G6XNI
FY4 4YF M1BUG
FY4 5BT G4SZB
FY4 5BX G6AOS
FY4 5DS G0BGR
FY4 5DW G6IFQ
FY4 5EZ G0SVQ
FY4 5HT G4DJY
FY4 5RE G6VRF
FY5 1BU G4BCT
FY5 1DL G3RFH
FY5 1EY M6KSA
FY5 1JD G3HKA
FY5 1JW G4KQC
FY5 1JY G4URG
FY5 1LP G0PIL
FY5 1LP G7URJ
FY5 1SP 2E0RWT
FY5 1SU G3ZGZ
FY5 2AR 2E0UHJ
FY5 2AR G3OSR
FY5 2AR M0UIJ
FY5 2AW M3PDH
FY5 2HX M3XGY
FY5 2JN M0EAL
FY5 2JS G1PPK
FY5 2LG G1NFE
FY5 2LL G3WBB
FY5 2NW G0VCN
FY5 2QX M3SRP
FY5 2RT G4ANE
FY5 2UG G4FRK
FY5 2ZA G0MWM
FY5 3AF G6FDS
FY5 3AQ G8XQI
FY5 3DH 2E0EDX
FY5 3DQ M3OYJ
FY5 3EY G6SRJ
FY5 3HZ G8ZSM
FY5 3QA M3YVW
FY5 3QB G0GER
FY5 3QR G1DEP
FY5 3QR M1AMZ
FY5 3QR M6RKW
FY5 3RQ G4RDY
FY5 3RY 2E0DDR
FY5 3RY M6UAR
FY5 3SU M6NZN
FY5 3UH G6HMX
[illegible]
FY5 4DR 2E0MKH
FY5 4DR M6KHB
FY5 4LD M6HTW
FY5 4NB G3SNH
FY5 4NG M3TWV
FY5 4PL 2E0TOG
FY5 4PL M6FAC
FY5 5AE G8MKQ
FY5 5AP G4APP
FY5 5AW G3YNG
FY5 5AY G4UHI
FY5 5BH M6DCH
FY5 5DR M3NXO
FY5 5HH 2E1ADR
FY5 5HP G3ZNG
FY5 5JD G4RKU
FY5 5NS M3NDF
FY6 0AJ 2E0CVG
FY6 0BS 2E0MYT
FY6 0BS M6CHC
FY6 0BT G7IED
FY6 0BT G7PDR
FY6 0DD G0JZT
FY6 0DF G0AJW
FY6 0DF G4RND
FY6 0EB M0IST
FY6 0EH M6GQG
FY6 0EW G7NYD
FY6 0HE G4RFJ
FY6 0HG 2E0DQB
FY6 0HG G8FFW
FY6 0HG M6FKQ
FY6 0LD G6KHD
FY6 0LZ M0HOI
FY6 0PB G7UWE
FY6 0PW G1OCK
FY6 0QE G4DRX
FY6 0QE G4YRX
FY6 0QW M6FUE
FY6 0RS G6NJE
FY6 7BL G6NLQ
FY6 7DB M3TVJ
FY6 7DJ G4KHJ
FY6 7DU M3UVD
FY6 7DY M3UUN
FY6 7EG G3UIS
FY6 7EW 2E1EIU
FY6 7EW M3UGA
FY6 7FY 2E0SEP
FY6 7FY M6FUQ
FY6 7HJ M6SLJ
FY6 7JZ G0AAY
FY6 7LW G7MUH
FY6 7NE G0LQN
FY6 7PB G6KOE
FY6 7PN G3BDT
FY6 7PW G3XCE
FY6 7QY G0RNA
FY6 7RD 2E0OMG
FY6 7RD M3XZN
FY6 7SR G6MQG
FY6 7UB G8KBH
FY6 7UB G8YOK
FY6 7UQ M3NVE
FY6 7UX G4LOR
FY6 8AA G0OSV
FY6 8AD G3MCE
FY6 8AD M6TGP
FY6 8BL G0BRX
FY6 8BL M6SJJ
FY6 8BZ G7FED
FY6 8EB G3VDO
FY6 8ED G4BVW
FY6 8ED G4VYL
FY6 8EH M3SJM
FY6 8EL G4TMA
FY6 9AP G4XKR
FY6 9AW G0RHF
FY6 9DR G0ETV
FY6 9EA G6UDA
FY6 9EB G4GPZ
FY6 9EH 2E0TDH
FY6 9EH M6TDP
FY7 6RG M0BAL
FY7 6DA 2E0FFS
FY7 6FG M0GNJ
FY7 6FG M0NAU
FY7 6JE G8GLS
FY7 6JR G8IMJ
FY7 6LJ M0JME
FY7 6LJ M6JLB
FY7 6QA G8YPY
FY7 6QA M3AYT
FY7 6QJ M3ZUA
FY7 6UE M6MRG
FY7 7BW G0GDT
FY7 7DU G3SSN
FY7 7EA G0CUX
FY7 7HA 2E1FUJ
FY7 7HA G0LRK
FY7 7HA G0NCY
FY7 7HH G4TTG
FY7 7HJ M0HMU
FY7 7HJ M0JFE
FY7 7HW M3FPS
FY7 7HY M0BLS
FY7 7JF 2E0XHG
FY7 7LF M0DLB
FY7 7LF M3MRN
FY7 7LJ M0HZP
FY7 7LJ M3IPY
FY7 7LJ M3TCY
FY7 7LX M6ELW
FY7 7LT M0UJE
[illegible]
FY7 7NE M6EFH
FY7 7NH G0IAE
FY7 7PY M3WWH
FY7 8BH 2E0KNF
FY7 8BQ G4GYF
FY7 8EG G4FWM
FY7 8EG G8RDP
FY7 8EG M3RAE
FY7 8NW M6BHE
FY7 8QH G0EPY
FY7 8RD G1EPL
FY8 1DW G4JFV
FY8 1EH G8LCZ
FY8 1HZ G7CWI
FY8 1LZ G1CDQ
FY8 1NZ G3FFR
FY8 1PG G2HFP
FY8 1PQ G1TJR
FY8 1PU G4OPT
FY8 1TN M3NZG
FY8 2DA 2E0BKZ
FY8 2DA M0GMN
FY8 2DA M3UGX
FY8 2ED M3NER
FY8 2HE 2E1DZV
FY8 2HE M3JMK
FY8 2HW G0HLB
FY8 2JF G7HJQ
FY8 2LT G4GFE
FY8 2QN G3WBI
FY8 2SG G7DIE
FY8 2SG M3SGS
FY8 3BG G0AIN
FY8 3DB G6KTO
FY8 3NH M6WIM
FY8 3NP M3ZOI
FY8 3NP M6SBH
FY8 3QF G0JBS
FY8 3RZ M6FRZ
FY8 3RZ M6UAE
FY8 3SL G3IEJ
FY8 3TL M6HWL
FY8 4AR G8LMF
FY8 4BJ G4NXW
FY8 4PF M1FHJ
FY8 4QR G4WLE
FY8 4UE G7CBY
FY8 4UE G7CBZ

## G

## (Glasgow)

G1 1HD MM6BPY
G1 1LH MM0IBO
G1 1NY MM3MMO
G11 5AP GM0EDR
G11 5EA GM8TXC
G11 6BL 2M0DKF
G11 6BL MM6ZWT
G11 6QP 2M0YYU
G11 7LG GM4ACM
G11 7PP GM7KVU
G11 7QX 2M0TLE
G12 0AS 2M0CTN
G12 0AX GM8ZGC
G12 0ER 2M0SCG
G12 0LE GM8DKG
G12 0PB GM0ILQ
G12 8TL GM3RUP
G12 9DZ GM7DLY
G12 9NX MM0AMY
G13 1BZ MM6BSK
G13 1DQ MM6SCG
G13 1JH MM6JOX
G13 2LA GM0EFH
G13 2RJ GM0EAH
G13 2YQ MM6KVI
G13 2YQ MM6RHQ
G13 3AQ GM0GYT
G13 3TT MM6YUJ
G13 3YE MM6CIJ
G13 4HL GM0FHJ
G13 4QE MM6CTL
G13 4QN GM4NWK
G15 6AU 2M0RCD
G15 6AU MM0HCO
G15 6AU MM6AOH
G15 7QE GM7NZI
G2 3AU MM6AXT
G20 0HJ 2M0UTH
G20 0HJ MM3UTH
G20 0HT 2M0NTY
G20 6AG MM0CXB
G20 6AG MM0PYG
G20 6XZ GM0FWY
G20 7YQ 2M0TTF
G20 7YQ MM0GUE
G20 7YQ MM6AYH
G20 8EB MM6NWH
G20 8LF GM0USI
G20 8LF GM4FVQ
G20 9JQ GM0KVD
G21 2AP MM3RDP
G21 2DE 2M0WFN
G21 2DE MM6AHJ
G21 2QF MM3UHK
G21 3HY GM0KUJ
G21 3JS MM6AKZ
G21 3SQ GM4ZGU
G21 3SQ MM0DXK
G21 4XY GM4ENN
[illegible]
G22 7RG 2M0GCF
G22 7HG MM0GUF
G23 5DJ GM0JHE
G3 6QD MM1EWA
G3 6QN GM3VBT
G31 3LJ GM0KTO
G31 4RT MM0TBH
G31 5RP GM3PUY
G32 0LP MM6LBS
G32 6TA GM3SPT
G32 7XR MM6WMK
G32 9BW GM3JKC
G33 1BU MM0SIL
G33 1LR MM6JBN
G33 1RS GM7GXI
G33 2AF MM6JSN
G33 2DD GM4PCT
G33 2DP MM3VZI
G33 2LD GM7BRL
G33 5HU GM1POA
G33 6NJ MM0XFA
G34 9AR 2M0FYF
G34 9AR MM0JRF
G4 0HF 2M0LEX
G4 0PH 2M1EZA
G4 0TQ MM3LFI
G4 9JX GM6AQR
G41 2AT GM3DIN
G41 3BS GM8PIV
G41 4QN GM0OAA
G41 5RL MM0TWX
G41 5SD MM6PKC
G42 0DW MM6FBU
G42 8DR GM3WUX
G42 8DW 2M0FRA
G42 8DW MM6CFH
G42 8RT 2M0IEC
G42 8RT MM3IEC
G42 9PZ MM6PND
G43 1BW 2M0HYZ
G43 1BW MM3UTU
G43 2BW MM3SQM
G43 2DU GM4WWU
G43 2DY MM0GYD
G43 2SY MM0GUW
G43 2YL MM3FYF
G44 3NQ MM6HIZ
G44 3XG MM6KHM
G44 4NA GM4REF
G44 4PA GM0WRR
G44 4TJ MM1BJT
G44 5JU 2M0MTO
G44 5JU MM0MTO
G44 5PF 2M0RRO
G44 5PF GM7WED
G44 5PF MM3LKV
G45 9QD 2M0PBC
G45 9UR MM6CHM
G46 6AA GM8MRW
G46 6BZ GM4JTA
G46 6DB GM3NGW
G46 6LA GM6FIK
G46 6QB GM4IYZ
G46 7AE GM3COB
G46 7LU GM0WRH
G46 8AB GM0LIM
G46 8HW MM6OCP
G46 8LP MM6VGS
G46 8NA MM6LGS
G46 8UR GM7UTD
G51 1QL MM6HFZ
G52 2BB 2M0DSY
G52 2BB MM6OAI
G52 2TP MM6FEX
G52 3AP 2M0SAX
G52 3AP MM0SAX
G52 3HA GM4BGS
G52 3HA MM6AHY
G52 3JY GM1JNC
G52 3PU MM0CDK
G52 4HJ MM3WHS
G53 6BS MM6CCY
G53 6QW GM4FFF
G53 6QW MM0DGR
G53 6QW MM0EPC
G53 7UJ GM0UTQ
G53 7XT 2M0AZW
G60 5DE MM6MCX
G60 5HR MM0TCQ
G60 5LF GM3YCB
G60 5LE GM6AQB
G60 5LJ MM0CZM
G61 1AP GM4AWB
G61 1EJ GM0GCO
G61 1EN GM3ZWG
G61 1RF GM3HQQ
G61 2LT GM7OOJ
G61 3HD 2M1HFE
G61 3HD 2M1MIC
G61 3HD 2M1OJD
G61 3HD GM3VTB
G61 3HD GM4VTB
G61 3HD MM1VTB
G61 3HD MM3CVB
G61 3HQ GM0ATL
G61 3JX GM3GTQ
G61 3LX GM0NUQ
G61 4HA MM1JAA
G61 4JP GM0RRK
G61 4JU GM0TOW
G61 4JU GM4CXM
G62 6JN 2M0GIL
G62 6JN MM0HLQ
G62 6QT GM4LSD
G62 7DT GM1XFR
G62 7HA 2M0IBW
[illegible]
G62 7JP 2M0VXL
G62 7JP MM3RXM
G62 7JP MM3VXL
G62 7RA GM4KAV
G62 7RL GM4JRF
G62 7RR GM3MAS
G62 8BE GM8ZKF
G62 8HD MM0GDL
G62 8NL GM8EJS
G63 0EX GM3ODP
G63 0NP GM0GIB
G63 0PF MM1FHO
G64 2HP GM0GMO
G64 2NS GM3HOM
G64 3AD GM8ZAK
G65 0EE 2M0HDA
G65 0EE GM0BRJ
G65 0EX 2M0RCZ
G65 0EX MM0POD
G65 0EX MM3RCZ
G65 0EX GM1AYT
G65 0QZ GM8IIH
G65 9EA MM3SAK
G65 9EJ MM1FZR
G65 9HJ MM0GFA
G65 9HJ MM1DHU
G65 9HQ MM6JJQ
G65 9UL GM0BUE
G65 9UT GM3GRG
G66 1JW MM1DEE
G66 1RS 2M0YEQ
G66 1RS MM0YEQ
G66 1RS MM3YEQ
G66 2BD GM4LYV
G66 2BH 2M1EKI
G66 2BH MM3XOK
G66 2PL MM0GUX
G66 2QE 2M0OMS
G66 2QE MM0OMS
G66 3AS GM3WYL
G66 3BW GM0VEK
G66 3HJ MM3HYG
G66 3JL GM1BTL
G66 3JW MM3RHT
G66 3NX GM7JDS
G66 3RY GM8WWY
G66 4BF GM7SPA
G66 4DF MM1AHL
G66 4DF MM3DIZ
G66 4EL GM6TIB
G66 4EN GM4RUP
G66 4RE GM3SER
G66 5HS 2M0LFS
G66 5HS 2M0TMZ
G66 5HS MM6LFS
G66 5NG GM0BEL
G66 5NG GM4ZRX
G66 7EP GM7FLG
G66 7HU GM0EEH
G66 8AY GM0HNV
G66 8AY GM0PEX
G66 8AY GM3JMM
G66 8BB GM0HJU
G66 8DT MM6HMB
G66 8ER GM7GBD
G66 8ET GM0HIM
G66 8ET GM0RED
G66 8ET GM7HIR
G66 8HG GM0ATA
G67 1JE MM3VBF
G67 1LR MM6ATR
G67 2BL GM1RRJ
G67 2BU GM0GFV
G67 2DA GM7FIS
G67 2NP 2M0UGL
G67 2NP MM0SEY
G67 2NP MM3UGL
G67 2NT MM1RIK
G67 2QW 2M0VKO
G67 2QW GM6FLL
G67 2QW MM3VKO
G67 3BN 2M0GLI
G67 3BN MM6GLI
G67 3LU GM4YLU
G67 4AD GM7RQK
G67 4AJ 2M0BYI
G67 4AJ MM0MYL
G67 4AJ MM3YUW
G67 4AJ MM3YUX
G67 4AW MM6HQK
G67 4ES MM0GXQ
G68 0EP MM3BSC
G68 0HX GM0NBM
G68 0JB GM0PLU
G68 0JB MM0LMC
G68 0JR 2M0SMM
G68 0JR MM0HUF
G68 0JR MM6CXA
G68 9BJ GM4EIW
G68 9DZ GM1VFR
G68 9DZ GM7JUX
G68 9EG GM7MZZ
G68 9ER MM1DEA
G68 9JD MM6DBN
G68 9NT MM1BJZ
G68 9NW GM4VWV
G68 9PE MM6GYR
G69 0JW GM1MRY
G69 0LZ MM0ODI
G69 0PH MM3CBO
G69 6JB MM6FTG
G69 6LQ GM4RPE
G69 6ND MM3DOP
G69 6QP MM0DNX
[illegible]
G69 7BH GM8YGI
G69 7HW GM3SAN
G69 7QZ MM6KCB
G69 8AG MM3DQV
G69 8AG MM3DQX
G69 8LE MM0OWL
G71 6ED GM7FGH
G71 6LG 2M0MBE
G71 6LG MM3VPK
G71 7BQ GM0IMW
G71 7ET GM7VYR
G71 7ET MM3VYR
G71 8AR GM1VBE
G72 0RQ GM0XFK
G72 7GS MM6YND
G72 7GS MM6ZDG
G72 7NN MM6JCL
G72 7PR 2M0OML
G72 7PR MM0OML
G72 7PR MM3OML
G72 7SQ MM3UMY
G72 7SQ MM3UMZ
G72 7TP GM7SFE
G72 7UW MM6GGE
G72 8DH MM6VLG
G72 8NL GM4NDV
G72 8RD GM4TTC
G72 9NX GM0AZU
G72 9UA MM0GPZ
G73 1JX MM6RXJ
G73 2PH GM4HYF
G73 3EN GM0LBN
G73 3EU MM3WJZ
G73 3QP GM7AYW
G73 3QP GM8SNB
G73 4AE GM0JVV
G73 4DX MM3CKP
G73 4EA 2M0DKV
G73 4EA MM6EGC
G73 5RG MM0AYE
G74 1BQ GM4DLU
G74 1DR MM0BMA
G74 2HU GM0TPI
G74 3AF GM8HBB
G74 3DL MM0ZXI
G74 3DN MM6CHY
G74 3HZ GM1BNA
G74 4QS GM4BOA
G74 4RS GM3ZDH
G74 4TG GM8NET
G74 4TZ GM7LOK
G75 0DL MM3YVU
G75 8DL 2M0YOY
G75 8DL MM6YOY
G75 8LH MM6EWR
G75 8RZ MM1EGS
G75 8SA MM1DME
G75 8TN GM1SQZ
G75 8XT GM7GDE
G75 8YG GM0UET
G75 8YG MM3CLA
G75 8YG MM3UET
G75 9LP MM6GOR
G75 9NR 2M0AKT
G75 9NR MM5AIR
G76 0DA GM4XOI
G76 0DG GM1PUR
G76 0EU GM0VIY
G76 0EU MM3PEY
G76 0HA GM0PHG
G76 7DU GM4VBE
G76 7HG GM0SEP
G76 7HG GM0UIG
G76 7HG GM4SRL
G76 7PL GM4XGY
G76 7XT GM0IVQ
G76 7XT GM6FOT
G76 8NR 2M0MCD
G76 8NR MM3NZX
G76 9BN GM0DVO
G77 5NF GM4JMU
G77 5PP GM1NEW
G77 5QJ GM7SKB
G77 5TQ GM4KLQ
G77 6EA MM6MML
G77 6HP GM3NEQ
G77 6JU GM0NBA
G77 6LQ GM1NET
G77 6LQ GM6OQN
G77 6PD GM0OPS
G77 6PZ MM6DVR
G77 6UJ 2M1EDM
G77 6UJ GM0UKZ
G77 6UZ MM0HFU
G77 6XX GM3KXQ
G77 6YG GM6LIN
G78 1TY GM0OPK
G78 2DH GM0GMI
G78 2DH GM1TFZ
G78 2LR MM6DAQ
G78 3JA GM6JIL
G78 3PZ GM4LCP
G78 3QJ MM3USV
G78 3QP GM4OSV
G81 1EH MM3OYL
G81 1ER GM4BLO
G81 2LL GM0GYM
G81 2PH 2M0NIT
G81 2PH MM6TNO
G81 2ST GM4PLI
G81 3EH MM6ZYZ
G81 3LH GM4KHE
G81 3NF MM6RPN
G81 3AH GM1DNG
[illegible]
G81 4LN GM7UMU
G81 5BS MM3YTB
G81 5EG GM4ZMK
G81 5HJ 2M1EDT
G81 5PD GM6KTP
G81 6AW GM3KCY
G81 6LW GM3RGU
G81 6LW MM0HJC
G81 6NR GM3ITN
G81 6PX GM4AGG
G81 6PX GM4TOQ
G82 2BN MM6NUP
G82 2BN MM6YUP
G82 2JA MM3ODV
G82 2PF MM6DHF
G82 2PN GM7NHU
G82 2RZ MM0CXZ
G82 3ER GM0KZX
G82 3ER GM4HEL
G82 3ER GM4URZ
G82 3ER MM0HLK
G82 3JU 2M0MSB
G82 3JU MM3MGK
G82 3NU 2M1HSG
G82 3NU MM3HSG
G82 3PB MM1JAC
G82 3QW GM8FFH
G82 4BG MM5AON
G82 4JL MM0LGT
G82 4NH MM6HAH
G82 4QA MM3FMB
G82 5HD GM1UIR
G82 5HD MM3DMZ
G82 5HP GM1MLW
G83 0BZ 2M0JBZ
G83 0BZ MM6JCU
G83 0DW GM7MYF
G83 0LL GM6JWF
G83 0RJ MM3OFE
G83 0RZ GM4LGM
G83 0UF MM1DMU
G83 0US MM1LJB
G83 7AD 2M0ZFG
G83 7AD MM0ZFG
G83 7AD MM6ZFG
G83 7DB 2M0OVV
G83 7DB MM3OVV
G83 8EX MM6EOE
G83 8JN MM0HDA
G83 8RP GM0RTY
G83 8RU MM3OCY
G83 9BJ MM3XDP
G83 9BU 2M0BIL
G83 9BU MM0ELF
G83 9DU MM3FCG
G83 9DB MM3XKH
G83 9DB MM6LAH
G83 9EB MM1BJP
G83 9LE MM6SWC
G83 9LG 2M0TXY
G83 9LG MM6FCA
G83 9NP 2M0LNF
G83 9NP MM3LNF
G83 9QL GM7OBM
G83 9QT MM3DDQ
G84 0AE 2M0DKU
G84 0AE MM6SUB
G84 0DW MM6BGV
G84 0EB 2M0VFV
G84 0EB MM6SMY
G84 0JN MM0EAI
G84 0QD GM1KWX
G84 0QR GM1FTZ
G84 0QX 2M0CTI
G84 0QX MM6CNV
G84 0QZ GM4MFO
G84 0RL 2M1HKA
G84 0RN MM6HOO
G84 7EE MM6HEZ
G84 7PL GM8VAM
G84 7SE MM3VXP
G84 8JP GM7OAF
G84 8NN 2M0BRH
G84 8NN MM0GPL
G84 8NN MM3FPI
G84 9DA 2M0BJU
G84 9DA MM0GON
G84 9DA MM3OZU
G84 9DN GM4RJX
G84 9DP 2M0RDR
G84 9DP MM6RDR
G84 9DX MM3ECO
G84 9DX MM3ERD
G84 9DX MM3YOL
G84 9JD GM1VYF
G84 9JZ MM0JWH
G84 9QP GM4ZWJ

## GL

## (Gloucester)

GL1 1GF G4WXF
GL1 2AR M6BBS
GL1 2PB G1ZSZ
GL1 2QZ M3TYI
GL1 2RX 2E0LGW
GL1 3DE 2E0HBE
GL1 3DE 2E1FXN

Postcode

**IMPORTANT NOTE**

**Revalidate licence to avoid revocation** – Ofcom has advised the Society that plans will be drawn up to revoke licences that have not been revalidated as required by the licence conditions. The quickest way to revalidate is to do so online via the Ofcom website: *https://services.ofcom.org.uk/* or by email: *amateur.validations@ofcom.org.uk* If you need assistance in the process, Ofcom staff are available to help, but please be patient during times of heavy workload.

GL1 3QE G2CIW
GL1 3QS M6SQF
GL1 5DD G3VMQ
GL1 5EL G0MPZ
GL1 5ER G3RKH
GL1 5HL 2E0BNF
GL1 5HL M0RGL
GL1 5HL M3OIV
GL1 5JU M6CZD
GL1 5QD G7BPX
GL10 2DG G1USZ
GL10 2DH G7MLW
GL10 2PZ G7MWC
GL10 2QH G4CMY
GL10 2QH G6XKV
GL10 3EG 2E0CRI
GL10 3EG M6JSP
GL10 3HS G4XWZ
GL10 3HX G4EXF
GL10 3JA 2E0CMC
GL10 3JA M6BBO
GL10 3JN G3TBF
GL10 3LD G4EDY
GL10 3LD G4YYR
GL10 3LD G8ILN
GL10 3LJ G0RUY
GL10 3NA G0PDE
GL10 3NL G4UBC
GL10 3PJ M6AKT
GL10 3QW G4SHB
GL10 3RT G4CRG
GL10 3RX G0TME
GL10 3SN G4FRR
GL10 3TU G4CIO
GL11 4AP G6HKL
GL11 4AS 2E0KLD
GL11 4EW G4JXC
GL11 4QB 2E0RBZ
GL11 4QB M6RAZ
GL11 5DA G4KYI
GL11 5EL G4VZR
GL11 5EW G7AHT
GL11 5JQ G8ZTM
GL11 5SW G1GDT
GL11 6DX G4YIC
GL11 6HB G4FQH
GL11 6HY G4ETS
GL11 6JE G0DQS
GL11 6LF G0BRW
GL11 6LT G4FOD
GL12 7BJ G4VCQ
GL12 7BL G0IWW
GL12 7LQ G0SYF
GL12 7QY M0HLV
GL12 7RH G7FEQ
GL12 8AS G1HXT
GL12 8DA G0MIG
GL12 8NB G1USW
GL12 8SG G7JWE
GL12 8TJ G0CBK
GL12 8TN G1VNL
GL13 9BU G0NVX
GL13 9BU G4HQX
GL13 9DF G8ZHN
GL13 9EB M6EAT
GL13 9LE G0RYV
GL13 9NP M6DZZ
GL13 9PL G6JWO
GL13 9PY G0UGR
GL13 9TE M6JJV
GL13 9TG G7OPB
GL13 9TQ G7FPU
GL13 9TQ M6DCL
GL13 9UA G6GLO
GL13 9UA M3YXF
GL13 9US G4RLT
GL13 9UT G0IHC
GL14 1JE M6HOF
GL14 1NB G6GUC
GL14 1QX G0XAE
GL14 2BH G4KRJ
GL14 2BH G6FGA
GL14 2DE G8PGH
GL14 2DW G0DZA
GL14 2DW G0WBS
GL14 2EB G0PBB
GL14 2EB G0SNB
GL14 2EB G7KXN
GL14 2EF G0ODN
GL14 2EF G6CHT
GL14 2QU G4THC
GL14 3DZ G0DAB
GL142ED M6APS
GL15 4AJ M3DCL
GL15 4HR G7VQI
GL15 4JX G4NIF
GL15 4NY G4ULG
GL15 4QD G4UHJ
GL15 5AZ 2E1IFM
GL15 5AZ 2E1IFN
GL15 5AZ 2E1IJL
GL15 5AZ 2E1KID
GL15 5AZ G7VHJ
GL15 5BS M6FOD
GL15 5JD G0ENF
GL15 5LP G0FDD
GL15 5LR G4NNJ
GL15 5NP 2E0FOD
GL15 5NP M0SSJ
GL15 5NP M6SWL
GL15 5QS M3YJM
GL15 5QS M3YJP
GL15 5QS M6BPG
GL15 5SL G4EPW
GL15 5TA 2E0NSS
GL15 5TA M3NSS
GL15 6BZ M1ZZA
GL15 6DN G4IKX
GL15 6HG 2E0DWP
GL15 6HG M3BRV
GL15 6JQ G0SDD
GL15 6LQ G3CZL
GL15 6NA 2E0NBC
GL15 6NB M6GSY
GL15 6NT G7EEG
GL15 6PE G8BXD
GL15 6TN G6UXY
GL15 6TN M3FTJ
GL15 6UN G4PWA
GL16 7AG G3NOC
GL16 7AQ G1EHX
GL16 7BE G0MJL
GL16 7LG 2E0CJP
GL16 7LG M3FHQ
GL16 7LR G8AOJ
GL16 7PU G0KWG
GL16 7QB G0OOF
GL16 7QB M0ACW
GL16 7QB M0ATX
GL16 7QD M1ART
GL16 7RG 2E1IFL
GL16 8AY M1AYN
GL16 8AY M1BKE
GL16 8AZ G3KTI
GL16 8BD M6RCP
GL16 8BG G1EDP
GL16 8BN M6CMZ
GL16 8BP M6RBU
GL16 8BY G7AEF
GL16 8BY M1EKM
GL16 8DE M6CNA
GL16 8DN G1BWP
GL16 8PQ G8WGD
GL16 8PT G3TLD
GL16 8PT G7GZU
GL17 0AU 2E0MIT
GL17 0DQ M3FHP
GL17 0JE G0NQI
GL17 0LP M6KGN
GL17 0PH G0FGZ
GL17 0SB G7MWW
GL17 9AU M0HIY
GL17 9AU M0RHR
GL17 9QD G1IZH
GL17 9QL G4HVD
GL17 9QS M1AVU
GL17 9SB G0RMX
GL17 9SB M6FUR
GL17 9UD G0ORD
GL17 9XR G7TUS
GL17 9XT G7IRW
GL17 9YN M3MYG
GL17 9YN M3MYI
GL18 1EH G3ZMD
GL18 1HN G6PMJ
GL18 1PS M6DDD
GL18 1PZ G3WVQ
GL18 1QB M3DGJ
GL18 1TL G1NRX
GL18 2BW G4WUH
GL18 2EJ G8SQH
GL19 3BN G8CQZ
GL19 3BP G1MHM
GL19 3NZ G0HVQ
GL19 3QY G8TTJ
GL19 4AH G4NNO
GL19 4BT G0ECJ
GL19 4BT G6VKA
GL19 4DE G6BDM
GL19 4DE G6BOQ
GL19 4NX G8DYG
GL19 4NY G4CIB
GL19 4NY G4RHK
GL2 0DP G4NVY
GL2 0DZ G4FLS
GL2 0EJ 2E0TOT
GL2 0EJ M6TOT
GL2 0ER M6LNX
GL2 0HA G1JMF
GL2 0JG M1DKA
GL2 0LX G4MGW
GL2 0NB M3NVO
GL2 0NH M6DLV
GL2 0NQ G3XMM
GL2 0PS M3CCJ
GL2 0RX G0EEA
GL2 4LZ G1GAT
GL2 4NP G4OIN
GL2 4PB G0DBM
GL2 4PY G7SKF
GL2 4PZ G8OTD
GL2 4QE G0JQX
GL2 4QJ G4IVD
GL2 4QQ G8RMI
GL2 4RT 2E0CMD
GL2 4RT M0HFY
GL2 4RT M6CBY
GL2 4SY G8MMG
GL2 4SY M0OMD
GL2 4UF M3NCQ
GL2 4US 2E0VFA
GL2 4US M6VFA
GL2 4WJ G0KVO
GL2 4YR M6MAH
GL2 4YS G1AXW
GL2 4YY G1DNT
GL2 5BJ M6DDT
GL2 5HH 2E0UPA
GL2 5HH M0UPA
GL2 5HH M6ULA
GL2 5NZ G7NGN
GL2 7DF G8IEW
GL2 7DJ G7JWQ
GL2 7ED G1WZO
GL2 7ET G4LZQ
GL2 7LH G3STZ
GL2 7LW G1XAL
GL2 7pt G3ILO
GL2 8EB G7DMZ
GL2 8EH G1ISY
GL2 8JP G4PJJ
GL2 8LJ G4ZYR
GL2 8NH G3PJQ
GL2 9BB 2E0CLZ
GL2 9BB M0IAJ
GL2 9BB M6IAJ
GL2 9ED G4DCK
GL2 9HB G1IDV
GL2 9NW G4BGW
GL2 9PS G7GVJ
GL2 9RB G0HTO
GL20 5DG G6VAR
GL20 5FB G7AEE
GL20 5FB G7VTL
GL20 5NH 2E0JCA
GL20 5NH M6JCA
GL20 5PD G1KNX
GL20 5RL G0MMA
GL20 5RX M5ADE
GL20 5TW G7AEC
GL20 5TZ G0VFZ
GL20 5TZ G8AZN
GL20 5TZ M6WUH
GL20 6BB G4CRN
GL20 6DW G0NXA
GL20 6JW G6AHX
GL20 7AH G0PTR
GL20 7AT G3FHG
GL20 7AU G3XGW
GL20 7EH G1XYF
GL20 7EP 2E0UHF
GL20 7EP G6LJU
GL20 7EP M0RAR
GL20 7NQ G8YMR
GL20 7QL G0HDB
GL20 7RS G0NUL
GL20 7RW G8WWC
GL20 7WL G0COZ
GL20 8AQ M3SZO
GL20 8AS M1AMA
GL20 8AT G4EAZ
GL20 8BA G4YGY
GL20 8BB M0WMR
GL20 8BT G1NFB
GL20 8ES G8TQP
GL20 8ES M0ADJ
GL20 8HS G3OLW
GL20 8NN 2E0CME
GL20 8NN M6BBU
GL20 8NT G7OHW
GL20 8PJ M3XIV
GL20 8PX G8JXS
GL20 8QP G8VSH
GL20 8QY G0FCM
GL20 8RB G1NVS
GL20 8RE G7CVC
GL20 8RP M6WRE
GL20 8TQ M6WIR
GL3 1AA G0JVH
GL3 1AT M6KCG
GL3 1BL M3WHG
GL3 1LL M3TMY
GL3 1NG G0AHU
GL3 1NT M6WTE
GL3 2AU G4ENZ
GL3 2BT G0VIG
GL3 2DS G3SZS
GL3 2DW G6XQO
GL3 2HT G1CMH
GL3 2LD G3SUA
GL3 2LZ G3ZKN
GL3 2PN G4FRI
GL3 2PQ 2E0ITU
GL3 2PQ M6ITU
GL3 2PU G0UHG
GL3 2PU G6UGW
GL3 2QE G4BCA
GL3 2QS G0EJF
GL3 2RY G6OTP
GL3 2TR G4MYW
GL3 3BX M6KZU
GL3 3DH G0VWH
GL3 3JE G4HBV
GL3 3JF G3UUL
GL3 3JH G3ZLM
GL3 3LF 2E0OLO
GL3 3LF M0OLO
GL3 3SX G8IWB
GL3 3TZ G6USD
GL3 3TZ G8DHF
GL3 4BD M6ECT
GL3 4DH M6RMI
GL3 4ER 2E0JIA
GL3 4ER M6IJB
GL3 4ES G1NVO
GL3 4GF M6TTN
GL3 4NP M3IXD
GL3 4PF G0FNF
GL3 4PF M0HWT
GL3 4PP 2E0PWC
GL3 4PP M6PWC
GL4 0AL G3XUC
GL4 0BX G0RGJ
GL4 0DA G3HXN
GL4 0HP G8WRI
GL4 0JT M3UEY
GL4 0NZ 2E0OKK
GL4 0NZ M0IRD
GL4 0QN G0HBB
GL4 0QY G1DIM
GL4 0RA G1IFF
GL4 0SR G6UER
GL4 0TD G4CLR
GL4 0TS G1AET
GL4 0TT G4TDG
GL4 0TW M6MIF
GL4 0XP G0GJR
GL4 0XW M0PCB
GL4 0XW M0VSQ
GL4 0YQ 2E0IXC
GL4 0YQ M3IXC
GL4 3AG G0UWU
GL4 3AG G0WUW
GL4 3AH G7RHM
GL4 3AX G1BWH
GL4 3ER G4MOH
GL4 3JL G1NJI
GL4 3TQ G4PTW
GL4 3YW M6AND
GL4 4AG G0ULH
GL4 4AG G2HX
GL4 4AG M6RYL
GL4 4DZ M6BRI
GL4 4NE 2E0GXK
GL4 4NE M6GXK
GL4 4RB G0GAJ
GL4 4RE M6YUL
GL4 4RN M6VED
GL4 4UA G7CSM
GL4 4WA G7BVS
GL4 4WH G0FBX
GL4 4WH G0FHK
GL4 4WH G1FHK
GL4 4WP G0JBV
GL4 4XD G4KXK
GL4 4XH G6BXT
GL4 4XJ G3TDT
GL4 5DG G3VTS
GL4 5FD G4BNW
GL4 5FJ G7AEA
GL4 5FJ M1AGK
GL4 5FQ 2E0PMA
GL4 5GD 2E0GDZ
GL4 5GD G7GQC
GL4 5GD G7JUP
GL4 5GD M0XAC
GL4 5GD M6GDZ
GL4 6DW G0MIE
GL4 6PB G0IMQ
GL4 6QU M0JJA
GL4 6SX G3JNP
GL4 6WE 2E1CAF
GL4 8AL G0NRZ
GL4 8DA M5AFX
GL4 8HB G1SCV
GL4 8LD G3YJE
GL5 1ES M3BGE
GL5 1HD M3UGV
GL5 1HS G6SQT
GL5 1PL G8VLY
GL5 1QE M6SHZ
GL5 1RD G0MZK
GL5 1RU 2E1GMT
GL5 1ST 2E0TPG
GL5 1ST G0NUN
GL5 1ST M0TPG
GL5 1ST M6TBG
GL5 1SY 2E0IND
GL5 1SY M0IND
GL5 1SY M6IND
GL5 1US G0FCO
GL5 2DG M0BJP
GL5 2DG M3CIJ
GL5 2EA G1NKF
GL5 2RF G3REB
GL5 2UG G8AER
GL5 3PJ G0OIS
GL5 3QR G4MQL
GL5 3QT M1FCV
GL5 3QZ G4GXB
GL5 3TS G4GWZ
GL5 3TS G4RNK
GL5 4DF G4RJG
GL5 4DJ G4KWW
GL5 4DQ G7HLD
GL5 4JW G1JCT
GL5 4PU M6UFC
GL5 4PX G0TNG
GL5 4PX G1OEM
GL5 4SA G0IUW
GL5 4UW G0FCJ
GL5 5JS G3ICA
GL5 5JY G3YIE
GL5 5LN G4ENA
GL50 2AW G3DNS
GL50 2LU G8JJK
GL50 2LX G4MQB
GL50 2NG G4GVZ
GL50 2QL 2E0DMI
GL50 2QL M6BGK
GL50 2RD M0WEB
GL50 3BW M6WWK
GL50 3NH M6KLA
GL50 4BY G8GKH
GL50 4HS G4PBY
GL50 4LL M6ORB
GL50 4NU G3SMD
GL50 4NW G3RMD
GL50 4NX G8DTA
GL50 4NY M3JIL
GL50 4PX G8XRS
GL50 4QB G4FZG
GL50 4RG G0NDU
GL50 4RG M6SMF
GL50 4RJ M0DCB
GL50 4SA G0UTR
GL50 4SA G0WJA
GL50 4SB G4VTA
GL50 4SE G6GJN
GL51 0EG 2E0GEL
GL51 0EG M6UHU
GL51 0JN 2E0OCH
GL51 0JN M0UCH
GL51 0LZ G8CYU
GL51 0NY G0SMM
GL51 0NZ G6AFE
GL51 0PP G6MBH
GL51 0QH M6FYK
GL51 0QT G3LHU
GL51 0QT M6ECJ
GL51 0TW G6UYN
GL51 0UP M0SYG
GL51 0UP M1SYG
GL51 0WN M3JQV
GL51 0WN M6XVC
GL51 0XE G8ZEE
GL51 3BB G6TRY
GL51 3EZ G4ISJ
GL51 3HH 2E0MDJ
GL51 3JF G4BZU
GL51 3LA G0EYP
GL51 3LL G6VQN
GL51 3LL M6EPL
GL51 3LX G4FUJ
GL51 3PD G4HQC
GL51 3QL G0ALA
GL51 3RA G3NKS
GL51 3RR G4SGI
GL51 3RZ M1ELN
GL51 3WA G6DPS
GL51 4GQ M3ZZX
GL51 4SB G0TPO
GL51 4SN G3KII
GL51 4SZ G7URT
GL51 4TG G4ERR
GL51 4UW G6RII
GL51 5RX G6XSC
GL51 6AA M6DZU
GL51 6AL G0EVX
GL51 6AL G4XDL
GL51 6BA G8TWS
GL51 6BU G6FPK
GL51 6GF M6HKK
GL51 6JE G4BSC
GL51 6JQ 2E1GKY
GL51 6JQ G4AYM
GL51 6JQ M3GKY
GL51 6JS G0WIG
GL51 6NP G1EDX
GL51 6NX G4WUJ
GL51 6QE G1WVK
GL51 6QP M0GPC
GL51 6QZ G8DOB
GL51 6RE G3IFB
GL51 6RL G3LVP
GL51 6RL G8APY
GL51 6RT G8JAY
GL51 6RT M3YAJ
GL51 6RW G4ILI
GL51 6SN G1AEB
GL51 7DQ G8PZD
GL51 8AF G3YEU
GL51 8DZ 2E0CLY
GL51 8DZ M0JLY
GL51 8DZ M6BBP
GL51 8HR G3JJT
GL51 8NX G6BHS
GL51 9DQ G1NMW
GL51 9RD M1AUO
GL51 9RF G3MZV
GL51 9RG M0WMT
GL51 9RW M6CJM
GL51 9SN G0UPB
GL51 9TG G0LRI
GL52 2NR G4NSZ
GL52 2NR G8NSZ
GL52 2NR M3NSZ
GL52 2QB G7STT
GL52 2QR G4YSF
GL52 3BB G3XPR
GL52 3DA M1BFR
GL52 3DF G8ZFT
GL52 3DG G3XKD
GL52 3DS 2E0JVV
GL52 3DS M0JVV
GL52 3DS M6JVV
GL52 3DT G4CWM
GL52 3DT G6UJI
GL52 3DU G3UYL
GL52 3DU G4LCM
GL52 3DU G4PKO
GL52 3DX G7NCW
GL52 3DY G0LFP
GL52 3DY G7DEI
GL52 3EH G3SNN
GL52 3EH G5BK
GL52 3ES G4MM
GL52 3ES G4PDQ
GL52 3HF G8LHP
GL52 3LB G6EEU
GL52 3PS G8ENW
GL52 3QB M0DQB
GL52 5AN G0COI
GL52 5BN G0ISP
GL52 5EG G3LME
GL52 5JQ G4KFT
GL52 5LX G6YCV
GL52 6JG M6KEF
GL52 6NN G0SOV
GL52 6NZ G1KYV
GL52 6RF G8UJF
GL52 6SX G3VKV
GL52 6TX G6DXD
GL52 6YA G7GQA
GL52 6YT M1MOD
GL52 7UZ M6FRG
GL52 7YR G8VVY
GL52 8AG G3XKH
GL52 8BG G6LTB
GL52 8BN G3PYI
GL52 8BY M1CBY
GL52 8BY M1CBZ
GL52 8NA G3AKI
GL52 8NQ G3WFL
GL52 8NU G7PTV
GL52 8NX G0IYO
GL52 8NZ G4GKK
GL52 8PA G4SYW
GL52 8PF M0RON
GL52 8SA G4ERP
GL52 8SS G0VSY
GL52 8SS G0VSZ
GL52 8TG G0MBZ
GL52 8TG G4MPJ
GL52 8TH G0MTW
GL52 8UL G4ZRD
GL52 8XP G4CKX
GL52 9HN 2E0DKZ
GL52 9HN M0HLI
GL52 9PY G3KGT
GL52 9QP G3NOI
GL52 9QS G3ZTI
GL52 9QU G4IZZ
GL52 9QX G0VCA
GL52 9TZ G8MGD
GL52 9UD G4BMP
GL525LG 2E0VHF
GL53 0AD G8CQX
GL53 0AZ G4RFU
GL53 0BH G0VCD
GL53 0HE G4NOE
GL53 0LU G4WGK
GL53 0NQ G6VVL
GL53 0NU G8BTV
GL53 0PQ G7AKI
GL53 0PU G6PYM
GL53 0PU M3GVY
GL53 7BQ G3ZQJ
GL53 7DB G3XTP
GL53 7DX G8KMR
GL53 7JJ G3CGD
GL53 7RT G3PCL
GL53 7RU M0DKP
GL53 7RY M0IBQ
GL53 8AR M6BCC
GL53 8NQ G4XXA
GL53 8NQ G4XXB
GL53 8NS G0LSQ
GL53 9AT M6TFE
GL53 9DQ G0VNJ
GL53 9EN G3MOE
GL53 9HU G4VQE
GL53 9LB G4INB
GL54 1DB G4CRB
GL54 1DJ M3RXD
GL54 1JD G3RWI
GL54 1LG G8FMZ
GL54 2NZ G1NAP
GL54 2QX G3NGZ
GL54 2QX G3TSO
GL54 3JJ M1BPS
GL54 3QJ G1VXS
GL54 4ET M6SFV
GL54 4HP G4VLH
GL54 4HP M0URF
GL54 4HZ G7IHN
GL54 4JX M6HNT
GL54 4LL M0AUW
GL54 5EP M6XBV
GL54 5JX G4MUW
GL54 5LQ G0TPD
GL54 5QR G3DPM
GL54 5XH M0BAM
GL55 6AG G3DWI
GL55 6BY G1TFY
GL55 6SZ G0WEO
GL55 6TD G0LSK
GL55 6XP G0TPA
GL56 0JF M1DCX
GL56 0JF M1DCY
GL56 0LL 2E0GVC
GL56 0LL M6GVC
GL56 9AE M0AFV
GL56 9JZ G8DCJ
GL56 9TE G4PQB
GL6 0AP G0FDE
GL6 0DR G3WMQ
GL6 0JR G4VSO
GL6 0LD G0DZM
GL6 0NJ G3ZTX
GL6 0PJ M0HMR
GL6 0PY G3OYX
GL6 0RT G6CYO
GL6 0TB G0PPM
GL6 6AD G7KUU
GL6 6AG G0ISH
GL6 6EY G4TCO
GL6 6NJ 2E0JYX
GL6 6NJ G4HXQ
GL6 6PH M0BAP
GL6 6RQ G4VLV
GL6 6SA G0VPT
GL6 6TJ G3OAD
GL6 7DA G6PKG
GL6 7EF G0PKJ
GL6 7NY G4SJN
GL6 7QA G3SPO
GL6 8AA G3KYZ
GL6 8AG 2E0IEI
GL6 8AG M6GZE
GL6 8DG G1VDE
GL6 8FB G3VBQ
GL6 8JN G4CIG
GL6 8LZ G3NQF
GL6 8ND G3TEV
GL6 8NX M0HNH
GL6 8NY G4OHA
GL6 9BA G4GUG
GL6 9BZ G4BSM
GL6 9BZ G7EXX
GL6 9HB G4IJV
GL6 9HF G3UDD
GL7 1AP G3SUG
GL7 1AU G0UIX
GL7 1BJ M1TCP
GL7 1BJ M6FXG
GL7 1BL M1BHZ
GL7 1BX G0AZD
GL7 1BX G7AEH
GL7 1GJ G4GPV
GL7 1HF 2E0ARJ
GL7 1HF G5HI
GL7 1JT M0AYX
GL7 1PR 2E0LLW
GL7 1PS M6NXA
GL7 1TG G0RYR
GL7 1TG M3GVJ
GL7 1TG M3WHA
GL7 1UG G8BAS
GL7 2EJ M0ANO
GL7 2LS G4EVE
GL7 2NG G1FRD
GL7 2PZ G0WIY
GL7 3AR G3UJZ
GL7 3NN M6CAW
GL7 3SD G4JTO
GL7 3SD G7AYE
GL7 4DZ M6NFO
GL7 4EQ G3ZVC
GL7 4EX G4XMR
GL7 4HN G6RPW
GL7 4HN M6JJG
GL7 4JU M0ONY
GL7 5BG G7IYI
GL7 5PR 2E0DIU
GL7 5PR M6EOY
GL7 5QS G1MYQ
GL7 5ST G2OU
GL7 5ST M1BEO
GL7 6BE G6EMB
GL7 7BA G7ORG
GL7 7BB G3UK
GL7 7DL G0IZP
GL7 7JU G3TA
GL7 7JY G4ABY
GL76PA 2E0IRE
GL76PA M6LFW
GL8 8BU 2E0XVZ
GL8 8BU M0GVQ
GL8 8BU M3LLN
GL8 8DR G3OIP
GL8 8HA G0RXQ
GL8 8JE G1TJW
GL8 8JL G8EMU
GL8 8JU 2E0AYX
GL8 8JU M0OSM
GL8 8LT G8NSO
GL8 8SN G4IHT
GL8 8YU G3MAV
GL9 1HH G1KKS
GL9 1HP G0XAY
GL9 1HP G4GCT
GL9 1HT 2E1IJK
GL9 1HT M0TFY

## GU (Guildford)

GU1 1BT 2E0BZH
GU1 1EP G0SWC
GU1 1EP G0SWE
GU1 1FR G7MZS
GU1 1HX G1MDS
GU1 1HX G8MAV
GU1 1NA 2E1CML
GU1 1PA M6HAC
GU1 1QL G0KTV
GU1 1RQ G4CDX
GU1 1TL 2E0OBC
GU1 1TL M6OBC
GU1 1UB G1ZDG
GU1 2FB 2E0RJO
GU1 2FB M0RCN
GU1 2FB M6RJO
GU1 2FJ G0LPG
GU1 2JE M6CUJ
GU1 2JQ G0EFO
GU1 2QF G1RNV
GU1 2RP M6IER
GU1 2RR G3NR
GU1 2TY G7UCR
GU1 3JR G4TJK
GU1 3NP G6AFK
GU1 3PZ G8ZAX
GU1 3PZ M0SBT
GU1 4DD M1TOD
GU1 4DN M6WEU
GU1 4NP G8TZN
GU1 4NQ G7VNL
GU10 1BY G0VYQ
GU10 1LE G4GNO
GU10 1RH G3ZSS
GU10 2JG G1RUG
GU10 2JG M0CJO
GU10 2QU 2E1JIM
GU10 2QU G4IFX
GU10 2QU M3CPX
GU10 4AX M6GYC
GU10 4AX M6HYJ
GU10 4BJ M0PSI
GU10 4DW G0LVR
GU10 4EX 2E0BOD
GU10 4JW G8RYJ
GU10 4LP G3FPQ
GU10 4LU G4ZKJ
GU10 4PQ M6FZG
GU10 4RJ M6AGQ
GU10 4RL G4AHN
GU10 4TP G6BOF
GU10 4UA G8WKA
GU10 5EL G4KWX
GU10 5HZ G0NFA
GU10 5LP G4ROM
GU10 5LS G4LES
GU10 5PA G8IXL
GU10 5QE M0VVV
GU11 1HA M0NGI
GU11 1HA M6NGI
GU11 1HA M6NGU
GU11 1YY G6RQZ
GU11 2HT M6RUE
GU11 3BW M0GPW
GU11 3DB M0AHJ
GU11 3DE M6FWB
GU11 3EL G4GVV
GU11 3EP M0PSE
GU11 3ET G4GMI
GU11 3JX G4GGZ
GU11 3PX G0BCH
GU11 3SL G4ZYM
GU12 4AG G1XIE
GU12 4AT G8YKM
GU12 4EU 2E1FHQ
GU12 4FQ G4MBZ
GU12 4HU M6JZY
GU12 4HY M6KIL
GU12 4HY M6PCE
GU12 4HZ G4YFU
GU12 4LD G8OLY
GU12 4RD G3KND
GU12 4SF G4SPD
GU12 5AY G7ITM
GU12 5HP G8PDP
GU12 5HR M6MGA
GU12 5HS G4WEL
GU12 5HS G4WEM
GU12 5JG G8ZAJ
GU12 5JP G1LKJ
GU12 5JP M3LKJ
GU12 5JT G8BCO
GU12 5JT M0TKJ
GU12 5NQ G6XNP
GU12 5PR 2E1HAC
GU12 5PR G8CXF
GU12 5QW G8NEF
GU12 5SF G4CDH
GU12 6HP 2E0DFI
GU12 6HP M6EHA
GU12 6LX G3VKI
GU12 6LX G8YCL
GU12 6LY G7PWA
GU12 6NZ G6TSX
GU12 6PL M0GEL
GU12 6PP M6KVK
GU12 6QB G1IYE
GU12 6QN G7CGT
GU12 6ST M1EGZ
GU124HY G1OQG
GU124HY M0COM
GU14 0DD G0EMR
GU14 0DW G6MRW
GU14 0ET 2E0WKT
GU14 0ET M0WKT
GU14 0ET M6GQX
GU14 0ET M6WKT
GU14 0HL M6AAO
GU14 0PB G1AAP
GU14 0RF G8IBC
GU14 0RW M6AOP
GU14 6AR 2E0TRX
GU14 6AR M6REX
GU14 6AX M6BQK
GU14 6DS G6RIZ
GU14 6JS 2E0DFL
GU14 6JS M6EHI
GU14 6JT M6BDM
GU14 6LX G1NTX
GU14 6NP G4OJR
GU14 6NU G4JFN
GU14 6NU G4WAH
GU14 6QA 2E0SEE
GU14 6QA G4KCC
GU14 6QA M6AZE
GU14 6RF G3WOS
GU14 6TH M0HAU
GU14 7AE G3SRR
GU14 7AR G3PQC
GU14 7BA 2E0CTQ
GU14 7DA G0DWM
GU14 7EU G3UHW
GU14 7EY G1OER
GU14 7HH G0HOD
GU14 7PP G7VTS
GU14 8AG 2E1CYZ
GU14 8AG M3MKB
GU14 8BJ M0RCD
GU14 8DG 2E0GFF
GU14 8DG M0YMA
GU14 8DG M6ADB
GU14 8DJ M6FKN
GU14 8NN G4PDR
GU14 8PH G1FVH
GU14 8QJ G3PQF
GU14 8RY M5AJP
GU14 8SR G7HGT
GU14 9AU 2E0CUL
GU14 9AU M0KUL
GU14 9EA G1KXQ
GU14 9JQ G0MPI
GU14 9PJ G0MCQ
GU14 9PW M3IHS
GU14 9RL G4BNK
GU14 9RY 2E0LHS
GU14 9RY M6FKO
GU14 9UH G0HUT
GU14 9XA G4HTL
GU14 9XU G4NSN
GU14 9XY M6ASQ
GU15 1BF G3YTX
GU15 1DE G8GQS
GU15 1DG G3HEJ
GU15 1DL G6FYC
GU15 1DL M6CHX
GU15 1EF G8BVB
GU15 1EQ M6JTU
GU15 1HP G3OKU
GU15 1HW G6UAW
GU15 1LF G4UQE
GU15 1NP G6IXM
GU15 1NZ G1POK
GU15 1NZ G2BOF
GU15 1PS G0RGX
GU15 1PS G8DXZ
GU15 1RD 2E0CIX
GU15 1RU G8EAD
GU15 2BH G3RKK
GU15 2BU M0HUI
GU15 2DE G0DSQ
GU15 2JQ G6JTO
GU15 2LD G4REE
GU15 2LT G4ELJ
GU15 2LY M6BTB
GU15 2NR G6LWA
GU15 2NT G0WZX
GU15 2RQ 2E1AFA
GU15 2SP G6ENU
GU15 2SP M6ENU
GU15 3EB M6EYA
GU15 3EN M6MNT
GU15 3HT M0KWY
GU15 3HT M0LKY
GU15 3HT M0RCK
GU15 3HT M6DBK
GU15 3HT M6ELI
GU15 3HT M6LFY
GU15 3HT M6LKY
GU15 3JL G4JDF
GU15 3NP G4GLP
GU15 3NR G3LPN
GU15 3XB G4YHK
GU15 4AR G8OUY
GU15 4AR M6SUN
GU15 4BW M0CMP
GU15 4JR M6BDQ
GU15 4LD M6PUS
GU15 4ZF M0CMF
GU16 6BZ G1OSO
GU16 6DT G6WZD
GU16 6ER G6BLU
GU16 6EY G0BXM
GU16 6EY M1CIS
GU16 6GA M6LCW
GU16 6HG M6BMJ
GU16 6JJ G4SYB
GU16 6PH G6MJB
GU16 6SD M6CTW
GU16 7DU 2E0SKZ
GU16 7DU 2E0XAI
GU16 7DU 2E0XPJ
GU16 7DU 2E0XPT
GU16 7DU M0XPJ
GU16 7DU M3MXA
GU16 7DU M3XAI
GU16 7DU M3XPJ
GU16 7DU M6XPT
GU16 7EB M6BQR
GU16 7RD M6LAC
GU16 8LP G7KNK
GU16 8LP M3KNK
GU16 8LR G3PPU
GU16 8NA G8XNC
GU16 8PS G7LTW
GU16 8RT G8IRM
GU16 8RW G4MEA
GU16 8SA G3KFU
GU16 8ST M6SWE
GU16 8XH M6RYA
GU16 8XQ G4DJB
GU16 8XR M6WJN
GU16 8XX G2DX
GU16 8XX G3TJI
GU16 8YN G7WDD
GU16 8YT M6ABB
GU16 9BL M6MHM
GU16 9BW M6SAC
GU16 9NY M6KDK
GU16 9QN M6POQ
GU16 9QU G3YAG
GU16 9RB G7AQL
GU17 0DH G3WCY
GU17 0DJ G0ADT
GU17 0DU G6KRY
GU17 0EN G1FOE
GU17 0EN M3NFW
GU17 0HB G8JZO
GU17 0JE G4VDF
GU17 0NJ G3TMU
GU17 9AY M6HXF
GU17 9BP G8HIO
GU17 9DL G4CGW
GU17 9DX G3ZYX
GU17 9DZ G3REL
GU17 9DZ G3SFU
GU17 9ET G4EMV
GU17 9HA G8HUF
GU17 9HH G0EFL
GU17 9HH G6AAB
GU17 9JH G0HEE
GU17 9JQ G3ZWK
GU18 5RZ G3TJS
GU18 5TA G0KDL
GU18 5TE G8JMP
GU18 5TP G3UKE
GU18 5TR G8HXD
GU18 5TS G8NYJ
GU18 5UJ G4SUO
GU18 5XH G4SJH
GU19 5DH M6HJQ
GU19 5DH M6HJU
GU19 5DP M6WEW
GU19 5JR M3XTA
GU19 5JX G4ZRB
GU19 5NU G8MZD
GU2 4AZ G2DBH
GU2 4EL G4PEA
GU2 4JF G8UHK
GU2 4LA G6FKN
GU2 4LF 2E0VHZ
GU2 4LF M6VHZ
GU2 5XH G7DQE
GU2 7JH M3RZU
GU2 7JL M3UJO
GU2 7JL M3VKT
GU2 7JL M6MFM
GU2 7JN M3ZRV
GU2 7JQ M6ILC
GU2 7QN G0SUL
GU2 7RT G7JVQ
GU2 7SA G1WTX
GU2 7SR G4OWL
GU2 7SR G8DPQ
GU2 7TP M6PZM
GU2 7TS M3MBF
GU2 7XE M6LXX
GU2 7XH 2E0LTS
GU2 7XH G3IGQ
GU2 7XH G8AHK
GU2 7XH M0LTS
GU2 7YW 2E0VNN
GU2 7YW M3VNN
GU2 8AP G8LOZ
GU2 8AR G8FUH
GU2 8AU M6ASD
GU2 8BL G1MGF
GU2 8JD M0SFI
GU2 8JU G4PMZ
GU2 8LX 2E0CZS
GU2 8LX G8IMS
GU2 8LX M0PVI
GU2 8LX M3PVI
GU2 8UT 2E1AOG
GU2 9NP G8TZU
GU2 9PJ G0HNL
GU2 9PJ G8NOB
GU2 9PL M6RJQ
GU2 9SB G7UPN
GU2 9TY M3UIU
GU2 9WA G1YJY
GU20 6BU G8WSP
GU20 6JT G7JFU
GU21 2DD G1OFW
GU21 2HY M6XRD
GU21 2JN G8FZV
GU21 2LD M6JXN
GU21 2LH G4HWI
GU21 2NG G0JXP
GU21 2PX M6XNL
GU21 2QP 2E0SSJ
GU21 2TA G1AOE
GU21 3BP G4PRW
GU21 3BZ M0RDB
GU21 3DB M0IRI
GU21 3DD G8JMU
GU21 3NZ G8BTL
GU21 3PJ G7WIY
GU21 3QS M0VPC
GU21 3QY 2E0XSG
GU21 3QY M0XSG
GU21 4BG G3USX
GU21 4JG G8SSY
GU21 4PW 2E0DRW
GU21 4PW M6KWG
GU21 5EG G0NIX
GU21 5JQ M3WSE
GU21 5PY G0NGI
GU21 6DQ M6FBT
GU21 7PN G7CSJ

GU21 7PR G7SZZ
GU21 7QT [illegible]
GU21 8AW M0PEB
GU21 8SS M0KFU
GU21 8SY M6CEI
GU22 0AA G4GGI
GU22 0DL 2E0MCQ
GU22 0DL M6MTW
GU22 0JL G6VBJ
GU22 0NQ G7HHI
GU22 0NT G3ZHB
GU22 0NT G6BQC
GU22 7BE 2E0KFK
GU22 7BE M6KFK
GU22 7JF M6HDW
GU22 7NE M6NZL
GU22 7NW 2E0CLH
GU22 7NW M6BGE
GU22 7NX G0MIC
GU22 7NX M0SRC
GU22 7TW 2E0BQN
GU22 7TW M3XLY
GU22 7UE G3GAQ
GU22 7UR G8LSA
GU22 8PE G0AAA
GU22 8PE G3WVG
GU22 8PE M0ZIP
GU22 8QW M0DGT
GU22 8QZ G0AFP
GU22 9AJ M6EMO
GU22 9AU G0NRJ
GU22 9AU M3YSC
GU22 9BQ G4YPC
GU22 9BQ G5RS
GU22 9PH G7PVU
GU22 0NU G4SLL
GU23 6DQ G0JOP
GU23 6EN G0DKN
GU23 7AD M0BUF
GU23 7AP G4EML
GU23 7ET G8AFU
GU23 7HR G3IUW
GU24 0HF 2E0MXC
GU24 0HF M0MXC
GU24 0HF M6MXC
GU24 8AR G3KMA
GU24 8PR G7CSI
GU24 8PS M6ISH
GU24 8RL M6AXS
GU24 9DG G0RVK
GU24 9NG G0TAN
GU24 9QE G0DER
GU25 4EZ G0TTG
GU25 4LD G8MLK
GU25 4NG M6ACC
GU26 6LR G0NIN
GU26 6NY G3ZFT
GU26 6PA 2E1JON
GU26 6PA G4NOT
GU26 6PW G7VBL
GU26 6PZ G7UEI
GU26 6PZ G7WBM
GU26 6QX G3VXF
GU26 6RP G4CMG
GU26 6SX G3WLH
GU27 1AZ M1CZX
GU27 1DF G7NDN
GU27 1JD G1OTZ
GU27 1JF G6TBT
GU27 1JF M6DFE
GU27 1JF M6XER
GU27 2FD 2E0CVQ
GU27 2FD M6CTN
GU27 2NY G1MAL
GU27 3AW G4TTJ
GU27 3DZ 2E0URJ
GU27 3JL G0XTL
GU27 3JS M0EBX
GU27 3ND G4DQZ
GU27 3RG G0NON
GU27 3SN 2E0SJI
GU28 0EQ G6NMQ
GU28 0EQ G8WSX
GU28 0EQ M3UQE
GU28 0NY G0HJZ
GU28 0PJ M6THA
GU28 0QE G1TLH
GU28 0QX M6XRX
GU28 9DH 2E0BQZ
GU28 9DH M0PUT
GU28 9DH M3UNB
GU28 9DH M6BDE
GU28 9DP M0CPD
GU29 0BP G4DCM
GU29 0DJ G4YEI
GU29 0NU M6WJJ
GU29 0NX M6LPI
GU29 9JG G0DDY
GU29 9QZ G0MWX
GU29 9TE M1STI
GU3 1AZ G0SJH
GU3 2AX 2E0ZNZ
GU3 2AX M0ZNZ
GU3 2AX M6KNC
GU3 2AX M6ZNZ
GU3 2EN G4HZV
GU3 2EU G0PET
GU3 2JW G3VCY
GU3 3AU G4PNB
GU3 3BE G6GPR
GU3 3BX G3TQC
GU3 3PQ M1EBL
GU3 3PR G4MPW
GU30 7BS G4VLF
GU30 7BS G6JQX
GU30 7BY M0RGI
GU30 7WW [illegible]
GU30 7HG G3TSR
GU30 7HR G4VKC
GU30 7SB G8OXS
GU30 7SH G8ISI
GU30 7XA G8UYW
GU30 7XD G4AER
GU31 4EU 2E0DFF
GU31 4EU 2E0INV
GU31 4EU 2E0JLB
GU31 4EU M6IVS
GU31 4EU M6OBJ
GU31 4EU M6TXT
GU31 4NX G1MGU
GU31 4PN G1LQH
GU31 4PN G1OXF
GU31 4QA G6SLZ
GU31 5AW M0NNL
GU31 5AW M3YQQ
GU31 5AX G3TZL
GU31 5HJ G0BUZ
GU31 5NZ 2E1JJM
GU31 5NZ G0WKL
GU31 5NZ M3JJM
GU31 5SB G1THA
GU32 1AG M0DAY
GU32 2AN G4VQT
GU32 2AX G6XJB
GU32 3AZ G4VRC
GU32 3BT G4SBU
GU32 3BW G0MBQ
GU32 3DP M0HXC
GU32 3LS G0BTU
GU32 3LS G8XDD
GU32 3PJ G4ACW
GU33 6DA M3TQP
GU33 6HG G8XJO
GU33 7BP 2E0MPA
GU33 7BP M3ZQF
GU33 7DB M1CQU
GU33 7DL G4ELM
GU33 7JE M0PDH
GU33 7JE M3EMX
GU33 7LR G4WKY
GU33 7SN M0AIR
GU34 1JA G4ASL
GU34 1NU G4FOY
GU34 1PW G0JMI
GU34 1QY G0JVL
GU34 1QY G1LAO
GU34 1RA G3RSI
GU34 1RR G8MOS
GU34 2BJ G1EML
GU34 2BQ 2E0BUF
GU34 2BQ M0VVQ
GU34 2ED G8XWR
GU34 2EE G0DBS
GU34 2EE M0TRO
GU34 2EW 2E0BTW
GU34 2EW M0GOK
GU34 2EW M3XRK
GU34 2HP G8YFH
GU34 2PE M3KOA
GU34 2PF G1JUP
GU34 2QT G0IBR
GU34 2QT G7LWH
GU34 2RD G4JOU
GU34 2RS G6YBN
GU34 2TL G8IRL
GU34 2TP G8UML
GU34 3EJ G6BXV
GU34 3NP G0DWR
GU34 4AN G8EAN
GU34 4AQ G0EFP
GU34 4AX G4BUW
GU34 4LJ G4DNK
GU34 4LJ G8DDH
GU34 5AX G8LES
GU34 5BG G7BCS
GU34 5BJ G3XVR
GU34 5BN 2E1GTD
GU34 5BX G0RMR
GU34 5BY G4IPI
GU34 5BY G8WSZ
GU34 5EF G3PRQ
GU34 5HZ M6SSO
GU34 5JF M0BDF
GU34 5LG G6UEQ
GU34 5NP G3YVI
GU34 5PB G0BHA
GU34 5PB G4ORN
GU34 5PR G6BHH
GU34 5PR G6BHI
GU34 5PF G8CKN
GU34 5PR G3SJX
GU35 0DG G0PLZ
GU35 0EF M6KGK
GU35 0HB G3ZRM
GU35 0PP G0OYN
GU35 0QB G8AZB
GU35 0RA M0EVF
GU35 0SG 2E0BII
GU35 0TB G4MQQ
GU35 0TB G8RYX
GU35 0TH G8NLF
GU35 0TL G0PVI
GU35 0XA G8BGI
GU35 8AJ G0PVJ
GU35 8BL G7TEP
GU35 8EP G1CBB
GU35 8HQ G6SNA
GU35 8HU G4ZEL
GU35 8NA G4LJB
GU35 9AJ 2E0SNJ
GU35 9AJ M6SWA
GU35 9BA M0MGA
GU35 9ED 2E0UUN
GU35 9ED M6BYZ
GU35 9EX 2E0DVD
GU35 9EX M0EUY
GU35 9EX M3EUY
GU35 9PJ G8SXB
GU35 9QN 2E0CCW
GU35 9TA G1TQY
GU4 7EQ 2E0PMZ
GU4 7EQ M0DLX
GU4 7EQ M3PDE
GU4 7FF M6FSD
GU4 7NP M1RJJ
GU4 7NQ M6MGV
GU4 7PD 2E1ICW
GU4 7PD G0KUF
GU4 7QB G6MQZ
GU4 7TJ M1ACK
GU4 7XR G3PIZ
GU4 7XR M0ECW
GU4 8AY M0DHN
GU4 8HZ G0JRE
GU4 8JG G8NTH
GU4 8JY M3MON
GU4 8LL G6ZAC
GU4 8LQ G1HTN
GU4 8PH G3WBQ
GU46 6BX G3WUL
GU46 6DW G4URP
GU46 6FA G8PJQ
GU46 6GZ 2E0PTG
GU46 6GZ M0URL
GU46 6NE G8IFH
GU46 6NT G6FTY
GU46 6PB G1WNL
GU46 6PE G7CFW
GU46 6YP G7AOA
GU46 6YW G0YOU
GU46 7AD G4RYV
GU46 7RD G8INZ
GU46 7SY G4ZHZ
GU46 7TT G4TTZ
GU47 0HH 2E0KAL
GU47 0HH G8NMJ
GU47 0HH M3HCL
GU47 0ST G7VHZ
GU47 0UP G8EII
GU47 0UT 2E0DCL
GU47 0UT M6CLK
GU47 0UT M6KSC
GU47 0XN M3BVQ
GU47 0YQ G8LMY
GU47 0YQ M5LMY
GU47 0YY G6ACJ
GU47 8HT G0FGA
GU47 8HT G3ZJQ
GU47 8HY M6SMR
GU47 8JH G1FNS
GU47 8JL G8FXU
GU47 8LD G8NMH
GU47 8PR G4JVV
GU47 8QS G4XYW
GU47 8QS G6GS
GU47 9AU G4SWM
GU5 0BL G0UNF
GU5 0JP G0DRR
GU5 0SA G4VQF
GU5 0SE G6ZPR
GU5 0SX G8WMF
GU5 9AR M6ZDZ
GU5 9LS M0HKP
GU5 9LS M6HKN
GU5 9RS G0FCU
GU51 1AU G8VIV
GU51 2TE G4KTZ
GU51 2TL M6SCA
GU51 3AH G7IIH
GU51 3AH M0CYF
GU51 3AH M3ZCM
GU51 3BS G7BQT
GU51 3DY G3MSL
GU51 3EB G4OQZ
GU51 3EL G7MCS
GU51 3LY G8YVM
GU51 3LY M0NAM
GU51 3NF G8UUG
GU51 3NF M3UUG
GU51 4AQ 2E0DOZ
GU51 4AQ M6DOZ
GU51 4HA G4EFY
GU51 4HB G8JNI
GU51 4HN G4FTK
GU51 5AH M3XSU
GU51 5DP G1CNV
GU51 5NJ 2E0BBS
GU51 5NR G7KIT
GU51 5SU M0DHO
GU51 5TZ M6MDM
GU52 0TE G6DPW
GU52 0TE M3WRJ
GU52 0UR G3KOB
GU52 6AS G3HUK
GU52 6AY G8SUJ
GU52 6AZ M0RHE
GU52 6AZ M3KFK
GU52 6BN G4AFI
GU52 6JD 2E0BBY
GU52 6JD M3KFR
GU52 6LJ G3HGI
GU52 6LQ G3WKW
GU52 6PW G3RCD
GU52 6QJ G3CYL
GU52 7LE G3LGT
GU52 7LN G1JPP
GU52 7UG G3BRQ
[illegible]
GU52 8NB 2E0ETE
GU52 8NS 2E0REB
GU52 8NS M6FFF
GU52 8NS M6REB
GU52 8XG G0NWL
GU6 7AA M3BSM
GU6 7DD M6VBP
GU6 7EN M0AWE
GU6 7ET G4CJT
GU6 7HR G4IBI
GU6 7HZ G6XN
GU6 7HZ M0GJH
GU6 7JB G3OHC
GU6 7JB G4GRG
GU6 8BZ G4YIF
GU6 8FD M6ONS
GU6 8NB M6TCH
GU6 8PQ G1DSM
GU6 8TP 2E0LSL
GU7 1QQ G8IBL
GU7 1RL G6WOT
GU7 1RL M0GKK
GU7 1SB 2E0DPO
GU7 1SB M0HZM
GU7 1SB M6UGE
GU7 1SY G4CWP
GU7 1YA 2E1IAC
GU7 1YB 2E0XEN
GU7 1YB M0NGY
GU7 1YY M0IAZ
GU7 2BS M6HXA
GU7 2LD G3TCU
GU7 2LJ 2E0ERK
GU7 2LJ 2E0YAP
GU7 2NE G1WTW
GU7 2NH M3BMI
GU7 2NT G4VRN
GU7 2QT 2E0DTN
GU7 2QT M0KRD
GU7 2QT M6KRD
GU7 2RW G1SWZ
GU7 3AQ M6WDV
GU7 3EU G8ZMC
GU7 3HG G3KZB
GU7 3NN G8YEF
GU7 3NT G3MBK
GU7 3NU G0MUQ
GU7 3NZ G1FJF
GU7 3QJ G6NAV
GU7 3SH M0IAM
GU8 4AB M3TGK
GU8 4BD G4JEF
GU8 4ND G0SUA
GU8 4NT G8EBT
GU8 4RG 2E1AFS
GU8 5AB G7CND
GU8 5AB M3FJB
GU8 5DN G4GIX
GU8 5JZ 2E0TMH
GU8 5JZ M0TMP
GU8 5JZ M6VMC
GU8 5NR G4XBF
GU8 5TU G4DUF
GU8 6DE G0GNE
GU8 6DE G6RLM
GU8 6DU G3YZN
GU8 6DZ 2E1AFR
GU9 0BS G6FRS
GU9 0BS G7NMT
GU9 0BS G8ATK
GU9 0HX G7GOK
GU9 0HY G8FQN
GU9 0NU M5AMN
GU9 0RZ G0FRS
GU9 0RZ G0YYY
GU9 7BP G4WUV
GU9 7BP G4WUW
GU9 7BT G7HOE
GU9 7BX G3VYI
GU9 7BX G4ALE
GU9 7DA M0RMW
GU9 7DH M0BSD
GU9 7RQ G4UEL
GU9 8BZ G3TAX
GU9 8JQ G8OGH
GU9 8LF G0CTZ
GU9 8RR 2E0KBG
GU9 8RR M6KBG
GU9 8SA M0PPA
GU9 8SA M6DPR
GU9 8TA G0CXU
GU9 8TW G0EYA
GU9 9AY G7JUR
GU9 9BA G7UWL
GU9 9EA G8UMA
GU9 9EY M3OSP
GU9 9HJ G6BEN
GU9 9HJ G7FAQ
GWYN EDD G7HLZ

## GY

## (Guernsey)

GY1 1FP GU7CMH
GY1 1FQ GU4RUK
GY1 1GD 2U0TKB
GY1 1HP GU7MXZ
GY1 1JB MU3MNG
GY1 1NT GU4WTN
GY1 1SF GU4WRP
GY1 1WJ GU0RAG
GY1 1WU GU4EON
GY1 1XJ 2U0NPT
GY1 1YL GU7PVI
GY1 1XJ MU3NTH
GY1 2BA GU4SXM
GY1 2PL GU6JQF
GY2 4AH 2U1EKH
GY2 4EW MU0GSY
GY2 4EX MU6GBG
GY2 4FL 2U0BGE
GY2 4FL MU0ZVV
GY2 4HJ GU4HUY
GY2 4HJ GU4XGB
GY2 4JS MU6CPV
GY2 4NS GU0VPA
GY2 4RN 2U0WGE
GY2 4RN MU3KBP
GY2 4RN MU6FXB
GY2 4XW 2U0EJL
GY2 4XW MU6GCI
GY2 4YW MU3UWX
GY3 5AF GU6NCZ
GY3 5AF GU6TKE
GY3 5EN GU4XEA
GY3 5JD GU4JHH
GY3 5JG GU4YBW
GY3 5NL MU6GUE
GY3 5PQ GU3ONJ
GY4 6AD MU6RAN
GY4 6AF GU6XCM
GY4 6JT GU7DSB
GY4 6JW GU4WMG
GY4 6NB GU6RWD
GY4 6NH GU1HYN
GY4 6QJ 2U1FNQ
GY4 6QJ GU4NYT
GY4 6QJ GU4SYQ
GY4 6SF GU3YVV
GY4 6UB GU0DXX
GY5 7BN GU7OYU
GY5 7DT GU3UMX
GY5 7DZ GU4YOX
GY5 7FQ GU0SUP
GY5 7FQ GU3HFN
GY5 7FQ MU7NCZ
GY5 7SA MU3ZHF
GY5 7XD GU4XIT
GY5 7XZ GU4CHY
GY5 7YB GU7NHX
GY5 7YS GU0GUX
GY5 7YS GU0NHD
GY6 8EZ 2U1EJF
GY6 8HS GU3ZOM
GY6 8LA GU4ASO
GY6 8LN GU1IIW
GY6 8NR MU0FAL
GY6 8RB GU3LYC
GY6 8RT GU6EFB
GY6 8RT MU3EFB
GY6 8RY 2U1EKE
GY6 8RY GU1HTY
GY6 8RY GU1WJA
GY6 8SJ GU0BDI
GY6 8TU GU3UOQ
GY6 8UF GU8ITE
GY6 8XL 2U0DWD
GY6 8XL MU0WLV
GY6 8XL MU6XGE
GY6 8XN MU6WZY
GY6 8XX GU2RS
GY6 8YJ GU7CNI
GY7 9DA GU7TQX
GY7 9HE GU1MUP
GY7 9LD GU0PSP
GY7 9NN GU7CQN
GY7 9NN MU0CHN
GY7 9PH 2U0EFR
GY7 9PH MU0EFR
GY7 9PH MU6GFR
GY7 9QF 2U0LRB
GY7 9YQ MU0FBO
GY7 9YQ MU6STK
GY79EL GU3WOW
GY8 0AB GU4LJC
GY8 0AJ GU8FSU
GY8 0JB GU1DWO
GY9 3JD GU3LPV
GY9 3TX GU1DWW
GY9 3TZ GU7APA
GY9 3UY GU0UVH
GY9 3XD GU3TUX
GY9 3XE MU0FDN
GY9 3YZ GU4IJF

H52 0EU GD0PKW

## HA

## (Harrow)

HA0 1AE GD6VIO
HA0 2NX 2D0LHR
HA0 3QC GD6MLV
HA0 3SG MD6SHL
HA1 1TD 2D0FGW
HA1 1TD MD6FGW
HA1 1XH 2D0SRP
HA1 1XH MD0SRP
HA1 1XH MD6PRS
HA1 3HT GD0UHK
HA1 4AJ GD3VFX
HA1 4AL GD4IQD
HA1 4AL GD4MSE
HA1 4BW MD0EGL
HA1 4DJ MD6CIX
HA1 4EF MD6EGA
HA2 0HJ MD6YOO
HA2 0PH GD4HAU
HA2 0PU GD7OZA
HA2 0QE 2D0GEE
HA2 0QE MD6OMG
HA2 0QJ GD7FSJ
HA2 0SX GD4TPM
HA2 6AP GD4JMG
HA2 6AQ GD1RYF
HA2 6DG GD6SSV
HA2 6HE GD0CAG
HA2 6HF GD3LGQ
HA2 6LG 2D0LBG
HA2 6LG MD0JNS
HA2 6LG MD6JNS
HA2 7AX GD6MFR
HA2 7EJ GD4PKV
HA2 7JJ MD6FFP
HA2 7NU GD4IUM
HA2 7RB GD4IRP
HA2 7RP MD3KHW
HA2 9EA GD4KEW
HA2 9EF GD8DKW
HA2 9JL GD4WVW
HA2 9LD GD4MMA
HA2 9LJ GD7IEF
HA2 9QF GD4UKD
HA20LW 2D0NSY
HA20LW MD6MPS
HA3 0PB GD3FKI
HA3 0XQ GD4KAB
HA3 6AJ GD0IPK
HA3 6BX GD3VHK
HA3 6JT 2D0DJX
HA3 6JT MD0ZDJ
HA3 6JT MD6XDJ
HA3 6QL 2D0HBT
HA3 6QL MD3HBT
HA3 6TN GD0BSP
HA3 7AX MD0GCI
HA3 7HU MD6ABT
HA3 7NF MD6HVA
HA3 8HA MD6FQF
HA3 8JB GD3BQE
HA3 8ND GD7DNP
HA3 8NF GD3ASR
HA3 8NF GD4KEP
HA3 8NQ GD4ZPR
HA3 9BD MD0GFK
HA3 9EJ GD8FAT
HA3 9ET MD6SFL
HA3 9LB GD7HIH
HA3 9NB GD1TMW
HA3 9QZ 2D0RDA
HA3 9QZ 2D0UTL
HA3 9QZ MD6OCZ
HA4 0AU GD3KRT
HA4 0BT GD4YGM
HA4 0DS GD0ACK
HA4 0DW GD0DVC
HA4 0LX 2D0EJR
HA4 0LX MD0RJE
HA4 0LX MD6RJE
HA4 6AJ GD0RNF
HA4 6BA MD3TLG
HA4 6ED GD0DZH
HA4 6ED GD1SWF
HA4 6HG GD0ULF
HA4 6ND MD0MAR
HA4 6QZ MD3PYB
HA4 6SW 2D0CBT
HA4 6SW MD6MHA
HA4 6SW GD0NRN
HA4 7JB GD3EFX
HA4 7JB GD4AUF
HA4 7LZ GD0JIM
HA4 7LZ GD0VTC
HA4 7LZ GD6JIM
HA4 7QD GD8UOZ
HA4 7RD GD6TDX
HA4 7TQ 2D1HVL
HA4 7TQ MD0XRA
HA4 7UL GD4LHT
HA4 7UL MD6GRR
HA4 7UN GD4UZE
HA4 8AJ GD4FFA
HA4 8PH GD8UMY
HA4 8RY GD3OEC
HA4 8SB GD0JCF
HA4 8SB GD1DES
HA4 8SD GD7KUG
HA4 8TA GD3BPG
HA4 8UA GD6SYB
HA4 9AG GD3JVM
HA4 9BY MD6FGZ
HA4 9HD GD6PLR
HA4 9JT GD1LQV
HA4 9LF GD6SPI
HA4 9SA MD0HJR
HA5 1JH GD0IRM
HA5 1NB GD4PXY
HA5 1NJ GD1RGT
HA5 1NL 2D0CGB
HA5 1NL MD3VUQ
HA5 1PH GD7HSO
HA5 1TL GD0KSL
HA5 1TN GD0CHQ
HA5 2AU MD0DLC
HA5 2DA GD8CAB
HA5 2EH GD3LFD
HA5 2HR 2D0ZYX
HA5 2HR GD0HGG
HA5 2HR MD6HGG
HA5 2NB GD8NTR
HA5 2NE MD6LXW
HA5 3LP GD1FBS
HA5 3UH [illegible]
HA5 3UF [illegible]
HA5 3YF 2D0XLJ
HA5 3YF GD4AZL
HA5 3YF GD6XUD
HA5 3YF GD8DAI
HA5 3YF MD3XLJ
HA5 4EP GD3ENV
HA5 4HB 2D0VYN
HA5 4HB MD6VYN
HA5 4LN GD3UAS
HA5 4NJ GD0WHV
HA5 4RA GD0NTA
HA5 4RA GD0NTB
HA5 4TX GD1VIO
HA5 5SP MD0SEW
HA6 1DW GD4MEX
HA6 1EZ GD4JZS
HA6 1EZ GD8LZE
HA6 1PF 2D0JTY
HA6 2BU 2D0SEO
HA6 2BU MD0AAV
HA6 2BU MD6SJX
HA6 2FY GD4AGM
HA6 2HN GD7NBF
HA6 2UJ GD8MAA
HA6 2YP GD7KWT
HA6 3AU GD4HMD
HA7 1JH MD3THQ
HA7 2HS MD6MIK
HA7 3HT 2D0UAY
HA7 3HT MD3XSG
HA7 3PB MD3WCA
HA7 3PL GD1DQQ
HA7 4EP GD1MYM
HA7 4FP GD0MOX
HA7 4LD GD3NDC
HA7 4PF MD6HUV
HA7 4SJ GD4SYI
HA7 4UQ MD0ALQ
HA8 0SA GD4DRO
HA8 0TW MD3BCS
HA8 0TW MD3OAK
HA8 5NQ 2D0PBW
HA8 5NQ MD6IVO
HA8 5RJ GD0SQL
HA8 6DL GD8RNM
HA8 6HL GD4JOO
HA8 6NT GD2TV
HA8 7SA MD6KEU
HA8 7UH 2D0CQT
HA8 7UH MD6FEA
HA8 8AE MD6EZM
HA8 8JU GD4XEW
HA8 8PS GD4GLM
HA8 8RH GD4RFI
HA8 8SH GD0STR
HA8 8TR GD0KCL
HA8 8UA GD4SCG
HA8 8XJ MD0HOF
HA8 9BZ 2D0IGL
HA8 9HE GD1GXC
HA8 9JD GD4IXL
HA8 9JD MD0CFM
HA8 9PW MD6GVF
HA8 9SP MD6MED
HA9 0EY 2D0CTX
HA9 0EY MD0TCX
HA9 0EY MD6CTU
HA9 7DQ MD0TAT
HA9 7QR GD4UDT
HA9 8NX GD4DNP
HA9 8TP GD3ZZF
HA9 9HA MD1FFR
HA9 9TQ GD3SRN

## HD

## (Huddersfield)

HD1 3SL G0BFJ
HD1 3SL G6LD
HD1 4LL 2E0PAD
HD1 4LL M3WUG
HD1 4NU M6SGM
HD1 4QF G4KDM
HD1 4UR G1MOZ
HD1 5DY G7BNS
HD2 1AS G1MSD
HD2 1BN G1JJR
HD2 1DA G1CYY
HD2 1DH G3WUI
HD2 1ED G8YNE
HD2 1EJ M0GDJ
HD2 1LB G0ODY
HD2 1LL 2E0ALJ
HD2 1LL M0OYZ
HD2 1LL M3SVZ
HD2 1QH G4YDI
HD2 1QH G6KKA
HD2 1QP G1AOR
HD2 1QY G8YOC
HD2 1RN G3XXR
HD2 1RU 2D0EDL
HD2 1RU M0RIU
HD2 1RU M6LDE
HD2 1TH M6BJO
HD2 1UY G3WLW
HD2 1XT 2E0LOG
HD2 1XT M0LAG
HD2 1YW 2D0RNR
HD2 1YW M6JCS
HD2 2EB G0PRF
HD2 2EH G4XTE
HD2 2HF M6SZP
[illegible]
HD2 2NF G0BWO
HD2 2NF G4BWC
HD2 2NF G7UBQ
HD2 2NF M0AGR
HD2 2NH G4BYW
HD2 2PE M0TKA
HD2 2PQ G4MEK
HD2 2YD 2E0WBH
HD3 3DB G3BKJ
HD3 3EJ G0IRY
HD3 3FE G4MSP
HD3 3QX G3TFO
HD3 3RJ G8MAR
HD3 3YW M3ZVU
HD3 4BN G8DTX
HD3 4GH G4LRD
HD3 4HQ 2D0BOR
HD3 4HQ M3SVJ
HD3 4LD G1AGA
HD3 4LD G1AGB
HD3 4RS G8RWN
HD3 4SW 2D0BWJ
HD3 4SW M3YLK
HD3 4SW M3YLL
HD4 5DR G6VBA
HD4 5DX 2D0CVB
HD4 5DX M3ZKU
HD4 5NS G4GNA
HD4 5RP M6JPT
HD4 6HN M3VHU
HD4 6PJ M3GDX
HD4 6PL M1DQX
HD4 6QX G6TGE
HD4 6RA G1GTH
HD4 6SS M0AFS
HD4 6SZ M0PIE
HD4 6TE G4TML
HD4 7AS M3LQE
HD4 7DA G0BWB
HD4 7DA G0LPV
HD4 7DA G4CDD
HD4 7EP G8TOT
HD4 7HE G7UID
HD4 7HS G1BGM
HD4 7JS G0NMH
HD4 7JS G8XSF
HD4 7JX G0SVU
HD4 7RA G3UEU
HD4 7RJ G0TPM
HD4 7SP 2E0CFX
HD4 7SP M0WIS
HD4 7SP M3ZIR
HD4 7SR G4JLO
HD5 0AT G4RLA
HD5 0ER G3KJO
HD5 0JB G7MHQ
HD5 0LJ G7UZO
HD5 0NJ M3CMI
HD5 8EX G4KMK
HD5 8LS 2D0RSB
HD5 8LS G4BQC
HD5 8LS M0RBG
HD5 8QB 2E0VET
HD5 8QB M6ARC
HD5 8RP M0AOB
HD5 8UJ G4ITV
HD5 8UP G4RAJ
HD5 8XT G7OOH
HD5 8XU G4OTE
HD5 9HG G1FYS
HD5 9HX G0BWO
HD5 9JW G0LUU
HD5 9RG G7FBT
HD5 9UW G6ZVU
HD5 9XT M3OAQ
HD5 9XT M3RTP
HD6 1DA G6NLW
HD6 1HH G0DIU
HD6 2AT G1GTQ
HD6 2AY G1GTR
HD6 2BJ G8AUL
HD6 2EP G4EEJ
HD6 2LE M1RIG
HD6 2LF G0GRR
HD6 2NA G7LBD
HD6 2NW M0ARK
HD6 2RS G8MLD
HD6 2RU G4SDX
HD6 2RU M3JQJ
HD6 3AH M3JLX
HD6 3JS G4WAP
HD6 3LD G7KXS
HD6 3NP G3WAH
HD6 3RF G1JFQ
HD6 3SR G4GMT
HD6 3XF G8RFF
HD6 3XF M3RFF
HD6 4AP G3JWN
HD6 4EQ G0TJQ
HD6 4FF M0EEP
HD6 4JQ G4IPH
HD6 4JZ G4REG
HD7 2SQ G7KPH
HD7 4BT G4FSQ
HD7 4BX G7AXL
HD7 4JR G1MQQ
HD7 4JR G4LYY
HD7 4JU G1HEP
HD7 4JU M6GWC
HD7 4JY G0HRF
HD7 4NS G4IDR
HD7 4NU 2E1CRI
HD7 4QP G4UFL
HD7 4QQ G4UFJ
HD7 4DA G3XWN
HD7 4HF G8VMF
HD7 5BW G0PLD
HD7 5DS G3CJD
HD7 5JS M6OTL
HD7 5LS G0JTM
HD7 5QU G7JMU
HD7 5QX 2E1BYI
HD7 5QX M0HVP
HD7 5TY M3ZNC
HD7 6BJ M3VQN
HD7 6JX M6BBT
HD8 0DE G3TSA
HD8 0GD 2D0SCJ
HD8 0GD M3SCJ
HD8 0JA 2E0RUD
HD8 0JA M3PBE
HD8 0JB G4AHJ
HD8 0JG G4XJG
HD8 0JG G7NDC
HD8 0QH 2E0BQM
HD8 0QH M0NCK
HD8 0QH M3VWY
HD8 0QW 2E0RJH
HD8 0QW M0RJX
HD8 0QW M6RJH
HD8 8AG G0ATC
HD8 8AG G3LUK
HD8 8AG M0ALT
HD8 8AP G4OPY
HD8 8DS G7UKK
HD8 8DS G7VBU
HD8 8HP G1DEN
HD8 8HP G3SDY
HD8 8HP G8KMK
HD8 8HP M3OEN
HD8 8JZ M3OUS
HD8 8LX G0PHI
HD8 8QF G6UJR
HD8 8QW G7EXD
HD8 8SG M1ELQ
HD8 8UA M3WYV
HD8 8UB 2E0MPB
HD8 8UB M6CNI
HD8 8XP G6VIQ
HD8 8YH 2D0HFT
HD8 8YH M0ODM
HD8 8YH M6FFT
HD8 8YH M6HFT
HD8 9AB M0RGC
HD8 9AH M0RTC
HD8 9BT G4HKY
HD8 9BW G4BYD
HD8 9DS G4TKO
HD8 9EH G4JCL
HD8 9ET G8PRN
HD8 9GY G4KGN
HD8 9PA M1DLR
HD8 9QP G4EMQ
HD9 1EH G0HNW
HD9 1EN G1CFA
HD9 1EN G1LFD
HD9 1ES G3RJT
HD9 1EU M0LUS
HD9 1EU M6AIN
HD9 1HL G7VDH
HD9 1LH M3CGP
HD9 1XP M1EGD
HD9 1XU G8MQK
HD9 2PQ G3YPE
HD9 2PX G1JOW
HD9 2PX G8LIK
HD9 2PZ M6WYD
HD9 2RF G0OHD
HD9 3BP 2E1IAE
HD9 3ES G8PUT
HD9 3ES M3PUT
HD9 3ET G6WSX
HD9 3XZ G4UFZ
HD9 4AG G8GFY
HD9 4AJ G3NAK
HD9 4AJ G7GKQ
HD9 4DR G3LDJ
HD9 4DY 2E0DAW
HD9 4DY M3VIJ
HD9 4ED M3XOW
HD9 5NL G1EDU
HD9 5PL M0GBK
HD9 6DD G4LLZ
HD9 6DS G0JTL
HD9 6EG 2E1FWA
HD9 6ER G0PLG
HD9 7ER G0SJB
HD9 7SG G1VLU
HD9 7TJ G8ZML

## HG

## (Harrogate)

HG1 1TY G0MKK
HG1 2AS G6HAT
HG1 2BW G0NHG
HG1 2DP 2E0UAB
HG1 2DP M6CBC
HG1 2HA G4CWB
HG1 2JF M6FJJ
HG1 2JN 2E0WAY
HG1 2JN M3WDZ
HG1 2QD M3YXR
HG1 2QD M6TTY
HG1 2QG M3EFV
HG1 2QG M3JYW
HG1 2QG M3KHT
HG1 2QG M3MIU
HG1 2QG M3OGC
HG1 2QG M3OHQ
HG1 2QG M3OHT
HG1 2QG M3OIY
HG1 2QG M3OSY
HG1 2QG M3SVF
HG1 2QG M3TNK
HG1 2QG M3TWK
HG1 2QG M3UCL
HG1 2QG M3UFE
HG1 2QG M3UYK
HG1 2QG M3VHC
HG1 2QG M3VHO
HG1 2QG M3WCY
HG1 2QG M3WSW
HG1 2QG M3WTO
HG1 2QG M3XOY
HG1 2QG M3XUF
HG1 2QG M3XZF
HG1 2QG M3YCK
HG1 2QG M3YCU
HG1 2QG M3YEE
HG1 2QG M3YHL
HG1 2QG M3YKF
HG1 2QG M3YKO
HG1 2QG M3YLM
HG1 2QG M3YTL
HG1 2QG M3YUH
HG1 2QG M3ZCJ
HG1 2QG M3ZJJ
HG1 2QG M3ZKO
HG1 2QG M3ZNO
HG1 2QG M6AYE
HG1 2QG M6BPW
HG1 2QG M6BTF
HG1 2QG M6CSW
HG1 2QG M6CYL
HG1 2QG M6FIA
HG1 2QG M6HDK
HG1 2QG M6HUH
HG1 2QG M6HYW
HG1 2QG M6JYL
HG1 2QG M6KKL
HG1 2QG M6KLK
HG1 2QG M6LBQ
HG1 2QG M6LCT
HG1 2QG M6LDH
HG1 2QG M6LLW
HG1 2QG M6LST
HG1 2QG M6LXH
HG1 2QG M6MMP
HG1 2QG M6NMT
HG1 2QG M6VAN
HG1 2QG M6VCM
HG1 2QG M6WUN
HG1 2QG M6ZAS
HG1 2QG M6ZPZ
HG1 2QL M3XOZ
HG1 2QL M3YHM
HG1 2QL M3YSZ
HG1 3AN G0UIW
HG1 3BW G1ZEA
HG1 3DF M6LHD
HG1 3EA G1VQK
HG1 3LT G8MJF
HG1 3NA G4KAT
HG1 4BP M6HDI
HG1 4DL 2E0PDO
HG1 4DL M0TUV
HG1 4DL M6PDO
HG1 4EH G6VKP
HG1 4ET G4ODS
HG1 4HR 2E0BWK
HG1 4HR M6KVB
HG1 4JN G7BBJ
HG1 4QR G0KIY
HG1 4RH 2E0WKZ
HG1 4RH M3WKZ
HG1 4SG G1JLB
HG1 4ST G7GCD
HG1 5HS G4LGX
HG1 5JU G1WUC
HG2 0AY G8ERQ
HG2 0BJ G4ZIU
HG2 0DA 2E0WFM
HG2 0DA M0VGA
HG2 0DA M6WFM
HG2 0DQ 2E0CTU
HG2 0DQ M6JFE
HG2 0ES G7FTS
HG2 0LL G3OGZ
HG2 7AJ 2E0LMH
HG2 7AJ G8XZC
HG2 7AJ M0LMH
HG2 7AJ M3YUV
HG2 7DH G3PWK
HG2 7EE G4RCD
HG2 7ER G4TJI
HG2 7ES 2E0FBM
HG2 7ES M0HYJ
HG2 7ES M0RDY
HG2 7HQ G1UIB
HG2 7LJ G1ARH
HG2 7LJ G1ROD
HG2 7NJ M0AVF
HG2 7QL M6RNC
HG2 7RT M6EIP
HG2 8HN M3YIT
HG2 8HW G4MNE
HG2 8LB G4EKJ
HG2 8LE G4KCR
HG2 8LX G4KEN



**IMPORTANT NOTE**

**Revalidate licence to avoid revocation** – Ofcom has advised the Society that plans will be drawn up to revoke licences that have not been revalidated as required by the licence conditions. The quickest way to revalidate is to do so online via the Ofcom website: *https://services.ofcom.org.uk/* or by email: *amateur.validations@ofcom.org.uk* If you need assistance in the process, Ofcom staff are available to help, but please be patient during times of heavy workload.

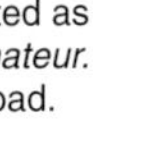

HG2 8PR G4LQG
HG2 8PR M6TJB
HG2 8QF M1DFW
HG2 9BS G1BPE
HG2 9HB G1CCX
HG2 9HP G3YHC
HG2 9LD M0JIL
HG2 9PG G8DMU
HG3 1BH 2E0PAE
HG3 1BH M0MCI
HG3 1HY M3URS
HG3 1JR G4IUF
HG3 1LY G1BHO
HG3 1LY G6WKO
HG3 1NF G3NXL
HG3 1NY M0DYO
HG3 1PB M1CTG
HG3 1QX G0JNG
HG3 1RJ G0DOZ
HG3 2AA G4MEM
HG3 2DG 2E0AGI
HG3 2DG 2E1BDV
HG3 2DG G1ZBO
HG3 2DG M5ACF
HG3 2DJ 2E1CJC
HG3 2DS 2E1CIX
HG3 2DS G3FMW
HG3 2QE G6EQT
HG3 2QF G3ZCY
HG3 2RF G0UFL
HG3 2SU M3UTP
HG3 2SU M3YHP
HG3 2UT G1JWL
HG3 2WE G1XKJ
HG3 2WE G6WTM
HG3 2WW M6MIG
HG3 3BX 2E0IVY
HG3 3BX G3RHR
HG3 3BX M6TUR
HG3 3DS G4PAH
HG3 3HE G5VO
HG3 3HE G6MC
HG3 3HE M0IPX
HG3 3HE M0WKR
HG3 3HE M3UZL
HG3 3HE M3YPG
HG3 3HE M6LJJ
HG3 3JR G4EEV
HG3 3JR G8IKS
HG3 3LA G4VNA
HG3 3PB G4AFT
HG3 4EW G3RYP
HG3 4HA 2E0JDM
HG3 5ET M6FXO
HG3 5LY G3NNO
HG3 5PG G1TBT
HG4 1NG M0TJW
HG4 1NW G7OFM
HG4 1PU G1WRE
HG4 1RG 2E0GEB
HG4 1RG M6HEE
HG4 1RZ G4DSC
HG4 1UA G3AB
HG4 1UJ G0ORJ
HG4 1UP M0DCD
HG4 1UP M3ZAZ
HG4 2LJ G4VMY
HG4 2LJ G6MGH
HG4 2LN G1VKC
HG4 2PG G3UVY
HG4 2QJ G3HTF
HG4 3LL G1PWM
HG4 4AU G0UFI
HG4 4AU G4VUN
HG4 4AY M6RCA
HG4 4LH G7LYN
HG4 4PW 2E0RBY
HG4 4PW 2E0SML
HG4 4PW M0RBY
HG4 4PW M6RGH
HG4 4PW M6SAN
HG4 5BU G1JYK
HG4 5DX M0ZAB
HG5 0BW G0EYL
HG5 0DX M0DAL
HG5 0LW M0AQH
HG5 0ND M6TTK
HG5 0NY G4VZS
HG5 0QA G8BNE
HG5 0SW 2E1EJX
HG5 0TH M0RMN
HG5 8EL M0CVG
HG5 8HD G0KGT
HG5 8HQ G7IPX
HG5 8HX G7HGD
HG5 8JN G6HEB
HG5 8JT G3DUW
HG5 9BJ G4YBD
HG5 9BZ G3XQJ
HG5 9DE G3XZV
HG5 9EL G3UNA
HG5 9EL G6CP
HG5 9HN 2E0DPL
HG5 9HN M0HYH
HG5 9LD G0GYU
HG5 9LS G3XPM

## HP (Hemel Hempstead)

HP1 1AS M6RTI
HP1 1LB G3GBN
HP1 1NS M6FTN
HP1 1XQ G8FXC
HP1 2BP M0ZJV
HP1 2BQ G0TIW
HP1 2HY 2E1EOI
HP1 2LN M1ABC
HP1 2PP M6ULY
HP1 2PP M6WTL
HP1 3BN G4WGA
HP1 3EW G7SGM
HP1 3HY M1EQO
HP1 3JU G7IPA
HP10 0HG M6YOS
HP10 0HJ G8ZRG
HP10 0PZ 2E1TWB
HP10 8BL G4XVP
HP10 8DL G1SHT
HP10 8HG M3MVO
HP10 8JJ G8FBW
HP10 8LN G4DDM
HP10 8LN G8FC
HP10 8LN G8RAF
HP10 9AX G3PWY
HP10 9DF G4VLL
HP10 9DW G4KCX
HP10 9DW G8NMT
HP10 9EQ G4PIE
HP10 9JY G0RNO
HP10 9LH G3WQG
HP10 9PL G3INZ
HP10 9RH G8JQV
HP11 1GL G3SRQ
HP11 1JL G7BOB
HP11 1JW G8XQT
HP11 1PA G4TJM
HP11 1QA M3OJW
HP11 1QT G8DJF
HP11 1TX M6FNE
HP11 1UB G1UCT
HP11 2JL G0OCY
HP11 2PL G1EUI
HP11 2UA G4CJY
HP12 3AS M0HCN
HP12 3DS G7GUO
HP12 3JQ G0IZC
HP12 3LP M6HXB
HP12 3NN G1LQX
HP12 3PA G8VJW
HP12 4AA G8WCH
HP12 4AL M0EAS
HP12 4AU G7OST
HP12 4BB G6BFP
HP12 4DA M0YOJ
HP12 4DD G0ANO
HP12 4DD G4ZOC
HP12 4DD G8ANO
HP12 4LG G4AOQ
HP12 4PG G8DLZ
HP12 4QD M1EBY
HP12 4RG G0IKE
HP12 4SP G3ZGQ
HP12 4TY G3PVJ
HP13 5BD 2E0RPY
HP13 5BD M0HWY
HP13 5BD M6NSE
HP13 5JN G3FNO
HP13 5JN G4TZQ
HP13 5JX 2E0KRK
HP13 5JX 2E0YGB
HP13 5JX G4EBY
HP13 5JX M0YGB
HP13 5JX M3YGB
HP13 5JX M6KRK
HP13 5LZ G6ZAM
HP13 5NG G8HDL
HP13 5NW G7RTI
HP13 5PX G8YRY
HP13 5QA G8AAU
HP13 5RH G0CTD
HP13 5RH G1EYG
HP13 5SP M1BDL
HP13 5SS G4LMM
HP13 5SS 2E0KNA
HP13 5SZ M6PTM
HP13 5TA G4MUI
HP13 5TZ G4TBG
HP13 5UD G6HBJ
HP13 5UN G8JAW
HP13 5XN G1CPC
HP13 5XT G4COS
HP13 5XY G0NCL
HP13 6HW M6PHC
HP13 6JG M6ELF
HP13 6JG M6YEY
HP13 6SN G3OUV
HP13 6SW M0HMB
HP13 6TJ G1GAR
HP13 6UD M0MIQ
HP13 6UD M6BQZ
HP13 6XL 2E0BKT
HP13 6XL M0GGW
HP13 6XL M3TUU
HP13 6XW G6GTZ
HP13 7EF G3GAA
HP13 7JF M6RKI
HP13 7PH G0OQK
HP13 7PH M6RVE
HP13 7PQ G0WEK
HP13 7UE G1MIY
HP13 7WZ M6UPE
HP13 7XN G4YAN
HP13 7YA G6HYJ
HP14 3BN G4KBB
HP14 3JN G4PUO
HP14 3JP M5BFL
HP14 3JW G3LYP
HP14 3LS G4KDN
HP14 3NT M6GEG
HP14 3PH G4HFS
HP14 3RP G0FUN
HP14 3RP G4WJS
HP14 3SJ G1OLY
HP14 3TF G0GGL
HP14 4JG G4YBH
HP14 4JJ G2BWW
HP14 4JJ G8DOR
HP14 4JN G4YKQ
HP14 4LN G3MGH
HP14 4QB G6XQB
HP14 4QD G1FEF
HP14 4RA G8GLV
HP14 4SG G0HND
HP14 4UH G3SRJ
HP14 4UY G0IOU
HP15 6BW G4XJN
HP15 6EG G6AHN
HP15 6EY G7MLX
HP15 6JR G3VCT
HP15 6LJ G3WNS
HP15 6LJ G8AKU
HP15 6SL G4PFA
HP15 6TN G0RBB
HP15 6UT G4GED
HP15 6XA G0TRE
HP15 6XD G7DZD
HP15 6XF G1RDX
HP15 7AT G3XZK
HP15 7BX M3YXC
HP15 7DT M6HRS
HP15 7ED G6UDF
HP15 7PH M6BEH
HP15 7RE G3TYG
HP15 7RP M6VVT
HP15 7TF G4PPK
HP15 7TF G6GUD
HP15 7TP G6EYJ
HP16 0BT M6ZSA
HP16 0DF G4SNQ
HP16 0HB G6YOZ
HP16 0HW M3HEI
HP16 0LF G4KGT
HP16 0LY 2E0DUU
HP16 0LY M6IJH
HP16 0NA G7TQA
HP16 0NJ G8XTJ
HP16 0NL M6FZK
HP16 0QD G8FKP
HP16 0SL M1CVT
HP16 9DS G3ZNU
HP16 9JH G6MJA
HP17 8AN G3OCP
HP17 8DL M6ERI
HP17 8DT G4CYR
HP17 8EU G1MZG
HP17 8EY G3WUN
HP17 8HD G3NPL
HP17 8HN G0IKP
HP17 8JG G1TRL
HP17 8RG G3YSR
HP17 8RG G7AAR
HP17 8SH G4IGK
HP17 8SH G7JLF
HP17 8SH G8DGW
HP17 8SH M0SOC
HP17 8XQ G8WDX
HP17 8XU M0SFD
HP18 0BL G6JRM
HP18 0BL G8BCL
HP18 0DN G6IFE
HP18 0HJ G6OIH
HP18 0LH G0MMI
HP18 0LY G0KMC
HP18 9AU M3BLF
HP18 9DG G3MAZ
HP18 9DP 2E1HNS
HP18 9DZ M6YMY
HP18 9HH G0RAS
HP18 9HP G7TPD
HP18 9JZ G1ULQ
HP18 9LD G6DAP
HP18 9QZ G0UUI
HP19 7BG G6YLV
HP19 7FY G1IJM
HP19 7HR M0TOL
HP19 7HR M1EIO
HP19 7HW G8UEI
HP19 7QJ G0MHZ
HP19 8GZ 2E0XLX
HP19 8GZ M0WDU
HP19 8GZ M6XLX
HP19 8JE M6HTK
HP19 8SW 2E0HAF
HP19 8WA G8AQP
HP19 9JQ M0PSH
HP19 9LN G0TGM
HP19 9QA 2E0PNP
HP19 9QA M6BEU
HP19 9QT 2E0TBQ
HP19 9QT G7HIU
HP19 9QY G7GQH
HP199QT 2E0PSX
HP199QT M0MBV
HP199QT M6LWT
HP2 4BA G0IAL
HP2 4BN G8RWU
HP2 4BP M6HTF
HP2 4HE G1URZ
HP2 4LZ M3AZF
HP2 4LZ M3RDH
HP2 4PB G8UCK
HP2 4PQ 2E0MMJ
HP2 4PQ M0MMJ
HP2 4PQ M6MMQ
HP2 4PX G1EIZ
HP2 4RP G7HCQ
HP2 4UG G1PAK
HP2 5AT 2E0MOB
HP2 5AT G0BLU
HP2 5AT M0RMO
HP2 5AT M6RMO
HP2 5EL G8VVB
HP2 5ES M6MBR
HP2 5HU G4DWZ
HP2 5JG G8CBE
HP2 5JL G1DDF
HP2 5JZ M0CQN
HP2 5LL G1JOA
HP2 5QF G8SYD
HP2 5QF M6TFQ
HP2 5SL M6HTZ
HP2 5WR G6FGV
HP2 5WR M0CZB
HP2 5YX 2E0VTV
HP2 5YX M6ITV
HP2 6DS G4BJN
HP2 6DX G8ZGQ
HP2 6LA G1PWF
HP2 7AR M3XHT
HP2 7JP G7WDO
HP20 1HW G7IMY
HP20 1JW G8ALQ
HP20 1QE M3VRL
HP20 1RB 2E0PNR
HP20 2AB M3PIA
HP20 2AJ 2E0MJH
HP20 2AJ M3CTT
HP20 2AS G4BKS
HP20 2EX M5ASR
HP20 2NU M3MMP
HP20 2QT G1NMP
HP21 7BD G4FXI
HP21 7EP G1FNA
HP21 7FE M6TBQ
HP21 7HP G3KGM
HP21 7NS G8VVZ
HP21 7RR G1ATY
HP21 7TN M0HZN
HP21 8BN G8MBD
HP21 8EH G1BHW
HP21 8JG G1MZW
HP21 8LN G7VFV
HP21 8NW 2E1PPM
HP21 8TW M0SVB
HP21 9AL M6GJC
HP21 9DS G8AYM
HP21 9DT G8MEM
HP21 9EF M3HER
HP21 9EF M3HIM
HP21 9QT 2E0DSV
HP21 9QT M6FIW
HP21 9RX G7ITO
HP21 9SL M6BFU
HP21 9UF G3IGH
HP22 4AZ G4GYO
HP22 4BX M3FKW
HP22 4EF G8BQH
HP22 4LH G3XOP
HP22 4LT G4HFU
HP22 4PA G0WZH
HP22 5AB M3UCC
HP22 5EG G1MRI
HP22 5HH G4VRS
HP22 5HH G6GDI
HP22 5HH M3OTB
HP22 5LA G1ZDU
HP22 5LD G8CPZ
HP22 5LD M0ALC
HP22 5LY G8AWY
HP22 5ND G8TJI
HP22 5RX M6CBH
HP22 5TX M0TIF
HP22 5UL M3OOE
HP22 6AR G1DDR
HP22 6AW G4PFR
HP22 6HN 2E0PJY
HP22 6HS G0BKP
HP22 6PH G1RAF
HP22 6PH G8ZGK
HP22 6QH G8ZME
HP23 4DS G6DZJ
HP23 4DS G7UYW
HP23 4EA G3NBZ
HP23 4HH G6MQD
HP23 4HZ G0SEY
HP23 4LX G4BYR
HP23 4TP G8MKX
HP23 5DG G1NUO
HP23 5HG G7JDE
HP23 5HG M3MGQ
HP23 5HL G7PWI
HP23 5PB G3WXM
HP23 5PB M0HXV
HP23 5PJ G3RWV
HP23 6ED G8GYY
HP23 6EN G3MEH
HP23 6HF G6XUJ
HP23 6HH G4PMG
HP23 6HH G8MKW
HP23 6HH M3PTG
HP23 6JB G1ODZ
HP23 6NE G6HGM
HP27 0BQ G0SNF
HP27 0EB G0SPB
HP27 0EB G1DAE
HP27 0EB M6BFQ
HP27 0EJ M6VOR
HP27 0HE G3UJK
HP27 0JA G7AYS
HP27 0LG G0NLJ
HP27 0LG G4HMC
HP27 0NB G3SNP
HP27 0PG G0MBB
HP27 0QY G7DPF
HP27 9AY G1THD
HP27 9HE G0WQQ
HP27 9JY G7CUD
HP27 9JZ G4JBE
HP27 9RP G3NBS
HP3 0BS G3VSQ
HP3 0BT M3UST
HP3 0BU M6NAS
HP3 0DS G6HIA
HP3 0EA G0VFW
HP3 0HG G1OEB
HP3 0QN G4VCO
HP3 0QR G8MFH
HP3 8AA 2E0NVS
HP3 8AA M0NVS
HP3 8AH 2E0WFC
HP3 8AH M0WTC
HP3 8AH M3WFC
HP3 8EQ G8OYQ
HP3 8HZ M6LGI
HP3 8JL 2E0HNC
HP3 8PR G6EAX
HP3 8QS G8LPC
HP3 8SH 2E0NDR
HP3 8SH G1BSZ
HP3 8SH G6EXX
HP3 8SH M0NGL
HP3 8SH M3NDR
HP3 9GS G6SVL
HP3 9HB G7JHM
HP3 9HS M3YYE
HP3 9JA M0SCS
HP3 9JA M1SCS
HP3 9JY G4MSW
HP3 9LW M6JFA
HP3 9NG 2E0BVU
HP3 9NG G6RXD
HP3 9NG M0GPE
HP3 9NG M3VUX
HP3 9PD G0GVN
HP3 9PW G2KG
HP3 9RP M0HSG
HP3 9UB M6PAG
HP3 9UE G1KOG
HP3 9UQ G0EVS
HP3 9WU G6HGE
HP4 1FG G1ABW
HP4 1HF 2E0CIJ
HP4 1HF M0HDT
HP4 1HF M6AZV
HP4 1JN G0TPK
HP4 1JN G0WIH
HP4 1QG G7WBY
HP4 1QQ G4CSI
HP4 2NZ G4ANZ
HP4 2PD G3RPA
HP4 3BS G7VJD
HP4 3BS M0AYG
HP4 3DR M6ZTP
HP4 3DW M0GVT
HP4 3PQ 2E0JCQ
HP4 3PQ M0JCQ
HP4 3PQ M6JCQ
HP4 3PW G1WMN
HP5 1EZ G4UXY
HP5 1RL M1BPU
HP5 1RW G0LZG
HP5 1SQ G8YDE
HP5 1SS G3WPD
HP5 1SS M1BAA
HP5 1TA G3XZG
HP5 2BU G6HPK
HP5 2BU G7PEU
HP5 2DD 2E0GDY
HP5 2DD M3GDY
HP5 2EB G0ABP
HP5 2EB G6LBG
HP5 2HF M6MRR
HP5 2HL G8BLB
HP5 2LH 2E0DSW
HP5 2LH M0MRL
HP5 2LH M6DSW
HP5 2PG M6EWZ
HP5 2QR M6BLF
HP5 2RY G7GLZ
HP5 2UQ G4XRV
HP5 2XL G3VZF
HP5 2XW G1CUZ
HP5 3AD G1MDG
HP5 3AD G3MDG
HP5 3AD G4HES
HP5 3DJ G3VRY
HP5 3JH G6CDV
HP5 3JQ 2E1LGJ
HP5 3LL G0BZN
HP5 3RJ G4CYY
HP6 5NT G6FUT
HP6 5PY G8ETV
HP6 5RW G6HUH
HP6 5RW G6XSL
HP6 6HG M0FJS
HP6 6HL G7MMC
HP6 6HP G7EAT
HP6 6JF G1JNG
HP6 6JF M0BTX
HP6 6LD G0NLX
HP6 6QL G0XDI
HP6 6QQ M1DIM
HP6 6QQ M3YWI
HP6 6QU M0GUL
HP6 6RJ G4GYN
HP6 6RP M3ZAI
HP6 6RR G4TNU
HP6 6SX M6DJG
HP6 6SX M6DWZ
HP7 0EF G0RTF
HP7 0PL M6HTI
HP7 0PN G4WQB
HP7 0PY G2SH
HP7 0RQ G1HPU
HP7 9AG M6MNS
HP7 9AU G3WLO
HP7 9BN M1DTG
HP7 9DZ G0VRV
HP7 9EP G1KTS
HP7 9EP G7STD
HP7 9HL G8GDI
HP8 4AR G3DTX
HP8 4BG M1DVT
HP8 4HQ G3LPY
HP8 4LG G3PBI
HP9 1AE M0GPY
HP9 1AW G0AFT
HP9 1AW M3MVR
HP9 1AW M6GFY
HP9 1BD G8FIG
HP9 1YB G0BKR
HP9 2AE G7ABQ
HP9 2BA G6DOV
HP9 2LA G3BII
HP9 2QU G3OHX
HP9 2XH G4LHE
HP9 2XP G3BEX
HP9 2YJ M6LGV

## HR (Hereford)

HR1 1AH M6JTA
HR1 1BY G8OHH
HR1 1DG G0JWJ
HR1 1DG G3YDD
HR1 1HE M6DUW
HR1 1HL 2E0EOK
HR1 1HL M6OEP
HR1 1HQ G7KVT
HR1 1HT G4BJD
HR1 1JJ M6GUO
HR1 1NH G1HWP
HR1 1NT M6WRS
HR1 1QL G6KWZ
HR1 1QX M1EIR
HR1 1QY G1YFC
HR1 1RQ G4DTB
HR1 1SS M1CAR
HR1 1TJ 2E1GFW
HR1 1UH G1CKR
HR1 1XA M1BDH
HR1 1XA M3BDH
HR1 1XY G8IVO
HR1 1XZ G0SDZ
HR1 2HB G4OJP
HR1 2QG M6XRF
HR1 2RQ G3LZM
HR1 2TY G3ESY
HR1 3BJ G1AYP
HR1 3BJ M3KAL
HR1 3DR G3YQC
HR1 3DU G7JUN
HR1 3EH G7LSG
HR1 3EP G0SBA
HR1 3ET G8RAC
HR1 3LP G3IXZ
HR1 3LR G1JWD
HR1 3NL G3RDC
HR1 3PS M3UKO
HR1 3RF G6PXQ
HR1 4AE G3WRA
HR1 4AF G0PMS
HR1 4BT M0MVL
HR1 4DL G6TID
HR1 4DQ G6PFX
HR1 4JS M0OTF
HR1 4LR G7LVG
HR1 4PR G8XMS
HR1 4RQ G1YBG
HR1 4UL G0LJP
HR2 0HP G4ASR
HR2 0JA G0NMC
HR2 0NX 2E0YAV
HR2 0NX 2E0YBL
HR2 0NX M0YAV
HR2 0NX M6WTG
HR2 0NX M6YAT
HR2 0QG G7KTP
HR2 0QR M1FDK
HR2 0SW M0DRL
HR2 6NG G8UVU
HR2 7DF G1YBB
HR2 7LJ M0JDP
HR2 7LS G0DYM
HR2 7LU G4MBJ
HR2 7NT 2E1MIB
HR2 7UG G1DYQ
HR2 7UG G7BTI
HR2 8EB M3RNL
HR2 8HJ G0SYT
HR2 8JT G4OGW
HR2 8NA G0VKC
HR2 8PU 2E0AFP
HR2 9AN G7RLO
HR2 9BE G4XTF
HR2 9EW M3VHB
HR2 9HD G4ZSY
HR2 9LT G1LSX
HR2 9NX M6DRW
HR2 9QN G0DHB
HR2 9QT G8RLH
HR2 9RH G1RLF
HR2 9RT M6IDD
HR2 9SE G4ZDU
HR3 5DW G0RHK
HR3 5JT M0ECX
HR3 5JT M6HRI
HR3 5NX M6URC
HR3 6BY G0IAH
HR3 6BY G1UMS
HR3 6HT G7IFR
HR3 6HY G4EYR
HR3 6LF G3IUC
HR3 6LX M6LBV
HR4 0JY G1DRW
HR4 0LP G4LNZ
HR4 0NE M6BDI
HR4 0PN G4MET
HR4 0SR 2E0PTA
HR4 0SR M3PTA
HR4 0TF M6OAW
HR4 7AZ G0RQF
HR4 7EN G0CXF
HR4 7PN G3YYC
HR4 7QP M0BHR
HR4 7SW G4GOG
HR4 8AJ G4FAD
HR4 8AN M6DVF
HR4 8EN G8ITX
HR4 8NN G4ZXQ
HR4 8NT 2E1HOF
HR4 8NT G6TVC
HR4 8QG G4RGB
HR4 8RQ G7HII
HR4 8SW G0UVE
HR4 8TA G3NPA
HR4 8TA G4ZA
HR4 9HP G4WZT
HR4 9NE G4HZT
HR4 9NJ G6DXC
HR4 9RW M6PFA
HR4 9TJ G3WAG
HR4 9TJ G6XPY
HR4 9TJ G8FFA
HR4 9TY M6JKM
HR4 9XF G0OGX
HR5 3EG M6VPS
HR5 3FJ 2E0ULY
HR5 3JS M6FTI
HR6 0AZ G4KOK
HR6 0BG G4EBL
HR6 0EF G7UUN
HR6 0HY G3NUG
HR6 0HY M0CDX
HR6 0JL M3HCP
HR6 0JZ G7AZC
HR6 0NG G3VST
HR6 0PF G8BPN
HR6 0SS G1JKN
HR6 8ER G4VMF
HR6 8ER M6XAO
HR6 8HE G4YPF
HR6 8HT G6OZU
HR6 8QQ M0PSK
HR6 8RX 2E0JRS
HR6 8RX G6UQI
HR6 8RX M6GFA
HR6 8RX M6YRS
HR6 8SA M6AUF
HR6 8SD G0BRH
HR6 8SL G6CYR
HR6 8SS G0JRV
HR6 8TD G3LFE
HR6 8TQ G6GQG
HR6 9BN G0ARF
HR6 9BZ M6GLZ
HR6 9DB 2E0DJH
HR6 9DB M0RNI
HR6 9DB M6DAW
HR6 9DN G4ASP
HR6 9EE G6FYE
HR6 9HB G4IIA
HR6 9HB G8NOD
HR6 9JQ G4VOY
HR6 9NW M1AGE
HR6 9QN G0OEA
HR6 9QR G4BOF
HR6 9UF G3XJP
HR6 9UU M6XLR
HR7 4BD G4CPL
HR7 4DZ G4ZWY
HR7 4EA G4XNK
HR7 4ES M1BPK
HR7 4EY 2E0TOP
HR7 4LB G1MUQ
HR7 4LY G3OGK
HR7 4UL G4FHQ
HR8 1BE G4FTX
HR8 1HG G3WPN
HR8 1ND G0TWT
HR8 1NX G3VZR
HR8 1PP G0WHN
HR8 1SG 2E0JAO
HR8 1SU G1JHY
HR8 1SU G1JHZ
HR8 2EG G0GFR
HR8 2FR G8AVB
HR8 2QB G8ZVZ
HR8 2UU G8EZE
HR8 2XD G1GCJ
HR9 5AU 2E0CUW
HR9 5AU M0HNK
HR9 5AU M6XRT
HR9 5BL G1ZND
HR9 5BL G8RZS
HR9 5JL G4EFH
HR9 5JY G3DYO
HR9 5NR M0NOS
HR9 5PN G0FGS
HR9 5QU G7VCZ
HR9 5RL G1OKP
HR9 5ST G4JNZ
HR9 5TA G6XLR
HR9 5TE M0GZF
HR9 6DH G0TRI
HR9 6LH G3BJC
HR9 7BE G0IMV
HR9 7BE G6TSL
HR9 7DN G7RVI
HR9 7DR M1AHN
HR9 7HR M6EQB
HR9 7HT G0RIJ
HR9 7JH G0GQI
HR9 7JP M0HZE
HR9 7JR G3EFL
HR9 7NF G7RIU
HR9 7PW G8PFR
HR9 7QW 2E0LTX
HR9 7QW M0RPI
HR9 7QW M6EQN
HR9 7TL G8CMU
HR9 7TW 2E0WYE
HR9 7TW M0TAY
HR9 7TW M1EJJ
HR9 7TW M6TAY
HR9 7UF G0EHT

## HS (Scottish Islands)

HS1 2DS GM0DRU
HS1 2DS GM6BGJ
HS1 2JN GM4PTQ
HS1 2NP GM0IPW
HS1 2PG GM7JED
HS1 2QN MM3IIT
HS1 2TL 2M0APB
HS1 2UP MM6CTQ
HS1 2UR MM6BIO
HS2 0DD GM6TYX
HS2 0DZ GM0LZE
HS2 0HD 2M1EOV
HS2 0HD GM0HMM
HS2 0HE GM0WZO
HS2 0HE MM0BPV
HS2 0HU GM0NHT
HS2 0LP MM6BIP
HS2 0LR GM7TTU
HS2 0LR MM6OEC
HS2 0PD MM0TYR
HS2 0PL MM0KNN
HS2 0QB GM1MQA
HS2 0QB GM3JFG
HS2 0QB GM3JIJ
HS2 0QB GM4KGK
HS2 0QW GM0FSY
HS2 0QX MM0IOL
HS2 0QX MM1RMS
HS2 0QX MM3SYB
HS2 0RT MM0TSB
HS2 0SW 2M0AKI
HS2 0SW GM7PBB
HS2 0SW MM3OHD
HS2 0XF GM7JFN
HS2 9AU GM7KUN
HS2 9HR MM0UIG
HS2 9JA GM6VHA
HS2 9PJ 2M0DNM
HS2 9PN 2M0OSK
HS2 9PN MM3OSK
HS2 9QE GM0TVT
HS3 3JA MM3WVP
HS6 5BP GM1MLY
HS6 5EU 2M1LPT
HS6 5EU GM0HBF
HS6 5EU MM0IRC
HS8 5SY GM7DMN
HS8 5TE MM3BCA
HS8 5TU MM0CWJ
HS8 5TU MM6TEW
HS8 5TX MM6SML
HS9 5XU MM1FHZ
HS9 5XX GM0OGN
HS9 5XX MM3BRR
HS9 5YD MM0IOB

## HU (Hull)

HU10 6AJ G8DCD
HU10 6AR G1RVH
HU10 6AW G3TEU
HU10 6BJ G0KLG
HU10 6DD G4NJB
HU10 6EU M6MKK
HU10 6HS G4LNR
HU10 6SF G0UGA
HU10 6SG M0DGU
HU10 6ST G0VQO
HU10 6ST G4BYG
HU10 6ST M0ARC
HU10 7AF 2E0KVK
HU10 7AF M0KVK
HU10 7AF M6EYR
HU10 7AF M6OGK
HU10 7DT G4CGG
HU10 7DT G4EKT
HU10 7JE G4AQA
HU10 7JJ G8JIU
HU10 7JN 2E0FZM
HU10 7JN M3FZM
HU10 7JU G8NIE
HU10 7UX G7PER
HU11 4AP G4IVB
HU11 4AX G0GBW
HU11 4BN G4FEJ
HU11 4BN G6ZEN
HU11 4DJ M6NIX
HU11 4DN M6HRQ
HU11 4ER G7CVA
HU11 4HB G0AJX
HU11 4NR G5GX
HU11 4PR M3KUH
HU11 4QH G6JFU
HU11 4RA M0BTN
HU11 4RN G4LZJ
HU11 4RN G7JZY
HU11 4RY G1UCZ
HU11 4SD G0NAA
HU11 4UW G6ZDV
HU11 4XF M0IOK
HU11 5BH G4YTV
HU11 5BH G6EKT
HU11 5BW M6OTR
HU11 5QP M0VEY
HU11 5RN 2E0HYC
HU11 5RN 2E0RIO
HU11 5RN G3BOK
HU11 5RN M0HYC
HU11 5RN M3UMR
HU12 0DP 2E0THS
HU12 0DP G0THS
HU12 0DS M6CZR
HU12 0HH G4PMJ
HU12 0JF M3XSV
HU12 0LE G0UMK
HU12 0NE G3ULD
HU12 0PJ G4KFA
HU12 0PT G1RON
HU12 0QE G4PYW
HU12 0RY G3UIF
HU12 0TE G6AWP
HU12 0TH M6PLH
HU12 0TJ M1DCE
HU12 0US 2E0EYC
HU12 0US M0PAY
HU12 0UX M1COV
HU12 8BY M6LDG
HU12 8ET G6AXC
HU12 8FP M6FPF
HU12 8GF G6ZJI
HU12 8GT M0MGF
HU12 8HQ G1BGO
HU12 8LB G7WDM
HU12 8LH G0VAY
HU12 8LX G0NYK
HU12 8NH G7SCT
HU12 8NH M0SCT
HU12 8NT G4CJJ
HU12 8QH G0LUP
HU12 8QH G4ZJC
HU12 8SG G4BDC
HU12 8TZ 2E0XKX
HU12 8TZ M6HUL
HU12 8UB G7MFO
HU12 8XF G1EFO
HU12 9AF G7RPW
HU12 9AX M1EHB
HU12 9BP M1VRC
HU12 9DY 2E0XYM
HU12 9DY M3XYM
HU12 9HP G4LHF
HU12 9LD M3BJJ
HU12 9LD M3JMJ
HU12 9NG G3YBU
HU12 9QG G1EQU
HU12 9RG G1EQF
HU13 0AS G7USB
HU13 0HB 2E0DVW
HU13 0JH G6GKT
HU13 0JW 2E0CIK
HU13 0LB G3NPT
HU13 0ND M1CBC
HU13 0NS G6MPK
HU13 0RL G4JIO
HU13 0RT G0KWE
HU13 0RT M3WJK
HU13 9BP G0CNV
HU13 9HE M6CKH
HU13 9HU G0OYX
HU14 3AE M3PLP
HU14 3AN G1EIB
HU14 3BD M6AYV
HU14 3BG G8FEK
HU14 3BX M6PBR
HU14 3DW G3YNO
HU14 3DW G3YQA
HU14 3DX G1REL
HU14 3HZ G0GVT
HU14 3JS G3RSB
HU14 3NL M0REC
HU14 3PR G7UJY
HU14 3RH G4XWA
HU15 1BE G0SWL
HU15 1BL G4VHM
HU15 1DA G7CLY
HU15 1EN 2E0SXJ
HU15 1EN G1SXJ
HU15 1EN G6KIA
HU15 1EN G6UQA
HU15 1EN M3SXJ
HU15 1FF G3VMV
HU15 1HY G4FGO
HU15 1JT M3HZW
HU15 1LA G6PIB
HU15 1NA G0ULN
HU15 1NL G8FDR
HU15 1NL M3UAG
HU15 1PE G8LZG
HU15 1QL G0GDS
HU15 1RJ 2E0DQP
HU15 1RJ G4DQP
HU15 1RJ M6DQP
HU15 1RQ 2E0CJA
HU15 2AL G4MJT
HU15 2AL G4SGU
HU15 2AL G4UVF
HU15 2HR G6POZ
HU15 2JG G3VLR
HU15 2LT G0WJK
HU15 2LT G4VNC
HU15 2NU 2E0DKI
HU15 2NU M0GWH
HU15 2QN G0LHV
HU15 2TY G4YEB
HU15 2XU G0JNQ
HU15 2ZU G3VDE
HU16 4AL M3PXL
HU16 4HN G0PAS
HU16 4HN G4KHT
HU16 4NF G0KBP
HU16 4SD G4EBT
HU16 5JG G3NOP
HU16 5RL G4GIY
HU16 5TF G0SXW
HU16 5TR G8EAH
HU16 5UG G4SEF
HU17 0AJ 2E1BDC
HU17 0BG G7TYT
HU17 0DF 2E1FIQ
HU17 0EY 2E0DYP
HU17 0EY M6WRH
HU17 0HP 2E0JMR
HU17 0HP M0BNR
HU17 0PA G8SGP
HU17 0PA M3NZG
HU17 0RU G4HBL
HU17 0TH G4EIM
HU17 5NR G7EFV
HU17 5NU G3RMX
HU17 7BP G4HYD
HU17 7BW M6DIO
HU17 7HD G6UGS
HU17 7HP G4GPY
HU17 7JN 2E0DXW
HU17 7JN M6GHS
HU17 7NX G0TAS
HU17 7NX M6NEG
HU17 7PN G8OEK
HU17 7RJ G7FDD
HU17 8GE G3MJM
HU17 8HY G4UOZ
HU17 8JL G4ESY
HU17 8LL M1EZK
HU17 8PF G0DMP
HU17 8PJ G4CMK
HU17 8PU G3NVX
HU17 8QA G4DEA
HU17 8SD G3ZRS
HU17 8SD M0GWA
HU17 8XB G6VCF
HU17 8YB M0HVQ
HU17 8YB M6BOP
HU17 9GL G4CQN
HU17 9LH G0KVR
HU17 9PG G6EUF
HU17 9QR 2E0VCP
HU17 9QR M6VCP
HU17 9RH G1AJD
HU17 9RH G1AJE
HU17 9RH G8UAI
HU18 1AF 2E0FLF
HU18 1AF M6FLF
HU18 1AL G6GHE
HU18 1BP G3OGE
HU18 1BX G0PSJ
HU18 1EF G4FCT
HU18 1EF G8PZX
HU18 1ES G0VET
HU18 1ES G6DKF
HU18 1EU 2E0LKE
HU18 1EU M0LKE
HU18 1EU M6LKT
HU18 1EW G4ADE
HU18 1JG G0IVP
HU18 1JU G7KHV
HU18 1LX G0CKF
HU18 1LX M0CZR
HU18 1PB G8LJQ
HU18 1PH 2E1HYT
HU18 1QZ G4CHH
HU18 1RE G0JWY
HU18 1TP G8KFK
HU18 1TT G0GXX
HU19 2AY M0MFP
HU19 2BB 2E0ENW
HU19 2BB M3EYH
HU19 2DB G6ZBT
HU19 2DB M0ZBT
HU19 2DS M6POP

HU19 2DT G2DPA
HU19 2DT G6YAS
HU19 2DT M0UDT
HU19 2DT M5EXY
[illegible] [illegible]
HU19 2EB G4HYY
HU19 2EB M0EGC
HU19 2EW G1OSG
HU19 2JA M6EUE
HU19 2LN M3ZHV
HU19 2LP 2E0MWW
HU19 2LZ G4SJI
HU19 2LZ G6YMI
HU19 2PB M6HGW
HU19 2QD G3RJM
HU19 2QU M0VES
HU2 8JR G8ZFX
HU2 9BL G1WFS
HU20 3UX G8HUG
HU20 3XA G0BEC
HU20 3XH G6KTK
HU3 1NS M0AYO
HU3 1NY M6HJE
HU3 1SA G0RUF
HU3 1SA G0UBY
HU3 1SA G6CTE
HU3 1UF G6WQH
HU3 1YD M6BYP
HU3 2HT M3MPM
HU3 2LT M0AAR
HU3 2SG M6SDA
HU3 2TL 2E0LKH
HU3 2TZ G0IRJ
HU3 5AS 2E0CNC
HU3 5AS M0GOC
HU3 5AS M6TKW
HU3 5LW M6GCP
HU3 5PP G1FPC
HU3 6AL M0TTL
HU3 6PH M6WRD
HU3 6QY G0UKS
HU3 6QY G3JDY
HU3 6RD M3OPX
HU3 6SL M3IDK
HU3 6SL M3PHR
HU4 6AP M1ADN
HU4 6EW G1LAN
HU4 6LJ M0BXB
HU4 6LU M0BEO
HU4 6PL G0IER
HU4 6PL M0HND
HU4 6RD M3UKH
HU4 6RW G1KNZ
HU4 6SQ G1PGX
HU4 6TL G3TKA
HU4 6TL M0GOY
HU4 6UG G1DQD
HU4 6XJ G6HVE
HU4 7AH G0POQ
HU4 7DA M3UNI
HU4 7HB G4ZPQ
HU4 7HL G0TRH
HU4 7QE M6ACF
HU4 7QG G4WCD
HU4 7QH G6VVZ
HU4 7SW 2E0LDR
HU4 7SW G3GJA
HU4 7SW G8EQZ
HU4 7SW M0BMB
HU4 7SW M0LDR
HU4 7SW M3LRZ
HU5 1LX M6YWG
HU5 1ND G0OVT
HU5 1QD 2E0OKP
HU5 2AB M3GIK
HU5 2AE G3AMW
HU5 2AE G8GBY
HU5 2TA M6PAE
HU5 3DW G3KDU
HU5 3JY G0UNC
HU5 3NZ M0ZYT
HU5 3NZ M3XWB
HU5 3PF M3PBR
HU5 3SQ G1PCQ
HU5 3TW M6APA
HU5 3TW M6EKA
HU5 4AA G1WQL
HU5 4AD M3KIU
HU5 4AQ M1EIE
HU5 4ED G1PLU
HU5 4ED G4HJD
HU5 4EQ M0ZIY
HU5 4HH M6JER
HU5 4HN G1MCI
HU5 4HN G1SQW
HU5 4JZ 2E0BWL
HU5 4JZ M0RWL
HU5 4JZ M3USZ
HU5 4LP G3RDP
HU5 4PF G1EYY
HU5 5DA 2E0SHW
HU5 5DJ G6VWF
HU5 5EG 2E0RVE
HU5 5HS G4CBL
HU5 5HW G0TGX
HU5 5HZ G0GXZ
HU5 5HZ G4CBA
HU5 5JW G7JGY
HU5 5LE G0VNH
HU5 5LN 2E0ZRB
HU5 5LN M3ZRB
HU5 5NX G6ABG
HU5 5QG M6ACX
HU5 5TP G7ONR
HU5 5YE G7BAC
HU5 5YY G4COT
HU6 7AG M6UUA
HU6 7AS G0CHP
HU6 7BA G4ERZ
[illegible] [illegible]
HU6 7DD G1LEZ
HU6 7DZ G3ZZP
HU6 7HJ M6INA
HU6 7JT G0GXF
HU6 7QB 2E0VCY
HU6 7QB M3VCY
HU6 7UN 2E0DMN
HU6 7UN M0ZUB
HU6 7UN M6ATB
HU6 7XE G4MPL
HU6 8DB M6JMR
HU6 8DG M6BZM
HU6 8DG M6BZN
HU6 8DJ 2E0SCA
HU6 8NJ M1FCW
HU6 8SA G3PQY
HU6 8SB G4AUQ
HU6 8SB G6LNV
HU6 8SB M0HFC
HU6 8SB M1DWT
HU6 9EB G0LPT
HU6 9ER G0RLB
HU6 9ER M3SBB
HU6 9EU G8VHG
HU6 9EZ G1TDO
HU6 9EZ M3TYS
HU6 9HA M3GZU
HU6 9HG M0DLT
HU6 9HG M0SRN
HU6 9JH M3VPH
HU6 9LS 2E0JRL
HU6 9RA G8ZFD
HU6 9RW 2E0TWZ
HU6 9SA G1TDN
HU6 9SA M6LEC
HU7 0DU G0UKS
HU7 0DU G0VTD
HU7 3AF G0OYQ
HU7 3AL G1XGE
HU7 3FA G4KXU
HU7 3FZ 2E1ICB
HU7 4AL M3MPT
HU7 4BX 2E0NYE
HU7 4HH 2E0NSQ
HU7 4HH M1BLO
HU7 4HH M3NSQ
HU7 4HN G0WEA
HU7 4NZ M1DRK
HU7 4PQ G0BAP
HU7 4PY G0EPA
HU7 4RF G1ARD
HU7 4RP M3VGP
HU7 4RU G7POW
HU7 4SP M6KMR
HU7 4ST 2E0DIX
HU7 4ST 2E0ELO
HU7 4ST 2E0HAL
HU7 4ST M3XHY
HU7 4ST M3ZZD
HU7 4ST M3ZZW
HU7 4UY G8WVB
HU7 4ZA M0BOM
HU7 5AQ 2E0OAH
HU7 5AQ M6OZY
HU7 5AS G7UCZ
HU7 5BU G0PTL
HU7 5DH 2E0GRH
HU7 5DH M6RRH
HU7 5FB M3RIE
HU7 5XD M0PLR
HU7 5XD M3XMQ
HU7 5YY G0VVP
HU7 6DE G0MKZ
HU7 6EE G0WDX
HU7 6EL G6WVR
HU8 0JG G6EZG
HU8 0QH G0HMS
HU8 0ST G0WFF
HU8 7RY 2E0BRQ
HU8 7RY M3OOQ
HU8 7RY M3SHW
HU8 8LQ G1FQI
HU8 8QR M3KLB
HU8 8TQ M3GWW
HU8 8TQ M3HME
HU8 9BN G1ADF
HU8 9EW M0TDK
HU8 9LY M3MWM
HU8 9NA 2E0LTC
HU8 9NA M6RWX
HU8 9RY 2E0FLA
HU8 9RY M6YLR
HU8 9UY M0EFE
HU9 1EJ G1CFB
HU9 1EJ G1LGW
HU9 1PL M6CKY
HU9 1ST 2E0GMW
HU9 2AS 2E0HKR
HU9 2PF 2E1XXX
HU9 2TE G0TUJ
HU9 3JQ G0PSL
HU9 3JU G7UGW
HU9 3LB G0NAU
HU9 3PN M0BYI
HU9 3PR M0GLV
HU9 3RG M1DHJ
HU9 4BT M6GGD
HU9 4DR 2E0KCW
HU9 4DR M0PKE
HU9 4DR M6BWP
HU9 4HH M3LUD
HU9 4PG 2E0UVZ
HU9 4PG M0HBO
HU9 4QH 2E0RWP
[illegible] [illegible]
HU9 4UJ M3DAY
HU9 5JZ 2E0HDX
HU9 5LH G0PCD
Hu94af M6BIG

## HX

## (Halifax)

HX1 1YP G8EKH
HX1 2BJ G8WFP
HX1 2NT G0FYQ
HX1 3EA G6FPF
HX1 3LE M0GQB
HX1 3LR 2E0PEC
HX1 3LR M6PEC
HX1 3RB G7SLP
HX1 4QG M3BXY
HX1 4TA G0CVC
HX1 5PU 2E0VRD
HX1 5PU M6HXI
HX1 5PU M6VRD
HX2 0DL G6SZP
HX2 0EF G3ONQ
HX2 0HT M3BXX
HX2 0LU G1NDQ
HX2 0NP G3TAY
HX2 0PA G7ELX
HX2 0PJ G0OZP
HX2 0PL G1RZZ
HX2 0QP M3DWA
HX2 0RB G8EXN
HX2 0RH G7OMF
HX2 0RL G4EHD
HX2 0SN M6MLR
HX2 0UL G4VOB
HX2 6EX G8RKX
HX2 6HQ M6DJN
HX2 6UX G6CNL
HX2 7HG G0INK
HX2 7HP G8SDE
HX2 7PN G6YGV
HX2 7RB G7RRC
HX2 7RB G7TDN
HX2 7RB M0CRG
HX2 7SG M3HTE
HX2 7TX G4NSH
HX2 8AA G3UI
HX2 8DL G1GWX
HX2 8DL G1VVB
HX2 8RE M1CMR
HX2 8RL M6BKN
HX2 8SG G8SJA
HX2 9AZ G4XYS
HX2 9EF M6RRR
HX2 9JD G6GGV
HX2 9JH G0JAQ
HX2 9JQ G2SU
HX2 9JQ G3TQA
HX2 9LY G3ZHP
HX2 9NR G0JBC
HX2 9PQ G0CBI
HX2 9PZ G6AVN
HX2 9SQ M6DKS
HX3 0AG G6COG
HX3 0AH M0HVI
HX3 0AL G4OYZ
HX3 0AL M3OOX
HX3 0AL M3OOY
HX3 0AL M3OPC
HX3 0AL M3OPD
HX3 0BD G7SWZ
HX3 0DL M6EMH
HX3 0HA M3IDA
HX3 0JQ 2E1SKR
HX3 0SR G0SPX
HX3 0SU G7IHX
HX3 0TH M0RWD
HX3 5FB M6VCU
HX3 5LU 2E1BOO
HX3 5LU 2E1BVJ
HX3 5LU G3UGF
HX3 5LU G6CQG
HX3 5NL M0DFJ
HX3 5PA 2E1AHK
HX3 5QF G8REO
HX3 6QG M6HEX
HX3 5QU G1WHU
HX3 5SZ G1HCU
HX3 6PL M0PDW
HX3 7AP G0CZL
HX3 7EB M0AHG
HX3 7EP G6MDC
HX3 7LB G8UTW
HX3 7NA G0PMU
HX3 7NE G8ZXZ
HX3 7NH G0HRJ
HX3 7NH M3KJK
HX3 7NY 2E0UPU
HX3 7NY M0UPU
HX3 7NY M6UPU
HX3 7PW G4KEX
HX3 7QY M0HFX
HX3 7QY M6LVW
HX3 7RN 2E0BAB
HX3 7RN M0TKD
HX3 7TS G4YRH
HX3 8HB G7FZJ
HX3 8NJ M3PNH
HX3 8QF G4BZS
HX3 8TJ G1ILO
HX3 8TJ M3JJU
HX3 8XP M6FYN
HX3 9BH M0FAX
HX3 9DD M0JRL
HX3 9DT G0GVB
HX3 9EE G6XTT
HX3 9LD 2E0BJQ
HX3 9PS G7DWY
HX3 9PS M0INB
HX3 9SE M3YEM
HX4 0BW 2E1HUE
HX4 0EW G8HTB
HX4 8HT M6JJE
HX4 8JF G6APB
HX4 8NU G3YGZ
HX4 8PA M3MQY
HX4 8QF G1MWT
HX4 9AS G4LIX
HX4 9DN M3JQT
HX5 0BB G4FMM
HX5 0BB M3LAG
HX5 0DR 2E0MWT
HX5 0LA 2E0HQJ
HX5 0PF 2E0PHU
HX5 0QA G3YCJ
HX5 0QH G6XNJ
HX5 0RN M6RIK
HX5 9JF 2E0ZXV
HX6 1AD G6DQU
HX6 1BX G8UTH
HX6 1EE M6UGX
HX6 1NL G4AES
HX6 1NS G0DUN
HX6 2EE M6JRC
HX6 2QW M3GSM
HX6 2RP G1MBM
HX6 2RU G0FOB
HX6 2SY M6JNC
HX6 3HZ M6GSO
HX6 4AG G0WFG
HX6 4AG G7NSN
HX6 4LQ G8EWN
HX6 4LS G8MKG
HX6 4NU M3JQS
HX6 4PA M1EYG
HX6 4RA G3VLO
HX6 4RZ G6AUC
HX7 5HS M6DTP
HX7 5NF G0CMQ
HX7 7AL 2E0KPT
HX7 7AL G3HZR
HX7 7ED G0TOB
HX7 7HD G0TVB
HX7 7HD G6TRG
HX7 7JA G4EFX
HX7 7JP 2E0JCK
HX7 7JP M0JCK
HX7 7NX G6XPZ
HX7 7PB G7CLX
HX7 7PG G3MLS
HX7 7PH G8BBZ

## IG

## (Ilford)

IG1 1TT M0ANB
IG1 2ER G6ZQJ
IG1 2UF G0IQK
IG1 3SL M6HQX
IG1 4RY M0NSP
IG1 4RY M6FDW
IG10 1HT 2E0DQL
IG10 1HT M6VTG
IG10 1PX M6SNH
IG10 1SB G0LWM
IG10 1SB G7EEE
IG10 1TS G0SGX
IG10 2AD G4WTQ
IG10 2AJ G4IMU
IG10 2LT M6GJH
IG10 2QL 2E0ZAU
IG10 2QL M6ZAX
IG10 2RE G0TOC
IG10 2RE G7KJV
IG10 2SA G3NQT
IG10 3BY M1EZC
IG10 3FP G0PFY
IG10 3JT G1DJI
IG10 3PL G1IZA
IG10 3PR G1IVP
IG10 3QY G7VKB
IG11 0DS G6MPJ
IG11 0LD M0BDQ
IG11 0NL G0POY
IG11 7EB M3ZBS
IG11 7QE M3WXW
IG11 7UW G1UAL
IG11 9DD G3OHS
IG11 9EE G0BIW
IG11 9HS M6PMB
IG11 9NX G7PRH
IG11 9NY M1EEY
IG11 9PS M3DVH
IG11 9PS M3EKY
IG11 9XB G0OVE
IG11 9XB G6JMJ
IG11 9XU M0KIB
IG2 6AS G3CAZ
IG2 6EQ G4PQW
IG2 6QA G7EMH
IG2 6QY G6RAH
IG2 6YG G1OYM
IG2 7DL G6EEF
IG2 7HR M6KNB
IG2 7NQ G4JRD
IG2 7SF G7JUZ
IG2 7UI M0JUZ
IG2 9JD G7UPQ
IG3 8NN G0VKH
IG3 8XE G7VBF
IG3 9TB G0MEW
IG4 5AE G3ALK
IG4 5HA M6OPJ
IG4 5HN G3WPR
IG4 5JX M0CTF
IG5 0AQ G4FXQ
IG5 0DL M1EMX
IG5 0HP G1ICK
IG5 0NP G8JFL
IG5 0RZ G6XSB
IG5 0TG G8VBK
IG5 0XN G8YAE
IG6 1BJ G0ATP
IG6 1PJ G4YUF
IG6 2LH G8PPA
IG6 2QE 2E0DDH
IG6 2QE M0MBD
IG6 2QE M6DLH
IG6 2QU G8HST
IG6 2QU G8PRJ
IG6 3AH G8BUF
IG6 3SR M1EDO
IG6 3TF G1HEQ
IG7 4DG 2E0CFT
IG7 4EA M1CPD
IG7 4HQ 2E0EER
IG7 4HQ M6AGD
IG7 5AU G2CD
IG7 5ED G4GMN
IG7 5HZ G4NZB
IG7 5QZ G7VGJ
IG7 6AD G8LGU
IG8 7LS 2E0DOW
IG8 7LS M0CEO
IG8 7LS M6LFE
IG8 7QU G4AJG
IG8 8DW 2E0RMM
IG8 8DW M6RML
IG8 8DX 2E0ALB
IG8 8DX M3XMA
IG8 8JN 2E1ICT
IG8 9AA G6SWT
IG8 9AA G8FRH
IG8 9HY G4BNB
IG8 9QZ M1ETW
IG9 5QE G4BMU
IG9 5RU G8MCR
IG9 5TZ G4YOA
IG9 6AQ G3VGR
IG9 6BY G0LWI

## IM

## (Isle of Man)

IM1 4BB GD7IEH
IM1 4EG GD0PLQ
IM1 4HQ GD3GBG
IM1 5EQ MD3WXS
IM2 1JE 2D0VMN
IM2 1JE MD3VMN
IM2 1JT MD3JTT
IM2 1NX MD6NSS
IM2 1QX MD3LEP
IM2 2EU MD6KBW
IM2 2LA MD3GAB
IM2 2LA MD3MAN
IM2 2NP GD3YEO
IM2 3RQ GD0JWR
IM2 4AH MD0RKI
IM2 4AQ MD6WFK
IM2 4AQ MD6ZKK
IM2 4AR 2D0IMN
IM2 4AR MD3FMN
IM2 4PE GD4FMB
IM2 4PE GD4ZAB
IM2 5BH GD0TEP
IM2 5BH GD7MAN
IM2 5BQ GD0PLR
IM2 5EH GD0IDU
IM2 5NG GD0TFO
IM2 6AG GD0KQE
IM2 6HH MD3MLB
IM2 6HQ 2D0YLX
IM2 6HQ MD3YLX
IM2 6HR 2D0GBM
IM2 6HR MD0VMD
IM2 6HR MD3VMD
IM2 6HW MD1DNT
IM2 6LL MD6RSV
IM2 6LX GD4EBA
IM2 6PB MD3CPK
IM2 7AW GD0KPN
IM2 7DX GD0JKA
IM3 3BU GD4RVQ
IM3 3BU GD6HCB
IM3 3GA GD0BFN
IM3 3GB 2D0CFZ
IM3 3GB MD0PWI
IM3 4AT 2D0JEA
IM3 4AT GD0OUD
IM3 4AT GD7HTG
IM3 4AT MD3WFJ
IM3 4BG MD3ZHD
IM3 4ET GD4DPK
IM3 4LE MD6TDU
IM3 4NR GD0BCN
IM3 4NR MD3OKG
IM3 4NR MD3OKH
IM3 4NU GD0IFU
IM4 1BJ GD7LAV
IM4 1EJ MD3LPW
IM4 1HQ GD7ELF
IM4 2BP GD3PLH
IM4 2BP GD3TNS
IM4 2DN GD0SFI
IM4 2DR 2D0DRM
IM4 3HE GD4EIP
IM4 3HE MD0DMV
IM4 3JB GD3MBC
IM4 3JB MD3UGY
IM4 3JP GD0AMD
IM4 3JP GD1LVY
IM4 3JP MD0GLK
IM4 3JP MD0HHH
IM4 3JP MD6IZI
IM4 3LZ GD4XWF
IM4 4AR MD6IKR
IM4 4ES GD4FWQ
IM4 4ES GD4GWQ
IM4 4EW GD0JGX
IM4 4LU MD3LEG
IM4 4NF 2D0BCR
IM4 4NF MD0MAN
IM4 4NG GD3FXN
IM4 4NZ MD6MGP
IM4 4QP MD3PER
IM4 4QZ GD0PDN
IM4 5EA GD4XOD
IM4 7AP GD3LSF
IM4 7AP GD7ESU
IM4 7DF GD3ZEX
IM4 7HH MD6LET
IM4 7JB GD3HFC
IM4 7JY GD0NFN
IM4 7PG GD4HPN
IM4 7PW GD0ELY
IM4 7PW GD4XTT
IM5 1BE GD4OEA
IM5 1BQ MD3LWQ
IM5 1GH MD6TSW
IM5 1HP GD4MCR
IM5 1JJ GD4NTR
IM5 1PJ GD4MDY
IM5 1PN GD4IOM
IM5 1PN GD6ICR
IM5 1PX GD4UQO
IM5 1UF MD3MCB
IM5 2AE 2D0WFH
IM5 2AE GD4UHB
IM5 2AE MD6WHG
IM5 3BA GD6TWF
IM6 1AF GD1GHK
IM6 1AF MD6AGF
IM6 1HU MD6KFH
IM7 1ED MD3ONP
IM7 1HE GD4ZUU
IM7 2AF GD4WOW
IM7 2AY GD3YTE
IM7 2EB GD4WBY
IM7 2EB GD8IOM
IM7 2EY GD4RGR
IM7 2HB GD1XMA
IM7 3BL GD4PTV
IM7 3DA GD0KEO
IM7 3EU GD7RVP
IM7 3HP GD3RFK
IM7 3HP GD4RFK
IM7 3HP MD3BZA
IM7 4AG GD4HOZ
IM7 4AQ GD1IOM
IM7 4AQ GD1MIP
IM7 4AQ MD6IOM
IM7 4HY 2D0STL
IM7 5AH GD3XNU
IM73EB GD1AQY
IM8 1JF GD0KWM
IM8 1LJ GD7BOJ
IM8 1LJ MD0BJM
IM8 1NA GD3GCE
IM8 1NF GD0HWA
IM8 1NF GD8ANU
IM8 2BH GD3SKH
IM8 2EG MD3OIS
IM8 2EX MD3EVY
IM8 2PT GD8BUE
IM8 3AN GD4AZJ
IM8 3DA 2D0JBE
IM8 3EB GD4IHC
IM8 3NZ MD6ZEE
IM8 3PU MD0FIX
IM8 3UF MD0DPG
IM9 1BL MD1RPC
IM9 1HP MD3KSN
IM9 1HR 2D0IOM
IM9 1HW 2D0VJK
IM9 1HY GD7DUZ
IM9 1HY MD3DUZ
IM9 1NN 2D0HEB
IM9 1NN MD0HEB
IM9 1NW MD3KCT
IM9 2DY MD3XUA
IM9 2ED GD1HIA
IM9 2EG GD6AFB
IM9 2EG MD6YBE
IM9 2ES GD0TFG
IM9 2ET MD3UMN
IM9 2EU MD1DXW
IM9 2EU MD3DXW
IM9 2EU MD6RAQ
IM9 2EW MD0PMN
IM9 2HF 2D0EDQ
IM9 2HF GD0BCJ
IM9 3AH GD3EFD
IM9 3AY 2D0JKW
IM9 3BA [illegible]
IM9 3BA GD4GNH
IM9 4DD GD0UHL
IM9 4EF GD6OXG
IM9 4EP GD6XHG
IM9 4ER 2D0RLA
IM9 4ER MD0RLA
IM9 4HA GD3SKZ
IM9 4HJ MD6SNV
IM9 4LN GD4TSO
IM9 4LQ 2D0RGW
IM9 4NB GD4VBA
IM9 4NH MD0IOM
IM9 4NL GD0HYM
IM9 4NL GD0MAN
IM9 4NX 2D0TRL
IM9 4PR MD6IUH
IM9 5AN GD1HVL
IM9 5ED GD0MWL
IM9 5ED MD3AAI
IM9 5HF 2D0FHG
IM9 5LP MD6RHN
IM9 5LX GD4RAG
IM9 5PR GD1USI
IM9 6DB GD4IZL
IM9 6DB GD8PPU
IM9 6DL GD1RHT
IM9 6EG MD3LJB
IM9 6EL GD0VIK
IM9 6EL MD0DTQ
IM9 6EL MD3EEW
IM9 6EN MD3JGS
IM9 6EP GD4FJI
IM9 6EU 2D0NOT
IM9 6HA MD6DYM
IM9 6HR GD8EUH
IM9 6LS GD3ZZN
IM9 6NB GD3JIU
IM9 6ND GD4ZZN
IM9 6PH MD3OED
IM9 6QU GD7KAM
IM9 6QZ MD3LJS
IM9 6TE GD8EXI
IM9 6TJ GD7ARS
IM9 6TJ GD7DPG
IM95DN MD3YUQ
IM99 2QF GD4TEM

IO9 9SU GD1MZJ

## IP

## (Ipswich)

IP1 2PH G7SMN
IP1 3QF G3XCO
IP1 3QU G8IIC
IP1 3QZ G0UBM
IP1 3SA G7LKY
IP1 3SD M6SVJ
IP1 4DP M3HGL
IP1 4JY G1YAB
IP1 4JY G4YUG
IP1 4LD G4LRB
IP1 4LU M3ZBR
IP1 4NZ G8MUF
IP1 5DR G0AKC
IP1 5DR G0JVT
IP1 5EA G7OCH
IP1 5HD G1WQY
IP1 5HS G8FTW
IP1 5JX G3YWM
IP1 5LD 2E0GCY
IP1 5LD M6GCK
IP1 5LR G7PLV
IP1 5NJ 2E0AXZ
IP1 5NL 2E0JSS
IP1 5NN G4XDK
IP1 6AB G7VLJ
IP1 6BD G1HGF
IP1 6BH G4BAV
IP1 6BH G4IRC
IP1 6DU G0SAR
IP1 6DU G8XOR
IP1 6ET M3PBK
IP1 6EW G7UQV
IP1 6JB G0HJK
IP1 6PA M6SFT
IP1 6PQ G8VNP
IP1 6PQ M0AKK
IP1 6PC M3YCZ
IP1 6RE M0JAJ
IP1 6RG G1YGV
IP1 6RL G8ZQA
IP1 6RL M3ZQA
IP1 6SS G7FNN
IP10 0DE G4LXX
IP10 0EU G4SYG
IP10 0EW G6RTY
IP10 0JX G4XQD
IP10 0NP G8MHA
IP10 0PF G4LYD
IP10 0PF G4RHR
IP10 0PL G4DYH
IP10 0PP G4GVW
IP10 0PP M0JSA
IP10 0PP M6ALX
IP10 0PP M6ISJ
IP10 0PZ G8BHC
IP10 0QF G0KBM
IP10 0QQ 2E0VAA
IP10 0QU G4DDK
IP10 0QU G8EMY
IP11 0RG G8BHK
IP11 0UU G3OJ
IP11 0UU [illegible]
IP11 0UY M6HKR
IP11 0XG [illegible]
IP11 0XL G8MXV
IP11 0XR G8BEH
IP11 0YQ M0DNJ
IP11 2BG M0OKB
IP11 2BZ M6MWT
IP11 2EA M3YKN
IP11 2NE G0CFT
IP11 2NS G8EAX
IP11 2NT M0LVW
IP11 2NY G1JRL
IP11 2PN G8BBV
IP11 2UH G4YQC
IP11 2UH G4ZFR
IP11 2UH G7OPS
IP11 2UL G7TUG
IP11 7AU 2E0IPX
IP11 7BE 2E0COH
IP11 7BE M0HLY
IP11 7BE M6SLN
IP11 7EF G3TRD
IP11 7EG G0BEU
IP11 7EG G3WDE
IP11 7JR M1EJS
IP11 7LG G4FAW
IP11 7NR G0UPD
IP11 7NX G3YHK
IP11 7QR 2E1HKB
IP11 7QR M0HKB
IP11 7RL G7ILA
IP11 7RP G3XFF
IP11 7RR G0PRU
IP11 7RR G4DNX
IP11 9AL G7MNS
IP11 9BU G6MKO
IP11 9DE M3BDQ
IP11 9DS M3JFA
IP11 9EE M6BFF
IP11 9HS G4IVC
IP11 9HS M3STQ
IP11 9JJ M0HQL
IP11 9JT G0GSL
IP11 9LR G4FBV
IP11 9NN G3ZKZ
IP11 9PJ M6FLX
IP11 9PS G8XRL
IP11 9SS G6MCG
IP11 9SU G8RRN
IP11 9TJ M0MWR
IP11 9TL G8XUL
IP11 9TL M6HJK
IP11 9TS M3MFG
IP11 9TX G0XEG
IP12 1AH G8VZZ
IP12 1BE M0JPS
IP12 1HB G8AXO
IP12 1JL 2E1GXU
IP12 1JQ G7JXR
IP12 1JS G0OZR
IP12 1LB 2E1FSR
IP12 1LD G1AGK
IP12 1LQ G0DDZ
IP12 1LW G0AOH
IP12 1PE M0BPW
IP12 1RL M6BHY
IP12 1RN G7DWN
IP12 2DD G0SBH
IP12 2ED G6FS
IP12 2ED M0BCT
IP12 2GA M6DON
IP12 2GL M3VDO
IP12 2HZ G4AVS
IP12 2JE G4HHA
IP12 2JH M6TCI
IP12 2PL G3ZYP
IP12 2PL G8BZJ
IP12 2PX M6PNC
IP12 2QA M6SWD
IP12 2QG G0JFM
IP12 2QG G1YOU
IP12 2TE M6AZN
IP12 2UD G1SAR
IP12 2UD G3ADZ
IP12 2UD G7CIY
IP12 2UR M0KTO
IP12 2UT G3RTB
IP12 3DZ M0LMC
IP12 3JT 2E0HOP
IP12 3JT M0NOC
IP12 3JY G0ILZ
IP12 3JY M6AOL
IP12 3LL G0THW
IP12 3QL M1AGP
IP12 3QU G1XWS
IP12 3QY G4DFD
IP12 3TP G0VWE
IP12 4DU 2E0YZC
IP12 4DU M0GQR
IP12 4DU M0VZO
IP12 4ED G4MRD
IP12 4HE 2E0TYL
IP12 4HE M3TYL
IP12 4HR G3PFH
IP12 4HR G7SCR
IP12 4HW G3TLY
IP12 4JN G3UYX
IP12 4JP G4FSG
IP12 4JP G7SJP
IP12 4JR G4AKW
IP12 4JW G8BRL
IP12 4NY M1CGM
IP12 4NZ G8ONH
IP12 4PT G1DIK
IP12 4PT G0WTQ
IP12 4SL G1MQS
[illegible] [illegible]
IP13 0AD G4NBP
IP13 0AT M6BID
IP13 0BP 2E0EVE
IP13 0BP 2E0MIG
IP13 0BP M0SDY
IP13 0BP M3EBZ
IP13 0BP M3KIT
IP13 0BP M3LDC
IP13 0BP M3MIG
IP13 0BP M3VDT
IP13 0ES G4TRE
IP13 0HE G0RZG
IP13 0LN 2E0BEW
IP13 0LN G4HUP
IP13 0LN M3IKR
IP13 0LR G3RPB
IP13 0LR G6KQ
IP13 0ND 2E1EJD
IP13 0RQ 2E0FIR
IP13 0RQ M0MJF
IP13 0RQ M6FIR
IP13 0SL G6AYX
IP13 6DP M0BVT
IP13 6DX 2E0IAF
IP13 6DX M0VCP
IP13 6DX M1DUD
IP13 6DX M5AEF
IP13 6EB G3RVC
IP13 6ES G8VCU
IP13 6JF G8XYQ
IP13 6LA 2E0SUF
IP13 6LA M0SUF
IP13 6ND M6ADY
IP13 6ND M6KYK
IP13 6PL G0VQS
IP13 6SE 2E0LFR
IP13 6SE 2E0OLE
IP13 6SU G0AOY
IP13 6TE M6MAG
IP13 6TH 2E0CQQ
IP13 6TH M0IAH
IP13 6UP 2E0CQR
IP13 6UP M6BTT
IP13 7AW M6EGZ
IP13 7JD G8VOC
IP13 7JP M6CCF
IP13 7NU M6EBX
IP13 7PP G4RSD
IP13 7RT M6NBW
IP13 8BB 2E0STQ
IP13 8BB M0LIE
IP13 8BB M6BZU
IP13 8DT G7JCF
IP13 8LZ G3PYW
IP13 9HA G7RMQ
IP13 9HA G8IQF
IP13 9HP G7MIP
IP13 9HQ M6SXC
IP13 9LY M3DFP
IP13 9NP M6KCK
IP13 9NP M6OLW
IP13 9PQ G3LQR
IP13 9SL 2E1GXV
IP13 9TE G6HTT
IP14 1BN G3KCF
IP14 1BT M1CGO
IP14 1DA M3NIC
IP14 1GH G0HEV
IP14 1LP G3TAQ
IP14 1RH G1HNH
IP14 1TD G3TZE
IP14 1TS 2E1BRC
IP14 1TS G6SUR
IP14 1UF 2E0XDD
IP14 1UF G0JJG
IP14 1UF M3XDD
IP14 1UQ G0NFE
IP14 2AB G3SUK
IP14 2DR M1ADV
IP14 2JT G6PGM
IP14 2LZ G4AWF
IP14 2NZ G0SCM
IP14 2RE G1UBH
IP14 0AD G7OND
IP14 3DJ G4AFX
IP14 3GA G3MXH
IP14 3HE M0EQV
IP14 3JT G7DME
IP14 3NW M1ALH
IP14 3PA G4BIY
IP14 3RQ M3XQX
IP14 4DB M6SCN
IP14 4EJ 2E0BBM
IP14 4EZ M5AEC
IP14 4LL M0JLN
IP14 4LP G0NMS
IP14 4NR G8SED
IP14 4NT M6DLA
IP14 4RF 2E0DIQ
IP14 4RF M6DIQ
IP14 4SP M6KNW
IP14 4TP G6XGT
IP14 5AN G4BJO
IP14 5AP G6JZV
IP14 5AX M6EVC
IP14 5BB G1YRJ
IP14 5DS G7BUL
IP14 5ET G3XLG
IP14 5ET G4THN
IP14 5GH 2E0DEH
IP14 5GH M6LMJ
IP14 5HB G3ZEQ
IP14 5HJ G8FUJ
IP14 5JL G3ZQU
[illegible] [illegible]
IP14 5LG M3IFD
IP14 5PE G0OZS
IP14 5SH M3XGL
IP14 5SN G6SYW
IP14 5UG M3VUH
IP14 6AJ G4GBA
IP14 6BU 2E1HWJ
IP14 6DJ G4VPA
IP14 6HD G8ZZS
IP14 6LB G3ONL
IP14 6LX G3VNT
IP14 6LX G4EVN
IP14 6RN G8FMI
IP15 5QB M6CZI
IP15 5QE 2E0ALD
IP16 4AR 2E1HBF
IP16 4AR G1YRF
IP16 4AR G3DBJ
IP16 4AT G4ZWB
IP16 4BY G3MYA
IP16 4DT G0UEA
IP16 4HJ M3KUM
IP16 4JP M6GLS
IP16 4QZ G0CFI
IP17 1BA M1DQE
IP17 1BU M6CUD
IP17 1DP M3DTD
IP17 1EA 2E0KOS
IP17 1EA M0IPS
IP17 1FB G0OIO
IP17 1HR M1ACB
IP17 1JY G0CJX
IP17 1LJ G1HIU
IP17 1UX G0GDJ
IP17 1UX M3TIL
IP17 1WB G4ZJH
IP17 2AA G0EGW
IP17 2AS G0KDR
IP17 2JA G8FIA
IP17 2JP G1MJV
IP17 2NW G4IIK
IP17 2PU M0BJR
IP17 2RA G4CXT
IP17 3AH M6EGL
IP17 3ED G4XVE
IP17 3EP M6MFG
IP17 3GZ G8BTX
IP17 3NY M6TWA
IP17 3PA M6CSR
IP18 6LR G6BPY
IP18 6RE G7BLK
IP18 6UJ G0BYK
IP18 6UL M1DNG
IP19 0BB 2E0FIB
IP19 0BB M3FIB
IP19 0BB M3OJU
IP19 0EA G4PUQ
IP19 0PY G0CFB
IP19 0RB M0LAY
IP19 8DB G8TBV
IP19 8EE 2E1CPJ
IP19 8EG G0JJE
IP19 8JF M6PDZ
IP19 8JT M6DRP
IP19 8RQ 2E0LRG
IP19 8RQ G7VKK
IP19 8RQ M0SAZ
IP19 8RQ M6LRG
IP19 8TJ G8KWN
IP19 8TU M6PFY
IP19 9BH M0ARY
IP19 9DY G4IHI
IP19 9DZ G8EUE
IP2 0EJ M6GPI
IP2 0NQ M0TMB
IP2 0QG G8LBS
IP2 8AZ G4BVI
IP2 8BQ G3XVL
IP2 8GA M0ULR
IP2 8GA M3ULQ
IP2 8HA 2E0BCM
IP2 8JZ G4AVL
IP2 8QB 2E0JBI
IP2 0QD M6JDG
IP2 9AS G3WJS
IP2 9BS M6CBQ
IP2 9JE M0IDK
IP2 9JF G4FSK
IP2 9PA G7PQX
IP2 9PD G8NYC
IP2 9QR G4RUR
IP2 9QZ G4CMQ
IP2 9SX G3SXV
IP2 9TA G4BQR
IP2 0UE M6CQD
IP2 9YQ M3UAQ
IP20 0NY G6IUF
IP20 0HY G3IRZ
IP20 9JP G4RAV
IP20 9JY G3ICG
IP20 9NQ G4PFG
IP20 9PU G6ZYM
IP21 4AG M6PCC
IP21 4DW G3UNU
IP21 4EE G4YFV
IP21 4EH 2E0EPM
IP21 4EH M6JLL
IP21 4LD M6XBB
IP21 4NG G4DML



**IMPORTANT NOTE**

**Revalidate licence to avoid revocation** – Ofcom has advised the Society that plans will be drawn up to revoke licences that have not been revalidated as required by the licence conditions. The quickest way to revalidate is to do so online via the Ofcom website: *https://services.ofcom.org.uk/* or by email: *amateur.validations@ofcom.org.uk* If you need assistance in the process, Ofcom staff are available to help, but please be patient during times of heavy workload.

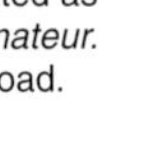

IP21 4PH M3YLZ
IP21 4RL M6JGQ
IP21 4TG 2E0HPJ
IP21 4TG M0HIJ
IP21 4TG M0HPJ
IP21 4TG M6JWW
IP21 4TQ G3IPG
IP21 4XP G3UCL
IP21 4XP G8OCV
IP21 4XP M0SNB
IP21 4XP M1GEO
IP21 4YJ 2E0XGW
IP21 4YJ M0XGW
IP21 4YJ M3XGW
IP21 5ER 2E0UAV
IP21 5ER M0UAV
IP21 5ER M6TYL
IP21 5JH M1DPX
IP21 5LS G0DFC
IP22 1AR G0OJJ
IP22 1BX M6ASW
IP22 1NQ G8DQZ
IP22 1NQ M0DQZ
IP22 1RN M3XIY
IP22 1RW G8IBR
IP22 1RZ M6FHO
IP22 2HS G8JPA
IP22 2JQ M6TFD
IP22 2PS G6WJW
IP22 2PS M3JNY
IP22 2PY M6CZL
IP22 2QR G3MFO
IP22 4DJ G0RPY
IP22 4GJ G4MMI
IP22 4HL G8BOB
IP22 4HR M3JLB
IP22 4HR M6VAZ
IP22 4NA G0WCJ
IP22 4NF G3XLL
IP22 4NP G6ILT
IP22 4PT M6EJL
IP22 4PW G8BLD
IP22 4PY M3XBF
IP22 4PY M6SAF
IP22 4QW 2E1LME
IP22 5RG G7OCQ
IP22 5RU G8RML
IP22 5SP 2E0TUF
IP22 5SP M3TUF
IP23 7AQ 2E1GRT
IP23 7DA M6GLU
IP23 7EE G3XAP
IP23 7JX M3PFM
IP23 8DY G1TIJ
IP23 8DY M3GFW
IP23 8HP M1BLX
IP24 1BZ G0GSZ
IP24 1DG M6FBW
IP24 1JJ 2E0PFG
IP24 1JJ M3VFE
IP24 1LG M6JAL
IP24 1LQ 2E0FCH
IP24 1LQ G0IFL
IP24 1LQ M6FCH
IP24 1NG G4LXD
IP24 1NP G3SXP
IP24 1PB G4RKK
IP24 1PF G0DCJ
IP24 1TQ G0CLT
IP24 2JH G7ACN
IP24 2LF M0CNM
IP24 2LY M3SPL
IP24 2QA 2E0GPS
IP24 2QA M3SKY
IP24 2QX G3ZZQ
IP24 2UU G4OQK
IP24 2YN M6PZA
IP24 2YN M6UES
IP24 3GA G0VGC
IP24 3GA G7MOY
IP24 3HQ G4VEL
IP24 3NF G1LRK
IP24 3PT M3UQL
IP25 6BZ G3GIB
IP25 6BZ G7JRP
IP25 6DB G4GRT
IP25 6DN G4OZY
IP25 6DW G1UKH
IP25 6EA G4AED
IP25 6EY G3DOV
IP25 6FF G0NMP
IP25 6FF G7DXN
IP25 6HE M6FHF
IP25 6HL M6DBH
IP25 6HL M6UCS
IP25 6LG G1NAN
IP25 6LL G8XOC
IP25 6NA 2E0GXX
IP25 6NA M3GXX
IP25 6NA M6RZO
IP25 6NJ M6ZXR
IP25 6NL G6YTV
IP25 6PF G7UNY
IP25 6SD M3ZVW
IP25 6SR G0ALQ
IP25 6SS M6EKO
IP25 6XA 2E0FLN
IP25 6XA M0NUX
IP25 6XA M6FLN
IP25 7DJ G4UEV
IP25 7EH G0KZI
IP25 7EH M6WWR
IP25 7EW G4TJY
IP25 7EZ G0SCT
IP25 7FD G1APL
IP25 7HW G4VBX
IP25 7LS M6KSS
IP25 7LT G0LGF
IP25 7LX M3LRW
IP25 7LX M5IMI
IP25 7LY M6SSY
IP25 7PJ G0FMI
IP25 7QL M3LGJ
IP25 7QN M0BHW
IP25 7QX M6MBK
IP25 7RA G6BLA
IP25 7RB M6GBB
IP26 4AR G6HGD
IP26 4BD G0CZR
IP26 4HZ G0HPM
IP26 4QJ G0CLH
IP26 5EG 2E0CNN
IP26 5EG M6BKM
IP26 5EQ 2E0BXS
IP26 5EQ M0GQP
IP26 5EQ M3YZI
IP26 5JA 2E0BCO
IP26 5JA M0FZX
IP26 5LL M6SGG
IP27 0BS G0WON
IP27 0DN G4KZK
IP27 0DX G8VMZ
IP27 0EN M6UAD
IP27 0GA M0JWM
IP27 0LJ G8ENY
IP27 0NR 2E0EAA
IP27 0NR M0SAA
IP27 0NR M3ZPW
IP27 0PU G7CNX
IP27 0PW 2E0GOL
IP27 0PW 2E0LJG
IP27 0PW M0LJD
IP27 0PW M0NKR
IP27 0PW M3UYG
IP27 0PW M6LJG
IP27 0QG 2E1IGK
IP27 0QG M0PBD
IP27 0TG M3SFK
IP27 9AJ M6JJD
IP27 9AU G8XQD
IP27 9ES G6LOJ
IP27 9EU M6EKL
IP27 9EU M6EMX
IP27 9EZ G1RFH
IP27 9EZ G6EUO
IP27 9EZ G6XYX
IP27 9EZ M3XYX
IP27 9HF G6DFR
IP27 9HS G6MGN
IP27 9SA G0BBN
IP27 9SA G0SDE
IP27 9SA G6TSP
IP270QH M6HZD
IP28 6DT M6NHS
IP28 6ER G3HGE
IP28 6ES G3PFJ
IP28 6TQ G6PCC
IP28 6UA G0FIU
IP28 6UG G4PXC
IP28 6UG G4XRK
IP28 6XF G3YFP
IP28 7BT G8DIY
IP28 7BZ G8ATS
IP28 7DP G4XTW
IP28 7HX G8BCA
IP28 7JU M6WVH
IP28 7LN G0DZY
IP28 7LS M6LSV
IP28 7PD M3LWV
IP28 7PD M3ZWK
IP28 7PD M6GEQ
IP28 7PD M6MFK
IP28 7PR G6VAZ
IP28 8LQ G4BWP
IP28 8LQ G4MBC
IP28 8PB G0BRM
IP28 8PB G1SXY
IP28 8QB G7OYF
IP28 8QD 2E0BJP
IP28 8SF M3PTB
IP29 4DL G7RKU
IP29 4PH G4KNO
IP29 4PL G3VTR
IP29 4SD M6AXM
IP29 4SS M6RIH
IP29 5AD G1XAM
IP29 5AD G8KMM
IP29 5AG G4CGV
IP29 5AP M0CWN
IP29 5DD G0WVE
IP29 5DX G4ERF
IP29 5HE G7WKC
IP29 5HR G4VBS
IP29 5QL G8CRM
IP29 5RD M3WOL
IP29 5RP G3JPM
IP29 5SE G0JRM
IP3 0BW M6BGB
IP3 0DU 2E0SLO
IP3 0DU M6WSW
IP3 0EE M0MCY
IP3 0EE M6HSA
IP3 0EJ M3FCS
IP3 0EW G4LVG
IP3 0LA G4LVD
IP3 0LF G7DNT
IP3 0LF M0XON
IP3 0LJ M3ESK
IP3 0LX M6LGF
IP3 0LZ M3KGJ
IP3 0LZ M3SMY
IP3 0LZ M3ULW
IP3 0PJ G4WAG
IP3 0PJ G6MAD
IP3 0QE G6RHK
IP3 0QH G1HSL
IP3 0RG M3TGP
IP3 0RR G7UWC
IP3 0SL G0IVV
IP3 8AT G6MMT
IP3 8BZ 2E0DWU
IP3 8BZ 2E0SER
IP3 8BZ M6MPZ
IP3 8BZ M6OAX
IP3 8EA M0CCX
IP3 8GG 2E0SKD
IP3 8GG M6SKD
IP3 8JW G0BPU
IP3 8NX M6CLV
IP3 8PD M3MXM
IP3 8PG G3PAI
IP3 8QW G8EHS
IP3 8SP G3TJU
IP3 8UT M6DME
IP3 9BY M3VCP
IP3 9DE G4POU
IP3 9GB M0ASJ
IP3 9LT 2E0MIY
IP3 9LT M6MIY
IP3 9NJ G3XKU
IP3 9PD M0CSQ
IP3 9QD M6CCZ
IP3 9QF M6AIB
IP3 9RE G4LBU
IP30 0AU M3FDB
IP30 0DG G4ICH
IP30 0JP M6FII
IP30 0QB 2E1BLP
IP30 0QD M6FYB
IP30 0SZ G7MPF
IP30 0TN M6WFE
IP30 0TS G7MLO
IP30 0TS M3XFG
IP30 9AF G1FTD
IP30 9BS G8XOM
IP30 9DQ M3TWS
IP30 9HJ G6AWO
IP30 9JB G0BXP
IP30 9JB G6TLS
IP30 9JB M0RTO
IP30 9PX G3MWO
IP30 9RL 2E0FBD
IP30 9RL M0STI
IP30 9RL M6FBD
IP30 9UF G4JPQ
IP31 1JJ G4UDD
IP31 1NG G4IXQ
IP31 1TB G0DUS
IP31 1TE G1FVU
IP31 2AY G0AHL
IP31 2BN G8YYC
IP31 2BZ M6KWX
IP31 2EE G0MEV
IP31 2EE G0MEZ
IP31 2EF G4SBW
IP31 2EP M3FRX
IP31 2EW G4LIG
IP31 2JQ M6HRN
IP31 2LB G8JSL
IP31 2LE G3DEN
IP31 2NJ G6TDW
IP31 2PL G1CFK
IP31 2QU G4BSA
IP31 2RP G0DUA
IP31 2UA M6XEE
IP31 2UQ M3SII
IP31 3EL G7UXD
IP31 3EN G2TO
IP31 3EN G3LPT
IP31 3HX G3GIH
IP31 3LG G3FOQ
IP31 3PD G4MID
IP31 3PF M3EGV
IP31 3PF M3EGY
IP31 3PF M6DSJ
IP31 3PF M6SJH
IP31 3PW M6NDN
IP31 3PY M6AWI
IP31 3RZ M0DFL
IP31 3SP G4UZF
IP31 3TR M6DSH
IP32 6DF G8VQJ
IP32 6ED 2E1FZH
IP32 6ED G7SDC
IP32 6PF M6AZD
IP32 6PU G6LPD
IP32 6QA G6VPK
IP32 6RJ M6KJY
IP32 6RR 2E0PUG
IP32 6RR M6CHL
IP32 6RU M6BLN
IP32 7AZ M6ERS
IP32 7HR M0PDA
IP32 7HR M6DDP
IP33 1JH G3OWQ
IP33 1RF G3TMX
IP33 1SW 2E0CHQ
IP33 1SW M6AUB
IP33 1YP M6AWO
IP33 2ES M1AKN
IP33 2GB M3VPD
IP33 2GB M6MRC
IP33 2GB M6NET
IP33 2GB M6SJA
IP33 2HZ G8XJK
IP33 2JA G6IGK
IP33 2LS G0DVT
IP33 2LT G3TQX
IP33 2LT M6FVE
IP33 2NJ M6FAK
IP33 2NS M6RGP
IP33 2QB M6AUQ
IP33 2QJ M6OCU
IP33 2QL 2E0KEA
IP33 2QL M3THY
IP33 2SY G4VSB
IP33 2UA M6PEX
IP33 3AT M6JAU
IP33 3DU G1UGH
IP33 3QF G0ARU
IP33 3QJ G1VGI
IP33 3SD M6AUP
IP33 3UB 2E0CWI
IP33 3UB M6CHU
IP33 3XQ G4XSM
IP4 1PJ M1BOP
IP4 1PJ M5BOP
IP4 1PU M6AWQ
IP4 2PJ G4SUU
IP4 2TL G8GCO
IP4 2TS G1SWK
IP4 2UB 2E0CZJ
IP4 2UB M0JBZ
IP4 2UB M0NBA
IP4 2UB M6CZJ
IP4 3AH G4KGO
IP4 3AS M3ZSA
IP4 3BT G3YUJ
IP4 3JH M3UOC
IP4 3LJ G3ZIN
IP4 3NG 2E1CWN
IP4 3PP G6HNN
IP4 4AD G1BEK
IP4 4AD M6EGN
IP4 4BS G8XKT
IP4 4BU M1NIZ
IP4 4EQ 2E0DPX
IP4 4HY M6KLN
IP4 4JJ G4ILN
IP4 4JN G8TPC
IP4 4JX G6PDE
IP4 4LP G4UCX
IP4 4QN G4TVT
IP4 4QP G8WVO
IP4 4QQ M6NCX
IP4 4SF G6FIL
IP4 5AX G4WMF
IP4 5BP M0CCZ
IP4 5DW G6BEH
IP4 5PP G4YVK
IP4 5PQ 2E0XAE
IP4 5PQ M6OXY
IP4 5RH G4ROH
IP4 5RZ G0CWW
IP4 5SA G4BRL
IP4 5SD G4FWA
IP4 5SR M6WUB
IP4 5SX G4ETC
IP4 5TF G7UWB
IP4 5UQ G8CJL
IP4 5UU G3PWB
IP5 1AQ G1YLE
IP5 1AS M6AAK
IP5 1AY G3WIU
IP5 1EB G4AEY
IP5 1ED G3UKW
IP5 1ED G3XDY
IP5 1ED G4FZZ
IP5 1EL G8VVR
IP5 1EN M1BMD
IP5 1JZ M1DLE
IP5 1LU G4VBQ
IP5 1NQ M3YCD
IP5 2DE M6DQD
IP5 2DQ M6EGM
IP5 2DQ M6ENJ
IP5 2DQ M6FCX
IP5 2EP M3YAW
IP5 2FB G0EBQ
IP5 2FB M3ZVF
IP5 2FX M6BFV
IP5 2GB 2E0VKQ
IP5 2GB G6GSG
IP5 2GB M1ADT
IP5 2GB M3ADT
IP5 2GB M3AKQ
IP5 2GP G8LQB
IP5 2GW G4LSQ
IP5 2GW M6CRG
IP5 2XF M3IBZ
IP5 2YN G4FNR
IP5 2YN M3XBL
IP5 2YN M3XPL
IP5 2YR M1DGY
IP5 2YT G0BQO
IP5 2YU M3MXG
IP5 2YU M3PNO
IP5 3QR G3NYK
IP5 3SD M6YAY
IP5 3SH G6XWK
IP5 3SN G8VQH
IP5 3ST G6KWH
IP5 3SU 2E1BIM
IP5 3SU G0NIK
IP5 3SU G3ZID
IP5 3SU G4DWF
IP5 3SY G4BTX
IP5 3SY M1ATJ
IP5 3SY M6EDN
IP5 3TZ M1CVB
IP5 3TZ M3BTZ
IP5 3TZ M3BUA
IP5 3UA G4EQE
IP5 3UF 2E1FWD
IP5 3UL G4ZVW
IP5 3UQ G4OPB
IP5 3UX G4PBE
IP6 0DH G8CZP
IP6 0DZ M6EFC
IP6 0GE G0PEY
IP6 0HQ G7VGE
IP6 0HZ G3TXE
IP6 0ND 2E0YKS
IP6 0ND M3YKS
IP6 0ND M5AEH
IP6 0PY G8ZYM
IP6 0RH M1CVG
IP6 0RJ M0DFQ
IP6 8DP G7VXS
IP6 8ER G4ZYZ
IP6 8JS 2E0FUD
IP6 8JS G3EKE
IP6 8JS M0NIL
IP6 8JS M6DQC
IP6 8RZ G8VHN
IP6 8TF 2E0ZBD
IP6 9AJ M0DNZ
IP6 9AR G0TCP
IP6 9AR M0HDR
IP6 9BE M0RLC
IP6 9ET M6HBP
IP6 9EU M6CDQ
IP6 9HG G1YRE
IP6 9LP G4MRS
IP6 9LP G4PIQ
IP6 9NG G3IRQ
IP6 9NX G7LNI
IP6 9PD G3RHP
IP6 9TD G0IIA
IP7 5BS G0MGN
IP7 5JL 2E0WBS
IP7 5JL M6WHA
IP7 5LJ G8HCZ
IP7 5LX G8UVG
IP7 5SQ G7JVE
IP7 6AW G1XVD
IP7 6FE G3OPB
IP7 6FE G8AAI
IP7 6JH G0PEC
IP7 6JH G0RNN
IP7 6NN G0ORG
IP7 6NN G0WTR
IP7 6PN G0NQK
IP7 7AL G0FJB
IP7 7BZ G4JIE
IP7 7BZ G4MGO
IP7 7DE G6HWT
IP7 7EJ 2E0CNG
IP7 7EJ M6AWN
IP7 7JH G7NGF
IP7 7LR G6MGQ
IP7 7LU 2E0XFH
IP7 7LU M3XFH
IP7 7RW G6UFI
IP8 3AP G0TJY
IP8 3DN G0RRC
IP8 3EY G0DVJ
IP8 3EY G0RGH
IP8 3EY M0GIM
IP8 3HZ G3IVH
IP8 3LF G4IJJ
IP8 3NH G4KYU
IP8 3RR G4TTB
IP8 3RX M6HJZ
IP8 4AU G3MYY
IP8 4HD G4FDF
IP8 4LR 2E0YSP
IP8 4PE G8FFU
IP8 4PP G1XZX
IP8 4PQ G3XOK
IP8 4PU G3JFW
IP8 4SP G3WJI
IP8 4SR G4DMT
IP9 1BA M3TWB
IP9 1BP M0GFJ
IP9 1DX G8PPD
IP9 1HS G0PJO
IP9 1HX G7VPN
IP9 1LL 2E0XKC
IP9 1NP 2E1HKC
IP9 1NW M3GLM
IP9 1QH G0DTC
IP9 1QH G4BXZ
IP9 1QP M1AOF
IP9 1RT M0HKC
IP9 1RW G3GOT
IP9 2DR M3TQD
IP9 2HT G4XOJ
IP9 2HW G8XRW
IP9 2JB G3WRT
IP9 2NF M3FDQ
IP9 2SW G3HMB
IP9 2TT G0LIN
IP9 2XD M1SRH

## IV

### (Inverness)

IV1 1NS GM7RJG
IV1 1XG GM7MWX
IV1 3XG GM3MUA
IV1 3XU GM4LXM
IV1 3XU MM6BDA
IV1 3YG MM0CMO
IV10 8RA 2M0ALS
IV10 8RA GM0UDL
IV10 8RA MM3UDL
IV10 8SQ GM6IYJ
IV10 8XA GM4KLN
IV11 8XF GM4OHY
IV12 4RH MM6YST
IV12 5AZ MM1EHO
IV12 5BX MM1HWB
IV12 5EW GM7BCC
IV12 5EW MM3BCC
IV12 5LF 2M0TAS
IV12 5LF MM3UBD
IV12 5PJ GM0RML
IV12 5PJ MM3SES
IV12 5RA GM1VAD
IV12 5SD 2M0RUP
IV12 5SE GM3PIL
IV12 5SE MM3YOC
IV12 5XX GM1TCN
IV13 7YE MM1MOY
IV13 7YQ MM6EPV
IV13 7YR GM3LVA
IV15 9NJ MM0EFJ
IV15 9PF GM4EWL
IV15 9PF GM8DRA
IV15 9PG GM8LEA
IV15 9RB GM4YGN
IV15 9RL 2M0AYU
IV16 9UZ GM4EKI
IV16 9XH MM6CCW
IV16 9XH MM6GBP
IV17 0QL MM6KNO
IV17 0SZ MM5ALX
IV17 0TR GM0JOL
IV17 0TR GM0SFQ
IV17 0TR GM8DFX
IV17 0TU GM4TTD
IV17 0XL MM3ZRF
IV17 0XN GM0CAD
IV17 0YQ GM6IDF
IV18 0BG GM3SAE
IV18 0EY GM0HNJ
IV18 0GF 2M0WTE
IV18 0PE GM3WOJ
IV18 0PR GM4FDT
IV18 0PR GM4MFL
IV19 1AZ MM3ONI
IV19 1ED MM0TWK
IV19 1LA GM4SUF
IV19 1NF GM3LKY
IV2 3AX MM0IPD
IV2 3DT GM0TNK
IV2 3ET GM4XWS
IV2 3EW GM7BOW
IV2 3HT GM0OMC
IV2 3HT GM1MYR
IV2 3HW MM6CEM
IV2 3LX GM0IQD
IV2 4AZ GM4JAE
IV2 4EN GM0GEE
IV2 4EX MM0BAG
IV2 4LD GM0SXP
IV2 4LD GM8DFC
IV2 4NL MM6DUV
IV2 4NL MM6GON
IV2 4XJ GM7RVR
IV2 5AQ MM0BGW
IV2 5DU GM4GKH
IV2 5EP MM0DHY
IV2 5EQ GM4OIJ
IV2 5EQ GM7IBM
IV2 5ER GM4FKD
IV2 5ES GM4UOD
IV2 5RG MM1ATR
IV2 5UE GM0NTI
IV2 5UE GM0OTI
IV2 5UE GM8RTI
IV2 6AX GM4VIK
IV2 6XJ MM6AAA
IV2 7HB MM1AEL
IV2 7HN GM4ZIT
IV2 7ND MM3MEH
IV2 7RW GM1BQP
IV2 7SR MM0BQL
IV2 7ST MM3IAG
IV20 1RX GM0CSZ
IV20 1XP MM6DJC
IV21 2BX MM6MJR
IV21 2DH GM4CHX
IV22 2HB GM0WXX
IV25 3PW MM0DVZ
IV25 3RT MM3GPL
IV26 2TB 2M0UAL
IV26 2TB MM6ULL
IV26 2TX GM6AJA
IV27 4AD GM7WHQ
IV27 4DG MM1YAM
IV27 4ED GM0IYP
IV27 4ED GM4VVX
IV27 4ED MM3MMB
IV27 4EG 2M1ECF
IV27 4JB GM6ZCX
IV27 4JB GM6ZCY
IV27 4JD GM4BYT
IV27 4JQ GM8IEM
IV27 4PQ GM0HLV
IV27 4QA GM6JOS
IV27 4RP MM0URN
IV27 4TG GM6JHH
IV27 4XD MM0WRL
IV27 4XQ MM3SCO
IV28 3XE MM1WKD
IV3 5AH MM3HJC
IV3 5BD 2M0HUD
IV3 5ES GM4UZR
IV3 5RE MM3PTS
IV3 8DN MM6HTS
IV3 8LB GM7HQW
IV3 8LH GM0AKJ
IV3 8PF GM7GVD
IV3 8SD GM0RIV
IV3 8SD GM0VPG
IV3 8SD GM4JNB
IV3 8SJ 2M0JAT
IV3 8SR GM8RMR
IV30 1SF GM6WLJ
IV30 1TB GM0UYZ
IV30 2YR GM0LOK
IV30 2YR GM7CPY
IV30 4BU MM3GLH
IV30 4DJ GM0MZD
IV30 4HJ MM1FAS
IV30 4HL GM7FPN
IV30 4HN GM7LSI
IV30 4HX 2M0CFB
IV30 4HX GM0MGO
IV30 4HX MM0GYX
IV30 4HX MM6AON
IV30 4JH GM0DQV
IV30 4JH GM8FFK
IV30 4LY GM4ILS
IV30 4NB GM8YKT
IV30 4NS GM0LPB
IV30 5PQ MM3FJA
IV30 5QN MM1GBS
IV30 5SU GM0OTS
IV30 5XY 2M0REH
IV30 5XY MM0TQH
IV30 5XY MM3TQH
IV30 5YD MM6CPE
IV30 5YE GM7LWA
IV30 5YN 2M0CXG
IV30 5YN MM6CXD
IV30 6EJ GM4EWM
IV30 6EJ MM0CJH
IV30 6ER GM1ASY
IV30 6GL MM6YGT
IV30 6JP 2M0BZB
IV30 6JY 2M0BPM
IV30 6JY MM0LER
IV30 6JY MM3WLA
IV30 6XF MM0JMB
IV30 8FE GM0GNY
IV30 8JU GM0ONN
IV30 8JZ GM0RYA
IV30 8NH GM0LDX
IV30 8NY 2M1HCP
IV30 8NY 2M1HCQ
IV30 8NY MM3BWV
IV30 8PE GM0AEG
IV30 8PE MM0HAR
IV30 8SY GM0EMC
IV30 8SY GM6TUE
IV30 8TB 2M0WMJ
IV30 8UA GM6IKN
IV30 8UR GM3VBY
IV31 6BA GM4PGM
IV31 6BA MM3ZSK
IV31 6JZ GM3UQU
IV31 6NA GM4HMN
IV31 6QH GM4MKU
IV31 6QQ MM6HFC
IV31 6RE 2M1HLE
IV31 6TP GM6PLG
IV32 7EF MM0EFW
IV32 7EH MM0DAT
IV32 7HF 2M0JVR
IV32 7JH MM3TVQ
IV32 7LU 2M0GYM
IV32 7LU MM3UBR
IV32 7NL MM0CQT
IV32 7NW 2M0DTJ
IV32 7NW MM6CYQ
IV32 7PX MM6IVP
IV32 7QS MM3XNP
IV327HJ 2M0BIK
IV36 1BQ 2M0MMM
IV36 1JB GM8DPV
IV36 1JL 2M0SWY
IV36 2SG GM0KNT
IV36 2UF GM0GHN
IV36 3FG 2M1DHG
IV36 3UA GM6NOO
IV36 3UA MM3XJW
IV4 7AB MM0LOZ
IV4 7AB MM3PSL
IV4 7AE GM7DXE
IV4 7AH 2M0XCT
IV4 7AW GM6LKQ
IV4 7EY GM4TJD
IV4 7EY GM4UMA
IV4 7GL 2M0DAC
IV4 7HT 2M0LWB
IV4 7HT MM0VSU
IV4 7HT MM6LWB
IV4 7HZ GM7ORJ
IV4 7HZ MM3SRK
IV4 7JQ GM4GZD
IV40 8BB MM6ZDY
IV42 8PY GM8RBR
IV44 8RD GM0HBK
IV44 8RF GM0MAC
IV44 8RQ GM3YAC
IV47 8SH MM6DTV
IV47 8SN GM0PRQ
IV49 9AB GM0JFL
IV49 9AN GM0MYV
IV49 9AQ GM1RLV
IV51 9DN GM3SWK
IV51 9EG GM4TRH
IV51 9JX MM0LUP
IV51 9NX MM6IAY
IV51 9PE GM0EWX
IV51 9PW GM0DXE
IV51 9YN MM3ZDI
IV52 8TN MM0GWO
IV53 8UX MM6TUG
IV55 8WF GM4AEK
IV55 8WL GM4RXD
IV55 8WP GM3MTW
IV55 8WS GM0CVD
IV55 8WS GM4WZD
IV56 8FJ GM4PUS
IV6 7PX MM3LQK
IV6 7QG MM0GKB
IV6 7RS GM3WZV
IV6 7SB GM0OGZ
IV6 7XN GM4RRP
IV63 6TN GM0BZS
IV63 7YJ GM6JJN
IV7 8AW 2M1DZX
IV7 8AW GM4SFW
IV7 8JS GM0NAI
IV7 8JS MM0APF
IV7 8LB GM3WED
IV7 8LH GM4FIZ
IV7 8LL GM0JFK
IV8 8PA GM7JYW
IV8 8PG GM0TQB
IV9 8PR 2M1EBJ
IV9 8QL GM1OVJ
IV9 8QL GM4MAI

## JE

### (Jersey)

JE2 3DE MJ0RGR
JE2 3GQ 2J0YAY
JE2 3GQ MJ6DEY
JE2 3JF GJ0NTD
JE2 3ND GJ7RIY
JE2 3RX MJ6ADQ
JE2 3XQ MJ0SIT
JE2 3ZA 2J0RZD
JE2 3ZA MJ0RZD
JE2 3ZA MJ3RZD
JE2 3ZB GJ4YMX
JE2 4GL GJ3YLI
JE2 4PA GJ1EXC
JE2 4PX GJ7DNI
JE2 4RS MJ0AQJ
JE2 4SA MJ3HWC
JE2 6FP GJ3XOJ
JE2 6FW GJ3XZE
JE2 6GF 2J0SZI
JE2 6GF MJ3SZI
JE2 6GH MJ3CMB
JE2 6GH MJ3GBJ
JE2 6NY GJ3DVC
JE2 6NY GJ3ECC
JE2 6NY MJ0JER
JE2 6NY MJ3IOJ
JE2 6SE GJ1YOT
JE2 7HP GJ3SND
JE2 7HQ GJ8PVL
JE2 7LT GJ0JSY
JE2 7LX MJ1CYD
JE2 7LX MJ3CDP
JE2 7PP MJ6SSF
JE2 7RL GJ6SNQ
JE2 7RT GJ7HTV
JE2 7RT MJ0BJU
JE2 7TA GJ7FGS
JE2 7TW 2J1CWG
JE2 7TW 2J1CWH
JE2 7TW GJ7UIT
JE2 7TX GJ7LJJ
JE2 7TZ GJ8RRP
JE2 7UD GJ4HSW
JE3 1EZ GJ4YBM
JE3 1FR GJ0KYZ
JE3 1GP GJ4CBQ
JE3 1GP MJ3JBQ
JE3 1GP MJ3KBQ
JE3 1GP MJ3MBQ
JE3 1GT GJ0PDJ
JE3 1GT GJ3IT
JE3 1GT GJ8RVT
JE3 1LA GJ3LFJ
JE3 1LE GJ0VJP
JE3 1LY MJ1COO
JE3 1NH MJ6THP
JE3 1NL MJ6RBI
JE3 2BJ GJ7AOG
JE3 3BQ GJ0FYB
JE3 5AA 2J0COQ
JE3 5AA GJ7DNJ
JE3 5AA MJ0LEL
JE3 5AA MJ6BDJ
JE3 6AS GJ4YLP
JE3 6AT GJ3AME
JE3 6ED GJ1TJP
JE3 6ER 2J0ODX
JE3 6ER MJ0PMA
JE3 6ER MJ6AUD
JE3 7AD GJ7SLU
JE3 7AQ GJ7DTA
JE3 7AZ GJ4BCC
JE3 7DT GJ6ENP
JE3 7ES GJ6WMZ
JE3 7YZ 2J0DXA
JE3 7YZ MJ0ULE
JE3 8AL MJ6VAA
JE3 8BL MJ1EPG
JE3 8BN 2J1EDR
JE3 8DP MJ3HMA
JE3 8GB GJ4JVP
JE3 8GP GJ4KBM
JE3 8GP GJ4ODX
JE3 8GP GJ6TPD
JE3 8GQ GJ7RWT
JE3 8HN GJ3YHU
JE3 8LS MJ6ORG
JE3 8NS MJ1CNB
JE3 9EF GJ8CEY
JE3 9EP GJ3YLN
JE3 9EP GJ8PCY
JE3 9HS GJ1KCB
JE4 9VI GJ4WRR

## KA

### (Kilmarnock)

KA1 1UF 2M0YAF
KA1 2HP GM0AAX
KA1 2HP MM3TYA
KA1 3AR MM3LWT
KA1 3DZ 2M0BSE
KA1 3LQ GM0ADX
KA1 3LQ GM3OZB
KA1 3QL MM3OJV
KA1 3RS GM0DYF
KA1 3RY GM0FKP
KA1 4EB MM3FET
KA1 4LQ GM4UYP
KA1 4PB GM0FQQ
KA1 4PB MM3VSC
KA1 4QN 2M0YFR
KA1 4QN MM3YFR
KA1 4QT GM0LYH
KA1 5AB GM7AAJ
KA1 5ND GM4CAM
KA10 6DA 2M0CMA
KA10 6DA MM0YET
KA10 6DA MM6AMA
KA10 6DS 2M0SOE
KA10 6NJ 2M0BMN
KA10 6NJ MM0CKF
KA10 6NJ MM3NNO
KA10 6SD GM4XRY
KA10 6TB GM3NIG
KA10 6TT GM7VXR
KA10 6TZ GM4SLY
KA10 6UG GM6BAO
KA10 6UP GM4IGS
KA10 6XE GM8GIQ
KA10 7AH MM0KJM
KA10 7DY GM6JEP
KA10 7EE GM4CXF
KA10 7HA GM4BIT
KA11 1BD GM0SEI
KA11 1BD GM7KSA
KA11 1BD MM0JJV
KA11 1BW GM7UAC
KA11 1BY MM3WNP
KA11 1EQ MM0BQJ
KA11 1HH MM6LIL
KA11 1JD GM0DWH
KA11 1LR GM6JOD
KA11 1LT MM6PDX
KA11 1NJ GM3JOB
KA11 1NW GM0NTY
KA11 1PN GM0JQE
KA11 2BF MM0UKW
KA11 2BY MM3PEV
KA11 2EB GM1VJD
KA11 2EU GM0DNG
KA11 3AL MM1DQW
KA11 4BN MM6JFU
KA11 4EB MM3OIZ
KA11 4ER GM0WUX
KA11 4EY 2M0WBJ
KA11 4EY MM0HYM
KA11 4EY MM6WBJ
KA11 4FB 2M0NLA
KA11 5AX MM6FPI
KA12 0QE GM7VFR
KA12 0QG GM7CZU
KA12 0QS MM1CXO
KA12 0SE GM3NYG
KA12 0SE GM4PGV
KA12 0TR GM7KHA
KA12 0XA MM3UOE
KA12 0XH 2M0BMF
KA12 8DR GM0FCI
KA12 8RZ MM0CTT
KA12 8RZ MM0CTU
KA12 8RZ MM3TNG
KA12 8TD GM6HGF
KA12 8TD MM0XPG
KA12 9DN GM1ASA
KA12 9DT MM3TWA
KA12 9ER MM3WUI
KA12 9JJ MM3LDR
KA12 9LU MM6DAF
KA12 9PA MM3UOS
KA13 6BE MM0DHQ
KA13 6BE MM0KLR
KA13 6EB GM1MSN
KA13 6JJ MM0HLN
KA13 6LU MM3YUU
KA13 6PL GM3YQK
KA13 6QN MM0BRG
KA13 6WH MM3YUY
KA13 7AR GM0TAE
KA13 7BA MM6WOC
KA13 7DT GM0UGG
KA13 7HE GM1MSO
KA13 7JN MM0CKK
KA13 7LQ MM1FHR
KA13 7ND MM1LBA
KA13 7PQ MM3VFK
KA13 7PT GM4KQS
KA14 3AJ GM8MMW
KA14 3BQ GM1AXI
KA15 1AR GM6OHF
KA15 1JE GM7DTC
KA15 2AJ GM4VHZ
KA15 2BA GM7DHA
KA15 2BE GM1FLQ
KA15 2DZ MM3OGS
KA15 2HG GM4LIS
KA15 2LN GM6UHE
KA15 2LN GM8NVG
KA16 9BJ GM6KPL
KA16 9HZ MM0OVK
KA17 0DQ GM4ZNS
KA17 0EE MM3OJE
KA17 0LP 2M0UEA
KA17 0LP GM0VBE
KA17 0LP GM1FNX
KA17 0LP MM3UEA
KA18 1HA MM0JOK
KA18 1HA MM3EJB
KA18 1HN MM3GQT
KA18 1NP MM3PHC
KA18 1PU GM0BKX
KA18 1PU GM0NET
KA18 1PU GM1MDO
KA18 2LL GM1TCP
KA18 2LL GM1YKE
KA18 2NZ MM1ANP
KA18 2RE GM0AYR
KA18 2RE GM7OKX
KA18 2RJ MM6EBZ
KA18 3EY GM4XFU
KA18 3GA GM0DBK
KA18 3HS GM4XMD
KA18 3TA GM0GRD
KA18 4HH GM7ANE
KA19 7AE 2M0BET
KA19 7AE MM3BOJ
KA19 7AU MM6JTG
KA19 7HF MM6ATU
KA19 7HZ 2M0GYN
KA19 7HZ MM6EBD
KA19 7RE MM0TFU
KA19 8AX MM6AAM
KA19 8AY MM6IMP
KA19 8EN MM6BNS
KA19 8ES GM0DHZ
KA19 8LR GM3YDN
KA197AE MM0LGR
KA2 0AY MM6PFT
KA2 0BZ GM7KIY
KA2 0DS MM6MHN
KA2 0ES 2M0DTZ
KA2 0ES MM0GKN
KA2 0LD MM3DHG
KA2 9DX MM3EJV
KA2 9EW GM7IUF
KA2 9EX GM4VKI
KA2 9EZ GM8JUY
KA2 9HZ MM3DHE
KA20 3BF MM6MNW
KA20 3DG GM4WNQ
KA20 3LG MM3RTH
KA20 3PQ MM6SGZ
KA20 4BB MM3UVF
KA20 4BG 2M0YIO
KA20 4BP MM6NAB
KA20 4EF MM3PXO
KA20 4HN MM0CWB
KA20 4HN MM6CCS
KA21 5DA MM6VFL
KA21 5DG MM6DHS
KA21 5NH GM0XAV
KA21 5PS MM3AQM
KA21 6AA MM0IBJ
KA21 6AA MM0ZIF
KA21 6AB 2M0JCG
KA21 6AB MM6AVE
KA21 6AQ GM0OTB
KA21 6AQ MM6CKC
KA21 6BE MM6GKE
KA21 6HY GM4UEH
KA21 6JX GM0KZG
KA22 7AJ MM3YBD
KA22 7BZ MM6BXT
KA22 7DY 2M0OVD
KA22 7DY MM0OVD
KA22 7ER MM3YFT
KA22 7LU MM3DKA
KA22 7NQ GM7VSB
KA22 7NQ MM6ARN
KA22 8LW MM3NMI
KA228HA MM3UOR
KA23 9BX GM0EFC
KA23 9BX GM0EFD
KA23 9JT GM1MQE
KA23 9LB GM0AYT
KA23 9LB MM6OOP
KA23 9LE GM3LLP
KA23 9LX GM3JIG
KA23 9NF GM0DEX
KA24 4AF GM0NHL
KA24 4DJ MM1EVJ
KA24 4HP MM0SVE
KA24 5HP MM6IIO
KA24 5HP MM6LJE
KA25 6AB MM3XXI
KA25 6EY GM0ODB

KA25 7JB MM6CCO
KA25 7LB GM0HQF
KA26 0AA 2M0NCM
KA26 0AA MM3GZG
KA26 0AA MM5DCM
KA26 0AA MM6CNC
KA26 0AE MM3GIR
KA26 0BP GM0FSZ
KA26 0BX GM0NBG
KA26 0BY MM6CTH
KA26 0EA GM0UDY
KA26 0EB MM1DWU
KA26 0EF MM0CBL
KA26 0NQ GM4ZTO
KA26 0PA GM4WEW
KA26 9AH GM0JMO
KA26 9AH GM7SAK
KA26 9AH MM0SAK
KA26 9AH MM3UDB
KA26 9DZ GM6OFB
KA26 9EL GM0KCY
KA26 9EL GM3VNW
KA26 9EU MM6ZUY
KA26 9JH GM0KWW
KA26 9LP GM4TPQ
KA27 8EW MM6RLD
KA27 8PG GM8LQL
KA27 8PR GM4BVZ
KA27 8QH GM3HEN
KA27 8QH GM4JFH
KA27 8QH GM4XJF
KA27 8RL GM3UA
KA28 0BL MM6WMM
KA28 0DP GM8BJJ
KA29 0AJ GM3PKV
KA29 0DG GM3DOD
KA29 0DP MM0JBS
KA3 1PZ GM0DJG
KA3 1PZ GM3USL
KA3 1TU GM4ZNC
KA3 1TU MM0TJT
KA3 2AS 2M0OOT
KA3 2AS MM3OOT
KA3 2DA GM0MZH
KA3 2DW GM4UYK
KA3 2EQ GM0JFH
KA3 2EZ GM6WTH
KA3 2GJ MM6VDP
KA3 2GN MM3LZU
KA3 2GP MM0GNS
KA3 2HU 2M0ONW
KA3 2JG GM0ONX
KA3 2JG MM3ONX
KA3 2JG MM6PGT
KA3 2RS 2M0VNW
KA3 2RS MM0ABB
KA3 2RS MM3VNW
KA3 3BP MM6SHM
KA3 3BT 2M0TSR
KA3 3DQ GM3AXX
KA3 3HG GM4VAY
KA3 3HT GM4OSS
KA3 4BP GM0GOV
KA3 4BP MM3LZD
KA3 4BP MM3YWZ
KA3 5AT MM3WGW
KA3 5DA MM6SUR
KA3 5JT GM0PDQ
KA3 5JT GM4HCO
KA3 5JT MM0MBH
KA3 6DB GM0OXS
KA3 6ES 2M0BEC
KA3 6ES MM3IOF
KA3 6FJ MM6GLO
KA3 6HJ GM0LYO
KA3 6JT GM0DVH
KA3 7DT GM4SQM
KA3 7DT GM6JNJ
KA3 7JB GM8ZEJ
KA3 7NT 2M0AAQ
KA3 7NT MM6WBP
KA3 7RB MM3JGT
KA30 8EF GM8EUG
KA30 8ER GM0BFW
KA30 8HU GM1RJS
KA30 8LE GM1LXM
KA30 8PG GM4ZZW
KA30 8QQ MM6JHH
KA30 8QU GM0NUI
KA30 8RQ MM3XIW
KA30 8SJ MM0RFA
KA30 8SN GM3MWX
KA30 8SN GM6XW
KA30 8TH MM0TCP
KA30 9BZ GM1BKR
KA30 9EA MM3WWQ
KA30 9EQ GM0DWY
KA30 9ER MM0MNS
KA30 9ET GM0GMN
KA30 9FT MM3YPH
KA30 9EX GM0RYD
KA4 8EA 2M0GCS
KA4 8EA MM0GHM
KA4 8EA MM0RAI
KA4 8EA MM3GDC
KA4 8EJ GM0KAZ
KA4 8HT GM8MNR
KA4 8NA GM4WZL
KA4 8NA MM3WZL
KA5 5AE GM1XBK
KA5 5JE GM0GRW
KA5 5QU GM7RPT
KA5 6DT GM7OPN
KA5 6HU MM0HSV
KA5 6HY MM3GSL
KA6 5BE MM0XCP
KA6 6BH GM8KXF
KA6 6EL GM1VPQ
KA6 6LD MM3FWF
KA6 6LR GM4PPT
KA6 6ND GM4SFA
KA6 6NL MM3WNH
KA6 6QG GM0POD
KA6 7AU 2M1FQI
KA6 7AU MM3FQI
KA6 7AU MM3WVQ
KA6 7DG MM0CNF
KA6 7EF GM1VXE
KA6 7LD GM6MD
KA6 7ND GM1OST
KA6 7PS GM0JBE
KA6 7RN GM0SDS
KA6 7SJ MM3JGR
KA6 7TX GM1VDZ
KA6 7UA MM6POV
KA7 2DJ GM4WFV
KA7 2LW GM4FGS
KA7 2NF 2M0LXX
KA7 3BU MM3DZG
KA7 3BZ GM7SXI
KA7 3DT GM4HSR
KA7 3JB GM0TBH
KA7 3JJ MM0MLO
KA7 3NF GM7OIN
KA7 3PE GM4SQO
KA7 3PE MM6OIR
KA7 3QJ MM0BPF
KA7 3RL MM6CPK
KA7 4EG GM3PMB
KA7 4EG GM5VG
KA7 4EQ GM0EPO
KA7 4ET MM0RAG
KA7 4JB GM7KXJ
KA7 4PB MM1JAS
KA7 4QF MM6RSP
KA7 4QN GM6BHR
KA7 4XA MM3HQL
KA8 8NX MM6BXR
KA8 9BW MM6BCF
KA8 9RD MM1DPH
KA9 1HY MM6CLM
KA9 1JW MM3WVN
KA9 1LT MM0GAI
KA9 1TT GM3WIL
KA9 2DB GM4TPX
KA9 2DL 2M0MOF
KA9 2DL MM6TCS
KA9 2ED GM0DEQ
KA9 2EQ MM3EDW
KA9 2HE GM3MQO
KA9 2HE GM4OOU
KA9 2HY GM4DOZ
KA9 2PW GM4LVW

## KT
### (Kingston Upon Thames)

KT1 2RU G3RLT
KT1 3EH G3MNB
KT1 3PS G8VXB
KT1 3QB M0BEJ
KT1 3RW G8SIK
KT10 0AZ G0EYT
KT10 0BJ M0XDV
KT10 0DT G4IUA
KT10 0DW G6FWO
KT10 0DW M6TLL
KT10 0HS G3RIM
KT10 0JZ G1PQT
KT10 0QE G4GPB
KT10 0RZ G8YXZ
KT10 8DL M0PLP
KT10 8PU G3YMN
KT10 8PY M6OKI
KT10 9EG 2E0KXX
KT10 9EG M6KXX
KT10 9PH G0IDB
KT11 1AU GCLIO
KT11 2AQ G1LKH
KT11 2BB G0OAS
KT11 2BH M1CNH
KT11 2LD G8ORX
KT11 2SX G4YKH
KT11 3HQ G8ETP
KT11 3HR G1UNB
KT12 1BB G4ARO
KT12 1LE G3LFX
KT12 1LW M6GZJ
KT12 1NF G8GTR
KT12 2AT G6AYS
KT12 2HS G0HJD
KT12 2LB G6CLK
KT12 2LY G3EAO
KT12 2SJ 2E0TIG
KT12 2SJ M6TIG
KT12 2SU M3FJE
KT12 3AW M1IAN
KT12 3AY 2E0TTL
KT12 3AY M0VLF
KT12 3BA G0UHQ
KT12 3BA G1OCY
KT12 3DA G4AWZ
KT12 3EQ G8DXP
KT12 3QH M1AEJ
KT12 3SQ G6INX
KT12 4HQ G1JKV
KT12 5DT M0AEJ
KT12 5HF G1KEV
KT12 5PH M6KSI
KT12 5PH M6NBY
KT12 5QG 2E1RJS
KT13 0BL G7SQH
KT13 0BS G3TXF
KT13 0JW G0CKV
KT13 0NR G4CXL
KT13 0NR G8HCL
KT13 0TB M6RDD
KT13 8AS G4BSV
KT13 8AS G4CPM
KT13 8PE G3NMH
KT13 8PU G6CUV
KT13 8UP G4LQD
KT13 9AR 2E0SKK
KT13 9AR M0RND
KT13 9AR M6AGG
KT13 9AT M3XSZ
KT13 9LS M0ZCO
KT13 9NU G8IPN
KT13 9RU G0JRH
KT14 6AE M0HMF
KT14 6DT G6JRZ
KT14 6DY M6AQW
KT14 7EF G6XAN
KT14 7HX G4EVA
KT14 7HY G3GLB
KT14 7NG G0PBL
KT14 7TG G0OBG
KT15 1DH G4WMP
KT15 1NX M6FDX
KT15 1SR G7HIN
KT15 2AX M6XYH
KT15 2DE G8HFW
KT15 2DN G0KEY
KT15 2UD M6DOG
KT15 3AA M6CUS
KT15 3AQ M3WQA
KT15 3DJ 2E0ABT
KT15 3DJ G4CCZ
KT15 3DL G1SMB
KT15 3DZ M0DJW
KT15 3DZ M1VPL
KT15 3ET G8JSN
KT15 3EY G1IDE
KT15 3EY G3UES
KT15 3EY M0SAR
KT15 3HS G0PVF
KT15 3LH 2E0BSM
KT15 3LH M0GPH
KT15 3LH M3YFI
KT15 3LX G3XSZ
KT15 3NB M0XMC
KT15 3SE G8CUG
KT15 3TU G6YHE
KT15 3TU G8KQH
KT15 3UA G0RMU
KT16 0BP G0EMT
KT16 0ER G1UMY
KT16 0HD G6KJE
KT16 0PH G7DUK
KT16 8AP M0DXP
KT16 8BU G8BWA
KT16 8EP G0PCZ
KT16 8HU G3ZNV
KT16 8JQ M6SEJ
KT16 8LE M0BSF
KT16 8QB G1AEA
KT16 8RA G7MYO
KT16 9DE 2E0BJX
KT16 9DE M0GPG
KT16 9DE M3UQJ
KT16 9ED G6YRI
KT16 9HP M6KSV
KT16 9HW G7WDN
KT16 9HX M6OIC
KT16 9JJ G8NYB
KT16 9PF G8CYL
KT16 9PF M3HJW
KT16 9PH G3OLH
KT17 1EJ 2E0WOL
KT17 1EJ M3XEI
KT17 1QP G8HPF
KT17 1BX G4VHJ
KT17 2HD G0DGF
KT17 2HS G4CPQ
KT17 2NG G7PBC
KT17 2NW G0CCK
KT17 2PP G6RHB
KT17 3LN G1ZVE
KT17 3NL G3LQP
KT17 3PU G1OXB
KT17 4EN G8EPS
KT17 4JP G4JSW
KT18 5JL G3TVL
KT18 5PB M6AQR
KT18 5PY G4IQV
KT18 5QT G4GHI
KT18 6HY G4YAR
KT18 7BU 2E0DQO
KT18 7BU M6EZS
KT18 7JE 2E0DZF
KT18 7JE G6KVG
KT18 7JE M3GQL
KT18 7RS G1NKN
KT18 7SL 2E1FYC
KT19 0AU G4ZSW
KT19 0BJ G8KWV
KT19 0HD M0AAF
KT19 0HF G8MTI
KT19 0LB G4ROI
KT19 0LS M0ZEY
KT19 0LU M6ITN
KT19 0PQ G8EPQ
KT19 0HJ G8YIJ
KT19 0SZ G0PFA
KT19 7EW G6REH
KT19 7LD G1DPW
KT19 8BW M3KQC
KT19 8HG M3IDB
KT19 8HH M6JHO
KT19 8JX G3YGG
KT19 8LU G0WZM
KT19 8RP G1LQC
KT19 8RP G4SYT
KT19 9DP G4WFL
KT19 9DP G6LFW
KT19 9EE G0KAS
KT19 9HD 2E0BJL
KT19 9HD M3ZLL
KT19 9HH G1IKG
KT19 9LB M0GDU
KT19 9LD M0TBS
KT19 9PQ G8NXB
KT19 9QR G0OXZ
KT19 9SY G0ROT
KT19 9TJ G0TNQ
KT19 9UL G4RSU
KT2 5GG G4PPN
KT2 5HH G6OBA
KT2 5NE M6MUF
KT2 5QB G8VOY
KT2 5TL M3XXE
KT2 5TU G6RFU
KT2 5TU M0ETA
KT2 6AH M3WEJ
KT2 6AQ 2E0MET
KT2 6PN G6JTD
KT2 6RA G8YJQ
KT2 7AL G6ZLS
KT2 7JT G0TRD
KT2 7QD G4GQA
KT20 5EW G6HVQ
KT20 5JF G4WNV
KT20 5JQ G7PHK
KT20 5PS M0JHM
KT20 5QQ G6AVP
KT20 5RZ G3SUX
KT20 5SQ G0XTM
KT20 5UA G4XXI
KT20 5UT G6ALG
KT20 6BS G3NPC
KT20 6BS M3URA
KT20 6ET G0SWN
KT20 6ET G6CJR
KT20 6ET G8WGP
KT20 6XE G4HVO
KT20 6XE G6GGY
KT20 7AD G7KWF
KT20 7BA G4BYZ
KT20 7LS G8GGS
KT20 7LZ G4NLB
KT20 7QE M6ZLP
KT20 7UD G8CCQ
KT21 1BE G4JUM
KT21 1EB G3KWJ
KT21 1HY 2E0DKP
KT21 1HY M0LUT
KT21 1HY M6LUT
KT21 1JL G1PGH
KT21 1NA G6TZE
KT21 1NE G1HWR
KT21 1NE G4KGE
KT21 1NN M0JSN
KT21 1PD G7LSB
KT21 1PY G6QN
KT21 1SG G7RSK
KT21 2EG G6DTW
KT21 2HB G8ZZR
KT21 2HU G4CLY
KT21 2HY G4XQE
KT21 2NS M3NIZ
KT21 2PG G4NHO
KT21 2SY G0CIT
KT21 2TL G8FXN
KT21 2UF G6BQQ
KT22 0HR G1BHS
KT22 0HR G0LYX
KT22 7EE M6PIC
KT22 7SL 2E0ZEH
KT22 7SL M0ZEH
KT22 7SL M3ZEH
KT22 8AR G7MAV
KT22 8JT G3ZQT
KT22 8LR M0CDJ
KT22 8LZ M1KOS
KT22 8NU G8MHN
KT22 8RD G4EYB
KT22 8UJ G4XYC
KT22 9AZ G3NFV
KT22 9AZ G8GWM
KT22 9EF M0CUB
KT22 9JR G0UZX
KT22 9NB G6BTC
KT22 9ND G3VVW
KT22 9NN G4CMR
KT22 9NT G3XKG
KT22 9QP G8OHP
KT22 9QT M6XTR
KT22 9TL M3FZE
KT22 9XA 2E0GBK
KT22 9XA M6BMB
KT23 3DA 2E0CDK
KT23 3DU G1FCU
KT23 3EY G4FDU
KT23 3JF M3CSI
KT23 3JY G4LHN
KT23 4BJ G1BJN
KT23 4BU G4AXN
KT23 4BS G8IWT
KT23 4BS G8KW
KT23 4DB G3UZW
KT23 4HP 2E0ESU
KT23 4JX G3SIA
KT23 4ND 2E0LTR
KT23 4ND M6ZLC
KT23 4RR G4CVN
KT24 5BU G3ENI
KT24 5BU G3TVS
KT24 5DY G6GEK
KT24 5SD G0RAL
KT24 6ED G5LK
KT24 6ED G7NJI
KT24 6ED G7RAT
KT24 6ED M1BFG
KT24 6QJ G3NIW
KT24 6QN G4PED
KT3 3BA G4OFO
KT3 3BX M6OMS
KT3 3DY G4JUZ
KT3 3HP G3TON
KT3 3HP G4FJP
KT3 3HZ G8BPY
KT3 3HZ M3BPY
KT3 3LB M3WZR
KT3 4HR G8GGI
KT3 4LD 2E0WNI
KT3 4LD M0WNI
KT3 4LD M3WNI
KT3 4LE G3XXH
KT3 5BZ G7UVF
KT3 5DE 2E0CEB
KT3 5DE M6HVS
KT3 6BS 2E0GRF
KT3 6BS M0HWS
KT3 6BS M3ZJV
KT3 6BY M1CEA
KT4 7DX G4ADM
KT4 7DX G7IQZ
KT4 7JQ G7LCS
KT4 7PH M3XWY
KT4 7RB 2E0WEE
KT4 7RB M6FRO
KT4 7SJ 2E0FMY
KT4 7SJ G0KRT
KT4 8EZ G4VLN
KT4 8NB G4BQY
KT4 8SN G7RBC
KT4 8UD M0BMZ
KT4 8XG G6VAX
KT4 8XU G7VNM
KT4 8XY G7OAJ
KT4 8YA G1LGB
KT4 8YA M0CLG
KT5 8JJ G6DKS
KT5 8JZ G8KEJ
KT5 8TS G0IRK
KT5 9BS G6SMJ
KT5 9DX G0JOS
KT5 9EA G7EKC
KT5 9EA M3EKC
KT5 9HA G3KQR
KT5 9JH G4PPU
KT5 9LJ 2E0OYN
KT5 9LJ G1KAG
KT5 9LJ M3OYN
KT5 9LW G7MLJ
KT5 9PH M3VXH
KT5 9RJ G0BQV
KT6 4DH G2SR
KT6 4DH G7JYQ
KT6 4DH M1SRC
KT6 4EX M6KFG
KT6 4SW M0SHA
KT6 5AF G0PCY
KT6 5JD G3JVC
KT6 5JN G1SKR
KT6 5RE 2E0MJO
KT6 6LJ G0SAC
KT6 6LJ G4WGE
KT6 6RJ G4LJI
KT6 7JT G0LFB
KT6 7LL M6SAL
KT6 7NA G1OEP
KT6 7NR G0KEB
KT6 7PP G8KFS
KT6 7QG G7BHH
KT6 7QH M6TCD
KT6 7SZ G7STL
KT6 7UN M0DWG
KT6 7UN M3HZK
KT7 0BL M0DHP
KT7 0DU M0IHN
KT7 0YH G7WEP
KT7 0YN G4CPV
KT8 1QF 2E0JAF
KT8 1QF M6CXF
KT8 1RR G3OJX
KT8 1SU 2E0GPH
KT8 1SU M0PGH
KT8 1SU M6GHT
KT8 1SU M6PHT
KT8 1TE G7HMU
KT8 2ET G1DZB
KT8 2ET G4YKK
KT8 2EX G8CLW
KT8 2HY G0KDI
KT8 2PY G1YVV
KT8 9AF G8WKT
KT8 9AN M0CGW
KT8 9DG 2E0LFO
KT8 9DG 2E0LJJ
KT8 9DG M3NCP
KT9 1BW G0DOE
KT9 1BY M6HLW
KT9 1EF M0AWN
KT9 1JY G0NCH
KT9 1NL G4HJY
KT9 1PL G6XKF
KT9 1PN G0WIT
KT9 2BN M6JFN
KT9 2BU G1SIP
KT9 2DE G1RMC
KT9 2DE G4LBM
KT9 2DE G7GLR
KT9 2DE G8RWH
KT9 2DT G6DFM
KT9 2EY G1JRR
KT9 2EY G1NYP
KT9 2HG G6KIE
KT9 2LA G0SDF
KT9 2NN G6HMG
KT9 2QD G3SXW

## KW
### (Kirkwall)

KW1 4DN GM7BNF
KW1 4DN GM7REF
KW1 4NT GM4JUE
KW1 4PF MM6EER
KW1 4PF MM6EEX
KW1 4PN MM6DRJ
KW1 4PN MM6JUE
KW1 4PN MM6WKC
KW1 4QT GM1GXH
KW1 4RD GM7FDS
KW1 4RG GM4MPR
KW1 4RX GM8EXU
KW1 4XP GM4DZX
KW1 4XP GM4OFI
KW1 4XX GM6KON
KW1 4XX MM6CXJ
KW1 4XX MM6SZA
KW1 5AS GM4XLN
KW1 5EY MM0HDW
KW1 5HQ GM4RAI
KW1 5HW GM8LFB
KW1 5NE MM6EFH
KW1 5NF 2M0WIC
KW1 5NF MM6NBI
KW1 5NL 2M0WTN
KW1 5NL MM6PSX
KW1 5NL MM6WCK
KW1 5SS GM7JGH
KW1 5TU MM0DUR
KW1 5TU MM5AJW
KW1 5TU MM6VRH
KW1 5UG MM6PKO
KW10 6TL 2M0SOP
KW10 6TT 2M0WTT
KW10 6TT MM6REV
KW12 6UZ MM3YLP
KW12 6XJ 2M0CEE
KW12 6XJ GM4NHX
KW12 6XJ MM0RJR
KW12 6XJ MM6AHK
KW12 6XT MM6HNQ
KW14 7ES GM6TMH
KW14 7ND GM6TJD
KW14 7NW GM6YQA
KW14 7QA MM3FMY
KW14 7RG GM3POT
KW14 7TB GM6LDG
KW14 7XB GM8YRE
KW14 7XB GM8YRX
KW14 7XH GM4RKH
KW14 7YJ GM8NYV
KW14 8JP MM6EWX
KW14 8SR GM8TTD
KW14 8SR MM0CYR
KW14 8SY MM3PPA
KW14 8TG GM6JRX
KW14 8UG GM6JNQ
KW14 8UT GM4EFR
KW14 8XN GM0TKB
KW14 8XE GM4JYB
KW14 8YT GM1DAN
KW14 8YT MM0PFH
KW14 8YT MM3YHA
KW14 8YT MM3YMU
KW14 8YT MM6FGR
KW15 1EE GM1RQD
KW15 1FF GM0GSG
KW15 1FF GM0HIG
KW15 1FP GM0OYU
KW15 1FP GM/FLZ
KW15 1FU MM0HDD
KW15 1LN GM4LNN
KW15 1NA GM4ZZH
KW15 1NA GM6WOF
KW15 1RL GM0MHS
KW15 1SL GM8NFG
KW15 1SX GM0WED
KW15 1SX MM0ORK
KW15 1SX MM3SUS
KW15 1SX MM3TLH
KW15 1SZ GM0OWM
KW15 1SZ GM3IBU
KW15 1TF MM6PHG
KW15 1UE MM0AUP
KW15 1UF 2M0IFS
KW15 1XF MM3XUI
KW15 1XG GM1VZU
KW15 1XH GM1MWK
KW16 3BU GM3PLO
KW16 3EH MM6HEQ
KW16 3EP GM4IOB
KW16 3EX GM0PMO
KW16 3LB GM6GJW
KW16 3NZ MM3JKX
KW16 3PQ MM3SDP
KW17 2AN GM1XPE
KW17 2AS MM1DXU
KW17 2BA GM0CVP
KW17 2BA GM4WUR
KW17 2EB GM7RDY
KW17 2EH MM0EAX
KW17 2EZ GM4WMM
KW17 2HR GM0HTT
KW17 2HW MM3VIV
KW17 2JA MM6MJG
KW17 2JE MM0SJH
KW17 2JS MM3LLU
KW17 2JU GM0HTH
KW17 2LE GM0EEY
KW17 2LE MM0SLB
KW17 2LQ GM7RDH
KW17 2NN GM0IDV
KW17 2NS MM3WJD
KW17 2NS MM5DWW
KW17 2NY GM7VKN
KW17 2PJ GM6BNS
KW17 2QL GM3POI
KW17 2QL MM0DGI
KW17 2QL MM0MWW
KW17 2QL MM3POI
KW17 2RB MM6ZRX
KW17 2RD GM1MXE
KW17 2RF GM0HQG
KW17 2RL GM8WJK
KW17 2RP 2M0ORK
KW17 2RP MM0MDH
KW17 2RP MM6ORK
KW17 2ST MM3GQY
KW17 2SU GM0KTH
KW3 6AS GM6YGW
KW3 6BX GM4XRT
KW6 6EA GM3YKP
KW8 6HH MM6BQG

## KY
### (Kirkcaldy)

KY1 2AT GM0TKE
KY1 2JG GM7DAJ
KY1 2XD 2M0ESL
KY1 2XD 2M0YIC
KY1 2XD MM0OMG
KY1 2XD MM6ESL
KY1 2XD MM6XSB
KY1 2XP GM7CNW
KY1 4LB GM0RSE
KY1 4LB GM4PSL
KY1 4LB MM0TGB
KY1 4PG MM6FSV
KY10 3AU GM8YUM
KY10 3HU GM4EOU
KY10 3UD GM0KDO
KY10 3UJ GM4FQE
KY11 1ET GM3KJZ
KY11 1LD GM4UKG
KY11 2AN GM7OQE
KY11 2EZ GM7PXJ
KY11 2HJ GM8YAQ
KY11 2LX GM0PUN
KY11 2QW GM0RLZ
KY11 2QW MM3GVE
KY11 2RW 2M0LRO
KY11 2SS MM0CZK
KY11 2SS GM0DYD
KY11 2SS MM0GBH
KY11 2TJ 2M0CLU
KY11 2TJ MM0TIE
KY11 2TJ MM6BKP
KY11 3RL GM3ZSH
KY11 3HS MM0QZS
KY11 3JG GM4YMI
KY11 3JY MM3ZCS
KY11 3LG GM3TAL
KY11 3LJ GM4JKT
KY11 4BW MM3RUQ
KY11 4NA GM0OVD
KY11 4PE GM3HYX
KY11 7HZ GM4FAU
KY11 8BF GM1ZVJ
KY11 8BH GM0TKC
KY11 8BT MM3LWO
KY11 8DH 2M0CSX
KY11 8DH MM0KZJ
KY11 8DH MM3CSX
KY11 8DH MM3UUT
KY11 8GE MM0MLB
KY11 8JL GM0SXO
KY11 8JN GM1YGW
KY11 8RA GM4HVM
KY11 9GT GM0VUY
KY11 9LF GM3OZJ
KY11 9LR GM8IXZ
KY11 9NG MM0KOZ
KY11 9SJ GM7IHJ
KY11 9TD MM1CHQ
KY11 9UA GM3SHR
KY11 9UQ GM4BRN
KY11 9UQ MM1CIR
KY11 9UW GM4GRC
KY11 9UW MM0DTX
KY11 9XH GM0LLJ
KY11 9XX 2M0JOK
KY11 9XX MM6JOK
KY11 9YE GM3VYJ
KY11 9YE GM4WZG
KY11 9YU GM3RXZ
KY12 0AH GM8VKN
KY12 0BY MM0TIR
KY12 0DD GM1ZCD
KY12 0DR 2M0RDH
KY12 0DR MM0RBR
KY12 0DR MM6JOH
KY12 0DR MM6ROH
KY12 0EP MM6HUE
KY12 0HN GM0IXO
KY12 0NL MM0TBY
KY12 0NL MM3TBY
KY12 0NL MM3XSF
KY12 0RL MM0KTE
KY12 7EF GM4FSF
KY12 7TB GM3NVQ
KY12 8PT GM3NCO
KY12 8QD GM8SNE
KY12 8QL GM4JJJ
KY12 8SS GM0VOU
KY12 8XL 2M0FLG
KY12 8XL MM6FLG
KY12 8XL MM6SHB
KY12 8YD GM0SRQ
KY12 8YL GM4ZNX
KY12 9BX GM4MPC
KY12 9JQ GM0RHP
KY12 9LP GM4NUU
KY12 9LU GM3YND
KY12 9TE MM0AOL
KY12 9UQ GM7LFT
KY12 9US 2M0RWZ
KY12 9US MM6RWZ
KY12 9XG MM6WSG
KY12 9YA GM0URD
KY12 9YD GM6CMQ
KY12 9YD MM5AHM
KY12 9YG 2M0BRD
KY12 9YG MM0XBD
KY13 8DN GM4HJO
KY13 8QY MM0GNH
KY13 8RT GM4VUG
KY13 8RU GM1ZOX
KY13 8TA GM0KVE
KY13 9JE MM6FXQ
KY13 9TB GM3OXA
KY13 9XG GM0VCN
KY13 9XG GM7FTK
KY14 7BE MM6DOC
KY14 7ES MM0JR
KY14 7HB 2M0EWY
KY14 7HB MM6EWY
KY15 4AE MM6AUI
KY15 4EF GM6GPH
KY15 4HS GM0SEF
KY15 4NA GM7STI
KY15 4NB MM3TOV
KY15 4PT GM3WKB
KY15 4SS GM0OFL
KY15 4UG GM3LAW
KY15 4UG GM4ENF
KY15 5BS GM6ZAK
KY15 5DA GM4ZCV
KY15 5JS MM0DWF
KY15 5LH GM0JRQ
KY15 5PJ MM0MUL
KY15 5QB GM0UZV
KY15 5RF GM4IEF
KY15 5SE 2M0IQU
KY15 5SE GM7CPL
KY15 5SE MM3IQU
KY15 5SH 2M0ZNQ
KY15 5SH MM3ZNQ
KY15 5SQ MM3ZJY
KY15 5UF GM4RGU
KY15 5WN GM4BUA
KY15 5YU MM6VAB
KY15 7BW GM3SJY
KY15 7PT MM0UAN
KY15 7PT MM0SGQ
KY15 7PT MM3SGQ
KY15 7QD MM6BMP
KY15 7QX GM0AIR
KY15 7ST GM6PGV
KY15 7TN GM0OFM
KY15 7UH GM3VPN
KY16 0FG MM0BGO
KY16 0UE 2M0WJS
KY16 0UE MM6WJS
KY16 0XD MM0IBL
KY16 0XE GM1MZZ
KY16 0JP MM0TAT
KY16 8NJ MM0IHL
KY16 8NS GM3YOI
KY16 8NT GM4FVO
KY16 8PP GM4DPC
KY16 8QP GM3FJA
KY16 8QR GM4FFP
KY16 8RA GM3VWY
KY16 8RB GM0WFA
KY16 8SB GM0HUO
KY16 8YA 2M0BYK
KY16 8YA MM0TRS
KY16 8YA MM6TRO
KY16 9DA GM4FPZ
KY16 9ND MM0LJA
KY16 9NE GM4ENP
KY16 9NG MM6IOR
KY16 9NQ GM4IIJV
KY16 9NQ MM0DQP
KY16 9UA GM0URI
KY16 9UA GM0GKR
KY2 5AD GM4XND
KY2 5DG MM6OMJ
KY2 5HS MM6BBK
KY2 5PH GM1BLX
KY2 5PX MM3CWO
KY2 5PZ MM0JSG
KY2 5RF MM0GSW
KY2 5SD 2M0NAN
KY2 5SD MM0YSK
KY2 5SD MM6CZH
KY2 5SE MM3DFZ
KY2 5SQ GM3YBQ
KY2 5SQ GM7UQM
KY2 5SR GM0HYW
KY2 5TU 2M0IWD
KY2 5TU GM4DOF
KY2 5TU GM7EEY
KY2 5TU MM8CZU
KY2 5TX MM6FJS
KY2 5XL 2M0IAH
KY2 5XL MM6UEN
KY2 5YX MM0SAH
KY2 6DR GM0NXO
KY2 6EL GM4LCJ
KY2 6ES GM0LKT
KY2 6GZ MM3PLC
KY2 6HR GM0SNT
KY2 6JE 2M0DSU
KY2 6JE MM3YKR
KY2 6XP GM0LOO
KY2 6YB MM8ZTV
KY2 6YB MM3MMI
KY2 6ZR GM0REW
KY3 0BT MM6DQG
KY3 0DQ GM0RDA
KY3 0JW GM0IJR
KY3 0RZ GM4TRZ
KY3 0XE 2M0ZBH
KY3 0XE MM0ZBH
KY3 0XH GM0CFC
KY3 0XJ GM4GUL
KY3 9EU GM0BUI
KY3 9JQ GM0VFD
KY3 9XS GM0CNP
KY4 0BG GM1SZM
KY4 0EQ 2M0GTR
KY4 0EQ GM8KSJ
KY4 0JF MM0DAA
KY4 0NL MM8ZOW
KY4 8AW GM8XUK
KY4 8AZ GM8NVE
KY4 8EU GM1ATW
KY4 8EU GM6OOA
KY4 9EP GM7KVB
KY5 0AX GM0KLP
KY5 0DG GM8JGB
KY5 0DG MM0WLL
KY5 0EN MM0OVV
KY5 0EN MM3NGJ
KY5 0ND GM3ZGH
KY5 0ND GM4NNH
KY5 0NW MM0RBN
KY5 0NW MM3YEC
KY5 8NS MM0PMW
KY5 8NS MM1DPC
KY5 9JS MM3VNT
KY5 9LT GM1COF
KY5 9QH MM6GIW
KY6 1DU GM0EZR
KY6 1HW GM4SBP
KY6 1JD MM6KCV
KY6 1JP GM0MRJ
KY6 1JR MM3IQA
KY6 1LZ 2M0KPE
KY6 1NA GM0CFK
KY6 2BQ GM4DEX
KY6 2JU MM3SXT
KY6 2PA GM0IJD
KY6 2QA GM1JZM
KY6 3AX GM7BOZ
KY6 3LH GM7CQQ
KY6 3PJ GM1CEJ
KY6 3QJ MM0JAW
KY6 3QR GM8ZQY
KY7 4ES GM3NMN
KY7 4HS GM4TNP
KY7 4HS GM7FRC
KY7 4HS MM3OYB
KY7 4PH GM4ZJI
KY7 4RF MM3KMX
KY7 4RU GM6AOJ
KY7 4RU MM3BHG
KY7 4RW GM0GTL
KY7 4SJ GM3VVF
KY7 4SS GM6HGW
KY7 4UE GM7FWA
KY7 5AJ GM1ADI
KY7 5AX GM4XAV
KY7 5DF 2M0HAY
KY7 5DF MM0HAY
KY7 5DF MM3EHM
KY7 5HX GM0IOA
KY7 5TA 2M0DSL
KY7 5TA MM0KDY
KY7 5TA MM6VXR
KY7 5TD GM7GKT
KY7 5TD MM3LYS
KY7 5TE GM3XMY
KY7 6BN GM7CTV
KY7 6FX 2M0ZAX
KY7 6FX MM0KFX
KY7 6FX MM0UIT
KY7 6HX MM3ATI
KY7 6LT GM0DDD
KY7 6LZ GM6SZJ
KY7 6TA MM6INN
KY7 6UP MM3XGP
KY7 6UT MM6HRZ
KY7 6XX GM0MNV
KY7 6YL 2M0AGS
KY7 6YL GM3KMF
KY74EA MM3VYY
KY8 1DJ MM6GPD
KY8 1DW MM6CFS
KY8 1EZ MM3KYO
KY8 2EQ GM4PWQ
KY8 2HA GM6NYT
KY8 3AR MM0HSB
KY8 3DH GM4HBQ
KY8 3QG MM6EHL
KY8 4DF GM7VQB
KY8 4HZ MM6GQP
KY8 4LH GM0GUJ
KY8 4QN GM3RUI
KY8 5BD MM6CHV
KY8 5BP MM3WWP
KY8 5DL MM3TWW
KY8 5DZ GM7KPE
KY8 5FA GM4FEO
KY8 5HA 2M1JKG
KY8 5JH MM3ZKQ
KY8 5LT GM1EAH
KY8 5QP GM4FYH
KY8 5QP MM1DJJ
KY8 5SW GM3AHR
KY8 5TL GM0IET
KY8 5TR GM4EJI
KY8 5XA GM0IWX
KY8 6HH GM6BGQ

## L
### (Liverpool)

L N12 G4NMP
L10 3LD G3XAN
L10 4XL G4VKV
L10 4YW M0RDC
L10 8LU M0EBD
L11 1AN G7SZF
L11 1AY G6NRK
L11 1AZ 2E0PUE
L11 1AZ M3PUE
L11 1BD M6BDR
L11 1BD M6SHV
L11 1DF G0WRE
L11 1DF M3XPS
L11 1EE M3DLP
L11 2YH 2E0JPC
L11 2YH M3JNQ
L11 2YH M6LLC
L11 2YH M6LLD
L11 2YH M6LLE
L11 3BQ G8WAW
L11 4UW G6OPV
L11 5AJ M3HMC
L11 7DJ M6XWB
L11 7DW M3ZYG
L11 8LR G0JSZ
L11 8LR G1TIQ
L11 9AG G4XAR
L11 9DP G4WWB
L12 0AW M6DKC
L12 0HX G8XTD
L12 0NF 2E0HLO
L12 0NF M6LHO
L12 0QN M6BQQ
L12 3HQ G6NFR
L12 4YR G4VYR
L12 6RE M0YVV
L12 7HE G6WIO
L12 8RN G0MXW
L12 9EQ 2E0BZE
L12 9EQ M3YVD
L12 9GY G0SMR
L12 9NE G8YGM
L13 2BA G1AVC
L13 3DY G6BFM
L13 4DH 2E0DKO
L13 4DH M6IQO
L13 7DJ G3OSI
L13 9AA M3SPY
L139DF M3UFY
L14 0LG G3RBD
L14 0NU G7DUY
L14 3LE M6SMQ
L14 3LF M3HSC
L14 3LH M3EFC
L14 3LT G7NHQ
L14 3LT M6RKA
L14 9PA G0UXA
L14 0PA M6CAB
L15 0HN M3GAP
L15 3JG G4KRN
L15 4PW G0STF
L15 5AX G0OSC
L15 6TG G6VNW
L16 3GE G0ARV
L16 3NE G6IMJ
L16 5ET G3ZIM
L16 5EU G8HTF



**IMPORTANT NOTE**

**Revalidate licence to avoid revocation** – Ofcom has advised the Society that plans will be drawn up to revoke licences that have not been revalidated as required by the licence conditions. The quickest way to revalidate is to do so online via the Ofcom website: *https://services.ofcom.org.uk/* or by email: *amateur.validations@ofcom.org.uk* If you need assistance in the process, Ofcom staff are available to help, but please be patient during times of heavy workload.

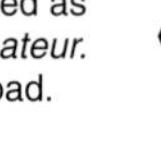

L16 6AD M6KFP
L16 7PA G0VMT
L16 7PQ G0KBS
L16 7QR G7GEE
L16 8NZ G3RWW
L17 0DP G6EQL
L17 1AB G4IHS
L17 5DB G8FXA
L17 6AD G0ADP
L17 7AL G1JHG
L17 8TT 2E0WJT
L17 8TT M0TWJ
L17 8UH G1VAA
L18 1EY G3PDC
L18 1JZ G0MIT
L18 1JZ G7ACK
L18 3EX G8IPK
L18 4PZ M5RMF
L18 4QE G1YEP
L18 4QP M3XSN
L18 5EX 2E0SNA
L18 5EX M0SET
L18 5EX M3YRO
L18 7HQ M0NVJ
L18 7HU G1YUX
L18 7JG G7VCN
L18 8AH M3XVU
L18 9SJ G0JUA
L18 9TB G0RBJ
L18 9TN G0MSO
L19 0LS G1RUZ
L19 1QJ M3XNT
L19 1RB 2E0CFD
L19 2HE M6IAV
L19 2QZ G1BES
L19 2QZ G1KOP
L19 3QX M6HMZ
L19 4TB G4KVL
L19 4TR M3PBB
L19 4UG 2E1PGA
L19 4UR M3OBD
L19 6PQ 2E0LKM
L19 6PQ M6LKM
L19 6PW M3LPJ
L19 6PZ M0CEW
L19 9AN G1ZBU
L19 9BX M0BAI
L19 9DG G2BGG
L19 9EA G0GQX
L20 0AL M3DYY
L20 0BY M3AKE
L20 4QR M6LPO
L20 5AN 2E0IOG
L20 5AN 2E0MTC
L20 5AN M3MTC
L20 5BG G7OLG
L20 5EA G1LOV
L20 6EJ G0OCC
L20 6ES M3GTT
L20 6NN M3JID
L20 9ET G3KVE
L20 9NH G3XLB
L21 2PE G3YXS
L21 4PJ G4BM
L21 7NN G0DBE
L21 7NN M3JIH
L21 7NN M3JII
L21 7QH G1DFP
L21 7QH G4VIL
L21 7RF M6MQK
L21 8HG G7OEA
L21 8NQ 2E0HGG
L22 1RJ M0CEC
L22 3YX G1PJK
L22 3YX G1SWE
L22 4RF 2E0ZOM
L22 4RF M3ZZS
L22 6QE M6RVA
L22 8QB G4XRX
L23 0RF G1MTJ
L23 0RG G6FDI
L23 0RQ G8JYV
L23 0SG M0CMW
L23 0TT M6EMV
L23 1US G3XMG
L23 2RD M1BTD
L23 2XF G4OJK
L23 3BN M6BFB
L23 3BZ G3IZF
L23 4TD 2E0DLK
L23 4TD M0HVE
L23 7UL G1KLK
L23 8SS M3IHR
L23 9SS 2E0PRO
L23 9SS M3AIR
L23 9UF G8WMG
L23 9XE M0TER
L23 9XE M1NIS
L23 9XE M3BMW
L23 9XJ G3VXK
L23 9XL G0WMJ
L23 9XS G6GOG
L23 9XY G0MJG
L23 9YU G3LNL
L24 2UB 2E0BCJ
L24 3TH G4CMP
L24 5RR M0NAC
L24 5RR M0NOW
L24 5RR M1SPW
L25 0NY G7RYW
L25 0QD G0AJL
L25 1PR M1CEW
L25 2RB G0MFH
L25 2RP M3NVL
L25 3PY G0KWD
L25 3QE G8NNX
L25 4SQ G3XSN
L25 4TJ G6BOK
L25 5PN M3XCS
L25 6DG M0BGT
L25 6DW G3KRX
L25 6EJ G4GSR
L25 7UH G1CZH
L25 7UJ M3AVF
L25 8QG 2E1KLT
L25 8QG M0KLT
L25 9QY M3PPZ
L25 9SN M6OAU
L26 1UN M0JYM
L26 7WG G0DXF
L26 7WR M6GKI
L26 7YW G7NHF
L26 9TL M1DHM
L26 9XP M0BJD
L28 3QB M0LPL
L28 3QB M3HAM
L28 5RJ M6KDC
L3 4BP M3SBQ
L3 6JF M3KXG
L30 0PH M6POB
L30 0PN 2E0GTM
L30 0PN M6CWA
L30 1PR M6MBX
L30 1PX M6BFP
L30 1PZ M6FYE
L30 1SG 2E0XTC
L30 1SG M0TNX
L30 1SG M0WYR
L30 5QD G7IKM
L30 7RN M0GSO
L31 0BW G8ZTF
L31 0DA G4OVD
L31 0DB M3LPR
L31 1DY G4WGN
L31 2JS G1ZEW
L31 2NN G0PWL
L31 3DN 2E1IIE
L31 3DN M3AZR
L31 3DX G8YPL
L31 5JJ G0DOR
L31 5NL G1DGW
L31 5PD G7FMW
L31 5PD M0AJJ
L31 6AS G1ACD
L31 6BS G4HDU
L31 6BY G3WDD
L31 7AN G8IIS
L31 8EG G8GTI
L31 9AL G4AGY
L31 9DE M1BWZ
L31 9DG M6CUX
L32 0TT M0TMC
L32 0TT M3FOD
L32 0TT M3OPM
L32 0TT M3XNB
L32 1TP M1BWH
L32 6QQ 2E0LAR
L32 6QQ M6KBY
L32 7QG 2E0PVW
L32 7QG M6XIP
L32 7QH G0XBL
L32 7QR G6OWS
L33 1UQ 2E0REG
L33 1UQ M6NYL
L33 1UW G0JIB
L33 4DW 2E0TXP
L33 4DW M0TXP
L33 4DW M6TXP
L34 1LW G3YCX
L34 1NH M3IJD
L34 1PX 2E0SJR
L34 1PX M0SJR
L34 1PX M3ITI
L34 2RS G4WCO
L34 2RY G6CKM
L34 9EH M6ARY
L35 0QH G1BEB
L35 0QR G1YCN
L35 0QR G6FIT
L35 1QG G8MKO
L35 1RJ M0FCB
L35 1RJ M6MNV
L35 1SF G4XNV
L35 2XX G1NVY
L35 2YG G0LPQ
L35 3JL G6MMG
L35 3PL 2E0PFH
L35 3PL M6PFH
L35 3PP G0IEW
L35 3RW G0ILC
L35 3SF G0RPD
L35 3SG G1CKV
L35 3UT M6JID
L35 3UX M6HCE
L35 4PA G3RJE
L35 5HL G4EMT
L35 6PJ G3WOH
L35 7NE M3LXA
L35 7NG G4VCP
L35 8NE G0SLK
L35 8PE G7KZN
L35 8PY 2E0YAR
L35 8PY M3ZVI
L35 9LX 2E0MJZ
L36 0XA M3RSN
L36 1TA M3OHR
L36 1TH G4EGM
L36 1UD G0NYR
L36 1XX 2E1CYD
L36 2NT 2E1CYS
L36 5RX G0CTC
L36 5SF G1ATG
L36 5TL M3CYS
L36 5TL M3KAY
L36 5TL M3TNY
L36 5TN G3TVN
L36 6ER M0ANN
L36 8DA M0HHR
L36 8EH G7AWW
L36 8JE M6PSV
L37 0AH G4PKP
L37 2DG G3MIP
L37 2JD G6OEM
L37 3JL G6CIP
L37 3LB M0CPL
L37 3LX G3POG
L37 3LY 2E0GMU
L37 3NH 2E0BZG
L37 3NH M0HUZ
L37 3NH M3ZIH
L37 4BP 2E0CYR
L37 4BP M6DTU
L37 6AD G8DUF
L37 6AY 2E0WFG
L37 6AY M3WFG
L37 6BF G7CVZ
L37 6BF M0DGA
L37 6BQ G4MZZ
L37 6DF G4SMT
L37 7ER M1DLM
L37 7EX M0COO
L38 0BG G0OWP
L38 3RP 2E0XDG
L38 3RP M0XZG
L38 3RP M6XDG
L39 0EB G6HIO
L39 0HW G6BZL
L39 1LR G0OXV
L39 1NN G3SZV
L39 1NU M6GXZ
L39 2BA 2E0ZMO
L39 2BA M3ZMO
L39 3LD G4ATU
L39 3NT G4LBX
L39 3PJ M0SPM
L39 3PX M6ZEB
L39 4RE G8TUN
L39 4TF G7LFC
L39 4TF M3LFC
L39 4TF M6BUB
L39 4TF M6COV
L39 4UD G8FUB
L39 4XP G4TAK
L39 4XP M1EQN
L39 5AP G6ODU
L39 5AT G0BLW
L39 5AY G3FIC
L39 5BG G3NNT
L39 5QJ G4UVB
L39 5QJ G6EXC
L39 6RF G8SAX
L39 6SE M6FAE
L39 6SY M6EEG
L39 7JL G7SWV
L39 7JU G7HSB
L39 7JY 2E0XJP
L39 7JY M6XJP
L39 7LD 2E0OLG
L39 7LD M3BHP
L39 8SX G3HWX
L39 9QA M6WIQ
L4 1UL 2E0VWL
L4 2QR M0WHC
L4 2UN G0OIU
L4 2UU M6KBT
L4 3SX M3KCC
L4 5RB 2E0MXM
L4 5RB M0HHA
L4 5RB M6MXM
L4 8UL M1SWB
L4 9UG G7UZA
L40 0RA G6ITU
L40 1SA G7EVY
L40 1TT G4CLE
L40 2QJ M0RDA
L40 2QS G0UCS
L40 2QS G6ZWZ
L40 3TF G0SGI
L40 3TG 2E0EUR
L40 3TG M6EKP
L40 5BE 2E0SSN
L40 5BE M0SSN
L40 5BE M6SSN
L40 5TJ G1EUH
L40 5TL G1JZZ
L40 5TN G0DDU
L40 5TU M0HNG
L40 5TU M6CUV
L40 5UA G6YGP
L40 5UP G4LBJ
L40 6HQ 2E0WHD
L40 6HQ M6SPT
L40 6JG G1XKL
L40 6JQ G6NJJ
L40 6JR G4LTI
L40 7RB G0AFQ
L40 7RP G0CBB
L40 9QG M0DWZ
L40 9RS G4IQJ
L41 8BX G7CQZ
L42 2BR G0DYG
L5 0TD G0ELZ
L5 0TD M0LBL
L5 2RF 2E0GNN
L5 2RF 2E0SMK
L5 2RF M3GNN
L5 2RF M6SMK
L5 3PD M0JCZ
L5 3PJ M0XSR
L5 6RH 2E0OFK
L5 6RJ M6LDK
L5 6SG 2E0KHI
L6 0BZ G1KON
L6 2NG M3FGX
L6 6BJ G7JCQ
L6 6BN G0EDF
L60 0BD G4XCM
L60 6BB G0KTY
L7 0JE M0RAT
L7 3EN M6BZB
L7L 1H3 G0EJI
L8 0UF G8DEY
L8 4XR G1MCW
L8 6QL G8ADA
L8 8NE M3WUN
L9 0NA G4XCY
L9 0NG G7VGK
L9 1AN G8CCD
L9 1JZ 2E0LPR
L9 1JZ M3MTL
L9 3BL M3ZUC
L9 7LG 2E0CLD
L9 7LG M0JPW
L9 7LG M0ODD
L9 7LG M0OGI
L9 7LG M6VMP
L9 9AY G0DHJ
L9 9BZ G0OIX
L9 9HH G7JZK

## LA
## (Lancaster)

LA1 2HA M3LCZ
LA1 2HE M6CPH
LA1 2NA M6SPH
LA1 2NW M1JHG
LA1 2QP M3KHM
LA1 2QP M3KIQ
LA1 2QP M3KNV
LA1 2QX M3UFT
LA1 2TZ G3VWJ
LA1 2UQ G1OHH
LA1 3AW G6ZHL
LA1 3BN M6IRU
LA1 3DS G7BBN
LA1 3DU G8EXJ
LA1 3DY G8UHO
LA1 3FA G0VGP
LA1 3HJ M0BUR
LA1 3ND G1HFA
LA1 3RJ M1FHQ
LA1 3SY 2E0FYE
LA1 3SY M3ZFY
LA1 4BD G1YYD
LA1 4ER G0AWM
LA1 4HX 2E0XYZ
LA1 4HX G1MBR
LA1 4HX G4YBS
LA1 4HX M1GDE
LA1 4HX M3GDE
LA1 4HX M3SNZ
LA1 4HX M3XNZ
LA1 4HX M5XYZ
LA1 4LQ G8TZJ
LA1 4LZ G4UGR
LA1 4QZ G4EKZ
LA1 4QZ M0RGS
LA1 4RF M6RDZ
LA1 4SJ G3SAQ
LA1 4SQ M6HLL
LA1 4TS M3BIO
LA1 4UU 2E1GPV
LA1 5BD G4DLP
LA1 5BY 2E1YES
LA1 5EB G3XEN
LA1 5EB G8KED
LA1 5EH G0FZA
LA1 5FS G6BJB
LA1 5JA M3VZH
LA1 5JD G0LLG
LA1 5LB G6PHF
LA1 5LY G6ZGH
LA1 5RS 2E0MBA
LA10 5AB M1ONE
LA10 5AB M3ONE
LA10 5AZ G1GVJ
LA10 5AZ G4MIS
LA10 5DE G1EKU
LA10 5JE 2E0CXJ
LA10 5JE M6CQV
LA10 5PJ G6NQL
LA10 5QH G3LUO
LA10 5TQ G6LZZ
LA10 5TQ G6SSQ
LA11 6DL G4ATB
LA11 6DP G3LZZ
LA11 6EX M0GKJ
LA11 6PH M0DCZ
LA11 6RB G8GEF
LA11 7BJ G4HLZ
LA11 7BX 2E0DTG
LA11 7BX M6DTG
LA11 7DY G3XTN
LA11 7HT G0KDG
LA11 7LJ G4CEJ
LA11 7PF G0WHD
LA12 0JB 2E0FZK
LA12 0JB G1DXM
LA12 0JF G1BMB
LA12 0LL M0RBE
LA12 0PZ 2E0BKJ
LA12 0PZ 2E0UUW
LA12 0PZ M0TES
LA12 0PZ M3UGJ
LA12 0PZ M3UUW
LA12 0QH G8ALE
LA12 0TB M3MEU
LA12 7DL 2E0DVV
LA12 7DL M6EYV
LA12 7ES G4UCL
LA12 7EU M0KYL
LA12 7JA M6DUI
LA12 8AA 2E0LCE
LA12 8AA M6IPL
LA12 8AW G0AOM
LA12 9AS M0KLJ
LA12 9EA G6NVF
LA12 9NU G7GUB
LA12 9PD G4EOB
LA12 9PD G6LKB
LA12 9QZ G4PXR
LA12 9RL M3WTA
LA13 0AU G0SKJ
LA13 0AU M6WAD
LA13 0BB M6MDJ
LA13 0BH 2E0TCC
LA13 0BH M6AMO
LA13 0BH M6NJD
LA13 0BX G4AFR
LA13 0EE G3ZFZ
LA13 0EQ M5RAG
LA13 0HE 2E0SLK
LA13 0HE M6BLZ
LA13 0HU G4ZEG
LA13 9AR G3VUS
LA13 9AR G4ARF
LA13 9HJ M6AGY
LA13 9HU M6RLX
LA13 9PX M6NEZ
LA13 9QG G7UKR
LA13 9QG M6WSU
LA13 9QY M1RDX
LA13 9RG G1GBF
LA13 9SF M3TXP
LA13 9TD M0MOL
LA14 1BQ G0MOM
LA14 1BQ M3FUR
LA14 1EZ M1FJF
LA14 1XU M6XAT
LA14 2AD M0FGB
LA14 2AD M1WDX
LA14 2BG 2E0HTC
LA14 2BG M6AUE
LA14 2ER M6MLH
LA14 2HH M6AHF
LA14 2JG M6FWO
LA14 2PZ M6RPL
LA14 2RX G6UMN
LA14 2SZ M1SKY
LA14 2UT M3PAU
LA14 3AN G4RQJ
LA14 3AY 2E0FAS
LA14 3AY M3CYW
LA14 3BP M1MIJ
LA14 3DE G1DNI
LA14 3DE G4NFR
LA14 3DX 2E0LBI
LA14 3DX M6LFB
LA14 3EJ G0GSJ
LA14 3JL M6WCN
LA14 3LP G1AGM
LA14 3LP G1CGP
LA14 3NU M3HQW
LA14 3NU M3UFU
LA14 3NU M3UQF
LA14 3PR M6DHV
LA14 3PS M1MCW
LA14 3PS M3YBL
LA14 3PY M6MRN
LA14 3QW 2E0NGC
LA14 3QW M0IGG
LA14 3QW M6LRA
LA14 3SF M6FWT
LA14 3TS M1AVV
LA14 3UD 2E0MYK
LA14 3UD M0MYK
LA14 3UD M6CET
LA14 3UZ M6TBT
LA14 4AH G0GJM
LA14 4JY 2E0AYW
LA14 4LE M0REJ
LA14 4LE M0SJJ
LA14 4LE M3GEN
LA14 4NH G4USW
LA14 4PA G0UXH
LA14 4PH G4BZR
LA14 5DW 2E0TEI
LA14 5DW M0TEB
LA14 5DW M6DTF
LA14 5DW M6GUM
LA14 5EL M6ZVD
LA14 5HW G4YXA
LA14 5LH M3MWV
LA14 5NZ G0SSO
LA14 5QQ G0LSU
LA14 5RL M3YNS
LA14 5RQ M6ANX
LA14 5SE M6EEO
LA14 5TP M3UDJ
LA14 5TX G7UAK
LA14 5XP G3IZD
LA14 5XP M3ZZV
LA15 8BW G7RNX
LA15 8EU M1CQF
LA15 8LZ G4VKE
LA15 8NL G4DKZ
LA15 8NQ M3UBE
LA15 8NR 2E0MND
LA15 8NR G7EGQ
LA15 8NR M3IJH
LA15 8NR M3MND
LA15 8NR M3TMX
LA15 8QA G4SPW
LA15 8QA M0SSD
LA15 8QA M6PDT
LA15 8SE G7MCE
LA16 7EW M3HVL
LA16 7EY G0HIK
LA16 7EY M3MGU
LA16 7JG 2E0CPG
LA16 7JG M0KPW
LA16 7JG M6AYY
LA17 7XJ G4TAZ
LA18 4JJ G1ZIM
LA18 4LS G7PKQ
LA18 5DB G1NBY
LA18 5EQ G0TUE
LA18 5HB G8JHG
LA18 5LE G6HEF
LA18 5LE G6HFB
LA19 5XL M1AOD
LA19 5XN M3TZI
LA2 0EG G4UJI
LA2 6AG G4TOT
LA2 6ER G0VPU
LA2 6HJ G0NTT
LA2 6HX G4ZJO
LA2 6LB M0BZR
LA2 6NL G7TOZ
LA2 6QB G0UWX
LA2 6QJ G4ZEZ
LA2 7AB M0DGK
LA2 7AG G0OIE
LA2 7BA M0DWK
LA2 7DD 2E0GBO
LA2 7DD G6OOK
LA2 7DD M6MUK
LA2 7EP G4LIY
LA2 7ET G1JYB
LA2 7ET M3YOG
LA2 7HX G8CSQ
LA2 7JN G1KLZ
LA2 7JZ 2E0CBL
LA2 7JZ M3ZMX
LA2 7LP 2E0RPO
LA2 7LP M6RPO
LA2 7LP M6XSS
LA2 8ER M0XPL
LA2 8NP G1LAT
LA2 9AN G4KWJ
LA2 9AN G4KWK
LA2 9BD M6AHE
LA2 9LF G0AXU
LA2 9LF G4NEQ
LA2 9NH 2E0MAF
LA2 9NH M0MOS
LA2 9NH M3WCZ
LA2 9NY G1CFJ
LA2 9PB G3UVQ
LA2 9PJ G4ZJE
LA2 9PJ M0CLR
LA20 6HS G7UGY
LA21 8EQ G8YGO
LA21 8EQ M0DCY
LA21 8ER 2E0COX
LA21 8ER M3MIO
LA22 9EX M3CQP
LA22 9EY M3HBB
LA23 1EU M3LQO
LA23 1PX 2E0NFS
LA23 2EU G3VGD
LA23 2LA G7JJW
LA23 2LB M6NOJ
LA23 3PW G8IRN
LA23 9BD G7OEW
LA3 1ET G0LRE
LA3 1FL M6GBK
LA3 1LR G7UBP
LA3 1SD G0TOO
LA3 1UH G0LQX
LA3 1UR G1RLK
LA3 2BB M0SED
LA3 2DJ G6BKY
LA3 2LR G3KOQ
LA3 2LT 2E0TFT
LA3 2LT M6JOG
LA3 2LX G3RCZ
LA3 2LY M1BSF
LA3 2PW M3LDF
LA3 2PW M3LQP
LA3 2QJ 2E0DXO
LA3 2TB G4ZJL
LA3 2UR 2E0TBF
LA3 2UR M6DBE
LA3 2YF G7IHD
LA3 3AA G4UME
LA3 3AJ G8BME
LA3 3EF G1HIB
LA3 3HU G6XVY
LA3 3JA G0LWU
LA3 3QJ M3APO
LA3 3RA M3OBM
LA3 3RA M3UAY
LA3 3SH G0JHG
LA3 3SN M6YPE
LA4 4EF M3IAE
LA4 4EW 2E0AZA
LA4 4EW 2E0AZB
LA4 4HS G4LDS
LA4 4LN M3HAZ
LA4 4LQ G6OUT
LA4 4LQ G7NLR
LA4 4NN 2E0BND
LA4 4NY G0IYU
LA4 4PZ G3CSY
LA4 4QD G0EBP
LA4 4QF M0LCY
LA4 4RE G1GRP
LA4 4RE G1ONH
LA4 4SZ G4DPT
LA4 4TJ G3VSE
LA4 4TU G7CAF
LA4 5BW M1TCQ
LA4 5LE 2E1DGL
LA4 5LJ G6ZJN
LA4 5NF M3REM
LA4 5PT M0CFZ
LA4 5QD G3KSP
LA4 5QR G1UKS
LA4 5RA G1PWY
LA4 5RU 2E0SBC
LA4 5RU M0SBC
LA4 5RU M3FRQ
LA4 5SR 2E0JOH
LA4 5SR M0YJO
LA4 5SR M6ALO
LA4 5SR M6LDL
LA4 5SU G6FKE
LA4 5UJ 2E1GNU
LA4 5UJ G0RDH
LA4 5UJ M0SCG
LA4 5XP G4XLC
LA4 6BN 2E0BQG
LA4 6BN M0GRR
LA4 6BW M6MWC
LA4 6DD G0FYW
LA4 6DQ M3ZTP
LA4 6EJ G0CHV
LA4 6HD G0GEU
LA4 6HS G0IFN
LA4 6JS 2E0VVJ
LA4 6JS M6TOD
LA4 6LA G1VTE
LA4 6LT G4IFT
LA4 6NU G4BWX
LA4 6PD G6HMA
LA4 6PS 2E0CKC
LA4 6PS M6APB
LA4 6QG G0AUF
LA4 6QL 2E1HER
LA4 6QL G8FJA
LA4 6QR G0ASQ
LA4 6TB G0FYH
LA5 0UG 2E0EET
LA5 0UG M0HPG
LA5 0UG M0PWD
LA5 0UG M6PEW
LA5 8AP G3XLS
LA5 8AQ G1PEE
LA5 8AW G4RHB
LA5 8DJ M0FSK
LA5 8EN G0CWP
LA5 8HA G4TRY
LA5 8HJ G0VLV
LA5 8HQ G0SCU
LA5 8HQ G6ZET
LA5 8HR G6IKE
LA5 8LY G3SZU
LA5 9HS G4ABE
LA5 9JQ G0NYD
LA5 9LD M3LCE
LA5 9LG M6AAG
LA5 9QR G4UGB
LA5 9QZ G4VAP
LA5 9SA M0RGB
LA6 1AX G8EUF
LA6 1DE 2E1GYR
LA6 1DJ G7IER
LA6 1HY G0KSS
LA6 1JE M6EOR
LA6 1PJ G8KGK
LA6 1PJ M3KGK
LA6 1QN G6POI
LA6 2AF G3HGD
LA6 2JL 2E1TIM
LA6 2JL M1JCB
LA6 2LJ G8JAI
LA6 3AN G4VPS
LA7 7BB G0BBE
LA7 7BB G0KRG
LA7 7BB G0KRS
LA7 7BB G0RLO
LA7 7BB G7KRC
LA7 7BE G8MGG
LA7 7DZ M0DZW
LA7 7JB G3JGP
LA7 7JP G1YMV
LA7 7ND M0PRN
LA7 7NL G6UPL
LA7 7NL G6UPM
LA7 7PT G4UXH
LA7 7QF G6NPC
LA8 0EN 2E0HOK
LA8 0EN M0KOH
LA8 0EN M6HOK
LA8 0NG M0PJX
LA8 8AT G8VTX
LA8 8DJ G4ZBW
LA8 8HP G3PLX
LA8 8LH G4ETD
LA8 9BT M1IKE
LA8 9HL G4GLH
LA8 9NJ M3OPS
LA9 4LN G3GEF
LA9 4QR G3VVT
LA9 4QR M3RIA
LA9 4UU G3HMR
LA9 5DU M3RBP
LA9 5HN M3ZOO
LA9 5QE G7OWV
LA9 5QE M0RIS
LA9 6AJ M6LWA
LA9 6AU G1IGC
LA9 6AU G6UYK
LA9 6AU G6YCM
LA9 6AU M0RMH
LA9 6BJ M1DBL
LA9 6BN G3VJI
LA9 6EA G3WRI
LA9 6HT G4ZBO
LA9 6LB M1KMC
LA9 7ED G7NJD
LA9 7EY 2E0BXM
LA9 7EY M3VHH
LA9 7HA G6IPN
LA9 7JB M3HSE
LA9 7JG G0KHA
LA9 7JG G0RTM
LA9 7JH 2E0TBT
LA9 7NA G0HIF
LA9 7NT M3ZWG
LA9 7PE G3XJI
LA9 7PL M3TYV
LA9 7PN G0RSL
LA9 7QB 2E0BNJ
LA9 7SN M3XVJ

## LD
## (Llandrindod)

LD1 5BH 2E0OZY
LD1 5BH M0OZI
LD1 5BH M6CQO
LD1 5LW G4XXJ
LD1 5NL G0RBH
LD1 5PD G4GWH
LD1 5PU G0GPZ
LD1 5RB 2E0XOT
LD1 5RB M0XOT
LD1 5RB M6ZYQ
LD1 5RD G4LUX
LD1 5SE G1PXM
LD1 5UR G7UNV
LD1 6BD G7JYJ
LD1 6BE M6RRW
LD1 6BU M0MVM
LD1 6DH 2E0MMX
LD1 6DH M6MMI
LD1 6DT G6OJK
LD1 6EW M3GUH
LD1 6PD G4YCJ
LD2 3PB G3PCY
LD2 3UL M6BUF
LD2 3UU G4MUJ
LD2 3UY M0LLY
LD3 0AT G6DQH
LD3 0AT G8YPR
LD3 0BP G0DIB
LD3 0DU G0BKJ
LD3 0HN G1SMJ
LD3 0HY M3RRU
LD3 0PW G7RFA
LD3 0UR G4NKR
LD3 7JF G0IVT
LD3 7NY G4BVT
LD3 7RT 2E0BOC
LD3 7RT M0GMH
LD3 7RT M3BOO
LD3 7UW G0ABT
LD3 8DE G3SB
LD3 8DH G6RUE
LD3 8DJ 2E0MCB
LD3 8DJ M3BTN
LD3 8PD G0XAP
LD3 8PD M6ELX
LD3 8UP G6JJX
LD3 9BR 2E0MFD
LD3 9BR M6HEA
LD3 9ES G1BXX
LD3 9HH 2E0HAS
LD3 9HH 2E0YLL
LD3 9HH 2E0ZJA
LD3 9HH M0HTO
LD3 9HH M0UAA
LD3 9HH M3VUJ
LD3 9HH M3YVK
LD3 9HH M6ZOL
LD3 9HT G0GIH
LD3 9LH 2E0RHI
LD3 9LH M3YWC
LD3 9NA M6GSS
LD3 9NG M6FVA
LD3 9PL 2E0GFX
LD3 9PL M6VGA
LD3 9PW G4KCV
LD3 9RL G6NSK
LD4 4DR G7KIV
LD6 5BN G0KQX
LD6 5HA G7BNC
LD6 5HA M6EQF
LD6 5LD G1XRD
LD6 5LG M0TJS
LD7 1AG G7HDC
LD7 1EF M3TMR
LD7 1HY M3RGD
LD7 1PT G0KIR
LD7 1RY M6YKI
LD7 1SW G0JEQ
LD7 1YL G8ARR
LD7 1YT G4GAF
LD8 2AN G3RMJ
LD8 2EG G1ENP
LD8 2EP G8CAK
LD8 2HL 2E0TLG
LD8 2HL M6PDP
LD8 2LB G0CTG
LD8 2LH G4VVF
LD8 2PA G7IZA

## LE
## (Leicester)

LE1 5XR G6ANA
LE1 5XS M1GUR
LE1 7RH G4ION
LE10 0DY G1CJI
LE10 0ED 2E0HTM
LE10 0ED M6AYN
LE10 0GB G8SUN
LE10 0HP M0BWF
LE10 0LA G0GSH
LE10 0LG G1CIV
LE10 0LY M6CJF
LE10 0NH G4CNZ
LE10 0PF M3OZN
LE10 0PH G0GGA
LE10 0PY M6WNE
LE10 0SE M3DZK
LE10 0TW G3XWD
LE10 0WU M3AQJ
LE10 1EB 2E0SCC
LE10 1EB M6HJR
LE10 1EQ M0HDS
LE10 1HE G1WJO
LE10 1JA 2E0MAQ
LE10 1JA M0MPF
LE10 1JA M3XOH
LE10 1LT G3MFK
LE10 1SJ G4NNI
LE10 1SL G4YFI
LE10 1SS M3IRQ
LE10 1TH M3IRP
LE10 1TJ 2E1GOP
LE10 1TJ G8JSE
LE10 1UA 2E0DRN
LE10 1UY G6UX
LE10 2AU G7FSD
LE10 2EG G8KFD
LE10 2EU 2E0TUT
LE10 2EU G8NEL
LE10 2EU M6OND
LE10 2LY G4JLB
LE10 2ND G3XPU
LE10 2NX G8CTJ
LE10 2PS 2E0BMJ
LE10 2SW G4PJP
LE10 2TN 2E0DTL
LE10 2TN M0LTD
LE10 2UD G1KFB
LE10 3ED M6AFL
LE10 3LF G4ZXA
LE10 3PR M6ATM
LE10 3PR M6SOG
LE11 1BA M3XMO
LE11 1BU M0WAI
LE11 1JW M0RWM
LE11 1LP 2E0FCZ
LE11 1LP M6FCZ
LE11 1NX M0JPP
LE11 1RD G7IQO
LE11 1SL G4YJT
LE11 1SN G4SGY
LE11 2AA G4IAQ
LE11 2AA G4IAR
LE11 2AA G4LAB
LE11 2JG G0VAU
LE11 2JL G6OUJ
LE11 2NU G3LTB
LE11 2PA G0LCU
LE11 2PB G0DMH
LE11 2PG 2E1HGJ
LE11 2PH M0CQV
LE11 2PH M3HZH
LE11 2QZ G7WIQ
LE11 2RU G6AK
LE11 2RU G8IXX
LE11 3AG G7NOF
LE11 3JP G1WLW
LE11 3JP G3MKE
LE11 3JS G0WTA
LE11 3JT G3OMK
LE11 3LT G7RAL
LE11 3LT G8SNF
LE11 3LT M3SNF
LE11 3PS G7LTU
LE11 3PT G3ZUB
LE11 3RA G7NII
LE11 3RL G4SBM
LE11 3RX G7DST
LE11 3SS G7WCP
LE11 3ST G8UIW
LE11 3TB G7BDR
LE11 3TU M3OOH
LE11 4EX 2E0JOG
LE11 4EX M6PPL
LE11 4LF G4AOP
LE11 4LF G8CPK
LE11 4LJ M3EKZ
LE11 4LL M3ROQ
LE11 4LQ G1ZAR
LE11 4ND M3AUC
LE11 4PG 2E1IBT
LE11 4PG G8RFY
LE11 4PP G4JQV
LE11 4PU G4CCI
LE11 4PX G7ACM
LE11 4QD G0MMJ
LE11 4QU 2E0DUW
LE11 4QU M6GGO
LE11 4QW G4OYP
LE11 4SN G4OKY
LE11 5AN 2E0MWJ
LE11 5EZ G4NTJ
LE11 5HB G4UAT
LE11 5JW G1VVT
LE11 5UJ G7OXA
LE11 5UU G8HMA
LE11 5UW G8LVL
LE11 5YB G7SCL
LE11 5YX M6NCM
LE11 5YZ G4VCN
LE11 5YZ G6VAW
LE12 5AP G4IOR
LE12 5DF G4MPH
LE12 5EE G1KBC
LE12 5HA M6FHV
LE12 5HQ G4DCI
LE12 5HW M6LKL
LE12 6LD G4ZMA
LE12 6NN G8HVF
LE12 6PP 2E0LEA
LE12 6PP G3ZJV
LE12 6PW G8JHE
LE12 6QL G7KNQ
LE12 6RH G3JHS
LE12 6ST G4BZP
LE12 6ST G8DTM
LE12 6UB G3XYC
LE12 6UZ G0INJ
LE12 7BP G0LMA
LE12 7FG M6GII
LE12 7HB G1IFX
LE12 7HY M3JNU
LE12 7ND G1DHQ
LE12 7NX M0BWU
LE12 7PX G4MIV
LE12 7PZ 2E0DGH
LE12 7RQ G6MOI
LE12 7SB G1HGA
LE12 7SH G0AUV
LE12 7SX G0JHJ
LE12 7TG M3OMU
LE12 7UH 2E0MFT
LE12 7UH M3MFT
LE12 7UX G0MCV
LE12 7UX M6MQV
LE12 8BB G4ERT
LE12 8BF G6TZO
LE12 8EJ G8IQT
LE12 8HT M3URH
LE12 8JZ G7UGA
LE12 8LA G4ZSP
LE12 8LG M6GAO
LE12 8LR 2E0MCW
LE12 8RP M3HAI
LE12 8SG G3THW
LE12 9AD G1HOD
LE12 9AT 2E0ZRT
LE12 9AT M0ZRR
LE12 9AT M6ZRT
LE12 9BZ M3ZTL
LE12 9DA M0TMM
LE12 9DG M6DJS
LE12 9DY G7HDW
LE12 9HP M0ZAK
LE12 9HY G4EUF
LE12 9LA G3KWY
LE12 9LS G0UGI
LE12 9LS G4HTH
LE12 9LW M0TAB
LE12 9LW M1CJT
LE12 9NL G7BMM
LE12 9PU G3MKU
LE12 9QH M6JLZ
LE12 9RW G4JCH
LE12 9SG M3OAB
LE12 9SH M6IAS
LE12 9SQ G1ETZ
LE12 9SS G8BUB
LE13 0BY G4LWB
LE13 0DS G0WYM
LE13 0DU M6CJI
LE13 0EJ M1TEM
LE13 0EU G1DPN
LE13 0EW G3XJW
LE13 0EW M3KSP
LE13 0GE G1JMC
LE13 0NB G3MLQ
LE13 0NN G8AKE
LE13 0RA G7PCT
LE13 0RA M0MKE
LE13 0SR G4LAU
LE13 0TF M6CXZ
LE13 1ES M3UOX
LE13 1HG G4NNZ
LE13 1HY G4YSP
LE13 1HZ M6HPS
LE13 1JZ G4FOX
LE13 1JZ G4PTK
LE13 1JZ G7FOX
LE13 1LE G3STG
LE13 1LE G4NRC

| | |
|---|---|
| LE13 1LE | M1NSC |
| LE13 1LJ | M0LBM |
| LE13 1QT | G1MZU |
| LE13 1QZ | G0SCO |
| LE13 1RZ | G4CAZ |
| LE13 1RZ | M6LGH |
| LE13 1SE | G8RBY |
| LE13 1UH | G6FDD |
| LE13 1UR | G8NYH |
| LE14 2AN | G6PZN |
| LE14 2AP | G8YEJ |
| LE14 2UB | M6BUU |
| LE14 3AF | G2UT |
| LE14 3DT | G4AMN |
| LE14 3EW | G7UDJ |
| LE14 3EX | G6PQI |
| LE14 3QA | G4CWC |
| LE14 3QE | M6IWA |
| LE14 3QG | G3UOD |
| LE14 3QH | G0PAG |
| LE14 3RY | G4HEE |
| LE14 3SA | M6WTD |
| LE14 4BU | 2E0MDT |
| LE14 4BU | M0VAR |
| LE14 4BU | M6CJQ |
| LE14 4DD | G4EVK |
| LE15 6BL | M6ELB |
| LE15 6EB | G3MCK |
| LE15 6GE | G4IYS |
| LE15 6HQ | M6MDC |
| LE15 6JQ | 2E0GPA |
| LE15 6JQ | M6GPA |
| LE15 6LT | G4FDP |
| LE15 6LZ | M1BAN |
| LE15 6PH | G6LEI |
| LE15 6QA | G0CWU |
| LE15 6RQ | M6OKM |
| LE15 6SE | G7BCW |
| LE15 6SJ | G3WMA |
| LE15 6SJ | M6NJE |
| LE15 6SL | G4PJY |
| LE15 7DB | G3PKD |
| LE15 7DZ | G3ZDW |
| LE15 7EU | M3HPF |
| LE15 7HP | M6ELS |
| LE15 7JL | G4PKZ |
| LE15 7LJ | G3KLT |
| LE15 7NJ | G4BOQ |
| LE15 7NX | G7FGR |
| LE15 7PS | G1VIG |
| LE15 7PX | 2E0HDG |
| LE15 7PX | M0HQC |
| LE15 7PX | M6CDO |
| LE15 7RL | G7JBD |
| LE15 7SD | M3PTQ |
| LE15 8AD | G3XQZ |
| LE15 8DH | 2E0OCW |
| LE15 8DH | M6WGJ |
| LE15 8EA | M6WFR |
| LE15 8HY | M3SNN |
| LE15 8JS | G3OKB |
| LE15 8PH | 2E0FYQ |
| LE15 8PH | M6FYQ |
| LE15 8PJ | M6GKN |
| LE15 8QB | M6EOT |
| LE15 8RH | G4OSJ |
| LE15 8SD | G3USR |
| LE15 8SU | G8TTE |
| LE15 9HU | G0ECI |
| LE15 9RL | M6EII |
| LE15 9TS | G0KAQ |
| LE16 7BQ | G4JOV |
| LE16 7BQ | M6GYG |
| LE16 7DD | G8GMB |
| LE16 7DD | M3GMB |
| LE16 7DE | G0SFJ |
| LE16 7EH | G0BHS |
| LE16 7JF | G4XRA |
| LE16 7JJ | G4WVR |
| LE16 7JJ | G8PAN |
| LE16 7LG | M3HZN |
| LE16 7LW | M3KXI |
| LE16 7NY | G4XEX |
| LE16 7PQ | G3YKS |
| LE16 7TA | M3HZA |
| LE16 7UU | 2E0XVX |
| LE16 7UU | M6XVX |
| LE16 7XE | 2E0GAP |
| LE16 7XE | G1IVG |
| LE16 7XE | M3YVF |
| LE16 7XE | M3YVG |
| LE16 7XE | M6AJP |
| LE16 8BF | G3VUH |
| LE16 8BH | G0VLR |
| LE16 8BH | G4ISN |
| LE16 8BH | G7VLR |
| LE16 8BH | G8LVM |
| LE16 8BQ | G4DIA |
| LE16 8EX | G6OFZ |
| LE16 8GJ | M0IAA |
| LE16 8LD | G1TTH |
| LE16 8LD | G4LXA |
| LE16 8QD | G1DUG |
| LE16 8QJ | G1FJH |
| LE16 8QJ | G3ZSU |
| LE16 8QJ | M3UXE |
| LE16 8QJ | M3UXM |
| LE16 8SJ | G1NQH |
| LE16 8SS | M6BCN |
| LE16 8XR | M3MRA |
| LE16 8XS | G7KRS |
| LE16 9AQ | M3ICF |
| LE16 9DP | M6BWL |
| LE16 9DX | G7PZB |
| LE16 9GA | 2E0OVB |
| LE16 9GA | M0OVB |
| LE16 9GA | M6OVB |
| LE16 9JL | M6LTD |
| LE16 9JS | M6BWJ |
| LE16 9JW | G1PIV |
| LE16 9LW | G3SFV |
| LE16 9NA | G4JYT |
| LE16 9NW | G1IVF |
| LE16 9NW | G1WVR |
| LE16 9RZ | G1ZHD |
| LE16 9SE | G1WYM |
| LE17 4AD | G6OGZ |
| LE17 4GR | G7GGF |
| LE17 4NW | G1ICI |
| LE17 4PG | M6PGA |
| LE17 4PS | G3ORY |
| LE17 4PS | G3SDC |
| LE17 4SP | 2E0EEU |
| LE17 4SP | M6EEU |
| LE17 4SQ | G0TTW |
| LE17 4TR | G0OLS |
| LE17 4TR | G7FFZ |
| LE17 4TU | M3RDV |
| LE17 4US | G1IRQ |
| LE17 4UT | M6HTQ |
| LE17 4XB | G3RIR |
| LE17 4XB | G8TKQ |
| LE17 4XF | G0JEW |
| LE17 4XF | G4APD |
| LE17 4XJ | G7DMG |
| LE17 4XJ | M5WAH |
| LE17 4YS | M6WOB |
| LE17 5AG | G6EDC |
| LE17 5AS | G4VSJ |
| LE17 5DE | G4VOZ |
| LE17 5DE | G8LM |
| LE17 5DL | G6TZT |
| LE17 5DL | G8ZUF |
| LE17 5EG | 2E0PBK |
| LE17 5EG | M6PBL |
| LE17 5HX | G4NGF |
| LE17 5HY | G4CLA |
| LE17 5QA | G8PGI |
| LE17 5RP | G1DPT |
| LE17 6AZ | G4ABX |
| LE17 6EQ | 2E1BKK |
| LE17 6EQ | G7LCK |
| LE17 6EQ | M3TDB |
| LE17 6HD | M6DMD |
| LE17 6JW | G4VTM |
| LE17 6NT | G6YQU |
| LE18 1BA | M6KRH |
| LE18 1BH | M3XFC |
| LE18 1BR | G6IFN |
| LE18 1DX | G0UVP |
| LE18 1FQ | G1UFT |
| LE18 1GD | G8SSX |
| LE18 1HU | G4OAI |
| LE18 1HY | G0HLL |
| LE18 1JZ | G4KKS |
| LE18 2EF | 2E0FOK |
| LE18 2EF | M6FOJ |
| LE18 2EP | G6LTR |
| LE18 2FU | G4XDX |
| LE18 2HQ | M3IHN |
| LE18 2JB | G1VNS |
| LE18 2JH | M0GTE |
| LE18 2QX | G0TCJ |
| LE18 3QS | G7DPU |
| LE18 3RL | G8EVD |
| LE18 3SX | G0BDE |
| LE18 3SX | G1HII |
| LE18 3TY | G4EMW |
| LE18 3WD | G4OKD |
| LE18 3XW | G1YQP |
| LE18 4LP | 2E0NLB |
| LE18 4LP | M6NDB |
| LE18 4LY | M0KVN |
| LE18 4LY | M1KVN |
| LE18 4NA | 2E0VOD |
| LE18 4NA | M3UFW |
| LE18 4NQ | 2E0LTD |
| LE18 4TH | 2E0APY |
| LE18 4WH | G6YII |
| LE18 4WL | M3ZPE |
| LE19 2AW | G0PBY |
| LE19 2FY | M6CBM |
| LE19 2HT | G0AIG |
| LE19 2RA | G7PPL |
| LE19 3PS | G0WBC |
| LE19 4AB | G4RNR |
| LE19 4NQ | G7VUU |
| LE19 4NQ | M0AZW |
| LE19 4QD | G8NTD |
| LE19 4QX | G1VIN |
| LE19 4QZ | G0TIZ |
| LE2 1PN | M3MBV |
| LE2 1PR | G1KOH |
| LE2 1WQ | G0TYF |
| LE2 1YD | M6BHJ |
| LE2 2AE | M6MJM |
| LE2 2EA | G1JAG |
| LE2 3EH | 2E0OTP |
| LE2 3EH | M6IKI |
| LE2 3PJ | G3TDX |
| LE2 3RJ | G6EHJ |
| LE2 3WR | M6FUD |
| LE2 4NT | G1IGP |
| LE2 4NY | 2E1FEC |
| LE2 4PB | 2E0RMJ |
| LE2 4PB | M6JEZ |
| LE2 4QA | 2E0BQE |
| LE2 4QA | M3XCU |
| LE2 4QD | 2E0BQF |
| LE2 4QD | M0OFL |
| LE2 4QD | M3UFG |
| LE2 4QN | G4EEL |
| LE2 4RJ | G0HNI |
| LE2 4UG | M3XMP |
| LE2 5FH | G7SEU |
| LE2 5HF | G6PFN |
| LE2 5PF | G3XKX |
| LE2 5PF | G5UM |
| LE2 5TR | G1HWO |
| LE2 5UE | G3HAN |
| LE2 6AD | G0TTE |
| LE2 6FF | M6EPN |
| LE2 6FN | G3MCP |
| LE2 6HQ | G3LQW |
| LE2 6JE | M6AER |
| LE2 6NT | M3CBY |
| LE2 6TS | 2E1CPC |
| LE2 7LN | M6OUI |
| LE2 8DB | G0LMO |
| LE2 8DH | G8FCQ |
| LE2 8DJ | G6WMU |
| LE2 8HW | G4NBH |
| LE2 8NJ | G1PPQ |
| LE2 8QA | G4UQS |
| LE2 8QA | G6ITW |
| LE2 8SF | M0HOP |
| LE2 8SF | M1HOP |
| LE2 8UH | G4MQS |
| LE2 8UH | M1CSL |
| LE2 8UP | G5VH |
| LE2 9DD | G1ZYN |
| LE2 9GA | G6WZM |
| LE2 9NS | G6HSI |
| LE2 9NS | G8RFE |
| LE2 9QJ | G0QQZ |
| LE2 9TH | G0PBP |
| LE2 9TJ | G4TQR |
| LE2 9TT | G6VLT |
| LE3 0LD | G6RRJ |
| LE3 0PB | G4OHF |
| LE3 0SA | G3ZCT |
| LE3 0TN | G0EVI |
| LE3 1AU | M0ZKA |
| LE3 1DD | M6BHI |
| LE3 1FG | G6EJU |
| LE3 1GR | G0FPU |
| LE3 1JF | M1NAS |
| LE3 1PA | 2E0DSQ |
| LE3 1PA | M0VSE |
| LE3 1PA | M6ESV |
| LE3 1RA | M3CGH |
| LE3 1RA | M6SBR |
| LE3 2BW | G3WQL |
| LE3 2EJ | G0FZC |
| LE3 2FN | M6KEH |
| LE3 2FN | M6TJX |
| LE3 2JB | M6HKJ |
| LE3 2JU | G0FRV |
| LE3 2RY | M1EZR |
| LE3 2SP | G0ORY |
| LE3 2UU | G1AEJ |
| LE3 2XQ | M1BSE |
| LE3 3AD | G4NUK |
| LE3 3AP | 2E0TPP |
| LE3 3BF | G7UOS |
| LE3 3DQ | G4DJK |
| LE3 3EY | G1CKY |
| LE3 3FF | G4SDZ |
| LE3 3GN | G7GLA |
| LE3 3HB | G3HYH |
| LE3 3LA | G7AYI |
| LE3 3PP | G1EBV |
| LE3 3PP | M6VBD |
| LE3 3PS | M3NNM |
| LE3 3SW | G0MTF |
| LE3 5RA | G6WSN |
| LE3 6BD | G6JOL |
| LE3 6FG | G0UHI |
| LE3 6FG | G7OCY |
| LE3 6LU | M3SAO |
| LE3 6NA | G4KGX |
| LE3 6NF | G4RLC |
| LE3 6PL | 2E0DLD |
| LE3 6PL | G1GEV |
| LE3 6PL | M3PIY |
| LE3 6PL | M3WBQ |
| LE3 6QQ | M6HMQ |
| LE3 8AF | G1YEZ |
| LE3 8AG | G6BBH |
| LE3 8AG | G6BBI |
| LE3 8FN | G7OBR |
| LE3 8GF | G8ZQG |
| LE3 8GH | M0GPO |
| LE3 8JH | G8SZX |
| LE3 8LF | M6PTO |
| LE3 8LU | 2E0KFH |
| LE3 8LU | M6CDK |
| LE3 9HG | M6DPA |
| LE3 9ND | M3PYG |
| LE3 9ND | M3UNL |
| LE3 9NH | M6FPD |
| LE3 9NH | M6FPE |
| LE3 9PS | M6CMW |
| LE4 0BB | 2E0HDY |
| LE4 0BB | M6TGH |
| LE4 0BJ | G3OCH |
| LE4 0GS | G3KQV |
| LE4 0LL | G4AGN |
| LE4 0PP | G7DHJ |
| LE4 0QY | 2E0BQC |
| LE4 0QY | M3XEA |
| LE4 0SF | M3PIK |
| LE4 0ST | M6LFL |
| LE4 0ST | M6TFO |
| LE4 0SU | G0AZM |
| LE4 0UG | M6FPG |
| LE4 0UR | G3UST |
| LE4 1BL | G0ICV |
| LE4 1BX | G8ATE |
| LE4 1EJ | M6XTP |
| LE4 2BH | 2E0XAY |
| LE4 2BH | M6SEV |
| LE4 2BJ | G4ITP |
| LE4 2HN | 2E0SVT |
| LE4 2HN | M6JLA |
| LE4 2HN | M6SVT |
| LE4 2LH | G4ILT |
| LE4 2PJ | M6POM |
| LE4 2PZ | M1SJE |
| LE4 3GW | M0GAH |
| LE4 3HB | G4ZCJ |
| LE4 3HQ | G4BJF |
| LE4 4DA | G4MEF |
| LE4 4EH | 2E1HMQ |
| LE4 4FU | G6TIW |
| LE4 4GS | G1DYT |
| LE4 5BQ | 2E0SWT |
| LE4 5BQ | M6TAA |
| LE4 5LH | M3BYX |
| LE4 5RB | M3NFA |
| LE4 7AE | M3FPT |
| LE4 7SU | M3EVP |
| LE4 8AY | G0VTL |
| LE4 8BD | G7DOA |
| LE4 8BP | G4VWI |
| LE4 8GL | M1BUJ |
| LE4 8JP | 2E1SDI |
| LE4 8NB | M6EIH |
| LE4 8NR | M3HVS |
| LE4 9DQ | M6WWF |
| LE4 9NR | M0LRG |
| LE5 0LF | G5MY |
| LE5 0PH | M3WZV |
| LE5 1EA | G4EOF |
| LE5 1ED | G6WAY |
| LE5 1FA | 2E0LDE |
| LE5 1PA | G3FJL |
| LE5 2DE | G7VDU |
| LE5 2DE | M3VDU |
| LE5 2EE | M6MDR |
| LE5 2EF | G1ECK |
| LE5 2EF | G4ZTD |
| LE5 2EF | G6MAM |
| LE5 2EF | M6NES |
| LE5 2GP | G3ZZW |
| LE5 2HQ | M6EIF |
| LE5 2LJ | M6VCN |
| LE5 2RL | G0ZIP |
| LE5 2RL | G7VIP |
| LE5 4LU | 2E0KOI |
| LE5 4QL | 2E0BIC |
| LE5 4QL | M3BIC |
| LE5 4WH | M6HCU |
| LE5 5UD | M0COQ |
| LE5 6AH | G1GKA |
| LE5 6EA | G6MJQ |
| LE5 6HL | M3UOZ |
| LE5 6JB | G4ZDQ |
| LE5 6PT | G4MGG |
| LE5 6SA | G1WZQ |
| LE5 6SY | G4IGL |
| LE5 6XT | G4YTF |
| LE6 0BA | G3UAA |
| LE6 0BN | G1ZQV |
| LE6 0BT | G4TFO |
| LE6 0EX | 2E0GSA |
| LE6 0LH | G6PGP |
| LE6 0LH | M5MUF |
| LE6 0YL | G8POS |
| LE6 1HT | G1XDS |
| LE6 1HT | G1YBK |
| LE65 1ES | G4WYN |
| LE65 1EU | G4LAI |
| LE65 1EW | G6EQB |
| LE65 1HT | 2E0RKK |
| LE65 1HT | M0GNO |
| LE65 1HT | M3RKK |
| LE65 1LY | G4SGD |
| LE65 1TR | G4JNH |
| LE65 1UL | M0BPS |
| LE65 1WA | G3UBB |
| LE65 2FN | G7SEK |
| LE65 2HL | G0NXT |
| LE65 2JQ | G3KTC |
| LE65 2JR | G7NBE |
| LE65 2JR | M3BBB |
| LE65 2JZ | G6CLP |
| LE65 2LW | 2E0SQN |
| LE65 2LW | M6SWS |
| LE65 2QQ | G1CWW |
| LE65 2QY | 2E1LGA |
| LE67 1BR | M0CEF |
| LE67 1GR | G1JOJ |
| LE67 2GL | G1ODQ |
| LE67 2GP | G8YGA |
| LE67 2HF | G4PLK |
| LE67 2HU | G1YQU |
| LE67 2ND | M1DHO |
| LE67 2NS | 2E0DFJ |
| LE67 2NS | M6RFJ |
| LE67 2QL | G0HME |
| LE67 2XD | 2E1HGF |
| LE67 3AF | G4AEO |
| LE67 3BE | G8YGT |
| LE67 3PB | M3KOF |
| LE67 3PL | G8VFM |
| LE67 3PW | G7WDG |
| LE67 4BF | G8ZZY |
| LE67 4DD | G1NNF |
| LE67 4DT | G0IXS |
| LE67 4JU | M6LAM |
| LE67 4TG | G4JDP |
| LE67 4TG | G0OZQ |
| LE67 5AY | G8JMG |
| LE67 5AZ | G1IWE |
| LE67 5BP | G4DCE |
| LE67 5BP | G6UZJ |
| LE67 5DF | G0FLG |
| LE67 5FD | G0PXK |
| LE67 5GF | M6UKM |
| LE67 5GN | 2E0FSK |
| LE67 5GN | M6GAH |
| LE67 5PA | M1AYC |
| LE67 5PA | M1BAC |
| LE67 5PF | M0DJB |
| LE67 5PT | G1KSC |
| LE67 5PT | G3RAL |
| LE67 5PT | G7SEG |
| LE67 6HP | G8YOG |
| LE67 6JT | G3ZJG |
| LE67 6JW | G0FZE |
| LE67 6LF | G4JKQ |
| LE67 6LL | M6MPW |
| LE67 6LP | G7NVS |
| LE67 6LW | M0JSD |
| LE67 6NX | G6ZVD |
| LE67 6QD | G0BAI |
| LE67 8LY | 2E0CGT |
| LE67 8LY | M3DFL |
| LE67 8LY | M6JBM |
| LE67 8NU | G6IHU |
| LE67 8RJ | G4IGC |
| LE67 9PY | M1FJB |
| LE67 9RJ | G6OWB |
| LE67 9RJ | G4RKD |
| LE67 9SN | M1EXO |
| LE67 9TX | G4ARI |
| LE67 9UJ | G1LTV |
| LE67 9UJ | G3SBF |
| LE67 9WA | G4CQQ |
| LE67 9WA | G4DLZ |
| LE67 9WH | G4BWF |
| LE7 1GQ | G1VBQ |
| LE7 1HJ | G8MZY |
| LE7 1HL | M1SCW |
| LE7 1HN | M6MFD |
| LE7 1HX | G0PVE |
| LE7 1LY | G0IPB |
| LE7 1PP | 2E0LAQ |
| LE7 1PP | M3LQL |
| LE7 2EH | M6KRR |
| LE7 2EN | G6FPO |
| LE7 2JH | 2E0ODF |
| LE7 2JN | G6NGV |
| LE7 2JS | G7IUI |
| LE7 2JW | G8TLC |
| LE7 3DN | G7OOT |
| LE7 3RJ | G8DBK |
| LE7 3TU | G3DAQ |
| LE7 3WZ | M1DMX |
| LE7 3ZJ | G0ORL |
| LE7 3ZR | M6PFR |
| LE7 4TJ | M3LQF |
| LE7 4WF | G1NPP |
| LE7 4YN | G6ABP |
| LE7 7BH | G7UYJ |
| LE7 7BQ | M5IAN |
| LE7 7DU | G3LRS |
| LE7 7DU | G4LRO |
| LE7 7DU | G6XRS |
| LE7 7EU | G6TMQ |
| LE7 7FA | G7NQU |
| LE7 7GN | M6OJM |
| LE7 7PU | G8LDB |
| LE7 7RL | G4NWS |
| LE7 7RL | G4WRC |
| LE7 9AN | M3MRZ |
| LE7 9DA | G4ZJR |
| LE7 9GW | G6ISM |
| LE7 9HD | 2E1CJZ |
| LE7 9HD | G7OEY |
| LE7 9HD | M0DBG |
| LE7 9HH | M6WKB |
| LE7 9HH | M6WVB |
| LE7 9JS | G3URQ |
| LE7 9PP | G3VOV |
| LE7 9PQ | G4DR |
| LE7 9UA | 2E0XZZ |
| LE7 9UA | M6KON |
| LE7 9UR | G4XXZ |
| LE8 0AP | G8DGH |
| LE8 0JJ | G0UFP |
| LE8 0JJ | M0BNS |
| LE8 0NP | M3JZX |
| LE8 3NE | G3RUO |
| LE8 4AB | M6DXS |
| LE8 4BE | G4WGU |
| LE8 4DL | G6KJH |
| LE8 4FQ | M0CDF |
| LE8 4FT | G1IAQ |
| LE8 4FU | G3KYF |
| LE8 4HF | G8OBP |
| LE8 5HT | G7WHI |
| LE8 5SB | M6ELN |
| LE8 5SU | G1IPP |
| LE8 5TB | G3MXV |
| LE8 5TG | G4YJU |
| LE8 5TL | G4FIE |
| LE8 5TP | G1HEN |
| LE8 5WJ | G8BTU |
| LE8 5WJ | G8CJA |
| LE8 6HY | 2E0KTV |
| LE8 6HY | M0KTV |
| LE8 6HY | M6BCH |
| LE8 6JQ | G6ZZE |
| LE8 6NF | G8ZQB |
| LE8 6NT | G7FMJ |
| LE8 6WB | G1OPV |
| LE8 8AP | G4EQL |
| LE8 8AX | M6GMY |
| LE8 8BF | 2E0LAA |
| LE8 8BP | G1PPU |
| LE8 8TR | M6BUR |
| LE8 8TX | G4IHR |
| LE8 8TX | G8RBI |
| LE8 9FH | G8HMJ |
| LE8 9FP | G8FWA |
| LE8 9GF | 2E1IDG |
| LE8 9GF | 2E1JMG |
| LE9 1SE | G8VEN |
| LE9 1SX | G1IUT |
| LE9 1UW | G1OOM |
| LE9 2BN | G4ECO |
| LE9 2BP | G4ZIZ |
| LE9 2DD | G0DMB |
| LE9 2DE | G4FSS |
| LE9 2DR | M0MRJ |
| LE9 2EN | G3ONV |
| LE9 2EU | G3OVH |
| LE9 3EB | G3WTD |
| LE9 3GE | G8XRG |
| LE9 3GX | G8PEA |
| LE9 4BT | G3RHZ |
| LE9 4BT | G4IQF |
| LE9 4BW | G4EPN |
| LE9 4BW | G4RCC |
| LE9 4DN | G4SHF |
| LE9 4DN | M0BAA |
| LE9 4DZ | G0CND |
| LE9 4DZ | G0JSE |
| LE9 4FX | G3TWY |
| LE9 4JZ | M6WAX |
| LE9 4LG | G8PGO |
| LE9 6EE | 2E1BAD |
| LE9 6HJ | G4OZD |
| LE9 6PT | G0LDO |
| LE9 6PY | M6OSU |
| LE9 6YU | G8OPX |
| LE9 7AF | G0BSN |
| LE9 7AW | G8EHM |
| LE9 7AY | M6CRO |
| LE9 7DA | G8SUM |
| LE9 7DU | G1LTK |
| LE9 7EL | G1MPD |
| LE9 7EY | M6BGL |
| LE9 7FP | 2E0SBM |
| LE9 7FY | G2AXO |
| LE9 7GT | G4VMM |
| LE9 7HJ | 2E0SHB |
| LE9 7HJ | M3NXE |
| LE9 7HJ | M6JAB |
| LE9 7HQ | G6MKL |
| LE9 7HW | G3TFV |
| LE9 7JF | G0HFL |
| LE9 7PD | 2E0RBU |
| LE9 7PH | M6BYL |
| LE9 7QG | G0LRG |
| LE9 7QG | G4AFJ |
| LE9 7QG | G4RMS |
| LE9 7TF | G3PVG |
| LE9 7TP | G6PHJ |
| LE9 8AE | M3TLJ |
| LE9 8AP | M1EYA |
| LE9 8BT | M0DJD |
| LE9 8DG | M6JHU |
| LE9 8DN | G8CGW |
| LE9 8DR | G6XXN |
| LE9 8EH | G4CAJ |
| LE9 8EW | G1WIW |
| LE9 8FS | G0UIF |
| LE9 8FS | G7TZU |
| LE9 8FS | G8LUV |
| LE9 8GP | G6FNQ |
| LE9 8JX | G7RXO |
| LE9 9JG | 2E1GXQ |
| LE9 9JJ | G6AYE |
| LE9 9JR | G0TPH |
| LE9 9LG | G0EDO |
| LE9 9LG | G8LMW |
| LE9 9LG | G8RYE |
| LE9 9LQ | G0HZG |
| LE9 9LQ | G1WSW |
| LE9 9LQ | G4UUG |
| LE9 9QF | G3LMR |

## LL

(Llandudno)

| | |
|---|---|
| LL11 2BG | G6WVD |
| LL11 2EU | G4AVC |
| LL11 2LU | G0NSG |
| LL11 2ST | 2E0SFB |
| LL11 2ST | M6CSS |
| LL11 2ST | M0SFD |
| LL11 3ED | M6EMK |
| LL11 3EN | G1MPR |
| LL11 3NN | G8UPJ |
| LL11 3PB | G7KNN |
| LL11 3PG | 2E0PWR |
| LL11 3PG | M3VSG |
| LL11 3PZ | M0ARL |
| LL11 3RY | M6FVT |
| LL11 3TW | G1IAW |
| LL11 3YT | G1LHV |
| LL11 4AF | G8ASD |
| LL11 4BY | G7UMW |
| LL11 4BY | M6DCN |
| LL11 4EB | M3WCE |
| LL11 4EE | M6EGD |
| LL11 4PF | M1VOD |
| LL11 4UE | 2E0VBK |
| LL11 4UE | M0ECF |
| LL11 5AD | 2E0PNX |
| LL11 5AD | M3OBL |
| LL11 5AD | M3PKU |
| LL11 5AH | G8KSE |
| LL11 5AH | G8KSF |
| LL11 5AS | M6DFC |
| LL11 5AW | G7CEA |
| LL11 5DH | G6IWC |
| LL11 5EP | 2E0MLG |
| LL11 5EP | M6GXU |
| LL11 5EU | G8VEE |
| LL11 5EU | M3SAI |
| LL11 5HH | G6JMC |
| LL11 5LH | G7MQE |
| LL11 5NG | 2E0OSG |
| LL11 5SH | M6EYU |
| LL11 5UF | G4TJN |
| LL11 5UH | G4GSS |
| LL11 5UN | G6MRO |
| LL11 5YB | M0ATI |
| LL11 5YF | G4OVH |
| LL11 5YP | G6NLP |
| LL11 6BP | M3IGZ |
| LL11 6DN | 2E0KGP |
| LL11 6DN | 2E0KGQ |
| LL11 6DN | G2HFR |
| LL11 6DN | M0KGP |
| LL11 6DN | M3KGP |
| LL11 6DN | M3KGQ |
| LL11 6EH | M0CCN |
| LL11 6HR | M3HQV |
| LL11 6HR | M3HRE |
| LL11 6NS | G0VMR |
| LL11 6RG | M3NDO |
| LL11 6RH | G1ZHI |
| LL12 0AU | G0MMY |
| LL12 0BP | G4IGF |
| LL12 0HN | G8HPL |
| LL12 0NW | G0KPU |
| LL12 0NW | M0RCH |
| LL12 0NW | M0YLS |
| LL12 0PW | 2E1VMR |
| LL12 0RT | G6MPX |
| LL12 0UW | G3YTL |
| LL12 7EP | M0GXC |
| LL12 7PD | G0ABE |
| LL12 7PP | G6SBD |
| LL12 7SG | M1DBA |
| LL12 7TW | 2E0TNB |
| LL12 7TW | M6BYX |
| LL12 7UG | G6GMF |
| LL12 7UH | 2E1DAO |
| LL12 8AF | G0EMB |
| LL12 8BE | G6FED |
| LL12 8BN | M1DAU |
| LL12 8DD | M3WPN |
| LL12 8HW | 2E0RBX |
| LL12 8LB | 2E0WWR |
| LL12 8LD | G8IJT |
| LL12 8RN | 2E0JCP |
| LL12 8RN | M3JWV |
| LL12 8RN | M3XLR |
| LL12 8SJ | G6JTX |
| LL12 8ST | G0CES |
| LL12 8YL | G0WXW |
| LL12 8YL | M0BNB |
| LL12 9PE | G0NOO |
| LL12 9PE | G0NOP |
| LL12 9PE | G7COB |
| LL12 9PJ | G6IGY |
| LL12 9PR | M6MJZ |
| LL12 9SE | G6PVK |
| LL13 0AY | G4VAG |
| LL13 0BL | G6ILY |
| LL13 0BZ | G6AJK |
| LL13 0LJ | G4XDR |
| LL13 0PQ | G3HEU |
| LL13 0UJ | G6FES |
| LL13 7AW | M3NIA |
| LL13 7EA | 2E0TYG |
| LL13 7EA | M0TBB |
| LL13 7EA | M6TYG |
| LL13 7QE | G0GWE |
| LL13 7QJ | G8MGF |
| LL13 7QW | 2E0ISZ |
| LL13 7QW | G6ZCS |
| LL13 7RW | M1BJB |
| LL13 9DA | M0CQR |
| LL13 9HR | 2E0AZM |
| LL13 9LY | G6ZHY |
| LL13 9NQ | G3XQO |
| LL13 9QH | G4TNF |
| LL13 9TA | G4ZYM |
| LL14 1BB | M0ARV |
| LL14 1NF | 2E0MDG |
| LL14 1NF | M0MDT |
| LL14 1PP | M0CSK |
| LL14 1ST | G4WVB |
| LL14 1UA | G6GTS |
| LL14 1UA | G8GGW |
| LL14 2DT | M6BYV |
| LL14 2EA | G6ZCR |
| LL14 2EA | M3JAP |
| LL14 2EY | 2E0JYN |
| LL14 2EY | M3WZZ |
| LL14 2EY | M6SHJ |
| LL14 2HT | G7PCX |
| LL14 2HT | M3PCX |
| LL14 2ND | M0ASL |
| LL14 2RL | G0WZZ |
| LL14 2RL | G8WYW |
| LL14 2RL | M0MRS |
| LL14 3EE | G0WLH |
| [illegible] | [illegible] |
| LL14 4DD | M3OXV |
| LL14 5BG | M6BYT |
| LL14 5DW | G7MHF |
| LL14 5NA | M6JTM |
| LL14 5PF | G1CTO |
| LL14 6AH | G0EHA |
| LL14 6AN | 2E0WXM |
| LL14 6AT | M3ZCI |
| LL14 6DD | G1MKV |
| LL14 6DP | 2E1PGL |
| LL14 6EG | G0MMB |
| LL14 6HS | G1GXQ |
| LL14 6TD | M6EAO |
| LL15 1AQ | G0GLI |
| LL15 1HB | 2E0PJM |
| LL15 1HB | M6BFM |
| LL15 1JA | M0GWY |
| LL15 1RR | M0ZUS |
| LL15 1YP | G3ODB |
| LL15 2DW | 2E1LCO |
| LL15 2EY | G6JWL |
| LL15 2YL | 2E1AED |
| LL16 3BE | M3AFR |
| LL16 3BE | M6FMW |
| LL16 3EU | G3RUE |
| LL16 3HE | G4CQZ |
| LL16 3PW | M3IIJ |
| LL16 3YA | G8FSN |
| LL16 4BD | G4IEZ |
| LL16 4BS | 2E0BLG |
| LL16 4BS | M3PNR |
| LL16 4BU | G0WWQ |
| LL16 4DT | G3GZX |
| LL16 4HH | G8RHP |
| LL16 4NN | G8SIE |
| LL16 4PQ | 2E0LJC |
| LL16 4PQ | M6CLJ |
| LL16 4RL | G4RWR |
| LL16 4YB | M0HSI |
| LL16 5PA | M3GQS |
| LL16 5UY | G3TKD |
| LL16 5YL | M3XHL |
| LL17 0AD | G6RNA |
| LL17 0BH | G7NIW |
| LL17 0BP | G6ONZ |
| LL17 0SX | G1TFB |
| LL17 0SX | M0YDX |
| LL17 0SX | M3GBD |
| LL17 0TD | M3CMG |
| LL17 0TH | 2E0GIX |
| LL17 0TH | M0OTH |
| LL17 0TH | M6GIX |
| LL17 0UP | M3DBF |
| LL18 1DA | G4ZPM |
| LL18 1TF | 2E1EPO |
| LL18 2AQ | 2E1EPL |
| LL18 2BX | M1EOK |
| LL18 2EN | M6GCN |
| LL18 2JJ | M3CEV |
| LL18 2LW | G4UJF |
| LL18 2NS | M6BVQ |
| LL18 2NU | G4XEF |
| LL18 2RS | G4XWN |
| LL18 2RY | G0DFY |
| LL18 2TP | G6IYP |
| LL18 2UL | G8WUM |
| LL18 2UL | M1EOO |
| LL18 2UL | M3TOB |
| LL18 3BG | G3OIN |
| LL18 3EE | G3UTG |
| LL18 3ER | G4NLD |
| LL18 3PE | G4ARC |
| LL18 3PE | G4HDR |
| LL18 3US | G0AYP |
| LL18 4AD | G1NGN |
| LL18 4AF | G0MOF |
| LL18 4DZ | G0MOH |
| LL18 4EH | 2E0FAP |
| LL18 4EH | M0HAP |
| LL18 4HH | G3FJI |
| LL18 4JF | G8OYT |
| LL18 4JH | G8XLL |
| LL18 4NF | M3CYM |
| LL18 4NF | M5CYM |
| LL18 4SH | M6EUC |
| LL18 4SU | G8TBY |
| LL18 4SU | M6NMG |
| LL18 4TN | G7VQA |
| LL18 4TY | M3HFO |
| LL18 5AG | G7EXH |
| LL18 5AG | G7NFY |
| LL18 5BN | M6TAK |
| LL18 5BP | M6ATG |
| LL18 5EW | G6VFN |
| LL18 5EY | G0HBZ |
| LL18 5HN | M0IDX |
| LL18 5HW | G4VLU |
| LL18 5JE | G4DMR |
| LL18 5JE | M0CCS |
| LL18 5LP | M3IXJ |
| LL18 5LS | G6XXY |
| LL18 5LS | M3XXY |
| LL18 5NH | G8FVI |
| LL18 5PP | M3CDL |
| LL18 5RB | M6FJQ |
| LL18 5RF | G6SLO |
| LL18 5RW | G0JSX |
| LL18 5TB | 2E1HTK |
| LL18 5TB | M3EGB |
| LL18 5TL | G0DSJ |
| LL18 5TL | G0NWR |
| LL18 5TT | 2E1BYK |
| LL18 5UP | G4JOT |
| [illegible] | [illegible] |
| [illegible] | G4TXH |
| LL18 6HS | M3SZD |
| LL18 6HT | G4UIR |
| LL19 7DF | 2E1CEE |
| LL19 7DS | G0ULP |
| LL19 7HH | M6TGF |
| LL19 7LP | G7RZN |
| LL19 7LP | G7TEO |
| LL19 7PF | G7AMS |
| LL19 7PF | M6FRW |
| LL19 7TH | M3HCW |
| LL19 7TS | G4DTQ |
| LL19 8DG | G7PEO |
| LL19 8EH | 2E1DIG |
| LL19 8HQ | M3FTP |
| LL19 8LP | G6RUO |
| LL19 8LU | G4XXP |
| LL19 8RD | 2E0CIV |
| LL19 8RD | M0MIE |
| LL19 8RD | M3YMY |
| LL19 8RN | G6YDT |
| LL19 8RN | M3YDT |
| LL19 8RU | 2E0ALZ |
| LL19 8SG | G3JGE |
| LL19 8TS | G6PMC |
| LL19 8YQ | M6EUK |
| LL19 9DT | G0ADC |
| LL19 9DU | G6ITJ |
| LL19 9HL | G3JGA |
| LL19 9HU | 2E0JPE |
| LL19 9HU | 2E0JWE |
| LL19 9HU | M3YYW |
| LL19 9HU | M6SAX |
| LL19 9NN | G0CNK |
| LL19 9NX | M6FXA |
| LL19 9PB | G0KZW |
| LL19 9SH | G4VHP |
| LL19 9SU | M1LLL |
| LL20 7AD | 2E1FJZ |
| LL20 7AR | 2E1EEP |
| LL20 7AS | 2E0CJJ |
| LL20 7AT | 2E1FJN |
| LL20 7AT | G7GHE |
| LL20 7BP | 2E0PFD |
| LL20 7BP | M0PDV |
| LL20 7BP | M6PDE |
| LL20 7DD | M3WAL |
| LL20 7UH | G0TBM |
| LL20 8AS | M0ICO |
| LL20 8AS | M3XQJ |
| LL20 8BE | 2E0HUU |
| LL20 8RD | 2E0XKL |
| LL20 8RD | M0XKL |
| LL20 8RD | M3XKL |
| LL20 X | G1UYW |
| LL21 0DL | G6WAG |
| LL21 0EH | M6BQM |
| LL21 0PE | G6NNB |
| LL21 0SB | G7KDI |
| LL21 0SB | M5DAD |
| LL21 9AA | G6UMU |
| LL21 9EG | G1ZKN |
| LL21 9NA | M0PPM |
| LL21 9NP | M3FLA |
| LL21 9NP | M3SLI |
| LL21 9TB | G6KAV |
| LL22 7AR | M3FLU |
| LL22 7BU | G1PYY |
| LL22 7BU | M3PYY |
| LL22 7DF | M3MRC |
| LL22 7DG | G4IDV |
| LL22 7DS | G4TWB |
| LL22 7HE | G7NTP |
| LL22 7JF | M0GWV |
| LL22 7LR | M0IBH |
| LL22 7LR | M6HHD |
| LL22 7UD | G0BCR |
| LL22 7UU | M1TAF |
| LL22 8DH | G0NSZ |
| LL22 8DH | G4ZWO |
| LL22 8DP | G7EVG |
| LL22 8EG | 2E0CEO |
| LL22 8EG | M0CDG |
| LL22 8EG | M6DIG |
| LL22 8EG | M6GOZ |
| LL22 8JA | G0SPY |
| LL22 8JD | G3WEQ |
| LL22 8PP | G6UNX |
| LL22 8QA | G1CDH |
| LL22 8UL | M3YHC |
| LL22 9LP | M1COJ |
| LL22 9ND | G3YIH |
| LL22 9NE | G4VHS |
| LL22 9YH | G4DLC |
| LL22 9YL | M6GFP |
| LL22 9YY | G8BIA |
| LL23 7NL | G4WPH |
| LL23 7PT | G6EWX |
| LL23 7SF | G4UFQ |
| LL23 7YY | M3WCS |
| LL26 0EP | M6ACZ |
| LL26 0EP | M6ADF |
| LL26 0RG | G0ETU |
| LL26 0RG | G4RRL |
| LL26 0RG | G4TSG |
| LL26 0TY | M6CVN |
| LL27 0JE | G0HUN |
| LL27 0JJ | M0BTI |
| LL27 0QA | G1IHB |
| LL28 4AJ | G0UTC |
| LL28 4AU | M6LFG |
| LL28 4NR | G8SXI |
| LL28 4QA | G4AMX |
| LL28 4RS | G6MLI |
| [illegible] | [illegible] |
| LL28 4TF | G6GKR |
| LL28 4TL | 2E0ENQ |
| LL28 4TL | M6ENQ |
| LL28 4TT | G3HGL |
| LL28 4UU | G1OIK |
| LL28 4UU | M3OIK |
| LL28 4YG | G7VHD |
| LL28 4YH | M3FVC |
| LL28 5HS | 2E0CVE |
| LL28 5NJ | G3DZJ |
| LL28 5NN | G8YDR |
| LL29 6AF | G4GJI |
| LL29 6DH | G7KGD |
| LL29 6DH | M6BAS |
| LL29 6DL | M0AFD |
| LL29 7AX | G8BQK |
| LL29 7BB | G3MDK |
| LL29 7BB | G6TM |
| LL29 7HB | G6STK |
| LL29 7NB | G0EVE |
| LL29 7NB | G0EVG |
| LL29 7ND | 2E0GGG |
| LL29 7ND | M0RKB |
| LL29 7ND | M3LTU |
| LL29 7ND | M6ECB |
| LL29 7SD | M3JUM |
| LL29 7TL | 2E0DFM |
| LL29 7TL | M6RMR |
| LL29 7TT | 2E0ITM |
| LL29 7TT | 2E0RWF |
| LL29 7TT | M0TMH |
| LL29 7YP | M0MFB |
| LL29 8EU | 2E1HNH |
| LL29 8EU | M0BXJ |
| LL29 8EU | M3RHI |
| LL29 8EU | M3UGW |
| LL29 8EX | G0PRM |
| LL29 8LE | G1SAM |
| LL29 8LG | G6HNW |
| LL29 8PD | G3WHU |
| LL29 8PW | G4NNL |
| LL29 8SA | G1CJJ |
| LL29 8TA | G0PNE |
| LL29 8UT | G0VOG |
| LL29 8UY | G0OGL |
| LL29 8YZ | M6WWZ |
| LL29 8ZA | G8VUG |
| LL29 9AJ | M6ZIZ |
| LL29 9DS | G8JJP |
| LL29 9EL | G0DYH |
| LL29 9HP | G4RCM |
| LL29 9LA | 2E1MWS |
| LL29 9LA | 2E1SWB |
| LL29 9LB | M3GPJ |
| LL29 9LB | M6PJN |
| LL29 9LD | 2E0CEF |
| LL29 9LD | M0HCA |
| LL29 9LJ | G0WNB |
| LL29 9LQ | G0VMW |
| LL29 9NL | G0SKC |
| LL29 9NL | G0TWI |
| LL29 9PY | G1KHH |
| LL29 9RH | M3DEL |
| LL30 1BL | G4WBT |
| LL30 1ES | G1BBH |
| LL30 1ES | G6HVA |
| LL30 1GD | G0NEC |
| LL30 1HQ | G0GST |
| LL30 1HZ | G1RVP |
| LL30 1JJ | 2E0FIE |
| LL30 1JJ | G4HBS |
| LL30 1JJ | M6MMF |
| LL30 1LP | G6WFW |
| LL30 1LT | M1AWT |
| LL30 1NG | 2E0AMB |
| LL30 1PF | G0EZB |
| LL30 1PF | G3UTL |
| LL30 1PY | G8WFS |
| LL30 1TU | G4PFL |
| LL30 1UH | G0AEZ |
| LL30 1UH | G7UUH |
| LL30 1UW | G7UIZ |
| LL30 2AA | G8VVX |
| LL30 2BU | G3FMR |
| LL30 2HL | G0LUA |
| LL30 2PQ | G2DNJ |
| LL30 2QA | G7TDQ |
| LL30 2RA | G3GJQ |
| LL30 2TY | M3EVN |
| LL30 2YE | G6NYR |
| LL30 2YP | G8FOY |
| LL30 2YQ | G7AHR |
| LL30 3EY | M0HCW |
| LL30 3HB | G1MIL |
| LL30 3LL | M0BET |
| LL30 3LN | G4PVU |
| LL30 3NL | 2E0DFN |
| LL30 3NL | M0HTG |
| LL30 3NL | M6EIK |
| LL30 3RB | M0PUF |
| LL31 0AN | M3MMJ |
| LL31 9BJ | G7RKQ |
| LL31 9BZ | G0SFP |
| LL31 9EL | G3LCQ |
| LL31 9ER | G1BDH |
| LL31 9HF | G6IVY |
| LL31 9HQ | G6IYA |
| LL31 9HS | 2E0DDZ |
| LL31 9HS | M6DAR |
| LL31 9LP | G4KFY |
| LL31 9PF | M1AHU |

Postcode

**IMPORTANT NOTE**

**Revalidate licence to avoid revocation** – Ofcom has advised the Society that plans will be drawn up to revoke licences that have not been revalidated as required by the licence conditions. The quickest way to revalidate is to do so online via the Ofcom website: *https://services.ofcom.org.uk/* or by email: *amateur.validations@ofcom.org.uk* If you need assistance in the process, Ofcom staff are available to help, but please be patient during times of heavy workload.

LL31 9PY G7KAX
LL31 9PY M3REJ
LL31 9QE 2E0CRB
LL31 9QE 2E0TYE
LL31 9QE M0HYK
LL31 9QF M0CPN
LL31 9RG G1PSW
LL31 9RR M3NFF
LL31 9UG G3XKB
LL31 9UG M0NAB
LL31 9UT G1KKJ
LL32 8EU G6ZDH
LL32 8NB M3NAQ
LL32 8NP G3YQP
LL32 8NR G8JKC
LL32 8NS G6SIX
LL32 8SB G6WKU
LL32 8SB M1JAN
LL32 8SB M3GNB
LL33 0EG G3MZY
LL33 0LL G0WUM
LL33 0RE M3XME
LL33 0SN G7JLG
LL33 0SR G0FPY
LL33 0SS G0LBA
LL330TN M3RAU
LL34 6ER G1ACV
LL34 6HB G0PQI
LL34 6LY G8XAS
LL34 6TE M0RXD
LL34 6UA G1SGG
LL35 0PT G0TWO
LL35 0RF M0HYA
LL36 0AG G4JPP
LL36 0AY G3XSR
LL36 0TA G6INF
LL36 9DE G4XXF
LL36 9EG M3MWE
LL36 9EG M3MWO
LL36 9EP G4KYK
LL36 9HG G6JDF
LL36 9LY G0COH
LL36 9ND G3LTX
LL36 9SB G4MSI
LL36 9SL M0ISF
LL36 9SL M6BXV
LL36 9UY G4XZJ
LL37 2JZ G1URD
LL37 2QB G6UWW
LL37 2UZ G6NXL
LL38 2AZ G1JPF
LL38 2BX M3WMI
LL40 1GA M5CKN
LL40 2AS M6WMW
LL40 2EE M3AXW
LL40 2EF M1CKK
LL40 2RP G0UCA
LL41 3DL G0WGM
LL41 3PU G1EKC
LL41 3SS 2E0ORT
LL41 3SS M0DOR
LL41 3SS M6YKS
LL41 4BH 2E0DUL
LL41 4BH M0WYN
LL41 4BH M6DUL
LL41 4BU 2E0OSH
LL41 4BU 2E0UPH
LL41 4BU M0UPH
LL41 4BU M3BSJ
LL41 4BU M6PZP
LL41 4BU M6UPH
LL41 4DD G8ZLT
LL41 4PN G7TTX
LL42 1DX G0HHD
LL42 1EN G6DDF
LL42 1PL M1KDP
LL42 1PT M3HGR
LL43 2AG 2E0LGG
LL43 2AG G8WNB
LL43 2AG M6LDM
LL43 2AL 2E0VTK
LL43 2AL M0VTK
LL43 2AL M3VTK
LL43 2AL M6LIZ
LL43 2AN 2E0LLL
LL43 2BB G0AYQ
LL43 2BB G4LZP
LL44 2BG M1WEJ
LL44 2BU M6CEG
LL44 2RQ G3FDZ
LL45 2HH G6JBN
LL45 2HT G0AGL
LL45 2LA M6DSU
LL45 2LU G0UKF
LL45 2ND G4ZTG
LL46 2SS G1KVI
LL46 2UW M6GHY
LL46 2UW M6PYH
LL46 2YW M6CHQ
LL47 6YA M3UKK
LL48 6AL G3JSG
LL48 6AY G4VWO
LL48 6AY G4VWP
LL48 6BH G8AAF
LL48 6BU M3DAO
LL48 6DU M6TLF
LL48 6EF M1BQO
LL48 6EG G6IMS
LL48 6LS G4PHB
LL48 6LS M3WMP
LL48 6SR G1EWY
LL48 6ST M3SZF
LL49 9AT M3RVN
LL49 9AT M5CAD
LL49 9BU G4HLO
LL49 9DF M6BOZ
LL49 9NB G6AZX
LL49 9UA M3DEI
LL49 9YA M6BQL
LL49 9YD G4OWQ
LL51 9SX M3WSC
LL51 9UJ G7KIO
LL51 9UQ G6VKI
LL52 0EF M3EOY
LL52 0EG M3ZHU
LL52 0SR G4IDC
LL52 0SR G8PKB
LL52 0SS M6VTA
LL53 5AP G0HGC
LL53 5NU 2E0VJL
LL53 5NU M3DEI
LL53 5NU M3VJL
LL53 5NU M6XSI
LL53 6DQ M0RHD
LL53 6DQ M3CYQ
LL53 6DQ M3CYU
LL53 6DQ M6RHD
LL53 6DW G6KLC
LL53 6EA G7UOH
LL53 6EG G3NNB
LL53 6RA M6EYI
LL53 6RQ G1ROE
LL53 6UB G0RMB
LL53 6UY G0IWD
LL53 7ED 2E0UUA
LL53 7PA 2E0BVK
LL53 7PF M0AEV
LL53 7PG G4UKU
LL53 7PG G4WKQ
LL53 7RL M3VVJ
LL53 8AE G0PZT
LL53 8AE G3KJW
LL53 8EH G8BZN
LL53 8ND G4UIL
LL53 8NG G7NJM
LL53 8NG M0NJM
LL53 8NW G1IEB
LL53 8UG G7KJO
LL54 5EF G1DKK
LL54 5ER M6SMD
LL54 5HG G4XRW
LL54 5HG G4XZP
LL54 5SF M3UWZ
LL54 6EW G0NDA
LL54 6HD G1OXJ
LL54 6LF M3ZKT
LL54 6RT G4CAT
LL54 6SN M3ZKY
LL54 6SS M3ZVD
LL54 7BH M3PZO
LL54 7DE M0VCC
LL54 7ED G0GLX
LL54 7HU M0LPG
LL54 7LB G8HQM
LL54 7NF M3YMI
LL54 7PU M3JQK
LL54 7YE G0ENT
LL55 1BE 2E0SVW
LL55 1EY 2E0CYM
LL55 1EY G6MUP
LL55 1HF G4MEI
LL55 1LL G0AQR
LL55 1UP G6TGR
LL55 1YB G1PIH
LL55 1YB G6REQ
LL55 1YL G4KAZ
LL55 1YT 2E0LLA
LL55 1YT M0KLW
LL55 1YT M6GBA
LL55 2AG G7CEQ
LL55 2EF 2E0OVT
LL55 2EF M3OVT
LL55 2RL G6BUW
LL55 2UR G4RLP
LL55 2UW M5DJO
LL55 3BN G4ZCY
LL55 3BN G4ZPL
LL55 3BN G6AKS
LL55 3BN G6UKO
LL55 3HD G1JIE
LL55 3LE 2E0AWW
LL55 3LU G1PCD
LL55 3LU G6NHL
LL55 3ND G0HHW
LL55 3NT 2E0GWK
LL55 3NT M6GWK
LL55 3PW G4PUX
LL55 3PW G7UQJ
LL55 3PW G8XUM
LL55 4AZ M3EWR
LL55 4LE 2E0GAQ
LL55 4LE M1EYU
LL55 4LE M6KTS
LL55 4RR M0PPO
LL55 4TW M0ANV
LL55 4YY M3DTC
LL56 4HQ M0LEF
LL56 4PB G4WKZ
LL56 4QZ 2E0KBO
LL56 4QZ M0IBT
LL56 4QZ M6SKV
LL57 1HF M3YNK
LL57 1LU 2E0JCN
LL57 1LU M0JWP
LL57 1LU M6CTB
LL57 1NA 2E0HMS
LL57 1NA M0RCW
LL57 1NG G0DXO
LL57 1NH 2E1AZU
LL57 1SH 2E0SJH
LL57 1SH M3KYG
LL57 2NS G4KTQ
LL57 2UD G0HGN
LL57 3AY M0EDS
LL57 3RE M6HRK
LL57 3SP G6MYY
LL57 3SP M0GOX
LL57 3SP M3AWI
LL57 3TD G8GOO
LL57 3YJ M3VQJ
LL57 4AS 2E0YLE
LL57 4AS M6TBC
LL57 4AX G7EWD
LL57 4DJ G0FQZ
LL57 4DP G3VFZ
LL57 4DR G0ETF
LL57 4LD G4FQU
LL57 4LX 2E0CZU
LL57 4LX M6BKS
LL57 4LX M6CTG
LL57 4NS G4HMR
LL57 4PA G6ORE
LL57 4PD G8PBX
LL57 4TG 2E0CHV
LL57 4TG M3SET
LL57 4UA M0BEA
LL57 4UR M0BYT
LL58 8AF M3GKI
LL58 8DB 2E0GVO
LL58 8HH G4EIE
LL58 8HW G0UWD
LL58 8LH M1DOU
LL58 8PG G4MBL
LL58 8PH G3GUX
LL58 8PT G8PZS
LL58 8RG G8TOX
LL58 8RW G6TOX
LL58 8SU G4JUI
LL58 8UE G4UCV
LL58 8YG G0BDW
LL59 5DA G8ZEI
LL59 5LR G3JBJ
LL59 5LR G4RQQ
LL59 5NB G4CFC
LL59 5NG G6DQB
LL59 5PB G4PNV
LL59 5PW G1SUK
LL59 5PW M5ADD
LL59 5RD G3XRM
LL59 5TH G4IGT
LL59 5UA G4TAU
LL59 5YA M3MEY
LL59 5YH G6KIW
LL59 5YH M0DNK
LL60 6BA G3PIO
LL60 6DG M3YQL
LL60 6DG M6WVC
LL60 6HD G3TLP
LL60 6HL M1BTA
LL60 6HL M3GWH
LL60 6HL M6FNK
LL60 6JN 2E0DER
LL60 6JN M0MON
LL60 6JN M3UUY
LL60 6JW G7BZR
LL60 6NW G2OG
LL61 5AX G4JXN
LL61 5JB G4GTC
LL61 5JB G8UTK
LL61 5JE G0CWG
LL61 5JF 2E0DIV
LL61 5JF 2E0SLP
LL61 5JF M5ADW
LL61 5JR G0IQZ
LL61 5JX G7OQO
LL61 5JY 2E0DAA
LL61 5JY G3VVC
LL61 5JY G4JKR
LL61 5JY G4TTA
LL61 5SZ G6IUK
LL61 5YT G6DOK
LL61 5YX G2DLK
LL61 6EQ M6TDB
LL61 6SY G0PZS
LL61 6TG M1DKM
LL61 6TG M6OTT
LL61 6TZ G3PRL
LL61 6TZ G4NBM
LL62 5AS 2E0MTD
LL62 5AS G7FNQ
LL62 5AS M3UPX
LL62 5AW G3EIZ
LL62 5BG 2E1FPK
LL62 5BG M1DAM
LL62 5LF G4HKX
LL63 5PQ G0GHG
LL63 5TW G0EAW
LL64 5JW G6WJM
LL64 5XB G3YTC
LL64 5XB M0JHC
LL64 5XB M3RWZ
LL65 1AL G0AVD
LL65 1BG G6VIC
LL65 1ES G4XAU
LL65 1EU G0LIS
LL65 1HT M6SPN
LL65 1LR M6GCQ
LL65 1LT M6KVO
LL65 1NW G4IEU
LL65 1SN G6JCK
LL65 1YR M0TMJ
LL65 1YR M1CFA
LL65 2AL 2E0RZL
LL65 2AL M6RZL
LL65 2AZ G1HHM
LL65 2DN G4WJO
LL65 2EF G0KPV
LL65 2EX G0NRW
LL65 2HD M0BLU
LL65 2LL 2E0RBW
LL65 2LL M6NLA
LL65 2LU M0AQZ
LL65 2NF G0IJY
LL65 2NS M3YQU
LL65 2PN G4MHR
LL65 2PN G4XYI
LL65 2UG M0REH
LL65 2WA M3GKB
LL65 2YH M0KGY
LL65 2YR M0CLB
LL65 3BT M3ZVB
LL65 3ES G0ENU
LL65 3ES M0GNF
LL65 3HS G7JKK
LL65 3LS M0DBB
LL65 3LU G0OGI
LL65 3LU M0MAH
LL65 3PL G0ONY
LL65 3PP M6OCJ
LL65 3RD G0VYG
LL65 3RD G4VEQ
LL65 3RR G6TVD
LL65 3SY G0MHK
LL65 3TB M6MFB
LL65 4AG G6CWZ
LL65 4AG G8JEI
LL65 4BG G4CWU
LL65 4EA M1EPI
LL65 4PY M3NDU
LL65 4SL G7JRT
LL65 4UY G4WZS
LL65 4YL G4DRR
LL65 4YL G8FOL
LL68 0RE G0FEM
LL68 0RE G8FDI
LL68 9AT M6DYF
LL68 9BG 2E0GIW
LL68 9BG M1EWJ
LL68 9BG M6PAM
LL68 9BT G3SRM
LL68 9DG G3RXD
LL68 9DL G6OAW
LL68 9DS M0SSB
LL68 9NE G8HYI
LL68 9NH 2E0GLV
LL68 9NH M0LLK
LL68 9NH M6GLV
LL68 9PD G1XAS
LL68 9RX G0ESK
LL69 9BZ G4ZWN
LL69 9YB G1WWW
LL71 7BU M0SEC
LL71 7DH M6RHR
LL71 8ED G0TPR
LL71 8EH 2E0GDB
LL71 8EH 2E0GMZ
LL71 8EH M0GMZ
LL71 8EJ M6FJX
LL72 8HB G1OKY
LL72 8HN G4EXE
LL73 8PE G8YUJ
LL73 8PP G0KRL
LL73 8PP G7APP
LL74 8RD 2E0VSW
LL74 8RD M3XQE
LL74 8RG G0MOI
LL74 8RW G6MWM
LL74 8SR G0BTB
LL75 8LN G7SBO
LL75 8NQ G0WEY
LL75 8UR 2E0CBZ
LL75 8UR M3ZVH
LL75 8UT 2E0SJT
LL75 8UT M0GCT
LL75 8YE M0JDW
LL75 8YG G1ISK
LL77 7DZ G6YMS
LL77 7EJ M1DNY
LL77 7EZ G0PJF
LL77 7JS M6FJT
LL77 7LJ G3TWN
LL77 7NA G6BMP
LL77 7QD 2E0ZZF
LL77 7QD M6ADU
LL77 7QE G3PRW
LL77 7RL G7BZY
LL77 7SJ M0BER
LL77 7SJ M0BTU
LL77 7SU G0ESU
LL77 7WD M3TOI
LL77 7YP G7CMF

## LN
(Lincoln)

LN1 1AY G1EUZ
LN1 2AS G4STO
LN1 2AS G5FZ
LN1 2AS G6COL
LN1 2BG G0RAF
LN1 2BG G3VCA
LN1 2DN G1AUQ
LN1 2DY G4ZKW
LN1 2EP G7OJU
LN1 2EP M6DDQ
LN1 2EX M6BGR
LN1 2GY 2E0ZLO
LN1 2GY M0ZLE
LN1 2GY M6ZAC
LN1 2HZ G7KIE
LN1 2JD M3EJR
LN1 2LH M6JHV
LN1 2LU G0BEN
LN1 2NH G0BUC
LN1 2QA G1WVO
LN1 2QA M3WVO
LN1 2QB G1HTM
LN1 2QL G3UPI
LN1 2QN G6BPH
LN1 2RG M3WBN
LN1 2SG G4GDC
LN1 2TL 2E0LEY
LN1 2TL 2E0VWG
LN1 2TL M6EGX
LN1 2XF M6JAG
LN1 3JS G4SSV
LN1 3LB G0VWW
LN1 3LU G4OEF
LN1 3NN G0RQQ
LN1 3QP G7KYJ
LN1 3QP M0XYZ
LN1 3SU G7KPM
LN10 5DY G6NKL
LN10 6RG G8VSX
LN10 6RT 2E0JHO
LN10 6RT M6DKF
LN10 6RW G1VMX
LN10 6TH G3OBZ
LN10 6UE G3MDI
LN10 6UR 2E0TPD
LN10 6UR M0VLC
LN10 6UR M3WUV
LN10 6UU G4IJA
LN11 0AP M3ZWI
LN11 0AZ 2E1GIK
LN11 0AZ M5COL
LN11 0BD G8SQY
LN11 0BP M6JPI
LN11 0DF 2E0PJS
LN11 0DN G7ENA
LN11 0DW M3CDV
LN11 0JD M3REL
LN11 0LD 2E1HWQ
LN11 0PB G0KKR
LN11 0PN G4MXI
LN11 0QD M6GOO
LN11 0XF G1VUG
LN11 0XL G4LVN
LN11 7AD G3PRR
LN11 7AG 2E1GZF
LN11 7HU M3WYL
LN11 7JR G3ZPM
LN11 7LN G6GZS
LN11 7QH G7VIG
LN11 7QH M0VIG
LN11 7QH M6AVS
LN11 7QU M3LRP
LN11 7QW G7MWH
LN11 7RL G4VFU
LN11 7SA G3NRQ
LN11 8AS 2E0EDS
LN11 8AS M6EDS
LN11 8EW G8OOS
LN11 8HL 2E0MBT
LN11 8HL M6MKB
LN11 8JD G6XXL
LN11 8LJ G3NHR
LN11 8RY G4HUD
LN11 8SG G1PRS
LN11 8SP M1BTR
LN11 8TA G0DPY
LN11 8UF M6RIT
LN11 8UJ G0VRU
LN11 8US G6IYS
LN11 8YN M6DIF
LN11 9AL 2E1FEF
LN11 9DG G8ECI
LN11 9HT G0GNW
LN11 9RQ G7AJP
LN11 9TF G4YZC
LN12 1BQ G0SWS
LN12 1BQ G6TAN
LN12 1BQ M0ERG
LN12 1BY 2E0BWQ
LN12 1BY G4ECE
LN12 1BY M6SIX
LN12 1DA G0FMG
LN12 1DA M3INQ
LN12 1JD 2E0JSJ
LN12 1JD M6TDF
LN12 1JF 2E0PGI
LN12 1JF M0PGI
LN12 1JF M6BHH
LN12 1JF M6RDW
LN12 1LL G3SLS
LN12 1NX G3WNQ
LN12 1PJ M3SCQ
LN12 1QB G0KED
LN12 1QE 2E1FEG
LN12 1QE G1XWD
LN12 1QR 2E0YYD
LN12 1QR M0XDX
LN12 2AJ 2E0GBH
LN12 2AJ G3VDV
LN12 2AJ G7BUK
LN12 2AJ M0DZA
LN12 2AJ M3FEY
LN12 2AS G1IEX
LN12 2AX M3YUB
LN12 2AX M6CTV
LN12 2AY G0DXK
LN12 2BF M6DRK
LN12 2BP M3FQT
LN12 2BS 2E0LRK
LN12 2BS M3NQL
LN12 2DF G7ANK
LN12 2EL M3RDY
LN12 2GY G3MNS
LN12 2HP G6NZG
LN12 2JU G8EWF
LN12 2NP G1BVV
LN12 2NU G1AAC
LN12 2NU G4BFV
LN12 2PT M3TMG
LN12 2QS G3LSA
LN12 2QY G4NJW
LN12 2QZ G4HHM
LN12 2RE 2E1FYZ
LN12 2RE G0CBM
LN12 2RT G0NQA
LN12 2RT M6PJX
LN12 2TX G1HEY
LN13 0JP G0DVY
LN13 0JP G0KUR
LN13 0JP G3SYB
LN13 0JP G4WEC
LN13 0JW G3ZUI
LN13 0NP 2E0KIM
LN13 0PN G0KBZ
LN13 9PH 2E1FFJ
LN13 9PH G7RAG
LN13 9RF M3MUI
LN13 9RL 2E0CDI
LN13 9TR G4VYF
LN2 1JD G8OPP
LN2 2AE G0WQA
LN2 2AY M6VAP
LN2 2JL G0CEG
LN2 2JL G4RUE
LN2 2JR G0PKN
LN2 2JU M1FBW
LN2 2JW G6BWM
LN2 2NE G3NIC
LN2 2PH G6BD
LN2 2PH G7FAR
LN2 2QU G2OT
LN2 2QU G3RRN
LN2 3JS G1LQB
LN2 3JX G0NXF
LN2 3SR G4YQK
LN2 3TR 2E0DAT
LN2 3TR M3NKL
LN2 3TT G7PMQ
LN2 3TU G0JOD
LN2 3US 2E1RWC
LN2 3US M3RWC
LN2 4BT 2E0XXO
LN2 4BT M6UGN
LN2 4LT 2E0LRJ
LN2 4LT M0TEF
LN2 4LT M1DYI
LN2 4LT M3LRJ
LN2 4LX G6EWP
LN2 4LX G6SWZ
LN2 4PT G4KSA
LN2 4QH M5RFD
LN2 4RE 2E0NDT
LN2 4RE M6NDT
LN2 4TX M0EJL
LN2 4WS G7JRD
LN2 5EY 2E0KWM
LN2 5EY M0KWM
LN2 5EY M6KWM
LN2 5LP G7RQO
LN2 5QF 2E0KVR
LN2 5QF M6KNG
LN2 5RU G3NGJ
LN3 4AN G6DKK
LN3 4AX G0VKF
LN3 4BE G3PTZ
LN3 4BG M0CZT
LN3 4BH M1AYR
LN3 4BW M5YEX
LN3 4DH G6RFR
LN3 4DQ M3CCQ
LN3 4DQ M3CTB
LN3 4EB G7NSK
LN3 4EG 2E0TMO
LN3 4EG 2E0VKW
LN3 4EG G0OSO
LN3 4EG M6TMO
LN3 4EG M6VKW
LN3 4EJ G0TQV
LN3 4HT G3NSL
LN3 4JQ G1FLL
LN3 4JS G4XMQ
LN3 4LS 2E0CTL
LN3 4LS M6BGA
LN3 4LS M6PPP
LN3 4NA G4SLG
LN3 5AB M0BKK
LN3 5TY G3VEV
LN3 5XS G7PHR
LN4 1AE G7GJT
LN4 1DB 2E0DKY
LN4 1DB M6SFQ
LN4 1DZ 2E1BME
LN4 1DZ G4URX
LN4 1EG G4DTL
LN4 1EY G4JQJ
LN4 1HG G4AWK
LN4 1LH G1SVI
LN4 1NB M6ICK
LN4 1NN G1OQF
LN4 1NP G3AWK
LN4 1NW G4YYE
LN4 1PG G4RYB
LN4 1PU G7EJH
LN4 1TT G4JNL
LN4 1TU G0GTV
LN4 2EX M0ONQ
LN4 2LE G3WOK
LN4 2PW 2E0TYC
LN4 2PW M6TYC
LN4 2PW M6VCK
LN4 2QG G7JBZ
LN4 2RD M6FYV
LN4 2RP G3WTT
LN4 2TD M6CDC
LN4 2TF M3GUU
LN4 2TS G3KRZ
LN4 2TY M6CXV
LN4 3AP 2E1GJP
LN4 3BQ M0RDS
LN4 3DS G7VSN
LN4 3DT G3NXT
LN4 3PD M0OCT
LN4 3RN G8IJC
LN4 3SL G3LAS
LN4 3YG 2E0DLH
LN4 3YG M0HIL
LN4 3YG M3LRI
LN4 4AQ G0PXB
LN4 4EA G3MON
LN4 4FN M6TFJ
LN4 4JJ M3GUF
LN4 4JJ M6GOG
LN4 4JJ M6YAR
LN4 4JQ 2E0SBL
LN4 4JQ M6SAY
LN4 4PB G0MPQ
LN4 4QA G1EFK
LN4 4QH 2E0WOW
LN4 4QH M6ROC
LN4 4QU G4DDI
LN4 4RG G6ZTL
LN4 4RG M0HAZ
LN4 4SP G4MWJ
LN4 4WR G6XFB
LN4 4XT M6RJP
LN5 0EE G6HTY
LN5 0ER G4PWM
LN5 0ER G4UNF
LN5 0FD G6PMW
LN5 0JA G7KPS
LN5 0RN G7JGI
LN5 0RN G7OVS
LN5 0RP 2E0MFF
LN5 0RP M6FFY
LN5 0SJ G3VHN
LN5 7PZ M6XDR
LN5 7QF M0MNV
LN5 7SJ G8TSV
LN5 7TH 2E0COF
LN5 7UT G0MRB
LN5 8DA G0UOZ
LN5 8DA M3VOZ
LN5 8EL G0LEN
LN5 8EL G7LEN
LN5 8QS M6CAJ
LN5 8RL 2E0BVG
LN5 8RL M0DIW
LN5 8RW G0EJQ
LN5 8RX G0FVI
LN5 8SH M0CES
LN5 8TG 2E0LUY
LN5 8TG G8YMW
LN5 8TG M0LUY
LN5 9DA 2E0GSJ
LN5 9DR M3ULU
LN5 9DT G4LQH
LN5 9FW G0WFV
LN5 9FW G7RAF
LN5 9JD G0KAU
LN5 9LJ M6SAV
LN5 9NE M3GEK
LN5 9QR G0MQD
LN5 9SW G4ZOX
LN5 9SX 2E0TRP
LN5 9UF M6AGR
LN5 9UT G0MEY
LN6 0AG M6NMD
LN6 0JB M6EBA
LN6 0LZ G1XZG
LN6 0NR 2E0EOD
LN6 0NR M6LJB
LN6 0SS G4OMD
LN6 0SY G8HMZ
LN6 0YF 2E0SJP
LN6 0YF M6MJP
LN6 0YZ M6DTO
LN6 3NR G7JQZ
LN6 3NU G0EUN
LN6 3RQ G6OYU
LN6 5TF G0EJV
LN6 5TW G4HNQ
LN6 5UF 2E0CJM
LN6 5UF M0KAN
LN6 5UF M6AVZ
LN6 5UU M0CQH
LN6 5UW G6KSR
LN6 7DR G7ERS
LN6 7DR M0ERS
LN6 7EW M6JGZ
LN6 7LQ G1ZBY
LN6 7NL G0VRE
LN6 7NL G8VEM
LN6 7NQ G4JES
LN6 7PA G6HUP
LN6 7PJ M6CKE
LN6 7PN G3PVU
LN6 7PY M6DKO
LN6 7RD G7UAV
LN6 7RQ M6CMG
LN6 7TG M3FKL
LN6 7TG M3FKM
LN6 7UD M3XCX
LN6 7UP M6CYW
LN6 8AP M6TCZ
LN6 8BD G4UAL
LN6 8BT M3GFZ
LN6 8BX G0LZS
LN6 8BY M6DUJ
LN6 8JU M6IMI
LN6 8LY G4OSB
LN6 8QS 2E0DDC
LN6 8QS M3XOE
LN6 8SD G6GLW
LN6 8SN G0IMP
LN6 8TZ M3OVA
LN6 8UB M1DJS
LN6 8UW 2E0GAL
LN6 8UW 2E0KAG
LN6 8UW M3YZV
LN6 8UW M6RVH
LN6 9AZ M6BXX
LN6 9BG G8BGL
LN6 9BS G4HIV
LN6 9DH 2E0OVR
LN6 9DH G7JFI
LN6 9DH M6PDV
LN6 9EX G3IYF
LN6 9HZ G3WZJ
LN6 9JE G8SGI
LN6 9LU G4OQR
LN6 9LW G7IRS
LN6 9PA 2E0IHH
LN6 9PA M6IHH
LN6 9PJ G7VTN
LN6 9QF G4LFE
LN6 9RG M3TPD
LN6 9RG M6LQO
LN6 9SB M0KED
LN6 9SP M6FXX
LN6 9SZ G0UND
LN6 9TB M3XBO
LN6 9TG 2E0DZV
LN7 6EL G0CVN
LN7 6GN G4JA
LN7 6HG G7OOE
LN7 6HZ G0EFZ
LN7 6JP G3ZRP
LN7 6NQ 2E0FML
LN7 6PA G4FQP
LN7 6RB G3RXP
LN8 2HE M6RCI
LN8 2HL G4CLL
LN8 3AA G7JXX
LN8 3AG G4UHZ
LN8 3BE M6DVZ
LN8 3DS 2E0RPD
LN8 3DS M0RPD
LN8 3DS M3RPD
LN8 3EE G0GMA
LN8 3EE G6MVD
LN8 3EF G3MZB
LN8 3EW G4OER
LN8 3EW M3OER
LN8 3LD G1YFQ
LN8 3LD G3KDY
LN8 3LE G4PWF
LN8 3PB M6VEN
LN8 3QP G4XFC
LN8 3UB M3GEZ
LN8 3US M0OOO
LN8 3US M3ZSW
LN8 3UT M0DBY
LN8 3XZ M1BRY
LN8 3YF M6EOX
LN8 3YL G8THR
LN8 3YL M3PCC
LN8 5RB M1DHV
LN8 5RB M3WRA
LN8 5RF G6TVP
LN8 5RF M5ZZZ
LN8 5RY G8KJI
LN8 5SL G7UEV
LN8 6AD G4KIZ
LN8 6AJ G3YFU
LN8 6AN M0BIC
LN8 6AZ G8UDI
LN8 6BJ G0TWE
LN8 6BN G3JKB
LN8 6DE G8JCS
LN8 6DH G4LAY
LN8 6DX 2E0EDG
LN8 6EW 2E0IJQ
LN8 6EW 2E0LJB
LN8 6EW 2E0ODL
LN8 6EW M3ODK
LN8 6EW M3ODL
LN8 6EW M6IJQ
LN8 6EX G4RBP
LN8 6JY G8RYO
LN9 5AH M6ARB
LN9 5AP M6RPZ
LN9 5AW G3VUE
LN9 5HF G0NDY
LN9 5JE M3VUV
LN9 5JF M1MSF
LN9 5NH G0SWW
LN9 5NH G7MES
LN9 5NW G6SWO
LN9 5PT G1JZG
LN9 6AF G6ZQS
LN9 6AG M3ZXH
LN9 6AW G3ZPU
LN9 6AW M0TCM
LN9 6BE G4FOT
LN9 6BH 2E0PKR
LN9 6JH M6HWT
LN9 6LB G3OUT
LN9 6LD G7NLA
LN9 6LF G4BZA
LN9 6NN M0AJD
LN9 6PQ G4WMO
LN9 6PQ G4ZWA
LN9 6QB G3SQA
LN9 6QP G3SCD
LN9 6RE G4AIE
LN9 6RR 2E0XOD
LN9 6RR M3XOD

## LS
(Leeds)

LS10 1LH 2E0CJQ
LS10 1LH M6TWL
LS10 3DT M6SDQ
LS10 3EJ M3IBJ
LS10 3LR 2E0RCR
LS10 3LR G4NOK
LS10 3LR M0RCX
LS10 3RB G1AUH
LS10 3TB 2E0CRM
LS10 3TB M3BZC
LS10 4EY 2E0JTI
LS10 4EY M3JTI
LS10 4JT M6NYY
LS10 4LH M6WDC
LS10 4NQ M6MWP
LS10 4QN 2E0SDT
LS10 4QN M0SDT
LS11 0HZ M3VBT
LS11 5RP G4RCH
LS11 5RX G4MPO
LS11 5SG G4IDT
LS11 5SG G8UAF
LS11 6NN G7HUJ
LS11 7LS 2E0BKY
LS11 7LS M3TQT
LS11 8AH 2E0MDA
LS11 8AH M6ARF
LS11 8BG M3BTG
LS11 8HR G7POS
LS11 8QJ M6OVE
LS11 8TP G1KSW
LS11 9RE G7OCX
LS12 1UA M1MIC
LS12 2PG G7ECA
LS12 2RE G0WXG
LS12 2RL M3XFI
LS12 3LB G7RME
LS12 3QJ G8SWK
LS12 3SN G8PKJ
LS12 4HL M6TFP
LS12 4HQ G4ICF
LS12 4HQ G8KZY
LS12 4JX G0RFA
LS12 4LA G6YMH
LS12 4RD 2E0ZRL
LS12 4RR G4IDD
LS12 4UW G6SFH
LS12 5BL M6RWB
LS12 5EA G4LXW
LS12 5LP G4YQW
LS12 5QS G4RYE
LS12 5SU G7HRP
LS12 5SZ 2E0GNU
LS12 5SZ M3XGI
LS12 6AY G1OUX
LS13 1BL G4SQG
LS13 1EA G7LBH
LS13 1ED M6EHO
LS13 2BJ 2E0JAQ
LS13 2BJ M3JTO
LS13 2BX G1DWT
LS13 2LE G4HLI
LS13 2LH G4OFA
LS13 3DF M0RSF
LS13 3DH G7BXA
LS13 3EH 2E0XPP
LS13 3EH M3XPP
LS13 3JX G0UOF
LS13 3RQ G6XGK
LS13 3RS M6PNX
LS13 3SG 2E1ODG
LS13 4QY G4OOJ
LS13 4RN M0AVZ
LS13 4SU 2E1RON
LS13 4TT 2E0ERM
LS14 1AP G0JUT
LS14 1BN M1EKL
LS14 1JP G0TGH
LS14 1LF M6OLS
LS14 1LJ 2E0KZH
LS14 1LJ M0KZH
LS14 1PL M0GIZ
LS14 2EP G0VTJ
LS14 2EQ G8TBX
LS14 3AU G1ZUB
LS14 5AS M6RGY
LS14 5BQ M3NPW
LS14 5PD G7VEB
LS14 6AE M0CVC
LS14 6JL G4JJS
LS14 6JZ G7PMX
LS14 6RH G4VRW
LS15 0EZ G7ELS
LS15 0EZ G7KJT
LS15 0HQ G7OZQ
LS15 0LW G4XZI
LS15 0NQ G7NXV
LS15 0NQ M5FOX
LS15 4EJ 2E0JOE
LS15 4EJ M3JOW
LS15 4EZ G3XVP
LS15 4HJ G4ZRF
LS15 4JD G3GXQ
LS15 4NY G4OOB
LS15 7DN M1CAE
LS15 7HA G6FFQ
LS15 7HD 2E1GLR
LS15 7HD M1DIE
LS15 7LF G4UYI
LS15 7QE G1IUZ
LS15 7QW G0HOB
LS15 7SQ G7DCT
LS15 8ED M0RAP
LS15 8JJ G8ZXT
LS15 8LW G4HSZ
LS15 8QZ M1ATB
LS15 8SD G7MTI
LS15 8SH G6INO
LS15 9JD G3RZY
LS16 5DN 2E0JAF
LS16 5DN M0JAF
LS16 5DY M3ZTN
LS16 5EG 2E0TMF
LS16 5EG M6AMT
LS16 5EG M6ZWP
LS16 6BU G8MDG
LS16 6BU M0DSI
LS16 6BX G0FWP
LS16 6DU 2E1ANQ
LS16 6EF G1SGM
LS16 6JA M0DBD
LS16 6NA M0APY
LS16 7AB G0PAN
LS16 7ES M0RCP
LS16 7ES M3OOL
LS16 7HF G4PHP
LS16 7PG G4TSF
LS16 7PN G0WRT
LS16 7PP G0SCL
LS16 7QX G4HAJ
LS16 7RB G7TEL
LS16 7SL G4BNL
LS16 8BL G3NRM
LS16 8DE G1SBK
LS16 8EJ G4CSZ
LS16 8JQ G1CXQ
LS16 9DQ G3MFJ
LS16 9DR 2E1CNO
LS16 9LE G4HGT
LS16 9LF G4BNM
LS17 5BH G8UHW
LS17 5HW M1AZM
LS17 5JP M1AHR
LS17 5NS G4THX
LS17 5PQ G3WNR
LS17 6DB M0SDU
LS17 6FD G4RYS
LS17 6HQ G4UZN
LS17 6HQ G8LCI
LS17 6LL M0BGS
LS17 6RF M1APL
LS17 6RY G3AAS
LS17 7DU G0SBZ
LS17 7DW G0CYL
LS17 7EP 2E1COM
LS17 7EP G7DHM
LS17 7NH G3KVJ
LS17 7PD G4LYM
LS17 7QB G4MSN
LS17 7RB G7RBA
LS17 8AR M6NYM
LS17 8BG 2E0LSE
LS17 8BG M0SME
LS17 8BG M0YKR
LS17 8BS G4XZG
LS17 8BY G4NCK
LS17 8DA M6NWO
LS17 8DF 2E0LBA
LS17 8DF M6LYA
LS17 8JW G3HDX
LS17 8LX G0PFT
LS17 8LX G8ASG
LS17 8TB G3HJP
LS17 9AB 2E0SFS
LS17 9AB M6PZW
LS17 9EG G1MPT
LS17 9ER G3TAF
LS17 9NL G1HHT
LS18 4BD M6XNO
LS18 4BE 2E0NCC
LS18 4BE M3NCT
LS18 4HL 2E0CUA
LS18 4HL M0ABO
LS18 4HL M3IDW
LS18 4HQ M3WTD
LS18 4HS G7RKJ
LS18 4PJ G0KVC
LS18 4RL G4OVT
LS18 4RL G4YDW
LS18 5AA G0BAK
LS18 5AW G7VHU
LS18 5DT M0ALD
LS18 5ES G3ZTB
LS18 5HB G4LAD
LS18 5HB G8WYR
LS18 5JL G1ELX
LS18 5JP G4XQV
LS18 5JS G4GYL
LS18 5LD G4GOP
LS18 5PP G4RVO

| Postcode | Callsign |
|---|---|
| LS18 5QE | G4OOH |
| LS18 5QU | M1DIB |
| LS18 5RN | G0LHU |
| LS18 5RQ | G4YBA |
| LS18 5SJ | G6OTS |
| LS18 5UN | G4FKY |
| LS19 6AD | G1BQV |
| LS19 6AR | G3ZNK |
| LS19 6BS | G4BZL |
| LS19 6EH | G7BHG |
| LS19 6NE | 2E0KDV |
| LS19 6NE | M0KDV |
| LS19 6NE | M3KDV |
| LS19 6PU | G4AYH |
| LS19 6RJ | G3RMQ |
| LS19 7AS | G4FBB |
| LS19 7AU | G4REC |
| LS19 7ED | G1PFZ |
| LS19 7NL | G0VXV |
| LS19 7QA | G0JUL |
| LS19 7SQ | G8RTK |
| LS19 7TE | G7KHT |
| LS19 7XE | G0HQU |
| LS20 8AY | G4GGR |
| LS20 8EJ | M0JSE |
| LS20 8HA | G1NEV |
| LS20 8NX | G3KKP |
| LS20 9AU | M1BZF |
| LS20 9DY | M0CSR |
| LS20 9EF | G7PFY |
| LS20 9EW | M6WWJ |
| LS20 9HN | M0JAM |
| LS20 9LW | M0TEX |
| LS21 1AH | G7VTW |
| LS21 1DH | G0CLD |
| LS21 1JZ | G6XOG |
| LS21 2AA | G8HPY |
| LS21 2AL | G7OXH |
| LS21 2BY | G0NIG |
| LS21 2DB | M1FHX |
| LS21 2DP | G0KDQ |
| LS21 2EJ | G0CLF |
| LS21 2FN | G4KUK |
| LS21 2RL | 2E0KBL |
| LS21 2RL | M6SKW |
| LS21 2RS | G1RRG |
| LS21 3DN | G7ELE |
| LS21 3DS | 2E0LAL |
| LS21 3DS | M3FZJ |
| LS21 3HY | M6FFK |
| LS21 3LE | G0SNW |
| LS21 3LE | M0HWL |
| LS21 3LN | G1TVW |
| LS21 3LW | G7RDJ |
| LS21 3NR | G4JVZ |
| LS21 3NZ | G4GDL |
| LS21 3NZ | G4LZT |
| LS21 3PN | G4MWQ |
| LS22 4BB | M3XVW |
| LS22 4EH | G7OAV |
| LS22 5BL | G1EGK |
| LS22 5BW | G4YZN |
| LS22 6NP | G1UDB |
| LS22 6RN | G1SBN |
| LS22 6TY | G3HRU |
| LS22 7PU | G3YTN |
| LS22 7QY | G4IEC |
| LS22 7RA | G4WWL |
| LS22 7TG | G0DJM |
| LS22 7UD | 2E0PUB |
| LS22 7UE | G1RPE |
| LS23 6AY | G4JVC |
| LS23 6EJ | M3HTO |
| LS23 6HU | G6RDO |
| LS23 6PU | G8XMU |
| LS23 6PX | G1VAG |
| LS23 6PX | M0VAG |
| LS23 6RN | G4XYY |
| LS23 6RP | G3VMW |
| LS23 7AL | G8JVS |
| LS23 7DZ | G3WDU |
| LS24 8AG | G0PEV |
| LS24 8JD | G1GCF |
| LS24 9BR | G4FFM |
| LS24 9DL | G0HXU |
| LS24 9HP | M3PWE |
| LS24 9HR | M3BLG |
| LS24 9HS | G0BCI |
| LS24 9PQ | G4TDC |
| LS24 9QW | M0EBR |
| LS25 1BZ | G4HKR |
| LS25 1DA | M0ASO |
| LS25 1EF | M6FVP |
| LS25 1EG | 2E0LDS |
| LS25 1EG | M0SNW |
| LS25 1EG | M6CHO |
| LS25 1EN | G4DTT |
| LS25 1HE | G4CRP |
| LS25 1HF | [illegible] |
| LS25 1HF | M6HGF |
| LS25 1HN | G8MWX |
| LS25 1NS | M0APC |
| LS25 1NS | M0NUK |
| LS25 1PW | M1BYT |
| LS25 2BP | G1PHS |
| LS25 2EG | G1VGP |
| LS25 2EN | G4DVZ |
| LS25 2EN | G7CCL |
| LS25 2EP | G7ADW |
| LS25 2HA | M1DGE |
| LS25 2JR | G1FDL |
| LS25 2LE | 2E0GAK |
| LS25 2LF | G0MYM |
| LS25 2NW | G4JKH |
| LS25 4AU | G4ZSS |
| LS25 5AY | G0FXY |
| LS25 5BT | G4ZSA |
| LS25 5PJ | G8INL |
| LS25 6AF | G4NCB |
| LS25 6BY | M3TFB |
| LS25 6BY | M3VXN |
| LS25 6LW | G6JWM |
| LS25 7BX | M3VSQ |
| LS25 7DY | G6JSR |
| LS25 7ET | G1UVI |
| LS25 7ET | G4TXK |
| LS25 7EX | 2E1AYS |
| LS25 7HU | G4AAU |
| LS25 7PB | G3KEP |
| LS25 7PB | G4DMP |
| LS25 7RN | M6EHR |
| LS26 0HL | G4OVR |
| LS26 0PP | G8TMD |
| LS26 0QZ | 2E0MEH |
| LS26 0QZ | 2E0VKG |
| LS26 0QZ | M0TCB |
| LS26 0QZ | M0VKG |
| LS26 0QZ | M6AXC |
| LS26 0QZ | M6BZT |
| LS26 0SW | 2E0JRZ |
| LS26 0SW | M0NOK |
| LS26 8DL | M3JEH |
| LS26 8EN | G7NZU |
| LS26 8ET | M3YDI |
| LS26 8QF | M3YTV |
| LS26 8QG | G0RDR |
| LS26 8SN | M1BNH |
| LS26 8SQ | G4JMT |
| LS26 8UE | M0NOA |
| LS26 8UN | M0NDO |
| LS26 8UW | G4PGD |
| LS26 9AF | G4MAK |
| LS26 9AJ | M3ZAA |
| LS26 9LE | 2E0GPF |
| LS26 9LE | 2E0VKY |
| LS26 9LE | M0HPP |
| LS26 9LE | M6GPF |
| LS26 9LE | M6VKY |
| LS27 0DP | G0BVT |
| LS27 0DS | 2E1HYD |
| LS27 0JA | G7NAP |
| LS27 0SG | G1MJT |
| LS27 7BT | G7AJN |
| LS27 7BT | M3OCL |
| LS27 7DJ | G0JVI |
| LS27 7DJ | G7PAF |
| LS27 7DJ | M6KSD |
| LS27 7DN | M0LGA |
| LS27 7RD | M0FJM |
| LS27 7RD | M3MSL |
| LS27 8NS | G6ZGF |
| LS27 8QB | M6SFS |
| LS27 8TD | G0PTI |
| LS27 8UG | M3PUI |
| LS27 8UH | 2E1GDK |
| LS27 8UL | M0NJW |
| LS27 8UL | M0SSW |
| LS27 8US | M3ZRH |
| LS27 9DS | M1EWV |
| LS27 9NP | 2E0JBA |
| LS27 9NP | M3FOQ |
| LS27 9NP | M3FOS |
| LS27 9PL | G4MBE |
| LS28 5AA | G6FMF |
| LS28 5BF | M0BNO |
| LS28 5DY | G4XZK |
| LS28 5DZ | G0DPS |
| LS28 5DZ | G4PTM |
| LS28 5HW | G7OQB |
| LS28 5LW | G4BXQ |
| LS28 5PZ | G7AQA |
| LS28 5QG | 2E1JRC |
| LS28 5QH | M0STS |
| LS28 5QH | M3BVL |
| LS28 5RE | G4NZN |
| LS28 6AA | G1III |
| LS28 6AA | M3WYR |
| LS28 6BG | G7RRO |
| LS28 6DJ | G4SKO |
| LS28 7AY | M6MNP |
| LS28 7EP | G1NSD |
| LS28 7PF | M3UNG |
| LS28 7SS | M0SGS |
| LS28 8AX | 2E0WJC |
| LS28 8AX | M3KFO |
| LS28 8EF | G3VXE |
| LS28 8EQ | 2E0HGR |
| LS28 8NH | 2E1ACZ |
| LS28 8NH | G0GGZ |
| LS28 8QD | G4MOT |
| LS28 9AA | G0ENO |
| LS28 9AA | G0GBC |
| LS28 9AD | 2E1DNB |
| LS28 9ND | [illegible] |
| LS28 9NQ | G8WCT |
| LS29 0LE | G8KSX |
| LS29 0QQ | G8HZJ |
| LS29 0QZ | G1RPP |
| LS29 6BY | G7LNT |
| LS29 6HE | G0SNX |
| LS29 6HH | G8SRN |
| LS29 7BX | 2E0DHK |
| LS29 7BX | M0HYN |
| LS29 7BX | M6DUZ |
| LS29 7EN | G3ZNR |
| LS29 7HS | G0LXR |
| LS29 7LL | M3HSV |
| LS29 7NP | G6MHR |
| LS29 7NR | G4BSS |
| LS29 7PA | G7VUH |
| LS29 7QB | G0KEV |
| LS29 7QB | G8AWN |
| [illegible] | [illegible] |
| LS29 7QH | G0CMU |
| LS29 7RG | G6NWK |
| LS29 7SR | G6TDH |
| LS29 8AH | G6JWM |
| LS29 8AH | M0TIN |
| LS29 8LU | 2E0KBJ |
| LS29 8LU | M3SBJ |
| LS29 8PH | G8AXR |
| LS29 8SX | M0DXN |
| LS29 8SX | M6XAB |
| LS29 9BP | G8WBK |
| LS29 9DB | G4TGJ |
| LS29 9JN | G4DCY |
| LS29 9SN | G0AJB |
| LS4 2LN | M6AYG |
| LS4 2SU | M6CBA |
| LS4 2TT | G0ETL |
| LS5 3HX | G4TFV |
| LS6 2AP | G1JTC |
| LS6 2RD | G0FXK |
| LS6 3DB | G4YES |
| LS6 3EX | M6PQR |
| LS6 3HB | G4MXM |
| LS6 3HB | G4OOI |
| LS6 3NX | G3FLV |
| LS6 3QB | G3TRE |
| LS6 4DN | M6GJB |
| LS6 4EX | G0TUM |
| LS6 4HD | G4IJI |
| LS6 4JT | G0JUI |
| LS6 4PZ | 2E0CSD |
| LS6 4SX | G0AEX |
| LS6 4SX | G3XEP |
| LS6 4SX | G8LVQ |
| LS7 1SL | G7VCK |
| LS7 2DY | G1EZF |
| LS7 2NN | G4UPD |
| LS7 3LY | G4TCB |
| LS7 3NX | M6WPI |
| LS7 4LH | G0SCV |
| LS7 4LH | G4RDL |
| LS7 4LH | G7KXT |
| LS8 1ED | M6XGF |
| LS8 1EW | G4ZOB |
| LS8 1NE | G7GUK |
| LS8 1NS | G3PKC |
| LS8 1RZ | G1DIR |
| LS8 2BS | G4KAX |
| LS8 2DR | 2E1EMN |
| LS8 2DR | M1APF |
| LS8 2EE | G4FTI |
| LS8 2LA | G3VFH |
| LS8 2PJ | M6XAV |
| LS8 2RH | G0KQP |
| LS8 2SX | G8HBQ |
| LS8 2TA | G6MTF |
| LS8 4DH | G7OOS |
| LS8 5BY | G1UVE |
| LS8 5DB | 2E1DLX |
| LS9 0AE | G4HJS |
| LS9 0LL | M3UEX |
| LS9 6BY | M1AUR |
| LS9 6DP | G3ZRL |
| LS9 6RG | M1AFV |
| LS9 7SY | G4FVV |

## LU

(Luton)

| Postcode | Callsign |
|---|---|
| LU1 4AN | G8ADC |
| LU1 4BU | M6UUU |
| LU1 4DJ | G8OBB |
| LU1 4DU | G6CQB |
| LU1 4DU | G6NDJ |
| LU1 4EN | G0GJA |
| LU1 4EN | G6WVS |
| LU1 4JA | G8LSS |
| LU1 5LB | G1OAM |
| LU1 5PE | G1AYH |
| LU1 5PE | M6VCS |
| LU1 5PE | M6ZLN |
| LU1 5PF | G8XSD |
| LU1 5PN | G3NNY |
| LU1 5QD | M6RPH |
| LU2 0DN | G6UUU |
| LU2 0FD | G7PJG |
| LU2 0JF | G0BWK |
| LU2 0PJ | M0GQU |
| LU2 0PL | M0HHM |
| LU2 0PR | G8GWK |
| LU2 0RE | G1OUA |
| LU2 0RF | G6OUA |
| LU2 0RE | G6UCQ |
| LU2 0RP | G6IAT |
| [illegible] | [illegible] |
| LU2 7BY | G4UVD |
| LU2 7DL | 2E0GQT |
| LU2 7DL | M6GQT |
| LU2 7EU | G3IWH |
| LU2 7EU | M0KIW |
| LU2 7JL | G4IET |
| LU2 7JY | 2E0GPB |
| LU2 7JY | M0PPG |
| LU2 7JY | M6GPB |
| LU2 7NL | G1FGC |
| LU2 7PP | G7BNW |
| LU2 7SA | G8NXS |
| LU2 7TE | G4ENB |
| LU2 7TE | G8DDC |
| LU2 7TX | G1PZD |
| LU2 7UH | G3NVL |
| LU2 7UU | G3TAZ |
| LU2 8AE | M0WHP |
| LU2 8BP | G1FAX |
| LU2 8EJ | G4NMY |
| LU2 8EP | G1BDI |
| LU2 8HQ | G8LXY |
| LU2 8JH | G0KUC |
| LU2 8PP | G0LQZ |
| LU2 8QP | G1JCC |
| LU2 8QP | M0JCC |
| LU2 8QP | M3KEJ |
| LU2 8QP | M6WEN |
| LU2 8RU | G8IZW |
| LU2 8TS | M0BHA |
| LU2 8TZ | G8FFM |
| LU2 9AN | G0VZH |
| LU2 9AX | G4BMM |
| LU2 9BL | G0TNO |
| LU2 9DP | G7SNW |
| LU2 9ET | G7RKW |
| LU2 9HL | G0KYJ |
| LU2 9HL | G4OIS |
| LU2 9JZ | G7VWA |
| LU2 9RA | G0ALB |
| LU2 9RA | G7VYB |
| LU3 1EF | G8IXK |
| LU3 1EF | M6JPO |
| LU3 1HB | G8JHM |
| LU3 1PY | G8UOJ |
| LU3 2AJ | 2E0XFF |
| LU3 2AJ | M0ZSS |
| LU3 2AJ | M6ESS |
| LU3 2AP | G4HPY |
| LU3 2DW | G3WBC |
| LU3 2HJ | G6PPV |
| LU3 2JJ | G1SCA |
| LU3 2LZ | G8CBU |
| LU3 2NY | G6DWM |
| LU3 2PY | G1SVL |
| LU3 2TH | 2E0YFZ |
| LU3 2TH | M3YFZ |
| LU3 2UA | G0DCS |
| LU3 2UR | G8CPQ |
| LU3 3BH | G8UYF |
| LU3 3JZ | G7HBN |
| LU3 3LA | G6MUJ |
| LU3 3PY | G4CPE |
| LU3 3SU | M0CAJ |
| LU3 3TU | M6SCQ |
| LU3 3TW | G1NWZ |
| LU3 3XQ | G3ZHJ |
| LU3 3XU | G6WRB |
| LU4 0BU | G0WKJ |
| LU4 0LP | M0ZMO |
| LU4 0QW | G7JAS |
| LU4 0UN | 2E0OTI |
| LU4 0UN | M3OTI |
| LU4 0XJ | M6JHP |
| LU4 8ER | M3HQB |
| LU4 8NU | G6RPD |
| LU4 8PY | G8EZV |
| LU4 9AL | 2E0DHA |
| LU4 9AL | M3JZL |
| LU4 9AS | G1XZW |
| LU4 9EN | G7BPI |
| LU4 9EN | G7DBT |
| LU4 9EN | M3XQT |
| LU4 9ER | M6IRJ |
| LU4 9HG | G3WLM |
| LU4 9JE | G6XFR |
| LU4 9LT | G1JIG |
| LU4 9TL | G4LZF |
| LU4 9UF | G7RIA |
| LU5 4AL | G1FXB |
| LU5 4AQ | 2E0CQC |
| LU5 4AQ | M0OBY |
| LU5 4AQ | M6RGC |
| LU5 4EA | G8NZD |
| LU5 4LF | G3ZNT |
| LU5 4LP | G3SGS |
| LU5 4PR | G1ANS |
| LU5 4PW | G4UOO |
| LU5 4QL | G6LGW |
| LU5 4SH | G7UAL |
| LU5 4SQ | G1NVV |
| LU5 5EP | M0SKM |
| LU5 5HU | M0JBF |
| LU5 5HU | M6JVB |
| LU5 5HZ | M0ICE |
| LU5 5JG | G1AYI |
| LU5 5NX | M3FMP |
| LU5 5NX | M3FMQ |
| LU5 5NZ | M6JLM |
| LU5 5PN | M0LTN |
| LU5 5QR | G1BNE |
| LU5 5TJ | M6ACP |
| LU5 6AL | G8HJK |
| LU5 6AZ | G4DUL |
| LU5 6BB | [illegible] |
| LU5 6EQ | G0MUC |
| LU5 6EL | G0WFT |
| LU5 6LE | G1PII |
| LU5 6LQ | [illegible] |
| LU5 6NF | G3NRW |
| LU5 6NT | G1ZNX |
| LU5 6QF | M0MBO |
| LU6 1AJ | M0MNO |
| LU6 1BN | M0KDH |
| LU6 1BN | M3ODH |
| LU6 1BQ | 2E1EOK |
| LU6 1BQ | G8ERV |
| LU6 1DN | M1BFI |
| LU6 1NQ | M6ALM |
| LU6 1NX | M0BZN |
| LU6 1PN | M0DZG |
| LU6 1QJ | G0ODU |
| LU6 1TS | M3WGM |
| LU6 1UG | M3WFO |
| LU6 2AD | 2E0TCK |
| LU6 2AD | M6TCK |
| LU6 2AG | G0WTL |
| LU6 2AH | G0HWK |
| LU6 2AW | G3ZFP |
| LU6 2BJ | G8DBH |
| LU6 2DD | G7TVT |
| LU6 2DF | G3SUL |
| LU6 2FH | G0SJR |
| LU6 2HL | 2E0ASW |
| LU6 2HL | 2E1GYH |
| LU6 2HL | G4CVS |
| LU6 2HL | M0CLW |
| LU6 2HL | M0UKC |
| LU6 2HR | G4XLG |
| LU6 2HR | M6MXA |
| LU6 2HR | M6OUS |
| LU6 2JQ | G3PTG |
| LU6 2JU | M3XJH |
| LU6 3AS | 2E0SBZ |
| LU6 3AS | M0SBZ |
| LU6 3DD | G6IAN |
| LU6 3JT | 2E0EAZ |
| LU6 3JT | G1HIJ |
| LU6 3JT | M6ETH |
| LU6 3JZ | 2E0NBZ |
| LU6 3JZ | M3NBZ |
| LU6 3LJ | G3NWH |
| LU6 3NB | G8ION |
| LU6 3NS | G7KGR |
| LU6 3PT | G4EKB |
| LU6 3RG | G8KGV |
| LU6 3RS | G4WYO |
| LU6 3RS | G8PJD |
| LU7 0AB | M6PEI |
| LU7 0ER | G4IVZ |
| LU7 0EX | G3XTQ |
| LU7 0HZ | G8GIK |
| LU7 0HZ | G8ZVI |
| LU7 0LB | 2E0HJJ |
| LU7 0LJ | G8GZX |
| LU7 0NT | G1HMW |
| LU7 0QE | G6JFN |
| LU7 0RY | 2E0BFG |
| LU7 0RY | M0PJD |
| LU7 0RY | M3LLZ |
| LU7 0SQ | G6FZW |
| LU7 0SY | G6NAG |
| LU7 1AH | M3MIV |
| LU7 1DJ | G8KKU |
| LU7 1FQ | 2E0OBI |
| LU7 1FQ | M3UHN |
| LU7 1HT | M6BFK |
| LU7 2NT | M0WAZ |
| LU7 2PE | M3MHP |
| LU7 2PE | M6KJM |
| LU7 2QQ | G0KRR |
| LU7 2QR | 2E0RZM |
| LU7 2QR | M3RZM |
| LU7 2QW | G0FRD |
| LU7 2QW | G4CAK |
| LU7 2RE | M6ACQ |
| LU7 2RH | G4CQA |
| LU7 2TS | G6POV |
| LU7 2TS | M6JNW |
| LU7 2TY | G3YZW |
| LU7 2XD | G4DDC |
| LU7 2XD | G8XTW |
| LU7 2YG | G3YYG |
| LU7 3AB | M6CMF |
| LU7 3AD | M6MNX |
| LU7 3DE | M0NTY |
| LU7 3DN | G1MMI |
| LU7 3DP | G6CHJ |
| LU7 3DS | G0AYC |
| LU7 3LH | G4ATQ |
| LU7 3LS | G4BUS |
| LU7 3LU | G3MBO |
| LU7 3NJ | M6VIT |
| LU7 3TT | G7OZJ |
| LU7 3UJ | G6IXH |
| LU7 4AT | M6IBF |
| LU7 4BS | M0POA |
| LU7 4HY | G0DYW |
| LU7 4JG | M6ADJ |
| LU7 4RF | M6CWO |
| LU7 4SU | M6CLZ |
| LU7 4TG | M6BQN |
| LU7 4UP | 2E0SIB |
| LU7 4UP | M6SBZ |
| LU7 4YU | 2E0FEB |
| LU7 4YU | M3SQP |
| LU7 0DZ | 2E0PGM |
| LU7 9DZ | M0PAO |
| LU7 9DZ | M6AZS |
| LU7 9JF | M6BAM |
| LU7 9JF | [illegible] |

## M

(Manchester)

| Postcode | Callsign |
|---|---|
| M11 1DP | M6RCF |
| M11 1EQ | M6OXB |
| M11 1HQ | M3VNH |
| M11 1HQ | M3YMH |
| M11 4FB | M3XGC |
| M12 5PU | G0WGL |
| M12 5RP | M3YPI |
| M13 9DA | M6TKG |
| M14 4UR | 2E0WLD |
| M14 4UR | M6WLD |
| M14 5JS | M6FML |
| M14 5RB | M1CVL |
| M14 6AH | G3YTZ |
| M14 6PE | G0BML |
| M14 6RW | M6EGU |
| M14 6RX | G4FFW |
| M14 7FN | M0TCC |
| M14 7FZ | G0EIZ |
| M14 7FZ | G4JS |
| M14 7WT | G1ZST |
| M15 4FZ | M6EQR |
| M15 5AF | M3NMR |
| M15 5RR | M0VAP |
| M15 5RR | M6SGV |
| M16 0HD | M0HXS |
| M16 8LX | M3NXZ |
| M16 8PW | G3SMN |
| M16 8QQ | M3OAG |
| M16 9GQ | M6CLQ |
| M18 8BP | G6YFL |
| M18 8LL | M6UCP |
| M18 8WW | G7NAL |
| M18 8WW | M6ISB |
| M19 1LE | 2E0TZD |
| M19 1LE | M6ZTD |
| M19 1QY | M0FVD |
| M19 2FA | G4GDG |
| M19 2LQ | G3JRK |
| M19 2NL | 2E0MQT |
| M19 2NL | G0PXZ |
| M19 2NP | G8OYF |
| M19 2UA | M6ZAQ |
| M19 3EH | G4BVQ |
| M19 3FL | M6EGW |
| M20 1DH | M0IGY |
| M20 1EN | 2E1HZI |
| M20 2SQ | 2E0CFM |
| M20 3DD | M3IYE |
| M20 4GH | G7AZW |
| M20 4WH | G3CCL |
| M20 5AB | M3XUW |
| M20 5GJ | M1DFO |
| M20 5LQ | 2E0BDV |
| M20 5LQ | M3MVM |
| M20 5PL | G0OVQ |
| M20 5QH | G0UXI |
| M20 5QS | M3PGO |
| M20 6BB | G7JME |
| M20 6BB | G7MIT |
| M20 6JA | G3YHQ |
| M20 6SU | 2E0BTB |
| M20 6TG | 2E1HLW |
| M20 6US | M6SQJ |
| M21 0QU | 2E0MWB |
| M21 0QU | M0UFC |
| M21 0QU | M6MWB |
| M21 0UX | G7EPL |
| M21 7DB | M1CXX |
| M21 7LE | M3OJK |
| M21 7QG | M3BLX |
| M21 7TG | G4RYT |
| M21 8FA | 2E0IZW |
| M21 8FA | M6IZW |
| M21 8XT | G6BHA |
| M21 9DP | G4OCR |
| M21 9JU | G0CRF |
| M21 9JU | G6CRF |
| M22 1AH | 2E1IHS |
| M22 1AH | G0PXG |
| M22 1AH | M3AHZ |
| M22 1GE | G7KFZ |
| M22 1GW | M6NKB |
| M22 1NN | G1UTM |
| M22 1UD | M3NHA |
| M22 1UD | M6NHA |
| M22 1UR | G0JYS |
| M22 4DP | G1OJB |
| M22 4DZ | G0PWA |
| M22 4GG | G1PNC |
| M22 4LW | M6MOJ |
| M22 4WB | M3XAO |
| M22 5AQ | G8OZD |
| M22 5DH | M3DAV |
| M22 5DJ | G7SYJ |
| M22 5FU | M6GTB |
| M22 5JY | G4RWN |
| M22 5QG | G7TEA |
| M22 5QG | M0TEA |
| M22 9PL | 2E1GNK |
| M22 9UX | M6BBL |
| M22 9UX | M6HNS |
| M22 9YE | 2E0SGY |
| M22 9YE | M6SGA |
| M23 1HF | G0DNY |
| M23 1JX | G0SAY |
| M23 1JY | G6BMY |
| M23 1LD | 2E0PLH |
| M23 1LE | M6XMN |
| M23 1LP | G0NID |
| M23 2AT | G7VJL |
| M23 2UL | G0LNA |
| M23 2YX | C0LNB |
| M23 9AW | G8LPX |
| M23 9BT | G8LUL |
| M23 9BW | M3YSF |
| M23 9NZ | G7PXS |
| M24 1HE | G6CVW |
| M24 1PN | G1XGN |
| M24 1TJ | G7VGX |
| M24 2EQ | G0TRK |
| M24 2NN | G6GCY |
| M24 2RA | M6VWP |
| M24 2TQ | G4VVT |
| M24 4ED | G4WLI |
| M24 4ED | G8IKW |
| M24 4JJ | G4WVP |
| M24 4RU | G4ZQL |
| M24 5JH | G0RRX |
| M24 5JH | G0UNL |
| M24 5LS | M3TRC |
| M24 5RL | G0FCZ |
| M24 6AX | M6RWT |
| M24 6BN | M1EKH |
| M25 0FZ | G7PYN |
| M25 0HE | G1HHU |
| M25 0HR | G8JWC |
| M25 0HX | G3ZVT |
| M25 1FN | M6RUN |
| M25 1FN | M6SAU |
| M25 1JY | M6AAX |
| M25 1NB | G4VUK |
| M25 2GP | G7TBM |
| M25 2RD | 2E0DIJ |
| M25 2RD | M6DIJ |
| M25 3BT | G6BVF |
| M25 3DY | M6AUC |
| M25 3HZ | G0CSU |
| M25 3JJ | G7VKJ |
| M25 9GW | G1PNB |
| M25 9GW | M3WWD |
| M25 9LB | M0MBS |
| M25 9NT | G0UAY |
| M25 9RE | G8OFI |
| M25 9RU | 2E0OAI |
| M25 9RU | M6DWS |
| M25 9TG | G1JZY |
| M26 1EP | M6SLE |
| M26 1HN | G1JUD |
| M26 1HN | G6RXF |
| M26 1JA | G1PKP |
| M26 1JA | G6CLW |
| M26 1YF | G3ZCH |
| M26 1YF | G6OTQ |
| M26 2GF | G6YGB |
| M26 2QG | M0GGQ |
| M26 2QS | G0WEF |
| M26 2RS | G0HEA |
| M26 2UP | 2E0GAO |
| M26 2UP | M0YYA |
| M26 2UP | M6GEF |
| M26 2UU | M6OVR |
| M26 2UX | G0ADL |
| M26 3GL | G4UAU |
| M26 3TB | G4UNB |
| M26 3TZ | M6ROY |
| M26 3UA | 2E0GPY |
| M26 3UJ | G1ESW |
| M26 4FA | M0BOQ |
| M26 4FS | G7BYS |
| M26 4FS | G8ZIC |
| M26 4NS | 2E0LCA |
| M26 4NS | M0XCT |
| M26 4NS | M6KFC |
| M26 4PT | G1ZOQ |
| M26 4QF | G6DGQ |
| M27 0YH | 2E0XBN |
| M27 0YH | G0MKA |
| M27 0YH | G6YIO |
| M27 0YH | M0XBN |
| M27 0YH | M3OTQ |
| M27 0YH | M6XBM |
| M27 4FA | G8RAX |
| M27 5GZ | G4IRB |
| M27 5GZ | G8SMR |
| M27 5NA | G0KLF |
| M27 5NJ | G4TSA |
| M27 5QY | M3OVE |
| M27 6AQ | G3STJ |
| M27 6PT | M1KCB |
| M27 6WH | G4FXA |
| M27 6WY | G1SYV |
| M27 8AF | M6MTV |
| M27 8FS | G0CVH |
| M27 8QS | G4VDJ |
| M27 8RE | G0MZY |
| M27 8RT | M6HGR |
| M27 8XP | G0LVG |
| M27 8XP | M6KMN |
| M27 9RD | G4EFD |
| M28 0SL | G8KOQ |
| M28 0SX | G3PMJ |
| M28 0TP | G4NTY |
| M28 1BE | M6FVL |
| M28 1DE | G7ELA |
| M28 1HW | G4SKG |
| M28 1LP | G7DPW |
| M28 1ND | G8DTF |
| M28 2JE | M6UIR |
| M28 2RJ | C0TEE |
| M28 2RJ | G3BSA |
| M28 2RJ | G4MWC |
| M28 2RW | G1VJQ |
| M28 2SL | G0RHY |
| M28 2TX | 2E0YJL |
| M28 2TX | M0YJL |
| M28 3DH | M6ASZ |
| M28 3DT | G0OHY |
| M28 3JW | M1VIP |
| M28 3NT | M6TBN |
| M28 3PW | G3OAX |
| M28 7EN | M0CAV |
| M28 7EX | 2E1WJB |
| M28 7EX | G4UPU |
| M28 7JD | G7FFK |
| M28 7JW | G4TJU |
| M28 7QF | G4JLG |
| M28 7TS | G8JOC |
| M28 7XS | G7FPS |
| M29 7AT | M6BSS |
| M29 7AX | 2E0REM |
| M29 7AX | M6AVT |
| M29 7BH | G4JHA |
| M29 7BJ | M6BST |
| [illegible] | [illegible] |
| M29 7DB | G0IZR |
| M29 7DB | M0SLR |
| M29 7DB | M3HXI |
| M29 7EJ | G8NSE |
| M29 7HH | G4GKT |
| M29 7HS | G4DZK |
| M29 7HS | G8FQZ |
| M29 7NW | M6ODA |
| M29 7PE | 2E1GSM |
| M29 7PH | G0TFP |
| M29 7PZ | M6ERH |
| M29 7RE | G7CQW |
| M29 7RT | G8SYM |
| M29 7WF | M3GAV |
| M29 8LQ | M0AVA |
| M29 8LS | M6SDH |
| M29 8LS | M6XZY |
| M29 8PG | M3RFI |
| M29 8PH | 2E1HZV |
| M29 8PH | M3EGC |
| M29 8PJ | M6MKX |
| M29 8WQ | M6AEQ |
| M29 9DN | M3LFH |
| M29 9EE | G4CFK |
| M30 0FS | M3LUZ |
| M30 0HW | G6YAQ |
| M30 0JF | 2E0OTE |
| M30 0JF | M0OTE |
| M30 0JF | M3OTE |
| M30 0JN | M6JII |
| M30 0LT | G1ASN |
| M30 0PZ | M6CPC |
| M30 0QU | M6WLA |
| M30 8AG | G4VPW |
| M30 8AJ | M1ABV |
| M30 8BP | G0BVM |
| M30 8DD | G0FPO |
| M30 8JR | 2E0FDD |
| M30 8JR | M6FDD |
| M30 8UW | G8FYK |
| M30 8WA | G8MOK |
| M30 8WB | M1NUS |
| M30 8WP | M3GHI |
| M30 8WP | M6GHI |
| M30 9DL | M6BGM |
| M30 9EE | G8YPH |
| M30 9FT | M6GIS |
| M30 9GW | M6TOR |
| M30 9LY | M0HTS |
| M31 4DL | G8BRU |
| M31 4DY | G3VYA |
| M31 4GW | G7GAK |
| M31 4NP | G0TRU |
| M31 4PT | G0LZL |
| M32 0AJ | M6RFK |
| M32 0AN | G7KRI |
| M32 0PG | M6UFR |
| M32 0RG | M3OVX |
| M32 0RY | M6APC |
| M32 8BE | M6PLI |
| M32 8DQ | G1JHX |
| M32 8NA | 2E0JED |
| M32 9AF | 2E0MUA |
| M32 9AF | M3MUA |
| M32 9BS | G7FKJ |
| M32 9BX | G4LVI |
| M32 9EF | 2E0KUH |
| M32 9EF | M6KUH |
| M32 9JB | G6YOP |
| M32 9QA | M0NTH |
| M32 9QF | G0BJK |
| M32 9RT | 2E0RDU |
| M32 9SJ | G1UFM |
| M33 2AR | M0WRA |
| M33 2BJ | G1NVL |
| M33 2DQ | G7FQY |
| M33 2DY | 2E1ICI |
| M33 2EA | G0FOU |
| M33 2EA | G4URV |
| M33 2EA | M0TGS |
| M33 2JT | M0DHE |
| M33 2LD | G8LQO |
| M33 2PJ | M6CGS |
| M33 2XF | G0OVY |
| M33 3GT | G2HW |
| M33 3GT | G3JVW |
| M33 3LD | 2E1IBP |
| M33 3LH | G0RFG |
| M33 3LQ | G0TKL |
| M33 3LQ | M3PSR |
| M33 3LW | 2E0CPF |
| M33 3LW | M6BQF |
| M33 3QP | G1BOO |
| M33 3TN | G0KKF |
| M33 3WL | G7NND |
| [illegible] | [illegible] |
| M33 4AE | G4JZZ |
| M33 4ER | 2E1IGP |
| M33 4FS | G4CDM |
| M33 4FW | M0IWV |
| M33 4HL | M3FSS |
| M33 4HP | G4KGU |
| M33 4HR | G7TUH |
| M33 4QL | M0HIH |
| M33 4RF | G4AUR |
| M33 4RG | G0NWJ |
| M33 4WG | G4MYB |
| M33 5DL | G8ILW |
| M33 5FB | G3KCB |
| M33 5HF | G0SSV |
| M33 5HR | 2E0ZVL |
| M33 5HR | M0LCR |
| M33 5HR | M0SII |
| M33 5HR | M6ZVL |
| M33 5NQ | M0QVM |
| M33 5NY | G0MRK |
| M33 5PR | G6ISA |
| M33 5QJ | G0BHP |
| M33 5QJ | G0TRC |
| M33 5RP | G7RUR |
| M33 5RP | M3XAU |
| M33 5UF | G0AOU |
| M33 5UF | G3FVA |
| M33 6EZ | G3ZDM |
| M33 6GS | M3VRU |
| M33 6HX | G4NOL |
| M33 6NF | G3WFT |
| M33 6NW | G3SUI |
| M33 6QW | G4FPA |
| M33 6WU | G0FRB |
| M33 7EN | G8RSI |
| M34 2DQ | G0ROW |
| M34 2ER | G0TBE |
| M34 2WP | G0WYP |
| M34 3GL | G7GXR |
| M34 3HU | 2E1YRK |
| M34 3NS | G7TYB |
| M34 3NX | G3XVG |
| M34 3NZ | M6LBI |
| M34 3QH | M6KKF |
| M34 3TE | G0BOH |
| M34 3TH | G4WGR |
| M34 3WP | M6ILS |
| M34 5BD | G8HTN |
| M34 5EJ | M6JRE |
| M34 5FB | M3IFG |
| M34 5GB | M3OWN |
| M34 5HE | G6IQC |
| M34 5LJ | G1JDT |
| M34 5QX | G0NHO |
| M34 5QX | G7DNR |
| M34 5TU | M0WBJ |
| M34 5UH | G7GTU |
| M34 6AF | G1BBK |
| M34 6EA | G6NVU |
| M34 6EA | G7HRZ |
| M34 6FT | G7DOY |
| M34 6JY | G0YRT |
| M34 6LR | G1EJK |
| M34 7LB | M3PBU |
| M34 7TF | M6EKD |
| M34 7TX | G1UTS |
| M34 7TX | M0GMA |
| M34 7TX | M3ZPM |
| M35 0LH | G0KTR |
| M35 0LR | M6GOX |
| M35 0PX | G7BAB |
| M35 9EJ | G3PYH |
| M38 0BU | M3WTN |
| M38 0FQ | G1SPM |
| M38 0LW | M0APZ |
| M38 9GL | G6MIU |
| M38 9GQ | 2E0SCD |
| M38 9GQ | M6FPC |
| M38 9HB | M3IRF |
| M38 9UH | M0KEJ |
| M38 9XU | G6PKV |
| M4 5LB | M0DAH |
| M4 7BB | 2E0NTW |
| M4 7BB | M3NTW |
| M4 7EB | M0SEB |
| M40 0DG | G6WGZ |
| M40 0DG | M6EAE |
| M40 0DG | M6HUN |
| M40 0DG | M6PHJ |
| M40 0GB | G0IUK |
| M40 1LF | M3GHS |
| M40 1QE | M0FRG |
| M40 2NP | M3JIE |
| M40 2SY | M3GAF |
| M40 3GG | 2E0AVS |
| M40 3GG | M3AJN |
| M40 3NL | G7TZV |
| M40 3SE | G6IRF |
| M40 3TD | M3VNP |
| M40 3WP | G7ORW |
| M40 5HL | G7VCG |
| M40 5HL | M3IAF |
| M40 5QD | G0VXW |
| M40 5QW | 2E0YES |
| M40 5QW | M0TLY |
| M40 5QW | M3GBC |
| M40 9BW | 2E0WCC |
| M40 9BW | M6LEJ |
| M40 9LF | G7USQ |
| M40 9ND | M3YPR |
| M41 0GF | G7HKQ |
| M41 0GW | M3IRJ |
| M41 0PY | G6PSA |
| M41 0RJ | G3JIS |
| M41 0SR | M3IRM |
| M41 0TY | G0XQE |
| M41 0TY | M0RML |
| M41 0XY | G0ONF |
| M41 0AT | 2E0DJY |
| M41 5RZ | G7LKL |
| M41 5HA | G7TIR |
| M41 6FJ | G1SCL |
| M41 6HP | G1NYJ |
| M41 6JL | M3RDS |
| M41 6LE | G3XXG |
| M41 6NQ | G7OTE |
| M41 6PU | M0XPD |
| M41 6PU | M6PAD |
| M41 6WA | G8YUK |
| M41 7BT | G4SDL |
| M41 7DY | 2E0FPA |
| M41 7DY | M0FPA |
| M41 7DY | M3FPA |
| M41 7ND | M1DYN |
| M41 7NR | G1UKZ |
| M41 7WH | G4WLJ |
| M41 7WX | C4MFI |
| M41 8PB | M0UUU |
| M41 8QT | 2E0DCS |
| M41 8RN | G8HXE |
| M41 8RN | M0TWC |
| M41 8RN | M3KMH |
| M41 8SB | G4XHT |
| M41 8UE | G4RYI |
| M41 8UE | M3OPV |
| M41 8UE | M3RYI |
| M41 8UT | G1VZB |
| M41 9JZ | G7HMW |
| M41 9NG | G7AGC |
| M41 9NJ | G1LWE |
| M41 9NT | G0SYP |
| M41 9PT | G4NCZ |
| M41 9QE | G0IMB |
| M43 6DS | G1WPL |
| M43 6DX | G1YCM |
| M43 6EF | G8CZE |
| M43 6FS | G4HBI |
| M43 6HL | G6KRS |
| M43 6QG | G6GXY |
| M43 7AN | M6LUI |
| M43 7EA | M6HEK |
| M43 7JX | G8OXX |
| M43 7LJ | M6BXN |
| M43 7NR | G1BZD |
| M43 7PL | 2E0IMS |
| M43 7SW | M6RCY |
| M43 7UQ | G4IFI |
| M43 7YG | M6EIL |
| M44 5AS | 2E0YDJ |
| M44 5AS | M0YDJ |
| M44 5AU | 2E0SGH |
| M44 5AU | G0GSK |
| M44 5AU | M0MBE |
| M44 5AU | M6SGH |
| M44 5JQ | G6LBQ |
| M44 6AQ | 2E0JIR |
| M44 6AQ | G8TXW |
| M44 6AQ | M3JIR |
| M44 6DD | M6PNT |
| M44 6EH | M3RYG |
| M44 6EH | M3YOZ |
| M44 6EN | G0JYK |
| M44 6HD | G6SPG |
| M44 6LR | G0IGT |
| M44 6QA | M0IPR |
| M44 6QA | M0SRA |
| M44 6TD | M3UJP |
| M44 6TE | G6MMA |
| M44 6WJ | M0JWE |
| M44 6WY | G0NPI |
| M45 6DP | G8PAL |
| M45 6EP | G8AOG |
| M45 6TJ | G8DUT |
| M45 7LR | G1COX |
| M45 7NT | G3GXI |
| M45 7NT | G8KRG |
| M45 7PR | M6BPL |
| M45 7RZ | G8GAJ |
| M45 7SS | G8NKJ |
| M45 7ST | G8OUT |
| M45 7TS | G4UGW |
| M45 7WH | M3OAZ |
| M45 8FR | G4FZP |
| M45 8FR | M3ZFJ |
| M45 8GG | M1CRZ |
| M45 8LH | 2E0GRX |
| M45 8NT | G1NSB |
| M45 8WY | G0AHR |
| M46 0EZ | 2E0LTT |
| M46 0EZ | M0LTT |
| M46 0EZ | M6MLA |
| M46 0LQ | G6SKM |
| M46 0PG | G6LCX |
| M46 0QE | G0FTO |
| M46 9AB | G6YBC |
| M46 9GZ | G0BOR |
| M46 9LN | G6TFV |
| M46 9LQ | G4HZJ |
| M46 9NQ | 2E0HBY |
| M46 9PQ | G0OMF |
| M46 9PY | M0GNY |
| M46 9QB | G0KBA |
| M46 9QB | G4GBK |
| M46 9RQ | 2E1IHB |
| M46 9RQ | M3WAC |
| M46 9RR | 2E1HLL |
| M46 9RR | M1HLL |
| M46 9TZ | G6HFW |
| M46 9XE | G4GWF |
| M5 3HU | M6GSH |
| [illegible] | [illegible] |
| M5 5LF | 2E0DMD |
| M50 3HB | 2E1MPB |
| M6 5YB | G0HON |
| M6 6BG | M6IKE |
| M6 6DW | G0GIF |
| M6 7AR | M6XBQ |
| M6 7LA | G0PLX |
| M6 7PQ | G6FEI |
| M6 7WP | M3UDC |
| M7 0QQ | G0IYM |
| M7 1LP | 2E0YWN |
| M7 2BZ | M5SJS |
| M7 2EU | M6CDY |
| M7 2GB | 2E0PDX |

**IMPORTANT NOTE**

**Revalidate licence to avoid revocation** – Ofcom has advised the Society that plans will be drawn up to revoke licences that have not been revalidated as required by the licence conditions. The quickest way to revalidate is to do so online via the Ofcom website: *https://services.ofcom.org.uk/* or by email: *amateur.validations@ofcom.org.uk* If you need assistance in the process, Ofcom staff are available to help, but please be patient during times of heavy workload.

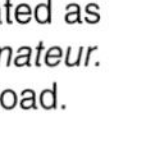

M7 3NX 2E1BLG
M7 3NZ G1JHB
M7 3TF M0LKS
M7 3TX G1ULB
M7 4EZ G7LDR
M7 4JA M0ASU
M7 4TH M3OSU
M7 4TH M3OSW
M8 0LS M6TGQ
M8 4GG M3VVQ
M8 4WQ G4YNI
M8 5AR M6ZAD
M8 5AS M0PGL
M8 5AW G5MS
M9 0RQ G6OGT
M9 4EX G4XBX
M9 5XY 2E0TRK
M9 5XY M3ZAV
M9 6HR G3KIQ
M9 7AR G3NSW
M9 7EQ G6MDG
M9 8AT G1XIH
M9 8AT G2ALN
M9 8JD G0SNZ

## ME

(Medway)

ME1 1DY G4NRT
ME1 1FB M3FPU
ME1 1HY G3NRU
ME1 1NB G7FZN
ME1 1QH G8WKX
ME1 1RN M0TEY
ME1 2BL G7VFU
ME1 2DL G0WFM
ME1 2HZ G0WYV
ME1 2NN G4SAW
ME1 2QF M3VJX
ME1 2RE M3JHW
ME1 2SS M3RMZ
ME1 2UA G2FSH
ME1 2UH G3ZJP
ME1 2UL M6MLM
ME1 2XD M0PLG
ME1 2XD M6SJQ
ME1 3AJ M6NST
ME1 3JU G0DKO
ME1 3NJ G1ERZ
ME10 1BY G3ISD
ME10 1DA 2E0BNH
ME10 1DA M0XFX
ME10 1DA M3SKV
ME10 1EB M3DSI
ME10 1EH G8LDJ
ME10 1ER M3ZKD
ME10 1ET M0HZW
ME10 1JF G0TNC
ME10 1JU G0JUW
ME10 1LB G3EHW
ME10 1QT G0UAS
ME10 1QY G3MXA
ME10 1QY G7GSR
ME10 1RD G3VPA
ME10 1RF G4TWC
ME10 1TJ M0ECS
ME10 1TJ M1DUB
ME10 1TJ M3PJS
ME10 1TJ M3PZX
ME10 1TS G6DUI
ME10 1TX G0UKJ
ME10 1UD G7RUC
ME10 1XF M6RQB
ME10 1XJ 2E1AUN
ME10 1XJ G0AXQ
ME10 1XJ M3PHF
ME10 1YL G1BOX
ME10 1YL G8ONY
ME10 2DP M6BAZ
ME10 2EY G4IBH
ME10 2HR G1JUO
ME10 2LR G1YTO
ME10 2LZ M0AZP
ME10 2PX M0PIA
ME10 2TE G7AEY
ME10 3AS G3ZAV
ME10 3BL 2E1JRS
ME10 3LG M6SDF
ME10 3NR G1VJN
ME10 3QY M6EWH
ME10 3TG 2E1CLG
ME10 4HS G0VTV
ME10 4HS G4LBS
ME10 4LF M3RXG
ME10 4NA G1SPX
ME10 4NA M6IAN
ME10 4NA M6RSD
ME10 4QD G0DRQ
ME10 4QD G1DLJ
ME10 4QE G4VEC
ME10 4RX G7BWV
ME10 4SL G0RNQ
ME10 5AN G7MEE
ME10 5AS G6PGT
ME11 5AG G8DPW
ME11 5AW G1JCL
ME11 5BU G6ILN
ME11 5DT 2E0RNF
ME11 5DT M6OEM
ME12 1BW M1WAW
ME12 1DR M6KYE
ME12 2AT M3TMQ
ME12 2BS G1AFW
ME12 2DJ G0PEH
ME12 2EJ M3ILB
ME12 2HG G0JYH
ME12 2HX M0CSU
ME12 2LH 2E0WGF
ME12 2LH M6GQJ
ME12 2NH G0DRT
ME12 2NL G7NJB
ME12 2QS G4KSU
ME12 2RA M6LFJ
ME12 3BB G7LZY
ME12 3BQ G3ZYQ
ME12 3DH G8XGB
ME12 3DH M0XGB
ME12 3EL G8XYU
ME12 3EL M0AKA
ME12 3EZ M3RAK
ME12 3FD M6NMS
ME12 3HX G3GDH
ME12 3JX G6NUI
ME12 3JX M1DJN
ME12 3LH G1JQH
ME12 3LH M0KCR
ME12 3LN G8JNZ
ME12 3LN M6KSP
ME12 3NJ G1SGS
ME12 3NR G2HKU
ME12 3PG 2E0ILN
ME12 3PG M0ILN
ME12 3PG M6ILN
ME12 3PG M6KCF
ME12 3PJ M6MZY
ME12 3PP M6DTN
ME12 3PT M6BRK
ME12 3PU M6NSW
ME12 3QR G7ECQ
ME12 3SA 2E0ELT
ME12 3SA M0KLK
ME12 3SA M6WJF
ME12 3SQ G7IAU
ME12 4EJ M3VMU
ME12 4JU G4WPB
ME12 4PU M3YGD
ME12 4PU M6MWA
ME12 4QY M3XIL
ME13 0EF M3MUX
ME13 7DL M1ARU
ME13 7ES G0PZX
ME13 7JG 2E0BRJ
ME13 7JZ G4VXB
ME13 7PD M3VWF
ME13 7TA M3MFF
ME13 8DE M6BOI
ME13 8DG M3NPS
ME13 8DZ G1HDK
ME13 8HP G1EUQ
ME13 8JQ M3YTP
ME13 8QL M3ZFB
ME13 8RU G7AFS
ME13 8XN G4UVX
ME13 9LF 2E1BAE
ME13 9NR M3XEQ
ME13 9SE G4AWG
ME14 1UT M6TLT
ME14 2BX M1BWJ
ME14 2EA G0REN
ME14 2ER G3YJS
ME14 2JN G3PCX
ME14 2QR G4UAQ
ME14 2QW M0CZN
ME14 3HP 2E1BBO
ME14 3HP G3YCN
ME14 3JN G3ZAL
ME14 4BH G0KYM
ME14 4EF G4VFQ
ME14 4HG G7MOX
ME14 4JB G0BUW
ME14 4NJ G0SPQ
ME14 5BH G6VVS
ME14 5HJ G8HLE
ME14 5PG G3WXU
ME14 5UX G4MXQ
ME15 0AJ G7DJT
ME15 0DR M6HIL
ME15 0QF M6HRX
ME15 0QH G4XMS
ME15 6FF G6XAK
ME15 6PX G7AZT
ME15 6TH G4CCQ
ME15 6UL M6FLB
ME15 6UU G8LEG
ME15 6YN M3XSP
ME15 7AU G7VOM
ME15 7AU G7VON
ME15 7PD 2E0WGP
ME15 7PD M3WZG
ME15 7PL M0HLD
ME15 7RR 2E0GTA
ME15 7RR M0CST
ME15 7RT G3WWI
ME15 7SP 2E0SVG
ME15 7SP M6SVG
ME15 8ED M0NPD
ME15 8EG G4IWN
ME15 8EP M0HBN
ME15 8JP G0VAR
ME15 8JZ G7GAZ
ME15 8RA G0DBY
ME15 8TN G8LGA
ME15 9AE G6ALJ
ME15 9BU G6PEA
ME15 9JB G8VPD
ME15 9JR G7GRJ
ME15 9PD G0NCW
ME15 9QR G7HON
ME15 9RA G3RJF
ME15 9RP G0LCH
ME15 9RS G4WBA
ME16 0DL G3ORP
ME16 0JL M0PMC
ME16 0JS G6RVS
ME16 0JS M0KMR
ME16 0JS M3RVS
ME16 0NF G3VNI
ME16 0NT G4AAW
ME16 0QP G0UED
ME16 8BS M1BGS
ME16 8DR G4AXD
ME16 8EN G8PWT
ME16 8NP G1XJK
ME16 8NX M0DWW
ME16 8PA G0FKK
ME16 8QP M1AJM
ME16 8QU 2E0JHG
ME16 8QU M0JHG
ME16 9HF 2E0HIZ
ME16 9HF 2E0WEL
ME16 9HF G4HIZ
ME16 9HF M0HIZ
ME16 9HF M1AEX
ME16 9HF M3SIZ
ME16 9HF M6SIZ
ME16 9HH G6HTH
ME16 9JG M0DHM
ME17 1AD G3YIF
ME17 1AY M1DDW
ME17 1BS G3YUU
ME17 1NP G1YSR
ME17 1QT G4GRR
ME17 1SR 2E0LAH
ME17 1SR M6BEC
ME17 1SX G1BLB
ME17 1TJ G3UOJ
ME17 1TZ G6KTG
ME17 2AN G6USR
ME17 2AY G0IJK
ME17 2EJ G3LRX
ME17 2ET G4XSI
ME17 2EU G8WHB
ME17 2NH 2E0PSR
ME17 2NH M0NAL
ME17 2NH M6GAE
ME17 2NS G1HQJ
ME17 3NP G1YQN
ME17 4DN G1ZMG
ME17 4JY G6HTS
ME17 4LG G4OPP
ME17 4QS M3XZK
ME18 5HT M6TFT
ME18 5LD G8MFI
ME18 5PR G3DEY
ME18 5PX G4LNM
ME18 5QH G8NVH
ME18 6DX M0GZH
ME19 4GT G4NHP
ME19 4PS G8LDU
ME19 4TT G0OLT
ME19 5BL M6GCX
ME19 5BW G6FLA
ME19 5EH G3PAG
ME19 5EN G3XJS
ME19 5EN M0VLP
ME19 5JX G0VUN
ME19 5QY G1AJY
ME19 5QY M0RWW
ME19 6ES G7LMT
ME19 6QS G1JJQ
ME19 6SJ M1CMW
ME2 1AX 2E0OFM
ME2 1AX M0OFM
ME2 1AX M6OFM
ME2 1AX M6TOG
ME2 1DJ G7DSU
ME2 1HB G3FTH
ME2 1HB G5MW
ME2 1HB G8MWA
ME2 1HW G7JJD
ME2 2NJ 2E0DCN
ME2 2QB 2E1IEK
ME2 2QY M0AZS
ME2 2TZ G7UTB
ME2 2YE G4ZFC
ME2 2YT M1DNJ
ME2 2YW G4TVW
ME2 3BA G0LAD
ME2 3HW G0LJD
ME2 3HW M0SKG
ME2 3JH G7LAN
ME2 3LW G1EWM
ME2 3NF G6CKK
ME2 3PN G0VMQ
ME2 3QG G0MIF
ME2 3RE M3BFX
ME2 3SN G4PPV
ME2 3SP G3FNZ
ME2 3TW G8ZRQ
ME2 4BB G6HIV
ME2 4BB M3LYZ
ME2 4EB G0VKL
ME2 4JD M0BPN
ME2 4LJ 2E0JAX
ME2 4LJ M6JHB
ME2 4NE G0GQT
ME2 4NL G4RFN
ME2 4NU G4CWV
ME2 4PG G1DOT
ME2 4PN G3NZR
ME2 4XF G8CAM
ME20 1AP M6PBP
ME20 6AE M1BJS
ME20 6BB M3VMQ
ME20 6EF G0PWX
ME20 6ET G3PYO
ME20 6LA G4RIU
ME20 6QZ G7LJQ
ME20 6RE M3MDY
ME20 7SF G1XUE
ME3 0BJ M6AAE
ME3 0BS M0ZMX
ME3 7BA G7AJT
ME3 7EH G0WAN
ME3 7JR 2E0BBA
ME3 7QX G7VQJ
ME3 7RN M3UIF
ME3 7SH G3OHP
ME3 7TN G1HRV
ME3 8BE G6KSK
ME3 8DW G8MIN
ME3 8HT G1GGI
ME3 8LR G3BSN
ME3 8NE G4FVB
ME3 8TR G4FJW
ME3 9AA 2E0KSG
ME3 9AA 2E0SGG
ME3 9AA M6KSG
ME3 9AA M6SDG
ME3 9BZ G0NCE
ME3 9DE G0PEK
ME3 9DE M6HLR
ME3 9DF 2E0BPG
ME3 9DF M3XFT
ME3 9EA 2E0BPY
ME3 9EA M3UIJ
ME3 9EA M3XBS
ME3 9EU G4TAM
ME3 9GF 2E0CYP
ME3 9GF M0HSU
ME3 9GF M6DVW
ME3 9HS G4EHQ
ME3 9JP G0AUN
ME3 9LH G7JYL
ME3 9PR G0PCA
ME3 9PR G0PCB
ME3 9QS G6ZEQ
ME3 9ST G1ALK
ME3 9TA G6IYD
ME3 9TG 2E0REY
ME3 9TG M3UVC
ME4 4XJ M0DKT
ME4 4XJ M3DKT
ME4 5BZ 2E0RAM
ME4 5LE G6TXP
ME4 5NN M3FOR
ME4 5SJ M3PQG
ME4 5TW G7CVY
ME4 6HL G1YXH
ME4 6QD 2E0CZR
ME4 6QD M6BYF
ME4 6RE G4YLT
ME4 6UU G8CJM
ME5 0BH G4HJE
ME5 0DD M1BIK
ME5 0DQ G1EDH
ME5 0DY G7MFW
ME5 0HJ 2E0KVB
ME5 0HJ M6ETA
ME5 0HJ M6WMA
ME5 0JS G2FJA
ME5 0JS G8VJU
ME5 0JX G4AKQ
ME5 0SB G8TGB
ME5 0UP M6WJB
ME5 7AA M3MDS
ME5 7BE M0JVT
ME5 7HH 2E0KLY
ME5 7HH M3SJY
ME5 7NG 2E0AXN
ME5 7NG M3DSU
ME5 7PD M0WDG
ME5 7PT G0ABN
ME5 7PZ G7ICV
ME5 7QD G4YTU
ME5 7QL M6WTM
ME5 8ER M3ZIM
ME5 8ER M5LRO
ME5 8JD G6FHK
ME5 8NE 2E0SEB
ME5 8NE M0STT
ME5 8NY G1LHD
ME5 8NY G7IBN
ME5 8QA G6XXQ
ME5 8SR G6TXB
ME5 8TN G0VWV
ME5 8TN G7KNM
ME5 8UL G3PSR
ME5 9AA G7MPZ
ME5 9AG M3MZN
ME5 9AR 2E0JSH
ME5 9AR M0FSH
ME5 9HA G8IXC
ME5 9HB M0DTK
ME5 9HE 2E0HRJ
ME5 9HE G0BIX
ME5 9HE M0GYI
ME5 9HE M0HRP
ME5 9HE M6HRJ
ME5 9JG G6BLK
ME5 9JG G6NAL
ME5 9JN G8CVS
ME5 9LA G4DQN
ME5 9LF G6SQS
ME5 9LU M6DAK
ME5 9QP 2E1GTB
ME5 9RD G0AKR
ME5 9SP G6YHP
ME5 9TD G3JHU
ME6 5HJ G6HXR
ME6 5HP 2E0MLE
ME6 5HP M6MLE
ME6 5PA G6ZMG
ME6 5PY G8CCJ
ME6 5RL G0JAR
ME7 1DZ G0GJW
ME7 1EW 2E0RNE
ME7 1JB 2E0WAK
ME7 1JB M0WKO
ME7 1JB M3WYQ
ME7 1ND G6OLV
ME7 1UE M3SML
ME7 1UG 2E1IJX
ME7 1UG M0PPS
ME7 2DN 2E0DCH
ME7 2HL M3MEI
ME7 2LN G3TCI
ME7 2RE 2E0BAY
ME7 2RE M3IQQ
ME7 2RE M6BAY
ME7 2RW M0SWF
ME7 2SW G0JLP
ME7 2SX G0VQB
ME7 2UN M0OOT
ME7 2UW G3JRD
ME7 2YH G5EDQ
ME7 3AY G6ITM
ME7 3DJ G0CWH
ME7 3EH M3NBK
ME7 3EX M1BDS
ME7 3JE G7AJK
ME7 3LS G4ZTF
ME7 3PT G6IVP
ME7 3QQ G0TAR
ME7 3QQ M0ECK
ME7 3SQ 2E1EVM
ME7 3TJ G7NJG
ME7 3TS 2E1EVJ
ME7 3TS G3VFC
ME7 3TS M6VFC
ME7 4BA M3EJL
ME7 4EP G3ZYV
ME7 4HB 2E1IJQ
ME7 4HB G6SXC
ME7 4HB M0FMY
ME7 4JB G6TBJ
ME7 4NN G1XIO
ME7 4RN M1ETC
ME7 4RR M0DYG
ME7 5EX M6OKR
ME7 5HY G8STY
ME7 5JB M0SAC
ME7 5QG 2E1HRY
ME7 5QG 2E1RIO
ME7 5QG M3ANE
ME7 5QG M3HRY
ME7 5QH M1ALM
ME8 0AQ G1IBJ
ME8 0DJ G7KAT
ME8 0DZ M3HWX
ME8 0JD G0VGR
ME8 0JG 2E0ICI
ME8 0JL M6VAG
ME8 0JX G7MNG
ME8 0LA G6OCO
ME8 0LZ G8VVM
ME8 0NR G6YLW
ME8 0PE G7MIM
ME8 0PE M1CQT
ME8 0RD 2E1DMU
ME8 0TH 2E1IJS
ME8 0TL G4HZI
ME8 6JE G0TCW
ME8 6LX M0PFW
ME8 6ND G1XNC
ME8 7PR G3VTT
ME8 7PR M0GVP
ME8 7QB G8MNL
ME8 7TA G1CBK
ME8 8EW 2E0HIP
ME8 8EW M0HLB
ME8 8EW M6KOM
ME8 8HE G3PQD
ME8 8JU G4ZRY
ME8 8JY G7AZM
ME8 8TA G0DWF
ME8 8TL 2E0WSS
ME8 8TL M3WSS
ME8 9EN 2E0RPA
ME8 9EN M0HNE
ME8 9EN M6REH
ME8 9ES G0APB
ME8 9HX G1ZBH
ME8 9JQ M6FSN
ME8 9NZ G6YNA
ME8 9PP G0BRC
ME8 9PP G7BRC
ME8 9PP M0AAK
ME9 0LR G1PGI
ME9 0PL G4MKG
ME9 0TY G4JJX
ME9 7AG G4TQS
ME9 7AS M3DWV
ME9 7BQ M6KRZ
ME9 7EX G3UIB
ME9 7JU G4IOG
ME9 7QA G4LUO
ME9 7RH G3PSS
ME9 7SE G8MVH
ME9 7SL M0LNE
ME9 7SL M6LNE
ME9 7SY G3UKL
ME9 8DX M6COL
ME9 8FR M3ZZF
ME9 8JZ 2E0IKV
ME9 8JZ M6VKA
ME9 8SB G7HTN
ME9 9HU G4MPA
ME9 9LH G6SIQ
ME9 9PL M0CUT
ME9 9SP G6HOS
ME9 9SW M3MXO

## MK

(Milton Keynes)

MK1 1BL G8ABB
MK1 1NG G0WFH
MK10 7BW G7UOD
MK10 7BW M6MGU
MK10 9AY 2E0ROY
MK10 9HG G8OSH
MK10 9HR 2E0CVV
MK10 9HR 2E0EBR
MK10 9HR M6EBR
MK10 9HR M6PNY
MK10 9HZ M0MPB
MK10 9JL G6KGK
MK11 1AT M3DXL
MK11 1ET G8XJL
MK11 1HX G6AZR
MK11 1LA M0CEU
MK11 1LD G0BOQ
MK11 1RB G4NNX
MK11 2AA 2E1CKQ
MK11 2AF G0GMB
MK12 5AY G6WXM
MK12 5BE G8AQB
MK12 5DN G6WZL
MK12 5DR M6WFG
MK12 5FJ M3LFZ
MK12 5HB M0BZK
MK12 5HP M0AFJ
MK12 5HZ 2E0RHO
MK12 5HZ M3LNH
MK12 6AL G0TGU
MK12 6AN G3NEH
MK12 6LR G3UQL
MK13 0PN G0RKV
MK13 0PN M3ROW
MK13 0QP G6INI
MK13 7AJ G8KFN
MK13 7AY G4LIJ
MK13 7AY G8CCV
MK13 7AZ M6JPW
MK13 7DA G1NRY
MK13 7DY G8NKN
MK13 7EP M0GPJ
MK13 7HS G3SHD
MK13 7LZ G4DNI
MK13 7PZ G7OOU
MK13 8AT G8EPQ
MK13 8DP M6ZXI
MK13 9BT 2E0EBP
MK13 9HS G0BYH
MK14 5AX G3YYN
MK14 5AX M0ALO
MK14 5BZ M1SHE
MK14 5BZ M3LEK
MK14 5EJ G7OSR
MK14 5JQ G4WBG
MK14 5JY G7AFL
MK14 5PA 2E0CKJ
MK14 5PA M3PWW
MK14 5QA G3JVR
MK14 6AY G6ALU
MK14 6BU 2E0XJL
MK14 6BU M3RZO
MK14 6BU M6LRW
MK14 6EJ G7UTY
MK14 6EJ M0CMS
MK14 6ER G6HKS
MK14 7AP M6SOU
MK14 7BB G0FMN
MK14 7BB M6EXE
MK14 7DY G4TOH
MK14 7LL G1DMR
MK14 7LU G1LRU
MK14 7PJ 2E0TFI
MK14 7PP 2E1HSP
MK14 7PP G1RNZ
MK14 7PP M3CMK
MK14 7PP M6MOG
MK14 7QN G7VOA
MK14 7QS G1SYU
MK14 7QS G7AUE
MK15 9HP 2E1HCT
MK15 9HP M3BEK
MK16 0EE G1BAQ
MK16 0LH G0FWD
MK16 0LL M0ATY
MK16 0LT G4MDR
MK16 0NG M3YJA
MK16 8BB M0HXN
MK16 8JB 2E0RTC
MK16 8JB G4GJR
MK16 8JB M6AIE
MK16 8JG G3WKL
MK16 8NR G8JWT
MK16 8PG M0GYK
MK16 8PL G4FIA
MK16 8PP G4VMX
MK16 8RF G1CYQ
MK16 8SR G6GCM
MK16 8ST G0TAI
MK16 8TZ M6BEI
MK16 9AR M6NPL
MK16 9DN G4MFT
MK16 9LZ G3ZCJ
MK16 9NQ G6ZQU
MK17 0BX G3ZLX
MK17 0DQ G1FXT
MK17 0DR G1YYU
MK17 0HP M0BZE
MK17 0LL M0RBD
MK17 0QR G0AGR
MK17 0QR G4NUG
MK17 0SB G7MFZ
MK17 0SB G8FAK
MK17 8AS G4XWM
MK17 8EA M0RPJ
MK17 8HX G0TUL
MK17 8HX G8RCK
MK17 8QG G0GQH
MK17 8TN G3VZV
MK17 9AG M3ZCN
MK17 9AJ G0NGQ
MK17 9AJ G4UQR
MK17 9ED G0LAX
MK17 9ED G6VNI
MK17 9JU M6KAG
MK18 1DA 2E0WDS
MK18 1DA M3UEZ
MK18 1DZ M1EUX
MK18 1LZ G4UCJ
MK18 1PL G3NPI
MK18 1UF G8ZUI
MK18 2FD G4HIF
MK18 2FG M3RHR
MK18 2HA 2E0EMP
MK18 2HA M6LCU
MK18 2JH G6JXC
MK18 2LZ G7PMW
MK18 2NY M3WRB
MK18 2PF G1UEO
MK18 2QD G0GGN
MK18 2QR G2PXQ
MK18 2QS G1MYO
MK18 3AB G8GLZ
MK18 3BD G1VHC
MK18 3BD G1ZWB
MK18 3BN 2E0KVA
MK18 3DY G4FYO
MK18 3HX G8FMC
MK18 3LZ M0THN
MK18 3NA G7JHV
MK18 3NT 2E0TAC
MK18 3NT M3TPI
MK18 3NT M5REV
MK18 4BX G8RDB
MK18 4EL G3SHZ
MK18 4HL G4UUF
MK18 4LH G0ICC
MK18 4LX M3KHI
MK18 5BJ G4YTL
MK18 7EQ G0BLQ
MK18 7HR G6TSJ
MK19 6BT M3JHR
MK19 6BY G4MDU
MK19 6BY M3LJF
MK19 6HX M3RKV
MK19 6JD 2E0ECI
MK19 6JD M6CIE
MK19 7AG M0JSZ
MK19 7AN G3LCS
MK19 7BE G3LFR
MK19 7BL G8CXT
MK19 7HF 2E0CMK
MK19 7HF M0IOT
MK19 7HF M6CPN
MK19 7JQ G4XJE
MK19 7LD G4KST
MK19 7LF 2E0FSG
MK19 7LF G4FZA
MK19 7LF M6AMJ
MK19 7LQ G4MGV
MK19 7LU G3XLN
MK19 7LX G0FQU
MK19 7NF G8UEZ
MK19 7NH G1WWP
MK19 7NY 2E0CFQ
MK19 7NY M0HCP
MK19 7PL G0TLE
MK2 2EZ G0EYZ
MK2 2HT G0UUF
MK2 2LA G0GQP
MK2 2PY G8GIL
MK2 2QD G0LRW
MK2 2RN G0OQE
MK2 2XP 2E1FCO
MK2 2XP 2E1GRG
MK2 3AD G3CPT
MK2 3EB M3IDQ
MK2 3FH G1GXX
MK2 3LS G4JGG
MK2 3NB G3YJD
MK2 3NG 2E1OUJ
MK3 5AF 2E0EAF
MK3 5AF M3UHY
MK3 5AF M3UXL
MK3 5AF M3ZYT
MK3 5DQ 2E0APN
MK3 5DQ G3PPO
MK3 5DR M6AYU
MK3 5EN G0TUO
MK3 6BP M0CEM
MK3 6BY G6ZJD
MK3 6HX M3ZFI
MK3 6JP 2E0MKT
MK3 6JP M0TSW
MK3 6JP M6TMK
MK3 6PE M3ELN
MK3 6PJ M3GGA
MK3 6PL 2E1GJO
MK3 7BG M3UMV
MK3 7BP 2E1BEV
MK3 7EU G3KQQ
MK3 7LD M0HFB
MK3 7PS M1ELI
MK3 7QP M6GCT
MK3 7RP G3UDC
MK3 7RQ G4GCJ
MK3 7TA G0EFN
MK3 7UN G4ZNY
MK4 1AF M6GRQ
MK4 1AR G6OIA
MK4 1AX M6MOU
MK4 1BJ 2E0BGP
MK4 1BQ G7NDO
MK4 1DP G1TTC
MK4 1EF G0UCK
MK4 1HY M0AEQ
MK4 1JH M1ESM
MK4 2AB M0VPE
MK4 2AU G0OGM
MK4 2BT G1PCN
MK4 2GF G7LXB
MK4 3AL G4BUL
MK4 3EQ G8XXM
MK4 4AX M3EQL
MK4 4AX M3SSI
MK4 4DH G4UFS
MK40 1FD 2E0MPR
MK40 1FD M3JZI
MK40 1TF M6CYD
MK40 1ZL M0LRS
MK40 2BE G6GCO
MK40 2DS 2E0ZPY
MK40 2DS M6ZPY
MK40 2HY M6OCB
MK40 2NX M6PVR
MK40 2TR G5BH
MK40 3BF G3USD
MK40 3RY M5DHB
MK40 3SA 2E0ACR
MK40 3UG G0BMG
MK40 3UG M3TIY
MK40 4BA G4BCS
MK40 4FZ M6HUP
MK40 4PZ G0ADB
MK41 0TF M0MKR
MK41 0TX M3XAK
MK41 0UP G0BKN
MK41 0UP G0MBR
MK41 0UP G6MBR
MK41 6BS 2E0OAK
MK41 6BS G0GBS
MK41 6BS M6JBU
MK41 6BS M6MBU
MK41 6DA G6UPA
MK41 7BS G4GDX
MK41 7BT G4KWH
MK41 7BT G4KYX
MK41 7DH G3UED
MK41 7HY G8CGM
MK41 7JP G4KJU
MK41 7LS G7RVC
MK41 7ST G3UBV
MK41 7ST G4EZQ
MK41 7TU G1BYT
MK41 7UH G6CVE
MK41 8AP G4UCY
MK41 8AS G4AQS
MK41 8AS G4GIR
MK41 8AW 2E1EMI
MK41 8BT G3RPL
MK41 8DR G4XDB
MK41 8DR G8HJH
MK41 8EL G0HER
MK41 8JS G6PAA
MK41 8JS G7RDA
MK41 8NX G6YNW
MK41 8QD G0VJJ
MK41 8QY G8HGL
MK41 9AL G8SDN
MK41 9AL G3XNG
MK41 9AS G7LEB
MK41 9DD G4YYC
MK41 9EP G1JZK
MK41 9ER M6SNP
MK41 9LD M1PGH
MK41 9LD M3PGH
MK41 9LG G8RTN
MK41 9NE G3WBP
MK41 9QR G1JZT
MK41 9QW M3UDU
MK41 9RB M6FLK
MK41 9TB G7MDY
MK42 0AF G3XDU
MK42 0AF G8HXW
MK42 0AT 2E0CEP
MK42 0AT M0HBX
MK42 0AT M3SDV
MK42 0LS 2E0DLG
MK42 0NA 2E0OAA
MK42 0NA M0TXS
MK42 0NA M6SRO
MK42 0PX G4IMH
MK42 0RS M3GHE
MK42 0SE G0GBI
MK42 0SJ G0OXY
MK42 0XB M6MUP
MK42 0XD M6GGW
MK42 6AH M6WCR
MK42 7BE G4OOQ
MK42 7DP M3DPQ
MK42 7DU M1MPA
MK42 7DU M3ZKA
MK42 7DU M6KGA
MK42 7EH G7MIS
MK42 7EN G4ZPH
MK42 7FR G1NRE
MK42 7JN 2E0GLD
MK42 7JN M3ORB
MK42 7PE G8MGP
MK42 7SX M6FAF
MK42 8BU G7SPE
MK42 8HD G0BDK
MK42 8HS M0ZAE
MK42 8HS M3ZIE
MK42 8NL G7SQM
MK42 8QS G4SKU
MK42 8RU G8WGQ
MK42 8RX G4OGG
MK42 9RG G0EYG
MK42 9TH G8OCH
MK42 9TZ 2E0RML
MK42 9UX G4NGD
MK42 9YS M0GRV
MK42 9YX G0VME
MK43 0DF G4DCW
MK43 0EL 2E0CQB
MK43 0EL M6CQB
MK43 0GN G7OKV
MK43 0HA G7BBD
MK43 0HS G7KLR
MK43 0HX M6BFI
MK43 0JA G4JXZ
MK43 0LW G4PMS
MK43 0PN G7UYB
MK43 0PN M0BZO
MK43 0QR G1MKR
MK43 0QR G3YQO
MK43 0QZ G0FGJ
MK43 0RU M6CYB
MK43 0RW G0RAV
MK43 0SD G2KER
MK43 7BB G4ABQ
MK43 7BG M0NGS
MK43 7BG M0PSR
MK43 7BG M5RPT
MK43 7DG M6JWK
MK43 7DR G4CEC
MK43 7ED G3INY
MK43 7HJ M6EMQ
MK43 7HN G8SZG
MK43 7HN M3HML
MK43 7JL G3LDG
MK43 7JT M3MSJ
MK43 7JX G3SVQ
MK43 7LP G8ADY
MK43 7QF G4PNK
MK43 7SG 2E0ACQ
MK43 7SG G0GPV
MK43 7SJ G0EKD
MK43 7TD G4XPJ
MK43 8JA G4MBA
MK43 8LF G4UMW
MK43 8QJ G3XYV
MK43 8QJ M3ZLE
MK43 9BB G0XIT
MK43 9BB G3RKP
MK43 9BB M6UBS
MK43 9BT 2E1SPY
MK43 9BT M3CLP
MK43 9BT M3OLI
MK43 9BT M5HOT
MK43 9DW G8BAK
MK43 9EX G1GAS
MK43 9EZ G0DIM
MK43 9JL G6SNV
MK43 9JW G3YUQ
MK43 9JW G6IEE
MK43 9LB G8XNN
MK43 9NG G6PHU
MK44 1DY G1BUV
MK44 1EN M1EBS
MK44 1EX G6HMF
MK44 1JP G8IIG
MK44 1NP G3WTP
MK44 1NP G3XNG
MK44 1NP M1BED
MK44 1NX G6EDT
MK44 1PL G0MOR
MK44 1SE G8ACQ
MK44 2AU G4FFC
MK44 2BA G0CZY
MK44 2DT G4TGV
MK44 2ER G4WDZ
MK44 2EW G4MSQ
MK44 2HP G4OCX
MK44 2JX G1PSH
MK44 2LA G4HUE
MK44 2LF 2E0HUR
MK44 2LF M6HUR
MK44 2RP G4HTX
MK44 2RP G6JJT
MK44 2RP M3TBF
MK44 3BH G4TXG
MK44 3BQ G4RLR
MK44 3NJ M6TVC
MK44 3NT M0DLM
MK44 3QW G4ACP
MK45 1AQ G8LLD
MK45 1BU G0COG
MK45 1JY G6JZW
MK45 1LN M1AFP
MK45 1LN M3ADL
MK45 1PJ G8CTB
MK45 1QA G4OWN
MK45 1QD G6PHZ
MK45 1RB G8OCF
MK45 1TB G0WZA
MK45 1TQ G7WAS
MK45 1TQ M0WAS
MK45 2AD G7PCF
MK45 2AE G0BVW
MK45 2AQ 2E1FDJ
MK45 2BT G4AHM
MK45 2DJ G0WAS
MK45 2EY G3VES
MK45 2QL G8VHX
MK45 2RS G7NBI
MK45 2RS M0CKA
MK45 2SP G0RNI
MK45 2SP G4FKI
MK45 2SP G8GPO
MK45 2TP G0JJK
MK45 3AY M6LBC
MK45 3BU G1WDQ
MK45 3DT G7IXM
MK45 3EF G6AJX
MK45 3LQ 2E0PSW
MK45 3LQ M0PSW
MK45 3LQ M3PPG
MK45 3LS G6JIF
MK45 3PJ G8HRW
MK45 3QB G4USN
MK45 3WE G4MGX
MK45 3WE G4XQQ
MK45 3WH G7STG
MK45 4AS G8LQM
MK45 4BG G3FJE
MK45 4BG G4LOO
MK45 4DA G4VLA
MK45 4FE G7HZQ
MK45 4LT M3IHQ
MK45 4NE G4DAQ
MK45 4NE M1EAJ
MK45 4NY M6BWF
MK45 4PF G4EIY
MK45 4QJ G0KOH
MK45 4TB G6BJO
MK45 5BW G7JSB
MK45 5JN G0TFI
MK45 5LB M3YQG
MK45 5LP 2E1GXE
MK45 5LP M3IEM
MK46 4AR M6GJG
MK46 4EY G7IMB
MK46 5BJ G4NEO
MK46 5ES G0HMF
MK46 5ES M6BJX
MK46 5FB G0SXY
MK46 5JA G4NPE
MK5 6AS M3JSQ
MK5 6AS M3LNQ
MK5 6AZ 2E0MEV
MK5 6AZ M6MHV
MK5 6HT G0UVG
MK5 6JE 2E0BZO
MK5 6JE M3XPY
MK5 6JL G4ETK
MK5 7AF G3HIU
MK5 7AF G3ZPA
MK5 7AF G8MKT
MK5 7AJ G4CLG
MK5 7AX G4LNC
MK5 7HA G6DOI
MK5 7PY G0CFD
MK5 8BL G4NIV
MK5 8EB G7HRH
MK57DE M1TLK
MK57DE M3TLK
MK6 2DS M6MPP
MK6 2ES G3XVA
MK6 2HT 2E1CJN
MK6 2HT G0FTI
MK6 2PZ G8JVA
MK6 3AY G3LMX
MK6 3EN G3EBP
MK6 3ES G6TTX
MK6 3LQ G7SYU
MK6 3LU G0RDG
MK6 4HL M6SAE
MK6 4HW M6DMQ
MK6 4HX M6GUX
MK6 4LA M6KOZ
MK6 5AE 2E1DAK
MK6 5DA M3HAW
MK6 5LB M6FIO
MK6 5NB 2E0HMD
MK7 6AA G0OUR
MK7 6AA M0ANS
MK7 7EG G6VTE
MK7 7TB G8XKH
MK7 7TT M6LGQ
MK7 8DH M3XKI
MK7 8LR M0MUZ
MK7 8LR M3YZU
MK7 8PL G1OWZ
MK7 8QB 2E0RLW
MK7 8QB M0UKA
MK7 8QB M6RLW
MK7 8QD G4VZN
MK7 8QS M6OSW
MK8 0AT G7THZ
MK8 0BB G1OWJ
MK8 0DB G4TZR
MK8 0EJ 2E0MGL
MK8 0EJ M3GLT
MK8 0EJ M3LTA
MK8 0EJ M3MGL
MK8 8AY G1WLX

| Postcode | Callsign |
|---|---|
| MK8 8AY | M0GOA |
| MK8 8BX | 2E0DHF |
| MK8 8BX | M0ELIM |
| MK8 8DJ | G4BRU |
| MK8 8NP | M3GHA |
| MK8 9EN | 2E0BQB |
| MK8 9EN | M3VPB |

## ML

(Motherwell)

| Postcode | Callsign |
|---|---|
| ML1 1LQ | MM3TRZ |
| ML1 2LB | GM0ARD |
| ML1 2RL | GM6OXL |
| ML1 3AS | GM3ULP |
| ML1 3JW | GM4UBJ |
| ML1 3NA | GM0WNR |
| ML1 3QU | GM0HWB |
| ML1 4JL | 2M0MMB |
| ML1 4JL | MM3BQK |
| ML1 4RD | 2M0FDZ |
| ML1 4RD | MM3BPR |
| ML1 4ST | GM3EIY |
| ML1 4ZL | 2M0CRQ |
| ML1 4ZL | GM1XOI |
| ML1 4ZL | MM0RKN |
| ML1 4ZL | MM0RWJ |
| ML1 4ZL | MM6CRQ |
| ML1 5EF | MM0SEK |
| ML1 5LB | MM3XJA |
| ML1 5LQ | GM0VXA |
| ML1 5NU | GM4LQR |
| ML10 6DL | GM0SYV |
| ML10 6FW | MM3SHT |
| ML10 6JT | 2M0YCG |
| ML10 6JT | MM3YCG |
| ML10 6SD | MM6MNE |
| ML11 0FF | GM0DQC |
| ML11 0HY | GM6YRH |
| ML11 0NZ | 2M0DSG |
| ML11 0NZ | MM0TDB |
| ML11 0QA | MM0MOT |
| ML11 7HL | MM6EEQ |
| ML11 7PS | GM4IIR |
| ML11 7SE | GM3YXY |
| ML11 8ES | GM0VYY |
| ML11 8HA | 2M0AIE |
| ML11 8LH | MM3GYU |
| ML11 8NB | MM0FCM |
| ML11 8NE | MM3XIA |
| ML11 8SU | MM1BJG |
| ML11 9AE | GM3NNZ |
| ML11 9QA | GM6BIG |
| ML11 9SR | GM4VGU |
| ML11 9XS | GM0GYN |
| ML12 6FW | GM4XHH |
| ML12 6GB | MM0MSH |
| ML12 6HQ | GM0SCA |
| ML12 6LB | GM7GIO |
| ML12 6PQ | MM0THE |
| ML12 6PU | GM4KKV |
| ML12 6RJ | GM0DXI |
| ML12 6RR | MM6EQW |
| ML12 6TP | MM6FPY |
| ML2 0AH | 2M0DQN |
| ML2 0LP | MM6JAW |
| ML2 7HG | GM0WRU |
| ML2 7SJ | GM0LIR |
| ML2 8BZ | MM6BIY |
| ML2 8JS | 2M0KVM |
| ML2 8JS | MM6KCM |
| ML2 8PG | 2M0SRX |
| ML2 8PG | MM6SNR |
| ML2 8QR | MM3VIS |
| ML2 8RA | 2M0NAX |
| ML2 8RA | MM6NAI |
| ML2 8SE | GM0VXZ |
| ML2 9BL | GM3TCW |
| ML2 9DB | MM6AHB |
| ML2 9QL | GM8KIQ |
| ML3 0HZ | MM6PIB |
| ML3 6BF | MM0KZA |
| ML3 6PD | MM6CWN |
| ML3 6PN | MM0RKT |
| ML3 6PP | GM4UBF |
| ML3 7DD | GM8NBV |
| ML3 7FB | GM0STB |
| ML3 7FB | GM4UQG |
| ML3 7FR | MM3JLS |
| ML3 7HB | GM0KDF |
| ML3 7HJ | 2M0ELP |
| ML3 7HJ | GM0ELP |
| ML3 7HJ | MM0ELP |
| ML3 7HN | 2M0RTD |
| ML3 7HN | MM0RTD |
| ML3 7JY | MM0GHN |
| ML3 7LL | GM8LBC |
| ML3 7LY | 2M0JHN |
| ML3 7PE | GM4ILE |
| ML3 7PN | 2M0JQD |
| ML3 7PN | MM3JSB |
| ML3 7PW | GM0MDX |
| ML3 7WS | MM6IZX |
| ML3 7WS | MM6VZA |
| ML3 7YG | MM6HQC |
| ML3 8AY | MM6LFN |
| ML3 8JS | 2M0VLF |
| ML3 8PH | MM6EVS |
| ML3 8PH | MM6YHO |
| ML3 8QH | GM0CSN |
| ML3 8TZ | MM3UQT |
| ML3 9AF | MM0HQD |
| ML3 9JR | 2M0OIC |
| ML3 9JR | MM6MCA |
| ML3 9QH | MM6RLL |
| ML3 9RQ | MM3VEE |
| ML3 9UU | MM6JJW |
| ML3 9UX | GM4RVU |
| ML4 1JB | MM0GGG |
| ML4 2BG | 2M0RMP |
| ML4 2HQ | 2M0RVF |
| ML5 4BJ | MM6PDA |
| ML5 4FN | 2M0CWV |
| ML5 4FN | MM6DKN |
| ML5 4JQ | GM4SZG |
| ML5 4NE | MM3VHM |
| ML5 5RD | 2M0TOR |
| ML54LL | MM0CEZ |
| ML6 0ES | MM6WBU |
| ml6 0nj | 2M0SSO |
| ml6 0nj | MM0SSG |
| ml6 0nj | MM3YNP |
| ML6 0QQ | GM4AUP |
| ML6 6ET | 2M0WLX |
| ML6 6ET | MM6OWL |
| ML6 6GP | MM6ZGS |
| ML6 6PA | GM0VXQ |
| ML6 6SP | MM6PYX |
| ML6 7DH | GM1MMK |
| ML6 7DH | GM8HBY |
| ML6 7DT | GM7NPR |
| ML6 7JF | GM7CPR |
| ML6 7JH | MM6KFJ |
| ML6 7NT | MM6NED |
| ML6 7QB | MM3VYA |
| ML6 7SZ | GM0ART |
| ML6 8BL | GM4WTS |
| ML6 8FX | GM0EEG |
| ML6 8LU | 2M0WXS |
| ML6 8LU | MM6WXS |
| ML6 8QG | GM0VWZ |
| ML6 8SF | GM4PRO |
| ML6 8XL | GM7SWX |
| ML6 9DF | GM3ZNC |
| ML6 9ND | MM0GHT |
| ML6 9ND | MM1XJS |
| ML6 9ND | MM3KNY |
| ML7 5AX | GM0PHW |
| ML7 5HW | MM6SFF |
| ML7 5PA | MM0BHX |
| ML7 5TJ | 2M0EGI |
| ML7 5TJ | MM0WEI |
| ML7 5TJ | MM3WEI |
| ML8 4AF | MM3VYU |
| ML8 4NR | GM4ARU |
| ML8 4QT | GM4XXO |
| ML8 4QZ | 2M0MUR |
| ML8 4QZ | MM0MUR |
| ML8 4QZ | MM0RRM |
| ML8 4QZ | MM6MUR |
| ML8 5GB | 2M0ZEE |
| ML8 5GB | MM6PJR |
| ML8 5HB | GM3UCI |
| ML8 5HB | MM3UCI |
| ML8 5HB | MM3VRI |
| ML8 5HR | GM4COX |
| ML8 5HR | GM4UXX |
| ML8 5JG | 2M0ISA |
| ML8 5JG | GM1OQT |
| ML8 5JG | MM0TPD |
| ML8 5JZ | GM6VVX |
| ML8 5LT | 2M0LAW |
| ML8 5LT | MM3SUV |
| ML8 5PH | GM4LAO |
| ML8 5RX | GM0OQV |
| ML8 5RX | MM3OQV |
| ML8 5TS | GM8FHK |
| ML8 5UW | 2M0ZBF |
| ML8 5UW | MM6EZO |
| ML9 1DN | GM0SYY |
| ML9 1EH | 2M0XXP |
| ML9 1EH | MM0XXP |
| ML9 1EH | MM3XXP |
| ML9 1JX | 2M0MMO |
| ML9 1JX | MM0MOB |
| ML9 1QT | GM3MXN |
| ML9 1RA | GM3HVK |
| ML9 1UR | GM1WYV |
| ML9 2DA | GM7AOM |
| ML9 2LG | GM3NKG |
| ML9 2TD | GM4ISM |
| ML9 3AJ | GM1DLS |
| ML9 3BT | 2M0VPU |
| ML9 3EN | GM6HFH |
| ML9 3JN | 2M0TGM |
| ML9 3JN | MM6FPX |
| MC0 0BD | M3JNJ |

## N

(North London)

| Postcode | Callsign |
|---|---|
| N1 0HN | M6PDR |
| N1 1QN | G1GBX |
| N1 1TN | G0TJD |
| N1 2BN | M0CYT |
| N1 2FP | M0ZAY |
| N1 2LW | G0DCP |
| N1 2PJ | G4FIH |
| N1 3NW | M3ZUT |
| N1 4NU | G0OOS |
| N1 4NU | M0ESP |
| N1 7AR | G0BQI |
| N1 7AY | 2E0JQW |
| N1 7AY | M0JQW |
| N1 7AY | M0THY |
| N1 7AY | M6JQW |
| N1 7AY | M6THY |
| N1 8LX | 2E1GPG |
| N1 0UA | G1GJV |
| N10 1FQ | M0TCU |
| N10 1LX | G3ASX |
| N10 2DE | G3KOD |
| N10 2HS | G4KPH |
| N10 2JU | M0PCS |
| N10 2PD | G8HHZ |
| N10 3AA | G4DFJ |
| N10 3EH | G0IPT |
| N10 3PB | G7FJU |
| N11 1AX | 2E1HAM |
| N11 1AX | M3GTA |
| N11 1RD | M3FWA |
| N11 2AD | 2E0DTY |
| N11 2AD | M0OTO |
| N11 2DE | G8YNC |
| N11 2JL | G6FDX |
| N11 2JY | G0NSO |
| N11 2LT | M0ULC |
| N11 2RL | G1IHA |
| N11 3HJ | M0PKL |
| N11 3NN | M6DLD |
| N11 3NQ | G3JPD |
| N12 0LS | G8XRP |
| N12 0NX | G6ABJ |
| N12 7JG | G4CCT |
| N12 8AF | G7VFY |
| N12 8QT | G3TLU |
| N12 8RB | 2E0KSO |
| N12 9AU | G1YSA |
| N12 9BG | G4RIH |
| N12 9EB | G6TQL |
| N13 4HD | 2E1HWV |
| N13 4RG | 2E1HXD |
| N13 5BS | G6NTW |
| N13 5BT | G3YTY |
| N13 5JS | M6SJF |
| N13 5JT | G3PRK |
| N14 4DH | G0KUX |
| N14 4DH | G1YJI |
| N14 4PG | G1BPU |
| N14 4RP | G4IEH |
| N14 4XD | G8FSL |
| N14 4XN | 2E1FMW |
| N14 4XN | G4AEZ |
| N14 4XN | G6VAL |
| N14 5AT | M6ORT |
| N14 6QU | G4BAN |
| N14 6RR | G1DEY |
| N14 7NJ | G6BLC |
| N15 4JN | G8OJR |
| N15 4LA | M6COI |
| N15 5RP | M0LPT |
| N16 0EL | M6BIU |
| N16 6DJ | 2E1PPK |
| N16 6DJ | M3GKG |
| N16 6EP | G0PIU |
| N16 8DD | 2E0LUK |
| N16 8DD | M6IVE |
| N16 8HW | G3NHS |
| N16 9PH | 2E0SDM |
| N16 9PH | M6REI |
| N17 0JD | G3YBS |
| N17 0PH | G8PHJ |
| N17 0TE | G0CEU |
| N17 6BP | G8ASC |
| N17 6EH | G0FWF |
| N17 6TG | G4KMM |
| N17 7HT | G3OVX |
| N18 1QD | M0TUK |
| N19 3DJ | M3FWJ |
| N19 3HF | G7LGI |
| N19 3SJ | G7VDI |
| N19 5BQ | G8WPA |
| N19 5NJ | M0HPZ |
| N19 5NJ | M6DYI |
| N19 5TB | G7RJO |
| N2 0HP | 2E0AQI |
| N2 0HP | G0UKA |
| N2 0HP | M0DAN |
| N2 0HP | M1DAN |
| N2 0QB | G7TWC |
| N2 0QB | M0HWC |
| N2 0SN | M0DJT |
| N2 8AY | M1KAZ |
| N2 8JT | G7LZB |
| N2 0JW | 2E0MCA |
| N2 9NS | G7ULM |
| N20 0AL | G4XYK |
| N20 0HT | G4LCR |
| N20 0HT | M6FKU |
| N20 0QN | G0EIQ |
| N20 8QH | G1FLX |
| N20 9PJ | G3CRS |
| N20 9PJ | G4BUO |
| N21 1EH | G3TIE |
| N21 1NP | G3SFG |
| N21 1NP | G1DFB |
| N21 1NP | G6QM |
| N21 1QP | G0LWF |
| N21 1QE | 2E0DTW |
| N21 2BE | G0RPM |
| N21 2BE | M6RPM |
| N21 2BE | M6RUT |
| N21 3AU | M5AGZ |
| N21 3LP | M0VKJ |
| N22 5BL | M6OPA |
| N22 5NB | M6NYA |
| N22 6LH | 2E1FVJ |
| N22 6NT | 2E1FVK |
| N22 6RG | M6FCQ |
| N22 6RR | 2E0VIS |
| N22 6RR | M6GIC |
| N22 7BN | G1WNZ |
| N22 7EX | G4KKT |
| N22 7YG | 2E0OJS |
| N22 8NN | M3HMI |
| N22 8NN | M3HGE |
| N3 1PT | G8LSH |
| N3 1QL | M1MRS |
| N3 1QL | M3SMR |
| N3 3HP | G7UNW |
| N3 3NJ | G0ODM |
| N3 3TX | 2E0NAZ |
| N4 1HN | G0PLC |
| N4 1RJ | M6KKG |
| N4 2LN | M3XLK |
| N4 3DW | G0JYU |
| N5 0AD | G4HOJ |
| N5 1NU | G0WZK |
| N5 2DE | G8ZQJ |
| N6 4BA | G4GRS |
| N6 4BA | G8MIC |
| N6 4BH | M6SIK |
| N6 4EU | G3XKY |
| N6 4QD | G7PZM |
| N6 5AU | G0RIC |
| N6 5PJ | G0OOD |
| N6 5QP | 2E0RUI |
| N6 5QP | M0RLM |
| N6 5TS | G4OJW |
| N6 5YT | G4VEA |
| N6 6JR | G4HDD |
| N6 6NB | M0NEU |
| N7 0EL | G1NWG |
| N7 0SH | M0RXB |
| N7 0SN | G0AKM |
| N7 7AY | M3MUQ |
| N7 7EL | M6LNS |
| N7 7FE | M6MZB |
| N7 7ND | G0VBI |
| N7 8HY | G4IZU |
| N7 9DQ | M6PRD |
| N8 0JB | M6ZIB |
| N8 0NA | G4DVG |
| N8 7RW | M6ZBP |
| N8 8NX | G4JNS |
| N8 9BP | G6RZR |
| N9 0BY | M6ATA |
| N9 0HJ | 2E1FNB |
| N9 0HJ | 2E1HRM |
| N9 0HJ | M3HRM |
| N9 0HR | G1ZBA |
| N9 7QG | G1ALD |
| N9 8HD | G4YGH |
| N9 8LJ | G4ZIS |
| N9 8QL | G1RFS |
| N9 9EQ | G7COQ |
| N9 9HP | G4UTR |
| N9 9HU | G8SZZ |

## NE

(Newcastle Upon Tyne)

| Postcode | Callsign |
|---|---|
| NE10 0DR | 2E0RLJ |
| NE10 0NG | M6DJK |
| NE10 8AY | G7MPJ |
| NE10 8EN | G7VDD |
| NE10 8QS | M6ETG |
| NE10 8UE | G7JVG |
| NE10 8WJ | M6GEE |
| NE10 9AB | M3YTZ |
| NE10 9BQ | 2E1DRX |
| NE10 9DH | G7EKG |
| NE10 9NB | G1OMX |
| NE10 9TZ | G4VTN |
| NE11 0BS | G0ESW |
| NE11 9BE | G7PUA |
| NE11 9LA | 2E1EDB |
| NE11 9XH | G0CXX |
| NE12 5BL | G7VKA |
| NE12 5XT | M6DYR |
| NE12 6BT | G3YRH |
| NE12 6YR | M0NTK |
| NE12 7EU | G1USF |
| NE12 7JP | G3LRI |
| NE12 7JR | M0AVU |
| NE12 8PH | M6GMM |
| NE12 8QG | M0JAO |
| NE12 8QG | M6DAM |
| NE12 8XQ | G0WUV |
| NE12 9AW | G3LYG |
| NE12 9QP | M6FHR |
| NE12 9RL | G1XXF |
| NE13 6EY | G6YIH |
| NE13 6QB | G0LNW |
| NE13 7HS | G0AOK |
| NE13 7HT | G1GDA |
| NE13 7HW | G0GYO |
| NE13 7JW | M0EUK |
| NE13 7NG | G7VLL |
| NE13 8AB | G4FFU |
| NE13 8AR | G4FFV |
| NE13 8AR | G8LCC |
| NE13 9AF | M0PGW |
| NE13 9AF | M1DPW |
| NE15 0EA | G8UZM |
| NE15 6LG | 2E0MCN |
| NE15 6LG | M6MPB |
| NE15 6QN | G1YUN |
| NE15 6RZ | G6SMI |
| NE15 7DN | M6FOE |
| NE15 7DQ | G1YDD |
| NE15 7LB | M0IDI |
| NE15 7LN | M6ADM |
| NE15 7LR | G1SKQ |
| NE15 8NB | G0WMX |
| NE15 8RL | G3KTT |
| NE15 8TW | G0DFT |
| NE15 9AT | M0NRS |
| NE15 9FR | G8XGS |
| NE15 9JU | G0EOY |
| NE15 9LH | M1DHC |
| NE16 3EZ | M6MSI |
| NE16 3JN | 2E1APX |
| NE16 3JN | G0PUK |
| NE16 4PF | G8DST |
| NE16 4PU | G3RAR |
| NE16 5PP | M0ERN |
| NE16 5PP | M1ETS |
| NE16 5PP | M5ERN |
| NE16 5QS | 2E0TSU |
| NE16 5QS | M3XQL |
| NE16 5QZ | G4GFV |
| NE16 5RX | G0WUI |
| NE16 5SZ | G6OSH |
| NE16 6LQ | G0RFN |
| NE16 6QF | M3WQL |
| NE17 7JR | G3OPE |
| NE19 1RH | G4HDS |
| NE19 1TA | G4AOS |
| NE2 1HQ | M6WDN |
| NE2 1XY | M6FTP |
| NE2 2HD | G4ETI |
| NE2 2JN | G0BEV |
| NE2 3NS | G2DJM |
| NE20 9AE | G0AWA |
| NE20 9AU | 2E0KCN |
| NE20 9AU | M6NQR |
| NE20 9AW | G4DSD |
| NE20 9EJ | G4SVE |
| NE20 9HQ | G8OEO |
| NE20 9PZ | G7VTT |
| NE20 9QQ | M0CIP |
| NE20 9QQ | M1DPI |
| NE20 9RA | G8KPD |
| NE20 9RA | G8RER |
| NE20 9RD | M0GPU |
| NE20 9RR | G4ITR |
| NE20 9RZ | G1SUM |
| NE21 4RR | G0SWB |
| NE21 5QL | M1CKU |
| NE21 6HJ | 2E0DZO |
| NE21 6HJ | M6HNV |
| NE21 6JU | G7LOG |
| NE22 5BU | G4TMQ |
| NE22 5DZ | G1DDS |
| NE22 5ER | G4UTQ |
| NE22 5HJ | G6WJD |
| NE22 5HJ | M3HLX |
| NE22 5TD | G0AXJ |
| NE22 5TD | M6AXJ |
| NE22 5TD | M6SRS |
| NE22 6DF | G4WNG |
| NE22 6DR | G3OTH |
| NE22 6HD | G4OAN |
| NE22 6HW | G0OWE |
| NE22 6JJ | G1DSB |
| NE22 6LD | G4AXF |
| NE22 6NE | G0DJO |
| NE22 6NS | M1AQY |
| NE22 6PB | G0KYR |
| NE22 7AF | G0JQR |
| NE22 7EH | G7SNP |
| NE22 7EJ | G0GLH |
| NE23 1HZ | G6OTV |
| NE23 1RQ | G7VBD |
| NE23 1TS | G1ZJQ |
| NE23 2BP | M1PJB |
| NE23 2BZ | G6WUD |
| NE23 2UE | G0TLZ |
| NE23 2UR | M6SEU |
| NE23 3FY | 2E0KNL |
| NE23 3FY | M6KNL |
| NE23 3LW | G0BAU |
| NE23 3TT | G4HJB |
| NE23 6AA | G6AVL |
| NE23 6AS | G0NHJ |
| NE23 6AS | G0NHK |
| NE23 6DN | G4HIH |
| NE23 6RG | M0ART |
| NE23 6SX | G1SLI |
| NE23 6TS | 2E0FIZ |
| NE23 6TU | G7MYT |
| NE23 6TW | M1DHG |
| NE23 6TW | G0JRX |
| NE23 6UB | G0LCE |
| NE23 6XQ | 2E0CKW |
| NE23 6XQ | M6BLH |
| NE23 7BH | G0PVT |
| NE23 7DE | G0JWO |
| NE23 7QX | M3WQN |
| NE23 7RT | G1TPO |
| NE23 7TZ | M0FTO |
| NE23 8HG | G7OVK |
| NE24 1NP | G8EXK |
| NE24 1PG | G0ECQ |
| NE24 2JR | G4YAV |
| NE24 2JW | M3PFE |
| NE24 2QW | 2E0XRG |
| NE24 2QW | M3XRG |
| NE24 2SU | M1BMQ |
| NE24 3DT | M6EAH |
| NE24 3DT | M6EFY |
| NE24 3EP | M1ARL |
| NE24 3HA | G0LFE |
| NE24 3HA | G6PNO |
| NE24 3HX | G1LMC |
| NE24 3JG | G1ATC |
| NE24 3JG | G7PYR |
| NE24 3QN | 2E0DTH |
| NE24 3QN | M6WBH |
| NE24 3QS | 2E0CWF |
| NE24 3QU | M0UNT |
| NE24 3QQ | M0CXF |
| NE24 3RD | M6YNY |
| NE24 3RL | G4VKY |
| NE24 4AZ | M3PTR |
| NE24 4DS | M0DAD |
| NE24 4DT | G0HXC |
| NE24 4EB | G0EVF |
| NE24 4JD | M3LXP |
| NE24 4LB | G0TWV |
| NE24 4PL | M0TWO |
| NE24 4QJ | 2E0MDV |
| NE24 4QP | M0CVU |
| NE24 5AS | M1EHZ |
| NE24 5ET | M3DFW |
| NE24 5HL | G4SIE |
| NE24 5RH | 2E0BMB |
| NE24 5RN | G7KDX |
| NE24 5TS | M6KVV |
| NE25 0AL | M0MLY |
| NE25 0BU | G0LNS |
| NE25 0DW | M1DZT |
| NE25 0DW | M3ILA |
| NE25 0EL | 2E0DCV |
| NE25 0EL | M6CLG |
| NE25 0FH | G1EXK |
| NE25 0NG | G0PKR |
| NE25 0NG | G0VRT |
| NE25 0NP | G3VYZ |
| NE25 0NP | G8ARF |
| NE25 0PA | G0PJU |
| NE25 0PZ | M0GXW |
| NE25 0SH | G4VOU |
| NE25 0ST | 2E0CMH |
| NE25 0ST | M0HLR |
| NE25 0ST | M6TEY |
| NE25 0TA | G4BEZ |
| NE25 8AU | G7PWL |
| NE25 8BJ | G6CEM |
| NE25 8JG | G4IZQ |
| NE25 8JG | M0IGB |
| NE25 8JQ | G7ELV |
| NE25 8RP | M6LHF |
| NE25 8XB | G4VPU |
| NE25 9AF | M0WJW |
| NE25 9QR | G3WDS |
| NE25 9XH | M3HSI |
| NE26 1AF | G1ODJ |
| NE26 1QR | M0USB |
| NE26 1SH | G7VTJ |
| NE26 1SH | M0VTJ |
| NE26 2JR | G0OBQ |
| NE26 2NR | G6PKM |
| NE26 3AD | G8FVT |
| NE26 3BA | G3GMS |
| NE26 3DJ | G1MHA |
| NE26 3DY | G0NWM |
| NE26 3DY | M0GAE |
| NE26 3EF | G5BW |
| NE26 3SY | M0LDX |
| NE26 4JH | 2E0UWT |
| NE26 4JH | M3UWT |
| NE26 4JQ | M3BMH |
| NE26 4RE | G3VUD |
| NE26 4RG | M0HTX |
| NE27 0EF | G1AIA |
| NE27 0UF | G4RUI |
| NE27 0UX | G6VEG |
| NE27 0UZ | 2E0KBE |
| NE28 6SN | M3ENS |
| NE28 7PG | M0HGV |
| NE28 7PH | G3ZTJ |
| NE28 7SQ | M6LBL |
| NE28 8DQ | M3JFF |
| NE28 8TL | M0ADR |
| NE28 9AA | M0CBK |
| NE28 9RX | M0NKY |
| NE28 9YW | M0IIM |
| NE29 0JS | M6REY |
| NE29 0QN | G4GWV |
| NE29 0QX | G4FRD |
| NE29 6XE | G0AIO |
| NE29 7AW | G1EYD |
| NE29 7HP | G1FNU |
| NE29 7JY | 2E0CIO |
| NE29 7JY | M0HIA |
| NE29 7JY | M6AZU |
| NE29 7QN | 2E0HWC |
| NE29 7QN | M6HWC |
| NE29 8DW | G0AXO |
| NE29 0HG | G0EDK |
| NE20 8LU | G1LNR |
| NE29 8NT | G3OZP |
| NE29 8QA | G3PCC |
| NE29 8SS | G2ARY |
| NE29 8SS | M0BQD |
| NE29 9DS | G1GYM |
| NE29 9BS | M0GYM |
| NE29 9EN | G0MFF |
| NE29 9EJ | G6FIV |
| NE29 9HJ | G0UBU |
| NE29 9NG | G4AXL |
| NE29 9NU | G8YCP |
| NE3 2DU | G6OQO |
| NE3 2FG | G4LIA |
| NE3 2QQ | 2E0LMK |
| NE3 2QQ | M6SAQ |
| NE3 2UL | 2E1HDB |
| NE3 2UL | 2E1HDC |
| NE3 2UL | M1MOB |
| NE3 2UL | M1TXT |
| NE3 2XU | M0PJF |
| NE3 2XU | M3ZKF |
| NE3 3TN | G1XYS |
| NE3 4BB | G8QM |
| NE3 4JH | G4WLA |
| NE3 4NH | G4LGH |
| NE3 4RN | G0CQK |
| NE3 4RN | G1GDM |
| NE3 4RZ | G4ILL |
| NE3 4TT | G1GAD |
| NE3 5BH | G3LIV |
| NE3 5BH | G8UEE |
| NE3 5BY | G3SQU |
| NE3 5DX | G1VOP |
| NE3 5QD | 2E1ICO |
| NE3 5TQ | G3URE |
| NE30 1NA | G8NLS |
| NE30 2BW | G8YFA |
| NE30 2JW | 2E0TDR |
| NE30 2JW | M6FVH |
| NE30 2TG | G6CCN |
| NE30 3AN | G8EZL |
| NE30 3DP | M1DSE |
| NE30 3JZ | G8GYN |
| NE30 4BJ | M3WHX |
| NE31 1DH | M6PKD |
| NE31 1HL | M6CYE |
| NE31 1XT | G7JTI |
| NE31 2AQ | M0OMT |
| NE31 2AQ | M3SRT |
| NE31 2BW | M3AHQ |
| NE31 2DW | 2E0NFC |
| NE31 2DW | M6NFC |
| NE31 2DY | 2E0HUT |
| NE31 2HN | 2E0GTE |
| NE31 2HN | M0TYN |
| NE31 2HN | M6GAK |
| NE31 2HN | M6MNC |
| NE31 2HN | M6NEE |
| NE31 2HN | M6SLU |
| NE31 2HR | 2E0MCM |
| NE31 2HT | G4SFN |
| NE31 2LJ | G0WZB |
| NE31 2LJ | M6WZB |
| NE31 2TB | 2E0GWC |
| NE31 2TB | M6CKQ |
| NE31 2XS | G1NZP |
| NE32 3QA | 2E0KPC |
| NE32 3QA | M0KPC |
| NE32 3QA | M3GAG |
| NE32 3SB | 2E0MTY |
| NE32 3SB | 2E0SBO |
| NE32 3SB | M6FDR |
| NE32 3SB | M6IRM |
| NE32 3SB | M6PRN |
| NE32 3SB | M6SBF |
| NE32 3TX | G0TFT |
| NE32 4AE | G0DIP |
| NE32 4DB | 2E0HAH |
| NE32 4DB | M0MBT |
| NE32 4DB | M6MRK |
| NE32 4DU | M6PFW |
| NE32 4EY | M6AEJ |
| NE32 4HX | G3JDO |
| NE32 4HX | M6BAG |
| NE32 4LP | G7MWU |
| NE32 4QN | M6JFJ |
| NE32 4QY | M6KAS |
| NE32 5DN | 2E0MBW |
| NE32 5QB | 2E0HHU |
| NE32 5QT | G4WUM |
| NE32 5UF | 2E0TBL |
| NE32 5UF | M6JWD |
| NE32 5YJ | M1FNE |
| NE33 2AN | G3SEG |
| NE33 2TH | M3EVV |
| NE33 3AU | G4NHC |
| NE33 3EH | G4LFG |
| NE33 3LQ | M0BVF |
| NE33 4HP | G0DSX |
| NE33 4LB | M6JAF |
| NE33 4UE | G3PRE |
| NE33 5AW | G7WLY |
| NE33 5HX | G1NZL |
| NE33 5NH | M1EEZ |
| NE34 0BQ | G4RNI |
| NE34 0BZ | M1BEX |
| NE34 0DR | G3KZZ |
| NE34 0EQ | G3JPO |
| NE34 0JB | 2E0KSW |
| NE34 0JB | M0BSZ |
| NE34 0JB | M6KVN |
| NE34 0NJ | G4XWR |
| NE34 0QH | M6RGS |
| NE34 0RA | M0RWI |
| NE34 0RT | G3HIF |
| NE34 6AZ | G0ZAT |
| NE34 6RY | M0DEO |
| NE34 6RY | M3VEX |
| NE34 7AQ | G6RIA |
| NE34 7DJ | M6KZR |
| NE34 7DY | M6FEU |
| NE34 7HG | M6CNN |
| NE34 7JP | G6UWI |
| NE34 7JJ | 2E0DID |
| NE34 7JQ | G4NS |
| NE34 7LP | M6REO |
| NE34 7PT | G7HSY |
| NE34 7QU | G0WUH |
| NE34 7RF | G1OHX |
| NE34 8DD | M6BMV |
| NE34 8EJ | G1XVR |
| NE34 8JT | M6SWO |
| NE34 8NL | 2E0PEM |
| NE34 8NL | M6PEM |
| NE34 8RU | G6LIK |
| NE34 9AE | 2E0TPR |
| NE34 9AE | M6PCU |
| NE34 9AJ | M0RPO |
| NE34 9DA | M6TUV |
| NE34 9BU | G8OIU |
| NE34 9JJ | G0WKQ |
| NE34 9JJ | G7PHG |
| NE34 9JJ | M3UJH |
| NE34 9JJ | M6XTA |
| NE34 9NQ | 2E0DZC |
| NE34 9NQ | M6LFR |
| NE35 9BS | M0NER |
| NE35 9EP | 2E0FAN |
| NE35 9EP | M6PGP |
| NE35 9HU | 2E0KBN |
| NE35 9HU | M6KBS |
| NE35 9LG | G8JUC |
| NE35 9LR | 2E0SID |
| NE35 9LR | M0ZID |
| NE35 9NG | G3XWU |
| NE35 9NH | M3IWN |
| NE36 0HJ | G4VSK |
| NE36 0JJ | M0MGK |
| NE36 0LE | G8HI |
| NE36 0RF | G0RCJ |
| NE37 1AG | M6WHN |
| NE37 1QD | G0DHS |
| NE37 1QQ | G4YEX |
| NE37 2BT | G4YDM |
| NE37 2JS | G7LOY |
| NE37 2JS | M6JIZ |
| NE37 2LZ | M6KKA |
| NE37 3BZ | G1LBU |
| NE37 3DE | G8SVR |
| NE37 3DJ | M6GBM |
| NE37 3LA | G0EPP |
| NE38 0NR | G1TWT |
| NE38 7AL | G4WNI |
| NE38 7BB | 2E0YYG |
| NE38 7BB | M6YYG |
| NE38 7DB | G8VYQ |
| NE38 7EF | G0SCI |
| NE38 7EF | M6YBV |
| NE38 7JN | G0NSP |
| NE38 7TA | G0OHK |
| NE38 7TQ | M0MCN |
| NE38 8BP | G4RTJ |
| NE38 8DR | G0UQP |
| NE38 8DR | G4YGW |
| NE38 8DS | G1HCC |
| NE38 8HS | M3MVI |
| NE38 8PF | G7NHE |
| NE38 8PF | M3ABY |
| NE38 8RX | G0ITM |
| NE38 9EG | G0HEU |
| NE38 9ET | G1JDV |
| NE38 9ET | M3ZDV |
| NE38 9HS | M3XRV |
| NE38 9JB | M0HIM |
| NE39 1EQ | G4NOX |
| NE39 1LT | G2DML |
| NE39 1LT | G3SEQ |
| NE39 2AD | G3TKB |
| NE39 2HD | G1VRC |
| NE39 2HF | 2E1BCF |
| NE39 2JP | 2E0RAK |
| NE39 2JP | M0PGX |
| NE39 2JP | M3PGY |
| NE39 2JQ | M6MLX |
| NE39 2JT | 2E1AII |
| NE39 2PN | 2E0AJK |
| NE4 6UX | M6NDI |
| NE4 7ER | G1YJQ |
| NE4 7RG | M6SQE |
| NE4 8BH | 2E0CAT |
| NE4 8BH | 2E0IGW |
| NE4 9AL | M3YSN |
| NE4 9EP | G3UKH |
| NE4 9QY | G8POG |
| NE4 9TB | M0CFB |
| NE40 3ED | G6PPD |
| NE40 3LF | G0SPL |
| NE40 3LF | G8AGB |
| NE40 3RU | G7KJP |
| NE40 4EU | G3ZVM |
| NE40 4QY | G8CYW |
| NE40 4SY | 2E0DUA |
| NE40 4SY | M6WDI |
| NE40 4TD | G4NSC |
| NE42 5DJ | M0OTP |
| NE42 5ER | M6KGS |
| NE42 5PN | G8SFA |
| NE42 6AT | 2E1BUI |
| NE42 6AT | G0NEE |
| NE42 6EH | G4STK |
| NE42 6JH | G0EQC |
| NE42 6PZ | G8RQF |
| NE42 6QE | 2E0GAG |
| NE42 6QE | 2E0JMD |
| NE42 6QE | M0XGD |
| NE42 6QE | M6JBO |
| NE42 6RB | G7JTK |
| NE43 7AF | M6LEH |
| NE43 7DX | G8YVR |
| NE43 7RQ | G3UVU |
| NE45 5AP | G1GLZ |
| NE45 5BA | G1SIO |
| NE45 5EX | G8POO |
| NE45 5EX | M5POO |
| NE45 5HB | G4OEU |
| NE45 5JH | G3TSS |
| NE46 2QB | G6DNV |
| NE46 2QD | G1UUF |
| NE46 3EP | M6MRP |
| NE46 3QG | M6ITM |
| NE46 4BE | M1SAB |
| NE46 4DE | G0CNU |
| NE46 4LE | G0LYC |
| NE46 4LU | G1TIK |
| NE46 4LW | G1HZI |
| NE46 4QF | G0TGK |
| NE46 4RX | G1PIF |
| NE46 4SS | G1MOB |
| NE47 6DE | G0OYL |
| NE47 6HG | G0GXO |
| NE47 9DR | G0AXZ |
| NE47 9EL | G4UEO |
| NE48 1AQ | G1SCB |
| NE48 1EG | G0AFU |
| NE48 2RY | G6ZIC |
| NE48 2RZ | G3TTI |
| NE48 2SL | G8JSR |
| NE48 2SQ | G1CGH |
| NE48 3EL | M6PHK |
| NE48 4BD | G1TIF |
| NE49 9BS | 2E0IBT |
| NE49 9BS | 2E0KFT |
| NE49 9BS | M0JKF |
| NE49 9BS | M3IBT |
| NE49 9BS | M3KFT |
| NE5 1BT | G8PHB |
| NE5 1SQ | G7SCV |
| NE5 1TL | 2E1IIP |
| NE5 2DL | 2E0MEG |
| NE5 2DL | M3TTK |
| NE5 2JB | G3XXQ |
| NE5 2JS | G4VMU |
| NE5 2NW | G0DMW |
| NE5 2NW | G0IOQ |
| NE5 2PA | 2E0CSV |
| NE5 2PA | M0HRW |
| NE5 2PA | M6BXS |
| NE5 2PJ | G3NIJ |
| NE5 3QB | 2E0MUW |
| NE5 3RE | M6SFJ |
| NE5 3SZ | G7UVN |
| NE5 3XB | G6JQE |
| NE5 3XL | 2E0GBJ |
| NE5 3XL | M6TXI |
| NE5 3YB | G8ECZ |
| NE5 4BG | M6MJS |
| NE5 4LB | G7HDZ |
| NE5 4PQ | G0RMO |
| NE5 5EA | G0WYY |
| NE5 5HS | 2E0WDH |
| NE5 5HS | M3WDH |
| NE5 5PR | M3YXP |
| NE6 1BH | M6SNE |
| NE6 3BB | M1DSU |
| NE6 3RY | M0YVG |
| NE6 4DH | 2E0LYN |
| NE6 4DR | G1EPD |
| NE6 4HA | G7TPG |
| NE6 4HL | G7UTG |
| NE6 4HL | M0UTG |
| NE6 4SR | G6JLL |
| NE6 4SR | G8GFW |
| NE6 4SY | G3ZWR |
| NE6 4TN | G7DAR |
| NE6 5AD | G4UOW |
| NE6 5HP | G3YVZ |
| NE6 5HP | G3ZQM |
| NE6 5JY | G7FSE |
| NE61 1XF | G8ONS |
| NE61 2AS | G0WAD |
| NE61 2HJ | G0DWO |
| NE61 2JT | M3ILZ |
| NE61 2PL | G3AGC |
| NE61 2PN | G4NFV |
| NE61 2RU | M6HHE |
| NE61 2SG | G0EVV |
| NE61 2SG | G6IIA |
| NE61 2SG | G7SPP |
| NE61 2TF | 2E0AOU |
| NE61 2TF | G1BNN |
| NE61 2TQ | G4ZTC |
| NE61 2XF | G7PSL |
| NE61 2XY | G7HHU |
| NE61 3LA | G4KBX |
| NE61 3RA | G6OTW |
| NE61 3RB | G4FXU |
| NE61 5ES | M6EJC |
| NE61 5HF | G7RWC |
| NE61 5HQ | M6BEN |
| NE61 5HR | 2E0DDF |
| NE61 5HR | M0ZVX |
| NE61 5HR | M6DDE |
| NE61 5JU | G4DGQ |
| NE61 5LN | M1JPS |
| NE61 5LT | M6JLY |
| NE61 5NI | M3KJK |
| NE61 5PU | G4ZTA |
| NE61 5QW | G0KNW |
| NE61 5RB | G4AAX |
| NE61 5RB | G8PNN |
| NE61 5RP | M6COR |
| NE61 5SP | M1DGW |
| NE61 5SR | M7ARM |
| NE61 5XP | G4INW |
| NE61 6DS | 2E0GCD |
| NE61 6LQ | G3BIK |
| NE61 6LT | G4DJJ |
| NE61 6YW | M3UAK |
| NE62 5DA | G4GWB |
| NE62 5DN | G1NOS |
| NE62 5LD | G4ZOY |
| NE62 5NG | M1ULD |
| NE62 5NN | M3GCT |
| NE62 5UE | G4URW |
| NE62 5UF | G0WZG |
| NE62 5UF | M3HJJ |
| NE62 5UL | G0KUI |
| NE62 5UT | 2E0SCM |
| NE62 5UT | M6SBW |
| NE62 5XA | M6SJN |
| NE62 9XF | G3ZKU |
| NE63 0EU | 2E0CVZ |
| NE63 0EU | M6CVE |
| NE63 0HT | M6ACJ |
| NE63 0LE | G0KGY |
| NE63 0LE | G4OIV |
| NE63 0PL | G7FFR |
| NE63 0PQ | M6KWP |
| NE63 0TH | M1DKK |
| NE63 8AB | M6EGS |
| NE63 8BQ | G0EFG |
| NE63 8EF | 2E0SCO |
| NE63 8EF | M3WLU |
| NE63 8HX | G1LMZ |
| NE63 8HX | M0GEC |
| NE63 8JF | G0CQH |
| NE63 8JF | G7UZS |
| NE63 8JT | G8VIB |
| NE63 8RZ | G0PJV |
| NE63 9BB | G7VDA |
| NE63 9BB | M0BMJ |
| NE63 9BB | M1EIZ |
| NE63 9BG | G4AVN |
| NE63 9EN | M6SIE |
| NE63 9GJ | 2E0WTQ |
| NE63 9GJ | M0HNN |
| NE63 9GJ | M6CPQ |
| NE63 9LU | M0WBB |
| NE63 9QJ | G6JXS |
| NE63 9RY | G0VRS |
| NE63 9RY | G0WYZ |
| NE63 9SD | G7VDV |
| NE63 9TJ | M0AHZ |
| NE63 9TJ | M3VQS |
| NE63 9TP | G3KLP |
| NE630TD | 2E0NHR |
| NE64 6BQ | 2E0VAI |
| NE64 6BQ | M0VHC |
| NE64 6BQ | M6TOM |
| NE64 6DQ | M3RPK |
| NE64 6ST | G0UDZ |
| NE64 6XN | G0VAD |
| NE64 6XT | G0LEF |
| ne646lh | 2E0LFM |
| NE65 0AP | G2FXV |
| NE65 0EL | G8GVN |
| NE65 0ER | G3OGH |
| NE65 0GB | G7OWP |
| NE65 0JS | G4BCP |
| NE65 0TF | 2E0BVD |
| NE65 0TF | M6ALB |
| NE65 7HF | M0CKG |
| NE65 7QN | 2E0NJE |
| NE65 7QN | M3NFG |
| NE65 7QS | G4WMQ |
| NE65 7RA | M1CCX |
| NE65 7RJ | 2E0SLN |
| NE65 7TU | G4VDZ |
| NE65 8JT | G3VKU |
| NE65 8QH | G1ODT |
| NE65 8UN | G1JKX |
| NE65 8UR | G1BBY |
| NE65 8UW | G4ADD |
| NE65 9DB | M1ZXG |
| NE65 9JL | G6CBL |
| NE65 9JN | G1RKD |
| NE65 9JW | G1DNA |
| NE65 9JW | G1HAB |
| NE65 9QW | G8MOG |
| NE65 9SP | M0SSE |
| NE65 9TF | G8YSH |
| NE65 9YA | M1VHT |
| NE66 1BS | G6KXB |
| NE66 1DW | M3WDN |
| NE66 1ET | M1CWO |
| NE66 2JD | 2E0ZED |
| NE66 2JD | M0ZDD |
| NE66 2JD | M3ZED |
| NE66 2QE | G7RKO |
| NE66 2QP | G7ANV |
| NE66 2QP | M3YOT |
| NE66 2YD | M0DSS |
| NE66 3AT | G1AHT |
| NE66 3AT | G3RNP |
| NE66 3AW | G4KRH |
| NE66 3JE | G3ZDU |
| NE66 3JQ | M6RVO |
| NE66 3JZ | G1YAE |
| NE66 3LS | M1CCY |
| NE66 3QN | G4EKF |
| NE66 3RG | M0VKX |
| NE66 4YD | M0FLC |
| NE67 5AW | G4FCC |
| NE68 7TB | M6TSM |
| NE68 7TD | G3AQB |
| NE68 7XP | G3TEP |
| NE68 7XR | G1VVU |
| NE7 7AP | G4KVK |
| NE7 7BX | G4WMA |
| NE7 7FB | G1XSA |
| NE7 7QE | G0BZB |
| NE70 7PA | M3NMD |
| NE71 6DB | G4LUF |
| NE71 6HU | G3HDT |
| NE71 6JL | G0AVU |
| NE8 1US | 2E0DZS |
| NE8 1US | G7MWB |
| NE8 1US | G7UUR |
| NE8 1US | M0MBA |
| NE8 1US | M6DZS |



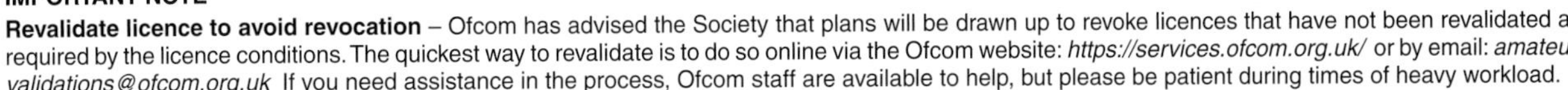
**IMPORTANT NOTE**

**Revalidate licence to avoid revocation** – Ofcom has advised the Society that plans will be drawn up to revoke licences that have not been revalidated as required by the licence conditions. The quickest way to revalidate is to do so online via the Ofcom website: *https://services.ofcom.org.uk/* or by email: *amateur.validations@ofcom.org.uk* If you need assistance in the process, Ofcom staff are available to help, but please be patient during times of heavy workload.

NE8 1XN 2E0DZQ
NE8 1XN M6HSI
NE8 2JB M0GRI
NE8 3RS M0LBD
NE8 4AJ G0NPQ
NE8 4DX G7SPN
NE8 4DX M0GGP
NE8 4DX M3NLA
NE8 4JA G7KLS
NE8 4UH G6JUP
NE8 4XH M3IBE
NE9 5DP G4WAX
NE9 5HG M0PHP
NE9 5HG M3UHP
NE9 5PY M3XWK
NE9 5TX G8XLZ
NE9 5XL M6RFL
NE9 5YN 2E0MWH
NE9 6DA G3IPD
NE9 6DH G0JHU
NE9 6JJ 2E0AOS
NE9 6JJ 2E1EYF
NE9 6NP M3XNN
NE9 6QN 2E0TTY
NE9 6QN M6ADE
NE9 6SN M6BGC
NE9 6TU 2E0EAU
NE9 6TU M6AQN
NE9 6TZ G0SQE
NE9 6UX G0HBO
NE9 7BN G1GEY
NE9 7DX 2E0XQK
NE9 7DX M3XQK
NE9 7EX G7VDK
NE9 7JD G7MKB
NE9 7NQ G1DZY
NE9 7PG G0SLQ
NE9 7TH G7CIT
NE9 7TR G0NPP
NE9 7TY 2E0CRN
NE9 7TY M6SRF

## NG
## (Nottingham)

NG10 1DR G4WHN
NG10 1DX G6DRH
NG10 1GB M6BYN
NG10 1JH M6GTU
NG10 1LS G8XAN
NG10 1NL G3REU
NG10 2BS 2E1HID
NG10 2BY G0CQR
NG10 2DZ G0FFQ
NG10 3BT G6AGR
NG10 3DG 2E0SMA
NG10 3DG M6WXN
NG10 3EW G1OPG
NG10 3FP M3IDO
NG10 3GF G0PSI
NG10 3GF G4AL
NG10 3GF G6MDR
NG10 3GG M3VBV
NG10 3GH G4OSR
NG10 3LF M6USM
NG10 3NL 2E0STT
NG10 3QE 2E0BQW
NG10 3QE M3SVP
NG10 3RG G8PTN
NG10 4DA G4IVO
NG10 4DD M6OOL
NG10 4DH M6TZA
NG10 4EB M3IZJ
NG10 4EW G6PVW
NG10 4GD 2E1GMQ
NG10 4GD M0BTG
NG10 4JA G7TZW
NG10 4JS G5OW
NG10 4LB 2E0SAK
NG10 4LB M0GSR
NG10 4LB M3OYC
NG10 4NZ G4TYP
NG10 5BS G0IVI
NG10 5EF G4HIC
NG10 5FF 2E0MLV
NG10 5FF M3YRB
NG10 5LQ 2E0XCH
NG10 5LQ M6BSV
NG10 5PD G4LOV
NG103GG 2E0NAF
NG103GG M3NAF
NG11 0HP G4BKQ
NG11 0JW G4NPT
NG11 6AA G3ZQH
NG11 6GF G0PIB
NG11 6GG G0JYE
NG11 6LB G0KUZ
NG11 6ND G8GWP
NG11 6QD G4NNY
NG11 6QE G7HZZ
NG11 6QE M0PHX
NG11 6QE M0VRG
NG11 6QE M0YHA
NG11 6QJ G1LOL
NG11 7AU 2E1GHI
NG11 7AU G4LPD
NG11 7AU G8HUR
NG11 7BE G0JPL
NG11 7BQ G7HUK
NG11 7BY M1ECT
NG11 7EB G8DD
NG11 7EB M0BWY
NG11 7FD G8LNG
NG11 7FD M6JRH
NG11 8FD G1PNX
NG11 8GF G0WXF
NG11 8GF G7SWE
NG11 8GN M6TLC
NG11 8JA M1BHP
NG11 8JG G7SMQ
NG11 8JN 2E0PJT
NG11 8JN M0PJT
NG11 8LG M6AAU
NG11 8SL G0JVW
NG11 9AL M0BCW
NG11 9ED M0BWV
NG11 9ET G7BGY
NG11 9JN G0LXX
NG11 9JZ M0BWW
NG12 1AY G0SGF
NG12 1AY G7ITW
NG12 1DG G4ZDF
NG12 2BU G6KDY
NG12 2BX G3YCE
NG12 2GA M1EJI
NG12 3DG M3UBG
NG12 3DJ G6OAN
NG12 3DP G1GSG
NG12 3FD G0MXX
NG12 3HT G7LFL
NG12 3JJ G6AFS
NG12 3JS G0MAY
NG12 3JY G3TNX
NG12 3NE G3FRE
NG12 4BW G4ABT
NG12 4DN G0NPC
NG12 4EA G4XRB
NG12 4ES M0BUY
NG12 4FW 2E1CEU
NG12 5BN G8ENB
NG12 5DA 2E1AEC
NG12 5DN G7ODB
NG12 5ET G0GWP
NG12 5HQ M1AZV
NG12 5JX G6PKY
NG12 5LQ G3KZX
NG12 5LQ M0RIA
NG12 5NX G3SJJ
NG12 5PY G1ZLB
NG12 5RA G4XZA
NG12 5RA M3WTB
NG13 0AR G4WFK
NG13 0BH G3YWO
NG13 0ED 2E0BUZ
NG13 0ED M3ZUB
NG13 0FP M6GJL
NG13 8NA M6EQJ
NG13 8QD G7JDI
NG13 8RL M0RMJ
NG13 8RL M1KEJ
NG13 8TG G8MZW
NG13 8TY G4FAB
NG13 8YR M6FVN
NG13 9AF G0FOG
NG13 9HL M6ZWB
NG13 9HY G4TTS
NG13 9HZ G0USJ
NG13 9JF G6XSS
NG14 5EH M3LVR
NG14 5FG 2E1CWQ
NG14 6FF G8DVN
NG14 6HL G4IJU
NG14 6HY 2E0LRA
NG14 6PH G4MJW
NG14 6QA G6SCG
NG14 7AR M5ROB
NG14 7DJ M0GDP
NG14 7EF G8FGZ
NG14 7FE G3VMK
NG14 7GW G1XOW
NG15 0EG G8YLS
NG15 6DQ G4KJA
NG15 6DQ G4XXS
NG15 6ED M6SPL
NG15 6EP G4WXR
NG15 6FF G6DDP
NG15 6FU M0HCZ
NG15 6FU M6CGY
NG15 6FU M6HUC
NG15 6FY 2E1GDM
NG15 6GE M6LHP
NG15 6GG G6PCN
NG15 6GN G4ZII
NG15 6HY G4EPL
NG15 6NS M0SCP
NG15 6RF M3MBH
NG15 7AH G0DLQ
NG15 7LU G4CMX
NG15 7PJ M6BHC
NG15 7QA G4JSM
NG15 7SJ M3EVC
NG15 7SL M6CCA
NG15 7SR G6NHY
NG15 7TF G6ORH
NG15 7UB G0RGE
NG15 8BG G0VNW
NG15 8EB G0CEV
NG15 8EU 2E1CJB
NG15 8JA 2E1GSC
NG15 9AD G8MMP
NG15 9DG G8EXS
NG15 9EA G0LUD
NG16 1AR G4XOU
NG16 1AU 2E0DME
NG16 1EG G0EVN
NG16 1FD M6VMG
NG16 1HE G0FJZ
NG16 1HE G1NBT
NG16 2AP M0GGH
NG16 2DP G8XPZ
NG16 2EN G3VYK
NG16 2FG 2E0NSC
NG16 2FG M0LAF
NG16 2JJ G0UUX
NG16 2LQ G1PMK
NG16 2PU M6BAN
NG16 2QF M1IDE
NG16 2QF M3EDI
NG16 2QX G7ODN
NG16 2QY G7UII
NG16 2RA G6MRN
NG16 2RH M0ANC
NG16 2TH 2E0FCS
NG16 2TH G6UFV
NG16 2TH M6FKZ
NG16 2UB M1EBW
NG16 3DL G6GCW
NG16 3DP G4JRJ
NG16 3DQ G8ZAU
NG16 3DR G0GYH
NG16 3DY G6HZX
NG16 3FR M6NTR
NG16 3FY G1WSD
NG16 3GW G0BXG
NG16 3GW G7HHM
NG16 3GY G7HGF
NG16 3LR M3AZH
NG16 3NL G4VAX
NG16 3QL G3RTO
NG16 3RB M6GTC
NG16 3RE M0ZCW
NG16 4GJ 2E1IJY
NG16 4GJ G6MAW
NG16 4GP G6ZAF
NG16 5BG G4UFC
NG16 5EH G1OJQ
NG16 5EH G6DJQ
NG16 5FN G0OBK
NG16 5GX G0LIQ
NG16 5HJ 2E1EVH
NG16 5JZ G0WXH
NG16 5JZ G7UVL
NG16 5PW G0WFO
NG16 6BQ G8JGF
NG16 6DX G4ROB
NG16 6DX G8RYK
NG16 6ET G1KGQ
NG16 6FA 2E1BJG
NG16 6FN G0CSS
NG16 6FP G0IZL
NG16 6FW M3UUL
NG16 6GP G4NXB
NG16 6JR G0OKD
NG16 6JR M0HLF
NG16 6JR M3LCW
NG16 6LQ G7BHU
NG16 6NH M6BYS
NG16 6NL G0UKL
NG16 6QH 2E0SBJ
NG16 6QR M0MRT
NG16 6RF G0IUV
NG16 6RJ 2E1GRA
NG17 1EP G4ZUC
NG17 1HA 2E0VCB
NG17 2BS M6FEC
NG17 2BU M3FBN
NG17 2DD M6WDL
NG17 2DG M6KCE
NG17 2DW G0PDM
NG17 2FB G4DFZ
NG17 2FH M3FWR
NG17 2GG M5ADF
NG17 2HS G3JFD
NG17 2HS G7TTY
NG17 2LL G6ZTH
NG17 2NP M6XCC
NG17 2QD G1XVF
NG17 2QF G0JDG
NG17 2QF G0RGU
NG17 2RA 2E0CLL
NG17 2RA 2E0XBX
NG17 2RA G4TSN
NG17 2RA M0TSN
NG17 2RA M3TSN
NG17 2SJ G8XGW
NG17 3AB G0UJD
NG17 3AP G4UDN
NG17 3AR M3VLN
NG17 3BW M0AJX
NG17 3DN M5ABH
NG17 3DP G0IUN
NG17 3EB M0AGA
NG17 3EJ G8EPH
NG17 3FF G6ROS
NG17 3FF M6NNC
NG17 3FT G0BCF
NG17 3HT G4REU
NG17 3JY M6GQR
NG17 4AY G1XNG
NG17 4AZ 2E0EOF
NG17 4BA M6AET
NG17 4BE G4HCD
NG17 4BX G7RXK
NG17 4BY G0UTN
NG17 4EH G0LHL
NG17 4EN G3RLJ
NG17 4EY M6TCF
NG17 4EZ G7PBV
NG17 4EZ M0HQG
NG17 4GA G1RBH
NG17 4GG M0XWD
NG17 4GG M3LIW
NG17 4GQ M6KTN
NG17 4HQ G6NWS
NG17 4HX G7PHL
NG17 4JA M3EZY
NG17 4LL M6KFF
NG17 4NL G6NWN
NG17 5AR M0GOH
NG17 5AX M3ZGI
NG17 5BA M6NOC
NG17 5BD G7INC
NG17 5BL M6CZA
NG17 5EH M3ZGF
NG17 5GH G3VDF
NG17 5HP G1SIU
NG17 5HU G0GZO
NG17 7EH G1VHY
NG17 7FH G1RJD
NG17 7HB M0DYQ
NG17 7HF G6OZH
NG17 7JW M0CWZ
NG17 7LY G3SQQ
NG17 7QB M3KHE
NG17 8AD G7JZC
NG17 8BA M6YJB
NG17 8BT G0JPZ
NG17 8DD 2E0UBN
NG17 8DD M0ORY
NG17 8DD M6UBN
NG17 8DP M0BNF
NG17 8EJ G7LHT
NG17 8FP M3HDL
NG17 8FR M6AGZ
NG17 8FU M3JZN
NG17 8FU M3KCU
NG17 8FX 2E0IQO
NG17 8GE M0SRO
NG17 8GH M0RMF
NG17 8HP M6FZA
NG17 8JJ G4VOG
NG17 8JT G0DMN
NG17 8LH G0NRA
NG17 8LH G0NZA
NG17 8NW G6UDB
NG17 8QB G7VJQ
NG17 8RL G4AJJ
NG17 8RS G3LNN
NG17 9AE 2E1CQM
NG17 9AH M3JZO
NG17 9AH M3JZP
NG17 9BG G6XNN
NG17 9BN G6XYR
NG17 9BR M0RBB
NG17 9BU M0BAE
NG17 9DN G1EZU
NG17 9DY G0CVB
NG17 9DY G3SDW
NG17 9ET G8ZUZ
NG17 9FE M6NEM
NG18 1QE M3SQE
NG18 2HP 2E0JLK
NG18 2HP G7TOY
NG18 2HP M6RRC
NG18 2LQ 2E1FTE
NG18 2LQ M1BVP
NG18 2NF M3NOY
NG18 2NF M6XDX
NG18 2PX G0HRO
NG18 2RN G0KIU
NG18 3AZ M3HNV
NG18 3BL M3FLE
NG18 3DD G0PGA
NG18 3EN G4NOR
NG18 3GR G1MMD
NG18 3HZ 2E1GYB
NG18 3HZ M0GYB
NG18 3HZ M3GYB
NG18 3JJ G4VWA
NG18 3JL G4LBY
NG18 3NP G0JRC
NG18 3NP G0KJC
NG18 3QP G0JVK
NG18 3RN M6PGU
NG18 3RZ G1DRR
NG18 3SP M1CBH
NG18 4EQ G4ZWI
NG18 4ER G4DFV
NG18 4FA G1EIV
NG18 4FB G7ROI
NG18 4HD G1LPQ
NG18 4HF G4AAH
NG18 4HF M3CSM
NG18 4HG G0VYT
NG18 4NB G0RWW
NG18 4PG G6RYW
NG18 4QD G3XDS
NG18 4RT G4GZU
NG18 5EE G3GQC
NG18 5EE M0DFX
NG18 5EE M1ARF
NG18 5JF 2E1EYC
NG18 5JF M3MKD
NG18 5LL M5DLA
NG18 5NB G0RDP
NG18 5NG G1ECS
NG18 5QS G4TGB
NG18 5RG G7VKG
NG18 5RG M6VKG
NG18 5SQ G4VUM
NG19 0AJ M6TLB
NG19 0BX G7GGG
NG19 0DS 2E0BSN
NG19 0DS M3ZEY
NG19 0DW M0CFD
NG19 0EJ G3VIY
NG19 0EL G0DAL
NG19 0EY G1HLT
NG19 0HZ G0ELB
NG19 0JP G4TUX
NG19 0LR 2E0GBV
NG19 0LR M0HOB
NG19 0LR M6CIU
NG19 0LS G4LQL
NG19 0PP G0TUP
NG19 0QJ 2E0DAO
NG19 0QJ M3NZR
NG19 6EG G1ZLY
NG19 6EL M6TRD
NG19 6HZ G7AVZ
NG19 6JU M3HNQ
NG19 6NB G4ANU
NG19 6NT 2E0EVA
NG19 6NT M6EVA
NG19 6QQ G4ZSD
NG19 7JR M3KAN
NG19 7JW M3LBY
NG19 7LG M6NHT
NG19 7LX M3CDE
NG19 7NP 2E1EIX
NG19 7NP M3OQG
NG19 7PY G6JFL
NG19 7QQ 2E0MLF
NG19 7RG G0RRZ
NG19 7RG G6EAH
NG19 7RG M6SXN
NG19 8AZ G1FEX
NG19 8BP G4KLX
NG19 8DJ M3CBV
NG19 8HT G0GAG
NG19 8QT G1IMY
NG19 8QT G1JGY
NG19 8QT G4BAS
NG19 8TL G4OIE
NG19 9AZ G0OYP
NG19 9DW G1WTB
NG19 9HA G7CKL
NG19 9HH G4WPW
NG19 9JR G6XMB
NG19 9NA G3VVE
NG19 9NB G4RPD
NG19 9PJ M3GJW
NG2 2EB M0DSN
NG2 3GD M0GLP
NG2 4AE G6RFJ
NG2 4AS M6PEG
NG2 4GF M0JOH
NG2 4LN G7FCC
NG2 4NY G0HMO
NG2 4QD 2E0BRP
NG2 4QD M3WWU
NG2 5AX M3EHJ
NG2 5BD G7HCJ
NG2 5DS G1TTB
NG2 5FU G1HXP
NG2 5GB G4VOW
NG2 5GB M3GCP
NG2 5GB M3JVP
NG2 5HF M6CZF
NG2 5LA M0AVH
NG2 5NA G4NDM
NG2 6DX M6WRG
NG2 6EB G3WNC
NG2 6FH G4ZSO
NG2 6FQ G3VLN
NG2 6GJ G4VUI
NG2 6GQ 2E0XIK
NG2 6GQ G3MOT
NG2 6GQ M0XIK
NG2 6GQ M3XIK
NG2 6HQ 2E0BBX
NG2 6HQ M0GWR
NG2 6HQ M3VJS
NG2 6HQ M3XHC
NG2 6LD M0SMH
NG2 6LD M6BMY
NG2 6LD M6SKL
NG2 6NA M6LOL
NG2 6PE G8ZSD
NG2 6PE M3ZSD
NG2 6QG G8STJ
NG2 6SG G0RMN
NG2 7AD M5ABR
NG2 7FD G0TCF
NG2 7FL G8KJJ
NG2 7GG G4NZU
NG2 7GN M1FDH
NG2 7LU G8GVL
NG2 7ND M3PMI
NG2 7QN G6RNF
NG2 7QR M6GIY
NG2 7RX M6JRW
NG2 7UQ G8TVM
NG20 0BN 2E0MIB
NG20 0BN M3VUZ
NG20 0BW G7BNO
NG20 0DJ 2E0SPN
NG20 0DJ G0TLT
NG20 0DJ M6HKA
NG20 0DN G7MNQ
NG20 0HJ 2E1FOU
NG20 0ND M6DEE
NG20 0PS 2E0DNJ
NG20 0PS M6CBN
NG20 0PS M6TAJ
NG20 0PT G0CAL
NG20 0PT G1OXT
NG20 0PT M6LCP
NG20 0PY M6NCB
NG20 8AZ G4PPH
NG20 8AZ G8UST
NG20 8AZ M6EJB
NG20 8BJ M0DHI
NG20 8EQ G6JCT
NG20 8EQ G6JCV
NG20 8JW 2E0SGI
NG20 8NH 2E0CAS
NG20 8NH M0TTY
NG20 8PJ G3RCW
NG20 8PJ M0ROY
NG20 8QH G4VLK
NG20 8XN 2E1WPW
NG20 9EB G4ODF
NG20 9PJ M0IFH
NG20 9PJ M3KBF
NG20 9PJ M3NHD
NG20 9PY G0ENV
NG20 9QN 2E0VKB
NG20 9QN M6WKW
NG20 9RP G1KBJ
NG21 0AR 2E1HGY
NG21 0AR G4FHK
NG21 0AR G8AOI
NG21 0AR M0BPY
NG21 0AT M6TRW
NG21 0AU G7KWA
NG21 0ED G0UYQ
NG21 0ED G8EHX
NG21 0FU M0JFB
NG21 0GX G4DFU
NG21 0HX G7LAK
NG21 0JJ G6SFE
NG21 0JZ G7BXG
NG21 0LW 2E0BAF
NG21 0NW G8KPV
NG21 0RN G7SVU
NG21 0RW 2E1IDE
NG21 0SJ M5TMG
NG21 0TL G7PMF
NG21 0TU G0PBW
NG21 0TU M6PBV
NG21 9ah M6HWE
NG21 9DS G1JXG
NG21 9EA G0CCV
NG21 9HJ G1SIX
NG21 9LE M1AEB
NG21 9LQ G4WHL
NG21 9LQ G4WHM
NG21 9NH G0SYS
NG21 9NJ M6JNE
NG21 9NZ M3KOR
NG21 9QG 2E0BWX
NG21 9QG M6TWW
NG21 9QT M1ANR
NG21 9QZ G0BYX
NG22 0HB M3BIR
NG22 0HF G4ODD
NG22 0JB G7NKE
NG22 0JR G6ZOB
NG22 0JR M3KBG
NG22 0PN G7PFD
NG22 0PP G4TDZ
NG22 8AP G4SCL
NG22 8BX G4HTO
NG22 8DQ G1WRO
NG22 8EA G8ITU
NG22 8LU G8BRG
NG22 8PB G6FLQ
NG22 8QF G3ZVV
NG22 9AS G7ACA
NG22 9BD 2E1HVB
NG22 9DG G4GFT
NG22 9DG M0AXX
NG22 9DX G8YPQ
NG22 9HP M0HHF
NG22 9LW G4EKL
NG22 9PQ 2E1FBA
NG22 9PS M6HVM
NG22 9PU 2E0AVA
NG22 9QX 2E0TBD
NG22 9QX M3UIS
NG22 9QX M6TRV
NG22 9QY 2E0SHA
NG22 9QY M6JLR
NG22 9RN 2E1FRZ
NG22 9RW M6DXV
NG22 9RZ M0ADB
NG22 9SG 2E0KOR
NG22 9SG M3KOU
NG22 9SX 2E0HUN
NG22 9SX 2E1DPQ
NG22 9SX M0MLJ
NG22 9SX M6BZK
NG22 9SX M6IAQ
NG22 9TJ G0BYQ
NG22 9TJ G4BZV
NG22 9TJ G6VGN
NG22 9TN M3IDF
NG22 9TN M3KBE
NG22 9TN M3KPL
NG22 9UL M0HXH
NG22 9UP G1DQF
NG22 9UU 2E1HIQ
NG22 9UU M0BDD
NG22 9UZ G4CCB
NG23 5BA G4BPE
NG23 5EA 2E0JMF
NG23 5EG G3RWP
NG23 5EG G8PY
NG23 5LB M6MXR
NG23 5NT M0BCF
NG23 5PQ 2E0RVV
NG23 5PQ 2E0SVV
NG23 5PX M3IKV
NG23 6QG 2E0FGA
NG23 6QG M6FGA
NG23 6ST G8TNB
NG23 7AA G1LVH
NG23 7ED G6MLH
NG23 7HL M1JTA
NG23 7HR M0VOS
NG23 7HR M6VOS
NG23 7LD G0DWB
NG23 7NT G7NDS
NG23 7PB M6SHH
NG23 7PR G7SYQ
NG23 7RA G3HLG
NG24 1DF G4TXO
NG24 1DF M0RHQ
NG24 1FN G0JAP
NG24 1FW G1VKN
NG24 2BU G7UXR
NG24 2FJ G7PRB
NG24 2FX M6JNM
NG24 2HA M6CJJ
NG24 2HT G7MUB
NG24 2JX M0DDI
NG24 2NL G3YWS
NG24 2NT G3XFU
NG24 2NT G7KFM
NG24 2NT G7KFN
NG24 2RX G4PCP
NG24 2SA G0ORP
NG24 2SU M0NWK
NG24 3AS 2E0DYK
NG24 3AZ G1KNQ
NG24 3DA M3MOQ
NG24 3FJ G4NSW
NG24 3HH G6PII
NG24 3LY G0BVU
NG24 3NS G8EGU
NG24 3NZ G0LBB
NG24 3RE 2E0FKH
NG24 3TZ M0KAW
NG24 4BZ G7DEH
NG24 4BZ M3MKY
NG24 4DT G4KFY
NG24 4DY G4ZHG
NG24 4EX M3OVM
NG24 4EX M3YHD
NG24 4JL M6CVU
NG24 4JL M6CVV
NG24 4NY M3OIC
NG24 4PP M6JAM
NG24 4QB G0MCP
NG24 4QB G7HOT
NG24 4RW G7BEJ
NG24 4UA G0PUY
NG25 0BG M3DQJ
NG3 2LN M6CWI
NG3 2LP 2E0TEW
NG3 2LP M0JSW
NG3 2LP M6BUD
NG3 2LS G0MLM
NG3 2PE G7SCN
NG3 3AN G1NKV
NG3 3EQ G0WYI
NG3 3LL 2E0CVD
NG3 3LL M6TBR
NG3 4PZ G1MQB
NG3 5EN G8CXV
NG3 5FD G4ZDX
NG3 5GF G6NLU
NG3 5HY G4PJZ
NG3 5QB G8FWH
NG3 5QW G8ZUU
Ng3 5RG M3LEF
NG3 6AD G3EKW
NG3 6AR G8BPQ
NG3 6DH G0JOX
NG3 6DY 2E0DLO
NG3 6DY M0SYS
NG3 6DY M6EYP
NG3 6EF G0MAD
NG3 6EF G4NRZ
NG3 6EQ G6ABU
NG3 6EU G6WHS
NG3 6FL G4ZTY
NG3 6FT G3TWB
NG3 6JA G8MIW
NG3 6LR M1JHL
NG3 7AP G3SEN
NG3 7BH G1IKH
NG3 7BY G4SGG
NG3 7FY 2E1HHG
NG3 7HD G4SJG
NG31 6JG G7ELG
NG31 6LE M6HET
NG31 6LT M3ZRP
NG31 6QW M0CVO
NG31 7BP M6KCS
NG31 7BX M6GJV
NG31 7EA G3YOQ
NG31 7EN 2E0KVF
NG31 7EN M0KVF
NG31 7EN M6KVF
NG31 7HH G0GRC
NG31 7HH G8WWJ
NG31 7JG M0SFA
NG31 7JL G6IPW
NG31 7LT 2E0CMF
NG31 7LT G4ENS
NG31 7LT G8ENS
NG31 7LT M0SDM
NG31 7LT M6AZT
NG31 7NH M6WDR
NG31 7NN G0RCI
NG31 7NN G1EUU
NG31 7PH G1UTZ
NG31 7RB 2E0TYT
NG31 7RB M6GBI
NG31 7RB M6MSJ
NG31 7RH 2E1DCV
NG31 7RH G7SJX
NG31 7RH M3SJX
NG31 7RH M3WPM
NG31 7XA 2E1AXL
NG31 7XG M0JHW
NG31 7XP M6SRG
NG31 7XR M3VUA
NG31 8AD G4GXI
Ng31 8aw M6HNX
NG31 8BN G4DZC
NG31 8DP M3SRY
NG31 8GA M0AUS
NG31 8GH G6UOH
NG31 8JR 2E0MDE
NG31 8LN G4WZU
NG31 8LP G4VUA
NG31 8QF M3OJX
NG31 8RL G6YMD
NG31 8RS G0FSP
NG31 8RX G4ATA
NG31 8RX G8DYT
NG31 9BL M0HNF
NG31 9BL M0SUR
NG31 9JD 2E1CSD
NG31 9JW G8SHE
NG31 9ND M3ZYX
NG31 9PF M1DDB
NG31 9PF M6ISD
NG31 9QA 2E1EFG
NG31 9QG 2E0XBM
NG31 9QG M0HLO
NG31 9QG M0XBM
NG31 9QS M0USY
NG31 9RD G3VJE
NG31 9RG 2E0ALF
NG32 1AU 2E0OBL
NG32 1AU G8OCO
NG32 1AU M0OBL
NG32 1ET M3XZJ
NG32 1HX G4MQM
NG32 1HX M5DJC
NG32 2BG 2E0MFM
NG32 2BG M6SUK
NG32 2EB M0TGX
NG32 2EB M6AFS
NG32 2ES M6FRR
NG32 2NS G3PTI
NG32 2PD G4RIO
NG32 2PD G6JQH
NG32 3DF G4HVC
NG32 3DX G6SDG
NG32 3EJ G3TBK
NG32 3EJ M0TBK
NG32 3EJ M3DVQ
NG32 3EJ M3TBK
NG32 3NR G0PJS
NG32 3RR 2E1EBR
NG32 3RU G0OJF
NG32 3SJ G0BUB
NG33 4HA G4FGY
NG33 4PZ M0STA
NG33 4RT G0KQK
NG33 4SB G3KHZ
NG33 5DX M3NQA
NG33 5HB G1GET
NG33 5HG G0BHT
NG33 5JH M3MFZ
NG33 5PU G0GRU
NG33 5PU G8STM
NG34 0DL 2E0YNT
NG34 0DL M6YNT
NG34 0LD M0HXF
NG34 0LF M0MLW
NG34 0QG G4RIP
NG34 0RS G6KVE
NG34 0SE G3KVP
NG34 7HG G0BTT
NG34 7JN 2E1FAN
NG34 7LQ G0CRE
NG34 7NW M6AQG
NG34 7QP G2VS
NG34 7TD G3LSJ
NG34 7TF G4ODG
NG34 7TF G6AWY
NG34 7US G1BPV
NG34 7WA G4SIJ
NG34 7WJ G7LPK
NG34 7WL M6JJX
NG34 7WL M6YBA
NG34 8AE G4GBE
NG34 8BJ G0BXU
NG34 8BJ G1LOW
NG34 8DB M1EGW
NG34 8HG G0IBI
NG34 8HZ 2E0ENN
NG34 8HZ M6ENN
NG34 8LR G4KQT
NG34 8NJ 2E1DPG
NG34 8NJ G3JFC
NG34 8NJ M3AYL
NG34 8QG 2E0CWM
NG34 8QG M0GUU
NG34 8QG M6GAC
NG34 8QQ M0XCR
NG34 8XB G6WYE
NG34 9AL M6OLT
NG34 9AT G0SGP
NG34 9BT G1PGN
NG34 9EQ G7UOL
NG34 9EU 2E0MPE
NG34 9EU M6MYN
NG34 9FH G6UCW
NG34 9GA G3WYH
NG34 9GU G4GUC
NG34 9HG G1XRF
NG34 9HG G3PXV
NG34 9HS G0NVY
NG34 9JE G0POM
NG34 9JG G0FLV
NG34 9JP G3ZUC
NG34 9JX M5AKT
NG34 9JZ G0DMJ
NG34 9PH G8YFP
NG34 9QY G4UYF
NG34 9RX G6JTW
NG34 9SA G3SRX
NG34 9TS G3VHI
NG4 1BU G6ZBO
NG4 1DA G6UZY
NG4 1DP G6JCI
NG4 1DR G4ZUS
NG4 1HF G4PFF
NG4 1LE G0IBY
NG4 1LY 2E1BHU
NG4 2JG M3HWS
NG4 2JG M3YHU
NG4 2LW M0BCI
NG4 2QJ M3CZE
NG4 3DA G7TMF
NG4 3DX M1EDW
NG4 3EH G4ZCW
NG4 3ET G6MBI
NG4 3LF G6IOE
NG4 3PE G1FYU
NG4 3QH G7SBK
NG4 3SF M3CBH
NG4 4AD G4EUJ
NG4 4AQ M1DJB
NG4 4AS M6FLQ
NG4 4AU G8LGY
NG4 4BB G8GCK
NG4 4BL G1EAB
NG4 4EA M6CCV
NG4 4FL G4OCS
NG4 4GF M3UEE
NG4 4GS G4CPG
NG4 4HZ M3VRV
NG4 4JP G6YCL
NG4 4JP G7SFS
NG4 4QE M1JCS
NG5 1BH G1GJT
NG5 1DW M0EPR
NG5 1DW M0GZD
NG5 1DW M6EFL
NG5 1EP G6SRZ
NG5 1EP M0WAB
NG5 1FB G6BMZ
NG5 1GP G6ONW
NG5 1JR G4EKW
NG5 1JR G6CW
NG5 1NH G3TVY
NG5 1NL G6SFF
NG5 1NW M1REJ
NG5 3DA G0DME
NG5 3GL M6TRI
NG5 3GS G1HRM
NG5 4AQ 2E0KAR
NG5 4AQ M0NPT
NG5 4BD G7LKV
NG5 4GQ G4SQV
NG5 4HU G4EDX
NG5 4JS G0JAC
NG5 4JX G4WDP
NG5 4LB G0KXZ
NG5 4NX 2E0CWE
NG5 4NX M6CIM
NG5 4PG G4ERY
NG5 5BL G8VCI
NG5 5DT G6UZA
NG5 5DU G6URR
NG5 5EE G8CQQ
NG5 5LT G1NXI
NG5 5NP 2E0LFK
NG5 5RX G7SOP
NG5 5US M6ANY
NG5 6BX M0DLZ
NG5 6DJ 2E0AXT
NG5 6DJ M0HLA
NG5 6DP M6SAH
NG5 6DP M6ZBD
NG5 6DX 2E0PHS
NG5 6DX M3XBH
NG5 6EB 2E0EMF
NG5 6EB M3JWQ
NG5 6EH M6RSJ
NG5 6FN G4PNX
NG5 6FT G1FDO
NG5 6GA 2E0SAU
NG5 6GA M3TKP
NG5 6GU G6MVF
NG5 6LQ G4EAN
NG5 6NH G8SSL
NG5 6QA G7WKH
NG5 6QL G1UDX
NG5 6QT 2E1DLT
NG5 6QZ G1HTL
NG5 6SY G2FGT
NG5 7ET G1YBA
NG5 7HF M3IVX
NG5 7HQ G4OOY
NG5 7LA G7DPV
NG5 7LW G8UGS
NG5 7NF 2E0SBK
NG5 7NF G4ITX
NG5 7NF M6SLA
NG5 8AG G4EGY
NG5 8BE M6HIK
NG5 8FQ G8RBU
NG5 8GE M0PDG
NG5 8GH M0KIR
NG5 8JS G6YLX
NG5 8NZ G7IWK
NG5 8QF G1GQQ
NG5 8QX G6IBN
NG5 8RG M3ROF
NG5 8SJ 2E0DSH
NG5 8SJ M0ZRD
NG5 9DQ G7ULJ
NG5 9LN G7MGV
NG5 9LN M0BCH
NG5 9QN G1HMZ
NG5 9QU G7IZE
NG6 0BH G0KZA
NG6 0BS G1XHO
NG6 0LS G0INA
NG6 7DL G4DIU
NG6 7FJ G1ZLA
NG6 8AY G8UQR
NG6 8DG G0GXH
NG6 8DL G1RHB
NG6 8NG G0NVS
NG6 8NL M3MDW
NG6 8PU G6VWV
NG6 8QY G0VFU
NG6 8SL G6NLD
NG6 8WD G6IQP
NG6 8XE G4PMM
NG6 8XN G0LCG
NG6 8XN G5RRH
NG6 8XN G7HRR
NG6 9DT G1GTF
NG6 9FB G8ZZV
NG6 9FU G7SYD
NG6 9FX G1YWN
NG6 9HN G8OHC
NG6 9JE G6GWY
NG7 1RG G6COZ
NG7 1TN G6JUT
NG7 3DF M6WMH
NG7 3HF 2E1HAL
NG7 5EB 2E1ECG
NG7 5FS G3WGH
NG7 6AD M6RIL
NG7 6HU G6ISB
NG7 7DR M3NQY
NG7 7LJ M6WAY
NG8 1AT 2E1IHY
NG8 1DE G1YEU
NG8 1JE M0GRH
NG8 1JU G6UAP
NG8 1LF M3MXH
NG8 1PU 2E0KNB
NG8 1PU M0KNB
NG8 2BH G6CIF
NG8 2BQ G8BVU
NG8 2BZ G4BWN
NG8 2EH G0FSJ
NG8 2ER G8UUS
NG8 2FB M6GHR
NG8 2NA G8BNG
NG8 2QR G4KLJ
NG8 2QR G8NWU
NG8 2RE G0FOE
NG8 2SB G8POL
NG8 3ES G3XBE
NG8 3FT M6MQP
NG8 3NE G4TYO
NG8 4AH M3JHL
NG8 4GY M0AUK
NG8 4LZ G1HRJ
NG8 4PU 2E1FDT
NG8 5EU G1XYG
NG8 5FH G1HSP
NG8 5FJ G1RCN
NG8 5LF G7DSO
NG8 5ND M3CIO
NG8 5PN G1DNZ
NG8 5QG M6GHN
NG8 6AD G6XJD
NG8 6AD G6XJE
NG8 6EX 2E0PJR
NG8 6EX M0LZM
NG8 6GP 2E0LZM
NG8 6GP M3XID
NG8 6GQ M0DAZ
NG8 6JT 2E0GAC
NG8 6LY M3TSG
NG8 6NE M1BSI
NG86LY G0TSG
NG86LY M0TVC
NG9 1AY M0DLR
NG9 1EU 2E0BCE
NG9 1FJ 2E0EEO
NG9 1FJ M0IOI
NG9 1FJ M6NOT
NG9 1GR G6JRE
NG9 1HE G6XOD
NG9 1HE G6XOE
NG9 1LY G0CQT
NG9 1NA M3VIO
NG9 1NE G6XXJ
NG9 1NS 2E1GYO
NG9 1PY 2E1EQQ
NG9 1PY G0OQQ
NG9 2AD 2E0PHS
NG9 2AD M6EET
NG9 2FJ G7OHD
NG9 2GU M6DZN
NG9 2QR G4JHN
NG9 2QZ M3RNU
NG9 3BB 2E0OBS
NG9 3BB M6CAH
NG9 3BY G7LNV
NG9 3FN M0DAC

NG9 3FP G0PXP
[illegible] [illegible]
NG9 3JN [illegible]
NG9 3JT G1PUZ
NG9 3JW [illegible]
NG9 3LW G8BFM
NG9 3PF G0PCK
NG9 3PX G4GQY
NG9 3QH G8JQW
NG9 3RB 2E0YZZ
NG9 4BB G3EJH
NG9 4BB G4VFK
NG9 4BB G8ZK
NG9 4ED M3UVQ
NG9 4ES G0UYV
NG9 4FG M3RUI
NG9 4FH G7NGX
NG9 4JB 2E1HGM
NG9 4JB M1ECB
NG9 5AE M0GHX
NG9 5DA M0BDS
NG9 5EB G8ZSZ
NG9 5EZ G4BNX
NG9 5FH G6BRP
NG9 5FJ G7TRB
NG9 5FU G4AQT
NG9 5FU M6STH
NG9 5GR 2E0GSW
NG9 5GR M0RMY
NG9 5GR M6CTX
NG9 5HS 2E0FBL
NG9 5HS M6ENH
NG9 5HX G0SPA
NG9 5HY 2E1CHX
NG9 5HY G7USM
NG9 5LA G7HIT
NG9 5LB 2E1SIS
NG9 5LB G3CIO
NG9 5LB G3SIG
NG9 5LB M0OIC
NG9 5NH G6IPC
NG9 5PB G3XAZ
NG9 6AB G6RCD
NG9 6BP G8IYZ
NG9 6EW G4KQP
NG9 6FW 2E1EAV
NG9 6FZ M5ADM
NG9 6GN 2E0DWB
NG9 6GN M3ZNR
NG9 6HP G0RSU
NG9 6HS G6UWO
NG9 6JW G4XEL
NG9 6JW G6MSC
NG9 6NH M6EDJ
NG9 6NX G7SHW
NG9 6RF M3KUB
NG9 7ET G0NVO
NG9 7FY G4OYO
NG9 7HN 2E0GIH
NG9 7HN M3GIH
NG9 8BN M1EWD
NG9 8DJ G4PMA
NG9 8HR G3JTO
NG9 8HR M1ERF
NG9 8LT G7LCV
NG9 8PQ G0SLZ
NG9 8QG G4TYN

## NN

(Northampton)

NN1 3BL G0PYI
NN1 3QL M3NNH
NN1 3QL M6EMA
NN1 3RR M6GEA
NN1 4HU G3YOV
NN1 4LU G1ICH
NN1 4NA M0ARZ
NN1 4PU M3KUG
NN1 4QR M3CET
NN1 5HP M6INW
NN1 5JS G4PXJ
NN1 5JY M3WUA
NN1 5LT G7KGP
NN1 5LT M3GTG
NN1 5LT M3NEC
NN1 5LT M3TTT
NN1 5LU M6LRB
NN1 5QP 2E0BDD
NN1 5ST 2E1BCC
NN1 5ST G3VMU
NN10 0AS M0DOK
NN10 0AT M0DOD
NN10 0DH M3NLI
NN10 0DJ G6VFC
NN10 0DT G4LJG
NN10 0DX G7PYQ
NN10 0EG M0JOT
NN10 0GE G6EQS
[illegible] [illegible]
NN10 0NF G7NBG
NN10 0NF M3NBG
NN10 0SJ G4BHT
NN10 0SN G8LII
NN10 0SW G4FEV
NN10 0SW G4WFT
NN10 0SY G3WDG
NN10 0SY G4KGC
NN10 0TZ M3UPJ
NN10 6BA G7TZD
NN10 6BG G3OMS
NN10 6BU M3CQW
NN10 6EH G7APQ

NN10 8DH G6NHO
NN10 8HF G1RTX
NN10 8HU [illegible]
NN10 8LE M3BJR
NN10 8NH G0WTO
NN10 8NH G7CMI
NN10 9ER G4GCQ
NN10 9EZ G6SDI
NN10 9HF G7LNP
NN10 9HH 2E0TSO
NN10 9HH M0TSA
NN10 9HL G1OOS
NN10 9LG G4NXV
NN10 9LN G6GOX
NN10 9LW 2E0LLI
NN10 9LW M6FIV
NN10 9NS G7OCC
NN10 9PF G0PYS
NN10 9QJ M6FPW
NN10 9RP 2E0ERG
NN10 9RP M6RAH
NN10 9YJ G6PUE
NN11 0GR G1ZJK
NN11 0GU G4BFT
NN11 0NZ G8KHF
NN11 0UN M3BXE
NN11 0XG M1AQX
NN11 0XH 2E0CTR
NN11 0XH M3PMY
NN11 0XT M3FKK
NN11 2JH G4SEV
NN11 3AD G1WVS
NN11 3AD M3HUX
NN11 3BW G4TBN
NN11 3ES G3YHI
NN11 3PR G4KQH
NN11 3QP 2E0LMR
NN11 3QP M6CNG
NN11 3YT G3MRQ
NN11 4AL G1DLH
NN11 4AL M3NUM
NN11 4AN M3FNY
NN11 4AQ G0HWT
NN11 4DQ G4DDH
NN11 4EY G7KRB
NN11 4GW G8VQQ
NN11 4GY 2E0AWS
NN11 4GY G6XLB
NN11 4GY M3BSI
NN11 4HJ G0DRE
NN11 4HJ G0FPM
NN11 4JQ 2E1FSV
NN11 4JQ M1CUX
NN11 4JR G6TVK
NN11 4NW 2E1AXD
NN11 4NW 2E1AXE
NN11 4NW 2E1CZF
NN11 4NW G5XX
NN11 4NW G6EGO
NN11 4NW G8GVV
NN11 4PR G7LII
NN11 4PT M3XTT
NN11 4QF M6FZR
NN11 4QU G1DIO
NN11 4QU G6MTE
NN11 4RX G7KJX
NN11 4SU G0NNG
NN11 4SX G3WTN
NN11 4TD 2E0BHC
NN11 6YJ 2E0BVT
NN11 6YJ M0GUC
NN11 6YJ M3UPF
NN11 7HE M6YAS
NN11 7HQ G8KTA
NN11 9AL M3KZI
NN11 9BT G7JDA
NN11 9WH M3IQS
NN12 6AG M3HEO
NN12 6AW G3UMT
NN12 6AW G8CHK
NN12 6DL G0EPI
NN12 6DN G1YMJ
NN12 6DN G6ZGA
NN12 6ED G6RHJ
NN12 6EZ M3NLQ
NN12 6JB G8EUX
NN12 6QQ G4GWT
NN12 6RA G8MAG
NN12 6RD 2E0YLP
NN12 6RD M6YLP
NN12 6RL 2E0TOY
NN12 6RL M0XER
NN12 6TH 2E0VLL
NN12 6TH M3VLL
NN12 6UP M3FFC
NN12 6UQ G1EBX
NN12 7AY G0IJI
NN12 7NA G8MQY
NN12 7NW M1DJC
NN12 7PH G4IUH
NN12 7PH G8AAT
NN12 7PP G1MZD
NN12 7QX M6GDQ
NN12 7RS G1TAI
NN12 7SA G1DJU
NN12 7TX G4KFC
NN12 7TY G4EQX
NN12 8AL G0VCB
NN12 8AL G4DMS
NN12 8JZ G8YJZ
NN12 8LU G4UXO
NN12 8NA G7ECG
NN12 8NG M1BXC
NN12 8NW G0FKW

NN12 8UY G4DQQ
NN12 8XJ M3KVG
[illegible] [illegible]
[illegible] [illegible]
NN13 5LX G4EBF
NN13 5QT G4KZT
NN13 5SN G4FCD
NN13 5TW M6HXU
NN13 6AQ 2E0BGZ
NN13 6AQ M0RGD
NN13 6AQ M3IIA
NN13 6BP M6SKX
NN13 6DA M3ULM
NN13 6HA G3NAN
NN13 6JH M1CLO
NN13 6JQ M0RCY
NN13 6JQ M6CKZ
NN13 6JS G6HKH
NN13 6LY G8JDC
NN13 6ND G4IYA
NN13 6NE 2E0CZN
NN13 6NE G1BZM
NN13 6NE M6DLW
NN13 6NS M6OYD
NN13 6PB G4DWC
NN13 7ET G8OZH
NN13 7NJ M3UBF
NN13 7TY G1EPO
NN14 1BG G7TZZ
NN14 1EE G8ZIH
NN14 1EL G0GRS
NN14 1HG G0NFL
NN14 1NG G4RPT
NN14 1RG M0CKP
NN14 1RQ G1NEN
NN14 1SL G4UYM
NN14 2FH G6GDR
NN14 2LB G0FFB
NN14 2LQ 2E0TJP
NN14 2LQ M3XDZ
NN14 2QB 2E1EFT
NN14 2QB G1LMN
NN14 2QY G4PNL
NN14 2RE G1OET
NN14 2RE G5KN
NN14 2RQ 2E0WPT
NN14 2RQ M6MPT
NN14 2RW 2E0HJB
NN14 2SQ M1PAH
NN14 2UD G1WPR
NN14 2UD G8JKB
NN14 2UY G7NBV
NN14 2XA G1IHY
NN14 2XB M6FOZ
NN14 2XG 2E0ZAH
NN14 2XG G6DHW
NN14 2XL G8DXI
NN14 3BT G6SWJ
NN14 3BT M6JTC
NN14 3DD 2E0PNG
NN14 3DZ G8ZPE
NN14 3JT M0NIG
NN14 3LJ 2E0XBG
NN14 3LJ M6SFW
NN14 3LQ G4YNG
NN14 4DD G4HLW
NN14 4EA G6LJH
NN14 4FL M6FKR
NN14 4JQ G8XQN
NN14 4JT G0CMP
NN14 4PQ G3YQJ
NN14 4PY G4PUZ
NN14 4QA 2E0GGM
NN14 4QA M0YGG
NN14 4QA M6GGM
NN14 4RE G3WIA
NN14 4TB 2E0GDT
NN14 4TB M0UKO
NN14 4TB M6UKH
NN14 4XA M3UFZ
NN14 4XB G7OGL
NN14 6AJ M6OAN
NN14 6DZ G1TPC
NN14 6EP M6UKB
NN14 6HT G4DCD
NN14 6HT G8HBZ
NN14 6HY G6BKD
NN14 6HY G8BKD
NN14 6LH G7JAV
NN14 6TN G1IQA
NN14 6YF G8EZG
NN14 6YG G3WKR
NN14 0YQ [illegible]
NN15 5BN G7AJJ
NN15 5BY G8WSV
NN15 5DF G7PQM
NN15 5EG G0NSA
NN15 5EQ G4VKX
NN15 5HB G8YKE
NN15 5HD G4NZZ
NN15 5HE G1VPS
NN15 5HE M1ECD
NN15 5HE M3JCA
NN15 5JB G7GMQ
NN15 5LH 2E0JLD
NN15 5LH M6JLD
NN15 5LJ M6SGP
NN15 5LS 2E0RBO
NN15 5LS M3YPW
NN15 5LS M6GIL
NN15 5ND G7WKV
NN15 5NT G6OLU
NN15 5PN M3LDM
NN15 5QB M6NDY
NN15 5QP M3XQO

NN15 5QW G4ASH
NN15 5RS M3ZRM
[illegible] [illegible]
[illegible] [illegible]
NN15 5TA G4HND
NN15 5YL G4HLL
NN15 6HF G0DEF
NN15 6HH G1YLM
NN15 6HL M6ENB
NN15 6JS M6SUM
NN15 6PS G4TTX
NN15 6QH G7SKA
NN15 6RE G6JTC
NN15 6RG G4TXM
NN15 6SF G8XPD
NN15 6TU 2E0VNK
NN15 6TU M0VNK
NN15 6TU M3VNK
NN15 6UW G4OYN
NN15 7AP G4NCA
NN15 7DR G0RDV
NN15 7DS G1XJN
NN15 7DZ M1AWN
NN15 7EF G1AZD
NN15 7HX G7CJG
NN15 7LL G0PSG
NN15 7LL G0WWT
NN15 7LL G7NTG
NN16 0HU M6MAL
NN16 0LA M6DAF
NN16 0LL G7VIH
NN16 0NG M6RUH
NN16 0PH M0AQP
NN16 0QB G7API
NN16 8EB G8FSJ
NN16 8PL G4LEM
NN16 8UF G3VJG
NN16 9BU G7AJS
NN16 9ET M6WOE
NN16 9EW G4KXG
NN16 9EW G8RAV
NN16 9HA G0PBU
NN16 9JA G7DMK
NN16 9JL G0LDR
NN16 9JR G8WSU
NN16 9JZ G6PEG
NN16 9LA G4NAC
NN16 9PF 2E0XJR
NN16 9RF M6PAV
NN16 9SB G6OXN
NN16 9TG 2E0WIZ
NN16 9TG M6WIZ
NN16 9TW M6AHP
NN17 1BJ G7TIK
NN17 1EG M3NVV
NN17 1EN G6BHX
NN17 1ER G7NJZ
NN17 1JD G0IKR
NN17 1SY G7SCX
NN17 2AF G1TQR
NN17 2BS G1IIY
NN17 2BZ G4YRX
NN17 2LA G4PDK
NN17 2LJ G1BKI
NN17 2LN G0PAD
NN17 2QJ M1BQS
NN17 2QP G6AFT
NN17 2RP G6RCY
NN17 3BT G4JCJ
NN17 3BU G7CIV
NN17 3DA M0AJI
NN17 3DN G4DNJ
NN17 3DN M3CCS
NN17 3HB G6VOE
NN17 3LR M6NYZ
NN18 0AE M1CDP
NN18 0AG M3JBW
NN18 0BN G0NRB
NN18 0LG G0RRO
NN18 0LZ G4LKF
NN18 0LZ G8JJE
NN18 0QU 2E0LGL
NN18 0QU M6LGL
NN18 0TA G4OMV
NN18 8DE G4XRD
NN18 8FZ G4NZE
NN18 8GS G5UI
NN18 8HX G3MTJ
NN18 8LL [illegible]
NN18 8NB M0RSH
NN18 9DE [illegible]
NN18 9EG G8SEV
NN18 9HH G0IYS
NN18 9JJ G4XKD
NN18 9JS G4LKU
NN18 9PG M3TLT
NN18 9PH G1JNY
NN2 6EP G6NYH
NN2 6ET G4DLD
NN2 6RP G3KLV
NN2 6RP M0OXZ
NN2 7DA [illegible]
NN2 7LP M0AIP
NN2 7PP 2E0YXZ
NN2 7PP G0JKZ
NN2 7PP M6SNO
NN2 7PP M6YYY
NN2 7PT G0GAQ
NN2 7QQ 2E0AAZ
NN2 7QX G1IPD
NN2 7QY M3FLP
NN2 7RA G4KPT
NN2 7RA M0DOL
NN2 7RR G3MJW
NN2 7RR M0EBU

NN2 7RY G0WYQ
NN2 7SP M0VQP
[illegible] [illegible]
[illegible] [illegible]
NN2 7TJ M3VIB
NN2 7TR G0NDG
NN2 7TR G8WPU
NN2 8BJ G7OGN
NN2 8BX G4FTL
NN2 8DZ G3ZJO
NN2 8EA G1GOY
NN2 8EB G7BFH
NN2 8EL M0GCR
NN2 8HD G0ONS
NN2 8HT G1ERY
NN2 8JR M0DUU
NN2 8NB G4YJP
NN2 8PE 2E1CCF
NN2 8PH G8PTH
NN2 8QU G6TVB
NN2 8QU G8OZY
NN2 8SX G7AAY
NN2 8TX G7TEG
NN2 8UU G8KNU
NN29 7AP M1TSU
NN29 7AT G8AMG
NN29 7DA G4PAV
NN29 7DH G6MKJ
NN29 7DP G0VOB
NN29 7DR G1ALL
NN29 7EA G8TFY
NN29 7EE G4RST
NN29 7EW 2E0KAK
NN29 7EW M0HAF
NN29 7EW M6AGK
NN29 7FD G0UBX
NN29 7HF M6RGR
NN29 7JU G0EAE
NN29 7JU G6NUS
NN29 7LP G7HBU
NN29 7LS G1ONQ
NN29 7ND G1RVF
NN29 7NS G0PSZ
NN29 7RS G3JRE
NN29 7TP G1EUG
NN3 2AF G0VFM
NN3 2AX G7UFW
NN3 2BU 2E0CCY
NN3 2BU M3NWH
NN3 2DD G4VWF
NN3 2HE G0MJK
NN3 2LP 2E0BXV
NN3 2LT 2E0WSJ
NN3 2LT M3WSJ
NN3 2PG M6BGS
NN3 2SY G0RTC
NN3 3AF G6AXH
NN3 3EQ 2E0SPZ
NN3 3FA G4MAB
NN3 3JE M3DYL
NN3 3JL G4MJF
NN3 3JX G1KMS
NN3 3ND M6NQT
NN3 3NX M6KKY
NN3 3PB G4JFC
NN3 5AD G1SLG
NN3 5AL M6WXP
NN3 5BH G1IRG
NN3 5BH G1NRG
NN3 5ET M6YPA
NN3 5EY G6VGS
NN3 5HA G7NIZ
NN3 5HY M6BOB
NN3 5JS G4OIG
NN3 5JS G8CXK
NN3 5LZ M6FEI
NN3 5NT M0DME
NN3 6DN M3XHW
NN3 6HP G4EUR
NN3 6JQ G4SCJ
NN3 6JQ G6NHV
NN3 7HU G6IVW
NN3 7LB 2E0GMO
NN3 7LB M0MAO
NN3 7LB M0POI
NN3 7LB M6GMO
NN3 7RD G8FZW
NN3 7RE G6CQR
[illegible] [illegible]
NN3 7UB M6BQJ
NN3 8AX M5AFY
NN3 8DE G3MBD
NN3 8DF G8INA
NN3 8HH M6MCR
NN3 8HY G7RFX
NN3 8HY [illegible]
NN3 8LD G0MHA
NN3 8LZ 2E0BKO
NN3 8LZ M3RGU
[illegible] 2E0NNH
NN3 8NE M0NCC
[illegible] M0NNH
[illegible] [illegible]
NN3 8QJ G4HGB
NN3 8QT 2E1DNF
NN3 8UR G0PTM
NN3 8UX 2E1CAH
NN3 8UX 2E1CBU
NN3 8UX G1BGF
NN3 8UX G4IRD
NN3 8XA 2E0PUN
NN3 8XA M3PUN
NN3 8XN G8TOP
NN3 9AX 2E1EET
NN3 9BW M6SBU

NN3 9DN G1VKJ
NN3 9NT M3RCD
NN3 9SN M1BTU
[illegible] [illegible]
NN4 0NJ G4LUW
NN4 0QQ G0NYE
NN4 0RQ G6LUU
NN4 0RS M0XDK
NN4 0SS G6ILZ
NN4 0TS M3TGW
NN4 0WE 2E0HAB
NN4 0WE M0HOU
NN4 0WF G7CIA
NN4 0XX G0KAK
NN4 5AZ G0FQI
NN4 6AH G6NPZ
NN4 6AL G0FSM
NN4 6AP M6ROJ
NN4 6AP M6VHA
NN4 6DJ M6KDW
NN4 6ER G4DHV
NN4 6EW M3TRP
NN4 6HE G7NEG
NN4 6LA G0TMW
NN4 7AF G3YNF
NN4 7BT M3NAL
NN4 7EQ 2E1EAS
NN4 7PT G8ZEW
NN4 8AU G8HTO
NN4 8AY G3KAN
NN4 8AZ G6FAL
NN4 8BY M0EAB
NN4 8LQ G6KIZ
NN4 8NR M6YYU
NN4 8PE G1TPN
NN4 8QB G1BHR
NN4 8QB M3BQT
NN4 8QB M3ZOR
NN4 8RU G4DTM
NN4 8RX G4RKR
NN4 8ST 2E0DNL
NN4 8TQ M0STN
NN4 8TT M6JBV
NN4 9QU 2E0VNL
NN4 9QU M6VNL
NN4 9QW G4YFX
NN4 9TX M6RGZ
NN4 9YR M3KFP
NN5 4AJ G4MLO
NN5 4WB 2E0XKD
NN5 4WB M0XKD
NN5 4WB M3XKD
NN5 5AR M6RDF
NN5 5LW G1UQF
NN5 6AY M6MIO
NN5 6BW G7IXK
NN5 6JW G0RDT
NN5 6LR M3MKH
NN5 6NH G7SYT
NN5 6NH M3EKA
NN5 6NL G7HLP
NN5 6NR G6FGC
NN5 6PU G8KST
NN5 6PY G4NVP
NN5 6PZ G4NUF
NN5 6PZ G8KIW
NN5 6QP G7ISE
NN5 6QX G6TJE
NN5 6RA G1OMI
NN5 6RA G3RLD
NN5 6RF G0VJK
NN5 6TU M0DAE
NN5 6TW 2E1FUQ
NN5 6XX G4ICC
NN5 6YW G0DLF
NN5 7ER 2E0PHX
NN5 7ER M6PHX
NN5 7ES G7SVT
NN6 0BD G4MTP
NN6 0BD G8SWL
NN6 0BD M0DON
NN6 0BY 2E0YXO
NN6 0HR G8SQA
NN6 0HX G8LOP
NN6 0NY G4LYC
NN6 0NZ G0JPF
NN6 0PB G4UHW
NN6 0PF G6OVC
NN6 0RQ G7RTC
NN6 0PX G7OXY
NN6 0QS M6BKJ
NN6 6AF 2E0TAY
NN6 6EW G4LRT
NN6 6EZ G6EGY
NN6 6EZ M3CZW
NN6 6LX G7DMH
NN6 7BH [illegible]
NN6 7GJ G6ARM
NN6 7GP 2E0JMY
NN6 7GP [illegible]
NN6 7PP G7BQM
NN6 7PQ M6AFU
NN6 7PQ M6ANM
NN6 7PQ M6ANO
NN6 7PU G6SJA
NN6 7QQ 2E1ECM
NN6 7QQ G4FMC
NN6 7RB G0OFB
NN6 7RR G4VMB
NN6 7RW G7HQH
NN6 7SR G0LHE
NN6 7ST G4EPA
NN6 7ST G8HYU
NN6 7TT G6YTB
NN6 7TX M0SPS

NN6 7YP M6OCR
NN6 8AF 2E0KRN
NN6 8AF 2E0PDH
NN6 8AF M0AOK
NN6 8AW [illegible]
NN6 8AW M3NKN
NN6 8QP G1TUS
NN6 8RP G6UNA
NN6 9BS G8GJW
NN6 9DS M6EJS
NN6 9JS G7NRR
NN6 9NE G4YKE
NN6 9SB M3ZST
NN6 9UE M6NRC
NN6 9XL G8JFX
NN7 1AE G7NDT
NN7 1BW G0DFV
NN7 1DA G6PHT
NN7 1DR 2E0RDN
NN7 1DR M6RFQ
NN7 1EQ G3TGE
NN7 1HB G6KPJ
NN7 1JW G1XXH
NN7 1NU G4FIN
NN7 2BW G6UVN
NN7 2DA G1AAD
NN7 2LF G0NRG
NN7 2NL 2E0SCH
NN7 2NL M0PIT
NN7 2NL M6ETC
NN7 2NR G8DHU
NN7 3AT G3TXQ
NN7 3DF G4XKM
NN7 3EL 2E0WBR
NN7 3EL G7PTA
NN7 3EL M6WBV
NN7 3EZ G4HWA
NN7 3JB G8EDX
NN7 3JU M6CRT
NN7 3NA 2E0SHV
NN7 3NA M0SHV
NN7 3QT G1VUY
NN7 3QU 2E0MOR
NN7 4BX G3IPL
NN7 4DD G3GWB
NN7 4DD G4CZB
NN7 4DD G8LED
NN7 4DN G4TGW
NN7 4DP G4IPL
NN7 4DR G8TMM
NN7 4LL G4PGY
NN7 4LS G0TML
NN7 4QL G1GJD
NN7 4RL M6MBS
NN7 4SB G4WOB
NN7 4SH G3ZAW
NN8 2DX G4ZIW
NN8 2DY G6WAS
NN8 2EN G6UWS
NN8 2ES M3LYG
NN8 2HR M6YPD
NN8 2HT G4HIQ
NN8 2JA 2E0PFF
NN8 2LZ G1IQE
NN8 2NQ G4HME
NN8 2PH M6NTY
NN8 2PQ M3CQT
NN8 2PU G0VRF
NN8 3BF M0WXF
NN8 3DB M3XLB
NN8 3DB M3XLW
NN8 3EW 2E1BFH
NN8 3JJ G0HBA
NN8 3JJ G1AUY
NN8 3PJ G0ING
NN8 3PJ G3KME
NN8 3PJ G5LP
NN8 3PJ G8PMA
NN8 3PS M3YIY
NN8 3PW G6ECT
NN8 3SH G4MOP
NN8 3ZA G6BBD
NN8 4AT G4WZB
NN8 4EB G7UPZ
NN8 4ET G8GOM
NN8 4RU G3ZAG
NN8 4RY G8NWZ
NN8 4RZ G8JPV
NN8 4SF G1IJJ
NN8 4UQ 2E0CRV
NN8 4UQ M6MSU
NN8 4UX G6CPX
NN8 4UZ G0FQA
NN8 5NX G4RDM
NN8 5PQ [illegible]
NN8 5QP G8OSZ
NN8 5WE G8EJQ
NN8 5WT G0DDJ
NN8 5WW G6STF
NN8 5YU M6ELJ
NN9 5BA M6JTL
NN9 5DA M6LTM
NN9 5BP M6YPG
NN9 5DP G0EHE
NN9 5LP G0HNG
NN9 5RB G3UQR
NN9 5RJ M6YDN
NN9 5RL M6MLN
NN9 5SY G4PPW
NN9 5TA G6ODT
NN9 5TL G0TLN
NN9 5TN G7EWL
NN9 5TY G3NXS
NN9 5XW G6DYU
NN9 5XW M3DYU

NN9 5YF G7PMO
NN9 5YS G6FEJ
NN9 6AL M3WBJ
[illegible] [illegible]
[illegible] [illegible]
NN9 6DF G7MKG
NN9 6DQ G0TES
NN9 6DQ G1JPC
NN9 6HE G0JXQ
NN9 6HE G0MEO
NN9 6HH G0VCW
NN9 6HN G6ONV
NN9 6LF M0MOI
NN9 6LS G7LPB
NN9 6LX G4LAM
NN9 6PY G6LSB
NN9 6QX G6TRO
NN9 6RD G1WWY

## NP

(Newport)

NP1 0BT G8IQC
NP1 4TB G3OVD
NP1 7PB G7MYD
NP1 8FY G0MTI
NP10 0BP G0ION
NP10 0BP G8MER
NP10 8ER G8WZR
NP10 8ET M6LHB
NP10 8JZ M6TJG
NP10 8LJ M6MBC
NP10 8LL G1MGI
NP10 8NF M3ZRK
NP10 8NY 2E0CVO
NP10 8NY M6SVZ
NP10 8PA G4JZY
NP10 8PD M0OPY
NP10 8PT M1DCF
NP10 8PT M6HPK
NP10 8RB G4COJ
NP10 8RB M6DGQ
NP10 8RF M6JEL
NP10 8RT G3YQM
NP10 8SL G4KHQ
NP10 8UP G7IAK
NP10 8WF 2E0ODS
NP10 8WF M6FNV
NP10 8WY M6NTW
NP10 9AH G7SMV
NP10 9AQ 2E0KEQ
NP10 9AQ M0KEQ
NP10 9AQ M6KEQ
NP10 9AT M6XSR
NP10 9EA 2E0OGY
NP10 9EA M6CVH
NP10 9FB G7RIB
NP10 9FT G6TUD
NP10 9FU G1JFT
NP10 9HN G4RYQ
NP10 9JG G4VCL
NP10 9JQ M0LZZ
NP11 3AE 2E0LBE
NP11 3AE M6BVU
NP11 3BE G5NF
NP11 3DF G0NZN
NP11 3DP G8NBF
NP11 3DW G4JJR
NP11 3EW M1CNN
NP11 3JQ 2E1IBN
NP11 3LJ 2E0DDP
NP11 3NT M1FDN
NP11 3NT M6DCI
NP11 4HJ 2E0KDR
NP11 4HJ M3RTU
NP11 4RE G0NUS
NP11 4RS G0JWF
NP11 5AU 2E0CYX
NP11 5AU M6SKR
NP11 5DL M0RZC
NP11 5GE 2E0KHK
NP11 5GE M3KHK
NP11 5GE M3XYW
NP11 5GE M6NON
NP11 5HN 2E0MAC
NP11 5JY 2E0MJC
NP11 5LA G0NPL
NP11 5LU 2E0END
NP11 6AW 2E0GEM
NP11 6AW M3XPM
NP11 6AW M6NUW
NP11 6BB 2E0BTQ
NP11 6BB M0TLR
NP11 6DN M3NZZ
NP11 6BP M3MXC
NP11 6GL M6CZV
NP11 6HJ 2E0BFC
NP11 6HJ M0GKV
NP11 6HJ M3LMZ
NP11 6HU 2E0BFD
NP11 6HU M3FDO
NP11 6NB 2E0RJL
NP11 6NB [illegible]
NP11 6NB M3LYQ
NP11 6PF 2E0EAD
NP11 6PF M3ZCU
NP11 6PF M6GVR
NP11 6QA M6XAD
NP11 6QH M3YLV
NP11 6RG M6CZU
NP11 7HY G1ZKE
NP11 7NR 2E0YAB
NP11 7NR M3YDS

NP11 7PF G7IAT
NP114HN M6NWM
NP12 0DA G8TRO
[illegible] [illegible]
[illegible] [illegible]
NP12 0EL M6SVV
NP12 0EL M6ZBE
NP12 0LU G0BZJ
NP12 0NQ G0DJX
NP12 0QE G0RQS
NP12 0QR G8OUM
NP12 0SF 2E1AVM
NP12 1DE M6GWO
NP12 1DF G3OAJ
NP12 1DZ G1PJP
NP12 1EE G4ROV
NP12 1EW G0LJW
NP12 1HQ G0MOW
NP12 1LX G0ARK
NP12 1QH 2E1ITI
NP12 1QH G8ITI
NP12 1SJ G0IXP
NP12 2AD G0JMJ
NP12 2DA M0CVT
NP12 2DA M0TTU
NP12 2ET M6YHR
NP12 2GA 2E0DGN
NP12 2GA G4JKV
NP12 2GE M3ZUF
NP12 2HP G4FCV
NP12 2HP G6BK
NP12 2JB G0SZN
NP12 2JU G4HBK
NP12 2JU G6GW
NP12 2LS M3VKM
NP12 2ND G0NPM
NP12 2NJ M3MNV
NP12 2PF M0LAO
NP12 2PF M3YLO
NP12 2PU G6OIO
NP12 3HB M0DNF
NP12 3NG G0EYH
NP12 3NR M0KEV
NP12 3NR M0MNX
NP12 3PG G3XVQ
NP12 3QS M3EAI
NP12 3RH G0ETM
NP12 3RH G1EHI
NP12 3UL M1EOR
NP13 1BS G3PFV
NP13 1DR 2E0SDO
NP13 1DR M0MDJ
NP13 1DR M6TBU
NP13 1EX M6PYP
NP13 1JE M6BWG
NP13 1NE M3XZB
NP13 2AD G7NTA
NP13 2AU M0LBB
NP13 2HU G0WVL
NP13 2HY 2E0VAW
NP13 2PJ 2E0ZAE
NP13 2PJ M3ZYE
NP13 2RT 2E1TDM
NP13 2RT M0BBM
NP13 2RY G4YDX
NP13 3AW M3IHX
NP13 3EQ G7RSE
NP13 3JS M3WIV
NP15 1AD G0DKF
NP15 1AT G6XBV
NP15 1BA M0XFU
NP15 1BA M6XFU
NP15 1HN G8AWM
NP15 1LG M6GWL
NP15 1QG G7KRY
NP15 1SP G3LDC
NP15 2BX G7HLZ
NP15 2BX M3HLZ
NP15 2EP M0CSC
NP15 2LF G8VKS
NP16 5AF 2E0HQD
NP16 5AF M3HQD
NP16 5BU M0OPS
NP16 5BX G1SXU
NP16 5EB M3NNA
NP16 5EH 2E0TSJ
NP16 5EH G4LWZ
NP16 5EH G8ZOE
NP16 5EH M3TSJ
NP16 5HP [illegible]
NP16 5JP G0HBD
NP16 5LA 2E1DCT
NP16 5LU G4KPD
NP16 5LY M3YLR
NP16 5RL G4REX
NP16 5TL G7TGB
NP16 6BN M0DRI
NP16 6DR 2E0ZXX
NP16 6DR M6ZXX
NP16 6EF G7IRD
NP16 6FA 2E0RCU
NP16 6FA M0RVC
NP16 6FA [illegible]
[illegible] [illegible]
NP16 6LA 2E0DWA
NP16 6LA 2E0PBU
NP16 6LA M6DWA
NP16 6LA M6PBO
NP16 6LE 2E0DNI
NP16 6LE M0NVY
NP16 6LE M6KGB
NP16 6PE G1ISR
NP16 6PN 2E0CXV
NP16 6PN M6DNP
NP16 6SE G4LDA

NP16 6TF 2E0BTZ
NP16 7AQ M6WOV
NP16 7DF 2E0JMH
[illegible] [illegible]
[illegible] [illegible]
NP16 7DR G7ERI
NP16 7DS G4RZE
NP16 7EX M6MGC
NP16 7HF 2E0DJC
NP16 7HF 2E0RDJ
NP16 7HF M6RDC
NP16 7LG G1VDT
NP16 7LZ G8OQV
NP18 1GT G4ITJ
NP18 1JF G6FLU
NP18 1LP G7NFT
NP18 1LT G0FBT
NP18 1RY G4BCZ
NP18 2AE M1EHW
NP18 2BY G7SNF
NP18 2JL M1BEQ
NP18 2JR G6GCK
NP18 2LS M0HID
NP18 2LT G7PIB
NP18 2NR G8GUJ
NP18 3BZ G4LCF
NP18 3EF G7BIL
NP18 3EU G4ZVQ
NP18 3EZ G4ESL
NP18 3PW M3ZGR
NP18 3SQ G8ZCV
NP19 0AN 2E0PAN
NP19 0AN M3PQE
NP19 0AN M3PQF
NP19 0EG 2E0SMB
NP19 0EG M3FIK
NP19 0ET M6LOD
NP19 0GX 2E0CJZ
NP19 0HU 2E0CHH
NP19 0HU M0LLO
NP19 0HU M6CJH
NP19 4NX G6EWQ
NP19 4RQ M3KSE
NP19 4TB M6PPD
NP19 7DB G7AVB
NP19 7ER 2E0BTF
NP19 7ER M0GXE
NP19 7ER M3XWE
NP19 7FJ M6KYN
NP19 7GF 2E0DWR
NP19 7HY G8SVN
NP19 7JU M0GOM
NP19 7PB 2E1ADO
NP19 7QB G4VNS
NP19 7QP G8MOZ
NP19 7QW G0DHG
NP19 7QY 2E0MNJ
NP19 7SA G1DLP
NP19 7SN M0ALN
NP19 8AG G3KWB
NP19 8NU G6KBD
NP19 9AP M3YKA
NP19 9AT M3GWZ
NP19 9ET G1TDV
NP19 9JN 2E1GLY
NP19 9NE G6RGT
NP19 9QA G7RNC
NP19 9QR G4TIW
NP19 9QS G4SEW
NP19 9RY G8OIJ
NP19 9SX 2E0HOT
NP19 9TA M3FYR
NP2 0LX G0LKJ
NP2 4JD G6EDP
NP20 2GN 2E1DXV
NP20 3BD M3FLK
NP20 3BJ M0YGJ
NP20 3DE G0MSY
NP20 3DJ G0FJE
NP20 3DJ G1NRS
NP20 3DJ G4EZW
NP20 3DJ G4SUE
NP20 3DN M3PUU
NP20 3DN M3ZCO
NP20 3DN M6PUT
NP20 3EE G0FXC
NP20 3HJ 2E0JDL
NP20 3HJ M3NEI
NP20 3HJ M6FFN
[illegible] [illegible]
NP20 3NR G8OKR
NP20 3NS G3KZT
NP20 3NT 2E1EID
NP20 3SP G4LOD
NP20 4DT M3KHC
NP20 4FN M0BRL
NP20 4JR [illegible]
NP20 4LP G7BSC
NP20 4LP M1AZR
NP20 5AA G4SDT
[illegible] [illegible]
NP20 5BA M6TBD
[illegible] [illegible]
[illegible] [illegible]
NP20 5EF G3XEJ
NP20 5EF G4BLE
NP20 5GG 2E0VVL
NP20 5GG M3SWJ
NP20 5HX 2E0NUC
NP20 5HX M3NUC
NP20 5PP 2E0CVM
NP20 5PP M6XAG
NP20 5XE 2E0CZZ
NP20 5XE M6CZS
NP20 6HE G3NIN



**IMPORTANT NOTE**

**Revalidate licence to avoid revocation** – Ofcom has advised the Society that plans will be drawn up to revoke licences that have not been revalidated as required by the licence conditions. The quickest way to revalidate is to do so online via the Ofcom website: *https://services.ofcom.org.uk/* or by email: *amateur.validations@ofcom.org.uk* If you need assistance in the process, Ofcom staff are available to help, but please be patient during times of heavy workload.

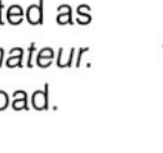

NP20 6JD 2E0JBJ
NP20 6JD M3JZV
NP20 6JN 2E0XAA
NP20 6JN G3WTZ
NP20 6JT G0OAJ
NP20 6LB 2E0MAO
NP20 6LG M3VWM
NP20 6LS G4ZCM
NP20 6NB 2E0ECZ
NP20 6NB M0ZCE
NP20 6NB M3ECZ
NP20 6QF G3YKZ
NP20 6QF G8CNF
NP20 6QN G4GFL
NP20 6QN G4YKW
NP20 7BZ G0KTL
NP20 7DJ M6FBJ
NP20 7DW M0USK
NP20 7DW M3USK
NP20 7EJ M3YJJ
NP20 7SQ M6GAV
NP20 7YA M3YJQ
NP22 3DD M6MPQ
NP22 3HE G0MQU
NP22 3HF 2E0IAO
NP22 3HF M6XAC
NP22 3HN M6SID
NP22 3NT M6DQQ
NP22 3PB 2E0TAR
NP22 3PB M6KAA
NP22 3PF 2E0LAC
NP22 3PF M6SAA
NP22 3RZ M0CPD
NP22 3SG G4WLT
NP22 3SN M6DBD
NP22 3TA 2E0SCB
NP22 3TA M6DBB
NP22 3TE M6EKT
NP22 3TH M6SFP
NP22 4JG G6TQH
NP22 4JG M3KSI
NP22 4PF M6IJW
NP22 4PL M3HIX
NP22 4PW M3TUO
NP22 5AR 2E0XBE
NP22 5AR M3UAP
NP22 5AR M3XAJ
NP22 5AR M3XBE
NP22 5BH G8TIX
NP22 5DW M6AOE
NP22 5EA M3VWO
NP22 5LS G8JWP
NP23 4AR G4KFI
NP23 4QY G0JTU
NP23 4QY M3HBC
NP23 4SD G7HTU
NP23 4TT 2E0DOE
NP23 4TT M6EYX
NP23 4UZ M6XRA
NP23 5AJ M3PMU
NP23 5BB M6RRG
NP23 5ES G0HCN
NP23 5FA M3UHO
NP23 5FE G1IOT
NP23 5FE M3WPH
NP23 5QR G7VEH
NP23 5SF 2E0CLJ
NP23 5SF M6EAD
NP23 5TF G0IRT
NP23 5TS 2E1CIP
NP23 5TS G0MNO
NP23 5TS G1HNF
NP23 6HT 2E0LCJ
NP23 6HT M0PWY
NP23 6HT M3VLJ
NP23 6JF G4CXK
NP23 6JR G0VSO
NP23 6LN G7VGB
NP23 6NE M3UDW
NP23 6NE M3UEG
NP23 6PJ M6EXL
NP23 6TR M3HKH
NP23 6UA G4UTS
NP23 7QY G0JTF
NP24 6AZ M3TLP
NP24 6BA M0HRD
NP24 6BS M3YBK
NP24 6BW M3DPV
NP24 6EY M3RZW
NP25 3HT G0MSW
NP25 3SD G3VFL
NP25 4LU M0ZEN
NP25 4LY G8WCA
NP25 4NB G1FJI
NP25 4QD G4FIC
NP25 5DE G8PTS
NP26 3AB G0GEV
NP26 3AG G3ORL
NP26 3AT G8YKS
NP26 3AX 2E0XWD
NP26 3AX M3LVF
NP26 3BZ 2E0LGE
NP26 3BZ G3NWS
NP26 3BZ G8GT
NP26 3BZ M0LGE
NP26 3BZ M6LGA
NP26 3FD M3ZMU
NP26 3FG 2E0CUJ
NP26 3JF M0GTN
NP26 3JU G8MZR
NP26 3PB G3MPP
NP26 3QG G4OGO
NP26 3SA M0HJG
NP26 3TB 2E1BOG
NP26 3UW G3WEZ
NP26 4DX M0DTH
NP26 4JD G4ZVL
NP26 4JL M3KRN
NP26 4PP G4FNO
NP26 5AF 2E0SYS
NP26 5BW G0EGH
NP26 5GB G1EPR
NP26 5RE G6NQU
NP26 5RE M3ESH
NP26 5RE M3HOY
NP26 5RE M3VEN
NP26 5RX M6RLE
NP26 5TB M6BSU
NP26 5TU M1FLY
NP26 5UW M6OTK
NP3 1AR G6BVS
NP3 2NQ G8LTV
NP3 4LZ M1AZI
NP4 0BN G1XVC
NP4 0BN M3HGO
NP4 0DG G0FJH
NP4 0HT 2E0REX
NP4 0HT M0LUK
NP4 0HT M6USK
NP4 0JD M3GHF
NP4 0NB G7LPM
NP4 0NJ G0DQT
NP4 0PT G4NXD
NP4 0RT M0AZN
NP4 5AB 2E1GMM
NP4 5BX G1VJB
NP4 5BZ G1SXT
NP4 5EA 2E0DDJ
NP4 5EZ G8CKJ
NP4 5JF 2E0TAX
NP4 5LT G4HYZ
NP4 5LU G4OCN
NP4 5SS 2E0BVV
NP4 5SS M3JQX
NP4 5XS G4ZQV
NP4 6DF G0MBW
NP4 6HE M6MZU
NP4 6HH M0AUH
NP4 6QZ M6HPC
NP4 6QZ M6RHY
NP4 6QZ M6WCT
NP4 7HJ 2E0IJL
NP4 7HJ M0KMS
NP4 7QF 2E0SLD
NP4 7QF M3SLO
NP4 7QF M6CAE
NP4 7QS G8ERA
NP4 7RZ M6PGC
NP4 7SN M6KTM
NP4 7UJ G1VDW
NP4 8DQ G4AYQ
NP4 8DQ M3DPF
NP4 8DQ M3LOI
NP4 8LL 2E0XVT
NP4 8LL M0YVT
NP4 8LL M6XVT
NP4 9EH M6IJJ
NP4 9ER G1WTL
NP4 9LQ G3XJA
NP4 9PA M0XRU
NP4 9QL G6STS
NP4 9QN 2E0CVL
NP4 9QN M3NAE
NP4 9SS G1GKV
NP44 1AT G4IZJ
NP44 1BJ G7RQV
NP44 1BQ 2E0UNY
NP44 1BQ M6UNY
NP44 1LH G0TTN
NP44 1LJ M1CVM
NP44 1LS G4UWR
NP44 1NB G0FGO
NP44 1NW M1DOO
NP44 1RB G7DIL
NP44 1RZ M3AVU
NP44 2AL G4IQB
NP44 2AN G0KAM
NP44 2AN G1YKY
NP44 2JW 2E0SEZ
NP44 2JW M0SEZ
NP44 2JW M3SEZ
NP44 2JW M6GZC
NP44 2JW M6SDE
NP44 2PF M6ERZ
NP44 3AE G4JBQ
NP44 3AE G8SBK
NP44 3AP 2E0CUK
NP44 3AP G8CRH
NP44 3AP M6CUR
NP44 3AP M6TRQ
NP44 3AZ G0HIR
NP44 3BX 2E1DEA
NP44 3BX G0OLZ
NP44 3EL G6RBZ
NP44 3HH G4YML
NP44 3LE M6RLA
NP44 3LS 2E1GDY
NP44 4LB G1YZF
NP44 4LD M3OSQ
NP44 4LF 2E0YAD
NP44 4LF M3NHN
NP44 4LG 2E0LFE
NP44 4LG M6KUD
NP44 4LW M3URO
NP44 4PQ M3LMV
NP44 4PX M0RPB
NP44 4QT M1FFE
NP44 4TE G0UKT
NP44 4TE G7VEL
NP44 5AS G1UXW
NP44 5HN G8JOY
NP44 5JE M0MTR
NP44 5TQ G0DQY
NP44 5TQ G1ZFX
NP44 5TY 2E0PRD
NP44 5TY M0PRD
NP44 5TY M3NQI
NP44 5UD G7SSQ
NP44 5UH M6OZB
NP44 5UN G0OZB
NP44 5UN G7DHG
NP44 5UU M6SOF
NP44 6EE G7UMS
NP44 6JH 2E1DNK
NP44 6JH 2E1DRB
NP44 6JH G0DHA
NP44 6JJ G7SSN
NP44 6UL M3MBG
NP44 6UL M6FAW
NP44 7AB G4CQT
NP44 7AH G0DXG
NP44 7AH G4RIB
NP44 7JG G0HDY
NP44 7JP 2E0MKG
NP44 7JP M0MKG
NP44 7JP M6MGY
NP44 7JX M0YAC
NP44 7JX M0YAD
NP44 7LL G7CBU
NP44 7LS M6EJG
NP44 8RJ G6BHQ
NP44 8UG G0DQW
NP44 8UG G6DYW
NP6 1PE G6BAH
NP7 0BP G6JVB
NP7 0BY G0JJF
NP7 0DX G6VED
NP7 0EL G1UVN
NP7 0EX 2E0GKS
NP7 0EX M0GUK
NP7 0EX M6GKS
NP7 5BQ G4JDE
NP7 5HQ G6JPC
NP7 5HQ M0RRD
NP7 5HW G4IHM
NP7 5JA G0UDH
NP7 5LG M0YVK
NP7 5TL G6BXU
NP7 5TL G6RAV
NP7 5TL M1FAT
NP7 6AF G6ZZF
NP7 6AP M6EYM
NP7 6BE M6ETW
NP7 6HB M3ETB
NP7 6HN G0UKG
NP7 6JY G6RAO
NP7 6PF G4XQH
NP7 7AL G4UHK
NP7 7AW G8KJK
NP7 8AA G3UXJ
NP7 8DH M6NFH
NP7 8DN G4IQA
NP7 8TG M6CYM
NP7 9HP G8KCH
NP7 9RS M6EJI
NP7 9SA G4EIN
NP7 9SD M0LDJ
NP8 1AP G0BWE
NP8 1DJ G0XYL
NP8 1DJ G4ZUW
NP8 1DQ G4FRH
NP8 1LN G6KGR
NP8 1LU 2E0MZM
NP8 1LU M6MJO
NP8 1NU G8IKH
NP8 1RU G8IAM
NP8 1SY G4UZC
NP9 7HW 2E1AYO

## NR (Norwich)

NR1 1ET G4FAI
NR1 1LW M6VOX
NR1 1QA G4MYQ
NR1 1TR M6NLR
NR1 2AL G7VZL
NR1 2JR G4RRX
NR1 2JX 2E1HOO
NR1 2JX G7UVB
NR1 2NB 2E0KFO
NR1 2NB M3WVC
NR1 2NL G0WJU
NR1 2NX G4VLS
NR1 3ED G0WFP
NR1 3HB M6HAE
NR1 3JB M6GCC
NR1 3QX 2E1HBG
NR1 3QX M6VDH
NR1 4BE G0MWH
NR1 4EN M6DPZ
NR1 4EP 2E0GDO
NR1 4HB G0BFM
NR1 4HQ G4JYG
NR1 4JX M1CRQ
NR1 4LN M3SDB
NR1 4LR 2E0CEY
NR1 4LR M6JET
NR1 4NJ M6BLI
NR1 4PP 2E0GFB
NR1 4PP M6RWV
NR1 4QB 2E0NLK
NR1 4QB M3NLK
NR1 4QB M6MLK
NR1 4QB M6SLK
NR10 3BH G4RSP
NR10 3EA G8BIG
NR10 3ES G0TMT
NR10 3HE G8YQH
NR10 3HE M1GPC
NR10 3HF G7ETC
NR10 3HF M3LJA
NR10 3HF M3UMA
NR10 3HS G7PDO
NR10 3HY M6COU
NR10 3LD G4WUG
NR10 3LF M6LAE
NR10 3LG G0SGT
NR10 3LP M6JWP
NR10 3LX G0DWV
NR10 3NN M6EOA
NR10 3PG G0AUT
NR10 3PP G1ESX
NR10 3QA G7JTZ
NR10 3QA M3JTZ
NR10 3QB G3VPT
NR10 3QB M0SPX
NR10 3QQ M3NGF
NR10 3QQ M6EKS
NR10 3QW G3TWX
NR10 3SE G0VRY
NR10 3SG M0BLH
NR10 3ST G0MJC
NR10 3SW 2E0SPA
NR10 3SW M6SPA
NR10 4AE M3EDC
NR10 4AH G3TNY
NR10 4AZ G0PFJ
NR10 4BG M3MJD
NR10 4BS G1WUY
NR10 4BS M3YGU
NR10 4HA G4WUU
NR10 4HJ 2E0CAW
NR10 4HJ 2E0YOP
NR10 4HJ M0SCO
NR10 4HJ M3CAW
NR10 4HJ M3YOP
NR10 4HL G7HXW
NR10 4HL M0NSR
NR10 4HL M3SOY
NR10 4LS G0IYK
NR10 4LU G8YJS
NR10 4PP 2E1HKE
NR10 5BB G3WCE
NR10 5BB M1BQW
NR10 5DJ M6FBE
NR10 5DJ M6HDM
NR10 5EX M3RZN
NR10 5EX M6SEN
NR10 5JT M6HVY
NR10 5JY 2E1GZY
NR10 5JY M1EPR
NR10 5JY M3EPR
NR10 5LE 2E1EDA
NR10 5LT G3VWQ
NR10 5PW G4SGX
NR10 5QE G3TOZ
NR10 5QW G3KMO
NR10 5QX G1NGE
NR11 6BD 2E0IWB
NR11 6BD M6IWB
NR11 6DJ 2E0DDL
NR11 6DJ M0LDV
NR11 6DJ M6DXN
NR11 6FZ 2E0CYV
NR11 6FZ M6DYC
NR11 6HW 2E0RDI
NR11 6HW M6OPL
NR11 6JD 2E0HAJ
NR11 6JD M0HET
NR11 6JF G7GQB
NR11 6LB G0NTJ
NR11 6RX G6ZRV
NR11 6UA M6DRT
NR11 6UT G1UQC
NR11 6UT G8RSQ
NR11 7AQ G0AZR
NR11 7AW G1GGN
NR11 7BE G4GQE
NR11 7DT G4PLZ
NR11 7DY G7VRK
NR11 7ED 2E0LSR
NR11 7ED M6EMT
NR11 7LQ 2E0WPZ
NR11 7PD M1BOA
NR11 8BE G3PPR
NR11 8DN G6NZY
NR11 8ED G0TAM
NR11 8ED G7RNN
NR11 8HJ G4AFQ
NR11 8JB G3BGF
NR11 8JD 2E0NAI
NR11 8JD M6CIK
NR11 8JF G6OQJ
NR11 8JF G7HFP
NR11 8JW G4SFY
NR11 8LP G0JLV
NR11 8NG G6KBZ
NR11 8PD G4RRN
NR11 8UW M6SNJ
NR12 0AL G3SGC
NR12 0HF G4SFQ
NR12 0JP 2E0NFK
NR12 0JP M6WNW
NR12 0LU M1EEN
NR12 0NE G4POG
NR12 0NJ G4MQK
NR12 0PD G3WQY
NR12 0PW M1DMB
NR12 0QE M6DXQ
NR12 0RB G6PYR
NR12 0RL M6ADT
NR12 0RL M6RKM
NR12 0RU G1ANZ
NR12 0RU G4CTT
NR12 0SF G7HXI
NR12 0SU M6SMG
NR12 0SX 2E1SBF
NR12 0TA G7TUM
NR12 0UX G7VAH
NR12 0YQ G3NMZ
NR12 7AB G0SMS
NR12 7AB G1KZD
NR12 7AB G1NJG
NR12 7AJ G0VDR
NR12 7AJ G3VZT
NR12 7AJ G7RPK
NR12 7AJ M6VDR
NR12 7BB 2E0NKI
NR12 7BB 2E0TZY
NR12 7BB M0TZY
NR12 7BB M6TKE
NR12 7BB M6TZY
NR12 7DL G4ILR
NR12 7DQ G4SOP
NR12 7DT G6ESQ
NR12 7EG G7RIE
NR12 7EL M6WTR
NR12 7ER 2E0SXC
NR12 7ER M6DBS
NR12 7EW G1XUW
NR12 7HD G7SNT
NR12 7HW G6GAC
NR12 7HZ G1EAJ
NR12 7JU G1HXN
NR12 7LG M6DEF
NR12 7NB G0DGU
NR12 8BA 2E1EIV
NR12 8BA M0HUG
NR12 8BA M0PUD
NR12 8DG G4VUR
NR12 8DG G1OSE
NR12 8DP M6VFN
NR12 8HP M0EAZ
NR12 8JH G4YDZ
NR12 8LQ G8UPD
NR12 8LZ G0AKL
NR12 8QR G0WJV
NR12 8SL M0THA
NR12 8SL M3MHN
NR12 8SQ G1YLV
NR12 8TU M0TVG
NR12 8UJ G0GHB
NR12 8XB G6IYY
NR12 8YL G7PRC
NR12 8YL G7VPD
NR12 9BE M6GUY
NR12 9BY G8IYK
NR12 9DA G8VSF
NR12 9EQ G1ONC
NR12 9PB G1TAY
NR12 9PJ G4HXK
NR12 9PZ G1LQM
NR12 9QN M6WZK
NR12 9RF G4PSH
NR12 9SA G4CDN
NR12 9SE G4CKB
NR12 9SE M6ABQ
NR12 9SE M6KLM
NR13 3AA G0ADA
NR13 3AB G4JNX
NR13 3DH G0OOB
NR13 3EY M6RGI
NR13 3HJ M6EJY
NR13 3LA G0KRU
NR13 3PL G4YGQ
NR13 3PP G4YYL
NR13 3PP G6WIT
NR13 3PU M6DYO
NR13 3PU M6POZ
NR13 3SW G3LNM
NR13 3TH M0HJI
NR13 4AH G4UUB
NR13 4AW 2E0BDB
NR13 4AW 2E0BSU
NR13 4AW M3NLF
NR13 4BA G4UAM
NR13 4BB G1MUC
NR13 4FA 2E0BCQ
NR13 4FA M0KXK
NR13 4JS G4WUO
NR13 4LZ G0KSD
NR13 4NB G4BFS
NR13 4NF G0GQV
NR13 4QF 2E1FNJ
NR13 4QF 2E1HPC
NR13 4QF 2E1HPD
NR13 4QF G7UVY
NR13 4RU G4FOS
NR13 5DG 2E0ZZC
NR13 5DG M6ZZC
NR13 5DL M6VXT
NR13 5HR G4MCA
NR13 5JQ 2E0NGR
NR13 5JQ M6NJG
NR13 5LG G0AKO
NR13 5LU M6WRJ
NR13 5PA G7RBS
NR13 5QG 2E0AWG
NR13 5QG M0EBJ
NR13 5RA 2E0YAO
NR13 5RA G0OOR
NR13 5RF G8UKO
NR13 6DP G0IRQ
NR13 6DY G6BXR
NR13 6LT G4KLM
NR13 6QD G0IMU
NR13 6QH G1BCU
NR13 6RQ G6WMG
NR13 6RR G3PDH
NR13 6SG M0HLZ
NR13 6SS M6GUW
NR134JX 2E0DJR
NR134JX M3UYV
NR135LD G7RTN
NR14 6BH G4IAO
NR14 6BQ G1MSK
NR14 6BY G3ITB
NR14 6LL M0NRW
NR14 6LL M0PTO
NR14 6NH G7HFE
NR14 6PG M0BIT
NR14 6PP G3VKM
NR14 6RE M3ZFO
NR14 6UT G4TVJ
NR14 6UZ G1MSG
NR14 7AH M6LBM
NR14 7AR G7DFP
NR14 7AS 2E0MYE
NR14 7AS M3MYE
NR14 7BU G4SOZ
NR14 7DN G0JIV
NR14 7DP G0RSS
NR14 7EB M3RMD
NR14 7EE G4GUS
NR14 7JE G0DEE
NR14 7LU G0VJH
NR14 7PB G4LOJ
NR14 7QN G4BID
NR14 7RY 2E1CPV
NR14 7WF 2E0BRF
NR14 7WF M0GMK
NR14 7WF M6CJD
NR14 8AL G4VCE
NR14 8AS G4LEP
NR14 8DH G0XAR
NR14 8HT G1PQJ
NR14 8HX G6VAD
NR14 8LQ G3LDI
NR14 8PH M0HJE
NR14 8RB G4ZOK
NR14 8RG G4UYR
NR15 1JB M3NKP
NR15 1NG G6ULJ
NR15 1NX G6WAO
NR15 1TH G7VNN
NR15 2AJ 2E0CQG
NR15 2AJ M0HIW
NR15 2DG M0TSM
NR15 2DS G3WDN
NR15 2HR 2E0VBW
NR15 2HR M0VBW
NR15 2NT G3VYN
NR15 2PH G8WSH
NR15 2ST 2E0JMB
NR15 2ST M0UKS
NR15 2TA M3LYA
NR15 2UH M6UDX
NR15 2WY M6PWG
NR16 1AW G8AKF
NR16 1DJ G1GSB
NR16 1DL G3MGX
NR16 1HH 2E1CYE
NR16 1NW M1CQK
NR16 1NW M1CQL
NR16 1RS G3SDT
NR16 2AN G3RXA
NR16 2BY 2E1CIR
NR16 2BY 2E1DLO
NR16 2BY 2E1DWM
NR16 2BY G7RVH
NR16 2DE G6XQP
NR16 2EQ M0RYB
NR16 2HL G0BOO
NR16 2JD G0ENB
NR16 2JD M1MNR
NR16 2JD M3HMK
NR16 2LW G7UFV
NR16 2NA G6HYP
NR16 2NE M6ABX
NR16 2PS G3JIE
NR16 2SZ M3PZZ
NR17 1BL G6AVI
NR17 1BW 2E0TFX
NR17 1BW M0TFX
NR17 1BW M3TFX
NR17 1DT M3SFZ
NR17 1EQ M6AKP
NR17 1JB G7URP
NR17 1JB M3DCP
NR17 1JT M3PLU
NR17 1LW 2E0DWZ
NR17 1LW M6RKD
NR17 1QU G4NRG
NR17 1UL G3UUR
NR17 1YF M1CVU
NR17 2AD G4RKL
NR17 2EW 2E0PPO
NR17 2EW M6RNS
NR17 2HU 2E0IXH
NR17 2HU M3IXH
NR17 2LA M3FUQ
NR17 2NG G1XVV
NR17 2NG M3MGJ
NR17 2NH G3JUU
NR17 2RL G0KXV
NR17 2RQ M3NGU
NR18 0AR 2E1DQT
NR18 0AR G7VZI
NR18 0BD 2E0YOK
NR18 0BD M6OKY
NR18 0DE G1FPK
NR18 0DN G3YIA
NR18 0EA G1EGL
NR18 0EN G4OJV
NR18 0EX G8DYA
NR18 0HS G6LQM
NR18 0HX G6YYU
NR18 0HY G7TIE
NR18 0NB M1EOV
NR18 0NT G3MPN
NR18 0NZ G6DAN
NR18 0QP G6YFZ
NR18 0TT 2E0CQH
NR18 0TT M6BVM
NR18 0TU M0RJB
NR18 0UN 2E0DWT
NR18 0UN M0WEL
NR18 0UN M6WEL
NR18 0WQ G0OPV
NR18 0XJ G0KYA
NR18 0XJ M6ELE
NR18 0XJ M6PUP
NR18 0XT M1BCU
NR18 9BH M6NVT
NR18 9BP M6PPV
NR18 9DD G7EOA
NR18 9DR 2E0XSW
NR18 9DR M0VSW
NR18 9DR M3VSW
NR18 9HQ M6ZZB
NR18 9RY G4VAO
NR18 9TB G7DIB
NR18 9TF G4VMG
NR19 1AG G0UOB
NR19 1AJ G6ULS
NR19 1EG G1OZR
NR19 1JB G3PKJ
NR19 1JB G8DQK
NR19 1JY G7HCT
NR19 1LU G0GEB
NR19 1LU M6LSE
NR19 1ND G0TYC
NR19 1ND G4EYA
NR19 1QE G6ZKX
NR19 1XF M6KJR
NR19 2BJ G8UUV
NR19 2BS M3KTH
NR19 2BS M3WKM
NR19 2LX G3YHO
NR19 2NS G4UVA
NR19 2SR M6EBY
NR19 2SR M6ODM
NR19 2SU M6NMR
NR19 2SY 2E0ZJB
NR19 2SY M0ZJB
NR19 2SY M3ZJB
NR19 2UB G0LGJ
NR2 2DW M3VGX
NR2 2JB G1AMN
NR2 2RN G4LFQ
NR2 2SN G4EXN
NR2 3DS 2E0CFG
NR2 3DS M6AOT
NR2 3NJ G7FUQ
NR2 3NJ G8GWJ
NR2 3PG M0GGT
NR2 3PG M3NKZ
NR2 3PT 2E1DRY
NR2 3RL G7NVB
NR2 3TL G6TSZ
NR2 3TP 2E0MAP
NR2 3TP 2E0PEW
NR2 3TP M0PEW
NR2 3TP M3PKZ
NR2 3TP M3WPO
NR2 3TP M3YYU
NR2 3TP M6HRW
NR2 4BL 2E0BSS
NR2 4BL M3OYE
NR2 4EH 2E0POI
NR2 4EH M0NFY
NR2 4EH M6POI
NR2 4SD 2E0NCG
NR2 4SD M0NCG
NR2 4SD M3NCG
NR20 3DG 2E0CAU
NR20 3DG G4LPW
NR20 3LX G0UYA
NR20 3QP G3YLA
NR20 3RE G4DYC
NR20 3RF G0UML
NR20 3SB G0VIJ
NR20 4AA G3ZIG
NR20 4AA G4ANT
NR20 4AY M3YCS
NR20 4AY M3ZCE
NR20 4BW G6YXB
NR20 4DW G7HJD
NR20 4ET M6TDI
NR20 4LR G4IPN
NR20 4LY M3TDT
NR20 4NQ G0TYQ
NR20 4SL G0UYC
NR20 5LL G8OO
NR20 5PS G0GHE
NR20 5QN M6GYV
NR20 5RU M6EIU
NR20 5SB M6JJK
NR20 5SZ G7EWK
NR20 5TA G4XCR
NR20 5TW G3JCK
NR21 0BU G4UKZ
NR21 0EJ G3XMP
NR21 0HX G0PPR
NR21 0LT G6LEK
NR21 0PT G0WCH
NR21 0QX G7OSJ
NR21 7BY G7PSK
NR21 7LG G7IJD
NR21 7NA G0FVF
NR21 7NA G6MDM
NR21 7QE G6YLR
NR21 8DW G1IJC
NR21 8HB G3PAX
NR21 8LR G0KYG
NR21 8NP M6AWA
NR21 8NZ M6GBD
NR21 8PH G4XQW
NR21 9ES G0KRD
NR21 9HW G4RRD
NR21 9JB G6IGO
NR21 9LH M6AIQ
NR21 9PD G0SQI
NR23 1BZ G4NRW
NR23 1HF G4DIC
NR23 1JY G4KOV
NR23 1JZ G2PK
NR23 1LR G0SRZ
NR23 1PA G4RAY
NR23 1QE G0DSN
NR23 1QE G0NWT
NR23 1QL G7BUS
NR23 1QL G7EVQ
NR24 2DE G4IWI
NR24 2JY G6VZG
NR24 2LB M3YPX
NR24 2LQ G0FVS
NR24 2LQ M6CYC
NR24 2PH G8AKP
NR24 2RD G6AUY
NR25 6DG G4BUF
NR25 6NW G0SIQ
NR25 6RU G3LQJ
NR25 6TG 2E0CTT
NR25 6TG M6BDX
NR25 6TU G7PTD
NR25 6TU M0SSF
NR25 6TX M3WZJ
NR25 6TZ G7OSK
NR25 7DQ G8YSJ
NR25 7HH 2E0JDE
NR25 7HH G0GFQ
NR25 7HH G1MIE
NR25 7HH M6JFV
NR25 7LL G4EBE
NR25 7UA G4UQA
NR25 7XJ G4OLF
NR26 8BJ M3ZNY
NR26 8EJ G7FSI
NR26 8EJ M6DPI
NR26 8HU G7CIK
NR26 8JY 2E0TBX
NR26 8JY M6CRP
NR26 8NQ G4ARX
NR26 8PT G7PHT
NR26 8RE G3VIC
NR26 8RR 2E0RBQ
NR26 8RR G8SAU
NR26 8RR M6RMT
NR26 8SH 2E0SLS
NR26 8SH M0SHK
NR26 8SH M0WPX
NR26 8SH M6HDS
NR26 8TG G4MIE
NR26 8UB M3BFU
NR26 8UN G3HRK
NR26 8UN G6DFA
NR26 8XP G4DMB
NR26 8YE M0VAW
NR26 8YJ M6ASE
NR27 0BG G2FT
NR27 0BP M6IFI
NR27 0BY M0DFF
NR27 0EE G0UIQ
NR27 0EQ M6FQK
NR27 0EQ M6GAD
NR27 0EQ M6LAQ
NR27 0HG 2E0BCW
NR27 0HG M0EBN
NR27 0HG M6BXA
NR27 0HQ G2BHG
NR27 0HX G0RHJ
NR27 0NY M6CFU
NR27 9BW G4NRE
NR27 9BW G6EHE
NR27 9BW G6HL
NR27 9DN G4RZN
NR27 9LQ G0BGX
NR27 9LQ G0SQK
NR27 9LW G0LYN
NR27 9LW G1SVD
NR27 9PH G3PND
NR27 9PH G4PQP
NR27 9PH M0NCA
NR27 9PW 2E0WBW
NR27 9PW M6WWD
NR27 9QT G3UDV
NR27 9RN G3ZRG
NR27 9RR M3UCA
NR28 0AU M6ACR
NR28 0BS M3YXV
NR28 0DH 2E0WOS
NR28 0DH M0WOS
NR28 0DH M3WOS
NR28 0EE 2E0UJD
NR28 0EE G8RZZ
NR28 0EE M0UJD
NR28 0EE M3UJD
NR28 0EE M3VTN
NR28 0EN G3WXW
NR28 0JP M3PGD
NR28 0LE 2E0FAY
NR28 0LE M6BLQ
NR28 0LR G4ZWM
NR28 0PJ G7TIN
NR28 0QG G3YOA
NR28 0QR 2E0WTG
NR28 0QR M0WTG
NR28 0QR M3WTG
NR28 0QU G0GRB
NR28 0RZ G1NST
NR28 0SX M3XSI
NR28 0SY G0IYD
NR28 0SY G1XYD
NR28 9BB G0AJJ
NR28 9BB G6VOV
NR28 9BS M0JMS
NR28 9HS G4RTH
NR28 9JE M6NWP
NR28 9LD G4ZAY
NR28 9LX M6ZAH
NR28 9PL M6AST
NR28 9RG G6JCY
NR28 9SD G7IIN
NR28 9SH G4TDQ
NR28 9SR M6RSB
NR28 9XA G6JCX
NR28 9XT 2E0CVY
NR28 9XT M0NAQ
NR28 9XT M6CFQ
NR28 9YA G6WME
NR29 3BG G7UDU
NR29 3HD G5NB
NR29 3HJ G0MYC
NR29 3HU G8VPE
NR29 3JB M6DAL
NR29 3LT G7DXQ
NR29 3NH G1CEU
NR29 3QD 2E0GFW
NR29 3QD M6GFW
NR29 3RJ G1LTH
NR29 3RJ G4SCS
NR29 3SP M3XUV
NR29 4AF G7BMY
NR29 4AQ G7TUK
NR29 4EA G3ZEB
NR29 4ES G1EBP
NR29 4JD G8PEN
NR29 4JL G8SYV
NR29 4JY G7HRF
NR29 4LT G3XAU
NR29 4PT G1RPV
NR29 4PU M6GZK
NR29 4QT M3XMU
NR29 4RJ G4MFX
NR29 4RY G4EMH
NR29 4SH G8UCZ
NR29 4TD G0FVT
NR29 4TW 2E1FQB
NR29 4UR M3JST
NR29 5BX M6NIL
NR29 5ED G1JBZ
NR29 5JY G4KCU
NR29 5LG G0APV
NR29 5LG G6TXQ
NR29 5NF G7VSG
NR29 5NU G0KCB
NR29 5NZ M6ALG
NR29 5PB G4AXA
NR29 5PB G4CMM
NR29 5PZ G0VSB
NR29 5QL G0OBH
NR3 1ND 2E0DRT
NR3 1ND M6MUD
NR3 1NY G0JAG
NR3 1PN M6FEE
NR3 1SE M6MBI
NR3 2AD M0TJL
NR3 2DL G7VUM
NR3 2HT 2E1HOS
NR3 2HT 2E1HOT
NR3 2HW G8AUN
NR3 2LF 2E0YTF
NR3 2LF M3YTF
NR3 2NG M3LXU
NR3 2NQ G0CWD
NR3 2NQ G0WJZ
NR3 2PR M0PPR
NR3 2PS M6EIM
NR3 2QN 2E0YRW
NR3 2QN M3YRW
NR3 2RL M6GVA
NR3 3DT G4YDQ
NR3 3LU M3JWH
NR3 3NN G0SSX
NR3 3NU G0MQI
NR3 3PY M1CRL
NR3 3QD M3WZN
NR3 3TB G4NFF
NR3 3TD M6DDO
NR3 3TD M6YRL
NR3 3TQ M6HDS
NR3 4BE M6ZOC
NR3 4EF G3ZTV
NR3 4QW G0PEQ
NR3 4RX G8XEN
NR3 4TS M1AFQ
NR30 1DU 2E0TWQ
NR30 1DU M6TWQ
NR30 1JG M3NNI
NR30 1JY 2E1BFX
NR30 1JY M1FJG
NR30 1JY M6LDY
NR30 1LY G0GGB
NR30 1QU G3YYQ
NR30 2AT G4TNY
NR30 2RU M0NPQ
NR30 2RW 2E1GVB
NR30 3JU G6ATW
NR30 4BB M3HVN
NR30 4BS 2E1HAU
NR30 4HL M6LBT
NR30 4JU 2E1HVU
NR30 4LD 2E1HVT
NR30 4LD M3RJH
NR30 4LR G4ALC
NR30 4LS M6CHJ
NR30 5AB M0WBR
NR30 5DT G8VQX
NR30 5HG M6AIO
NR30 5PH G4VSD
NR30 5PQ 2E0BNI
NR30 5PQ M3ORZ
NR30 5PQ M3TFZ
NR30 5RH M1BUC
NR30 5RW 2E0GAV
NR30 5RW M6AEG
NR30 5UN 2E0BNT
NR30 5UN M0ZOR
NR30 5UN M3RZJ
NR31 0BU M1CDV
NR31 0HR G4YZF
NR31 0HR G6ZG
NR31 0PW G6VZU
NR31 6EF M0DLG
NR31 6LZ M6DBA
NR31 6PR M6MIL
NR31 6PX M3YQR
NR31 6PX M6MKW
NR31 6PZ M3XEO
NR31 6QY G0SHJ
NR31 6RQ G6FQL
NR31 6RT M6JKS
NR31 7AS 2E1AGQ
NR31 8DN G3WPG
NR31 8DN G4GBT
NR31 8DR G1HNU
NR31 8DR G7LNM
NR31 8JG G3VSV
NR31 8LD M6LGT
NR31 8NT M3MJY
NR31 8PB G7ATW
NR31 8QJ G1GER
NR31 8RG G7RPJ
NR31 8SX G8TAE
NR31 9AR M6CBJ
NR31 9HT G3UPS
NR31 9JT G0FEI
NR31 9JU M1CQR
NR31 9LZ G6KZI
NR31 9NT M6FBZ
NR31 9NX G0GGE
NR31 9NX G0VMK
NR31 9NX G3YRC
NR31 9NX G7MKV
NR31 9PD M6SVY
NR31 9PH G8PIO
NR31 9PP 2E0CBH
NR31 9PP G0BAW
NR31 9PP G0CLG
NR31 9PP M6WOO
NR31 9PQ G7WHM
NR31 9QY M6NTB
NR31 9SB G4VYN
NR31 9TY 2E0JDF
NR31 9TY M0ABG
NR31 9TY M6EKK
NR31 9TY M6SGJ
NR32 1EH G7PWK
NR32 1HZ M6JPK
NR32 1JY 2E1BPN
NR32 1RY G0WCO
NR32 1TP M1ATP
NR32 2AF G0RFI
NR32 2AN M3OYS
NR32 2BZ G3JRM
NR32 2BZ G4RLS
NR32 2BZ G8PIR
NR32 2HW M1SSB
NR32 2JS 2E0PAP
NR32 2JS G3ETP
NR32 2JS M3UGQ
NR32 2LA M6KWC
NR32 2NW M3NNJ
NR32 2QG M3GIF
NR32 2RY G4RMT
NR32 3DN M6BAW
NR32 3DS M0XGK
NR32 3EL G1MPL
NR32 3HS G0VMF
NR32 3JF G3AMK
NR32 3LQ M1CRF
NR32 3NF G3XGK
NR32 3QL 2E0PKY
NR32 3QL M6MRL
NR32 4AU G7TJZ
NR32 4DR G4CPW
NR32 4DU 2E0GGO
NR32 4JU M6HCG
NR32 4LW G8JLW
NR32 4PZ 2E0OBB
NR32 4PZ M3OBB
NR32 4PZ M6KPT
NR32 4QB G3AER

NR32 4QB M0DMR
NR32 4QW M0JUO
NR32 4SU G4HSG
NR32 4UB G6NGA
NR32 5AS G0MEE
NR32 5AS G8RHQ
NR32 5BE 2E0FDG
NR32 5BE M6XBP
NR32 5DJ G4TAD
NR32 5DW G8JUK
NR32 5HJ G4UKX
NR32 5LB G4RKP
NR32 5LL G4FCZ
NR33 0DX 2E0BFT
NR33 0DX M1FRH
NR33 0DX M3AGE
NR33 0LG G8HAU
NR33 0LY G8HRF
NR33 0QN 2E1LJL
NR33 0QN M0LJL
NR33 0TZ G4KDL
NR33 7BQ 2E1OZO
NR33 7BT G7JRU
NR33 7BX G0VDQ
NR33 7DW G0GBV
NR33 7EY M6IBC
NR33 7HH G7MJX
NR33 7PN G6NKS
NR33 7PS M0DCU
NR33 7RR M3SKB
NR33 7SQ G3PGQ
NR33 7TF 2E0NAQ
NR33 7TF M6CPP
NR33 7TR 2E1WCD
NR33 7TR G4CKH
NR33 7TU M6BLP
NR33 8BQ G8FGY
NR33 8BY M6VFR
NR33 8EG 2E0JPW
NR33 8JZ G3WDI
NR33 8NE M3KPQ
NR33 8NN 2E0SJF
NR33 8NN M0SLF
NR33 8PJ G6MCB
NR33 8PJ G7BYI
NR33 8QG G1ZCS
NR33 8QN G0WLS
NR33 8QQ M0DFY
NR33 8SD G7SDM
NR33 8TB 2E0VRT
NR33 8TB M6MPK
NR33 8TP G0JDL
NR33 8WJ M0ALX
NR33 8WP G2DWB
NR33 9AQ G3JMX
NR33 9AW G4ZMH
NR33 9BB M0JTH
NR33 9BB M3NMV
NR33 9BB M6PLO
NR33 9BG M3AGA
NR33 9DH G4HRC
NR33 9DH G4ZST
NR33 9DH G8HRC
NR33 9HA G1HQK
NR33 9HA M3TIK
NR33 9HB G7USP
NR33 9HL G1IAD
NR33 9JE G0HSN
NR33 9LZ M3FVX
NR33 9NX G0ULS
NR33 9PG G7PQB
NR33 9PJ G4ZSX
NR33 9QL G7MWM
NR33 9RH M5TAM
NR34 0AU M0TDK
NR34 0EX G6PSC
NR34 7BJ 2E1DDJ
NR34 7DJ G7RXI
NR34 7PE 2E0CQJ
NR34 7PE M0NAG
NR34 7PE M6NAL
NR34 7PG M6MOP
NR34 7PQ M6LVN
NR34 7QL M1RJG
NR34 7RD G0WQW
NR34 7TH G0LRP
NR34 7UW 2E0VKK
NR34 7UW M0VKK
NR34 7UW M6VKK
NR34 8AQ M3NOS
NR34 8HR M6DAS
NR34 8HR M6PSG
NR34 8RE G1XWX
NR34 8RE M0DSB
NR34 8RN M6GUD
NR34 9AS G7FWE
NR34 9DW M1VLS
NR34 9HN G8BJQ
NR34 9JJ G4XAE
NR34 9NN M0ZKK
NR34 9NN M0DCQ
NR34 9PL G6OAU
NR34 9QT 2E1BLA
NR34 9QT M3BXZ
NR34 9RH 2E0CEH
NR34 9RT G4KFS
NR34 9RT M6GVT
NR34 9YW G3KIJ
NR34 9YW M0DEB
NR35 1BY G6VZF
NR35 1HE G4JUV
NR35 1JU G6LSW
NR35 1LE M1TES
NR35 1PP M6JAR
NR35 1RE G0JKI
NR35 2LZ M0MRO
NR35 2LZ M0NON
NR35 2PH M6LIH
NR35 2QP M1DHT
NR35 2SF 2E0XDI
NR35 2SF M3XDI
NR35 2SF M6PUD
NR4 6AA G3CIM
NR4 6AF M6FHE
NR4 6HA G8ZGS
NR4 6JD G3GLA
NR4 6LT G4NZQ
NR4 6LT G7MKP
NR4 6PS M0SWO
NR4 6QB G0PFN
NR4 6RA M0RLW
NR4 6TZ G4BDW
NR4 6XH M0EVK
NR4 7EA M6GNO
NR4 7ET M0CKL
NR4 7EZ M1ACT
NR4 7HP G8VLL
NR4 7HP M1BKF
NR4 7JA G3SCV
NR4 7LS M6LWP
NR4 7LT G0FDT
NR4 7LZ M6HTJ
NR4 7PD G4VUV
NR4 7PY G0IYY
NR4 7QY M3IGA
NR4 7RY G0CKH
NR4 7SE M0HBL
NR4 7SF G4KAU
NR4 7TJ 2E0CHU
NR4 7TJ M0HEJ
NR4 7TJ M6ATL
NR5 0EH 2E1GTE
NR5 0EH G4ARN
NR5 0EH M3AEZ
NR5 0ES G3HUL
NR5 0EZ M3DKC
NR5 0HB G4ONF
NR5 0HG G6KCG
NR5 0HL 2E0FIJ
NR5 0HL M6FIJ
NR5 0HZ 2E0OCL
NR5 0HZ 2E0ODT
NR5 0HZ 2E0OEE
NR5 0HZ G0NNS
NR5 0HZ G0ODR
NR5 0HZ M0OCL
NR5 0JA G3PEN
NR5 0JA G6TAF
NR5 0JA G7DTR
NR5 0JB G4KQY
NR5 0JD G8UJO
NR5 0LP 2E0IMW
NR5 0LP M0IMW
NR5 0LP M6IMW
NR5 0NE G0UUT
NR5 0NH G6KAI
NR5 8AR G8TAY
NR5 8AS G0SMH
NR5 8AS G7MLC
NR5 8BJ 2E0BWA
NR5 8BJ M6NNE
NR5 8DQ M3BVY
NR5 8ED 2E0GCO
NR5 8ED M3TJL
NR5 8ED M6EXY
NR5 8HW M6XTD
NR5 8RA G4RMN
NR5 8TE 2E0AZF
NR5 9AX G4WTD
NR5 9BH G3THV
NR5 9LB G1DXQ
NR6 5EP G4GPZ
NR6 5HF G4GVR
NR6 5HU 2E0DBS
NR6 5HU M3DBS
NR6 5NN G6UPI
NR6 5PD M6FIG
NR6 5PE G4YEJ
NR6 5PQ G0CUL
NR6 5PQ M6EIZ
NR6 5RQ G0ULI
NR6 5SJ G4VWF
NR6 5SW G4YGD
NR6 6DG M1DPU
NR6 6JF 2E0KNE
NR6 6JF 2E0LSX
NR6 6JF M0LSX
NR6 6JF M3LSX
NR6 6JF M3NKW
NR6 6JF M3UYY
NR6 6LN M3ILY
NR6 6PD G4XNP
NR6 6PL G8SUD
NR6 6PU G0ANN
NR6 6RB G0FKG
NR6 6RZ 2E0JKB
NR6 6RZ M0HDC
NR6 6RZ M0JKB
NR6 6RZ M6JKB
NR6 6RZ M6ZLA
NR6 6TD M3PEQ
NR6 6UA G7DVO
NR6 6UX G7WEW
NR6 6UX M1AEH
NR6 6XD G3IOR
NR6 6XF G4EYT
NR6 7BB G6TGQ
NR6 7BG G6EJM
NR6 7BL G0VDO
NR6 7DP G4WEE
NR6 7DR G4YOA
NR6 7HA 2E0QOT
NR6 7HA M0GSZ
NR6 7HA M6GST
NR6 7HA M6RND
NR6 7HE G4JSD
NR6 7HE G4UIY
NR6 7HR G8EEY
NR6 7HR M0RGF
NR6 7HR M0RGY
NR6 7HR M0WHQ
NR6 7HZ G0UEM
NR6 7LG 2E0JZU
NR6 7LG G4AUV
NR6 7LG M6JZU
NR6 7LJ M1JLM
NR6 7NB M0XWS
NR6 7NL G1NCN
NR6 7PH G0ULG
NR6 7QN G0CLR
NR6 7QN G0NFV
NR6 7RQ G4LSG
NR6 7RQ G4TWS
NR6 7RQ G4TWT
NR7 0AS G6TLP
NR7 0BQ G1KJH
NR7 0DS M1AUW
NR7 0DW G7NFW
NR7 0GY M6DYD
NR7 0JJ G3HQS
NR7 0JJ G4VYH
NR7 0JS M6GUJ
NR7 0LY 2E1IKM
NR7 0NA M1ADX
NR7 0PD M6OOJ
NR7 0PH 2E0HAW
NR7 0PH M3CJX
NR7 0RR 2E0BMI
NR7 0RR M3RQY
NR7 0SF 2E0TAZ
NR7 0SF M3IPD
NR7 0UD M3XEG
NR7 0US M6VST
NR7 0XZ M6NME
NR7 8AA G7AJX
NR7 8AD G3HKD
NR7 8EE G6OUM
NR7 8EG G0UEB
NR7 8EJ G4AJO
NR7 8NQ G4LGB
NR7 8QE G7HHL
NR7 8RG M6LXM
NR7 8TF 2E0YWP
NR7 8TF M3YWP
NR7 8XH G4PCZ
NR7 9AD G7SNC
NR7 9DQ 2E0SYW
NR7 9DQ M6SYW
NR7 9DZ M3WYG
NR7 9HW M3IXM
NR7 9LG G6AIO
NR7 9LL G0BBV
NR7 9LL G4OLP
NR7 9RR 2E0NNE
NR7 9RR M6NNE
NR7 9TW 2E0CIS
NR7 9TW M6BAV
NR7 9TW M6UAS
NR7 9UH M6KDD
NR7 9XU M6XYL
NR8 5AT M6TCE
NR8 5AU G3PZX
NR8 5AU G3YIR
NR8 5AX G6LZM
NR8 5DP G3EZB
NR8 5DT M3PFN
NR8 5EH M6VWN
NR8 6AB G4SVI
NR8 6AB G4XVI
NR8 6BD G7IGR
NR8 6BU M6XAI
NR8 6GA G4URS
NR8 6GA M6YLE
NR8 6GG M6JSJ
NR8 6LD G7NIX
NR8 6LT G7AYQ
NR8 6NF 2E0WJA
NR8 6NF M0HHW
NR8 6NF M6RFV
NR8 6NN G0CBO
NR8 6SL M0MDI
NR8 6TR G7LBO
NR8 6UU M6SBK
NR8 6UZ M3VDZ
NR8 6XZ G7SQW
NR9 3AR G3WHG
NR9 3EP G0DJI
NR9 3EZ G0AHO
NR9 3HQ M6CHT
NR9 3HS M6XTL
NR9 3JP G4NST
NR9 3JP G0WQN
NR9 3JY G4YKZ
NR9 3NR G4TUK
NR9 3NR G6JDW
NR9 4AB G3PTB
NR9 4AD M0TNE
NR9 4AF M3ULB
NR9 4AF M6FJZ
NR9 4HA 2E0JRR
NR9 4JB G3SEM
NR9 4NY 2E1FDM
NR9 4QJ M1EXQ
NR9 4QS G0PSY
NR9 4QS G4ZDT
NR9 4QE G7ENR
NR9 4QZ G8VYH
NR9 5BB M3RWK
NR9 5BY G0TCO
NR9 5EJ M1CQS
NR9 5QT 2E0KPL
NR9 5QT M6LPK
NR9 5SH G3FEW
NR9 5TF G3KVT

## NW (North West London)

NW1 0LJ 2E0BGC
NW1 0LJ M3MUP
NW1 2JL G1YPZ
NW1 5ND M0MLV
NW1 6UE G3LAU
NW1 7ES M6VAR
NW1 8TH G1BNV
NW1 8TL M1EXW
NW1 8UB M5AHF
NW1 9BX G4RPK
NW1 9JH G4MGI
NW1 9TL G3AFT
NW1 9TL G3ZKE
NW1 9TL G8DWL
NW10 0DQ M3UQD
NW10 1BU G0NLN
NW10 1PA M3XAY
NW10 1PL G3THQ
NW10 6QY M0GPQ
NW11 6UP G4PLX
NW11 7AA G4DBQ
NW11 7DL G3GAF
NW11 7DL G4HXX
NW11 8AA M0HFW
NW11 8AA M0YGW
NW11 8SG G3RDG
NW11 8TE G1INJ
NW2 1BP M3OXY
NW2 1DY G6IWZ
NW2 2AR G0VPV
NW2 3SN 2E0NQU
NW2 3SN M0NQU
NW2 3SN M3NQU
NW2 3TL G1DWC
NW2 4PG 2E0TAV
NW2 4PG M0RVT
NW2 4PG M6RPD
NW2 5EH G4PEF
NW2 6RS 2E0EZK
NW2 6RS M6EZK
NW3 1EJ G8PDM
NW3 2DW M0PCX
NW3 2JJ M0GXK
NW3 2LD G4DFX
NW3 4JP G3UGX
NW3 5TY M6TXX
NW3 7UP M0NPA
NW4 1JR G4GHZ
NW4 1LS 2E0CVW
NW4 1LS M0HWD
NW4 1LS M6CWG
NW4 1NP G1GIA
NW4 2NS M3MML
NW4 2PN G3UML
NW4 3NB G3XVH
NW4 3TR 2E0FMX
NW4 3TR M6CZB
NW4 4EN G6CKD
NW4 4EN G8JGE
NW4 4RR G3RDR
NW5 1JL 2E1GSX
NW5 2PE G7TKW
NW5 2RD M6FHD
NW5 4BX M6TUT
NW5 4JB G7UZN
NW5 4SA M6HBG
NW6 1LJ M6KFA
NW6 1NT M6LMT
NW6 2DT G3VCK
NW6 4DG G0KVS
NW6 4EJ 2E0UAU
NW6 4EJ M0SHN
NW6 4EJ M6SHK
NW6 4LY 2E0GMN
NW6 4LY M0ZEE
NW6 4LY M6USD
NW6 5LR M1FIP
NW6 5SL M6EKC
NW6 6PY 2E0HWD
NW6 6PY M0INF
NW6 6PY M6FSJ
NW6 7LA M6MCW
NW7 1JL G3MYG
NW7 1NH 2E0NAR
NW7 1NH M0NRA
NW7 1NS G0JUE
NW7 2AD G1YJH
NW7 2EH G6HIU
NW7 2RE G3WZO
NW7 3LH M0HJW
NW7 3LH M6CJX
NW7 4NH G6LPT
NW7 4QP G1NRM
NW8 7EP M0SEO
NW8 8HR 2E0KBP
NW8 8HR M0KBP
NW8 8SX G3KCT
NW9 0DJ 2E0JSO
NW9 0DJ M0JEO
NW9 5AZ G1IAG
NW9 5BS M0KWB
NW9 5HF G7G3F
NW9 5JP 2E0RVI
NW9 5JP M0RVI
NW9 5JP M6RVI
NW9 5RE M3TKN
NW9 5UQ M3CXX
NW9 6EU G0FAB
NW9 6HT M6CYP
NW9 7PG M6FHB
NW9 8ES M6ESY
NW9 8JL G1AWF
NW9 9NG G1DMS
NW9 9NG G6XLL

## OL (Oldham)

OL1 2JS G0FZF
OL1 2QG 2E0WBG
OL1 2QG M0WBS
OL1 2QG M6WBS
OL1 2UA G0OCF
OL1 4NP M6ZEW
OL1 4PB M3MUO
OL1 4QB M0SSO
OL1 4QU M3MDI
OL10 1AS 2E0YMT
OL10 1AS M6UMC
OL10 2AU G1NXR
OL10 2LH M3EBG
OL10 2LN M5DZH
OL10 3AF G6FJP
OL10 3JB G8JCL
OL10 3JG M6NTL
OL10 3RR G4GSL
OL10 3RR G8TML
OL10 3RS G0RFE
OL10 4DH M6MVN
OL10 4DH M6MZN
OL10 4PW 2E0LJS
OL10 4PW M3YBT
OL10 4QD G7NJE
OL10 4RQ G4TQL
OL10 4ST 2E0MAN
OL11 1DT G7NIH
OL11 1QS 2E1HWI
OL11 1QS M0CXQ
OL11 2BN G3ZRB
OL11 2BN G8EEA
OL11 2EB G0JSF
OL11 2EB M6JCB
OL11 2EL M6XTY
OL11 2LN G4TDW
OL11 2NA 2E0TCO
OL11 2NA M6TKC
OL11 2QB G1ITS
OL11 2XZ 2E0DRG
OL11 2XZ M0ICS
OL11 2XZ M6GRD
OL11 3BU G0EUP
OL11 3JG G4TDU
OL11 3LF G4HHS
OL11 3LG 2E1GGL
OL11 3LG G7BRJ
OL11 3QG 2E0FAC
OL11 3QG M0PEB
OL11 3QG M3SQO
OL11 4AX 2E0DTX
OL11 4AX M0LDZ
OL11 4BT G3TOJ
OL11 4DD G8HDS
OL11 4HW G3DTP
OL11 4PP G3XPI
OL11 4QF M3XGD
OL11 5AW G6CSN
OL11 5JD G0WHL
OL11 5NB G3RLE
OL11 5QA G0LSP
OL11 5QS G6GXK
OL11 5QW G0TFL
OL11 5RT G7TOB
OL11 5TG 2E1DEN
OL11 5UP 2E1OJO
OL11 5XA 2E1GKB
OL11 5XB G6EFO
OL11 5XE 2E1MDC
OL11 5XF M0TVL
OL11 5XN 2E0DHB
OL11 5XW G0WTB
OL11 5YG 2E0AWK
OL11 5YG M5ALA
OL12 0AS G4XKC
OL12 0HY M0UZX
OL12 0QR M0DOO
OL12 0SH G1ZUH
OL12 0SQ M0IEO
OL12 0DJ G1OQJ
OL12 6LZ G1CIA
OL12 6UH M5ADI
OL12 7JN G4GAI
OL12 7JQ 2E1FIX
OL12 7NH G0SIY
OL12 7NJ M3XFZ
OL12 7NX G0OXP
OL12 7RT G7LKR
OL12 7SU G7UCP
OL12 8PX G4TMV
OL12 8QB M0SPK
OL12 8RU M3UBQ
OL12 8SU 2E1TKD
OL12 8SU G8TKD
OL12 8UX M0VUM
OL12 8UX M6XVJ
OL12 9AJ G0VRX
OL12 9LT 2E0FTL
OL12 9NH G6BWK
OL12 9NY M6TYN
OL12 9QP G7OAI
OL12 9SW M6ZCL
OL12 9TY G3RIK
OL12 9UW M6THI
OL12 9UX G4TOZ
OL129PX M3ROU
OL13 8BH 2E0RBC
OL13 8GE M0DVW
OL13 8PX M0CFX
OL13 8QJ 2E0CWB
OL13 8QJ M0TXK
OL13 8QJ M6CQH
OL13 8QL 2E0XCM
OL13 8QL M0LEX
OL13 8RN G8TYY
OL13 9LS 2E1AHU
OL13 9LS G7KLT
OL13 9PH M0CEB
OL13 9RL G7JQF
OL14 5BU G0TVO
OL14 5ET G6LLU
OL14 5JE G7RFO
OL14 5NX G3OLP
OL14 5PH M6MLF
OL14 5PH M6OOC
OL14 5RU G8WWO
OL14 6AX G4KDU
OL14 6AX G4RCJ
OL14 6AX M3FQA
OL14 6BB 2E0GPK
OL14 6BB M6GPK
OL14 6HB M6BZV
OL14 6ND G4TYW
OL14 6NP G3ZMX
OL14 7ET G1JBW
OL14 7NE 2E0SCV
OL14 7NE M6BKI
OL14 7NF G4HSX
OL14 8DS G8IC
OL14 8RJ M6GQB
OL15 0AZ G6NGR
OL15 0BP 2E1DKU
OL15 0DS G0OWH
OL15 0DS G3RJV
OL15 0HQ G7BPM
OL15 0NG G8NQI
OL15 8EB G3DNN
OL15 8EN G0BLM
OL15 8EN G6NCL
OL15 9EF G4OUJ
OL15 9HU M0CQD
OL15 9JE G8TTU
OL15 9LJ 2E0JHP
OL15 9QQ G7LHV
OL16 2DD G0IPN
OL16 2DD M3ULL
OL16 2EG M6EXT
OL16 2PB G0WPO
OL16 2YH M3SMK
OL16 3PJ G0JFP
OL16 3QZ G1FHI
OL16 3UJ G4PJE
OL16 3UN M3BVX
OL16 4AY G0WSY
OL16 4BA G0HXQ
OL16 4EL M6GZR
OL16 4EW M3VZP
OL16 4EZ G7EQK
OL16 4RJ G6TVE
OL16 4SE M5AFE
OL16 4SH G7LFM
OL16 4TY M0FDX
OL16 4UB G0UFD
OL16 4XG M3XHK
OL16 4XP 2E0ZGA
OL16 4XP M0ZAM
OL16 4XP M3ZGA
OL16 4YD G3SAO
OL16 5BB G0FNM
OL16 5BB G0GVS
OL16 5BB G0KZH
OL2 5AS G3WXN
OL2 5HL G8YKO
OL2 5SD G6PBI
OL2 5TQ 2E0MEQ
OL2 5TQ M6MHQ
OL2 6BH G1UTP
OL2 6JY 2E0CHB
OL2 6JY M0EAM
OL2 0NJ G7MXQ
OL2 6PH G1GKR
OL2 6RW G0BJR
OL2 6RW G1ORC
OL2 6RW G4ORC
OL2 6SR 2E0PUS
OL2 6SR M0GIE
OL2 6SR M3LQA
OL2 6TP G0KUY
OL2 6TW G1UKW
OL2 7AA G6EJI
OL2 7AY G4GNG
OL2 7DA G8CZI
OL2 7EG M0AUG
OL2 7JT M0CRZ
OL2 7JT M3SZC
OL2 7NQ M6ATY
OL2 7QA 2E1OBI
OL2 7TX M3AUB
OL2 7YN 2E0JCX
OL2 7YN M0FTA
OL2 7TQ M3OFN
OL2 8BG G6OVX
OL2 8DA G6TCD
OL2 8DR G0MMX
OL2 8EA G7MCT
OL2 8JF G0PVO
OL2 8JJ G7LMI
OL2 8NE M1FEW
OL3 5AP G0GTC
OL3 5DS G8ZZW
OL3 5NU G0EOX
OL3 5PL G1GZK
OL3 5PL G1KFG
OL3 5PW G7ENC
OL3 6EB G4AIW
OL3 7AW G0HFC
OL3 7EJ G3SNA
OL3 7HD 2E0TZM
OL3 7HD M3TMZ
OL4 1AS G4HFG
OL4 1LX G3TJX
OL4 2DJ G1JZU
OL4 2PE G7EBF
OL4 2PN M6MIC
OL4 2PU G0COX
OL4 2PW G4UFG
OL4 2QQ G7ICE
OL4 2TH G1XGW
OL4 3LW G4AVF
OL4 4AZ G0EYU
OL4 4AZ G4EBQ
OL4 4ED G4SSC
OL4 4RY G6DAD
OL4 4SJ G6TAI
OL4 5LF G4PLV
OL4 5SH G0RKE
OL4 5SH G4GLV
OL5 0SN G3RKM
OL5 9DE 2E0SBN
OL5 9DE M6HBB
OL5 9JN G0FCX
OL5 9LQ G6EJT
OL5 9LQ G8GVZ
OL5 9PD G1JAA
OL6 6SD G3WGQ
OL6 6TX G6YBV
OL6 7DU G0SNQ
OL6 7HG 2E0RGY
OL6 7HG 2E0ZEN
OL6 8BP G4RGM
OL6 8BP G6GMR
OL6 8BP M0NCZ
OL6 8BU G6IML
OL6 8DS 2E0OQZ
OL6 8DS M0GMQ
OL6 8DS M3OQZ
OL6 8DT M3KAQ
OL6 8QN G0AQS
OL6 8RJ G7MRO
OL6 8RL M6GNU
OL6 8RP G6BRD
OL6 8SX M3OBN
OL6 8UA M3OHC
OL6 8UT 2E0TJI
OL6 8UT M6ETR
OL6 9AP G3BPQ
OL6 9DU G7DGC
OL6 9EG M6PVC
OL6 9RH G7TSQ
OL7 0DU G1GWJ
OL7 0DU G7NHC
OL7 0JD G4GHB
OL7 9AN G0MRD
OL7 9HN G3WI
OL7 9NL M0AAD
OL7 9QY G3NTF
OL8 2AN M6DKL
OL8 2EZ M6MDB
OL8 2HG 2E0MCK
OL8 2HG M0MBM
OL8 2HG M3UXF
OL8 2HG M6CLA
OL8 2HG M6LEA
OL8 2HL G1HAC
OL8 2LP M3HAC
OL8 2NS M0CLK
OL8 2PD M1BGF
OL8 2XE 2E0FIT
OL8 2XE M0NVQ
OL8 2XE M0RCL
OL8 3AD M6REE
OL8 3BA G0MGL
OL8 3BA G3LGV
OL8 3JH M6CIG
OL8 3LQ M1CZZ
OL8 3NF G7WAE
OL8 3NP M0NTI
OL8 3NP M3KMW
OL8 3RN M3XHD
OL8 3RY M6ULE
OL8 3TX 2E0BZU
OL8 3TX M6PMF
OL8 3TZ M6KCI
OL8 4DJ M0LYN
OL8 4JB M0KGM
OL9 0EJ G8JRZ
OL9 0NT M3NYQ
OL9 0PH M5BGR
OL9 0QX 2E0JSN
OL9 0RF G1SLA
OL9 0RW M1FET
OL9 8AP G4TLW
OL9 8DA M0ZOV
OL9 8HH G3SMZ
OL9 8QD G1MKI
OL9 8QJ 2E0KDE
OL9 8QJ M0KDE
OL9 8RG G6MMS
OL9 9QR G3IGC

## OX (Oxford)

OX1 3BW G7VDJ
OX1 3BW M6FID
OX1 3UN 2E0SQL
OX1 3UN M3PHP
OX1 4EW M3NUB
OX1 5HW G7OGO
OX1 5JP G3SLI
OX1 5LH G8FED
OX1 5LL 2E1ABE
OX1 5LS G4AZA
OX1 5LX G3KCZ
OX1 5LY G0RFS
OX1 5NF G3JFR
OX1 5NH 2E0ADA
OX1 5NH G0REL
OX1 5NH G0RJX
OX1 5NH G0RKG
OX1 5NU M6YFX
OX1 5NZ G6YTW
OX1 5PE G0SPS
OX1 5PE G7KGH
OX1 5PE M6BLA
OX1 5PG 2E0DKG
OX1 5PG M0KLB
OX1 5PG M6ATD
OX1 5RX G4AZT
OX10 0HN G7VIA
OX10 0PT G6NLN
OX10 0QG G1BZW
OX10 0QT G1XNN
OX10 0SD M0WGA
OX10 6AE G3RDF
OX10 6HP G0MUZ
OX10 6HW G0MUJ
OX10 6LX M3LUK
OX10 6RR G3OEB
OX10 6RY G4SOH
OX10 6SH G0SPC
OX10 6SJ G4YSQ
OX10 6SJ G6IMW
OX10 6SL G4ATL
OX10 7DB G6YQT
OX10 7JE M0PIB
OX10 7LB G4ITC
OX10 7LF 2E0BCG
OX10 7LF M0DYU
OX10 7NL G1URJ
OX10 7PJ G0IDF
OX10 7PZ 2E0PWI
OX10 7QB G7IVF
OX10 7RA G1ULR
OX10 8AD M6MYC
OX10 8BN 2E0SFC
OX10 8BN M3UMX
OX10 8HP G0ADH
OX10 8LU G7TNQ
OX10 9ER G4DAT
OX10 9JD G8FKL
OX10 9NE 2E0SBH
OX10 9NE M6SJR
OX10 9PA G7TIB
OX10 9PE G6UXM
OX10 9PE M0BWL
OX10 9PY G6IMQ
OX10 9QB G4CUR
OX10 9QT M1BQT
OX11 0AD G1AIO
OX11 0AD M6CGJ
OX11 0AE G8SMH
OX11 0BP M3ZQC
OX11 0DQ G8HKS
OX11 0DQ M0HKS
OX11 0DX G4MFP
OX11 0ES G0FRO
OX11 0HB G1GGT
OX11 0LU G0FLB
OX11 0PE M3KVW
OX11 0SA G8BFK
OX11 0SB G8ATP
OX11 0SU G1TRF
OX11 0AD M0PQU
OX11 7AX G4PWI
OX11 7DB G4ZKI
OX11 7DD G6VSY
OX11 7DF G0DPK
OX11 7DT G6VMB
OX11 7FG 2E0TTI
OX11 7FG M0MII
OX11 7FG M6TTI
OX11 7JP G1HKI
OX11 7JP M3HKI
OX11 7JQ G3GQQ
OX11 7NG M6FIB
OX11 7SQ G1HQH
OX11 7TZ 2E0GDH
OX11 7TZ 2E0ZOZ
OX11 7TZ M0SXH
OX11 7TZ M6AVM
OX11 7TZ M6DEC
OX11 7TZ M6EMW
OX11 7TZ M6RGO
OX11 7UT G6VLC
OX11 8BP G1SZK
OX11 8BP G4SYL
OX11 8HD G1OWI
OX11 8HD G1SSI
OX11 8HD G7AKP
OX11 8HD M3SSL
OX11 8HP G0BSA
OX11 8HP G0MQM
OX11 8HY M6CRH
OX11 8LG G4EQM
OX11 8SW G1UPX
OX11 8TE G4XYM
OX11 8TX 2E0ZMM
OX11 8TX M3ZMM
OX11 8UD M3XHV
OX11 9BB G8TYX
OX11 9DY M6YHV
OX11 9JX G8CUL
OX11 9JX G8NVI
OX11 9JX M0CUL
OX11 9JX M6YHS
OX11 9NS G4IYC
OX11 9PU G4SAJ
OX11 9PU G6DPI
OX11 9RB 2E0CKQ
OX11 9RB M0HDU
OX11 9RB M6BKO
OX12 0AT G8PTY
OX12 0BD G1NBP
OX12 0BN G3VHE
OX12 0DA M1EWM
OX12 0DA M1JIM
OX12 0HA G3RRS
OX12 0HA G3VPW
OX12 0HA M6HFA
OX12 0HT G1YDG
OX12 0HU M1EHV
OX12 0LD G0ANX
OX12 0LD G3YSQ
OX12 0LJ G0JEK
OX12 0LJ G6LRY
OX12 0NY G0LCB
OX12 0PY G1NDV
OX12 7BG 2E0MTL
OX12 7BG M0TGF
OX12 7BG M6NXY
OX12 7DF G0OAB
OX12 7DF M6XDB
OX12 7DN G4DPA
OX12 7EP 2E0LNU
OX12 7EP G6LNU
OX12 7EP M3LNU
OX12 7EQ M6ROI
OX12 7EW G0HLS
OX12 7HB M0CLH
OX12 7HP G4BGH
OX12 7JA G4LRQ
OX12 7NQ G6UUQ
OX12 7PN G1IYB
OX12 7QF G4XIW
OX12 8DF M6CGZ
OX12 8DJ G6SRY
OX12 8FE G4HLX
OX12 8QW G0CEI
OX12 9AY 2E0RTW
OX12 9DN M1RJL
OX12 9DN M1SFS
OX12 9GD M6PGY
OX12 9GS M0ADN
OX12 9JP M3NTR
OX12 9RG G7WCG
OX12 9RG M0CIY
OX13 5BX G3YFM
OX13 5BZ G4IXW
OX13 5DB G3SSW
OX13 5ER G4VXE
OX13 5HL G3YIK
OX13 5HL G3ZHL
OX13 5HL M6BVV
OX13 5HS G0AOZ
OX13 5JJ G0SXG
OX13 5LN G1FBQ
OX13 5LR G3VHS
OX13 5PF G4BRK
OX13 6BL G7WJZ
OX13 6BX G8UDJ
OX13 6DU G6ZKY
OX13 6LB M0BRE
OX13 6LP G0KMW
OX13 6NU G4AEV
OX13 6PP M3CHU
OX13 6QJ G0LMX
OX13 6RZ G4JLV
OX13 6TF G1KXX
OX14 1DW G1KEI
OX14 1HG M6GYQ
OX14 1JL M3FSG
OX14 1JQ M6PIA
OX14 1NZ M6RST
OX14 1PH G1HJW
OX14 1PR G0IWW
OX14 1PT M1FII
OX14 1QE G0FJI
OX14 1QT G0UCQ
OX14 1QU M0CIW
OX14 1XA G3ZSX
OX14 1XB M0GRA
OX14 1XB M0HEY
OX14 2AB G4DNA
OX14 2BH M6UKI
OX14 2BJ 2E0KHO
OX14 2BJ M0KHO
OX14 2BL G1IVK
OX14 2BL M6HUK
OX14 2DQ M0BRT
OX14 2EP G1OVY
OX14 2ES G7CAG
OX14 2HL G4WXC
OX14 2ND M0AEN
OX14 2NQ G0HOO
OX14 3JF G3HKH
OX14 3NF M0WPT
OX14 3NF M6EPT
OX14 3NL M3PRA
OX14 3PE G6ZHB
OX14 3RX 2E0KEP
OX14 3RX M0KEP
OX14 3RX M6GYY
OX14 3SD G4RKF
OX14 3SR G0OYS
OX14 3SW G7VGA
OX14 3SX G0NPA
OX14 3TD G6MRP
OX14 3TE 2E0CFH
OX14 3TE G8NRP
OX14 3TE M0NRP
OX14 3TE M6YMA
OX14 3TR G4SQI
OX14 3TU G0SLL
OX14 3XB 2E1HGA
OX14 3XB G8KIG
OX14 3XB M3HGA
OX14 3YN M1SLH
OX14 4AY M6KTH
OX14 4DA M3ZAE
OX14 4DF M3XAG
OX14 4EG G3RTM
OX14 4HY 2E1BMJ
OX14 4JX G1BDP
OX14 4LG M1HFM
OX14 4NL G3MZC
OX14 4NN 2E0ZSA
OX14 4NN M3ZSI
OX14 4PU G7LUK
OX14 4QB 2E0MHM
OX14 4QB M6MLQ
OX14 5LW G4JEO
OX14 5NE 2E0UBH
OX14 5NE M6UBH
OX14 5NG G6PBZ
OX14 5NW G0RWJ
OX14 5PP G0OMZ
OX14 5RF 2E1HNN
OX14 5RL G0INZ
OX14 5RN G1OQB
OX14 5RN G7DOE
OX14 5RN G7DOF
OX14 5RQ G6PEP
OX15 0SR G6STD
OX15 4AX M6JFW
OX15 4BB G8JPW
OX15 4DY M6ZZK
OX15 4JE G1PUY
OX15 4JL 2E0FKS
OX15 4JL M3FKS
OX15 4NT 2E0HAN
OX15 4NT M3UHH
OX15 4QU M6SJU
OX15 5AP G0UWB
OX15 5HW 2E0YBS
OX15 5JB M0ZZA
OX15 5LH G7KYH
OX15 6BJ M1BMV
OX15 6BJ M1EDA
OX15 6DS G6SLG
OX16 0BS M3WGK
OX16 0DB M1CNJ
OX16 0LE M3SBA
OX16 0LU G1XRQ
OX16 0QW M6DNL
OX16 0RR 2E0SMS
OX16 0RR 2E0UKG
OX16 0RR G0LDB
OX16 0RR M3SYL
OX16 0RR M3UKP
OX16 0RZ M6CPS
OX16 0SB G7CEB
OX16 0UG G8ZRU
OX16 0XF M3YCJ
OX16 1AA G8ZLN
OX16 1BQ M6FUM
OX16 1BX G7LNK
OX16 1DJ M6SKH
OX16 1EZ M6WTW
OX16 1FA 2E1EHF
OX16 1FA M0RVJ
OX16 1PQ G4XPY
OX16 1UE M1AOG
OX16 1XA M6GMC
OX16 2NF G7UUW
OX16 2PA M0HVK
OX16 3QS 2E1IBJ
OX16 3WS 2E0FTH
OX16 3WS M3SZM
OX16 3WX 2E0DWV
OX16 3WX M0UXO
OX16 3WX M3ITT
OX16 4SD M3YEU
OX16 4SD M6JMG
OX16 5DR 2E0LGO
OX16 5DR M6ZLD
OX16 5DW 2E0TEN
OX16 5DW M0RDI
OX16 5DW M6AMR
OX16 5HQ M6ETJ
OX16 9AJ 2E0TAO
OX16 9AJ M0GUF
OX16 9AJ M3UQI
OX16 9AP G0BJI
OX16 9AR G8KSH



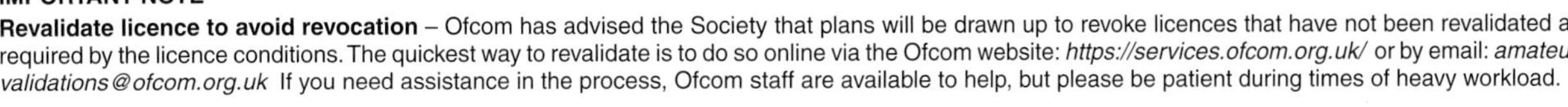

**IMPORTANT NOTE**

**Revalidate licence to avoid revocation** – Ofcom has advised the Society that plans will be drawn up to revoke licences that have not been revalidated as required by the licence conditions. The quickest way to revalidate is to do so online via the Ofcom website: *https://services.ofcom.org.uk/* or by email: *amateur.validations@ofcom.org.uk* If you need assistance in the process, Ofcom staff are available to help, but please be patient during times of heavy workload.

OX16 9AR M3FUD
OX16 9AR M3KSH
OX16 9BQ G1IIO
OX16 9BQ G7ILJ
OX16 9DH G8DCX
OX16 9DP G3MXK
OX16 9EL G6TAS
OX16 9EL M6LVJ
OX16 9ES M1CJZ
OX16 9JU G0BRA
OX16 9JU M0BJN
OX16 9LB G4RWV
OX16 9RZ M0AXN
OX16 9SR G1HXZ
OX16 9ST M6KGE
OX16 9TA G7LKC
OX16 9TW G0MEC
OX16 9TY G0PRT
OX17 1BE G1KKE
OX17 1BU M1CZF
OX17 1DH G6CFU
OX17 1GJ M6CKF
OX17 1JA G1KIB
OX17 1LU M0BRI
OX17 1NG G0MWV
OX17 1NR 2E0TMY
OX17 1NR M0TTH
OX17 1NR M3TTH
OX17 2AB M1AHY
OX17 2DZ M6DMN
OX17 2EH G8CIJ
OX17 2QQ G4DKD
OX17 2RX M6BFX
OX17 3AE G1AHQ
OX17 3AG G8SXQ
OX17 3DG G7VYT
OX17 3FE M6NBX
OX17 3HA M3RZP
OX17 3HR G3YZV
OX17 3LB M3RQJ
OX17 3PF G3OPH
OX17 3RG G6KFR
OX17 3RJ G1BJE
OX17 3RJ M3RUR
OX17 3RS M6ZEK
OX17 3XB G0BIV
OX18 1DX G3KIL
OX18 1ED G7SRB
OX18 1FF G6HEJ
OX18 1TT 2E0CHT
OX18 2HX G0NZE
OX18 2LH 2E1ESK
OX18 2NN G4RJY
OX18 2NW G3LPC
OX18 2NW G3LQC
OX18 3AZ G7KDM
OX18 3LS G1WYD
OX18 3PR G8KOD
OX18 3QZ G6OUI
OX18 3XT G0MFV
OX18 3YB 2E0BHA
OX18 3YB M3SHZ
OX2 0AS 2E0DFO
OX2 0AS M0HWV
OX2 0BB 2E0EDM
OX2 0BB M0TEK
OX2 0BE G3BLS
OX2 0HN 2E0YXB
OX2 0HN M3YXB
OX2 0HS M0MLG
OX2 0HS M3LNY
OX2 0HS M3MLG
OX2 6AY G1OWM
OX2 6HY M6GOC
OX2 6QE G0AXI
OX2 6TX G4ZHE
OX2 6UD M0HDN
OX2 7EN G0FGP
OX2 7EN G7CUB
OX2 7QG G6SRT
OX2 8AX 2E0HNK
OX2 8AX M6HNB
OX2 8BB 2E0RNM
OX2 8BB M3RNM
OX2 8BY G3XGC
OX2 8EA G3LPU
OX2 8EA G8PX
OX2 8PE G0AZP
OX2 8PE G6YSQ
OX2 8PX G4ZSV
OX2 8QT G1JXS
OX2 9DN G1VSO
OX2 9DS G6RXV
OX2 9HZ G3UOM
OX2 9JG G6ASA
OX2 9PA G1LVW
OX2 9PE 2E0FZJ
OX2 9PE 2E0RQN
OX2 9PE M0JHF
OX2 9PE M0RQN
OX2 9PE M6FZJ
OX2 9PE M6RQN
OX2 9PR M0OXD
OX2 9PR M3OXD
OX2 9QA G4MQR
OX2 9SN G0CFN
OX2 9SW 2E0FGH
OX2 9SW M6FGH
OX20 1JT M6GBJ
OX20 1JY G4POL
OX20 1RZ G6HRA
OX25 1PX G1FZR
OX25 2AH G7AHB
OX25 2PS G0WSA

OX25 4AW G3NSP
OX25 4SQ M0DKU
OX25 5TH M6PLR
OX25 6LP G0REV
OX26 1TZ G4ATG
OX26 2AR G7ILP
OX26 2BE G8EKN
OX26 2DZ G1CEO
OX26 2GS G4SYV
OX26 2LP 2E0CYT
OX26 2LP M6CWZ
OX26 2LR G4BKB
OX26 3FD M3CVW
OX26 3GA 2E0JCU
OX26 3SR 2E0DAB
OX26 3SR M3DAB
OX26 3YB G6HPT
OX26 3YL G0GLG
OX26 3ZJ G3UTE
OX26 4HB G0TSK
OX26 4UH G6PLF
OX26 4XA M3GRF
OX26 4XJ 2E0GBE
OX26 4XX M1TAZ
OX26 4YA 2E0PSC
OX26 4YA M0PSC
OX26 4YA M6PSC
OX26 5DW G6FJO
OX26 5DX G0LUQ
OX26 6SR G0JRY
OX26 6UQ G6FPP
OX26 6YG M6DTH
OX26 6YP G1BYQ
OX26 6YR G8UWD
OX27 0AP G7DBN
OX27 0EY G1FXC
OX27 0HJ G4IHX
OX27 0HP G6PHH
OX27 7PA 2E0BPN
OX27 7PA M3ZVX
OX27 8DF G4FTA
OX27 8DX G1WMK
OX27 8FL M0GKA
OX27 8FQ M0BWS
OX27 8TU M0CJN
OX27 9AZ G4BII
OX28 1HQ G4IOK
OX28 1NS G4EGN
OX28 2EG M0BLV
OX28 2EG M0BRP
OX28 2JB G4GUN
OX28 3HH M6NRM
OX28 5JQ G7CKP
OX28 5JQ G8LKB
OX28 5JT G3BNP
OX28 5NG G0CQZ
OX28 5NL G4YUA
OX28 6EG G8RDA
OX29 0RQ G6UPQ
OX29 0RT G6ZNT
OX29 4BU 2E0BMH
OX29 4BU G2KQ
OX29 4BU M3SSO
OX29 4EW G1RCX
OX29 4EW G6LDO
OX29 4LU G7OQQ
OX29 4NT G6ANV
OX29 4QJ G0BQZ
OX29 4QR M3CUK
OX29 4QY G8JVI
OX29 5RH G3ZYR
OX29 6SQ G1DOL
OX29 6UW G8EWT
OX29 7PA G4DSE
OX29 7PB M3NGE
OX29 7TH G3UKC
OX29 7TH M1FRB
OX29 7YD G1VBL
OX29 8HH G1ZBG
OX29 8JP M1ENJ
OX29 8JP M1ENK
OX29 8JP M3ENJ
OX29 9UW G7VIB
OX3 0EX G0BPX
OX3 0JG G6DVE
OX3 0JJ G1OYF
OX3 0NW 2E0CDN
OX3 0NW M6HMU
OX3 0RD G0AGJ
OX3 0RJ G3UGJ
OX3 0SJ M0BJE
OX3 7EE G0LBZ
OX3 7EE M3FHV
OX3 7RR M0DDT
OX3 8AA 2E0ETV
OX3 8AA M0ZRX
OX3 8AA M6FMZ
OX3 8AX M1FFY
OX3 8ED G4KSQ
OX3 8HX G0KPY
OX3 8JA G4AYR
OX3 8JB G3KZU
OX3 8LP 2E0OXF
OX3 8LP M6CFW
OX3 8LT M6EYT
OX3 8LT M6RXE
OX3 8PD G3UJO
OX3 8TA G3UMF
OX3 9LQ 2E1IGG
OX3 9PB M3EGU
OX3 9QF G0AFY
OX3 9RZ G4FON
OX33 1EN G8KNJ
OX33 1RQ M3JMW

OX33 1SQ G6DBU
OX33 1TJ G0HZQ
OX33 1TR M0LTK
OX33 1YL G4LLL
OX39 4AA G0ODQ
OX39 4AE G0HHN
OX39 4DZ G4FRL
OX39 4EA G4RWI
OX39 4EU G1ZRR
OX39 4EW G0IKI
OX39 4HB G0IIB
OX39 4HP 2E0PCC
OX39 4HP M6EHP
OX39 4JY G1GQJ
OX39 4JY G7AGR
OX39 4JY G8GQJ
OX39 4PE M1BXQ
OX39 4PL G3AKN
OX39 4PL G3UNI
OX39 4RA G3ZRY
OX39 4TP G0FRN
OX39 4UD G7IKS
OX4 1BY G0CPP
OX4 1EB 2E0LLM
OX4 1EB M3LLM
OX4 1LT G3LOF
OX4 2HF M6MTA
OX4 2JW 2E0OKS
OX4 2JW M6BHM
OX4 3BX M0OXR
OX4 3HT G1PCU
OX4 3NU G7IMH
OX4 3NW G0HEN
OX4 3QU M0ACU
OX4 3RB G1YDI
OX4 3SA M6WHZ
OX4 3SW G3HJS
OX4 4AR G0XBG
OX4 4BB M6NMW
OX4 4NE G6BTB
OX4 4NS M6RLC
OX4 4PW G8OMW
OX4 4XG M3WVB
OX4 6DL G7TLD
OX4 6DX G4KON
OX4 6ER G6ZPL
OX4 6ER M6ZPL
OX4 6RA M3MSY
OX4 6SS M3LVM
OX4 7FD G0TYZ
OX4 7GU G4OBB
OX44 7LH G3ZKO
OX44 7RZ M3ZFH
OX44 7SJ 2E0DBH
OX44 7SJ M3ZFN
OX44 7TA M1GGG
OX44 7TA M6VKE
OX44 7TJ G3SBM
OX44 7TN G4DAP
OX44 7YH M6BOQ
OX44 7YN G0MCO
OX44 9DL G1YSX
OX44 9DL G6ZTR
OX44 9EX 2E1GSW
OX44 9HP G8ECG
OX49 5AD G8LGM
OX49 5LW G7NTO
OX49 5PR M6HHH
OX49 5RA 2E0CBN
OX49 5RA M0HBW
OX49 5RA M6MBE
OX49 5RF G0DUF
OX5 1DE M3VQG
OX5 1HF M3IAP
OX5 1HH G0THY
OX5 1NX M6RAR
OX5 1TR G0LXW
OX5 2EQ G7OIT
OX5 2EU G0EDY
OX5 2EU G7KGI
OX5 2LL G0LUN
OX5 2UB G3MSO
OX5 2US G8JOX
OX5 2XP G8BNB
OX5 2XZ G0XGM
OX6 4JR G1XTA
OX7 3HB G6GEV
OX7 5DZ G0NYH
OX7 5JS G0JZE
OX7 5PZ M6AGI
OX7 5QH M6CUN
OX7 5SW G4TLS
OX7 5XS G1RLA
OX7 6AD G6KCJ
OX7 6EA M6EZY
OX7 6ER G8BII
OX7 6GE G8DJT
OX7 6JN G8GJG
OX7 6LS G0BXS
OX7 6QA G6NLG
OX7 7BP 2E0DKJ
OX8 6PZ G0MKY
OX8 6PZ G0PYF
OX8 8EU G8PKG
OX9 2AX G0THH
OX9 2EE M6KIE
OX9 2EJ M6OXF
OX9 2NE G0MVW
OX9 3BJ G8AAD
OX9 3DG G6GHU
OX9 3EP M6RPP
OX9 3JF G8NRR
OX9 3JS G0FOK
OX9 3NH G4HXU
OX9 3NJ 2E0SSD

OX9 3NJ M6YPX
OX9 3NQ G0UUR
OX9 3NQ G4ADR
OX9 3NQ M6TXU
OX9 3PL G0VLI
OX9 3QL G1GGK
OX9 3TE 2E0BIO
OX9 3XN M6AZR
OX9 3YD M6NOK
OX9 3ZH M6AMW
OX9 4JY G1ECE

## PA

## (Paisley)

PA1 1GT GM0SCW
PA1 1RH 2M0HCF
PA1 2JE GM6PCW
PA1 2QT MM6WER
PA1 2SX GM0UKD
PA1 3NG GM4VLX
PA1 3RT GM4MPY
PA1 3SR GM1SXX
PA10 2EL MM6DCT
PA11 3NB MM3NGV
PA11 3PQ 2M0JSF
PA12 4EG GM4FLX
PA12 4HL GM0KMA
PA12 4NB MM0WOA
PA13 4BB GM0JVC
PA13 4DE 2M0FSF
PA13 4DE MM0GTG
PA13 4HP MM6WHW
PA13 4JJ GM7ALI
PA13 4PY GM7ADU
PA13 4PY GM7ADY
PA14 5AT MM3UNQ
PA14 5LH 2M0ROT
PA14 5LH MM6ROT
PA14 5QG GM3SUZ
PA14 5UT MM3YIH
PA14 5XF GM0WDF
PA14 6DT 2M0MOK
PA14 6DT MM6WUL
PA14 6EF MM6MVM
PA14 6JE MM0RHH
PA14 6JE MM3RHH
PA14 6LF GM0CZM
PA14 6PL MM6SGO
PA14 6XH MM1GAR
PA14 6XP GM4HRJ
PA15 2EE GM4EGD
PA15 2HJ MM6IAB
PA15 2JD MM6KIW
PA15 4DW 2M0OLK
PA15 4HH GM4FBU
PA15 4HW MM3MZX
PA15 4NF GM7VCV
PA15 4SX MM0XPZ
PA15 4SX MM3SWG
PA15 4TA MM3UXO
PA16 0DS MM0CDW
PA16 0EL MM0STU
PA16 0EW GM0GNK
PA16 0EW GM3YOR
PA16 0HX MM1AUG
PA16 0QF MM6LCL
PA16 0YL MM3YIG
PA16 0YL MM3YIQ
PA16 7AL MM3WYI
PA16 7BJ MM3IBM
PA16 7BL 2M0BOS
PA16 7BL MM3MHQ
PA16 7QE GM3KJI
PA16 8AS GM6LEZ
PA16 8NP MM0KNE
PA16 8NP MM0SRX
PA16 8QB MM6OOK
PA16 8QG GM3YLB
PA16 9AZ GM0KYU
PA16 9BG 2M0NFZ
PA16 9DH GM3ZXG
PA16 9DN MM6BFH
PA16 9JG MM6ROV
PA16 9JG MM0JET
PA16 9JG MM3YCI
PA17 5DX GM0MDD
PA17 5DX GM0VKG
PA18 6DX GM0ADF
PA18 6DX GM0EUM
PA18 6DX GM0WSR
PA18 6DX MM6AHN
PA18 6EB MM3XLQ
PA19 1DY MM3WYM
PA19 1EJ MM3YQX
PA19 1EW GM4SVW
PA19 1HF 2M1EJI
PA19 1HF MM1AWV
PA19 1HY GM0HPU
PA19 1HY GM8YUI
PA19 1HY MM1AUF
PA19 1JE MM3YUS
PA19 1JS GM4CEA
PA19 1SG 2M0BEL
PA19 1SG MM3KSV
PA19 1UW MM6YNG
PA19 1XE GM0ATQ
PA19 1XG 2M0IOB
PA19 1XG MM3WWV
PA19 1XS MM3UIX
PA19 1XU MM3YSJ
PA19 1YA GM0IOY
PA2 0AR GM0KTJ

PA2 0BD GM1TDT
PA2 0BE MM3GEW
PA2 0DP MM6KHA
PA2 0NG MM3EKL
PA2 0RP 2M0OXQ
PA2 0RP GM0PYM
PA2 0RP MM0PAZ
PA2 0RP MM3OXQ
PA2 0RP MM6LPT
PA2 0RP MM6SWT
PA2 6DB MM1SYD
PA2 6ER MM0ZZO
PA2 6JT GM0UOU
PA2 6RA GM4EQY
PA2 6SD MM6WGT
PA2 6UJ MM3HQC
PA2 7NU GM4YAA
PA2 7QY GM0BCY
PA2 7RL 2M0BIN
PA2 7RL MM3VVZ
PA2 7RP GM4BRM
PA2 8AW GM3DDL
PA2 8BY GM4LHQ
PA2 8EB GM3RPM
PA2 8LG MM0PJQ
PA2 8QW GM0NAE
PA2 9AJ 2M0KZI
PA2 9JW MM3YMM
PA2 9NY MM6CEL
PA2 9RD GM8TVV
PA20 0DY GM6OLM
PA20 0QT 2M0CGE
PA20 9EU GM0ELL
PA20 9HT MM3FGI
PA20 9LP MM6KSJ
PA20 9NW GM0FGI
PA20 9NW GM8SBH
PA21 2DH GM0PXR
PA23 7DP MM3IMC
PA23 7EW GM4EAW
PA23 7HX MM6IPP
PA23 7JJ MM1HMV
PA23 7JJ MM3MHS
PA23 7UA GM7KFS
PA23 7UP GM8PZR
PA23 8AX GM3YLD
PA23 8DG MM6DSD
PA23 8EA 2M0FSB
PA23 8EA MM6FSB
PA23 8FG GM7TFN
PA23 8FG MM3CJP
PA23 8JR MM3DCN
PA23 8LQ GM4NSL
PA23 8NA GM6HLT
PA23 8PQ GM4SFT
PA23 8QT GM4HPK
PA23 8SG GM8RGO
PA24 8AD GM7NVG
PA27 8BX MM3ROV
PA27 8BZ GM0UPE
PA27 8DB GM6NLB
PA27 8DE GM1USN
PA27 8DE MM3USN
PA28 6EN GM0GON
PA28 6GY MM0BED
PA28 6NZ MM0GTX
PA28 6PG MM6GHH
PA28 6PL GM1PKN
PA28 6PL MM0GHG
PA28 6RX GM7BAS
PA28 6RZ GM4HUL
PA28 6ST MM0EAR
PA28 6ST MM0KRC
PA28 6ST MM3RCX
PA29 6TF GM7OSQ
PA29 6XY GM4ZVF
PA29 6YL MM6PHO
PA3 1BE MM3LVT
PA3 1JW 2M0FDN
PA3 2LR 2M0EDY
PA3 3RT MM0RLN
PA3 3ST GM3WNB
PA3 3SU GM7UFN
PA3 3SU GM7UVS
PA3 4LL MM6KHI
PA3 4LT MM0LEN
PA3 4ST MM6COF
PA30 8FB GM4VXA
PA30 8HH GM4VYQ
PA31 8AG GM1PWL
PA31 8LG 2M0ONS
PA31 8LG MM6MMG
PA31 8QA GM4JLD
PA31 8QA GM4OHT
PA31 8QA MM5DOG
PA31 8QL GM0BPF
PA31 8QL GM4HNK
PA31 8RQ MM1FEO
PA31 8SQ GM8ICC
PA32 8TW MM0WAX
PA34 4BG GM0NTR
PA34 4HZ MM3HMM
PA34 4QT GM0EWU
PA34 4RP MM3GQR
PA34 4RX MM3IMK
PA34 4SF 2M0WWM
PA34 4SF MM0WWM
PA34 4SF MM6WWM
PA34 4YB GM0ERV
PA34 4YL GM0MNW
PA34 5AR MM0TOB
PA34 5ED MM6RWI
PA34 5EF MM3PKZ
PA34 5JQ GM4MTI
PA34 5JR MM3IJI

PA34 5NA MM0DCC
PA34 5NS MM3XUY
PA34 5PB MM0WRO
PA35 1HD GM0LRA
PA35 1HD GM3RTJ
PA35 1HG 2M1ELU
PA35 1HY GM0EQW
PA35 1JJ MM3IKS
PA35 1JQ GM0FHS
PA37 1PJ MM6MVI
PA37 1PY GM1FPD
PA37 1QJ MM3FUG
PA37 1QQ MM3FGH
PA37 1QS MM0NDM
PA37 1SG MM0AHC
PA37 1SG MM6CLC
PA37 1SR GM4AXS
PA4 0EG GM0LKS
PA4 0NE MM6ABO
PA4 0NP 2M0TXA
PA4 0SL GM8IID
PA4 0XA GM7MBB
PA4 8FB 2M0STV
PA4 8FB MM0WRX
PA4 8XS MM3NYY
PA4 9DD GM7VHQ
PA43 7HZ MM6EBJ
PA44 7PZ GM1DCB
PA44 7PZ GM3AEI
PA44 7PZ GM3PAK
PA49 7UN 2M0UMH
PA49 7UN MM0UMH
PA49 7UN MM6LMH
PA5 0EX GM0EDJ
PA5 8AR GM3VCR
PA5 8HR MM3TNF
PA5 8LH 2M0RDT
PA5 8LH MM6RDT
PA5 8RZ MM6CBI
PA5 8TA GM8JET
PA5 8UB GM7DFI
PA5 9AD GM4FDM
PA5 9BE MM5TUW
PA5 9BH GM1JHU
PA5 9LJ MM6TGB
PA5 9NE GM0CHM
PA5 9NR 2M0JAL
PA5 9NR MM3AQW
PA5 9NS MM5AII
PA5 9RG MM3SWU
PA6 7DA GM4GZQ
PA6 7DX MM5AII
PA6 7HJ 2M0BZZ
PA6 7HJ MM3XWS
PA60 7XG MM6JDN
PA65 6BG GM0PRO
PA7 5DT GM7OAW
PA7 5DT GM7WOS
PA7 5DT MM3OAW
PA7 5ES GM8OAH
PA7 5HE GM3UWX
PA7 5HW GM3NLB
PA70 6HB GM4PMK
PA72 6JU GM0MHE
PA75 6GN GM7NYB
PA75 6PX GM4EHB
PA75 6QL MM6SEI
PA77 6TW MM3GOX
PA77 6TW MM3GOY
PA77 6TW MM3GPB
PA77 6UA MM3GMP
PA77 6UE 2M0GOE
PA77 6UE MM3GOE
PA77 6UL MM3GOT
PA77 6UT GM3PGY
PA78 6TB MM0JUL
PA8 6BJ MM1JWF
PA8 6BP MM3WZH
PA8 6EE MM6BLY
PA8 6HG GM0GDD
PA8 6JN MM6CMY
PA8 7DH GM1JDJ
PA8 7HX GM4AEL
PA80 5XS GM4RKM
PA80 5XW MM0GFR
PA9 1BJ MM1HMZ
PA9 1BP GM1VYG

## PE

## (Peterborough)

PE1 2EJ M3YSS
PE1 3DS G3PQB
PE1 3DT G4BVM
PE1 3DZ 2E0SAZ
PE1 3DZ M3OWZ
PE1 3JJ G4DSP
PE1 3JJ M1CDL
PE1 4DR G1FSX
PE1 4EE G6UST
PE1 4HA G4FKQ
PE1 4LR M6PHZ
PE1 4LT G8CKV
PE1 4RA G4BBA
PE1 4SN G3RDZ
PE1 4UR G1ZHL
PE1 5HF G4FKG
PE1 5HG G1XVT
PE1 5XD M1HQX
PE1 5XE G0FHO
PE10 0BZ G7RFE
PE10 0EE G4WJM
PE10 0ES G3MMS
PE10 0FG M6HOV

PE10 0HT G4NIZ
PE10 0HZ G4GMS
PE10 0LS G1IXV
PE10 0NT G8NWM
PE10 0NW G6URT
PE10 0QE G6VEY
PE10 0RB G8VVC
PE10 0RN G7RFD
PE10 0SG G6SSN
PE10 0SG G7GRC
PE10 0SR G7NUM
PE10 0SR M1DYK
PE10 0SR M1DYL
PE10 0TU G4DHF
PE10 0TU G6LI
PE10 0UR G7AHO
PE10 0UY G4HLB
PE10 0XR G4LEX
PE10 0XW M6DKU
PE10 9AU G0TIS
PE10 9BX G3HCQ
PE10 9BX G4FHU
PE10 9HE M3MJE
PE10 9HE M3SKC
PE10 9JJ G0HSD
PE10 9JX G0OTE
PE10 9NS M3RCE
PE10 9PY M6CEU
PE10 9QT G6SKS
PE10 9QT G6SKT
PE10 9QY G0ERW
PE10 9SB G0BHU
PE10 9SQ G4EMK
PE10 9SQ G8ILU
PE10 9TT G1CPM
PE10 9XL G8PUE
PE10 9XL G8TXX
PE11 1GZ G4CEY
PE11 1HD M0CTM
PE11 1HQ G1LSB
PE11 1JQ M0DNF
PE11 1PD G8NWC
PE11 1QB G4ZXZ
PE11 1QF 2E0JEE
PE11 1UU M0PAC
PE11 2BJ G4XWW
PE11 2FG G0RNV
PE11 2HE G0JUR
PE11 2NA 2E0BOZ
PE11 2NA M3WUE
PE11 2RT M0MFC
PE11 2UN G0UOQ
PE11 2UW G0HGH
PE11 2YU G7FML
PE11 3AB G8VQK
PE11 3AF G4NQS
PE11 3BD M6UUE
PE11 3FA G7NGQ
PE11 3LL G0NHF
PE11 3NJ G4BMH
PE11 3PN G6UHS
PE11 3PU 2E0GRR
PE11 3PU G8BYB
PE11 3QW G0HXR
PE11 3QW G0JPI
PE11 3SG M1BMW
PE11 3TG M0WQR
PE11 4AX G1DSP
PE11 4AX G4NBR
PE11 4AX G4OO
PE11 4BA M0DJA
PE11 4BA M1DJA
PE11 4BQ G8KGR
PE11 4DA G3UYE
PE11 4DH M0HZA
PE11 4DQ G7CWM
PE11 4EU G3VFF
PE11 4HG G3SJR
PE11 4HG M3DIS
PE11 4ND G8FMX
PE11 4NE G7RTL
PE11 4NE M0NEL
PE11 4QB G1ZJP
PE11 4QB M1MHZ
PE11 4QQ G3UDP
PE11 4RF G1ELK
PE11 4XH G3NHF
PE11 4YA G7THF
PE12 0DE M3PYI
PE12 0EE M3WPP
PE12 0EG G1TUL
PE12 0EG M3SHN
PE12 0EZ G3REH
PE12 0HP G0YKC
PE12 0JG G0OMM
PE12 0NP G1KLI
PE12 0PQ 2E0DHW
PE12 0PQ M0ZLH
PE12 0PQ M6AWS
PE12 0QR 2E0LCN
PE12 0RU G1JKO
PE12 0TY G8FPW
PE12 0XF M0GTS
PE12 0XN G0FDJ
PE12 0XN G6NVC
PE12 6AD G7AKV
PE12 6AD G8PQB
PE12 6AH G6UKC
PE12 6AH M0PLX
PE12 6BG G3XDA
PE12 6BX 2E0BGX
PE12 6BX M3HZO
PE12 6DA G3VPR
PE12 6DH G4ODA
PE12 6LD M0KRU

PE12 6PN G1PAT
PE12 6PS G4SJV
PE12 6PS G7CUP
PE12 6PT G0LWC
PE12 7AU M0KCE
PE12 7BT G1WRC
PE12 7BT M0CKE
PE12 7BT M3DEJ
PE12 7BT M5ARC
PE12 7BT M6ZTO
PE12 7DQ M3KKX
PE12 7HR G1WYC
PE12 7JS G1JOL
PE12 7JS G7RRD
PE12 7NG G8MEE
PE12 7PP G7HCR
PE12 7QD G6DDU
PE12 8BA G0SHC
PE12 8BA G1SCT
PE12 8DH 2E0TUB
PE12 8DH 2E0TUS
PE12 8DH M6TJN
PE12 8DH M6TJO
PE12 8JT G8ETD
PE12 8NA G6VPN
PE12 8QQ G4PST
PE12 8RW M0GTR
PE12 8RW M3SKQ
PE12 9AR G6HEE
PE12 9BT M6XMX
PE12 9DS G8LZK
PE12 9DS M6OAD
PE12 9EG G1SRD
PE12 9EG M3JNT
PE12 9EN G3YPZ
PE12 9HS G4ZID
PE12 9LQ G0WKN
PE12 9LQ G3RKQ
PE12 9NT M0BFT
PE12 9PF G8XND
PE12 9PJ G4RMJ
PE12 9PJ M6XCO
PE12 9QU G4TPS
PE12 9RG G4KPE
PE12 9RG G6DHT
PE12 9RH G7FYV
PE12 9YF G0BEJ
PE12 9YF M0CKI
PE12 9YF M3DLE
PE12 9YF M3OJP
PE12 9YR G7BZC
PE13 1PZ G1VYB
PE13 1SW G4UQN
PE13 2EG G8HQO
PE13 2EL G1HJP
PE13 2JD G6NNK
PE13 2JD M0CNX
PE13 2JD M3RGN
PE13 2JR G7TFL
PE13 2JU M0EAY
PE13 2QA G8NIL
PE13 2RJ 2E0MAB
PE13 2RJ G8NED
PE13 2RJ M0DUQ
PE13 2RJ M3MBC
PE13 3ED M0BII
PE13 3EJ M3SNO
PE13 3EJ M3WPS
PE13 3HF M1BIG
PE13 3JR M0CMK
PE13 3LF 2E0RGB
PE13 3LQ 2E0USC
PE13 3LQ M3USC
PE13 3NQ G6XMU
PE13 3PA 2E0URF
PE13 3PA M3URF
PE13 3QH G8BIJ
PE13 3RH M1AZA
PE13 3RP G2AXQ
PE13 3RP G8GIN
PE13 3RR 2E0DEP
PE13 3RR M3NBH
PE13 3RR M3WUJ
PE13 3TU M6BKK
PE13 3UD M1KGL
PE13 3UN G1FRM
PE13 3UT M1AMJ
PE13 4AR M0RAU
PE13 4AR M6AVJ
PE13 4ER G4NFY
PE13 4HS M0BNP
PE13 4HU G1ATQ
PE13 4JU G8RZN
PE13 4LB 2E0WNM
PE13 4LB M6WNM
PE13 4LJ G0SXM
PE13 4QQ M0CNW
PE13 4RG G7HHQ
PE13 4RT G7INY
PE13 4RX M3OVZ
PE13 4SQ 2E0LTJ
PE13 4TL 2E1HSD
PE13 5AE G8YXJ
PE13 5AS G1ULP
PE13 5AS M3LDE
PE13 5BD G1HQW
PE13 5BP G0OPC
PE13 5BP M3TXJ
PE13 5DR G0XTA
PE13 5DW M1AYU
PE13 5ET G4KKR
PE13 5ET M3RJP
PE13 5HH G3EUG
PE13 5HH G7AIR
PE13 5HH G7EUG

PE13 5HT G4OKH
PE13 5LG G0MFT
PE13 5PQ G8HDK
PE13 5QR G6LTN
PE13 5QR M0BSZ
PE14 0BA G7LWU
PE14 0DF G4FCI
PE14 0ND G8MLA
PE14 0ND M3MLA
PE14 0NR M3KWF
PE14 7BJ M6HTM
PE14 7DF G1KSI
PE14 7DN G6OEJ
PE14 7EJ G7JJX
PE14 7HU G7LPW
PE14 7JF G4ORJ
PE14 7LT G3RCQ
PE14 7LT M3DLC
PE14 7LT M3RCQ
PE14 7LZ G0WDK
PE14 7LZ M6BAA
PE14 7NP M1AEV
PE14 7PQ G7HDU
PE14 8AY M1CZM
PE14 8AY M3CZM
PE14 8DQ G1MHN
PE14 8ES G7UFI
PE14 8JB G8MXQ
PE14 8JR G6PJE
PE14 8JT M0MSX
PE14 8JT M3PYD
PE14 8RQ M3GSQ
PE14 9JR G0EYM
PE14 9JT G0FND
PE14 9NY M3AKR
PE14 9PU M0MVB
PE14 9QE G0SDM
PE14 9QE M3VZZ
PE14 9RF G8EEK
PE14 9RG G4NVL
PE14 9RG G8SWW
PE14 9RG M3HQU
PE14 9RG M3TUD
PE15 0AJ G6NHK
PE15 0AU G1JEA
PE15 0FE G1EXM
PE15 0GA M6DGR
PE15 0LN M3GZW
PE15 0PE G7IFD
PE15 0PJ 2E0JDK
PE15 0PJ M6JHZ
PE15 0QF G1OQX
PE15 0QF G6SXB
PE15 0RR G3YLY
PE15 0RW G4AJE
PE15 0SA M0GAA
PE15 0SB G8OXE
PE15 0SE G7NCG
PE15 0SE M0BRM
PE15 0SP G7ADS
PE15 0SP M0BLR
PE15 0SR 2E0OVL
PE15 0TW G3PYE
PE15 0TW G6PYE
PE15 0TW G8IER
PE15 0UH M6DDM
PE15 0XQ G0INO
PE15 0YF G7TBX
PE15 8DD G7PJW
PE15 8EL G3PMH
PE15 8EL G8PHS
PE15 8HF G7EVP
PE15 8JH G6BOP
PE15 8QA G7NEM
PE15 8RY G0PER
PE15 8SN G0FLP
PE15 9AG G0GJL
PE15 9AH G6JVX
PE15 9AP G3VQG
PE15 9DH G6MMB
PE15 9DL G1VTO
PE15 9DT G0OYZ
PE15 9EA G0BXJ
PE15 9EQ G3VMI
PE15 9NN M3YGQ
PE15 9QA 2E0TGF
PE15 9QA M3BIB
PE16 6AY M0HTA
PE16 6BD G4GPQ
PE16 6BJ M6JWE
PE16 6BW M6DPP
PE16 6DZ G3ESK
PE16 6JF G0GLJ
PE16 6JN G0MQL
PE16 6LG G4ZUH
PE16 6NB M3RKZ
PE16 6QE G6RBR
PE16 6RN 2E0OPU
PE16 6RN M3OPU
PE16 6RX G1YFE
PE16 6RX M6LYY
PE16 6SF G7MHV
PE16 6SG M6TFK
PE16 6TP G6OHM
PE16 6TP M3FTV
PE16 6TP M6NIE
PE16 6UR G7JXT
PE16 6UR G7VQC
PE16 6UX G0HKE
PE17 2TB G1HJD
PE17 6YB G1OSJ
PE19 1AE G1CSO
PE19 1DT 2E0OTY
PE19 1DT M6ODY
PE19 1HA M6YWO

PE19 1JU G8WRV
PE19 1LD G7HMZ
PE19 1NP M1FGO
PE19 1UF G8FDE
PE19 1UF G8FFC
PE19 2DT 2E0TMC
PE19 2DT M6TMC
PE19 2EN 2E0BRS
PE19 2EN M3PYW
PE19 2NN G8BKG
PE19 2NN M0DFW
PE19 2NN M0OLG
PE19 2QF G7JUC
PE19 2QF M3JUC
PE19 2QF M6JUC
PE19 2QY G6OXI
PE19 4UE G0ABW
PE19 5DQ G7LFZ
PE19 5RD G6JAK
PE19 5SE M0SRS
PE19 5SF G1YVS
PE19 5SR M5AEE
PE19 5SS G8CHC
PE19 5TZ M0AXZ
PE19 5UB G6FJG
PE19 5UY G8AKL
PE19 5XR G4DPF
PE19 5YA G0CJQ
PE19 5YJ M6TYS
PE19 6AL 2E0BIS
PE19 6BD 2E0KHW
PE19 6QW G8TQI
PE19 6RX G3WCD
PE19 6SD G4LIQ
PE19 6ST G8KDF
PE19 6TJ G3UKB
PE19 7AN G6TNW
PE19 7LE G8ELG
PE19 7LF G4MZV
PE19 7LS 2E0YJW
PE19 7LS M0YJW
PE19 7LS M3YJW
PE19 7RH G6KVK
PE19 7RH M5KVK
PE19 8DF G3PNP
PE19 8GJ G7NRV
PE19 8HY G3LWJ
PE19 8PD 2E0DIP
PE19 8PD M6DWE
PE19 8PE G4ULM
PE19 8PU G8LWO
PE19 8QL 2E0CZG
PE19 8QL M0ZSM
PE19 8QL M6GEK
PE19 8TZ G0JIW
PE2 5EN G7CQG
PE2 5ER G1FWU
PE2 5LG G3EHQ
PE2 5LY G6AVS
PE2 5NT G4UGK
PE2 5PG M1AKH
PE2 5PR G7IBX
PE2 5PS G7AMW
PE2 5SB G4PYR
PE2 5SH 2E0DZY
PE2 5SH M6MTU
PE2 5SL G4XGR
PE2 5SL G7JWD
PE2 5SL G8HTZ
PE2 5XS G3NSF
PE2 5YW G0HYN
PE2 5YW M0VTG
PE2 5YW M3VTG
PE2 6XN M3MXP
PE2 6YB G4LMY
PE2 6YB M0AOG
PE2 6YL G4OIR
PE2 6YP G8ILG
PE2 6YY G1NIT
PE2 6YZ M3YYV
PE2 7AH G0LEH
PE2 8LH G0HZE
PE2 8LS G4CNW
PE2 8LS G4FVK
PE2 8PF G6CZS
PE2 8PL G6GYM
PE2 8QH M3VZU
PE2 8UP G8XLH
PE2 8UP M5ATR
PE2 9DN 2E0DBI
PE2 9DN G4SQA
PE2 9DN M0HYX
PE2 9DN M0WAF
PE2 9DN M3UEF
pe2 9dn M6HYX
PE2 9EG G8HXR
PE2 9FE 2E0TEK
PE2 9HL 2E0HLF
PE2 9HL M6HLF
PE2 9JE G7HUG
PE2 9JG M0BSV
PE2 9JG M3NWK
PE2 9JN G8VWU
PE2 9PD G4DXW
PE2 9PD G4EHW
PE2 9PD M1PRC
PE2 9RQ 2E0WBT
PE2 9RQ M3WBT
PE20 1JA G0MJJ
PE20 1JA G8SKG
PE20 1JT G0MRA
PE20 1JW M3TSI
PE20 1LB M6BOT
PE20 1LR 2E0TNV
PE20 1LR M0TNV

PE20 1LR M3TNY
PE20 1LR M3TOJ
PE20 1LR M6AHJ
PE20 1LR M6DAC
PE20 1LR M6RYN
PE20 1LZ 2E0TQB
PE20 1LZ M1GOH
PE20 1LZ M3TQB
PE20 1NU G3OYL
PE20 1QN M0NDY
PE20 1SG M1DXG
PE20 2AT G3VC
PE20 2BU M1DKF
PE20 2EG M0DCH
PE20 2LQ G6DJY
PE20 2LU 2E0ZTL
PE20 2LU G7STM
PE20 2LU M6CAI
PE20 2NS M6JKH
PE20 2PF M0NAE
PE20 2PF M6AKI
PE20 3AP 2E0WTY
PE20 3AP M0WTY
PE20 3AP M3WTY
PE20 3AP M6BEK
PE20 3AP M6JCC
PE20 3AP M6MOM
PE20 3DF M1CNX
PE20 3DF M3GPX
PE20 3EF 2E0CBU
PE20 3EF M6HBV
PE20 3ES G0PBO
PE20 3LH G0MKG
PE20 3PG G0BGY
PE20 3PZ G7HNM
PE20 3PZ M3MIS
PE20 3QG G7JSV
PE20 3QT G3KLC
PE21 0AU G6ADG
PE21 0EJ M3NJY
PE21 0EN M0PAG
PE21 0EZ M0KOG
PE21 0LL M6EHH
PE21 0QS G8UNP
PE21 0RD 2E0TRA
PE21 0RD G7SZA
PE21 0RD M3OGU
PE21 0SF M0WBD
PE21 6DP 2E0RSG
PE21 6DS M6BXB
PE21 6JN 2E0CDF
PE21 6JN M6AEC
PE21 6PW G7LSP
PE21 6QW M3ZLR
PE21 7BZ G6NUZ
PE21 7BZ M0NUZ
PE21 7EJ 2E0BCB
PE21 7EJ M0GDT
PE21 7HL G3DFY
PE21 7HL M0AZV
PE21 7JG 2E1FRQ
PE21 7JG G4OID
PE21 7NQ M3DPB
PE21 7PN G4GZL
PE21 7PN M3KRQ
PE21 7RU G7MDT
PE21 7TJ 2E0BNW
PE21 7TJ M0WFK
PE21 7TJ M3WFK
PE21 8BJ G8ZTR
PE21 8BW M0HQZ
PE21 8DA G8OYL
PE21 8DL G7MTE
PE21 8EU G3WZA
PE21 8HD G1RAE
PE21 8HD M6GAT
PE21 8HY G6SKF
PE21 8LL 2E0KRD
PE21 8LL M3KRD
PE21 8NG G7ZRT
PE21 8NG M0TLC
PE21 8NL M6JWT
PE21 8NQ 2E0CON
PE21 8NQ M3WUH
PE21 8PU M0EAO
PE21 8SX M3WGB
PE21 8UF G0HPV
PE21 8XB M6EUZ
PE21 9AE G3JRY
PE21 9HA G0DAF
PE21 9HY G7UTE
PE21 9JJ 2E0BFS
PE21 9LE 2E0JBD
PE21 9LE M3JND
PE21 9LR G7LCW
PE21 9PD G4RPI
PE21 9PN G7WID
PE22 0BU G7IBH
PE22 0BU G2FFD
PE22 0BZ G0UFC
PE22 0DH 2E0BBQ
PE22 0DZ M0GVL
PE22 0DZ M6AHL
PE22 0JD G7DYD
PE22 0JG M3DIT
PE22 0NA 2E0MJL
PE22 0NA M0JWL
PE22 0NA M3MJL
PE22 0PA M6IOG
PE22 0PA M6KPX
PE22 0YD G4FI
PE22 0YD G7FUM
PE22 7AB G1EBZ
PE22 7BS G8LDW

PE22 7EA 2E0MVD
PE22 7EA M0MOU
PE22 7HJ 2E1FDE
PE22 7JU G3XLE
PE22 7LW 2E1CJJ
PE22 7LW G0KYD
PE22 7LW M0LIO
PE22 7LW M6DAX
PE22 7RA 2E0HLM
PE22 7RA M0HLM
PE22 7RA M3ICA
PE22 8EX G7GZV
PE22 8EY G8IEV
PE22 8HA G8UMO
PE22 8LX 2E0MID
PE22 8LX 2E0VBR
PE22 8LX M0MID
PE22 8LX M0VBR
PE22 8LX M6VBF
PE22 8LX M6VBR
PE22 8QJ G7EXT
PE22 8RD M6EON
PE22 8RR 2E0BDS
PE22 8RR M0LFC
PE22 8RU G1RNL
PE22 8RU G7HZS
PE22 9LS G0KOO
PE22 9NY M0VPK
PE22 9PR M6PAY
PE22 9PW G7KYX
PE22 9QX G4DYV
PE22 9RA G0KZN
PE23 4AU M0DXJ
PE23 4BE M1CNI
PE23 4BY M6BUL
PE23 4EJ G0ESO
PE23 4JB G4UBM
PE23 5AG 2E0NPS
PE23 5BD G7IVW
PE23 5BE M3JQN
PE23 5DB G4UNW
PE23 5EP 2E0ITV
PE23 5EX G4DYJ
PE23 5HH G6HYF
PE23 5SE G6IDL
PE24 4DL G0LAG
PE24 4JJ G1MJN
PE24 4JP M3UEN
PE24 4PJ M3FSE
PE24 4QN M3ZNF
PE24 5AT G8HWJ
PE24 5DQ G7GQO
PE24 5ES G6MQU
PE24 5HQ G0OHF
PE24 5HT G4LOP
PE24 5HT G7BNN
PE24 5JQ M0CCW
PE24 5LG 2E0ZRQ
PE24 5LG M3ZRQ
PE24 5NF M0KFB
PE24 5NJ G4CPI
PE24 5NN G0JPQ
PE24 5NZ G0WCU
PE24 5PU G8SFU
PE24 5QU 2E0MCC
PE24 5QZ G4ITB
PE24 5RQ M6AFW
PE24 5RY G1LLZ
PE24 5SL G0JMZ
PE24 5TZ M6CRE
PE24 5YQ G4PRF
PE24 5YQ G7JCX
PE24 5YQ M3YCN
PE24 5YS M6SRN
PE25 1BN G1MHB
PE25 1EL G0HNZ
PE25 1HE G4XTX
PE25 1HE M3NQN
PE25 1HL G0ECN
PE25 1HQ M1ERY
PE25 1SE G3NPY
PE25 1TQ G4ZQT
PE25 2BP G0OTH
PE25 2BP M0FRC
PE25 2DD G0EPE
PE25 2EE M0BBE
PE25 2EW G4ZBZ
PE25 2JA 2E0ROS
PE25 2JA M3FFI
PE25 2LN M6AWU
PE25 2NB M6HSO
PE25 2QQ M3KVD
PE25 2QZ G4GOR
PE25 2RA M3KVI
PE25 2TF M6AFX
PE25 2TH G0NDV
PE25 2TX 2E0YSU
PE25 2TX M0WNO
PE25 2TX M3YSU
PE25 3DE M5PWR
PE25 3ER G4MVZ
PE25 3JS M6IPI
PE25 3LF G0CHB
PE25 3LJ M0GHA
PE25 3PU 2E0BAK
PE25 3QU G1YBT
PE25 3RX G3XZY
PE25 3RZ G3OTD
PE26 1BZ G4RRH
PE26 1EY 2E0CUB
PE26 1EY M6BRQ
PE26 1JP G8JWE
PE26 1LU M6FDK
PE26 1LX G1ZBB

PE26 1LZ G0OPL
PE26 1NB 2E0CYP
PE26 1NB 2E1AQK
PE26 1NB G3CXP
PE26 1NB G3MQI
PE26 1NB G3NKJ
PE26 1NB G3ROQ
PE26 1NB M1CXP
PE26 1SF G1XIY
PE26 1SH G8TVC
PE26 2NH G4LQJ
PE26 2QE G4KPZ
PE26 2QE G6RHA
PE26 2SN G4SSO
PE26 2SY M0VVT
PE26 2TD 2E0MKC
PE26 2TF G0BNR
PE26 2YW G0FJR
PE27 3DQ G4NKW
PE27 3FN G1UNG
PE27 3HF M6CKD
PE27 3HF M6INV
PE27 3NL G0RSA
PE27 3TJ G8XIY
PE27 3TZ G4YDD
PE27 3XZ G6THP
PE27 3YZ G0KHY
PE27 4SH 2E0BVQ
PE27 4SN G0GXS
PE27 4SW G7JAE
PE27 4TB G4XGT
PE27 4UD 2E1EJU
PE27 5FX G4KJJ
PE27 5FX G7JXJ
PE27 5JN G3ASE
PE27 5XG G6NBM
PE27 6HW G7CSL
PE27 6SS G6VBQ
PE27 6UB G0IUM
PE28 0AJ G8GRT
PE28 0BB G0IWB
PE28 0BP G7GJS
PE28 0BP M0MAW
PE28 0BS G3NKQ
PE28 0DZ G0AXY
PE28 0DZ G4JZV
PE28 0PE M3ZUW
PE28 0PF M6BTY
PE28 0RY G4NZG
PE28 0UU G1MVF
PE28 2AG M5BRY
PE28 2DQ G3NID
PE28 2FP G1LMS
PE28 2NU G4CTI
PE28 2QG G7FKF
PE28 2SD G8FZT
PE28 2UQ M6MEN
PE28 2UT M6FDI
PE28 3AH 2E0FRF
PE28 3AH 2E0FRK
PE28 3AH M3KUY
PE28 3AT 2E0YSO
PE28 3BJ M1FEZ
PE28 3DA G8HYP
PE28 3DJ M6WMZ
PE28 3DL G4DLT
PE28 3DL G8LEB
PE28 3EG G4GVN
PE28 3ER M3TBG
PE28 3EY 2E1ECN
PE28 3JB 2E0HIG
PE28 3JB M6BGH
PE28 3LE G0RLS
PE28 3LE G1KGV
PE28 3LE G7TSB
PE28 3LY G3NSD
PE28 3QG G8YUR
PE28 3QH G4BIK
PE28 3QH G6GJY
PE28 3XA G4FCF
PE28 3YJ G7NBL
PE28 4DZ G3ECP
PE28 4EW G7ODT
PE28 4FH G4GWU
PE28 4JX M6LRF
PE28 4PX G8UJV
PE28 4RP G1OVH
PE28 4TQ G8RSA
PE28 4US G3TCL
PE28 4US G4ZEW
PE28 4WS G4RKO
PE28 5AW 2E1AQH
PE28 5AW G1NOO
PE28 5SN G4SOQ
PE28 5SY G7EHH
PE28 5TZ M6PPY
PE28 5UA G0HSR
PE28 5UA G7KJW
PE28 5UX 2E0DOJ
PE28 5UX M0MCX
PE28 5WD G0MND
PE28 5YE 2E1IGI
PE28 5YE G7ELC
PE28 5YE M3CMW
PE28 9AN G4YUL
PE28 9BS G3WOT
PE28 9EH G3HSU
PE28 9JL G4KLE
PE28 9JR G3SIT
PE28 9LP G1KHM
PE28 9NA M0ETE
PE29 1QL 2E0EML
PE29 1QL M6CSK
PE29 1QL M6TXR

PE29 1RP G4LHI
PE29 1RZ G3MKV
PE29 1SR G3HPV
PE29 1TA 2E1AHG
PE29 1TA G7OJX
PE29 1TJ M3ZGG
PE29 1TL G1JGT
PE29 1TS M3OWU
PE29 1WT G7JAO
PE29 1XE 2E0BQH
PE29 1XE M6GTR
PE29 1XF G0UFB
PE29 1XP G7MJD
PE29 1XY G3YQB
PE29 1YD G3AAZ
PE29 1YL M0ALK
PE29 1YW G8YCI
PE29 1YZ G1KWF
PE29 2AF G3XJE
PE29 2AL G3ZGG
PE29 2AL G4ZGG
PE29 2LU G1SMC
PE29 2UX G0CCL
PE29 6UF G7DIU
PE29 7BP G6YHF
PE29 7HJ G6PRL
PE29 7JA G6TGW
PE29 7XE G3UXV
PE3 6BB G3YYW
PE3 6FB M0BSW
PE3 6LB G3ZJW
PE3 8BA G4KEY
PE3 8ES G8NGZ
PE3 8LG G4UQF
PE3 8NA M6FPU
PE3 8SH M0PXY
PE3 8SH M6PXY
PE3 9AU G6MGA
PE3 9FS M0RHG
PE3 9PA M3EUP
PE3 9TS M6CWF
PE3 9UD G1PJO
PE3 9UH G4FFS
PE3 9UH G6GNC
PE3 9YD G6GLB
PE30 2ED G1HYU
PE30 2EL M6AJF
PE30 2HX M6MJF
PE30 2NW 2E0EVX
PE30 2PX G7UGR
PE30 2QG M3GIE
PE30 2QL 2E1HYE
PE30 2QL M3HYE
PE30 3BG 2E1HXC
PE30 3BS G3RSV
PE30 3BS G3VET
PE30 3DE G0IJU
PE30 3DP G8NSK
PE30 3DX G1KLP
PE30 3EX 2E1TOM
PE30 3EZ G3ZCA
PE30 3EZ G8MAY
PE30 3HB G8GKL
PE30 3LY G8KOC
PE30 3NA G3YII
PE30 3NJ G6REC
PE30 3PW 2E0TEZ
PE30 3PW M0TAJ
PE30 3PW M3TQJ
PE30 3PW M3UZK
PE30 3PX G4CBO
PE30 3QQ G8CBB
PE30 3UZ G0VHH
PE30 3XD G8WAP
PE30 4AB G1XYZ
PE30 4AB G3HRX
PE30 4AB G3XYZ
PE30 4AB G4OZG
PE30 4AT G6IIP
PE30 4DN G1LOK
PE30 4EJ M3RQQ
PE30 4QA G0WWU
PE30 4QE G8STE
PE30 4XD 2E1IHT
PE30 4XT M6GRZ
PE30 4YH G1XDK
PE30 4YY G6WAN
PE30 5BB M6HSP
PE30 5BB M6NRX
PE30 5BQ G7OUZ
PE30 5DT G8BQT
PE30 5DY G6IWU
PE30 5NH M3UEC
PE30 5PA G8RES
PE31 6AX M3ZOH
PE31 6BT G4JNQ
PE31 6DP M6HNG
PE31 6HQ G0AED
PE31 6LH G8HHU
PE31 6PR G0MTB
PE31 6PR G6TVI
PE31 6QN G7WGL
PE31 7AR G1SCQ
PE31 7AR G7RNA
PE31 7BS G6WDC
PE31 7DX G8JXU
PE31 7EB G6RQA
PE31 7LA M1CCN
PE31 7LQ G4DDT
PE31 7QF G4EOJ
PE31 7RE 2E1TMB
PE31 7SA G6DAC
PE31 7UG G3YLW
PE31 8AQ G0OFD
PE31 8AQ G8MBJ

PE31 8BP M3LCU
PE31 8BS G1YMA
PE31 8DA G7UDQ
PE31 8LN G0CQD
PE31 8NJ M0ZLP
PE31 8RL G4DCJ
PE31 8RL G8FTP
PE31 8RS G0LRU
PE31 8RS G4ZQC
PE31 8SF G3OVL
PE31 8SF M3NHZ
PE32 1AW G7RSA
PE32 1DQ G7NJP
PE32 1HY G0WIW
PE32 1HY M1CFG
PE32 1NY G7OVE
PE32 1QU M3GKE
PE32 1RJ G7DQL
PE32 1RL G6CPS
PE32 1SF G8TUH
PE32 1SS G1LMQ
PE32 1SY G4CXE
PE32 1TG M0BQB
PE32 1TG M0BYD
PE32 1WA G4GXL
PE32 2DT M6KRM
PE32 2DU M6JHC
PE32 2EA G0TZZ
PE32 2LJ G7MFY
PE32 2LZ M6OTH
PE32 2NA G7BUN
PE32 2QX M3KAU
PE32 2TE G1CSY
PE32 2TE M0DZB
PE32 2UA G1ICX
PE33 0HE G0JWG
PE33 0HS G0BPL
PE33 0JG G8CWE
PE33 0LB G4RKN
PE33 0PG G0UMM
PE33 0PR G8ULL
PE33 0PR M6JOW
PE33 0UA G8PHQ
PE33 9BD G1EMW
PE33 9DB M0JPM
PE33 9HP G4IXD
PE33 9JJ G3THS
PE33 9JQ 2E0RZB
PE33 9JQ M6RZB
PE33 9PX M3YXJ
PE33 9PX M3YXK
PE33 9RP G0FQO
PE33 9RP G1RHE
PE33 9SX G3PRU
PE33 9TQ G0CCS
PE34 3BD M0GVI
PE34 3BN M3YFX
PE34 3BX G0UWW
PE34 3HF G3JKM
PE34 3HF M1CCQ
PE34 3HL G7MOD
PE34 3LS G4LBQ
PE34 3PF G7UEL
PE34 3PP 2E0XXM
PE34 3PP M3XXM
PE34 4DE G6MQK
PE34 4DE M3JZK
PE34 4DY G6FSK
PE34 4EA G4NJJ
PE34 4HQ G0EUD
PE34 4HZ G4MOC
PE34 4JB M3GHO
PE34 4JP M3KRO
PE34 4JX M0TBJ
PE34 4JX M6WGC
PE34 4PR M3YCB
PE34 4QJ G7PTB
PE34 4QS M3GJN
PE36 5BD G0KHF
PE36 5BP M3VDY
PE36 5BS G1NJV
PE36 5BZ G0XBO
PE36 5DJ G4VUF
PE36 5EA M0HQA
PE36 5EJ G4RQU
PE36 5PJ G8MRI
PE36 6AP G7APU
PE36 6BS G1LSK
PE36 6BX 2E1EBX
PE37 7BT G0HZA
PE37 7JB G1JBT
PE37 7SP G7MOK
PE37 7TP M0LAI
PE37 0EE G4FZL
PE37 0ET G1KIW
PE37 8HA G8XOU
PE37 8HA M0XOU
PE37 8HY M6NEC
PE37 8LU G0NQN
PE37 8NN G1UGJ
PE37 8NN G7VPS
PE38 0BY G6THA
PE38 0DH G4HCP
PE38 0DP G1VIG
PE38 0DP G6NMK
PE38 0DY G1HBV
PE38 0DY M0ERJ
PE38 0EN G4LMX
PE38 0EU M3HPT
PE38 0EU M3KBB
PE38 9AX G3UPN
PE38 9EU G3XVN
PE38 9HA G7MUN
PE38 9HA M3SMI
PE38 9NJ G8VCO

PE38 9PG G8JAN
PE38 9QP G8FFF
PE38 9QU G4TUO
PE38 9QY G4RAU
PE38 9RQ G3ASG
PE38 9TZ M6OST
PE4 5AF G7FTD
PE4 5AU G3GMW
PE4 5BJ G1DRI
PE4 5DD G3WRL
PE4 5DF G7DWO
PE4 5ED G1MLC
PE4 6AQ G6AYU
PE4 6JS G0CNL
PE4 6LY G4EOD
PE4 6QY G1EUN
PE4 6RL G4ZKM
PE4 6RP M3FVE
PE4 6SA M6DLJ
PE4 6SB G8EQY
PE4 6ZH G4DJZ
PE4 7BJ M3JGI
PE4 7EN 2E0ROD
PE4 7EN 2E1IGN
PE4 7EN M0RHB
PE4 7EN M5AEI
PE4 7PY G7JJP
PE4 7PY M3LPT
PE4 7TN G4DEW
PE4 7TT M3DRA
PE4 7TW G3RED
PE4 7UP M0CEG
PE4 7ZD G3VYU
PE4 7ZD G7SQY
pe46nt M6HWO
PE5 7BH G4ULI
PE6 0BA M0AWB
PE6 0BA M1BXJ
PE6 0DG G4NCP
PE6 0HA 2E0GRP
PE6 0HA M0GRP
PE6 0HA M6GRP
PE6 0JB G8YOX
PE6 0LH G0TFR
PE6 0QG G0NUU
PE6 7EU G1ESC
PE6 7HZ G1ZDX
PE6 7LG G0EVU
PE6 7NW G3YOY
PE6 7QD G1JBJ
PE6 7RG 2E0BEF
PE6 7RG G4FXY
PE6 7RG G4NRK
PE6 7RG G6CZO
PE6 7UB G0IAG
PE6 7UB G1XXW
PE6 7UP G1OAU
PE6 7XG G0BLV
PE6 7YF G8KQZ
PE6 8AG G6IDW
PE6 8BS G1AUM
PE6 8DA G7OOB
PE6 8DJ G4FIQ
PE6 8EH G8YVS
PE6 8HP G7OKT
PE6 8HR G1FJD
PE6 8JS G3ZCV
PE6 8JS G4MZM
PE6 8NW G4VJN
PE6 8NW G6KEZ
PE6 8NW G6MTY
PE6 8NY G6CVV
PE6 8PQ M0HTU
PE6 8QJ G7GWA
PE6 8QJ M5ALU
PE6 8RD G1HOJ
PE6 8RD G6NYL
PE6 8RF G1AGW
PE6 8SF M0NMH
PE6 8SF M3MQH
PE6 8SH G1RVK
PE6 8SN G7UHL
PE6 9DN M0DBH
PE6 9LP G7SOE
PE6 9NP G8OAD
PE6 9RB G6ENN
PE7 1BP G0RCH
PE7 1BX G4KIY
PE7 1DL G0MTA
PE7 1HJ G4NLO
PE7 1RF G1NLS
PE7 1RL G0IHK
PE7 1RR G0PGY
PE7 1RT G4WJB
PE7 1TS M6GGU
PE7 1TY 2E0GPU
PE7 1TY M3YEK
PE7 1UE 2E1GDG
PE7 1UF G4FFN
PE7 1XX G7VPA
PE7 2BT M3JCE
PE7 2DD G0JJW
PE7 2DW M0ZWD
PE7 2HN G4YIT
PE7 2PD G1JIH
PE7 2RG G3XRK
PE7 3AA M3YGC
PE7 3AX M6CUQ
PE7 3AY M3NWD
PE7 3BU G6JAS
PE7 3FA G6EUY
PE7 3GE M0DYR
PE7 3JW G1PLE
PE7 3PG M3LSF
PE7 3PR G1KQN

PE7 3RG M3JDX
PE7 3RS G7IVU
PE7 3SH G8UFO
PE7 3SS G0HOF
PE7 3SU G4PPJ
PE7 3TX M0RKH
PE7 3UA G3DQW
PE7 3UA G3TGO
PE7 3UU M1SAC
PE7 3YD M3MEP
PE7 3YQ 2E0DVO
PE7 3YQ M6FDH
PE7 3ZF G0XAH
PE7 3ZR G7AFO
PE7 8FL 2E0CCQ
PE8 4DQ G4JPB
PE8 4JQ G3VZL
PE8 4LT G7NDQ
PE8 4NX 2E0OTZ
PE8 4NX M3OTZ
PE8 4QR G6DYK
PE8 5AN G7HLU
PE8 5AW G8YDI
PE8 5HP G4HCG
PE8 5PS G0UIH
PE8 6AU 2E0HEP
PE8 6AU 2E0ZAF
PE8 6AU M0KDT
PE8 6AU M0ZAV
PE8 6AU M3ZWL
PE8 6LP 2E0KNT
PE8 6LP M0KWR
PE8 6LP M3KNT
PE9 1DS 2E0BVR
PE9 1HF G4NHL
PE9 1QQ G4ZBG
PE9 1SN G6HNP
PE9 1UP G3TBG
PE9 2BH G3HEE
PE9 2DU G3OJQ
PE9 2FB G6ENZ
PE9 2JY G1FSW
PE9 2NX G4YNT
PE9 2SG G1DCX
PE9 2TS G1JGR
PE9 2UB G3THC
PE9 2YR M0YDK
PE9 2YX G7IGV
PE9 3AG G4VAM
PE9 3LL 2E0CNP
PE9 3LL 2E0CNS
PE9 3LL M6FBH
PE9 3LL M6MSH
PE9 3LN G8IOA
PE9 3LZ G7DGP
PE9 3NA G4OZM
PE9 3PF 2E0JBX
PE9 3PF M0THB
PE9 3PF M6THB
PE9 3PW G3MFG
PE9 3QD G4RQK
PE9 3SY M1REC
PE9 3UQ G8VDJ
PE9 4EB G1CUG
PE9 4JJ G4PEL
PE9 4NQ G0WUU
PE9 4RJ 2E0ZAL

## PH

## (Perth)

PH1 1DT GM4HJK
PH1 1JD GM4BVD
PH1 1JD MM1FHS
PH1 1LE GM8UGO
PH1 1QB GM6JKU
PH1 2DU GM6OFO
PH1 2DU MM3KVN
PH1 2NF GM6UJG
PH1 2NS MM3XUX
PH1 2QJ MM6NAA
PH1 2SE MM6HBY
PH1 3DD GM1AHF
PH1 3DD GM1AHG
PH1 3LW GM1DSK
PH1 3YX MM6AFQ
PH1 4EY MM3VKR
PH1 4QS GM4EAF
PH1 4QS GM4ZRH
PH1 5FN MM3YGI
PH10 6PF 2M1EPV
PH10 6QJ GM7ESM
PH10 6QJ MM3IOM
PH10 6RX GM7MTQ
PH10 6RX MM3MTQ
PH10 6TH GM4NUN
PH10 6UF MM6HIG
PH10 6XE GM7GUL
PH10 7JL GM0VIT
PH10 7NY GM3WFI
PH10 7RP GM4LMG
PH11 8AS GM1ZQF
PH11 8BU GM0LYT
PH11 8DW GM3WAP
PH11 8HQ 2M0RJJ
PH11 8HQ MM0RJJ
PH11 8HQ MM6GHX
PH12 8SL 2M0BMK
PH12 8SL 2M0CLN
PH13 9HS MM6RGT
PH15 2BE GM6YRN
PH15 2EB MM6CBH

PH15 2QY GM4FOZ
PH15 2QY GM8KJO
PH16 5HH GM8KPH
PH16 5JF 2M0IMI
PH16 5JF MM0ZVO
PH16 5JJ 2M0DGI
PH16 5JL MM0ZOG
PH16 5JL MM6BFR
PH16 5JS GM4ODW
PH16 5JS GM6BGL
PH16 5LA GM0CWR
PH18 5UG MM6PRH
PH2 0AE MM3SYQ
PH2 0AR GM1PFU
PH2 0BL MM6OAG
PH2 0GY GM4MSL
PH2 0HH GM0FTX
PH2 0LD GM6UWF
PH2 0LD MM6FSQ
PH2 6QA 2M0TGN
PH2 6QA MM0WXE
PH2 6QA MM6TGN
PH2 6QE GM8YRT
PH2 6RR GM1BUY
PH2 6RR GM3EOB
PH2 6SD GM4DQJ
PH2 6SP GM0HCQ
PH2 6TD GM4EVS
PH2 6TQ GM4AWA
PH2 6TQ MM0EMC
PH2 7BY GM1GHZ
PH2 7BY GM3UYR
PH2 7BY MM0MHZ
PH2 7HJ GM4YZT
PH2 7LL GM1FBM
PH2 7NF GM0CEA
PH2 7QA GM0IST
PH2 7QA GM0NYP
PH2 7QA GM1RGM
PH2 7QP MM3CNS
PH2 7QQ GM0LVI
PH2 7QU MM3SYO
PH2 7QU MM6CEW
PH2 7RB GM6VRC
PH2 7TB GM0IMH
PH2 7TS GM7SDP
PH2 7XD MM0MBC
PH2 8HR GM0FVJ
PH2 8PT GM3THI
PH2 8SA 2M1VFO
PH2 9HS GM0MXZ
PH2 9LW GM0PRG
PH2 9LW GM3YEW
PH2 9PF GM4ZET
PH20 1BS GM8CEA
PH21 1JS GM0KEQ
PH21 1JS GM6NUL
PH21 1NY GM4EBX
PH21 1NY GM4JDK
PH23 3AA GM0PWS
PH26 3LX GM1SRP
PH26 3PA GM4CCN
PH26 3PA GM4LPG
PH26 3PX MM6CJC
PH3 1BZ GM8TCG
PH3 1DD GM1BXI
PH3 1EW GM4JZB
PH3 1JS GM4SKB
PH3 1LX MM6YCB
PH32 4DW GM3VKN
PH33 6HB GM0ANG
PH33 6HX GM4YAT
PH33 6JS GM3RFA
PH33 6NX GM1YGV
PH33 6NY GM4PWR
PH33 6PY GM7REY
PH33 6SJ GM7PKT
PH33 6SZ GM0UMJ
PH33 6UG GM8AOB
PH33 6UH GM4NFI
PH33 7AB MM3XFX
PH33 7AL GM6AES
PH33 7AL MM0VSG
PH33 7ER GM0ERB
PH33 7LE GM4OPU
PH33 7LR GM4TFJ
PH33 7PF GM4UGN
PH33 7PQ GM1YPJ
PH34 4DZ GM7PXL
PH34 4EQ GM4GXH
PH34 4EX GM4TJL
PH39 4NX 2M0SBP
PH39 4NX MM3YBQ
PH4 1QE GM0GQH
PH40 4PD GM4OAS
PH41 4QF GM4XJY
PH41 4QF GM6VYZ
PH41 4QY GM6VYY
PH41 4RQ 2M0LUG
PH41 4RQ MM6INV
PH49 4HN GM3ZRT
PH5 2DD GM0WEZ
PH5 2BD MM6SOH
PH5 2BD MM6YEZ
PH6 2BB GM4YRO
PH6 2HB MM0GKT
PH7 3FB GM7SAQ
PH7 3RP GM0AOF
PH7 3RP GM1LTM
PH7 4DH GM4OFC
PH7 4LE GM4YDC
PH8 0ET GM4LFK
PH9 0LG GM4NGJ
PH9 0NH GM6IQH

## PL

## (Plymouth)

PL1 2HY M3MWG
PL1 3JF G1KXJ
PL1 3JN M3PXP
PL1 3LB G7CNC
PL1 3PS G0OJW
PL1 3QR 2E0AZD
PL1 3RN M6FGC
PL1 3RZ M6GUZ
PL1 4GQ G6RFS
PL1 4GU 2E1GMD
PL1 4GU M3YOX
PL1 4HL G4WDR
PL1 4HL G6BJJ
PL1 4HU 2E0NJK
PL1 4HU M3PXQ
PL1 4RE G0ESY
PL1 4ST M3PYT
PL1 5HD M6ABG
PL1 5NB M6HLA
PL10 1AF M6HFK
PL10 1DA G7AWG
PL10 1DA G8UDA
PL10 1DH M6GJO
PL11 2BZ 2E0LAD
PL11 2BZ M3UXG
PL11 2HJ G8XOX
PL11 2LP M6OLD
PL11 2LT M3TBQ
PL11 2NE 2E1ABQ
PL11 2PX G4ZGZ
PL11 2PZ G4ZZD
PL11 3AQ G0LFQ
PL11 3DX G7FGD
PL11 3JA M0PJY
PL11 3LY G1PRE
PL11 3LY G4LHY
PL12 4AL M3ZGY
PL12 4BJ G0AKH
PL12 4BJ G4GXK
PL12 4BJ G7WAB
PL12 4BJ G8SAL
PL12 4BN G4YXJ
PL12 4BX 2E0OMI
PL12 4BX M0HTK
PL12 4BX M6DZH
PL12 4DL M1CQC
PL12 4DY G4HZE
PL12 4JH G3UBY
PL12 4JZ G8PVR
PL12 4LF M3KFL
PL12 4NG G0VSM
PL12 4NG G3ZYY
PL12 4NG G4KYY
PL12 4NG M3ZYY
PL12 4NJ M0BHG
PL12 4PA G7TLK
PL12 4PA M0RKA
PL12 4PA M6WKI
PL12 4PP G0XAW
PL12 4RG G0KDW
PL12 4SH 2E0CXA
PL12 4SH M6BKH
PL12 4SW 2E0MCJ
PL12 4SW M6ZEJ
PL12 4TJ G1OMZ
PL12 5AE 2E0CGP
PL12 5AE M0HBU
PL12 5AE M6AKW
PL12 5NH G6SEK
PL12 5NQ G6ELG
PL12 6BJ G3EXL
PL12 6DR G4YGZ
PL12 6DU G8HVZ
PL12 6EF G7FTF
PL12 6EL G3XCS
PL12 6EN G6YNV
PL12 6EN G7VBZ
PL12 6EN M6RRD
PL12 6JE M1BZK
PL12 6LJ G6KQK
PL12 6NY 2E0RBA
PL12 6NY M0RAZ
PL12 6PH G3SGV
PL12 6PS M0BHK
PL12 6RH G0DAV
PL12 6SA G0ION
PL12 6SP G4ALY
PL12 6TD G3ZHK
PL12 6XP M6ZEL
PL13 1PN G4MXP
PL13 1PR G8TXT
PL13 2HT G1YMC
PL13 2JP 2E0PDQ
PL13 2JP G1YDQ
PL13 2JP G4BRF
PL13 2JS G0PWW
PL13 2JY G8TEF
PL13 2LF 2E0SJG
PL13 2ND G4DHC
PL13 2ND G0HNO
PL13 2NX G0HRX
PL13 2NX M1SAN
PL13 2PU G7CLO
PL13 2SD G6YXV
PL14 3AQ M6KBJ
PL14 3FN G0CAY
PL14 3HX G0GUT
PL14 3LW 2E0TUH
PL14 3LW M6TUH
PL14 3NS 2E0TGL
PL14 3NS M6SOC

PL14 3PT G3UAE
PL14 3PU M0GGM
PL14 3QG G1NSV
PL14 3RL G4TIQ
PL14 3SQ G7PHH
PL14 3TD M0BEV
PL14 3TQ G3ZHO
PL14 3TS G1ZRQ
PL14 4NJ G1STP
PL14 4NJ G6SHF
PL14 4NW G3WOA
PL14 4PA G0LYR
PL14 4PT 2E0DCY
PL14 4PT M0KRR
PL14 4PT M3ZHX
PL14 4PT M6DCY
PL14 4QP G7WCF
PL14 4RS G8BCG
PL14 5BW G4MFE
PL14 5DP G0DAU
PL14 5EL G4VCA
PL14 5EN G4TYD
PL14 5EY G6OWT
PL14 5LE G8BIX
PL14 5NP G0EIA
PL14 5NQ G1KTZ
PL14 5PP 2E0MLJ
PL14 5PP M6TMJ
PL14 5PW G7LJA
PL14 5QT G0EFR
PL14 5QT M6MND
PL14 5RA G7JLK
PL14 5RF G7OAH
PL14 5RU G7BIM
PL14 5SA M6RFR
PL14 6EP G8GLI
PL14 6JS G1NTV
PL14 6PZ G8VZR
PL14 6RY G4EOG
PL15 7HB 2E0VCC
PL15 7LQ G0BNU
PL15 7NB G6FOH
PL15 7PQ M3ZZH
PL15 7PS G4RZM
PL15 7PW G1TAZ
PL15 7PW G3LNW
PL15 7QD G3ZYL
PL15 7RQ M3IIV
PL15 8UB G1MBG
PL15 8UF G3SYM
PL15 8UZ G0EES
PL15 9BB G4KKZ
PL15 9EB G0ANM
PL15 9EB G6ZMO
PL15 9NA G7HEP
PL15 9PH M0ACI
PL15 9QL G6AAC
PL15 9RL G6NPP
PL15 9RR G0AAM
PL15 9SB G4PEK
PL15 9SY G0UFV
PL15 9SY G1SOX
PL15 9TN M3IME
PL16 0AH M0ESW
PL16 0AH M3PSS
PL16 0AS G3LUW
PL16 0EH G4NIA
PL16 0HH G3SBP
PL16 0HJ 2E1EOR
PL17 7AQ M6JMM
PL17 7BX M6AKO
PL17 7EU G7NHW
PL17 7EU M0CDS
PL17 7EU M1APB
PL17 7GB M3RYY
PL17 7HP G1KBG
PL17 7JJ G1SCY
PL17 7LW G1IPI
PL17 7PA G0IVZ
PL17 7PE G1YTV
PL17 7PL M3XWX
PL17 7PT G1FMU
PL17 7PT G1XIC
PL17 7PT G7UDX
PL17 7PT M0SCR
PL17 7QH M3VQV
PL17 7QH M6BMT
PL17 7QH M6VQV
PL17 7QL G1VQG
PL17 8DL M2FAC
PL17 8DE M3YMS
PL17 8DF G0MIH
PL17 8GL 2E0VAF
PL17 8GL M3PCQ
PL17 8GL M3VJE
PL17 8JX G6BER
PL17 8NR G6IJW
PL17 8PA M6GAR
PL17 8PZ G8AWB
PL18 9AZ M0XGG
PL18 9BD G1KYE
PL18 9BL G7CQK
PL18 9BN G4UMS
PL18 9BU M6BFD
PL18 9BY M6FZV
PL18 9DA M6OPO
PL18 9DA M6POD
PL18 9DA M6ROP
PL18 9DH G3XIB
PL18 9DX G4PBN
PL18 9HH M1BBB
PL18 9JB 2E0NMC
PL18 9JB M0NMC
PL18 9JB M1APC
PL18 9JB M3HXZ



**IMPORTANT NOTE**

**Revalidate licence to avoid revocation** – Ofcom has advised the Society that plans will be drawn up to revoke licences that have not been revalidated as required by the licence conditions. The quickest way to revalidate is to do so online via the Ofcom website: *https://services.ofcom.org.uk/* or by email: *amateur.validations@ofcom.org.uk* If you need assistance in the process, Ofcom staff are available to help, but please be patient during times of heavy workload.

PL18 9JB M6BTS
PL18 9JB M6NEW
PL18 9NG G4FWN
PL18 9PB G4BVB
PL18 9QN M6YRW
PL18 9RY M0BSC
PL19 0DP 2E0MDH
PL19 0DP M6BWO
PL19 0EA 2E0PAA
PL19 0EE G1RFQ
PL19 0ER G3ZNE
PL19 0ET G8ONR
PL19 0EU M0YDW
PL19 0EU M3YDW
PL19 0NJ G7VUP
PL19 0NR G3FHT
PL19 0PN G0GSR
PL19 8DP G0LSG
PL19 8DR G1KKA
PL19 8DX G1OJL
PL19 8EY 2E0YJF
PL19 8EY M3YJF
PL19 8HA G3MWZ
PL19 9AJ G7IZU
PL19 9DA G3TGN
PL19 9DG G4OJB
PL19 9DJ G6FOV
PL19 9DL G7IIB
PL19 9DN G3TYO
PL19 9DQ G3XOU
PL19 9LJ G0NUO
PL19 9LQ G7VYQ
PL19 9NQ G8KSM
PL19 9PR G0LEV
PL19 9QD G8AKC
PL2 1DT M6TGR
PL2 1EA M3LOE
PL2 1EA M3SCX
PL2 1HS M6CPW
PL2 1JD 2E0OLF
PL2 1JD M0MLZ
PL2 1JD M6OLF
PL2 1WT M3TOR
PL2 2BR G4PKM
PL2 2BU G7LUL
PL2 2BY M6HCA
PL2 2ER G0WGB
PL2 2JL M0RSM
PL2 2QY 2E0BJD
PL2 2QY M3MQX
PL2 3AP M3YQM
PL2 3AX 2E0DVN
PL2 3AX M6BNM
PL2 3AZ G3XLZ
PL2 3ET G8EHD
PL2 3LF 2E0ITN
PL2 3LF M6ITN
PL2 3QG 2E0SPR
PL2 3QG M3PYJ
PL2 3QZ G7ANY
PL2 3RS G1SQI
PL2 3RS G7LLD
PL2 3RS M3IOT
PL20 6AT G7MOH
PL20 6EA M3BKB
PL20 6HP G6IIZ
PL20 6LJ G3TXL
PL20 6LJ M3TXL
PL20 6ND M6LJK
PL20 6NG G4GUY
PL20 6PT G0IAI
PL20 6SY G4DND
PL20 7AH G4HPX
PL20 7AR G3SGZ
PL20 7AZ 2E0RPH
PL20 7BZ M6WBR
PL20 7DD G1SQG
PL20 7JS G8MWN
PL20 7JS M3TNJ
PL20 7LH G0PGI
PL20 7LR G3ZXO
PL20 7NA G3VVG
PL20 7PJ M6JCG
PL20 7QB M6TGY
Pl206nn M6MQM
PL21 0AS M6OKO
PL21 0AX G7HIC
PL21 0ET G4VFG
PL21 0ET G6NSU
PL21 0HT G0TQR
PL21 0LD M0VFG
PL21 0LT G4NDL
PL21 0RQ G3UZI
PL21 0WD M5DAP
PL21 0YW G3TSE
PL21 9BD G4XFM
PL21 9BH G7LNG
PL21 9BT G4ONC
PL21 9DH G0EOZ
PL21 9EU G4OUZ
PL21 9EU G4RIM
PL21 9JE G4ETP
PL21 9JX 2E0KMF
PL21 9PP M3ZBI
PL21 9QT 2E0GWD
PL21 9SN G3PRC
PL21 9SN G8PRC
PL21 9SN G8XTE
PL21 9TE 2E0XCP
PL21 9TE M0ZCP
PL21 9TE M3XCP
PL21 9XA 2E0PGH
PL21 9XA M6PGH
PL22 0ET G3XSC
PL22 0JB G0MFQ
PL22 0QH G3HUB
PL23 1ET G3MTG
PL23 1JH G3SXR
PL23 1NB G4AIR
PL23 1NB G7RNB
PL23 1QQ M3UIB
PL24 2AT G0AEW
PL24 2DE M6OSO
PL24 2DG G8XNA
PL24 2DL M6PCP
PL24 2LD G0CIG
PL24 2LH G4ZEB
PL24 2RL G1JXP
PL25 3AU G0KTD
PL25 3BB M1EWP
PL25 3BZ M6GBY
PL25 3DH G4RLN
PL25 3DN M0PIP
PL25 3DR G0NNO
PL25 3DY M3UMM
PL25 3EB G4LTY
PL25 3EX G3HTO
PL25 3HB G8WBG
PL25 3QE G4UON
PL25 3QG G4JOD
PL25 3TJ G4XBW
PL25 4DS G8TNA
PL25 4HR G1XMI
PL25 4HR G8NYR
PL25 4HR M1HZR
PL25 4HS M1ALX
PL25 4HT G4XOP
PL25 4JA G4XBC
PL25 4QF M0AGY
PL25 4RJ G4MFQ
PL25 4UD M0DOA
PL25 4UW G6GAB
PL25 5EA G4TRV
PL25 5TA G8MNC
PL254EA M3XTG
PL26 6BN 2E1EAK
PL26 6BN G4LED
PL26 6JA G3GHS
PL26 6NU M0HDP
PL26 6NZ G4MLI
PL26 6QS G8HYL
PL26 6QZ G3YIY
PL26 6TG G7JIB
PL26 6TZ G6MNI
PL26 6TZ G6MNJ
PL26 6XD G8TCP
PL26 6XD M1BRZ
PL26 7AR G7VTC
PL26 7BH 2E0KUC
PL26 7BH M0ORS
PL26 7BW G4OPL
PL26 7EH G4HFO
PL26 7ER G8ZQM
PL26 7ER M3ZQM
PL26 7NN G8VRV
PL26 7PF G4KNI
PL26 7PN G7FLX
PL26 7PT G4ZGQ
PL26 7PW M0GTH
PL26 7TL G2KF
PL26 7TP G3KYM
PL26 7TY G0TRP
PL26 7XN M1AEG
PL26 7XQ G0FCB
PL26 7XS G0HZD
PL26 8BE M0HKI
PL26 8DB G6CNW
PL26 8DS M6TGK
PL26 8EJ M3TEL
PL26 8GY G4VSY
PL26 8HD M6HVK
PL26 8HW M6CJR
PL26 8JA M1AZJ
PL26 8JH G1TTK
PL26 8JN 2E0BXD
PL26 8JN M0PSB
PL26 8JN M6JOE
PL26 8PH G0AHM
PL26 8PS M6RSI
PL26 8QL G4JYF
PL26 8QL M3OBQ
PL26 8QN 2E1CYI
PL26 8TL G0VDU
PL26 8UA 2E1CWE
PL26 8UB G0RJI
PL26 8UH G0GAW
PL26 8UH G3WKF
PL26 8UQ G4OKS
PL26 8UX 2E0NAM
PL26 8UX M0NAW
PL26 8UX M6NKC
PL26 8YE G4ZZY
PL27 6AN M6HVO
PL27 6NW G8LOU
PL27 7LL M3UWM
PL27 7PG G7DUB
PL27 7QA M6HEB
PL27 7TB G7TLC
PL27 7TS 2E0XTM
PL27 7TS M3XTM
PL28 8ES G0MWW
PL29 3RU G4UZO
PL3 4BR M3UHQ
PL3 4DN M6BWK
PL3 4HE M1FCG
PL3 4HS 2E0DCF
PL3 4HS M0WCL
PL3 4HS M6LJR
PL3 4LN G8PYE
PL3 4PD G0OYO
PL3 4QN M3OUV
PL3 4RB G6GEX
PL3 4RB M0GEX
PL3 5BS G0FCG
PL3 5DA G8RMP
PL3 5DU G4BCX
PL3 5HQ G6ALR
PL3 5NP G3SQN
PL3 5NP M6RMK
PL3 5TX G0KPQ
PL3 6BY M6DIZ
PL3 6DB M3ZWD
PL3 6DE M6DIB
PL3 6DH G0GVX
PL3 6HD M1AGY
PL3 6HE M0WDC
PL3 6HL M3OJJ
PL3 6JF G1LOE
PL3 6JZ G0LSJ
PL3 6JZ G7HCN
PL3 6LT G1OGB
PL3 6LX G4XZS
PL3 6LX G6ZGK
PL3 6NQ G7DIR
PL3 6PB G1KCW
PL3 6PT G3SVZ
PL3 6PX G6GFO
PL3 6PX M3NJA
PL3 6QY M5PLY
PL3 6SZ M3RNK
PL30 3AU G3IWE
PL30 3BS G4WQU
PL30 3PH G4AXC
PL30 3PN G3XKS
PL30 5BE M3HJU
PL30 5ED G1JBB
PL30 5EP G4UBK
PL30 5HD G8TWA
PL30 5JL G3MCD
PL30 5LU G1OFG
PL30 5PJ G1FXD
PL31 1BE 2E1AFC
PL31 1BH G3IGV
PL31 1EL M0PHM
PL31 1PY G4WVD
PL31 1QJ 2E0DSI
PL31 1QJ M6DQO
PL31 2BP G8XNH
PL31 2BZ M3ZNX
PL31 2FB 2E0BHN
PL31 2FB M3RKJ
PL31 2FE M6AKV
PL31 2FP G0UFF
PL31 2NU G0EEJ
PL31 2QP G4GPD
PL32 9RZ G1VWU
PL32 9UB G4YME
PL32 9UP G4YVB
PL32 9UX G0TZD
PL32 9YN G0ISJ
PL33 9AT G4WAV
PL33 9BN G6GWX
PL33 9BX 2E0JJR
PL33 9BX M3PXK
PL33 9DP M6CGB
PL33 9DT 2E0BJV
PL33 9DT M3PXF
PL34 0BH G0EDH
PL34 0BH G0EWR
PL34 0DT M0HJO
PL34 0EH M3KKG
PL34 0EL G3LOV
PL34 0EL M3MXZ
PL34 0HH G0ATS
PL4 0EF G8BOI
PL4 0EZ M3YGT
PL4 0PL M6EQQ
PL4 6AZ G6URM
PL4 6NW G3RRW
PL4 6PR G1ZSK
PL4 6PZ G0JIA
PL4 6QA G4EKV
PL4 6QA M3LUW
PL4 6RD M3BHK
PL4 7HB M0UNJ
PL4 7LA M6NDE
PL4 8AA G0UOP
PL4 8AA G3YQF
PL4 8DS M6VEL
PL4 8TA M0VRT
PL4 8TA M3KYD
PL4 9DF M6JIY
PL4 9EL G0BDB
PL4 9EN 2E0CSO
PL4 9EN M6BKE
PL4 9ET G0TQT
PL4 9EZ G7KII
PL5 1AB G0JCA
PL5 1DS G7DQC
PL5 1NN M6MLO
PL5 1PD G0NCS
PL5 1SB G8PSC
PL5 1SR G3YJQ
PL5 1UH M3YYM
PL5 1UX G0WOU
PL5 1UX M0WFF
PL5 2AL M3ZZE
PL5 2DB G0VKY
PL5 2LN M3ZZI
PL5 2NU G7MME
PL5 2NW G8IDE
PL5 2PJ G4EJQ
PL5 2QS M3OMX
PL5 2SN 2E0GLS
PL5 2SN M6GPS
PL5 3DJ 2E0GNI
PL5 3RB 2E0DTC
PL5 3RB M6BNL
PL5 3TL 2E0ONO
PL5 3TL 2E0OSO
PL5 3UW M0CCA
PL5 4AZ 2E0BIM
PL5 4HX 2E0OPS
PL5 4HX M3LCF
PL5 4JQ 2E0IOS
PL5 4LT G0KIK
PL5 4PU 2E0BNV
PL5 4PU 2E0NGG
PL5 4PU M3NGG
PL5 4PU M3PXE
PL5 4QE G3RYZ
PL6 5HH M0PDB
PL6 5NR G0JNZ
PL6 5PU G3JFS
PL6 5PU G7UTI
PL6 5QE 2E0STI
PL6 5QE M3XXL
PL6 5QN M0CPW
PL6 5RJ G4BZF
PL6 5SD G7NIA
PL6 5SD M0BWN
PL6 5SN G0KYE
PL6 5TR G7BXS
PL6 5TR M3BXS
PL6 5TX G4WLS
PL6 5TZ M3ZZJ
PL6 5XL G4IML
PL6 6AH G3JTJ
PL6 6AY 2E0SKE
PL6 6AY M6FGE
PL6 6DW 2E0PLY
PL6 6JP G4CGM
PL6 6LS M0BLO
PL6 6RE G6IPQ
PL6 6TN M6CDN
PL6 7BY G1HHC
PL6 7BY G1HHD
PL6 7BY G1RCD
PL6 7DT G7OOV
PL6 7HS M6DGN
PL6 7HX G3ZZS
PL6 7JA G4YDR
PL6 7JA M3BAS
PL6 7LN M3YLQ
PL6 7SA G7DQA
PL6 7SP G7RTO
PL6 7SY G4UTX
PL6 7UB G4GHR
PL6 8QD M6VKN
PL6 8RQ G8WZJ
PL6 8SB G0LCP
PL6 8SP M3KZT
PL6 8TD G7ESO
PL6 8TL M0DWT
PL6 8UU M3SQQ
PL6 8XF G0OBP
PL7 1HG M0RFH
PL7 1JJ M1AHF
PL7 1JR G3ARE
PL7 1JY G3LCY
PL7 1PU G6CYH
PL7 1PZ M0GBA
PL7 1PZ M1CXV
PL7 1QZ G3VCN
PL7 1SN M3YYS
PL7 1TX G3TZQ
PL7 2BX G6IBU
PL7 2DG G8MLI
PL7 2DY G4LOI
PL7 2EJ G0FSD
PL7 2EQ G3IVP
PL7 2EQ G7LPT
PL7 2EY G4NLU
PL7 2GT G7HHW
PL7 2HP G0KML
PL7 2QP M6FGI
PL7 2RU G7NHB
PL7 2SU M1SAM
PL7 2YF G0MMW
PL7 2ZA G1SYZ
PL7 4BL G3HPC
PL7 4BP M3YYJ
PL7 4DA M0AVS
PL7 4EG 2E0ZRM
PL7 4HH 2E0TZR
PL7 4HY G0UMI
PL7 4HZ M6EQZ
PL7 4JB 2E1HSL
PL7 4JB G1CKT
PL7 4JB M0UOK
PL7 4JE G0NAP
PL7 4NU M3YYR
PL7 4PF G1LTC
PL7 4QY 2E0MGC
PL7 4QY G8CMG
PL7 4QY M0WMB
PL7 4QY M6CQG
PL7 4RW G7ART
PL8 1BP G3MMX
PL8 1BP G8ADX
PL8 1BZ G4BVS
PL8 2ED G6FBJ
PL8 2JQ M0DZL
PL8 2NT G0ESL
PL9 0AW 2E0CYZ
PL9 0AW M6AYZ
PL9 0DS G4PRL
PL9 0EU G0MQK
PL9 0JZ 2E0OZI
PL9 0JZ M6OZI
PL9 7DQ G7BAV
PL9 7LA G3VNG
PL9 7LD G4NDD
PL9 7LF G3KHU
PL9 7LU G7UWO
PL9 7NU G4KFZ
PL9 7PG M0BHV
PL9 8DB G1BMP
PL9 8DB G3XZX
PL9 8HU G0VZX
PL9 8NW M3LVX
PL9 8PJ G3ULN
PL9 8QZ G0LRJ
PL9 8RB G3RMZ
PL9 8TW G0RMC
PL9 8UR G3SPI
PL9 9AG M0AKR
PL9 9HJ G7BYW
PL9 9LU G7FXW
PL9 9NN G8VLZ
PL9 9PT M0ASI
PL9 9RR G4SJD
PL9 9TX M3LTT
PL9 9UU 2E0DQU
PL9 9UU M6IAX

## PO (Portsmouth)

PO1 1QN 2E1AKW
PO1 1QN G6ASJ
PO1 2PB G4NMD
PO1 2PB G6APJ
PO1 2PB M3CIS
PO1 2PB M3KVU
PO1 2SN 2E0HYM
PO1 2SN M6HYM
PO1 3RD M0GYS
PO1 5AR G0LFI
PO1 5AR M6UBI
PO1 5LL M3XDH
PO1 5RR M6PWB
PO10 7JY G0BBJ
PO10 7NH G4WFF
PO10 7NS G8RRC
PO10 7QP G6NUX
PO10 7RA G0USE
PO10 7RP 2E0DMB
PO10 7TR G4OZX
PO10 8BN G4XMJ
PO10 8BN M0GZN
PO10 8HS G1MDJ
PO10 8JG G3HCO
PO10 8LB 2E0HEX
PO10 8LB M0HEX
PO10 8LB M6HEX
PO10 8RN G8WUS
PO10 8XD G0UFY
PO11 0AW G6RGI
PO11 0AZ G3VNU
PO11 0DT G3JVL
PO11 0ER G3MCV
PO11 0HL G4VKX
PO11 0JW G0AYI
PO11 0JW G7KQT
PO11 0LX G4NKX
PO11 0NR M6WAJ
PO11 0QE G0KCF
PO11 0QE G8DLL
PO11 0QE M3TEV
PO11 0QR G8VMY
PO11 0RL G3NDO
PO11 0RL G6VXZ
PO11 9BT G0HIN
PO11 9HY G7ARK
PO11 9LG 2E0DYX
PO11 9LT G2JL
PO11 9LT G3DIT
PO11 9NE G4ZMY
PO11 9PS G8ZJK
PO11 9RA 2E1IHO
PO11 9RA G3YPW
PO11 9RA M1FFN
PO11 9SN 2E0JIG
PO11 9SN M6IFO
PO11 9ST M6ZZY
PO12 1DH G0SFG
PO12 1EN G4BEQ
PO12 1HE G4JOA
PO12 1JX M6BBB
PO12 1PW M3UWR
PO12 1QY G4XQX
PO12 1RB M6CPT
PO12 2HB G0MPJ
PO12 2HB G1BHG
PO12 2HB M6KEV
PO12 2HX 2E1CNM
PO12 2HX G0AVP
PO12 2HX G6XHJ
PO12 2HX M0MIT
PO12 2HX M0RCV
PO12 2LG 2E0CWJ
PO12 2LG M0HTE
PO12 2LG M6CNR
PO12 2NE G8EIE
PO12 2NL G6JGT
PO12 2NL G7AUR
PO12 2NN M0BEC
PO12 2PA G4NEJ
PO12 2QU G8BSD
PO12 2RA G7SAX
PO12 2SZ G7TUQ
PO12 2UP G6UGG
PO12 3BY 2E0TPE
PO12 3BY M6GQC
PO12 3DR G4VVL
PO12 3DS M6AIH
PO12 3EB G6ERI
PO12 3EU G6TQF
PO12 3HF M0KPK
PO12 3JJ G0GIA
PO12 3JY 2E0AYK
PO12 3JY M0EHL
PO12 3LD G7RWN
PO12 3QN G6NHA
PO12 3QY G0NCX
PO12 3SX G1ZTN
PO12 4BU G0HZY
PO12 4BX G4POW
PO12 4DB G1WXU
PO12 4EP G1EDA
PO12 4ES 2E0CGH
PO12 4ES M6WFA
PO12 4EU G1OAW
PO12 4EW G8GBE
PO12 4GG G0JFD
PO12 4GL G4CXJ
PO12 4GS G6FKA
PO12 4HN M6LFM
PO12 4JF G0LPP
PO12 4LS G0SGM
PO12 4NS G1UID
PO12 4NX M6WBX
PO12 4UE M5AGB
PO13 0DG G3UVC
PO13 0DG G4MWW
PO13 0DP G4UHS
PO13 0EQ G0VEP
PO13 0JG M3NMM
PO13 0JS 2E0EFA
PO13 0NF G0DGW
PO13 0PT G3MZZ
PO13 0QS G8MRN
PO13 0RB G1YTL
PO13 0SE 2E0BXX
PO13 0SE M6HAS
PO13 0SJ M0EEL
PO13 0SP G0GLW
PO13 0TN 2E1ESN
PO13 0TN M1ANC
PO13 0TT G7TXW
PO13 0TU M6NAQ
PO13 0TY 2E0DOF
PO13 0TY M3SVB
PO13 0YN G0DDW
PO13 0YX G0JYV
PO13 0ZD 2E0DBA
PO13 0ZD M0LOW
PO13 0ZD M3ERR
PO13 0ZU M6BLX
PO13 8GP 2E0GLL
PO13 8GP M6TIU
PO13 8HD 2E0YZM
PO13 8HD M6YZM
PO13 8JY M6FUW
PO13 9AU G4MHQ
PO13 9AU G7USV
PO13 9BB M6WYY
PO13 9DH G7MFR
PO13 9EN G4TKW
PO13 9EU 2E0LHC
PO13 9EY G8GUS
PO13 9HU G3NPZ
PO13 9NA G6APD
PO13 9NJ G8OUH
PO13 9RD G4BMQ
PO13 9UJ G1CLJ
PO139DU 2E0EMG
PO139DU M0MAQ
PO139DU M3VTQ
PO14 1AS M6HST
PO14 1EF G6AIQ
PO14 1EG G0VIX
PO14 1PH M3FNO
PO14 1QF 2E0BQK
PO14 1QF M3VOU
PO14 1SL G7ILI
PO14 1SW G6GPV
PO14 2AQ G8POQ
PO14 2BQ G6AVT
PO14 2BZ M0BQZ
PO14 2HA G4XVV
PO14 2JF G3GVM
PO14 2PX G4PZV
PO14 2QS G0RHV
PO14 2QX G0PPH
PO14 2SQ G4SPS
PO14 3AA G6BAT
PO14 3AD G1KNI
PO14 3AF G7NYF
PO14 3AH M6OSL
PO14 3BS G8APL
PO14 3DR G8PGF
PO14 3HW G3IBI
PO14 3LB G6HGX
PO14 3LB M0DDA
PO14 3LB M3ADJ
PO14 3LF 2E0ZFV
PO14 3LF M3ZFV
PO14 3RX M0GLL
PO14 3SY M1FCE
PO14 4BH G3HKT
PO14 4BX M1FIR
PO14 4EN 2E0LLD
PO14 4EN M6LDD
PO14 4JP M1EMP
PO14 4LQ G0ERI
PO14 4NB M0NRG
PO14 4NP G0GFD
PO14 4QX G8RNV
PO14 4SL G0MAT
PO14 4SL G0TPB
PO14 4SY G8RLF
PO15 5AJ G1NCK
PO15 5BL G6HHE
PO15 5DU M0RIK
PO15 5EQ G3YTQ
PO15 5HP G3RDA
PO15 5JJ G0AMS
PO15 5LG G0JYQ
PO15 5NE G4VNM
PO15 5NF 2E0JTN
PO15 5NF M0JTN
PO15 5NF M6JTN
PO15 5NQ M6JEG
PO15 5PF G1MPP
PO15 5QB G3UWE
PO15 5QH G0OBM
PO15 5RS G0XGL
PO15 6HF 2E0JRD
PO15 6HF M0JDL
PO15 6HF M3YJD
PO15 6HT M6AVH
PO15 6JU G3KLF
PO15 7EA G0SHT
PO15 7EA G4CAA
PO15 7LE G3UPZ
PO16 0RA G3VEF
PO16 0RA G4JLP
PO16 0RX 2E0ABL
PO16 0SQ G0DDA
PO16 0TR G4ZMP
PO16 7DP 2E0MOY
PO16 7DP M0YSR
PO16 7DP M6FFR
PO16 7HB M0HBE
PO16 7JF G4FCL
PO16 7LU G0HZC
PO16 7NL G0RNM
PO16 7NL G4UOR
PO16 7NW G0MST
PO16 7QL G4ITG
PO16 7QW G3YIW
PO16 7RR G6KTX
PO16 7TB 2E0ATB
PO16 7XA G0JYZ
PO16 7XR G1WKZ
PO16 7XR G1WLD
PO16 8DS G8XEZ
PO16 8DU M6DRG
PO16 8DY G0FAD
PO16 8JW G4PWG
PO16 8JY G7MDV
PO16 8LB G4CYC
PO16 8LF G0LYI
PO16 8LF G3XUF
PO16 8PB G8IOJ
PO16 8PE M6GSZ
PO16 8QX M0DQO
PO16 8TL G1OJS
PO16 8UF G6ORL
PO16 8UF G8TXK
PO16 9AA M6ZAU
PO16 9AP G7DWV
PO16 9BQ G6JSI
PO16 9DN G4YPA
PO16 9DX M0CAA
PO16 9HP G4IOJ
PO16 9LA M6EKX
PO16 9NJ G4MRW
PO16 9NR G3UFF
PO16 9PA G0KCG
PO17 5AZ 2E0NCF
PO17 5AZ M0SNT
PO17 5AZ M6FBK
PO17 5EP G6FLE
PO17 5EX 2E0SMO
PO17 5GS G6LYA
PO17 6DG G4GWJ
PO17 6DG G4HKZ
PO17 6DG M0XML
PO17 6EY 2E0CFE
PO17 6HS G3XUX
PO17 6JB G1WID
PO17 6JJ G8JXV
PO18 0AY M1BTO
PO18 0EX G3NXO
PO18 8LF G6XZA
PO18 8QQ M0GNL
PO18 8QW M3XYC
PO18 8SR G7MIN
PO18 8SR M3MIN
PO18 8SU G3LSQ
PO18 8SU G4JDG
PO18 8TH 2E0CKE
PO18 8TH 2E0XAO
PO18 8TH M3UJZ
PO18 8TH M6SXO
PO18 8TZ G0SDC
PO18 9JJ G0IOP
PO18 9LS G7ISD
PO18 9NP G0LNX
PO19 1QZ G4ETX
PO19 3AE G1UFS
PO19 3AN 2E0CGK
PO19 3JL M6GRT
PO19 3LD G4SQJ
PO19 3LY G7BVL
PO19 3NL M6VOL
PO19 3QF G8KJT
PO19 3QU G0WSD
PO19 5RL G4EHG
PO19 5UA G3ZEN
PO19 5UA G4VQZ
PO19 6BY G3MVZ
PO19 6GG M0CXY
PO19 6GL G4ZTQ
PO19 6TD M0OCK
PO19 6TD M6OCK
PO19 7NE 2E1GCF
PO19 7NW 2E0CSH
PO19 7NW M0OSB
PO19 7NW M6BYU
PO19 7UY M6KTK
PO19 7XE G0ISL
PO19 8AU G0CRG
PO19 8AU G4WIR
PO19 8BP M3SYW
PO19 8DE M6CJU
PO19 8QR G1BZU
PO19 8QR G3BZU
PO19 8QR G3CRS
PO19 8QR G3ZDF
PO19 8QR G7DOL
PO19 8TP G8NYK
PO2 0EX G7SCE
PO2 0JZ 2E0KBK
PO2 0JZ M0KWN
PO2 0JZ M6KWN
PO2 0QS M0DND
PO2 0TW G0HCI
PO2 7AY G0SMB
PO2 7DD 2E0BUN
PO2 7DF 2E0BUP
PO2 7HB G7OKI
PO2 7HB M3WCM
PO2 7JW G0DAE
PO2 7LB G8UYY
PO2 7LX M3ZNG
PO2 7PG G4EFB
PO2 7SN M6HCR
PO2 8EX M3AMA
PO2 8LX 2E1HSB
PO2 8NA M0GIF
PO2 8NF G6HTK
PO2 9HE M1DMR
PO2 9RB 2E0EFH
PO2 9RB M0TFH
PO2 9RB M3UYQ
PO20 0EB G8HPZ
PO20 0EE G4KEC
PO20 0HY G4MVS
PO20 0JB M1BCY
PO20 0JG G7SZW
PO20 0JS G4ZPP
PO20 0NA M6SXM
PO20 0RG M5DNK
PO20 0SF M3WLG
PO20 1JY G3KDE
PO20 1JZ G0POU
PO20 1JZ G8CBO
PO20 1NY G0WBR
PO20 1NY M6LUK
PO20 1PA G3NFW
PO20 1PE G4BZB
PO20 2HX 2E0FVL
PO20 2WB M3DSW
PO20 3SJ G4ZJP
PO20 3TH M6VTB
PO20 3TL 2E0WSX
PO20 3TL M0HIX
PO20 3TL M6ARP
PO20 3YR M0BXQ
PO20 7JJ G0LMJ
PO20 7RR G0WVM
PO20 8AH G7BKJ
PO20 8DW G4VRG
PO20 8EX 2E0RXC
PO20 8EX M6RCS
PO20 8NT 2E0YAS
PO20 8NT M6YTS
PO20 8PB G6VLV
PO20 8RG G0BVV
PO20 8RJ G3SFB
PO20 9AX G4YJD
PO20 9AY G6XTK
PO20 9DT M0GOO
PO20 9HL G7HJT
PO21 1AB M1FCC
PO21 1HQ G1GKW
PO21 2DG 2E1FIE
PO21 2EJ G0SZK
PO21 2ET G8EWP
PO21 2JU M6XCZ
PO21 2PY M0MLM
PO21 2QA G0CHE
PO21 2RB G4XPT
PO21 2SF G6FAX
PO21 3AH G3SYG
PO21 3DH G3GUR
PO21 3EL G8OCM
PO21 3EQ G0CIR
PO21 3EZ G0TXU
PO21 3HQ G7IMT
PO21 3ND G0AKK
PO21 3ND M3TUQ
PO21 3SL G4RPA
PO21 4AW M0CME
PO21 4DY G2ARU
PO21 4ET G3LTM
PO21 4HB M0GSC
PO21 4HT G3RJS
PO21 4LH M6GMV
PO21 4LW G7EPR
PO21 4NB G8TGH
PO21 4NN G6HTB
PO21 4PS G8REF
PO21 4QT 2E0NUG
PO21 4QT M0VPL
PO21 4RN G0RQH
PO21 4TB G8YAS
PO21 4TJ G4ITY
PO21 4TN G1XIV
PO21 4UR G6FDU
PO21 4XN G3IJS
PO21 5AD M6DHB
PO21 5LL G0KJU
PO21 5TD G0OSU
PO21 5TW G8ZTD
PO22 0AD G6AII
PO22 0AR M3DOA
PO22 0HF M0KPT
PO22 0JN G1ITL
PO22 0LH G7BWW
PO22 0LR M3DKK
PO22 6AH G7TPH
PO22 6BX G0AFN
PO22 6ED G6UTT
PO22 6HG G1ORB
PO22 6JU M6EIC
PO22 6QH G0ONW
PO22 7NW G4PDY
PO22 7QG G4ECF
PO22 7SE 2E0FAH
PO22 7SE M3FPH
PO22 7SL G0DFA
PO22 8DP G6XJN
PO22 8JJ G3YLL
PO22 8PH 2E0SCK
PO22 8PH M0HJB
PO22 8PH M6BXJ
PO22 9DG G1LHE
PO22 9EX G3SFE
PO22 9HF 2E0BUI
PO22 9LA G0VCJ
PO22 9LY G7SQC
PO3 5DE M3YZE
PO3 5EL M0DDU
PO3 5HG M3CZB
PO3 5LN M6DNJ
PO3 5PU M6JSS
PO3 5TN 2E0JCE
PO3 5TR G8KQV
PO3 5TR G8NVZ
PO3 6AU M1SKA
PO3 6BE M0GMI
PO3 6BG M0ASE
PO3 6HA G6NNS
PO3 6LR G3ZBP
PO30 1AE M6NGK
PO30 1AF G1DEO
PO30 1DG G4MBD
PO30 1DG G7IRH
PO30 1DR G6OOH
PO30 1DT G3XOC
PO30 1HA G4FYI
PO30 1HG G6ORJ
PO30 1HN 2E0BUK
PO30 1LN G0MHN
PO30 1NR 2E1IWD
PO30 1NR M0IWA
PO30 1NR M3DIW
PO30 1PA G0EHR
PO30 1PZ G0KQO
PO30 1QD 2E0UYB
PO30 1QD M3CIG
PO30 1QT G7NAO
PO30 1RE M6HQW
PO30 1RJ G1FMC
PO30 1XN M0GQJ
PO30 2BH M3AYC
PO30 2BH M3BRL
PO30 2BH M3KLS
PO30 2BH M3SYZ
PO30 2DB 2E0AAJ
PO30 2DB M6MPD
PO30 2DP G8AZM
PO30 2JU M0PDL
PO30 2LA M0ZZI
PO30 2LL G8RWM
PO30 2LL M6KFW
PO30 3DD G1KAK
PO30 3HQ G4MXX
PO30 3JP 2E0JIW
PO30 3JP M0PBN
PO30 3JP M3JIW
PO30 3JY M1FBL
PO30 4AE G4BIM
PO30 4BA G8NHM
PO30 4BG G0GMY
PO30 4BG M6NBV
PO30 4DJ G0PXM
PO30 4HH M0CZP
PO30 4JS G1YXG
PO30 4LZ M6FQE
PO30 5BS 2E0DPF
PO30 5BS M6TYD
PO30 5GU 2E0EJK
PO30 5GU M6CUC
PO30 5JJ M3FBR
PO30 5JR M1IOW
PO30 5JR M3IWG
PO30 5NG 2E0DUS
PO30 5NG M6YEG
PO30 5QZ G0RUT
PO30 5QZ G7GGH
PO30 5SF G0DWE
PO30 5SF G1FHB
PO30 5SF M6NKY
PO30 5SG M0OIW
PO30 5SJ M3LJZ
PO30 5SJ M3ZBJ
PO30 5TL M0TVR
PO30 5TL M3MEE
PO30 5TL M6BRT
PO30 5TP G3IMX
PO30 5TY G6CUT
PO30 5UF M6DAZ
PO30 5UH G0LFV
PO30 5UR G7MID
PO30 5XS M6FPJ
PO30 5ZF 2E0BSK
PO30 5ZF M3WHV
PO31 7HF 2E0AGQ
PO31 7HF M3AYJ
PO31 7HW M6PBM
PO31 7JN G6ZUZ
PO31 7ND M6CUW
PO31 7ND M6SPU
PO31 7NF G4GRK
PO31 7NX M6LTS
PO31 7PP M6FQH
PO31 7PS 2E0WKG
PO31 7PS M6WKG
PO31 7PY M3BBF
PO31 7SG M0DKJ
PO31 7SR M3RGP
PO31 8AD G4ZFQ
PO31 8AL G3WXC
PO31 8AS 2E0ENG
PO31 8AS M3ORQ
PO31 8AS M3WKL
PO31 8DP G3XYB
PO31 8DT G3YZK
PO31 8DW 2E0AKQ
PO31 8DW M0JCH
PO31 8DW M3NNV
PO31 8DX G4SCB
PO31 8HQ G3XZB
PO31 8JP G4ZBH
PO31 8JQ M6DYN
PO31 8LB M0TFN
PO31 8NE 2E1KJB
PO31 8NR G0HDH
PO31 8PE G0CWX
PO31 8PN G3PZB
PO31 8PN G3SKY
PO31 8PN G8MBU
PO31 8PT G8RKH
PO31 8PZ G3IW
PO31 8QP G4ZOH
PO31 8RL G4VYC
PO32 6AX G7GZJ
PO32 6EA G7VGY
PO32 6EJ G4UCZ
PO32 6GL M6CYN
PO32 6HR G0PQW
PO32 6HT M3RKR
PO32 6JE M6FWE
PO32 6JE M6LAS
PO32 6JT G7TQC
PO32 6LG M6FDU
PO32 6LS G0PEB
PO32 6LS G7ATJ
PO32 6LS G7RAU
PO32 6LS M0TAM
PO32 6NG G4IKI
PO32 6NT G0JHQ
PO32 6NT G6RTE
PO32 6NT G6TVX
PO32 6PS G0KQR
PO32 6QN G7SVF
PO32 6QN M0GAQ
PO32 6QW M6SGT
PO32 6RU G1TUZ
PO32 6RZ G7RCC
PO32 6SS G3PQJ
PO32 6TD 2E0ZMB
PO32 6TD M3SQG
PO33 1AB G8MYF
PO33 1BX G0BAR
PO33 1BX M0DSF
PO33 1ED G6LXP
PO33 1EL G0MWU
PO33 1HA 2E0CCR
PO33 1HA M3BQM
PO33 1JD G0MIZ
PO33 1NT G4TAT
PO33 1PR G4RSN
PO33 1QY M6DQW
PO33 1TA M0BTP
PO33 1XE G4RGE
PO33 1YF G0SEB
PO33 2BG G1WQC
PO33 2BG M0CGO
PO33 2BH 2E1CDK
PO33 2BH M3AIL
PO33 2HD M3TCU
PO33 2JP G1FPP
PO33 2NY G3VDZ
PO33 2QF G1HRQ
PO33 2QQ G0DRW
PO33 2SS G6LVS
PO33 2UH G3XHM
PO33 2UP G7MAR
PO33 2UX G3KPO
PO33 2UX M0TIW
PO33 3BJ G7UHT
PO33 3BU G6SQL
PO33 3DL G4SBN
PO33 3EL M1OBR
PO33 3EN 2E1EBN
PO33 3HU M3YPP
PO33 3JU G1HCM
PO33 3LH G3XLP
PO33 3NX G1VGM
PO33 3NX G2CNN
PO33 3QS M3JIU
PO33 3SF G4TFW

| | |
|---|---|
| PO33 3TA | G4KZI |
| PO33 3TL | [illegible] |
| PO33 3TL | G0UFK |
| PO33 3TL | G4SHH |
| PO33 3UX | G6HZG |
| PO33 4BB | G4OAG |
| PO33 4ED | G3UHX |
| PO33 4EU | 2E0DUI |
| PO33 4EU | M6FLZ |
| PO33 4JE | M0GUN |
| PO33 4JJ | G4KKB |
| PO33 4JR | G0RSY |
| PO33 4LG | G4BCH |
| PO33 4LG | G4CQO |
| PO33 4LQ | M1DBM |
| PO33 4LU | G6DOD |
| PO33 4NQ | G4YNC |
| PO34 5JE | G0GNI |
| PO35 5QS | 2E0XXB |
| PO35 5QS | M3XXB |
| PO35 5QW | G3JLN |
| PO35 5RA | 2E0MFA |
| PO35 5RA | G3YEG |
| PO35 5RA | M0MFA |
| PO35 5RA | M6IOW |
| PO35 5RY | 2E1PHW |
| PO35 5TN | M0HKK |
| PO35 5TS | G4LUY |
| PO35 5UW | 2E0ZML |
| PO35 5XU | G4MZC |
| PO35 5YJ | G0VZV |
| PO36 0BA | M6JBA |
| PO36 0DX | G1RIR |
| PO36 0JD | G4ANW |
| PO36 0JT | G3GEG |
| PO36 0JY | G8TAQ |
| PO36 0LH | G4OYR |
| PO36 8BA | G0VPO |
| PO36 8BE | 2E0DEI |
| PO36 8BG | M3BKV |
| PO36 8DU | G8IPA |
| PO36 8DZ | 2E0MRD |
| PO36 8DZ | 2E0SUS |
| PO36 8DZ | M6DEL |
| PO36 8HE | G4RTW |
| PO36 8QE | G0WVD |
| PO36 8QL | G4ADK |
| PO36 9BX | G6ETC |
| PO36 9BY | G8FRY |
| PO36 9DS | G4UNM |
| PO36 9HF | G0WLX |
| PO36 9HQ | G6WUR |
| PO36 9HW | G0MPA |
| PO36 9JA | G4JLX |
| PO36 9JA | G4ULT |
| PO36 9JL | G1SMY |
| PO36 9NS | 2E0CZW |
| PO36 9NS | M6SGS |
| PO37 6DB | G8PPN |
| PO37 6EJ | G0NTH |
| PO37 6NN | G8JBM |
| PO37 6NX | G4SVY |
| PO37 7BU | G6EVX |
| PO37 7EJ | G4NOU |
| PO37 7NA | G4WNZ |
| PO37 7NJ | M0JAG |
| PO37 7NZ | G4RTY |
| PO37 7PA | G4UUJ |
| PO38 1AA | G1JYZ |
| PO38 1AL | G4CZP |
| PO38 1AP | G0DFU |
| PO38 1BD | 2E0GZT |
| PO38 1BD | M6BRL |
| PO38 1BH | G1CPO |
| PO38 1BT | 2E0IIT |
| PO38 1DQ | G4ZEN |
| PO38 1NX | G4MPI |
| PO38 1QL | G0LGZ |
| PO38 1RZ | G0RMJ |
| PO38 1TH | G7ONE |
| PO38 2DE | G8DDY |
| PO38 2DZ | M1BOD |
| PO38 2HY | M6SSM |
| PO38 2JN | G4EWE |
| PO38 2NE | M6SKY |
| PO38 2QW | G4XIU |
| PO38 2RD | G8FWE |
| PO38 3BU | M6YEB |
| PO38 3DB | 2E0RSD |
| PO38 3DB | M6BRR |
| PO38 3EF | G6KRN |
| PO38 3EL | M3FBG |
| PO38 3EL | M3FPG |
| PO38 3EQ | M1BWR |
| PO38 3HB | M6DRI |
| PO38 3HR | G0PEF |
| PO38 3HW | M6MYD |
| PO38 3HZ | G3LYD |
| PO38 3NH | G4RZQ |
| PO38 3NT | G4RQP |
| PO38 3PH | 2E0PRB |
| PO38 3PH | M3PRD |
| PO39 0AH | G4NCD |
| PO39 0AL | G4CSM |
| PO39 0BL | G7BZD |
| PO39 0BL | M3BDA |
| PO39 0DN | M3GRI |
| PO39 0DX | M6DRH |
| PO39 0EF | M1CYX |
| PO39 0JL | G0PTT |
| PO4 0BA | G0AXS |
| PO4 0NT | G6ZKM |
| PO4 0RL | M6ROU |
| PO4 8AS | G4YCG |
| PO4 8AS | G6NZ |
| PO4 8AU | G4VFX |
| PO4 8AU | G0OFJ |
| PO4 8HH | G6CUA |
| PO4 8HH | [illegible] |
| PO4 8JG | G4LRH |
| PO4 8JS | G1ERF |
| PO4 8JU | G0IWN |
| PO4 8NF | M3JRN |
| PO4 8NP | G3VXM |
| PO4 8NX | 2E0CSY |
| PO4 8NX | M6BYY |
| PO4 8PB | M1DXO |
| PO4 8RH | 2E0CRU |
| PO4 8RH | M6ASV |
| PO4 8UF | 2E1ILH |
| PO4 8UF | M3HAD |
| PO4 9BG | G3CAJ |
| PO4 9EF | G6DWO |
| PO4 9EL | G4CKW |
| PO4 9HP | M6BRN |
| PO4 9JE | G7UPL |
| PO4 9JU | M3ETI |
| PO4 9LR | G7ARJ |
| PO4 9PZ | G6FDP |
| PO4 9QU | G1PGQ |
| PO4 9XZ | G6SGW |
| PO40 9DL | G1RFB |
| PO40 9ES | G6ENY |
| PO40 9HB | G4NAK |
| PO40 9JH | G7AXM |
| PO40 9LF | G4EDN |
| PO40 9LQ | M3IUC |
| PO40 9NH | G1JGS |
| PO40 9QS | G7DZY |
| PO40 9TG | G7OPY |
| PO40 9UA | G3IIN |
| PO40 9UR | G7RES |
| PO41 0PS | M3LMB |
| PO41 0PY | G0GEZ |
| PO41 0RX | G0DKS |
| PO41 0SA | G3VPK |
| PO41 0SL | G0GPK |
| PO41 0TA | G3OMZ |
| PO41 0TL | G4AZC |
| PO41 0XS | 2E0BHB |
| PO5 1RU | M6MTN |
| PO5 2AJ | G3SQD |
| PO5 2AZ | 2E0HPL |
| PO5 2AZ | M6MSN |
| PO5 2JJ | M6IRC |
| PO5 2NL | G0LFN |
| PO5 3AU | M3JVR |
| PO5 3HP | 2E0IAN |
| PO5 3JA | G0DRL |
| PO5 4AL | M6CXC |
| PO6 1DB | G4IQO |
| PO6 1DU | G3CNO |
| PO6 1DU | M0GWD |
| PO6 1EW | 2E0PWK |
| PO6 1EW | M3PWK |
| PO6 1LB | M3MUB |
| PO6 1LZ | G4GZO |
| PO6 1NB | G6HJV |
| PO6 1NG | G4OVM |
| PO6 1NR | G4ZPA |
| PO6 1PY | 2E0AWE |
| PO6 2ES | G7OWQ |
| PO6 2JE | G8SBQ |
| PO6 2JU | G7DRT |
| PO6 2NL | G4FOW |
| PO6 2PS | G4HUM |
| PO6 2PU | G6UXW |
| PO6 2RL | M0TYW |
| PO6 2TJ | G6XDY |
| PO6 2TX | G0RPX |
| PO6 3DG | G3OQC |
| PO6 3DG | G7HQP |
| PO6 3EU | G0OBA |
| PO6 3HD | G1JAB |
| PO6 3JP | M3HSS |
| PO6 3JP | M3SSU |
| PO6 3PE | 2E0GTT |
| PO6 3PE | M6MSM |
| PO6 3QT | G1OVG |
| PO6 3QY | M6TON |
| PO6 3RD | M0ALF |
| PO6 3RH | 2E0KIT |
| PO6 3RU | M0ADQ |
| PO6 3SB | G4AVX |
| PO6 3SH | M0DNR |
| PO6 4AE | G1XZQ |
| PO6 4AF | M3AUK |
| PO6 4BB | G4PYS |
| PO6 4EZ | M6LBU |
| PO6 4LS | 2E0BUO |
| PO6 4LS | M0LBJ |
| PO6 4LS | M3EOL |
| PO6 4QH | G3RCE |
| PO6 4QL | G4LIO |
| PO6 4QL | G6GYF |
| PO6 4TA | 2E0OEU |
| PO6 4TA | M0NEH |
| PO6 4TA | M6NEH |
| PO7 4QU | 2E0RFM |
| PO7 4QU | M6RFM |
| PO7 4RU | G8CKS |
| PO7 4SP | G0SSY |
| PO7 4SW | M1AHA |
| PO7 5BL | 2E0NFB |
| PO7 5BL | M0NFB |
| PO7 5BL | M6ATF |
| PO7 5BT | G1TDP |
| PO7 5DY | G7DUE |
| PO7 5EB | G6BHB |
| PO7 5ED | G0ABB |
| PO7 5HH | M5KZI |
| PO7 5HJ | G0IFY |
| PO7 5HJ | G0IUY |
| [illegible] | [illegible] |
| PO7 5QF | M0CYM |
| PO7 5QR | G1WXW |
| PO7 5QW | G6ISY |
| PO7 5SF | G4GUA |
| PO7 5TB | G7IUE |
| PO7 5TW | M3GTK |
| PO7 6AA | G1UFA |
| PO7 6AH | M0TAP |
| PO7 6AH | M3XYI |
| PO7 6AQ | M3TQY |
| PO7 6BG | G4CRM |
| PO7 6BT | G0JWL |
| PO7 6BX | M0JAK |
| PO7 6DD | G0RPV |
| PO7 6DP | G0ASZ |
| PO7 6EB | G4JIH |
| PO7 6ED | G3KOJ |
| PO7 6EG | M3AEE |
| PO7 6HE | G4TST |
| PO7 6HH | M1EXJ |
| PO7 6LJ | G6XBG |
| PO7 6PE | G8FXV |
| PO7 6PR | G4WQZ |
| PO7 6PR | G6RST |
| PO7 6PR | G8IRS |
| PO7 6SH | G6CZZ |
| PO7 6UA | G4DCP |
| PO7 6UE | G3RTP |
| PO7 6YJ | G8LNU |
| PO7 7BA | 2E0SJN |
| PO7 7BD | G3YZY |
| PO7 7BX | M0SSP |
| PO7 7BX | M3SHI |
| PO7 7EW | 2E0CKI |
| PO7 7EW | M6BCG |
| PO7 7HX | G0ATG |
| PO7 7JE | G1OGR |
| PO7 7LG | G8VOI |
| PO7 7LG | M3VOI |
| PO7 7LN | M5ACT |
| PO7 7NA | M3KYZ |
| PO7 7NY | G0DNF |
| PO7 7PB | G3TVI |
| PO7 7PE | M6DTM |
| PO7 7PG | G3LKW |
| PO7 7QB | G0PSF |
| PO7 7QQ | G3WYT |
| PO7 7RG | G1FMT |
| PO7 7RP | M3ZQJ |
| PO7 7SB | G0MUK |
| PO7 7UB | G8PIQ |
| PO7 7XP | G4VIQ |
| PO7 8AG | G4MKQ |
| PO7 8AL | M6TID |
| PO7 8BP | M0GMU |
| PO7 8BP | M0JDS |
| PO7 8BX | 2E0TRW |
| PO7 8BX | G4FBS |
| PO7 8BX | M0KTT |
| PO7 8BX | M6BSA |
| PO7 8JN | G4SAC |
| PO7 8ND | G8KOS |
| PO7 8QD | G8OKE |
| PO7 8RS | G6BQE |
| PO8 0HF | 2E0FMA |
| PO8 0HF | M6TAU |
| PO8 0JF | M6BNU |
| PO8 0JX | G4BQV |
| PO8 0JX | M6EYZ |
| PO8 0LD | G0GDV |
| PO8 0LJ | G0VOJ |
| PO8 0LQ | G7EYS |
| PO8 0NF | G1PPD |
| PO8 0NQ | 2E0HES |
| PO8 0NQ | M6GEV |
| PO8 0NU | G4VNR |
| PO8 0PD | G0EDM |
| PO8 0PJ | G4JCX |
| PO8 0PJ | G8TZE |
| PO8 0RH | 2E0REC |
| PO8 0RH | M6MCU |
| PO8 0RH | M6TUD |
| PO8 0TX | G7IWA |
| PO8 8BG | G3RRY |
| PO8 8DY | M3WPJ |
| PO8 8EW | G0NPE |
| PO8 8HS | G3ZFF |
| PO8 8HX | G3WLY |
| PO8 8JE | G0NHZ |
| PO8 8JE | M3MXI |
| PO8 8JJ | G8LNC |
| PO8 8PN | G3VCR |
| PO8 8QH | G3TKN |
| PO8 8RP | 2E1COC |
| PO8 8RU | G8YTR |
| PO8 8RX | 2E1BOM |
| PO8 8SF | G0JRN |
| PO8 8SG | G3XWM |
| PO8 8SQ | G3LIK |
| PO8 8SQ | G4FOC |
| PO8 8TS | 2E0BSX |
| PO8 8UB | G6YSO |
| PO8 8UB | M3GVC |
| PO8 9BE | G6EHL |
| PO8 9DA | M3KUQ |
| PO8 9DA | M3YDB |
| PO8 9EW | G6WXK |
| PO8 9HE | G0MGH |
| PO8 9JL | 2E0BVO |
| PO8 9JL | M3YIV |
| PO8 9NX | M0WAH |
| PO8 9QH | G0KUA |
| PO8 9QU | G8UVF |
| PO8 9RD | G0JDH |
| [illegible] | [illegible] |
| PO8 9TJ | G0UHM |
| PO8 9UB | 2E0BRC |
| PO8 9UB | G7GNA |
| PO8 9UB | M3TXR |
| PO8 9UY | M3CWA |
| PO8 9XF | G3MYI |
| PO9 1AG | M3YIE |
| PO9 1AG | M6KLH |
| PO9 1HZ | G1JXL |
| PO9 1LQ | M1DPJ |
| PO9 1QS | G0LUB |
| PO9 1RL | G6IOV |
| PO9 1RN | G0VKX |
| PO9 2BS | M3FRS |
| PO9 2ET | G4TLO |
| PO9 2HR | G1FFU |
| PO9 2NQ | M0HNO |
| PO9 2PU | 2E0JAZ |
| PO9 2PU | M0JAZ |
| PO9 2PU | M6JRS |
| PO9 2QX | G6TQZ |
| PO9 2RD | G0BXV |
| PO9 2RP | G8RUX |
| PO9 2RW | G8LWC |
| PO9 2RZ | G4XQZ |
| PO9 2TN | G6WBX |
| PO9 2UW | G0JEZ |
| PO9 2UX | M1BIY |
| PO9 2XE | 2E1FPP |
| PO9 2XE | M3LDQ |
| PO9 2XE | M3OQQ |
| PO9 3AA | G8VEZ |
| PO9 3AZ | G3YYK |
| PO9 3BW | G6XRH |
| PO9 3DL | M3EYS |
| PO9 3JZ | G7VDN |
| PO9 3JZ | M1BSX |
| PO9 3LX | 2E0TEE |
| PO9 3LX | M0YEE |
| PO9 3LX | M6AXI |
| PO9 3NJ | 2E1GZV |
| PO9 3NJ | G4UXJ |
| PO9 3NS | M6FEL |
| PO9 3PL | G0FYX |
| PO9 3RA | G4RGO |
| PO9 3RS | M0HRH |
| PO9 3RS | M3FEA |
| PO9 4HR | G7WGI |
| PO9 4JA | M6KAQ |
| PO9 4LG | G7UCL |
| PO9 4LG | M3IVY |
| PO9 4NS | M6PJD |
| PO9 4PT | 2E0GMS |
| PO9 4PT | M6GHB |
| PO9 4QY | M0CBG |
| PO9 5AR | M3EDS |
| PO9 5BJ | G7EYV |
| PO9 5DZ | G8XNO |
| PO9 5ED | M6SUI |
| PO9 5HU | G0DOK |
| PO9 5LA | M0RWS |
| PO9 5LA | M3ZRS |
| PO9 5LH | G0EOI |
| PO9 5LS | 2E0LGT |
| PO9 5PW | G0ERS |
| PO9 5RY | G0BSJ |
| PO9 6AG | G3TSM |
| PO9 6DA | G6XCV |
| PO9 6DQ | G0BAG |
| PO9 6ED | 2E0ZKD |
| PO9 6HN | M3YNB |

## PR

(Preston)

| | |
|---|---|
| PR1 0BN | G6LOC |
| PR1 0DT | 2E0BYW |
| PR1 0EE | M0DOM |
| PR1 0JL | M3OHO |
| PR1 0LL | G4LKM |
| PR1 0TD | M3DXN |
| PR1 0UR | G1PUK |
| PR1 0UX | G4RFA |
| PR1 0XN | M0BKS |
| PR1 0XN | M3AUL |
| PR1 0XQ | G6EPX |
| PR1 0YE | G0EIF |
| PR1 1QW | M5KEN |
| PR1 1RY | M3OEE |
| PR1 2YJ | M6GBW |
| PR1 2YL | G1VTQ |
| PR1 3NA | 2E0NAP |
| PR1 3NA | M6DOJ |
| PR1 4NJ | G6PFZ |
| PR1 4TS | 2E0RAF |
| PR1 4UD | G3NQX |
| PR1 4UD | G6NKI |
| PR1 4YB | G7RCK |
| PR1 5HJ | G3KUE |
| PR1 5HJ | G4WYH |
| PR1 5TA | G7ING |
| PR1 5TB | M6SJW |
| PR1 5TP | G8RIP |
| PR1 5TR | G4WXI |
| PR1 5UY | G6WXS |
| PR1 5XE | M3WHB |
| PR1 6HN | M6FBV |
| PR1 6HP | 2E0ZDX |
| PR1 6HP | M6DCB |
| PR1 6NS | M6UDS |
| PR1 6QN | G4THA |
| PR1 7LL | M6CCB |
| [illegible] | [illegible] |
| PR1 8EL | G6ORT |
| PR1 8PJ | M0LDI |
| PR1 8TP | G0JEH |
| PR1 9DD | G7JZJ |
| PR1 9DR | G1HMY |
| PR1 9EL | G3BWI |
| PR1 9EQ | G4WQT |
| PR1 9EQ | G4ZKA |
| PR1 9HA | M6CNK |
| PR1 9HD | M3CMM |
| PR1 9NG | M0HIQ |
| PR1 9RH | G4FSJ |
| PR1 9RP | G7TCB |
| PR1 9SY | M6FZE |
| PR1 9TB | M0UWS |
| PR2 1BH | M3ZIX |
| PR2 1JD | G6EUG |
| PR2 1JP | G7VAS |
| PR2 1PB | 2E0ACV |
| PR2 1RX | M0NDU |
| PR2 1RX | M6JFK |
| PR2 1SH | G6MCC |
| PR2 1TY | M6LFZ |
| PR2 2AS | M3KEC |
| PR2 2HH | 2E0FJD |
| PR2 2HH | M6FJD |
| PR2 2HH | M6FLL |
| PR2 2PX | M6TCX |
| PR2 2YW | M1AMW |
| PR2 3EX | 2E1AFH |
| PR2 3EX | G4UQI |
| PR2 3FQ | G0SBY |
| PR2 3GA | G7TRL |
| PR2 3HS | G7ILX |
| PR2 3JL | 2E0ZPN |
| PR2 3LP | G7AQD |
| PR2 3RU | M3JDN |
| PR2 3RU | M3MDN |
| PR2 3RY | 2E1AFI |
| PR2 3RY | G0LCR |
| PR2 3RY | G4RPW |
| PR2 3RY | G7CLR |
| PR2 3RY | G7PZE |
| PR2 3SX | G0KSN |
| PR2 3UU | G7IFM |
| PR2 3UU | G7LPF |
| PR2 3YR | G4LHR |
| PR2 3YS | M0AKQ |
| PR2 3YY | M0WCM |
| PR2 3ZT | G4UTC |
| PR2 6AN | M6AQY |
| PR2 6AN | M6RRA |
| PR2 6DA | 2E1IJM |
| PR2 6DH | G3SYA |
| PR2 6EX | M0DMD |
| PR2 6EY | G4HQA |
| PR2 6TH | 2E1FJP |
| PR2 7BE | 2E0FMB |
| PR2 7DA | 2E0RFD |
| PR2 7DA | M0RDZ |
| PR2 7DA | M3OSS |
| PR2 7DA | M3URD |
| PR2 8GT | G3OIH |
| PR2 8NY | G7KXV |
| PR2 8XN | G7PSZ |
| PR2 8XN | M3PSZ |
| PR2 9AW | M0RWH |
| PR2 9FP | 2E0CXW |
| PR2 9FP | M6AYH |
| PR2 9QA | G7OFU |
| PR2 9QX | G1VZW |
| PR2 9QX | G7OET |
| PR2 9RF | 2E0HMM |
| PR2 9RF | M3UCO |
| PR2 9SQ | G3RSM |
| PR2 9SS | G4PNH |
| PR2 9TP | M3OEB |
| PR2 9YJ | M0KHS |
| PR25 1AR | M3ULI |
| PR25 1BH | G3ZRE |
| PR25 1BL | M6CGL |
| PR25 1HT | 2E0BBN |
| PR25 1JA | M6HHA |
| PR25 1JD | G3UPY |
| PR25 1JH | M6JTD |
| PR25 1HJ | G4TWG |
| PR25 1RN | 2E0BSB |
| PR25 1RN | G3WSW |
| PR25 1RN | M1CND |
| PR25 1RN | M3TPW |
| PR25 1YB | G0JSL |
| PR25 1YB | G0JSM |
| PR25 2DD | G0MAH |
| PR25 2LJ | M0HPR |
| PR25 2XA | M6TZU |
| PR25 2XW | M6AQF |
| PR25 2YL | G0IDE |
| PR25 3AA | G3GGS |
| PR25 3AP | G4SBQ |
| PR25 3AR | 2E0OII |
| PR25 3AR | G0NGE |
| PR25 3AR | G6VBK |
| PR25 3AR | M0PQI |
| PR25 3AR | M6AQK |
| PR25 3BD | G6LUF |
| PR25 3HA | M6PCD |
| PR25 3NR | G1IQF |
| PR25 3NS | G1RBZ |
| PR25 3UH | G0FDX |
| PR25 3UH | G0GVA |
| PR25 4QZ | G4YIA |
| PR25 4QZ | M6HBU |
| PR25 4UX | G0FQN |
| PR25 4XL | M0CYE |
| PR25 4XP | M1EYQ |
| PR25 4XT | G3KQY |
| [illegible] | [illegible] |
| PR25 4YJ | G4AYU |
| PR25 4ZR | G0HKW |
| PR25 5PA | G1FKT |
| PR25 5PD | G1TTL |
| PR25 5PD | M0FWO |
| PR25 5PJ | M3SHQ |
| PR25 5RQ | 2E0ZMR |
| PR25 5RQ | M3ZMR |
| PR25 5SP | G3XII |
| PR25 5SX | 2E0CAR |
| PR26 6QA | M3DKZ |
| PR26 6QS | 2E0HPF |
| PR26 7AJ | 2E0GUY |
| PR26 7AJ | M3XXS |
| PR26 7QJ | G0KLT |
| PR26 7QJ | M3KLT |
| PR26 7XJ | 2E0BDJ |
| PR26 7XT | G0JHC |
| PR26 7XT | G6ZGO |
| PR26 8LB | G0VHO |
| PR26 8NP | G7VOQ |
| PR26 9AP | G0PFU |
| PR26 9HP | G0JSJ |
| PR3 0BB | G6LNS |
| PR3 0JY | M3ZRA |
| PR3 0RH | 2E0IBI |
| PR3 0RH | M6IBI |
| PR3 0XD | M6HUM |
| PR3 1AA | 2E0PNC |
| PR3 1AA | M3LJI |
| PR3 1AD | G8CWQ |
| PR3 1BA | G7CUA |
| PR3 1FJ | G0LXP |
| PR3 1FS | G4GVG |
| PR3 1LH | G4JCG |
| PR3 1NL | G4AMY |
| PR3 1NQ | M0DKL |
| PR3 1PD | G4TVN |
| PR3 1PL | G4IAL |
| PR3 1QF | G0PMY |
| PR3 1RD | G1HKR |
| PR3 1RD | G4BSD |
| PR3 1RD | G7FNM |
| PR3 1RD | M3GDI |
| PR3 1RD | M3TAW |
| PR3 1RF | G6AIZ |
| PR3 1YL | G0VWX |
| PR3 1YQ | M0SHM |
| PR3 1YQ | M3PRY |
| PR3 2BQ | G8CQV |
| PR3 2JX | M3RZF |
| PR3 2LH | M3BTI |
| PR3 2NA | M3WKK |
| PR3 2XB | G0SDJ |
| PR3 3EL | G3WGK |
| PR3 3SL | G4VOJ |
| PR3 3SY | G3NNA |
| PR3 3TB | M3FSQ |
| PR3 3TH | M6BTK |
| PR3 3TH | M6IFH |
| PR3 3TQ | G6MAC |
| PR3 3TQ | G8XJT |
| PR3 3TX | G3LZO |
| PR3 3TX | G7NOI |
| PR3 3TY | G3NKL |
| PR3 3UA | G0HJB |
| PR3 3WD | G1CFG |
| PR3 3WN | G6RTD |
| PR3 3YS | M3ZZQ |
| PR3 5HB | G0KMP |
| PR3 5JN | G6DDR |
| PR3 6AB | G1HJO |
| PR3 6AB | M1AXP |
| PR3 6AP | G3LPL |
| PR3 6BD | M3NPX |
| PR3 6BN | G6HCF |
| PR3 6SS | G0IYT |
| PR4 0HH | G7WEM |
| PR4 0HH | M3TDH |
| PR4 0NP | M6DFZ |
| PR4 0PA | M3EPC |
| PR4 0TT | M6GNN |
| PR4 1AJ | M3HVY |
| PR4 1DY | G1TBN |
| PR4 1DN | G4COQ |
| PR4 1EG | G7NOQ |
| PR4 1EN | 2E0TLD |
| PR4 1HG | G4WIM |
| PR4 1JL | M3EWZ |
| PR4 1JN | G8HEU |
| PR4 1PT | M6CEB |
| PR4 1RG | G4RNF |
| PR4 1RL | G4RCF |
| PR4 1RQ | M3TVK |
| PR4 1RX | G0CPJ |
| PR4 1SB | 2E0UBW |
| PR4 1UD | G7CUL |
| PR4 1UG | G4MFF |
| PR4 1UQ | G7GVP |
| PR4 1WA | 2E0BLL |
| PR4 1XT | M3EWY |
| PR4 1XU | M3EWW |
| PR4 1XU | M3MHD |
| PR4 1XY | 2E0NHM |
| PR4 1YD | G7LPD |
| PR4 2AY | G1GQY |
| PR4 2AY | G1GQZ |
| PR4 2DS | G8OTZ |
| PR4 2EL | G4AHZ |
| PR4 2NQ | M1AKF |
| PR4 2XA | G7JGW |
| PR4 2XA | M3JGW |
| PR4 2ZA | G1TUI |
| PR4 3AG | M1FHT |
| [illegible] | [illegible] |
| PR4 3BB | M3VXY |
| PR4 3BQ | M6TSI |
| PR4 3DN | M1ACJ |
| PR4 3PN | G3TKK |
| PR4 3SU | M3HVX |
| PR4 3SX | 2E1ECL |
| PR4 3SX | 2E1FJL |
| PR4 3SX | M0ABK |
| PR4 3SX | M0GJA |
| PR4 3TU | G4MRX |
| PR4 3TX | G8MED |
| PR4 3UD | G0HIJ |
| PR4 3UD | G1UCG |
| PR4 3UQ | 2E0KMZ |
| PR4 3UQ | M1HZZ |
| PR4 3UQ | M6EQE |
| PR4 4JD | G3VBL |
| PR4 4JX | 2E0AFL |
| PR4 4RQ | G0LEE |
| PR4 4XJ | G0LQK |
| PR4 5AX | 2E0JAR |
| PR4 5BB | 2E0CDE |
| PR4 5BH | G1MBN |
| PR4 5BX | G1IPY |
| PR4 5HB | G4ZCG |
| PR4 5NP | G4WAL |
| PR4 5NP | G6TNA |
| PR4 5QD | G3PMO |
| PR4 6AA | G0TUC |
| PR4 6AT | G0EHK |
| PR4 6HD | G4MGB |
| PR4 6JS | M6HOU |
| PR4 6JX | G0VAV |
| PR4 6JX | M1AUZ |
| PR4 6LY | G6TMN |
| PR4 6RB | G4OWS |
| PR4 6SX | M3LNM |
| PR4 6TD | G6EWH |
| PR4 6TR | G0JJD |
| PR4 6UD | G6PLT |
| PR4 6UL | G3MPF |
| PR4 6US | M3VOY |
| PR5 0BB | M3CJH |
| PR5 0DT | G3PS |
| PR5 0JR | M0DBT |
| PR5 0LA | M0PVP |
| PR5 0LX | M0FYA |
| PR5 4BP | G4BGP |
| PR5 4JX | 2E0PCA |
| PR5 4JX | M0NED |
| PR5 4TT | M3BAN |
| PR5 4UT | G3XUH |
| PR5 5HH | G3ZOC |
| PR5 5LA | M6BXD |
| PR5 5RA | G0EHW |
| PR5 5TY | M0ETS |
| PR5 5UP | G3AZI |
| PR5 5UU | G1BTN |
| PR5 5UW | G0DPG |
| PR5 5UW | M3DPG |
| PR5 5YB | G3ZVQ |
| PR5 6EP | G0FQC |
| PR5 6FT | G4TUF |
| PR5 6LS | M3NSM |
| PR5 6RA | 2E0DZR |
| PR5 6RA | M6CKV |
| PR5 6TD | M6MLY |
| PR5 6UY | G0ASH |
| PR5 6XR | G1DER |
| PR5 8DD | G0EIG |
| PR5 8DS | 2E1HQY |
| PR5 8DU | G3UCA |
| PR5 8EN | 2E0VMA |
| PR5 8EN | G0VMA |
| PR5 8EN | M6VMA |
| PR5 8HJ | M1DMH |
| PR5 8HT | G4GOM |
| PR5 8JX | G0MXU |
| PR6 0AG | M6SRZ |
| PR6 0DB | M0AQE |
| PR6 0DB | M0ISN |
| PR6 0DG | G0USM |
| PR6 0JD | M1CEC |
| PR6 0LL | G0JMN |
| PR6 0NJ | M6ERW |
| PR6 0PU | 2E0MEO |
| PR6 0PY | G7SYY |
| PR6 0PY | M3SYY |
| PR6 7AN | 2E0DEX |
| PR6 7AQ | 2E1BVQ |
| PR6 7AU | G0HBX |
| PR6 7BG | 2E0NYF |
| PR6 7BG | M6NYF |
| PR6 7BJ | G0CUN |
| PR6 7DN | 2E0GBI |
| PR6 7DN | M6LSB |
| PR6 7HF | G8COR |
| PR6 7TT | 2E0LON |
| PR6 7TT | M0ZWT |
| PR6 7TT | M6DFT |
| PR6 7TZ | G6JJI |
| PR6 8EJ | G1TSV |
| PR6 8PJ | G4BOB |
| PR6 8PX | G7KTQ |
| PR6 8UE | G8JZI |
| PR6 9DA | M1CGI |
| PR6 9DA | M1CGJ |
| PR6 9DU | G6AHO |
| PR6 9ET | M0HGD |
| PR6 9LA | G0LBT |
| PR6 9LA | M3KEY |
| PR6 9LA | M3NEL |
| PR6 9NQ | 2E0ZRG |
| PR6 9NQ | M5SQV |
| [illegible] | [illegible] |
| PR6 9PD | G3GSL |
| PR6 9RS | G1KVC |
| PR6 9SB | G6MEI |
| PR7 1LX | G4WYZ |
| PR7 1PH | G4TZK |
| PR7 1RE | 2E0DHT |
| PR7 1RE | G7KUR |
| PR7 1RE | M6GCE |
| PR7 1RH | M0CGA |
| PR7 1UH | G0DAI |
| PR7 2HL | M0DMI |
| PR7 2JA | M0DNW |
| PR7 2JA | M6NNA |
| PR7 2JB | 2E1IKA |
| PR7 2JB | M1SMF |
| PR7 2JG | M1EHI |
| PR7 2JW | M6RKO |
| PR7 2NT | M6RKL |
| PR7 2YB | 2E0SUD |
| PR7 2YB | 2E0WOZ |
| PR7 3AP | M3JQW |
| PR7 3HS | G0UZF |
| PR7 3NH | G4IYP |
| PR7 3QS | M6GIP |
| PR7 4AJ | G3RYY |
| PR7 4AN | M6SEC |
| PR7 4JU | G3URK |
| PR7 4NL | G7PYW |
| PR7 4NS | G1DAK |
| PR7 4PH | M0GED |
| PR7 4PT | G0NGK |
| PR7 5AE | M6KPG |
| PR7 5EH | G4WGT |
| PR7 5HH | G3JMZ |
| PR7 5NR | M6COP |
| PR7 5NY | G0OWA |
| PR7 5NY | M6OWA |
| PR7 5NZ | G1MBE |
| PR7 5PW | G1AHM |
| PR7 5PY | G0KDX |
| PR7 5QW | G8SNQ |
| PR7 5RF | M6LSH |
| PR7 5SN | M6BFZ |
| PR7 5TW | 2E0YPJ |
| PR7 5TW | M0YPJ |
| PR7 5TW | M3YPJ |
| PR7 6AS | G7DKY |
| PR7 6BA | G8GFB |
| PR7 6BE | 2E0IDA |
| PR7 6BP | G4PAT |
| PR7 6BS | G1ZTG |
| PR7 6BU | G3ANG |
| PR7 6JW | G6ZOL |
| PR7 6LY | G7GYR |
| PR7 6PD | G0WTD |
| PR7 6PD | G7NER |
| PR7 6PL | G0SHU |
| PR7 6PN | G4PQM |
| PR7 6PP | M6AQI |
| PR7 6PT | G0CUB |
| PR7 6PT | G3RPO |
| PR7 6PW | G0WTM |
| PR7 6PW | G7SSJ |
| PR7 7AG | M3FTW |
| PR7 7DN | G7UAY |
| PR8 1JA | G4ZYN |
| PR8 1JH | G6EZY |
| PR8 1LG | 2E0PAJ |
| PR8 1NQ | G0RLT |
| PR8 1RS | G3TTY |
| PR8 1RT | G4IQK |
| PR8 2FB | G7VJU |
| PR8 2FY | M6CJA |
| PR8 2HF | G2ART |
| PR8 2JJ | G3WTB |
| PR8 2LW | G4BEU |
| PR8 2NS | G0EKX |
| PR8 2QF | G7MJS |
| PR8 2QF | G7OAA |
| PR8 2QW | G4JYQ |
| PR8 2RR | G6JUQ |
| PR8 2RS | G6ILX |
| PR8 3DB | G3ZII |
| PR8 3DU | 2E0NEL |
| PR8 3DU | M3WFY |
| PR8 3DW | M6JHQ |
| PR8 3EQ | M0GZS |
| PR8 3HE | M3WDU |
| PR8 3NP | G8CIX |
| PR8 3QF | 2E0TDF |
| PR8 3QF | M0MVD |
| PR8 3RP | G1DFT |
| PR8 3RS | G1UCI |
| PR8 3SZ | M1CWW |
| PR8 3UA | M3WFL |
| PR8 4BJ | G7CVF |
| PR8 4DT | M3JGX |
| PR8 4EQ | 2E0VAM |
| PR8 4EQ | M3XMH |
| PR8 4EQ | M6AJH |
| PR8 4EQ | M6ALH |
| PR8 4HA | G0MPW |
| PR8 4HA | G7CCH |
| PR8 4HH | G3HWS |
| PR8 4NH | 2E0YTT |
| PR8 4NH | M3LPQ |
| PR8 4SF | 2E0OIL |
| PR8 4SF | M0MKO |
| PR8 4SF | M6OHL |
| PR8 5HB | M6SEA |
| PR8 5HS | G4SBE |
| PR8 5LP | G0JCD |
| PR8 6HF | M3FNH |
| [illegible] | [illegible] |
| PR8 6JP | M6CER |
| PR8 6JW | G0PNA |
| PR8 6NA | G7TIM |
| PR8 6NR | M3UYU |
| PR8 6PY | 2E0BUJ |
| PR8 6PY | M3WJV |
| PR8 6PY | M6ARU |
| PR8 6PY | M6EDI |
| PR8 6SQ | M3LBG |
| PR8 6XH | G3ORK |
| PR9 0QT | 2E0FAB |
| PR9 0TW | G3YIB |
| PR9 7AA | G3STT |
| PR9 7BE | G0LZX |
| PR9 7BE | G1VWL |
| PR9 7BE | G6FVM |
| PR9 7BY | G1EPS |
| PR9 7DX | M6AIW |
| PR9 7EZ | M3KDO |
| PR9 7HR | M1BUL |
| PR9 7HX | G4EVC |
| PR9 7JU | G6DDC |
| PR9 7JX | G3OCR |
| PR9 7JX | G8ILJ |
| PR9 7JX | M3YUD |
| PR9 7PQ | G3HAA |
| PR9 8DG | G0HRT |
| PR9 8ET | G6CIO |
| PR9 8LS | M3UGK |
| PR9 8NL | G0JCQ |
| PR9 8NL | G2OA |
| PR9 8NL | G4VYP |
| PR9 8NY | 2E0CXD |
| PR9 8NY | G4NMU |
| PR9 8NY | M0HOQ |
| PR9 8NY | M6JEY |
| PR9 8PS | G0OFY |
| PR9 8QR | 2E0VEF |
| PR9 8QW | G4UPK |
| PR9 9BT | G4DRA |
| PR9 9FA | M3TCX |
| PR9 9GA | G4SCO |
| PR9 9GJ | G0PVU |
| PR9 9GJ | G1ZMJ |
| PR9 9QY | G7HRQ |
| PR9 9RN | G8RON |
| PR9 9TW | M3HQN |
| PR9 9XE | G4TUP |
| PR9 9XF | 2E0CWO |
| PR9 9XF | M6CWD |
| PR9 9XW | G3TMB |
| PR9 9XZ | G4FMQ |
| PR9 9YX | G4YPH |

## RG

(Reading)

| | |
|---|---|
| RG1 2RE | M3FVA |
| RG1 3LP | G0RSR |
| RG1 3LP | G6ZTZ |
| RG1 3QQ | 2E0KGB |
| RG1 3QQ | M0TLN |
| RG1 3QQ | M3WNX |
| RG1 4QD | 2E0GUT |
| RG1 4QD | M6GUT |
| RG1 4RF | G7IYF |
| RG1 5DU | M6PBX |
| RG1 5LR | G6FBA |
| RG1 5LW | G4RSC |
| RG1 5QP | 2E0PIK |
| RG1 5QP | M0PIK |
| RG1 5QP | M3ZUY |
| RG1 5RD | G3WNP |
| RG1 5SD | M1XTN |
| RG1 5SD | M3DKG |
| RG1 5SD | M3VSL |
| RG1 6AS | 2E0PIX |
| RG1 6AS | M6LOK |
| RG1 6QD | G6YLN |
| RG1 6QE | M0HAO |
| RG1 7HT | G6SBN |
| RG1 7PA | M6DUD |
| RG1 7TY | G4JPE |
| RG1 7UG | G4FLY |
| RG1 8DL | M0NGC |
| RG1 8EN | G8WBN |
| RG1 8QS | 2E0DOP |
| RG1 8QS | M6ENK |
| RG10 0AU | M0OUL |
| RG10 0AY | M3BBS |
| RG10 0BL | 2E0TSM |
| RG10 0BL | M0XSM |
| RG10 0BL | M3UTK |
| RG10 0JB | G1ZQR |
| RG10 0SF | G8LZG |
| RG10 8BH | M1FAI |
| RG10 8BJ | G8KZG |
| RG10 8BN | G3WPH |
| RG10 8DH | 2E0JBL |
| RG10 8DR | M0OAA |
| RG10 8DR | M6BPQ |
| RG10 9AX | 2E0BPT |
| RG10 9AX | M0BSJ |
| RG10 9AX | M3VYM |
| RG10 9AY | G3LRQ |
| RG10 9BN | G8NXJ |
| RG10 9BN | M3NXJ |
| RG10 9BT | 2E0CJB |
| RG10 9BT | M0TVA |
| RG10 9BT | M3ZAW |
| RG10 9BT | M6RUB |
| RG10 9ED | G7GEF |
| RG10 9JG | G6AWF |
| RG10 9LJ | 2E0UDA |
| RG10 9LJ | M0UDA |
| RG10 9LU | M0UUB |
| RG10 9PY | G3VKQ |
| RG10 9PY | M6MCE |
| RG10 9QD | M1BFO |
| RG10 9QF | G7REH |
| RG10 9QT | M6LSG |
| RG10 9SJ | M6EFU |
| RG10 9TS | G8EPZ |
| RG10 9YD | G4SYE |
| RG10 9YT | G8KOM |
| RG12 0TB | M6MBY |
| RG12 0TN | M6PCM |
| RG12 0TN | G1HLQ |
| RG12 0TR | G6YLA |
| RG12 0TW | M3JBM |
| RG12 0UA | 2E0WWS |
| RG12 0UA | M3YXW |
| RG12 0UD | G7IBU |
| RG12 0UR | 2E1AFN |
| RG12 0XE | G3ZIR |
| RG12 0XU | G4DDL |
| RG12 2JR | G4MXY |
| RG12 2JW | M3YNJ |
| RG12 2JW | M6SER |
| RG12 2LU | G8RKO |
| RG12 2LU | M3XBZ |
| RG12 2PT | G0NEF |
| RG12 2PT | M6NEF |
| RG12 2QG | G3RXM |
| RG12 2QP | M3WVI |
| RG12 2RA | G4YBG |
| RG12 2SE | G0OVA |
| RG12 7BE | M6EXZ |
| RG12 7DU | G4BZJ |
| RG12 7ET | G7FSR |
| RG12 7LD | G4CQH |
| RG12 7PS | G6NFC |
| RG12 7QA | G8AMK |
| RG12 7QG | G4DDN |
| RG12 7QG | M0BRA |
| RG12 7QQ | G0IRH |
| RG12 7RX | G1BBI |
| RG12 7TF | G4AUC |
| RG12 7WF | G3TKS |
| RG12 7WG | G4KNZ |
| RG12 7WH | G0CGE |
| RG12 7WL | G0SCY |
| RG12 7WZ | M6PEO |
| RG12 7YX | G0FCT |
| RG12 7YX | G6BRA |
| RG12 8AP | G1MHP |
| RG12 8QU | G0UIS |
| RG12 8QX | 2E0TVD |
| RG12 8QX | M0HQH |
| RG12 8QX | M6SCP |
| RG12 8QY | 2E0INT |
| RG12 8QY | M6CTA |
| RG12 8UD | 2E0XKT |
| RG12 8UD | M0KJT |
| RG12 8UD | M6XKT |
| RG12 8UP | G0AQU |
| RG12 8UQ | G1CPX |
| RG12 8XE | G7JDF |
| RG12 8XP | G1RXV |
| RG12 8XU | G6TNR |
| RG12 8XY | G4PCF |
| RG12 8YD | 2E0APZ |
| RG12 8YD | M6YPZ |
| RG12 8YG | M3NAH |
| RG12 8ZJ | G6DNL |
| RG12 9BY | 2E1JKP |
| RG12 9EF | G1GUI |
| RG12 9EF | M1KIP |
| RG12 9ES | G4WYC |
| RG12 9EX | G3YMC |
| RG12 9EY | G6VJU |
| RG12 9HT | G7USG |
| RG12 9JE | M1CAX |
| RG12 9JL | G3OFW |
| RG12 9NP | G8HZL |
| RG12 9PA | G7SEJ |
| RG12 9PS | M6AFU |
| RG12 9PX | G1VXX |
| RG12 9QH | G3ZWP |
| RG12 9QX | G0NQN |
| RG12 9TY | M6AQV |
| RG12 9YL | G0VQX |
| RG14 1LL | G0CKU |
| RG14 1RI | G8NHG |
| RG14 1TR | G6DIC |
| RG14 1UH | G3OPG |
| RG14 2FE | M0AAP |
| RG14 2HA | G6OEW |
| RG14 2HA | G6XZS |
| RG14 2HR | G1DAV |
| RG14 2JL | M3YXQ |
| RG14 2LS | 2E0POQ |
| RG14 2LS | M0POQ |
| RG14 2LO | M0POQ |
| RG14 2ND | G3OUC |
| RG14 2PN | G6RBP |
| RG14 2QD | G6AZG |
| RG14 2RT | G7VNJ |
| RG14 2TH | G7HUO |
| RG14 2TH | G7HUP |
| RG14 3AJ | M6GMR |
| RG14 3JA | 2E0NGB |
| RG14 5JA | M0JEC |
| RG14 5JA | M0NHK |
| RG14 5JA | M3NBL |



**IMPORTANT NOTE**

**Revalidate licence to avoid revocation** – Ofcom has advised the Society that plans will be drawn up to revoke licences that have not been revalidated as required by the licence conditions. The quickest way to revalidate is to do so online via the Ofcom website: *https://services.ofcom.org.uk/* or by email: *amateur.validations@ofcom.org.uk* If you need assistance in the process, Ofcom staff are available to help, but please be patient during times of heavy workload.

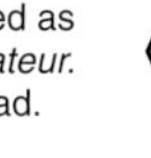

RG14 5JE M1CIJ
RG14 5JF 2E0FAE
RG14 5JF M6ASF
RG14 5JJ G1FBU
RG14 5JN G7SYC
RG14 5JP G4RUW
RG14 5NR G7SLL
RG14 5QW G7REJ
RG14 6AZ 2E0JOF
RG14 6AZ M3JOF
RG14 6BA 2E0GOW
RG14 6BA 2E0UAO
RG14 6BA M0GOW
RG14 6BA M3UAO
RG14 6DD G3UVM
RG14 6DN G3URI
RG14 6DN G4MKF
RG14 6EE G4YMY
RG14 6HD M1BRU
RG14 6HP G0HBJ
RG14 6HP G6ZSF
RG14 6JG G0VFE
RG14 6JX G4WEV
RG14 6PY G8KHU
RG14 6PZ 2E0JHC
RG14 6PZ M0IDC
RG14 6PZ M3JHC
RG14 6RU G3ZGC
RG14 6RU M3ZGC
RG14 6RY G1WTS
RG14 6RY G8AKM
RG14 6SX G8AYC
RG14 7AL G4ZDP
RG14 7DJ G8MWU
RG14 7FX G7KXZ
RG14 7RA M1DLG
RG14 7RB G4TKS
RG14 7RR G6LAE
RG14 7TL G8JUS
RG14 7TT G1DFI
RG17 0AZ 2E0BJM
RG17 0AZ M0NKS
RG17 0AZ M6OAJ
RG17 0BZ G3ZPK
RG17 0JE G1OQV
RG17 0JE M6CUE
RG17 0JR G0LJJ
RG17 0JR G1FVP
RG17 0LA G3LWT
RG17 0LJ M3TZN
RG17 0LL G6GCJ
RG17 0NE G4TPH
RG17 7DG G0DTQ
RG17 7ED G6WLE
RG17 7JL M6MPX
RG17 7TS G4DNH
RG17 7UN 2E0DYN
RG17 7UN M0ICZ
RG17 7UN M6FWX
RG17 8RF G7BWI
RG17 8YQ G6EES
RG17 8YQ M3XEL
RG17 9QE G0LTX
RG17 9QE M1ABX
RG17 9UE G4FNK
RG17 9UW G7PGY
RG17 9UY G6WEI
RG18 0RR 2E0BPS
RG18 0RR M0JCM
RG18 0RR M3XGZ
RG18 3AF M3HSR
RG18 3BA G4BOO
RG18 3BF G7TOO
RG18 3BF M0PJC
RG18 3BN G6AMF
RG18 3BP G3RVM
RG18 3DH G4UET
RG18 3DT G3VMT
RG18 3EB G7JLT
RG18 3EG G0POT
RG18 3EG G5XV
RG18 3PD M1PTT
RG18 3PD M3TVD
RG18 3UH M3JHV
RG18 4DL G4ORX
RG18 4DQ G1PCG
RG18 4DQ G4WLG
RG18 4DS M1CEY
RG18 4EE 2E0SXP
RG18 4EE M3SXP
RG18 4LA 2E0TVR
RG18 4LA M3PRN
RG18 4LQ G1JKP
RG18 4LS M3NXF
RG18 4LS M3SXF
RG18 4LS M6PXF
RG18 4NP G0ORH
RG18 4NP G7HNN
RG18 9HQ G3VOW
RG18 9HT G8LTN
RG18 9HZ 2E0HQO
RG18 9HZ M6HQO
RG18 9PB G0MIA
RG18 9PD G0PUB
RG18 9PD M3PUB
RG18 9PD M3RBX
RG18 9PH G4LMW
RG18 9PH G8XCW
RG18 9PH M6LMW
RG18 9PU G8URB
RG18 9QP G3KJC
RG18 9QW M6DGU
RG18 9RJ G6SOZ
RG18 9RJ G7DRR
RG18 9TG G6HUN
RG19 3LE G6ZVL
RG19 3LE G8TSC
RG19 3PF G0OBJ
RG19 3PF G4YMH
RG19 3RL G4FWR
RG19 3RS G3ICB
RG19 3SD 2E0ESO
RG19 3SF G0KQT
RG19 3SH G7RJW
RG19 3XE G6PNG
RG19 3XW G3DBV
RG19 3XW G7UXK
RG19 3XX G4BWE
RG19 4DY G8XEC
RG19 4FD G3WYW
RG19 4FD M1PGT
RG19 4FN 2E0GBG
RG19 4FN M6DMG
RG19 4FQ G8JIP
RG19 4GJ G4RTQ
RG19 4LX G7VGL
RG19 4WA 2E0MAD
RG19 4WA G6IZA
RG19 8BE G3MEV
RG19 8DD G3LLK
RG19 8EY 2E0RMZ
RG19 8EY M0RMZ
RG19 8SH M6DHT
RG19 8SS M0SBF
RG2 7AH M6YYM
RG2 7BG M0ALB
RG2 7DX G4YBX
RG2 7HU M3UYW
RG2 7JR G4OMN
RG2 7JU G4JXH
RG2 7TD M0GQS
RG2 8DN 2E0FUZ
RG2 8DN M3ZTK
RG2 8DN M3ZVK
RG2 8HJ G1ASD
RG2 8JB G6NVY
RG2 8LJ G1NYZ
RG2 8QP G4DBF
RG2 8QP G7BXU
RG2 8SD 2E1HJE
RG2 9HA G8IJG
RG2 9HR M6FHN
RG2 9PD M3VME
RG2 9PU G7KWN
RG20 0AT G8OID
RG20 0BW G6UJC
RG20 0LY G0LUK
RG20 5ER G4HDE
RG20 5RJ G3BFL
RG20 5RT G4UEF
RG20 5SD G8IAR
RG20 5SL G8HJF
RG20 5SL G8HJG
RG20 7BE G3NAQ
RG20 7BT G8JRN
RG20 7EH M0KKA
RG20 7EZ M0CUK
RG20 7JS 2E0FBN
RG20 7JS M6FYG
RG20 8QL M3EHK
RG20 8RU M6BEM
RG20 8SD G1YOS
RG20 8SE G1NAB
RG20 8TZ M0HPS
RG20 9BN G6GXZ
RG20 9BW G3SVD
RG20 9BW M3SVD
RG20 9BW M3UKN
RG20 9DY M6NHZ
RG20 9EY G7VFE
RG20 9HY M6NAC
RG20 9TB M3TYU
RG20 9TS 2E1ETJ
RG20 9TS G4DOQ
RG20 9TS G6PFF
RG20 9UY G7DXC
RG20 9UY M3PCW
RG21 3AS M6SEK
RG21 3HG G4WIZ
RG21 3JH G7WCN
RG21 3JW G8AOO
RG21 3NF G7FFB
RG21 5HL G6BBW
RG21 5NN G1POJ
RG21 5NR G7SNB
RG21 5NR M6NIS
RG21 5NZ G8YEQ
RG21 5SR G8PIY
RG21 5SX 2E0AVP
RG21 5UA G4CSD
RG21 7TG G7SKX
RG21 8XT M0NQB
RG22 4BG M0EDR
RG22 4EL G8AOK
RG22 4HN G4JIQ
RG22 4HR G8JLB
RG22 4JN M6KTZ
RG22 4JR G8GOS
RG22 4LJ M3JSK
RG22 4LZ 2E1LIS
RG22 4LZ M0LIS
RG22 4NB G6JOR
RG22 4NP G6UZO
RG22 4NP M3BRU
RG22 4PH G7LYL
RG22 4RF M1CJX
RG22 4RQ G8AGJ
RG22 4TY 2E0GUV
RG22 4TY G7PKD
RG22 4TY M0CJJ
RG22 4UB G4BHE
RG22 4UJ G6OPK
RG22 4UL G3WXD
RG22 4UL G8DUV
RG22 4UX G4PND
RG22 4WU G0UVQ
RG22 4XD G6MDS
RG22 4XT G8VML
RG22 4XT M0DRG
RG22 5BA G7LJB
RG22 5BQ G1CBS
RG22 5DN G0LEP
RG22 5JP G1JHM
RG22 5JY G8GTZ
RG22 5LY G6KQZ
RG22 5NN G0LQD
RG22 5NN G6PMF
RG22 5NX M0SAB
RG22 5QD G3UCF
RG22 5QF G6ZZR
RG22 5RA 2E0CBO
RG22 5RA M0HHD
RG22 5RA M3YQC
RG22 6AX G0PTA
RG22 6BN G7PFL
RG22 6BQ G7MAT
RG22 6BQ M3HYO
RG22 6DF M0CBF
RG22 6EP M0RAB
RG22 6JA G1PEU
RG22 6JL G1WMV
RG22 6JL M0TAD
RG22 6NU 2E0BHD
RG22 6NU G0KQA
RG22 6NZ 2E1FTH
RG22 6QG M6MGQ
RG22 6QP G3OAZ
RG22 6QQ M0HNX
RG22 6TD M1CQI
RG23 7AQ G3JTQ
RG23 7BB 2E0PGP
RG23 7BB 2E1FSF
RG23 7BB 2E1FSG
RG23 7BB M6KIK
RG23 7BL G4JGS
RG23 7DD G3HVA
RG23 7DJ G6PJP
RG23 7JP M1CQP
RG23 7JX M0LAS
RG23 7LB G6CFA
RG23 7LD M0CJM
RG23 8AD G0XBA
RG23 8EX G4YPK
RG23 8JD G4EDH
RG23 8JF G7KAK
RG23 8NG G0JSR
RG23 8NH G8JYN
RG23 8QL 2E0NBM
RG24 7EA M3RLS
RG24 7EF G8JHH
RG24 7EH M0GWC
RG24 7JE G4XIP
RG24 8AA G7PTH
RG24 8EU G6VFI
RG24 8GL M3BVM
RG24 8RB G4XMP
RG24 8RG 2E0BTV
RG24 8RG M0GOQ
RG24 8RG M3XWV
RG24 8SB G8JSF
RG24 8SL M3NHW
RG24 8SU G0NWH
RG24 8SU M1ENA
RG24 8UJ G7PAG
RG24 8WN M1KPW
RG24 9BE M3NRI
RG24 9DD G7FFC
RG24 9GH 2E0ISO
RG24 9GH M3ISO
RG24 9HE G4YNH
RG24 9HX M0CWY
RG24 9LR G8UBN
RG24 9PQ M0DOP
RG24 9PQ M3DMY
RG24 9PQ M3OOU
RG24 9PQ M6DMY
RG24 9PY M3GCJ
RG24 9RR G7TAE
RG25 2BP G8JMY
RG25 2BZ G7NDB
RG25 2NH G3CEI
RG25 2RN G7EVF
RG25 3EJ M6CEE
RG25 3HP M0FSN
RG25 3LD G0NJG
RG25 3LZ G4GFM
RG25 3NL G7VZR
RG25 3NL M0VZR
RG26 3ED G0CAK
RG26 3EL G8AKA
RG26 3HP 2E0MPX
RG26 3HP M0PXM
RG26 3HP M6KMG
RG26 3LF M6XST
RG26 3NJ G1WKK
RG26 3SH M1BCH
RG26 3TL G8DVU
RG26 3UQ G6AOH
RG26 3UR G6MNN
RG26 3YH G4LUA
RG26 3YH M6JJL
RG26 3YJ 2E0VVC
RG26 3YJ M0VVC
RG26 3YJ M6TTL
RG26 4EW M0MEI
RG26 4HF G6IOX
RG26 4HH G8MSY
RG26 5AN 2E0KER
RG26 5AN M0KER
RG26 5AN M3WXY
RG26 5BX M3XLC
RG26 5DG G1EHF
RG26 5DG M0HNA
RG26 5DG M3EHF
RG26 5EP G4LXJ
RG26 5HD G8FMD
RG26 5JX M0PPZ
RG26 5NY G4XMO
RG26 5PJ G1JWG
RG26 5UE M6AII
RG26 5UU G8VKO
RG26 5XH M1ECW
RG27 0DQ G0FEJ
RG27 0ES G3UQW
RG27 8BF M0SVV
RG27 8JS G3YGE
RG27 8NJ G0AMZ
RG27 8QX G3YFL
RG27 8SW M0AET
RG27 8SW M3MOF
RG27 9EY G4SFH
RG27 9HW G4YEG
RG27 9JS G6AHH
RG27 9NP M1CAO
RG27 9PU G6OFD
RG27 9QJ G4SXX
RG27 9QZ G3RXI
RG27 9RA M0VEC
RG27 9RP G6ETX
RG27 9SG G0RSV
RG28 7JG 2E0ERV
RG28 7NF G3NYS
RG28 7SE G4ASY
RG29 1AE M0SOX
RG29 1BH 2E0EJM
RG29 1BH 2E0ICU
RG29 1BH 2E0PIP
RG29 1BH G1ZBL
RG29 1BH M0AEU
RG29 1BH M3OFS
RG29 1BH M3OUI
RG29 1BH M3TIZ
RG29 1EJ G7VSP
RG29 1JZ G7BRB
RG29 1ND G8WAM
RG29 1NN 2E0NWE
RG29 1NN M6YNL
RG29 1PG G3NVM
RG29 1SX 2E0LSV
RG29 1SX M0LSV
RG3 4UL G7SJK
RG30 2EJ M3NOW
RG30 2HA G6GBL
RG30 2NT G1ANQ
RG30 2NX G0AFR
RG30 2PE G1DSJ
RG30 2PE G1ZSY
RG30 2RN G4YFB
RG30 2SF G0ODK
RG30 2TH G6JUI
RG30 2UL M6ETP
RG30 3EN G4BTI
RG30 3NQ G7UET
RG30 3NQ M3AGI
RG30 3NQ M5ALG
RG30 4HU M3MSQ
RG30 4JT 2E1EPQ
RG30 4PD 2E0DII
RG30 4PD M6IAO
RG30 4QP G0GBR
RG30 4QP G0RZM
RG30 4YJ G7KWD
RG30 4YP G0VHK
RG30 6AG G7ITU
RG30 6DT 2E1CAJ
RG30 6EH M6GYD
RG30 6EP M6XBS
RG30 6TP 2E0AKY
RG30 6TP M0LUV
RG30 6TP M3PQS
RG30 6UE M6EML
RG306EH 2E0MGT
RG31 4US 2E0CGL
RG31 4US M0EGN
RG31 4XP 2E0FUN
RG31 4XP M0RXX
RG31 4XP M3VHZ
RG31 5DZ G4EFE
RG31 5HJ G8SMA
RG31 5HU M1AUY
RG31 5JU G7VHX
RG31 5JY G0IZI
RG31 5WE G4XKA
RG31 6DE G7JJG
RG31 6FZ M5ACR
RG31 6HN 2E0BJS
RG31 6HN M0RBJ
RG31 6HN M3VFP
RG31 6NP G4JOB
RG31 6PY 2E1FGB
RG31 6RH G8PQZ
RG31 6RH M0GMC
RG31 6RR M6BNV
RG31 6SR G8LCA
RG31 7AT G7MFP
RG31 7DD M0SMS
RG31 7DN G0TFU
RG31 7JR 2E0BZX
RG31 7JR M6RFC
RG31 7RX M0TBR
RG31 7ZL G3VUK
RG31 7ZU 2E0SHP
RG31 7ZU M6SHP
RG4 5AL G4AWY
RG4 5AY G4BXY
RG4 5LE M6WOT
RG4 6DB 2E0VJO
RG4 6DB M0JSX
RG4 6DB M3VJO
RG4 6PL G4VSQ
RG4 6PY 2E0CFI
RG4 6PY M0HAH
RG4 6PY M6CFO
RG4 6QA M3IEP
RG4 6QB G8NQC
RG4 6RT M6KCY
RG4 6SF G8ROG
RG4 6SZ G0KIA
RG4 7DT 2E0DNE
RG4 7HH G0CCC
RG4 7HH G4CCC
RG4 7HR G3AKF
RG4 7HR G4JTR
RG4 7HR M0AAA
RG4 7NE G4JNU
RG4 7NR 2E0IET
RG4 7NR M0IET
RG4 7NR M3IET
RG4 7NR M3IEV
RG4 7NR M3SGF
RG4 7NT G7ENS
RG4 7RP G4EJK
RG4 8EN G3TBJ
RG4 8HH G0OIW
RG4 8HH G3WKX
RG4 8HJ G4ZSZ
RG4 8HJ G7KSS
RG4 8JH G3OYN
RG4 8LE G4TXA
RG4 8PA G3YZZ
RG4 8PH 2E0FSM
RG4 8PH M6RFP
RG4 8PL G4OPK
RG4 8TT G0MQW
RG4 9AD G0LGG
RG4 9LS M6KBX
RG4 9NT 2E0WWF
RG4 9NT M6XTT
RG4 9NT 2E0DWM
RG4 9RU M3TXG
RG4 9RY G6NEA
RG4 9RY G6RKJ
RG4 9RY M3AZP
RG4 9SA G8PDY
RG4 9SP G7RZQ
RG4 9TF G0JMS
RG40 1AW M5AGS
RG40 1DE M6EFE
RG40 1DE M6FNX
RG40 1DG G8PJC
RG40 1PL G8VZY
RG40 1QG G8CIT
RG40 1QG G8IEI
RG40 1QW 2E0PEP
RG40 1QW M6PEP
RG40 1RL G1DNP
RG40 1RY M3ZRW
RG40 1TW G4ORU
RG40 1XS G8YRL
RG40 1XX G8FBF
RG40 1YE M6PHE
RG40 2HT G2DD
RG40 3EB G4LRY
RG40 3HT G4OKM
RG40 3JU 2E0ZDA
RG40 3LB G7JTV
RG40 3LD G8DUO
RG40 3LG G3XPC
RG40 3QB G8NZK
RG40 3RL G3WME
RG40 4EN M0GRW
RG40 4JA G1SEO
RG40 4PA M6DUM
RG40 4PA M6SSS
RG40 4PF G4HRE
RG40 4PY 2E0WGI
RG40 4PY M0WGI
RG40 4PY M3WGI
RG40 4RA 2E0CKO
RG40 4RA G1ZSF
RG40 4RA G4NIP
RG40 4RD G4DYO
RG40 4TS M0ZEQ
RG40 4TS M6SUZ
RG40 4UD G4HOU
RG40 4UJ M1VHF
RG40 4UJ M3WOK
RG40 4UN 2E0CDU
RG40 4UN M0GVY
RG40 4UN M6DBX
RG40 5PE G4ZRZ
RG40 5QU M1AXG
RG40 5YB G3ZBS
RG40 5YE G4XAL
RG40 5YL G4KVD
RG41 1JE G3ZZH
RG41 1LJ M6POA
RG41 1NN M6HML
RG41 1NR G6CGC
RG41 1PH 2E1AUQ
RG41 1PH G3YHG
RG41 1QH G3XZJ
RG41 2RJ G0JHW
RG41 2TL M3IWK
RG41 3AG G8GJM
RG41 3HG G3MGU
RG41 3UE M3XPR
RG41 4AJ M6EFF
RG41 4AR G6TLX
RG41 4AW G8IBP
RG41 4AX G4LJF
RG41 4BA G8FIE
RG41 4BD G6IJK
RG41 4BX G4JXU
RG41 4DH G1KTF
RG41 4ED G3TUY
RG41 4SY G4AEP
RG41 4TA M6HRM
RG41 4UR G8VPO
RG41 4UY G7SEO
RG41 5JF G4GBI
RG41 5JG M0GXB
RG41 5JZ G7COP
RG41 5LX M6CMB
RG41 5NJ G6WXN
RG41 5NW G6IVB
RG41 5PE G1HBD
RG42 1RT G6XPB
RG42 1RT M6XPB
RG42 1UE G8OLK
RG42 2AD M6RCE
RG42 2BT G4WZJ
RG42 2HJ 2E0PTS
RG42 2HJ M0PTS
RG42 2HJ M6PGB
RG42 2HL G3NCN
RG42 2LD G3VG
RG42 2LE G3TAI
RG42 2LH G4NOC
RG42 2QX G6DKI
RG42 3TQ 2E0ZKT
RG42 3TQ M3UER
RG42 3TR 2E0XDF
RG42 3TR 2E0XNF
RG42 3TR M0XDF
RG42 3TR M3XJF
RG42 3UN G0ATZ
RG42 3UN M3UTQ
RG42 3XA G4HLF
RG42 3XU 2E0RTP
RG42 3XU M0RTP
RG42 3XU M6RTP
RG42 3XX G4ZFE
RG42 4DF G4DQA
RG42 4DS G4ORB
RG42 4UR G7CJO
RG42 5JG G4PKE
RG42 5LG G0GJV
RG42 5LG G4BRA
RG42 5LG G6XSY
RG42 5QZ M3NPK
RG42 6AS G3IKN
RG42 6HU G0BBB
RG45 6AL G8EYM
RG45 6BT G6AXY
RG45 6DU G8XML
RG45 6EF G6LLG
RG45 6EN G3GRY
RG45 6EX G4EPX
RG45 6HE 2E1DIH
RG45 6HE G6LKZ
RG45 6NL G6ETZ
RG45 6NR G3DVF
RG45 6PE G3NDI
RG45 6QE M1FEY
RG45 6QF G1WGM
RG45 6QF G3KBS
RG45 6QR G0FPT
RG45 6SE M0DZU
RG45 7EG G7VWO
RG45 7ER G4DHY
RG45 7HR G6FXR
RG45 7JP G8STR
RG45 7NE G8ZLL
RG45 7NR G0WCS
RG45 7NW 2E0PZK
RG45 7NW M0OPK
RG45 7NW M6DGI
RG45 7PD G8GSU
RG45 7QR G4DGF
RG5 3AD G3RAC
RG5 3AD G4HNF
RG5 3AX 2E0JCZ
RG5 3AX M0JRZ
RG5 3AX M6FJL
RG5 3BG M6HDD
RG5 3BL M6JWX
RG5 3DA G3WTV
RG5 3DU M0DNP
RG5 3DY M6CCM
RG5 3HA G0DWD
RG5 3HF M3NXH
RG5 3JA G0IZV
RG5 3LQ 2E0ZIP
RG5 3LQ M3XHH
RG5 3LQ M6PEN
RG5 3LR G0AHI
RG5 3QG G8DZC
RG5 3SE 2E0PLL
RG5 3SE M6BBZ
RG5 4AP G0VQR
RG5 4AP M0CYB
RG5 4AW G4ELY
RG5 4BL G2NF
RG5 4HB 2E0SZZ
RG5 4HB M0XMS
RG5 4HB M6SZZ
RG5 4HH G4IZS
RG5 4LA G8GZR
RG5 4LD 2E0JCW
RG5 4LD M6JCW
RG5 4LJ G4JVD
RG5 4LR G0FIJ
RG5 4LR G8MBQ
RG5 4NA M3MZW
RG5 4PA. M6GJS
RG5 4QB G7UOQ
RG5 4QT G3UUB
RG5 4RT G3PBR
RG5 4UD M0MCG
RG5 4UN G0VPE
RG5 4UN G4KWT
RG5 4XE M6PGX
RG5 4XR G8CSK
RG6 1AP G0ILT
RG6 1AS M0IZS
RG6 1AT G0RKD
RG6 1HD G7RFS
RG6 1HS 2E0JPM
RG6 1HS M6JMO
RG6 1HW G3VMY
RG6 1LH 2E0GFE
RG6 1LH M6GFE
RG6 1QE G4OKA
RG6 3AH G4CDF
RG6 3DG G6HCT
RG6 4AB G7UZY
RG6 4AZ G3TEB
RG6 4AZ G4VTQ
RG6 4DB G7FQE
RG6 4EP M6LHH
RG6 4HN 2E0XUU
RG6 4HN M0XUU
RG6 4HN M3XUU
RG6 4NY 2E0DLV
RG6 4NY G8FIF
RG6 4NY M0HQU
RG6 4NY M6DPU
RG6 4XA G6IEI
RG6 5GX G6YIE
RG6 5PR 2E1CXE
RG6 5PW G7NDI
RG6 5PY G8NPZ
RG6 5QX G8LCL
RG6 5QZ G4FXT
RG6 5RY G8ADQ
RG6 5SL 2E0BPW
RG6 5SL M0JNX
RG6 5SL M3KYK
RG6 5SN G8VRW
RG6 5SR 2E0JLN
RG6 5SR G3UAX
RG6 5SR G7NZO
RG6 5SR M6JIL
RG6 5TG G0CEC
RG6 5TP M3EZJ
RG6 5XG G0MZN
RG6 5XG G4AOL
RG6 5XG G8VWH
RG6 5YQ G6JVK
RG6 7JU M3ZLW
RG6 7LH G4RDC
RG6 7LJ G7LWY
RG6 7LQ G7AOK
RG6 7PA G4OQJ
RG6 7PD G0JTN
RG6 7PE G4OAE
RG6 7RT G4HLT
RG6 7RT G4MUT
RG7 1BB G8VWV
RG7 1BQ G7FBE
RG7 1DD M3KVC
RG7 1DW M6ZRZ
RG7 1HP G7MTW
RG7 1QY G8MVY
RG7 1SG G7OMQ
RG7 1TJ M0SCB
RG7 1TT G8YVQ
RG7 1XS G4ZAC
RG7 2LH G4XXH
RG7 2PZ G3SCZ
RG7 2QB M0GVN
RG7 3BU G4CLD
RG7 3ES G3YGR
RG7 3EZ G4IWS
RG7 3JA M1DGQ
RG7 3QE G1COW
RG7 3QU M1BNG
RG7 3TH M0SEL
RG7 3TL G1AWD
RG7 3TR G3WND
RG7 3TU G3ZOI
RG7 3XW 2E0PHM
RG7 3XW M0MPM
RG7 3XW M0PAQ
RG7 3XW M6PHM
RG7 4DE G8ZWN
RG7 4PA M0OJO
RG7 4TH G8DGR
RG7 4TR G4EEE
RG7 5DN G3LXQ
RG7 5NS G7TKP
RG7 5RY 2E0DEU
RG7 6HA M6KDQ
RG7 6NN G0OBV
RG7 6NU G3KIW
RG7 6NU M3KXY
RG7 6RA G4ZSG
RG7 6TN G0KPE
RG8 0DJ 2E0CCJ
RG8 0EB G4CEI
RG8 0HX 2E0SSG
RG8 0HX M3SSG
RG8 0JL G3NGX
RG8 0PL M0AZG
RG8 0SE G3LLG
RG8 7BD M3LJJ
RG8 7PX G1GMH
RG8 7QG G3TRY
RG8 7QG G4CUE
RG8 7QG G4MDF
RG8 7QG G8NMO
RG8 7QL G0LHZ
RG8 7QL G3ULT
RG8 8BU M6SIF
RG8 8DG G7KSV
RG8 8DQ 2E0LAX
RG8 8DQ M6AWC
RG8 8LN 2E0LAV
RG8 8LN G7TXR
RG8 8LN M0XYX
RG8 8LT 2E0SRJ
RG8 8LT M0MLE
RG8 8LT M6RJS
RG9 1AP G8VOB
RG9 1JT 2E0HIT
RG9 1JT M6HIT
RG9 1LT G3TFL
RG9 1PH G0NCO
RG9 1UU G4WBI
RG9 1UU G7LSF
RG9 2BG M6SII
RG9 2LX G3UYY
RG9 3JS G8FXX
RG9 3JS M0FXX
RG9 4DH G6CHA
RG9 5HJ G3XTT
RG9 5PS G3PZL
RG9 5SG M0BTY
RG9 5WE M3PXT
RG9 6NR M0XVI

## RH (Redhill)

RH1 1AF G4VXN
RH1 2BN G6YAH
RH1 2BZ G3VGE
RH1 2DY G8ZFQ
RH1 2EZ G0RHG
RH1 2HA G0LXV
RH1 2HH G3MPB
RH1 2JB G1WIS
RH1 2JB G7LZM
RH1 2LA G8XOV
RH1 3AA G3YSX
RH1 3BH G7ODV
RH1 3BN G1YRR
RH1 3ER G4WGD
RH1 3EY G8UNO
RH1 3LL 2E0BXZ
RH1 3LL M0JCR
RH1 3LL M6JCR
RH1 3PB 2E0UOJ
RH1 3PB M3UOJ
RH1 3PX G0KCH
RH1 4AS M6DPW
RH1 4AY G3NIQ
RH1 4DE G3TRC
RH1 4QT G7KGV
RH1 5BJ G4LJU
RH1 5DP 2E0ZFX
RH1 5DP M0HUD
RH1 5DP M6ZFX
RH1 5JB G7LUO
RH1 5JB M0DBO
RH1 5LY M6BZY
RH1 6AW G8HMG
RH1 6BJ M0DVR
RH1 6LP 2E0JRN
RH1 6LP M0JRL
RH1 6LW M6AOB
RH1 6PB G1YXJ
RH1 6PB G4PFX
RH1 6PQ G8LAM
RH1 6PQ M0WAO
RH1 6TB M1APH
RH10 3AL G7OMM
RH10 3DW G3CTP
RH10 3DW M6WCG
RH10 3EG G8MQF
RH10 3HX G6NPJ
RH10 3JP G4XHF
RH10 3NG G0DSU
RH10 3NG G4PBF
RH10 3PS G3VJM
RH10 3QS M6WPN
RH10 3QT G0KKS
RH10 3RG M0GZE
RH10 3TA M6CSP
RH10 3TB 2E1HGT
RH10 3TG G0ORX
RH10 3UQ 2E0ZLM
RH10 3UQ M3RXO
RH10 3XG G4ANN
RH10 3XG G6RC
RH10 4EU M0GAD
RH10 4HQ G7ITZ
RH10 4HS M6JPR
RH10 4JB M6MPO
RH10 4JQ G0PYJ
RH10 4LN G6NQO
RH10 4LN M3APQ
RH10 4LU M6RXL
RH10 4NG G3RMK
RH10 4UB G0GML
RH10 4UJ G3YTR
RH10 4UL G4TYY
RH10 4UP G0BKW
RH10 4XA G3VLH
RH10 4XY G6LEB
RH10 4YY G3NZP
RH10 5AR 2E1AVT
RH10 5DQ G3ZON
RH10 5DW 2E1BMF
RH10 5DW G4UMJ
RH10 5LD G6FPH
RH10 6BG 2E0MZB
RH10 6BJ G4FYY
RH10 6BP G6CKZ
RH10 6DL M0HQM
RH10 6HQ G0VYN
RH10 6HQ G3WKI
RH10 6HW G8MGZ
RH10 6JS G4MKD
RH10 6JS G7OMI
RH10 6LX G0IOE
RH10 6LX G6XTG
RH10 6PB G6XAV
RH10 6PN G4SFP
RH10 6QF G8INS
RH10 6RL 2E1YAP
RH10 6RL M3GQM
RH10 6RL M3YAP
RH10 6SE G3SYD
RH10 6TH G3MGL
RH10 7AE G1AYU
RH10 7BS G4UBT
RH10 7DB M6EWQ
RH10 7DB M6EXW
RH10 7DD G8PQH
RH10 7ED M0PMM
RH10 7FL G1XST
RH10 7JH M0VPG
RH10 7JR G3FRV
RH10 7NU G0IPE
RH10 7RQ G0VIQ
RH10 7SQ G3ZSJ
RH10 7SW M0LPK
RH10 7WQ M6MRY
RH10 7XG G1BGH
RH10 7YW G7RUN
RH10 8BZ G0FPI
RH10 8EH G4TVC
RH10 8GD G6TUG
RH10 8HA G8IZY
RH10 8JR G6GGW
RH10 8JX M6GYE
RH10 8NB G6DOW
RH10 8XG M0GLU
RH11 0BS G4ONJ
RH11 0HW G0EGT
RH11 0JJ G8XHK
RH11 0NA G1GHY
RH11 0NA G3KAA
RH11 0RB G6OCM
RH11 0UF G6AUE
RH11 7BD G3IPP
RH11 7BW G8RRR
RH11 7EB M0CWT
RH11 7JB G6SHQ
RH11 7JE G3UNS
RH11 7JP G0RGM
RH11 7PE G4LEG
RH11 7PU M6RRU
RH11 7RF G4TZF
RH11 7UH G3UUC
RH11 8BA M6ECM
RH11 8EH G1LTI
RH11 8EU G7LYB
RH11 8EY 2E0BTO
RH11 8EY 2E0DDW
RH11 8EY M0GUZ
RH11 8EY M3YFJ
RH11 8EY M6DDW
RH11 8HW G1XFX
RH11 8LQ G4PFW
RH11 8SQ G8AVZ
RH11 8XJ G6YPM
RH11 9AF G3LET
RH11 9AN M6SXT
RH11 9BZ M6BLS
RH11 9DT M3XHZ
RH11 9EG G6ECE
RH11 9EG G6EDR
RH11 9ES G6ZHO
RH11 9HG 2E1IGJ
RH11 9HN M6SIU
RH11 9HN M6TOV
RH11 9JH G0PVQ
RH11 9LH M6HNJ
RH11 9NS G1PLV
RH11 9NY G3ZNW
RH11 9QT G8PUH
RH11 9QZ G7VZQ
RH12 1DQ G0VPZ
RH12 1LD G8YRF
RH12 1LF M3WOI
RH12 1NA G8GHL
RH12 1PZ G4HAY
RH12 1SB 2E0RKR
RH12 1SB G7DMQ
RH12 1SB M0RKR
RH12 1SB M6RKR
RH12 1UB M6MPO
RH12 1UY M6ARD
RH12 2AF G4LJR
RH12 2AF M6FLU
RH12 2DA G4CCA
RH12 2HH 2E0ZMT
RH12 2HH M0ZMT
RH12 2HH M3ZMT
RH12 2HU M6MON
RH12 2LB 2E0NKC
RH12 2LB M3NKC
RH12 2NH 2E0OVI
RH12 2NH M0OVI
RH12 2PY G4JHI
RH12 2QL G4LLI
RH12 3DU G3TQY
RH12 3DU G3ZTU
RH12 3ET G6EWJ
RH12 3HD G3SWC
RH12 3HE G4ZEJ
RH12 3LH G7EYL
RH12 3ND G4PEY
RH12 3NH G0RBQ
RH12 3QR 2E0TJT
RH12 3QR M3YZQ
RH12 4AR G4TMC
RH12 4GR G6DMM
RH12 4GX G8FQS
RH12 4HR G6XTJ
RH12 4JB G8LHS
RH12 4JB M6GQH
RH12 4JE G1CKF
RH12 4LN M6PLS
RH12 4RE G4YJA
RH12 4SH M3SYN
RH12 4TP M0LTA
RH12 4UF G0URO
RH12 5EU G1XFM
RH12 5EU G4MWD
RH12 5EU M6TZR
RH12 5EW G8RSK
RH12 5FR G6NQQ
RH12 5PJ G1ALU
RH12 5PN G7SRV
RH12 5PZ G7MQP
RH12 5UB G0JGI
RH12 5WA G4KDR
RH12 5WD G8IBE
RH12 5XL G6FYU
RH12 5XW G3VQO
RH12 5XX 2E1IGA
RH13 0AL G6PGQ
RH13 0PJ G7AVF
RH13 0RQ 2E0CSJ
RH13 5AR M6MTM
RH13 5DB M6NXZ
RH13 5HB G0DJS
RH13 5HG G6SBI
RH13 5HH G4EUG
RH13 5JS 2E0BXW
RH13 5JS M6EAA
RH13 5JS M6SPJ
RH13 5LA M6SLD
RH13 5LH M6END
RH13 5ND G7UEC
RH13 5NL 2E0DPR
RH13 5NL M0SMN
RH13 5NL M6CGK
RH13 5NZ G4TPO
RH13 5PE G1ODN
RH13 6AE G4MKW
RH13 6AJ 2E0WEK
RH13 6AJ M6WEK
RH13 6AL M6GPJ
RH13 6AX G3ZBU
RH13 6AX M3EUR
RH13 6AX M3GCR
RH13 6BQ M0GRU
RH13 6BS M6IHC
RH13 6BS M6LPS
RH13 6DD G4NWW
RH13 6DG G6MQJ
RH13 6DT G8BAD
RH13 6DW M3FTK
RH13 6ED G4FQR
RH13 6EJ G3WSC
RH13 6EJ G4PEO
RH13 6LZ G8YLA
RH13 6LZ M0YLA
RH13 6ND G7HWM
RH13 8AZ G3XEI
RH13 8EF M6RJK
RH13 8EH G4JCY
RH13 8EQ G3ZQW
RH13 8ER G4TPW
RH13 8GD M6EXD
RH13 8HR 2E0BQU
RH13 8HR G3RAM
RH13 8HR M6CSL
RH13 8HZ G7DFV
RH13 8HZ M3DFV
RH13 8JB G8FBM
RH13 8JF G4OLZ
RH13 8JX 2E0PML
RH13 8JX M0VUE
RH13 8JX M6EGK
RH13 8LG G1VVY
RH13 8LT G3WZT
RH13 8LT G4HRS
RH13 9BB G0LOC
RH13 9BB M6AUZ
RH13 9BG G4BVP
RH13 9DQ G4VFC
RH13 9HP G1MPW
RH13 9HZ 2E0AED
RH13 9HZ G3PVH
RH13 9HZ G4HMM
RH13 9HZ M6CJV
RH13 9TF G4AZQ
RH13 9TF G8TQJ
RH13 9XR G0TAO
RH13 9XX G6DID
RH13 9XY G4LRP
RH14 0DY G3PUX
RH14 0PJ 2E0FTC

RH14 0RJ G4NUX
RH14 0[illegible] M6FSU
RH14 0SR M6FLR
RH14 0TF G8SLU
RH14 9BH G3OGP
RH14 9GL G0ERF
RH14 9HB G0FQD
RH14 9LJ G3RDN
RH14 9LT 2E0RGC
RH14 9NG G4EFO
RH14 9NG M6NCD
RH14 9NP G6ZDP
RH14 9PU G7MXT
RH14 9RJ M0HWH
RH14 9TU 2E0VCU
RH14 9TU M6PIP
RH15 0AE M6NAW
RH15 0DE M1BQU
RH15 0EU G3YBM
RH15 0HB M0TPJ
RH15 0JN M1CSI
RH15 0LA G3IZW
RH15 0LD M3TFS
RH15 0LJ M1EBU
RH15 0NB G3RMY
RH15 0ND G0WGP
RH15 0NF M6BRO
RH15 0NH M6JOR
RH15 0NJ M1CSE
RH15 0NR G4PLT
RH15 0NZ M0MIG
RH15 0PH G6ZQA
RH15 0PT G0DDE
RH15 0PT G6DBX
RH15 0QA G6IHG
RH15 0QN G0UUP
RH15 0QU M5JWS
RH15 0QY G1NYS
RH15 0QY M6LAP
RH15 0RP G1AZA
RH15 0RQ G0GNV
RH15 0RQ G1TDL
RH15 0RW G7BLD
RH15 0RW G7TPW
RH15 0RW M3HZM
RH15 0RW M3HZP
RH15 0TD M1CNG
RH15 8AW M1CYL
RH15 8BD 2E1IAS
RH15 8BD G1DZD
RH15 8BL G0IDP
RH15 8BW G3XUP
RH15 8DL G0LFF
RH15 8EU G4AKG
RH15 8JQ G4ZXO
RH15 8LE G4ULZ
RH15 8NP 2E0RJA
RH15 8QF G0KYX
RH15 8QW G7ZMS
RH15 8TF G3XUD
RH15 8TR G3XQM
RH15 8UR G4ORS
RH15 8UY M5BTB
RH15 9HA 2E0DRA
RH15 9HA M0IKD
RH15 9HA M6IKD
RH15 9HR G1PJM
RH15 9HR G1POD
RH15 9HZ G8KPE
RH15 9JA G4WEH
RH15 9PL 2E1BRD
RH15 9PL G7OIA
RH15 9PY G0RPL
RH15 9RF M6HWJ
RH15 9RR M3ZGS
RH15 9SP G3VAK
RH15 9ST G0SQF
RH15 9SZ G7TMR
RH15 9UL G8GJV
RH15 9UT M6WYX
RH15 9XA G4ZCS
RH15 9XA M6TPE
RH15 9XG G0THD
RH16 1EJ G7TNU
RH16 1HJ G4KKO
RH16 1NB G4DAY
RH16 1[illegible] M1CTO
RH16 1TF G3ZEK
RH16 1UZ 2E0DEE
RH16 1UZ 2E0LIV
RH16 1UZ G8ZOV
RH16 1UZ M3MZA
RH16 1UZ M3MZC
RH16 1UZ M3ZOU
RH16 2AB G3FSX
RH16 2DQ G4IOV
RH16 2LJ 2E0WVE
RH16 2LJ M6WVE
RH16 2LQ G3HUA
RH16 2PH M6AKK
RH16 2QG G4OKB
RH16 2QG M0WGS
RH16 2QJ M6OJT
RH16 3AX 2E0CBQ
RH16 3HW G3ASV
RH16 3JS G8MFM
RH16 3LG M6FOW
RH16 3PE G3YTU
RH16 3PE M6YTU
RH16 3QW G0AUI
RH16 4DH 2E1FQF
RH16 4HL G7FHV
RH16 4JR G1SKI
RH17 5AW G0XAA
RH17 5AW G3VKW
RH17 5AW M3HJE
RH17 5[illegible] G0WYN
RH17 5HB M0TD[illegible]
RH17 5HH G4JST
RH17 5QQ G6SOY
RH17 6AF G8VKQ
RH17 6DQ G7WBR
RH17 6JA G4ZXB
RH17 6NG G7IAS
RH17 6NJ G0HXF
RH17 6SR 2E0PMB
RH17 6SR M6PBK
RH17 6UL M6COE
RH17 7HE G4MLR
RH17 7JU G4HYI
RH17 7LP M6GVD
RH17 7PG G1ACA
RH17 7PY G0ENA
RH17 7QS G4RBQ
RH17 7RA M6NIC
RH17 7RN M0JTJ
RH17 7RW G8ZAT
RH18 5AF 2E0TNC
RH18 5BX M0HBM
RH18 5BX M6AGO
RH18 5GD G5DJW
RH19 1DQ 2E0XDZ
RH19 1DQ M6XDZ
RH19 1EE 2E1JGW
RH19 1JG G3VPS
RH19 1JR G3KOA
RH19 1LP M3LUO
RH19 1SG G7KBR
RH19 1TA 2E0BQV
RH19 1TA M0GQD
RH19 1TA M0UQH
RH19 2DD G3YQW
RH19 2ER G0EID
RH19 3BS G4KIU
RH19 3HP M6SXB
RH19 3QE G0TSQ
RH19 3QL M1ENX
RH19 3RB M3NGY
RH19 3SE G1AML
RH19 3TN 2E0AZU
RH19 3TN 2E0BQX
RH19 3TN M3ICN
RH19 3TN M3MYT
RH19 3UX M3WIC
RH19 3XF G0RCF
RH19 4BZ G1YXT
RH19 4EA G4TRG
RH19 4JT 2E1HGR
RH19 4TF G4PFU
RH2 0QA G6YRV
RH2 0RE G8PZI
RH2 0US G8TIO
RH2 7BS G7CVM
RH2 7HJ G4AVE
RH2 7JX M0KDX
RH2 8DQ G0VGT
RH2 8DS G8ZRV
RH2 8EL M3WOQ
RH2 8NA G3ZIY
RH2 8NA G1TPA
RH2 8NE G6VMI
RH2 8QJ G4WBF
RH2 9DG G7OBF
RH2 9DH 2E0VBT
RH2 9DH M0VBT
RH2 9DH M6KVS
RH2 9HA M3SGJ
RH2 9NG M6EOQ
RH20 1AH G6OSO
RH20 1AW 2E1BSC
RH20 1AW G7PTX
RH20 1AZ M0EPX
RH20 1AZ M1LTS
RH20 1BZ G0INQ
RH20 1HS G1EAE
RH20 1LA G1PMA
RH20 1PS M3IYX
RH20 2AT G0LFH
RH20 2EE 2E0BSF
RH20 2EE M6MST
RH20 2HJ G4BUE
RH20 2HJ M0RUF
RH20 2JL M6FEP
RH20 2PR G0HWS
RH20 2PZ 2E0ZYL
RH20 2PZ M0TCD
RH20 2PZ M0[illegible]
RH20 3BD G8IPF
RH20 3GL M6RPX
RH20 3HG M1AIM
RH20 3HJ 2E0JGD
RH20 3HZ M6GNM
RH20 3JW G3XJN
RH20 3NS M3PTX
RH20 4AR M0BAH
RH20 4AU 2E1DZH
RH20 4AU G4OAK
RH20 4BS G4LCU
RH20 4JU M6DMP
RH20 4LL G6AAR
RH20 4LT G4JCA
RH20 4LU G3POQ
RH20 4LW G0AGB
RH20 4QL G6LBE
RH20 4QL G6YCZ
RH20 4QX G7TMU
RH20 4RF G4YPG
RH4 1EY G7APO
RH4 1HD M1DDF
RH4 1LP G0GNA
RH4 1NA M6SLR
RH4 [illegible] [illegible]
RH4 [illegible] G3IKV
RH4 2DG G0GQY
RH4 2HT G4YFK
RH4 2JE M3WLX
RH4 2NR G4VUW
RH4 3BZ G6EQP
RH4 3BZ M3EQP
RH4 3DS G3YWX
RH4 3DZ G6DTH
RH4 3HX G3AFB
RH4 3JH M0DSX
RH5 4AD G3RDH
RH5 4AE M6FWG
RH5 4EY 2E0NNB
RH5 4EY M0NNB
RH5 4EY M6NNB
RH5 4NN G4FTQ
RH5 4PE M6DCD
RH5 4PP G6YPF
RH5 4QB G4VTC
RH5 4QH G1THW
RH5 4RH G1VNU
RH5 4TN 2E0DKE
RH5 4TN M6FCM
RH5 4TN M6GWI
RH5 5AQ G4CMU
RH5 5AQ G7DOR
RH5 5BS G1ABQ
RH5 5BX G4OBT
RH5 5DZ G0RKS
RH5 5HT G6KNE
RH5 5JB M6ATH
RH5 5LL G6OBJ
RH5 5SG M0OTT
RH5 5TR M6RKE
RH6 0BL G8OLP
RH6 0DR M1AZB
RH6 0HU 2E0IBN
RH6 7BU G1TRI
RH6 7BX M3JXV
RH6 7JF G1HLY
RH6 7LN M6BJT
RH6 7NH M3TAG
RH6 8BS G8AMU
RH6 8HW 2E0TZO
RH6 8HW G1ZQN
RH6 8HW M0HTJ
RH6 8HW M0TZO
RH6 8HW M3TZO
RH6 8JG G0SOG
RH6 8JG G7KYW
RH6 8JR G3JHP
RH6 8LF G6DCH
RH6 8QT 2E0HAP
RH6 8QT M6AVO
RH6 8RE G6VKX
RH6 8RX G4EIV
RH6 8RZ M6RQB
RH6 9BY M6GRU
RH6 9DX G0VQD
RH6 9EA G6MGZ
RH6 9NA G3ZIY
RH6 9QA G4PFT
RH6 9QP G0HUF
RH6 9RB M6SJO
RH6 9SY G3OKS
RH6 9TX G4FBI
RH6 9UE M0MIM
RH7 6BT M6UTT
RH7 6DG G3OYU
RH7 6EX 2E0CSQ
RH7 6EY G8OYY
RH7 6JX 2E0JXN
RH7 6JX M3JXN
RH7 6LL G3LHZ
RH7 6NQ G7TWJ
RH7 6PS 2E0TZZ
RH7 6PS M0TZZ
RH7 6PS M6TZZ
RH7 6QX G6AHR
RH7 6RL G3IOM
RH8 0EW G7VPU
RH8 0JA M3WNV
RH8 0JE G3IKB
RH8 0PD 2E0IPW
RH8 9AU [illegible]
RH8 9AU G7BKN
RH8 9AU M5ALC
RH8 9DJ M6OHN
RH8 9LT 2E0GPD
RH8 9LT M6OBZ
RH8 9PF M0JEP
RH9 8ED G7VIO
RH9 8ER G0ALE
RH9 8ER G3RQZ
RH9 8JW G0FTU
RH9 8LD M0JCJ
RH9 8LD M3ZUS

## RM

## (Romford)

RM1 2AD M3NHS
RM1 2NP G6YCF
RM1 2SR G8JBV
RM1 4AU M0VIN
RM1 4BZ G6BTX
RM1 4HD G7MMU
RM1 4LA G3XLR
RM1 4SP G6TGM
RM1 4XR G8TQZ
RM10 7BT G0RFL
RM10 7EX G4SHN
RM10 7YR M6LDR
RM10 [illegible] G0GGH
RM10 8BS G1JQH
RM10 8BT G1YKI
RM10 8DE G0WXE
RM10 8ES G1WUH
RM10 9BS 2E0YAL
RM10 9BS M0YOL
RM10 9BS M3TBV
RM10 9PP M1GDH
RM10 9QD M0GRT
RM11 1BW G7IYN
RM11 1FL G3LXJ
RM11 2BA G0BSF
RM11 2BS G0CBU
RM11 2NF M3DSA
RM11 2ST 2E1RBA
RM11 3EN G0RIQ
RM11 3LD 2E0LBL
RM11 3LD M0LBY
RM11 3LD M3RQW
RM11 3PD G0PIA
RM11 3PY 2E0EBV
RM11 3PY G3JHI
RM11 3PY M0TAZ
RM11 3QD G1VSH
RM11 3QH G8WUO
RM11 3SG G0TJN
RM11 3SG G4VIX
RM12 4EY G3SFK
RM12 4HZ M0HQE
RM12 4JZ M0MAC
RM12 4JZ M1CUC
RM12 4LL G0IAP
RM12 4NR G3JWH
RM12 4SE G6MVS
RM12 4YF G4VHR
RM12 5BL G0AJH
RM12 5DR G7GJY
RM12 5EU G6BEL
RM12 5HH 2E0AWI
RM12 5LS G8PMR
RM12 5QS G8UZW
RM12 5RD G3XYA
RM12 6AT G4MXF
RM12 6TF G4EEF
RM13 7AS 2E0UKD
RM13 7AS M3UKD
RM13 7EJ G4RZZ
RM13 7PD 2E0WJI
RM13 7PH G8ZDT
RM13 8AA M6GAI
RM13 8AJ G4JRB
RM13 8AL G3SFM
RM13 8BB G0IEO
RM13 8ND G8NWI
RM13 9LF M3NYI
RM13 9LW G0PBN
RM13 9LW G1BTF
RM13 9RL M6AMI
RM13 9UU G7KHF
RM13 9UU G8FJG
RM14 1AL G6TRX
RM14 1AT G0KSJ
RM14 1HN G0IQN
RM14 1HP G4AZX
RM14 1HT M1CZO
RM14 1NA G3RJI
RM14 1NB G0DOM
RM14 1QU 2E0AVQ
RM14 1QU M3AQF
RM14 1XR G8XQZ
RM14 2EZ 2E1EVK
RM14 2EZ M6HDU
RM14 2HJ G8GGO
RM14 2HR 2E0MES
RM14 2HR M0SEM
RM14 2HR M3ELS
RM14 2HY G3ZWF
RM14 2JH G4UNS
RM14 2YX M3DFU
RM14 3AA G4VIW
RM14 3AH G7GES
RM14 3AU G7DXX
RM14 3AZ G8VVG
RM14 3EA G4XZC
RM14 3[illegible] [illegible]
RM14 3YT G4LQX
RM15 4AA M3VTH
RM15 4NR M0CJR
RM15 5BZ G0PNN
RM15 5DJ 2E0PFX
RM15 5DJ M0PFX
RM15 5DJ M0XYD
RM15 5DJ M6PFX
RM15 5DS M6FZN
RM15 5ES M6PST
RM15 5ET 2E0JGG
RM15 6HP 2E0BYN
RM15 6HP M6ILP
RM15 6TS G1DXH
RM15 6XE G1TJT
RM16 2GB G3SCT
RM16 2GB G4HCK
RM16 2GB G4HKO
RM16 2QL G1WEF
RM16 2RS G1ZAY
RM16 3AU G6TYF
RM16 3DW M0MOR
RM16 3HB G1VSE
RM16 3LT G0DEK
RM16 4BS G7HMA
RM16 4HL 2E0KGJ
RM16 4JB G8WTN
RM16 4JF M6TTM
RM16 4RR G6NVD
RM16 4[illegible] [illegible]
RM16 [illegible] G8DOJ
RM16 6FT M5BAZ
RM16 6LT 2E0BWT
RM16 6LT M6AEW
RM16 6NP M6EIR
RM16 6NS M6JYB
RM16 6QT M6KBA
RM16 6QT M6MES
RM16 6RE M6GNJ
RM16 6RE M6GNS
RM17 5AB G0FDS
RM17 5AB M1MOG
RM17 5HA G4EPM
RM17 5LW M1DPE
RM17 5QR G0BKL
RM17 5RG G3KMD
RM17 5RG G6EFE
RM17 5RG G6OBO
RM17 5RG M1YOW
RM17 5RH 2E0ELI
RM17 5RH M6WJL
RM17 5SF G4LUT
RM17 5UW G8VZJ
RM17 5YN G1NOR
RM17 6DJ M6URS
RM17 6LD M0TIL
RM17 6LD M1DUC
RM17 6PX M6GPM
RM17 6SG M6WRO
RM18 8BA G3PHD
RM18 8DT G1EHM
RM18 8HF G7FCJ
RM18 8HQ G7VQL
RM18 8SX G4OVG
RM18 8XP 2E0PDM
RM18 8XP M6MTR
RM19 1LE G8TND
RM19 1RQ M1TDD
RM2 5PP G3ZQY
RM2 5PS M6EDO
RM2 5XJ G4XXT
RM2 6AP G3TPJ
RM2 6DX G4JDT
RM2 6LD G4GBW
RM2 6QX G8EDH
RM20 3DA M6GDJ
RM20 3JA M0TIU
RM20 4AB G1SID
RM20 4YD M6BKV
RM20 4YR G6LCU
RM3 0AW M6HKU
RM3 0BQ G8VXR
RM3 0DL G4TTN
RM3 0EB G0BCZ
RM3 0EJ G3SKV
RM3 0JD M1JEC
RM3 0RH M3BFJ
RM3 0XN M0BOB
RM3 0YT G3SVK
RM3 0ZA M3TXM
RM3 7DX G0CBT
RM3 7DX G3XBF
RM3 7EQ 2E0SLH
RM3 7EQ M0HXE
RM3 7QL M3CGJ
RM3 7SP M6WCE
RM3 7UB G1WZK
RM3 8DB M6OOX
RM3 8HJ G6MDN
RM3 8HN G7VVL
RM3 8HN G8TPP
RM3 8HQ G6WKI
RM3 9DE M3MIX
RM3 9LA M1BFX
RM3 9NA G1CSN
RM3 9NL G4YOT
RM3 9PT 2E0PSN
RM3 9PT M6PSN
RM3 9XR M6MTC
RM4 1AX G6SVV
RM4 1DL G0NFJ
RM4 1DR M6LRL
RM4 1DT G4XVH
RM4 1NH G6DIM
RM5 2HF G6EBO
RM5 2SD G4ZMS
RM5 2SL M0PAX
RM5 3ND G0OVC
RM5 3QA M0ONP
RM5 3RX G0BLO
RM5 3SX G8OPO
RM5 3TJ G4MWG
RM6 4BF M6ING
RM6 4EB G0FFK
RM6 4EB G1HKS
RM6 4NU M1BDY
RM6 4RL G3JSR
RM6 4SL G4YGL
RM6 5AT M3EDU
RM6 5EJ G7VVK
RM6 5JP M6HFI
RM6 5NR M3MRO
RM6 5QL 2E0OKC
RM6 5QP M0CMH
RM6 6AA M0HUA
RM6 6PY G1NIV
RM6 6RX G0PIS
RM6 6UH 2E1GAD
RM7 0JP M3ZNL
RM7 0LA G8YIN
RM7 0NJ G0BOF
RM7 7LH M6JYO
RM7 7LJ 2E0ROV
RM7 7L[illegible] [illegible]
RM7 7L[illegible] M6ROW
RM7 8AZ M6ZXB
RM7 8BD G7WJE
RM7 8BE 2E0TMA
RM7 8BE M6RMN
RM7 8HD G0GFZ
RM8 1DS G7TAX
RM8 1JY G0TJE
RM8 1PP G6XDG
RM8 1XD 2E0DPD
RM8 1XD M6FCR
RM8 2EW 2E0TPN
RM8 2LR G8SLB
RM8 2QL M0BLY
RM8 2RA G7UYT
RM8 2RA M0HEM
RM8 2TU G4CUQ
RM8 3ST G6AIU
RM9 4ET G7CXT
RM9 4ST G4HOI
RM9 4ST G8ZZL
RM9 4TJ M3SJH
RM9 5RS M0DPV
RM9 5RS M3ARB
RM9 5XH G7UVW
RM9 6LD 2E0WWN
RM9 6LD G0TOD
RM9 6LD M3WWN

## S

## (Sheffield)

S1 3JD G3UOS
S10 1DB G7JAF
S10 1HT G0IYV
S10 1PF G8PNM
S10 1WQ G8YXI
S10 2BA M6HUX
S10 3BU G4SGF
S10 3PT G4SOM
S10 3RF G0BSX
S10 3RZ G6HYI
S10 3TR M1GWA
S10 4DN G4CUI
S10 4ED G6YFY
S10 4EU G3YBA
S10 4EX M1AGR
S10 4GF G3WDM
S10 4LF G7SSB
S10 4LL G8AGN
S10 5NX G7LQO
S10 5RE G1JVN
S10 5RE G1JVO
S11 7EE G4CYA
S11 7EF M6HSY
S11 7GG 2E0SSY
S11 7GG M6MHO
S11 7LX G0IIF
S11 8UA M1AAC
S11 9AA G0PFI
S11 9BR G0HHA
S11 9BR G0LNV
S11 9BU M0AXE
S11 9DE M6UAJ
S11 9DR G0WLR
S11 9FA G7PFI
S11 9LH 2E0ZSB
S11 9LH M3ZSB
S11 9LQ G4IJB
S11 9PX G8AKQ
S11 9RA G0HGI
S11 9RP G4HIA
S11 9RR G3NHE
S11 9SB G8RFP
S11 9SP G3RVI
S11 9SP M3NSO
S12 2AY G3UXM
S12 2BL M5ASK
S12 2BU G4VLZ
S12 2FN G1HFK
S12 2GL M6NLF
S12 2JH 2E0PCG
S12 2JP M3ZBF
S12 2NB 2E0GNS
S12 2NB M3HNU
S12 3DT M6MPM
S12 3HB G1GPM
S12 3HN M6SCH
S12 3LP 2E0KXD
S12 3LP M3KXD
S12 3LR G7GHH
S12 4ER 2E0SPD
S12 4ER M0SZD
S12 4ER M6AXD
S12 4HU 2E0TCX
S12 4HU M6HBH
S12 4HU M6IGW
S12 4JB M0CBT
S12 4LF G4GPF
S12 4ND M0PBO
S12 4SE M3SZT
S12 4XE G0BQW
S12 4XH G7UEX
S12 4XS 2E0MJG
S12 4XS M0NCN
S12 4XS M3NCN
S12 4YF G0NRM
S13 7LS G0WPC
S13 7RH G3JCJ
S13 8EH M3DYR
S13 8EL G0JJR
S13 8NJ G4VBP
S13 8TH G0HYS
S13 [illegible] M0TPM
S13 9HX G0VDN
S14 1DF M1ADP
S14 1GD M3DDZ
S14 1RJ G1RZJ
S17 3DN G4UWP
S17 3EG G8BLP
S17 3ER G3SXT
S17 3EZ G0VNE
S17 3GF G0TVW
S17 3GZ G4AYO
S17 3HE G0LVA
S17 3LL M6AYM
S17 3NB M0BAK
S17 3QF G0TKK
S17 4PF G4KPU
S17 4QX M6RWS
S17 4RJ G0OUK
S18 1ND 2E1HLO
S18 1ND G0LUM
S18 1ND M3CAZ
S18 1PH M6DPF
S18 1PW 2E0BMD
S18 1PW M0UCD
S18 1RF G6CTA
S18 1RJ G0XVL
S18 1RP G0UGJ
S18 1SD G4ULN
S18 1UQ G3KNJ
S18 1UQ G3VKK
S18 1UZ G4JUH
S18 1WH G3YON
S18 1XX G4HOP
S18 2EA 2E0HAV
S18 2EA M0ZEL
S18 2EA M6DQZ
S18 2EH G7BUF
S18 2EJ G1EQL
S18 2EL G4PXF
S18 2ER G0MVC
S18 2FF G4KXW
S18 2HQ G4PPR
S18 2LE G1DFW
S18 3BB G8RFW
S18 3BQ G1KPZ
S18 3BW M5SAV
S18 3BY M3EVJ
S18 4DE M6DDI
S18 7WA G8UHM
S18 8NA G0XVS
S18 8YW G0WYN
S2 1GH M3XAH
S2 1HE G0BJL
S2 1RG G0JJV
S2 1RH M6XXX
S2 1UU M3DFS
S2 2AG 2E0WST
S2 2AG M0VGC
S2 2AG M6WST
S2 2AR G8YQU
S2 2EA 2E0IUK
S2 2EA M0IUK
S2 2EA M6IUK
S2 2GB G1LWY
S2 2NZ G1ASG
S2 2PJ 2E0ZEV
S2 2PJ M3ZEV
S2 3AE M6KJP
S2 3BJ M3NGZ
S2 3EH G7SWQ
S2 3EH M6WNC
S2 3LB G1YED
S2 3LQ G1DBZ
S2 3NF 2E0KSH
S2 3NF M6KSH
S2 3QJ 2E0AJT
S2 3SY M6RUK
S2 3UF M6EEZ
S2 4AH M6ICR
S2 4BP M6JDQ
S2 5FD 2E0ZMZ
S2 5FD M6MFJ
S2 5QW G4MWL
S20 1AD G7KDQ
S20 1AW G4RQF
S20 1AY 2E0PCF
S20 1[illegible] [illegible]
S20 1FW M0JCS
S20 1FW M6ICO
S20 4GB 2E0UKX
S20 4HL G6XLC
S20 6QJ M6JUB
S20 6SD G6YAI
S20 6SG G0FZD
S20 6SU M6BTE
S20 8EF G0OUG
S20 8EZ 2E0EZX
S20 0LE 2E0USA
S21 1BU M6TSP
S21 1QA M[illegible]
S21 1GR M3HGT
S21 1TF G0PBH
S21 1TG M0AFW
S21 2BX G3HQG
S21 3UL M6ZCB
S21 4AR M3NXC
S21 4BR G8IQA
S21 4FY 2E1HXB
S21 4FY M3HXB
S21 4JX G7SCP
S21 5FD G7TZQ
S21 5RS G6XUX
S21 5RW 2E0YZA
S21 5RW M0YZA
S25 1YP G8UHV
S25 1YT G8WVH
S25 2N[illegible] [illegible]
S25 2NY G6XUX
S25 2RP G4NUB
S25 2SW G3YYE
S25 2SX M0YZV
S25 2XT M6RIS
S25 3RG M0JWJ
S25 3RG M6GZW
S25 4BJ G0VUH
S25 4BR 2E0MKE
S25 4BR M6MKE
S25 4BW G7FTM
S25 4DQ G0DZX
S25 4EP 2E0USV
S25 4EP M0USV
S25 4EP M6RIE
S25 4EP M6USV
S25 4FG G8EMH
S25 4FL G4SRX
S25 4GD G6YZR
S25 4HB G4GQS
S25 4HB M3IXE
S25 4JP G4RMX
S25 4JY G7GFR
S25 5BA G6TNI
S25 5BA G7IOI
S25 5DR G0JZL
S25 5EG G1DPI
S25 5EG M0CKX
S25 5GL G7BCI
S25 5JG G1HEW
S26 1HJ G7TBF
S26 1HJ G7TDR
S26 1HL G3NEO
S26 1JG G1YGP
S26 1JN G3HKF
S26 1JN G7HAR
S26 1JQ G1WRD
S26 2AA 2E0DCA
S26 2AA M3UKB
S26 2AS G0CCN
S26 2BR G4YSG
S26 2DH M0TWG
S26 2EP G4ZDD
S26 2FJ G6BCM
S26 2GA 2E0TIF
S26 2GA M6TIF
S26 2GL M3VPP
S26 2GR 2E0GBP
S26 2GX M3VII
S26 3RB M0TRV
S26 3XS G0DCF
S26 4NQ 2E0CFK
S26 4NQ M6PER
S26 4RQ M6DNS
S26 4TA G7LQN
S26 4TH M6PJB
S26 4TL G1UDE
S26 4UP 2E0PFO
S26 4UP M0PFO
S26 4UP M6PFO
S26 4UY M1CLW
S26 4WB G0BPM
S26 5LZ G8OCE
S26 5PH M3KXS
S26 5QS M0ONI
S26 5RG M1TVR
S26 5RG M3HIT
S26 5RG M3MTR
S26 5RU G0SVX
S26 6RA M0CUI
S26 6SJ 2E0ZPA
S26 6SJ M0PJA
S26 6SJ M6PJA
S26 6SJ M6ZCA
S3 7AH 2E0OKH
S3 7AH M6HOM
S3 7AH M6OKH
S3 7JH M6CGC
S3 7NL G8PLI
S3 7PB G0FTR
S3 9GX M3IVN
S3 9JF G4MRB
S3 9JF G6DIZ
S32 1AS G4KWE
S32 1[illegible] G7ODR
S32 1DU G0WEX
S32 2JL G0MFB
S32 5QB G0NOU
S32 5QX M1CRP
S33 6SB 2E0HDM
S33 6SB M3WBA
S33 8WB G4RNA
S33 8WB G6HBZ
S33 9JQ M6SSR
S33 9JX G4NYW
S35 0AF G0FLI
S35 0BB G4PHL
S35 0DQ M6RKS
S35 0EA G7BPQ
S35 0GD G1OKB
S35 0HR G0CCM
S35 1AH G0KNL
S35 1AL M0AZB
S35 1EA G6ZWL
S35 1QG M1EOU
S35 1SQ M6HFG
S35 1SY 2E0CVX
S35 1TX M0BVN
S35 1UR 2E1CJD
S35 1YS 2E0ZSJ
S35 1YS M0ZSJ
S35 1YS M3ZSJ
S35 2TD G0UAC
S35 2UG G4TQT
S35 [illegible] G1YLU
S35 [illegible] 2E1DJD
S35 3HS M0GCC
S35 3HZ G0VUM
S35 3LP M6KVX
S35 3PA G0LQC
S35 4HF G6OFV
S35 4NS M0CBN
S35 4NS M0DEI
S35 4NT M0CKS
S35 7BS M1AKT
S35 7DH M6SFM
S35 7DQ G3LZI
S35 8PE G4CTE
S35 9YW M6HBD
S36 1AY G2CFC
S36 1BZ 2E0RGS
S36 1BZ M0RBQ
S36 1BZ M6AOV
S36 2NA M6KTP
S36 2ND 2E0LMG
S36 2NH G8UCN
S36 2NP G0TKF
S36 2PT G7DGE
S36 2QX M0IBY
S36 2QX M0TGW
S36 2QX M3VCZ
S36 2SG 2E0TKF
S36 2SG M3UCY
S36 2SG M3UEL
S36 2TF G1YKL
S36 2TN G0UNY
S36 6AF G7MZJ
S36 6AU 2E0BIB
S36 6AU M3GPN
S36 6BH G0EWD
S36 6EA G7MZK
S36 6EH M6MIN
S36 8HP G0BVK
S36 8HP G4MWF
S36 8WW G4VRT
S36 8ZS M6VMJ
S36 9DB G4DFS
S36 9EF M6ZXL
S36 9NE G6ZTD
S36 9RX M3LGM
S4 7DD M3BUH
S4 7DE G4DJD
S4 8AA 2E0TWS
S4 8AA M0TWS
S4 8HL M6SAI
S40 1DG G0UUC
S40 1HQ M0EME
S40 1HU 2E0JIF
S40 1HU M0OND
S40 1HU M6GIM
S40 1HZ G4LNG
S40 1NJ M6NGR
S40 2HT M6EDM
S40 2LL 2E1EAZ
S40 2LT G7SFF
S40 2RH G1IOR
S40 2RL 2E0GIG
S40 2RL 2E0SLE
S40 2RL M0XAI
S40 2TN M3ZQV
S40 2UF G1VPE
S40 2UR M3XQG
S40 3BT G3ZLF
S40 3BX 2E0MAZ
S40 3BX M0MAZ
S40 3BX M3XIO
S40 3DF G4BFR
S40 3DR M3OOP
S40 3EY G3GLX
S40 3HF G4NKI
S40 3HG G3OOP
S40 3HT G7KUM
S40 3LW G1UGL
S40 3NB M3AJV
S40 3NN G8OKI
S40 3NT G6DPE
S40 3PJ 2E0BNK
S40 3PJ M0ZIG
S40 3PJ M3SRQ
S40 4DB G8YTF
S40 4DB M0VAA
S40 4EA G4GSD
S40 4ES G6NZA
S40 4HN M0APL
S40 4LE G4DDI
S40 4LG G6OKU
S40 4LG M0OKT
S40 4LQ G7PUP
S40 4NE M3SMN
S40 4NW M0EJF
S40 4PQ 2E0STA
S40 4PW G6EYD
S40 4RJ 2E0DXK
S40 4RJ M6JWG
S40 4RS M3GAE
S40 4RX 2E0BNO
S40 4SJ M6MBO
S40 4TF G0RDF
S40 4UT M3JCU
S40 4XD G4VHX
S40 4XJ G0EWI
S41 0AT M6EYG
S41 0BB M3VJI
S41 0BD M3LGF
S41 0BS M3XAN
S41 0GF M0KBW
S41 0JP M6MPV
S41 0NJ G0UEH
S41 0NN G7HIQ
S41 0[illegible] [illegible]
S41 0RY [illegible]
S41 0RX M6HIM
S41 0SU G4HQH
S41 7HA G0TRN
S41 7JF M6NDO
S41 7QE M0RRR
S41 8AU M3OHL
S41 8BN M6KZE
S41 8LX G8TFU
S41 8LZ M0DRE
S41 8PE G0LJI
S41 8PE M6SSA
S41 8RU G8VQS
S41 8SF G6FTH
S41 9DA G2AS
S41 9DA G3PHO
S41 9DA G5TO
S41 9JB M3YZK
S41 9JD M6MJD
S41 9JU G6ZOT
S41 [illegible] G6NZO
S42 5AR M0GXO
S42 5BH G1EUT
S42 5GA G1CBY
S42 5HH G6SZB
S42 5HH G8NSD
S42 5JA M0FOX
S42 5JP G0JOM
S42 5NS G6URK
S42 5NT G6PRP
S42 5PU 2E0MKV
S42 5PU M6MKV
S42 5PX M6VEC
S42 6BD M3UXU
S42 6BE M3ESS
S42 6BE M3LBM
S42 6BE M3NST
S42 6BG M3SAA
S42 6BW 2E0CAK
S42 6BW M6ARM
S42 6HD G3WAM
S42 6HS G4XFF
S42 6JD G3CTZ
S42 6JF G4ROP
S42 6LB G3KMY
S42 6LZ G7HNG
S42 6NR M3YJN
S42 6PL 2E0WNT
S42 6PL M3WNT
S42 6RX G4VSR
S42 6SP G6SVH
S42 6SP M3SVH
S42 6TF M3HUB
S42 6TY G3YBO
S42 6UR G4CVO
S42 6XR M3UXH
S42 7AN G1WAB
S42 7AN G6PBW
S42 7AN M3JWW
S42 7DR G6DQA
S42 7EA M3EVR
S42 7JQ G1FTU
S42 7LR G3RLL
S42 7NE M0CRW
S42 7NH G6OYV
S43 1HB 2E0UCV
S43 1HB M3UCV
S43 1HG M3WAV
S43 1HP M6GOB
S43 1HP M6PAX
S43 1HP M6SSP
S43 1JA G8YMT
S43 1JU 2E0MTN
S43 1JU M0VTR
S43 1JU M3BPN
S43 1LJ G4PBC
S43 1LJ M0APH
S43 1NU M3RGG
S43 2AP G3ZSZ
S43 2BG M3CKM
S43 2BG M3CKO
S43 2DF M6REF
S43 2DZ G4RIQ
S43 2EA G3OVK
S43 2HN M6RUP
S43 2LH M1BVI
S43 2NR M3RNY
S43 2NW 2E0GRY
S43 2NW M0GRY
S43 3AG G7SMC
S43 3DH G0VRH
S43 3HJ G7VND
S43 3HS M0IKE
S43 3HS M0JOY
S43 3HU 2E0MCS
S43 3HU M6LSA
S43 3HU M6UPS
S43 3NJ 2E0DBM
S43 3NJ M3UDZ
S43 3PQ 2E0YNI
S43 3PQ M3ICO
S43 3PQ M3IXT
S43 3PY G7SLY
S43 3QB G4OQU
S43 3QZ 2E0SYM
S43 3RA M6SAS
S43 3SY M3FVJ
S43 3UH G3BNF
S43 3XH G4VWT
S43 3XH M3UHV
S43 3XH M6SLI



**IMPORTANT NOTE**

**Revalidate licence to avoid revocation** – Ofcom has advised the Society that plans will be drawn up to revoke licences that have not been revalidated as required by the licence conditions. The quickest way to revalidate is to do so online via the Ofcom website: *https://services.ofcom.org.uk/* or by email: *amateur.validations@ofcom.org.uk* If you need assistance in the process, Ofcom staff are available to help, but please be patient during times of heavy workload.

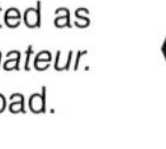

| Postcode | Callsign |
|---|---|
| S43 4AP | G2FVL |
| S43 4AQ | M3KKB |
| S43 4BD | G4VDB |
| S43 4BS | G0KUW |
| S43 4DH | 2E0FIF |
| S43 4DH | M0MKH |
| S43 4DW | G7MOO |
| S43 4EG | G6DAO |
| S43 4EH | G3WEU |
| S43 4NB | M6DVE |
| S43 4NG | M3KOJ |
| S43 4NG | M3LBN |
| S43 4NJ | G4RTA |
| S43 4PF | M3KQD |
| S43 4PG | M3VUP |
| S43 4PQ | M3DLZ |
| S43 4PQ | M3LDJ |
| S43 4QP | 2E1FDF |
| S43 4RF | G6YFH |
| S43 4RU | 2E0CAG |
| S43 4RZ | M3ZIO |
| S43 4SA | G0RUV |
| S43 4SR | G4BJG |
| S43 4ST | G4RSS |
| S43 4SX | G0EOJ |
| S43 4TR | G8KDM |
| S44 5AE | G3GGU |
| S44 5AL | M6DEK |
| S44 5AP | M3FRT |
| S44 5BA | M3VDA |
| S44 5ET | G0MBS |
| S44 5HE | M1CLI |
| S44 5LU | G6EZH |
| S44 5LX | M6PFK |
| S44 5NG | 2E1SAZ |
| S44 5NG | G0THF |
| S44 5NG | M0RSC |
| S44 5NG | M3SAZ |
| S44 5PH | M6AWZ |
| S44 5QA | 2E0GUN |
| S44 5QA | M0MTX |
| S44 5QA | M3ORL |
| S44 5QD | M6ASO |
| S44 5UE | G0TJG |
| S44 6AQ | M6EYB |
| S44 6BH | G0IXZ |
| S44 6DH | M3LEX |
| S44 6DY | G6TEX |
| S44 6EG | 2E0SUE |
| S44 6EG | M3PUZ |
| S44 6EJ | M6GGT |
| S44 6ER | G8YBR |
| S44 6EW | G0DJA |
| S44 6EY | G4UPA |
| S44 6HF | M3DGR |
| S44 6HS | G7WAF |
| S44 6HT | G1SLE |
| S44 6HU | G4AGE |
| S44 6HU | G8AVC |
| S44 6LL | M3MYK |
| S44 6LS | M6FBP |
| S44 6NX | G7GSX |
| S44 6NX | M0CWX |
| S44 6NX | M3CFP |
| S44 6PS | G0CTS |
| S44 6SE | G4FEM |
| S44 6TX | G4KSY |
| S44 6XN | G7DJN |
| S44 6XR | G8NRC |
| S45 0AL | G1ISN |
| S45 0DA | M6VRR |
| S45 0EA | G3MME |
| S45 0LL | M1WWW |
| S45 8AH | M0CJD |
| S45 8BQ | G6XZC |
| S45 8DH | 2E0JUL |
| S45 8DH | G3VNH |
| S45 8DH | M0JHD |
| S45 8DH | M3VNG |
| S45 8EJ | 2E0IBU |
| S45 8EJ | M6WBM |
| S45 8ET | G6POJ |
| S45 8ET | M6SKF |
| S45 9BG | M6NUR |
| S45 9FA | G8RBW |
| S45 9HH | M6PWH |
| S45 9HH | M6SIY |
| S45 9JT | 2E1FKM |
| S45 9LY | G3UAF |
| S45 9NH | G0MNC |
| S45 9PL | G0IVB |
| S45 9RE | G1HFT |
| S45 9RE | G3THT |
| S5 0DH | M6DCA |
| S5 0DY | M6BGJ |
| S5 0GX | M3DTH |
| S5 0JL | 2E1FDD |
| S5 0JP | G4TCE |
| S5 0JY | G7WGX |
| S5 0JY | M0SAY |
| S5 0NN | 2E0JOX |
| S5 0NN | 2E0TJX |
| S5 0NN | M3KUK |
| S5 0QE | M6FJO |
| S5 0TY | G6ADD |
| S5 6AA | M6MVT |
| S5 6HB | G8BMI |
| S5 6LX | G7DMP |
| S5 6SG | M3LEB |
| S5 6SQ | G0UHF |
| S5 7JX | G7ACR |
| S5 7TL | G0OLL |
| S5 7WE | G0UQZ |
| S5 7WU | G0BWG |
| S5 8AX | 2E0OMV |
| S5 8AX | M0OMV |
| S5 8DT | M0DMA |
| S5 8GL | 2E0DBN |
| S5 8GL | M6DMB |
| S5 8GN | 2E0SNU |
| S5 8LL | M6DWW |
| S5 8NE | M6GVS |
| S5 8WF | 2E1AGV |
| S5 8WF | M1FJQ |
| S5 9AH | M6JFS |
| S5 9DY | G3UXH |
| S5 9GA | G0SEW |
| S5 9GS | G8AOZ |
| S5 9QR | G0RNB |
| S6 1BQ | M0GAV |
| S6 1HU | G7DEE |
| S6 1JE | G3LLV |
| S6 1JP | M6TBX |
| S6 1LA | 2E0YDB |
| S6 1LA | M0SDB |
| S6 1LA | M6YDB |
| S6 1XA | M6MPR |
| S6 2AT | M6SLO |
| S6 2TY | M3XWF |
| S6 3NG | G4BOJ |
| S6 3TH | G7NUC |
| S6 3TL | M1FFV |
| S6 4AH | G4VYI |
| S6 4AY | G0BKE |
| S6 4BE | G3OZL |
| S6 4FG | G0CGT |
| S6 4FN | G3VCQ |
| S6 4FN | M0SDC |
| S6 4FN | M3VCQ |
| S6 4NB | G0JKE |
| S6 4QG | M0EMR |
| S6 4QP | G7NNZ |
| S6 4SU | M3PMK |
| S6 4WG | M1EQW |
| S6 5BA | G0RNC |
| S6 5BH | G6NVT |
| S6 5DD | G4YWJ |
| S6 5DQ | M0DPJ |
| S6 5GQ | M6YSM |
| S6 5HQ | M3UJL |
| S6 5LN | G4TFZ |
| S6 5ND | M6XAW |
| S6 5PA | G0JTT |
| S6 6AB | G1MLV |
| S6 6DJ | G0JKH |
| S6 6EW | G8FDJ |
| S6 6GL | G4FAL |
| S6 6GL | M0HDG |
| S6 6HX | M6MBG |
| S6 6LJ | G1WWI |
| S6 6SG | G8KB |
| S6 6TG | M1JWM |
| S60 2UZ | G1NFO |
| S60 3AZ | G3VSK |
| S60 3ER | G1MDE |
| S60 3EW | G4RGH |
| S60 3HB | M0GUD |
| S60 3JJ | G1HWA |
| S60 3JN | G1IDF |
| S60 3JZ | M3CPN |
| S60 4DZ | G6BKL |
| S60 4EA | M6AYW |
| S60 4EF | G4LFS |
| S60 4NH | G4NLC |
| S60 5ES | G0ISK |
| S60 5HG | G1DEZ |
| S60 5JU | G0CGS |
| S60 5PU | M3FZO |
| S60 5RE | G6CPY |
| S60 5TH | G8AHN |
| S61 1BJ | G0UOM |
| S61 1HW | M3PJI |
| S61 1RY | G1RGG |
| S61 2BL | G6LXW |
| S61 2DU | 2E1BKF |
| S61 2EX | G1OHU |
| S61 2HB | G8EQD |
| S61 2HB | M0EQD |
| S61 2JW | G6MLS |
| S61 2JZ | M3FFK |
| S61 2LT | 2E0DBO |
| S61 2LT | M6IDX |
| S61 2NP | M3FJC |
| S61 2QS | M6CLW |
| S61 2RP | 2E1ENZ |
| S61 2RP | 2E1ETB |
| S61 2RP | G4ZVN |
| S61 2SS | G4NXS |
| S61 2SS | G6XGJ |
| S61 2TQ | 2E0EVP |
| S61 2TQ | M6ATJ |
| S61 2TT | G4EBG |
| S61 2UH | G1TYU |
| S61 2UN | G0OUN |
| S61 3HN | G7OFV |
| S61 3NP | G4GET |
| S61 3RQ | M6FNU |
| S61 4DL | G0MTT |
| S61 4LP | G4HVW |
| S61 4PA | M0BEQ |
| S61 4PD | G7OGR |
| S61 4PL | M6EXX |
| S62 5BD | M3TKU |
| S62 5BH | M3IFA |
| S62 5ED | M0UGD |
| S62 5JR | G1GBR |
| S62 5LW | G0VBR |
| S62 5LW | G7JSE |
| S62 5NH | G7BQS |
| S62 5NT | 2E0GMA |
| S62 5NT | M6AXX |
| S62 5NT | M6GMA |
| S62 5QJ | G7AOQ |
| S62 6DB | 2E0BKB |
| S62 6DB | M3ZYO |
| S62 6HJ | 2E0BNE |
| S62 6HJ | 2E0SDA |
| S62 6HJ | M3OBX |
| S62 6LN | G7MST |
| S62 7BX | G6JDC |
| S62 7EE | M3UGD |
| S62 7EN | G3USW |
| S62 7FE | M3MGN |
| S62 7JA | M6NOL |
| S62 7QF | G6HGU |
| S63 0JA | M1EUN |
| S63 0LZ | G0TLA |
| S63 0LZ | G7KMF |
| S63 0PX | 2E0BMY |
| S63 0PX | M0GVX |
| S63 0PX | M0HKT |
| S63 0PX | M3PGK |
| S63 0QT | M3UOD |
| S63 0RZ | G3WLV |
| S63 0SA | 2E0GAR |
| S63 0SA | M3PGU |
| S63 0TG | 2E0ZTG |
| S63 0TG | M0ZTG |
| S63 0TG | M3OTG |
| S63 0TN | G4VNX |
| S63 0TY | G4XSB |
| S63 6AA | 2E0CDW |
| S63 6AA | M0KCW |
| S63 6AA | M6CDW |
| S63 6DD | 2E0DLC |
| S63 6DD | M6DLX |
| S63 6EL | 2E0DYJ |
| S63 6EL | M6EZC |
| S63 6ET | G4UZG |
| S63 6EY | G0JKG |
| S63 6HW | G8XUH |
| S63 6HZ | M6MGN |
| S63 6JN | M6EUU |
| S63 6JY | 2E0EXX |
| S63 6JY | M6BSP |
| S63 6JY | M6OMZ |
| S63 6NQ | M6KJQ |
| S63 6NU | G1KZA |
| S63 7GT | 2E1HTE |
| S63 7JB | G4TZL |
| S63 7JU | G7EVT |
| S63 7LR | G3YOC |
| S63 7ND | M6PSR |
| S63 7NF | M0KRL |
| S63 7NF | M3NBI |
| S63 7ST | G1YPR |
| S63 7TD | M6HCB |
| S63 8BZ | G6IIK |
| S63 8DA | G0TZC |
| S63 8HQ | M3FJD |
| S63 8HR | G1EQM |
| S63 8HR | M0EQM |
| S63 8JL | G0UAD |
| S63 9BA | G0GPR |
| S63 9BY | G6WSZ |
| S63 9DT | M6ZEF |
| S63 9DX | M3ZYI |
| S63 9DZ | 2E0CMO |
| S63 9EA | M6KMS |
| S63 9JW | G0JDC |
| S63 9LB | 2E0TAU |
| S63 9LB | M3LJK |
| S63 9LB | M6JAJ |
| S64 0AD | G8CXA |
| S64 0DG | G8HPJ |
| S64 0DG | M6DBG |
| S64 0DJ | G7BJR |
| S64 0DT | M0DKX |
| S64 0JD | G4KWM |
| S64 0JG | G4PJT |
| S64 0JZ | M0CTJ |
| S64 0NL | G3VJR |
| S64 5TY | G0VPS |
| S64 5UG | G6VIN |
| S64 5UQ | G7MUE |
| S64 5UQ | M3ILR |
| S64 5UQ | M6HOQ |
| S64 5UU | G0VXC |
| S64 8AH | 2E1FCE |
| S64 8DQ | M6PKN |
| S64 8DS | G6MFU |
| S64 8DU | G8BIW |
| S64 8EF | 2E0KLV |
| S64 8EF | M3VAF |
| S64 8LF | G0PBF |
| S64 8NF | M3KSK |
| S64 8NF | M3PSD |
| S64 8NL | M6FLW |
| S64 8NU | G0VJY |
| S64 8NX | G0CSK |
| S64 8NX | G6ZFX |
| S64 8PT | 2E0BQQ |
| S64 8PT | M3UPK |
| S64 8PY | M3UPL |
| S64 8QR | 2E0BQP |
| S64 8QR | M0KVM |
| S64 8QR | M3UOF |
| S64 8QW | M6PZB |
| S64 8RF | M6FJF |
| S64 8SQ | G0FUO |
| S64 8SQ | G4BTS |
| S64 8SQ | M6FUO |
| S64 8TE | 2E1FTF |
| S64 8UH | G6OET |
| S64 8UL | G4IDL |
| S64 9EH | G1UVJ |
| S64 9EH | M3ZPR |
| S64 9EZ | 2E1FTV |
| S64 9EZ | M3ACY |
| S64 9EZ | M3EIJ |
| S64 9JX | 2E0HNX |
| S64 9PF | G4RPL |
| S64 9SG | 2E0EBD |
| S64 9SG | M0IBR |
| S64 9SG | M6EBO |
| S65 1LH | G8UCR |
| S65 1LT | M3ZHI |
| S65 1NB | M6JED |
| S65 2HQ | M3OAM |
| S65 2JR | M6LYS |
| S65 2JU | M6UZZ |
| S65 2LR | M6EOS |
| S65 2NT | M6FVO |
| S65 2QP | M0CTI |
| S65 2UA | G4NJI |
| S65 2UE | 2E0DXJ |
| S65 2UE | M6AWE |
| S65 2UX | 2E0DKB |
| S65 2UX | M6ICQ |
| S65 2XA | 2E0TAL |
| S65 2XA | M3ZXN |
| S65 3JX | 2E0DZE |
| S65 3JX | M0CHU |
| S65 3LG | M6BZE |
| S65 3NH | G1RWR |
| S65 3NL | M6RGE |
| S65 3NN | M0CJY |
| S65 3NX | M3EWQ |
| S65 3PQ | M0MTD |
| S65 3RW | M6BRY |
| S65 4AG | M6KBH |
| S65 4HH | G8NVX |
| S65 4LD | G1IFH |
| S65 4NZ | G4APO |
| S65 4NZ | G8VJP |
| S65 4NZ | M3NGO |
| S65 4QR | G4SKM |
| S65 4QR | G8DRQ |
| S66 1AF | M6RBV |
| S66 1DL | G4CGP |
| S66 1NN | 2E1BGQ |
| S66 1UH | M6DMW |
| S66 2BN | G4ENC |
| S66 2BN | G8EUV |
| S66 2HZ | G4WFW |
| S66 2HZ | G7RIJ |
| S66 2LJ | 2E0DID |
| S66 2LJ | M3UAX |
| S66 2LU | G8PUK |
| S66 2NL | M6WNN |
| S66 2NS | M6VTT |
| S66 2NT | M3JDF |
| S66 2PQ | M6JIS |
| S66 2SJ | G1DTF |
| S66 2SP | G0WXD |
| S66 2SS | G8HUO |
| S66 2SW | 2E1IJO |
| S66 2SW | M3BBL |
| S66 3WA | G6PMR |
| S66 3YW | G4UHQ |
| S66 3ZY | 2E0DOM |
| S66 7AJ | M3YXT |
| S66 7DQ | M6NOH |
| S66 7DQ | M6ZAB |
| S66 7EN | 2E0ODO |
| S66 7EN | M3ODO |
| S66 7EU | G1BHQ |
| S66 7HA | G0EIB |
| S66 7HA | G0GOO |
| S66 7HB | G6PCX |
| S66 7HF | G0EPU |
| S66 7HJ | G7OFI |
| S66 7HJ | G4BVV |
| S66 7HX | G1POC |
| S66 7JZ | M1DBF |
| S66 7JZ | M6JJH |
| S66 7LU | G1ILC |
| S66 7LY | M6AVF |
| S66 7NR | M6COX |
| S66 7PA | M6FBF |
| S66 7PD | G7TKM |
| S66 7SG | 2E0YLH |
| S66 7SH | G3AMH |
| S66 8AZ | G0HDG |
| S66 8AZ | M6BVW |
| S66 8BD | M0MCT |
| S66 8BD | M6TBO |
| S66 8BL | G6TJC |
| S66 8DL | M3XQH |
| S66 8DS | M6BVX |
| S66 8EE | 2E0ZOR |
| S66 8JT | M1CYJ |
| S66 8JT | M3CYJ |
| S66 8JU | G1OJT |
| S66 8NA | M6GIZ |
| S66 8NN | M6LWJ |
| S66 8QZ | M6WBF |
| S66 9DL | G4JMC |
| S66 9HT | M6MQN |
| S66 9LG | 2E0JRA |
| S66 9LG | M0JRA |
| S66 9LG | M3JRA |
| S66 9NJ | M1EKA |
| S66 9PD | M3KUN |
| S66 9QP | M6SUF |
| S7 1FE | M6ZYH |
| S7 1NE | G7GFH |
| S7 1PB | M3HEE |
| S7 1PB | M3XBT |
| S7 1SJ | G4ILX |
| S7 2EB | G0WFL |
| S7 2QN | 2E0DRE |
| S7 2QN | M3MIJ |
| S70 1NA | G1BAL |
| S70 1NA | M0DRM |
| S70 1NA | M0MHY |
| S70 1PJ | M6WJA |
| S70 1PL | G3ESP |
| S70 1PL | G3WRS |
| S70 1PL | M0TNG |
| S70 1QE | G1KXZ |
| S70 1UA | 2E1HAF |
| S70 3AA | G1ANF |
| S70 3AA | G6ZXO |
| S70 3DW | M6BAD |
| S70 3EW | G3TEH |
| S70 3FG | G0NSC |
| S70 3FG | G1XUH |
| S70 3RH | G4DYG |
| S70 4AU | M3LXR |
| S70 4BW | M6XLG |
| S70 4DW | G3EAE |
| S70 4HH | G3ZQV |
| S70 4LR | 2E0EWS |
| S70 4QN | M6TEV |
| S70 5AY | M6BRU |
| S70 5DU | G6ZEW |
| S70 5JB | G6PDJ |
| S70 5NR | G4IAY |
| S70 5QR | G0OHR |
| S70 5QY | 2E1ACS |
| S70 5QY | G8VDP |
| S70 5QZ | G7MRZ |
| S70 5RL | G0JEA |
| S70 5SN | G8EGM |
| S70 5SU | G4KUD |
| S70 5TL | G1IWH |
| S70 6JY | G4TMZ |
| S70 6JY | G6YOR |
| S70 6LG | G0DUM |
| S70 6PQ | G6AUP |
| S70 6QQ | G6XQT |
| S70 6RG | G6TVA |
| S70 6SU | M6WLC |
| S71 1JR | G8BRK |
| S71 1SB | M3TRO |
| S71 1SS | M6MDG |
| S71 1SU | M0GAG |
| S71 1XQ | G4POI |
| S71 2BA | M0RTL |
| S71 2BA | M1OXR |
| S71 2BA | M3ITM |
| S71 2BA | M3SPG |
| S71 2BQ | M6EDL |
| S71 2ES | G8ISE |
| S71 2EY | G6DER |
| S71 2HP | G4DSA |
| S71 2HS | M3ZME |
| S71 2HS | M6KJD |
| S71 2JY | M3HUY |
| S71 2LL | G7VYN |
| S71 2PP | G1WTN |
| S71 2RA | G4JUR |
| S71 3DS | 2E0ETP |
| S71 3JJ | M3VYV |
| S71 3JJ | M6CBD |
| S71 3NR | M3VCK |
| S71 3QD | M1DZM |
| S71 3RH | G0SZE |
| S71 3SJ | G0ANE |
| S71 4AA | G1ANI |
| S71 4DH | G1SYP |
| S71 4NG | M0OCC |
| S71 4NG | M0OXO |
| S71 4NG | M0UKI |
| S71 4NG | M3ZLZ |
| S71 4NG | M3ZYZ |
| S71 4NH | G1BQR |
| S71 4NL | M0GEY |
| S71 4NL | 2E0KAF |
| S71 4NL | M3KRY |
| S71 5AL | M3KAE |
| S71 5BS | G4UFR |
| S71 5BS | G6LWK |
| S71 5DH | G4IHZ |
| S71 5PT | M3VWG |
| S71 5QX | G0JUK |
| S71 5RB | G3AMH |
| S71 5RB | G6KOB |
| S71 5RB | M0HFE |
| S71 5SD | M6CPR |
| S72 0AU | G0PQG |
| S72 0AU | G1JDO |
| S72 0AU | G6CND |
| S72 0DW | M3WZF |
| S72 0HG | 2E1HUQ |
| S72 7NA | M3ICH |
| S72 8BE | M1ERN |
| S72 8DE | G1EFF |
| S72 8DY | G6NXW |
| S72 8DY | M3RYR |
| S72 8DY | M3VPN |
| S72 8EW | M6SEB |
| S72 8JB | M1SWS |
| S72 8JP | G4EPD |
| S72 8NQ | M0DMS |
| S72 8NQ | M6BNJ |
| S72 8PZ | 2E1HXR |
| S72 8RN | G4LUE |
| S72 8RN | G8PLJ |
| S72 8SE | M0SYR |
| S72 8SE | M6URG |
| S72 8XG | G0RUX |
| S72 9AN | 2E0FYL |
| S72 9AN | 2E0TJC |
| S72 9DL | 2E1HPM |
| S72 9DL | G4XDV |
| S72 9DL | M3XDV |
| S72 9EG | M6FLY |
| S72 9JG | M3NTQ |
| S72 9JX | 2E1HPZ |
| S72 9JX | M3WOC |
| S72 9LL | M0JOR |
| S73 0EF | M3IDJ |
| S73 0EF | M3SPQ |
| S73 0RW | G8VHB |
| S73 0TN | M6PRW |
| S73 0YJ | G1BCN |
| S73 8LD | G0VMN |
| S73 8PP | 2E0CKP |
| S73 8PP | M6CKP |
| S73 8PS | G0SKD |
| S73 8QB | 2E0FWR |
| S73 8QB | M0WFR |
| S73 8QB | M6FRK |
| S73 8RX | G1VGO |
| S73 8SF | G6YJD |
| S73 8SG | M6GGV |
| S73 8SJ | M3ULT |
| S73 9AB | M1LAP |
| S73 9AE | G0UQF |
| S73 9BJ | M3UJF |
| S73 9DU | G4BHL |
| S73 9DU | M3ZXA |
| S73 9ED | G0TKJ |
| S73 9ED | G1YTX |
| S73 9HA | M3ZPY |
| S73 9HQ | G0CQO |
| S73 9JA | M0BZH |
| S73 9NB | M3UBB |
| S73 9NF | M0CLJ |
| S73 9PE | G1YUU |
| S73 9PL | M3XRH |
| S73 9PP | G4NSO |
| S73 9QE | M3HVH |
| S74 0AL | G4PEN |
| S74 0AZ | M3ITU |
| S74 0AZ | M3LSS |
| S74 0BU | M0GXH |
| S74 0BU | M3SNY |
| S74 0EJ | M3YFG |
| S74 0LB | 2E0PVT |
| S74 0PS | M6RTM |
| S74 0PS | M6SLT |
| S74 8BW | M0PAF |
| S74 8BW | M0PPP |
| S74 8DP | M0CSZ |
| S74 8DR | G0CUI |
| S74 8DS | M3HHX |
| S74 8DT | G6UQO |
| S74 8JR | 2E0PJE |
| S74 8JR | M6PJE |
| S74 9AX | G6NVW |
| S74 9EQ | G4YTM |
| S74 9HF | G7GQD |
| S74 9HF | M0CNG |
| S74 9HS | M3XJO |
| S74 9HU | M0BJK |
| S74 9HU | M0FCG |
| S74 9HU | M1KEY |
| S74 9HZ | M3KXF |
| S74 9PW | G4YAP |
| S74 9RF | M3KZV |
| S749EE | G4YER |
| S75 1EE | 2E1HJO |
| S75 1EE | M0HYD |
| S75 1LX | 2E0TTS |
| S75 1LX | M3TTS |
| S75 1PS | M3XZD |
| S75 1PX | G4OUS |
| S75 1PZ | G4HBA |
| S75 1PZ | G8KRT |
| S75 1QG | G4MLQ |
| S75 2AF | M3EVE |
| S75 2EF | 2E1HOK |
| S75 2EQ | M3EGF |
| S75 2EW | M6FJC |
| S75 2JW | M5BAD |
| S75 2ND | G7BZE |
| S75 2QB | 2E0BLF |
| S75 2QB | M3UXC |
| S75 2QB | M6BFC |
| S75 2QJ | G8MQX |
| S75 2RX | M3BCQ |
| S75 2TQ | M1GCS |
| S75 2TX | G3OXR |
| S75 2TX | G6AJ |
| S75 3DD | M3KBZ |
| S75 3JE | 2E0DNO |
| S75 3JE | M6DCO |
| S75 4HD | G1UQT |
| S75 4HT | 2E0RLR |
| S75 4HT | M6RLR |
| S75 4JS | G1VOC |
| S75 4NN | G0RUZ |
| S75 4QQ | G4DCF |
| S75 4QQ | M6CWW |
| S75 4SA | M0GOI |
| S75 5DF | G7JRC |
| S75 5EL | M3VGF |
| S75 5ER | M0RCI |
| S75 5JQ | G7TGF |
| S75 5LG | M0JLT |
| S75 5NS | M3IOQ |
| S75 5PG | M6BJE |
| S75 5PW | 2E0BRI |
| S75 5PW | G1NAQ |
| S75 5QP | M6KHS |
| S75 6ET | M3UMD |
| S75 6EU | 2E1HJS |
| S75 6EX | M6DDV |
| S75 6EX | M6JGU |
| S75 6HB | M0LOU |
| S75 6HB | M1LOU |
| S75 6HB | M3OCP |
| S75 6HB | M3ODN |
| S75 6HB | M3SLF |
| S75 6HD | M3NSX |
| S75 6HD | M3NTJ |
| S75 6HH | G0SBO |
| S75 6JA | G1LES |
| S75 6JP | G7ULW |
| S75 6LE | G8GDH |
| S75 6LY | G0TOQ |
| S8 0BN | G0JPE |
| S8 0EY | M6XMH |
| S8 0GA | M0GDX |
| S8 0RA | 2E0BFZ |
| S8 0RA | G0NFY |
| S8 0RW | M0WTV |
| S8 4BA | M1CGF |
| S8 7BW | G0PEW |
| S8 7DP | G1LJL |
| S8 7DR | G6FGL |
| S8 7DZ | G6SWW |
| S8 7FW | 2E0MDZ |
| S8 7FW | M6MDZ |
| S8 7UF | 2E0PGL |
| S8 7UF | M6PGL |
| S8 7UF | M6TAL |
| S8 8DA | M3NPA |
| S8 8DA | M3NPE |
| S8 8DA | M3NRQ |
| S8 8DQ | G8LIH |
| S8 8DS | M6BYH |
| S8 8EZ | 2E0GCH |
| S8 8EZ | M3GCH |
| S8 8FT | G8YVW |
| S8 8GE | G0HSA |
| S8 8GE | G3RCM |
| S8 8HE | G4RVS |
| S8 8JJ | G4DYT |
| S8 8LB | G3XSI |
| S8 8PJ | G0NQY |
| S8 8QR | G0NJD |
| S8 8RD | G8OFN |
| S8 8SE | G3RKL |
| S8 9JL | G3KVG |
| S8 9NG | G1HGB |
| S8 9RL | G7DFX |
| S80 1RZ | 2E0MNU |
| S80 1RZ | M0MNU |
| S80 1RZ | M3MNU |
| S80 1SU | 2E11EXP |
| S80 1SW | 2E0NVB |
| S80 1TD | 2E0MZZ |
| S80 1TD | M6ELR |
| S80 1TT | G4YRZ |
| S80 1UZ | G0MZP |
| S80 1YA | G3WTY |
| S80 1YG | M0KAD |
| S80 2DL | G3UVB |
| S80 2EB | M3AZS |
| S80 2EW | M0CMN |
| S80 2HN | 2E0RPR |
| S80 2HN | M0RPR |
| S80 2HN | M6HEP |
| S80 2HN | M6RPR |
| S80 2LL | 2E1GJE |
| S80 2NQ | G0DKQ |
| S80 2NT | 2E0PRS |
| S80 2NT | M0YES |
| S80 2RB | M0MIR |
| S80 2RR | 2E0BRY |
| S80 2RR | M3MYW |
| S80 2SA | 2E0MZU |
| S80 2SA | 2E0WAJ |
| S80 2SB | G0KVG |
| S80 2SD | 2E1GJT |
| S80 2ST | M3KUE |
| S80 2TT | G7MQW |
| S80 2TW | M6LVE |
| S80 2TW | M6NNJ |
| S80 2TW | M6NNU |
| S80 3DB | M0GKR |
| S80 3DF | G7TNZ |
| S80 3DF | M1FEX |
| S80 3DF | M3RDW |
| S80 3DF | M6REW |
| S80 3DL | M6SRA |
| S80 3EB | M0DOB |
| S80 3HF | 2E0CHA |
| S80 3HF | G7OBP |
| S80 3HG | G0CEJ |
| S80 3HR | G7SOH |
| S80 3NF | M0MRK |
| S80 3NQ | M0WEN |
| S80 3QB | M1LAN |
| S80 3QG | M0BMT |
| S80 3QG | G6JBL |
| S80 3QJ | G7BJD |
| S80 3QZ | G4PYV |
| S80 4AN | M1DAB |
| S80 4BB | 2E0BEV |
| S80 4BN | M3OHZ |
| S80 4DE | G0AHD |
| S80 4DL | M3WTP |
| S80 4DY | G1EVA |
| S80 4HN | 2E0XRM |
| S80 4HN | M0XRM |
| S80 4HS | 2E0AVD |
| S80 4HS | M0JMJ |
| S80 4JR | G1KIT |
| S80 4NP | G6OZT |
| S80 4NS | M6MUJ |
| S80 4PY | G8SRZ |
| S80 4SN | G3VRU |
| S80 4TN | 2E0ITJ |
| S80 4TN | M0HEW |
| S80 4TN | M6TJZ |
| S80 4UD | G1OIZ |
| S80 4UF | 2E0RXN |
| S80 4UF | M6RXN |
| S80 4XT | M3JHJ |
| S81 0AX | G4YJY |
| S81 0AX | M0WDP |
| S81 0BS | G0AGD |
| S81 0DH | M6WSM |
| S81 0DP | M6GTQ |
| S81 0EE | G3OZN |
| S81 0HD | 2E0NCK |
| S81 0HD | 2E0PNK |
| S81 0HD | M6CAR |
| S81 0HD | M6NCK |
| S81 0HD | M6PNK |
| S81 0JS | G0WTK |
| S81 0JS | G8TBF |
| S81 0LP | G3XXO |
| S81 0QN | G0BDR |
| S81 0QN | G0CJD |
| S81 0QN | G4ZUN |
| S81 0QN | G6EXG |
| S81 0RJ | 2E1SKY |
| S81 0TB | G0HMX |
| S81 0UY | M3HOV |
| S81 0XH | M3MSH |
| S81 0XJ | 2E0CAJ |
| S81 0XJ | M0TAL |
| S81 0XJ | M6CMT |
| S81 0XY | M0EEB |
| S81 7BH | G1VBP |
| S81 7DS | 2E0CTJ |
| S81 7DS | M6CTJ |
| S81 7ED | G0FMP |
| S81 7HL | M6KDB |
| S81 7JS | 2E0RYN |
| S81 7JS | M6DCK |
| S81 7LE | G4CRE |
| S81 7LJ | 2E0ZTD |
| S81 7LJ | M3UHC |
| S81 7LJ | M3ZTD |
| S81 7PG | G8TWT |
| S81 7PS | M6PRB |
| S81 7QE | G0TKG |
| S81 7QW | G6TLB |
| S81 7RT | 2E0GEF |
| S81 7RT | G6CNX |
| S81 7RT | M3YHT |
| S81 8NH | G8RSV |
| S81 8NQ | M0JSP |
| S81 8PH | M0JAV |
| S81 8TD | G4IBZ |
| S81 9BD | 2E0RSH |
| S81 9BD | G3ZMM |
| S81 9BD | M0ZMM |
| S81 9BD | M3RSH |
| S81 9JT | G4MWX |
| S81 9LE | M3LRK |
| S81 9NW | G3XXN |
| S81 9NX | G6DRG |
| S81 9PT | M3BXN |
| S81 9PX | 2E0LTV |
| S81 9PX | M6FDT |
| S81 9RA | G0NFB |
| S81 9RR | G0DMA |
| S81 9RR | M6SBM |
| S81 9SB | M3YRC |
| S81 9SH | M6SDL |
| S81 9SL | G0NGW |
| S81 9SS | 2E0TRU |
| S81 9SS | M6TRU |
| S81 9SS | M6TWI |
| S9 1AG | M6GOE |
| S9 1AR | 2E0JBS |
| S9 1AY | 2E1IDM |
| S9 1BA | G7TBC |
| S9 1LU | G6VUE |
| S9 1PZ | 2E0HRY |
| S9 1PZ | M3UQG |
| S9 4AU | G4RBU |
| S9 5EH | M0AFY |
| S9 5FQ | G6DCT |
| S9 5FQ | M0RCU |
| S9 5GJ | G8SVT |

## SA

(Swansea)

| Postcode | Callsign |
|---|---|
| SA1 1NZ | GW0JFQ |
| SA1 1RZ | GW6TYO |
| SA1 2HQ | MW0SBX |
| SA1 3SN | GW7RLS |
| SA1 3SN | GW7SKC |
| SA1 3UX | 2W0LBN |
| SA1 5DQ | 2W0IMD |
| SA1 5DQ | MW0UNU |
| SA1 5DQ | MW6KPA |
| SA1 6EN | GW3PPB |
| SA1 6HN | GW1XBG |
| SA1 6LS | GW8DSO |
| SA1 6NH | MW1AUV |
| SA1 6PD | GW4PHT |
| SA1 6TZ | GW3NDB |
| SA1 6TZ | GW8ASA |
| SA1 6UW | MW6PMN |
| SA1 7AD | GW4VBV |
| SA1 7EF | GW4WFM |
| SA1 7FJ | 2W0HAC |
| SA1 7FJ | MW6HTT |
| SA1 7JY | GW4YID |
| SA1 8BG | MW6PED |
| SA1 8EJ | GW6MLL |
| SA1 8HQ | GW4HDB |
| SA10 6AA | 2W0DAP |
| SA10 6AA | MW6THW |
| SA10 6BQ | MW3JEK |
| SA10 6DG | GW0KZK |
| SA10 6DL | GW0CBL |
| SA10 6DS | GW0HPC |
| SA10 6ET | MW6FAM |
| SA10 6JR | MW6FIX |
| SA10 6SD | GW6KRK |
| SA10 6SP | GW0VSW |
| SA10 6SP | MW0NSC |
| SA10 7BU | GW4YJI |
| SA10 7DD | GW8DOA |
| SA10 7DY | MW0CIA |
| SA10 7LB | 2W1CPS |
| SA10 7LB | MW6SWN |
| SA10 7NR | 2W0EPE |
| SA10 7NR | MW6EPE |
| SA10 7PL | GW1WGR |
| SA10 7PL | GW4WWN |
| SA10 7RW | 2W0JJW |
| SA10 7RW | MW6RGW |
| SA10 7RX | GW7IBT |
| SA10 7TW | GW7EHD |
| SA10 7UG | GW0CYG |
| SA10 8AL | MW6PNZ |
| SA10 8BG | GW4FOI |
| SA10 8DR | 2W0ZWR |
| SA10 8DR | MW0ZWR |
| SA10 8DR | MW3ZWR |
| SA10 8DW | GW1SGH |
| SA10 8HD | GW8VCA |
| SA10 8HT | GW4FXF |
| SA10 9BU | GW6APK |
| sa108nd | MW0ABH |
| SA11 1DJ | GW4IUK |
| SA11 1JL | GW0VFF |
| SA11 1JL | MW3CNL |
| SA11 1PP | MW3GTM |
| SA11 2AY | 2W0TAI |
| SA11 2BJ | MW6PWS |
| SA11 2JG | MW6CBL |
| SA11 2JU | GW6NXH |
| SA11 2LU | MW6MSE |
| SA11 2NG | GW4DWN |
| SA11 2PG | 2W0RRY |
| SA11 2PG | MW6BWR |
| SA11 2PU | 2W0JYC |
| SA11 2PU | MW0HNM |
| SA11 2PU | MW0JYC |
| SA11 2TE | GW4PCO |
| SA11 3AL | GW1EWW |
| SA11 3FD | MW3GQE |
| SA11 3JW | GW0NKJ |
| SA11 3NJ | GW2ABJ |
| SA11 3PJ | GW0HNT |
| SA11 3QE | MW0CNC |
| SA11 3RN | MW0HYP |
| SA11 3SS | GW4JQQ |
| SA11 3SS | GW8SZC |
| SA11 3SS | MW3GRC |
| SA11 3XA | MW3LHA |
| SA11 3YQ | GW0TWR |
| SA11 4AA | GW7JHK |
| SA11 4HS | GW7TZG |
| SA11 4NN | 2W0PEE |
| SA11 4NN | MW6NRT |
| SA11 5BG | GW4SRE |
| SA11 5BG | GW4UYT |
| SA11 5NT | GW4WYX |
| SA12 6BS | GW4RKI |
| SA12 6DF | 2W0BVS |
| SA12 6QJ | GW8JJZ |
| SA12 6TF | 2W0CED |
| SA12 6TF | MW6CAN |
| SA12 7DE | GW4RML |
| SA12 7ED | 2W0RMY |
| SA12 7ED | MW6ADZ |
| SA12 7HG | GW1NBW |
| SA12 8AP | GW0RLQ |
| SA12 8AS | GW4XWC |
| SA12 8EU | 2W0ACD |
| SA12 8EU | GW0KTE |
| SA12 8EY | GW1RQM |
| SA12 8LE | GW4TVQ |
| SA12 8QH | GW0RCG |
| SA12 8TU | MW6ESA |
| SA12 8YE | GW3TKG |
| SA12 9EA | MW0POB |
| SA12 9TP | GW6UHY |
| SA12 9TP | MW0CLU |
| SA12 9TW | GW8DUY |
| SA12 9YE | GW3SCX |
| SA13 1BR | MW1GLD |
| SA13 1DQ | MW1BAJ |
| SA13 1HA | GW4DOO |
| SA13 1LD | MW6AHV |
| SA13 1NE | GW7TJM |
| SA13 1SG | GW0TWF |
| SA13 1TD | 2W0CLT |
| SA13 1TD | MW6BOC |
| SA13 1YD | GW4PRP |
| SA13 2HL | MW0CDO |
| SA13 2LB | GW3NKM |
| SA13 2LF | GW0KPD |
| SA13 2US | GW3XHD |
| SA13 2YD | MW0LEW |
| SA13 2YE | GW3TMJ |
| SA13 3AN | GW4RDW |
| SA13 3SD | GW4TVU |
| SA13 3UU | MW6IMT |
| SA13 3YE | MW6FGN |
| SA14 6AP | MW3YNA |
| SA14 6BD | GW0RUD |
| SA14 6BP | 2W0MXT |
| SA14 6BP | MW0MXT |
| SA14 6BR | 2W0DSP |
| SA14 6BR | MW3WLS |
| SA14 6DT | GW0SRF |
| SA14 6LY | GW7RQI |
| SA14 6NT | MW1DUJ |
| SA14 7AR | GW6FBV |
| SA14 7HT | MW6TFL |
| SA14 7HY | GW1LFN |
| SA14 7NA | GW0LNM |
| SA14 7PR | 2W0TRD |
| SA14 7PU | 2W0CDJ |
| SA14 7PU | MW0HMV |
| SA14 7PY | MW1FJK |
| SA14 7RJ | GW4BYA |
| SA14 7RW | GW0AKV |
| SA14 7SA | MW3UNZ |
| SA14 8AE | GW4PUC |
| SA14 8AT | 2W0SCP |
| SA14 8AT | MW0GIN |
| SA14 8DB | MW6HDT |
| SA14 8DN | GW4KFD |
| SA14 8EL | GW3OPC |
| SA14 8PU | MW0ATR |
| SA14 8PW | MW6JCH |
| SA14 8QU | MW0CUA |
| SA14 8UB | GW4APF |
| SA14 9AH | GW4OH |
| SA14 9AH | MW0JZE |
| SA14 9AH | MW3ZAQ |
| SA14 9LH | 2W0FOG |
| SA14 9LT | GW4RXO |
| SA14 9SS | GW0EZQ |
| SA14 9SS | MW0DBV |
| SA14 9UP | GW0KJZ |
| SA15 1DG | GW0DZL |
| SA15 1NQ | 2W0LDX |
| SA15 1NQ | MW3SNJ |
| SA15 1NT | MW3XVR |
| SA15 1NZ | MW0PJR |
| SA15 1PB | MW0EAN |
| SA15 1SR | 2W0DQT |
| SA15 1SR | MW6EFK |
| SA15 1SW | GW6PDR |
| SA15 1SW | MW6CCG |
| SA15 1SW | MW6ZOD |
| SA15 1TR | MW6FQD |
| SA15 2AB | MW0CXH |
| SA15 2BG | MW0CJB |
| SA15 2LB | GW0RHE |
| SA15 2RB | GW6VFH |
| SA15 2RH | GW4XJK |
| SA15 3AE | GW0JQT |
| SA15 3AE | GW1OII |
| SA15 3ED | GW4UIE |
| SA15 3NF | GW0BCL |
| SA15 3NS | GW0IXQ |
| SA15 3RA | MW0DFN |
| SA15 3RU | GW0OFH |
| SA15 3RW | MW0JAW |
| SA15 3SN | GW4TFS |
| SA15 3TF | GW7FXX |
| SA15 4BR | GW0RTR |
| SA15 4HN | MW0RGM |
| SA15 4LA | 2W1ETW |
| SA15 4RR | MW6FIY |
| SA15 5AQ | GW4LJS |
| SA15 5DJ | GW8VUV |
| SA15 5EN | MW3GDL |
| SA15 5HF | GW0NLB |
| SA15 5HN | GW0KJT |
| SA15 5HN | MW6FHK |
| SA15 5HP | GW7EXQ |
| SA15 5NY | MW3HNP |
| SA15 5RT | GW0TMU |
| SA15 5SF | GW0SRE |
| SA15 5SG | GW0IVG |
| SA15 5SH | MW6PLP |
| SA15 5SU | 2W0CKV |
| SA15 5SU | GW0TDA |
| SA15 5TF | GW8KCY |
| SA15 5TP | GW7UXY |
| SA15 5UB | MW0CTX |
| SA15 5UG | GW6HUD |
| SA15 5UH | MW6OCT |
| SA15 5UH | MW6SCD |
| SA16 0EG | MW3VMY |
| SA16 0HF | GW0DWQ |
| SA16 0HJ | GW6XGA |
| SA16 0HP | GW0DIX |
| SA16 0LD | GW0WGE |
| SA16 0NF | 2W0MMD |
| SA16 0NF | GW0JDW |
| SA16 0RH | GW8WXP |
| SA16 0TE | MW6NAX |
| SA16 0UR | MW3IKC |
| SA16 0XT | MW3IFZ |
| SA17 4AE | MW6GAA |
| SA17 4BQ | GW8XMW |
| SA17 4EB | GW1ENG |
| SA17 4EY | MW0CWS |
| SA17 4EY | MW3DWZ |
| SA17 4HF | 2W0BKM |
| SA17 4HF | MW3VFN |
| SA17 4NW | MW1BNY |
| SA17 4RA | MW6RDL |
| SA17 4RX | MW6HEI |
| SA17 5EJ | GW3PLB |
| SA17 5EN | GW3YAF |

| | |
|---|---|
| SA17 5TQ | GW3UAY |
| [illegible] | [illegible] |
| SA18 1AF | MW6IRD |
| SA18 1AN | MW6MTG |
| SA18 1AU | GW0WRI |
| SA18 1BB | MW6RXZ |
| SA18 1BD | GW0LDZ |
| SA18 1EN | 2W0RKF |
| SA18 1EN | MW6RKF |
| SA18 1PE | GW0HNE |
| SA18 1RF | GW0RAD |
| SA18 1SB | GW0JDS |
| SA18 1TR | MW3UIQ |
| SA18 1TR | MW3UIY |
| SA18 1TR | MW6TLN |
| SA18 2AY | MW3OLX |
| SA18 2DB | MW1FGV |
| SA18 2DB | MW3FGV |
| SA18 2EP | 2W0JMK |
| SA18 2GE | GW1LRN |
| SA18 2HF | GW4IMC |
| SA18 2LQ | MW0ANX |
| SA18 2LX | MW6AIC |
| SA18 2NG | GW1ANW |
| SA18 2TY | GW1JOV |
| SA18 3HA | GW4UPG |
| SA18 3HD | MW3XDB |
| SA18 3LN | GW4JPC |
| SA18 3QL | GW0IXM |
| SA18 3QW | GW7HJN |
| SA18 3RD | MW0GTY |
| SA18 3RY | GW1TJK |
| SA18 3SF | 2W0TCM |
| SA18 3SF | MW0XTZ |
| SA18 3TG | GW3WMP |
| SA18 3UA | GW4BVJ |
| SA19 6EB | MW0AMI |
| SA19 6UL | MW3VNV |
| SA19 7HD | 2W0GAY |
| SA19 7HD | MW0SIP |
| SA19 7LY | GW0GPQ |
| SA19 7UA | GW0AIY |
| SA19 7UL | GW4FZM |
| SA19 7YR | MW0CIS |
| SA19 7YR | MW0CIT |
| SA19 7YT | GW6WOB |
| SA19 8AD | GW1SPW |
| SA19 8DF | GW0AWT |
| SA2 0AT | MW3TAF |
| SA2 0FQ | 2W0RGA |
| SA2 0FQ | MW6RGA |
| SA2 0FZ | GW3OMN |
| SA2 0GB | GW4SRI |
| SA2 0PJ | GW1YHL |
| SA2 0PR | GW0SAJ |
| SA2 0QE | GW3TYI |
| SA2 0RL | MW0KRS |
| SA2 0YD | MW0CGP |
| SA2 7DF | GW1FBL |
| SA2 7EH | GW4BNJ |
| SA2 7EQ | GW6UFH |
| SA2 7ES | MW0RDF |
| SA2 7HW | GW1BFB |
| SA2 7NB | GW7NVM |
| SA2 7PQ | GW8VFQ |
| SA2 7PR | MW6HUY |
| SA2 7QE | GW6VKY |
| SA2 7RP | GW8HDH |
| SA2 7RW | GW0HNS |
| SA2 7SG | GW7RRM |
| SA2 7TH | GW1DPL |
| SA2 7TH | MW6VNP |
| SA2 7UH | GW8TVX |
| SA2 7UJ | MW3FTY |
| SA2 7XQ | GW0FYO |
| SA2 8BE | GW3XIS |
| SA2 8DP | MW0BBL |
| SA2 8JL | MW0RBA |
| SA2 8LL | GW6AAG |
| SA2 8LQ | GW6MMM |
| SA2 8LT | GW3TSQ |
| SA2 8LX | GW3KGI |
| SA2 8PP | MW0HGK |
| SA2 9BW | GW3UWS |
| SA2 9BW | GW4ADL |
| SA2 9DF | 2W1TBD |
| SA2 9DT | GW4JGU |
| SA2 9DZ | MW3TKI |
| SA2 9EQ | GW0INN |
| SA2 9GR | GW4JUC |
| SA2 9HY | GW3TMS |
| SA2 9LG | GW1TFI |
| SA2 9LG | MW3FTB |
| SA2 9LG | MW3FTC |
| SA2 9LY | GW4HAT |
| SA20 0ED | GW6REF |
| SA20 0LD | GW0LLD |
| SA20 0NT | GW8HYT |
| SA20 0UP | GW7VJK |
| SA20 0UP | MW3LDY |
| SA20 0US | GW3PGJ |
| SA3 1AE | GW4KTT |
| SA3 1BA | GW4AZI |
| SA3 1BR | GW6OTD |
| SA3 1LB | GW8VGB |
| SA3 2EQ | MW0FRY |
| SA3 3JR | GW0VRL |
| SA3 3JW | GW1AXU |
| SA3 3JW | GW4RKX |
| SA3 3JW | MW6EDR |
| SA3 3LA | GW0BNO |
| SA3 4BQ | MW6LHW |
| SA3 4EW | GW7PMA |
| SA3 4HF | GW4EPF |
| SA3 4JD | GW3LJS |
| SA3 4LZ | GW0CCF |
| SA3 4PD | GW1WWE |
| SA3 4PD | GW3SRG |
| SA3 4PD | GW3YQU |
| SA3 4QW | GW0KWA |
| SA3 4RJ | GW8DYR |
| SA3 4SA | GW1AUT |
| SA3 4SW | GW6WEU |
| SA3 4UB | GW0GJD |
| SA3 4UR | MW0EYE |
| SA3 5AN | GW4PEX |
| SA3 5BU | MW0JGE |
| SA3 5BU | MW3ORY |
| SA3 5DL | MW0HGM |
| SA3 5HS | GW4TGA |
| SA3 5HT | MW0AJH |
| SA3 5LA | MW0DCT |
| SA3 5NL | GW4LDP |
| SA3 5NQ | GW4CC |
| SA3 5NQ | GW4HSH |
| SA3 5PE | GW3SIY |
| SA3 5PR | GW4EVL |
| SA3 5QJ | GW1PFK |
| SA3 5QL | GW0RHC |
| SA31 1BS | GW6YUC |
| SA31 1HG | MW6HBM |
| SA31 1JJ | GW4RVA |
| SA31 1LR | GW0PUM |
| SA31 1NN | MW0COZ |
| SA31 1SY | GW0CVY |
| SA31 1SZ | 2W0CZP |
| SA31 1SZ | MW6DVQ |
| SA31 1TS | GW4XUE |
| SA31 2DT | 2W1EAN |
| SA31 2JB | GW1ADY |
| SA31 2JL | GW0HGP |
| SA31 2LH | GW0IXK |
| SA31 2NL | GW8PFT |
| SA32 7SA | GW7OIK |
| SA32 8AA | GW8THL |
| SA32 8DQ | MW0AMJ |
| SA32 8DX | GW3NXR |
| SA32 8LJ | GW0AJU |
| SA32 8LJ | GW0HYL |
| SA32 8LJ | GW1IIZ |
| SA32 8LP | GW0HCK |
| SA32 8LP | GW1SGE |
| SA32 8PR | GW4TGL |
| SA32 8SA | GW0ADY |
| SA33 4AR | GW7JUB |
| SA33 4ES | GW0TVX |
| SA33 4ET | MW0CFQ |
| SA33 4ET | MW3GVU |
| SA33 4LG | MW3VWE |
| SA33 4PD | GW2OP |
| SA33 4PD | GW3XJQ |
| SA33 4PD | MW0RLD |
| SA33 5AH | GW0TSE |
| SA33 5BL | MW3VNR |
| SA33 5DR | GW0TXP |
| SA33 5HA | GW3YVN |
| SA33 5NX | GW0GFN |
| SA33 5NX | GW1NED |
| SA33 6BL | MW6BHO |
| SA33 6BN | GW4PCJ |
| SA33 6BN | GW4YCT |
| SA33 6ES | GW0LYK |
| SA33 6UD | MW6IGC |
| SA33 6XJ | MW3WJP |
| SA34 0AF | GW0DKG |
| SA34 0EX | MW0VWC |
| SA34 0HL | 2W0HDC |
| SA34 0HL | 2W0KJO |
| SA34 0HL | 2W0RMO |
| SA34 0HL | MW0WSD |
| SA34 0HL | MW6OKJ |
| SA34 0HL | MW6TJH |
| SA34 0JD | GW3WWH |
| SA34 0QX | GW1UHF |
| SA35 0BD | GW1GSW |
| SA37 0EH | GW3KAX |
| SA37 0HZ | 2W0BLA |
| SA37 0HZ | MW3JTJ |
| SA38 9AS | MW3VJN |
| SA38 9BA | GW7IPS |
| SA38 9BE | 2W0LLT |
| SA38 9DE | MW3OLT |
| SA38 9JS | GW0JCT |
| SA38 9LS | GW8SBO |
| SA38 9QB | GW1CIY |
| SA38 9RL | GW3WCA |
| SA39 9AZ | GW1UOV |
| SA39 9BY | GW0MWN |
| SA39 9DH | GW4VPX |
| SA4 3AY | MW3UZH |
| SA4 3DZ | GW8NBI |
| SA4 3GY | GW7LDP |
| SA4 3HD | MW1AAH |
| SA4 3LJ | MW0CLT |
| SA4 3LL | MW3UYX |
| SA4 3PE | MW6GIU |
| SA4 3PU | GW4UCK |
| SA4 3PU | GW7MMH |
| SA4 3SE | GW4EGS |
| SA4 4BJ | GW7PQS |
| SA4 4DT | 2W0KPN |
| SA4 4DT | MW6KPN |
| SA4 4GE | GW3JBZ |
| SA4 4GW | GW1WTZ |
| SA4 4GZ | GW0DIQ |
| SA4 4GZ | GW0DRI |
| SA4 4GZ | GW0NHE |
| SA4 4PN | MW3UZP |
| SA4 4PR | GW4VWY |
| SA4 4UX | MW0LMW |
| SA4 6SW | MW6BOW |
| SA4 6TR | 2W0GYB |
| SA4 6TR | MW6GYR |
| SA4 6UF | 2W0HUL |
| SA4 6UF | MW3CBS |
| SA4 6UH | GW7SDE |
| SA4 8EE | GW7AFC |
| SA4 8HU | GW3XYW |
| SA4 8HX | 2W0APT |
| SA4 8HX | GW0VEW |
| SA4 8HZ | MW0WZX |
| SA4 8JF | GW4JDZ |
| SA4 8QF | GW6VBR |
| SA4 8RD | GW0RNK |
| SA4 9AB | MW0EAT |
| SA4 9EB | GW0MGQ |
| SA4 9WH | GW3OQK |
| SA40 9RD | 2W0DNV |
| SA40 9RD | MW6CTS |
| SA40 9SR | 2W0BFV |
| SA40 9SR | MW3WFH |
| SA40 9ST | MW3BGP |
| SA40 9SU | MW6TDC |
| SA40 9TY | 2W0NRA |
| SA40 9TY | MW3JVH |
| SA41 3QG | MW3SFP |
| SA41 3QJ | MW0MJB |
| SA41 3RN | 2W0RFX |
| SA41 3RN | MW0VGH |
| SA41 3RN | MW3RFX |
| SA41 3RR | GW0TMV |
| SA42 0PL | GW4UEP |
| SA42 0QS | GW6CJJ |
| SA42 0XF | MW0ARD |
| SA42 0XS | GW4OQB |
| SA43 1AF | GW4OUU |
| SA43 1BU | GW0PNI |
| SA43 1DR | MW0CXW |
| SA43 1DR | MW0WWR |
| SA43 1DW | GW6LHF |
| SA43 1LT | GW6DGU |
| SA43 1PG | GW7GWO |
| SA43 1PG | MW6BTC |
| SA43 1RF | GW7WCR |
| SA43 1SN | MW1BDV |
| SA43 2AE | GW6RWJ |
| SA43 2AS | MW0LEA |
| SA43 2BD | GW3SIK |
| SA43 2DA | GW6AMK |
| SA43 2DH | GW0CMI |
| SA43 2HR | GW4HGJ |
| SA43 2HR | GW4SZV |
| SA43 2JE | GW0PLP |
| SA43 2JG | GW3KZO |
| SA43 2JH | GW4SJO |
| SA43 2JH | GW4TVE |
| SA43 2LA | GW7EUL |
| SA43 2NY | GW3EJR |
| SA43 2PE | GW0MBN |
| SA43 2RJ | GW4JRK |
| SA43 2RT | MW0WRP |
| SA43 3EF | MW0CRI |
| SA43 3HZ | GW4XMV |
| SA43 3LH | GW6HRL |
| SA44 4HS | MW3TUB |
| SA44 4NA | GW3UEP |
| SA44 4RJ | GW0PDB |
| SA44 4SJ | GW0HYH |
| SA44 4SJ | GW8JDB |
| SA44 4SJ | MW3URG |
| SA44 4TU | MW6BZX |
| SA44 5AT | MW0DXX |
| SA44 5HE | GW1VMA |
| SA44 5JA | GW6JSO |
| SA44 5JA | MW0GUV |
| SA44 5NZ | GW4JPJ |
| SA44 5PN | GW3WXA |
| SA44 5RF | MW3LSL |
| SA44 5XJ | MW6ICM |
| SA44 5YB | 2W0ZZU |
| SA44 5YB | GW0EWY |
| SA44 5YB | MW3ZZU |
| SA44 5YG | GW8PSJ |
| SA44 6AG | 2W0HOH |
| SA44 6AG | MW6HOH |
| SA44 6BT | GW3RYE |
| SA44 6EA | GW4TQD |
| SA44 6ES | GW4JKK |
| SA44 6EY | MW6FMV |
| SA44 6JE | GW0POZ |
| SA44 6NP | GW3KDB |
| SA44 6NQ | MW0WEE |
| SA44 6QY | MW0AIZ |
| SA45 9QR | GW0DDL |
| SA45 9RJ | GW7SXU |
| SA45 9RL | GW3WVV |
| SA45 9RR | GW3ZJS |
| SA45 9SY | MW3JQC |
| SA45 9TH | MW0ALG |
| SA46 0DL | GW3DRV |
| SA46 0ED | GW3LHK |
| SA46 0JS | GW4XSX |
| SA47 0QN | GW0HXS |
| SA48 7LG | GW0ADS |
| SA48 7PA | GW0GEI |
| SA48 7SB | GW8GQE |
| SA48 7SG | GW0ARA |
| SA48 8DT | GW3KCQ |
| SA48 8LH | GW0PNC |
| SA48 8NH | MW3SNH |
| SA48 8RP | GW4VNK |
| SA5 4RE | MW6BMM |
| SA5 5NX | MW3UYJ |
| SA5 7BP | MW0CNB |
| SA5 7DH | MW0UNA |
| SA5 7DW | GW7ODP |
| SA5 7DW | GW4MII |
| SA5 7EG | 2W0BTE |
| SA5 7EN | GW4WOV |
| SA5 7JJ | MW6DGW |
| SA5 7PU | GW0FZY |
| SA5 8AU | GW7TZI |
| SA5 8AW | MW6GIV |
| SA5 8BT | MW6CEX |
| SA5 8DU | MW0CND |
| SA5 8QN | GW4KUS |
| SA5 9AU | MW6HDP |
| SA5 9BS | GW4URB |
| SA5 9DG | GW0AZW |
| SA5 9EH | GW4MTD |
| SA5 9NH | GW0NCU |
| SA5 9PG | GW8NXK |
| SA6 5AD | MW1FGB |
| SA6 5AX | GW0COU |
| SA6 5DX | 2W1DGM |
| SA6 5DX | GW7MMG |
| SA6 5HF | GW4SPL |
| SA6 5HR | MW0OUC |
| SA6 5HR | MW0SDD |
| SA6 5HR | MW0SGX |
| SA6 5HR | MW3OUC |
| SA6 5HR | MW3SGX |
| SA6 5JF | GW7OJT |
| SA6 5JP | GW4EVJ |
| SA6 5LA | 2W1EIN |
| SA6 5RE | MW6ULX |
| SA6 5RN | GW0DXZ |
| SA6 5RN | GW4ZRK |
| SA6 5TB | MW3UEU |
| SA6 6AP | GW7PBP |
| SA6 6AS | GW4TFX |
| SA6 6BA | GW0UIZ |
| SA6 6BJ | MW3ZKW |
| SA6 6BY | 2W1CNN |
| SA6 6BY | 2W1EYZ |
| SA6 6BY | GW1NWF |
| SA6 6BY | GW6ZUS |
| SA6 6DA | GW4HZH |
| SA6 6ER | GW4ZQY |
| SA6 6LD | GW0BNN |
| SA6 6LL | MW6RYD |
| SA6 6LU | GW1DTA |
| SA6 6NW | 2W0AUC |
| SA6 6NX | GW0NKH |
| SA6 6NX | GW1YFP |
| SA6 6QB | GW0MVS |
| SA6 6QD | GW0LDQ |
| SA6 6SW | GW0KYY |
| SA6 6TB | 2W1FGR |
| SA6 6TF | GW6YPA |
| SA6 7AG | GW7AOE |
| SA6 7LN | 2W1GAC |
| SA6 7PB | GW3NUO |
| SA6 7PH | 2W0JMX |
| SA6 7PN | GW1EOI |
| SA6 7PQ | GW4MVY |
| SA6 7PS | GW4WPA |
| SA6 8HU | GW0CKX |
| SA6 8LE | 2W0REJ |
| SA6 8LE | MW0WRY |
| SA6 8LQ | MW1DCI |
| SA6 8LT | GW8KSL |
| SA61 1BW | GW0EJE |
| SA61 1BW | GW0GUY |
| SA61 1HJ | GW0UXX |
| SA61 1LB | MW6XRO |
| SA61 2HA | 2W0VVO |
| SA61 2HA | MW0VVO |
| SA61 2HA | MW3VVO |
| SA61 2HL | MW0CAB |
| SA61 2HL | MW1SAS |
| SA61 2JA | GW8KXW |
| SA61 2RD | MW6NPW |
| SA61 2RE | MW3EPJ |
| SA61 2SB | GW8YJN |
| SA61 2UE | MW6BGO |
| SA62 3BG | GW0ZDL |
| SA62 3ET | MW6VOW |
| SA62 3HX | GW4PCX |
| SA62 3LS | MW3YBX |
| SA62 3QU | MW6VLH |
| SA62 3SZ | GW0UQH |
| SA62 3TD | GW8WRC |
| SA62 3TH | GW7NGU |
| SA62 3TS | 2W0DOX |
| SA62 3TS | MW0GOV |
| SA62 3TS | MW3SQA |
| SA62 3XH | GW0SXS |
| SA62 4HJ | MW0GDM |
| SA62 4JB | GW6IZZ |
| SA62 4LR | GW1MV[illegible] |
| SA62 4PR | GW1HEV |
| SA62 5DY | GW0KQV |
| SA62 5NT | MW0AMN |
| SA62 5NT | MW0AMQ |
| SA62 5RZ | MW0KAK |
| SA62 5XD | MW6XAE |
| SA62 6EN | GW8NAC |
| SA62 6EU | MW3UAA |
| SA62 6HF | MW6COD |
| SA62 6UA | GW4HXO |
| SA64 0AZ | MW6JFI |
| SA64 0EX | 2W0HDB |
| SA64 0EX | MW0HVB |
| SA64 0EX | MW6BDS |
| SA64 0EX | MW6HDB |
| SA64 0LL | MW0RLJ |
| SA65 9LN | MW6AGS |
| [illegible] | [illegible] |
| SA65 9PT | GW8UOC |
| SA65 9QJ | MW0BEL |
| SA65 9RL | GW4UEJ |
| SA65 9SL | GW8YLK |
| SA65 9TP | 2W1IHN |
| SA66 7HH | GW3PPQ |
| SA66 7JP | GW1ALV |
| SA66 7LB | GW0RWM |
| SA66 7PX | 2W0BMM |
| SA66 7QS | GW3RKV |
| SA67 7DY | MW0HUU |
| SA67 7EE | GW1MNU |
| SA67 7EZ | GW8PKV |
| SA67 8JE | GW3VEW |
| SA67 8JH | MW6JTR |
| SA67 8JL | MW0BEY |
| SA67 8NR | GW4MIP |
| SA67 8SP | GW3LEW |
| SA67 8ST | GW0CCO |
| SA67 8TQ | MW6BEG |
| SA68 0RN | 2W0XDT |
| SA68 0RN | MW0XDT |
| SA68 0SY | GW4WMD |
| SA68 0XS | GW6TDE |
| SA68 0XS | MW6HED |
| SA7 9HS | GW4LKS |
| SA7 9JA | MW0TCJ |
| SA7 9JR | GW4UVN |
| SA7 9LD | 2W0SRD |
| SA7 9QH | GW1PAV |
| SA7 9QU | MW6HLU |
| SA7 9QX | GW4JGW |
| SA7 9SF | 2W0BJR |
| SA7 9SF | MW3OSI |
| SA7 9ST | GW4XES |
| SA7 9ST | GW4ZBU |
| SA7 9SX | GW8DUP |
| SA70 7DP | GW0JRF |
| SA70 7EB | MW0ECY |
| SA70 7EB | MW3NTE |
| SA70 8DB | GW7GKN |
| SA70 8EB | MW6GQD |
| SA70 8LT | GW4NOO |
| Sa708en | M6HNN |
| SA71 4EP | MW0CHI |
| SA71 4EP | MW1BXX |
| SA71 4ET | MW0DSV |
| SA71 4HX | MW6HLQ |
| SA71 4JL | MW6FGJ |
| SA71 4NB | GW4WMK |
| SA71 4PX | MW6DDR |
| SA71 4QL | 2W0BGI |
| SA71 4QL | MW3NUP |
| SA71 5BD | GW7FBV |
| SA71 5HJ | GW6TEO |
| SA71 5HY | GW4OZU |
| SA71 5JH | GW4VTG |
| SA71 5JQ | MW6RGK |
| SA71 5QX | 2W0MCT |
| SA71 5QX | MW3KNE |
| SA72 6DF | GW6NLO |
| SA72 6DU | GW0WGN |
| SA72 6EZ | GW4RGI |
| SA72 6FB | GW4OXL |
| SA72 6HL | GW7VMT |
| SA72 6JP | GW4TGT |
| SA72 6NE | MW1BLE |
| SA72 6NW | MW0AIE |
| SA72 6PN | GW4REI |
| SA72 6SH | GW4VRO |
| SA72 6TP | GW4YVN |
| SA72 6XQ | MW0BRO |
| SA73 1EA | GW3CR |
| SA73 1HD | GW0LIK |
| SA73 1HD | GW4ZYV |
| SA73 1HR | MW0BYS |
| SA73 1HR | MW3ESE |
| SA73 1HR | MW3ESF |
| SA73 1LG | GW4ERB |
| SA73 1RA | GW0JDY |
| SA73 1RE | 2W0JPF |
| SA73 1RE | MW3JPF |
| SA73 1SA | GW0VND |
| SA73 1TB | GW7HGU |
| SA73 1TB | MW1DFQ |
| SA73 1TG | GW0VTF |
| SA73 1TL | MW0HWU |
| SA73 1TL | MW0IBZ |
| SA73 1TR | GW4HGS |
| SA73 1TR | MW0COB |
| SA73 1TR | MW6FGQ |
| SA73 2AU | 2W0OZO |
| SA73 2AU | MW6OZO |
| SA73 2BW | MW6DFX |
| SA73 2DS | GW4OEJ |
| SA73 2ED | GW4OKE |
| SA73 2EN | MW0SGD |
| SA73 2NU | GW0TLJ |
| SA73 2NU | MW0BBU |
| SA73 3AY | MW0XDN |
| SA73 3HN | GW4ODN |
| SA73 3LW | GW0LVH |
| SA73 3PF | GW1PND |
| SA73 3RS | GW4CGE |
| SA73 3TF | MW3VFB |
| SA8 3DT | GW4KCQ |
| SA8 3EP | GW8SBN |
| SA8 3EP | GW8TBG |
| SA8 3ER | MW6JDY |
| SA8 4AA | GW1RZE |
| SA8 4AH | MW3ZWS |
| SA8 4AL | GW6VBN |
| SA8 4DB | GW6UGC |
| [illegible] | GW4IYT |
| SA8 4NB | 2W0BBQ |
| SA8 4NB | 2W0BBT |
| SA8 4NB | MW0GCS |
| SA8 4PH | GW7ORB |
| SA8 4QL | GW3UCJ |
| SA8 4QN | MW3BMF |
| SA8 4QT | GW4ONI |
| SA9 1AT | GW4YWM |
| SA9 1BP | GW4GLU |
| SA9 1SP | MW3EPK |
| SA9 1XF | GW7MLN |
| SA9 2FT | MW3SKW |
| SA9 2NZ | MW3UIN |
| SA9 2RY | GW1PKM |
| SA9 2XQ | GW3YQH |

## SE (South East London)

| | |
|---|---|
| SE1 0QB | 2E0JWG |
| SE1 0QB | M0JXG |
| SE1 0QB | M6JHG |
| SE1 1EL | 2E1HFH |
| SE1 1EL | M3HFH |
| SE1 2JW | M6VRE |
| SE1 2XN | M6YLD |
| SE1 2YD | 2E0CPR |
| SE1 2YD | M6BZH |
| SE1 3LD | M0ATD |
| SE1 4DT | M3RZL |
| SE1 5NU | M6EOM |
| SE1 7AN | M0HOH |
| SE1 7DX | 2E0HFA |
| SE1 7DX | M0JKN |
| SE1 7DX | M6HFA |
| SE1 8DU | G1UFX |
| SE1 8LJ | 2E0ESC |
| SE1 8LJ | M6MBB |
| SE10 8AX | 2E1DBP |
| SE11 5UL | G3RGM |
| SE12 0EG | G1IYO |
| SE12 0UJ | 2E0CIT |
| SE12 0UJ | M6BDD |
| SE12 8AE | M0HUS |
| SE12 8AG | 2E0FSN |
| SE12 8AG | M0SBR |
| SE12 8AG | M6HGH |
| SE12 8JN | G7GLW |
| SE12 9AB | G0LRM |
| SE12 9BG | G3GDB |
| SE12 9EX | G8IBO |
| SE12 9LX | G4XUI |
| SE12 9RS | G1XFR |
| SE13 5QW | G4JOI |
| SE13 6QP | M0BYM |
| SE13 6QW | M3WUO |
| SE13 7HP | G0ZIG |
| SE13 7PA | M6JKX |
| SE13 7PL | M3RJK |
| SE13 7PR | 2E0SNX |
| SE13 7PR | M0SNX |
| SE13 7PR | M6SNX |
| SE13 7SP | M6LTU |
| SE14 5DN | M6HJX |
| SE14 5NJ | 2E1GKP |
| SE14 6DU | M6URX |
| SE15 2DB | 2E0UAK |
| SE15 2DB | M6YAV |
| SE15 2TX | M6FMY |
| SE15 3EG | G3VBU |
| SE15 4AN | G3XBH |
| SE15 4DZ | G1BYO |
| SE15 4JX | G7KYD |
| SE15 4PY | G3HMO |
| SE16 2AP | M3JYG |
| SE16 3DQ | M0IVE |
| SE16 4BN | G0NVJ |
| SE16 4BY | M3IYG |
| SE16 4TT | G7NIL |
| SE16 5XN | G0BUD |
| SE16 6PU | 2E0HOU |
| SE16 6PU | M0OSY |
| SE16 6PU | M0EYD |
| SE16 6YP | 2E1CAQ |
| SE16 7AU | M6JCX |
| SE18 1LN | M3HIN |
| SE18 1SL | M6HJH |
| SE18 2PH | M0JRJ |
| SE18 2RT | G1CJK |
| SE18 3HF | G8GJO |
| SE18 5BY | G0EPR |
| SE18 5JL | 2E0WHF |
| SE18 5JL | M0YXO |
| SE18 5JL | M6WHF |
| SE18 5LE | G7PCW |
| SE18 5PU | M0TDD |
| SE19 2DA | G1PKS |
| SE19 2PY | G1FBI |
| SE19 2SA | G4EBI |
| SE19 3AD | G3TAO |
| SE19 3PZ | M0ICL |
| SE19 3QT | G7HUC |
| SE2 0AJ | M1FFS |
| SE2 0DN | G4MLY |
| SE2 0DW | G4WNF |
| SE2 0PD | G4JGQ |
| SE2 0PP | G0EFA |
| SE2 0QX | G3SPJ |
| SE2 0UY | M6IDM |
| SE2 9HL | G0ATB |
| [illegible] | [illegible] |
| SE2 9RU | M0SKO |
| SE20 7BW | G1XVW |
| SE20 7JG | G0VZH |
| SE20 8NU | G4JHW |
| SE21 8JY | G0PAR |
| SE22 0SB | G0FAH |
| SE22 8RH | M0NTN |
| SE22 9AH | M3TMM |
| SE22 9JH | G6ICC |
| SE23 1BW | G3SXA |
| SE23 1HG | G4PRQ |
| SE23 1RH | G6BJR |
| SE23 1SL | G1MKS |
| SE23 2QP | G0DIA |
| SE23 2RU | G3TPO |
| SE23 3HT | 2E0HGO |
| SE23 3HT | M6HGO |
| SE23 3SJ | 2E0EUI |
| SE23 3SJ | M6EUI |
| SE24 0BE | M0BXG |
| SE25 4NG | 2E0EBK |
| SE25 4NG | M6EUH |
| SE25 4TQ | G2HBA |
| SE25 5HB | G7PWJ |
| SE25 6HY | G7TQE |
| SE25 6SY | G3WRR |
| SE25 6TA | 2E0OPO |
| SE25 6TA | M6DKY |
| SE26 5DF | M6GGQ |
| SE26 5DP | G0RTZ |
| SE26 5RJ | 2E0CIG |
| SE26 5RJ | M0JAD |
| SE26 5RJ | M6BBG |
| SE26 6BP | M3YRS |
| SE26 6JQ | M3XBN |
| SE27 0LG | M6AMN |
| SE27 9NA | 2E0DKW |
| SE27 9NA | M6DKW |
| SE28 0NJ | M3YSY |
| SE28 8NF | 2E0ATY |
| SE28 8NF | M0MYC |
| SE28 8PG | M1BPN |
| SE28 8PU | G1NYI |
| SE3 0EZ | G0CUA |
| SE3 0QG | 2E0MUS |
| SE3 0QG | M6MUS |
| SE3 7AH | G1UKG |
| SE3 7BG | G4FUG |
| SE3 7BH | M0UOG |
| SE3 7NN | G8WBU |
| SE3 7PE | 2E0CQS |
| SE3 7PE | M0WXU |
| SE3 7PE | M6BZD |
| SE3 7QS | G3GHN |
| SE3 7QS | G4RFC |
| SE3 7QX | G7UQQ |
| SE3 7RQ | G0ILD |
| SE3 8HF | G4ZXV |
| SE3 8LR | G6IRE |
| SE3 8QY | M6SLZ |
| SE3 8RP | M6LHN |
| SE3 8TT | G1NTP |
| SE3 9HL | M0BGR |
| SE3 9LJ | M0RAC |
| SE3 9QN | G0BSH |
| SE3 9QP | G1KLW |
| SE4 1AS | 2E0CMQ |
| SE4 1AS | M6BCQ |
| SE4 2AX | M0IAG |
| SE4 2DN | M6BVK |
| SE4 2JQ | G0HBW |
| SE4 2NL | 2E0TXJ |
| SE4 2NL | G6DJH |
| SE4 2NL | M6TXJ |
| SE5 0DJ | 2E0BWM |
| SE5 8QQ | G4WUQ |
| SE6 1EZ | M3WNM |
| SE6 1JF | 2E0LSW |
| SE6 1NQ | G8WZO |
| SE6 1NR | M3WRQ |
| SE6 1SP | M6BHU |
| SE6 1UR | 2E0LBZ |
| SE6 1UR | M6BPD |
| SE6 2EG | M6WRB |
| SE6 3TG | M3VNI |
| SE6 3TW | G6ZWM |
| SE6 4DH | G4DKC |
| SE6 4RN | G7PKY |
| SE6 4UW | G0TCA |
| SE7 7AD | M3YKI |
| SE7 7DW | G1HBR |
| SE7 7EU | M1JWS |
| SE7 7LH | 2E0GLT |
| SE7 7LH | M0HPF |
| SE7 8SH | G1FAD |
| SE8 3AS | M1MAL |
| SE8 3EX | M6GUS |
| SE8 5AJ | G0VXY |
| SE9 1PP | M3UJR |
| SE9 1QH | G8XDR |
| SE9 1QJ | 2E0RCV |
| SE9 1QJ | M0MCV |
| SE9 1QJ | M3RCV |
| SE9 1SA | 2E1GUC |
| SE9 1SA | G7SCU |
| SE9 1SE | G0DCI |
| SE9 2EX | G8IPY |
| SE9 2LP | G6WCX |
| SE9 2NZ | G0FDP |
| SE9 3BN | M6FWQ |
| SE9 3BW | G3XCJ |
| SE9 3EB | G0OWV |
| SE9 3EP | G4VFH |
| SE9 3HL | G0OKX |
| SE9 3JJ | M6VSR |
| SE9 3JX | G3IGZ |
| SE9 3SJ | G3MIQ |
| SE9 3XD | G0ULL |
| SE9 4JL | G0JAJ |
| SE9 4LF | G0LYG |
| SE9 5RF | M6FDG |
| SE9 6EX | 2E0HTV |
| SE9 6EX | 2E0UKK |
| SE9 6EX | M6UKX |
| SE9 6QP | M0JHB |
| SE9 6RA | M6ALW |

## SG (Stevenage)

| | |
|---|---|
| SG1 1HE | 2E0DGL |
| SG1 1HE | M6ENV |
| SG1 1LS | G3MHX |
| SG1 1RP | G6XSZ |
| SG1 1RR | M0GBZ |
| SG1 1TE | G2BKZ |
| SG1 1TE | G3SAD |
| SG1 1TN | M3VRB |
| SG1 1UJ | G7DNV |
| SG1 1UP | M6TZO |
| SG1 2AS | M6GKC |
| SG1 2NE | M3BMU |
| SG1 2NE | M3BYS |
| SG1 2QL | M3RUW |
| SG1 2RS | M0KPB |
| SG1 3DF | M6BSZ |
| SG1 3EZ | G7CYQ |
| SG1 3JL | G3UFB |
| SG1 3SL | M6JJM |
| SG1 3SL | M6MLP |
| SG1 4AY | G8WWI |
| SG1 4EP | G4SPV |
| SG1 4HD | G8IUG |
| SG1 4JP | M6JSV |
| SG1 4LF | G3INU |
| SG1 4NQ | M0NEV |
| SG1 4NU | G1FWS |
| SG1 4PH | M6ZCR |
| SG1 4PJ | M3IJV |
| SG1 4PL | M3PCP |
| SG1 4PW | 2E0PYM |
| SG1 4PW | 2E0YUD |
| SG1 4RN | G6ANR |
| SG1 4SD | G4DDX |
| SG1 4TB | M1EMB |
| SG1 5EA | M3ZII |
| SG1 5HF | G7DRG |
| SG1 5JA | G1SOY |
| SG1 5LH | M3JMU |
| SG1 5LH | M3NDC |
| SG1 5LQ | G7MSG |
| SG1 5NF | G4BYE |
| SG1 5PH | M6EJW |
| SG1 5QS | G7HCL |
| SG1 5QS | M0EBG |
| SG1 5SF | G4IKY |
| SG1 5SF | G8FMA |
| SG1 5TB | M3MBR |
| SG1 5TT | G0RAM |
| SG1 6DS | M0MPI |
| SG10 6AX | G4XVO |
| SG10 6BD | M6TRS |
| SG10 6BN | G0KTW |
| SG10 6JG | G3OMD |
| SG11 1HT | G0KGR |
| SG11 1JH | G4VMR |
| SG11 1JH | G4VSL |
| SG11 1LH | M6FMF |
| SG11 1QF | M6AHH |
| SG11 1QS | 2E0CIR |
| SG11 1QS | M3IJS |
| SG11 1RN | G6UEG |
| SG11 1TG | G0DCW |
| SG11 1TG | G4YOS |
| SG11 2DN | G3OTR |
| SG11 2EH | M0SHF |
| SG12 0HP | G7VRY |
| SG12 0JQ | G3OJG |
| SG12 0NL | G4YFC |
| SG12 0SY | G4PHV |
| SG12 7DT | M6CSU |
| SG12 7FA | M1BYG |
| SG12 7EJ | 2E1GCC |
| SG12 7JS | G4GNK |
| SG12 7PD | G0OJY |
| SG12 8AX | G7UAH |
| SG12 8AZ | G4VXD |
| SG12 9JN | G3TIK |
| SG12 9LS | G4MAS |
| SG12 9PG | G8NDH |
| SG12 9QL | G6WYF |
| SG13 7DP | M6MOX |
| SG13 7EJ | G8OIY |
| SG13 7JD | G8OVZ |
| SG13 7QT | M3PJJ |
| SG13 7QU | G3SZF |
| SG13 7RB | G4UDZ |
| SG13 7RR | G0AHB |
| SG13 8AD | G8OPC |
| SG13 8BN | M6LKA |
| SG13 8DE | G0HCC |
| SG13 8HR | M6GBQ |
| SG13 8RB | G1IMS |
| SG14 1NE | M6NTT |
| SG14 2BZ | G4EEZ |
| SG14 2EU | G3IOJ |
| SG14 2HH | G0NSH |
| SG14 2QG | M6IQD |
| SG14 3AF | 2E0[illegible] |
| SG14 3AF | M6LOG |
| SG14 3DL | G1BLQ |
| SG14 3JU | G0OSK |
| SG14 3SF | 2E0LFT |
| SG14 3SF | M0WWV |
| SG14 3SF | M6WKZ |
| SG14 3SN | G7VLD |
| SG14 3TF | G0JGV |
| SG15 6RP | G0TUX |
| SG15 6RX | 2E0PLV |
| SG15 6RX | M0PLV |
| SG15 6RX | M6PLV |
| SG15 6TS | G4XBD |
| SG15 6UF | G6UMX |
| SG15 6UY | G3WKA |
| SG16 6BN | G3ORG |
| SG16 6BQ | M0RJG |
| SG16 6EY | M3HYQ |
| SG16 6JS | G1SJO |
| SG17 5AA | G3NII |
| SG17 5AD | G6RHL |
| SG17 5AE | M0CPE |
| SG17 5AU | G8UFX |
| SG17 5AU | G8XZQ |
| SG17 5BN | G3JNB |
| SG17 5EA | G0GGQ |
| SG17 5GZ | G0RLL |
| SG17 5GZ | G1RIV |
| SG17 5HB | G6MIS |
| SG17 5HB | G6VBE |
| SG17 5HW | G1CIT |
| SG17 5LU | G4JBD |
| SG17 5NH | G0PQO |
| SG17 5NH | G7CUY |
| SG17 5NY | M1AIN |
| SG17 5RA | M3HPO |
| SG17 5RZ | G7MYN |
| SG17 5RZ | M0MYN |
| SG17 5UP | G6JLU |
| SG18 0DA | M0JXM |
| SG18 0DG | G7ABF |
| SG18 0ED | M6DAU |
| SG18 0HL | G4GYP |
| SG18 0HT | 2E0BNB |
| SG18 0HT | M0HAS |
| SG18 0HT | M3JZM |
| SG18 0JJ | G7OZU |
| SG18 0JW | M3PKM |
| SG18 0NL | G0AKI |
| SG18 0NL | G3VHF |
| SG18 0NL | M3FAY |
| SG18 0NX | G8MCY |
| SG18 0PE | M3PIH |
| SG18 0PS | G4NNB |
| SG18 0PX | M3TXS |
| SG18 8BL | M6SCI |
| SG18 8BT | G0DEZ |
| SG18 8DX | G0PHY |
| SG18 8EE | M0WNV |
| SG18 8EE | M1GXL |
| SG18 8FS | M6BFL |
| SG18 8GG | 2E1GBM |
| SG18 8JR | M0SLP |
| SG18 8JU | G6ZEM |
| SG18 8LX | M0PLT |
| SG18 8PB | G3XIZ |
| SG18 8PF | M3NHP |
| SG18 8QS | G6XDK |
| SG18 8RF | M6MII |
| SG18 8SR | G1FYF |
| SG18 9AH | G8PQN |
| SG18 9BJ | G4KUY |
| SG18 9BX | G4FGJ |
| SG18 9DD | G4ETG |
| SG18 9DF | G2DPQ |
| SG18 9DF | G4YRF |
| SG18 9DL | G3CCO |
| SG18 9JL | G1JVY |
| SG18 9JL | M3AWS |
| SG18 9NE | M1EPK |
| SG18 9NY | G0ECZ |
| SG18 9PQ | 2E0RFI |
| SG18 9PQ | M6UTC |
| SG18 9PS | G7LUR |
| SG18 9PT | G0ETA |
| SG18 9QP | G1GSN |
| SG18 9QP | G4XIM |
| SG18 9RS | G3XCK |
| SG19 1AU | 2E1CWJ |
| SG19 1AU | G0TMJ |
| SG19 1AU | G8GHR |
| SG19 1AX | G6REA |
| SG19 1DT | G4CEP |
| SG19 1ED | M6BKT |
| SG19 1EU | G4MEO |
| SG19 1HE | G4ZDN |
| SG19 1HF | G7SVF |
| SG19 1HP | 2E1DEP |
| SG19 1HP | 2E1FDU |
| SG19 1JJ | 2E0KMD |
| SG19 1JJ | M6KMD |
| SG19 1JL | 2E1CDZ |
| SG19 1PQ | M0BIK |
| SG19 1RH | 2E0KKJ |
| SG19 1RH | M6MKJ |
| SG19 1TX | G3KDD |
| SG19 2AX | G8BXM |
| SG19 2JU | 2E1EFQ |
| SG19 2JY | G3ZWM |
| SG19 2LB | G0IKZ |
| SG19 2PB | M0YLG |
| SG19 2PS | G7JGQ |
| SG19 2PS | M3JGQ |
| [illegible] | [illegible] |
| SG19 2QD | G4MKR |
| SG19 2QQ | G1YRM |
| SG19 2QQ | G3OXG |
| SG19 2QR | G3WSD |
| SG19 2UR | G0OQI |
| SG19 3AN | G7SOZ |
| SG19 3AP | G6CTP |
| SG19 3BA | M1FFF |
| SG19 3BT | 2E1WEB |
| SG19 3DX | G4MKP |
| SG19 3LG | G0FOT |
| SG193NY | G1ZPU |
| SG193NY | M0ZPU |
| SG2 0BP | 2E0AWM |
| SG2 0DE | M6SDN |
| SG2 0EB | M3IVA |
| SG2 0EJ | M6EGT |
| SG2 0EQ | G3JLZ |
| SG2 0HD | G7HCB |
| SG2 0JH | G7PNE |
| SG2 0JJ | G0IOP |
| SG2 0JN | 2E0ZBZ |
| SG2 0JN | M0ZBZ |
| SG2 0JN | M3ZBZ |
| SG2 0NZ | G6BYK |
| SG2 0QE | M6RIZ |
| SG2 0QG | G0BXH |
| SG2 7BH | G4UUM |
| SG2 7DR | M3YSL |
| SG2 7DS | G0KUQ |
| SG2 7DT | G4XUW |
| SG2 7DW | M3XTK |
| SG2 7EF | G4TFH |
| SG2 7JL | 2E0OSX |
| SG2 7JL | M0OSX |
| SG2 7JR | M3DLU |
| SG2 7QS | G4CZZ |
| SG2 7ST | G7JSS |
| SG2 8DB | G0MHR |
| SG2 8DB | G3WGC |
| SG2 8DB | G4KGP |
| SG2 8JA | M6AJM |
| SG2 8JD | G4KFL |
| SG2 8NA | M1DGS |
| SG2 8PX | 2E1IID |
| SG2 8PY | M6IVR |
| SG2 8QZ | M3KCG |
| SG2 8RJ | G6UXE |
| SG2 8RW | M3YCM |
| SG2 8RX | M6NCR |
| SG2 8RY | G8ZFI |
| SG2 8SH | G4LPP |
| SG2 8UG | G0UVN |
| SG2 9DW | G8UEY |
| SG2 9DW | M3NEA |
| SG2 9DX | M0CUH |
| SG2 9ER | M3XTP |
| SG2 9LZ | M6FKF |
| SG2 9NJ | G4NMA |
| SG2 9NL | G8HGP |
| SG2 9NP | G6YMA |
| SG2 9NR | G7NKI |
| SG2 9RN | G0EVZ |
| SG2 9RR | G8KHI |
| SG2 9TP | G4NEA |
| SG2 9TT | 2E0LKS |
| SG2 9TT | M0LDC |
| SG2 9TT | M6LKS |
| SG2 9TY | G0MFY |
| SG3 6BQ | G1HBC |
| SG3 6DE | 2E0MJD |
| SG3 6DE | M0XJP |
| SG3 6DF | G0TAG |
| SG3 6DN | M0AZZ |
| SG3 6NU | G0FRM |
| SG4 0AN | G3LQO |
| SG4 0BU | M6SWP |
| SG4 0DP | G4OXD |
| SG4 0DP | G8PUR |
| SG4 0DP | M6JUK |
| SG4 0JH | 2E0RCT |
| SG4 0JH | M0LGC |
| SG4 0JH | M0ROC |
| SG4 0JH | M3RCT |
| SG4 0LG | G4KDW |
| SG4 0QP | G6WEH |
| SG4 0RA | M0RLO |
| SG4 7DR | M0ARH |
| SG4 7NT | G2YT |
| SG4 7PD | G6YIQ |
| SG4 7PD | M5AIQ |
| SG4 7QZ | M0RRC |
| SG4 7QZ | M1ECY |
| SG4 7RA | G4GXM |
| SG4 7RQ | G0KJN |
| [illegible] | M0JVZ |
| SG4 8BE | 2E0CBF |
| SG4 8EQ | 2E0HUB |
| SG4 8EQ | G7HMV |
| SG4 8EQ | M0KSG |
| SG4 8EQ | M6HMV |
| SG4 8EQ | M6HOT |
| SG4 8EQ | M6MDH |
| SG4 8ER | G4HSO |
| SG4 8JH | G4PLW |
| SG4 8NP | G3PZE |
| SG4 8NT | G4NGL |
| SG4 8PG | G8WKE |
| SG4 8PR | G0CRK |
| SG4 8QD | M6STY |
| SG4 8RA | M6IAC |

Postcode

**IMPORTANT NOTE**

**Revalidate licence to avoid revocation** – Ofcom has advised the Society that plans will be drawn up to revoke licences that have not been revalidated as required by the licence conditions. The quickest way to revalidate is to do so online via the Ofcom website: *https://services.ofcom.org.uk/* or by email: *amateur.validations@ofcom.org.uk* If you need assistance in the process, Ofcom staff are available to help, but please be patient during times of heavy workload.

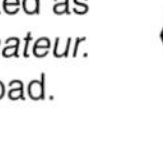

SG4 8RR M6CSN
SG4 8SD G8TRR
SG4 8TJ M3VQQ
SG4 9AN G3WRJ
SG4 9EP G4CTM
SG4 9LT M0JMC
SG4 9NP G4KNT
SG4 9NW G3FVR
SG4 9PJ G0GXU
SG4 9TB G7URR
SG4 9TL G1VQI
SG4 9TL M6UTI
SG5 1PA G6CFC
SG5 1PN G8BFH
SG5 1PT G6LSC
SG5 1QE M6RXS
SG5 1UA M3XZE
SG5 1UA M6GHW
SG5 2BE M3HPN
SG5 2HP G6XCK
SG5 2JH G4VXU
SG5 2QZ G7LXH
SG5 3AZ 2E0CPX
SG5 3AZ M6SOE
SG5 3DR G0EUV
SG5 3DR G0UEW
SG5 3LS G4PSO
SG5 3LS G7PQP
SG5 3LS M3GGV
SG5 3NB G0IYJ
SG5 3NS G4CML
SG5 3RG G6BRW
SG5 3RX G8FMT
SG5 3RX M0FMT
SG5 3SL G4YDH
SG5 3UP G3UWM
SG5 3XY G6CRC
SG5 4EH 2E0JKT
SG5 4EH M6TEM
SG5 4EL G0UVX
SG5 4HH M1CAH
SG5 4HT G4THF
SG5 4JE 2E0BYJ
SG5 4JX 2E0MIH
SG5 4JX M6MHU
SG5 4RU G4ALR
SG5 4RU G4KQD
SG6 1PH G8EDS
SG6 1PP G7HRJ
SG6 1RF M6FKE
SG6 1UE M1WHO
SG6 1UE M3EMS
SG6 2BL G4SWH
SG6 2DE 2E0ASU
SG6 2DQ 2E0PJN
SG6 2DQ M3PJN
SG6 2EU G6JBQ
SG6 2JA G8DKK
SG6 2LU G0EVD
SG6 2NP 2E0EPT
SG6 2NZ 2E0BYG
SG6 2NZ M6EGO
SG6 2TY 2E0XMC
SG6 2TY M6XMC
SG6 3PY M0ARQ
SG6 4BH 2E0HVL
SG6 4BH M6HVL
SG6 4DB G4DKQ
SG6 4HG G1UFH
SG6 4HY G1BYJ
SG6 4JQ M6KAZ
SG6 4JZ G0AMX
SG6 4LA 2E1RAF
SG6 4LA G0TAK
SG6 4LA G3ELV
SG6 4LA M0KAC
SG6 4LF G1OKV
SG6 4NQ 2E0ISS
SG6 4PL G0MJP
SG6 4RZ G0PJC
SG7 5AF M0SFR
SG7 5HL G0NZJ
SG7 5LJ G0RRV
SG7 5QS G0RGW
SG7 5QS G8EXF
SG7 5RN M6CIP
SG7 6DA M6LCC
SG7 6DG G8ABX
SG7 6HJ G6WKZ
SG7 6LT G4HIE
SG7 6LT G6KBS
SG7 6RX M6ERC
SG7 6RZ G1XXR
SG7 6RZ G8NHO
SG7 6TD 2E0STP
SG7 6TW G6MAY
SG8 0DJ G0CMB
SG8 0DJ G1HND
SG8 0JU G3WAB
SG8 0NE G3WMD
SG8 0PF G6PCE
SG8 0QG M3LUA
SG8 0RB G6VIK
SG8 5BB G4TEU
SG8 5BP G4MES
SG8 5BW G4HPE
SG8 5BW G6XQY
SG8 5ET 2E0BKQ
SG8 5ET M0ZDG
SG8 5ET M3UYC
SG8 5HX G7BIX
SG8 5JX 2E0NKM
SG8 5JX M6NKM
SG8 5LR G0DCU
SG8 5NG G1HDR
SG8 5PT G0LZD
SG8 6BZ G0FMT
SG8 6DD G1VRA
SG8 6HD M6VET
SG8 6JD G8DJU
SG8 6JX G8KDD
SG8 6UH M6FTY
SG8 7AD M5AGR
SG8 7DJ G4NIX
SG8 7ED G1GIJ
SG8 7ER M6CIQ
SG8 7ES G4NIY
SG8 7ES G6XEF
SG8 7EU G0TIL
SG8 7PZ M6FVS
SG8 7XU G0AUJ
SG8 8LT G0KFT
SG8 8RP G4FOH
SG8 9BN G0SII
SG8 9DE G0FLT
SG8 9DQ G8RZL
SG8 9EA G7VEX
SG8 9EA M3TNW
SG8 9HP M0BDW
SG8 9JF G6EDD
SG8 9QL G1HQE
SG8 9QX G0AQI
SG8 9SE G1YZI
SG8 9SE G1YZJ
SG8 9UQ G1OKF
SG9 0QR M3LHZ
SG9 0SU 2E0CIQ
SG9 0SU M3IWR
SG9 9BQ 2E0HPR
SG9 9BQ M6HPR
SG9 9DZ G6BWE
SG9 9DZ M1NHR
SG9 9EX M1ETN
SG9 9HJ M0DWP
SG9 9RA G1RPO

## SK

## (Stockport)

SK1 2QE G0OGP
SK1 3HD G6UQC
SK1 3NL M3AHU
SK1 4AB M1ANQ
SK1 4BL G8WNK
SK1 4BP 2E0RFU
SK1 4BP M0RFU
SK1 4BP M6BYR
SK1 4DS G0ELX
SK10 1QP G3HUR
SK10 2DA M3JLD
SK10 2DQ M3YFL
SK10 2EL G4XRG
SK10 2HJ G1AKA
SK10 2JB G7NNU
SK10 2NS G6CGO
SK10 2PF G4JNE
SK10 2PN G1NUS
SK10 2PS 2E0NSR
SK10 2PS 2E1AEQ
SK10 2PS G0DMV
SK10 2PS G0JNJ
SK10 2PS G4MWS
SK10 2PS M6NSR
SK10 2PW G0IUI
SK10 2RN G8JQG
SK10 2TU G6AKK
SK10 3BG M0TEZ
SK10 3BR G3JQ
SK10 3BS G1SHI
SK10 3DA G0DRO
SK10 3DB G0WUZ
SK10 3DJ G0DCO
SK10 3HB M6XCP
SK10 3JA G3CWI
SK10 3JG G6TDJ
SK10 3JL G0GWA
SK10 3LN G3LIO
SK10 3LN G8RDJ
SK10 3NQ G6XFU
SK10 3PE G1IZD
SK10 3RE G0IKB
SK10 4BG 2E0SYN
SK10 4BG M6CXO
SK10 4HZ G0AHJ
SK10 4QW G3UJA
SK10 5AH G6VSQ
SK10 5DT G6ECN
SK10 5HJ G8BPU
SK10 5HT G4XTO
SK10 5JG G1JDF
SK10 5JZ G3SXI
SK10 5LJ G0NYS
SK10 5SD G4GMK
SK11 0BU 2E1FLW
SK11 0EP G8GXO
SK11 6RW G0KJK
SK11 7EN G3VKF
SK11 7ES G0KVU
SK11 7ES G4WOL
SK11 7GE 2E0ZSH
SK11 7GE M3ZSH
SK11 7NJ G1NTR
SK11 7PL G7POV
SK11 7QA M3MMG
SK11 7SF M6CQK
SK11 7YG 2E0BAX
SK11 7YG M0GIA
SK11 7YG M1BYH
SK11 7YG M3HVP
SK11 7YH 2E0RXX
SK11 7YH M0TXX
SK11 7YH M3ZRY
SK11 8EX G6LFG
SK11 8JD M6RJM
SK11 8LL G3RWE
SK11 8LL M3DUR
SK11 8QA G3GKG
SK11 8QL G8XQH
SK11 8QP G0DMU
SK11 8RH 2E0EYP
SK11 8RH M0HGY
SK11 8RH M1EYP
SK11 8RH M3EYP
SK11 8RJ G1NPC
SK11 8RS 2E0MUD
SK11 8RS M6FAY
SK11 8RS M6HRG
SK11 8SD G7MSH
SK11 8TH G7OQG
SK11 9AP G7NKV
SK11 9AS G1NXV
SK11 9BF G3HFM
SK11 9BW M3JPI
SK11 9BX G8BVF
SK11 9LQ M3FLJ
SK11 9RR 2E1HNB
SK11 9RR M3CBN
SK11 9SW G3VDB
SK12 1AR G6VKL
SK12 1BG G6XRL
SK12 1BG M3AJS
SK12 1BL G0PJW
SK12 1BU G4ZGP
SK12 1EF G1AFI
SK12 1HA G3SHF
SK12 1HA G6UQ
SK12 1HA G8SRS
SK12 1HA M5MDX
SK12 1HG M0AGW
SK12 1HN G0ONG
SK12 1PU G4IJJ
SK12 1PU G8BRF
SK12 1PW M0GWF
SK12 1QH M1PTR
SK12 1RW G6RJH
SK12 1RW G7UJC
SK12 1SP G3THF
SK12 1TG G4MEH
SK12 1TH G6FUD
SK12 1XB G6HJU
SK12 1YE G0EVP
SK12 1YX M6LBX
SK12 2AE G0CQU
SK12 2AY G8CET
SK12 2BY G0JYD
SK12 2DB G6YCE
SK12 2RG G0NLL
SK12 2RZ G8MHT
SK13 1AR G1XCB
SK13 1EN G7JUJ
SK13 1JZ 2E0MPC
SK13 1LF G0DTT
SK13 1LH M0AIC
SK13 1LR M1EYO
SK13 1NX 2E0PSP
SK13 1PD G7JNM
SK13 2AT M3RUO
SK13 2BN G0UPO
SK13 2BN G4YLQ
SK13 5HL M3XAV
SK13 5HP G6BIX
SK13 6NF G4KWF
SK13 6XY G6YRK
SK13 7AH G6EKS
SK13 7AJ G4GNQ
SK13 7AJ G4LMR
SK13 7BG G8BEQ
SK13 7RL M0SAQ
SK13 7TL G4OQH
SK13 8EL 2E0EBN
SK13 8EL M6EBN
SK13 8NW G1YKK
SK13 8NW G6LLL
SK13 8NZ G0BAY
SK13 8RG G4IFJ
SK13 8RJ G0WJS
SK13 8RZ G6ARR
SK13 8UD G4TJC
SK13 8UD M6HMK
SK14 1DT 2E1INC
SK14 1DT G8INC
SK14 1DT M3NXA
SK14 1QX G0KLD
SK14 1RP M6YBB
SK14 2SW G8OKD
SK14 3BW G4HLA
SK14 3QW M6AOG
SK14 3QX 2E0DVM
SK14 3QX M6FCJ
SK14 3SN G0RRR
SK14 3SN G7KKW
SK14 4AF 2E0SWZ
SK14 4AF M0SWZ
SK14 4BS G0RGN
SK14 4DJ G3VDS
SK14 4DW M3LPI
SK14 4DW M3LPK
SK14 4JS G4BLL
SK14 4RT G6PZS
SK14 4SY 2E0MBV
SK14 4SY M3WPU
SK14 4TD M6VAK
SK14 5AB G0URT
SK14 5AU G4ING
SK14 5AW M1TZR
SK14 5BB G8LCS
SK14 5DD G4USX
SK14 5JX G4PYQ
SK14 5LJ G1UTN
SK14 5NS G0WVV
SK14 5RE G7KEI
SK14 5RU 2E0CGX
SK14 5RU M3NBU
SK14 5ST M6MYS
SK14 6LE G4EZF
SK14 6SE 2E1BVS
SK14 8LE G0NPK
SK14 8PE M3PVX
SK14 8PL G7JGF
SK14 8QH G6SLY
SK145QZ M6BIX
SK15 1AJ M6TMM
SK15 1BU M6NWI
SK15 1EX 2E0OFF
SK15 1EX M6GRK
SK15 1HD 2E0MKK
SK15 1HD G7FEG
SK15 1HD M0MRN
SK15 1HD M3ATC
SK15 1HD M3MKK
SK15 1HG G0NIL
SK15 1HL G0KYB
SK15 1HU G0PVR
SK15 1HU M3SHX
SK15 1LE M0NSI
SK15 1US G8YIG
SK15 2EA M6RNV
SK15 2HR G1VTP
SK15 2LT G0WFD
SK15 2QQ G6LBO
SK15 2QZ G7WFD
SK15 2TH G6YSZ
SK15 2TR G4NHW
SK15 3DZ G1TYP
SK15 3EN 2E0MXW
SK15 3EN M0SAD
SK15 3EN M3MXW
SK15 3HJ G0TIX
SK15 3HU G0ITZ
SK15 3JX M6LEH
SK15 3LS G6DQK
SK15 3RL M6JRL
SK16 4AB G0HRQ
SK16 4AU M6ZEP
SK16 4BE G0JNV
SK16 4EW G1BSY
SK16 4JJ G7BCK
SK16 4NU M6GYT
SK16 4SJ M6GSC
SK16 4UD M1DJO
SK16 5AB M3PQT
SK16 5BU 2E1SUE
SK16 5BU G4ZPZ
SK16 5DN G0NAJ
SK16 5DP M0AHF
SK16 5EG G4KFJ
SK16 5HR 2E1CWP
SK16 5HR 2E1CWX
SK16 5HR G0XAU
SK16 5HS G4LTM
SK16 5JA M1EEW
SK16 5JL G1HBE
SK16 5JL G1HTF
SK16 5NW G0LOE
SK16 5QS M6STP
SK16 5RT G0CKM
SK17 0DH G0JWD
SK17 0LU 2E0SCF
SK17 0LU M0TGT
SK17 0LU M6CRF
SK17 0LU M6JCF
SK17 0LU M6LEF
SK17 0LU M6SCF
SK17 0NF 2E1AZQ
SK17 6HH G4BJT
SK17 6HN G1OSA
SK17 6HQ G1LVV
SK17 6NQ G8DAM
SK17 6QX 2E0BSG
SK17 6QX M0GOB
SK17 6QX M6ARS
SK17 6RD G7TMM
SK17 6ST M0DZM
SK17 6UA M6NCL
SK17 6WJ 2E0WAT
SK17 6WJ M3AJW
SK17 6XX M1CGQ
SK17 7DD 2E0NIB
SK17 7DD G6DBJ
SK17 7DD M6NIB
SK17 7DX G8YTX
SK17 7EA 2E0DHS
SK17 7EA M3DHS
SK17 7HW M6KWL
SK17 7JF M3MQM
SK17 7PH M3LIB
SK17 7PU G0MUR
SK17 7TA 2E0CFV
SK17 7TF 2E0CWR
SK17 7TF M6CWR
SK17 7TJ G0WSP
SK17 7TQ M3VMA
SK17 7TW 2E0TCG
SK17 7TW G1INK
SK17 7TW M6TCG
SK17 8AY G1OHD
SK17 8DP M3JQG
SK17 8DW G4MRQ
SK17 8DW G4SPA
SK17 8ET G6LXE
SK17 8JT G4HGN
SK17 8LP G0CWS
SK17 8NP G0IOK
SK17 8PX M3ENO
SK17 8RJ M1WVS
SK17 8RJ M3TOT
SK17 8SW G1TBI
SK17 9AG M3YXL
SK17 9HG M0AVK
SK17 9HG M3MSZ
SK17 9JS G0XKK
SK17 9JS G4BUX
SK17 9JS G4IHO
SK17 9JS G8BUX
SK17 9JU G7TZN
SK17 9NP G7NFK
SK17 9PL G7NHL
SK17 9PL M3NFK
SK17 9RE G6MIF
SK17 9RY G0RKT
SK17 9SE G0RYQ
SK17 9SG 2E0GAU
SK17 9SG M0OKK
SK17 9SG M6OKK
SK17 9TS G4KNQ
SK2 5AZ 2E0JLR
SK2 5AZ M6FCF
SK2 5BS G4DSR
SK2 5BS G8MFR
SK2 5EP G4WHF
SK2 5RN G7ILS
SK2 5SF M1ANT
SK2 5SF M1SIM
SK2 5UR G1FFR
SK2 6AA M0SAV
SK2 6BN G8CHY
SK2 6BX G0KHR
SK2 6DW 2E0HSB
SK2 6DW 2E0RHM
SK2 6DW M3UVM
SK2 6DW M6ACE
SK2 6DX M6HGN
SK2 6EP G7IOC
SK2 6EP M3IOC
SK2 6HA G4ORV
SK2 6JS G4FRM
SK2 6PY 2E0DWV
SK2 6PY M6FVZ
SK2 6SP 2E0TJU
SK2 6SP M0TJU
SK2 6SP M3TJU
SK2 7AR M3UKY
SK2 7BH M6AAL
SK2 7EB G1JVF
SK2 7EB G3UHF
SK2 7EB G4HON
SK2 7EB M0ECR
SK2 7HH G0ASI
SK2 7JD 2E0DJZ
SK2 7JD M6FFO
SK2 7LQ G3RZJ
SK22 2JG G7OJA
SK22 3AY G7KUB
SK22 3DH G3TPI
SK22 4DP G4YAB
SK22 4HR G1IMD
SK22 4HU M6NVB
SK22 4JT G4XDT
SK23 0EZ 2E0FOL
SK23 0EZ M6FOL
SK23 0HY 2E0GYZ
SK23 0JF G0SMP
SK23 0JF G7KRZ
SK23 0LF G7CJW
SK23 0PX G6YDN
SK23 0QF G3NZV
SK23 0QF G3ZVS
SK23 0TA G1ABJ
SK23 6BD M6PTZ
SK23 7AY G1OXH
SK23 7BD G0VIE
SK23 7BP M0WLH
SK23 7DP G4IXF
SK23 7DT G0BDN
SK23 7HD G0GCJ
SK23 7LH G4CDZ
SK23 7NF G6UOO
SK23 7NH G8MUX
SK23 7QU G1HQO
SK23 9TU M3OXN
SK23 9UN G3TDF
SK3 0JJ G7SRH
SK3 0PX G3JZT
SK3 0QL G0RXA
SK3 0QL M1BAR
SK3 0QY G6RSU
SK3 0TX G0MLC
SK3 0UJ G1HWJ
SK3 0UN G7TQT
SK3 8HQ G8ZAD
SK3 8PA G7VFQ
SK3 8PG 2E0UTX
SK3 8PG M0UTA
SK3 8QU G3NBY
SK3 8SL M3JGU
SK3 9DR 2E0GXE
SK3 9DR M6GCY
SK3 9ET 2E0CTF
SK3 9ET M0HNJ
SK3 9ET M6PRE
SK3 9HF G8UWM
SK3 9JN G7DSV
SK3 9JU G6PWQ
SK3 9NJ 2E0GCL
SK3 9NJ M6GBV
SK3 9QG M0BIH
SK3 9QL G0OQR
SK3 9RZ 2E0REU
SK3 9RZ M6PFP
SK4 1HZ 2E0BDI
SK4 1HZ M3LQD
SK4 1NH M3IRS
SK4 1NH M6TGV
SK4 2AA G6LQE
SK4 2DB G8YTP
SK4 2DQ M3UTJ
SK4 2PE G3TFR
SK4 3AD M0CGF
SK4 3HD G1JFU
SK4 3HD G4RPJ
SK4 3JB G4DVI
SK4 3JS G4HNO
SK4 3NB G8UBP
SK4 3PG G8NRU
SK4 3PU G1CQT
SK4 3QF G0AMY
SK4 4AX G3RRG
SK4 4BX G3WFW
SK4 4BX G7KSP
SK4 4EB M6EJR
SK4 4NE G3YFD
SK4 4NL G4HWW
SK4 4PY G6KSO
SK4 4RL G0WOP
SK4 4RP G7PRD
SK4 5AW G1FMA
SK4 5BR M6AVA
SK4 5DJ M1BSM
SK4 5HP 2E0XYA
SK4 5HS M1BBU
SK4 5NW M3ZIA
SK5 6JD M6BTA
SK5 6LE M6DLL
SK5 6SH G0IES
SK5 6TD 2E1HQV
SK5 6TD 2E1HQW
SK5 6UH G8WPL
SK5 6UU G0LJF
SK5 7EU G1JQK
SK5 7EX G3RNV
SK5 7HU 2E0RTN
SK5 7HU M0RUM
SK5 7HU M6KBI
SK5 7LB G0IQC
SK5 7LL G8ZLU
SK5 7NA 2E0TLC
SK5 7NA M0ZPZ
SK5 8BG 2E0NLY
SK5 8BG M6RBB
SK5 8DB G1JIR
SK5 8JY G0TIP
SK5 8NQ M6ANP
SK6 1BL G1YZH
SK6 1EE G6LKJ
SK6 1EF G0GSM
SK6 1EW G4WMV
SK6 1HA M6VBZ
SK6 1HR G0WMZ
SK6 1JE M3UQP
SK6 1LH 2E1CXP
SK6 1PG G4KUC
SK6 1PG G8VHF
SK6 1PT G6GUT
SK6 1PY G4OZC
SK6 2ED G1TNK
SK6 2HQ G8ILD
SK6 3EE M0ATV
SK6 3EN G8IAN
SK6 3EY G0EOM
SK6 3JT G0PUD
SK6 4JF G3PTX
SK6 4PT G6RIC
SK6 4QE M0ABW
SK6 5BB M1BMC
SK6 5BR G3PMV
SK6 5BT G3XGH
SK6 5EB G7GGN
SK6 5PG G4NPU
SK6 5PN G8OOC
SK6 5PR 2E1WVF
SK6 5PR G4WVF
SK6 6DF G4MUL
SK6 6HJ G3MUO
SK6 6HJ G4BEV
SK6 6JL M0BEX
SK6 6LL G7BYU
SK6 6NA G0EQH
SK6 6PJ G3SNG
SK6 6PT 2E0JEK
SK6 6PT M6BMZ
SK6 7BG G0PDH
SK6 7DT G8NOS
SK6 7HP M3EAE
SK6 7JS 2E0JEH
SK6 7JS M6FRI
SK6 7PB 2E0TTG
SK6 7PB M6TMH
SK6 7PW M3EXY
SK6 7RJ M6TMA
SK6 8AB G4FBG
SK6 8AE 2E0DOD
SK6 8AE M0PAI
SK6 8AE M3TUL
SK6 8BH G0NKU
SK6 8BJ G6WWA
SK6 8JE G3NXX
SK7 1HR G4BIC
SK7 1LE G8OBK
SK7 1LG G3TDH
SK7 1PJ M3ZVS
SK7 1PP G4GRU
SK7 2AB 2E0HPB
SK7 2AE 2E0RPE
SK7 2AE M6RPE
SK7 2BY M3NMP
SK7 2DT G3NUQ
SK7 2ER G3KAF
SK7 2JR G0ADJ
SK7 2QW M3FFU
SK7 3DT 2E1HLU
SK7 3JS G7EAH
SK7 3JU G8VLR
SK7 3JW G0KZO
SK7 3JW G4SYC
SK7 4BJ G1GYC
SK7 4HW M1AQJ
SK7 4QE G0FLQ
SK7 4QG G4GEY
SK7 4QP G0CJA
SK7 4QW G0HEX
SK7 4RL M1VSR
SK7 4RU M6MYL
SK7 5JB M0KJC
SK7 5JB M3KJC
SK7 5JN G4OQP
SK7 5LR G0LDU
SK7 5PW M0AFF
SK7 5QY G4VSW
SK7 6EJ G8YUP
SK7 6EJ M3YUP
SK7 6EL M6XDW
SK7 6HG G6DBQ
SK7 6HP 2E0MIU
SK7 6HP M0JFM
SK7 6HP M6MIU
SK7 6HU G8WPV
SK7 6JB G8EVR
SK7 6NR G8RIC
SK7 6NR M5KJM
SK8 1BA G4CSV
SK8 1DX G4DQL
SK8 1HY G7TKT
SK8 1QY G3MTR
SK8 2AQ G6TRN
SK8 2AQ M6UAP
SK8 2EE M0BMW
SK8 2LZ M6CMK
SK8 3AA M3FJQ
SK8 3AJ G4FAS
SK8 3AS 2E1INW
SK8 3BG G0WOA
SK8 3DL 2E0LKC
SK8 3DL 2E0LMD
SK8 3DL M6LCH
SK8 3DL M6YEH
SK8 3DZ G0HVT
SK8 3DZ G0KCE
SK8 3ET G3HJK
SK8 3JA M0AGV
SK8 3LF G1TAU
SK8 3LL G3ZBZ
SK8 3NH G7PIG
SK8 3NR G8BHX
SK8 3PF G4SVV
SK8 3PJ G0DOU
SK8 3RH M6MTJ
SK8 3RH M6NHJ
SK8 3RR M3EEJ
SK8 3RW M3GIY
SK8 3UD G8ZCJ
SK8 4AE M0SGW
SK8 4AL G3UYG
SK8 4NA G3NYE
SK8 5DD M3DRM
SK8 5HP G6AXK
SK8 5PF G0NKM
SK8 5PN 2E0ETN
SK8 5PN M0TKT
SK8 5PN M6ETN
SK8 5RU 2E1GGT
SK8 6AX G3VOU
SK8 6DG G1WTY
SK8 6EQ G4NFA
SK8 6EX G8TMJ
SK8 6JG G4KJK
SK8 6LS M0ARA
SK8 6PW G0BAA
SK8 6PW G0OZJ
SK8 6PW G1NCR
SK8 7AJ G8GRP
SK8 7AJ M6HKZ
SK8 7AL G8LDC
SK8 7DS G4FRW
SK8 7EH G6YDO
SK8 7EP G8GHO
SK8 7HU G6TKV
SK8 7HX M3XMS
SK8 7LG G4EUC
SK8 7PB G6FOI
SK8 7PZ G0IGB
SK9 1LU G4IRU
SK9 1QE G4CQV
SK9 2AD G1JDQ
SK9 2BB G3RKF
SK9 2BJ M0SMV
SK9 2EY G1MWS
SK9 2EY M0GMG
SK9 2HQ M3RWV
SK9 3AF G6OER
SK9 3AR G0OPG
SK9 3AR G7IOB
SK9 3JT G4IRG
SK9 3NN G4GXQ
SK9 3UF G8JHL
SK9 4EP G8IXP
SK9 4HE G3WPF
SK9 5BN M0DFD
SK9 5BN M3CIE
SK9 5EN G3ONI
SK9 5JA G1HFH
SK9 5NQ G3TSZ
SK9 5PX G1GYH
SK9 5QE M0AAS
SK9 6BG G3VOM
SK9 6EG G0AXE
SK9 6HD G1ORS
SK9 6HD G7CMN
SK9 6HD G7RMZ
SK9 6HD M3ORT
SK9 6JD 2E1INT
SK9 6LG G3USO
SK9 6LP G3WTQ
SK9 6LS G4WKD
SK9 7HW G7TNT
SK9 7HZ G6TSM
SK9 7PN G3BNW
SK9 7PQ M3NAW
SK9 7RE 2E0PNA

## SL

## (Slough)

SL0 0DT 2E1DRC
SL0 0NJ G6XRK
SL0 9LF G3OZK
SL0 9NJ G3WCB
SL0 9NJ G4SAT
SL0 9RB G0VSG
SL0 9RD G6MKZ
SL0 9RJ M6PZY
SL1 1TS 2E0RNX
SL1 1TS M6RNX
SL1 2DA M6ETS
SL1 2HX G7HID
SL1 2HY M1ESV
SL1 2LQ G3RZF
SL1 2XY G7PEX
SL1 2YR 2E0DFS
SL1 2YR M6DBL
SL1 3JU G4FKE
SL1 3XG G4XGD
SL1 5BB 2E0MWN
SL1 5BB M6FJN
SL1 5DF G4CIJ
SL1 5QZ M6RBO
SL1 5RE G6GEN
SL1 5UR G8WQC
SL1 6AH G4HIN
SL1 6AY G8NNP
SL1 6HD G4RAA
SL1 6JU M1ACN
SL1 7BQ G3YFO
SL1 7EW G6DRP
SL1 7LY G6WZC
SL1 7NA G4ZGC
SL1 7NF 2E1AGE
SL1 8AT G8TEQ
SL1 8AW M1JJS
SL1 9HB G0WYR
SL2 1HG G6POC
SL2 1LG G7WGE
SL2 1PF G0BON
SL2 1SL G4XTZ
SL2 1SL M1DSQ
SL2 1SU M6CUU
SL2 2AD 2E0EBZ
SL2 2AD M3OJJ
SL2 2JY G6XJI
SL2 3HE G4LLN
SL2 3HW M6BFS
SL2 3LZ G0SKA
SL2 3NT G6OCF
SL2 3QT M6VVB
SL2 3QY 2E0LDZ
SL2 3QY M0TDZ
SL2 3QY M6LDZ
SL2 3XL 2E1AXI
SL2 3XT G0GUN
SL2 3XT G1DSA
SL2 4AX G3PFO
SL2 4DQ G4PXA
SL2 4ER G6VUF
SL2 5BG 2E0EBX
SL2 5BG M6KGR
SL2 5EZ M0MRY
SL2 5PN G1SPJ
SL2 5XH M3FZB
SL3 0LJ G0JAM
SL3 6HJ M1AEZ
SL3 6RQ M6SLC
SL3 7BB G1LMI
SL3 7DA G4XAN
SL3 7ET G6TSF
SL3 7ET M3TSF
SL3 7FQ G0HZK
SL3 7FQ M0GPX
SL3 7HN G4HTB
SL3 7JY M3CJE
SL3 7RL M6FEZ
SL3 8AU G0TGQ
SL3 8AU G6FCJ
SL3 8AX G1MDG
SL3 8ED G8GGM
SL3 8JH G0HXM
SL3 8JH G3VHH
SL3 8RJ G3VKB
SL3 8UU M6DFQ
SL3 9BA G4CRW
SL3 9DJ G3BXS
SL3 9NF M3MUU
SL3 9NJ 2E0BHZ
SL3 9NJ M6TBM
SL4 2AN G3ZCD
SL4 2LT G6PWF
SL4 2NQ 2E0USQ
SL4 2NQ M3USQ
SL4 2PD G8IUM
SL4 2QY M6HND
SL4 3LP G8FUO
SL4 3RD G1HDG
SL4 3SQ G8UYK
SL4 4AE G0HTK
SL4 4JP G7WFZ
SL4 4YU G7LTT
SL4 4YZ G0VHT
SL4 5BX G4WPR
SL4 5DX G8DPH
SL4 5PS G4PUB
SL4 5PS G4SVG
SL4 5TD G4WQS
SL4 6BN G7VME
SL4 6BW G4ZWX
SL4 6HL 2E1IJD
SL4 6HL 2E1IJE
SL4 6HL G7KYI
SL4 6HL M0HWN
SL4 6HL M1BEC
SL4 6LE G1NWO
SL4 6LE M3ABQ
SL4 6NF G7CYD
SL5 0HU G3PLY
SL5 0NR G1FXX
SL5 7SG 2E0DMQ
SL5 8HQ G8LF
SL5 8LL G7PYB
SL5 8NB G1PSS
SL5 8QB G6EIZ
SL5 8RS G3UBC
SL5 9GF G4EYV
SL5 9QH M1CSZ
SL6 0PB G6ZEZ
SL6 1DD G6CJB
SL6 1NX G0OGE
SL6 1XE G6GFR
SL6 2BJ G0TNY
SL6 2ED M6JPX
SL6 2GZ G0VKE
SL6 2JS G0MMQ
SL6 2US G3WHB
SL6 2YJ M6OAF
SL6 2YP G3ONR
SL6 2YU M3BJZ
SL6 3DD M3TGT
SL6 3EP M1HVJ
SL6 3JA 2E0CJV
SL6 3JA M6AYL
SL6 3JY G0BMU
SL6 3NR G4XOW
SL6 3PY M3MVJ
SL6 3UZ G8GQF
SL6 3XH G0SVN
SL6 3XL G7PVZ
SL6 3XL M3JAC
SL6 3YS G3VMR
SL6 4ED G7REC
SL6 4GY G4USK
SL6 4NF G4FTG
SL6 4NG G4XYN
SL6 4NR M0IBN
SL6 4QT G1MRX
SL6 4SA G8UPO
SL6 5AR G3UXY
SL6 5BJ M1TUG
SL6 5BW M6IEM
SL6 5DN 2E1BPV
SL6 5DW M6OJB
SL6 5EG G4SZA
SL6 5JD G8GHT
SL6 5JP G3OTN
SL6 5JW M1BAD
SL6 6AX G8SRV
SL6 6BE M0PGM
SL6 6DA G4XDU
SL6 6DF G8MYV
SL6 6EG 2E0TVM
SL6 6HN G4KMH
SL6 6HN G8TFR
SL6 6LG G0GLA
SL6 6LT M0RFK
SL6 6RX G0IIP
SL6 6RY G6NBI
SL6 6SD M6PTX
SL6 6SS G1PCR
SL6 7DT M3HXW
SL6 7EF G8AJM
SL6 7EJ 2E1ANN
SL6 7EZ M6ZMC
SL6 7HE M3HCG
SL6 7LF G0PMZ
SL6 7SH G0FFL
SL6 7TZ G6BJY
SL6 7UH G0BXC
SL6 7UH G0BQE
SL6 8EU G3YDT
SL6 8HT G0CXV
SL6 8JZ G4BOV
SL6 8QN G7FQP
SL6 9DH G7EBR
SL6 9DN G3PQA
SL6 9JF G0CMW
SL6 9JT G3WYK
SL6 9LS M3UCZ
SL6 9NF G1LIG
SL6 9SN G0NJQ
SL7 1DQ G3TOP
SL7 1DR G3DXD
SL7 1JW G7JDN
SL7 1NX M5SSB
SL7 1SA G7TWW
SL7 1TX G6GIF
SL7 1UW G4KCD
SL7 2PL G5BCO
SL7 2QU G6VF
SL7 2RE G4IDJ
SL7 3BZ G4YSH
SL7 3HA 2E1JOY
SL7 3LU G1SNO
SL7 3PY G3LVW
SL7 3PY M6LHS
SL7 3QW G4BEB
SL7 3RB G0UYM
SL7 3RH G3IQF
SL8 5AA 2E0IWF
SL8 5AA M6IWF
SL8 5AZ G3INQ
SL8 5TJ 2E0GCM
SL8 5TJ M0VCE
SL8 5TJ M6GCM
SL8 5TJ M6IOS
SL8 5TY G0MGM
SL9 0DD G0RAU
SL9 0DD G6XDB
SL9 0EJ M1SJA
SL9 0ES G4CLB
SL9 0HD G8TLT
SL9 0HE G4UNE
SL9 0HH M3ZGH
SL9 0LR G4MDN
SL9 7EB 2E1GXS
SL9 7HX G7CZF
SL9 7LA G4KGA
SL9 7QR G8CZQ
SL9 8LJ G3NZW
SL9 8PX M6LGM
SL9 8QT M5AIQ
SL9 9JZ M6WBK
SL9 9NX M0WAQ
SL9 9PA G1LIK

## SM

## (Sutton)

SM1 1JE G3WYB
SM1 1QH M6EXN
SM1 2BL G4HSD
SM1 2EB M0WLS
SM1 2EB M6WLS
SM1 2PA G0BWV
SM1 2PA G2XP
SM1 2PA G7SAC
SM1 2SQ G7KIN
SM1 3EA 2E0JHF
SM1 3EA M6JHF
SM1 3JX M0RFY
SM1 3QB G3NHX
SM1 3SL G8GLC
SM1 4BG G0SWU
SM2 5BA M6LOF
SM2 5EJ G6CTV
SM2 5ER G1XKD
SM2 5HP G1GBI
SM2 5HP M3NLP
SM2 5HW G8NFP
SM2 5HW G8TQK
SM2 5JA G4PLH
SM2 5JH G0PMM
SM2 5ND 2E0VAV
SM2 5NN G1EAM
SM2 5RP G7JDB
SM2 6RL 2E0VLB
SM2 6RW G1DQL
SM2 6RW M0DPF
SM2 6UA G1UBV
SM2 7QA G0KBL
SM3 8ES G3XTC
SM3 8NA M0BWB
SM3 8NN M1ABY
SM3 8QH G0XRA
SM3 9EG G4DSY
SM3 9HY G4PGS
SM3 9ND 2E0EXL
SM3 9ND M0ZPG
SM3 9NE G6SGM
SM3 9NL G8VCL
SM3 9PP G8YEP
SM3 9QJ G7UZI
SM3 9RF G6XGV
SM3 9RH M1ROD
SM3 9RL M3PUQ
SM3 9SG G1HWK
SM3 9TT M3VBZ
SM3 9TZ G8CVQ
SM3 9UB G0EAU
SM4 4BS G1SHN
SM4 4DL G0VXX
SM4 4LF G0BXC
SM4 4QG G0PNT
SM4 4QN G4ZHA
SM4 4RJ G3DCZ
SM4 4SX M6AVV
SM4 4TD G4CQR
SM4 5PR 2E0UVO
SM4 5PR M3UVO
SM4 6HL M3GYI
SM5 1QT G6ZTM

SM5 2PB G1EGZ
[illegible] M1MHD
[illegible]
SM5 3AB 2E0LEM
SM5 3DZ G7BDK
SM5 3DZ G8VUK
SM5 3EJ G4VAV
SM5 3HA G7NKH
SM5 3LS G0PUQ
SM5 3NG G4FDN
SM5 3NG G8PAT
SM5 3QH G7ELZ
SM5 3RA 2E0MJS
SM5 3RA M0SSM
SM5 3RA M6MCM
SM5 3SF G8HIG
SM5 3SJ M3RUL
SM5 3SU 2E0FTM
SM5 3SU M6TMF
SM5 3SW G7AQN
SM5 4QH G4BVG
SM6 0AZ 2E0MFV
SM6 0AZ M6MFV
SM6 0PB G7CRQ
SM6 0QY G6XZP
SM6 0TN G3BMQ
SM6 7AG G4XSA
SM6 7JU G7NHY
SM6 8AE G3SRC
SM6 8AE G4CCY
SM6 8AE G4DDY
SM6 8BE M0DWB
SM6 8EP G6SAQ
SM6 8EP G8ZOY
SM6 8EW G8XDM
SM6 8EW M0BEH
SM6 8HL G0TXL
SM6 8LE 2E0WBE
SM6 8LE M6WBE
SM6 8PT G6ZLD
SM6 8PY M3LRN
SM6 8QB 2E0JJK
SM6 8QB M0JJK
SM6 8QB M6JAK
SM6 9GL G6UCY
SM6 9GX 2E0XSL
SM6 9GX M3RHL
SM7 1AB 2E0NVK
SM7 1AB M3NVK
SM7 1AJ G3OLX
SM7 1BU M3UQO
SM7 1HG G0ARG
SM7 1HG G4DFA
SM7 1HG G8GYX
SM7 1HG G8OTS
SM7 1JW 2E0TTA
SM7 1LB G6XZM
SM7 1LE 2E0JLX
SM7 1LE 2E0LEG
SM7 1LE M0WOJ
SM7 1LE M3XJL
SM7 1LE M3YCZ
SM7 1LE M6EMC
SM7 2BA G7DRD
SM7 2DG G1DPX
SM7 2ER M3NUO
SM7 2ER M3ZON
SM7 2HG 2E0GTZ
SM7 2HG M6AXB
SM7 2HZ M0NQJ
SM7 2HZ M3NOE
SM7 2QJ G0IUH
SM7 3JR G3KTA
SM7 3NA M6UTB
SM7 3PN G8GNX

## SN

(Swindon)

SN1 2AN M6EWK
SN1 2AX 2E0GSF
SN1 2GL M6USO
SN1 2JR G7LET
SN1 2JU G4KEZ
SN1 2QT M3KHJ
[illegible] G0BNH
SN1 3AE M3GKK
SN1 3HW 2E0LAY
SN1 3HW M6LAY
SN1 3NJ G0TLI
SN1 3PY G2BUJ
SN1 3QA G1ERM
SN1 4AY G0DPK
SN1 4DP G3YKC
SN1 4EY G1KMN
SN1 4GE G1PJC
SN1 4GU G7JGE
SN1 4HH M6FVJ
SN1 4HX G1NAU
SN1 4LL G7PKK
SN1 4NB M3YFM
SN1 5EE G8PAB
SN1 5ES M1EGN
SN10 1BF G8LGP
SN10 1LY 2E0MBS
SN10 1LY M0MSZ
SN10 1LY M6MMS
SN10 1RW G4RQL
SN10 2EN G6NTE
SN10 2FB G6IBW
SN10 2FY G6PYF
SN10 2LD G3WAE
SN10 2RH G6EYA
SN10 2UB G7FWD
[illegible]
SN10 3AN G4GRX
SN10 3AP G3WZH
SN10 3BJ G3PYP
SN10 3BJ G4NWR
SN10 3BJ G4TIX
SN10 3BJ G6EVY
SN10 3DG G6KHW
SN10 3PQ M3PVV
SN10 3SL G4SHA
SN10 4AG G7LPE
SN10 4ED G6SND
SN10 4EL G6FNJ
SN10 4EU M1FHC
SN10 4JY G0VNB
SN10 4LP G3ZJJ
SN10 4PL 2E0SBS
SN10 4PL M3UZZ
SN10 4PP 2E1GQD
SN10 4RR G3MQD
SN10 4RR G3PZV
SN10 5AJ G1JMK
SN10 5AJ G6ZFA
SN10 5AZ M1FHA
SN10 5BA G8XYA
SN10 5DH G0WHY
SN10 5HD G4VKJ
SN10 5LE G7LWF
SN10 5NJ 2E0MPN
SN10 5NJ M3PQV
SN10 5SR G3XKL
SN10 5TD G4EES
SN10 5TP G0RIK
SN11 0AQ 2E0GKA
SN11 0AQ M0XCX
SN11 0AQ M6WKL
SN11 0EP G6HTZ
SN11 0LQ G7VEY
SN11 0NE M6FCS
SN11 0PZ G4SXH
SN11 0PZ M6TBS
SN11 0QT G8DNH
SN11 8EZ G4XYB
SN11 8PW 2E0ZGX
SN11 8PW M3ZGX
SN11 8QS G0ROV
SN11 8SQ G4NHQ
SN11 8UR 2E0EME
SN11 8UY G7TPS
SN11 8YD G0MLE
SN11 9AB G0OIM
SN11 9AT M6EIJ
SN11 9BG M1AZF
SN11 9BT G7RGG
SN11 9DU 2E0CRH
SN11 9DU G8LRD
SN11 9DU M0HKV
SN11 9DU M6BSQ
SN11 9EA G3WIW
SN11 9EW G4NKU
SN11 9FN 2E0VRE
SN11 9FN M3ZHC
SN11 9FN M6DTA
SN11 9LD G1JRZ
SN11 9LR G4UJL
SN11 9PA G4SUX
SN11 9QH 2E0LRD
SN11 9QH M3ZWP
SN11 9QQ G7TIV
SN12 6AG G4DMC
SN12 6FD 2E0DIM
SN12 6FD M0VAI
SN12 6FD M6DIM
SN12 6FX 2E0DRQ
SN12 6FX M6EEM
SN12 6HN G0HAD
SN12 6TH 2E1EYL
SN12 6TQ G4SRD
SN12 6UL G1PWU
SN12 7AB G3RVY
SN12 7DT G0KTP
SN12 7HL M0ECZ
SN12 7PF M3ZMZ
SN12 7PG G7TKO
SN12 7PT M6HJF
SN12 7QW M0OTL
SN12 7SW M0DGJ
SN12 7TE G6EJD
SN12 7TE M3IZB
SN12 8DQ M6XBX
SN12 8LA G4NDC
SN12 8NR G1AJC
SN13 0JD G4GUK
SN13 0JR G3ORX
SN13 0JR G4DIE
SN13 0JT M1EIJ
SN13 0LD G7BNX
SN13 0LS M3IGN
SN13 0QR G3JSK
SN13 0QW [illegible]
SN13 8AN M0TRK
SN13 8AN M6AIJ
SN13 8DF G3KRW
SN13 9AP G0FGR
SN13 9AU M3YPA
SN13 9JL M1EGC
SN13 9LE G4SEZ
SN13 9NH G0KLJ
SN13 9NJ G4NLA
SN13 9NS G3MEY
SN13 9QS M6CDU
SN13 9SH G3MBN
SN13 9SH G4SKN
SN13 9ST 2E0BZI
SN13 9TN G8DY
SN13 9TN M0PHO
[illegible]
SN13 9TX G0IUE
SN13 9TX M3KIR
SN13 9UL M6BYA
SN13 9UT 2E0GET
SN14 0DN G8MPM
SN14 0HT G3LWF
SN14 0JB G0RBD
SN14 0PS G4THG
SN14 0RB M3UXR
SN14 0TH M3NAT
SN14 0TN G7PSS
SN14 0UJ G1OKD
SN14 0UJ G1SOG
SN14 0UJ M3SOG
SN14 0UT M0RXM
SN14 0XP 2E0JWP
SN14 0XP M6CHH
SN14 6DE M0PYE
SN14 6EG M6BPA
SN14 6EL G1YIQ
SN14 6LU M6FWP
SN14 6XJ G1JRP
SN14 6YH M3MMN
SN14 7BB M3JYO
SN14 7EA 2E0NEY
SN14 7EA G8NEY
SN14 7EA M0GHZ
SN14 8DN G3PIJ
SN14 8QU G0RCP
SN14 8SA G8WMC
SN15 1AE M6DXW
SN15 1AR G0FGW
SN15 1BQ 2E0PLS
SN15 1BQ M0PLS
SN15 1BQ M6BXZ
SN15 1DD M1GTI
SN15 1NU M0CTR
SN15 1PZ G6FNY
SN15 1QQ G1USK
SN15 1QQ M6XJW
SN15 2AY G0IAY
SN15 2BD G8HAM
SN15 2DJ G4TKF
SN15 2EE G8FLL
SN15 2EF G4ASG
SN15 2JP G3NFP
SN15 2JW G0LJS
SN15 2LB G0MVM
SN15 3DZ G0HPN
SN15 3DZ G4MLG
SN15 3EF 2E1FBK
SN15 3EG M0MJD
SN15 3EL G4OQG
SN15 3FY 2E0SCS
SN15 3FY M6WSB
SN15 3GU G3UUV
SN15 3JZ M3HEC
SN15 3LX G0GJE
SN15 3RX G4SFD
SN15 3UA G4BBY
SN15 3UA G4KPI
SN15 4AR G6MFB
SN15 4AS G7OPI
SN15 4EL 2E0BPL
SN15 4EL M3WDV
SN15 4ER G0OID
SN15 4LG M6SDZ
SN15 4PJ M3EIA
SN15 4RL G4GWR
SN15 4TR G6VHE
SN15 5BA M3WBS
SN15 5EX 2E0CXH
SN15 5EX M6BXY
SN15 5HG G4KJV
SN15 5LD G8TTI
SN15 5NN G0AEV
SN15 5NT G6HUI
SN15 5QF G4CDW
SN16 0BL G6UYM
SN16 0DA M3WEF
SN16 0DR G4TLY
SN16 0EA M3WSH
SN16 0ES G4TRA
[illegible]
SN16 0PZ G4SWN
SN16 0PZ G7IBL
SN16 0RN 2E0EMX
SN16 0RN M6AGH
SN16 9DL G8DKI
SN16 9HF 2E1ASF
SN16 9NR G8KKD
SN16 9QB G8WLV
SN16 9QT G7GPU
SN16 9QZ G8POP
SN16 9SR 2E0GMG
SN16 9SR M3VGT
SN16 9UA G4SWO
SN2 1BA M1CZI
SN2 1BH M0SPN
SN2 1HE M6TGM
SN2 1NE G8CPA
SN2 1NL M0BFA
SN2 1NL M0PBZ
SN2 1RQ M3TSV
SN2 1RX G7FEA
SN2 1RX M6NDR
SN2 2AF M6EPH
SN2 2EL M3VST
SN2 2HG G7VJG
SN2 2PE G6OMH
SN2 2SG G4VWG
SN2 2SL G7UZG
SN2 5DF G0VCY
SN2 5ED G8XWI
[illegible]
SN2 5JJ M1EMC
SN2 7HP 2E0CYS
SN2 7HP M6VDX
SN2 7JN G4MNB
SN2 7LE G8SRC
SN2 7LE M0ACM
SN2 7LJ G7PYV
SN2 7LL G0JTD
SN2 7LQ G0VQW
SN2 7NJ G8MGO
SN2 7QA G4MDT
SN2 7SN G7CGB
SN2 8BT G7HPI
SN2 8DL M1ECQ
SN25 1PY 2E0ZVG
SN25 1WG G4SOA
SN25 1WG G6KAW
SN25 2AZ M6XTB
SN25 2BD 2E0FUR
SN25 2BL M6PWM
SN25 2EE G1WYP
SN25 2GQ M6ZYK
SN25 2NJ G8CZJ
SN25 3AZ G6XML
SN25 3BT G4ZAM
SN25 3DB G6ARO
SN25 3EZ M1DBW
SN25 3EZ M3PBW
SN25 3LZ M1AOU
SN25 3PP G8ELH
SN25 3QB G0UWS
SN25 3QB G7NWR
SN25 3QB M6UWS
SN25 4GX M1SHA
SN25 4TP G3YMV
SN25 4YD M3VAG
SN25 4YJ G1HYC
SN25 4YR G1LHQ
SN26 7AB M6LVX
SN26 7AQ G3GYQ
SN26 7AR G8NTS
SN26 7BE 2E0IMM
SN26 7BE M6IMM
SN26 7DE G4HGV
SN26 7DE G8LYG
SN26 7EA G4SHK
SN3 1AZ G6VDK
SN3 1BX G7MGY
SN3 1BZ G7CRM
SN3 1DJ G0FCH
SN3 1DN 2E0LXD
SN3 1DN 2E1EQI
SN3 1DN G0VTA
SN3 1DN G0WDQ
SN3 1DR G8VJY
SN3 1EA G1NCG
SN3 1ER G8IWO
SN3 1HZ G7MTX
SN3 1NB G8FWK
SN3 1NJ G7KIW
SN3 1PT M1AGH
SN3 2AL M5ECX
SN3 2DR M1ACC
SN3 2EA G1FTV
SN3 3EW G1MHZ
SN3 3NB G3NPM
SN3 3NW M5CBS
SN3 3PJ G4ZYH
SN3 3TB 2E0RJP
SN3 3TB M0OZD
SN3 3TB M3OZD
SN3 4AA G3LLZ
SN3 4AB G0PSO
SN3 4AY G3SIR
SN3 4EE M6ESR
SN3 4EQ M3RLH
SN3 4HB G0BQK
SN3 4HU G3HSV
SN3 4JA 2E0EXJ
SN3 4JA M0WXY
SN3 4JA M6EXJ
SN3 4LA G3MAI
SN3 4NF G3IDW
SN3 4EE G6BFG
SN3 4SF G1DCI
SN3 4SN G5WQ
SN3 4SN M0BCG
SN3 4SR M0LEY
SN3 4WE 2E0BNZ
SN3 4WE M3WDI
SN3 5AG G8ETI
SN3 5AU G3WEF
SN3 5AX G4LDL
SN3 5AX G8KWD
SN3 5BN G0CPA
SN3 5BN M3CSN
SN3 5DE 2E0PMV
SN3 5DE G0SNM
SN3 5DE M0PMV
SN3 5DE M6PMZ
SN3 6AR G3XYD
SN3 6JB G4JVJ
SN3 6JB G6NYC
SN3 6JB G7BPO
SN3 6JR M0BDJ
SN3 6LA G1YGY
SN3 6LS G8UKY
SN3 6LT 2E0POZ
SN3 6NF G4ENR
SN3 6NL G4AQK
SN4 0AE 2E0BZT
SN4 0AE M3ZPB
SN4 0DF 2E0FEC
[illegible]
SN4 0DF M3YOW
SN4 0DF M6IDB
SN4 0JF G4NSB
SN4 0LU G4NHD
SN4 0LW G3IRA
SN4 0RL M1CDJ
SN4 7AU G4NCF
SN4 7DG G8JWK
SN4 7DN G0OWR
SN4 7DR G4RZF
SN4 7DU G8IJM
SN4 7DX G7KRH
SN4 7EU M6DNB
SN4 7EZ G8IYJ
SN4 7FN 2E0BQJ
SN4 7FN M0TGN
SN4 7FN M0WCB
SN4 7FN M6DAN
SN4 7JG G4OIQ
SN4 7RG M3YOQ
SN4 7SH G7TIW
SN4 8AS G0LTP
SN4 8DJ G4MDH
SN4 8HJ M3YPS
SN4 8JY M3GRY
SN4 8LL G4FTZ
SN4 8LP 2E0HYK
SN4 8LW G8HCW
SN4 8NQ G4YAL
SN4 9AP 2E0CTW
SN4 9AP M0ZGB
SN4 9AP M6HBS
SN4 9AR G3JOT
SN4 9AT G0HOJ
SN4 9AX 2E0VAU
SN4 9AX M0TTE
SN4 9AX M6PZF
SN4 9AX M6VAU
SN4 9BB M1DHW
SN4 9BP M6FWM
SN4 9DR G0WGI
SN4 9DR M0BKN
SN4 9DR M1EZP
SN4 9EP G7VHG
SN4 9HY G0DVB
SN5 0AD G3HCT
SN5 0AD G3RZP
SN5 0AD G4FNC
SN5 3LE G3APS
SN5 3LX G0JSU
SN5 4AD G3YBY
SN5 4BA G4YEF
SN5 4BS G3TPQ
SN5 4DE G4MQP
SN5 4ET G8VGQ
SN5 4EX G4MGH
SN5 4HQ G8IYH
SN5 4JN M1KDJ
SN5 5AL G0DHL
SN5 5BH G1MUM
SN5 5DB M6BKA
SN5 5QT G4OED
SN5 5SA G8EFK
SN5 5SH G6UZM
SN5 5TL G4DSF
SN5 5TS G3OMA
SN5 5UP M6BLG
SN5 5UQ G4DOA
SN5 6AA M3PFU
SN5 6BB G7WIC
SN5 6BQ G6WAU
SN5 6EB M0PRT
SN5 6EQ M6BJQ
SN5 7AF G1VRJ
SN5 7AG G6UAJ
SN5 7AQ M6FSG
SN5 7AW G6LBR
SN5 7BB G1TOL
SN5 7BG G3KUD
SN5 7BN M3NTI
SN5 7BX G4GXW
SN5 7DR 2E0GKM
SN5 7DR M0GKG
SN5 7DR M0HUM
SN5 7DR M3RMY
SN5 7EF G4KIM
SN5 7EG G6BTR
SN5 8AH 2E1FJV
SN5 8AJ M6RGX
SN5 8AR M3FGR
SN5 8DB M6EVU
SN5 8HL M3XFS
SN5 8PF G4WSB
SN5 8QX M6BMF
SN5 8RT G7MTF
SN6 6AJ 2E0ZPT
SN6 6AJ M0MTA
SN6 6AJ M0MEA
SN6 6AW M1WIN
SN6 6BN G8OFR
SN6 6JG G0TDY
SN6 6RD 2E0EAV
SN6 6RD M0NMI
SN6 6RD M6EIW
SN6 7BA 2E0KGV
SN6 7BA M0KGV
SN6 7BA M0TBG
SN6 7BA M3KGV
SN6 7BU G0NUH
SN6 7BZ G4AJA
SN6 7BZ G4BPO
SN6 7BZ M0ATC
SN6 7EB G4GDR
SN6 7ER M3BWF
[illegible]
SN6 7HN M1ATC
[illegible]
SN6 7HS G8EOZ
SN6 7HU M1SKI
SN6 7LA G0NPN
SN6 7NU 2E0VPW
SN6 7NU M1EJD
SN6 7NU M6VPW
SN6 7PD G8JHC
SN6 7PG G8JRF
SN6 7RB M3EHY
SN6 7RL M3JIA
SN6 8AJ M0DAW
SN6 8AZ G3XEY
SN6 8ER G3ONU
SN7 7BB M0ABI
SN7 7DG G0JBJ
SN7 7EY G3NNG
SN7 7EY G3PIA
SN7 7LB G8BGM
SN7 7NG G4RUZ
SN7 7RL 2E0EJA
SN7 7SE M3CVH
SN7 7SS G8BCF
SN7 7YT G8GIF
SN7 8AT G4VLW
SN7 8EY G0SCQ
SN7 8LE G4PFY
SN7 8LE G6HIX
SN7 8LN M6GOI
SN7 8LR G1HYO
SN7 8LX 2E1HFX
SN7 8LZ M6JAZ
SN7 8ND G0MYX
SN7 8ND G3ZEF
SN7 8NP G3XVB
SN7 8PX G3ZUS
SN7 8QR G0ITO
SN7 8QR G6ZMX
SN7 8RP G0VBG
SN8 1NQ M0DSW
SN8 2AS G4XVW
SN8 2AZ G6EPN
SN8 2DH G1FKJ
SN8 2JQ G6EDJ
SN8 2PX G8JWD
SN8 3AF 2E0AAX
SN8 3AN M1EBH
SN8 3AS G3MFL
SN8 3DZ G4DHT
SN8 3HN G8SXA
SN8 3HN G8SXD
SN8 3HQ 2E0JOC
SN8 3HQ M3JOC
SN8 3NS M3AXZ
SN8 3QB M0TDW
SN8 3RP M3ANW
SN8 3SS M0GQM
SN8 3TD G4LMA
SN8 4HT M1CJE
SN9 5ES M0BXU
SN9 5HP G7GEA
SN9 5HU G8PMJ
SN9 5HU M6CFV
SN9 5LE 2E0GWR
SN9 5LE M0SWL
SN9 5NW 2E1EMK
SN9 6AE G4TOY
SN9 6AF M0BUT
SN9 6EN M6IIG

## SO

(Southampton)

SO14 0LB G6NJT
SO14 3FE M6DLU
SO14 3TY G4HXE
SO14 6FR M6LYP
SO14 6GR M0XGT
SO15 1JJ 2E0CZK
SO15 1JJ M6CYF
SO15 3EW G8SBS
SO15 3FF 2E0XEE
SO15 4HU 2E1ESQ
SO15 5DS G4MTS
SO15 5EA M6KFS
SO15 7NG M0RBH
SO15 8NY G6JVP
SO15 8PT M3VWJ
SO15 8PT M3VXG
SO15 8QZ M3YAO
SO16 2NU M0TWR
SO16 2NZ G3IXN
SO16 3EB G0SWY
SO16 3EF G0VNO
SO16 3GZ G4VAS
SO16 3HY G4ZAS
SO16 3LF G7UDD
SO16 3NY G4DIV
SO16 3TN G6UTL
SO16 3UJ M6LPF
SO16 4BN G1XBR
SO16 4NR M0RLP
SO16 4QT M6PHA
SO16 5DW G3CGE
SO16 5FD G4ZFY
SO16 5FG 2E0AJQ
SO16 5FG G8KVN
SO16 5FG M0APN
SO16 5GJ M6VRO
SO16 5HB G4WYW
SO16 5JX M3UVK
SO16 5NQ G3VOF
SO16 5SW G7GIJ
[illegible]
SO16 6PG M3KLF
SO16 6QB M6PSO
SO16 6TY M0BIJ
SO16 7DE G1CEI
SO16 7DR M6ZSH
SO16 7EL G3VXY
SO16 7FB M1ARI
SO16 7GJ M0MPT
SO16 7PE G3ZIL
SO16 8AH 2E0BGJ
SO16 8AH M0NXP
SO16 8AY G6VDW
SO16 8BE G4WFR
SO16 8DW G1XRM
SO16 8EH G6CRG
SO16 8FX G7VZK
SO16 8GU M0DVF
SO16 8LR G8LRS
SO16 9AU 2E1EUE
SO16 9AW M6ZOO
SO16 9EZ M6HTG
SO16 9JF G0XAZ
SO16 9JQ M6PCV
SO16 9LB 2E0DGR
SO16 9LB M6DZX
SO16 9WA M3XVQ
SO17 1DY 2E0GDG
SO17 1EN M6EJJ
SO17 1NU G0DHR
SO17 1NU G0MSR
SO17 1PF M6AFG
SO17 1SD G2HNI
SO17 2LJ M0EBO
SO17 2LJ M1CWB
SO17 3AE M6GHC
SO17 3AP 2E0DYG
SO17 3AP M0YAY
SO17 3AP M6YGD
SO17 3RT G6KYE
SO17 3RU M6FZQ
SO17 3SJ M6EEF
SO17 3SY G3KMI
SO17 3SY M0DNY
SO17 3TG M6KOH
SO18 1HJ M0CYJ
SO18 1JR M6DBF
SO18 1JS G1VGA
SO18 1LS G3TUF
SO18 1QR G3OFX
SO18 2AP 2E0YEP
SO18 2AP M0MMR
SO18 2FP G7UYI
SO18 2HQ M0GCH
SO18 2NU 2E0NWA
SO18 2NU M6CDG
SO18 2NU M6NWA
SO18 2PQ G1LWF
SO18 2QE G3IDB
SO18 3AG 2E1JJN
SO18 3AG M1JJN
SO18 3AG M3JJN
SO18 3RB G3KDW
SO18 4EH 2E1HQP
SO18 4FN 2E0ZBE
SO18 4FN M6SFC
SO18 5FE 2E0CZI
SO18 5FE M6DQV
SO18 5FS G3VCL
SO18 5GY 2E0NJC
SO18 5GY M3VQZ
SO18 5QJ G4PTF
SO18 5QL G3KXE
SO18 6BB G0UJP
SO19 0JG M0HME
SO19 1BX G6XYS
SO19 1DD M6JLW
SO19 1DP G8BJB
SO19 1FW G4UDB
SO19 2EQ G0CAJ
SO19 2HF G6HHH
SO19 2HU M0JBD
SO19 2NU G1OQI
SO19 4DE G3NZL
[illegible]
SO19 5JP G1LDN
SO19 5LF G0LCN
SO19 5LN G0ILA
SO19 6FB M3VPA
SO19 6FU G4BDQ
SO19 6HB G7HAB
SO19 6NG M6PHP
SO19 6PS G7BLX
SO19 6QX G0EMV
SO19 7DD 2E0PAU
SO19 7DD M3YXU
SO19 7GG G8RNT
SO19 7GG M5PSW
SO19 7JN G7MTA
SO19 7JN M0DDB
SO19 8AF G7BZU
SO19 8EY G4UEN
SO19 8FF G0IOI
SO19 8LD G1GVM
SO19 8LH 2E0KBX
SO19 8LN G0JBH
SO19 8NT G4OLY
SO19 8NT G8WWM
SO19 8SJ M6AKE
SO19 9AT G6YVJ
SO19 9AW G4BYY
SO19 9DA 2E0TAQ
SO19 9DA M3UUF
SO19 9FF G4WDH
SO19 9GB G2SZ
[illegible]
SO19 9HY G8FXZ
SO19 9LJ 2E0PKK
SO19 9PE G8GBP
SO19 9SS M3OCR
SO20 6AH G8IUP
SO20 6BA G3RDQ
SO20 6BA G6MCN
SO20 6EY M6AEL
SO20 6EY M6GTH
SO20 6HE G8CDA
SO20 6NY 2E0UKM
SO20 6NY M3UKU
SO20 6PR G4NMS
SO20 6PR M1BPD
SO20 8DB G3LGA
SO20 8DB G8RJM
SO20 8EA M1BDJ
SO21 1AF G3JRH
SO21 1BA 2E0VBN
SO21 1BA M3VBN
SO21 1BQ G6LKW
SO21 1TT M0WZM
SO21 1US G3UPD
SO21 2BX G8TEC
SO21 2DE G3HRH
SO21 2EG G8IDJ
SO21 2EP G3AJD
SO21 2EP G4NZC
SO21 2LG G3ZEO
SO21 3DU G0LMD
SO21 3EB G0GWG
SO21 3EB G3ZHV
SO21 3EE M3MKJ
SO21 3EQ G0JLX
SO21 3EQ M0HHT
SO21 3EY G1NBU
SO21 3EY G3LMH
SO21 3EZ G3VEH
SO21 3HP G3SJK
SO21 3HP G4XRJ
SO21 3HY 2E0FNY
SO21 3HY M6FNY
SO21 3JG G4EDW
SO21 3RU G3GXG
SO21 3RU M6YOG
SO22 4DN 2E0SSC
SO22 4DN M6SJC
SO22 4DU M6CNS
SO22 4EE M0AYV
SO22 4EE M0REX
SO22 4ET G3NVB
SO22 4EY G3MCL
SO22 4HS G3VPG
SO22 4HS G4JXE
SO22 4JL G0UMD
SO22 4JY G3PTS
SO22 4LD G8PGE
SO22 4LF G4SVC
SO22 4LQ G4MCM
SO22 4NU G8DGC
SO22 4PQ 2E1FKZ
SO22 4PQ G6AAZ
SO22 4PQ M3SFC
SO22 4PZ G4AFA
SO22 4QH G3XJM
SO22 4QL G8NEH
SO22 5AT G3RQR
SO22 5AX G0DZV
SO22 5BU G4HDY
SO22 5BX G1DAU
SO22 5DJ G6BZE
SO22 5HU G3RUZ
SO22 5JX G4CJO
SO22 5ND G0JMK
SO22 5ND G3UYK
SO22 6BA M0BTR
SO22 6BA M3KER
SO22 6EU G0BES
SO22 6FE G4FKR
SO22 6HE G3YSK
SO22 6HE M0AYY
SO22 6JH G0EBK
SO22 6LT M1RGW
[illegible] M6IRI
SO22 6SG G4AXO
SO23 0NU G4AXY
SO23 0PP G0IHA
SO23 0PX G4AZU
SO23 7DA M3KMT
SO23 7EQ G3ROG
SO23 7HT G1HRA
SO23 7ND G0TAH
SO23 7ND G1NCM
SO23 7QQ G4DKH
SO23 7TP M6SEE
SO23 7XH G3HQX
SO23 9DT G1EHE
SO23 9RG M3HND
SO23 9SR G4LYL
SO24 0DU G4YDE
SO24 0DU G6VXR
SO24 0DY G7JKY
SO24 0HP G4LQW
SO24 9HH 2E0CGS
SO24 9HH M3ETH
SO24 9NJ G8UDZ
SO24 9PE G6SDC
SO24 9PL G0JHT
SO24 9PP G4DMG
SO24 9RP G8JCB
SO24 9SG G7DNF
SO24 9TX G7JTR
SO3 4BA G6XIR
SO3 4NH G0PDL
[illegible]
SO30 0FD 2E0YAW
SO30 0FD M6MZZ
SO30 0GR 2E0JVM
SO30 0LZ M3TZF
SO30 0PH G4JNT
SO30 2AX M0JWA
SO30 2BJ M0HBC
SO30 2NY G3YOM
SO30 2RU G1PVD
SO30 2UD M3UVX
SO30 2UQ M6RGQ
SO30 3EQ G3IBY
SO30 4FT G3RQF
SO30 4HE M6CNQ
SO31 1AA 2E0DKT
SO31 1AA M0IAX
SO31 1AA M6FKY
SO31 1AA M6FLJ
SO31 1AD M6GXV
SO31 1AP G3PDP
SO31 1BJ G3SED
SO31 1DA G4WED
SO31 1EA G3YSG
SO31 1EA G8YUO
SO31 4HD M1ABG
SO31 4QG M6BQB
SO31 4RT G6IXE
SO31 5FE G0WSB
SO31 5GR G0HAE
SO31 6EP G1WXT
SO31 6EP G4UDY
SO31 6LJ M6DNN
SO31 6PE G0UTU
SO31 6PE G6CMF
SO31 6RN G0UUS
SO31 6RR G8HER
SO31 6SP G1GSY
SO31 6SY G4KWY
SO31 6WX 2E0IED
SO31 6WX M0IED
SO31 6WZ M3UUS
SO31 6XF 2E0HWG
SO31 6XF M6HWG
SO31 6XX G0TCQ
SO31 7BX 2E1IIC
SO31 7BX G0SCX
SO31 7BX M3IEQ
SO31 7BZ G0FBC
SO31 7EY 2E1FKT
SO31 7HQ G8ACL
SO31 8AT G8IOK
SO31 8AY G3MNJ
SO31 8AY G4NHN
SO31 8EL G0MYJ
SO31 8EN 2E0BXQ
SO31 8EN G6XQR
SO31 8EN M6LJD
SO31 8HE M6BGF
SO31 9FG G3HQT
SO31 9FU G3NNW
SO31 9GD M6ETZ
SO31 9HG G4PXH
SO31 9JU M6BPR
SO31 9TB M0UOO
SO31 9TB M3UOO
SO32 1AQ G3JMK
SO32 1BP 2E0NAJ
SO32 1BP M0NAF
SO32 1BP M6NAJ
SO32 1EW G4EWT
SO32 1EY 2E1IWG
SO32 1JR G1GXB
SO32 1JS G4NYD
SO32 1RQ G1JBG
SO32 1SG 2E1CXI
SO32 1SJ G4CFS
SO32 2AA M3FOV
SO32 2AR G4COM
SO32 2AR G4IDW
SO32 2BD G1UTC
SO32 2BD M1CCA
SO32 2DH G6HNJ
SO32 2LG M0GFM
SO32 2NP 2E0CSK
SO32 2NP M0HPU
SO32 2NP M6BZA
SO32 2NR G8KZO
SO32 2TL G4ZRT
SO32 2TR G6NXM
SO32 3LE M6CSA
SO32 3NP M6UKG
SO32 3QN G1PRP
SO4 4WQ G0ORV
SO40 2NA G0COC
SO40 2NL 2E0SJQ
SO40 2NL M0BJL
SO40 2NL M6SCX
SO40 2NW G6ANI
SO40 2RF G7HOK
SO40 2SD G7FOH
SO40 2UP G6MEW
SO40 2UY G0LDJ
SO40 2WD G3GLW
SO40 3BN G6XYF
SO40 3HP G0BOC
SO40 3HP G0OSD
SO40 3LP G4VII
SO40 3NX 2E0WHH
SO40 3NX M3WHH
SO40 3QP M6TIY
SO40 4SL G1KSE
SO40 4UT G4IXE
SO40 4UT M3YZJ
[illegible]
SO40 7AL G3VSL
SO40 7AL G0UPA
SO40 7AT 2E0CZI
SO40 7AT M0RHO
SO40 7AT M6MQD
SO40 7AT M6RFG
SO40 7AY 2E0GJE
SO40 7AY M6GJE
SO40 7GP G4BIZ
SO40 7GY G0FOH
SO40 7JH G4YWZ
SO40 7LA G0VNI
SO40 7LA G0WIL
SO40 8JB G4PGW
SO40 8US G1LGY
SO40 8WB M0DYA
SO40 9AX M3PFF
SO40 9BT G0PGZ
SO40 9DR M0WSR
SO40 9DR M3WSR
SO40 0LZ G0WFQ
SO40 9LZ G3SOU
SO40 9LZ G4SJW
SO40 9LZ G8FAB
SO40 9UZ G8KYK
SO41 0GG G4TOG
SO41 0GP 2E0KGC
SO41 0GP M0GTL
SO41 0GP M6KGC
SO41 0GU G3GRV
SO41 0HG 2E0DCD
SO41 0HG M0LMB
SO41 0HG M6BOF
SO41 0HQ G1RAX
SO41 0HZ G3JAF
SO41 0JD G4HYW
SO41 0LQ G4RFP
SO41 0PB G7JWV
SO41 0PP G4LUN
SO41 0PR M0LKD
SO41 0QT M3OBU
SO41 0QU 2E0NHS
SO41 0QU M3NHE
SO41 0QW G4DNE
SO41 0RL 2E1JLC
SO41 0RR 2E0VDM
SO41 0RR M0VDM
SO41 0RR M6BOJ
SO41 0SG G3KLH
SO41 0TE G1UFL
SO41 0TE M3OTU
SO41 0TF G3ZJY
SO41 0UL G4BMC
SO41 0UX G2HCG
SO41 0XG G8YFK
SO41 0XS 2E0DAY
SO41 0XS M0VBD
SO41 0XS M3SJD
SO41 0XS M3VBD
SO41 0ZB G8FAS
SO41 0ZG M6WOK
SO41 3PB G0COE
SO41 3SD G0TXN
SO41 3TF M6NGW
SO41 5PA M6CJG
SO41 5PA M6LYN
SO41 6AB G1OCR
SO41 6AL G0PCW
SO41 6AX G7OGT
SO41 6BB G4SBB
SO41 6BQ G0SNK
SO41 6DH M3ZIN
SO41 6DY G4XNA
SO41 8BD G1KFH
SO41 8BG G3ZIE
SO41 8DN G4ZAX
SO41 8DT M3ECF
SO41 8DY G0BCS
SO41 8DY G3TEI
SO41 8DY G4UAA
SO41 8ES G8KTV
SO41 8GQ M6GKH
SO41 8HN M6DPY
SO41 9BL M0ZDU
SO41 9BP M0NEC
SO41 9DX 2E0UXJ
SO41 9DX M0UXS
SO41 9DX M3UXS
SO41 9GT M6VCC
SO41 9LB G0JZW
SO41 9LB M0ULD
SO42 7TT G4FDX
SO42 7WQ 2E0DGT
SO42 7WQ M6DPN
SO43 7AA G6JYN
SO40 7DN M3FQM
SO43 7DN M3FQN
SO43 7JB G3NOA
SO43 7JP M0FWT
SO45 1BH G1RET
SO45 1BH G6UAN
SO45 1BJ G0HDI
SO45 1BN G7KDN
SO45 1BN M6STA
SO45 1BR G4CLF
SO45 1DA G0EUJ
SO45 1EG G6LVJ
SO45 1FH 2E0TVZ
SO45 1FH M3TVZ
SO45 1FX 2E0WPH
SO45 1FX M0PMH

Postcode

**IMPORTANT NOTE**

**Revalidate licence to avoid revocation** – Ofcom has advised the Society that plans will be drawn up to revoke licences that have not been revalidated as required by the licence conditions. The quickest way to revalidate is to do so online via the Ofcom website: *https://services.ofcom.org.uk/* or by email: *amateur.validations@ofcom.org.uk* If you need assistance in the process, Ofcom staff are available to help, but please be patient during times of heavy workload.

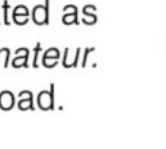

SO45 1FX M6PPH
SO45 1WF G6NXV
SO45 1WJ G7RUH
SO45 1WX G0OSG
SO45 1XD G6XWZ
SO45 1XP G3TPV
SO45 1XW G0BPA
SO45 1XW G6TGB
SO45 1YN G1UEV
SO45 1YX G6UMT
SO45 2EY G7AYA
SO45 2HP G0SBV
SO45 2JH 2E0GYP
SO45 2JH M6JAS
SO45 2JT G6MNL
SO45 2JT G6XMA
SO45 2JZ G4KNN
SO45 2NQ G4JYN
SO45 2NQ G4YVY
SO45 2PP G4SBF
SO45 3DJ M6GAG
SO45 3GB 2E0CXQ
SO45 3HJ M0GZB
SO45 3LG G6MZF
SO45 3LG M0LCC
SO45 3LJ G1UDW
SO45 3LJ M6CRA
SO45 3LS G3KSF
SO45 3QF 2E0MBG
SO45 3QF M6PWT
SO45 3QG 2E0GAF
SO45 3QG M6GDS
SO45 4BY M6LFO
SO45 4HS G3OZT
SO45 4HU M6HYT
SO45 4LE G0WCB
SO45 4LR G8CLK
SO45 4LS 2E0JHD
SO45 4LS M0IKT
SO45 4LS M6JHD
SO45 4RH G7GLH
SO45 4RP G3KCD
SO45 5AU G0OSP
SO45 5BP G8APM
SO45 5DL G1JRU
SO45 5EL G7AFT
SO45 5ER G3NJG
SO45 5FP G3WJM
SO45 5FQ G1UBN
SO45 5QQ G0WSI
SO45 5QS G4SGJ
SO45 5QW G6MUX
SO45 5TU 2E0PBN
SO45 5TU M6BPS
SO45 5TW G4NOP
SO45 5UD G1RCW
SO45 5UW G4GCI
SO45 6AY G4JAX
SO45 6BN 2E0EXC
SO45 6BP G6DLJ
SO45 6DL G4MOL
SO45 6DW G8BMQ
SO45 6EW M3ZNJ
SO45 6JT G4BPN
SO45 6LA 2E0BSQ
SO45 6LA M0GMD
SO45 6LA M6EMG
SO50 4BX 2E0DMW
SO50 4BX M3DMW
SO50 4LW 2E0IOU
SO50 4NS M6PDR
SO50 4NZ M3NSH
SO50 4PP G0TBC
SO50 4PP M1FDO
SO50 4PQ G0OFX
SO50 4QY G3VHW
SO50 4RJ M3HBG
SO50 4RW G3NML
SO50 5AD M0MTN
SO50 5AD M3TYM
SO50 5AS M3FEL
SO50 5EN G4KMP
SO50 5EN M6BBH
SO50 5GD M6NCE
SO50 5HA M5MDH
SO50 5HZ M1CNK
SO50 5NE M3DCJ
SO50 6AY 2E0MDK
SO50 6AY M0HHX
SO50 6AY M6IDE
SO50 6BD G1TJH
SO50 6DG G3VRB
SO50 6FX 2E0GSC
SO50 6FX M0OHI
SO50 6FX M3OHI
SO50 7HB G3YGF
SO50 7JJ M1MAB
SO50 7NY 2E1EEK
SO50 8AA M3VIA
SO50 8AG M1RST
SO50 8FL G3XIV
SO50 8HJ 2E0DHG
SO50 8ND M1FCF
SO50 8PF M6FCW
SO50 8PH G8VQA
SO50 8PU M3RBU
SO51 0AX G8IPG
SO51 0GQ G0RLA
SO51 0HG M3ECD
SO51 0JL 2E0CHL
SO51 0JL M6ASP
SO51 0JP M0TOG
SO51 0LF G4GSK
SO51 0LW G1GYT
SO51 0RA G8LLJ
SO51 5PQ G1TWW
SO51 5PY G0OGW
SO51 5RJ G6FVZ
SO51 5SP G7VZY
SO51 5ST G0IVR
SO51 5ST G4EOW
SO51 5ST G6IVR
SO51 5SW G8JNJ
SO51 6AR G8DXF
SO51 6BT 2E0GHZ
SO51 6BT M0XIG
SO51 6EE G0FHC
SO51 6EX G3NAE
SO51 6FU G0DZU
SO51 6FU G0HLA
SO51 7HU M0BTZ
SO51 7JW M3ECU
SO51 7JY G6CPE
SO51 7JZ G8AUJ
SO51 7JZ G8OFX
SO51 7LE G4ZCD
SO51 7LH G3FWI
SO51 7LL G4YVE
SO51 7RR M3ECJ
SO51 7RU G4MNP
SO51 7RU G8ZMM
SO51 7TJ G1NTN
SO51 8BR 2E0CGR
SO51 8DF G3MDR
SO51 8DP G8OQP
SO51 8DY M3ECQ
SO51 8EQ G6GCI
SO51 8FJ M6BWB
SO51 8FN G2DSY
SO51 8HG G7OOF
SO51 8LH G4XBZ
SO51 8JT G4YEE
SO52 9DR 2E0BHY
SO52 9EU G6TRW
SO52 9FD 2E0DJU
SO52 9FD M6EPA
SO52 9GB 2E0KWC
SO52 9GB M3YKC
SO52 9GB M3YMC
SO52 9HF M1FJA
SO52 9JG G3XSD
SO52 9JT G7AJR
SO52 9NS M1PVF
SO53 1GD G4DZS
SO53 1LE G7UDM
SO53 1LN G0UKB
SO53 1LN M0ACL
SO53 1LN M1LMJ
SO53 1NA 2E0KCO
SO53 1NA M0KCO
SO53 1NA M3KCO
SO53 1SA M3FFO
SO53 1SA M3HOE
SO53 1SF M3YJT
so53 2ay G1UDR
SO53 2BP 2E0SIM
SO53 2BP M3ALZ
SO53 2DJ G4EZC
SO53 2FE G8ZBN
SO53 2FY M0AYU
SO53 2GX G7AJE
SO53 2LJ G3UYD
SO53 2LL M6SRD
SO53 2PU M0MEH
SO53 3BS G4BGT
SO53 3BS G6JGR
SO53 3DX G0TSM
SO53 3EB G4LDC
SO53 3GD G0TAA
SO53 3GJ G6AMW
SO53 3PA 2E0GBT
SO53 3TP G7POC
SO53 4HP G6MCX
SO53 4HQ 2E0FMG
SO53 4HQ M6FMG
SO53 4HU M3SKT
SO53 4LD G7VLH
SO53 4LN 2E0NIF
SO53 4LN M0NIF
SO53 4QF G7BOH
SO53 4SW M3SEE
SO53 4SW G0DQH
SO53 4TT G3WII
SO53 5AT G4MZU
SO53 5AX G4HHZ
SO53 5BX 2E1ECV
SO53 5PA G4HWM
SO53 5PB G8LVC
SO53 5QD 2E0AJP
SO53 5QP G6IQI
SO53 5RP G1NPI
SO53 5RP G8IPQ
SO53 5RP M3IPQ

## SP (Salisbury)

SP1 1NU G0DJV
SP1 1PU M0NJP
SP1 1PU M3VOW
SP1 1QL G3MXP
SP1 1SA G2AQJ
SP1 2JH G8WJB
SP1 2JX 2E1MEL
SP1 3AU G4YFJ
SP1 3BH G7DNG
SP1 3DN G6CZB
SP1 3HS G4MOE
SP1 3JN M0DYV
SP1 3JZ G6BBG
SP1 3LW G4YVM
SP1 3PR G7WAA
SP1 3PR M0CLI
SP1 3PZ G4TRD
SP1 3RT M6DPX
SP10 1HH G4ABL
SP10 1PL G0UJW
SP10 2AH G0OMD
SP10 2BU 2E0BZJ
SP10 2BU M3CEB
SP10 2EN G8CPJ
SP10 2HE G4YSB
SP10 2PZ M0ALE
SP10 2QT G7PVE
SP10 2RB G2FSJ
SP10 3BN G6FIB
SP10 3BU M3FWS
SP10 3DL 2E0EPR
SP10 3DL M6ESP
SP10 3DS G6YEY
SP10 3EP G7VRJ
SP10 3HL G0JBY
SP10 3JT M0BMF
SP10 3JY 2E0DUM
SP10 3JY M6IVN
SP10 3PE 2E0CLW
SP10 3PE M0VZS
SP10 3PE M3VZS
SP10 3SS M6GAB
SP10 3TD M6SPC
SP10 3UH M3SVC
SP10 3XD M6DQN
SP10 4AR G1HPS
SP10 4EN G0AMO
SP10 4EN M3DEA
SP10 4EZ M1EYS
SP10 5DB G7CFX
SP10 5HT G8CEP
SP10 5NA G6ARC
SP10 5NA G8NDN
SP10 5NJ 2E0NTJ
SP10 5NJ M6NTJ
SP11 0QF M6GNC
SP11 0QT G8ARH
SP11 0RE G6SDE
SP11 0RS G8FHI
SP11 6EA G3LTF
SP11 6FD G6VTA
SP11 6JY G6SJG
SP11 6RH G4XYG
SP11 6RR M6AIL
SP11 7BJ 2E0LBK
SP11 7BJ M0LBK
SP11 7BJ M3LBK
SP11 7BN G4FYM
SP11 7BX G6YTY
SP11 7EH G3XVS
SP11 7ER G4OZL
SP11 7ER G6WBG
SP11 7ES G4WEN
SP11 7JA G4YHQ
SP11 7LL M1CFW
SP11 7RN 2E0MUZ
SP11 7RN M0SWT
SP11 7RN M6MUZ
SP11 8LG G6JSF
SP11 8NE G8GYS
SP11 8NS G3ZWN
SP11 9AN G4NNS
SP11 9JD G4CIZ
SP11 9PG G1XSV
SP11 9SJ M6RQE
SP2 0AX G4WDA
SP2 0LW G4RLF
SP2 0NF G4SXQ
SP2 7BY G7MJV
SP2 7BY M3YCV
SP2 7DD M0BBR
SP2 7EX G8CWJ
SP2 7SU G0MLJ
SP2 7TW G0EET
SP2 8DD G4DUA
SP2 8DL M6ZCM
SP2 8DN G0BDP
SP2 8DU G8WBO
SP2 8EF M0BWC
SP2 8EP G3KOZ
SP2 8FE G0VMC
SP2 8LY G7BSO
SP2 8NR G3OIL
SP2 8PB G0TGW
SP2 8QS G4HQM
SP2 8RN G4LWU
SP2 8RN G6RLG
SP2 9BY G8NDV
SP2 9EJ G7EXZ
SP2 9HJ M6THJ
SP2 9HN 2E1MEP
SP2 9LD M3EQW
SP2 9LD M3UED
SP2 9LF G1BMT
SP2 9LU M0VSR
SP2 9LU M0WSC
SP2 9PA G1DGL
SP2 9PE M0RED
SP2 9PJ 2E0LWT
SP2 9PJ M3RMU
SP3 4AH G3KTM
SP3 4BQ M3VBM
SP3 5JL G4FRB
SP3 5PE M3BIZ
SP3 5PW M0TRB
SP3 6EE G3UJU
SP3 6JL M0ISW
SP4 0AA G6ABM
SP4 0BX G4NWJ
SP4 0BY G4AEI
SP4 0EP G7WKW
SP4 0ER G8ALR
SP4 0ER M0BVO
SP4 0LE 2E0XKO
SP4 0LE M0XKO
SP4 0LE M3XKO
SP4 6AU G8HUV
SP4 6DJ G6JGP
SP4 6EE 2E0LZE
SP4 6HW M0MWS
SP4 6LS G8OFA
SP4 6LS G8RHC
SP4 6LX G0HSV
SP4 6NB G1SSZ
SP4 6PX M6FHS
SP4 7AD G4WPI
SP4 7AS G6YUX
SP4 7EE G3YFE
SP4 7FS M6FQL
SP4 7HW M6CCN
SP4 7NN G4UED
SP4 7PJ 2E1CMZ
SP4 7PJ G4SXR
SP4 7PP G4SXT
SP4 7QS G6UXK
SP4 7RE G7GKD
SP4 7RG G1WRY
SP4 7RG G4YRV
SP4 7TU M3ZYR
SP4 8AB M1AFF
SP4 8DB 2E0HKC
SP4 8DB G0HKC
SP4 8DB M6HKC
SP4 8DJ G6JRS
SP4 9QG M6NIQ
SP4 9QZ 2E0TEF
SP4 9QZ G3TBL
SP4 9QZ M6TEF
SP5 1JJ G0RTN
SP5 1PL M3VKB
SP5 1PP G4YIS
SP5 1RB G4LDR
SP5 1RE G8PBY
SP5 1RJ M6WFH
SP5 1RS G1ITJ
SP5 1SH 2E0KDI
SP5 1SH G3MDM
SP5 1SZ 2E0RFL
SP5 1SZ M6HBN
SP5 2AL G4VQJ
SP5 2AW M0GWQ
SP5 2BP M6KMZ
SP5 2BZ G7WBA
SP5 2ED G0OSW
SP5 2HT G6ILC
SP5 2HU G3OSQ
SP5 2JA 2E0HMC
SP5 2JA M0JJC
SP5 2JA M6DYU
SP5 2JA M6HAU
SP5 2LN G0RZB
SP5 2LQ G8UXW
SP5 2LQ M3HCB
SP5 2LR M0CBM
SP5 2LR M3DZN
SP5 2NQ G3SHK
SP5 2NR G4TEK
SP5 2PR G3NZK
SP5 2SE G4KCM
SP5 2SZ M0GSP
SP5 3AR G1ELZ
SP5 3AT G3YWT
SP5 3AZ G0CDZ
SP5 3LB 2E1TON
SP5 3LB G6ZHJ
SP5 3LG M3MGD
SP5 3LX G4NKP
SP5 3QH M6TOW
SP5 3TH 2E0HEA
SP5 3TH M3FQG
SP5 4JT 2E1NAC
SP5 4JT 2E1PEC
SP5 4LH M6HMM
SP5 5BH G6FPC
SP5 5ED G4UNG
SP5 5NZ M6OVI
SP5 5QJ 2E0DZJ
SP5 5QJ M6HFO
SP5 5QL G7IRU
SP5 5QL M3CAQ
SP5 5SA M6CPO
SP6 1AY 2E0KZC
SP6 1AY M0KZC
SP6 1AY M3KZC
SP6 1BJ M3HZD
SP6 1BP G7OZH
SP6 1BW 2E1FRY
SP6 1EG M3UQA
SP6 1EQ G6CEZ
SP6 1JF M0WDZ
SP6 1LW G4POF
SP6 1NW G4IWU
SP6 1NW M3NMX
SP6 2AX G8OFO
SP6 2HT G4LTR
SP6 2LJ G1MVG
SP6 2NR G4GBP
SP6 3DJ G0TQZ
SP6 3EE G7VOX
SP6 3EP 2E0RAH
SP6 3EW G8BIH
SP6 3HA G1VOQ
SP6 3LD 2E0RFK
SP6 3LD M3RFK
SP6 3RB G1JMH
SP6 3RB G6LVI
SP63bl 2E0CFP
SP63bl M6AFN
SP7 0BJ M3HEV
SP7 0DP G7ESE
SP7 0DP M6DGS
SP7 0HS M3SQI
SP7 0LB G0MFR
SP7 0NY G0USF
SP7 0NY G7SFL
SP7 8AL G0YBU
SP7 8EY G3XNK
SP7 8HX G3LGF
SP7 8LQ G8BSP
SP7 8NF G3BVB
SP7 8NQ G3FWU
SP7 8QS M0TMO
SP7 8QX M0CBI
SP7 8RX G4OVJ
SP7 9HD G8LBG
SP7 9HZ G0UOD
SP7 9NX G1RJA
SP7 9PA 2E1EGU
SP7 9PE G4CK
SP7 9PF G1NWT
SP7 9QB M3JLH
SP8 4EL G4FUY
SP8 4EU G3ZBB
SP8 4GN M6RFI
SP8 4HH G0VNA
SP8 4LZ G4OVI
SP8 4NR G8PQJ
SP8 4PE G7SUS
SP8 4QJ G4ILM
SP8 4RB G8DDN
SP8 4RR G4VRM
SP8 4SS G1SNI
SP8 4TW G4FQV
SP8 4UP G0YLO
SP8 4UP G0ZEE
SP8 4UP G6TER
SP8 4UP M3UZE
SP8 4WE M6ROA
SP8 5AL G3LCL
SP8 5EL M0AKI
SP8 5ET 2E0RIH
SP8 5ET M6JGH
SP8 5EW G4FDI
SP8 5JY G3NRH
SP8 5LB G0MPM
SP8 5LW G8SYA
SP8 5NB G1THG
SP8 5QR M5BJC
SP8 5RL G7JIF
SP8 5RN M3TBH
SP8 5SJ G8HPN
SP9 7FN M6ZBW
SP9 7JX M6YSD
SP9 7SB M6NHP
SP9 7TR 2E0DUJ

## SR (Sunderland)

SR1 2DP M3XYH
SR2 0BP G4DGB
SR2 0JU G0SLN
SR2 7EZ G6GGT
SR2 7HB G7FFV
SR2 7RX M3REP
SR2 7TS G4GTX
SR2 8NF M0MDP
SR2 8QB G7PQD
SR2 8QT 2E0ZDM
SR2 8RS G6LMR
SR2 8RX G6INK
SR2 9BB G0KFY
SR2 9DU M0SWD
SR2 9DX G0ASM
SR2 9EE G4MSJ
SR2 9EJ G4TOI
SR2 9HQ G3YJG
SR2 9LQ G0KVJ
SR2 9LQ G0NSK
SR2 9LQ M6NSK
SR3 1AN G4HPS
SR3 1AN G4OBX
SR3 1HJ G0BNK
SR3 1HJ M0IMD
SR3 1HJ M3HYD
SR3 1JF G0TAX
SR3 1JW 2E0PYN
SR3 1JW M6AVG-
SR3 1LG G7KOF
SR3 1LX G4MTW
SR3 1QS M1CQX
SR3 1UJ G0BWJ
SR3 2RF M0NGB
SR3 2RG G6SGZ
SR3 3AL M1DPQ
SR3 3AN M6JIH
SR3 3PX G3ZOG
SR3 3SF M0ATA
SR3 4AA G0AOE
SR3 4AT 2E0HHE
SR3 4BG M3UKJ
SR3 4DJ M6IOL
SR3 4EE M6HAM
SR3 4EZ G4VLT
SR3 4JN 2E0PAX
SR3 4JN M6UVF
SR3 4PA G7PUK
SR4 0AE G4HIX
SR4 0BA M0KLL
SR4 0DD G0MCT
SR4 0HN 2E0EGR
SR4 0HN M0XGR
SR4 0NA M3XYP
SR4 0QZ 2E0UKA
SR4 0QZ M3XJX
SR4 6AU M3JFB
SR4 6BA 2E0PSH
SR4 6BA M0HTI
SR4 6BA M6DVV
SR4 6EY M6MOB
SR4 6XG G7MXM
SR4 7LL M6RNQ
SR4 7RY M0GFN
SR4 7SA G1YUL
SR4 7SU M1CPB
SR4 7TB M0RKW
SR4 8AU G0EHX
SR4 8BG M6GNX
SR4 8BS 2E0YDT
SR4 8HT G8HPW
SR4 8JB G3WOM
SR4 9DW G1WEV
SR4 9EN G8NPP
SR4 9NQ G0SRG
SR4 9NQ G4EKM
SR5 1DY G4XYP
SR5 1LG 2E0OAP
SR5 1LG M6OAP
SR5 1LL M6UNS
SR5 2BU G6VTH
SR5 3PG M0GOL
SR5 3RE M5GHT
SR5 4BU G1CSA
SR5 4DF G0OBN
SR5 4LA M3XAW
SR5 4LH M6MMY
SR5 4ND M0KAE
SR5 4PD G7SZB
SR5 5AH 2E0HYE
SR5 5AH M0HYE
SR5 5AH M6BYE
SR5 5AW 2E0SNE
SR5 5AW M6FIT
SR5 5ET M6CWQ
SR5 5LH M6ELL
SR5 5QA G0UNE
SR5 5RD G7CKQ
SR5 5RD M3XQB
SR5 5SA M6OKH
SR6 0AQ G0WKZ
SR6 0JP G0BAN
SR6 0NL M6LND
SR6 0NT G4TMX
SR6 7AL G7PHI
SR6 7BW 2E0TLX
SR6 7BW M0TLX
SR6 7BW M3TLX
SR6 7LN 2E0CUC
SR6 7XE G8PDE
SR6 8BD G3TEC
SR6 8ER 2E0ERD
SR6 8ER M6ERD
SR6 8ET G4MRK
SR6 9HJ G8EQB
SR6 9HP G0IID
SR6 9NT 2E0DJJ
SR6 9NT M0KCF
SR6 9NT M6GPM
SR7 0AN G8TDP
SR7 0BD G6AGY
SR7 0BD G6AGZ
SR7 0JT G0MJV
SR7 0LP G0WTW
SR7 7BQ G4GSO
SR7 7DJ G0MKC
SR7 7DJ G4PPL
SR7 7EU G6LNL
SR7 7NG 2E0WRY
SR7 7NG M6WRY
SR7 7SA G1ERU
SR7 7SA M3SMM
SR7 7UF M6OYZ
SR7 8DG G0DVP
SR7 8DZ G4NMF
SR7 8EE M6GTQ
SR7 8HW 2E0JRB
SR7 8JG M6EQG
SR7 8JT 2E0SNM
SR7 8JT M6OOB
SR7 8JT M6SMA
SR7 8JZ G0UFN
SR7 8NL G0NWY
SR7 8PY M6DTJ
SR7 8RS M0MED
SR7 8RZ M6SQC
SR7 9PQ G4NMK
SR8 1DD G0FBW
SR8 1DG 2E0HYG
SR8 1DG G0CWF
SR8 1LN G0TOK
SR8 1LP G4PLU
SR8 1PZ M0GZW
SR8 2AN G6NXP
SR8 2DJ G4WKT
SR8 2DT G4RXR
SR8 2EA G0RDD
SR8 2HB G6VHG
SR8 2HE M6HWN
SR8 2JS M0WUS
SR8 2JS M3AUP
SR8 2JW 2E1DMI
SR8 2NN G0NDD
SR8 3AJ G6RKS
SR8 3DF G7DBV
SR8 3HU G3ZMG
SR8 3SA M6PQD
SR8 4BW G0NXC
SR8 4EN M0IGR
SR8 4NE M0TEN
SR8 4RD 2E0YZX
SR8 4RD M6TMG
SR8 5EG M3OTM
SR8 5PF M0BLI
SR8 5QZ G3NSI
SR8 5RS G0MKE
SR8 5UD G4RVY

## SS (Southend on Sea)

SS0 0AN 2E0SET
SS0 0AN M6NFW
SS0 0EU 2E0SCE
SS0 0EU M6BMN
SS0 0EU M6CSE
SS0 0LE G1POV
SS0 0NL M6ZPS
SS0 0NY G6CVB
SS0 0QL G6WCI
SS0 0RA G0ENN
SS0 7AZ M1FIB
SS0 7DR G1PVT
SS0 7DR G6AXE
SS0 7LA G0MBP
SS0 7PU G2BBI
SS0 8BD M0DRS
SS0 8BD M3VNS
SS0 8BN G4DVJ
SS0 8BS M6HJC
SS0 8NL G0FNB
SS0 9JN G1KPV
SS0 9PR M0INP
SS0 9RA G8IWI
SS0 9SU G3MJN
SS0 9SZ 2E0TWL
SS0 9SZ 2E0TWW
SS0 9SZ M0TWL
SS0 9SZ M0TWW
SS1 1HG G0RSW
SS1 1NH M3XMZ
SS1 1QB M1KES
SS1 1QE G0VCY
SS1 2QN M3OLW
SS1 2RP G1VJJ
SS1 2SX G1HNN
SS1 2SX G4UAI
SS1 2SX G4UTF
SS1 2TZ M0KUR
SS1 3AD G1IQU
SS1 3DF G6XNK
SS1 3DF G8LYW
SS1 3DG G8ZPO
SS1 3HD G1JLG
SS1 3LE M6MEO
SS1 3NW G0TIG
SS1 3PX G4YAK
SS1 3QU G3YVK
SS1 3SS G6TLN
SS11 7DG M6MVA
SS11 7DN G7DAH
SS11 7EH G4FKX
SS11 7HX M0JRW
SS11 7JE G8BPW
SS11 7LN G0DZQ
SS11 7ND 2E0CUU
SS11 7ND M6CHI
SS11 7PF G1EUM
SS11 7PP 2E0CNA
SS11 7PP M0PSD
SS11 7PP M6BMI
SS11 7PT G1HHS
SS11 7PT G1LCS
SS11 8QT G4GUJ
SS11 8QT G8OIV
SS11 8TZ M3IVD
SS11 8XB 2E0DVF
SS11 8XB M6KOA
SS12 0AR G3IOI
SS12 0AX 2E1FZO
SS12 0AX M1CKA
SS12 0DX M1DFC
SS12 0EN G7MBY
SS12 0ER G3LRL
SS12 0HE G3PZZ
SS12 0HS G7HGI
SS12 0QB 2E0KGT
SS12 0QB M0ICG
SS12 0QB M6GTK
SS12 9DY G1KOR
SS12 9ED M3LHX
SS12 9EQ G4YIH
SS12 9FA G6WKN
SS12 9JX G7HCO
SS13 1BQ G6WLQ
SS13 1HA G7EIE
SS13 1JS 2E0FMS
SS13 1JU G6AJT
SS13 1NR G8WSS
SS13 1PJ M6ILM
SS13 1QU G0SIV
SS13 2BH 2E0RCF
SS13 2BH M3OBO
SS13 2EJ M6PMP
SS13 3EW G7JRK
SS13 3NB 2E1EOZ
SS13 3QH 2E1IJU
SS14 1LX G7PFG
SS14 1NF G1GYQ
SS14 1RS M6EWL
SS14 1UA G1PXQ
SS14 2DA M0TTF
SS14 2FH M6HSH
SS14 2JD G8CUN
SS14 2JQ G7KDR
SS14 2NS G1YKX
SS14 2TN G0UCH
SS14 3JN G0NPO
SS14 3LZ 2E0SSX
SS14 3LZ M0SSK
SS14 3NA G6SNN
SS14 3QS 2E0NCE
SS14 3QS G7OED
SS14 3QS M0LMR
SS14 3QS M6EBQ
SS15 4DE G0NGA
SS15 4EH G1FCW
SS15 4EH G4ZUL
SS15 4EH M6YYL
SS15 4EJ M6HKT
SS15 4EW 2E0TKV
SS15 4EW M3TKV
SS15 4JE G6TNE
SS15 5EA G4IWO
SS15 5FY M6HCV
SS15 5GT 2E0JPS
SS15 5RH G7UZX
SS15 5RH M1EAW
SS15 5YN G0RTH
SS15 5YN G7KCN
SS15 6AL G3UPM
SS15 6JR G1AUR
SS15 6PR G6LZX
SS15 6QB M6MRH
SS15 6RA M6UKJ
SS16 4NE G1CHV
SS16 4PQ G3VYF
SS16 4PQ M1BHC
SS16 4PQ M6GJN
SS16 4RU 2E0BRK
SS16 4TL G4EPU
SS16 5BN M1ECC
SS16 5HD G8KIH
SS16 5HX G4IEG
SS16 5JH G8OSG
SS16 5QQ 2E0KPR
SS16 5RA 2E0GQD
SS16 5RA M3GQD
SS16 5RB M6WXY
SS16 5RW 2E0LEZ
SS16 5RW M6LTO
SS16 5TW G7ALR
SS16 6AQ G1RCQ
SS16 6AQ G4NVT
SS16 6DU 2E0WKV
SS16 6DU M3WKV
SS16 6NA G6RAQ
SS16 6NU 2E0CTA
SS16 6NU M0ONZ
SS16 6NU M6ORC
SS16 6SD G7BNZ
SS16 6SJ G0PAE
SS16 6SJ G8JCV
SS16 6SP G1JGD
SS17 0BB 2E0TOL
SS17 0BB M0STO
SS17 0BB M6GRX
SS17 0BE 2E0EEB
SS17 0BE M6EEB
SS17 0DF G4HXY
SS17 0EF G0NGG
SS17 0NH G7JVB
SS17 7BD M0AKE
SS17 7BQ G4KIQ
SS17 7BZ M6RFW
SS17 7EW G1PQK
SS17 7EW M6XSZ
SS17 7HE M1BZR
SS17 7JU G0WKH
SS17 7LB G4JIU
SS17 7LD G8EXQ
SS17 7NH M6EBU
SS17 7QL G4LTH
SS17 7QZ G4LNT
SS17 7SB 2E0CJO
SS17 7SB M0LEH
SS17 7SS M6ITX
SS17 7SX 2E1EXA
SS17 8DD M1DKP
SS17 8DE G1ERQ
SS17 8DT G1YAH
SS17 8HL G7FEL
SS17 8NL G6WXZ
SS17 8NS G6DXP
SS17 8PN G8LWA
SS17 8QU G7DXV
SS17 9AJ G3SMF
SS2 4AZ M6IIL
SS2 4DH M0RLI
SS2 4ED G3UUI
SS2 4HR G0SKI
SS2 4JR G0FVB
SS2 4LJ G7JLO
SS2 4LU G0XOX
SS2 4LW G0TUI
SS2 4LZ M3IAC
SS2 4NN G1YCR
SS2 4NW G0UKP
SS2 4PA M6PEQ
SS2 4PX G0UXN
SS2 4QA G8OQR
SS2 4RD M3VHQ
SS2 4RD M6RDK
SS2 4SP G0DPC
SS2 5EJ M6LHG
SS2 5JG 2E0UEH
SS2 5JG M6UEH
SS2 6HB M6CSJ
SS2 6HR G1DXD
SS2 6LW M3YLB
SS2 6QP G8GYH
SS2 6QU G0KEI
SS2 6QY G8RAU
SS2 6RW 2E0FTX
SS2 6RW M6HRT
SS2 6SQ G1FEV
SS2 6TB 2E0VZL
SS2 6TB M3VZL
SS2 6TF G0TTI
SS2 6UA G0TZP
SS3 0AR G1HPV
SS3 0BE 2E0VNO
SS3 0BE M0VNO
SS3 0BE M3VNO
SS3 0DR G0NXN
SS3 0EX G1EXR
SS3 0EZ M3POW
SS3 0JG G7BLJ
SS3 0JS 2E0BJB
SS3 0JS M0GGZ
SS3 0JS M3SYC
SS3 0LS M6EDD
SS3 8AN G4ZMU
SS3 8BL 2E0DVX
SS3 8BL M6DVX
SS3 9AR G7CQA
SS3 9FA G0CXW
SS3 9JG M0KVA
SS3 9JG M3VAR
SS3 9JP G0PCF
SS3 9LR G7PMK
SS3 9LU M0ABA
SS3 9NT M3UFA
SS3 9NY G1HQQ
SS3 9PX M3YLN
SS3 9SB G3SCY
SS3 9SG G6BHE
SS3 9SG G7RYM
SS3 9SL G4CEU
SS3 9TL M0WBK
SS3 9TL M3WBK
SS3 9TN M6DNZ
SS3 9TP G4HTZ
SS3 9YE G8OPY
SS3 9YS G1XXV
SS4 1DP M1AHT
SS4 1EA G3CQL
SS4 1EH M6GZO
SS4 1GH M3UPQ
SS4 1JE G6HNI
SS4 1LH M3KLU
SS4 1QH G7MFX
SS4 1QH M6HLP
SS4 1QJ G8SGH
SS4 1RS G0KTC
SS4 1SE M6HPY
SS4 1SH G8GHK
SS4 1SH M1BHW
SS4 2ER G1HML
SS4 3AH G1RAP
SS4 3AQ 2E0TRF
SS4 3BX G7FFI
SS4 3DQ G6KFD
SS4 3EP M6JTP
SS4 3FJ 2E0KFB
SS4 3FJ M6KFB
SS4 3HE 2E0WHU
SS4 3HE M6WHU
SS4 3PY G1GCY
SS4 3PY G7PUZ
SS5 4BX G1RYY
SS5 4DL M0IDG
SS5 4DL M6IDG
SS5 4DN G1KHS
SS5 4EL M6HFP
SS5 4EY G0PEP
SS5 4EY G3OJV
SS5 4GN G1KVO
SS5 4JE M3IWJ
SS5 4JN G0DFE
SS5 4JY G0DTP
SS5 4NT G7WHZ
SS5 4PJ G4TWH
SS5 4QG G6LYM
SS5 4QS G6XYU
SS5 4QS M3JJS
SS5 4SW G4MCU
SS5 5AT G7GBN
SS5 5BP G0PEJ
SS5 5DY G4HKQ
SS5 5EE G7ENT
SS5 5EL G3ZJZ
SS5 5HD 2E1DOA
SS5 5HJ G3VQY
SS5 5HN G7WHP
SS5 5PF M6MVW
SS5 6AA G4KDH
SS5 6BG G0EBG
SS5 6BH G4ZFJ
SS5 6BN G7TBW
SS5 6DD G4RDS
SS5 6LR G4MUS
SS5 6LT G8RAN
SS5 6LU G0MGT
SS5 6LU G1OOG
SS5 6LZ G3HWM
SS5 6NE 2E0JTW
SS5 6NE M6BPK
SS5 6NE M6GEJ
SS6 7JR G1TWS
SS6 7LB G0DRV
SS6 7NS G4FSE
SS6 7QF G6DOF
SS6 7QH G8FAX
SS6 7RG G0DRH
SS6 7TD G0EFI
SS6 8AB G0MLO
SS6 8AR G0HSK
SS6 8AR G6XAT
SS6 8BN G3OGX
SS6 8BP G0FKS
SS6 8EB G4KDE
SS6 8HP G1KQU
SS6 8JP M6MKU
SS6 8LQ G7PPS
SS6 8LW G4GNU
SS6 8SG 2E0SIS
SS6 8SG M6WYZ
SS6 8UA G0WIX
SS6 8UY M0DDK
SS6 8YF G4KIH
SS6 9AL G4PWB
SS6 9EJ G7LAX
SS6 9JZ G4XEO
SS6 9LY G3ZHA
SS6 9ND M6KTW
SS6 9PD G6AVK
SS6 9PH G8LHW
SS6 9TU G8OXU
SS6 9UH G3TRH
SS7 1AL M0PGC
SS7 1DN G1BAR
SS7 1JL G4EZP
SS7 1JL G4JAR
SS7 1QB G4FCX
SS7 1SS G6MVW
SS7 2BG 2E0XLM
SS7 2BG M0SCX
SS7 2DL G4WAK
SS7 2HA M1EGP
SS7 2JP G3UTA
SS7 2LN G3LUZ
SS7 2LP G3SRA
SS7 2ST M0AOK
SS7 2TY G4ZMN
SS7 3BB M0DQK
SS7 3DU G1KJG
SS7 3HE G7MPH
SS7 3LD M6RMU
SS7 3NA G8FAR
SS7 3PL M0PYD
SS7 3RJ G1FBZ
SS7 3SG M6MQJ
SS7 3TU G0TTM
SS7 3TU G5QK
SS7 3UU 2E0TCY
SS7 3UU G8UWI
SS7 3YL G4AJY
SS7 4DT G6IDU
SS7 4EE G4TPK
SS7 4EF G4UMP
SS7 4EN G8WUU
SS7 4LA G6TKW
SS7 4LS G1ZHN
SS7 4NR 2E0YPK
SS7 4NT G7EGU
SS7 4NW M3BFB
SS7 5BQ G8EOM
SS7 5DT G1GBV
SS7 5EN G0EAG
SS7 5ES G4GDS
SS7 5JQ G8VFI
SS7 5LH G7IIO
SS7 5NU G6ZUE
SS7 5PH G8YPK
SS7 5RD G4PBO
SS7 5RL G0LYX
SS7 5SJ G1LAW
SS8 0BB G7MXL
SS8 0BP G0JXJ
SS8 0BP G0LTO
SS8 0BU G1BGJ
SS8 0DN G7HQF
SS8 0EP G4RSE
SS8 0EP G4UVJ
SS8 0EP G6RSE
SS8 0EW 2E0LFX
SS8 0EW M6LFX
SS8 0EX G0WAX
SS8 0EX G0WGA
SS8 0GX M6GTV
SS8 0JB M6EYW
SS8 0JB M6XXI
SS8 0JG G4MAN
SS8 7DN M6SDX
SS8 7EA G1FBW
SS8 7EE G7TFA
SS8 7EH M3MSP
SS8 7EL G1IEO
SS8 7HL G1ZSG
SS8 7JD 2E0FME
SS8 7JD M3FME
SS8 7PB M3IKI
SS8 7RD M6HCT
SS8 7TJ G0KSC
SS8 7TS G4BQF
SS8 8HX G0JAN

SS8 8HX G1OXO
SS8 8LG M6KEB
SS8 8LG M6UCY
SS8 8NZ G0THG
SS8 8NZ M0FZW
SS8 8NZ M6FZW
SS8 9AB G6CNF
SS8 9BL G7PDU
SS8 9DJ G4ZQJ
SS8 9DS G8ZTB
SS8 9DZ G4YQJ
SS8 9EJ 2E0EVZ
SS8 9EJ M6EVZ
SS8 9HB G6LUO
SS8 9HL 2E0IEO
SS8 9HL M0IEO
SS8 9HL M6CXS
SS8 9HL M6JNX
SS8 9HL M6NFR
SS8 9LP 2E0NNQ
SS8 9LP M3NNQ
SS8 9QL G6SPH
SS8 9QP G0NCT
SS8 9RQ M6HDO
SS8 9TW 2E0RMT
SS8 9TW M6RKC
SS8 9YB G1SOB
SS9 1HQ M0BDB
SS9 1HQ M1ANK
SS9 1PT 2E0JTQ
SS9 1PT M0JTQ
SS9 1PT M6JTQ
SS9 1RU G0KMF
SS9 2DT M3KSS
SS9 2HT G0JAO
SS9 2JX G7CDO
SS9 2NL G3MVU
SS9 2PB G4SNV
SS9 2PX G0JYI
SS9 2QS M1DKL
SS9 2SY M3UCH
SS9 2UW M3YSV
SS9 2XD G4OQ
SS9 3BD G0KVZ
SS9 3EB M6TGC
SS9 3JT G0JDD
SS9 3LF 2E0MMH
SS9 3LF M6FFB
SS9 3NT G3KGN
SS9 3NT M0SXA
SS9 3PJ M6FSU
SS9 3QF 2E0IMZ
SS9 3QF M6IMZ
SS9 3QS G1GXF
SS9 3SG G0GBY
SS9 4AZ G6OAV
SS9 4DA G8WSW
SS9 4DE G0DCN
SS9 4DE G0FPV
SS9 4HA G1YAF
SS9 4HA M6BMK
SS9 4HG G6GGZ
SS9 4NB G6MMR
SS9 4PW G6DAH
SS9 4PW G6NLS
SS9 4PY G4UQY
SS9 4QY G3PHL
SS9 4RY G3SVI
SS9 4SZ G1NCO
SS9 5AB G8CEX
SS9 5AR 2E0ESX
SS9 5AR M6LOS
SS9 5AX G1ZHM
SS9 5DN G4ENW
SS9 5EL G3RPZ
SS9 5HR M1RKB
SS9 5NF 2E0KPX
SS9 5NN G7EGX
SS9 5NW G6OLY
SS9 5NX G4KEG
SS9 5RF G7DTS
SS9 5SW G3RCX
SS9 5XR G6WFS

## ST

## (Stoke on Trent)

ST1 2DF M6LOZ
ST1 2NE 2E0YYY
ST1 2NE M6MMM
ST1 3BA G8ZES
ST1 3BA M0GAN
ST1 3GX M6MVK
ST1 3HS 2E1SOB
ST1 3NE G6ING
ST1 5QP M6CEQ
ST1 6BY G6KKN
ST1 6DD G1XLN
ST1 6DE G8NTG
ST1 6EF 2E0BLD
ST1 6EF 2E0PTI
ST1 6EF M3PTI
ST1 6EP M3PXU
ST1 6PQ M3GVT
ST1 6SL G4FMJ
ST10 1AT G8VSR
ST10 1DN G0ODH
ST10 1DP M1AMB
ST10 1DT G4SXK
ST10 1LU G6UEH
ST10 1LW G4DSQ
ST10 1LZ M3YWH
ST10 1PY G6GA
ST10 1QG G4CHG
ST10 1QQ G7PIJ
ST10 1QS G0TPY
ST10 1RU 2E0OEJ
ST10 1RU 2E1IHF
ST10 1RU G7MWS
ST10 1RU G7PBH
ST10 1RU M3MHL
ST10 1RU M3MIP
ST10 1RX M6EWV
ST10 1SA 2E0PBH
ST10 1SA M3PHZ
ST10 1XB G1MAD
ST10 1XB G4NHT
ST10 1XB G4OUG
ST10 1XB G4PGG
ST10 1XB G7PHB
ST10 1YU G0KQY
ST10 2AE G6FFU
ST10 2AG G6KTE
ST10 2DW M3GBA
ST10 2EG G6YIW
ST10 2JQ M6BME
ST10 2PA G3UNM
ST10 2PT G6YCN
ST10 3BW M3OFA
ST10 3EA G4OTX
ST10 3ER G1OGE
ST10 3HD G0DJQ
ST10 4AN 2E0TVW
ST10 4AN M6AVW
ST10 4BH G4YYO
ST10 4DZ G0VBT
ST10 4EG M1RKY
ST10 4FE 2E1FOW
ST10 4HL G8FHC
ST10 4LD 2E0WJE
ST10 4LD M3DPY
ST10 4LT G7MGA
ST10 4LY G6GVZ
ST10 4NJ M0KDU
ST10 4PE G0UUA
ST11 9AA G7CEY
ST11 9AP G0VHY
ST11 9AU G6IXN
ST11 9AZ M3AAQ
ST11 9AZ M3JDJ
ST11 9BH G0VRK
ST11 9DA G6OAS
ST11 9DR M6YAW
ST11 9EN 2E0YYF
ST11 9EN 2E0YYT
ST11 9EN M6YYF
ST11 9EN M6YYT
ST11 9HA G7VLA
ST11 9HQ M0DSR
ST11 9HU 2E0IDK
ST11 9HU M0IDK
ST11 9HU M6IDK
ST11 9LY 2E0DVK
ST11 9NT G1DOG
ST11 9NX G4YYG
ST11 9NX G4YYP
ST11 9NZ G7UKN
ST11 9PL 2E0AAO
ST11 9PL G4YRR
ST11 9PL M6BWZ
ST11 9PP G4WUX
ST11 9QT G1BHF
ST11 9RH 2E0MZL
ST11 9RH M6MZL
ST11 9RN G4HUO
ST12 9BD M3XKM
ST12 9BL M6HOG
ST12 9EQ G4DUB
ST12 9JA G6ITO
ST12 9JQ G0RJT
ST13 5DB M3KMO
ST13 5DB M5JAO
ST13 5JU 2E11DLR
ST13 5LA M3XMY
ST13 5LS 2E0MJJ
ST13 5LS M3TJJ
ST13 5NZ 2E0SXY
ST13 5NZ M3SIM
ST13 5PW 2E0SAA
ST13 5PW M3RRZ
ST13 5RZ M1RIY
ST13 5RR G7HIJ
ST13 5SZ G4MDJ
ST13 6BU M0RMP
ST13 6ES G8YQA
ST13 6PX M3BVP
ST13 7AA G3VDQ
ST13 7AN G4XIZ
ST13 7ED G8FWD
ST13 7EX G0LAZ
ST13 7HH M3HVA
ST13 7HJ G0WOM
ST13 7JP G6KPT
ST13 7LW G7SSK
ST13 8BL 2E1SGK
ST13 8BL M0SGK
ST13 8DD M1ALE
ST13 8DP G3FKY
ST13 8EQ 2E1HRN
ST13 8EQ M0FAZ
ST13 8EQ M3FCN
ST13 8EQ M3FCO
ST13 8EQ M3HRN
ST13 8EQ M3SKD
ST13 8ES G4TGS
ST13 8JA G8JGL
ST13 8JN M1ABF
ST13 8LD G7MQF
ST13 8LL G7CFT
ST13 8LN G4UQM
ST13 8QL M0GBF
ST13 8XF G0VMQ
ST14 5AG 2E0KMS
ST14 5AG M0SRJ
ST14 5DE M0DNU
ST14 5DH G3VSB
ST14 5DP 2E0SRC
ST14 5DP M0RRN
ST14 5DP M6YUK
ST14 5JU G7LGS
ST14 5LE 2E0AUM
ST14 5LF M3HEJ
ST14 5LT G0LJH
ST14 7AX M3MTP
ST14 7BY 2E1DLM
ST14 7DZ G4EVW
ST14 7EN 2E1BMV
ST14 7EQ M0VCS
ST14 7ET 2E1CLM
ST14 7ET G6VGC
ST14 7FE G1RKJ
ST14 7FE G7PAY
ST14 7HL 2E0LIZ
ST14 7HL 2E1LIZ
ST14 7HL M3ZIL
ST14 7LB G6SVJ
ST14 7NB G7STC
ST14 7NF G7UCN
ST14 7QG G6CUY
ST14 7QY G6UOX
ST14 8BB G8YAT
ST14 8BT G0CAP
ST14 8EF M0WUL
ST14 8LT 2E0PBB
ST14 8PU G3XJY
ST14 8QS G8FGQ
ST14 8XG M6UTX
ST15 0DH G0KJP
ST15 0DW 2E1BRA
ST15 0DX G0SKQ
ST15 0EB G0TED
ST15 0EG G1LAP
ST15 0EG G7GJM
ST15 0EH 2E0VDP
ST15 0EH M3VDP
ST15 0EP G1DSF
ST15 0EP G7KEE
ST15 0JA G3JUX
ST15 0JF G8HWI
ST15 0JN G1HSJ
ST15 0LF M0CDN
ST15 0PX G4TTM
ST15 0QH 2E0PSD
ST15 0RD M6MGO
ST15 0RH G4NJR
ST15 0RP M3CZL
ST15 8BL M6BUX
ST15 8LA G3XMK
ST15 8LF G0VWT
ST15 8LQ M0CZA
ST15 8LW G4CRK
ST15 8PR G6JAF
ST15 8TG M3ZGD
ST15 8TZ M6KGL
ST15 8XL G4EJD
ST15 8YP G8MZZ
ST16 1FJ G8RFV
ST16 1HA 2E0HHK
ST16 1HA M6ZZQ
ST16 1HJ G3ZHS
ST16 1HJ M1EUM
ST16 1LD G4TMD
ST16 1NL G6YLZ
ST16 1NL M6MTS
ST16 1PG G0ITS
ST16 1PZ G7BJG
ST16 1QP G6ZLJ
ST16 1QR M6HQR
ST16 1SD 2E0HBB
ST16 1SD M6HRL
ST16 1TB G4YFF
ST16 1TB M0RWR
ST16 1TB M6VIX
ST16 1TE G1UDS
ST16 1XA G0FXS
ST16 2DZ G3XPD
ST16 3HD G0IAC
ST16 3HQ G0EYX
ST16 3HS G7PFT
ST16 3NX 2E1EDT
ST16 3NX G0GAP
ST16 3NX M6FFE
ST16 3PJ M6TPG
ST16 3PL G0GOZ
ST16 3PL G3SBL
ST16 3RE 2E0RPF
ST16 3RE M0RPF
ST16 3RE M3RPF
ST16 3TY M0DFL
ST16 3WG G1SPU
ST17 0AJ G0OIQ
ST17 0AQ G4NVH
ST17 0AQ M3NLW
ST17 0AQ M3NVH
ST17 0HJ G4PKF
ST17 0HS M1EOP
ST17 0NH M0SJD
ST17 0NU G6KQD
ST17 0PA G3LOE
ST17 0PA G6LXF
ST17 0RE G6IRZ
ST17 0SJ G4RWQ
ST17 0SL 2E1AZK
ST17 0TW G0TFD
ST17 0TW G1UUJ
ST17 0WE G1JNH
ST17 4AN M0DHK
ST17 4BP G6RFM
ST17 4BX G2HNA
ST17 4EH G7VLB
ST17 4HB G6TBV
ST17 4HZ 2E0MQA
ST17 4HZ M3MQR
ST17 4NR G1XWO
ST17 4QJ 2E0VTS
ST17 4QJ M0POG
ST17 4QJ M0VTS
ST17 4QJ M3VTS
ST17 4QP G0GAR
ST17 4QR M6KIX
ST17 4QS G4PWV
ST17 4RY 2E0RDQ
ST17 4RY M6BVD
ST17 4YA G6JKF
ST17 9BE G8KUA
ST17 9DH M6PCW
ST17 9DS 2E0AVK
ST17 9EF M3GCS
ST17 9FR G6OYF
ST17 9FR M1LIP
ST17 9FT G6KJK
ST17 9HP G1SJG
ST17 9HP G1UDT
ST17 9HP G3OSP
ST17 9NA G3GYC
ST17 9QJ G3VUL
ST17 9RL G4RSW
ST17 9SB M3NOF
ST17 9TW M3XZY
ST17 9YA G3JDM
ST18 0DR 2E0PRC
ST18 0DR M6SFC
ST18 0EW G8JVU
ST18 0GQ G0GMS
ST18 0HJ G4ZMM
ST18 0NJ G0VGK
ST18 0NN G4RJM
ST18 0QE G0GUF
ST18 0QH G6UZL
ST18 0QW G0AXR
ST18 0QZ G4OUT
ST18 0RD G4PET
ST18 0RD G4PEU
ST18 0SP 2E1ELE
ST18 0UR G4THY
ST18 0XB G6KGA
ST18 9DA G6DGX
ST18 9DQ 2E0EVB
ST18 9DQ G4KQK
ST18 9DQ M6EVB
ST18 9HS M6MRM
ST18 9QR G4UKA
ST19 5AH G3JBF
ST19 5BX G3PIN
ST19 5EH G0VJC
ST19 5HD G0GJG
ST19 5HF M0UXB
ST19 5HP 2E1EAW
ST19 5QH G0BFK
ST19 5SP M3VOR
ST19 9AB G4WAF
ST19 9DX G0NEN
ST19 9HZ G3XFB
ST19 9JZ G4IDG
ST19 9LX M0RSY
ST19 9NQ 2E1GHX
ST19 9NQ M3PLN
ST19 9NU 2E1GVJ
ST2 0AQ M3DPP
ST2 0BX G1YQL
ST2 0HB M6EYS
ST2 0QY G4GRZ
ST2 0RQ G7RFH
ST2 0SS M6YYD
ST2 0SX M0LAA
ST2 0SX M0LEK
ST2 0SX M1DOA
ST2 7AR G4BEM
ST2 7AR G4DPV
ST2 7BZ G4CKT
ST2 7HJ G4OSI
ST2 7JJ G1STO
ST2 7LD M1MLM
ST2 7LR G0UZE
ST2 7LR G7RTX
ST2 7NF G3ISX
ST2 7NG G6CCQ
ST2 7PA G1SQC
ST2 7PF G0TKR
ST2 8AQ G4DVA
ST2 8DQ G0LJO
ST2 8DQ G7VSW
ST2 8DS 2E1STO
ST2 8DU M3PIM
ST2 8EH M6YAH
ST2 8HU 2E1MAZ
ST2 8HU M0DVT
ST2 8HU M0FTR
ST2 8HU M3LVA
ST2 8HX M3SRH
ST2 8JN M1DXQ
ST2 8JZ 2E1DZP
ST2 8LP M3KEL
ST2 9BZ M6PAN
ST2 9DR G7VUL
ST2 9JA M0CTL
ST20 0BA M6HMR
ST20 0BN G0BMB
ST20 0DT G0DRN
ST20 0HL G1WMO
ST20 0JD G1OTD
ST20 0LG M6AJB
ST20 0QH M3KQF
ST20 0RP 2E0LME
ST20 0RP M3LME
ST21 6EX G6EKM
ST21 6LE G4LSA
ST21 6LT G4DDZ
ST21 6RG G0UUM
ST21 6RG G0UUN
ST3 1AD 2E1IVT
ST3 1SJ G0TTO
ST3 1SJ G7KDJ
ST3 2AJ 2E1GVS
ST3 2EG 2E0ESJ
ST3 2EG M3HNM
ST3 2HA G0CWO
ST3 2RS M6MEP
ST3 2RZ M1IRM
ST3 3BX M0OBU
ST3 3JQ M3JZD
ST3 3LX M3OLE
ST3 4NF G6TPI
ST3 5DX G7EWX
ST3 5EG G3VTE
ST3 5JA G7NEE
ST3 5JD G0JGB
ST3 5JL G6AYH
ST3 5LA M3AOP
ST3 5PN G7BMP
ST3 5QY G6XJF
ST3 5RP G7FSA
ST3 5RQ G8NAI
ST3 5ST G4WJX
ST3 5UB M3BIK
ST3 5UD G1FHH
ST3 5UG G0SKM
ST3 5XW G0DUI
ST3 6AH G7SWR
ST3 6EQ M3YBF
ST3 6HA 2E0RDW
ST3 6HA G0SMN
ST3 6HA M0RDW
ST3 6HY G1XMP
ST3 6HY M0ODS
ST3 6JY G0PSH
ST3 6NS G6THM
ST3 6PN 2E0VTR
ST3 6PQ G4RQG
ST3 6RG G1UYT
ST3 7AN M6HVI
ST3 7AP G8UWL
ST3 7DP M6VGR
ST3 7EN G0WLC
ST3 7EW 2E0PHB
ST3 7HN G0UPV
ST3 7JA M0DPW
ST3 7JY G3MFH
ST3 7LH M3YMD
ST3 7LJ M3YSD
ST3 7QF G7MMW
ST3 7RZ M0FOG
ST3 7SS G0NED
ST3 7UG M0EUI
ST3 7WH G6LQG
ST3 7YE G8DZJ
ST4 1ED G6UDI
ST4 2EX G0NMY
ST4 2RQ 2E1HCG
ST4 3DT M6SAM
ST4 3HH G1YPH
ST4 3HQ G7JWO
ST4 4DY G1EZJ
ST4 4NG G0VVT
ST4 4PH G4CJM
ST4 4PL G0VKI
ST4 4QP G3UD
ST4 5AN G0JZS
ST4 5AZ G7FPJ
ST4 5DH M3HQQ
ST4 5DW G6WFF
ST4 5EE G1CHE
ST4 5EF G1OQW
ST4 5HA M6PAW
ST4 5HE M6LZT
ST4 5HR 2E0BZC
ST4 5HR M0ZPA
ST4 5HR M6PMD
ST4 5PU M3WZY
ST4 6ED G7HQJ
ST4 6LA G0JGF
ST4 6LD G0DTW
ST4 6PR G6HFK
ST4 6QZ G3UJI
ST4 6RE M3YHN
ST4 6RU 2E0TQO
ST4 6RU M1DCK
ST4 7DZ G0YDJ
ST4 7HE G1UCC
ST4 7HF G1XZA
ST4 7HF G1XZB
ST4 8AB G4EJM
ST4 8AZ M6PXL
ST4 8EL G8SAR
ST4 8FR M6BWH
ST4 8HP G6HBQ
ST4 8LP M6PNG
ST4 8NJ G3SLX
ST4 8NJ G4MDK
ST4 8NJ G4MJU
ST4 8NW G7EVK
ST4 8PD 2E0COY
ST4 8PL M3PQL
ST4 8PQ G7WGZ
ST4 8PQ G7WIA
ST4 8RQ M0ACA
ST4 8RT G7NHZ
ST4 8TF G0PXL
ST4 8UZ G0NBE
ST4 8XY G6AZP
ST4 8YQ G8DYI
ST46DB G3GBU
ST5 0EX G3KLY
ST5 0JS M6GAS
ST5 0JY G1WVW
ST5 0JY M0CRJ
ST5 0LB G8KUZ
ST5 0LE G4DQB
ST5 0LH M0PWC
ST5 0NG G4VFJ
ST5 0NL G6IFS
ST5 0PB G0OQP
ST5 0PD G1TMF
ST5 0SN 2E0NUL
ST5 0SN M6WIX
ST5 1JH G7IDH
ST5 1NH G4IMV
ST5 2EF G7JXU
ST5 2HA 2E0IBB
ST5 2HA M6OBB
ST5 2HU G6VTX
ST5 2JL G0WDT
ST5 2LG G0PSD
ST5 2PQ G7BJN
ST5 2QH M3MIF
ST5 2QT G6IXS
ST5 2SD G8WKH
ST5 2TB M0BSH
ST5 3BS G6HCI
ST5 3DF G3TJP
ST5 3DF G6CXN
ST5 3DG G8BMG
ST5 3DL G7BPG
ST5 3DX M3AUN
ST5 3EA M3KFE
ST5 3EF G4TQB
ST5 3EW G4UAY
ST5 3EY M0HIO
ST5 3LT G7RSM
ST5 3LT G7UQG
ST5 3NY G4MDM
ST5 3NY G4UMG
ST5 3NZ M3FMK
ST5 3PQ G1JJE
ST5 3PQ G7MMK
ST5 3QR 2E0XXX
ST5 3RW G1JIW
ST5 4BH G3LBS
ST5 4BN G8NPD
ST5 4DA G8MTB
ST5 4EG G6LQP
ST5 4EN G1DDA
ST5 4HS G4TMR
ST5 4LA G1STQ
ST5 5EU G4NFL
ST5 5HP G3USF
ST5 6AR G0GOI
ST5 6BB G7PNF
ST5 6DT 2E0SAL
ST5 6DT G1LMT
ST5 6DT M0MVK
ST5 6DT M3MIB
ST5 6EY 2E0DMM
ST5 6HP 2E0XCB
ST5 6LS 2E0SDQ
ST5 6LS M3GDK
ST5 6QP G7GRR
ST5 6RL M0OMA
ST5 6TB 2E0PLC
ST5 6TB M0TKS
ST5 6TB M3TYG
ST5 6TB M3WSV
ST5 7AH G4CBW
ST5 7EP M3ZJH
ST5 7JR M6ECH
ST5 7QD G1VBY
ST5 7QP 2E0SCX
ST5 7QP M6ALQ
ST5 7RQ G7KJE
ST5 7ST G6KDU
ST5 7ST M3YCO
ST5 7TB G4UDH
ST5 8DR G6IKH
ST5 8EA G0SOK
ST5 8EL G4GNW
ST5 8EL G7MTV
ST5 8EL M3PIQ
ST5 8EY G7UAT
ST5 8HH M3WXI
ST5 8LD M6YCA
ST5 8ND G0ZHU
ST5 8QA G0VZE
ST5 9AJ G0GZT
ST5 9AJ M3XIE
ST5 9BP M3YHZ
ST5 9ET G7RBL
ST5 9HE G7OKO
ST5 9JZ G4DPW
ST5 9LH M6LBY
ST5 9NL G1JFL
ST5 9NY G1XYN
ST5 9NY M6CEP
ST5 9PS G7MRF
ST5 9PU G6LLF
ST6 1HB M1ACL
ST6 1PS M1PYE
ST6 1PX G7TJQ
ST6 1RA M6USE
ST6 2ER M6LJS
ST6 2BW M0QBF
ST6 2BY G7SSW
ST6 3DJ M6HKX
ST6 3JB M6HUS
ST6 3PN M6EDZ
ST6 4DZ M6BCJ
ST6 4ES M6MIR
ST6 4ET M3MOH
ST6 4EX M6JLH
ST6 4JB M6LLL
ST6 4NN M6CYA
ST6 4NW M3YUR
ST6 5HS 2E0JTB
ST6 5HS M3KJV
ST6 5LP M0EOT
ST6 5PP G7JWI
ST6 5PZ M6CLY
ST6 6DZ G4WEP
ST6 6EB M6HCD
ST6 6EX M3JIV
ST6 6HE G8WBP
ST6 6HW 2E1HQZ
ST6 6LX G6DON
ST6 6PB M3FNR
ST6 6PE G0UDO
ST6 6RA G3ZRQ
ST6 6TJ 2E0NGF
ST6 6TJ M6NGF
ST6 6TZ M0CDZ
ST6 6UR G0MGJ
ST6 6XF 2E0HAG
ST6 6XF M3XHU
ST6 7AL G7MSQ
ST6 7BA M6SWH
ST6 7DL M6SOT
ST6 7JY G7EUF
ST6 8BZ G3ICZ
ST6 8DX G7HMB
ST6 8JF M1MDP
ST6 8LX M0RDX
ST6 8PL G1VYS
ST6 8QH G0BRL
ST6 8RN 2E0MWA
ST6 8RT G8RWZ
ST6 8RX G0UTZ
ST6 8SD M3SDJ
ST6 8TG G3UHV
ST6 8TQ G6OBD
ST6 8TX M0INY
ST7 1AR M3DBU
ST7 1AR M3EBU
ST7 1AT G6YQN
ST7 1AW 2E0KDB
ST7 1AW M6KWK
ST7 1DN 2E0JSM
ST7 1DN M6FFJ
ST7 1EZ 2E0IQT
ST7 1JN 2E0BOF
ST7 1JN M3LRF
ST7 1JY M3BTJ
ST7 1LW G6NQY
ST7 1NP G4WBW
ST7 1NQ G6NGN
ST7 1PF G4EQZ
ST7 1PT M6AHA
ST7 1RB G7VCT
ST7 1RH G4VJB
ST7 1SG G0MVU
ST7 1SG G4VEW
ST7 1SQ G6SGA
ST7 1TA G0RDS
ST7 1UB G1TBX
ST7 2BA G4ERQ
ST7 2BN G4WHT
ST7 2BW G8OSJ
ST7 2BY G7FZB
ST7 2DS M0CLL
ST7 2EF G6GFC
ST7 2HX G6NVH
ST7 2HY G4GHT
ST7 2LP G3VXS
ST7 2ND G0OFW
ST7 2NH G4UFU
ST7 2NQ G0EIR
ST7 2NR G0UKK
ST7 2RL G0WLJ
ST7 2SH 2E0DFB
ST7 2SU 2E0NCN
ST7 2TQ M0CVJ
ST7 3BL G6LMJ
ST7 3LD M0RJS
ST7 3ND G8MKS
ST7 3PB 2E0MKX
ST7 3PB M0MJK
ST7 3PH G3OHH
ST7 3RB G4DLA
ST7 3RY G0NAI
ST7 3SE G4UKP
ST7 3TJ 2E0ICY
ST7 3TJ M0RZY
ST7 3TJ M6ORE
ST7 3TJ G6TWB
ST7 3TL G8UVN
ST7 3TQ G8LIY
ST7 4AL G7IAM
ST7 4DQ G1CKJ
ST7 4DY G0KBJ
ST7 4HF G0UDG
ST7 4HP G4UDG
ST7 4HR M6BVZ
ST7 4HW G0UWK
ST7 4JF G0HWV
ST7 4JU G8ILP
ST7 4JX 2E0TAK
ST7 4JX M3TBZ
ST7 4LW G7TXX
ST7 4RT G8LZO
ST7 4RQ G4YUN
ST7 4RS G1ELQ
ST7 4RS G6BQM
ST7 4TA G0FZU
ST7 4TA G6USG
ST7 4TH M1AUP
ST7 4UR M6EGF
ST7 4YA G8NYZ
ST7 4YL G0RDK
ST7 4YL G1PUV
ST7 8BB M3DOO
ST7 8HB M0SSY
ST7 8HU G7OSO
ST7 8JQ M3TLZ
ST7 8LA G0MVT
ST7 8LP G0LSX
ST7 8NG G0PTD
ST7 8NH G8GLD
ST7 8PA G0FWU
ST7 8PE G6UKQ
ST7 8PL G6YCO
ST7 8QB M3WXU
ST7 8QF G6TCV
ST8 6BP G6PAO
ST8 6EH M3SJQ
ST8 6LS M3FIZ
ST8 6NT G8KSC
ST8 6RD M0GIP
ST8 6RD M1PAS
ST8 6SE G0BYU
ST8 6SQ 2E0DDK
ST8 6SQ M6DUY
ST8 6SZ M3EOT
ST8 6TJ G6CYV
ST8 7AH G6DSG
ST8 7AS G1ZNZ
ST8 7HL G6AKX
ST8 7LN G6UKM
ST8 7LY G4WAM
ST8 7NG G4UQW
ST8 7NW G0CHL
ST8 7PF G6PJC
ST8 7SW G7LYS
ST8 7SW G8APB
ST8 7SZ G4GVE
ST8 7SZ G4RXB
ST8 7UF G0DBC
ST9 0BD G3LHG
ST9 0DG 2E0CIA
ST9 0DG G0RKN
ST9 0DG M0ASR
ST9 0EP G4SCY
ST9 0ER M1FIL
ST9 0HU 2E0CAA
ST9 0JN G4DVN
ST9 0LF G1EWC
ST9 0LR 2E0DDD
ST9 0PF G8NSS
ST9 0PQ G1DRP
ST9 9BW G4KME
ST9 9JB G8WIM
ST9 9JB M3YHF
ST9 9JB M3YHH
ST9 9LW G4LCL
ST9 9QD M0GYN
ST9 9QG G8JDD

## SW

## (South West London)

SW10 9BJ G7GBJ
SW10 9NJ G3JHH
SW11 1RH G4CKS
SW11 2JE M6HEG
SW11 2RA G1HRD
SW11 2SU G1VVH
SW11 3NY G8JKD
SW11 3TN 2E0LVR
SW11 3TN M0LVR
SW11 3TN M6ODP
SW11 4AA G1EGB
SW11 4DU G8NXA
SW12 0NA G1ULG
SW12 8EG M6UMM
SW12 8TQ 2E0BVH
SW12 8TQ M0MEN
SW12 8TQ M3YIX
SW12 8UQ M3YNX
SW12 9LU G7IWU
SW12 9SS G6CKY
SW13 9LH G0WLD
SW13 9NB M0FCA
SW14 7AT 2E0TTT
SW14 7AT M6GGG
SW14 7NL 2E0RSI
SW14 7NL M3NNG
SW14 8JH G4AVK
SW14 8JJ G3KWW
SW14 8JJ G6JMX
SW14 8JU G8ANN
SW15 1LZ M6VIE
SW15 6EX G1SAK
SW16 1AA M1BJC
SW16 1SN G8ZFU
SW16 2LT 2E0BWY
SW16 2LT M3ZLV
SW16 2UU M3NLN
SW16 3LS M0BIN
SW16 3SN M0HHP
SW16 4HL G7IRG
SW16 4HT G0WLF
SW16 6JY G6RBN
SW17 0PG M0KLH
SW17 0QQ G1LAR
SW17 7DH M1ECI
SW17 7DW M0AGP
SW17 8PR G7NLF
SW17 8SF G8TQV
SW17 8SF M0TQV
SW17 9EN 2E0BWD
SW17 9EN M6OJO
SW18 1SD G0TEO
SW18 2AJ G3PAQ
SW18 2EB 2E1RWN
SW18 3BB G4PZJ
SW18 3JU M0CON
SW18 3RQ M0FCR
SW18 5AN G4PKK
SW18 5PB 2E0TRE
SW18 5PB M3TNN
SW18 5TN 2E0HAT
SW18 5TN M6HAT
SW19 1EB M6HSX
SW19 1EB M6ZIA
SW19 1QU M1PFS
SW19 1XF G7RHU
SW19 3JJ G4ZHT
SW19 3JJ M0JCL
SW19 3JJ M1JCL
SW19 4PR G4EZG
SW19 4UR M0ARM
SW19 5AZ G6LKV
SW19 6JA G4UNI
SW19 6LW G3VPQ
SW19 6QU G4CGD
SW19 6SJ G8ILZ
SW19 7LW G3NXN
SW19 7RQ G6GKG
SW19 8DQ G4ILP
SW19 8JP M0TNB
SW19 8NZ G8WUF
SW19 8RY 2E0KRB
SW19 8RY M0SOT
SW19 8SF M0HJJ
SW19 8TU 2E1HJA
SW1P 4PS M3LBJ
SW1P 4RW G3ZDY
SW1V 1JJ M0HTQ
SW1V 2AD 2E0GSL
SW1V 2AD M0LAE
SW1V 2AD M6LXC
SW1V 2DB G4MRL
SW1V 2QS G1TQT
SW1V 2RT G4JKE
SW1V 4AS G6ZFK
SW1W 8JW M3AJC
SW2 1DD M6BXM
SW2 1NQ G4ILW
SW2 1PA G7IHV
SW2 3AA M6ATZ
SW2 5LU G3KHQ
SW20 0BA M6LXP
SW20 0JN G6WIM
SW20 0UD G3DRN
SW20 8AP G0CRU
SW20 9AB M3LWX
SW20 9BQ 2E0JGH
SW20 9BQ M0JGH
SW20 9HG G8IAK
SW20 9HX G8BUZ
SW20 9JT G0OVT
SW20 9LB M3UFL
SW3 3LB G1XGZ
SW3 3TS M1DNQ
SW3 4LA G6LVB
SW3 4QE G8ECR
SW3 4SR M6YDG
SW3 6BU G8DOH
SW4 0AN M6TAH
SW4 0ET G4PIA
SW4 6BE M6DUE
SW4 7AF M3ZFL
SW4 9HD G8LHQ
SW4 9PB G8OOF
SW4 9RR 2E0KJI
SW4 9RR M6KJI
SW5 9EB 2E0CWK
SW6 1BS G0HFK
SW6 2RB G6BUH
SW6 2SX 2E0EFO
SW6 3SA M6BDG
SW6 5RG G0ADG
SW6 5DW 2E0LCW
SW6 6QB G4OXG
SW6 6SA M6FHH
SW6 7PP G7IUB
SW7 1DL G0RZC
SW7 3DG M3TND
SW7 3DQ G0RFX
SW7 5QF G4MFW
SW7 4AZ M6NOA
SW7 4ET G4UAF
SW7 5NS G4WQO
SW8 1BH G7VQR
SW8 2PD G0PIY
SW8 2TF 2E0DYV
SW8 2TF M6GQZ
SW8 3DD M6GYI
SW8 4DP 2E0ULH
SW9 7UE G0VPH
SW9 9RH G6NAD
SW9 9RZ M6DUH

## SY

## (Shrewsbury)

SY1 1RD G8ZWF
SY1 1ST G6EAM
SY1 1UH G0GTN
SY1 2AB M6FNF
SY1 2PY G8DXM
SY1 3HF 2E0OCM
SY1 3PY G0EML
SY1 3PZ G4CBT
SY1 3RA M3ENY
SY1 3RH G4XBI
SY1 4BF M1BCM
SY1 4ER G0IRI
SY1 4JE G7NBP
SY1 4LD 2E1FZU
SY1 4PL 2E0BMP
SY1 4RB G8SIU
SY1 4RP G8RTB
SY1 4RU M0KZB
SY10 0AN G6KLQ
SY10 0LH M6FEH
SY10 0NS G2HCA
SY10 7AE G7LXI
SY10 7AS G6YGI
SY10 7AS M3XRI
SY10 7BX G4TQE
SY10 7DX G7CQB
SY10 7HL G3OKT
SY10 7HP G3SMT
SY10 7JY M3IHB
SY10 7NF M6LWF
SY10 7PQ G1PLJ
SY10 7PQ M3WQV
SY10 7PQ M6BNF
SY10 7QB G3LYU
SY10 7QB G6XPO
SY10 7RQ G4TEQ
SY10 7TU G0MDQ
SY10 7TU G6DBP
SY10 8HQ M6OLL
SY10 8HU G1HAX
SY10 8PS G7OTQ
SY10 8QH G4JUW
SY10 9AU G8AJA
SY10 9AZ G0DLW
SY10 9BA G4DFQ
SY10 9BS G4UDE
SY10 9DF G1VIR
SY10 9DP M6FST
SY10 9DP M6NJM
SY10 9DP M6WEE
SY10 9HN G1JVH
SY10 9HQ G4IOQ
SY10 9NZ M6TBL
SY10 9QR G1APU
SY10 9RB M0DSZ
SY11 1EP M6PIH
SY11 1LX M0RRY
SY11 1TP G7JHC
SY11 2QW M6FTK
SY11 2SG M3NFZ
SY11 2UA G7NNA
SY11 2UF G7CEC
SY11 2UN M6HCY
SY11 2XA G6AUS
SY11 2YD G4WVK
SY11 3AH M6HBK
SY11 3AH M6NCA
SY11 3BU M1EYH
SY11 3DH G0CWZ
SY11 3JF 2E1HXT
SY11 3LD 2E0TDF
SY11 3LD M0TDF
SY11 3LD M3XTF
SY11 3NR M6KQL
SY11 3PJ G1OIB
SY11 4AH 2E1MSC
SY11 4AH M3PVC
SY11 4LF G0VEU
SY11 4PA G0EXD
SY12 0EG M0YCQ
SY12 0EX M0CCQ
SY12 0PF 2E1MAR
SY12 0QJ G3XGD
SY12 0QZ G0TLS
SY12 9ER M0COT
SY12 9NA G6ZMD
SY12 9PA G0AJA
SY12 9QA G1VXY
SY13 1BS M0NDC
SY13 1BZ M1BKL
SY13 1EX G4HRH
SY13 1PH M0SIN
SY13 1PH M1SIN
SY13 1QT M6NBQ
SY13 1RQ G3TZU
SY13 1UD M2IHC
SY13 2AR G1GZZ
SY13 2HD G8CTD
SY13 2HZ G1JOP
SY13 2LW G3YLV
SY13 2ND G8WDG
SY13 2NJ G4VQH
SY13 2PT G4UJS
SY13 2PX G1GSJ
SY13 2QL 2E0ZSU
SY13 2QL M3ZSU
SY13 2RA G4ANY
SY13 3AD M6SHI
SY13 3LT G0MXD
SY14 7AA M0VKC
SY14 7AR G4RRR

Postcode

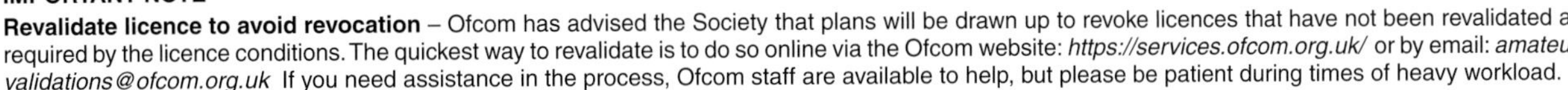

**IMPORTANT NOTE**

**Revalidate licence to avoid revocation** – Ofcom has advised the Society that plans will be drawn up to revoke licences that have not been revalidated as required by the licence conditions. The quickest way to revalidate is to do so online via the Ofcom website: *https://services.ofcom.org.uk/* or by email: *amateur.validations@ofcom.org.uk* If you need assistance in the process, Ofcom staff are available to help, but please be patient during times of heavy workload.

SY14 7AX G3GIZ
SY14 7AX G3TZO
SY14 7DB G4GOO
SY14 7JJ M3NOR
SY14 8AY M3ZLA
SY14 8JB G6XTC
SY14 8JQ G4XVR
SY14 8JQ G4XVS
SY14 8LL M0ROJ
SY14 8NE M6MEC
SY147DN M0PZC
SY15 6AB G7DRX
SY15 6AL G0IQP
SY15 6EB M6IMH
SY15 6JH G4RYK
SY15 6NQ M6RBH
SY15 6RS G0RJV
SY15 6RS M3GEX
SY15 6RS M3HFT
SY15 6RT M3MYQ
SY15 6SB 2E1CEZ
SY15 6SB G4NQJ
SY15 6TJ G4WND
SY15 6TJ G4YNL
SY16 1DW G7TIX
SY16 1PR G7CAH
SY16 1QQ G0TWL
SY16 1QQ G1PVN
SY16 1QT G7DJL
SY16 1RH M6RQM
SY16 2HN 2E0GWM
SY16 2HN M6GWM
SY16 2HU 2E0MTE
SY16 2HU M6MTE
SY16 2JN M6WCU
SY16 3ER M5TLE
SY16 3HA M6EIB
SY16 3HX G7NNM
SY16 3JY G3LXE
SY16 4DB G3IWM
SY17 5BU 2E0CPO
SY17 5BU M6TEL
SY17 5JG G3FXI
SY17 5JG M5AMV
SY17 5JP G4ZUD
SY17 5PU M6EDV
SY17 5PU M6EDW
SY18 6AD M6KPF
SY18 6AR G4CWG
SY18 6DQ G3KAJ
SY18 6HZ G1UOY
SY18 6NP G4DTU
SY18 6PQ 2E0CYE
SY18 6PQ M6CKM
SY18 6PS M0BLM
SY18 6SN G3XPK
SY19 7DJ G0EHS
SY19 7DJ G1JNI
SY2 5EF G3TSV
SY2 5LA M6FYT
SY2 5LQ G0IGM
SY2 5PF G3SRT
SY2 5PF G3VZG
SY2 5QL 2E1JCM
SY2 5SW M1EQV
SY2 5TA 2E0EBG
SY2 5TA M0MEB
SY2 5TA M6EBB
SY2 5TS M6FNM
SY2 5UG G4RLO
SY2 5UZ 2E1EQR
SY2 5YB M0GSL
SY2 5YD G0EBD
SY2 6EE M6OEN
SY2 6HH M6BUS
SY2 6HQ G3IDY
SY2 6ND M3ZXE
SY2 6SJ G0RQX
SY20 8DL 2E0RDZ
SY20 8DL M0RCZ
SY20 8DL M6RFS
SY20 8EJ G0JAI
SY20 8NY G8BFO
SY20 8RA 2E0IGN
SY20 8RA M6GKQ
SY20 9BQ G7VBY
SY20 9EP G7NYP
SY20 9EZ G4LPU
SY20 9PR 2E0XUL
SY20 9RD G7GAH
SY20 9RU G7JUV
SY21 0AE G1JNR
SY21 0EP G4AUD
SY21 0JT G4DEP
SY21 0PW G3KGV
SY21 0RH 2E0UTT
SY21 0RH M0UTT
SY21 0RH M6EHX
SY21 7BB G4DYY
SY21 7BB G8HOS
SY21 7NQ G4DWX
SY21 7RD G3JPT
SY21 7RH G8HEB
SY21 7TP M1ABT
SY21 7UL G4TFM
SY21 8AH M0DDE
SY21 8AU G3JSV
SY21 8BJ G6EUR
SY21 8DY M6CYX
SY21 8ER G1ORP
SY21 8ER M3ORP
SY21 8EZ M5LMG
SY21 8HL G4MZB
SY21 8LA M1BEW
SY21 8LF G7DLD
SY21 8LF G7KMD
SY21 8NJ G4SGQ
SY21 8RT G0NNE
SY21 8SG G3TCV
SY21 8TS G6EUT
SY21 9JE G0KGD
SY21 9JY G4BVE
SY21 9JY G8DHT
SY21 9LH M6WLY
SY21 9PN G1KTW
SY21 9PX G1EAV
SY21 9PX G4GNY
SY22 5AF M0DAX
SY22 6BE G3KVX
SY22 6EW G0AGZ
SY22 6HB G6HUR
SY22 6NF G4UBQ
SY22 6NQ M3JKB
SY22 6PZ G7FYG
SY22 6PZ G8WTB
SY22 6PZ M3NVQ
SY22 6PZ M3OFH
SY22 6QL G1YQM
SY22 6QL M0EDX
SY22 6QL M6CYU
SY22 6QL M6MFN
SY22 6RJ G3YDX
SY22 6SL G4CRH
SY22 6TW G3YRP
SY22 6UJ M6MGM
SY22 6XL G3UBH
SY226LL M6VLS
SY23 1AT G8SIT
SY23 1BT M3ROX
SY23 1BW M6TXZ
SY23 1DW 2E0FLI
SY23 1DW M6FNA
SY23 1EU 2E0OLT
SY23 1EU M0OLE
SY23 1EU M6LRO
SY23 1HH G4YAW
SY23 1JU M6HRR
SY23 1QB G4TUD
SY23 1RJ M0GLS
SY23 1RY G1URF
SY23 1SS 2E0LJD
SY23 1SS G7AGG
SY23 1SS M6EFV
SY23 1SS M6XIE
SY23 2EL M6NFN
SY23 2EL M6WOD
SY23 2EU M6NBU
SY23 2JA G7HAE
SY23 2JA M3HAE
SY23 2JA M3WTR
SY23 2JU G1HIN
SY23 3AB M6XFM
SY23 3BL G7OZP
SY23 3BQ M6IJD
SY23 3EA M3LFL
SY23 3EZ G4CTV
SY23 3GZ M6ORW
SY23 3HF G8LZY
SY23 3HQ 2E0FFL
SY23 3HQ G8SFT
SY23 3HQ M0LBR
SY23 3HQ M6FLA
SY23 3HU M6SVM
SY23 3LH M6GZZ
SY23 3LS 2E0DVP
SY23 3LS M0PRI
SY23 3LS M6DVP
SY23 3PD M6ZXO
SY23 3PL M6ZXG
SY23 3QU M6ZCT
SY23 3RP M6TFB
SY23 3TE 2E0TRI
SY23 3TE M6NRO
SY23 3TR G7HSW
SY23 3TX 2E0DPU
SY23 3TX M0ZXY
SY23 3TX M6ZXY
SY23 4BQ G1FWC
SY23 4DX 2E0DNR
SY23 4DX M6TWH
SY23 4PP G4LHL
SY23 4RF G0WAY
SY23 4RN M6AEN
SY23 4TP 2E0GIA
SY23 4TP M6GIA
SY23 5BW M3WKC
SY23 5BZ M6FQA
SY23 5HG 2E0CPD
SY23 5HG M0RBV
SY23 5HG M6BXI
SY23 5HG M6CGV
SY23 5HP M3UYH
SY23 5HS 2E0DCK
SY23 5LF M6RIY
SY23 5NA G8UKZ
SY23 5NB G0OBB
SY23Â £DT G7JSH
SY24 5AE G8ONP
SY24 5BE G2CQX
SY24 5BS M6LJP
SY24 5DF M6MLI
SY24 5DJ G3BV
SY24 5HD M6GKT
SY24 5JD G6JWD
SY24 5LT G8NCU
SY24 5LX M3NZV
SY24 5LX M3VRD
SY24 5NP G0PUW
SY24 5NP G7GWT
SY25 6AE M3XOT
SY25 6AJ G4ZHI
SY25 6EJ 2E0TOF
SY25 6EJ M0TOF
SY25 6EJ M6TOF
SY25 6PE G0CEP
SY25 6QU G4YMZ
SY25 6QZ G7LUB
SY25 6TT M0CBD
SY3 0AG G1KQP
SY3 0BE G4JSZ
SY3 0EQ G7WGK
SY3 0ER G7JTD
SY3 0EX G4HAC
SY3 0LA G1SCR
SY3 0LA M1DQI
SY3 0LE G7NLY
SY3 0QF G1CWZ
SY3 5AN G0EIY
SY3 5AS 2E0THZ
SY3 5AS M0XMH
SY3 5AS M6MGH
SY3 5BJ G6TGJ
SY3 5BW G3UDA
SY3 5DF M0AMP
SY3 5PG G4ZPC
SY3 6AB G3NRX
SY3 6AW G3OWJ
SY3 7JN G3JPU
SY3 7NB G0RCS
SY3 7NB G8OFZ
SY3 7QB G3WVL
SY3 7QU G7LRB
SY3 7TG M6XXD
SY3 8AD 2E1MJF
SY3 8AD M5WJF
SY3 8QQ M3XEJ
SY3 8RU G4ZZP
SY3 8RX G1EBT
SY3 8SU G4AZS
SY3 8TJ 2E0NWB
SY3 8TJ M6NWB
SY3 8UX M1FMC
SY3 9DB G3VWH
SY3 9PH M0PDP
SY3 9PH M6OBS
SY4 1BJ G6WNG
SY4 1DT G0ARP
SY4 1DT G6UDG
SY4 1EE G8RIK
SY4 1EE M0DFA
SY4 1JD G3NSS
SY4 1JH G0DNI
SY4 1LX G0BXZ
SY4 2BB G8DSG
SY4 2EH G4SDU
SY4 2HT G0WDW
SY4 2HY G6DQY
SY4 2LG G1CLD
SY4 2LH G7CJC
SY4 3AZ G4EDC
SY4 3JL G4VVD
SY4 3LY G4CBM
SY4 3QX G3LEK
SY4 3YL G4YKX
SY4 4JX G0RVE
SY4 4LA G0JLS
SY4 4PQ 2E0ETC
SY4 4PQ M3ZHZ
SY4 4SN G3SOA
SY4 5BX G1WVV
SY4 5DE G1YJB
SY4 5DE M0BAV
SY4 5NB G3VAO
SY4 5NB G4PNF
SY4 5NB G7BSP
SY4 5NW G7MHL
SY4 5NX G3TUL
SY4 5QZ 2E0SUT
SY4 5QZ 2E0TGG
SY4 5QZ M6RDS
SY4 5QZ M6TGG
SY4 5SL 2E0TVS
SY4 5SL M0KZP
SY4 5SL M6TVS
SY4 5YD G3PLW
SY5 0BG 2E0SIX
SY5 0EH G8DIR
SY5 0NG M3WKC
SY5 0NS M0PWT
SY5 6DS G7DCM
SY5 6DS G7JWH
SY5 7AP G6FHM
SY5 7AP M0FHM
SY5 7AT M3SUW
SY5 7AX G3ZQC
SY5 8AN G7DWX
SY5 8AN M0DEV
SY5 8AU 2E1DEM
SY5 8EP G4MYZ
SY5 8LE M1DQG
SY5 8ND M0ZPM
SY5 8NH G8DIQ
SY5 8RA G2CQX
SY5 9AS G4FRX
SY5 9AT G4XDC
SY5 9DJ G3YFK
SY5 9EG 2E0GQW
SY5 9EG M0GQW
SY5 9EG M3GQW
SY5 9EU G3VVL
SY5 9JA 2E1UTD
SY5 9JQ G3LNP
SY5 9JS G8NGF
SY5 9JY M1DQH
SY5 9LF M6JUX
SY5 9QR G7VYI
SY6 6DD G3LRH
SY6 6DP M1DTS
SY6 6EE M6GVL
SY6 6HT M6GVM
SY6 6JW G4SMA
SY6 6LX G7UMF
SY6 6NJ G4UST
SY6 6QD G0WPX
SY6 6QD G3BJ
SY6 6QD G4JKS
SY6 6RB M0COP
SY6 7AB M1CHU
SY6 7BN G8WPF
SY7 0EF M6FNB
SY7 0EP M3JLR
SY7 0NB G0NGN
SY7 8BA M6DUX
SY7 8DG M3MRM
SY7 8EN M6FSW
SY7 8QR M3YZO
SY7 9DU 2E0CHZ
SY7 9DU G7RHF
SY7 9DU M6RAL
SY7 9DW M3SCF
SY7 9DW M6XCF
SY7 9DX G7EYM
SY7 9ER G3ZZX
SY7 9ET M6ZJH
SY7 9HX M1CMM
SY7 9LF M6ZJP
SY7 9PG G0CGI
SY7 9PS M0DUV
SY7 9RF M6PNR
SY7 9RL M6AUA
SY8 1BH M3ZXC
SY8 1BQ G6CRD
SY8 1BQ M3FRJ
SY8 1DW G3VWX
SY8 1EY G3SEZ
SY8 1JE M3YAI
SY8 1JR M6BQC
SY8 1LE M6AOW
SY8 1NB 2E0DRI
SY8 1NB M6XKG
SY8 1NB M6YDA
SY8 1NH G1VVX
SY8 1NS M3RLM
SY8 1UJ M6CSF
SY8 1XN G4HQB
SY8 1XN G7HSA
SY8 1XN M3XLM
SY8 1XP M0SJL
SY8 1XZ G3KZE
SY8 2DH M6JBX
SY8 2DT G8EZD
SY8 2NS M3JWM
SY8 2PH 2E0XTL
SY8 2PH G4OYX
SY8 2PH G8XYJ
SY8 2PH M3XTL
SY8 2QD M6GGJ
SY8 3AE 2E0CDS
SY8 3AE G6LXL
SY8 3AE M0MTS
SY8 3AW M0CBA
SY8 3DR M3EBF
SY8 3DR M3OMF
SY8 3EB 2E0PKL
SY8 3EB M6PKL
SY8 3ED M1DKZ
SY8 3JX M6ZGZ
SY8 3NJ G8HMV
SY8 3NZ G1AJS
SY8 3NZ G8GUN
SY8 4BX M3NFH
SY8 4EP M1AXD
SY8 4HD M6AUH
SY8 4LF G4AIJ
SY8 4LF M0GZU
SY8 4LF M6EGG
SY8 4LX G1JOD
SY8 4ND M1MST
SY8 4NE G0EXA
SY8 4NQ G1KGE
SY9 5AF G3NUB
SY9 5BS M5DND
SY9 5EF 2E1HUB
SY9 5EL M0GGU
SY9 5ES G0RPW
SY9 5HB 2E1HTY

## TA
## (Taunton)

TA1 1LB G4XUR
TA1 1PH G0UJU
TA1 1UW M6AXP
TA1 1UW M6KGV
TA1 2AF 2E0SBB
TA1 2AF M0SBB
TA1 2AF M6AES
TA1 2AF M6TBJ
TA1 2BD 2E0BBZ
TA1 2BS G1YLN
TA1 2JQ G1NZK
TA1 2LT M3LQI
TA1 2NA G7CCV
TA1 2NJ G0PSE
TA1 2PL G8DUI
TA1 2QJ G4ASK
TA1 2QX 2E0WDI
TA1 2QX M3ONB
TA1 2RA 2E0AKN
TA1 2RA M0GSN
TA1 2RA M3LQV
TA1 2RS M0HQB
TA1 2TA G0PNB
TA1 2XF G0WKU
TA1 2XN G5JJ
TA1 2XN M0CIF
TA1 3DA G4RWF
TA1 3EH G0EYR
TA1 3EJ G4RNT
TA1 3EL G0PGL
TA1 3HQ G0UIL
TA1 3JE M0CIE
TA1 3PA G4BJX
TA1 3RR G4VVM
TA1 3RW G7SLZ
TA1 3RW M3SLZ
TA1 3XN G0UFU
TA1 4JE G4MYE
TA1 4JX G3JKE
TA1 4NL G7RAZ
TA1 4QH M3PVP
TA1 4RE G8JXK
TA1 5DS G4WTA
TA1 5EH G0RIE
TA1 5EP 2E0ZSE
TA1 5EP M0SEV
TA1 5EP M6ZSE
TA1 5ES G8XHU
TA1 5HH 2E0HJZ
TA1 5JE M6GXH
TA1 5JH 2E1BFW
TA1 5JH G7KIQ
TA1 5NS M3XNO
TA1 5PQ G4VVS
TA1 5PQ G8ZSP
TA1 5PW G3TJH
TA1 5QT M3PDD
TA1 5QZ 2E0BPI
TA1 5QZ M3WEQ
TA10 0HH G3RSU
TA10 0JH G0PNF
TA10 0NG G6AYD
TA10 0NX G4BWL
TA10 0PP G6JNS
TA10 9AE G0DGA
TA10 9AE G0DGE
TA10 9AX 2E1JKL
TA10 9DD 2E1LGE
TA10 9DH G1HQN
TA10 9EZ G8UIG
TA10 9HW G6YTZ
TA10 9JX 2E0PDB
TA10 9JX M6PEF
TA10 9NJ G3PCJ
TA10 9NL G3USC
TA10 9NP 2E0GKR
TA10 9NP M0VKR
TA10 9NP M3OLZ
TA10 9PR G0RGC
TA10 9RH G0SZT
TA10 9ST M6PGN
TA10 9TA M3XSR
TA10 9TF G0FZH
TA10 9TH G7LPO
TA11 6BW G0HDJ
TA11 6ER G1PUQ
TA11 6HP G3YYR
TA11 6JY M6AXU
TA11 6LG G7IHP
TA11 6PS G3TZA
TA11 6PX G1DCZ
TA11 7AP M0BJO
TA11 7DN G3TCT
TA11 7HD G6HIQ
TA12 6BG G8GVW
TA12 6DE M6FDF
TA12 6DG G0NKC
TA12 6DT G1AKD
TA12 6HG 2E0LQW
TA12 6HG M0RGE
TA12 6HG M3LQW
TA12 6JU G3AIK
TA12 6NY G0PNS
TA12 6NY G3LWM
TA13 5AD G3OTK
TA13 5AP G4PDG
TA13 5AQ G8VBI
TA13 5BN M1AIK
TA13 5BY 2E0ZXG
TA13 5BY M0ZXG
TA13 5BY M3ZXG
TA14 6QN G6EER
TA14 6QN G8LKP
TA14 6SH M1BSN
TA16 5NF M6NBG
TA16 5NP G3TEL
TA16 5QX M6KMF
TA18 7AH G6AFL
TA18 7AP M3UBH
TA18 7AQ G0NPF
TA18 7AX G1PVZ
TA18 7BE G4UTY
TA18 7PB M0GJD
TA18 7SG G4NLI
TA18 8BL G4FVW
TA18 8DB M1RMW
TA18 8DF 2E0BFJ
TA18 8DF M3JYH
TA18 8DF M3TGD
TA18 8ET G1AWU
TA18 8HS M6XAA
TA18 8HT M0AGT
TA18 8JA M0BCL
TA18 8JB M0BBO
TA18 8JJ G3WMX
TA18 8LY G4NCU
TA18 8PL G0JCY
TA18 8QE G0EGC
TA19 0EA M3FGQ
TA19 0HH G0MEQ
TA19 0NT M0WDL
TA19 0NZ G4OXR
TA19 9AH G6AOV
TA19 9AH G6LVG
TA19 9DG G7LNY
TA19 9ER M0ILT
TA19 9NH G4KJD
TA19 9NJ G4FDG
TA19 9NW G4HTD
TA19 9QH G6PQP
TA19 9QL G0WSC
TA19 9QN M0AOD
TA19 9QS M6TVH
TA19 9QU M0BHE
TA19 9RJ G4NBN
TA2 6LE G1TKQ
TA2 6RR G4SZS
TA2 6SR G8HNM
TA2 6TA 2E0BQY
TA2 6TF G6EED
TA2 6TN M6CTT
TA2 7DT G1CNZ
TA2 7EF G4WSF
TA2 7EN G1AJZ
TA2 7PB G4XKK
TA2 7SP G7LDD
TA2 7SP M3TJQ
TA2 7TG M6SDS
TA2 8DX G0PWV
TA2 8DX G0RWA
TA2 8DZ G7KYG
TA2 8LB G7CWT
TA2 8NA G4UTM
TA2 8NA G8BTY
TA2 8PP G6EIO
TA2 8QB G1VVE
TA2 8RL M0FBM
TA2 8RP G3PCG
TA20 1BX 2E0TAW
TA20 1BX M3MNT
TA20 1BZ G3HAL
TA20 1DG M1DGP
TA20 1DZ G4IRS
TA20 1EZ G8DLH
TA20 1FH G1ZYS
TA20 1JU G6WZA
TA20 1NJ G0LNK
TA20 1NT M6OCO
TA20 2BL G4VHB
TA20 2BU G1ECV
TA20 2EB G8BVR
TA20 2EG M0FAK
TA20 2EG M3AIG
TA20 2EG M3BXG
TA20 2EW 2E1IJR
TA20 2HH M3UVV
TA20 2JB G8GZC
TA20 2ND G8LKK
TA20 2PY G8GFZ
TA20 2RX M3UTN
TA20 2SX G7MFH
TA20 2TJ M3LBZ
TA20 3BH M1FAY
TA20 3HJ G0KJF
TA20 3HJ G8HOI
TA20 3PH 2E0DZT
TA20 3PH M6DZT
TA20 4DN G0AEN
TA20 4PJ G4TIA
TA21 0AY G8UXY
TA21 0DX 2E0PDU
TA21 0DX M0PDU
TA21 0DX M3PDK
TA21 0DX M3PDU
TA21 0RH G8FZI
TA21 8BB G0MJF
TA21 8DH G8ZVK
TA21 8EP M6JST
TA21 8ER G1CQG
TA21 8EX G4RGA
TA21 8HZ G4YXR
TA21 8HZ M0MAX
TA21 8JB M5WSS
TA21 8LG G7BSG
TA21 8PY G4OLA
TA21 8TF M6RPQ
TA21 9AL 2E0AKL
TA21 9AL M0GFO
TA21 9AL M3NLX
TA21 9AR G6TWA
TA21 9AT G4RGF
TA21 9AX G4AIU
TA21 9DY M3SNL
TA21 9ED M5SLC
TA21 9HT M0CVB
TA21 9HW M6TEJ
TA22 9BY M6MRW
TA22 9EL G7KWP
TA22 9EN G1TKE
TA22 9EN M6SES
TA22 9PT G0JFA
TA22 9RU G8CCF
TA23 0AG G6BWJ
TA23 0BP 2E0EAI
TA23 0BP M6SMB
TA23 0HZ G6KRG
TA23 0PZ M6NRG
TA24 5HS G4PHC
TA24 5JU G3NFY
TA24 5NZ G8YZY
TA24 5NZ M0AOJ
TA24 5TD G7PBO
TA24 6AD M3LSE
TA24 6AD M3LSK
TA24 6BH G1CGU
TA24 6BT M3GQI
TA24 6BY G6HOR
TA24 6EE 2E0ONV
TA24 6EE M3ONV
TA24 6HJ G4HOW
TA24 6JH M6RJW
TA24 6PW M3BLR
TA24 6SD 2E0WSR
TA24 6SD M6DUN
TA24 6UL G8CPF
TA24 7UL M3BXH
TA24 8AF G0WMN
TA24 8AW G0ACA
TA24 8AX M1AFZ
TA24 8BZ G0OUO
TA24 8EB G4FNJ
TA24 8HL G1ONV
TA24 8JR M6ARV
TA24 8JR M6ASI
TA24 8QF G4TIC
TA24 8TA G8BCI
TA246AD 2E0AKO
TA3 5AU G6PPU
TA3 5BX G4LZU
TA3 5BY M3PBP
TA3 5DW G4AVJ
TA3 5EW G4IBC
TA3 5EW M0AID
TA3 5JW G1NTK
TA3 5LP M6AYQ
TA3 5PW G4MIB
TA3 5QP G0MZQ
TA3 5TE G4JZL
TA3 5TE G8BDM
TA3 6AD M6USG
TA3 6DX M0HMX
TA3 6HA G1DRY
TA3 6HA G1KNK
TA3 6PU G8XOE
TA3 6QN G8XET
TA3 6QN M3LVK
TA3 6TN G7DKZ
TA3 7AQ G4JKZ
TA3 7EG G4UVV
TA3 7HB G4NIL
TA3 7HH G4MFD
TA3 7HS G8WXV
TA3 7LG G4KIK
TA3 7NR G3THG
TA3 7RE G6TOY
TA3 7SD G4UVZ
TA4 1AQ G1XQP
TA4 1EU M6OPS
TA4 1EX G4ZTS
TA4 1JA G4FDA
TA4 1NY G7HKZ
TA4 2BU G0GWS
TA4 2DP G3ULL
TA4 2DP M3BAO
TA4 2JY G0FUS
TA4 2LZ M0GHR
TA4 2QF G4ETN
TA4 2SB G0HMG
TA4 3AE M6NPN
TA4 3AQ G3UWH
TA4 3AU M6HFU
TA4 3LA G3VFB
TA4 3TR G4BQB
TA4 3TX M0GFX
TA4 4DE G0NKZ
TA4 4DE G0SZO
TA4 4JR M6BGI
TA4 4JZ G4RZR
TA4 4LA G0WIE
TA4 4NZ M6CJE
TA4 4PD G3HMG
TA4 4PD G3POM
TA4 4PG G4AJU
TA4 4SL G0NRI
TA5 1LZ G8KOE
TA5 1QG G8AYV
TA5 1SF G4OEC
TA5 2AP G3OJK
TA5 2BQ G7UIO
TA5 2GG M6FQM
TA5 2GT G0MRR
TA5 2HW G3TTP
TA5 2JN M6CTY
TA5 2PY G7UUL
TA5 2PZ G4WMY
TA6 3HZ M6YOX
TA6 3LP G4BRW
TA6 3QE G7SYS
TA6 3TD 2E0PAH
TA6 3TL M3SZK
TA6 4DW G3RHW
TA6 4DW M6JNQ
TA6 4ET M6XBE
TA6 4HL M3TOE
TA6 4HQ G4SMQ
TA6 4JL G6JTT
TA6 4LD G6UNN
TA6 4NA 2E0WEC
TA6 4NA G7GMZ
TA6 4NA M0WFN
TA6 4NA M3CSK
TA6 4PN G0OYA
TA6 4RY G1BKL
TA6 4SH G0MIK
TA6 4SH G7FAD
TA6 4SL G0VOE
TA6 4UB M3HET
TA6 4UT G7KHZ
TA6 4XD M6MXB
TA6 4XP M1FFC
TA6 5NS G0LCV
TA6 5PF G0COM
TA6 5PH G0CNA
TA6 5SA M3WSO
TA6 6AX G0OXB
TA6 6BW G4DII
TA6 6DZ G8PVG
TA6 6GR M1GRA
TA6 6HF M6MHL
TA6 6JU G1MCY
TA6 6LF M3JRF
TA6 6LJ M6AZF
TA6 6QW 2E0VCM
TA6 6QW G6MYH
TA6 6QW M3VCM
TA6 6RN G8FDF
TA6 6SJ G8ZOJ
TA6 6UR G1XOZ
TA6 7AJ G0WRQ
TA6 7LJ G1BTI
TA6 7NZ G7WBL
TA6 7PD G1HLP
TA6 7QG G0UQT
TA6 7QY G7URS
TA6 7RF G8YPV
TA7 0DE G3FSA
TA7 0EZ 2E0SST
TA7 0EZ M6SST
TA7 0HD G8VHO
TA7 0HD M3DBY
TA7 0HD M3PQJ
TA7 0HJ G4ELW
TA7 0JR M6FUL
TA7 0NP M6IRL
TA7 0RH G6OBB
TA7 0SB G4WGX
TA7 8AL G4ZBS
TA7 8AP M0BZV
TA7 8AS G6YGH
TA7 8BS 2E0TDI
TA7 8EA G7VWM
TA7 8EP G3AJW
TA7 8ER G3MDD
TA7 8ER G4RGY
TA7 8HJ G0GCA
TA7 8HL G0LHX
TA7 8HQ G1NMQ
TA7 8JB G4UNO
TA7 8JP G0BDJ
TA7 8QR G6RRV
TA7 9AR G4BSK
TA7 9BT G1HSF
TA7 9BX G3HZW
TA7 9DY G4GHM
TA7 9HB G4VHQ
TA7 9LR G8SUW
TA7 9LS G4NQQ
TA8 1BE M3KJS
TA8 1HE G4SJJ
TA8 1HE M1DIL
TA8 1HG G0BBK
TA8 1HY G4HLN
TA8 1JA G4DVK
TA8 1JU 2E1IIM
TA8 1LD G1MOK
TA8 1LL 2E0MIJ
TA8 1LL M0VLN
TA8 1LL M6PMH
TA8 1LN G1KVQ
TA8 1LT G4EJW
TA8 1LW G7VWG
TA8 1LW M3IOX
TA8 1LW M6IFT
TA8 1RE M3UHS
TA8 2EJ G6TAH
TA8 2EL G6TKR
TA8 2HP G0FYP
TA8 2JT G3WUZ
TA8 2LE G7NME
TA8 2LN G4CGH
TA8 2LT M6ADD
TA8 2NL G3GZZ
TA8 2NN G0PQD
TA8 2NW G3WLA
TA8 2PH G3NYD
TA8 2PN G6PJT
TA8 2PS G3UBP
TA8 2QA M3PHG
TA8 2QJ M0NDE
TA8 2QJ M3SGG
TA8 2TT G6SEE
TA8 2TT G6SEF
TA9 3DQ M1DSZ
TA9 3EH 2E0VOK
TA9 3EH 2E0ZUX
TA9 3EH M0VOK
TA9 3LA M6GPP
TA9 3LD G4UOS
TA9 3LE G4NFO
TA9 3NF M0FEY
TA9 3QU G3XBI
TA9 3QZ G0VYV
TA9 3RF G7GJZ
TA9 4BS G7FEP
TA9 4BU G8HUH
TA9 4DQ G1JGM
TA9 4DT G4TBO
TA9 4DU G4WSM
TA9 4DU G7VIV
TA9 4DY G0GOB
TA9 4HE M6GLM
TA9 4LZ M0NDL
TA9 4NG G3VLW
TA9 4PH M3LDX
TA9 4QT G3SXQ

## TD
## (Tweed)

TD1 1HD G7GIS
TD1 1QL G8KOF
TD1 1RG M3RQG
TD1 1SQ G6IZU
TD1 2BA G4JPG
TD1 2EE G1JLP
TD1 2HZ G7GPG
TD1 2LB G8UUW
TD1 2RB G8CJW
TD1 2RJ 2E1GEZ
TD1 2RJ 2E1GXX
TD1 2RJ M0BPX
TD1 3JW G4YEQ
TD1 3JW G6WRY
TD1 3ND G7LUN
TD1 3RF G0HQT
TD1 3SB 2E1BBY
TD10 6XN M6CES
TD11 3EE M6KFE
TD11 3HG G0TUS
TD11 3HQ G0RDZ
TD11 3HS M3URQ
TD11 3HW G1DVO
TD11 3LG 2E0TIN
TD11 3LG M0XXL
TD11 3LG M3ZFW
TD11 3PP M6BNX
TD11 3TQ G0EUL
TD11 3UF M6KWA
TD12 4BS G0HNP
TD12 4ED 2E0ICB
TD12 4ED M0ICB
TD12 4EY 2E0CDO
TD12 4JQ G4SYF
TD12 4NG G8BDX
TD12 4SP M6SBN
TD14 5AU M0WZZ
TD14 5DL G0BRS
TD14 5DX G4ZXJ
TD14 5EU G1JFF
TD14 5EU M1DTU
TD14 5JF M6BPM
TD14 5LG 2E0DZZ
TD14 5LG M0GZZ
TD14 5LG M3UQN
TD14 5LS G7SRJ
TD14 5NJ 2E0BXY
TD14 5NJ M3ZFK
TD14 5PB 2E0XDX
TD14 5PB M0XAB
TD14 5PB M6FRD
TD14 5PB M6LRK
TD14 5RR G4FVM
TD14 5SA M3LSO
TD14 5TB G7LNO
TD15 1BZ G3VAJ
TD15 1SU G4WDO
TD15 1XB G7ROC
TD15 1XW G6VGO
TD15 2BB G4XVF
TD15 2BS G0JCP
TD15 2BS G1IVI
TD15 2DX G4GCL
TD15 2EB G3KML
TD15 2EE M0GHK
TD15 2JH M3INO
TD15 2JN G6IBP
TD15 2LA 2E0KLN
TD15 2LA M0KLN
TD15 2RW 2E0DDO
TD15 2RW M6CMI
TD15 2TY G0HBV
TD2 6TT G0SRO
TD3 6ND M6TWE
TD3 6NF G4YSN
TD4 6AS G0RWU
TD4 6BN G3RXU
TD4 6HS 2E0YBR
TD4 6HS M6BIH
TD5 7AS G0ALW
TD5 7AS G0ALX
TD5 7AS G4KHS
TD5 7EU G0SIA
TD5 7EU G1FMV
TD5 7NB G6ZFI
TD5 7NJ G4CXP
TD5 7NN G6PTX
TD5 7NP G6TBE
TD5 7PB G6RTN
TD5 7RU G0CDV
TD5 7SZ G3PPE
TD5 8BB G3VLB
TD5 8LB G4NLJ
TD5 8SA G8TCH
TD6 0QS M0BPP
TD6 0RY G4CID
TD6 0RY M3OXB
TD6 9BH M0RMM
TD6 9EP G0RDC
TD6 9JB G4PNM
TD7 4BG G7GRH
TD7 4HS M0HTL
TD7 5BH M3XXO
TD7 5BP G0KCN
TD7 5DD G0FTJ
TD7 5EY G4UFP
TD7 5JB G4VYU
TD7 5LS G3VAL
TD7 5NF M3PAE
TD7 5NF M3YDH
TD7 5NF M3YJH
TD8 6AG G4UPX
TD8 6DU G1ZTB
TD8 6RB G0VHR
TD8 6SA G0FSH
TD8 6SA G0IQI
TD8 6SD G7WLO
TD8 6SU M6BUG
TD8 6TZ G6MCV
TD9 0QQ 2E0EBU
TD9 0QQ M6FZT
TD9 0SN G4LPJ
TD9 0TT 2E0XRZ
TD9 0TT M0XRZ
TD9 7BH G0FTK
TD9 7PR G0IGJ
TD9 7QE M1APS
TD9 8BA G8GUX
TD9 8EN G6CKN
TD9 8SG G1KWA

## TF
## (Telford)

TF1 1JH G1VIT
TF1 2AJ 2E1HKS
TF1 2AS G8VZT
TF1 2AS M5AFG
TF1 2BY G1BHV
TF1 2JF G8BKH
TF1 2JP G0RQI
TF1 2LE G4AUY
TF1 2ND M3YWF
TF1 2PQ G0ASP
TF1 3DA G3MHY
TF1 3DA M6BVC
TF1 3DB G1JJK
TF1 3ED G6HKP
TF1 3GA G6NRM
TF1 3HU M0CFT
TF1 5FL M3IJT
TF1 5LF G0UCN
TF1 6FZ G1ZPQ
TF1 6PQ G1IRX
TF1 6PQ G6NOL
TF1 6QR G2RSA
TF1 6TD G8RXZ
TF1 6XT M0CPK
TF1 6XU G6ZFU
TF1 6YL M1SWR
TF10 7BS 2E0PNN
TF10 7BS M0PNN
TF10 7DT G7KZG
TF10 7DX G7ACD
TF10 7EF G4IFR
TF10 7EG G1AJT
TF10 7HH G7KJA
TF10 7JD G0KBK
TF10 7LS G0UCI
TF10 7NN M6IMA
TF10 7RG G0MKP
TF10 7RU G4ZAQ
TF10 8AH G4JOW
TF10 8DA G0GQK
TF10 8JL G1KRU
TF10 9AQ G4OLO
TF10 9EN 2E0BAJ
TF10 9EN M0EAK
TF10 9EZ M0TAW
TF10 9EZ M3OGH
TF10 9JQ G0WJH
TF11 8EB M3YBW
TF11 8HZ G0VSJ
TF11 8RD G0EXN
TF11 9BF M3FAL
TF12 5PE G6APE
TF12 5PE M3SXK
TF12 5SB G4YDT
TF12 5SH G6DBZ
TF13 6BW G1HMI
TF13 6BZ G6KJY
TF13 6ED G8AQA
TF13 6ED M0HMO
TF13 6EN 2E0YPW
TF13 6EN M0YPW
TF13 6EN M6YPW
TF2 0AS 2E0DTB
TF2 0DT G4OMI
TF2 0DX G1OAR
TF2 0DX M0OAR
TF2 0DY M0VZT
TF2 6EZ G7NIB
TF2 6HD M6HBE
TF2 6JR M6AAD
TF2 6NA M1BGL
TF2 6RF G0GUV
TF2 6RR 2E0APG
TF2 7DA 2E1ACW
TF2 7LJ G0VXJ
TF2 7LQ G1SKW
TF2 7NP M3XNU
TF2 8AR M3DLB
TF2 8DD M3ZXX
TF2 8HY G0HDC
TF2 8JU G4UMM
TF2 8NA G4YVD
TF2 8SQ M3GZT

| Postcode | Callsign |
|---|---|
| TF2 9EN | M6DCJ |
| TF2 9HD | 2E0DCM |
| TF2 9HD | M0ECM |
| TF2 9HD | M0EMM |
| [illegible] | [illegible] |
| TF2 9LU | M6ANN |
| TF2 9LX | 2E0KLS |
| TF2 9PT | G6APX |
| TF2 9QJ | G0BCP |
| TF2 9RB | G3JKX |
| TF2 9SH | M6FDN |
| TF2 9SL | G4KLD |
| TF2 9SR | 2E1HTV |
| TF3 1BW | 2E0TIL |
| TF3 1BW | M6TIL |
| TF3 1ED | 2E1DZJ |
| TF3 1ED | G8ZIP |
| TF3 1EG | G0UFE |
| TF3 1EG | G6UDX |
| TF3 1LA | G0BST |
| TF3 1LJ | 2E0BAH |
| TF3 1NH | G8ZXU |
| TF3 1NY | M3UWE |
| TF3 1PT | M6LEU |
| TF3 1SS | M6YVN |
| TF3 2AH | M0HBI |
| TF3 2AQ | G7SFI |
| TF3 2BU | M0TYK |
| TF3 2BY | 2E0TTB |
| TF3 2BY | M6LMB |
| TF3 2BY | M6TTB |
| TF3 2DP | M1IHM |
| TF3 2EE | 2E1IHJ |
| TF3 2ES | M6MGG |
| TF3 2HX | 2E1GKE |
| TF3 2JB | M6KZM |
| TF3 2JJ | M6HEW |
| TF3 2JW | M0TBQ |
| TF3 2JX | G4FBZ |
| TF3 2NH | M0IRS |
| TF3 2NQ | G6FFR |
| TF3 5DA | G0HQK |
| TF3 5GQ | 2E1GJJ |
| TF3 5HD | 2E0UDE |
| TF3 5HD | M0RKY |
| TF3 5HD | M6RLB |
| TF3 5HD | M6TZE |
| TF3 5HE | G0GAL |
| TF3 5HN | M6KCA |
| TF4 2AR | 2E0ZGS |
| TF4 2EB | 2E0CVC |
| TF4 2EB | M6CSZ |
| TF4 2EW | G0MQX |
| TF4 2GA | M0EBT |
| TF4 2HS | G6NWT |
| TF4 2NX | 2E0MCL |
| TF4 2NX | M3LML |
| TF4 2PP | M3NZW |
| TF4 2QA | 2E1HTU |
| TF4 2QE | G0VNO |
| TF4 2RW | 2E0PLX |
| TF4 3BJ | G7SEY |
| TF4 3DD | G8JVM |
| TF4 3HP | 2E0PWF |
| TF4 3HP | M6EGV |
| TF4 3JF | G0VPW |
| TF4 3LB | M6BRB |
| TF4 3NG | 2E1FNX |
| TF5 0LL | G8OYB |
| TF5 0NQ | M1DGX |
| TF5 0NQ | M1DLX |
| TF5 0PG | G7BCO |
| TF6 5AU | 2E0TRO |
| TF6 5AU | M3UWB |
| TF6 5ER | G6SGD |
| TF6 6BA | M3PHQ |
| TF6 6DH | G0JBO |
| TF6 6HH | G4WRK |
| TF6 6HQ | G3UKV |
| TF6 6HQ | G3ZME |
| TF6 6HQ | G8UKV |
| TF6 6HQ | M3UKV |
| TF7 4AB | M1ACA |
| TF7 4AQ | M0DNV |
| TF7 4AS | G0GQO |
| TF7 4BX | 2E1CFB |
| TF7 4DP | M1GIZ |
| TF7 4DT | G4VZL |
| TF7 4DW | G7VDS |
| TF7 4FA | M6HDA |
| TF7 4HP | M1EZJ |
| TF7 4HT | G6ZMF |
| TF7 4HT | G8UGL |
| TF7 4HT | M0UGL |
| TF7 4NH | G4VZK |
| TF7 4QE | G0LCO |
| TF7 4QE | G4TQZ |
| [illegible] | [illegible] |
| TF7 5HB | M3UXN |
| TF7 5JJ | M6OLY |
| [illegible] | [illegible] |
| TF7 5QH | 2E0KDF |
| TF7 5QH | M6BVF |
| TF7 5QN | M3KXB |
| TF7 5QR | G0JAL |
| TF7 5RX | M1NMG |
| TF7 5SF | 2E0IVR |
| TF7 5SN | G0HCT |
| TF7 5SU | G4EIX |
| TF8 7AU | 2E0FPO |
| TF8 7BS | G7OQT |
| TF8 7EW | M0AXV |
| TF8 7ND | 2E1ESW |
| TF8 7ND | G0VXG |
| TF8 7ND | M3BHI |
| TF9 1AR | G8EWD |
| TF9 1EA | G0JNA |
| TF9 1FN | 2E1FYI |
| [illegible] | [illegible] |
| TF9 1NP | G1DVH |
| TF9 2BA | G3OKD |
| TF9 2BB | G6JKV |
| TF9 2BD | G0PLB |
| TF9 2DG | G0MDJ |
| TF9 2DU | G0DHM |
| TF9 2HG | M6GMU |
| TF9 2HG | M6SNU |
| TF9 2QG | G0GHL |
| TF9 2QX | G0AGC |
| TF9 2SF | G3ZGU |
| TF9 2SX | 2E0CBA |
| TF9 2SX | M6HRC |
| TF9 2TF | M3WXG |
| TF9 2TF | M3XKJ |
| TF9 2TG | M3MQJ |
| TF9 2TH | M1MDE |
| TF9 3BD | G4MZQ |
| TF9 3HJ | G6COB |
| TF9 3LP | G6WPL |
| TF9 3NP | G8EOH |
| TF9 3QB | G0CER |
| TF9 3QB | G6VSG |
| TF9 3QF | G0CGA |
| TF9 3QH | M3THE |
| TF9 3SY | 2E0UNI |
| TF9 3SY | M0UNI |
| TF9 3UB | 2E0SHG |
| TF9 3UH | G8PSZ |
| TF9 3UJ | G0CDS |
| TF9 4BE | G3JPB |
| TF9 4DD | 2E0SNP |
| TF9 4DN | M0INI |
| TF9 4DN | M6INI |
| TF9 4ED | G4JKF |
| TF9 4LQ | 2E0MKF |
| TF9 4QE | G6GTJ |
| TF9 4QR | G3WEI |
| TF9 4RJ | M1AOL |

## TN
(Tonbridge)

| Postcode | Callsign |
|---|---|
| TN1 2LH | G4ZFP |
| TN1 2SH | G3WTR |
| TN10 3HB | M3XST |
| TN10 3HD | M6TVE |
| TN10 3HG | G0DJC |
| TN10 3HG | G7TYR |
| TN10 3HG | G8WZK |
| TN10 3HQ | M3TWP |
| TN10 3LW | G1OGV |
| TN10 3LW | M3OGV |
| TN10 3NN | G1ZZL |
| TN10 4AJ | M6CBU |
| TN10 4BJ | M6YOU |
| TN10 4DG | G4FFY |
| TN10 4HD | G7KNS |
| TN10 4NG | G4FYG |
| TN10 4NN | M3ZKL |
| TN10 4QS | G0VGD |
| TN11 0HU | G4HGR |
| TN11 0NL | 2E0RXW |
| TN11 8HP | G1WKS |
| TN11 8JH | M0RBT |
| TN11 8LE | G3YJW |
| TN11 9DA | G6IUS |
| TN11 9DR | G6MXE |
| TN11 9ED | G7THJ |
| TN11 9ED | M3THJ |
| TN11 9HB | G4AUL |
| TN11 9HD | G8CAA |
| TN11 9HQ | G4TPJ |
| TN11 9LG | G4JED |
| TN11 9LG | M6JHL |
| TN11 9PL | G3ZDT |
| TN12 0BT | G7GMU |
| TN12 0BX | G4VEG |
| TN12 0LD | G0EIH |
| TN12 0LQ | G0IWJ |
| TN12 0LR | G0UXG |
| TN12 0LR | M0KWA |
| TN12 0NA | G7GGA |
| TN12 0NE | M0ORQ |
| TN12 0PA | M3PWZ |
| TN12 0PG | M1CGR |
| TN12 0RH | G1NDK |
| TN12 0TE | 2E0KCX |
| TN12 5BW | G0EBS |
| TN12 5DB | M6VKH |
| TN12 5EQ | G4YIM |
| [illegible] | [illegible] |
| TN12 6BG | M3IMP |
| TN12 6DX | M1ATA |
| [illegible] | G0MID |
| TN12 6LS | G1YLB |
| TN12 7EA | G8CDD |
| TN12 7HA | G3KOM |
| TN12 8EF | M6MPY |
| TN12 9AX | G0WBV |
| TN12 9BS | G8YBH |
| TN12 9JD | G0RHO |
| TN12 9LS | G7FKX |
| TN12 9NB | G3XBQ |
| TN12 9SG | M3RTE |
| TN12 9TQ | G3UUG |
| TN12 9TW | M3XPF |
| TN13 1EL | G8KVO |
| TN13 1EU | 2E0RJX |
| TN13 1EU | M0RMI |
| TN13 1EU | M6EYJ |
| TN13 1TJ | G0HAU |
| TN13 1TJ | M0HAU |
| TN13 2AH | G3THM |
| TN13 2DZ | 2E0DJP |
| TN13 2DZ | M0PRV |
| TN13 2DZ | M0RWK |
| TN13 2DZ | M6DJP |
| TN13 2HE | G1TGZ |
| TN13 2UA | G4ASZ |
| TN13 2UG | 2E1EAX |
| TN13 2UG | G4AKY |
| TN13 3DR | M3PPR |
| TN13 3PJ | 2E0MEK |
| TN13 3PJ | M6TEO |
| TN13 3SB | M6YJK |
| TN13 3XA | G0CDQ |
| TN13 3XN | M3KDM |
| TN14 5AS | M1DUA |
| TN14 5AT | G7MYY |
| TN14 5BT | G7RHE |
| TN14 5JF | 2E0KJJ |
| TN14 5JF | G8PHM |
| TN14 5LD | G0SLU |
| TN14 5PN | G8OMQ |
| TN14 5PT | G6BNJ |
| TN14 5QP | G0WQC |
| TN14 5RR | G6KLF |
| TN14 6BX | 2E0YEW |
| TN14 6BX | M6KTC |
| TN14 6BX | M6PRA |
| TN14 6HT | G3SAR |
| TN14 6LY | G1TTX |
| TN14 6NH | M1RSB |
| TN14 7AU | G7IET |
| TN14 7EY | G1KVW |
| TN14 7HW | M1NTY |
| TN14 7JU | G7JYG |
| TN14 7NA | G7KTR |
| TN14 7SW | 2E0JDA |
| TN14 7SW | M0JDA |
| TN14 7SW | M6JDA |
| TN14 7TD | G7PVF |
| TN15 0LZ | M1JDW |
| TN15 6BB | G1FHR |
| TN15 6BL | G3GJW |
| TN15 6BL | G3ZRX |
| TN15 6BL | G4BRC |
| TN15 6BP | G1MNY |
| TN15 6DT | G4RVV |
| TN15 6EZ | G7MNZ |
| TN15 6HP | G8ZZK |
| TN15 6JJ | G8TLP |
| TN15 6LL | M1DMN |
| TN15 6NU | G0KFW |
| TN15 6PG | G6EDM |
| TN15 6PT | G6GLH |
| TN15 6QL | G8PWO |
| TN15 6TG | G3WBS |
| TN15 6YF | G4MGY |
| TN15 7JU | G8YDB |
| TN15 7NR | G3RGS |
| TN15 7NR | G4CTC |
| TN15 8DZ | M0WAR |
| TN15 8EA | G3IAR |
| TN15 8HT | G0IPH |
| TN15 8LL | G8ZYI |
| TN15 8LQ | G4DFY |
| TN15 8RG | M0MEA |
| TN15 8RS | G3EWM |
| TN15 8RS | G4FQT |
| TN15 8SD | 2E1FDP |
| TN16 1DX | G0POC |
| TN16 1LS | 2E0DIT |
| TN16 1LS | M0KOT |
| TN16 1LS | M6DIT |
| TN16 2AT | G1BMW |
| TN16 2BQ | G6JIY |
| TN16 2BT | G4NSD |
| TN16 2JE | G3KEQ |
| TN16 2LA | G1FKP |
| TN16 3AU | M0WHY |
| TN16 3BD | G3ZVN |
| TN16 3BN | G4ASI |
| TN16 3BX | G3OXS |
| TN16 3HH | G6UNR |
| TN16 3HX | 2E0DWG |
| TN16 3HY | M6GVJ |
| TN16 3QX | M6TBV |
| TN16 3SA | G1XPF |
| TN16 3SG | G8JQS |
| TN16 3UU | G4GLN |
| TN16 3UX | G6THC |
| TN16 3XE | G8AXN |
| TN16 3XY | M3TLW |
| [illegible] | [illegible] |
| TN17 3DZ | 2E1FVY |
| TN17 3JR | 2E0IAK |
| [illegible] | [illegible] |
| TN17 4DB | M0OOD |
| TN18 4JE | G3XWL |
| TN18 4RA | G0SLD |
| TN18 4RW | 2E0IXI |
| TN18 4RW | M6IXI |
| TN18 4RW | M6MOC |
| TN18 5DU | 2E0ZJO |
| TN18 5DU | 2E0ZJQ |
| TN18 5DU | M0ZJO |
| TN18 5DU | M0ZJQ |
| TN18 5DU | M3ZJO |
| TN18 5DU | M3ZJQ |
| TN18 5DU | M6YLC |
| TN18 5LW | 2E0OCB |
| TN18 5LW | M3VXO |
| TN18 5LX | G0IPO |
| TN18 5NR | M0GHW |
| TN18 5PN | G3URA |
| TN19 7DN | [illegible] |
| TN19 7DB | 2E0KWW |
| TN19 7DD | M0KWW |
| TN19 7DB | M6LUD |
| TN19 7PJ | G4NVM |
| TN19 7QY | G7IGF |
| TN2 3BL | M3LCP |
| TN2 3BS | M6MTH |
| TN2 3DA | G4OSH |
| TN2 3HF | G8SQP |
| TN2 3NA | M6HXK |
| TN2 3NY | 2E0XFR |
| TN2 3RX | 2E0PMC |
| TN2 3RX | M0NLP |
| TN2 3RX | M6BFG |
| TN2 3RX | M6FTW |
| TN2 3RX | M6TBH |
| TN2 4BT | G6CRR |
| TN2 4BX | G0VQA |
| TN2 4BX | G3AIO |
| TN2 4ER | G0VYK |
| TN2 4HG | G4KGF |
| TN2 4LF | G6AHK |
| TN2 4NP | G7PMG |
| TN2 4NS | G8JXG |
| TN2 4SU | M3AKW |
| TN2 5DD | G4JFD |
| TN2 5HG | G0AOA |
| TN2 5HT | M0WTW |
| TN2 5LU | G3IIW |
| TN2 5NB | G4CTY |
| TN2 5NN | G4LXC |
| TN2 5RG | G6XPF |
| TN2 5UN | G8HGI |
| TN2 5XU | M6SVN |
| TN20 6UD | G0BUX |
| TN21 0AD | G3MUJ |
| TN21 0EF | G3WWS |
| TN21 0HW | G7CSX |
| TN21 0JH | 2E0AVZ |
| TN21 0JH | M0EDQ |
| TN21 0JH | M3BPZ |
| TN21 0SR | M3YIL |
| TN21 8AS | G6FPQ |
| TN21 8AS | M0FPQ |
| TN21 8BB | G1ITE |
| TN21 8BB | M5ITE |
| TN21 8ED | G4YDB |
| TN21 8EF | M6ANE |
| TN21 8EY | G4KEI |
| TN21 8HE | G4AWJ |
| TN21 8HG | M6IMR |
| TN21 8HG | M6LHR |
| TN21 8HP | 2E0HDE |
| TN21 8HP | M0MDE |
| TN21 8HP | M6HDE |
| TN21 8NR | G8VHK |
| TN21 8NS | 2E0ATF |
| TN21 8NS | 2E1GOM |
| TN21 8NS | G3YNN |
| TN21 8NS | M1EXL |
| TN21 8NS | M3WOW |
| TN21 8NS | M5FUN |
| TN21 8PF | G2BHY |
| TN21 8PF | G3TGF |
| TN21 8PG | M3EMU |
| TN21 8SA | G8KQA |
| TN21 8SJ | G0KOC |
| TN21 8SJ | G4BJC |
| TN21 8SJ | M1SWL |
| TN21 8TG | G1PJT |
| TN21 8TW | G0FDV |
| TN21 8YS | G0MSA |
| TN21 8YS | M0WAJ |
| TN21 8YW | 2E0OPM |
| TN21 8YW | M0WRD |
| TN21 8YW | M6WHS |
| TN21 9AL | G3BBK |
| TN21 9NG | M0RYA |
| TN21 9NR | 2E1LAM |
| TN21 9PP | 2E1GIZ |
| TN21 9PP | M0FCD |
| TN22 1BY | M3NYA |
| TN22 1BZ | G4ZOQ |
| TN22 1DW | 2E0ZCP |
| TN22 1DW | M6ZCP |
| TN22 1HU | G7DQZ |
| TN22 1NN | M1XRC |
| TN22 1RY | G0VZN |
| TN22 1TT | M6SXA |
| TN22 1TU | G8BTC |
| TN22 2AT | G7KID |
| TN22 2AY | M6DYX |
| [illegible] | [illegible] |
| TN22 2BX | M6MTK |
| TN22 2DJ | 2E0MKI |
| [illegible] | [illegible] |
| TN22 2DJ | M3IKM |
| TN22 2DP | M3IZD |
| TN22 3AP | G0BUK |
| TN22 3BE | M3YZH |
| TN22 3DT | G1IFV |
| TN22 3HX | G4GEN |
| TN22 3LJ | M6KAE |
| TN22 3LP | 2E0IDN |
| TN22 3LP | M6IDN |
| TN22 3LS | G4FMA |
| TN22 4HT | G4NAJ |
| TN22 4JA | G4UDW |
| TN22 4JU | G6GSF |
| TN22 4LB | G3YMH |
| TN22 4NG | M3OII |
| TN22 4QB | G4MMH |
| TN22 5AH | G1SHH |
| TN22 5BU | G0MAR |
| [illegible] | [illegible] |
| TN22 5JH | G7VAD |
| TN22 5JU | G6BRV |
| TN22 5LA | G7HQY |
| TN22 5TG | M3TJT |
| TN23 1LN | G7NUG |
| TN23 1LN | M6VWE |
| TN23 3AG | G0PDP |
| TN23 3BG | 2E0YQC |
| TN23 3BG | M3TGO |
| TN23 3DY | G1EJA |
| TN23 3EG | G6EXU |
| TN23 3EX | M6EWW |
| TN23 3FJ | M0DMX |
| TN23 3LF | M3HFA |
| TN23 3LQ | M6IFF |
| TN23 3NH | M3VBL |
| TN23 3NJ | G4AAR |
| TN23 3NJ | G6ZAA |
| TN23 3NJ | M0ZAA |
| TN23 3NJ | M3YAA |
| TN23 3NR | 2E0FAA |
| TN23 3NR | M0TMX |
| TN23 3NR | M3FAA |
| TN23 3QY | M3RIU |
| TN23 5DF | 2E0SJA |
| TN23 5EH | M3BGN |
| TN23 5LN | G6NFE |
| TN23 5NX | G7WJW |
| TN23 5PA | 2E0XXT |
| TN23 5PA | M6TEZ |
| TN23 5QR | M6HJG |
| TN23 5SA | G0TJI |
| TN23 5UR | G6AJW |
| TN23 5UR | M6RRJ |
| TN23 5UW | G4UUI |
| TN23 5YF | M3ZRG |
| TN23 6JL | G4UKO |
| TN23 7HL | G4EWV |
| TN23 7UE | M3WXD |
| TN23 7UP | 2E0KEZ |
| TN23 7UP | M3KEZ |
| TN24 0BJ | M6GXP |
| TN24 0DN | 2E0XBW |
| TN24 0DN | M0XBW |
| TN24 0DN | M6XBW |
| TN24 0DZ | 2E1KCC |
| TN24 0FX | 2E0RBP |
| TN24 0FX | M6RPC |
| TN24 0HJ | 2E0TED |
| TN24 0HJ | 2E1ITE |
| TN24 0HJ | M3WHY |
| TN24 0JH | M3PHO |
| TN24 0TA | M3YOO |
| TN24 0XA | G8EWL |
| TN24 8HN | G7BYV |
| TN24 8NF | G0HVP |
| TN24 9AE | G3ZPI |
| TN24 9BD | G0KGE |
| TN24 9EU | G7HJJ |
| TN24 9EU | G8YNF |
| TN24 9JW | 2E1FWM |
| TN24 9JY | G7KLJ |
| TN24 9LR | M3CAA |
| TN24 9LU | M0AVY |
| TN24 9PT | M1PAB |
| TN24 9QL | M3ILJ |
| TN24 9RS | M3NKX |
| TN25 4DF | G3ZAJ |
| TN25 4DW | G8MFV |
| TN25 4NX | M3WAF |
| TN25 4PQ | G6SRE |
| TN25 5AB | M6IPH |
| TN25 5BD | M0WYE |
| TN25 5JB | G4FPG |
| TN25 5LR | M6JNJ |
| TN25 6EB | G1BXT |
| TN25 6EP | M1DRL |
| TN25 6HG | M1SEM |
| TN25 6NE | G0JIR |
| TN25 6RA | G0PEG |
| TN25 6RA | G8WLB |
| TN25 6RW | G7EVC |
| TN25 6UA | G3YBE |
| TN26 1HR | M3YPU |
| [illegible] | [illegible] |
| TN26 1HW | G3MMN |
| TN26 1LS | G0MTJ |
| TN26 2AH | M3YBR |
| TN26 2EF | M6PDD |
| TN26 2HL | G0CRL |
| TN26 2LU | 2E0BVB |
| TN26 2LU | M6RIO |
| [illegible] | [illegible] |
| TN26 3HA | G3DT |
| TN26 3LY | G6HXZ |
| [illegible] | [illegible] |
| TN26 3QS | G0CHG |
| TN26 3QU | G8PFZ |
| TN27 0AQ | G6LYD |
| TN27 0BX | M0EUS |
| TN27 0DD | G6JPG |
| TN27 0EE | M3SNX |
| TN27 0HA | 2E0SPT |
| TN27 0HA | M0DRQ |
| TN27 0HB | G8FYX |
| TN27 0JA | G6KEN |
| TN27 0JN | G6BME |
| TN27 0JQ | G3KAP |
| TN27 0JR | G7HKT |
| TN27 0JT | G0EZU |
| TN27 0NG | M0MSF |
| TN27 0NG | M1DDY |
| TN27 0PE | M0KIM |
| TN27 0RR | [illegible] |
| [illegible] | [illegible] |
| TN27 8EA | G7FKZ |
| TN27 8EA | M3YWE |
| TN27 8ER | G1AOQ |
| TN27 8PZ | M0DKN |
| TN27 9DW | G8ISM |
| TN27 9NH | 2E0CMN |
| TN27 9QQ | G4ZXI |
| TN27 9QR | G0ICP |
| TN27 9UF | 2E0CUE |
| TN27 9UF | M0HNI |
| TN27 9UF | M6RCW |
| TN28 8BJ | G8BIS |
| TN28 8JB | G4ZTW |
| TN28 8JL | G4TGK |
| TN28 8JY | G7PWS |
| TN28 8LB | G1HJS |
| TN28 8NF | G1UTF |
| TN28 8NL | G0DCG |
| TN28 8NX | G0BGH |
| TN28 8PB | G8IIK |
| TN28 8QA | M0DXS |
| TN28 8RD | M0DAS |
| TN28 8RL | G8GZW |
| TN28 8RX | G4YAZ |
| TN28 8SB | 2E1KAJ |
| TN28 8SF | G6JQD |
| TN28 8SU | M6PUY |
| TN28 8SW | G0KRH |
| TN28 8SZ | M6BQO |
| TN28 8UJ | M3OMZ |
| TN28 8UL | G0IMD |
| TN28 8XP | G0OIN |
| TN29 0LA | G0LNN |
| TN29 0PW | G1VBO |
| TN29 0PQ | G4LJT |
| TN29 0QX | G0IEE |
| TN29 0QX | G6ZVB |
| TN29 0RD | G6DZT |
| TN29 0RD | G6OJZ |
| TN29 0RU | G4TZX |
| TN29 0SF | 2E0ZIV |
| TN29 0SF | M3ZIV |
| TN29 0TY | G7UBK |
| TN29 0UA | G1EVI |
| TN29 9BN | G4LMN |
| TN29 9EJ | G6NLZ |
| TN29 9EL | M6JTE |
| TN29 9ES | 2E0SDJ |
| TN29 9ES | M0SDJ |
| TN29 9JP | 2E0SVK |
| TN29 9JP | M0LMI |
| TN29 9JP | M6LMI |
| TN29 9LE | G8XCL |
| TN29 9LF | G1HSX |
| TN29 9NL | M0DTI |
| TN29 9NP | M0LYD |
| TN29 9NS | G1KPU |
| TN29 9RG | G1HSI |
| TN29 9SN | 2E0AKU |
| TN29 9SN | M3KQY |
| TN29 9UU | G0CPZ |
| TN29 9YF | G4RLU |
| TN3 0AN | 2E0LSS |
| TN3 0RB | M1KWH |
| TN3 8DY | M6ONO |
| TN3 9QU | G0CHY |
| TN30 6DQ | 2E1GKF |
| TN30 6DY | M6RCB |
| TN30 6EL | G4KPS |
| TN30 7BA | 2E1GJD |
| TN30 7BA | G8VGU |
| TN30 7NB | G8GFS |
| TN30 7NT | G0GCQ |
| TN30 7NT | M3JCQ |
| TN31 6BX | 2E0IEE |
| TN31 6DL | M6FLH |
| TN31 6DN | G6IPH |
| TN31 6DS | M3ULE |
| TN31 6EJ | M6BSB |
| TN31 6EJ | M6MCP |
| TN31 6EP | G0OZM |
| TN31 6EX | M0EDU |
| TN31 6HQ | G8VOH |
| TN31 6NB | G0FLA |
| TN31 6NJ | G3ROC |
| TN31 6TA | M6LTB |
| TN31 6UL | G4TLE |
| TN31 6UR | G0TFV |
| TN31 7BG | G6LPC |
| TN31 7FH | G4PDU |
| TN31 7HJ | M3MZR |
| TN31 7HT | M3KPG |
| TN31 7NY | G0DYF |
| TN31 7PJ | G6XII |
| TN31 7RU | G0ISM |
| TN32 5AU | G4DIA |
| TN32 5JZ | M6EWI |
| TN32 5NP | G1VUK |
| TN32 5QS | G2HIX |
| TN32 5QS | G6KGU |
| TN32 5QS | M3URX |
| TN32 5UG | G4JWK |
| TN33 0DT | M6KBO |
| TN33 0DT | M6WSC |
| TN33 0DU | G7THK |
| TN33 0EU | G7CKS |
| TN33 0HD | M3KPU |
| TN33 0NL | G8AJP |
| TN33 0NL | G8CLZ |
| TN33 0QS | G3UYB |
| TN33 0SF | G0LTV |
| TN33 0TP | G3VFO |
| TN33 9BP | M3ESN |
| [illegible] | [illegible] |
| [illegible] | [illegible] |
| TN33 9LP | M6JHR |
| TN34 1TG | M3KQT |
| TN34 1TS | G8FET |
| TN34 1UU | G0OOU |
| TN34 2AP | 2E0KES |
| TN34 2AP | M6RNO |
| TN34 2AT | G1LZS |
| TN34 2AU | G1BFS |
| TN34 2AW | M0NUC |
| TN34 2AW | M0SSR |
| TN34 2BD | G1HHH |
| TN34 2BD | G2LL |
| TN34 2BD | G4FET |
| TN34 2BD | G6HH |
| TN34 2BG | G4PWE |
| TN34 2BT | M3MRQ |
| TN34 2DN | M3TPZ |
| TN34 2DZ | G3VUR |
| TN34 2ES | G8ELW |
| TN34 2ET | M6RBA |
| TN34 2HA | M6HXG |
| TN34 2JA | G7PIP |
| TN34 2JE | G0GKL |
| TN34 2JZ | M6HXN |
| TN34 2NQ | G4UWF |
| TN34 2PJ | G6TFE |
| TN34 2PS | G7WWW |
| TN34 2QT | G4KMJ |
| TN34 2SF | G3UFI |
| TN34 2UL | 2E0SAY |
| TN34 3BJ | G6HVX |
| TN34 3JN | G0SYQ |
| TN34 3LQ | G7KMA |
| TN34 3LQ | M3AXB |
| TN34 3LY | M3BKI |
| TN34 3LY | M3BKJ |
| TN34 3SL | G8RWJ |
| TN35 4AS | M6EHE |
| TN35 4BL | G3FHN |
| TN35 4DN | G3SYZ |
| TN35 4DP | G3OZZ |
| TN35 4DP | M0AHY |
| TN35 4DP | M0CCV |
| TN35 4DP | M1CPL |
| TN35 4EP | G3BDQ |
| TN35 4JH | G8ZGM |
| TN35 4LG | G8THE |
| TN35 4LG | M3XYZ |
| TN35 4PA | M6NJK |
| TN35 4PB | G3TCG |
| TN35 4PB | M6PWD |
| TN35 4QU | G4NER |
| TN35 4QZ | G1IKV |
| TN35 4SL | G1GFZ |
| TN35 5DH | 2E0EAO |
| TN35 5DH | M0ROK |
| TN35 5DH | M6BBM |
| TN35 5DJ | M1AKL |
| TN35 5EP | G0FUU |
| TN35 5HH | 2E1GWX |
| TN35 5HZ | G4NVQ |
| TN35 5JN | G0CJO |
| TN35 5LZ | M3BRQ |
| TN355BS | 2E0PSK |
| TN36 4AS | G6TFJ |
| TN37 6EL | G1HSM |
| TN37 6HF | G1FTK |
| TN37 6JA | 2E1HEO |
| TN37 6JY | G7OJY |
| TN37 6LA | G7ERC |
| TN37 6PF | G4KLF |
| TN37 6PN | G1ORK |
| TN37 6QR | G1ZPA |
| TN37 6RX | 2E0SPH |
| TN37 6RX | M0ZPL |
| TN37 6RX | M6FRJ |
| TN37 6SD | G1IFW |
| TN37 6SJ | G8FEJ |
| TN37 7AJ | G3XXM |
| TN37 7AL | G7WGA |
| TN37 7BS | G3ZFX |
| TN37 7BS | G6CLU |
| TN37 7DF | 2E0LPW |
| TN37 7DF | M0LPW |
| TN37 7DF | M6LPW |
| TN37 7DF | M6NXT |
| TN37 7EG | M3YXH |
| TN37 7JY | G7VQO |
| TN37 7PS | G1DVU |
| TN37 7PS | M0BMD |
| TN37 7QX | G0JHK |
| TN37 7UA | M3XVC |
| [illegible] | G1AJN |
| TN38 0BT | G0TDJ |
| [illegible] | G7MQQ |
| [illegible] | 2E1DHC |
| [illegible] | [illegible] |
| TN38 0JN | G0NJO |
| TN38 0LP | M6BQS |
| TN38 0PS | G3DQY |
| TN38 0PS | G3WQK |
| TN38 0PU | 2E0RAJ |
| TN38 0PU | M3MFE |
| TN38 0QA | G8CMK |
| TN38 0QA | G8SBJ |
| TN38 0QP | G1GFF |
| TN38 0SE | G1OHL |
| TN38 0TR | G0AQT |
| TN38 0UA | G4EOA |
| TN38 0XL | G4WOE |
| TN38 0XP | M6ZZS |
| TN38 8WD | 2E0WJR |
| TN38 9BU | G1EHS |
| [illegible] | [illegible] |
| [illegible] | G4VBK |
| TN38 9HP | M3BKU |
| TN38 9LQ | G0ILK |
| TN38 9RS | G7JVN |
| TN38 9RX | G4BOZ |
| TN38 9UA | G0OFZ |
| TN39 3AQ | M6WIB |
| TN39 3AX | G8GQG |
| TN39 3EE | M6CWU |
| TN39 3HP | G1WLO |
| TN39 3LU | G4MMG |
| TN39 3NJ | G8MYG |
| TN39 3PB | G3WUH |
| TN39 3PD | M0BDE |
| TN39 3PH | M6TJY |
| TN39 3PS | M3ZBH |
| TN39 3RS | M6FKS |
| TN39 3SS | G6MDF |
| TN39 3TD | G7TAJ |
| TN39 3TF | G7OOP |
| TN39 3TS | M6HHC |
| TN39 3UA | G3JOR |
| TN39 4AS | G7LEL |
| TN39 4BN | G1PDS |
| TN39 4DY | G4HGK |
| TN39 4HE | M6IBO |
| TN39 4HZ | G0THV |
| TN39 4JH | M0OMI |
| TN39 4JJ | G8EYQ |
| TN39 4JL | M0BPU |
| TN39 4LT | G1ZQO |
| TN39 4PP | 2E1GTW |
| TN39 4QB | 2E0WCO |
| TN39 5BX | M6TOY |
| TN39 5EQ | G1MTB |
| TN39 5HH | 2E0JND |
| TN39 5HH | M6JND |
| TN39 5HN | G0JBM |
| TN39 5HU | G7GHP |
| TN39 5HY | 2E0GHX |
| TN39 5JH | G4MPK |
| TN4 0DX | G1EUD |
| TN4 0EQ | G4MIK |
| TN4 0NX | M3JCT |
| TN4 8ED | G4GTN |
| TN4 8PS | 2E0CMP |
| TN4 8PS | M0NAY |
| TN4 8PS | M6BOX |
| TN4 8PY | G7RUX |
| TN4 8TH | 2E0DGS |
| TN4 8TH | M0VIR |
| TN4 8TH | M6GBZ |
| TN4 9BL | 2E0CQL |
| TN4 9BL | M0KAO |
| TN4 9BL | M6BQW |
| TN4 9DH | 2E0TWP |
| TN4 9DH | M6PJM |
| TN4 9DJ | G0BMN |
| TN4 9DR | G3KIP |
| TN4 9JG | G7DAB |
| TN4 9JU | G6TXH |
| TN4 9JY | G0ABM |
| TN4 9PA | M6GAX |
| TN4 9RF | M0RYK |
| TN4 9RF | M0XKX |
| TN4 9RP | G0VPY |
| TN40 1EG | G6SGV |
| TN40 1EG | M6GQV |
| TN40 1QE | G0TMA |
| TN40 1QH | G8KPE |
| TN40 1TE | G6HIB |
| TN40 1TU | G3NMJ |
| TN40 1UE | G6NGM |
| TN40 1UG | G8TUU |
| TN40 2AZ | G3GGH |
| TN40 2AZ | G4SIF |
| TN40 2BQ | G6ULD |
| TN40 2EA | M3TNO |
| TN40 2EL | G4LPA |
| TN40 2HN | M3LGI |
| TN40 2HU | G6MWB |
| TN40 2NT | G4YZL |
| TN40 2NT | G4ZSR |
| TN40 2PG | G1VNZ |
| TN40 2PW | M1FFX |
| TN40 2SD | G3MVX |
| TN40 2SH | G0ILN |
| TN40 2SH | G7TAF |
| TN40 2TA | M6PFB |
| TN40 2TB | M1EQD |
| TN40 2UL | M6WDQ |
| TN5 6BT | M6LVK |
| TN5 6DG | M6HXF |
| TN5 6DL | G6VGA |
| TN5 6RJ | G8GYB |
| [illegible] | M0NTA |
| TN5 6UT | G1NZH |
| TN5 7AF | G1EHB |
| TN5 7EG | G4XST |
| TN5 7JA | G7IRN |
| TN5 7PS | G7MFE |
| TN6 1BS | G4LZK |
| TN6 1EN | M1CNE |
| TN6 1JF | G3TXZ |
| TN6 1QQ | G7JAN |
| TN6 1TF | G4EZE |
| TN6 2AD | G0GPE |
| TN6 2AD | G4XCE |
| TN6 2AG | G8LKS |
| TN6 2BG | G3WKS |
| TN6 2BG | G4OTV |
| TN6 2BG | G4UPI |
| TN6 2BG | M0SSX |
| TN6 2BN | G6GWE |
| [illegible] | [illegible] |
| TN6 2DU | M0MOF |
| TN6 2EN | M1ARX |
| TN6 2ES | 2E0MDU |
| TN6 2ES | M6NAU |
| TN6 2ET | G0WTC |
| TN6 2ET | G6UIF |
| TN6 2HA | 2E0TLB |
| TN6 2NJ | G8UBD |
| TN6 2NY | G3CYU |
| TN6 2RY | G3RST |
| TN6 2SB | G8LSD |
| TN6 2UU | 2E0GCP |
| TN6 3BH | G0XBV |
| TN6 3BQ | G4LDJ |
| TN6 3BY | M3TLB |
| TN6 3EB | M3UAR |
| TN6 3LG | M6YLS |
| TN6 3LY | G7MZL |
| TN6 3QT | M0CPC |
| TN6 3TA | 2E0HKK |
| TN7 4DH | G0NAR |
| TN7 4EN | M1PVC |
| TN8 5BB | G7RXJ |
| TN8 5BX | M6JRB |
| TN8 5DB | M3WYZ |
| TN8 5DF | G0TKZ |
| TN8 5DS | M6MVV |
| TN8 5EL | G3JMJ |
| TN8 5EN | G6MXV |
| TN8 5PS | M6CSC |
| TN8 5PW | G6DTT |
| TN8 6AJ | M3ZHM |
| TN8 6AJ | M3ZHQ |
| TN8 6AJ | M6GPE |
| TN8 6AJ | M6GRJ |
| TN8 6AU | M6SNC |
| TN8 6LN | M0GZC |
| TN8 6LN | M3ZWM |
| TN8 6SX | M6SRC |
| TN8 7AU | M3UJM |
| TN9 1UX | G1CPA |
| TN9 1UX | M0AFZ |
| TN9 2AN | 2E1IJZ |
| TN9 2AN | M0IJZ |
| TN9 2AN | M3ANX |
| TN9 2EJ | G7GDV |
| TN9 2NQ | G3YPY |
| TN9 2PG | M3LHE |
| TN9 2PG | M3LHG |
| TN9 2TR | M3RPE |
| TN9 2UY | G0JUY |
| TN9 2UY | M3MOP |
| TN9 2YR | M6SHC |
| TN9 2YS | G4EPC |

## TQ
(Torquay)

| Postcode | Callsign |
|---|---|
| TQ1 1EQ | G6LPS |
| TQ1 1JU | M6CON |
| TQ1 1JY | 2E0PCZ |
| TQ1 1JY | M0PCZ |
| TQ1 1JY | M6PCZ |
| TQ1 1QR | G0KEC |
| TQ1 1QR | G1XDJ |
| TQ1 1TJ | G7TVQ |
| TQ1 2ED | G0CRJ |
| TQ1 2HH | G3VOF |
| TQ1 2LQ | M0IME |
| TQ1 2ND | G0PQA |
| TQ1 2NL | G0RBV |
| TQ1 3JN | G1KPI |
| TQ1 3LD | G6TEQ |
| TQ1 3LJ | G6MTB |
| TQ1 3LZ | M3YSA |
| TQ1 3NZ | G4FKU |
| TQ1 3PD | G4YTY |
| TQ1 3TP | G3KPV |
| TQ1 3TX | 2E0PLR |
| TQ1 3TX | M0LAI |
| TQ1 3TX | M6LBN |
| TQ1 4AA | M0HSH |
| TQ1 4EA | M3LKE |
| TQ1 4HZ | G0EDF |
| TQ1 4LD | G0SUU |
| TQ1 4NJ | G4OVO |
| TQ1 4PU | M6PHW |
| TQ1 4PX | G1EFT |
| TQ1 4QY | G8XST |
| TQ1 4RD | G8GPF |
| TQ1 4UN | G1TNP |
| TQ10 0DT | M1DKW |
| TQ10 9EJ | G4TCP |
| TQ10 9EU | G4XTR |
| [illegible] | [illegible] |
| TQ10 9JF | M6FQZ |
| TQ10 9LL | G0PCT |
| TQ11 0AF | G8YZA |
| TQ11 0BP | G7RYL |
| TQ11 0DL | M0LED |
| TQ11 0EA | G1GMG |
| TQ11 0ER | G4XWP |
| TQ11 0HB | G6NWC |
| TQ12 1AA | M3IDC |
| TQ12 1DX | M3JKA |
| TQ12 1DZ | G0EKH |
| TQ12 1ER | G0OFA |
| TQ12 1FY | 2E0CVN |
| TQ12 1GQ | G0MUN |
| TQ12 1HS | G3TAA |
| TQ12 1HT | M6DXG |
| TQ12 1LH | M1DVV |
| TQ12 1RQ | G3YAR |
| [illegible] | [illegible] |
| [illegible] | [illegible] |
| TQ12 1YJ | G1WQN |
| TQ12 2ND | G0RDO |
| TQ12 2NN | M3PZF |
| TQ12 2PT | G4IHY |
| TQ12 2PU | G0WAE |
| TQ12 2PX | G7AQV |
| TQ12 2TL | M3GFH |
| TQ12 2TP | M3GFE |
| TQ12 3BP | M1AIY |
| TQ12 3HD | G1NNU |
| TQ12 3JE | G0WWD |
| TQ12 3LE | G1WUU |
| TQ12 3LY | M3DHA |
| TQ12 3NJ | M6ERU |
| TQ12 3NZ | G4GZM |
| TQ12 3TE | G4DTW |
| TQ12 3TJ | M1AKV |
| TQ12 3YS | G6YJO |
| TQ12 3YT | 2E0YLY |
| TQ12 3YT | M0YLY |
| TQ12 3YU | M3GHD |
| TQ12 4EH | G7AFV |
| TQ12 4HA | G1DBH |
| TQ12 4HE | G4VUD |
| TQ12 4JG | G4LAK |
| TQ12 4JZ | G4MNA |
| TQ12 4LF | G3LHJ |
| TQ12 4LF | G3NJA |
| TQ12 4LF | G8NJA |
| TQ12 4LF | M0LTW |
| TQ12 4NE | G0IGK |
| TQ12 4NH | G0NXI |
| TQ12 4NJ | G4ELZ |
| TQ12 4NS | G3XXE |
| TQ12 4PG | M3IEU |
| TQ12 4QS | G0CWQ |
| TQ12 4QS | G4XJL |
| TQ12 4RE | G0SDL |
| TQ12 4SB | G3TLK |
| TQ12 5BZ | G6FIN |
| TQ12 5DZ | M3UMJ |
| TQ12 5EW | G0BNJ |
| TQ12 5JG | G4NRH |
| TQ12 5NX | G8YEN |
| TQ12 5PZ | 2E0GPL |
| TQ12 5PZ | M3UAW |
| TQ12 5PZ | M3ZDT |
| TQ12 5QJ | G8VSV |
| TQ12 5QS | G0SBM |
| TQ12 5QS | G4FCN |
| TQ12 5QS | G6CLD |
| TQ12 5QS | G7MNL |
| TQ12 5QS | G8GCS |
| TQ12 5QS | M1EPU |
| TQ12 5RX | M0UAC |
| TQ12 5RX | M6GBU |
| TQ12 5XB | M3GDQ |
| TQ12 6FA | M6JAQ |
| TQ12 6GX | G1ZBJ |
| TQ12 6HA | G1XBX |
| TQ12 6HE | G3PVB |
| TQ12 6HE | G8SLM |
| TQ12 6HE | G8YBY |
| TQ12 6HJ | 2E0CKY |
| TQ12 6HJ | M6AIV |
| TQ12 6SB | 2E0WSZ |
| TQ12 6SB | M6WNB |
| TQ12 6TX | M6KAR |
| TQ12 6UP | M3YYG |
| TQ12 6YA | G4DMM |
| TQ13 0BB | G1BJN |
| TQ13 0EE | G4VTO |
| TQ13 0EW | G4RYH |
| TQ13 0EW | M6CCX |
| TQ13 0GB | G4TND |
| TQ13 0JH | G4NRR |
| TQ13 0JH | G6DHD |
| TQ13 0JN | G7OZI |
| TQ13 0NW | M0NAA |
| TQ13 0NW | M6NMN |
| TQ13 0PL | M3RXW |
| TQ13 7AN | M3DBA |
| TQ13 7BS | G7CUU |
| TQ13 7DU | G7RKX |
| TQ13 7DX | M3XYU |
| TQ13 7HT | G1BGK |
| TQ13 7HY | G8KEO |
| TQ13 7RU | G3FYF |
| TQ13 8AR | G3UUU |
| TQ13 8EJ | G0BYF |
| TQ13 8QA | 2E0FWD |
| TQ13 8QA | M0EYZ |
| TQ13 8SD | 2E0MLH |
| TQ13 8SD | 2E0RIW |
| [illegible] | [illegible] |
| TQ13 9DE | 2E0EZL |
| TQ13 9EP | G6ICH |
| TQ13 9EP | M0LHK |
| TQ13 9FF | G0RMP |
| TQ13 9HP | M0CLE |
| TQ13 9HR | G3RPD |
| TQ13 9JY | M0ECQ |
| TQ13 9YL | M3UOY |
| TQ14 0DG | G3WLT |
| TQ14 8AQ | 2E0DOX |
| TQ14 8AQ | M6ADX |
| TQ14 8BJ | G0JZA |
| TQ14 8BX | G0CEL |
| TQ14 8EF | G8KTC |
| TQ14 8HE | G7LJN |

Postcode

**IMPORTANT NOTE**

**Revalidate licence to avoid revocation** – Ofcom has advised the Society that plans will be drawn up to revoke licences that have not been revalidated as required by the licence conditions. The quickest way to revalidate is to do so online via the Ofcom website: *https://services.ofcom.org.uk/* or by email: *amateur.validations@ofcom.org.uk* If you need assistance in the process, Ofcom staff are available to help, but please be patient during times of heavy workload.

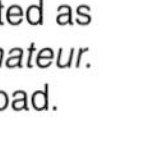

TQ14 8LE G1MKE
TQ14 8NH G4CDJ
TQ14 8PN G4YTT
TQ14 8QG G6XD
TQ14 8QX G4BG
TQ14 8RJ G0VEB
TQ14 8RJ G1GHU
TQ14 8RS G3LQX
TQ14 8UF G3YHH
TQ14 9DT G3KFP
TQ14 9HG G3MOA
TQ14 9JP G0KDT
TQ14 9NF G0PWU
TQ14 9NJ G1EUA
TQ14 9NW M3FZS
TQ14 9QA M3HYI
TQ14 9QA M3IZW
TQ14 9QR M0HHG
TQ14 9QR M3TYO
TQ14 9RF G0PNA
TQ14 9RP G8WBL
TQ2 6BS G1ZZG
TQ2 6BZ G8TYD
TQ2 6DX G4OTU
TQ2 6EA G0TMH
TQ2 6EA M3WFE
TQ2 6EA M3YVE
TQ2 6HS M6IDP
TQ2 6HS M6RWJ
TQ2 6LL G0OXT
TQ2 6PN G1NIC
TQ2 6RG M3IPZ
TQ2 6UT G3YDE
TQ2 7AG G7JHE
TQ2 7DS G7MEA
TQ2 7EH G3AH
TQ2 7EH M0RIV
TQ2 7GJ G3ZYN
TQ2 7GJ G4JIN
TQ2 7JX 2E0PBT
TQ2 7JX M0PBT
TQ2 7JX M6PBT
TQ2 7NR G0HSW
TQ2 7QX G0JDO
TQ2 7QX M6GTO
TQ2 8AJ G0HGV
TQ2 8DH G4ZQM
TQ2 8EF M0BKD
TQ2 8HB G0MRZ
TQ2 8HU G7SME
TQ2 8LB M0PAW
TQ2 8LE G0DDF
TQ2 8LF G8GRB
TQ2 8LR G6FTA
TQ2 8LR M3FTA
TQ2 8PA M6DFY
TQ2 8QA G4PCK
TQ2 8QE G8UKI
TQ2 8QH G4UIH
TQ2 8QH G4UII
TQ2 8QQ G7HIK
TQ3 1AB G7DKB
TQ3 1AE G0AZX
TQ3 1AE G0LQM
TQ3 1AG G3HTX
TQ3 1EE M5ABN
TQ3 1EL G7FVA
TQ3 1HN M6PNJ
TQ3 1JX G0GWC
TQ3 1LE G7DMX
TQ3 1NH G4OYC
TQ3 1NH G6XXB
TQ3 1SH G0NOB
TQ3 1SQ G6UPR
TQ3 1TA G4AQR
TQ3 1TD G7IIC
TQ3 2AB 2E0HOO
TQ3 2AB M3UNF
TQ3 2BN G4CLN
TQ3 2DP M0DPQ
TQ3 2EN G8BKQ
TQ3 2HT G0BAJ
TQ3 2HZ G0UWF
TQ3 2JP G6ZZS
TQ3 2LR G8ZEV
TQ3 2NX G0SUB
TQ3 2NX G1GRT
TQ3 2PB G6JAR
TQ3 2PE G4OFP
TQ3 2PT G4LPY
TQ3 2QF M0BKL
TQ3 2QS G7MCK
TQ3 2RE G0ORK
TQ3 2RG M0IAT
TQ3 2SW G7AHP
TQ3 2SY M6AJJ
TQ3 3BT G3GCW
TQ3 3BU M6DPG
TQ3 3EQ M1DXB
TQ3 3JJ G0TQC
TQ3 3JJ G3YCH
TQ3 3NA G0CGM
TQ3 3NU G0CDB
TQ3 3NU G4SSD
TQ3 3NX M3AXX
TQ3 3QN G1XKQ
TQ3 3QW 2E0WAU
TQ3 3QW M6VAL
TQ3 3TD M3GDV
TQ3 3TU G1TUU
TQ3 3UD G1JKL
TQ4 5AX 2E0DBQ
TQ4 5AX M0ZYD
TQ4 5AX M6AYK
TQ4 5ES G0KPA
TQ4 5EY M6HNH
TQ4 5HU G4UUT
TQ4 5HZ G4NNP
TQ4 5JQ G3CLW
TQ4 5JQ G4KFB
TQ4 5JW G7EBI
TQ4 5LF G4WVM
TQ4 5LH 2E0PDL
TQ4 5LH M0LOB
TQ4 5LH M3UVT
TQ4 5LH M3VBG
TQ4 5LH M6DFI
TQ4 5NA G7CHC
TQ4 5NG G0SPF
TQ4 6EG G6AOB
TQ4 6ES 2E0OJE
TQ4 6HA G0UAV
TQ4 6HD 2E0JNP
TQ4 6HD M0JNP
TQ4 6HD M6GSW
TQ4 6HT G4AWO
TQ4 6JF 2E0RJE
TQ4 6JF M3URV
TQ4 6JN G7JFM
TQ4 6JU G4FLW
TQ4 7HG G0UQI
TQ4 7HZ G8XLG
TQ4 7LW G0HZB
TQ4 7ND G0KTU
TQ4 7NZ M3NYX
TQ4 7RL G0OSH
TQ4 7RL G4DZH
TQ4 7RL G7OSH
TQ4 7RL M0BKG
TQ4 7RN M0BHJ
TQ4 7RN M1ANO
TQ4 7RN M1MUM
TQ4 7RN M3UNR
TQ4 7RU 2E0TAG
TQ4 7RU M3TQU
TQ4 7RW G0BTH
TQ4 7SD G4VQI
TQ4 7SD G6UIM
TQ5 0AN M6DFJ
TQ5 0AY G0OJS
TQ5 0DD G7DLE
TQ5 0DG G1SDX
TQ5 0JZ M0DQN
TQ5 0LW M1DVJ
TQ5 0LZ G8JMS
TQ5 0PB G0RGP
TQ5 0PB G7LMR
TQ5 0PB G8ETU
TQ5 0RJ G3RVA
TQ5 8JN G4WZN
TQ5 8NP G1OBM
TQ5 8QG G6ZOJ
TQ5 8QP M3TQG
TQ5 8QR G3XNX
TQ5 8RF G4UXP
TQ5 8RN G4NGR
TQ5 8RS G0PTK
TQ5 9AG M0PLH
TQ5 9BT G6JZN
TQ5 9DJ M1EJE
TQ5 9ED G0TDQ
TQ5 9EE G0KEK
TQ5 9EL G0VQY
TQ5 9EL M3GCD
TQ5 9LJ G1EZI
TQ5 9LS G3REW
TQ5 9LS M1EIW
TQ5 9NA M1DJX
TQ5 9PB G0LCD
TQ5 9PF G4OJD
TQ5 9PJ G0TSB
TQ5 9PJ G1AEU
TQ5 9QZ G3ZUO
TQ5 9RL G4JXK
TQ5 9SW 2E0XOL
TQ5 9SW M0XOL
TQ5 9SW M6XON
TQ5 9TG G4UAW
TQ5 9UD G4OKC
TQ5 9UN G0NPY
TQ5 9UR G7VSL
TQ6 0DP G4XNO
TQ6 0DQ G3PCT
TQ6 9BN M3TQA
TQ6 9HN G8FAW
TQ6 9LQ M3TQN
TQ6 9NZ G7XPC
TQ7 1DP G0MJU
TQ7 1JU M3RPH
TQ7 1LL G8YBT
TQ7 1LP G0WHZ
TQ7 1LR G8KWJ
TQ7 1NR G8EMA
TQ7 1QB G6STF
TQ7 1QU G4RYO
TQ7 1RW M6AFB
TQ7 2AA G8KOV
TQ7 2DJ M3PRZ
TQ7 2DP G6SBG
TQ7 2NB 2E0POB
TQ7 2NB M0HXG
TQ7 2NB M6BWQ
TQ7 3JZ G7IMV
TQ7 3LX G7OSB
TQ7 4NE G3XLW
TQ7 4PF G3PFX
TQ7 4QH G0RIY
TQ7 4QJ M1DWQ
TQ7 4QP G6TOI
TQ8 8AN M3WNZ
TQ8 8DP M0GJL
TQ8 8JP G3UBL
TQ8 8NU G6STJ
TQ8 8PJ G8TQH
TQ8 8PW G3HST
TQ9 5AN G8KRB
TQ9 5EF G3PNO
TQ9 5FH G6JPM
TQ9 5GN M3VTO
TQ9 5LH G3OPX
TQ9 5QY G0UBG
TQ9 5RF M0AVN
TQ9 5RF M3NSB
TQ9 6AN M3DBG
TQ9 6BD M1BXD
TQ9 6DA G7JLS
TQ9 6DB G8LXS
TQ9 6DJ G3SNU
TQ9 6LS M6MFO
TQ9 6RL 2E0PVQ
TQ9 6RL M3PVQ
TQ9 6SR G7KCE
TQ9 7BA G4VHL
TQ9 7BD M6TUK
TQ9 7EY G0DIH
TQ9 7NQ G7DHW
TQ9 7PU G4FBN
TQ9 7TL G1WZG

## TR

(Truro)

TR1 1AT G1ZWH
TR1 1BW G0NDC
TR1 1DS G0RIZ
TR1 1EF G1SZD
TR1 1JD G1UWD
TR1 1QR 2E0LTH
TR1 1QR M3SNQ
TR1 1RA G3MFW
TR1 1RU G0NKQ
TR1 1SS G0WXL
TR1 1WL 2E1ADQ
TR1 1WL G0NFI
TR1 1WL G8VCJ
TR1 1WL M3CKH
TR1 1XJ G0WID
TR1 1YS G0SHY
TR1 2BS G3XFL
TR1 2BW G4PGD
TR1 2BY G1ZRP
TR1 2DN M0WYM
TR1 2JZ M0BIZ
TR1 3DG 2E1CIT
TR1 3JR G8KJP
TR1 3LU M1EWT
TR1 3LX G0UWI
TR1 3LX G1DTS
TR1 3PA M0JLM
TR1 3PE G0FHX
TR1 3PE G0FHY
TR1 3PT G7FLI
TR1 3QD G0CUH
TR1 3TX G3WKP
TR1 3XB G0KVX
TR1 3YJ 2E0EHT
TR1 3YJ M0IUM
TR1 3YJ M6MPL
TR10 8BH G3UFX
TR10 8GF M0SEA
TR10 8HQ M3URE
TR10 8QA G7ERQ
TR10 8QF G4UIT
TR10 8SJ G8UOL
TR10 9DT G0EOH
TR10 9DZ G1EGI
TR10 9HB G0BSK
TR10 9HS G3NVJ
TR10 9JW G4PCW
TR10 9WS M1EQA
TR11 2DX G7FTA
TR11 2ER M0CNU
TR11 2HE G6CZX
TR11 2RN G0ETQ
TR11 2RW G4FNP
TR11 2SW G4RUA
TR11 2TD G1DDK
TR11 2TE 2E0PFA
TR11 2TE M0PBX
TR11 2TE M3SOV
TR11 3DZ 2E0MRP
TR11 3DZ M6MAP
TR11 3JW M6WRN
TR11 3YL M6AWJ
TR11 4AQ G3LAI
TR11 4BE G0HJW
TR11 4BL G8LCE
TR11 4HR G6YHK
TR11 4PD M1AVW
TR11 4RF G4HVF
TR11 4RG G7RKE
TR11 4SG G8AYJ
TR11 4SR G3VQS
TR11 4SW G4MYY
TR11 4SW G6DGW
TR11 5AP M0DYH
TR11 5BJ 2E0XZI
TR11 5BW G4PXN
TR11 5DT G8CTR
TR11 5EJ G6IMN
TR11 5HE G0OCS
TR11 5HE G0OCT
TR11 5HF G4VMW
TR11 5LR G4ELI
TR11 5NA 2E1ESM
TR11 5NA G7WDS
TR11 5QG M0RFM
TR11 5QN G4NBC
TR11 5SF 2E0BYH
TR11 5SF M3XUE
TR11 5SL G7LKZ
TR11 5SR G3NHL
TR11 5TE M1FIG
TR11 5TJ G3LZN
TR11 5TT M3XOA
TR11 5TT M6EME
TR11 5UG G3NWW
TR12 6LY G4WIA
TR12 6QD G7FOT
TR12 6QL G3UYN
TR12 6TB G8IKK
TR12 6UB G0XPD
TR12 6UW G6ZKC
TR12 7AG G6IWK
TR12 7BJ M1ERA
TR12 7BJ M1ERD
TR12 7BW G3PLE
TR12 7BZ M6NIK
TR12 7DN G3MPD
TR12 7DN G4YQP
TR12 7DS M0BNZ
TR12 7DY 2E0TDP
TR12 7DY M0TDP
TR12 7DY M3IXK
TR12 7DY M3MNQ
TR12 7ET G0RIT
TR12 7HZ 2E0ZSY
TR12 7HZ M3SYV
TR12 7HZ M6RLQ
TR12 7JD M0CNZ
TR12 7JH G1ZPC
TR12 7JJ M6NDK
TR12 7JL G0GUO
TR12 7LF G0XAO
TR12 7LU G3NJV
TR12 7NN G3KJK
TR12 7NU 2E0LLO
TR12 7NU G3WGZ
TR12 7NU M0SOU
TR12 7NU M6LZP
TR12 7NZ G4WSH
TR12 7PA G1JHN
TR12 7QR 2E0RER
TR12 7QR M0RRX
TR12 7QR M6RER
TR12 7QW G0HWU
TR13 0BY G4JGX
TR13 0DT G6FQP
TR13 0DZ M1EZT
TR13 0HE G8ZVM
TR13 0LD G0FKX
TR13 0PE 2E0VJB
TR13 0PE M3VJJ
TR13 0PQ M0BSI
TR13 0PR G4LOH
TR13 0SN G4GFY
TR13 8BP G0HFA
TR13 8JZ G3KGP
TR13 8LU 2E0LDM
TR13 8LU M3YAX
TR13 8PQ G0WYS
TR13 8PQ G0FGX
TR13 8PQ G1OZV
TR13 8PY G0UDB
TR13 8QJ G1ZPJ
TR13 8RX M0BUA
TR13 8SR G7VFA
TR13 8WF 2E0CTD
TR13 8XH M3VJC
TR13 9AW G4WZH
TR13 9BL G4ZYO
TR13 9LR 2E0KTW
TR13 9LR M0EMW
TR13 9NB G4ATR
TR13 9NB G8PPF
TR13 9PB G7MSN
TR14 0AR M6LDW
TR14 0AS M3WRM
TR14 0ET M6REJ
TR14 0EY G8SEE
TR14 0HD 2E1HSR
TR14 0HF G1OQU
TR14 0JF G4KES
TR14 0JX G4POT
TR14 0JX G8XCR
TR14 0LL 2E0PCR
TR14 0LL G0TRJ
TR14 0LL M6PCB
TR14 0LL M6PUQ
TR14 0NQ M0DUY
TR14 0QE G8WWW
TR14 7AW G4BHD
TR14 7BP M3UML
TR14 7EW M6RIU
TR14 7HB G6IZK
TR14 7HN G0WYU
TR14 7JN G0PGJ
TR14 7PH 2E0CUH
TR14 7RN M0JIU
TR14 7RR M6ZXZ
TR14 7SU G8YWJ
TR14 7TT G0ILI
TR14 7XN G4BHC
TR14 8AW 2E0BOI
TR14 8AW G4ZTZ
TR14 8AW M0GIY
TR14 8AW M3UWW
TR14 8JB G7GIG
TR14 8QF G1HFE
TR14 8QF G1LNA
TR14 8RQ G4XMA
TR14 8RQ M0RSJ
TR14 8RW G3WJP
TR14 8ST G8ZDS
TR14 8TW M6ASJ
TR14 8UP G4ZUI
TR14 8UP G6EON
TR14 9AA G4YYM
TR14 9AT G1BAX
TR14 9DE G4DEO
TR14 9ER G4YYH
TR14 9JT G4ZKN
TR15 1AX 2E0BON
TR15 1AX 2E0NKT
TR15 1AX M3XNA
TR15 1AX M6TRC
TR15 1EP M6LGD
TR15 1EW G4LOG
TR15 1NX G8IXN
TR15 1PA 2E0RBR
TR15 1PA M3WOY
TR15 1QR G6XKO
TR15 1SE G0FKI
TR15 2LL 2E0SCQ
TR15 2LL M6SCV
TR15 2NY G0FHT
TR15 2RJ G6HGR
TR15 2TP G7VIR
TR15 3AR 2E0MDN
TR15 3AR M0MBZ
TR15 3AR M6MEJ
TR15 3AW G3PUQ
TR15 3AW G4DBZ
TR15 3BP G0IDH
TR15 3BZ G4GDU
TR15 3DF G6KTB
TR15 3HJ G3JQK
TR15 3JQ M3YVY
TR15 3NJ M3IJO
TR15 3UD M6WDO
TR15 3YG 2E0TCV
TR15 3YJ G7DUC
TR16 4AQ G1NRF
TR16 4AQ M6KIR
TR16 4AQ M6NRF
TR16 4AY M6MYM
TR16 4DQ M3PHJ
TR16 4DY G6HGK
TR16 4SG G7ADP
TR16 4SH G7VOH
TR16 4SH M3YKT
TR16 5AH M6JCJ
TR16 5DT G4WQL
TR16 5JL 2E0BOJ
TR16 5JL M3TNB
TR16 5JY M0IAF
TR16 5LG 2E0CWS
TR16 5LG M0KDR
TR16 5LT G3MZN
TR16 5LT G7RIO
TR16 5PN G0FKF
TR16 5PT G0CAM
TR16 5QA G4ZCT
TR16 5QZ G0FIC
TR16 5QZ G4CRC
TR16 5RJ 2E0JGW
TR16 5RJ M6WBB
TR16 5TQ G0MYR
TR16 6DQ G1WLN
TR16 6HN G7WJJ
TR16 6HN M0DWS
TR16 6HP G4ZKH
TR16 6LR G8WNQ
TR16 6LS G8NXD
TR16 6NH M3RRN
TR16 6NT G0VZB
TR16 6PB G4VQP
TR16 6PG 2E0JBW
TR16 6PQ M6EVY
TR16 6SF G6UHD
TR166PN M3URW
TR17 0BP M6KRL
TR17 0HF M0DBA
TR18 2AH 2E0CJS
TR18 2AH M0PZR
TR18 2AH M6ANI
TR18 2DN G7SBP
TR18 2EX M6GBO
TR18 2HH G4PEM
TR18 2PL 2E1CCN
TR18 2QE G7CCS
TR18 2QE M0GOP
TR18 3BA G1JAH
TR18 3BB G3NRD
TR18 3LD G1IVO
TR18 3NA G3RID
TR18 3PD G8GRO
TR18 3PW G0WZC
TR18 3QL G8EJC
TR18 4BH G0PGB
TR18 5AS 2E0BKN
TR18 5AS M0MAY
TR18 5AS M3RPZ
TR18 5DQ G4BPJ
TR18 5NG M0SBL
TR18 5NR 2E0MEI
TR18 5NR M6DZB
TR18 5QA M0ABZ
TR18 5QR M0SBD
TR18 5QW G0MWZ
TR19 6BB 2E0CAQ
TR19 6BB M0LDQ
TR19 6BB M6SUV
TR19 6DT G1ZME
TR19 6LX G3XRJ
TR19 6TT G4RRQ
TR19 7BG M6YHQ
TR19 7BS G0THQ
TR19 7BT G4AMT
TR19 7DS G4XCV
TR19 7DZ G3OYB
TR19 7ET G4YAF
TR19 7SP M1BKS
TR19 7ST G3UUZ
TR19 7TJ M6FVC
TR19 7UT G3PZN
TR2 4AG M3HBS
TR2 4AU M6YEO
TR2 4BE G6OJX
TR2 4BS G3ZZV
TR2 4DS G4JZA
TR2 4NY G7KMP
TR2 4PQ G3VWK
TR2 4TD 2E0BBI
TR2 5AR G0REA
TR2 5DD G4NBF
TR2 5DD M0TOP
TR2 5JP G0JWV
TR2 5JX G0ENZ
TR2 5NP G4EIK
TR20 8AJ G8ARM
TR20 8AJ M0BMX
TR20 8BQ G4SEN
TR20 8DJ G3YSN
TR20 8JY G7TBJ
TR20 8UL G7TJD
TR20 8XA G1GKF
TR20 8XD G3GYE
TR20 8XT 2E1GNN
TR20 8XT G8HZN
TR20 8XT M3PHS
TR20 8XW G0AIX
TR20 9AR G7SUT
TR20 9BW G3ZPJ
TR20 9BX M3ZDK
TR20 9HN G4SOK
TR20 9LY M0PNZ
TR20 9LY M3PNZ
TR20 9PT G4ZLF
TR20 9SX G3ZPW
TR20 9TE M0TNT
TR21 0JD M1IOS
TR21 0NR G8GHQ
TR26 1AX G8JAB
TR26 1BL G3KDP
TR26 1ER G3YGM
TR26 2DU G1UAY
TR26 2SP M3KRZ
TR26 3AL G4YQS
TR26 3DX G8UJQ
TR26 3HL G4DZJ
TR26 3LE G0ISY
TR27 4AE M0XEK
TR27 4AG G4TXD
TR27 4AJ M3KGE
TR27 4EX G0PZR
TR27 4EX M0WIN
TR27 4JQ G1ZOB
TR27 4LY G1CHM
TR27 4NA 2E0WVS
TR27 4NJ G4KYO
TR27 4PJ G1LCY
TR27 4PS G3UCQ
TR27 4QT G4HUG
TR27 4RL G8UXD
TR27 5AB G4GKY
TR27 5BD M0FFS
TR27 5DT M6EEH
TR27 5ET G6LGM
TR27 5HA G3YNK
TR27 5HA G4MSV
TR27 5HA G4RVP
TR27 5HL 2E0JYA
TR27 5JD G4HFI
TR27 5JF G6YVD
TR27 5LJ M6FWV
TR27 6BA G8UTY
TR27 6HJ G4TXE
TR27 6PJ G4KYH
TR3 6BB G3PPT
TR3 6BJ G3UFJ
TR3 6BJ M5GUS
TR3 6BQ G1OEQ
TR3 6DB 2E1ERJ
TR3 6DY G4WKW
TR3 6EB G3MRT
TR3 6EN G0CQJ
TR3 6EN G6MOT
TR3 6ES M6ZXQ
TR3 6LJ G8PBI
TR3 6LN 2E1FNY
TR3 6LN G4LJY
TR3 6NW G8END
TR3 6PQ G0NNR
TR3 6PQ G0PGX
TR3 7AD 2E0BUD
TR3 7AD G3WPP
TR3 7AD M3AKH
TR3 7AT G0AWR
TR3 7BU M0MMT
TR3 7BU M6MIE
TR3 7DT 2E0OEM
TR3 7DT M6WDW
TR3 7EN G1XJT
TR3 7FG G3OCB
TR3 7HJ G8SLC
TR3 7JP G0UWO
TR3 7NW G0WGV
TR3 7RB M3ZAY
TR4 8DZ M6YFT
TR4 8ER M6IEW
TR4 8HH M0VAU
TR4 8QL G0DHT
TR4 8QL G4ZZZ
TR4 8RJ G8RBS
TR4 8SU 2E0PCO
TR4 8SU M6PCX
TR4 8TS G7WER
TR4 8TS M0AWP
TR4 8TS M0BDH
TR4 8TT G0LIA
TR4 9LT G4OOL
TR4 9PF G6WLA
TR4 9QE G8KLE
TR4 9RB G8ZYR
TR5 0TR 2E1ADT
TR5 0TU G6WMT
TR5 0XQ G3TDM
TR6 0AJ G3IKX
TR6 0DZ G1ZEI
TR6 0EF G0GQJ
TR6 0EY G3TYA
TR7 1AS M3AFS
TR7 1EN G4RMG
TR7 1PF G4SSL
TR7 1QQ G0DJL
TR7 1TY G4ADV
TR7 1TY M0BFB
TR7 2DE 2E0NWR
TR7 2DE M0RHW
TR7 2DG G0LCJ
TR7 2EJ G3WUA
TR7 2EJ G6ZSQ
TR7 2JN G8SCY
TR7 2LE G3JVN
TR7 2QE M0CJE
TR7 2RB G4VPJ
TR7 2RH M1AWX
TR7 2RW G0FLU
TR7 2SQ G6ZWI
TR7 2TB G0NWS
TR7 2TE G4WOI
TR7 3AE M3KUZ
TR7 3AE M3KVK
TR7 3AG G1LQT
TR7 3AW G0HEW
TR7 3BN G6CEP
TR7 3EB G7CRN
TR7 3EJ G8GOR
TR7 3JT G6AFA
TR7 3LB 2E0DXL
TR7 3LB M6HBQ
TR8 4AW G4CYI
TR8 4PL 2E1FOX
TR8 4PL G7KFQ
TR8 4PL M3KFQ
TR8 5HH G8YOE
TR9 6ER 2E0KTD
TR9 6ER G1PGV
TR9 6ER G4PSU
TR9 6ER M3KTD
TR9 6ER M3PSU
TR9 6FH M3ZLO
TR9 6NB M6DPH
TR9 6PD G4TGG
TR9 6PD G7OQL
TR9 6PT M3AEJ
TR9 6TW G0PNM

## TS

(Teeside)

TS1 4PR G7VOT
TS10 1JA G0HPQ
TS10 2BP G0GKY
TS10 2BP G4XXG
TS10 2BP M0BYV
TS10 2DG 2E1DMZ
TS10 2DL G0MBU
TS10 2EN G7CBR
TS10 2EQ G7CRA
TS10 2HW G6ZGC
TS10 2JZ G3UZB
TS10 2LJ M0BWO
TS10 2LQ G6IVD
TS10 2LT G0AJZ
TS10 2RU M6DST
TS10 2RU M6EMZ
TS10 2RY G8ZNL
TS10 4AG G3NVP
TS10 4HD M6SEG
TS10 4NJ G3VZO
TS10 4PS M3MII
TS10 4PS M6TST
TS10 4SG M3PYS
TS10 5DP M3JGN
TS10 5EB G0VLK
TS11 6BD 2E1BLT
TS11 6BD G3DAV
TS11 6DF G4WNA
TS11 6JX G4NUO
TS11 6NW G0VZT
TS11 7HJ M0RIG
TS11 7HJ M6RNE
TS11 7HT G0NUT
TS11 7LP G4HWC
TS11 8AG G8GFA
TS11 8AU G4RHX
TS11 8BP G1GMF
TS11 8BT G4GCU
TS11 8DB 2E1CIO
TS11 8DJ M0CIC
TS11 8DU G4CRS
TS11 8DU G4OLK
TS11 8EB M1SPY
TS11 8JJ M6YBD
TS12 1AL G8JLA
TS12 1DP G4IJO
TS12 1JP 2E0GLA
TS12 1JP M3NUH
TS12 2DQ G3XAG
TS12 2JZ G4FCU
TS12 2NJ M3RNO
TS12 2TJ G3KBI
TS12 2XG G0EWT
TS12 2YN M0BTL
TS12 3AP G4KIR
TS12 3AW 2E1FFZ
TS12 3AW M3AOC
TS12 3DX M6KYC
TS12 3EE G0RJA
TS12 3EW M6KAV
TS12 3JZ G6HHK
TS12 3LE G0FSG
TS12 3LS G0PZM
TS13 4DT G7EBX
TS13 4DW G4WUS
TS13 4JB G6DIR
TS13 4JD M6DJY
TS13 4NX M1FCZ
TS13 4PB G0SVJ
TS13 4UG G4JCS
TS13 5LF G1EDE
TS14 6DJ G4HEB
TS14 7BY 2E0DRJ
TS14 7BY M0NDT
TS14 7DP G7ONL
TS14 7LQ M0STF
TS14 7LZ G1AAG
TS14 7LZ M1AMP
TS14 7NB G4HRU
TS14 7PE M0MJT
TS14 8DL G6MVQ
TS14 8JG M0AGO
TS14 8JG M6GSX
TS14 8JY G1YHP
TS14 8LD G7ION
TS14 8LN G4YMB
TS14 8LT G0MBV
TS14 8LW G0VAP
TS15 0BA 2E0ECD
TS15 0BA M6ECD
TS15 0HT G4JLJ
TS15 0JQ G3BPP
TS15 9EF G1AHW
TS15 9EZ M1AFU
TS15 9JQ G1FIZ
TS15 9JQ G4ZXT
TS15 9LB M0DTS
TS15 9ND G7RAJ
TS15 9NL G8KIK
TS15 9NL G8MBK
TS15 9NQ G4JJM
TS15 9RZ G3VGZ
TS15 9RZ G4EEH
TS15 9RZ G8AOE
TS15 9TG M3KNF
TS16 0AQ G6KQJ
TS16 0JB G7NQZ
TS16 9EH G3FDW
TS16 9HP G3RUG
TS16 9HU M3XPH
TS17 0LT M0AVW
TS17 0NL G0IBW
TS17 0TB M3VPX
TS17 5BG G6ICZ
TS17 5DJ G1HSH
TS17 5HQ M6BHQ
TS17 8LX G6CKH
TS17 9AN M3LBD
TS17 9AU M3VDV
TS17 9QJ G0JOC
TS18 1JY G0MLB
TS18 4AZ M6KIT
TS18 4JA G3WWG
TS18 5AR M0HOK
TS18 5LH G4JXR
TS18 5NH 2E0MLX
TS19 0BD M6MMX
TS19 0DE M3FNA
TS19 0ER 2E0KBD
TS19 0ER M0KBD
TS19 0ER M6PII
TS19 0RA G3UFV
TS19 0SL G4YOV
TS19 7AP M3EVI
TS19 7JL G0ELC
TS19 7JT G1ZRS
TS19 7LY G6MPE
TS19 7SH M0DNN
TS19 8ES 2E0DBZ
TS19 8ES M3HSJ
TS19 8SZ M3ASC
TS19 8XF G8KSA
TS19 8XF M0DSC
TS19 9JS M3YNN
TS20 1JG M3XAC
TS20 1JQ G7CLG
TS20 1JU M0CYD
TS20 1JU M6HVX
TS20 1LE M3RFO
TS20 1LG M0NBK
TS20 1LW G8ILB
TS20 1PZ G7FFM
TS20 1SJ G7CLH
TS20 1SJ G7SUQ
TS20 1SZ M3JOA
TS20 2DF M6MOQ
TS20 2EX G0BZC
TS20 2QB M6MKF
TS20 2SP G4LIM
TS21 1DZ G4DXP
TS21 1HT G8LTD
TS21 1LQ G3NBL
TS21 2DH M3DGP
TS21 2DS G0WXC
TS21 2EZ G0SJU
TS21 3BZ 2E1JRB
TS21 3LT G8ZIA
TS22 5HA G1RYM
TS22 5JY G7NRO
TS23 1DL G4PVN
TS23 1HX G7CUO
TS23 1QN G3YVY
TS23 2AN G0HJM
TS23 2BQ G4ABM
TS23 2HX G6LDJ
TS23 2PJ G7LOA
TS23 2QH M0ASN
TS23 2RF M3UBL
TS23 3GP M6SWI
TS23 3HG G8OKS
TS23 3QN G4YFD
TS23 3SY G0JON
TS23 3UA G4MYN
TS24 0DX 2E0HLC
TS24 0DX M6AXG
TS24 0UF M6KQJ
TS24 0XF M0WFA
TS24 7HY 2E0YSF
TS24 7HY M6IEO
TS24 7NW M6NUF
TS24 8EU G7GRQ
TS24 8NN G7VGM
TS24 8RB G3IJV
TS24 9BQ G4VCJ
TS24 9BQ G6CGY
TS25 1GQ M6MQE
TS25 1HF M6TJI
TS25 1LP M6FQP
TS25 1LW M6GDO
TS25 1RN M6BHN
TS25 1RW G0SBX
TS25 2LA G1DFZ
TS25 2PY M3NFQ
TS25 2RD G7VGO
TS25 2RG G4ZCN
TS25 3DP G3UUF
TS25 3QY M1BNK
TS25 4NZ G1KBE
TS25 5HX G3NWY
TS25 5HZ G0VGB
TS25 5LB G1GTP
TS25 5NG M6OTB
TS25 5PA G1ETQ
TS25 5RQ G0FZZ
TS26 0NW 2E0FAU
TS26 0NW M6EVW
TS26 0PJ G3NUA
TS26 0TP 2E0AZS
TS26 0TP M0VED
TS26 0XG 2E0DGA
TS26 0XG M0VRW
TS26 0XG M6EOK
TS26 8ND G7SNJ
TS26 9AN 2E0HPI
TS26 9AN M6HPL
TS26 9BJ M0HXK
TS26 9ES M0BKF
TS26 9PD G4PDM
TS26 9PN M0HPL
TS26 9PR G4SHJ
TS27 3QR G7JXY
TS27 4AB M6MKM
TS27 4AX M0BZB
TS27 4DF M0DEL
TS27 4EB M6DLC
TS27 4RT G0GKO
TS28 5BP M6PIT
TS28 5BT G0HJX
TS28 5EJ G4XPP
TS28 5JE G0GWD
TS28 5JF M6LAZ
TS28 5JW M6RDQ
TS29 6DA G0MHC
TS29 6EE M0ZEM
TS29 6EF M3SHB
TS3 0NN M6SOV
TS3 0RL 2E0JKD
TS3 6PE M6CFD
TS3 7DR 2E0BJK
TS3 7NQ M6IBH
TS3 7PU M6TNA
TS3 7RX 2E0KOM
TS3 7RX M0KOM
TS3 7SL 2E0FJP
TS3 7SL M3FJP
TS3 8LX G4IIN
TS3 8NX G4OOK
TS3 9HD M6MQL
TS3 9NT M6MFC
TS4 2HG M6LCY
TS4 3HJ M6RTA
TS4 3HR M6KRI
TS4 3HU M6WCC
TS4 3HX G3RGB
TS5 4AJ G4UTV
TS5 4HG M0SMA
TS5 5BN M3CWV
TS5 5JP M3YRZ
TS5 5LD M0BJX
TS5 5NQ G4HSB
TS5 6DP G4VZC
TS5 6QE G3MXZ
TS5 6SQ M3KRR
TS5 8BT G1NTL
TS5 8DR G3KXV
TS5 8EB G8WKZ
TS5 8NT G4IJM
TS5 8NT G8HDM
TS5 8RE M0GQV
TS5 8RY M6LLM
TS6 0BN G0NYZ
TS6 0BQ G7TWU
TS6 0BS G4MCF
TS6 0PL G1DAT
TS6 0PP M0BZA
TS6 0QQ G0FXR
TS6 0QQ G0LWE
TS6 0QQ M3TVN
TS6 0RW M6CLI
TS6 7EZ G7NQR
TS6 7EZ M0HMS
TS6 7NA G1CLT
TS6 7ND 2E0CNX
TS6 7ND M3ITL
TS6 8AF G1CCM
TS6 8EH G7FVH
TS6 9LY G4ORQ
TS6 9QW G0PDK
TS6 9SZ G7NQX
TS7 0EZ G7WEN
TS7 0JL M0DIQ
TS7 0LB 2E0LGB
TS7 0LB M0LGB
TS7 0LT G1IGW
TS7 0QL G8YDC
TS7 8AN G7UMA
TS7 8EH G4LVY
TS7 8HJ G4PPS
TS7 8SE 2E0CPK
TS7 8SE M0HHC
TS7 8SE M6BKL
TS7 9BB 2E0PEL
TS7 9BB M0SMP
TS8 0RU G0AOO
TS8 0SU G7RXB
TS8 0UJ G7PTC
TS8 9AB G4POD
TS8 9BU G4ZML
TS8 9HH G0BQP
TS8 9LJ G0UTP
TS8 9LJ M6UTP
TS8 9QQ G8LIE
TS8 9SL G7IZC
TS9 5AG G4HWV
TS9 5BU G0TYM
TS9 5EL M0CSD
TS9 5HU G0UYG
TS9 5HX G7RNQ
TS9 5PQ G4OIW
TS9 6DW 2E0FAU
TS9 6DW M3FYV
TS9 6EN M6BHP
TS9 6JF G7TFG
TS9 7DA G8ADH

## TW

(Twickenham)

TW1 1AG M0ICI
TW1 1PY G4RBR
TW1 1QL G3CPC
TW1 1RS M6MDN
TW1 2AX 2E0JAB
TW1 2DD G4CXZ
TW1 2DF G4CBD
TW1 3AU G6CIA
TW1 3EW G1LRR
TW1 4BQ G1TOB
TW1 4QZ G8DSU
TW1 4SW M0MDC
TW10 5DU 2E1DQM
TW10 5EF G4FZY
TW10 6DS 2E0RYS
TW10 6DS M0RYS
TW10 6DS M6RSY
TW10 7ED G1PHA
TW10 7EQ 2E0ZWE
TW10 7EQ M3RHO
TW10 7NQ G0SLI
TW10 7YG G8CYE
TW11 0AW G4RWA
TW11 0BG G4LGO
TW11 0DH G7PMV
TW11 0SB G7TYH
TW11 8AS G3PPC
TW11 8DE G8BAJ
TW11 8JB G6RHV
TW11 8LT G1OPT
TW11 8SH G8ZWA
TW11 9DW G8WGE
TW11 9HA G0DEH
TW11 9HY G8JUG
TW11 9LN G8AUU
TW11 9PD G8DEL
TW11 9QS G0JME
TW11 9RS G3ZLD
TW12 1DE M0LHR
TW12 1DW M0BYY
TW12 1HX G4PPE
TW12 2BL M6ZES
TW12 2JH G8SNV
TW12 2LU G3RCB
TW12 2NE M6JXD
TW12 2PZ G7DWM
TW12 2QG G4ZOU

TW12 2SR M0NAI
TW12 2TR G7OMA
TW12 3BW G4UXB
[illegible]
TW13 2YN G6NIX
TW13 4HX G1KCU
TW13 4LX M0GFZ
TW13 4QG M3XHB
TW13 5DJ G7GFQ
TW13 6LZ G0EEZ
TW13 6PE G0KXG
TW13 6PX G1YVZ
TW14 0AH 2E0RUS
TW14 0AH M3SQS
TW14 0JR M0KGA
TW14 0JY G1TFM
TW14 8SJ G8MWD
TW14 9ED G1LKL
TW14 9JE G0DEO
TW14 9LW G7UWS
TW14 9QP G4ALA
TW14 9XG G7OXK
TW15 1AB G8FVJ
TW15 1AL M1ASS
TW15 1DG G0CPF
TW15 1DP G1IBS
TW15 1DP G4ZVU
TW15 1DR 2E0UFM
TW15 1DR G4ESG
TW15 1DR M6YON
TW15 1HF G0ONA
TW15 1JB 2E0NOC
TW15 1PQ G0RJE
TW15 1PW G0NBJ
TW15 1PW G0OSX
TW15 1QE G1WWB
TW15 1SQ G8ZZG
TW15 1UT M6HKS
TW15 2AP G8CAH
TW15 2EP M3YXE
TW15 2LP G3XTZ
TW15 2LU G8LPA
TW15 2LU M0LPA
TW15 2RF M1CSU
TW15 2SJ G1WZB
TW15 2SL M5CBR
TW15 2TD G1ZDT
TW15 2TD G7AYP
TW15 3ET G2HS
TW15 3PF M0AHS
TW15 3QA G3JUL
TW16 5NB G4LDW
TW16 5PT G3PSW
TW16 6HF G3HTC
TW16 6HF M0CFR
TW16 6SG G3ZXA
TW16 7NA G4YAS
TW16 7NL G7PAK
TW16 7PL 2E0NBE
TW16 7PL M0NCE
TW16 7PL M6FUF
TW16 7QU M6NPF
TW16 7TL 2E1AOF
TW16 7UA G0HYT
TW16 7UA G7EAR
TW17 0DN G3UPW
TW17 0EN G0VDZ
TW17 0JB G6IFR
TW17 0RP G6KLH
TW17 0SH G4XGI
TW17 8AY G4GNS
TW17 8EU G0LGA
TW17 8QQ M6GAN
TW17 8RR G7MII
TW17 8RX G4CKQ
TW17 9DQ G8BQZ
TW17 9HE M6HSG
TW18 1DG G0MIJ
TW18 1DJ G0JSP
TW18 1EE G1SXB
TW18 1JB 2E1AVX
TW18 1JB M0JFP
TW18 1JB M0MXO
TW18 1NE G7ECE
TW18 2AP G4HKS
TW18 2AP M6GIQ
TW18 2AP M6SUC
TW18 2DD G6PVC
TW18 2LE M6PDU
TW18 2QE G3ZLR
TW18 3DW G0NIF
TW18 3EE 2E0VBX
TW18 3EE M6VBX
TW18 3HH G0OEK
TW18 3HQ M6NVG
TW18 3NQ G4GSC
TW18 4NR G3NTM
TW19 5NF M6ZJW
TW19 6AX G4AKA
TW19 6DT G4UKO
TW19 7AJ M3OPT
TW19 7EU 2E0WKH
TW19 7EU M6HAR
TW19 7JE G0TID
TW19 7JE M3CJA
TW19 7JE M3EMN
TW19 7LF G4MTH
TW19 7LG M6CTI
TW2 5BY G1NSQ
TW2 5BY G6AWM
TW2 5JJ G7TOF
TW2 5JP G1FKS
TW2 5JT M3JXY
TW2 5LS G0EZL
TW2 5PD M0DLL
TW2 5PE M0XDY
TW2 6AA G2AIW
TW2 6BL G8WRL
[illegible]
TW2 6EJ G3ZSP
TW2 6HW G1LVZ
TW2 6SP G6GVF
TW2 6SR 2E0WEJ
TW2 6SR M0GBC
TW2 6SW G6VUG
TW2 7AL G7GZC
TW2 7BL 2E0WMD
TW2 7BL G0TSU
TW2 7BL M0DMJ
TW2 7BL M3ESQ
TW2 7DF G0DNU
TW2 7DT G7GAP
TW2 7EA G1OVO
TW2 7EU M6RZP
TW2 7EX G4OQN
TW2 7JE G0OFN
TW2 7JG G7HOV
TW2 7NH G0MRF
TW2 7NP M0DCW
TW2 7SN G6LKH
TW2 7SN G8RHZ
TW20 0BT G7MVN
TW20 0HQ M3GXG
TW20 0HQ M3VOA
TW20 0JF G8NIU
TW20 0RZ M0HJY
TW20 0RZ M6BRJ
TW20 8AN G0LSE
TW20 8NL G6PVV
TW20 9AH M0HMI
TW3 1XG G3WKE
TW3 1YH G4CWE
TW3 2DW M3NFE
TW3 2HH 2E0JXX
TW3 2HH M0XXJ
TW3 2HH M6JXX
TW3 3QJ 2E0RBH
TW3 3QJ M3VYS
TW3 4NL M0RJZ
TW4 5BB G1EIH
TW4 5EW 2E0BLQ
TW4 5LZ G4TSH
TW4 5PF G0RHB
TW4 5QH G3UIJ
TW4 6AG G3AHE
TW4 6NA G1CRN
TW4 7BH G3YAS
TW4 7JG M0WBC
TW4 7JQ G0WMD
TW4 7JQ M0HIC
TW4 7NH G0IIK
TW4 7NU G7GBE
TW4 7RA G0USK
TW5 0ND G6YTX
TW5 0NF G4IXB
TW5 0PZ M0TCE
TW5 9AW G1JRW
TW5 9EX G0VLQ
TW7 4AJ G0LUH
TW7 4LS M0EMD
TW7 4PF G7RNF
TW7 4PQ G4DUO
TW7 5DP G7RPP
TW7 5HB G8SJO
TW7 6AD G3LWR
TW7 6HW G4LTC
TW7 6HX G4PWS
TW7 6LF G4NJN
TW7 7DJ M3HVO
TW7 7NL G6AGA
TW8 0EB G4JMB
TW8 0PL G3XKV
TW8 9PT G6SUK
TW8 9RD 2E0JDC
TW8 9RD 2E0YVR
TW8 9RD 2E0ZBW
TW8 9RD M3YVR
TW8 9RD M3ZBW
TW9 1AX M3NAO
TW9 1LL 2E0SWE
TW9 1LL M6BIV
TW9 1YD 2E0AKR
TW9 1YD M0GSV
TW9 1YD M3MIH
TW9 2DZ 2E0CEN
TW9 2DZ M3ZEI
TW9 2HA G1CRT
TW9 2TJ G0ACD
TW9 3AY G0LYJ
TW9 3BD 2E1GHZ
TW9 4AS G0OOI
TW9 4AS G3CIK
TW9 4FF G0UHW
TW9 4JB G4FGW
TW9 4JP M0YDU

## UB
## (Uxbridge)

UB1 1DG 2E0UMR
UB1 1DG M6UMR
UB1 2SA M0GBO
UB1 2UP G1YES
UB1 3JP G3KLK
UB1 3NT G1YRY
UB1 3QD G6ESM
UB10 0BS G6KNK
UB10 0DN G1ECY
UB10 0HH G1DYR
UB10 0HP G6IRY
UB10 0HW 2E0CXE
UB10 0HW M0HQR
UB10 0HW M6CKU
UB10 0QH G4DIG
UB10 0LD G4DVI
UB10 8HS G4CTD
UB10 8LN G4IPJ
UB10 8LS G1PAF
UB10 8QA G8LIU
UB10 8QA G8UBF
UB10 8TA G6LEY
UB10 8UF M3MST
UB10 8UF M3WCR
UB10 9HZ M6ODL
UB2 4RG G7OPG
UB2 5AN G7IVN
UB2 5HN G6VAE
UB3 1PZ G0LQW
UB3 1ST G3SUN
UB3 2BT M0NYW
UB3 2BT M1PRO
UB3 2FJ M6FPP
UB3 2JE G8HOU
UB3 2QX G6XDI
UB3 2TP G8GYP
UB3 2TQ G6STI
UB3 3PA G1VTS
UB3 3PP G7JSW
UB3 4AD G0EFS
UB3 4QJ G1JLM
UB3 5ET G3PKR
UB4 0AE G0CHR
UB4 0AQ M0MBB
UB4 0DZ M6XJS
UB4 0EF G0RGL
UB4 0EF G0VGY
UB4 0JH G1LLQ
UB4 8ET G4OHQ
UB4 9QT M6GQU
UB4 9RB G1LVR
UB4 9YF G0KYN
UB5 4AH M0MDG
UB5 4AQ 2E1IDC
UB5 4HF M6FPN
UB5 4NL G1XLT
UB5 4QS G4ZKT
UB5 4SG 2E0JGP
UB5 4SG M6JGP
UB5 4SW M6WIK
UB5 4TL M6DTC
UB5 5BX G1DLA
UB5 5HN G6VXE
UB5 5JQ M6RHU
UB5 6AR G1KOX
UB5 6EU G0SLH
UB5 6HP G7GDC
UB5 6PX G1SAJ
UB6 0BH G1VSM
UB6 0BH M0DBI
UB6 0HQ G3PSC
UB6 0JZ G0MXY
UB6 0RF G4AWM
UB6 0SE G7KYF
UB6 0SS M3CGO
UB6 7AD G4FJX
UB6 7HR G6PGJ
UB6 7QJ 2E0EBA
UB6 7QJ M0TMT
UB6 7QJ M6EST
UB6 8AJ 2E1HFW
UB6 8LN M1AZQ
UB6 9BY G7ABE
UB6 9LS G1UBL
UB6 9NN G0JJQ
UB6 9NN G6UIT
UB6 9NT G0HRK
UB7 0LE M0AAC
UB7 0LS 2E0LJR
UB7 0LS M6ZZV
UB7 7AN G0FFK
UB7 7AP 2E0TTW
UB7 7AP M6EJF
UB7 7AP M6UAF
UB7 7TZ M6BBR
UB7 8BU M3OTR
UB7 8HN G6JOV
UB7 8PF G6BSS
UB7 9DR G2FDN
UB7 9PE G4ZRC
UB8 1BL G8FKH
UB8 1BL M0DGB
UB8 1QX G3JVP
UB8 2FX G7BRC
UB8 2NY 2E0CRC
UB8 2NY M0KBA
UB8 2NY M6CCR
UB8 2PQ G7NFD
UB8 2UL G4CVF
UB8 3EL M0HZV
UB8 3HB G7VUN
UB8 3HS G7VHO
UB8 3HX M0LTP
UB8 3JQ M0LEB
UB8 3QU G6JHG
UB8 3SR G0OLD
UB8 3TE G7IYG
UB8 3TE G7IYH
UB8 3WS G7LCD
UB9 5DJ G0VOV
UB9 5ED M3MES
UB9 5HJ G7LVE
UB9 5LL G0PBS
UB9 5PB G6FGJ
UB9 6LZ M6TGI
UB9 6QB G3VXA
UB9 6QB G6SSM
UB94AG G4YEO

## W
## (West London)

W10 5TA M0TUR
W10 6QL M6FUH
W11 1EP M0IAK
W11 4BT M1DGK
W11 4HE M6EFG
W11 4SQ M5OOO
W11 4TG G8EZT
W12 0AP G7AAS
W12 0LP G3SEF
W12 7BL G3IJL
W12 7HQ G3XBX
W12 7PY 2E0TBH
W12 8JN 2E1IFA
W12 9LU 2E0AVB
W12 9TF M3NKO
W13 0AE G6GIU
W13 0ED G0PIN
W13 0HP G1ARF
W13 0NT G7JHZ
W13 0TF M0GHC
W13 8AJ 2E0PMP
W13 8AJ M6PRP
W13 8JZ G3RGP
W13 9DT G0AIS
W13 9EN G8LVF
W13 9HS M0CPB
W13 9JY M6ZBS
W13 9LA G6OJN
W13 9QA M6CWP
W13 9QU G8VDQ
W13 9RA G6JEU
W13 9TN M0FZU
W13 9TY G1HGT
W14 0JX G4HTY
W14 9BS G1EIG
W14 9JS M6ENW
W1H 7DP M0KYR
W1J 8PE M0CER
W1U 3PY G3UXI
W2 1TT 2E0CAO
W2 2RJ M3YGY
W2 2TH M0SYM
W2 3RL M6RVJ
W2 4LA M3MOC
W2 5HA M0HWW
W2 5PB 2E0CLX
W2 6DD 2E0JLM
W2 6DD M0XJM
W2 6DW 2E0ZDC
W2 6DW M0ZDC
W2 6DW M6ZDC
W3 0DE G3YMM
W3 0HH G1CNN
W3 0HR G3HHU
W3 0RP G8LWS
W3 7AQ G0ZHP
W3 7NP G0UCC
W3 7PA M1ELS
W3 9AE G8SPE
W3 9EJ G6HXB
W3 9RQ G4ZZV
W4 2HR M6COY
W4 2JH G4LIC
W4 2LL G4SJL
W4 2SF 2E0ZHN
W4 3LR G1SAT
W4 3LR G3VKT
W4 3NH G4ZJD
W4 4EH 2E1HUJ
W4 5DN G4TZA
W4 5EN G0OYJ
W4 5JS G3SQX
W5 1HL G0VDN
W5 1LS G4CEK
W5 1PY M0BCZ
W5 2JA 2E0RCC
W5 3LL M3SDQ
W5 3SL G8MVC
W5 3XH M6PFM
W5 4BJ M6XKN
W5 4DR G7ANG
W5 4JW G0SCG
W5 4XA 2E0JAJ
W5 4XA G7BRZ
W5 4XA G8ROJ
W5 5BH 2E0JFS
W5 5BH M0JXS
W5 5BU M0JIO
W5 5HS G6FVB
W5 5LG 2E0HCW
W5 5LG M0HSW
W5 5PD G6YJJ
W5 5QD G8RVO
W5 5QT 2E1DTE
W5 5QT 2E1DTF
W6 0HY M6ZDF
W6 7LD G3YFV
W6 8BD G7EEJ
W6 8HN G7VLF
W6 8RN 2E0CUO
W6 8RN M0HYG
W6 8RN M6CNL
W7 1BT G6TJY
W7 1JQ M6TVX
W7 2JH G7VEI
W7 2JS M6MQH
W7 3AG G8RBX
W7 3DJ G6TNK
W7 3EJ G3TIG
W8 6PJ M6RIQ
W8 7AX G0AVQ
W8 7BS G3SKR
W8 7SL G4EHN
W8 7SX G0HTX
W9 2BJ M6HEJ
W9 2QU M6VLD
W9 3LZ M3DZQ

## WA
## (Warrington)

WA1 3AY M1AUH
WA1 3EN G4FMI
WA1 3EN G8XVJ
WA1 3EN M0SDA
WA1 3HB G4XQA
WA1 3HX G0RFM
WA1 3TN 2E0RYP
WA1 3TZ M3LFU
WA1 4BJ G1NWA
WA1 4BJ G7JOA
WA1 4DY G0ANL
WA1 4EB G1JHL
WA1 4ED G4TZT
WA1 4EZ G6LFT
WA1 4EZ M0CUQ
WA1 4EZ M3PQN
WA1 4LU G0CDA
WA1 4NW G0PZP
WA1 4NY G1NXT
WA1 4PE M0CSE
WA10 2HA G0KQI
WA10 3JH M0BWP
WA10 3JH M0TLR
WA10 3JH M3UGT
WA10 3NE M0AQK
WA10 3QL M6BJA
WA10 3RR 2E0MVH
WA10 3RX 2E0PBO
WA10 3RX M6PPK
WA10 3XT M1BWS
WA10 4BL M6ZLM
WA10 4DN M3MSB
WA10 4GY G4IFU
WA10 4JG G0JBR
WA10 4JW 2E0IAZ
WA10 4JW M6IAZ
WA10 4LN G1OSL
WA10 4LU 2E0KNM
WA10 4LU M6KNM
WA10 4NL M3OEM
WA10 4RH M0ACK
WA10 5AH G0EZI
WA10 5DW M3DUO
WA10 5PB G8MHE
WA10 6BP G4FQN
WA10 6QG G0BCX
WA10 6SH G4MWO
WA10 6SH M0SHR
WA10 6TP G1NVE
WA11 0BL G8CXZ
WA11 0DY G1XAP
WA11 0ES G0DZI
WA11 0ES G6DZI
WA11 0GN G4WJR
WA11 0LY G4YKV
WA11 0NB G4WGJ
WA11 0PP M3VXM
WA11 0PY G0AFJ
WA11 0QD M0NZR
WA11 0QS G1ZKZ
WA11 0RH M3NJQ
WA11 0RS G0SSL
WA11 0SD 2E0YGH
WA11 0SD M6TTO
WA11 0SQ G1WAE
WA11 0SX G0NWC
WA11 0UB G4XIE
WA11 0YQ G0HBU
WA11 0YS 2E0RCC
WA11 0YS M0RCC
WA11 7AG M6FME
WA11 7LD G4KIN
WA11 7LD G8TMR
WA11 8AG G4KIP
WA11 8AT G3JIR
WA11 8BH M3DJW
WA11 8BW G0BHH
WA11 8DJ G4FKP
WA11 8ER G8JSM
WA11 8JR G4EST
WA11 8JW M0AOO
WA11 8LA M0DZC
WA11 8NR G0GGO
WA11 8SW M3JGH
WA11 0DQ M0EZZ
WA11 9EL G4DIY
WA11 9EL G4WXZ
WA11 9HU G6YSN
WA11 9JD M6ZZE
WA11 9JJ M3KYQ
WA11 9NH M3VOL
WA11 9RE G0NEB
WA11 9RW G0EAM
WA12 0AS M3OWQ
WA12 0AU 2E0BGM
WA12 0AU M3ISI
WA12 0DR G0VVQ
WA12 0DU G4KHG
WA12 0JT G4XOL
WA12 0JW 2E0NLW
WA12 0JW M6NLW
WA12 0LN G7RUN
WA12 0LN M3PQH
WA12 0LN M0TLN
WA12 0LY G1HIO
WA12 0LY G1HIP
WA12 0LY M0NRC
WA12 0NE 2E0IUH
WA12 0NE M3IUH
WA12 0NN G6OBG
WA12 8BY M0JVW
WA12 8HJ M3HHQ
WA12 8HP M3IRR
WA12 8JE G1HSA
WA12 8JF 2E1FSH
WA12 8JF M0ZLK
WA12 8JF M6BWV
WA12 8JF M6YZX
WA12 8LT G3XRI
WA12 8LT M0NLW
WA12 8LZ G6IKM
WA12 8SQ G1JPT
WA12 8SQ G6JPT
WA12 9DB M3ISJ
WA12 9EY 2E0IUI
WA12 9GB M3HGZ
WA12 9LS G0VAM
WA12 9LT G7MFA
WA12 9TH 2E0LDJ
WA12 9TH M0NEX
WA12 9TH M6LDJ
WA12 9UD G1MMA
WA12 9YG G3YVH
WA12 9YL M3NBQ
WA13 0JS G3VJV
WA13 0JT G7BGO
WA13 0QR G1ANK
WA13 0RD G3NKH
WA13 0SN G3YHD
WA13 0SY 2E0DGU
WA13 9BA G8VMQ
WA13 9EY G0AMU
WA13 9EY G6XKK
WA13 9LY G3NKW
WA13 9NL G8TYH
WA13 9NN M1DQQ
WA13 9NW G1DVA
WA13 9NW G6WRC
WA13 9NW M6KDM
WA13 9QA G4VSX
WA14 1JL G4UGM
WA14 2EQ G8JGM
WA14 3ND G3WFH
WA14 3HF G0LNO
WA14 3HY G0KTT
WA14 4HW 2E0WHA
WA14 4HW M0WIA
WA14 4HW M3WIA
WA14 4JE G0CBJ
WA14 4QT M3VOJ
WA14 4RQ G6HLU
WA14 4UE G4PYH
WA14 5AZ G7UCT
WA14 5AZ G4NAV
WA14 5JF G7AGA
WA14 5JY M6AAV
WA14 5LJ 2E0NDG
WA14 5LJ M6NDG
WA14 5LT G4HUF
WA14 5LT G7EYR
WA14 5NQ M1DXL
WA14 5YF M0OKS
WA15 0AS G4FYB
WA15 0LT G4CNI
WA15 6AB G0WEV
WA15 6AB M1CTM
WA15 6BG 2E0KCZ
WA15 6BG G3SMM
WA15 6BG M0KCZ
WA15 6BG M6KCZ
WA15 6BW G3EOO
WA15 6BX G6RGN
WA15 6BX G7MVY
WA15 6DY G4VBJ
WA15 6HQ G4WQD
WA15 6HR G6MCE
WA15 6RS G6DOQ
WA15 6SZ M1BTI
WA15 6UE G4TFU
WA15 7DB M6JCN
WA15 7HH G1GLS
WA15 7HY M3VCV
WA15 7NP G1KXP
WA15 7NP M3OEO
WA15 7PA G7RHD
WA15 7YU G0UAL
WA15 8DN G6GTH
WA15 8EA 2E0CXI
WA15 8ET G1OVR
WA15 8HA G7VCF
WA15 8LY G6MNB
WA15 8PD G0UXR
WA15 8PW G0GWI
WA16 0BN G7EPN
WA16 0JX M6EOB
WA16 0NE G0AKF
WA16 0TU M5ACS
WA16 0TZ M1ROE
WA16 6PS G6EXE
WA16 7AX G3UJE
WA16 7AX M0TTG
WA16 7ET G4HZW
WA16 7HB 2E0CSU
WA16 7HE G0AUB
WA16 7HF G0FOY
WA16 7RD G8CJQ
WA16 8BB G3LFQ
WA16 8BB G4LFQ
WA16 8BG M0LHU
WA16 8BH G8WUY
WA16 8JG 2E0KNU
WA16 8JG M6KNU
WA16 8JR G6FWK
WA16 8LH M6JMP
WA16 8TR G4GEO
WA16 9AW G0IFX
WA16 9DE G4MGK
WA16 9DZ G3ISB
WA16 9SB G1OQO
WA2 0AG G3YRQ
WA2 0AG M3WIJ
WA2 0BE G1KSK
WA2 0BL G3NFB
WA2 0BL M1DOT
WA2 0EZ G6OSV
WA2 0GL M3RZI
WA2 0HN 2E0MMP
WA2 0NQ M3ISN
WA2 0QE M3KPB
WA2 0SQ 2E1GMO
WA2 0SQ 2E1HEF
WA2 0SQ G4YPI
WA2 0SQ G1OND
WA2 0SQ M0DWQ
WA2 0SQ M1GMO
WA2 0UF M6KEG
WA2 0UH M0RAN
WA2 0UJ 2E1HEK
WA2 0UJ G6FKL
WA2 0WR M6WAA
WA2 7AT M1CXW
WA2 7AX M1FAX
WA2 7DN M3ARU
WA2 7JQ 2E0BXA
WA2 7JQ M6JON
WA2 7RW M6JEP
WA2 7SE G1FOW
WA2 8AJ M6TJJ
WA2 8AW M3PVB
WA2 8BQ M1EFT
WA2 9BQ G0RFT
WA2 9JA 2E0DHO
WA2 9JA G6LXV
WA2 9JA M6EWO
WA2 9NQ M6NBN
WA2 9PS G0VQK
WA2 9PS G7TOA
WA2 9SD G7PMB
WA2 9TW G7HKN
WA3 1DG M0NTA
WA3 1EB M1BRX
WA3 1EU G6AHF
WA3 2DU G4UPO
WA3 2EE G4VDX
WA3 2EP G1EFU
WA3 2EP G1THF
WA3 2ES G0LVJ
WA3 2QL G4ONG
WA3 2RN G4RNC
WA3 3DF G4JPX
WA3 3HH 2E0COA
WA3 3HH M0HKJ
WA3 3HH M0LGL
WA3 3HH M6CFN
WA3 3HH M6EEL
WA3 3HH M6LTL
WA3 3JJ M6DOX
WA3 3LA M6JHI
WA3 3NA G0TQG
WA3 3QX G0RPO
WA3 3TZ G0LVY
WA3 3TZ G6TPG
WA3 3UX G6WVL
WA3 4DF G4LWY
WA3 4ES G1IQK
WA3 4JF 2E0COD
WA3 4LD G0RPG
WA3 4LD M0MRC
WA3 4LD M1CNP
WA3 4LR G8YKG
WA3 4NW G0WJX
WA3 4NW G7GZB
WA3 4PD G8NDE
WA3 5AP M6MQA
WA3 5JJ G0IZN
WA3 5QY G3XDP
WA3 5RU G6HKN
WA3 5RW G6DSP
WA3 6AQ G4YKR
WA3 6DJ G4VBD
WA3 6LB M3EUF
WA3 6NE G1YLK
WA3 6NY 2E0SLM
WA3 6NY M3YHG
WA3 6PD G0ILL
WA3 6PS M6RJB
WA3 6PT M0HVN
WA3 6PT M6GUV
WA3 6QD G0JXI
WA3 6QP G6KTN
WA3 6RR 2E0ISK
WA3 6RS M3MHR
WA3 6TH G1VSK
WA3 6TP G1ZOY
WA3 6TR M6CFL
WA3 6TZ G0BCU
WA3 6UY 2E0FXP
WA3 6UY G0CBN
WA3 6UY G0CPD
WA3 7JG G0TDC
WA3 7LA G8NXY
WA3 7LR G0JPC
WA3 7LR G6MKD
WA3 7LS G1NXS
WA3 7LT G1OOO
WA3 7NU 2E0KTG
WA3 7NU M0XDJ
WA3 7NU M3YGL
WA3 7NU M3ZLP
WA3 7PD G1VPH
WA3 7QA M1MKL
WA4 1AX G7NPT
WA4 1DW G0YSS
WA4 1EX G7CED
WA4 1HX M3ZMN
WA4 1LY G8SQK
WA4 1NF G6DOZ
WA4 1PY 2E0DTO
WA4 1PY M0HFF
WA4 1QN G1PIY
WA4 1RJ G1KOD
WA4 1RU G0MWY
WA4 1RW M3OQS
WA4 1TU G7VAG
WA4 2AE G3OGQ
WA4 2ET M6WCY
WA4 2HE G0KXW
WA4 2HQ G4YPI
WA4 2HY M0PDX
WA4 2JA 2E0MFG
WA4 2JA M6TTP
WA4 2PF G0SSJ
WA4 2RE G4PBZ
WA4 2RX G0RPF
WA4 2TR G6MQI
WA4 3AE M3TRJ
WA4 3BG G0IHF
WA4 3BJ G6GEL
WA4 3ES M3MXJ
WA4 3JA G7SKR
WA4 3LE 2E0NDP
WA4 3LE M0NDP
WA4 3LE M6NDP
WA4 4AF M3TPG
WA4 4BY G3JDT
WA4 4BY G3ZRN
WA4 4EE 2E0TBI
WA4 4EE M0WAU
WA4 4EE M6JPL
WA4 4NG M6JWO
WA4 4PY 2E0WVM
WA4 4PY M3ISY
WA4 4QB G1GNX
WA4 4QB G4UKV
WA4 5AW G4EAQ
WA4 5DL M6JBZ
WA4 5EJ G0MYN
WA4 5QD G4LLG
WA4 5QJ G7LHS
WA4 5RD G8KBB
WA4 5RD M6TLP
WA4 6AF G0KXY
WA4 6DA M3GIX
WA4 6DE 2E0GGI
WA4 6DE M6CPX
WA4 6ES M3RVK
WA4 6GY G4OEX
WA4 6LF G7GPL
WA4 6QU G4BQJ
WA4 6UA G3MUX
WA5 0AG G6YHL
WA5 0GB G7KHW
WA5 0GD M1EGV
WA5 0GJ M6SPZ
WA5 0HP M6ABE
WA5 0HP M6TLX
WA5 0LW M6LRH
WA5 0LY M3HIG
WA5 1BL 2E0TGS
WA5 1BL M3TGS
WA5 1BQ 2E0XJJ
WA5 1GU 2E0CDT
WA5 1GU M0HLC
WA5 1GU M3ZRR
WA5 1HN G0CWA
WA5 1HP M1FHP
WA5 1JF G7LPZ
WA5 1JH G6DPH
WA5 1JY M0LOG
WA5 1SY M0CRN
WA5 1TQ 2E0LOL
WA5 1TQ M3LXS
WA5 1TQ M3MQA
WA5 1XG G0CVM
WA5 1XN G7CFS
WA5 2AQ G3ZHE
WA5 2AD G0UCI
WA5 2BY G0NVT
WA5 2DB G0KKU
WA5 2DD G0ZKD
WA5 2HX G4TGO
WA5 2HY G8PXI
WA5 2JN M3MIQ
WA5 2PA G7GGM
WA5 2QE G0SLR
WA5 2QY G3GAH
WA5 2RB G8WAJ
WA5 2SG G4HSS
WA5 2SR G8NRF
WA5 2TW G3RRM
WA5 3AJ M0DCG
WA5 3AN G0RBI
WA5 3AN G3XUB
WA5 3DA G0APY
WA5 3EE G0OON
WA5 3EH 2E0ELK
WA5 3ET G8VVP
WA5 3LU G3YZU
WA5 3NU G1NHN
WA5 3NW G0XHE
WA5 3RX G8RCL
WA5 4AX G4WNW
WA5 4DS 2E1GNR
WA5 4ES M3JNB
WA5 4HJ 2E0MGW
WA5 4NE M3JAZ
WA5 4NS G4OMZ
WA5 4NW M6GFZ
WA5 4PW M6JIQ
WA5 7XT 2E0WRK
WA5 8GB G8NHD
WA5 8QL G0WRS
WA5 8QL G4VSS
WA5 8QL M3CFI
WA5 8QL M3FIW
WA5 8QL M3FIY
WA5 8WU 2E0DKA
WA5 8WU M6EWN
WA5 9PL M3ZGU
WA5 9QE M3YRJ
WA5 9SB M5AGI
WA5 9UU G4GPJ
WA6 0NW G6IRW
WA6 0QX G0LZJ
WA6 0QX G8KTE
WA6 6DN G3GKS
WA6 6PT 2E0JDI
WA6 6PT M0JLR
WA6 6PT M6XJR
WA6 6PY M0DYB
WA6 6QQ G3VBA
WA6 6SG G8WQE
WA6 6SG M3MHV
WA6 7JR M3LLB
WA6 7JR M6FMU
WA6 7PG 2E0LYR
WA6 7PG M3LYR
WA6 7QJ G8DVF
WA6 7RU G4OTI
WA6 8AE G4VYJ
WA6 8DA M1KTY
WA6 8DA M1WRX
WA6 8HP 2E1FKD
WA6 8HP 2E1HEE
WA6 8HP G1IVV
WA6 8HP G7NCP
WA6 8HP M3IVV
WA6 8JJ G6ZHS
WA6 8JS G8MMN
WA6 9LB 2E0OPC
WA6 9LB 2E0XMP
WA6 9LB M6OPC
WA6 9LB M6YTC
WA6 9LL G3INP
WA7 1DH G8YAZ
WA7 1NY G6AGT
WA7 1QZ M0TFS
WA7 1UH G1LCC
WA7 1XE G4YZP
WA7 2AP G7EOK
WA7 2DX 2E1BZH
WA7 2FP M3DAE
WA7 2FR G0OJG
WA7 2GR G7ICD
WA7 2JF M6MZI
WA7 2JR G0MQH
WA7 2LG G0NWE
WA7 2LR M3GTB
WA7 2NS G6FPX
WA7 2QS G1PIX
WA7 2QS M0PIX
WA7 2QS M3GZD
WA7 2UE 2E1DTR
WA7 2UH G0BSD
WA7 2UH G4DIT
WA7 3AQ G0VQL
WA7 3AQ G8HLQ
WA7 3BH M3JPM
WA7 3DY G7JJJ
WA7 3EG 2E0ZMI
WA7 3EG M6ZMI
WA7 3ER G1SJT
WA7 3JA G1DYN
WA7 3JA M6LFC
WA7 3JH G4LUQ
WA7 3JH G8IEL
WA7 3JH M3IEL
WA7 3JR M3NCD
WA7 4AD G0MKD
WA7 4BG G8SIM
WA7 4EH G4WPG
WA7 4QZ 2E0UHC
WA7 4QZ M3NBH
WA7 4RN 2E1HDZ
WA7 4RN M0TLM
WA7 4RP M3KIN
WA7 4TW G7SXG
WA7 4TX G7VCP
WA7 4XL G4HOD
WA7 5AR M0GIB
WA7 5JA 2E1IIA
WA7 5JA M3GZE
WA7 5JJ M3FIP
WA7 5LR G4OAB
WA7 5RF G0HEF
WA7 5RJ G0NSL
WA7 5RW G1MSY
WA7 5ST M3NMB
WA7 5XE M3MOV
WA7 5XW M3ELP
WA7 5XW M3GZJ
WA7 5YN M3XRY
WA7 5YU G0RLF
WA7 5YU G1SES
WA7 6AN G0YNP
WA7 6BJ M3CVL
WA7 6BN G0RBM
WA7 6DL G6UMH
WA7 6DQ G0MSF
WA7 6DT G1JMP
WA7 6DW G0MIX
WA7 6DW G7IAE
WA7 6EH G0WHQ
WA7 6JF M0KOI
WA7 6LB M6AML
WA7 6LB M6HZL
WA7 6RJ G1ZUZ
WA8 0BX 2E1DTK
WA8 0QX 2E0YDM
WA8 0QX M0YZF
WA8 0QX M6CAP
WA8 3BA G7MEU
WA8 3JE 2E0HCL
WA8 3JE M6WID
WA8 4GA M6LHM
WA8 4NT G1LFM
WA8 4TB G8BJQ
WA8 5AS 2E0AIL
WA8 5BT G6URF
WA8 5BU G7ERH
WA8 5QG G6CYF
WA8 6ET G4YJS
WA8 6HR G1AAK
WA8 6HR G7UXU
WA8 7BN G0ICD
WA8 7LD 2E1EBL
WA8 7LD M3DIG
WA8 7LR M6JCP
WA8 7NB M6JDD
WA8 7NQ G1HGC
WA8 7PL M0CMT
WA8 7UW G0SVZ
WA8 8BG G0FQP
WA8 8QE M3WID
WA8 8SQ M3VUO
WA8 8SS G8RIB
WA8 9DP G0HKZ
WA8 9WX G1SBW
WA9 1BL M3IQP
WA9 1DY M6TKP
WA9 1EN 2E0MRA
WA9 1EN M6GMF
WA9 1EN M6OKC
WA9 1QB 2E0ZAJ
WA9 1QB M3OCJ
WA9 1QY M3MLK
WA9 1SQ M3KLY
WA9 2AQ M3ZWQ
WA9 2AR G7KZY
WA9 2BD 2E0VWX
WA9 2BD M3VNQ
WA9 2BU M3LWP
WA9 2DP M3SFJ
WA9 2JD M6XIT
WA9 2PL M6DRO
WA9 2QG 2E0JCY
WA9 2QG M3JCY
WA9 3LW M3OCA
WA9 3NF G7OXP
WA9 3NF G8HLH
WA9 3RE 2E0KZJ
WA9 3RE M3KZJ
WA9 3SD M3ZYW
WA9 3SH M3MZT
WA9 3SX M3LXF
WA9 3WT M6DWG
WA9 3XQ 2E0NCB
WA9 3XQ M0GGK
WA9 3XQ M3NCB
WA9 4AN M3EJX
WA9 4AP M0JGM
WA9 4AY M6AXF
WA9 4BD M6UNA
WA9 4BJ M6HVR
WA9 4BS G1OMY
WA9 4DL M3RCS
WA9 4DN M3KZP
WA9 4DQ G7RMD
WA9 4HD M3RJF
WA9 4HW G7IFO
WA9 4NA G3REV
WA9 4PW G8OUI
WA9 4RY G1ZNT
WA9 4XH G4ROU
WA9 5AR G0STH
WA9 5AR G4WGB
WA9 5EQ G1EVV
WA9 5HL 2E0EDP
WA9 5HL M6FSZ
WA9 5LH M0WHL
WA9 5LZ G2ABV
WA9 5LZ G4KZQ
WA9 5NA 2E0PKU
WA9 5TD M0CHR

## WC
## (West Central London)

WC1N 1AS M6XYY
WC1N 3XX G7HMK
WC1X 0BG G1UCR

Postcode

**IMPORTANT NOTE**

**Revalidate licence to avoid revocation** – Ofcom has advised the Society that plans will be drawn up to revoke licences that have not been revalidated as required by the licence conditions. The quickest way to revalidate is to do so online via the Ofcom website: *https://services.ofcom.org.uk/* or by email: *amateur.validations@ofcom.org.uk* If you need assistance in the process, Ofcom staff are available to help, but please be patient during times of heavy workload.

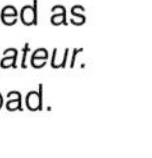

## WD (Watford)

WD17 2LN G6RXY
WD17 2RQ G6JGF
WD17 3AQ G3LGR
WD17 3BB G4BYS
WD17 3DA G4IVT
WD17 3DD G4TJA
WD17 3DP G8LDY
WD17 3DU 2E0NKR
WD17 3DU M6MYR
WD17 3DU M6TQW
WD17 3DX M0ZRG
WD17 3DX M6ZRG
WD17 4FJ M6SBB
WD17 4PJ G0PAU
WD17 4QT G3YKB
WD17 4SU G3SHY
WD17 4SW G4KUJ
WD17 4SZ G3WSB
WD17 4TA M3IFJ
WD18 0AY G7VEF
WD18 0FA M3WVX
WD18 0FA M3WVY
WD18 0HA M6NAM
WD18 0HQ G8IIZ
WD18 6LB G7LXP
WD18 6LB M0SAT
WD18 6NX G0OYC
WD18 7JD G0HTM
WD18 7JD G0NUR
WD18 7PT G6LZB
WD19 4AS M5DIK
WD19 4DA 2E0DTQ
WD19 4DA 2E0ZLA
WD19 4DA M6TPI
WD19 4DH M6ZLR
WD19 4DH G3WDX
WD19 4LW G8BXH
WD19 4PA G3MED
WD19 5AA G3NDK
WD19 5DF M6ICB
WD19 5EH G0DDV
WD19 5ET G6IUD
WD19 6DH 2E0JSK
WD19 6JH G0OJT
WD19 6LF M1BCR
WD19 6LN M0ESB
WD19 6TR G8SUG
WD19 6YL M6ROM
WD19 7AY M6SQU
WD2 5HY 2E1HCB
WD23 1EX M0SSH
WD23 1FB G4SWY
WD23 1PE G3VNB
WD23 1PZ G3OMR
WD23 2BU M0CIO
WD23 2HA G4IID
WD23 2HF M6PQF
WD23 3AR G0LZV
WD23 3BP G1XPB
WD23 3BQ 2E0MUT
WD23 3BQ M0HSZ
WD23 3BQ M6ZBA
WD23 3SS G4GYS
WD23 4GH G4RNW
WD23 4GL M6MLU
WD23 4NP G8XHN
WD23 4QT G8XXV
WD24 4RL M0BLD
WD24 5HB M0HZH
WD24 5LT G4WMN
WD24 6DA M0CJG
WD24 6RP G1IGA
WD24 6RU 2E0WTA
WD24 6SY M6GHJ
WD24 7EE 2E0ERP
WD24 7EE M0TNC
WD24 7EE M6SQL
WD24 7LN M0HYL
WD25 0DB M6ILY
WD25 0EH G0MDM
WD25 0HG G6FAF
WD25 0HX M6IZA
WD25 7AN M0CMC
WD25 7EW G6OBU
WD25 9AR 2E0PBR
WD25 9AR M0PAJ
WD25 9AR M6PAO
WD25 9DZ G4CNH
WD25 9PS G6MHF
WD25 9PS M3DSE
WD25 9PS M3LHM
WD25 9PX M3UXK
WD25 9QH G4RMC
WD25 9QQ M0TTB
WD25 9QQ M6BBC
WD25 9RB 2E0GMD
WD25 9RB M6CHS
WD3 1HL G0WBL
WD3 1NA M1CIG
WD3 1PQ M6WAQ
WD3 3AU G1XEP
WD3 3AU G6AJG
WD3 3DW 2E0LWR
WD3 3EE G6ZFV
WD3 3EG G7HHT
WD3 3FE G3TVH
WD3 3JQ G3EBV
WD3 3JT M3JLE
WD3 3NE G1HKU
WD3 3NH G1MPC
WD3 3PH G8MCJ
WD3 4EB G4JHU
WD3 4EB M6YEQ
WD3 4EE G8MM
WD3 5BD G3VOS
WD3 5BJ G4COV
WD3 5HQ 2E0BST
WD3 5HQ M0SKC
WD3 5HQ M3OAJ
WD3 5JD G4ORE
WD3 5JJ G0VJI
WD3 5JJ G1DNY
WD3 5JT M0LAL
WD3 5JT M6CFT
WD3 5NH G8NIK
WD3 5PX G6YJR
WD3 5QH G3WJG
WD3 5QQ G4TAO
WD3 5RG G4AEM
WD3 7AR M0GFF
WD3 7DY G1ISX
WD3 7EN G3ZER
WD3 7ES 2E0ERE
WD3 7ES M0ITI
WD3 7ES M6HNZ
WD3 7NR G8VTV
WD3 8FH G6NLX
WD3 8GW G6IBD
WD3 8GW M0DAB
WD3 8HT G8OQT
WD3 8LN M6UNI
WD3 8QA G7HGQ
WD3 8QG G1XET
WD3 9TZ G6LFA
WD3 9YZ M6TDJ
WD4 8DY G0SGV
WD4 8JE G7GTG
WD4 8JE G7GTH
WD4 8NJ G7TMH
WD4 8PP G6WPR
WD4 9HF G3XYJ
WD5 0BJ G4FRZ
WD5 0DA G4KUF
WD5 0DH G3JCR
WD5 0EG G3XLI
WD5 0EU G3OYT
WD5 0PJ G0TVD
WD5 0QG G4DPP
WD5 0ST G7PDH
WD6 1BT M0OSE
WD6 1EP G7MKJ
WD6 1HH M3SQX
WD6 1NT 2E0SHY
WD6 1NT M6ACA
WD6 2DA G4TEP
WD6 2HG G6APH
WD6 2HG M6MID
WD6 2LQ G7AIH
WD6 2PB G7DFW
WD6 2RB G8XJN
WD6 2RP G6DAU
WD6 3DH M6DZA
WD6 3JH G6JMO
WD6 3LH G4GPL
WD6 3NJ G4TIH
WD6 3PT G1INI
WD6 3PU 2E0GSK
WD6 3PU M0GSK
WD6 3PU M6GSK
WD6 4HY 2E0WEF
WD6 4JD G3JPJ
WD6 4JD M0TDG
WD6 4NB 2E0EDE
WD6 4NB M0HIE
WD6 4NB M6EDY
WD6 4QT 2E0SSK
WD6 4QT M3UON
WD6 5AD M3OYQ
WD6 5BJ M1BCB
WD6 5DG G0PQB
WD6 5LR G1XVL
WD6 5PE G6RHN
WD6 5QE G8RCO
WD7 7DU G4WIS
WD7 7NF 2E0CQO
WD7 7NF M0MBR
WD7 7NF M6BTG
WD7 8BA M3XQQ
WD7 8BA M3YUK
WD7 8HJ G6JMB
WD7 8JD G6OKA
WD7 9JW G7OIR
WD7 9JW M0ADY

## WF (Wakefield)

WF1 2AL G7VDT
WF1 2DU M3WSQ
WF1 2LF G4VQR
WF1 2LS G4OVL
WF1 2LS M6MEB
WF1 2RA G3OMJ
WF1 3AB M1MCL
WF1 3DN G4LXV
WF1 3DY G0CYU
WF1 3DY M6FYH
WF1 3NW G4VRJ
WF1 3QD M6ZTA
WF1 3RL M6MGW
WF1 3SZ 2E0LAI
WF1 3TL G8LGE
WF1 3TX M6BRH
WF1 4EZ M3OUH
WF1 4JL M6MXO
WF1 4PW M3EVB
WF1 5AH M3XNX
WF1 5HE M3CJI
WF1 5LB M3OUG
WF1 5LG M3XWR
WF1 5NT G6NYF
WF1 5TD G7BSL
WF10 1LN M6ASC
WF10 1LN M6MRS
WF10 1PZ M6MEV
WF10 2AA G4FEQ
WF10 2AP M6GOL
WF10 2BS M3TVV
WF10 2DN G0GYA
WF10 2QA M6BLM
WF10 2QF M6PZZ
WF10 2RB G4TCG
WF10 2RN M6KTI
WF10 2SB G8OXD
WF10 3EY 2E0UAM
WF10 3EY M3UAM
WF10 3HN G3HNC
WF10 3HT G4RQI
WF10 3HY G7MTG
WF10 3QS M6RIN
WF10 3QZ M3VRA
WF10 4AG 2E0BEP
WF10 4AG M0EDE
WF10 4JT M3YNC
WF10 4LF G3OAR
WF10 4LN M6JEQ
WF10 4PH G0PMB
WF10 4QL G7MVE
WF10 4SE G1YNH
WF10 4SE M3YNH
WF10 5AF M6FXY
WF10 5DN 2E0KMA
WF10 5DN M6KMA
WF10 5UJ M6YOB
WF103QN 2E0CCF
WF103QN M3TZQ
WF11 0HG G0VXD
WF11 0JH G0NQE
WF11 0JH G3FYQ
WF11 0NQ 2E0WRF
WF11 8DH G6KIH
WF11 8JF M6RGD
WF11 8JL G0TVL
WF11 8LH G4SOL
WF11 8RH 2E0JBF
WF11 8RH M6BWE
WF11 8SR G0NER
WF11 8TD G0VXM
WF11 9AT G4AAQ
WF11 9HR G4TLM
WF11 9NN G8NDF
WF11 9PA G1PZA
WF12 0BB G4CLI
WF12 0BD M1BAV
WF12 0BP 2E0SLT
WF12 0EQ M6SON
WF12 0HB 2E1FMC
WF12 0HT G8VOQ
WF12 0JZ G7LBM
WF12 0NF M6GCB
WF12 0NL G0CBW
WF12 0PJ M6ZMR
WF12 0QH M6UZH
WF12 0QX G4CLJ
WF12 0RH M6BCB
WF12 7AS 2E0EOP
WF12 7AS M3PAX
WF12 7AW 2E1FFL
WF12 7DE M6ZXC
WF12 7LA G7NPL
WF12 7LG 2E0DQD
WF12 7LG M6FHP
WF12 7LS M1AJG
WF12 7PD G3CPG
WF12 7PU M0EDO
WF12 7SN M6AIA
WF12 7SQ G8POK
WF12 8AJ G6FTJ
WF12 8AQ G6ZJV
WF12 8BY M0WEC
WF12 8NR G1NYN
WF12 8PL M1BHE
WF12 8PY M3LZL
WF12 8PZ 2E0GEG
WF12 8PZ M3VBP
WF12 9AY M3YHJ
WF12 9DN G4GHQ
WF12 9PU M3YGS
WF12 9QF M3HTR
WF13 3LZ G4YAJ
WF13 3SR G6ZUO
WF13 4DQ 2E0LEV
WF13 4EN G6WGE
WF13 4HT G1VIF
WF14 0AJ G6ITV
WF14 0AQ G6XTZ
WF14 0LF 2E0BCS
WF14 0LF M3JUF
WF14 0NH G7VNE
WF14 0NR G3HPD
WF14 8JP 2E0JTH
WF14 8JP M3ZTX
WF14 8JW 2E0RFE
WF14 8LG M5AIB
WF14 8NW M6MBQ
WF14 8PN M6AYC
WF14 8PX G6PHX
WF14 9AN M6NDC
WF14 9AW G2FCP
WF14 9BJ G8NZR
WF14 9ED G3YDL
WF14 9ED M0GWG
WF14 9ED M3YDL
WF14 9HZ 2E0LBB
WF14 9LJ M6BJY
WF14 9NL M6JRJ
WF14 9PB G4PHR
WF14 9PB G7CSV
WF14 9PW G1BGQ
WF14 9PY M6MIA
WF14 9QS G3ZXZ
WF14 9SA 2E0CYL
WF15 6DY G1DEX
WF15 6QE G8DZW
WF15 7BW G1IKF
WF15 7DN G0GMJ
WF15 7DP G4GOX
WF15 7HW M3SGI
WF15 7LP G0PMP
WF15 7NJ G4EIL
WF15 7QH G4OTL
WF15 8DG G8NYM
WF15 8JU G7MHD
WF15 8JW 2E0IPC
WF16 0BE G3JQC
WF16 9HS 2E0CNQ
WF16 9HS M0HRM
WF16 9HS M6CSG
WF16 9JL G0TAL
WF17 0DX G4MLV
WF17 0JL G4FPE
WF17 0LJ G4EOC
WF17 0RG G7LTO
WF17 6DB M6GNH
WF17 6DZ G4XZM
WF17 6EH G4SEQ
WF17 6HG G1NEG
WF17 6HG M3NEG
WF17 7NT G1UVD
WF17 9BX G0IAX
WF17 9DL G6XHK
WF17 9JF M0JBW
WF17 9JF M3JKM
WF17 9JF M3JSM
WF17 9JF M3LMC
WF17 9JF M3SPR
WF17 9QF G6JLI
WF17 9RG 2E0OOC
WF17 9RG M3OOC
WF2 0BH M6EVL
WF2 0BJ M6YSB
WF2 0FQ M3MXA
WF2 0JB M6PPS
WF2 0NZ M3VSZ
WF2 0PA 2E0NER
WF2 0PA M6FMC
WF2 0PD M6IGK
WF2 0PE M3PNV
WF2 0PP G4RCG
WF2 0PP G6LCP
WF2 0QR G6XXE
WF2 0RU G4ROS
WF2 0SE G0CEW
WF2 0SP G3TRV
WF2 0TJ G6LDP
WF2 0UT M6JOU
WF2 0UU M6FMP
WF2 6HJ M0AKD
WF2 6JG M3LDI
WF2 6LQ M6BNC
WF2 6NP M6SSW
WF2 6PL G0PPQ
WF2 6PL G3NJB
WF2 6PL M1CRA
WF2 6RN G4CPC
WF2 6RT M6FPV
WF2 6SE G4BLT
WF2 6SE G8DVS
WF2 6SR G7JTH
WF2 6SR M1DDR
WF2 7DA M6CXI
WF2 7DE 2E0DJB
WF2 7DE G1WRS
WF2 7DE M0LDI
WF2 7DE M3WLV
WF2 7DH M3XYK
WF2 7EF G8IJI
WF2 7EG M6SUP
WF2 7HU G0DIS
WF2 7HW G1OOU
WF2 7JN 2E0WHN
WF2 7JN M3TXF
WF2 7JN M3WHN
WF2 7JT M0CUF
WF2 7LS M6NIV
WF2 7LS M6SNW
WF2 7PR G7SWH
WF2 8BN 2E0NKF
WF2 8BN M3NKF
WF2 8BX M3EOZ
WF2 8BY M3OFV
WF2 8EB G0EVA
WF2 8EB M3YNY
WF2 8EL G0MVR
WF2 8EL M3LYC
WF2 8ET M6DDL
WF2 8EX M3KZB
WF2 8HA M3XRW
WF2 8JZ 2E0APJ
WF2 8LD M3TCT
WF2 8LY M6DRR
WF2 8UP G7UQA
WF2 8YB M3YXD
WF2 9DR M3XRO
WF2 9JS M6FCE
WF2 9JZ G6AJS
WF2 9NZ M6RLK
WF2 9QB M3HTA
WF2 9QG 2E0KTX
WF2 9QG 2E0PIA
WF3 1AJ G6ERZ
WF3 1DL 2E1KYQ
WF3 1HT G0IKD
WF3 1HZ G4DZU
WF3 1JB G0PXF
WF3 1JL M6MHE
WF3 1PF M0AOI
WF3 1QB G4CNX
WF3 1RE G8JHA
WF3 1TA M6HJD
WF3 1UD G6NPW
WF3 1UG M3LOX
WF3 2AX G0VTI
WF3 2AX G1RHW
WF3 2BL M1TRC
WF3 2DD G0CQD
WF3 2JG G1FOM
WF3 3NS M6EZV
WF3 3PX 2E0JAA
WF3 3PX 2E1JAA
WF3 3PX M3BGT
WF3 3PX M3MAA
WF3 3PX M3SCA
WF3 3TG M6GDV
WF3 4JJ G0EVT
WF4 1EF G8OJK
WF4 1EH G1XBE
WF4 1HS G7HYZ
WF4 1HX G8JQH
WF4 1NU 2E0KYI
WF4 1NU M0KYI
WF4 1NU M3JZA
WF4 1PF G4AYL
WF4 1RN G0TYS
WF4 1RN M3UEW
WF4 1RN M3UXI
WF4 1RN M3VEW
WF4 1TJ G1YPQ
WF4 2BG G3LRP
WF4 2BU M6HRO
WF4 2DU M0SZQ
WF4 2DU M6SUJ
WF4 2EJ G4PIR
WF4 2EJ M3WXX
WF4 2ER 2E0WFD
WF4 2NF G0UOS
WF4 3AJ 2E0VKN
WF4 3AJ M0IAA
WF4 3AJ M3VKN
WF4 3AS M6WND
WF4 3BL G0HDP
WF4 3BY M1FBN
WF4 3EF 2E0OPB
WF4 3EF M3OPB
WF4 3EG M1CBU
WF4 3JA G8GOT
WF4 3JZ G0EVR
WF4 3JZ G0FEV
WF4 3NH G7IOF
WF4 3PG G4IZH
WF4 3QD M3TLD
WF4 4AX M6JBB
WF4 4ED G0BQB
WF4 4HN G4JSS
WF4 4QF M6EXG
WF4 4RU G7PNM
WF4 4TE G0COA
WF4 5AN G4IAU
WF4 5HH M6FFV
WF4 5HH M6JFR
WF4 5JF G3PXF
WF4 6AE G1AHS
WF4 6BX M6RIJ
WF4 6EQ G4DWO
WF5 0PP G1WXK
WF5 0RU G0MJZ
WF5 0TJ G1HYA
WF5 8EN G4TDI
WF5 8GN M0LUD
WF5 8LF G3SEY
WF5 8LH G1ISP
WF5 8LJ M3OUF
WF5 8NA M0LAB
WF5 8QP G4MVE
WF5 8RQ M6LPN
WF5 8RY G1VAO
WF5 9EW G7NTI
WF5 9JL G4LJK
WF5 9RE M3SVO
WF5 9RQ G4IOD
WF5 9RQ G4KFP
WF5 9SJ G6AAK
WF6 1AW M3LYX
WF6 1BL M6HKI
WF6 1BN M6MMB
WF6 1DA G7MJP
WF6 1EE G0PHS
WF6 1ER M3HBX
WF6 1NY M3OOA
WF6 1SN G4SBG
WF6 1SS G4SAV
WF6 2AA G6MID
WF6 2AA G6XEX
WF6 2LB 2E0GBF
WF6 2LB G0FMO
WF6 2LB M6AHZ
WF6 2PL M6WBO
WF6 2RZ G7PZL
WF6 2TT G1IIX
WF7 5AB G1PHK
WF7 5AG G1ZPO
WF7 5DP M1NTV
WF7 5EA M6MAD
WF7 5EA M6STV
WF7 5EB 2E1GLS
WF7 5EB G4OOC
WF7 5LH G4NRF
WF7 5LW G7JJY
WF7 6AA 2E0IJK
WF7 6AA M6EOW
WF7 6AH G1ZGH
WF7 6BL G0HPA
WF7 6LQ G1SJZ
WF7 7HE G0KNH
WF7 7HQ G8TKY
WF7 7HU M0AGU
WF7 7JG M6TCY
WF7 7LA G0KMN
WF7 7LA G1LFI
WF7 7LQ G0RAE
WF7 7NA G0PAZ
WF8 1LA G6DUH
WF8 1NH G4IBN
WF8 1QA M6EHW
WF8 1SB G4ISU
WF8 1TD M3UDD
WF8 1TD M6TDV
WF8 2AB M6PTF
WF8 2AJ G0RLN
WF8 2AJ G1FYQ
WF8 2BY M3FOK
WF8 2BY M3OLD
WF8 2BY M3SUI
WF8 2DJ G7HQC
WF8 2EJ 2E0AOZ
WF8 2HW M6KEP
WF8 2JW M3WUW
WF8 2LX M3KBL
WF8 2PN M3UDK
WF8 2QN M6EYK
WF8 2QQ G4KMW
WF8 2RT 2E0KDA
WF8 2UP G0FVN
WF8 3ES 2E0LUL
WF8 3ES G5VZ
WF8 3ES M0JRQ
WF8 3ES M3JRQ
WF8 3NJ G6WBT
WF8 3NT G0TVU
WF8 3QD G4ICB
WF8 3QD G8TEB
WF8 4BU M3WYA
WF8 4BX G6VZS
WF8 4LG G0SNS
WF8 4ND G4YNU
WF8 4RP G0TPN
WF8 4SJ G3VTD
WF9 1AF M0WYC
WF9 1AZ 2E0BSI
WF9 1AZ G0BPK
WF9 1AZ M6CHF
WF9 1ED G8YCQ
WF9 1EU M6XLP
WF9 1LG M3AOQ
WF9 2AR 2E0VAT
WF9 2AR M0ZIM
WF9 2AR M6DEV
WF9 2BT G8HRA
WF9 2BT M0CIR
WF9 2BZ 2E0DYQ
WF9 2BZ M6SYK
WF9 2DD G0CYX
WF9 2RD M6HPH
WF9 2TL 2E1FSX
WF9 2TL G1DKV
WF9 2TL G6OXZ
WF9 3EP 2E0MJP
WF9 3JF M6CXM
WF9 3JT 2E1CIY
WF9 3NJ G4OSY
WF9 3QE G1WGO
WF9 3QE G4ZVB
WF9 3RJ M6FBQ
WF9 3RY G4NDP
WF9 3SJ M3XYO
WF9 4AP 2E1XDJ
WF9 4AY G0GEH
WF9 4EA M0RTV
WF9 4EQ 2E0DWJ
WF9 4EQ M6DWJ
WF9 4ES 2E0MAH
WF9 4JA G1PPX
WF9 4NW M3EXJ
WF9 4RL G1CDN
WF9 4TA 2E0ICT
WF9 4TA M0ICT
WF9 4TQ M3KAX
WF9 5AJ G4PLL
WF9 5BX G8GXS
WF9 5DP M3IUK
WF9 5DS G1CDO
WF9 5DT M6WRX
WF9 5HE G1FFH
WF9 5HE M0JQK
WF9 5HE M3MLF
WF9 5HY G0JLL
WF9 5JL M0JUK

## WN (Wigan)

WN1 2AT 2E0WIG
WN1 2BA G1SUH
WN1 2DL G1INU
WN1 2HX 2E0DMP
WN1 2QJ G1VYA
WN1 2RF G1JMS
WN1 2RR G3YNU
WN1 2SS G4WGF
WN1 3DQ M6CYS
WN1 3PP 2E0CHY
WN1 3PP M6AMV
WN1 3UE 2E0JKR
WN1 3UQ G7ADF
WN1 3XT 2E0YYZ
WN1 3XT M3VIW
WN2 1HW G8OMC
WN2 1JE 2E1LEN
WN2 1LT G0IYX
WN2 1LT G4KKN
WN2 1NA G1SWH
WN2 1NA M0AWX
WN2 1RL M1AWS
WN2 1SZ M0UNN
WN2 2BZ 2E0LTF
WN2 2HJ M3UCU
WN2 2HJ M6JSM
WN2 2NA G0FYE
WN2 2PS M6BZF
WN2 3BN G6ZIY
WN2 3DP 2E0VXI
WN2 3DP M6VXI
WN2 3EE 2E1WNA
WN2 3EQ M6HTC
WN2 3HP G0AYF
WN2 3HR G1YMP
WN2 3SB G1TAR
WN2 3XD G4HPH
WN2 4AF M3KUO
WN2 4EH 2E1BYY
WN2 4EH G6DHI
WN2 4EJ G0AVH
WN2 4EJ G4YYB
WN2 4ET G1BDU
WN2 4ET G4WIL
WN2 4HJ 2E0DPY
WN2 4HJ M6PYE
WN2 4HL 2E0MLS
WN2 4HL G1ECI
WN2 4HL M0ICK
WN2 4HL M3ITZ
WN2 4JA G4ZVK
WN2 4LD G4EHY
WN2 4PL G1EIO
WN2 4PL G8ALD
WN2 4PT 2E0FJZ
WN2 4QS G0CBD
WN2 4RG G7JZI
WN2 4SZ G4PGQ
WN2 4TY G7LVS
WN2 5JN G7GAG
WN2 5QD M3PZL
WN2 5QD M3PZN
WN2 5QR M3KZR
WN2 5TF G7FMB
WN2 5XA 2E1EHB
WN2 5XA M0GRE
WN2 5XA M1AIX
WN2 5XG G6ZJM
WN2 5XR 2E0CRX
WN2 5XR M3LBR
WN2 5XR M3XFN
WN2 5YL M6OBO
WN2 5YL M6XXB
WN3 4NS G1CZN
WN3 4PQ M6SDU
WN3 4QA 2E0MNP
WN3 4QA M0XOC
WN3 4QA M6INT
WN3 5PL 2E0BTX
WN3 5PL M6GBR
WN3 5PR G8DPE
WN3 5PS G4OVV
WN3 5QJ G4DEN
WN3 5QN G0DPI
WN3 5QN G0HRW
WN3 5QN G7EDK
WN3 5QN G7IXC
WN3 5UT M3XDQ
WN3 6AA G3BPK
WN3 6AA G4GWG
WN3 6AJ G0KHH
WN3 6EN G4GSM
WN3 6JQ G6CHD
WN3 6JY G6LST
WN3 6LD G4WDC
WN3 6LD G6TYB
WN3 6TQ G4HSC
WN4 0AZ G4WXX
WN4 0DW G7SXB
WN4 0EF M3WJN
WN4 0EQ 2E0ZRX
WN4 0EQ M3ZRX
WN4 0JT M1XCG
WN4 0NJ G0SJS
WN4 0QF G1LWX
WN4 0QX G8XLI
WN4 0SJ G8CXW
WN4 8AY G7IHE
WN4 8QT G0SIW
WN4 8RX M1BZJ
WN4 8ST G0UNG
WN4 8SU G7LTR
WN4 8SU M3BRT
WN4 8SU M3LTR
WN4 8TG 2E0JCT
WN4 8TG M6JGI
WN4 9DY G6OCB
WN4 9JN G4XCQ
WN4 9LD G3HUX
WN4 9PJ G6NTM
WN4 9SE M0HTR
WN4 9SE M0RGN
WN4 9SE M6CLB
WN4 9UP 2E0WES
WN4 9UZ 2E0RAG
WN4 9XD 2E0BQO
WN4 9XD M3IQF
WN5 0AJ G1DON
WN5 0EB 2E0WEZ
WN5 0EB M3RMQ
WN5 0PU G0DSR
WN5 7AF 2E0ARV
WN5 7AH 2E0NRB
WN5 7AH M6NRB
WN5 7AX G0NEO
WN5 7BG G4HAP
WN5 7EB G3PJW
WN5 7HF G0FZN
WN5 7HS 2E0ISQ
WN5 7HS M0ISQ
WN5 7HS M3ISQ
WN5 7HT G4EII
WN5 7JA M1DYU
WN5 7LH 2E0SYY
WN5 7LH M0SYY
WN5 7LH M6OFF
WN5 7PX G4LON
WN5 7QL G0OQS
WN5 7TF G4VFR
WN5 7UA G6RZJ
WN5 8LX G7SOV
WN5 8NG 2E0ROI
WN5 8NG 2E1LED
WN5 8NG 2E1WRC
WN5 8NG M3ROI
WN5 8RH G7CWE
WN5 9BH G6ORO
WN5 9DL 2E1DFE
WN5 9EF M3JZF
WN5 9HZ G0TRY
WN5 9JQ M6ATX
WN5 9NQ 2E1HZY
WN5 9NQ 2E1TCP
WN5 9PA G4KTU
WN5 9PJ M6AIF
WN5 9PR M6MTL
WN5 9SB M6HTB
WN5 9TG G0UVL
WN5OEN 2E0UOG
WN5OEN M3UOG
WN6 0AQ 2E0ICK
WN6 0AQ 2E1HDE
WN6 0AQ M0HDE
WN6 0AQ M3POP
WN6 0AZ 2E0VCE
WN6 0BE M3NJK
WN6 0LW 2E0BJT
WN6 0LW G4PPG
WN6 0LW M3LBT
WN6 0NP 2E0TAJ
WN6 0NR G8EHF
WN6 0QP M3TYQ
WN6 0QR G1MSR
WN6 0RY 2E0BMO
WN6 0RY M3RXH
WN6 0SA G0EOK
WN6 0UA G6XKY
WN6 7NH 2E0AWR
WN6 7PA M3RRJ
WN6 7PU M0SCU
WN6 7RF G4EOT
WN6 8AU G0PBE
WN6 8DT G0GBP
WN6 8EH 2E0WPE
WN6 8EH M3HXO
WN6 8EY G7RGV
WN6 8HR G4KCP
WN6 8LH G0PPK
WN6 8NJ G1EPF
WN6 8NJ G4PPB
WN6 8NS 2E0SAF
WN6 8PJ M6PFZ
WN6 8PL 2E0SYE
WN6 8QA G4SFJ
WN6 8QA G6PUV
WN6 9AG 2E0WWZ
WN6 9AG M3WWZ
WN6 9DJ G3RSW
WN6 9DS 2E0BUE
WN6 9DS M3URT
WN6 9HU G8XXC
WN6 9JF G4EHK
WN6 9JL M0HPT
WN6 9JN M3SJL
WN6 9LF G8JGU
WN7 1HN 2E0DMU
WN7 1HN G0SQX
WN7 1HN M0HZK
WN7 1HN M6FHY
WN7 1HU M6BSY
WN7 1HW M6TOE
WN7 1JZ G0MNY
WN7 1NB G3VGK
WN7 1NB G6HNQ
WN7 1SG G0KMB
WN7 1TF M0JJD
WN7 1TP 2E0TUR
WN7 1TP M3UAE
WN7 1TS G1HFS
WN7 2HH M0KDM
WN7 2HH M6MRD
WN7 2JJ G0SSK
WN7 2NG G6TET
WN7 2NH 2E0GLW
WN7 2NH M3IIW
WN7 2PL M6CYV
WN7 2RG G4UDF
WN7 2UU G4JXI
WN7 2YR M1TET
WN7 3BU G3WIS
WN7 3DS M3HCA
WN7 3EB G4COE
WN7 3ET M6TFV
WN7 3LD M6GOS
WN7 3NE G8PUN
WN7 3NU G4GFD
WN7 3PQ 2E1CJF
WN7 3PQ G6VPU
WN7 3PQ M0HOY
WN7 4EQ M6LEY
WN7 4HY 2E0DRZ
WN7 4HY M6DRZ
WN7 4HY M6DWT
WN7 4HY M6JPS
WN7 4SX G0LGC
WN7 4TA G1FLV
WN7 5BT G1VKT
WN7 5DG G6LDY
WN7 5DG M3FNT
WN7 5EU M6KML
WN7 5HG M1SJH
WN7 5PN G4VXW
WN7 5PN G6HPL
WN7 5PN M0NAR
WN7 5PW G1DUS
WN7 5PW G4NTT
WN7 5QX 2E0HVN
WN7 5QX M6HVN
WN7 5SW M6JGM
WN8 0AA G0MRM
WN8 0AA G6TKY
WN8 0AE G4ACI
WN8 0DE G4FEU
WN8 0JG G0UMY
WN8 0QT G3VWA
WN8 6AA M6HTU
WN8 6AT G4WWG
WN8 6AX G8UXL
WN8 6ED G6IKU
WN8 6PA G6OMN
WN8 6PE G4PJS
WN8 6RD M3NSG
WN8 6RJ G4JTP
WN8 6TA G3VSR
WN8 6TA G6HXL
WN8 7AR G3PNQ
WN8 7DA M0SBH
WN8 7LA M6JFY
WN8 7NB 2E0ESS
WN8 7NB M0TAA
WN8 7NB M3XQW
WN8 7NS G0ELM
WN8 7NT G6ORS
WN8 7NT M3CDY
WN8 7RA G1DMW
WN8 8BD M3UHG
WN8 8EG G6YRB
WN8 8EN G6WEW
WN8 8HN M6PLB
WN8 8NP M3DLT
WN8 8NW M3FRD
WN8 8NW M6HHO
WN8 8QS M6GOF
WN8 8QZ M6DDH
WN8 9BP G0LAK
WN8 9BP G4SME
WN8 9BQ G8XEI
WN8 9EF M3MJN
WN8 9JZ G1UOR
WN8 9NB G6UGE
WN8 9QQ G3KTJ

## WR (Worcester)

WR1 1NR G3RMF
WR1 1PA 2E1MFC
WR1 1PA M3CER
WR1 1QE G3VXH
WR1 3NY G8XMO
WR10 1HW G0FFF
WR10 1JH G1IPE
WR10 1JY G7TGK
WR10 1LL M6GQS
WR10 1LW G1AFJ
WR10 1LW G4TXF
WR10 1NQ G7MYM
WR10 1PW G3UEY
WR10 1RE G7CRU
WR10 2AX G4OZQ
WR10 2JE G4YIG
WR10 2JW G6DEA
WR10 2LA G4RMV
WR10 2NY G4NIJ
WR10 2PL G1UQK
WR10 2RJ G7RRJ
WR10 3BB 2E1EOD
WR10 3EF G8LQP
WR10 3EL G8BGT
WR10 3EL M5AGY
WR10 3EZ G0NXE
WR10 3HJ G4TDR
WR10 3HL M0CUU
WR10 3HS M6ZNF
WR10 3JP M0DLP
WR10 3JZ G3ZKH
WR10 3NA G7WEB
WR11 1BU G4YJB
WR11 1DE G6NBL
WR11 1EQ G4EKG
WR11 1XZ G8OWO
WR11 1YW G4NRD
WR11 2AH 2E0RKX
WR11 2AH M0RKX
WR11 2AH M6RKX
WR11 2NB G7JSC
WR11 2NE M0PDY
WR11 2NE M3PDY
WR11 2QA G6XKX
WR11 2QJ G1JLQ
WR11 2QZ G7JWW
WR11 3BJ G3XCW
WR11 3BS M1PLC
WR11 3EA G0IBE
WR11 3FF G4XQG
WR11 3HE G3KLZ
WR11 3HE G8SQZ
WR11 4AH G4AXW
WR11 4NL G6DRC
WR11 4NL G6YXY
WR11 4QY G8NMK
WR11 4RJ M6WYR
WR11 4SL M3ZID
WR11 5UF G4MFK
WR11 7AA 2E0IDR
WR11 7AA M0IDR
WR11 7AA M6IDR
WR11 7EU G1VNV
WR11 7GH G0UPY
WR11 7GQ G1SHU
WR11 7HQ G0GBL
WR11 7PT M3JKI
WR11 7QG G1UNQ
WR11 7RP G4UXC
WR11 7RP G4WET
WR11 7TA G1IME
WR11 7UE 2E0HDU
WR11 7UE M3CKD
WR11 7UZ G3CUF
WR11 7XX 2E0BQL
WR11 7XX G1JMD
WR11 7XX M3MWQ
WR11 8JP G6YMY
WR11 8JU G3YKI
WR11 8QW M0TZT
WR11 8SN G1XWZ
WR11 8TL M3JFP
WR11 8XN G0KIN
WR12 7AX M3DVM
WR12 7EP G8HFL
WR12 7HB G3IXI
WR12 7HS G6HAA
WR12 7NQ G7OAS
WR12 7PF G3KAR
WR12 7QB G7EBL
WR12 7QB M5GJO
WR13 5AA 2E0MGR
WR13 5AA M6MGR
WR13 5DP G6FLR
WR13 5HX 2E0NON
WR13 5HX M0PYG
WR13 5HX M6PYG
WR13 5LA G4FCA
WR13 6AA G1RFX
WR13 6DL G3WIK
WR13 6ER G8EAJ
WR13 6ET G0RWY
WR13 6NN G8AXV
WR13 6QW M1CYM
WR13 6RA G6EOA
WR13 6SE G4VHV
WR14 1AD G4BVY
WR14 1AP 2E0GWS
WR14 1AP M6GWS
WR14 1BD M0OPG
WR14 1BU G1CBL
WR14 1DH G3YPU
WR14 1DT G3RNP
WR14 1FU G3OOW
WR14 1HA G4ZAI
WR14 1HX G6CMV
WR14 1HX G7EME
WR14 1JX G1VRP
WR14 1JX M0GEF
WR14 1LP G4FNZ
WR14 1NB G0GFI
WR14 1NX G7TTH
WR14 1PD G8ZKG
WR14 1PH G7WIG
WR14 1PJ G6NAP
WR14 1PU G4IKJ
WR14 1PU G7JLC
WR14 1RQ G8IDK
WR14 1SB M3MXF
WR14 1SG M3AIZ
WR14 1TU M6NVR
WR14 1TY G4BZM
WR14 1UY M6ATK
WR14 2BQ G4FFY
WR14 2BQ G8LLS
WR14 2DQ G2CKR
WR14 2DS M0XVX
WR14 2HU G3UIK
WR14 2LE G4EYJ
WR14 2ND M1CAK
WR14 2NF M0PAR
WR14 2NG G0FXD
WR14 2NJ G4BFC
WR14 2NN G8YTU
WR14 2SD G3CVK
WR14 2SR 2E0CQM
WR14 2SR M0XCH

WR14 2SR M3MVN
WR14 2SW 2E0DJI
WR14 2SW M0JJM
WR14 2SW M6DJL
WR14 2TE G7HAS
WR14 2TU G8VUS
WR14 2UL G3WIJ
WR14 2UY 2E0JVP
WR14 2UY M6CDR
WR14 2WB M3RJI
WR14 2YD 2E1BZI
WR14 3BH G4RNX
WR14 3EA G4GHL
WR14 3JZ G3NQZ
WR14 3NJ M3RVJ
WR14 3NP G0WLG
WR14 3PU M0GYL
WR14 3QP G3WGY
WR14 3QP G8CMD
WR14 3QW G4IPY
WR14 4AH G1MFK
WR14 4AX G7PRW
WR14 4BX G0ROB
WR14 4DW G3OAH
WR14 4EF G1GDS
WR14 4HT G3FHL
WR14 4JR G6JBY
WR14 4LJ G6TEB
WR14 4LU G4VWX
WR14 4LX G1IDZ
WR14 4NL G1FXS
WR14 4NL G3YPM
WR14 4PL G6BUY
WR14 4XE M3OAC
WR142ST 2E0BSW
WR15 8DD M6EIG
WR15 8JX M6LNC
WR15 8LX G7RVW
WR15 8QN M3NPC
WR15 8QY G0DAZ
WR15 8SP G3BPF
WR15 8SR G0CCQ
WR15 8TW G1ENR
WR2 4DJ G4RPC
WR2 4DJ G8SSE
WR2 4DP 2E1DLD
WR2 4DP G4ROR
WR2 4DP M3AQG
WR2 4DP M3KVH
WR2 4ES G8LJU
WR2 4HE G6FRB
WR2 4JQ G6CBP
WR2 4RB G7LUF
WR2 4SE G3TGD
WR2 4SF G4NXP
WR2 4SR G8URZ
WR2 4TE G3OCW
WR2 5EQ 2E1HLS
WR2 5ND G6EQF
WR2 5PX G1EIP
WR2 5QR M0BXD
WR2 5QY G0BAM
WR2 5RH M3JJT
WR2 5RR G1IVL
WR2 5RW G1AQF
WR2 5RW M0DEK
WR2 5SU G3TQZ
WR2 5TE G0RWS
WR2 6AA 2E0JEM
WR2 6AA M3JEM
WR2 6DA G1ICQ
WR2 6PQ G1JMN
WR2 6PQ M0JMN
WR2 6RW G0AQB
WR3 7AN G4DEV
WR3 7BH M0WFO
WR3 7DQ M6CEN
WR3 7EH G4VZH
WR3 7HY G0PNR
WR3 7LR G4BYB
WR3 7ND G8PHV
WR3 7PJ 2E0CLE
WR3 7PJ 2E0SAN
WR3 7PJ M6EXI
WR3 7PP G4NUZ
WR3 7RX M6AZK
WR3 7SR G3RWQ
WR3 7TT M3NEP
WR3 7TZ 2E0EAN
WR3 7TE M0PNA
WR3 7TZ M6ABN
WR3 7UP G4YZK
WR3 8BQ G0AOC
WR3 8EZ G1VKB
WR3 8NU M6DUF
WR3 8PJ G6TYT
WR3 8QT M6CEO
WR3 8RL M3YSY
WR3 8RL M3ZGT
WR3 8XA M6RPU
WR4 0HY G0WJV
WR4 0HY M0CZF
WR4 0HY M6BSG
WR4 0JW 2E0RGO
WR4 0JW M0VNG
WR4 0PE G6DMG
WR4 9AH M6LGB
WR4 9HS G3SPP
WR4 9JD M1ESI
WR4 9LB G0HUV
WR4 9NT M3RFW
WR4 9RD M3UZB
WR4 9RF M1FJH
WR4 9TU M3SBE

WR4 9UG G3VQW
WR4 9XL M6TIC
WR4 9YU G0DEP
WR5 1AG 2E0URU
WR5 1AG M6WRW
WR5 1HH G0WXJ
WR5 1HH M0RAD
WR5 1HH M0ZOO
WR5 1HH M3WRZ
WR5 1JJ G4IDF
WR5 1JJ G4MHC
WR5 1LL G1ZFS
WR5 1QB G4ADJ
WR5 1QF M6EXF
WR5 1QX G1FQD
WR5 1SD G4KPL
WR5 2AA 2E0DGV
WR5 2AL G6PMO
WR5 2BD G1JGF
WR5 2BD G8STI
WR5 2BG G0WJN
WR5 2DU 2E0RNA
WR5 2DU M0XPB
WR5 2DU M6RNA
WR5 2DU M6ZPB
WR5 2ET G4URN
WR5 2HL G4HNZ
WR5 2HL G8PIN
WR5 2JL 2E0CAL
WR5 2JL 2E0DAJ
WR5 2JL M3VAQ
WR5 2JT G1CSS
WR5 2JT G7SLV
WR5 2JU G4ELL
WR5 2LN G4GIM
WR5 2NG G0SMZ
WR5 3AL G6UHL
WR5 3AY 2E0RHE
WR5 3AY M0TBW
WR5 3AY M6CNY
WR5 3BW G7FAS
WR5 3HD G1WPH
WR5 3HD G6NFJ
WR5 3HT G0AAT
WR5 3JG G4LVV
WR5 3NX G8ASJ
WR5 3QB G1XBL
WR5 3SH G7CRR
WR5 3SJ G6UXF
WR5 3SZ G1EME
WR5 3SZ G6FZV
WR5 3UA M0TXD
WR6 5LZ G4KTW
WR6 5NE G3GNA
WR6 5NG 2E0BTS
WR6 5NG M6GJT
WR6 5RD G3RYH
WR6 5SR G6CXV
WR6 5SR M0ELA
WR6 6AY G1MAC
WR6 6DW G6SOA
WR6 6EB G6JJP
WR6 6HQ G3WLG
WR6 6ND G4OPD
WR6 6QA G4AAL
WR6 6QX G4OWK
WR6 6TQ M3IQJ
WR7 4AP G6VVE
WR7 4BT G0CRX
WR7 4PR G6ZHF
WR7 4QJ G7HNL
WR7 4RF G1XWK
WR7 4RH G8LPN
WR8 0AT M6KVD
WR8 0DS G4TQY
WR8 0ET G6GWU
WR8 0LQ G6UZT
WR8 0LR M3ZLI
WR8 0PD G4FAT
WR8 0QQ G1PQO
WR8 0SJ G8XQL
WR8 9AN G0EMS
WR8 9BE G1AKB
WR8 9EH G6XLG
WR8 9JR G3URZ
WR8 9JR G4JKA
WR8 9LP G8LCM
WR9 0AU G1DYC
WR9 0DF G6FOF
WR9 0DN G0UTU
WR9 0DX G4HPD
WR9 0EZ 2E0TTH
WR9 0EZ M6PMT
WR9 0RP G1YBI
WR9 0RP M0XYL
WR9 7AF G3RLF
WR9 7AY G3KTH
WR9 7AZ G3TRB
WR9 7DQ G4RQO
WR9 7FR G0WFK
WR9 7HT G0XUU
WR9 7NI G6WPO
WR9 7QE G0BLS
WR9 7QS G1DYL
WR9 7QU G4THU
WR9 7RX G0WIS
WR9 7RY G7BBY
WR9 7SE G4HDO
WR9 7SP G7CTT
WR9 8HG G3VPH
WR9 8NR 2E0NRW
WR9 8NR M6NRW
WR9 8QR G4TID
WR9 8RA G1MLK

WR9 8SS G7VDX
WR9 8TQ G3VGG
WR9 8TQ G7VJM
WR9 8TQ M0BQE
WR9 8TQ M3IMM
WR9 8UD G0BDM
WR9 8WA G4KXP
WR9 9BZ G8HHR
WR9 9ED 2E1IAY
WR9 9ED 2E1IAZ
WR9 9EG G0AXB
WR9 9HU G7RVT
WR9 9LA 2E1CBH
WR9 9LA G4LVO
WR9 9LG G4BYM

## WS
## (Walsall)

WS1 2DA G0FSL
WS1 2DA G3PJY
WS1 2DA G4GUW
WS1 2DH M6GWZ
WS1 2PJ G4EJU
WS1 3AL G4KBA
WS1 3HB G0HUD
WS1 3HL M3BJW
WS1 3LB G1MMN
WS1 4DZ G0ENM
WS1 4DZ G4YVF
WS1 4HF G4FCB
WS10 0BH M3NXK
WS10 0DN G7EKT
WS10 0EU 2E0WDB
WS10 0EU M3IPT
WS10 0EW M6ISA
WS10 0HL M3STJ
WS10 0HQ M3NZN
WS10 0JF G0EAN
WS10 0LW G0JAA
WS10 0PN G0WJJ
WS10 0TB G0CMR
WS10 7HY G7CNZ
WS10 7RH M3DAM
WS10 7RH M3RYA
WS10 7RQ G1RBA
WS10 7RR G6RTG
WS10 8LE M0WRS
WS10 8NS 2E0VGB
WS10 8NS M0JGB
WS10 8NS M6FQW
WS10 8NU M6DLI
WS10 8RJ M3INC
WS10 8RN M3WCI
WS10 8RQ M6UXB
WS10 8SH M1PTE
WS10 8UB G6ZYX
WS10 8XY G1WRU
WS10 9BQ 2E0SOZ
WS10 9BT G1PQY
WS10 9HH M0CRU
WS10 9LH G6IDO
WS10 9NF M0JPB
WS10 9PH G4TDP
WS10 9RQ G6CSR
WS10 9UA G6JAM
WS108HJ 2E0MBK
WS108HJ M6MEL
WS11 0AF G7TCW
WS11 0AQ M3WDC
WS11 0AR M6UKC
WS11 1AR G1BRD
WS11 1AR G7PCV
WS11 1AZ G1UUL
WS11 1BB G3PSU
WS11 1LJ G3URL
WS11 1NQ G4ICE
WS11 1NS M3RHG
WS11 1PE 2E0OWH
WS11 1PE G7ASY
WS11 1PW M3SLQ
WS11 1RU G8GRC
WS11 2SE G0UYH
WS11 3SP M3NCO
WS11 4NN 2E1FQO
WS11 4PT M3NOM
WS11 6EH M3MXX
WS11 6EN M6DHG
WS11 6LX 2E0ZXQ
WS11 6LX M0ZXQ
WS11 6LX M3ZXQ
WS11 7FR G7DWH
WS11 7YX 2E0IQX
WS11 7YX M0IQX
WS11 8ES 2E0OUK
WS11 0QZ G4YTK
WS11 9TN G7IZN
WS12 0FR M6DVH
WS12 0PR G0VSK
WS12 0QL G1LFH
WS12 0QW 2E0MGX
WS12 0QW M0WFM
WS12 0SX G0MVV
WS12 1BG G1NZQ
WS12 1BG M0SST
WS12 1BS G4KWQ
WS12 1BS M1FJC
WS12 1BS M3STR
WS12 1BW M6CDE
WS12 2AW G4RJD
WS12 2DR G6JNZ
WS12 2DR M3NVA
WS12 2DR M3WTC

WS12 2DW G8NTJ
WS12 2EZ M3KUJ
WS12 2GF M3NPI
WS12 2GH M3FYZ
WS12 2GR 2E0DFP
WS12 2GR M0GYO
WS12 2GR M6DFP
WS12 2RN G0CCF
WS12 2RN G7WBH
WS12 3DS G1SIG
WS12 3HA M1CYR
WS12 3TG 2E0VRC
WS12 3TG M6VRC
WS12 3YA M3IXY
WS12 3YJ G0XAT
WS12 4BD 2E0FWN
WS12 4BL M3JMX
WS12 4DL M3OVF
WS12 4DP 2E0DVU
WS12 4DP M6DVU
WS12 4DQ 2E1EXK
WS12 4DQ G0DUQ
WS12 4DQ M3EXK
WS12 4DS 2E0DCJ
WS12 4DS M6CQE
WS12 4EN 2E0RGR
WS12 4JT G7AMD
WS12 4LF M3YXS
WS12 4LR G4JVT
WS12 4LT G4VYE
WS12 4LT G6AJV
WS12 4QA G8VPR
WS12 4SF 2E0KBA
WS12 4SF G4ZBF
WS12 4SF G6POP
WS12 4SN G3ZVK
WS12 4SS 2E0WTD
WS12 4SS M0NEG
WS12 4SS M6AAP
WS12 4SU G0ICW
WS12 4TA 2E0FAQ
WS12 4TP M3NNY
WS13 6AU G0RRM
WS13 6BF G4GGH
WS13 6BH G6KTC
WS13 6DA 2E0UVP
WS13 6DA M0HPB
WS13 6DA M6BIS
WS13 6DB G6OHQ
WS13 6DQ G0HQX
WS13 6EP M6THO
WS13 6EP M6TYB
WS13 6SD M3YJY
WS13 6ST G3PFT
WS13 7ED G0OYM
WS13 7ET G4EHT
WS13 7LW G8XXA
WS13 7LZ G0DRA
WS13 7NQ G0PHE
WS13 7PW G0EVJ
WS13 7SN G8NXE
WS13 8AA G8BFL
WS13 8EF G8GSL
WS13 8JE G8EQC
WS13 8NJ G1JLX
WS13 8NY 2E0DCX
WS13 8NY M0VAD
WS13 8PW G7IXG
WS13 8SA G4YBP
WS13 8UZ G6LWZ
WS14 0JH G8NTY
WS14 9DA G4NPY
WS14 9DP 2E0JPX
WS14 9DP M6ELC
WS14 9HH G4OUM
WS14 9LN G0DKZ
WS14 9NH G0RIF
WS14 9PF G7BMC
WS14 9QT 2E0KRT
WS14 9QT M3KRP
WS14 9RF G1SMP
WS14 9RF G7JXQ
WS14 9RJ G6NEZ
WS14 9SF G6HZH
WS14 9SN G8HYK
WS14 9SZ G3YPD
WS14 9US G1OBC
WS14 9US G8KHV
WS14 0XF G8ZID
WS15 1AB M0PWY
WS15 1AT M6SLG
WS15 1BB G6GOW
WS15 1BJ M6TNC
WS15 1EJ 2E0JWH
WS15 1EP G4NLK
WS15 1EW M1AYA
WS15 1EZ G8TNS
WS15 1JF G4LAN
WS15 1JJ 2E0HPC
WS15 1JJ M3XCF
WS15 1LP [illegible]
WS15 1LP M3GGF
WS15 1QA G2HKS
WS15 2AR G7ACG
WS15 2AR G7JMB
WS15 2DD G3TJA
WS15 2EE G7ULC
WS15 2EE G8OTC
WS15 2ES M6SJB
WS15 2GY M6JJF
WS15 2JL G1IDQ
WS15 2LL M3DIY
WS15 2LL M3TOY
WS15 2NH M0DJF
WS15 2NP M0EWW

WS15 2NS G7VJY
WS15 2PE M0TTO
WS15 2QB 2E0BXK
WS15 2QB M6HQP
WS15 2QP M6JSH
WS15 2QW G7KBH
WS15 2QY G1WVZ
WS15 2RE M6FHM
WS15 2XH G1KQH
WS15 2YG G0JJO
WS15 2YG G3TCY
WS15 2YG G4PWD
WS15 3EF G4FMO
WS15 3HJ G8MIA
WS15 3LD G7HZU
WS15 3QU M3TKQ
WS15 3QX 2E1FKJ
WS15 3QX G4DBR
WS15 4AF G8YNP
WS15 4DQ M6KEA
WS15 4EF G8JNR
WS15 4ER M0NDZ
WS15 4HF M6XWG
WS15 4QG G8XGK
WS15 4RG G3NSO
WS15 4RG M3XAR
WS15 4TH G6NVS
WS2 0AY 2E0ALA
WS2 0AY M3NRK
WS2 0EG G0CLX
WS2 0EW M0IAW
WS2 0HJ G0JVN
WS2 0HZ G8ZBJ
WS2 0JD 2E0VPT
WS2 0JD M3VPT
WS2 0NH G0MPR
WS2 0NT M3AGH
WS2 7BB G4EFG
WS2 7BG G8SCI
WS2 7EJ G7DHQ
WS2 7EN M3RMX
WS2 8AT M3UFK
WS2 8QL 2E0WCM
WS2 8QL M6MLL
WS2 8RE G7TGG
WS2 8RR G4RVK
WS2 8RX G0HWP
WS2 9PZ 2E0YAX
WS2 9PZ M6YAF
WS2 9QB G6ZJS
WS2 9QX M0AVL
WS3 1AL G7CST
WS3 1AL M0PET
WS3 1DF M3JRR
WS3 1DL 2E0FRO
WS3 1DL M3LEY
WS3 1DL M6BJF
WS3 1EW G8ZHA
WS3 1JX M6CAS
WS3 1JX M6CEZ
WS3 1NQ G1GZM
WS3 1NW M3CKU
WS3 1PJ M6UKO
WS3 1PY M3LLT
WS3 1RF M6GTD
WS3 1TT G4PPP
WS3 2AL M3BJH
WS3 2AQ G6FOX
WS3 2AQ G8DEJ
WS3 2AR M6JJS
WS3 2AT M3CWZ
WS3 2EG M3TLU
WS3 2EN M6WBG
WS3 2EZ G4PPC
WS3 2EZ G7PPC
WS3 2HT G0MKW
WS3 2NG M1AGW
WS3 2QD M1EVH
WS3 2RJ M6LBR
WS3 2SS G0VFB
WS3 2UQ M6HPF
WS3 3AU M0TMF
WS3 3AU M1TMF
WS3 3BZ M1EBI
WS3 3DG G7GCU
WS3 3DX M3RET
WS3 3EN M6DGO
WS3 3LT M6CKI
WS3 3NB G1LUF
WS3 3NB G0UDL
WS3 3PG G1SJB
WS3 3QA G0KFS
WS3 3QD G0CXO
WS3 3QD G1YWI
WS3 3RF G8IJE
WS3 3XB G0CNG
WS3 4AG 2E0DEC
WS3 4AG M6DDC
WS3 4AH M6HGA
WS3 4AH M6HGD
WS3 4AW G1XLL
WS3 4DH G6RYI
WS3 4EZ M0LRD
WS3 4HR G8PWE
WS3 4LS G0DAC
WS3 4LW G3GGR
WS3 4PB G4OKE
WS3 4PU G1WQH
WS3 4PU G1XLW
WS3 4QN G7BNK
WS4 1AP 2E0LEF
WS4 1AP M0SPA
WS4 1AP M0VSP
WS4 1AP M6APG
WS4 1AP M6ENF

WS4 1AP M6LIN
WS4 1BB M6OYA
WS4 1DQ G0IMX
WS4 1DS G0NFQ
WS4 1DS G4GRM
WS4 1EQ G6UEU
WS4 1HP G1KVP
WS4 1HS G6EQD
WS4 1HT G4YZM
WS4 1JB G0DQO
WS4 1JX G7IIZ
WS4 1NG M3ZUL
WS4 1NH 2E0ZYK
WS4 1NH M3ZYK
WS4 1QP M0CSB
WS4 1QU 2E0DUZ
WS4 1QU G0UYP
WS4 1SA 2E0PPA
WS4 1SA M6CKL
WS4 1XA G4UDV
WS4 2AB M3XTR
WS4 2EN G6HZI
WS4 2HB G4GJE
WS5 3AD M0AST
WS5 3AW G0KRY
WS5 3DH G3YHN
WS5 3DH G8MLW
WS5 3DT G1RLR
WS5 3DT G1UBC
WS5 3ES G4GKC
WS5 3ES G4HLL
WS5 3NQ G4SEL
WS5 3PN G8NQY
WS5 3QJ G6JVA
WS5 3QX G3ZHC
WS5 4BX M3UTZ
WS5 4DJ M0ZDO
WS5 4DN 2E1EXI
WS5 4DN G3TIN
WS5 4DN G8HTM
WS5 4HE G4TCM
WS5 4LU M0GRX
WS5 4NH G8XXU
WS6 6DF M0BHQ
WS6 6DT G4CJK
WS6 6HA G8IHC
WS6 6HE G4PFO
WS6 6JA M3WIX
WS6 6LD G4JUK
WS6 6LQ G7TCQ
WS6 6LS G7GDA
WS6 6PE G8MFU
WS6 7BE M3VTL
WS6 7BX G0GPB
WS6 7DP G4TQC
WS6 7EJ G6RIJ
WS6 7EJ G8TRQ
WS6 7EP G7IPI
WS6 7EU G6RIG
WS6 7EZ M3UXX
WS6 7EZ M3VEY
WS6 7HD G1TNR
WS6 7LE 2E0JPT
WS6 7LE M3MQC
WS7 0BB G0HKF
WS7 0DJ 2E0OWC
WS7 0DJ M6OWC
WS7 0ED G4POR
WS7 0EE G6JJB
WS7 0EQ G6PXN
WS7 0EQ G6UCO
WS7 0ES G0PJR
WS7 0ES G6PAP
WS7 0EW G8SYS
WS7 0HA G1GKH
WS7 0LQ 2E0CAP
WS7 0LQ 2E0SDV
WS7 0LQ M3XIF
WS7 0LQ M3XIH
WS7 0LQ M6GDI
WS7 1FA G0GFC
WS7 1FA G6DFC
WS7 1NQ G7DOS
WS7 1PW G0AGU
WS7 2AS G4ZBE
WS7 2AS G8RDN
WS7 2DL G6PUR
WS7 2DL G7HJX
WS7 2HS G0NPE
WS7 2HU M3LUD
WS7 2LL M3IZI
WS7 2PA G1FSE
WS7 3GD 2E0BRT
WS7 3GD 2E0KGD
WS7 3GD M3VPJ
WS7 3GD M6KGD
WS7 3QL G0PTL
WS7 4QU G0VEL
WS7 4QQ M0KIG
WS7 4QQ M6GHO
WS7 4QU G1JWO
WS7 4QS G7WRG
WS7 4QU G3PET
WS7 4QU G8MII
WS7 4SU G4ZYL
WS7 4UJ G7EKL
WS7 4XY M6ERP
WS7 4YD G3RTY
WS7 8BT G3LXB
WS7 8TX G0EOG
WS7 9AB G4TSD
WS7 9AQ G0DAY
WS7 9AT G0MCM
WS7 9BT G1PGD
WS7 9DS G6OTZ

WS7 9EA G3NLY
WS7 9EA G3WAS
WS7 9EJ G8BMP
WS7 9JG G4FOO
WS8 6AG G8OLJ
WS8 6BB M0XCJ
WS8 6BY 2E0CDL
WS8 6BY M0RUK
WS8 6BY M3KDL
WS8 6DL G8KXO
WS8 6EZ M6JKD
WS8 6HR G4DDE
WS8 6JF G0MPO
WS8 6LN M3LQY
WS8 7AY G0OEI
WS8 7BZ G0FBO
WS8 7BZ G1JMW
WS8 7DH G0JJP
WS8 7EB M6XMS
WS8 7LH G4XOG
WS8 7LL M6NNK
WS8 7ND 2E0CVP
WS8 7ND M6CYO
WS8 7PL G3OHN
WS9 0BP M6MPE
WS9 0DB G4JSQ
WS9 0DU G7IXP
WS9 0QA G4GJU
WS9 0QE G4XFT
WS9 0TG G1BQH
WS9 8HG G6ODF
WS9 8HN M0DWM
WS9 8JA G4FGP
WS9 8JE G4EWI
WS9 8JG G4FAP
WS9 8JG M3IEF
WS9 8LA G4PJK
WS9 8SN G1MJA
WS9 8XE G0OEB
WS9 9BD G7WCB
WS9 9EA M1CHF
WS9 9HU G8KSZ
WS9 9JR G0DSL
WS9 9LS G7APS
WS9 9PD G4FAJ
WS9 9PL G1MZP
WS9 9RD G8LPI

## WV
## (Wolver-hampton)

WV1 1QH G4MJI
WV1 1QQ G0MCE
WV1 2AR G4SGE
WV1 2LD M0TCR
WV1 2LD M1TCR
WV1 4PL G4JUD
WV1 4PL G4PVZ
WV10 0HR G6ZSH
WV10 0JB G7UHY
WV10 0NA G0EGP
WV10 0SH G4ZBC
WV10 6BQ G0GEL
WV10 6DW 2E0DAI
WV10 6DW M0YDH
WV10 6DW M6EMP
WV10 6DW M6GGZ
WV10 6DW M6WOW
WV10 6DX G4IGG
WV10 6EN G4OSU
WV10 6EZ G1AEF
WV10 6NN G1WAM
WV10 6NW G8ZWU
WV10 6QS M6TTV
WV10 6RU G7MEG
WV10 6RU M3MEG
WV10 6SL M3UJV
WV10 6TH M6FDL
WV10 6XH G3FWD
WV10 7BB G4CHJ
WV10 7HS G7NZM
WV10 7HS M3PQQ
WV10 7JS G4URM
WV10 7JS M6TPR
WV10 7JT M3YDM
WV10 7NP G7RCU
WV10 7NP M1DFK
WV10 7TY 2E0XCV
WV10 7TY M3XCV
WV10 7TY M3XIG
WV10 8AR M0FIW
WV10 8AY G1MSB
WV10 8HJ M3UFJ
WV10 8JX G6ZFZ
WV10 0JZ M0EQY
WV10 8NB M3XKN
WV10 8NG G0ULO
WV10 8NS 2E0UHL
WV10 8NS M6BQE
WV10 8SL M6PDJ
WV10 8TH G6UVU
WV10 8UN M6NTN
WV10 9BU G4HVR
WV10 9HN M0AXG
WV10 9HN M3SAB
WV10 9LZ G0SOY
WV10 9PP 2E1HHY
WV10 9PP M0SRB
WV10 9PQ M3UEP
WV10 9QE G0SMO
WV10 9QE M6FZX

WV10 9RJ M0ZUI
WV10 9SJ M3WDB
WV107HS M3YYK
WV11 1AU M3LLO
WV11 1AP G6MJM
WV11 1BD G6LWD
WV11 1BG G0ILH
WV11 1BH G8FVM
WV11 1DD M6BIJ
WV11 1DL M0GTP
WV11 1EG 2E0PPR
WV11 1NW M3WDK
WV11 1PR G7CFC
WV11 1RF G6JYR
WV11 1YD G4PBP
WV11 1YD G8BHH
WV11 2DB G6ALZ
WV11 2DW G4WAS
WV11 2HR M0WAY
WV11 2HR M6TDZ
WV11 2JS M3UKR
WV11 2ND 2E0FRY
WV11 2ND M6FRY
WV11 2PD G1BCE
WV11 2PP M0CVK
WV11 2PZ M0EAF
WV11 2PZ M3GLA
WV11 2PZ M3JLA
WV11 2PZ M3RDA
WV11 2RQ G4FGR
WV11 3AQ M0ZTE
WV11 3AW G4CGL
WV11 3EG G0KYK
WV11 3EW G6CKJ
WV11 3HR 2E0WJL
WV11 3NL G7BIY
WV11 3NU G0BET
WV11 3QJ M3KJD
WV11 3RT G0EGG
WV12 4AN G4JZF
WV12 4JA M6CRZ
WV12 4PX 2E0MYH
WV12 4QF G4MAR
WV12 4SL G1XII
WV12 5AU G4CFH
WV12 5RD M0BQF
WV12 5RD M0JOD
WV12 5RF G0BJZ
WV12 5TD G4VIT
WV12 5YJ G1YLJ
WV13 1AP G3TQL
WV13 1AW 2E0MAY
WV13 1AW M0DCM
WV13 1JQ G8GMA
WV13 2RY G6HRX
WV13 3BD M3UWY
WV13 3DG G0KRK
WV13 3DQ G0OVK
WV13 3HJ M3SBS
WV13 3PB G1PBF
WV13 3PB G4BXC
WV13 3PB M6EXR
WV13 3QA G8JBC
WV13AP M6ITI
WV14 0HT M3ZYM
WV14 0JT 2E0YDA
WV14 0JT M3VTV
WV14 0SF G7UWV
WV14 0TN 2E0LST
WV14 0TN M3YLT
WV14 6EW 2E0TWT
WV14 6NP G8FMW
WV14 6NZ G6YWL
WV14 6NZ M0GRF
WV14 6PU M3XOU
WV14 6RX G7VYW
WV14 7BD M6YMN
WV14 7BN G8JHO
WV14 8AP 2E0OTB
WV14 8AP M6FNP
WV14 8AU M3GKX
WV14 8BQ 2E0XAG
WV14 8BQ M3WQF
WV14 8DA G4ZAD
WV14 8DB M6PAS
WV14 8DE M0AXO
WV14 8DE M6GJM
WV14 8EA M6MCDR
WV14 8EA M6MVB
WV14 8EW 2E0OKY
WV14 8EW M3OKY
WV14 8HX G4LUB
WV14 8JE M6HWQ
WV14 8SB M0GII
WV14 8SB M0SHY
WV14 8XD G0WEB
WV14 8YH G1RBY
WV14 0YW G1KED
WV14 9AN G0JCN
WV14 9AY 2E0SBB
WV14 9AX M6SPB
WV14 9AX M3PPY
WV14 9BE 2E0GCW
WV14 9EU G4YVX
WV14 9TB 2E0WLY
WV14 9TB M6WOL
WV14 9UH G0WCI
WV14 9UT M0BHN
WV14 9XF G4ETW
WV14 9XF M0MSG
WV15 5BS M3OPA
WV15 5DJ M3SRV
WV15 5DS G0UYE
WV15 5DU G4TCX
WV15 5HH M1KDH

WV15 5PB G0BXD
WV15 5PH M3GHL
WV15 5PH M3KCP
WV15 6BU [illegible]
WV15 6FR M6ROI
WV15 6JL 2E0BLC
WV15 6LT G0IZE
WV15 6NU G4OBA
WV16 4HU G8UPF
WV16 4JD G6JMG
WV16 4JD G8RDQ
WV16 4JS G0RQG
WV16 4JW G1SVR
WV16 4JW G3SVR
WV16 4JW G3TVR
WV16 4NW G4NKC
WV16 4SJ G6GUH
WV16 4SP G7LEX
WV16 5AH G6SXD
WV16 5JQ 2E0TDV
WV16 6JY G4EQR
WV16 6LD M3XDM
WV16 6LQ G4YGT
WV16 6LQ M6FCC
WV16 6PR G8CBA
WV16 6RP G0GXT
WV2 1JA M3RZB
WV2 1JA M3XMB
WV2 2AU G7BZQ
WV2 2HF M3VOB
WV2 2QZ 2E0EOU
WV2 2QZ M0EOU
WV2 2QZ M3SPJ
WV2 3ET G1RRU
WV2 3HU G4DGM
WV2 4PS 2E0GRW
WV2 4PS M0RNW
WV2 4PS M6MMR
WV3 0PY G1LEX
WV3 7DN G1GVP
WV3 7HA M6NGA
WV3 7HZ G7GJA
WV3 7LZ G3UBX
WV3 7NJ G7CXB
WV3 8AS 2E0MGI
WV3 8AS M0MGI
WV3 8AS M6EKQ
WV3 8DN M0HJQ
WV3 8HJ G3ZLJ
WV3 8HR M6RNM
WV3 8HY G8BWP
WV3 8JT G0TQP
WV3 9AR G1BKB
WV3 9AR M3NXQ
WV3 9BL G4WRB
WV3 9HN G4BTE
WV3 9HZ G0CYO
WV3 9JF G8EBD
WV3 9JG G8JJF
WV3 9LW G3URJ
WV3 9PX G6ZFG
WV3 9RJ M1BSU
WV4 4AY G3RQX
WV4 4JJ G4LWC
WV4 4LN 2E0FWY
WV4 4LN M6FWY
WV4 4PF 2E0KAX
WV4 4PF G0NWV
WV4 4PF M6NAO
WV4 4RQ G7ACJ
WV4 4ST G0BKQ
WV4 4TR G1ELJ
WV4 4XP G7GEP
WV4 5AQ G6UFZ
WV4 5HH M6FQJ
WV4 5LN G4XNE
WV4 5QD 2E0DGM
WV4 5QD M6HBT
WV4 5QY G4XEZ
WV4 5RF 2E0AQU
WV4 5RF G1RLI
WV4 5RF M0BMN
WV4 5RP G7DDQ
WV4 6DA G0EEN
WV4 6NW M0TBA
WV4 6QY G7SVM
WV4 6RJ 2E1EUY
WV5 0BD G3ZGY
WV5 0BQ G6MZN
WV5 0EA G0TUN
WV5 0EA G1ZAK
WV5 0HQ G0BGB
WV5 0HW G4WLK
WV5 0HY M3AZF
WV5 0JX G1DFF
WV5 0LF M6GFV
WV5 0LG G1BAA
WV5 0LH G0OOE
WV5 0LX M1AUN
WV5 7HF G6AKN
WV5 7HT 2E0TGB
WV5 7HT M0WBF
WV5 7HT M6AZY
WV5 8AF G0OGS
WV5 8AZ M3GWC
WV5 8BH G3TPP
WV5 8BL G4EAJ
WV5 8DL M6OWM
WV5 8EF G1ZTK
WV5 8HQ G4FIF
WV5 8HQ G4WRX
WV5 8HT 2E0BJY
WV5 8HT M0VKY
WV5 8HT M3NFB
WV5 8JJ M3UAJ

WV5 9AJ G0CRO
WV5 9AW G4OTS
WV5 9BP 2E0BLY
WV5 9BP 2E0KMB
WV5 9BP M0KMB
WV5 9BP M3JTU
WV5 9BP M3WUX
WV5 9ER M3TKW
WV5 9HX G4MZI
WV5 9HX G8JVW
WV6 0NT M3ULX
WV6 0SF G4TDO
WV6 0SF G6MZT
WV6 7AQ G1GTK
WV6 7DU G4CYO
WV6 7LX G1DKI
WV6 7NQ 2E0RZX
WV6 7NQ M0RZX
WV6 7NQ M6RZX
WV6 7PH M0MTJ
WV6 7RR G0OKT
WV6 7SB G6HDF
WV6 7SX G7RXW
WV6 7XH M6PYO
WV6 7YH G8CZM
WV6 7YY G3IOB
WV6 8EZ G4DFG
WV6 8HT G0VSP
WV6 8LA G1SGA
WV6 8PT M6WWB
WV6 8RQ G7LPY
WV6 8TR G0HDD
WV6 8TS G4HRG
WV6 8UZ G8ZIK
WV6 9AZ G8GXF
WV6 9HB G4DHL
WV6 9JJ G6VDY
WV6 9LF G8UYR
WV6 9NQ G6JJG
WV7 3AA 2E0BYA
WV7 3AA M0MYA
WV7 3AA M3BYA
WV7 3DQ G4VYA
WV7 3DZ M3WFB
WV7 3EN G0ISI
WV7 3HL G7JPN
WV7 3HW G0RNX
WV7 3LF 2E0JDP
WV7 3LF G7FEF
WV7 3LF M0NJJ
WV7 3LF M6NJP
WV7 3LS G4EAB
WV8 1ES G8IMH
WV8 1JP G8XUE
WV8 1NT G8RF
WV8 1PG G1LGJ
WV8 1PG G4EVP
WV8 1PG G6NBK
WV8 1SA G6DET
WV8 1SG M0HKE
WV8 1TS M0WEV
WV8 1XZ M0AFQ
WV8 2AW G3KNG
WV8 2BE G6YIU
WV8 2BY G0MRP
WV8 2DJ G8JBT
WV8 2DS G0LDY
WV8 2HA G8JIS
WV8 2JT G7ETM
WV9 5DE G4GZK
WV9 5PH G1BWG
WV9 5PJ G6PVT
WV9 5RN G4FAQ

XX99 1AA G4CQX

## YO
## (York)

YO1 6DB 2E1CIK
YO1 7LW M3XYT
YO10 3AW M3LVL
YO10 3AY M1CJN
YO10 3HE G0PMI
YO10 3JX G0WUY
YO10 3JX G4YRC
YO10 3LJ M6FSA
YO10 3LR G4IIX
YO10 3PD 2E1BRG
YO10 3PD M1AQI
YO10 3QB G4KCT
YO10 3QB G8SFI
YO10 3QH M1CJM
YO10 3QJ G6MCQ
YO10 3TJ G1SKV
YO10 4BQ G4KDX
YO10 4DH G0KFV
YO10 4PB G3OZE
YO10 5HH 2E0DMO
YO10 5HH M0HNL
YO11 1QD 2E0SCN
YO11 1QB M6INX
YO11 1TL M0VOL
YO11 2AW G4IAJ
YO11 2BJ G0OII
YO11 2DA G3NZY
YO11 2DT M3UEJ
YO11 2HF G0PHD
YO11 2SP G4OVW
YO11 2TP G4HDL
YO11 2UQ M1CHS
YO11 2XD 2E0OEZ
YO11 2XD M3SPA



**IMPORTANT NOTE**

**Revalidate licence to avoid revocation** – Ofcom has advised the Society that plans will be drawn up to revoke licences that have not been revalidated as required by the licence conditions. The quickest way to revalidate is to do so online via the Ofcom website: *https://services.ofcom.org.uk/* or by email: *amateur.validations@ofcom.org.uk* If you need assistance in the process, Ofcom staff are available to help, but please be patient during times of heavy workload.

| | |
|---|---|
| YO11 3AE | G8HYM |
| YO11 3AN | G3ACQ |
| YO11 3AP | G3YZR |
| YO11 3AP | G4VDH |
| YO11 3AP | G7PHC |
| YO11 3AP | G8BVL |
| YO11 3AP | M6EMY |
| YO11 3AQ | G0AOL |
| YO11 3AQ | G4UHM |
| YO11 3BN | G0IEB |
| YO11 3DE | 2E0BLN |
| YO11 3DE | M3MEO |
| YO11 3DR | G0FBM |
| YO11 3EP | 2E0OTT |
| YO11 3EP | 2E0ZWA |
| YO11 3EP | M0WMO |
| YO11 3LQ | G1TBE |
| YO11 3LY | 2E0GOQ |
| YO11 3LY | M6GIB |
| YO11 3NR | 2E0GOO |
| YO11 3NR | M6JIW |
| YO11 3PB | G6AIB |
| YO11 3PZ | G8WVZ |
| YO11 3RE | G8AZA |
| YO11 3TA | M6YLH |
| YO11 3TN | 2E0NCS |
| YO11 3TN | M6NCS |
| YO11 3TS | G4AKR |
| YO11 3UB | 2E0IKW |
| YO11 3UB | M0IKW |
| YO11 3UB | M6GMT |
| YO11 3UD | M0OLD |
| YO11 3UU | 2E0YDX |
| YO11 3UU | G0NXX |
| YO11 3UU | M1NXX |
| YO11 3XA | G3MYZ |
| YO112LZ | 2E0BLJ |
| YO112LZ | M3SWQ |
| YO12 4AS | 2E0ERS |
| YO12 4AS | M6ERN |
| YO12 4AT | G0VAG |
| YO12 4DX | G7CAS |
| YO12 4EF | 2E0LKT |
| YO12 4EF | M0LKT |
| YO12 4EF | M3TUJ |
| YO12 4ES | 2E0CNH |
| YO12 4EW | G4JJQ |
| YO12 4EY | G7OWX |
| YO12 4JE | G0RPU |
| YO12 4JJ | 2E0PLE |
| YO12 4JR | G0NNZ |
| YO12 4LA | G0FXT |
| YO12 4LL | G8WYB |
| YO12 4QT | G0VBM |
| YO12 4RJ | G3RIX |
| YO12 4RN | 2E0OOO |
| YO12 4RN | G0OOO |
| YO12 4RN | G4SSH |
| YO12 4RN | G7OOO |
| YO12 4RN | G7ROY |
| YO12 4RQ | 2E0ZZZ |
| YO12 4RQ | G4YSS |
| YO12 4RQ | M1NNN |
| YO12 4SR | 2E0PAV |
| YO12 4SR | M0PAV |
| YO12 5BT | G4WXJ |
| YO12 5HG | 2E0KWT |
| YO12 5HG | G4OOE |
| YO12 5HG | M3YWR |
| YO12 5HW | G1PPO |
| YO12 5HW | M3PPO |
| YO12 5HX | M0GME |
| YO12 5JA | G0PRI |
| YO12 5LB | M1ESW |
| YO12 5NB | M6BYM |
| YO12 5NJ | G0WHO |
| YO12 5NQ | G3TVW |
| YO12 5PU | G0UUU |
| YO12 5QJ | G3PEJ |
| YO12 5QJ | G3XIH |
| YO12 5RG | G0DOA |
| YO12 5RQ | G0JNR |
| YO12 6AN | M0GZX |
| YO12 6DF | G6DUT |
| YO12 6DQ | G0TOS |
| YO12 6DQ | G8VAR |
| YO12 6EP | 2E0BLS |
| YO12 6EP | M3XLS |
| YO12 6HL | G4ZAO |
| YO12 6JT | G4BP |
| YO12 6JT | M0MXX |
| YO12 6JW | G6AFZ |
| YO12 6LE | G0KDA |
| YO12 6LJ | G8PIC |
| YO12 6QY | M6HCO |
| YO12 6RA | G4EGB |
| YO12 6RQ | G3HKO |
| YO12 6SB | G0SHM |
| YO12 6SF | G4NSE |
| YO12 6SP | G0VXE |
| YO12 6SP | G2CP |
| YO12 6TB | G4BON |
| YO12 6TF | G3WOD |
| YO12 6TG | G8ETS |
| YO12 6TW | G0KFG |
| YO12 6UF | G4DWU |
| YO12 6UN | 2E0XAR |
| YO12 6UN | M6TOK |
| YO12 7AU | M6OAV |
| YO12 7DH | G3ZJT |
| YO12 7HL | G0PFE |
| YO12 7HU | 2E0RJR |
| YO12 7HU | M6FFZ |
| YO12 7JT | G1OSP |
| YO12 7JZ | M6HGE |
| YO12 7NP | M3CSV |
| YO12 7SD | M6DSE |
| YO13 0AZ | 2E0PIT |
| YO13 0AZ | M6FVW |
| YO13 0DQ | G0OPQ |
| YO13 0HN | G0VFV |
| YO13 0LW | G4FNG |
| YO13 0QH | G8IKG |
| YO13 0RU | G2API |
| YO13 0RX | G0COL |
| YO13 0RX | M3SEJ |
| YO13 0SG | G0CDR |
| YO13 0SN | G3KEV |
| YO13 9AR | G4MGP |
| YO13 9AR | G4MGQ |
| YO13 9EL | G4UUU |
| YO13 9ER | G0NUP |
| YO13 9ER | G0SIG |
| YO13 9ET | G4FLM |
| YO13 9ET | G4JAQ |
| YO13 9EU | G0FUE |
| YO13 9HB | 2E0BAU |
| YO13 9HB | G7SGK |
| YO13 9HB | M0IKB |
| YO13 9HB | M3IJZ |
| YO13 9HL | G3KAE |
| YO13 9JA | G6SDY |
| YO13 9JQ | G4OPI |
| YO13 9JW | G0PSK |
| YO13 9PA | G0UVR |
| YO14 0AN | G8YQN |
| YO14 0DL | M6DGP |
| YO14 0DQ | G4WCY |
| YO14 0FD | G7PKG |
| YO14 0JA | M6OZT |
| YO14 0LB | G8ZFS |
| YO14 0NB | G4KZZ |
| YO14 0PY | 2E0CYO |
| YO14 9AY | 2E0OKZ |
| YO14 9AY | M0MCA |
| YO14 9AY | M3OKZ |
| YO14 9DU | 2E0AUI |
| YO14 9ER | G7MZE |
| YO14 9NL | G7LOV |
| YO14 9NY | G4EDR |
| YO14 9NY | G8HWQ |
| YO14 9QE | G6TKH |
| YO14 9QJ | G0LAN |
| YO14 9QJ | G4ETM |
| YO14 9RT | 2E0BLK |
| YO14 9RT | M3NUL |
| YO14 9UZ | 2E0TQF |
| YO14 9UZ | M3JLF |
| YO14 9UZ | M3TQF |
| YO15 1AE | G3UEN |
| YO15 1AP | G0IAD |
| YO15 1HL | G8NJI |
| YO15 1JF | G0LES |
| YO15 1JT | G0FLD |
| YO15 1LJ | G4XBU |
| YO15 1LU | M0EGV |
| YO15 1LW | G8PWU |
| YO15 1LY | M0ABP |
| YO15 2DS | G0VEX |
| YO15 2ED | M3FCR |
| YO15 2HF | M3PFL |
| YO15 2JW | G4CVA |
| YO15 2NA | G7VAY |
| YO15 3DX | M6BWW |
| YO15 3EE | M6KBV |
| YO15 3HS | M6GPU |
| YO15 3JT | G1ORT |
| YO15 3JU | 2E0JJB |
| YO15 3JU | M0JJB |
| YO15 3JU | M3JJB |
| YO15 3NA | G4GKU |
| YO15 3NP | G0RMP |
| YO15 3NT | M0FUN |
| YO15 3NX | M1ECV |
| YO15 3NX | M3ECV |
| YO16 4AT | M6EZE |
| YO16 4AU | M3MGI |
| YO16 4AZ | 2E1GHF |
| YO16 4BA | 2E0JME |
| YO16 4HL | G6VYK |
| YO16 4HN | G4LOB |
| YO16 4JD | G6JEF |
| YO16 4JN | 2E0IJW |
| YO16 4JN | M3OQK |
| YO16 4JZ | M3FDV |
| YO16 4JZ | M3PFY |
| YO16 4ND | M6MUB |
| YO16 4NL | 2E0ASX |
| YO16 4NL | G0DEB |
| YO16 4NL | G0SGR |
| YO16 4QS | M6HGM |
| YO16 4RG | M6OBR |
| YO16 4XU | 2E0BRU |
| YO16 4XU | 2E0HLH |
| YO16 6ES | 2E0PCL |
| YO16 6ES | G1HAH |
| YO16 6ES | M6PPT |
| YO16 6FB | G8XFK |
| YO16 6FY | M1AEK |
| YO16 6GW | G3ZTR |
| YO16 6HQ | G1UCN |
| YO16 6JA | G6XDZ |
| YO16 6RW | M0PDQ |
| YO16 6TB | M0AHV |
| YO16 6TB | M6IDF |
| YO16 6UA | G3PWN |
| YO16 6XR | G0LRO |
| YO16 6YP | G7OHD |
| YO16 7DB | G4KHR |
| YO16 7DZ | G7TGN |
| YO16 7GB | M0DPH |
| YO16 7HL | G4KNR |
| YO16 7HR | 2E1IIG |
| YO16 7NH | M1BLW |
| YO16 7NL | M0KXQ |
| YO16 7NL | M6TEE |
| YO16 7PN | G0MNN |
| YO16 7PN | G0NUA |
| YO16 7PZ | G6WNB |
| YO16 7SA | G3SWW |
| YO17 6NJ | G8HCK |
| YO17 6QX | G3VVR |
| YO17 6SY | G3TPW |
| YO17 7BE | G3UWP |
| YO17 7BQ | G0WZJ |
| YO17 7DF | 2E0EAY |
| YO17 7HE | G8OPE |
| YO17 7LB | 2E0NPN |
| YO17 7LB | G1VDO |
| YO17 7LB | M3SQZ |
| YO17 7LE | G0FNS |
| YO17 7YN | G4HNW |
| YO17 7YP | G7KMH |
| YO17 8LZ | G4CVG |
| YO17 9AE | G8JYS |
| YO17 9AE | M1AVM |
| YO17 9AR | G6OJV |
| YO17 9DZ | 2E0CYU |
| YO17 9DZ | M0JLP |
| YO17 9DZ | M6EGE |
| YO17 9SJ | G0KOE |
| YO18 7HB | G4OKW |
| YO18 7HF | G7PBK |
| YO18 7HJ | G4JYW |
| YO18 7HJ | G4OBK |
| YO18 7HN | 2E0PEF |
| YO18 7HN | G3EEH |
| YO18 7HN | G3TQQ |
| YO18 7HN | M0HQO |
| YO18 7HZ | G4BNS |
| YO18 7LE | G0FNP |
| YO18 7PG | M6XGB |
| YO18 7TD | G6LBL |
| YO18 8DA | 2E0ZYG |
| YO18 8DA | M6ZYG |
| YO18 8HH | G3UBI |
| YO18 8PS | G7IID |
| YO18 8TA | G4LLQ |
| YO19 4QQ | G0IEH |
| YO19 4QQ | G7COG |
| YO19 4RS | M0CYU |
| YO19 5PH | G0LIY |
| YO19 5UL | G0TUV |
| YO19 6DE | G0AWH |
| YO19 6GE | G6JRI |
| YO19 6PB | M6HNF |
| YO19 6PB | M6SUH |
| YO19 6RG | G1IBX |
| YO21 1JD | G0ANH |
| YO21 1JJ | M0CKU |
| YO21 1JS | M0AZC |
| YO21 1NA | G3NTA |
| YO21 1PN | 2E1FHO |
| YO21 1QX | G6HTA |
| YO21 1UE | G1VOY |
| YO21 2BL | G0WDC |
| YO21 2DE | 2E1IJN |
| YO21 2DE | M1PXB |
| YO21 2DE | M3EWN |
| YO21 2JN | 2E1NII |
| YO21 2JN | M3NII |
| YO21 3DW | G7RRY |
| YO21 3DW | G8SGX |
| YO21 3DX | 2E0PFR |
| YO21 3JB | G7RTJ |
| YO21 3LR | G0UBK |
| YO21 3LY | 2E0NYX |
| YO21 3LY | M3LNR |
| YO21 3PD | 2E0ONE |
| YO21 3SD | G4QA |
| YO21 3UE | 2E0NAS |
| YO21 3UE | M6NSJ |
| YO21 3UR | G4SAB |
| YO22 4DE | G0WNF |
| YO22 4DY | G8LQN |
| YO22 4EE | M0CLN |
| YO22 4HL | G0EQI |
| YO22 4JD | 2E0LAB |
| YO22 4JD | 2E0TLY |
| YO22 4JD | M0KEG |
| YO22 4JD | M3TLY |
| YO22 4JD | M3UNY |
| YO22 4JU | G0EGE |
| YO22 4PB | G0DSO |
| YO22 4QG | G4HHH |
| YO22 4QN | G4EQS |
| YO22 4RQ | M1EZH |
| YO22 4RQ | M3EZH |
| YO22 4RW | G6ZRO |
| YO22 4RW | M3ZRO |
| YO22 4RW | M6GYU |
| YO22 5AN | G0EBL |
| YO22 5AN | G4DAX |
| YO22 5AN | G8MZQ |
| YO22 5AN | G8XLA |
| YO22 5BG | G6HWA |
| YO22 5EP | G0AOP |
| YO22 5QQ | G0WNJ |
| YO23 1JN | 2E0VBB |
| YO23 1LE | G4YEK |
| YO23 2RX | 2E0WYZ |
| YO23 2RX | M6AQL |
| YO23 2UL | M1AXX |
| YO23 3PS | M0GHY |
| YO23 3SH | G4FUO |
| YO23 3SH | M3CEN |
| YO23 3XU | M0KOO |
| YO23 3XU | M6FUU |
| YO23 3YD | G7KJD |
| YO23 7DA | G3PRO |
| YO24 1DL | 2E0JMW |
| YO24 1LX | M1EUL |
| YO24 1UF | G0RQZ |
| YO24 2QX | M0EHA |
| YO24 2RF | G6FJA |
| YO24 2RT | M0RNC |
| YO24 3FB | G0MVE |
| YO24 3FB | M0AQW |
| YO24 3HG | G4MLW |
| YO24 3LF | G0PWO |
| YO24 4EG | 2E0RXT |
| YO24 4EG | M6YBT |
| YO24 4EY | M3RUK |
| YO24 4JU | M3VQL |
| YO24 4LE | G7SUA |
| YO24 4NP | G8KWH |
| YO25 3BJ | G7JVJ |
| YO25 3QW | M0CCU |
| YO25 3TD | G8MVJ |
| YO25 3TJ | G7ONF |
| YO25 4DE | 2E0ADL |
| YO25 4DE | G3XYF |
| YO25 4JZ | G1IBF |
| YO25 4PA | G0AZQ |
| YO25 4PD | G4YTD |
| YO25 4QA | G3WOV |
| YO25 4QE | 2E0CGI |
| YO25 4SF | 2E0WLK |
| YO25 4SF | M0WLK |
| YO25 4SF | M3WLY |
| YO25 4ST | G0VIA |
| YO25 5AT | G7FSH |
| YO25 5AY | G1PJZ |
| YO25 5BG | G6FMS |
| YO25 5BG | G7TXF |
| YO25 5BS | G7POT |
| YO25 5DJ | M1BKW |
| YO25 5ES | M0GVZ |
| YO25 5EZ | 2E0DMS |
| YO25 5EZ | M3ZTU |
| YO25 5HX | G6CPF |
| YO25 5HZ | 2E0LBJ |
| YO25 5HZ | M6LEQ |
| YO25 5NB | G0LYZ |
| YO25 5PF | M6RSR |
| YO25 6QT | G0HTG |
| YO25 6TX | M1ETX |
| YO25 6YE | M0DMB |
| YO25 8BH | 2E1IJF |
| YO25 8DX | G1IKT |
| YO25 8DX | G7JZS |
| YO25 8HL | 2E0OZE |
| YO25 8HL | M0VOZ |
| YO25 8HL | M6OZZ |
| YO25 8JJ | G1IMI |
| YO25 8LE | G7VGT |
| YO25 8LS | G6BUV |
| YO25 8PG | M3XHQ |
| YO25 8PP | G1BWZ |
| YO25 8QJ | G4PAA |
| YO25 8SJ | M1ASR |
| YO25 9DR | G1ILJ |
| YO25 9HE | G0KGA |
| YO25 9HE | G4SMB |
| YO25 9PF | M6CXN |
| YO25 9SH | G6ENR |
| YO25 9UL | G3ZOU |
| YO25 9UN | 2E0GDL |
| YO25 9XL | G1VWZ |
| YO26 4SH | M3ZUG |
| YO26 4SH | M3ZUU |
| YO26 4TU | G4NFE |
| YO26 4TX | M0XZX |
| YO26 4TX | M6XQX |
| YO26 4XP | G4TOM |
| YO26 5EN | G4VXG |
| YO26 5HQ | G7FGA |
| YO26 5QS | G4EJP |
| YO26 5QZ | G1BYP |
| YO26 5RP | G4LKP |
| YO26 6AW | G0EVO |
| YO26 6JB | G4FDD |
| YO26 6PS | G3ORD |
| YO26 7LW | G1ANV |
| YO26 7ND | M6FWC |
| YO26 7PY | G0LPX |
| YO26 7QD | G0GYP |
| YO26 7QN | G4OVX |
| YO26 8BG | G3XWH |
| YO26 8DD | G0NHM |
| YO26 8DE | 2E1STK |
| YO26 9RQ | G4GTU |
| YO26 9SF | M0RWG |
| YO26 9SF | M6RMG |
| YO3 9UA | G0AWZ |
| YO30 1AA | G4YCS |
| YO30 2DF | G6NIZ |
| YO30 4SU | M6DOW |
| YO30 5PT | M0LKL |
| YO30 5PT | M6LLB |
| YO30 5PZ | G4EMA |
| YO30 5QG | G3HWW |
| YO30 5QG | G4ESU |
| YO30 5RP | G0UDP |
| YO30 5SU | G0DLK |
| YO30 5SU | G0GYJ |
| YO30 5ZH | 2E0PMD |
| YO30 6DZ | G0VWP |
| YO30 6EQ | 2E0WWT |
| YO30 6EQ | M6WWT |
| YO30 6HL | G6SYX |
| YO30 6NA | G0LOP |
| YO30 6PE | G4HEV |
| YO30 6PX | G1MTP |
| YO31 0LU | G0HVH |
| YO31 0PX | G0FDH |
| YO31 0PZ | G1DRG |
| YO31 0TD | M6DOF |
| YO31 0UR | G1VIZ |
| YO31 1AL | G3HWF |
| YO31 1BL | M3YUC |
| YO31 1BR | G7SBZ |
| YO31 1DF | 2E1ENN |
| YO31 1ED | G6YXO |
| YO31 1JD | G4FRA |
| YO31 1JW | G4CTZ |
| YO31 7EJ | G4VRU |
| YO31 8SQ | G0ULQ |
| YO31 9BD | G0TTS |
| YO31 9DB | G1KAO |
| YO31 9HL | 2 E0YYK |
| YO31 9HL | M6YYK |
| YO31 9PE | M6FZS |
| YO32 2DG | G0UKO |
| YO32 2FL | G1LRM |
| YO32 2FL | G6MED |
| YO32 2GU | M6YPS |
| YO32 2GX | M0DPY |
| YO32 2PB | M0STV |
| YO32 2PR | G1YRC |
| YO32 2PR | G8IMZ |
| YO32 2QE | G4JQF |
| YO32 2WT | 2E0JYM |
| YO32 2YE | M6PXG |
| YO32 2ZZ | G1PJV |
| YO32 3DT | G0WPF |
| YO32 3EG | M1VGH |
| YO32 3ET | M6HPG |
| YO32 3GG | M1EHD |
| YO32 3NL | G4PBJ |
| YO32 3NQ | M6KJB |
| YO32 3NS | G0XAB |
| YO32 3NS | M6HQZ |
| YO32 3NW | 2E0FNQ |
| YO32 3RP | G8INO |
| YO32 3RR | G0VZI |
| YO32 3RR | M0BWQ |
| YO32 3YW | G3PHJ |
| YO32 3YY | M0LET |
| YO32 5PA | G3IJA |
| YO32 5TE | G0TXY |
| YO32 5TL | G4BNT |
| YO32 5XW | M0RSA |
| YO32 9LY | G7BXJ |
| YO32 9NJ | G3BMO |
| YO32 9QG | M1BHN |
| YO32 9QU | 2E0PWD |
| YO32 9SD | G8RLW |
| YO32 9SD | M1BMU |
| YO32 9SH | G3VGH |
| YO32 9TF | G4DBP |
| YO32 9UJ | G3TEE |
| YO32 9YG | G1GHG |
| YO41 1AQ | G4VXV |
| YO41 1DP | G4XHX |
| YO41 1DZ | M6MFS |
| YO41 1EZ | 2E1BRT |
| YO41 1LX | G4USC |
| YO41 1PR | G0SNG |
| YO41 1PR | M1AFX |
| YO41 5PL | 2E0GCE |
| YO41 5PL | G4XBJ |
| YO41 5PL | M3HWY |
| YO41 5RW | M0RCL |
| YO42 1RX | M0EBV |
| YO42 1YJ | G1JMY |
| YO42 1YQ | G3YRU |
| YO42 2BX | G0FYU |
| YO42 2BX | G4KCF |
| YO42 2ET | M6PDN |
| YO42 2FQ | G1KUG |
| YO42 2GX | G0AOQ |
| YO42 2HB | G4BBZ |
| YO42 2HJ | G7BGM |
| YO42 2JF | G0LIW |
| YO42 2PA | G4FBO |
| YO42 2WA | M0AOA |
| YO42 4JG | M0YDF |
| YO43 3BR | M6WIC |
| YO43 3JR | M0CTP |
| YO43 3NB | G7FKP |
| YO43 3ND | G0FZO |
| YO43 3RB | 2E0JIX |
| YO43 3RB | M6JIX |
| YO43 4BG | M1AEQ |
| YO43 4DF | M0XBR |
| YO43 4DQ | M0BVD |
| YO43 4HJ | G0EAT |
| YO43 4SD | G0UFZ |
| YO43 4TT | G0VRM |
| YO43 4TT | G4CMT |
| YO43 4TT | M0GYR |
| YO51 9BU | M6PHU |
| YO51 9BZ | G4KFH |
| YO51 9DP | G4KWZ |
| YO51 9DY | G0RHI |
| YO51 9EA | G4CPD |
| YO51 9GN | G4TEW |
| YO51 9LP | G0MVX |
| YO51 9LP | G1WCY |
| YO51 9LP | G4SJM |
| YO60 6QB | G1NBK |
| YO60 7DX | 2E1ATV |
| YO60 7DZ | 2E1HBS |
| YO60 7DZ | G0ODS |
| YO60 7HH | M0AGL |
| YO60 7QT | G2FM |
| YO60 7QT | G2YL |
| YO60 7QT | G3QI |
| YO60 7QT | G5KC |
| YO60 7RJ | M6ZEN |
| YO61 1HW | G1AVZ |
| YO61 1JR | G1TLE |
| YO61 1PS | M1FCH |
| YO61 2NH | G7AXN |
| YO61 2NH | M0RCM |
| YO61 3AB | M0HAN |
| YO61 3AB | M0WPA |
| YO61 3BB | G7KBZ |
| YO61 3DE | G1JKE |
| YO61 3DE | M0MUC |
| YO61 3DE | M0TDE |
| YO61 3DE | M3JKE |
| YO61 3DE | M3YTE |
| YO61 3ER | 2E0GHP |
| YO61 3ER | M3JKT |
| YO61 3JS | M3MKO |
| YO61 3QQ | G4IRV |
| YO61 3RE | G3ZNB |
| YO61 4PF | G6JCM |
| YO61 4PX | G1FGI |
| YO61 4QA | G4JKY |
| YO61 4QA | G4LXU |
| YO62 4DG | G0GXI |
| YO62 5DY | 2E1CYU |
| YO62 5EZ | G7HAF |
| YO62 5LG | G4KNV |
| YO62 5YJ | G7CRV |
| YO62 6DQ | 2E0ZMS |
| YO62 6DQ | M0ZMS |
| YO62 6DQ | M3ZMS |
| YO62 6EL | G1SZT |
| YO62 6EL | M0YAL |
| YO62 6LN | M3RQB |
| YO62 7JX | G4ZXF |
| YO62 7SH | M3MWT |
| YO7 1AA | M6PUN |
| YO7 1AP | G6JAP |
| YO7 1BQ | G1JST |
| YO7 1FH | M0CXO |
| YO7 1FJ | G4JBK |
| YO7 1FW | G8RJZ |
| YO7 1FW | M3UQB |
| YO7 1FW | M3ZXB |
| YO7 1JN | G4SPC |
| YO7 1QD | M6JTI |
| YO7 1RT | G6JTI |
| YO7 1UD | G4ZNZ |
| YO7 2DJ | G7SKH |
| YO7 2LL | G3HHD |
| YO7 3AN | G6FTE |
| YO7 3BU | M3ZSV |
| YO7 3PF | G6NJO |
| YO7 3QR | M0SGH |
| YO7 3QR | M3HAL |
| YO7 3QW | M0SWE |
| YO7 4HX | G4VMA |
| YO7 4LR | G4YGP |
| YO7 4PS | M0HQJ |
| YO7 4RT | G7VBN |
| YO8 3RD | G0DIR |
| YO8 3TB | M3SXV |
| YO8 4AZ | G1KNA |
| YO8 4BY | 2E0GHK |
| YO8 4BY | M6SBV |
| YO8 4HA | M3VDH |
| YO8 4HA | M3VTX |
| YO8 4HA | M6DDU |
| YO8 4NY | G4MWH |
| YO8 4XX | M6CIW |
| YO8 5HF | G3WQ |
| YO8 5HP | G4VSV |
| YO8 6LN | G1UVK |
| YO8 6NH | M0DCS |
| YO8 6PE | 2E0UIP |
| YO8 6PW | G0DWC |
| YO8 6QL | G3ODD |
| YO8 6QP | 2E1DRU |
| YO8 6QP | G0OLE |
| YO8 6QP | G6YYN |
| YO8 6QW | G7AUP |
| YO8 6QX | G0FLX |
| YO8 6RA | G6JYX |
| YO8 6TG | G3ZIV |
| YO8 8ES | G0GFA |
| YO8 8HD | 2E0XJT |
| YO8 8HD | G4YPV |
| YO8 8HD | G6XJT |
| YO8 8HZ | M1CCG |
| YO8 8JU | M6BQD |
| YO8 8NQ | M0UTX |
| YO8 8NQ | M0VQJ |
| YO8 8QT | G8JZX |
| YO8 9AR | 2E0BIU |
| YO8 9AR | M0ICA |
| YO8 9AR | M3PYR |
| YO8 9DW | G1CMC |
| YO8 9EU | 2E0NDY |
| YO8 9EU | M0JOO |
| YO8 9EU | M3PNY |
| YO8 9EU | M5TLA |
| YO8 9HW | G8EEM |
| YO8 9LL | G0BGV |
| YO8 9ND | M1AJA |
| YO8 9PD | G3PSM |
| YO8 9PD | G4JYL |
| YO8 9PF | G4OJN |
| YO8 9PF | G8LUP |
| YO8 9QD | G6JRL |
| YO8 9QJ | 2E0RCA |
| YO8 9QJ | M0KLM |
| YO8 9RR | M6JPC |
| YO8 9XH | G4LYE |
| YO8 9YB | G8BQF |
| YO8 9YB | G8ZZB |

## ZE

## (Zetland)

| | |
|---|---|
| ZE1 0BR | GM4PXG |
| ZE1 0NE | 2M0ZET |
| ZE1 0NE | MM3ZET |
| ZE1 0PA | GM4WXQ |
| ZE1 0PJ | GM3ZET |
| ZE1 0PJ | GM4LER |
| ZE1 0UQ | MM6BZQ |
| ZE2 9DA | GM8MMA |
| ZE2 9DD | MM0ZRC |
| ZE2 9DL | GM4GQM |
| ZE2 9EF | GM4GPP |
| ZE2 9ER | GM6RQW |
| ZE2 9GJ | GM7GWW |
| ZE2 9GY | 2M0CPN |
| ZE2 9GY | MM0ZAW |
| ZE2 9GY | MM6ACW |
| ZE2 9GY | MM6ZDW |
| ZE2 9HG | MM6ZBG |
| ZE2 9HH | GM3WHT |
| ZE2 9HL | GM0EKM |
| ZE2 9JD | MM6HHB |
| ZE2 9JG | GM0GFL |
| ZE2 9JG | MM0XAU |
| ZE2 9JG | MM0ZCG |
| ZE2 9JG | MM5YLO |
| ZE2 9JX | GM1KKI |
| ZE2 9LA | 2M0BDR |
| ZE2 9LA | 2M0BDT |
| ZE2 9LA | MM0LSM |
| ZE2 9LA | MM0ZAL |
| ZE2 9LD | GM8YEC |
| ZE2 9LG | GM8LNH |
| ZE2 9LX | GM4SLV |
| ZE2 9QF | GM7AFE |
| ZE2 9QN | GM0ILB |
| ZE2 9QS | GM0JDB |
| ZE2 9RH | MM6SJK |
| ZE2 9RS | GM4SSA |
| ZE2 9RS | MM0ZET |
| ZE2 9SX | GM1ZNR |
| ZE2 9TD | GM1FGN |
| ZE2 9TF | MM6IKB |
| ZE2 9TF | MM6IMB |
| ZE2 9TH | GM4ZHL |
| ZE2 9TX | MM1FJM |
| ZE2 9UL | GM4LBE |
| ZE3 9JN | MM0HUQ |
| ZE3 9JS | 2M0GFC |
| ZE3 9JS | GM4IPK |
| ZE3 9JS | MM0NQY |
| ZE3 9JS | MM6PTE |
| ZE3 9JS | MM6YLO |
| ZE3 9JW | 2M0SEG |
| ZE3 9JW | MM6MFA |
| ZE3 9JX | MM5PSL |
| ZZ1 9FR | G4EXU |
| ZZ2 3TP | G3WUK |
| | G0FMX |
| | G6JPE |
| | G6REM |